CHILTON'S
IMPORT CAR REPAIR
MANUAL 1993-97

Publisher & Editor-In-Chief	Kerry A. Freeman, S.A.E.
Executive Editors	Dean F. Morgantini, S.A.E., W. Calvin Settle Jr., S.A.E.
Managing Editor	Nick D'Andrea
Senior Editors	Jacques Gordon, Michael L. Grady, Kevin M. G. Maher, Debra McCall, Richard J. Rivele, S.A.E., Richard T. Smith, Jim Taylor, Ron Webb
Project Managers	Larry Braun, S.A.E., A.S.C., Thomas P. Browne III, Joseph L. DeFrancesco, A.S.E., Robert E. Doughten, Ben Greisler, S.A.E., Martin J. Gunther, Craig P. Nangle, A.S.E., S.A.E., Ernest H. Ralph, A.S.E., S.A.E., Richard Schwartz
Editorial Staff	Jaffer A. Ahmad, Robert Chabot, William Cottman, A.S.E., Leonard Davis, A.S.E., Michael DiFurio Jr., S.A.E., Sam Fiorani, Matthew E. Frederick, William C. Friedauer, Edward Giacomucci, A.S.E., S.A.E., Al Gibbs, Herbert Guie Jr, George B. Heinrich III, Dawn M. Hoch, Daniel Howells, A.S.E., David E. Jester, A.S.E., Lori L. Johnson, A.S.E., Will Kessler, A.S.E., Kenneth F. Konzelman, Neil J. Leonard, A.S.E., James R. Marotta, Robert McAnally, Thomas A. Mellon, Raymond K. Moore, A.S.E., Norman D. Norville, A.S.E., Christine L. Nuckowski, Eric S. Peterson, A.S.E., Charles Ramsey, A.S.E., William L. Renn, A.S.E., Roy Ripple, A.S.E., George E. Ritter, Robert Saxton, A.S.E., S.A.E., Paul Shanahan, Larry E. Stiles, Gordon L. Tobias, S.A.E., Albert A. Wood, A.S.E.
Production Manager	Andrea M. Steiger
Assistant Production Manager	Marsha Park-Herman
Production Specialists	Christina Davis, Kimberly T. Hayes, Joseph C. McGinty, Liz Thompson
Director of Manufacturing	Mike D'Imperio
Manufacturing Manager	Robin Norman
OFFICERS	
Senior Vice President	Ronald A. Hoxter

CHILTON BOOK COMPANY

ONE OF THE **DIVERSIFIED PUBLISHING COMPANIES,**
A PART OF **CAPITAL CITIES/ABC,INC.**

Manufactured in
© 1996 Chilton Book Company
Chilton Way, Radnor, PA 19089
ISBN 0-8019-7920-X
ISSN

CH   7/97

CAR MODELS

Table of Contents

HOW TO USE THIS MANUAL

HOW TO USE THIS MANUAL

Car Section

Car sections are grouped by manufacturer and arranged in alphabetical order. The text and illustrations that comprise the service procedures in each Car Section are arranged in the following order of systems and components: Firing Orders, Engine Electrical, Chassis Electrical, Engine Cooling, Fuel System, Emission Controls, Engine Mechanical, Engine Lubrication, Transmission/Transaxle, Drive Axle, Steering, Front Suspension, Rear Suspension, Brakes.

All illustrations are located as close as possible to the pertinent text. Procedures are for all models in the particular section unless specifically noted otherwise.

Locating Information

The Table of Contents, at the front of the book, lists the beginning of each Car Section in the manual.

To find where a particular Car Section is located in the book, you need only look in the Table of Contents. Once you have found the proper section, you may wish to find where specific procedures are located in that section. Turn to the Index at the front of the section. At the upper left-hand side is a listing of the main topics within the section and the page number they will be found on. Following the main topics is an alphabetical listing of all the procedures within the section and their page numbers.

Safety Notice

Proper service and repair procedures are vital to the safe, reliable operation of all motor vehicles, as well as the personal safety of those performing repairs. This manual outlines procedures for servicing and repairing vehicles using safe effective methods. The procedures contain many NOTES and CAUTIONS which should be followed along with standard safety procedures to eliminate the possibility of personal injury or improper service which could damage the vehicle or compromise its safety.

It is important to note that repair procedures and techniques, tools and parts for servicing vehicles, as well as the skill and experience of the individual performing the work vary widely. It is not possible to anticipate all of the conceivable ways or conditions under which vehicles may be serviced, or to provide cautions as to all of the possible hazards that may result. Standard and accepted safety precautions and equipment should be used when handling toxic or flammable fluids, and safety goggles or other protection should be used during cutting, grinding, chiseling, prying, or any other process that can cause material removal or projectiles.

Some procedures require the use of tools specially designed for a specific purpose. Before substituting another tool or procedure, you must be completely satisfied that neither your personal safety, nor the performance of the vehicle will be endangered.

Part Numbers

Part numbers listed in this book are not recommendations by Chilton for any product by brand name. They are references that can be used with interchange manuals and aftermarket supplier catalogs to locate each brand supplier's discrete part number.

Although information in this manual is based on industry sources and is as complete as possible at the time of publication, the possibility exists that some car manufacturers made later changes which could not be included here. Information on very late models may not be available in some circumstances. While striving for total accuracy, Chilton Book Company cannot assume responsibility for any errors, changes, or omissions that may occur in the compilation of this data.

Copyright Notice

ACURA 1

Integra • Legend • Vigor • 2.5TL • 3.2TL

FIRING ORDERS

NOTE: To avoid confusion, always replace spark plug wires one at a time.

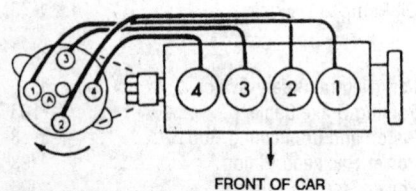

FRONT OF CAR

323652

1.7L, and 1.8L Engines
Engine Firing Order: 1-3-4-2
Distributor Rotation: Clockwise

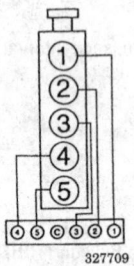

327709

2.5L Engine
Firing Order:
1-2-4-5-3
Distributorless
ignition

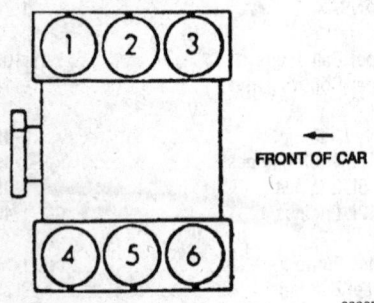

FRONT OF CAR

333879

3.2L Engine
Engine Firing Order: 1-2-3-4-5-6
Distributorless ignition

ENGINE ELECTRICAL

NOTE: Disconnecting the negative battery cable on some vehicles may interfere with the functions of the on board computer systems and may require the computer to undergo a relearning process, once the negative battery cable is reconnected.

Distributor

REMOVAL AND INSTALLATION

1.7L and 1.8L (B18A1, B17A1, B18B1, B18C1) Engines

NOTE: The radio may contain a coded anti-theft circuit. Make sure you have the security code number before disconnecting the battery.

1. Disconnect the negative battery cable.
2. Disconnect engine wire harness and connectors from distributor.
3. Disconnect the spark plug wires from distributor cap.
4. If removing the ignition coil, remove the distributor cap, rotor, and cap seal, then remove the leak cover.
5. Remove the 2 screws to disconnect the wires from the coil.
6. Remove the 2 screws and slide the ignition coil out of the distributor housing.
7. Remove distributor hold-down bolts, and remove distributor from cylinder head.
   To install:
8. Use new O-ring on distributor housing. Coat new O-ring with engine oil before installation.
9. Slip the distributor into position.

NOTE: Lugs on the end of the distributor and the matching grooves in the camshaft end are offset to eliminate any possibility of installing the distributor 180 degrees out of time.

10. Install hold-down bolts, hand tighten.
11. Slide the ignition coil into the distributor housing and install the 2 mounting screws.
12. Reconnect the 2 wires to the coil and install the 2 screws. Install the leak cover, rotor, cap seal, and cap.
13. Connect engine wire harness and connector to distributor.

14. Connect the spark plug wires. Reconnect the negative battery cable.
15. Set timing, using a timing light, and then torque the hold-down bolts to 16 ft. lbs. (22 Nm).

2.5L (G25A1, G25A4) Engines

NOTE: The radio contains a coded anti-theft circuit. Make sure you have the security code number before disconnecting the battery.

1. Disconnect the negative battery cable.
2. Disconnect the spark plug and coil wires from the distributor cap and mark their positions.
3. Remove the distributor mounting bolts. Remove the distributor from the cylinder head.
   To install:
4. Install a new O-ring on the distributor housing. Coat the O-ring with engine oil before installation.
5. Install the distributor into position, verifying that the lugs on the distributor shaft end fit into the grooves on the camshaft end.
6. Install the mounting bolts. For Vigor, torque the mounting bolts to 16 ft. lbs. (22 Nm). For 2.5TL, torque to 13 ft. lbs. (18 Nm).
7. Connect the spark plug and coil wires. Connect the negative battery cable.
8. Check the ignition timing with a timing light. The timing marks are located on the crankshaft pulley and lower timing cover.

Ignition Timing

ADJUSTMENT

1.7L and 1.8L (B17A1, B18A1, B18B1, B18C1) Engines

1. If equipped with an automatic transaxle, place the shifter in Park. If equipped with a manual transaxle place the shifter in Neutral. Set the parking brake and block the drive wheels.
2. Start the engine and hold the engine speed at 3000 rpm, until the radiator fan comes on. The engine will be at normal operating temperature. Make sure all electrical systems (radio, air conditioning, lights, etc.,) are shut OFF.
3. Pull the service check connector located behind the right kick panel. Connect the BRN/WHT and BLK terminals with the SCS service connector or equivalent device.

4. Connect a timing light to No. 1 ignition wire and point the light toward the pointer on the timing belt cover.

5. The red mark on the crankshaft pulley should be aligned with the pointer no the timing belt cover. The ignition timing specification should be: 16° ±2° BTDC (red mark) at 700–800 rpm.

NOTE: The red mark on the crank pulley is 16° BTDC.

6. Adjust the ignition timing by: loosening the distributor mounting bolts, turn the distributor housing counterclockwise to advance the timing, turn the distributor housing clockwise to retard the timing.

7. Torque the distributor bolts to 17 ft. lbs. (24 Nm) and recheck the timing.

8. Remove the SCS service connector from the service check connector.

2.5L (G25A1, G25A4) Engines

Vigor

— CAUTION —

All Supplemental Restraint System (SRS, air bag system) electrical wiring harnesses are covered with yellow outer insulation for easy identification. To avoid the possibility of personal injury, if it is necessary to disconnect the SRS harness, install the shorting connector on the air bag, then disconnect the wire harness. Replace the entire affected SRS harness assembly if there is an open circuit or damage to the wiring.

1. Start the engine and allow it to warm up until the cooling fan comes on.

2. Pull out the service connector, located under the middle of the dash. Connect the WHT/GRN and BRN terminals with a jumper wire.

3. Check the idle speed and adjust as required.

4. Connect the timing light to the No. 1 plug wire; while engine idles, point the light toward the pointer on the timing belt cover.

5. Inspect ignition timing at idle. Timing should be 15 ± 2 degrees BTDC, using the RED timing mark, at 700±50 rpm in N.

6. If timing adjustment is necessary, complete the following:

a. Remove the control box upper cover and the ignition timing adjuster from control box.

b. Drill 2 rivets off with a ⅛ in. drill bit, then separate the stay cover from the adjuster.

— CAUTION —

To avoid the possibility of personal injury, always wear eye protection when drilling. Do not damage the adjuster when removing rivets.

c. Adjust the timing as necessary by turning the adjusting screw on the adjuster; turn the adjusting screw counterclockwise to retard timing, or clockwise to advance timing.

d. After adjustment, reinstall stay cover to the ignition timing adjuster with new rivets, then reinstall adjuster to control box. Reinstall the control box cover.

2.5TL

— CAUTION —

All Supplemental Restraint System (SRS, air bag system) electrical wiring harnesses are covered with yellow outer insulation for easy identification. To avoid the possibility of personal injury, if any SRS component or wiring harness must be disconnected, the air bags must first be disabled. Replace the entire affected SRS harness assembly if there is an open circuit or damage to the wiring.

NOTE: This vehicle's ignition timing is not adjustable. The timing is controlled by the PCM.

1. Start the engine and allow it to idle at 3000 rpm with all electrical accessories off and the transmission in N or P. Allow the engine to warm up until the cooling fan comes on.

2. Pull the service check connector out from under the glove box. Connect the WHT/BLU and ORN terminals with a service connector, No. 07PAZ–0010100.

3. Check the idle speed and adjust if necessary:

a. Connect a test tachometer to the test tachometer connector on the right side of the engine compartment to the rear of the right shock tower.

b. Idle speed must be 700 ± 50 rpm with the transmission in N or P and all electrical accessories off.

4. Connect a timing light to the No. 1 plug wire. While engine idles, point the light toward the pointer on the timing belt cover.

5. Inspect ignition timing at idle. Timing should be 15 ± 2 degrees BTDC, using the red timing mark, at 700 ± 50 rpm in N or P.

6. If the ignition timing is incorrect, it cannot be adjusted. The PCM must be replaced.

NOTE: All mechanical and electrical systems should checked for proper operation before replacing the PCM. Only replace the PCM as a last resort.

7. Remove the timing light.

8. Remove the service connector. Reconnect the check connector, and tuck it under the glove box.

3.2L (C32A1) Engine

Legend

1. Start the engine and allow it to warm up until the cooling fan comes on.

2. Pull up the upper right edge of the front passenger's floor carpeting. Pull the 2–P service check connector out from under the right side of the dash.

3. Connect the WHT and BLK terminals with a jumper wire.

4. Connect an inductive timing light to the service loop located on the right shock tower. Point light toward the pointer on timing belt cover while the engine is idling.

5. Check the idle speed by connecting a test tachometer to the test connector on the right shock tower. Adjust the idle speed if necessary.

6. Ignition timing should be 15 ± 2 degrees BTDC, using the RED timing mark, at 600±50 rpm in N or P for vehicles with automatic transmissions or 650 ± 50 rpm in N for vehicles with manual transmissions.

7. If timing adjustment is necessary, follow this procedure:

a. Remove the control box cover from the firewall. Be careful not to damage the vacuum hoses when removing the cover.

b. Drill the two cover rivets off with a ⅛ in. drill bit. Then, remove the cover from the adjuster.

— CAUTION —

To avoid personal injury, always wear eye protection when drilling. Do not damage the adjuster body when removing rivets.

c. Turn the adjusting screw counterclockwise to retard the timing, or clockwise to advance the timing.

d. After the adjustment, install the cover to the ignition timing adjuster with new rivets.

e. Reinstall the control box cover.

f. Remove the jumper wire from the service check connector.

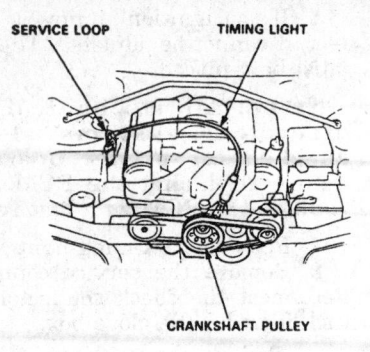

Ignition timing check — Legend

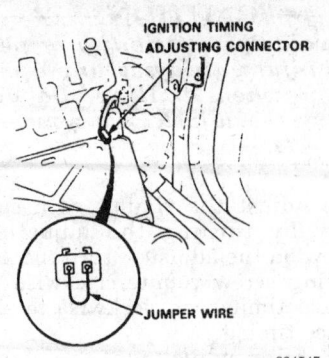

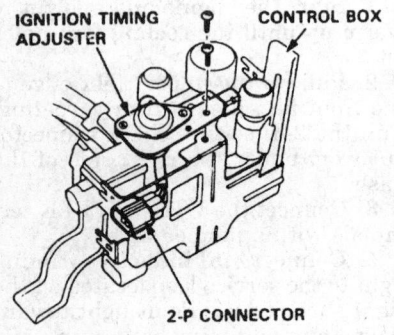

Control box and timing adjuster — Legend

8. Recheck the ignition timing and remove the timing light.

3.2TL

This procedure is for inspection only. The ignition timing is not adjustable on these vehicles.

1. Start the engine and hold the engine at 3000 rpm with no load (shift lever in **P** or **N**) until the radiator fan comes on, then let the engine idle.

2. Locate the service connector under the glove box. Connect the GRN/BLU and ORN wire terminals with the special tool (SCS service connector Part # 07PAZ–0010100).

3. Check the idle speed, the engine should idle at 640 ±50 rpm in park or neutral.

4. Connect a timing light to the service loop; with the engine idling, point the light toward the pointer on the timing belt cover.

5. Inspect the timing: the timing should be 15±2° (indicated by the red mark on the pulley) at 640±50 rpm (engine at idle with no load).

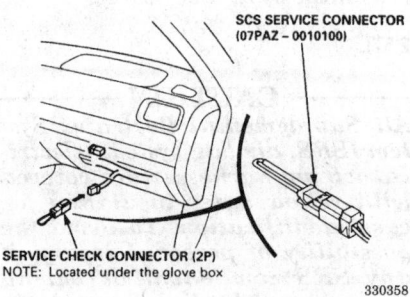

Service check connector — 3.2TL

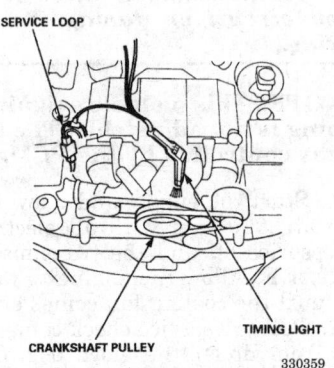

Timing light attachment — 3.2TL

6. If the ignition timing is incorrect, replace the PCM (there is no way to adjust the ignition timing).

NOTE: All mechanical and electrical systems should checked for proper operation before replacing the PCM. Only replace the PCM as a last resort.

7. Remove the timing light.
8. Disconnect the special tool (SCS service connector) from the service check connector.

Alternator

PRECAUTIONS

Several precautions must be observed with alternator equipped vehicles to avoid damage to the unit.

• If the battery is removed for any reason, make sure it is reconnected with the correct polarity. Reversing the battery connections may result in damage to the 1-way rectifiers.

• When utilizing a booster battery as a starting aid, always connect the positive to positive terminals and the negative terminal from the booster battery to a good engine ground on the vehicle being started.

• Never use a fast charger as a booster to start vehicles.

• Disconnect the battery cables when charging the battery with a fast charger.

• Never attempt to polarize the alternator.

• Do not use test lights of more than 12 volts when checking diode continuity.

• Do not short across or ground any of the alternator terminals.

• The polarity of the battery, alternator and regulator must be matched and considered before making any electrical connections within the system.

• Never separate the alternator on an open circuit. Make sure all connections within the circuit are clean and tight.

• Disconnect the battery ground terminal when performing any service on electrical components.

• Disconnect the battery if arc welding is to be done on the vehicle.

REMOVAL AND INSTALLATION

NOTE: The radio may have a coded theft protection circuit. Obtain the code from the owner before disconnecting the battery, removing the radio fuse, or removing the radio.

1.7L and 1.8L (B17A1, B18A1, B18B1, B18C1) Engines

1993 Integra

1. It is necessary to remove the halfshaft to remove the alternator. Disconnect the negative battery cable.

2. Raise the locking tab on the left front spindle nut and loosen the nut with a socket.

3. Raise and safely support the vehicle and remove the left front wheel.

4. Remove the damper fork nut and damper pinch bolt. Remove the damper fork.

5. Remove the knuckle-to-lower arm castle nut and separate the lower ball joint using a suitable puller with the pawls applied to the lower arm.

6. Pull the knuckle outward and remove the halfshaft outboard CV-joint from the knuckle using a plastic mallet.

7. Carefully pry the inner CV-joint out of the intermediate shaft and remove the halfshaft.

NOTE: Do not pull on the driveshaft, as the CV-joint may come apart. Use care when prying out the assembly and pull it straight to avoid damaging the intermediate shaft seals.

8. Disconnect and tag the alternator wire connection from the alternator. Remove the terminal nut and the white wire from the **B** terminal.

9. Loosen the adjusting nut and remove the alternator nut. Remove the alternator belt from the alternator pulley. Remove the lower through-bolt and raise the alternator.

10. Remove the 3 mounting bracket bolts and mounting brackets. Remove the adjusting nut and upper through-bolt, pull out the alternator.

To install:

11. Position the alternator and install the through-bolt and adjusting nut. Install the mounting brackets and mounting bracket bolts and torque to 33 ft. lbs. (45 Nm).

12. Lower the alternator and install the lower through-bolt. Install the alternator belt and adjusting nut. Tension the belt and torque the adjusting nut to 17 ft. lbs. (24 Nm).

13. Reconnect all the wiring.

14. Use a new set ring on the end of the inner CV-joint and slide it into the intermediate shaft. Use the plastic mallet to tap in on the halfshaft and set the ring.

15. Slide the outer CV-joint into place and position the ball joint in the lower arm. Install the nut and torque to 40 ft. lbs. (55 Nm) and tighten as necessary to install a new cotter pin.

16. Install the damper fork, pinch bolt and fork nut. Torque the pinch bolt to 32 ft. lbs. (44 Nm) and the fork nut to 47 ft. lbs. (65 Nm).

17. Install a new self-locking spindle nut and wheel and place the vehicle on the ground before torquing the spindle nut. Attempting to torque the spindle nut while the vehicle is on jack stands or a lift may cause the vehicle to fall.

18. With the vehicle on the ground, torque the spindle nut to 134 ft. lbs. (185 Nm).

19. Reconnect the negative battery cable.

1994–97 Integra

1. Disconnect both the negative and positive battery cables.

2. Label and disconnect the alternator connector. Disconnect the white wire from the B terminal on the rear of the alternator.

3. Loosen the adjusting nut, then remove the mounting nut.

4. Remove the belt from the pulley.

5. Remove the lower mount bolt. Lift the alternator upwards.

6. Remove the bolts from the upper and lower mounting brackets, Remove the brackets.

7. Remove the upper mounting bolt and adjusting nut; remove the alternator.

To install:

8. Position the alternator into the upper bracket and install the alternator adjusting nut and lockbolt.

9. Position the alternator into the vehicle and install the lower and upper mounting brackets. Torque the lower and upper bracket mounting bolts to 33 ft. lbs. (44 Nm).

10. Install the lower mounting bolt.

11. Position alternator belt into pulley. Adjust the alternator belt tension and tighten the adjusting bolt. Tighten the locknut on the belt adjuster. Tighten the lower mounting nut to 33 ft. lbs. (44 Nm).

12. Reconnect the wires to the rear of the alternator.

13. Reconnect the positive and negative cables to the battery and enter the radio security code.

2.5L (G25A1, G25A4) Engines

1. Disconnect both the negative and positive battery cables.

2. Disconnect the four-pin alternator connector. Disconnect the black wire from the B terminal on the rear of the alternator.

3. Loosen the adjusting lockbolt and the adjusting nut. remove the belt from the pulley.

4. Remove the alternator mounting bolts. Remove the alternator.

To install:

5. Position the alternator into the brackets and install the alternator mounting bolt, then install the adjusting nut and lockbolt.

6. Position alternator belt onto the pulley. Adjust the alternator belt tension and tighten the adjusting bolts. Tighten the locknut on the belt adjuster.

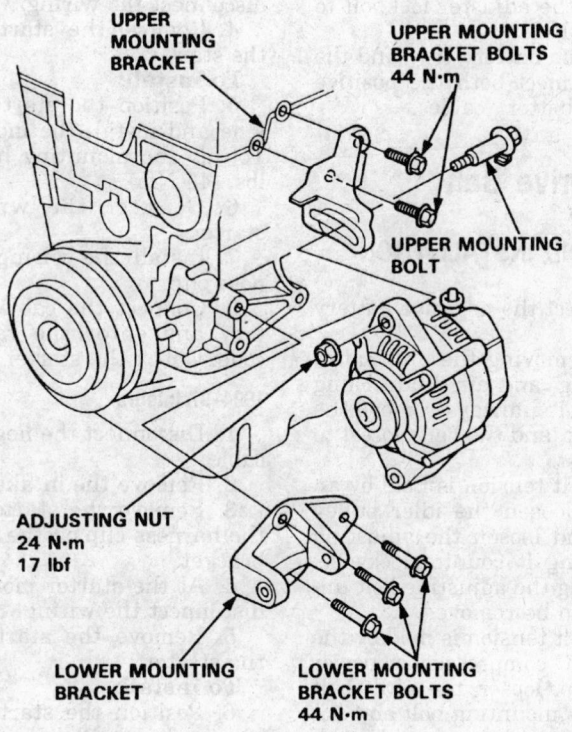

UPPER MOUNTING BRACKET

UPPER MOUNTING BRACKET BOLTS 44 N·m

UPPER MOUNTING BOLT

ADJUSTING NUT 24 N·m 17 lbf

LOWER MOUNTING BRACKET

LOWER MOUNTING BRACKET BOLTS 44 N·m

314519

Alternator mounting — 1994–97 Integra

7. Torque the mounting bolt to 33 ft. lbs. (45 Nm).

8. Torque the adjustment lockbolt to 16 ft. lbs. (22 Nm) for Vigor models, or 18 ft. lbs. (25 Nm) for 2.5TL models. Torque the adjusting nut to 18 ft. lbs. (25 Nm) for Vigor models, or 6 ft. lbs. (8 Nm) for 2.5TL models.

9. Reconnect the wires to the rear of the alternator. Install alternator harness cover.

10. Reconnect the positive and negative cables to the battery.

3.2L (C32A1) Engine

1. Disconnect both battery cables. If necessary, remove the battery and battery tray.

2. Remove the adjusting lockbolt and the lower mounting bolt and slip the alternator belt off the pulley.

3. Rotate the alternator counterclockwise far enough to release it from the mounting bracket; then pull the alternator straight forward.

4. Tilt the alternator and disconnect the wiring harnesses. Remove the alternator.

To install:

5. Reconnect the wiring to the alternator terminals; then install the alternator along with the alternator mounting bolt and adjusting nut. Position the belt onto the pulley.

6. Adjust the belt tension. Torque the lower mounting bolt to 33 ft. lbs. (45 Nm) and the adjuster lockbolt to 16 ft. lbs. (22 Nm).

7. Install the battery tray and the battery. Reconnect both the positive and negative battery cables.

Drive Belt

REMOVAL AND INSTALLATION

1. Disconnect the negative battery cable.

2. When removing the alternator, power steering, and air conditioning compressor belts, always remove the outer belt first, and the belt closest to the engine last.

3. If the belt tension is held by an idler pulley, loosen the idler pulley center nut and loosen the adjusting bolt by turning it counterclockwise. Keep loosening the adjusting bolt until the belt can be removed.

4. If the belt tension is held by the alternator, A/C compressor, or power steering pump, loosen the pivot bolt first, then the mounting bolt and adjuster bolt. Keep loosening the adjuster bolt until the belt can be removed.

To install:

5. Install the new belts in the order of the belt closest to the engine first, and the outer belt on last.

6. Make sure that each belt is seated in each groove, on each pulley properly.

7. Adjust the tension of each belt. Tighten the mounting bolts and pivot bolts. Tighten the idler pulley center nut, if equipped.

8. Reconnect the negative battery cable. Before starting the engine double check to make sure that the belts were installed properly.

Starter

REMOVAL AND INSTALLATION

NOTE: The radio may have a coded theft protection circuit. Obtain the code from the owner before disconnecting the battery, removing the radio fuse, or removing the radio.

1.7L (B17A1), 1.8L (B18A1, B18B1, B18C1) Engines

1993 Integra

1. Disconnect the negative battery cable.

2. Remove the starter wiring from the harness clip.

3. At the starter motor, label and disconnect the wiring.

4. Remove the starter bolts and the starter.

To install:

5. Position the starter to the engine and install the mounting bolts. Torque the mounting bolts to 32 ft. lbs. (45 Nm).

6. Connect the wiring to the starter.

7. Install the wiring to the harness clip.

8. Connect the cables to the battery and enter the radio security code. Check the starter operation.

1994–97 Integra

1. Disconnect the negative battery cable.

2. Remove the intake air duct.

3. Remove the starter cable from the harness clip on the starter motor bracket.

4. At the starter motor, label and disconnect the wiring.

5. Remove the starter bolts and the starter.

To install:

6. Position the starter to the engine and install the mounting bolts. Torque the mounting bolts to 33 ft. lbs. (44Nm).

7. Connect the wiring to the starter.

8. Install the intake air duct.

9. Connect the cables to the battery and enter the radio security code. Check the starter operation.

2.5L (G25A1, G25A4) Engines

1. Disconnect the negative battery cable.

2. Remove the intake manifold rear bracket.

3. Disconnect the cable from the B terminal on the solenoid.

4. Disconnect the black/white wire from the solenoid terminal connector.

5. Remove the starter upper and lower mounting bolts. Remove the starter.

To install:

6. Position the starter onto the engine and torque the upper bolts to 32–33 ft. lbs. (44–45 Nm). Torque the lower bolts on Vigor to 54 ft. lbs. (75 Nm), or 2.5TL to 47 ft. lbs. (64 Nm).

7. Connect the wiring to the starter. Torque the B terminal nut to 7.2 ft. lbs. (10 Nm).

8. Install the intake manifold bracket and tighten it to 16 ft. lbs. (22 Nm).

9. Connect the cables to the battery. Check the operation of the starter.

3.2L (C32A1) Engine

Legend

1. Disconnect the negative battery cable.

2. Remove the starter cable from the starter mounting bracket.

3. Label and disconnect the starter motor wiring.

4. Remove the starter bolts and the starter.

To install:

5. Position the starter on the engine. Tighten the upper and lower mounting bolts to 54 ft. lbs. (75 Nm), and the lower bolts are tightened to 33 ft. lbs. (45 Nm).

6. Install the starter mounting bracket and reattach the starter cable.

7. Connect the black/white wire and the starter cable to the starter terminals.

8. Connect the cables to the battery. Check the starter operation.

3.2TL

1. Disconnect the negative then the positive battery cables.

2. Raise and safely support the vehicle.

3. Remove the left halfshaft from the vehicle.

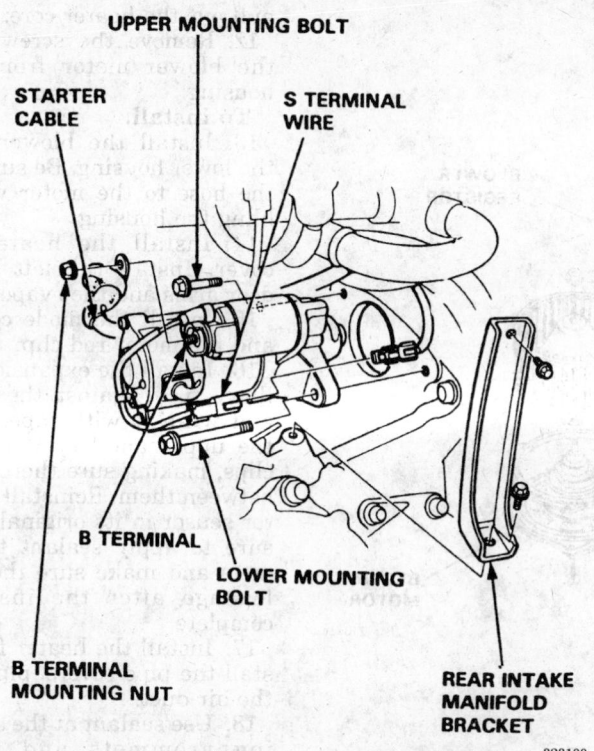

UPPER MOUNTING BOLT

STARTER CABLE

S TERMINAL WIRE

B TERMINAL

LOWER MOUNTING BOLT

B TERMINAL MOUNTING NUT

REAR INTAKE MANIFOLD BRACKET

323100

Starter mounting — Vigor and 2.5TL

4. Remove the heat shields from exhaust pipe A.

5. Remove the bolts attaching the catalytic converter to exhaust pipe A. Discard the nuts and gasket.

6. Remove the nuts attaching exhaust pipe A to the exhaust manifolds. Remove exhaust pipe A from the vehicle then discard the gaskets and attaching nuts.

7. Disconnect the black/white wire and the cable from the terminal on the starter solenoid.

8. Remove the two starter mounting bolts then remove the starter.

To install:

9. Position the starter in the vehicle and install the mounting bolts. Torque the mounting bolts to 32 ft. lbs. (44 Nm).

10. Connect the cable to the solenoid with the crimped side facing out. Connect the black/white wire to the solenoid.

11. Install exhaust pipe A into the vehicle with new gaskets and new attaching nuts. Torque the nuts attaching exhaust pipe A to the manifolds to 40 ft. lbs. (54 Nm). Torque the nuts attaching exhaust pipe A to the catalytic converter to 16 ft. lbs. (22 Nm).

12. Install the heat shields to exhaust pipe A and torque the attaching nuts to 8.7 ft. lbs. (12 Nm).

13. Install the left halfshaft into the vehicle.

14. Lower the vehicle.

15. Connect the positive then the negative battery cables and enter the radio security code.

CHASSIS ELECTRICAL

Blower Motor

REMOVAL AND INSTALLATION

NOTE: The radio may have a coded theft protection circuit. Obtain the code from the owner before disconnecting the battery, removing the radio fuse, or removing the radio.

1993 Integra

1. Disconnect the negative battery cable.

2. Properly discharge the refrigerant from the air conditioning system into recovery equipment. Disconnect the receiver line and suction hose

from the evaporator assembly. Be sure to cap the open fitting to prevent moisture from entering the system.

3. Remove the passenger side lower dashboard cover.

4. Remove the glove box assembly.

5. Remove the front console assembly.

6. Remove the passenger side knee bolster panel, located under the glove box frame.

7. Remove the 2 self tapping screws and the air conditioning bands from around the evaporator assembly.

8. Disconnect the wire connector from the thermostat switch and pull off the wire harness from the clamps. Remove the evaporator.

9. Remove the self tapping screws and remove the heater duct assembly.

10. Remove the heater blower motor mounting bolts.

11. Disconnect the electrical connectors from the blower motor, resistor and recirculation control motor. Remove the blower motor from the blower motor housing.

To install:

12. Install the blower motor, resistor and recirculation control motor. Torque the bolts to 7 ft. lbs. (10 Nm)

13. Install the blower motor housing into the car.

14. When reattaching the actuator, make sure its positioning will not allow the air door to be pulled too far.

15. Attach the actuator and all linkage, then apply battery voltage and watch the door movement. If necessary, loosen the holding screw and move the actuator up or down.

16. When adjusting the control rod, connect the recirculation control motor connection to the main wire harness, push **RECIRC** and open the air doors. Then connect the control rod to the arm while holding the air doors open.

17. Install the heater duct, or A/C band, if equipped.

18. Install automatic shoulder seat belt control unit. Make sure that all wiring connections are reconnected

19. Install the knee bolster, kick panel, glove box, and lower dashboard cover.

20. Recharge the air conditioning system.

21. After installation, make sure the recirculation control motor operates smoothly.

1994–97 Integra

1. Disconnect the negative battery cable.

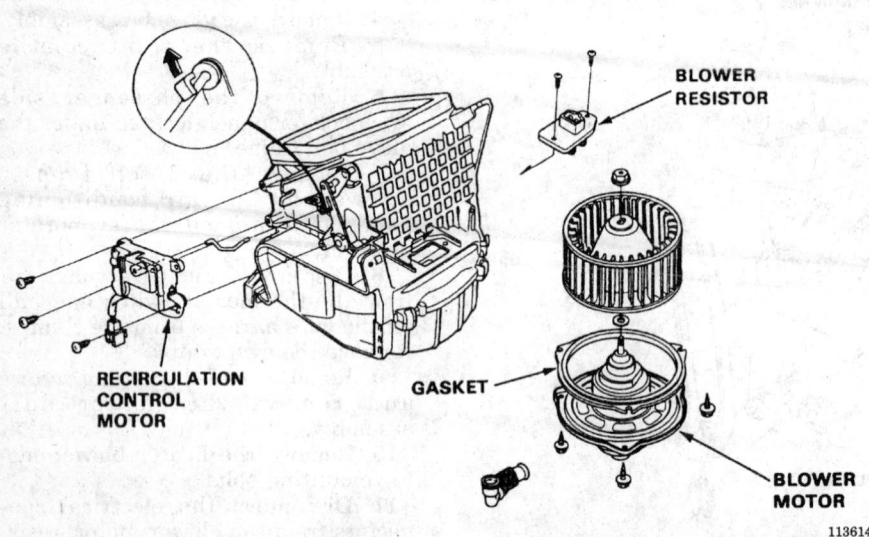

Blower motor — 1993 Integra

113614

2. Working under the passenger side of the dash, disconnect the electrical connections from the blower motor.

3. Remove the three screws attaching the blower motor to the blower unit and remove the blower motor.

To install:

4. Position the blower motor in the blower unit and install the three attaching screws.

5. Connect the electrical connections to the blower motor.

6. Connect the negative battery cable to the battery and enter the radio security code.

7. Turn the ignition **ON** and test the blower motor at all speeds.

Vigor

1. Disconnect the negative battery cable.

2. Remove the dashboard.

3. Drain the coolant from the system and recover the refrigerant from the air conditioning system. The wiring harness for the supplemental restraint system is routed near the heater–evaporator, so use caution.

--- **CAUTION** ---

The Supplemental Restraint System (SRS) must be disarmed before removing the blower motor. Failure to do so may cause accidental deployment of the air bag, resulting in unnecessary system repairs and/or personal injury.

4. Disconnect the heater hoses from the heater at the firewall. Catch the coolant that will run out.

5. Remove the bolts holding the air conditioner lines to the evaporator at the firewall. Remove the mounting nuts at the firewall.

6. Remove the duct and connector from the assembly. Remove the mounting nuts and the heater/evaporator assembly.

7. Remove the duct, pipe covers and pipe clamp.

8. Remove the blower motor relay.

9. Disconnect the clip from the mode control motor rod. Remove the mounting screws and the mode control motor.

10. Remove the clips and the screws to separate the housing covers. Remove the evaporator from the case and remove the expansion valve.

11. Remove the evaporator sensor, mode door arms and the left cover.

Remove the heater core cover and pull out the heater core.

12. Remove the screws to remove the blower motor from the lower housing.

To install:

13. Install the blower motor into the lower housing. Be sure to connect the hose to the motor when assembling the housing.

14. Install the heater core and cover. Install the left cover, mode door arms and the evaporator sensor.

15. Install the mode control motor and the motor rod clip.

16. Install the expansion valve capillary tube against the suction line and wrap it with tape. Reassemble the upper and lower housings with clips, making sure there are no gaps between them. Reinstall the evaporator sensor in its original position. Be sure to apply sealant to the grommets and make sure there is no air leakage after the installation is complete.

17. Install the heater fan relay. Install the pipe covers, pipe clamp and the air duct.

18. Use sealant at the air conditioning grommets and install the heater/evaporator assembly. Torque the mounting bolts to 7 ft. lbs. (10 Nm).

19. Install the firewall located mounting nut and torque to 16 ft. lbs. (22 Nm).

20. Install the air conditioner lines and torque the mounting nut to 16 ft. lbs. (22 Nm).

21. Install the heater hoses and control valve. Install the dashboard. Fill and bleed the cooling system. Recharge the air conditioning system.

22. Reconnect the negative battery cable.

2.5TL

--- **CAUTION** ---

The Supplemental Restraint System (SRS, air bag system) wiring harness is routed around the heater unit, evaporator, and blower motor. The harness is wrapped in yellow insulation. Be careful not to damage the insulation or SRS wiring. If any SRS component must be disconnected or serviced, the system must first be disarmed.

1. Disconnect the negative and positive battery cables.

2. Evacuate and recover the air conditioning refrigerant using a R134a certified recovery system.

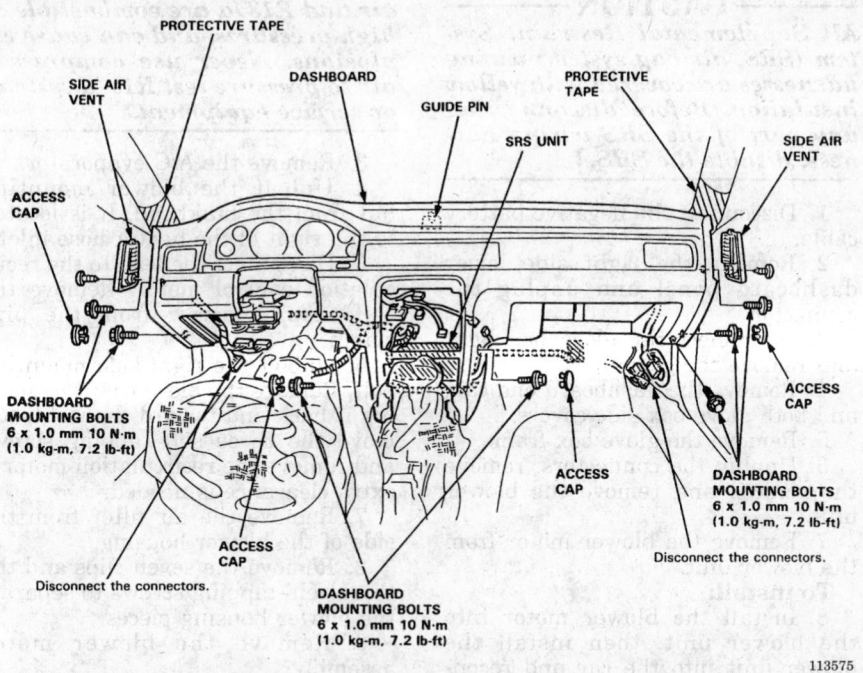

Dashboard removal — Vigor

113575

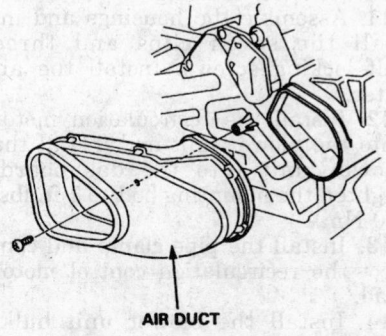

Air duct — Vigor

113580

— CAUTION —

Exposure to air conditioning refrigerant and lubricant can irritate your eyes, nose, and throat. Do not breathe refrigerant or lubricant vapor or mist. Mixtures of air and R134a are combustible at high pressures, and can cause explosions. Never use compressed air to pressure test R134a systems or service equipment.

3. Remove the A/C evaporator.
4. Unbolt the blower mounting nut from the bulkhead. It is located to the right of the heater hose inlets.

5. Disconnect the lead to the recirculation control motor. Remove the self–tapping screw from the pipe clamp.

6. Unbolt the right-side mounting bolt. Remove the blower unit by moving it back and toward the right. Remove the three self–tapping screws and remove the recirculation motor if extra clearance is needed.

7. Remove the air filter from the side of the blower housing.

8. Remove the seven clips and the three self–tapping screws to separate the blower housing pieces.

9. Remove the blower motor assembly.

To install:

10. Install the blower motor into the lower housing. Make sure that the blower motor air hose is fitted into the lower housing.

11. Assemble the housings and install the seven clips and three self–locking screws. Install the air filter.

12. Install the recirculation motor onto the blower unit and install the blower unit into the dashboard. Tighten the mounting bolt to 7 ft. lbs. (10 Nm).

13. Install the pipe clamp and connect the recirculation control motor lead.

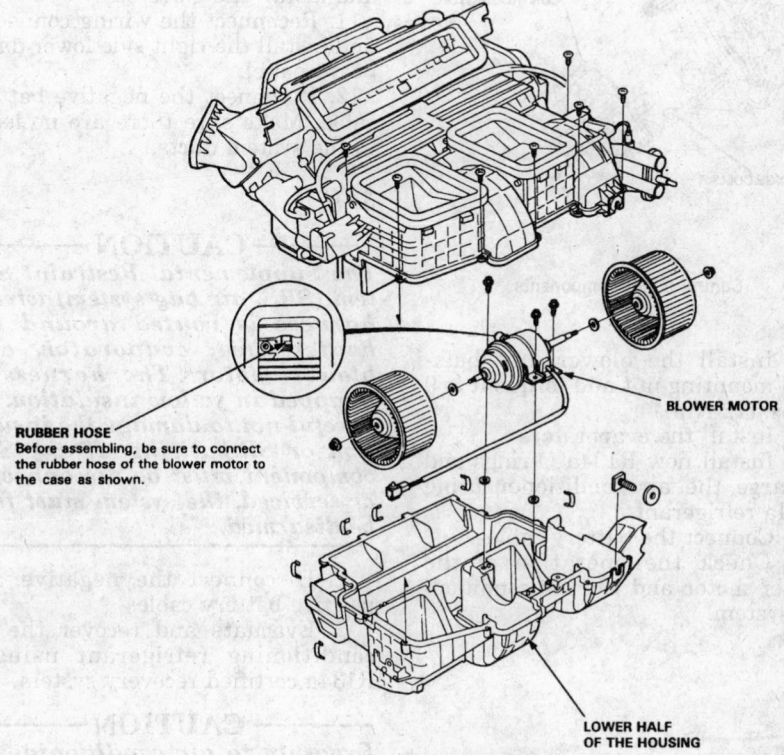

Blower motor — Vigor

113588

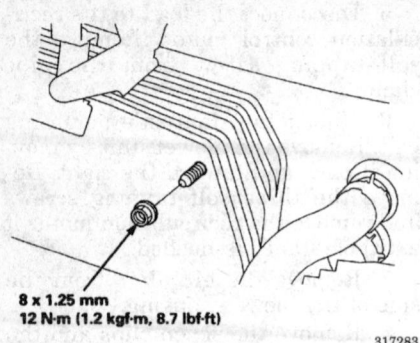

8 x 1.25 mm
12 N·m (1.2 kgf·m, 8.7 lbf·ft)

317283

Bulkhead mounting nut

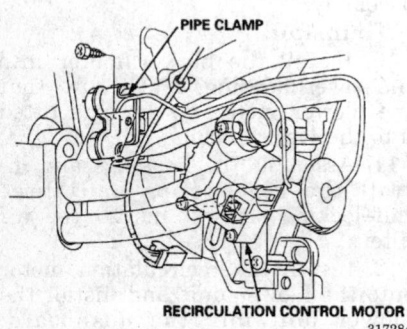

PIPE CLAMP

RECIRCULATION CONTROL MOTOR

317284

Recirculation control motor

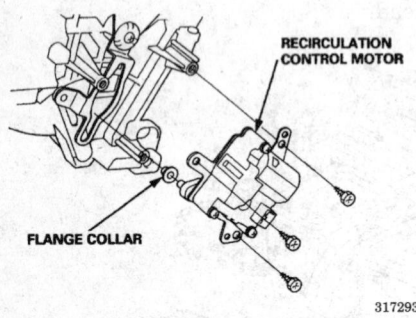

RECIRCULATION CONTROL MOTOR

FLANGE COLLAR

317293

Control motor components

14. Install the blower unit bulkhead mounting nut and torque it to 9 ft. lbs. (12 Nm).
15. Install the evaporator.
16. Install new R134a O-rings and recharge the air conditioner using R134a refrigerant.
17. Connect the battery cables.
18. Check the operation of the blower motor and the air conditioning system.

Legend

------ CAUTION ------
All Supplemental Restraint System (SRS, air bag system) wiring harnesses are covered with yellow insulation. Before disconnecting any part of the SRS wiring harness, disable the SRS.

1. Disconnect the negative battery cable.
2. Remove the right side lower dashboard panel and unplug the connector.
3. Disconnect the glove box light and remove the glove box.
4. Remove the dashboard end cap, and both glove box side covers.
5. Remove the glove box frame.
6. Unplug the connectors, remove the screws, and remove the blower unit.
7. Remove the blower motor from the blower unit.
To install:
8. Install the blower motor into the blower unit, then install the blower unit into the car and reconnect the wiring connectors. Torque the mounting bolts to 7 ft. lbs. (10 Nm).
9. Install the glove box frame, the glove box side covers, and the dashboard end caps.
10. Reconnect the glove box light and install the glove box.
11. Reconnect the wiring connector and install the right side lower dashboard panel.
12. Reconnect the negative battery cable. Make sure there are no leaks in the system ducts.

3.2TL

------ CAUTION ------
The Supplemental Restraint System (SRS, air bag system) wiring harness is routed around the heater unit, evaporator, and blower motor. The harness is wrapped in yellow insulation. Be careful not to damage the insulation or SRS wiring. If any SRS component must be disconnected or serviced, the system must first be disarmed.

1. Disconnect the negative and positive battery cables.
2. Evacuate and recover the air conditioning refrigerant using a R134a certified recovery system.

------ CAUTION ------
Exposure to air conditioning refrigerant and lubricant can irritate your eyes, nose, and throat.

Do not breathe refrigerant or lubricant vapor or mist. Mixtures of air and R134a are combustible at high pressures, and can cause explosions. Never use compressed air to pressure test R134a systems or service equipment.

3. Remove the A/C evaporator.
4. Unbolt the blower mounting nut from the bulkhead. It is located to the right of the heater hose inlets.
5. Disconnect the lead to the recirculation control motor. Remove the self–tapping screw from the pipe clamp.
6. Unbolt the right-side mounting bolt. Remove the blower unit by moving it back and toward the right. Remove the three self–tapping screws and remove the recirculation motor if extra clearance is needed.
7. Remove the air filter from the side of the blower housing.
8. Remove the seven clips and the three self–tapping screws to separate the blower housing pieces.
9. Remove the blower motor assembly.
To install:
10. Install the blower motor into the lower housing. Make sure that the blower motor air hose is fitted into the lower housing.
11. Assemble the housings and install the seven clips and three self–locking screws. Install the air filter.
12. Install the recirculation motor onto the blower unit and install the blower unit into the dashboard. Tighten the mounting bolt to 7 ft. lbs. (10 Nm).
13. Install the pipe clamp and connect the recirculation control motor lead.
14. Install the blower unit bulkhead mounting nut and torque it to 9 ft. lbs. (12 Nm).
15. Install the evaporator.
16. Install new R134a O-rings and recharge the air conditioner using R134a refrigerant.
17. Connect the battery cables.
18. Check the operation of the blower motor and the air conditioning system.

Windshield Wiper Motor

REMOVAL AND INSTALLATION

NOTE: The radio may have a coded theft protection circuit. Obtain the code from the owner before disconnecting the battery, removing the radio fuse, or removing the radio.

1993 Integra

Front

1. Disconnect the negative battery cable.

2. Open the hood, remove the cap nuts, and carefully remove the wiper arms.

3. Remove the cowl cover plate and hood seal by prying out their trim clips and removing the screws.

4. Pry the wiper linkage off the motor arm. Disconnect the electrical connector.

5. Remove the mounting bolts and the wiper motor assembly.

To install:

6. Install the motor. Torque the mounting bolts to 7 ft. lbs. (10 Nm). Connect the electrical plug.

7. Apply grease to the moving joints and press the wiper linkage onto the wiper arm.

8. Install the cowl cover and trim clips.

9. Install the wiper arms and torque the cap nuts to 10 ft. lbs. (14 Nm), then adjust so the tips are 0.8–1.2 in. (20–30mm) above the cowl cover. Connect the negative battery cable.

10. Test the wipers for proper operation.

Rear

1. Remove the hatch trim panel.

2. Remove the nut cover, wiper arm, cap, special nut, special washer and the cushion rubber.

3. Disconnect the 4–pin connector from the wiper motor.

4. Remove the bolts for the motor and remove the wiper motor.

To install:

5. Install the wiper motor.

6. Reconnect the wiring.

7. Install the washers, wiper arm and the nut and cover. Check for water leakage in the rear wiper arm.

8. Install the hatch trim panel.

1994–97 Integra

Front

1. Disconnect the negative cable at the battery.

2. Remove the cap nuts and carefully remove the wiper arms without damaging the hood.

3. Remove the hood seal and air scoop by prying out their trim clips.

4. Disconnect the wiring connector from the wiper motor.

5. Remove the 4 mounting bolts and wiper linkage assembly.

6. Remove the wiper harness from the wiper linkage.

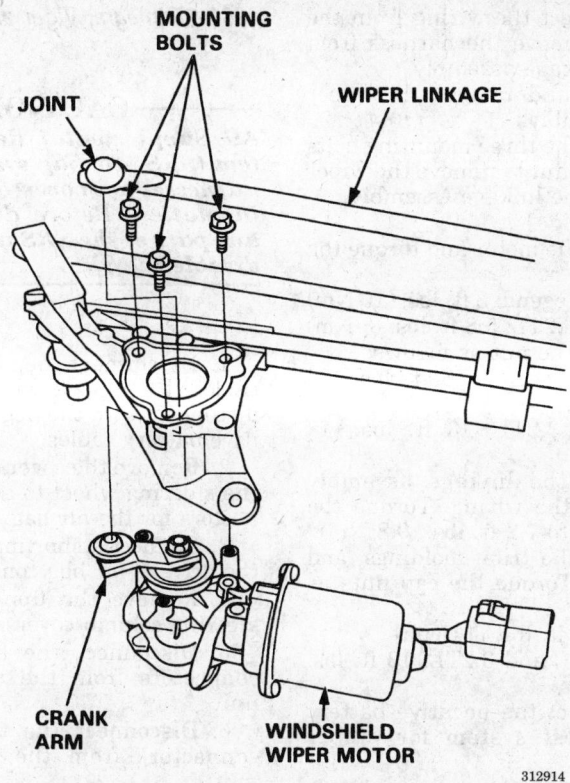

Wiper motor replacement — 1994–97 Integra

7. Separate the wiper linkage and crank arm at the joint, then remove the 3 mounting bolts and wiper motor.

To install:

8. Reconnect the wiper motor to the wiper linkage then mount to the wiper harness. Torque the mounting bolts to 7 ft. lbs. (10 Nm). Lightly coat the linkage joints with grease and make sure the linkage moves smoothly.

9. After connecting the wiper linkage to the wiper harness, install the full wiper linkage assembly and 4 mounting bolts.

10. Reconnect the wiring connector to the wiper motor and install the hood seal and air scoop.

11. Carefully install wiper arms and adjust so the tips are 0.8 to 1.2 in. (20 to 30mm) above the cowl cover, then mount the cap nuts.

12. Reconnect the negative battery cable and enter the radio security code. Test the wiper assembly for proper operation.

Rear

1. Disconnect the negative battery cable.

2. Remove the cover, mounting nut, and wiper arm.

3. Remove the rubber seal, special nut, and the washer.

4. Open the tailgate and remove the tailgate trim panels.

5. Disconnect the 4–pin wire connector from the wiper motor.

6. Remove the three mounting bolts and the wiper motor assembly.

To install:

7. Install the wiper motor assembly and the mounting bolts. Torque the mounting bolts to 7.2 ft. lbs. (9.8 Nm).

8. Reconnect the wiring.

9. Install the tailgate trim panels.

10. Install the rubber seal, special nut, and the washer. Torque the nut to 7 ft. lbs. (9 Nm)

11. Install the wiper arm, mounting nut, and the cover. Torque the nut to 10 ft. lbs. (14 Nm)

12. Connect the negative battery cable and enter the radio security code.

Vigor, 2.5TL, Legend and 3.2TL

1. Disconnect the negative battery cable.

2. Open the hood. Remove the wiper arm cap nuts and carefully remove the wiper arms so that they don't scratch the hood.

3. Remove the hood seal, lower windshield molding, and the air scoops by prying out the trim clips and removing the screws.

4. Disconnect the wiring from the motor and remove the harness from the wiper linkage assembly.

5. Unbolt and remove the wiper linkage assembly.

6. Unbolt the three mounting bolts and the shaft nut to remove the wiper motor from the linkage assembly.

To install:

7. Install the motor and torque the bolts as follows:

Vigor and Legend: 8 ft. lbs. (10 Nm)
2.5TL and 3.2TL: 5.8 ft. lbs. (8 Nm)

8. Torque the nut as follows:

Vigor and Legend: 16 ft. lbs. (22 Nm)
2.5TL and 3.2TL: 13 ft. lbs. (18 Nm).

9. Install the linkage assembly and connect the wiring. Torque the linkage bolts to 7.2 ft. lbs. (9.8 Nm).

10. Install the trim, moldings, and wiper arms. Torque the cap nuts as follows:

Legend: 10 ft. lbs. (14 Nm)
Vigor, 2.5TL and 3.2TL: 13 ft. lbs. (18 Nm)

11. Reconnect the negative battery cable and test system for proper operation.

Combination Switch

REMOVAL AND INSTALLATION

NOTE: The radio may have a coded theft protection circuit. Obtain the code from the owner before disconnecting the battery, removing the radio fuse, or removing the radio.

1993 Integra

1. Disconnect the negative battery cable.

2. Remove the steering wheel.

3. Remove the column covers and disconnect the wiring. If equipped with cruise control, remove the slip ring.

4. Remove the screws and slide the lighting and wiper/washer switch out to the side.

To install:

5. Install the lighting and the wiper/washer switch on the side, install screws.

6. Install slip ring if it was removed. Reconnect the wiring and install column covers.

7. Install the steering wheel. Replace the self locking nut and tighten to 50 Nm (36 ft. lbs.).

8. Reconnect the negative battery cable. Test the combination switch.

1994–97 Integra, Vigor and 2.5TL

—————— **CAUTION** ——————

All Supplemental Restraint System (SRS, air bag system) wiring harnesses are covered with yellow insulation. Before disconnecting any part of the SRS wire harness, disable the SRS.

1. Disconnect negative battery cable. If equipped with an air bag, disconnect both the negative and positive battery cables.

2. Remove the cover on the back of the steering wheel to access the connectors for the air bag.

3. Connect a shorting connector to the driver's air bag connector.

4. Remove the upper and lower steering column covers.

5. Disconnect the two electrical connectors from the wiper control unit.

6. Disconnect the two electrical connectors from the light control unit.

7. Remove the lighting and the wiper switch mounting screws, then remove the switches.

To install:

8. Install the the lighting and wiper switches and install the mounting screws.

9. Connect the electrical connectors to the wiper and light control units.

10. Install the steering column covers.

11. Remove the shorting connector and arm the air bag system.

12. Connect the battery and turn the ignition switch to the **ON** position. The instrument panel air bag light should go on for approximately 8 seconds and then go off.

13. Enter the radio security code.

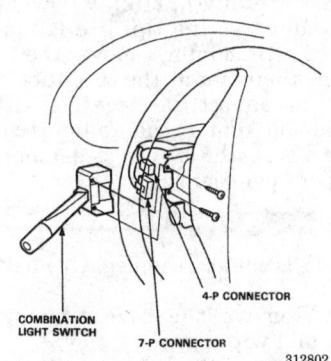

Light switch — 1994–97 Integra

Legend and 3.2TL

—————— **CAUTION** ——————

All Supplemental Restraint System (SRS, air bag system) wiring harnesses are covered with yellow insulation. Before disconnecting any part of the SRS wire harness, disable the SRS.

1. Disconnect negative battery cable. If equipped with an air bag, disconnect both the negative and positive battery cables

2. Remove the cover on the back of the steering wheel to access the connectors for the air bag.

3. Connect a short connector to the driver's air bag connector.

4. Gently pry the courtesy light controller and the TCS switch (if equipped) out of the dash board lower cover.

—————— **WARNING** ——————

On vehicles equipped with traction control, do not remove the steering angle sensor.

5. Disconnect the electrical connectors from the courtesy light controller and the TCS switch (if equipped), then remove the courtesy light controller and the TCS switch.

6. Remove the dashboard lower cover.

7. Remove the steering column tilt cover.

8. Remove the upper and lower steering column covers.

9. Disconnect the electrical connector for the wiper switch.

10. Disconnect the electrical connector for the light switch.

11. Turn the steering wheel toward the switch being removed then remove the mounting screws and the switch.

To install:

12. Install the the lighting and wiper switches and install the mounting screws.

13. Connect the electrical connectors to the wiper and light switches.

14. Install the steering column covers.

15. Install the steering column tilt cover.

16. Install the dash board lower cover.

17. Connect the electrical connectors to the courtesy light controller and the TCS switch (if equipped), then install the courtesy light controller and the TCS switch into the dash board lower cover.

18. Remove the short connector and arm the air bag system.

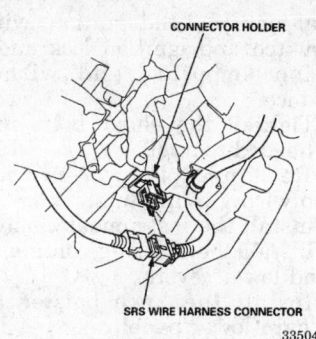

SRS wiring harness and connector holder — 3.2TL

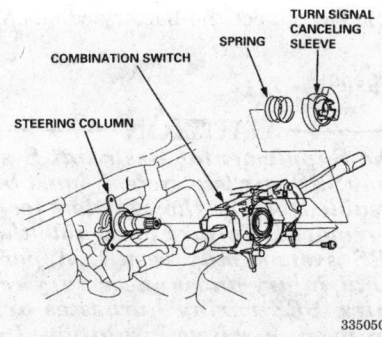

Turn signal sleeve, spring, and combination switch — 3.2TL

19. Connect the battery and turn the ignition switch to the on position. The air bag lamp, on the instrument panel, should illuminate for approximately 8 seconds and then extinguish.

20. Enter the radio security code.

Ignition Lock Cylinder

REMOVAL AND INSTALLATION

NOTE: The radio may have a coded theft protection circuit. Obtain the code from the owner before disconnecting the battery, removing the radio fuse, or removing the radio.

1993 Integra

Electrical Switch Replacement

1. Disconnect the negative battery cable.
2. Remove the steering wheel; remove the steering column covers.
3. Remove lower dashboard panel and knee bolster.
4. Disconnect the 4-pin connector at the fuse/relay box and disconnect the 5-pin connector from the main wiring harness.
5. Insert the key and turn it to **O**.

6. Remove the 2 mounting bolts and remove the switch.
To install:
7. Install the switch and 2 mounting bolts.
8. Reconnect the wiring connector to the main wire harness. Reconnect the wiring connector to the under–dash fuse/relay box.
9. Install the knee bolsters and dashboard panels.
10. Install the steering column covers.
11. Install the steering wheel and tighten the center nut to 36 ft. lbs. (50 Nm).
12. Reconnect the negative battery cable.

Lock Cylinder Replacement

NOTE: This procedure is for vehicles with manual transaxles only. For vehicles with automatic transaxles, the steering lock assembly must be replaced.

1. Disconnect the negative battery cable.
2. Remove the steering wheel; remove the steering column covers.
3. Insert the key and turn it to **I**.
4. Remove the screw from the lock body.
5. Use a small tool to push in the pin on the lock body; remove the lock cylinder from the body.
To install:
6. Turn the key to the **O** position. Align the lock cylinder with the lock body.
7. Turn the key almost to the **I** position; insert the lock cylinder until the pin touches the body.
8. Turn the key fully to the **I** position. Push in the pin and insert the lock cylinder into the body until the pin clicks into place.
9. Install the screw into the lock body.
10. Install the steering wheel; tighten the center nut to 36 ft. lbs. (50 Nm).
11. Install the steering column covers.
12. Connect the negative battery cable.

Steering Lock Replacement

1. Disconnect the negative battery cable.
2. Remove the steering wheel. Remove the steering column covers.
3. Remove lower dashboard panel and knee bolster.
4. Remove the nuts and bolts holding the steering column; lower the column.
5. Center punch the two shear bolts; use a ³/₁₆-inch bit to drill out the

bolts. Take care not to damage the switch when drilling. Remove the bolts from the switch body.
To install:
6. Install the new ignition switch and lock assembly without the key inserted. Install and hand-tighten the new shear bolts.
7. Make certain the projection on the ignition switch is aligned with the hole in the steering column. Adjust the lock assembly as needed.
8. Insert the key; check for proper operation of the steering lock and normal motion of the ignition switch. The switch and ignition lock should function smoothly and without resistance.
9. Tighten the shear bolts until the heads break off.
10. Install the steering column to the lower dash. Tighten the nuts and bolts to 9.5 ft. lbs. (13 Nm).
11. Install the steering wheel.
12. Install the steering column covers.
13. Install the knee bolsters and dashboard panels.
14. Reconnect the negative battery cable.

1994–97 Integra

——— WARNING ———
Always disarm the air bag Supplemental Restraint System (SRS, air bag system) before performing any work on the steering column or dashboard. Air bag wiring is encased in yellow insulation; never disconnect this wiring without disarming the system. Failure to disarm the system may result in unintentional air bag deployment.

1. Disconnect the negative battery cable.
2. Disarm the air bag (SRS) system.
3. Remove lower dashboard panel and knee bolster.
4. Disconnect the 5–pin connector at the fuse/relay box and disconnect the 7–pin connector from the main wiring harness.
5. Remove steering column covers.
6. Remove the nuts and bolts holding the steering column; lower the column.
7. Center punch the two shear bolts then use a ³/₁₆ inch (5mm) bit to drill the heads of the bolts off. Take care not to damage the switch when drilling. Remove the bolts from the switch body.
To install:
8. Install the new ignition switch and lock assembly without the key

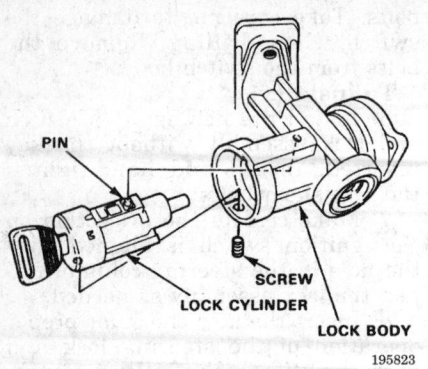

PIN

SCREW

LOCK CYLINDER

LOCK BODY

195823

Key cylinder removed — 1993 Integra

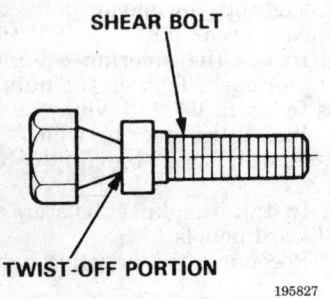

SHEAR BOLT

TWIST-OFF PORTION

195827

New shear bolt — 1993 Integra

inserted. Install and hand–tighten the new shear bolts.

9. Insert the key; check for proper operation of the steering lock and normal motion of the ignition switch. The switch and ignition lock should function smoothly and without resistance.

10. Tighten the shear bolts until the heads break off.

11. Install the steering column to the lower dash. Tighten the nuts to 5 ft. lbs. (13 Nm) and the bolts to 16 ft. lbs. (22 Nm).

12. Reconnect the wiring connector to the main wire harness. Reconnect the wiring connector to the under–dash fuse/relay box.

13. Install the steering column covers.

14. Install the knee bolsters and dashboard panels.

15. Arm the air bag (SRS) system.

16. Reconnect the negative battery cable.

Vigor, 2.5TL and 3.2TL

─────── **WARNING** ───────
Always disarm the Supplemental Restraint System (SRS, air bag system) before performing any work on the steering column or dashboard. Air bag wiring is en-

cased in yellow insulation; never disconnect this wiring without disarming the system. Failure to disarm the system may result in unintentional air bag deployment.

Electrical Switch Replacement

1. Disconnect the negative and positive battery cables.

2. Remove the steering wheel access cover.

3. Disarm the air bag (SRS) system and install a shorting connector.

4. Remove lower dashboard panel and knee bolster.

5. Disconnect the 7-pin connector from the under–dash fuse box.

6. Remove the steering column covers.

7. Insert the key and turn it to **O**.

8. Remove the two mounting bolts to remove the electrical switch.

To install:

9. Install the switch and the two mounting bolts.

10. Reconnect the wiring connector to the under–dash fuse box.

11. Install the steering column covers, knee bolster, and lower dash panel.

12. Arm the air bag (SRS) system by removing the shorting connectors.

13. Install the steering wheel access panel.

14. Reconnect the battery cables.

15. Check the operation of the ignition switch.

Steering Lock Replacement

1. Disconnect the negative and positive battery cables.

2. Remove the steering wheel access panel.

3. Disarm the air bag (SRS) system and install a shorting connector.

4. Remove lower dashboard panel and knee bolster.

5. Disconnect the 7-pin and the 8-pin connectors from the under–dash fuse box.

6. Remove steering column covers.

7. Remove the instrument cluster for easier access to the lock assembly.

8. Center punch the two shear bolts holding the switch body. Use a 3/16 in. (5mm) bit to drill out the bolt heads.

9. Remove the shear bolt from the switch body.

10. Remove the ignition switch.

To install:

11. Install the new ignition switch without the key inserted.

12. Loosely tighten the new shear bolts.

13. Insert the key; check for proper operation of the steering lock and

normal motion of the ignition switch. The switch and ignition lock should function smoothly and without resistance.

14. Tighten the shear bolts until their heads break off.

15. Reconnect the 17–pin and 18–pin wiring connectors.

16. Install the instrument cluster.

17. Install the steering column upper and lower covers.

18. Install the knee bolster and dashboard lower panel.

19. Arm the air bag (SRS) system by removing the shorting connectors.

20. Install the steering wheel access panel.

21. Reconnect the battery cables.

Legend

─────── **CAUTION** ───────
The Supplemental Restraint System (SRS, air bag system) must be disabled before the ignition lock is removed. Failure to disable the SRS system may result in personal injury and unnecessary repairs. SRS wiring harnesses are wrapped in yellow insulation. Do not damage the wiring harness or its insulation.

1. Disconnect the negative and positive battery cables. Wait at least three minutes before working around the air bags.

2. Disable the driver's side air bag:

   a. Remove the lower steering wheel access cover.

   b. Uncouple the air bag connector from the cable reel connector.

   c. Remove the red shorting connector from the access cover and install it onto the air bag connector.

   d. Connect the SRS shorting connector A (tool No. 07MAZ–SP0020A) to the cable reel connector.

3. Carefully pry the dimmer switch controller and the traction control switch out of the driver's side lower dashboard panel. Disconnect the wiring connectors from both switches.

4. Remove the driver's side lower dashboard panel and the knee bolster.

5. Disconnect the 7-pin and the 8-pin connector from the under–dash fuse box.

6. Remove the upper and lower steering column covers.

7. Remove the steering column mounting nuts and bolts and lower the steering column to the floor of the vehicle.

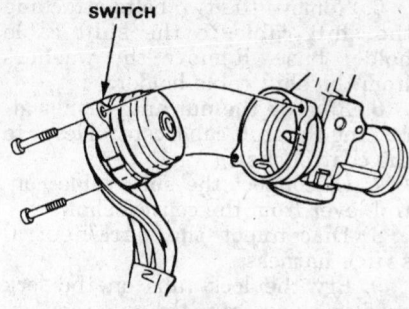

Electrical switch — Integra shown, others similar

--- **WARNING** ---

Wrap a large towel around the steering wheel to protect the air bag. Be careful not to damage the yellow SRS wiring harness when removing the shear bolts.

8. Notch the shear bolt head. Using a suitable tool to tap the bolt counterclockwise until it is loose. Work carefully to avoid damaging the lock body and steering column.

9. Remove the shear bolt from the switch body.

10. Insert the key and turn it to the **I** position.

11. Push in on the lock pin and pull the ignition lock assembly out of the steering column.

To install:

12. Turn the key to the **I** position, push in on the lock pin and insert the ignition lock assembly into the steering column until it clicks into place.

13. Install the knee bolster bracket which is held in place by the shear bolt. Only tighten the shear bolt enough to hold the bracket in place.

14. Insert the key; check for proper operation of the ignition lock. The ignition lock and switch should function smoothly and without resistance.

15. Verify that the ignition lock body will fit properly into the dashboard lower panel.

16. After the ignition lock assembly has been properly installed, tighten the shear bolts until their heads break off.

17. Raise the steering column into position. Verify that no wiring harnesses are pinched or kinked by the steering column or its brackets. Tighten the nuts to 12 ft. lbs. (16 Nm) and the bolts to 16 ft. lbs. (22 Nm).

18. Reconnect the wiring connectors to the under–dash fuse/relay box.

19. Install the upper and lower steering column covers.

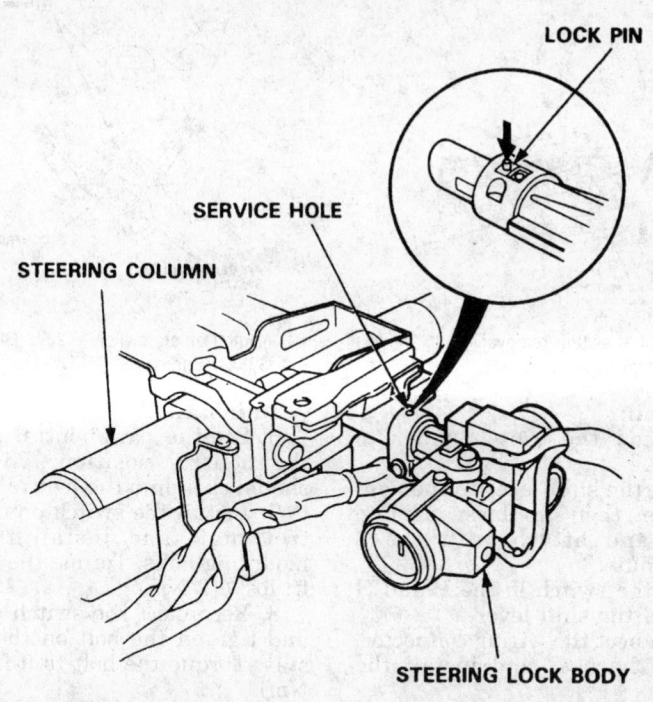

Steering lock pin and service hole — Legend

20. Install the knee bolster and driver's side dashboard lower panel.

21. Install the dimmer switch and the traction control switch.

22. Enable the driver's side air bag:

a. Remove the shorting connector from the air bag harness.

b. Remove the SRS shorting connector A from the cable reel connector.

c. Reconnect the air bag and cable reel connectors.

d. Install the shorting connector back into its holder on the lower access cover. Install the access cover onto the steering wheel.

23. Reconnect the positive and negative battery cables.

24. Turn the ignition switch to the **ON** position and check the operation of the SRS indicator light. The SRS indicator light should come on for six seconds and then turn off. This sequence indicates that the SRS system is enabled and functioning normally. If the indicator light doesn't turn on, or turn off, the system fault must be corrected.

25. Check the operation of the dashboard switches.

Ignition Switch

REMOVAL AND INSTALLATION

Except on 1993 Integra, the ignition switch is integral with the ignition lock cylinder. The removal and installation procedures for these components are found under "Ignition Lock Cylinder."

Park/Neutral Safety Switch

REMOVAL AND INSTALLATION

NOTE: The radio may have a coded theft protection circuit. Obtain the code from the owner before disconnecting the battery, removing the radio fuse, or removing the radio.

Integra and Legend

1. Disconnect negative cable at the battery.

2. Remove the console, then disconnect the wiring connector from the switch.

3. Remove the 2 console switch mounting nuts, then remove the switch.

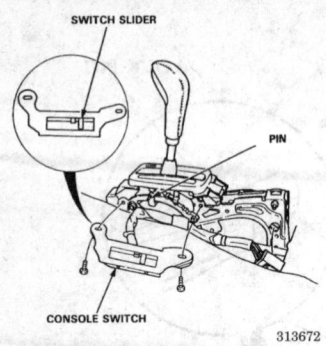

Park/neutral switch removal —
Integra and Legend

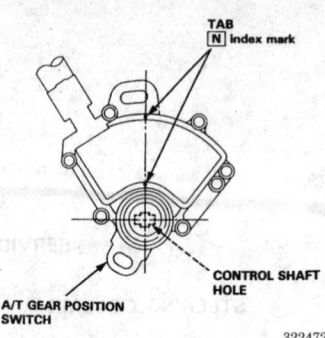

Neutral safety switch — 2.5L (G25A1,
G25A4) Engines

To install:

4. Position the switch slider to neutral.

5. Move the shift lever to the neutral position, then slip the switch into position, and attach with the 2 mounting nuts.

6. Test the switch in the **P** and **N** positions of the shift lever.

7. Reconnect the wiring connector, clamp the harness and install the console.

8. Connect the negative cable at battery and enter the radio security code.

9. Move the shift lever through all the gears, and verify that the gear position indicator follows the gear position switch.

10. Start the engine and check the shift lever through all the gears again.

Vigor and 2.5TL

1. Raise and safely support the vehicle.

2. Shift the select lever to the neutral position.

3. Remove the transmission under–guard and switch cover.

4. Disconnect the park/neutral switch connector, then remove the harness stay.

5. Remove the mounting bolt from the rear transmission cover. Remove the two mounting bolts from the switch body, then remove the switch from the control shaft.

To install:

6. Set the park/neutral switch to the neutral position. The switch should click into the neutral position.

7. Install the switch onto the control shaft and install the three mounting bolts. Torque the bolts to 9 ft. lbs. (12 Nm).

8. Reconnect the switch connector and tighten the bolt on the harness stay. Torque the bolt to 9 ft. lbs. (12 Nm).

9. Install the transmission under–guard. Torque the mounting bolts to 9 ft. lbs. (12 Nm).

10. Check the switch and the gear position indicator. Move the shift lever through all the gears and verify that the gear position indicator follows the switch.

11. Lower the vehicle.

12. Start the engine and check the shift lever operation through all the gears.

3.2TL

NOTE: The radio may contain a coded theft protection circuit. Always obtain the code number before disconnecting the battery.

1. Set the parking brake and chock the drive wheels. Place the transmission selector in **N** (neutral).

2. Disconnect the negative battery cable.

3. Raise and safely support the vehicle.

4. Remove the shift cable cover mounting bolts and remove the cover.

5. Remove the two bolts attaching the shift cable to the shift cable holder base. Remove the washers from the shift cable holder.

6. Remove the nut and washer attaching the shift cable control lever to the control shaft.

7. Disconnect the shift cable control lever from the control shaft.

8. Disconnect the park/neutral switch harness.

9. Pry the lock tabs on the lock washer away from the nut then remove the nut and washer.

10. Remove the bolts attaching the park/neutral switch and remove the switch.

To install:

11. Set the switch to the **N** position by aligning the mark in the switch body and the slot in the control shaft hole.

12. Install the switch to the transaxle with the mounting bolts. Do not fully tighten the mounting bolts.

13. Install a new lock washer and the locknut to the control shaft, the pointer on the lock washer should align with the neutral mark on the switch body.

14. Torque the locknut to 8.7 ft. lbs. (12 Nm), then bend the lock tabs against the locknut.

15. Torque the 6mm switch mounting bolts to 8.7 ft. lbs. (12 Nm).

16. Connect the park/neutral switch harness.

17. Connect the shift cable control lever to the control shaft, then install the washer and nut.

18. Install the shift cable holder to the shift cable holder base with the mounting washers. Install the attaching bolts and torque the bolts to 8.7 ft. lbs. (12 Nm).

19. Install the shift cable cover and torque the mounting bolts to 8.7 ft. lbs. (12 Nm).

20. Lower the vehicle and place the transmission selector in **P** (park).

21. Connect the negative battery cable.

22. Test the operation of the park/neutral switch, by making sure the engine will only start in park or neutral and the back–up lights work properly.

23. Enter the radio security code.

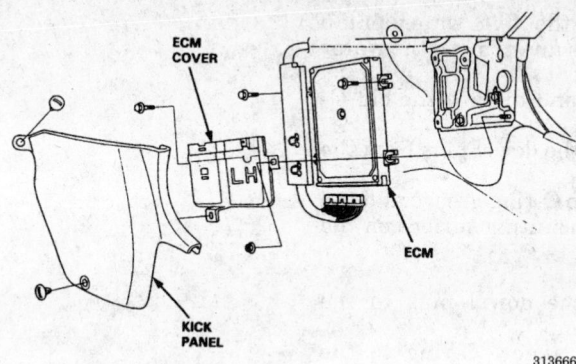

Powertrain control module removal — 1994–97 Integra shown, others similar

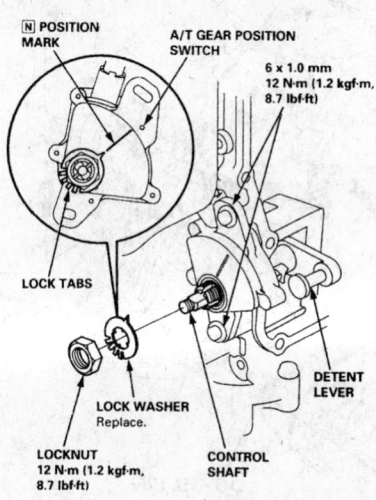

Park/neutral switch — 3.2TL

Powertrain Control Module

REMOVAL AND INSTALLATION

NOTE: The radio may contain a coded theft protection circuit. Always obtain the code number before disconnecting the battery, radio fuse, or removing the radio.

1. Disconnect the negative battery cable.
2. Remove the right door sill molding and the small cover on the right kick panel.
3. Pull the carpet back to expose the engine control module.
4. Unbolt the cover and disconnect the wiring connectors.
5. Unbolt and remove the control module.
To install:
6. Install the engine control module and mounting bolts.
7. Connect the wiring connectors, install the module cover and mounting bolts.
8. Place the carpet back to cover the module, then install the small cover on the right kick panel, and the right door sill molding.
9. Connect the negative battery cable.

ENGINE COOLING

Radiator

REMOVAL AND INSTALLATION

NOTE: The original radio may contain a coded anti-theft circuit. Make sure you have the security code number before disconnecting the battery.

1. Disconnect the negative battery cable.
2. Drain the cooling system into a sealable container.
3. On some models, it may be necessary to remove the front splash guard to gain access to the lower radiator and shroud mounting bolts.
4. Disconnect the thermo–switch wire, the fan motor wire, and the air conditioning fan motor wire.

NOTE: The radiator can be removed from the vehicle with the fan shrouds attached.

5. Disconnect the upper and lower hoses at the radiator. If equipped with an automatic transaxle, disconnect and plug the transmission cooling lines.
6. Remove the hoses to the coolant reservoir.
7. Remove the radiator bracket bolts and lift the radiator and fan assembly from the vehicle.
8. Remove the cooling fan assemblies from the radiator.
To install:
9. Install the radiator and fan assembly into the engine bay. Install the radiator bracket, cushion, and mounting bolts. Torque the mounting bolts to 5–7 ft. lbs. (7–10 Nm).
10. Reconnect the upper and lower hoses at the radiator. Unplug and reconnect the transmission cooling lines at the radiator. Be sure not to overtighten the cooling lines.
11. Reconnect the hose to the coolant reservoir.
12. Reconnect the fan motor wiring connectors and thermo–switch wire connection.
13. Install the splash apron if it was removed.
14. Fill the cooling system. When filling the system, open the bleeder on the thermostat housing. On Vigor and 2.5TL vehicles, there is a second bleeder near the fuel pressure regulator.
15. Connect the negative battery cable and enter the radio security code. Start the engine and watch the coolant level in the radiator. Top the coolant to the proper level as needed.

Water Pump

REMOVAL AND INSTALLATION

NOTE: The radio may have a coded theft protection circuit. Obtain the code from the owner before disconnecting the battery, removing the radio fuse, or removing the radio.

1.7L (B17A1), 1.8L (B18A1, B18B1, B18C1) Engines

1. Disconnect the negative battery cable.
2. If applicable, remove the front under panel.
3. Gradually release the system pressure by slowly and carefully removing the radiator cap. Be sure to protect your hands with gloves or a shop rag.
4. Drain the engine coolant into a sealable container.
5. Remove the timing belt from the engine.
6. Remove the camshaft pulleys and remove the back cover.

7. Remove the five water pump mounting bolts and remove the water pump.

8. Remove and discard the old O-ring.

9. Remove the dowel pins from the oil water pump.

10. Clean the O-ring groove and the water pump mounting surface on the engine.

To install:

11. Install the dowel pins to the new water pump.

12. Position a new O-ring to the new water pump, Apply a small amount of sealant to the O-ring to hold it in position.

13. Place the new water pump on the engine and install the mounting bolts. Torque the mounting bolts to 8.7 ft. lbs. (12 Nm).

14. Install the back cover and the camshaft pulleys.

15. Install the timing belt.

16. Fill the engine with coolant and bleed the air from the cooling system.

17. Connect the negative battery cable and enter the radio security code.

18. Run the engine and check for cooling system leaks.

2.5L (G25A1, G25A4) and 3.2L (C32A1) Engines

Legend

NOTE: Perform this service operation with the engine cold.

1. Disconnect the negative battery cable.

2. Remove the front splash panel and release the system pressure by slowly removing the radiator cap.

3. Drain the cooling system.

4. Remove the timing belt. Inspect the timing belt for any signs of damage or oil and coolant contamination. Replace the timing belt if there is any doubt about its condition.

5. If extra clearance is required, remove the camshaft pulleys and the timing belt rear cover.

6. Remove the two mounting bolts from the thermostat housing.

7. Remove the water pump bolts. Then, remove the water pump and sprocket assembly from the engine block. Remove the O-rings from the water passage.

To install:

8. Before installation, make sure all gasket and O-ring groove surfaces are clean.

9. Install the water pump with a new O-ring. Use liquid gasket if it was present on the sealing surfaces of the water pump that was removed.

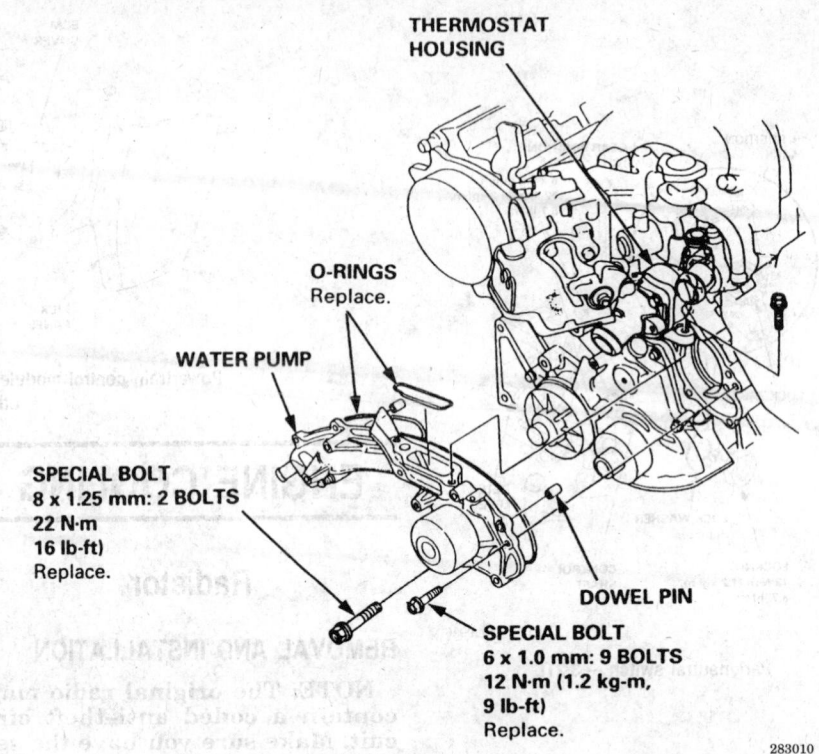

Water pump — Legend

Use new 6mm mounting bolts and evenly tighten them to 9 ft. lbs. (12 Nm). Use new 8mm bolts and tighten them to 16 ft. lbs. (22 Nm).

10. Install the timing belt rear cover and camshaft pulleys. Tighten the pulley bolts for 2.5L (G25A1, G25A4) engines to 33 ft. lbs. (44 Nm). For 3.2L (C32A1) engines, tighten the pulley bolts to 23 ft. lbs. (32 Nm).

11. Install the thermostat housing and the two mounting bolts. Use a new O-ring.

12. Install the timing belt and timing belt covers.

13. Install and adjust the tension of the accessory drivebelts.

14. Close the cooling system drain plug. Refill and bleed the cooling system.

15. Connect the negative battery cable.

16. Start the engine, allow it to reach normal operating temperature, and check for leaks.

Thermostat

REMOVAL AND INSTALLATION

NOTE: The radio may have a coded theft protection circuit. Obtain the code from the owner before disconnecting the battery, removing the radio fuse, or removing the radio.

1. Disconnect the negative battery cable.

2. When the radiator is cool, remove the radiator cap and drain plug. Drain the radiator coolant into a clean container.

3. Remove the upper radiator hose from the thermostat housing.

4. Remove the thermostat housing bolts, the housing, and the thermostat.

5. Clean the gasket mounting surfaces.

To install:

6. Install new gaskets and the thermostat with the spring end toward the engine. Torque the housing bolts to 9 ft. lbs. (12 Nm).

7. Install the radiator hose and tighten the hose clamp.

8. Make sure the coolant drain plug is closed.

9. Connect the negative battery cable.

10. Refill the cooling system, then start the engine and bleed the system. Allow the engine to reach normal operating temperature and check for leaks.

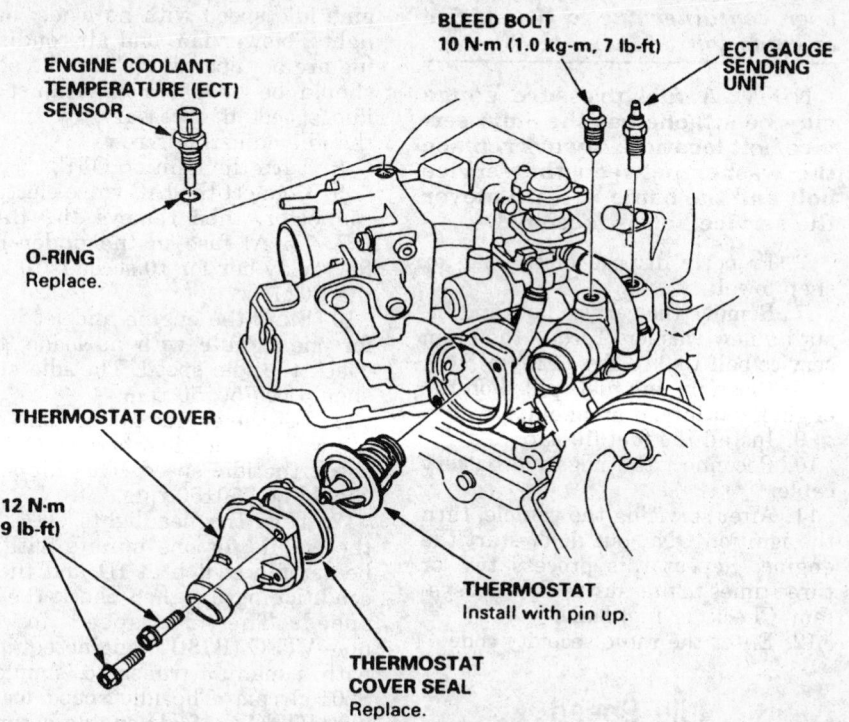

ENGINE COOLANT
TEMPERATURE (ECT)
SENSOR

O-RING
Replace.

BLEED BOLT
10 N·m (1.0 kg-m, 7 lb-ft)

ECT GAUGE
SENDING
UNIT

THERMOSTAT COVER

12 N·m
9 lb-ft)

THERMOSTAT
Install with pin up.

THERMOSTAT
COVER SEAL
Replace.

194493

Thermostat — all models similar

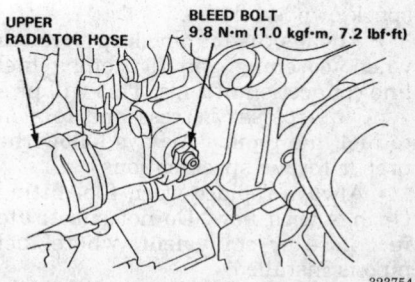

UPPER
RADIATOR HOSE

BLEED BOLT
9.8 N·m (1.0 kgf·m, 7.2 lbf·ft)

323754

Coolant bleed bolt — 1994–97 1.8L (B18B1, B18C1) Engines

4. Refill the radiator to the base of the filler neck. Put the radiator cap on the radiator, and tighten the cap only to the first stop.

5. Start the engine and let it run until the engine warms up (the radiator fan should come on at least twice).

6. Turn the engine **OFF** and check the level of the coolant in the radiator. Add coolant as necessary then install the radiator cap and fully tighten it.

7. Fill the reservoir to the **MAX** mark then install the reservoir cap.

8. Check for any coolant leaks.

Electric Cooling Fan

REMOVAL AND INSTALLATION

NOTE: The radio may have a coded theft protection circuit. Obtain the code from the owner before disconnecting the battery, removing the radio fuse, or removing the radio.

1. Disconnect the negative battery cable.

2. To gain extra space to remove the lower fan shroud bolts, the vehicle's front splash guard may be removed.

3. Disconnect the fan sensor control wire and the cooling fan motor connector.

4. Disconnect the air conditioning fan motor connector.

5. Remove the radiator fan shroud retaining bolts and remove the shroud assembly from the vehicle.

6. Remove the cooling fan mounting nuts and remove the cooling fan from the shroud.

To install:

7. Install the fan onto the shroud and install the shroud onto the radiator. Tighten the shroud bolts to 5–7 ft. lbs. (7–10 Nm).

8. Reconnect the fan motor wiring connectors and the fan sensor control wire connection.

9. Install the front splash guard if it was removed.

10. Connect the negative battery cable.

Cooling System Bleeding

PROCEDURE

— **CAUTION** —

Servicing the cooling system or removing the radiator cap while the engine is hot can cause the coolant to spray out, causing serious personal injury. Always let the engine and radiator cool — usually for a minimum of an hour — before servicing the cooling system or removing the radiator cap.

1. Mix a recommended antifreeze/coolant with an equal amount of water in a clean container.

2. Add the coolant mixture to the radiator, fill the system up to the radiator neck.

3. Loosen the bleed bolt on top of the engine. Tighten the bleed bolt when coolant comes out in a steady stream with no bubbles.

FUEL SYSTEM

Fuel System Service Precautions

Safety is the most important factor when performing not only fuel system maintenance but any type of maintenance. Failure to conduct maintenance and repairs in a safe manner may result in serious personal injury or death. Maintenance and testing of the vehicle's fuel system components can be accomplished safely and effectively by adhering to the following rules and guidelines.

• To avoid the possibility of fire and personal injury, always disconnect the negative battery cable unless the repair or test procedure requires that battery voltage be applied.

• Always relieve the fuel system pressure prior to disconnecting any fuel system component (injector, fuel rail, pressure regulator, etc.), fitting or fuel line connection. Exercise extreme caution whenever relieving fuel system pressure to avoid exposing skin, face and eyes to fuel spray.

Please be advised that fuel under pressure may penetrate the skin or any part of the body that it contacts.

• Always place a shop towel or cloth around the fitting or connection prior to loosening to absorb any excess fuel due to spillage. Ensure that all fuel spillage (should it occur) is quickly removed from engine surfaces. Ensure that all fuel soaked cloths or towels are deposited into a suitable waste container.

• Always keep a dry chemical (Class B) fire extinguisher near the work area.

• Do not allow fuel spray or fuel vapors to come into contact with a spark or open flame.

• Always use a backup wrench when loosening and tightening fuel line connection fittings. This will prevent unnecessary stress and torsion to fuel line piping. Always follow the proper torque specifications.

• Always replace worn fuel fitting O-rings with new. Do not substitute fuel hose or equivalent, where fuel pipe is installed.

Fuel System Pressure

RELIEVING

NOTE: The radio may have a coded theft protection circuit. Obtain the code from the owner before disconnecting the battery, removing the radio fuse, or removing the radio.

——————— CAUTION ———————
Fuel injection systems remain under pressure after the engine has been turned OFF. Properly relieve fuel pressure before disconnecting any fuel lines. Failure to do so may result in fire or personal injury.
———————————————————————

1. Disconnect the negative battery cable.
2. Remove the fuel fill cap.
3. Use a box wrench on the 6mm service bolt on the fuel rail while holding the special banjo bolt with another wrench.
4. Place a rag or shop towel over the 6mm service bolt.
5. Slowly loosen the 6mm service bolt one complete turn.

——————— CAUTION ———————
Do not allow fuel spray or fuel vapors to come in contact with a spark or open flame. Keep a dry chemical fire extinguisher

nearby. Never store fuel in an open container due to risk of fire or explosion.
———————————————————————

NOTE: A fuel pressure gauge may be attached at the 6mm service bolt location. Always replace the washer between the service bolt and the banjo bolt whenever the service bolt is loosened.

6. Properly dispose of the rag or shop towel.
7. Remove the service bolt and install a new washer. Torque the 6mm service bolt to 9 ft. lbs. (12 Nm).
8. Clean up any fuel spilled on the engine and intake manifold.
9. Install the fuel fill cap.
10. Reconnect the negative battery cable.
11. After servicing the vehicle, turn the ignition **ON** , but don't start the engine. Repeat this process two or three times to pressurize the fuel system. Check for fuel leaks.
12. Enter the radio security code.

Idle Speed

ADJUSTMENT

1994–97 1.8L (B18B1, B18C1) Engines

1. Check the following items before setting the idle speed:
   a. The MIL has not been reported on
   b. Ignition timing
   c. Spark plugs
   d. Air cleaner
   e. PCV system
2. Start the engine. Hold the engine at 3000 rpm with no load (in **N** or **P**) until the radiator fan comes on, then let it idle.
3. Shut the engine off and connect a tachometer to the engine.
4. Disconnect the Idle Air Control (IAC) valve electrical connector.
5. Start the engine with the accelerator pedal slightly depressed. Stabilize the engine at 1000 rpm, then slowly release the pedal until the engine idles.
6. Check the engine idle speed with no loads: headlights, blower fan, and air conditioning are not operating. The idle speed should be 480±50 rpm. Adjust the idle speed, if necessary, by turning the idle adjusting screw.
7. Check the ignition timing. If the timing is out of specification shut the engine off then start the engine with the accelerator pedal slightly depressed. Stabilize the engine at 1000 rpm, then slowly release the pedal

until the engine idles. Check the engine idle speed with no loads: headlights, blower fan, and air conditioning are not operating. The idle speed should be 480±50 rpm. Adjust the idle speed, if necessary, by turning the idle adjusting screw.
8. Turn the ignition OFF.
9. Connect the IAC valve electrical connector, then remove the BACK UP (7.5 A) fuse in the under–hood fuse/relay box for 10 seconds to reset the PCM.
10. Start the engine and let it idle for one minute with no loads, then check the idle speed. The idle speed should be 750±50 rpm.
11. Let the engine idle for one minute with the headlights (low) ON and check the idle speed. The idle speed should be 750±50 rpm.
12. Turn the headlights OFF. Idle the engine for one minute with the heater fan switch at HI and the air conditioning on, then check the idle speed. The idle speed for the non–VTEC (B18B1) engine equipped with a manual transaxle, should be 820±50 rpm. The idle speed for the non–VTEC (B18B1) engine equipped with a automatic transaxle, should be 840±50 rpm. The idle speed for the VTEC (B18C1) engine should be 850±50 rpm.

2.5L (G25A1, G25A4) Engines

NOTE: Idle is not adjusted at the throttle plate stop screw. That screw is set at the factory and is not adjustable. All idle adjustments are made by turning the idle adjusting screw on the throttle body. All idle test conditions must be met and the idle must remain stable under all test conditions.

1. Start engine and warm up to normal operating temperature: the cooling fan must come on at least once. Switch the ignition off.
2. Connect a test tachometer.
3. Disconnect the 2–pin connector from the IAC valve.
4. Start the engine with the accelerator pedal slightly depressed. Use the pedal to stabilize the engine speed at 1000 rpm. Slowly release the throttle until the engine idles. Check the idle with air conditioning, cooling fan and all electrical loads off.
5. The idle should be 550±50 rpm with the vehicle in **P** or **N**.
6. If the rpm is not within specifications, adjust as needed by turning the idle adjusting screw on the throttle body.
7. Switch the ignition off. Reconnect the connector to the IAC valve.

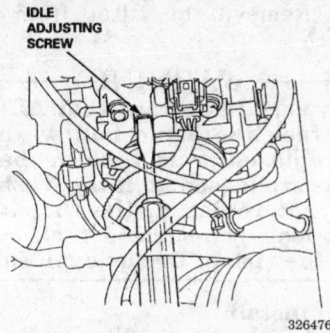

Idle adjusting screw — 1994–97 1.8L (B18B1, B18C1) engines

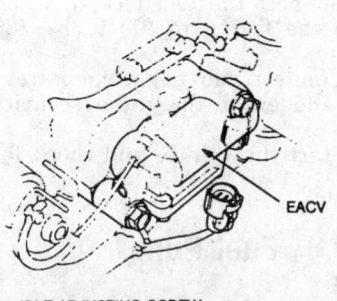

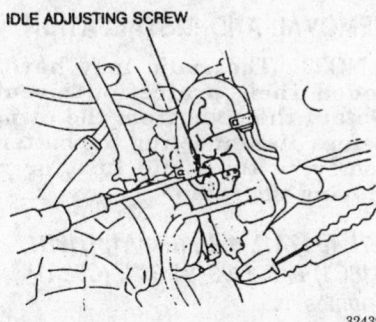

Idle speed adjustment — 2.5L (G25A1, G25A4) engines

Then, reset the PCM by removing the Back up/Radio 10A fuse from the underhood fuse box for at least 10 seconds.

8. Restart the engine, allowing it to idle in **N** or **P** for one minute. With all electrical loads including the air conditioning and cooling fan off, idle speed should be 700±50 rpm.

9. Allow the engine to idle for one minute with the headlight high beams and rear defogger on. The idle speed should be 770±50 rpm.

10. Turn the headlights and defogger off. Switch the air conditioner on and the blower fan to high speed. After one minute, the idle should be 770±50 rpm.

11. Enter the radio security code.

3.2L (C32A1) Engine

Legend

1. Start the engine and allow it to warm up to normal operating temperature. The cooling fan should come on.

2. Securely block the vehicle's wheels.

3. Connect a test tachometer to the test connector located at the rear of the right shock tower, or to the inductive service loop on the right shock tower. Consult your test equipment's instructions.

4. Hold the engine at 3000 rpm with the transmission in **P** or **N**. All electrical accessories must be off. On vehicles equipped with daytime running lights, pull the parking brake lever up to turn the lights off.

5. Check the vehicle's idle speed:
• GS and Coupe models: M/T 680 ± 50 rpm; A/T 630 ± 50 rpm
• L & LS models: M/T 650 ± 50 rpm; A/T 600 ± 50 rpm

6. Pull back the front edge of the passenger's side carpet. Pull the 2–P service check connector out from under the glove box. Connect the

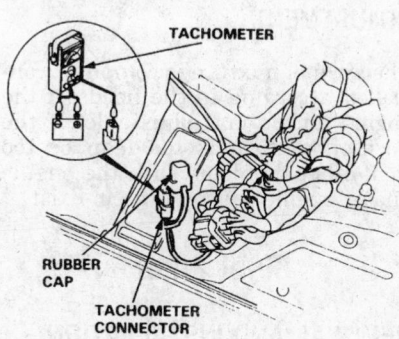

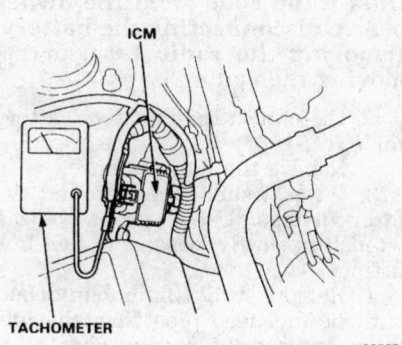

Tachometer test connections — Legend

check connector's terminals with a jumper wire.

7. Check the yellow LED display on the cover of the PCM. Use the LED display to determine the amount of idle speed adjustment.
   a. If the yellow LED is off, no idle speed adjustment is needed.
   b. If the yellow LED is blinking, turn the idle speed adjusting screw only 1/4 turn clockwise.
   c. If the yellow LED is on, turn the idle speed adjusting screw only 1/2 turn counterclockwise.

8. Check that the LED turns off after approximately 30 seconds. If the LED doesn't turn off, slowly turn the idle screw another 1/4 turn in the required direction of the adjustment. Repeat this procedure until the LED turns off.

9. Check the idle speed with the headlights on high beam, the rear window defogger and air conditioner on. Also check the idle speed while the steering wheel is being turned. Idle speed should be as follows:
• GS and Coupe models: M/T 680 ± 50 rpm; A/T 630 ± 50 rpm
• L & LS models: M/T 650 ± 50 rpm; A/T 600 ± 50 rpm

10. Remove the jumper wire and tuck the service check connector back under the glove box. Put the carpet edge back into place.

11. Remove all test equipment and install any connector covers.

3.2TL

NOTE: Idle is not adjusted at the throttle plate stop screw. That screw is set at the factory and is not adjustable. All idle adjustments are made by turning the idle adjusting screw on the throttle body. All idle test conditions must be met and the idle must remain stable under all test conditions.

1. Start engine and warm up to normal operating temperature: the cooling fan must come on at least once. Switch the ignition off.

2. Connect a OBDII scan tool or a tachometer.
• Connect a tachometer to the loop of the igniter unit (ignition module) secondary, or...
• Remove the rubber cap from the tachometer connector and connect a tachometer.

3. Disconnect the 2–pin connector from the IAC valve.

4. Start the engine with the accelerator pedal slightly depressed. Use the pedal to stabilize the engine speed at 1000 rpm. Slowly release the throttle until the engine idles. Check

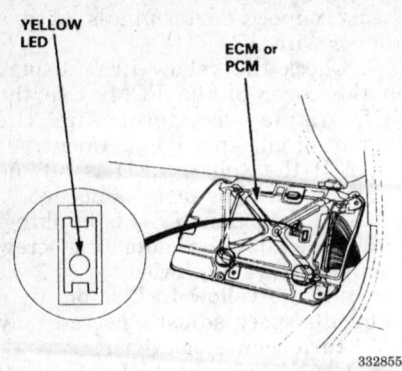

Yellow LED on PCM cover — Legend

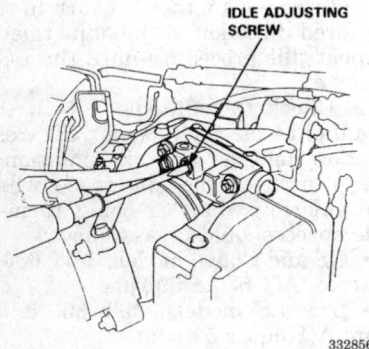

Idle adjusting screw — Legend

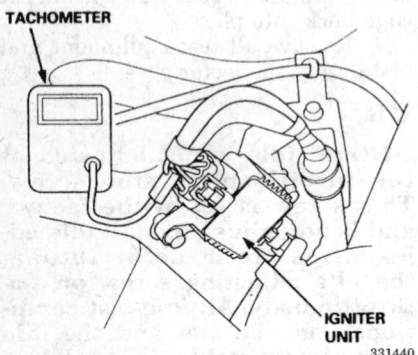

Tachometer connection — 3.2TL

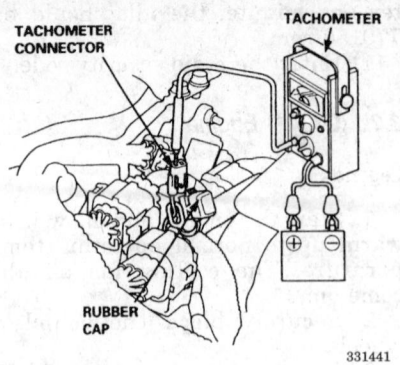

Tachometer connection — 3.2TL

the idle with air conditioning, cooling fan and all electrical loads off.

5. The idle should be 510±10 rpm with the vehicle in **P** or **N**.

6. If the rpm is not within specifications, adjust as needed by turning the idle adjusting screw on the throttle body.

7. Check the ignition timing: if the timing is out of specification, shut the engine OFF then start the engine and adjust the idle.

8. Switch the ignition off. Reconnect the connector to the IAC valve. Then, reset the PCM by removing the Back up/Radio 10A fuse from the un-

derhood fuse box for at least 10 seconds.

9. Restart the engine, allowing it to idle in **N** or **P** for one minute. With all electrical loads including the air conditioning and cooling fan off, idle speed should be 640±50 rpm.

10. Allow the engine to idle for one minute with the headlight low on. The idle speed should be 650±50 rpm.

11. Turn the headlights off. Switch the air conditioner on and the blower fan to high speed. After one minute, the idle should be 650±50 rpm.

12. Enter the radio security code.

Mixture

ADJUSTMENT

The air/fuel mixture is computer controlled according to the needs of the engine and is not adjustable. If the air/fuel mixture is too lean or too rich, other problems with the engine and/or engine control system exist.

Fuel Filter

REMOVAL AND INSTALLATION

NOTE: The radio may have a coded theft protection circuit. Obtain the code from the owner before disconnecting the battery, removing the radio fuse, or removing the radio.

1. Disconnect the negative battery cable.
2. Relieve the fuel pressure.
3. Wrap a shop towel around the filter fittings. Use a properly-sized wrench to slowly loosen the fuel line fittings.
4. Remove the 12mm banjo bolt and the fuel feed pipe from the fuel filter. Discard the used washers.
5. Remove the fuel filter clamp retaining bolt and open the clamp.

6. Remove the filter from the vehicle.

--- **WARNING** ---

It is very important that ALL of the fuel line banjo bolt washers be replaced every time the banjo bolts are loosened. If the washers are not replaced, the fuel lines will leak pressurized fuel, causing the risk of fire or explosion.

To install:

7. Install the new filter in position and tighten the clamp mounting bolt to 7 ft. lbs. (10 Nm).

8. Install the banjo bolts with new washers. Torque the bolt to 25 ft. lbs. (33 Nm).

9. Connect the fuel feed line and torque the fitting to 27 ft. lbs. (37 Nm).

10. Connect the negative battery cable and enter the radio security code.

11. Start the vehicle and check for leaks.

Fuel Pump

REMOVAL AND INSTALLATION

NOTE: The radio may have a coded theft protection circuit. Obtain the code from the owner before disconnecting the battery, removing the radio fuse, or removing the radio.

1.7L (B17A1), 1.8L (B18A1, B18B1, B18C1) and 1993–94 3.2L (C32A1) Engines

--- **CAUTION** ---

Do not smoke while working on the fuel system. Keep open flames away from your work area.

1. Disconnect the negative battery cable.
2. Relieve the fuel system pressure.
3. Remove the rear seat to gain access to the fuel pump access panel. Remove the maintenance access cover.
4. Disconnect the electrical connector from the fuel pump.
5. If equipped with quick connect fittings, hold the fuel line connector with one hand and press down the retainer tabs with the other hand, then pull the connector off. Check the contact area of the pipe for dirt or damage, clean or replace the pipe or pump as required. Remove the old retainer from the pipe and discard. Cover the connector and pipe with

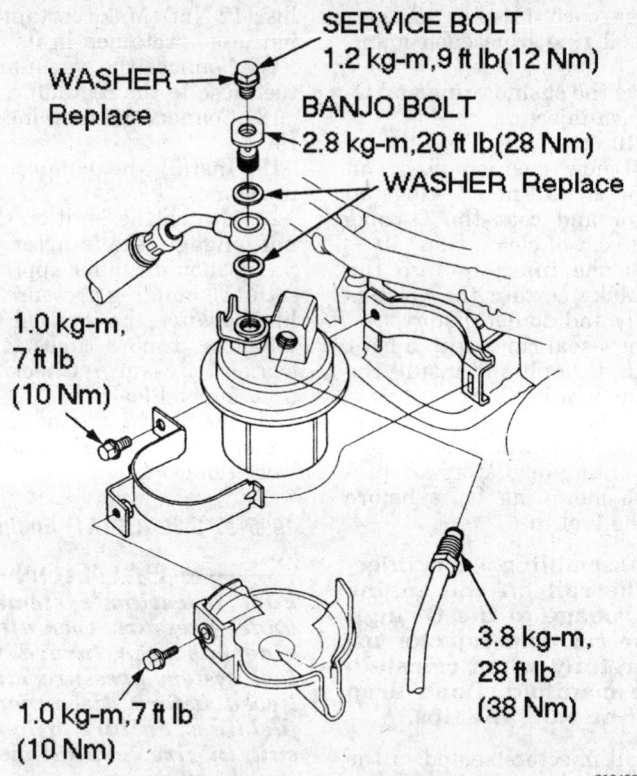

SERVICE BOLT
1.2 kg-m, 9 ft lb (12 Nm)

WASHER
Replace

BANJO BOLT
2.8 kg-m, 20 ft lb (28 Nm)

WASHER Replace

1.0 kg-m, 7 ft lb (10 Nm)

1.0 kg-m, 7 ft lb (10 Nm)

3.8 kg-m, 28 ft lb (38 Nm)

213499

Exploded view of the fuel filter — Legend

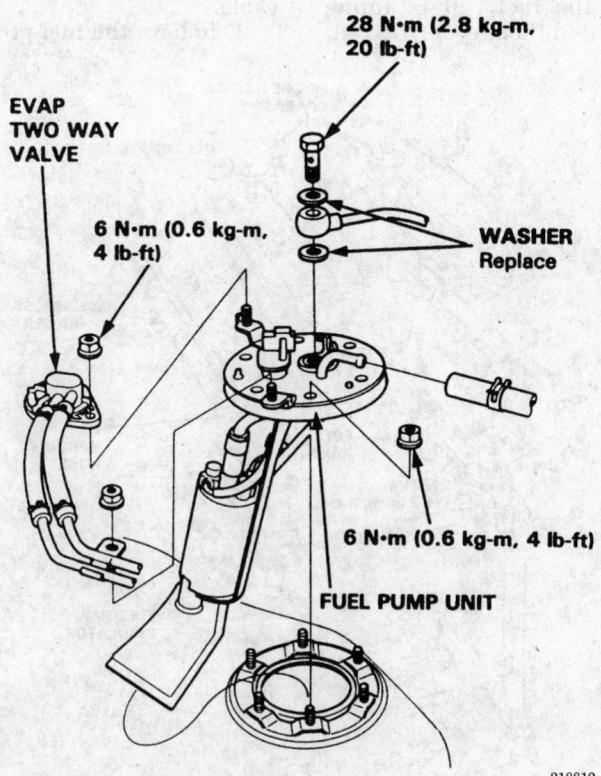

28 N·m (2.8 kg-m, 20 lb-ft)

EVAP TWO WAY VALVE

6 N·m (0.6 kg-m, 4 lb-ft)

WASHER Replace

6 N·m (0.6 kg-m, 4 lb-ft)

FUEL PUMP UNIT

218812

Fuel pump components — 1993 1.7L and 1.8L engines

plastic bags to prevent damage and keep foreign material out.

6. Remove the fuel pump mounting nuts then remove the fuel pump from the fuel tank.

To install:

7. Install the fuel pump into the fuel tank. Torque the mounting nuts to 4 ft. lbs. (6 Nm).

8. Reconnect the fuel lines and electrical connectors. Use new sealing washers when connecting the fuel pressure hose. Torque the fuel line banjo bolt to 21 ft. lbs. (27 Nm).

9. Install a new retainer into the fuel tube connector and press the tube on to the pipe. The retainer pawls should lock with a clicking sound.

10. Install fuel pump access cover.

11. Install the rear seat.

12. Reconnect the negative battery cable and enter the radio security code.

13. Turn the ignition switch on and off two or three times to pressurize the system and check for leaks.

2.5L (G25A1, G25A4) and 1996–97 3.2L (C32A1) Engines

CAUTION

Observe all applicable fuel precautions when working around fuel. Do not allow fuel spray or fuel vapors to come in contact with a spark or open flame. Keep a dry chemical (Class B) fire extinguisher near the work area. Never drain or store fuel in an open container due to the possibility of fire or explosion.

1. Disconnect the negative battery cable. Raise and safely support the vehicle. Remove the left rear wheel.

2. Remove the tank drain bolt and drain the fuel into an approved container.

3. Disconnect the pump and float wiring connectors located under the trunk floor.

4. Remove the fuel hose and pipe covers from the inside of the quarter panel.

5. Support the tank with a transmission jack, remove the straps and lower the tank out of the vehicle. If it sticks on the undercoating, carefully pry it free using a blunt or wooden instrument as a lever.

6. Disconnect the fuel line by removing the banjo bolt or uncoupling the quick-connect fittings.

7. Remove the fuel pump mounting nuts. Remove the fuel pump from the fuel tank.

8. Install fuel pump and mounting nuts. Torque the mounting nuts to 4 ft. lbs. (6 Nm).

9. Use a jack to place the fuel tank in position. Reconnect the fuel hoses with new washers, or reconnect the quick-connect fittings with new retainers. Reconnect the pump and float wiring connectors. Torque the fuel line to 28 ft. lbs. (38 Nm).

10. Tighten and torque the strap bolts on Vigor to 29 ft. lbs. (40 Nm), or 2.5TL and 3.2TL to 28 ft. lbs. (38 Nm).

11. Use a new sealing washer on the drain plug and tighten the plug to 36 ft. lbs. (49–50 Nm).

12. Refill the tank. Check for leaks.

13. Connect the negative battery cable and enter the radio security code.

Fuel Injector

REMOVAL AND INSTALLATION

NOTE: The radio may have a coded theft protection circuit. Obtain the code from the owner before disconnecting the battery, removing the radio fuse, or removing the radio.

Except 1995–97 3.2L (C32A1) Engines

——— CAUTION ———
Fuel injection systems remain under pressure, even after the engine has been turned OFF. The fuel system pressure must be relieved before disconnecting any fuel lines. Failure to do so may result in fire, explosions, or personal injury.

1. Disconnect the negative battery cable.

2. Relieve the fuel pressure.

3. Remove the engine harness cover, if so equipped.

4. Label and disconnect the electrical harnesses from the injectors.

5. Place a rag below the fuel pressure regulator. Disconnect the vacuum hose and disconnect the fuel line. Plug the fuel line to prevent spillage and the entry of dirt.

6. Disconnect the fuel hose from the fuel rail.

7. Remove the nuts holding the fuel pipe (rail) and the wiring harness.

8. Remove the fuel rail from the injectors, leaving the injectors in the manifold.

9. Remove each injector and remove the seal ring from each manifold port.

10. Remove the cushion ring and O-ring from each injector.

To install:

11. Install new cushion rings on each injector. Install new O-rings on each injector and coat the O-rings with a light coat of clean, thin oil.

12. Install the injectors into the fuel pipe. Make certain the O-rings seat properly and do not distort.

13. Coat new seal rings with a light coat of clean, thin oil and install the rings into the manifold.

14. Install the fuel rail and injectors to the intake manifold. Make certain the mounting insulators are present on the mounting bolts before installing the fuel rail.

NOTE: Assembling each injector into the rail off the engine prevents damage to the O-rings. Handle the rail and injector assembly carefully when reinstalling to the manifold. Don't drop an injector or bang the tips.

15. With all injectors seated in the manifold, align the center mark on each injector electrical connector with the mark on the fuel rail.

16. Install the fuel rail retaining nuts and tighten them evenly to 9 ft.

lbs. (12 Nm). Make certain the wiring harness is retained in its clips.

17. Connect the vacuum hose and fuel hose to the regulator.

18. Connect the fuel line to the fuel rail.

19. Install the connectors to the injectors.

20. Switch the ignition ON but do not engage the starter. The fuel pump should run for approximately 2 seconds, building pressure within the lines. Switch the ignition OFF, then ON 2 or 3 more times to build full system pressure. Check the work area for fuel leaks.

21. Install the engine harness covers or intake manifold covers if they were removed.

1995–97 3.2L (C32A1) Engines

——— CAUTION ———
Fuel injection systems remain under pressure, even after the engine has been turned OFF. The fuel system pressure must be relieved before disconnecting any fuel lines. Failure to do so may result in fire, explosions, or personal injury.

1. Disconnect the negative battery cable.

2. Relieve the fuel pressure.

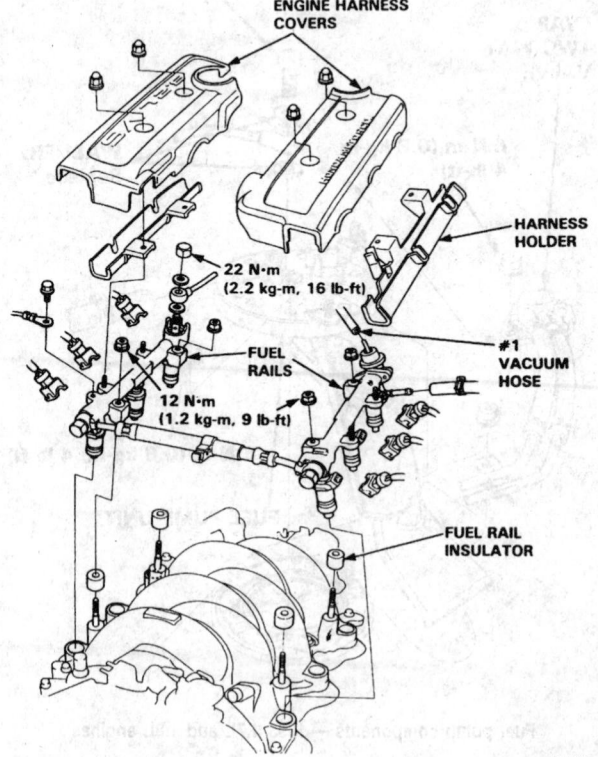

Fuel supply system — except 1995–97 3.2L (C32A1) engines

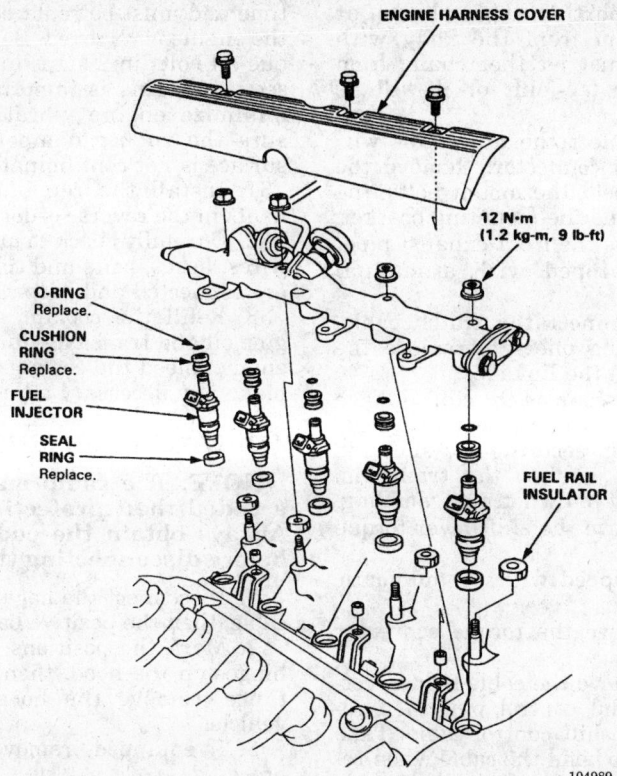

ENGINE HARNESS COVER

12 N·m
(1.2 kg-m, 9 lb-ft)

O-RING
Replace.

CUSHION
RING
Replace.

FUEL
INJECTOR

SEAL
RING
Replace.

FUEL RAIL
INSULATOR

104989

Fuel supply system — except 1995–97 3.2L (C32A1) engines

3. Remove the engine harness covers.

4. Label and disconnect the electrical harnesses from the injectors.

5. Place a rag below the fuel pressure regulator. Disconnect the vacuum hose and the fuel return hose. Plug the fuel line to prevent leaks.

6. Disconnect the fuel line from the fuel rail.

7. Remove the nuts holding the fuel rail and the wiring harness.

8. Remove the fuel rail from the injectors, leaving the injectors in the manifold.

9. Remove each injector, noting its position, and remove the seal ring from each manifold port.

10. Remove the cushion ring and O-ring from each injector.

To install:

11. Install new cushion rings on each injector.

12. Coat new O-rings with clean engine oil and install them on the fuel injectors.

13. Install the injectors into the fuel rail. Make certain the O-rings seat properly and are not distorted.

NOTE: Assembling each injector into the fuel rail prevents damage to the O-rings. Handle the rail and injector assembly

carefully when reinstalling it to the manifold. Don't drop the injectors or bang their tips.

14. Coat new seal rings with a light coat of clean, thin oil and install them into the manifold.

15. Install the fuel rail and injectors to the intake manifold. Make sure the mounting bolt insulators are in place before installing the fuel rail.

16. With all injectors seated in the manifold, align the matching mark on each injector with the matching mark on the fuel rail.

17. Install the fuel rail retaining nuts and tighten them evenly to 9 ft. lbs. (12 Nm). Make certain the wiring harness is retained in its clips.

18. Connect the vacuum hose and fuel return hose to the regulator.

NOTE: Always use new sealing washers when reconnecting the fuel lines to prevent fuel leaks.

19. Connect the fuel line to the fuel rail using new washers. Tighten the nut to 16 ft. lbs. (22 Nm).

20. Connect the electrical harness to the injectors.

21. Reconnect the negative battery cable.

22. Switch the ignition **ON** but do not engage the starter. The fuel pump should run for approximately 2 seconds, building pressure within the

lines. Switch the ignition **OFF**, then **ON** 2 or 3 more times to build full system pressure. Check for fuel leaks.

23. Install the engine harness covers.

ENGINE MECHANICAL

Engine Assembly

REMOVAL AND INSTALLATION

1993 1.7L and 1.8L (B17A1, B18A1) Engines

1. Raise and safely support the vehicle. Remove the engine and wheel well splash shields.

2. Disconnect the battery cables from the battery, negative cable first. Remove the battery and the battery tray from the engine compartment.

3. It is not necessary to remove the hood. Raise it to a full vertical position and support it properly.

4. Drain the oil from the engine, the coolant from the radiator and the fluid from the transaxle.

5. Remove the intake air duct and air cleaner assembly.

6. Remove the fuel filler cap and loosen the service bolt on the fuel filter banjo bolt to relieve the fuel pressure. Disconnect the fuel feed hose from the fuel filter and the fuel return hose from the pressure regulator.

— CAUTION —
The fuel system may be under pressure and fuel will be sprayed. Be sure there is good ventilation and take the appropriate fire safety precautions.

7. Disconnect the charcoal canister hose from the throttle body. Vehicles with automatic transaxle have an emissions control equipment box on the firewall. Rather than disconnecting the vacuum hoses, it may be easier to unplug the connectors, remove the box from the firewall and lay it on the engine.

8. Remove the ground cable from the transaxle. Disconnect the 2 distributor electrical connectors, make an alignment mark on the distributor body and the engine, and remove the distributor.

9. Remove the throttle cable by loosening the locknut. Slip the cable end out of the throttle bracket and accelerator linkage. Take care not to bend the cable when removing it. Do not use pliers to remove the cable from the linkage.

10. Remove the mounting bolts and the V-belt for the power steering pump. Without disconnecting the hoses, secure the pump out of the way.

11. Near the brake booster, disconnect the engine wiring harness connectors and remove the wire from the clamp.

12. Remove the brake booster vacuum hose from the intake manifold. Disconnect the upper and lower radiator hoses and heater hoses.

13. Without disconnecting the hydraulic hoses or cable, remove the bolt and remove the speed sensor from the transaxle as an assembly.

14. Disconnect the transaxle cooling hoses, if equipped, and unplug the radiator fan connectors. Remove the radiator and fans as an assembly.

15. If not already removed, remove the driver's side splash shield at the wheel well.

16. Remove the air conditioning belt. Remove the compressor with the hoses still attached and secure the compressor to the front beam.

NOTE: Do not loosen or disconnect the air conditioning lines. Do not vent refrigerant into the air.

17. To disconnect the halfshafts, the lower damper fork and ball joint must be disconnected, allowing the steering knuckle to move away from the transaxle.

   a. With the front wheels removed, remove the damper fork nut and pinch bolt. Remove the damper fork from the lower arm.

   b. Remove the lower ball joint castle nut press the ball joint out of the lower arm using a suitable puller, with the puller jaws grasping the lower arm.

   c. Carefully pry the inner CV-joint away from the transaxle to force the set ring at the inner end past the groove.

   d. Pull the inboard CV-joint, not the halfshaft, and remove the CV-joint out of the intermediate shaft.

NOTE: Do not pull on the halfshaft, as the CV-joint may come apart. Use care when prying out the assembly and pull it straight to avoid damaging the intermediate shaft seals.

   e. Support the halfshafts or hang them from the body with wire. Do not let them hang from the outer CV-joint or it will be damaged.

18. Disconnect the alternator wiring harness connector. Remove the alternator belt, the mount bolts, the alternator and the mounting bracket.

19. Remove the front exhaust pipe.

20. If equipped with a manual transaxle:

   a. Disconnect the clutch cable. Avoid using pliers to remove the cable from the linkage.

   b. To disconnect the shift linkage rod, push back the boot and drive out the pin securing the shift rod universal joint to the transaxle. Use a new pin when reassembling.

   c. Remove the shift lever torque rod.

21. If equipped with an automatic transaxle:

   a. Remove the torque converter cover.

   b. Remove the cable holder, cotter pin and control pin, then remove the shift control cable. Take care not to bend the cable when removing it. Avoid using pliers to remove the cable from the linkage.

22. Attach a suitable lift to the engine. Raise the engine slightly to remove all the slack from the hoist.

23. Remove the rear transaxle mount and rear transaxle mounting bracket.

24. Remove the front transaxle mount. Remove the side transaxle mounting bracket. Remove the side engine mount. The transmission and engine mounts can be removed from the engine at this time. It is not necessary to remove the engine if only the transmission and motor mounts need to be removed. However, the engine assembly must be safely and firmly supported by an engine hoist to take the weight off the mounts.

25. Check that the engine/transaxle assembly is completely free of all vacuum hoses and electrical wires.

26. Slowly raise the engine approximately 6 inches and check again.

27. Raise the engine/transaxle assembly all the way and remove it from the vehicle. Place the engine on a suitable engine stand.

   To install:

28. Slowly lower the engine and transaxle assembly into the vehicle.

29. Check that the engine and transaxle are clear of any hoses or electrical connectors.

30. Install the mounts and bolts. Some of the transaxle mount bolts are designed to be torqued only one time and must be replaced whenever the engine is removed. Be sure to torque all bolts in 2 steps in the correct sequence. This is important to help minimize engine vibrations. Make sure the rubber damper mounting surface is not contaminated with oil.

31. Install the remaining components in the reverse order of removal.

32. Carefully check to make sure all wires, hoses, belts and control cables are connected and properly adjusted.

33. Refill the coolant system, engine oil and transaxle fluid. Start the engine, bleed the cooling system and make any necessary adjustments.

1994–97 1.8L (B18B1, B18C1) Engines

NOTE: The radio may contain a coded theft protection circuit. Always obtain the code number before disconnecting the battery.

1. Disconnect the negative battery cable, then the positive battery cable.

2. Mark the positions of the hood hinges on the hood, then with assistance, remove the hood from the vehicle.

3. If equipped, remove the strut brace.

4. Disconnect the battery cables from the under–hood fuse/relay box and under–hood ABS fuse/relay box.

5. Remove the intake air duct, air cleaner housing assembly and mounting bracket.

6. Remove the evaporative emission (EVAP) control canister hose and vacuum hose from the intake manifold.

7. Disconnect the engine wire harness connectors on the right side of engine compartment.

8. Relieve the fuel pressure by loosening the service bolt on the fuel filter about one turn. Place a shop towel over the fuel filter to prevent pressurized fuel from spraying over the engine.

——— CAUTION ———
Fuel injection systems remain under pressure after the engine has been turned OFF. Properly relieve fuel pressure before disconnecting any fuel lines. Failure to do so may result in fire or personal injury.

9. Disconnect the fuel feed hose, brake booster vacuum hose, and fuel return hose.

10. Remove the throttle cable by loosening the locknut, then slip the cable end out of the accelerator linkage. Be careful not to bend the cable when removing it. Replace the cable if it gets kinked.

11. Remove the engine wire harness connectors, terminal, and clamps on the left side of engine compartment.

12. Remove the cruise control actuator, and engine ground cable at the body end.

13. Remove the adjusting bolt and mounting bolt, then remove the power steering belt and pump. Do not disconnect the power steering hoses.

14. Loosen the idler pulley bolt and adjusting bolt, then remove the air conditioning compressor belt.

15. On manual transmission only, remove the clutch slave cylinder and pipe/hose assembly. Do not disconnect the pipe/hose assembly.

16. Remove the transmission ground cable and hose clamp. Remove the radiator cap.

17. Safely raise and support the vehicle, then remove the front wheels and splash shield.

18. Drain the engine coolant, engine oil, and transmission fluid. Reinstall the drain plugs using new washers. Be careful not to overtighten the drain plugs.

19. Disconnect the upper and lower radiator hoses and the heater hoses from the engine.

20. If equipped with a automatic transmissions, disconnect the ATF (automatic transmission fluid) cooler hoses.

21. Remove the radiator assembly.

22. Remove the air conditioning compressor mounting bolts and position the compressor out of the way. Suspend the compressor on a wire, do not let it hang by its hoses. Do not disconnect the hoses.

23. Disconnect the heated oxygen sensor (HO2S) connector.

24. Remove the nuts and bolts connecting exhaust pipe A to the catalytic converter. Discard the gasket and the locknuts.

25. Remove and discard the nuts attaching exhaust pipe A to the exhaust hanger.

26. Remove and discard the locknuts attaching exhaust pipe A to the exhaust manifold, then remove exhaust pipe A from the vehicle. discard the exhaust gaskets.

27. If equipped with a manual transaxle, disconnect the shift rod and extension rod from the transaxle.

28. If equipped with a automatic transaxle, remove the shift cable cover, then disconnect the shift cable from the transaxle.

29. Remove the right damper fork bolt, discard the nut.

30. Remove the right damper pinch bolt then remove the damper fork.

31. Disconnect the suspension lower arm ball joints using a ball joint remover.

32. Carefully pry the inner CV-joint away from the transaxle to force the set ring at the inner end past the groove. Remove the other CV-joint out of the intermediate shaft. Do not let the halfshafts hang down. Support the halfshafts or hang them from the body with wire and cover the halfshaft ends with plastic bags.

33. Attach a hoist to the engine.

34. Remove the left and right front mounts and brackets, then remove the rear mount bracket.

35. Remove the side engine mount, then remove the transmission mount.

36. Check that the engine is completely clear of vacuum hoses, fuel and engine coolant hoses, and electrical wiring.

37. Slowly raise the engine approximately 6 in. (150mm). Check one more time to make sure that all hoses and wires are disconnected from the engine.

38. Raise the engine and transaxle assembly all the way and remove it from the car.

39. Separate the engine and transaxle.

To install:

40. Install the transaxle to the engine assembly. If equipped with a manual transaxle, torque the transaxle housing mounting bolts to 47 ft. lbs. (64 Nm), the two bolts and new washers to the rear mounting bracket to 87 ft. lbs. (118 Nm), and torque the upper mounting bolts to 47 ft. lbs. (64 Nm). If equipped with a automatic transaxle, torque the transaxle housing mounting bolts to 43 ft. lbs. (59 Nm), the two bolts and new washers to the rear mounting bracket to 87 ft. lbs. (118 Nm), and torque the upper mounting bolts to 54 ft. lbs. (74 Nm).

41. If equipped with a automatic transaxle torque the bolts attaching the torque converter to the drive plate to 8.7 ft. lbs. (12 Nm).

42. Install the torque converter/clutch cover.

43. Install the rear engine stiffener, torque the bolts attaching the stiffener to the engine to 17 ft. lbs. (24 Nm). If equipped with a manual transaxle, torque the bolts attaching the stiffener to the transaxle to 42 ft. lbs. (57 Nm). If equipped with a automatic transaxle, torque the bolts attaching the stiffener to the transaxle to 32 ft. lbs. (43 Nm).

44. If equipped with the VTEC (B18C1) engine, install the front engine stiffener. Torque the bolt attaching the stiffener to the engine to 17 ft. lbs. (24 Nm), torque the bolts attaching the stiffener to the transaxle to 42 ft. lbs. (57 Nm).

45. Install the engine and transaxle into the engine compartment. Install the transmission mount, then tighten the bolt/nuts on the transmission side. Leave the mount bolt loose.

46. Install the engine side mount, then tighten the bolt/nuts on the engine side. Leave the mount bolt loose. Tighten the mount bolt on the transmission mount and then tighten the mount bolt on the side engine mount.

47. Install the rear mount bracket, then tighten the bolts in the proper sequence.

48. Install the right front mount/bracket, then tighten the bolts in the proper sequence.

49. Install the left front mount, then tighten the bolts in the proper sequence.

50. The remaining components are installed in the reverse order of removal.

51. Reconnect the battery cables to the fuse/relay boxes and the battery. Connect the positive cable then the negative cable to the battery.

52. Turn on the ignition switch (do not operate the starter) so that the fuel pump operates for approximately 2 seconds and the fuel line pressurizes. Repeat this operation 2 or 3 times and check for fuel leakage.

53. Enter the radio security code then test drive the vehicle.

2.5L (G25A1, G25A4) Engines

NOTE: A hydraulic lift is very helpful for removing this engine. The recommended engine removal and installation procedure requires the vehicle to be raised and lowered to unbolt the transmission mounts before the engine may be lifted out.

1. On vehicles with automatic transmissions, make sure the selector is in **P**. On manual transmissions, put the selector into first gear.

2. Open the hood and secure it in its fully opened position (vertical). The hood may be removed if more clearance and working room is desired.

3. Disconnect the battery and remove it. Remove the battery tray.

4. Remove the splash shield.

5. Drain the engine oil and the differential and transmission oil.

6. Remove the engine ground cables and ignition coil wire.

7. Unbolt the ABS relay and move it out of the way. Remove the battery heat shield.

8. Remove the battery cables from the under-hood fuse/relay box.

9. Disconnect the engine harness connectors on the right side of the engine compartment.

10. Remove the intake air cleaner duct and the air cleaner housing.

11. Loosen the component adjusting and mounting bolts and remove the power steering pump and air conditioning compressor belts.

12. Without disconnecting the hoses, remove the power steering pump and the air conditioner compressor and secure them out of the way.

NOTE: Do not loosen or disconnect the air conditioning refrigerant lines.

13. Loosen the service bolt on the fuel filter to relieve the fuel system pressure. Remove the banjo bolt to remove the fuel feed hose from the fuel filter. Remove the fuel return hose from the pressure regulator.

--------- CAUTION ---------
Fuel injection systems remain under pressure after the engine has been turned OFF. Properly relieve fuel pressure before disconnecting any fuel lines. Failure to do so may result in fire or personal injury.
---

14. Remove the throttle cable by loosening the locknut, then slip the cable end out of the throttle bracket and accelerator linkage. Do not bend the cable when removing it. Unbolt the throttle cable clamp and move the cable aside.

15. Label and disconnect the engine wiring harness connectors on the left side of the engine compartment.

16. Disconnect the charcoal canister hoses, fuel return hose, brake booster hose and the emission control vacuum hoses.

17. Disconnect the transmission wiring connector that is located near the firewall.

18. The distributor may be removed for extra access to the upper transmission case bolts (Vigor). Disconnect the wiring and remove the two bolts to remove the distributor. Do not lose the collar that fits on the distributor shaft.

19. Remove the upper transmission bolts and the 26mm differential shim. Note the location of the shim for installation.

20. Remove the power steering speed sensor from the differential case without disconnecting the hydraulic hoses. Disconnect the wiring and secure the sensor out of the way.

21. Drain the engine coolant.

22. Disconnect the heater hoses.

23. Disconnect the transmission cooler hoses from the radiator tank. Disconnect the upper and lower radiator hoses and the fan wiring.

24. Remove the two upper brackets and lift the radiator and cooling fan assembly from the engine compartment.

25. Raise and support the vehicle safely.

26. On vehicles with automatic transmissions, remove the small torque converter cover. Rotate the crankshaft to remove the torque converter bolts.

27. Disconnect the halfshafts. The lower damper fork and ball joint must be disconnected so that the inner CV-joint can be separated from the intermediate shaft.

  a. Remove the front wheels. Remove the damper fork nut and pinch bolt. Remove the damper fork from the lower arm.

  b. Remove the lower ball joint castle nut. Press the ball joint out of the lower arm using a suitable puller.

  c. Carefully pry the inner CV-joint away from the transaxle to force the set ring at the inner end past the groove.

  d. Pull the inner CV-joint, not the halfshaft, and remove the CV-joint from the intermediate shaft.

NOTE: Do not pull on the halfshaft, the CV-joint may come apart. Use care when prying out the assembly and pull it straight to avoid damaging the intermediate shaft seals.

  e. Hang the halfshafts from the body with wire. Do not let them hang from the outer CV-joint or it will be damaged.

28. Disconnect the oxygen sensor lead from the exhaust system and remove the front exhaust pipe and its brackets.

29. Remove the transmission side mount and bracket.

30. Make sure the transmission is in **P** or first gear. Remove the extension shaft sealing cap on the lower left side of transmission housing.

31. Use an extension shaft puller tool, Acura tool number 07LAC–PW50101, to disengage the extension shaft from the differential. The differential is removed with the engine.

32. Remove the transmission mid-mounts and the mid-mount spacer.

33. Support the transmission with a jack and remove the transmission case bolts.

34. Install the transmission mid-mounts and spacer to hold the transmission in the vehicle. Make sure the engine will separate from the transmission, and that the transmission will be supported by the mounts after the engine has been removed.

35. Lower the vehicle and install a chain hoist onto the engine lifting hooks.

36. Raise the hoist just enough to take up the weight of the engine.

37. Unbolt the front engine mounts.

38. If only the engine mounts need to be removed, it is not necessary to remove the engine. However, the weight of the engine and transaxle must be safely supported by a floor jack or engine hoist before the mounts may be removed.

39. Slowly raise the engine a few inches to separate it from the transmission. On manual transmission vehicles, make sure the engine clears the mainshaft. Verify that all wiring harnesses, fuel and coolant lines, and vacuum hoses are disconnected.

40. Lift the engine out of the vehicle. Make sure the engine clears the mounts, the transmission case, and the differential extension shaft.

To install:

NOTE: Use new mounting bolts when installing the transmission side mount and bracket.

41. Install the front engine mounts into the engine compartment.

42. Install the engine in the vehicle. Keep the lifting chain attached to hold the weight of the engine. Install the mounting nuts to hold the engine in place. Do not torque the nuts and bolts at this time. Make sure the differential lines up with the extension shaft. Make sure the mainshaft is aligned in the clutch pressure plate, or the torque converter is flush against the drive plate and mounted on the mainshaft.

43. Raise the vehicle.

44. Install new snap and set rings onto the extension shaft. Apply high-temperature molybdenum grease to its splines and the shaft mating surface in the differential.

45. Support the transmission with a jack and remove the mid-mounts. Carefully fit the engine into position and start the upper engine bolts. Slowly tighten two bolts on opposite sides just enough to draw the engine and transmission together. Install all the remaining bolts except for the differential bolt and its shim. Do not fully tighten the bolts yet.

46. If either the engine, transmission or differential is being replaced, the space between the differential and transmission housings must be measured and the correct shim installed. Shims are available in increments of 0.004 in. (0.1mm). Measure the space between the housings with a feeler gauge. Install the largest shim possible that does not exceed the measurement. If the wrong shim is installed, the differential or transmission housing will be out of alignment and could develop cracks.

47. Install the shim and torque the bolts to 54 ft. lbs. (75 Nm). Torque the transmission case bolts to 54 ft. lbs. (75 Nm).

48. Install the mid-mount and spacer. Loosely install the nuts and bolts for all the mounts and set the engine into place. Remove the engine lifting equipment.

49. Tighten the mounting nuts and bolts in the to the correct torque in the proper sequence. This step is important to preload the engine and transmission mounts. Following the proper sequence will minimize engine vibration and premature mount failure. Make sure the rubber damper mounting surface is not contaminated with oil.

- Left and right front mount nut: 54 ft. lbs. (75 Nm)
- Left front damper bolt: 28 ft. lbs. (39 Nm).
- Left front damper bracket bolt: 40 ft. lbs. (55 Nm).
- Vigor front mount-to-subframe bolts: 43 ft. lbs. (60 Nm)
- 2.5TL front mount-to-subframe bolts: 28 ft. lbs. (38 Nm).
- Mid-mount nuts: 32 ft. lbs. (43 Nm)
- Mid-mount bolts: 28 ft. lbs. (38 Nm)

50. After the engine and transmission have been bolted together, install the extension shaft. Make sure the set ring snaps firmly into place. Coat the threads of the 33mm sealing cap with a sealing compound, install the cap and torque it to 58 ft. lbs. (80 Nm).

51. On vehicles with automatic transmissions, install the torque converter bolts and torque them to 9 ft. lbs. (12 Nm). Do not over torque these bolts or the drive plate will warp. Install the torque converter cover.

52. Install the transmission side mount and bracket. Torque the bracket bolts to 40 ft. lbs. (54 Nm). Torque the mount bolts to 47 ft. lbs. (64 Nm).

53. The balance of the installation is the reverse of the removal procedure.

54. Fill the engine with fresh oil. Refill the differential and transmission. Fill the radiator with a coolant mixture containing no more than 50–60 percent antifreeze.

55. Verify that all components have been reinstalled and connected properly. Check for loose or disconnected wires and lines.

56. Connect the negative and positive battery cables.

57. Bleed the cooling system. Check fluid levels. Run the engine and check its operation.

3.2L (C32A1) Engines

Legend

NOTE: The engine and transaxle are removed as an assembly.

1. Do not remove the hood. Disconnect the hood stay strut and reconnect it to hold the hood in a vertical position.

2. Disconnect the negative battery cable, then the positive battery cable. Remove the battery and the battery box.

3. Remove the radiator cap.

4. Remove the strut bar and its bracket from the bulkhead.

5. Working underneath the vehicle, remove the splash shield. Drain the engine coolant and oil, the transaxle fluid, and the differential oil.

6. Label and disconnect the starter and battery wiring from the main fuse/relay box. Remove the ground cable from the engine block. Label and disconnect the main engine wiring harness.

7. Remove the throttle cable cover. Without turning the adjusting nut, loosen the locknut, which is closer to the throttle, and disconnect the throttle cable from the throttle and bracket.

8. Remove the air cleaner assembly and intake duct.

9. Disconnect the igniter unit located on the right shock tower and remove the wiring harness clamp. Disconnect the engine ground cable.

10. On the firewall behind the right cylinder head is a control box containing emission control equipment. Unplug the electrical connectors and remove the control box from the firewall without disconnecting the vacuum lines. Lay the box on top of the engine.

11. Label and disconnect the main engine wiring harness connectors and remove the bracket.

12. Relieve the fuel system pressure by slowly loosening the service bolt one turn.

— CAUTION —
Fuel injection systems remain under pressure after the engine has been turned OFF. Properly relieve fuel pressure before disconnecting any fuel lines. Failure to do so may result in fire or personal injury.

---

13. Remove the fuel supply hose and disconnect the return hose from the pressure regulator.

14. Disconnect the vacuum hose to the brake booster at the check valve.

15. The transaxle wiring harness is located at the left rear of the engine compartment. Disconnect the harness and remove the clamp.

16. Disconnect and plug the transaxle cooling hoses at the radiator. Disconnect the upper and lower hoses and the fan and sensor wiring. Remove the radiator and fans as an assembly.

17. Remove the bypass solenoid valve assembly, vacuum pipes, and air tank. The valve assembly is located near the power steering reservoir.

18. On GS models with traction control, remove the TCS control valve and its mounting bracket. The valve assembly is located on the right side of the engine compartment behind the air intake duct.

19. Remove the power steering pump without disconnecting the hoses, and secure it out of the way.

20. Raise and safely support the vehicle and remove the front wheels.

21. Remove the lower damper forks and remove the nut from the lower ball joint. Use a suitable ball joint removal tool to disconnect the ball joint from the lower control arm.

22. Carefully pry the inner CV-joints from the differential. Hang the halfshafts out of the way with wire to avoid damaging the outer CV-joint. Cover the inner joints with plastic bags to protect the splines.

23. Remove the steering rack lower cover plate from the rear beam.

24. Without disconnecting any hoses, remove the power steering speed sensor from the differential housing.

25. Disconnect the front exhaust pipe from the catalyst and remove it from the manifolds. On GS models, remove the twin warm-up catalytic converters

26. Leaving the lines connected, remove the air conditioning compressor and hang it from the body with wire.

NOTE: Do not loosen or disconnect the air conditioning refrigerant lines. Do not vent refrigerant into the air.

27. On vehicles with manual transaxles, remove the clutch slave cylinder without disconnecting the hydraulic line. Disconnect the shift lever torque rod and disconnect the shift linkage by driving out the 8mm roll pin.

28. On vehicles with automatic transaxles, remove the converter heat shields. Unbolt the shift cable from its mounting bracket. Disconnect the the shift cable from the input shaft and hang it from the underbody of the vehicle.

29. Remove the engine mid-mounts and the transaxle rear mount and bracket.

30. Working from above, remove one of the EGR valve passage bolts and install a lifting hook. Attach a chain hoist and take up the slack.

31. Remove all engine and transaxle mounting nuts and bolts. Raise the engine/transaxle slightly and check to make sure that all hoses and wires have been disconnected. As the unit is raised, allow it to tilt up so the transaxle will clear the rear beam.

32. The engine and transaxle mounts can be unbolted, then removed from the engine compartment and engine assembly at this time.

33. If removing and replacing the engine mounts, it is not necessary to remove the engine. However, the weight of the engine must be off of the mount and firmly supported by an engine hoist or floor jack before performing this procedure.

To install:

NOTE: Portions of the sub-frame are made of aluminum alloy. Using normal steel bolts

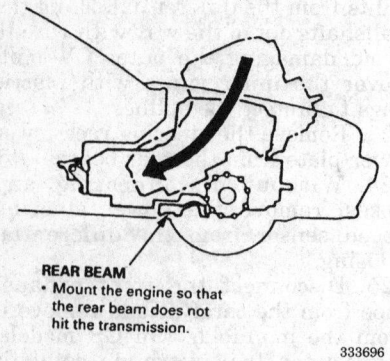

REAR BEAM
• Mount the engine so that the rear beam does not hit the transmission.

333696

Engine installation — Legend

will cause corrosion and looseness. Only use bolts that are specifically designed for this application. These parts are available from the dealer.

34. When installing the engine/transaxle, rotate the unit up in front to avoid hitting the rear beam with the transaxle.

35. Check carefully to make sure the rubber mounts are not twisted or offset. Start all the mounting nuts and bolts before torque tightening them. This step is important to help minimize engine vibrations.

36. Torque the front engine mounts to 28 ft. lbs. (39 Nm). The damper mounting bolts are tightened to 16 ft. lbs. (22 Nm).

37. Torque the mid mount to subframe nut to 35 ft. lbs. (49 Nm).

38. Torque the rear transaxle mount and bracket bolts to 28 ft. lbs. (39 Nm).

39. Remove the lifting hook and install the EGR valve passage bolt. Make sure the rubber damper mounting surface is not contaminated with oil.

40. On vehicles with manual transaxles, reconnect the shift linkage and install the clutch slave cylinder.

41. On vehicles with automatic transaxles, reconnect the shift cable and its mounting bracket. Install the heat shield.

42. When installing the exhaust pipe and catalytic converter, use new gaskets and self-locking nuts. Torque the manifold nuts to 40 ft. lbs. (55 Nm) and the catalyst flange nuts to 16 ft. lbs. (22 Nm).

43. The balance of the removal is the reverse of the installation. Refer to the appropriate portions of this section for tightening torque specifications needed for the various components to be installed.

44. Verify that all engine wiring, fluid lines, and vacuum lines have been connected properly.

45. When the battery is connected, turn the ignition switch **ON** and **OFF** a number of times to pressurize the fuel system and check for leaks.

46. After checking carefully and making sure everything is properly connected, start the engine and bleed the cooling system. When the engine is at operating temperature, stop the engine and adjust the throttle cable.

47. To adjust the throttle cable, loosen both nuts and take up the slack in the cable. Back the adjusting nut away from the bracket so there is 0.120 in. (3.0mm) gap between the nut and bracket. Make sure the

throttle opens and closes fully with pedal movement.

48. Start the engine and check for leaks and proper operation.

3.2TL

NOTE: The engine and transaxle are removed as an assembly.

1. Move the front passenger's seat forward.

2. Do not remove the hood. Disconnect the hood support strut and reconnect it to hold the hood in a vertical position.

NOTE: The radio may contain a coded theft protection circuit. Always obtain the code number before disconnecting the battery.

3. Disconnect the negative battery cable, then the positive battery cable. Remove the battery and the battery box.

4. Remove the engine cover.

5. Remove the air cleaner assembly and intake duct.

6. Remove the throttle cable cover. Without turning the adjusting nut, loosen the locknut, which is closer to the throttle, and disconnect the throttle cable and cruise control cable from the throttle and bracket.

7. Disconnect the engine wire harness connector on the left side of the engine compartment.

8. Remove the engine ground cable and engine wire harness clamps.

9. Disconnect the vacuum hoses, then remove the clamp from the under-hood fuse/relay box.

10. Label then disconnect the battery cables from the under-hood fuse/relay box, then remove the under-hood fuse/relay box.

11. Disconnect the engine wire harness connector, located by the under-hood fuse/relay box.

12. Raise the power steering fluid reservoir, then disconnect the vacuum hoses and remove the vacuum pipe and vacuum tank.

NOTE: Do not disconnect the power steering hoses.

13. Disconnect the ignition control module (igniter) located on the right shock tower and remove the wiring harness clamp. Disconnect the engine ground cable.

14. Disconnect the engine wire harness connectors on the right side of the engine compartment.

15. Remove the ground cable and wire harness clamp.

16. Disconnect the wire harness from the control box and solenoid valve, then remove the control box.

17. Disconnect the brake booster vacuum hose.

18. Remove the two bolts mounting the heater valve.

19. Properly relieve the fuel pressure.

--- **CAUTION** ---

Do not allow fuel spray or fuel vapors to come in contact with a spark or open flame. Keep a dry chemical fire extinguisher nearby. Never store fuel in an open container due to risk of fire or explosion.

20. Disconnect the engine fuel feed hose from the fuel filter and disconnect the fuel return hose from the fuel regulator.

--- **CAUTION** ---

Fuel injection systems remain under pressure after the engine has been turned OFF. Properly relieve fuel pressure before disconnecting any fuel lines. Failure to do so may result in fire or personal injury.

21. Disconnect the evaporative emissions (EVAP) control canister hose and the vacuum hose.

22. Disconnect the transaxle sub–harness connector, and remove the wire harness clamp.

23. Loosen the alternator mounting bolt, lockbolt and adjusting rod, then remove the drive belt.

24. Loosen the A/C idler pulley center nut and adjusting bolt then remove the drive belt.

25. Disconnect the power steering pressure switch connector.

26. Remove the power steering pump adjusting bolt, locknut and mounting bolt then remove the drive belt and pump.

27. Pull the carpet back under the passenger seat to expose the secondary heated oxygen sensor (HO2S) connector, then disconnect the connector.

28. Remove the radiator cap.

29. Raise and safely support the vehicle.

30. Remove the front wheels and the splash guard.

31. Drain the engine coolant into a sealable container.

32. Drain the transaxle fluid into a proper container, then install the drain plug with a new washer.

33. Drain the oil from the differential, then install the drain plug with a new washer.

34. Drain the engine oil into a proper container, then install the drain plug with a new washer.

35. Remove the front suspension damper forks.

36. Disconnect the lower ball joints from the steering knuckles.

37. Disconnect the halfshafts from the differential and the intermediate shaft. Support the halfshafts with wire out of the way and cover the inner CV–joints with plastic bags.

38. Disconnect the A/C compressor clutch connector. Remove the compressor, without disconnecting the hoses.

39. Disconnect the vehicle speed sensor connector, then remove the VSS/power steering sensor. Do not disconnect the fluid hoses.

40. Remove the heat shields from exhaust pipe A.

41. Remove the nuts attaching exhaust pipe A to the exhaust manifolds and the catalytic converter. Remove exhaust pipe A and discard the gaskets.

42. Remove the oxygen sensor wire harness cover and grommet, then remove the catalytic converter. Discard the nuts and gasket.

43. Remove the exhaust heat shield from the floor of the vehicle.

44. Disconnect the transaxle cooler hoses, then plug the hoses and pipes.

45. Remove the shift cable cover mounting bolts and remove the wire harness clamps from the cover. Remove the shift cable cover from the transaxle.

46. Remove the shift cable holder from the holder base, do not lose the washers.

47. Remove the locknut attaching the shift cable to the control lever, then remove the shift cable.

48. Lower the vehicle.

49. Remove the upper and lower radiator hoses.

50. Remove the radiator assembly.

51. Remove the heater hoses.

52. Attach a hoist to the engine lifting points.

53. Remove the center bracket from the front engine mounts.

54. Remove the center mount from the front beam.

55. Remove the nuts and bolts attaching the left and right engine mount brackets to the left and right brackets.

56. Working under the vehicle, Remove the shift cable guide bracket.

57. Remove the bolts attaching the transmission beam to the body, and loosen the three bolts on the transmission beam.

58. Remove the stop holder, the mid mount stops and the mid mounts.

59. Verify that the engine and transaxle assembly is completely free of vacuum hoses, fuel and coolant lines and electrical wiring.

60. Slowly raise the engine and transaxle. Remove the left end right brackets from the front engine mounts.

61. Raise the engine all of the way and remove it from the vehicle. Separate then engine and the transaxle assembly.

To install:

62. Carefully install the engine and transaxle into the engine compartment, take care to not hit the rear beam.

NOTE: Check carefully to make sure that the rubber mounts are not twisted or offset. Start all of the mount nuts and bolts before torquing them in place. This is important to help minimize engine vibrations.

63. Install the center mount to the front beam and torque the bolts to 40 ft. lbs. (54 Nm).

64. Install the left bracket to the left front mount and the engine mount bracket. Do not tighten the nuts and bolts at this time.

65. Install the right bracket to the right front mount and the engine mount bracket. Do not tighten the nuts and bolts at this time.

66. Install the center bracket. Torque the bolts attaching the brackets to 40 ft. lbs. (54 Nm) then torque the mount through-bolt to 40 ft. lbs. (54 Nm).

67. Install the transmission beam mounting bolts, do not tighten the bolts at this time. Loosen the bolts attaching the mount to the beam and the mount through-bolt.

68. Install the mid mounts, then the mid mount stops. Tighten the 8mm bolts loosely then install the stop holder.

69. Torque the mid mount 10mm bolts to 28 ft. lbs. (38 Nm) and torque the 8mm bolts to 16 ft. lbs. (22 Nm). Torque the new nuts attaching the midmounts to 35 ft. lbs. (48 Nm) and torque the nuts attaching the stop holder to 40 ft. lbs. (54 Nm).

70. Torque the nut and bolt attaching the left bracket and the left engine mount bracket to 28 ft. lbs. (38 Nm). Torque the left bracket through-bolt to 40 ft. lbs. (54 Nm).

71. Torque the nut and bolt attaching the right bracket and the right engine mount bracket to 28 ft. lbs. (38 Nm). Torque the right bracket through-bolt to 40 ft. lbs. (54 Nm).

72. Torque the bolts attaching the transmission beam to the vehicle to 28 ft. lbs. (38 Nm), then torque the bolts attaching the mount to the beam to 40 ft. lbs. (54 Nm). Torque the three bolts attaching the bracket to the transaxle to 28 ft. lbs. (38 Nm) then torque the mount through-bolt to 40 ft. lbs. (54 Nm).

73. Remove the engine hoist.

74. Connect the heater hoses to the heater core at the bulkhead.

75. Install the radiator assembly.

76. Install the upper and lower radiator hoses.

77. Raise and safely support the vehicle.

78. Connect the shift cable control lever to the control shaft, then install the washer and nut. Torque the nut to 8.7 ft. lbs. (12 Nm).

79. Install the shift cable holder to the shift cable holder base with the mounting washers. Install the attaching bolts and torque the bolts to 8.7 ft. lbs. (12 Nm).

80. Install the shift cable cover and torque the mounting bolts to 8.7 ft. lbs. (12 Nm).

81. The balance of the removal is the reverse of the installation. As necessary, refer to the appropriate portions of this section for tightening torque specifications needed for the various components to be installed.

82. Fill the engine with engine oil.

83. Fill the transmission with transmission oil.

84. Connect the positive then the negative battery cables and enter the radio security code.

85. Switch the ignition **ON** but do not engage the starter. The fuel pump should run for approximately 2 seconds, building pressure within the lines. Switch the ignition **OFF**, then **ON** 2 or 3 more times to build full system pressure. Check for fuel leaks.

86. Fill the cooling system and bleed the air from the cooling system.

87. Run the engine and check for leaks, check the transmission oil level and add if necessary.

Engine Mounts

REMOVAL AND INSTALLATION

NOTE: The radio may have a coded theft protection circuit. Obtain the code from the owner before disconnecting the battery, removing the radio fuse, or removing the radio.

1.7L (B17A1), 1.8L (B18A1, B18B1, B18C1) Engines

Side Engine Mount

1. Secure the hood as far open as possible.

2. Disconnect the negative battery cable then the positive battery cable.

3. Attach an engine hoist to the engine lifting points and raise the hoist to remove all slack from the chain.

4. Remove the nuts and bolt attaching the side engine mount to the engine.

5. Remove the through-bolt from the side engine mount.

To install:

6. Install the side engine mount. Do not tighten the mount through-bolt. Torque the nuts and bolt attaching the mount to the engine to 38 ft. lbs. (52 Nm).

7. Torque the nut and bolt attaching the side engine mount to the engine to 47 ft. lbs. (64 Nm).

8. Remove the hoist equipment from the engine.

9. Connect the positive then the negative battery cables and enter the radio security code.

Left Front Engine Mount

1. Secure the hood as far open as possible.

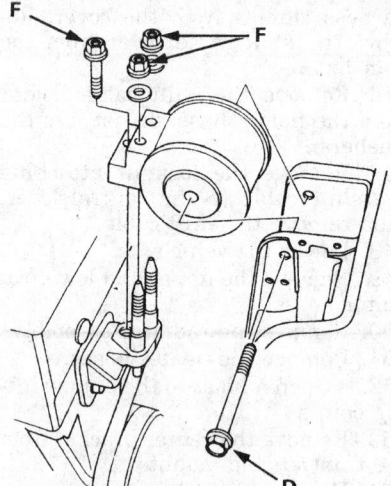

Torque Specifications:
D: 12 x 1.25 mm
   74 N·m (7.5 kgf·m, 54 lbf·ft)
F: 10 x 1.25 mm
   52 N·m (5.3 kgf·m, 38 lbf·ft)
Replace.

323550

Side engine mount — 1.7L (B17A1), 1.8L (B18A1, B18B1, B18C1) engines

2. Disconnect the negative battery cable then the positive battery cable.

3. Attach an engine hoist to the engine lifting points and raise the hoist to remove all slack from the chain.

4. Remove the nut and washer from the mount bolt, discard the nut.

5. Remove the bolts attaching the front engine mount to the vehicle and remove the mount.

6. Remove and discard the mount bolt.

To install:

7. Install a new mount bolt to the engine and torque it to 61 ft. lbs. (83 Nm).

8. Install the left front mount and the bolts attaching the mount to the vehicle. Torque the bolts to 33 ft. lbs. (44 Nm).

9. Install the washer and new nut to the mount bolt then torque the nut to 43 ft. lbs. (59 Nm).

10. Remove the hoist equipment from the engine.

11. Connect the positive then the negative battery cables and enter the radio security code.

Right Front Engine Mount

1. Secure the hood as far open as possible.

2. Disconnect the negative battery cable then the positive battery cable.

3. Attach an engine hoist to the engine lifting points and raise the hoist to remove all slack from the chain.

4. Remove the nut and washer from the mount bolt, discard the nut.

5. Remove the bolts attaching the right front engine mount to the vehicle.

To install:

6. Install the left front mount and the bolts attaching the mount to the vehicle. Torque the bolts to 33 ft. lbs. (44 Nm).

7. Install the washer and new nut to the mount bolt then torque the nut to 43 ft. lbs. (59 Nm).

8. Remove the hoist equipment from the engine.

9. Connect the positive then the negative battery cables and enter the radio security code.

Rear Engine Mount

1. Secure the hood as far open as possible.

2. Disconnect the negative battery cable then the positive battery cable.

3. Attach an engine hoist to the engine lifting points and raise the hoist to remove all slack from the chain.

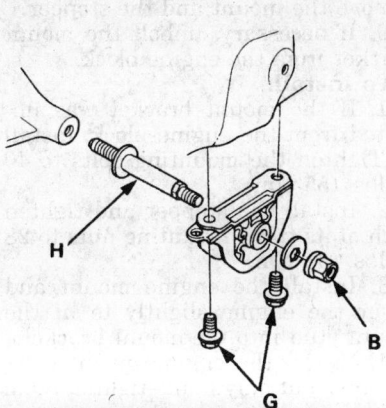

M/T:

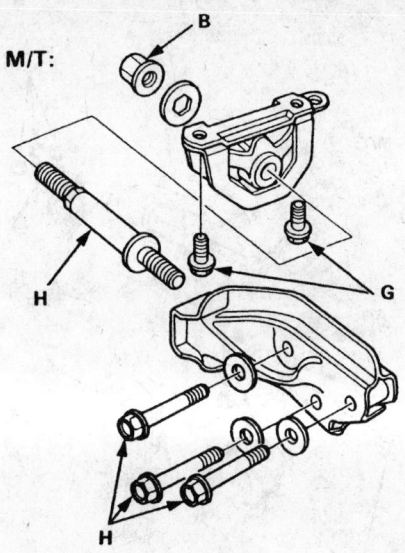

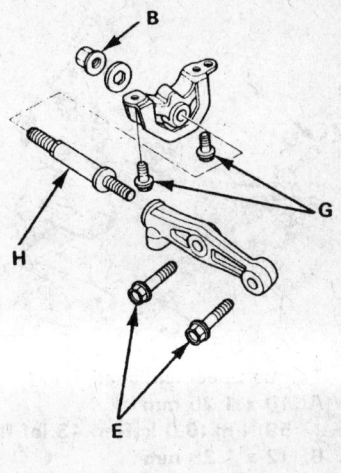

Torque Specifications:

B: 12 x 1.25 mm
 59 N·m (6.0 kgf·m, 43 lbf·ft)
 Replace.

G: 10 x 1.25 mm
 44 N·m (4.5 kgf·m, 33 lbf·ft)

H: 12 x 1.25 mm
 83 N·m (8.5 kgf·m, 61 lbf·ft)

323551

Left front engine mount — 1.7L (B17A1), 1.8L (B18A1, B18B1, B18C1) engines

Torque Specifications:

B: 12 x 1.25 mm
 59 N·m (6.0 kgf·m, 43 lbf·ft)
 Replace.

E: 12 x 1.25 mm
 64 N·m (6.5 kgf·m, 47 lbf·ft)

G: 10 x 1.25 mm
 44 N·m (4.5 kgf·m, 33 lbf·ft)

H: 12 x 1.25 mm
 83 N·m (8.5 kgf·m, 61 lbf·ft)

323552

Right front engine mount — 1.7L (B17A1), 1.8L (B18A1, B18B1, B18C1) engines

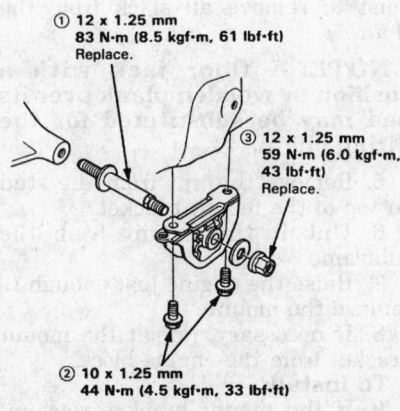

① 12 x 1.25 mm
83 N·m (8.5 kgf·m, 61 lbf·ft)
Replace.

③ 12 x 1.25 mm
59 N·m (6.0 kgf·m, 43 lbf·ft)
Replace.

② 10 x 1.25 mm
44 N·m (4.5 kgf·m, 33 lbf·ft)

223498

Left front mount torque sequence — 1994–97 1.8L (B18B1, B18C1) engines

A/T:

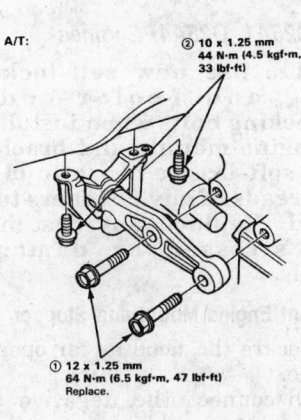

② 10 x 1.25 mm
44 N·m (4.5 kgf·m, 33 lbf·ft)

① 12 x 1.25 mm
64 N·m (6.5 kgf·m, 47 lbf·ft)
Replace.

M/T:

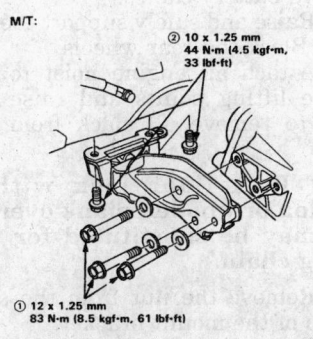

② 10 x 1.25 mm
44 N·m (4.5 kgf·m, 33 lbf·ft)

① 12 x 1.25 mm
83 N·m (8.5 kgf·m, 61 lbf·ft)

223497

Right front mount torque sequence — 1994–97 1.8L (B18B1, B18C1) engines

4. Remove and discard the through-bolt from the rear engine mount.

5. Remove the bolts attaching the mount to the rear beam.

To install:

6. Install the rear engine mount to the rear beam. Torque the bolts attaching the mount to the beam to 43 ft. lbs. (59 Nm).

7. Install a new rear engine mount through-bolt. Torque the new through-bolt to 43 ft. lbs. (59 Nm).

8. Remove the hoist equipment from the engine.

9. Connect the positive then the negative battery cables and enter the radio security code.

Transaxle Mount

1. Secure the hood as far open as possible.

2. Disconnect the negative battery cable then the positive battery cable.

3. Attach an engine hoist to the engine lifting points and raise the hoist to remove all slack from the chain.

4. Remove the nuts and bolt attaching the mount to the transaxle.

5. Remove the through-bolt from the transaxle mount.

6. Remove the transaxle mount from the mounting bracket.

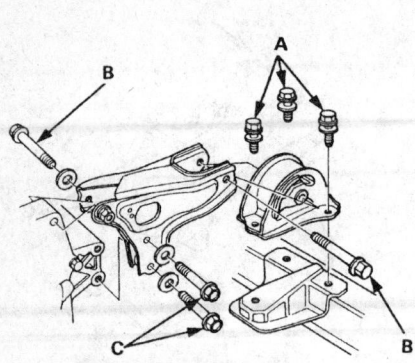

Torque Specifications:
A: 10 x 1.25 mm
   59 N·m (6.0 kgf·m, 43 lbf·ft)
B: 12 x 1.25 mm
   59 N·m (6.0 kgf·m, 43 lbf·ft)
   Replace.
C: 14 x 1.5 mm
   118 N·m (12.0 kgf·m, 86.8 lbf·ft)
   Replace.

323553

Rear engine mount — 1.7L (B17A1), 1.8L (B18A1, B18B1, B18C1) engines

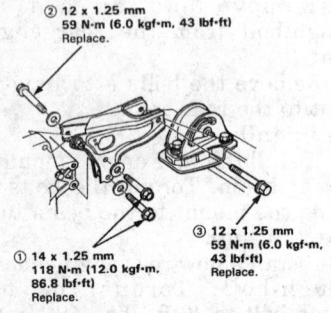

② 12 x 1.25 mm
   59 N·m (6.0 kgf·m, 43 lbf·ft)
   Replace.

① 14 x 1.25 mm
   118 N·m (12.0 kgf·m, 86.8 lbf·ft)
   Replace.

③ 12 x 1.25 mm
   59 N·m (6.0 kgf·m, 43 lbf·ft)
   Replace.

223496

Rear mount torque sequence — 1994–97 1.8L (B18B1, B18C1) engines

To install:

7. Install the transaxle mount and the attaching nuts and bolts.
8. Torque the nuts and bolts attaching the transaxle mount to the transaxle to 38 ft. lbs. (52 Nm).
9. Torque the transaxle mount through-bolt to 54 ft. lbs. (74 Nm).
10. Remove the hoist equipment from the engine.
11. Connect the positive then the negative battery cables and enter the radio security code.

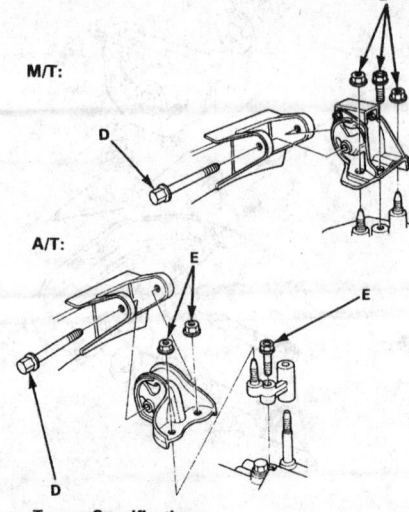

Torque Specifications:
D: 12 x 1.25 mm
   74 N·m (7.5 kgf·m, 54 lbf·ft)
E: 12 x 1.25 mm
   64 N·m (6.5 kgf·m, 47 lbf·ft)

323554

Transaxle mount — 1.7L (B17A1), 1.8L (B18A1, B18B1, B18C1) engines

2.5L (G25A1, G25A4) Engines

NOTE: Use new self-locking nuts and color-coded self-locking bolts when installing the engine mounts and brackets Use a soft-bristle brush to clean the threads of any fasteners to be reused. Replace any fasteners with stressed or damaged threads.

Left Front Engine Mount and Stopper

1. Secure the hood as far open as possible.
2. Disconnect the negative and positive battery cables.
3. Raise and safely support the vehicle. Block the rear wheels.
4. Attach an engine hoist to the engine lifting points and raise the hoist to remove all slack from the chain.

NOTE: A floor jack with a cushion or wooden plank over its pad may be substituted for the lifting chain.

5. Remove the nut from the stud on top of the mount bracket.
6. Remove the through-bolt from the front stopper bracket.
7. Unbolt the mount from the subframe.
8. Remove the stopper from the subframe.

9. Raise the engine just enough to remove the mount and the stopper.
10. If necessary, unbolt the mount bracket from the engine block.

To install:

11. If the mount bracket was unbolted from the engine block, install it. Tighten the mounting bolts to 40 ft. lbs. (55 Nm).
12. Install the stopper and tighten each of its three mounting nuts to 28 ft. lbs. (38 Nm).
13. Install the engine mount and lower the engine slightly to fit the mount stud into the mount bracket.
14. Install the engine mount bolts and nut, but only hand-tighten them at this point.
15. Lower the engine so its weight is on the mounts. First, tighten the four mount bracket bolts to 43 ft. lbs. (60 Nm) each. Next, tighten the stopper through-bolt to 40 ft. lbs. (55 Nm). Finally, tighten the stud nut to 54 ft. lbs. (74 Nm).
16. Remove the hoist equipment from the engine.
17. Lower the vehicle.
18. Reconnect the positive and negative battery cables.

Right Front Engine Mount

1. Secure the hood as far open as possible.
2. Disconnect the negative and positive battery cables.
3. Raise and safely support the vehicle. Block the rear wheels.
4. Attach an engine hoist to the engine lifting points and raise the hoist to remove all slack from the chain.

NOTE: A floor jack with a cushion or wooden plank over its pad may be substituted for the lifting chain.

5. Remove the nut from the stud on top of the mount bracket.
6. Unbolt the mount from the subframe.
7. Raise the engine just enough to remove the mount.
8. If necessary, unbolt the mount bracket from the engine block.

To install:

9. If the mount bracket was unbolted from the engine block, install it. Tighten the mounting bolts to 40 ft. lbs. (55 Nm).
10. Install the engine mount and lower the engine slightly to fit the mount stud into the mount bracket.
11. Install the engine mount bolts and nut, but only hand-tighten them at this point.
12. Lower the engine so its weight is on the mounts. First, tighten the four mount bracket bolts to 43 ft. lbs.

Mount Locations:

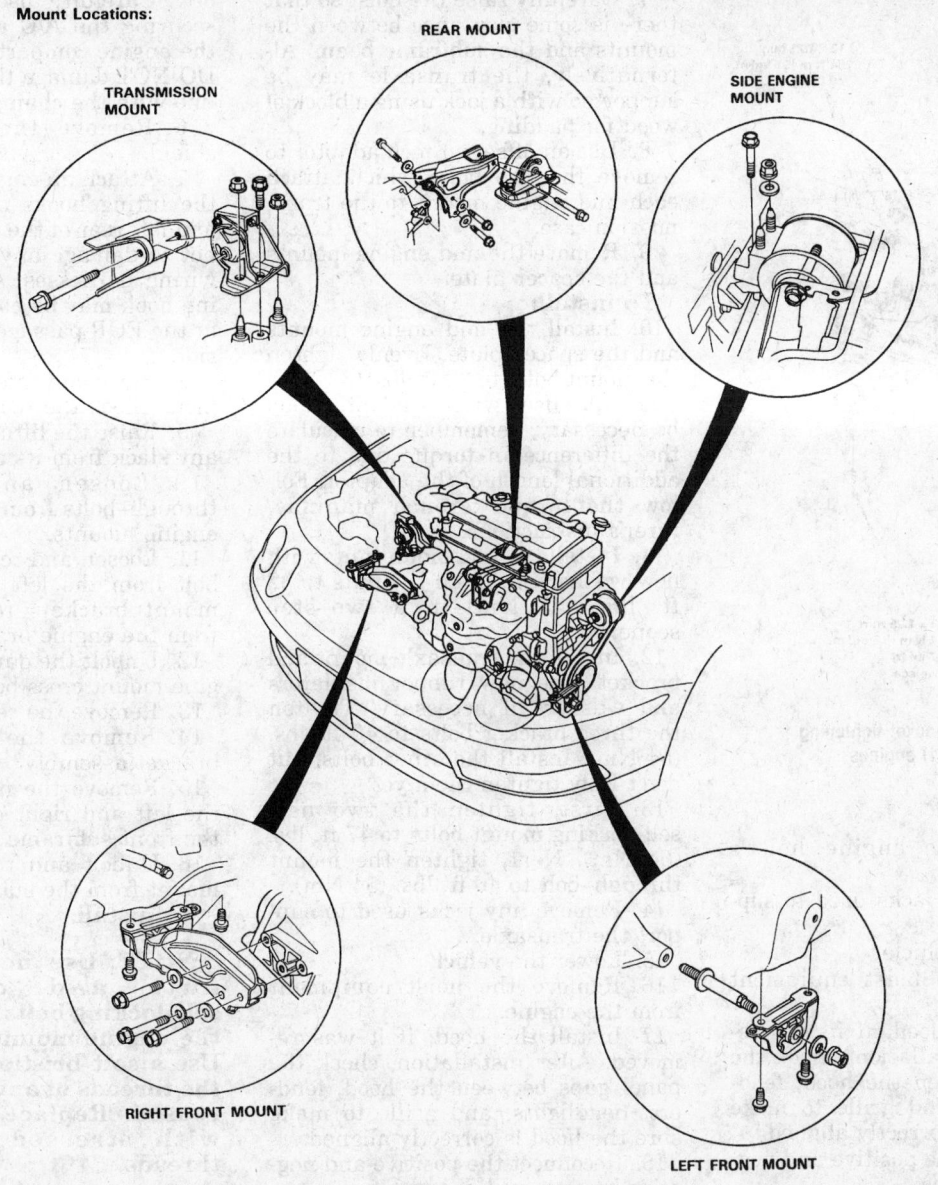

TRANSMISSION MOUNT

REAR MOUNT

SIDE ENGINE MOUNT

RIGHT FRONT MOUNT

LEFT FRONT MOUNT

323549

Engine mount locations — 1.7L (B17A1), 1.8L (B18A1, B18B1, B18C1) engines

(60 Nm) each. Next, tighten the stud nut to 54 ft. lbs. (74 Nm).

13. Remove the hoist equipment from the engine.

14. Lower the vehicle.

15. Reconnect the positive and negative battery cables.

Transaxle Mount and Bracket

1. Support the hood as far open as possible. If necessary, matchmark the hood hinges with a felt–tipped marker and remove the hood for extra hoist clearance.

2. Disconnect the negative and positive battery cables.

3. Attach an engine hoist to the engine lifting points and raise the hoist to remove all slack from the chain.

4. Carefully raise the hoist so that there is some clearance between the mounts and the subframe beam. Alternatively, the transaxle may be supported with a jack using a block of wood for padding.

5. Remove the self–locking bolts which attach the mount to the subframe. Remove the rubber isolators and washers. Inspect the isolators and washers for distortion and disintegration and replace them if necessary.

6. Remove the through–bolt and the tab–locking nut from the mount. Remove the mount.

7. If necessary, the mount bracket may be removed by removing the three bolts.

To install:

8. Install the transaxle mount bracket and tighten each of the three bolts to 40 ft. lbs. (54 Nm).

9. Install the transaxle mount. Use new rubber insulators and washers if necessary. Install the through–bolt and two new mount bolts, but don't fully tighten them yet.

10. First, tighten the two new self–locking mount bolts to 47 ft. lbs. (64 Nm). Next, tighten the mount through–bolt to 40 ft. lbs. (54 Nm).

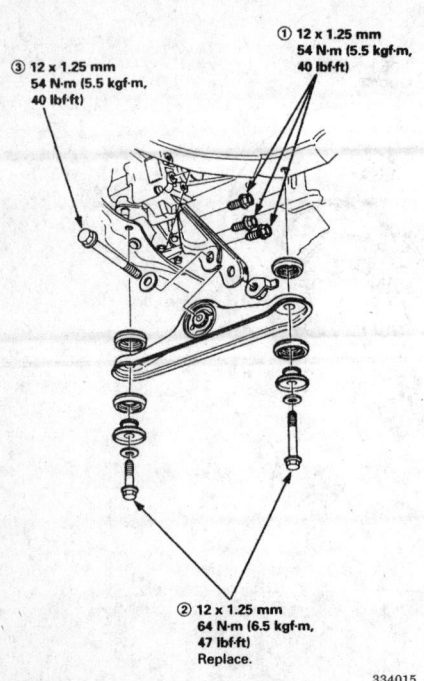

① 12 x 1.25 mm
54 N·m (5.5 kgf·m,
40 lbf·ft)

③ 12 x 1.25 mm
54 N·m (5.5 kgf·m,
40 lbf·ft)

② 12 x 1.25 mm
64 N·m (6.5 kgf·m,
47 lbf·ft)
Replace.

334015

Transaxle mount and bracket tightening sequence — 2.5L (G25A1) engines

11. Release the engine hoist's tension.

12. Remove any jacks used to support the transaxle.

13. Lower the vehicle.

14. Remove the hoist equipment from the engine.

15. Install the hood, if it was removed. After installation, check the panel gaps between the hood, fenders, headlights, and grille to make sure the hood is correctly aligned.

16. Reconnect the positive and negative battery cables.

Mid Engine Mount and Spacer

1. Support the hood as far open as possible. If necessary, matchmark the hood hinges with a felt–tipped marker and remove the hood for extra hoist clearance.

2. Disconnect the negative and positive battery cables.

3. Raise and safely support the vehicle.

4. Attach an engine hoist to the engine lifting points and raise the hoist to remove all slack from the chain.

5. Remove the transaxle mount and bracket.

6. Remove the two nuts which secure the mid engine mounts to the subframe beam.

7. Carefully raise the hoist so that there is some clearance between the mounts and the subframe beam. Alternatively, the transaxle may be supported with a jack using a block of wood for padding.

8. Use an offset wrench adaptor to remove the two bolts which attach each mid engine mount to the transmission case.

9. Remove the mid engine mounts and the spacer plate.

To install:

10. Install the mid engine mounts and the spacer plate. Evenly tighten the mount bolts to 28 ft. lbs. (38 Nm). Since an offset wrench adaptor may be necessary, remember to calculate the difference in torque due to the additional length of the adaptor. Follow the torque wrench manufacturer's instructions.

11. Install the mid mount nuts with new washers. Tighten the nuts to 32 ft. lbs. (43 Nm) in a two–step sequence.

12. Install the transaxle mount and bracket. Use new rubber insulators and washers if necessary. Tighten the three bracket bolts to 40 ft. lbs. (54 Nm). Install the other bolts, but don't fully tighten them yet.

13. First, tighten the two new self–locking mount bolts to 47 ft. lbs. (64 Nm). Next, tighten the mount through–bolt to 40 ft. lbs. (54 Nm).

14. Remove any jacks used to support the transaxle.

15. Lower the vehicle.

16. Remove the hoist equipment from the engine.

17. Install the hood, if it was removed. After installation, check the panel gaps between the hood, fenders, headlights, and grille to make sure the hood is correctly aligned.

18. Reconnect the positive and negative battery cables.

1993–95 3.2L (C32A1) Engines

Left and Right Front Engine Mounts, Bracket, Center Mount, Beam, and Damper

1. Support the hood in the fully–open position. Use a felt–tipped marker to matchmark the hood brackets and hinge plates.

2. Disconnect the hood struts, and then unbolt the hinge plates and remove the hood. Work carefully to avoid damaging the paint.

3. Disconnect the negative and positive battery cables and then remove the battery. If desired, the battery tray can be removed.

4. Remove the air intake duct and the air cleaner assembly.

5. Carefully disconnect the clamp securing the A/C refrigerant line to the engine compartment strut brace. DO NOT damage the A/C refrigerant line with the chain hoist.

6. Remove the engine splash shield.

7. Attach an engine lifting hoist to the lifting hooks at the front right and left rear of the engine. Be careful not to damage any control cables or wiring harnesses. An additional lifting hook may be bolted to the engine at the EGR passage on the right rear side.

8. Raise and safely support the vehicle. Block the rear wheels.

9. Raise the lifting hoist to remove any slack from its chains.

10. Loosen and remove the through–bolts from the left and right engine mounts.

11. Loosen and remove the nut and bolt from the left and right engine mount brackets to separate them from the engine brackets.

12. Unbolt the damper from the engine mount cross beam.

13. Remove the center mount nut.

14. Remove the crossbeam and bracket assembly.

15. Remove the nuts which attach the left and right engine mounts to the front subframe beam.

16. Unbolt and remove the center mount from the subframe beam.

To install:

NOTE: Use new self–locking nuts and color–coded self–locking bolts when installing the engine mounts and brackets Use a soft–bristle brush to clean the threads of any fasteners to be reused. Replace any fasteners with stressed or damaged threads.

17. Install the center mount. Tighten its bolts to 28 ft. lbs. (38 Nm).

18. Install the left and right engine mounts and tighten the new self–locking nuts to to 28 ft. lbs. (38 Nm).

19. Install the crossbeam and bracket assembly. If the four 10mm crossbeam bolts were loosened or removed, they must be replaced. Don't fully tighten the crossbeam and bracket assembly fasteners at this point.

20. Install the left and right engine mount through–bolts. Don't fully tighten them yet.

21. Reconnect the left and right engine mounts and mount brackets to the engine brackets. Don't fully tighten the fasteners at this point.

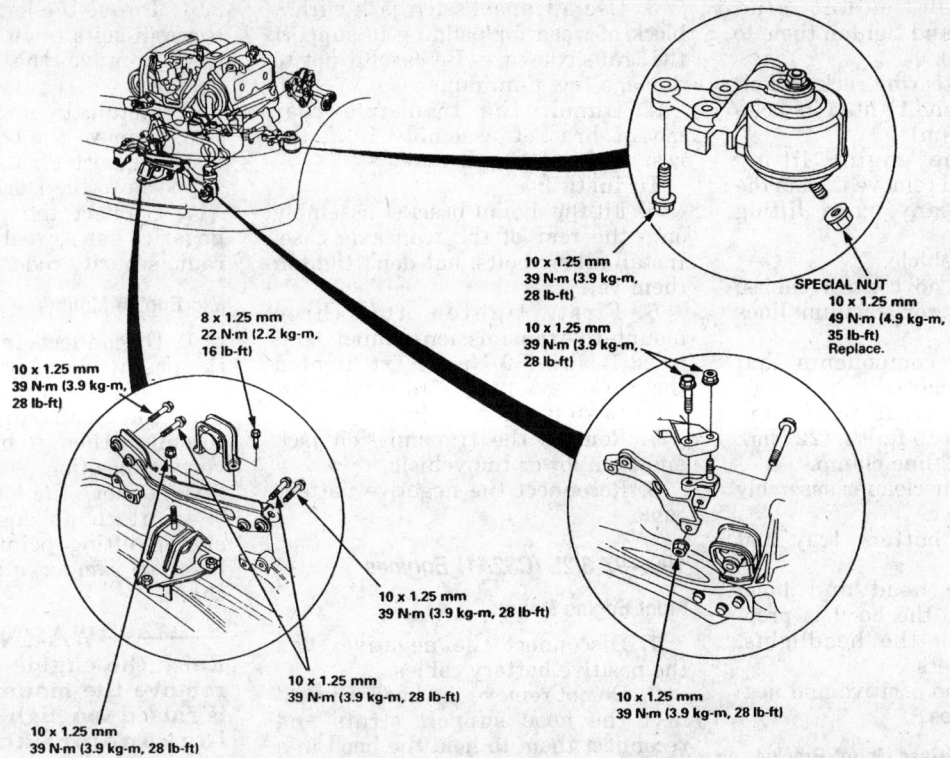

10 x 1.25 mm
39 N·m (3.9 kg-m,
28 lb-ft)

8 x 1.25 mm
22 N·m (2.2 kg-m,
16 lb-ft)

10 x 1.25 mm
39 N·m (3.9 kg-m,
28 lb-ft)

SPECIAL NUT
10 x 1.25 mm
49 N·m (4.9 kg-m,
35 lb-ft)
Replace.

10 x 1.25 mm
39 N·m (3.9 kg-m,
28 lb-ft)

10 x 1.25 mm
39 N·m (3.9 kg-m, 28 lb-ft)

10 x 1.25 mm
39 N·m (3.9 kg-m, 28 lb-ft)

10 x 1.25 mm
39 N·m (3.9 kg-m, 28 lb-ft)

10 x 1.25 mm
39 N·m (3.9 kg-m, 28 lb-ft)

333797

Front engine mount beam, bracket, and damper components — 1993–95 3.2L (C32A1) engines

22. Install the center mount nut. Don't fully tighten it, or install the damper yet.

23. Tighten the left and right engine mount and crossbeam fasteners to their final torque specifications using the following sequence:

 a. Tighten the left and right mount lower self–locking nut to 28 ft. lbs. (39 Nm). This nut secures the mount to the subframe.

 b. Tighten the left and right mount bracket nut and bolt to 28 ft. lbs. (39 Nm).

 c. Tighten the outer two 10mm crossbeam bolts to 28 ft. lbs. (39 Nm).

 d. Tighten the inner two 10mm crossbeam bolts to 28 ft. lbs. (39 Nm).

 e. Tighten the center mount nut to 28 ft. lbs. (39 Nm).

 f. Tighten the left and right mount through–bolts to 28 ft. lbs. (39 Nm).

 g. Install the damper and tighten its mounting bolts to 16 ft. lbs. (22 Nm).

24. Release the engine lifting hoist's tension and remove it from the engine. Remove any extra lifting hooks.

25. Install the splash shield and lower the vehicle.

26. Verify that no control cables, wiring harnesses, or vacuum lines have be damaged.

27. Install any components that may have been removed.

28. Install the strut brace and tighten its bolts to 16 ft. lbs. (22 Nm). Reconnect the A/C line clamp.

29. Install the air cleaner assembly and air intake duct.

30. Install the battery tray and battery.

31. Install the hood and hood struts. Make sure the hood is properly aligned with the headlights, bumper, and fenders.

32. Reconnect the positive and negative battery cables.

Left and Right Mid Engine Mounts

1. Support the hood in the fully–open position. Use a felt–tipped marker to matchmark the hood brackets and hinge plates.

2. Disconnect the hood struts, and then unbolt the hinge plates and remove the hood. Work carefully to avoid damaging the hood or scratching the paint.

3. Disconnect the negative and positive battery cables and then remove the battery. If desired, the battery tray can be removed.

4. Remove the air intake duct and the air cleaner assembly.

5. Carefully disconnect the clamp securing the A/C refrigerant line to the engine compartment strut brace. DO NOT damage the A/C refrigerant line with the chain hoist.

6. Attach an engine lifting hoist to the lifting hooks at the front right and left rear of the engine. Be careful not to damage any control cables or wiring harnesses. An additional lifting hook may be bolted to the engine at the EGR passage on the right rear side.

7. Raise and safely support the vehicle. Block the rear wheels.

8. Raise the lifting hoist to remove any slack from its chains.

9. Unbolt and remove the guard plate from underneath the steering rack and pinion.

10. Remove the mount stud's self–locking nut to release the mount from the subframe beam.

11. Remove the three bolts attaching each mid mount assembly to the transaxle case.

12. Remove the mounts.

 To install:

13. Install the left and right mid engine mounts.

14. Install the three mount bolts and tighten each to 28 ft. lbs. (39 Nm).

15. Install new mount stud self–locking nuts and tighten them to 35 ft. lbs. (49 Nm).

16. Install the steering rack and pinion guard plate and tighten its bolts to 28 ft. lbs. (39 Nm).

17. Release the engine lifting hoist's tension and remove it from the engine. Remove any extra lifting hooks.

18. Lower the vehicle.

19. Verify that no control cables, wiring harnesses, or vacuum lines have be damaged.

20. Install any components that may have been removed.

21. Install the strut brace and tighten its bolts to 16 ft. lbs. (22 Nm). Reconnect the A/C line clamp.

22. Install the air cleaner assembly and air intake duct.

23. Install the battery tray and battery.

24. Install the hood and hood struts. Make sure the hood is properly aligned with the headlights, bumper, and fenders.

25. Reconnect the positive and negative battery cables.

Automatic Transaxle Rear Mount Bracket

1. Disconnect the negative battery cable.

2. Raise and safely support the vehicle.

3. Use a transmission jack with a block of wood for padding to support the transaxle case. Be careful not to damage the aluminum case.

4. Remove the bolts which attach the mount beam to the mount bracket. Unbolt and remove the mount beam.

5. Unbolt the transaxle rear mount bracket assembly from the rear of the transaxle case.

To install:

6. Fit the mount bracket assembly onto the rear of the transaxle case. Install the three bolts and tighten each one to 28 ft. lbs. (39 Nm).

7. Install the mount beam. Install all seven mounting bolts, but don't tighten them yet.

8. First, tighten the four beam–to–transmission–tunnel bolts to 28 ft. lbs. (39 Nm). Next, tighten the three mount bolts to 28 ft. lbs. (39 Nm).

9. Remove the transmission jack and lower the vehicle.

10. Reconnect the negative battery cable.

Manual Transaxle Rear Mount Bracket

1. Disconnect the negative battery cable.

2. Raise and safely support the vehicle.

3. Use a transmission jack with a block of wood for padding to support the transaxle case. Be careful not to damage the aluminum case.

4. Unbolt the transaxle rear mount bracket assembly from the rear of the transaxle case.

To install:

5. Fit the mount bracket assembly onto the rear of the transaxle case. Install all six bolts, but don't tighten them yet.

6. First, tighten the three mount–to–transmission–tunnel bolts to 28 ft. lbs. (39 Nm). Next, tighten the three mount–to–transaxle–case bolts to 28 ft. lbs. (39 Nm).

7. Remove the transmission jack and then lower the vehicle.

8. Reconnect the negative battery cable.

1996–97 3.2L (C32A1) Engines

Front Engine Mounts

1. Disconnect the negative then the positive battery cables.

2. Do not remove the hood. Disconnect the hood support struts and reconnect them to hold the hood in a vertical position.

3. Remove the intake air duct.

4. Attach an engine hoist to the engine lifting points and raise the hoist to remove all slack from the chain.

5. Remove the through-bolts from the center, left and right front engine mounts.

6. Raise the engine off of the front engine mounts.

―――――― WARNING ――――――
Raise the engine only enough to remove the mount, if the engine is raised too high the engine wiring and/or hoses may be damaged.

7. If removing the center mount, remove the bolts attaching the mount to the front beam then remove the mount.

8. If removing the left or right mount, remove the nut attaching the mount to the beam then remove the mount.

To install:

9. If the left or right mount was removed, install the mount and torque the nut to 28 ft. lbs. (38 Nm).

10. If the center mount was removed, install the mount and install the bolts. Torque the bolts to 40 ft. lbs. (54 Nm).

11. Lower the engine on to the front mounts and install the through-bolts.

12. Torque the center mount through-bolt to 40 ft. lbs. (54 Nm).

13. Torque the left and right mount through-bolts to 40 ft. lbs. (54 Nm).

14. Remove the hoist from the engine.

15. Install the intake air duct.

16. Remove the bolts attaching the hood support struts and install the supports to their original positions.

17. Connect the positive then the negative battery cables and enter the radio security code.

Mid Engine Mounts

1. Disconnect the negative then the positive battery cables.

2. Do not remove the hood. Disconnect the hood support struts and reconnect them to hold the hood in a vertical position.

3. Remove the intake air duct.

4. Attach an engine hoist to the engine lifting points and raise the hoist to remove all slack from the chain.

―――――― WARNING ――――――
Raise the engine only enough to remove the mount, if the engine is raised too high the engine wiring and/or hoses may be damaged.

5. Remove the through-bolts from the center, left and right front engine mounts.

6. Loosen the bolts attaching the transmission mount to the beam, loosen the transmission mount through-bolt, then loosen the bolts attaching the transmission mount bracket to the transaxle.

7. Remove the nuts attaching the stop holder, if equipped.

8. Remove and discard the nut attaching the mid mount to the beam.

9. Remove the three bolts attaching the midmount bracket to the engine, then remove the mount and bracket.

10. Remove the bracket from the mount.

To install:

11. Install the bracket to the midmount, make sure that the mount is properly positioned with the guide pin through the bracket. Torque the midmount nut to 28 ft. lbs. (38 Nm).

12. Install the midmount and bracket to the engine, loosely install the bolts.

13. Install the nuts to the stop holder, if equipped.

14. Torque the mid mount 10mm bolts to 28 ft. lbs. (38 Nm) and torque the 8mm bolts to 16 ft. lbs. (22 Nm). Torque the new nuts attaching the midmounts to 35 ft. lbs. (48 Nm) and torque the nuts attaching the stop holder to 40 ft. lbs. (54 Nm).

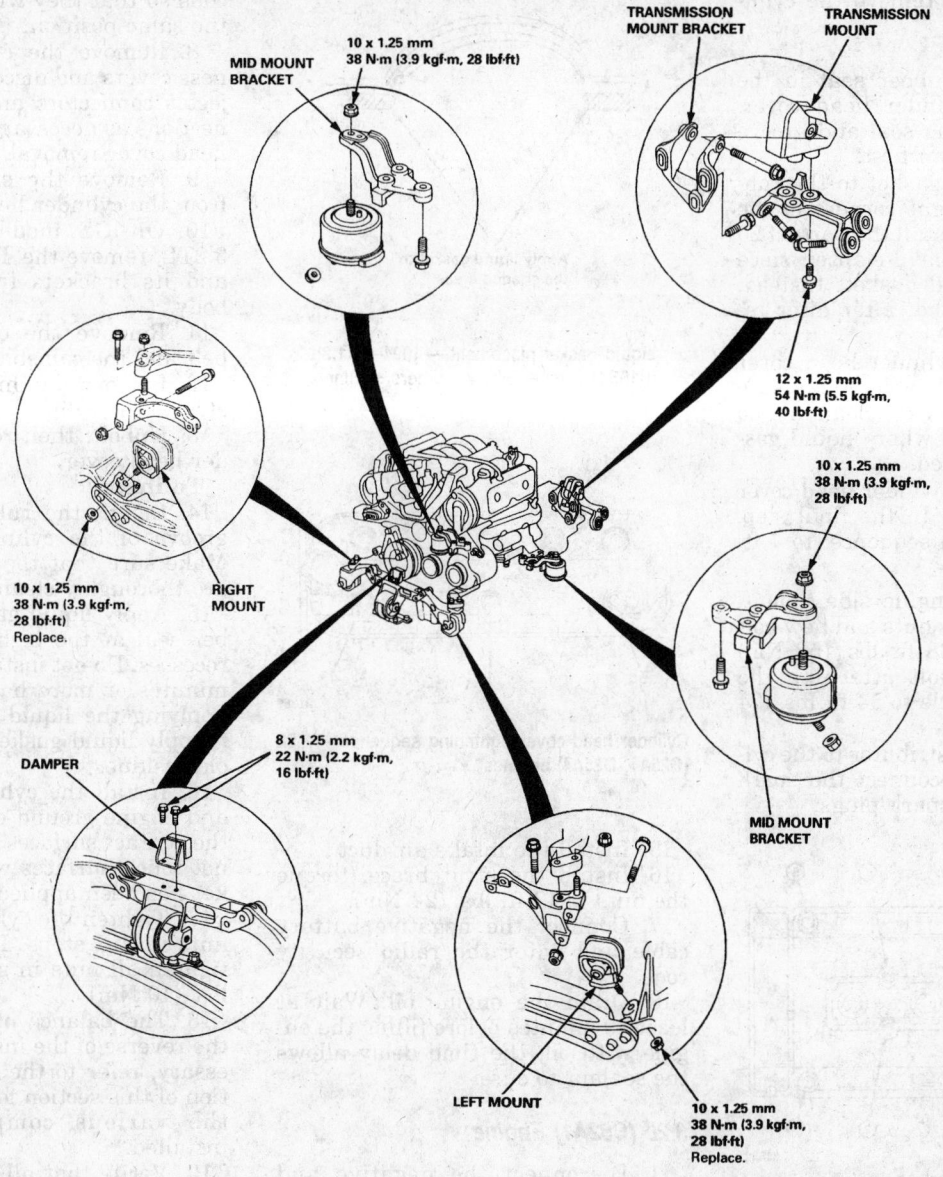

MID MOUNT BRACKET

10 x 1.25 mm
38 N·m (3.9 kgf·m, 28 lbf·ft)

TRANSMISSION MOUNT BRACKET   TRANSMISSION MOUNT

12 x 1.25 mm
54 N·m (5.5 kgf·m, 40 lbf·ft)

10 x 1.25 mm
38 N·m (3.9 kgf·m, 28 lbf·ft)

RIGHT MOUNT

10 x 1.25 mm
38 N·m (3.9 kgf·m, 28 lbf·ft)
Replace.

MID MOUNT BRACKET

DAMPER

8 x 1.25 mm
22 N·m (2.2 kgf·m, 16 lbf·ft)

LEFT MOUNT

10 x 1.25 mm
38 N·m (3.9 kgf·m, 28 lbf·ft)
Replace.

334426

Engine mounts — 1996–97 3.2L (C32A1) engines

15. Install the through-bolts to the three front mounts. Torque the center mount through-bolt to 40 ft. lbs. (54 Nm). then torque the left and right mount through-bolts to 40 ft. lbs. (54 Nm).

16. Torque the bolts attaching the transmission mount to the beam to 40 ft. lbs. (54 Nm) then torque the three bolts attaching the bracket to the transaxle to 28 ft. lbs. (38 Nm). Torque the mount through-bolt to 40 ft. lbs. (54 Nm) last.

17. Remove the hoist from the engine.

18. Install the intake air duct.

19. Remove the bolts attaching the hood support struts and install the supports to their original positions.

20. Connect the positive then the negative battery cables and enter the radio security code.

Cylinder Head Cover

REMOVAL AND INSTALLATION

NOTE: The radio may have a coded theft protection circuit. Obtain the code from the owner before disconnecting the battery, removing the radio fuse, or removing the radio.

1.7L (B17A1), 1.8L (B18C1, B18A1, B18B1), 2.5L (G25A1, G25A4) Engines

1. Disconnect the negative battery cable.

2. Remove the strut brace.

3. If necessary, remove the intake air duct.

4. Remove the spark plug wires and distributor from the cylinder head.

5. Disconnect the engine ground cable.

6. If necessary, remove the engine side mount (4-cylinder engines).

7. Remove anything else that obstructs the cylinder head cover.

8. Unbolt, then remove the cylinder head cover.

To install:

9. Install the rubber seal in the groove of the cylinder head cover. Make sure that the seal and groove are thoroughly clean first.

10. Apply liquid gasket to the rubber seal at the eight corners of the recesses. Do not install the parts if 20 minutes or more have elapsed since applying the liquid gasket. Instead, reapply liquid gasket after after removing old residue.

11. Install the cylinder head cover and engine ground cable. Make sure the contact surfaces are clean and do not touch surfaces where liquid gasket has been applied.

12. Tighten the cylinder head cover nuts in 2–3 steps. In the final step, tighten all nuts in sequence, to 7 ft. lbs. (10 Nm).

13. Install the engine side mount, torque the two new nuts and new bolt to the engine to 38 ft. lbs. (52 Nm) and torque the bolt attaching the mount to the vehicle to 54 ft. lbs. (74 Nm).

14. Install the distributor to the cylinder head and reconnect the spark plug wires to the spark plugs.

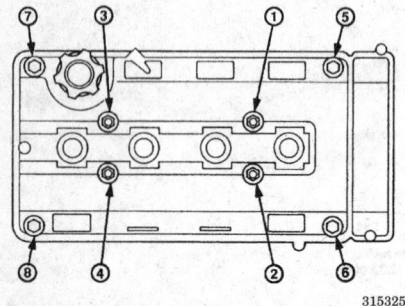

Cylinder head cover torque sequence — 1.7L (B17A1), 1.8L (B18C1) engines

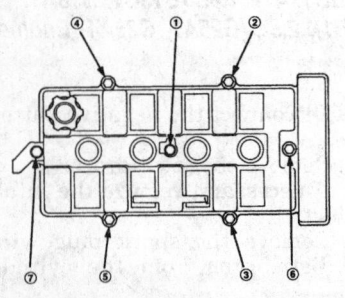

Cylinder head cover torque sequence — 1.8L (B18A1, B18B1) engines

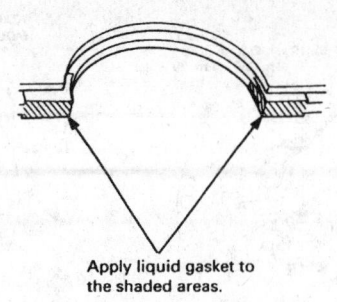

Apply liquid gasket to the shaded areas.

Liquid gasket placement — 1994–97 1.8L (B18B1) engines shown, others similar

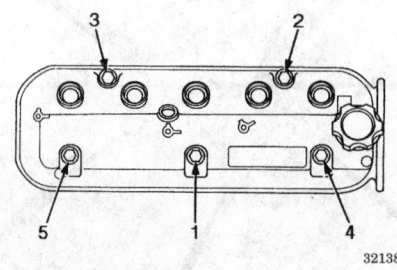

Cylinder head cover tightening sequence — 2.5L (G25A1, G25A4) engines

15. Install the intake air duct.

16. Install the strut brace, torque the nuts to 17 ft. lbs. (24 Nm).

17. Connect the negative battery cable and enter the radio security code.

18. Drain the engine oil. Wait at least 20 minutes before filling the engine with oil; the time delay allows the sealant to cure.

3.2L (C32A1) Engine

1. Disconnect the negative and then the positive battery cables.

2. Remove the secondary ground cable from the cylinder head and block. Be sure to mark all wiring and emission hoses before disconnecting them.

3. Disconnect the air cleaner and ducting as needed.

4. Remove the cover at the throttle body and disconnect the throttle and cruise control cables by loosening their locknuts. Slip the cables out of the throttle linkage.

5. Remove the strut bar and bracket.

6. Remove the injector resistor and the connector.

7. Unplug the connectors and remove the ignition coils. Mark the coils so that they will be installed in the same position.

8. Remove the engine wire harness covers and disconnect the six injector connectors and all other connections as necessary for the cylinder head cover removal.

9. Remove the six ignition coils from the cylinder heads.

10. On GS model Legends and 3.2TL, remove the TCS control valve and its brackets from the throttle body.

11. Remove the camshaft timing belt and the camshaft pulleys.

12. Remove anything else that obstructs the cylinder head covers.

13. Unbolt, then remove the cylinder head cover.

To install:

14. Install the rubber seal in the groove of the cylinder head cover. Make sure that the seal and groove are thoroughly clean first.

15. Apply liquid gasket to the rubber seal at the eight corners of the recesses. Do not install the parts if 20 minutes or more have elapsed since applying the liquid gasket. Instead, reapply liquid gasket after removing old residue.

16. Install the cylinder head cover and engine ground cable. Make sure the contact surfaces are clean and do not touch surfaces where liquid gasket has been applied.

17. Tighten the cylinder head cover nuts in 2–3 steps. In the final step, tighten all nuts in sequence, to 7 ft. lbs. (10 Nm).

18. The balance of the removal is the reverse of the installation. If necessary, refer to the appropriate portion of this section for information on the various components to be installed.

19. Verify that all wires and hoses are properly connected, then reconnect the battery.

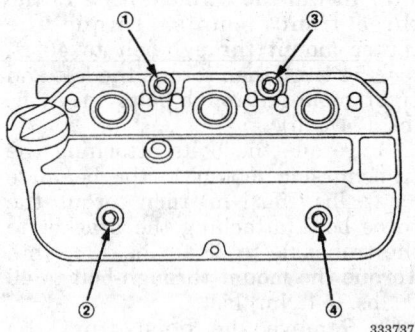

Cylinder head cover torque sequence — 3.2TL

Cylinder Head

REMOVAL AND INSTALLATION

NOTE: The radio may have a coded theft protection circuit. Obtain the code from the owner before disconnecting the battery, removing the radio fuse, or removing the radio.

1.7L(B17A1), 1.8L (B18A1) Engines

1. Before performing this procedure, the cylinder head must be below 100 degrees F. (38 degrees C.). The procedure is best done on a fully-cooled engine.
2. Disconnect the negative battery cable, then the positive battery cable.
3. Drain the cooling system.
4. Remove the air cleaner:
   a. Remove the air cleaner cover and filter.
   b. Disconnect the hot/cold air intake ducts and remove the air chamber hose.
   c. Remove the air cleaner.
5. Relieve the fuel pressure by slowly loosening the service bolt on the top of the fuel filter about one turn. Wrap the fitting in a shop rag to catch fuel spray.
6. Be sure to mark all connectors and vacuum hoses before disconnecting them.
7. Disconnect the fuel feed line. Remove the vacuum hose, breather hose and air intake hose.
8. Remove the water bypass hose from cylinder head. Remove the charcoal canister hose from the throttle body.
9. Remove the brake booster vacuum hose from the intake manifold. Remove the fuel return hose. Remove the PCV hose.
10. Remove the throttle cable from the throttle body. Take care not to bend the cable when removing it. Do not use pliers to remove the cable from the linkage. Always replaced a kinked cable with a new one.
11. Disconnect the ignition coil connector, TDC and crankshaft/cylinder sensor connector from the distributor.
12. Remove and tag the spark plug wires.
13. Remove the emission control equipment bracket but do not disconnect the emission hoses.
14. Disconnect the 3 engine harness connectors on the left side of the engine compartment.
15. Disconnect the engine sub-harness connectors from the cylinder

CYLINDER HEAD BOLTS TORQUE SEQUENCE

NOTE: Put longer bolts here.

110438

Cylinder head bolt torque sequence — 1.7L (B17A1), 1.8L (B18A1) engines

head and intake manifold. The connectors are as follows:
   a. Four injector connectors.
   b. TA sensor connector.
   c. Throttle angle sensor connector.
   d. EGR valve lift sensor; automatic transaxle only.
   e. Ground cable terminal.
   f. TW sensor ground.
   g. Coolant temperature gauge sender terminal.
   h. Oxygen sensor terminal.
   i. EACV connector.
16. Remove the upper radiator hose and heater inlet hose from the cylinder head.
17. Remove the power steering belt and power steering pump. Do not disconnect the hoses from the pump.
18. Raise and safely support the vehicle.
19. Remove the left front wheel and then remove the left splash shield.
20. Remove the intake manifold bracket bolts and remove the exhaust manifold upper shroud.
21. Remove the exhaust manifold bracket. Remove front exhaust pipe and remove the exhaust manifold.
22. Remove the cylinder head cover and engine ground cable.
23. Remove the timing belt.
24. Remove the camshaft driven pulleys. Loosen all the camshaft

holder bolts 1 full turn at a time to release valve spring pressure evenly. Remove the camshaft holder bolts, then remove the camshaft holders, camshafts and rocker arms.
25. Remove the cylinder head bolts and remove the cylinder head. To prevent warpage, unscrew all the cylinder head bolts in the reverse of the torque sequence, 1/3 turn at a time. Repeat this sequence until all the bolts are loosened.
26. Remove the intake manifold from the cylinder head.

To install:
27. Use new gaskets and install the intake manifold onto the cylinder head and tighten the nuts in a crisscross pattern in 2–3 steps, beginning in the middle. Torque the nuts to 17 ft. lbs. (23 Nm).
28. Install the cylinder head onto the engine block, after making sure the mating surface was cleaned and a new gasket was installed. Be sure to pay attention to the following points:
   a. Be sure the No. 1 cylinder is at TDC and the camshaft pulley UP mark is on the top before positioning the head in place.
   b. The cylinder head dowel pins and oil control jet must be aligned.
   c. Torque the cylinder head bolts, in 2 progressive steps: First to 22 ft. lbs. (30 Nm), in sequence, then to 61 ft. lbs. (85 Nm), in the same sequence. This sequence is the same as on earlier Integra engines.
   d. Apply engine oil to the cylinder head bolts and washers. Use the longer bolt in the No. 1 and No. 2 positions.
29. Make sure the keyways on the camshafts are facing up. The valve locknuts should be loosened and the adjusting screw backed off before installation. Replace the rocker arms in their original position.
30. Place the rocker arms on the pivot bolts and the valve stems.
31. Install the camshafts and the camshaft seals with the open spring side facing in.
32. Be sure to note the I and E marks that are stamped on the camshaft holders. Do not apply oil to the seal mating surface of the camshaft holders.
33. Apply a liquid gasket to the head of the mating surfaces of the No. 1 and No. 6 camshaft holders then install them, along with No. 2, 3, 4 and 5. Tighten each bolt 1 turn at a time to insure that the rockers do not bind on the valves.
34. Torque the camshaft holder bolts and make sure the rocker arms

are properly positioned on the valve stems. Start at the center holders and work out towards the ends, torque the bolts to 9 ft. lbs. (12 Nm).

35. Press in the camshaft seal securely with a suitable seal driver.

36. Install the keys into their groves in the camshafts. To set the No. 1 piston at TDC, align the holes on the camshaft with the holes in the No. 1 camshaft holders and drive 5.0mm pin punches into the holes.

37. Push the camshaft pulleys onto the camshafts, then torque the retaining bolts to 27 ft. lbs. (38 Nm).

38. Install the timing belt and adjust the tension. Install the lower and middle timing belt covers and bolts.

39. The balance of the removal is the reverse of the installation.

40. After installation, check to see that all hoses and wires are installed correctly.

41. Refill the coolant system.

42. Adjust the valve clearance. Make all other necessary adjustments. After completing the cylinder head removal and installation, it is recommended that the engine oil be changed.

1994–97 1.8L (B18B1) Engines

1. Before removing the cylinder head, make sure the engine temperature is below 100 degrees F. (38 degrees C.); a fully cooled engine is best.

2. Disconnect the negative battery cable.

3. Make sure the crankshaft is at TDC on No. 1 cylinder by aligning the white mark on the crankshaft pulley with the pointer on the lower timing belt cover.

4. Drain the engine coolant. Remove the radiator cap to speed draining.

5. Remove the intake air duct.

6. Relieve the fuel pressure. Exercise proper safety precautions.

――――― CAUTION ―――――

Do not allow fuel spray or vapors to come in contact with a spark or open flame. Never store fuel in an open container due to risk of fire or explosion.

7. Disconnect the fuel feed hose. Be sure to mark all connectors and vacuum hoses before disconnecting them.

8. Disconnect the breather hose, water bypass hose and the evaporative emission control canister hose.

9. Remove the PCV hose and the fuel return hose.

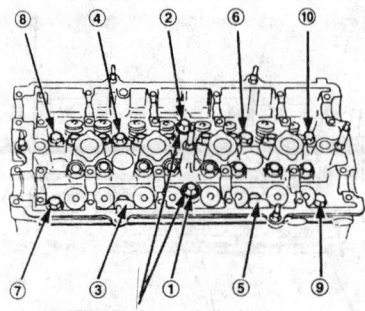

NOTE: Put longer bolts here.

315412

Cylinder head bolts torque sequence — 1994–97 1.8L (B18B1) engines

10. Remove the brake booster vacuum hose, water bypass and EVAP (Evaporative emissions) purge control solenoid vacuum hose.

11. Remove the throttle cable. Remove the throttle control cable (automatic transaxle only). Be careful not to bend the cables when removing them.

12. Remove the wire harness clamps and disconnect the following:

• Four fuel injector connectors
• Intake Air Temperature (IAT) sensor connector
• Engine Coolant Temperature (ECT) sensor connector
• TDC/CKP/CYP sensor connector
• Ignition coil connector
• ECT gauge sending unit connector
• Throttle Position (TP) sensor connector
• Manifold Absolute Pressure (MAP) sensor connector
• Idle Air Control (IAC) valve connector
• EVAP purge control solenoid valve connector
• Crankshaft Speed Fluctuation (CKF) sensor connector, if equipped

13. Remove the upper radiator hose, heater hose and water bypass hose.

14. Remove the splash shield.

15. Remove the power steering adjusting and mounting bolts, then remove the power steering pump and belt. Do not disconnect the power steering hoses.

16. Remove the air conditioning compressor and alternator belts, then remove the cruise control actuator.

17. Remove the engine side mount.

18. Remove the cylinder head cover, timing belt cover, and timing belt.

19. Remove the camshaft pulleys and back cover.

20. Remove the exhaust manifold cover, bracket, and exhaust manifold.

21. Remove the bolts attaching the intake manifold to the support bracket.

22. Remove the nuts attaching the intake manifold to the cylinder head. Remove the nuts in a crisscross pattern, beginning from the center and moving out to both ends.

23. Remove the manifold and the old gasket.

24. Loosen the locknuts and adjusting screws, then remove the camshaft holder bolts. Remove the camshaft holders, camshafts and rocker arms.

25. Remove the cylinder head bolts, then remove the cylinder head. To prevent warpage, unscrew the bolts in the reverse of the torque sequence, 1/3 turn at a time. Repeat the sequence until all bolts are loosened.

To install:

26. Install the cylinder head onto the engine block, after making sure the mating surface was cleaned and a new gasket was installed. Be sure to pay attention to the following points:

• Be sure the No. 1 cylinder is at top dead center and the camshaft pulley UP mark is on the top before positioning the head in place.
• The cylinder head dowel pins and oil control orifice must be cleaned and aligned.
• Replace the washer when damaged or deteriorated.
• Apply engine oil to the cylinder head bolts and the washers.
• Use the longer cylinder head bolts at the No. 1 and No. 2 positions.

27. Tighten the cylinder head bolts in two steps. In the first step tighten all bolts in sequence to 22 ft. lbs. (29 Nm), then in the second step tighten all bolts in the same sequence to 63 ft. lbs. (85 Nm).

28. Use new gaskets and install the intake manifold onto the cylinder head and tighten the nuts in a crisscross pattern in 2–3 steps, beginning in the middle. Torque the nut nuts to 17 ft. lbs. (23 Nm).

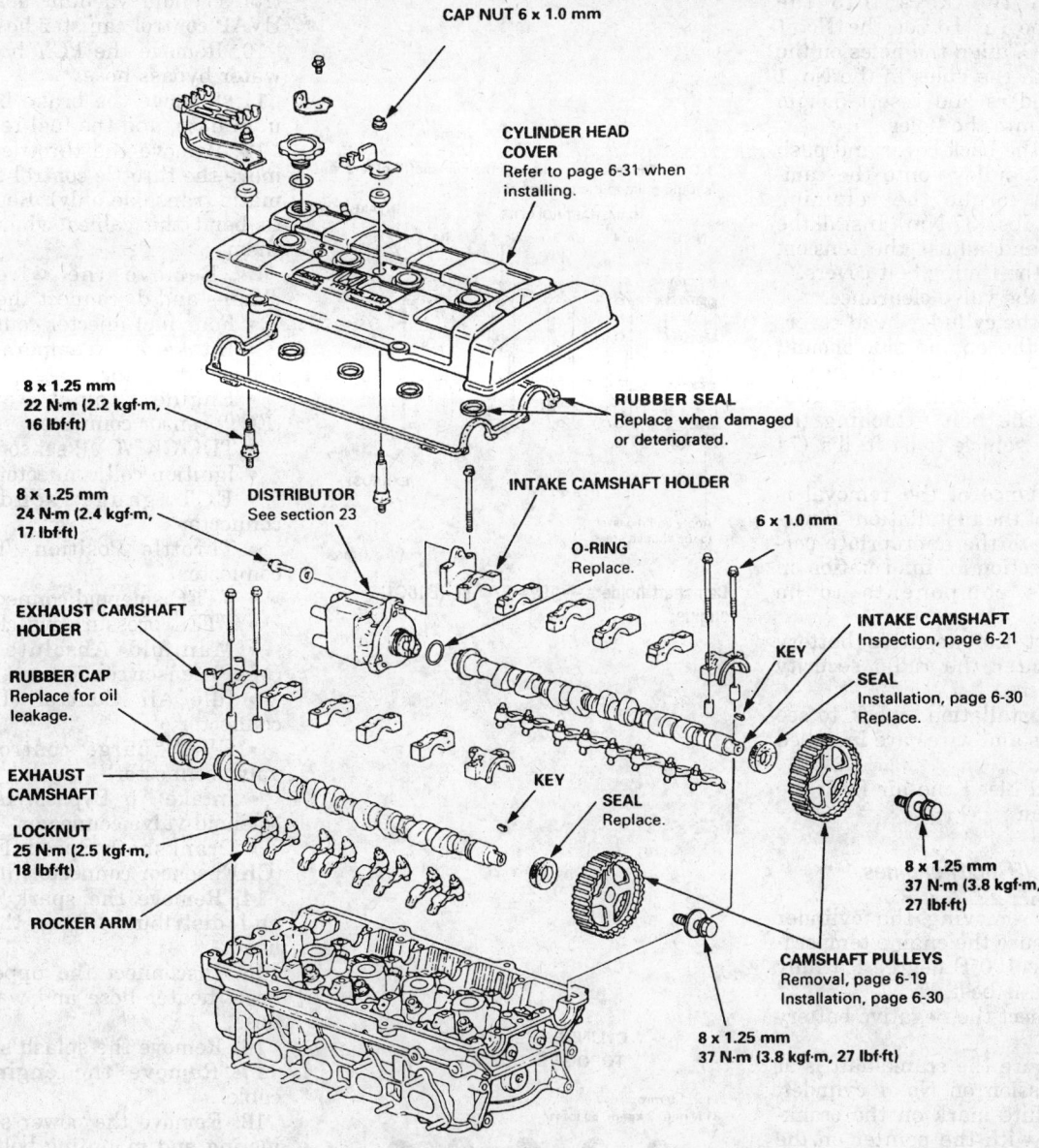

CAP NUT 6 x 1.0 mm

CYLINDER HEAD COVER
Refer to page 6-31 when installing.

8 x 1.25 mm
22 N·m (2.2 kgf·m, 16 lbf·ft)

RUBBER SEAL
Replace when damaged or deteriorated.

8 x 1.25 mm
24 N·m (2.4 kgf·m, 17 lbf·ft)

DISTRIBUTOR
See section 23

INTAKE CAMSHAFT HOLDER

O-RING
Replace.

6 x 1.0 mm

INTAKE CAMSHAFT
Inspection, page 6-21

KEY

EXHAUST CAMSHAFT HOLDER

SEAL
Installation, page 6-30
Replace.

RUBBER CAP
Replace for oil leakage.

EXHAUST CAMSHAFT

KEY

SEAL
Replace.

8 x 1.25 mm
37 N·m (3.8 kgf·m, 27 lbf·ft)

LOCKNUT
25 N·m (2.5 kgf·m, 18 lbf·ft)

ROCKER ARM

CAMSHAFT PULLEYS
Removal, page 6-19
Installation, page 6-30

8 x 1.25 mm
37 N·m (3.8 kgf·m, 27 lbf·ft)

315401

Cylinder head assembly — 1994–97 1.8L (B18B1) engines

29. Install and tighten the intake manifold bracket bolts to 17 ft. lbs. (24 Nm).

30. Install the exhaust manifold and tighten the new self–locking nuts in a crisscross pattern in 2–3 steps, beginning with the inner nuts. Torque the nuts to 23 ft. lbs. (31 Nm). Install a new exhaust pipe gasket and torque the new nuts to 40 ft. lbs. (54 Nm).

31. Make sure that the keyways on the camshafts are facing up and that the rocker arms are in their original position. The valve locknuts should be loosened and the adjusting screw backed off before installation

32. Place the rocker arms on the pivot bolts and the valve stems.

33. Install the camshafts, then install the camshaft seals with the open side facing in. Install the rubber cap with liquid gasket applied. If the rubber cap has two horizontal marks, align the marks with the cylinder head upper surface.

34. Apply liquid gasket to the head of the mating surfaces of the No. 1 and No. 6 camshaft holders then install them, along with No. 2, 3, 4, and

5. Be sure to pay attention to the following points:
• ”I“ or ”E“ marks are stamped on the camshaft holders.
• Do not apply oil to the holder mating surface of camshaft seals.
• The arrows marked on the camshaft holders should point to the timing belt.

35. Tighten the camshaft holders temporarily. Make sure that the rocker arms are properly positioned on the valve stems.

36. Tighten each bolt in 2 steps to ensure that the rockers do not bind on the valves. Torque the bolts to 7 ft. lbs. (10 Nm) working from the middle outward.

37. Install the keys into the camshaft grooves. To set the No. 1 piston at TDC, align the holes on the camshaft with the holes in the No. 1 camshaft holders and insert 5.0mm pin punches into the holes.

38. Install the back cover and push the camshaft pulleys onto the camshafts, then torque the retaining bolts to 27 ft. lbs. (37 Nm). Install the timing belt and adjust the tension, then install the timing belt covers.

39. Adjust the valve clearance.

40. Install the cylinder head cover.

41. Install the engine side mount, torque the two new nuts and new bolt to the engine to 38 ft. lbs. (52 Nm) and torque the bolt attaching the mount to the vehicle to 54 ft. lbs. (74 Nm).

42. The balance of the removal is the reverse of the installation. If necessary, refer to the appropriate portion of this section for information on the various components to be installed.

43. Connect the negative battery cable and enter the radio security code.

44. After installation, check to see that all hoses and wires are installed correctly.

45. Fill and bleed the air from the cooling system.

1994–97 1.8L (B18C1) engines

1. Before removing the cylinder head, make sure the engine temperature is below 100°F degrees; a fully cooled engine is best.

2. Disconnect the negative battery cable.

3. Make sure the crankshaft is at TDC/compression on No. 1 cylinder. Align the white mark on the crankshaft pulley with the pointer on the lower timing belt cover.

4. Drain the engine coolant into a sealable container. Remove the radiator cap to speed draining.

5. Remove the strut brace.

6. Remove the intake air duct.

7. Relieve the fuel pressure. Exercise proper safety precautions.

CAUTION

Do not allow fuel spray or vapors to come in contact with a spark or open flame. Never store fuel in an open container due to risk of fire or explosion.

8. Disconnect the fuel feed hose.

9. Be sure to mark all connectors and vacuum hoses before disconnecting them. Disconnect the EVAP (Evaporative emissions) purge con-

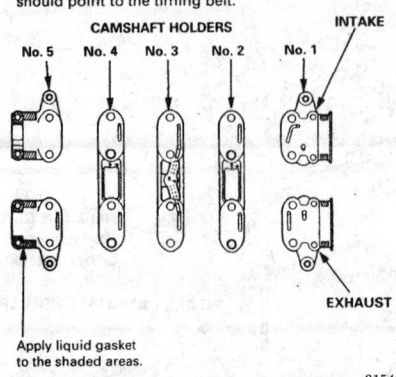

NOTE: The arrows marked on the camshaft holders should point to the timing belt.

CAMSHAFT HOLDERS

No. 5   No. 4   No. 3   No. 2   No. 1   INTAKE

EXHAUST

Apply liquid gasket to the shaded areas.

315452

Camshaft holders — 1994–97 1.8L (B18C1) engines

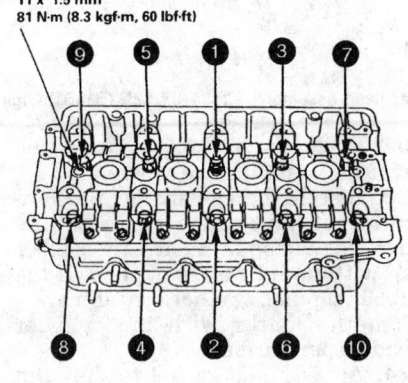

CYLINDER HEAD BOLT TORQUE SEQUENCE

11 x 1.5 mm
81 N·m (8.3 kgf·m, 60 lbf·ft)

9   5   1   3   7

8   4   2   6   10

315451

Cylinder head bolt torque sequence — 1994–97 1.8L (B18C1) engines

trol solenoid vacuum hose and the EVAP control canister hose.

10. Remove the PCV hose and the water bypass hose.

11. Remove the brake booster vacuum hose, and the fuel return hose.

12. Remove the throttle cable. Remove the throttle control cable (automatic transaxle only). Be careful not to bend the cables when removing them.

13. Remove the wire harness clamps and disconnect the following:
- Four fuel injector connectors
- Intake Air Temperature (IAT) sensor connector
- Engine Coolant Temperature (ECT) sensor connector
- TDC/CKP/CYP sensor connector
- Ignition coil connector
- ECT gauge sending unit connector
- Throttle Position (TP) sensor connector
- VTEC solenoid connector
- VTEC pressure switch connector
- Manifold Absolute Pressure (MAP) sensor connector
- Idle Air Control (IAC) valve connector
- EVAP purge control solenoid valve connector
- Intake Air Bypass (IAB) control solenoid valve connector
- Crankshaft Speed Fluctuation (CKF) sensor connector, if equipped

14. Remove the spark plug wires and distributor from the cylinder head.

15. Disconnect the upper radiator hose, heater hose and water bypass hose.

16. Remove the splash shield.

17. Remove the engine ground cable.

18. Remove the power steering adjusting and mounting bolts, then remove the power steering pump and belt. Do not disconnect the power steering hoses.

19. Remove the heat shield from the power steering bracket.

20. Remove the air conditioning compressor and alternator belts.

21. Remove the cruise control actuator.

22. Remove the engine side mount.

23. Remove the cylinder head cover, timing belt cover, and timing belt.

24. Remove the camshaft sprockets and back cover.

25. Remove the exhaust manifold cover, bracket, and exhaust manifold.

26. Remove the bolts attaching the intake manifold to the support bracket.

27. Remove the nuts attaching the intake manifold to the cylinder head.

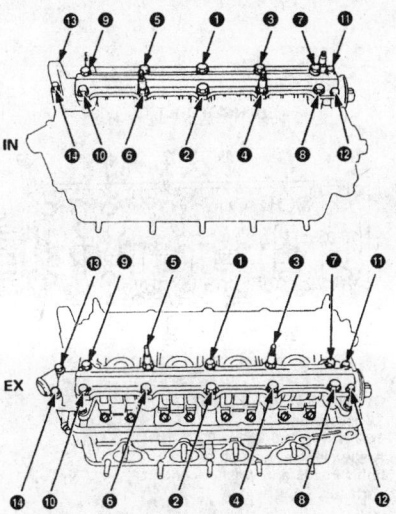

● – ●: 8 x 1.25 mm 27 N·m (2.8 kgf·m, 20 lbf·ft)
● – ●: 6 x 1.0 mm 9.8 N·m (1.0 kgf·m, 7.2 lbf·ft)

315453

Camshaft holder plates torque sequence —
1994–97 1.8L (B18C1) engines

Remove the nuts in a crisscross pattern, beginning from the center and moving out to both ends.

28. Remove the manifold and the old gasket.

29. Remove the VTEC solenoid from the cylinder head.

30. Loosen the rocker arm locknuts and adjusting screws.

31. Remove the camshaft holder bolts, then remove the camshaft holder plates, the camshaft holders, and camshafts.

32. Remove the cylinder head bolts, then remove the cylinder head. To prevent warpage, loosen the bolts in the reverse of the torque sequence 1/3 turn at a time. Repeat this sequence until all bolts are loosened.

To install:

33. Install the cylinder head onto the engine block, after making sure the mating surface was cleaned and a new gasket was installed. Be sure to pay attention to the following points:

• Be sure the No. 1 cylinder is at top dead center and the camshaft pulley UP mark is on the top before positioning the head in place.

• The cylinder head dowel pins and oil control orifice must be cleaned and aligned.

• Replace the washer when damaged or deteriorated.

• Apply engine oil to the cylinder head bolts and the washers.

• Use the longer cylinder head bolts at the No. 1 and No. 2 positions.

34. Tighten the cylinder head bolts in two steps. In the first step tighten all bolts in sequence to 22 ft. lbs. (29 Nm). In the second step tighten all the bolts in the same sequence to 60 ft. lbs. (81 Nm) for 1994 and 1995 models, and to 63 ft. lbs. (85 Nm) for 1996 models.

35. Use new gaskets and install the intake manifold onto the cylinder head; tighten the nuts in a crisscross pattern in 2–3 steps, beginning in the middle. Torque the nuts to 17 ft. lbs. (23 Nm).

36. Install and tighten the intake manifold bracket bolts to 17 ft. lbs. (24 Nm).

37. Install the VTEC solenoid with a new filter, torque the attaching bolts to 17 ft. lbs. (24 Nm).

38. Install the exhaust manifold and tighten the new self–locking nuts in a crisscross pattern in 2–3 steps, beginning with the inner nuts. Torque the nuts to 23 ft. lbs. (31 Nm). Install a new exhaust pipe gasket and torque the new nuts to 40 ft. lbs. (54 Nm).

39. Install the exhaust manifold bracket and cover. Torque the bracket attaching bolts to 33 ft. lbs. (44 Nm) and the cover bolts to 17 ft. lbs. (24 Nm).

40. Make sure that the keyways on the camshafts are facing up and that the rocker arms are in their original position. The valve locknuts should be loosened and the adjusting screw backed off.

41. Install the camshafts, then install the camshaft seals with the open side facing in. Install the rubber cap with liquid gasket applied. If the rubber cap has two horizontal marks, align the marks with the cylinder head upper surface.

42. Install a new O–ring and the dowel pin to the oil passage of the No. 3 camshaft holder.

43. Apply liquid gasket to the head of the mating surfaces of the No. 1 and No. 5 camshaft holders then install them, along with No. 2, 3, and 4. Be sure to pay attention to the following points:

• Do not apply oil to the holder mating surface of camshaft seals.

• The arrows marked on the camshaft holders should point to the timing belt.

44. Tighten the camshaft holders temporarily. Make sure that the rocker arms are properly positioned on the valve stems.

45. Tighten each bolt in 2 steps to ensure that the rockers do not bind on the valves. Torque the 8·1.25mm bolts to 20 ft. lbs. (27 Nm), and the 6·1.0mm bolts to 7.2 ft. lbs. (9.8 Nm).

46. Install the back cover and torque the attaching bolt to 7.2 ft. lbs. (9.8 Nm). Install the keys into the camshaft grooves, then push the camshaft pulleys onto the camshafts, then torque the retaining bolts to 41 ft. lbs. (56 Nm).

47. Install the timing belt and adjust the tension, then install the timing belt covers.

48. Adjust the valve clearance.

49. Install the rubber seal in the groove of the cylinder head cover. Make sure that the seal and groove are thoroughly clean first.

50. Apply liquid gasket to the rubber seal at the eight corners of the recesses. Do not install the parts if 20 minutes or more have elapsed since applying the liquid gasket. Instead, reapply liquid gasket after after removing old residue.

51. Install the cylinder head cover and engine ground cable. Make sure the contact surfaces are clean and do not touch surfaces where liquid gasket has been applied.

52. Tighten the cylinder head cover nuts in 2–3 steps. In the final step, tighten all nuts in sequence, to 7 ft. lbs. (10 Nm).

53. Install the engine side mount, torque the two new nuts and new bolt to the engine to 38 ft. lbs. (52 Nm) and torque the bolt attaching the mount to the vehicle to 54 ft. lbs. (74 Nm).

54. The balance of the removal is the reverse of the installation. If necessary, refer to the appropriate portion of this section for information on the various components to be installed.

55. Connect the negative battery cable and enter the radio security code.

56. After installation, check to see that all hoses and wires are installed correctly.

57. Fill and bleed the air from the cooling system.

58. Change the engine oil. Wait at least 20 minutes before filling the engine with oil; the time delay allows the sealants to cure.

2.5L (G25A1, G25A4) Engines

1. Disconnect the negative battery cable and drain the coolant.

2. Disconnect the wiring from the ignition coil and the ground wire.

3. Remove the ABS motor relay box and the battery heat shield.

CAUTION

Fuel injection systems remain under pressure after the engine has been turned OFF. Properly relieve fuel pressure before disconnecting any fuel lines. Failure to do so may result in fire or personal injury.

---

4. Remove the fuel filler cap and loosen the service bolt on the fuel filter banjo bolt to relieve the fuel system pressure. Remove the banjo bolt to disconnect the fuel feed hose from the fuel filter. Disconnect the fuel return hose from the pressure regulator.

5. Remove the intake air duct and air cleaner assembly.

6. Loosen the air conditioner compressor and alternator adjustment bolts. Remove the drive belts.

7. Remove the throttle cable by loosening the locknut, then slip the cable end out of the throttle bracket and accelerator linkage. Do not bend the cable when removing it. Unbolt the throttle cable clamp and move the cable aside.

8. Label and disconnect the fuel and vacuum hoses from the intake manifold. Be sure to mark all electrical connectors and vacuum hoses before disconnecting them.

9. Disconnect the upper radiator hose, the heater hoses, and the water bypass hoses and unbolt the wire harness clips.

10. Disconnect the brake booster hose, canister hose, and the two vacuum hoses from the rear of the cylinder head.

11. Remove the two distributor mounting bolts. Remove the distributor, ignition wires, and ground cables from the cylinder head.

12. Label and disconnect the wiring harness holder that is routed across the front of the cylinder head.

13. Disconnect and label the fuel injector leads; throttle position sensor; IAC, EGR, EVAP, and air bypass solenoid valves, and the ECT connections. Disconnect the TDC, crankshaft, and camshaft position sensors.

14. Unbolt the intake manifold support brackets. The manifold may be removed after removing the cylinder head.

15. Disconnect the oxygen sensor wire .

16. Remove the exhaust manifold heat shields and disconnect the exhaust pipe from the manifold.

17. Remove the support bracket and remove the exhaust manifold.

18. Remove the cylinder head cover and upper timing belt cover.

19. Remove the timing belt. Replace the belt if it shows any signs of stress or damage.

20. Remove the camshaft position sensor and the camshaft sprocket.

21. Remove the back cover and unbolt the TDC/CKP sensor.

22. Loosen each cylinder head bolt about 1/3 turn at a time. Follow the reverse of the torque sequence to prevent warping the head. Repeat until all bolts are loose and can be removed.

23. If the cylinder head is stuck to the block, there are pry points at each end of the cylinder head. Do not pry against the gasket surfaces.

24. Carefully remove the cylinder head from the vehicle. If the intake and exhaust manifolds are still attached, have an assistant help in removal since the head assembly will be very heavy.

25. After removing the cylinder head, the intake manifold and exhaust manifolds maybe removed.

To install:

26. Make sure the cylinder head and the engine block sealing surfaces are flat and clean. Resurface the head if it is warped. Clean all gasket surfaces and run a tap through the bolt holes in the block to clean the threads. Make sure the bolt holes are clean and dry so the head can be torqued properly.

27. Install a new O-ring onto the oil control orifice and install the orifice and dowel pins onto the block. Lay the new head gasket in place.

28. Install the intake manifold onto the cylinder head with a new gasket. Torque the nuts in a crisscross pattern in 2 steps to 16 ft. lbs. (22 Nm).

29. Verify that the crankshaft and camshaft are both at TDC for number one piston.

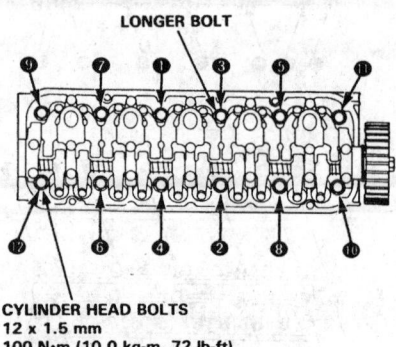

CYLINDER HEAD BOLTS
12 x 1.5 mm
100 N·m (10.0 kg-m, 72 lb-ft)
Apply clean engine oil to the bolt threads and washer contact surfaces.

321387

Cylinder head bolt torque sequence — Vigor (G25A1 engine)

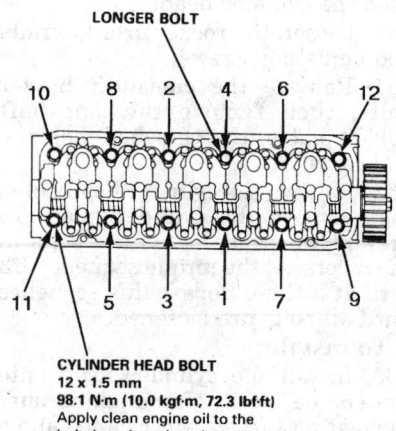

CYLINDER HEAD BOLT
12 x 1.5 mm
98.1 N·m (10.0 kgf·m, 72.3 lbf·ft)
Apply clean engine oil to the bolt threads and washer contact surfaces.

321388

Cylinder head bolt torque sequence — 2.5TL (G25A4 engine)

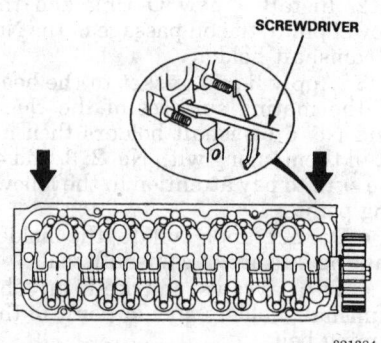

Cylinder head prying points — 2.5L (G25A1, G25A4) engines

321384

30. Carefully fit the cylinder head to the block. Make sure the oil control orifice is properly aligned.

31. Lightly oil the threads and washer surfaces of the cylinder head bolts and install them. Torque the bolts in 3 steps to 72 ft. lbs. (100 Nm) in the correct sequence.

32. Install the intake manifold brackets.

33. Loosely install the exhaust manifold bracket onto the manifold. Install the exhaust manifold with a new gasket and new self-locking nuts and torque the nuts to 23 ft. lbs. (32 Nm).

34. Connect the exhaust pipe and install the manifold shields.

35. Reconnect the oxygen sensor.

36. Install the timing belt and covers.

37. Adjust the valves.

38. Apply silicone sealant to the ends of the cylinder head near the camshaft holders. Install the cylinder head cover with new rubber seals as required.

39. The balance of the removal is the reverse of the installation. If necessary, refer to the appropriate portion of this section for information on the various components to be installed.

40. Add fresh engine oil and a new filter.

41. Verify that all wiring, grounds, hoses, and cables are properly connected.

42. Connect the battery cable. Run the engine to bleed the cooling system and check for leaks. Check for proper cooling system and engine operation.

3.2L (C32A1) Engines

Legend

1. Disconnect the negative and then the positive battery cables.

2. Remove the battery and battery tray.

3. Engine temperature must be below 100 degrees F. (38 degrees C.) before performing this procedure. Turn the crankshaft so that the No. 1 cylinder is at top dead center.

4. Drain the cooling system.

5. Remove the vacuum hose from the brake booster.

6. Remove the secondary ground cable from the cylinder head and block. Be sure to mark all wiring and emission hoses before disconnecting them.

7. Remove the air cleaner and ducting.

CAUTION

Fuel injection systems remain under pressure, even after the engine has been turned OFF. The fuel system pressure must be relieved before disconnecting any fuel lines. Failure to do so may result in fire and/or personal injury.

8. Relieve the fuel pressure by loosening the service bolt on the top of the fuel filter about a turn. Disconnect the fuel return hose from the pressure regulator. Remove the special nut and the fuel hose.

9. Remove the cover at the throttle body and disconnect the throttle and cruise control cables by loosening their locknuts. Slip the cables out of the throttle linkage.

10. Disconnect the charcoal canister hose from the throttle body.

11. Remove the intake manifold.

12. Remove the upper timing belt covers.

13. Do not remove the timing belt adjuster bolt. Loosen it ½ turn, relieve the belt tension and tighten the bolt.

14. Remove the camshaft timing belt and the camshaft pulleys.

15. Remove the timing belt cover plates from the heads and remove the crank angle/cylinder sensor from the left head.

16. Remove the cylinder head covers.

17. Remove the bolts from the alternator and power steering pump brackets as required.

18. Working under the vehicle, unbolt the twin three-way catalytic converters and disconnect them from the exhaust manifolds.

19. Loosen each head bolt about ½ turn in the opposite of the installation sequence. This is important to prevent warping the heads. Repeat until all bolts are loose and the head can be removed.

20. Remove the heads, and then remove the exhaust manifolds from the heads.

To install:

21. It is easier to install the exhaust manifolds and their covers onto the heads before installing the heads to the engine. Use new gaskets and self-locking nuts. Torque the nuts to 22 ft. lbs. (31 Nm).

22. Install the heads with new gaskets and O-rings, making sure the dowel pins and control orifices are properly positioned. Oil the threads and washers on the head bolts and torque in two steps in the sequence shown to 56 ft. lbs. (78 Nm).

23. Apply liquid gasket to the corners of the camshaft holders and install the cylinder head covers.

24. Install the crankshaft/cylinder sensor to the left cylinder head, then install both timing belt cover plates.

25. Install the camshaft pulleys and torque the bolts to 23 ft. lbs. (32 Nm). The left and right pulleys are different; the left one goes with the crankshaft/cylinder sensor.

26. Align the timing marks on the crankshaft and camshaft pulleys, install the timing belt and adjust the belt tension. Install the timing belt covers.

27. The balance of the removal is the reverse of the installation. If necessary, refer to the appropriate portion of this section for information on the various components to be installed.

28. Refill the cooling system. Whenever cylinder heads are removed and installed, it is always recommended to change the oil.

29. Verify that all wires and hoses are properly connected and install the battery. Before starting the engine, turn the ignition switch **ON** and **OFF** a number of times to pressurize the fuel system. Check for leaks.

30. After starting the engine, bleed the cooling system.

3.2TL

1. Disconnect the negative then the positive battery cables.

2. Drain the engine coolant into a sealable container.

3. Remove the engine cover.

4. Remove the air intake dust and the air cleaner housing.

5. Remove the throttle cable cover. Without turning the adjusting nut, loosen the locknut, which is closer to the throttle, and disconnect the throttle cable and cruise control cable from the throttle and bracket.

NOTE: Take care to not bend the throttle cables, replace the cables if they become bent or kinked.

6. Remove the upper and lower radiator hoses.

7. Remove the battery and the battery base.

8. Disconnect the vacuum hoses, then remove the clamp from the under-hood fuse/relay box.

9. Label then disconnect the battery cables from the under-hood fuse/relay box, then remove the under-hood fuse/relay box.

10. Disconnect the engine wire harness connector, located by the under-hood fuse/relay box.

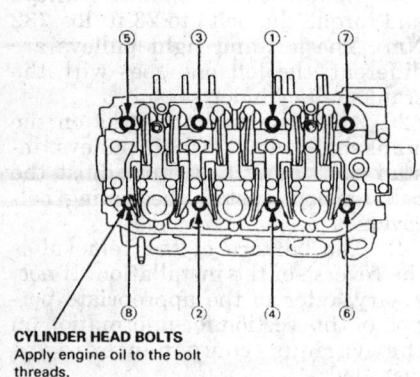

CYLINDER HEAD BOLTS
Apply engine oil to the bolt threads.

333579

Cylinder head bolt torque sequence — 3.2L engines

11. Disconnect the wire harness from the control box and solenoid valve, then remove the control box.

12. Properly relieve the fuel pressure.

CAUTION

Do not allow fuel spray or fuel vapors to come in contact with a spark or open flame. Keep a dry chemical fire extinguisher nearby. Never store fuel in an open container due to risk of fire or explosion.

13. Disconnect the engine fuel feed hose from the fuel filter and disconnect the fuel return hose from the fuel regulator.

CAUTION

Fuel injection systems remain under pressure after the engine has been turned OFF. Properly relieve fuel pressure before disconnecting any fuel lines. Failure to do so may result in fire or personal injury.

14. Disconnect the evaporative emissions (EVAP) control canister hose and the vacuum hose.

15. Disconnect the bypass hoses and the heater hose.

16. Loosen the alternator mounting bolt, lockbolt and adjusting rod, then remove the drive belt.

17. Loosen the A/C idler pulley center nut and adjusting bolt then remove the drive belt.

18. Remove the power steering pump drive belt.

19. Remove the bolt attaching the engine ground cable to the intake manifold. Remove the engine wire harness clamps.

20. Remove the Traction Control System (TCS) control valve assembly upper and lower brackets.

21. Disconnect the TCS throttle sensor connector and the TCS throttle actuator connector, then remove the TCS control valve assembly.

22. Remove the oil pressure switch connector, engine ground cable and the engine wire harness cover.

23. Remove the EGR pipe, then remove the intake manifold and water passage.

24. Remove the covers from the exhaust manifolds, then remove the exhaust manifolds.

25. Remove the timing belt from the engine.

26. Remove the left and right camshaft sprockets, then remove the timing belt back covers.

27. Remove the left and right cylinder head covers.

28. Remove the three bolts attaching the alternator bracket to the cylinder head.

29. Remove the two bolts attaching the power steering pump bracket to the cylinder head.

30. Remove the cylinder head bolts 1/3 at turn at a time in the reverse of the torque sequence. Repeat until all the bolts are loosened.

31. Remove the cylinder heads from the cylinder block.

To install:

32. Remove the oil control orifices, then clean them and install new O–rings.

33. Clean the cylinder block and the head mating surfaces.

34. Install the oil control orifices and the dowel pins to the cylinder block.

35. Install new gaskets to the cylinder block then install the cylinder heads.

36. Torque the cylinder head bolts in two or three steps following the proper sequence to 56 ft. lbs. (76 Nm). The manufacturer recommends using a beam–type torque wrench.

37. Clean the groove in the cylinder head cover and install the gasket into the groove. Seat the recesses for the camshaft first, then work it into the

groove around the outside edges. Apply liquid gasket to the cylinder head cover gasket at the four corners of the recesses.

38. Install the cylinder head cover, slide the cover back and forth to seat the cylinder head cover gasket. Replace the washer when damaged or deteriorated. Torque the nuts in two or three steps following the proper sequence, to 8.7 ft. lbs. (12 Nm).

39. Install the two bolts used to attach the power steering pump bracket to the cylinder head, then torque the bolts to 33 ft. lbs. (44 Nm).

40. Install the three bolts used to attach the alternator bracket to the cylinder head, then torque the bolts to 16 ft. lbs. (22 Nm).

41. Install the timing belt back covers and torque the bolts to 8.7 ft. lbs. (12 Nm).

42. Install the left and right camshaft sprockets and torque the attaching bolts to 23 ft. lbs. (31 Nm).

43. Install the timing belt and timing belt covers.

44. The balance of the removal is the reverse of the installation. If necessary, refer to the appropriate portion of this section for information on the various components to be installed.

45. Install the engine cover and torque the nuts to 8.7 ft. lbs. (12 Nm).

46. Install the battery base and torque the bolts to 16 ft. lbs. (22 Nm), then install the battery.

47. Fill and bleed the air from the cooling system.

48. Connect the positive then the negative battery cables. Enter the radio security code.

Lash Adjusters

BLEEDING

3.2L (C32A1) Engines

1. Remove the hydraulic lash adjusters (tappets) from the rocker arms. Note the position of each rocker arm and its lash adjuster for reassembly.

2. Fill a clean container with clean 10W–30 engine oil.

3. Submerge the lash adjuster in the oil and insert a thin, stiff piece of wire into its plunger to depress the check ball.

4. Depress the check ball and slowly pump the plunger to bleed any air from the lash adjuster. Continue to bleed the lash adjuster until it stops releasing air bubbles into the oil.

5. Lubricate new O–rings with clean engine oil and then install them onto the lash adjuster.

6. Clean and lubricate the rocker arms and then reinstall the lash adjusters. Make sure the sets of rocker arms and lash adjusters are not mismatched.

Valve Lash

ADJUSTMENT

1.7L (B17A1), 1.8L (B18A1, B18B1, B18C1) Engines

NOTE: While all valve adjustments must be as accurate as possible, it is better to have the valve adjustment slightly loose rather than too tight. Burned valves may result from overly–tight adjustments. Perform the valve adjustment for each cylinder in the same sequence as the firing order: 1–3–4–2.

1. Make sure the engine is cold; cylinder head temperature must be below 100°F (38°C). Overnight cold is best.

2. Remove the cylinder head cover and the upper timing belt cover.

3. Set the No. 1 cylinder to Top Dead Center (TDC). The word **UP** should appear at the top and the TDC grooves on the pulley should align with the cylinder head surface or the mark on the rear belt cover.

4. Valve clearances are:

   a. B17A1 and B18A1 (1993) engines: Intake — 0.003–0.005 in. (0.08–0.12mm) Exhaust — 0.006–0.008 in. (0.16–0.20mm)

   b. B18B1 engine: Intake — 0.003–0.005 in. (0.08–0.12mm) Exhaust — 0.006–0.008 in. (0.16–0.20mm)

   c. B18C1 (VTEC) engine: Intake — 0.006–0.007 in. (0.15–0.19mm) Exhaust — 0.007–0.008 in. (0.17–0.20mm)

5. With the No. 1 cylinder at TDC, adjust the valves of the No. 1 cylinder by performing the following procedures:

   a. Hold the rocker arm against the valve and place the feeler gauge between the rocker arm and the camshaft lobe. There should be a slight drag on the feeler gauge.

   b. If adjustment is required, loosen the valve adjusting the screw locknut.

   c. Turn the adjusting screw to obtain the proper clearance.

   d. Hold the adjusting screw and torque the locknut(s) to 18 ft. lbs.

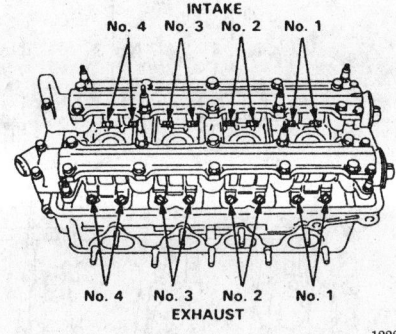

Valve arrangement — 1993–94 1.7L (B17A1), 1.8L (B18A1, B18B1, B18C1) engines

   e. Recheck the clearance.

6. Turn the crankshaft 180 degrees counterclockwise; the cam pulley will turn 90 degrees. With the No. 3 cylinder at TDC, the **UP** marks should be at the exhaust side. Adjust the valves on the No. 3 cylinder.

7. Turn the crankshaft 180 degrees counterclockwise; the cam pulley will turn 90 degrees. With the No. 4 cylinder at TDC, both **UP** marks should be at the bottom. Adjust the valves on the No. 4 cylinder.

8. Turn the crankshaft 180 degrees counterclockwise. The No. 2 cylinder will now be on TDC and the **UP** marks should be at the intake

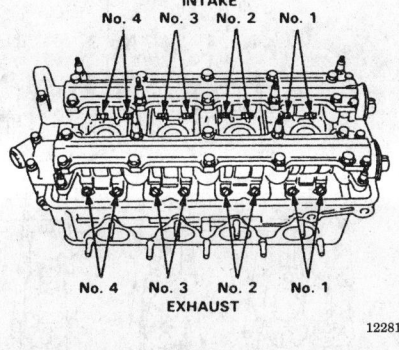

Crankshaft pulley timing mark — 1.8L (B18B1) engines

side. Adjust the valves on the No. 2 cylinder.

9. Install the cylinder head cover and upper timing belt cover.

2.5L (G25A1, G25A4) Engines

1. Disconnect the negative battery cable.

2. Remove the cylinder head cover and the upper timing belt cover.

3. Rotate the crankshaft to align the white TDC on the crankshaft pulley with the pointer on the cover. Make sure the **UP** mark on the camshaft sprocket is up and the TDC marks align with the edge of the cylinder head.

4. Align the No. 1 mark on the back of the camshaft sprocket with the notch in the camshaft holder.

5. Hold a No. 1 cylinder rocker arm against the camshaft and use a feeler gauge to check the clearance at the valve stem. Intake valve clearance should be 0.010 in. (026mm), exhaust valve clearance should be 0.012 in. (0.30mm). The service limit for both intake and exhaust valves is plus or minus 0.0008 in. (0.02mm). Loosen the locknut and turn the adjusting screw to adjust the clearance. Tighten the locknut and recheck the clearance.

6. Rotate the crankshaft counterclockwise to align the TDC marks for each piston with the notch. Adjust the valves of each cylinder. The adjustment order is 1–2–4–5–3.

7. Install the cylinder head and timing belt covers.

8. Reconnect the negative battery cable.

3.2L (C32A1) Engines

These engines are equipped with hydraulic valve lash adjusters on the rocker arms. No valve clearance adjustments are possible or necessary.

Rocker Arms and Shafts

REMOVAL AND INSTALLATION

NOTE: The radio may have a coded theft protection circuit. Obtain the code from the owner before disconnecting the battery, removing the radio fuse, or removing the radio.

1.7L and 1.8L (B17A1, B18C1) Engines

1. Disconnect the negative battery cable.

2. Remove the cylinder head from the vehicle.

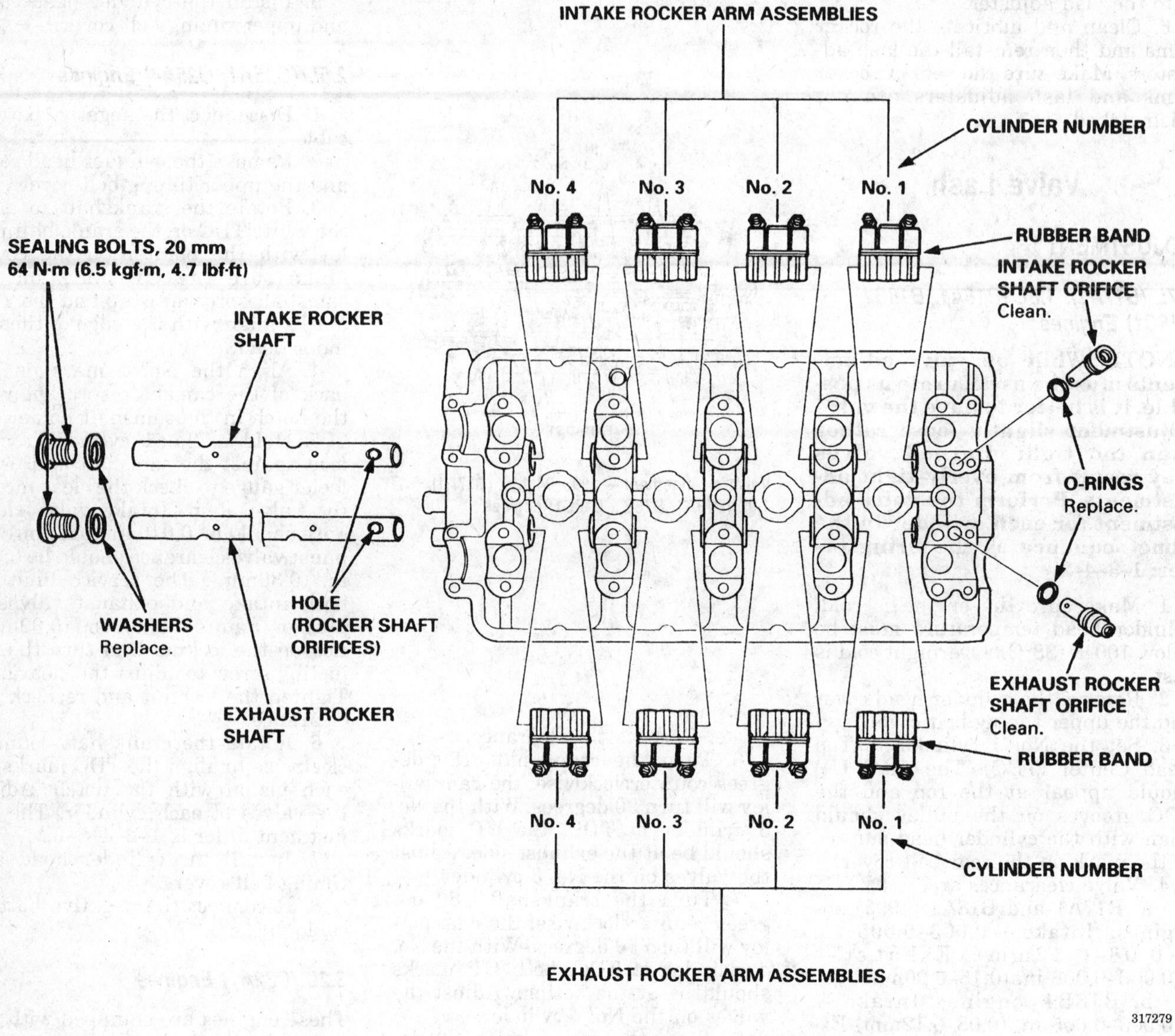

INTAKE ROCKER ARM ASSEMBLIES

CYLINDER NUMBER

No. 4   No. 3   No. 2   No. 1

RUBBER BAND

INTAKE ROCKER SHAFT ORIFICE
Clean.

SEALING BOLTS, 20 mm
64 N·m (6.5 kgf·m, 4.7 lbf·ft)

INTAKE ROCKER SHAFT

O-RINGS
Replace.

WASHERS
Replace.

HOLE (ROCKER SHAFT ORIFICES)

EXHAUST ROCKER SHAFT

EXHAUST ROCKER SHAFT ORIFICE
Clean.

RUBBER BAND

No. 4   No. 3   No. 2   No. 1

CYLINDER NUMBER

EXHAUST ROCKER ARM ASSEMBLIES

317279

Rocker arms and shafts — 1.7L (B17A1), 1.8L (B18A1, B18B1, B18C1) engines

3. Hold each rocker arm assembly together with a rubber band to prevent them from separating.

4. Remove the intake and exhaust rocker shaft orifices from the cylinder head. The rocker shaft orifices are different and should be identified when removed. Discard the O–rings on the orifices.

5. Remove the VTEC solenoid from the cylinder head and discard the filter.

6. Remove the rocker arm shaft sealing bolts, discard the washers.

7. Insert 12mm bolts into the rocker arm shafts. Remove each

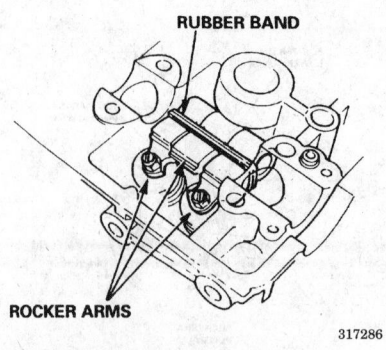

RUBBER BAND

ROCKER ARMS

317286

Rocker arms with rubber band installed — 1.7L (B17A1), 1.8L (B18A1, B18B1, B18C1) engines

rocker arm set while slowly pulling out the rocker arm shaft.

NOTE: Tag each rocker arm set to to assure installation in their original locations.

8. Inspect the rocker arm pistons. If they do not move smoothly, replace the rocker arm assembly.

9. Remove the lost motion assembly from the cylinder head. Inspect the lost motion assembly by pushing the plunger with your finger. Replace the lost motion assembly if it does not move smoothly.

To install:

10. Install the lost motion assembly to the cylinder head.

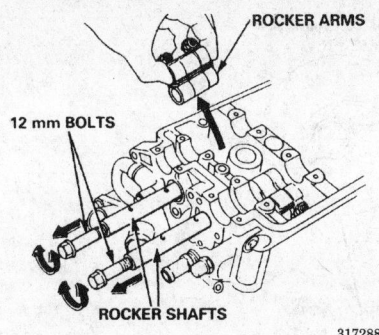

Removing the rocker arms — 1.7L (B17A1), 1.8L (B18A1, B18B1, B18C1) engines

11. Apply engine oil to the rocker arm pistons, then bundle the rocker arms with a rubber band. Apply a light coat of clean engine oil to the rocker arms.

12. Position the rocker arms in their original locations, if they are being reused. If new assembles are being used place them in the cylinder head.

13. Lightly coat the rocker arm shafts with clean engine oil, then install the rocker arm shafts into the cylinder head. A 12mm bolt can be installed into the end of the rocker arm shafts to aid in their installation. Be sure to install the shafts in the proper positions. Remove the 12mm bolts from the rocker arm shafts, if used.

14. Clean and install the rocker arm shaft orifices with new O–rings. If the holes in the rocker arm shafts are not aligned screw a 12mm bolt into the end of the shaft to position the shaft.

15. Install the sealing bolts with new washers, torque the bolts to 47 ft. lbs. (64 Nm).

16. Install the cylinder head into the vehicle.

1.8L (B18A1, B18B1) Engines

1. Disconnect the negative battery cable.

2. Remove the spark plug wires.

3. Remove the cylinder head cover and timing belt cover.

4. Rotate the crankshaft to TDC, compression of No. 1 piston and remove the timing belt.

5. Remove the distributor from the cylinder head.

6. Install 5.0mm pin punches to the No.1 camshaft holders then remove the camshaft sprockets.

7. Loosen the valve adjusters to remove as much spring tension as possible.

8. Remove the pin punches from the camshaft holders.

9. To check camshaft end-play:

  a. Loosen the end bearing cap bolts 1 turn.

  b. Install the dial indicator.

  c. Push the camshaft fully towards the back of the head, zero the dial indicator and push the camshaft fully the other way to read end–play.

  d. End–play on a new camshaft should be 0.002–0.006 in. (0.05–0.15mm), 0.020 in. (0.5mm) is the service limit.

10. To remove the camshaft bearing caps, loosen each bolt 2 turns at a time in a crisscross pattern to avoid damage to the valves or rockers. Mark the caps so they can be replaced in their original position.

11. Lift the camshafts from the cylinder head, wipe them clean and inspect the lift ramps. Replace the camshafts and rockers if the lobes are pitted, scored or excessively worn.

12. Tag or label the rocker arms before removing to install them to their original locations.

13. Use Plastigage ®to check bearing clearance: for 1993, 0.002–0.004 in. (0.050–0.089mm) is standard clearance, with 0.006 in. (0.15mm) service limit. For 1994 the standard clearance is 0.0015–0.0027 in. (0.039–0.069mm), and absolute service limit is 0.006 in. (0.15mm). For 1995–96 the standard clearance is 0.0012–0.0027 in. (0.030–0.069mm), and absolute service limit is 0.006 in. (0.15mm).

To install:

14. Check the following before installing the camshafts:

  a. Be certain the keyways on the camshafts are facing UP (No. 1 cylinder at TDC).

  b. The valve adjuster locknuts should be loosened and the adjusting screws backed off before installation.

15. Lubricate the rocker arms and camshafts with clean oil.

16. Place the rocker arms on the pivot bolts and the valve stems, making sure that the rocker arms are in their original positions.

17. Install the camshaft seals with the open side (spring) facing in. Lubricate the lip of the seal.

18. Make sure the keyways on the camshafts are facing up and install the camshafts to the cylinder head.

19. Apply liquid gasket to the head mating surfaces of the No. 1 and No. 6 camshaft holders, then install them along with No. 2, 3, 4 and 5 camshaft holders. The arrows stamped on the holders should point toward the timing belt. Do not apply oil to the holder

mating surface where the camshaft seals are housed.

20. Tighten the camshaft holders temporarily and make sure that the rocker arms are properly positioned.

21. Press the oil seals into the No.1 camshaft holders with a seal driver.

22. On 1993–95 models, torque the bolts in a crisscross pattern to 7 ft. lbs. (10 Nm). Check that the rockers do not bind on the valves.

23. Install the cylinder head plug to the end of the cylinder head. If the plug has alignment marks, align the marks with the cylinder head upper surface.

24. If equipped with a timing belt back cover, install the cover and torque the bolts to 7.2 ft. lbs. (9.8 Nm).

25. Install 5.0mm pin punches to the No.1 camshaft holders then install the camshaft pulley keys onto the grooves in the camshafts.

26. Push the camshaft pulleys onto the camshafts, then torque the retaining bolts to 27 ft. lbs. (38 Nm).

27. Install the timing belt and timing belt covers. Remove the pin punches from the camshaft holders.

28. Adjust the valves and pour oil over the camshafts and rocker arms.

29. Apply liquid gasket to the rubber seal at the eight corners of the recesses. Do not install the parts if 20 minutes or more have elapsed since applying the liquid gasket. Instead, reapply liquid gasket after after removing old residue.

30. Install the cylinder head cover and engine ground cable. Make sure the contact surfaces are clean and do not touch surfaces where liquid gasket has been applied.

31. Tighten the cylinder head cover nuts in 2–3 steps. In the final step, tighten all nuts in sequence, to 7 ft. lbs. (10 Nm).

32. Install the distributor to the cylinder head and reconnect the spark plug wires to the spark plugs.

33. Connect the negative battery cable and enter the radio security code.

34. Change the engine oil. Wait at least 20 minutes before filling the engine with oil. The time interval allows the gasket sealants to cure properly.

2.5L (G25A1, G25A4) and 3.2L (C32A1) Engines

1. Disconnect the negative battery cable. Remove the timing belt covers and cylinder head covers.

2. Rotate the crankshaft to TDC compression of No. 1 piston and remove the timing belt.

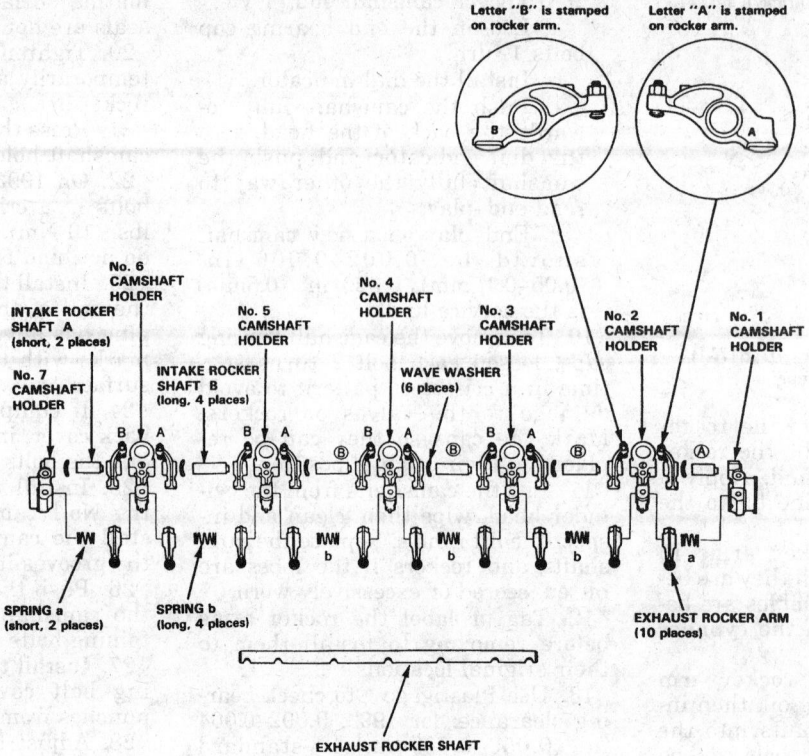

Rocker arm and shaft assembly — 2.5L (G25A1, G25A4) engines

arms are not all the same length. Carefully note their positions during disassembly.

4. Remove the camshaft sprocket.

5. Remove the cylinder head from the vehicle.

6. Loosen the rocker shaft holder bolts 1 turn at a time in the opposite of the installation sequence. Following this procedure will prevent the camshafts and rocker assemblies from warping.

7. After all bolts are loose, remove the rocker arm shafts as an assembly with the bolts still in the holders.

8. If the rocker shafts are to be disassembled, note that each rocker arm has a letter **A** or **B** stamped into the side. Before disassembling the rocker arms, make a note of the position of each letter so the arms can be reassembled the same way.

9. For 3.2L (C32A1) engine, do not remove the hydraulic tappets from the rocker arms unless they are to be replaced. Handle the rocker arms carefully so the oil does not drain out of the tappets.

10. Lift the camshafts from the cylinder head, wipe them clean and inspect the lift ramps. Replace the camshafts and rockers if the lobes are pitted, scored, or excessively worn.

To install:

11. Lubricate the camshaft and its journals with fresh engine oil.

12. Place a new camshaft seal on the end of the camshaft. The spring side of the seal must face in. Lubricate the journals and set the camshaft in place on the head.

13. Install the camshaft onto the cylinder head with the keyway pointed up.

14. Apply liquid gasket to the mounting surfaces of the camshaft end holders.

15. Set the rocker arm assemblies in place and start all the cam holder bolts. Make sure the rocker arms are properly positioned and turn each bolt in sequence two turns at a time until the holders are seated on the head. Follow this procedure to avoid damaging the camshaft and rocker assemblies.

16. When all the camshaft and rocker holders are seated, torque the bolts in the same sequence. Torque the 8mm bolts to 16 ft. lbs. (22 Nm) and the 6mm bolts to 9 ft. lbs. (12 Nm).

17. Install the cylinder head.

18. Install the camshaft sprocket and torque the bolts for 2.5L (G25A1, G25A4) engines to 51 ft. lbs. (70 Nm). Torque the bolts for 3.2L (C32A1) engine to 23 ft. lbs. (32 Nm).

Specified torque:
8 mm bolts: 22 N·m (2.2 kg-m, 16 lb-ft)
6 mm bolts: 12 N·m (1.2 kg-m, 9 lb-ft)

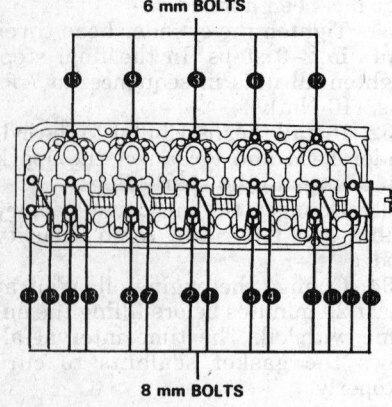

Rocker arm assembly holder bolt torque sequence — 2.5L (G25A1, G25A4) engines

3. For 2.5L (G25A1, G25A4) engines, the springs between the rocker

19. Install the timing belt, adjust the valves and oil the camshaft before completing the assembly.

20. Install the cylinder head cover and timing cover.

21. Install the distributor.

22. Reconnect the negative battery cable.

23. Check for proper engine and valve train operation.

Intake Manifold

REMOVAL AND INSTALLATION

NOTE: The radio may have a coded theft protection circuit. Obtain the code from the owner before disconnecting the battery, removing the radio fuse, or removing the radio.

1.7L (B17A1), 1.8L (B18A1, B18B1, B18C1) Engines

1. Disconnect the negative battery cable.

2. Drain the cooling system into a sealable container. Remove the strut brace if equipped.

3. Remove the air intake duct.

4. Relieve the fuel pressure.

——————— **CAUTION** ———————
Do not allow fuel spray or vapors to come in contact with a spark or open flame. Never store fuel in an open container due to risk of fire or explosion.

5. Disconnect the fuel feed hose.

6. Remove the breather hose, the water bypass hose and the EVAP control canister hose from the throttle body.

7. Remove the fuel return hose.

8. Disconnect the PCV hose.

9. Remove the brake booster vacuum hose, the water bypass hose and the vacuum hose from the manifold.

10. Remove the throttle cable from the throttle body. Take great care not to kink or damage the cable.

11. If necessary, remove the throttle body.

12. For 1994, and later engines, if equipped with automatic transaxle, disconnect the throttle control cable.

13. Label and disconnect all the emission vacuum hoses from the intake manifold.

14. Label and disconnect the wiring connected to the intake manifold. Disconnect sensors as needed; release wiring retainers and clips.

15. Disconnect the water bypass hoses from the manifold.

16. Remove the bolts attaching the intake manifold to the support bracket.

17. Remove the nuts attaching the intake manifold to the cylinder head. Remove the nuts in a crisscross pattern, beginning from the center and moving out to both ends.

18. Remove the manifold and the old gasket.

19. Clean the intake manifold mating surfaces. Inspect the manifold for cracks, flatness and/or damage; replace the parts, if necessary. If the intake manifold is to be replaced, transfer all the necessary components to the new manifold. On B18C1 engines, the intake manifold may be removed from the air bypass valve body. If the manifold is removed, always install a new gasket before reassembly.

To install:

20. If the manifold was removed from the air bypass valve body, reassemble the components before installation. Tighten the through-bolts to 17 ft. lbs. (24 Nm).

21. Use new gaskets when installing the intake manifold. Torque the nuts, in a crisscross pattern, in 2–3 steps, starting with the inner nuts, to 17 ft. lbs. (23 Nm).

22. Install the bolts to the manifold support bracket. Tighten the bolts to 17 ft. lbs. (24 Nm).

23. The remainder of the procedure is the reverse of the removal. When connecting the fuel feed hose to the filter, use new washers and torque the banjo bolt to 25 ft. lbs. (33 Nm) and the service bolt to 11 ft. lbs. (15 Nm). If applicable, when installing the strut brace, torque the attaching nuts to 17 ft. lbs. (24 Nm). If removed, use a new gasket when installing the throttle body and tighten the nuts to 14 ft. lbs. (20 Nm).

24. After all removed components and connections have been re-installed, refill and bleed the air from the cooling system.

25. Connect the negative battery cable and enter the radio security code. Start the engine and allow it to reach normal operating temperature.

26. Check for leaks and proper engine operation. Top off the engine coolant as necessary.

2.5L (G25A1, G25A4) Engines

1. Disconnect the negative battery cable.

2. Remove the fuel filler cap and loosen the service bolt on the fuel filter to relieve the fuel system pressure. Remove the banjo bolt to remove the fuel feed hose from the fuel

filter. Remove the fuel return hose from the pressure regulator.

——————— **CAUTION** ———————
Fuel injection systems remain under pressure after the engine has been turned OFF. Properly relieve fuel pressure before disconnecting any fuel lines. Failure to do so may result in fire or personal injury. Do not allow fuel spray or fuel vapors to come in contact with a spark or open flame. Keep a dry chemical fire extinguisher nearby. Never store fuel in an open container due to risk of fire or explosion.

3. Remove the intake air duct and air cleaner assembly.

4. Remove the throttle cable by loosening the locknut, then slip the cable end out of the throttle bracket and throttle linkage. Take care not to bend the cable when removing it. Move the cable aside.

5. Remove the engine harness cover. Label and disconnect the vacuum hoses and all wiring from the intake manifold.

6. To avoid having to drain the cooling system, remove the fast idle valve and the IAC valve without disconnecting the coolant hoses. Move these components out of the work area so that they will not be damaged.

7. Remove the EGR pipe and the vacuum pipe.

8. Remove the fuel rail. Remove the fuel injectors from the manifold. Handle the injectors and fuel rail carefully to avoid damaging them or contaminating them with dirt.

9. Unbolt the top bolts on the intake manifold brackets.

10. Remove the nuts that secure the manifold to the head. Remove the intake manifold from the engine.

To install:

NOTE: Use new O-rings when installing the IAC and fast idle valves.

11. Inspect the manifold and its components for any signs of damage.

12. Fit the manifold to the engine with a new gasket and torque the nuts to 16 ft. lbs. (22 Nm). Torque the manifold bracket bolts to 16 ft. lbs. (22 Nm).

13. Install the fuel injectors into the rail and install the assembly onto the manifold with new sealing rings and cushion rings to prevent noise and leakage.

14. Connect the fuel injector harnesses and install the harness cover.

15. Install the IAC, fast idle, EGR, and EVAP valves. Use new O-rings.

16. Connect the wiring, vacuum hoses, and fuel lines. Use new sealing washers when connecting the fuel lines.

17. Install the throttle cable into its bracket and linkage. The throttle cable deflection is the measured by pressing down on the cable between the rubber boot and the linkage. The deflection must be 0.39–0.47 in. (10–12mm). Adjust the throttle cable as required.

18. Verify that all wiring and vacuum hoses are installed correctly.

19. Connect the negative battery cable. Run the engine and check for leaks.

3.2L (C32A1) Engines

1. Disconnect the negative battery cable.

2. Drain the cooling system.

3. Remove the air intake duct from the throttle body.

4. On GS models, remove the TCS control valve assembly and its brackets from the throttle body.

5. Relieve the fuel system pressure by loosening the service bolt on the fuel filter about 1 turn, then disconnect the fuel supply and return lines from the manifold.

--- CAUTION ---

Fuel injection systems remain under pressure, even after the engine has been turned OFF. The fuel system pressure must be relieved before disconnecting any fuel lines. Failure to do so may result in fire and/or personal injury.

6. Remove the engine harness covers, tag and disconnect the wiring harnesses from the fuel injectors.

7. Remove the vacuum pipe harness, air inlet pipe, and EGR pipe.

8. Remove the pulsed air injection pipe and valve.

9. Remove the intake manifold nuts and bolts in a crisscross pattern, beginning from the center and moving out to both ends of the manifold.

10. Verify that all vacuum lines are disconnected and remove the intake manifold and throttle body as a unit.

11. Remove the water passage and clean the gasket mounting surfaces.

12. Inspect the manifold for cracks, flatness, or other damage; replace any damaged parts. If the intake manifold is to be replaced, transfer all the necessary components to the new manifold.

To install:

NOTE: Always use new gaskets and O-rings during installation.

13. Install the water passage.

14. Install the intake manifold and torque the nuts/bolts, in a crisscross pattern in 2–3 steps, starting with the inner nuts. Torque the 8mm bolts to 16 ft. lbs. (22 Nm) and the 6mm bolts to 9 ft. lbs. (12 Nm).

15. Reconnect the vacuum lines and the air inlet, and the EGR pipes.

16. Reconnect the fuel supply and return lines to the manifold. Tighten the fuel system service bolt.

17. Reconnect all intake manifold wiring connectors.

18. Install the TCS control valve assembly and brackets.

19. Install the air duct to the throttle body. Refill the cooling system.

20. Reconnect the negative battery cable. Start the engine, allow it to reach normal operating temperature and check for leaks and proper engine operation.

Exhaust Manifold

REMOVAL AND INSTALLATION

NOTE: The radio may have a coded theft protection circuit. Obtain the code from the owner before disconnecting the battery, removing the radio fuse, or removing the radio.

1.7L (B17A1), 1.8L (B18A1, B18B1 and B18C1) Engines

--- WARNING ---

This procedure should only be performed on a cold engine.

1. Disconnect the negative battery cable.

2. Remove the exhaust manifold cover.

3. Raise and safely support the vehicle.

4. Remove the three nuts attaching the front exhaust pipe to the exhaust manifold. Discard the nuts. Separate the exhaust pipe from the manifold and discard the gaskets.

5. Lower the vehicle.

6. Disconnect the bracket attaching the exhaust manifold to the engine.

7. Remove the exhaust manifold attaching nuts and discard the nuts.

8. Remove the exhaust manifold from the engine. Clean any old gasket material from the engine and the exhaust manifold mating surfaces.

9. Remove the rear cover from the exhaust manifold.

To install:

10. Install the rear cover to the exhaust manifold and torque the mounting bolts to 17 ft. lbs. (24 Nm).

11. Install a new exhaust manifold gasket to the cylinder head.

12. Install the exhaust manifold to the engine and install new attaching nuts. Torque the nuts to 23 ft. lbs. (31 Nm).

13. Install the bracket to the exhaust manifold and the engine. Torque the bolts to 33 ft. lbs. (44 Nm).

14. Raise and safely support the vehicle.

15. Install new gaskets to the front exhaust pipe where it connects to the exhaust manifold.

16. Connect the front exhaust pipe to the exhaust manifold. Install new nuts and torque the nuts to 40 ft. lbs. (54 Nm).

17. Lower the vehicle.

18. Install the exhaust manifold cover and torque the bolts to 17 ft. lbs. (24 Nm).

19. Connect the negative battery cable and enter the radio security code.

20. Run the engine and check for exhaust leaks.

2.5L (G25A1, G25A4) Engines

1. Disconnect the negative battery cable.

2. Remove the outer manifold heat shields.

3. Disconnect the wire and remove the oxygen sensor from the manifold.

4. Disconnect the manifold from exhaust pipe.

5. Remove the mounting bracket and remove the nuts to remove the manifold.

To install:

6. Make sure the gasket mating surfaces are clean. Install the bracket loosely and install the manifold with new gaskets and self-locking nuts. Torque the nuts to 23 ft. lbs. (31–32 Nm), then tighten the bracket bolts. Be sure not to bend or damage the contact surface of the metal gasket.

7. Coat the threads of the oxygen sensor with an anti-seize compound. Be careful not to get any on the head of the sensor. Install the sensor and carefully torque to 33 ft. lbs. (44–45 Nm). Connect the sensor wire.

8. Install a new gasket and connect the manifold to the exhaust pipe. Torque the nuts to 40 ft. lbs. (54–55 Nm). Install the outer manifold heat shields and tighten the bolts to 22 ft. lbs. (29 Nm).

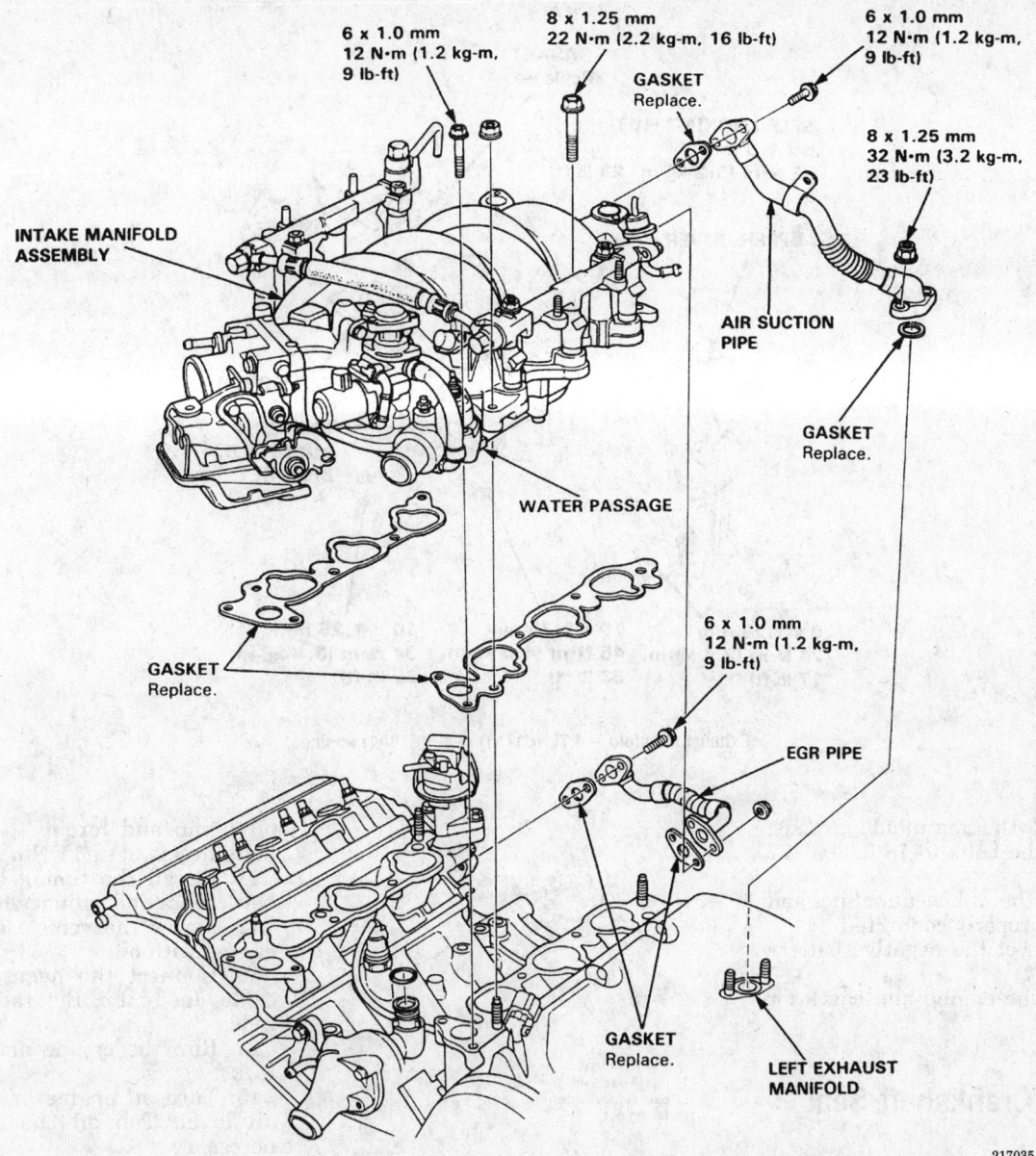

6 x 1.0 mm
12 N·m (1.2 kg-m, 9 lb-ft)

8 x 1.25 mm
22 N·m (2.2 kg-m, 16 lb-ft)

GASKET
Replace.

6 x 1.0 mm
12 N·m (1.2 kg-m, 9 lb-ft)

8 x 1.25 mm
32 N·m (3.2 kg-m, 23 lb-ft)

INTAKE MANIFOLD ASSEMBLY

AIR SUCTION PIPE

WATER PASSAGE

GASKET
Replace.

GASKET
Replace.

6 x 1.0 mm
12 N·m (1.2 kg-m, 9 lb-ft)

EGR PIPE

GASKET
Replace.

LEFT EXHAUST MANIFOLD

217035

Intake manifold assembly — 3.2L (C32A1) Engines (Legend shown, 3.2TL similar)

9. Connect the negative battery cable. Start the engine and check for exhaust leaks.

3.2L (C32A1) Engines

NOTE: This operation should be performed with the engine and exhaust cold.

1. Disconnect the negative battery cable. Make sure the engine is not hot or warm before performing this operation. Remove the exhaust manifold shrouds.
2. If applicable, remove the two small heat shields from the cylinder heads.
3. Remove the exhaust pipe nuts.

4. Remove the oxygen sensors.
5. Remove the air suction tube.
6. Remove the exhaust attaching nuts in a crisscross pattern starting from the center of the manifold.
7. Clean the gasket mounting surfaces. Inspect the manifold for cracks, flatness and/or damage; replace the parts, if necessary.

To install:

8. To install, use new gaskets and self–locking nuts. Make sure all mating surfaces are clean before installing exhaust manifold. Torque the manifold nuts in a crisscross pattern starting from the center, for Legend models to 25 ft. lbs. (34 Nm). For

3.2TL, torque the manifold nuts to 22 ft. lbs. (30 Nm).
9. If applicable, install the two small heat shields and torque the attaching bolts to 16 ft. lbs. (22 Nm).
10. Use new gaskets when installing the exhaust pipe to the manifold and torque the nuts to 40 ft. lbs. (55 Nm).

NOTE: On GS series Legends, the exhaust manifold is bolted to twin three-way catalytic converters.

11. Install the air suction tube, then install the oxygen sensors. Torque the oxygen sensors to 33 ft. lbs. (45 Nm).

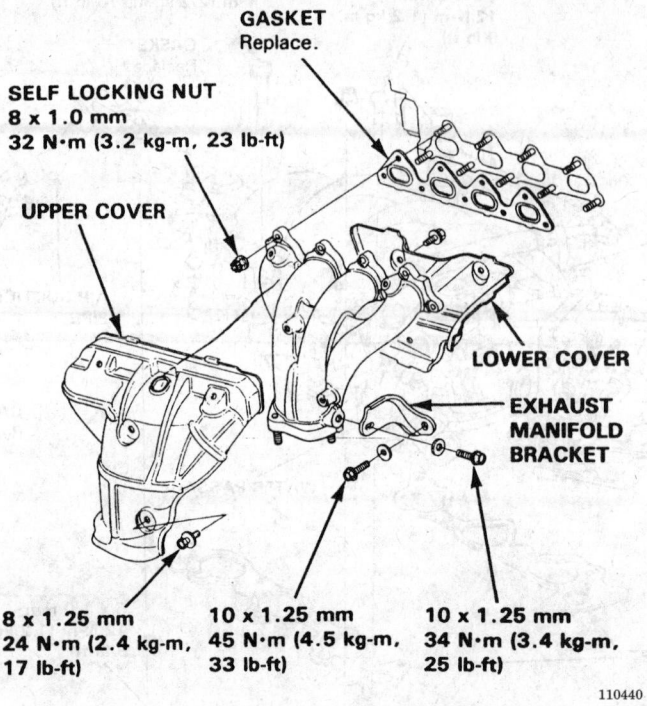

GASKET
Replace.

SELF LOCKING NUT
8 x 1.0 mm
32 N·m (3.2 kg-m, 23 lb-ft)

UPPER COVER

LOWER COVER

EXHAUST
MANIFOLD
BRACKET

8 x 1.25 mm
24 N·m (2.4 kg-m, 17 lb-ft)

10 x 1.25 mm
45 N·m (4.5 kg-m, 33 lb-ft)

10 x 1.25 mm
34 N·m (3.4 kg-m, 25 lb-ft)

110440

Exhaust manifold — 1.7L (B17A1), 1.8L (B18A1) engines

12. Install the manifold shrouds, tightening the bolts to 16 ft. lbs. (22 Nm).

13. Verify that all vacuum lines and wiring are properly connected.

14. Reconnect the negative battery cable.

15. Start the engine and check for leaks.

Front Crankshaft Seal

REMOVAL AND INSTALLATION

NOTE: The radio may have a coded theft protection circuit. Obtain the code from the owner before disconnecting the battery, removing the radio fuse, or removing the radio.

1. Disconnect negative cable at the battery.

2. Raise and safely support the vehicle. Drain the engine oil and properly dispose of it.

3. Make sure the crankshaft is at TDC on No. 1 cylinder by aligning the white mark on the crankshaft pulley with the pointer on the lower timing belt cover.

4. Remove the cylinder head cover, timing belt cover and timing belt.

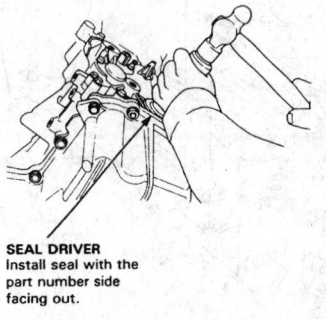

SEAL DRIVER
Install seal with the part number side facing out.

334059

Installing the seal

5. If equipped with a Crankshaft Speed Fluctuation (CKF) sensor, remove the sensor form the oil pump.

6. Remove the timing belt gear from the crankshaft.

7. Using a small prying tool, remove the seal from the oil pump.

To install:

8. Apply a light coat of oil to the seal lip.

9. Position the seal on the oil pump then using a seal driver, install the seal into the oil pump.

10. Install the timing belt pulley to the crankshaft. If equipped with a CKF sensor, install the sensor to the

oil pump and torque the attaching bolts to 8 ft. lbs. (11 Nm).

11. Install the timing belt, timing belt covers and cylinder head cover.

12. Lower the vehicle and fill the engine with oil.

13. Connect the negative battery cable and enter the radio security code.

14. Run the engine and check for leaks.

15. Turn off engine and check the oil level. Top off the oil level if necessary.

Timing Belt, Sprockets, Tensioner and Front Cover

REMOVAL AND INSTALLATION

NOTE: The radio may have a coded theft protection circuit. Obtain the code from the owner before disconnecting the battery, removing the radio fuse, or removing the radio.

1.7L (B17A1), 1.8L (B18A1) Engines

1. Disconnect the negative battery cable. Turn the crankshaft pulley until cylinder No. 1 is set to TDC, compression. The white crankshaft pulley mark should be aligned with the

pointer on the cover and the camshaft sprockets **UP** mark should be facing upward, sprocket timing marks aligned.

2. Raise and safely support the vehicle. Remove the left front wheel. Remove the wheel well splash guard.

3. Remove the power steering belt and power steering pump. Do not disconnect the power steering fluid lines.

4. Remove the air conditioning belt and the alternator belt.

5. Remove the left side engine mount.

6. Remove the cylinder head cover.

7. Remove the middle timing belt cover.

8. Remove the special crankshaft pulley bolt and the crankshaft pulley.

9. Remove the lower timing belt cover.

NOTE: Do not use the timing belt covers to store small parts. Grease or oil can transfer from the parts to the cover, then to the belt. Clean the covers thoroughly before installation.

10. Loosen, but do not remove, the tensioner adjusting bolt. Push the tensioner to slacken the timing belt, then retighten the bolt. If the timing

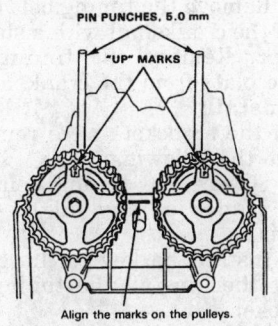

PIN PUNCHES, 5.0 mm
"UP" MARKS
Align the marks on the pulleys.
110436

Timing marks — 1.7L(B17A1), 1.8L (B18A1) engines

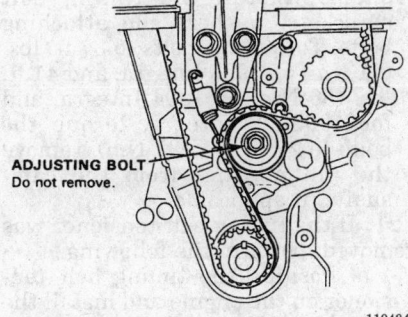

ADJUSTING BOLT
Do not remove.
110434

Timing belt tensioner — 1.7L (B17A1), 1.8L (B18A1) engines

belt is to be reinstalled, mark the direction of rotation.

11. Remove the timing belt.

12. If necessary, remove the timing belt tensioner by performing the following:

  a. Remove the timing belt tensioner spring.

  b. Remove the bolt from the timing belt tensioner and remove the tensioner.

To install:

13. If the timing belt tensioner was removed, perform the following:

  a. Position the timing belt tensioner on the engine and install the attaching bolt loosely.

  b. Install the timing belt tensioner spring.

  c. Push the tensioner down then snug the tensioner bolt to hold this position.

14. Recheck that No. 1 piston is at TDC on its compression stroke. To set the camshafts to top dead center for No. 1 cylinder, align the hole in the camshafts with the holes in the No. 1 camshaft holders; push 5.0mm pin punches into the holes.

15. If necessary, remove the timing belt sprockets by performing the following:

  a. Remove the timing belt sprocket–to–camshaft bolts, the washers, and the sprockets. Remove the sprocket with a pulley remover or a brass hammer. Be careful not to lose the Woodruff key.

NOTE: The camshaft oil seal can be replaced without removing the camshaft holders. To check camshaft end–play, the rocker arm adjusting screws must be loosened.

  b. If equipped with a Crankshaft Speed Fluctuation (CKF) sensor, remove the sensor mounting bolts and remove the sensor.

  c. Remove the timing belt pulley from the crankshaft with a suitable puller. Remove the timing belt guide plate from the crankshaft.

16. If the sprockets were removed, perform the following:

  a. Clean the timing belt gear and guides of any oil or other debris.

  b. Install the key to the crankshaft then install the timing belt sprocket.

  c. If equipped with a CKF sensor, install the sensor. Torque the mounting bolts to 8 ft. lbs. (11 Nm).

  d. Install the keys to the camshafts. Install the timing belt sprockets, washers and attaching bolts. Torque the bolts to 37 ft. lbs.

(51 Nm) on 1993 Integra, and 41 ft. lbs. (56 Nm) on 1994 Integra, and for 1995–97 Integra, torque the bolts to 27 ft. lbs. (37 Nm), remove the pin punches from the camshafts, if applicable.

17. Install the timing belt. If installing the old belt, make sure it is turning the same direction.

18. Make sure the timing belt is properly installed on the crankshaft and that the front oil seal does not leak. Replace the cover and crankshaft pulley and carefully oil the threads of the pulley bolt without getting oil on the washer or the head of the bolt. If installing a new crankshaft or new bolt, torque the bolt in 3 steps: 145 ft. lbs. (200 Nm), then loosen the bolt completely, then retorque to 130 ft. lbs. (180 Nm).

19. Make sure the camshaft sprockets and crankshaft pulley are properly aligned with the timing marks. Loosen the tensioner bolt about ½ turn, then torque to 40 ft. lbs. (55 Nm). Tension adjustment is automatically accomplished with the spring on the tensioner. Check timing mark alignment again.

20. Install the middle timing belt cover.

21. Remove the pin punches from the camshafts. Install the cylinder head cover.

22. Install the side engine mount. The through-bolt holding the mount to the engine should be tightened to 40 ft. lbs. (55 Nm). The nut and bolt holding the engine to the mount should be tightened to 54 ft. lbs. (75 Nm).

23. Install and adjust the alternator, air conditioning and power steering belts.

24. Install the splash shield and wheel.

25. Connect the negative battery cable.

1.8L (B18B1, B18C1) Engines

1. Disconnect the negative battery cable.

2. Turn the crankshaft pulley until cylinder No. 1 is set to TDC, compression. The white crankshaft pulley mark should be aligned with the pointer on the lower timing belt cover.

3. Raise and safely support the vehicle. Remove the left front wheel. Remove the wheel well splash guard.

4. Remove the power steering belt and power steering pump. Do not disconnect the power steering fluid lines.

5. Remove the air conditioning belt and the alternator belt.

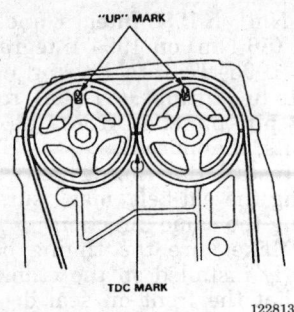

Camshaft pulleys aligned at TDC of No. 1 piston — 1993–94 1.7L (B17A1), 1.8L (B18A1, B18B1, B18C1) engines

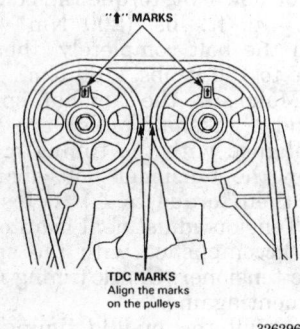

Camshaft pulley position for No. 1 cylinder at TDC — 1995–97 1.8L (B18C1) engines

Number 1 piston at TDC:

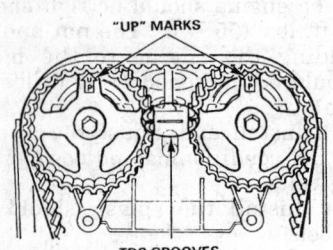

Camshaft pulley position for No. 1 piston at TDC — 1.8L (B18B1) engines

6. Remove the cruise control actuator.
7. Support the engine.
8. Remove the side engine mount.
9. Tag, then disconnect the spark plug wires from the spark plugs.
10. Remove the cylinder head cover.
11. Remove the crankshaft pulley bolt and the crankshaft pulley. To remove the crankshaft pulley bolt a pulley holder (holder attachment tool part # 07MAB–PY3010A and holder handle tool part # 07JAB–001020A,

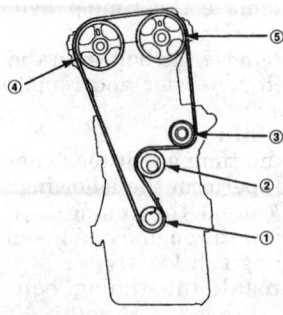

2. Install the timing belt tightly in the sequence shown.
① Timing belt drive pulley (crankshaft) → ②
Adjusting pulley → ③ Water pump pulley → ④
Exhaust camshaft pulley → ⑤ Intake camshaft pul-

Timing belt installation order — 1.8L (B18B1, B18C1) engine

or equivalent) will be needed to keep the crankshaft from turning.
12. Remove the middle timing belt cover.
13. Remove the lower timing belt cover.

NOTE: Do not use the timing belt covers to store small parts. Grease or oil can transfer from the parts to the cover, then to the belt. Clean the covers thoroughly before installation.

14. Recheck that No. 1 piston is at TDC on its compression stroke. Align the groove on the tooth side of the crankshaft timing belt drive sprocket to the arrow pointer on the oil pump.
15. To set the camshafts to top dead center for No. 1 cylinder, align the hole in each camshaft with the holes in the No. 1 camshaft holders; push 5.0mm pin punches into the holes. Make sure that the **UP** arrows are pointing up and the TDC marks on the intake and exhaust sprockets are aligned.
16. Loosen the tensioner adjusting bolt 180 degrees. Push the tensioner to remove the tension on the timing belt, then retighten the bolt. If the timing belt is to be reinstalled, mark the direction of rotation.

17. Remove the timing belt.

NOTE: Make sure the water pump pulley turns counter–clockwise freely. Check for signs of seal leakage; a small amount of weeping from the bleed hole is normal.

18. If necessary, remove the timing belt tensioner by performing the following:
a. Remove the timing belt tensioner spring.
b. Remove the bolt from the timing belt tensioner and remove the tensioner.
19. If necessary, remove the timing belt sprockets by performing the following:
a. Remove the timing belt sprocket–to–camshaft bolts, the washers, and the sprockets. Remove the sprocket with a pulley remover or a brass hammer. Be careful not to lose the Woodruff key.

NOTE: The camshaft oil seal can be replaced without removing the camshaft holders. To check camshaft end–play, the rocker arm adjusting screws must be loosened.

b. If equipped with a Crankshaft Speed Fluctuation (CKF) sensor, remove the sensor mounting bolts and remove the sensor.
c. Remove the timing belt pulley from the crankshaft with a suitable puller. Remove the timing belt guide plate from the crankshaft.
To install:
20. If the sprockets were removed, perform the following:
a. Clean the timing belt gear and guides of any oil or other debris.
b. Install the key to the crankshaft then install the timing belt sprocket.
c. If equipped with a CKF sensor, install the sensor. Torque the mounting bolts to 8 ft. lbs. (11 Nm).
d. Install the keys to the camshafts. Install the timing belt sprockets, washers and attaching bolts. Torque the bolts to 37 ft. lbs. (51 Nm) on 1993 Integra, and 41 ft. lbs. (56 Nm) on 1994 Integra, and for 1995–97 Integra, torque the bolts to 27 ft. lbs. (37 Nm), remove the pin punches from the camshafts, if applicable.
21. If the timing belt tensioner was removed, perform the following:
a. Position the timing belt tensioner on the engine and install the attaching bolt loosely.
b. Install the timing belt tensioner spring.

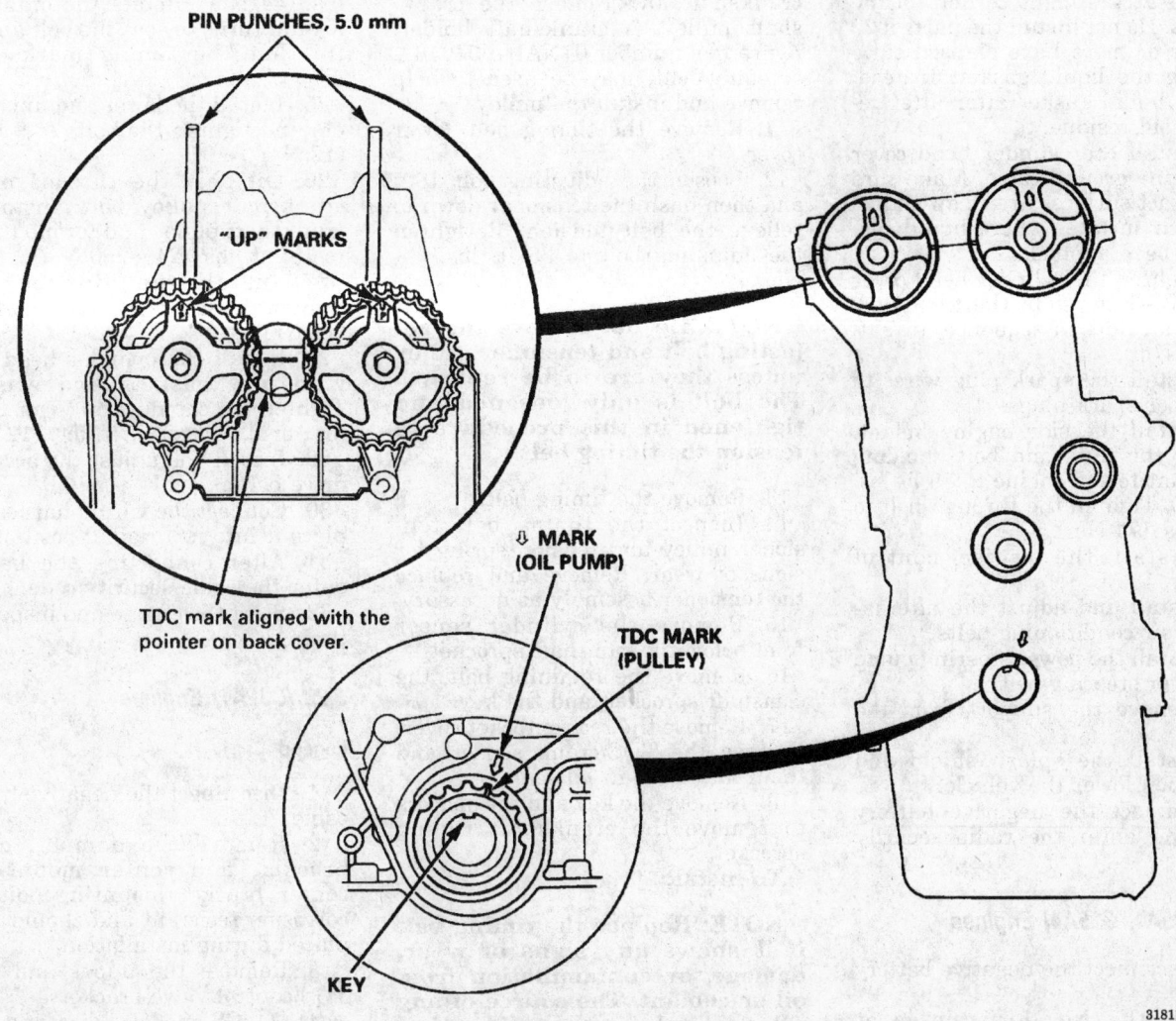

PIN PUNCHES, 5.0 mm

"UP" MARKS

TDC mark aligned with the pointer on back cover.

⇩ MARK (OIL PUMP)

TDC MARK (PULLEY)

KEY

318113

Timing marks — 1995–97 1.8L (B18B1, B18C1) engines

c. Push the tensioner down then snug the tensioner bolt to hold this position.

NOTE: Before reinstallation, check every component for cleanliness. All covers, pulleys, shields, etc. must be completely free of grease and oil.

22. Install the timing belt in the correct sequence. If installing the old belt, make sure it is turning the same direction. Install the belt first to the crankshaft pulley, then to the adjuster, then to the water pump pulley, the exhaust camshaft and finally to the intake camshaft pulley.

23. Install the lower belt cover. Install the crankshaft pulley, tightening the bolt to 130 ft. lbs. (177 Nm). Lubricate the threads and the flange of the bolt with engine oil before installation.

24. Loosen the adjusting bolt, allowing the adjuster to tension the belt. Retighten the bolt to 40 ft.lbs. (54 Nm).

25. Remove the pin punches from the camshafts.

26. Rotate the crankshaft 4 to 6 turns counterclockwise. This allows the belt to equalize tension across all the pulleys.

27. Once again, set the engine to TDC compression for cylinder No. 1.

Check that all timing marks for the cam and crankshafts align. If any mark is out of alignment, remove the timing belt and reinstall it.

28. Loosen the adjusting bolt 180 degrees. Rotate the crankshaft counterclockwise until the camshaft pulleys have moved 3 teeth. Retighten the adjusting bolt to 40 ft.lbs. (54 Nm).

29. Check the torque of the crankshaft pulley bolt.

30. Install the middle belt cover, torque the attaching bolts to 7.2 ft. lbs. (9.8 Nm).

31. Install the rubber seal in the groove of the cylinder head cover.

Make sure that the seal and groove are thoroughly clean first.

32. Apply liquid gasket to the rubber seal at the eight corners of the recesses. Do not install the parts if 20 minutes or more have elapsed since applying the liquid gasket. Instead, reapply liquid gasket after after removing old residue.

33. Install the cylinder head cover and engine ground cable. Make sure the contact surfaces are clean and do not touch surfaces where liquid gasket has been applied.

34. Tighten the cylinder head cover nuts in 2–3 steps. In the final step, tighten all nuts in sequence, to 7 ft. lbs. (10 Nm).

35. Install the spark plug wires to the proper spark plugs.

36. Install the side engine mount. Tighten the nuts and bolts holding the mount to the engine to 38 ft. lbs. (52 Nm). Tighten the through-bolt to 54 ft. lbs. (74 Nm).

37. Install the cruise control actuator.

38. Install and adjust the alternator and air conditioning belts.

39. Install the power steering pump and power steering belt.

40. Remove the support from the engine.

41. Install the splash shield and wheel then lower the vehicle.

42. Connect the negative battery cable and enter the radio security code.

2.5L (G25A1, G25A4) Engines

1. Disconnect the negative battery cable.

2. Set the No. 1 piston is at TDC/compression. The white TDC mark on the crankshaft pulley must line up with the pointers on the lower belt cover.

3. Label and disconnect the engine wiring harness that is routed over the upper belt cover to the alternator and temperature sending unit. Move the harness out of the work area.

4. Loosen the component adjusting bolts and remove the accessory drive belts.

5. Remove the dipstick tube.

6. Mark the sparkplug wires and remove the cylinder head cover.

7. Remove the timing belt upper cover.

8. Make sure the UP mark and the TDC marks on the camshaft sprocket are correctly positioned.

9. Rotate the crankshaft to align the white timing mark on the crankshaft pulley with the pointer on the lower cover. There are similar align-

ment marks on the crankshaft sprocket and oil pump housing

10. Remove the center bolt from the crankshaft and remove the crankshaft pulley. A crankshaft holder, Acura tool number 07NAB–001040A, or equivalent, may be used to help remove and install the pulley

11. Remove the timing belt lower cover.

12. Loosen the adjusting bolt 180° and then push the tensioner down to relieve the belt tension. Retighten the adjusting bolt to 33 ft. lbs. (44 Nm).

NOTE: Do not remove the adjusting bolt and tensioner pulley unless they are to be replaced. The bolt is only loosened and tightened in this procedure to tension the timing belt

13. Remove the timing belt.

14. Inspect the timing belt tensioner pulley and tension spring for signs of wear. Remove and replace the tensioner assembly as necessary.

15. Remove the cylinder sensor from below the camshaft sprocket.

16. Remove the retaining bolt, the camshaft sprocket, and the key.

17. Remove the rear sprocket cover. Remove the TDC/crank sensor and the front camshaft oil seal.

18. Remove the key and the spacers to remove the crankshaft timing sprocket.

To install:

NOTE: Replace the timing belt if it shows any signs of wear, damage, or contamination from oil or coolant. The source of any oil or coolant contamination must be determined and corrected before the new timing belt may be installed.

19. If removed, install the crankshaft timing sprocket making sure the spacers have their concave surfaces out. Install the key. The groove must align with the pointer on the oil pump.

20. Install the front camshaft oil seal, the TDC/crank sensor, and the rear sprocket cover.

21. Install the key, timing sprocket and retaining bolt. Torque the retaining bolt to 54 ft. lbs. (75 Nm).

22. Verify that the crankshaft and camshaft sprockets are properly aligned.

23. Install the timing belt in the following order: first onto the crankshaft sprocket, then the tensioner pulley, the water pump sprocket, and finally the camshaft sprocket.

24. Loosen the tensioner adjusting bolt to allow the spring to set the tension. Then, tighten the bolt to 33 ft. lbs. (44 Nm). Rotate the crankshaft six full turns to seat the belt and verify that the timing marks align properly.

25. Install the lower and upper covers and tighten the bolts to 9 ft. lbs. (12 Nm).

26. Oil only the threads on the crankshaft pulley bolt. Install the crankshaft pulley and torque the bolt to 181 ft. lbs. (245–250 Nm).

27. Install the dipstick tube, tighten the mounting bolt to 9 ft. lbs. (12 Nm).

28. Install the cylinder head cover with new gaskets and washers. Tighten the cap nuts on Vigor to 7 ft. lbs., or 2.5TL to 8.7 ft. lbs. (12 Nm).

29. Install and adjust the accessory drive belts.

30. Connect the wiring harness and place it into its original position.

31. After connecting the battery, enter the radio security code.

32. Start the engine and inspect the timing.

3.2L (C32A1) Engines

Legend

1. Disconnect the negative battery cable.

2. Remove the damper, center bracket, and center mount. The center bracket mounting bolts are corrosion resistant and should be replaced during installation.

3. Remove the upper and lower TCS control valve brackets.

4. On GS models, disconnect the TCS throttle sensor connector and TCS throttle actuator connector, then remove the TCS control valve assembly from the throttle body.

5. Remove the engine wiring harness covers and the wiring harness from the front of the engine.

6. Remove the oil pressure switch connector and the engine ground cable.

7. Remove the alternator, power steering, and air conditioning compressor belts.

8. Remove the drive belts for the alternator, air conditioner compressor, and power steering pump.

9. Remove the upper timing belt covers.

10. Rotate the crankshaft to TDC/compression for the No. 1 piston. The white mark on the crankshaft pulley will be aligned with the pointer on the lower cover, and the camshaft sprocket marks will be al-

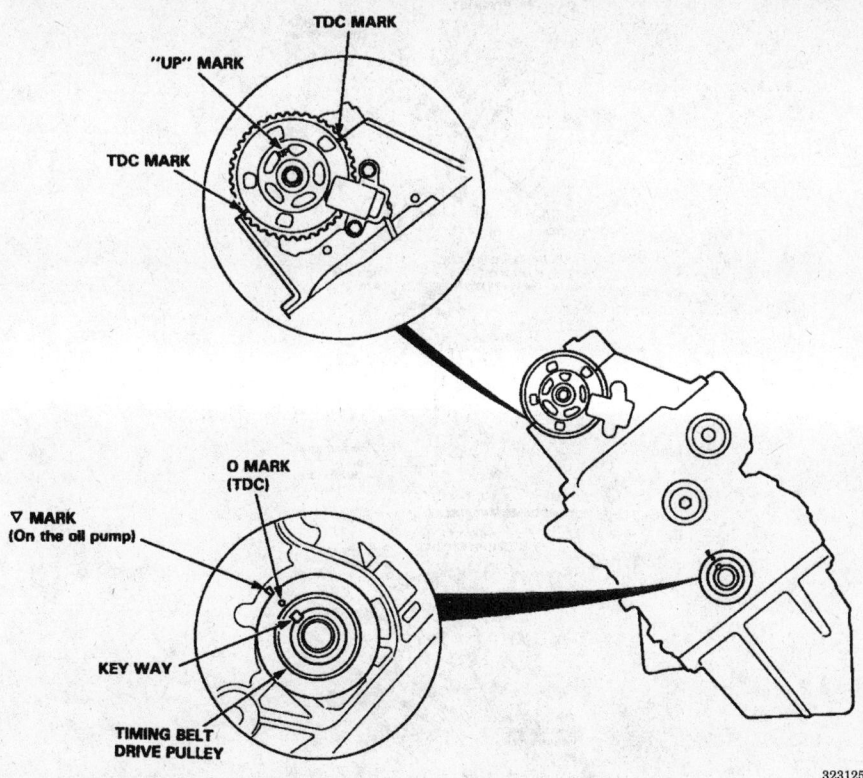

Timing belt TDC mark alignment — 2.5L (G25A1, G25A4) engines

323125

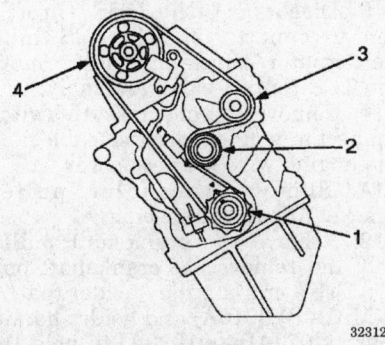

323127

Timing belt installation sequence — 2.5L (G25A1, G25A4) engines

igned with the yellow marks on the rear covers.

NOTE: Do not rotate the crankshaft or camshafts with the belt removed. The pistons will contact the valves and cause engine damage. If it is necessary to move the camshaft(s) for proper alignment, first advance the crankshaft by 15 degrees from TDC. Adjust the camshaft position as needed, then return the crankshaft 15 degrees to TDC.

11. Remove the crankshaft pulley and the air conditioner belt tensioner pulley.

12. Remove the dipstick tube. Remove the lower timing belt cover.

13. Loosen the timing belt tensioner pulley bolt about ½ turn (180 degrees) and push the pulley to slacken the belt tension. Tighten the bolt and remove the belt. If the belt is to be reinstalled, mark the direction of rotation.

14. Using a camshaft holding tool or equivalent to secure the timing belt sprockets, remove the sprocket bolts and the sprockets. On the rear camshaft, remove the bolt opposite the locating pin last.

To install:

15. Install the timing belt sprocket and mounting bolts. Torque the bolts to 23 ft. lbs. (32 Nm).

NOTE: Replace the belt if it is worn, cracked, or oil soaked. Find and repair the source of the oil leak before installing a new belt. Inspect the water pump. If there is any doubt about the condition of the water pump, it should be replaced now that the timing belt is removed.

16. Install the belt in sequence on the crankshaft, adjuster pulley, the left camshaft, water pump, then the right camshaft.

17. To adjust the tension, loosen the tensioner pulley bolt about ½

turn (180 degrees). The spring will automatically set the proper tension.

18. Install the lower cover and the crankshaft pulley. Apply oil to the pulley bolt threads and washer, and tighten the pulley to 174 ft. lbs. (240 Nm).

19. Rotate the crankshaft 5-6 turns clockwise and check that the timing marks on the crankshaft and camshafts align properly. Adjust the timing belt tension again by rotating the crankshaft to align the blue mark on the pulley with the pointer.

20. Loosen and retorque the tensioner pulley bolt to 31 ft. lbs. (43 Nm).

21. Install the dipstick pipe with a new O-ring and tighten the mounting bolt to 9 ft. lbs. (12 Nm). Install the A/C idler pulley and tighten the mounting bolt to 16 ft. lbs. (22 Nm).

22. Install the upper covers, then install the alternator, power steering, and compressor belts. Adjust the belts to the proper tension. Tighten the cover bolts to 9 ft. lbs. (12 Nm).

23. Reconnect and install the wiring harness. Reconnect the oil pressure switch connector and the engine ground cable.

24. On GS models, install the TCS control valve assembly and brackets. Reconnect the wiring.

25. Install the center mount, center bracket, and damper. Torque the center mount and bracket bolts to 28 ft. lbs. (39 Nm) and the damper bolts to 16 ft. lbs. (22 Nm).

26. Reconnect the negative battery cable.

3.2TL

1. Disconnect the negative battery cable.

2. Remove the air intake duct.

3. Turn the engine to align the timing marks and set cylinder No.1 to TDC. The white mark on the crankshaft pulley should align with the pointer on the timing belt cover. Remove the inspection caps on the upper timing belt covers to check the alignment of the timing marks. The pointers for the camshafts should align with the marks on the camshaft sprockets.

4. Support the engine with a jack fitted with a cushion (block of wood). Remove the center bracket from the front engine mounts.

5. Remove the alternator belt.

6. Remove the air conditioner compressor belt.

7. Remove the power steering pump belt.

8. Remove the Traction Control System (TCS) control valve assembly upper and lower brackets.

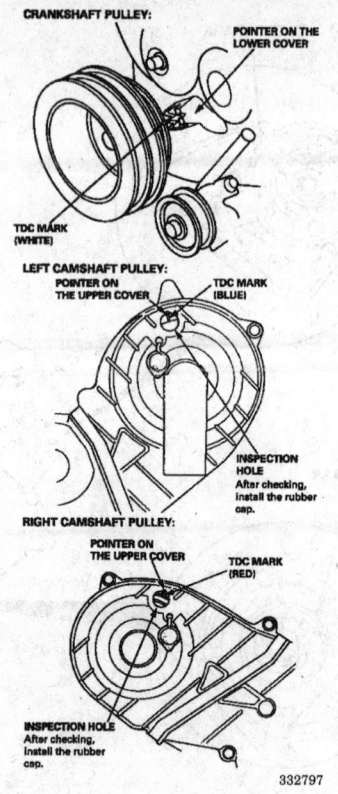

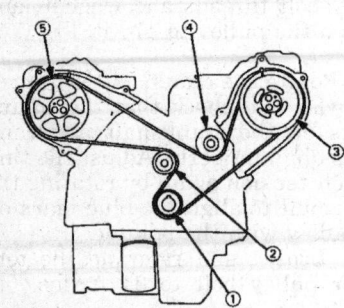

Install the timing belt tightly in the sequence shown.

①Timing belt drive pulley (crankshaft) → ②Adjusting pulley → ③Left camshaft pulley → ④Water pump pulley → ⑤Right camshaft pulley.

• For easy installation, advance the right camshaft pulley by about a half tooth from the TDC position.

216606

Timing belt placement order — Legend

Timing marks alignment — 3.2TL

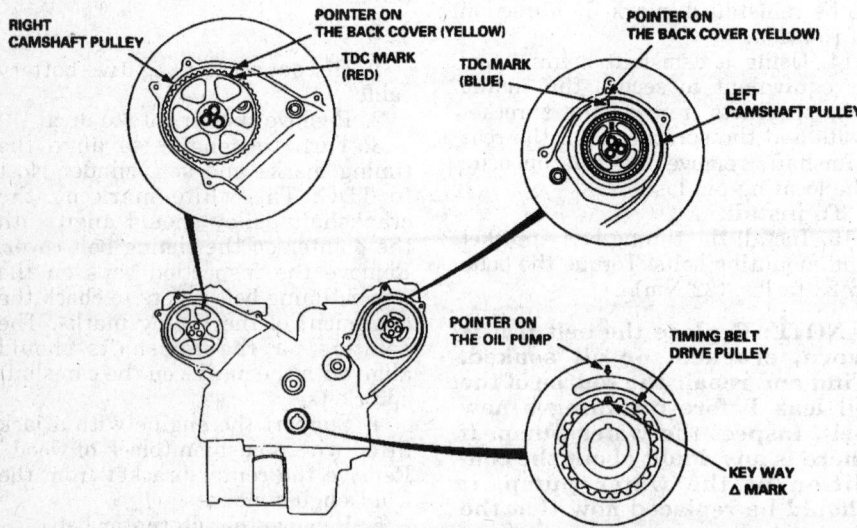

216607

Timing marks — Legend

9. Disconnect the TCS throttle sensor connector and the TCS throttle actuator connector, then remove the TCS control valve assembly.

10. Remove the oil pressure switch connector, engine ground cable and the engine wire harness cover.

11. Remove the idler pulley bracket, dipstick and pipe.

12. Remove the crankshaft pulley bolt and remove the crankshaft pulley. Use a crank pulley holder (part # 07MAB–PY3010A) and holder handle (part # 07JAB–001020A) to hold the crankshaft pulley in place while removing the bolt.

13. Remove the upper and lower timing belt covers. Clean any dirt, oil or grease from the covers. Do not use the covers for storing removed items.

14. Loosen the adjusting bolt 180°. Push the tensioner to remove tension from the timing belt, then tighten the adjusting bolt.

15. Remove the timing belt.

16. Remove the crankshaft timing belt sprocket, do not lose the crankshaft key.

17. Remove the three bolts attaching each of the camshaft sprockets and remove the sprockets.

18. Remove the timing belt tensioner by performing the following:

a. Remove the spring from the tensioner.

b. Remove the bolt attaching the tensioner and remove the tensioner.

To install:

— **CAUTION** —

Do not rotate the crankshaft pulley or camshaft pulleys with the timing belt removed. The pistons may hit the valves and cause damage.

19. Install the timing belt tensioner by performing the following:

   a. Install the tensioner and the attaching bolt.

   b. Move the tensioner its full deflection to the left and tighten the bolt.

   c. Install the spring to the tensioner.

20. Install the camshaft sprocket with the blue TDC mark to the left camshaft and install the camshaft sprocket with the red TDC mark to the right mark. Torque the sprocket mounting bolts to 23 ft. lbs. (31 Nm).

21. Install the timing belt sprocket to the crankshaft and the key. Make sure that the timing belt guide plates are properly installed with the concave surface facing the sprocket.

22. Remove the spark plugs.

23. Set the timing belt drive sprocket so that the No. 1 piston is at

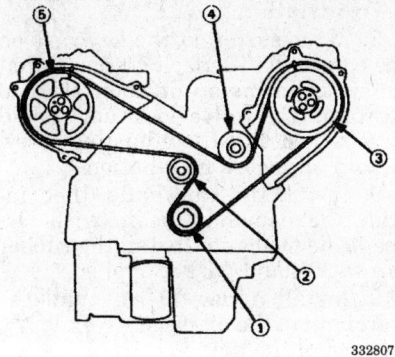

Timing belt installation sequence — 3.2TL

top dead center (TDC). Align the TDC mark on the tooth of the timing belt drive sprocket with the pointer on the oil pump.

24. Set the camshaft pulleys so that the No. 1 piston is at TDC. Align the TDC mark on the camshaft pulley to the pointers on the back covers.

25. Install the timing belt in the following sequence: First, install the timing belt to the drive sprocket (crankshaft). Second to the adjusting pulley, third to the left camshaft sprocket, Fourth to the water pump pulley, and last to the right camshaft sprocket.

26. Loosen then retighten the timing belt adjuster bolt to tension the timing belt.

27. Install the lower timing belt cover and torque the attaching bolts to 8.7 ft. lbs. (12 Nm).

28. Install the crankshaft sprocket and the crankshaft pulley bolt. Torque the bolt to 174 ft. lbs. (235 Nm) with the aid of the crank pulley holder.

29. Rotate the crankshaft five or six turns clockwise so that the timing belt positions on the sprockets.

30. Set cylinder No. 1 to TDC by aligning the timing marks. If the timing marks do not align, remove the timing belt then adjust the components and reinstall the timing belt.

31. Rotate the crankshaft clockwise enough to move nine teeth on the camshaft pulley (the blue mark on the crankshaft pulley should line up with pointer on the lower cover).

32. Loosen the timing belt adjusting bolt 180° then torque the bolt to 31 ft. lbs. (42 Nm).

33. Install the upper timing belt covers. Torque the timing belt covers attaching bolts to 8.7 ft. lbs. (12 Nm).

34. Install the spark plugs and torque them to 13 ft. lbs. (18 Nm).

35. Install the dipstick pipe and the dipstick. Install a new O–ring and

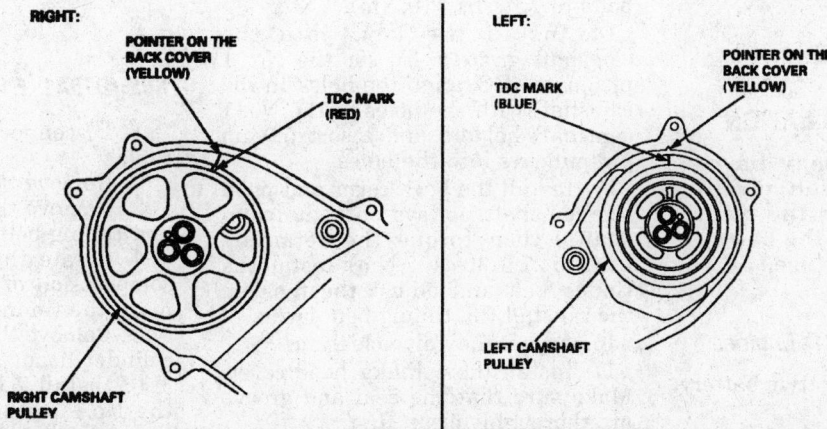

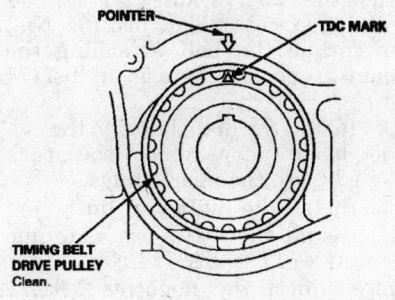

Timing sprockets aligned — 3.2TL

torque the attaching bolts top 9 ft. lbs. (12 Nm).

36. Install the A/C belt idler pulley and torque the mounting bolt to 16 ft. lbs. (22 Nm).

37. Install the engine wire harness cover then the engine ground cable and the oil pressure switch connector.

38. Install the TCS control valve assembly to the throttle body. Connect the TCS throttle actuator connector and the TCS throttle sensor connector.

39. Install the TCS control valve assembly lower and upper brackets. Torque the lower bracket bolts to 16 ft. lbs. (22 Nm) and torque the upper bracket bolts to 8.7 ft. lbs. (12 Nm).

40. Install and adjust the power steering drive belt.

41. Install and adjust the A/C compressor drive belt.

42. Install and adjust the alternator drive belt.

43. Install the center bracket. Torque the bolts attaching the brackets to 40 ft. lbs. (54 Nm) then torque the mount through-bolt to 40 ft. lbs. (54 Nm).

44. Remove the jack from under the engine.

45. Install the air intake duct.

46. Connect the negative battery cable and enter the radio security code.

Camshaft

REMOVAL AND INSTALLATION

NOTE: The radio may have a coded theft protection circuit. Obtain the code from the owner before disconnecting the battery, removing the radio fuse, or removing the radio.

1.7L (B17A1), 1.8L (B18C1) Engines

1. Disconnect the negative battery cable.

2. Make sure the crankshaft is at TDC/compression on No. 1 cylinder by aligning the white mark on the crankshaft pulley with the pointer on the lower timing belt cover.

3. Remove the strut brace.

4. Remove the cylinder head cover, timing belt cover, and timing belt.

5. Remove the camshaft pulleys and back cover.

6. Loosen the rocker arm locknuts and adjusting screws.

7. Remove the camshaft holder bolts, then remove the camshaft holder plates, the camshaft holders, and camshafts.

To install:

8. Make sure that the keyways on the camshafts are facing up and that the rocker arms are in their original position. The valve locknuts should be loosened and the adjusting screw backed off before installation

9. Install the camshafts, then install the camshaft seals with the open side facing in. Install the rubber cap with liquid gasket applied.

10. Install a new O–ring and the dowel pin to the oil passage of the No. 3 camshaft holder.

11. Apply liquid gasket to the head of the mating surfaces of the No. 1 and No. 5 camshaft holders then install them, along with No. 2, 3, and 4. Be sure to pay attention to the following points:

• Do not apply oil to the holder mating surface of camshaft seals.

• The arrows marked on the camshaft holders should point to the timing belt.

12. Tighten the camshaft holders temporarily. Make sure that the rocker arms are properly positioned on the valve stems.

13. Torque the camshaft holder bolts in two steps, following the proper sequence, to ensure that the rockers do not bind on the valves. Torque the 8·1.25mm bolts to 20 ft. lbs. (27 Nm). Torque the 6·1.0mm bolts to 7 ft. lbs. (10 Nm).

14. Install the keys into the camshaft grooves. To set the No. 1 piston at TDC, align the holes on the camshaft with the holes in the No. 1 camshaft holders and insert 5.0mm pin punches into the holes.

15. Install the back cover and push the camshaft pulleys onto the camshafts, then torque the retaining bolts to 27 ft. lbs. (37 Nm). Install the timing belt and adjust the tension, then install the timing belt covers.

16. Adjust the valve clearance.

17. Install the cylinder head cover. Make sure that the seal and groove are thoroughly clean first.

18. Install the engine side mount, torque the two new nuts and new bolt to the engine to 38 ft. lbs. (52 Nm) and torque the bolt attaching the mount to the vehicle to 54 ft. lbs. (74 Nm).

19. Install the distributor to the cylinder head and reconnect the spark plug wires to the spark plugs.

20. Install the intake air duct.

21. Install the strut brace, torque the nuts to 17 ft. lbs. (24 Nm).

22. Connect the negative battery cable and enter the radio security code.

23. Drain the engine oil. Wait at least 20 minutes before filling the engine with oil; the time delay allows the sealant to cure.

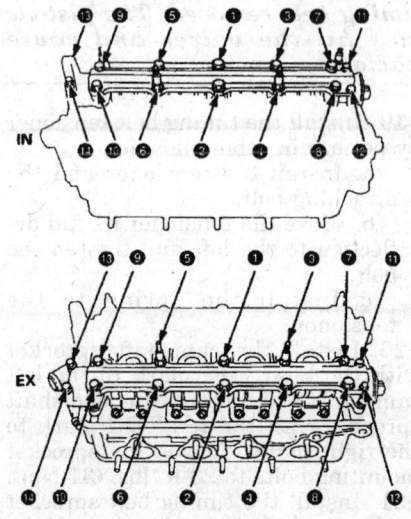

0 – ⑩: 8 x 1.25 mm 27 N·m (2.8 kgf·m, 20 lbf·ft)
⑪ – ⑭: 6 x 1.0 mm 9.8 N·m (1.0 kgf·m, 7.2 lbf·ft)

315324

Camshaft holder plates torque sequence — 1.7L (B17A1), 1.8L (B18C1) engines

1.8L (B18A1, B18B1) Engines

1. Disconnect the negative battery cable.

2. Remove the spark plug wires.

3. Remove the cylinder head cover and timing belt cover.

4. Rotate the crankshaft to TDC, compression of No. 1 piston and remove the timing belt.

5. Remove the distributor from the cylinder head.

6. Install 5.0mm pin punches to the No.1 camshaft holders then remove the camshaft sprockets.

7. Loosen the valve adjusters to remove as much spring tension as possible.

8. Remove the pin punches from the camshaft holders.

To install:

9. Check the following before installing the camshafts:

   a. Be certain the keyways on the camshafts are facing UP (No. 1 cylinder at TDC).

   b. The valve adjuster locknuts should be loosened and the adjusting screws backed off before installation.

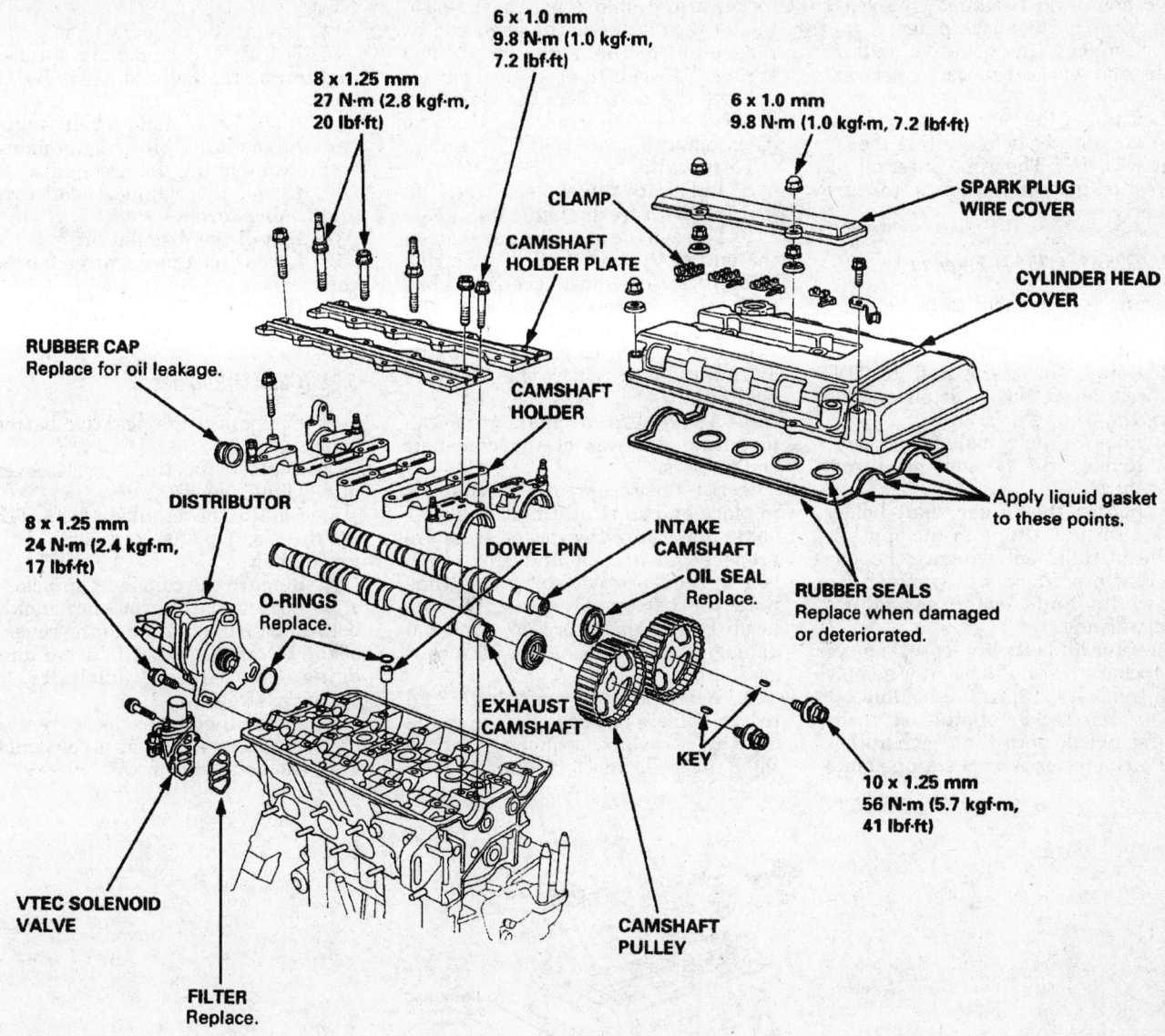

6 x 1.0 mm
9.8 N·m (1.0 kgf·m, 7.2 lbf·ft)

8 x 1.25 mm
27 N·m (2.8 kgf·m, 20 lbf·ft)

6 x 1.0 mm
9.8 N·m (1.0 kgf·m, 7.2 lbf·ft)

SPARK PLUG WIRE COVER

CLAMP

CAMSHAFT HOLDER PLATE

CYLINDER HEAD COVER

RUBBER CAP
Replace for oil leakage.

CAMSHAFT HOLDER

DISTRIBUTOR

Apply liquid gasket to these points.

8 x 1.25 mm
24 N·m (2.4 kgf·m, 17 lbf·ft)

INTAKE CAMSHAFT

DOWEL PIN

OIL SEAL
Replace.

RUBBER SEALS
Replace when damaged or deteriorated.

O-RINGS
Replace.

EXHAUST CAMSHAFT

KEY

10 x 1.25 mm
56 N·m (5.7 kgf·m, 41 lbf·ft)

VTEC SOLENOID VALVE

CAMSHAFT PULLEY

FILTER
Replace.

315318

Cylinder head components — 1.7L (B17A1), 1.8L (B18C1) engines

10. Lubricate the rocker arms and camshafts with clean oil.

11. Place the rocker arms on the pivot bolts and the valve stems, making sure that the rocker arms are in their original positions.

12. Install the camshaft seals with the open side (spring) facing in. Lubricate the lip of the seal.

13. Make sure the keyways on the camshafts are facing up and install the camshafts to the cylinder head.

14. Apply liquid gasket to the head mating surfaces of the No. 1 and No. 6 camshaft holders, then install them along with No. 2, 3, 4 and 5 camshaft holders. The arrows stamped on the holders should point toward the tim-

ing belt. Do not apply oil to the holder mating surface where the camshaft seals are housed.

15. Tighten the camshaft holders temporarily and make sure that the rocker arms are properly positioned.

16. Press the oil seals into the No.1 camshaft holders with a seal driver.

17. On 1993–95 models, torque the bolts in a crisscross pattern to 7 ft. lbs. (10 Nm). Check that the rockers do not bind on the valves.

18. Install the cylinder head plug to the end of the cylinder head. If the plug has alignment marks, align the marks with the cylinder head upper surface.

19. If equipped with a timing belt back cover, install the cover and torque the bolts to 7.2 ft. lbs. (9.8 Nm).

20. Install 5.0mm pin punches to the No.1 camshaft holders then install the camshaft pulley keys onto the grooves in the camshafts.

21. Push the camshaft pulleys onto the camshafts, then torque the retaining bolts to 27 ft. lbs. (38 Nm).

22. Install the timing belt and timing belt covers. Remove the pin punches from the camshaft holders.

23. Adjust the valves and pour oil over the camshafts and rocker arms.

24. Install the cylinder head cover and engine ground cable.

25. Install the distributor to the cylinder head and reconnect the spark plug wires to the spark plugs.

26. Connect the negative battery cable and enter the radio security code.

27. Change the engine oil. Wait at least 20 minutes before filling the engine with oil. The time interval allows the gasket sealants to cure properly.

2.5L (G25A1, G25A4) Engines

1. Disconnect the negative battery cable. Remove the timing belt covers and cylinder head covers.

2. Rotate the crankshaft to TDC compression of No. 1 piston and remove the timing belt.

3. Remove the camshaft sprocket.

4. Remove the cylinder head from the vehicle.

5. Loosen the rocker shaft holder bolts 1 turn at a time in the opposite of the installation sequence. Following this procedure will prevent the camshafts and rocker assemblies from warping.

6. After all bolts are loose, remove the rocker arm shafts as an assembly with the bolts still in the holders.

7. If the rocker shafts are to be disassembled, note that each rocker arm has a letter **A** or **B** stamped into the side. Before disassembling the rocker arms, make a note of the position of each letter so the arms can be reassembled the same way. The springs between the rocker arms are not all the same length. Carefully note their positions during disassembly.

To install:

8. Lubricate the camshaft and its journals with fresh engine oil.

9. Place a new camshaft seal on the end of the camshaft. The spring side of the seal must face in. Lubricate the journals and set the camshaft in place on the head.

10. Install the camshaft onto the cylinder head with the keyway pointed up.

11. Apply liquid gasket to the mounting surfaces of the camshaft end holders.

12. Set the rocker arm assemblies in place and start all the cam holder bolts. Make sure the rocker arms are properly positioned and turn each bolt in sequence two turns at a time until the holders are seated on the head. Follow this procedure to avoid damaging the camshaft and rocker assemblies.

13. When all the camshaft and rocker holders are seated, torque the bolts in the same sequence. Torque the 8mm bolts to 16 ft. lbs. (22 Nm) and the 6mm bolts to 9 ft. lbs. (12 Nm).

14. Install the cylinder head.

15. Install the camshaft sprocket and torque the bolts to 51 ft. lbs. (70 Nm).

16. Install the timing belt, adjust the valves and oil the camshaft before completing the assembly.

17. Install the cylinder head cover and timing cover.

18. Install the distributor.

19. Reconnect the negative battery cable.

20. Check for proper engine and valve train operation.

3.2L (C32A1) Engines

1. Disconnect the negative battery cable.

2. Remove the timing belt covers and cylinder head covers.

3. Rotate the crankshaft to TDC for the No. 1 piston and remove the timing belt.

4. Remove the camshaft sprockets.

5. Loosen the rocker shaft holder bolts one turn at a time in the reverse of the torque sequence to avoid damaging the valves, camshafts, or rocker assemblies.

6. After all bolts are loose, remove the rocker arm shafts as an assembly with the bolts still in the holders.

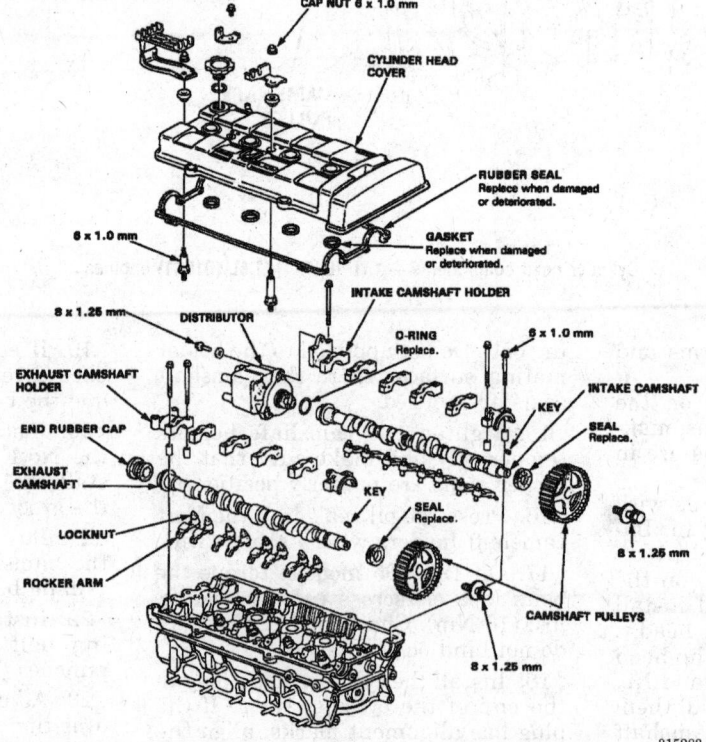

315283

Camshaft and rocker arm assembly — 1.8L (B18A1, B18B1) engines

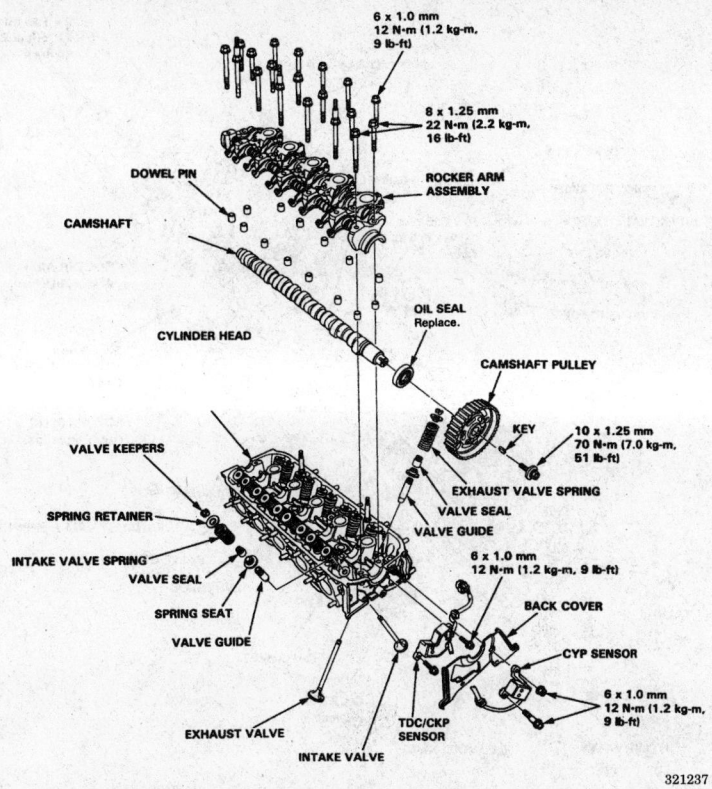

Camshaft and rocker arm assembly — 2.5L (G25A1, G25A4) engines

7. If the rocker shafts are to be disassembled, note that each rocker arm has a letter **A** or **B** stamped into the side. Before disassembling the rocker arms, make a note of the position of each letter so that the arms can be reassembled in the same position.

8. Do not remove the hydraulic tappets from the rocker arms unless they are to be replaced. Handle the rocker arms carefully so the oil does not drain out of the tappets.

9. Lift the camshafts from the cylinder head, wipe them clean and inspect the lift ramps. Replace the camshafts and rockers if the lobes are pitted, scored, or excessively worn.

To install:

10. Place a new seal on the end of the camshaft, lubricate the journals and set the camshaft in place on the head.

NOTE: The pin hole in the front of the camshaft designates the top position.

Specified torque:
8 mm bolts: 22 N·m (2.2 kg-m, 16 lb-ft)
6 mm bolts: 12 N·m (1.2 kg-m, 9 lb-ft)

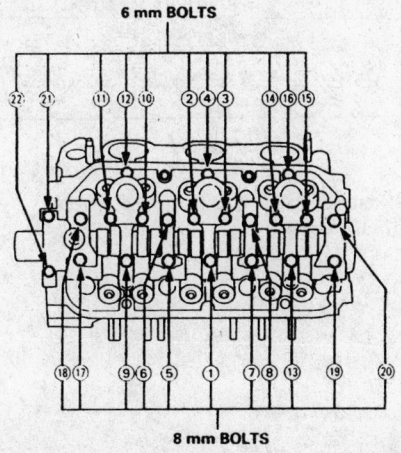

Camshaft holder bolt torquing sequence — Legend

Specified torque:
8 mm bolts: 22 N·m (2.2 kgf·m, 16 lbf·ft)
6 mm bolts: 12 N·m (1.2 kgf·m, 8.7 lbf·ft)
6 mm bolts: ②, ③, ④, ⑩, ⑪, ⑫, ⑭, ⑮, ⑯, ㉑, ㉒

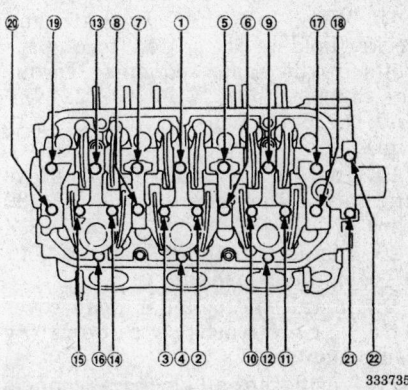

Rocker arm assembly/camshaft holders torque sequence — 3.2TL

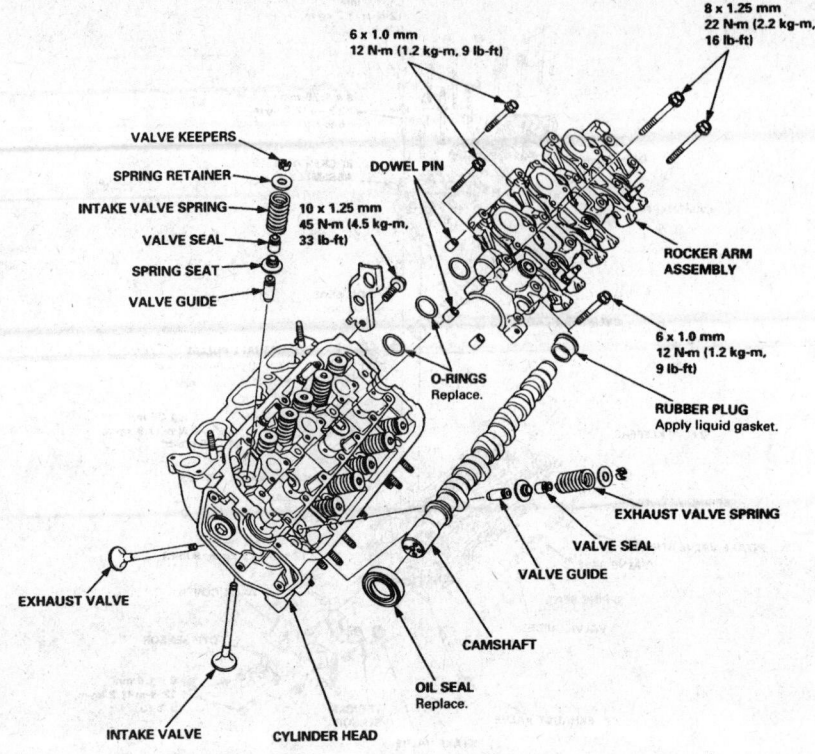

6 x 1.0 mm
12 N·m (1.2 kg-m, 9 lb-ft)

8 x 1.25 mm
22 N·m (2.2 kg-m, 16 lb-ft)

VALVE KEEPERS
SPRING RETAINER
INTAKE VALVE SPRING
VALVE SEAL
SPRING SEAT
VALVE GUIDE

DOWEL PIN

10 x 1.25 mm
45 N·m (4.5 kg-m, 33 lb-ft)

ROCKER ARM ASSEMBLY

6 x 1.0 mm
12 N·m (1.2 kg-m, 9 lb-ft)

O-RINGS
Replace.

RUBBER PLUG
Apply liquid gasket.

EXHAUST VALVE SPRING
VALVE SEAL
VALVE GUIDE

EXHAUST VALVE

CAMSHAFT

INTAKE VALVE

CYLINDER HEAD

OIL SEAL
Replace.

217054

Camshaft and rocker arm assembly — 3.2L engines

11. Apply liquid gasket to the mounting surfaces of the camshaft end holders.

12. Set the rocker arm assemblies in place and start all of the camshaft holder bolts. Make sure the rocker arms are properly positioned and turn each bolt in sequence two turns at a time until the holders are seated on the head to avoid damaging the valves or rocker assemblies.

13. When all the camshaft and rocker holders are seated, torque the bolts in the same sequence. Torque the 8mm bolts to 16 ft. lbs. (22 Nm) and the 6mm bolts to 9 ft. lbs. (12 Nm).

14. Install the camshaft pulleys and torque the bolts to 23 ft. lbs. (32 Nm).

15. Install the timing belt and pour oil over the camshafts.

16. Install the cylinder head cover and reassemble accessory components.

17. Verify that all electrical connections and vacuum lines are connected.

18. Reconnect the negative battery cable.

19. Run the engine and check for leaks and proper operation.

Piston and Connecting Rod

POSITIONING

PISTON INSTALLATION DIRECTION

EXHAUST

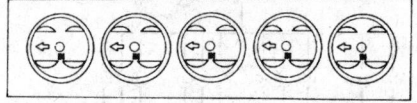

INTAKE

332295

Piston installation direction — 2.5L (G25A1, G25A4) engines

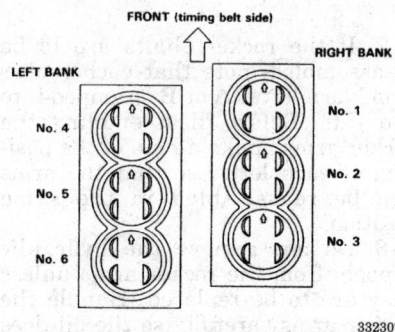

FRONT (timing belt side)

LEFT BANK   RIGHT BANK

No. 4   No. 1
No. 5   No. 2
No. 6   No. 3

332307

Piston numerical positions — 3.2L (C32A1) engines

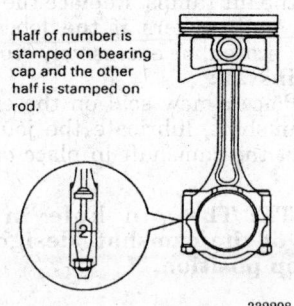

Half of number is stamped on bearing cap and the other half is stamped on rod.

332298

Connecting rod code number location — 2.5L (G25A1, G25A4) engines

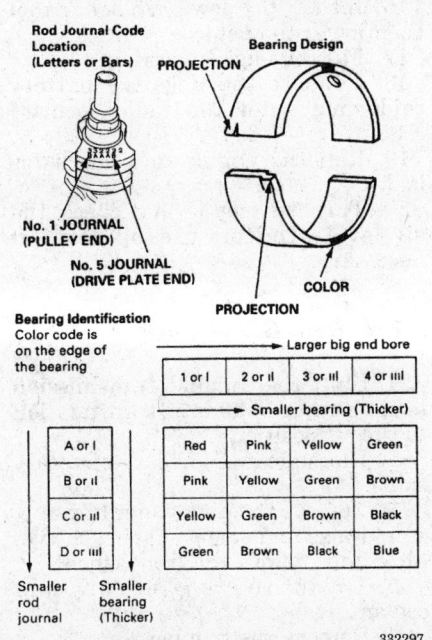

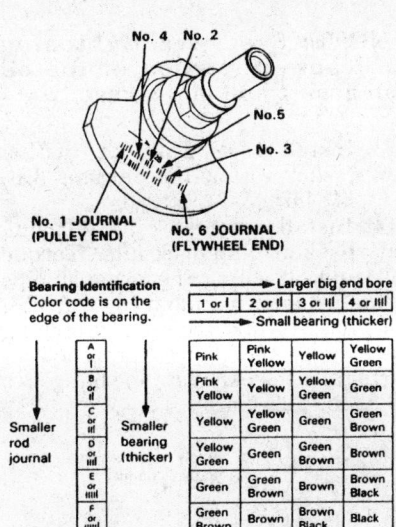

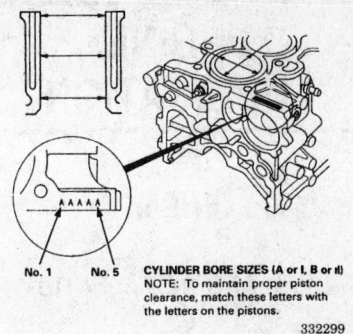

CYLINDER BORE SIZES (A or I, B or II)
NOTE: To maintain proper piston clearance, match these letters with the letters on the pistons.

332299

Cylinder bore size code location — 2.5L (G25A1, G25A4) engines

Bearing Identification
Color code is on the edge of the bearing

	Larger big end bore			
	1 or I	2 or II	3 or III	4 or IIII
	Smaller bearing (Thicker)			
A or I	Red	Pink	Yellow	Green
B or II	Pink	Yellow	Green	Brown
C or III	Yellow	Green	Brown	Black
D or IIII	Green	Brown	Black	Blue

Smaller rod journal / Smaller bearing (Thicker)

332297

Rod bearing identification color codes and thicknesses — 2.5L (G25A1, G25A4) engines

Bearing Identification
Color code is on the edge of the bearing.

	Larger big end bore			
	1 or I	2 or II	3 or III	4 or IIII
	Small bearing (thicker)			
A or I	Pink	Pink Yellow	Yellow	Yellow Green
B or II	Pink Yellow	Yellow	Yellow Green	Green
C or III	Yellow	Yellow Green	Green	Green Brown
D or IIII	Yellow Green	Green	Green Brown	Brown
E or IIII	Green	Green Brown	Brown	Brown Black
F or IIII	Green Brown	Brown	Brown Black	Black

Smaller rod journal / Smaller bearing (thicker)

NOTE: When using bearing halves of different colors, it does not matter which color is used in the top or bottom.

332310

Rod bearing size and thickness identification codes and colors — 3.2L (C32A1) engines

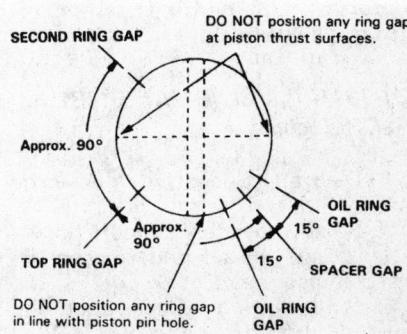

DO NOT position any ring gap at piston thrust surfaces.

SECOND RING GAP
Approx. 90°
TOP RING GAP
Approx. 90°
OIL RING GAP
15°
15°
SPACER GAP
OIL RING GAP

DO NOT position any ring gap in line with piston pin hole.

332301

Ring end gap positions

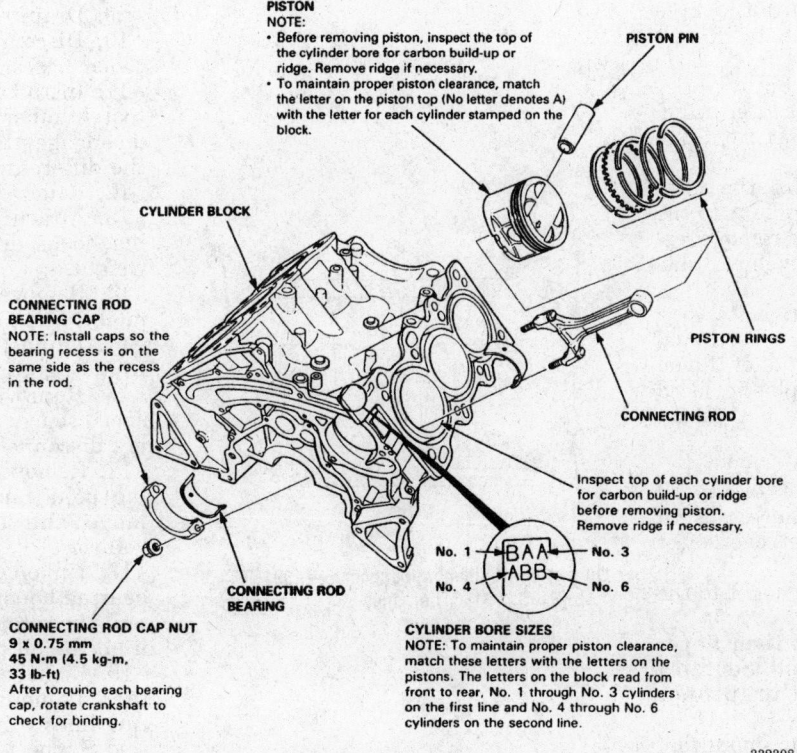

PISTON
NOTE:
• Before removing piston, inspect the top of the cylinder bore for carbon build-up or ridge. Remove ridge if necessary.
• To maintain proper piston clearance, match the letter on the piston top (No letter denotes A) with the letter for each cylinder stamped on the block.

PISTON PIN

CYLINDER BLOCK

PISTON RINGS

CONNECTING ROD BEARING CAP
NOTE: Install caps so the bearing recess is on the same side as the recess in the rod.

CONNECTING ROD

Inspect top of each cylinder bore for carbon build-up or ridge before removing piston. Remove ridge if necessary.

CONNECTING ROD BEARING

CONNECTING ROD CAP NUT
9 x 0.75 mm
45 N·m (4.5 kg-m, 33 lb-ft)
After torquing each bearing cap, rotate crankshaft to check for binding.

No. 1   No. 3
B A A
A B B
No. 4   No. 6

CYLINDER BORE SIZES
NOTE: To maintain proper piston clearance, match these letters with the letters on the pistons. The letters on the block read from front to rear, No. 1 through No. 3 cylinders on the first line and No. 4 through No. 6 cylinders on the second line.

332308

Engine block cylinder bore sizes and identification marks — 3.2L (C32A1) engines

ENGINE LUBRICATION

Oil Pan

REMOVAL AND INSTALLATION

NOTE: The radio may have a coded theft protection circuit. Obtain the code from the owner before disconnecting the battery, removing the radio fuse, or removing the radio.

1.7L (B17A1), 1.8L (B18A1, B18B1, B18C1) Engines

1. Disconnect negative cable at the battery.
2. Raise and safely support the vehicle. Drain the oil and remove the lower splash panel.
3. If equipped, disconnect the heated oxygen sensor (HO2S) connector.
4. Remove the nuts and bolts connecting exhaust pipe A to the catalytic converter. Discard the gasket and the locknuts.
5. Remove and discard the nuts attaching exhaust pipe A to the exhaust hanger.
6. If applicable, remove the mounting bolts from the center beam. Remove the center beam from the subframe.
7. Remove and discard the locknuts attaching exhaust pipe A to the exhaust manifold, then remove exhaust pipe A from the vehicle. discard the exhaust gaskets.
8. Loosen the oil pan bolts in a crisscross pattern. To remove the oil pan, lightly tap the corners of the oil pan with a rubber or plastic faced mallet. Clean off all the old gasket material.

To install:

9. Apply liquid gasket to the oil pan mating surface where the oil pump and the right side cover meet the engine block.
10. Install the oil pan gasket to the oil pan.
11. Install the oil pan, then finger tighten the center and end mounting nuts and bolts in the proper sequence.
12. Tighten the oil pan mounting nuts and bolts starting from the center bolt next to the oil drain plug (bolt # 1) and work clockwise, tight-

ening the bolts in three steps. Torque the bolts to 10 ft. lbs. (14 Nm).

NOTE: Excessive tightening can cause distortion of the oil pan gasket and oil leakage.

13. Install the oil drain plug with a new gasket, torque the plug to 33 ft. lbs. (44 Nm).
14. Install exhaust pipe A using new gaskets and locknuts. Torque the nuts attaching the exhaust pipe to the exhaust manifold to 40 ft. lbs. (54 Nm). Torque the nuts attaching the exhaust pipe to the catalytic converter and the exhaust pipe hanger to 16 ft. lbs. (22 Nm).

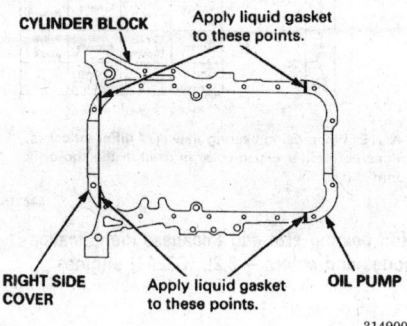

Application of liquid gasket — 1994–97 1.8L (B18B1, B18C1) engines

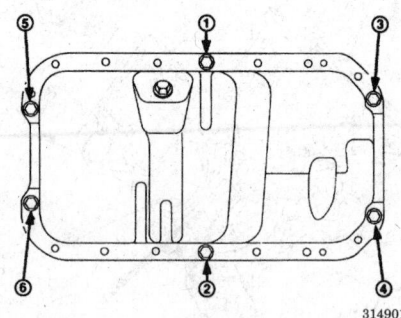

Oil pan bolt tightening sequence — 1994–97 1.8L (B18B1, B18C1) engines

15. Connect the heated oxygen sensor (HO2S) connector.
16. Install the lower splash panel then lower the vehicle.
17. Fill the engine with oil.
18. Connect the negative battery cable and enter the radio security code.
19. Run the engine and check for leaks.
20. Turn off engine and check the oil level. Top off the oil level if necessary.

2.5L (G25A1, G25A4) Engines

1. Shift the manual transmission to 1st gear or automatic transmission to the **P** position.
2. Disconnect the negative battery cable.
3. Remove the air cleaner housing.
4. Raise and safely support the vehicle and remove the front wheels.
5. Drain the engine oil and coolant.
6. Remove the damper forks.
7. Remove the lower ball joint nut. Use a ball joint remover to disconnect the ball joint from the control arm.
8. Carefully pry the inner CV-joints out of their sockets. Wrap them in plastic to keep them clean. Do not let the driveshafts hang by the outer CV-joint.
9. Drain the differential oil.
10. Disconnect the differential oil cooler hoses.
11. Install the shaft puller, Acura tool number 07LAC–PW50101, and disengage the extension shaft from the differential.
12. Remove the side splash shield.
13. Attach a chain hoist to the lifting hooks and take up the engine's weight.
14. Remove the transmission side mount and bracket.
15. Unbolt and remove the left front engine mount bracket.
16. Remove the power steering speed sensor from the differential. Do not disconnect the hoses.
17. Remove the differential mounting bolts and the 26mm shim. Remove the differential from the vehicle.
18. Unbolt the intermediate shaft bearing housing from the oil pan and pull the intermediate shaft assembly from the oil pan pipe.
19. Remove the A/C compressor and then its mounting bracket. Leave the A/C lines connected to the compressor. Support the compressor with a piece of wire to move it out of the work area and take the weight off the A/C lines.

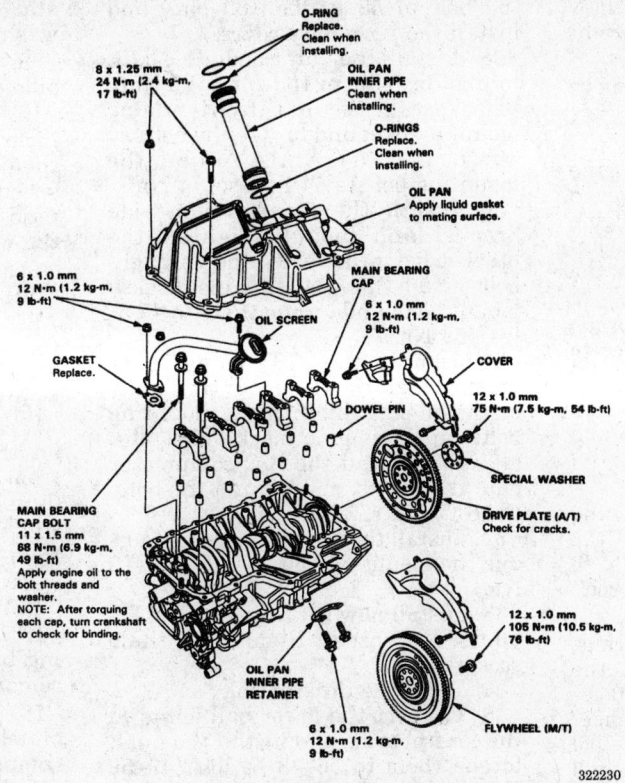

Oil pan components — 2.5L engines

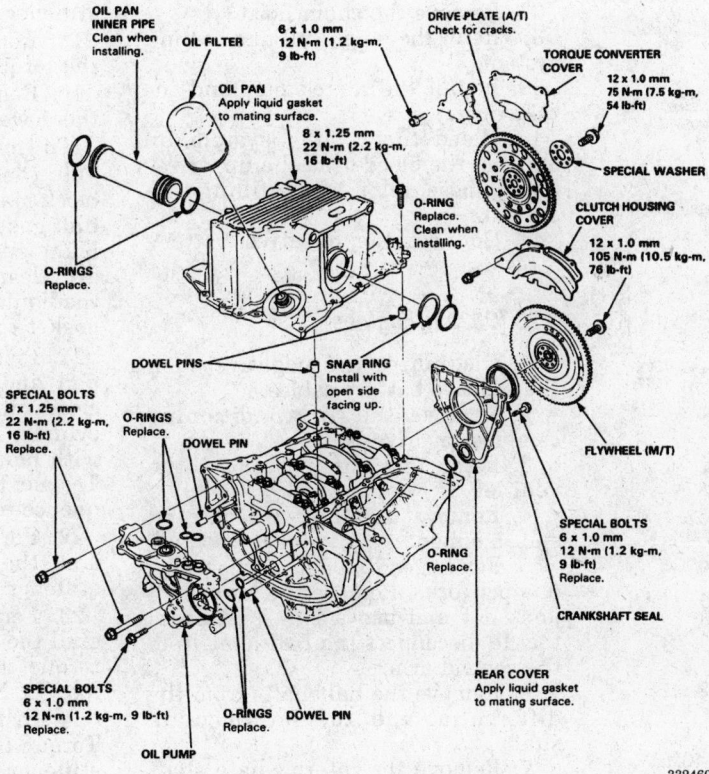

Oil pan components — 3.2L engines

20. Remove the set plate that holds the oil pan inner pipe from the right side of the engine.

21. Unbolt the oil pan and remove it from the vehicle.

To install:

22. Clean the oil pan and engine block mating surfaces. Apply an even bead of liquid gasket to the engine block sealing surface. Apply some liquid gasket to the inner threads of the bolt holes.

23. Install the oil pan and torque the bolts in the correct sequence to 16–17 ft. lbs. (22–24 Nm).

24. Install new O-rings on the oil pan inner pipe. Install the pipe and tighten the set plate bolts to 9 ft. lbs. (12 Nm).

25. Install the differential, making sure the original shim is in the proper position. Torque the bolts to 54 ft. lbs. (75 Nm). Connect the cooling hoses.

26. Install new set and snaprings on the extension shaft. Coat the splines and their mating surfaces with high temperature grease. Thread the special installation tool into the transmission case to install the extension shaft.

27. Pack the extension shaft cavity with high temperature grease and install the 33mm sealing bolt. Torque

the bolt to 58 ft. lbs. (80 Nm) and install the secondary cover.

28. Install the intermediate shaft, torque the bolts to 16 ft. lbs. (22 Nm).

29. Install the left front engine mount bracket and torque the bolts to 40 ft. lbs. (54 Nm). Torque the mounting bolt to 54 ft. lbs. (74 Nm).

30. Install the transmission side bracket and mount. Tighten the bracket mounting bolts and through-bolt to 40 ft. lbs. (54 Nm). Install new mount bolts and torque them to 47 ft. lbs. (64 Nm).

31. Install the A/C compressor mount. Tighten the mounting bolts on the oil pan and then the mounting bolts on the engine block, 36 ft. lbs. (49 Nm). Install the A/C compressor onto the mount and tighten the bolts to 16 ft. lbs. (22 Nm).

32. Install the speed sensor and torque the mounting bolt to 7 ft. lbs. (10 Nm).

33. Install new set rings on the CV-joints and press them into their sockets.

34. Refill the differential.

35. Connect the lower ball joints to the control arms and install the nuts, torque them to 36–43 ft. lbs. (49–59 Nm) and install a new cotter pin. Torque the damper fork bolts to 47 ft. lbs. (64 Nm). Install the front wheels.

36. Lower the vehicle.

37. Remove the chain hoist.

38. Refill the engine oil and cooling system.

39. Install the air cleaner and intake duct.

40. Bleed the cooling system by opening the bleeder on the upper radiator hose inlet when filling the system.

41. Connect the negative battery cable.

3.2L (C32A1) Engines

1. Disconnect the negative then the positive battery cables.

2. Remove the air conditioning compressor drive belt.

3. Raise and safely support the vehicle.

4. Remove the front wheels and splash shield.

5. For Legend models, remove the damper forks. Remove the lower ball joint nut and use a ball joint press tool to disconnect the ball joint from the control arm.

6. Remove the halfshafts from the differential and the intermediate shaft.

7. Remove the intermediate shaft from the oil pan.

8. Drain the oil from the differential into a sealable container the in-

stall the drain plug with a new washer.

9. Drain the engine oil into a sealable container.

10. If equipped, disconnect the Vehicle speed Sensor (VSS) harness, then remove the VSS/power steering speed sensor. Do not disconnect the fluid hoses, support the sensor out of the way.

11. Remove the right front beam bridge.

12. Remove the lower plate from the rack and pinion, install the two rack and pinion mounting bolts that were removed.

13. Remove the 36mm sealing bolt on the transaxle. Ensure that the transaxle is in 1st gear (manual) or **P** (automatic).

14. Disconnect the extension shaft from the differential with the extension shaft puller (part # 07LAC–PW50101).

15. Remove the differential mounting bolts and the 26mm shim, then remove the differential.

16. Disconnect the A/C compressor clutch connector, then remove the compressor. Do not disconnect the A/C hoses from the compressor and do not let the compressor hang by the hoses.

17. Remove the rear engine stiffener.

18. Remove the flywheel cover or the torque converter covers.

19. Remove the oil pan. Do not lose the dowel pins from the oil pan

To install:

20. Clean the oil pan and cylinder block mating surfaces then apply liquid gasket to the cylinder block. Make sure that the mating surfaces are clean and dry before installing the liquid gasket. Do not apply liquid gasket to the O-ring grooves.

21. Install the dowel pins to the oil pan and new O-rings coated with clean oil. Install the oil pan to the cylinder block. Coat the oil pan bolts with liquid gasket then install them. Torque the bolts in the proper sequence to 16 ft. lbs. (22 Nm).

22. For manual transmission, install the flywheel cover and engine stiffener.

23. For automatic transmission, install the torque converter covers and torque the mounting bolts to 8.7 ft. lbs. (12 Nm)

24. Install the rear engine stiffener. Torque the bolt attaching the engine stiffener to the transaxle first, to 47 ft. lbs. (64 Nm), then torque the bolts to the engine block to 16 ft. lbs. (22 Nm).

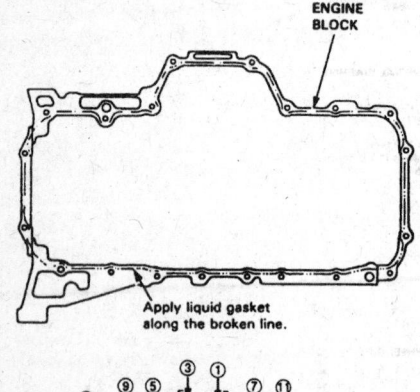

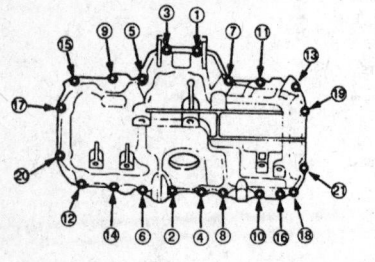

322232

Oil pan bolt torque sequence — 2.5L (G25A1, G25A4) engines

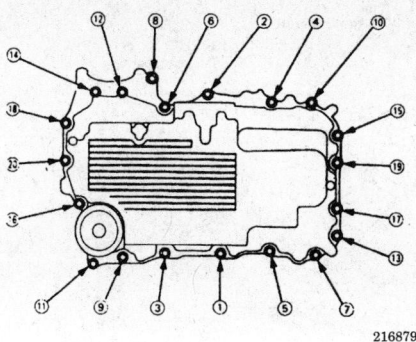

Torque sequence — 3.2L engines

216879

25. Install the A/C compressor to the engine block and torque the mounting bolts to 16 ft. lbs. (22 Nm).

26. Connect the A/C clutch connector.

27. Install the dowel pins to the differential then install the differential to the engine. Install the mounting bolts loosely and install the 26mm shim. Torque all of the mounting bolts to 47 ft. lbs. (64 Nm).

28. Install a new set ring to the extension shaft. Using the extension shaft installer (part # 07MAF–PY40100 or 07MAF–PY40101) install the shaft to the differential.

29. Fill the secondary gear with super high temperature grease (Part # 08798–9002). Applying sealer to the threads of the 36mm sealing bolt then install the bolt and torque to 58 ft. lbs.(78 Nm).

30. Remove the two bolts from the rack and pinion necessary to install the lower plate, then install the lower plate and the attaching bolts. Torque the lower plate attaching bolt to 28 ft. lbs. (38 Nm) and torque the rack and pinion bolts to 43 ft. lbs. (59 Nm).

31. For Legend models, when installing the lower ball joint nuts, torque them to 51–58 ft. lbs. (70–80 Nm) and install a new cotter pin. Torque the damper fork bolts to 51 ft. lbs. (70 Nm).

32. Install the right beam bridge and torque the attaching bolts to 28 ft. lbs. (38 Nm).

33. Install the VSS and torque the attaching bolt to 8.7 ft. lbs. (12 Nm). Connect the VSS harness to the VSS sensor.

34. Install the intermediate shaft and the halfshafts.

35. Fill the differential with oil.

36. Install the engine splash shield and torque the bolts to 7.2 ft. lbs. (9.8 Nm).

37. Install the front wheels.

38. Lower the vehicle.

39. Install the A/C compressor drive belt.

40. Fill the engine with oil.

41. Connect the positive then the negative battery cables and enter the radio security code.

42. Run the engine and check for leaks.

43. Check the front wheel alignment.

Oil Pump

REMOVAL AND INSTALLATION

NOTE: The radio may have a coded theft protection circuit. Obtain the code from the owner before disconnecting the battery, removing the radio fuse, or removing the radio.

1. Disconnect the negative battery cable. Raise and safely support the vehicle. Drain the oil and remove the lower splash panel, if necessary.

2. Make sure the crankshaft is at TDC on No. 1 cylinder and remove the timing belt cover, timing belt, and the gear off the crankshaft.

3. For Legend models, remove the oil pump cap.

NOTE: On GS model Legends, an engine oil cooler is installed in place of the oil pump cap. Remove this assembly and disconnect the oil cooler lines.

4. Remove the oil pan. Remove the pick–up screen.

5. Remove the oil filter assembly, if necessary.

6. Remove the oil pump from the front of the engine. Any time the oil pump is removed, the front oil seal should be replaced.

To install:

7. Install the oil pump, using new O–rings and liquid gasket applied to a clean pump mounting face. For all engines, except 1994–97 1.8L (B18B1, B18C1) engines, torque the 6mm bolts to 9 ft. lbs. (12 Nm) and the 8mm bolts to 16 ft. lbs. (22 Nm). For 1994–97 1.8L (B18B1, B18C1) engines, torque the 8·1.25mm bolts to 17 ft. lbs. (24 Nm), torque the 6·1.0mm bolts to 8 ft. lbs. (11 Nm).

8. Install the oil pump cap or oil cooler unit, as applicable. Replace the cooler hoses if they show signs of damage. Torque the center bolt to 30 ft. lbs. (42 Nm).

— WARNING —
The B18B1 and B18C1 engines use different oil pumps, make sure that you have the correct oil pump. Match the crankshaft tim-

ing mark on the new oil pump with the timing mark on the old oil pump, because the timing marks are in different locations. If an oil pump is used with the timing mark in the wrong position the pistons may contact the valves.

9. Install the pick–up screen, then the oil pan. Torque the oil pan bolts to 9 ft. lbs. (12 Nm).

10. Install the oil filter assembly, exhaust pipe, center beam, and lower splash panel, if necessary.

11. Wait at least 30 minutes after completion of procedure before refilling the engine with oil. The waiting period is to allow a curing period for the silicone sealant. Refill the engine with oil and connect the negative battery cable. Start the engine and check the engine for leaks.

12. Turn off engine and check the oil level. Top off the oil level if necessary.

TRANSAXLE

Manual Transaxle Assembly

REMOVAL AND INSTALLATION

NOTE: The radio may have a coded theft protection circuit. Obtain the code from the owner before disconnecting the battery, removing the radio fuse, or removing the radio.

1993–94 Integra

1. Disconnect the negative battery cable first, then the positive cable and remove the battery.

2. Drain the transaxle oil. Reinstall the drain plug with a new washer.

3. Remove the air cleaner case complete with the air intake tube. Disconnect the transaxle ground cable.

4. On 1994 Integra models, remove the lower radiator hose clamp from transaxle hanger B.

5. Loosen the clutch cable adjusting nut and disconnect the clutch cable at the release arm, then disconnect from the clutch cable bracket.

6. Disconnect the electrical connectors for the back-up light switch, oxygen sensor and the starter motor

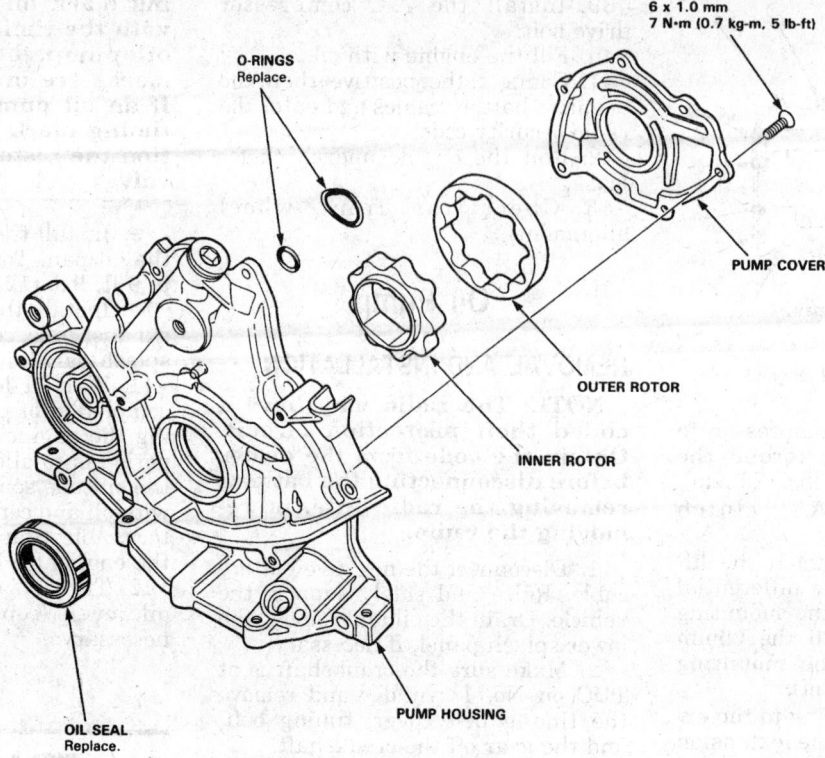

O-RINGS
Replace.

6 x 1.0 mm
7 N·m (0.7 kg-m, 5 lb-ft)

PUMP COVER

OUTER ROTOR

INNER ROTOR

PUMP HOUSING

OIL SEAL
Replace.

316911

Oil pump components — 1.8L engines

OIL PUMP HOUSING

Apply liquid gasket
along the broken line.

322433

Oil pump sealing surface — 2.5L (G25A4) engine

cables and wire harness clamp from the starter.

7. Remove the power steering speed sensor without disconnecting the sensor hose.

8. On 1994 Integra models, remove the clutch pipe bracket and slave cylinder. Do not operate the clutch pedal once the slave cylinder has been removed.

9. Disconnect the distributor connectors and remove the distributor mounting bolts. Before removing the distributor, be sure to make some alignment marks on the distributor housing and engine to aid in the installation procedure.

10. Raise and safely support the vehicle.

11. Remove the starter.

12. Remove the right front splash shield and splash guard. Remove the center beam bolts and remove the center beam.

13. Remove the cotter pin from the lower right ball joint castle nut, remove the nut and using a ball joint separator, remove the ball joint from the lower arm.

14. Remove the right damper fork. On 1993 Integra models, remove the right radius rod locknut, then the bolts and remove the right radius arm.

15. Remove the right halfshaft assembly.

16. Remove the cotter pin from the lower left ball joint castle nut, remove the nut and using a ball joint separator, remove the ball joint from the lower arm.

17. Remove the left halfshaft from the intermediate shaft. Remove the intermediate shaft bolts and remove the intermediate shaft. Be sure to coat all precision finished surfaces with clean engine oil or grease and tie plastic bags over the halfshaft ends.

18. Remove the shift rod and shift lever torque rod.

19. Remove the front engine stiffener and the rear engine stiffener. Remove the 4 bolts from the clutch housing cover and remove the cover.

20. Remove the 2 transaxle mount bolts from the engine side.

21. Remove the 2 transaxle mount bolts from the rear engine mount bracket.

22. Remove the side transaxle mount bolt from the underside. Remove the front transaxle mount bolts and mount.

23. Install the bolts into the cylinder head and attach a suitable lifting device or chain hoist to the bolts. Lift the engine slightly to take the load off of the engine mounts.

24. Place a suitable transaxle jack under the transaxle and raise it enough to take the weight off of the transaxle mounts. Remove the bolts and nuts that attach the brackets to the side transaxle mounts.

25. Remove the 3 transaxle mount bolts from the transaxle side.

26. Pull the transaxle away from the clutch pressure plate until it clears the mainshaft, then remove the transaxle by lowering the jack.

To install:

27. Install the transaxle on a transaxle jack. Clean and lubricate the clutch release bearing surfaces.

28. Make sure both 14mm dowel pins are installed in the clutch housing.

29. Loosely install the transaxle mount bolts, then torque them to 49 ft. lbs. (68 Nm).

30. Secure the transaxle to the engine with the engine side mounting bolt and torque it to 50 ft. lbs. (68 Nm).

31. Install the transaxle to side transaxle mount. Install the transaxle to the front transaxle mount.

32. Install the transaxle to the rear engine mount bracket.

33. Loosely install the bolt in the front stiffener and then torque then to 17 ft. lbs. (24 Nm).

34. Loosely install the bolt in the rear stiffener and then torque then to 17 ft. lbs. (24 Nm).

35. Remove the transaxle jack. Remove the lifting device by removing the hoist bolts from the cylinder head.

36. The remaining components are installed in the reverse order of removal.

37. Install the battery base. Refill the transaxle with the recommended oil.

38. Install the battery and connect the battery cables.

39. Adjust the clutch free-play. Check the ignition timing and road test the vehicle to be sure the transaxle is operating properly.

1995–97 Integra

1. Disconnect the negative battery cable, then the positive battery cable.

2. Drain the transaxle oil. Install the drain plug with a new washer.

3. Remove the air cleaner case and the air intake tube.

4. Disconnect the back–up light switch connector and the transaxle ground wire.

5. Remove the lower radiator hose clamp from the transaxle hanger.

6. Remove the wire harness clips.

7. Disconnect the starter motor cables and the Vehicle Speed Sensor (VSS) connector.

8. Remove the clutch pipe bracket and slave cylinder. Do not operate the clutch pedal once the slave cylinder has been removed.

9. Remove the three upper transaxle mounting bolts and the lower starter mounting bolt.

10. Safely raise and support the vehicle.

11. Remove the engine splash shield.

12. Disconnect the Heated Oxygen Sensor (HO2S) connector.

13. Remove and discard the two nuts attaching exhaust pipe A to the hanger bracket.

14. Remove and discard the nuts attaching exhaust pipe A to the exhaust manifold, discard the exhaust gaskets.

15. Remove and discard the three nuts attaching the exhaust system to the catalytic converter, discard the exhaust gasket. Remove exhaust pipe A from the vehicle.

16. Remove the cotterpins and castle nuts from the front lower ball joints. Separate the ball joints from the lower control arms.

17. Remove the right damper fork pinch bolt and lower nut and bolt, then remove the damper fork from the vehicle.

18. Pry the right halfshaft out of the differential, discard the set ring on the inner joint.

19. Pry the left halfshaft out of the intermediate shaft, discard the set ring on the inner joint.

20. Tie plastic bags over the halfshaft joints to keep the splines of the joints clean.

21. Remove the intermediate shaft mounting bolts, and remove the intermediate shaft.

22. Remove the set ring from the intermediate shaft and install a new set ring.

23. Disconnect the extension rod and the shift rod:

   a. If equipped with a VTEC (B18C1) engine, remove the heat shield.

   b. Disconnect the shift extension rod from the transaxle case.

   c. Slide the boot on the shift rod back to expose the clip and spring pin. Remove the clip from the shift rod.

   d. Drive out the spring pin with a punch and disconnect the shift control rod. Note that on reassembly, install the clip back into place after driving the spring pin in.

24. Remove the rear engine stiffener, and if equipped with a VTEC (B18C1) engine remove the front engine stiffener.

25. Remove the clutch cover.

26. Remove the right front mount/bracket. Discard the long self-locking bolt.

27. Place a transmission jack under the transaxle and a jack stand under the engine.

28. Remove the transaxle mount.

29. Remove the transaxle mounting bolts and the bolts from the rear mounting bracket. Discard the self-

locking bolts from the rear mounting bracket.

30. Pull the transaxle assembly away from the engine until it clears the main shaft, then lower it on the transmission jack.

To install:

31. Install the dowel pins to the clutch housing.

32. Apply super high temperature grease to the following components:

   a. The release fork bolt.

   b. The spline of the transaxle input shaft.

   c. The inside of the release bearing and the sleeve on the transaxle input shaft where the release bearing rides.

   d. The tips of the release fork and where the slave cylinder pin rides on the release fork.

33. Install the release fork boot.

34. Place the transaxle assembly on the transmission jack and raise it to engine level.

35. Install the transaxle mounting bolts and new rear mount bracket bolts. Torque the transaxle mounting bolts to 47 ft. lbs. (64 Nm) and the rear mount bolts to 87 ft. lbs. (118 Nm).

36. Raise the transaxle and install the transaxle mount. First torque the mounting nuts and bolt to the transaxle to 47 ft. lbs. (64 Nm), then torque the mounting bolt to 54 ft. lbs. (74 Nm).

37. Install the three upper transaxle mounting bolts and the lower starter bolt. Torque the upper transaxle mounting bolts to 47 ft. lbs. (54 Nm), torque the lower starter bolt to 33 ft. lbs. (44 Nm).

38. Install the right front mount/bracket. Torque the new self locking bolt to 61 ft. lbs. (83 Nm), and torque the other mounting bolts to 33 ft. lbs. (44 Nm).

39. Install the clutch cover. Torque the 6·1mm bolts to 9 ft. lbs. (12 Nm), torque the 8·1.25mm bolts to 17 ft. lbs. (24 Nm). If equipped with a NON–VTEC (B18B1) engine torque the 12·1.25mm bolt to 42 ft. lbs. (57 Nm).

40. Install the rear engine stiffener and the front engine stiffener, if equipped. Torque the bolts attaching the stiffener(s) to the transaxle to 42 ft. lbs. (57 Nm). Torque the bolts attaching the stiffener(s) to the engine to 17 ft. lbs. (24 Nm).

41. Remove the transmission jack and the jack stand from the engine.

42. Install the shift rod to the transaxle and install the spring pin and clip. Install the shift rod boot to the

shift rod, making sure the drain hole is facing down.

43. Install the extension rod and torque the attaching bolt to 16 ft. lbs. (22 Nm).

44. If equipped with a VTEC (B18C1) engine, install the heat shield. Torque the mounting bolts to 7 ft. lbs. (9.8 Nm).

45. Install new set ring to the intermediate shaft and the halfshafts.

46. Install the intermediate shaft, torque the mounting bolts to 29 ft. lbs. (39 Nm).

47. Turn the right steering knuckle fully outward and slide the halfshaft into the differential unit until you feel the spring clip engage. Turn the left steering knuckle fully outward and slide the halfshaft onto the intermediate shaft until you feel the spring clip engage.

48. Install the right damper fork. Torque the pinch bolt to 32 ft. lbs. (43 Nm) and torque the fork lower nut and bolt to 47 ft. lbs. (64 Nm).

49. Connect the lower ball joints to the lower control arms, torque the castle nuts to 36–43 ft. lbs. (49–59 Nm). Install new cotter pins.

50. Install exhaust pipe A using new gaskets and locknuts. Torque the nuts attaching the exhaust pipe to the exhaust manifold to 40 ft. lbs. (54 Nm). Torque the nuts attaching the exhaust system to the catalytic converter to 25 ft. lbs. (33 Nm). Torque the exhaust pipe hanger nuts to 12 ft. lbs. (16 Nm).

51. Install the engine splash shield.

52. Apply super high temperature grease to the tip of the slave cylinder and install the slave cylinder to the transaxle. Torque the mounting bolts to 16 ft. lbs. (22 Nm). Install the clutch bracket pipe and torque the mounting bolts to 7 ft. lbs. (9.8 Nm).

53. Install the front wheels and lower the vehicle.

54. Connect the VSS sensor and the starter motor connectors.

55. Connect the lower radiator hose clamp to the transaxle hanger.

56. Connect the transaxle ground cable and the back–up light switch connector.

57. Install the air cleaner housing assembly with the air intake tube.

58. Refill the transaxle with 4.6 pints of 10W–30 or 10W–40, SF or SG grade oil.

59. Connect the positive then the negative battery cables.

60. Check the operation of the clutch and smooth operation of the shifter.

61. Check the front wheel alignment and road test the vehicle.

1993–94 Vigor

1. Disconnect both battery cables. Remove the battery and the battery tray.

2. Remove the ABS relay box, but do not disconnect the harness.

3. Remove the heat shield, distributor and the control box. Do not disconnect the vacuum lines from the control box.

4. Disconnect the transaxle ground wire and the back-up light switch connector.

5. Remove the clutch slave cylinder and the transaxle housing mounting bolts with the 26mm shim. Do not operate the clutch pedal once the slave cylinder has been removed.

6. Remove the mount beam and bracket.

7. Remove the secondary cover and the 33mm sealing bolt. Pull the extension shaft out using a puller that threads into the shaft bore.

8. Remove the exhaust and disconnect the shift and extension rods.

9. Remove the transaxle mount nuts and support the housing with a jack. Remove the mounts and the housing bolt.

10. Remove the clutch cover and the housing bolt.

11. Remove the transaxle.

To install:

12. Lubricate the release bearing, fork and guide with molybdenum grease and make sure the dowel pins are properly placed in the clutch housing.

13. While fitting the transmission to the engine, turn the release lever up and make sure the fork engages the release bearing on the clutch.

14. Make sure the transmission is properly fitted and install the lower transmission bolts. Torque the bolts to 47 ft. lbs. (65 Nm).

15. If the transmission or differential is being replace, measure the gap between them with a feeler gauge and select the correct 26mm shim.

16. Install the front transmission mounts.

17. Install a new set ring onto the extension shaft and lubricate the shaft with a high temperature molybdenum grease. Install the extension shaft.

18. Pack the shaft area with grease but keep the threads clean. Apply a liquid gasket compound to the threads and install the 33mm sealing bolt. Torque the bolt to 58 ft. lbs. (80 Nm) and install the secondary cover.

19. Connect the shift linkage and the extension rod.

20. Use new gaskets and install the exhaust pipe.

21. Install the rear transmission mount and bracket.

22. Install the 26mm shim and the remaining transmission bolts. Torque the shim bolt to 55 ft. lbs. (75 Nm) and the rest to 47 ft. lbs. (65 Nm).

23. Install the slave cylinder.

24. Connect the backup light switch connector and the transmission ground wire.

25. Install the distributor.

26. Install the heat shield and the ABS relay box.

27. Install the battery and connect the battery cables.

28. Refill the transmission with oil and check the clutch operation.

29. Check the ignition timing and test drive to check the transmission smoothness.

1993–94 Legend

1. Disconnect both battery cables.

2. Unbolt the strut bar from the bulkhead and shock towers.

3. Drain the transaxle fluid and replace the plug with a new washer.

4. Remove the control box, but do not disconnect the vacuum lines.

5. Disconnect the neutral switch, back-up light switch, and the reverse lockout solenoid connector.

6. Remove the transaxle housing bolts and the clutch hose bracket from the rear engine hanger.

7. Remove the front exhaust pipe and catalytic converter. On Legends with 6–speed transaxle, remove the twin three-way catalytic converters and their brackets.

8. Remove the converter heat shield and bracket.

NOTE: An extension shaft puller, Acura #07LAC-PW50100 or equivalent, should be used to remove the extension shaft before removing the transaxle.

9. Lock the transaxle by shifting it into first gear.

10. Remove the extension shaft secondary cover and the 36mm sealing bolt. Use an extension shaft puller to remove the extension shaft from the rear of the transaxle case. The differential is not removed with the transmission.

11. Disconnect the transaxle linkage extension and the shift rod.

12. Disconnect the oil cooler lines at the oil pump pipes.

13. Remove the release fork cover and the clutch slave cylinder.

14. Support the steering rack with a jack, and remove the rack cover plate. Reinstall the steering rack bolts to hold the rack in the vehicle.

15. Remove the exhaust pipe bracket.

16. Remove the transaxle rear mount and bracket assembly.

17. Pull out on the clutch fork to release it from the throw-out bearing. Don't remove it from the clutch housing.

18. Use a jack to take up the transaxle's weight. Remove the transaxle mid mounts.

19. Remove the transaxle housing mounting bolts.

20. Remove the engine stiffener and the clutch housing cover.

21. Remove the transaxle mounting bolts and the 26mm shim under the transaxle.

22. Verify that all linkages, vacuum lines, and wiring harnesses have been disconnected.

23. Slide the transaxle back and off the input shaft, and lower it from the vehicle.

To install:

24. Make sure that the transaxle mounting dowel pins are seated in the clutch housing.

25. Clean and lightly lubricate the input shaft and release fork contact points with molybdenum grease and install the fork.

26. Install the extension shaft in place. Use a new set ring on the shaft and lightly lubricate the splines with molybdenum grease. Install the sealing bolt and secondary cover.

27. Install the transaxle and start all of the bolts. Install the 26mm transaxle shim. Torque the 12mm bolts to 55 ft. lbs. (75 Nm).

28. Install the clutch cover and engine stiffener bolts. Torque to stiffener bolts to 16 ft. lbs. (22 Nm), and the clutch cover bolts to 9 ft. lbs. (12 Nm).

29. Install the mid mounts and the exhaust bracket. Torque the 10mm bolts to 29 ft. lbs. (39 Nm), 10mm nuts to 36 ft. lbs. (49 Nm).

30. Install the transaxle rear mount and bracket assembly. Torque the mount bolts to 29 ft.lbs. (39 Nm).

31. Install the shift linkage and extension rod. Make sure the hole in the shift rod boot is facing down.

32. Install the release fork and the slave cylinder.

33. Install the release fork cover and connect the oil cooler hoses.

34. With the transaxle in gear, install the extension shaft using the special tool. Coat the extension shaft with molybdenum grease and use a new set ring. Make sure the shaft snaps into place on the set ring.

35. Pack the shaft area with molybdenum grease, but keep the thread area clean. Apply liquid gasket to the sealing bolt threads and install the bolt and cover.

36. Support the steering rack with a jack and remove the bolts. Install the rack cover plate and torque the bolts to 28 ft. lbs. (39 Nm). These bolts thread into aluminum and must have the special Dacro® coating to avoid corrosion.

37. On six-speed Legends, install the twin three-way catalytic converters and brackets.

38. Install the heat shield and the exhaust pipe and catalytic converter. Use new locking nuts and gaskets. Torque the exhaust flange nuts to 40 ft. lbs. (55 Nm) and the catalyst flange nuts to 26 ft. lbs. (34 Nm).

39. Install the clutch hose bracket.

40. Install the upper transaxle mounting bolts and torque to 55 ft. lbs. (75 Nm).

41. Reconnect the neutral position switch, backup light switch, and the reverse lockout solenoid connectors.

42. Install the control box.

43. Install the strut bar.

44. Verify that all linkages, vacuum lines, and wiring harnesses have been reconnected.

45. Refill the transaxle fluid.

46. Connect the battery cables.

47. Check the clutch operation and adjust if necessary.

48. Shift the transaxle through the gear range and check for smooth operation.

Clutch Assembly

REMOVAL AND INSTALLATION

NOTE: The radio may have a coded theft protection circuit. Obtain the code from the owner before disconnecting the battery, removing the radio fuse, or removing the radio.

1993–94 Integra and 1993–94 Vigor

1. Disconnect the negative battery cable.

2. Remove the transaxle.

3. On Integra, remove the release shaft retaining bolt and remove the release shaft and release bearing assembly.

4. Remove the slave cylinder with the hydraulic hose still connected. Remove the boot from the clutch case and remove the release fork with the bearing.

5. Matchmark the flywheel and pressure plate for easy reassembly. Remove the pressure plate bolts in a crisscross pattern 2 turns at a time to prevent warping the plate.

6. Inspect the flywheel for scoring and wear. Use a dial indicator to make sure it is flat and reface or replace, as necessary.

To install:

7. Make sure the flywheel and the end of the crankshaft are clean before assembly. Torque the flywheel bolts to 76 ft. lbs. (103 Nm). Torque the bolts in a criss–cross pattern.

8. Apply grease to the splines of the clutch disc, and install the clutch disc using a clutch alignment shaft.

9. Install the release bearing on the pressure plate, then install the pressure plate. When installing the pressure plate, align the mark on the outer edge of the flywheel with the alignment mark on the pressure plate. Failure to align these marks will result in imbalance. After installing the pressure plate, make sure the release bearing does not come off.

10. Torque the pressure plate bolts, using a pilot shaft to center the friction disc. After centering the disc, tighten the bolts 2 turns at a time, in a crisscross pattern to avoid warping the diaphragm springs; torque to 19 ft. lbs. (26 Nm).

11. Install the transaxle, make sure the mainshaft is properly aligned with the disc spline and the aligning pins are in place, before torquing the case bolts.

1995–97 Integra

1. Disconnect the negative battery cable.

2. Remove the manual transaxle assembly from the vehicle.

3. Insert the clutch alignment shaft (part # 07NAF–PR30100) with the clutch alignment disc (part # 07JAF–PM7011A) and handle (part # 07936–3710100). Use a feeler gauge and measure the clearance between the pressure plate spring fingers and the clutch alignment disc. There should be a maximum of 0.02 in. (0.6mm) of clearance for a new pressure plate with 0.03 in. (0.8mm) limit for a used pressure plate.

4. Remove the clutch alignment disc.

5. Install a flywheel holder (part # 07LAB–PV00100 or 07924–PD20003) to aid in the removal of the pressure plate and clutch disc.

6. Matchmark the flywheel and pressure plate for easy reassembly. Remove the pressure plate bolts in a crisscross pattern 2 turns at a time to prevent warping the plate.

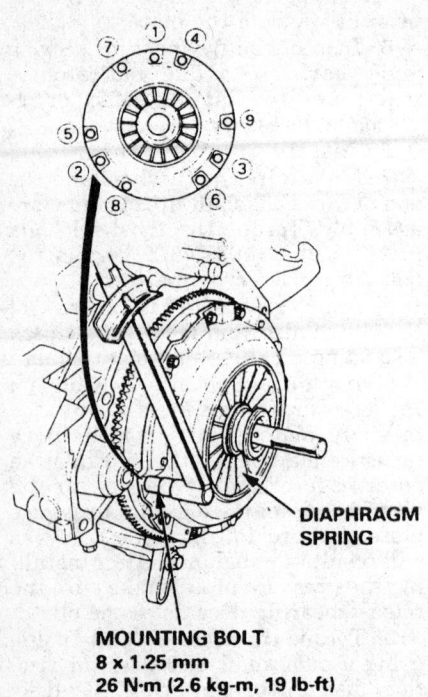

DIAPHRAGM SPRING

MOUNTING BOLT
8 x 1.25 mm
26 N·m (2.6 kg-m, 19 lb-ft)

121002

Torquing pressure plate mounting bolts — 1993–94 Integra and 1993–94 Vigor

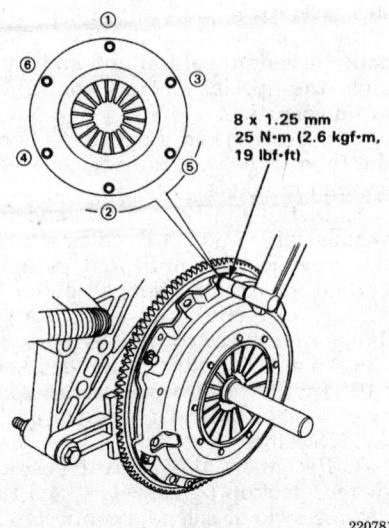

8 x 1.25 mm
25 N·m (2.6 kgf·m, 19 lbf·ft)

220781

Pressure plate bolt torque sequence — 1995 Integra

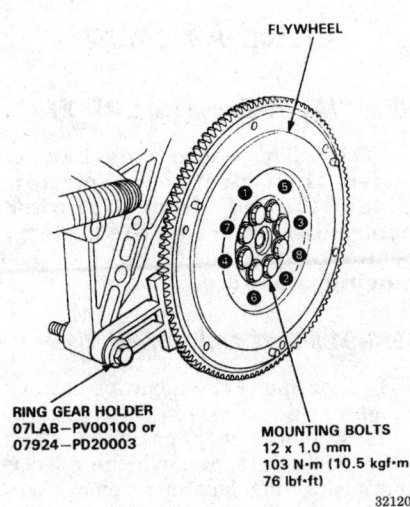

FLYWHEEL

RING GEAR HOLDER
07LAB–PV00100 or
07924–PD20003

MOUNTING BOLTS
12 x 1.0 mm
103 N·m (10.5 kgf·m, 76 lbf·ft)

321206

Flywheel bolt torque sequence — 1996 Integra

7. Remove the pressure plate then the clutch disc with the alignment shaft.

To install:

8. If the flywheel was removed, align the hole in the flywheel with the crankshaft dowel pin and install the mounting bolts finger tight.

9. Install the flywheel holder then torque the flywheel mounting bolts in a crisscross pattern in several steps. The mounting bolts final torque should be 76 ft. lbs. (103Nm).

10. Apply high temperature grease (part # 08798–9002) to the spline of the clutch disc, then install the clutch disc using the clutch alignment shaft.

11. Install the pressure plate, torque the mounting bolts in the proper sequence to 19 ft. lbs. (25 Nm).

12. Remove the flywheel holding tool and the clutch alignment shaft.

13. Insert the clutch alignment shaft with the clutch alignment disc and handle. Use a feeler gauge and measure the clearance between the pressure plate spring fingers and the clutch alignment disc. There should be a maximum of 0.02 in. (0.6mm) of clearance for a new pressure plate with 0.03 in. (0.8mm) limit for a used pressure plate.

14. Install the transaxle assembly.

15. Connect the negative battery cable and enter the radio security code.

Legend

1. Disconnect the negative battery cable.

2. Raise and support the vehicle.

3. Remove the transaxle.

4. Matchmark the flywheel and pressure plate for reassembly. Remove the pressure plate bolts in a crisscross pattern 2 turns at a time to prevent warping the plate.

5. Inspect the pressure plate and clutch disk for signs of wear.

6. Inspect the flywheel for scoring and wear. Use a dial indicator to make sure it is flat and resurface or replace, as necessary.

To install:

7. Make sure the flywheel and the end of the crankshaft are clean before assembly. Torque the flywheel bolts to 76 ft. lbs. (105 Nm) on all others. Torque the bolts in a criss–cross pattern.

8. Apply grease to the splines of the clutch disc, and install the clutch disc using a clutch alignment shaft.

9. Install the release bearing on the pressure plate, then install the pressure plate. When installing the pressure plate, align the mark on the outer edge of the flywheel with the alignment mark on the pressure plate. Failure to align these marks will result in imbalance.

10. Torque the pressure plate bolts using an alignment shaft to center the friction disc. After centering the disc, tighten the bolts 2 turns at a time, in a crisscross pattern to avoid warping the diaphragm springs; torque to 19 ft. lbs. (26 Nm).

11. Jack the transaxle into place and make sure the mainshaft is properly aligned with the disc spline and that the aligning pins are in place.

12. Install the transaxle and torque the mounting bolts to 55 ft. lbs. (75 Nm).

13. Install and connect the slave cylinder and its hydraulic line. Fill the reservoir with fluid.

14. Verify that all wiring harnesses, vacuum lines, and linkages are connected properly.

15. Reconnect the negative battery cable.

16. Check the clutch adjustment and road test the vehicle.

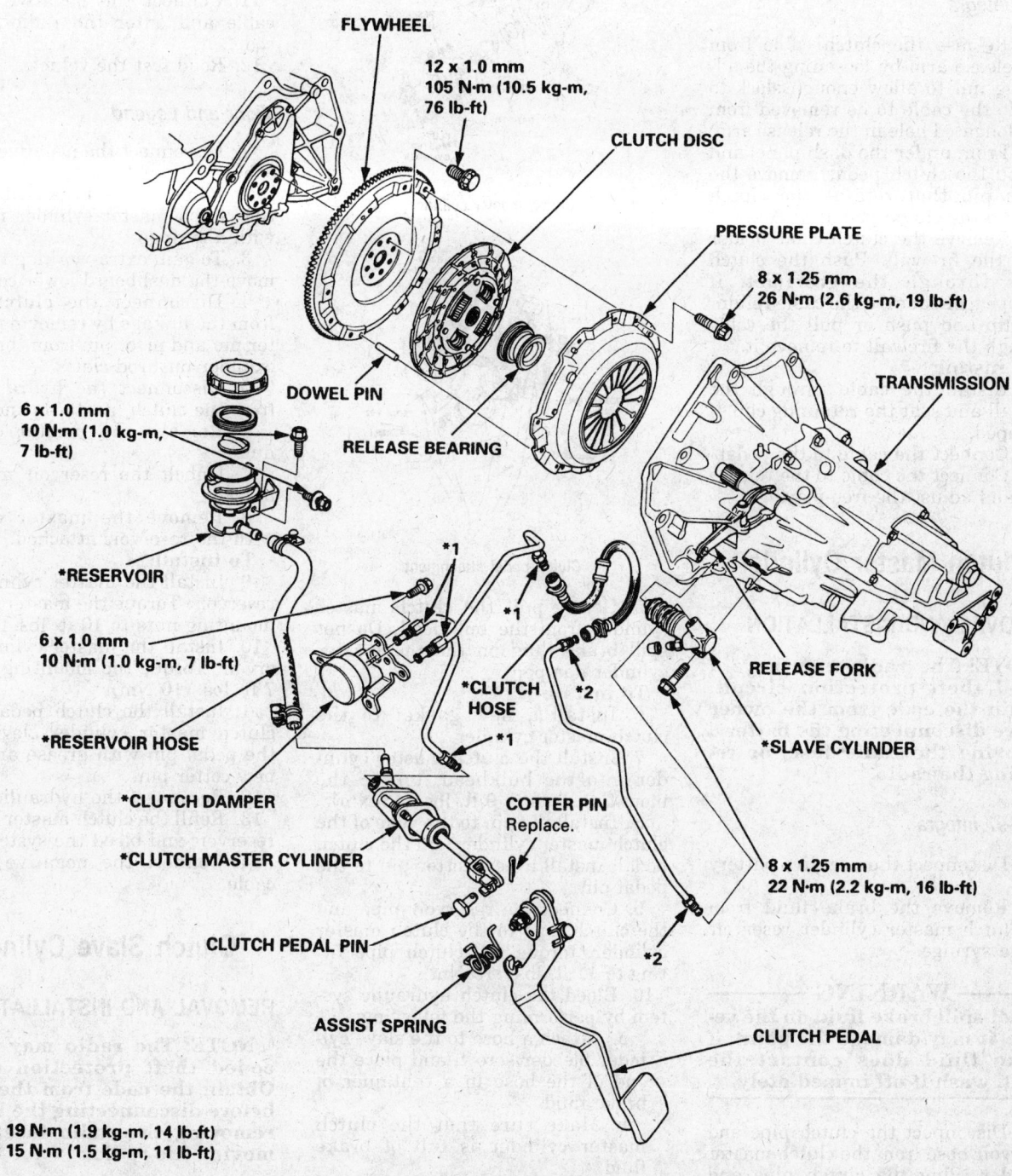

FLYWHEEL

12 x 1.0 mm
105 N·m (10.5 kg-m, 76 lb-ft)

CLUTCH DISC

PRESSURE PLATE

8 x 1.25 mm
26 N·m (2.6 kg-m, 19 lb-ft)

TRANSMISSION

DOWEL PIN

RELEASE BEARING

6 x 1.0 mm
10 N·m (1.0 kg-m, 7 lb-ft)

*RESERVOIR

6 x 1.0 mm
10 N·m (1.0 kg-m, 7 lb-ft)

*RESERVOIR HOSE

*CLUTCH DAMPER

*CLUTCH MASTER CYLINDER

CLUTCH PEDAL PIN

*CLUTCH HOSE

*CLUTCH HOSE

COTTER PIN
Replace.

RELEASE FORK

*SLAVE CYLINDER

8 x 1.25 mm
22 N·m (2.2 kg-m, 16 lb-ft)

ASSIST SPRING

CLUTCH PEDAL

*1: 19 N·m (1.9 kg-m, 14 lb-ft)
*2: 15 N·m (1.5 kg-m, 11 lb-ft)

79201g01

Clutch components — Legend shown, others similar

Clutch Cable

ADJUSTMENT

1993 Integra

1. Release the clutch cable from the release arm by loosening the adjusting nut to allow enough slack to enable the cable to be removed from the elongated hole in the release arm.
2. From under the dash panel and behind the clutch pedal, remove the clevis pin that retains the clutch cable to the clutch pedal.
3. Remove the clutch cable holder from the firewall. Push the clutch cable through the grommet, if equipped, or squeeze the cable retaining clip and push or pull the cable through the firewall to remove it.

To install:

4. Install the cable through the firewall and seat the retaining clip, if equipped.
5. Connect the cable to the pedal.
6. Connect the cable to the release arm and adjust the free-play.

Clutch Master Cylinder

REMOVAL AND INSTALLATION

NOTE: The radio may have a coded theft protection circuit. Obtain the code from the owner before disconnecting the battery, removing the radio fuse, or removing the radio.

1994–97 Integra

1. Disconnect the negative battery cable.
2. Remove the brake fluid from the clutch master cylinder reservoir with a syringe.

── WARNING ──
Do not spill brake fluid on the vehicle, it may damage the paint; if brake fluid does contact the paint, wash it off immediately.

3. Disconnect the clutch pipe and reservoir hose from the clutch master cylinder. Plug the clutch pipe and reservoir hose to prevent brake fluid from coming out.
4. Working under the dashboard, remove and discard the cotterpin from the clutch pedal pin. Remove the pin from the yoke then remove the clutch master cylinder mounting nuts.

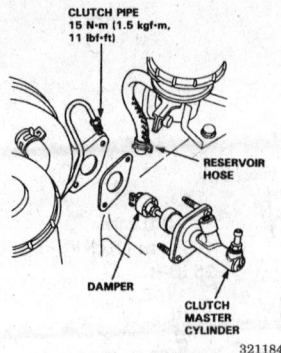

CLUTCH PIPE
15 N·m (1.5 kgf·m, 11 lbf·ft)

RESERVOIR HOSE

DAMPER

CLUTCH MASTER CYLINDER

321184

Clutch master cylinder

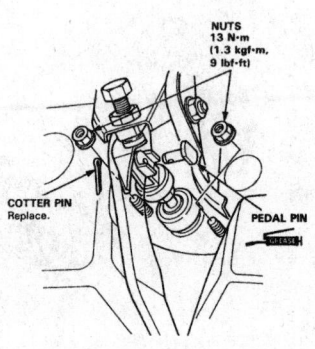

NUTS
13 N·m (1.3 kgf·m, 9 lbf·ft)

COTTER PIN
Replace.

PEDAL PIN

321185

Clutch pedal attachment

5. Gently pull the clutch master cylinder from the bulkhead. Do not spill brake fluid on the the master cylinder damper.

To install:

6. Install a new gasket to the clutch master cylinder.
7. Install the clutch master cylinder into the bulkhead, torque the mounting nuts to 9 ft. lbs. (13 Nm).
8. Install the pin to the yoke of the clutch master cylinder and the clutch pedal. Install a new cotter pin to the pedal pin.
9. Connect the reservoir pipe and the clutch pipe to the clutch master cylinder, torque the clutch pipe fitting to 11 ft. lbs. (15 Nm).
10. Bleed the clutch hydraulic system by performing the following:
   a. Attach a hose to the slave cylinder bleeder screw, and place the end of the hose in a container of brake fluid.
   b. Make sure that the clutch master cylinder is full of brake fluid.
   c. Pump the clutch pedal slowly, until no more bubbles appear at the bleeder hose.
   d. Tighten the slave cylinder bleeder screw to 7 ft. lbs. (9.8 Nm).

e. Refill the clutch master cylinder with clean brake fluid.

NOTE: Use only DOT 3 or 4 brake fluid.

11. Connect the negative battery cable and enter the radio security code.
12. Road test the vehicle.

Vigor and Legend

1. Disconnect the negative battery cable.
2. Remove the brake fluid from the clutch master cylinder reservoir with a syringe.
3. To gain extra working room, remove the dashboard lower cover.
4. Disconnect the clutch pedal from the linkage by removing the cotter pin and pivot pin from the master cylinder pushrod clevis.
5. Disconnect the hydraulic line from the clutch master cylinder.
6. Remove the master cylinder nuts.
7. Unbolt the reservoir mounting bracket.
8. Remove the master cylinder with the reservoir attached.

To install:

9. Install the master cylinder and reservoir. Torque the master cylinder mounting nuts to 10 ft. lbs. (13 Nm).
10. Install the master cylinder reservoir. Torque the mounting bolts to 7 ft. lbs. (10 Nm).
11. Install the clutch pedal to the clutch master cylinder clevis. Coat the pedal pin with grease and use a new cotter pin.
12. Reconnect the hydraulic lines.
13. Refill the clutch master cylinder reservoir and bleed the system.
14. Connect the negative battery cable.

Clutch Slave Cylinder

REMOVAL AND INSTALLATION

NOTE: The radio may have a coded theft protection circuit. Obtain the code from the owner before disconnecting the battery, removing the radio fuse, or removing the radio.

1994–97 Integra

1. Disconnect the negative battery cable.
2. Disconnect the clutch pipe from the slave cylinder. Plug the clutch pipe to prevent brake fluid from coming out.

WARNING

Do not spill brake fluid on the vehicle, it may damage the paint; if brake fluid does contact the paint, wash it off immediately.

3. Remove the slave cylinder mounting bolts and remove the slave cylinder.

To install:

4. Install the boot to the slave cylinder, if not installed. Apply high temperature grease (part # 08798–9002) to the tip of the slave cylinder rod and apply brake assembly lubricant to the slave cylinder where the boot contacts the slave cylinder.

5. Position the slave cylinder on the transaxle and install the mounting bolts. Torque the mounting bolts to 16 ft. lbs. (22 Nm).

6. Connect the clutch pipe to the slave cylinder and torque the fitting to 11 ft. lbs. (15 Nm).

7. Bleed the clutch hydraulic system by performing the following:

a. Attach a hose to the slave cylinder bleeder screw, and place the end of the hose in a container of brake fluid.

b. Make sure that the clutch master cylinder is full of brake fluid.

c. Pump the clutch pedal slowly, until no more bubbles appear at the bleeder hose.

d. Tighten the slave cylinder bleeder screw to 7 ft. lbs. (9.8 Nm).

e. Refill the clutch master cylinder with clean brake fluid.

NOTE: Use only DOT 3 or 4 brake fluid.

8. Connect the negative battery cable and enter the radio security code.

9. Road test the vehicle.

Vigor and 1993–94 Legend

1. On 1993–94 Legend models, remove the release fork cover.

2. Disconnect and plug the hydraulic line at the slave cylinder.

3. Remove the clutch hose on the Vigor model.

4. Release the pushrod from the throwout arm.

5. Remove the slave cylinder bolts and the slave cylinder. Check for leaks or deterioration and replace, if necessary.

To install:

6. Install the clutch slave cylinder onto the clutch housing. Torque the slave cylinder bolts to 16 ft. lbs. (22 Nm).

7. Reconnect the end of the pushrod to the throwout arm. Be sure

to first apply super high temperature urea grease to the end of the pushrod.

8. Reconnect the hydraulic line fitting to the slave cylinder. On Vigor models, reconnect the clutch hose and install the banjo bolt.

9. Install the release fork cover, if one was removed.

10. Refill the clutch master cylinder reservoir and bleed the hydraulic system. After the hydraulic system has been bled, top off the master cylinder reservoir.

1995 Legend

1. Disconnect the negative battery cable.

2. Remove the release fork cover.

3. Drain the reservoir using a syringe.

4. Disconnect the hydraulic line at the slave cylinder.

5. Disconnect the end of the pushrod from the release fork.

6. Remove the slave cylinder from the clutch housing.

To install:

7. Install the clutch slave cylinder onto the clutch housing. Torque the slave cylinder bolts to 16 ft. lbs. (22 Nm).

8. Apply high temperature grease to the end of the pushrod. Reconnect the end of the pushrod to the release fork.

9. Reconnect the hydraulic line fitting to the slave cylinder.

10. Install the release fork cover. Torque the bolts to 9 ft. lbs. (12 Nm).

11. Refill the clutch master cylinder reservoir and bleed the hydraulic system.

12. Attach a hose to the slave cylinder bleeding screw and suspend it into a drain container.

13. Pump the clutch pedal until no bubbles appear in the bleeder hose.

14. After the hydraulic system has been bled, top off the master cylinder reservoir.

15. Connect the negative battery cable.

Hydraulic Clutch System Bleeding

PROCEDURE

NOTE: Use DOT 3 or 4 brake fluid in the clutch master and slave cylinders. Brake fluid will damage the vehicle's paint — immediately clean up any spills.

1. Fit a flare or box-end wrench onto the slave cylinder bleeder screw.

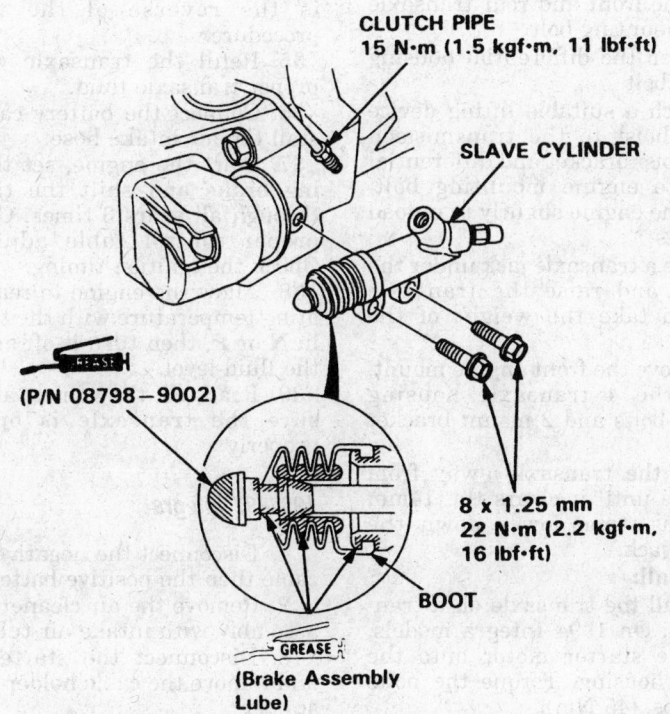

CLUTCH PIPE
15 N·m (1.5 kgf·m, 11 lbf·ft)

SLAVE CYLINDER

GREASE
(P/N 08798–9002)

8 x 1.25 mm
22 N·m (2.2 kgf·m, 16 lbf·ft)

GREASE
(Brake Assembly Lube)

BOOT

321120

Clutch slave cylinder — 1994–97 Integra

2. Attach a rubber tube to the slave cylinder bleeder screw and suspend it into a clear drain container partially filled with brake fluid.

3. Fill the clutch master cylinder with brake fluid.

4. Open the bleeder screw and pump the clutch pedal until no more bubbles appear in the tube.

5. Close the bleeder screw.

6. Refill the clutch master cylinder reservoir with brake fluid.

Automatic Transaxle Assembly

REMOVAL AND INSTALLATION

NOTE: The radio may have a coded theft protection circuit. Obtain the code from the owner before disconnecting the battery, removing the radio fuse, or removing the radio.

1993 Integra

1. Disconnect the negative first and then the positive battery cables.

2. Remove the 4 battery mounting bolts and remove the battery.

3. Remove the air cleaner case complete with air intake tube. Disconnect the transaxle ground cable.

4. Remove the speed sensor and the connectors, but leave its hoses connected. Be careful not to bend the speedometer cable.

5. Disconnect the speed pulser connector. Disconnect the lock-up control solenoid valve wire connectors.

6. Disconnect the vacuum hose from the vacuum modulator valve. Drain the transaxle fluid into a suitable drain pan.

7. Disconnect the transaxle cooler lines at the joint pipes. Be sure to turn the ends up so as to prevent fluid from flowing out. Be sure to check for leakage at the hose joints at this time.

8. Remove the center beam, except on GSR models. Remove the header pipe, and disconnect the heated oxygen sensor (HO2S) connector, if necessary.

9. Remove the cotter pins from the lower ball joint castle nuts and remove the lower ball joints nuts. Separate the ball joints from the lower arms with a suitable ball joint separator tool.

10. Remove the damper fork. On 1993 Integra models, remove the radius rod.

11. Pry the right and left halfshafts out of the differential and the inter-mediate shaft. Pull on the inboard CV-joint and remove the right and left halfshafts. Coat all precision finished surfaces with clean engine oil and tie plastic bags over the halfshaft ends.

12. Remove the 3 mounting bolts and lower the bearing support. Remove the intermediate shaft from the differential.

13. Remove the engine splash shield and the right wheel well splash shield.

14. Remove the right damper pinch bolt, then separate the damper fork and damper.

15. Remove the bolts and nut from the right radius arm and remove the radius arm.

16. Remove the front and rear engine stiffeners. Remove the torque converter cover and cable holder.

17. Remove the shift control cable by removing the cotter pin, control pin and control lever roller from the control lever.

18. Remove the shift control cable guide, take care not to bend the control cable.

19. Remove the plug, then remove the driveplate bolts 1 at a time while rotating the crankshaft pulley. Remove the mounting bolt from the front engine mount.

20. Remove the 2 mounting bolts from the rear engine mount bracket. Remove the front and rear transaxle housing mounting bolt.

21. Loosen the differential housing mounting bolt.

22. Attach a suitable lifting device or chain hoist to the transmission housing hoist bracket and differential housing to engine mounting bolt, then lift the engine slightly to unload the mounts.

23. Place a transaxle jack under the transaxle and raise the transaxle enough to take the weight of the mounts.

24. Remove the front engine mount. Remove the 4 transaxle housing mounting bolts and 2 mount bracket bolts.

25. Pull the transaxle away from the engine until it clears the 14mm dowel pins, then lower down the transaxle jack.

To install:

26. Install the transaxle on a transaxle jack. On 1994 Integra models, install the starter motor onto the transaxle housing. Torque the bolts to 33 ft. lbs. (45 Nm).

27. Make sure both 14mm dowel pins are installed in the torque converter housing.

28. Raise the transaxle high enough to align the dowel pins with the matching holes in the block. Align the torque converter match-mark and the bolt heads with holes in the driveplate.

29. Install the 4 transaxle housing mounting bolts, then install the transaxle to the engine block. Torque the mounting bolts to 47 ft. lbs. (65 Nm) for 1993 Integra, and 54 ft. lbs. (74 Nm) for 1994 Integra.

30. Install the front engine mount to the front beam. Install the transaxle to the front engine mount.

31. Install the transaxle to the transaxle mount bracket. Remove the transaxle jack.

32. Install the 2 transaxle housing mounting bolts engine side and rear engine mount bracket bolts. Torque the transaxle housing mounting bolts (engine side) to 47 ft. lbs. (65 Nm) for 1993 Integra, and 43 ft. lbs. (59 Nm) for 1994 Integra. Torque the rear engine mount bracket bolts to 43 ft. lbs. (60 Nm) for 1993 Integra, and 87 ft. lbs. (118 Nm) for 1994 Integra.

33. Attach the torque converter to driveplate with 8 (6 x 1 x 12mm) bolts and torque to 9 ft. lbs. (12 Nm). Rotate the crankshaft, as necessary, to tighten the bolts half torque, then the final torque in a crisscross pattern. Check for free rotation after tightening the last bolt.

34. The balance of the installation is the reverse of the removal procedure.

35. Refill the transaxle with the proper transaxle fluid.

36. Connect the battery cables. Install the air intake hose.

37. Start the engine, set the parking brake and shift the transaxle through all gears 3 times. Check for proper control cable adjustment. Check the ignition timing.

38. Allow the engine to reach operating temperature with the transaxle in **N** or **P**, then turn it off and check the fluid level.

39. Road test the vehicle and make sure the transaxle is operating properly.

1994–97 Integra

1. Disconnect the negative battery cable then the positive battery cable.

2. Remove the air cleaner housing assembly with intake air tube.

3. Disconnect the starter cables and remove the cable holder from the starter.

4. Disconnect the transaxle ground cable from the transaxle hanger.

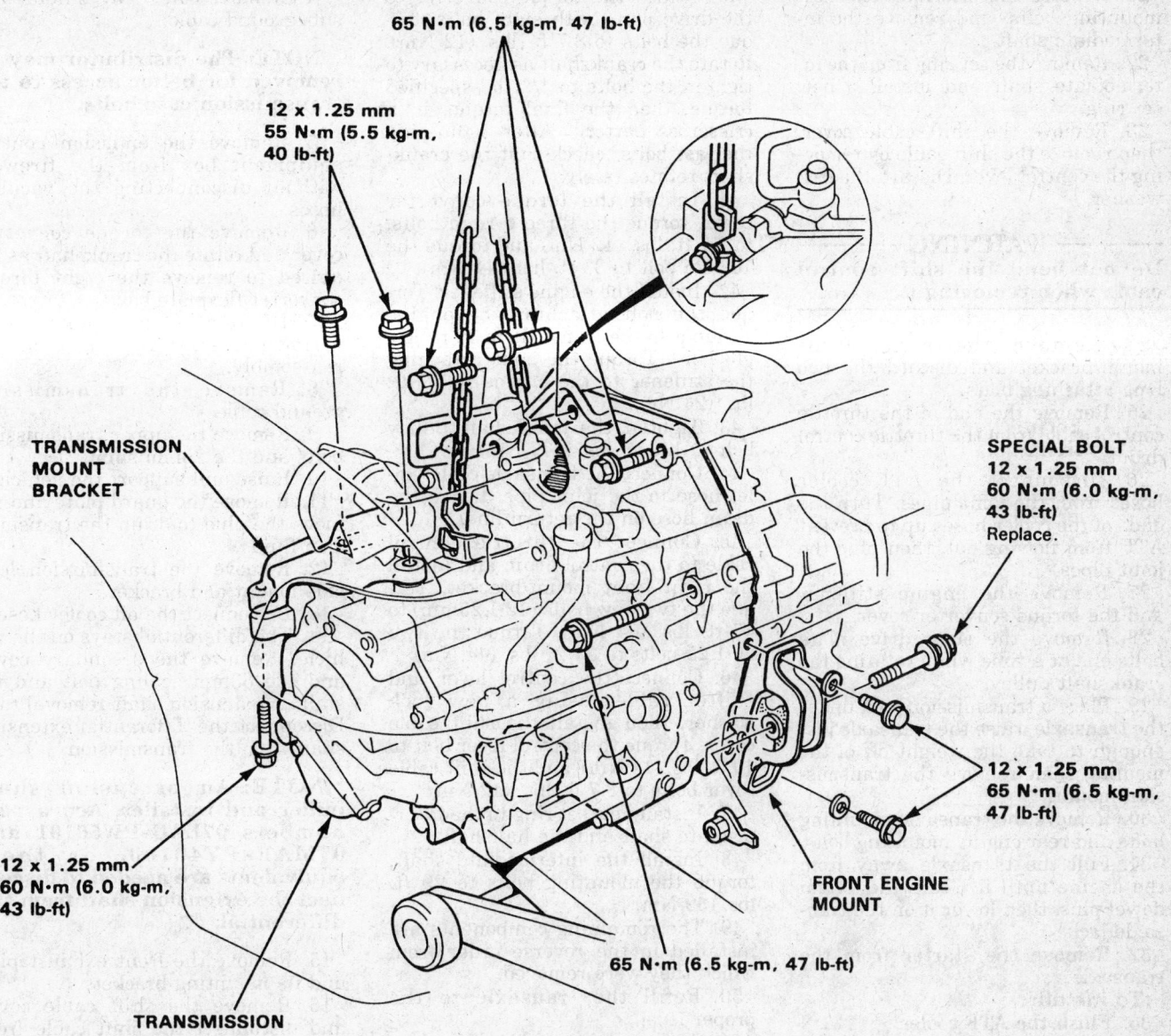

65 N·m (6.5 kg-m, 47 lb-ft)

12 x 1.25 mm
55 N·m (5.5 kg-m, 40 lb-ft)

TRANSMISSION MOUNT BRACKET

SPECIAL BOLT
12 x 1.25 mm
60 N·m (6.0 kg-m, 43 lb-ft)
Replace.

12 x 1.25 mm
60 N·m (6.0 kg-m, 43 lb-ft)

TRANSMISSION

12 x 1.25 mm
65 N·m (6.5 kg-m, 47 lb-ft)

FRONT ENGINE MOUNT

65 N·m (6.5 kg-m, 47 lb-ft)

220704

Transaxle mounts — 1993 Integra

5. Disconnect the lock–up control solenoid valve connector and the shift control solenoid valve connector. Remove the harness clamp on the lock–up control solenoid harness from the harness stay.

6. Disconnect the vehicle speed sensor (VSS), main shaft speed sensor and the counter shaft speed sensor connectors.

7. Remove the the upper transaxle mounting bolts.

8. Remove the drain plug from the transaxle and drain the used fluid into a sealable container. Properly dispose of the used fluid. Reinstall the drain plug with a new sealing washer.

9. Remove the splash shield.

10. Remove the front wheels.

11. Remove the cotter pins and castle nuts from the front lower ball joints. Separate the ball joints from the lower control arms.

12. Remove the right damper fork bolt, discard the nut.

13. Remove the right damper pinch bolt then remove the damper fork.

14. Pry the right halfshaft out of the differential, discard the set ring on the inner joint.

15. Pry the left halfshaft out of the intermediate shaft, discard the set ring on the inner joint.

16. Tie plastic bags over the half-shaft joints to keep the splines of the joints clean.

17. Disconnect the heated oxygen sensor (HO2S) connector.

18. Remove the nuts and bolts connecting exhaust pipe A to the catalytic converter. Discard the gasket and the locknuts.

19. Remove and discard the nuts attaching exhaust pipe A to the exhaust hanger.

20. Remove and discard the locknuts attaching exhaust pipe A to the exhaust manifold, then remove exhaust pipe A from the vehicle. discard the exhaust gaskets.

21. Remove the intermediate shaft mounting bolts, and remove the intermediate shaft.

22. Remove the set ring from the intermediate shaft and install a new set ring.

23. Remove the shift cable cover, then remove the shift cable by removing the control lever. Discard the lock washer.

——— WARNING ———
Do not bend the shift control cable when removing it.

24. Remove the right front mount/bracket and discard the two long attaching bolts.

25. Remove the end of the throttle control cable from the throttle control drum.

26. Disconnect the ATF cooler hoses from the joint pipes. Turn the ends of the cooler hoses up to prevent ATF from flowing out, then plug the joint pipes.

27. Remove the engine stiffener and the torque converter cover.

28. Remove the eight drive plate bolts one at a time while rotating the crankshaft pulley.

29. Place a transmission jack under the transaxle, raise the transaxle just enough to take the weight off of the mounts, then remove the transmission mount.

30. Remove the transaxle mounting bolts and rear engine mounting bolts.

31. Pull the transaxle away from the engine until it clears the 14mm dowel pins, then lower it on the transaxle jack.

32. Remove the starter from the transaxle.

To install:

33. Flush the ATF cooler.

34. Install the starter to the transaxle, torque the bolts to 33 ft. lbs. (45 Nm). Install the 14mm dowel pins to the torque converter housing.

35. Place the transaxle on a transmission jack, and raise to engine level.

36. Fit the transaxle to the engine, then install the transaxle housing mounting bolts and the two rear engine mounting bolts with new washers. Torque the transaxle housing mounting bolts to 43 ft. lbs. (59 Nm) and the rear engine mounting bolts to 86.8 ft. lbs. (118 Nm).

37. Install the transmission mount. Torque the bolt to 54 ft. lbs. (74 Nm) and the nuts to 47 ft. lbs. (64 Nm).

38. Install the three transaxle upper mounting bolts, torque the bolts to 54 ft. lbs. (74 Nm).

39. Remove the transmission jack.

40. Attach the torque converter to the drive plate with eight bolts, torque the bolts to 8.7 ft. lbs. (12 Nm). Rotate the crankshaft as necessary to tighten the bolts to 1/2 the specified torque, then the final torque, in a crisscross pattern. After tightening the last bolts, check that the crankshaft rotates freely.

41. Install the torque converter cover, torque the three 6·1mm bolts, to 8.7 ft. lbs. (12 Nm) and torque the 10·1.25 bolt to 33 ft. lbs. (44 Nm).

42. Install the engine stiffener. Torque the bolt attaching the engine stiffener to the transaxle to 32 ft. lbs. (43 Nm). Torque the bolts attaching the stiffener to the engine to 17 ft. lbs. (24 Nm).

43. Tighten the crankshaft pulley bolt to 130 ft. lbs. (177 Nm).

44. Connect the transaxle cooler inlet hose to the joint pipe. Leave the drain hose on the return line.

45. Connect the throttle control cable to the control drum and install the right front mount/bracket. Torque the two new bolts (12·1.25mm) to 47 ft. lbs. (64 Nm). Torque the two 10·1.25 bolts to 33 ft. lbs. (44 Nm)

46. Connect the control lever and shifter cable, using a new lock washer, then install the shift cable cover. Torque the control lever bolt to 10 ft. lbs. (14 Nm), and the shift cable cover bolts to 8.7 ft. lbs. (12 Nm).

47. Install new set ring to the intermediate shaft and the halfshafts.

48. Install the intermediate shaft, torque the mounting bolts to 29 ft. lbs. (39 Nm).

49. The remaining components are installed in the reverse order from which they were removed.

50. Refill the transaxle to the proper level.

51. Start the engine, with the parking brake set, and shift the transaxle through all gears three times.

52. Check and adjust the shift cable as necessary.

53. Let the engine reach operating temperature (the cooling fan comes on) with the transaxle in **P** or **N**, then turn the engine off and check the fluid level.

54. Road test the vehicle.

Vigor and 2.5TL

1. Shift the transmission into **P**.

2. Disconnect both battery cables and remove the battery and battery tray.

3. Without disconnecting the wires, remove the ABS relay box and set it aside.

4. Remove the heat shield and sub-ground cable.

NOTE: The distributor may be removed for better access to the transmission case bolts.

5. Remove the emission control equipment box from the firewall without disconnecting the vacuum hoses.

6. Remove the torque converter cover and rotate the crankshaft as required to remove the eight torque converter flexplate bolts.

7. Disconnect the transmission wiring harnesses and tag them for reassembly.

8. Remove the transmission ground cable.

9. Remove the upper transmission bolts and the 26mm shim.

10. Raise and support the vehicle.

11. Remove the guard plate and remove the plug to drain the transmission fluid.

12. Remove the transmission left-side mount and bracket.

13. Disconnect the oil cooler hoses.

14. The differential stays on the vehicle. Remove the secondary cover and the 33mm sealing bolt and install an extension shaft removal tool. Disconnect the differential extension shaft from the transmission.

NOTE: An extension shaft puller and installer, Acura part numbers 07LAC–PW50101 and 07MAF–PY40100, or their equivalents are needed to disconnect the extension shaft from the differential.

15. Remove the front exhaust pipe and its mounting brackets.

16. Remove the shift cable cover and disconnect the shift cable from the control shaft. Remove the cable mounting bracket and wire the cable up and out of the way.

17. Place a transmission jack securely under the transmission and raise it to take the weight off the mounts.

18. Use an offset wrench to remove the mid-mounts and the mid-mount spacer.

19. Remove the transmission case bolts. Remove the torque converter cover mounting bolt located on the converter housing. Do not remove the torque converter cover.

20. Slide the transmission back and away from the engine. Carefully lower it from the vehicle.

To install:

21. Flush the ATF cooler lines.

22. Install the torque converter onto the mainshaft using a new O-

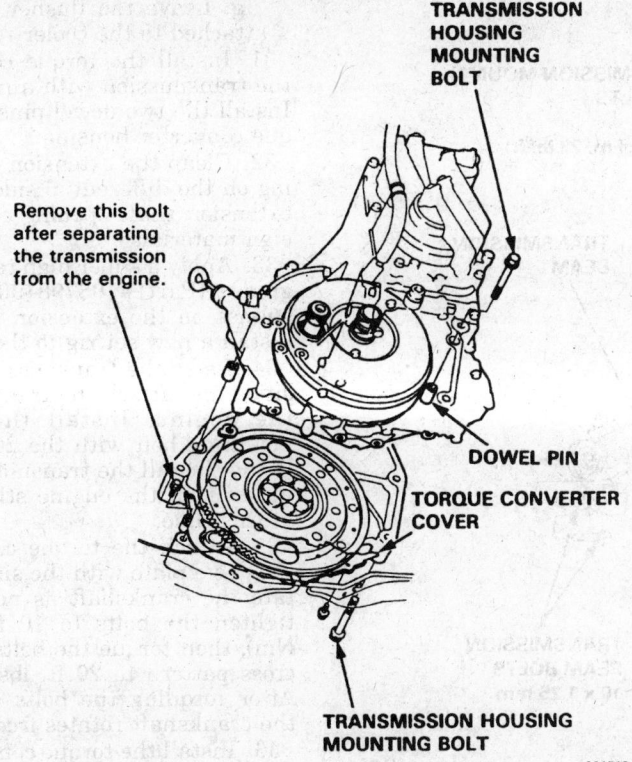

Automatic transaxle — 2.5L (G25A1, G25A4) engines

TRANSMISSION
HOUSING
MOUNTING
BOLT

Remove this bolt
after separating
the transmission
from the engine.

DOWEL PIN

TORQUE CONVERTER
COVER

TRANSMISSION HOUSING
MOUNTING BOLT

321519

ring. Install the mounting pins into the transmission case.

23. Install a new set ring on the extension shaft and lightly lubricate the splines with high temperature molybdenum grease. Pack the opening in the drive pinion with high temperature molybdenum grease.

24. Install the transmission and start all of the bolts. Don't forget the 26mm shim between the transmission and differential. Torque the 12mm bolts to 54 ft. lbs. (75 Nm). Torque the torque converter cover bolt to 9 ft. lbs. (12 Nm).

25. Install the mid-mounts and brackets. Torque the bolts to 28 ft. lbs. (39 Nm) and the nuts to 32 ft. lbs. (43–44 Nm).

26. Install a new set ring, and install the extension shaft using an extension shaft installer. Make sure the shaft snaps into place.

27. Pack the shaft area with molybdenum grease, but keep the thread area clean. Apply liquid gasket to the sealing bolt threads and torque the bolt to 58 ft. lbs. (78–80 Nm). Install the cover.

28. Install the transmission left-side mount and bracket. Torque the mount bolts to 47 ft. lbs. (65 Nm). Torque the bracket bolts to 40 ft. lbs. (54 Nm).

29. Install the torque converter bolts and torque them in 2 steps in a crisscross pattern to 9 ft. lbs. (12 Nm). Install the torque converter cover and tighten the bolts to 9 ft. lbs. (12 Nm).

30. The remaining components are installed in the reverse order from which they were removed.

31. Connect all the wiring harnesses and the battery cables. Refill the transmission with fresh fluid.

32. When all parts have been installed, start the engine and shift through all the gears 3 times to fill all the passages with fluid, then check the shift cable and adjust as needed. When the engine is fully warmed up, stop the engine and check the fluid level.

Legend and 3.2TL

1. Disconnect the negative then the positive battery cables.
2. Shift the transmission into **P** (Park).
3. Remove the control box from the bulkhead without disconnecting the vacuum hoses. Place the control box out of the way.
4. Disconnect the transmission sub–harness connectors, and remove the sub–harness clamp.

5. If applicable, remove the three bolts securing the ATF dipstick pipe bracket.
6. Remove the upper transmission mounting bolts.
7. Drain the fluid from the transmission into a sealable container. Install the drain plug with a new washer and torque the plug to 36 ft. lbs. (49 Nm).
8. Pull the carpet back under the passenger seat to expose the secondary heated oxygen connector (HO2S sensor 2) connector. Disconnect the connector and push it out from the inside of the vehicle.
9. Remove the heat shields from exhaust pipe A.
10. Remove the nuts attaching exhaust pipe A to the exhaust manifolds and the catalytic converter. Remove exhaust pipe A and discard the gaskets.
11. Remove the oxygen sensor wire harness cover and grommet, then remove the catalytic converter. Discard the nuts and gasket.
12. Remove the exhaust heat shield from the floor of the vehicle.
13. Disconnect the transmission cooler hoses, then plug the hoses and pipes.
14. Remove the shift cable cover mounting bolts and remove the wire harness clamps from the cover. Remove the shift cable cover from the transmission.
15. Remove the shift cable holder from the holder base, do not lose the washers.
16. Remove the locknut attaching the shift cable to the control lever, then remove the shift cable.
17. Remove the ATF dipstick pipe from the torque converter housing.
18. Remove the lower plate from under the rack and pinion, then install the two rack and pinion mounting bolts.
19. Remove the shift cable guide bracket from the transmission beam.
20. Remove the transmission beam, rear transmission mount racket/mount and exhaust pipe hanger.
21. Make sure that the transmission is in park then remove the 36mm sealing bolt from the transmission.
22. Install the extension shaft puller (Part # 07LAC–PW50100 or 07LAC–PW50101) onto the end of the extension shaft. Using the extension shaft puller, disconnect the extension shaft from the differential. Pull the

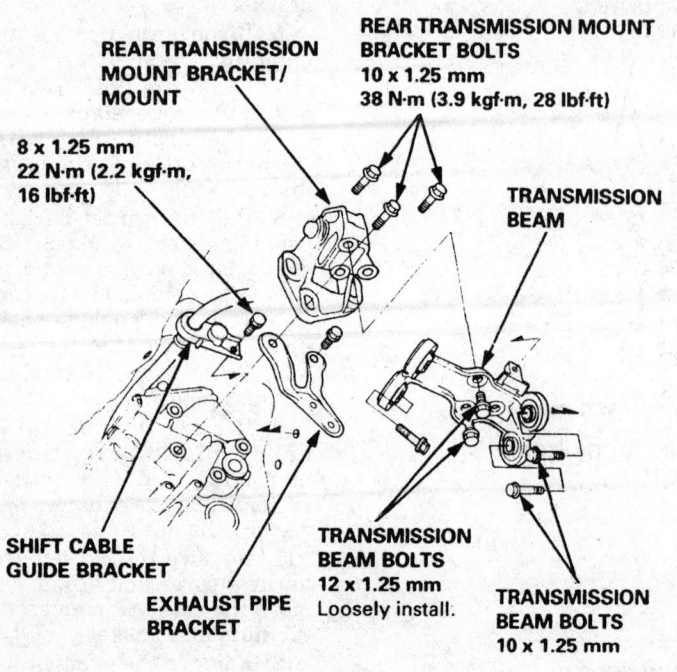

REAR TRANSMISSION MOUNT BRACKET/MOUNT

8 x 1.25 mm
22 N·m (2.2 kgf·m, 16 lbf·ft)

REAR TRANSMISSION MOUNT BRACKET BOLTS
10 x 1.25 mm
38 N·m (3.9 kgf·m, 28 lbf·ft)

TRANSMISSION BEAM

SHIFT CABLE GUIDE BRACKET

EXHAUST PIPE BRACKET

TRANSMISSION BEAM BOLTS
12 x 1.25 mm
Loosely install.

TRANSMISSION BEAM BOLTS
10 x 1.25 mm

334743

Rear transmission mount — 3.2TL

extension shaft out enough to remove the set ring.

NOTE: Do not try to remove the extension shaft, it cannot be removed from the transmission this way.

23. Place a transmission jack under the transmission and raise the transmission to take the weight off of the mounts.
24. Remove the stop holder, the mid mount stops and the mid mounts.
25. Remove the engine stiffener.
26. Remove the torque converter covers.
27. Remove the six drive plate bolts one at a time while rotating the crankshaft.

NOTE: If necessary, remove the spark plugs while removing the drive plate bolts.

28. Remove the transmission mounting bolts.
29. Pull the transmission away from the engine until it clears the dowel pins, then lower it on the transmission jack.

To install:
30. Flush the transmission cooling lines before installing the transmission. Use a pressurized flushing canister, such as Honda tool No. J38405–A, or its equivalent. Use only biodegradable flushing fluid, Honda part No. J35944–20. Other types of flushing fluid may damage the A/T cooling system.

a. Fill the flusher with 21 ounces of fluid (canister is 2/3 full). Pressurize the flusher to 80–120 PSI (560–845 kpa), following the procedure on the fluid container and flusher.

b. Clamp the discharge hose of the flusher to the cooler return line. Clamp the drain hose to the cooler inlet line and route it into a bucket or drain tank.

c. Connect the flusher to air and water lines. Open the flusher water valve and flush the cooler for ten seconds. The air line should be equipped with a water trap to keep the system dry.

d. Depress the flusher trigger to mix flushing fluid with the water. Flush for two minutes, turning the air valve on and off for five seconds every 15–20 seconds to create a surging action.

e. After finishing one flushing cycle, reverse the hose and flush in the opposite direction following the same steps.

f. Dry the cooler lines with compressed air for two minutes, or until flushing agent stops draining from the system.

g. Leave the flusher drain hose attached to the cooler return line.

31. Install the torque converter to the transmission with a new O–ring. Install the two dowel pins to the torque converter housing.

32. Clean the extension shaft opening on the differential side. Keep the extension shaft opening clean of foreign material.

33. Apply a super high temperature grease (Part # 08798–9002) to the splines on the extension shaft, then install a new set rig to the groove.

34. Raise the transmission into position and attach the transmission to the engine. Install the housing mounting bolt with the 26mm shim. Do not install the transmission housing bolt on the engine stiffener side at this time.

35. Attach the torque converter to the drive plate with the six bolts. Rotate the crankshaft as necessary to tighten the bolts to 10 ft. lbs. (13 Nm), then torque the bolts in a criss cross pattern to 20 ft. lbs. (26 Nm). After torquing the bolts check that the crankshaft rotates freely.

36. Install the torque converter covers and torque the bolts to 8.7 ft. lbs. (12 Nm).

37. Install the engine stiffener. Tighten the 8mm bolts loosely and torque the transmission housing bolt to 16 ft. lbs. (22 Nm), then torque the rest of the bolts to 16 ft. lbs. (22 Nm).

38. Install the transmission housing mounting bolts from the transmission side. Torque the bolts to 47 ft. lbs. (64 Nm).

39. Install the mid mounts, then the mid mount stops. Tighten the 8mm bolts loosely then install the stop holder.

40. Torque the mid mount 10mm bolts to 28 ft. lbs. (38 Nm) and torque the 8mm bolts to 16 ft. lbs. (22 Nm). Torque the new nuts attaching the mid mounts to 35 ft. lbs. (48 Nm) and torque the nuts attaching the stop holder to 40 ft. lbs. (54 Nm).

41. Remove the transmission jack from the transmission.

42. Install the transmission beam to the rear transmission mount bracket/mount. Tighten the two bolts loosely, then install them with the exhaust pipe bracket on the rear cover and body.

43. Torque the three transmission beam bolts to 28 ft. lbs. (38 Nm).

44. Torque the three rear transmission mount bracket bolts to 28 ft. lbs. (38 Nm).

45. Torque the two bolts attaching the mount it the beam to 40 ft. lbs. (54 Nm).

46. Install the shift cable guide to the transmission beam and torque the bolt to 7.2 ft. lbs. (9.8 Nm).

47. Install the extension shaft using the extension shaft installer (Part # 07MAF-PY40100 or 07MAF-PY40101). Make sure that the extension shaft locks into the secondary gear and the differential.

48. Fill the secondary gear with super high temperature grease (Part # 08798-9002). Applying sealer to the threads of the 36mm sealing bolt then install the bolt and torque to 58 ft. lbs.(78 Nm).

49. Remove the two bolts from the rack and pinion necessary to install the lower plate, then install the lower plate and the attaching bolts. Torque the lower plate attaching bolt to 28 ft. lbs. (38 Nm) and torque the rack and pinion bolts to 43 ft. lbs. (59 Nm).

50. Connect the shift cable control lever to the control shaft, then install the washer and nut. Torque the nut to 8.7 ft. lbs. (12 Nm).

51. Install the shift cable holder to the shift cable holder base with the mounting washers. Install the attaching bolts and torque the bolts to 8.7 ft. lbs. (12 Nm).

52. Install the shift cable cover and torque the mounting bolts to 8.7 ft. lbs. (12 Nm).

53. Connect the cooler feed hose to the pipe.

54. Install the ATF dipstick pipe with a new O-ring on the torque converter housing.

55. Install the transmission sub-harness clamp to the harness.

56. Install the exhaust heat shield to the floor of the vehicle and torque the bolts to 7.2 ft. lbs. (9.8 Nm).

57. Install the catalytic converter with a new gasket. Install the wire harness and the grommet to the vehicle then install the harness cover. Torque the cover bolts to 7.2 ft. lbs. (9.8 Nm).

58. Install exhaust pipe A with new gaskets and new nuts. Torque the nuts attaching the pipe to the manifolds to 40 ft. lbs. (54 Nm) and torque the nuts attaching the exhaust pipe to the catalytic converter to 16 ft. lbs. (22 Nm). Torque the catalytic converter rear attaching nuts to 24 ft. lbs. (32 Nm).

59. Install the heat shields to exhaust pipe A and torque the nuts to 8.7 ft. lbs. (12 Nm).

60. Connect the secondary heated oxygen sensor (HO2S) connector, located under the passenger front seat.

61. Install the transmission upper mounting bolts and torque the bolts to 47 ft. lbs. (64 Nm).

62. Install the ATF dipstick pipe bracket bolts and torque the bolts to 8.7 ft. lbs. (12 Nm).

63. Connect the transmission sub-harness connectors.

64. Install the control box and torque the mounting bolts to 8.7 ft. lbs. (12 Nm).

65. Connect the positive then the negative battery cable.

66. Fill the transmission with ATF. Use only Honda Premium ATF or an equivalent DEXRON®II ATF.

   a. Leave the flusher drain hose attached to the cooler return line.

   b. With the transmission in park, run the engine for 30 seconds, or until approximately one quart of fluid is discharged. As soon as one quart of fluid drains, shut off the engine. This completes the cooler flushing process.

   c. Remove the drain hose and reconnect the cooler return line.

   d. Refill the transmission to the proper level with ATF.

67. Let the engine reach proper operating temperature (the radiator fan comes on) with the transmission in **N** (neutral) or **P** (park). Turn the engine **OFF** and check the fluid level.

68. Enter the radio security code.

DRIVE AXLE

Halfshaft

REMOVAL AND INSTALLATION

NOTE: The outer CV-joint cannot be removed from the halfshaft, the boot is serviceable, if the joint requires replacement the shaft must be replaced as an assembly.

1. With the vehicle on the ground, raise the locking tab on the spindle nut and loosen it with a suitable socket.

2. Disconnect the negative battery cable.

3. Raise and safely support the vehicle and remove the spindle nut and front wheels.

4. For Integra and Vigor, if removing the right side halfshaft, drain the transaxle or differential oil.

5. For 2.5TL, drain the differential oil if the left halfshaft is to be removed.

6. Remove the damper fork nut and damper pinch bolt. Remove the damper fork.

7. Remove the lower ball joint nut and separate the lower ball joint using a ball joint remover.

8. Pull the knuckle outward and remove the halfshaft outboard CV-joint from the knuckle using a plastic mallet.

9. Using a small prybar with a 3.5 x 7mm tip, carefully pry out the inboard CV-joint approximately ½ in. (13mm) in order to force the spring clip out of the groove in the differential side gears.

NOTE: Be careful not to damage the oil seal. Do not pull on the inboard CV-joint, it may come apart.

10. Pull the halfshaft out of the differential or the intermediate shaft. Replace the spring clip on the end of the inboard joint.

11. Be sure to mark the roller grooves during disassembly to ensure proper positioning during reassembly.

12. Remove the front and rear boot retaining bands, then separate the inboard joint from the halfshaft assembly.

13. Mark the spider gear and the driveshaft so they can be installed in their original positions.

14. Remove snapring, spider gear, then remove the stopper ring.

15. Be sure to mark the position on the shaft where the dynamic damper goes, to ensure it will be reinstalled in its original position. Remove the inboard CV-joint boot, dynamic damper, then the outboard CV-joint boot.

To install:

16. Wrap the spline with vinyl tape to prevent damage to the boots. Install the outboard boot, dynamic damper, and inboard boot, then remove the vinyl tape.

17. Install the stopper ring onto the halfshaft groove, then install the spider gear in its original position by aligning the marks.

18. Fit the snapring into the halfshaft groove.

19. Pack the outboard joint boot with CV-joint grease only. Do not use a substitute or mix types of grease.

20. Fit the rollers to the spider gear with their high shoulders facing outward. Reinstall the rollers in their original positions on the spider gear.

21. Pack the inboard joint boot with CV-joint grease.

22. Fit the inboard joint onto the halfshaft. Hold the halfshaft assembly so the inboard joint points up to prevent it from falling off.

23. With the boots installed, adjust the CV-joints in or out to place the

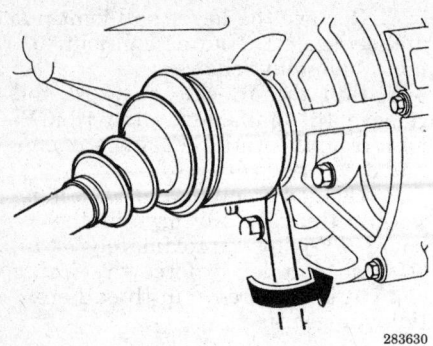

Removal of the inboard joint

283630

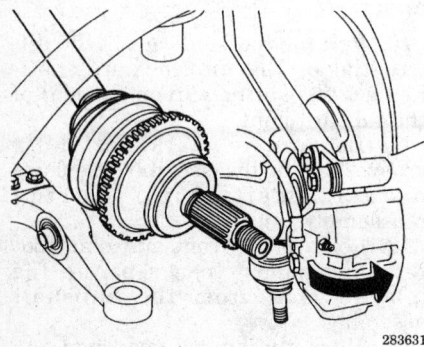

Removal of the outboard joint

283631

inner boot ends in the original positions.

24. Install the new boot bands on the boots and bend both sets of locking tabs. Lightly tap on the locking tabs to ensure a good fit.

25. Always use a new set ring whenever the driveshaft is being installed. Be sure the driveshaft locks in the differential side gear groove and that the CV-joint subaxle bottoms in the differential or the intermediate shaft.

26. Torque the ball joint nut to:
1993–94 Integra: 29 ft. lbs. (40 Nm)
1995–97 Integra and 2.5TL: 36–43 ft. lbs. (49–59 Nm)
1993–94 Vigor: 40 ft. lbs. (55 Nm)
Legend and 3.2TL: 54 ft. lbs. (75 Nm)

27. Install and torque the lower damper nut and bolt for all models except Legend and 3.2TL to 47 ft. lbs. (65 Nm). For Legend and 3.2TL, torque the lower damper nut and bolt to 50 ft. lbs. (70 Nm).

28. Install and torque the upper pinch bolt for all models except Legend and 3.2TL to 32 ft. lbs. (44 Nm). For Legend and 3.2TL, torque the upper pinch bolt to 36 ft. lbs. (49 Nm).

29. With the vehicle on the ground, torque the spindle nut, then stake the nut.
Integra: 134 ft. lbs. (182 Nm)
Vigor and 2.5TL: 181 ft. lbs. (245 Nm)
Legend and 3.2TL: 242 ft. lbs. (355 Nm)

CV-Joint Boot

REPLACEMENT

NOTE: The outer CV-joint cannot be removed from the half-shaft, the boot is serviceable, if the joint requires replacement the shaft must be replaced as an assembly.

1. With the vehicle on the ground, raise the locking tab on the spindle nut and loosen it with a suitable socket.
2. Raise and safely support the vehicle and remove the spindle nut and front wheels.
3. Remove the halfshaft from the vehicle.
4. Remove the front and rear boot retaining bands, then separate the inboard joint from the halfshaft assembly.
5. Be sure to mark the roller grooves during disassembly to ensure proper positioning during reassembly.
6. Mark the spider gear and the halfshaft so they can be installed in their original positions.
7. Remove snapring, spider gear, then remove the stopper ring.
8. Be sure to mark the position on the shaft where the dynamic damper goes, to ensure it will be reinstalled in its original position. Remove the inboard CV-joint boot, dynamic damper, then the outboard CV-joint boot.

To install:
9. Wrap the spline with vinyl tape to prevent damage to the boots. Install the outboard boot, dynamic damper, and inboard boot, then remove the vinyl tape.
10. Install the stopper ring onto the halfshaft groove, then install the spider gear in its original position by aligning the marks.
11. Fit the snapring into the halfshaft groove.
12. Pack the outboard joint boot with CV-joint grease only. Do not use a substitute or mix types of grease.
13. Fit the rollers to the spider gear with their high shoulders facing outward. Reinstall the rollers in their original positions on the spider gear.

14. Pack the inboard joint boot with CV-joint grease.
15. Fit the inboard joint onto the halfshaft. Hold the halfshaft assembly so the inboard joint points up to prevent it from falling off.
16. With the boots installed, adjust the CV-joints in or out to place the inner boot ends by measuring the length of the halfshaft. The shafts (measured from the outer ends of the CV-joint housing) should be as follows:
Integra: 18.7–18.9 in. (475–480mm)
Vigor: Right: 21.2–21.4 in. (539–544mm). Left: 19.4–19.6 in. (494–499mm).
2.5TL: Right: 19.8–20.0 in. (503–508mm). Left: 19.5–19.7 in. (496–501mm).
Legend: Right: 19.7–19.9 in. (500–505mm). Left: 23.1–23.3 in. (586–591mm).
3.2TL: Right: 19.5–19.7 in. (495–500mm). Left: 22.9–23.1 in. (581–586mm).
17. Install the new boot bands on the boots and bend both sets of locking tabs. Lightly tap on the locking tabs to ensure a good fit.
18. Install the halfshaft into the vehicle.
19. Lower the vehicle.
20. With the vehicle on the ground, torque the spindle nut to the proper specification, then stake the nut.

STEERING

Air Bag

— CAUTION —
Some vehicles are equipped with an air bag system, also known as the Supplemental Restraint System (SRS). The system must be disabled before performing service on or around system components, steering column, instrument panel components, wiring and sensors. Failure to follow safety and disabling procedures could result in accidental air bag deployment, possible personal injury and unnecessary system repairs.

PRECAUTIONS

Several precautions must be observed when handling the inflator

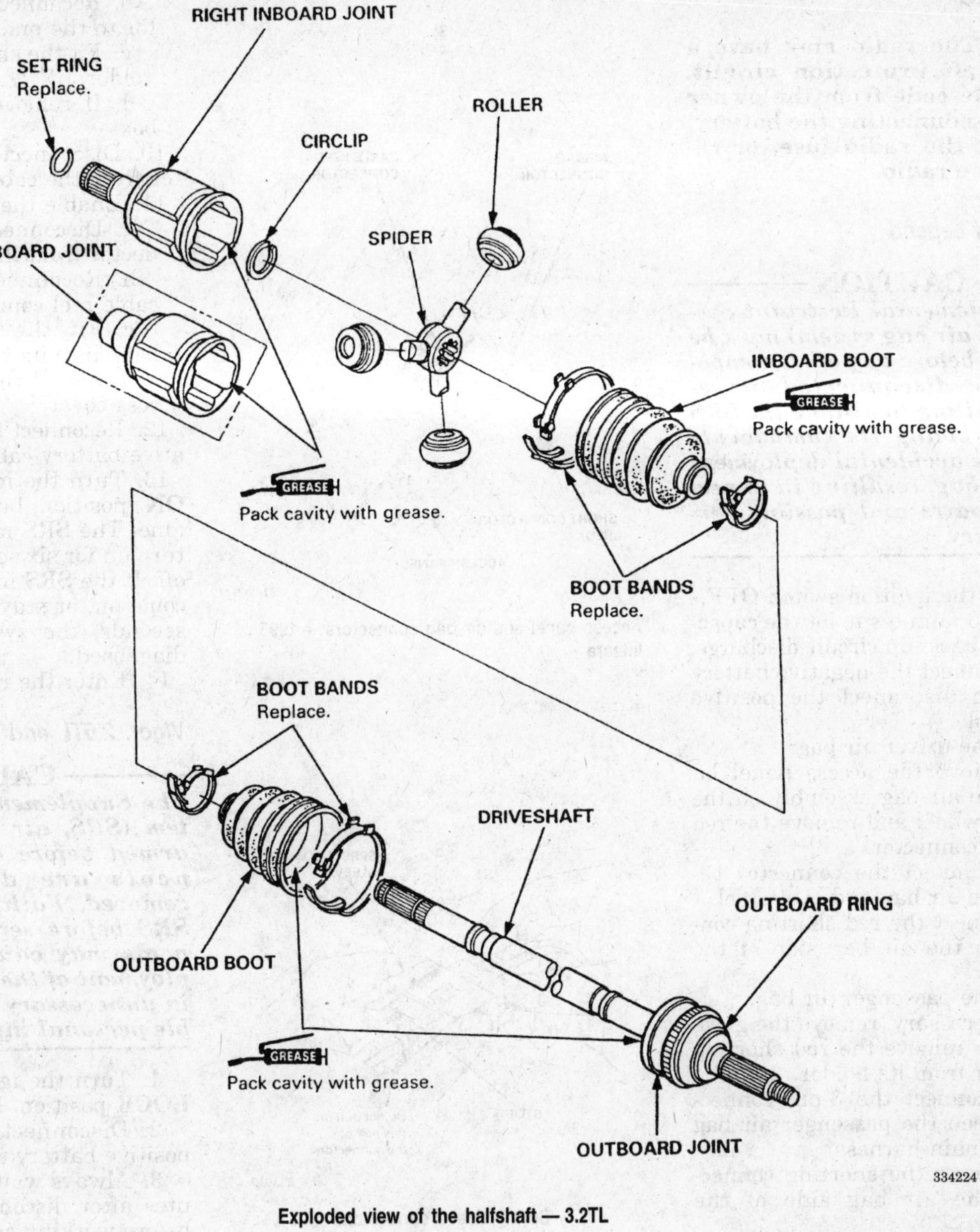

SET RING
Replace.

RIGHT INBOARD JOINT

CIRCLIP

ROLLER

SPIDER

LEFT INBOARD JOINT

INBOARD BOOT
GREASE
Pack cavity with grease.

GREASE
Pack cavity with grease.

BOOT BANDS
Replace.

BOOT BANDS
Replace.

DRIVESHAFT

OUTBOARD RING

OUTBOARD BOOT

GREASE
Pack cavity with grease.

OUTBOARD JOINT

334224

Exploded view of the halfshaft — 3.2TL

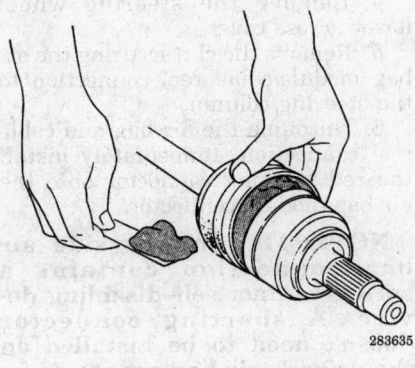

283635

Packing the boot with grease

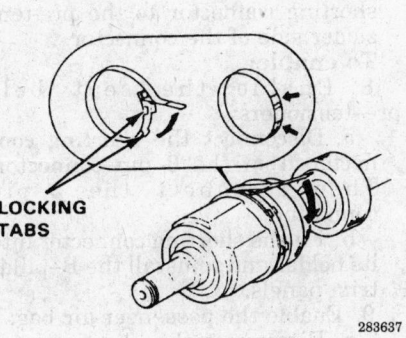

LOCKING TABS

283637

Retaining band installation

module to avoid accidental deployment and possible personal injury.

• Never carry the inflator module by the wires or connector on the underside of the module.

• When carrying a live inflator module, hold securely with both hands, and ensure that the bag and trim cover are pointed away.

• Place the inflator module on a bench or other surface with the bag and trim cover facing up.

• With the inflator module on the bench, never place anything on or close to the module which may be thrown in the event of an accidental deployment.

DISARMING

NOTE: The radio may have a coded theft protection circuit. Obtain the code from the owner before disconnecting the battery, removing the radio fuse, or removing the radio.

Integra and Legend

——————— **CAUTION** ———————

The Supplemental Restraint System (SRS, air bag system) must be disarmed before any of its components are disconnected or removed. Failing to disable the SRS before servicing its components may cause accidental deployment of the air bag, resulting in unnecessary repairs and possible personal injury.

1. Turn the ignition switch **OFF**.
2. Wait 3 minutes to let the capacitor in the back–up circuit discharge.
3. Disconnect the negative battery cable, then disconnect the positive battery cable.
4. For the driver air bag:
   a. Remove the access panel lid below the air bag assembly on the steering wheel and remove the red shorting connector.
   b. Disconnect the connector between the air bag and cable reel.
   c. Connect the red shorting connector to the air bag side of the connector.
5. For the passenger air bag:
   a. If necessary, remove the glove box, then remove the red shorting connector from its holder.
   b. Disconnect the 3-pin connector between the passenger air bag and the main harness.
   c. Connect the shorting connector to the air bag side of the connector.
6. After installing the shorting connectors on the air bags, connect shorting connector 07–MAZ–SP0020A, or equivalent, on the cable reel connector and another on the main harness connector of the passenger air bag to prevent static electricity from setting off the seat belt pre-tensioners before you disconnect them.
7. For the seat belt pre-tensioners, disarm them one side at a time:
   a. Remove the B-pillar trim panels.
   b. Remove the red shorting connector from the short connector holder.
   c. Disconnect the pre-tensioner 3-pin connector, then install the

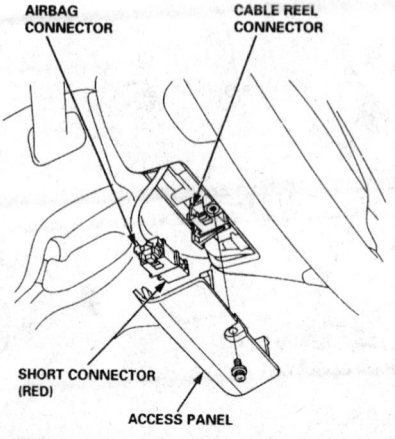

Access panel and air bag connectors — 1995 Integra

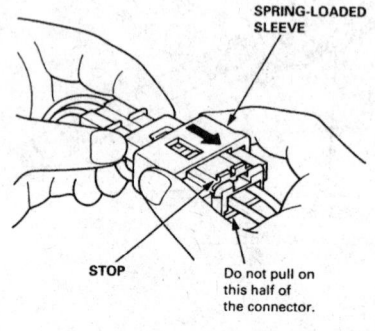

Spring-sleeve connectors

shorting connector to the pre-tensioner side of the connector.
To enable:
8. Enable the seat belt pre–tensioners:
   a. Disconnect the shorting connector from the 3–pin connector. Then reconnect the 3–pin connector.
   b. Fit the shorting connector into its holder and reinstall the B–pillar trim panels.
9. Enable the passenger air bag:
   a. Disconnect the shorting connectors from the air bag and main harness connectors.

b. Reconnect the air bag connector to the main harness.
   c. Fit the short connector into its holder.
   d. If removed, install the glove box.
10. Disconnect the shorting connector from the cable reel connection.
11. Enable the driver's air bag:
   a. Disconnect the shorting connector from the air bag connector.
   b. Reconnect the air bag and cable reel connectors.
   c. Fit the shorting connector back into its holder.
   d. Install the steering wheel access cover.
12. Reconnect the positive and negative battery cables.
13. Turn the ignition switch to the **ON** position, but don't start the engine. The SRS indicator light should turn on for six seconds and then turn off. If the SRS indicator light doesn't come on, or stays on longer than six seconds, the system fault must be diagnosed.
14. Enter the radio security code.

Vigor, 2.5TL and 3.2TL

——————— **CAUTION** ———————

The Supplemental Restraint System (SRS, air bag) must be disarmed before any of its components are disconnected or removed. Failing to disable the SRS before servicing its components may cause accidental deployment of the air bag, resulting in unnecessary repairs and possible personal injury.

1. Turn the ignition switch to the **LOCK** position. Remove the key.
2. Disconnect the negative and positive battery cables.
3. Always wait at least three minutes after disconnecting the battery before working around the air bags.
4. Remove the steering wheel lower access cover.
5. Remove the clip securing the air bag module/cable reel connection to the steering column.
6. Uncouple the air bag and cable reel connection. Immediately install the red shorting connector onto the air bag module connector.

NOTE: The driver's side air bag connection contains a spring–contact self–disabling device. A shorting connector doesn't need to be installed on the driver's air bag connector.

7. After servicing has been completed, couple the air bag and cable reel connectors.

8. Install the clip securing the air bag/cable reel connection to the steering column.

9. Install the access cover.

10. Reconnect the positive and negative battery cables.

11. Turn the ignition switch to the **ON** position, but don't start the engine. The SRS indicator light should turn on for six seconds and then turn off. If the SRS indicator light doesn't come on, or stays on longer than six seconds, the system fault must be diagnosed.

12. Enter the radio security code.

Steering Wheel

REMOVAL AND INSTALLATION

1993 Integra

1. Disconnect the negative battery cable.

2. Remove the steering wheel center pad.

3. Remove the steering wheel shaft nut.

4. Remove the steering wheel by rocking it slightly from side to side and pull steadily with both hands.

5. Remove the right and left lower instrument panel covers. Remove the front console.

6. Remove the driver's side knee bolster from the steering hanger. Remove the steering joint cover.

7. Remove the steering joint bolts and move the joint toward the steering column. Remove the upper and lower steering column covers.

8. Disconnect each wire coupler from the combination switch. Remove the turn signal canceling sleeve and combination switch assembly.

9. Disconnect each wire coupler from the fuse box under the left side of the dash.

10. Remove the steering column holder. Remove the attaching nuts and bolts, then remove the steering column assembly.

To install:

11. Install the column in the vehicle. Torque the steering column support bracket bolts to 9 ft. lbs. (13 Nm) and the nuts to 16 ft. lbs. (22 Nm). Torque the lower bolt on the steering joint to 16 ft. lbs. (22 Nm). If equipped with a tilt steering column, use the following procedures.

12. Connect each wire coupler to the fuse box under the left side of the dash.

13. Install the combination switch assembly and the turn signal canceling sleeve.

14. Connect the wires to the combination switch.

15. Loosely install the steering joint on the steering gearbox pinion. Make sure that the lower bolt is securely in the groove in the steering gearbox pinion.

16. Install the steering joint cover with the clamps and the clip.

17. Install the upper and lower column covers.

18. Install the driver's knee bolster and the front console.

19. Install the right and left lower covers.

20. Install the steering wheel. Be sure the steering wheel shaft engages the canceling sleeve.

21. Attach the cruise control set/resume switches connector to the steering wheel clip (if equipped).

22. Connect the horn connector.

23. Make sure the steering wheel nut is torqued to 36 ft. lbs. (50 Nm) and install the center pad.

24. Reconnect the battery. Check the operation of horn buttons and cruise control switches.

1994–97 Integra

1. Disconnect the negative and positive cables from the battery.

2. Disable the SRS air bag system.

3. Remove lid B (located on the left rear side of the steering wheel) and the cruise control set/resume switch cover.

4. Carefully remove the air bag assembly from the steering wheel.

───── **CAUTION** ─────
Store a removed air bag assembly with the pad surface facing up. If the air bag is improperly stored face down, accidental deployment could propel the unit with enough force to cause serious injury.

5. Disconnect the electrical connectors from the horn and cruise control switches.

6. Remove the steering wheel nut.

7. Remove the steering wheel by rocking it slightly from side to side, as you pull steady with both hands.

To install:

8. Before installing the steering wheel the cable reel must be centered. Center the cable reel as follows:

    a. Rotate the cable reel clockwise until it stops.

    b. Then rotate the cable reel counterclockwise (approximately two turns) until, the yellow gear tooth lines up with the mark on the cover and the arrow on the cable reel label points straight up.

9. Feed the cable reel electrical wiring throughout the steering wheel. Align the notches in the steering wheel shaft with the canceling sleeve, and align the pins on the cable reel with the holes in the steering wheel. Install the steering wheel to the steering shaft making sure that the cable reel and canceling sleeve engage.

10. Install the steering wheel nut and torque the nut to 36 ft. lbs. (49 Nm).

11. Attach the cruise control switches electrical connector to the steering wheel clip.

12. Connect the horn electrical connector.

13. Carefully install the air bag assembly, use new TORX® bolts. Torque the bolts to 7 ft. lbs. (10 Nm).

───── **WARNING** ─────
Carefully inspect the air bag assembly before installing. Do not install an air bag assembly that shows signs of being dropped or improperly handled, such as dents, cracks or deformation. Always keep the shorting connector on the air bag connector when the harness is disconnected. Do not disassemble or tamper with the air bag assembly.

───── **CAUTION** ─────
Confirm that the air bag assembly is securely attached to the steering wheel; otherwise severe personal injury could result during air bag deployment.

14. Install the cruise control switch cover and lid B to the steering wheel.

15. Disconnect the short connector from the air bag connector.

16. Connect the air bag electrical connector to the cable reel electrical connector.

17. Attach the short connector to the access panel, and install the access panel to the steering wheel.

18. Connect the positive cable to the battery and then connect the negative cable to the battery. Enter the radio security code.

19. After installing the air bag assembly, confirm proper system operation.

• Turn the ignition to the on position, without starting the vehicle: the SRS indicator light should come on for about 6 seconds and then go off.

• Confirm the operation of the horn buttons.

• Confirm the operation of the cruise control set/resume switches.

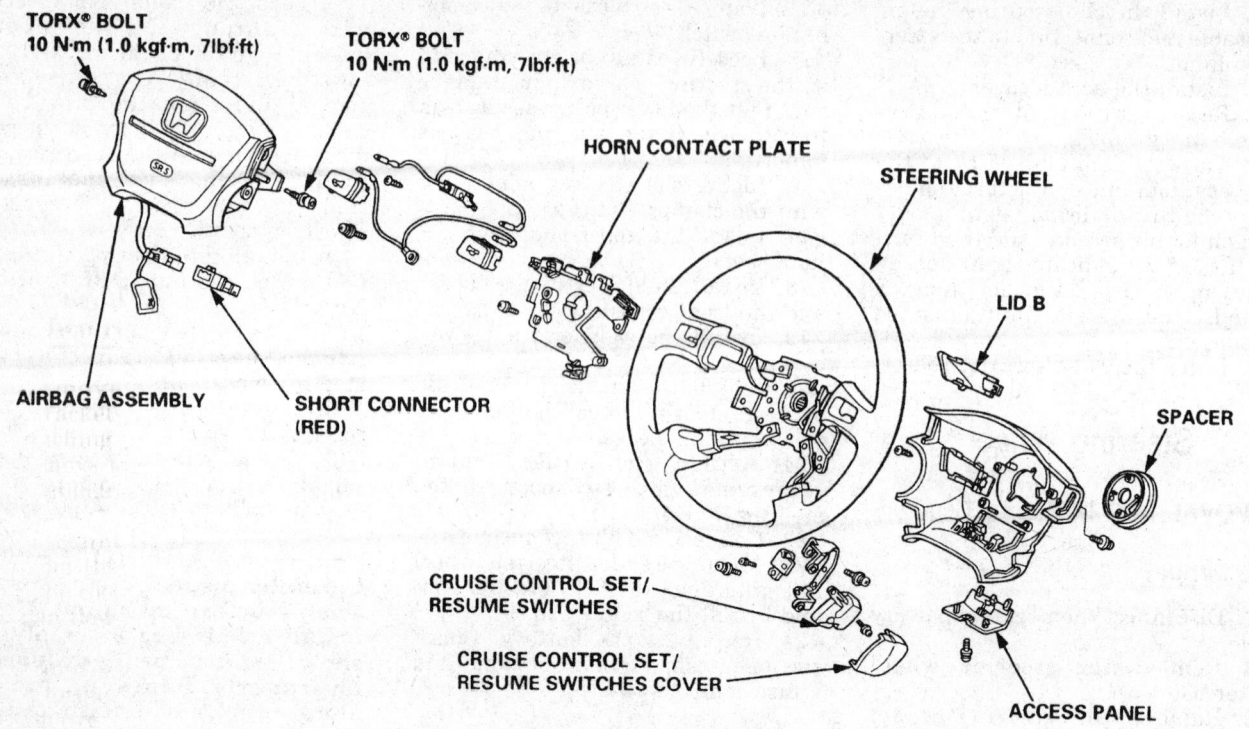

TORX® BOLT
10 N·m (1.0 kgf·m, 7lbf·ft)

TORX® BOLT
10 N·m (1.0 kgf·m, 7lbf·ft)

HORN CONTACT PLATE

STEERING WHEEL

LID B

SPACER

AIRBAG ASSEMBLY

SHORT CONNECTOR (RED)

CRUISE CONTROL SET/
RESUME SWITCHES

CRUISE CONTROL SET/
RESUME SWITCHES COVER

ACCESS PANEL

216214

Exploded view of the steering wheel components — 1995 Integra

• Turn the steering wheel counterclockwise and make sure the yellow gear tooth still lines up with the alignment mark.

1993–94 Vigor and 1993–94 Legend

1. Disconnect both the negative and positive battery cables from the battery.
2. The air bag system and pre-tensioner system must be disabled before performing any service on or around the steering column, air bag, or instrument panel components or wiring.
3. Remove the lower maintenance lid below the air bag and then remove the short connector.
4. Disconnect the connector between the air bag and the cable reel.
5. Connect the short connector to the air bag side of the connector.
6. Remove the left side maintenance lid and the cruise control/set resume switch cover.
7. Insert a T30 Torx® bit and remove the Torx® bolts. Remove the air bag assembly.

— CAUTION —
The air bag system (SRS) must be disarmed before removing the steering wheel. Failure to do so may cause accidental deploy-ment, property damage or personal injury. Always carry an air bag assembly with the bag and cover facing away from your body. Store the assembly with the bag and cover facing up; never place the assembly face down on the floor or workbench.

8. Remove the steering wheel retaining nut.
9. Remove the steering wheel by rocking it slightly from side to side as you pull steadily with both hands.
10. Remove the upper and lower steering column covers, the driver's side knee bolster and the yoke joint cover at the base of the column.
11. Unplug the wiring connectors on the underside of the column. Remove the screws to remove the cable reel and combination switch.
12. Remove the rear steering column holder. Remove the steering column mounting nuts.
13. Remove the steering yoke joint bolts and disconnect the yoke joint, then remove the column assembly.
   To install:
14. Install the column and torque the bolts to 16 ft. lbs. (22 Nm) and the nuts to 12 ft. lbs. (16 Nm).
15. Install the upper bolt in the steering joint and torque the bolt to 16 ft. lbs. (22 Nm)

16. Install the cable reel and combination switch and connect all the wiring.
17. Center the cable reel, by rotating the cable reel clockwise until it stops. Then rotate it counterclockwise (approximately 2 turns) until the yellow gear tooth lines up with the mark on the cover. The arrow on the cable reel label points straight up.
18. Before installing the steering wheel, the front wheels should be aligned straight forward. Torque the steering wheel nut to 36 ft. lbs. (50 Nm).
19. After reassembly confirm that the wheels are still straight ahead and that the steering wheel spoke angle is correct. If minor spoke angle adjustment is necessary, do so only by adjustment of the tie rods, not by removing and repositioning the steering wheel.
20. Install the air bag assembly with new Torx® bolts.
21. Disconnect the short connector from the air bag and connect the air bag and cable reel connectors.
22. Reconnect the battery cables.
23. Make sure that all wiring is reconnected and all components reassembled.
24. Turn the ignition switch **ON**; the instrument panel SRS light

should turn **ON** for about 8 seconds and turn **OFF**.

2.5TL and 3.2TL

— **CAUTION** —

The air bag system (SRS) must be disarmed before removing the steering wheel. Failure to disarm the system may cause accidental deployment of the air bag, resulting in unnecessary repairs and possible personal injury.

1. Disconnect the negative and positive battery cables.
2. Remove the steering wheel lower access covers.
3. Disconnect the air bag–to–cable reel connector. First, pull the connector out of its holder. Next, pull the spring loaded sleeve toward the stop. Separate the connectors by pulling on the sleeve, not the wires.

NOTE: The driver's side air bag is shorted automatically when disconnected.

4. Remove the steering wheel side covers.
5. Remove the Torx® bolts; then, remove the air bag module.

— **CAUTION** —

Always carry and store a live air bag module with the trim cover pointed away from your body or the work surface. Following this procedure will reduce the chance of personal injury if the air bag is accidentally deployed.

6. Disconnect the horn and cruise control switches.
7. Remove the steering wheel bolt.
8. Loosely reinstall the steering wheel bolt. Install a steering wheel puller onto the steering wheel.

— **WARNING** —

To avoid damage to the cable reel assembly, do not thread the puller bolts more than five threads into the hub. Install a set of nuts five threads up on the puller bolts to act as stops.

9. Remove the steering wheel and bolt from the column.

To install:

NOTE: Use new Torx® bolts when installing the air bag module. These are available at a Honda or Acura dealer.

10. Make sure that the front wheels are aligned straight ahead before installing the steering wheel.
11. Verify the position of the cable reel before installing the steering

wheel. The hole in the white gear must line up with the window in the steering column cover.

Tie Rod Ends

REMOVAL AND INSTALLATION

1. Raise and safely support the vehicle. Remove the front wheels.
2. Loosen and remove the tie rod end castle nut.
3. Using a press-type ball joint removal tool, separate the tie rod end from the steering knuckle. Take care to not damage the threads on the joint.
4. While supporting the power steering rod, remove the tie rod end. Be sure to count revolutions required to remove the tie rod end.

To install:

5. Install the tie rod to the steering rod, but do not tighten the locknut at this time.
6. Install the tie rod end to the steering knuckle.

NOTE: Use a new cotter pin when installing the tie rod castle nut. Tighten the castle nut to the lower torque specification, and then tighten it only enough to slide the cotter pin through the hole.

7. Torque the castle nut to the following specifications:
Integra: 29–35 ft. lbs. (39–47 Nm)
Legend: 36–43 ft. lbs. (50–60 Nm)
Vigor: 36–43 ft. lbs. (50–60 Nm)
2.5TL: 36–43 ft. lbs. (49–59 Nm)
3.2TL: 36–43 ft. lbs. (49–59 Nm)
8. Install the wheels and check the wheel alignment.
9. Torque the tie rod locknut to the following specifications:
1993–94 Integra: 40 ft. lbs. (55 Nm)
Legend: 33 ft. lbs. (45 Nm)
Vigor: 33 ft. lbs. (45 Nm)
2.5TL: 33 ft. lbs. (44 Nm)
3.2TL: 33 ft. lbs. (44 Nm)

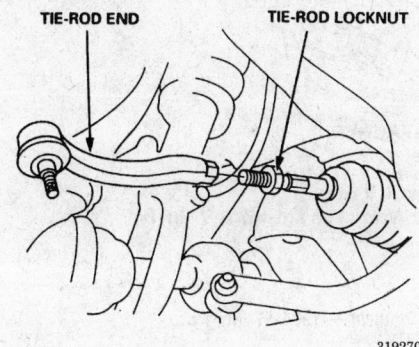

319270

Tie rod end

10. Road test the vehicle.

Power Rack and Pinion

REMOVAL AND INSTALLATION

NOTE: The radio may have a coded theft protection circuit. Obtain the code from the owner before disconnecting the battery, removing the radio fuse, or removing the radio.

Integra

1. Lift the power steering reservoir and disconnect the return hose that goes to the oil cooler.
2. Connect a hose of suitable diameter to the disconnected return hose and place the end of the hose in a container to collect the power steering fluid.

— **CAUTION** —

Take care not to spill the fluid on the body and engine assembly. Wipe off any spilled fluid at once.

3. Start the engine, let it run at idle, and turn the steering wheel from lock to lock several times. When fluid stops running out of the hose, shut off the engine. Connect the return hose to the reservoir. Properly dispose of the used fluid.
4. Disconnect the negative battery cable.
5. Using cleaning solvent and a brush, wash any oil and dirt off the valve body unit and its lines, and the end of the rack.
6. Remove the steering joint cover.
7. Remove the ignition key and lock the steering wheel in the straight forward position.
8. Remove the steering joint lower bolt and pull the joint toward the column.
9. Raise and safely support the vehicle.
10. Remove the front wheels.
11. Remove the cotter pins and unscrew the tie rod end ball joint nuts halfway.
12. Break the tie rod ball joints loose from the steering knuckles, by using a tie rod end removal tool (part # 07MAC–SL00200) or equivalent.
13. Remove the nuts and lift the tie rod ends out of the steering knuckles.
14. If equipped with a manual transaxle, disconnect the extension rod and the shift rod by performing the following:
  a. If equipped with a VTEC engine, remove the heat shield.
  b. Disconnect the shift extension rod from the transaxle case.

c. Slide the boot on the shift rod back to expose the clip and spring pin. Remove the clip from the shift rod.

d. Drive out the spring pin with a punch and disconnect the shift control rod. Note that on reassembly, install the clip back into place after driving the spring pin in.

15. If equipped with an automatic transaxle, disconnect the shift cable from the transaxle by performing the following:

a. Remove the shift cable holder from the vehicle.

b. Remove the shift cable cover from the transaxle.

c. Remove the bolt attaching the control lever to the control shaft, then remove the control lever from the control shaft. Discard the lock washer.

d. Position the cable out of the way without bending the cable.

16. Disconnect the oxygen sensor electrical connector.

17. Remove the catalytic converter front attaching nuts and bolts, then remove the rear attaching nuts. Remove the catalytic converter from the vehicle. Discard the locknuts and old exhaust gaskets.

18. Remove the return clamp from the left side of the rear beam, and

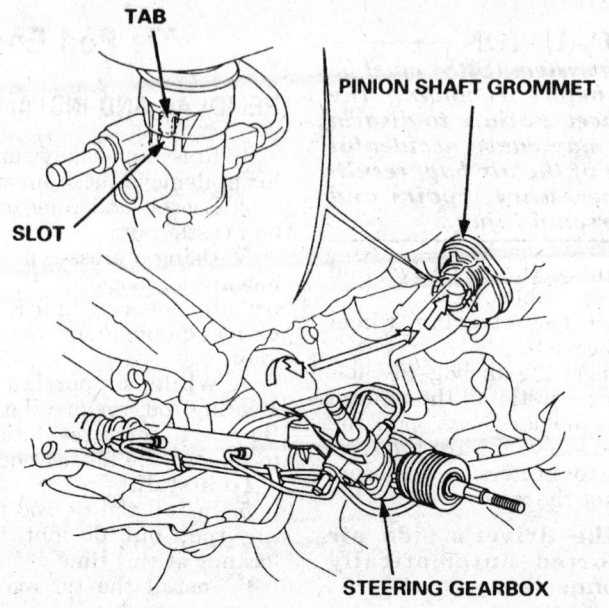

Installing the rack and pinion — 1994-97 Integra

8 x 1.25 mm
22 N·m (2.2 kgf·m, 16 lbf·ft)

SHIFT CABLE HOLDER
SHIFT CABLE

SHIFT CABLE COVER

CONTROL SHAFT
CONTROL LEVER
LOCK WASHER
Replace.

6 x 1.0 mm
14 N·m (1.4 kgf·m, 10 lbf·ft)

6 x 1.0 mm
12 N·m (1.2 kgf·m, 8.7 lbf·ft)

319236

Automatic transaxle shift cable attachment — 1994-97 Integra

move the return pipe above the rack and pinion.

19. Remove the rear beam brace.

20. Remove the left tie rod end, then slide the inner tie rod all of the way to the right.

21. Disconnect the two lines from the valve body unit on the rack and pinion. Place the pipe and hose to the rear side of the rack and pinion, so they do not hinder in the removal of the rack and pinion.

CAUTION
After disconnecting the hose and pipe, plug or cap the hose and pipe to prevent foreign materials from entering the valve body unit.

NOTE: Do not loosen the cylinder pipes between the valve body unit and the cylinder.

22. Remove the gearbox mounting bolts.

23. Pull the rack and pinion all the way down to clear the pinion shaft from the bulkhead, and remove the pinion shaft grommet.

24. Holding the rack and pinion assembly, slide the rack all of the way to the right, then place the left rack end below the rear beam.

25. Move the rack and pinion assembly to the left, and tilt the left

side down to remove it from the vehicle.

To install:

26. Gently push the left inner tie rod end into the rack and pinion assembly until it reaches the end of its travel.

27. Place the right side of the rack and pinion over the rear beam and move the assembly completely to the right. Lift the left side of the rack and pinion over the rear beam.

28. Install the pinion shaft grommet, then slide the rack and pinion to the left and up into position. Make sure the pinion shaft fits the pinion shaft grommet properly by aligning the tab on the grommet with the slot in the valve body.

29. With the rack and pinion properly positioned install the mounting bolts. Torque the left side mounting bolts to 28 ft. lbs. (38 Nm), and the right side mounting bolts to 43 ft. lbs. (58 Nm).

NOTE: After installing the rack and pinion, check the air hose connections for interference with adjacent parts.

30. Center the steering rack within its stroke.

31. If applicable, make sure that the cable reel in the steering column is cantered by performing the following:

 a. Turn the steering wheel left approximately 150°, to check the cable reel position with the indicator.

 b. If the cable reel is centered, the yellow gear tooth lines up with the alignment mark on the cover.

 c. Return the steering wheel right approximately 150°, to position the steering wheel in the straight ahead position.

——————— CAUTION ———————
Do not connect the steering joint to the pinion without the cable reel being centered. Damage to the SRS system components and personal injury may occur.

32. Slip the lower end of the steering joint onto the pinion shaft (line up the bolt hole with the groove around the shaft), and tighten the lower bolt. Torque the bolt to 16 ft. lbs. (22 Nm).

NOTE: Be sure that the lower steering joint is securely in the groove in the steering pinion. If the steering wheel and rack and pinion are not centered, reposition the serrations at the lower end of the steering joint.

33. Install the steering joint cover with the clamps and clips.

34. Connect the feed pipe to the rack and pinion valve body unit, torque the fitting to 27 ft. lbs. Connect the return hose to the rack and pinion valve body unit and tighten the hose clamp.

35. Install the rear beam brace rod and the return pipe clamp on the rear beam. Torque the rear beam brace rod bolts to 28 ft. lbs. (38 Nm).

36. Install the catalytic converter with new gaskets and new locknuts. Torque the rear nuts to 25 ft. lbs. (33 Nm). Torque the front nuts and bolts to 16 ft. lbs. (22 Nm). Connect the oxygen sensor electrical connector.

37. If equipped with a manual transaxle connect the shift linkage by performing the following:

 a. Install the shifter rod and install the spring pin to attach the shifter rod to the transaxle.

 b. Install the clip over the spring pin, then cover the clip and spring pin with the boot.

 c. Install the extension rod and torque the bolt to 16 ft. lbs. (22 Nm).

 d. Install the heat shield, if equipped and torque the bolts to 7.2 ft. lbs. (9.8 Nm).

38. If equipped with an automatic transaxle connect the shift cable by performing the following:

 a. Install the shift control lever to the control shaft, use a new lock washer to secure the bolt. Torque the bolt to 10 ft. lbs. (14 Nm).

 b. Install the shift cable cover and torque the bolts to 16 ft. lbs. (22 Nm).

 c. Attach the shift control cable holder, torque the bolt to 8.7 ft. lbs. (12 Nm).

39. Thread the right and left tie rod ends on to the rack and pinion an equal number of turns.

40. Connect the tie rods to the steering knuckles and install the castle nuts. Torque the castle nuts to 29–35 ft. lbs. (39–47 Nm), tighten the nuts enough to install new cotter pins. Do not loosen the castle nuts to install the cotter pins.

41. Install the front wheels.

42. Fill the power steering reservoir to the upper line.

43. Connect the negative battery cable and enter the radio security code.

44. Start the engine and run at a fast idle, then turn the steering lock to lock several times to bleed the air from the system.

45. Check the power steering fluid again and add, if necessary. Check the system for leaks.

46. Check and adjust the front end alignment.

Vigor, 2.5TL, Legend and 3.2TL

1. Disconnect the fluid return hose from the rack and put the end in a container. Start the engine and turn the steering wheel lock–to–lock several times. When fluid stops coming out, stop the engine.

2. Disconnect the negative battery cable, then disconnect the positive battery cable.

3. Raise and safely support the vehicle and remove the front wheels.

4. Remove the cotter pins and disconnect the tie rod end joints using a separator tool. Be careful to not damage the threads on the joints.

5. Loosen the steering joint bolt but do not remove it yet.

6. Remove the splash guard. The 2 long bolts also hold the rack in place, and the rack will now be partially hanging on the steering joint.

7. Carefully clean all the hydraulic fitting connections with solvent and a brush and blow them dry.

8. For Vigor and 2.5TL, disconnect the 8mm sensor line from the valve body by removing the 14mm flare nut.

9. Disconnect the hydraulic fittings and hoses.

10. Remove the hydraulic line mounting clamps from the rack.

11. Place a jack under the rack and remove the steering joint bolt. Remove the rack assembly.

To install:

NOTE: Several bolts thread into aluminum. When replacing fasteners, be sure to use bolts that have a Dacro® coating specifically designed for such applications. Using normal steel bolts could cause corrosion and loosening of the bolt.

——————— WARNING ———————
Use ONLY genuine Honda power steering fluid. Using ANY other type or brand of fluid will damage the power steering system.

12. For Vigor and 2.5TL, loosely install the right mounting bracket to hold the rack in the vehicle. The arrow stamped on the bracket should face the front of the vehicle. Install the 8mm sensor line in its clip to secure it to the rack cylinder tube.

13. For Legend and 3.2TL, fit the pinion into the steering joint and install the right side mounting rubber

and bracket. Do not tighten the bolts yet.

14. Loosely connect the hydraulic lines. Install the hydraulic line cushions and clamps, then tighten the line connections.

15. Connect the four lines to the control unit. Torque the bolts to 8 ft. lbs. (11 Nm).

16. Connect the 6mm return line to 9 ft. lbs. (13 Nm) and the 10mm line to 21 ft. lbs. (29 Nm)

17. Install the hoses and the hose clamps.

18. Make sure that the air bag system cable reel is centered. Turn the steering wheel left until the yellow gear tooth is visible through the lower left inspection hole. The yellow gear tooth should align with the mark on the inspection cover. Do not bolt the steering joint until the marks match.

19. Install the steering joint bolts, make sure the joint does not bind when turned, then torque the bolts to 16 ft. lbs. (22 Nm).

20. Torque the right side mount bolts to 28 ft. lbs. (39 Nm).

21. Install the splash guard and torque the short bolts to 28 ft. lbs. (39 Nm). Torque the long bolts to 43 ft. lbs. (60 Nm).

22. For Vigor and 2.5TL, connect the feed line, the inlet hose, and the outlet hose to the valve body unit. Torque the bolts to 8 ft. lbs. (11 Nm). Install the 8mm sensor line to the valve body unit. Torque the line to 18 ft. lbs. (25 Nm).

23. Connect the tie rod ends and torque the nuts to 36–43 ft. lbs. (50–60 Nm), then tighten them enough to install a new cotter pin.

24. When installation is complete, refill the hydraulic reservoir with new steering fluid, reconnect the battery, start the engine, and turn the steering wheel lock–to–lock several times to bleed the system. Check fluid level again.

25. Check the system for leaks.

26. Check and adjust the front wheel alignment.

Power Steering Pump

BLEEDING

Integra

1. Fill the power steering reservoir to the upper line.

2. Start the engine and run at a fast idle, then turn the steering lock to lock several times to bleed the air from the system.

3. Check the power steering fluid again and add, if necessary. Do not fill the reservoir beyond the upper line.

Vigor, 2.5TL, Legend and 3.2TL

— WARNING —
Use only genuine Honda power steering fluid. Any other type or brand of fluid will damage the power steering pump.

1. Raise the front of the vehicle and support it with safety stands. Block the rear wheels.

2. Lift the power steering reservoir off of its mount. Disconnect the return hose from the steering rack at the reservoir. Immediately plug the reservoir inlet to prevent fluid loss and contamination. Don't disconnect the hose that connects the pump to the reservoir.

3. Insert a length of rubber tubing into the return hose and route the tubing into a drain container.

4. With the engine running at idle, turn the steering wheel lock–to–lock several times until fluid stops running out of the hose. Immediately shut off the engine.

5. After servicing, reconnect the reservoir return line. Fill the reservoir to the upper line with genuine Honda power steering fluid.

6. Run the engine at idle and turn the steering wheel lock–to–lock several times to bleed air from the system and fill the rack valve body.

7. Recheck the fluid level and add more if necessary. Don't overfill the reservoir.

8. Check the power steering system for leaks.

9. Lower the vehicle.

REMOVAL AND INSTALLATION

NOTE: The radio may have a coded theft protection circuit. Obtain the code from the owner before disconnecting the battery, removing the radio fuse, or removing the radio.

Vigor, 2.5TL, Legend and 3.2TL

1. Disconnect the fluid return hose from the rack and put the end in a container. Start the engine and turn the steering wheel lock–to–lock several times. When fluid stops coming out, stop the engine.

2. Disconnect the negative battery cable.

3. Remove the air cleaner cover and duct.

4. Disconnect the hydraulic lines from the pump and plug them.

5. Loosen the adjustment bolt and remove the belt.

6. Remove the mounting bolt and nut and remove the pump.

To install:

— WARNING —
Use ONLY genuine Honda power steering fluid. Using ANY other type or brand of fluid will damage the power steering system.

7. Check pump preload by mounting the pump into a bench vise and using a torque wrench. Preload should be 6 ft. lbs. (8 Nm).

8. Install the pump. On Integra, torque the mounting bolts to 17 ft. lbs. On Legend, Vigor and 3.2TL, torque the mounting bolts to 33 ft. lbs. (45 Nm), on 2.5TL –36 ft. lbs. (39 Nm). Torque the mounting nut on all vehicles to 16 ft. lbs. (22 Nm).

9. Install a new O–ring on the outlet hose and connect the hoses to the pump. Torque the outlet hose bolts to 8 ft. lbs. (11 Nm).

10. Install and adjust the belt and refill the reservoir.

11. Connect the negative battery cable.

12. Run the engine and turn the steering wheel lock–to–lock several times to bleed the air from the power steering system. Check the fluid level and add as required.

BRAKES

Anti-Lock Brake System Service

PRECAUTIONS

• Certain components within the Anti-Lock Brake System (ABS) are not intended to be serviced or repaired individually. Only those components with removal and installation procedures should be serviced.

• Do not use rubber hoses or other parts not specifically specified for and ABS system. When using repair kits, replace all parts included in the kit. Partial or incorrect repair may lead to functional problems and require the replacement of components.

• Lubricate rubber parts with clean, fresh brake fluid to ease assembly. Do not use lubricated shop air to clean parts; damage to rubber components may result.

6. Install the red cap.

Vigor and Legend

---- **CAUTION** ----

The hydraulic accumulator contains brake fluid and nitrogen gas at extremely high pressures. Certain portions of the hydraulic system also contain brake fluid at high pressure. The system must be depressurized before disconnecting any hoses, lines, or fittings; or personal injury may result.

1. Disconnect the negative battery cable.
2. Clean any dirt or grime off of the modulator reservoir and bleeder screw.
3. Remove the red service cap from the bleeder screw on the body of the ABS modulator.
4. Fit a bleeder T-wrench, tool No. 07HAA–SG00101 or equivalent, squarely onto the bleeder screw.
5. Slowly turn the T-wrench 90° to release the high-pressure fluid from the modulator. To completely drain the brake fluid from the modulator, continue turning the T-wrench until it has been rotated one complete turn.
6. If necessary, a suction pump or syringe may be used to suck the brake fluid from the modulator reservoir.
7. After servicing, use the T-wrench to tighten the bleeder screw to 6.5 ft. lbs. (9 Nm). Reinstall the red service cap.
8. Refill and bleed the brake system. Use only clean DOT 3 or 4 fluid. Don't mix brands or types of brake fluid.

2.5TL and 3.2TL

---- **CAUTION** ----

The hydraulic accumulator contains brake fluid and nitrogen gas at extremely high pressures. Certain portions of the hydraulic system also contain brake fluid at high pressure. Do not loosen the relief plug on the accumulator. The system must be depressurized before disconnecting any hoses, lines, or fittings; or personal injury may result.

Relieving brake system pressure

1. Remove the cap from the modulator maintenance bleeder.
2. Connect a flare or box-end wrench to the maintenance bleeder.

PUMP MOUNTING NUT
22 N·m
(2.2 kg-m, 16 lb-ft)

PUMP MOUNTING BOLT
45 N·m
(4.5 kg-m, 33 lb-ft)

P/S PUMP

ADJUSTING BOLT

323293

Power steering pump mounting

- Use only specified brake fluid from an unopened container.
- If any hydraulic component or line is removed or replaced, it may be necessary to bleed the entire system.
- A clean repair area is essential. Always clean the reservoir and cap thoroughly before removing the cap. The slightest amount of dirt in the fluid may plug an orifice and impair the system function. Perform repairs after components have been thoroughly cleaned; use only denatured alcohol to clean components. Do not allow ABS components to come into contact with any substance containing mineral oil; this includes used shop rags.
- The Anti-Lock control unit is a microprocessor similar to other computer units in the vehicle. Ensure that the ignition switch is **OFF** before removing or installing controller harnesses. Avoid static electricity discharge at or near the controller.
- If any arc welding is to be done on the vehicle, the control unit should be unplugged before welding operations begin.

DEPRESSURIZING

Integra

---- **CAUTION** ----

The hydraulic accumulator and other components contain brake fluid under extremely high pressure. To avoid injury, the pressure MUST be properly relieved before loosening any lines, fittings, or components.

1. Drain the fluid from the modulator reservoir by sucking the fluid out with a syringe or disconnect the pump hose from the pump joint.
2. Drain the brake fluid from the master cylinder by loosening the bleed screw and and pumping the brake pedal.
3. Remove the red cap from the bleeder on top of the power unit.
4. Install the bleeder t-wrench and turn it out slowly 90° to collect the high pressure fluid into the reservoir. Turn the bleeder one complete turn to drain the brake fluid thoroughly.
5. For 1993 models, tighten the bleeder to 4 ft. lbs. (5.5 Nm). For 1994–97 torque the maintenance bleeder to 8 ft. lbs. (11 Nm).

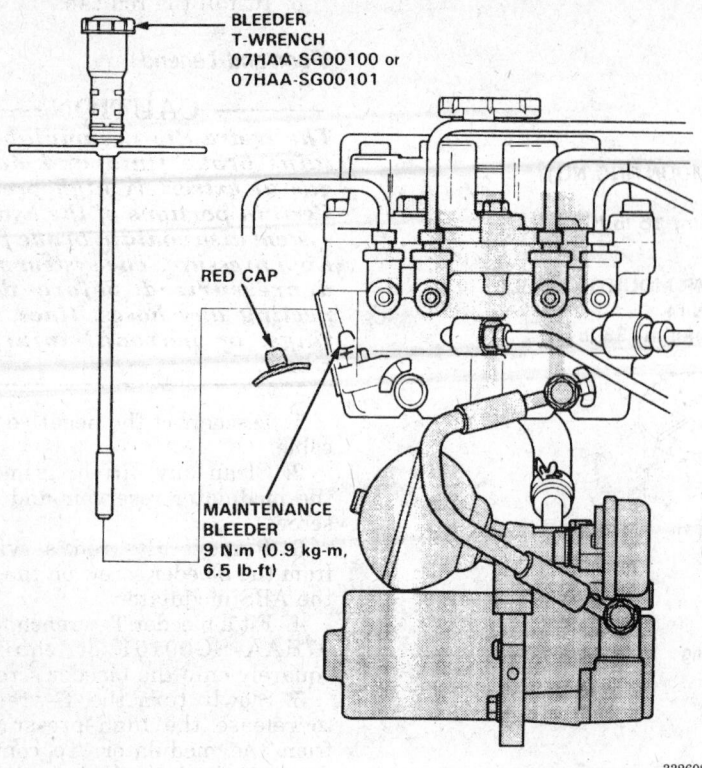

Bleeder T-wrench and bleeder location — Vigor and Legend

WARNING

Brake fluid will damage the vehicle's paint. Immediately clean up any spills.

3. Attach a snug–fitting rubber tube to the bleeder and route the tube's end into a clear container partially filled with brake fluid.

4. Hold the tube with one hand and slowly loosen the maintenance bleeder about 1/8 or 1/4 turn. Do not loosen the maintenance bleeder too much, as the highly–pressurized fluid may burst out.

5. After relieving the pressure, tighten the bleeder to 8 ft. lbs. (11 Nm).

6. After servicing, start the engine and make sure that the ABS indicator light turns off. Check the performance of the brake system.

Draining the ABS modulator

1. Relieve the brake system pressure. Then, retighten the maintenance bleeder.

2. Start the engine and allow it to idle for one minute. Then, shut the engine off.

3. Check to see that the brake fluid in the modulator reservoir is below the MAX line.

4. To drain the remaining brake fluid from the modulator, repeat steps 1, 2, and 3. The total fluid capacity of the modulator is 5 fl. oz. (150 ml); about 1.3–1.5 fl. oz. (40–45 ml) of fluid is drained during each cycle.

NOTE: The modulator should be bled if air enters it during the draining and brake fluid replacement processes.

5. Remove the modulator reservoir cap and fill the reservoir to the MAX line with clean brake fluid. Recap the reservoir.

6. Repeat steps 1, 2, and 3 two more times. Then, refill the reservoir with clean brake fluid.

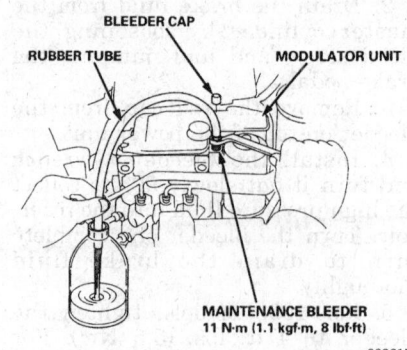

Modulator bleeding — 2.5TL and 3.2TL

7. Tighten the bleeder to 8 ft. lbs. (11 Nm).

8. After servicing, start the engine and make sure that the ABS indicator light turns off.

Bleeding the ABS modulator

1. Fill the modulator reservoir to the MAX line.

2. Attach a snug–fitting rubber tube to the maintenance bleeder and route the end of the tube into a clean container partially filled with brake fluid.

3. Loosen the maintenance bleeder. Then start the engine to active the modulator's pump motor.

4. Tighten the maintenance bleeder when fluid starts to flow out of the bleeder.

5. Shut the engine off after the modulator pump motor stops.

NOTE: If the ABS indicator light turns on and the pump motor stops, restart the engine and repeat steps 3, 4, and 5.

6. Make sure the brake fluid level is at the reservoir's MAX line.

7. After servicing, start the engine and make sure that the ABS indicator light turns off.

Master Cylinder

REMOVAL AND INSTALLATION

NOTE: The radio may have a coded theft protection circuit. Obtain the code from the owner before disconnecting the battery, removing the radio fuse, or removing the radio.

1. Disconnect the negative battery cable.

2. Disconnect the brake fluid level switch connectors.

3. Remove the reservoir cap and pump the fluid from the master cylinder.

4. Use a flare wrench to disconnect the brake lines from the master cylinder. Plug the ports and the ends of the lines to avoid spilling fluid.

5. For Vigor, 2.5TL and 3.2TL, remove the four-way joint mounting bolt. Disconnect the check valve and the throttle cable from the four-way joint bracket.

6. Remove the master cylinder mounting nuts.

7. For Vigor, 2.5TL and 3.2TL, remove the four-way joint bracket.

8. Remove the master cylinder from the brake booster.

9. Remove and replace the rod seal between the brake booster and the master cylinder.

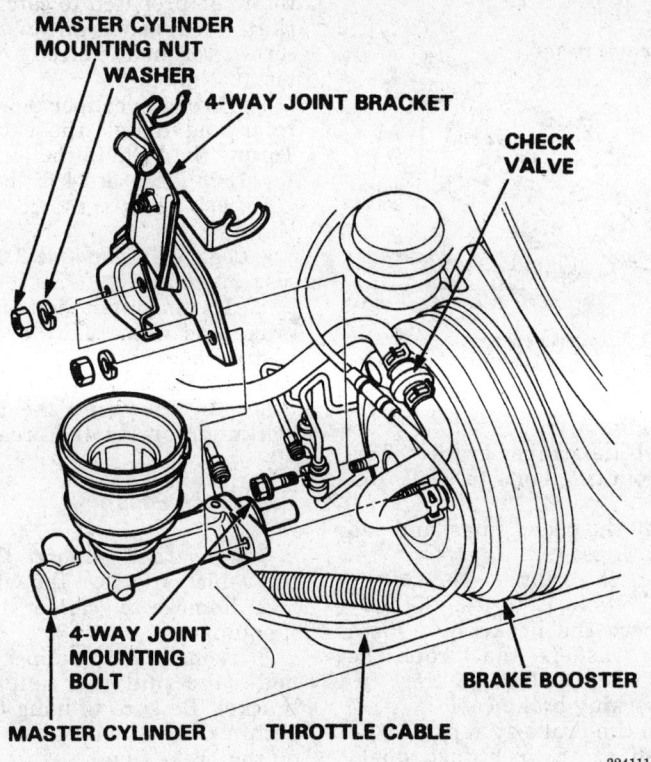

Master cylinder and related components — Vigor, 2.5TL and 3.2TL

324111

To install:

10. Set the pushrod adjustment gauge (tool part # 07JAG–SD40100) on the master cylinder. After making sure the gauge is properly seated, adjust the center shaft. The center shaft should contact the end of the master cylinder piston.

11. Without disturbing the center shaft's position, install the adjustment gauge upside down on the brake booster.

12. Install the master cylinder mounting nuts and torque them to 11 ft. lbs. (15 Nm).

13. Connect a vacuum gauge in line with the vacuum line for the booster. Run the engine to produce about 20 in. Hg. and maintain the engine speed to give a steady vacuum reading.

14. Use a feeler gauge and measure the clearance between the adjustment gauge and the adjusting nut. The clearance should be 0–0.02 inches (0–0.4mm), if adjustment is necessary the preferred clearance is 0–0.01 inches (0–0.2mm).

15. Adjustment of the pushrod is done by loosening the star locknut and then turning the adjuster. Vacuum must be applied to the booster while adjusting the pushrod. Hold the clevis while adjusting the pushrod.

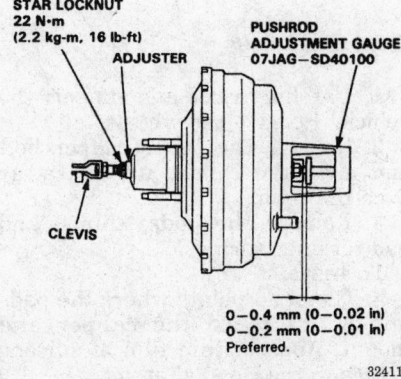

Brake booster and components with the pushrod installed Vigor, 2.5TL and 3.2TL

324113

16. Tighten the star locknut securely, then remove the pushrod adjustment gauge.

17. Install the master cylinder (for Vigor, 2.5TL and 3.2TL, install the four-way joint to the booster). Install the mounting nuts and torque them to 11 ft. lbs. (15 Nm).

18. For Vigor, 2.5TL and 3.2TL, connect the check valve and the throttle cable to the four-way joint bracket. Install the four-way joint bracket mounting bolt, torque the bolt to 11 ft. lbs. (15 Nm).

19. Connect the brake lines and torque the fittings to 14 ft. lbs. (19 Nm).

20. Reconnect the brake fluid level switch connector.

21. Fill and bleed the brake system.

22. Check the brake pedal height and the free-play.

23. Connect the negative battery cable.

Brake Caliper

REMOVAL AND INSTALLATION

Front Brake Caliper

1. Raise and safely support the vehicle, then remove the left and right front wheels.

2. Remove the banjo bolt and disconnect the brake line from the caliper. Be sure to plug the brake line to avoid brake fluid loss or damage to the finish.

3. Remove the caliper mounting bolts and remove the caliper.

4. Remove the brake pads and shims.

5. If there is a pad spring, remove it from the caliper body.

6. Remove the caliper bracket mounting bolts and remove the bracket.

To install:

7. Install the bracket and torque the bolts to:
   Integra: 80 ft. lbs. (110 Nm)
   Vigor: 80 ft. lbs. (110 Nm)
   2.5TL: 80 ft. lbs. (108 Nm)
   3.2TL: 80 ft. lbs. (108 Nm)

8. Install the pad spring, brake pads, shims, caliper and slide mounting bolts.

9. Torque the caliper slide mounting bolts:
   1993 Integra: 24 ft. lbs. (33 Nm)
   1994–96 Integra: 23 ft. lbs. (31 Nm)
   Vigor: 36 ft. lbs. (50 Nm)
   2.5TL: 36 ft. lbs. (49 Nm)
   3.2TL: 36 ft. lbs. (49 Nm)

10. Connect the brake line and the banjo bolt. Replace the crush washers and torque the banjo bolt to 25 ft. lbs. (35 Nm).

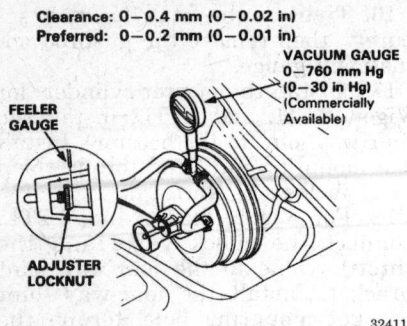

Measuring the booster pushrod clearance Vigor, 2.5TL and 3.2TL

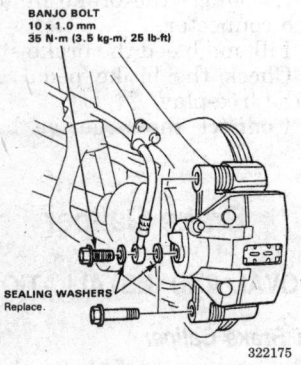

Front brake caliper

11. Be sure to properly bleed the brake system. Torque the bleed screws to 7 ft. lbs. (9 Nm).

12. Install the front wheels and tighten the wheel nuts to 80 ft. lbs. (110 Nm).

Rear Brake Caliper

1. Carefully raise and support the rear of the vehicle. Remove the wheels.

2. Remove the caliper dust shield. Disconnect the parking brake cable from the caliper arm by removing the lock pin and clevis pin. Remove the cable clip, and disconnect the cable from the arm.

3. Disconnect the brake line from the caliper. Avoid loss of brake fluid by plugging the brake line.

4. Remove the caliper mounting bolts and pull the caliper off the bracket.

5. Remove the pads, shim, and pad retainer spring. Clean all points where the shoes and shim touch the caliper and mount. Apply a thin film of silicone grease to the cleaned areas before installation.

6. Remove the caliper bracket mounting bolts. Remove the bracket from the rotor.

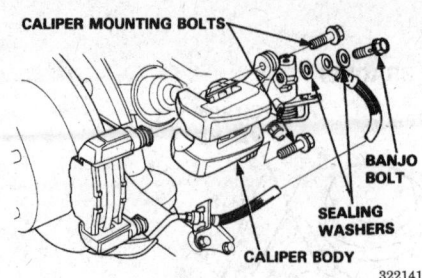

Rear brake caliper

To install:

7. Install the caliper bracket. Torque the mounting bolts to 28 ft. lbs. (39 Nm).

8. Install the pads, shims, and pad retainer springs.

9. Install the caliper. Torque the mounting bolts to 17 ft. lbs. (23 Nm).

10. Connect the brake hose with new crush washers and torque the banjo bolt to 25 ft. lbs. (35 Nm). Install the parking brake cable.

11. Bleed the brake system.

12. Install the caliper dust shield and tighten the bolts. Torque the bolts to 7 ft. lbs. (10 Nm).

13. Install the rear wheels, torque the wheel nuts to 80 ft. lbs. (110 Nm).

Disc Brake Pads

REMOVAL AND INSTALLATION

Front Brake Pads

1. Carefully raise and support the vehicle. Remove the wheels.

2. Remove the lower caliper bolt and pivot the caliper up and away from the rotor.

3. Remove the pads, shims, and pad retainer springs.

To install:

4. Clean all points where the pads and shims touch the caliper and mount. Apply a thin film of silicone grease to the cleaned areas.

5. Place the pad retainers in position on the caliper bracket.

6. Apply a high temperature brake grease to the back side of the pads and both sides of shims and wipe off the excess. Install the pads and shims.

7. Install the inner brake pad with the wear indicator facing upward.

8. Loosen the bleed screw slightly and push in the caliper piston to allow mounting of the caliper over the

rotor. Be prepared to catch any fluid that escapes. Tighten the bleed screw. Torque the bleed screw to 7 ft. lbs. (9 Nm).

9. Pivot the caliper down over the rotor and install the caliper bolts. Torque the bolts to the following:
• 1993 Integra: 24 ft. lbs. (33 Nm)
• 1994–97 Integra: 23 ft. lbs. (31 Nm)
• Legend, Vigor and 3.2TL: 36 ft. lbs. (50 Nm)

10. If disconnected, install the brake pad wear indicator connector. Install the wheels.

11. Depress the brakes several times to make sure the brakes are working properly, then road test.

Rear Brake Pads

1. Raise and support the rear of the vehicle. Remove the wheels.

2. Remove the caliper dust shield, if equipped.

3. Remove both caliper mounting bolts and pull the caliper off the bracket. Be sure to hang the caliper with a piece of wire so no tension is on the brake line.

4. Remove the pads, shim and pad retainer spring. Clean all points where the pads and shims touch the caliper and mount. Apply a thin film of silicone grease to the cleaned areas.

To install:

5. Apply a high temperature brake grease to the pads and shims. Install the pads and shims, making sure the inner pad has the wear indicator facing downward.

6. Rotate the brake caliper piston clockwise into the the cylinder, using a locknut wrench (part # 07916–6390001), or equivalent. Align the cutout in the piston with the tab on the inner pad by turning the piston back.

——— **WARNING** ———
Lubricate the piston boot with grease to avoid twisting the piston boot. If the piston boot is twisted, back the piston out so it sits properly.

7. Install the caliper on the bracket and torque the bolts to 17 ft. lbs. (23 Nm)

8. Install the parking brake cable and the dust shield, if necessary.

9. Install the wheels and lower the vehicle.

10. Depress brakes several times to make sure the brakes work properly. Road test the vehicle.

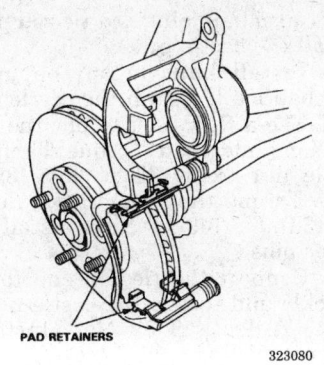

PAD RETAINERS

323080

Front caliper body pivoted up

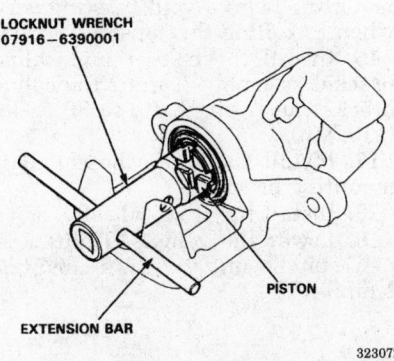

LOCKNUT WRENCH
07916–6390001

PISTON

EXTENSION BAR

323072

Rotating the piston into the bore

Brake Rotor

REMOVAL AND INSTALLATION

Integra and Legend

Front Brake Rotor

1. Safely raise and support the vehicle. Remove the front wheels.
2. Remove the caliper bracket mounting bolts. It is not necessary to disconnect the brake line. Hang the caliper with a piece of wire so that no tension is on the brake line.
3. Remove the screws holding the brake rotor on the hub. Sometimes a light tap with a punch and hammer will loosen the screws to make removal easier. Be careful not to damage the surfaces of the rotor.
4. If the rotor does not pull off, screw two 8mm bolts into the holes near the studs to push it away from the hub. Turn each bolt two turns at a time to prevent cocking the rotor.
5. Remove the brake rotor from the knuckle.
To install:
6. Before installing the brake rotor, clean the mating surfaces of the front hub and the brake rotor.
7. Install the brake rotor with the 6mm brake rotor retaining screws.

Torque retaining screws to 7 ft. lbs. (10 Nm).
8. Install the brake caliper and bracket assembly with the caliper bracket mounting bolts.
9. Torque mounting bolts to 80 ft. lbs. (110 Nm)
10. Install the front wheel, tighten the wheel nuts to 80 ft. lbs. (110 Nm).

Rear Brake Rotor

1. Release the parking brake.
2. Raise and safely support the vehicle. Remove the rear wheels.
3. Remove the caliper, pads, and mounting bracket assembly. It is not necessary to disconnect the brake line. Hang the caliper with a piece of wire so no tension is on the brake line.
4. Remove the screws holding the brake rotor on the hub. Sometimes a light tap with a punch and hammer will loosen the screws to make removal easier.
5. If the rotor does not pull off, thread two 8mm bolts into the holes near the studs. To remove the rotor, turn each bolt 2 turns at a time to prevent cocking the brake rotor excessively.
6. Remove brake rotor.
To install:
7. Be sure to clean the hub before installing the rotor. Install the brake

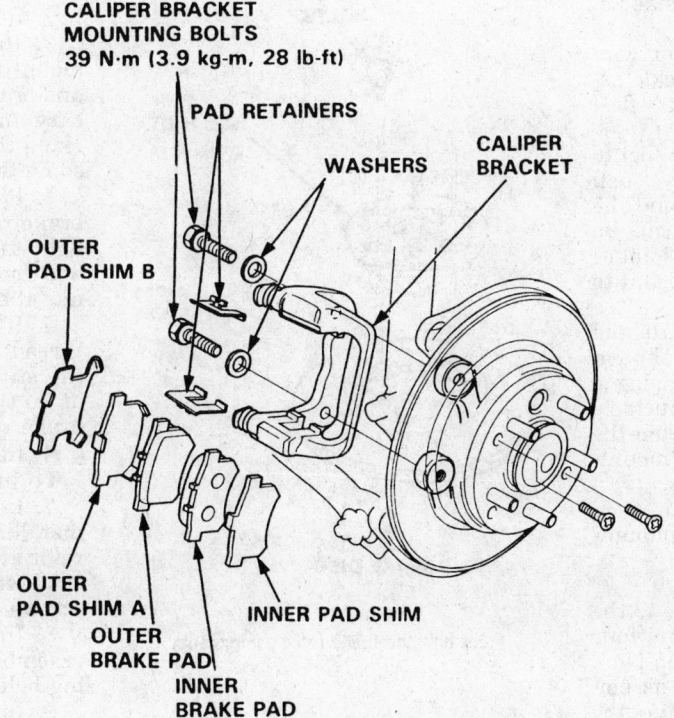

CALIPER BRACKET
MOUNTING BOLTS
39 N·m (3.9 kg-m, 28 lb-ft)

PAD RETAINERS

WASHERS

CALIPER BRACKET

OUTER PAD SHIM B

OUTER PAD SHIM A

OUTER BRAKE PAD

INNER BRAKE PAD

INNER PAD SHIM

323141

Exploded view of the rear pads — Legend

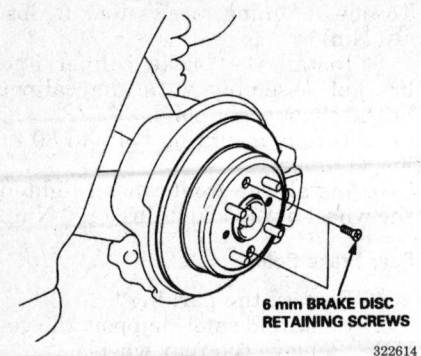

6 mm BRAKE DISC RETAINING SCREWS

322614

Rotor retaining screws

rotor and tighten the two brake rotor retainer screws. Torque retainer screws to 7 ft. lbs. (10 Nm).

8. Install caliper, pads, and bracket assembly. Torque mounting bracket bolts to 28 ft. lbs. (39 Nm).

9. Install the rear wheels and tighten the wheel nuts to 80 ft. lbs. (110 Nm).

Vigor, 2.5TL and 3.2TL

Front Brake Rotor

1. Raise and support the vehicle. Remove the front wheels.

2. Remove the brake hose mounting bolt. Leave the brake line connected to the caliper.

3. Remove the caliper bracket mounting bolts and hang the caliper and bracket assembly to one side using a piece of wire.

4. Remove the wheel sensor wire and its bracket from the knuckle assembly. Do not disconnect the wheel sensor wire.

5. Use a ball joint removal tool to separate the upper and lower ball joints from the control arms and the tie rod end from steering arm. Be careful not to damage the ball joint or its boot. Use penetrating lubricant to help loosen the ball joints.

6. Pull the knuckle outward and remove the halfshaft outboard CV–joint from the knuckle using a plastic mallet. Remove the knuckle.

7. Separate the hub unit from the knuckle by removing the four mounting bolts.

8. Separate the brake rotor from the hub by removing the four mounting bolts.

To install:

9. Install the hub assembly to the brake rotor and torque the four mounting bolts to 40 ft. lbs. (55 Nm).

10. Install the hub and rotor assembly to the knuckle and torque the bolts to 33 ft. lbs. (45 Nm).

11. Install the knuckle on the halfshaft. Install the knuckle on the

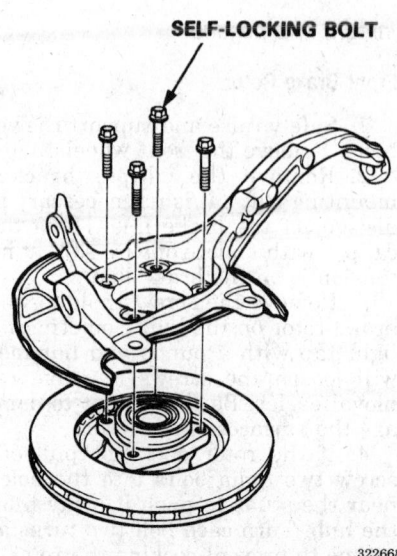

SELF-LOCKING BOLT

322668

Steering knuckle and hub components

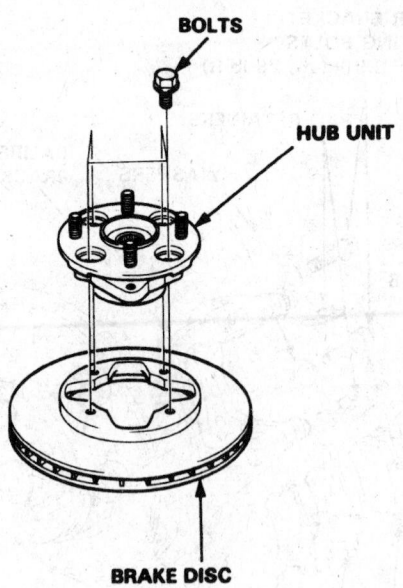

BOLTS

HUB UNIT

BRAKE DISC

322669

Front hub and brake rotor components

lower arm ball joint and tie rod, then install the castle nuts.

12. Install knuckle on the upper arm ball joint. Install the castle nut.

13. Use a floor jack to put the suspension under load. Torque the upper castle nut to 29–35 ft. lbs. (39–47 Nm). Torque the lower castle nut to 36–43 ft. lbs. (49–59 Nm). Install new cotter pins.

14. Connect the tie rod end to the knuckle and torque the castle nut to 36–43 ft. lbs. (49–59 Nm). Install a new cotter pin.

15. Install the knuckle protector, wheel sensor, wheel sensor wire, and mounting bolts. Avoid twisting wires when installing the sensors.

16. Install the caliper and caliper bracket assembly. Torque the caliper bracket mounting bolts to 80 ft. lbs. (110 Nm).

17. Install the brake hose and its mounting bracket.

18. Install the front wheels.

19. Lower the vehicle. Tighten the new spindle nut to 181 ft. lbs. (245 Nm).

Rear Brake Rotor

1. Carefully raise and safely support the vehicle. Remove the rear wheels.

2. Remove the caliper dust shield.

3. Remove the caliper bracket mounting bolts. Remove the caliper and bracket assembly. It is not necessary to disconnect the brake line. Hang the caliper with a piece of wire so no tension is on the brake line.

4. Remove the screws holding the brake rotor on the hub. Sometimes a light tap with a punch and hammer will loosen the screws to make removal easier.

5. If the rotor does not pull off, thread two bolts into the holes near the studs. Turn each bolt two turns at a time to prevent cocking the brake rotor excessively.

6. Remove the brake rotor.

To install:

7. Be sure to clean the hub before installing the rotor. Install the brake rotor and tighten the two brake rotor retainer screws. Torque the retainer screws to 7 ft. lbs. (10 Nm).

8. Install the caliper and bracket assembly. Torque the bracket mounting bolts to 28 ft. lbs. (39 Nm).

9. Install the caliper dust shield.

10. Install the rear wheels and tighten the wheel nuts to 80 ft. lbs. (110 Nm). Lower the car.

Parking Brake Cable

ADJUSTMENT

Integra

NOTE: Pull the parking brake handle with about 44 lbs. (200 N) of force. The parking brake lever should come up 6–10 notches. Adjust the parking brake if the brake is not applied within these notches.

1. If the rear brakes were serviced, make sure that the parking brake arm on the rear brake caliper contacts the brake caliper pin.
2. Pull the parking brake lever up one notch.
3. Slide the driver and passenger front seats forward.
4. Remove the rear console access cap located below the parking brake handle.

NOTE: When prying with a flat bladed tool, wrap the blade with tape to prevent damage to the vehicles interior trim.

5. Remove the two machine screws from the front of the rear console and the two screws from the rear of the console.
6. Lift up on the parking brake handle to aid in the removal of the rear console. Lift the front of the rear console, then slide it rearward to detach the hooks on the front of the rear console. Remove the rear console from the vehicle.
7. Remove the cap from the parking brake cable end.
8. Safely raise and support the rear wheels.
9. Tighten the parking brake adjusting nut until the rear wheels drag slightly when turned.
10. Release the parking brake lever and verify that the rear wheels do not drag when turned. Readjust the brake if necessary.

Lever Locked Notches: 6–10

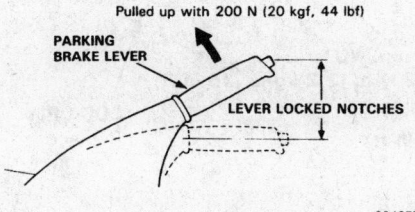

Pulled up with 200 N (20 kgf, 44 lbf)

PARKING BRAKE LEVER

LEVER LOCKED NOTCHES

334857

Parking brake lever

11. Make sure that the parking brake is fully applied when the lever is pulled 6–10 notches.
12. Install the parking brake cable end cap.
13. Place the rear console on the parking brake handle and slide the hooks on the rear console into the front console. Push the console down into place and install the two rear screws and the two front machine screws.
14. Install the access cap to the rear console.
15. Lower the vehicle.

Legend

Minor adjustment

1. Remove the console rear cover located below the rear cigarette lighter. It is held in place by two clips.
2. Pull the parking brake lever up one click only.
3. Raise the rear wheels off of the ground. Support the vehicle safely and block the front wheels.
4. Turn the parking brake cable adjusting nut clockwise until the rear wheels drag slightly when you turn them by hand.
5. Release the parking brake lever. Check that the rear wheels do not drag when you turn them by hand.
6. With the parking brake cable adjusted properly, the parking brake should be fully applied when the lever is pulled up six to eight clicks.
7. Install the console rear cover.

Major adjustment

NOTE: Make sure the parking brake shoe linings are not worn beyond the service limit of 0.04–0.10 in. (1.0–2.5mm).

1. Remove the console rear cover located below the rear cigarette lighter. It is held in place by two clips.
2. Raise the rear wheels off of the ground. Support the vehicle safely and block the front wheels.
3. Remove the rear wheels.
4. Release the parking brake lever. Loosen the adjusting nut by turning it counterclockwise.
5. Insert a brake adjusting tool into the rear brake shoe adjusting hole. Turn the adjuster up until the shoes lock, then back the adjuster off eight stops.
6. Check to make sure that the parking brake is fully applied when the lever is pulled up six to eight clicks.

7. If the number of clicks is out of range, turn the adjusting nut until brake cable is adjusted.
8. Install the rear wheels and lower the vehicle.
9. Install the console rear cover.

Vigor, 2.5TL and 3.2TL

NOTE: The radio may contain a coded theft protection circuit. Always obtain the code number before disconnecting the battery.

1. Disconnect the negative battery cable.
2. Remove the console screws and lift the console up to remove it. The console trim is held in place by spring clips. On Vigor models, the adjusting nut is accessible by removing the rear ashtray. Turn the nut with a socket and extension.
3. Pull the parking brake lever up one click only.
4. Raise the rear wheels off of the ground. Support the vehicle safely and block the front wheels.
5. Make sure that the parking brake arm on the caliper contacts the caliper pin.
6. Turn the parking brake cable adjusting nut until the rear wheels drag slightly when you turn them by hand.
7. Release the parking brake lever. Check that the rear wheels do not drag when they are turned by hand.
8. With the parking brake cable adjusted properly, the parking brakes should be fully applied when the lever is pulled up seven to eleven clicks.
9. Install the console and any console components that may have been removed.
10. Reconnect the negative battery cable.
11. Enter the radio security code.

REMOVAL AND INSTALLATION

NOTE: The radio may have a coded theft protection circuit. Obtain the code from the owner before disconnecting the battery, removing the radio fuse, or removing the radio.

Integra, Vigor, 2.5TL and 3.2TL

1. Disconnect the negative battery cable.
2. Remove the console trim panel by removing the screws and carefully prying up the securing clips.
3. Set the parking brake lever to a fully released position.

4. Remove the adjusting nut from the equalizer at the rear of the brake lever and separate the cable from the equalizer. The equalizer is held by an adjusting nut, or a clip and clevis pin.

5. Raise and support the vehicle safely.

6. Remove the front section of the exhaust system and the heat shields

7. Remove the rear caliper covers.

8. Remove the lock pin and pull out the clevis pin which connects the cable and the caliper lever.

9. Detach the cable retaining clips.

10. Remove the cables from their guides on the underbody and suspension members of the vehicle.

11. Remove the cables from the vehicle. Be careful not to tear the grommets that seal the cable entrances to the body.

To install:

12. Apply silicon grease to the cable and the guides before installing the cable.

13. Install the cable. Install the cable guides working from the rear of the vehicle forward. Torque the cable guide nuts and bolts to 16 ft. lbs. (22 Nm).

14. Feed the connecting ends of the cables through the grommets and into the body.

15. Install the cable onto the caliper lever and secure it with new clips.

Install the clevis pins with new cotter pins.

16. Install the rear caliper covers.

17. Verify that all the cable mounting brackets are installed and tightened.

18. Install the exhaust heat shields and any exhaust system components that were removed. Use new self-locking nuts on the heat shields and exhaust pipe flanges.

19. Lower the vehicle.

20. Grease the sliding surfaces of the parking brake lever.

21. Reconnect the front end of the cable to the equalizer with the adjusting nut, or the clip and clevis pin.

22. Test the operation of the parking brake cable.

23. Check the adjustment of the parking brake.

24. Connect the negative battery cable and enter the radio security code.

Legend

1. Set the parking brake lever to a fully released position. Block the front wheels, raise and support the rear of the vehicle safely.

2. On some vehicles, it may be necessary to remove the front exhaust heat shield and the fuel tank heat shield.

3. Remove the adjusting nut from the equalizer at the brake lever and separate the cables from the equalizer.

4. Remove the guides and brackets which mount the left and right parking brake cables to the body.

5. Remove the parking brake cable brackets from the radius rod assembly.

6. Remove the caliper, the brake disc, the hub, and the parking brake shoe assembly.

7. Disconnect the parking brake cable from the parking brake lever on the shoe assembly.

8. Use a 12mm offset box wrench to remove the parking brake cable from the backing plate.

To install:

9. Apply grease to the cable and guides before the cable is installed.

10. Install the cable, the cable guides, and brackets. Torque the cable guide nuts and bolts to 17 ft. lbs. (23 Nm).

11. Insert the cable through the backing plate and connect it to the parking brake arm. Install the brake shoe assembly. Make sure the retainer springs and return springs are in the correct positions.

12. Install the hub and torque the nut to 206 ft. lbs. (285 Nm) when the vehicle is lowered. Install the hub

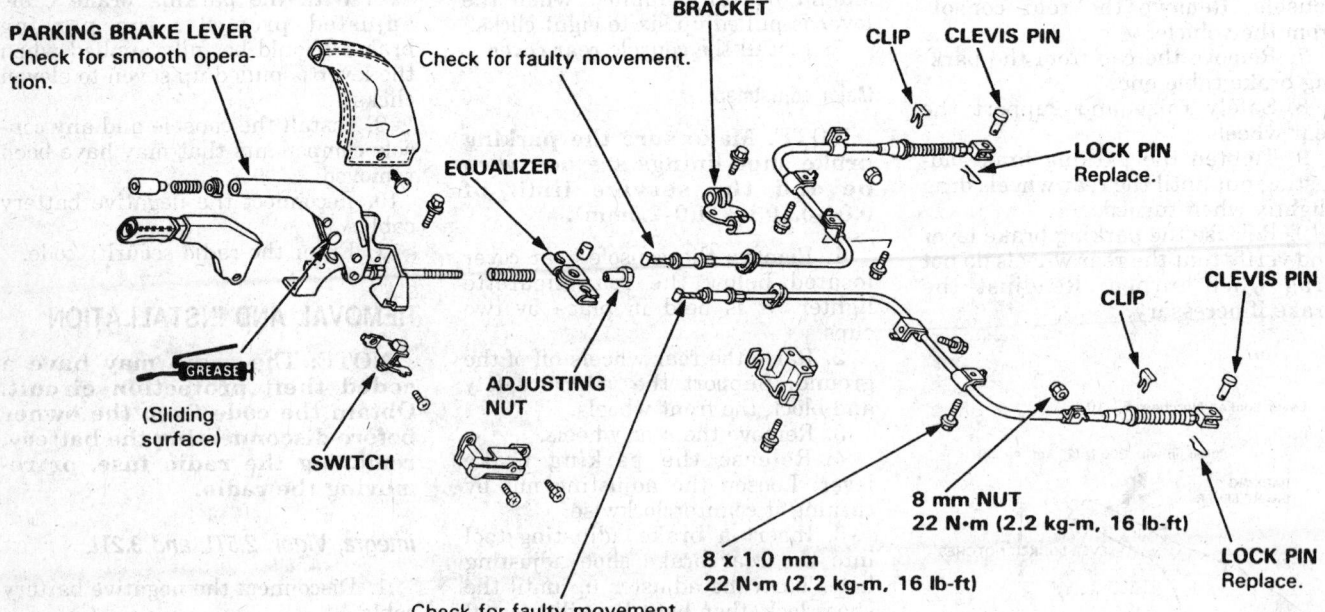

PARKING BRAKE LEVER
Check for smooth operation.

Check for faulty movement.

BRACKET

CLIP **CLEVIS PIN**

LOCK PIN
Replace.

EQUALIZER

GREASE

(Sliding surface)

SWITCH

ADJUSTING NUT

Check for faulty movement.

8 x 1.0 mm
22 N·m (2.2 kg-m, 16 lb-ft)

8 mm NUT
22 N·m (2.2 kg-m, 16 lb-ft)

CLIP **CLEVIS PIN**

LOCK PIN
Replace.

321875

Cable diagram — Integra, Vigor, 2.5TL and 3.2TL

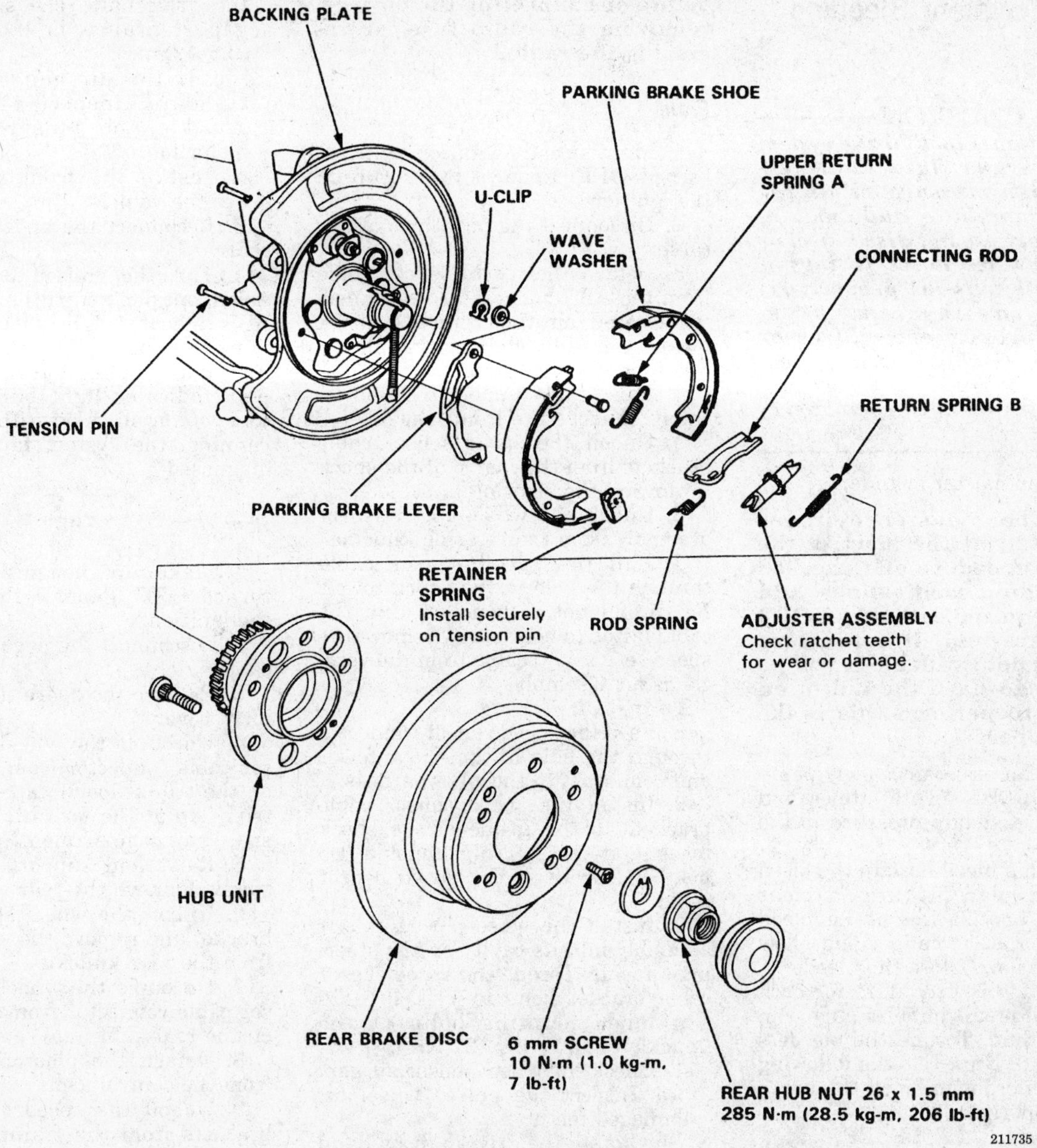

BACKING PLATE

PARKING BRAKE SHOE

UPPER RETURN
SPRING A

U-CLIP

WAVE
WASHER

CONNECTING ROD

TENSION PIN

RETURN SPRING B

PARKING BRAKE LEVER

RETAINER
SPRING
Install securely
on tension pin

ROD SPRING

ADJUSTER ASSEMBLY
Check ratchet teeth
for wear or damage.

HUB UNIT

REAR BRAKE DISC

6 mm SCREW
10 N·m (1.0 kg-m,
7 lb-ft)

REAR HUB NUT 26 x 1.5 mm
285 N·m (28.5 kg-m, 206 lb-ft)

211735

Parking brake components — Legend

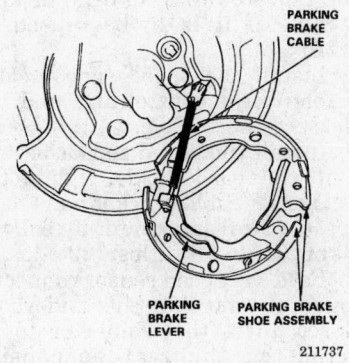

PARKING
BRAKE
CABLE

PARKING
BRAKE
LEVER

PARKING BRAKE
SHOE ASSEMBLY

211737

Cable removal — Legend

cap. Install the brake disc and mounting screws. Torque the screws to 7 ft. lbs. (10 Nm).

13. Install the caliper, torque the mounting bolts to 80 ft. lbs. (110 Nm).

14. Reconnect the front end of the cable to the equalizer with the adjusting nut.

15. Install the heat shields if they were removed.

16. Replace the rear wheels, tightening the wheel nuts to 80 ft. lbs. (110 Nm).

17. Adjust the parking brake cable if necessary.

Brake System Bleeding

PROCEDURE

---------- CAUTION ----------
The Acura anti-lock brake system contains brake fluid under extremely high pressure within the pump, accumulator and modulator assembly. Do not disconnect or loosen any lines, hoses, fittings or components without properly relieving the system pressure. Improper procedures or failure to discharge the system pressure may result in severe personal injury and/or property damage.

1. Fill the master cylinder.

NOTE: The master cylinder must be full at the start of the bleeding procedure and checked after bleeding each caliper. Add fluid as required. Use only DOT 3 or 4 brake fluid. If a pressure bleeder is not available it will be necessary to have the aid of an assistant to perform this brake bleeding operation.

2. Have an assistant slowly pump the brake pedal several times and then apply a steady pressure to the brake pedal.
3. Attach a bleed hose to the bleed screw and place it into a clear container. Loosen the brake bleed screw at the brake caliper and allow the fluid to flow. Close the bleeder.
4. Repeat this procedure for each wheel, until no air bubbles appear in the brake fluid. Torque the bleeders to 7 ft. lbs. (9 Nm). Use the following sequence in order to bleed the brake system properly:
- Integra: RR, LF, LR, RF
- Legend: RR, LF, LR, RF
- Vigor: RR, LF, LR, RF
- 2.5TL: RR, LF, LR, RF
- 3.2TL: RR, LF, LR, RF

NOTE: Bleeding sequence: LF (Left Front), RF (Right Front), LR (Left Rear), RR (Right Rear).

5. Check the fluid level in the master cylinder and add, if necessary. Road test the vehicle and check the brake performance.

Wheel Speed Sensor

REMOVAL AND INSTALLATION

NOTE: The radio may have a coded theft protection circuit. Obtain the code from the owner before disconnecting the battery, removing the radio fuse, or removing the radio.

Front

1. Make sure the ignition switch is turned **OFF**. Remove the key from the ignition.
2. Disconnect the negative battery cable.
3. Detach the wheel sensor cable from the ABS harness at its connector located on the vehicle's shock tower. Carefully detach the plastic clip from the shock tower.
4. Raise and support the vehicle safely. Remove the front wheels.
5. Unbolt the speed sensor cable brackets from the inside of the shock tower and the steering knuckle.
6. Unbolt the wheel speed sensor assembly from the steering knuckle.
7. Pull the speed sensor cable through the hole in the shock tower. Be careful not to bend or kink the cable if it is to be reused. Remove the speed sensor and cable from the vehicle as an assembly.
   To install:
8. Feed the speed sensor connector through the hole in the shock tower and connect it to the ABS harness.
9. Install the speed sensor cable brackets to the inside of the shock tower and the steering knuckle. Do not bend or twist the speed sensor cable.
10. Install the wheel speed sensor assembly into its cavity on the steering knuckle. Torque the speed sensor cable bracket bolts to 7 ft. lbs. (10 Nm) and mounting bolts at the knuckle to 16 ft. lbs. (22 Nm).
11. Check the wheel sensor air gap:
    a. Inspect the pulser wheel for damaged teeth.
    b. Rotate the driveshaft by hand and measure the clearance between the wheel speed sensor and the pulser wheel. Measure the clearance for one full rotation of

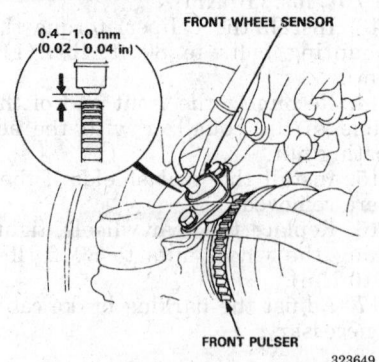

Speed sensor air gap — Integra

the driveshaft. The standard air gap clearance is 0.02–0.04 in. (0.5–1.0mm).
    c. If the air gap exceeds the standard, inspect the steering knuckle and pulser wheel for distortion.
12. Install the front wheels and lower the vehicle.
13. Reconnect the negative battery cable.
14. Turn the ignition to the **ON** position, but don't start the engine. The ABS indicator light should turn on. Start the engine and verify that the ABS indicator light turns off. If the ABS indicator light blinks, doesn't come on, or stays on with the engine running, the system fault must be diagnosed.

Rear

1. Make sure the ignition switch is turned **OFF**. Remove the key from the ignition.
2. Disconnect the negative battery cable.
3. Remove the spare tire compartment cover.
4. Uncouple the wheel speed sensor cable connector from its junction on the trunk floor located under the trim flap at the forward edge of the spare tire compartment.
5. Raise and support the vehicle safely. Remove the rear wheels.
6. Unbolt the wheel speed sensor bracket and remove the speed sensor from the rear knuckle.
7. Uncouple the wheel speed sensor cable connector from its junction on the rear subframe.
8. Detach the plastic cable clips from the control arm.
9. Unbolt the speed sensor cable brackets from the trailing arm and the underbody of the vehicle.
10. Remove the speed sensor and cable. Be careful not to bend or kink the cable if it is to be reused. **To install:**
11. Install the speed sensor cable brackets to the control arm and underbody of the vehicle.
12. Install the speed sensor assembly into the rear knuckle. Tighten the speed sensor cable bracket bolts to 10 ft. lbs. (16 Nm) and mounting bolts at the knuckle to 16 ft. lbs. (22 Nm).
13. Feed the speed sensor connector through the hole in the underbody and reconnect the connector to its junction on the rear suspension beam. Fasten the plastic cable clips.
14. Check the wheel sensor air gap:
    a. Inspect the pulser wheel for damaged teeth.

Front:

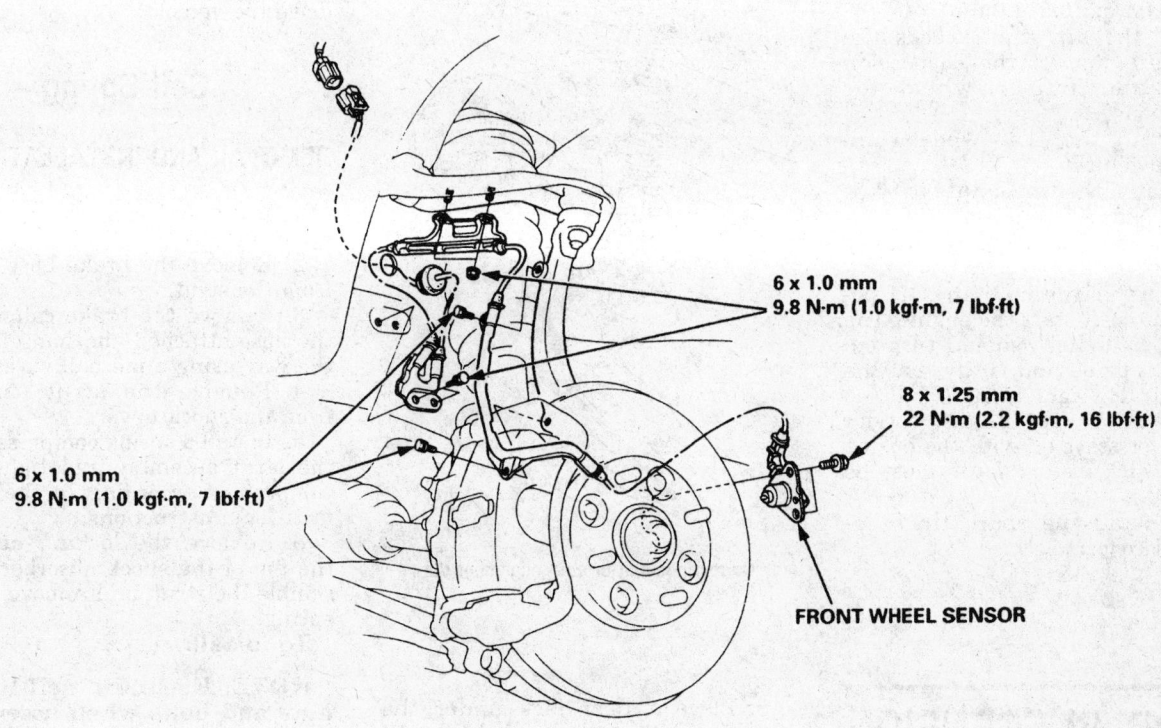

6 x 1.0 mm
9.8 N·m (1.0 kgf·m, 7 lbf·ft)

8 x 1.25 mm
22 N·m (2.2 kgf·m, 16 lbf·ft)

6 x 1.0 mm
9.8 N·m (1.0 kgf·m, 7 lbf·ft)

FRONT WHEEL SENSOR

Rear:

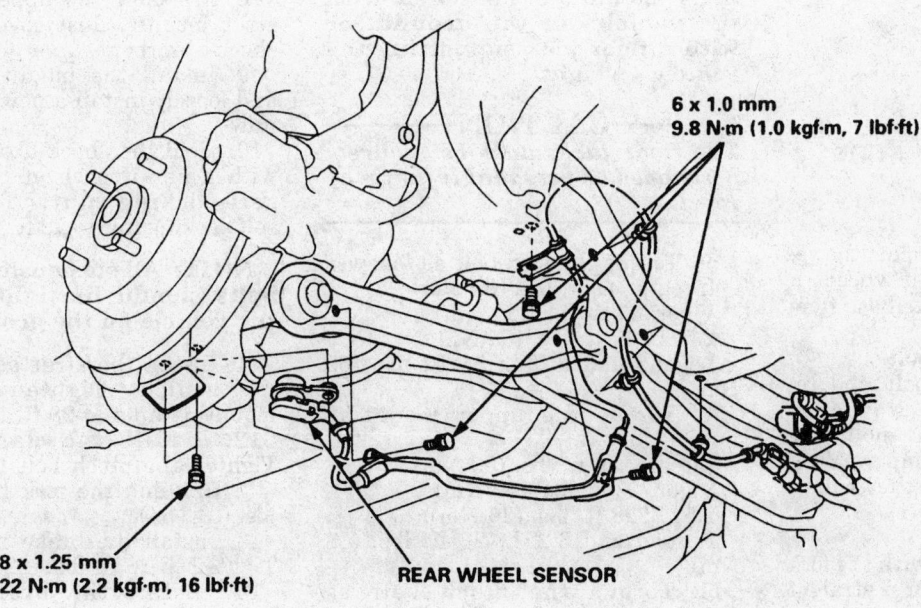

6 x 1.0 mm
9.8 N·m (1.0 kgf·m, 7 lbf·ft)

8 x 1.25 mm
22 N·m (2.2 kgf·m, 16 lbf·ft)

REAR WHEEL SENSOR

332118

Front and rear wheel speed sensor components — 2.5TL

b. Rotate the hub by hand and measure the clearance between the wheel speed sensor and the pulser wheel. Measure the clearance for one full rotation of the hub. The standard air gap clearance is 0.02–0.04 in. (0.5–1.0mm).

c. If the air gap exceeds the standard, inspect the suspension arms and pulser wheel for distortion.

15. Install the rear wheels and lower the vehicle.

16. Tighten the wheel nuts to 80 ft. lbs. (110 Nm).

17. Reconnect the negative battery cable.

18. Turn the ignition to the **ON** position, but don't start the engine. The ABS indicator light should turn on. Start the engine and verify that the ABS indicator light turns off. If the ABS indicator light blinks, doesn't come on, or stays on with the engine running, the system fault must be diagnosed.

19. Reinstall the spare tire compartment cover.

FRONT SUSPENSION

Strut

REMOVAL AND INSTALLATION

1. Raise and safely support the vehicle and remove the front wheels.

2. Disconnect the brake hose from the damper.

3. Remove the pinch bolt.

4. Remove the fork bolt and remove the fork.

5. Remove the upper mounting nuts and remove the damper. Mark the left and right sides so they will be installed correctly.

To install:

6. Install the damper with the fork alignment tabs aligned in a straight line when both units are installed in the vehicle. Loosely install the upper mount nuts.

7. Install the fork on the lower arm and the damper and loosely install the pinch bolt and fork bolt.

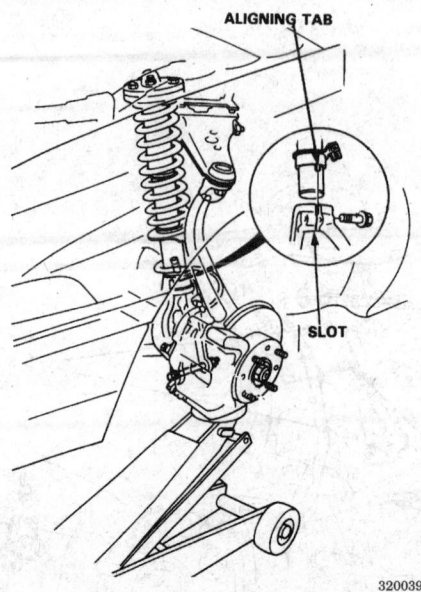
Correct position of strut alignment tab

320039

8. Place a floor jack under the lower ball joint and raise it just until the vehicle raises off the jack stand.

NOTE: All suspension nuts and bolts should be tightened with the vehicle on the ground, or with a floor jack supporting the vehicle's weight.

——— **CAUTION** ———
The floor jack must be securely positioned or personal injury may result.

9. Torque the fork bolt as follows:
Integra: 47 ft. lbs. (65 Nm)
Vigor: 47 ft. lbs. (65 Nm)
2.5TL: 47 ft. lbs. (65 Nm)
Legend and 3.2TL: 51 ft. lbs. (69 Nm)

10. Torque the upper mounting nuts as follows:
Integra: 36 ft. lbs. (50 Nm)
Vigor: 28 ft. lbs. (39 Nm)
2.5TL: 28 ft. lbs. (39 Nm).
Legend and 3.2TL: 28 ft. lbs. (38 Nm)

11. Torque the pinch bolt as follows:
Integra: 32 ft. lbs. (44 Nm)
Vigor: 32 ft. lbs. (44 Nm)
2.5TL: 32 ft. lbs. (43 Nm)
Legend and 3.2TL: 37 ft. lbs. (50 Nm)

12. Connect the brake hose to the damper and torque the bolts to 16 ft. lbs. (22 Nm).

13. Install the front wheels and lower the vehicle.

14. Check the alignment and test drive the vehicle.

Coil Spring

REMOVAL AND INSTALLATION

1. Raise and support the vehicle, remove the front wheels.

2. Remove the brake hose clamps from the strut.

3. Remove the brake caliper with the hose attached and hang it out of the way using a piece of wire.

4. Remove the strut assembly from the shock tower.

5. Install a spring compressor onto the strut assembly and tighten the compressor according to the manufacturer's instructions.

6. Remove the locking nut from the top of the shock absorber, disassemble the strut, and remove the coil spring.

To install:

NOTE: Use new self-locking nuts and bolts when assembling the strut.

7. Install a spring compressor onto the coil spring.

8. Assemble the upper and lower strut mounts, dust covers, and the shock absorber.

9. Install the mounting washer, and loosely install a new self-locking nut.

10. Hold the shock absorber piston with a hex wrench and tighten the self-locking nut. Torque the self-locking nut to 22 ft. lbs. (30 Nm).

NOTE: All suspension nuts and bolts should be tightened with the vehicle on the ground.

11. Install the strut assembly into the vehicle. Tighten the upper mounting nuts to 28 ft. lbs. (39 Nm).

12. Install the damper fork. Tighten the pinch bolt to 32 ft. lbs. (43 Nm), and the fork bolt to 47 ft. lbs. (64 Nm).

13. Install the brake hose clamps. Tighten to 16 ft. lbs. (22 Nm).

14. Install the brake caliper. Tighten the mounting bolts to 80 ft. lbs. (108 Nm).

15. Install the wheel, and tighten the wheel nuts to 80 ft. lbs. (110 Nm).

16. Check and adjust the vehicle's front wheel alignment.

10. Install the front wheels, lower the vehicle, and torque the nuts connecting the upper control arm to the chassis to 47 ft. lbs. (65 Nm).

11. Check the wheel alignment and road test the vehicle.

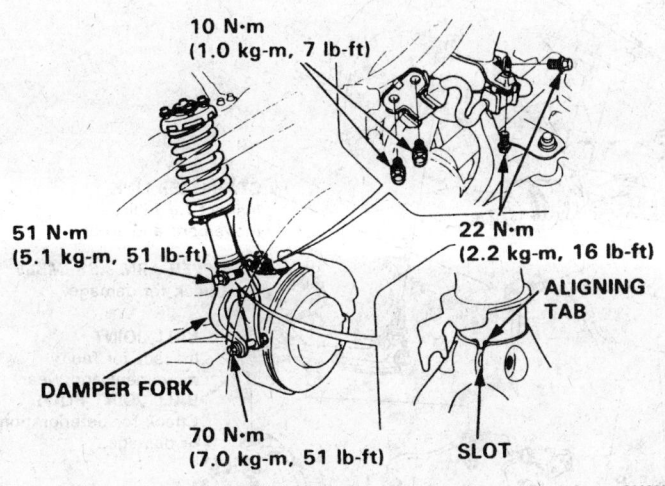

10 N·m (1.0 kg-m, 7 lb-ft)

51 N·m (5.1 kg-m, 51 lb-ft)

22 N·m (2.2 kg-m, 16 lb-ft)

ALIGNING TAB

DAMPER FORK

70 N·m (7.0 kg-m, 51 lb-ft)

SLOT

218020

Strut alignment mark — Legend

Upper Ball Joints

REMOVAL AND INSTALLATION

Vigor, 2.5TL and Legend

NOTE: The upper ball joint cannot be removed from the control arm. If the ball joint is damaged, the upper arm assembly must be replaced.

1. Raise and safely support the vehicle. Remove the front wheel.

2. If equipped, remove the ball joint nut cover. Remove the cotter pin and the nut connecting the upper control arm to the steering knuckle.

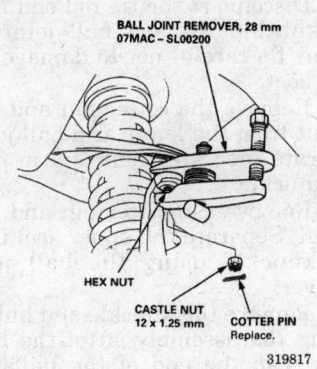

BALL JOINT REMOVER, 28 mm 07MAC - SL00200

HEX NUT

CASTLE NUT 12 x 1.25 mm

COTTER PIN Replace.

319817

Separating the upper ball joint

3. Support the lower control arm assembly with a floor jack.

4. Using a ball joint removal tool, separate the upper control arm from the steering knuckle.

5. Remove the upper control arm nuts, washers and the upper control arm from the vehicle.

To install:

6. Replace all self–locking nuts upon installation. Clean off any dirt, oil or grease off of the threads of the fasteners. Do not tighten any nuts on any rubber mounts or bushings until the vehicle is lowered onto the ground.

7. Install the upper control arm and mounting bolts to the chassis. The upper control arms are not interchangeable.

8. Raise the steering knuckle up with a floor jack, just enough to install the upper control arm ball joint into the steering knuckle. Torque the ball joint nut as follows:

Integra: 29–35 ft. lbs. (39–47 Nm)
Legend: 32 ft. lbs. (44 Nm)
Vigor: 29 ft. lbs. (40 Nm)
2.5TL: 32 ft. lbs. (44 Nm)
3.2TL: 29–35 ft. lbs. (39–47 Nm)

9. Tighten the castle nut enough to install a new cotter pin. If removed, install the ball joint nut cover.

Lower Ball Joints

REMOVAL AND INSTALLATION

Integra, Vigor, and 1993–94 Legend

1. Raise and support the vehicle safely. Remove the front wheel assemblies.

2. Remove the steering knuckle.

3. Remove the boot by prying off the snapring. Remove the 40mm clip. Check the boot for deterioration and damage, replace if necessary.

4. Install the special ball joint with a removal/installation tool 07965-SB00100 or equivalent, on the ball joint and tighten the ball joint nut.

5. Position the ball joint in this special tool and set this assembly in a vise. Press the ball joint out of the steering knuckle.

To install:

6. Place the ball joint in position by hand. Install the ball joint into the special tool and press in the new ball joint in the vise.

─── **WARNING** ───
After installing the boot, check the ball joint pin tapered section for grease contamination and wipe it if necessary.

7. Install the 40mm circlip. Adjust the special tool with the adjusting bolt until the end of the tool aligns with the groove on the boot. Slide the clip over the tool and into position.

8. Install the knuckle.

9. Check the front wheel alignment and adjust if necessary.

2.5TL

NOTE: The lower ball joint is pressed into the steering knuckle and cannot be removed.

1. Pry up the lock tab and loosen the spindle nut. Slightly loosen the lug nuts.

2. Raise and safely support the vehicle. Remove the front wheel and spindle nut.

3. Remove the brake caliper mounting bolts and remove the caliper from the knuckle. Hang the caliper out of the way with a length of wire.

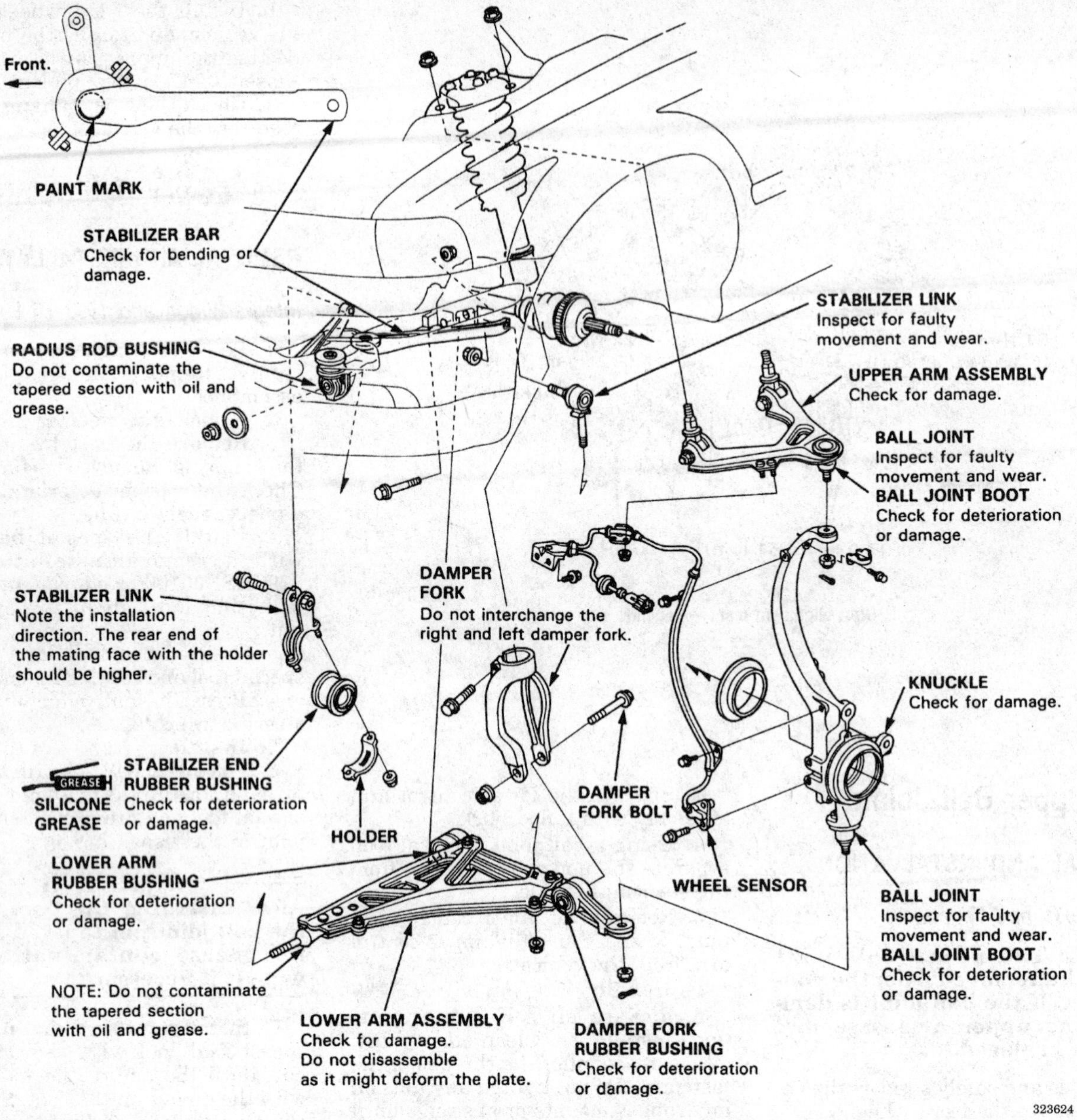

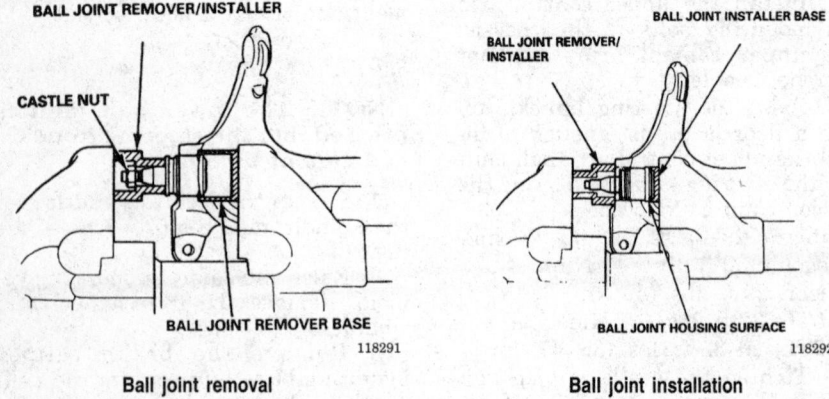

PAINT MARK

STABILIZER BAR
Check for bending or damage.

RADIUS ROD BUSHING
Do not contaminate the tapered section with oil and grease.

STABILIZER LINK
Inspect for faulty movement and wear.

UPPER ARM ASSEMBLY
Check for damage.

BALL JOINT
Inspect for faulty movement and wear.

BALL JOINT BOOT
Check for deterioration or damage.

STABILIZER LINK
Note the installation direction. The rear end of the mating face with the holder should be higher.

DAMPER FORK
Do not interchange the right and left damper fork.

STABILIZER END RUBBER BUSHING
Check for deterioration or damage.

SILICONE GREASE

HOLDER

LOWER ARM RUBBER BUSHING
Check for deterioration or damage.

DAMPER FORK BOLT

WHEEL SENSOR

KNUCKLE
Check for damage.

BALL JOINT
Inspect for faulty movement and wear.

BALL JOINT BOOT
Check for deterioration or damage.

NOTE: Do not contaminate the tapered section with oil and grease.

LOWER ARM ASSEMBLY
Check for damage. Do not disassemble as it might deform the plate.

DAMPER FORK RUBBER BUSHING
Check for deterioration or damage.

323624

A common upper control arm and ball joint assembly

BALL JOINT REMOVER/INSTALLER

CASTLE NUT

BALL JOINT REMOVER BASE

118291

Ball joint removal

BALL JOINT INSTALLER BASE

BALL JOINT REMOVER/INSTALLER

BALL JOINT HOUSING SURFACE

118292

Ball joint installation

4. Remove the ABS speed sensor from the knuckle.

5. Disconnect the tie rod end from the knuckle using a ball joint remover. Be careful not to damage the joint boot.

6. Remove the cotter pin and castle nut from the lower arm ball joint. Separate the lower control arm from the knuckle.

7. Remove the cotter pin and castle nut. Separate the upper arm from the knuckle using the ball joint remover.

8. Remove the knuckle and hub by sliding the assembly off of the halfshaft. Tap the end of the halfshaft

with a plastic mallet to release it from the knuckle.

9. To remove the hub and rotor assembly from the knuckle, remove the four self-locking bolts from the back of the knuckle.

10. Remove the four bolts from the hub to separate it from the brake disc.

11. The bearing can be pressed off the hub with a hydraulic press. The inner race will stay on the hub and can be removed with a bearing puller. Any time the hub and bearing are separated, the wheel bearing must be replaced with a new one.

To install:

12. Clean all the parts and examine them for wear. A worn or damaged hub will cause premature bearing failure and should be replaced.

NOTE: When pressing on a new bearing, be sure to press only on the inner race or the bearing will be damaged.

13. Install the brake disc and torque the bolts to 40 ft. lbs. (55 Nm).

14. Install the hub assembly and torque the self-locking bolts to 33 ft. lbs. (45 Nm).

NOTE: Make sure that all the hub bolts are properly torqued to avoid warpage of the brake disc.

15. Install the knuckle and hub assembly onto the halfshaft.

16. Install the knuckle on the tie rod, and the upper and lower control arms. Torque the lower ball joint nut to 40 ft. lbs. (54 Nm) and tighten as required to install a new cotter pin.

17. Torque the upper ball joint nut to 32 ft. lbs. (44 Nm) and tighten as required to install a new cotter pin. Torque the tie rod end to 36 ft. lbs. (50 Nm) and tighten as required to install a new cotter pin. Install the knuckle protector.

18. Install the speed sensor, sensor wire, and mounting bolts. Be careful to avoid twisting the wires.

19. Install the brake caliper, brake hoses, and mounting bolts.

20. With the wheel installed and the vehicle on the ground, torque the spindle nut to 180 ft. lbs. (250 Nm) and stake it in place. Tighten the wheel nuts to 80 ft. lbs. (110 Nm).

1995 Legend and 3.2TL

NOTE: These special tools, or their equivalents are recommended by Acura for removing and replacing wheel bearings:

- #07749-0010000: driver
- #07HAD-SG00100: driver attachment
- #07746-0010500: wheel bearing driver attachment
- #07GAF-SD40700: hub assembly base
- #07GAF-SE00100: hub assembly tool

1. Pry the lock tab away from the spindle and loosen the nut. Slightly loosen the lug nuts.

2. Raise and safely support the vehicle. Remove the front wheel and spindle nut.

3. Remove the wheel sensor from the knuckle, but do not disconnect it.

4. Remove the caliper mounting bolts. Hang the caliper out of the way with a piece of wire.

5. Remove the brake rotor retaining screws. Screw both 12mm bolts into the disc brake removal holes and turn the bolts to press the rotor from the hub. Only turn each bolt 2 turns at a time to prevent cocking the disc.

6. Remove the tie rod from the knuckle using a tie rod end removal tool. Use care not to damage the ball joint seals.

7. Remove the cotter pin from the lower arm ball joint and remove the castle nut.

NOTE: The lower ball joints cannot be separated from the steering knuckle

8. Remove the lower control arm from the knuckle using the ball joint removal tool.

9. Remove the cotter pin from the upper arm ball joint and remove the castle nut.

10. Remove the upper arm from the knuckle using the ball joint remover.

11. Remove the knuckle and hub by sliding the assembly off of the halfshaft. Be sure to clean any dirt or grease off of the ball joints.

12. The hub can be removed with a slide hammer. Clamp the knuckle in a vise and secure the slide hammer to the wheel studs.

13. Remove the splash guard and snaprings.

14. Support the knuckle and press the bearing out towards the wheel side.

15. If the inner bearing race stayed on the hub, use a puller to remove it.

To install:

16. Clean all parts and examine for wear and damage.

17. When pressing in a new bearing, install the inner snapring first and press the bearing in from the wheel side. Be sure to press only on the outer race or the bearing will be damaged.

18. Install the outer snapring and the splash guard.

19. Properly support the knuckle and press the hub into the bearing. Do not press on the wheel studs or they will press out of the hub. Support the knuckle by the inner race or the bearing will be damaged. Be sure to lubricate the bearings.

20. Install the knuckle/hub/bearing assembly onto the halfshaft and reassemble the knuckle to the upper and lower control arms.

21. Torque the lower ball joint nut to 51-58 ft. lbs. (70-80 Nm) and tighten only enough to install a new cotter pin.

22. Torque the upper ball joint nut to 29-35 ft. lbs. (40-48 Nm) and tighten only enough to install a new cotter pin. Torque the tie rod end to 36-43 ft. lbs. (50-60 Nm) and tighten only enough to install a new cotter pin.

23. Install the disc brake rotor, caliper, mounting bolts, and brackets. Reconnect the wheel sensor bracket to the knuckle.

24. With the wheel installed and all 4 wheels on the ground, torque the spindle nut to specification and stake it in place.

Lower Control Arms

REMOVAL AND INSTALLATION

Integra, Vigor, 2.5TL and 1993–94 Legend

1. Raise and safely support the vehicle. Remove the front wheels.

2. Remove the lower bolt from the damper fork. Discard the used locknut.

3. Disconnect the stabilizer bar from the arm. Discard the used locknut.

4. Remove the lower ball joint nut. Using a ball joint removal tool (part # 07MAC–SL00200 or equivalent), separate the ball joint from the steering knuckle.

5. Remove the flange bolt from the front of the control arm.

6. Remove the nut and washer from the rear of the control arm. Discard the used locknut.

7. Remove the lower control arm from the vehicle.

To install:

8. Be sure to replace the self–locking nuts after removal. Wipe off all dirt, oil or grease on the threads before tightening the fasteners. Use a silicone grease on the rubber bushings upon installation.

9. Reconnect the lower control arm to the radius rod/arm and to the chassis. Install the mounting bolts but do not tighten until the vehicle is on the ground.

10. Reconnect the lower arm ball joint to the steering knuckle.

11. For 1993–94 Legend, torque the ball joint nut to 54 ft. lbs. (75 Nm) and tighten as required to insert a new cotter pin.

12. For 1993–94 Integra and Vigor, torque the lower ball joint nut to 40 ft. lbs. (55 Nm) and tighten as required to install a new cotter pin.

13. For 1994–97 Integra and 2.5TL, torque the lower ball joint to 36–43 ft. lbs. (49–59 Nm).

14. Reconnect the stabilizer bar to the arm and install the damper fork and connecting bolt. Do not tighten the stabilizer bar bolts until the vehicle is on the ground.

15. Install the front wheels and carefully lower the vehicle.

16. Torque the radius rod–to–control arm bolts as follows:
   1993 Integra: 76 ft. lbs. (105 Nm)
   1994–97 Integra: 61 ft. lbs. (85 Nm)
   1993–94 Legend: 76 ft. lbs. (105 Nm)
   Vigor and 2.5TL: 76 ft. lbs. (105 Nm)

17. Torque the chassis bolts as follows:
   1993 Integra: 43 ft. lbs. (60 Nm)
   1994–97 Integra: 47 ft. lbs. (64 Nm)
   1993–94 Legend: 90 ft. lbs. (125 Nm)
   Vigor and 2.5TL: 39 ft. lbs. (55 Nm)

1995 Legend and 3.2TL

1. Raise and safely support the vehicle. Remove the front wheels.

2. Remove the damper fork lower bolt.

3. Support the lower control arm assembly with a floor jack.

4. Disconnect the stabilizer bar from the lower arm.

5. Remove the lower arm ball joint castle nut. Separate the ball joint from the steering knuckle.

6. Remove and discard the nut and bolt from the rear of the control arm.

7. Remove and discard the nut and washer which attach the control arm to the radius rod bushing.

8. Move the assembly downward and toward the rear of the vehicle to remove it.

To install:

NOTE: Always use new self-locking nuts and corrosion resistant bolts during reassembly.

9. Clean off all dirt, oil, or grease on the threads before tightening the fasteners. Use a silicone grease to lubricate the rubber bushings .

NOTE: All final torque adjustments should be made with the vehicle on the ground, or by using a floor jack to hold the vehicle's weight on the suspension.

10. Install the lower control arm to the radius rod bushing, and then the subframe. Install a new mounting bolt and nuts but do not tighten them until the vehicle is on the ground.

11. Reconnect the lower arm ball joint to the steering knuckle.

12. Torque the ball joint castle nut to 51–58 ft. lbs. (69–78 Nm) and tighten as required to insert a new cotter pin.

13. Reconnect the stabilizer bar to the arm, do not tighten the stabilizer bar bolts until the vehicle is on the ground.

14. Position the damper fork to the lower control arm then install the connecting bolt. Do not tighten the damper fork bolt until the vehicle is on the ground. Tighten damper fork bolt to 51 ft. lbs. (70 Nm).

15. Install the front wheels and lower the vehicle.

16. Torque the control arm flange bolt to 90 ft. lbs. (123 Nm). Torque the radius rod nut to 76 ft. lbs. (103 Nm).

17. Torque damper fork bolt to 51 ft. lbs. (70 Nm).

18. Torque the stabilizer bar link mounting nuts to 28 ft. lbs. (39 Nm).

Sway Bar

REMOVAL AND INSTALLATION

1. Raise and safely support the vehicle.

2. Remove the nuts attaching the sway bar to the lower control arms. Discard the used nuts.

3. Remove the nuts and bolts attaching the sway bar to the under body of the vehicle.

4. Remove the sway bar from the vehicle, noting the position of the bushings.

5. Disconnect the sway bar links from the sway bar, then remove the bushings. Discard the nuts attaching the links to the sway bar.

To install:

6. Before installation, inspect all parts and replace if worn or damaged. Apply silicone grease to the rubber bushings before reassembly. Do not tighten any nuts or bolts that

are for rubber mounts or bushings unless the car is on the ground.

7. Install the rubber mounts onto the stabilizer bar with the stabilizer mark aligned with the end of the bushing. Install the stabilizer bar into the vehicle along with the stabilizer brackets and mounting bolts. Be careful to ensure that the stabilizer bar is installed in the same way that it was removed.

8. Torque the stabilizer bracket bolts to 16 ft. lbs. (22 Nm), except on Vigor models, torque the bracket bolts to 29 ft. lbs. (40 Nm).

9. Install the stabilizer links between the control arms and stabilizer bar. Do not torque.

10. Install the wheels, then lower the vehicle to the ground.

11. Torque the stabilizer link fasteners as follows:
   1993–94 Integra: 16 ft. lbs. (22 Nm)
   Integra w/VTEC engine: 22 ft. lbs. (29 Nm)
   Vigor and 2.5TL: 14 ft. lbs. (19 Nm)
   Legend and 3.2TL, BOLTS: 28 ft. lbs. (39 Nm); NUTS 16 ft. lbs. (22 Nm)

Front Wheel Bearings

ADJUSTMENT

The wheel bearings are not adjustable or repairable and should be replaced if found defective.

REMOVAL AND INSTALLATION

Integra

1. Raise and safely support the vehicle.

2. Remove the front wheels, then pry the lock tab away and loosen the spindle nut.

3. Remove the brake hose mounting bolts.

4. Remove the brake caliper bolts and remove the caliper from the knuckle. Do not allow the caliper to hang by the brake hose, support it with a length of wire.

5. Remove the disc brake rotor.

6. If equipped with ABS, remove the wheel sensor wire bracket, then remove the wheel sensor from the knuckle. Do not disconnect the wheel sensor connector.

7. Remove the lower ball joint.

8. Remove the upper ball joint using a suitable ball joint removal tool (tool part # 07MAC–SL00200 or equivalent).

9. Pull the knuckle outward and remove the halfshaft outboard joint

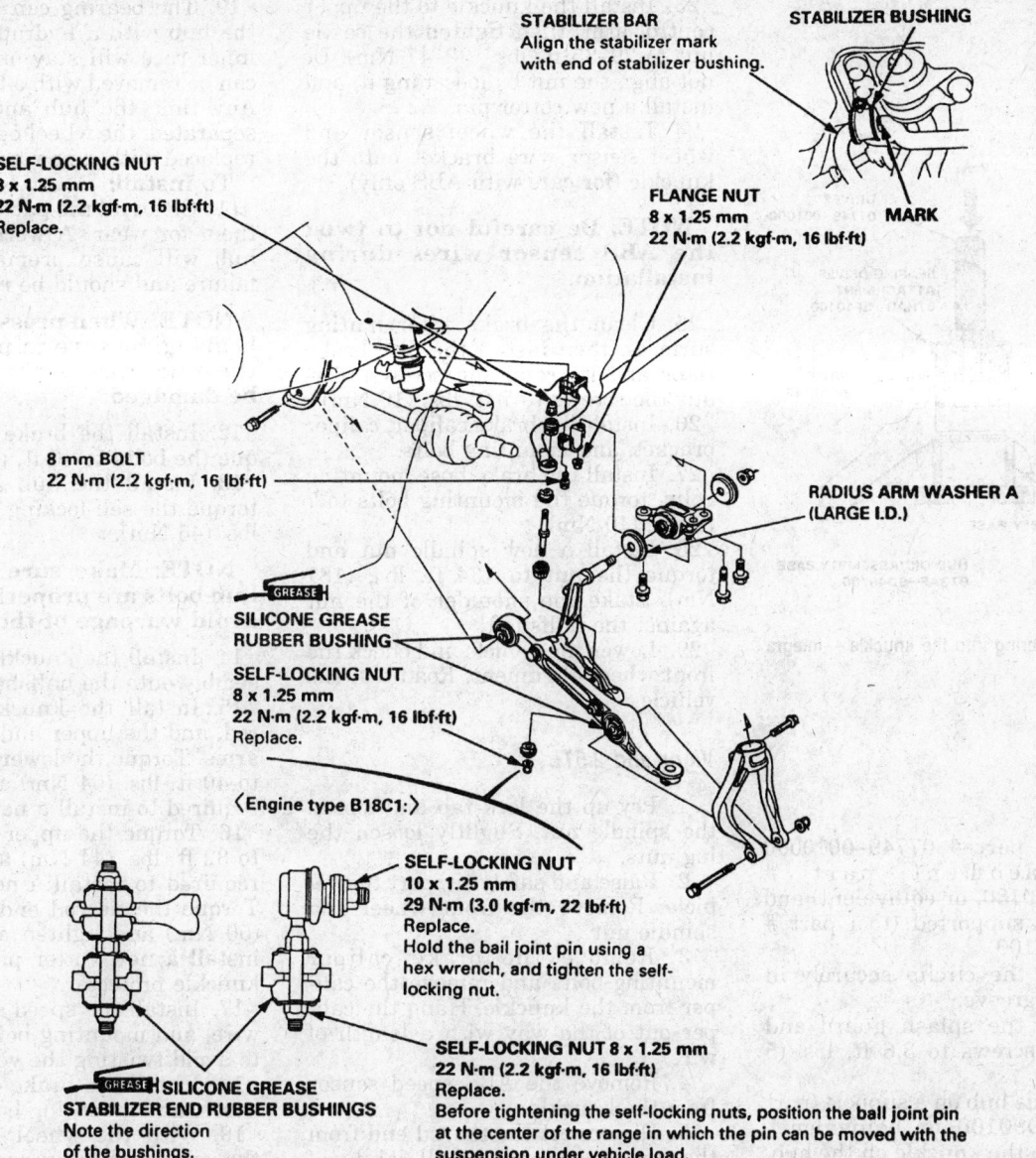

STABILIZER BAR
Align the stabilizer mark with end of stabilizer bushing.

STABILIZER BUSHING

MARK

SELF-LOCKING NUT
8 x 1.25 mm
22 N·m (2.2 kgf·m, 16 lbf·ft)
Replace.

FLANGE NUT
8 x 1.25 mm
22 N·m (2.2 kgf·m, 16 lbf·ft)

8 mm BOLT
22 N·m (2.2 kgf·m, 16 lbf·ft)

**RADIUS ARM WASHER A
(LARGE I.D.)**

GREASE

**SILICONE GREASE
RUBBER BUSHING**

SELF-LOCKING NUT
8 x 1.25 mm
22 N·m (2.2 kgf·m, 16 lbf·ft)
Replace.

⟨Engine type B18C1:⟩

SELF-LOCKING NUT
10 x 1.25 mm
29 N·m (3.0 kgf·m, 22 lbf·ft)
Replace.
Hold the ball joint pin using a hex wrench, and tighten the self-locking nut.

GREASE **SILICONE GREASE
STABILIZER END RUBBER BUSHINGS**
Note the direction of the bushings.

SELF-LOCKING NUT 8 x 1.25 mm
22 N·m (2.2 kgf·m, 16 lbf·ft)
Replace.
Before tightening the self-locking nuts, position the ball joint pin at the center of the range in which the pin can be moved with the suspension under vehicle load.

118954

Stabilizer bar and links — 1993-94 Integra and Vigor

**OIL SEAL DRIVER
07947–6340000**

Press

**HUB DIS/ASSEMBLY BASE
07GAF–SE00401**

218010

Separating the bearing from the knuckle — Integra

from the knuckle using a plastic hammer, then remove the knuckle.

10. Place the knuckle on a base (tool part # 07GAF–SD40700 or equivalent), take care not to distort the splash shield. Insert a disassembly tool into the hub (tool part # 07GAF–SE00100 or equivalent), then using a press, remove the hub from the knuckle. Hold onto the hub to keep it from falling when pressed clear.

11. Remove the knuckle ring from the rear of the knuckle.

12. Remove the circlip from the knuckle, then remove the splash guard.

13. Place the knuckle on the disassembly base and install a driver (driver tool part # 07749–0010000 and attachment tool part # 07746–0010500, or equivalent) to the bearing. Using a press, remove the bearing from the knuckle.

14. Press the wheel bearing inner race from the hub using the hub disassembly tool (part # 07GAF–SE00100) and a bearing separator.

To install:

15. Remove the old grease from the hub and knuckle and thoroughly dry and wipe clean all components.

16. Press a new wheel bearing into the knuckle with a suitable driver

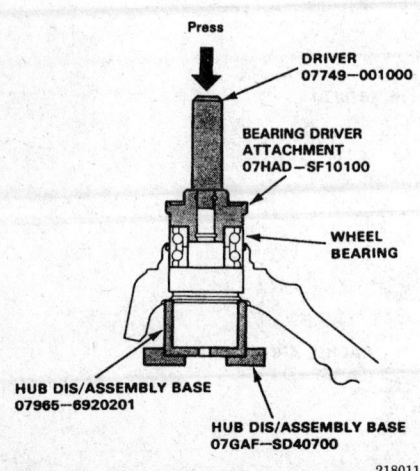

Pressing the bearing into the knuckle — Integra

(Driver tool part # 07749-0010000 and attachment part # 07HAD–SF10100, or equivalent) and the knuckle supported (tool part # 07965–SD90100).

17. Install the circlip securely in the knuckle groove.

18. Install the splash guard and tighten the screws to 3.6 ft. lbs. (5 Nm).

19. Place the hub on a support (part # 07965–SD90100 or equivalent), then position the knuckle on the hub. Use a suitable driver (Driver tool part # 07749-0010000 and attachment part # 07HAD–SF10100, or equivalent) to press the knuckle onto the hub.

20. Install the knuckle ring to the rear of the knuckle.

21. Install the knuckle/hub assembly onto the halfshaft.

22. Install the knuckle to the lower control arm and the tie rod end. Torque the lower ball joint nut to 36–43 ft. lbs. (49–59 Nm) and torque the tie rod end castle nut to 29–35 ft. lbs. (39–47 Nm). Do not align the nuts by loosening them, and install new cotter pins.

——— WARNING ———
Be careful not to damage the ball joint boots.

23. Install the knuckle to the upper control arm, then tighten the castle nut to 29–35 ft. lbs. (39–47 Nm). Do not align the nut by loosening it, and install a new cotter pin.

24. Install the wheel sensor and wheel sensor wire bracket onto the knuckle (for cars with ABS only).

NOTE: Be careful not to twist the ABS sensor wires during installation.

25. Clean the brake rotor mating surfaces, then install the disc brake rotor and its retaining screws. Torque the screws to 7 ft. lbs. (10 Nm).

26. Install the brake caliper, caliper bracket, and mounting bolts.

27. Install the brake hose mounting bolts, torque the mounting bolts to 7 ft. lbs. (10 Nm).

28. Install a new spindle nut and torque the nut to 134 ft. lbs. (181 Nm). Stake the shoulder of the nut against the halfshaft.

29. Lower the vehicle and check the front wheel alignment. Road test the vehicle.

Vigor and 2.5TL

1. Pry up the lock tab and loosen the spindle nut. Slightly loosen the lug nuts.

2. Raise and safely support the vehicle. Remove the front wheel and spindle nut.

3. Remove the brake caliper mounting bolts and remove the caliper from the knuckle. Hang the caliper out of the way with a length of wire.

4. Remove the ABS speed sensor from the knuckle.

5. Disconnect the tie rod end from the knuckle using a ball joint remover. Be careful not to damage the joint boot.

6. Remove the cotter pin and castle nut from the lower arm ball joint. Separate the lower control arm from the knuckle using a ball joint remover.

7. Remove the cotter pin and castle nut. Separate the upper arm from the knuckle using the ball joint remover.

8. Remove the knuckle and hub by sliding the assembly off of the halfshaft. Tap the end of the halfshaft with a plastic mallet to release it from the knuckle.

9. To remove the hub and rotor assembly from the knuckle, remove the four self-locking bolts from the back of the knuckle. Remove the four bolts from the hub to separate it from the brake disc.

10. The bearing can be pressed off the hub with a hydraulic press. The inner race will stay on the hub and can be removed with a bearing puller. Any time the hub and bearing are separated, the wheel bearing must be replaced with a new one.

To install:

11. Clean all the parts and examine them for wear. A worn or damaged hub will cause premature bearing failure and should be replaced.

NOTE: When pressing on a new bearing, be sure to press only on the inner race or the bearing will be damaged.

12. Install the brake disc and torque the bolts to 40 ft. lbs. (55 Nm).

13. Install the hub assembly and torque the self-locking bolts to 33 ft. lbs. (45 Nm).

NOTE: Make sure that all the hub bolts are properly torqued to avoid warpage of the brake disc.

14. Install the knuckle and hub assembly onto the halfshaft.

15. Install the knuckle on the tie rod, and the upper and lower control arms. Torque the lower ball joint nut to 40 ft. lbs. (54 Nm) and tighten as required to install a new cotter pin.

16. Torque the upper ball joint nut to 32 ft. lbs. (44 Nm) and tighten as required to install a new cotter pin. Torque the tie rod end to 36 ft. lbs. (50 Nm) and tighten as required to install a new cotter pin. Install the knuckle protector.

17. Install the speed sensor, sensor wire, and mounting bolts. Be careful to avoid twisting the wires.

18. Install the brake caliper, brake hoses, and mounting bolts.

19. With the wheel installed and the vehicle on the ground, torque the spindle nut to 180 ft. lbs. (250 Nm) and stake it in place. Tighten the wheel nuts to 80 ft. lbs. (110 Nm).

Legend and 3.2TL

NOTE: These special tools, or their equivalents are recommended by Acura for removing and replacing wheel bearings:

- #07749-0010000: driver
- #07HAD–SG00100: driver attachment
- #07746-0010500: wheel bearing driver attachment
- #07GAF-SD40700: hub assembly base
- #07GAF-SE00100: hub assembly tool

1. Pry the lock tab away from the spindle and loosen the nut. Slightly loosen the lug nuts.

2. Raise and safely support the vehicle. Remove the front wheel and spindle nut.

3. Remove the wheel sensor from the knuckle, but do not disconnect it.

4. Remove the caliper mounting bolts. Hang the caliper out of the way with a piece of wire.

5. Remove the brake rotor retaining screws. Screw both 12mm bolts into the disc brake removal holes and turn the bolts to press the rotor from the hub. Only turn each bolt 2 turns at a time to prevent cocking the disc.

6. Remove the tie rod from the knuckle using a tie rod end removal tool. Use care not to damage the ball joint seals.

7. Remove the cotter pin from the lower arm ball joint and remove the castle nut.

NOTE: The lower ball joints cannot be separated from the steering knuckle

8. Remove the lower control arm from the knuckle using the ball joint removal tool.

9. Remove the cotter pin from the upper arm ball joint and remove the castle nut.

10. Remove the upper arm from the knuckle using the ball joint remover.

11. Remove the knuckle and hub by sliding the assembly off of the half-

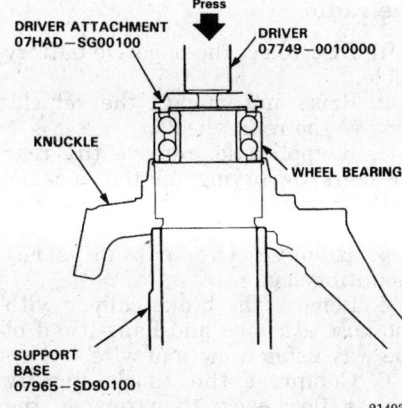

Installing the wheel bearing — Legend and 3.2TL

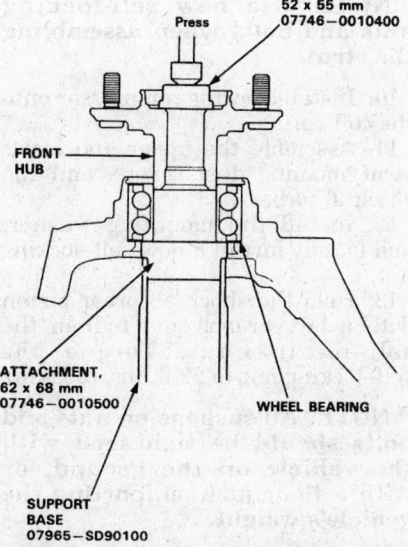

Pressing out the wheel bearing — Legend and 3.2TL

Hub installation — Legend and 3.2TL

shaft. Be sure to clean any dirt or grease off of the ball joints.

12. The hub can be removed with a slide hammer. Clamp the knuckle in a vise and secure the slide hammer to the wheel studs.

13. Remove the splash guard and snaprings.

14. Support the knuckle and press the bearing out towards the wheel side.

15. If the inner bearing race stayed on the hub, use a puller to remove it.

To install:

16. Clean all parts and examine for wear and damage.

17. When pressing in a new bearing, install the inner snapring first and press the bearing in from the wheel side. Be sure to press only on the outer race or the bearing will be damaged.

18. Install the outer snapring and the splash guard.

19. Properly support the knuckle and press the hub into the bearing. Do not press on the wheel studs or they will press out of the hub. Support the knuckle by the inner race or the bearing will be damaged. Be sure to lubricate the bearings.

20. Install the knuckle/hub/bearing assembly onto the halfshaft and reassemble the knuckle to the upper and lower control arms.

21. Torque the lower ball joint nut to 51-58 ft. lbs. (70-80 Nm) and tighten only enough to install a new cotter pin.

22. Torque the upper ball joint nut to 29-35 ft. lbs. (40-48 Nm) and tighten only enough to install a new cotter pin. Torque the tie rod end to 36-43 ft. lbs. (50-60 Nm) and tighten only enough to install a new cotter pin.

23. Install the disc brake rotor, caliper, mounting bolts, and brackets. Reconnect the wheel sensor bracket to the knuckle.

24. With the wheel installed and all 4 wheels on the ground. For Legend models, torque the spindle nut to 242 ft. lbs. (335 Nm). For 3.2TL, torque the spindle nut to 181 ft. lbs. (245 Nm). Stake the nut in place.

REAR SUSPENSION

Strut

REMOVAL AND INSTALLATION

Integra

1. Raise and safely support the vehicle and remove the rear wheels.
2. Remove the upper strut mount cover from the rear panel, just below the speaker. On sedans, remove the trunk side panel.
3. Remove the damper cover, then remove the upper mount nuts.
4. On cars with ABS, remove the wheel sensor wire brackets but do not disconnect the wheel sensor connector.
5. Remove the lower strut mounting bolt.
6. Remove the flange bolt the at connects the lower arm to the trailing arm.
7. Lower the rear suspension and remove the strut assembly from the vehicle.
8. Use a spring compressor to remove the spring from the strut assembly.

To install:

9. Reassemble the strut absorber and coil spring assembly. Torque the damper self–locking nut to 22 ft. lbs. (30 Nm).
10. Lower the rear suspension and position the strut assembly in the vehicle. The nut welded to the lower strut mounting should face the front of the vehicle.
11. Loosely install the upper mounting nuts.
12. Raise the rear suspension to install the lower strut mounting bolt and the bolt connecting the lower arm to the trailing arm.
13. Raise the vehicle until the vehicle just lifts off the safety stand and tighten the lower strut bolt and lower control arm bolt. Torque the lower strut bolt and the control arm bolt to 40 ft. lbs. (54 Nm).
14. Install the wheel sensor wire bracket on cars with ABS. Torque the mounting bolts to 7 ft. lbs. (10 Nm).
15. Torque the upper mounting nuts to 36 ft. lbs. (49 Nm).
16. Install the rear wheels then lower the vehicle.
17. Install the access panel, or the trunk side panel.
18. Check the vehicle's alignment.

Legend

NOTE: The radio may contain a coded theft protection circuit. Make sure you have the security code before disconnecting the battery, radio fuse, or removing the radio.

1. Disconnect the negative battery cable.
2. Raise and support the vehicle, remove the rear wheels.
3. If applicable, remove the rear speakers by prying off the speaker grilles and removing the four retaining screws.
4. Remove the rubber strut mounting cap.
5. Remove the brake caliper with the hose attached and hang it out of the way using a piece of wire.
6. Compress the shock slightly with a floor jack; then remove the shock mounting bolt from the knuckle.
7. Unbolt the three flange bolts and remove the strut assembly from the shock tower.
8. Install a spring compressor onto the strut assembly and tighten the compressor according to the manufacturer's instructions.
9. Remove the locking nut from the top of the shock absorber, disassemble the strut, and remove the coil spring.

To install:

NOTE: Use new self-locking nuts and bolts when assembling the strut.

10. Install a spring compressor onto the coil spring.
11. Assemble the upper and lower strut mounts, dust covers, and the shock absorber.
12. Install the mounting washer, and loosely install a new self–locking nut.
13. Hold the shock absorber piston with a hex wrench and tighten the self–locking nut. Torque the self–locking nut to 22 ft. lbs. (30 Nm).

NOTE: All suspension nuts and bolts should be tightened with the vehicle on the ground, or with a floor jack supporting the vehicle's weight.

14. Install the strut assembly into the vehicle. Tighten the upper mounting nuts to 28 ft. lbs. (39 Nm).
15. Install the shock mounting bolt at the knuckle and tighten to 76 ft. lbs. (105 Nm).
16. Install the brake caliper. Tighten the mounting bolts to 28 ft. lbs, (39 Nm).

17. Install the wheel, and tighten the wheel nuts to 80 ft. lbs. (110 Nm).
18. Install the strut mount caps and the rear speakers. Reconnect the negative battery cable.

Vigor, 2.5TL and 3.2TL

1. Raise and safely support the vehicle and remove the rear wheels.
2. Remove the rear seat:
   a. Remove the lower cushion bolt located under the armrest.
   b. Pull the rear of the lower cushion up and lift it forward to release it from the clips.
   c. Pull down the trunk bulkhead trim and release the armrest lid clips.
   d. Remove the three bolts from the back cushion, and then lift it up and forward to disengage the securing hooks.
3. Place a floor jack under the lower arm and slightly compress the spring.
4. Remove the upper mounting nuts and the lower flange bolt.
5. Lower the jack to remove the strut. Be sure and mark the right and left struts so they can be reinstalled on the proper sides.
6. Use a spring compressor to remove the spring from the struts.

To install:

7. Reassemble the spring and strut assembly. Torque the damper self–locking nut to 22 ft. lbs. (30 Nm).
8. Install the struts into the vehicle. Loosely install the mounting nuts and mounting bolt, but do not torque them until the weight of the car is on the suspension.
9. Raise the rear suspension with a floor jack until the weight of the car is on the strut. Torque the upper mounting nuts to 28 ft. lbs. (39 Nm), then torque the lower mounting bolts to 40 ft. lbs. (55 Nm). Be careful not to pinch the ABS speed sensor wire between the damper and bracket.
10. Install the rear wheels and lower the vehicle.
11. Install the rear seat cushions.

Coil Spring

REMOVAL AND INSTALLATION

Vigor, 1993–94 Integra and Legend

1. Remove the damper unit and make a note of the spring seat and bracket positions for reassembly.
2. Install the damper into a spring compressor and tighten the compressor according to manufacturer's instructions.

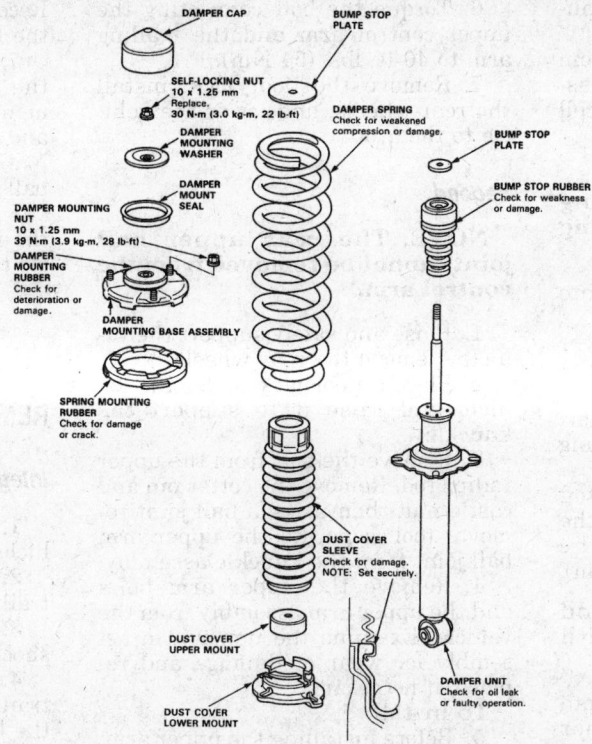

Rear strut - exploded view — Legend

218030

Labels within the exploded view:

DAMPER CAP

BUMP STOP PLATE

SELF-LOCKING NUT
10 x 1.25 mm
Replace.
30 N·m (3.0 kg-m, 22 lb-ft)

DAMPER SPRING
Check for weakened compression or damage.

BUMP STOP PLATE

DAMPER MOUNTING WASHER

DAMPER MOUNT SEAL

BUMP STOP RUBBER
Check for weakness or damage.

DAMPER MOUNTING NUT
10 x 1.25 mm
39 N·m (3.9 kg-m, 28 lb-ft)

DAMPER MOUNTING RUBBER
Check for deterioration or damage.

DAMPER MOUNTING BASE ASSEMBLY

SPRING MOUNTING RUBBER
Check for damage or crack.

DUST COVER SLEEVE
Check for damage.
NOTE: Set securely.

DUST COVER UPPER MOUNT

DUST COVER LOWER MOUNT

DAMPER UNIT
Check for oil leak or faulty operation.

NOTE: Mark the right and left dampers or store them separately so they are reinstalled on the proper sides.

3. Remove the locking nut from the top of the shock absorber and disassemble the damper and spring as required.

To install:

4. Install the spring into a spring compressor and compressor according to the manufacturer's instructions.

5. Install the damper mounting rubber and the damper mounting washer, and loosely install a new 10mm locking nut.

6. Torque the self–locking damper mounting nut to 22 ft. lbs. (30 Nm).

1995–97 Integra

1. Remove the strut assembly from the vehicle. Make a note of the spring seat and bracket positions for reinstallation.

2. Install the strut into a spring compressor and tighten the compressor according to the manufacturer's instructions.

3. Remove the locknut from the top of the strut and discard.

4. Remove the strut assembly from the compressor.

5. Remove the upper mounting components and spring as required.

6. Remove the dust cover components and the bump stop components.

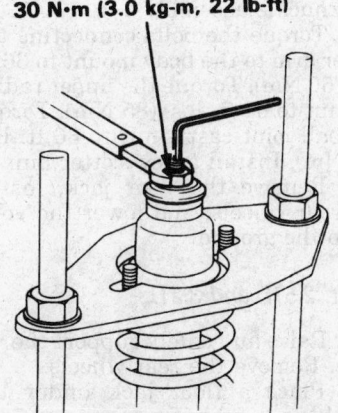

30 N·m (3.0 kg-m, 22 lb-ft)

116744

Self-locking damper nut — 1993–94 Integra, Vigor (All) and 1993–94 Legend

To install:

7. Install the bump stop components and the dust cover components to the strut.

8. Position the upper strut mounting components in their proper sequence, then install the assembly into the spring compressor.

9. Compress the strut assembly with the spring compressor.

10. Install the damper mounting washer, and loosely install a new self–locking nut.

11. Hold the damper shaft with a hex wrench and tighten the self–locking nut. Torque the self–locking nut to 22 ft. lbs. (30 Nm).

12. Remove the strut assembly from the spring compressor and install in the vehicle.

1995 Legend, 2.5TL and 3.2TL

1. Remove the rear seat cushions.

2. Raise and safely support the vehicle, remove the rear wheels.

3. Use a jack under the knuckle to compress the coil spring.

4. Remove the shock mounting bolt from the knuckle.

5. Unbolt the three flange bolts and remove the strut assembly from the shock tower.

6. Install a spring compressor onto the strut assembly and tighten the

compressor according to the manufacturer's instructions.

7. Remove the locking nut from the top of the shock absorber, disassemble the strut, and remove the coil spring.

To install:

NOTE: Use new self-locking nuts and bolts when assembling the strut.

8. Install a spring compressor onto the coil spring.

9. Assemble the upper and lower strut mounts, dust covers, and the shock absorber.

10. Install the mounting washer, and loosely install a new self–locking nut.

11. Hold the shock absorber piston with a hex wrench and tighten the self–locking nut. Torque the self–locking nut to 22 ft. lbs. (30 Nm).

NOTE: All suspension nuts and bolts should be tightened with the vehicle on the ground.

12. Install the strut assembly into the vehicle. Tighten the upper mounting nuts to 28 ft. lbs. (39 Nm).

13. Install the shock mounting bolt at the knuckle and tighten to 40 ft. lbs. (54 Nm).

14. Install the wheel, and tighten the wheel nuts to 80 ft. lbs. (110 Nm).

15. Install the rear seat cushions.

16. Check the vehicle's rear wheel alignment.

Upper Control Arms

REMOVAL AND INSTALLATION

Integra

1. Raise and safely support the vehicle. Remove the rear wheels.

2. Place a floor jack under the vehicle and raise it to support the trailing arm assembly.

3. Remove the upper arm from the trailing arm assembly, then remove the bolt connecting the upper control arm to the body. Remove the upper control arm from the vehicle.

To install:

4. Before installation, inspect the upper arm for wear or damage and replace, if necessary. Apply a light coating of silicone grease to the bushings.

5. Install the upper arm into the vehicle. With the floor jack raised and placing a load on the suspension, torque the bolt connecting the upper control arm and the body to 29 ft. lbs. (39 Nm)

6. Torque the bolt connecting the upper control arm and the trailing arm to 40 ft. lbs. (54 Nm).

7. Remove the floor jack, install the rear wheels, and lower the vehicle to the ground.

Legend

NOTE: The rear upper ball joint cannot be removed from the control arm.

1. Raise and safely support the vehicle. Remove the rear wheels.

2. Place a floor jack under the vehicle and raise it to support the knuckle.

3. Remove the bolt from the upper radius rod. Remove the cotter pin and castle nut, then using a ball joint removal tool, separate the upper arm ball joint from the knuckle assembly.

4. Remove the upper arm bolts and the upper arm assembly from the vehicle. Examine the upper arm assembly for wear or damage and replace, if necessary.

To install:

5. Before installing the upper arm, apply a light coating of silicone grease to each rubber bushing.

6. Install the upper arm into the vehicle and the upper radius rod into the upper arm. Install the upper arm bolts.

7. Install the upper arm ball joint into the knuckle and install the castle nut. Apply a load to the suspension by raising the floor jack under the knuckle assembly.

8. Torque the bolts connecting the upper arm to the body mount to 36 ft. lbs. (50 Nm). Torque the upper radius rod nut to 61 ft. lbs. (85 Nm). Torque the ball joint castle nut to 36 ft. lbs. (50 Nm). Install a new cotter pin.

9. Remove the floor jack, install the rear wheels, and lower the vehicle to the ground.

Vigor, 2.5TL and 3.2TL

1. Raise and safely support the vehicle. Remove the rear wheels.

2. Place a floor jack under the knuckle assembly for support.

3. Remove the upper arm ball joint cap.

4. Remove the cotter pin and ball joint castle nut. Position a ball joint puller between the knuckle and upper arm; then, separate the knuckle from the upper arm.

5. Unbolt the upper arm and remove it from the vehicle.

To install:

6. Install the upper arm and its mounting bolts onto the vehicle.

Reconnect the upper arm ball joint to the knuckle.

7. With the floor jack supporting the knuckle assembly, torque the mounting bolts to 28 ft. lbs. (39 Nm), and the ball joint nut to 32 ft. lbs. (44 Nm). Install a new cotter pin. Install ball joint cover.

8. Remove the floor jack, install the rear wheels, and lower the vehicle to the ground.

Lower Control Arms

REMOVAL AND INSTALLATION

Integra

1. Raise and safely support the vehicle. Remove the rear wheels.

2. Place a floor jack under the trailing arm and support it.

3. Unbolt the stabilizer link and shock absorber from the lower arm.

4. Unbolt the lower arm from the trailing arm assembly, then unbolt the lower arm from the body bracket and remove the lower arm from the vehicle.

5. If the bushings in the lower arm are to be replaced, they must be removed using a press tool suitable for bushing replacement.

To install:

6. Before installing the new bushings, apply a light coating of silicone grease on each bushing. Install the bushings with a proper bushing installation tool.

7. Install the lower arm into the body bracket and tighten the bolt. Do not torque the bolt until the vehicle is under load.

8. Install the lower arm into the trailing arm assembly, but do not torque the bolt.

9. Reconnect the stabilizer link and shock absorber to the lower arm. Install the bolts but do not torque.

10. Apply vehicle load to the suspension by raising the floor jack under the trailing arm assembly. Torque the lower arm bolts to 40 ft. lbs. (54 Nm). Torque the stabilizer link bolt 29 ft. lbs. (39 Nm). Torque the shock absorber bolt to 40 ft. lbs. (54 Nm).

11. Install the rear wheels, remove the floor jack, and lower the vehicle.

Vigor and 1993–94 Legend

1. Raise and safely support the vehicle. Remove the rear wheels.

2. Place a floor jack under and support the rear knuckle.

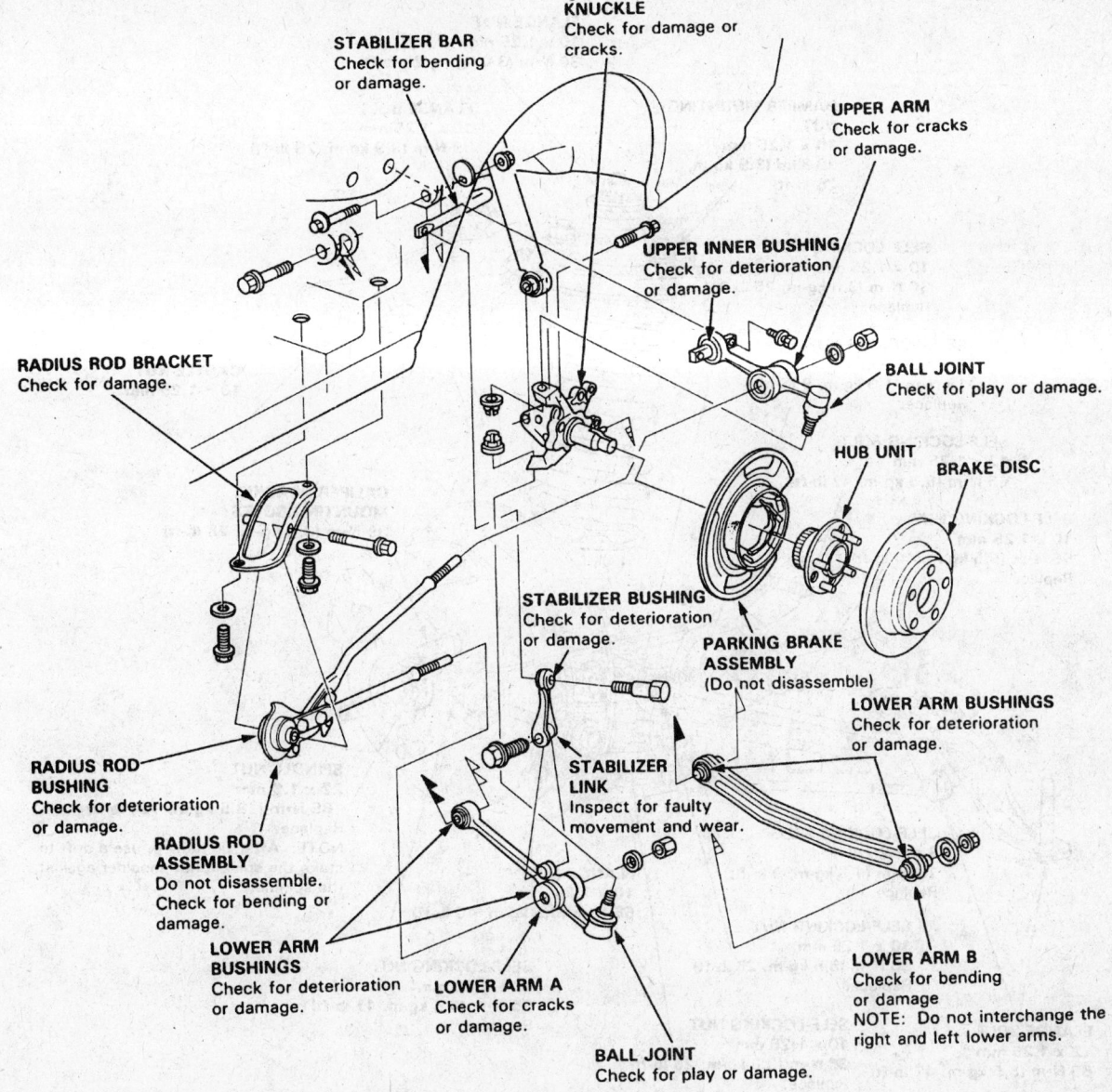

KNUCKLE
Check for damage or
cracks.

STABILIZER BAR
Check for bending
or damage.

UPPER ARM
Check for cracks
or damage.

UPPER INNER BUSHING
Check for deterioration
or damage.

RADIUS ROD BRACKET
Check for damage.

BALL JOINT
Check for play or damage.

HUB UNIT

BRAKE DISC

STABILIZER BUSHING
Check for deterioration
or damage.

**PARKING BRAKE
ASSEMBLY**
(Do not disassemble)

LOWER ARM BUSHINGS
Check for deterioration
or damage.

**RADIUS ROD
BUSHING**
Check for deterioration
or damage.

**STABILIZER
LINK**
Inspect for faulty
movement and wear.

**RADIUS ROD
ASSEMBLY**
Do not disassemble.
Check for bending or
damage.

**LOWER ARM
BUSHINGS**
Check for deterioration
or damage.

LOWER ARM A
Check for cracks
or damage.

LOWER ARM B
Check for bending
or damage
NOTE: Do not interchange the
right and left lower arms.

BALL JOINT
Check for play or damage.

250894

Rear suspension assembly — Legend

3. Unbolt the stabilizer link and radius rod from lower arm A on 1993–94 Legend models.

4. Unbolt lower arm A and lower arm B from the knuckle assembly. On 1993–94 Legend models, you must remove the cotter pin, castle nut and use a suitable ball joint removal tool to separate lower arm A from the knuckle.

5. Remove lower arm A and lower arm B from both body brackets. Remove the bushings with a proper bushing removal tool.

To install:

6. Install the new bushings into the lower arms using a suitable bushing installation tool. Apply a light coating of silicone grease to the bushing before installation.

7. Install both lower arms into the body brackets. Tighten the body bracket bolts but do not torque until the vehicle is on the ground.

8. Install lower arm A and lower arm B on the knuckle assembly. Tighten the bolt, but do not torque. On the 1993–94 Legend models, install the lower arm A ball joint into the knuckle assembly and torque the castle nut to 36 ft. lbs. (50 Nm), then install a new cotter pin.

9. On the 1993–94 Legend models, install the stabilizer link and radius rod to lower arm A and tighten the nuts, but do not torque them.

10. Lower the floor jack and remove it from under the vehicle. Install the rear wheels and lower the vehicle to the ground.

11. Torque the lower arm A bolts to 40 ft. lbs. (55 Nm), except on 1991–94 Legend, torque the body bracket bolt to 76 ft. lbs. (105 Nm).

12. Torque the bolt for the lower arm B to the body to 47 ft. lbs. (65 Nm), except on 1993–94 Legend, torque the body bracket bolt to 40 ft. lbs. (55 Nm). Torque the knuckle bolts from the lower arm A and lower arm B to 47 ft. lbs. (65 Nm).

13. On the 1993–94 Legend models, torque the radius rod nut to 61 ft. lbs.

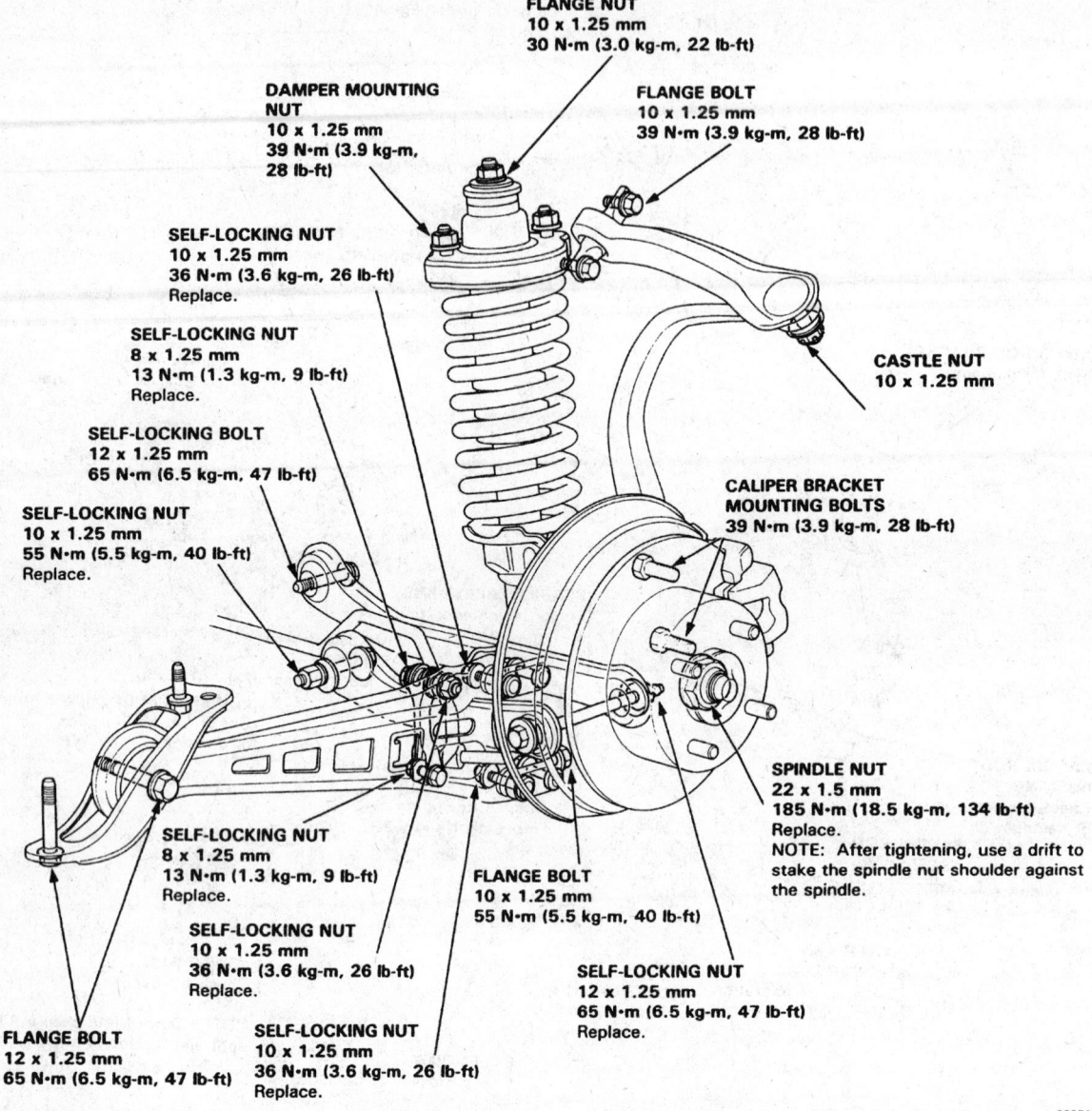

FLANGE NUT
10 x 1.25 mm
30 N·m (3.0 kg-m, 22 lb-ft)

DAMPER MOUNTING NUT
10 x 1.25 mm
39 N·m (3.9 kg-m, 28 lb-ft)

FLANGE BOLT
10 x 1.25 mm
39 N·m (3.9 kg-m, 28 lb-ft)

SELF-LOCKING NUT
10 x 1.25 mm
36 N·m (3.6 kg-m, 26 lb-ft)
Replace.

SELF-LOCKING NUT
8 x 1.25 mm
13 N·m (1.3 kg-m, 9 lb-ft)
Replace.

CASTLE NUT
10 x 1.25 mm

SELF-LOCKING BOLT
12 x 1.25 mm
65 N·m (6.5 kg-m, 47 lb-ft)

CALIPER BRACKET MOUNTING BOLTS
39 N·m (3.9 kg-m, 28 lb-ft)

SELF-LOCKING NUT
10 x 1.25 mm
55 N·m (5.5 kg-m, 40 lb-ft)
Replace.

SPINDLE NUT
22 x 1.5 mm
185 N·m (18.5 kg-m, 134 lb-ft)
Replace.
NOTE: After tightening, use a drift to stake the spindle nut shoulder against the spindle.

SELF-LOCKING NUT
8 x 1.25 mm
13 N·m (1.3 kg-m, 9 lb-ft)
Replace.

SELF-LOCKING NUT
10 x 1.25 mm
36 N·m (3.6 kg-m, 26 lb-ft)
Replace.

FLANGE BOLT
10 x 1.25 mm
55 N·m (5.5 kg-m, 40 lb-ft)

SELF-LOCKING NUT
12 x 1.25 mm
65 N·m (6.5 kg-m, 47 lb-ft)
Replace.

FLANGE BOLT
12 x 1.25 mm
65 N·m (6.5 kg-m, 47 lb-ft)

SELF-LOCKING NUT
10 x 1.25 mm
36 N·m (3.6 kg-m, 26 lb-ft)
Replace.

323519

Upper arm components — Vigor, 2.5TL and 3.2TL

(85 Nm), and the stabilizer link to 16 ft. lbs. (22 Nm).

1995 Legend

1. Raise and safely support the vehicle. Remove the rear wheels.
2. Place a floor jack under the rear knuckle for support.
3. Unbolt the wheel sensor from the lower arm.
4. Unbolt the stabilizer link and radius rod from lower arm A.
5. Unbolt lower arm A and lower arm B from the knuckle assembly. Remove the castle nut and use a suitable ball joint removal tool to separate lower arm A from the knuckle.

6. Remove lower arm A and lower arm B from both body brackets. Remove the bushings with a proper bushing removal tool.

To install:

7. Install the new bushings into the lower arms using a suitable bushing installation tool. Apply a light coating of silicone grease to the bushing before installation.
8. Install both lower arms into the body brackets. Tighten the body bracket bolts but do not torque until the vehicle is on the ground.
9. Install the lower arm A ball joint into the knuckle assembly and torque the castle nut to 36 ft. lbs. (50 Nm), then install a new cotter pin.

10. Install the stabilizer link and radius rod to lower arm A and tighten the nuts, but do not torque them.
11. Install the wheel sensor, tighten the bolts to 7 ft. lbs. (10 Nm).
12. Lower the floor jack and remove it from under the vehicle. Install the rear wheels and lower the vehicle to the ground.
13. Torque the lower arm A body bracket bolt to 76 ft. lbs. (105 Nm).
14. Torque the lower arm B body bracket bolt to 40 ft. lbs. (55 Nm).
15. Torque the lower ball joint castle nuts to 36-43 ft. lbs. (50-60 Nm). Torque the arm B to knuckle bolt to 47 ft. lbs. (65 Nm).

16. Torque the radius rod nut to 61 ft. lbs. (85 Nm), and the stabilizer link to 16 ft. lbs. (22 Nm).

2.5TL and 3.2TL

1. Raise and safely support the vehicle. Remove the rear wheels.
2. Place a floor jack under the rear knuckle for support.
3. Unbolt lower arm A and lower arm B from the knuckle assembly.
4. Remove lower arm A and lower arm B from their body brackets.

To install:

NOTE: Use new self-locking nuts when installing the lower control arms. Reinstall the toe adjuster cams (on the lower arm A body bracket) into their original positions.

5. Apply a light coating of silicon grease to the bushing before installing the control arms.
6. Install both lower arms into their body brackets. Tighten the body bracket bolts but do not torque them until the vehicle is on the ground.
7. Install lower arm A and lower arm B on the knuckle assembly. Tighten the bolts.
8. Lower the floor jack and remove it from under the vehicle.
9. Install the rear wheels and lower the vehicle to the ground.
10. Torque the control arm A flange bolt to 47 ft. lbs. (64 Nm). Torque the control arm B flange bolt to 40 ft. lbs. (54 Nm).
11. Torque the control arm knuckle bolt to 47 ft. lbs. (64 Nm).

Sway Bar

REMOVAL AND INSTALLATION

1. Raise and safely support the vehicle. Remove the rear wheels.
2. Unbolt the stabilizer link from the trailing arm.
3. Remove the mounting bolts from the stabilizer brackets and remove the brackets from the stabilizer bar.
4. Remove the stabilizer bar from the vehicle. Remove the stabilizer links from the stabilizer bar. Make sure that the stabilizer bar is installed in the same position that it was in before it was removed. Examine the components for damage and replace if necessary.

To install:

5. Apply silicone grease to the rubber bushings, and clean all components before installation.

6. Assemble the stabilizer links to the stabilizer bar and loosely install the nuts, but do not torque.
7. Place a floor jack under the lower arm of the rear suspension and raise the rear, putting the weight of the car on the rear suspension. Connect the stabilizer link to the trailing arm.
8. Torque the stabilizer bar bracket bolts to 16 ft. lbs. (22 Nm).
9. Torque the nut connecting the stabilizer link and the trailing arm as follows:
   Integra: 29 ft. lbs. (39 Nm)
   Legend: 16 ft. lbs. (22 Nm)
   Vigor, 2.5TL and 3.2TL: 9 ft. lbs. (13 Nm)
10. Torque the nut connecting the stabilizer link to the stabilizer bar as follows:
   Integra: 16 ft. lbs. (22 Nm)
   Legend: 22 ft. lbs. (30 Nm).
   Vigor, 2.5TL and 3.2TL: 9 ft. lbs. (13 Nm)

Wheel Bearings

ADJUSTMENT

The wheel bearings are not adjustable or repairable and should be replaced if found defective.

REMOVAL AND INSTALLATION

Integra and Legend

1. With the vehicle on the ground, remove the hub grease cap and pry the spindle nut lock tab away from the spindle. Loosen the nut.
2. Make sure the emergency brake is disengaged.
3. Raise and safely support the vehicle and remove the rear wheels.
4. Remove the brake hose mounting bolt from the knuckle.
5. Remove the caliper with the brake hose connected, and hang it out of the way with wire.
6. Remove the caliper bracket.
7. Remove the two brake disc retaining screws, and remove the disc by pressing it off with a pair of 8mm bolts threaded into the holes between the studs. Turn each bolt 2 turns at a time.
8. Remove the spindle nut and remove the hub assembly from the knuckle. The bearing is part of the hub and the assembly is replaced as one piece.
9. Be sure to wash the bearing and spindle thoroughly in solvent before reassembly.

To install:

10. Install the hub and bearing assembly onto the spindle and install the spindle nut. Tighten the spindle nut, but do not torque it until the vehicle is on the ground.
11. Install the brake disc and torque the disc retaining screws to 7 ft. lbs. (10 Nm).
12. Install the caliper mounting bracket and tighten the bolts to 28 ft. lbs. (39 Nm).
13. Install the brake caliper and the brake hose mounting bolts. For Integra, torque the mounting bolts to 28 ft. lbs. (38 Nm). For Legend, torque the caliper mounting bolts to 17 ft. lbs. (23 Nm).
14. Install the brake hose mounting bolt, torque the bolt to 16 ft. lbs. (22 Nm).
15. If applicable, install the brake caliper shield. Torque the mounting bolts to 7 ft. lbs. (10 Nm).
16. Tighten the brake disc retaining screws to 7 ft. lbs. (10 Nm).
17. Install the rear wheels and lower the vehicle to the ground.
18. Install a new hub nut and torque the nut for Integra models to 134 ft. lbs. (181 Nm). For Legend models, torque it to 206 ft. lbs. (285 Nm). Stake the spindle nut, and install the grease cap.

Vigor, 2.5TL and 3.2TL

1. With the vehicle on the ground, remove the hub cap and pry the spindle nut lock tab away from the spindle. Loosen the spindle nut.
2. Engage the parking brake to provide leverage to help with loosening the brake disc retaining screws.
3. Raise and safely support the vehicle and remove the rear wheels.
4. Remove the brake hose mounting bolt. Remove the caliper without disconnecting the hydraulic hose. Support the caliper with a piece of wire.
5. Remove the two disc retaining screws. Remove the disc by pressing it off with a pair of 8mm bolts threaded into the holes between the studs. Turn each bolt two turns at a time. Release the parking brake after removing the disc brake retaining screws, and before removing the brake disc.
6. Remove the spindle nut and remove the hub and bearing assembly from the knuckle. The wheel bearing is part of the hub assembly and the components are replaced as one unit.
7. Clean the bearing and spindle with solvent before reassembly. Clean the mating surfaces of the hub and brake disc.

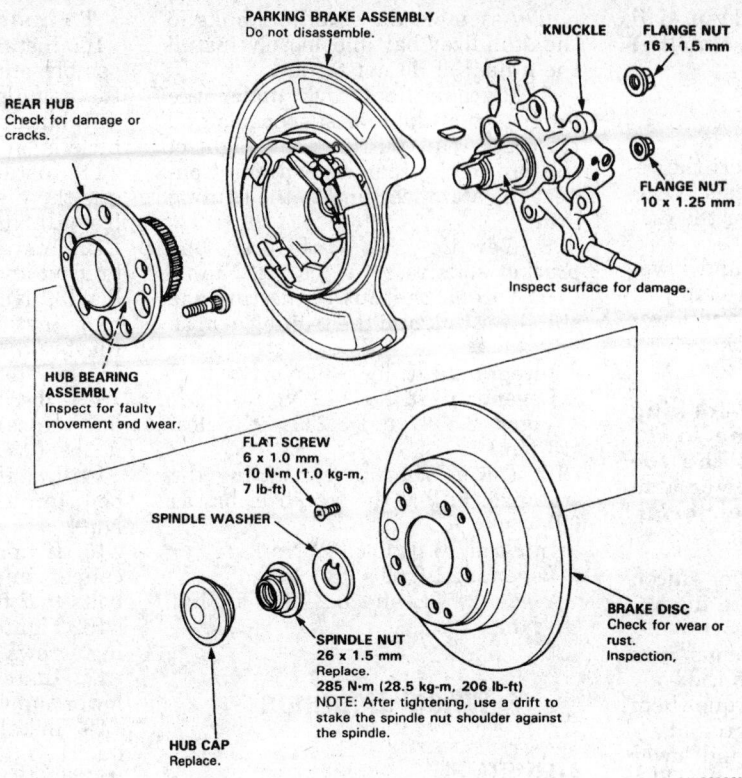

PARKING BRAKE ASSEMBLY
Do not disassemble.

KNUCKLE

FLANGE NUT
16 x 1.5 mm

REAR HUB
Check for damage or
cracks.

FLANGE NUT
10 x 1.25 mm

Inspect surface for damage.

HUB BEARING
ASSEMBLY
Inspect for faulty
movement and wear.

FLAT SCREW
6 x 1.0 mm
10 N·m (1.0 kg-m,
7 lb-ft)

SPINDLE WASHER

BRAKE DISC
Check for wear or
rust.
Inspection,

SPINDLE NUT
26 x 1.5 mm
Replace.
285 N·m (28.5 kg-m, 206 lb-ft)
NOTE: After tightening, use a drift to
stake the spindle nut shoulder against
the spindle.

HUB CAP
Replace.

220869

Hub and bearing assembly — Legend

To install:

8. Install the hub and bearing assembly onto the spindle and install a new spindle nut. Tighten the spindle nut, but do not torque it until the vehicle is on the ground.

9. Install the brake disc and retaining screws. Install the brake caliper and the brake hose mounting bolts. Torque the retaining screws to 7 ft. lbs. (10 Nm). Torque the caliper mounting bolts to 28 ft. lbs. (39 Nm). Torque the hose mounting bolts to 16 ft. lbs. (22 Nm).

10. Install the rear wheels and lower the vehicle to the ground. Torque the spindle nut for the Integra to 134 ft. lbs. (181–185 Nm). For the Vigor and 2.5TL, 181 ft. lbs. (245 Nm), and for the 3.2TL, tighten to 242 ft. lbs. (355 Nm). Stake the nut to the spindle with a punch. Install the hub cap.

AUDI 2

80 • 90 • 100 • 200 • Coupe • S4 • V8

FIRING ORDERS

NOTE: To avoid confusion, always replace spark plug wires one at a time.

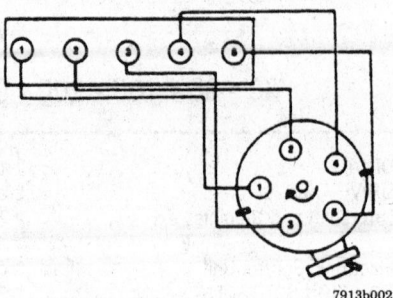

7913b002

2.2L and 2.3L Engines
Engine Firing Order: 1–2–4–5–3
Distributor Rotation: Clockwise

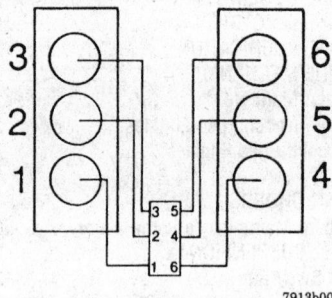

7913b004

2.8L Engine
Engine Firing Order: 1–4–3–6–2–5
Distributorless Ignition

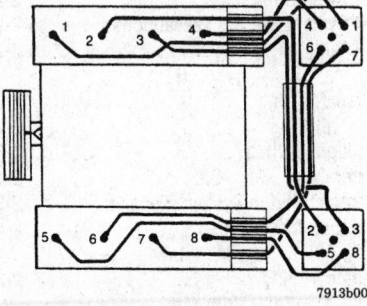

7913b003

3.6L and 4.2L Engines
Engine Firing Order: 1–5–4–8–6–3–7–2
Distributor Rotation: Clockwise

SERIAL NUMBER IDENTIFICATION

Vehicle Identification Plate

The Vehicle Identification Number (VIN) is located on a plate on top of the instrument panel. The VIN number is visible from outside through the left side of the windshield. The VIN number is also stamped into the upper right corner of the firewall. The vehicle identification plate is mounted on the right front wheel housing.

Engine Number

4– 5–Cylinder Engines

The engine serial number is stamped into the left rear side of the engine block. In addition to the serial number, an engine code number is stamped into the starter end of the engine block, below the cylinder head mounting surface. This number indicates the original cylinder bore size of the engine.

6–Cylinder Engine

The engine serial number is stamped on the right hand side inside of the engine block between the cylinder head and hydraulic pump.

8–Cylinder Engine

The engine serial number is stamped into the left side of the engine block, just above the power steering hydraulic pump.

Transaxle Number

To identify a transaxle there are 2 sets of numbers. The first set is located at the top of the bell housing. This set contains the transaxle code letters (first 3 digits or letters) and date of production. The second set is located at the side of the case. This set of numbers is the transaxle type.

NOTE: The transaxle code is also listed on the Vehicle Identification Label.

Model Data Sticker

On the inside of the luggage compartment lid is a sticker indicating the model type, chassis number, body type, paint number, engine and transaxle codes and the options package.

Body Panel Identification

The major body panels of all 1991 and later vehicles are marked with the complete VIN and the Audi logo. This is done in accordance with the Motor Vehicle Theft Law Enforcement Act to discourage theft and resale of the vehicle for parts. All authorized replacement parts will also have the Audi logo and a label with the "R DOT" designation in place of the VIN for that vehicle.

ENGINE MECHANICAL

NOTE: Disconnecting the negative battery cable on some vehicles may interfere with the functions of the on-board computer systems and may require the computer to undergo a relearning process, once the negative battery cable is disconnected. Most vehicles are equipped with theft protected radios, which cannot be operated if power to the radio is interrupted. Before disconnecting the battery cables, obtain the security code.

Engine Assembly

REMOVAL AND INSTALLATION

5–Cylinder Engine

WITH TURBOCHARGER

NOTE: Tag all hoses and wiring during removal, to use as reference during reassembly.

1. Disconnect the negative battery cable. Relieve fuel pressure.
2. Open the heater control valve all the way and drain the cooling system.
3. Remove the fuel injector cooling fan blower motor and intake hose from the engine.
4. On 80/90, remove the upper radiator cover, grille, bumper strip. Disconnect the wiring harness in bumper for turn signals and headlights. Remove the bumper.
5. Disconnect the electrical connector from the coolant fan. Remove the upper radiator hose from the engine. Remove the radiator-to-expansion tank hose from the tank and the bleeder hose from the auxiliary radiator.
6. Disconnect the wire from the thermo-switch. Remove the radiator mounting bolts, right-side radiator cover and bottom radiator cover.
7. Remove the windshield washer reservoir from the mount and support it aside.

──────── **CAUTION** ────────
The compressed refrigerant used in the air conditioning system expands and evaporates into the atmosphere at a temperature of −21.7°F or less. This will freeze any surface it comes in contact with, including eyes. In addition, the refrigerant decomposes into a poisonous gas in the presence of flame. To avoid injury, always wear eye protection and gloves when discharging the system.

8. Properly discharge the air conditioning system and disconnect the refrigerant hoses from the air conditioning condenser.
9. Remove the radiator and air conditioning condenser together. Remove the air conditioning compressor and mounting bracket from the engine.
10. Remove the power steering pump drive belt from the pump. Remove the pump from the mounts. Leaving the hoses attached, support it aside.
11. Disconnect the coolant hose from the thermostat housing and dis-

connect the wires from the oil pressure switch and temperature sender. Disconnect the wire plugs from the control pressure regulator.
12. Remove the control pressure regulator from the engine but leave the fuel lines connected. Support it aside.
13. Remove the throttle rod clips and remove the rod from the engine. Remove the injector line holder and remove the fuel injectors from the cylinder head.
14. Disconnect the electrical connector from the cold start valve and remove the valve from the intake manifold. Leave the fuel line connected.
15. At the throttle body, disconnect the electrical connectors from the throttle valve switches and intake air temperature switch.
16. Disconnect the air intake hose. Disconnect the wire from the auxiliary air regulator, pull off the vacuum hoses and disconnect the breaker hose from the engine.
17. At the 2-way valve, remove and tag the vacuum hoses. Remove the thermo-pneumatic valve. Leave the vacuum lines connected and remove the rpm sensor.
18. Disconnect the speedometer cable from the transaxle.
19. Remove the distributor from the engine.
20. Disconnect and tag the thermo-time switch and overheating warning lamp connectors. Disconnect the heater hoses from the engine.
21. At the left engine mount, disconnect the brake booster and reservoir from the firewall and leave the lines connected. On Quattro vehicles, disconnect the differential lock control lights connector. Disconnect the backup light switch wires.
22. Disconnect the tie rods from the steering rack. Disconnect the steering linkage.
23. If equipped with a manual transaxle, remove the clutch slave cylinder from the bell housing. Leave the line attached. Remove the bracket and pin under the transaxle bracket.
24. Disconnect the left engine mount ground strap. Disconnect the vacuum hose from the auxiliary air valve.
25. Remove the air duct from the intercooler and remove the intercooler.
26. Disconnect and tag the electrical connectors from the alternator. Remove the oil cooler. Leave the lines attached. Disconnect and tag the starter wiring.

27. Disconnect the exhaust pipe at the turbocharger. Remove the transaxle cover plates and the right side transaxle mount. Disconnect the halfshafts from the transaxle. On Quattro vehicles, disconnect the driveshaft from the rear of the transaxle.

28. On Quattro vehicles at the transaxle, disconnect the differential lock, remove the front and rear circlips and push back the boot. Disconnect the cable.

29. Remove the left-side transaxle mounting bolt and mounts from both sides.

30. At both front wheels, remove the ball joint pinch bolts. At the subframe, remove the mounting bolts and subframe. Separate the ball joints from the steering knuckle.

31. Install an engine lifting device on the engine. Raise the engine slightly and remove the left and right engine mounts. Lower the engine/transaxle assembly from the vehicle.

32. Raise the front of the vehicle and slide the engine/transaxle assembly from under the vehicle.

33. Separate the engine from the transaxle.

To install:

34. Install the engine assembly and temporarily secure the engine mounts.

35. Install the steering joints to the steering knuckles and torque to 22 ft. lbs. (30 Nm). Install the ball joints with the pinch bolts and torque to 44 ft. lbs. (65 Nm).

36. Install the exhaust and 4WD differential lock clips, if equipped.

37. Reconnect the exhaust pipe at the turbocharger.

38. Install the halfshafts to the transaxle.

39. Install the right side transaxle mount and transaxle cover plates.

40. On Quattro vehicles, reconnect the driveshaft to the rear of the transaxle.

41. Lower the vehicle and install all electrical connections.

42. Install the oil cooler.

43. Install the intercooler assembly and air duct.

44. Reconnect the left engine mount ground strap.

45. Install the clutch slave cylinder, if equipped.

46. Install the tie rods to the steering rack assembly.

47. Install the brake booster and reservoir. On Quattro vehicles, connect the differential lock control

lights connector. Reconnect the backup light switch wires.

48. Install the heater hoses.

49. Install the distributor.

50. Install the speedometer cable and all vacuum lines.

51. Install the fuel injectors from the cylinder head.

52. Connect the throttle body connectors, throttle linkage and fuel injection pressure regulator.

53. Install the power steering pump and all drive belts.

54. Install the radiator and air conditioning condenser. Connect the hoses.

55. Install the windshield washer reservoir.

56. On 80/90, install the upper radiator cover and grille. Reconnect the wiring harness in bumper for turn signals and headlights. Install the bumper.

57. Install the cooling fan blower motor and intake hose to the engine.

58. Refill and bleed the cooling system. Connect the battery negative cable.

59. To minimize vibration, loosen all the engine and subframe mounting bolts, then torque to 25 ft. lbs. (34 Nm) while the engine is running at idle.

60. Check all fluid levels, road test for proper operation.

WITHOUT TURBOCHARGER

NOTE: Tag all hoses and wiring during removal to use as reference during reassembly.

1. Disconnect the negative battery cable. Relieve fuel pressure.

2. Drain the cooling system.

3. Disconnect the radiator and heater hoses from the engine.

4. Remove the control pressure regulator from the engine, without disconnecting the fuel lines.

5. Remove the cold start valve from the intake manifold, without disconnecting the fuel lines.

6. Pull out the fuel injectors from the cylinder head and support the injectors and fuel lines aside.

NOTE: Protect the fuel injectors and the cold start valve with caps.

7. Loosen the air duct and vacuum hoses from the throttle valve assembly.

8. Remove the air box cover and filter.

9. At the top of the grille, pull the hood latch cable guide off its bracket.

10. If equipped with air conditioning, perform the following procedures:

   a. Remove the 2 clips from the top of the grille and the screw from the bottom. Remove the grille.

   b. Remove the condenser mounting bolts.

   c. Remove the air duct to auxiliary air regulator hose and remove the air duct from the throttle valve housing.

   d. Remove the fuel distributor, air flow sensor, fuel injectors and air box, as a unit.

NOTE: When removing the fuel injectors, leave all of the lines connected and cover the fuel injectors with caps.

   e. Remove the accessory drive belts.

CAUTION

The compressed refrigerant used in the air conditioning system, expands and evaporates into the atmosphere at a temperature of -21.7°F or less. This will freeze any surface it comes in contact with, including eyes. In addition, the refrigerant decomposes into a poisonous gas in the presence of flame. To avoid injury, always wear eye protection and gloves when discharging the system.

---

   f. Properly discharge the refrigerant from the air conditioning system. Remove and plug the air conditioning hoses, move them away from the engine.

   g. Remove the upper/lower compressor mounting bolts and remove the compressor from the engine.

11. Remove the power steering pump from the engine, leaving the hoses connected.

12. Remove the vacuum amplifier.

13. Remove the EGR control valve.

14. Remove the windshield washer reservoir from its holder.

15. Remove the distributor cap and ignition wires. Remove the distributor vacuum hose(s).

NOTE: Tape the distributor dust cap on to prevent it from falling off.

16. Disconnect the throttle linkage from the engine.

17. If equipped with an automatic transaxle, remove the throttle pushrod.

18. Disconnect the oil pressure and water temperature sensor wiring.

19. Remove the exhaust pipe to manifold nuts.

20. Remove the exhaust pipe support bracket from the transaxle.

21. Remove the front engine mount bolts and remove the mount. Disconnect the ground strap on left engine mount, if equipped.

22. Tag and disconnect all wires from the starter and remove the starter.

23. Tag and disconnect all wires leading from the alternator and remove the alternator.

24. If equipped with an automatic transaxle, work through the starter mounting hole to remove the torque converter mounting bolts.

25. Remove the lower engine to transaxle mounting bolts.

26. Support the transaxle and lower the vehicle.

27. Remove the upper engine to transaxle mounting bolts.

28. Remove the left engine support bracket.

29. Loosen the right engine bracket from the right engine mount.

30. Lift the engine until the crankshaft V-belt pulley is behind the grille opening.

31. Carefully detach the engine from the transaxle.

32. Remove the engine assembly by turning it to the right while lifting it out.

To install:

33. Install the engine assembly and secure the mounts.

34. On vehicles with automatic transaxle, install the torque converter bolts. On 087 and 089 units, torque the bolts to 22 ft. lbs. (30 Nm). On 097 units, torque the bolts to 44 ft. lbs. (60 Nm).

35. Install the alternator and all wiring.

36. Install the starter and all necessary wiring.

37. Install the engine ground strap and engine mount.

38. Install the exhaust system and brackets.

39. Reconnect all engine compartment wiring including the oil pressure and water temperature sensor wiring.

40. Reconnect the throttle linkage.

41. Install the distributor cap and ignition wires.

42. Install the washer reservoir.

43. Install EGR valve and vacuum amplifier.

44. Install the power steering pump and all drive belts.

45. Install the air conditioning compressor and condenser.

46. Install the fuel injectors, control pressure regulator and cold start valve.

47. Install the hood latch cable guide onto the bracket.

48. Install the air box cover and all necessary hoses.

49. Install the radiator, refill and bleed the cooling system. Connect the battery negative cable.

50. Torque the engine-to-transaxle mounting bolts to 43 ft. lbs. (58 Nm), the starter bolts to 14 ft. lbs. (19 Nm), the air conditioner mounting bolts to 29 ft. lbs. (39 Nm) and the power steering pump and the control pressure regulator mounting bolts to 14 ft. lbs. (19 Nm).

51. Check all fluid levels, road test for proper operation.

NOTE: To minimize vibration, loosen all the engine and subframe mounting bolts, then tighten while the engine is running at idle. Tighten the engine and subframe mounting bolts to 32 ft. lbs. (43 Nm).

8–Cylinder Engine

The engine is taken out towards the front without the transaxle. All cable ties which have to be released or cut open when removing the engine must be replaced in the same position when the engine is installed.

1. Disconnect the battery negative cable. The battery is under the rear seat.

2. Under the dashboard, remove the retainers, remove left dashboard end cap and the knee protector.

3. Remove the heater duct for the driver-side area, under the dash panel, by removing the right-hand screw and loosening the left-hand clip.

4. Unclip the floor lamp and push it through the opening.

5. Remove the control units with brackets.

6. Open the locking mechanisms on the control unit connectors and disconnect the harness connectors. Lock the connectors so they don't become tangled when removing the cable harness. Pull off connectors 11, 13 and 15. The numbers are marked on the wiring harness.

7. Remove the connector panel by removing the lower screw, loosening the upper screw and pulling the panel downwards. Press the latch on the butterfly connector lock and slide it sideways to release the connector from the panel.

8. Under the hood, remove the plenum chamber cover and lift the rubber grommet on the middle wiring harness. Cut the cable tie and pull the wiring harness carefully out of the passenger compartment and plenum chamber. Open the expansion tank cap for coolant.

9. Raise and safely support vehicle. Remove the sound insulation or under pan.

10. Remove the bolts and nuts holding bumper and bracket, remove the electrical connectors and pull off the bumper towards the front.

11. Drain the coolant from the radiator. Also open the block drains on both sides.

12. Remove the oil cooler hoses from the oil filter housing and the hose from the bottom of the transaxle oil cooler. Remove the bracket for the line on the air conditioner and disconnect the engine ground cable at the engine.

13. Separate the harness for the headlight washing system and air conditioner harness. Disconnect the coolant temperature sensor harness, the cool air duct from the alternator, the coolant hose and hose for headlight washer system.

14. Disconnect the outer half of the air intake elbow for the alternator. Disconnect the wiring from the alternator and starter motor. Unscrew oil filter.

15. Remove the upper bolt on the starter by guiding a 10mm Allen® socket, with extension and flex fitting, through the opening on the transaxle housing over the final drive. Remove starter.

16. Locate and remove the bolts from the front of the subframe on left and right sides. Remove the bolts securing the exhaust system on the left and right sides.

17. Remove the 4 bolts connecting the bottom of the engine and transaxle.

18. Remove the bolts above and below the long member on the left and right sides.

19. Remove the harness connector from the temperature sensor at the front of the air conditioning condenser and disconnect the hood release cable.

20. Open the fuse box on the left side behind the hydraulic reservoir. Disconnect the wire to the fan and the ground wire at the top of the suspension strut. Expose the wiring.

21. Remove the air conditioning dryer with the bracket.

22. Remove the intake manifold bracket on the left and right. Pull out the exchanger. Remove the bar-shaped reinforcement strut.

23. Remove the air conditioning condenser bolts and swivel the condenser downward. Wiring should remain connected.

24. Remove the wiring bracket from the top of the transaxle cooler. Remove the water hose from both sides of the engine. Remove the bleeder hose from the expansion tank on the radiator. Remove the front apron.

25. Remove the screws securing the upper part of the air cleaner housing. There are 4 screws on the housing and 3 screws at the rear of the housing.

26. Remove the screws securing the lower part of the air cleaner housing. Press toward the rear and lift out. Remove the right-hand stud from the lower air cleaner housing.

27. Disconnect the carbon canister hose from the front of the engine. Cut both cable ties on the fuel injector. Remove the fuel supply and return lines.

28. Remove the coolant hose to expansion tank. Disconnect the high tension wire and connector on the ignition coil at the left and right. Remove the bolts holding the left coil, cable housing and retaining clip. Watch for the spacer sleeve.

29. Remove the left and right heatshields. Remove the housing for the ignition wire retainer. Disconnect the vacuum line.

30. Remove the screws securing the left and right distributor caps. Wiring remains connected. Using a cable tie, secure the 2 distributor caps aside against the PCV hose.

31. Disconnect the throttle cable by unclipping both retaining clips. Remove the screws securing the throttle cable support bracket. Unscrew both coolant hoses. Disconnect the supporting clamp.

32. Separate both harness connectors from the oxygen sensor. Disconnect the vacuum line to the cruise control system.

33. Remove the transaxle oil fill tube bolt from the engine. Pull the wiring harness through the contact plate and place on the engine.

34. Remove the 6 torque converter bolts through the starter opening.

35. Release the tension on the ribbed drive belt and remove the belt in a downward direction by placing a 13mm box wrench on the hexagon guide of tensioner and pressing wrench slowly upwards.

36. Remove the bolts holding the air conditioning compressor. Lift the compressor over the strut. The lines remain connected.

37. Remove the bolts for the air conditioning compressor bracket and the hydraulic pump. Note the guide sleeves. Place the bracket with the hydraulic pump on the long member.

The lines remain connected. At installation, fit the lower bolts with the guide sleeves in position and tighten lightly. Then the upper bolts can be installed.

38. Remove the nuts from the left and right engine mountings. Remove the bolts securing the engine support at the front. Take note of the shims. The same thickness shims must be used at installation.

39. Disconnect the engine and transaxle at the top by removing 3 of the 4 bolts. Loosen the 4th bolt but do not remove.

40. Install a suitable lifting sling to the front left-hand side and rear right-hand side.

41. Move in hoist, being careful of the air conditioning compressor. Lift engine carefully. Remove last bolt from the top of the engine and transaxle. Unscrew the engine mounting from the left and right sides and pull the engine out from the front. Lift carefully to prevent damage to the transaxle mainshaft, clutch and body.

42. Use care when selecting a suitable engine repair stand. Attaching to some types of stands could cause the engine block to distort and cause any cylinder bore measurements to be inaccurate.

To install:

43. Note that there are guide sleeves in the engine block for centering the engine and transaxle. Make sure they are installed.

44. Install the engine assembly and secure to the transaxle and engine mounts.

45. Install the 6 torque converter bolts.

46. Install the air conditioning compressor and hydraulic pump. Fit the lower bolts with the guide sleeves first. Tighten lightly and then install the top bolts.

47. Install and adjust the ribbed drive belt.

48. Install all electrical connectors, hoses and sheet metal heatshields.

49. Reconnect the throttle cable.

50. Install the distributor caps.

51. Install the coolant hose to the expansion tank.

52. Reconnect all ignition coil wiring.

53. Install the fuel supply and return lines.

54. Install the air cleaner housing assembly.

55. Install the front apron.

56. Install the bleeder hose to the expansion tank on the radiator. Install all water hoses.

57. Install the radiator and air conditioning condenser and the small brackets that were removed.

58. Install the air conditioning dryer with bracket.

59. Install and route the hood release cable.

60. Install the starter and wiring.

61. Install a new oil filter. Reconnect all necessary wiring and hoses.

62. Install the engine ground cable.

63. Install the bumper bracket, bumper and all wiring.

64. Install the sound insulator or underpan.

65. Install the control units and interior parts removed. Reconnect all necessary wiring.

66. Connect battery negative cable.

67. Always replace all self-locking nuts. Make sure the exhaust system is installed free of strain. Check all fluid levels before starting the engine Check the fluid level in the automatic transaxle.

68. Road test the vehicle for proper operation.

Cylinder Head

NOTE: Before removing or installing the cylinder head, align the engine timing marks at TDC. Rotate the crankshaft mark away about 1/4 turn (BTDC). This will prevent the valves from hitting the piston heads. Be sure to turn the crankshaft to the proper position after cylinder head installation.

REMOVAL AND INSTALLATION

NOTE: Cylinder head removal should not be attempted unless the engine is cold.

4– 5-Cylinder Engines

1. Disconnect the negative battery cable.

2. Drain the cooling system.

3. Disconnect the air duct from the throttle valve assembly on all vehicles except the Turbo and Quattro. On the Turbo and Quattro, remove the hose which runs between the air duct and the turbocharger.

4. Disconnect the throttle cable from the throttle valve assembly.

5. Remove the air duct for the injector cooling fan on the Turbo and Quattro.

6. Clean and remove the fuel injectors and all other fuel lines.

NOTE: Protect the fuel injectors and the cold start valve with caps.

7. Tag and disconnect all vacuum and PCV lines.

8. Remove the hose which runs from the intake manifold to the turbocharger on the Turbo and Quattro.

9. Tag and disconnect all electrical lines leading to the cylinder head.

10. Remove the intake manifold.

11. Disconnect all radiator and heater hoses where they are attached to the cylinder head. Position them aside.

12. Tag and remove all spark plug wires.

13. Remove the distributor. To aid installation, scribe a mark on the body of the distributor and the cylinder head.

14. Separate the exhaust manifold from the exhaust pipe.

NOTE: Exhaust pipe detachment differs slightly on the Turbo and Quattro. First the exhaust pipe must be unbolted from the turbocharger. Second, it must be unbolted from the wastegate at the rear of the engine.

15. Disconnect the EGR valve and oxygen sensor from the exhaust manifold.

16. Remove the heat deflector shield.

17. Remove the oil lines (2) from the turbocharger.

18. Remove the exhaust manifold.

NOTE: When removing the exhaust manifold on the Turbo and Quattro, the manifold, turbocharger and wastegate should all be removed as a unit.

19. Remove the air hose cover from the back of the alternator.

20. Tag and disconnect all wires coming from the back of the alternator and remove the alternator from the engine.

21. Disconnect and plug the hoses coming from the power steering pump.

22. Remove the power steering pump and the V-belt.

23. Remove the timing belt cover and belt.

24. Remove the valve cover.

25. Loosen the cylinder head bolts in the reverse order of the tightening sequence.

26. Remove the bolts and lift the cylinder head off the engine.

To install:

27. Clean the cylinder head and engine block mating surfaces thoroughly and install the new gasket without any sealing compound. Make sure the words **TOP** or **OBEN** are facing up, when the gasket is installed.

28. Place the cylinder head on the engine block and install bolts No. 8 and 10 first. These holes are smaller and will properly locate the gasket and the head on the engine block.

29. Install the remaining bolts. Torque them in sequence in 3 stages as follows:

Step 1 — 29 ft. lbs. (39 Nm)
Step 2 — 43 ft. lbs. (58 Nm)
Step 3 — Tighten ½ turn more (180 degrees).

30. Install the valve cover.

31. Install timing belt and timing belt cover.

32. Install the power steering pump and drive belt.

33. Install the alternator and all wiring.

34. Install the exhaust manifold.

35. Reconnect the 2 oil lines from the turbocharger assembly.

36. Install the heat deflector shield.

37. Reconnect the EGR valve and oxygen sensor to the exhaust manifold.

38. Reconnect the exhaust system.

39. Install the distributor and spark plug wires.

40. Install the intake manifold.

41. Install the radiator and heater hoses.

42. Install the fuel injectors. Reconnect the air duct.

43. Reconnect the throttle cable.

44. Connect the negative battery cable.

45. Refill and bleed the cooling system. Check all fluid levels.

46. Road test the vehicle check for proper operation.

6–Cylinder Engine

This procedure is for the left cylinder head assembly with the engine installed in the vehicle. Modify the service procedure as necessary for the right side.

1. Disconnect the negative battery cable. Drain engine coolant. Relieve fuel pressure.

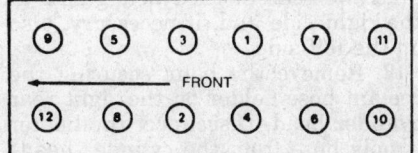

Cylinder head torque sequence — 5–cylinder engines

2. Remove the ribbed V-belt.

3. Remove the timing belt. The vibration damper should remain installed.

4. Remove the exhaust pipe from the manifold.

5. Remove the EGR valve hose at manifold.

6. Remove the air guide hose between air mass sensor and intake manifold.

7. Remove and tag all spark plug wires and injector connectors.

8. Remove the crankcase breathers on the left and right cylinder head covers.

9. Remove the fuel feed and return lines. Remove the silencer.

10. Remove the left side cover for fuel line.

11. Disconnect the throttle cable.

12. Mark and remove all vacuum hoses from vacuum pump and intake manifold.

13. Disconnect connectors on idling stabilization valve and throttle valve potentiometer.

14. Disconnect vacuum hose on vacuum control unit.

15. Disconnect connectors on oil pressure sender and oil pressure switch.

16. Disconnect connector for Hall sender sensor.

17. Remove the EGR valve from the intake manifold.

18. Remove the intake manifold assembly.

19. Remove the coolant pipe at the rear of the cylinder head.

20. Remove the oxygen sensor.

21. Remove the heatshield on the exhaust manifold.

22. Remove the cylinder head cover.

23. Remove the timing belt rear belt guard. Remove the hydraulic line from reservoir to pump.

24. Reverse the installation torque sequence and remove the cylinder head assembly from the engine.

To install:

25. Clean all sealing surfaces. Check cylinder head for distortion. Measure at several locations. The maximum permissible distortion is 0.1mm.

26. Install cylinder head gasket. The lettering must face upwards.

27. Install the cylinder head assembly, check centering pins in the cylinder block.

28. Install cylinder head bolts by hand.

29. Tighten the cylinder head bolts in sequence in 2 steps. Step 1 — 44 ft. lbs. and Step 2 — ½ turn (180 degrees). It is not necessary to retighten

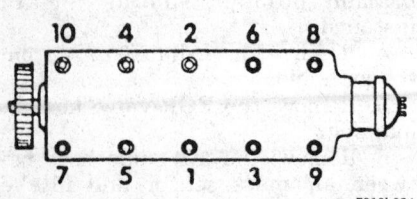

Cylinder head torque sequence — V8 engine

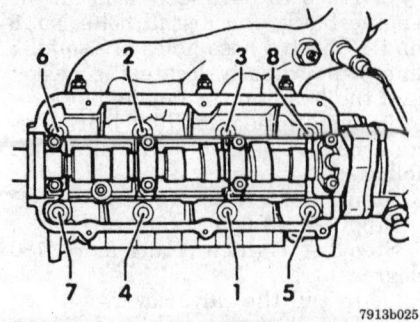

Cylinder head torque sequence — V6 engines

the cylinder head bolts after repairs or as part of inspection service.

30. Install the timing belt rear belt guard.

31. Install the cylinder head cover.

32. Install the oxygen sensor. Install the heatshield on the exhaust manifold.

33. Install the intake manifold assembly.

34. Install the EGR valve to the intake manifold.

35. Reconnect the hall sender sensor, oil pressure sender and oil pressure switch.

36. Reconnect connectors on idling stabilization valve and throttle valve potentiometer.

37. Install all vacuum hoses to vacuum pump and intake manifold.

38. Reconnect the throttle cable.

39. Reconnect the fuel feed and return lines. Install the silencer.

40. Install the crankcase breather on the cylinder head cover.

41. Install all spark plug wires and injector connectors.

42. Install the exhaust manifold.

43. Install the timing belt and V-belt.

44. Refill and bleed the cooling system.

45. Check all fluid levels. Operate the engine at normal operating temperatures and check for leaks.

46. Road test the vehicle for proper operation.

8–Cylinder Engine

1. Disconnect the negative battery cable. The battery is under the rear seat.

2. Remove the toothed cam drive belt. It should not be necessary to loosen or remove the vibration damper.

3. Open the left and right block drains.

4. Remove the bolts holding the exhaust pipe on both sides.

5. Remove the supply line at the fuel manifold. Remove the fuel return line. Remove the bolt holding the pressure regulator to the fuel manifold. At assembly, press the regulator directly into the seat of the O-ring seal. Do not pull in with bolts.

6. Disconnect the breather hose at the rear of the intake manifold, the vacuum hose and the bolt holding the upper part of the supporting clamp for the engine wiring harness. Remove the mounting bracket for ignition wire holder.

7. Disconnect the linkage for the cruise control. Release the throttle cable by unclipping both retaining clips.

8. Disconnect the heater supply hose for coolant at the rear of the cylinder heads. Release the hose under the engine wiring harness clamp.

9. Remove the screws securing the 2 spark plug covers, disconnect the spark plug connectors and remove the ignition cables complete with the distributor caps. Disconnect the harness connector for the left and right knock sensors. Disconnect the air mass sensor harness connector.

10. Disconnect the connector for the throttle valve harness connector potentiometer. Disconnect the idle stabilizer valve harness connector. Remove the idle stabilizer valve. Remove the hose from the carbon canister.

11. Remove the air temperature sensor. Remove the wiring holder behind the rear toothed belt guard at the right side and, if necessary, also on the left side.

12. Remove the bolts securing the coolant hose holder on the right rear cylinder head. Disconnect the heater supply hose from the cylinder heads by pulling towards the rear.

13. Disconnect the harness connector for the Hall sender on the right cylinder head. If the right cylinder head needs to be removed, disconnect

the temperature sensor harness connector on the right cylinder head.

14. Remove the breather hose. Turn the hose with pliers until the retaining lug unlatches. Remove the breather hose on the cylinder head, on the left side at the rear.

15. Remove the bolts securing the fuel rails and lift out complete with injectors and place on plenum chamber. At installation, make sure the O-rings are not damaged. Moisten slightly with fuel.

16. Remove the bolts securing the intake manifold and lift out. Watch the front breather hose under the intake manifold.

17. If it is the left hand cylinder head being removed, unscrew the dipstick guide and pull out. Remove the transaxle oil filler tube bolt.

18. Remove the cylinder head cover. Note that the 2 center mounting bolts are different. The bolt with the longer hexagon fitting must be installed at the rear.

19. If the left cylinder head is being removed, remove the bolts securing the cruise control vacuum unit and bracket. Do not forget the deflector plate during installation.

20. Remove the cylinder head bolts in reverse sequence of tightening. Remove the cylinder head.

To install:

21. Observing the following during the installation. Check that the gasket surface is not distorted. Maximum permissible distortion is 0.004 in. (0.1mm).

22. Make sure the gasket surface is clean. Look for the word **OBEN** on the gasket. This is the top, facing the cylinder head. When installing the cylinder head, watch the centering dowels in the block.

23. Insert the cylinder head bolts hand-tight. Torque in sequence in 3 steps. First to 30 ft. lbs. (41 Nm), then to 44 ft. lbs. (60 Nm). The 3rd step is to give the bolts an additional 1/2 turn. Use a regular ratchet handle or breaker bar and turn in one smooth motion without stopping. It should not be necessary to retighten the cylinder head bolts during maintenance or after repairs.

24. Install the cylinder head cover.

25. Install the intake manifold assembly.

26. Install the fuel rails with fuel injectors. Make sure the O-rings are not damaged. Moisten slightly with fuel.

27. Install the breather hose.

28. Reconnect the coolant hose holder and install heater hose.

29. Install the air temperature sensor. Reconnect all wiring.

30. Reconnect the connector for the throttle valve harness connector potentiometer.

31. Install the idle stabilizer valve. Install the hose from the carbon canister.

32. Install spark plug wires, air mass sensor harness connector and distributor caps. Install the spark plug covers.

33. Reconnect the cruise control linkage and throttle cable.

34. Press the regulator directly into the seat of the O-ring seal. Install the bolt holding the pressure regulator to the fuel manifold.

35. Install the supply line and fuel return line.

36. Install the toothed cam drive belt.

37. Refill and bleed the cooling system.

38. Check all fluid levels. Operate the engine at normal operating temperatures and check for leaks.

39. Road test the vehicle for proper operation.

Valve Lifters

REMOVAL AND INSTALLATION

Use care when handling valve lifters. Always place a removed valve lifter on a clean surface with the contact surface or camshaft side facing downward. The camshaft bearing caps are marked at the top for correct installation.

1. Disconnect the negative battery cable. Remove the toothed camshaft drive belt.

2. Remove the cylinder head cover.

3. Remove the camshaft sprocket.

4. On the 8-cylinder engine, on the exhaust side, remove the intermediate flange and bearing cap for the distributor. Loosen the bearing cap in front of the chain, plus caps 2 and 3. Loosen bearing caps 1 and 4 alternately and in a diagonal sequence.

5. On the 8-cylinder engine, on the intake side, remove bearing caps 6 and 7. Loosen bearing caps 5 and 8 alternately and in a diagonal sequence.

6. Remove the cam and lift out the valve lifter. If it is to be reused, it must go in the bore from which it was removed.

To install:

7. Installation is the reverse of removal. Before assembly coat the moving parts with clean oil.

8. On the 8-cylinder engine, install the camshafts with the chain so the markings on the chain sprockets are in alignment. Install the bearing caps so the stamped-on numbers can be read from the intake side.

9. Make sure the bearing caps are installed properly. They will cause cam shaft failure if installed backwards. Tighten bearing caps alternately, in a diagonal sequence to 11 ft. lbs. (15 Nm).

Valve Lash

ADJUSTMENT

All engines are equipped with hydraulic valve lash adjusters that eliminate the need for routine valve lash adjustments. Intermittent valve noise is normal when the engine is cold. If valve noise persists, check the camshaft lobes and/or camshaft followers for wear. Replace if necessary. Do not attempt valve lifter repair. If worn or damaged, replace the complete assembly.

After working on the valve train, carefully turn the engine by hand at least 2 turns to make sure the valves do not strike the pistons when the engine is started. Do not start the engine for 30 minutes after installing new valve lifters or the valves may strike the pistons. The lifter must be allowed to bleed down to proper adjustment.

To check a suspect lifter, use the following procedure:

1. Warm the engine to operating temperature until the radiator fan comes ON at least once.

2. Bring the engine to approximately 2500 rpm for 2 minutes. If a lifter is still noisy, shut OFF engine and remove the cylinder head cover.

3. Turn the crankshaft pulley bolt clockwise until the cam lobes of the cylinder to be checked point upward.

4. Push down against the valve lifter with light pressure using a suitable tool. If the valve lifter can be pushed down more than 0.004 in. (0.1mm), replace the lifter.

5. Do not start the engine for 30 minutes after installing new valve lifters or the valves may strike the pistons. The lifter must be allowed to bleed down to proper adjustment.

Intake Manifold

REMOVAL AND INSTALLATION

5-Cylinder Engine

1. Disconnect the negative battery cable.

2. Relieve the fuel system pressure.

3. On non-turbocharged engines, disconnect the air duct from the throttle valve assembly. On turbocharged engines, remove the hose between the air duct and turbocharger.

4. Disconnect the throttle cable and rod from the throttle valve assembly.

5. On turbocharged engines, remove the air duct for the injector cooling fan.

6. Remove the fuel injectors from the cylinder head, with the fuel lines attached.

7. Disconnect the cold start valve wiring and remove the fuel line from the valve.

NOTE: Protect the fuel injectors and cold start valve with caps.

8. Tag and disconnect all vacuum and PCV lines.

9. Tag and disconnect all electrical lines leading to the cylinder head.

10. On turbocharged engines, remove the hose which runs from the intake manifold to the turbocharger. On the Quattro vehicle, remove the hose which runs from the intake manifold to the intercooler.

11. Remove the auxiliary air regulator. Remove the air box cover and filter element.

12. Remove the intake manifold mounting nuts and remove the manifold from the engine.

To install:

13. Clean the gasket mating surfaces on the manifold and engine.

14. Using a new gasket, install the manifold on the cylinder head and tighten the nuts to 15 ft. lbs. (20 Nm).

15. Install the auxiliary air regulator.

16. On turbocharged engines, install the hose which runs from the intake manifold to the turbocharger. On the Quattro vehicle, install the hose which runs from the intake manifold to the intercooler.

17. Reconnect all vacuum and electrical connections.

18. Install the fuel line to the cold start valve and reconnect the cold start valve wiring.

19. Install the fuel injectors.

20. Connect the throttle cable and rod to the throttle valve assembly. Install all air ducts.

21. Check all fluid levels. Operate the engine at normal operating temperatures and check for leaks.

22. Road test the vehicle for proper operation.

8–Cylinder Engine

1. Disconnect the negative battery cable.

2. Relieve the fuel system pressure.

3. Remove the 7 bolts retaining the upper part of the air cleaner housing and the 2 bolts securing the lower part. Press the housing towards the rear and lift out.

4. Remove the fuel supply and return lines. Remove the fuel pressure regulator from the fuel rail.

5. Remove the breather hose at the rear of the intake manifold, remove the vacuum hose and the bolt securing the upper part of the engine wiring harness clamp.

6. Disconnect the harness connectors for the left and right knock sensor, the air mass sensor, the potentiometer, the thermo-switch and idle stabilizer valve. Remove the idle stabilizer valve and disconnect the hose to the carbon canister.

7. Remove the air temperature sensor bolts. Remove the breather hose at the top of the cam cover by using pliers to turn the hose until the retaining lug unlatches. Remove the breather on the cylinder head at the left rear.

8. Remove the bolts securing the fuel rails and lift out the rails with the injectors. Place on the plenum chamber.

9. Remove the intake manifold bolts and lift out the manifold. Watch the front breather hose under the intake manifold. There are 2 oil retention valves that must be replaced if they sound noisy on short drives, although the noise may go away on longer drives. Replace these valves as follows:

    a. After the intake manifold has been removed, remove the right hand knock sensor and the bolts securing the engine breather cover.

    b. Lift out the cover along with the bulkhead panel. It may be easier to lift the right side first.

    c. Locate the oil retention valves and remove the circlips.

    d. Screw a M6 x times; 50mm bolt with a large washer into the oil retention valve. Remove the valve by prying evenly under the washer.

To install:

10. Use care when installing the intake manifold. First install the front breather hose to the engine under the intake manifold. Install the intake manifold assembly.

11. Install the fuel rails with fuel injectors. When reinstalling the fuel system parts, use care not to damage any O-rings.

12. Install the air temperature sensor, breather and hose.

13. Install the idle stabilizer valve and reconnect the hose to the carbon canister.

14. Reconnect the harness connectors for the left and right knock sensor, the air mass sensor, the potentiometer, the thermo-switch and idle stabilizer valve.

15. Install the fuel pressure regulator, by pressing it directly into the seat of the O-ring seal.

16. Install the fuel supply and return lines.

17. Check all fluid levels. Operate the engine at normal operating temperatures and check for leaks.

18. Road test the vehicle for proper operation.

Exhaust Manifold

REMOVAL AND INSTALLATION

Non-turbocharged Engines

EXCEPT 8 CYLINDER ENGINE

NOTE: Although not necessary, more working clearance will be found by removing the intake manifold before removing the exhaust manifold. Before beginning, soak the manifold studs with lubricant to aid in the removal.

1. Disconnect the negative battery cable. Raise and support the vehicle safely. Disconnect the exhaust pipe from the exhaust manifold.

2. Disconnect the EGR valve and oxygen sensor from the manifold.

3. Disconnect the CO probe receptacle tube.

4. Remove the exhaust manifold mounting nuts and remove the manifold from the engine.

To install:

5. Clean the gasket mating surfaces of the manifold and engine.

6. Using a new gasket, install the manifold on the engine and tighten the nuts to 22 ft. lbs. (30 Nm).

NOTE: Always replace the old mounting nuts with new brass nuts. Check the condition of the

studs before installation. The oxygen sensor, EGR tube, bolts and nuts exposed to high temperatures should receive a light coating of anti-seize compound on the threads before assembly.

7. Connect the exhaust pipe to the manifold, using a new gasket and tighten nuts to 26 ft. lbs. (35 Nm).

8. Install and tighten the CO measuring tube to 22 ft. lbs. (30 Nm).

9. Install EGR valve and oxygen sensor to the manifold.

10. Install the heat deflector shield, if equipped.

11. Reconnect the battery. Operate the engine at normal operating temperatures and check for leaks.

8 CYLINDER ENGINE

1. Disconnect negative battery cable.

2. Disconnect the front exhaust pipe at the manifolds.

3. For the left side manifold, unscrew the guide for the dipstick tube and remove.

4. For both the left and right manifolds, remove the bolts securing the exhaust manifolds to the cylinder heads.

5. Remove the manifolds one at a time. Loosen the engine mounts. Lift the engine at the edge of the oil pan on the side of the manifold to be removed.

To install:

6. Clean the gasket mating surfaces of the manifolds and engine.

7. Using a new gasket, install the manifold on the engine and tighten the nuts to 18 ft. lbs. (24 Nm). Lower the engine.

8. Install the dipstick guide and replace the sealing ring, if necessary.

9. Reconnect the front exhaust pipe at the manifold.

10. Reconnect the battery. Operate the engine at normal operating temperatures and check for leaks.

Turbocharged Engine

1. Disconnect the negative battery cable. Remove the hose which runs between the air duct and the turbocharger.

2. If the intake manifold has not been removed, disconnect the hose which runs from the intake manifold to the turbocharger or intercooler.

3. Disconnect the exhaust pipe from the turbocharger.

4. Disconnect the exhaust pipe from the wastegate on the rear of the manifold.

5. Disconnect the EGR valve and the oxygen sensor, if necessary, from the manifold.

6. Remove the oil lines from the turbocharger.

7. Remove the line from the bottom of the turbocharger to the intercooler, if equipped.

NOTE: The manifold, turbocharger and wastegate are removed as a unit.

8. Remove the manifold assembly.
To install:
9. Clean the gasket mating surfaces of the manifolds and engine.
10. Install the exhaust manifold. Tighten the exhaust manifold mounting nuts to 26 ft. lbs. (35 Nm). Always use new gaskets and O-rings where necessary.

NOTE: The oxygen sensor, EGR tube, bolts and nuts exposed to high temperatures should receive a light coating of anti-seize compound on the threads before assembly.

11. Install the line from the bottom of the turbocharger to the intercooler, if equipped.
12. Reconnect the oil lines from the turbocharger.
13. Install the EGR valve and oxygen sensor, if necessary.
14. Install the exhaust system.
15. Install all hoses or air ducts.
16. Reconnect the battery. Operate the engine at normal operating temperatures and check for leaks.

Turbocharger

REMOVAL AND INSTALLATION

1. Disconnect the negative battery cable. Spray all mounting bolts with a lubricant.
2. Remove the vacuum tube between the intake air boot and turbocharger.
3. Remove the intake boot and crankcase ventilation hose. Remove the hose assembly between the intake manifold and throttle housing.
4. Remove the air box cover and remove the filter element.
5. Remove the right side engine mount heatshield.
6. Remove the oil supply pipe from the turbocharger. Remove the exhaust pipe from the corrugated pipe. Loosen the exhaust pipe at the transaxle mount and catalytic converter.
7. Remove the retaining clamp from the starter housing and sensor air hose.
8. Remove the exhaust pipe from the turbocharger.

9. Remove the alternator support bolt and position the alternator to the side.
10. Remove the oil return pipe from the turbocharger. Remove mounting bolts and turbocharger.
To install:
11. Install the turbocharger with new gaskets and torque the mounting nuts to 43 ft. lbs. (60 Nm).
12. Connect the oil supply and return lines using new gaskets.
13. Install the alternator assembly.
14. Connect the exhaust pipe and torque the nuts to 25 ft. lbs. (34 Nm). Install all exhaust brackets.

NOTE: Bolts and nuts exposed to high temperatures should receive a light coating of anti-seize compound on the threads before assembly. After servicing the turbocharger, always replace the engine oil along with the turbocharger filter and engine oil filter.

15. Connect the outlet air hose and breather hoses.
16. Install heatshield, filter element and air box cover.
17. Reconnect the battery. Operate the engine at normal operating temperatures and check for leaks.

Turbocharger Wastegate

REMOVAL AND INSTALLATION

NOTE: Although not necessary, more working clearance will be found by removing the intake manifold before removing the wastegate. Before starting, soak the studs with lubricant to aid in the removal.

1. Disconnect the negative battery cable. Remove the wastegate to exhaust pipe connecting tube. There are 3 bolts on the top and on the bottom.
2. Remove the mounting bolt for the tube leading from the wastegate to the exhaust manifold.
3. Remove the vacuum line from the end of the wastegate.
4. Remove the 4 mounting bolts and remove the wastegate from the exhaust manifold.
5. Installation is in the reverse order of removal. Torque the mounting nuts/bolts to 18 ft. lbs. (25 Nm).

NOTE: Bolts and nuts exposed to high temperatures should receive a light coating of anti-seize compound on the threads before assembly.

Timing Belt Front Cover

REMOVAL AND INSTALLATION

5–Cylinder Engine

UPPER COVER

1. Disconnect the negative battery cable. Loosen the alternator adjusting bolts and remove the drive belt.
2. Loosen the power steering pump adjusting bolts and remove the drive belt.
3. Remove the retaining nuts and remove the timing belt cover. Take care not to lose any of the washers or spacers.
4. Installation is the reverse of the removal procedure.
5. Adjust the drive belt tension when finished.

LOWER COVER

1. Disconnect the negative battery cable. Remove the upper timing belt cover.
2. Loosen the air conditioning compressor mounting bolts and remove the drive belt.
3. Remove the crankshaft balancer center bolt.

NOTE: To remove the crankshaft balancer bolt, on manual transaxle vehicles, place the vehicle in 5th gear and have an assistant apply the brake. The will stop the engine from rotating while loosening the bolt. On automatic transaxle vehicles, remove the starter and hold the flywheel from turning, using a flywheel holding tool VW 10–201 or equivalent. This bolt is extremely tight.

4. Remove the lower timing belt cover bolts and remove the cover.
To install:
5. Installation is the reverse of the removal procedure.
6. Use the same procedure to install the crankshaft center bolt as when removing the bolt. Apply a locking compound on the bolt threads and tighten the bolt to 258 ft. lbs. (350 Nm) in several steps.
7. Adjust the drive belt tension when finished.

8–Cylinder Engine

1. Disconnect the negative battery cable. The battery is under the rear seat.
2. Remove the coolant expansion tank cap. Raise and safely support vehicle. Remove the sound insulator or lower pan. Drain coolant from radiator.

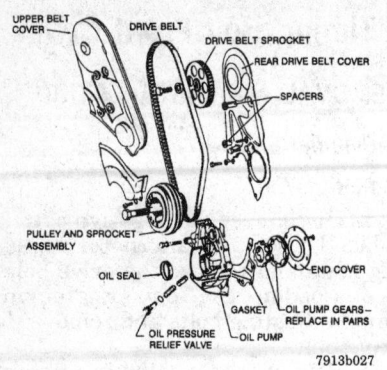

Timing belt — 5-cylinder engine

3. Remove the alternator cooling air duct. Loosen the clip and disconnect the coolant temperature sensor harness connector. Remove the coolant hose.

4. Disconnect the outer half of the air duct at the alternator. Remove the screws securing the wiring at the alternator. Note that the alternator wiring must be brought out at the side, not downward. Otherwise, the alternator air duct cannot be mounted.

5. Release tension on the poly-ribbed drive belt and remove the belt in a downward direction. Place a 13mm box wrench on the hexagon guide of the tensioner and pressing the wrench slowly upwards.

6. Remove the alternator mounting bolts and remove the alternator.

7. Working from below, remove the 3 bolts for the toothed belt guard.

8. Remove the bolts securing the bracket for the air intake ducts on the left and right and remove the ducts. Remove the bar-shaped strut brace.

9. Remove the 7 screws securing the upper part of the air cleaner housing, then remove the bolts holding the lower part of the housing. Press towards the rear and lift out.

10. Remove the upper radiator hose and clips.

11. Disconnect the electric fan bolts, lift the fan assembly out and lay to one side, wiring still connected.

12. Remove the bolts securing the supporting clamp for the lower radiator hose and take off the upper part. Disconnect the radiator hose at the thermostat housing. Swing the radiator hose to right at rear.

13. Remove the fan shroud by unscrewing the bolts for the viscous fan at the top. Remove the bolts securing the viscous fan. A spanner wrench or equivalent may be needed to hold the fan hub. Note that this is a left hand

thread. Turn to the right to loosen. Lift out the fan and shroud together.

14. Disconnect the engine support at the front. Note any shims. The same thickness shims must be used at installation. Remove the radiator hose. At installation, install the radiator hose first.

15. Loosen the center bolt of the vibration damper by one turn. A special tool may be needed to hold the damper from turning. This bolt was installed to over 250 ft. lbs. (340 Nm) torque and will be difficult to remove.

16. Remove the bolts securing the left side toothed belt cover or guard.

17. Remove the lower part of the supporting clamp for the lower radiator hose. Turn the tensioner for the poly-ribbed accessory drive belt in the loosening direction and insert an appropriate size holding pin in the hole provided.

18. Remove the bolts securing the right side toothed belt cover, with the exception of the top bolt. Remove the tensioner holding pin. Remove the belt cover top screw and remove the cover or guard. Carefully lift the guard from the bottom to avoid damaging the radiator.

To install:

19. Install the right cover or guard.

20. Install tensioner holding pin. Install the left cover or guard.

21. Install crankshaft damper. The crankshaft damper center bolt must be torque to 258 ft. lbs. (350 Nm). A holding tool may be needed to keep the crankshaft from turning.

22. Install radiator hoses. Reconnect the engine support.

23. Install the viscous fan hub and shroud assembly. This uses a left hand thread. Turn right to loosen, left to tighten.

24. Install the electric fan and reconnect the wiring.

25. Install the alternator and wiring.

26. Install drive belt.

27. Install the sound insulator or lower pan.

28. Install the coolant expansion tank.

29. Refill and bleed the cooling system.

30. Install the air cleaner housing, air ducts and brace.

31. Check all fluid levels. Operate the engine at normal operating temperatures and check for leaks.

Oil Seal Replacement

Camshaft Seal

EXCEPT 8 CYLINDER ENGINE

1. Disconnect the negative battery cable. Set engine to TDC. Make sure crankshaft and cam timing marks are aligned.

2. Remove timing belt.

3. Hold camshaft from turning and remove cam drive sprocket by tapping from behind. Take care not to lose the cam drive key.

4. Pry out the old oil seal. In some cases it may be easier to remove the front camshaft bearing cap.

To install:

5. Installation is the reverse of the removal procedure. Lubricate the seal lip with clean engine oil. Reinstall the front bearing cap if removed. Torque hold-down nuts to 15 ft. lbs. (20 Nm).

6. Use a suitable driver and tap the new seal into position.

7. Install the cam drive sprocket with the drive key. Check the timing marks. Torque center bolt to 60 ft. lbs. (81 Nm).

8. Install cam belt and cover.

8 CYLINDER ENGINE

1. Disconnect the negative battery cable. Align timing marks cylinder No. 1 TDC of the compression stroke. Remove the toothed timing belt.

2. Remove the camshaft sprockets.

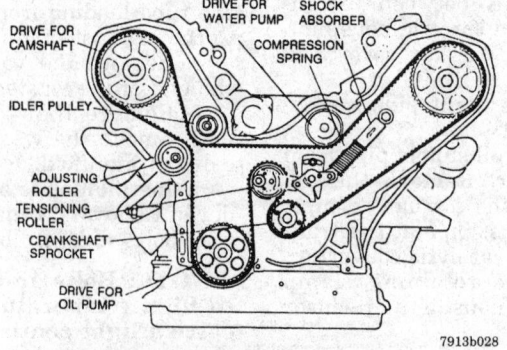

Timing belt layout — V8 engine

3. Remove the left and right belt guards.

NOTE: The toothed belt guard on the left-hand side contains 2 sealing rings and 1 shaft seal. The toothed belt guard on the right side contains 1 sealing ring and 1 shaft seal. If one shaft seal is leaking, the seals on both sides must be replaced.

4. Drive out the shaft seal.
To install:
5. When driving in the new seals, make sure the shaft seal is flush with the front edge of the belt guard. Replace the sealing rings.
6. Guide the camshaft sprocket into place.
7. Install the cam belt. Check timing marks on crankshaft and cam sprockets. Align timing marks cylinder No. 1 TDC of the compression stroke, if necessary.
8. Clean the sealing surface at the rear of the guard and on the cylinder head. After installing the new sealing rings, apply a thin coat of sealant to both surfaces. Install the rear toothed belt cover and torque the bolts to 7 ft. lbs. (10 Nm).

Crankshaft Seal

5 CYLINDER ENGINE

The oil seal is a part of the oil pump.
1. Disconnect the negative battery cable. Remove the timing belt.
2. Using a small prybar, pry the oil seal from the oil pump.
3. Clean out the seal seat. Using a new seal, lubricate the lip with engine oil. Using a suitable socket, drive the seal in to the seal seat.

NOTE: When installing a new seal, be careful not to damage the lip of the seal.

4. Reinstall the timing belt. Torque the crankshaft pulley bolt to 253 ft. lbs. (343 Nm).

8 CYLINDER ENGINE

1. Disconnect the negative battery cable. Remove the toothed cam drive belt.
2. Remove the vibration damper.
3. A special oil seal extractor is recommended to pull the seal from the front cover flange.
4. Install the new seal after lubricating it with engine oil. Press the seal in only until flush. Note that if the crankshaft shows signs of scoring, press the sealing ring in completely.
5. Reinstall the crankshaft sprocket and timing belt.

Timing Belt and Tensioner

Adjustment

5–Cylinder Engine

1. Remove the timing belt cover.
2. Loosen the water pump adjusting bolts and rotate the pump clockwise to tighten and counterclockwise to loosen.
3. The belt is correctly tensioned when it can be twisted 90 degrees midway between the camshaft drive sprocket and the water pump.

8–Cylinder Engine

1. Remove the timing belt covers.
2. Perform the basic setting of the toothed belt tensioner by turning the idler pulley eccentric clockwise. A special turning tool may be needed. Turn the eccentric and measure the damper length (the damper is shaped like a small shock absorber). Measure the overall length, not counting the mounting eyes. Measure the length of the barrel. Turn the eccentric until the damper length is 5.11–5.23 in. (130–133mm). Tighten the idler pulley eccentric to 18 ft. lbs. (24 Nm).
3. Remove any camshaft and crankshaft locking tool previously installed. Turn engine at least 2 turns. Tighten the vibration damper center bolt to 258 ft. lbs. (350 Nm). Check the damper length and if necessary, readjust the idler pulley.
4. Reinstall the timing belt covers.

REMOVAL AND INSTALLATION

5–Cylinder Engine

1. Disconnect the negative battery cable. Using the large bolt on the crankshaft sprocket, rotate the engine until the No. 1 cylinder is at TDC of the compression stroke. Align the TDC mark **0** with the cast mark on the bell housing. If the belt hasn't jumped teeth, the timing mark on the rear face of the camshaft sprocket should be aligned with the upper left edge of the valve cover.
2. Remove the alternator and air conditioner compressor drive belts.
3. Remove the upper and lower timing belt covers.
4. Loosen the water pump bolts only enough to turn the pump clockwise.

NOTE: By loosening the water pump bolts, the coolant may drain from the engine at the water pump. If necessary, drain

the cooling system, remove the water pump and reinstall it with a new O-ring.

5. Slide the timing belt off the sprockets.
To install:
6. If necessary, turn the camshaft until the notch on the back of the sprocket is in line with the left side edge of the cylinder head gasket surface.
7. If necessary, align the TDC **0** mark with the with the lug cast on the bell housing.
8. Install the timing belt and turn the water pump counterclockwise to tighten the belt. Tighten the water pump bolts to 15 ft. lbs. (20 Nm).

NOTE: The timing belt is correctly tensioned when it can be twisted 90 degrees along the straight run between the camshaft sprocket and water pump. The belt must not be jammed between the oil pump and sprocket when installing the vibration damper.

9. Install the timing belt covers and tighten the bolts to 7 ft. lbs. (10 Nm).
10. Install the alternator and air conditioning compressor belts. These belts are correctly tensioned when they can be depressed ⅜ in. along their longest straight run.

8–Cylinder Engine

1. Disconnect the negative battery cable. The battery is under the rear seat.
2. Remove the coolant expansion tank cap. Raise and safely support vehicle. Remove the sound insulator or lower pan. Drain coolant from radiator.
3. Remove the alternator cooling air duct. Loosen the clip and disconnect the coolant temperature sensor harness connector. Remove the coolant hose.
4. Disconnect the outer half of the air duct at the alternator. Remove the screws securing the wiring at the alternator.
5. Release tension on the polyribbed drive belt and remove the belt in a downward direction. Do this by placing a 13mm box wrench on the hexagon guide of the tensioner and pressing the wrench slowly upwards.
6. Remove the alternator mounting bolts and remove the alternator.
7. Working from below, remove the 3 bolts for the toothed belt guard.
8. Remove the bolts securing the bracket for the air intake ducts on the left and right and remove the

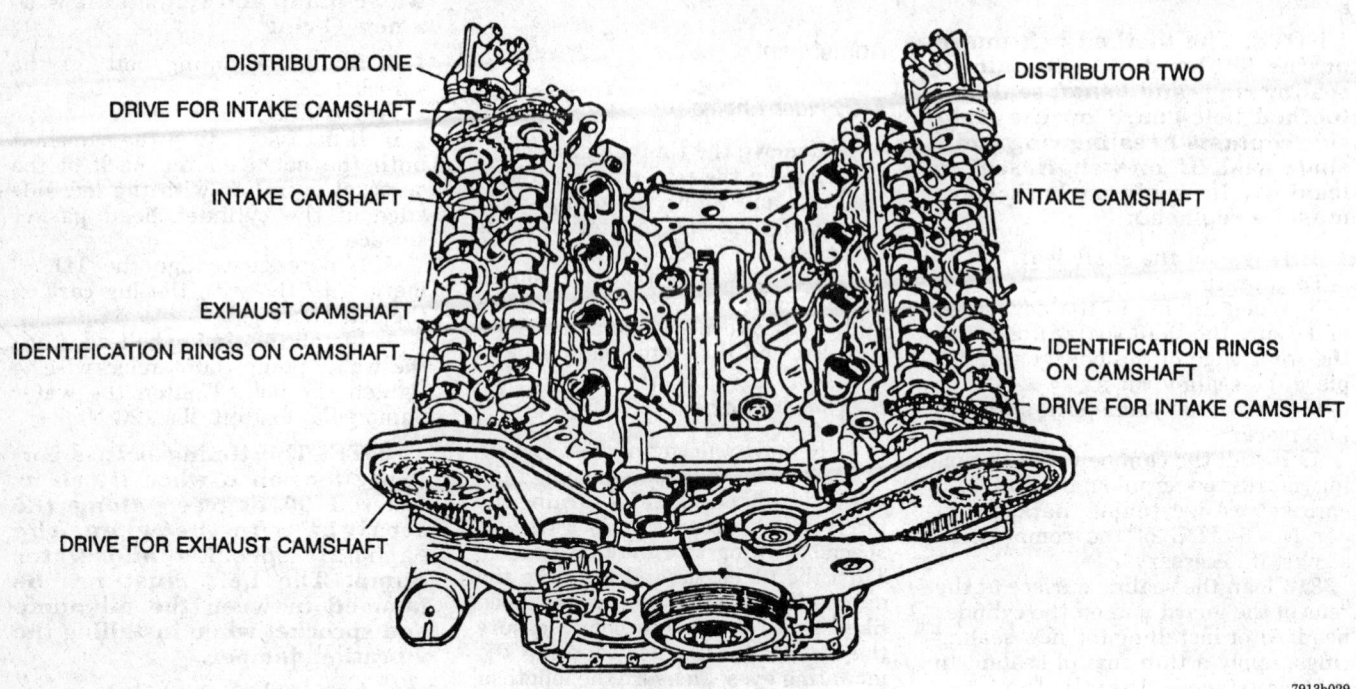

DISTRIBUTOR ONE
DRIVE FOR INTAKE CAMSHAFT
INTAKE CAMSHAFT
EXHAUST CAMSHAFT
IDENTIFICATION RINGS ON CAMSHAFT
DRIVE FOR EXHAUST CAMSHAFT

DISTRIBUTOR TWO
INTAKE CAMSHAFT
IDENTIFICATION RINGS ON CAMSHAFT
DRIVE FOR INTAKE CAMSHAFT

Camshaft layout train — V8 engine

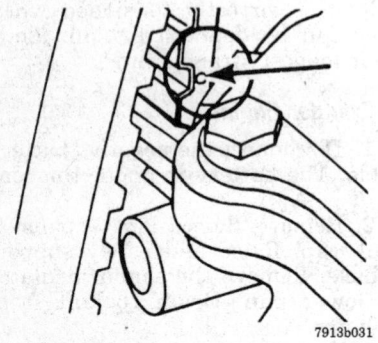

Camshaft sprocket alignment with cylinder head — 4– 5-cylinder engines

Crankshaft pulley alignment marks — 5-cylinder engine

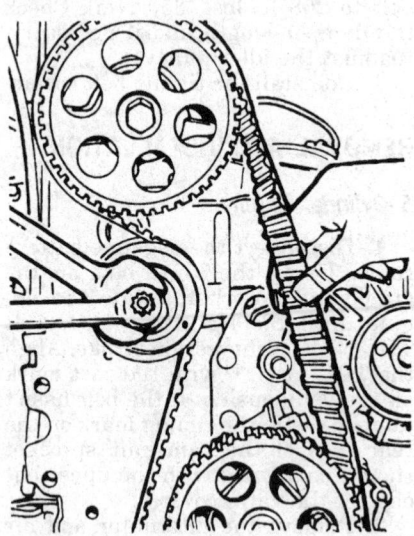

The timing belt on all models is correctly tensioned when the belt can be twisted 90° with thumb and 1 finger

ducts. Remove the bar-shaped strut brace.

9. Remove the 7 screws securing the upper part of the air cleaner housing, then remove the bolts holding the lower part of the housing. Press towards the rear and lift out.

10. Remove the upper radiator hose and clips.

11. Disconnect the electric fan bolts, lift the fan assembly out and lay to one side, wiring still connected.

12. Remove the bolts securing the supporting clamp for the lower radiator hose and take off the upper part. Disconnect the radiator hose at the thermostat housing. Swing the radiator hose to the right.

13. Remove the fan shroud by unscrewing the bolts for the viscous fan at the top. Remove the bolts securing the viscous fan. A spanner wrench or equivalent may be needed to hold the fan hub. Note that this is a left hand thread. Turn to the right to loosen. Lift out the fan and shroud together.

14. Disconnect the engine support at the front. Note any shims. The same thickness shims must be used at installation. Remove the radiator hose. At installation, install the radiator hose first.

15. Loosen the center bolt of the vibration damper by one turn. A special tool may be needed to hold the

damper from turning. This bolt was installed to over 250 ft. lbs. (340 Nm) torque and will be difficult to remove.

16. Remove the bolts securing the left side toothed belt cover or guard.

17. Remove the lower part of the supporting clamp for the lower radiator hose. Turn the tensioner for the poly-ribbed accessory drive belt in the loosening direction and insert an appropriate size holding pin in the hole provided.

18. Remove the bolts securing the right side toothed belt cover, with the exception of the top bolt. Remove the tensioner holding pin. Remove the belt cover top screw and remove the cover or guard. Carefully lift the guard from the bottom to avoid damaging the radiator.

19. Disconnect the ignition cables at the coils. Remove both distributor caps. Turn the engine to TDC. It may be necessary to temporarily install the damper to align the timing marks. Check that the distributor rotor is pointing to the mark on the housing. If not, turn the crankshaft 1 additional turn. Remove both distributors.

20. Remove the stop plate at the toothed belt tensioner. Disconnect the shock-absorber shaped damper at the top bolt.

21. Remove the belt from the tensioning idler pulley (with eccentric). Pulley is on right side of engine. Take the belt off both camshaft sprockets.

22. A special holding tool is available that is installed on the back of the camshafts. It fits the locating pin on the distributor flanges. If necessary, use a special hook wrench tool to turn the camshaft until the pins latch into the special holding tool. Secure the special tool with the distributor mounting bolts.

23. At the camshaft sprocket end, loosen the mounting bolts 2 turns. Using a plastic hammer, tap the edge of the camshaft sprockets loose.

NOTE: After the sprockets are removed, note the grooves machined in the camshaft ends. Woodruff keys must not be installed in the camshaft sprocket/camshaft connection. Unlike the other engines, this engine does not use cam keys.

24. Remove the vibration damper which was previously temporarily reinstalled to line up TDC timing marks. A puller can be used. Unscrew 2 opposing bolts of the 4 bolts connecting the vibration damper and

toothed belt sprocket. Use a puller in these holes.

25. Remove the toothed belt.

To install:

26. Before installing a new belt, the rollers and tensioners must be checked for dirt, rough running and ease of rotation. Clean or replace rollers and tensioners, as necessary.

27. Fit the toothed belt at the crankshaft and install the vibration damper with the belt on the crankshaft.

28. Apply thread locking compound to the center bolt and tighten to 332 ft. lbs. (450 Nm) using hand wrench. A holding tool may be required to keep the crankshaft from turning. It is a must to ensure that the TDC mark on the vibration damper is aligned with the TDC pointer before and after tightening the camshaft sprockets.

29. Guide the toothed belt into position and install the idler pulley with eccentric. Snug the nut enough to hold the eccentric in place but do not fully tighten it yet.

30. Tighten the damper top bolt to 18 ft. lbs. (24 Nm). Engage the damper to the tensioner lever by pressing the lever downward.

31. Perform the basic setting of the toothed belt tensioner by turning the idler pulley eccentric clockwise.

32. A special turning tool may be needed. Turn the eccentric and measure the damper length. Measure the overall length of the barrel not counting the mounting eyes. Turn the eccentric until the damper barrel length is 5.11–5.23 in. (130–133mm).

33. Tighten the idler pulley eccentric to 18 ft. lbs. (24 Nm). Tighten the camshaft sprockets to 33 ft. lbs. (45 Nm).

34. Remove any camshaft and crankshaft locking tool previously installed.

35. Turn engine at least 2 turns. Check the damper length and if necessary, readjust the idler pulley.

36. Reconnect the ignition cables at the coils. Install both distributor caps.

37. Install the right side cover or guard.

38. Install the left side cover or guard.

39. Reconnect the engine support. Install any shims, if necessary.

40. Install the fan shroud. Note this left-handed thread.

41. Install all coolant hoses.

42. Install the electric fan.

43. Install the alternator and electrical wiring.

44. Install the drive belt and adjust.

45. Install the air filter housing. Install all air ducts. Install strut brace.

46. Install the coolant expansion tank.

47. Install the sound insulator or lower pan.

48. Refill and bleed the cooling system.

49. Check all fluid levels. Operate the engine at normal operating temperatures and check for leaks.

50. Road test the vehicle for proper operation.

Timing Sprockets

REMOVAL AND INSTALLATION

Except 8–Cylinder Engines

All timing belt sprockets, are located by keys on their respective shafts. Each sprocket is retained by a bolt. To remove any or all of the sprockets, first remove the timing belt cover(s) and timing belt.

1. Disconnect the negative battery cable. Remove the center retaining bolt for the sprocket.

2. Pull the sprocket off the shaft.

3. If the sprocket is sticking on the shaft, use a gear puller or tap lightly with a plastic mallet. Do not hammer on the sprocket or damage may occur.

4. Remove the sprocket, being careful not to lose the key.

5. Installation is in the reverse order of removal.

NOTE: Always check valve timing after removing the drive sprockets.

8–Cylinder Engines

1. Disconnect the negative battery cable. Remove timing cover, toothed timing belt and both distributors. Note that the timing belt drives the exhaust camshaft. A chain from the exhaust cam drives the intake cam.

2. A special holding tool is available that is installed on the back of the camshafts. It fits the locating pin on the distributor flanges. If necessary, use a special hook wrench to turn the camshaft until the pins latch into the special holding tool. Secure the special tool with the distributor mounting bolts.

3. At the camshaft sprocket end, loosen the mounting bolts 2 turns. Using a plastic hammer, tap the edge

2 AUDI 80/90/100/200/COUPE/S4/V8

of the camshaft sprockets slightly loose.

NOTE: After the sprockets are removed, note the grooves machined in the camshaft ends. Woodruff keys must not be installed in the camshaft sprocket/camshaft connection.

4. When assembling the sprockets to the camshafts, make sure the timing marks found on the backside of the chain sprockets are still aligned. They should align next to each other at the 3 o'clock and 9 o'clock positions. Tighten the camshaft sprockets to 33 ft. lbs. (45 Nm).

Camshaft

REMOVAL AND INSTALLATION

5–Cylinder Engine

80 and 90

1. Disconnect the negative battery cable. Remove the upper drive belt cover, valve cover and upper part of intake manifold, if necessary.
2. Using the large bolt on the crankshaft sprocket, rotate the engine until the No. 1 cylinder is at TDC of the compression stroke. Align the TDC mark **0** with the cast mark on the bell housing. If the belt hasn't jumped the timing mark on the rear face of the camshaft sprocket should be aligned with the upper left edge of the valve cover.
3. Remove the timing belt from the camshaft sprocket. Remove the camshaft sprocket.
4. Remove bearing caps No. 1 and 3. Bearing caps are marked on the top.
5. Diagonally loosen bearing caps No. 2 and 4 and remove the bearing caps.
6. Lift the camshaft out of the cylinder head.

To install:
7. When installing, lightly oil the camshaft and bearing journals with clean engine oil.

8. Position the caps on the same journals from which they were removed.
9. Install bearing caps No. 2 and 4. Tighten alternately and diagonally to 15 ft. lbs. (20 Nm).
10. Install bearing caps No. 1 and 3 and tighten to 15 ft. lbs. (20 Nm).
11. Install the camshaft sprocket and timing belt. Install the valve cover. The camshaft sprocket bolt is tightened to 59 ft. lbs. (80 Nm).

100 and 200

1. Disconnect the negative battery cable. Remove the upper drive belt cover, valve cover and upper part of intake manifold, if necessary.
2. Using the large bolt on the crankshaft sprocket, rotate the engine until the No. 1 cylinder is at TDC of the compression stroke. Align the TDC mark **0** with the cast mark on the bell housing. If the belt hasn't jumped, the timing mark on the rear face of the camshaft sprocket should be aligned with the upper left edge of the valve cover.
3. Remove the timing belt from the camshaft sprocket. Remove the camshaft sprocket.
4. Diagonally loosen bearing caps No. 2 and 4 and remove the bearing caps. Bearing caps are marked on the top.
5. Diagonally loosen bearing caps No. 1 and 3 and remove the bearing caps.
6. Lift the camshaft out of the cylinder head.

To install:
7. When installing, lightly oil the camshaft and bearing journals with clean engine oil.
8. Position the caps on the same journals from which they were removed.
9. Tighten the nuts of caps 2 and 4 until snug.
10. Tighten all nuts to 15 ft. lbs. (20 Nm).
11. Install the camshaft sprocket and timing belt. Install the valve cover. The camshaft sprocket bolt is tightened to 58 ft. lbs. (79 Nm).

8–Cylinder Engine

1. Disconnect the negative battery cable. Remove the toothed camshaft drive belt.
2. Remove the cylinder head cover.
3. Remove the camshaft sprocket.
4. On the exhaust side, remove the intermediate flange and bearing cap for the distributor. All bearing caps that are removed should be marked or identified so they can be installed in the same position. Do not mix bearing caps or install then backwards. Loosen the bearing cap in front of the chain, plus caps No. 2 and 3. Loosen bearing caps No. 1 and 4 alternately and in a diagonal sequence. Bearing caps are marked on the top for correct installation.
5. On the intake side, remove bearing caps No. 6 and 7. Loosen bearing caps No. 5 and 8 alternately and in a diagonal sequence.
6. Remove the cam. If a lifter is removed and is to be reused, it must go in the bore from which it was removed.

To install:
7. Installation is the reverse of removal. Before assembly coat the moving parts with clean oil.
8. Install the camshafts with the chain so the markings on the chain sprockets are in alignment. The marks are on the back side of the chain sprocket. They should face each other at the 3 o'clock and 9 o'clock positions. Install the bearing caps so the stamped-on numbers can be read from the intake side.
9. Make sure the bearing caps are installed properly. They will cause parts failure if installed backwards. Torque the bearing caps to 11 ft. lbs. (15 Nm) in the following sequence:
Caps 5 and 8 in a diagonal sequence
Caps 1 and 4 in a diagonal sequence
Caps 2 and 3 in a diagonal sequence
Caps 6 and 7 in a diagonal sequence

2-16

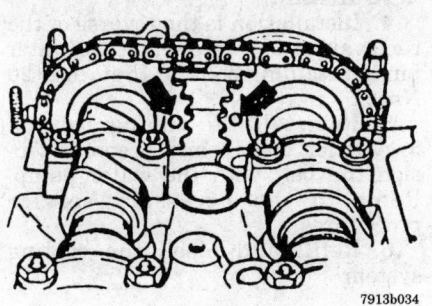

Cam gear alignment marks — V8 engine

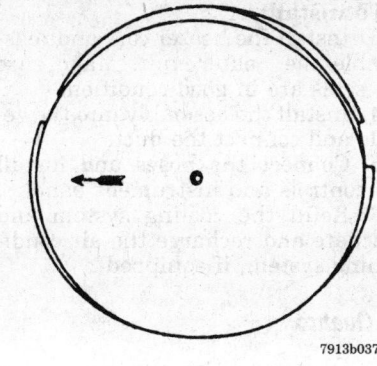

Arrow on piston faces front of vehicle

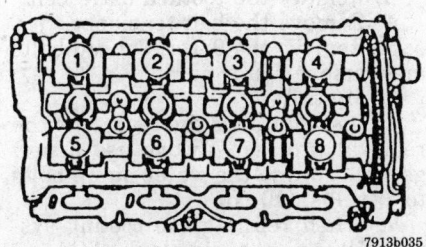

Cam bearing caps are numbered — V8 engine

Piston and Connecting Rod

Positioning

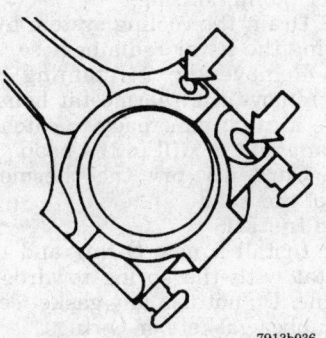

Align the forge marks when
assembling the connecting rod caps

ENGINE LUBRICATION

Oil Pan

REMOVAL AND INSTALLATION

1. Disconnect the negative battery cable. Raise and support the vehicle safely.

2. Drain the oil from the crankcase. Remove the cover plate from under the engine, if equipped.

3. If necessary, remove the 4 bolts from the subframe and lower the subframe. Remove the oil pan bolts while supporting the pan.

4. Lower the pan from the engine. Discard the gasket. Note that the 8–cylinder engine uses a 2 piece oil pan as well as a honeycomb baffle insert. Both an upper and lower pan gasket will be required.

   To install:

5. Coat both sides of a new gasket with sealer and install the gasket and oil pan.

6. Tighten the pan bolts to 15 ft. lbs. (20 Nm) on 5–cylinder engines. On 8–cylinder engines, tighten the lower pan-to-upper pan bolts to 14 ft. lbs. (19 Nm) and the upper pan-to-block bolts to 18 ft. lbs. (24 Nm).

Oil Pump

REMOVAL AND INSTALLATION

5–Cylinder Engine

1. Disconnect the negative battery cable. Loosen and remove the crankshaft pulley bolt.

2. Remove the timing belt covers.

3. Loosen the water pump bolts and turn the pump body clockwise.

4. Remove the timing belt and V-belt pulley with the timing belt sprocket.

5. Remove the dipstick. Raise and safely support vehicle. Drain the engine oil.

6. Remove the front bolts on the subframe and remove the oil pan.

7. Remove the oil suction pipe from the base of the oil pump and bracket to the engine block.

8. Remove the oil pump bolts and remove the oil pump from the front of the engine.

9. Installation is the reverse of the removal procedure. Refill with oil to correct level.

8–Cylinder Engine

1. Disconnect the negative battery cable. Pull out dipstick. Raise and safely support vehicle.

2. Drain engine oil

3. Remove oil pan bolts and pull off pan assembly.

4. Remove bolts securing oil pump and pull out, disengaging from drive.

5. Installation is the reverse of the removal procedure. Refill with oil to the correct level.

Rear Main Bearing Oil Seal

REMOVAL AND INSTALLATION

The rear main oil seal is located at the rear of the engine block. It can be found in a housing or flange behind the flywheel/flexplate. To replace the seal, remove the transaxle or pull the engine.

1. Disconnect the negative battery cable. Remove the transaxle.

2. Remove the flywheel/flexplate.

3. Using a suitable tool, pry the old seal out of its housing.

4. To install, lightly oil the replacement seal and press it into place.

   NOTE: Be careful not to damage the seal or score the crankshaft.

5. Install the flywheel/flexplate and the transaxle.

ENGINE COOLING

Radiator

REMOVAL AND INSTALLATION

1. Drain the cooling system.
2. Remove the 3 pieces of the radiator cowl and the fan motor assembly. Take care in removing the fan motor connectors to avoid bending them.
3. Remove the upper and lower radiator hoses and the coolant tank supply hose.
4. Disconnect the coolant temperature switch located on the lower right side of the radiator.
5. Remove the radiator mounting bolts and lift out the radiator.
6. Installation is the reverse of removal. Torque radiator mounting bolts to 14 ft. lbs. (19 Nm) and cowl bolts to 7 ft. lbs. (10 Nm). Refill and bleed the cooling system.

Heater Core

REMOVAL AND INSTALLATION

Except V8 Quattro

1. Disconnect the negative battery cable.
2. At the radiator, pull off the bottom hose and drain the coolant into a container for reuse.
3. Remove the heater hoses from the heat exchanger.
4. At the heater assembly control valve, disconnect the control wire.
5. Remove the console. Remove the left and the right heater covers from below the dashboard. On 80 and 90 model vehicles, remove the instrument panel.
6. Properly discharge the air conditioning system and remove the refrigerant lines from the evaporator.
7. At the heater control unit, pull off the control knobs.
8. Remove the trim plate from the heater control unit.
9. At the heater control unit, remove the retaining screws and the center cover.
10. Remove the heater air ducts and the heater assembly retaining springs.
11. Remove the air plenum from the cowl and the heater assembly from the vehicle.
12. Separate the heater unit and remove the heater core.

To install:
13. Install the heater core and reassemble the heater unit. Make sure the seals are in good condition.
14. Install the assembly into the vehicle and connect the ducts.
15. Connect the hoses and install the controls and instrument panel.
16. Refill the cooling system and evacuate and recharge the air conditioning system, if equipped.

V8 Quattro

1. Disconnect the negative battery cable.
2. Matchmark hood hinges and remove hood.
3. Remove the windshield wiper assembly.
4. Remove the cap from the engine coolant overflow bottle.
5. Remove the heater retaining band.
6. Remove the vacuum hoses from the vacuum servo motors.
7. Clamp the heater hoses and disconnect them from the heater core.
8. Remove the retainers between the body and heater and remove the heater box.
9. Remove the silicone rubber sealant from the heater core inlet/outlet area and separate the housing halves. Remove the heater core from the housing.

To install:
10. Installation is the reverse of the removal procedure. Before installing the fresh air blower guides, lubricate with petroleum jelly.
11. Apply gasket compound around the heater box before installing the heater core to seal box.
12. After installing the heater core, fill the opening between the heater core and the housing with silicone sealant.
13. When connecting the water hoses, make sure the lower connection on the heater core is connected to the hose going to the water pump.
14. Refill the cooling system.

Water Pump

REMOVAL AND INSTALLATION

5-Cylinder Engines

1. Drain the cooling system.
2. Remove the V-belts and the timing belt covers. On 5-cylinder engine, remove the timing belt from the water pump.
3. Always replace the old gasket or O-ring.

To install:
4. Installation is the reverse of the removal procedure. Torque the water pump retaining bolts to 15 ft. lbs. (20 Nm).
5. Reinstall the timing belt on 5-cylinder engine and properly tension the belt with the water pump. Refer to the necessary service procedures.
6. Refill and bleed the cooling system.

8-Cylinder Engine

1. Disconnect the negative battery cable.
2. Drain cooling system.
3. Remove the toothed drive belt.
4. Remove the belt tensioner.
5. Remove the 9 Torx® head bolts and remove water pump.

To install:
6. Installation is the reverse of the removal process. Always use a new gasket. Torque the water pump bolts to 7 ft. lbs. (10 Nm).
7. When refilling the cooling system, fill the expansion tank with new coolant to the maximum mark. Close the expansion tank and warm the engine until the radiator cooling fan cycles.
8. Check the coolant level and if necessary, top off. When the engine is at normal temperature, the level should be slightly over the maximum mark. When the engine is cold, the fluid should be between the maximum and minimum marks.

Thermostat

REMOVAL AND INSTALLATION

5-Cylinder

The thermostat is located in the lower radiator hose neck, on the left side of the engine block, behind the water pump housing.
1. Drain the cooling system by removing the lower radiator hose.
2. Remove the 2 retaining bolts and remove the thermostat housing. Have a catch pan ready to catch the coolant that is still in the head.
3. Carefully pry the thermostat out of the head.

To install:
4. Install a new O-ring and thermostat with the spring towards the engine. Do not use any gasket sealer on rubber gaskets or O-rings.
5. When installing the housing, be careful to properly seat the O-ring. Torque the bolts to 7 ft. lbs. (10 Nm).

6. Reconnect the hose. Refill and bleed the system.

6–Cylinder Engine

It is not necessary to remove the ribbed V-belt for access to thermostat and gasket.

1. Drain the cooling system.
2. Remove the drive (timing) belt cover.
3. Remove the thermostat housing from under the timing belt. Do not get coolant on timing belt.
4. Remove the thermostat and gasket.
5. Installation is the reverse of the removal procedure. Refill and bleed the cooling system.

8–Cylinder Engine

The thermostat is located at the front right-hand side of the engine below the intake manifold.

1. Drain the cooling system.
2. Remove the 2 retaining bolts from the thermostat housing.

NOTE: It is not necessary to disconnect the lower radiator hose. Removing the hose from the thermostat yoke may ease installation.

3. Remove the thermostat housing.
4. Carefully, pry the thermostat from the engine.
To install:
5. Install the new thermostat with the breather valve at the top or 12 o'clock position. Use a new gasket or O-ring.
6. Install the radiator hose, if removed. Refill and bleed the cooling system.

Cooling System Bleeding

After working on the cooling system, even to replace the thermostat, the system should be bled. Air trapped in the system will prevent proper filling and leave the radiator coolant level low, causing a risk of overheating.

1. To bleed the system, start with the system cool, the radiator cap off and the radiator filled to about an 1 in. below the filler neck.
2. Start the engine and run it at slightly above normal idle speed. This will insure adequate circulation. If air bubbles appear and the coolant level drops, fill the system with an antifreeze/water mixture to bring the level back to the proper level.
3. Run the engine this way until the thermostat opens. When this happens, coolant will move abruptly across the top of the radiator and the temperature of the radiator will suddenly rise.
4. At this point, air is often expelled and the level may drop quite a bit. Keep refilling the system until the level is near the top of the radiator and remains constant.
5. Fill the radiator right up to the filler neck. Replace the radiator filler cap. Refill the overflow tank.

ENGINE ELECTRICAL

NOTE: Most vehicles are equipped with theft protected radios, which cannot be operated if power to the radio is interrupted. Before disconnecting the battery cables, obtain the security code. Never disconnect any electrical connector with the ignition key ON unless specified in the repair procedure, or damage to electronic components may result.

Distributor

REMOVAL

1. Disconnect the wiring harness connector from the distributor cap.

NOTE: V8 Quattro vehicles have 2 distributors, driven off the exhaust camshafts.

2. Unclip and remove the distributor cap and static shield with the spark plug wires still attached.
3. Disconnect and tag the vacuum lines at the distributor, if equipped.
4. Note the position of the rotor in relation to the distributor housing. Scribe a mark on the distributor and engine block or cylinder head for installation. Matchmark the tip of the rotor to the engine. Note the approximate position of the vacuum advance unit in relation to the engine.
5. Remove the distributor holddown bolt and clamp.
6. Lift the distributor assembly from the engine.

INSTALLATION

Timing Not Disturbed

1. With the rotor pointing in the same direction as when removed, insert the distributor into the engine.

2. Once the distributor is seated into the engine, line up the marks on the distributor and engine with the metal tip of the rotor.
3. Make sure the vacuum advance unit, if equipped, is pointed in the same direction as it was pointed originally. If the marks on the distributor and the engine are lined up properly, this will be done automatically.
4. Install the distributor holddown clamp and bolt.
5. Install the distributor cap and static shield.
6. Install the vacuum lines, if equipped.
7. Install the distributor wiring harness connector.
8. Start the engine. Adjust the ignition timing.

Timing Disturbed

NOTE: If the engine has been turned or disturbed in any manner while the distributor was removed or if the marks were not drawn, it will be necessary to initially time the engine. Follow the procedure given below.

1. It is necessary to place the No. 1 cylinder in the firing position (TDC) to correctly install the distributor. To locate this position, the ignition timing marks on the flywheel and the clutch housing are used.
2. Remove the spark plug from the No. 1 cylinder. Turn the crankshaft until the piston in the No. 1 cylinder is moving up on the compression stroke. This can be determined by placing a finger over the spark plug hole and feeling the air being forced out of the cylinder. Stop turning the engine when the timing mark on the flywheel is aligned with the lug on the flywheel housing.
3. Remove the timing belt cover.
4. Align the mark on the camshaft sprocket with the upper edge of the drive belt cover or with the upper edge of the valve cover gasket mounting surfaces.
5. Oil the distributor housing lightly where it bears on the cylinder block.
6. Install the distributor so the rotor tip points to the mark on the distributor housing for the No. 1 cylinder.
7. Clean the distributor cap and check for signs of cracking or carbon tracks. Install the cap and continue the installation procedure.

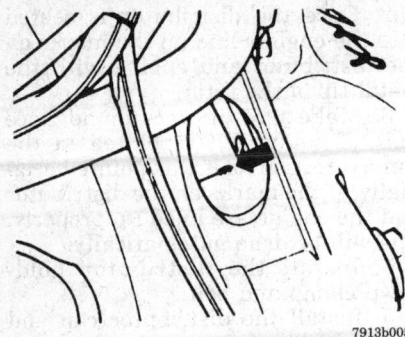

7913b005

The timing mark is on the flywheel and aligns with a pointer on the bell housing

Ignition Timing

ADJUSTMENT

Some tachometers, dwell-meters and oscilloscopes will not work with these ignition systems. Some test equipment may be damaged. Consult the manufacturer of the test equipment if there is any doubt.

All engines are timed by aligning the distributor housing with reference marks, no dynamic adjustment is possible. The electronic control unit will retard or advance the timing for each cylinder as required.

NOTE: V8 Quattro vehicles have 2 ignition coils with power stages and 2 distributors. They are both controlled by the Motronic ECU. Each coil and distributor is responsible for providing spark to 4-cylinders. One distributor is mounted on the back of each cylinder head. Both distributors are driven by lugs on the exhaust camshafts. A Hall sender is installed in the distributor mounted on the right cylinder head. The signal from this Hall unit identifies cylinder No. 1 for the start of the sequential fuel injection and cylinder selective

knock regulation. Ignition timing is determined by the electronic control unit and cannot be adjusted.

Alternator

PRECAUTIONS

Several precautions must be observed with alternator-equipped vehicles to avoid damage to the unit.
• If the battery is removed for any reason, make sure it is reconnected with the correct polarity. Reversing the battery connections may result in damage to the one-way rectifiers.
• When utilizing a booster battery as a starting aid, always connect the positive to positive terminals and the negative terminal from the booster battery to a good engine ground on the vehicle being started.
• Never use a fast charger as a booster to start vehicles.
• Disconnect the battery cables when charging the battery with a fast charger.
• Never attempt to polarize the alternator.
• Do not use test lamps of more than 12 volts when checking diode continuity.
• Do not short across or ground any of the alternator terminals.
• The polarity of the battery, alternator and regulator must be matched and considered before making any electrical connections within the system.
• Never separate the alternator as an open circuit. Make sure all connections within the circuit are clean and tight.
• Disconnect the battery ground terminal when performing any service on electrical components.
• Disconnect the battery if arc welding is to be done on the vehicle.

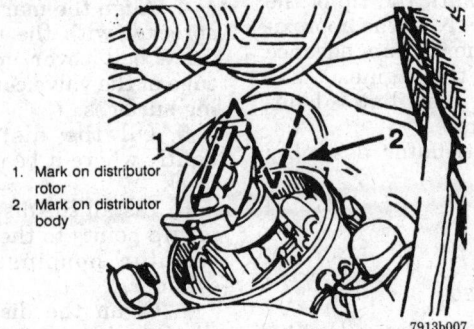

1. Mark on distributor rotor
2. Mark on distributor body

7913b007

With the cap removed, turn the engine to align the rotor with the mark on the distributor body

BELT TENSION ADJUSTMENT

The drive belts are correctly tensioned when the longest span of belt between pulleys can be depressed $1/8$–$1/2$ in. (3–13mm) using moderate thumb pressure. To adjust, loosen the slotted adjusting bracket bolt on the alternator. If the alternator hinge bolts are very tight, it may be necessary to loosen them slightly to move the alternator. Move the alternator in or out to obtain the correct tension. Tighten the adjusting bolt when finished.

V-belts under 39 inches in length should deflect about $1/8$ in. (3mm). Belts over 40 inches long should deflect about $1/2$ in. (13mm).

Poly-ribbed belt alignment is especially important on 8-cylinder engines. Check that the poly-ribbed belt between the air conditioner compressor and the hydraulic pump is in alignment front-to-back to prevent damage to the belt. If the 2 pulleys are not in alignment, remove the bolts securing the ribbed belt pulley for the hydraulic pump. Using shims, available in sizes 0.020, 0.040 and 0.060 in. (0.5, 1.0 and 1.5mm) adjust the pulleys until they are in alignment. Note that the poly-ribbed belt used on 8-cylinder engines is designed to last the life of the engine.

REMOVAL AND INSTALLATION

NOTE: If necessary, remove and install the alternator from below the vehicle.

1. Disconnect the negative battery cable.
2. Disconnect and tag the alternator wiring. On turbocharged engines, the cold air housing must be removed from the back of the alternator.
3. Remove the pivot bolt from the adjusting bracket.
4. Remove the drive belt.
5. Unbolt and remove the alternator.
To install:
6. Hold the alternator in position and install the pivot bolts.
7. Install the drive belt and adjusting bolt.
8. Adjust the belt tension.
9. Connect the electrical connections, making sure they are installed in their original locations.
10. Connect the negative battery cable.

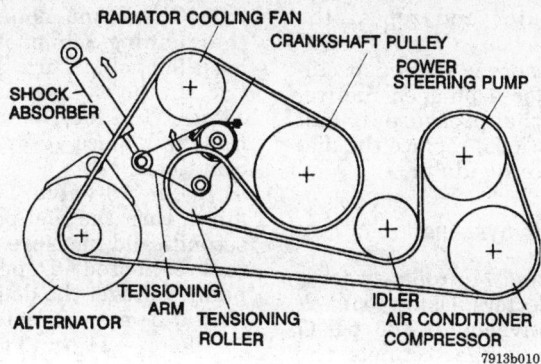

RADIATOR COOLING FAN
CRANKSHAFT PULLEY
POWER STEERING PUMP
SHOCK ABSORBER
TENSIONING ARM
ALTERNATOR
TENSIONING ROLLER
IDLER
AIR CONDITIONER COMPRESSOR
7913b010

A single belt is used — V8 Quatro

Starter

REMOVAL AND INSTALLATION

1. Disconnect both battery cables.
2. Raise and safely support the vehicle.
3. Disconnect and tag the starter wiring.
4. Remove the starter support bracket bolts. Remove the starter mounting bolts from the rear of the starter.
5. On some engines, 1 bolt goes through the transaxle with a nut on the end of the bolt.
6. Remove the starter from the engine.
7. Installation is the reverse of the removal procedure. Check system for proper operation.

EMISSION CONTROLS

Emission Warning Lamps

RESETTING

Oxygen Sensor Reminder

Every 30,000 miles a maintenance reminder light in the dashboard will come ON. This is an indication that the emission system should be checked and that the oxygen sensor should be replaced.

1. To reset the non-turbocharged vehicles, remove the instrument panel cluster. Remove the switch cover near the **OXS** button. Push the switch to reset the light.
2. On turbocharged vehicles, lift the rear seat and push the button marked **OXS** on the reset box.

FUEL SYSTEM

Fuel System Service Precautions

Safety is the most important factor when performing not only fuel system maintenance but any type of maintenance. Failure to conduct maintenance and repairs in a safe manner may result in serious personal injury or death. Maintenance and testing of the vehicle's fuel system components can be accomplished safely and effectively by adhering to the following rules and guidelines.

• To avoid the possibility of fire and personal injury, always disconnect the negative battery cable unless the repair or test procedure requires that battery voltage be applied.

• Always relieve the fuel system pressure prior to disconnecting any fuel system component (injector, fuel rail, pressure regulator, etc.), fitting or fuel line connection. Exercise extreme caution whenever relieving fuel system pressure to avoid exposing skin, face and eyes to fuel spray. Please be advised that fuel under pressure may penetrate the skin or any part of the body that it contacts.

• Always place a shop towel or cloth around the fitting or connection prior to loosening to absorb any excess fuel due to spillage. Ensure that all fuel spillage is quickly removed from engine surfaces. Ensure that all fuel soaked cloths or towels are deposited into a suitable waste container.

• Always keep a dry chemical (Class B) fire extinguisher near the work area.

• Do not allow fuel spray or fuel vapors to come into contact with a spark or open flame.

• Always use a backup wrench when loosening and tightening fuel line connection fittings. This will prevent unnecessary stress and torsion to fuel line piping. Always follow the proper torque specifications.

• Always replace worn fuel fitting O-rings with new. Do not substitute fuel hose or equivalent where fuel pipe is installed.

RELIEVING FUEL SYSTEM PRESSURE

Modern fuel injection systems operate under high pressure. This makes it necessary to first relieve the system of pressure before servicing. The pressurized fuel, when released, may ignite or cause personal injury.

1. Disconnect the power to the fuel pump by removing the relay or the fuel pump fuse. Check the list on the fuse box lid to make sure. The fuse can be removed to stop the fuel pump from running. With the engine operating at idle, wait until the engine stalls from fuel starvation.
2. Switch the ignition **OFF** and remove the negative battery cable.
3. Carefully loosen the fuel line on the control pressure regulator or component to be serviced.
4. Wrap a clean rag around the connection, while loosening, to catch any fuel.
5. After service is complete, discard the fuel soaked rag in the proper manner and reconnect negative battery cable, relay or fuses.

Fuel Tank

REMOVAL AND INSTALLATION

1. Disconnect the negative battery cable.
2. Relieve the fuel system pressure. Empty fuel tank and remove expansion tank if equipped.
3. Remove the rubber boot and overflow line from the body.
4. Remove the trunk panel and disconnect the harness connector from sending unit.
5. Remove all lines from the tank assembly. Loosen fuel tank straps or fuel tank retaining bolts and remove the fuel tank from the body.
6. Installation is the reverse of the removal procedure. Tighten fuel tank straps or retaining bolts evenly and install sound deadening strips in the the same position, as necessary.

Fuel Filter

REMOVAL AND INSTALLATION

Most vehicles use a fuel filter mounted under the vehicle, below the fuel tank. An arrow should be on the filter indicating fuel flow direction. Install with arrow pointing to engine. Use care not to mix up fuel supply or return lines. Fuel pressure applied to the return side of the system will cause damage.

In addition, some vehicles use a filter in the engine compartment near the fuel distributor. If equipped, use the following procedure:

1. Make certain to follow precautions and relieve fuel pressure.
2. Disconnect the fuel lines leading into and out of the fuel distributor.
3. Unscrew the filter retaining bracket and remove the filter.
4. Install a new filter in the bracket and reattach the bracket. Make sure the arrows are pointing in the direction of the fuel flow to the distributor.
5. Reconnect the fuel lines, start the engine and check for leaks.

Electric Fuel Pump

PRESSURE TESTING

NOTE: The fuel tank is pressurized. Open the filler cap carefully. Fuel system pressure is not adjustable.

CIS Systems

1. Using tool VW 1318 or an equivalent 0–100 psi (7 BAR) pressure gauge, connect the gauge in the line to the cold start valve. If using the special tool, position the lever so the valve is closed.
2. Remove the fuel pump relay from the main relay panel and plug a long jumper wire into terminal 52. If special tool US 4480/3 or equivalent is available, connect it in place of the fuel pump relay with the switch OFF.
3. Remove the electrical connector from the differential pressure regulator, if equipped.
4. Apply 12 volts to the jumper wire or turn the special tool switch ON to run the fuel pump. The pressure should be 84–96 psi (5.8–6.6 Bar). If the pressure is low, check the fuel pump delivery quantity.
5. If the pressure is higher than specifications, disconnect the fuel tank return line from the diaphragm

pressure regulator and repeat the test.
6. If the pressure is within specifications, check for a plugged fuel return line. If the pressure is not within specifications, replace the diaphragm pressure regulator.

MPI and Motronic Systems

1. The MPI and Motronic systems uses electric injectors. Using tool VW 1318 or an equivalent 0–100 psi (7 BAR) pressure gauge, connect the gauge in the system at a convenient place to read the pressure in the supply rail. This can be either at the inlet line or before the pressure regulator. If using the special tool, position the lever so the valve is open.
2. Disconnect the vacuum line to the pressure regulator.
3. Remove the fuel pump relay from the main relay panel and plug a long jumper wire into terminal 52. If special tool US 4480/3 or equivalent is available, connect it in place of the fuel pump relay with the switch OFF.
4. Apply 12 volts to the jumper wire or turn the special tool switch ON to run the fuel pump. The pressure should be 55–61 psi (3.8–4.2 BAR). If the pressure is low, check the fuel pump delivery quantity.
5. If fuel comes out of the regulator at the vacuum hose connection, replace the regulator.
6. Install the fuel pump relay and start the engine. At idle with the engine warm, connect the vacuum line to the pressure regulator. The pressure should decrease by about 8 psi (0.6 BAR). If not, check for blockage in the return lines, no vacuum to the pressure regulator or a bad regulator.

DELIVERY TESTING

1. Remove the fuel pump relay from the main relay panel and plug a long jumper wire into terminal 52. If special tool US 4480/3 or equivalent is available, connect it in place of the fuel pump relay with the switch OFF.
2. Remove the fuel pump cover from the floor of the trunk. On 80 and 90 vehicles, raise and safely support vehicle to access the pump.
3. At the fuel pump connector, pull back the rubber cover to expose the terminals, leave the plug connected to the pump.
4. Using a voltmeter, connect the probes across the terminals and apply 12 volts to the jumper wire or turn the special tool switch ON to run the pump.

5. Check and note the voltage of the running pump, it should be at least 9.0 volts. Turn the pump OFF.
6. In the engine compartment, disconnect the fuel line return connection and place it into a graduated container.
7. On 10-valve, 5–cylinder engines, turn the fuel pump ON for 30 seconds and measure the quantity of fuel collected. Depending on the pump voltage, the delivered quantity should be approximately:
   9 volts — 11 oz. (335cc)
   10 volts — 15 oz. (450cc)
   11 volts — 20 oz. (600cc)
   12 volts — 26 oz. (760cc)
8. On turbocharged, V8 and 5–cylinder, 20-valve engines, run the pump for 15 seconds. Depending on the pump voltage, the delivered quantity should be approximately:
   8 volts — 10 oz. (295cc)
   9 volts — 12 oz. (355cc)
   10 volts — 16 oz. (480cc)
   11 volts — 19 oz. (560cc)
   12 volts — 22 oz. (660cc)
   13 volts — 25 oz. (750cc)
   14 volts — 28 oz. (835cc)

REMOVAL AND INSTALLATION

All Except 80/90

1. Make certain to follow precautions and relieve fuel pressure. The fuel pump is located in the fuel tank. Remove the floor cover from the luggage compartment.
2. Disconnect the negative battery cable and the electrical connector from the fuel gauge sender.
3. Mark and remove the hoses from the fuel gauge sender.
4. Loosen the fuel gauge sender-to-fuel tank retaining ring. Pull out the fuel gauge/fuel pump assembly.
5. From inside the assembly housing, pull off the fuel hoses, detach the electrical connections and remove the gravity vent valve.
6. To install, reverse the removal procedures. Start the engine and check for leaks.

80 and 90

The fuel pump is located under the vehicle on a bracket in front of the fuel tank. The fuel pump assembly is located on the right side of front wheel drive vehicles and on the left side on Quattro vehicles. The 80 and 90 vehicles do not use a separate fuel pump filter. The filter is expected to be a lifetime unit unless the fuel was contaminated.

1. Make certain to follow precautions and relieve fuel pressure.

2. Disconnect the negative battery cable.

3. Raise and safely support vehicle.

4. Carefully loosen fuel line at fuel pump. Catch excess fuel in a container.

5. Remove fuel pump electrical connectors and remove the fuel pump.

To install:

6. Install fuel pump. Connect the fuel lines.

7. Connect the fuel pump electrical connectors.

8. Lower vehicle and connect the negative battery cable.

9. Replace any relays or fuses, that had been removed. Start engine and inspect for fuel leakage.

Fuel Injector

REMOVAL AND INSTALLATION

4– 5–Cylinder Engines

1. Disconnect the battery negative cable. Relieve the fuel pressure.

2. Remove the wiring harness support clip attaching bolts and position the wiring harness aside.

3. Disconnect the fuel return and supply lines, then the pressure regulator vacuum hose.

4. Remove the fuel rail attaching bolts, then the fuel rail and injectors as an assembly.

To install:

5. Use new O-rings and make sure the injectors fit properly onto the fuel rail. Install them as an assembly and torque the bolts to 15 ft. lbs. (20 Nm).

6. Use new gaskets and connect the fuel supply and return lines. Torque the fittings to 18 ft. lbs. (25 Nm).

7. Reconnect all wiring. Install pressure regulator vacuum hose.

8. Start engine and check for fuel leaks.

6– 8–Cylinder Engines

1. Disconnect the battery negative cable. Relieve the fuel pressure.

2. Remove the left and right support braces, then the intake air duct.

3. Remove the engine compartment support brace.

4. Remove the upper air cleaner attaching bolts, then the upper air cleaner.

5. Remove the lower air cleaner housing attaching bolts, then push the housing back and lift outward.

6. Remove the upper ventilation hose, then the right fuel rail cable tie.

NOTE: When assembling, install a new cable tie in the same position and location as the original.

7. Remove the engine wiring harness support clip attaching bolts and position the wiring harness aside.

8. Disconnect the throttle valve potentiometer, thermo-switch and idle stabilizer valve electrical connectors.

9. Remove the idle stabilizer valve, then disconnect the vacuum hoses to the carbon canister.

10. Remove the bolts attaching the intake air temperature sensor, then position aside.

11. Disconnect the air mass sensor electrical connectors, then the 2 knock sensor plug connections.

12. Disconnect the fuel return and supply lines, then the pressure regulator vacuum hose.

13. Remove the fuel rail attaching bolts, then the fuel rail and injectors as an assembly.

To install:

14. Use new O-rings and make sure the injectors fit properly onto the fuel rail. Install them as an assembly and torque the bolts to 7 ft. lbs. (10 Nm).

15. Use new gaskets and connect the fuel supply and return lines. Torque the fittings to 18 ft. lbs. (25 Nm).

16. Reconnect all wiring and vacuum hoses.

17. Install the air cleaner housing and ducting and properly secure the wires.

18. Install the support brace. Start engine and check for fuel leaks.

DRIVE AXLE

Front Halfshaft

REMOVAL AND INSTALLATION

Coupe, 80 and 90

NOTE: When loosening or tightening axle nut or bolt, make sure the vehicle is on the ground. Axle nut torque is high enough that attempting to loosen it may cause the vehicle to fall off the support.

1. Loosen the axle nut or bolt. Raise and safely support the vehicle.

2. Unbolt and remove the half-shaft-to-transaxle drive flange bolts.

3. Mark the position of the ball joint on the control arm, remove the 2 retaining nuts and disconnect the ball joint.

4. Remove the ball joint-to-steering knuckle bolt and separate the knuckle from the ball joint. Remove the mounting bolts for the control arm/stabilizer and push control arm downward, if necessary.

5. Pivot the strut outward and remove the halfshaft.

To install:

6. When installing the right half-shaft, take care not to damage the boot on the cover plate.

7. Tighten the ball joint-to-control arm/knuckle nuts/bolt to 47 ft. lbs. (64 Nm). Tighten the halfshaft flange bolts to 33 ft. lbs. (45 Nm).

8. Install the wheel and snug the axle nut or bolt. Place the vehicle on the ground and torque the axle nut or bolt to 200 ft. lbs. (270 Nm).

9. Check and adjust wheel alignment when finished.

100, 200 and V8

EXCEPT QUATTRO

NOTE: When loosening or tightening axle nuts, make sure the vehicle is on the ground. Axle nut torque is high enough that attempting to loosen it may cause the vehicle to fall off the support. A puller is required for this procedure.

1. Remove the halfshaft end nut.

2. Raise and support the vehicle safely and remove the wheels. If equipped with ABS, slide the speed sensor partly out of its mount.

3. On the right side, remove the halfshaft skid plate.

4. Disconnect the halfshaft from the transaxle. Using wire, support the halfshaft.

5. Using a 4-armed puller mounted on the wheel hub, press the halfshaft out of the hub.

6. Guide the inside end of the shaft up over the transaxle and out of the hub.

7. If equipped with an automatic transaxle, perform the following:

   a. Remove the stabilizer bar clamps.

   b. Remove the ball joint-to-hub bolt. Remove the ball joint from the hub.

   c. Press the halfshaft from the hub.

   d. Swing the suspension strut outward and press the halfshaft from the hub.

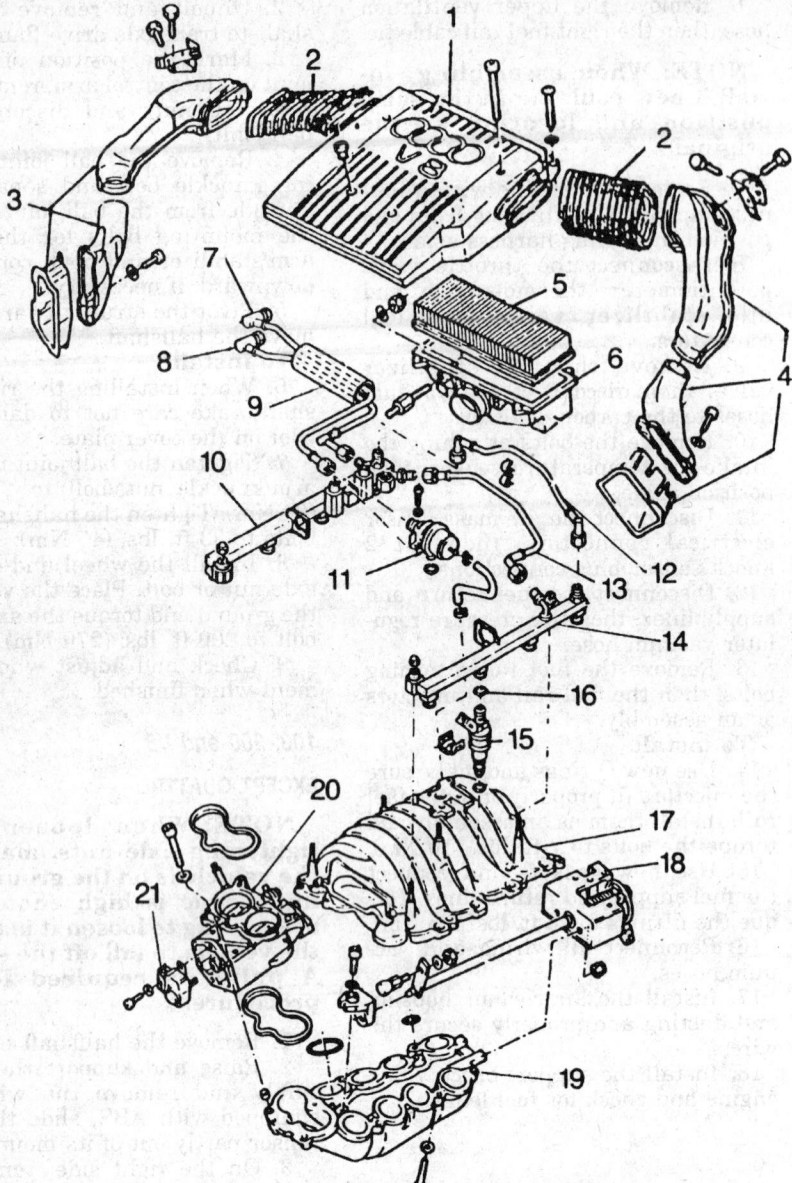

1. Upper air cleaner housing
2. Flexhose
3. Right air intake
4. Left air intake
5. Air cleaner filter element
6. Lower air cleaner housing
7. Return fuel line
8. Supply fuel line
9. Insulator
10. Right fuel manifold
11. Fuel pressure regulator
12. Supply line crossover
13. Return line crossover
14. Left fuel manifold
15. Fuel injector
16. O-ring
17. Upper intake manifold
18. Throttle shaft housing
19. Lower intake manifold
20. Throttle housing
21. Idle and full throttle switch

7913b021

Injectors and intake system — V8 engine

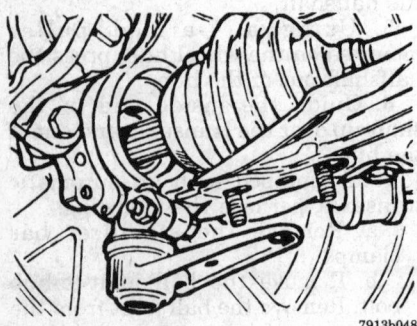

7913b048

Remove the front halfshaft by pivoting the strut out from the bottom

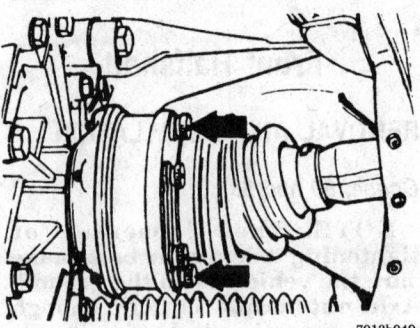

7913b049

Halfshaft bolts at the transaxle drive flange

To install:

8. When installing, make certain that the splines are clean and free of grease. Apply a ¼ in. (3mm) bead of RTV silicone sealant around the leading edge of the splines. Allow it to set at least 1 hour after installation.

9. Install the shaft-to-transaxle flange bolts and torque to 32 ft. lbs. (43 Nm).

10. Install the skid plate. If equipped with ABS, install the speed sensor.

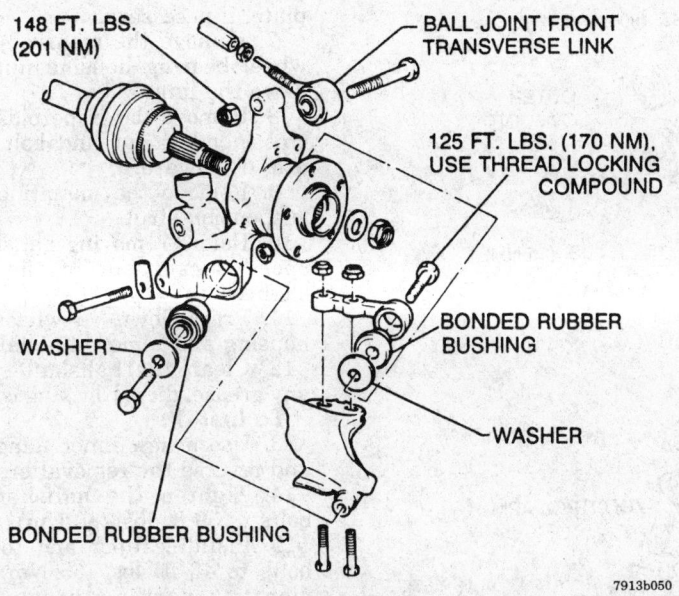

148 FT. LBS. (201 NM)

BALL JOINT FRONT TRANSVERSE LINK

125 FT. LBS. (170 NM), USE THREAD LOCKING COMPOUND

WASHER

BONDED RUBBER BUSHING

WASHER

BONDED RUBBER BUSHING

7913b050

Rear suspension

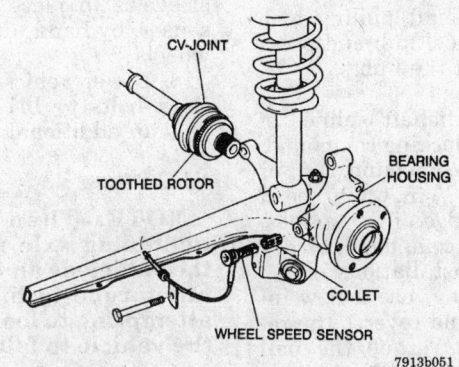

CV-JOINT

TOOTHED ROTOR

BEARING HOUSING

COLLET

WHEEL SPEED SENSOR

7913b051

Rear axle speed sensor for ABS — 200 shown

11. With the wheels installed and the vehicle on the ground, torque the axle nut to 203 ft. lbs. (275 Nm).

QUATTRO

1. Remove the wheel cover and loosen the lug nuts. Remove the dust cover and the halfshaft nut or through bolt, if equipped.
2. Raise and support the vehicle safely. Remove the wheel. On vehicles equipped with ABS, slide the speed sensor partly out of its mount.

Remove the right backing plate, if necessary.
3. Disconnect the halfshaft at the transaxle flange and position it aside.
4. Using a suitable puller, press out the stub axle from the hub. Use only a mechanical or hydraulic puller to remove the stub axle. Never use hot air blower or a flame to heat the stub axle.

To install:
5. Replace the gasket on the inner CV-joint.

6. Make sure the splines on the stub axle and the wheel hub are free of oil, grease and old locking compound. Apply a bead of suitable locking compound approximately $3/64$ in. wide around the splines and install the stub halfshaft. Allow at least 1 hour for the locking compound to harden after installation.
7. Install and torque the halfshaft to transaxle bolts to 58 ft. lbs. (79 Nm) and install the wheel.
8. With the vehicle on the ground, torque the halfshaft end nut to 207 ft. lbs. (280 Nm).

V8 QUATTRO

NOTE: When loosening or tightening axle nuts, make sure the vehicle is on the ground. Axle nut torque is high enough that attempting to loosen it may cause the vehicle to fall off the support.

1. Raise and safely support the vehicle. Remove the wheel cover and remove the hexagon through bolt and washer from the halfshaft end.
2. Slightly pull back the wheel speed sensor. Remove both stabilizer bar brackets.
3. Remove the clamp bolt at the steering knuckle and press out the ball joint pivot pin for the lower control arm. Be careful not to damage the joint boot or seal. Never widen the slit in the steering knuckle housing. Use penetrating oil as required.
4. Remove the halfshaft from the transaxle drive flange.
5. Push the strut outward and remove the halfshaft.
To install:
6. Always use a new gasket at the transaxle drive flange. Pull off the protective film from the replacement gasket and stick to CV-joint.
7. Install the halfshaft to the transaxle. Torque halfshaft-to-transaxle drive flange bolts to 59 ft. lbs. (80 Nm).
8. Install halfshaft through bolt, always use a new bolt and tighten when the vehicle's weight is on the floor. Torque to 148 ft. lbs. (200 Nm) plus an additional ¼ turn.
9. After installing the halfshaft, insert the wheel speed sensor into the housing as far as possible.
10. Install the ball joint. When installing the ball joint, install the lock bolt with the head facing to the rear of the vehicle. torque lock bolt to 48 ft. lbs. (65 Nm).
11. Install both stabilizer bar brackets.
12. Install wheel cover.

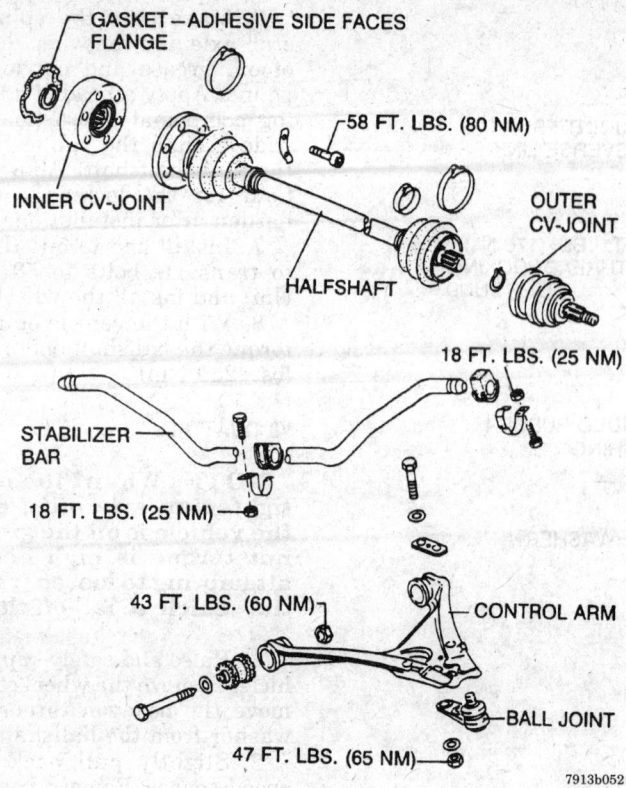

GASKET—ADHESIVE SIDE FACES
FLANGE

58 FT. LBS. (80 NM)

INNER CV-JOINT

OUTER
CV-JOINT

HALFSHAFT

18 FT. LBS. (25 NM)

STABILIZER
BAR

18 FT. LBS. (25 NM)

43 FT. LBS. (60 NM)   CONTROL ARM

BALL JOINT

47 FT. LBS. (65 NM)

7913b052

Rear drive axle — Quattro

Rear Halfshaft

REMOVAL AND INSTALLATION

80 and 90 Quattro

NOTE: When loosening or tightening axle nuts, make sure the vehicle is on the ground. Axle nut torque is high enough that attempting to loosen it may cause the vehicle to fall off the support.

1. With the vehicle resting on the ground, loosen the halfshaft nut.
2. Raise and support the vehicle safely.
3. Remove the halfshaft nut, wheel bolts and wheel assembly.
4. Remove the ball joint nut. Using a ball joint removal tool, separate the ball joint from the strut.
5. Using a suitable tool, pry downward on the lower control arm to remove the ball joint from the control arm. If necessary, loosen lower control arm mounting bolts.
6. Pull the brake hose and parking brake cable, with grommets, from the holding fixture.
7. Remove the inner halfshaft flange bolts. Separate the shaft from the flange and support it.

8. Using a halfshaft pulling tool, attach it to the wheel hub and press the halfshaft out of the hub.
To install:
9. Clean the halfshaft splines of any grease, dirt or locking compound. Using the locking compound D-6, or equivalent, apply a ¼ in. (3mm) bead around the outer edge of the splines. Allow the locking compound to dry for an hour after installation.
10. When installing, use a new inner flange gasket and reverse the removal procedures. Torque the ball joint nut to 47 ft. lbs. (64 Nm).
11. Tighten the inner halfshaft flange bolts to 58 ft. lbs. (79 Nm) and install the wheel.
12. With the vehicle on the ground, torque the halfshaft to hub nut to 238 ft. lbs. (322 Nm).

100 and 200 Quattro

1. With the vehicle weight on the ground, loosen the halfshaft end nut.
2. Raise and support the vehicle safely.
3. Remove the halfshaft nut, wheel bolts and wheel assembly.
4. Remove the brake caliper to strut retaining bolts and remove the caliper, without disconnecting the hydraulic line. Using wire, support the caliper.

5. Remove the brake rotor. Remove the inner halfshaft flange bolts and support the halfshaft.
6. Remove the fuel tank cover plate, if necessary.
7. Remove the transverse link-to-wheel bearing housing nut and remove the link.
8. Remove the trapezoidal arm-to-crossmember nut and bolt. Pry the arm downward.
9. Remove the mounting bolt for suspension strut.
10. Before removing halfshaft, pull speed sensor out of the housing slightly.
11. Press down on wheel bearing housing and remove the halfshaft.
12. Clean the halfshaft splines of any grease, dirt or locking compound.
To install:
13. Use a new inner flange gasket and reverse the removal procedures.
14. Tighten the halfshaft flange bolts to 59 ft. lbs. (80 Nm).
15. Install caliper and torque the bolts to 48 ft. lbs. (65 Nm). Adjustment of parking brake may be necessary.
16. Install the halfshaft bolt and washer assembly, tighten until just snug.
17. Make certain speed sensor sleeve is in place and install speed sensor, by hand, until seated. Install wheels.
18. Lower vehicle and torque halfshaft bolts to 147 ft. lbs. (200 Nm) plus an additional ¼ turn.

V8 Quattro

NOTE: When loosening or tightening axle nuts, make sure the vehicle is on the ground. Axle nut torque is high enough that attempting to loosen it may cause the vehicle to fall off the support.

1. Remove the center cap from the wheel and remove the axle bolt and washer.
2. Raise and safely support the vehicle and remove the wheel.
3. Slide the speed sensor out of the holder and remove the brake caliper without disconnecting the hydraulic line. Hang the caliper with wire.
4. Remove the brake disc.
5. Disconnect the transverse link from the wheel bearing housing.
6. Remove the halfshaft bolts from the differential drive flange and support the axle.
7. Remove the lower strut mount bolt and push the suspension down to remove the halfshaft.
To install:
8. Always use a new gasket at the transaxle drive flange. Pull off the

protective film from the replacement gasket and stick to CV-joint.

9. Install rear halfshaft to drive flange. Torque the halfshaft-to-drive flange bolts to 59 ft. lbs. (80 Nm).

10. Install halfshaft through bolt, always use a new bolt and tighten when the vehicle's weight is on the floor. Torque to 148 ft. lbs. (200 Nm) plus an additional 1/4 turn.

11. Use a new self-locking nut on the lower strut mount bolt and torque to 66 ft. lbs. (90 Nm).

12. Install a new self-locking nut on the transverse link bolt and torque to 148 ft. lbs. (200 Nm).

13. Install the brake disc and caliper. Torque the bolts to 48 ft. lbs. (65 Nm) and return the speed sensor its normal position.

14. Install the wheel and center cap.

CV-Boot

REMOVAL AND INSTALLATION

NOTE: On some vehicles, the entire halfshaft must be replaced, not serviced. Always check parts availability before removing CV-boots. If boot kits are available, use the following procedure.

1. Raise and safely support the vehicle and remove the halfshaft. Always loosen the halfshaft end locking nut or through bolt with the vehicle on the floor.

2. On the inner joint, remove the circlip from the inner CV-joint stub shaft. On a shop press, support the ball hub and press out the shaft. Remove the boot.

3. On the outer joint, spread the circlip and drive the joint off the shaft by tapping lightly with a soft copper or brass drift against the hub.

To install:

4. Installation is the reverse of the removal process. Press the joint onto the shaft until the circlip can be pressed into the groove. The chamfer on the inside diameter of the ball hub splines must face the halfshaft.

5. After applying the correct amount of special lube, usually supplied with the replacement boots, install the new boot band and tighten it according to the instructions in the kit.

Driveshaft and U-Joints

REMOVAL AND INSTALLATION

Quattro

1. Raise and support the vehicle safely.

2. Using a scribing tool, mark the position of the driveshaft to the transaxle flange and the rear differential.

3. Remove the driveshaft flange mounting bolts from both ends and remove the driveshaft from the vehicle. Remove the center bearing bolts.

To install:

4. To install, reverse the removal procedures. Note that the universal joints are not replaceable. If a universal joint is damaged or worn out, replace the driveshaft assembly.

5. Align scribe marks and install driveshaft.

6. Tighten the driveshaft-to-transaxle/differential flange bolts to 39 ft. lbs. (53 Nm) on 80 and 90 Quattro, or 33 ft. lbs. (45 Nm) on 100, 200 and V8 Quattro. Torque the driveshaft center bearing to frame bolts to 14 ft. lbs. (19 Nm).

Front Wheel Hub, Knuckle and Bearings

REMOVAL AND INSTALLATION

80 and 90

NOTE: 80 and 90 vehicles use 2 types of front wheel bearing housing assemblies. They are a single-piece unit that cannot be separated from the strut and a bearing housing that is removable from the strut for service. The repair procedures are similar for both with the exception that the single piece unit housing, if defective, must be replaced as a strut assembly.

1. Raise and safely support vehicle. Remove the halfshafts.

2. Remove the strut housing to body nuts and remove the strut/hub assembly from the vehicle.

3. Remove the brake disc and splash shield.

4. Using an arbor press with suitable drivers, press the wheel hub from the strut housing.

5. Using snapring pliers, remove the snaprings from both sides of the wheel bearing.

6. Using an arbor press with suitable drivers, press the wheel bearing from the strut housing.

7. Using a wheel puller, pull the wheel bearing race from the wheel hub.

To install:

8. Lightly grease inside the strut housing before installing the new bearing.

9. Be sure to press only on the outer race when pressing the new bearing into the hub. When installing the snaprings, make sure they are properly seated.

10. Install the brake plate onto the bearing housing. Be sure to press only on the inner bearing race when pressing the bearing over the hub.

11. To install the strut and wheel hub assembly, reverse the removal procedures. Torque the upper strut to body nut to 44 ft. lbs. (60 Nm). On all other models, torque the 3 upper strut nuts to 22 ft. lbs. (30 Nm). Check front wheel alignment.

100, 200, S4 and V8 Quattro

1. Raise and safely support vehicle. Remove the halfshafts.

2. Remove the strut to vehicle nuts and remove the strut from the vehicle.

3. Remove the disc brake rotor and splash shield.

4. Using an arbor press with suitable drivers, press the wheel hub from the strut housing.

5. Using snapring pliers, remove the snaprings from both sides of the wheel bearing.

6. Using an arbor press with suitable driver, press the wheel bearing from strut housing.

7. Using a suitable puller, remove the bearing race from the wheel hub.

To install:

8. Install the outer snapring into the strut housing.

9. Be sure to press only on the outer race when pressing the new bearing into the hub. When installing the snapring, make sure it is properly seated.

10. Install the brake plate onto the bearing housing. Be sure to press only on the inner bearing race when pressing the bearing over the hub.

11. To install the strut and wheel hub assembly, reverse the removal procedures. Tighten the strut to body nuts to 44 ft. lbs. (60 Nm). Check front wheel alignment.

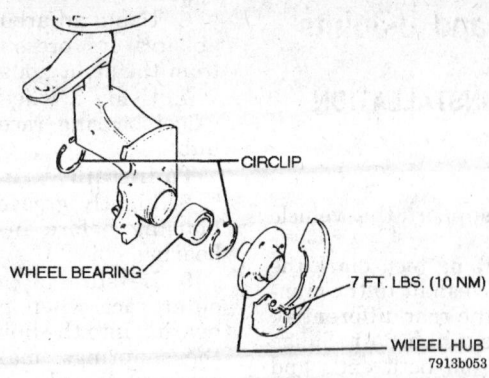

Front hub and bearing assembly

Differential Carrier

REMOVAL AND INSTALLATION

80 and 90 Quattro

1. Raise and safely support the vehicle. Matchmark and disconnect the axle shafts from the flanges and disconnect the driveshaft. Hang them from the body with wire.
2. Remove the differential lock servo and bracket. Disconnect the wiring.
3. Remove the nut from the rear mount bushing.
4. Place a transaxle jack under the differential and take the weight of the unit off the mounts. Remove the right side and rear mounts.
5. Pull the differential slightly forward and lower it from the vehicle.
   To install:
6. Installation is the reverse of removal. Torque the mount–to–sub frame nuts or bolts to 18 ft. lbs. (25 Nm). Torque the differential–to–mount nuts or bolts to 33 ft. lbs. (45 Nm).
7. Torque the axle flange bolts to 33 ft. lbs. (45 Nm) and the driveshaft bolts to 40 ft. lbs. (55 Nm).

100 and 200 Quattro

1. Raise and safely support the vehicle and remove the cover plate.
2. Engage the differential lock and block a wheel or set the parking brake.
3. Matchmark and disconnect the axle shafts and driveshaft and hang them from the body with wire.
4. Release the parking brake and disconnect the cables at the caliper and at the front.
5. Unbolt the left parking brake retainer and remove the heatshield.
6. Support the differential with a transaxle jack and take the weight off the mounts.

7. Remove the crossmember and the exhaust pipe.
8. Disconnect the mounts and lower the unit slightly.
9. Note the color coding of the vacuum hoses so they can be properly connected during installation. Disconnect the wiring and vacuum lines and carefully lower the unit out of the vehicle.
   To install:
10. Installation is the reverse of removal. Make sure to properly connect the vacuum hoses.
11. Torque the crossmember-to-body bolts and the suspension-to-crossmember bolts to 33 ft. lbs. (45 Nm).
12. Torque the axleshaft 8mm bolts to 33 ft. lbs. (45 Nm) and the 10mm bolts to 60 ft. lbs. (80 Nm).
13. Torque the driveshaft bolts to 40 ft. lbs. (55 Nm).

V8 Quattro

NOTE: Any time the differential is removed from this vehicle, the driveshaft must be properly aligned. This requires the special alignment tool 3139 or equivalent.

1. Raise and safely support the vehicle and remove the rear wheels.
2. Remove the rear muffler system and remove the heatshields, fuel tank shield and the axle shaft joint shield.
3. Disconnect the parking brake cables and the exhaust system support from the crossmember.
4. Matchmark the axle shafts and driveshaft and disconnect them from the differential. Hang them from the body with wire.
5. Detach the parking brake cables from the differential and disconnect them from the calipers.
6. Disconnect the brake hydraulic hoses from the distributor valve on the differential.
7. Use a 15mm socket to remove the mounting nut at the hole in the rear crossmember.
8. Support the rear crossmember and remove the crossmember mounting nuts.
9. At the front crossmember, loosen the left mounting bolt and remove the right mounting bolt to lower the crossmember slightly on the right side. Reposition the parking brake cables to the front.
10. Carefully support the differential with a transaxle jack. Separate the 2 halves of the front crossmember and unbolt the differential. Carefully lower the differential out of the vehicle.

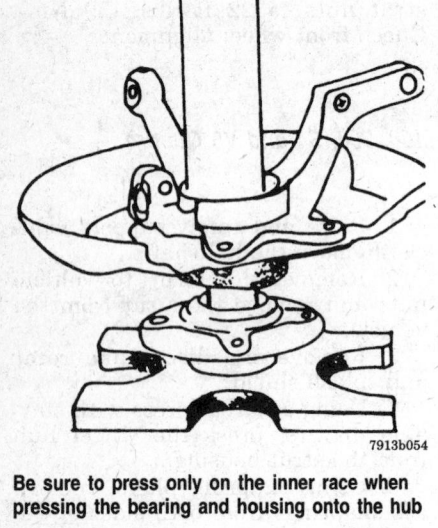

Be sure to press only on the inner race when pressing the bearing and housing onto the hub

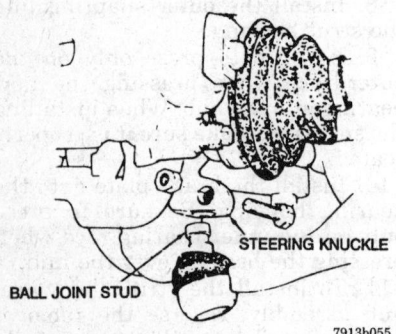

Remove the bolt to separate the ball joint from the knuckle

To install:

11. Install the differential assembly to the vehicle with transmission jack.

12. Loosely install the differential and all crossmember bolts before torquing any of them. Torque the front crossmember bolts, front crossmember-to-body bolts and the differential–to–front and rear crossmember bolts to 33 ft. lbs. (45 Nm).

13. Install the rear crossmember. Torque the rear crossmember-to-suspension bolts to 37 ft. lbs. (50 Nm).

14. Reconnect the brake cables and hydraulic lines.

15. Clean the threads in the differential flanges and use new self sealing bolts when connecting the driveshaft. Connect the axleshafts and driveshaft, making sure to align the matchmarks. Torque the axleshaft bolts to 60 ft. lbs. (80 Nm) and the driveshaft bolts to 40 ft. lbs. (55 Nm).

16. To align the driveshaft, install the driveshaft alignment tool and remove the bolts from the center support. Measure the gap between the support bolt holes and the body and install the proper spacers to make the gap even on both sides. Make sure the support is centered front to rear and install the bolts. Torque the bolts to 15 ft. lbs. (20 Nm) and remove the alignment tool.

17. Install the exhaust system. Install all hangers and align the exhaust components before tightening the clamps.

18. Install the heatshields, fuel tank shield and the axle shaft joint shield.

19. Install wheels. Bleed the brakes. Road test the vehicle.

MANUAL TRANSAXLE

Transaxle Assembly

REMOVAL AND INSTALLATION

NOTE: If the flywheel has been removed from the crankshaft for any reason, tighten the mounting bolts to: bolt without shoulder — 72 ft. lbs. (98 Nm); bolt with shoulder — 54 ft. lbs. (73 Nm). Coat all threads with a locking compound.

80 and 90

EXCEPT QUATTRO

1. Disconnect the negative battery cable.

2. Unplug the 2 electrical connectors for the backup lights. They can be found between the ignition coil and the fuel distributor filter.

3. Remove the upper engine-to-transaxle bolts.

4. Detach the speedometer cable from the transaxle.

5. Detach the clutch cable from the clutch lever.

6. Unbolt the exhaust pipe from the exhaust manifold.

7. Unscrew the 3 mounting bolts and remove the center engine mount.

8. Unbolt the front exhaust pipe from the support bracket and unbolt it from the catalytic converter.

9. Unscrew the 6 screws and remove the left halfshaft from the transaxle. Wire the halfshaft up and aside. Repeat the procedure for the right halfshaft.

10. Remove the clutch cover plate.

11. Tag and disconnect all wires leading to the starter and remove the starter.

12. Remove the bolt from the shift rod coupling.

13. Pry off the linkage coupling with a small prybar.

14. Pull the shift rod coupling off the shift rod. Place a transaxle jack under the transaxle and support it by lifting slightly.

15. Loosen the left chassis bolt on the rear transaxle support. Remove the bolts from the right or transaxle side of the support and pivot the support aside.

16. Remove the rubber mounting block.

17. Unscrew 3 bolts and remove the front transaxle support.

18. Remove the lower engine-to-transaxle bolts.

19. Carefully pry the transaxle apart from the engine and remove it.

To install:

20. Carefully fit the transaxle to the engine and torque the bolts to 40 ft. lbs. (54 Nm).

21. Install the front transaxle support and torque the bolt to 18 ft. lbs. (24 Nm). Torque the rubber mount-to-body bolts to 80 ft. lbs. (108 Nm).

22. Install the subframe bolts and torque to 51 ft. lbs. (69 Nm). Install the rubber mount-to-transaxle bolts and torque to 40 ft. lbs. (54 Nm).

23. Connect the shift linkage and adjust as required. Secure the bolt with safety wire.

24. Install the starter and connect all electrical wiring. Install the clutch cover.

25. Install the halfshaft bolts and torque to 33 ft. lbs. (45 Nm).

26. Reassembly the exhaust system using new self-locking nuts. Torque to 25 ft. lbs. (34 Nm).

27. Connect the speedometer and clutch cables and adjust as required.

28. Road test the vehicle for proper operation.

QUATTRO

1. Disconnect the negative battery cable.

2. Remove the 3 upper engine-to-transaxle bolts.

NOTE: Tag all bolts during removal so all bolts can be replaced in their correct locations, as they are not all the same size.

3. Disconnect the ground strap from the transaxle.

4. Remove the wiring connectors from the speedometer sender and the multi-function switch.

5. Disconnect the wiring for the oxygen sensor and oxygen sensor heating element.

6. Remove the engine protection plate.

7. Disconnect the exhaust pipe from the manifold.

8. Separate the exhaust pipe behind the catalyst and remove the pipe and catalyst.

9. Matchmark and remove the driveshaft.

10. Remove the rear crossmember.

11. Remove the shift rod securing bolt at the transaxle and let the shift rod hang.

12. Remove the transaxle cover plate.

13. Remove the right halfshaft shield.

14. Disconnect the left and right halfshafts, turn the steering to the right lock and tie both shafts up.

15. Remove the clutch slave cylinder.

16. Remove the tie rod coupling from the steering rack and turn wheel to the left.

17. Support the engine.

18. Support the transaxle.

19. Remove the transaxle strut at the left rear and front engine mount.

20. Remove the heatshield from the bonded rubber bushing.

21. Remove the bonded rubber bushing support bracket from the transaxle.

22. Remove the bonded rubber bushing.

23. Remove the bolt from the seat belt tension in cable guide at the left

rear of the transaxle. Position the cables and guide aside.

24. Lower the right rear subframe by loosening the mounting bolts.

25. Remove the remaining transaxle to engine bolts.

26. Remove the transaxle.

To install:

27. Install the transaxle, make certain that the alignment bushings are in the cylinder block before reassembly.

28. Tighten transaxle retaining bolts as follows: Torque the 8mm bolts to 18 ft. lbs. (24 Nm), the 10mm bolts to 33 ft. lbs. (45 Nm) and the 12mm bolts to 48 ft. lbs. (65 Nm).

29. Install the subframe. Tighten subframe mounting bolts to 25 ft. lbs. (34 Nm), plus an additional 90 degree turn.

30. Install the seatbelt tension cables and guides.

31. Install transaxle mount and engine mount. Support engine and transaxle as necessary.

32. Install the tie rod to the steering rack assembly. Torque the tie rod coupling–to–steering rack to 33 ft. lbs. (45 Nm).

33. Install the clutch slave cylinder.

NOTE: A replacement bolt for mounting the clutch slave cylinder is available from Audi, with a pointed tip for easier installation.

34. Install the right and left halfshafts. Torque the halfshaft–to–flange bolts to 33 ft. lbs. (45 Nm).

35. Install shift rod.

36. Install rear crossmember.

37. Install the driveshaft. Torque the driveshaft to transaxle and final drive bolts to 40 ft. lbs. (54 Nm).

38. Install the exhaust system.

39. Install all necessary wiring, transaxle cover plate, halfshaft shield and engine protection plate.

40. Install ground strap to the transaxle.

41. Reconnect the battery. Bleed system. Road test the vehicle for proper operation.

100

1. Disconnect the negative battery cable.

2. Remove the upper engine-to-transaxle bolts.

NOTE: Tag all bolts during removal, so all bolts can be replaced in their correct locations, as they are not all the same size.

3. Disconnect the ground strap from the transaxle, if equipped.

4. Remove the wiring connectors from the speedometer sender and the multi-function switch.

5. Support the engine.

6. Disconnect the wiring for the oxygen sensor and oxygen sensor heating element.

7. Raise and safely support vehicle. Remove the splash shield, if equipped.

8. Disconnect the exhaust pipe from the manifold.

9. Separate the exhaust pipe behind the catalyst and remove the pipe and catalyst.

10. Remove the bolt for the shift rod at the transaxle and separate.

11. Remove the heatshield from the right inner CV-joint.

12. Remove the halfshafts from the flanges and tie aside.

13. Remove the heatshield for the bonded rubber bushing on the right side.

14. Support the transaxle, with a suitable holding fixture.

15. Remove the strut at the rear of the transaxle.

16. Remove the clutch slave cylinder. Do not remove the hydraulic line from the slave cylinder.

17. Remove the subframe assembly.

18. Remove the lower transaxle-to-engine bolts.

19. Pry transaxle back and lower assembly.

20. Remove the transaxle.

To install:

21. Install the transaxle, make certain alignment bushings are in the cylinder block before reassembly.

22. Tighten transaxle retaining bolts as follows: Torque the 8mm bolts to 18 ft. lbs. (24), the 10mm bolts to 33 ft. lbs. (45 Nm) and the 12mm bolts to 48 ft. lbs. (65 Nm).

23. Install the subframe.

24. Tighten subframe mounting bolts to 25 ft. lbs. (34 Nm), plus an additional 90 degree turn.

25. Install the strut at the rear of the transaxle. Support the transaxle as necessary.

26. Install the clutch slave cylinder.

NOTE: A replacement bolt for mounting the clutch slave cylinder is available from Audi, with a pointed tip for easier installation.

27. Install the halfshafts. Torque the halfshaft-to-flange bolts to 33 ft. lbs. (45 Nm) and the halfshaft-to-transaxle to 40 ft. lbs. (54 Nm).

28. Install the shift rod at the transaxle.

29. Install the exhaust system.

30. Install all necessary wiring, halfshaft shield and engine protection plate.

31. Install ground strap to the transaxle, if equipped.

32. Reconnect the battery. Bleed system. Road test the vehicle for proper operation.

200

1. Disconnect the negative battery cable.

2. Remove the upper engine to transaxle bolts.

3. Remove the connector for the speedometer sender by pressing in the clips.

4. Remove the clip from the clutch slave cylinder and drive out spring pin, if equipped. Remove the bolt securing the clutch slave cylinder to the transaxle and remove the cylinder. Leave the hydraulic line connected.

5. Support the engine. Tie up coolant hoses and cables, as needed.

6. Remove the right side guard plate.

7. Disconnect the halfshafts from the flanges and rest both halfshafts on top of the subframe.

8. Tag and disconnect the wire from the backup light switch. Disconnect vacuum hoses at the servo if so equipped.

9. Pry off the shift and adjusting rods.

10. Remove the lower engine-to-transaxle bolts.

11. Remove the starter.

12. Remove the guard plate from the subframe.

13. With suitable jack, lift transaxle slightly.

14. Remove both rear subframe mounting bolts.

15. Remove both transaxle support bolts from the subframe.

16. Remove the bracket from the transaxle, push tension system cable and bracket off the retainer on transaxle. The retainer can only be removed with the transaxle out of the vehicle.

17. Remove the right side transaxle bracket.

18. Pull transaxle off dowel sleeves.

19. Lower transaxle and take out from below.

To install:

20. Installation is the reverse of the removal procedure. Before installing the transaxle, rest both halfshafts on top of the subframe.

21. Lubricate mainshaft splines.

22. Install transaxle onto dowels and install the lower bolts.

23. Install the tensioning system bracket and cable to the transaxle.

24. Tighten the transaxle bracket and subframe upper bolts to 29 ft. lbs. (39 Nm).

25. Check alignment of transaxle and torque transaxle-to-engine bolts to 40 ft. lbs. (54 Nm).

26. Torque subframe-to-body bolts to 80 ft. lbs. (108 Nm).

27. Torque halfshaft-to-drive flange bolts to 58 ft. lbs. (79 Nm).

V8 QUATTRO

NOTE: Any time the transaxle is removed from this vehicle, the driveshaft must be properly aligned. This requires the special alignment tool 3139 or equivalent.

1. Disconnect the negative battery cable, located under the rear seat.

2. Remove the strut between the front shock towers.

3. Remove the air cleaner assembly and the ducts.

4. Remove the bracket for the ignition wires. This is held with self-locking bolts, which should be replaced.

5. Remove the wiring harness bracket and the wiring retainers from the transaxle. Remove the 2 upper transaxle-to-engine bolts.

6. Remove the bolts for the right side engine mount.

7. Raise and safely support the vehicle and remove the front wheels.

8. Remove the splash shield under the engine and the body crossmember.

9. Remove the front exhaust pipe with the catalytic converter and the transaxle heatshield.

10. Install the driveshaft alignment tool. Matchmark the driveshaft and remove the driveshaft and heatshield.

11. Remove the transaxle mount-to-frame bolts.

12. Lower the vehicle and install an engine support bridge VAG 10-222A or equivalent across the inner fenders. Connect the bridge to the left engine mount and raise the engine just enough to take the weight off the mounts.

13. Remove the halfshaft shields and disconnect the halfshafts from the flanges. Support the shafts so they do not hang by the outer CV-joints.

14. Disconnect the hydraulic line from the clutch slave cylinder and plug the fittings to keep them clean.

15. Remove the clamp and lock sleeve to disconnect the shift linkage rods.

16. Disconnect the wiring for the backup lights and speedometer sensor and remove the wiring from the brackets.

17. Remove the cable guide for the seat belt tensioning system and position the cable/guide up out of the way.

18. At the left transaxle mount, pry the wire bracket open to release the oxygen sensor wire.

19. Disconnect the engine oil lines near the transaxle and plug the fittings.

20. Support the transaxle and subframe with a suitable transmission jack and remove the subframe mounting bolts. Lower the subframe to give clearance but make sure the radiator fan still turns freely; loosen the engine mounts if needed.

21. Tie the subframe in place so the jack can be removed. Secure the halfshafts to the subframe.

22. Disconnect the transaxle mounts and remove the engine–to–transaxle bolts. Move the unit back and carefully lower it out of the vehicle.

To install:

23. Check the condition of the dowel sleeves in the engine block and replace if necessary.

24. Install the transaxle assembly. Install all the mounting bolts loosely before torquing any of them. Observe the following engine-to-transaxle bolt torques:

12mm — 48 ft. lbs. (65 Nm)
10mm — 32 ft. lbs. (45 Nm)
8mm — 18 ft. lbs. (20 Nm)

25. Install the subframe. Torque the subframe mounting bolts to 48 ft. lbs. (65 Nm) plus an additional ¼ turn. Reconnect the engine mounts if necessary.

26. Install the cable guide seat belt tensioning system. Torque the seat belt tensioning system bolts to 30 ft. lbs. (40 Nm).

27. Connect and adjust the shift linkage, if necessary.

28. Reconnect the line to the clutch slave cylinder.

29. Install the halfshafts. Torque the halfshaft flange bolts to 58 ft. lbs. (80 Nm). Install halfshaft shields.

30. Remove the engine support equipment or equivalent. Install transaxle mount.

31. Install the driveshaft with the matchmarks aligned and torque the flange bolts to 41 ft. lbs. (55 Nm). The shaft must be aligned.

32. To align the driveshaft, install the driveshaft with the alignment tool attached but do not install the bolts for the center support. Measure the gap between the support bolt holes and the body and install the proper spacers to make the gap even on both sides. Make sure the support is centered front to rear and install the bolts. Torque the bolts to 15 ft. lbs. (20 Nm) and remove the alignment tool.

33. Install the exhaust system. When installing the exhaust, tighten the clamps after making sure everything is properly positioned to minimize vibration.

34. Install engine mount, body crossmember and strut between the front shock towers.

35. Install air cleaner and ducts.

36. Install all necessary wiring and engine protection plate.

37. Reconnect the battery. Bleed system. Road test the vehicle.

Linkage Adjustments

Except V8 Quattro, Quattro Turbo, Coupe 100 and 200

1. Place the shift lever into the **N** position.

2. Raise and safely support the vehicle and loosen the clamp nut on the shift rod. Be certain that the shift finger slides freely on the shift rod.

3. Working inside the vehicle, remove the gear shift lever knob and the boot.

4. Loosen the shifter base plate bolts slightly. Align the holes in the plate with the holes in the bearing housing and tighten the bolts.

5. Using the alignment tool 3057 or equivalent, slip it over the gearshift lever and make sure the locating pin is in the front centering hole.

6. Position the shift lever to the right detent cut-out for 5th and **R** and tighten the lower knurled nut of the tool.

7. At the top of the tool, move the slide with the gear shift lever to the right stop. Tighten the upper knurled nut of the tool.

8. Position the gear shift lever into the left cut-out of the slide. Adjust the shift rod and the shift finger with the transaxle in **N** and tighten the clamp nut.

9. Remove the tool and check the shifting of the gears for smoothness.

V8 Quattro, Quattro Turbo and Coupe

1. Place the shift lever in **N**. Adjust the length of the adjusting rod so the distance between the center point of the end holes is 5.275 in. (134mm).

2. Loosen the clamp nut, making sure the shift rod moves freely. Loosen the bolts slightly, align the centering holes of the gear shift lever

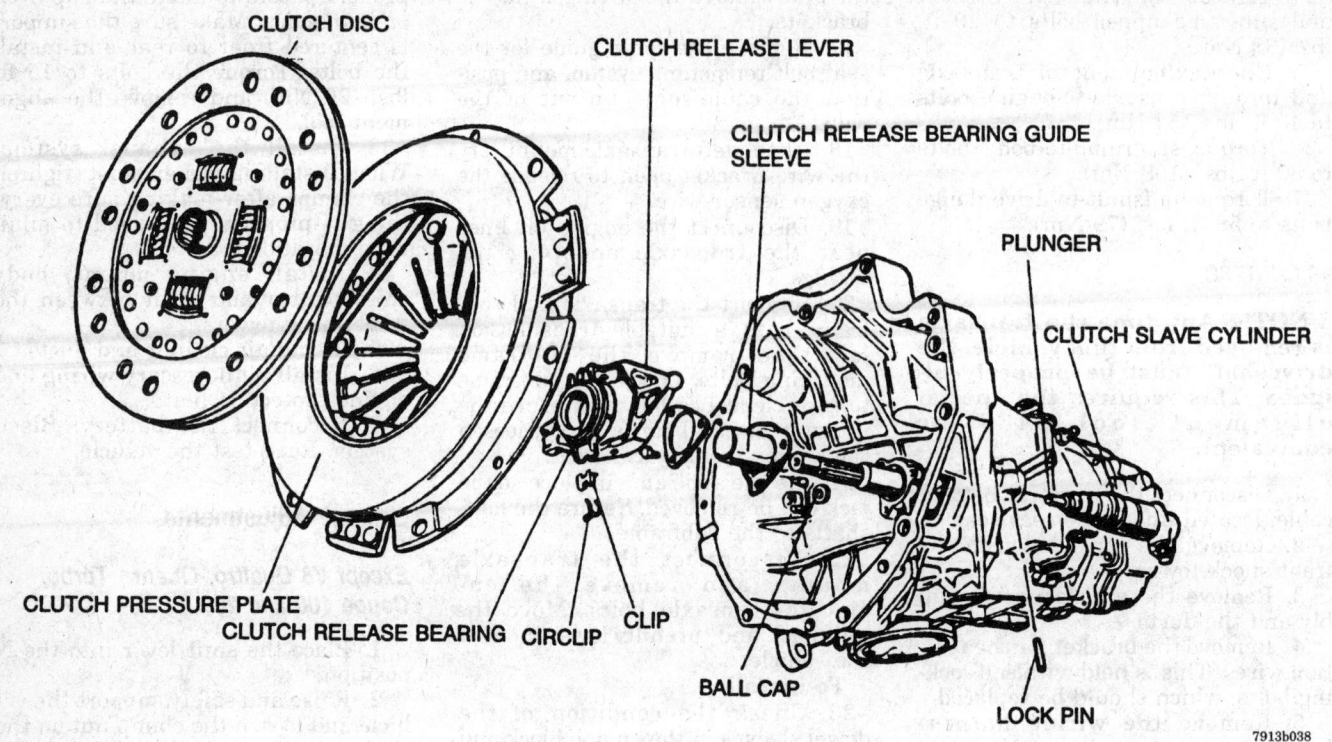

Clutch assembly and hydraulic release system

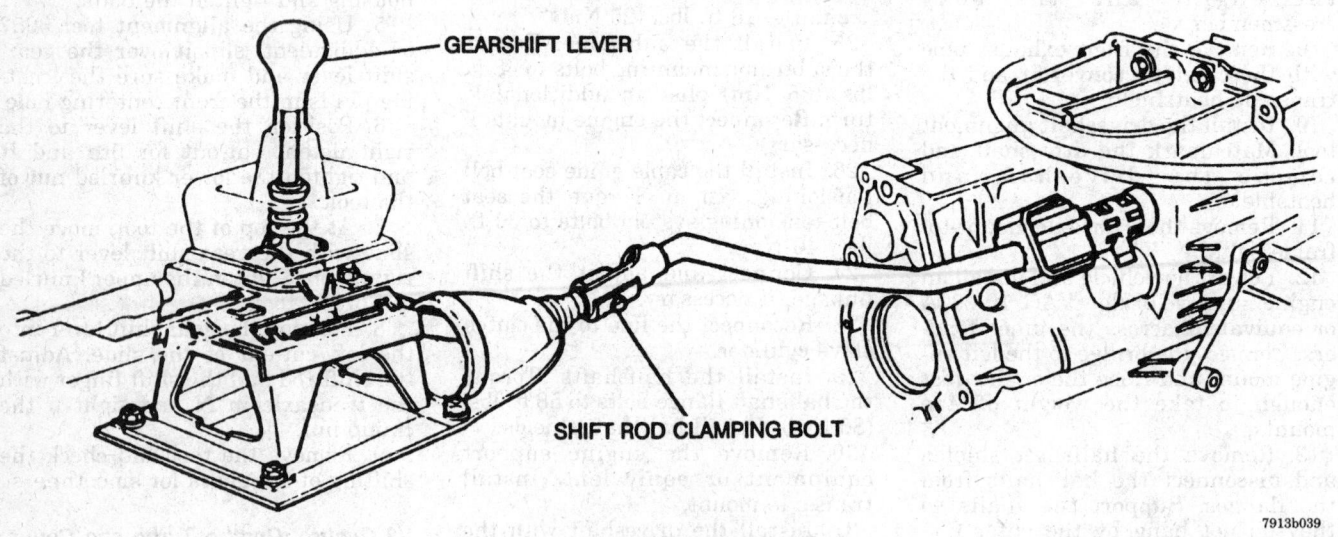

Manual shift linkage — 80/90 and 100

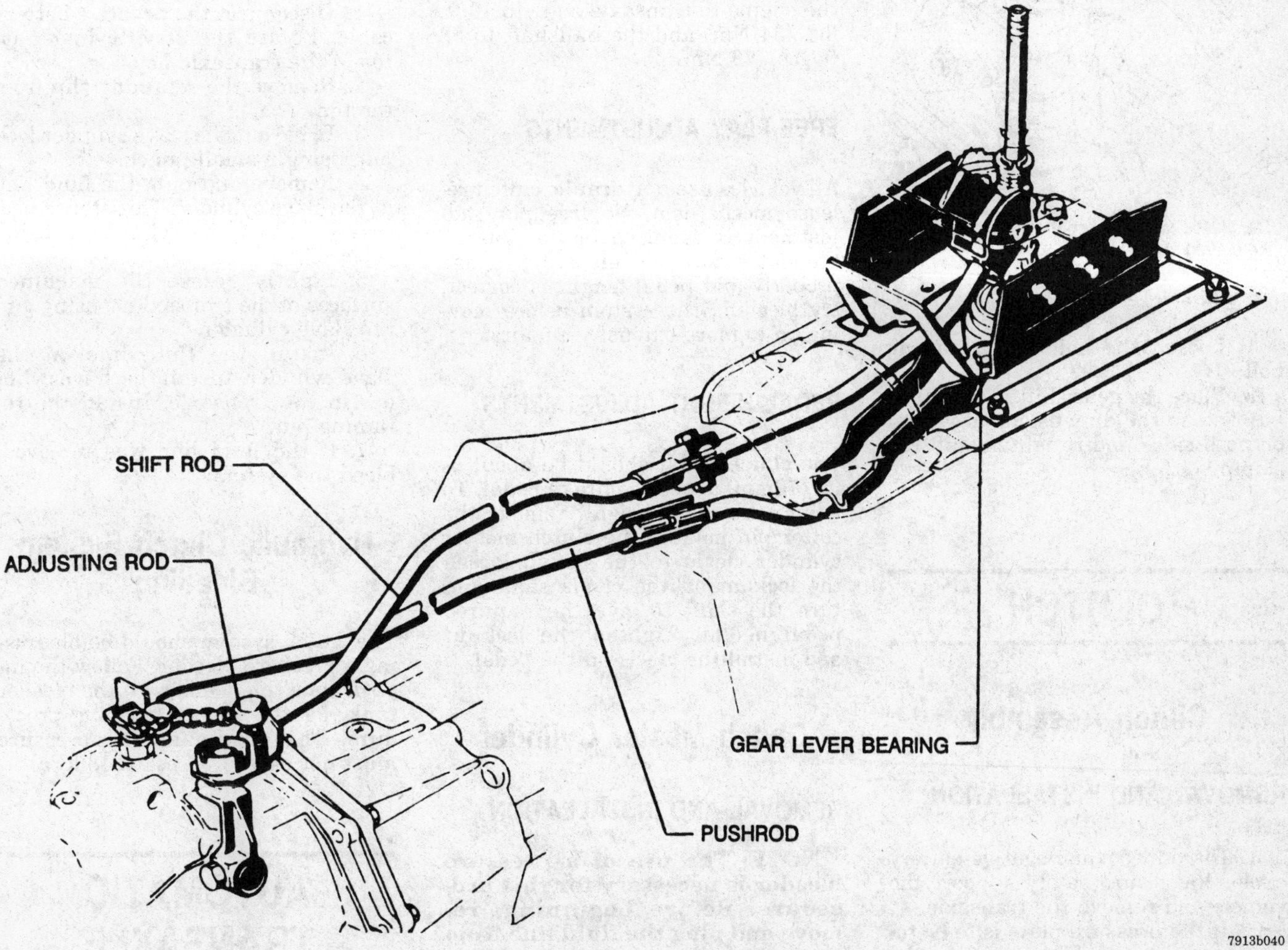

SHIFT ROD

ADJUSTING ROD

GEAR LEVER BEARING

PUSHROD

7913b040

Manual shift linkage — 200 and V8 Quattro

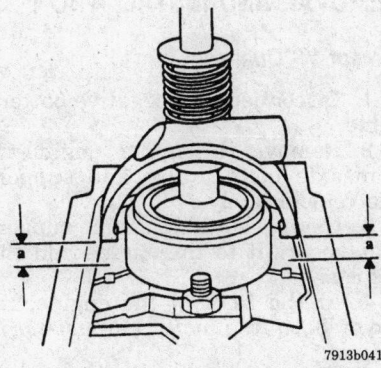

7913b041

Keep shifter centered for proper adjustment

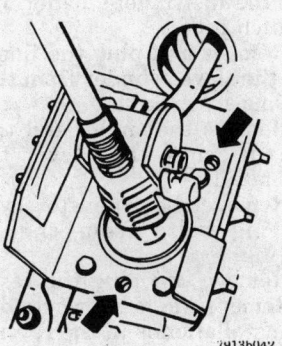

7913b042

Align the centering holes of the stop lever bearing, then tighten the bolts

housing and stop plate and tighten the bolts.

3. Install tool 3048 or equivalent, tighten the clamp nut and remove the tool. Engage 1st gear, press the shift lever to the left, stop and release the shift lever.

4. Engage 5th gear, press the shift lever to the right, stop and release the shift lever.

5. If the lever does not spring back approximately the same distance as in Steps 3 and 4, move the gear shift lever housing slightly in the slots sideward.

6. Make sure all gears engage easily without jamming.

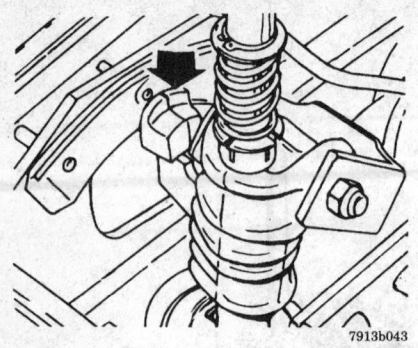

The plastic stop bracket should align with the curved stop plate

100 and 200

1. Loosen the shift rod clamping bolt.
2. Place the gear shift in a vertical position so the dimensions are equal on both sides and retighten shift rod clamp bolt.

CLUTCH

Clutch Assembly

REMOVAL AND INSTALLATION

1. Disconnect the negative battery cable. Raise and safely support the vehicle and remove the transaxle.
2. If the pressure plate is to be reused, mark its relationship to the flywheel.
3. Using a suitable tool, lock the flywheel. Unbolt the pressure plate from the flywheel, loosening the bolts alternately, a little at a time, to prevent warpage.

To install:

4. Install the clutch with the driven plate on the pressure plate so the spring cage is facing the pressure plate.
5. Hold the clutch assembly against the flywheel, aligning the marks made in Step 2 and the dowel pins on the flywheel with the pressure plate. Insert an alignment shaft tool through the pressure plate and the driven plate into the crankshaft pilot bearing.
6. Install the pressure plate bolts finger tight. Tighten the bolts evenly, in a diagonal pattern, to avoid distortion. Torque the bolts to 18 ft. lbs. (24 Nm). Remove the alignment shaft.

7. The clutch release bearing in the front of the transaxle should be checked before reassembly. It is retained by 2 springs.
8. Replace the transaxle. Torque the engine-to-transaxle bolts to 40 ft. lbs. (54 Nm) and the halfshaft to 58 ft. lbs. (79 Nm).

FREE-PLAY ADJUSTMENTS

All vehicles use a hydraulic clutch release mechanism. No free play adjustment is required or possible. If the clutch does not release or engage properly and pedal height is correct, try bleeding the system before moving on to more extensive repairs.

PEDAL HEIGHT ADJUSTMENTS

The clutch pedal should be about ⅜ in. (10mm) above the brake pedal. To adjust the pedal height, remove the cotter pin holding the clutch master cylinder clevis to the pedal, loosen the locknut on the clevis shaft and turn the shaft to give the required pedal height. Tighten the locknut and install the clevis on the pedal.

Clutch Master Cylinder

REMOVAL AND INSTALLATION

NOTE: The use of a pressure bleeder is necessary for this procedure. Before beginning, remove and plug the fluid line from the reservoir to the master cylinder. Empty the fluid in the line into a container.

1. Disconnect the negative battery cable. Locate the master cylinder under the instrument panel, behind the clutch pedal.
2. Remove and plug the line leading to the slave cylinder from the end of the master cylinder.
3. Remove the circlip and the pin which attaches the clevis to the clutch pedal.
4. Remove the 2 master cylinder mounting bolts from the pedal mounting.
5. Remove and plug the reservoir line. Remove the master cylinder.
6. Installation is in the reverse order of removal. Tighten the master cylinder mounting bolts to 15 ft. lbs. (20 Nm).
7. Bleed the clutch system when finished.

Clutch Slave Cylinder

REMOVAL AND INSTALLATION

1. Disconnect the negative battery cable. Locate the slave cylinder on top of the transaxle housing.
2. Remove the retaining clip from the pin.
3. Drive out the slave cylinder lock pin, using a small punch.
4. Remove and plug the fluid line at the slave cylinder. This step is necessary only if the cylinder is being replaced.
5. Lightly grease the machined surfaces of the transaxle housing and the slave cylinder.
6. Install the fluid line on the slave cylinder. Install the slave cylinder in the transaxle. Install the retaining pin.
7. If the fluid line was removed, bleed the system.

Hydraulic Clutch System Bleeding

The clutch system should be bled using a pressure bleeder. Follow the instructions that come with the bleeder tank, for the proper bleeding procedure. The maximum line pressure must not exceed 36 psi. (248 kPa).

AUTOMATIC TRANSAXLE

Transaxle Assembly

REMOVAL AND INSTALLATION

Except V8 Quattro

1. Disconnect the negative battery cable.
2. Remove the upper engine-to-transaxle bolts. Raise and support the vehicle safely.
3. Using a suitable engine support tool, secure it to the engine and the vehicle.
4. At the front of the engine, remove both top bolts. Remove the starter.
5. Through the starter opening, remove the torque converter to drive plate bolts and remove torque converter cover plate.

6. Clamp off the coolant hoses at the ATF cooler and remove the hoses from the cooler.

7. Remove the speedometer cable from the transaxle.

8. Remove the inner halfshaft-to-transaxle bolts. Using a wire, tie up the halfshafts.

9. At the left control arm, mark the position of the ball joint and remove the ball joint and the support.

10. Place an oil catch pan under the transaxle, remove the oil filler tube from the oil pan and drain the fluid.

11. Remove the exhaust pipe-to-transaxle bracket.

12. Remove the selector cable bracket from the transaxle. At the transaxle shift lever, remove the selector cable circlip and the cable.

13. At the transaxle, remove the accelerator cable bracket and the cable from the operating lever.

14. Remove the center bolt, from the transaxle mount. Using the engine support tool, lift the engine slightly.

15. Remove the throttle cable bracket bolts and the bracket.

16. Support the transaxle and lift it slightly. Remove the lower transaxle-to-engine bolts.

17. Separate the engine from the transaxle and lower it from the vehicle. Be sure to secure the torque converter.

To install:

18. When installing the transaxle, should the torque converter slip off the one-way clutch support, the oil pump shaft could be pulled from the oil pump. This may cause severe damage when bolting the transaxle to the engine. Make sure the torque converter is properly positioned before installing the bolts.

19. Torque the engine-to-transaxle bolts to 41 ft. lbs. (56 Nm), the subframe bolts to 52 ft. lbs. (71 Nm) and the transaxle mount center bolt to 30 ft. lbs. (40 Nm).

20. Install the torque converter-to-driveplate bolts and torque to 22 ft. lbs. (30 Nm). Install the cover plate.

21. Install the halfshaft-to-transaxle bolts and torque to 33 ft. lbs. (45 Nm).

22. Connect the ball joint to the control arm and torque the bolts to 48 ft. lbs. (65 Nm).

23. Install the exhaust system. When installing the exhaust, tighten the clamps after making sure everything is properly positioned to minimize vibration.

24. Connect the hoses to the oil cooler. Install the oil filler tube and refill the transaxle.

25. Connect and adjust the selector cable as required.

26. Connect and adjust the accelerator linkage and align the engine-to-transaxle mounts, if necessary.

V8 Quattro

1. Disconnect the negative battery cable. The battery is under the rear bench seat.

2. Raise and safely support vehicle. Remove the front wheels.

3. Remove the cross struts for the spring strut domes.

4. Remove the air cleaner housing and air cleaner assembly.

5. Remove the supports for the ignition cables and the left and right distributor caps. Remove the throttle cable and support.

6. Disconnect the oxygen sensor and probe heater. Remove the cable clamp for the electrical cables running alongside.

7. Disconnect the 2-pin plug connection at the transaxle bell housing. Disconnect the cable clamp for the wire harness next to the 2-pin plug.

8. Disconnect the retaining strap for the ventilation hose at the firewall.

9. Loosen the transaxle filler pipe on the cylinder head

10. Remove the radiator fan. The left fan can be set aside while still connected.

11. Remove the right engine mounting bolts on the body.

12. Raise and safely support vehicle. Remove the lower engine cover, the crossmember, the front exhaust pipe with catalytic converter and the heat deflector for the driveshaft.

13. Remove the driveshaft by loosening the driveshaft mounting bolts on the transaxle, rear final drive and body.

NOTE: A special driveshaft assembly device or aligning jig is required, tool 3213, or equivalent. This tool keeps the multi-piece driveshaft straight and in proper alignment. Attach this jig to the driveshaft and tighten the nuts. Remove the bolts on the transaxle and rear final drive. Support the driveshaft and alignment jig, remove the mounting bolts from the body and carefully lower the the driveshaft from the vehicle. Take note of any shims. Always move and store the driveshaft flat.

14. Detach the selector lever cable and support bracket at the transaxle

and remove the oil filter. Remove the upper bolt on the starter by guiding a 10mm Allen socket with extension and flex fitting through the opening, on the transaxle housing, over the final drive. Remove the bolt and take out the starter.

15. Working through the starter opening, remove the 6 torque converter bolts.

16. Remove the mounting bolts on both sides of the transaxle mounting.

17. Attach a lifting hoist to the engine mounting on the left side and lift slightly.

18. Under the vehicle, support the subframe using a suitable transaxle jack. Remove the mounting bolts on the subframe and lower the subframe carefully until it hangs freely. Disconnect the halfshafts from the transaxle and tie up.

19. Drain the transaxle and remove the filler pipe. Unscrew the cooling lines, the retaining clip that holds both lines together and tie to the subframe, aside.

20. Remove the left and right transaxle mounts.

21. Disconnect the 8-pin electrical connector on the transaxle by turning counterclockwise.

22. Disconnect the plug on the speed sensor. Push the locking lever down and remove the plug from the multi-function transaxle switch. Remove the retainers for the electrical wires from the transaxle and unclamp the ventilation hose.

23. Remove the tabs on the seat belt tensioning cables and disconnect the cables.

24. Remove the speed and TDC sensor with heat deflector plate on the engine. Remove the 2 upper engine-to-transaxle bolts.

25. Support the transaxle with a suitable transaxle jack. Remove the remaining engine-to-transaxle connecting bolts and remove the transaxle from the engine. Lower the transaxle carefully and secure the torque converter to keep it from falling out. Use caution not to damage the halfshafts, bolt-on parts or the multi-function switch.

To install:

26. If a new replacement transaxle is being installed, the following must be carried out.

   a. Transfer the catalytic converter mountings on the left and right side of the housing.

   b. Transfer the pipes for the transaxle cooler and the support for the seat belt tensioning cables.

   c. Check that both guide sleeves are installed in the engine block at the 2 o'clock and 8 o'clock positions, viewed from the flywheel.

27. Make sure the torque converter is secured and install the transaxle. Note that if only the torque converter is replaced, it should be carefully positioned on the free-wheel support and should not be tilted. To engage the splines of the pump shaft, rotate the torque converter forward and backward slightly. The torque converter must be inserted onto the free-wheel support up to the stop.

28. Torque the converter-to-driveplate to 26 ft. lbs. (35 Nm).

29. Torque transaxle-to-engine to 44 ft. lbs. (60 Nm).

30. Install the halfshafts. Torque halfshaft-to-transaxle to 59 ft. lbs. (80 Nm).

31. Install the speed and TDC sensor with heat deflector.

32. Install the seat belt tensioning cables. Torque support for seat belt cables to 30 ft. lbs. (45 Nm).

33. Install the left and right transaxle mounts. Torque transaxle supports to 30 ft. lbs. (45 Nm).

34. Install automatic transaxle filler pipe.

35. Install the subframe assembly. Torque subframe-to-body to 48 ft. lbs. (65 Nm) plus ¼ turn.

36. Install the starter all necessary wiring.

37. Install oil filter, support bracket and selector lever cable.

38. Install and adjust the driveshaft. Align with care. Use the following procedure:

a. With the alignment jig holding the driveshaft straight, set the driveshaft in place.

b. Carefully, measure the distance between the body pan bolt holes and the mounting ears of the center support bracket. This distance should be the same left to right.

c. Five different shims are available: 2, 4, 6, 8 and 10mm. Determine the thickness of the required shims.

d. Push the driveshaft all the way to the rear up to the stop. Mark the position of the center mounting bracket on the floor pan.

e. Push the driveshaft forward all the way and park the position of the center mounting bracket on the floor pan. The center mounting must be midway between these 2 marks.

f. Insert the bolts and shims as determined beforehand and tighten to 15 ft. lbs. (20 Nm). Remove the alignment jig.

39. Install the heat deflector for the driveshaft.

40. Install the exhaust system. When installing the exhaust, tighten the clamps after making sure everything is properly positioned to minimize vibration.

41. Install the crossmember.

42. Install the radiator fan assembly.

43. Reconnect all necessary wiring and vacuum hoses.

44. Install throttle cable support and throttle cable.

45. Install the left and right distributor caps.

46. Install the air cleaner housing assembly.

47. Install the cross-strut support braces. Torque cross brace to suspension strut domes to 17 ft. lbs. (23 Nm).

48. Install front wheels.

49. Fill the unit with automatic transaxle fluid.

50. Adjust the selector cable. Check the throttle cable linkage and, if necessary, adjust.

51. Install and tighten all bolts for the lower engine cover.

52. Check engine oil level. Align front end. Road test the vehicle for proper operation.

Automatic Shift Linkage

All vehicles with automatic transaxle are equipped with the shiftlock transaxle control system. The system must be properly adjusted to make sure the transaxle is fully engaged in each shift position. If this is not done, the transaxle may be only partially engaged in a certain range position, causing severe damage due to slippage. Improper adjustment may also make it impossible to shift into or out of **P**.

The shiftlock system is designed to make it impossible to shift out of **P** or **N** unless certain parameters are met. The system depends on proper operation or condition of fuse S12, the brake light switch circuit, the interior light relay control unit, the driver's door switch circuit and vehicle speed signal.

SYSTEM FUNCTION

With ignition switch **ON**, the selector cannot be shifted out of **P** or **N** unless the brake pedal is depressed. At speeds under 3.7 mph, when shifted into **N** the shifter should lock in **N** after 1 second, unless the brake pedal

is depressed. At speeds above 3.7 mph the shifter should not lock.

SYSTEM ADJUSTMENT

1. Adjust the solenoid switch, using a 1mm gauge between the selector lever and solenoid switch. With lever in **R**, push the solenoid against the gauge and tighten to 7 ft. lbs. (9 Nm).

2. Center the lower bore of fork piece and supply voltage to switch. The solenoid pin locks the fork piece.

3. Position the gear shift lever housing so selector lever is in the **N** position relative to the housing.

4. Position the shift arm with a 4mm aligning pin through the housing bore.

5. Shift the selector lever to the **N** position. Remove aligning pin.

6. Check for correct operation.

7. If shifter does not function properly, it may be necessary to check the electronic control systems.

KICKDOWN SWITCH ADJUSTMENT

087 Series

The accelerator control is to be adjusted so at closed throttle the operating lever on the transaxle is at the no-throttle position. If adjustment is incorrect, shift speeds will be too high at part throttle and main pressure will be too high at idle.

1. Place selector in **P**.

2. Apply parking brake.

3. Adjust accelerator control in idle position with closed throttle.

4. Disconnect the locks on ball sockets and and disconnect the pull rod from the levers of the routing guide.

5. Disconnect the rods for cruise control.

6. Loosen locknut on the pull rod.

7. Position lever for pull rod approximately 0.040 in. (1mm) before stop.

8. Install pull rod, without tension. Ball socket must be twisted to be in line with ball and throttle lever must contact the stop.

NOTE: Turbocharged vehicles have 2 pull rods.

9. Loosen pushrod length adjusting bolt (loosen back locknut).

10. Push operating lever into the no-throttle position. The throttle valve must contact the idle stop.

11. Tighten the pushrod length adjusting bolt.

12. Adjust the pushrod length by shifting the adjusting plate. The

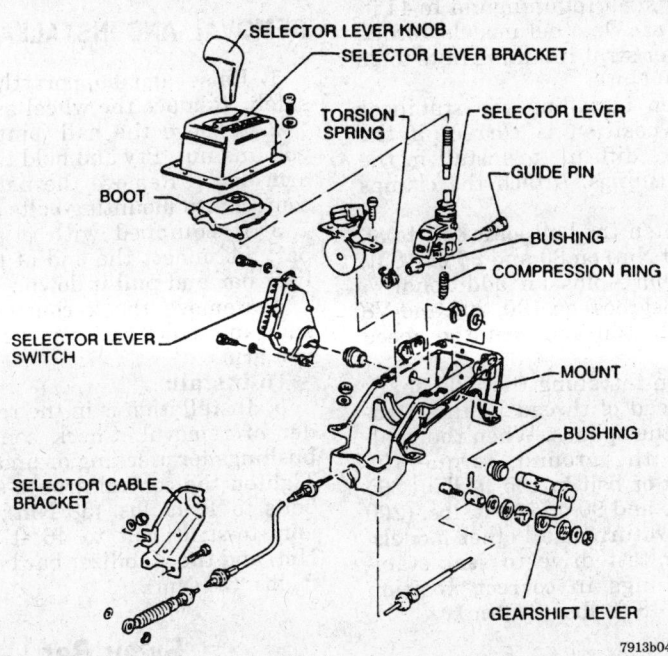

SELECTOR LEVER KNOB
SELECTOR LEVER BRACKET
TORSION SPRING
SELECTOR LEVER
GUIDE PIN
BOOT
BUSHING
COMPRESSION RING
SELECTOR LEVER SWITCH
MOUNT
BUSHING
SELECTOR CABLE BRACKET
GEARSHIFT LEVER

7913b044

Shiftlock II automatic transaxle shifter assembly

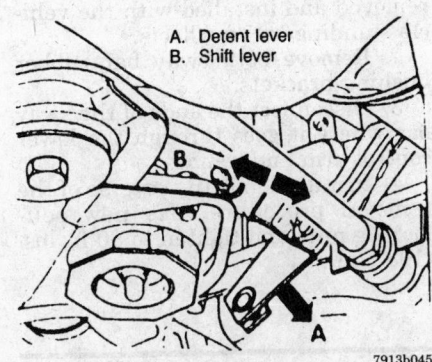

A. Detent lever
B. Shift lever

7913b045

Adjusting transaxle pushrod

pushrod must install on the operating lever without tension.

13. Remove the pushrod from the operating lever.

14. Have an assistant depress the accelerator pedal to the stop.

15. Depress the operating lever to the kickdown stop.

16. Using pliers, pull the accelerator pedal cable back and fasten.

17. Check that the operating lever is in contact with the kickdown stop, readjust the pedal cable, if necessary.

18. Install the pushrod onto operating lever and secure.

19. Check throttle lever operation.

20. Push accelerator cable through full throttle position to kickdown stop.

21. Transaxle lever must be in contact with stop.

22. Over center spring must be compressed approximately 0.320 in. (8mm).

23. Release accelerator pedal and install rod to cruise control. Rod must be tension free, adjust as necessary.

089 Series

1. Remove covering for throttle control.

2. Loosen the 2 cable locking nuts.

3. Turn the throttle to the full throttle position and hold.

4. Using tool 3004 or equivalent, hold the throttle cable brackets at full open throttle. Attach tool end on the lever lower cable bracket and tool end on the end of the hood gas strut.

5. Insert a $^{11}/_{16}$ in. (17mm) spacer between the accelerator pedal and pedal stop.

6. An assistant is needed to push the pedal down to the stop.

7. Pull accelerator cable and install locking clip.

8. Pull cable until pressure from transaxle kickdown position is felt.

9. Tighten the nut on cable side against the bracket, next tighten the nut on pivot side against bracket.

10. Remove tool.

11. Throttle lever must rest against the idle stop when the accelerator is released.

12. Press accelerator to full throttle position, not kickdown.

13. The pressure point for full throttle position of the accelerator pedal must be approximately $^{3}/_{4}$ in. (19mm) away from the pedal stop.

14. Press the accelerator to the pedal (kickdown) stop.

15. Transaxle operating lever must contact the kickdown stop.

16. The spring between the cable brackets must be stressed.

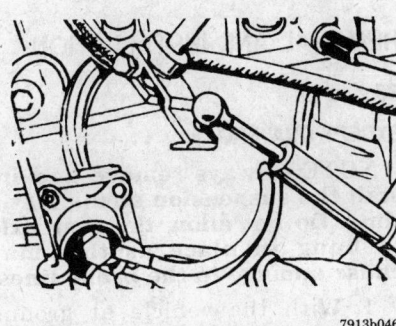

7913b046

Kickdown detent linkage

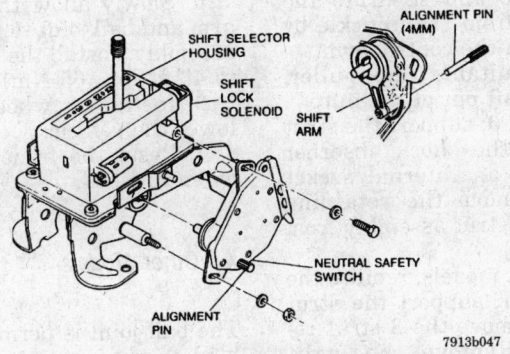

SHIFT SELECTOR HOUSING
ALIGNMENT PIN (4MM)
SHIFT LOCK SOLENOID
SHIFT ARM
NEUTRAL SAFETY SWITCH
ALIGNMENT PIN

7913b047

Adjusting neutral safety switch

17. For vehicles with cruise control, adjust the coupling rod by moving the ball end 0.039–0.059 in. (1–1.5mm).

18. For vehicles with cruise control, the air conditioning switch must only switch **ON** at kickdown and not at wide-open throttle. Approximately 5/64 in. (2mm) between the switch and cable bracket at wide-open throttle (not kickdown).

FRONT SUSPENSION

MacPherson Strut

REMOVAL AND INSTALLATION

1. Disconnect the negative battery cable. With the vehicle on the ground, remove the front axle nut or bolt and loosen the wheel bolts.

2. Raise and support the vehicle safely. Remove the wheel assembly.

3. Remove the brake caliper mounting bolts and disconnect the brake line from the bracket without disconnecting the line from the caliper. Remove speed sensor, if equipped.

4. Remove the brake caliper with the line still attached and support it aside.

5. Remove disc brake rotor.

6. Remove the ball joint clamp bolt and nut.

7. Remove the tie rod end nut and separate the tie rod end from the strut.

8. If equipped with a stabilizer bar, remove the retaining bolt and remove the stabilizer bar end clamps. Pivot the stabilizer bar downward.

9. Remove the 2 center stabilizer bar clamps and unbolt stabilizer bar from the lower control arm.

10. Remove the pinch bolt from the steering knuckle and separate the lower ball joint from the knuckle by pushing down on the control arm.

11. Using a suitable hub puller, press the halfshaft out of the hub.

12. On 80 and 90, support the strut assembly, hold the shock absorber piston rod with an internal socket wrench and remove the retaining nut. Remove the strut assembly from the vehicle.

13. On all other models, remove the upper strut cover, support the strut assembly and remove the 3 strut retaining nuts. Remove the strut assembly.

To install:

14. On 80 and 90 models, torque the upper strut retaining nut to 44 ft. lbs. (60 Nm). On other models, torque the 3 upper strut retaining nuts to 22 ft. lbs. (30 Nm).

15. When installing the stabilizer bar, the position is correct if the clamps are difficult to install in the rubber bushings. Attach the clamps loosely.

16. Tighten the ball joint bolt to 36 ft. lbs. (49 Nm) on 80 and 90 or 48 ft. lbs. (65 Nm), plus an additional 1/4 turn (90 degrees) on 100, 200 and V8 Quattro. Install and seat the speed sensor.

17. When installing the axle shaft, apply a bead of thread locking compound to the splines. When the vehicle is on the ground, torque the center nut or bolt to 195 ft. lbs. (265 Nm) on 80 and 90 or 148 ft. lbs. (200 Nm) plus 1/4 turn on all other models.

18. After test drive to seat stabilizer bushings in correct position tighten to 18 ft. lbs. (24 Nm).

Lower Ball Joint

REMOVAL AND INSTALLATION

Coupe, 80 and 90

1. Raise and safely support the vehicle. Remove the wheel assembly.

2. Remove the lower ball joint clamp nut and bolt and pry the control arm down to disengage the ball joint from the steering knuckle.

3. Remove the ball joint to control arm mounting bolts and remove the ball joint.

To install:

4. Install the new ball joint on the control arm and tighten the mounting bolts to 46 ft. lbs. (62 Nm).

5. Slowly allow the lower control arm and ball joint to fit into the strut assembly. Install the bolt and tighten to 48 ft. lbs. (65 Nm).

6. Install the wheel assembly and lower the vehicle.

7. Reset the front end alignment when finished.

All Other Models

The ball joint is permanently assembled to the lower control arm and cannot be replaced separately.

Lower Control Arms

REMOVAL AND INSTALLATION

1. Raise and support the vehicle safely. Remove the wheel assembly.

2. Remove the ball joint to strut bolt and nut. Pry and hold the control arm down. Remove the ball joint to control arm mounting bolts and nuts.

3. If equipped with a stabilizer bar, disconnect the end of the stabilizer bar and pull it down.

4. Remove the 2 control arm to subframe bolts and remove the control arm.

To install:

5. Installation is in the reverse order of removal. Check control arm bushings for cracking or undue wear. Tighten the control arm-to-subframe bolts to 43 ft. lbs. (58 Nm); the ball joint-to-strut bolt to 46 ft. lbs. (62 Nm) and the stabilizer bar bolts to 18 ft. lbs. (24 Nm).

Sway Bar

REMOVAL AND INSTALLATION

1. The sway bar or stabilizer bar is removed and installed with the vehicle standing on its wheels.

2. Remove both sway bar rubber bushing brackets.

3. Disconnect the ends of the sway bar where it goes through the lower control arm and remove.

4. Installation is the reverse of the removal procedure. Use new self-locking nuts and tighten to 80 ft. lbs. (108 Nm).

REAR SUSPENSION

MacPherson Struts

REMOVAL AND INSTALLATION

80 and 90

EXCEPT QUATTRO

NOTE: Always remove and install the suspension struts 1 at a time. Do not allow the rear axle to hang in place as this may cause damage to the brake lines.

1. With the vehicle at ground level, open the trunk and remove the trim from around the shock tower.

2. Remove the rubber cap.

3. Hold the strut rod and remove the strut mounting nut.

4. Raise and support the vehicle safely.

5. Remove the lower strut mounting bolt from the axle beam and remove the strut.

6. Installation is the reverse of removal. Torque the upper strut mounting bolt to 14 ft. lbs. (19 Nm) and the lower strut mounting bolt to 43 ft. lbs. (58 Nm).

80 AND 90 QUATTRO WITH SINGLE PIECE STRUT

1. With the vehicle on the ground, remove the axle nut.

2. Raise and safely support the vehicle and remove the rear wheels.

3. Unbolt the halfshaft from the differential flange.

4. Remove the self-locking lower ball joint nut and press the ball joint out of the wheel bearing housing. Use the nut to protect the threads.

5. Loosen the control arm mounting bolts and allow the arm to pivot down out of the way.

6. Remove the tie rod nut and press the tie rod joint out of the bearing housing arm. Use the nut to protect the threads.

7. Pull the parking brake cable out of the strut bracket. Remove the brake caliper with its bracket without disconnecting the hydraulic line. Hang the caliper from the body with wire.

8. Remove the brake disc and slide the wheel speed sensor out of its mounting, if equipped. Press the halfshaft out of the hub.

9. Support the strut from below. In the luggage compartment, remove the trim and the rubber cap at the top of the strut.

10. Hold the top of the strut with an Allen wrench and remove the self-locking nut. Lower the strut from the vehicle.

To install:

11. Replace all self-locking nuts. Install the strut into the upper mount and torque the new upper nut to 48 ft. lbs. (60 Nm).

12. Make sure the axle shaft splines are clean and apply a bead of threat locking compound to the outer end of the splines. Install the axle shaft but do not torque the center nut until the vehicle is on its wheels. Torque the axle flange bolts to 35 ft. lbs. (45 Nm).

13. Install the brake disc and caliper. Torque the caliper bracket mounting bolts to 92 ft. lbs. (125 Nm).

14. Attach the ball joint to the control arm and the tie rod to the bearing housing and install new self-locking nuts. Torque the tie rod nut to 29 ft. lbs. (40 Nm) and the ball joint nut to 54 ft. lbs. (75 Nm).

15. When installation is complete and the vehicle is on the ground, torque the center axle nut to 236 ft. lbs. (320 Nm).

80 AND 90 QUATTRO WITH 2-PIECE STRUT

1. In the luggage compartment, remove the trim and the cap from the top of the strut. With the vehicle on its wheels, hold the strut rod from turning and remove the upper strut nut.

2. Place a block of wood between the axle shaft and the frame. Carefully raise and safely support the vehicle and remove the rear wheels.

3. Remove the bolts securing the lower strut to the wheel bearing housing and remove the strut. These are stretch bolts that cannot be reused and must be replaced with the newer type.

To install:

4. Install the strut to the bearing housing with new bolts. Torque the new stretch bolts to 59 ft. lbs. (80 Nm), plus an additional ½ turn.

5. Install the wheel and lower the vehicle until the wood block can be removed. Be sure to carefully guide the strut into the upper mount.

6. Install a new self-locking upper strut mounting nut and torque to 44 ft. lbs. (60 Nm). Check the wheel alignment.

100 and 200

EXCEPT QUATTRO AND V8 QUATTRO

NOTE: The struts must be removed with the weight of the vehicle on the rear wheels. If not, a spring compressor must be used on the rear springs.

1. If the vehicle is not on its wheels, install the spring compressor and compress the spring. Do not attempt to remove the shock with the rear wheels raised without a compressor.

2. Remove the upper strut mounting nut.

3. Remove the lower strut mounting nut.

4. Remove the shock absorber.

5. Installation is the reverse of removal. Torque the lower mounts to 66 ft. lbs. (89 Nm) and the upper to 14 ft. lbs. (19 Nm).

QUATTRO AND V8 QUATTRO

1. Raise and support the vehicle safely. Remove the wheel assembly.

2. Open the trunk and remove the shock absorber covers, the remove the shock absorber-to-body nuts/bolts.

3. Remove the shock absorber-to-rear wheel knuckle assembly. Remove the shock absorber from the vehicle.

4. To install, reverse the removal procedures. Torque the shock absorber-to-body nuts/bolts to 15 ft. lbs. (20 Nm) and the shock absorber-to-rear wheel knuckle assembly bolt to 66 ft. lbs. (89 Nm).

Rear Control Arms

REMOVAL AND INSTALLATION

80 and 90 Quattro

1. Raise and support the vehicle safely, under the frame and differential.

2. Using a scribing tool, mark the position of the ball joint carrier with the control arm.

3. Remove the ball joint carrier-to-control arm nuts and the lock plate. Separate the ball joint carrier from the control arm.

4. Remove the control arm-to-subframe bolts and the control arm from the vehicle.

5. To install, reverse the removal procedures. Always use new replacement nuts. Torque the control arm-to-subframe bolts to 43 ft. lbs. (58 Nm) and the ball joint nut to 54 ft. lbs. (75 Nm). Check the rear wheel alignment.

100 and 200 Quattro

On these vehicles, the control arm is called the trapezoidal arm. It is connected to the wheel bearing housing and to 2 separate crossmembers. Always use new self-locking nuts on all applications.

1. Raise and support the vehicle safely, under the frame and differential.

2. Remove the wheel. Along the trapezoidal arm, remove the speed sensor wiring bracket nuts/bolts and the guide.

3. Remove the wheel bearing housing-to-trapezoidal arm front and rear bolts.

4. Remove the trapezoidal arm-to-rear crossmember bolt.

5. At the brake pressure regulator, disconnect the spring.

6. Remove the trapezoidal arm-to-front crossmember nut and the trapezoidal arm from the vehicle.

7. To install, reverse the removal procedures. Torque the trapezoidal arm-to-front crossmember nut to 44 ft. lbs. (60 Nm), the trapezoidal arm-to-rear crossmember bolt to 63 ft. lbs. (85 Nm), the trapezoidal arm-to-wheel bearing housing bolts to 125 ft. lbs. (169 Nm) and the speed sensor guide nut/bolts to 7 ft. lbs. (10 Nm). Adjust the rear wheel alignment.

NOTE: Before installing the trapezoidal arm, be sure to coat the fasteners with locking compound.

Rear Wheel Bearings

REMOVAL AND INSTALLATION

1. Raise and support the vehicle safely. Remove the wheel assembly.
2. Without disconnecting the hydraulic line, remove caliper assembly from rotor. Suspend caliper from the body with wire, do not let it hang by brake hose.
3. Pry off the grease cap and remove the cotter pin, nut and washer.
4. Remove the outer bearing.
5. Remove the brake rotor.
6. Remove the bearing inner bearing and seal from the rotor hub, using a soft drift or press.
7. Remove the bearing inner and outer race(s) from the rotor, using a soft drift or press.
   To install:
8. Clean and inspect mating surfaces for bearing races.
9. Install new races, using soft drift or press.
10. Pack the new bearing with grease and set it into the inner race.
11. Install seal, making sure it is square in the rotor hub.
12. Install rotor, outer bearing, washer, nut and adjust bearing play.
13. Install cotter pin and dust cap.
14. Install caliper assembly.
15. If hydraulic lines had be removed, install and bleed brakes.
16. If parking brake cable has been remove, install and adjust as necessary.
17. Install the wheel assembly.
18. Lower vehicle and check brakes for proper operation.

ADJUSTMENT

1. Raise and support the vehicle safely.
2. Remove the grease cap.

3. Remove the cotter pin and the locking nut.
4. While turning the wheel, so the wheel bearing does not jam, tighten the adjusting nut firmly.
5. Back the nut off slightly. The nut is properly adjusted when it is possible to pry the thrust washer side to side with some drag but using light pressure on the tool.
6. Install the locking nut and a new cotter pin. When installing the cap, make sure it is securely in place.

Rear Axle Assembly

REMOVAL AND INSTALLATION

Front Wheel Drive

1. Raise and support the vehicle safely. Remove the wheel assembly. Detach the muffler hanger bands. Lower and support the muffler and tail pipe.
2. Remove the parking brake cable to equalizer nut. Pry the cable sleeve from the bracket. Remove both parking brake cables at the brackets and disconnect the brake hoses at the brake line brackets. Cap all hoses and lines. Vehicles equipped with anti-lock brakes, disconnect the speed sensor.
3. Remove the nuts from the bolts attaching the trailing arms to the body. Do not remove the bolts at this time. On the right side, disconnect the spring from the brake pressure regulator.
4. Remove the bolts attaching the diagonal arms to the axle and remove the bolts attaching the strut to the axle. Slide out the trailing arm to the body attaching bolts and carefully remove the axle from the vehicle.

NOTE: All bolts through rubber bushings should be tightened with the weight of the vehicle on its wheels. This is done to preset the bushings in a level, nonstressed position to avoid poor handling or tire wear.

   To install:
5. After positioning the axle in the vehicle, install both trailing arm bolts finger tight. Attach the lower strut mounts and install the wheel and tire assemblies. Lower the vehicle to the ground.
6. Torque the trailing arm attaching bolts to 72 ft. lbs. (98 Nm) and the strut bolts to 66 ft. lbs. (90 Nm).
7. Install the diagonal arm and torque the bolt at the axle end to 70 ft. lbs. (95 Nm). Torque the bolt at the body end to 66 ft. lbs. (90 Nm).

Raise the vehicle again to install the remaining components.
8. After the installation procedure has been completed, install the speed sensor, if equipped, bleed the brake system and adjust the parking brake, as necessary.

STEERING

Steering Wheel

---- **CAUTION** ----
On vehicles equipped with an air bag, the negative battery cable and reserve power supply must both be disconnected before working on the vehicle. Failure to do so may result in deployment of the air bag and personal injury.

REMOVAL AND INSTALLATION

Without Air Bag

1. Center the steering wheel. Disconnect the negative battery cable.
2. Pull off the center horn pad and disconnect the wire. Mark the relationship of the steering wheel to the steering shaft.
3. Remove the steering wheel mounting nut and remove the steering wheel.
4. To install, align the matchmarks and tighten the nut to 30 ft. lbs. (41 Nm).

NOTE: Never strike or pound on the steering wheel. The collapsible steering column may be damaged.

With Air Bag

1. Center the steering wheel. Disconnect the negative battery cable.
2. Remove the side trim from center console, disconnect the red power supply connector to the air bag.

---- **CAUTION** ----
The reserve power supply can trigger the air bag even with the battery disconnected. The power connector to the air bag must be disconnected.

3. Remove the screws for the upper steering column trim and remove the upper trim.
4. Separate the connector for the air bag spiral spring.

5. Remove the air bag Torx® head retaining bolts.

6. Unhook the air bag unit, lift safety clamp and remove the air bag wiring at the terminal. Place the removed air bag unit face up in a safe place where it will not be disturbed.

7. Remove the steering wheel mounting nut and remove the steering wheel. A steering wheel puller may be necessary.

8. If removing spiral spring, wheel must be in the straight-ahead position. Do not twist spring after removing it.

To install:

9. To install, align the matchmarks and tighten the nut to 30 ft. lbs. (41 Nm).

NOTE: Never strike or pound on the steering wheel. The collapsible steering column may be damaged.

10. Reinstall air bag unit, air bag connector and Torx® head screws. Torque the Torx® head screws to 53 inch lbs. (6 Nm).

11. Install steering column upper trim.

12. Connect air bag system power connector.

13. Install the side trim on center console.

14. Connect the negative battery cable.

Power Steering Rack

REMOVAL AND INSTALLATION

80 and 90

1. Raise and safely support the vehicle.

2. Remove the lower left instrument panel cover, the steering column-to-steering rack clamp bolt and the steering column-to-dash bolts. Remove the steering column from the vehicle.

3. Using a pair of locking pliers, clamp off the fluid return line to the reservoir. Disconnect the fluid pressure line from the steering gear.

4. At the steering column boot, press in on the clips and remove the boot from the panel. From inside the vehicle, remove the fluid return line from the control valve body. On 5-cylinder vehicles, push off the dash panel boot and push the boot into the passenger compartment to access the pressure and return line.

5. At the left wheel housing, disconnect the steering rack from the frame.

6. At the steering rack, remove the tie rod coupling locknuts/bolts and the tie rods from the rack. Push the rack back into the steering housing.

7. Disconnect the steering assembly from the firewall. Turn the wheels to the right. Remove the assembly between the left wheel housing and the control arm.

To install:

8. Install the rack assembly and torque the left side bolts and nuts to 14 ft. lbs. (20 Nm). Use new self-locking nuts to secure the rack to the firewall and torque to 35 ft. lbs. (45 Nm).

9. Use new self-locking nuts and secure the tie rod coupling to the rack. Torque the nuts to 35 ft. lbs. (45 Nm).

10. Install the steering column, connect the hydraulic lines and bleed the system.

100, 200 and V8 Quattro

1. Raise and safely support the vehicle.

2. Pry off the lock plate and remove both tie rod mounting bolts from the steering rack, inside the engine compartment. Pry the tie rods out of the mounting pivot.

3. Remove the lower instrument panel trim.

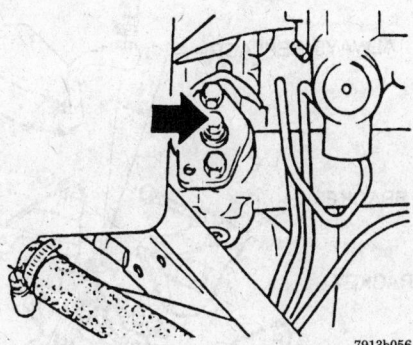

Steering rack adjustment bolt

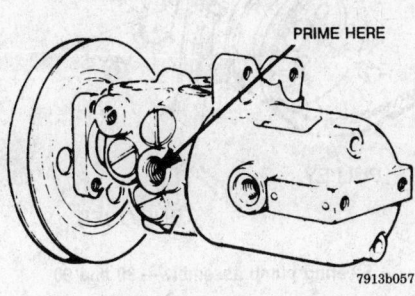

Prime the pump with fluid before installation

4. Remove the pressure and return lines from the steering rack control valve body.

5. Remove the shaft clamp bolt, pry off the clip and drive the shaft toward the inside of the vehicle with a brass drift.

6. Remove the steering gear mounting bolts at both ends. There is a single bolt at the right end.

7. Turn the wheels all the way to the right and remove the steering gear through the opening in the right wheel housing.

To install:

8. Temporarily install the tie rod mounting pivot to the rack with both mounting bolts. Remove 1 bolt, install the tie rod and replace the bolt. Do the same on the other tie rod. Make sure to install the lock plate.

9. Torque the following bolts:
Except the 100 and 200
Tie rod to 39 ft. lbs. (53 Nm)
Mounting pivot bolt to 15 ft. lbs. (20 Nm)
Steering gear-to-body mounting bolts to 15 ft. lbs. (20 Nm)
100 and 200
Tie rod to 44 ft. lbs. (60 Nm)
Pivot bolt to 30 ft. lbs. (41 Nm)
Gear-to-body bolts to 15 ft. lbs. (20 Nm)

10. Install hose lines with new O-rings and torque to 30 ft. lbs. (41 Nm).

11. Bleed the hydraulic system.

Power Steering Pump

REMOVAL AND INSTALLATION

Except V8 Quattro

1. Disconnect the negative battery cable. Remove the hoses from the pump. Plug the openings.

2. Remove the belt adjusting bolt or tensioner locknut, push the pump to 1 side and remove the belt.

3. Support the pump, remove the mounting bolts and lift out the pump.

4. Installation is the reverse of the removal. Torque the power steering mounting bolts to 14 ft. lbs. (20 Nm) and banjo bolts to 35 ft. lbs. (45 Nm). Adjust the belt, torque the tensioner locknut to 14 ft. lbs. (20 Nm). Be sure to fill the pump suction chamber with hydraulic fluid before attaching lines or the pump may be damaged.

V8 Quattro

1. Disconnect the battery negative cable.

2. Remove the air duct tube bolt near the engine oil dipstick.

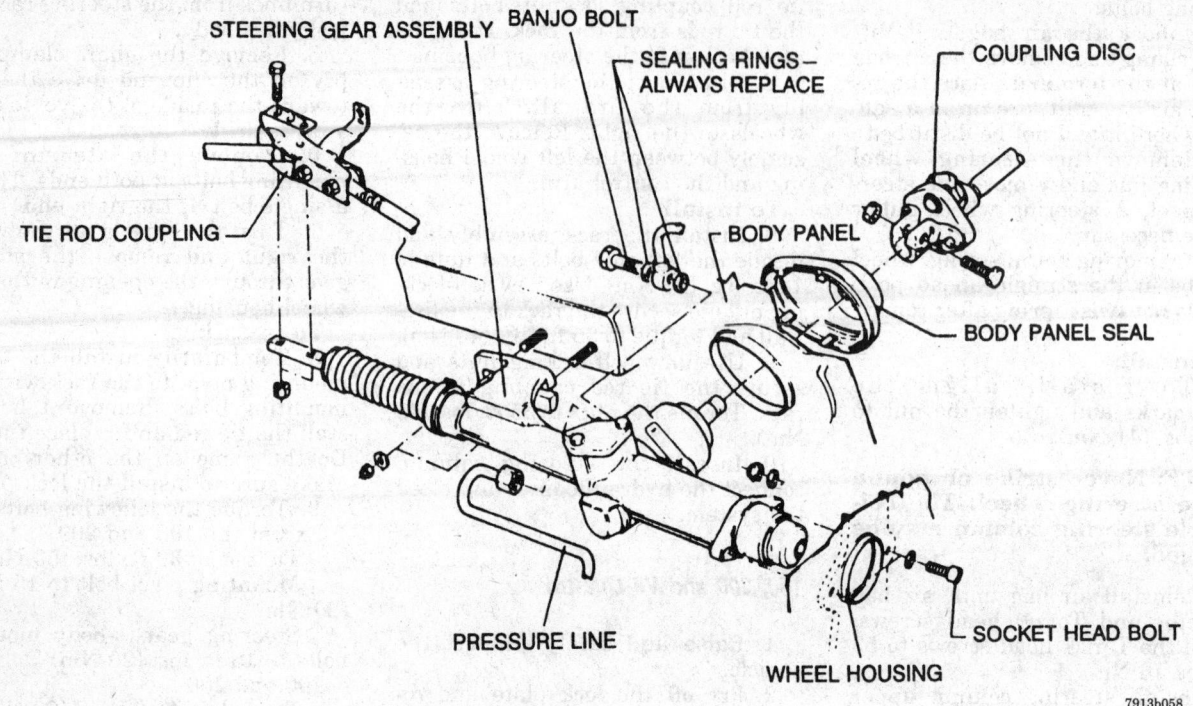

STEERING GEAR ASSEMBLY

BANJO BOLT

SEALING RINGS— ALWAYS REPLACE

COUPLING DISC

TIE ROD COUPLING

BODY PANEL

BODY PANEL SEAL

PRESSURE LINE

SOCKET HEAD BOLT

WHEEL HOUSING

7913b058

Steering rack assembly removal — 80 and 90

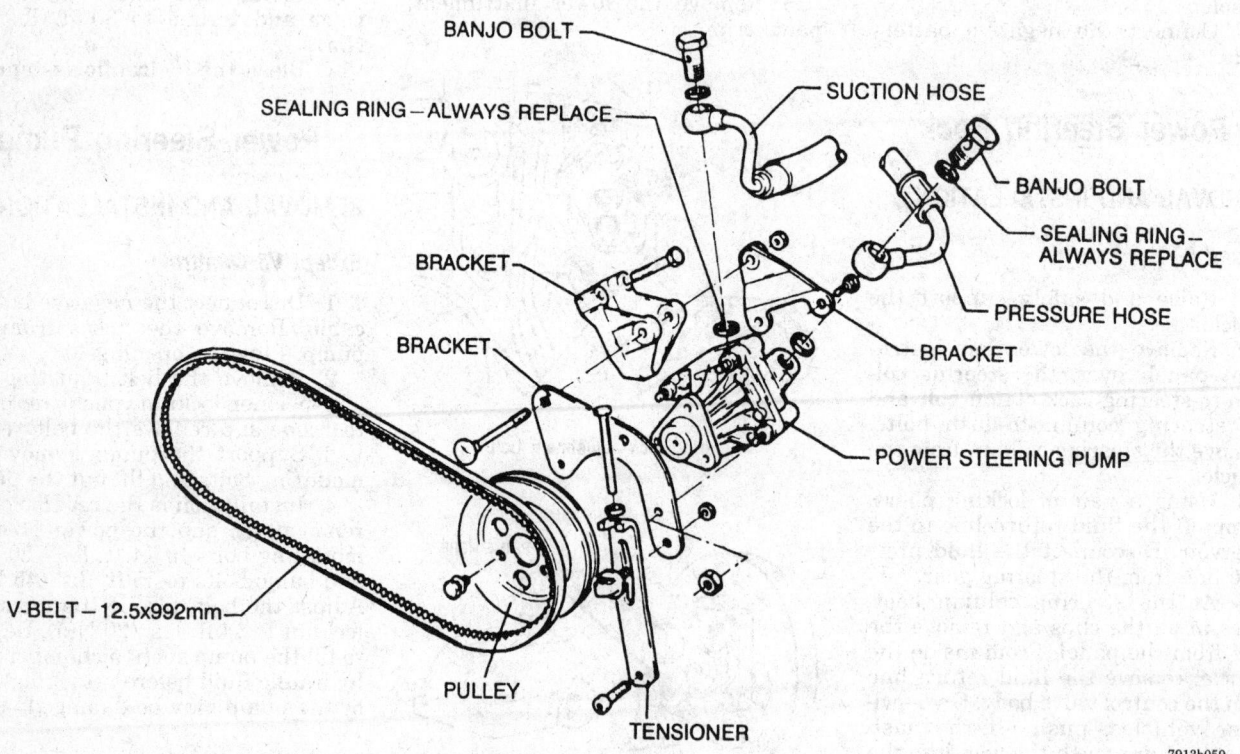

BANJO BOLT

SUCTION HOSE

SEALING RING—ALWAYS REPLACE

BANJO BOLT

SEALING RING— ALWAYS REPLACE

BRACKET

PRESSURE HOSE

BRACKET

BRACKET

POWER STEERING PUMP

V-BELT—12.5x992mm

PULLEY

TENSIONER

7913b059

Steering pump assembly — 80 and 90

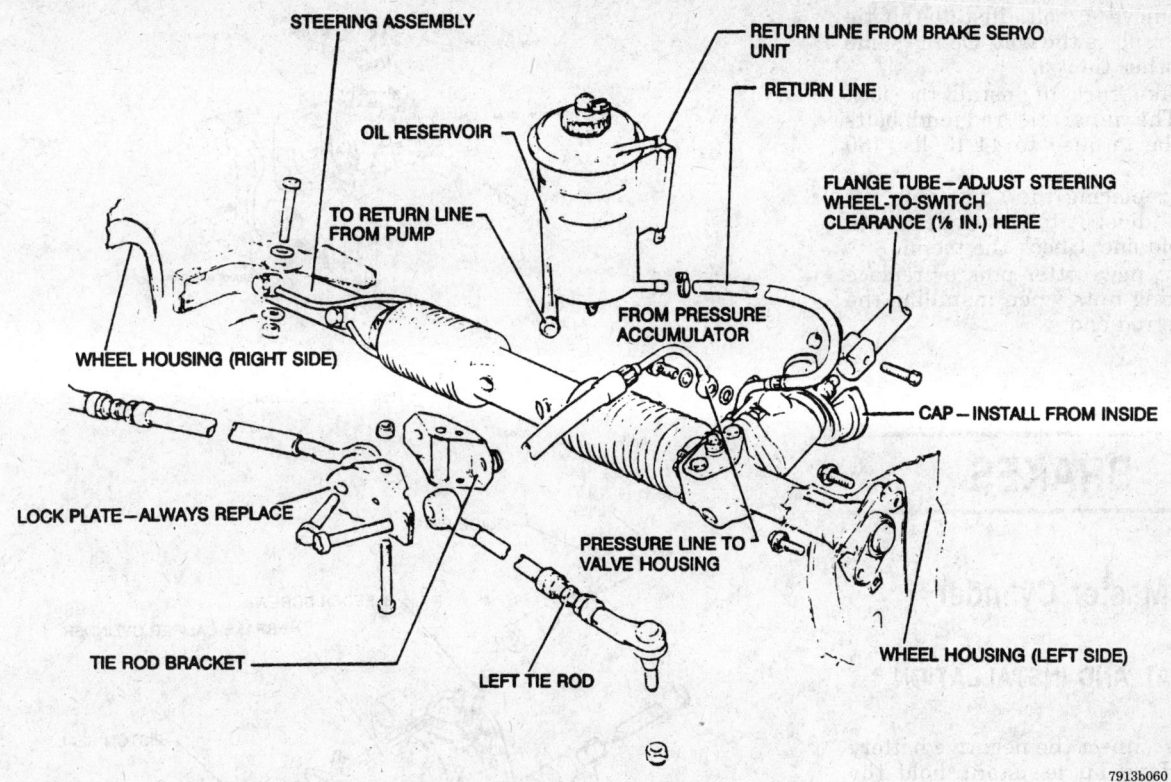

Steering rack assembly — 100, 200 and V8 Quattro

3. Remove the retaining screws and remove the air duct elbow.

4. Remove the retaining bolts and move the radiator fan and motor assembly to one side.

5. Loosen the pump pulley mounting bolts.

6. Place a 13mm wrench on the tensioner bolt and move the wrench upward slowly and firmly until tension on the ribbed belt is released. Remove the drive belt from the idler pulley and take the pulley off the pump. Save any adjustment shims.

7. Clamp off the fluid intake and return lines, remove the lines from the pump and seal the openings. Before removing the banjo bolt take note of the proper installation of the pipe, over the corner of the pump support.

8. Remove the mounting bolts and the front support bracket, then remove the rear pump bolts. Push the pump forward from the support and twist the pump for clearance as it is removed.

To install:

9. Install the pump in support.

10. Loosely install bolts on pump at rear. Loosely install bolts on pump at front.

11. Torque all pump mounting bolts to 15 ft. lbs. (20 Nm).

12. Before installing the banjo bolt for the hydraulic line, make sure the pipe is properly aligned with the pump support. Use new seals. Torque the banjo bolts to 30 ft. lbs. (40 Nm).

13. Check the belt and pulley alignment. Shim the pulley, if necessary. Install the pulley and torque pulley retaining bolts to 15 ft. lbs. (20 Nm).

14. Adjust drive belt.

15. Install the radiator fan motor assembly.

16. Check and refill fluid level in reservoir.

17. Install air duct. Reconnect the battery. Check system for proper operation.

SYSTEM BLEEDING

1. Fill the reservoir to the **FULL** mark.

2. Raise and safely support the vehicle.

3. Turn the steering wheel with the engine not running from lock to lock several times to remove the air from the system.

4. Add fluid to the reservoir until the level is maintained at 1 3/16 in. (30mm) below the **FULL** mark.

5. Start the engine. As the fluid in the reservoir continues to drop, add fluid to maintain the 1 3/16 in. (30mm) level.

NOTE: When turning the steering wheel, do not use more force than necessary to turn it.

6. Keep bleeding the system until no more air bubbles appear in the reservoir.

7. Turn OFF the engine and pump the brake pedal at least 20 times.

8. Replenish the fluid to the proper level.

Tie Rod Ends

REMOVAL AND INSTALLATION

NOTE: A puller or press is required for this procedure.

1. Raise and support the vehicle safely. Remove the front wheels.

2. Disconnect the outer end of the steering tie rod from the steering knuckle by removing the cotter pin and nut and pressing out the tie rod end. A small puller or press is required to free the tie rod end.

3. Under the hood, pry off the lock plate and remove the mounting bolts from both tie rod inner ends. Pry the tie rod out of the mounting pivot.

4. Install the mounting pivot to the rack with both mounting bolts.

5. Remove 1 bolt, install the tie rod and replace the bolt. Do the same on the other tie rod.

6. Make sure to install the lock plate. The inner tie rod end bolts should be torqued to 44 ft. lbs. (60 Nm).

7. If replacing the adjustable left tie rod, adjust it to the same length as the old one. Check the toe-in.

8. Use new cotter pins or replace self-locking nuts when installing the outer tie rod end.

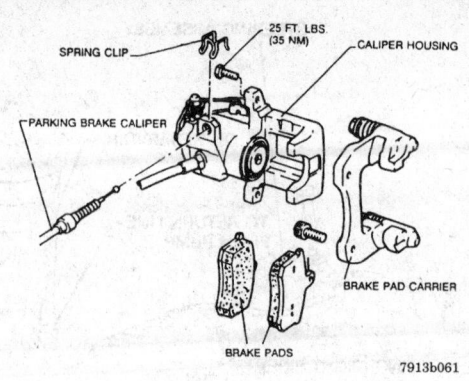

Girling rear brake caliper

BRAKES

Master Cylinder

REMOVAL AND INSTALLATION

1. Disconnect the negative battery cable. Have an assistant hold the brake pedal down about 1½ in. Disconnect the brake lines nearest the firewall.

2. Hold a container under the fitting disconnected in Step 1 and have the assistant release the pedal. The contents of the reservoir will drain into the container. Discard the used fluid.

3. Disconnect the other brake line.

4. Disconnect the stoplight switch and any warning switches from the master cylinder.

5. Remove the master cylinder from the power brake unit. Be careful not to lose the sealing ring between the 2 units.

To install:

6. Install the new master cylinder and and torque the mounting bolts to 17 ft. lbs. (23 Nm). Install the switches.

7. Transfer the reservoir from the old master cylinder to the new unit.

NOTE: Bench bleeding master cylinder will speed the on-vehicle bleeding procedure. Raise the front or rear of the vehicle, if necessary, to maintain bleeding locations at the highest point in the hydraulic system.

8. Fill and bleed the system. There should be a pedal free-play of 0.2 in. (5mm). Free-play can be adjusted with the linkage, inside.

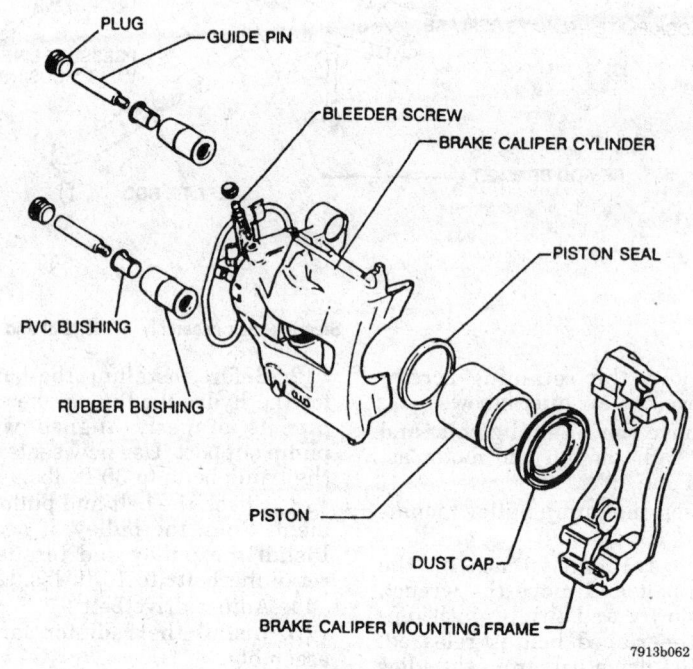

Teves front brake caliper

Proportioning Valve

REMOVAL AND INSTALLATION

This device is not used on vehicles with ABS.

1. Disconnect the negative battery cable. Remove and plug the 4 brake lines leading from the proportioning valve.

2. Disconnect the spring which is attached to the valve and the axle beam.

3. Remove the 2 mounting bolts and remove the valve.

NOTE: Do not disassemble the valve.

4. Installation is in the reverse order of removal.

5. Bleed the brake system when finished.

Power Brake Booster

REMOVAL AND INSTALLATION

1. Remove the master cylinder. Do not disconnect the brake lines.

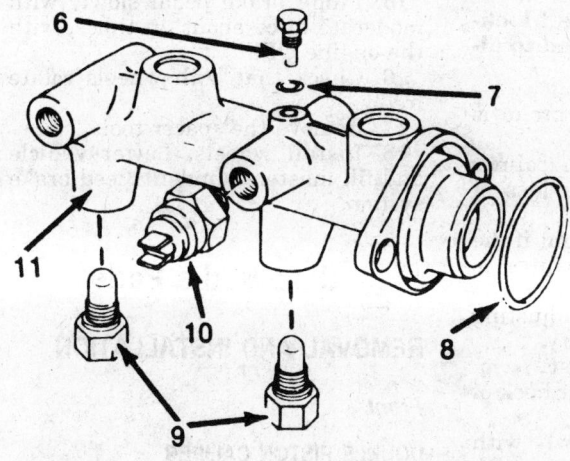

1. Reservoir Cap
2. Washer
3. Filter screen
4. Reservoir
5. Master cylinder plugs
6. Stop screw
7. Stop screw seal
8. Master cylinder seal
9. Residual pressure valves
10. Warning light sender unit
11. Brake master cylinder housing

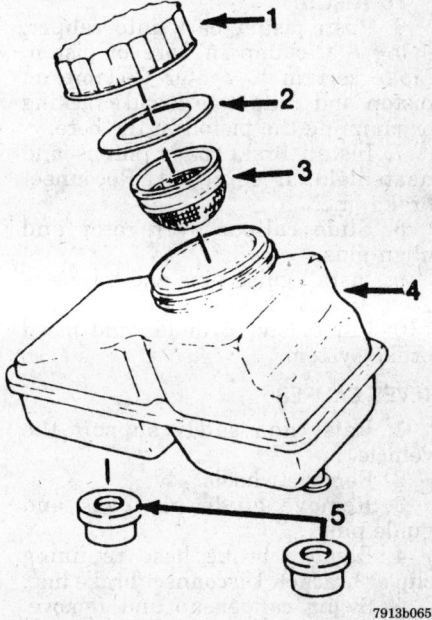

7913b065

Master cylinder assembly

2. Disconnect the vacuum hose from the power brake booster.

3. From under the dash, disconnect the pushrod from the brake pedal, remove the power brake booster-to-firewall nuts and remove the booster from the vehicle.

4. To install, reverse the removal procedures. Torque the master cylinder nuts to 17 ft. lbs. (23 Nm).

Brake Caliper

REMOVAL AND INSTALLATION

Front

DOUBLE PISTON CALIPER

1. Raise and safely support the vehicle.

2. Remove wheels.

3. Remove the lower caliper bolt, hold guide pin with an open end wrench while loosening. Disconnect the wear indicator, if equipped.

4. Swing brake caliper up and remove brake pads, taking note of spacer shims and heat-shield locations, if equipped.

5. Disconnect brake line and remove caliper.

To install:

6. Push pistons back into caliper. Place an old disc pad on piston side of

caliper, using a C-clamp centered on old pad across both piston, push pistons back into bore. Make certain to center C-clamp on pad and caliper to avoid cracking or jamming the pistons in their bores.

7. Install brake pads, shims and heatshield, if equipped. Install brake line.

8. Slide caliper over rotor and align pins. Make sure caliper moves freely on the pins.

9. Install guide pins and torque to 26 ft. lbs. (35 Nm).

10. Connect the wear indicator and install wheels.

11. Lower vehicle and fill master cylinder. Bleed brake system.

INTERNAL CALIPER

This type of brake system, used on the V8 Quattro, has the caliper mounted in the inside the rotor. The rotor is mounted with an external hat.

1. Raise and safely support the vehicle. Remove the wheels.

2. Press and unhook the brake pad retention spring to release the pads.

3. Press the piston back into its bore by pushing on the caliper housing. Once the piston is bottomed, push the entire housing towards the center of the vehicle to gain access for

removal of the inside pad. Remove the pad.

4. Remove the retaining bolt for the rotor. Remove the rotor by turning the rotor on an angle.

5. Disconnect the brake hose.

6. Remove the brake caliper from the wheel bearing housing. Brake caliper cannot be repaired, replace if necessary.

To install:

7. Install the brake caliper to the wheel bearing housing. Torque the retaining bolts to 92 ft. lbs. (125 Nm).

8. Install the brake hose.

9. Replace the rotor and torque the retaining or locating bolt to 44 inch lbs. (5 Nm). Install the inside pad and retaining spring.

10. Bleed the brakes.

11. Mount the wheels and depress the brake pedal to seat the pads before road test. Check brake fluid.

GIRLING CALIPER

1. Raise and safely support the vehicle.

2. Remove wheels.

3. Remove the lower caliper bolt, hold guide pin with open end wrench, while loosening. Disconnect the wear indicator, if equipped.

4. Swing brake caliper up and remove brake pads, taking note of

spacer shims and heat-shield locations, if equipped. Remove brake line.

5. Remove the caliper.

To install:

6. Push piston back into caliper, using a C-clamp in bore of piston. Make certain to center C-clamp on piston and caliper to avoid cracking or jamming the piston in the bore.

7. Install brake pads, shims and heatshield, if equipped. Reconnect brake line.

8. Slide caliper over rotor and align pins.

9. Install guide pins and torque to 25 ft. lbs. (35 Nm).

10. Fill master cylinder and bleed brake system.

TEVES CALIPER

1. Raise and safely support the vehicle.

2. Remove wheels.

3. Remove guide pin caps and guide pins.

4. Remove brake hose retaining clip or bracket. Disconnect brake line.

5. Swing caliper up and remove. Remove brake pads, taking note of spacer shims and heat-shield locations. Disconnect the wear indicator, if equipped.

To install:

6. Installation is the reverse of the removal process. Push piston back into caliper, using a C-clamp in bore of piston. Make certain to center C-clamp on piston and caliper to avoid cracking or jamming the piston in the bore.

7. Install brake pads, shims and heatshield, if equipped.

8. Slide caliper over rotor and align pins. Install guide pins and torque to 18 ft. lbs. (25 Nm).

9. Install brake hose and clip. Fill the master cylinder and bleed the brake system.

REAR

GIRLING CALIPER

1. Raise and safely support the vehicle.

2. Remove wheels.

3. Remove rear brake caliper housing, hold guide pin with open end wrench while loosening bolts. Disconnect the wear indicator, if equipped.

4. Remove brake pads, taking note of spacer shims and heat-shield locations. Disconnect brake line.

To install:

5. Screw piston into housing by turning it clockwise with a socket head wrench while pushing in firmly.

6. Install brake pads, shims and heatshield, if equipped.

7. Install brake line. Install caliper on housing. Install new bolts and torque to 25 ft. lbs. (35 Nm).

NOTE: The bolts are self-locking type; it is recommended to always use new bolts.

8. Make certain parking brake is free of tension.

9. Use a prybar to push caliper lever against stop on both sides of vehicle.

10. Parking brake is too tight if the lever of the opposite side caliper is pulled away from the stop.

11. Loosen parking brake adjusting nut and position, as necessary.

12. Push a tool of at least ¼ in. (6mm) diameter between rear hook of spring and roller.

13. Pump brake pedal slowly with moderate force about 40 times, with the engine OFF.

14. Check that both wheels rotate freely.

15. Remove spacer tool.

16. Install wheels. Lower vehicle and fill master cylinder. Bleed brake system.

TEVES CALIPER

1. Raise and safely support the vehicle.

2. Remove rear wheels.

3. Remove both protective caps. Loosen both guide pins but do not pull out of the rubber boots.

4. Pull caliper housing toward the outside of vehicle by hand. Disconnect brake line.

5. Swing the caliper housing to the rear and remove. Disconnect the wear indicator, if equipped.

6. Remove brake pads, taking note of spacer shims and heat-shield locations.

To install:

7. Push piston back into caliper, using a C-clamp in bore of piston. Make certain to center C-clamp on piston and caliper to avoid cracking or jamming the piston in the bore.

8. Install brake pads, shims and heatshield, if equipped.

9. Slide caliper over rotor and align pins. Install guide pins and torque to 18 ft. lbs. (25 Nm). Install the protective caps. Reconnect brake line.

10. Make certain parking brake is free of tension.

11. Use a prybar to push caliper lever against stop on both sides of vehicle.

12. Parking brake is too tight, if lever of opposite side caliper is pulled away from the stop.

13. Loosen parking brake adjusting nut and position, as necessary.

14. Push a tool of at least ¼ in. (6mm) diameter between rear hook of spring and roller.

15. Pump brake pedal slowly with moderate force about 40 times, with the engine OFF.

16. Check that both wheels rotate freely.

17. Remove the spacer tool.

18. Install wheels. Lower vehicle and fill master cylinder. Bleed brake system.

Disc Brake Pads

REMOVAL AND INSTALLATION

Front

DOUBLE PISTON CALIPER

1. Siphon a sufficient quantity of brake fluid from the master cylinder reservoir to prevent the brake fluid from overflowing the master cylinder when installing pads.

2. Raise and safely support the vehicle.

3. Remove wheels.

NOTE: Change the pads on 1 side at a time and use other side for reference. Do not disconnect the brake hoses, unless the caliper is to be serviced.

4. Remove the lower caliper bolt, hold guide pin with open end wrench, while loosening. Disconnect the wear indicator, if equipped.

5. Swing brake caliper up and remove brake pads, taking note of spacer shims and heat-shield locations, if equipped.

6. Push pistons back into caliper. Place old disc pad on piston side of caliper, using a C-clamp centered on old pad across both piston, push pistons back into bore. Make certain to center C-clamp on pad and caliper to avoid cracking or jamming the pistons in their bores.

To install:

7. Install brake pads, shims and heatshield, if equipped.

8. Slide caliper over rotor and align pins.

9. Install guide pins and torque to 26 ft. lbs. (35 Nm).

10. Connect the wear indicator and install wheels.

11. Lower the vehicle and fill master cylinder.

12. Pump the brake pedal slowly several time to force pads against the rotors.

13. Check master cylinder level again and add new fluid, if needed.

INTERNAL CALIPER

This type of brake system, used on the V8 Quattro, has the caliper mounted in the rotor. The rotor is mounted with an external hat. This is a unique system, but poses no real difficulties in servicing.

1. Raise and safely support the vehicle. Remove the wheels.
2. Press and unhook the brake pad retention spring to release the pads.
3. Press the piston back into its bore by pushing on the caliper housing. Once the piston is bottomed, push the entire housing towards the center of the vehicle to gain access for removal of the inside pad. Remove the pad.
4. Remove the retaining bolt for the rotor. Remove the rotor by turning the rotor on an angle.
5. Remove the brake wear sensor wire from the holder. Remove the outer pad and wear sensor.

To install:
6. Install the outer pad and wear sensor. Clip the sensor wire into the holder.
7. Replace the rotor and torque the retaining bolt to 44 inch lbs. (5 Nm). Install the inside pad and retaining spring.
8. Mount the wheels and depress the brake pedal to seat the pads.

GIRLING CALIPER

1. Siphon a sufficient quantity of brake fluid from the master cylinder reservoir to prevent the brake fluid from overflowing the master cylinder when installing pads.
2. Raise and safely support the vehicle.
3. Remove wheels.

NOTE: Change the pads on 1 side at a time and use other side for reference. Do not disconnect the brake hoses, unless the caliper is to be serviced.

4. Remove the lower caliper bolt, hold guide pin with open end wrench, while loosening. Disconnect the wear indicator, if equipped.
5. Swing brake caliper up and remove brake pads, taking note of spacer shims and heat-shield locations, if equipped.
6. Push piston back into caliper, using a C-clamp in bore of piston. Make certain to center C-clamp on piston and caliper to avoid cracking or jamming the piston in the bore.

To install:
7. Install brake pads, shims and heatshield, if equipped.
8. Slide caliper over rotor and align pins.

9. Install guide pins and torque to 25 ft. lbs. (35 Nm).
10. Install wheels.
11. Lower vehicle and fill master cylinder.
12. Pump the brake pedal slowly several time to force pads against the rotors.
13. Check master cylinder level again and add new fluid, if needed.

TEVES CALIPER

1. Siphon a sufficient quantity of brake fluid from the master cylinder reservoir to prevent the brake fluid from overflowing the master cylinder when installing pads.
2. Raise and safely support the vehicle.
3. Remove wheels.

NOTE: Change the pads on 1 side at a time and use other side for reference. Do not disconnect the brake hoses, unless the caliper is to be serviced.

4. Remove guide pin caps.
5. Remove guide pins.
6. Remove brake hose retaining clip or bracket.
7. Swing caliper up and secure in position, using a wire. Do not allow caliper to hang from brake hose.
8. Remove brake pads, taking note of spacer shims and heat-shield locations. Disconnect the wear indicator, if equipped.
9. Push piston back into caliper, using a C-clamp in bore of piston. Make certain to center C-clamp on piston and caliper to avoid cracking or jamming the piston in the bore.

To install:
10. Install brake pads, shims and heatshield, if equipped.
11. Slide caliper over rotor and align pins.
12. Install guide pins and torque to 18 ft. lbs. (25 Nm).
13. Install brake hose clip.
14. Install wheels.
15. Lower vehicle and fill master cylinder.
16. Pump the brake pedal slowly several time to force pads against the rotors.
17. Check master cylinder level again and add new fluid if needed.

Rear

GIRLING CALIPER

1. Siphon a sufficient quantity of brake fluid from the master cylinder reservoir to prevent the brake fluid from overflowing the master cylinder when installing pads.
2. Raise and safely support the vehicle.

3. Remove wheels.

NOTE: Change the pads on 1 side at a time and use other side for reference. Do not disconnect the brake hoses, unless the caliper is to be serviced.

4. Remove brake caliper housing, hold guide pin with open end wrench while loosening bolts. Disconnect the wear indicator, if equipped.
5. Remove brake pads, taking note of spacer shims and heat-shield locations.
6. Screw piston into housing by turning it clockwise with a socket head wrench while pushing in firmly.

To install:
7. Install brake pads, shims and heatshield, if equipped.
8. Install caliper on housing.

NOTE: The bolts are self-locking type; it is recommended to always use new bolts.

9. Install new bolts and torque to 25 ft. lbs. (35 Nm).
10. Make certain parking brake is free of tension.
11. Use a prybar to push caliper lever against stop on both sides of vehicle.
12. Parking brake is too tight if lever of opposite side caliper is pulled away from the stop.
13. Loosen parking brake adjusting nut and position, as necessary.
14. Push a tool of at least ¼ in. (6mm) diameter between rear hook of spring and roller.
15. Pump brake pedal slowly with moderate force about 40 times, with the engine OFF.
16. Check that both wheels rotate freely.
17. Remove the spacer tool.
18. Install wheels.
19. Lower vehicle and fill master cylinder.

TEVES CALIPER

1. Siphon a sufficient quantity of brake fluid from the master cylinder reservoir to prevent the brake fluid from overflowing the master cylinder when installing pads.
2. Raise and safely support the vehicle.
3. Remove wheels.

NOTE: Change the pads on 1 side at a time and use other side for reference. Do not disconnect the brake hoses, unless the caliper is to be serviced.

4. Remove both protective caps.
5. Loosen both guide pins but do not pull out of the rubber boots.

6. Pull caliper housing toward the outside of vehicle by hand.

7. Swing the caliper housing to the rear and remove. Do not allow caliper to hang from brake hose. Disconnect the wear indicator, if equipped.

8. Remove brake pads, taking note of spacer shims and heat-shield locations.

9. Push piston back into caliper, using a C-clamp in bore of piston. Make certain to center C-clamp on piston and caliper to avoid cracking or jamming the piston in the bore.

To install:

10. Install brake pads, shims and heatshield, if equipped.

11. Slide caliper over rotor and align pins.

12. Install guide pins and torque to 18 ft. lbs. (25 Nm).

13. Install the protective caps.

14. Make certain parking brake is free of tension.

15. Use a prybar to push caliper lever against stop on both sides of vehicle.

16. Parking brake is too tight if lever of opposite side caliper is pulled away from the stop.

17. Loosen parking brake adjusting nut and position, as necessary.

18. Push a tool of at least ¼ in. (6mm) diameter between rear hook of spring and roller.

19. Pump brake pedal slowly with moderate force about 40 times, with the engine OFF.

20. Check that both wheels rotate freely.

21. Remove the spacer tool.

22. Install wheels. Refill master cylinder.

Brake Rotor

REMOVAL AND INSTALLATION

Except Internal Caliper Type

1. Raise and safely support vehicle.
2. Remove wheels.
3. Remove brake caliper.
4. Remove disc rotor.
5. When replacing rotors, always replace in pairs. If machining, watch wear limit. Always machine both sides, never one side only.

Internal Caliper Type

1. Raise and safely support vehicle.
2. Remove wheels.
3. Press and unhook the brake pad retention spring to release the pads.

4. Press the piston back into its bore by pushing on the caliper housing. Once the piston is bottomed, push the entire housing towards the center of the vehicle to gain access for removal of the inside pad. Remove the pad.

5. Remove the retaining bolt for the rotor. Remove the rotor by turning the rotor on an angle.

To install:

6. Replace the rotor and torque the retaining or locating bolt to 44 inch lbs. (5 Nm). Install the inside pad and retaining spring.

7. Mount the wheels and depress the brake pedal to seat the pads before road test. Check brake fluid.

Brake System Bleeding

To bleed the system, use a bottle with transparent hose attached so brake fluid can be checked for air bubbles. Do not re-use fluid removed from reservoir.

1. Raise and safely support vehicle. Start the engine.

2. Connect bleeder hose and bleed calipers. Have an assistant press the pedal. When in the down position, open the bleeder screw. Close the bleeder screw and press the pedal again. Open the bleeder screw and again allow the air to come out. Repeat at each wheel until the fluid in the container shows no sign of air bubbles.

3. Bleed in the following sequence:
   a. Right rear caliper
   b. Left rear caliper
   c. Right front caliper
   d. Left front caliper

4. After bleeding, refill brake fluid reservoir.

Anti-Lock Brake System Service

PRECAUTIONS

The following precautions should be observed when working on the Anti-lock Brake System (ABS).

• Electrical testing should be done using the factory LED tester. This tester must be used to check the hydraulic modulator, ABS control unit, wheel speed sensors and ABS wiring harness. It is also necessary to perform the test procedure if the brake lines or brake pressure regulators are replaced because of accident damage.

• Switch the ignition OFF before connecting or disconnecting the ABS control unit connector.

• Disconnect the ABS control unit connector before using electrical welding equipment on the vehicle.

• Disconnect the battery connections before charging the battery or replacing the hydraulic modulator.

• Remove the ABS control unit before drying paint repairs in an oven if the temperature will be more than 185°F. for more than 2 hours.

• Do not use mini-spare tires on vehicles equipped with ABS. Use wheels and tires of matching size on ABS equipped vehicles.

• Do not drive the vehicle with the anti-lock brake tester connected.

• Do not fabricate brake lines. Use only original equipment parts. Brake line flare nuts are to be tightened to 11 ft. lbs. (15 Nm).

• Do not repair the hydraulic modulator except when replacing relays. If the hydraulic modulator is defective, it must be replaced.

RELIEVING ANTI-LOCK BRAKE SYSTEM PRESSURE

A special factory tool is recommended for ABS system bleed-down. It is a set of high and low pressure gauges that is also used for system pressure testing. It is used as follows:

1. Raise and safely support vehicle. Remove one front wheel.

2. Make sure the ignition switch is **OFF**. Remove the bleeder screw from the caliper.

3. Connect the special tool and install a brake pedal depressor between the brake pedal and the driver's seat.

4. Press the brake pedal until the pressure gauge drops down indicating system pressure has been relieved.

SYSTEM TEST

The operational ABS system test must be performed after every repair to the service brake components. These repairs include replacement of linings and/or discs, hoses, booster or master cylinder, cables and parking brake components.

This short test insures that nothing within the ABS components was disturbed during the repair. To perform the test:

1. Turn the ignition switch **ON**; the ABS warning lamp should light.

2. Start the engine; the warning lamp should turn OFF.

3. While the engine is running, switch the ABS **OFF**; the warning lamp(s) should come ON.

4. Turn the ignition switch **OFF** and restart it; the ABS warning lamp

should go out, showing that the manual switch position has been overridden and the system reset.

5. Drive the vehicle over 20 mph (30 kph); the ABS warning lamp should not come ON. The differential locks on Quattro vehicles must not be engaged during this test.

6. In a safe location and under controlled conditions, perform at least one test stop from about 20 mph which engages the anti-lock system. Check system function and vehicle stability.

Hydraulic Modulator

REMOVAL AND INSTALLATION

1. Disconnect negative battery cable.
2. Bleed down system pressure.
3. Remove the modulator cover.
4. Remove the harness retainer.
5. Disconnect the hydraulic modulator from the mounting.
6. Unclip the hydraulic fluid reservoir from the mounting.
7. Unscrew the brake lines. Mark the lines so they will be reinstalled in their proper locations. Seal any openings immediately with plugs.

To install:
8. Connect the hydraulic lines to the modulator but do not tighten them yet.
9. Install the modulator, then tighten the lines. Install the reservoir and connect the wiring.
10. Refill with DOT 4 brake fluid only. Bleed and test the system.

Wheel Speed Sensors

REMOVAL AND INSTALLATION

Front

The front wheel speed sensor is hand-pressed into the wheel bearing housing, also called a steering knuckle. A sleeve is inserted into a hole in the housing which retains the speed sensor.
1. Raise and safely support vehicle.
2. Remove wheel and tire assembly.
3. Locate the sensor in the side of the housing and pull out. Left and right are identical.
4. To disconnect the wiring, remove the engine lower cover.
5. Unclip the connector from its mount and disconnect sensor.

To install:
6. Installation is the reverse of the removal procedure. Before inserting the retainer sleeve, the opening in the wheel bearing housing should be greased.
7. Press the sleeve in as far as possible.
8. Push the sensor in to stop by hand.

Rear

The rear wheel speed sensor is hand pressed into the rear wheel bearing housing. A sleeve is inserted into a hole in the front side of the housing which retains the speed sensor.
1. Raise and safely support vehicle.
2. Remove wheel and tire assembly.
3. Locate the sensor in the front side of the rear wheel bearing housing and pull out. Left and right are identical. Unclip the connector and remove sensor.

To install:
4. Installation is the reverse of the removal process. Before inserting the retainer sleeve, the opening in the rear wheel bearing housing should be greased.
5. Press the sleeve in as far as possible.
6. Push the sensor in to stop by hand.

Acceleration Switch

REMOVAL AND INSTALLATION

4WD Vehicles

The ABS acceleration switch is located under the left rear seat bench. It is a mercury switch activated during deceleration. When braking with ABS the switch helps the braking system provide additional stabilization during braking.
1. Switch the ignition **OFF**.
2. Remove the mounting screws.
3. Disconnect the wiring.
4. At installation, note that the arrow on the cover must point forward, in the direction of driving.

Electronic Control Unit

REMOVAL AND INSTALLATION

The ABS electronic control unit is located under the left side of the rear passenger seat.
1. Switch ignition **OFF**.

2. Raise rear seat and locate ABS control unit. Remove hold-down nuts.
3. The ABS control unit plug has locks to secure it. Disconnect by pressing the spring on the narrow end of the plug. To connect, insert the lug of the connector into the adapter and push the connector against the spring. The connector should snap into the lock with a click.
4. Installation is the reverse of the removal process.

CHASSIS ELECTRICAL

— CAUTION —
On vehicles equipped with an air bag, the negative battery cable and reserve power supply must both be disconnected before working on the system. Failure to do so may result in deployment of the air bag and possible personal injury. Never disconnect any electrical connector with the ignition switch turned ON or damage to electronic controlling devices could occur.

NOTE: All vehicles are equipped with theft protected radios, which cannot be operated if power to the radio is interrupted. Before disconnecting the battery cables, obtain the security code.

Air Bag

DISARMING

1. Disconnect the battery and the red air bag power supply connector. This connector is labeled "Air bag". On 80 and 90, the connector is behind a panel under the driver side dashboard. On all other models, the connector is at the front of the center console.
2. Remove the screws for the upper steering column trim and remove the upper trim.
3. Separate the connector for the air bag spiral spring.
4. If required, a memory saver can now be safely used to preserve the radio code while the battery is disconnected.

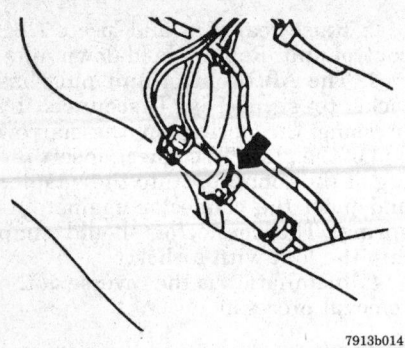

Power supply connector must be disconnected first

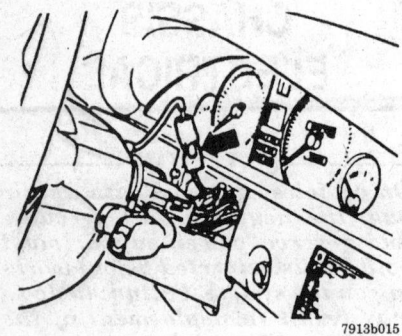

The air bag can be disconnected at the steering column

Heater Blower Motor

REMOVAL AND INSTALLATION

Coupe, 80 and 90

1. The blower is removed from under the hood. Disconnect the negative battery cable.
2. Remove the air plenum from the cowl.
3. Remove the ballast resistor.
4. At the blower motor, disconnect the electrical connector.
5. Remove the blower mounting bolts and the blower from the heater assembly.
6. To install, reverse the removal procedures.

100, S4 and 200

NOTE: Blower or core removal requires removal and disassembly of the heater/air conditioning unit.

1. Disconnect the negative battery cable.

2. Drain the cooling system.

NOTE: If equipped with air conditioning, the system must be discharged.

3. Properly discharge the air conditioning system.
4. Disconnect the following components:
   a. Temperature sensor connector
   b. Evaporator/heater connector clamp
   c. Temperature control cable
   d. Fresh air door vacuum hose
5. Disconnect the main harness connector.
6. Loosen the case retaining strap.
7. Remove the coolant hoses at the heater core tubes.
8. Label and remove the yellow, green and red vacuum hoses from the heater case.
9. Remove the air duct hoses.
10. Remove the heater case mounting screws, 2 in the passenger compartment, 1 in the engine compartment. Remove the 4 evaporator housing mounting screws in the passenger compartment.
11. Support the heater/evaporator unit and pull it away from the firewall.
12. Remove the control cable grommet to facilitate case removal.
13. The case halves may be separated by removing the clips at the top and bottom with a small prybar.
14. Remove the blower motor and the heater core from the unit.
   To install:
15. Install the blower motor and heater core into the case. Make sure the case gasket is in good condition and assemble the case. Install the clips.
16. Position the case at the firewall, install the control cable grommet and secure the case in place with the screws.
17. Attach the ducts, wires, control cable, vacuum lines and coolant hoses.
18. Refill the cooling system and leak test. Recharge and leak test the air conditioner.

─────── CAUTION ───────
Refrigerant will freeze any surface it contacts, including skin and eyes. It also turns into a poisonous gas in the presence of an open flame. Wear eye protection and suitable gloves when working on or around the air conditioning system.

V8 Quattro

1. Disconnect the negative battery cable.
2. Matchmark hood hinges and remove hood.
3. Remove the windshield wiper assembly.
4. Remove the cap from the engine coolant overflow bottle.
5. Remove the heater retaining band.
6. Label and remove the vacuum hoses from the vacuum servo motors.
7. Clamp off the heater hoses to the heater core and remove the hoses.
8. Remove the retainers between the body and heater. Remove the heater box.
9. After removing the heater box, remove the blower cooling hose.
10. Remove the lockring, stop washer and grommet. Remove the blower from the housing.
   To install:
11. Before installing the fresh air blower guides, lubricate with petroleum jelly. Install the blower motor using the black electrical connection area only.
12. Install the heater unit into the vehicle and connect the vacuum and coolant hoses. When connecting the water hoses, the lower connection on the heater core is connected to the hose going to the water pump.
13. Run the engine and check for leaks before completing the assembly.

Windshield Wiper Motor

REMOVAL AND INSTALLATION

1. Disconnect the negative battery cable and disconnect the wiring harness connector at the motor. Remove the wiper arm retaining nut. Pry off the wiper arms and remove the nuts from the studs in the cowl.
2. Remove the brace-to-body screws. While holding the crank, remove the nut securing the crank to the wiper motor and remove the crank.
3. Remove the bolts securing the wiper motor to the support and remove the motor.
4. Connect the new motor to the wiring harness, run the motor 2 revolutions and turn the wiper switch to the OFF position. The wiper motor should stop in the park position.
5. To install, reverse the removal procedures. Make sure the crank is installed in the proper position.

FIRING ORDERS

NOTE: To avoid confusion, always replace spark plug wires one at a time.

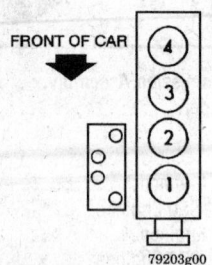

M42/M44 Engines
Engine Firing Order:
1–3–4–2
Distributorless Ignition System

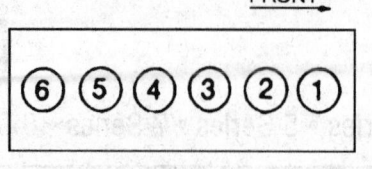

M50/M52/S50 Engines
Engine Firing Order: 1–5–3–6–2–4
Distributorless Ignition System

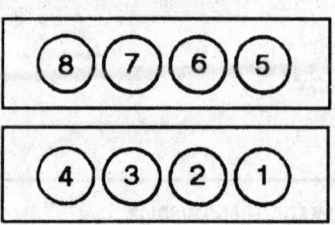

M60/M63 Engines
Engine Firing Order: 1–5–4–8–6–3–7–2
Distributorless Ignition System

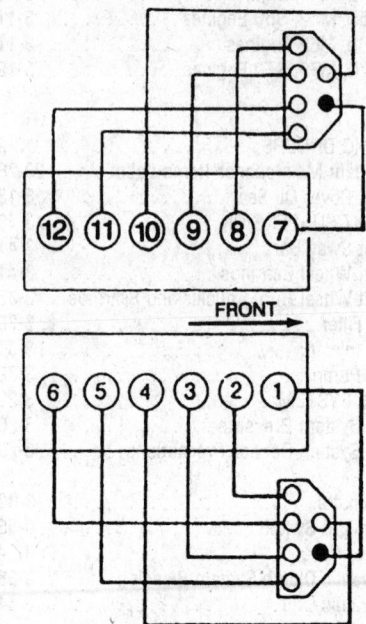

M70/M73/S70 Engines
Engine Firing Order: 1–7–5–11–3–9–6–12–2–8–4–10
Distributor Rotation: Clockwise

SERIAL NUMBER IDENTIFICATION

Vehicle Identification Plate

The manufacturer's plate is located in the engine compartment, on the right side inner fender panel or support.

Engine Number

The engine number is located on the left rear side of the engine, above the starter motor.

Vehicle Identification Number

The vehicle identification number is located on a plate, on the drivers side of the instrument panel.

Chassis Number

The chassis number can be found in the engine compartment on the right front inner fender support or facing forward on the right side of the bulkhead. A label is also attached to the upper steering column cover, inside the vehicle.

ENGINE MECHANICAL

NOTE: Disconnecting the negative battery cable on some vehicles may interfere with the functions of the on-board computer systems and may require the computer to undergo a relearning process.

Engine Assembly

REMOVAL AND INSTALLATION

M42/M44 Engines

1. Disconnect the battery ground cable. Remove the transmission and remove the engine splash guard. Disconnect the gas spring and prop rod and support hood safely in the fully open position.

2. Remove the fan cowl by turning the expansion rivets on the left and right sides. Lift the cowl up and out of the engine compartment.

3. Hold the fan pulley while unscrewing the fan nut from the shaft. The shaft uses left hand threads; turn the nut counterclockwise to unscrew.

4. Drain the coolant from the engine block. Disconnect the bottom hose from the radiator expansion tank, the engine coolant hoses and the heater hoses from the splash wall. Drain all coolant into clean containers for reuse or proper disposal.

5. Disconnect the air flow meter electrical plug and loosen the hose clamp and mounting screws. Lift the air sensor with the air cleaner up and out of the engine compartment.

6. Unclip the throttle cable and pull the cable out with the rubber holder.

7. Disconnect the fuel lines taking note of their positions. Pull off the vent hose to the filter for tank venting.

8. Disconnect the vacuum fitting at the brake booster.

9. Remove the ignition leads from the coil. Unscrew the connections at the alternator and starter. Disconnect the 2 plugs from the electrical duct. Remove the plug from the throttle valve potentiometer located at the throttle neck. Pull off the tank venting valve plug located next to the air cleaner. Disconnect the fuel injector plug located at the end of the electrical duct near to the fuel pipes. Pull off the idle speed control connector at the rear of the intake manifold. Disconnect the oil pressure switch electrical connection.

10. Unscrew the front and rear intake manifold supports.

11. Remove the electrical duct from the engine. Disconnect the coolant temperature senders for the gauge and the DME.

12. Disconnect the electrical duct and wiring harness on the engine and lay it off to the side of the engine.

13. Use a suitable lifting yoke to attach to the engine lifting eyes. Unscrew the motor mounts and the engine ground strap. Lift out the engine.

To install:

14. Lower engine into engine compartment. Fasten the motor mounts and the ground strap.

15. The balance of installation is the reverse of the removal procedure.

16. Install the fan using tool 11 5 040 or equivalent. Torque the nut to 29 ft. lbs. (40 Nm). If using the fan tool, set the torque wrench to 22 ft. lbs. (30 Nm); the additional length of the tool multiplies the torque to achieve 29 ft. lbs. (40 Nm) at the nut.

17. Add the proper coolant mixture and bleed the cooling system.

18. Connect the battery leads and check all fluid levels before starting the engine.

M50/M52/S50 Engines

3 Series

1. Disconnect the battery ground cable.

2. Remove the transmission from the vehicle.

3. Remove the engine splash guard.

4. Press the hinge so it goes over center and support hood safely in the fully open position.

5. Disconnect the air mass sensor plug and loosen the air intake duct hose clamp. Remove the air cleaner assembly.

6. Disconnect the hoses for the idle speed control and the crankcase breather.

7. Unscrew and remove the ducting for the alternator. Pull out the fan cowl expansion rivets and remove the cowl upwards.

8. Hold the pulley with tool 11-5-030 or equivalent, and unscrew the fan clockwise. Remove the fan and keep upright.

9. Drain the cooling system. The engine block drain plug is accessible through the exhaust manifold.

10. Remove the upper and lower radiator hoses from the radiator. Disconnect the coolant level switch and the automatic transmission cooler lines, if equipped. Plug the cooler lines.

11. Disconnect the right side hose and the temperature sensor.

12. Insert a tool into the radiator support clips and press down on the tab. Pull back on the radiator to release it. Remove the radiator.

13. Disconnect the heater hoses from the heater and heater valve.

14. Remove the grill from the air intake cowl at the base of the windshield.

   a. Remove the electrical lead tray.

   b. Remove the screws on the right side cowl holder bracket and the screw on the left side.

   c. Remove the cowl from the engine compartment.

15. Unscrew the fastener from the throttle cable cover and pull the cover

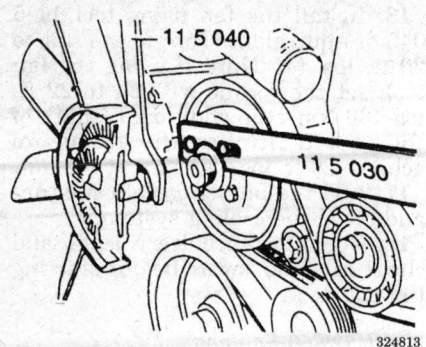

Cooling fan removal

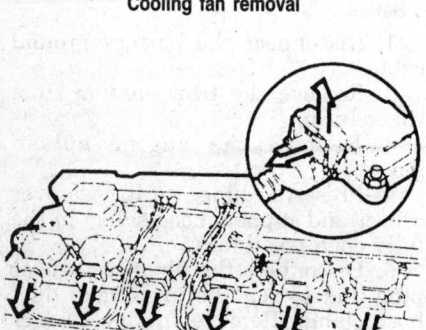

Ignition leads on coil — 3 Series with M50/M52/S50 engine

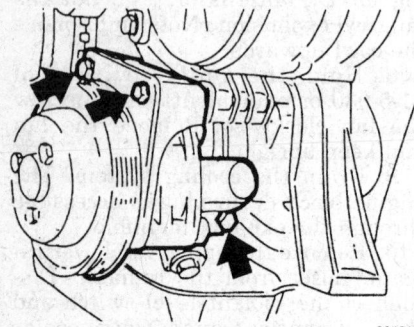

Power steering pump mounting bolts — 3 Series with M50/M52/S50 engine

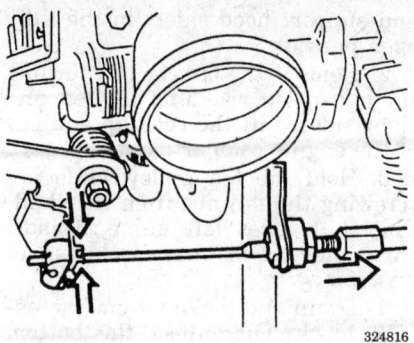

Throttle cable — 3 Series with M50/M52/S50 engine

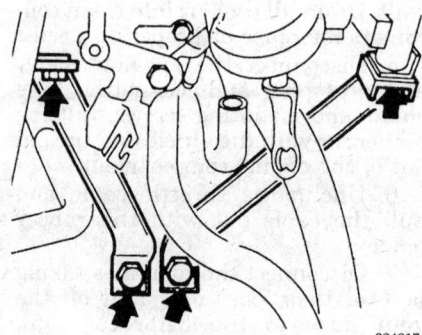

Front and rear intake manifold supports — 3 Series with M50/M52/S50 engine

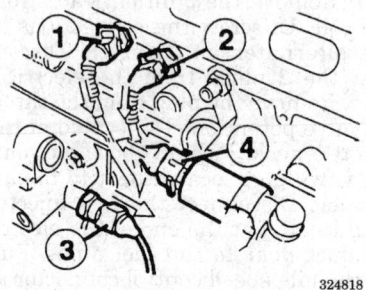

Temperature sensor (1), temperature gauge (2), oil pressure switch (3) and idle speed control valve (4) electrical connector locations — 3 Series with M50/M52/S50 engine

forward and off. Unclip the cable and pull the cable out with the rubber holder.

16. Pull the vacuum fitting from the brake booster and plug the openings.

17. Remove the engine and intake manifold covers. Unscrew the bolt holding the ground strap on the front lifting eye. Replace the bolt before lifting the engine.

18. Unscrew the 2 bolts holding the plug plate and pull off the plug plate. Be careful not to damage the rubber seals. Take off the ignition coil elec-

trical plugs. Remove the plug plate complete with the electrical leads.

19. Remove the cylinder head vent hose and pull off the air temperature sensor plug.

20. Remove the tank venting hose and the throttle heating hoses from the throttle body.

21. Remove the throttle valve switch plug.

22. Unclip the idle speed control valve mounted on the manifold.

23. Disconnect the fuel hoses from the pipes.

24. Unscrew the hardware holding the intake manifold to the cylinder head. Remove the intake manifold taking care not to drop anything into the exposed ports.

25. Disconnect the plugs from the temperature sensor, temperature gauge, the oil pressure switch and the idle speed control valve.

26. Disconnect the cylinder identifying sender plug (black) and the pulse sender plug (gray) for the DME. Unscrew the oxygen sensor plug in the holder.

27. Remove the electric leads from the alternator and the starter. Unscrew the electrical lead tray and place the engine wiring harness to the side.

28. Loosen the drive belt for the power steering pump and the air conditioner compressor by turning their respective tensioners clockwise. This will release the tension on the belt and allow the belt to be removed.

29. Unbolt the power steering pump and place to the side without disconnecting the hoses.

30. Unbolt the air conditioner compressor and place to the side without disconnecting the lines.

31. Attach a lifting fixture to the engine lifting hooks.

32. Unscrew the engine mounts and ground strap.

33. Lift the engine out of the vehicle being careful of the front radiator mount.

To Install:

34. Lower the engine into the vehicle and attach the motor mounts and ground strap. Torque the engine mount: 8mm bolts to 16 ft. lbs. (22 Nm), and the 10mm bolts to 31 ft. lbs. (42 Nm).

35. Install the power steering pump and the air conditioner compressor. Install the drive belts.

36. Reposition the wiring harness and electrical lead tray on the engine. Connect the leads to the starter and alternator.

37. Screw in the plug for the oxygen sensor holder.

38. Connect the leads for the cylinder identifying sender, DME pulse sender, temperature sensor, temperature gauge sender, oil pressure switch and idle speed control valve.

39. Install the intake manifold making sure that the intake seals are intact. Replace the intake seals if any signs of deterioration are evident.

40. Attach the fuel lines. The upper line is the return and the lower is the feed.

41. Attach the idle speed control valve hose located on the manifold.

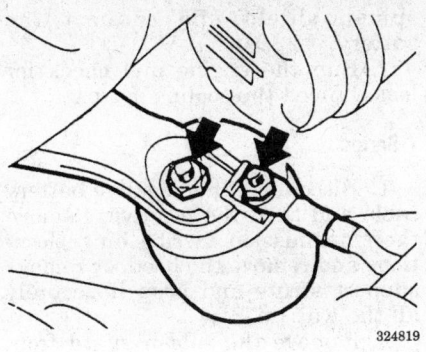

324819

Right engine mount — 3 Series with M50/M52/S50 engine

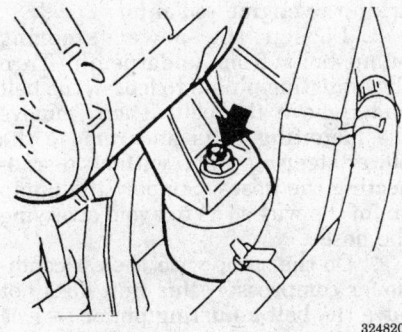

324820

Left engine mount — 3 Series with M50/M52/S50 engine

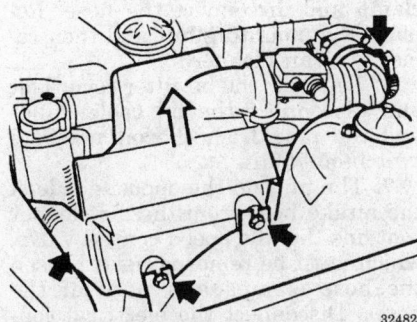

324821

Air cleaner and air mass sensor — 3 Series with M50/M52/S50 engine

42. Connect the throttle valve switch plug, the throttle valve heating lines and the tank vent line.

43. Connect the air temperature sensor plug and attach the cylinder head venting hoses.

44. Reconnect the plugs for the ignition coils and mount the plug plate. Attach the ground strap to the front lifting eye.

45. Replace the engine and manifold covers. Connect the line to the brake booster.

46. Reconnect the throttle cable and cover. Install the air intake cowl at the base of the windshield.

47. Connect the heater hoses to the valve and inlet.

48. Remount the radiator by pressing down on the mounting clips to fasten. Check that the lower mounts are in place.

49. Connect the temperature switch plug for the air conditioner and replace the trim panel.

50. Connect the cooling system hoses and the automatic transmission lines. Use new seals on the transmission lines and tighten to 13–15 ft. lbs. (18–21 Nm).

51. Install and tighten the engine block drain plug.

52. Install the fan using tool 11-5-040 or equivalent wrench and holding too 11-5-030 or equivalent. Torque the nut to 29 ft. lbs. (40 Nm). If using the tool set the torque wrench to 22 ft. lbs. (30 Nm): the additional length of the tool multiplies the torque to achieve 29 ft. lbs. (40 Nm) at the nut.

53. Replace the radiator cowling into its mounting slots and press in the rivets.

54. Install the alternator air ducting.

55. Connect the idle speed and crankcase breather hose to the air intake duct. Replace the air cleaner assembly and connect the electrical plug.

56. Install the transmission to the vehicle.

57. Refill and bleed the cooling system.

58. Install the splash shield.

59. Connect the negative battery cable.

60. Check all fluids before starting engine.

5 Series

1. Disconnect the battery terminals, negative side first and remove the battery. Unscrew and remove the battery tray. Remove the transmission.

2. Loosen the clamp on the cooling duct to the alternator and remove the duct.

3. Disconnect the plug to the air flow meter and loosen the clamps to the air cleaner duct. Unscrew the mounting bolts and remove the air cleaner assembly.

4. Pull out the expansion rivets that hold the fan cowl. Remove the cowl by pulling up out of the engine compartment.

5. Hold the fan pulley while unscrewing the fan nut from the shaft. The shaft uses left hand threads; turn the nut counterclockwise to unscrew.

6. Drain the coolant from the block. The drain plug is located between the exhaust manifolds. Disconnect the coolant hoses from the radiator and remove the coolant level switch plug. On automatic transmission equipped vehicles, remove the oil lines to the radiator and plug.

7. Disconnect the bottom radiator hose and remove the trim panel from the right side of the engine compartment to expose the side of the radiator and the air conditioner condenser.

8. Pull the plug off of the air conditioner temperature switch.

9. Remove the radiator supporting clips by inserting a small prybar down from above into the slot and pulling back. Pull the radiator free from the clip. Remove the radiator from the vehicle.

10. Disconnect the heater hoses from the heater valve and the heater.

11. Unscrew the fastener from the throttle cable cover and pull the cover forward and off. Unclip the cable and pull the cable out with the rubber holder.

12. Pull the vacuum fitting from the brake booster and plug the openings.

13. Remove the engine and intake manifold covers. Unscrew the bolt holding the ground strap on the front lifting eye. Replace the bolt before lifting the engine.

14. Unscrew the 2 bolts holding the plug plate and pull off the plug plate. Be careful not to damage the rubber seals. Take off the ignition coil electrical plugs. Remove the plug plate complete with the electrical leads.

15. Remove the cylinder head vent hose and pull off the air temperature sensor plug. Remove the tank venting hose and the throttle heating hoses from the throttle body. Remove the throttle valve switch plug. Unclip the idle speed control valve mounted on the manifold. Disconnect the fuel hoses from the pipes.

16. Unscrew the hardware holding the intake manifold to the cylinder head. Remove the intake manifold taking care not to drop anything into the exposed ports.

17. Disconnect the plugs from the temperature sensor, temperature gauge, the oil pressure switch and the idle speed control valve. Disconnect the cylinder identifying sender plug (black) and the pulse sender plug (gray) for the DME. Unscrew the oxygen sensor plug in the holder.

18. Remove the electric leads from the alternator and the starter. Unscrew the electrical lead tray and

place the engine wiring harness to the side.

19. Loosen the drive belt for the power steering pump and the air conditioner compressor by turning their respective tensioners clockwise. This will release the tension on the belt and allow the belt to be removed.

20. Unbolt the power steering pump and place to the side without disconnecting the hoses. Unbolt the air conditioner compressor and place to the side without disconnecting the lines.

21. Attach a lifting fixture to the engine lifting hooks. Unscrew the engine mounts and ground strap. Lift the engine out of the vehicle being careful of the front radiator mount.

To install:

22. Lower the engine into the vehicle and attach the motor mounts and ground strap.

23. Install the power steering pump and the air conditioner compressor. Install the drive belts.

24. Install the remaining components in the reverse order of removal.

25. Install the fan using tool 11 5 040 or equivalent. Torque the nut to 29 ft. lbs. (40 Nm). If using the fan tool, set the torque wrench to 22 ft. lbs. (30 Nm); the additional length of the tool multiplies the torque to achieve 29 ft. lbs. (40 Nm) at the nut. Replace the radiator cowling.

26. Replace the air cleaner assembly and connect the electrical plug. Install the transmission and fill and bleed the cooling system. Install the battery tray and battery. Check all fluids before starting engine.

M60/M62 and M70/M73/S70 Engines

5 Series

1. Allow the engine to cool. Disconnect the battery cable from the battery, negative side first. Open the hood to the widest position possible and secure in place.

2. Remove the splash shield from under the car. Drain the coolant and remove the radiator.

3. Remove the transmission.

4. Remove the nut holding the transmission oil cooler lines to the engine oil pan.

5. Loosen and remove the drive belts to the power steering pump and the air conditioner compressor. Remove the bolts holding the pump and compressor to the engine and remove them from the engine, keeping the lines connected. Wire the pump and compressor out of the way and without any tension on the hoses.

6. Disconnect the hoses from the coolant expansion tank. remove the

screws on the side of the expansion tank and remove the expansion tank from the engine compartment.

7. Disconnect the heater hoses from the heater control valve and the heater inlet pipe.

8. Pull off the connections to the ignition coil. Remove the air cleaner assembly. Disconnect and remove the idle speed control from the intake duct.

9. Disconnect the harness to the air flow meter. Disconnect the ducting to the air flow meter and remove along with the crankcase breather vacuum line.

10. On non-ASC equipped cars, disconnect the cruise control cable and the throttle cable at the throttle. Remove the cable mounting bracket. If equipped with ASC, remove the connector to the throttle control unit as there will be no throttle cable to disconnect.

11. Disconnect the leads to the starter. Disconnect the 2 electrical connectors in the starter area. Disconnect the oil level sender leads and the alternator connections. Remove the air duct to the alternator.

12. Disconnect the tank venting valve and the hose to the carbon canister.

13. Mark the feed and return fuel lines. Disconnect the lines and catch any spilled fuel.

14. Disconnect the vacuum line from the brake booster and plug the opening. Disconnect the harness connections to the temperature sensors and DME sensors.

15. Disconnect the ground strap and make a check for any remaining lines or electrical leads still attached.

16. Attach a lifting sling to the engine. Remove the engine mount nuts and bolts and lift the engine from the engine bay.

To install:

17. Install the engine into the engine compartment. Torque the engine mounts to 32.5 ft. lbs. (45 Nm). Connect the ground strap.

18. Connect the brake booster vacuum line, the temperature sensors and the DME sensors.

19. Connect the fuel lines to the proper locations as previously marked. Connect the tank venting valve and the carbon canister line. Attach the alternator leads and the cooling duct.

20. The remaining components are installed in the reverse order of removal.

21. Install the transmission and the radiator. Fill the cooling system and check the engine fluids. Install the

splash shield and connect the battery.

22. Run the engine and check for leaks. Bleed the cooling system.

7 Series

1. Disconnect the negative battery cable and then the positive. Remove the transmission. Scribe hinge locations and remove the hood, or remove support struts and prop it securely all the way up.

2. Remove the splash guard from underneath the engine. Then, with the engine cool, remove the drain plugs in the radiator and block and drain the engine coolant.

3. Loosen the power steering pump bolts from underneath. Turn the adjusting pinion to loosen the belt and remove the belt. Then, remove the mounting bolts and remove the power steering pump without disconnecting the hoses. Support the pump out of the way so as to avoid stressing the hoses.

4. Do the same with the air conditioner compressor, this unit does not have the belt adjusting pinion — it is necessary only to loosen all the bolts and push the compressor toward the engine to remove the belt.

5. Loosen the air intake hose clamp and disconnect the hose. Remove the mounting nut and then remove the air cleaner(s).

6. Unscrew the oil filter cover bolt and disconnect the oil cooler lines and the plug from the oil pressure switch on 750iL.

7. The unit on the opposite side of the intake hose from the air cleaner contains the idle speed control valve, which must be removed next. Loosen the hose clamps and pull off the hoses. Disconnect the electrical connector. Remove the mounting nut and then pull the idle speed control out of the air intake hose.

8. Pull off the retainers for the air flow sensor, and then pull the unit off its mounts, disconnecting the vacuum hose from the PCV system at the same time.

9. Working on the coolant expansion tank, disconnect the electrical connector. Remove the nuts on both sides. Loosen their clamps and then disconnect the hoses and remove the tank.

10. Disconnect the heater hoses at both the control valve and at the heater core.

11. Disconnect the throttle and cruise control cables at the throttle lever. Unbolt the cable housing retainer and remove the housing and cables.

12. Pull off the low amperage starter connectors and disconnect the high amperage connector coming from the battery.

13. Loosen its clamp and then disconnect the coolant hose that runs to the alternator.

14. Disconnect the connecting plug for the oxygen sensor, as well as the other plugs.

15. Loosen the clamps and then disconnect the fuel supply and return pipes.

16. Disconnect the fuel pipe at the injector supply manifold. Disconnect the plug. Disconnect the electrical connector at the throttle body. Lift off the protective caps and then remove the attaching nuts for the protective cover for the wiring harness for the injectors and remove it.

17. Disconnect the ground strap at the block. Remove the engine mount nut from the top on both sides.

18. Attach a lifting sling to the engine and support the assembly. Disconnect the ground lead. Carefully lift the engine out of the compartment, tilting the front of the engine upward for clearance.

To install:

19. Keep these points in mind during installation:

   a. Torque the engine mounting bolts to 32.5 ft. lbs. (43 Nm).

   b. Adjust the belt tension for the air conditioning compressor and power steering pump drive belts to give 1/2–3/4 in. deflection.

   c. Torque the oil cooler line flare nuts to 25 ft. lbs. (34 Nm).

   d. When reconnecting the intake manifold to the throttle necks, inspect and, if necessary, replace the O-rings. Torque the mounting nuts to 6.5 ft. lbs. (8 Nm).

20. Lower the engine into the engine compartment. When the engine is positioned, the guide pin must fit in the bore of the axle carrier. Torque the mounting bolts on the front axle carrier (small bolt) to 18–20 ft. lbs. (25–27 Nm); the larger bolt to 31–35 ft. lbs. (40–47 Nm). The mount-to-bracket bolts are torqued to 31–35 ft. lbs. (40–47 Nm).

21. Connect the fuel lines, use new hose clamps to connect the fuel lines to the fuel filter. Connect all of the multi-prong plugs and all vacuum hoses.

22. The balance of installation is the reverse of the removal procedure.

23. Make sure all fluid levels are correct before starting the engine. Bleed air from the cooling system.

8 Series

1. Disconnect the negative battery cable and then the positive. Remove the transmission. Scribe hinge locations and remove the hood, or remove support struts and prop it securely all the way up.

2. Remove the mass air flow sensor from the engine. Remove the windshield washer tank.

3. Loosen the oil filter cap to permit the oil to drain back into the oil pan. Then, remove the oil filter lines.

4. Remove the radiator and expansion tank.

5. Unclip the diagnostic clip. Disconnect the wires at the left and right coils. Unscrew the right ignition coil.

6. Disconnect the D+ (thin lead) from the alternator. Disconnect the oil sender.

7. Disconnect the hoses and wire connections from the tank venting valves.

8. Remove the oil catching tray. Disconnect the wire connectors from the sensors, injectors and throttle body.

9. Disconnect the wiring cover at the rear of the engine.

10. Disconnect hoses at pressure regulator, noting there arrangement.

11. Disconnect the temperature sensors and remove the injector wiring harness.

12. Disconnect the alternator main feed wires, at the B+ connection point. Disconnect the starter wire connections.

13. Remove the air conditioning compressor, leaving the hoses connected.

14. Remove the cool air duct for the alternator.

15. Disconnect and plug the fuel lines.

16. Remove the heat shields located under the vehicle on the thrust struts.

17. Remove the heat shields on the right exhaust manifold and remove the right exhaust pipe from the manifold.

18. Drain the power steering fluid and remove the hose from the supply tank.

19. Remove the starter assembly. Disconnect any necessary connections.

20. Disconnect the heater hoses.

21. Attach a suitable lifting device to the engine and unscrew the ground strap and engine mounts.

22. Remove the guide tube for the oil dipstick.

23. Lift the engine slightly and remove the right engine bracket. Turn the rear of the engine to the right to clear the left exhaust pipe past the steering spindle.

24. Remove the engine and place on a suitable holding fixture.

To install:

25. Keep these points in mind during installation:

   a. Torque the engine mounting bolts to 32.5 ft. lbs. (43 Nm).

   b. Adjust the belt tension for the air conditioning compressor and power steering pump drive belts to give 1/2–3/4 in. deflection.

   c. Ensure all hose and wires are connected as prior to removal.

26. Lower the engine into the engine compartment. The mount-to-bracket bolts are torqued to 31–35 ft. lbs. (40–47 Nm).

27. Install the remaining components in the reverse order of removal.

28. Make sure all fluid levels are correct before starting the engine. Bleed air from the cooling system.

Engine Mounts

REMOVAL AND INSTALLATION

1. Raise and safely support the vehicle.

2. Support the engine using a suitable lifting device. Disconnect the mounting bolts.

3. Remove the ground strap, if equipped. Remove the engine mounts.

To install:

4. Install the mount onto the mounting bracket and replace the bolts.

5. Replace the ground strap, if equipped. Remove the lifting device.

6. Lower the vehicle.

Cylinder Head

REMOVAL AND INSTALLATION

M42/M44 Engines

1. Disconnect the negative battery cable.

2. Remove the ignition coil cover and pull off the spark plug connectors.

3. Remove the complete ignition tackle. Remove the cylinder head cover.

4. Disconnect the coolant hoses and unscrew the temperature sensor.

5. Remove the thermostat housing and thermostat. Unscrew the upper timing case cover.

6. Rotate the engine in the direction of the rotation until the camshaft peaks of the intake and exhaust cam-

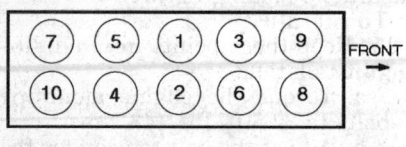

293746

Cylinder head tightening sequence — M42/M44 engines

shafts for cylinder No. 1 face each other. The arrows on the sprocket face up.

7. Remove the chain tensioner. Remove the upper chain guide, chain guide bolt on the right side and the sprockets.

8. Remove the cylinder head bolts from the outside to the inside in several steps using the proper tool.

9. Remove the cylinder head. Clean the sealing surfaces on the cylinder head and the crankcase.

To install:

10. Install the cylinder head onto the engine with a new gasket.

11. Tighten the cylinder head in 3 steps by following the sequence shown as follows:

   a. Step 1 — 24 ft. lbs. (32.5 Nm)
   b. Step 2 — 90–95 degree turn.
   c. Step 3 — plus an additional 90–95 degree turn.

12. The balance of installation is the reverse of the removal procedure.

13. Connect the negative battery cable, then start the engine and inspect for any fuel, vacuum or coolant leaks.

M50/M52/S50 Engines

1. If engine is not already removed from the vehicle, disconnect the negative battery cable and drain the engine coolant. Remove the intake manifold and throttle valve. Disconnect the exhaust pipes and the oxygen sensor wire. Remove the exhaust manifolds. Remove the thermostat housing and engine lifting eye.

2. Pull off the connectors for the ignition coils and remove the coils. Unscrew the cylinder heads cover and remove. Remove the sender from the head and the electrical lead duct.

3. Remove the upper timing case cover and the camshaft cover. Crank the engine in the direction of rotation so the intake and exhaust camshaft peaks for cylinder No. 1 face each

other. Hold the camshafts in place with tool 11 3 240 or equivalent. With the camshafts in this alignment, the arrows on the sprockets will be facing up. Remove the valve cover mounting studs. Lock the flywheel in place to prevent movement of the crankshaft.

4. Unscrew the chain tensioner and carefully remove. There is a spring contained within the tensioner and may eject if care is not taken.

5. Press down on the upper chain tensioner and lock it into place using tool 11 3 290 or equivalent. Unscrew the transfer timing chain sprockets and pull the 2 off together with the chain. Remove the upper chain tensioner and the lower chain guide. Pull off the main timing chain sprocket along with the chain. Use a bent piece of wire to hold the chain from falling down into the engine. Do not rotate the engine after this point or the valve timing will be disturbed when the engine is reassembled.

6. Unscrew the bolts on the head at the ends of the cams. Using a proper sized Torx® bit or tool 11 2 250, loosen the cylinder head bolts in several steps. Use an outside to inside pattern to prevent warpage. On production heads the bolt washers are locked into place while on replacement heads the washers are loose. Keep track of the bolt washers.

To install:

7. If the camshafts have been removed and reinstalled a waiting period dependent on the ambient temperature is necessary before mounting the cylinder head on the engine. At room temperature wait 4 minutes to allow the lifters to compress fully. At temperatures down to 50°F (10°C) wait 11 minutes. At temperatures lower than 50°F (10°C) wait 30 minutes. This is to prevent contact between the valves and the piston tops. The engine may not be cranked under the same condition for a period of 10 minutes at room temperature; 30 minutes for temperatures down to 50°F (10°C); 75 minutes for temperatures below 50°F (10°C).

8. Clean all mounting surfaces and check the head for warpage. Take care not to drop any pieces of gasket or dirt into the oil or coolant passages. Check the condition of the head locating dowel sleeves.

9. Place a new head gasket on the engine block over the locating dowels and gently place the head on the engine. Align the head with the dowel sleeves and check that the head sits flat on the engine.

10. Cylinder head bolts may only be used once. Lightly oil the threads of the new cylinder head bolts. Check that the head bolt washers are in place and install the bolts. Torque the head bolts in 3 steps; Step 1 to 24 ft. lbs. (33 Nm), Step 2 and 3 to 93 degree torque angle. Torque the center bolts first and go out in a diagonal pattern.

11. Align main timing chain and sprocket on the can so the arrow faces up. The bolt holes in the camshaft should be on the left sides of the sprocket slots. This will allow the tensioner to take up the slack in the chain and rotate the gear to the counterclockwise position.

12. The balance of installation is the reverse of the removal procedure.

13. Connect the negative battery cable, start the engine and check for any leaks.

M60/M62 Engines

1. Disconnect negative battery cable.

2. Remove both exhaust manifolds from each side of the engine. Remove the heat shields from the front axle carrier.

3. Drain the engine coolant, and remove the coolant expansion tank.

4. Remove the upper timing case cover.

5. Remove the oil pipes on the cylinder head.

6. Remove the intake manifold.

7. Remove the cylinder head cover.

8. Remove the engine vent pipe together with the O-ring. Disconnect all the coolant hoses on the coolant collector. Remove the coolant collector mounting bolts, and remove the coolant collector.

9. Remove all eight spark plugs.

10. Remove the camshaft sprockets, and the timing chain tensioner.

11. Remove the bolts retaining the guide rail on the cylinder head's left hand side.

12. Remove the cyclone oil trap.

13. Remove the cylinder head bolts from the outside to inside. Lift off the cylinder head.

NOTE: The cylinder head bolts must be replaced.

To install:

14. Thoroughly clean all mounting surfaces and check the head for warpage. Take care not to drop any pieces of gasket or dirt into the oil or coolant passages. Check the condition of the head locating dowel sleeves and clean out the bolt threads with a tap.

15. Mount the cylinder head and new bolts. Torque the bolts in the proper sequence in three steps:
- Step 1–Torque each bolt to 24 ft. lbs. (32 Nm)
- Step 2–Wait 10 to 20 minutes, then turn each bolt an additional 80°
- Step 3–Wait 10 to 20 minutes, then turn each bolt an additional 80°

16. Install the cyclone oil trap.

17. Install the bolts retaining the guide rail on the cylinder head's left hand side.

18. Install the camshaft sprockets and the timing chain tensioner. Torque the sprocket mounting bolts to 11 ft. lbs. (15 Nm).

19. The remaining components are installed in the reverse order from which they were removed.

20. Connect negative battery cable.

M70/M73/S70 Engines

1. Unbolt the exhaust pipe connections at the manifold and at the transmission pipe clamp. Disconnect the negative battery cable.

2. Remove the splash shield from under the engine. With the engine cool, remove the drain plugs from the bottom of the radiator and block. Drain the engine oil.

3. Remove the fan. Lift out the expansion rivets on either side and remove the fan shroud.

4. Loosen the hose clamp and disconnect the air inlet hose. Remove the mounting nut and remove the air cleaner.

5. The unit on the opposite side of the intake hose from the air cleaner contains the idle speed control valve, which must be removed next. Loosen the hose clamps and pull off the hoses. Disconnect the electrical connector. Remove the mounting nut and then pull the idle speed control out of the air intake hose.

6. Pull off the retainers for the air flow sensor, and then pull the unit off its mounts, disconnecting the vacuum hose from the PCV system at the same time.

7. Working on the coolant expansion tank, disconnect the electrical connector. Remove the nuts on both sides. Loosen their clamps and then disconnect all hoses and remove the tank.

8. Disconnect the heater hoses at both the control valve and at the heater core. Remove the valve, if needed.

9. Disconnect the throttle and cruise control cables at the throttle lever. Unbolt the cable housing retainer and remove the housing and cables.

10. Disconnect the plugs near the thermostat housing. Loosen the hose clamps and pull off the coolant hoses.

11. Disconnect the plug in the line leading to the oxygen sensor. Disconnect the other plugs.

12. Disconnect the fuel supply and return lines, collecting fuel in a metal container for safe disposal.

13. Disconnect the fuel pipe running along the cylinder head, near the manifold. Pull off the electrical connector at the throttle body. Remove the caps, then remove the attaching bolts and remove the wiring harness carrier and harness for the fuel injectors.

14. Disconnect the coil high tension lead. Disconnect the high tension wires at the plugs. Then, remove the mounting nuts and remove the carrier for the high tension wires from the head.

15. Remove the attaching nuts for the camshaft cover and remove it.

16. Turn the engine until the timing marks are at TDC and the No. 6 valves are at overlap, both slightly open.

17. Remove the upper timing case cover. Remove the timing chain tensioner piston.

18. Remove the upper timing chain sprocket bolts and pull the sprocket off, holding it upward and then supporting it securely so the relationship between the chain and sprockets top and bottom will not be lost.

19. Disconnect the upper radiator hose at the thermostat housing. Remove the bolts and remove the support for the intake manifold.

20. Remove the cylinder head bolts in the opposite of numbered order. Then, install 4 special pins part 11 1 063 or equivalent. This is necessary to keep the rocker arm shafts from moving. Then, lift off the head.

To install:

21. Make checks of the lower cylinder head and block deck surface to make sure they are true. Install a new head gasket, making sure all bolt, oil and coolant holes line up. Use a gasket marked M30B35. Use a 0.3mm thicker gasket, if the head has been machined.

22. Apply a very light coating of oil to the head bolts. Don't let oil get into the bolt holes or apply excessive amounts of oil, or torque could be incorrect and the block could crack. Use the type of bolt without a collar. Install the bolts finger-tight.

23. Torque bolts 1–6 in the correct order to 42–44 ft. lbs. (57–60 Nm). Remove the pins holding the rocker shafts in place. Now, complete the first stage of torquing by torquing bolts 7–14 in the correct order, to the same specification. Adjust the valves after a 15 minute wait. Tighten the bolts, in the correct order, with a torque angle gauge 30–36 degrees, using special tool 11 2 110 or equivalent.

24. Reinstall the timing sprocket to the camshaft. Make sure the camshaft is in proper time, that new lockplates are used and that nuts are properly torqued.

25. When reinstalling the timing cover, make sure to apply a liquid sealer to the joints between upper and lower timing covers. The remainder of installation is the reverse of removal. Note these points:

a. Adjust throttle, speed control and accelerator cables. Inspect and if necessary replace the exhaust manifold gasket.

b. When reinstalling the cylinder block coolant plug, coat it with sealer. Make sure to refill the cooling system and bleed it. Make sure to refill the oil pan with the correct amount of oil.

c. Install the timing chain so the down pin on the camshaft sprocket is at the 8 o'clock when its tapped bores are at right angles to the engine. Torque the sprocket bolts to 6.5–7.5 ft. lbs. (8–10 Nm).

d. Check the camshaft cover gasket, replacing, as necessary. Tighten camshaft cover bolts in the order shown. Torque the bolts to 6.5–7.5 ft. lbs. (8–10 Nm).

e. When reinstalling the fan shroud, make sure all guides are located properly.

f. Coat the tapered portion of the exhaust pipe connection flange with the proper sealant. Torque the attaching nuts to 4.5 ft. lbs. (6 Nm) and loosen 1½ turns.

26. Start the engine and run it until hot (25 minutes). Then, again remove the valve cover and turn the head bolts, in the correct order, 30–40 degrees.

Valve Lash

ADJUSTMENT

All engines are equipped with hydraulic valve lash adjusters. This design does not require adjustments nor are adjustments possible.

Intake Manifold

REMOVAL AND INSTALLATION

M42/M44 Engines

1. Disconnect the negative battery cable. Unscrew the upper manifold section.
2. Disconnect the rear mounting bracket and remove the coolant hose.
3. Loosen the front mounting bracket and disconnect the holder for the preheater.
4. Remove the mounting bolts and lift off the upper manifold section. Pull the hose off the fuel pressure regulator at the same time.
5. Pull the plug plate off the fuel injectors and remove the wire holding clamp.
6. Remove the injection pipe with the fuel injectors attached and remove the lower manifold section.

To install:

7. Install the lower manifold section onto the engine after cleaning the mating surfaces of both components. Tighten the manifold bolts evenly from the middle out to either end.
8. Clean the fuel injector mounting holes, then install the injection pipe and fuel injectors onto the lower intake manifold.
9. The remaining components are installed in the reverse order from which they were removed.
10. Connect the negative battery cable, cycle the ignition **ON** and **OFF** several times to allow fuel pressure to build. Each time allow the ignition to remain **ON** for 5–7 seconds.
11. Check for fuel leaks.

M50/M52/S50 Engines

1. Disconnect the negative battery cable and drain the coolant to a level below that of the throttle housing. Unscrew the fastener from the throttle cable cover and pull the cover forward and off. Unclip the cable and pull the cable out with the rubber holder.
2. Pull the vacuum fitting from the brake booster and plug the openings.
3. Remove the engine and intake manifold covers. Unscrew the bolt holding the ground strap on the front lifting eye. Replace the bolt before lifting the engine.
4. Unscrew the 2 bolts holding the plug plate and pull off the plug plate. Be careful not to damage the rubber seals. Take off the ignition coil elec-

trical plugs. Remove the plug plate complete with the electrical leads.
5. Remove the cylinder head vent hose and pull off the air temperature sensor plug. Remove the tank venting hose and the throttle heating hoses from the throttle body. Remove the throttle valve switch plug. Unclip the idle speed control valve mounted on the manifold. Disconnect the fuel hoses from the pipes.
6. Unscrew the hardware holding the intake manifold to the cylinder head. Remove the intake manifold taking care not to drop anything into the exposed ports.

To install:

7. Install the lower manifold section onto the engine after cleaning the mating surfaces of both components. Tighten the manifold bolts evenly from the middle out to either end.
8. Install the remaining components in the opposite order from which they were removed.
9. Connect the negative battery cable, cycle the ignition **ON** and **OFF** several times to allow fuel pressure to build. Each time allow the ignition to remain **ON** for 5–7 seconds.
10. Check for fuel leaks.

M60/M62 Engines

1. Relieve the fuel system pressure. Disconnect the negative battery cable.
2. Remove the center cover from the cylinder head cover.
3. Loosen the hose clamps on the idle speed control and the throttle valve assembly.
4. Disconnect the plug on the mass air flow sensor.
5. Unclip and remove the upper section of the air cleaner assembly along with the mass air flow sensor.
6. Remove the right hand cover from the cylinder head cover.
7. Raise and safely support the vehicle. Disconnect the plug for the oil level switch. Lower the vehicle.
8. Disconnect the plugs for the ignition coils.
9. Disconnect both knock sensors (For cylinders 1 and 2 along with 3 and 4), and the pulse sensor.
10. Disconnect the intake air temperature sensor, throttle valve potentiometer and the idle speed control.
11. Disconnect the diagnosis plug and the engine plug.
12. Unscrew the ignition coil ground wire, located near the rear engine lifting eye. Disconnect the temperature sensor (black) for the temperature gauge, and the temperature sensor (white) for the DME.

13. Remove the four bolts for the holder of the intake manifold cover. Remove the holder.
14. Disconnect and remove the throttle cable.
15. Remove the left hand cover from the cylinder head cover.
16. Disconnect the ignition coil electrical connectors.
17. Disconnect both knock sensors (For cylinders 5 and 6 along with 7 and 8), and the camshaft sender.
18. Disconnect the coolant expansion tank plug and spill hose. Remove the two mounting bolts, and move the tank aside.
19. Disconnect the oil pressure switch electrical connector and remove the wiring.
20. Remove the screws for the wiring ducts on the cylinder heads.
21. Disconnect the vacuum hoses on the radiator, and loosen the hose clamp. Pull the vacuum hose off of the brake booster.
22. Remove the tank vapor venting hose off of the throttle valve assembly.
23. Disconnect the fuel feed and return lines.
24. Remove the hose off of the end cover on the back of the manifold.
25. Remove the seven mounting bolts, and pull off the end cover together with the pressure regulating valve straight back to prevent damaging the vent pipe.
26. Remove the five intake manifold bolts, and remove the intake manifold upwards.

To install:

27. Scrape the gasket off of the manifold and the cylinder head. Replace the gasket, and position the intake manifold. Install the mounting bolts. Torque the bolts to 14–17 ft. lbs. (20–24 Nm).
28. Check and replace the seal and gasket for the end cover, if necessary. Position the end cover and install the mounting bolts. Torque the M8 bolts to 14–17 ft. lbs. (20–24 Nm), and the M6 bolts to 7 ft. lbs. (10 Nm).
29. Install the hose onto the end cover on the back of the manifold.
30. Connect the fuel feed and return lines.
31. Install the tank vapor venting hose onto the throttle valve assembly.
32. The balance of installation is the reverse of the removal procedure.

——WARNING——
Mixing up the knock sensor connectors will lead to engine damage.

33. Connect negative battery cable.

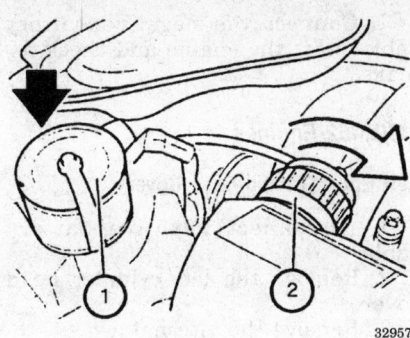

Connectors: 1– Diagnosis plug, 2– Engine plug — M60/M62 engines

M70/M73/S70 Engines

1. Disconnect the negative battery cable. Loosen the clamps for the fuel lines.

2. Pull off the vacuum hoses for the pressure regulators. Lift out the injection pipes with the injectors attached.

3. Remove the distributor caps and the throttle valve necks on the manifolds.

4. Disconnect the spark plug wires and remove the ignition lead pipes.

5. Disconnect the crankcase breather hose and loosen the manifold support nuts.

6. Disconnect the nose guard and remove the intake manifold, using the proper tool.

To install:

7. Scrape the gasket off of the manifold and the cylinder head. Replace the gasket, and position the intake manifold. Install the mounting bolts to 14–17 ft. lbs. (19–23 Nm) from the center of the manifold out to the ends.

8. Install the remaining components in the opposite order from which they were removed.

9. Connect the negative battery cable, cycle the ignition **ON** and **OFF** several times to allow fuel pressure to build. Each time allow the ignition to remain **ON** for 5–7 seconds.

10. Check for fuel leaks.

Exhaust Manifold

REMOVAL AND INSTALLATION

M42/M44 Engine

1. Disconnect the negative battery terminal.

2. With the exhaust system cool, disconnect the exhaust pipe from the manifold. Remove the 4 nuts on the flange connection and lower the exhaust pipes. Support the exhaust sys-

tem. Make sure the oxygen sensor wire is not being stretched.

3. Remove the nuts securing the manifold to the cylinder head. Remove the manifolds.

To install:

4. Clean the mounting surfaces on the manifolds and the cylinder head. Check the condition of the studs and replace if necessary.

5. Install the new exhaust manifold gaskets with the graphite side towards the cylinder head and install the manifolds. Torque the nuts to 16–18 ft. lbs. (22–25 Nm). Use new nuts and anti-seize.

6. Connect the exhaust pipe to the manifolds. Connect the negative battery terminal. Start engine and check for leaks.

7. After 1200 miles, loosen and then tighten each nut to 10 ft. lbs. (12 Nm).

M50/M52/S50 Engine

1. Disconnect the negative battery terminal.

2. Remove the mounting nuts on each flange connection and separate the exhaust pipes from the manifold. Support the exhaust system. Make sure the oxygen sensor wire is not being stretched.

3. Remove the nuts securing the manifold to the cylinder head. Remove the manifolds.

To install:

4. Clean the mounting surfaces on the manifolds and the cylinder head. Check the condition of the studs and replace if necessary.

5. Install the new exhaust manifold gaskets with the graphite side towards the cylinder head and install the manifolds. Torque the nuts to 14 ft. lbs. (19 Nm). Use new nuts and anti-seize.

6. Install the exhaust pipe to the manifolds using new mounting nuts.

7. Connect the negative battery terminal.

8. Start engine and check for exhaust leaks.

M60/M62 Engine

Left Exhaust Manifolds

1. Disconnect negative battery cable.

2. Disconnect the oxygen sensor plug and remove the exhaust assembly.

3. Remove the alternator.

4. Remove the left cylinder head cover.

5. Remove the complete air cleaner upper section along with the mass air flow sensor.

6. Remove the bolts from the left and right engine mounts at the bottom.

7. Remove the rear engine splash guard.

8. Remove the bolts of the center of gravity mount to front axle carrier. Remove the left heat shields on the front axle carrier.

9. Lift the engine, with a suitable tool, at the front eye. Ensure clearance between the engine and the firewall.

10. Remove the manifold bolts and remove the manifolds downwards.

To install:

11. Scrape the old gasket off of the cylinder head and exhaust manifold and replace the gasket. The gasket beads face the exhaust manifolds.

12. Coat the upper row of the exhaust bolts with locking fluid. Position the exhaust manifold and install the bolts. Torque the mounting bolts to 16 ft. lbs. (22 Nm).

13. Lower the engine to its original position. Install the left heat shields and the center of gravity mount-to-front axle carrier bolts.

14. Install the rear engine splash guard.

15. Install the bolts for the left and right engine mounts at the bottom. Torque the 10mm bolts to 31 ft. lbs. (42 Nm) and the 8mm bolts to 16 ft. lbs. (22 Nm).

16. Install the complete air cleaner upper section along with the mass air flow sensor.

17. Install the left cylinder head cover and replace the gasket. Torque the mounting bolts in a criss-cross pattern to 11 ft. lbs. (15 Nm)

18. Install the alternator. Connect the oxygen sensor plug, and install the exhaust assembly.

19. Connect negative battery cable.

Right Exhaust Manifold

1. Disconnect negative battery cable.

2. Disconnect the oxygen sensor plug and remove the exhaust assembly.

3. Remove the right heat shields on the front axle carrier.

4. Remove the rear engine splash guard.

5. Remove the washing fluid tank.

NOTE: Remove the manifold for cylinders two and four first.

6. Remove the manifold bolts and remove the manifolds upwards.

To install:

7. Scrape the old gasket off of the cylinder head and exhaust manifold and replace the gasket. The gasket beads face the exhaust manifolds.

8. Coat the upper row of the exhaust bolts with locking fluid. Position the exhaust manifold and install the bolts. Torque the mounting bolts to 16 ft. lbs. (22 Nm).

9. Install the washing fluid tank.

10. Install the rear engine splash guard.

11. Install the right heat shields.

12. Connect the oxygen sensor plug and install the exhaust assembly.

13. Connect negative battery cable.

M70/M73/S70 Engine

1. Disconnect the negative battery terminal.

2. Remove the left side upper section of the air cleaner assembly along with the air mass sensor.

3. Remove the clamp on the left and right split pipes.

4. Remove the heat shields on the left manifold and on the steering gear.

5. Remove the manifold/split pipe bolts on the left hand side.

6. On the left hand side, remove the front and rear manifolds along with the gaskets.

7. Remove the nuts on the staybolts. Remove the staybolts in the cylinder head for the left manifold.

8. Remove the right side upper section of the air cleaner assembly along with the air mass sensor.

9. Remove the windshield washing fluid tank.

10. Remove the oil dipstick guide tube.

11. Remove the heat shields on the right manifold.

12. On the right hand side, remove the front and rear manifolds along with the gaskets.

13. Remove the nuts on the staybolts. Remove the staybolts in the cylinder head for the right manifold.

To install:

14. Clean the mounting surfaces on the manifolds and the cylinder head. Check the condition of the studs and replace if necessary.

15. Install the staybolts in the cylinder head for the right manifold. Install the nuts onto the staybolts.

16. On the right hand side, install the new exhaust manifold heat shield gaskets and install the manifolds. Torque the nuts to 16–18 ft. lbs. (22–25 Nm). Use new self-locking nuts.

17. Install the heat shields on the right manifold.

18. Install the oil dipstick guide tube.

19. Install the windshield washing fluid tank.

20. Install the right side upper section of the air cleaner assembly along with the air mass sensor.

21. Install the staybolts in the cylinder head for the left manifold. Install the nuts onto the staybolts.

22. On the left hand side, install the new exhaust manifold heat shield gaskets and install the manifolds. Torque the nuts to 16–18 ft. lbs. (22–25 Nm). Use new self-locking nuts.

23. Install the manifold/split pipe bolts on the left hand side.

24. Install the heat shields on the left manifold and on the steering gear.

25. Install the clamp on the left and right split pipes.

26. Install the left side upper section of the air cleaner assembly along with the air mass sensor.

27. Connect the negative battery terminal. Start engine and check for leaks.

Timing Chain Front Cover

REMOVAL AND INSTALLATION

M42/M44 and M50/M52/S50 Engines

1. Disconnect the negative battery cable. Drain the cooling system and remove the radiator and fan assembly.

2. Remove the drive belts and any accessories that block access to the timing cover. Remove the engine splash shield, if necessary.

3. Remove the vibration damper using the proper tool. Unscrew the central bolt and remove the vibration damper hub.

4. Remove the timing case cover bolts and remove the timing cover.

NOTE: The timing case cover can be removed without removing the water pump.

To install:

5. Clean the mounting flanges of the timing chain cover and the engine block. Install the timing chain cover with a new manifold. Tighten the bolts evenly in a crisscross manner.

6. Install the remaining components in the opposite order from which they were removed. Make sure to tighten the central hub bolt to 224 ft. lbs. (304 Nm) on the M42/M44 engines or to 295 ft. lbs. (400 Nm) on the M50/M52/S50 engines. Torque the vibration damper bolts to 17 ft. lbs. (24 Nm).

7. Connect the negative battery cable, start the engine and check for leaks.

M60/M62 Engines

Left Upper Timing Case Cover

1. Disconnect negative battery cable.

2. Remove the left cylinder head cover.

3. Remove the alternator.

4. Remove the cover from the oil filter. Unbolt the return pipe at the oil filter housing. Remove the full flow oil filter housing retaining nuts and remove the oil filter housing.

5. Remove the battery positive wire from the alternator. Remove the protective tube mounting screws and move the wire aside.

6. Remove the nine timing case mounting bolts and remove the timing case.

To install:

7. Check for the correct seating of the dowel sleeves. Clean the sealing surfaces to remove any oil or old gasket. Position a new gasket.

8. Mount the timing case cover together with the inserted bolt. This bolt cannot be installed with the cover in place. Install the rest of the mounting bolts and screw in the vertically mounted bolts until the cover just contacts the cylinder head. Do not tighten the bolts yet.

9. Install the horizontally mounted bolts, then tighten the vertically mounted bolts in two steps. After the vertically mounted bolts are tight, tighten the horizontally mounted bolts in two steps. Torque the 6mm bolts to 7 ft. lbs. (10 Nm), 8mm bolts to 16 ft. lbs. (22 Nm), and 10mm bolts to 35 ft. lbs. (47 Nm).

10. Position the battery positive wire for the alternator and install the protective tube mounting screws. Connect the wire to the alternator.

11. Install the oil filter housing. Install the return pipe and replace the housing cover.

12. Install the alternator and the cylinder head cover.

13. Fill the engine oil and connect the negative battery cable.

Right Upper Timing Case Cover and Tensioner

1. Disconnect negative battery cable.

2. Remove the right cylinder head cover.

3. Remove the air cleaner upper section along with the mass air flow sensor.

4. Unscrew the timing chain tensioner mounting element from the side of the cover. Remove the complete mounting element and hydraulic tensioner.

5. Remove the camshaft sender screw.

6. Remove the upper mounting bolt for the oil dipstick guide tube. Remove the lower mounting nut for the tube and remove the tube.

7. Remove the nine right timing case mounting bolts and remove the cover from the cylinder head.

To install:

8. Replace the hydraulic tensioner oil seal in the timing case cover.

9. Check for the correct seating of the dowel sleeves. Clean the sealing surfaces to remove any oil or old gasket. Position a new gasket.

10. Mount the timing case cover. Screw in the vertically mounted bolts until the cover contacts the cylinder head. Do not tighten the bolts yet.

11. Install the horizontally mounted bolts, then tighten the vertically mounted bolts in two steps. After the vertically mounted bolts are tight, tighten the horizontally mounted bolts in two steps. Torque the 6mm bolts to 7 ft. lbs. (10 Nm), 8mm bolts to 16 ft. lbs. (22 Nm), and 10mm bolts to 35 ft. lbs. (47 Nm).

12. Install the oil dipstick guide tube, making sure to replace the O-ring.

13. Install the camshaft sender screw and the chain tensioner.

14. Install the air cleaner upper section along with the mass air flow sensor.

15. Install the right cylinder head cover.

16. Connect negative battery cable.

Lower Timing Case Cover and Seal

1. Disconnect negative battery cable.

2. Remove the intake manifold cover. Remove the intake hose between the throttle body and the air volume meter.

3. Remove the cooling fan and the drive belt.

4. Remove the pulley on the water pump.

5. Remove the eight mounting bolts for the vibration damper and remove the damper.

6. Raise and safely support the vehicle. Remove the front engine splash shield. Remove the cooling air guide for the alternator, located on the engine carrier. Lower the vehicle.

7. Position special tool 11 2 450, or equivalent, on the center bolt for the vibration damper hub. Remove the

bolt. Install a suitable puller on the hub and remove the hub.

8. Press out the oil seal using special tool 11 2 380 and 11 2 383, or equivalent.

9. Remove the 15 mounting bolts for the lower timing case cover and remove the cover.

To install:

10. Check for the correct seating of the dowel sleeves. Clean the sealing surfaces and remove all pieces of gasket. Position a new gasket on the lower cover.

11. Cut off the protruding ends of the gasket, making sure the cutting tool is level. Do not allow the pieces to fall into the engine.

12. Position the lower cover and install the mounting bolts with an even distribution of pressure. Torque the 6mm bolts to 7 ft. lbs. (10 Nm), 8mm bolts to 16 ft. lbs. (22 Nm) and 10mm bolts to 35 ft. lbs. (47 Nm).

13. Install the oil seal in the timing case cover using special tool 11 1 220 or an appropriate seal install tool and the crankshaft damper hub bolt. Make sure the seal is flush with the cover.

14. Install the vibration damper hub and install the bolt. Torque the hub bolt in three steps:
- Step 1: 74–81 ft. lbs. (100–110 Nm)
- Step 2: turn an additional 60 degrees
- Step 3: turn an additional 60 degrees
- Step 4: turn an additional 30 degrees

15. Raise and safely support the vehicle. Install the cooling air guide for the alternator, located on the engine carrier. Install the front engine splash shield. Lower the vehicle.

16. Position the vibration damper. Install the mounting bolts on the vibration damper.

17. Install the pulley on the water pump.

18. Install the drive belt and the cooling fan.

19. Install the intake hose between the throttle body and the air volume meter. Install the manifold cover.

20. Connect negative battery cable.

M70/M73/S70 Engines

UPPER

1. Disconnect the negative battery cable. Drain the cooling system and remove the fan assembly.

2. Remove both intake manifolds and distributor housings.

3. Disconnect the round rubber mounts, bolts and nuts. Remove both cylinder head covers.

4. Remove the mounting bolts and lift out the timing cover.

To install:

5. Clean the timing cover and engine block mounting surfaces, then position the timing cover on the engine. Install the mounting bolts.

6. Install both cylinder head covers.

7. The remaining components are installed in the reverse order from which they were removed.

8. Connect the negative battery cable, start the engine and check for leaks.

LOWER

1. Disconnect the negative battery cable. Drain the cooling system and remove the fan assembly.

2. Remove the drive belts and the engine splash shield. Remove the tensioning bolt.

3. Unscrew the bolts but do not remove the vibration damper. Remove the central hub bolt with the proper tool.

4. Remove the vibration damper using the proper tool to pull the vibration damper hub from the crankshaft.

5. Drain the engine oil and remove the lower section of the oil pan. Remove the bottom mounting screws from the timing case cover and loosen the adjacent oil pan bolts on both sides.

6. Remove the timing belt tensioner and reference mark sender.

7. Remove the mounting screws and take off the timing case cover.

To install:

8. Clean the timing chain cover and the mounting area on the engine. Install the timing chain cover and mounting bolts.

9. Install the reference mark sender and the timing belt tensioner.

10. Install the remaining components in the reverse order of removal. Make sure to tighten the central hub bolt to 318 ft. lbs. (430 Nm) and the vibration damper mounting bolts to 17 ft. lbs. (25 Nm).

11. Connect the negative battery cable, start the engine and check for leaks.

Front Cover Oil Seal

REPLACEMENT

1. Disconnect the negative battery cable.

2. Remove the vibration damper and hub assembly.

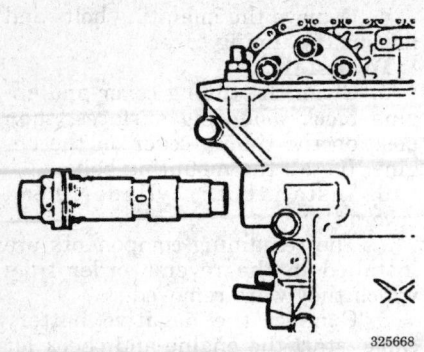

Timing chain tensioner — M60/M62 engines

325668

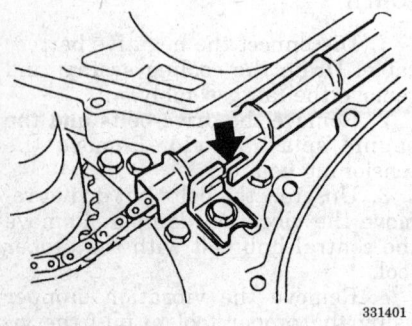

331401

With the timing case cover fitted correctly, the retaining tab is not visible — M70/M73/S70 engines

3. Press out the radial oil seal, using the proper tool.

To install:

4. Install the new seal flush in conjunction with the central bolt and washer, using the proper seal installer.

5. Install the vibration damper and hub assembly.

6. Connect the negative battery cable.

Timing Chain and Sprockets

REMOVAL AND INSTALLATION

M42/M44 and M50/M52/S50 Engines

1. Disconnect the negative battery cable. Remove the vibration damper and hub assembly.

2. For M42/M44 engines, remove the lower timing case cover. Remove the timing chain tensioner.

3. For M50/M52/S50 engines, remove the upper and lower timing case covers. Compress and lock the upper timing chain tensioner. Unbolt the upper timing chain sprockets and

remove from the camshafts along with the chain.

4. Unscrew the upper chain guide and top bolt on the right chain guide.

5. Remove the timing chain sprockets and the lift out the chain. Remove the timing chain guide.

6. Remove the tensioning rail, if necessary. Remove the crankshaft sprocket with the proper tool and lift out the Woodruff key.

7. Remove the reversing roller, if needed.

NOTE: The reversing roller can only be replaced complete with bearings.

To install:

8. Install the Woodruff key into the channel in the crankshaft. Slide the crankshaft sprocket over the end of the crankshaft with the Woodruff key aligning with the channel in the crankshaft. Use the central mounting bolt to draw the sprocket entirely into position.

9. The remaining components are installed in the reverse order from which they were removed. Make sure to tighten the camshaft sprocket bolts to 16 ft. lbs. (22 Nm), the lower timing cover bolts to 6.5–8.0 ft. lbs. (9–11 Nm) for M6 bolts and 16 ft. lbs. (22 Nm) for M8 bolts, the vibration damper hub bolt to 295 ft. lbs. (410 Nm) and the vibration damper bolts to 17 ft. lbs. (23 Nm). Use sealer at the intersections of the timing cover and the pan.

10. Connect the negative battery cable, start the engine and check for leaks.

M60/M62 Engines

1. Disconnect the negative battery cable.

2. Remove the upper and lower timing chain covers. The timing chain tensioner has to come off with the cover.

3. Turn the engine in the direction of rotation and set cylinder number 1 to TDC. The arrows on the sprockets should face up in the cylinder axis. Use a crankshaft holder to keep the TDC position.

4. Loosen and remove the camshaft sprocket bolts from both banks of cylinders. Compress the hydraulic tensioning element to loosen the timing chain. Lock the element with special tool 11–3–420, and remove the sprockets with the chain. Do not rotate the engine with the timing chain removed.

5. Guide the chain out of the tensioner rails and off the lower sprocket.

6. To remove the guide rail:

a. On the left side (cylinder bank 1–4), remove the lower mounting bolt and remove the tensioning rail. Pull off the spacer with oil supply for the tensioning rail.

b. On the left side, remove the two mounting bolts for the guide rail. Do not mix up the two bolts, it is important to install the same bolt in the same hole. Remove the sliding rail.

c. On the right side (cylinder bank 5–8), remove the two mounting bolts on the tensioning rail, and the two bolts on the guide rail. Remove the rails.

To install:

7. On the right cylinder bank, position the guide rail and install the mounting bolts. Position the tensioning rail, and install the mounting bolts.

8. On the left cylinder bank, check the seal for the spacer. Position the guide rail, and install the mounting bolts in the correct holes. Install the spacer. Position the tensioning rail, and install the lower mounting bolt.

9. Inspect the sprockets for wear and replace if necessary.

10. Install the chain in position.

11. Be sure No. 1 piston remains at the top of its firing stroke and the key on the crankshaft is in the 12 o'clock position.

12. Position the chain on the guide rail and swing the chain inward and to the left.

13. Engage the chain on the crankshaft gear and install the camshaft sprockets into the chain.

14. The sprocket on the intake camshaft for cylinder bank 1–4 has a sender pin. With the arrow pointing up, align the pin in the middle of the slots. Install the camshaft sprockets. Remove special tool 11–3–420. Remove the crankshaft holder.

15. Install the chain tensioner piston, spring and cap plug, but do not tighten.

16. To bleed the chain tensioner, fill the oil pocket, located on the upper timing housing cover, with engine oil and move the tensioner back and forth with a suitable prybar until oil is expelled at the cap plug. Tighten the cap plug securely.

17. Install the timing chain covers. Torque the 6mm bolts to 7 ft. lbs. (10 Nm), the 8mm bolts to 16 ft. lbs. (22 Nm), and the 10mm bolts to 35 ft. lbs. (47 Nm).

18. Connect negative battery cable.

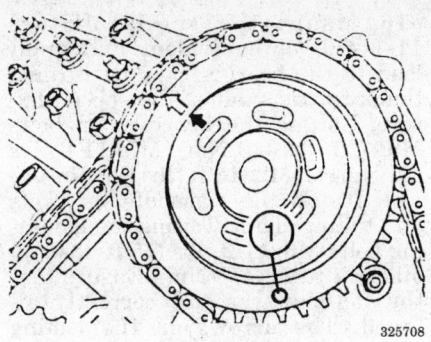

Camshaft sprocket sender pin — M60/M62 engines

M70/M73/S70 Engines

1. Disconnect the negative battery cable. Drain the cooling system and remove the fan assembly.

2. Remove the drive belts and the engine splash shield. Remove the tensioning bolt.

3. Remove the timing covers.

4. Press out the front cover oil seal using special tool 11–1–210 or equivalent.

5. Set the engine to TDC. Install a holder in the crankshaft. The valves of No. 1 and No. 7 cylinders are

closed. The dowel pins in the camshafts should face in.

6. Remove the mounting bolts on the camshaft sprockets and carefully remove the sprockets with the timing chain.

7. Remove mounting bolts for the timing chain guide and remove the guide.

To install:

8. Position and install the timing chain guide.

9. Position the timing chain on the sprockets and install the camshaft sprockets. Verify that the timing chain is correctly aligned on all the sprockets and remove the crankshaft holder.

10. Lubricate the sealing lip of the shaft seal with oil. Install the new seal using a suitable seal installer. The seal should be flush with the cover.

11. Install the timing chain covers.

12. Install the engine splash shield. Install the fan assembly.

13. Connect negative battery cable.

14. Before starting the engine, change the engine oil. Refill and bleed the cooling system.

15. Start the engine and check for proper operation.

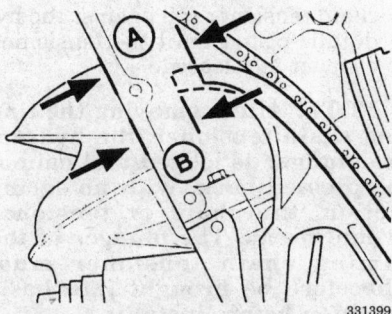

Timing chain tensioner adjustment: dimension "B" from "A" should be 0.216–0.256 in. (5.48–6.5mm) — M70/M73/S70 engines

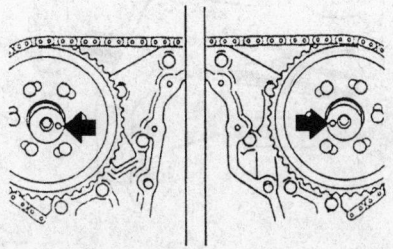

Camshaft dowel pins with engine at TDC — M70/M73/S70 engines

1. Adjusting screw
2. Lock nut
3. Screw plug
4. Replace sealing ring
5. Replace o-ring
6. Dowel sleeve
7. Compression spring
8. Chain tensioning piston

Timing chain tensioner components — M70/M73/S70 engines

Camshaft

REMOVAL AND INSTALLATION

M42/M44 Engines

1. Disconnect the negative battery cable.

2. Unscrew the ignition coil cover and remove the spark plug connectors and spark plugs. Remove mounting nuts and lift out complete ignition assembly.

3. Disconnect and tag all wiring and hoses which may interfere with cylinder head cover removal. Position aside.

4. Unscrew and remove the cylinder head cover.

5. Rotate the crankshaft in normal direction of rotation until the cam lobes of intake and exhaust cams of No. 1 cylinder faces each other. The arrows on the timing chain sprockets face UP.

NOTE: If the camshaft must be rotated with the timing chain and sprockets removed, the crankshaft must first be rotated approximately 90 degrees away from TDC position, in the direction of engine rotation. In this position contact between valves and pistons will be avoided.

6. To avoid lost of ignition timing, mount the crankshaft holding tool 11–2–300 or equivalent into position.

7. Matchmark the timing chain and sprockets relative to each other.

8. Remove the timing chain tensioner.

9. Unbolt and remove the upper chain guide.

10. Unbolt and remove the timing chain and sprockets. Suspend the timing chain and sprockets after removing.

11. Mount the special fixture 11–3–260 or equivalent in spark plug holes and tighten to 17 ft. lbs (23 Nm).

──── **WARNING** ────
The camshafts can be damage or broken when removing/installing without the fixture.

12. Apply load to the bearing caps by turning the eccentric shaft of the special fixture. Loosen and remove the bearing cap bolts.

13. Unscrew and remove the special fixture. Lift out the bearing caps and camshafts. Camshafts are marked with "A" for exhaust and "E" for intake. Bearing caps are marked with "A1...A5" for the exhaust side and "E1...E5" for the intake side.

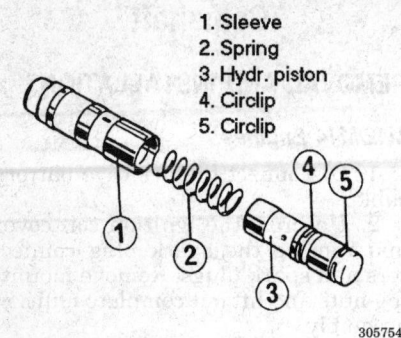

1. Sleeve
2. Spring
3. Hydr. piston
4. Circlip
5. Circlip

Hydraulic lash adjuster components — M42/M44 engines

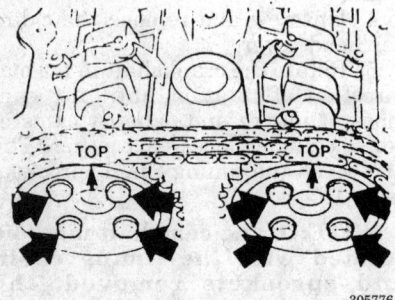

The camshaft sprockets should be positioned as shown when the engine is at TDC on the No. 1 cylinder — M42/M44 engines

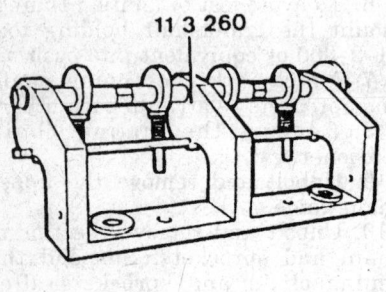

11 3 260

Camshaft removal tool 11-3-260 — M42/M44 engines

To install:

14. Fit the camshafts and bearing caps into position. Note the correct location of the intake and exhaust camshafts. Also note the position of each bearing caps. Do not tighten the bearing caps at this time.

15. Align the camshafts so that lobes of the intake and exhaust cams face each other in no. 1 cylinder (see note above). The camshafts can be turned on the hexagon with a 27mm wrench. The camshafts can be held in

Bearing cap bolt locations — M42/M44 engines

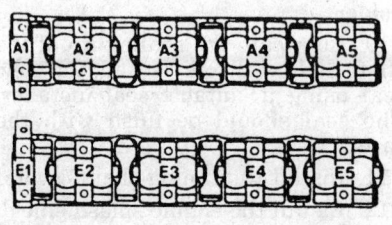

Bearing caps are marked with A1...A5 for exhaust side and E1...E5 for intake side — M42/M44 engines

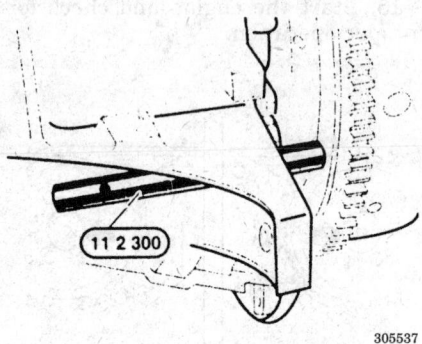

11 2 300

Hold crankshaft in TDC position with tool 11-2-300 or equivalent — M42/M44 engines

position using special holding tool 11-3-240 or equivalent.

NOTE: The valve clearance compensating elements expand whenever no load is applied by the camshaft (camshaft removed) and require a certain amount of time after installation before they compress again. Always wait approximate 10 minutes (at room temperature 68° (20°C) after installation of the camshaft, before the engine is cranked. Allow a longer waiting time at lower temperature.

16. Mount the special fixture 11-3-260 or equivalent and apply load to the bearing caps by turning the eccentric shaft of the special fixture. Torque the bearing cap bolts M6: 7 ft. lbs (10 Nm), M7: 11 ft. lbs (15 Nm) or M8: 15 ft. lbs (20 Nm).

17. Install the camshaft sprockets and timing chain. Torque the mounting bolts M6: 11 ft. lbs (15 ft. lbs), all others 7 ft. lbs (10 Nm). Ensure that the matchmarks are correctly aligned. The arrows on the timing chain sprockets face UP.

18. Before installing the timing chain tensioner, adjust the basic position:

a. Knock the outside sleeve of the chain tensioner on a hard surface. This will cause the piston to jump of the lock.

b. Re-assemble the parts and clamp in a vice fitted with soft jaws.

c. Push the chain tensioner together and fit the circlip in the slip bevel of the sleeve.

d. Push the chain tensioner together even further until the circlip is heard engage.

e. Remove the chain tensioner form the vice and install the chain tensioner to the engine and torque to 29 ft. lbs (40 Nm). Using a suitable tool, reach down and push the chain tensioner rail against the hydraulic piston until the tensioning element is released.

NOTE: After removing the timing chain tensioner, the hydraulic plunger is locked and cannot be pressed back. With no spring action, the chain or tensioner would break. The plunger of the timing chain tensioner must therefore be brought into basic position before installing.

19. Install the upper timing chain guide.

20. Check the cylinder head cover gasket and replace if necessary. Install the cylinder head cover and check for correct seating of the gasket at the rear of the cylinder head. Torque the cover bolts to 7-11 ft. lbs (10-15 Nm).

21. Connect all wiring and hoses. Install complete ignition assembly, spark plugs, spark plug connectors and ignition coil cover.

WARNING
Remove the crankshaft holding tool before operating the engine.

22. Connect the negative battery cable.

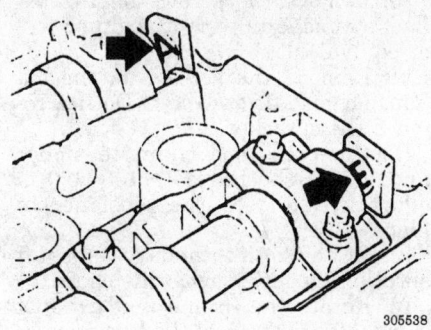

Camshafts are marked with "A" for exhaust and "E" for intake — M42/M44 engines

305538

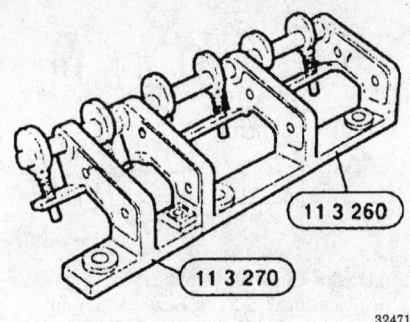

Camshaft removal tools 11 3 260 and 11 3 270 — M50/M52/S50 engines

324716

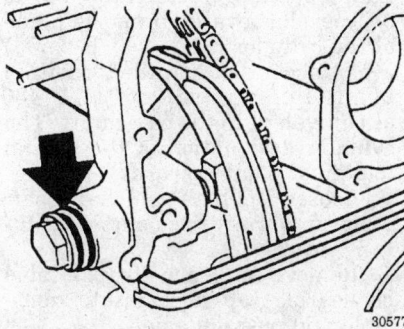

Chain tensioner location — M42/M44 engines

305774

Bearing cap bolt locations — M50/M52/S50 engines

324717

Upper chain guide location — M42/M44 engines

305775

Bearing cap markings — M50/M52/S50 engines

324718

23. Start the engine and check for proper operation.

NOTE: A new or disassembled timing chain tensioner is without oil. To assure correct functioning, the engine must be operated the first time at 3500 rpm for approximately 20 seconds.

M50/M52/S50 Engines

1. Disconnect the negative battery cable.

2. Remove the cylinder head.

NOTE: Special tools are required to perform this operation. BMW tools 11–3–260/270/250 or equivalent are required for proper removal and installation of the camshafts and for retention of the valve lash compensators. Without these tools the camshafts will be damaged during removal or installation.

3. Remove the spark plugs and attach the 11–3–260 (plus addition 11–3–270) camshaft removal fixture. Torque the hold down bolts in the spark plug bores to 17 ft. lbs. (23 Nm).

4. Apply load to the bearing caps by rotating the eccentric shaft. This relieves the tension on the bearing cap bolts. Loosen and remove the bearing cap bolts.

5. Remove the camshaft removal fixture after releasing the tension from the eccentric shaft.

6. Remove the camshafts and the bearing caps. Note that the intake camshaft is marked "E" and the exhaust camshaft is marked "A". The camshaft bearing are consecutively numbered with "A" or "E" to designate intake or exhaust side.

7. Hold the valve lash compensators in place using tool 11–3–250 or equivalent, and remove the bearing plate along with the valve plungers.

To install:

8. Inspect the camshafts and valve lash compensators for damage and wear and replace as necessary.

9. Install the camshafts with the cylinder number 1 intake and exhaust cam peaks pointing at each other. The flats on the sprocket ends of the camshafts should be parallel. The exhaust camshaft is marked with a notch on the flange.

10. Install the fixture. Place the bearing caps into position and press the caps down with the tool. Torque the bolts to 10–12 ft. lbs. (13–17 Nm).

11. When the camshafts have been removed and reinstalled a waiting period dependent on the ambient temperature is necessary before mounting the cylinder head on the engine. At room temperature wait 4 minutes to allow the lifters to compress fully. At temperatures down to 50° F (10° C) wait 11 minutes. At temperatures lower than 50° F (10° C) wait 30 minutes. This is to prevent contact between the valves and the piston tops.

12. The engine may not be cranked under the same conditions as above for a period of 10 minutes at room temperature; 30 minutes for temperatures down to 50° F (10° C); 75 minutes for temperatures below 50° F (10° C).

M60/M62 Engines

Left Camshaft (Cylinder Bank 5–8)

1. Disconnect negative battery cable.

2. Remove the left and right cylinder head covers.

3. Remove all spark plugs.

4. Remove the top left timing case cover.

5. Remove the splash guard.

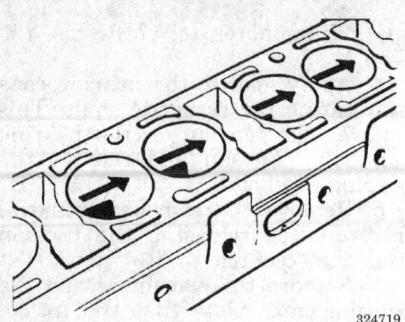

324719

Check bearing surfaces of valve clearance compensators for scoring — M50/M52/S50 engines

324720

Bearing plate markings — M50/M52/S50 engines

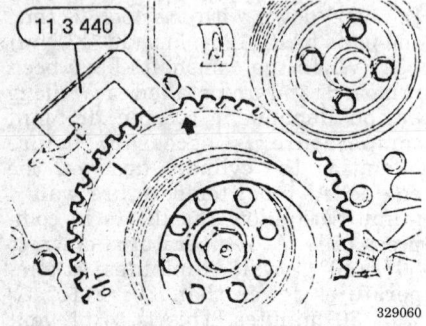

329060

Gap in increment gear must fit in special tool 11–3–440 — M60/M62 engines

6. Remove all oil lines to the left and right cylinder head.

7. Rotate the crankshaft in direction of rotation until the first cylinder is in TDC position.

8. Brace the camshaft on the hex head with a suitable open-end wrench and loosen the 3 accessible screws on each right sprocket approximately ½ a turn.

  a. Turn the engine over once and loosen the remaining 3 screws on each right sprocket approximately ½ a turn.

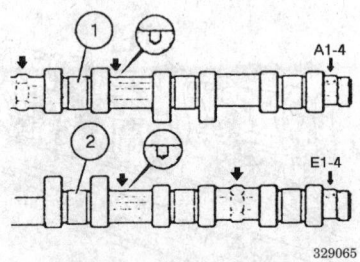

329064

Left-side camshaft identification (cylinder bank 5 to 8): Hex-head (3) on intake camshaft between cylinder 7 and 8. Hex-head (4) on exhaust camshaft between cylinder 5 and 6 — M60/M62 engines

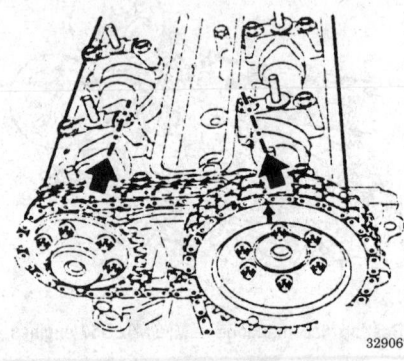

329065

Right-side camshaft identification (cylinder bank 1 to 4): Hex-head (2) on intake camshaft between cylinder 3 and 4. Hex-head (1) on exhaust camshaft between cylinder 1 and 2 — M60/M62 engines

329067

Camshaft positioning — M60/M62 engines

  b. Unscrew and remove the primary sprocket from the left-hand intake camshaft (cylinder bank 5–8). Secure the chain to prevent it from dropping.

9. Rotate the engine to 45 degrees BTDC setting position. Rotate the crankshaft against direction of rotation until the gap in the increment gear fits in the special tool 11–3–440 or equivalent.

10. Remove the screws on the exhaust camshafts with a spanner tool or equivalent.

11. Remove the screws on the exhaust camshaft sprocket. Do not remove the sprocket.

12. Compress the chain tensioner and install special tool 11–3–420 or equivalent to lock the tensioner in place.

13. Lift off both secondary camshaft sprockets together with the chain.

14. Rotate the intake and exhaust camshafts to the installed position:

  a. Using special tool 11–3–430 or equivalent, rotate the camshafts until the recess in both camshaft flange points approximately 30–40 degrees downwards from the plane of the cylinder head.

  b. Check the installed position by installing special tool 11–2–430 or equivalent to the camshafts. The cylinder designation of the special tool must point upwards.

15. Loosen the both camshaft bearing caps uniformly from outside to inside ½ turn.

16. Remove all bearing caps. Label each bearing cap to facilitate re-assemble and position aside.

17. Remove the camshafts noting their locations.

18. To remove the hydraulic valve lifters use special tool 11–3–250 to pull them out of the cylinder head. Make sure that no damage occurs to the guides in the head. Inspect the bearing surfaces of the bucket tappets (lifters) for wear and scoring.

To install:

19. If the lifters were removed, install them with special tool 11–32–250 or equivalent.

20. Lubricate and install the camshafts in their correct position.

NOTE: The intake camshaft will have a hexagon between cylinders 7 and 8. The exhaust camshaft will have the hexagon between cylinders 5 and 6.

21. Rotate the intake and exhaust camshafts to the installed position:

  a. Using special tool 11–3–430 or equivalent, rotate the camshafts until the recess in both camshaft flange points approximately 30–40 degrees downwards from the plane of the cylinder head.

  b. Check the installed position by installing special tool 11–2–430 or equivalent to the camshafts. The cylinder designation of the special tool must point upwards.

22. Install the bearing caps. Tighten the bearing caps from outside to inside in 1/2 turn incre-

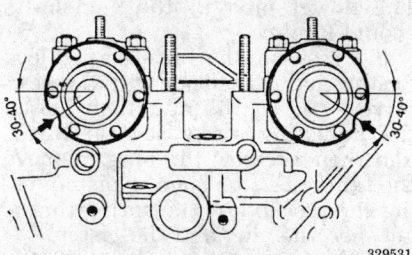

Install left-hand camshafts: Recesses in camshaft point downwards approximately 30–40 degrees from plane of cylinder head — M60/M62 engines

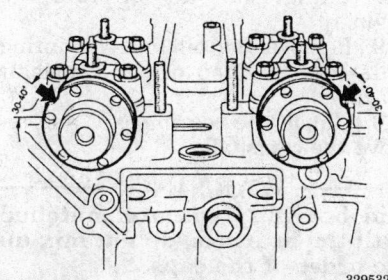

Install right-hand camshafts: Recesses in camshaft point upwards approximately 30–40 degrees from plane of cylinder head — M60/M62 engines

ments. Torque the bolts to 9–13 ft. lbs. (12–17 Nm).

NOTE: Do not confuse camshaft bearing caps of cylinders No. 1–4 and 5–8. The exhaust camshaft bearing caps are marked with A1–A5 from intake side. The intake camshaft bearing caps are marked with E1–E5 from intake end.

23. Fit the special tool 11–3–430 or equivalent to the camshaft. Rotate the camshaft until the marker bores face upwards.

   a. Install special tools 11–2–442/446 or equivalent to the camshaft on cylinder bank 5–8.

   b. Install special tools 11–2–441/445 or equivalent to the camshaft on cylinder bank 1–4.

   c. Using a suitable open-end wrench, align all camshafts in such a way that the special tools fit on the cylinder heads without any gaps.

   d. Fit special tools 11–2–443 or equivalent to special tools 11–2–441/442/445/446 and secure them with special tools 11–2–444 using spark plug threads.

24. Install the secondary sprockets together with chain to the camshafts on cylinder bank 5–8.

25. Install the screws on the exhaust camshaft sprocket and tighten snug.

26. Remove the special tool used to lock the chain tensioner in position.

27. Rotate the engine from 45 degrees BTDC in direction of rotation as far as TDC setting. Install special tool 11–2–300 at the flywheel to lock the crankshaft in TDC position.

28. Assemble the primary sprocket and chain to the intake camshaft with the arrow pointing upwards (in cylinder axis) and the long bores centrally aligned. Install the screws snug.

29. Install the special tool 11–3–390 or equivalent in the right timing case cover and with a suitable torque wrench, tension the tool to 1.3 Nm.

30. Tighten the sprockets to 11 ft. lbs (15 Nm) in the following order:

• All screws on the left exhaust camshaft

• 3 screws in the right exhaust camshaft

• All screws on the left intake camshaft

• 3 screws in the right intake camshaft

31. Remove special tools 11–2–444/443/441/445.

32. Remove special tools 11–2–444/443/442/446.

33. Remove special tool 11–2–300 used to locked the crankshaft in TDC position.

34. Turn the engine over once.

35. Tighten the remaining 3 screws on right exhaust camshaft and remaining 3 screws on right intake camshaft to 11 ft. lbs. (15 Nm).

36. Relieve the load and remove the special tool 11–3–390 or equivalent from the right timing case cover.

37. The balance of installation is the reverse of the removal procedure.

38. Start the engine. Check for leaks and proper operation.

Right Camshaft (Cylinder Bank 1–4)

1. Disconnect negative battery cable.

2. Remove the left and right cylinder head covers.

3. Remove all spark plugs.

4. Remove the fan assembly.

5. Remove the top right timing case cover.

6. Remove the splash guard.

7. Remove all oil lines to the left and right cylinder head.

8. Rotate the crankshaft in direction of rotation until the first cylinder is in TDC position.

9. Brace the camshaft on the hex head with a suitable open-end wrench and loosen the 3 accessible screws on each left sprocket approximately ½ a turn.

   a. Turn the engine over once and loosen the remaining 3 screws on each left sprocket approximately ½ a turn.

   b. Unscrew and remove the primary sprocket from the right-hand intake camshaft (cylinder bank 1–4). Secure the chain to prevent it from dropping.

10. Rotate the engine to 45 degrees BTDC setting position. Rotate the crankshaft against direction of rotation until the gap in the increment gear fits in the special tool 11–3–440 or equivalent.

11. Remove the screws on the exhaust camshaft sprocket. Do not remove the sprocket.

12. Compress the chain tensioner and install special tool 11–3–420 or equivalent to lock the tensioner in place.

13. Lift off both secondary camshaft sprockets together with the chain.

14. Rotate the intake and exhaust camshafts to the installed position:

   a. Using special tool 11–3–430 or equivalent, rotate the camshafts until the recess in both camshaft flange points approximately 30–40 degrees upwards from the plane of the cylinder head.

   b. Check the installed position by installing special tool 11–2–430 or equivalent to the camshafts. The cylinder designation of the special tool must point upwards.

15. Loosen the both camshaft bearing caps uniformly from outside to inside ½ turn.

16. Remove all bearing caps. Label each bearing cap to facilitate re-assemble and position aside.

17. Remove the camshafts noting their locations.

18. To remove the hydraulic valve lifters use special tool 11–3–250 to pull them out of the cylinder head. Make sure that no damage occurs to the guides in the head. Inspect the bearing surfaces of the bucket tappets (lifters) for wear and scoring.

To install:

19. If the lifters were removed, lubricate and install them with special tool 11–32–250 or equivalent.

20. Lubricate and install the camshafts in their correct position.

NOTE: The intake camshaft will have a hexagon between cylinders 3 and 4. The exhaust camshaft will have the hexagon between cylinders 1 and 2.

21. Rotate the intake and exhaust camshafts to the installed position:

   a. Using special tool 11–3–430 or equivalent, rotate the camshafts until the recess in both camshaft flange points approximately 30–40 degrees upwards from the plane of the cylinder head.

   b. Check the installed position by installing special tool 11–2–430 or equivalent to the camshafts. The cylinder designation of the special tool must point upwards.

22. Install the bearing caps. Tighten the bearing caps from outside to inside in 1/2 turn increments. Torque the bolts to 9–13 ft. lbs. (12–17 Nm).

NOTE: Do not confuse camshaft bearing caps of cylinders No. 1–4 and 5–8. The exhaust camshaft bearing caps are marked with A1–A5 from intake side. The intake camshaft bearing caps are marked with E1–E5 from intake end.

23. Fit the special tool 11–3–430 or equivalent to the camshaft. Rotate the camshaft until the marker bores face upwards.

   a. Install special tools 11–2–442/446 or equivalent to the camshaft on cylinder bank 5–8.

   b. Install special tools 11–2–441/445 or equivalent to the camshaft on cylinder bank 1–4.

   c. Using a suitable open-end wrench, align all camshafts in such a way that the special tools fit on the cylinder heads without any gaps.

   d. Fit special tools 11–2–443 or equivalent to special tools 11–2–441/442/445/446 and secure them with special tools 11–2–444 using spark plug threads.

24. Install the secondary sprockets together with chain to the camshafts on cylinder bank 1–4.

25. Install the screws on the exhaust camshaft sprocket and tighten snug.

26. Remove the special tool used to lock the chain tensioner in position.

27. Rotate the engine from 45 degrees BTDC in direction of rotation as far as TDC setting. Install special tool 11–2–300 at the flywheel to lock the crankshaft in TDC position.

28. Assemble the primary sprocket/chain with sensor pin to the intake camshaft with the arrow pointing upwards (in cylinder axis) and the long bores centrally aligned. Install the screws snug.

29. Install the special tool 11–2–400 or equivalent to the right cylinder head (cylinder bank 1–4). Install the special tool 11–3–390 to special tool 11–2–400. Using a suitable torque wrench, tension the tool to 1.3 Nm.

30. Tighten the sprockets to 11 ft. lbs (15 Nm) in the following order:

   • 3 screws on the left exhaust camshaft
   • All screws on the right exhaust camshaft
   • 3 screws on the left intake camshaft
   • All screws in the right intake camshaft

31. Remove special tools 11–2–444/443/441/445.

32. Remove special tools 11–2–444/443/442/446.

33. Remove special tool 11–2–300 used to locked the crankshaft in TDC position.

34. Turn the engine over once.

35. Tighten the remaining 3 screws on left exhaust camshaft and remaining 3 screws on left intake camshaft to 11 ft. lbs (15 Nm).

36. Relieve the load and remove the special tool 11–3–390 and 11–2–400.

37. Install the remaining components in the reverse order of removal.

38. Start the engine. Check for leaks and proper operation.

M70/M73/S70 Engines

1. Disconnect the negative battery cable. Drain the cooling system and remove the fan assembly.

2. Remove both intake manifolds and distributor housings.

3. Disconnect the round rubber mounts, bolts and nuts. Remove both cylinder head covers.

4. Remove the mounting bolts and lift out the upper timing cover.

5. Set the engine to TDC. Install a holder in the crankshaft. The valves of cylinders one and seven are closed.

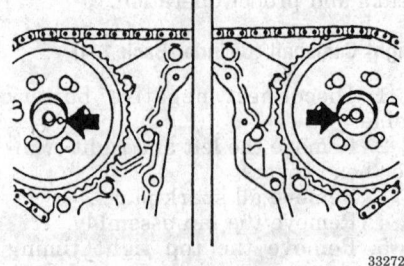

At TDC, the dowel pins in the camshaft sprockets should face each other — M70/M73/S70 engines

The dowel pins in the camshafts should face in.

6. Press off the anti-tamper lock for the chain tensioner with a screwdriver. Loosen the nut, then loosen the adjusting screw several turns, and then unscrew the plug. Remove the timing belt tensioning piston, using care not to lose the spring that is between the plug and the piston.

7. Remove mounting bolts for the timing chain guide, and remove the guide. Remove the tensioning rail.

8. Remove the mounting bolts on the camshaft sprockets, and carefully remove the sprockets. Do not allow the timing chain to fall into the engine.

9. Remove the oil pipe mounting bolts from the top of the camshaft bearings.

10. Unbolt the bearing caps and remove the camshaft.

WARNING
The bearing caps are matched with the bearings, do not mix up the order of the caps.

To install:

11. With the crankshaft positioned at TDC, install the camshaft with the dowel pin facing the center of the engine. Position the bearing caps and install the mounting bolts from inside to outside. Torque the bolts to 11 ft. lbs. (15 Nm).

12. Hold both camshafts in position with special tool 11 3 190.

13. Mount the oil pipes with the oil outlet bores facing the camshaft. Install the hollow union bolt in the bearing cover. Install the mounting bolts and torque to 9 ft. lbs. (12 Nm).

14. Install the camshaft sprockets mounting bolts finger tight. Position the timing chain on the sprockets in the opposite direction of engine rotation, beginning at the crankshaft. Verify that the timing chain is correctly aligned on all the sprockets, and remove the crankshaft holder.

15. Position and install the timing chain guide, and tensioning rail.

16. Install the timing chain tensioner.

17. Tighten the camshaft sprocket bolts to 7 ft. lbs. (10 Nm).

18. Install the upper timing cover and a new gasket.

19. Install the cylinder head covers.

20. Install both intake manifolds and distributor housings.

21. Install the fan assembly. Fill and bleed the cooling system.

22. Connect negative battery cable.

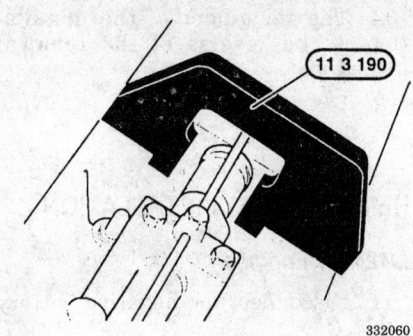

Camshaft alignment gauge — M70/M73/S70 engines

Piston and Connecting Rod

POSITIONING

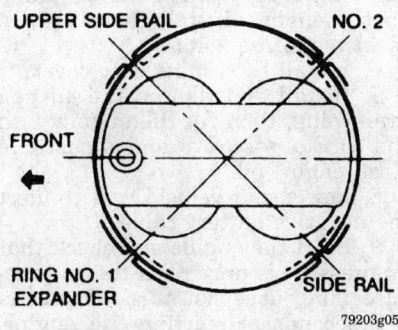

Location of piston in the the cylinder bore with ring gaps located 180 degrees apart

ENGINE LUBRICATION

Oil Pan

REMOVAL AND INSTALLATION

M42/M44 Engines

1. Disconnect the negative battery cable. Raise and safely support the vehicle.
2. Drain the engine oil.
3. Disconnect the exhaust pipe, if necessary.
4. Remove the lower oil pan mounting bolts and take off the lower oil pan. Remove the upper section oil pan bolts and remove the upper oil pan.

To install:

5. Clean the mounting surfaces and install new gaskets.

6. Position the oil pan against the engine and install the mounting bolts. Tighten the mounting bolts to 6 ft. lbs. (9 Nm).
7. Install the exhaust pipe, if removed.
8. Install and tighten the oil pan drain plug, then fill the engine with the correct viscosity and amount of clean engine oil.
9. Lower the vehicle and connect the negative battery cable.
10. Start the engine and check that oil pressure is present; if the oil pressure lamp does not turn off within 5–7 seconds of starting the engine, turn the engine **OFF**. Check for any oil leaks.

M50/M52/S50 Engines

3 Series

1. Disconnect the negative battery cable. Raise the vehicle and support it. Drain the engine oil.
2. Remove the front lower splash guard, if necessary.
3. Disconnect the electrical terminal from the oil sending unit.
4. Remove the power steering gear from the front axle carrier, if necessary.
5. Remove the flywheel cover.
6. Remove the oil pan bolts and lower the oil pan. Remove the oil pump bolts and take out the oil pump and oil pan.

To install:

7. Before installing the oil pan, clean the gasket surfaces and install a new gasket on the oil pan.
8. Coat the joints on the ends of the front engine cover with a universal sealing compound.
9. Install the flywheel cover.
10. If the power steering gear was removed, make sure to refill and bleed this system.
11. Connect the electrical wiring harness to oil sending unit.
12. Install the front lower splash guard, if removed.
13. Install and tighten the oil pan drain plug, then fill the engine with the correct viscosity and amount of clean engine oil.
14. Lower the vehicle and connect the negative battery cable.
15. Start the engine and check that oil pressure is present; if the oil pressure lamp does not turn off within 5–7 seconds of starting the engine, turn the engine **OFF**. Check for any oil leaks.

5 Series

1. Disconnect the negative battery terminal and raise and safely support the vehicle. Drain the engine oil.

2. Loosen the holding bolt for the oil dipstick guide pipe and remove the clamp. Pull the guide tube free of the pan.
3. Remove all the oil pan bolts and remove the pan. Raise the engine slightly if needed for clearance.

To install:

4. Apply sealer to the joint between the pan, front cover and block.
5. Install new gaskets and install the pan. Torque mounting bolts to 6.5–8.0 ft. lbs. (9–11 Nm).
6. Install the dipstick guide tube using a new base seal and tighten the holding bolt.

M60/M62 Engines

1. Disconnect the negative battery cable.
2. Remove the intake manifold cover. Remove the top clips on the radiator.
3. Remove the cooling fan.
4. Remove the guide tube for the oil dipstick.
5. Remove the engine splash guards.
6. Remove the cover for the oil filter so the oil will run back to the pan. Drain the engine oil. Disconnect the plug for the level switch.
7. Remove the lower oil pan bolts and remove the lower oil pan. Remove the gasket and clean the mounting surfaces.
8. Disconnect the left and right engine mounts at the bottom.
9. Unbolt the power steering pump at the holder. On automatic transmissions, remove the oil pipes at the power steering pump.
10. Remove the banjo bolt for the oil return pipe from the oil filter at the oil pan.
11. Remove the mounting bolt on the sprocket for the oil pump and remove the sprocket along with the chain.
12. Remove the three oil pump mounting bolts and remove the oil pump. Remove the oil pipes out of the crankcase.
13. Lift the engine by the front eye hook. Observe the distance between the engine and the firewall while lifting the engine.
14. Unscrew the upper oil pan bolts and remove the upper oil pan.

To install

15. Clean the mounting surfaces and install a new gasket.
16. Install the upper oil pan and torque the bolts to 7–8 ft. lbs. (9–11 Nm).
17. Lower the engine.

18. Check the seals on the oil pipes and replace it if necessary. Lubricate the seals with oil and the oil pipes.

19. Check the seal in the oil pump and replace it if necessary. Screw the hexagon adapter back into the oil pump until it stops.

20. Position the oil pump and install the two right side oil pump mounting bolts. Torque the bolts to 14–17 ft. lbs. (20–24 Nm). Position the chain on the pump and the sprocket and install the sprocket. Torque the bolt to 35 ft. lbs. (47 Nm). Verify that the chain is positioned correctly.

21. Adjust the chain sag to 0.315–0.472 in. (8–12mm) by turning the hexagon adapter in the oil pump. Install the left side mounting bolt.

22. Install the banjo bolt for the oil return pipe from the oil filter at the oil pan.

23. Install the power steering pump and connect the oil lines (if equipped).

24. Making sure to connect the ground strap. Connect the left and right engine mounts at the bottom and tighten to 32 ft. lbs. (43 Nm).

NOTE: If replacing the lower oil pan, remove the level switch from the old pan and install it in the new pan with a new O-ring.

25. Position the lower oil pan with a new gasket. Install the bolts and torque to 7–8 ft. lbs. (9–11 Nm), beginning in the middle and working to the outside.

26. Connect the plug for the level switch, making sure to replace the O-ring.

27. Install the engine splash guards.

28. Install the oil dipstick guide tube, making sure to replace the O-ring. Install the cooling fan.

29. Install the intake manifold cover. Install the top clips on the radiator.

30. Fill the engine oil. Connect negative battery cable.

M70/M73/S70 Engines

7 Series

1. Disconnect the negative battery cable. Raise and safely support the vehicle.

2. Remove the transmission and the oil pump assembly. Lower the vehicle.

3. Disconnect and remove the windshield washer tank and the coolant expansion tank.

4. Remove the guide tube for the oil dipstick. Disconnect the oil pipe on

the tandem pump. Remove the mounting bracket.

5. Unscrew the belt tensioner and remove the oil drain hose.

6. Crank the engine to TDC and unscrew the flywheel using the proper tool.

7. Disconnect the left and right engine mounts at the bottom. Pull off the pipe adapter for oil extraction.

8. Remove the oil pump consoles. Unscrew the oil pan bolts and remove the oil pan.

To install

9. Clean the mounting surfaces and install a new gasket.

10. Install the oil pan and tighten the mounting bolts to 7 ft. lbs. (11 Nm).

11. Connect the left and right engine mounts at the bottom and tighten to 32.5 ft. lbs. (43 Nm).

12. Replace the oil consoles and tighten to 25 ft. lbs. (34 Nm).

13. Install the flywheel and tighten the bolts to 72 ft. lbs. (97 Nm).

14. The remainder of the installation is the reverse of the removal procedure.

8 Series

1. Disconnect the negative battery cable. Raise and safely support the vehicle.

2. Remove the transmission and the oil pump assembly. Lower the vehicle.

3. Disconnect and remove the windshield washer tank and the coolant expansion tank.

4. Remove the guide tube for the oil dipstick. Disconnect the oil pipe on the tandem pump. Remove the mounting bracket.

5. Unscrew the belt tensioner and remove the oil drain hose.

6. Crank the engine to TDC and unscrew the flywheel using the proper tool.

7. Disconnect the left and right engine mounts at the bottom. Pull off the pipe adapter for oil extraction.

8. Remove the oil pump consoles. Unscrew the oil pan bolts and remove the oil pan.

To install

9. Clean the mounting surfaces and install a new gasket.

10. Install the oil pan and tighten the mounting bolts to 7 ft. lbs. (11 Nm).

11. Connect the left and right engine mounts at the bottom and tighten to 32.5 ft. lbs. (43 Nm).

12. Replace the oil consoles and tighten to 25 ft. lbs. (34 Nm).

13. Install the flywheel and tighten the bolts to 72 ft. lbs. (97 Nm).

14. The remainder of the installation is the reverse of the removal procedure.

Oil Pump

REMOVAL AND INSTALLATION

M42/M44 Engines

1. Disconnect the negative battery cable.

2. Raise and safely support the vehicle. Drain the engine oil.

3. Remove the timing case cover.

4. Disconnect the oil pump cover mounting bolts and remove the oil pump assembly.

To install:

5. Clean the oil pump mounting surfaces, then position the oil pump on the engine. Install the oil pump cover mounting bolts.

6. Install the timing case cover.

7. Install and tighten the oil pan drain plug, then fill the engine with the correct viscosity and amount of clean engine oil.

8. Lower the vehicle and connect the negative battery cable.

9. Start the engine and check that oil pressure is present; if the oil pressure lamp does not turn off within 5–7 seconds of starting the engine, turn the engine **OFF**. Check for any oil leaks.

M50/M52/S50 Engines

1. Raise and safely support vehicle. Disconnect the negative battery cable. Drain the oil from the engine. Remove the oil pan to access the oil pump drive sprocket.

2. Remove the oil pump drive sprocket nut. Note that it is a left-hand thread. Remove the oil pump drive sprocket from the oil pump shaft. Check the shaft splines.

3. Unbolt the oil pump body from the block and remove. Check the condition of the dowel sleeves.

To install:

4. Clean the oil pump mounting surfaces, then position the oil pump on the engine. Install the oil pump body mounting bolts to 16 ft. lbs. (22 Nm).

5. Install the oil pump drive sprocket onto the oil pump shaft. Install the oil pump drive sprocket nut to 18 ft. lbs. (25 Nm). The sprocket nut must be tightened in a counterclockwise direction; it has a reverse, or left-hand thread.

6. Install the oil pan.

7. Install and tighten the oil pan drain plug, then fill the engine with

the correct viscosity and amount of clean engine oil.

8. Lower the vehicle and connect the negative battery cable.

9. Start the engine and check that oil pressure is present; if the oil pressure lamp does not turn off within 5–7 seconds of starting the engine, turn the engine **OFF**. Check for any oil leaks.

M60/M62 Engines

1. Disconnect the negative battery cable.

2. Remove the lower oil pan.

3. Disconnect the left and right engine mounts at the bottom.

4. Unbolt the power steering pump at the holder. On automatic transmissions, remove the oil pipes at the power steering pump.

5. Remove the banjo bolt for the oil return pipe from the oil filter at the oil pan.

6. Remove the mounting bolt on the sprocket for the oil pump and remove the sprocket along with the chain.

7. Remove the three oil pump mounting bolts and remove the oil pump. Remove the oil pipes out of the crankcase.

To install

8. Check the seals on the oil pipes and replace it if necessary. Lubricate the seals with oil and the oil pipes.

9. Check the seal in the oil pump and replace it if necessary. Screw the hexagon adapter back into the oil pump until it stops.

10. Position the oil pump and install the two right side oil pump mounting bolts. Torque the bolts to 14–17 ft. lbs. (20–24 Nm). Position the chain on the pump and the sprocket and install the sprocket. Torque the bolt to 35 ft. lbs. (47 Nm). Verify that the chain is positioned correctly.

11. Adjust the chain sag to 0.315–0.472 in. (8–12mm) by turning the hexagon adapter in the oil pump. Install the left side mounting bolt.

12. Install the banjo bolt for the oil return pipe from the oil filter at the oil pan.

13. Install the power steering pump and connect the oil lines (if equipped).

14. Install the lower oil pan.

15. Fill the engine oil. Connect negative battery cable.

M70/M73/S70 Engines

1. Disconnect the negative battery cable and remove the oil pan.

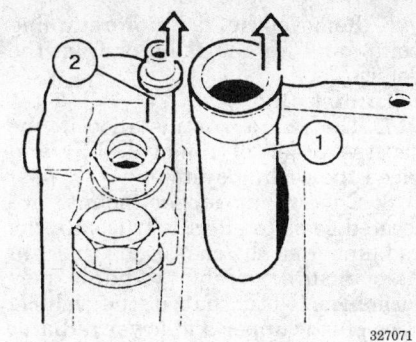

327071

Fresh oil pipe (1), and pure oil pipe (2) — M60/M62 engines

2. Remove the bolts retaining the sprocket to the oil pump shaft and remove the sprocket.

3. Remove the oil pump retaining bolts and lower the oil pump from the engine block. There are 3 bolts at the front and 2 bolts attaching the rear of the oil pickup to the lower end of a support bracket. It is necessary to remove all 5 bolts.

4. Do not loosen the chain adjusting shims from the 2 mounting locations.

To install:

5. Add or subtract shims between the oil pump body and the engine block to obtain a slight movement of the chain under light thumb pressure.

6. Install the oil pump in position.

NOTE: When used, the 2 shim thicknesses must be the same. Tighten the pump holder at the pickup end after shimming is completed to avoid stress on the pump.

7. After the main pump mounting bolts are torqued, loosen the bolts at the bracket on the rear of the pickup, allowing the pickup to assume its most natural position. This will relieve tension on the bracket.

8. The balance of installation is the reverse of the removal procedure.

9. Install the oil pan.

10. Install and tighten the oil pan drain plug, then fill the engine with the correct viscosity and amount of clean engine oil.

11. Lower the vehicle and connect the negative battery cable.

12. Start the engine and check that oil pressure is present; if the oil pressure lamp does not turn off within 5–7 seconds of starting the engine, turn the engine **OFF**. Check for any oil leaks.

Rear Main Bearing Oil Seal

REMOVAL AND INSTALLATION

The rear main bearing oil seal can be replaced after the transmission and clutch/flywheel or the converter/flywheel has been removed from the engine.

1. Raise and safely support the vehicle. Drain the engine oil and loosen the oil pan bolts. Carefully use a sharp bladed tool to separate the oil pan gasket from the lower surface of the end cover housing.

2. Remove the 2 rear oil pan bolts.

3. Remove the bolts around the outside of the cover housing and remove the end cover housing from the engine block. Remove the gasket from the block surface.

4. Remove the seal from the housing.

To install:

5. Coat the sealing lips of the new seal with oil. Install a new seal into the end cover housing with a special seal installer tool. On the 6–cylinder engines, press the seal in until it is about 0.039–0.079 in. (0.991–2.070mm) deeper than the standard seal, which was installed flush.

6. While the cover is off, check the plug in the rear end of the main oil gallery. If the plug shows signs of leakage, replace it with another, coating it with the proper sealant to keep it in place.

NOTE: Fill the cavity between the sealing lips of the seal with grease before installing.

7. Coat the mating surface between the oil pan and end cover with sealer. Using a new gasket, install the end cover on the engine block and bolt it into place.

8. Install the remaining components in the opposite order from which they were removed.

9. Install and tighten the oil pan drain plug, then fill the engine with the correct viscosity and amount of clean engine oil.

10. Lower the vehicle and connect the negative battery cable.

11. Start the engine and check that oil pressure is present; if the oil pressure lamp does not turn off within 5–7 seconds of starting the engine, turn the engine **OFF**. Check for any oil leaks.

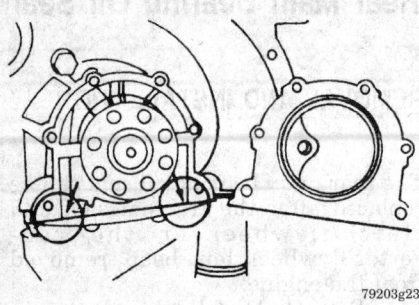

79203g23

Apply sealer to the joints as marked during rear main seal housing installation

ENGINE COOLING

Radiator

REMOVAL AND INSTALLATION

1. Disconnect the negative battery cable. Drain the cooling system. On some engines, this requires removing the plug from the bottom radiator tank.

2. If equipped with a coolant expansion tank, remove the cap, disconnect the hose at the radiator and drain the coolant into a clean container. If equipped with a splash guard, remove it.

3. Remove the coolant hoses and disconnect the automatic transmission oil cooler lines and plug their openings as well as the openings in the cooler.

4. Disconnect any of the temperature switch wire connectors.

5. Remove the shroud from the radiator. On some vehicles, this is done by simply pressing plugs toward the rear of the vehicle. On others, there are metal slips that must be pulled upward and off to free the shroud from the radiator. The shroud will remain in the vehicle, resting on the fan on most vehicles. On 7 Series models, remove the fan and shroud together. Make sure to store the fan in a vertical position. The fan must be held stationary with some sort of flat blade cut to fit over the hub and drilled to fit over 2 of the studs on the front of the pulley. Then, unscrew the retaining nut at the center of the fluid drive hub turning it clockwise to remove it because it has left hand threads.

6. Remove the radiator retaining bolt(s) and lift the radiator from the vehicle.
   To install:

7. The radiator is installed in the reverse order of removal. Fill and bleed the cooling system.

8. Check that rubber mounts are located so as to effectively isolate the radiator from the chassis, as this will help ensure reliable radiator performance. Note that if the vehicle uses plastic upper and lower radiator tanks and has a radiator drain plug, be careful not to over torque the plugs.

9. Torque engine oil cooler pipes to 18–21 ft. lbs. (23–28 Nm) and transmission cooler pipes to 13–15 ft. lbs. (17–20 Nm).

10. Torque the thermostatic fan hub on the 7 Series models to 29–36 ft. lbs. (40–48 Nm).

Heater Core

REMOVAL AND INSTALLATION

3 Series

1. Disconnect the negative battery cable. Remove the package tray. Remove bolts and remove the left/lower dish trim panel.

2. Drain the coolant, loosen the bolt and remove the clamp bracing the 2 lines going to the heater core.

3. Remove the left side duct carrying air from the heater to the rear seat duct.

4. Unscrew the bolts and remove the lower heater discharge duct.

5. Unscrew the bolts fastening the water lines from the engine compartment to the lines coming down from the heater core. Remove and discard the O-ring seals.

6. Unscrew the bolts, separate the halves of the core housing and pull the core out of the housing.
   To install:

7. Install the core into the housing, position the two halves of the core housing together, then install the bolts until tight.

8. Install new O-ring seals on the water lines.

9. Install the remaining components in the reverse order from which they were removed.

10. Fill the engine with the correct amount and mixture of antifreeze and water. Connect the negative battery cable, then start the engine and check for coolant leaks.

5 and 7 Series

1. Disconnect the negative battery cable. Drain coolant from the cooling system. Remove the center console.

2. Remove the mounting bolts and remove the right core mounting bracket. Lift out the front blower motor on the 7 Series.

3. Remove the core cover screws. Loosen the wire straps and clips and remove the cover.

4. Disconnect the temperature sensor(s).

5. Unscrew the mounting bolts and lift out the heater pipes. Replace the O-rings. Then, lift out the core from the right side.
   To install:

6. Install the core in from the right side. Install the heater pipes with new O-rings, then tighten the mounting bolts.

7. Attach the wiring connector to the temperature sensor(s).

8. The remaining components are installed in the reverse order from which they were removed.

9. Fill the engine with the correct amount and mixture of antifreeze and water. Connect the negative battery cable, then start the engine and check for coolant leaks.

8 Series

1. Disconnect the negative battery cable. Drain coolant from the cooling system. Remove the instrument dashboard.

2. Remove the heater core hoses.

3. Disconnect the temperature sensors.

4. Remove the core cover screws. Loosen the wire straps and clips and remove the cover.

5. Unscrew the mounting bolts and lift out the heater pipes. Replace the O-rings. Then, lift out the core from the right side.
   To install:

6. Install the core in from the right side. Install the heater pipes with new O-rings, then tighten the mounting bolts.

7. Attach the wiring connector to the temperature sensor(s).

8. The remaining components are installed in the reverse order from which they were removed.

9. Fill the engine with the correct amount and mixture of antifreeze and water. Connect the negative battery cable, then start the engine and check for coolant leaks.

Water Pump

REMOVAL AND INSTALLATION

M42/M44 Engines

1. Disconnect the negative battery cable. Drain the cooling system.
2. Remove the drive belt and the water pump pulley.
3. Remove the pump mounting bolts.
4. Screw 2 bolts into the tapped bores and press the water pump out of the cover uniformly.

To install:

5. Lubricate and install a new O-ring.
6. Install the water pump and tighten the bolts to 6 ft. lbs. (9 Nm).
7. The remaining components are installed in the reverse order from which they were removed.
8. Start the engine and check for coolant leaks.

M60/M62 Engines

1. Disconnect the negative battery cable.
2. Drain the cooling system.
3. Remove the heat shields at the left and right hand sides of the front axle carrier.
4. Remove the front and rear engine splash guards.
5. Remove the fan. The fan must be held stationary with tool 11–5–030 or some sort of flat blade cut to fit over the hub and drilled to fit over 2 of the studs on the front of the pulley. Remove the fan coupling nut; left hand thread–turn clockwise to remove.
6. Remove the drive belt tensioner, and the serpentine drive belt.
7. Remove the vibration damper and hub. There are eight mounting bolts, and a central bolt. Use a suitable holding tool 11–2–230 or equivalent for the central bolt.
8. Disconnect the coolant hose at the cover of the thermostat housing. Remove the thermostat.
9. Remove the water pump pulley by counterholding the pulley with the drive belt and removing the four pulley mounting bolts.
10. Disconnect the hoses from the water pump. Remove the 6 mounting bolts and remove the water pump.

To install:

11. Clean the gasket surfaces and use a new gasket.
12. Check for the correct seating of the dowel sleeves. Install the water pump in position. Torque the M8 mounting bolts to 16 ft. lbs. (22 Nm),

and the M6 mounting bolts to 7 ft. lbs. (10 Nm). Connect the hoses.
13. Install the vibration damper and hub. Torque the eight mounting bolts to 17 ft. lbs. (24 Nm). Torque the central mounting bolt in four steps as follows:
• An initial torque of 74–81 ft. lbs. (100–110 Nm)
• Add an additional 60 degrees torque
• Add an additional 60 degrees torque
• Then add an additional 30 degrees torque.
14. Install the thermostat. Connect the coolant hose.
15. Install the drive belt tensioner and the drive belt.
16. Install the pulley and tighten the bolts to 6–7 ft. lbs. (8–10 Nm). Install the belt and tighten. Install the fan.
17. Install the heat shields and splash guards.
18. Connect the negative battery terminal.
19. Refill and bleed the cooling system.

M50/M52/S50 and M70/M73/S70 Engines

1. Disconnect the negative battery cable. Drain the cooling system.
2. Remove the fan cowl and fan, if necessary.
3. Remove the drive belt and the pulley. Disconnect the bracket, if necessary.
4. Remove the air cleaner with the air flow sensor, if needed.
5. Disconnect the cooling hoses and remove the water pump.
6. The installation is the reverse of the removal procedure. Torque the M8 bolts to 16 ft. lbs. (22 Nm) and the M6 bolts to 6.5 ft. lbs. (9 Nm).

Thermostat

REMOVAL AND INSTALLATION

The thermostat is located near the water pump, either on the cylinder head or intake manifold on some vehicles and is located between 2 coolant hose sections on some vehicles.
Always drain some coolant out and save it in a clean container before removing the thermostat.

Cooling System Bleeding

WITH BLEEDER SCREW

Set the heat valve in the **WARM** position, start the engine and bring it to normal operating temperature. Run the engine at fast idle and open the venting screw on the thermostat housing until the coolant comes out free of air bubbles. Close the bleeder screw and refill the cooling system.

WITHOUT BLEEDER SCREW

Fill the cooling system, place the heater valve in the **WARM** position, close the pressure cap to the second (fully closed) position. Start the engine and bring to normal operating temperature. Carefully release the pressure cap to the first position and squeeze the upper and lower radiator hoses in a pumping action to allow trapped air to escape through the radiator. Recheck the coolant level and close the pressure cap to its second position.

ENGINE ELECTRICAL

NOTE: Disconnecting the negative battery cable on some vehicles may interfere with the functions of the on board computer systems and may require the computer to undergo a relearning process.

Distributorless Ignition

All ignition and fuel injection functions are controlled by the Digital Motor Electronics (DME) control unit. Ignition timing is fully electronically controlled; there is no vacuum advance or manual adjustment. Ignition functions are calculated from internal maps and from the same sensors used for the fuel injection system. Vehicles with an automatic transmission, the control unit will retard ignition timing briefly when the transmission is about to shift up or down. For this reason, there is a data link between the DME control unit and the transmission control unit.
Variations of this ignition system are used on different engines. The M42B18 engine uses distributorless ignition with a coil pack mounted on

the inner fender. The M50B25 engine uses distributorless ignition with a coil pack mounted above each spark plug.

Ignition Timing

ADJUSTMENT

The ignition timing is controlled by the DME. As a result, checking and adjusting the timing is impossible. There is no method of setting dynamic or static timing.

Alternator

PRECAUTIONS

Several precautions must be observed with alternator equipped vehicles to avoid damaging the unit. They are as follows:

• If the battery is removed for any reason, make sure it is reconnected with the correct polarity. Reversing the battery connections may result in damage to the one-way rectifiers.

• When utilizing a booster battery as a starting aid, always connect it as follows: positive to positive, and negative (booster battery) to a good ground on the engine.

• Never use a fast charger as a booster to start vehicles with alternating-current (AC) circuits.

• When servicing the battery with a fast charger, always disconnect the battery cables.

• Never attempt to polarize an alternator.

• Avoid long soldering times when replacing diodes or transistors. Prolonged heat is damaging to alternators.

• Do not use test lamps of more than 12 volts for checking diode continuity.

• Do not short across or ground any of the terminals on the alternator.

• The polarity of the battery, alternator and regulator must be matched and considered before making any electrical connections within the system.

• Never operate the alternator on an open circuit. Make sure all connections within the circuit are clean and tight.

• Turn OFF the ignition switch and then disconnect the battery terminals when performing any service

on the electrical system or charging the battery.

• Disconnect the battery ground cable if arc welding is to be done on any part of the vehicle.

BELT TENSION ADJUSTMENT

On serpentine belt equipped engines, belt tension is automatically adjusted by the tensioner idler. If not equipped with a single serpentine belt, the fan belt tension is adjusted by moving the alternator on the slack adjuster bracket. The belt tension is adjusted to a deflection of approximately 1/2 in. under moderate thumb pressure in the middle of its longest span. On many engines, the position of the top of the alternator is adjusted via a bolt that is geared to the bracket. This bolt is turned to position the alternator and determine tension and then is locked in position with a lock bolt.

REMOVAL AND INSTALLATION

M42/M44 Engines

1. Disconnect the battery ground cable.
2. Remove the air cleaner.
3. Remove the protective cap. Disconnect the wires (B+, D+, and ground strap) from the rear of the alternator, marking them for later installation.
4. Loosen the adjusting and pivot bolts. Loosen the lockbolt, turn the tensioning bolt so as to eliminate belt tension and then remove the belt. Remove the bolts and remove the alternator.
5. If the voltage regulator needs to be replaced, unfasten the two cover screws on the back of the alternator and lift out the voltage regulator

To install:
6. If the voltage regulator was removed, clean the contact surfaces and check the contact spring tension. The contacts should extend at least 5 mm from the voltage regulator body. Install the voltage regulator, replace the cover and install the two cover screws.
7. Position the alternator, connect the belt and loosely install the bolts.
8. To adjust belt tension, turn the tension adjusting bolt with a torque wrench. When the torque is 60 inch lbs. (7 Nm), hold the adjustment bolt steady and tighten the locknut at the rear of the unit.
9. Tighten the remaining bolts.
10. Connect the wires to the back of the alternator.

11. Install and connect the air cleaner.
12. Connect negative battery cable.

M50/M52/S50 Engines

1. Disconnect the battery ground cable.
2. Remove the air cleaner and air mass sensor.
3. Remove the fan and cowl.
4. Loosen the hose clamps on the intake manifold.
5. Unclip the upper section of the filter housing and place to one side.
6. Loosen the hose clamp on the vent hose and remove the vent hose.
7. Remove the covers of the tensioning roller and the reversing roller. Loosen and remove the drive belt.
8. Remove the protective cap. Disconnect the wires from the rear of the alternator, marking them for later installation.
9. Remove the bolts and remove the alternator.
10. If the voltage regulator needs to be replaced unfasten the two cover screws on the back of the alternator and lift out the voltage regulator

To install:
11. If the voltage regulator was removed, clean the contact surfaces and check the contact spring tension. The contacts should extend at least 5 mm from the voltage regulator body. Install the voltage regulator, replace the cover and install the two cover screws.
12. Position the alternator and install the bolts. Tighten the bolts to 2.5 ft. lbs. (3.5 Nm).
13. The tab on the tensioning roller must be engaged in the slot/groove .
14. Connect the wires to the back of the alternator.
15. Route the drive belt and tighten. Replace the covers on the rollers.
16. Install the vent hose and tighten the hose clamp.
17. Position the upper section of the filter housing and clip the fasteners.
18. Tighten the hose clamps on the intake manifold.
19. Install the fan and cowl.
20. Install and connect the air cleaner and air mass sensor.
21. Connect negative battery cable.

M60/M62 Engines

1. Disconnect the battery ground cable.
2. Remove the holder of the vacuum hoses on the radiator cowl panel.
3. Hold the fan pulley with tool 11 5 030 or equivalent. The tool is de-

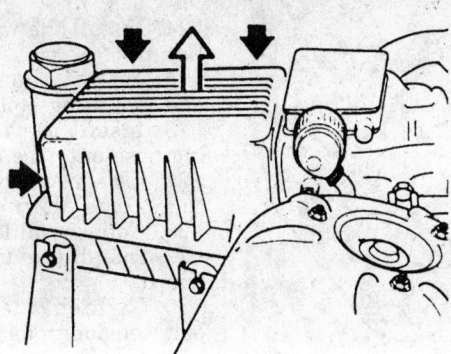

Air cleaner assembly — M42/M44 engine

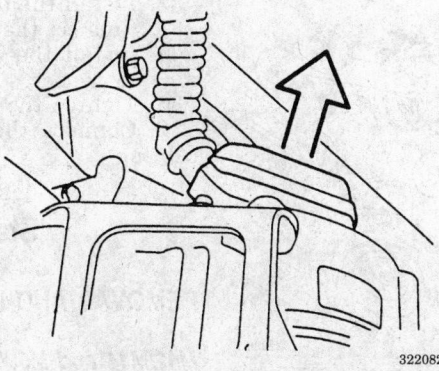

Alternator wires — M42/M44 engine

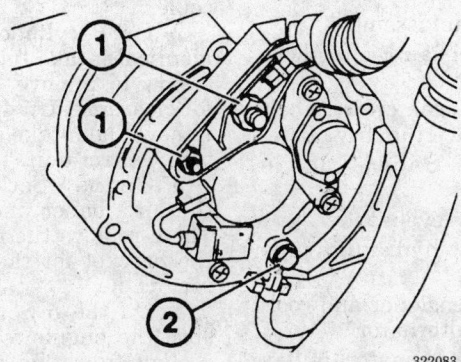

Mounting bolts (1), and ground strap (2) — M42/M44 engine

signed to be bolted to 2 of the pulley bolts and to provide the resistance to allow the fan to be unscrewed from the pulley.

4. Unscrew the fan bolts clockwise. Remove the fan and keep upright. Be careful of the radiator fins.

5. Remove the fan cowl by disconnecting the left and right clips. Take off upwards.

6. Loosen the drive belt tensioner nuts to slacken the drive belt. Remove the drive belt. Remove the drive belt tensioner nuts and the tensioner pulley bolt. Remove the tensioner.

7. Loosen the oil filter nuts and pull the oil filter forward about 5/32 inch (4mm).

8. Remove the top and side alternator mounting bolts.

9. Lift and safely support the vehicle. Remove the splash guards.

10. Disconnect the vent hose of the alternator at the alternator.

11. Remove the bottom alternator mounting bolt.

12. Pull the alternator forward. Remove the protective cap. Disconnect the wires (B+ and D+) from the rear of the alternator, marking them for later installation.

13. Lower the vehicle.

14. Disengage the radiator at the left and right sides by pushing in with a screwdriver.

15. Protect the radiator with a sheet metal plate to prevent damage while removing the alternator. Unscrew the pulley on the water pump.

16. To remove the alternator: the alternator mounting flange must be turned clockwise out from underneath the oil filter mounting flange. Pull the alternator forward. Remove the alternator upwards.

17. To remove the voltage regulator, remove the two bolts from the rear of the alternator and carefully remove the voltage regulator. Clean the contact surfaces and check the tension of the spring contacts.

To install:

18. Mount the voltage regulator and install the two bolts.

19. Lower and rotate the alternator into position making sure to align the mounting holes.

20. Install the pulley on the water pump and remove the sheet metal protective plate from the radiator. Torque the pulley nut to 40 ft. lbs. (55 Nm).

21. Install the clips on the radiator and squeeze together so that the splines are heard to engage several times.

22. Raise and safely support the vehicle.

23. Connect the wires to the back of the alternator.

24. Install the bottom alternator mounting bolt.

25. Connect the vent hose of the alternator and tighten the hose clamp.

26. Install the splash guards. Lower the vehicle.

27. Install the top and side alternator mounting bolts.

28. Return the oil filter to its original position and tighten the nuts.

29. Route the drive belt. Install the drive belt tensioner and pulley. As shown in illustration, preload the adjusting plate on the hexagon nut (1) up to the end of the slot (2). Tighten the nuts (3).

30. Install the fan cowl by connecting the left and right clips.

31. Install the fan using tool 11 5 040 or equivalent wrench and holding tool 11 5 030 or equivalent. Torque the nut to 29 ft. lbs. (40 Nm). If using the tool set the torque wrench to 22 ft. lbs. (30 Nm). The additional length of the tool multiplies the tor-

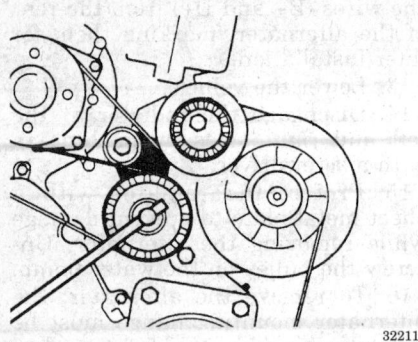

Remove the drive belt — M50/M52/S50 engine

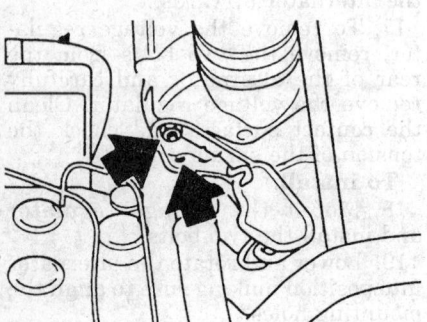

Alternator wires — M50/M52/S50 engine

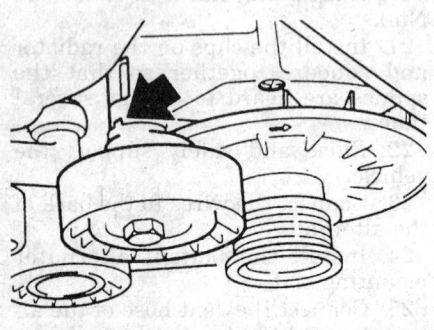

Tensioner tab alignment — M50/M52/S50 engine

que to achieve 29 ft. lbs. (40 Nm) at the nut.

32. Install the holder of the vacuum hoses on the radiator cowl panel.

33. Connect negative battery cable.

M70/M73/S70 Engines

1. Disconnect the negative battery cables.

2. Loosen the oil filter cap.

3. Remove the air filter and mass air flow sensor.

4. Disconnect the ignition coil wires.

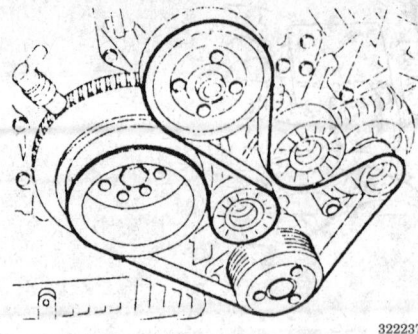

Drive belt routing — M60/M62 engines

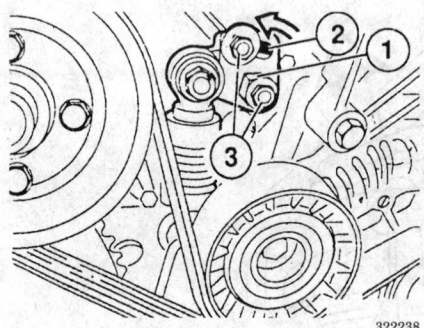

Drive belt tensioner — M60/M62 engines

5. Remove the fan assembly.

6. Loosen clamp of air cooling hose on alternator.

7. Loosen the alternator drive belt tensioner and remove the drive belt.

8. Remove the under vehicle splash guard.

9. Disconnect oil level sender.

10. Unscrew the hydraulic pipe clamp on alternator.

11. Remove the tensioner and cooling hose from the alternator.

12. Remove the alternator pulley and mounting bolts.

13. Remove the upper hydraulic line clamp.

14. Remove the oil filter assembly including hoses.

15. Disconnect the alternator wires.

16. Remove the distributor cover and loosen the wire strap.

17. Remove the heat shield.

18. Remove the alternator assembly.

   To install:

19. Install the alternator assembly.

20. Install the heat shield.

21. Install the distributor cover and tighten the wire strap.

22. Connect the alternator wires.

23. Install the oil filter assembly and hoses.

24. Install the upper hydraulic line clamp.

25. Install the alternator pulley and mounting bolts.

26. Install the tensioner and cooling hose onto the alternator.

27. Unscrew the hydraulic line clamp onto alternator.

28. Connect oil level sender.

29. Install the under vehicle splash guard.

30. Tighten the alternator drive belt tensioner and install the drive belt.

31. Tighten clamp of air cooling hose onto alternator.

32. Install the fan assembly.

33. Connect the ignition coil wires.

34. Install the air filter and mass air flow sensor.

35. Tighten the oil filter cap.

36. Connect the negative battery cables.

Starter

REMOVAL AND INSTALLATION

M42/M44 and M50/M52/S50 Engines

1. Disconnect the negative battery cable.

2. On 6–cylinder engines with 6 identical intake tubes, it may be necessary to remove No. 6 intake tube for clearance. On 4–cylinder vehicles, remove the intake cowl from the mixture control unit.

3. On the 3 Series models, remove the air cleaner and air flow sensor. Then, remove the mounting bolts for the bracket for the air collector and remove it.

4. On the 5 Series models, make sure the engine is cool. Drain some coolant from the cooling system and then remove the expansion tank.

5. Operate the brake pedal hard 20 times. Disconnect the power steering line that would otherwise prevent access to the starter.

6. Cut off the straps and remove the solenoid switch insulating cover, located right near the solenoid.

7. Cut off the straps and remove the solenoid switch insulating cover, located right near the solenoid.

8. Remove the starter solenoid wire leads, marking them for later installation, unless they have already been removed. On 4–cylinder vehicles, disconnect the mounting bracket at the block.

9. On the 3 Series models, drain coolant out of the engine and then disconnect the heater hose located near the starter; also unscrew and re-

move the coolant pipe if necessary for clearance.

NOTE: Remove the accelerator cable holder on automatic transmission equipped vehicles.

10. Unbolt and remove the starter.

NOTE: On the 5 Series models, it may be necessary to use a box wrench with an angled handle to unscrew the main starter mounting bolts. On the 3 Series models, the starter must be pulled out from above.

To install:

11. Install the starter and install the retaining bolts. Install all removed components on all vehicles.

12. Make sure to reconnect all hoses and refill and bleed the cooling system or power steering system.

13. Where the solenoid switch cover has been unstrapped, reinstall it with new straps to locate it properly for electrical safety.

M60/M62 Engines

1. Disconnect negative battery cable. Lift and safely support the vehicle.

2. Remove the splash guard.

3. Remove the heat baffle plates on the right side of the engine, on the starter motor, and from in front of the front axle carrier.

4. Remove the electrical leads from the starter motor.

5. Remove the mounting bolts for the starter motor. Remove the starter motor by lifting out from the back of the ring gear, and remove the starter downwards.

To install:

6. Install the starter and tighten the mounting nuts and bolts to 34–36 ft. lbs. (47–50 Nm).

7. Connect the wires to the original locations on the solenoid. If the battery cable connection is all metal, torque to 3.6–4.4 ft. lbs. (5–6 Nm). If the battery cable connection is plastic, torque to 0.7–1.1 ft. lbs. (1.0–1.5 Nm).

8. Install the heat baffles that were previously removed, and the splash guard.

9. Lower the vehicle, and connect the negative battery cable.

M70/M73/S70 Engines

1. Disconnect the negative battery cable.

2. Remove the exhaust system. Remove the heat shields on steering linkage.

3. Remove headlight washing tank and pump, position aside.

4. Remove the heat shields on the intake manifold. Unscrew the lines on the starter connection point.

5. Remove the exhaust manifolds, front manifold first.

6. Remove the starter mounting bolts. Remove the brackets and heat shields. Remove the starter assembly.

To install:

7. Situate the starter motor, brackets and heat shield in position against the engine, then install the mounting bolts until tight.

8. Using new gaskets and after having satisfactorily cleaned the mounting flanges, install the rear exhaust manifold, then the front exhaust manifold onto the engine.

9. The balance of installation is the reverse of the removal procedure.

10. Connect the negative battery cable. Start the engine and check for exhaust leaks.

EMISSION CONTROLS

Emission Warning Lamps

RESETTING

Service Interval Reminder Lights

The on-board computer is used to evaluate mileage, average engine speed and engine and coolant temperatures as well as other computer input factors that determine maintenance intervals. There are 5 green, a yellow and a red **LED** used to remind the driver of oil changes and other maintenance services.

The green LED'S will be illuminated when the ignition is in the **ON** position and the engine OFF. There will not be as many green LED'S illuminated when the maintenance time gets closer. A yellow LED that is illuminated when the engine is running, will indicate maintenance is now due. The red LED will be illuminated when the service interval has been exceeded by approximately 1000 miles. This is the computers way of saying this is your last warning.

There is a service interval reset tool manufactured by the Assenmacher Tool Company (tool 62–1–100). This tool is used to reset BMW 6-cylinder and 4-cylinder vehi-

cles with the aid of an additional adapter tool.

1. Locate the diagnostic connector near the thermostat housing.

2. Plug the special reset tool into the diagnostic connector and place the ignition switch in the **ON** position.

3. Depress the reset button on the tool until all 5 green LED'S are illuminated, showing that the reset has completed.

FUEL SYSTEM

Fuel System Service Precautions

Safety is the most important factor when performing not only fuel system maintenance but any type of maintenance. Failure to conduct maintenance and repairs in a safe manner may result in serious personal injury or death. Maintenance and testing of the vehicle's fuel system components can be accomplished safely and effectively by adhering to the following rules and guidelines.

• To avoid the possibility of fire and personal injury, always disconnect the negative battery cable unless the repair or test procedure requires that battery voltage be applied.

• Always relieve the fuel system pressure prior to disconnecting any fuel system component (injector, fuel rail, pressure regulator, etc.), fitting or fuel line connection. Exercise extreme caution whenever relieving fuel system pressure to avoid exposing skin, face and eyes to fuel spray. Fuel under pressure may penetrate the skin or any part of the body that it contacts.

• Always place a shop towel or cloth around the fitting or connection prior to loosening to absorb any excess fuel due to spillage. Ensure that all fuel spillage (should it occur) is quickly removed from engine surfaces. Ensure that all fuel soaked cloths or towels are deposited into a suitable waste container.

• Always keep a dry chemical (Class B) fire extinguisher near the work area.

• Do not allow fuel spray or fuel vapors to come into contact with a spark or open flame.

• Always use a backup wrench when loosening and tightening fuel line connection fittings. This will prevent unnecessary stress and torsion to fuel line piping. Always follow the proper torque specifications.

• Always replace worn fuel fitting O-rings with new. Do not substitute fuel hose or equivalent where fuel pipe is installed.

RELIEVING FUEL SYSTEM PRESSURE

To relieve the pressure in the system, first find the fuel pump relay plug, located on the cowl. Unplug the relay, leaving it in a safe position where the connections cannot ground. If necessary, tape the plug in place or tape over the connector prongs with electrical tape. Then, start the engine and operate it until it stalls. Crank the engine for 10 seconds after it stalls to remove any residual pressure.

Fuel Filter

REMOVAL AND INSTALLATION

On filters that are located near the fuel tank, it is necessary to clamp the fuel lines closed before disconnecting them, or fuel will run out continuously.

1. Disconnect the negative battery cable. Relieve fuel system pressure. Clamp the lines closed if the filter is mounted low, near the fuel tank. Then, loosen the clamps and disconnect the inlet and outlet hoses. Remove the hose clamps or slide them back, well off the connections to make it easier to pull off the hoses, if necessary.

2. The filters will usually be attached to a frame, floor pan or wheel well by a bracket. Loosen the bracket and remove the filter. Note the direction of flow and then remove the filter.

To install:

3. Situate the new fuel filter, while observing the direction of flow markings on the filter, into position on the frame, floor pan or wheel well (depending on the particular model). Install the mounting bracket until snug.

4. Install the fuel lines onto the correct fuel filter fittings. Tighten the fuel line clamps until tight, but not to the point where the fuel lines become excessively pinched or damaged.

5. Connect the negative battery cable and cycle the ignition **ON** and **OFF** several times to build fuel pressure.

6. Inspect the fuel filter and fuel lines for any fuel leaks.

Electric Fuel Pump

PRESSURE TESTING

Except 3 Series

1. Relieve fuel system pressure. Tee a pressure gauge into the fuel feed line in front of the pressure regulator. Plug the fuel return hose.

2. Pull off the pump relay. Jumper terminals **87** and **30**. Measure the delivery pressure. It should be 43 psi on vehicles except the 7 Series. On these vehicles, it should be 48 psi.

3 Series

1. Relieve fuel system pressure. Tee a pressure gauge into the fuel feed line in front of the pressure regulator.

2. Disconnect the fuel pump relay.

3. Connect a remote starter switch between terminals **KL30** and **KL87** of the relay. Close the switch and check the pressure. It should be 43 psi. If not, the filter is severely clogged or the fuel pump is defective.

REMOVAL AND INSTALLATION

3 Series

1. Relieve fuel system pressure. Disconnect the negative battery cable. Going to the pump, which is under the vehicle and near the fuel tank, push back any protective caps, note the routing and disconnect the electrical connector(s).

2. Securely clamp the suction hose (coming from the tank) and plug the discharge hose so no fuel can escape.

3. Open the hose clamp connecting the suction hose to the pump and disconnect it.

4. Remove the attaching nuts which mount the pump and bracket to the floor pan and remove both as an assembly.

5. Remove the bolt passing through the 2 parts of the bracket and also mounting the hose attaching strap to the bracket. Then, pull the pump out of the bracket.

6. Loosen the hose clamp for the discharge hose and disconnect it at the pump. Pull the rubber ring off the pump.

7. Note the code number on the pump and make sure to replace it with one of the same number. Inspect all the rubber mounts on the pump

mounting bracket and replace any that are cracked or crushed.

To install:

8. Attach the discharge hose to the fuel pump, then slide the pump into the mounting bracket. Install the hose strap mounting bolt to the pump mounting bracket.

9. Install the remaining components in the opposite order from which they were removed.

10. Connect the negative battery cable. Make sure to remove the clamps from the hoses and then run the engine and check for leaks. Check the fuel system pressure.

5 Series

The fuel pump is an electrical unit, delivering fuel through a pressure regulator, to a fuel distributor or a ring-line for the injection valves. The fuel pump is mounted under the vehicle, in the fuel tank, or in the engine compartment.

1. Relieve fuel system pressure. Disconnect the negative battery connector. Push back any protective caps and disconnect the electrical connector(s).

2. If the fuel lines are flexible, pinch them closed with an appropriate tool. Disconnect the fuel lines and plug the ends.

3. Remove the retaining bolts and remove the pump and expansion tank as an assembly.

4. The pump can be separated from the expansion tank after removal.

To install:

5. Install the pump in the correct position, be sure to use similar types of hose clamps, if any need replacing. The wrong type clamp can damage the pressure lines.

6. Run the engine and check the fuel lines for leakage. Check the fuel system pressure.

7 Series

The pump on this vehicle is mounted in the top of the tank along with the fuel level sending unit.

1. Disconnect the negative battery cable. Drain the fuel tank, enough to prevent spillage when removing the pump.

2. Relieve fuel system pressure. Remove the trim panels from the trunk. Then, remove the screws from the cover for the pump/sending unit assembly.

3. Label the fuel hoses connecting at the top of the pump/sending unit assembly. Unclamp and disconnect the fuel hoses and then plug them.

4. Slide the collar for the electrical connector to one side and then unplug the connector.

5. Remove the mounting screws and remove the pump/sending unit assembly. Replace the gasket.

6. Press the retaining locks for the pump unit inward and slide the pump out of the pump/sending unit assembly.

7. Note the routing of the fuel and electrical lines to the pump from the top of the pump/sending unit assembly. Loosen the hose clamp screws and the screws attaching the electrical connectors to the pump. Disconnect the hose and connector.

8. Unscrew the pressure regulator from the top of the check valve. Then, unscrew the check valve from the top of the pump.

9. Pull the insulating sleeve off the pump. Then, loosen the retaining screw and slide the filter off the pump.

To install:

10. Slide a new filter onto the pump, then install and tighten the retaining screw.

11. Install the insulating sleeve on the pump.

12. Install the remaining components in the reverse order of removal.

13. Be careful to ensure that the 2 retaining locks fasten the pump in

1. Fuel level transmitter
2. Gasket
3. Inlet line
4. Return line
5. Pressure damper
6. Check valve
7. Fuel pump
8. Pump insulating sleeve
9. Fuel intake filter
10. Pump holder

79203g07

View of the in-tank fuel pump — 7 Series

place in a secure manner. Operate the engine and check for leaks.

8 Series

The pump on this vehicle is mounted in the top of the tank along with the fuel level sending unit.

1. Disconnect the negative battery cable. Drain the fuel tank, enough to prevent spillage when removing the pump.

2. Relieve fuel system pressure. Remove the rear seat. Then, remove the screws from the cover for the pump/sending unit assembly.

3. Label the fuel hoses connecting at the top of the pump/sending unit assembly. Unclamp and disconnect the fuel hoses and then plug them.

4. Slide the collar for the electrical connector to one side and then unplug the connector.

5. Remove the coupling nut and remove the pump/sending unit assembly. Replace the gasket.

6. Remove the filter screens and rubber liner.

7. Mark the routing of the fuel and electrical lines to the pump from the top of the pump/sending unit assembly. Loosen the hose clamp screws and the screws attaching the electrical connectors to the pump. Disconnect the hose and connector.

To install:

8. Attach the hose and connector to the fuel pump. Tighten the hose clamps and electrical connector retaining screws until snug.

9. Install the filter screens and rubber liner onto the pump.

10. The remaining components are installed in the reverse order from which they were removed.

11. Be careful to ensure that the 2 pumps are connected as previously marked.

12. Connect the negative battery cable. Operate the engine and check for leaks.

Fuel Injector

REMOVAL AND INSTALLATION

M42/M44 Engines

1. Disconnect the negative battery cable. Remove the upper section of the collector.

2. Disconnect the rear brace and remove the injector hose.

3. Remove the holder for the preheater.

4. Remove the screws and lift off the upper section of the intake manifold. Disconnect the hose from the

fuel pressure regulator at the same time.

5. Remove the plug plate from the fuel injectors and unscrew the clamp.

6. Remove the injection pipe with the injectors attached. Lift off the retainer and remove the fuel injector.

To install:

7. Install the fuel injector onto the injection pipe, then fasten the retainer in place.

8. Install the injection pipe, with the injectors installed, onto the engine.

9. Install the remaining components in the reverse order of removal.

10. Connect the negative battery cable.

11. Start the engine and check for leaks.

M50/M52/S50 Engines

1. Relieve the fuel pressure and disconnect the negative battery cable.

2. Remove the injector valley cover and disconnect and remove the plug plate.

3. Pull off the vacuum hose from the pressure regulator. Disconnect the fuel feed and return hoses from the fuel rail.

4. Unscrew the fasteners holding the fuel rail and remove the fuel rail along with the injectors.

5. Pull off the retainers and remove the injectors. If replacing an injector use the correct style and code number for the application.

To install:

6. Inspect the O-rings and replace as necessary. Replace the injectors on the fuel rail and fit the retainers in place.

7. Lubricate the O-rings with Vaseline or gear oil prior to installation and replace the fuel rail.

8. The remaining components are installed in the reverse order from which they were removed.

9. Connect the negative battery cable, start the engine and check for fuel leaks.

M60/M62 Engines

1. Relieve the fuel system pressure. Disconnect the negative battery cable.

2. Remove the oil filler cap. Unclip the four caps and remove the nuts. Remove the manifold cover.

3. Remove the cable securing straps and disconnect the throttle cable on the throttle valve lever.

4. Unscrew and remove the guards for the cylinder head covers on both sides.

5. Remove the electrical connectors for the ignition coils.

6. Unclip and move aside the throttle cable. Unscrew and move aside the wiring ducts. Remove the fuel injector connectors.

7. Plug the fuel feed and return lines with special tool 13 3 010, or equivalent.

8. Unbolt the injection pipe rail and remove the rail together with the fuel injectors.

9. Press off the clip and remove the fuel injector.

To install:

10. Check the O-rings on the injector and replace if necessary. Check that all the injectors are of like types. Lightly grease the O-rings with petroleum jelly and install on the fuel rail.

11. Install the fuel rail and tighten the bolts to 6–8 ft. lbs. (9–11 Nm). Unplug and connect the fuel lines.

12. Connect the fuel injector connectors. Install the wiring ducts. Position and clip the throttle cable.

13. Install the electrical connectors for the ignition coils.

14. Position the guard for the cylinder head cover and install the mounting bolt. Repeat the process on the other side.

15. Connect the throttle cable on the throttle valve lever and connect the cable securing straps.

16. Install the manifold cover, replace the caps and replace the oil filler cap.

17. Connect negative battery cable.

M70/M73/S70 Engines

1. Disconnect the negative battery cable. Remove the hose and plastic caps, if necessary.

2. Remove the plugs, plate and screws from the injector pipe.

3. Push up on the injector pipe until the injectors have cleared the guides on the intake manifold or the throttle valve housing.

4. Pull of the plugs and lift up the retainer. Remove the injectors.

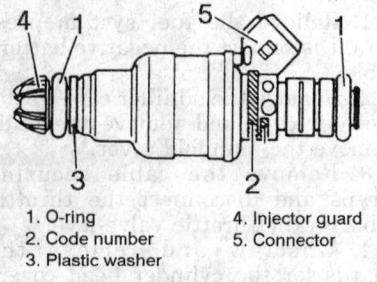

1. O-ring
2. Code number
3. Plastic washer
4. Injector guard
5. Connector

298177

Fuel injector — M60/M62 engines

To install:

5. Inspect the O-rings and replace as necessary. Replace the injectors on the fuel rail and fit the retainers in place.

6. Lubricate the O-rings with Vaseline or gear oil prior to installation and replace the fuel rail.

7. The remaining components are installed in the reverse order from which they were removed.

8. Connect the negative battery cable, start the engine and check for fuel leaks.

DRIVE AXLE

Halfshafts

REMOVAL AND INSTALLATION

M3, Z3 and 3 Series

1. Raise and safely support the vehicle.

2. Remove the rear tire and wheel assembly.

3. Disconnect the output shaft at the outer flange and suspend it with wire.

4. Unbolt the caliper and suspend it with the brake line connected. Unbolt and remove the rear disc.

5. Remove the large nut and remove the lock plate. If equipped with ABS, disconnect and then remove the ABS speed sensor by unscrewing it.

6. Unscrew the collar nut. Then, pull off the drive flange with the proper tool(s).

7. Screw on the collar nut until it is just flush with the end of the shaft and use a suitable hammer to knock out the shaft.

8. Remove the snap ring. Pull out the wheel bearings, using the proper tool.

9. Pull the inner bearing race off the axle shaft with special tool 00 7 500 or equivalent.

To install:

10. Install the new bearing assembly using the proper tools. Then, reinstall the snap ring.

11. Install the rear axle shaft with special tools: 23 1 300, 33 4 080 and 33 4 020 or equivalent.

12. Lubricate and install the collar nut, and drive in the lock plate with the proper tool(s). Torque the collar nut to 148 ft. lbs. (200 Nm).

13. Install and connect the ABS speed sensor.

14. Remount the brake disc and caliper.

15. Reconnect the output shaft.

16. Install the rear tire and wheel assembly.

17. Lower the vehicle.

5 and 7 Series

1. Raise and safely support the vehicle. Remove the rear tire and wheel assembly.

2. Lift out the lock plate and if equipped, remove the ABS sensor.

3. remove the retaining nut from the output flange. Remove the flange.

4. Disconnect the output (half) shaft from the final drive (differential carrier) by pressing out with the proper tool and suspend it.

5. Pull out the output shaft from the drive flange hub with a special tool.

6. Drive out the rear axle shaft with the proper tool.

7. Lift out the snap ring. Then, pull out the wheel bearings, using the proper tool.

8. Pull out the seal with a suitable tool.

9. If the inner bearing shell is damaged, pull it off with a puller and thrust pad.

To install:

10. Using an appropriate bearing installer, pull in the wheel bearing assembly, pull in the seal, insert the snap ring and then pull in the rear axle shaft, all in reverse of the removal procedure. Install the axle shaft seal.

11. To install the output shaft, screw the threaded spindle into the shaft all the way and then use the nut and washer against the outside of the bridge.

12. Reconnect the output shaft to the final drive. Torque the mounting bolts to 42–46 ft. lbs. (58–63 Nm).

13. Lubricate the bearing surface of the outer nut with oil and install the nut. Torque the nut to 169–188 ft. lbs. (234–260 Nm).

14. Install the ABS sensor, if removed.

15. Install the rear tire and wheel assembly and lower the vehicle.

8 Series

1. Raise and safely support the vehicle. Remove the rear tire and wheel assembly.

2. Remove the ABS sensor.

3. Remove the retaining nut from the output flange. Remove the drive flange hub.

4. Remove the half-shaft from the vehicle by unscrewing the shaft from the final drive output flange and by

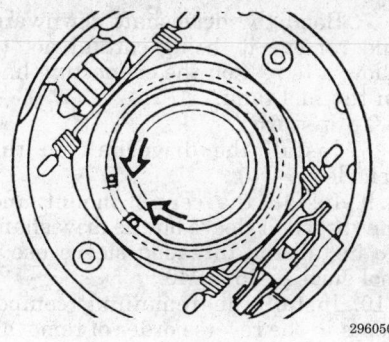

296050

Remove the snap ring from the hub — M3, Z3 and 3 Series

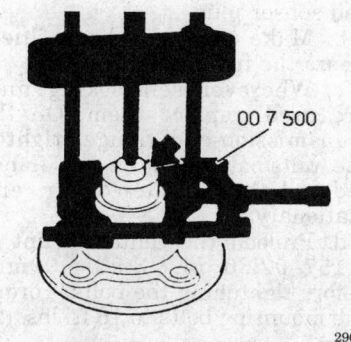

00 7 500

296051

Special Tool 00 7 500: Bearing puller — M3, Z3 and 3 Series

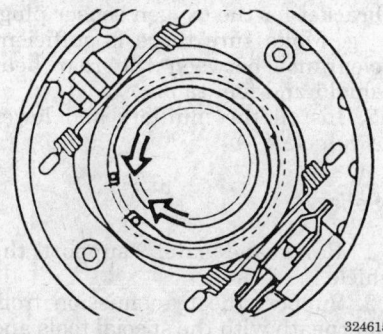

324615

Remove the snapring from the hub — 5 and 7 Series

pressing the half-shaft out of the drive flange hub with special tools 33–2–111, 116 and 117.

5. Press out the drive flange hub with special tools 33–3–250 and 33–2–105.

6. Lift out the snap-ring. Then, pull out the wheel bearings, using the special tools 33–3–261, 262, and 263.

7. Pull out the seal with a suitable tool.

8. If the bearing inner race is damaged, pull it off of the drive flange hub with special tool 33–3–240.

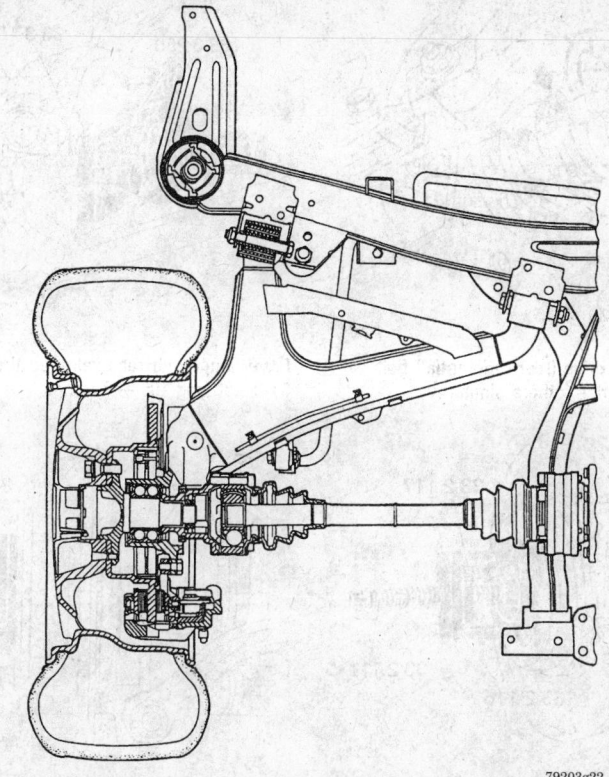

79203g22

Cut-away schematic of the rear halfshaft and suspension system — 5 Series

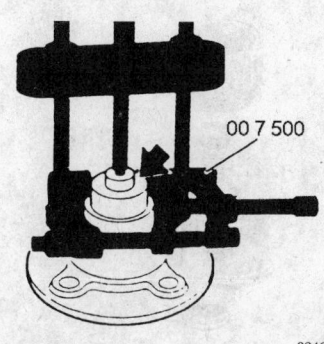

00 7 500

324616

Special Tool 00 7 500: Bearing puller — 5 and 7 Series

To install:

9. Using an appropriate bearing installer, pull in the wheel bearing assembly, pull in the seal, insert the snap-ring and then pull in the drive flange hub with special tools 33–3–261, 263, and 264. Install the axle shaft seal.

10. To install the output shaft, pull in half-shaft with special tools 33–2–116 and 118.

11. Reconnect the output shaft to the final drive. Torque the M10 mounting bolts to 61 ft. lbs. (83 Nm).

12. Lubricate the bearing surface of the outer nut with oil and install. Torque the nut to 221 ft. lbs. (300 Nm).

13. Install the ABS sensor.

14. Install the rear tire and wheel assembly and lower the vehicle.

CV-Joint Boot

REMOVAL AND INSTALLATION

1. Raise and safely support the vehicle.

2. Remove the output shaft and the snapring. Press off the cap and the dust cover.

3. Press the output shaft out of the CV-joint. Clean and remove the grease from the splines of the joint.

To install:

4. Install the dust cover with the inside cover on the output shaft.

5. Press on the joint with the cap and install the snapring. Pack the joint and dust cover with the proper grease.

6. Mount he dust cover and install new clamps.

7. Install the output shaft and lower the vehicle.

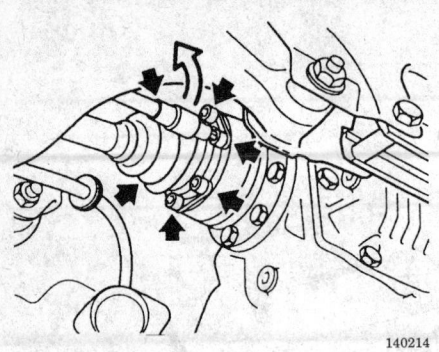

Halfshaft-to-final drive (rear differential) bolt locations — 8 Series, others similar

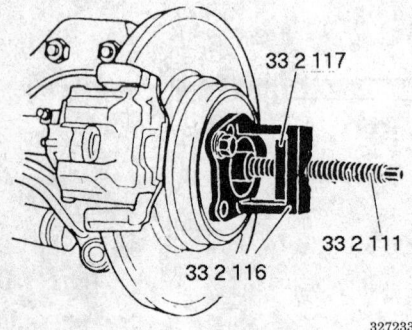

Halfshaft special tools: 33-2-111, 116, and 117 — 8 Series

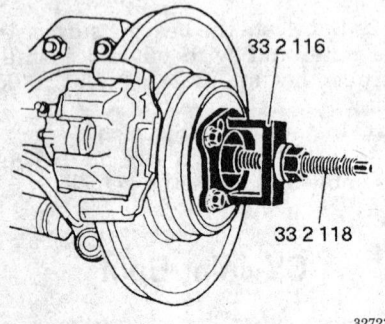

Halfshaft special tools: 33-2-116 and 118 — 8 Series

Driveshaft and U-Joints

REMOVAL AND INSTALLATION

M3, Z3 and 3 Series

1. Raise and safely support the vehicle. Remove the mufflers. Unscrew and remove the exhaust system heat shield near the fuel tank.
2. Unbolt and remove the cross brace that runs under the driveshaft.
3. Support the transmission. The automatic transmission must be supported by the case and not the pan.

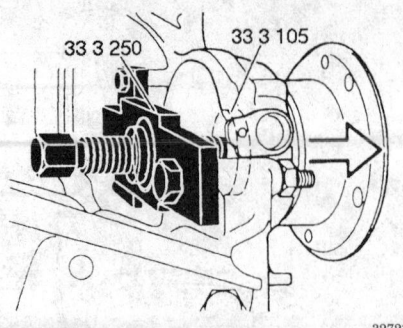

Drive flange hub removal special tools: 33-3-250 and 105 — 8 Series

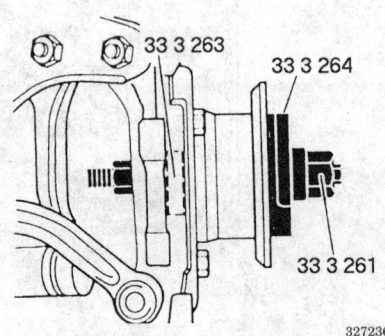

Drive flange hub installation special tools: 33-3-261, 263, and 264 — 8 Series

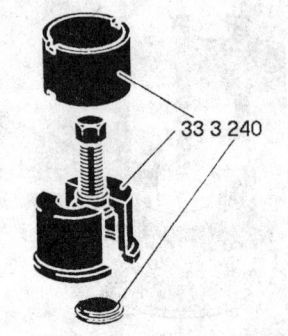

Special tool: 33-3-240 — 8 Series

Loosen all transmission support bolts and remove. Remove the transmission rear support crossmember.
4. Lower the manual transmission for clearance. Remove the driveshaft bolts from the front coupling.

NOTE: Make sure the drive axle does not rest on the fuel line that runs across under it.

5. Unscrew and remove bolts at the coupling near the final drive.
6. Loosen the threaded sleeve on the driveshaft with a tool such as tool 26 1 040 or equivalent. Unbolt and remove the center mount.

7. Bend the driveshaft downward and remove it, being careful not to allow it to rest on the connecting line on the fuel tank.

To install:
8. Install the driveshaft in the vehicle.
9. Install the center mount and the threaded sleeve on the driveshaft. To install the threaded sleeve use a tool such as 26 1 040.
10. Install the remaining components in the reverse order of removal.
11. Upon installation keep the following in mind:
   a. Mount the holder for the oxygen sensor plug.
   b. Make sure the heat shield clears the fuel tank.
   c. Wherever self-locking nuts are used, replace them. On the transmission-end flange, tighten the nuts/bolts only on the flange side, holding the other end stationary.
   d. Preload the center mount to 0.157–0.236 in. (3.98–5.99mm) before tightening the bolts. Torque the mounting bolts to 16 ft. lbs. (21 Nm).
   e. Lubricate the center bearing with the proper lubricant.
   f. Make sure to reinstall the bracket for the oxygen sensor plug.
   g. Make sure there is sufficient clearance between the rear heat shield and fuel tank.
12. Install the mufflers and lower the vehicle.

5 Series

1. Raise and safely support the vehicle.
2. Support the transmission from underneath with the special tools and a floor jack. Remove the nuts and washers from the transmission mounts on top of the rear transmission mounting crossmember. Loosen but do not remove the nuts located underneath which fasten the crossmember to the body. Then, slide this crossmember as far to the rear as it will go.
3. Unscrew the fastening nuts on the forward end of the CV-joint and then discard them.
4. Using a prybar to keep the driveshaft from turning, remove the self-locking nuts and bolts fastening the rear of the driveshaft to the final drive.
5. Remove the bolts fastening the center mount to the body. Bend the driveshaft down and pull the CV-joint off the transmission flange. Cover the joint to keep it clean.

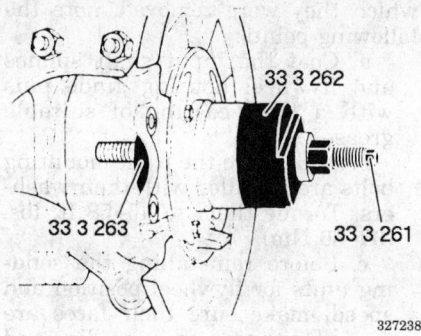

Bearings special tools: 33-3-261, 262 and 263 — 8 Series

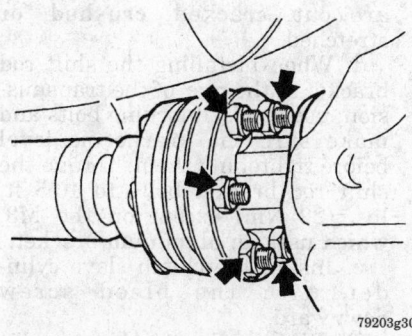

Driveshaft-to-differential flange bolt locations — M3, Z3 and 3 Series, other models similar

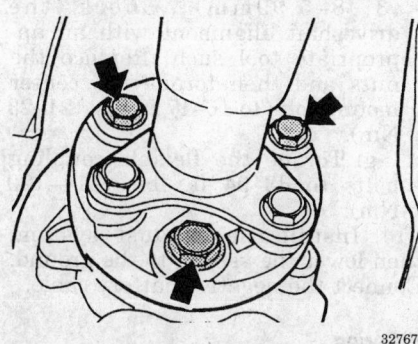

Mounting bolts for the front driveshaft joint (Guibo) disc — 5 Series, other models similar

To install:

6. Replace the gasket that fits between the joint bolts.

7. The remaining components are installed in the reverse order from which they were removed. Make sure to replace the self-locking nuts used at either end of the shaft. Preload the center mount forward by forcing the bracket 0.157-0.197 in. (3.98-5.00mm) forward from the neutral position.

8. Reposition the rear transmission mounting crossmember, install the mounting bolts until tight, then secure the transmission to the rear crossmember.

9. Lower the vehicle.

7 and 8 Series

NOTE: If equipped with a front universal joint, use tools 24 0 120 and 00 2 020 or equivalent to support the transmission during this operation.

1. Raise and safely support the vehicle. Remove the exhaust system. Remove the heat shield from the floorpan. Remove the nuts and bolts fastening the driveshaft to the transmission at the flexible coupling. Replace the self-locking nuts.

2. If equipped with a front U-joint, support the transmission from underneath with the proper tools. When the transmission is securely supported, remove the 6 bolts and remove the rear transmission mounting crossmember.

3. Remove the self-locking nuts and then the bolts fastening the driveshaft to the final drive. Replace the self-locking nuts. Remove the driveshaft, taking care to keep it protected from dirt.

4. Remove the bolts from the crossbrace and remove the center driveshaft mount. Then, bend the shaft at the middle and remove it from the vehicle by pulling it off the centering pin on the forward end.

To install:

5. Install the driveshaft into the vehicle. Install the center driveshaft mount and mounting bolts until tight.

6. Install the remaining components in the reverse order of removal. Make sure to repack the CV-joint with approved grease and replace the gasket, if necessary. Check the center bearing for lubrication and if it's dry, lubricate with the proper lubricant. If the vibration damper at the forward end of the driveshaft must be replaced, turn it 60 degrees to remove it. When remounting the center mount, preload it forward from its most natural position 0.157-0.236 in. (3.98-5.99mm).

7. Torque U-joint bolts to 52 ft. lbs. (70 Nm) and CV-joint bolts to 51 ft. lbs. (69 Nm).

8. Install the exhaust system, then lower the vehicle.

Front Wheel Hub, Knuckle and Bearings

REMOVAL AND INSTALLATION

M3, Z3, 3 Series and 5 Series

1. Raise and safely support the vehicle. Remove the front tire and wheel assembly.

2. Disconnect and suspend the brake caliper from the body without disconnecting the brake line.

3. Remove the setscrew with an Allen® wrench. Pull off the brake disc and pry off the dust cover with a small prybar.

4. Using a chisel, knock the tab on the collar nut away from the shaft. Unscrew and discard the nut.

5. Pull off the bearing with the proper puller set and discard it.

6. If the inside bearing inner race remains on the stub axle, unbolt and remove the dust guard. Bend back the inner dust guard and pull the inner race off with a special tool capable of getting under the race. Reinstall the dust guard.

NOTE: Do not reuse the bearing unit if removed.

To install:

7. If the dust guard has been removed, install a new one.

8. Install a new inner race using a race installation tool, or equivalent.

9. Install a new bearing.

10. Install the remaining components in the reverse order of removal. Make sure to tighten the wheel hub collar nut to 188 ft. lbs. (255 Nm). Lock the collar nut by bending over the tab.

11. Install the front brake caliper.

12. Install the front tire and wheel assembly, then lower the vehicle to the ground.

7 and 8 Series

1. Raise and safely support the vehicle. Remove the front tire and wheel assemblies. Remove the attaching bolts and remove and suspend the brake caliper, hanging it from the body so as to avoid putting stress on the brake line.

2. Remove the setscrew with an Allen® wrench. Pull off the brake disc and pry off the dust cover with a small prybar.

3. Using a chisel, knock the tab on the collar nut away from the shaft. Unscrew and discard the nut.

4. Install a puller collar such as 31 2 105 or equivalent to the bearing housing with 3 bolts. Install a puller

such as 31 2 102 and 312 2 106 or equivalent and pull off the bearing and discard it.

5. If the inside bearing inner race remains on the stub axle, unscrew and remove the dust guard, using a socket extension. Bend back the inner dust guard and pull the inner race off with a special tool capable of getting under the race.

To install:

6. Reinstall the dust guard and install a new dust cover.

7. Then install a special tool over the stub axle and screw it in for the entire length of the guide sleeve's threads. Slide the bearing on and follow it with 31 2 100 or equivalent, and use this tool to press the bearing on.

8. Reverse the remaining removal procedures to install the disc and caliper. Torque the wheel hub collar nut to 210 ft. lbs. (285 Nm). Lock the collar nut by bending over the tab.

9. Install a new grease cap coated with the proper lubricant.

10. Install the front tire and wheel assemblies, then lower the vehicle to the ground.

MANUAL TRANSMISSION

Transmission Assembly

REMOVAL AND INSTALLATION

M3, Z3 and 3 Series

1. Disconnect the negative battery cable. Raise and safely support the vehicle. Remove the exhaust system. Remove the cross brace and heat shield.

2. Hold the nuts on the front with one wrench and remove bolts from the rear with another to disconnect the flexible coupling at the front of the driveshaft. Some vehicles have a vibration damper at this point in the drivetrain. This damper is mounted on the transmission output flange with bolts that are pressed into the damper. On these vehicles, unscrew and remove the nuts located behind the damper.

3. Loosen the threaded sleeve on the driveshaft. Get a special tool to hold the splined portion of the shaft while turning the sleeve.

4. Remove its mounting bolts and remove the center driveshaft mount. Then, bend the driveshaft down at the center and pull it off the transmission output flange. Keep the sections of the driveshaft from pulling apart and suspend it from the vehicle with wire.

5. Remove the retainer and washer and pull out the shift selector rod.

6. Use a hex-head wrench to remove the self-locking bolts that retain the shift rod bracket at the rear of the transmission and then remove the bracket. If equipped with a shift arm, use a suitable prybar to pry the spring clip up off the boss on the transmission case and swing it upward. Then, pull out the shift shaft pin.

7. Unscrew and remove the clutch slave cylinder and support it so the hydraulic line can remain connected.

8. The transmission incorporates sending units for flywheel rotating speed and position. Remove the heat shield that protects these from exhaust heat and then remove the retaining bolt for each sending unit. Note that the speed sending unit, which has no identifying ring goes in the bore on the right, and that the reference mark sending unit, which has a marking ring, goes in the bore on the left. If the sending units are installed in reverse positions, the engine will not run at all. Pull these units out of the flywheel housing.

9. Disconnect the wiring connector going to the backup light switch and pull the wires out of the harness.

10. Support the transmission from underneath in a secure manner. Remove mounting bolts and remove the crossmember holding the rear of the transmission to the body. Then, lower the transmission onto the front axle carrier.

11. Using the proper tool, remove the bolts holding the transmission flywheel housing to the engine at the front. Make sure to retain the washers with the bolts. Pull the transmission rearward to slide the input shaft out of the clutch disc and then lower the transmission and remove from the vehicle.

To install:

12. Install the transmission in position under the vehicle. Align the input shaft and install the transmission.

13. The remaining components are installed in the reverse order from

which they were removed, note the following points:

a. Coat the input shaft splines and flywheel housing guide pins with a light coating of suitable grease.

b. Make sure the front mounting bolts are installed with their washers. Torque them to 46–58 ft. lbs. (62–80 Nm).

c. Before reinstalling the sending units for flywheel position and speed, make sure their faces are free of either grease or dirt and then coat them with a light coating of a suitable lubricant. Inspect the O-rings and replace them if they are cut, cracked, crushed, or stretched.

d. When installing the shift rod bracket at the rear of the transmission, use new self-locking bolts and make sure the bracket is level before tightening them. Torque the shift rod bracket bolts to 16.5 ft. lbs. (22 Nm) except on the M3, which uses an aluminum bracket.

e. Install the clutch slave cylinder with the bleed screw downward.

f. When installing the driveshaft center bearing, preload it forward 0.157–0.236 in. (3.98–5.99mm). Check the driveshaft alignment with an appropriate tool such. Replace the nuts and then torque the center mount bolts to 16–17 ft. lbs. (21–23 Nm).

g. Torque the flexible coupling bolts to 83–94 ft. lbs. (114–129 Nm).

14. Install the exhaust system, then lower the vehicle to the ground. Connect the negative battery cable.

5 Series

1. Disconnect the negative battery cable. Raise and safely support the vehicle. Disconnect and lower the exhaust system to provide clearance for transmission removal. Remove the heat shield brace and transmission heat shield.

2. Support the driveshaft and then unscrew the driveshaft coupling at the rear of the transmission. Use a wrench on both the nut and the bolt.

3. Working at the front of the driveshaft center bearing, unscrew the screw-on ring type connector which attaches the driveshaft to the center bearing. Then, unbolt the center bearing mount. Bend the driveshaft down and pull it off the centering pin. If equipped with a vibration damper, turn it and pull it back over the output flange before

pulling the driveshaft off the guide pin. Suspend it from the vehicle.

4. Pull off the wires for the backup light switch. Unscrew the passenger compartment console to disconnect it from the top of the transmission by removing the self-locking bolts. Discard and replace.

5. Pull out the locking clip and disconnect the shift rod at the rear of the transmission. Take care to keep all the washers.

6. If the transmission is linked to the shift lever with an arm, use a small prybar to lift the spring out of the holder on the bracket and then raise the arm. Pull out the shift shaft bolt.

7. If equipped with a flywheel housing cover (semi-circular in shape), remove the mounting bolts and remove the cover.

8. The speed sensor and reference mark sensor on the flywheel housing must be disconnected. Note their locations. The speed sensor goes in the upper bore, marked D. The reference mark sensor, which has a ring, goes in the lower bore, marked B. Check the O-rings for the sensors and install new ones if they are damaged.

9. Support the transmission securely. Then, unbolt and remove the rear transmission crossmember.

10. Remove the upper and lower attaching nuts and remove the clutch slave cylinder, supporting it so the hydraulic line need not be disconnected. Disconnect the reverse gear backup light switch and pull the wires out of the holders.

11. Unscrew the bolts fastening the transmission to the bell housing, using the proper tool. On some vehicles there are Torx® bolts used ; use a Torx® wrench for these. Pull the transmission rearward until the input shaft has disengaged from the clutch disc and then lower and remove it.

To install:

12. Place the transmission in gear. Insert the guide sleeve of the input shaft into the clutch pilot bearing carefully. Turn the output shaft to rotate the front of the input shaft until the splines line up and it engages the clutch disc.

13. Perform the remaining portions of the procedure in reverse of removal, observing the following points:

   a. Make sure the arrows on the rear crossmember point forward.

   b. Preload the center bearing mount forward of its most natural position 0.079–0.157 in. (2.07–3.99mm).

   c. In tightening the driveshaft screw on ring, use tool 26 1 040 or equivalent.

   d. When reconnecting the nuts and bolt at the transmission coupling, replace the nuts with new ones and turn only the nut, holding the bolts stationary.

   e. Make sure DME sensor faces are clean. Coat the sensor outside diameters with the proper lubricant.

   f. If equipped with a shift arm, lubricate the bolt with a light layer of a suitable lubricant.

   g. Observe these torque figures:
   Transmission-to-bell housing — 52–58 ft. lbs. (70–80 Nm).
   Rear/top transmission Torx® bolts — 46–58 ft. lbs. (62–80 Nm).
   Center mount-to-body — 16–17 ft. lbs. (21–23 Nm).
   Front joint-to-transmission — 83–94 ft. lbs. (114–129 Nm).

14. Install the exhaust system, then lower the vehicle to the ground. Connect the negative battery cable.

7 and 8 Series

1. Disconnect the negative battery cable. Raise and safely support the vehicle. Remove the exhaust system. Remove the attaching bolts and remove the heat shield mounted just to the rear of the transmission on the floorpan.

2. Support the transmission securely from underneath. Then, remove the crossmember that supports it at the rear from the body by removing the mounting bolts on both sides.

3. Using wrenches on both the bolt heads and on the nuts, remove the bolts passing through the vibration damper and front universal joint at the front of the driveshaft.

4. Remove its mounting bolts and remove the center driveshaft mount. Then, bend the driveshaft down at the center and pull it off the transmission output flange. Keep the sections of the driveshaft from pulling apart and suspend it from the vehicle with wire.

5. Pull out the circlip, slide off the washer and then pull the shift selector rod off the transmission shift shaft. Disconnect the backup light switch.

6. Lower the transmission slightly for access. Then, use a small prybar to lift the spring out of the holder on the bracket and then raise the arm. Pull out the shift shaft bolt.

7. Remove the upper and lower attaching nuts and remove the clutch slave cylinder, supporting it so the

hydraulic line need not be disconnected.

8. Unscrew the bolts fastening the transmission to the bell housing. Use a Torx® wrench to remove the bolts. Make sure to retain the washer with each bolt to ensure that they can be readily removed later, if necessary. Pull the transmission rearward until the input shaft has disengaged from the clutch disc and then lower and remove the transmission.

To install:

9. Install the transmission in position under the vehicle. Align the input shaft and install the transmission.

10. Preload the center bearing mount forward of its most natural position 0.157–0.236 in. (3.98–5.99mm).

11. When reconnecting the nuts and bolt at the transmission coupling, replace the nuts with new ones and turn only the nut, holding the bolts stationary. Tighten the center mount-to-body nut to 16–17 ft. lbs. (21–23 Nm). Tighten the front joint-to-transmission nut to 58 ft. lbs. (80 Nm).

12. Install the remaining components in the reverse order of removal.

13. Reconnect the shift arm, if equipped, and lubricate the bolt with a light layer of a suitable lubricant, then check the O-ring for crushing, cracks or cuts, replacing it, if damaged.

14. When installing the clutch slave cylinder, make sure the bleeder screw faces downward.

15. Install the exhaust system. Connect the negative battery cable, then lower the vehicle to the ground.

CLUTCH

Clutch Assembly

REMOVAL AND INSTALLATION

1. Disconnect the negative battery cable. Raise and safely support the vehicle. Remove the heat shield and then the mounting bolts. Disconnect the speed and reference mark sensors at the flywheel housing. Mark the plugs for reinstallation.

2. Remove the transmission and clutch housing.

3. On vehicles with 6–cylinder engines, a Torx® socket is required. If equipped with a 265/6 transmission

(without an integral clutch housing), remove the clutch housing.

4. Prevent the flywheel from turning, using a locking tool.

5. Loosen the mounting bolts one after another gradually, 1–1½ turns at a time, to relieve tension from the clutch.

6. Remove the mounting bolts, clutch and driveplate. Coat the splines of the transmission input shaft with Molykote® Long-term 2, Microlube® GL 2611 or equivalent. Make sure the clutch pilot bearing, located in the center of the crankshaft, turns easily.

7. Check the clutch driven disc for excess wear or cracks. Check the integral torsional damping springs, used with lighter flywheels only, for tight fit. Inspect the rivets to make sure they are all tight. Check the flywheel to make sure it is not scored, cracked, or burned, even at a small spot. Use a straightedge to make sure the contact surface is true. Replace any defective parts.

To install:

8. To install, fit the new clutch plate and disc in place and install the mounting bolts.

9. When installing the clutch retaining bolts turn them in gradually to evenly tighten the clutch disc and to prevent warpage.

10. Install the transmission and the clutch housing.

11. If equipped, install the speed and reference mark sensors. Install the heat shield.

12. Note that on vehicles with 6–cylinder engines, the clutch pressure plate must fit over dowel pins. Torque the clutch mounting bolts to 16–19 ft. lbs. (21–26 Nm).

Clutch Master Cylinder

REMOVAL AND INSTALLATION

1. Remove the necessary trim panel or carpet.

2. Disconnect the pushrod at the clutch pedal.

3. Remove the cap on the reservoir tank. On some vehicles, there is a clutch master cylinder reservoir, while on others there is a common reservoir shared with the brake master cylinder. Remove the float container, if equipped. Remove the screen and remove enough brake fluid from the tank until the level drops below the refill line or the connection for the filler pipe, if there is one.

4. Disconnect the coolant expansion tank without removing the hoses on the 7 Series models.

5. Remove the lower/left instrument panel trim. Then, remove the retaining nut from the end of the master cylinder actuating rod where the bolt passes through the pedal mechanism.

6. Disconnect the line to the slave cylinder and the fluid fill line going to the top of the master cylinder. Remove the retaining bolts and remove the master cylinder from the firewall.

To install:

7. Install the clutch master cylinder in position. The piston rod bolt should be coated with the proper lubricant. Make sure all bushings remain in position.

8. Install the remaining components in the reverse order of removal.

9. Connect the pushrod to the clutch pedal, then install any trim panel or carpet pieces removed earlier.

10. Bleed the system and adjust the pedal travel with the pushrod to 6 in.

Clutch Slave Cylinder

REMOVAL AND INSTALLATION

1. Remove enough brake fluid from the reservoir until the level drops below the refill line connection.

2. Remove the snapring or retaining bolts and pull the unit down.

3. Disconnect the line and remove the slave cylinder.

To install:

4. Install the slave cylinder on the transmission. On the 3 Series, if the engine uses the 2–section flywheel, make sure a larger cylinder with a diameter of 0.874 in. (22.2mm) is used instead of the usual cylinder with a diameter of 0.809 in. (20.5mm). Make sure to install the cylinder with the bleed screw facing upward. When installing the front pushrod, coat it with the proper anti-seize compound.

5. Refill the clutch system reservoir to the correct level.

6. Bleed the clutch hydraulic system.

Hydraulic Clutch System Bleeding

1. Fill the reservoir.

2. Connect a bleeder hose from the bleeder screw to a container filled with brake fluid so air cannot be drawn in during bleeding procedures.

3. Pump the clutch pedal about 10 times and then hold it down.

4. Open the bleeder screw and watch the stream of escaping fluid. When no more bubbles escape, close the bleeder screw and tighten it.

5. Release the clutch pedal and repeat the above procedure until no more bubbles can be seen when the screw is opened.

6. If this procedure fails to produce a bubble-free stream:

 a. Pull the slave cylinder off the transmission without disconnecting the fluid line.

NOTE: Do not depress the clutch pedal while the slave cylinder is dismounted.

 b. Depress the pushrod in the cylinder until it hits the internal stop. Then, reinstall the cylinder.

AUTOMATIC TRANSMISSION

Transmission Assembly

REMOVAL AND INSTALLATION

Except 3 Series

NOTE: To perform this operation, the following tools or equivalents are required. Special transmission support tools 24 0 120 and 00 2 020 and driveshaft locking ring tool 26 1 040 or equivalent.

1. Disconnect the battery ground cable. Loosen the throttle cable adjusting nuts, release the cable tension and disconnect the cable at the throttle lever. Then, remove the nuts and pull the cable housing out of the bracket.

2. Disconnect the exhaust system at the manifold and hangers and lower it out of the way. Remove the hanger that runs across under the driveshaft. Remove the exhaust heat shield from under the center of the vehicle.

3. Support the transmission a suitable lifting device. Remove the crossmember that supports the transmission at the rear.

4. Remove the driveshaft coupling through bolts and nuts or the CV-joint through bolts and nuts. Either type is located right at the rear of the transmission. Discard used self-lock-

ing coupling nuts. Keep the CV-joint clean and replace its gasket.

5. Unscrew the transmission locking ring at the center mount, if equipped. Then, remove the bolts and remove the center mount. Bend the driveshaft downward and pull it off the centering pin. Suspend it with wire from the underside of the vehicle.

6. Drain the transmission oil and discard it. Remove the oil filler neck. Disconnect the oil cooler lines at the transmission by unscrewing the flare nuts and plug the open connections.

7. If equipped, remove the converter cover by removing the Torx® bolts from behind and the regular bolts from underneath.

8. Remove the bolts fastening the torque converter to the driveplate, turning the flywheel as necessary to gain access from below.

9. If equipped, remove the guard for the speed and reference mark sensors. Remove the attaching bolt for each and remove each sensor. Keep the sensors clean.

10. Disconnect the shift cable by loosening the locknut fastening it to the shift lever and disconnecting the cable at the cable housing bracket.

11. If the transmission has an electrical connection, turn the bayonet fastener to the left to release the connection, disconnect it and pull the wire out of the ties.

12. Lower the transmission as far as possible. Then, remove all the Torx® or standard type bolts attaching the transmission to the engine.

13. Remove the small grill from the bottom of the transmission. Then press the converter off with a large prybar through this opening while sliding the transmission out.

To install:

14. Install the transmission under the vehicle and raise it into position. Slide the torque converter and the transmission together before installing the transmission completely to the engine.

15. Install the small grille onto the transmission.

16. Raise the transmission to install it against the engine block.

17. The remaining components are installed in the reverse order from which they were removed. Make sure the converter is fully installed onto the transmission — so the ring on the front is inside the edge of the case.

 a. When reinstalling the driveshaft, tighten the lockring with a special tool. If the driveshaft has a simple coupling, rather than a CV-joint, make sure to replace

the self-locking nuts and to hold the bolts still while tightening the nuts to keep from distorting the coupling.

 b. When installing the center mount, preload it forward from its most natural position 0.157–0.236 in. (3.98–5.99mm).

18. Connect the negative battery cable, then adjust the throttle cables.

3 Series

NOTE: To perform this operation, a support for the transmission, BMW tool 24 0 120 and 00 2 020 or equivalent and a tool for tightening the driveshaft locking ring, BMW tool 26 1 040 or equivalent, are required.

1. Disconnect the battery ground cable. Loosen the throttle cable adjusting nuts, release the cable tension and disconnect the cable at the throttle lever. Then, remove (and retain) the nuts and pull the cable housing out of the bracket.

2. Disconnect the exhaust system at the manifold and hangers and lower it aside. Remove the hanger that runs across under the driveshaft. Remove the exhaust heat shield from under the center of the vehicle.

3. Drain the transmission oil and discard it. Remove the oil filler neck. Disconnect the oil cooler lines at the transmission by unscrewing the flare nuts and plug the open connections.

4. Support the transmission with the proper tools. Separate the torque converter housing from the transmission by removing the Torx® bolts with the proper tool from behind and the regular bolts from underneath. Retain the washers used with the Torx® bolts.

5. Remove bolts attaching the torque converter housing to the engine, making sure to retain the spacer used behind one of the bolts. Then, loosen the mounting bolts for the oil level switch just enough so the plate can be removed while pushing the switch mounting bracket to one side.

6. Remove the bolts attaching the torque converter to the driveplate. Turn the flywheel as necessary to gain access to each of the bolts, which are spaced at equal intervals around it. Make sure to re-use the same bolts and retain the washers.

7. To remove the speed and reference mark sensors, remove the attaching bolt for each and remove each sensor. Keep the sensors clean.

8. Turn the bayonet type electrical connector counterclockwise and then pull the plug out of the socket. Then,

lift the wiring harness out of the harness bails.

9. Support the transmission using the proper jack. Then, remove the crossmember that supports the transmission at the rear.

10. Disconnect the transmission shift rod. Then, remove the nuts and then the through bolts from the damper-type U-joint at the front of the transmission.

11. Unscrew the transmission locking ring at the center mount, if equipped, using the special tool designed for this purpose. Then, remove the bolts and remove the center mount. Bend the driveshaft downward and pull it off the centering pin. Suspend it with wire from the underside of the vehicle.

12. Lower the transmission as far as possible. Then, remove all the Torx® or standard type bolts attaching the transmission to the engine.

13. Remove the small grill from the bottom of the transmission. Then press the converter off with a large prybar passing through this opening while sliding the transmission out.

To install:

14. Install the transmission in position under the vehicle and install the torque converter onto the transmission.

15. Make sure the converter is fully installed onto the transmission — so the ring on the front is inside the edge of the case. Install the small grille onto the bottom of the transmission.

16. Raise the transmission up and bring it together with the engine, then install the engine-to-transmission attaching bolts until tight.

17. The remaining components are installed in the reverse order from which they were removed. When reinstalling the driveshaft, tighten the lockring with the proper tool. Make sure to replace the self-locking nuts on the driveshaft flexible joint and to hold the bolts still while tightening the nuts to keep from distorting it. When installing the center mount, preload it forward from its most natural position 0.157–0.236 in. (3.98–5.99mm). When reconnecting the bayonet type electrical connector, make sure the alignment marks are aligned after the plug it twisted into its final position. When reinstalling the speed and reference mark sensors, inspect the O-rings used on the sensors and install new ones, if necessary. Make sure to install the speed sensor into the bore marked **D** and the reference mark sensor, which is marked with a ring, into the bore

marked **B**. Torque the crossmember mounting bolts to 16–17 ft. lbs. (21–23 Nm). If O-rings are used with the transmission oil cooler connections, replace them.

18. Install the negative battery cable, then adjust the throttle cables.

SHIFT LINKAGE ADJUSTMENT

1. Move the selector lever to **P** position. Loosen the nut until the cable is free.

2. Push the transmission lever to the **D** or **P** position. Then push the cable rod in the opposite direction.

3. Clamp down the cable rod without tension.

4. Tighten the nut to 7.0–8.5 ft. lbs. (9–11 Nm).

NOTE: Do not bend the cable.

THROTTLE LINKAGE ADJUSTMENT

1. On the injection system throttle body, loosen the 2 locknuts at the end of the throttle cable and adjust the cable until there is a play of 0.010–0.030 in. (0.254–0.762mm).

2. Loosen the locknut and lower the kickdown stop under the accelerator pedal. Have someone depress the accelerator pedal until the transmission detent can be felt. Then, back the kickdown stop back out until it just touches the pedal.

3. Check that the distance from the seal at the throttle body end of the cable housing is at least 1.732 in. (43.9mm) from the rear end of the threaded sleeve. If this dimension checks out, tighten all the locknuts.

FRONT SUSPENSION

MacPherson Strut

REMOVAL AND INSTALLATION

M3, Z3 and 3 Series

1. Disconnect the negative battery cable.

2. Raise and safely support the vehicle. Remove the tire and wheel assembly.

3. Disconnect the brake pad wear indicator plug and ground wire. Pull the wires out of the holder on the strut. Remove the ABS pulse sender, if equipped.

4. Unbolt the caliper and pull it away from the strut, suspending it with a piece of wire from the body. Do not disconnect the brake line.

5. Remove the attaching nut and then detach the pushrod on the stabilizer bar at the strut.

6. Unscrew the attaching nut and press off the guide joint with the proper tool.

7. Unscrew the nut and press off the tie rod joint.

8. Press the bottom of the strut outward and push it over the guide joint pin, using the proper tool. Support the bottom of the strut.

9. Unscrew the nuts at the top of the strut, from inside the engine compartment, then remove the strut.

To install:

10. Position the strut in the vehicle, then tighten the upper strut mounting nuts 16–17 ft. lbs. (21–23 Nm). The upper strut mounting nuts must be replaced with new self-locking nuts.

11. The remaining components are installed in the reverse order from which they were removed. Tie rod and guide joints must have both pins and both bores clean for reassembly. Replace both self-locking nuts. Torque the control arm to spring strut attaching nut to 43–51 ft. lbs. (59–69 Nm).

12. Install the front tire and wheel assemblies, then lower the front of the vehicle.

13. Connect the negative battery cable.

5, 7 and 8 Series

1. Disconnect the negative battery cable.

2. Raise and safely support the vehicle. Remove the tire and wheel assembly.

3. Disconnect the brake pad wear indicator plug and ground wire. Pull the wires out of the holder on the strut. Remove the ABS pulse sender, if equipped.

4. Disconnect the stabilizer pushrod with the proper tool.

5. Disconnect the lower strut bolts at the control arm.

6. Support the bottom of the strut and unscrew the nuts at the top of the strut, from inside the engine compartment. Remove the strut.

7. The installation is the reverse of the removal procedure.

Lower Control Arms

REMOVAL AND INSTALLATION

M3, Z3 and 3 Series

1. Raise and safely support the vehicle. Remove the front tire and wheel assembly.

2. Disconnect the rear control arm bracket where it connects to the body by removing the bolts.

3. Remove the nut and disconnect the thrust rod on the front stabilizer bar where it connects to the center of the control arm.

4. Unscrew the nut which attaches the front of the stabilizer bar to the crossmember and remove the nut from above the crossmember. Then, use a plastic hammer to knock this support pin out of the crossmember.

5. Unscrew the nut and press off the guide joint where the control arm attaches to the lower end of the strut, using the proper tool.

To install:

6. Keep these points in mind:

   a. Replace the self-locking nut that fastens the guide joint to the control arm.

   b. Make sure the support pin and the bore in the crossmember are clean before inserting the pin through the crossmember. Replace the original nut with a replacement nut and washer.

   c. Torque the control arm-to-spring strut nut to 43–51 ft. lbs. (58–69 Nm). Torque the control arm support to crossmember nut to 29–34 ft. lbs. (40–46 Nm). Torque the pushrod on the stabilizer bar to 29–34 ft. lbs. (40–46 Nm).

5, 7 and 8 Series

1. Raise and safely support the vehicle. Remove the tire and wheel assembly.

2. Remove the mounting bolts that fasten the bottom of the strut to the steering knuckle.

3. Remove the cotter pin and castellated nut. Use a suitable ball joint remover to press the ball joint end of the control arm off the steering knuckle.

4. Remove the self-locking nut. Then, remove the through bolt and the washers, slide the inner end of the strut and bushing out of the front suspension crossmember.

To install:

5. Note the following points:

   a. Make sure both washers are replaced to cushion the bushing where it contacts the suspension crossmember.

b. Replace the bushing if it is worn or cracked.

c. Use a new self-locking nut on the bolt fastening the inner end of the strut.

d. Align the bottom of the strut with the steering knuckle so the tab on the arm fits into the notch on the bottom of the strut. Install the bolts with a locking type sealer.

e. When installing the arm ball joint onto the steering knuckle, tighten the nut until the cotter pin hole lines up and then use a new cotter pin in the nut.

f. Final tighten the through bolt for the inner end of the arm after the vehicle is on the ground at normal ride height.

Sway Bar

REMOVAL AND INSTALLATION

1. Raise and safely support the vehicle.
2. Disconnect the push/thrust rod on both sides.
3. Disconnect the left side control arm bracket on the 3 Series.
4. Disconnect the left and right stabilizer mounts. Remove the stabilizer bar.

To install:
5. Position the stabilizer bar in the vehicle, then install the left and right stabilizer mounts. Tighten the stabilizer mount bolts to 16 ft. lbs. (21 Nm).
6. On the 3 Series models (including the M3 and Z3), install the left side control arm bracket.
7. Install the push/thrust rod on both sides.
8. Lower the vehicle to the ground.

Front Wheel Bearings

REMOVAL AND INSTALLATION

M3, Z3 and 3 Series

1. Raise and safely support the vehicle. Remove the tire and wheel assembly.
2. Remove the attaching bolts and remove and suspend the brake caliper, hanging it from the body so as to avoid putting stress on the brake line.
3. Remove the setscrew with an Allen® wrench. Pull off the brake disc and pry off the dust cover with a small prybar.

4. Using a chisel, knock the tab on the collar nut away from the shaft. Unscrew and discard the nut.
5. Pull off the bearing with a suitable bearing puller and discard it.
6. If the inside bearing inner race remains on the stub axle, unbolt and remove the dust guard. Bend back the inner dust guard and pull the inner race off with a special tool capable of getting under the race. Reinstall the dust guard.

To install:
7. If the dust guard has been removed, install a new one. Install a special tool over the stub axle and screw it in for the entire length of the guide sleeve's threads. Press the bearing on.
8. Reverse the remaining removal procedures to install the disc and caliper. Torque the wheel hub collar nut to 188 ft. lbs. (255 Nm). Lock the collar nut by bending over the tab.

5, 7 and 8 Series

1. Raise and safely support the vehicle. Remove the tire and wheel assembly.
2. Remove the attaching bolts and remove and suspend the brake caliper, hanging it from the body so as to avoid putting stress on the brake line.
3. Remove the setscrew with an Allen® wrench. Pull off the brake disc and pry off the dust cover with a small prybar.
4. Using a chisel, knock the tab on the collar nut away from the shaft. Unscrew and discard the nut.
5. Using the proper tool, pull off the bearing and discard it.

To install:
6. If the inside bearing inner race remains on the stub axle, unscrew and remove the dust guard, using a socket extension. Bend back the inner dust guard and pull the inner race off with a special tool capable of getting under the race. Reinstall the dust guard and install a new dust cover.
7. Then install a special tool over the stub axle and screw it in for the entire length of the guide sleeve's threads. Slide the bearing on and follow it with the proper tool and use this tool to press the bearing on.
8. Reverse the remaining removal procedures to install the disc and caliper. Torque the wheel hub collar nut to 210 ft. lbs. (285 Nm). Lock the collar nut by bending over the tab.
9. Install a new grease cap coated with a suitable sealer.

REAR SUSPENSION

Shock Absorbers

REMOVAL AND INSTALLATION

M3, Z3 and 3 Series

1. Remove the trunk trim panel to expose the upper shock mounts.
2. Raise the rear of the vehicle and support safely.
3. Support the trailing arm and remove the lower mounting bolt.

WARNING
The support must not be removed until the new shock absorber is installed, and the vehicle must not be raised since this could damage the halfshafts.

4. Pull off the cap and remove the upper mounting nuts and remove the shock from the vehicle.
To install:
5. Place the shock into position with new seals fitted between the shock absorber and body. Renew the upper self-locking nuts and torque to 11 ft. lbs. (15 Nm) for the Z3 and 318ti, and to 16 ft. lbs. (22 Nm) for all other 3 Series vehicles (including the M3).
6. Install the trunk trim panel.
7. Install the lower shock mounting to the rear axle assembly. The thrust washer on the rubber mount must face the screw head.
8. Lower the vehicle. With the vehicle resting at standard ride height, torque the mounting bolt to 63 ft. lbs. (87 Nm), or to 94 ft. lbs. (130 Nm) if marked with 10.9, for the Z3 and 318ti models, or to 74 ft. lbs. (100 Nm) for all other 3 Series models (including the M3).

5 and 7 Series

Standard Suspension

1. Raise and support the rear of the vehicle.
2. Remove the rear seat cushion and back rest. Remove the trim panel over the strut mount.
3. Support the control arm, pull off rubber cap and remove the nuts at the top of the strut mount.
4. Remove the lower mounting bolt and lower the spring/shock assembly. Remove the assembly from the vehicle.
5. Use a spring compressor and compress the spring. Remove the top

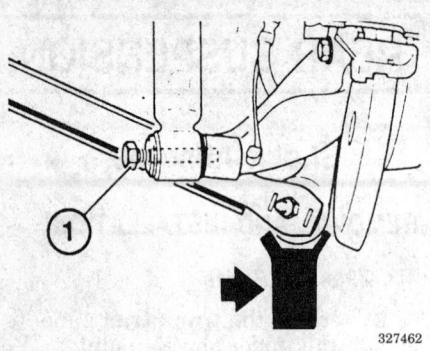

Support the trailing arm and remove bolt (1) — M3, Z3 and 3 Series

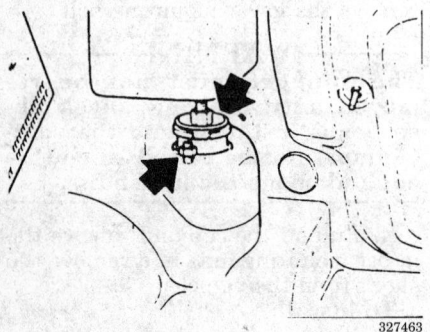

Upper mounting nut locations — M3, Z3 and 3 Series

nut and pull the top mount off. Remove the spring.

To install:

6. Compress the new spring or replace the old spring on the shock. Install the mount and washers. Use a new locknut and torque to 18 ft. lbs. (25 Nm). Release the spring.

7. Install the shock and torque the upper mount nuts to 16 ft. lbs. (21.5 Nm). Loosely install the lower mounting bolt.

8. With the vehicle lowered to the ground and at standard riding height, torque the lower mount to 94 ft. lbs. (130 Nm).

9. Install the trim and seat cushions.

Ride Level Height Control Suspension

1. Raise and support the rear of the vehicle.

2. Remove the rear seat cushion and back rest. Remove the trim panel over the strut mount.

NOTE: The coil spring, shock absorber assembly acts as a strap so the control arm should always be supported.

3. Disconnect the low pressure switch electrical connection and turn on the ignition.

4. Disconnect the control rod nut, holding the collar with an 8mm

wrench against torque. Don't disconnect the rod at the ball joint.

5. Operate the lever on the control switch in the "discharge" direction for about 20 seconds to discharge fluid from the lines.

6. Disconnect the hydraulic line on the strut and turn off the ignition.

7. Support the control arm, pull off rubber cap and remove the nuts at the top of the strut mount.

8. Remove the lower mounting bolt and lower the spring strut assembly. Remove the assembly from the vehicle.

9. Use a spring compressor and compress the spring. Remove the top nut and pull the top mount off. Remove the spring.

To install:

10. Compress the new spring or replace the old spring on the strut. Install the mount and washers. Use a new locknut and torque to 18 ft. lbs. (25 Nm). Release the spring.

11. Install the spring strut and torque the upper mount nuts to 16 ft. lbs. (21.5 Nm). Loosely install the lower mounting bolt.

12. Connect the hydraulic line on the strut.

13. Connect the control rod nut, holding the collar with an 8mm wrench against torque.

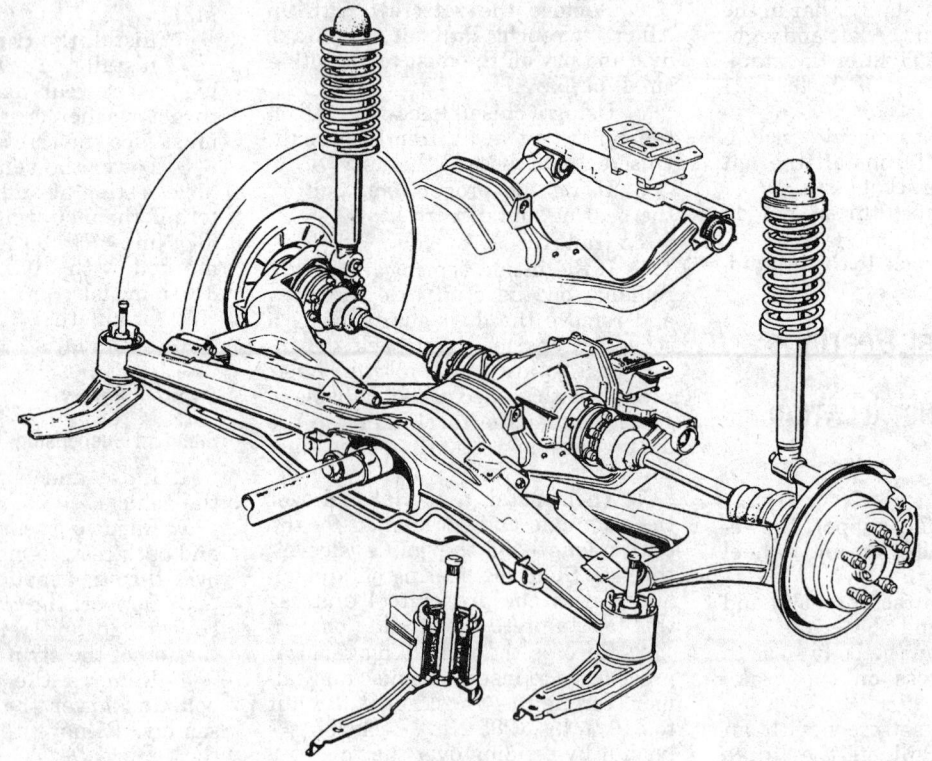

Rear suspension system — 5 Series

REAR SPRING STRUT LAYOUT DRAWING

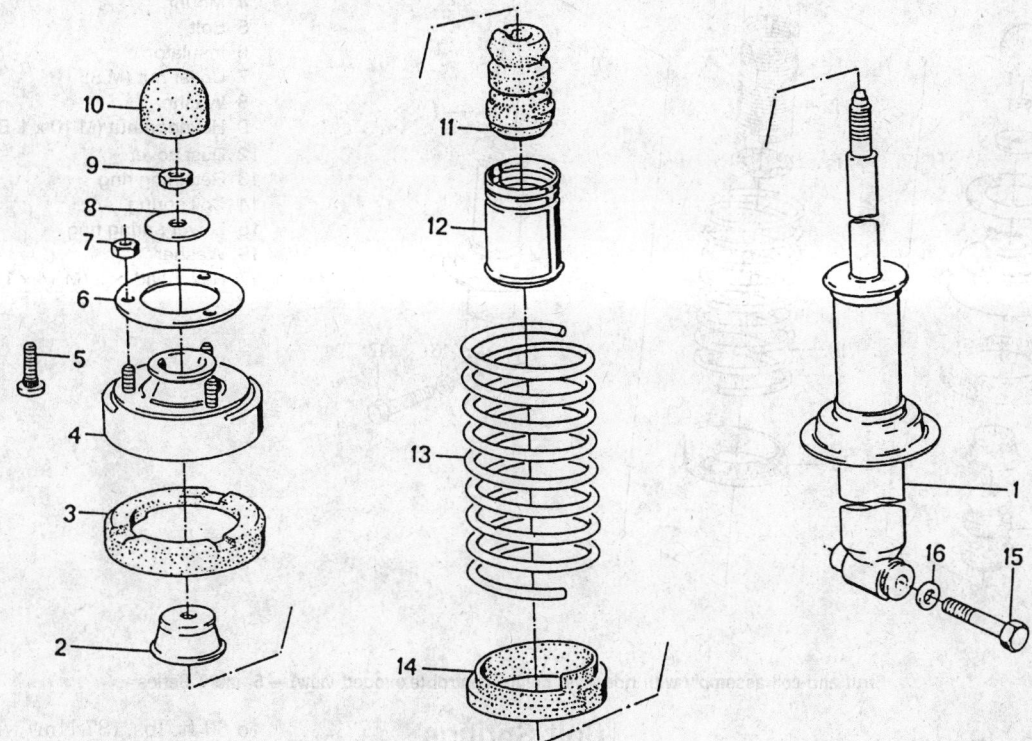

3 Upper spring ring	12 Protective tube
4 Mount	13 Coil spring
5 Bolt	14 Lower spring ring
6 Insulator	15 Bolt M 14 x 1.5 x 85
7 Collar nut M 8	16 Washer
8 Disc	
9 Hexagon nut M 10 x 1.8 ZN	

323210

Shock and coil spring assembly (exploded view) — 5 and 7 Series

14. Connect the low pressure switch electrical connection.

15. With the vehicle lowered to the ground and at standard riding height, torque the lower mount to 94 ft. lbs. (130 Nm).

16. Install the trim and seat cushions.

8 Series

————— **WARNING** —————
The support must not be removed until the new shock absorber is installed, and the vehicle must not be raised since this could damage the halfshafts.

1. Raise and safely support the vehicle.

2. Properly support the trailing arms.

3. Compress the coil springs safely, using a suitable tool.

4. Remove the trunk mat and remove upper shock mount bolts.

5. Remove the lower shock mount bolts and remove shock absorber.

To install:

6. Exchange or replace the upper shock mount and torque the upper shock nut to 11 ft. lbs. (15 Nm).

7. Replace the gasket between the shock mount and the body. Install the shock and torque the new self-locking nuts to 16 ft. lbs. (22 Nm). Install the trunk mat.

8. The thrust disk on the rubber mount must face the screw head. Install the lower shock mounting bolt and replace the cap.

9. Release the coil spring, and remove the support for the trailing arm. Lower the vehicle. With the vehicle resting at standard ride height, torque the mounting bolt to 85 ft. lbs. (115 Nm).

REAR SPRING STRUT ASSEMBLY DRAWING (WITH RIDE LEVEL HEIGHT CONTROL)

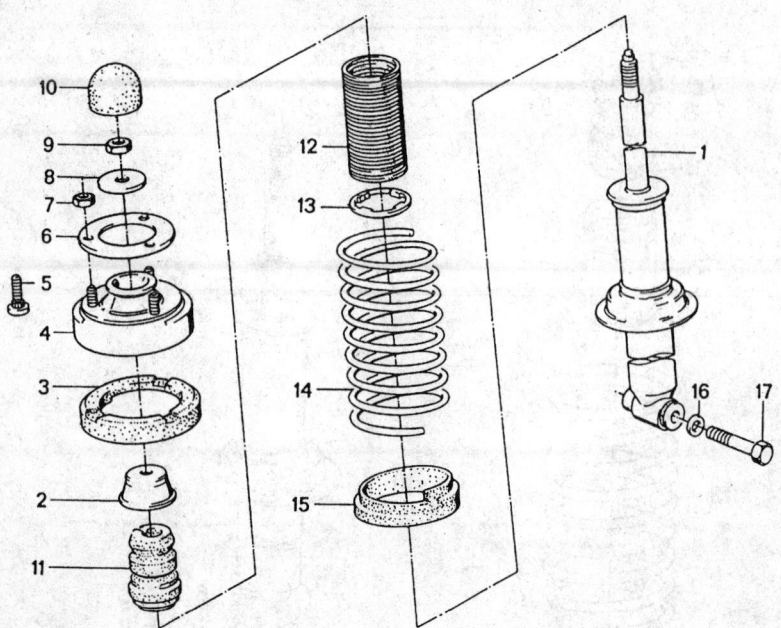

1 Shock absorber
2 Support
3 Upper spring ring
4 Mount
5 Bolt
6 Insulator
7 Collar nut (M 8)
8 Washer
9 Hexagon nut (M 10 x 1-BZN)
12 Dust cover
13 Retaining ring
14 Coil spring
15 Lower spring ring
16 Washer
17 Hex. head bolt (M 14 x 1.5 x 85)

323211

Strut and coil assembly with ride level height control (exploded view) — 5 and 7 Series

Compress the coil spring — 8 Series
323208

Lower mounting nut location — 8 Series
323209

Coil Springs

REMOVAL AND INSTALLATION

M3, Z3 and 3 Series

Except Z3 and 318ti

1. Raise the rear of the vehicle and support securely. Do not support on the suspension parts.
2. Remove the tire and wheel assembly.
3. Support the lower trailing arm at the hub and disconnect the stabilizer bar at the control arm and the subframe.
4. Remove the shock absorber lower mounting bolt.
5. Lower the trailing arm slowly and remove the spring to the side.
To install:
6. Install the spring with the bushing in place and the top of the upper spring ring lubricated.
7. Raise the trailing arm to a level where the bolt can be replaced in the lower shock mount. Connect the stabilizer bar. Do not tighten any bolts yet.
8. Install the tire and wheel assembly.
9. Lower the rear of the vehicle.
10. Torque the stabilizer bolt to 16 ft. lbs. (21.5 Nm) and the shock bolt

to 63 ft. lbs. (87 Nm) with the control arm in the normal ride position.

Z3 and 318ti

1. Disconnect the rear portion of the exhaust system and hang it from the body.
2. Disconnect the final drive rubber mount, push it down, and hold it down with a wedge.
3. Remove the bolt that connects the rear stabilizer bar to the strut on the side being worked on. Be careful not to damage the brake line.

NOTE: Support the lower control arm securely with a jack or other device that will permit it to be lowered gradually, while maintaining secure support.

4. Then, to prevent damage to the output shaft joints, lower the control arm only enough to slip the coil spring off the retainer.
To install:
5. Make sure, in replacing the spring, that the same part number, color code, and proper rubber ring are used. Install the spring, making sure that the spring is in proper position.
6. Keep the control arm securely supported while raising and replace the shock bolt. Install the bolts in the final drive rubber mount and torque to 69 ft. lbs. (95 Nm).

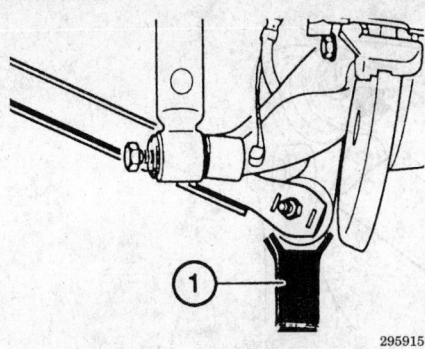

Support the trailing arm (1) — 3 Series, except Z3 and 318ti

295915

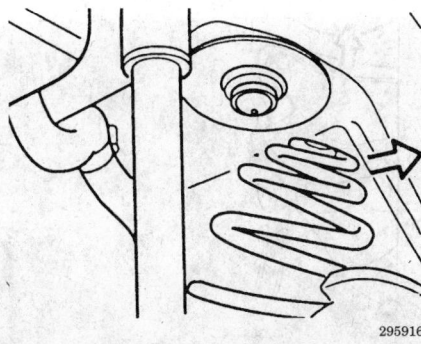

Remove the coil spring — 3 Series, except Z3 and 318ti

295916

7. Torque the stabilizer bolt to 16 ft. lbs. (21.5 Nm), and the shock bolt to 63 ft. lbs. (87 Nm) with the control arm in the normal ride position. Install the exhaust system.

5 and 7 Series

The coil spring is removed along with the shock absorber. The 5 and 7 Series use a "coil over" type shock absorber where the spring is mounted to the shock in one compact unit. Once the shock is removed from the vehicle, the spring can be compressed and separated from the shock absorber.

8 Series

1. Raise and safely support the vehicle.
2. Position the coil spring compressor tool onto the coil spring.
3. Place spring tensioning plate in the middle of the coil spring. Turn up upper spring tensioning plate completely, and turn down lower spring tensioning plate completely.
4. Slide in and turn the tensioning shaft until crosshead is inserted perfectly in the opening of the upper spring retainer.
5. Compress the spring.
6. Properly support the trailing arm.

7. Remove the spring assembly.
To install:
8. Make sure, in replacing the spring, that the same part number, color code, and proper rubber ring are used. Install the spring, making sure that the spring is in proper position.
9. Keep the control arm securely supported. Check for correct position of the rubber liners.
10. Release the spring.
11. Remove the support and lower the vehicle.

Upper Control Arm

REMOVAL AND INSTALLATION

M3 and 3 Series — Except Z3 and 318ti

NOTE: Make sure to performing a chassis alignment after completing this procedure.

1. Raise and safely support the vehicle.
2. Remove the stabilizer support bolt, and remove the stabilizer support.
3. Remove the wheel carrier-to-upper control arm bolt.
4. Disconnect the rear axle differential. Do not detach the half-shafts.
5. Retract the rear axle differential as far as possible towards the rear using special tool 33–04–390 or equivalent.
6. Unclip the power cable for brake lining wear indicator and the ABS sensor.
7. Remove the bolt on the rear axle carrier and remove the upper control arm.
To install:
8. Position the upper control arm. Install a screw and washer assembly in place of the expansion bolt removed from the rear axle carrier. Torque the mounting bolt to 57 ft. lbs. (77 Nm).
9. Clip the power cable in position.
10. Reposition the rear axle carrier to its normal place and connect.
11. Install the wheel carrier-to-upper control arm bolt. Torque the mounting bolt to 94 ft. lbs. (127 Nm).
12. Position the stabilizer support and install the retaining bolt.
13. Lower the vehicle.
14. Check rear axle for correct wheel alignment.

Z3 and 318ti Models

The Z3 and 318ti models, although considered an E36 chassis, utilize the rear suspension of the E30 chassis. The E30 rear suspension setup only

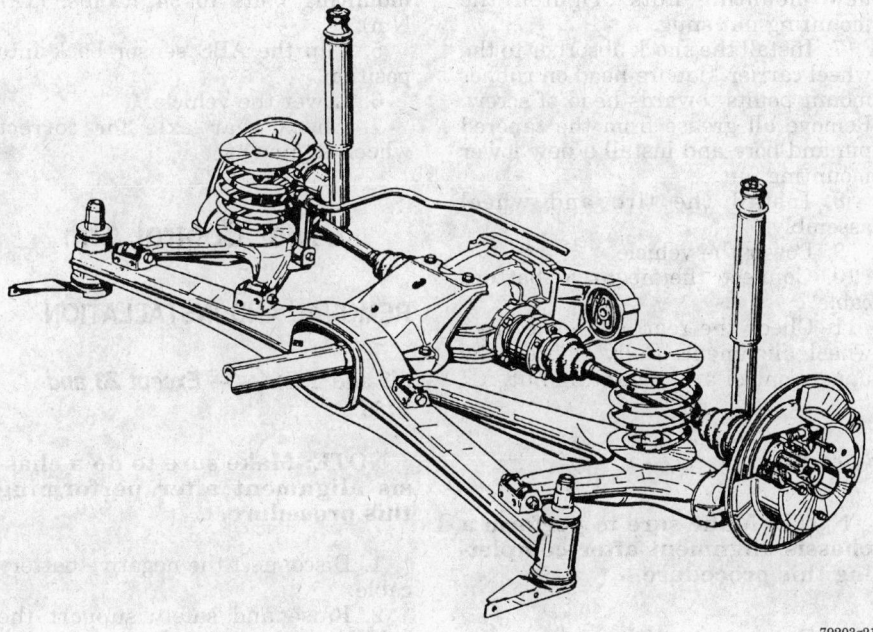

Rear suspension setup on the Z3 and 318ti models

79203g21

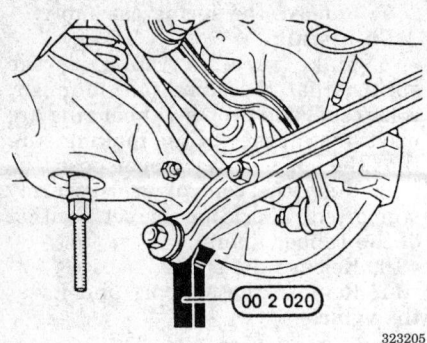

Support the trailing arm (1) — 8 Series

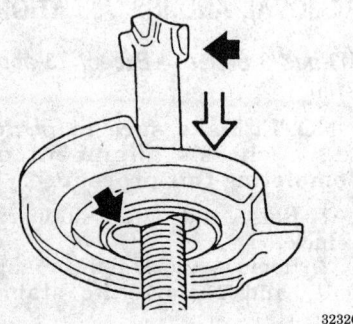

Tensioning shaft — 8 Series

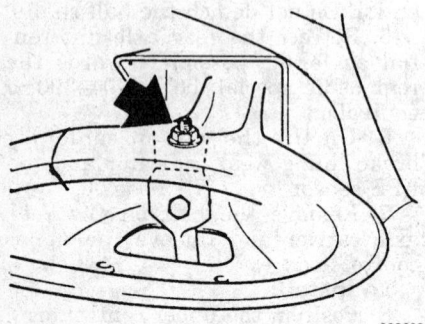

Stabilizer support — 3 Series, except Z3 and 318ti

consists of a lower control arm — no upper control arm is used.

7 Series

1. Disconnect the negative battery cable.
2. Raise and support the vehicle safely.
3. Remove the rear wheel and tire assembly.
4. Unfasten the lower shock absorber mounting nut and pull the shock absorber from the wheel carrier.

5. Unfasten and remove the upper control arm from the vehicle.

To install:

6. Install the upper control arm to the vehicle. Secure the arm using new mounting nuts. Tighten the mounting nut snug.
7. Install the shock absorber to the wheel carrier. Square-head on rubber mount points towards head of screw. Remove all grease from the tapered pin and bore and install a new lower mounting nut.
8. Install the tire and wheel assembly.
9. Lower the vehicle.
10. Connect the negative battery cable.
11. Check the rear axle for correct wheel alignment. Fully tighten the upper control arm mounting nut.

8 Series

NOTE: Make sure to perform a chassis alignment after completing this procedure.

1. Raise and safely support the vehicle.
2. Remove the three mounting bolts for the upper control arm.
3. Unclip the ABS sensor wire.

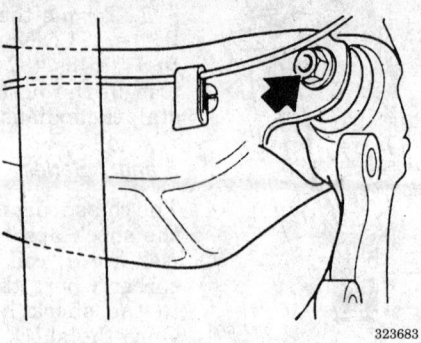

Wheel carrier-to-upper control arm bolt location — 3 Series, except Z3 and 318ti

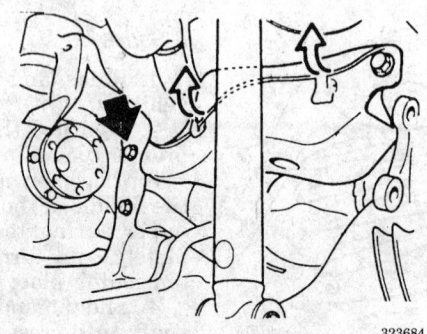

Rear axle carrier-to-upper control arm bolt location — 3 Series, except Z3 and 318ti

To install:

4. Position the upper control arm. Install the three mounting bolts. Torque the mounting bolt rear sub-frame to 57 ft. lbs. (77 Nm), and torque the mounting bolts to 94 ft. lbs. (127 Nm).
5. Clip the ABS sensor back into position.
6. Lower the vehicle.
7. Check rear axle for correct wheel alignment.

Lower Control Arm

REMOVAL AND INSTALLATION

M3 and 3 Series — Except Z3 and 318ti

NOTE: Make sure to do a chassis alignment after performing this procedure.

1. Disconnect the negative battery cable.
2. Raise and safely support the vehicle.
3. Disconnect the rear axle differential. Do not detach the halfshafts.
4. Retract the rear axle differential as far as possible towards the

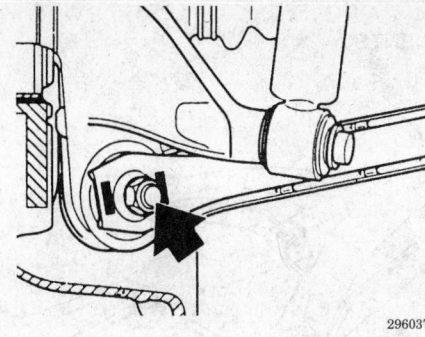

Unfasten nut at control arm — 7 Series

323725

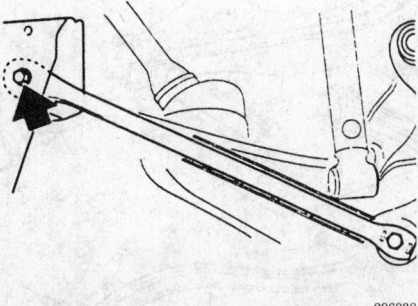

Trailing arm bolt location — 3 Series, except Z3 and 318ti

296037

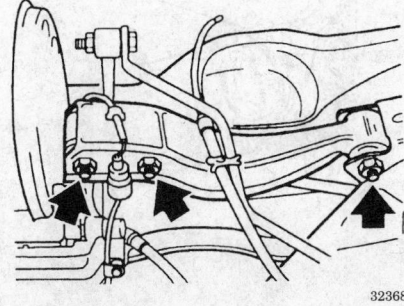

Upper control arm mounting bolt locations — 8 Series

323680

Rear axle carrier-to-lower control arm bolt location — 3 Series, except Z3 and 318ti

296038

rear using special tool 33 04 390 or equivalent.

5. Support the trailing arm. Remove the bolt on the trailing arm.

6. Remove the bolt on the rear axle carrier and remove the lower control arm.

To install:

7. Position the lower control arm. Install a bolt and washer assembly in place of the expansion bolt removed from the rear axle carrier. Torque the bolt to 81 ft. lbs. (110 Nm).

8. Install the trailing arm bolt. Torque the bolt to 81 ft. lbs. (110 Nm), and remove the support.

9. Reposition the rear axle carrier to its normal place and connect. Torque the bolts to 57 ft. lbs. (77 Nm).

10. Lower the vehicle.

11. Check rear axle for correct wheel alignment.

Z3 and 318ti Models

1. Raise the vehicle and remove the rear wheel. Apply the parking brake and disconnect the output shaft at the rear axle shaft. Remove the parking brake lever.

2. Remove the brake fluid from the master cylinder reservoir. To do this, it will be necessary to remove

the strainer at the top of the reservoir.

3. Disconnect the brake line connection on the rear control arm. Plug the openings.

4. Remove the rear brake components, then remove the halfshafts.

5. Support the control arm securely. Disconnect the shock absorber at the control arm. Lower the control arm slowly and remove the spring.

6. Remove the nuts and then slide the bolts out of the mounts where the control arm is mounted to the axle carrier.

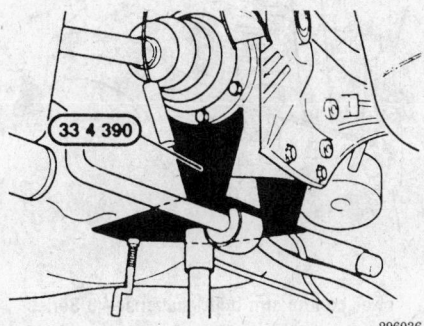

Special tool: 33-4-390 — 3 Series, except Z3 and 318ti

296036

To install:

7. Install the control arm into position. Install the bolt that goes into the inner bracket first.

8. Tighten the bolts holding the control arm to the axle carrier to 48–54 ft. lbs. (62–70 Nm).

9. Make sure the spring is positioned properly top and bottom. Tighten the shock absorber bolt to 52–63 ft. lbs. (68–82 Nm).

10. Install the halfshafts and rear brake components.

11. Reconnect the parking brake cable and adjust. Then apply the brake and reconnect the output shaft.

12. Reconnect the brake line, replenish with the proper brake fluid, and bleed the system.

8 Series

NOTE: Make sure to perform a chassis alignment after completing this procedure.

1. Disconnect the negative battery cable.

2. Raise and safely support the vehicle.

3. Carefully support the rear axle with a service jack.

4. Compress and remove the coil spring.

5. Remove the bottom shock absorber bolts.

6. Remove the 3 lower support arm bolts.

To install:

7. Raise the lower support arm into position. Install the 3 bolts. Torque the bolt to the rear sub-frame to 94 ft. lbs. (127 Nm), and the bolts to the wheel carrier to 111 ft. lbs. (150 Nm).

8. Install the bottom shock absorber bolt. Torque the bolts to 85 ft. lbs. (115 Nm).

9. Install and slowly release the coil spring.

10. Lower the vehicle.

11. Connect the negative battery cable.

12. Check rear axle for correct wheel alignment.

Rear Stabilizer Bar

REMOVAL & INSTALLATION

1. Disconnect the stabilizer from the trailing arm on either side by removing the connecting bolt from the lower end of the link.

2. Disconnect the stabilizer on the crossmember.

3. Check the rubber bushings for wear and replace as necessary.

REAR AXLE COMPONENTS

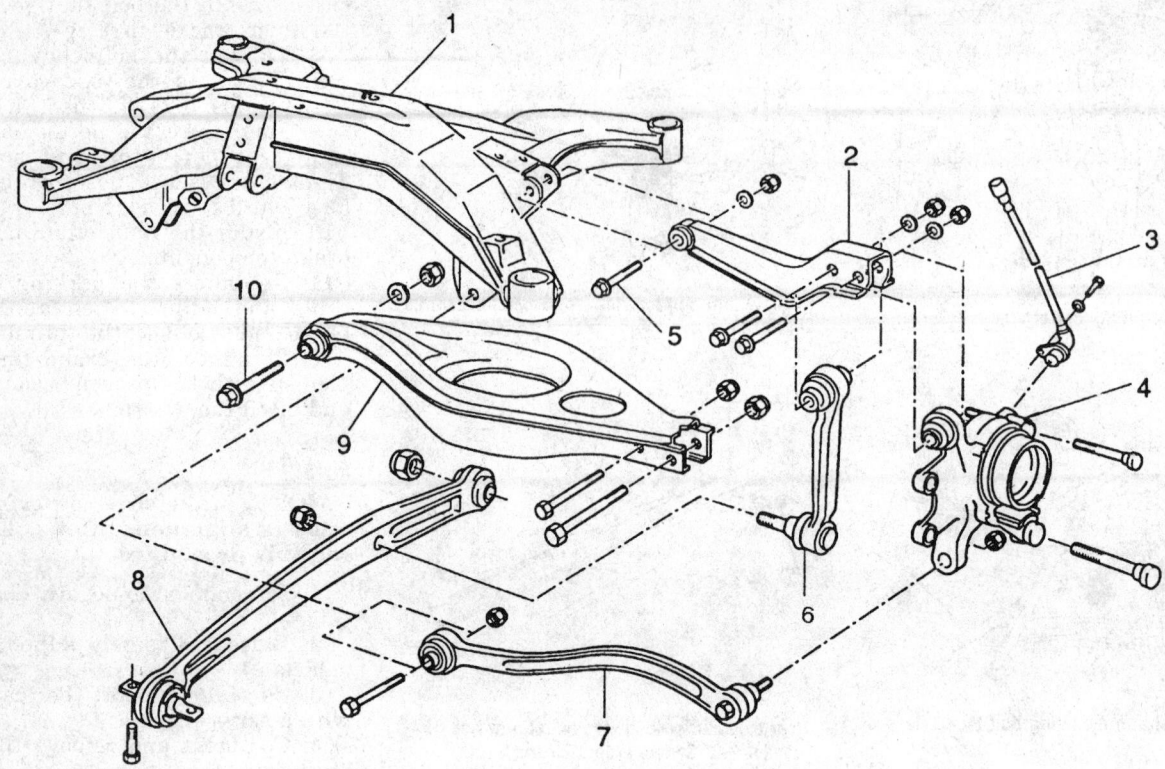

1 Rear axle carrier
2 Upper control arm
3 ABS sensor
4 Wheel carrier
5 Camber adjusting bolt
6 Push rod / integral
7 Guide arm
8 Trailing arm
9 Lower support arm
10 Toe adjusting bolt

332471

Rear axle components (exploded view) — 8 Series

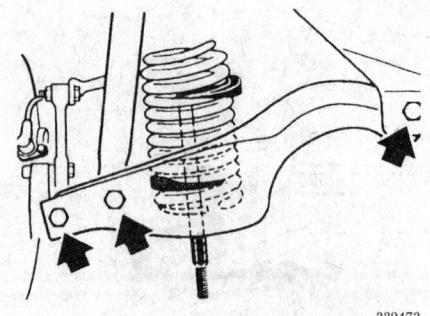

332472

Bottom shock absorber bolt location (1) — 8 Series

332473

Lower control arm bolt locations — 8 Series

To install:

4. Install the stabilizer bar and tighten the mounting bolts securely.

5. Attach the stabilizer bar to the trailing arm on either side.

STEERING

Steering Wheel

CAUTION

On vehicles equipped with an air bag, the negative battery cable must be disconnected, before working on the system. Failure to do so may result in deployment of the air bag and possible personal injury.

Disarming the Airbag System

1. Place the ignition switch in the **OFF** position.
2. Disconnect the negative battery terminal and cover the battery terminal to prevent accidental contact.
3. Once the battery has been disconnected, wait for a period of approximately 10 minutes allowing the capacitor in the control unit to discharged. Once the capacitor is discharged, a trigger pulse cannot be generated inadvertently.
4. Disconnect the crash sensors in the engine compartment.
5. Remove the lower cover of the steering column and disconnect the orange connector. The E30 3 Series has a small panel on the bottom of the steering column that when pulled down, holds the connector.
6. On the E34 5-Series, disconnect the seatbelt tensioner connectors.

Enabling the Airbag System

1. Place the ignition switch in the **OFF** position.
2. Connect the sensors, the steering column connector and the seatbelt tensioner connectors.
3. Connect the negative battery terminal.
4. Place the ignition switch in the **ON** position. Check that the SRS light illuminates for 6 seconds and extinguishes. If it illuminates in any other pattern, there is a problem that needs to be rectified by a qualified BMW technician.

REMOVAL AND INSTALLATION

With AIR BAG 1 System

NOTE: The Air Bag 1 system can be identified by the air bag telltale lamp located on the instrument panel.

CAUTION

All precautions must be adhered to or premature deployment of the air bag may occur causing personal injury. If the air bag system is damaged or rendered inoperative, the additional protection an air bag provides in a frontal collision will not be available and may lead to greater injury in an accident.

1. Disconnect the negative battery terminal. Wait for a period of approximately 10 minutes allowing the capacitor in the control unit to discharged.
2. Disarm the SRS: Remove the lower steering column trim or the small panel and disconnect the orange connector.
3. Use a Torx® T30 socket to remove the air bag module bolts from behind the steering wheel.
  a. Remove the air bag module and disconnect the wire on the back.
  b. Place the air bag module in the trunk with the pad facing up.
4. Turn the steering wheel to straight-ahead position. Marks on steering gear and steering spindle should be aligned.
5. Hold the steering wheel and remove the nut from the steering shaft.
6. Mark the relationship of the steering wheel to the shaft and pull the steering wheel off the shaft.

WARNING

Once the steering wheel is removed, the torsion spring securing the contact ring in its center position becomes effective. On no account should the steering be moved. The contact ring may be damaged.

To install:

7. If the contact ring has moved, it must be adjusted to its center position. Determine the center position of the contact ring by pressing on the spring and rotating the ring completely left or right to the stop. Return the ring half the total number of turns and align the marks on the contact ring.
8. Before installing the steering wheel, coat the slip ring with grease.
9. Install the steering wheel in its original position with the lock pin in the bore. Install the washer and a new locknut. Torque the nut to 58 ft. lbs. (80 Nm). Do not use the steering column lock to hold the steering wheel against the torque. Hold the wheel during torquing.
10. Install the air bag module and connect the wires. Torque the mounting bolts to 6 ft. lbs. (8 Nm), right side first. Check for smooth rotation of the wheel and operation of the horn.
11. Plug in the orange connector and install the lower steering column trim or panel.
12. Connect the negative battery cable.

AIR BAG 2 SYSTEM

NOTE: The Air Bag 2 system can be identified by the air bag telltale lamp located on the steering wheel.

CAUTION

All precautions must be adhered to or premature deployment of the air bag may occur causing personal injury. If the air bag system is damaged or rendered inoperative, the additional protection an air bag provides in a frontal collision will not be available and may lead to greater injury in an accident.

1. Disconnect the negative battery terminal. Wait for a period of approximately 10 minutes allowing the capacitor in the control unit to discharged.

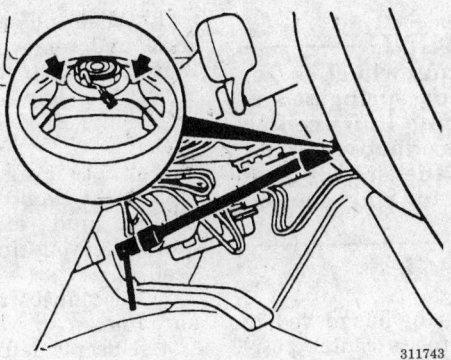

311743

Air bag 1: retaining screw locations

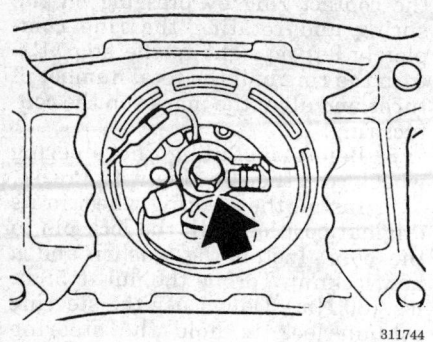

Air bag 1: Steering wheel retaining nut

311744

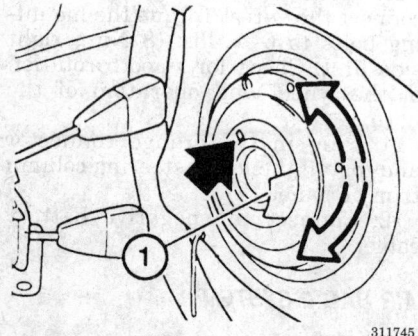

311745

Air bag 1: adjusting contact ring (1- release spring)

2. Remove the lower steering column trim or the small panel and disconnect the orange connector.

3. Remove the three retaining bolts from the air bag module.

a. Remove the air bag module and disconnect the wire on the back.

b. Place the air bag module in the trunk with the pad facing up.

4. Turn the steering wheel to straight-ahead position. Marks on steering gear and steering spindle should be aligned.

5. Hold the steering wheel and remove the nut from the steering shaft.

6. Mark the relationship of the steering wheel to the shaft and pull the steering wheel off the shaft.

——— WARNING ———
Once the steering wheel is removed, the torsion spring securing the contact ring in its center position becomes effective. On no account should the steering be moved. The contact ring may be damaged.

To install:

7. If the contact ring has moved, it must be adjusted to its center position. Determine the center position of the contact ring by pressing on the

311746

Air bag 2: retaining screw locations

spring and rotating the ring completely left or right to the stop. Return the ring half the total number of turns and align the marks on the contact ring.

8. Before installing the steering wheel, coat the slip ring with grease.

9. Install the steering wheel in its original position with the lock pin in the bore. Install the washer and a new locknut. Torque the nut to 58 ft. lbs. (80 Nm). Do not use the steering column lock to hold the steering wheel against the torque. Hold the wheel during torquing.

10. Install the air bag module and connect the wires. Torque the mounting bolts to 6 ft. lbs. (8 Nm), right side first. Check for smooth rotation of the wheel and operation of the horn.

11. Plug in the orange connector and install the lower steering column trim or panel.

12. Connect the negative battery cable.

Power Steering Rack

REMOVAL AND INSTALLATION

M3, Z3 and 3 Series

1. Raise and safely support the vehicle and remove front wheels. Remove the pinch bolt and loosen bolt. Press the spindle off the steering gear.

2. Use a syringe to empty the power steering fluid reservoir. Loosen the clamp and pull off the hydraulic fluid return line from the power steering unit. Discard drained fluid.

3. Disconnect and plug the pressure line.

4. Unscrew left and right side nuts and press off the tie rods where they connect to the spring struts.

5. Remove the bolts attaching the steering unit to the front cross-member and remove it.

To install:

6. Install the steering unit to the front crossmember, then install and tighten the mounting bolts.

7. Install in reverse order, keeping the following points in mind:

a. The steering unit bolts to the rear holes of the axle carrier. Use new self-locking nuts and torque them to 29–34 ft. lbs. (40–46 Nm).

b. When reconnecting tie rods to the spring struts, make sure tie rod pins and strut bores are clean. Replace self-locking nut and torque to 40–48 ft. lbs. (54–66 Nm).

c. Replace the seals on the power steering pump connection and torque the bolt to 29–32 ft. lbs. (40–43 Nm).

8. Refill the fluid reservoir with specified fluid. Idle the engine and turn the steering wheel back and forth until it has reached right and left lock 2 times each. Then, turn the engine **OFF** and refill the reservoir.

Power Steering Gear

REMOVAL AND INSTALLATION

5, 7 and 8 Series

1. Disconnect the negative battery cable.

2. Remove the steering wheel, if equipped with an air bag (SRS).

3. Discharge the pressure reservoir by pushing in on the brake pedal about 10 times. Draw off hydraulic fluid in the supply tank.

4. Unscrew the bolt and press the tie rod off the steering drop arm with the proper tool.

5. Remove the heat shield on the steering gear and disconnect the ride level height control pipes on the 750iL.

6. Remove the bolt and push the U-joint from the steering gear. Disconnect and plug the hydraulic lines.

7. Unscrew the steering gear mounting bolts and remove the steering gear.

NOTE: If necessary, move the steering drop arm by turning the steering stub to enable the removal of the gear assembly.

To install:

8. Install the steering gear and tighten the mounting bolts.

9. Connect the hydraulic lines, using new seals.

10. Turn the steering wheel counterclockwise or clockwise against the

stop and then back about 1.7 turns until the marks are aligned.

11. Connect the U-joint to the steering gear making sure the bolt is in the locking groove of the steering stub.

12. Install the tie rod to the steering drop arm and replace the self locking nut.

13. Replace the heat shield on the steering gear and connect the ride level height control pipes on the 750iL.

14. Refill the hydraulic fluid and replace the steering wheel, if equipped with an air bag (SRS).

15. Connect the negative battery cable.

Power Steering Pump

REMOVAL AND INSTALLATION

NOTE: If the vehicle is equipped with a tandem pump, the removal and installation procedure is the same.

1. Disconnect the negative battery cable. Release the pressure from the reservoir.

2. Draw the hydraulic fluid from the pump reservoir. Disconnect and plug the hydraulic lines.

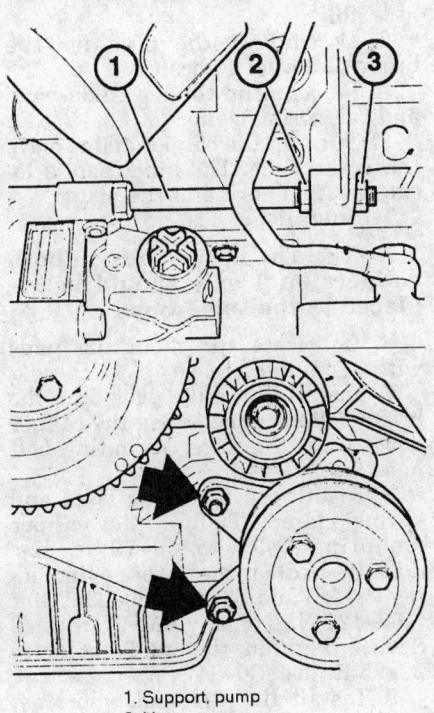

1. Support, pump
2. Nut
3. Nut

313682

Secure the nut (2) hand-tight against the engine support — 3 Series

3. Disconnect and plug the hydraulic lines. Remove the bolts and loosen the nuts to turn the adjusting pinion.

4. Remove the drive belt.

5. Remove the bolts from the brackets holding the pump and remove the pump assembly.

To install:

6. Situate the pump in position, then install the retaining bolts.

7. Install the drive belt.

8. The balance of installation is the reverse of the removal procedure. Tighten the adjusting pinion to 6 ft. lbs. (8 Nm).

9. Connect the negative battery cable, start the engine and inspect the power steering system for correct operation and for fluid leaks.

BELT ADJUSTMENT

Tighten the drive belt so when pressure is applied to the belt, the distance between both belt pulleys is 0.2–0.4 in. (5–10mm) of deflection.

1. Disconnect the negative battery cable. Loosen the nuts on the adjusting pinion.

2. Tighten the belt to the recommended specification and tighten the adjusting pinion nuts.

3. Connect the negative battery cable.

SYSTEM BLEEDING

1. Fill the reservoir to the **MAX** mark on the oil stick.

2. Rotate the steering in both directions fully, to each stop, until all the air is removed from the fluid.

3. Check the oil level and fill to the specified mark, if necessary.

Tie Rod Ends

REMOVAL AND INSTALLATION

1. Raise and safely support the vehicle. Loosen the clamping bolt that retains the toe-in adjustment by keeping the tie rod end from turning in relation to the tie rod.

2. Remove the cotter pin and castellated or self-locking nut from the bottom of the tie rod end. Then, press the tie rod end out of the steering knuckle with the proper tool. Then, unscrew the tie rod end from the tie rod and remove it, counting the number of turns required.

To install:

3. Install in reverse order, using a new cotter pin or castellated nut.

Recheck the front alignment and reset the toe-in, if necessary.

4. Torque the castellated or self-locking nut to 26 ft. lbs. (36 Nm) and the clamping screw to 10 ft. lbs. (14 Nm). Final torque the clamping bolt with the vehicle resting on its wheels.

BRAKES

Master Cylinder

REMOVAL AND INSTALLATION

1. Disconnect the negative battery cable. Draw off the brake fluid from the reservoir.

2. Remove the plug and disconnect the hydraulic lines. Remove the hose for the hydraulic clutch, if needed.

3. Remove the reservoir. Remove the mounting bolts and lift out the master cylinder.

To install:

4. Install the master cylinder, making sure the rubber ring is making a good seal.

5. Install the reservoir.

6. Attach the brake system hydraulic lines to the master cylinder. Also install the hydraulic clutch line, if removed.

7. Bleed the brake system and connect the negative battery cable.

Proportioning Valve

REMOVAL AND INSTALLATION

1. Disconnect the negative battery cable. Draw off hydraulic fluid from the master cylinder with a syringe or hose used only with clean brake fluid.

2. Disconnect the brake lines at the top and bottom of the proportioning valve.

3. Remove the clamp from the valve and disconnect the pressure connection at the union.

To install:

4. Check day/year codes, reduction factor and switch-over pressure to make sure the new valve is identical.

5. Attach the pressure connection to the proportioning valve. Install the clamp.

6. Install the brake system hydraulic lines to the proportioning valve.

7. Bleed the hydraulic brake system.

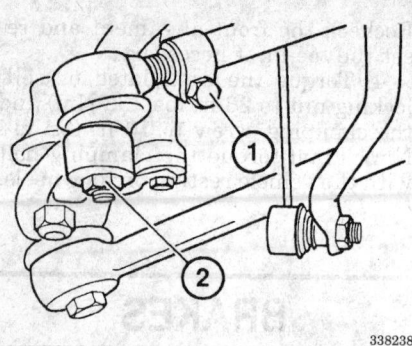

338238

Tie rod end clamping bolt (1) and tie rod end-to-steering knuckle nut (2)

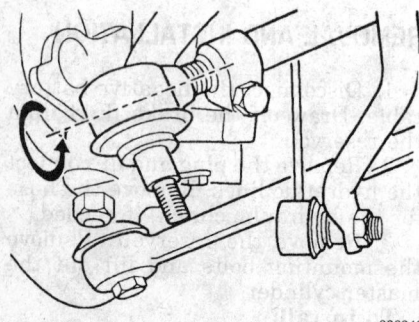

338240

Unscrew the tie rod end from the tie rod

Power Brake Booster

REMOVAL AND INSTALLATION

1. Disconnect the negative battery cable. Draw off brake fluid in the reservoir and discard.

2. Remove the reservoir and disconnect the clutch hydraulic hose.

3. Disconnect all brake lines from the master cylinder.

4. Remove the instrument panel trim from the bottom/left inside the passenger compartment.

5. Remove the return spring from the brake pedal. Press off the clip and remove the pin which connects the booster rod to the brake pedal.

6. Remove the 4 nuts and pull the booster and master cylinder off in the engine compartment.

7. If the filter in the brake booster is clogged, it will have to be cleaned. To do this, remove the dust boot, retainer, damper and filter, and clean the damper and filter. Make sure when reinstalling that the slots in the damper and filter are offset 180 degrees.

To install:

8. Inspect the rubber seal between the master cylinder and booster and replace it, if necessary.

9. Situate the brake booster and master cylinder in position on the firewall in the engine compartment. Tighten the 4 nuts in a crisscross fashion until tight.

10. Connect the booster rod to the brake pedal by installing a new retaining clip. Install the return spring to the brake pedal.

11. Adjust the stoplight switch for a clearance of 0.197–0.236 in. (5.00–5.99mm).

12. Install the remaining components in the reverse order of removal.

13. Connect the negative battery cable, then bleed the brake hydraulic system.

Brake Caliper

REMOVAL AND INSTALLATION

Front

1. Disconnect the negative battery cable. Draw off brake fluid with a suitable syringe.

2. Disconnect the hydraulic brake lines.

3. Raise and safely support the vehicle. Remove the front tire and wheel assembly.

4. Remove the caliper mounting bolts and disconnect the brake pad wear indicator plug.

5. Remove the caliper assembly.

To install:

6. Install the brake caliper onto the steering knuckle; tighten the caliper mounting bolts to 63–79 ft. lbs. (85–108 Nm) on the 3 Series (including the M3 and Z3) or to 80–89 ft. lbs. (109–121 Nm) on any BMW but the 3 Series. Tighten the guide bolts to 22–25 ft. lbs. (30–34 Nm).

NOTE: Make sure the brake wear indicator wire is held in the correct position by the tab of the dust cap.

7. Install the front tire and wheel assemblies, then lower the vehicle.

8. Connect the hydraulic brake system lines.

9. Connect the negative battery cable, then bleed the brake system.

Rear

1. Disconnect the negative battery cable. Draw off brake fluid with a suitable syringe.

2. Disconnect the hydraulic brake lines.

3. Raise and safely support the vehicle. Remove the rear tire and wheel assembly.

4. Remove the caliper mounting bolts and disconnect the brake pad wear indicator plug.

5. Remove the caliper assembly by pulling to the rear.

To install:

6. Install the brake caliper onto the steering knuckle; tighten the caliper mounting bolts to 42–48 ft. lbs. (56–66 Nm). Tighten the guide bolts to 22–25 ft. lbs. (30–34 Nm).

NOTE: Make sure the brake wear indicator wire is held in the correct position by the tab of the dust cap.

7. Install the front tire and wheel assemblies, then lower the vehicle.

8. Connect the hydraulic brake system lines.

9. Connect the negative battery cable, then bleed the brake system.

Disc Brake Pads

REMOVAL AND INSTALLATION

Front

1. Raise and safely support the vehicle.

2. Remove the tire and wheel assembly.

3. Disconnect the plug for the brake pad wear indicator

4. Remove the caliper guide bolts and the spring clamp.

5. Turn up the caliper and remove the brake pads. The inner pad is located with a spring in the piston.

To install:

NOTE: The brake pads on both calipers on 1 axle should be replaced at the same time.

6. Lubricate the mounting pads with a suitable grease.

7. Install the brake pads onto the brake caliper, then swing the caliper down until the lower mounting bolt holes are aligned.

8. Install the mounting bolts and spring clamp. Tighten the caliper mounting bolts to 63–79 ft. lbs. (85–108 Nm) on the 3 Series (including the M3 and Z3) or to 80–89 ft. lbs. (109–121 Nm) on any BMW but the 3 Series. Tighten the guide bolts to 22–25 ft. lbs. (30–34 Nm).

9. Install the tire and wheel assemblies, then lower the vehicle to the ground.

10. Bleed the brake system if any of the brake system lines were disconnected.

FIST CALIPER (FRONT)

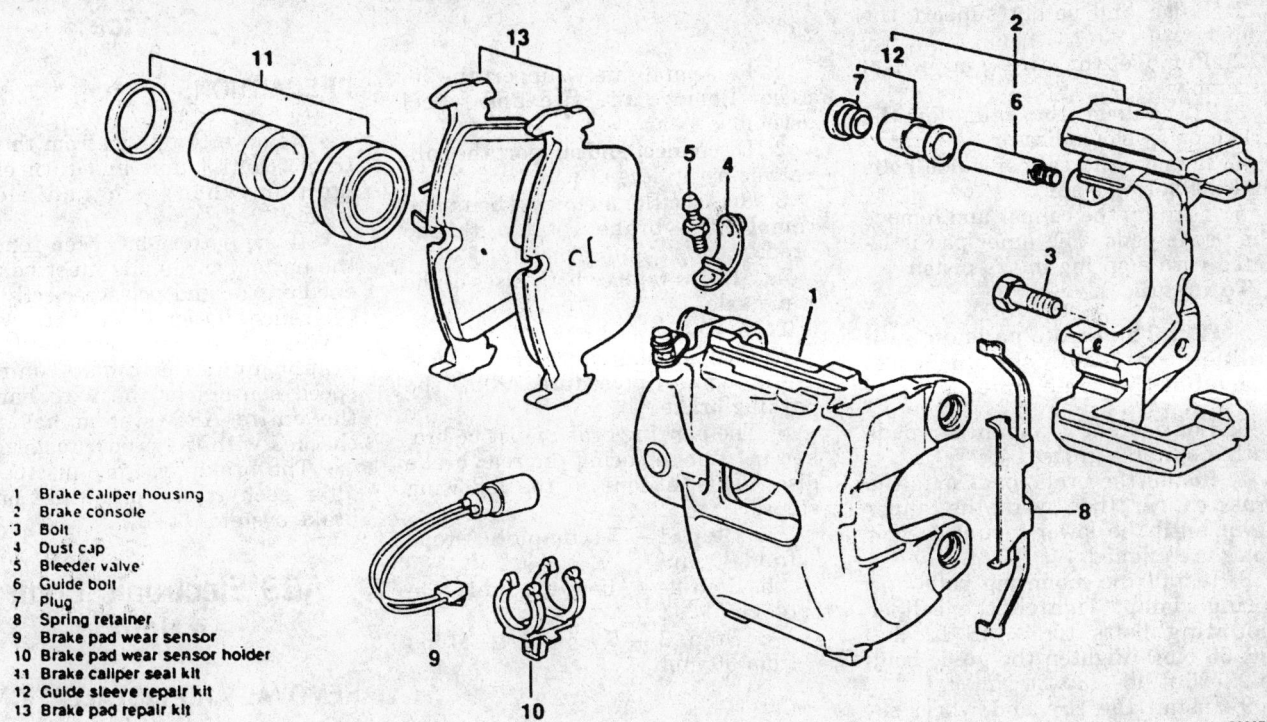

1 Brake caliper housing
2 Brake console
3 Bolt
4 Dust cap
5 Bleeder valve
6 Guide bolt
7 Plug
8 Spring retainer
9 Brake pad wear sensor
10 Brake pad wear sensor holder
11 Brake caliper seal kit
12 Guide sleeve repair kit
13 Brake pad repair kit

300070

Exploded view of the front brake caliper — M3, Z3 and 3 Series

FIST CALIPER (REAR)

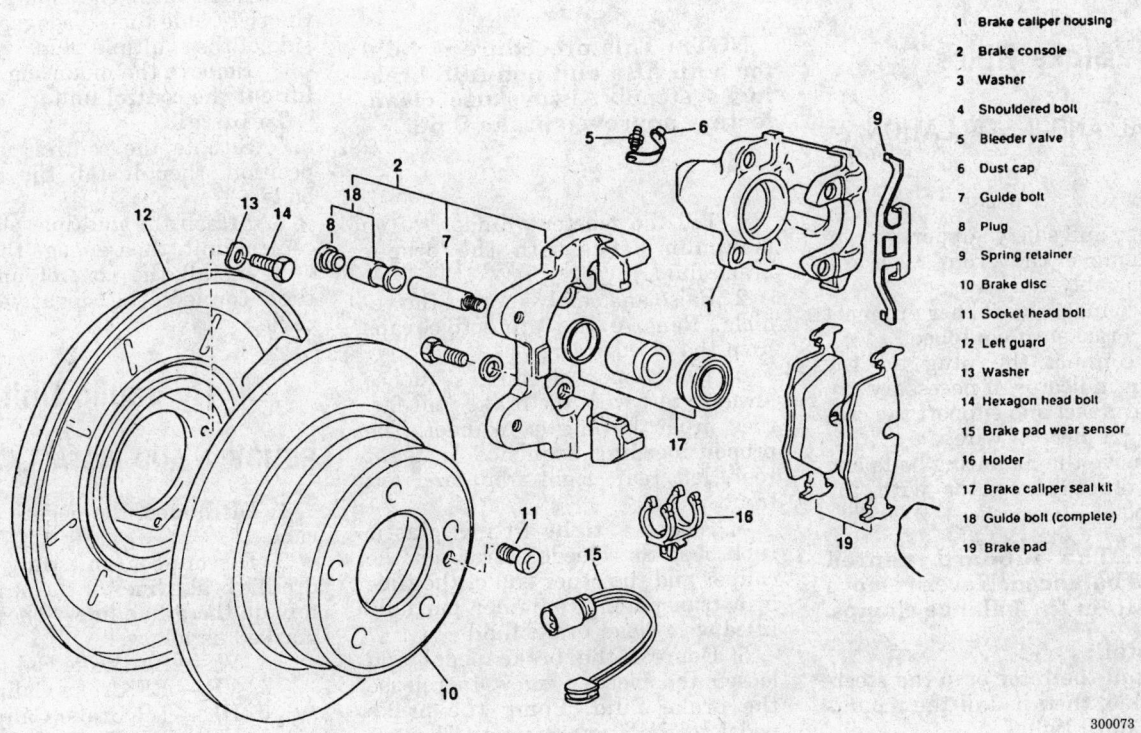

1 Brake caliper housing
2 Brake console
3 Washer
4 Shouldered bolt
5 Bleeder valve
6 Dust cap
7 Guide bolt
8 Plug
9 Spring retainer
10 Brake disc
11 Socket head bolt
12 Left guard
13 Washer
14 Hexagon head bolt
15 Brake pad wear sensor
16 Holder
17 Brake caliper seal kit
18 Guide bolt (complete)
19 Brake pad

300073

Exploded view of the rear brake caliper — M3, Z3 and 3 Series

Rear

1. Raise and safely support the vehicle.
2. Remove the tire and wheel assembly.
3. Disconnect the plug for the brake pad wear indicator
4. Remove the caliper guide bolts and the spring clamp.
5. Turn up the caliper and remove the brake pads. The inner pad is located with a spring in the piston.

To install:

NOTE: The brake pads on both calipers on 1 axle should be replaced at the same time.

6. Lubricate the mounting pads with a suitable grease.
7. Install the brake pads onto the brake caliper, then swing the caliper down until the lower mounting bolt holes are aligned.
8. Install the mounting bolts and spring clamp. Tighten the caliper mounting bolts to 42–48 ft. lbs. (56–66 Nm). Tighten the guide bolts to 22–25 ft. lbs. (30–34 Nm).
9. Install the tire and wheel assemblies, then lower the vehicle to the ground.
10. Bleed the brake system if any of the brake system lines were disconnected.

Brake Rotor

REMOVAL AND INSTALLATION

Front

1. Raise and safely support the vehicle. Remove the front tire and wheel assembly.
2. Disconnect the rubber grommet from the bracket, if equipped.
3. Disconnect the plug for the brake pad indicator, if necessary.
4. Disconnect and support the caliper, using a piece of wire.
5. Remove the mounting bolts and remove the brake rotor with the proper tool.

NOTE: The inboard vented discs are balanced. Never remove or reposition the balance clamps.

To install:

6. Install the rotor onto the steering knuckle, then install the mounting bolts until tight.
7. Install the brake caliper.
8. Install the front tire and wheel assembly, then lower the vehicle to the ground.

Rear

1. Raise and safely support the vehicle. Remove the tire and wheel assembly.
2. Disconnect and support the caliper, using a piece of wire.
3. Remove the mounting bolts and remove the brake rotor with the proper tool.
4. Always replace both discs of the same axle.

To install:

5. The installation is the reverse of the removal procedure. Adjust the parking brake.
6. The parking brake must be broken in after replacing the rear brake discs. This is done in the following steps:
   a. Step 1 — 5 complete stops from 30 mph
   b. Step 2 — Allow the brakes to cool off
   c. Step 3 — 5 complete stops from 30 mph

Brake System Bleeding

NOTE: This procedure is valid for both ABS and non-ABS braking systems. Always use clean, factory approved brake fluid.

1. Fill the master cylinder to the maximum level with the proper brake fluid.
2. Raise and safely support the vehicle. Remove the protective caps from the bleeder screws.
3. The proper bleeding sequence always start with the brake unit farthest from the master cylinder. The proper bleeding sequence is: right rear, left rear, right front and left front.
4. Insert a tight fitting plastic tube over the bleeder screw on the caliper and the other end of the tube in a transparent container partially filled with clean brake fluid.
5. Depress the brake pedal and loosen the bleeder screw to release the brake fluid. Pump the brake pedal to the stop 12 times. Tighten the bleeder screw when the escaping brake fluid is free of air bubbles.
6. Repeat this step on all 4 wheels. Lower the vehicle.

Anti-Lock Brake System Service

PRECAUTIONS

• Remove the plugs from the electronic control unit and turn off the ignition when using an electric welder.
• If the battery has been removed, the battery terminals must be tightened on the end poles perfectly after the reinstallation of the battery.
• After the replacement of the hydraulic unit, the control unit, the speed sensors or the wire harness, the entire ABS system has to be checked with the proper tester.
• The brake system must be bled after each repair procedure on the brake system.

ABS Electronic Control Unit

REMOVAL AND INSTALLATION

1. Disconnect the negative battery cable.
2. Remove the cover in the engine compartment on the right side.
3. Push back the clamp. Pull off the right side then disengage the left side of the multiple plug.
4. Remove the mounting bolts and lift out the control unit.

To install:

5. Situate the control unit in its position, then install the mounting bolts.
6. Attach the multiple plug to the control unit, then engage the clamp.
7. Install the control unit cover, then connect the negative battery cable.

Hydraulic Unit

REMOVAL AND INSTALLATION

1. Disconnect the negative battery cable.
2. Disconnect and plug the hydraulic brake lines at the unit. Do not mix up the brake lines. The lines are marked as follows:
   a. VL — LF Brake Caliper
   b. VR — RF Brake Caliper
   c. HL — LR Brake Caliper
   d. HR — RR Brake Caliper
3. Remove the cover mounting bolts and lift off the cover.
4. Disconnect the electrical connections and plugs.

5. Loosen the mounting nuts, pull up and remove the hydraulic control unit.

To install:

6. Position the hydraulic control unit in place, then install the mounting nuts.

7. Attach the electrical connections and plugs to the control unit.

8. Install the cover.

9. Attach the hydraulic lines to the hydraulic control unit. Make sure attach the proper line to its terminal on the control unit.

10. Connect the negative battery cable, then bleed the brake system.

CHASSIS ELECTRICAL

CAUTION

On vehicles equipped with an air bag, the negative battery cable must be disconnected, before working on the system. Failure to do so may result in deployment of the air bag and possible personal injury.

Windshield Wiper Motor

The electric wiper motor assembly is located under the engine hood, at the top of the cowl panel. A few vehicles have covers over the wiper motor assembly, while others have the motors exposed. Link rods operate the left and right wiper pivot assemblies from a drive crank bolted to the wiper motor output shaft.

REMOVAL AND INSTALLATION

M3, Z3 and 3 Series

1. Disconnect the negative battery cable. Remove the heater motor, as described above. Remove the bracket bracing the windshield wiper motor, which is now visible.

2. Disconnect the electrical connector for the motor.

3. Lift out the grill located at the top of the cowl and disconnect the linkages to both wiper arms at the left side shaft mounts.

4. Disconnect both wiper arms from their shafts by lifting the cover, unscrewing the nut and pulling the arm off. Then, remove the cover, nut and washer surrounding the shafts

and holding the console in place. Now remove the entire console.

5. With the motor still mounted, remove the nut retaining the linkage to the motor shaft. Then, unbolt and remove the motor from the console.

To install:

6. Install the motor in the vehicle and tighten the mounting bolts, then attach the linkage to the motor shaft.

7. Install the wiper motor cover console.

8. Install the remaining components in the reverse order of removal.

9. Connect the negative battery cable, then check the wipers for proper function.

5 Series

1. Disconnect the negative battery cable. Remove the cowl cover to expose the wiper motor, if equipped.

2. Disconnect the wiper motor crank arm from the motor output shaft by removing the nut and pulling off the crank arm.

3. Remove the motor retaining screws and disconnect the electrical connector.

4. Remove the wiper motor from the vehicle.

To install:

5. Install the wiper motor into the vehicle.

6. Attach the electrical connector to the wiper motor, then install and tighten the retaining screws.

7. Attach the wiper motor linkage to the motor.

8. Install the cowl cover and connect the negative battery cable.

7 and 8 Series

1. Disconnect the negative battery cable. Make sure the wipers are in the parked position. Remove the heater blower. Take off the cover near the blower.

2. Disconnect the heater cable and lift out the linkage. Disconnect the temperature sensor.

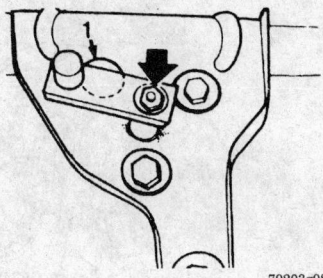

79203g08

On the 7 Series, install the motor crank in the position shown, with the bolt (1) half hidden by the upper edge of the crank

3. Disconnect the clips, lift the cowl cover slightly and then remove the fresh air inlet cowls on either side. Then, remove the cover.

4. Unscrew bolts and remove the mounting bracket for the wiper housing. Remove the left wiper arm by pulling up the cover, loosening the pinch bolt and removing it. Remove the right wiper by pulling up the cover, removing the through bolt and then pulling it off.

5. Lift out the clips and remove the cover for the linkage.

6. Unscrew and remove the nuts fastening the linkage to the cowl. Pull the linkage arms downward and out of the cowl.

7. Mark the relationship between the linkage lever and the motor. Remove the nut and disconnect the linkage at the motor shaft. Disconnect the electrical connector and remove the linkage.

8. Remove the mounting bolts and remove the wiper motor. If installing a new motor, connect the motor and operate it until it reaches parked position; then install the linkage so the shaft lever and linkage link are in a straight line.

To install:

9. Install the motor, then attach the linkage to the motor shaft.

10. Attach the electrical connector to the motor.

11. The balance of installation is the reverse of the removal procedure. Make sure the wiper arms are pressed all the way onto the linkage shafts so the contact pressure control will work. Make sure the inlet cowling is installed in proper relation to the blower housing and fresh air flap.

12. Install the negative battery cable, then inspect the wiper system for proper function.

Windshield Wiper Switch

REMOVAL AND INSTALLATION

1. Disconnect the negative battery cable. Remove the steering wheel. Remove the lower/left instrument panel trim.

2. Remove the screws and remove the lower steering column cover.

3. Push the locking hook for the flasher back and remove the relay, socket facing downward.

4. Take off the upper steering column cover. If equipped with air bags, drive out the pins and lift out the expansion rivet first.

5. Press the retaining hooks inward on both sides, pull the switch

out, and then disconnect the electrical connector.

To install:

6. Attach the electrical connector to the switch, then push the switch into place.

7. Make sure that the retaining hooks engage the switch unit in position.

8. The remaining components are installed in the reverse order from which they were removed.

9. Install the steering wheel and negative battery cable.

Instrument Cluster

REMOVAL AND INSTALLATION

1. Disconnect the negative battery cable. Remove the attaching screws and remove the lower instrument panel trim from under the steering column.

2. Remove the mounting nuts for the trim just under the instrument carrier and remove it.

3. Unscrew the 4 screws underneath and 2 above the instrument carrier and remove trim that surrounds the instrument carrier.

4. Remove the 2 screws at the top of the carrier, lift it out of the instrument panel, and then disconnect the plugs. To disconnect the combination plug, first pull the sliding clamp off the center.

5. To replace the speedometer, pull the speedometer from the instrument carrier.

To install:

6. Install the speedometer into the instrument carrier.

7. Position the instrument carrier close to the instrument panel, attach all electrical connectors to it, then slide the carrier into the instrument panel. Install the 2 retaining screws at the top of the panel.

8. Install the remaining components in the reverse order of removal.

9. Connect the negative battery cable.

Headlight Switch

REMOVAL AND INSTALLATION

1. Disconnect the negative battery cable. Remove the lower/left trim panel screws and remove the panel.

2. Unscrew the knob from the switch.

3. Pull off the connector plug from behind the dash panel. Pull out the switch from behind and remove it.

To install:

4. Push the switch onto the back of the dashpanel — make sure that the terminal connectors are installed in the proper holes.

5. Screw the knob back onto the switch.

6. Install the trim panels and connect the negative battery cable.

Combination Switch

REMOVAL AND INSTALLATION

――――― **CAUTION** ―――――
On vehicles equipped with an air bag, the negative battery cable must be disconnected, before working on the system. Failure to do so may result in deployment of the air bag and possible personal injury.

1. Disconnect the negative battery cable. Remove the steering wheel.

2. Remove the lower instrument panel on the left side.

3. Remove the steering column casing lower section.

4. If equipped with an air bag, drive out the pins and lift out the expansion rivet. Remove the upper section of the steering column casing.

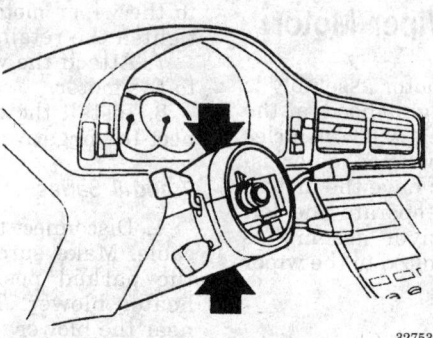

Remove lower section of the steering column casing by unscrewing the fasteners

327539

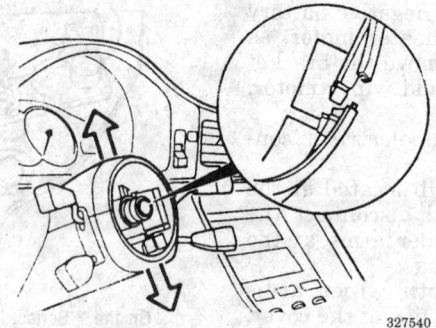

Remove upper section of steering column casing after unclipping the catches

327540

5. Remove the plug and disconnect the electrical connectors.

6. Push in the retaining hooks on both sides and pull out the switch.

To install:

7. Press the switch in position until the retaining hooks engage it in place.

8. Attach all electrical connectors and install the plug.

9. The remaining components are installed in the reverse order from which they were removed.

10. Install the steering wheel and connect the negative battery cable.

FIRING ORDERS

NOTE: To avoid confusion, always replace spark plug wires one at a time.

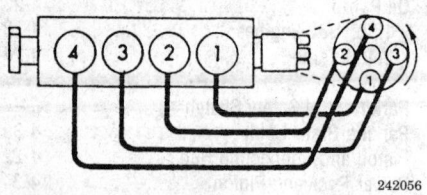

242056

1.5L, 1.8L and 2.4L Engines
Engine Firing Order: 1–3–4–2
Distributor Rotation: Counterclockwise

ENGINE ELECTRICAL

NOTE: Disconnecting the negative battery cable on some vehicles may interfere with the functions of the on board computer systems and may require the computer to undergo a relearning process, once the negative battery cable is reconnected.

Distributor

REMOVAL

Before removing the distributor, position No. 1 cylinder at TDC on the compression stroke and align the timing marks.

1. Disconnect the negative battery cable.
2. Disconnect the spark plug wires from the distributor cap.
3. Remove the vacuum hose from the advance unit, if equipped.
4. Remove the cap from the distributor.
5. Verify the rotor points to the No. 1 cylinder position and the timing marks on the crankshaft pulley and the timing tab are aligned at TDC.
6. Mark the distributor body to the exact place the rotor points. Matchmark both the distributor

mounting flange and the cylinder head.
7. Loosen and remove the retaining nut from the mounting stud. Lift the distributor from the cylinder head. The rotor may turn slightly from the mark on the distributor body. Make note of how far. When the distributor is reinstalled, this is the point to position the rotor.
To install:

Engine Undisturbed

1. Position the distributor into the engine while aligning the matchmarks made during removal.
2. Verify the rotor points to the No. 1 cylinder position and the timing marks on the crankshaft pulley and the timing tab are aligned at TDC.
3. Install the distributor retaining nut on the mounting stud and tighten.
4. Install the cap on the distributor.
5. Connect the spark plug wires to the distributor cap.
6. Connect the negative battery cable to the battery.
7. Start the engine and check the ignition timing whenever the distributor has been removed.

Engine Disturbed

1. With the distributor removed from the engine, turn the crankshaft so the No. 1 piston is on the TDC of the compression stroke and the timing marks are aligned.

NOTE: With the distributor properly installed, the rotor will be pointed toward the No. 1 terminal of the distributor cap.

2. Insert the distributor; if resistance is met, slight wiggling of the rotor shaft will help seat the distributor.
3. When the distributor seats against the head, align the matchmarks and install the retaining nut. Do not tighten the retaining nut all the way, as the timing must be checked.
4. Reinstall the rotor, cap, plug wires, primary lead or harness and connect the vacuum hoses.
5. Connect the negative battery cable. Start the engine, allow it to reach operating temperature and check the ignition timing. Adjust the timing by turning the distributor as needed. Tighten the retaining nut.

Ignition Timing

ADJUSTMENT

Colt and Summit

1. Attach the timing light according to the manufacturer's instructions.
2. Locate the timing tab line on the front of the engine and the notch on the crankshaft pulley. Mark them so they are easily recognizable with the timing light. Connect a tachometer to the engine.
3. Start the engine and allow it to reach operating temperature.
4. Check that the engine idle is at specifications.
5. Turn the engine OFF.
6. Locate the ignition timing adjustment connector (brown) and connect a jumper wire to from the connector to a good ground connection.

NOTE: Grounding the ignition timing adjustment connector will set the engine to basic ignition timing.

7. Point the timing light at the crankshaft pulley marks and inspect the ignition timing.
8. If the timing is not within specifications, loosen the distributor mounting nut and rotate the distributor slowly, in either direction, to align the timing marks.
9. Tighten the distributor mounting nut when the ignition timing is correct. Stop the engine and remove the timing light.
10. Adjust engine idle if needed.

Summit Wagon and Vista

1. Set the parking brake, start and run the engine until normal operating temperature is obtained. Keep all lights and accessories OFF and the front wheels straight-ahead. Place the transaxle in **P** or automatic transaxle or neutral for manual transaxle.

NOTE: On Canadian vehicles the lights will remain on when the vehicle is running, this will not be a problem.

2. Locate the wire connector on the ignition coil connector. Insert a paper clip behind the TACH terminal connector to act as a tachometer adapter. Connect a tachometer to the paper clip. If not at specification, set the idle speed at the correct level.
3. Turn the engine **OFF** and remove the water-proof cover from the ignition timing adjusting connector. This connector is a brown connector

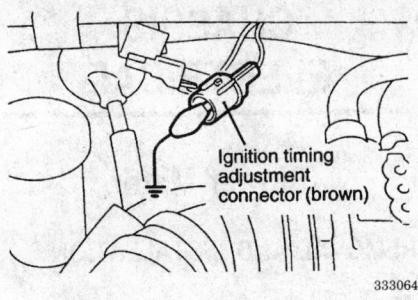

Timing connector identification — Colt and Summit models

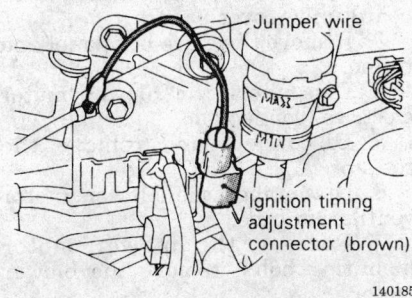

Ignition timing connector location — Summit Wagon and Vista models with 2.4L (VIN G) engines

located near the center of the firewall. Connect a jumper wire from this terminal to a good ground.

4. Connect a conventional power timing light to the No. 1 cylinder spark plug wire. Start the engine and run at idle.

5. Aim the timing light at the timing scale located near the crankshaft pulley.

6. Loosen the distributor holddown nut just enough so the housing can be rotated.

7. Turn the housing in the proper direction until the specified timing is reached. Tighten the hold-down nut and recheck the timing. Turn the engine **OFF**.

8. Remove the jumper wire from the ignition timing adjusting terminal and install the water-proof cover.

9. Start the engine and check the actual timing without the terminal grounded. This reading should be approximately 5 degrees more than the basic timing. Actual timing may increase according to altitude. Also, actual timing may fluctuate because of slight variation accomplished by the ECU. As long as the basic timing is correct, the engine is timed correctly.

10. Turn the engine **OFF**. Disconnect the timing equipment and tachometer.

Alternator

PRECAUTIONS

Several precautions must be observed with alternator equipped vehicles to avoid damage to the unit.

• If the battery is removed for any reason, make sure it is reconnected with the correct polarity. Reversing the battery connections may result in damage to the 1-way rectifiers.

• When utilizing a booster battery as a starting aid, always connect the positive to positive terminals and the negative terminal from the booster battery to a good engine ground on the vehicle being started.

• Never use a fast charger as a booster to start vehicles.

• Disconnect the battery cables when charging the battery with a fast charger.

• Never attempt to polarize the alternator.

• Do not use test lights of more than 12 volts when checking diode continuity.

• Do not short across or ground any of the alternator terminals.

• The polarity of the battery, alternator and regulator must be matched and considered before making any electrical connections within the system.

• Never separate the alternator on an open circuit. Make sure all connections within the circuit are clean and tight.

• Disconnect the battery ground terminal when performing any service on electrical components.

• Disconnect the battery if arc welding is to be done on the vehicle.

REMOVAL AND INSTALLATION

1.5L (VIN A) Engine

1. Disconnect the negative battery cable.

2. Remove the left side cover panel under the vehicle.

3. Remove the drive belts.

4. Remove the water pump pulleys.

5. Remove the alternator upper bracket/brace.

6. Disconnect the alternator wiring connectors and remove the alternator.

To install:

7. Position the alternator on the lower mounting fixture and install the lower mounting bolt and nut. Tighten nut just enough to allow for movement of the alternator.

8. Install the alternator upper bracket/brace and connect the alternator electrical harness.

9. Install the water pump pulleys.

10. Install the drive belts and adjust to the proper tension.

11. Install the left side cover panel under the vehicle as required.

12. Connect the negative battery cable and check for proper operation.

1.8L (VIN C) Engine

1. Disconnect negative battery cable.

2. Remove the accessory drive belts.

3. Disconnect the electrical harness from the alternator.

4. Remove the alternator mounting nut, bolt and upper brace assembly from the vehicle.

To install:

5. Install the alternator and secure using mounting nuts. Make sure the upper brace assembly is in place.

6. Install and adjust drive belts to the proper tension. Secure all mounting hardware.

7. Reconnect the negative battery cable and check system operation.

2.4L (VIN G) Engine

1. Disconnect the negative battery cable.

2. Remove the left side cover panel under the vehicle.

3. Remove the drive belts.

4. Remove both water pump pulleys.

5. Remove the alternator upper bracket/brace.

6. Disconnect the alternator electrical connectors and remove alternator.

To install:

7. Position the alternator on the lower mounting fixture and install the lower mounting bolt and nut. Tighten nut just enough to allow for movement of the alternator.

8. Install the alternator upper bracket/brace and connect the alternator electrical harness.

9. Install the water pump pulleys.

10. Install the drive belts and adjust to the proper tension.

11. Install the left side cover panel under the vehicle as required.

12. Connect the negative battery cable and check for proper operation.

Drive Belt

REMOVAL AND INSTALLATION

All Engines

Excessive belt tension will cause damage to the alternator and water pump pulley bearings, while loose belt tension will produce slip and premature wear on the belt. Be sure to adjust the belt tension to the proper level.

To adjust the tension on a drive belt, loosen the adjusting bolt or fix-

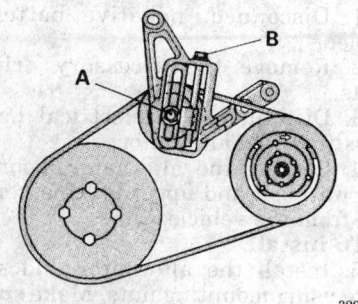

Example of air conditioner belt tensioner adjustment procedure on all engines — (A) securing bolt and (B) adjusting bolt

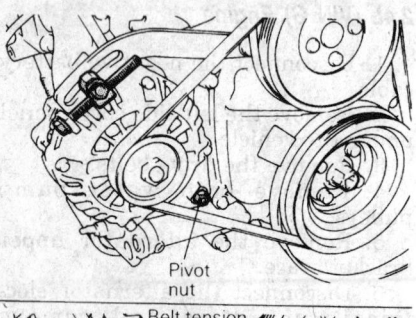

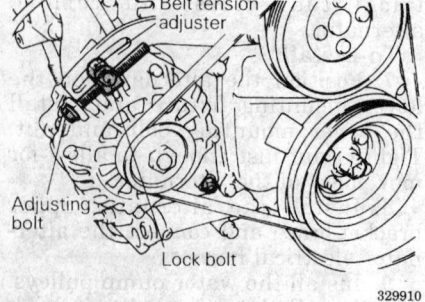

Example of alternator adjustment and pivot points — All engines

ing bolt locknut on the desired component, bracket or tension pulley. Then move the component or turn the adjusting bolt to adjust belt tension. Once the desired value is reached, secure the bolt or locknut and recheck tension.

Belt replacement is similar to adjustment, with the exception the belt will have to be properly routed around the pulleys. It is important to note the routing of the belt before removal. For individual belt replacement, start with the outer most belt. If a removed belt is to be reused, be certain to mark the direction of rotation on the belt, to extend belt life.

Starter

REMOVAL AND INSTALLATION

1.5L (VIN A) and 1.8L (VIN C) Engines

1. Disconnect the speedometer cable connector at the transaxle end.
2. Disconnect the starter motor electrical connections.
3. Remove the starter motor mounting bolts and remove the starter.

To install:
4. The installation is the reverse of the removal procedure.
5. Tighten starter mounting bolts to 22 ft. lbs. (31 Nm.)
6. Connect the negative battery cable and check the starter for proper operation.

2.4L (VIN G) Engine

1. Disconnect the negative battery cable.
2. Disconnect and remove the air cleaner assembly as required.
3. Disconnect the starter motor electrical connections.
4. Remove the starter motor mounting bolts and remove the starter.

To install:
5. Install the starter.
6. Tighten starter mounting bolts to 20–25 ft. lbs. (27–34 Nm.)
7. Connect the starter motor electrical connections.
8. Connect the negative battery cable and check the starter for proper operation.

CHASSIS ELECTRICAL

Blower Motor

REMOVAL AND INSTALLATION

All Models

1. Disconnect the negative battery cable.
2. Remove the right side dashboard undercover panel.
3. Remove the glove box panel and frame.
4. Disconnect the blower motor electrical connection.
5. Disconnect and remove the resistor.
6. Disconnect the blower motor ventilation tube.
7. Remove the blower motor mounting bolts, remove the blower motor.

To install:
8. Replace the blower motor and install the mounting bolts.
9. Connect the blower motor electrical connection.
10. Connect the blower motor ventilation tube.
11. Install the resistor and the glove box assembly.
12. Install the right side dashboard undercover panel.
13. Connect the negative battery cable.

Windshield Wiper Motor

REMOVAL AND INSTALLATION

Colt and Summit

1. Disconnect the negative battery cable.
2. Remove the windshield wiper arms by unscrewing the cap nuts and lifting the arms from the linkage posts.
3. Remove the front deck garnish panel.
4. Remove both windshield holders.
5. Remove the clips that hold the deck cover. If they are the pin type, they may be removed using the following procedure:
   a. Remove the clip by pressing down on the center pin with a suitable blunt pointed tool. Press down a little more than $1/16$ in. (2mm).

This releases the clip. Pull the clip outward to remove it.

b. Do not push the pin inward more than necessary because it may damage the grommet or if pushed too far, the pin may fall in. Once the clips are removed, use a plastic trim stick to pry the deck cover loose.

6. Remove the air intake screen.

7. Loosen the wiper motor assembly mounting bolts and remove the windshield wiper motor. Disconnect the linkage from the motor assembly. If necessary, remove the linkage from the vehicle.

NOTE: The installation angle of the crank arm and motor has been factory set. Do not remove unless necessary. If arm must be removed, remove only after marking mounting positions.

To install:

8. Install the windshield wiper motor and connect the linkage. Connect the electrical harness to the motor.

9. When installing the trim and garnish pieces and reusing pin type clips, use the following procedure:

a. With the pin pulled out, insert the trim clip into the hole in the trim.

b. Push the pin inward until the pin's head is flush with the grommet.

c. Check that the trim is secure.

10. Install the wiper arms and tighten nuts.

11. Connect the negative battery cable and check the wiper system for proper operation.

Summit Wagon and Vista

Front

1. Disconnect the negative battery cable.

2. Remove the windshield wiper arms by unscrewing the cap nuts and lifting the arms from the linkage posts.

3. Remove the front deck garnish panel.

4. Loosen the wiper motor assembly mounting bolts, then using a flat tipped screwdriver to pry the linkage from the motor.

5. Disconnect the motor wiring and remove the motor assembly.

NOTE: The installation angle of the crank arm and motor has been factory set, do not remove them unless it is necessary to do so. If arm must be removed, remove them only after marking their mounting positions.

To install:

6. Install the windshield wiper motor mounting bolts and connect the linkage.

7. Connect the electrical harness to the motor.

8. Install the front deck garnish panel.

9. Install the wiper arms and tighten nuts to 17 ft. lbs. (24 Nm).

10. Connect the negative battery cable and check the wiper system for proper operation.

Rear

1. Disconnect the negative battery cable.

2. Remove the rear wiper arm by removing the cap nut cover, unscrewing the cap nut and lifting the arm from the linkage post.

3. Remove the large interior trim panel. Use a plastic trim stick to unhook the trim clips of the liftgate trim. There will be a row of metal liftgate clips across the top. There will be two rows of trim clips that retain the rest of the panel.

4. Disconnect the electrical harness at the wiper motor.

5. Remove the rear wiper assembly mounting bolt.

NOTE: Do not loosen the grommet for the wiper post.

To install:

6. Install the motor and torque mounting bolt to 7 ft. lbs. (9 Nm). If the grommet was removed, ensure that it is installed with the arrow pointing downward.

7. Install the wiper arm and tighten mounting nut to 6 ft. lbs. (8 Nm).

8. Connect the negative battery cable and check rear wiper system for proper operation.

9. If operation is satisfactory, fit the tabs on the upper part of the liftgate trim into the liftgate clips and secure the liftgate trim.

Combination Switch

REMOVAL AND INSTALLATION

Colt and Summit

――― CAUTION ―――
The air bag system (SRS or SIR) must be disarmed before removing the steering wheel. Failure to do so may cause accidental deployment, property damage or personal injury.

1. Disconnect the negative battery cable.

2. If equipped with an air bag, disarm as follows:

a. Position the front wheels in the straight-ahead position and place the key in the **LOCK** position. Remove the key from the ignition lock cylinder.

b. Disconnect the negative battery cable and insulate the cable end with high-quality electrical tape or similar non-conductive wrapping.

c. Wait at least 1 minute before working on the vehicle. The air bag system is designed to retain enough voltage to deploy the air bag for a short period of time even after the battery has been disconnected.

3. Remove the steering wheel as follows:

a. Remove the air bag module mounting nut from behind the steering wheel.

b. To disconnect the connector of the clockspring from the air bag module, press the air bag's lock toward the module to spread the lock open. While holding lock in this position, use a small tipped prying tool to gently pry the connector from the module.

c. Store the air bag module in a clean, dry place with the pad cover facing up.

d. Remove the steering wheel retaining nut and use a steering wheel puller to remove the wheel. Do not use a hammer or the collapsible mechanism in the column could be damaged.

4. Remove the knee protector panel under the steering column, then the upper and lower column covers.

5. Disconnect all connectors, remove the wiring clip and remove the column switch assembly.

To install:

6. Install the switch assembly and secure all harness connectors with clips if needed. Make sure the wires are not pinched or out of place.

7. Install the column covers and knee protector.

8. Confirm that the front wheels are in a straight-ahead position. Center the clockspring by aligning the **NEUTRAL** mark on the clockspring with the mating mark on the casing. Then install the steering wheel and torque the retaining nut to 29 ft. lbs. (40 Nm).

9. Install the air bag module.

10. Connect the negative battery cable, turn the key to the **ON** position, the SRS warning light should illuminate for seven seconds and go

out. If the warning light is not functioning properly, refer to SRS system diagnosis.

11. Check all functions of the combination switch for proper operation.

Summit Wagon and Vista

------ CAUTION ------

On vehicles equipped with an air bag, be sure to disarm the system before attempting repairs on the vehicle. Failure to do so could result in severe personal injury and damage to vehicle.

1. To disarm air bag, perform the following steps:

a. Position the front wheels in the straight-ahead position and place the key in the **LOCK** position. Remove the key from the ignition lock cylinder.

b. Disconnect the negative battery cable and insulate the cable end with high-quality electrical tape or similar non-conductive wrapping.

c. Wait at least one minute before working on the vehicle. The air bag system is designed to retain enough voltage to deploy the air bag for a short period of time even after the battery has been disconnected.

2. To gain access to clockspring wiring, perform the following steps:

a. Remove the four meter hood mounting screws and lift the hood off the three retaining clips.

b. Remove the two screws securing the hood release to the lower instrument panel. Remove the lower instrument panel under the steering column, then the upper and lower column covers.

3. Remove the steering wheel as follows:

a. Remove the air bag module mounting nut from behind the steering wheel.

b. To disconnect the connector of the clockspring from the air bag module, press the air bag's lock toward the module to spread the lock open. While holding lock in this position, use a small tipped prying tool to gently pry the connector from the module.

c. Store the air bag module in a clean, dry place with the pad cover facing up.

d. Remove the steering wheel retaining nut and use a steering wheel puller to remove the wheel. Do not use a hammer or the collapsible mechanism in the column could be damaged.

4. Disconnect all connectors and remove the wiring clip. Remove the three retaining screws and remove the column switch assembly.

To install:

5. Install the switch assembly and secure all harness connectors with clips if needed. Make sure the wires are not pinched or out of place.

6. Confirm that the front wheels are in a straight-ahead position. Center the clockspring by aligning the **NEUTRAL** mark on the clockspring with the mating mark on the casing. Then install the steering wheel and torque the retaining nut to 29 ft. lbs. (40 Nm).

7. Install the air bag module.

8. Install the column covers and lower instrument panel.

9. Connect the hood release to the lower instrument panel.

10. Install the meter hood assembly.

11. Connect the negative battery cable, turn the key to the **ON** position, the SRS warning light should illuminate for seven seconds and go out. If the warning light is not functioning properly, refer to SRS system diagnosis.

12. Check all functions of the combination switch for proper operation.

Ignition Lock Cylinder

REMOVAL AND INSTALLATION

All Models

------ CAUTION ------

If vehicle is equipped with an air bag, disarm the system before attempting to perform repair.

1. Disconnect the negative battery cable. Remove the hood lock release lever from the lower panel.

2. Remove the lower instrument panel knee protector.

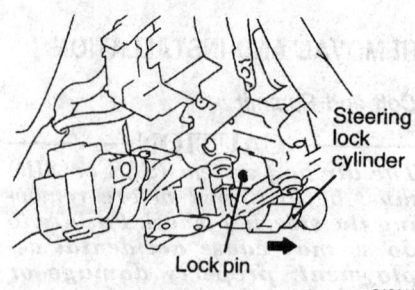

Location of the lock pin for ignition lock removal — all models

3. Remove ductwork with lower knee panel and remove the steering wheel assembly.

NOTE: Use proper steering wheel puller equipment when removing the steering wheel. The use of a hammer for removal could damage the collapsible mechanism within the column.

4. Remove the lower steering column cover.

5. Remove the upper steering column cover.

6. Remove the clip that holds the wiring harness against the steering column.

7. Insert the key into the steering lock cylinder and turn to the **ACC** position.

8. With a small pointed tool, push the lock pin of the steering lock cylinder inward and pull the lock cylinder out.

9. Remove the key reminder switch, if equipped.

To install:

10. With the key in the **ACC** position of the lock cylinder, install the cylinder to the housing of the steering column.

11. Connect harness connections and install the wiring clip.

12. Install the steering column upper and lower covers.

13. Install the knee protector.

14. Install the steering wheel assembly.

15. Connect the negative battery cable and check the ignition switch and lock for proper operation.

Ignition Switch

REMOVAL AND INSTALLATION

Colt and Summit

------ CAUTION ------

If vehicle is equipped with an air bag, disarm the system before attempting to perform repair.

1. Disconnect the negative battery cable. Remove the hood lock release lever from the lower panel.

2. Remove the lower instrument panel knee protector.

3. Remove ductwork with lower knee panel and remove the steering wheel assembly.

NOTE: Use proper steering wheel puller equipment when removing the steering wheel. The use of a hammer for removal could damage the collapsible mechanism within the column.

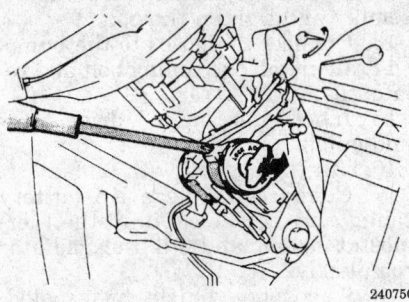

Push lock pin inward to release the lock cylinder — Summit Wagon and Vista models

4. Remove the lower steering column cover.

5. Remove the upper steering column cover.

6. Remove the clip that holds the wiring harness against the steering column.

7. Unplug the ignition switch harness connector. Remove the ignition switch mounting screws and pull the switch from the steering lock cylinder.

To install:

8. Install the ignition switch into the rear of the lock cylinder housing. Be sure to align the keyway of the ignition switch with interlock cylinder.

9. Connect harness connections and install the wiring clip.

10. Install the steering column upper and lower covers.

11. Install the knee protector.

12. Connect the negative battery cable and check the ignition switch and lock for proper operation.

Summit Wagon and Vista

1. Disconnect the negative battery cable.

2. Remove the hood lock release lever from the lower panel.

3. Remove the lower instrument panel knee protector.

NOTE: Use proper steering wheel puller equipment when removing the steering wheel. The use of a hammer for removal could damage the collapsible mechanism within the column.

4. Remove the instrument panel hood.

5. Remove the lower steering column cover.

6. Remove the upper steering column cover.

7. Remove the clip that holds the wiring harness against the steering column.

8. Insert the key into the steering lock cylinder and turn to the **ACC** position. With a small pointed tool, push the lock pin of the steering lock cylinder inward and pull the lock cylinder out.

9. Remove the key reminder switch, if equipped.

10. Unplug the ignition switch harness connector. Remove the ignition switch mounting screws and pull the switch from the steering lock cylinder.

To install:

11. Install the ignition switch into the rear of the lock cylinder housing. Be sure to align the keyway of the ignition switch with interlock cylinder.

12. Connect harness connections and install the wiring clip.

13. Install the steering column upper and lower covers.

14. Install the instrument panel hood.

15. Install the knee protector.

16. Connect the negative battery cable and check the ignition switch and lock for proper operation.

Park/Neutral Safety Switch

REMOVAL AND INSTALLATION

All Models

1. Disconnect the negative battery cable.

2. Disconnect the selector cable from the lever.

3. Remove the two retaining screws and lift off the switch.

To install

4. Mount and position new switch. Do not tighten the bolts until the switch is adjusted.

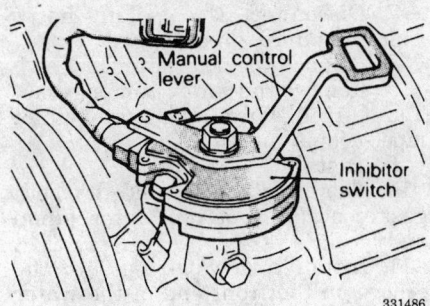

Park/Neutral switch location — all models

5. Connect selector cable and adjust switch.

6. After installation and adjustment make sure the engine only starts in the **P** and **N** selections. Also check that the reverse lights operate only in the **R** selection.

Powertrain Control Module

REMOVAL AND INSTALLATION

Colt and Summit

NOTE: The electronic control unit is located above the passenger side kickpanel.

1. Disconnect the negative battery cable.

2. Remove the glove box, right side kickpanel and lower panel assemblies.

3. Unplug the connectors and remove fasteners. Remove the control unit.

To install:

4. Position the control unit into the vehicle.

5. The balance of installation is the reverse of the removal procedure.

6. Connect the negative battery cable.

Colt Wagon and Summit Wagon

The engine control module is located behind glove box.

1. Disconnect negative battery cable.

2. Locate and access the module.

3. Unplug wiring connector and remove mounting hardware. Slide out control unit.

To install:

4. Position the control unit into the vehicle.

5. The balance of installation is the reverse of the removal procedure.

6. Connect the negative battery cable.

ENGINE COOLING

Radiator

REMOVAL AND INSTALLATION

All Models

1. Disconnect the negative battery cable.

CAUTION

Allow the cooling system to completely cool before attempting any repair or draining the system. Injury from scalding could result if radiator cap or hose connections are removed while system is hot.

---

2. Drain the cooling system when safe.

3. Disconnect the overflow tube. Some vehicles may also require removal of the overflow tank.

4. Disconnect upper and lower radiator hoses.

5. Disconnect electrical connectors for cooling fan and air conditioning condenser fan, if equipped. Remove the fan assembly.

6. Disconnect thermo sensor wires.

7. Disconnect and plug automatic transaxle cooler lines, if equipped with automatic transaxle.

8. Remove the upper radiator mounts and lift out the radiator assembly.

9. Service the lower mounts, as required.

To install:

10. Install the radiator and fan assembly, if removed as an assembly.

11. Connect the automatic transaxle cooler lines, if disconnected.

12. Install the remaining components in the reverse order of removal.

13. Fill the system with coolant.

14. Connect the negative battery cable, run the vehicle until the thermostat opens, fill the radiator completely and check the automatic transaxle fluid level, if equipped.

15. Once the vehicle has cooled, recheck the coolant level.

Water Pump

REMOVAL AND INSTALLATION

1.5L (VIN A) and 1.8L (VIN C) Engines

1. Disconnect the negative battery cable.

2. Rotate the engine and position the No. 1 piston to TDC of its compression stroke.

3. Drain the cooling system.

4. Remove the engine undercover.

5. Disconnect the clamp bolt from the power steering hose.

6. Support the engine with the appropriate equipment and remove the engine mount bracket.

7. Remove the timing belt from the front of the engine.

8. Remove the timing belt rear cover.

9. Remove the power steering pump bracket.

10. Remove the alternator brace if necessary.

NOTE: The water pump mounting bolts are different in length, note their positioning for reassembly.

11. Remove the water pump mounting bolts and remove the pump.

To install:

12. Thoroughly clean and dry both mating surfaces of the water pump and block.

13. Apply a 0.09–0.12 inch (2.5–3.0mm) continuous bead of sealant to water pump and install the pump assembly.

NOTE: Install the water pump within 15 minutes of the application of the sealant. Wait 1 hour after installation of the water pump to refill the cooling system or starting the engine.

14. Properly position the bolts and tighten the bolts to 18 ft. lbs. (24 Nm).

15. The remaining components are installed in the reverse order from which they were removed.

16. Fill the system with coolant.

17. Connect the negative battery cable, run the vehicle until the thermostat opens and fill the radiator completely.

18. Once the vehicle has cooled, recheck the coolant level.

2.4L (VIN G) Engine

1. Disconnect the negative battery cable.

2. Drain the cooling system.

3. Remove the engine undercover.

4. Disconnect the clamp bolt from the power steering hose.

5. Support the engine with the appropriate equipment and remove the engine mount bracket.

6. Remove the timing belt from the front of the engine.

7. Disconnect the coolant hoses from the pump, if equipped.

8. Remove the alternator brace.

9. Remove the water pump, gasket and O-ring where the water inlet pipe(s) joins the pump.

To install:

10. Thoroughly clean and dry both gasket surfaces of the water pump and block.

11. Install a new O-ring into the groove on the front end of the water inlet pipe. Do not apply oils or grease to the O-ring. Wet with clean antifreeze only.

12. Install the gasket and pump assembly and tighten the bolts.

13. Connect the hoses to the pump.

14. Reinstall the timing belt and related parts.

15. Install the engine drive belts and adjust.

16. Fill the system with coolant.

17. Connect the negative battery cable, run the vehicle until the thermostat opens and fill the radiator completely.

18. Once the vehicle has cooled, recheck the coolant level.

Thermostat

REMOVAL AND INSTALLATION

1.5L (VIN A) Engine

1. Disconnect the negative battery cable.

2. Drain the cooling system.

3. Remove the air cleaner assembly.

4. Disconnect the upper radiator hose from the thermostat housing.

5. Remove the thermostat housing and gasket.

6. Remove the thermostat taking note of its original position in the housing.

To install:

NOTE: In order to prevent leakage, make sure both mating surfaces are clean and free of all old gasket material.

7. Install the thermostat so its flange seats tightly in the machined recess in the thermostat housing. Refer to its location prior to removal.

8. Use a new gasket and reinstall the thermostat housing. Torque the housing mounting bolts to 16 ft. lbs. (22 Nm).

9. Connect the hoses and fill the system with coolant.

10. Install the air cleaner assembly.

11. Connect the negative battery cable, run the vehicle until the thermostat opens and fill the radiator completely.

12. Bleed the cooling system.

13. Once the vehicle has cooled, recheck the coolant level.

1.8L (VIN C) and 2.4L (VIN G) Engines

Except 1995–97 Summit Wagon and Vista Models

1. Disconnect the negative battery cable.

2. Drain the cooling system.

3. Disconnect the lower radiator hose from the thermostat housing.

4. Remove the thermostat housing and gasket.

5. Remove the thermostat taking note of its original position in the housing.

To install:

6. In order to prevent leakage, make sure both mating surfaces are clean and free of all old gasket material.

NOTE: Be sure to position the thermostat with the jiggle valve facing straight up.

7. Install the thermostat so its flange seats tightly in the machined recess in the thermostat housing. Refer to its location prior to removal.

8. Use a new gasket and reinstall the thermostat housing. Torque the housing mounting bolts to 14 ft. lbs. (19 Nm).

9. Connect the lower hose and fill the system with coolant.

10. Connect the negative battery cable, run the vehicle until the thermostat opens and fill the radiator completely.

11. Bleed the cooling system.

12. Once the vehicle has cooled, recheck the coolant level.

1995–97 Summit Wagon and Vista Models

1. Disconnect the negative battery cable.

2. Drain the cooling system.

3. Disconnect the lower radiator hose from the water outlet.

4. Remove the water outlet.

5. Remove the thermostat and gasket.

To install:

6. Use a new rubber gasket and install the thermostat with the jiggle valve facing up. Torque the water outlet bolts to 14 ft. lbs. (19 Nm).

7. Connect the lower radiator hose to the water outlet. Install the clamp in the original position.

8. Refill the cooling system and connect the negative battery cable.

9. Start the engine and check for leaks.

Electric Cooling Fan

REMOVAL AND INSTALLATION

1993–94 1.5L (VIN A), 1.8L (VIN C) and All 2.4L (VIN G) Engines

1. Disconnect the negative battery cable.

2. Disconnect the electrical connectors from the cooling fan.

3. Remove the fan and shroud-to-radiator bolts. Remove the fan

shroud assembly. It may be necessary to remove the condenser fan during this procedure.

4. With fan removed from the vehicle, remove the center fan nut and the three attaching bolts.

To install:

5. Install the fan blade to the motor, then connect the motor to the shroud.

6. Align the fan assembly to the radiator and secure with mounting bolts.

7. Connect all fan wiring and secure away from abrasion.

8. Connect the negative battery cable. Start the engine and check fan operation.

1995–97 1.5L (VIN A) and 1.8L (VIN C) Engines

1. Disconnect the negative battery cable.

2. Drain the coolant from the radiator.

3. Disconnect the overflow hose.

4. Remove the upper radiator hose.

5. If equipped with an automatic transaxle, remove the fluid cooler hoses.

6. Disconnect the fan motor harness connector.

7. Remove the fan shroud mounting bolts and remove the fan assembly.

To install:

8. Install the fan and shroud assembly.

9. Connect the fan motor harness connector.

10. Connect the transaxle fluid cooler hoses if removed.

11. Install the upper radiator hose.

12. Connect the radiator overflow hose.

13. Connect the negative battery cable.

14. Refill and bleed the cooling system.

Cooling System Bleeding

PROCEDURE

All Engines

1. Disconnect the negative battery cable.

2. Set the heater control lever to the HOT position.

3. With the coolant drained, close the radiator drain plug and install the engine drain plugs.

4. Slowly fill the radiator to the brim with the proper mixture of coolant and water. Fill the reservoir bot-

tle to the FULL line and install the radiator cap.

5. Connect the negative battery cable.

6. Start the engine and allow it to reach normal operating temperature. Watch the coolant temperature gauge for signs of overheating. Race the engine two or three times with no-load.

7. Shut the engine off and let it cool down completely.

8. After the engine has cooled down, remove the radiator cap and fill the radiator to the brim. Fill the reservoir to the FULL line on the bottle.

9. Check the radiator drain plug and the engine drain plugs for leakage.

10. Install the radiator cap.

FUEL SYSTEM

Fuel System Service Precautions

Safety is the most important factor when performing not only fuel system maintenance but any type of maintenance. Failure to conduct maintenance and repairs in a safe manner may result in serious personal injury or death. Maintenance and testing of the vehicle's fuel system components can be accomplished safely and effectively by adhering to the following rules and guidelines.

• To avoid the possibility of fire and personal injury, always disconnect the negative battery cable unless the repair or test procedure requires that battery voltage be applied.

• Always relieve the fuel system pressure prior to disconnecting any fuel system component (injector, fuel rail, pressure regulator, etc.), fitting or fuel line connection. Exercise extreme caution whenever relieving fuel system pressure to avoid exposing skin, face and eyes to fuel spray. Please be advised that fuel under pressure may penetrate the skin or any part of the body that it contacts.

• Always place a shop towel or cloth around the fitting or connection prior to loosening to absorb any excess fuel due to spillage. Ensure that all fuel spillage (should it occur) is quickly removed from engine surfaces. Ensure that all fuel soaked cloths or towels are deposited into a suitable waste container.

• Always keep a dry chemical (Class B) fire extinguisher near the work area.

• Do not allow fuel spray or fuel vapors to come into contact with a spark or open flame.

• Always use a backup wrench when loosening and tightening fuel line connection fittings. This will prevent unnecessary stress and torsion to fuel line piping. Always follow the proper torque specifications.

• Always replace worn fuel fitting O-rings with new. Do not substitute fuel hose or equivalent, where fuel pipe is installed.

Fuel System Pressure

RELIEVING

———— **CAUTION** ————

Fuel injection systems remain under pressure after the engine has been turned OFF. Properly relieve the fuel pressure before disconnecting any fuel lines. Failure to do so may result in fire or personal injury.

1. Loosen the fuel filler cap to release fuel tank pressure.
2. Remove rear seat for access and disconnect the fuel pump harness connector.
3. Start the vehicle and allow it to run until it stalls from lack of fuel. Turn the key to the **OFF** position.
4. Disconnect the negative battery cable, then reconnect the fuel pump connector and reinstall the fuel filler cap.

———— **WARNING** ————

Always wrap shop towels around a fitting that is being disconnected to absorb residual fuel in the lines.

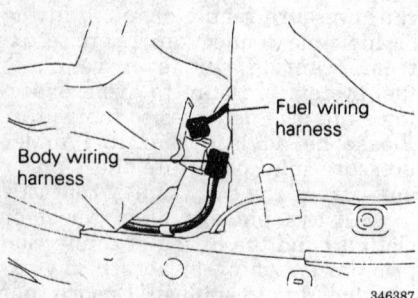

Fuel pump harness location — all models

Idle Speed

ADJUSTMENT

Colt and Summit

NOTE: The idle speed is controlled electronically and adjustment is usually unnecessary. However, the idle speed may be checked using the following procedures.

1. Warm the engine to operating temperature, leave lights, electric cooling fan and accessories **OFF**. The

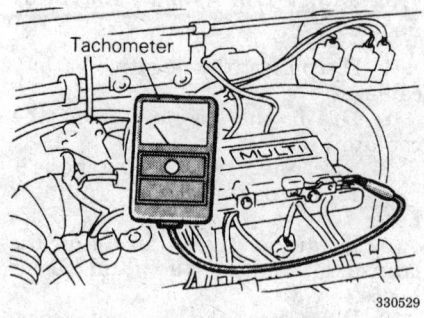

Early style tachometer pickup location — Colt and Summit models

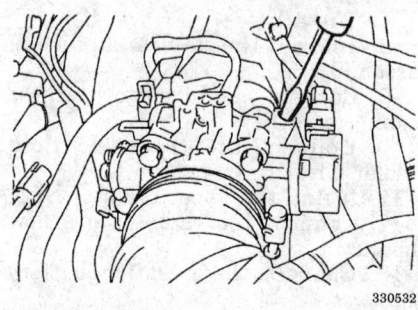

Engine speed adjusting screw location — Colt and Summit models

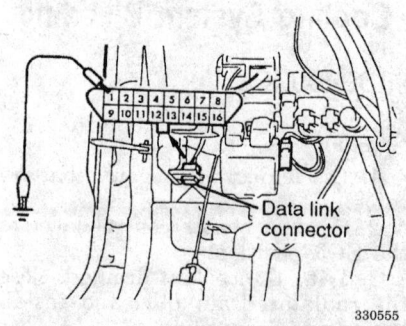

Diagnostic connector and terminal identification — all models except 1993 Colt and Summit

transaxle should be in **N** or **P** for automatic transaxle. The steering wheel in a neutral position for vehicles with power steering.

2. Insert the paper clip into the single terminal rpm connector in the engine compartment, and connect the primary voltage detection type tachometer to the paper clip.

3. Locate the self-diagnostic connector next to the fuse box and for 1993 vehicles ground terminal number 10 of the diagnostic connector with a jumper wire or ground terminal number 1 for 1994–97 vehicles.

4. Remove the waterproof female connector from the ignition timing adjustment connector. Ground the ignition timing adjustment terminal.

5. Start the engine and run at idle. Check the basic idle speed, the desired value is 700–800 rpm.

6. If the value is not within specifications, turn the Speed Adjusting Screw (SAS) to make the necessary adjustment.

NOTE: If the idle speed is higher than the standard value, inspect the SAS screw for evidence of movement. If there is evidence that the SAS screw has been adjusted, readjust to the proper setting. If the screw does not look as though it has been adjusted, it is possible that there is leakage as a result of deterioration of the Fast Idle Air Valve (FIAV) and, if so the throttle body should be replaced.

7. Turn the ignition **OFF**. Disconnect and remove the jumper wires from the diagnosis control terminal and the ignition timing adjustment terminal.

8. Start the engine and let run at idle speed for about 10 minutes, check to be sure the idling condition is normal.

Summit Wagon and Vista

NOTE: The idle speed is controlled electronically and adjustment is usually unnecessary. However, the idle speed may be checked using the following procedures.

1. Warm the engine to operating temperature, leave lights, electric cooling fan and accessories **OFF**. The transaxle should be in **N** or **P** for automatic transaxle. The steering wheel in a neutral position for vehicles with power steering.

2. Insert the paper clip into the single terminal rpm connector in the engine compartment, and connect the

primary voltage detection type tachometer to the paper clip.

3. Locate the self-diagnostic connector next to the fuse box. Ground number 1 terminal of the diagnostic connector with a jumper wire.

4. Remove the waterproof female connector from the ignition timing adjustment connector. Ground the ignition timing adjustment terminal.

5. Start the engine and run at idle. Check the basic idle speed, the desired value is 700–800 rpm.

6. If the value is not within specifications, turn the Speed Adjusting Screw (SAS) to make the necessary adjustment.

NOTE: If the idle speed is higher than the standard value, inspect the SAS screw for evidence of movement. If there is evidence that the SAS screw has been adjusted, readjust to the proper setting. If the screw does not look as though it has been adjusted, it is possible that there is leakage as a result of deterioration of the Fast Idle Air Valve (FIAV) and, if so the throttle body should be replaced.

7. Turn the ignition **OFF**. Disconnect and remove the jumper wires from the diagnosis control terminal and the ignition timing adjustment terminal.

8. Start the engine and let run at idle speed for about 10 minutes, check to be sure the idling condition is normal.

Mixture

ADJUSTMENT

All Models

Air/Fuel mixture is controlled by the ECU and is not adjustable.

Fuel Filter

REMOVAL AND INSTALLATION

All Engines

A replaceable fuel filter is located in the engine compartment.

——————— CAUTION ———————

Do not use conventional fuel filters, hoses or clamps when servicing fuel injection systems. They are not compatible with the injection system and could fail, causing personal injury or damage to

the vehicle. Use only hoses and clamps specifically designed for fuel injection.*

——————— CAUTION ———————

Fuel injection systems remain under pressure, after the engine has been turned OFF. Properly relieve fuel pressure before disconnecting any fuel lines. Failure to do so may result in fire or personal injury.

1. Relieve the fuel system pressure.

NOTE: Wrap shop towels around the fitting that is being disconnected to absorb residual fuel in the lines.

2. Hold the fuel filter nut securely with a backup or spanner wrench. Cover the hoses with shop towels and remove the eye bolt. Discard the gaskets.

3. Separate the flare nut connection at the filter. Discard the gaskets.

4. Remove the mounting bolts and the fuel filter from the vehicle.

To install:

5. If equipped with flare fitting, tighten the fitting by hand before installing the filter to the vehicle.

6. Install the filter to its bracket only finger-tight. Movement of the filter will ease attachment of the fuel lines.

7. Install new gaskets and connect the high pressure hose and eye bolt, then the main pipe. While holding the fuel filter nut, tighten the eye bolts to 22 ft. lbs. (30 Nm). Tighten the flare nut to 27 ft. lbs. (37 Nm).

8. Tighten the filter mounting bolts fully.

9. Install the air cleaner assembly, if removed.

10. Connect the negative battery cable, install the fuel filler cap, turn the key to the **ON** position to pressurize the fuel system and check for leaks.

11. Release the fuel pressure and repair leaks as required.

Fuel Pump

REMOVAL AND INSTALLATION

Colt and Summit

1. Relieve the fuel system pressure using proper procedures. Disconnect negative battery cable.

NOTE: Wrap shop towels around the fitting that is being disconnected to absorb residual fuel in the lines.

——————— CAUTION ———————

Fuel injection systems remain under pressure after the engine has been turned OFF. Properly relieve fuel pressure before disconnecting any fuel lines. Failure to do so may result in fire or personal injury.

2. Raise the vehicle and support safely.

3. Drain the fuel from the fuel tank into an approved container.

4. Disconnect the return hose, high pressure hose and vapor hoses from the fuel pump.

5. Disconnect the electrical connectors at the pump/sending unit.

——————— CAUTION ———————

Cover all fuel hose connections with a shop towel, prior to disconnecting, to prevent splash of fuel that could be caused by residual pressure remaining in the fuel line.

6. Disconnect the filler and vent hoses.

7. Place a transmission jack under the center of the fuel tank and apply a slight upward pressure. Remove the fuel tank strap retaining nut.

8. Lower the tank slightly and disconnect any remaining electrical or hose connectors at the fuel tank.

9. Remove the fuel tank from the vehicle.

10. Remove the five nuts securing the access plate to the fuel tank and remove the pump assembly.

To install:

11. Install fuel pump into fuel tank, with new packing gasket, and tighten mounting nuts.

12. Install the fuel tank onto the transmission jack. Raise the tank in position under the vehicle. Leave enough clearance to attach the electrical and hose connections to the top of the fuel pump.

13. Attach all connections to the top of the tank.

14. Raise the tank completely and position the retainer straps around the fuel tank. Install new fuel tank self-locking nuts and tighten to 22 ft. lbs. (31 Nm).

15. Connect the return hose and high pressure hoses.

16. Install the vapor hose and the filler hose. Install the filler hose retainer screws to the fender, if removed.

17. Lower the vehicle and pour the drained fuel into the gas tank.

18. Connect the negative battery cable. Check the fuel pump for proper pressure and inspect the entire system for leaks.

Summit Wagon and Vista

1. Relieve fuel system pressure. Remove the fuel filler cap.

———— **CAUTION** ————
Fuel injection systems remain under pressure after the engine has been turned OFF. Properly relieve fuel pressure before disconnecting any fuel lines. Failure to do so may result in fire or personal injury.

2. Disconnect the negative battery cable.

3. Raise and safely support the vehicle.

4. The fuel pump is located in the fuel tank. Drain the fuel from the fuel tank.

———— **CAUTION** ————
Do not allow fuel spray or fuel vapors to come in contact with a spark or open flame. Keep a dry chemical fire extinguisher nearby. Never store fuel in an open container due to risk of fire or explosion.

5. On vehicles equipped with AWD, remove the rear propeller shaft from the vehicle as follows:

   a. Remove the center exhaust pipe bracket.

   b. Matchmark the differential companion flange to the propeller flange yoke.

   c. Remove the bolts, washers and nuts from the center support. Remove the propeller shaft assembly in a straight and level manner to avoid damage to the boot caused by pinching.

   d. Install cover into the rear end of the transfer case to prevent the entry of foreign materials.

6. Disconnect the return hose, high pressure hose and all other hoses and connectors connected to the pump and sending unit.

7. Disconnect the filler and vent hoses. Place a support under the tank and remove the retaining nuts. Lower the tank from vehicle.

8. Remove retaining nuts and remove the fuel pump assembly from tank.

To install:

9. Install the replacement pump using a new gasket. Be certain the pump is installed in the same location, facing the same direction as before.

10. Install the fuel tank and secure the retainer nuts. Connect all electrical harness connectors. Reconnect all vent hoses, fuel supply and fuel return hoses securing with the proper clamps.

11. On vehicles equipped with AWD, install the propeller shaft aligning the matchmarks prior to installation. Tighten the rear yoke nuts to 22–25 ft. lbs. (30–35 Nm) and the center support self-locking nuts to 22 ft. lbs. (30 Nm).

12. Install the exhaust pipe center bracket. Check that electrical connectors are properly installed and all fuel hose connections are tight.

13. Connect the negative battery cable and check the entire fuel system for proper operation and leaks. If repairing of a fuel leak is required, release the fuel system pressure prior to repairing system.

Fuel Injector

REMOVAL AND INSTALLATION

All Engines

1. Relieve the fuel system pressure following proper procedures.

———— **CAUTION** ————
Fuel injection systems remain under pressure after the engine has been turned OFF. Properly relieve fuel pressure before disconnecting any fuel lines. Failure to do so may result in fire or personal injury.

2. Disconnect the PCV hose from the valve cover. Also disconnect the breather hose at the opposite end of the valve cover.

3. Remove the bolts holding the high pressure fuel line to the fuel rail and disconnect the line. Be prepared to contain fuel spillage; plug the line to keep out dirt and debris.

4. Remove the vacuum hose from the fuel pressure regulator.

5. Disconnect the fuel return hose from the pressure regulator. Remove the fuel pressure regulator mounting bolts and remove from the fuel rail.

6. Label and disconnect the electrical connector from each injector.

7. Remove the bolt(s) holding the fuel rail to the manifold. Carefully lift the rail up and remove it with the injectors attached. Take great care not to drop an injector. Place the rail and injectors in a safe location on the workbench; protect the tips of the injectors from dirt and/or impact.

8. Remove and discard the injector insulators from the intake manifold. The insulators are not reusable.

9. Remove the injectors from the fuel rail by pulling gently in a straight outward motion. Make certain the grommet and O-ring come off with the injector.

To install:

10. Install a new insulator in each injector port in the manifold.

11. Remove the old grommet and O-ring from each injector. Install a new grommet and O-ring; coat the O-ring lightly with clean, thin oil.

12. If the fuel pressure regulator was removed, replace the O-ring with a new one and coat it lightly with clean, thin oil. Insert the regulator straight into the rail, then check that it can be rotated freely. If it does not rotate smoothly, remove it and inspect the O-ring for deformation or jamming. When properly installed, align the mounting holes and tighten the retaining bolts to 7 ft. lbs. (9 Nm). This procedure must be followed even if the fuel rail was not removed.

13. Install the injector into the fuel rail, constantly turning the injector left and right during installation. When fully installed, the injector should still turn freely in the rail. If it does not, remove the injector and inspect the O-ring for deformation or damage.

14. Install the delivery pipe and injectors to the engine. Make certain that each injector fits correctly into its port and that the rubber insulators for the fuel rail mounts are in position.

15. Install the fuel rail retaining bolts and tighten them to 9 ft. lbs. (12 Nm).

16. Connect the wiring harnesses to the appropriate injector.

17. Connect the fuel return hose to the pressure regulator, then connect the vacuum hose.

18. Install a new O-ring on the high pressure fuel line, coat the O-ring lightly with clean, thin oil and install the line to the fuel rail. Tighten the mounting bolts to specifications.

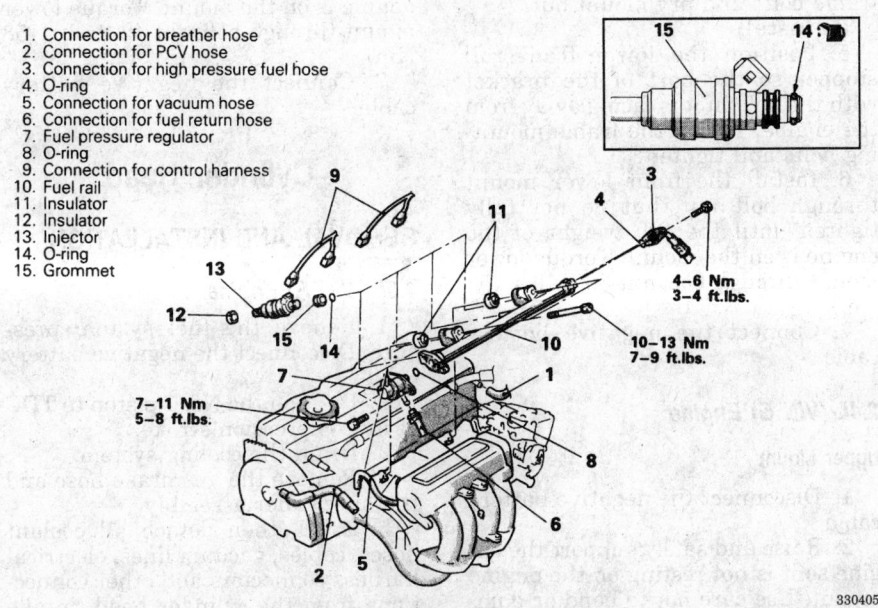

1. Connection for breather hose
2. Connection for PCV hose
3. Connection for high pressure fuel hose
4. O-ring
5. Connection for vacuum hose
6. Connection for fuel return hose
7. Fuel pressure regulator
8. O-ring
9. Connection for control harness
10. Fuel rail
11. Insulator
12. Insulator
13. Injector
14. O-ring
15. Grommet

4–6 Nm
3–4 ft. lbs.

10–13 Nm
7–9 ft. lbs.

7–11 Nm
5–8 ft. lbs.

330405

Fuel injector and related parts — all models

19. Connect the negative battery cable. Pressurize the fuel system and inspect all connections for leaks.

ENGINE MECHANICAL

Engine Assembly

REMOVAL AND INSTALLATION

All Engines

1. Relieve fuel system pressure.
2. Disconnect the negative battery cable. Remove the undercover if equipped.
3. Matchmark the hood and hinges and remove the hood assembly. Remove the air cleaner assembly and all adjoining air intake duct work.
4. Drain the engine coolant and remove the radiator assembly, coolant reservoir and intercooler.
5. Remove the transaxle assembly.
6. Disconnect the ground cable, accelerator cable, breather hose and heater hose connections from the engine.
7. Note locations and remove vacuum hoses from engine. Be sure to disconnect brake booster vacuum supply.
8. For 2.4L (VIN G) engines, disconnect the heater hoses at the cylinder head and coolant inlet pipe.
9. Disconnect fuel feed and return hoses.
10. Disconnect the crankshaft and camshaft sensor wiring.
11. Disconnect oxygen sensor connections, coolant temperature gauge and coolant temperature sensor connections.
12. Disconnect the oil pressure switch connection.
13. On models with automatic transmissions, disconnect the thermo switch.
14. Disconnect harness connections for the idle speed control motor and throttle position sensor.
15. Disconnect harness connections for the intake air temperature sensor.
16. Disconnect EGR temperature sensor (California).
17. Note locations for reassembly and disconnect injector harness plugs.
18. Disconnect power transistor and the ignition coil connections.

19. Disconnect alternator and power steering switch wiring.
20. Remove the air conditioner drive belt and the air conditioning compressor. Leave the hoses attached. Do not discharge the system. Wire the compressor aside.
21. Remove the power steering pump and wire aside.
22. Remove the alternator and starter harness clamps.
23. Remove the exhaust manifold to head pipe nuts. Discard the gasket.
24. Attach a hoist to the engine and support the engine weight. Remove the engine mount bracket. Remove any torque control brackets (roll stoppers).
25. Remove the engine assembly from the vehicle.

To install:
26. Install the engine and secure in position. The front lower mount through bolt nut should not be tightened until the full weight of the engine is on the mount. Tighten the bolts to the following specifications:
  • 1.5L (VIN A) and 1.8L (VIN C) through bolt — 72 ft. lbs. (100 Nm)
  • 1.5L (VIN A) and 1.8L (VIN C) bracket mounting bolts — 42 ft. lbs. (58 Nm)
  • 1.5L (VIN A) and 1.8L (VIN C) bracket mounting nut — 38 ft. lbs. (53 Nm)
  • 2.4L (VIN G) through bolt — 51 ft. lbs. (69 Nm)
  • 2.4L (VIN G) bracket mounting bolts — 42 ft. lbs. (58 Nm)
27. Using a new gasket, position exhaust pipe onto the manifold and tighten the flange nuts to 36 ft. lbs. (50 Nm).
28. Install the remaining components in the opposite order from which they were removed.
29. Connect negative battery cable and run engine.
30. Inspect all connections and check all fluid levels.

Engine Mounts

REMOVAL AND INSTALLATION

1.5L (VIN A) and 1.8L (VIN C) Engines

Upper Mount

1. Disconnect the negative battery cable.
2. Raise and safely support the engine so it is not resting on the engine mount. One suggested way is a block of wood between a floor jack and the oil pan. Use care not to bend or damage any components.

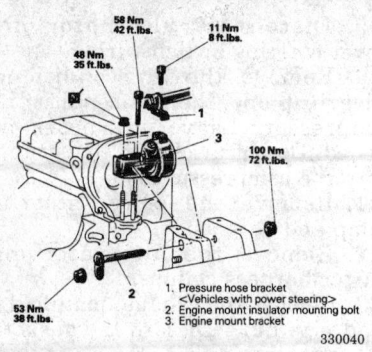

1. Pressure hose bracket <Vehicles with power steering>
2. Engine mount insulator mounting bolt
3. Engine mount bracket

330040

Engine mount identification — 1.5L (VIN A) and 1.8L (VIN C) engines

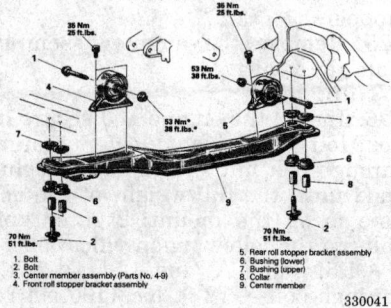

1. Bolt
2. Bolt
3. Center member assembly (Parts No. 4-9)
4. Front roll stopper bracket assembly
5. Rear roll stopper bracket assembly
6. Bushing (lower)
7. Bushing (upper)
8. Collar
9. Center member

330041

Engine roll stoppers and related components — 1.5L (VIN A) and 1.8L (VIN C) engines, 2.4L (VIN G) similar

3. Remove the retainer bolt from the clamp securing the power steering pressure hose and the air conditioning low pressure hose, if equipped.

4. Remove the engine mount bracket and body connection through bolt.

5. Disconnect the engine mounting bracket, by removing the two nuts and bolts securing it to the engine and remove.

To install:

6. Install the engine mounting bracket and stopper plate. Align the through bolt hole to the notch in the body bracket, make sure they are installed properly. Torque upper mount to engine nuts to 35 ft. lbs., bolts to 42 ft. lbs. (58 Nm) and the upper mount through bolt nut to 72 ft. lbs. (100 Nm). There is a second nut that secures the through bolt, tighten the nut to 38 ft. lbs. (53 Nm).

Lower Mount

1. Disconnect the negative battery cable.

2. Raise and safely support the engine so it is not resting on the engine mount. One suggested way is a block of wood between a floor jack and the oil pan. Use care not to bend or damage any components.

3. Remove the stopper through bolt.

4. Remove the stopper mount frame bolts and pry mount out.

To install

5. Position the lower front roll stopper so the part of the bracket with the hole in it is facing away from the engine. Install the frame mounting bolts and tighten.

6. Install the front lower mount through bolt nut, but do not fully tighten until the full weight of the engine is on the mount. Torque lower mount through bolt nut 38 ft. lbs. (53 Nm).

7. Connect the negative battery cable.

2.4L (VIN G) Engine

Upper Mount

1. Disconnect the negative battery cable.

2. Raise and safely support the engine so it is not resting on the engine mount. Use care not to bend or damage any components.

3. Remove the retainer bolt from the clamp securing the power steering pressure hose and the air conditioning low pressure hose, if equipped.

4. Remove the engine mount bracket and body connection through bolt.

5. Disconnect the engine mounting bracket, by removing the two nuts and bolts securing it to the engine and remove.

To install:

6. Install the engine mounting bracket and stopper plate. Align the through bolt hole to the notch in the body bracket, make sure they are installed properly. Torque upper mount to engine nut and bolt to 42 ft. lbs. (58 Nm) and the upper mount through bolt nut to 51 ft. lbs. (70 Nm).

Lower Mount

1. Disconnect the negative battery cable.

2. Raise and safely support the engine so it is not resting on the engine mount. Use care not to bend or damage any components.

3. Remove the stopper through bolt.

4. Remove the stopper mount frame bolts and pry mount out.

To install

5. Position the lower front roll stopper so the part of the bracket with the hole in it is facing away from the engine. Install the frame mounting bolts and tighten.

6. Install the front lower mount through bolt nut, but do not fully tighten until the full weight of the engine is on the mount. Torque lower mount through bolt nut 42 ft. lbs. (58 Nm).

7. Connect the negative battery cable.

Cylinder Head

REMOVAL AND INSTALLATION

1.5L (VIN A) Engine

1. Relieve the fuel system pressure. Disconnect the negative battery cable.

2. Position the No. 1 piston to TDC of its compression stroke.

3. Drain the cooling system.

4. Remove the air intake hose and the air cleaner assembly.

5. Label, then detach all coolant hoses, cables, vacuum lines, electrical harness connectors and other connections from the cylinder head, intake manifold, exhaust manifold and all related components.

6. Remove the upper radiator hose, throttle body hoses, bypass hose and heater hose connections.

7. Place a shop towel around the high pressure fuel line to absorb any residual fuel remaining in the system. Disconnect the high pressure fuel line.

8. Disconnect and plug the fuel return line.

9. Remove the spark plug cables.

10. Remove the clamp that holds the power steering pressure hose to the engine mounting bracket.

11. Place a jack and wood block under the oil pan and carefully lift just enough to take the weight off the engine mounting bracket, remove the bracket.

12. Remove the valve cover and gasket.

13. Remove the timing belt front upper cover.

14. If not already done, rotate the crankshaft clockwise and align the timing marks.

15. While securing the sprocket, remove the sprocket bolt and remove the sprocket with the timing belt attached. Be sure to keep the sprocket and belt wired or tied together as one unit.

16. Remove the timing belt rear upper cover.

——— CAUTION ———
The crankshaft must always be rotated clockwise. Do not rotate engine by turning the camshaft.

17. Remove the exhaust pipe self-locking nuts and separate the exhaust pipe from the exhaust manifold. Discard the gasket.

18. Loosen the cylinder head mounting bolts in the reverse sequence as shown in the tightening illustration, using three steps. Lift off the cylinder head assembly and remove the head gasket.

To install:

19. Thoroughly clean and dry the mating surfaces of the head and block. Check the cylinder head for cracks, damage or engine coolant leakage. Remove scale, sealing compound and carbon. Clean oil passages thoroughly. Check the head for flatness. End to end, the head should be within 0.002 inch with 0.008 inch the maximum allowed out of true. The total thickness allowed to be removed from the head and block is 0.008 in. maximum.

20. Place a new head gasket on the cylinder block with the identification marks facing upward. Make sure the gasket has the proper identification mark for the engine. Do not use sealer on the gasket.

NOTE: The cylinder head torque specifications are for a cold cylinder head.

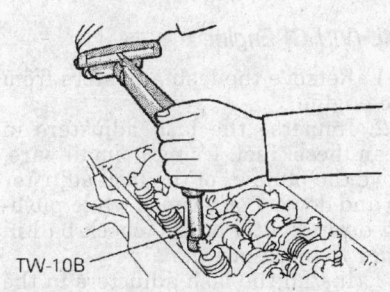

TW-10B

← Front of engine

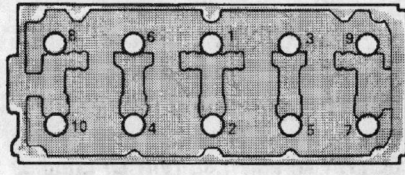

329646

Cylinder head bolt torque sequence — 1.5L (VIN A) engine

21. Carefully install the cylinder head on the block. Using 3 even steps, torque the head bolts in sequence to 53 ft. lbs. (73 Nm).

22. The remaining components are installed in the reverse order from which they were removed. Tighten the camshaft sprocket retaining bolt to 51 ft. lbs. (70 Nm) and the valve cover retaining bolts to 16 inch lbs. (2 Nm).

23. Fill the system with coolant.

24. Connect the negative battery cable, run the vehicle until the thermostat opens, fill the radiator completely.

25. Check and adjust the idle speed and ignition timing.

26. Once the vehicle has cooled, recheck the coolant level.

1.8L (VIN C) Engine

1. Relieve fuel system pressure. Disconnect the negative battery cable.

2. Position the No. 1 piston to TDC of its compression stroke.

3. Remove the air cleaner assembly.

4. Drain the cooling system.

5. Label, then detach all coolant hoses, cables, vacuum lines, electrical harness connectors and other connections from the cylinder head, intake manifold, exhaust manifold or related components.

6. Remove the upper radiator hose, overflow tube and the water hose from the thermostat to the throttle body.

7. Wrap the connection with a shop towel and disconnect the high pressure fuel line at the fuel rail. Discard the O-ring.

8. Disconnect the fuel return hose from the fuel pressure regulator.

9. Remove the thermostat housing, thermostat and the thermostat case with O-ring from the engine.

10. Remove the rocker cover.

11. Remove the timing belt upper cover.

12. If not already done, rotate the crankshaft in the clockwise direction to align the camshaft timing marks. Matchmark the camshaft sprocket and the timing belt. Tie the camshaft sprocket and the timing belt together so the sprocket will not move with respect to the timing belt.

13. While holding the camshaft sprocket in position using the appropriate wrench, remove the camshaft sprocket and with the belt attached. Wire the sprocket and belt aside making sure constant tension is maintained on the belt. Do not allow

the belt to slacken or engine timing may be altered.

NOTE: When removing the camshaft sprocket, do not allow the crankshaft to rotate. Confirm proper engine timing during installation.

14. Loosen the cylinder head bolts in two or three steps in the reverse of the proper tightening sequence.

15. Remove the cylinder head from the engine.

──── **CAUTION** ────
When removing the cylinder head, take care not to bend or damage the plug guide. The plug guide can not be replaced.
────────────

16. Remove the cylinder head gasket from the block.

To install:

17. Thoroughly clean and dry the mating surfaces of the head and block. Check the cylinder head for cracks, damage or engine coolant leakage. Remove scale, sealing compound and carbon. Clean oil passages thoroughly. Check the head for flatness end to end, the head should be within 0.002 inch with 0.008 inch the maximum allowed out of true. The total thickness allowed to be removed from the head and block is 0.008 in. maximum.

18. Place a new head gasket on the cylinder block with the identification marks facing upward. Make sure the gasket has the proper identification mark for the engine. Do not use sealer on the gasket.

19. Carefully install the cylinder head on the block.

20. Inspect the cylinder head bolt prior to installation. The length below the head of the bolts should not exceed 3.795 inches (96.4mm). If bolt shank length exceeds limit, bolt must be replaced. New bolts are always recommended.

21. Apply a small amount of engine oil to the thread section of the bolt and install so the chamfer of the washer faces upward.

22. Tighten the cylinder head bolts in the proper order as follows:

    a. In the proper tightening sequence, torque bolts to 54 ft. lbs. (75 Nm).

    b. In the reverse order of the tightening sequence, fully loosen bolts.

    c. In the proper tightening sequence, torque bolts to 14 ft. lbs. (20 Nm).

    d. In the proper tightening sequence, tighten bolts an additional ¼turn (90 degrees).

e. In the proper tightening sequence, tighten bolts an additional ¼turn (90 degrees).

23. Install the remaining components in the opposite order from which they were removed. Tighten the camshaft sprocket mounting bolt to 65 ft. lbs. (90 Nm), the valve cover retaining bolts to 29 inch lbs. (3 Nm), the thermostat case assembly bolts to 16 ft. lbs. (22 Nm) and the thermostat housing bolts to 10 ft. lbs. (14 Nm).

24. Fill the system with coolant.

25. Connect the negative battery cable, run the vehicle until the thermostat opens, fill the radiator completely.

26. Check and adjust the idle speed and ignition timing.

27. Check all systems for leaks. Allow the engine to cool and recheck the coolant level.

2.4L (VIN G) Engine

1. Relieve fuel system pressure. Disconnect the negative battery cable.

2. Drain the cooling system.

3. Label, then detach all coolant hoses, cables, vacuum lines, electrical harness connectors and other connections from the cylinder head, intake manifold, exhaust manifold or related components.

4. Remove the radiator.

5. Wrap the connection with a shop towel and disconnect the high pressure fuel line at the fuel rail.

6. Disconnect the fuel return hose and remove the O-ring.

7. Remove the bolt retaining the power steering hose and air conditioner hose clamp.

8. Remove the coolant reservoir. Remove the bolt holding the ground wire to the manifold.

9. Place a jack and wood block under the oil pan and carefully lift just enough to take the weight off the engine mounting bracket. Then remove the engine mounting bracket taking note of the position of the mount stopper.

10. Remove the valve cover, gasket and half-round seal.

11. Remove the timing belt front upper cover.

12. If possible, rotate the crankshaft clockwise until the timing marks on the cam sprocket and belt align. Matchmark the timing sprocket to the belt. Remove the sprocket bolt and remove the sprocket with the timing belt attached. Attach a flexible cord to the hood and suspend the sprocket so it cannot turn and there is no slack in

the belt. Remove the timing belt rear upper cover.

13. Loosen the head bolts in the reverse of the correct tightening sequence in 2 or 3 steps. Remove the cylinder head bolts and head assembly from the block.

To install:

14. Thoroughly clean and dry the mating surfaces of the head and block. Remove scale, sealing compound and carbon. Clean oil passages thoroughly.

15. Perform the following checks before reassembly:

• Check the cylinder head for cracks, damage or engine coolant leakage.

• Check the head for flatness. End to end, the head should be within 0.002 in. (0.051mm) normally with 0.008 in. (0.200mm) the maximum allowed out of true. The total thickness allowed to be removed from the head and block is 0.008 in. (0.200mm) maximum.

• Check the cylinder head bolts for stretching or necking. The shank length should not exceed 3.91 in. (99.4 mm).

16. Place a new head gasket on the cylinder block with the identification marks at the top (upward) position. Make sure the gasket has the proper identification mark for the engine. Do not use sealer on the gasket. Replace the turbo gasket and ring, if equipped.

17. Carefully install the cylinder head on the block. Torque the cylinder head using the following procedure:

a. Torque all bolts in sequence to 58 ft. lbs. (78 Nm).

b. Using loosening sequence, fully loosen all bolts.

c. Torque all bolts in sequence to 14 ft. lbs. (20 Nm).

d. Tighten all bolts in sequence ¼ turn (90 °).

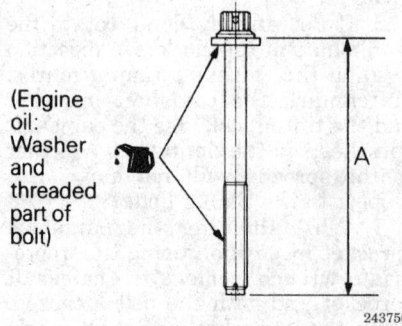

(Engine oil: Washer and threaded part of bolt)

A

243750

Checking cylinder head bolt for necking — 2.4L (VIN G) engines

e. Once again, tighten all bolts in sequence ¼ turn (90 °).

NOTE: Install the head bolt washer so the sagging side made by tapping out the washer is facing upward.

18. Install the camshaft sprocket and tighten bolt to 65 ft. lbs. (90 Nm), while holding the sprocket in place using the appropriate wrench. Confirm proper timing mark alignment.

19. Apply sealer to the perimeter of the half-round seal and to the lower edges of the half-round portions of the belt-side of the new gasket. Install the valve cover.

20. Install the remaining components in the reverse order of removal.

21. Firmly set the parking brake. Start the engine and allow to idle until the thermostat opens, add coolant as required to fill system to the appropriate level.

22. Check all systems for leaks. Allow the engine to cool and recheck the coolant level.

Lash Adjusters

BLEEDING

1.5L (VIN A) and 1.8L (VIN C) Engines

The 1.5L (VIN A) and 1.8L (VIN C) engines do not utilize hydraulic lash adjusters. These engines use solid lash adjusters, which do not require bleeding.

2.4L (VIN G) Engine

1. Remove the lash adjusters from the engine.

2. Immerse the lash adjusters in clean diesel fuel. Using a small wire, move the plunger of the lash adjuster up and down 4 or 5 times while pushing down lightly on the check ball in order to bleed out the air.

3. Install the lash adjusters in the rocker arms and secure with lash retainers.

Valve Lash

ADJUSTMENT

1.5L (VIN A) and 1.8L (VIN C) Engines

NOTE: Incorrect valve clearances will cause unsteady engine operation, excessive noise and reduced engine output. Check the valve clearances and adjust as required while the engine is hot.

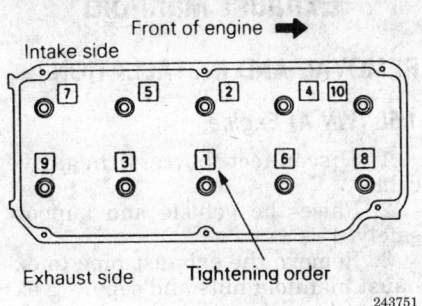

Cylinder head bolt tightening sequence — 1.8L (VIN C) and 2.4L (VIN G) engines

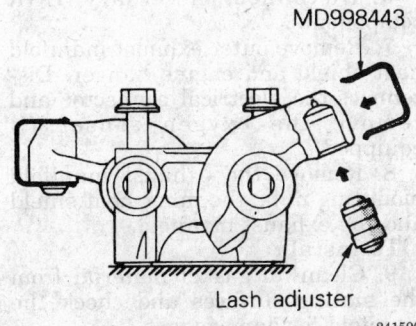

Valve lash adjuster identification — 2.4L (VIN G) engines

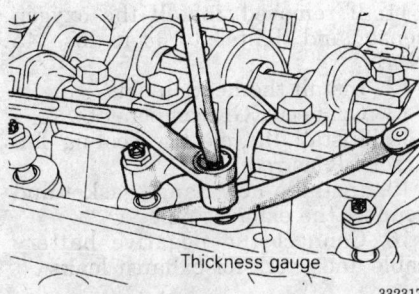

Proper method of adjusting valve clearance — 1.5L (VIN A) and 1.8L (VIN C) engines

1. Warm the engine to operating temperature, turn **OFF** and disconnect the negative battery cable.
2. Remove all spark plugs so engine can be easily turned by hand.
3. Remove the valve cover.
4. Turn the crankshaft clockwise to position the No. 1 cylinder at TDC of its compression stroke. The notch on the crankshaft pulley will be al-

igned with the **T** mark on the timing belt lower cover.

NOTE: When the No. 1 cylinder at TDC of its compression stroke, the No. 1 cylinder intake and exhaust valves will have valve lash.

5. Check the valve lash at cylinder No. 1 intake, cylinder No. 1 exhaust, cylinder No. 2 intake and cylinder No. 3 exhaust valves.
6. Rotate the crankshaft clockwise 1 complete turn and align the **T** mark. This will position the No. 4 cylinder at TDC of its compression stroke.

NOTE: When the No. 4 cylinder at TDC of its compression stroke, the No. 4 cylinder intake and exhaust valves will have valve lash.

7. Check the valve lash at cylinder No. 2 exhaust, cylinder No. 3 intake and cylinder No. 4 intake and exhaust valves.
8. If the valve clearances are out of specification, loosen the rocker arm locknut and adjust the clearance using a feeler gauge while turning the adjusting screw. Be sure to hold the screw to prevent it from turning when tightening the locknut.
9. After adjusting the valves, install the valve cover with new gasket and spark plugs, and connect the negative battery cable.

2.4L (VIN G) Engine

The 2.4L (VIN G) engines utilize hydraulic lash adjusters, which automatically retain the correct valve lash. These engines do not require manual valve lash adjustments.

Rocker Arms and Shafts

REMOVAL AND INSTALLATION

1.5L (VIN A) Engine

1. Disconnect the negative battery cable.
2. Rotate the engine and position the No. 1 piston to TDC of its compression stroke.
3. Disconnect the accelerator cable, breather hose and PCV hose connections.
4. Remove the valve cover and discard the gasket.
5. Loosen both rocker arm assemblies gradually and evenly and remove the rocket shafts from the vehicle.
6. If disassembly is required, keep all parts in the exact order of removal. Inspect the roller surfaces of the rockers. Replace if there are any

signs of damage or if the roller does not turn smoothly. Check the inside bore of the rockers and the adjuster tip for wear.
   To install:
7. Lubricate the rocker shaft with clean engine oil and install the rockers and springs in their proper places.
8. Install the rocker shaft assemblies. Torque the bolts gradually and evenly to 23 ft. lbs. (32 Nm).
9. Check valve adjustment and install the valve cover with a new gasket. Tighten the valve cover bolts to 16 inch lbs. (1.8 Nm).
10. Connect the accelerator cable, breather hose and PCV hose.
11. Connect the negative battery cable and check the engine adjustments.

1.8L (VIN C) Engine

1. Disconnect the negative battery cable.
2. Rotate the engine and position the No. 1 piston to TDC of its compression stroke.
3. Label and disconnect the spark plug cables.
4. Disconnect the air flow sensor connector and remove the air cleaner case cover.
5. Disconnect the accelerator cable, breather hose and PCV hose connections.
6. Remove the rocker cover and discard the gasket.
7. Loosen both rocker arm shaft assemblies gradually and evenly and remove the rocket shafts from the vehicle. Do not disassembly rocker arms and rocker arm shaft assemblies.
8. If disassembly is required, keep all parts in the exact order of removal. Inspect the roller surfaces of the rockers. Replace if there are any signs of damage or if the roller does not turn smoothly. Check the inside bore of the rockers and the adjuster tip for wear.
   To install:
9. Lubricate the rocker shaft with clean engine oil and install the rockers and springs in their proper places.
10. Install the rocker arm and shaft assemblies. Tighten the rocker arm shaft retainer bolts to 23 ft. lbs. (32 Nm).
11. Check valve adjustment and install valve cover with a new gasket. Tighten the valve cover bolts to 29 inch lbs. (3.3 Nm).
12. Connect the spark plug cables.
13. Connect the accelerator cable, breather hose and PCV hose.

14. Connect the air flow sensor connector and install the air cleaner case cover.

15. Connect the negative battery cable. Run the engine at idle until normal operating temperature is reached. Check idle speed and ignition timing and adjust as required.

2.4L (VIN G) Engine

1. Disconnect the negative battery cable.

2. Disconnect the PCV and breather hoses. Remove the valve cover.

3. Install lash adjuster retainer tools MD998443 or equivalent, to prevent the auto-lash adjuster from falling out of the rocker arm.

4. Loosen rocker arm and shaft assembly evenly in several steps. Remove the rocker arm and shaft assembly as a complete unit.

NOTE: If any parts are to be reused, it is essential that all parts be kept in the same order and orientation for reinstallation. Be sure to mark and separate parts, so parts won't be mixed during reassembly.

5. Carefully disassemble the shaft assembly. Visually inspect the rocker arm roller and replace if damage or seizure is evident. Check the roller for smooth rotation. Replace if excess play or binding is present. Also, inspect valve contact surface for possible damage or seizure. It is recommended that all rocker arms and lash adjusters be replaced together.

To install:

6. Immerse the lash adjusters in clean diesel fuel. Using a small wire, move the plunger of the lash adjuster up and down 4 or 5 times while pushing down lightly on the check ball in order to bleed out the air. Install the lash adjusters in the rocker arms.

7. Assemble the rocker assembly components in their original locations, lubricate the rocker shaft with heavy engine oil and position on the cylinder head.

8. Install the shaft mounting bolts and tighten all bolts evenly and gradually to 21–25 ft. lbs. (28–34 Nm).

9. Remove the lash adjuster retainers.

10. Install the valve cover, with a new gasket and semi-circular packing in place.

11. Connect the negative battery cable.

Intake Manifold

REMOVAL AND INSTALLATION

All Models

1. Relieve the fuel system pressure.

——— CAUTION ———
Fuel injection systems remain under pressure after the engine has been turned OFF. Properly relieve fuel pressure before disconnecting any fuel lines. Failure to do so may result in fire or personal injury.

2. Disconnect battery negative cable and drain the cooling system.

3. Label, then detach all coolant hoses, cables, vacuum lines, electrical harness connectors and other connections from the intake manifold and related components.

4. Disconnect the high pressure fuel line and the fuel return hose.

5. Remove the fuel rail, fuel injectors, pressure regulator and insulators.

6. Remove the intake manifold support bracket.

7. If the thermostat housing is preventing removal of the intake manifold, remove it.

8. Remove the intake manifold mounting bolts/nuts and remove the intake manifold assembly.

To install:

9. Clean all gasket material from the cylinder head intake mounting surface and intake manifold assembly. Check both surfaces for cracks or other damage. Check the intake manifold water passages and jet air passages for clogging. Clean if necessary.

10. Using a straight edge, measure the distortion of the intake manifold-to-cylinder head. Total distortion or warpage should be 0.006 inches (0.15mm or less).

11. Install a new intake manifold gasket to the head and install the manifold. Torque the manifold in a crisscross pattern, starting from the inside and working outwards to 14 ft. lbs. (20 Nm).

12. Install the remaining components in the opposite order from which they were removed.

13. Connect the negative battery cable, run the vehicle until the thermostat opens, fill the radiator completely.

14. Check and adjust the idle speed and ignition timing.

15. Once the vehicle has cooled, recheck the coolant level.

Exhaust Manifold

REMOVAL AND INSTALLATION

1.5L (VIN A) Engine

1. Disconnect battery negative cable.

2. Raise the vehicle and support safely.

3. Remove the exhaust pipe to exhaust manifold nuts and separate exhaust pipe. Discard gasket.

4. Lower vehicle.

5. Remove electric cooling fan assembly, if necessary.

6. Disconnect necessary EGR components.

7. Remove outer exhaust manifold heat shield and engine hanger. Disconnect the electrical connector and remove the oxygen sensor (if equipped).

8. Remove the exhaust manifold mounting nuts, the inner heat shield and the exhaust manifold.

To install:

9. Clean all gasket material from the mating surfaces and check the manifold for damage.

10. Install a new gasket and install the manifold. Tighten the nuts to in a crisscross pattern to 13 ft. lbs. (18 Nm).

11. If removed, install the oxygen sensor and tighten to 33 ft. lbs. (45 Nm).

12. Install the heat shields.

13. Connect EGR components.

14. Install the electric cooling fan assembly as required.

15. Install a new flange gasket and connect the exhaust pipe.

16. Connect the negative battery cable and check for exhaust leaks.

1.8L (VIN C) Engine

1. Disconnect battery negative cable.

2. Raise the vehicle and support safely.

3. Remove the exhaust pipe to exhaust manifold nuts and separate exhaust pipe. Discard gasket.

4. Lower vehicle.

5. Remove electric cooling fan assembly, if necessary.

6. If the oxygen sensor is located in the manifold, remove the sensor.

7. Disconnect necessary EGR components.

8. Remove outer exhaust manifold heat shield and engine hanger. Disconnect the electrical connector and remove the oxygen sensor.

9. Remove the exhaust manifold mounting bolts, the inner heat shield and the exhaust manifold.

To install:

10. Clean all gasket material from the mating surfaces and check the manifold for damage.

11. Using a new gasket and install the manifold. Tighten the inner nuts to in a crisscross pattern to 13 ft. lbs. (18 Nm) and tighten the two outer (larger) nuts to 22 ft. lbs. (30 Nm).

12. Install the heat shields.

13. Connect EGR components.

14. If removed, install the oxygen sensor.

15. Install the electric cooling fan assembly as required.

16. Install a new flange gasket and connect the exhaust pipe.

17. Connect the negative battery cable and check for exhaust leaks.

2.4L (VIN G) Engine

1. Disconnect battery negative cable.

2. Raise the vehicle and support safely.

3. Leaving hoses connected, disconnect the power steering pump and position aside.

4. Remove the exhaust pipe to exhaust manifold nuts and separate exhaust pipe. Discard gasket.

5. Lower the vehicle.

6. Remove the electric cooling fan assembly.

7. Remove the outer exhaust manifold heat shield and engine hanger.

8. Disconnect the electrical connector and remove the oxygen sensor.

9. Remove the exhaust manifold mounting bolts, the inner heat shield and the exhaust manifold.

To install:

10. Clean all gasket material from the mating surfaces and check the manifold for damage.

11. Install a new gasket and install the manifold. Tighten the nuts to in a crisscross pattern to 22 ft. lbs. (30 Nm).

12. Install the heat shields.

13. Install the electric cooling fan assembly as required.

14. Install a new flange gasket and connect the exhaust pipe.

15. Install the power steering pump and adjust the drive belt for proper tension.

16. Connect the negative battery cable and check for exhaust leaks.

Front Crankshaft Seal

REMOVAL AND INSTALLATION

All Models

1. Disconnect the negative battery cable.

2. Remove the timing belt.

3. Drain the engine oil.

4. Remove the crankshaft pulley retainer bolts and remove the pulley.

5. On 1.5L (VIN A) and 1.8L (VIN C) engines, remove the vibration damper retainer bolt and washer and remove damper. If difficult to remove, the appropriate puller may be used.

6. Remove the crankshaft sprocket retainer bolt and washer from the sprocket, if used, and remove the sprocket. If no bolts are used on the sprocket, use the appropriate puller to remove.

7. Pry out the oil seal from front of engine.

To install:

8. Using proper size driver, install new front seal.

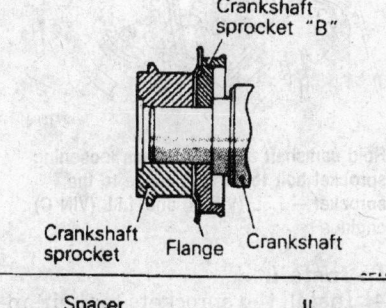

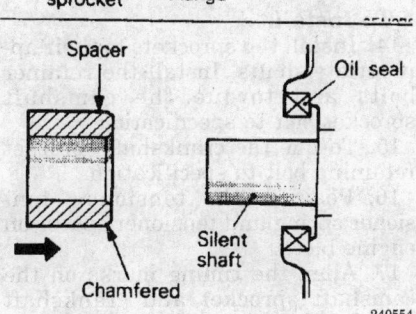

Front crankshaft oil seal — 2.4L (VIN G) engines

WARNING

Small nicks and burrs on crankshaft surface will damage oil seal. Use care when installing oil seal not to damage crankshaft surface.

9. Install the crankshaft sprocket and torque the retaining bolt to 80–94 ft. lbs. (110–130 Nm).

10. Install the timing belt, timing covers, valve cover and remaining components.

11. Install the engine undercover and connect the negative battery cable.

12. Fill engine oil, start the engine and check for leaks.

Timing Belt

REMOVAL AND INSTALLATION

1.5L (VIN A) and 1.8L (VIN C) Engines

1. Disconnect the negative battery cable. Remove the engine undercover.

2. Rotate crankshaft clockwise and position engine at TDC compression stroke.

3. Raise and safely support the weight of the engine using the appropriate equipment. Remove the A/C clamp, front engine mount bracket and accessory drive belts.

4. Remove the crankshaft pulley.

5. Remove timing belt upper and lower covers.

6. Make a mark on the back of the timing belt indicating the direction of rotation so it may be reassembled in the same direction if it is to be reused.

7. For 1.5L (VIN A) engines, loosen the timing belt tensioner and move the tensioner to provide slack to the timing belt. Tighten the tensioner in this position.

8. For 1.8L (VIN C) engines, loosen the timing belt tensioner, insert a screwdriver into the tensioner and release tension by prying screwdriver against spring tension. Temporarily tighten tensioner bolt to provide slack and remove the timing belt.

9. Remove the timing belt.

NOTE: Coolant and engine oil will damage the rubber in the timing belt, drastically reducing its life. Do not allow engine oil or coolant to contact the timing belt, the sprockets or tensioner assembly.

10. Remove the tensioner spacer, tensioner spring and tensioner assembly.

NOTE: It is recommended that the timing belt is replaced every 60,000 miles (96,000 km).

11. Inspect the timing belt for cracks or wear. Check the tensioner pulley for smooth rotation.

12. Remove the crankshaft sprocket retainer bolt and washer from the sprocket, if used, and remove sprocket. If sprocket is difficult to remove, the appropriate puller may be used. If no bolts are used on the sprocket, use the appropriate puller to remove.

13. Remove the camshaft sprocket using the appropriate spanner wrench if available. If not, note that most camshafts are made with a hexagon-shaped piece near the forward cam lobes so that a wrench can be used to hold the camshaft to keep it from turning when the sprocket bolt is being loosened or tightened. In no case should a tool be inserted through the sprocket to jam the sprocket to keep it from turning. This will damage the sprocket so that it should not be reused.

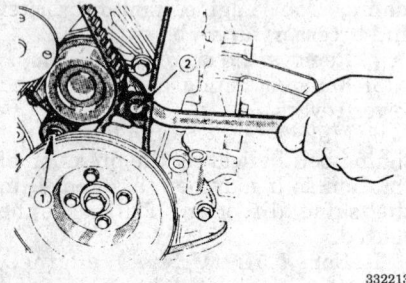

332213

Loosen pivot bolt No. 1 then slot side bolt No. 2. Tighten No. 2 first, then No. 1 — 1.5L (VIN A) engine

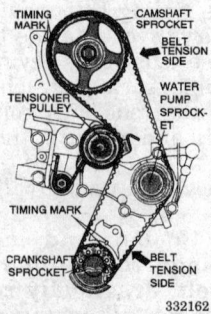

332162

Timing belt routing and mark identification — 1.8L (VIN C) engine

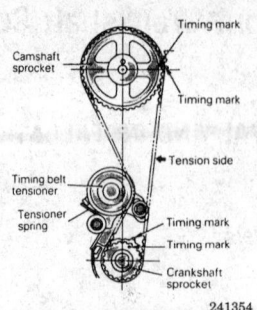

241354

Timing marks and belt installation — 1.5L (VIN A) engines

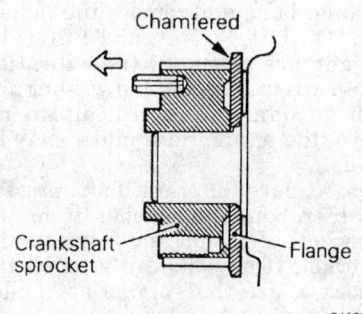

241355

Proper installation of crankshaft sprocket — 1.5L (VIN A) engines

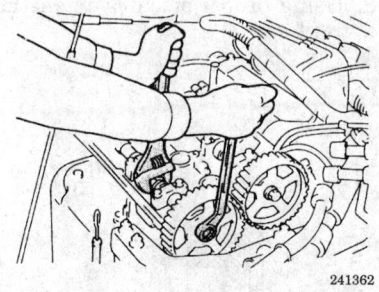

241362

Hold camshaft as shown when loosening sprocket bolt to avoid damage to the sprocket — 1.5L (VIN A) and 1.8L (VIN C) engines

To install:

14. Install the sprockets to their appropriate shafts. Install the retainer bolts and torque the camshaft sprocket bolt to specification.

15. Torque the crankshaft sprocket retaining bolt to specification.

16. Position the tensioner, tensioner spring and tensioner spacer on engine block.

17. Align the timing marks on the camshaft sprocket and crankshaft sprocket. This will position No. 1 piston on TDC on the compression stroke.

18. For 1.5L (VIN A) engines, position the timing belt on the crankshaft sprocket and keeping the tension side of the belt tight, set it on the camshaft sprocket and then the tensioner.

19. For 1.8L (VIN C) engines, position the timing belt on the crankshaft sprocket, water pump sprocket, camshaft sprocket and the tensioner keeping the tension side of the belt tight.

20. Apply slight counterclockwise force to the camshaft sprocket to give tension to the belt and make sure all timing marks are aligned.

21. Loosen the pivot side tensioner bolt and the slot side bolt. Allow the spring to remove the slack.

22. For 1.5L (VIN A) engines, tighten the slot side tensioner bolt and then the pivot side bolt. If the pivot side bolt is tightened first, the tensioner could turn with bolt, causing over tension.

23. For 1.8L (VIN C) engines, turn the crankshaft clockwise two rotations and tighten the adjuster bolt to 18 ft. lbs. (24 Nm) and tighten the pivot (spring) bolt to 35 ft. lbs. (45 Nm).

24. Turn the crankshaft clockwise. Loosen the pivot side tensioner bolt and then the slot side bolt to allow the spring to take up any remaining slack. Tighten the slot bolt and then the pivot side bolt to 17 ft. lbs. (24 Nm).

25. Install the timing belt covers and torque the cover bolts to 96 inch lbs. (11 Nm).

26. Install the crankshaft pulley and torque the retaining bolt to 83 ft. lbs. (113 Nm) for 1.5L (VIN A) engines, and to 134 ft. lbs. (185 Nm) for 1.8L (VIN C) engines.

27. Install the front engine mount bracket and mount.

28. Connect the A/C hose clamp and install the accessory drive belts.

29. Install the engine undercover.

30. Connect the negative battery cable.

31. Install the timing belt covers and all related items.

32. Connect the negative battery cable.

2.4L (VIN G) Engine

1. Position the engine so the No. 1 piston is at TDC.

2. Disconnect the negative battery cable.

3. Remove the timing belt covers.

4. To loosen the timing (outer) belt tensioner, install special tool MD998738 or equivalent, to the slot and screw inward to move tensioner

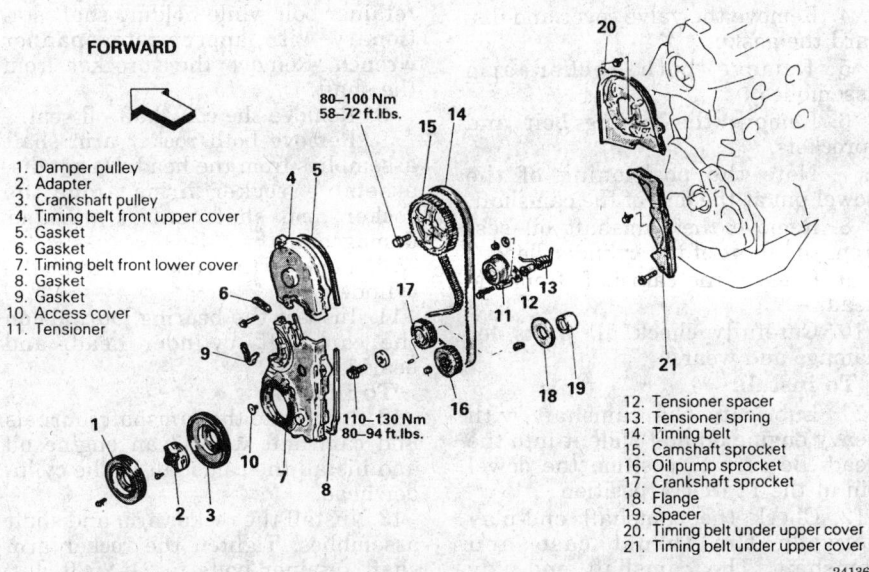

FORWARD

1. Damper pulley
2. Adapter
3. Crankshaft pulley
4. Timing belt front upper cover
5. Gasket
6. Gasket
7. Timing belt front lower cover
8. Gasket
9. Gasket
10. Access cover
11. Tensioner

80–100 Nm
58–72 ft.lbs.

110–130 Nm
80–94 ft.lbs.

12. Tensioner spacer
13. Tensioner spring
14. Timing belt
15. Camshaft sprocket
16. Oil pump sprocket
17. Crankshaft sprocket
18. Flange
19. Spacer
20. Timing belt under upper cover
21. Timing belt under upper cover

241363

Timing belt and related components — 1.8L (VIN C) engines

toward the water pump. Once the tension has been relieved, remove the outer timing belt.

NOTE: If timing belts are going to be reused, mark the direction of rotation on the belt. This will ensure the belt is reinstalled in same direction, extending belt life.

5. Remove the outer crankshaft sprocket and flange.
6. Loosen the silent shaft (inner) belt tensioner and remove the belt.
7. Remove the crankshaft sprocket retainer bolt and washer from the sprocket, if used, and remove sprocket. If sprocket is difficult to remove, the appropriate puller may be used. If no bolts are used on the sprocket, use the appropriate puller to remove.
8. Remove the camshaft sprocket using the appropriate spanner wrench; hold the shaft in position while removing the bolt.

To install:
9. Install the sprockets to their appropriate shafts. Install the retainer bolts and torque the camshaft sprocket bolt to 65 ft. lbs. (88 Nm).
10. Torque the crankshaft sprocket retaining bolt to 80–94 ft. lbs. (110–130 Nm).

11. Turn both tensioner pulleys and check for any signs of bearing wear.
12. Align the timing marks of the silent shaft sprockets and the crankshaft sprocket with the timing marks on the front case. Wrap the timing belt around the sprockets so there is no slack in the upper span of the belt and the timing marks are still aligned.
13. Install the tensioner pulley and move the pulley by hand so the long side of the belt deflects about ¼ in. (6mm).
14. Hold the pulley tightly so the pulley cannot rotate when the bolt is tightened. Tighten the bolt to 14 ft.

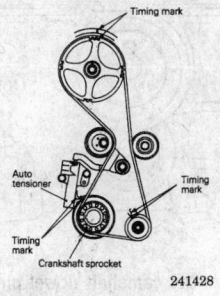

Timing mark

Auto tensioner

Timing mark

Timing mark

Timing mark

Crankshaft sprocket

241428

Camshaft timing belt timing mark identification — 2.4L (VIN G) engines

lbs. (19 Nm) and recheck the deflection amount.

15. Align the timing marks of the camshaft, crankshaft and oil pump sprockets with their corresponding marks on the front case or rear cover.

NOTE: There is a possibility to align all timing marks and have the oil pump sprocket and silent shaft out of time, causing an engine vibration during operation. If the following step is not followed exactly, there is a 50 percent chance that the silent shaft alignment will be 180 degrees off.

16. Before installing the timing belt, ensure that the left side (rear) silent shaft (oil pump sprocket) is in the correct position as follows:

a. Remove the plug from the rear side of the block and insert a tool with shaft diameter of 0.31 inches. (8mm) into the hole.

b. With the timing marks still aligned, the shaft of the tool must be able to go in at least 2 1/2 inches. If the tool can only go in about 1 in., the shaft is not in the correct orientation and will cause a vibration during engine operation. Remove the tool from the hole and turn the oil pump sprocket 1 complete revolution. Realign the timing marks and insert the tool. The shaft of the tool must go in at least 2 1/2 inches.

c. Recheck and realign the timing mark.

d. Leave the tool in place to hold the silent shaft while continuing.

17. Install the belt to the crankshaft sprocket, oil pump sprocket, then camshaft sprocket, in that order. While doing so, make sure there is no slack between the sprocket except where the tensioner is installed.

18. To adjust the timing (outer) belt perform the following steps:

a. Turn the crankshaft ¼ turn counterclockwise, then turn it clockwise to move No. 1 cylinder to TDC.

b. Loosen the center bolt. Using tool MD998752 or equivalent and a torque wrench, apply a torque of 2.6 ft. lbs. (3.6 Nm). Tighten the center bolt.

c. Screw the special tool into the engine left support bracket until its end makes contact with the tensioner arm. At this point, screw the special tool in some more and remove the set wire attached to the auto tensioner, if the wire was not previously removed. Then remove the special tool.

d. Rotate the crankshaft two complete turns clockwise and let it

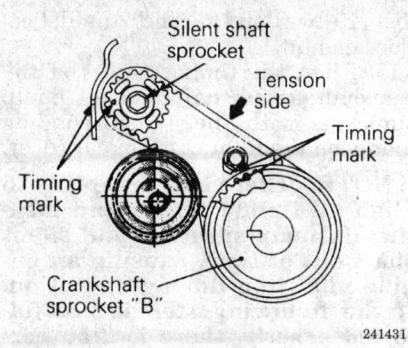

Oil pump timing belt timing mark identification — 2.4L (VIN G) engines

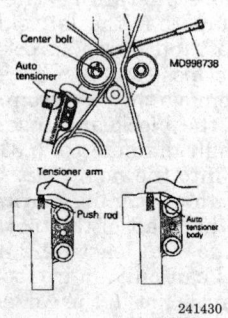

Auto tensioner adjustment — 2.4L (VIN G) engines

sit for approximately 15 minutes. Then, measure the auto tensioner protrusion (the distance between the tensioner arm and auto tensioner body) to ensure that it is within 0.15–0.18 inch (3.8–4.5mm). If out of specification, repeat Step 1–4 until the specified value is obtained.

NOTE: Do not manually overtighten the belt or it will howl.

19. Install the timing belt covers and all related items.
20. Connect the negative battery cable.
21. Run the engine until the thermostat opens. Check and adjust ignition timing.

Camshaft

REMOVAL AND INSTALLATION

1.5L (VIN A) and 1.8L (VIN C) Engines

1. Disconnect the negative battery cable.

2. Disconnect the accelerator cable, breather hose and PCV hose connections.
3. Remove the distributor.
4. Remove the valve cover and discard the gasket.
5. Remove both rocker arm assemblies.
6. Remove the timing belt and sprockets.
7. Note the positioning of the dowel pin at the end of the camshaft.
8. Remove the camshaft oil seal from the front of the cylinder head.
9. Remove the camshaft from the head.
10. Carefully check all parts for damage and wear.
   To install:
11. Lubricate the camshaft with heavy engine oil and slide it into the head. Be sure to position the dowel pin at the 12 o'clock position.
12. Check the camshaft end-play between the thrust case and camshaft. The camshaft end-play should be 0.002–0.008 in. (0.05–0.20mm). If the end-play is not within specification, replace the camshaft thrust bearing.
13. Install a new camshaft oil seal. Be sure to lubricate the lips of the seal with clean engine oil.
14. The balance of installation is the reverse of the removal procedure.
15. Connect the negative battery cable and check the ignition timing.

2.4L (VIN G) Engine

1. Disconnect the negative battery cable.
2. Remove the breather hose. Disconnect the PCV hose.
3. Matchmark and remove the distributor.
4. Remove the timing belt.

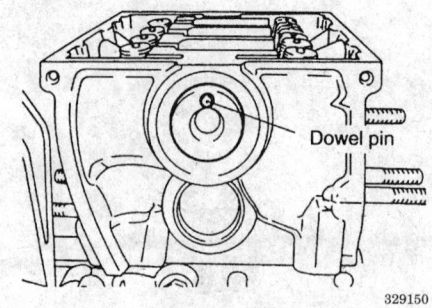

Positioning of the camshaft dowel pin — 1.5L (VIN A) engine

5. Disconnect and tag the spark plug wires.
6. Remove the rocker cover.
7. Remove the camshaft sprocket retainer bolt while holding shaft stationary with appropriate spanner wrench. Remove the sprocket from the shaft.
8. Remove the camshaft oil seal.
9. Remove both rocker arm shaft assemblies from the head. Do not disassemble rocker arms from the rocker arm shaft unless worn or damaged.
10. Remove the camshaft from the cylinder.
11. Inspect the bearing journals on the camshaft, cylinder head, and bearing caps.
   To install:
12. Lubricate the camshaft journals and camshaft with clean engine oil and install the camshaft in the cylinder head.
13. Install the rocker arm and shaft assemblies. Tighten the rocker arm shaft retainer bolts to 21–25 ft. lbs. (29–35 Nm).
14. Install new camshaft oil seal.
15. Install camshaft sprocket and retainer bolt tightening to 65 ft. lbs. (90 Nm).
16. The remaining components are installed in the reverse order from which they were removed.
17. Connect the negative battery cable. Run the engine at idle until normal operating temperature is reached. Check idle speed and ignition timing and adjust as required.

Piston and Connecting Rod

POSITIONING

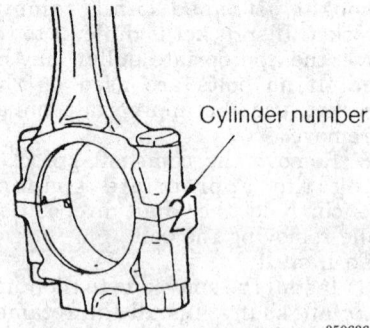

Connecting rod matchmarks — all engines

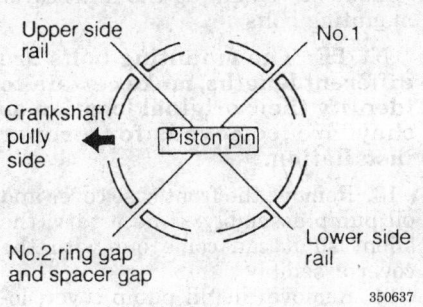

Piston ring positioning — all engines

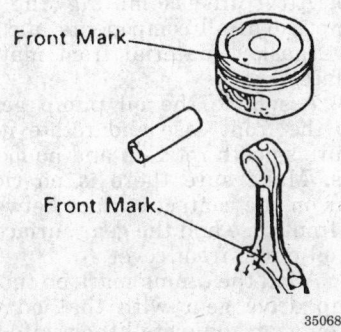

Aligning the piston and the connecting rod — all engines

ENGINE LUBRICATION

Oil Pan

REMOVAL AND INSTALLATION

1.5L (VIN A) and 1.8L (VIN C) Engines

1. Disconnect the negative battery cable.
2. Raise the vehicle and support safely.
3. Remove the oil pan drain plug and drain the engine oil.
4. Disconnect and lower the exhaust pipe from the engine manifold.
5. Remove the bell housing lower cover.
6. Remove the oil pan retainer bolts. Tap the oil pan with a rubber mallet to break the seal.

NOTE: Do not use a chisel, screwdriver or similar tool when removing the oil pan. Damage to engine components may occur. If

available, oil pan remover tool MD998727 or equivalent may be used break the seal.

7. Inspect the oil pan for damage and cracks. Replace if faulty. While the pan is removed, inspect the oil screen for clogging, damage and cracks. Replace if faulty.

To install:

8. Using a wire brush or other tool, scrape clean all gasket surfaces of the cylinder block and the oil pan so that all loose material is removed. Clean sealing surfaces of all dirt and oil.
9. Apply sealant around the gasket surfaces of the oil pan in such a manner that all bolt holes are circled and there is a continuous bead of sealer around the entire outside edge of the oil pan.

NOTE: The continuous bead of sealer should be applied in a bead approximately 0.16 in. (4mm) in diameter.

10. Install the oil pan onto the cylinder block within 15 minutes after applying sealant. Install the fasteners and tighten to 60 inch lbs. (7 Nm).
11. Install the bell housing cover.
12. Connect the exhaust pipe from the engine manifold with new gasket in place. Tighten the exhaust pipe to manifold flange nuts to 33 ft. lbs. (45 Nm). Install and tighten the support bolt to 18 ft. lbs. (25 Nm).
13. Install the oil drain plug and tighten to 29 ft. lbs. (40 Nm).
14. Lower the vehicle and fill the crankcase to the proper level with clean engine oil.
15. Connect the negative battery cable. Start the engine and check for leaks.

2.4L (VIN G) Engine

1. Disconnect the negative battery cable.
2. Raise the vehicle and support safely.
3. Remove the oil pan drain plug and drain the engine oil.
4. Disconnect and lower the exhaust pipe from the engine manifold.
5. On FWD models, remove the left side axle shaft.
6. On AWD models, remove the transfer assembly.
7. Remove the oil pan retainer bolts. Tap in thin prybar between the engine block and the oil pan.

NOTE: Do not use a chisel, screwdriver or similar tool when removing the oil pan. Damage to engine components may occur.

8. Inspect the oil pan for damage and cracks. Replace if faulty. While the pan is removed, inspect the oil screen for clogging, damage and cracks. Replace if faulty.

To install:

9. Using a wire brush or other tool, scrape clean all gasket surfaces of the cylinder block and the oil pan so that all loose material is removed. Clean sealing surfaces of all dirt and oil.
10. Apply sealant around the gasket surfaces of the oil pan in such a manner that all bolt holes are circled and there is a continuous bead of sealer around the entire perimeter of the oil pan.

NOTE: The continuous bead of sealer should be applied in a bead approximately 0.16 in. (4mm) in diameter.

11. Install the oil pan onto the cylinder block within 15 minutes after applying sealant. Install the fasteners and tighten to 4–6 ft. lbs. (6–8 Nm).
12. If applicable, install the transfer assembly and check fluid level.
13. Connect the exhaust pipe from the engine manifold with new gasket in place. Tighten the exhaust pipe to manifold flange nuts to 29 ft. lbs. (40 Nm). Install and tighten the support bolt to 29 ft. lbs. (40 Nm).
14. Install the oil drain plug and tighten to 29 ft. lbs. (40 Nm), If not already done.
15. Lower the vehicle and fill the crankcase to the proper level with clean engine oil.
16. Connect the negative battery cable. Start the engine and check for leaks.

Oil Pump

REMOVAL AND INSTALLATION

1.5L (VIN A) and 1.8L (VIN C) Engines

NOTE: Whenever the oil pump is disassembled or the cover removed, it is suggested the gear cavity is filled with petroleum jelly for priming purposes. Do not use grease.

1. Disconnect the negative battery cable.
2. Remove the front engine mount bracket and accessory drive belts.
3. Remove timing belt upper and lower covers.
4. Remove the timing belt and crankshaft sprocket.

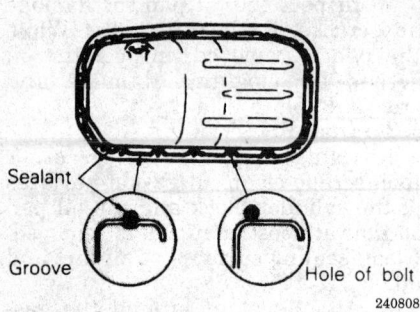

Sealant

Groove

Hole of bolt

240808

Applying sealant to oil pan — 1.8L (VIN C) and 2.4L (VIN G) engines

5. Remove the oil pan and remove the oil screen.

6. Remove and tag the front cover mounting bolts. Note the lengths of the mounting bolts as they are removed for proper installation.

7. Remove the front case assembly and oil pump assembly.

8. Remove the oil pump cover.

9. Remove the inner and outer gears from the front case.

NOTE: The outer gear has no identifying marks to indicate direction of rotation. Clean the gear and mark it with an indelible marker.

10. Check the front case for damage or cracks. Replace the front seal. Replace the oil screen O-ring. Clean all parts thoroughly with a safe solvent. Check the pump gears for wear or damage, and ensure that the relief valve can slide freely in the case.

To install

11. Remove all gasket material from the mating surfaces and clean all parts.

12. Thoroughly coat both oil pump gears with clean engine oil and install them in the correct direction of rotation.

13. Install the pump cover and torque the bolts to 84 inch lbs. (10 Nm).

14. Coat the relief valve and spring with clean engine oil, install them and tighten the plug to 33 ft. lbs. (45 Nm).

15. Install a new front crankshaft seal and coat the lips of the seal with clean engine oil.

16. Install the front case and oil pump assembly to the engine block using a new gasket. Use the noted locations of the mounting bolts for proper positioning and tighten the bolts to 10 ft. lbs. (14 Nm).

17. Install the oil screen with new gasket. Torque the screen bolts to 14 ft. lbs. (19 Nm).

18. Install the oil pan.

19. Install the sprocket, timing belt and pulley.

20. Connect the battery cable refill the engine oil, run the engine and check for leaks.

2.4L (VIN G) Engine

NOTE: Whenever the oil pump is disassembled or the cover removed, the gear cavity must be filled with petroleum jelly. This seals the pump and acts lime a prime so the oil pump draws oil as soon as the engine starts to turn. Do not use grease.

1. Disconnect the negative battery cable. Rotate the engine so No. 1 cylinder is on Top Dead Center (TDC) of its compression stroke. The timing marks should be aligned at this point.

2. Raise and safely support the vehicle.

3. Drain the engine oil. Lower the vehicle.

4. Using the proper equipment, support the weight of the engine. Remove the front engine mount bracket and accessory drive belts.

5. Remove timing belt upper and lower covers.

6. Remove the timing belt and crankshaft sprocket.

7. Disconnect the electrical connector from the oil pressure sending unit and remove the oil pressure sensor. Remove the oil filter and the oil filter bracket.

8. Remove the oil pan, oil screen and gasket.

9. Using special tool MD998162, remove the plug cap in the engine front cover.

10. Remove the plug on the side of the engine block. Insert a Phillips screwdriver with a shank diameter of 0.32 in. (8mm) into the plug hole. This will hold the silent shaft.

11. Remove the driven gear bolt that secures the oil pump driven gear to the silent shaft.

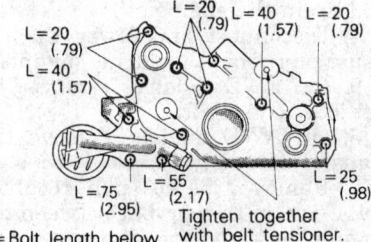

L = 20 (.79) L = 40 (1.57) L = 20 (.79)

L = 20 (.79)

L = 40 (1.57)

L = 75 (2.95) L = 55 (2.17) L = 25 (.98)

Tighten together with belt tensioner.

L = Bolt length below head [mm (in.)]

325180

Front cover bolt identification and location — 2.4L (VIN G) engine

12. Remove and tag the front cover mounting bolts.

NOTE: The mounting bolts are different lengths, make certain to identify their original location as they are removed, for proper installation.

13. Remove the front case cover and oil pump assembly. If necessary, the silent shaft can come out with the cover assembly.

14. Remove the oil pump cover, located on the back of the engine front cover. Remove the oil pump drive and driven gears.

To install

15. After disassembling the oil pump, clean all components and remove gasket material from mating surfaces.

16. Assemble the oil pump gears into the front case and rotate it to ensure smooth rotation and no looseness. Make sure there is no ridge wear on the contact surface between the front case and the gear surface of the oil pump front cover.

17. Align the timing mark on the oil pump drive gear with that on the driven gear and install them into the engine front case. Apply engine oil to the gears.

18. Install the oil pump cover and tighten the retainer bolts to 17 ft. lbs. (24 Nm).

19. Using the appropriate driver, install a new crankshaft seal into the front case.

20. Position a new front case gasket in place. Set seal guide tool MD998285 on the front end of the crankshaft to protect the seal from damage. Apply a thin coat of oil to the outer circumference of the seal pilot tool.

21. Install the front case assembly through a new front case gasket and temporarily tighten the flange bolts.

22. Mount the oil filter on the bracket with new oil filter bracket gasket in place. Install the 3 bolts with washers and tighten to 16 ft. lbs. (22 Nm).

23. Insert a Phillips screwdriver into a hole in the left side of the engine block to lock the silent shaft in place.

24. Secure the oil pump drive gear onto the left silent shaft by installing and tightening the driven gear bolt to 29 ft. lbs. (40 Nm).

25. Install new O-ring to the groove in the front case and install the plug cap. Using the special tool MD998162, tighten the cap to 20 ft. lbs. (27 Nm).

26. Install the oil screen in position with new gasket in place.

27. Clean both mating surfaces of the oil pan and the cylinder block. Apply sealant in the groove in the oil pan flange, keeping towards the inside of the bolt holes. The width of the sealant bead applied is to be about 0.016 in. (4mm) wide.

NOTE: After applying sealant to the oil pan, do not exceed 15 minutes before installing the oil pan.

28. Install the oil pan to the engine and secure with the retainers. Tighten bolts to 6 ft. lbs. (8 Nm).
29. Install the oil pressure gauge unit and the oil pressure switch. Connect the electrical harness connector.
30. Refill crankcase with oil. Install new oil filter.
31. Install the timing belts and timing belt covers. Assemble the remaining components to the front of the engine.
32. Connect the negative battery cable and start the engine. Verify oil pressure. Inspect for leaks.

TRANSAXLE

Manual Transaxle Assembly

REMOVAL AND INSTALLATION

Colt and Summit

NOTE: If the vehicle is going to be rolled while the halfshafts are out of the vehicle, obtain 2 outer CV-joints or proper equivalent tools and install to the hubs. If the vehicle is rolled without the proper torque applied to the front wheel bearings, the bearings will no longer be usable.

NOTE: The suspension components should not be tightened until the vehicles weight is resting on the ground.

1. Disconnect the negative battery cable.
2. Remove the front wheels and the inner wheel panels.
3. Remove the air cleaner assembly and vacuum hoses.
4. Note the locations and disconnect the shifter cables.
5. Disconnect the backup lamp switch connector, speedometer cable connection and remove the starter motor.

6. Remove the upper transaxle-to-engine mounting bolts.
7. Raise the vehicle and support safely.
8. Remove the undercover and splash pan.
9. Drain the transaxle oil.
10. Support the engine and remove the crossmember.
11. Remove the upper transaxle mounting bolt and bracket.
12. Disconnect the stabilizer bar, tie rod ends and the lower ball joint connections.
13. Remove the clutch release cylinder and clutch oil line bracket. Do not disconnect the fluid lines and secure the slave cylinder with wire. Disconnect the clutch cable, if equipped with cable controlled clutch system.
14. Remove the halfshafts by inserting a prybar between the transaxle case and the driveshaft and prying the shaft from the transaxle. Do not pull on the driveshaft. Doing so damages the inboard joint. Do not insert the prybar so far the oil seal in the case is damaged.

NOTE: It is not necessary to disconnect the halfshafts from the steering knuckle. Remove the shaft with the hub and knuckle as an assembly. Tie the shafts aside. Note the circle clip on the end of the inboard shafts should not be reused.

15. Remove the bell housing lower cover.
16. Remove the transaxle to engine bolts and lower the transaxle from the vehicle.

To install:

NOTE: When installing the transaxle, be sure to align the splines of the transaxle with clutch disc.

17. Install the transaxle to the engine and install the mounting bolts. Torque the bolts to specifications.
18. Install the bell housing cover.
19. Install the remaining components in the opposite order from which they were removed.
20. Connect the negative battery cable and check the transaxle for proper operation. Make sure the reverse lights operate when in reverse.

Vista and Summit Wagon

FWD Models

1. Disconnect the battery cables, negative cable first. Remove the battery and tray.
2. Remove the coolant reservoir.
3. Remove the air cleaner.

4. Disconnect the clutch cable, speedometer cable and backup light wiring from the transaxle.
5. Remove the upper engine-to-transaxle bolts.
6. Disconnect the select control lever and switch harness.
7. Remove the starter.
8. Disconnect and tag all wiring from the transaxle.
9. Raise and support the vehicle safely.
10. Remove the front wheels.
11. Drain the transaxle fluid.
12. Remove the extension and shift rod from the engine compartment.
13. Remove the stabilizer and strut bar from the lower control arm.
14. Remove the left and right halfshafts.
15. Support the transaxle with a suitable floor jack, taking care to avoid damaging the pan.
16. Remove the bell housing cover.
17. Remove the remaining transaxle-to-engine bolts.
18. Remove the transaxle mounting bolt.
19. Lower the jack and slide the transaxle from under the vehicle.

To install:

20. Secure the transaxle on a transaxle jack and position it to the engine.
21. Carefully guide the transaxle input shaft into the clutch assembly. Make sure the transaxle is seated properly and is flush to the engine flange. Install two transaxle-to-engine bolts.
22. Install the transaxle mounting bolt. Torque the bolt to 29–36 ft. lbs. (40–50 Nm).
23. Install the lower transaxle-to-engine bolts. Torque the bolts to 32–39 ft. lbs. (43–53 Nm).
24. Remove the floor jack from the transaxle.
25. Install the remaining components in the reverse order of removal.
26. Tighten the upper engine-to-transaxle bolts to 32–39 ft. lbs. (43–53 Nm).
27. Install the coolant reservoir and refill with coolant.
28. Install the battery tray and battery. Connect the battery cables, positive cable first.

AWD Models

1. Disconnect the battery cables, negative cable first. Remove the battery.
2. Remove the coolant reserve tank.
3. Disconnect the speedometer cable, shift control cable and backup light harness at the transaxle.

4. Remove the range select control valves and connectors.

5. Tag and disconnect all other wiring attached to the transaxle.

6. Remove the clutch slave cylinder.

7. Remove the vacuum reservoir tank.

8. Disconnect the starter wiring and remove.

9. Remove the upper engine-to-transaxle bolts.

10. Raise and support the vehicle safely.

11. Remove the front wheels, lower engine cover and skid plate.

12. Drain the transaxle and transfer case.

13. Remove the driveshaft.

14. Remove the transfer case extension housing.

15. Remove the left and right halfshafts.

16. Disconnect the right strut from the lower arm.

17. Remove the right fender liner.

18. Take up the weight of the transaxle with a suitable floor jack.

19. Remove the bell housing cover bolts and remove the cover.

20. Remove the remaining engine-to-transaxle bolts.

21. Remove the transaxle mount insulator bolt.

22. Remove the transaxle mounting bracket attaching bolts.

23. Move the transaxle/transfer case assembly to the right. Tilt the right side of the transaxle down, until the transfer case is about level with the upper part of the steering rack tube, then turn it to the left and lower the assembly.

To install:

24. Secure the transaxle/transfer case assembly to a transaxle jack.

25. Raise the assembly in position to the engine. It may be necessary to tilt or angle the assembly in and around the steering rack tube.

26. Once the transaxle/transfer assembly is positioned to the engine, carefully guide the input shaft into the clutch assembly.

27. Install the transaxle mounting bracket attaching bolts. Torque the bolts to 40–43 ft. lbs. (55–60 Nm).

28. Install the transaxle mount insulator bolt. Torque to 40–43 ft. lbs. (55–60 Nm).

29. Install the lower engine-to-transaxle bolts. Torque to 31–40 ft. lbs. (43–55 Nm).

30. Install the bell housing cover bolts and install the cover.

31. The remaining components are installed in the reverse order from which they were removed.

32. Install the battery. Connect the battery cables, positive cable first.

Clutch Assembly

REMOVAL AND INSTALLATION

All Models

1. Disconnect the negative battery cable. Raise and safely support the vehicle.

2. Remove the transaxle assembly from the vehicle.

3. Install a dummy shaft through the clutch disc to support the clutch during the removal procedure.

4. Remove the pressure plate attaching bolts, pressure plate and clutch disc. If the pressure plate is to be reused, loosen the bolts in a diagonal pattern, 1 or 2 turns at a time. This will prevent warping the clutch cover assembly.

5. Remove the return clip and the release bearing from the transaxle. Do not use solvent to clean the bearing.

6. Inspect the clutch release fork and fulcrum for damage or wear. If necessary, remove the release fork and unthread the fulcrum from the transaxle.

7. Carefully inspect the condition of the clutch components and replace any worn or damaged parts.

To install:

8. Inspect the flywheel for heat damage or cracks. Resurface or replace the flywheel as required. Install the flywheel using new bolts.

9. Install the fulcrum and tighten to 25 ft. lbs. (35 Nm). Install the release fork. Apply a coating of multipurpose grease to the point of contact with the release bearing. Apply a coating of multi-purpose grease to the end of the release cylinder's

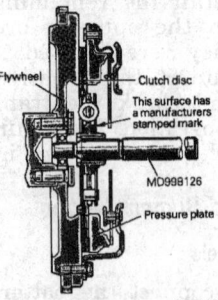

329219

Clutch plate and disk alignment procedure — all models

pushrod and the pushrod hole in the release fork.

NOTE: When installing the clutch be careful not to apply excessive grease. Excessive grease will cause clutch slippage and shudder.

10. Apply multi-purpose grease to the clutch release bearing. Pack the bearing inner surface and the groove with grease. Do not apply grease to the resin portion of the bearing. Place the bearing in position and install return clip.

11. Apply a coating of grease to the clutch disc splines and then use a brush to rub it in the grooves. Using a universal clutch disc aligner, position the clutch disc on the flywheel. Install the retainer bolts and tighten gradually in a diagonal sequence. Tighten them to a final torque of 15 ft. lbs. (21 Nm). Remove the aligning tool.

12. Install the transaxle assembly and check fluid level.

13. Connect the negative battery cable.

14. Check for proper clutch operation and adjust if necessary.

Clutch Cable

ADJUSTMENT

Colt and Summit

1. Measure the clutch pedal height (measurement A). The specification is 6.38–6.50 inches (162–165mm).

NOTE: The clutch pedal height is not adjustable. If not within specifications, part replacement is required.

2. Depress clutch pedal several times and check the pedal free-play (measurement B).

3. If measurement is not 0.67–0.87 inches (17–22mm), adjustment is required.

4. To adjust turn the outer cable adjusting nut, located at the firewall, until free-play is within range.

5. Depress clutch pedal several times and recheck measurement.

Clutch Master Cylinder

REMOVAL AND INSTALLATION

Colt and Summit

1. Disconnect the negative battery cable.

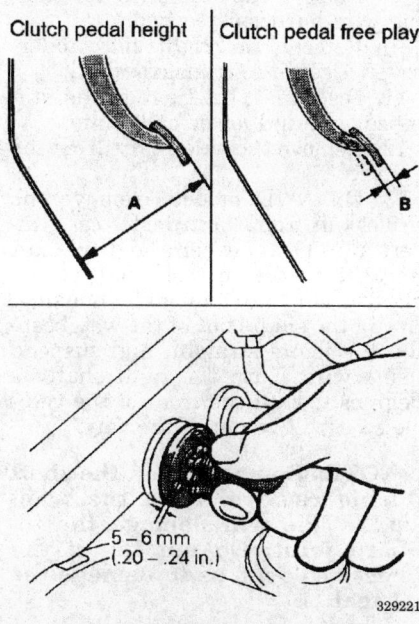

Proper method of adjusting clutch pedal free-play — Colt and Summit models

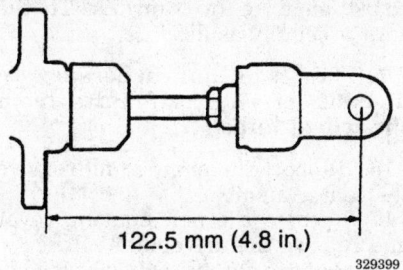

Clutch cylinder pushrod adjustment — Colt and Summit models

2. Remove necessary underhood components in order to gain access to the clutch master cylinder.

— WARNING —
The clutch hydraulic system uses brake fluid. Use care; brake fluid is harmful to painted surfaces.

3. Loosen the clutch fluid line at the cylinder and plug the openings.

4. Remove the clevis pin retainer at the clutch pedal and remove the washer and clevis pin.

5. Remove the two nuts that secure the master cylinder to the firewall and pull the cylinder from the firewall. A seal should be found between the mounting flange and firewall. This seal should always be replaced, not reused.

To install

NOTE: The length of the clutch master cylinder pushrod should be 4.8 inches (122.5mm) from the flange to the center of the clevis pin hole.

6. Mount the master cylinder on the firewall studs using a new seal. Torque the retaining nuts to 9 ft. lbs. (13 Nm).

7. Lubricate all pivot points with grease.

8. Connect the clevis pin under the dash and secure it with a new cotter pin.

9. Connect hydraulic line and fill the reservoir with fresh DOT 3 brake fluid. Bleed the system at the slave cylinder.

10. Check the adjustment of the clutch pedal for proper height and free-play.

Vista and Summit Wagon

1. Remove necessary underhood components in order to gain access to the clutch master cylinder.

— WARNING —
Clutch hydraulic system uses brake fluid which is harmful to painted surfaces.

2. Loosen the bleeder screw on the slave cylinder and drain the system.

3. Disconnect the pushrod from the clutch pedal.

4. Disconnect the clutch pedal from the pedal bracket.

5. Disconnect the fluid line and reservoir tube from the master cylinder.

6. Remove the reservoir and bracket on models with externally mounted fluid reservoirs.

7. Remove the two nuts and pull the cylinder from the firewall. A seal should be between the mounting flange and firewall. This seal should be replaced.

NOTE: On the AWD models, the lower master cylinder mounting nut is accessed from inside the vehicle.

To install

8. Mount master cylinder on firewall studs, using new seal, and torque nuts to 9 ft. lbs. (13 Nm).

9. Lubricate all pivot points with grease.

10. Connect the pushrod to the clutch pedal.

11. Install the reservoir and bracket, if removed.

12. Connect hydraulic line and bleed the system at the slave cylinder using fresh DOT 3 brake fluid.

13. Check the adjustment of the clutch pedal for proper free-play.

Clutch Slave Cylinder

REMOVAL AND INSTALLATION

1. Disconnect the negative battery cable. Remove necessary underhood components in order to gain access to the clutch release cylinder.

2. Remove the hydraulic line and plug the openings.

3. Remove the mounting bolts and the cylinder from the transaxle housing.

To install

4. Lubricate all pivot points with grease.

5. Mount the slave cylinder to the transaxle and tighten bolts to 11 ft. lbs. (15 Nm).

6. Connect hydraulic line and fill the system with clean brake fluid meeting DOT 3 specifications.

7. Bleed the system and adjust the clutch pedal height and the clevis pin play.

Hydraulic Clutch System Bleeding

— CAUTION —
The clutch hydraulic system uses brake fluid. Use care; brake fluid is harmful to painted surfaces.

1. Fill the reservoir with clean brake fluid meeting DOT 3 specifications.

2. Press the clutch pedal to the floor then open the bleeder screw on the slave cylinder.

3. Tighten the bleed screw and release the clutch pedal.

4. Repeat the procedure until the fluid is free of air bubbles.

NOTE: It is suggested that a hose be attached to the bleeder with the other end immersed in a container at least half full of brake fluid during the bleeding operation. Do not allow the reservoir to run out of fluid during bleeding.

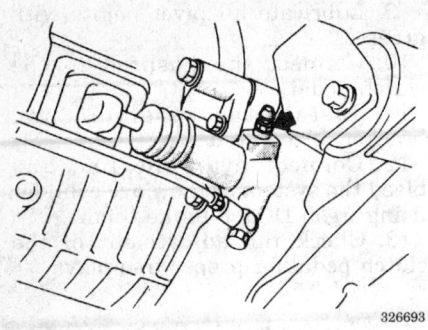

326693

Clutch hydraulic system bleeder screw — all models

Automatic Transaxle Assembly

REMOVAL AND INSTALLATION

Colt and Summit

NOTE: If the vehicle is going to be rolled on its wheels while the halfshafts are out of the vehicle, obtain two outer CV-joints or proper equivalent tools and install to the hubs. If the vehicle is rolled without the proper torque applied to the front wheel bearings, the bearings will no longer be usable.

1. Disconnect the negative battery cable.
2. Remove the battery and battery tray.
3. Remove the air hose and air cleaner assembly.
4. Raise the vehicle and support safely.
5. Remove the under guard pan.
6. Drain the transaxle oil.
7. If equipped with 1.6L engine, remove the tension rod.
8. Disconnect the control cable and cooler lines.
9. Disconnect the shift control solenoid valve connector.
10. Disconnect the inhibitor switch, kickdown servo switch, the pulse generator and oil temperature sensor, if equipped.
11. Disconnect the speedometer cable and remove the starter.
12. Remove the transaxle mounting bolts and bracket.
13. Disconnect the stabilizer bar from the lower control arm.
14. Disconnect the steering tie rod end and the ball joint from the steering arm.

15. Remove the halfshafts at the inboard side from the transaxle. Tie the joint assembly aside.

NOTE: It is not necessary to disconnect the halfshafts from the wheel hubs.

16. Support the engine and remove the center member.
17. Remove the bell housing cover and remove the driveplate bolts.
18. Remove the transaxle assembly lower connecting bolt, located just over the halfshaft opening.
19. Properly support the transaxle assembly and lower it moving it to the right for clearance.

To install:

20. After the torque converter has been mounted on the transaxle, install the transaxle assembly on the engine. Install the mounting bolts and torque to specifications.
21. Tighten the driveplate bolts to 33–38 ft. lbs. (46–53 Nm). Install the bell housing cover.
22. Install the center member and torque the bolts to specifications.
23. The balance of installation is the reverse of the removal procedure.
24. Refill with Dexron® II, Mopar ATF Plus type 7176 or equivalent, automatic transaxle fluid.
25. Start the engine and allow to idle for two minutes. Apply parking brake and move selector through each gear position, ending in **N**. Recheck fluid level and add if necessary. Fluid level should be between the marks in the **HOT** range.

Vista and Summit Wagon

NOTE: On both Front Wheel Drive (FWD) and All Wheel Drive (AWD) vehicles, the transaxle and converter must be removed and installed as an assembly.

1. Disconnect negative battery cable.
2. Remove the air cleaner assembly.
3. Disconnect the transaxle control lever. Disconnect and plug the oil cooler lines.
4. Disconnect the pulse generator connector, oil temperature connector, kickdown servo switch connector, inhibitor switch connector and solenoid valve connection.
5. Disconnect the speedometer cable connection. Remove the oil level dipstick and tube.
6. Install holding fixture to the top of the engine to support engine weight.
7. Remove the top transaxle upper coupling bolts.

8. Raise and safely support the vehicle.
9. Remove the starter motor leaving wire harness attached.
10. Remove the right side undercover. Drain the transaxle fluid.
11. Disconnect the tie rod ends, stabilizer bar and lower ball joints.
12. Remove the axle shafts from the vehicle.
13. On AWD models, remove the driveshaft from the transfer case, insert a prybar between the driveshaft and the transaxle case and pry the shaft from the transaxle housing. Swing the shafts out of the way keeping the joints straight, and suspend using wire. Turn the right shaft 90 degrees toward the front of the vehicle so it will be out of the way.

NOTE: Do not pull on the shaft during removal from the transaxle. This will damage the inboard joint. Do not insert the prybar so deep as to damage the oil seal.

14. Remove the lower bell housing cover. Scribe a mark on the driveplate and transaxle converter face using chalk. Remove the driveplate connecting bolts while turning the crankshaft.
15. Support the transaxle using a transmission jack. Remove the center support.
16. Remove the transaxle mount bolt and bracket.
17. On AWD models, disconnect the front exhaust pipe, then drain and remove the transfer assembly.
18. Remove the lower transaxle case coupling bolts, press the torque converter towards the transfer case to prevent separation during removal and lower the transfer case from the vehicle.

To install:

19. Install the transaxle into the vehicle and secure using the lower case coupling bolts.
20. Install the transaxle mount bolt and bracket, torque through bolt nut to 51 ft. lbs. (70 Nm).
21. Align the scribe marks on the converter and the driveplate. Install the driveplate connecting bolts tightening to 33–38 ft. lbs. (46–53 Nm).
22. Install the transfer assembly, if applicable, and the center crossmember. Remove the transmission jack.
23. The remaining components are installed in the reverse order from which they were removed.
24. Refill with DEXRON® II, Mopar ATF Plus type 7176, or equivalent automatic transaxle fluid.

Fill the transfer case to proper level GL-4 or higher, SAE 75W-90W.

25. Start the engine and allow to idle for 2 minutes. Apply parking brake and move selector through each gear position, ending in **N**. Recheck fluid level and add if necessary. Fluid level should be between the marks in the **HOT** range. Check operation of all gauges and meters.

DRIVELINE

Driveshaft

REMOVAL AND INSTALLATION

1. Raise the vehicle and support safely.
2. Drain the transfer case.
3. Matchmark the differential companion flange to the driveshaft flange yoke, and the two driveshaft sections to one another to ensure proper phasing on reassembly.
4. Unbolt the driveshaft from the differential flange.
5. Remove the two center bearing attaching nuts.

NOTE: Do not confuse the flat washer and the adjusting spacer. Keep them separate for assembly.

6. Pull the driveshaft from the transfer case. Be careful to avoid damaging the transfer case oil seal.
To install:
7. Install the driveshaft to the vehicle and align the matchmarks at the rear yoke.
8. Install the bolts at the rear differential flange and torque to 22–25 ft. lbs. (30–35 Nm).
9. Install the center support bearing with all spacers in place. Torque the retaining nuts to 22–29 ft. lbs. (30–40 Nm).
10. Check the fluid levels in the transfer case and rear differential case.

U-Joints

REMOVAL AND INSTALLATION

1. Raise and properly support vehicle.

NOTE: Matchmark the rear yoke to shaft and/or the center yoke to yoke for proper installation reference.

2. Remove driveshaft and secure in vise.
3. Remove the snaprings which retain the bearing caps in the slip yoke and the driveshaft.
4. Use a large punch or an arbor press and drive one of the bearing caps in toward the center of the universal joint. The joint will be forced through the opposite side of the yoke.
5. As the opposite side bearing cap is forced from the yoke, grip it with a pair of pliers and pull it, in a twisting motion, out of the yoke.
6. Press the spider cross toward the side you just pushed to force the cap back into the yoke. When the bearing cap starts to clear the yoke, pull it free with a pair of pliers. Repeat the procedure with the other side bearing caps.
7. After removing the bearing caps, lift the bearing (spider) cross from the yoke. Thoroughly clean all dirt and foreign matter from the yoke area on both ends of the driveshaft.
To install:

NOTE: When installing new bearing caps within the yokes, it is advisable to use an arbor press. However, if a press is not available, the bearings should be driven into position with extreme care. A heavy jolt on the needle bearing, in the cap, can easily damage or misalign the bearings. A large vise and correct

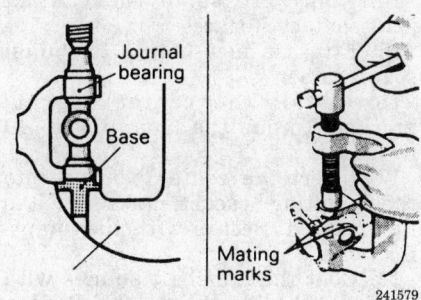

Pressing out universal joint — Summit Wagon and Vista models

size drivers and spacers can sometimes be used, in place of a punch, to push the bearing caps in or out.

8. Start a bearing cap into the yoke bore.
9. Position the spider into the yoke and into the bearing cap. Push the cap the rest of the way into the yoke bore until it is about 6mm below the outside surface of the yoke. Install a new snapring.
10. Start a bearing cap into the yoke on the opposite side of the one just installed. Carefully press it into the yoke while aligning the spider cross with the bearing center.
11. Continue to pry the cap in until the opposite side bearing cap contacts the snapring. Install a new snapring on the side just installed. Using a feeler gauge, check the clearance between the bearing cap face and the snapring. Clearance should be between 0.0008 — 0.0024 in. (0.02 — 0.06 mm), if not within specification, install appropriate size snapring. Once proper clearance is obtained, check the joint for free movement. The replacement snap-ring sizes and identification colors are as follows:
• 0.0503 in. (1.28mm) — standard thickness
• 0.0516 in. (1.31mm) — Yellow
• 0.0528 in. (1.34mm) — Blue
• 0.0539 in. (1.37mm) — Purple
• 0.0551 in. (1.40mm) — Brown
12. Complete the installation of the yoke and bearing caps.
13. Position the driveshaft and work on the other end if service is required.

NOTE: After service is completed, check the assembled joints and yokes for freedom of movement. If misalignment of any part causes it to bind, a sharp rap on the side of the yoke with a brass hammer should seat the needle bearings, and provide the desired freedom of movement. Care should be exercised to firmly support the shaft end during this operation, as well as to prevent blows to the bearing caps themselves. Under no circumstances should a driveshaft be installed in a vehicle if there is any bind in the U-joints. If the binding remains, disassemble the yoke and joint a check the needle bearings for correctly vertical alignment.

14. Align the matchmarks and install the driveshaft.

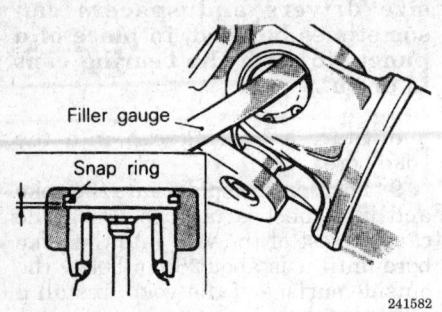

Filler gauge

Snap ring

241582

Checking universal joint installation — Summit Wagon and Vista models

Halfshaft

REMOVAL AND INSTALLATION

Front

Except Left Side of AWD Vista and Summit Wagon

NOTE: If the vehicle is going to be rolled while the halfshafts are out of the vehicle, obtain 2 outer CV-joints or proper equivalent tools and install to the hubs. If the vehicle is rolled without the proper torque applied to the front wheel bearings, the bearings will no longer be usable.

1. Disconnect the negative battery cable.
2. Remove the cotter pin, halfshaft nut and washer.
3. Raise the vehicle and support safely. Disconnect the lower ball joint and the tie rod end from the steering knuckle.
4. On vehicles with an inner shaft, remove the center support bearing bracket bolts and washers.
5. Remove the halfshaft by setting up a puller on the outside wheel hub and pushing the halfshaft from the front hub. After pressing the outer shaft, insert a prybar between the transaxle case and the halfshaft and pry the shaft from the transaxle. Do not pull on the shaft; doing so damages the inboard joint. Do not insert the prybar too far or the oil seal in the case may be damaged.
   To install:
6. Inspect the halfshaft boot for damage or deterioration. Check the ball joints and splines for wear.
7. Replace the circlips on the ends of the halfshafts.
8. Insert the halfshaft into the transaxle. Make sure it is fully seated.

9. Pull the strut assembly out and install the outer end of the shaft to the hub.
10. Install the washer so the chamfered edge faces outward. Install the nut and tighten temporarily.
11. Install the tie rod end and ball joint.
12. Install the wheel and lower the vehicle to the floor. Tighten the axle nut with the brakes applied. Tighten the nut to a maximum torque of 188 ft. lbs. (260 Nm). Install the cotter pin and bend to secure.

Left Side of AWD Vista and Summit Wagon

1. Remove the center wheel hub. Loosen the wheel lugs and the center halfshaft nut.
2. Raise and safely support the front of the vehicle with the suspension hanging.
3. Remove the front wheels.
4. Drain the transaxle fluid.
5. Disconnect the lower ball joint from the steering knuckle.
6. Remove the strut bar and the stabilizer bar from the lower arm.
7. Remove the center bearing mount snapring/bolts from the bracket.
8. Lightly tap the double-offset joint outer race with a wooden mallet and disconnect the Cardan joint.
9. Disconnect the halfshaft from the center bearing bracket.
10. Use a pusher/puller tool mounted to the wheel studs and press the halfshaft from the drive hub.
11. Unbolt and remove the bearing bracket.
12. Use a wooden mallet and lightly tap the Cardan joint yoke and remove it from the transaxle. DO NOT pry the Cardan joint from the transaxle, damage can be caused to the joint and boot.
   To install:
13. Service the halfshaft as required. Install the Cardan joint.
14. Apply grease to the Cardan joint contact surfaces.
15. Attach a new O-ring to the oil seal retainer.
16. Install the center bearing bracket. Torque the mounting to 40 ft. lbs.
17. Insert the center bearing into the mounting bracket, make sure it is fully seated. Secure with the snapring bolts.
18. Coat the halfshaft splines with grease and slide it into the Cardan joint.
19. Slide the halfshaft into the drive hub. Install the suspension

components and wheel. Lower the vehicle and tighten the wheel lugs and the center halfshaft nut. Torque the nut to 188 ft. lbs.

Rear

AWD Vista and Summit Wagon

1. Raise the vehicle and support safely.
2. Remove the bolts that attach the rear halfshaft to the companion flange.
3. Remove the cotter pin and axle nut from the outer shaft.
4. Remove the rear driveshaft from the vehicle.
5. If the differential companion shaft is to be removed, connect a slide hammer to the flange and pull the shaft from the differential.
   To install:
   NOTE: If the companion shaft is being replaced or both shafts were removed together, it is important to properly identify the companion shaft. The right and left side companion shafts are different, as are limited slip differentials, which use a two stage serration on the companion shaft.
6. Replace the circlip and install the companion flange to the differential case. Make sure it snaps in place. On limited slip differentials, ensure that both serrations are fully engaged to the differential.
7. Install the axle shaft through the hub and install the axle nut. Torque the nut to 145 to 188 ft. lbs. (200 to 260 Nm) and secure with cotter pin.
8. Install the companion flange bolts and tighten to 40 to 47 ft. lbs. (55 to 65 Nm).
9. Check the fluid level in the rear differential.

CV-Joint Boot

REPLACEMENT

These vehicles use several different types of joints. Engine size, transaxle type, whether the joint is an inboard or outboard joint, even which side of the vehicle is being serviced could make a difference in joint type. Be sure to properly identify the joint before attempting joint or boot replacement. Look for identification numbers at the large end of the boots and/or on the end of the metal retainer bands.

The types of joints used are the Tripod Joint (T.J.), Birfield Joint, (B.J.),

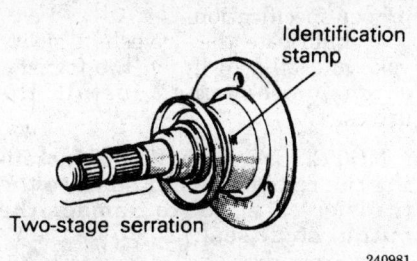

Identification stamp

Two-stage serration

240981

Companion shaft identification — Summit Wagon and Vista models

the Rzeppa Joint (R.J.) and the Double Off-set Joint (D.O.J.). Both the outer joint styles are part of the axle assembly, and can not be removed from the axle. In addition, some left side shafts will have a round dynamic damper installed on the shaft. Special grease is generally used with these joints and is often supplied with the replacement joint and/or boot. Do not use regular chassis grease.

In most cases, there is a specified distance between the large and small boot bands. This is so the boot will not be installed either too loose or too tight, which could cause early wear and cracking, allowing the grease to get out and water and dirt in leading to early joint failure.

Colt and Summit

1. Raise and properly support the vehicle. Remove the halfshaft.
2. Remove the T.J. boot bands from the boot. Side cutter pliers can be used to cut off the metal retaining bands. Remove the T.J. case from the halfshaft.
3. Remove the snapring next to the tripod joint spider assembly from the halfshaft with snapring pliers and remove the spider assembly from

the shaft. Do not disassemble the spider and use care in handling.

NOTE: Both of the halfshaft boots are going to be removed from the T.J. case side of the halfshaft.

4. If the boot is be reused, wrap vinyl tape around the spline part of the shaft so the boot will not be damaged when removed. Remove the dynamic damper, if used, and boots from the shaft.
To install:
5. Clean old grease from joint with solvent and blow dry.
6. Double check that the correct replacement parts are being installed. Wrap vinyl tape around the splines to protect the boot and install the boots and damper, if used, in the correct order.
7. Install the tripod joint spider assembly to the shaft and secure with snapring.
8. Fill the inside of the boot and case with the specified grease. Often the grease supplied in the replacement parts kit is meant to be divided in half, with half being used to lubricate the joint and half being used inside the boot. Keep grease off the rubber part of the dynamic damper (if used).
9. Install the T.J. case to the spider assembly.
10. Secure the boot bands with the halfshaft in a horizontal position. Make sure the boot span on the halfshaft is 3.25 ±0.12 in. (80 ±3mm) in length.
11. Check dynamic damper for proper positioning and adjust if necessary.
12. Install the halfshaft.

Vista and Summit Wagon

Except Double Off-Set Joint

Although joint types vary, the basic procedures are the same, with the ex-

ception of the Double Offset Joint. The following is a general procedure which should apply to most applications.
1. Raise and properly support the vehicle. Remove the halfshaft.
2. Remove the T.J. boot bands from the boot. Side cutter pliers can be used to cut off the metal retaining bands. Remove the T.J. case from the halfshaft.
3. Remove the snapring next to the tripod joint spider assembly from the halfshaft with snapring pliers and remove the spider assembly from the shaft. Do not disassemble the spider and use care in handling.

NOTE: Both of the halfshaft boots are going to be removed from the T.J. case side of the halfshaft.

4. If the boot is be reused, wrap vinyl tape around the spline part of the shaft so the boot will not be damaged when removed. Remove the dynamic damper, if used, and boots from the shaft.
To install:
5. Clean old grease from joint with solvent and blow dry.
6. Double check that the correct replacement parts are being installed. Wrap vinyl tape around the splines to protect the boot and install the boots and damper, if used, in the correct order.
7. Fill the inside of the boot with the specified grease. Often the grease supplied in the replacement parts kit is meant to be divided in half, with half being used to lubricate the joint and half being used inside the boot. Keep grease off the rubber part of the dynamic damper (if used).
8. Secure the boot bands with the halfshaft in a horizontal position. Make sure the boot span on the halfshaft is the proper length.
9. Install the halfshaft.

Double Off-Set Joint

1. Remove the halfshaft. The Double Off-set Joint (D.O.J.) is bigger than other joints and in these applications, is only used as an inboard joint.
2. Side cutter pliers can be used to cut the metal retaining bands.
3. Locate and remove the large circlip at the base of the joint. Remove the outer race (the body of the joint).
4. Matchmark the shaft, D.O.J. inner race and cage. Remove the joint balls and the small snapring from the shaft. With a brass drift pin, tap lightly and evenly around the inner

Items	1.5L Engine	1.8L Engine	
		L.H.	R.H.
A mm (in.)	351 ± 3 (13.82 ± .12)	365 ± 3 (14.37 ± .12)	200.5 ± 3*² (7.89 ± .12)
B mm (in.)	481 ± 3*¹ (18.94 ± .12)	—	—

NOTE
*¹: A/T
*²: M/T

329602

Dynamic damper installation position — Colt and Summit models

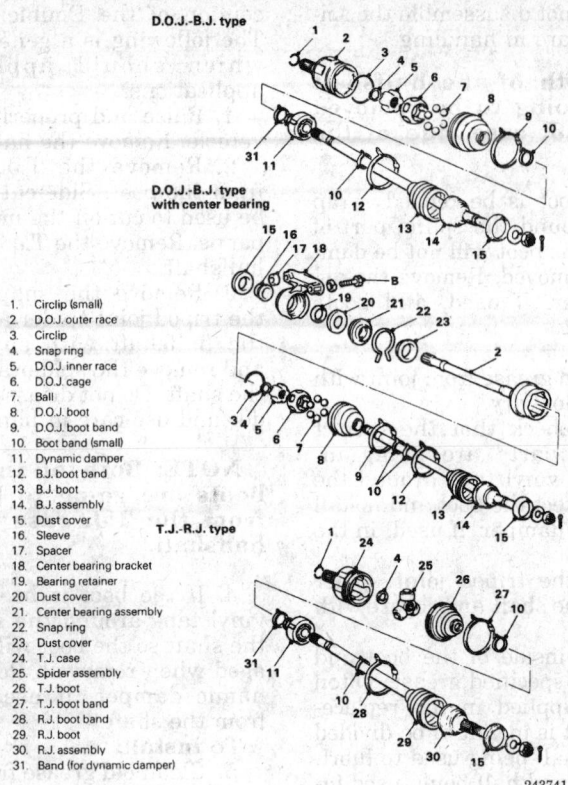

D.O.J.-B.J. type

D.O.J.-B.J. type
with center bearing

T.J.-R.J. type

1. Circlip (small)
2. D.O.J. outer race
3. Circlip
4. Snap ring
5. D.O.J. inner race
6. D.O.J. cage
7. Ball
8. D.O.J. boot
9. D.O.J. boot band
10. Boot band (small)
11. Dynamic damper
12. B.J. boot band
13. B.J. boot
14. B.J. assembly
15. Dust cover
16. Sleeve
17. Spacer
18. Center bearing bracket
19. Bearing retainer
20. Dust cover
21. Center bearing assembly
22. Snap ring
23. Dust cover
24. T.J. case
25. Spider assembly
26. T.J. boot
27. T.J. boot band
28. R.J. boot band
29. R.J. boot
30. R.J. assembly
31. Band (for dynamic damper)

243741

CV-joint identification and exploded view — Summit Wagon and Vista models

race to remove the race and the inner cage from the shaft.

NOTE: Both of the halfshaft boots are going to be removed from the D.O.J. case side of the halfshaft.

5. If the boot is to be reused, wipe the grease from the splines and wrap the splines in vinyl tape before sliding the boot from the shaft.

To install:

6. Be sure to tape the shaft splines before installing the boots. Fill the inside of the boot with the specified grease. Often the grease supplied in the replacement parts kit is meant to be divided in half, with half being used to lubricate the joint and half being used inside the boot.

7. Install the cage onto the halfshaft so the small diameter side of the cage is installed first. Align the matchmarks made at disassembly on the inner race and shaft. With a brass drift pin, tap lightly and evenly around the inner race to install the race until it comes into contact with the rib of the shaft. Apply the specified grease to the inner race and cage and fit them together aligning the matchmarks. Insert the balls into the cage.

8. Install the outer race (the body of the joint) after filling with the specified grease. The outer race should be filled with this grease.

9. Secure the boot bands with the halfshaft in a horizontal position. Make sure the boot span on the halfshaft is the proper length.

10. Install the halfshaft.

Transfer Case Assembly

REMOVAL AND INSTALLATION

1. Disconnect the battery negative cable.

2. Raise and properly support vehicle. Drain the transfer assembly.

3. Disconnect the front exhaust pipe and hanger.

4. Matchmark and remove the driveshaft.

5. Unbolt the transfer case assembly and remove by sliding out from the transaxle. Cover the opening in the transaxle and transfer case to keep oil from dripping and to keep dirt out.

To install:

6. Install the transfer case assembly to the transaxle. Tighten the

transfer case to transaxle bolts to proper specification.

7. Lubricate the driveshaft sleeve yoke and oil seal lip on the transfer extension housing. Install the driveshaft.

NOTE: Use care when installing the rear propeller shaft to the transfer case, not to damage the output shaft seal.

8. Connect the exhaust pipe, using a new gasket.

9. Refill the transfer case with gear oil of correct classification. Check fluid level in transaxle and add as required.

10. Lower the vehicle and connect the negative battery cable.

STEERING

Air Bag

---— CAUTION ———

Some vehicles are equipped with an air bag system, also known as the Supplemental Inflatable Restraint (SIR) or Supplemental Restraint System (SRS). The system must be disabled before performing service on or around system components, steering column, instrument panel components, wiring and sensors. Failure to follow safety and disabling procedures could result in accidental air bag deployment, possible personal injury and unnecessary system repairs.

PRECAUTIONS

Several precautions must be observed when handling the inflator module to avoid accidental deployment and possible personal injury.

• Never carry the inflator module by the wires or connector on the underside of the module.

• When carrying a live inflator module, hold securely with both hands, and ensure that the bag and trim cover are pointed away.

• Place the inflator module on a bench or other surface with the bag and trim cover facing up.

• With the inflator module on the bench, never place anything on or close to the module which may be thrown in the event of an accidental deployment.

DISARMING

All Models

1. Position the front wheels in the straight-ahead position and place the key in the **LOCK** position. Remove the key from the ignition lock cylinder.
2. Disconnect the negative battery cable and insulate the cable end with high-quality electrical tape or similar non-conductive wrapping.
3. Wait at least one minute before working on the vehicle. The air bag system is designed to retain enough voltage to deploy the air bag for a short period of time after the battery has been disconnected.

To arm:

4. Connect the negative battery cable, turn the ignition switch to the **ON** position and check the SRS warning light for proper operation.

Steering Wheel

REMOVAL AND INSTALLATION

1993 Models

1. Disconnect the negative battery cable.
2. Remove the horn pad from the steering wheel by, pulling the lower end of the pad upward. Disconnect horn button connector.
3. Remove steering wheel retaining nut.
4. Matchmark the steering wheel to the shaft.
5. Use a steering wheel puller to remove the steering wheel.

------- **WARNING** -------
Do not hammer on steering wheel to remove it. The collapsible column mechanism may be damaged.

To install:

6. Line up the matchmarks and install the steering wheel to the shaft.
7. Torque the steering wheel attaching nut to 29 ft. lbs. (40 Nm).
8. Reconnect the horn connector and install the horn pad.
9. Connect the negative battery cable.

1994–97 Models

------- **CAUTION** -------
If equipped with an air bag, be sure to disarm it before starting repairs on the vehicle. Failure to do so could result in severe personal injury and damage to vehicle.

1. Disarm the air bag.
2. Remove the air bag module mounting nut from behind the steering wheel.
3. To disconnect the connector of the clockspring from the air bag module, press the air bag's lock toward the module to spread the lock open. While holding lock in this position, use a small tipped prying tool to gently pry the connector from the module.
4. Remove the air bag module and store in a clean, dry place with the pad cover facing up.
5. Matchmark the steering wheel to the shaft.
6. Remove the steering wheel retaining nut and use a steering wheel puller to remove the wheel. Do not use a hammer or the collapsible mechanism in the column could be damaged.

To install:

7. Confirm that the front wheels are in a straight-ahead position. Center the clockspring by aligning the **NEUTRAL** mark on the clockspring with the mating mark on the casing. Then install the steering wheel and torque the new retaining nut to 29 ft. lbs. (40 Nm).
8. Install the air bag module.
9. Connect the negative battery cable, turn the key to the **ON** position, the SRS warning light should illuminate for seven seconds and go out. If the warning light is not functioning properly, refer to SRS system diagnosis.

Tie Rod Ends

REMOVAL AND INSTALLATION

All Models

Outer

1. Raise the front of the vehicle and support it on jackstands. Remove the wheel.
2. Remove the cotter pin and the tie rod ball joint stud nut. Note the position of the steering linkage.
3. Wire brush the threads on the tie rod shaft and lubricate with penetrating oil.
4. Using a suitable ball joint separator tool, remove the tie rod ball joint from the steering knuckle.
5. Loosen the locknut and remove the outer tie rod end from the tie rod. Count the number of complete turns it takes to completely remove it.

To install:

6. Install the new tie rod end, turning it in exactly as many turns as it was to remove the old one. Make sure it is correctly positioned in relationship to the steering linkage.
7. Connect the outer tie rod end to the steering knuckle and install the castle nut. Torque the nut to 25 ft. lbs. (34 Nm).
8. Install a new cotter pin to the castle nut.
9. Tighten the tie rod end locking nut to 30 ft. lbs. (42 Nm).
10. Install the wheel and tire assembly.
11. Lower the vehicle and perform a front end alignment.

Inner

1. Raise the front of the vehicle and support it on jackstands. Remove the wheel.
2. Remove the cotter pin and the outer tie rod ball joint stud nut. Note the position of the steering linkage.
3. Wire brush the threads on the tie rod shaft and lubricate with penetrating oil.
4. Using a suitable ball joint separator tool, remove the tie rod ball joint from the steering knuckle.
5. Loosen the locknut and remove the tie rod end from the tie rod. Count the number of complete turns it takes to completely remove it.
6. Remove the tie rod-to-steering gear locknut.
7. Remove the clamps that secure the flexible boot to the steering gear.
8. Slide the boot from the inner tie rod and remove the boot.
9. Bend the lock plate tabs from the inner tie rod end nut.
10. Loosen the inner tie rod end nut from the steering gear and remove the inner tie rod end.

To install:

11. Using a new lock plate, install the tie rod end and torque the tie rod to 65 ft. lbs. (90 Nm).
12. Bend the tabs of the new lock plate to secure the inner tie rod end.
13. Slide the boot onto the steering gear and secure it with new clamps.
14. Install the outer tie rod end to the steering gear locknut.
15. Install the outer tie rod end, turning it in exactly as many turns as it was to remove the old one. Make sure it is correctly positioned in relationship to the steering linkage.
16. Connect the outer tie rod end to the steering knuckle and install the castle nut. Torque the nut to 25 ft. lbs. (34 Nm.).

17. Install a new cotter pin to the castle nut.

18. Tighten the tie rod end locking nut to 30 ft. lbs. (42 Nm).

19. Install the wheel and tire assembly.

20. Lower the vehicle and perform a front end alignment.

Manual Rack and Pinion

REMOVAL AND INSTALLATION

All Models

NOTE: If equipped with air bag, prior to removal of the steering rack, center the front wheels and remove the ignition key. Failure to do so may damage the SRS clockspring and render SRS system inoperative, risking serious driver injury. Be sure to properly disarm the air bag system.

1. Disconnect the battery negative cable. Raise the vehicle and support safely and remove the wheels.

2. Disconnect the oxygen sensor and remove the front exhaust pipe.

3. Properly support the engine. Remove both roll stopper mounting bolts and the four center member installation bolts.

4. Remove the center member.

NOTE: Matchmark the pinion input shaft of the rack to the lower steering column joint for installation purposes.

5. Remove the pinch bolt holding the lower steering column joint to the rack and pinion input shaft.

6. Remove the cotter pins and disconnect the tie rod ends from the steering knuckle.

7. Remove the rack and pinion steering assembly and its rubber mounts from the right side of the vehicle.

To install:

8. Align the matchmarks of the input shaft and install the rack to the vehicle.

9. Secure the rack using the retainer clamps and bolts. Torque the bolts to 51 ft. lbs. (70 Nm).

10. Torque the steering column pinch bolt to 13 ft. lbs. (18 Nm).

11. Install the center member.

12. Install the front exhaust pipe.

13. Connect the tie rod ends to the steering knuckles and torque the castle nuts to 25 ft. lbs. (34 Nm). Install new cotter pins.

14. Install the wheels and connect the negative battery cable.

15. Perform a front end alignment.

Power Rack and Pinion

REMOVAL AND INSTALLATION

Colt and Summit

NOTE: If equipped with air bag, prior to removal of the steering gear box, center the front wheels and remove the ignition key. Failure to do so may damage the SRS clockspring and render SRS system inoperative, risking serious driver injury.

1. Drain power steering system:

    a. Disconnect the return hose at the reservoir and place into a suitable container.

    b. Disable the ignition system. While cranking the engine, turn the wheels several times, until system has been drained.

2. Disconnect the battery negative cable. Raise the vehicle and support safely.

3. Disconnect the oxygen sensor and remove the front exhaust pipe.

NOTE: It may be easier on some vehicles, to disconnect all hangers and lower the complete exhaust system.

4. Properly support the engine. Remove both roll stopper mounting bolts and the four center member installation bolts. Remove the center member.

5. Remove the center member.

NOTE: Matchmark the pinion input shaft of the rack to the lower steering column joint for installation purposes.

6. Remove the pinch bolt holding the lower steering column joint to the rack and pinion input shaft.

7. Remove the cotter pins and disconnect the tie rod ends from the steering knuckle.

8. Disconnect the power steering fluid pressure pipe and return hose from the rack fittings.

9. Remove the rack and pinion steering assembly and its rubber mounts from the right side of the vehicle.

To install:

10. Align the matchmarks of the input shaft and install the rack to the vehicle.

11. Secure the rack using the retainer clamps and bolts. Torque the bolts to 51 ft. lbs. (70 Nm).

12. Torque the steering column pinch bolt to 13 ft. lbs. (18 Nm).

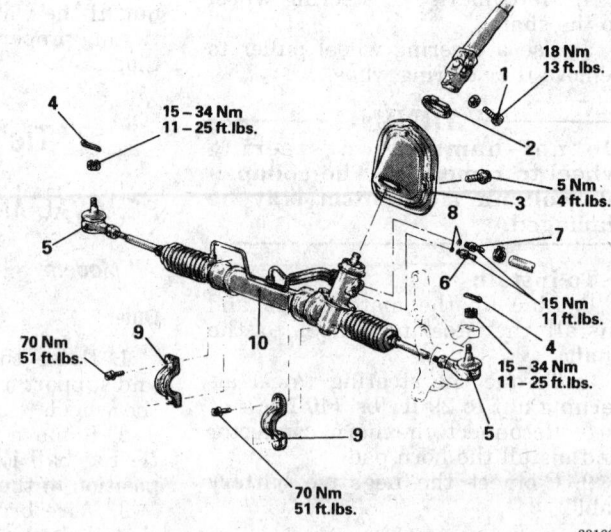

1. Joint assembly and gear box connecting bolt
2. Band
3. Steering cover
4. Cotter pin
5. Connection for tie-rod end and knuckle
6. Pressure pipe
7. Return pipe
8. O-ring
9. Clamp
10. Gear box assembly

18 Nm
13 ft.lbs.

15 – 34 Nm
11 – 25 ft.lbs.

5 Nm
4 ft.lbs.

15 Nm
11 ft.lbs.

15 – 34 Nm
11 – 25 ft.lbs.

70 Nm
51 ft.lbs.

70 Nm
51 ft.lbs.

331603

Power assisted steering rack and pinion and related parts — Colt and Summit models

13. Using new O-rings, connect the power steering fluid lines to the rack fittings.

14. Install the center member.

15. Install the front exhaust pipe.

16. Connect the tie rod ends to the steering knuckles and torque the castle nuts to 25 ft. lbs. (34 Nm). Install new cotter pins.

17. Install the wheels and connect the negative battery cable.

18. Refill the reservoir and bleed the system.

19. Perform a front end alignment.

Vista and Summit Wagon

———— WARNING ————

If equipped with air bag, prior to removal of the steering gear box, center the front wheels and remove the ignition key. Failure to do so may damage the SRS clockspring and render SRS system inoperative, risking serious driver injury.

1. Drain power steering system:

   a. Disconnect the return hose at the reservoir and place into a suitable container.

   b. Disable the ignition system. While cranking the engine, turn the wheels several times, until system has been drained.

2. Disconnect the battery negative cable. Raise the vehicle and support safely.

3. On AWD vehicles, properly support the engine and remove the rear engine mount bracket.

4. Remove the pinch bolt holding the lower steering column joint to the rack and pinion input shaft.

5. Remove the cotter pins and disconnect the tie rod ends from the steering knuckle.

6. Disconnect the power steering fluid pressure pipe and return hose from the rack fittings.

7. Remove the rack and pinion steering assembly and its rubber mounts from the right side of the vehicle.

To install:

8. Align steering shaft and install the steering gear into the vehicle. Secure rack assembly using the retainer clamps and bolts.

9. Install the pinch bolt and torque to 13 ft. lbs. (18 Nm).

10. Install the engine mount bracket, if removed.

11. Connect the power steering fluid lines to the rack fittings.

12. Connect the tie rod ends to the steering knuckles.

13. Connect the negative battery cable. Refill the reservoir and bleed the system.

14. Perform a front end alignment.

Power Steering Pump

BLEEDING

1. Raise and support the front of the vehicle safely.

2. Check and add fluid to the reservoir, if necessary.

3. Disconnect the ignition coil wire so the engine will not start.

4. Turn the steering wheel (all the way), right and left, just touching the stops while cranking the key. Continue turning the steering wheel until the fluid level in the reservoir no longer decreases. Do not crank the key for longer than 15–20 seconds.

5. Check the fluid level, add fluid as required.

6. Attach the coil wire.

7. Start the engine and continue turning the steering wheel from right to left, lightly touching the stops until the fluid level in the reservoir no longer decreases.

8. Stop the engine and check the fluid level, add fluid as required.

9. Repeat the steps until all of the air is bled from the system.

10. If the air cannot be bled from the system, turn and hold the steering wheel at each stop for at least 5 seconds but never more than 15 seconds.

REMOVAL AND INSTALLATION

Colt and Summit

1. Drain power steering system:

   a. Disconnect the return hose at the reservoir and place into a suitable container.

   b. Disable the ignition system. While cranking the engine, turn the wheels several times, until system has been drained.

2. Disconnect the battery negative cable.

3. Remove the pressure switch connector from the side of the pump.

4. If the alternator is located under the oil pump, cover it with a shop towel to protect it from oil.

5. Disconnect the pressure line.

6. Remove the steering pump drive belt and the water pump pulley drive belt.

7. Remove the water pump pulley.

8. Remove the bolts that secure the oil pump the remove the pump from its bracket.

To install:

9. Install the power steering pump and torque the mounting bolts to specifications.

10. Install the water pump pulley and torque the mounting bolts 14 ft. lbs. (19 Nm).

11. Install and adjust the drive belts.

12. Replace the O-rings and connect the pressure line. Connect the pressure line so the notch in the fitting aligns and contacts the pump's guide bracket. Tighten the nut that secures the pressure line to 13 ft. lbs. (18 Nm).

13. Connect the return line. Connect the pressure switch connector.

14. Refill the reservoir and bleed the system.

Vista and Summit Wagon

1. Disconnect the battery negative cable.

2. Loose and remove the alternator drive belt.

3. Remove the pressure switch connector from the side of the pump.

NOTE: If the alternator is located under the oil pump, cover it with a shop towel to protect it from oil.

4. Disconnect the return fluid line. Remove the reservoir cap and allow the return line to drain the fluid from the reservoir. If the fluid is contaminated, disconnect the ignition high tension cable and crank the engine several times to drain the fluid from the gearbox.

5. Disconnect the pressure line.

6. Remove the pump drive belt and unbolt the pump from its bracket.

To install:

7. Install the pump, wrap the belt around the pulley and tighten the mounting bolts.

8. Replace the O-rings and connect the pressure line. Connect the pressure line so the notch in the fitting aligns and contacts the pump's guide bracket.

9. Connect the return line. Connect the pressure switch connector.

10. Adjust both the power steering and alternator belts for proper tension and tighten the adjusting bolts.

11. Refill the reservoir and bleed the system.

BRAKES

Anti-Lock Brake System Service

PRECAUTIONS

• Certain components within the Anti-Lock Brake System (ABS) are not intended to be serviced or repaired individually. Only those components with removal and installation procedures should be serviced.

• Do not use rubber hoses or other parts not specifically specified for and ABS system. When using repair kits, replace all parts included in the kit. Partial or incorrect repair may lead to functional problems and require the replacement of components.

• Lubricate rubber parts with clean, fresh brake fluid to ease assembly. Do not use lubricated shop air to clean parts; damage to rubber components may result.

• Use only specified brake fluid from an unopened container.

• If any hydraulic component or line is removed or replaced, it may be necessary to bleed the entire system.

• A clean repair area is essential. Always clean the reservoir and cap thoroughly before removing the cap. The slightest amount of dirt in the fluid may plug an orifice and impair the system function. Perform repairs after components have been thoroughly cleaned; use only denatured alcohol to clean components. Do not allow ABS components to come into contact with any substance containing mineral oil; this includes used shop rags.

• The Anti-Lock control unit is a microprocessor similar to other computer units in the vehicle. Ensure that the ignition switch is **OFF** before removing or installing controller harnesses. Avoid static electricity discharge at or near the controller.

• If any arc welding is to be done on the vehicle, the control unit should be unplugged before welding operations begin.

Master Cylinder

REMOVAL AND INSTALLATION

All Models

1. Disconnect the negative battery cable.
2. Disconnect the fluid level sensor connector, if equipped.
3. Disconnect the brake lines from the master cylinder. If a separate fluid reservoir is used, plug the lines to prevent drainage.
4. Remove the two nuts securing the master cylinder to the brake booster and remove the master cylinder.

To install:

5. Fill the reservoir to the proper level with clean DOT 3 brake fluid. Bleed the master cylinder.
6. Install master cylinder to the mounting studs and install the mounting nuts. Tighten mounting nuts to 7–9 ft. lbs. (10–12 Nm).
7. Connect reservoir hoses to master cylinder and secure with clamps.
8. Fill the reservoir to the proper level with clean DOT 3 brake fluid. Bleed the master cylinder.
9. Connect the brake lines to the master cylinder.
10. Apply the brake pedal and check for firmness. If the pedal is spongy, air is present in the system. If air remains in the system, bleeding the entire system is required.
11. Check the brakes for proper operation and leaks.

Brake Caliper

REMOVAL AND INSTALLATION

Except Colt and Summit Rear Disc Brakes

Unlike most rear disc brake designs, the Vista and Summit Wagon's system does not incorporate the parking brake system into the rear brake caliper. Therefore, the rear brake system is serviced the same as the front system found on all models.

1. Raise the vehicle and support safely.
2. Remove the appropriate tire and wheel assembly.

NOTE: Do not let air into the master cylinder by allowing the reservoir to empty or the complete system bleeding will be required.

3. To disconnect the front brake hose, hold the nut on the brake hose side and loosen the flared brake line nut. Remove the brake hose from the caliper.
4. Remove the caliper guide and lock pins and lift the caliper assembly from the caliper support.

To install

NOTE: Be sure the tab on the disc brake pad is in alignment with the cutout of the caliper piston.

5. Position caliper onto the caliper support. Install the guide pin and lock pin. Tighten to specification.
6. Reconnect brake hose.

NOTE: Use caution not to twist brake hose during installation.

7. Bleed the brake system.
8. Ensure proper operation and no brake fluid leakage.
9. Install tire and wheel assembly. Torque lug nuts to 87–101 ft. lbs. (120–140 Nm).

Colt and Summit Rear Disc Brakes

1. Disconnect the battery negative cable.
2. Raise the vehicle and support safely.
3. Remove the appropriate tire and wheel assemblies. Loosen the parking brake cable adjustment from inside the vehicle.
4. Disconnect the parking brake cable end installed to the rear brake caliper assembly.
5. Remove the caliper lock and guide pins. Lift the caliper assembly from the caliper support.
6. Remove the rear brake hose from the caliper. Remove the caliper from the vehicle.

To install:

NOTE: Be sure the tab on the disc brake pad is in alignment with the cutout of the caliper piston.

7. Install the rear brake hose onto the caliper with new washers in place. If equipped with brake hose retainer bolt, tighten bolt to 25 ft. lbs. (35 Nm) torque. If no bolt is used, tighten the brake hose fitting to 12 ft. lbs. (17 Nm).

NOTE: Do not twist the brake hose during installation.

8. Install the caliper over the brake pads, making sure stopper grove lines up with pad projection.
9. Lubricate and install the lock pin and tighten to 23 ft. lbs. (32 Nm). Install the guide pin and tighten to 23 ft. lbs. (32 Nm).
10. Bleed the brake system.
11. Inspect the brake system for leaks and ensure proper operation.
12. Install tire and wheel assemblies. Torque wheels to 87–101 ft. lbs. (120–140 Nm).
13. Properly adjust parking brake cable.

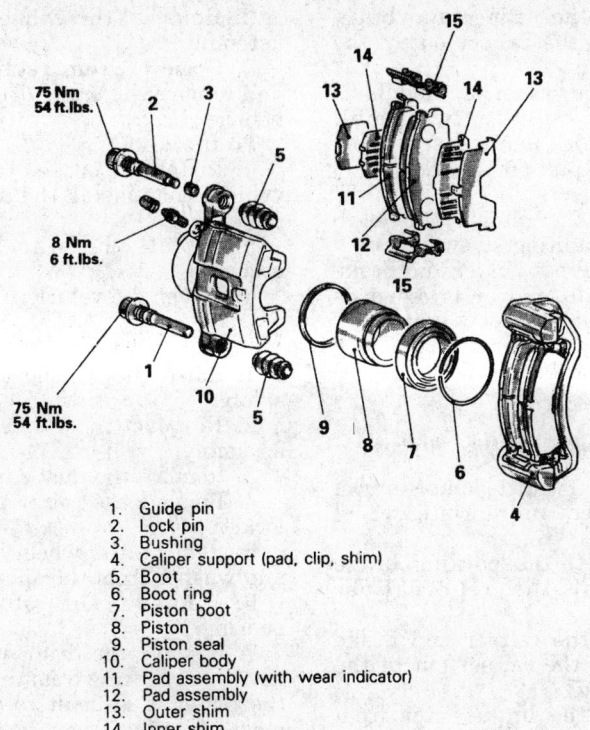

75 Nm
54 ft.lbs.

8 Nm
6 ft.lbs.

75 Nm
54 ft.lbs.

1. Guide pin
2. Lock pin
3. Bushing
4. Caliper support (pad, clip, shim)
5. Boot
6. Boot ring
7. Piston boot
8. Piston
9. Piston seal
10. Caliper body
11. Pad assembly (with wear indicator)
12. Pad assembly
13. Outer shim
14. Inner shim
15. Clip

239359

Front brake caliper and related components — Summit Wagon and Vista models

Disc Brake Pads

REMOVAL AND INSTALLATION

———— **CAUTION** ————
Brake pads and shoes contain asbestos, which has been determined to be a cancer causing agent. Never clean the brake surfaces with compressed air! Avoid inhaling any dust from brake surfaces! When cleaning brakes, use commercially available brake cleaning fluids.

Except Colt and Summit Rear Disc Brakes

Unlike most rear disc brake designs, the Vista and Summit Wagon's system does not incorporate the parking brake system into the rear brake caliper. Therefore, the rear brake system is serviced the same as the front system on all four models.

1. Remove some of the brake fluid from the master cylinder reservoir. The reservoir should be no more than ½ full. When the pistons are depressed into the calipers, excess fluid will flow up into the reservoir.

2. Remove some of the brake fluid from the master cylinder reservoir. The reservoir should be no more than half full. When the pistons are depressed into the calipers, excess fluid will flow up into the reservoir.
3. Raise the vehicle and support safely.
4. Remove the appropriate tire and wheel assemblies.
5. Remove the caliper guide and lock pins and lift the caliper assembly from the caliper support. Tie the caliper out of the way using wire. Do not allow the caliper to hang by the brake line.

NOTE: On some vehicles, the caliper can be flipped up by leaving the upper pin in place and using it as a pivot point.

6. Remove the brake pads, spring clip and shims. Take note of positioning to aid installation.
7. Install the wheel lug nuts onto the studs and lightly tighten. This is done to hold the disc on the hub.

To install:

NOTE: The piston in the disc caliper has a cutout that aligns with the tab on the disc brake pad.

8. Use a large C-clamp to compress piston back into caliper bore.
9. Lubricate slide points and install the brake pads, shims and spring clip onto the caliper support. Install the caliper over the brake pads.

NOTE: Be careful that the piston boot does not become caught when lowering the caliper onto the support. Do not twist the brake hose during caliper installation.

10. Lubricate and install the caliper guide and lock pins in their original positions. Tighten guide and locking pins to specification.
11. Install the tire and wheel assemblies. Lower the vehicle.

———— **WARNING** ————
Pump brake pedal several times, until firm, before attempting to move vehicle.

12. Road test the vehicle and check brakes for proper operation.

Colt and Summit Rear Disc Brakes

1. Remove some of the brake fluid from the master cylinder reservoir. The reservoir should be no more than half full. When the pistons are depressed into the calipers, excess fluid will flow up into the reservoir.
2. Raise the vehicle and support safely.
3. Remove the appropriate tire and wheel assemblies. Loosen the parking brake cable adjustment from inside the vehicle.
4. Disconnect the parking brake cable end installed to the rear brake caliper assembly.
5. Remove the caliper lock and guide pins and lift the caliper assembly from the caliper support. Tie the caliper out of the way using wire. Do not allow the caliper to hang by the brake line.
6. Remove the outer shim, brake pads and spring clips from the caliper support. Take note of positioning of each to aid in installation.
7. Install the wheel lug nuts onto the studs and lightly tighten. This is done to hold the disc on the hub.
8. Clean the caliper piston. Using rear disc brake driver tool MB990652 or equivalent, thread the piston into the caliper bore clockwise. Be sure at this point, that the groove of the piston correctly fits into the projection on the replacement brake pads rear surface.

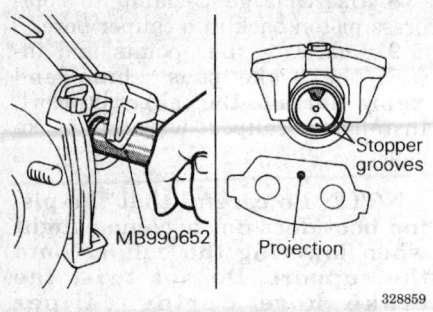

Retracting brake caliper piston and aligning pad to piston — Colt and Summit models

To install:

NOTE: The piston in the disc caliper has a cutout that aligns with the tab on the disc brake pad.

9. Lubricate all sliding points. Install the brake pads, shims and spring clip to the caliper support. Install the caliper over the brake pads, making certain that the stopper groove lines up with the projection on the brake pad.

NOTE: Be careful that the piston boot does not become caught when lowering the caliper onto the support. Do not twist the brake hose during caliper installation.

10. Lubricate and install the caliper guide and lock pins. Tighten the pins to 20 ft. lbs. (27 Nm). Attach the parking brake cable to the rear brake assembly.

11. Start the engine and forcefully depress the brake pedal 5–6 times. Apply the parking brake and make sure the adjustment is within specifications. Adjust the parking brake cable, as required.

12. Install the tire and wheel assemblies. Lower the vehicle.

13. Test the brakes for proper operation.

Brake Rotor

REMOVAL AND INSTALLATION

Except Colt and Summit Rear Disc Brakes

The following procedure is applicable to both the front and rear brake systems of the Summit Wagon and Vista, but to only the front system on the Colt and Summit models.

1. Raise the vehicle and support safely. Remove appropriate wheel assembly.

2. Remove the caliper and brake pads. Support the caliper out of the way using wire.

3. The rotor on most models is held to the hub by two small threaded screws. Remove screws, if equipped, and pull off the rotor.

To install

4. Position the rotor on the hub and install mounting screws.

5. Install caliper holder and brake pads. Slide caliper over brake pads and tighten guide pins.

6. Install wheel and torque lug nuts. Check brake pedal before attempting to move vehicle.

Colt and Summit Rear Disc Brakes

1. Raise the vehicle and support safely. Remove appropriate wheel assembly.

2. Disconnect the parking brake connection at the rear caliper assembly.

3. Remove the caliper and brake pads. Support the caliper out of the way using wire.

4. Remove the brake rotor from the rear hub assembly.

To install

5. Position the rotor on the hub. Install a couple of lug nuts and lightly tighten to hold rotor on hub.

6. Install the caliper holder and place brake pads in holder. Slide caliper over brake pads and install guide pins. Once caliper is secured, lug nuts can be removed.

7. Reconnect parking brake cable and install wheel(s).

Brake Drums

REMOVAL AND INSTALLATION

All Models

With Rear Hub Assembly

1. Raise the vehicle and support safely.

2. Remove the wheel and tire assembly.

3. Remove drum retaining screw and remove the brake drum from the vehicle.

To install:

4. Install the brake drum onto the vehicle, then install the drum retaining screw.

5. Install the wheel and tire assembly.

6. Lower the vehicle.

Without Rear Hub Assembly

1. Raise the vehicle and support safely.

2. Remove the wheel and tire assembly.

3. Remove the dust cap.

4. Remove the cotter pin and nut lock.

5. Remove the wheel bearing nut and washer from the spindle.

6. Remove the outer wheel bearing.

7. Remove the drum with the inner wheel bearing from the spindle. If the drum is difficult to remove, remove the plug from the rear of the backing plate and push the self adjuster lever away from the star wheel. Rotate the star wheel with an upward motion to retract the shoes and remove the drum. Remove the grease seal.

To install:

8. For Colt and Summit models, determine if the self-locking nut is reusable as follows:

   a. Screw in the self-locking nut until about ⅛ in. of the spindle is showing.

   b. Measure the torque required to turn the self-locking nut counterclockwise.

   c. The lowest allowable torque is 48 inch lbs. (5.5 Nm). If the measured torque is less than the specification, replace the nut.

9. Lubricate and install the inner wheel bearing. Install a new grease seal.

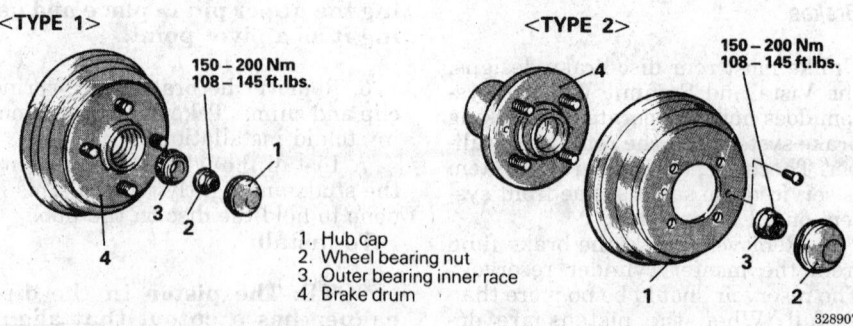

1. Hub cap
2. Wheel bearing nut
3. Outer bearing inner race
4. Brake drum

Rear brake drum type identification — Colt and Summit models

10. Install the drum to the spindle.

11. Lubricate and install the outer wheel bearing, washer and nut.

12. For Summit Wagon and Vista models, tighten the wheel bearing nut to 20–25 ft. lbs. (27–34 Nm) while rotating the drum. Back off the adjusting nut, then tighten it to 7 ft. lbs. (10 Nm). Install the nut lock and a new cotter pin.

13. For Colt and Summit models, torque the self-locking nut to 108–145 ft. lbs. (150–200 Nm).

14. Install the wheel assembly and lower the vehicle.

Brake Shoes

REMOVAL AND INSTALLATION

All Models

------- **WARNING** -------
Brake shoes contain asbestos, which has been determined to be a cancer causing agent. Never clean the brake surfaces with compressed air! Avoid inhaling any dust from brake surfaces! When cleaning brakes, use commercially available brake cleaning fluids.

1. Raise vehicle and support safely. Remove the appropriate tire and wheel assembly.

2. Remove the brake drum.

3. Remove the shoe to shoe spring.

4. Remove the shoe to lever spring and remove the adjuster assembly.

NOTE: Note the location of all springs and clips for proper reassembly.

5. Remove the shoe hold-down clips and the brake shoes.

6. Disconnect the parking brake cable from the rear shoes by spreading the horseshoe clip apart.

To install

7. Thoroughly clean and dry the backing plate. Ensure the backing plate bosses are smooth, so not to cause binding. If shoes are being replaced due to contamination from brake fluid. repair the leaking wheel cylinder as required.

8. Lubricate backing plate bosses, anchor pin, and parking brake actuating mechanism with a lithium-based grease.

9. Remove, clean and dry all remaining parts. Apply anti-seize to the star wheel threads and transfer all parts to the new shoes.

10. Connect the parking brake arm to the appropriate brake shoe. Attach shoes to backing plate and install all

remaining hardware in the reverse order it was removed.

11. Pre-adjust the shoes so the drum slides on with a light drag and install brake drum. Properly adjust the wheel bearings, if required.

12. Adjust rear brake shoes and install the wheel assemblies.

Wheel Cylinder

REMOVAL AND INSTALLATION

All Models

NOTE: It is important not let the master cylinder reservoir run dry, at any time during this procedure, or the entire system will have to be bleed.

1. Raise the vehicle and support it safely.

2. Remove the wheel and the brake drum.

3. Remove the shoe–to–lever spring and the upper shoe-to-shoe spring. Spread the upper portion of the brake shoes slightly.

4. Remove and plug the brake line from the wheel cylinder.

5. Remove the wheel cylinder retaining bolts and remove the cylinder from the backing plate.

To install:

6. Apply a very thin coating of silicone sealer to the cylinder mounting surface, install the cylinder to the backing plate and install the retaining bolts.

7. Connect the brake line to the wheel cylinder.

8. Install brake springs and the brake drum.

9. Install the tire and wheel assembly.

10. Fill the system with clean brake fluid and bleed the rear brakes.

Parking Brake Cable

ADJUSTMENT

Colt and Summit

NOTE: If the vehicle is equipped with rear drum brakes, make certain that the brake shoes are properly adjusted before attempting to adjust the parking brake.

1. Make sure the parking brake cable is free and is not frozen or sticking.

2. Apply the parking brake with 45 lbs. (200 N) of force while counting the number of notches. The desired

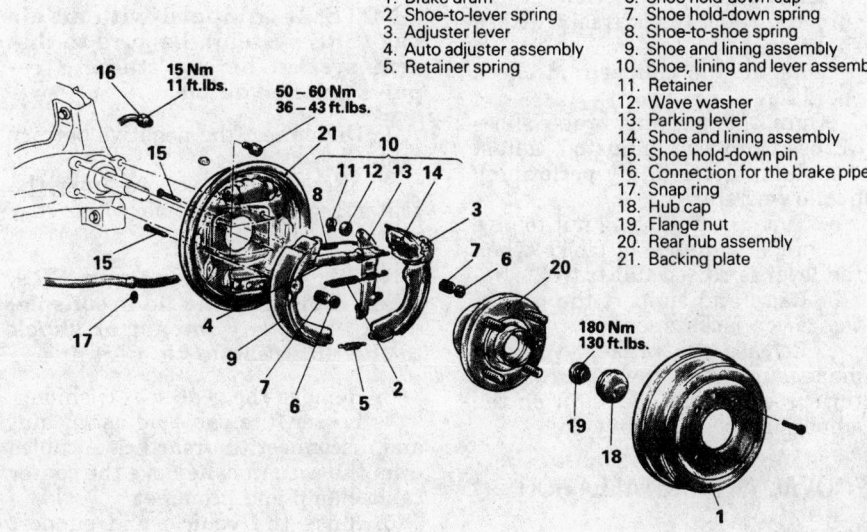

1. Brake drum
2. Shoe-to-lever spring
3. Adjuster lever
4. Auto adjuster assembly
5. Retainer spring
6. Shoe hold-down cup
7. Shoe hold-down spring
8. Shoe-to-shoe spring
9. Shoe and lining assembly
10. Shoe, lining and lever assembly
11. Retainer
12. Wave washer
13. Parking lever
14. Shoe and lining assembly
15. Shoe hold-down pin
16. Connection for the brake pipe
17. Snap ring
18. Hub cap
19. Flange nut
20. Rear hub assembly
21. Backing plate

15 Nm 11 ft.lbs.
50 – 60 Nm 36 – 43 ft.lbs.
180 Nm 130 ft.lbs.

329004

Brake shoes and related components with rear wheel hub — Colt and Summit shown, others models similar

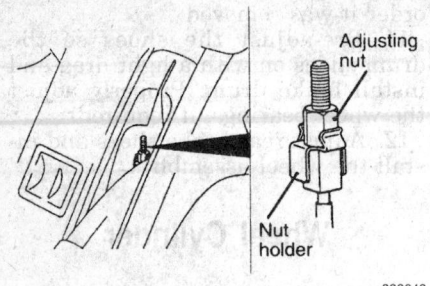

333043

Parking brake adjusting nut — Colt and Summit models

parking brake stroke should be 5–7 notches.

3. If adjustment is required, access the adjusting nut from inside the floor console.

4. Loosen the locknut on the cable rod. Rotate the adjusting nut to adjust the parking brake stroke to the 5–7 notch setting. After making the adjustment, check there is no looseness between the adjusting nut and the parking brake lever, then tighten the locknut.

NOTE: Do not adjust the parking brake too tight. If the number of notches is less than specification, the cable has been pulled too much and the automatic adjuster will fail or the brakes will drag.

5. After adjusting the lever stroke, raise the rear of the vehicle and safely support. With the parking brake lever in the released position, turn the rear wheels to confirm that the rear brakes are not dragging.

6. Check that the parking brake holds the vehicle on an incline.

Summit Wagon and Vista

With Drum Brakes

NOTE: Make certain that the brake shoes are properly adjusted before attempting to adjust the parking brake.

1. Pull the parking brake lever up with a force of about 45 lbs. If that value cannot be determined, just pull it up as far as possible. The total number of clicks heard should be 5–7.

2. If the number of clicks was not within that range, release the lever and back off the cable adjuster locknut at the base of the lever and tighten the adjusting nut until there is no more slack in the cable.

3. Operate the lever and brake pedal several times, until no more clicks are heard from the automatic adjuster.

4. Turn the adjusting nut to give the proper number of clicks when the lever is raised full travel.

5. Raise and support the rear of the car on jackstands.

6. Release the brake lever and make sure that the rear wheels turn freely. If not, back off on the adjusting nut until they do.

With Disc Brakes

1. Pull the parking brake lever up with a force of about 45 lbs. (61 N). The total number of clicks heard should be 3–5. If the number of clicks was not within that range, system requires adjustment.

NOTE: The parking brake shoes must be adjusted before attempting to adjust the cable mechanism

2. To adjust the parking brake shoes perform the following steps.

   a. remove the floor console, release the lever and back off the cable adjuster locknut at the base of the lever.

   b. Raise the vehicle, support safely and remove the wheel. Remove the hole plug in the brake rotor.

   c. Remove the brake caliper and hang out of the way with wire.

   d. Use a suitable prybar to pry up on the self-adjuster wheel until the rotor will not turn.

   e. Return the adjuster 5 notches in the opposite direction. Make sure the rotor turns freely with a slight drag.

   f. Install the caliper and check operation.

3. Once the parking brake shoes have been properly adjusted, adjust the cable mechanism, by performing the following steps:

   a. Turn the adjusting nut to give the proper number of clicks when the lever is raised full travel.

   b. Raise and support the rear of the car on jackstands.

   c. Release the brake lever and make sure that the rear wheels turn freely. If not, back off on the adjusting nut until they do.

REMOVAL AND INSTALLATION

Colt and Summit

With Rear Drum Brakes

NOTE: If equipped with an air bag (SRS system), be sure to disarm system before starting repairs on the vehicle.

1. Disconnect the negative battery cable.

2. Remove the screws from the center section and remove the rear part of the console.

NOTE: If equipped with SRS, when removing the floor console, don't allow any impact or shock to the SRS diagnostic unit.

3. Remove the rear seat cushion.

4. Remove the center cable clamp and grommet.

5. Raise the vehicle and support safely.

6. At the rear wheel, remove the brake drum and shoes. Disconnect the cable end from the parking brake strut lever. Compress the retaining strips to remove the cable from the backing plate.

7. Unfasten any other frame retainers and remove the cables.

To install

8. The parking brake cables may be color coded to indicate side. Check the parking brake cables for an identification mark.

9. Install the cable to the rear actuator. Secure in place with the parking brake cable clip and retainer spring.

10. The balance of installation is the reverse of the removal procedure.

11. Connect the negative battery cable.

12. Check that the parking brake holds the vehicle on an incline.

With Rear Disc Brakes

NOTE: If equipped with an air bag (SRS system), be sure to disarm system before starting repairs on the vehicle.

1. Disconnect the negative battery cable.

2. Remove the screws from the center section and remove the rear part of the console.

NOTE: If equipped with SRS, when removing the floor console, don't allow any impact or shock to the SRS diagnostic unit.

3. Remove the rear seat cushion.

4. Loosen the cable adjusting nut and disconnect the rear brake cables from the actuator. Remove the center cable clamp and grommet.

5. Raise the vehicle and support safely. Remove the parking brake cable clip and retainer spring. Disconnect the cable end from the parking brake assembly.

6. Unfasten any remaining frame retainers and remove the cables from the vehicle.

To install:

7. The parking brake cables may be color coded to indicate side. Check the parking brake cables for an identification mark.

8. Install the cable to the rear actuator. Secure in place with the parking brake cable clip and retainer spring.

9. Position the cable in the under the vehicle and install retainers loose.

10. Reattach the parking brake cables to the actuator inside the vehicle.

11. Tighten the adjusting nut until the proper tension is placed on the cable. Adjust the parking brake stroke using appropriate method.

12. Install the remaining components in the reverse order of removal.

13. Road test the vehicle and check for proper brake operation. Check that the parking brake holds the vehicle on an incline.

Summit Wagon and Vista

With Rear Drum Brakes

1. Disconnect the negative battery cable.

2. Remove the shifter knob, boot and cover assembly.

NOTE: If equipped with SRS, when removing the floor console, don't allow any impact or shock to the SRS diagnostic unit.

3. Remove the interior rear seat and carpet, in order to gain access to the cables.

4. Loosen the cable adjuster nut and then remove the parking brake cable, by pulling it from the passenger compartment.

5. Raise the vehicle and support safely.

6. At the rear wheel, remove the brake drum and shoes.

7. Disconnect the cable end from the parking brake strut lever. Compress the retaining strips to remove the cable from the backing plate.

8. Unfasten any other frame retainers and remove the cables.

To install

9. Install the cable to the rear actuator. Secure in place with the parking brake cable clip and retainer spring.

10. Install the brake shoes and drum.

11. Position the cable under the vehicle and install retainers loose.

12. Reattach the parking brake cables to the actuator inside the vehicle. Tighten the adjusting nut until the proper tension is placed on the

cable. Adjust the parking brake stroke.

13. Install the remaining components in the opposite order from which they were removed.

14. Check that the parking brake holds the vehicle on an incline.

With Rear Disc Brakes

Unlike conventional rear disc brake systems, the parking brake operation is **not** incorporated into the brake caliper. This system, uses a separate set of brake shoes, located behind the brake rotor.

1. Disconnect the negative battery cable.

2. Remove the shifter knob, boot and cover assembly.

NOTE: If equipped with SRS, when removing the floor console, don't allow any impact or shock to the SRS diagnostic unit.

3. Remove the interior rear seat and carpet, in order to gain access to the cables.

4. Loosen the cable adjuster nut and then remove the parking brake cable, by pulling it from the passenger compartment.

5. Raise the vehicle and support safely.

6. At the rear wheel, remove the brake caliper and rotor.

7. Remove the parking brake shoes, following the same procedures as conventional drum brake shoes.

8. Disconnect the cable end from the parking brake strut lever. Compress the retaining strips to remove the cable from the backing plate.

9. Unfasten any other frame retainers and remove the cables.

To install

10. Install the cable to the rear actuator. Secure in place with the parking brake cable clip and retainer spring.

11. Install the parking brake shoes.

12. Install the brake rotor and caliper assembly.

13. Position the cable in the under the vehicle and install retainers loose.

14. Reattach the parking brake cables to the actuator inside the vehicle. Tighten the adjusting nut until the proper tension is placed on the cable. Adjust the parking brake stroke.

15. Secure all cable retainers. Apply and release the parking brake a number of times once all adjustments have been made.

16. Assemble the interior components which were removed.

17. Adjust the parking brake shoes and parking brake cables.

18. Connect the negative battery cable and check the rear wheels to confirm that the rear brakes are not dragging.

19. Check that the parking brake holds the vehicle on an incline.

Brake System Bleeding

Bleeding the brake system is required anytime the normally closed system has been opened to the atmosphere. When bleeding the system, keep the brake fluid level in the master cylinder reservoir above half full. If the reservoir is empty, air will be pushed through the system. Hydraulic brake systems must be totally flushed if the fluid becomes contaminated with water, dirt or other corrosive chemicals. To flush, bleed the entire system until all fluid has been replaced with the correct type of new fluid.

NOTE: If using a pressure bleeder, follow the instructions furnished with the unit and choose the correct adapter for the application. Do not substitute an adapter that "almost fits" as it will not work and could be dangerous.

MASTER CYLINDER

Due to the location of the fluid reservoir, bench bleeding of the master cylinder is not recommended. The master cylinder is to be bled while mounted on the brake booster. If the fluid reservoir runs dry, bleeding of the entire system will be necessary. Two people will be required to bleed the brake system.

1. Fill the brake fluid reservoir with clean brake fluid. Disconnect the brake tube from the master cylinder.

2. Have a helper slowly depress the brake pedal. Once depressed, hold it in that position. Brake fluid will be expelled from the master cylinder.

——————— **CAUTION** ———————
When bleeding the brake system, keep your face away from the area. Spewing brake fluid may cause facial and/or visual damage. Do not allow brake fluid to spill onto the car's finish; it will remove the paint.
———————————————————————————

3. While the pedal is held down, use a finger to close the outlet port of the master cylinder. While the port is

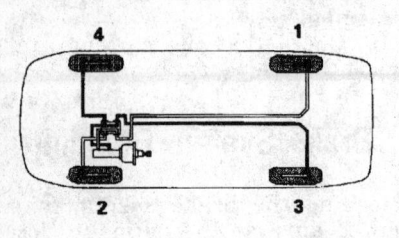

313356

Brake bleeding sequence — all models

closed, have the helper release the brake pedal.

4. Repeat this procedure until all air is bled from the master cylinder. Check the brake fluid in the reservoir every 4–5 times, making sure the reservoir does not run dry. Add clean DOT 3 brake fluid to the reservoir as needed. All air is bled from the master cylinder when the fluid expelled from the port is free of bubbles.

5. Connect the brake tube to the port on the master cylinder and torque to 10 ft. lbs.(13.5 Nm). Add clean fluid to fill the reservoir to the appropriate level.

6. Pressurize system and check for leaks. On ABS cars, turn key on until system warning light goes off. Then remainder of system can be bled with key **OFF**, following normal bleeding procedure.

CALIPERS

1. Fill the master cylinder with fresh brake fluid. Check the level often during this procedure. Raise and safely support the vehicle.

NOTE: ABS cars, system must be bleed with the key in the OFF position.

2. Starting with the wheel farthest from the master cylinder, remove the protective cap from the bleeder and place where it will not be lost. Clean the bleeder screw.

3. Start the engine and run at idle.

——————— CAUTION ———————
When bleeding the brakes, keep your face away from the brake area. Spewing fluid may cause physical and/or visual damage. Do not allow brake fluid to spill onto the car's finish; it will remove the paint.

4. If the system is empty, the most efficient way to get fluid down to the wheel is to loosen the bleeder about

½–¾turn, place a finger firmly over the bleeder and have a helper pump the brakes slowly until fluid comes out the bleeder. Once fluid is at the bleeder, close it before the pedal is released inside the vehicle.

NOTE: If the pedal is pumped rapidly, the fluid will churn and create small air bubbles, which are almost impossible to remove from the system. These air bubbles will accumulate and a spongy pedal will result. Also note, it is important not to exceed normal pedal travel during bleeding procedure. This will prevent possible master cylinder piston(s) damage, due to build-up on the bore walls.

5. Once fluid has been pumped to the caliper, open the bleed screw again, have a helper press the brake pedal, lock the bleeder and have the helper slowly release the pedal. Wait 15 seconds and repeat the procedure (including the 15 second wait) until no more air comes out of the bleeder upon application of the brake pedal. Remember to close the bleeder before the pedal is released inside the vehicle each time the bleeder is opened. If not, air will be introduced into the system.

6. If a helper is not available, connect a small hose to the bleeder, place the end in a container of brake fluid and proceed to pump the pedal from inside the vehicle until no more air comes out the bleeder. The hose will prevent air from entering the system.

7. Repeat the procedure on the remaining calipers in the following order:
   a. Left front caliper
   b. Left rear caliper
   c. Right front caliper
8. Pressurize the system and check for fluid leaks. Install the bleeder cap on the bleeder to keep dirt out.

9. Always road test the vehicle after brake work of any kind is done.

Wheel Speed Sensor

REMOVAL AND INSTALLATION

Front Speed Sensor

1. Disconnect the negative battery cable.

——————— CAUTION ———————
Wait at least 90 seconds after the negative (-) battery cable is disconnected to prevent possible deployment of the air bag.

2. Raise and safely support the vehicle. Remove the necessary tire and wheel assembly.

3. Remove the fender splash shield.

4. Disconnect the ABS speed sensor connector.

5. Remove the sensor harness clamp bolts and clamps.

6. Remove the ABS speed sensor mounting bolt and the sensor.

To install:

7. Install the ABS speed sensor with its mounting bolt.

NOTE: The clearance between the wheel speed sensor and the rotor's toothed surface is not adjustable, but measure the distance between the sensor installation surface and the rotor's toothed surface. Standard value is: 0.012–0.035 in. If not within specifications, replace the speed sensor or the toothed rotor.

8. Reinstall the sensor harness with its clamps and bolts.

9. Reconnect the speed sensor connector.

10. Install the fender splash shield.

11. Reinstall the tire and wheel, safely lower the vehicle, and reconnect the negative battery cable.

Rear Speed Sensor

1. Disconnect the negative battery cable.

——————— CAUTION ———————
Wait at least 90 seconds after the negative (-) battery cable is disconnected to prevent possible deployment of the air bag.

2. Raise and safely support the vehicle. Remove the necessary tire and wheel assembly.

3. Disconnect the ABS speed sensor connector.

4. Remove the sensor harness clamp bolts and clamps.

5. Remove the ABS speed sensor mounting bolt and the sensor.

To install:

6. Install the ABS speed sensor with its mounting bolt.

NOTE: The clearance between the wheel speed sensor and the rotor's toothed surface is not adjustable, but measure the distance between the sensor installation surface and the rotor's toothed surface. Standard value is: 0.008–0.028 in. for the FWD and 0.012–0.035 in. for the AWD. If not within specifications, replace the speed sensor or the toothed rotor.

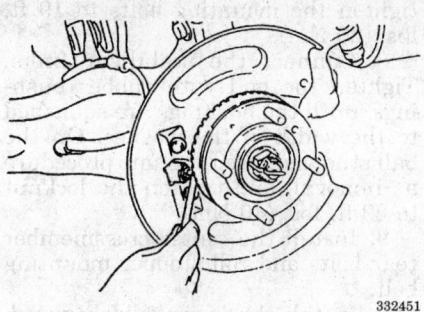

332451

Adjusting the rear speed sensor air gap — Colt and Summit models

7. Reinstall the sensor harness with its clamps and bolts.

8. Reconnect the speed sensor connector.

9. Reinstall the tire and wheel, safely lower the vehicle, and reconnect the negative battery cable.

FRONT SUSPENSION

Strut

REMOVAL AND INSTALLATION

All Models

1. Disconnect the negative battery cable.

2. If applicable, remove the daytime running lamp relay mounting bracket and position relay assembly aside.

3. Raise and safely support the vehicle.

4. Remove the brake hose and tube bracket retainer bolt and bracket from the front strut. Do not pry the brake hose and tube clamp away when removing.

5. If equipped with ABS, disconnect the front speed sensor mounting clamp from the strut.

6. Support the lower arm using floor jack or equivalent. Remove the lower strut to knuckle bolts. Once the mounting bolts have been removed, jack up the lower arm. Use a piece of wire to attach the brake hose, tube and driveshaft to the knuckle and to help keep the weight off. These components are not to be pulled.

7. Before removing the top bolts, make matchmarks on the body and the strut insulator for proper reassembly. If this plate is installed improperly, the wheel alignment will be

wrong. Remove the strut upper mounting bolts. Remove the strut assembly from the vehicle.

8. To remove the coil spring from the strut assembly, perform the following:

   a. Hold the spring upper seat with a spring compressor.

— CAUTION —
Do not remove the nut unless the spring is held by a spring compressor. Failure to do so may result in personal injury.

   b. Compress the spring and then remove the self–locking nut holding the strut insulator.

   c. Remove the spring.

To install:

9. To install the spring onto the strut, perform the following:

   a. With the spring being held in the spring compressor, align the spring in the grooves in the upper and lower seats.

   b. Install the self-locking nut and tighten to 43–51 ft. lbs. (60–70 Nm).

10. Install the strut to the vehicle and install the top mounting bolts. Make sure the insulator is installed so the matchmarks made during disassembly are in alignment. Tighten the mounting bolts to 33 ft. lbs. (45 Nm) for Summit Wagon and Vista models, or to 29 ft. lbs. (40 Nm) for Colt and Summit models.

11. Position the strut on the knuckle and install the mounting bolts. While holding the head of the lower mounting bolt, tighten the nuts to 78 ft. lbs. (108 Nm) for Summit Wagon and Vista models, or to 80–94 ft. lbs. (110–130 Nm) for Colt and Summit models.

12. Install the brake hose bracket and the ABS clamp.

13. Install the relay mounting bracket and connect the negative battery cable.

14. Install the wheel and tire assembly. Perform a front end alignment.

Lower Ball Joints

REMOVAL AND INSTALLATION

All Models

The lower ball joint is an integral part of the lower control arm assembly, and can not be serviced separately. A worn or damaged ball joint, requires replacement of lower control arm assembly.

Lower Control Arms

REMOVAL AND INSTALLATION

All Models

NOTE: The suspension components should not be tightened until the vehicle's weight is resting on its wheels.

1. Raise the vehicle and support safely.

2. Remove the wheel and tire assembly.

3. For Colt and Summit models, remove sway bar links or mounting nuts and bolts from lower control arm. Remove the joint cups and bushings.

4. For Vista and Summit Wagon models, disconnect the sway link by holding the ball stud with a hex wrench and removing the self-locking nut with a box wrench.

5. Disconnect the ball joint stud from the steering knuckle.

6. Remove the inner lower arm mounting bolt and nut.

7. Remove the rear mount bolts from the retaining clamp. Remove the rear retainer clamp if equipped.

8. Remove the arm from the vehicle.

To install:

9. Install the control arm to the vehicle and install the inner mounting bolt. Install new nut and torque to 78 ft. lbs. (108 Nm).

10. Install the rear mount clamp and bolts. Torque the clamp mounting bolts to 65 ft. lbs. (90 Nm) for Colt and Summit models, or to 51 ft. lbs. (70 Nm) for Summit Wagon and Vista models.

11. Connect the ball joint stud to the knuckle. Install a new nut and torque to 49 ft. lbs. (68 Nm).

12. Install the sway bar and links.

13. Lower the vehicle to the floor for the final tightening of the inner frame mount bolt.

14. Inspect all suspension bolts, making sure they all have been fully tightened.

15. Install the wheel and tire assembly.

Sway Bar

REMOVAL AND INSTALLATION

Colt and Summit

1. Disconnect the negative battery cable.

2. Raise and safely support vehicle.

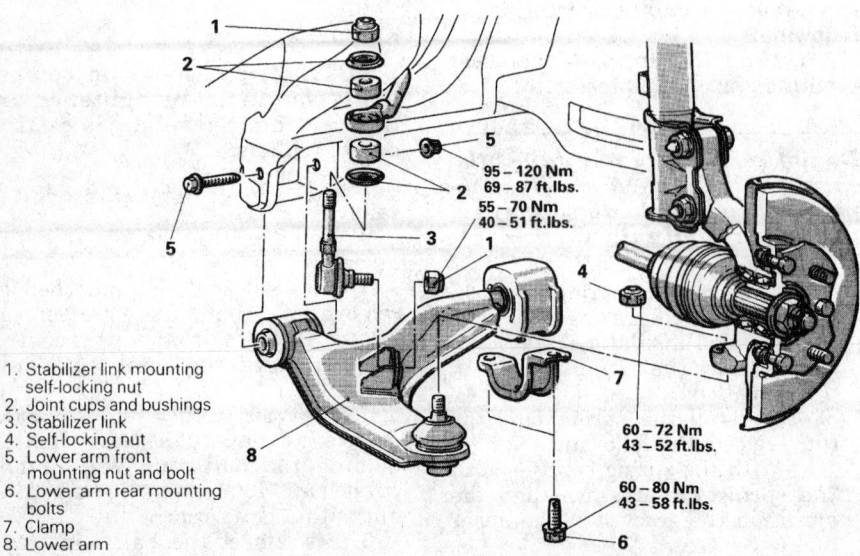

1. Stabilizer link mounting self-locking nut
2. Joint cups and bushings
3. Stabilizer link
4. Self-locking nut
5. Lower arm front mounting nut and bolt
6. Lower arm rear mounting bolts
7. Clamp
8. Lower arm

95 – 120 Nm
69 – 87 ft.lbs.
55 – 70 Nm
40 – 51 ft.lbs.

60 – 72 Nm
43 – 52 ft.lbs.

60 – 80 Nm
43 – 58 ft.lbs.

330852

Lower control arm and related parts — Colt and Summit models

3. Disassemble the links, remove the locknut, joint cup, bushing, and collar. Remove the stabilizer link bolts.

4. It will be necessary to remove the center crossmember in order to remove the sway bar. The following steps are required to remove the crossmember.

   a. Remove the front exhaust pipe.

   b. Properly support the engine, remove the engine roll stopper bolts. Remove the four center member mounting bolts and remove the center member assembly.

   c. Remove both steering rack mounts.

   d. Disconnect the lower control arm from the crossmember.

   e. Support the crossmember, remove the mounting bolts and lower the crossmember for access.

5. Remove the stabilizer bar mounts and remove the bar from the vehicle.

To install

NOTE: Note that the bar brackets are marked left and right.

6. Position the stabilizer bar in the vehicle, install the crossmember in the reverse order it was removed.

7. Install the sway bar mount brackets, and tighten the mounting bolts to 16 ft. lbs. (22 Nm).

8. Connect the stabilizer links and tighten the bolts with rubber bushings, until the amount of bolt protrusion at the end of link mounting bolt is 0.87 inches (22 mm).

9. Lower the vehicle and connect the negative battery cable.

Summit Wagon and Vista

FWD Vehicles

1. Disconnect the negative battery cable.

2. Raise and safely support vehicle.

3. Remove the splash guard, if equipped.

4. Remove the center crossmember rear installation bolts and roll stopper mounting bolt.

5. Remove the stabilizer link bolts.

6. Remove the stabilizer bar mounts and remove the bar from the vehicle.

To install

NOTE: Lubricate all rubber parts when installing. Note that the bar brackets are marked left and right.

7. Position the stabilizer bar in the vehicle, install the brackets, and

tighten the mounting bolts to 19 ft. lbs. (26 Nm).

8. Connect the stabilizer links. Tighten the end with rubber bushings, until the bushings are squashed to the width of the washer. On the ball stud end, use the same procedure as removal, and tighten the locknut to 29 ft. lbs. (40 Nm).

9. Install the center crossmember rear bolts and roll stopper mounting bolt.

10. Install the lower splash guard, if removed.

11. Lower the vehicle and connect the negative battery cable.

AWD Vehicles

1. Disconnect the negative battery cable.

2. Raise and safely support vehicle.

3. Remove the splash guard, if equipped.

4. Matchmark and remove the driveshaft.

5. Disconnect the exhaust at the manifold connection. Remove the front hangers and lower the exhaust.

6. To disconnect the sway link, hold ball stud with a hex wrench and remove the self-locking nut with a box wrench.

7. Remove the stabilizer bar mounts and remove the bar from the vehicle.

To install

NOTE: Lubricate all rubber parts when installing. Note that the bar brackets are marked left and right.

8. Position the stabilizer bar in the vehicle, install the brackets, and tighten the mounting bolts to 16 ft. lbs. (22 Nm).

9. Connect the stabilizer links and tighten the locknut, using the same procedure as removal.

10. Align the driveshaft and install.

11. Using a new gasket, connect the exhaust pipe.

12. Install the lower splash guard, if removed.

13. Lower the vehicle and connect the negative battery cable.

Front Wheel Bearings

ADJUSTMENT

All Models

1. Remove the hub, knuckle and bearing assembly from the vehicle.

2. Using pressing tool MB990998 or equivalent, mount the front hub assembly into the knuckle. Tighten

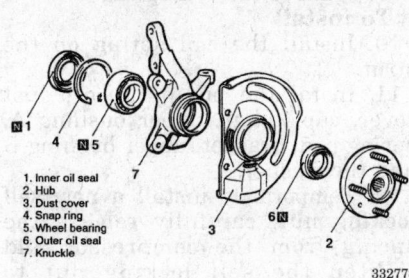

1. Inner oil seal
2. Hub
3. Dust cover
4. Snap ring
5. Wheel bearing
6. Outer oil seal
7. Knuckle

332778

Wheel bearing assembly exploded view — Colt and Summit models

the nut of the pressing tool to 144–188 ft. lbs. (200–260 Nm). Rotate the hub to seat the bearing.

3. Mount the knuckle assembly in a vise. Check the hub assembly turning torque and end-play as follows:

    a. Using a torque wrench and socket MB990998 or equivalent, turn the hub in the knuckle assembly. Note the reading on the torque wrench and compare to the desired reading of 16 inch lbs. (1.8 Nm) or less. This is known as the breakaway torque.

    b. Check for roughness when turning the bearing.

    c. Mount a dial indicator on the hub so the pointer contacts the machined surface on the hub.

    d. Check the end-play.

    e. Compare the reading to the limit of 0.002 in. (0.05mm).

4. If the starting torque or the hub end-play are not within specifications while the nut is tightened to 144–188 ft. lbs. (200–260 Nm), the bearing, hub or knuckle have probably not been installed correctly. Repeat the disassembly and assembly procedure and recheck starting torque and end-play.

5. Install the hub and knuckle assembly onto the vehicle.

REMOVAL AND INSTALLATION

All Models

1. Disconnect the negative battery cable.

2. Remove the cotter pin from the driveshaft nut. With the brakes applied, loosen the halfshaft nut.

3. Raise the vehicle and support safely. Remove the halfshaft nut.

4. If equipped with ABS, remove the front wheel speed sensor.

5. If equipped with Active-ECS, disconnect the height sensor from the lower control arm.

6. Remove the caliper assembly and brake pads. Suspend the caliper with a wire.

7. Using tool MB991113 or equivalent, disconnect the ball joint and tie rod end from the steering knuckle.

NOTE: It is important to use proper methods of joint separation. Use of unapproved techniques can result in damage to joint and possible failure.

8. Remove the halfshaft by setting up a puller on the outside wheel hub and pushing the halfshaft from the front hub. After pressing the outer shaft, insert a prybar between the transaxle case and the halfshaft and pry the shaft from the transaxle.

9. Unbolt the lower end of the strut and remove the hub and steering knuckle assembly from the vehicle.

10. Install the hub/knuckle assembly in a vise. Using puller MB991056 or equivalent, remove the hub from the knuckle.

NOTE: Do not use a hammer to accomplish this or the bearing will be damaged.

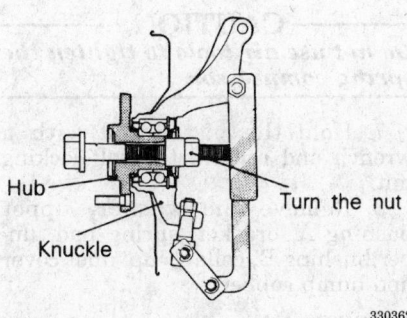

Hub   Turn the nut

Knuckle

330362

Use of press tool for hub removal — all models

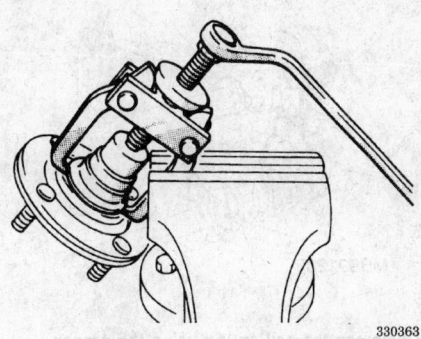

330363

Removing inner race from hub — all models

11. Remove the oil seal from the axle side of the knuckle using a small prying tool.

12. Remove the wheel bearing inner race from the front hub using a puller.

NOTE: Be careful that the front hub does not fall when the inner race is removed.

13. Remove the snapring from the axle side of the knuckle. Remove the bearing from the knuckle using a puller.

14. Once the bearing is removed, the bearing outer race can be removed by tapping out with a brass drift pin and a hammer.

    To assemble:

15. Fill the wheel bearing with multipurpose grease. Apply a thin coating of multipurpose grease to the knuckle and bearing contact surfaces.

16. Press the wheel bearing into the knuckle using appropriate pressing tool. Once the bearing is installed, install the inner race using the proper driving tool.

17. Drive the oil seal into the knuckle by using the proper size driver. Drive seal into knuckle until it is flush with the knuckle end surface.

18. Using pressing tool MB990998 or equivalent, mount the front hub assembly into the knuckle. Tighten the nut of the pressing tool to 144–188 ft. lbs. (200–260 Nm). Rotate the hub to seat the bearing.

19. Mount the knuckle assembly in a vise. Check the hub assembly turning torque and end-play as follows:

    a. Using a torque wrench and socket MB990998 or equivalent, turn the hub in the knuckle assembly. Note the reading on the torque wrench and compare to the desired reading of 16 inch lbs. (1.8 Nm) or less. This is known as the breakaway torque.

    b. Check for roughness when turning the bearing.

    c. Mount a dial indicator on the hub so the pointer contacts the machined surface on the hub.

    d. Check the end-play.

    e. Compare the reading to the limit of 0.002 in. (0.05mm).

20. If the starting torque or the hub end-play are not within specifications while the nut is tightened to 144–188 ft. lbs. (200–260 Nm), the bearing, hub or knuckle have probably not been installed correctly. Repeat the disassembly and assembly procedure and recheck starting torque and end-play.

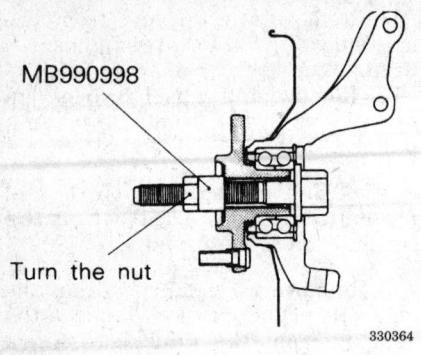

MB990998

Turn the nut

330364

Pressing new bearing assembly into knuckle — all models

21. Install the hub and knuckle assembly onto the vehicle. Install the lower ball joint stud into the steering knuckle and install new nut. Tighten to 52 ft. lbs. (72 Nm).

22. Install the halfshaft into the transaxle extension housing and guide the outer end through the hub/knuckle assembly.

23. Install the two front strut lower mounting bolts and tighten to 80–94 ft. lbs. (110–130 Nm).

24. Install the connection for the tie rod end and tighten nut to 25 ft. lbs. (34 Nm). Install new cotter pin and bend to locknut in position.

25. Install the brake disc and caliper assembly.

26. If equipped with Active-ECS, connect the height sensor and tighten the mounting bolt to 15 ft. lbs. (20 Nm).

27. Install the front speed sensor, if removed.

NOTE: When installing front speed sensor, make sure harness is routed in the original position and that it is not twisted.

28. Install the washer and new locknut to the end of the halfshaft. Tighten the locknut snugly.

29. Install the tire and wheel assembly onto the vehicle. Lower the vehicle to the ground.

30. With the weight of the vehicle on the ground and the brakes applied, tighten the locknut to 144–188 ft. lbs. (200–260 Nm).

31. Install the cotter pin in the first matching holes and bend it securely.

REAR SUSPENSION

Strut

REMOVAL AND INSTALLATION

Colt and Summit

NOTE: The strut assembly is a load bearing component, therefore the vehicle chassis and axle weight must be supported separately, requiring the use of two separate lifting devices.

NOTE: Matchmark the upper spring plate to the vehicle chassis for reassembly.

1. Remove the trunk interior trim to gain access to the top mounting nuts.

2. Remove the top cap and upper shock mounting nuts.

3. Raise and support vehicle chassis.

4. Raise and support the trailing arm assembly slightly.

5. Remove the shock lower mounting bolt and remove the assembly from the vehicle.

6. Compress the coil spring using the proper spring compressor.

── CAUTION ──
Do not use air tools to tighten the spring compressor.

7. Hold the piston rod with a wrench and remove the self locking nut.

8. Remove the washer, upper bushing A, bracket, spring pad, upper bushing B, collar, cup, dust cover and bump rubber.

NOTE: Align the stepped part of the spring pad with the end of the spring.

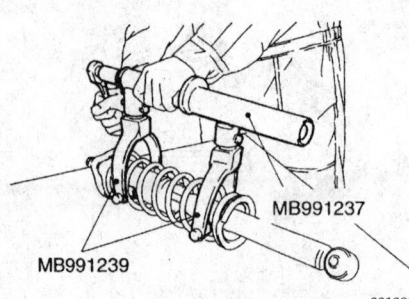

MB991237

MB991239

331881

Compress the coil spring using the proper tool — Colt and Summit models

9. Remove the coil spring.
To install
10. Install the coil spring on the strut.

11. Instal the bump rubber, dust cover, cup, collar, upper bushing A, spring pad, bracket, upper bushing B and the washer.

12. Temporarily install a new self locking nut, carefully release the spring from the compressor and tighten the self locking nut to specifications.

13. Position strut assembly so that lower mounting bolt can be installed and lightly tightened.

14. Use jack to raise or lower the axle assembly so that top strut plate studs aligns through body. Raise jack to hold strut assembly in position. Be sure to properly position the upper spring plate.

15. Install top plate nuts on studs. Tighten the upper shock mounting nuts to 20 ft. lbs. (28 Nm) and the lower mounting bolt to 65 ft. lbs. (90 Nm).

16. Lower the vehicle. Install top cap and interior trim.

Shock Absorber

REMOVAL AND INSTALLATION

Summit Wagon and Vista

NOTE: The strut assembly is a load bearing component, therefore the vehicle chassis and axle weight must be supported separately, requiring the use of two separate lifting devices.

1. Raise and support vehicle chassis.

2. Raise and support arm assembly slightly.

3. Remove the trunk interior trim to gain access to the top mounting nuts.

4. Remove the top cap and upper shock mounting nuts.

5. Remove the shock lower retaining nut and remove the assembly from the vehicle.
To install
6. Position strut assembly so that lower mounting nut can be installed and lightly tightened.

7. Use jack to raise or lower arm, so that top strut plate studs aligns through body. Raise jack to hold strut assembly in position.

8. Install top plate nuts on studs. Tighten the upper shock mounting nuts to 33 ft. lbs. (45 Nm).

9. With the full weight of the vehicle on the suspension, torque the

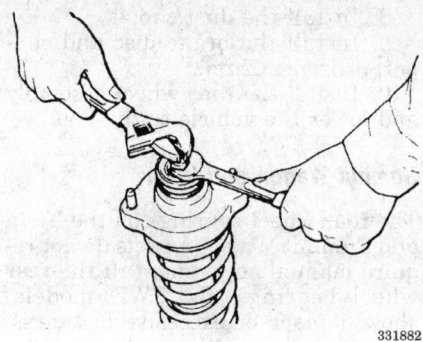

331882

Hold the piston rod and remove the self locking nut — Colt and Summit models

lower mounting bolt to 72 ft. lbs. (100 Nm).

10. Install top cap and interior trim.

Coil Spring

REMOVAL AND INSTALLATION

Summit Wagon and Vista

1. Raise and properly support the vehicle.
2. Remove the rear stabilizer bar.

NOTE: Perform the following steps, working on one side at a time.

3. On AWD models, remove the rear driveshaft mounting bolts at the carrier flange and hang the driveshaft from the vehicle body using wire.
4. Using a jack to support the lower arm, remove the rear shock absorber lower mounting bolt.
5. If equipped with ABS, remove the speed sensor clamp bolt and relocate out of the way. Do not apply tension to the wire harness of the connector.
6. Scribe mating marks on the lower control arm shaft (inner mounting bolt) and the crossmember. To remove the coil spring, loosen the shaft assembly nut and flange bolt nut (outer mounting bolt), then slowly lower the rear end of the lower arm. It is not necessary to remove the nuts, only to loosen them.

To install:

7. Install the coil spring into the seats making sure both ends of the spring are correctly aligned with the spring seat groove.
8. Slowly raise the rear the rear end of the lower arm and align the scribe marks made during disassembly. Once the full weight of the vehicle is on the ground, tighten shaft

and flange mounting nuts to 69 ft. lbs. (95 Nm).

9. Install the speed sensor clamp to it's original location and secure the wire harness making.
10. Reconnect the lower portion of the shock and tighten the retaining bolt to 72 ft. lbs. (100 Nm).
11. On AWD models, install the rear driveshaft to the flange and secure tightening mounting bolts to 40–47 ft. lbs. (55–65 Nm).
12. Install the stabilizer bar.
13. Lower the arm and remove the jack.
14. Check rear alignment.

Lower Control Arms

REMOVAL AND INSTALLATION

Summit and Colt

1. Raise and safely support the rear of the vehicle on jackstands.
2. Remove the rear wheels.
3. Remove the control link-to-trailing arm attaching bolts.
4. Support the lower arm with a hydraulic floor jack, then remove the lower arm-to-trailing arm mounting bolt.
5. Remove the rear strut assembly lower mounting bolt from the lower control arm.
6. Scribe a mating mark on the toe-in or camber adjusting bolt, then remove the control link and lower control arm from the vehicle together.

To install:

7. Before reinstalling the lower control arm, check the bushings for wear or deterioration, the control link upper link and lower arm for bending or breakage, and all bolts for straightness or other damage.
8. Position the lower control arm in the vehicle, and install the inboard attaching bolts. Before tightening the bolts, make sure that the toe-in or camber scribe marks are aligned. Temporarily tighten the nuts.
9. Swing the lower control arm up until the lower strut and arm holes are aligned, then install the lower strut-to-lower control arm mounting bolt and nut. Temporarily tighten the nut snugly.
10. Install the lower control arm-to-trailing arm attaching bolt, then temporarily tighten the nut until snug.
11. Install the control link-to-trailing arm bolts. Tighten the bolts to 18 ft. lbs. (25 Nm).
12. Remove the hydraulic floor jack so that the rear suspension is allowed

to hang free, then tighten the various bolts to the following values:
• Lower control arm-to-frame nuts — 51 ft. lbs. (70 Nm)
• Lower control arm-to-trailing arm nut — 65 ft. lbs. (90 Nm)
• Lower control arm-to-lower strut mounting nut — 65 ft. lbs. (90 Nm)
13. Install the rear wheels.
14. Lower the rear of the vehicle.

Summit Wagon and Vista

1. Disconnect negative battery cable.
2. Remove the rear stabilizer bar.
3. If equipped with AWD, remove the rear axle shaft.
4. Remove the rear brake drum.
5. If equipped with ABS, remove the rear caliper assembly and brake disc.
6. Remove the rear hub assembly. If equipped with ABS, take care not to damage the rotor teeth during hub removal.
7. Disconnect the parking brake cable from the rear brake shoe.
8. If equipped with ABS, disconnect and remove the rear wheel sensor.

NOTE: The speed sensor has a pole piece projecting from it. This exposed tip must be protected from impact or damage. Do not allow the pole piece to contact the toothed wheel during removal or installation.

9. Remove the rear shock and coil spring.
10. Remove the brake line and parking brake mounting bolts from the lower control arm.
11. Matchmark and remove the inboard lower arm pivot bolt. Remove the flange bolt and the arm from the vehicle.

To install:

12. Install the arm on the vehicle and secure with the flange bolt, temporarily tighten the nut. Install the arm pivot bolt and temporarily tighten the nut.
13. Install the rear shock and coil spring.
14. Install the brake line and parking brake mounting bolts to the lower control arm.
15. Connect the parking brake cable to the rear brake shoe.
16. Install the rear hub assembly.
17. Install the rear brake drum or, if equipped with ABS, install the rear caliper assembly and brake disc.
18. Install the rear axle shaft.
19. Install and connect the rear wheel speed sensor. Use a brass or other non-magnetic feeler gauge to

check the air gap between the tip of the pole piece and the toothed wheel. Correct gap is 0.012–0.035 in. (0.3–0.9mm). Tighten the two sensor bracket bolts to 10 ft. lbs. (14 Nm) with the sensor located so the gap is the same at several points on the toothed wheel. If the gap is incorrect, it is likely that the toothed wheel is worn or improperly installed.

20. Lower the vehicle and tighten the lower arm flange bolt nut and the arm pivot bolt to 69 ft. lbs. (95 Nm).

21. Install the rear stabilizer bar and reconnect the negative battery cable.

22. Bleed the brake system if any lines where opened. Adjust the parking brake and perform a rear wheel alignment.

Sway Bar

REMOVAL AND INSTALLATION

Summit Wagon and Vista

1. Raise and support the vehicle safely.

2. Remove the self-locking nuts at the sway link. Once the stabilizer bar nut is removed, remove the joint cups and stabilizer rubber bushings.

3. Remove the retainer bolts and the stabilizer bar brackets. Remove the bushing.

4. Remove the stabilizer bar.

5. Inspect the bar for damage, wear and deterioration and replace as required.

To install:

6. Install the stabilizer bar into the vehicle.

7. Install the center stabilizer bar bushings, brackets and bolts. Tighten the bolts to 17 ft. lbs. (23 Nm).

8. Assemble the joint cups and stabilizer rubber to the link. Install a new self-locking nut onto the link. Tighten the self-locking nut so the protrusion of the stabilizer link from

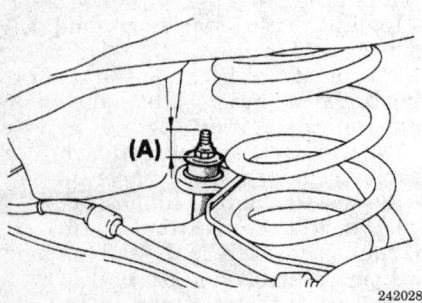

Rear stabilizer link protrusion measurement point — Summit Wagon and Vista models

242028

the top of the joint cup is within 0.98 to 1.06 in. (25 to 27 mm).

9. Lower the vehicle.

Wheel Bearings

ADJUSTMENT

Colt and Summit

NOTE: Never disassemble the rear hub bearing. The wheel bearing is serviced by replacement of the hub.

1. Raise and safely support the vehicle.

2. Remove the rear wheel.

3. Remove the caliper and brake disc or brake drum.

4. Remove the dust cap and torque the flange nut to 130 ft. lbs. (180 Nm).

5. Using a dial indicator, measure wheel bearing end-play. The maximum limit for end-play is 0.0020 inches (0.05mm).

6. Using a spring scale and a rope wrapped around the bolts, measure the rotary sliding resistance of the bearing/hub. The maximum limit for resistance is 4 lbs. (19 N).

7. If any of the readings exceed the specifications, replacement of the hub is required.

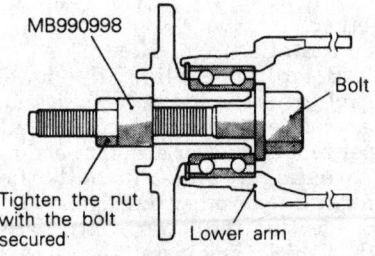

MB990998

Bolt

Tighten the nut with the bolt secured

Lower arm

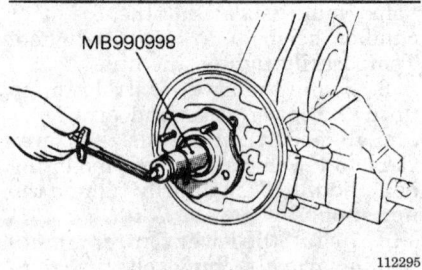

MB990998

112295

Wheel bearing preload adjustment method — Summit Wagon and Vista models

8. Install the dust cap.

9. Install the brake disc and caliper, or brake drum.

10. Install the rear wheel assembly and lower the vehicle to the floor.

Summit Wagon and Vista

The rear wheel bearings on the Vista and Summit Wagon models do not require manual adjustment. If the rear wheel bearings (on FWD models) show damage or excessive looseness, the rear hub must be replaced.

REMOVAL AND INSTALLATION

Colt and Summit

NOTE: Some vehicles may be equipped with a non-serviceable bearing/hub assembly, and are identified as a Type II assembly. If the vehicle has this design, the bearing/hub is serviced as an assembly.

1. Loosen the lug nuts. Raise the vehicle and support it safely.

2. Remove the wheel and tire assemblies.

3. Remove the grease cap.

4. Remove the nut.

5. Pull the drum off. If equipped with disc brakes, remove the caliper assembly, then remove the disc rotor.

  a. The outer bearing will fall out while the drum is coming off. Do not drop it. Remove the hub assembly.

  b. Pry out the oil seal. Discard it.

  c. Remove the inner bearing.

  d. Check the bearing races. If any scoring, heat checking or damage is noted, they should be replaced.

NOTE: When bearing or races need replacement, replace them as a set.

  e. Inspect the bearings. If wear or looseness or heat checking is found, replace them.

  f. If the bearings and races are to be replaced, drive out the race with a brass drift.

To install:

6. Before installing new races, coat them with wheel bearing grease. Drive into place with proper size driver. Make sure they are fully seated.

7. Thoroughly pack the bearings and lubricate the hubs with wheel bearing grease. Install the inner bearing and coat the lip and rim of the grease seal with grease. Drive the seal into place with a seal driver.

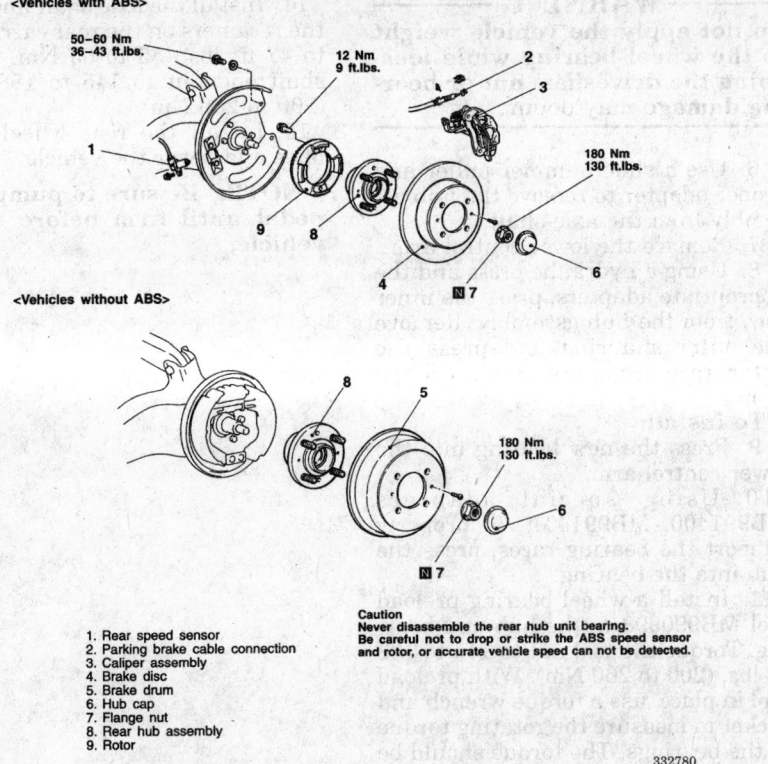

1. Rear speed sensor
2. Parking brake cable connection
3. Caliper assembly
4. Brake disc
5. Brake drum
6. Hub cap
7. Flange nut
8. Rear hub assembly
9. Rotor

Caution
Never disassemble the rear hub unit bearing.
Be careful not to drop or strike the ABS speed sensor and rotor, or accurate vehicle speed can not be detected.

332780

Rear axle hub removal — Colt and Summit models

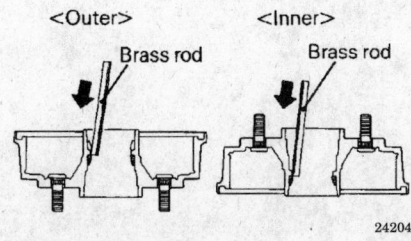

242049

Proper method of removing bearing race — Colt and Summit models

8. To determine if the self-locking nut is reusable:
   a. Screw in the self-locking nut until about 1/10 in. (2.54mm) of the spindle is showing.
   b. Measure the torque required to turn the self-locking nut counterclockwise.
   c. The lowest allowable torque is 48 inch lbs. (5.5 Nm). If the measured torque is less than the specification, replace the nut.
9. Place the drum or rotor on the shaft and install the outer bearing.

10. Torque the self-locking nut to 108–145 ft. lbs. (150–200 Nm).
11. If brake caliper was removed, reinstall.
12. Install the wheel and lower the vehicle.

Summit Wagon and Vista

FWD Vehicles

1. Raise the vehicle and support safely.
2. Remove the tire and wheel assembly.
3. Remove the bolt(s) holding the speed sensor bracket to the knuckle and remove the assembly from the vehicle.

NOTE: The speed sensor has a pole piece projecting from it. This exposed tip must be protected from impact or scratches. Do not allow the pole piece to contact the toothed wheel during removal or installation.

4. Remove the brake drums. If equipped with rear disc brakes, remove the caliper from the brake disc and suspend with a wire. Remove the brake rotor.
5. Remove the grease cap, locking nut and tongued washer.

6. Remove the rear hub and bearing assembly.

NOTE: The rear hub assembly can not be disassembled. If bearing replacement is required, replace the assembly as a unit.

To install:

7. Install the hub and bearing assembly.
8. Install the tongued washer and a new locking nut. Torque the locknut to 166 ft. lbs. (230 Nm). Once the locknut has been properly torqued, crimp the nut flange over the slot in the spindle.
9. Install the grease cap and brake parts.
10. Temporarily install the speed sensor to the knuckle; tighten the bolts only finger-tight.
11. Route the speed sensor cable correctly and loosely install the clips and retainers. All clips must be in their original position and the sensor cable must not be twisted. Improper installation may cause cable damage or system failure.

NOTE: The wiring in the harness is easily damaged by twisting and flexing. Use the white stripe on the outer insulation to keep the sensor harness properly placed.

12. Use a brass or other non-magnetic feeler gauge to check the air gap between the tip of the pole piece and the toothed wheel. Correct gap is 0.012–0.035 in. (0.3–0.9mm). Tighten the 2 sensor bracket bolts to 10 ft. lbs. (14 Nm) with the sensor located so the gap is the same at several points on the toothed wheel. If the gap is incorrect, it is likely that the toothed wheel is worn or improperly installed.
13. Install the tire and wheel assembly. Be sure to pump brake pedal until firm before moving vehicle.

AWD Vehicles

1. Raise and safely support the rear of the vehicle, with the suspension hanging free.
2. Remove the rear wheels.
3. Remove the brake drums. If equipped with rear disc brakes, remove the caliper and rotor assemblies.
4. Remove the bolts that attach the rear halfshaft to the rear carrier.
5. Remove the cotter pin, driveshaft nut cover and nut from the rear driveshaft.

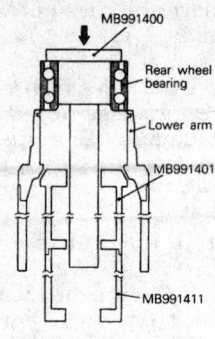

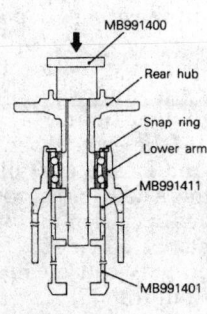

Proper method of assembling bearing and hub — Summit Wagon and Vista models

---

———WARNING———
Do not apply the vehicle weight to the wheel bearing while loosening the driveshaft nut or bearing damage may occur.

6. Use a slide hammer puller and proper adapter to remove the hub assembly from the axle-shaft.

7. Remove the lower control arm.

8. Using a hydraulic press and the appropriate adapters, press the inner race from the hub assembly. Remove the outer snapring and press the outer race from the lower control arm.

To install:

9. Press the new bearing into the lower control arm.

10. Using special adapters MB991400, MB991401 to properly support the bearing races, press the hub into the bearing.

11. Install a wheel bearing preload tool MB990998 to the hub and bearing. Torque the tool nut to 145 to 188 ft. lbs. (200 to 260 Nm). With preload tool in place, use a torque wrench and socket to measure the rotating torque of the bearings. The torque should be 9 inch lbs. (1.1 Nm) or less.

12. Install the lower control arm.

13. Install the rear brake components.

14. Install the axle shaft and torque the retainers on the rear carrier to 40 to 47 ft. lbs. (55 to 65 Nm) and the shaft end nut to 145 to 188 ft. lbs. (200 to 260 Nm).

15. Install the rear wheel assemblies and lower the vehicle.

NOTE: Be sure to pump brake pedal until firm before moving vehicle.

HONDA 5

Accord • Civic • del Sol • Prelude

FIRING ORDERS

NOTE: To avoid confusion, always replace spark plug wires one at a time.

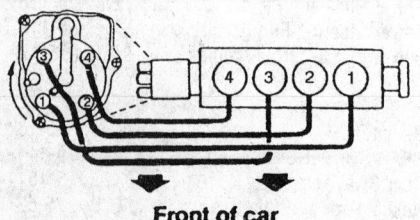

1.5L, 1.6L, 2.2L (F22B1 only) Engine
Engine Firing Order: 1–3–4–2
Distributor Rotation Clockwise

308855

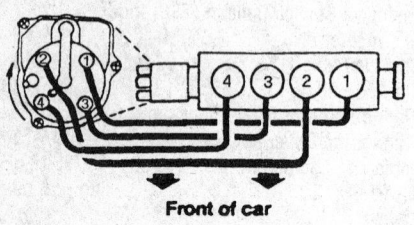

2.3L, 2.2L (Except F22B1) Engine
Engine Firing Order:1–3–4–2
Distributor Rotation: Clockwise

302173

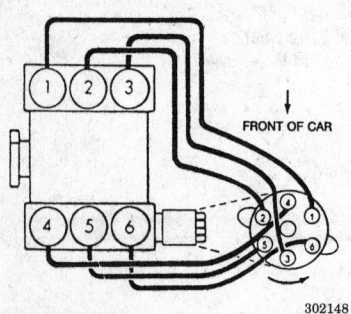

2.7L Engine
Engine Firing Order: 1–4–2–5–3–6
Distributor Rotation: Counterclockwise

302148

ENGINE ELECTRICAL

NOTE: Disconnecting the negative battery cable on some vehicles may interfere with the functions of the on board computer systems and may require the computer to undergo a relearning process, once the negative battery cable is reconnected.

Distributor

REMOVAL AND INSTALLATION

NOTE: The radio may contain a coded theft protection circuit. Always obtain the code number before disconnecting the battery. If the vehicle is equipped with 4WS, the steering control unit is shut down when the battery is disconnected. After connecting the battery, turn the steering wheel lock–to–lock to reset the steering control unit.

1. Disconnect the negative battery cable.
2. Rotate the crankshaft to bring No. 1 cylinder to TDC, and align the white mark on the crankshaft pulley with the pointer on the timing belt cover.
3. Remove the distributor cap with the ignition wires attached.
4. Disconnect the electrical connectors from the distributor.
5. Mark the direction the ignition rotor is pointing on the distributor housing to aid in installation.
6. Match mark the distributor housing with the cylinder head to aid in installation.
7. Remove the three distributor mounting bolts and remove the distributor.
8. Remove and discard the O–ring from the distributor housing.
To install:
9. Coat a new O–ring with clean engine oil and install it to the distributor housing.
10. Align the ignition rotor with the mark made on the distributor housing.The drive lugs are off-set so the distributor cannot be installed incorrectly. Fit the distributor into place and turn the rotor until the drive lugs engage and the distributor seats in the cylinder head.

NOTE: The lugs on the end of the distributor and their mating grooves in the camshaft end, are offset to eliminate the possibility of installing the distributor 180° out of time.

11. Align the matchmark on the distributor housing and the cylinder head and install the mounting bolts snugly.
12. Install the distributor cap with the ignition wires.
13. Connect the distributor electrical connectors.
14. Connect the negative battery cable and enter the radio security code.
15. If equipped with 4WS, start the engine and turn the steering wheel lock–to–lock to reset the 4WS control unit.
16. Adjust the ignition timing.
17. Torque the distributor mounting bolts to 16 ft. lbs. (22 Nm), except on 1996–97 Civics. On 1996–97 Civics, tighten to 13 ft. lbs. (18 Nm).

Ignition Timing

ADJUSTMENT

1.5L and 1.6L Engines

1. Set the parking brake and block the front wheels.
2. Connect a timing light to the No. 1 spark plug wire.
3. Start the engine and allow it to warm up.
4. Pull out the service check connector located behind the right kick panel. On the 2–P connector, connect the WHT/BGN or BRN and BLK terminals with service connector 07PAZ–0010100, or equivalent. Don't connect a jumper wire to the 3–P data link connector.
5. Shift the transaxle to neutral. All electrical accessories must be off. If equipped with Daytime Running Lights (DRL's), turn them off by engaging the parking brake lever.
6. Connect a test tachometer to the test tachometer connector located on the left shock tower. Check the idle speed.
7. While the engine idles, point the timing light at the mark on the timing belt cover.
8. Timing specifications: D15Z1 engine
• M/T: 16° BTDC at 600 rpm (USA) and 700 rpm (Canada)
9. Timing specifications: D15B8 engine
• M/T: 12° BTDC at 650 rpm (USA) and 750 rpm (Canada)

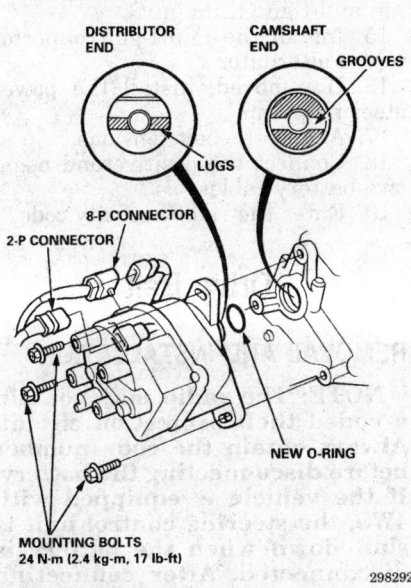

Distributor components — 1993–95 Civic and del Sol

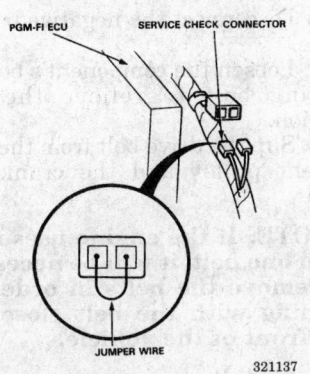

Service check connector — 1.6L engines

10. Timing specifications: D15B7 and D16Z6 engines
- M/T: 16° BTDC at 650 rpm (USA) and 750 rpm (Canada)
- A/T: 16° BTDC at 700 rpm (USA) and 750 (Canada)

11. Timing specifications: B16A2 and B16A3 engines
- M/T: 16° BTDC at 700±50 rpm (USA) and 750±50 rpm (Canada)

12. Timing specifications: D16Y5 engine
- M/T: 12±2° BTDC at 670±50 rpm (USA only)
- A/T and CVT: 12±2° BTDC at 700±50 rpm (USA only)

13. Timing specifications: D16Y7 and D16Y8 engines
- M/T: 12±2° BTDC at 670±50 rpm (USA) or 750±50 rpm (Canada)
- A/T: 12±2° BTDC at 700±50 rpm (USA) or 750±50 rpm (Canada)

14. If adjustment is needed, loosen the distributor adjusting bolts and turn the distributor counterclockwise to advance the timing or clockwise to retard the timing.

15. Tighten the distributor adjusting bolts to 17 ft. lbs. (24 Nm) and recheck the timing and the idle.

16. After everything has been rechecked, remove the service connector from the service check connector. Tuck the service check connector back behind the kick panel.

2.7L Engines

The ignition timing is only adjustable by the PCM, but the ignition base timing can be checked by performing the following:

1. Connect a timing light to the number 1 spark plug wire.

NOTE: Set the parking brake and block the front wheels.

2. Start the engine and allow it to warm up.
3. Pull out the service check connector located behind the glove box. Connect the GRN/BLU and RED terminals with the SCS service connector.
4. While the engine idles, point the timing light toward the pointer on the timing belt cover.
5. Timing specifications:
- 15°± 2° BTDC (Red mark on the crankshaft pulley) at 700 ± 50 rpm in neutral
6. After everything has been rechecked, remove the SCS service connector from the check connector.

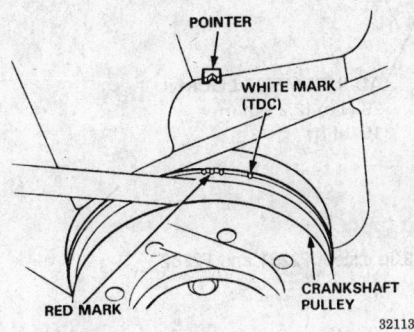

Crankshaft pulley timing mark locations

Alternator

PRECAUTIONS

Several precautions must be observed with alternator equipped vehicles to avoid damage to the unit.

- If the battery is removed for any reason, make sure it is reconnected with the correct polarity. Reversing the battery connections may result in damage to the 1-way rectifiers.
- When utilizing a booster battery as a starting aid, always connect the positive to positive terminals and the negative terminal from the booster battery to a good engine ground on the vehicle being started.
- Never use a fast charger as a booster to start vehicles.
- Disconnect the battery cables when charging the battery with a fast charger.
- Never attempt to polarize the alternator.
- Do not use test lights of more than 12 volts when checking diode continuity.
- Do not short across or ground any of the alternator terminals.
- The polarity of the battery, alternator and regulator must be matched and considered before making any electrical connections within the system.
- Never separate the alternator on an open circuit. Make sure all connections within the circuit are clean and tight.
- Disconnect the battery ground terminal when performing any service on electrical components.
- Disconnect the battery if arc welding is to be done on the vehicle.

REMOVAL AND INSTALLATION

NOTE: The radio may contain a coded anti–theft circuit. Obtain the security code number before disconnecting the battery cables.

NOTE: The voltage regulator is an internal part of all of the Nissan alternators in this manual. If it is faulty, the alternator must be replaced or rebuilt.

All 1.5L and 1.6L Engines

1. Disconnect the negative and the positive battery cables.
2. Disconnect the engine harness connector from the alternator. Remove the nut to disconnect the white wire from the B terminal.
3. Loosen the alternator adjusting bolt and through-bolt. Slip the belt off

the alternative pulley — 1.5L and 1.6L engines

4. Remove the alternator adjusting bolt and through-bolt. Remove the alternator. If necessary, the mounting bracket bolts and the upper and lower mounting brackets may be removed.

To install:

5. Install the alternator into its mounting brackets. Slip the belt over the alternator pulley.

6. Adjust the alternator belt tension. Tighten the alternator through-bolt nut to 33 ft. lbs. (45 Nm). Tighten the alternator adjusting bolt to 17 ft. lbs. (24 Nm). If the upper mount bracket was loosened, tighten its bolt to 33 ft. lbs. (45 Nm).

7. Reconnect the positive and negative battery cables.

2.2L, 2.3L and 2.7L Engines

1. With the ignition **OFF**, disconnect the negative battery cable, and then the positive battery cable.

2. If necessary, remove the power steering pump.

3. Disconnect the multi-pin electrical connector from the alternator.

4. Remove the terminal nut and remove the wire from the B terminal.

5. Loosen the through-bolt, then loosen the adjustment locknut, and then the adjusting bolt.

6. Remove the belt from the alternator pulley.

7. Remove the adjustment bolt and nut.

8. Support the alternator. Remove the through-bolt and remove the alternator.

To install:

9. If the alternator brackets were removed, reinstall them. Coat the bolts with a thread sealer or liquid thread lock. Install the bracket bolts and tighten them to 36 ft. lbs. (50 Nm).

10. Fit the alternator into place. Install the through-bolt but do not tighten the bolt at this time.

11. Install the adjustment bolt and locknut then the adjusting bolt. Make certain the adjusting bolt is properly installed into the adjustment through-bolt.

12. Install the belt and adjust the belt tension. Tighten the alternator adjustment locknut to 16 ft. lbs. (22 Nm).

13. Tighten the through-bolt to 33 ft. lbs. (44 Nm).

14. Connect the wire to the B terminal and tighten the nut.

15. Install the multi-pin connector to the alternator.

16. If removed, install the power steering pump.

17. Adjust the belt tensions.

18. Connect the positive and negative battery cables.

19. Enter the radio security code.

Drive Belt

REMOVAL AND INSTALLATION

NOTE: The radio may contain a coded theft protection circuit. Always obtain the code number before disconnecting the battery. If the vehicle is equipped with 4WS, the steering control unit is shut down when the battery is disconnected. After connecting the battery, turn the steering wheel lock–to–lock to reset the steering control unit.

1.5L (D15B7, D15B8 and D15Z1), 1.6L (D16Z6 and B16A3) Engines

1. Disconnect the negative battery cable.

2. Loosen the component's belt adjusting bolt to relieve the belt tension.

3. Slip the drive belt from the component pulley and the crankshaft pulley.

NOTE: If the engine uses more than one belt, it will be necessary to remove the belts in order beginning with the belt closest to the front of the vehicle.

To install:

4. If there are cracks or damage on the belt, replace it with a new one.

5. Install the drive belt onto the crankshaft pulley and then onto the accessory drive pulley.

6. Move the component into position and tighten the mounting bolts.

7. Tighten the component's belt adjusting bolt to bring the belt to the specified tension or deflection.

8. After the drivebelt is tensioned, tighten the mounting bolts to 33 ft. lbs. (44 Nm). Tighten the adjusting bolts to 16–17 ft. lbs. (24–26 Nm).

9. Reconnect the negative battery cable. Run the engine and check for proper operation of the belt-driven components.

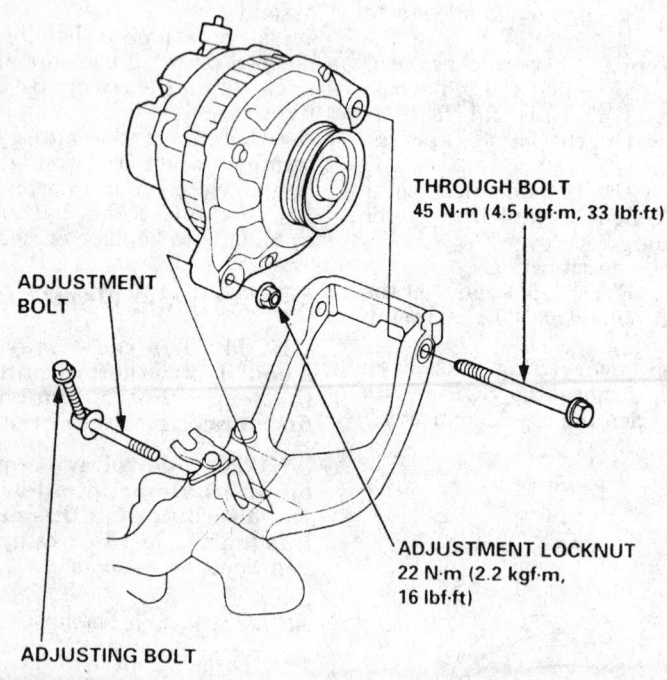

ADJUSTMENT BOLT

THROUGH BOLT
45 N·m (4.5 kgf·m, 33 lbf·ft)

ADJUSTMENT LOCKNUT
22 N·m (2.2 kgf·m, 16 lbf·ft)

ADJUSTING BOLT

296228

Alternator mounting bolts — all 2.2L and 2.3L except F22B1 and F22B2 engines

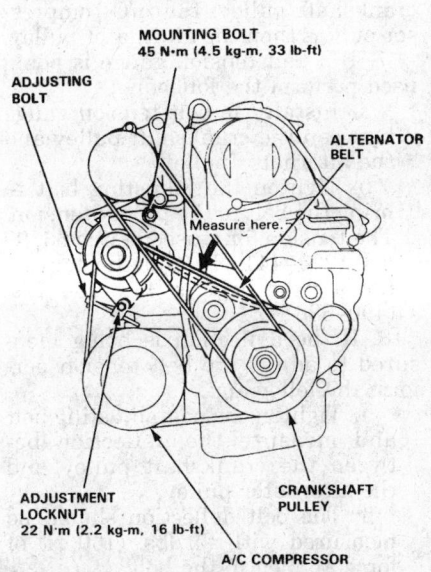

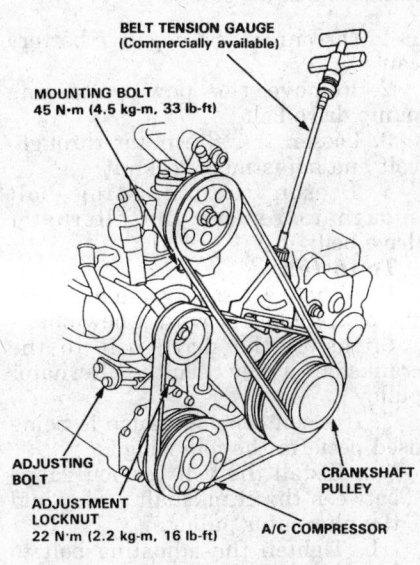

Measuring the alternator belt tension by checking deflection and using a belt gauge

I.6L (D16Y5, D16Y7, D16Y8 and B16A2) Engines

———— **WARNING** ————

When using a new belt, first adjust the deflection or tension to the values for a new belt, then readjust the deflection or tension to the values for the used belt after running the engine for five minutes. If this procedure is not followed component damage and possibly personal injury may occur.

Power Steering Belt

1. Disconnect the negative battery cable.
2. Loosen the power steering pump mounting bolts to remove the drive belt.
To install:
3. If there are cracks or damage on the belt, replace it with a new one.
4. Install the drive belt to the crankshaft and power steering pump.
5. If equipped with a square leverage point on the pump bracket, use a 1/2 inch breaker bar to lever the pump up or down to achieve the correct tension. Or, tighten the pump belt adjusting bolt.

6. Adjust the belt to the proper tension:
  a. Deflection for a new belt is 0.30–0.39 inches (7.5–10mm) and a used belt is 0.41–0.55 inches (10.5–14.0mm). For the B16A2 engine; new belt deflection is 0.20–0.28 inches (5.0–7.0mm), used belt deflection is 0.30–0.43 inches (7.5–11.0mm)
  b. Tension gauge is used the tension for a new belt is 143–176 lbs. (640–780 N), and the tension for a used belt is 77–110 lbs. (340–490 N). For the B16A2 engine; new belt tension is 170–200 lbs. (740–880 N), used belt tension is 88–120 lbs. (390–540 N).
7. Tighten the mounting bolts to 17 ft. lbs. (24 Nm). Remove the breaker bar.
8. Connect the negative battery cable and enter the radio security code.

Air Conditioning Belt

1. Disconnect the negative battery cable.
2. Remove the power steering belt.
3. Loosen the idler pulley adjusting bolt.
4. Loosen the idler pulley center nut.
5. Remove the two mounting bolts from the left front engine mount. Remove the air conditioning belt through the gap between the body and the left front engine mount.
To install:
6. If there are cracks or damage on the belt, replace it with a new one.
7. Install the belt to the crankshaft and the air conditioning compressor.
8. Install the two bolts to the left front engine mount. Torque the bolts to 33 ft. lbs. (44 Nm).
9. The air conditioning belt deflection and tension is measured between the crankshaft and the air conditioning compressor pulley.
  a. Deflection for a new belt is 0.2–0.3 inches (5.0–6.5mm). Used belt deflection, is 0.3–0.4 inch (7.5–9.5mm). For the B16A2 engine; new belt deflection is 0.2–0.4 inches (6.0–9.5mm), and used belt deflection is 0.18–0.26 inches (4.5–6.5mm).
  b. Tension for a new belt is 150–190 lbs. (690–830 N). The tension for a used belt is 77–110 lbs. (340–490 N). For the B16A2 engine; new belt tension is 170–200 lbs. (740–880 N), and used belt tension is 88–120 lbs. (390–540 N).
10. Tighten the idler pulley adjusting bolt.
11. Tighten the idler pulley center nut.
12. Install the power steering belt.
13. Connect the negative battery cable and enter the radio security code.

Alternator Belt

1. Disconnect the negative battery cable.
2. Remove the power steering then the air conditioning belts.
3. Loosen the alternator lower mounting nut and upper mounting bolt.
4. Remove the alternator belt.
To install:
5. If there are cracks or damage on the belt, replace it with a new one.
6. Install the alternator belt to the crankshaft pulley, then the alternator pulley.
7. The alternator belt deflection and tension is measured between the crankshaft and the alternator pulley.
  a. Deflection for a new belt is 0.26–0.33 inches (6.0–8.5mm) and a used belt is 0.31–0.41 inch (8.0–10.5mm). For the B16A2 engine; new belt deflection is 0.20–0.28 inches (5.0–7.0mm), used belt tension is 0.28–0.41 inch (5.0–7.0mm).
  b. Tension for a new belt is 121–165 lbs. (540–740 N), and the tension for a used belt for all en-

gines is 77–110 lbs. (340–490 N). For the B16A2 engine, new belt tension is 144–199 lbs. (70–90 N).

8. Move the alternator to reach the correct belt tension.

9. Tighten the mounting nut to 17 ft. lbs. (24 Nm).

10. Tighten the mounting nut to 33 ft. lbs. (44 Nm).

11. Install the air conditioning and power steering belts.

12. Connect the negative battery cable and enter the radio security code.

2.2L (H22A1 and F22A1) and 2.3L (H23A1) Engines

Power Steering Belt

1. Disconnect the negative battery cable.

2. Loosen the power steering pump mounting bolt and nut.

3. Loosen the power steering pump adjusting bolt enough to remove the drive belt and remove the belt.

To install:

4. If there are cracks or damage on the belt, replace it with a new one.

5. Install the belt to the crankshaft and the power steering pump pulley.

6. If a belt tension gauge is being used perform the following:

a. Install the belt tension gauge between the crankshaft pulley and the power steering pump pulley.

b. Tighten the adjusting bolt to bring the belt to the proper tension. The tension for a used belt is 77–110 lbs. (350–500 N), and the tension for a new belt is 154–198 lbs. (700–900 N).

7. If the deflection is being measured to adjust the belt tension perform the following:

a. Tighten the belt adjusting bolt and measure the deflection between the crankshaft pulley and the power steering pump pulley.

b. The belt deflection should be measured with 22 lbs. (98 N) of force applied to the belt.

c. The belt deflection should be 0.53–0.65 inches (13.5–16.5mm) for a used belt and 0.37–0.45 inches (9.5–11.5mm) for a new belt.

8. Torque the mounting nuts and bolt to 16 ft. lbs. (22 Nm) and the adjusting bolt locknut to 11 ft. lbs. (15 Nm).

9. Recheck the belt tension.

10. Connect the negative battery cable and enter the radio security code.

11. Start the engine and turn the steering lock–to–lock several times,

then stop the engine and recheck the belt tension.

Alternator Belt without A/C

1. Disconnect the negative battery cable.

2. Remove the power steering pump drive belt.

3. Loosen the alternator through-bolt and adjustment locknut.

4. Loosen the adjusting bolt enough to remove the alternator drive belt.

To install:

5. If there are cracks or damage on the belt, replace it with a new one.

6. Install the drive belt to the crankshaft pulley then the alternator pulley.

7. If a belt tension gauge is being used perform the following:

a. Install the belt tension gauge between the crankshaft pulley and the alternator pulley.

b. Tighten the adjusting bolt to bring the belt to the proper tension. The tension for a used belt is 66–99 lbs. (300–450 N), and the tension for a new belt is 99–143 lbs. (450–650 N).

8. If the deflection is being measured to adjust the belt tension perform the following:

a. Tighten the belt adjusting bolt and measure the deflection between the crankshaft pulley and the alternator pulley.

b. The belt deflection should be measured with 22 lbs. (100 N) of force applied to the belt.

c. The belt deflection should be 0.42–0.51 inches (10.5–12.5mm) for a used belt and 0.32–0.40 inches (8–11mm) for a new belt.

9. Torque the adjustment locknut to 16 ft. lbs. (22 Nm) and the through-bolt to 33 ft. lbs. (45 Nm).

10. Recheck the belt tension.

11. Install power steering belt.

12. Connect the positive then the negative battery cable and enter the radio security code.

13. If equipped with 4WS, turn the steering wheel lock–to–lock to reset the 4WS control unit.

Alternator Belt with A/C

1. Disconnect the negative battery cable.

2. Remove the power steering pump drive belt.

3. Loosen the alternator through-bolt and adjustment locknut.

4. Loosen the adjusting bolt enough to remove the drive belt from the alternator and the A/C compressor.

To install:

5. If there are cracks or damage on the belt, replace it with a new one.

6. Install the drive belt to the crankshaft pulley, the A/C compressor pulley, then the alternator pulley.

7. If a belt tension gauge is being used perform the following:

a. Install the belt tension gauge between the crankshaft pulley and the alternator pulley.

b. Tighten the adjusting bolt to bring the belt to the proper tension. The tension for a used belt is 66–99 lbs. (294–441 N), and the tension for a new belt is 110–154 lbs. (490–686 N).

8. If the deflection is being measured to adjust the belt tension perform the following:

a. Tighten the belt adjusting bolt and measure the deflection between the crankshaft pulley and the alternator pulley.

b. The belt deflection should be measured with 22 lbs. (100 N) of force applied to the belt.

c. The belt deflection should be 0.39–0.47 inches (10–12mm) for a used belt and 0.33–0.43 inches (8.5–11mm) for a new belt.

9. Torque the adjustment locknut to 16 ft. lbs. (22 Nm) and the through-bolt to 33 ft. lbs. (45 Nm).

10. Recheck the belt tension.

11. Install power steering belt.

12. Connect the positive then the negative battery cable and enter the radio security code.

13. If equipped with 4WS, turn the steering wheel lock–to–lock to reset the 4WS control unit.

2.2L (F22A6) Engine

Power Steering Belt

1. Disconnect the negative battery cable.

2. Loosen the power steering pumps mounting bolt and nut.

3. Loosen the power steering pump adjusting bolt enough to remove the drive belt and remove the belt.

To install:

4. If there are cracks or damage on the belt, replace it with a new one.

5. Install the belt to the crankshaft and the power steering pump pulley.

6. If a belt tension gauge is being used perform the following:

a. Install the belt tension gauge between the crankshaft pulley and the power steering pump pulley.

b. Tighten the adjusting bolt to bring the belt to the proper tension. The tension for a used belt is 77–110 lbs. (350–500 N), and the tension for a new belt is 110–154 lbs. (500–700 N).

7. If the deflection is being measured to adjust the belt tension perform the following:

　a. Tighten the belt adjusting bolt and measure the deflection between the crankshaft pulley and the power steering pump pulley.

　b. The belt deflection should be measured with 22 lbs. (100 N) of force applied to the belt.

　c. The belt deflection should be 12.5–16mm for a used belt and 9.5–11.5mm for a new belt.

8. Torque the mounting bolt to 33 ft. lbs. (45 Nm) and the nut to 16 ft. lbs. (22 Nm).

9. Recheck the belt tension.

10. Connect the negative battery cable and enter the radio security code.

11. Start the engine and turn the steering lock–to–lock several times, then stop the engine and recheck the belt tension.

Alternator Belt without A/C

1. Disconnect the negative battery cable.

2. Remove the power steering pump drive belt.

3. Loosen the alternator mounting bolt and adjustment locknut.

4. Loosen the adjusting bolt enough to remove the alternator drive belt.

To install:

5. If there are cracks or damage on the belt, replace it with a new one.

6. Install the drive belt to the crankshaft pulley then the alternator pulley.

7. If a belt tension gauge is being used perform the following:

　a. Install the belt tension gauge between the crankshaft pulley and the alternator pulley.

　b. Tighten the adjusting bolt to bring the belt to the proper tension. The tension for a used belt is 66–99 lbs. (300–450 N), and the tension for a new belt is 99–143 lbs. (450–650 N).

8. If the deflection is being measured to adjust the belt tension perform the following:

　a. Tighten the belt adjusting bolt and measure the deflection between the crankshaft pulley and the alternator pulley.

　b. The belt deflection should be measured with 22 lbs. (100 N) of force applied to the belt.

　c. The belt deflection should be 0.39–0.47 inches (10–12mm) for a used belt and 0.33–0.43 inches (8.5–11mm) for a new belt.

9. Torque the adjustment locknut to 16 ft. lbs. (22 Nm) and the mounting bolt to 33 ft. lbs. (45 Nm).

10. Recheck the belt tension.

11. Install power steering belt.

12. Connect the negative battery cable.

Alternator Belt with A/C

1. Disconnect the negative battery cable.

2. Remove the power steering pump drive belt.

3. Loosen the alternator mounting bolt and adjustment locknut.

4. Loosen the adjusting bolt enough to remove the drive belt from the alternator and the A/C compressor.

To install:

5. If there are cracks or damage on the belt, replace it with a new one.

6. Install the drive belt to the crankshaft pulley, the A/C compressor pulley, then the alternator pulley.

7. If a belt tension gauge is being used perform the following:

　a. Install the belt tension gauge between the crankshaft pulley and the alternator pulley.

　b. Tighten the adjusting bolt to bring the belt to the proper tension. The tension for a used belt is 99–132 lbs. (450–600 N), and the tension for a new belt is 209–253 lbs. (950–1150 N).

8. If the deflection is being measured to adjust the belt tension perform the following:

　a. Tighten the belt adjusting bolt and measure the deflection between the crankshaft pulley and the alternator pulley.

　b. The belt deflection should be measured with 22 lbs. (100 N) of force applied to the belt.

　c. The belt deflection should be 0.39–0.47 inches (10–12mm) for a used belt and 0.18–0.28 inches (4.5–7mm) for a new belt.

9. Torque the adjustment locknut to 16 ft. lbs. (22 Nm) and the mounting bolt to 33 ft. lbs. (45 Nm).

10. Recheck the belt tension.

11. Install power steering belt.

12. Connect the negative battery cable.

2.2L (F22B1 and F22B2) Engines

Power Steering Belt

1. Disconnect the negative battery cable.

2. Loosen the power steering pumps mounting bolt and nut.

3. Loosen the power steering pump adjusting bolt enough to remove the drive belt and remove the belt.

To install:

4. If there are cracks or damage on the belt, replace it with a new one.

5. Install the belt to the crankshaft and the power steering pump pulley.

6. If a belt tension gauge is being used perform the following:

　a. Install the belt tension gauge between the crankshaft pulley and the power steering pump pulley.

　b. Tighten the adjusting bolt to bring the belt to the proper tension. The tension for a used belt is 88–120 lbs. (390–540 N), and the tension for a new belt is 170–200 lbs. (740–880 N).

NOTE: A new belt is a belt that has been run less that five minutes.

7. If the deflection is being measured to adjust the belt tension perform the following:

　a. Tighten the belt adjusting bolt and measure the deflection between the crankshaft pulley and the power steering pump pulley.

　b. The belt deflection should be measured with 22 lbs. (98 N) of force applied to the belt.

　c. The belt deflection should be 0.51–0.63 inches (13–16mm) for a used belt and 0.43–0.49 inches (11–12.5mm) for a new belt.

8. Torque the mounting nuts to 17 ft. lbs. (24 Nm).

9. Recheck the belt tension.

10. Connect the negative battery cable and enter the radio security code.

11. Start the engine and turn the steering lock–to–lock several times, then stop the engine and recheck the belt tension.

Alternator Belt without A/C

1. Disconnect the negative battery cable.

2. Remove the power steering pump drive belt.

3. Loosen the alternator mounting bolt and adjustment locknut.

4. Loosen the adjusting bolt enough to remove the alternator drive belt.

To install:

5. If there are cracks or damage on the belt, replace it with a new one.

6. Install the drive belt to the crankshaft pulley then the alternator pulley.

7. If a belt tension gauge is being used perform the following:

　a. Install the belt tension gauge between the crankshaft pulley and the alternator pulley.

　b. Tighten the adjusting bolt to bring the belt to the proper tension.

The tension for a used belt is 66–99 lbs. (294–441 N), and the tension for a new belt is 121–165 lbs. (540–535 N).

8. If the deflection is being measured to adjust the belt tension perform the following:

a. Tighten the belt adjusting bolt and measure the deflection between the crankshaft pulley and the alternator pulley.

b. The belt deflection should be measured with 22 lbs. (100 N) of force applied to the belt.

c. The belt deflection should be 0.41–0.49 inches (10.5–12.5mm) for a used belt and 0.32–0.39 inches (8–10mm) for a new belt.

9. Torque the adjustment locknut to 16 ft. lbs. (22 Nm) and the mounting bolt to 33 ft. lbs. (44 Nm).

10. Recheck the belt tension.

11. Install power steering belt.

12. Connect the negative battery cable.

Alternator Belt with A/C

1. Disconnect the negative battery cable.

2. Remove the power steering pump drive belt.

3. Loosen the alternator mounting bolt and adjustment locknut.

4. Loosen the adjusting bolt enough to remove the drive belt from the alternator and the A/C compressor.

To install:

5. If there are cracks or damage on the belt, replace it with a new one.

6. Install the drive belt to the crankshaft pulley, the A/C compressor pulley, then the alternator pulley.

7. If a belt tension gauge is being used perform the following:

a. Install the belt tension gauge between the crankshaft pulley and the alternator pulley.

b. Tighten the adjusting bolt to bring the belt to the proper tension. The tension for a used belt is 99–132 lbs. (441–588 N), and the tension for a new belt is 209–254 lbs. (932–1123 N).

8. If the deflection is being measured to adjust the belt tension perform the following:

a. Tighten the belt adjusting bolt and measure the deflection between the crankshaft pulley and the alternator pulley.

b. The belt deflection should be measured with 22 lbs. (100 N) of force applied to the belt.

c. The belt deflection should be 0.32–0.41 inches (8–10.5mm) for a used belt and 0.20–0.28 inches (5–7mm) for a new belt.

9. Torque the adjustment locknut to 16 ft. lbs. (22 Nm) and the mounting bolt to 33 ft. lbs. (44 Nm).

10. Recheck the belt tension.

11. Install power steering belt.

12. Connect the negative battery cable.

2.7L Engines

Air Conditioning Belt

1. Disconnect the negative battery cable.

2. Loosen the adjusting pulley nut.

3. Loosen the adjusting bolt enough to remove the drive belt.

To install:

4. If there are cracks or damage on the belt, replace it with a new one.

5. Install the drive belt to the crankshaft pulley, the A/C compressor pulley, then the adjusting pulley.

6. If a belt tension gauge is being used perform the following:

a. Install the belt tension gauge between the crankshaft pulley and the adjusting pulley.

b. Tighten the adjusting bolt to bring the belt to the proper tension. The tension for a used belt is 110–143 lbs. (490–640 N), and the tension for a new belt is 209–254 lbs. (930–1130 N).

7. If the deflection is being measured to adjust the belt tension perform the following:

a. Tighten the belt adjusting bolt and measure the deflection between the crankshaft pulley and the adjusting pulley.

b. The belt deflection should be measured with 22 lbs. (98 N) of force applied to the belt.

c. The belt deflection should be 0.26–0.31 inches (6.5–8mm) for a used belt and 0.16–0.22 inches (4–5.5mm) for a new belt.

8. Torque the adjustment pulley nut to 33 ft. lbs. (44 Nm).

9. Recheck the belt tension.

10. Connect the negative battery cable and enter the radio security code.

Alternator belt

1. Disconnect the negative battery cable.

2. Remove the A/C compressor drive belt.

3. Loosen the alternator through-bolt and adjustment locknut.

4. Loosen the adjusting bolt enough to remove the alternator drive belt.

To install:

5. If there are cracks or damage on the belt, replace it with a new one.

6. Install the drive belt to the crankshaft pulley then the alternator pulley.

7. If a belt tension gauge is being used perform the following:

a. Install the belt tension gauge between the crankshaft pulley and the alternator pulley.

b. Tighten the adjusting bolt to bring the belt to the proper tension. The tension for a used belt is 77–110 lbs. (345–490 N), and the tension for a new belt is 154–200 lbs. (685–880 N).

8. If the deflection is being measured to adjust the belt tension perform the following:

a. Tighten the belt adjusting bolt and measure the deflection between the crankshaft pulley and the alternator pulley.

b. The belt deflection should be measured with 22 lbs. (100 N) of force applied to the belt.

c. The belt deflection should be 0.55–0.65 inches (14–16.5mm) for a used belt and 0.35–0.45 inches (9–11.5mm) for a new belt.

9. Torque the adjustment locknut to 16 ft. lbs. (22 Nm) and the through-bolt to 33 ft. lbs. (44 Nm).

10. Recheck the belt tension.

11. Install A/C compressor belt.

12. Connect the negative battery cable.

Power Steering Belt

1. Disconnect the negative battery cable.

2. Remove the belts from the A/C compressor and the alternator.

3. Loosen the power steering pump mounting nuts.

4. Loosen the power steering pump adjusting nut enough to remove the drive belt and remove the belt.

To install:

5. If there are cracks or damage on the belt, replace it with a new one.

6. Install the belt to the crankshaft and the power steering pump pulley.

7. If a belt tension gauge is being used perform the following:

a. Install the belt tension gauge between the crankshaft pulley and the power steering pump pulley.

b. Tighten the adjusting nut to bring the belt to the proper tension. The tension for a used belt is 88–120 lbs. (390–540 N), and the tension for a new belt is 170–200 lbs. (740–880 N).

8. If the deflection is being measured to adjust the belt tension perform the following:

a. Tighten the belt adjusting nut and measure the deflection between the crankshaft pulley and the power steering pump pulley.

b. The belt deflection should be measured with 22 lbs. (98 N) of force applied to the belt.

c. The belt deflection should be 0.57–0.67 inches (14.5–17mm) for a used belt and 0.41–0.49 inches (105–12.5mm) for a new belt.

9. Torque the mounting nuts to 17 ft. lbs. (24 Nm).

10. Recheck the belt tension.

11. Install and adjust the A/C compressor and the alternator drive belts.

12. Connect the negative battery cable and enter the radio security code.

13. Start the engine and turn the steering lock–to–lock several times, then stop the engine and recheck the power steering belt tension.

Starter

REMOVAL AND INSTALLATION

NOTE: The radio may contain a coded theft protection circuit. Always obtain the code number before disconnecting the battery. If the vehicle is equipped with electronic 4WS, the steering control unit is shut down when the battery is disconnected. After connecting the battery, start the engine and turn the steering wheel lock–to–lock to reset the steering control unit.

1.5L, 1.6L, 2.2L and 2.3L Engines

1. Disconnect the negative and positive battery cables.

2. Certain Honda engines have a bracket on the starter motor housing for the engine wiring harness and upper radiator hose. If equipped with a bracket, unclamp it and move the wiring harness or radiator hose out of the way.

3. Disconnect the starter cable from its terminal.

4. Disconnect the black/white wire from the S terminal on the starter solenoid.

5. Remove the two starter mounting bolts. Remove the starter.

6. Inspect the starter terminals and clean off any corrosion to ensure good electrical contact. Inspect the cables for brittle or cracked insulation and damaged or loose terminals and replace as necessary.

To install:

7. Install the starter. Install the mounting bolts and tighten them to 32 ft. lbs. (44–45 Nm).

8. Reconnect the starter cable with the crimp side facing up and tighten the nut to 7 ft. lbs. (9 Nm). Reconnect the black/white wire.

9. Reconnect the positive and negative battery cables.

10. Test the operation of the starter.

11. On vehicles equipped with electronic 4WS, start the engine and turn the steering wheel lock–to–lock to reset the control unit.

12. If equipped, enter the radio security code.

2.7L Engines

1. Disconnect the negative and positive battery cables.

2. Unclamp the engine wiring harness from its bracket on the starter motor. Move the wiring harness out of the way.

3. Move the upper radiator hose as far out of the way as you can and secure it. If necessary: drain the coolant to a level below the upper radiator hose. Then, disconnect the hose and move it out of the way.

4. Disconnect the starter cable from its terminal.

5. Disconnect the black/white wire from the S terminal on the starter solenoid.

6. Remove the two starter mounting bolts. A long extension may be helpful, since the starter is mounted below the distributor.

7. Remove the starter.

To install:

8. Install the starter and it's mounting bolts, torque the bolts to 47 ft. lbs. (64 Nm).

9. Connect the starter cable and torque the nut to 7 ft. lbs. (9 Nm). Connect the black/white wire.

10. Install the engine wiring harness into its clamp. Fasten the clamp.

11. Move the upper radiator hose back into position. If the hose was disconnected; connect it, refill and bleed the cooling system.

12. Connect the positive and negative battery cables.

13. Test the operation of the starter.

14. Enter the radio security code.

CHASSIS ELECTRICAL

Blower Motor

REMOVAL AND INSTALLATION

— CAUTION —
The air bag main wiring harness is routed near the blower motor housing. The air bag wiring harness is encased in yellow insulation. Be careful not to damage the air bag wiring. If any air bag component must be serviced, the air bag system must first be disabled.

Civic, del Sol, Prelude and 1994–97 Accord

1. Disconnect the negative and positive battery cables. Wait at least three minutes before working around the air bags.

2. Remove the glove box door and glove box frame.

3. Disconnect the blower motor lead.

4. Remove the three self–tapping screws from the bottom of the blower housing. Lower the blower motor out of the housing.

To install:

5. Fit the blower motor into the housing and install the three self–tapping screws. Make sure the edges of the blower fan rotor do not rub the edge of the housing.

6. Reconnect the blower motor lead.

7. Install the glove box frame and glove box door.

8. Reconnect the negative and positive battery cables. Start the vehicle.

9. Test the operation of the blower motor at all of its speed settings.

10. On vehicles equipped with 4WS, turn the steering wheel lock–to–lock to reset the steering control module.

11. Enter the radio security code.

1993 Accord

1. Disconnect the negative and positive battery cables.

2. If equipped, evacuate the air conditioning system using certified R–12 recovery/charging equipment. Carefully disconnect the evaporator suction and receiver lines. Plug the receiver and suction lines to keep out moisture.

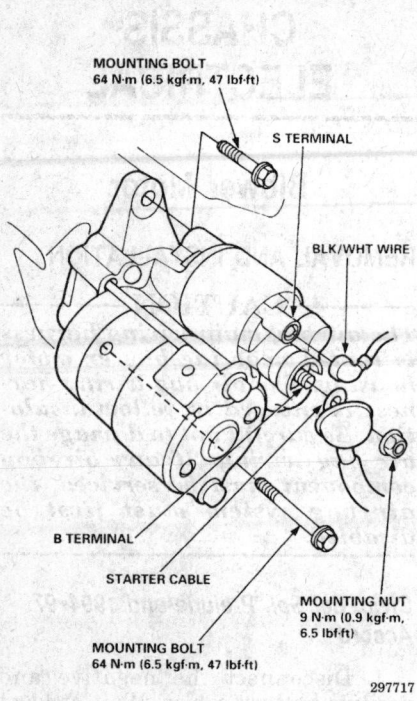

MOUNTING BOLT
64 N·m (6.5 kgf·m, 47 lbf·ft)

S TERMINAL

BLK/WHT WIRE

B TERMINAL

STARTER CABLE

MOUNTING NUT
9 N·m (0.9 kgf·m, 6.5 lbf·ft)

MOUNTING BOLT
64 N·m (6.5 kgf·m, 47 lbf·ft)

297717

Starter components — 2.7L engine

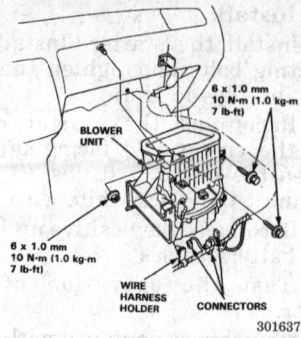

6 x 1.0 mm
10 N·m (1.0 kg·m, 7 lb·ft)

BLOWER UNIT

6 x 1.0 mm
10 N·m (1.0 kg·m, 7 lb·ft)

WIRE HARNESS HOLDER

CONNECTORS

301637

Blower unit components and location in dashboard — Civic and del Sol

— WARNING —
Do not discharge air conditioning refrigerant into the atmosphere. Use only certified recovery/charging equipment and follow all applicable safety precautions when working around refrigerant.

3. Remove the glove box and the glove box frame. Two screws secure the glove box hinges, and the frame is held in place by four screws. If the vehicle is not equipped with A/C, remove the heater duct at this point.

4. If equipped with an air bag, remove the red passenger's air bag

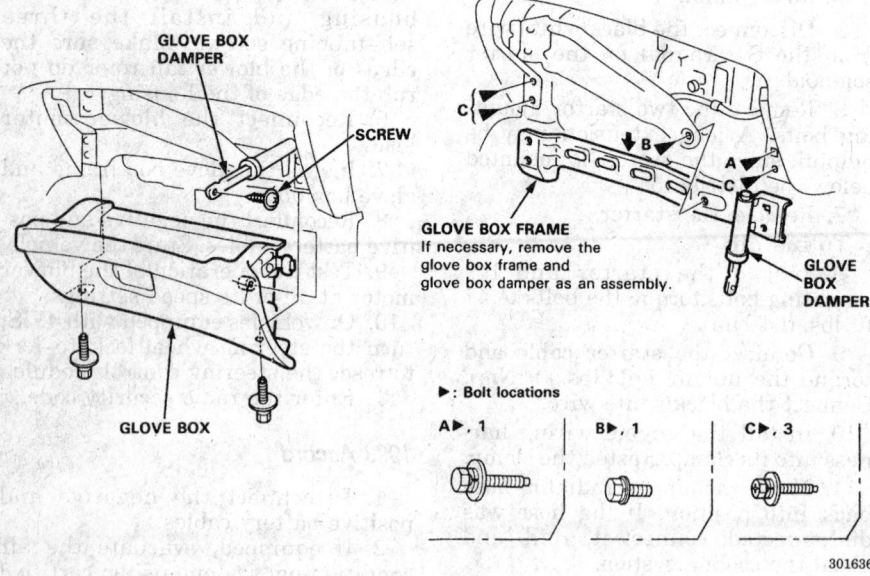

GLOVE BOX DAMPER

SCREW

GLOVE BOX

GLOVE BOX FRAME
If necessary, remove the glove box frame and glove box damper as an assembly.

GLOVE BOX DAMPER

▶: Bolt locations

A▶, 1	B▶, 1	C▶, 3

301636

Glove box — del Sol

shorting connector from its holder. Disconnect the passenger's air bag module 3–P connector from the air bag main harness. Install the shorting connector onto the the air bag module connector.

5. Remove the A/C evaporator brackets which are held in place by two screws.

6. Disconnect the A/C thermostat connector and the drain hose.

7. Remove the evaporator, it is held in place by three nuts.

8. Remove the right kick panel. It is secured by three clips.

9. Pull back the carpet to expose the PCM.

10. Remove the PCM bracket mounting nuts. Disconnect five PCM connectors and remove the PCM and its bracket.

11. Remove the clip from the front of the blower case, then remove the two lower blower case cover pieces.

12. Remove the two retaining bands. Then, remove the two screws and the three blower case mounting nuts.

13. Remove the recirculation control arm cover. Carefully remove the clips and screws; and then, separate the two pieces of the blower case.

14. Unbolt and remove the blower motor from the lower half of the blower case.

To install:
15. Install the blower motor into the lower half of the blower case.

16. Assemble the two halves of the blower case and install the clips and screws. Install the recirculation control arm cover.

17. Install the blower case into the dashboard and tighten the three mounting nuts to 7 ft. lbs. (10 Nm). Install the two screws.

18. Install the retaining bands and the two lower blower case cover pieces.

19. If equipped, install the A/C evaporator, reconnect the thermostat lead and the drain hose. Tighten the mounting nuts to 6 ft. lbs. (10 Nm) and install the retaining bands. If not equipped with A/C install the heater duct.

20. Remove the shorting connector from the passenger's air bag module. Reconnect the passenger's air bag module to the air bag main harness. Place the shorting connector back into its holder. Place the air bag connection back into its clip.

21. Install the glove box frame and glove box.

22. Reconnect the receiver and dryer lines using new O–rings. Use only components designed for use

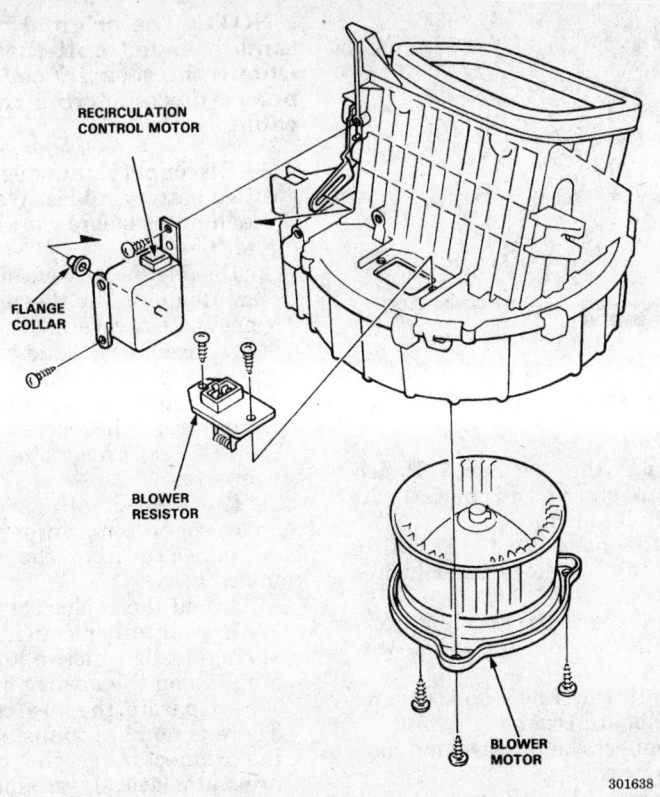

Blower motor components — Civic and del Sol

301638

with R–12 systems. Carefully tighten the receiver line coupling to 10 ft. lbs. (14 Nm), and the suction line coupling to 12 ft. lbs. (16 Nm) for ABS–equipped vehicles, or 24 ft. lbs. (33 Nm) for vehicles without ABS.

23. Evacuate the A/C system to remove any moisture.

24. Leak test the air conditioning system.

25. Recharge the A/C system with refrigerant and the proper refrigerant oil.

26. Reconnect the negative and positive battery cables.

27. Test the performance of the heater and air conditioning systems and the recirculation control. Make sure there are no air leaks.

28. If equipped, enter the radio security code.

Windshield Wiper Motor

REMOVAL AND INSTALLATION

Front

1. Disconnect the negative battery cable.

2. Remove the wiper arm cap nuts. Then, remove the wiper arms.

NOTE: Be careful not to scratch the vehicle's paint, glass, or trim.

3. Remove the hood seal and air scoop by carefully prying out their clips.

4. Disconnect the 5–P wiper motor connector.

5. Unbolt and remove the wiper linkage assembly.

6. Remove the wiper motor linkage nut. Matchmark the wiper linkage to the wiper motor mount.

7. Unbolt the wiper motor and remove it from the linkage assembly.

To install:

8. Install the wiper motor onto the linkage assembly. Tighten the bolts to 6–7 ft. lbs. (8–10 Nm). Connect the linkage to the motor and tighten the nut to 13–16 ft. lbs. (18–22 Nm).

9. Install the linkage assembly into the vehicle and tighten the mounting bolts to 7 ft. lbs. (10 Nm).

10. Reconnect the 5–P wiper motor connector.

11. Install the air scoop and hood seal. Make sure the clips are fully seated.

12. Install the wiper arms and cap nuts.

13. Reconnect the negative battery cable. Check the operation of the wiper motor and arms at all of the wiper switch speed settings.

Rear

1. Disconnect the negative battery cable.

2. Remove the nut cover and retaining nut from the rear window wiper arm. Then, remove the rear wiper arm and its base cap. Be careful not to scratch the vehicle's paint or rear window glass.

3. Carefully pry out the clips to remove the hatch interior trim panel. Be careful not to scratch the trim panel or break the clips.

4. Disconnect the wiper motor wiring harness. Unbolt and remove the wiper motor.

To install:

5. Install the wiper motor and tighten the three bolt to 7 ft. lbs. (10 Nm). Reconnect the wiring harness.

6. Install the hatch interior trim panel. Carefully press all the clips into position.

7. Install the base cap and wiper arm. In the parked position, the wiper arm should be approximately 0.39 in. (10mm) from the rear window trim.

8. Install the wiper arm nut and tighten it to 7 ft. lbs. (10 Nm). Then, install the nut cover.

9. Reconnect the negative battery cable and test the operation of the rear wiper.

Combination Switch

REMOVAL AND INSTALLATION

—————— CAUTION ——————
The air bag cable reel is located in the steering column near the combination switch. Air bag wiring harnesses are wrapped with yellow insulation. Do not damage these wires. If any air bag system component must be serviced, the air bag system must be disabled to prevent possible accidental air bag deployment.

NOTE: The original radio contains a coded anti–theft circuit. Obtain the security code number before disconnecting the battery cables.

1. Don't lock the steering column.

2. Disconnect the negative and positive battery cables.

3. Remove the lower dashboard cover and knee bolster. On some vehicles, the air duct routed behind the

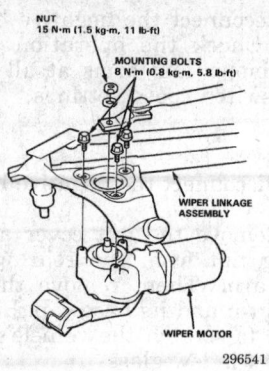

Front wiper motor components

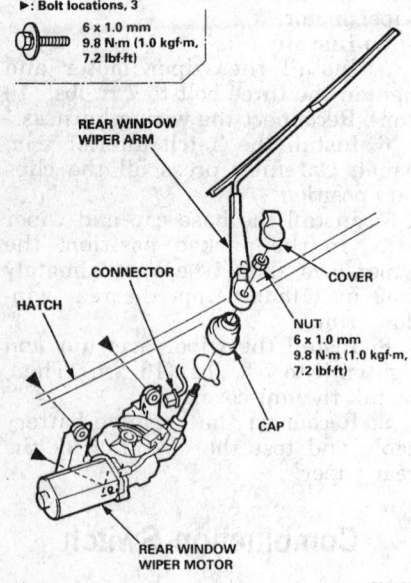

Rear motor and arm components — Civic and del Sol

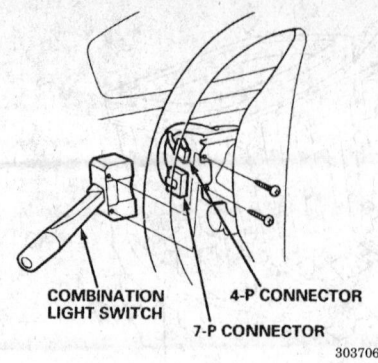

Combination switch components

knee bolster may be removed to access the lower steering column cover.

4. Remove the upper and lower steering column covers.

5. Disconnect the 4–P and 7–P electrical connectors from the headlight/dimmer/turn signal combination switch.

6. Turn the steering wheel 90° to the left.

7. Remove the two switch retaining screws. Remove the switch from its mount.

To install:

8. Connect the 4–P and 7–P electrical connectors, and install the switch into its mount.

9. Install the two retaining screws. Verify that no electrical wires are pinched.

10. Install the steering column covers. Install the air duct if it was removed.

11. Install the knee bolster and lower dashboard cover.

12. Reconnect the positive and negative battery cables.

13. If equipped with an air bag, turn the ignition switch to the **ON** position, but don't start the engine. The air bag indicator light should turn on for six seconds, and then turn off. This light sequence indicates that the air bag system is functioning normally. If the light stays on longer, there is a system malfunction which must be diagnosed.

14. On vehicles with 4WS, turn the steering wheel lock–to–lock to reset the steering control unit.

15. Check the operation of the headlights and turn signals, then enter the radio security code.

Ignition Switch and Lock Cylinder

REMOVAL AND INSTALLATION

Civic and del Sol

—— **CAUTION** ——
The supplemental restraint system (air bag) must be disabled before the ignition lock is removed. Failure to disable the air bag system may result in personal injury and unnecessary repairs. Air bag wiring harnesses are wrapped in yellow insulation. Do not damage the wiring harness or its insulation.

NOTE: The original radio contains a coded anti–theft circuit. Obtain the security code number before disconnecting the battery cable.

1. Disconnect the negative and positive battery cables. Wait at least three minutes before working around the air bags.

2. Disable the driver's side air bag:

a. Remove the lower steering wheel access cover.

b. Uncouple the air bag connector from the cable reel connector.

c. Remove the red shorting connector from the access cover and install it onto the air bag connector.

3. If equipped with spring–loaded air bag connectors, uncouple the air bag connector from the cable reel connector:

a. Hold the cable reel connector. With your other hand, slide the spring–loaded sleeve toward the stop tab on the air bag connector.

b. Separate the two connectors. There is no need to install a shorting connector, as the connectors are automatically grounded when they are uncoupled.

4. Remove the dashboard lower cover by removing the screws and detaching the clips. Unbolt and remove the knee bolster.

5. Disconnect the 5–P connector from the fuse/relay box and the 7–P connector from the main harness.

6. Remove the upper and lower steering column covers.

7. Unbolt the ignition electrical switch from the lock cylinder.

8. Remove the steering column mounting nuts and bolts and lower the column to the floor of the vehicle.

—— **WARNING** ——
Wrap a large towel around the steering wheel to protect the air bag. Be careful not to damage the yellow air bag wiring harness when drilling out the shear bolts.

9. Center punch the two shear bolts and then drill them out with a 3/16 in. (5mm) bit. Remove the shear bolts and the ignition lock body.

To install:

10. Install the ignition switch without the key inserted. Install new shear bolts, but tighten them only enough to hold the switch in place.

11. Insert the key and check for proper operation of the steering lock. The key must turn freely.

12. Tighten the shear bolts until their heads twist off.

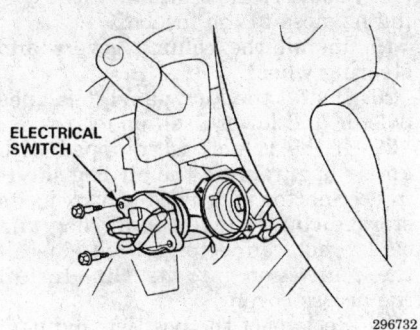

Ignition electrical switch

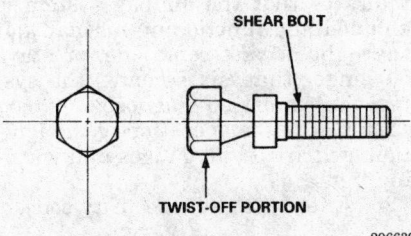

New shear bolt

13. Raise the steering column into position. Install the mounting bracket, nuts, and bolts. Then, tighten the mounting nuts to 9 ft. lbs. (13 Nm), and the mounting bolts to 16 ft. lbs. (22 Nm).

14. Install the upper and lower steering column covers.

15. Reconnect the fuse/relay box and main harness connectors.

16. Install the knee bolster and dashboard lower cover.

17. Turn the ignition key to the **OFF** position and remove the key.

18. Enable the driver's side air bag:

a. Remove the shorting connector from the air bag harness.

b. Reconnect the air bag and cable reel connectors.

c. Install the shorting connector back into its holder on the lower access cover. Then, install the access cover.

19. If equipped with spring–loaded connectors, reconnect the air bag and cable reel connectors. Make sure the connectors fit squarely together. Then, press the connectors to couple them. The spring–loaded sleeve will lock into place as the two connectors are coupled.

20. Reconnect the positive and negative battery cables.

21. Turn the ignition switch **ON** and check for proper operation. The air bag indicator light should come on for six seconds and then turn off. This sequence indicates that the air bag system is enabled and functioning normally.

Prelude

— CAUTION —

The air bag system must be disarmed before removing the ignition switch and steering lock. Failure to disarm the air bag system may cause accidental deployment of the air bag resulting in unnecessary repairs and personal injury.

NOTE: The radio may contain a coded theft protection circuit. Always obtain the code number before disconnecting the battery. If the vehicle is equipped with 4WS, the steering control unit is shut down when the battery is disconnected. After connecting the battery, turn the steering wheel lock–to–lock to reset the steering control unit.

1. Disconnect the negative and then the positive battery cables.

2. Remove the steering wheel access cover. Uncouple the air bag connector from the cable reel connector. Install the red shorting connector onto the air bag connector.

3. Remove the dashboard lower cover, knee bolster, and the left kick panel.

4. Remove the upper and lower steering column covers.

5. Disconnect the 5–P terminal from the under–dash fuse box. Disconnect the 3–P terminal from the main wiring harness.

6. Insert the key and turn the ignition to the **O** position.

7. Remove the cover from the ignition switch.

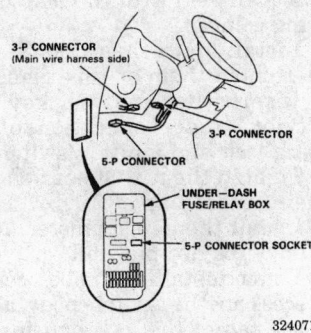

5-P and 3-P terminals and under-dash fuse box — Prelude

8. Remove the two screws and remove the ignition switch.

9. If the ignition lock needs to be replaced perform the following:

a. Center punch each of the two shear bolts. Use a 3/16 in. (5mm) bit to drill out the shear bolts. Be careful not to damage the switch body or the air bag harness.

b. Remove the shear bolts and the steering lock assembly.

To install:

10. If the ignition lock was removed perform the following to install it:

a. Install the ignition switch onto the steering column without its key inserted.

b. Install new shear bolts, but only hand–tighten them at this point.

c. Insert the ignition key and check for proper operation of the ignition switch and steering lock.

d. Tighten the new shear bolts until their heads snap off.

11. Install the new ignition switch.

12. Install the switch cover.

13. Reconnect the 5–P and 3–P terminals.

14. Install the steering column covers, kick panel, knee bolster, and the dashboard lower cover.

15. Remove the red shorting connector and couple the air bag and cable reel connectors. Install the connection back into its clip, and put the shorting connector back into its holder.

16. Reconnect the positive and negative battery cables.

17. Turn the ignition switch to the **ON** position. The air bag indicator light should turn on for six seconds, and then turn off. This light sequence indicates that the air bag system is enabled and functioning normally. If the air bag light doesn't come on, or stays on longer than six seconds, the system fault must be diagnosed.

18. On vehicles equipped with 4WS, start the engine and turn the steering wheel lock–to–lock to reset the steering control unit.

19. Check the operation of the ignition switch. Enter the radio security code.

1993 Accord

— CAUTION —

The air bag system must be disarmed before removing the ignition switch. Failure to disarm the air bag system may cause accidental deployment of the air bag, resulting in unnecessary air bag system repairs and personal injury.

NOTE: The radio may contain a coded theft protection circuit. Always obtain the code number before disconnecting the battery.

Manual transaxle

1. Disconnect the negative and then the positive battery cables. Wait three minutes before servicing the air bag.

2. If the vehicle is equipped with an air bag; remove the steering wheel access cover. Uncouple the air bag and cable reel connectors and install the red shorting connector onto the air bag connector.

3. Remove the lower dash panel, the left knee bolster, and the left kick panel.

4. If equipped, remove the driver's air bag.

5. Remove the steering wheel and the steering column covers.

6. At the fuse box, disconnect the 7-pin ignition switch connector.

7. Insert the key and turn it to the O position.

8. Remove the two screws holding the switch and remove the switch.

9. Turn the bulb socket 45° to remove it from the key light shroud.

10. Remove the setscrew and then remove the light housing from the lock cylinder.

11. Turn the key to the I position. Push the pin down and slide the lock cylinder out of the switch body.

To install:

12. Turn the key to the O position; then, align the lock cylinder with the lock body.

13. Turn the key almost to the I position, the key must NOT click into this position.

14. Insert the lock cylinder into the switch body until the pin touches the body of the switch.

15. Now, turn the key so that it clicks into the I position. Insert the the lock cylinder into the lock body until the pin clicks into place.

16. When installing, make certain the recess in the switch body aligns with the tab of the lock housing. Insert the switch and install the retaining screws.

17. Connect the ignition switch wiring harness to the fusebox.

18. Install the column covers and steering wheel.

19. Install the kick panel, left knee bolster and lower dash panel.

20. If the vehicle is equipped with an air bag; remove the air bag shorting connector and place it back in its storage clip. Couple the air bag and cable reel connectors; then, install the connection into its clip. Install the access cover.

21. Reconnect the positive and negative battery cables.

22. Turn the ignition switch to the ON position. The air bag indicator light should turn on for six seconds, and then turn off. This light sequence indicates that the air bag system is enabled and functioning normally. If the light doesn't come one, or stays on longer than six seconds, the system fault must be diagnosed.

23. Check the operation of the ignition switch and all gauges and warning lights.

24. Enter the radio security code.

Automatic Transaxle

1. Disconnect the negative and then the positive battery cables. Wait three minutes before servicing the air bag.

2. If the vehicle is equipped with an air bag; remove the steering wheel access cover. Uncouple the air bag and cable reel connectors and install the red shorting connector onto the air bag connector.

3. Remove the lower dash panel, the left knee bolster, and the left kick panel.

4. If equipped, remove the driver's air bag.

5. Remove the steering wheel and the steering column covers.

6. At the fuse box, disconnect the 7-pin ignition switch connector.

7. Insert the key and turn it to the O position.

8. Remove the two screws holding the switch and remove the switch.

9. Remove the dashboard trim panel.

10. Remove the gauge cluster.

11. Center punch each of the two shear bolts. Use a 3/16 in. (5mm) bit to drill out the shear bolts. Be careful not to damage the switch body or the air bag harness.

To install:

12. Install the new ignition switch without the key installed. Make sure that the projection on the ignition switch is aligned with the hole in the steering column.

13. Install new shear bolts, only hand-tighten them at this time.

14. Insert the ignition key and check for proper operation of the steering lock and ignition switch.

15. Tighten the new shear bolts until their heads snap off.

16. Install the gauge cluster. Make sure no wires are pinched.

17. When installing, make certain the recess in the switch body aligns with the tab of the lock housing. Insert the switch and install the retaining screws.

18. Connect the ignition switch wiring harness to the fusebox.

19. Install the column covers and steering wheel.

20. Install the kick panel, left knee bolster and lower dash panel.

21. If the vehicle is equipped with an air bag; remove the air bag shorting connector and place it back in its storage clip. Couple the air bag and cable reel connectors; then, install the connection into its clip. Install the access cover.

22. Reconnect the positive and negative battery cables.

23. Turn the ignition switch to the ON position. The air bag indicator light should turn on for six seconds, and then turn off. This light sequence indicates that the air bag system is enabled and functioning normally. If the light doesn't come one, or stays on longer than six seconds, the system fault must be diagnosed.

24. Check the operation of the ignition switch and all gauges and warning lights.

25. Enter the radio security code.

1994–97 Accord

——————— **CAUTION** ———————

The air bag system must be disarmed before removing the ignition switch and steering lock. Failure to disarm the air bag system may cause accidental deployment of the air bag resulting in unnecessary repairs and personal injury.

NOTE: The radio may contain a coded theft protection circuit. Always obtain the code number before disconnecting the battery.

1. Disconnect the negative and then the positive battery cables.

2. Remove the steering wheel access panel and remove the air bag connector from the connector holder. Pull the spring loaded sleeve, on the connector, toward the stop while holding the opposite half of the connector and pull the connector apart.

3. Remove the dashboard lower cover and knee bolster.

4. Remove the upper and lower steering column covers.

5. Disconnect the two connectors from the ignition switch.

6. Remove the steering column holder bolts and mounting nuts. Lower the steering column.

7. Center punch each of the two shear bolts. Use a 3/16 in. (5mm) bit to drill out the shear bolts. Be careful not to damage the switch body or the air bag harness.

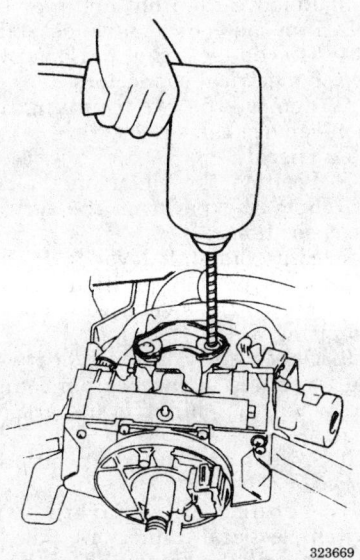

Drill out the shear bolts for the lock cylinder — 1993 Accord with automatic transaxle

8. Remove the ignition switch assembly.

9. Remove the three screws which attach the electrical switch to the ignition switch.

To install:

10. Install the new electrical switch onto the ignition switch.

11. Install the ignition switch onto the steering column without its key inserted.

12. Install new shear bolts, but only hand–tighten them at this point.

13. Insert the ignition key and check for proper operation of the ignition switch and steering lock.

14. Tighten the new shear bolts until their heads snap off.

15. Raise the steering column into position and install the holder bolts and nuts. Tighten the bolts to 16 ft. lbs. (22 Nm), tighten the nuts to 9.4 ft. lbs. (13 Nm).

16. Connect the two connectors to the ignition switch.

17. Install the upper and lower steering column covers, the knee bolster, and the dashboard lower cover.

18. Connect the driver's air bag connector to the cable reel connector and place this coupling back into its clip. Install the access cover to the steering wheel.

19. Connect the positive then the negative battery cables.

20. Turn the ignition switch to the **ON** position. The air bag indicator light should turn on for six seconds, and then turn off. This light sequence indicates that the air bag system is enabled and functioning normally. If the air bag light doesn't come on, or stays on longer than six seconds, the system fault must be diagnosed.

21. Enter the radio security code.

Park/Neutral Safety Switch

REMOVAL AND INSTALLATION

Civic

——— **CAUTION** ———

The air bag system main wiring harness is routed near the console. The air bag wiring is encased in yellow insulation. Do not damage the wiring or insulation. If any air bag system component must be serviced, the air bag system must first be disabled.

NOTE: The original radio contains a coded anti–theft circuit. Obtain the security code number before disconnecting the battery cables.

1. Disconnect the negative and positive battery cables. If equipped with an air bag, wait at least three minutes before working around the air bags.

2. Remove the center console. The rear part of the console is removed first, and then the front part.

3. Disconnect the 14–P connector from the park/neutral switch, which is located on the right side of the shift lever assembly.

4. Remove the switch mounting nuts. Remove the switch.

To install:

5. Position the slider on the switch to the neutral position.

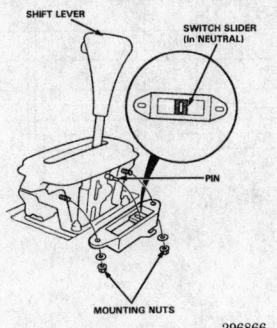

Park/neutral switch components — Civic and del Sol

6. Shift the shift lever to neutral; then, install the switch onto the shift lever assembly. Evenly tighten the mounting nuts.

7. Test the switch in the **P** and **N** positions. The engine should start when the shift lever is in either position.

8. Connect the 14–P connector.

9. Install the center console components.

10. Reconnect the positive and negative battery cables.

Prelude

NOTE: The radio may contain a coded theft protection circuit. Always obtain the code number before disconnecting the battery. If the vehicle is equipped with 4WS, the steering control unit is shut down when the battery is disconnected. After connecting the battery, turn the steering wheel lock–to–lock to reset the steering control unit.

1. Disconnect the negative battery cable.

——— **CAUTION** ———

Air bag wire harnesses are routed near the front console. All air bag wire harnesses are covered with yellow insulation. Before any part of the air bag wire harness is disconnected the air bag system must be disabled.

2. Use a flat tipped tool wrapped with tape and pry the shift indicator trim ring from the front console. Cover the shift lever with a shop towel to prevent damaging the shift lever knob.

3. Remove the front console attaching screws then remove the front console.

4. Disconnect the park/neutral switch electrical connectors.

5. Remove the two mounting nuts then remove the switch.

To install:

6. Position the slider on the new switch to the park, or the forward most position.

7. Shift the shift lever to **P**, then slip the switch into position.

8. Attach the switch with the two mounting nuts.

9. Check for continuity between the terminals in each switch position according to the gear position switch table.

10. If the switch needs adjustment perform the following:

   a. Shift the shift lever to **P** and loosen the switch nuts.

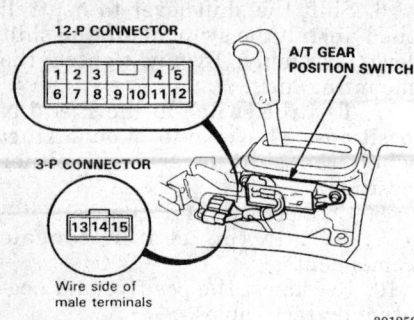

Gear position switch connector pin identification — Prelude

b. Slide the switch rearward up to 0.08 in. (2.0mm) until there is continuity between No. 8 and No. 11 terminals, within the range of freeplay of the shift lever.

c. Recheck for continuity between each of the other terminals.

NOTE: If adjustment is not possible, check for damage to the shift lever and bracket. If there is no damage to the bracket replace the switch.

11. Connect the park/neutral switch electrical connectors.

12. Install the front console and its attaching screws.

13. Remove the shop towel from the shift lever and install the shift indicator trim ring.

14. Connect the negative battery cable and enter the radio security code.

15. If equipped with 4WS, turn the steering wheel lock–to–lock to reset the 4WS control unit.

16. Test the operation of the switch by starting the vehicle.

NOTE: The engine should start when the shift lever is in park, anywhere in the range of freeplay.

1993 Accord

NOTE: The radio may contain a coded theft protection circuit. Always obtain the code number before disconnecting the battery.

1. Disconnect the negative battery cable.

--- **CAUTION** ---

Air bag wire harnesses are routed near the front console. All air bag wire harnesses are covered with yellow insulation. Before any part of the air bag wire harness is disconnected the air bag system must be disabled.

2. Remove the screws attaching the front and rear consoles to the vehicle.

3. Remove the front and rear console from the vehicle as an assembly.

4. Disconnect the park/neutral switch electrical connectors.

5. Remove the two mounting nuts then remove the switch.

To install:

6. Position the slider on the new switch to the park, or the forward most position.

7. Shift the shift lever to **P**, then slip the switch into position.

8. Attach the switch with the two mounting nuts.

9. Check for continuity between the terminals in the two pin connector, when the shifter is in park and neutral.

10. Adjust the switch position if necessary.

11. Connect the park/neutral switch electrical connectors.

12. Install the front and rear console assembly and install the screws that attach the consoles to the vehicle.

13. If equipped with a manual transaxle, install the shift lever knob.

14. Connect the negative battery cable and enter the radio security code.

15. Test the operation of the switch by starting the vehicle.

NOTE: The engine should start when the shift lever is in park, anywhere in the range of freeplay.

1994–97 Accord

NOTE: The radio may contain a coded theft protection circuit. Always obtain the code number before disconnecting the battery.

1. Disconnect the negative battery cable.

--- **CAUTION** ---

Air bag wire harnesses are routed near the front console. All air bag wire harnesses are covered with yellow insulation. Before any part of the air bag wire harness is disconnected the air bag system must be disabled.

2. Open the armrest and remove the inner panel from the bottom of the compartment.

3. Remove the access lid from the bottom of the armrest compartment, then remove the two screws.

4. Remove the beverage holder and attaching screw.

5. Lift the parking brake lever then lift the front of the rear console

	A/T Gear Position Switch								Back-up Light Switch	Neutral Position Switch		
Terminal Position	8	1	2	3	4	5	6	7	11	9	10	13	15
1	○				○								
2	○		○			○							
D3	○			○		○							
D4	○		○			○							
N	○						○					○	○
R	○							○		○	○		
P	○								○			○	○

301260

Gear position switch table — Prelude

to release the hooks. Slide the rear console back then lift and remove it from the vehicle.

6. Remove the ashtray and the console panel mounting screws.

7. If equipped with a manual transaxle, remove the shift lever knob.

8. Lift the console panel to detach the clips. Disconnect the cigarette lighter electrical connector and the hazard warning switch electrical connector, then remove the console panel.

9. Remove the radio mounting bolts then pull the radio from the dash and disconnect the electrical connectors.

10. Remove the coin pocket from the dashboard lower cover, then remove the screw from where the coin pocket was.

11. Pull the lower cover from the dash to release the clips and set the lower cover aside.

12. Open the glove box and remove the screw attaching the damper to the glove box door.

13. Remove the two bolts attaching the glove box door to the dash and set the door aside.

14. Remove the screws attaching the front console to the dash.

15. Cover the shift lever with a shop towel then remove the front console.

16. Disconnect the park/neutral switch electrical connectors.

17. Remove the two mounting nuts then remove the switch.

To install:

18. Position the slider on the new switch to the park, or the forward most position.

19. Shift the shift lever to **P** (park), then slip the switch into position.

20. Attach the switch with the two mounting nuts.

21. Check for continuity between the terminals in the two pin connector, when the shifter is in park and neutral.

22. Adjust the switch position if necessary.

23. Connect the park/neutral switch electrical connectors.

24. Install the front console and its attaching screws, then remove the shop towel from the shift lever.

25. Position the glove box in the dash and install the two mounting bolts.

26. Install the screw attaching the damper to the glove box door.

27. Push the lower dash cover into place to engage the retaining clips.

28. Install the screw to the lower dash cover, then install the coin pocket.

29. Connect the electrical connectors to the radio then place the radio in the dash. Install the radio mounting bolts and torque the bolts to 7 ft. lbs. (10 Nm).

30. Connect the cigarette lighter electrical connector and the hazard warning switch electrical connector, then push the console panel into place to engage the retaining clips.

31. If equipped with a manual transaxle, install the shift lever knob.

32. Install the console mounting screws and the ashtray.

33. Position the rear console and engage the hooks on the front of the console.

34. Install the beverage holder and the beverage holder attaching screw.

35. Install the two screws to the rear console and install the access lid to the bottom of the armrest compartment.

36. Install the inner panel to the armrest compartment.

37. Connect the negative battery cable and enter the radio security code.

38. Test the operation of the switch by starting the vehicle.

NOTE: The engine should start when the shift lever is in park, anywhere within the range of freeplay.

Powertrain Control Module

REMOVAL AND INSTALLATION

Civic and del Sol

NOTE: The original radio contains a coded anti–theft circuit. Obtain the security code number before disconnecting the battery cables.

1. Disconnect the negative battery cable.
2. Remove the right door sill moulding.
3. Pull the carpeting back and away from the right kick panel and floor to expose the PCM.

4. Remove the PCM cover and unbolt the PCM from its mounting bracket.

5. Remove the PCM from its mounting bracket and disconnect it from the wiring harness.

To install:

6. Connect the PCM to its wiring harness and install it onto the mounting bracket.

7. Install the PCM cover and its mounting bolts.

8. Fit the carpet back into its original position. Install the right door sill moulding.

9. Reconnect the negative battery cable.

Prelude and Accord

NOTE: The radio may contain a coded theft protection circuit. Always obtain the code number before disconnecting the battery. If the vehicle is equipped with 4WS, the steering control unit is shut down when the battery is disconnected. After connecting the battery, turn the steering wheel lock–to–lock to reset the steering control unit.

1. Disconnect the negative battery cable.
2. Remove the small cover from the right kick panel.
3. Pull the right door sill moulding from its retaining clips and place out of the way.
4. Pull the carpet back enough to expose the PCM.
5. Remove the PCM cover nuts and remove the cover and the PCM.

To install:

6. Install the PCM and the PCM cover, then install the attaching nuts.
7. Push the carpet back into place.
8. Position the right door sill moulding in the vehicle then push into place to seat the moulding in the retaining clips.
9. Install the small cover to the right kick panel.
10. Connect the negative battery cable and enter the radio security code.
11. If equipped with 4WS, start the engine then turn the steering wheel lock–to–lock to reset the 4WS control unit.

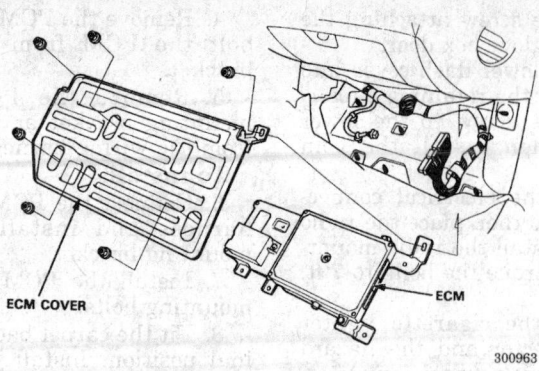

ECM COVER

ECM

300963

PCM mounting — Prelude

ENGINE COOLING

Radiator

REMOVAL AND INSTALLATION

NOTE: The radio may contain a coded theft protection circuit. Always obtain the code number before disconnecting the battery. If the vehicle is equipped with 4WS, the steering control unit is shut down when the battery is disconnected. After connecting the battery, turn the steering wheel lock–to–lock to reset the steering control unit.

1. Disconnect the negative battery cable.
2. Drain the coolant into a sealable container.
3. Remove the upper and lower radiator hoses.
4. Disconnect the reservoir hose from the radiator.
5. If equipped with a automatic transaxle, disconnect the ATF coolant hoses from the radiator.
6. Disconnect the cooling fans electrical connectors and the relay connector, if equipped.
7. Remove the upper radiator brackets then lift the radiator from the vehicle.
8. Remove the cooling fan assemblies from the radiator.
   To install:
9. Install the cooling fans to the radiator, torque the radiator cooling fan mounting bolts to 7 ft. lbs. (10 Nm).
10. Install the radiator into the vehicle and install the upper radiator brackets. Make sure that the radiator is secure in the upper and lower radiator cushions. Torque the upper radiator bracket bolts to 7 ft. lbs. (10 Nm).

11. Connect the cooling fans electrical connectors and the relay connector, if equipped.
12. If equipped with a automatic transaxle, connect the ATF cooler lines.
13. Connect the reservoir hose to the radiator.
14. Install the upper and lower radiator hoses.
15. Fill and bleed the air from the cooling system.
16. Connect the negative battery cable and enter the radio security code.
17. If equipped with 4WS, start the engine and turn the steering wheel lock–to–lock to reset the 4WS control unit.

Water Pump

REMOVAL AND INSTALLATION

1.5L, 1.6L, 2.2L and 2.3L Engines

NOTE: The original radio contains a coded anti–theft circuit. Obtain the security code number before disconnecting the battery cables.

1. Disconnect the negative battery cable.
2. Drain the cooling system.
3. Remove the accessory drive belts, the valve cover, and the upper timing belt cover.
4. Set the timing at TDC/compression for No. 1 piston.
5. Remove the crankshaft pulley and lower timing belt cover.
6. Remove the timing belt. Replace the timing belt if it is contaminated with oil or coolant or shows any signs of wear and damage.
7. If equipped with a Crankshaft Speed Fluctuation (CKF) sensor at the crankshaft sprocket, unbolt the sensor bracket and move the sensor out of the way. Cover the sensor with a shop towel to keep coolant off of it.

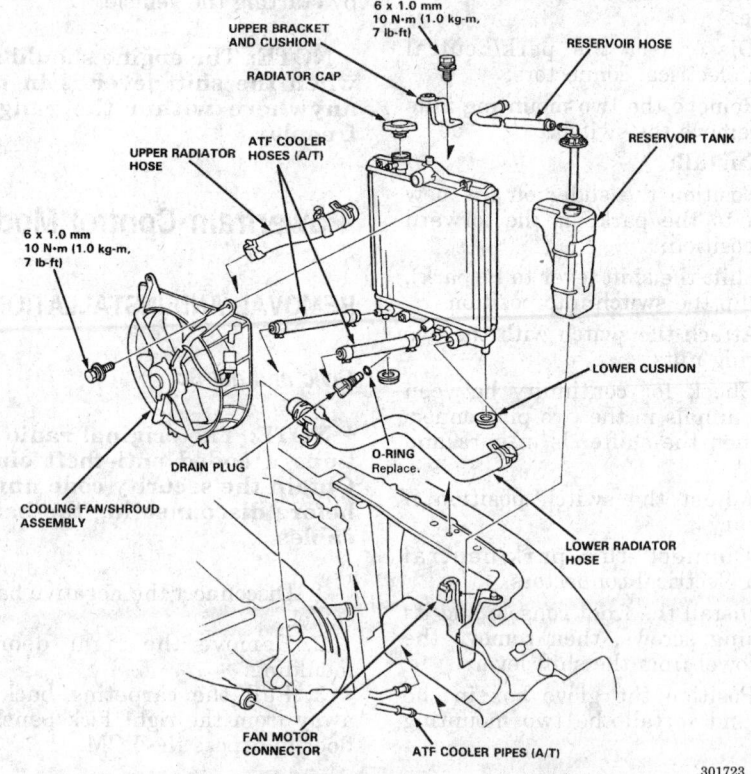

UPPER BRACKET AND CUSHION

6 x 1.0 mm 10 N·m (1.0 kg-m, 7 lb-ft)

RADIATOR CAP

RESERVOIR HOSE

UPPER RADIATOR HOSE

ATF COOLER HOSES (A/T)

RESERVOIR TANK

6 x 1.0 mm 10 N·m (1.0 kg-m, 7 lb-ft)

LOWER CUSHION

DRAIN PLUG

O-RING Replace.

COOLING FAN/SHROUD ASSEMBLY

LOWER RADIATOR HOSE

FAN MOTOR CONNECTOR

ATF COOLER PIPES (A/T)

301723

Radiator components — 1.5L and 1.6L engines

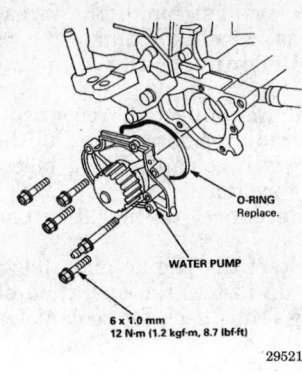

Water pump — 2.2L and 2.3L engines

8. Unbolt the water pump and remove it from the engine block. On 1.5L and 1.6L engines, the top right water pump mounting bolt also secures the alternator adjusting bracket. Leave the bracket attached to the alternator.

To install:

9. Clean the water pump and O-ring mating surfaces before installation.

10. Install the water pump with a new O-ring. Coat only the bolt threads with liquid gasket and torque them to 9 ft. lbs. (12 Nm). On 1.5L and 1.6L engines, tighten the bracket bolt to 33 ft. lbs. (44 Nm).

11. Install the timing belt. Make sure it is fitted and adjusted properly.

12. If equipped, install the CKF sensor and tighten the bracket bolts to 9 ft. lbs. (12 Nm).

13. Install the lower belt cover and crankshaft pulley.

14. Install the upper timing belt cover, the valve cover, and the accessory drive belts.

15. Be sure the cooling system drain plug is closed. Refill and bleed the cooling system.

16. Connect the negative battery cable and enter the radio security code.

17. Start the engine, allow it to reach normal operating temperature, and check for coolant leaks Check the tensions of the accessory belts.

18. If equipped with 4WS, turn the steering wheel lock-to-lock to reset the 4WS control unit.

2.7L Engine

1. Disconnect the negative battery cable.

2. Drain the coolant into a sealable container.

3. Remove the timing belt covers and the timing belt.

4. Remove the timing belt tensioner.

5. Remove the nine water pump bolts, take note of their locations for reinstallation.

6. Remove the water pump from the engine and discard the O-ring. Remove the dowel pins.

To install:

7. Clean the water pump mounting surface and O-ring groove, then install the dowel pins to the engine.

8. Install a new O-ring to the engine then install the water pump, be careful not to pinch the O-ring. Install the mounting bolts to their original locations and when tightening the bolts, make sure that the O-ring does not bulge out of the groove. Torque the six 1.0mm bolts to 9 ft. lbs. (12 Nm), and torque the eight 1.25mm bolts to 16 ft. lbs. (22 Nm).

9. Inspect the water pump, making sure that the pump turns freely.

10. Install the timing belt tensioner.

11. Install the timing belt and timing belt covers.

12. Refill and bleed the air from the cooling system.

13. Connect the negative battery cable and enter the radio security code.

Thermostat

REMOVAL AND INSTALLATION

NOTE: The radio contains a coded anti-theft circuit. Obtain the security code number before disconnecting the battery cables.

1. Disconnect the negative battery cable.

2. Drain the coolant to a level below the upper radiator hose.

3. Disconnect the coolant temperature switch lead from the thermostat cover.

4. Disconnect the upper radiator hose from the thermostat cover.

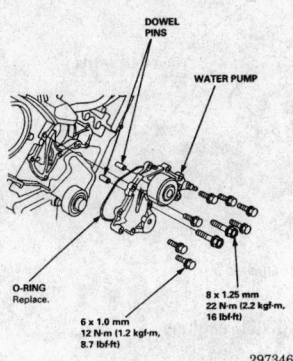

Water pump — 2.7L engine

5. Unbolt and remove the thermostat cover.

6. Remove the thermostat and its rubber gasket.

7. Clean any traces of sealant from the thermostat housing.

To install:

NOTE: Install the thermostat with a new rubber gasket. Replace any O-rings that where removed.

8. Install the new thermostat with its pin facing up. The pin fits into a notch in the thermostat housing. Install a new rubber gasket. Apply sealant to the thermostat cover if it was present when the cover was removed.

9. Install the thermostat cover and torque the mounting bolts to 9 ft. lbs. (12 Nm).

10. Reconnect the upper radiator hose.

11. Reconnect the coolant temperature switch lead.

12. Refill the radiator with a coolant mixture that contains antifreeze. Use only antifreeze formulated to prevent the corrosion of aluminum parts.

13. If equipped with 4WS, turn the steering wheel lock-to-lock to reset the 4WS control unit.

Electric Cooling Fan

REMOVAL AND INSTALLATION

1.5L and 1.6L Engines

NOTE: The original radio contains a coded anti-theft circuit. Obtain the security code number before disconnecting the battery cables.

Radiator Cooling Fan

1. Disconnect the negative battery cable.

2. For extra working room, disconnect the upper radiator hose and move it out of the way. Catch any spilled coolant with a drip fan.

3. Uncouple the fan motor connector.

4. Unbolt the fan shroud. The lower splash shield may be removed to access the lower mounting bolts.

5. Remove the fan shroud from the radiator.

6. Unbolt the fan blade and motor from the fan shroud.

To install:

7. Install the fan motor and bland into the fan shroud.

8. Install the fan shroud onto the radiator. Carefully tighten the mounting bolts to 5–7 ft. lbs. (7–10

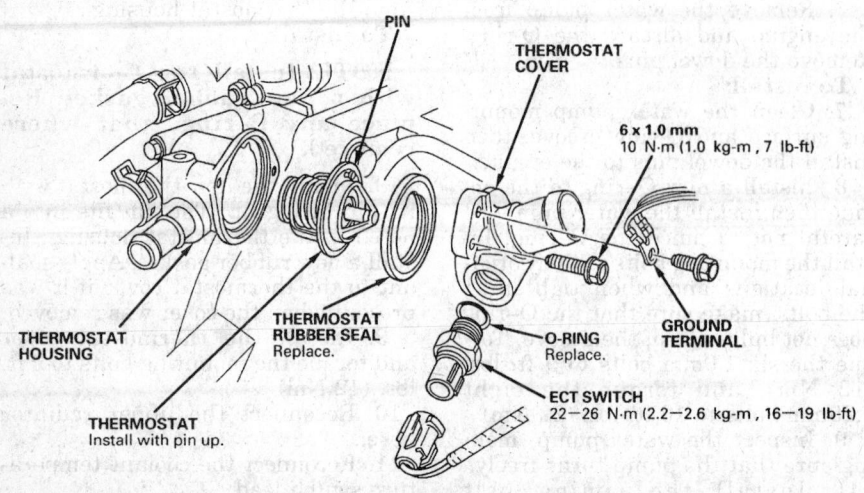

Thermostat components — 1.5L and 1.6L engines

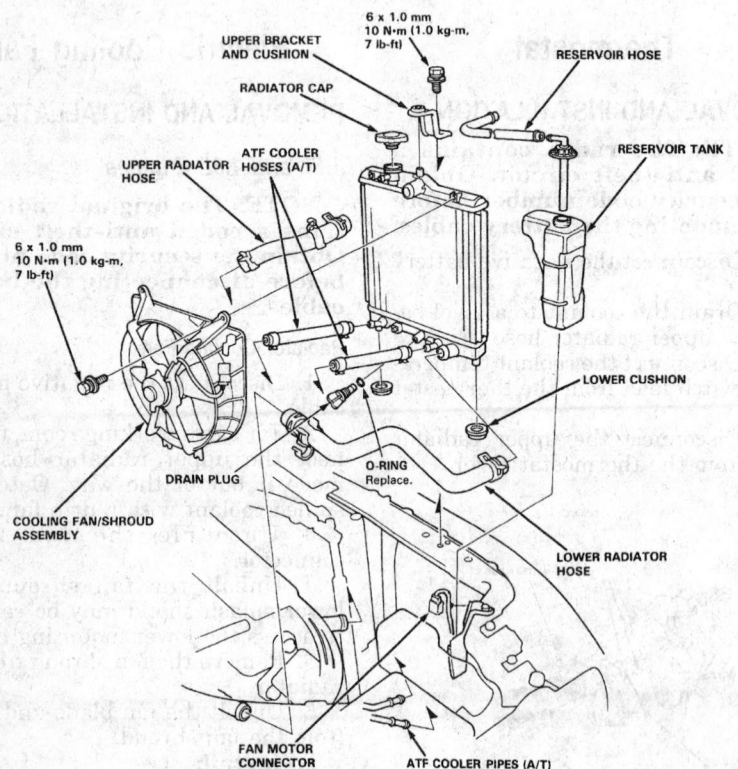

Cooling fan components — 1.5L and 1.6L engines

Nm) to avoid stripping the threads of the plastic radiator tank.

9. Reconnect the fan motor connectors.

10. If disconnected, reconnect the upper radiator hose and refill the radiator with coolant. Then, bleed the cooling system.

11. Reconnect the negative battery cable.

12. Start the engine and allow it to warm up to test the operation of the cooling fan. Check for coolant leaks.

A/C Condenser Fan

1. Disconnect the negative battery cable.

2. Uncouple the condenser fan connector.

3. Unbolt the upper fan shroud mount brackets.

4. Remove the condenser fan from the vehicle.

To install:

5. Install the condenser fan into the vehicle. Fit the lower fan mounts into their rubber grommets.

6. Install the upper and lower mounting bolts.

7. Reconnect the wiring.

8. Reconnect the negative battery cable. Allow the engine to reach operating temperature; then, turn the air conditioning on, and test the operation of the fan.

Except 1.5L and 1.6L Engines

NOTE: The radio may contain a coded theft protection circuit. Always obtain the code number before disconnecting the battery. If the vehicle is equipped with 4WS, the steering control unit is shut down when the battery is disconnected. After connecting the battery, turn the steering wheel lock–to–lock to reset the steering control unit.

1. Disconnect the negative battery cable.

2. Drain enough coolant from the radiator to remove the upper radiator hose. Drain the coolant into a sealable container.

3. Disconnect the cooling fan electrical connectors.

4. Remove the cooling fan mounting bolts then lift the cooling fan out of the vehicle. Be careful not to damage the radiator.

To install:

5. Position the cooling fan into the vehicle and install the mounting bolts. Torque the radiator cooling fan mounting bolts to 5 ft. lbs. (7 Nm), and the condenser cooling fan mounting bolts to 7 ft. lbs. (10 Nm).

6. Connect the cooling fan electrical connectors.

7. Add the drained coolant to the radiator.

8. If equipped with 4WS, turn the steering wheel lock–to–lock to reset the 4WS control unit.

9. Connect the negative battery cable and enter the radio security code.

10. Test the operation of the cooling fan.

Cooling System Bleeding

PROCEDURE

Without Bleed Valve

1. Fill the reservoir tank to the MIN line with water. Then, fill the reservoir tank to the MAX line with antifreeze.

2. Fill the radiator up to the base of the filler neck with the coolant mixture.

3. Set the heater to maximum heat.

4. Loosely install the radiator cap. Then, start the engine and let it run until warmed up (the radiator fan comes on at least twice).

5. Shut the engine off. Check the coolant level in the radiator and add more coolant mixture if necessary.

6. Install the radiator cap securely and run the engine to check for coolant leaks.

With Bleed Valve

1. Loosen the air bleed bolt in the thermostat housing, then fill the radiator to the bottom of the filler neck with the mixed coolant. Tighten the bleed bolt as soon as coolant starts to run out in a steady stream without bubbles. Torque the bleed bolt to 7 ft. lbs. (10 Nm).

—— **WARNING** ——

When pouring engine coolant, be sure to shut the relay box lid and not to let coolant spill on the electrical components or the paint. If any coolant spills, rinse it off immediately.

2. Fill the reservoir to the MAX mark with the coolant mixture.

3. Connect the negative battery cable and enter the radio security code.

4. With the radiator cap removed, start the engine and let it run until warmed up (the radiator fan comes on at least twice). Then, if necessary, top off the radiator with the coolant mixture.

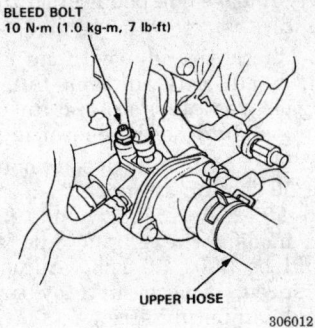

BLEED BOLT
10 N·m (1.0 kg-m, 7 lb-ft)

UPPER HOSE

306012

Bleed bolt — 2.2L (H22A1, F22A1, F22A6) and 2.3L engines

5. If equipped with 4WS, turn the steering wheel lock–to–lock to reset the 4WS control unit.

6. Install the radiator cap securely and run the engine to check for coolant leaks.

FUEL SYSTEM

Fuel System Service Precautions

Safety is the most important factor when performing not only fuel system maintenance but any type of maintenance. Failure to conduct maintenance and repairs in a safe manner may result in serious personal injury or death. Maintenance and testing of the vehicle's fuel system components can be accomplished safely and effectively by adhering to the following rules and guidelines.

• To avoid the possibility of fire and personal injury, always disconnect the negative battery cable unless the repair or test procedure requires that battery voltage be applied.

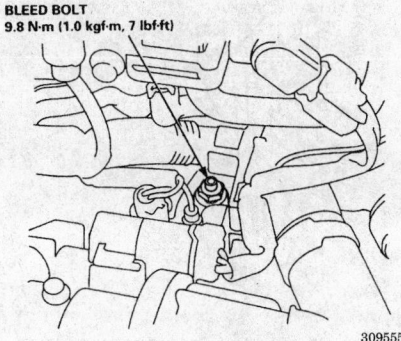

BLEED BOLT
9.8 N·m (1.0 kgf·m, 7 lbf·ft)

309555

Bleed bolt — 2.7L engines

• Always relieve the fuel system pressure prior to disconnecting any fuel system component (injector, fuel rail, pressure regulator, etc.), fitting or fuel line connection. Exercise extreme caution whenever relieving fuel system pressure to avoid exposing skin, face and eyes to fuel spray. Please be advised that fuel under pressure may penetrate the skin or any part of the body that it contacts.

• Always place a shop towel or cloth around the fitting or connection prior to loosening any excess fuel due to spillage. Ensure that all fuel spillage (should it occur) is quickly removed from engine surfaces. Ensure that all fuel soaked cloths or towels are deposited into a suitable waste container.

• Always keep a dry chemical (Class B) fire extinguisher near the work area.

• Do not allow fuel spray or fuel vapors to come into contact with a spark or open flame.

• Always use a backup wrench when loosening and tightening fuel line connection fittings. This will prevent unnecessary stress and torsion to fuel line piping. Always follow the proper torque specifications.

• Always replace worn fuel fitting O-rings with new. Do not substitute fuel hose or equivalent, where fuel pipe is installed.

Fuel System Pressure

RELIEVING

—— **CAUTION** ——

Fuel injection systems remain under pressure after the engine has been turned OFF. Properly relieve fuel pressure before disconnecting any fuel lines. Failure to do so may result in fire or personal injury.

NOTE: The radio may contain a coded theft protection circuit. Always obtain the code number before disconnecting the battery. If the vehicle is equipped with 4WS, the steering control unit is shut down when the battery is disconnected. After connecting the battery, turn the steering wheel lock–to–lock to reset the steering control unit.

1. Disconnect the negative battery cable.

2. Remove the fuel filler cap.

3. Use a box wrench to loosen the 6mm service bolt while holding the special banjo bolt with another

wrench. On 1.5L and 1.6L engines, it is located on the fuel filter. On other engines, it is found on the fuel rail.

4. Place a rag or shop towel over the 6mm service bolt.

5. Slowly loosen the 6mm service bolt one complete turn.

─────── **CAUTION** ───────

Do not allow fuel spray or fuel vapors to come in contact with a spark or open flame. Keep a dry chemical fire extinguisher nearby. Never store fuel in an open container due to risk of fire or explosion.

NOTE: A fuel pressure gauge may be attached at the 6mm service bolt location. Always replace the washer between the service bolt and the banjo bolt whenever the service bolt is loosened.

6. Remove the service bolt and install a new washer. Torque the 6mm service bolt to 9 ft. lbs. (12 Nm). Don't overtighten the service bolts, their threads may strip and cause leaks.

7. Clean up any fuel spilled on the engine and intake manifold.

8. Install the fuel filler cap.

9. Reconnect the negative battery cable.

10. Turn the ignition **ON**, but don't start the engine. Repeat this two or three times to pressurize the fuel system. Check for fuel leaks.

11. Enter the radio security code.

12. If equipped with 4WS, turn the steering wheel lock-to-lock to reset the 4WS control unit.

Idle Speed

ADJUSTMENT

Prelude and Accord

1. Before setting the engine idle speed check the following items:
• The MIL has not been reported on.
• Ignition timing
• Spark plugs
• Air cleaner
• PCV system

2. Start the engine. Hold the engine at 3000 rpm with no load (A/T in N or P, M/T in neutral) until the radiator fan comes on, then let it idle.

3. Shut the engine OFF and connect a tachometer to the engine.

4. Disconnect the connector from the Idle Air Control (IAC) valve.

5. Start the engine with the accelerator pedal slightly depressed. Stabilize the idle at 1000 rpm then

slowly release the pedal until the engine idles.

6. Check the idle with no engine load: headlights, blower fan, rear defogger, radiator fan, and air conditioning should not be operating.

7. The engine idle speed should be 550 ±50 rpm on all models except the 1996–97 2.7L model. On 1996–97 2.7L models the engine idle speed should be 600± 50 rpm. Adjust the idle speed, if necessary, by turning the idle adjusting screw.

NOTE: After adjusting the idle speed in this step, check the ignition timing. If it is out of specification shut the engine OFF and readjust the idle speed.

8. Shut the engine OFF.

9. Connect the electrical connector to the IAC valve, then remove the CLOCK RADIO (10A) fuse in the under-hood fuse/relay box for 10 seconds to reset the PCM/PCM.

10. Restart and idle the engine with no load conditions for one minute, then check the idle speed. The engine idle speed should be 700 ±50 rpm.

11. Idle the engine for one minute with the headlights on HIGH and the rear defogger ON then check the idle speed.
• The engine idle speed for the H23A1 engine should be 780 ±50 rpm.
• The engine idle speed for the H22A1 engine should be 790 ±50 rpm.
• The engine idle speed for the F22A1 engine should be 770 ±50 rpm.

12. Turn the headlights and rear defogger off. Idle the engine for one minute with the heater fan switch at

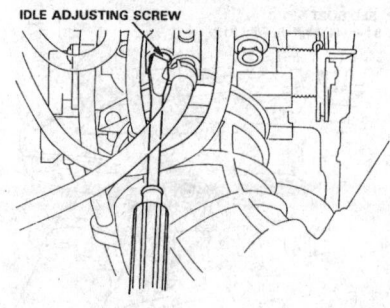

IDLE ADJUSTING SCREW

311607

Idle adjusting screw — 2.2L (H22A1, F22A1) and 2.3L (H23A1) engines

HI and the air conditioning on, then check the idle speed.
• The engine idle speed for the H23A1 engine should be 780 ±50 rpm.
• The engine idle speed for the H22A1 engine should be 790 ±50 rpm.
• The engine idle speed for the F22A1, F22B1 and F22B2 engines should be 770 ±50 rpm.

1996–97 Civic and del Sol

NOTE: The radio contains a coded anti-theft circuit. Disconnecting the battery or removing the 7.5A back-up fuse cancels the radio and clock presets. Obtain the security code.

1.6L (D16Y5, D16Y8, B16A2) Engines with Manual Transaxles

1. Check the following items before beginning the idle speed inspection and adjustment procedure:
a. Make sure the MIL is not on. Check any stored codes.
b. Ignition timing
c. Spark plug type and condition
d. Air cleaner and PCV system.

2. Connect a test tachometer to the test connector on the left shock tower.

3. Start the engine and warm it up to normal operating temperature at 3000 rpm with the transaxle in neutral and no electrical load. The cooling fan must come on at least once. After warming up, allow the engine to idle.

4. Uncouple the IAC valve connector.

5. Start the engine with the accelerator pedal slightly depressed. Use the pedal to stabilize the engine speed at 1000 rpm. Slowly release the pedal until the engine idles. Check the idle with air conditioning, cooling fan, and all electrical loads off. On vehicles equipped with daytime running lights (DRL's), pull the parking brake lever up to turn the headlights off.

6. The idle speed should be: 450±50 rpm.

7. If the rpm is not within specifications, adjust the idle speed as needed. Remove the cap and turn the idle adjusting screw on the throttle body.

8. After turning the idle adjusting screw, check the ignition timing. Then reset the idle speed if necessary.

9. Switch the ignition **OFF**. Reconnect the connector to the IAC valve. Then, reset the PCM by removing the Back up/Radio 7.5 A fuse

from the underhood fuse box for at least 10 seconds.

10. Restart the engine, allowing it to idle in neutral for one minute. With all electrical loads including the air conditioning and cooling fan off, idle speed should be as follows:
- D16Y5 engine: 670 ± 50 rpm (USA only)
- B16A2 engine: 700 ± 50 rpm (USA); 750±50 rpm (Canada)
- D16Y8 engine: 670 ± 50 rpm (USA); 750±50 rpm (Canada)

11. Allow the engine to idle for one minute with the headlight low beams on. The idle speed should be 750±50 rpm for all engines.

12. Turn the headlights off. Switch the air conditioner on and the blower fan to high speed. After one minute, the idle should be 810±50 rpm for all engines.

13. Remove the test equipment.

14. Verify that the MIL is not on. Reset the PCM if necessary.

D16Y7 engines with M/T or A/T; D16Y5 and D16Y8 engines with A/T or CVT

NOTE: If the vehicle is equipped with an Automatic or Continuously Variable Transaxle, the idle speed can be checked and adjusted with the Honda PGM-Tester or an equivalent OBD-II scan tool. When working with A/T or CVT vehicles, don't disconnect the IAC valve connector.

1. Check the following items before beginning the idle speed inspection and adjustment procedure:
  a. Make sure the MIL is not on. Check any stored codes.
  b. Ignition timing
  c. Spark plug type and condition
  d. Air cleaner and PCV system.

2. Connect a test tachometer to the test connector on the left shock tower.

3. Start the engine and warm it up to normal operating temperature at 3000 rpm in the **P** or **N** position with no electrical load. The cooling fan must come on at least once. After warming up, allow the engine to idle.

4. Start the engine with the accelerator pedal slightly depressed. Use the pedal to stabilize the engine speed at 1000 rpm. Slowly release the pedal until the engine idles. Check the idle with air conditioning, cooling fan, and all electrical loads off. On vehicles equipped with daytime running lights (DRL's), pull the parking brake lever up to turn the headlights off.

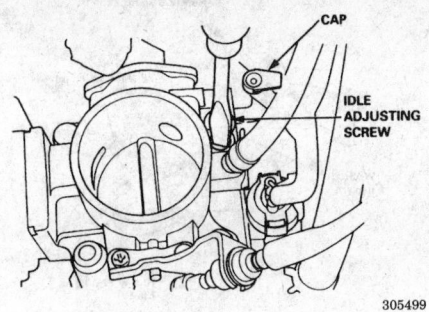

Idle adjusting screw — 1996–97 Civic and del Sol (D16Y7) engine

5. The idle speed with the vehicle in **P** or **N** should be as follows:Vehicles for the USA:
- D16Y7 engine with M/T: 670±50 rpm
- D16Y5 engine with CVT: 700±50 rpm
- D16Y7 & D16Y8 engine with A/T: 700±50 rpm

Vehicles for Canada:
- D16Y7 engine with M/T: 750±50 rpm
- D16Y7 & D16Y8 engines with A/T: 750±50 rpm

6. On the D16Y7 engine, do not disconnect the IAT sensor when removing the air cleaner housing from the throttle body.

7. If the rpm is not within specifications, the idle speed may be adjusted. Remove the cap, and turn the idle adjusting screw on the throttle body 1/2 turn clockwise or counterclockwise.

8. After turning the idle adjusting screw, recheck the idle speed. If the idle speed doesn't fall within the specified range, turn the adjusting screw an addition 1/2 turn. Then, recheck the idle speed.

——— WARNING ———
Don't turn the idle adjusting screw when the air conditioner is on.

9. Allow the engine to idle for one minute with the air conditioner on and the blower fan set on high speed. The idle speed should now be 810±50 rpm for all engine and transaxle combinations.

10. Remove the test equipment.

11. On D16Y7 engines, install the air cleaner assembly.

12. Verify that the MIL is not on. Reset the PCM if necessary.

Mixture

ADJUSTMENT

All Models

The air/fuel mixture is computer controlled according to the needs of the engine and is not adjustable. If the air/fuel mixture is too lean or too rich, other problems with the engine and/or engine control system exist.

Fuel Filter

REMOVAL AND INSTALLATION

Civic and del Sol

——— CAUTION ———
Fuel injection systems remain under pressure, even after the engine has been turned OFF. The fuel system pressure must be relieved before disconnecting any fuel lines. Failure to follow this procedure may result in fire or explosion.

NOTE: The original radio contains a coded anti–theft circuit. Obtain the security code number before disconnecting the battery.

1. Disconnect the negative battery cable.

2. Place a rag under the fuel filter to catch fuel spray.

3. Relieve the fuel pressure by first loosening the fuel filler cap. Use a 6mm flare wrench to hold the banjo bolt. Then, loosen the service bolt one complete turn with a box–end wrench or socket.

4. Use two flare wrenches to disconnect the fuel inlet line from the bottom of the filter. Plug the fuel line to keep out dirt.

5. Unbolt and remove the fuel filter clamp. Remove the filter from its bracket.

To install:

NOTE: Use new sealing washers when installing the fuel filter to prevent fuel leaks and the possibility of fire.

6. Clean the fuel line fittings before installing the filter.

7. Install the fuel filter and its clamp. Tighten the clamp bolt to 7 ft. lbs. (10 Nm).

8. Connect the fuel inlet line and carefully torque its fitting to 27 ft. lbs. (38 Nm).

9. Connect the fuel line with new washers and install the banjo bolt. Tighten the banjo bolt to 16 ft. lbs.

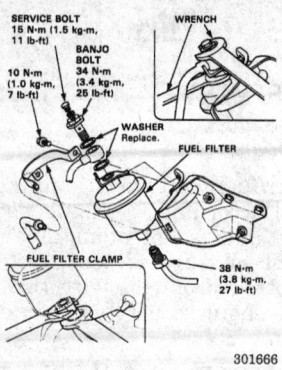

Fuel filter components — 1.5L and 1.6L engines

301666

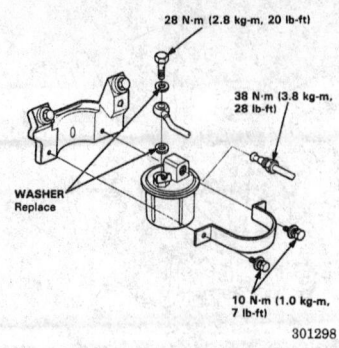

Fuel filter — Prelude with 2.2L (F22A1) and 2.3L (H23A1) engines

301298

(22 Nm). Install the service bolt and torque it to 9 ft. lbs. (12 Nm).

10. Connect the battery cable. Tighten the fuel filler cap.

11. Turn the ignition on and off several times to pressurize the fuel system. Start and run the engine and check for fuel leaks.

Prelude and Accord

NOTE: The radio may contain a coded theft protection circuit. Always obtain the code number before disconnecting the battery. If the vehicle is equipped with 4WS, the steering control unit is shut down when the battery is disconnected. After connecting the battery, turn the steering wheel lock–to–lock to reset the steering control unit.

1. Disconnect the negative battery cable.

2. Place a shop towel under and around the fuel rail, then relieve the fuel pressure.

────── **CAUTION** ──────

Do not allow fuel spray or fuel vapors to come in contact with a spark or open flame. Keep a dry chemical fire extinguisher nearby. Never store fuel in an open container due to risk of fire or explosion.

3. Remove the 12mm banjo bolt and the fuel feed pipe from the fuel filter. Discard the washers.

4. Remove the fuel filter clamp and the fuel filter.

To install:

5. Position the fuel filter on the bracket and install the filter clamp. Torque the clamp bolts to 7 ft. lbs. (10 Nm).

NOTE: Clean the fuel fittings thoroughly before reconnecting them.

6. Connect the fuel feed pipe to the filter, torque the fitting to 28 ft. lbs. (38 Nm).

7. Connect the fuel outlet pipe to the filter using new gaskets around the fitting. Torque the banjo bolt to 20 ft. lbs. (28 Nm) on all Prelude models and 16 ft. lbs. (22 Nm) on all other models.

8. Connect the negative battery cable and enter the radio security code.

9. If equipped with 4WS, turn the steering wheel lock–to–lock to reset the 4WS control unit.

10. Turn the ignition **ON** and check for fuel leaks.

Fuel Pump

REMOVAL AND INSTALLATION

NOTE: The radio may contain a coded theft protection circuit. Always obtain the code number before disconnecting the battery. If the vehicle is equipped with 4WS, the steering control unit is shut down when the battery is disconnected. After connecting the battery, turn the steering wheel lock–to–lock to reset the steering control unit.

Civic and del Sol

────── **CAUTION** ──────

Fuel injection systems remain under pressure, even after the engine has been turned OFF. The fuel system pressure must be relieved before disconnecting any fuel lines. Failure to follow this procedure may result in fire, explosion, or personal injury.

1. Disconnect the negative battery cable.

2. Loosen the fuel filler cap. Then, loosen the fuel filter service bolt to relieve the fuel pressure.

3. Remove the rear seat cushions (Civic), or the rear compartment trim (del Sol).

4. Remove the fuel pump access panel.

5. Disconnect the two–wire fuel pump harness.

6. Clean the fuel line fittings before disconnecting them.

7. Disconnect the fuel line and the hose from the fuel pump.

8. Unbolt the fuel pump and lift it out the fuel tank. Allow the fuel in the pump drain into the tank before removing the pump from the vehicle.

9. Disconnect and remove the fuel pump motor from its bracket.

To install:

NOTE: Use new sealing washers when reconnecting the fuel line banjo bolt.

10. Install the fuel pump into the fuel tank with a new O–ring. Then, tighten the mounting nuts to 4 ft. lbs. (6 Nm).

11. Reconnect the hose and the fuel line. Carefully tighten the banjo bolt to 20 ft. lbs. (28 Nm). Reconnect the fuel pump harness.

12. Tighten the fuel filler cap. Tighten the fuel filter service bolt to 11 ft. lbs. (15 Nm).

13. Connect the battery cable and turn the ignition switch **ON** and **OFF** several times to pressurize the fuel system.

14. Check the connections at the fuel pump for any leaks. Check the fuel filter service bolt for leaks.

15. Install the fuel pump access cover.

16. Install the rear seat cushions or rear compartment trim. Make sure the clips are properly seated.

Prelude

────── **CAUTION** ──────

Fuel injection systems remain under pressure after the engine has been turned OFF. Properly relieve fuel pressure before disconnecting any fuel lines. Failure to do so may result in fire or personal injury.

1. Disconnect the negative battery terminal.

2. Relieve the fuel pressure.

3. Lift or reposition the carpet in the luggage area. Remove the fuel pump maintenance access cover in the floor.

4. Disconnect the electrical connector at the pump unit.

5. Label and disconnect the fuel lines. Discard the washers from the fuel feed connection.

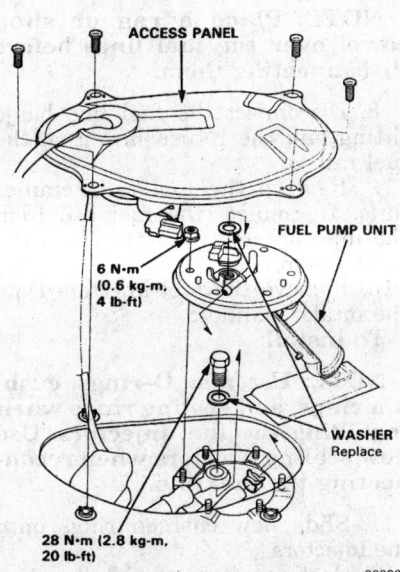

ACCESS PANEL

FUEL PUMP UNIT

6 N·m
(0.6 kg-m,
4 lb-ft)

WASHER
Replace

28 N·m (2.8 kg-m,
20 lb-ft)

303933

Fuel pump components–1993–95 Civic and
1993–97 del Sol

6. Carefully remove the retaining nuts holding the pump. When all are removed, lift the pump up and out of the tank.

NOTE: The pump sits on an angle and may require some manipulation to remove. If the pump still won't come out, loosen the fuel tank mounting nuts under the car; slide the tank downward a bit to give more clearance at the top.

To install:
7. Using a new sealing ring, reinstall the pump making certain it is correctly seated and not wedged or jammed. Install the retaining nuts, tightening them evenly and alternately to 4 ft. lbs. (6 Nm).
8. Install the fuel lines. Make certain the clamp is secure; use new ones if necessary. Install new washers to the fuel feed connection before installing the attaching bolt. Torque the fuel feed attaching bolt to 20 ft. lbs. (28 Nm).
9. Connect the fuel pump connector.
10. Connect the negative battery cable and enter the radio security code.
11. Switch the ignition **ON** but do not engage the starter. The fuel pump should run for approximately 2

seconds, building pressure within the lines. Switch the ignition **OFF**, then **ON** 2 or 3 more times to build full system pressure. Check for fuel leaks.
12. If equipped with 4WS, turn the steering wheel lock–to–lock to reset the 4WS control unit.
13. Install the maintenance access cover and seal or gasket, if used.
14. Reposition the carpeting in the luggage compartment.

Accord

1. Disconnect the negative battery cable.
2. Relieve the fuel pressure.

------ CAUTION ------
Fuel injection systems remain under pressure after the engine has been turned OFF. Properly relieve fuel pressure before disconnecting any fuel lines. Failure to do so may result in fire or personal injury.

3. Remove the fuel tank from the vehicle.
4. Disconnect the electrical connector from the fuel pump.
5. Remove the fuel feed line attaching bolt, and discard the washers. Disconnect the fuel return line from the fuel pump.
6. Remove the fuel pump mounting nuts.
7. Carefully remove the fuel pump from the fuel tank.
To install:
8. Clean the fuel pump mounting surface and install a new gasket.
9. Install the fuel pump into the tank, being careful not to damage the pickup screen.
10. Install the mounting nuts; torque the nut s to 4 ft. lbs. (6 Nm).
11. Connect the fuel return hose to the pump and make sure that the clamp is secure.
12. Connect the fuel feed line to the pump with new washers, torque the bolt to 21 ft. lbs. (27 Nm).
13. Connect the fuel pump electrical connector.
14. Install the fuel tank into the vehicle.
15. Connect the negative battery cable and enter the radio security code.
16. Switch the ignition **ON** but do not engage the starter. The fuel pump should run for approximately 2 seconds, building pressure within the lines. Switch the ignition **OFF**, then **ON** 2 or 3 more times to build full system pressure. Check for fuel leaks.

Fuel Injector

REMOVAL AND INSTALLATION

NOTE: The radio may contain a coded theft protection circuit. Always obtain the code number before disconnecting the battery. If the vehicle is equipped with 4WS, the steering control unit is shut down when the battery is disconnected. After connecting the battery, turn the steering wheel lock–to–lock to reset the steering control unit.

1993–95 Civic and del Sol

------ CAUTION ------
Fuel injection systems remain under pressure even after the engine has been turned OFF. The fuel system pressure must be relieved before disconnecting any fuel lines. Failure to do so may result in fire and/or personal injury.

1. Disconnect the negative battery cable.
2. Relieve the fuel pressure.
3. Disconnect the electrical connectors from the fuel injectors. On B16A3 engines (del Sol VTEC), disconnect the IAT sensor harness.
4. Disconnect the vacuum hose and fuel return hose from the fuel pressure regulator.

NOTE: Place a rag or shop towel over the fuel lines before disconnecting them.

5. Disconnect the fuel line and pulsation damper from the fuel rail.
6. Remove the fuel rail retainer nuts. Disconnect the fuel rail from the injectors and remove it.
7. Remove the injectors from the intake manifold.
To install:

NOTE: Use new O-rings, cushion rings, and sealing rings when installing the fuel injectors. Use new sealing washers when reconnecting the fuel line.

8. Slide new cushion rings onto the injectors.
9. Coat new O-rings with clean engine oil and install them on the injectors.
10. Insert the injectors onto the fuel rail.
11. Coat new seal rings with clean engine oil and press them into the intake manifold.

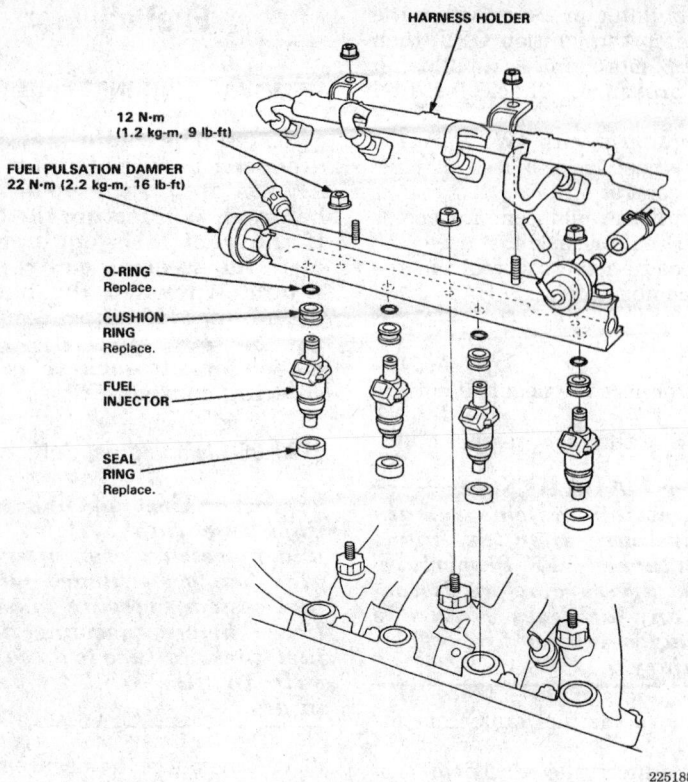

HARNESS HOLDER

12 N·m
(1.2 kg-m, 9 lb-ft)

FUEL PULSATION DAMPER
22 N·m (2.2 kg-m, 16 lb-ft)

O-RING
Replace.

CUSHION RING
Replace.

FUEL INJECTOR

SEAL RING
Replace.

225188

Fuel rail components and fuel injectors — 1993–95 Civic and del Sol

12. Install the injectors and fuel rail assembly onto the intake manifold.

NOTE: To prevent damage to the O-rings, install the injectors in the fuel rail first, then install the assembly onto the intake manifold.

13. On vehicles equipped with the D15Z1 engine (Civic VX), align the center line marking on the fuel injector with the mark on the fuel pipe.

14. Install and tighten the fuel pipe retainer nuts to 9 ft. lbs (12 Nm).

15. Connect the fuel line and pulsation dampener to the fuel rail, using new sealing washers. Connect the vacuum hose and fuel return line to the pressure regulator. Carefully tighten the pulsation dampener to 16 ft. lbs. (22 Nm).

16. Connect the electrical connectors to the injectors and the IAT sensor. Install the harness holder bolts.

17. Connect the negative battery cable and turn the ignition switch **ON** for 2 seconds, but do not start the engine. Repeat 2–3 times and check for fuel leaks.

18. Road test the vehicle.

1996–97 Civic and del Sol

—————— **CAUTION** ——————
Fuel injection systems remain under pressure even after the engine has been turned OFF. The fuel system pressure must be relieved before disconnecting any fuel lines. Failure to do so may result in fire and/or personal injury.

1. Disconnect the negative battery cable.

2. Relieve the fuel pressure.

3. Clean up any fuel spilled on the engine or intake manifold.

4. Detach the fuel injector wiring harness holder from the two tabs on the fuel rail.

5. Lift the harness holder up and label and uncouple the four fuel injector connectors. Then, disconnect the harness and the vacuum line from the EVAP solenoid valve. Move the harness holder out of the way.

6. Clean the fuel hose connections and the bodies of the injectors before disconnecting and removing them. This will help keep dirt from entering the fuel lines or intake manifold.

7. Disconnect the vacuum hose and fuel return hose from the fuel pressure regulator.

NOTE: Place a rag or shop towel over the fuel lines before disconnecting them.

8. Disconnect the fuel line banjo fitting and the return line from the fuel rail.

9. Remove the fuel rail retainer nuts. Disconnect the fuel rail from the injectors and remove it from the intake manifold.

10. Remove the fuel injectors from the intake manifold.

To install:

NOTE: Use new O-rings, cushion rings, and sealing rings when installing the fuel injectors. Use new sealing washers when reconnecting the fuel line.

11. Slide new cushion rings onto the injectors.

12. Coat new O-rings with clean engine oil and install them on the injectors.

13. Insert the injectors onto the fuel rail.

14. Coat new seal rings with clean engine oil and press them into the intake manifold.

15. Install the injectors and fuel rail assembly onto the intake manifold.

NOTE: To prevent damage to the O-rings, install the injectors in the fuel rail first, then install the assembly onto the intake manifold.

16. Align the any centering marks on the fuel injector with the mark on the fuel rail during installation.

17. Install and tighten the fuel rail retainer nuts to 9 ft. lbs (12 Nm).

18. Connect the fuel line banjo fitting and return line to the fuel rail using new sealing washers. Connect the vacuum hose and fuel return line to the pressure regulator. Carefully tighten the banjo bolt to 16 ft. lbs. (22 Nm) for the D16Y8, D16Y8, B16A2 engines, or 21 ft. lbs. (28 Nm) for the D16Y5 engine.

19. Connect the electrical connectors to the injectors. Connect the vacuum line and harness connector to the EVAP solenoid valve. Fit the harness holder securely onto its mounting tabs.

20. Install a new sealing washer on the fuel filter service bolt, tighten the bolt to 9 ft. lbs. (12 Nm).

21. Connect the negative battery cable and turn the ignition switch **ON** for two seconds, but do not start

the engine. Repeat this step two or three times and check for fuel leaks.

Prelude

1. Disconnect the negative battery cable.

---------------- CAUTION ----------------
Fuel injection systems remain under pressure after the engine has been turned OFF. Properly relieve fuel pressure before disconnecting any fuel lines. Failure to do so may result in fire or personal injury.

2. Relieve the fuel pressure.
3. Label and disconnect the electrical harnesses from the injectors.
4. Disconnect the IAC valve connector.
5. Place a rag below the fuel pressure regulator. Disconnect the vacuum hose and disconnect the fuel line. Plug the fuel line to prevent spillage and the entry of dirt.
6. Disconnect the fuel hose from the fuel rail.
7. Disconnect the vacuum hose from the EGR valve.
8. Remove the nuts holding the fuel pipe (rail).
9. Disconnect the PCV valve from the engine.
10. Remove the fuel rail from the injectors, leaving the injectors in the manifold.
11. Remove each injector and remove the seal ring from each manifold port.
12. Remove the cushion ring and O-ring from each injector.
   To install:
13. Install new cushion rings on each injector. Install new O-rings on each injector and coat the O-rings with a light coat of clean, thin oil.
14. Install the injectors into the fuel pipe. Make certain the O-rings seat properly and do not distort.
15. Coat new seal rings with a light coat of clean, thin oil and install the rings into the manifold.
16. Install the fuel rail and injectors to the intake manifold. Make certain the insulators are present on the mounting bolts before installing fuel rail.

   NOTE: Assembling each injector into the rail off the engine prevents damage to the O — rings. Handle the rail and injector assembly carefully when reinstalling to the manifold. Don't drop an injector or bang the tips.

17. With all injectors seated in the manifold, align the center mark on each injector electrical connector with the mark on the fuel rail.
18. Install the fuel rail retaining nuts and tighten them evenly. Torque the nuts to 9 ft.lbs. (12 Nm). Make certain the wiring harness is retained in its clips.
19. Connect the vacuum hose and fuel hose to the regulator.
20. Connect the fuel line to the fuel rail, use new washers at the connection. Torque the fuel line attaching nut to 16 ft. lbs. (22 Nm).
21. Install the connectors to the injectors.
22. Connect the IAC valve connector.
23. Connect the EGR valve vacuum hose.
24. Install the PCV valve to the engine.
25. Connect the negative battery cable and enter the radio security code.
26. Switch the ignition **ON** but do not engage the starter. The fuel pump should run for approximately 2 seconds, building pressure within the lines. Switch the ignition **OFF**, then **ON** at least twice to build full system pressure. Check for fuel leaks.
27. If equipped with 4WS, start the engine and turn the steering wheel lock–to–lock to reset the 4WS control unit.

Accord

2.2L Engines

1. Disconnect the negative battery cable.

---------------- CAUTION ----------------
Fuel injection systems remain under pressure after the engine has been turned OFF. Properly relieve fuel pressure before disconnecting any fuel lines. Failure to do so may result in fire or personal injury.

2. Relieve the fuel pressure.
3. Label and disconnect the electrical connectors from the fuel injectors. Position the harness out of the way.
4. Place a rag below the fuel pressure regulator. Disconnect the vacuum hose and disconnect the fuel line. Plug the fuel line to prevent spillage and the entry of dirt.
5. Label and disconnect the vacuum lines from the fuel rail.
6. Disconnect the fuel hose from the fuel rail. Discard the washers.

7. Remove the nuts holding the fuel rail.
8. Remove the fuel rail from the injectors, leaving the injectors in the manifold.
9. Remove each injector and remove the seal ring from each manifold port.
10. Remove the cushion ring and O-ring from each injector.
   To install:
11. Install new cushion rings on each injector. Install new O-rings on each injector and coat the O-rings with a light coat of clean, thin oil.
12. Install the injectors into the fuel rail. Make certain the O-rings seat properly and do not distort.
13. Coat new seal rings with a light coat of clean, thin oil and install the rings into the manifold.
14. Install the fuel rail and injectors to the intake manifold. Make certain the insulators are present on the mounting bolts before installing the fuel rail.

   NOTE: Assembling each injector into the rail off the engine prevents damage to the O-rings. Handle the rail and injector assembly carefully when reinstalling to the manifold. Don't drop an injector or bang the tips.

15. Install the fuel rail retaining nuts and tighten them evenly. Torque the nuts to 9 ft. lbs. (12 Nm). Make certain the wiring harness is retained in its clips.
16. Connect the fuel hose to the fuel rail with new washers. Torque the attaching nut to 16 ft. lbs. (22 Nm).
17. Connect the vacuum hose and fuel hose to the regulator.
18. Connect the electrical harness to the injectors.
19. Connect the vacuum hoses to the fuel rail.
20. Install the service bolt to the fuel rail with a new washer. Torque the service bolt to 9 ft.lbs. (12 Nm).
21. Connect the negative battery cable and enter the radio security code.
22. Switch the ignition **ON** but do not engage the starter. The fuel pump should run for approximately 2 seconds, building pressure within the lines. Switch the ignition **OFF**, then **ON** at least twice to build full system pressure. Check for fuel leaks.

2.7L Engines

1. Disconnect the negative battery cable.

CAUTION

Fuel injection systems remain under pressure after the engine has been turned OFF. Properly relieve fuel pressure before disconnecting any fuel lines. Failure to do so may result in fire or personal injury.

---

2. Relieve the fuel pressure.

3. Remove the intake manifold covers.

4. Label and disconnect the electrical harnesses from the injectors.

5. Remove the wiring harness holders and position them out of the way.

6. Disconnect the vacuum hose from the fuel pressure regulator.

7. Remove the fuel rail attaching nuts.

8. Remove the fuel rail from the injectors, leaving the injectors in the manifold.

9. Remove each injector, noting its position, and remove the seal ring from each manifold port.

10. Remove the cushion ring and O-ring from each injector.

To install:

11. Install new cushion rings on each injector.

12. Coat new O-rings with clean engine oil and install them on the fuel injectors.

13. Install the injectors into the fuel rail. Make certain the O-rings seat properly and are not distorted.

NOTE: Assembling each injector into the fuel rail prevents damage to the O-rings. Handle the rail and injector assembly carefully when reinstalling it to the manifold. Don't drop the injectors or bang their tips.

14. Coat new seal rings with a light coat of clean oil and install them into the manifold.

15. Install the fuel rail and injectors to the intake manifold.

16. With all injectors seated in the manifold, make sure each injector is positioned properly.

17. Install the fuel rail retaining nuts and tighten them evenly to 9 ft. lbs. (12 Nm).

18. Connect the vacuum hose to the regulator.

19. Install the wiring harness to the fuel rail, torque the mounting bolts to 9 ft. lbs. (12 Nm).

20. Connect the electrical harness to the injectors.

21. Connect the negative battery cable and enter the radio security code.

22. Switch the ignition **ON** but do not engage the starter. The fuel pump should run for approximately 2 seconds, building pressure within the lines. Switch the ignition **OFF**, then **ON** 2 or 3 more times to build full system pressure. Check for fuel leaks.

23. Install the intake manifold covers, torque the mounting bolts to 9 ft. lbs. (12 Nm).

EMISSIONS CONTROLS

Service Interval Lamp

RESETTING

1993–95 Accord

Vehicles equipped with a Maintenance Reminder indicator will indicate it is time for scheduled maintenance. When it is near 7,500 miles (12,000 km) since the last maintenance, the indicator will turn yellow. If you exceed 7,500 miles (12,000 km), the indicator will turn red. The indicator can be reset by inserting the ignition key or other similar object into the slot below and to the left of the indicator. This will extinguish the indicator for the next 7500 miles.

ENGINE MECHANICAL

Engine Assembly

REMOVAL AND INSTALLATION

NOTE: The original radio contains a coded anti–theft circuit. Obtain the security code number before disconnecting the battery cables.

1993–95 Civic and 1993–97 del Sol

1. Disconnect the negative and positive battery cables.

2. Raise the vehicle and support it safely.

3. Remove the radiator cap.

NOTE: The engine and transaxle are removed from the vehicle as one unit.

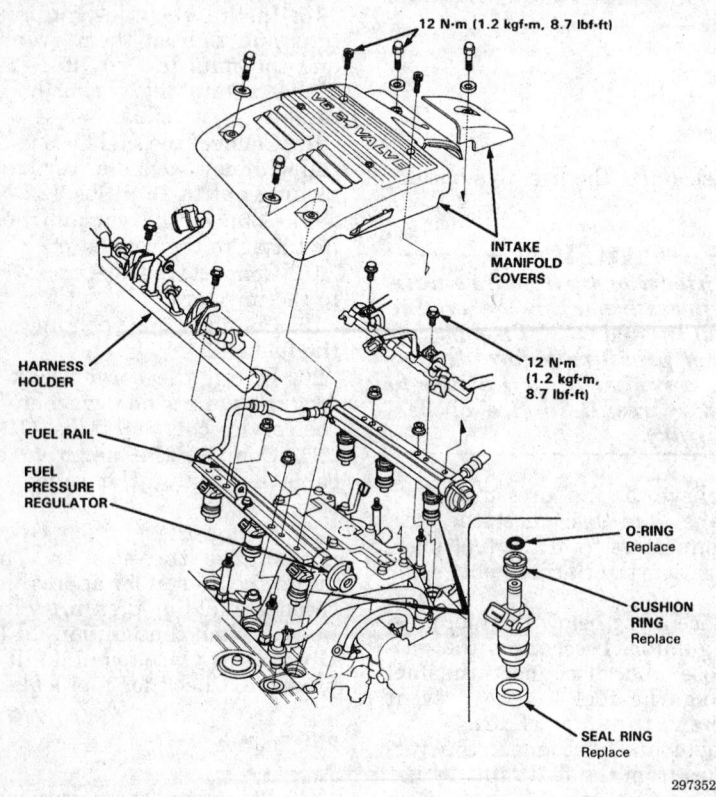

Fuel injectors and related components — 2.7L engines

297352

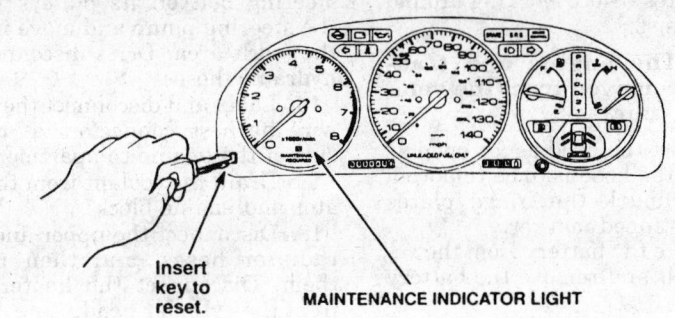

Insert key to reset.

MAINTENANCE INDICATOR LIGHT

276483

Resettting the Maintenance Required indicator — 1993 Accord

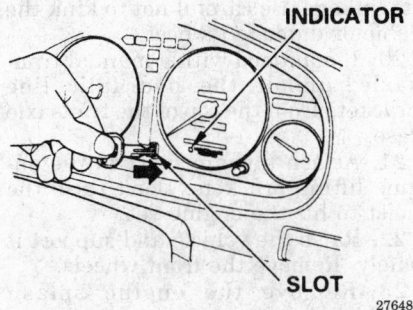

INDICATOR

SLOT

276484

Resetting the Maintenance Required indicator — 1994–95 Accord

4. Remove the engine splash shield.

5. Drain the coolant, transaxle oil and engine oil.

6. Lower the vehicle, Secure the hood as far open as possible.

7. Unbolt the underhood ABS fuse/relay box and move it out of the way.

8. Remove the air intake hose, resonator and air cleaner assembly.

9. Relieve the system fuel pressure by turning the fuel filter service bolt one turn.

10. Remove the fuel feed hose and charcoal canister hose from the intake manifold.

11. Remove the throttle cable by loosening the locknut and slipping the cable end out of the accelerator linkage.

12. Disconnect the engine wire harness connectors at the left side of the engine compartment.

13. Remove the fuel return hose and brake booster vacuum hose.

14. Disconnect the engine wire harness connectors, terminal, and clamps on the right side of the engine compartment.

15. Remove the battery/starter cable from the underhood fuse/relay box. Remove the ABS power cable from the battery terminal.

16. Remove the engine ground cable from the cylinder head.

17. Remove the power steering pump and belt, but do not disconnect the power steering hoses.

18. If equipped with A/C, unbolt the left front engine mount bracket from the body.

19. Remove the air conditioning belt and compressor and disconnect the electrical connector, but do not disconnect the air conditioning hoses.

20. Remove the transaxle ground cable. On vehicles with automatic transaxles, disconnect the ATF cooler lines.

21. On vehicles with manual transaxles, remove the slave cylinder without disconnecting the hydraulic line.

22. Raise and support the vehicle safely. Remove the front wheels.

23. Remove the upper and lower radiator hoses and heater hoses.

24. Remove the front exhaust pipe and stay.

25. On automatic transaxle, disconnect the shift cable.

26. On manual transaxle, disconnect the shift rod and extension rod from the transaxle.

27. Remove the strut damper fork. Disconnect the suspension lower arm ball joint using a ball joint separator.

28. Remove the driveshafts. Tie plastic bags over the inboard CV–joints to prevent damage to the boots.

29. Attach a chain hoist to the engine lifting brackets and raise it slightly to take up the weight of the engine/transaxle assembly.

30. At this point the engine mounts can be removed individually if any need to be replaced without the removal of the entire engine.

31. Remove the left and right engine stopper rubbers and brackets.

32. Remove the rear engine mounting bracket. Remove the engine support nuts.

33. Loosen the mount bolt and pivot the engine side mount out of the way.

34. Remove the transaxle mount nuts and pivot the mount out of the way.

35. Raise the chain hoist so it is tight.

36. Verify that all electrical, vacuum, coolant and fuel lines have been disconnected.

37. Remove engine/transaxle assembly from the vehicle.

To install:

NOTE: Use new self–locking nuts and gaskets when installing the front exhaust pipe and when assembling the front suspension. Use new set rings on the inboard CV–joint shaft.

38. Lower the engine into the vehicle.

39. Install and connect the engine and transaxle mounts and brackets. At this point only tighten the mounting nuts and bolts by hand.

NOTE: Failure to tighten the bolts in the proper sequence can cause excessive noise and vibration and reduce bushing life. Be sure to check that the bushings are not twisted or offset.

40. The engine and transaxle mount and bracket fasteners must be torqued in the proper sequence with the weight of the engine resting upon them. This step is important for engine mount preloading. Torque the engine mount bolts in the following sequence:

   a. Side transaxle mount nuts — 47 ft. lbs. (65 Nm).

   b. Engine side mount nuts — 47 ft. lbs. (65 Nm).

   c. Side transaxle mount bolt — 54 ft. lbs. (75 Nm).

   d. Engine side mount bolt — 54 ft. lbs. (75 Nm).

   e. Rear engine mount–to–engine bracket — 61 ft. lbs. (85 Nm).

   f. Rear engine mount through-bolt — 43 ft. lbs. (60 Nm).

   g. Right front stopper bracket — 47 ft. lbs. (65 Nm). Stopper–to–body bolts:33 ft. lbs. (45 Nm).

   h. Left front mount — stud: 61 ft. lbs. (85 Nm), nut:43 ft. lbs. (60 Nm), bolts: 33 ft. lbs. (45 Nm).

41. Check that the spring clip on the end of each driveshaft clicks into place. Be sure to use new spring clips on installation.

42. Install the damper fork and reconnect the lower ball joint. With the vehicle on the ground, torque the pinch bolt to 32 ft. lbs. (44 Nm), and the the fork bolt to 47 ft. lbs. (65 Nm).

43. Install the slave cylinder and connect the shift rod and extension rod to the transaxle on manual transaxle vehicles. On automatic transaxle vehicles, connect the shift cable and install its cover.

44. Install the front exhaust pipe. Use new self–locking nuts. Torque the converter flange nuts to 16 ft. lbs. (22 Nm). Torque the exhaust manifold nuts to 40 ft. lbs. (55 Nm).

45. Reconnect the radiator and heater hoses. Reconnect the ATF cooler lines.

46. Install the A/C compressor and torque its mounting bolts to 16 ft. lbs. (22 Nm).

47. Reconnect the engine and transaxle ground cables. Reconnect the stater cables and the ABS power cable.

48. Reconnect the engine wiring harnesses.

49. Reconnect the throttle cable and adjust its deflection to be 10–12mm (0.39–0.47 in.).

50. Reconnect the fuel lines to the fuel rail and fuel filter. Use new sealing washers. Tighten the fuel filter banjo bolt to 25 ft. lbs. (34 Nm) and the service bolt to 11 ft. lbs. (15 Nm).

51. Install the relay box, resonator, air cleaner, and intake duct.

52. Install the accessory belts and adjust their tensions.

53. Install the splash shield.

54. Refill the cooling system.

55. Refill the engine with fresh oil.

56. Refill the transaxle with the proper fluid.

57. Install and reconnect the battery.

58. Verify that all fuel and vacuum lines and electrical harnesses and ground cables have been reconnected properly.

59. After assembling the fuel line parts, turn the ignition switch to the ON, but don't start the engine. Then, turn the ignition switch OFF. Repeat this procedure two or three times to pressurize the fuel system and check for leak.

60. Bleed the air from the cooling system at the bleed bolt with the heater valve open.

61. Adjust the clutch cable free-play and check that the transaxle shifts into gear smoothly.

62. Check the ignition timing.

63. Check the and adjust the front wheel alignment.

64. Road test the vehicle.

1996–97 Civic

1. Disconnect the negative and positive battery cables. Wait at least three minutes before working around the air bags.

NOTE: The engine and transaxle are removed from the vehicle as one unit.

2. Support the hood as far open as possible. If the hood is to be removed, first matchmark the hinge plates with a felt–tipped marker.

3. Remove the battery from the vehicle. Unbolt and remove the battery tray.

4. Disconnect the battery and alternator cables from the underhood fuse and relay box on the right shock tower.

5. Remove the lower right kick panel to expose the PCM.

6. Label and disconnect the five wiring harness connections from the PCM.

7. Unbolt the main wiring harness retainer from the rear of the fuse and relay box on the right side of the bulkhead. Carefully pull the grommet out of its bulkhead opening. Next, pull the PCM harness and connectors through the opening. Be careful not to damage the wiring, insulation, or connectors.

8. Relieve the fuel pressure:
   a. Loosen the fuel filler cap.
   b. Use a box–end wrench and a flare wrench to hold the fuel filter banjo fitting.
   c. Place a shop towel over the fuel filter to catch the fuel spray.
   d. Slowly loosen the fuel filter service bolt one full turn.
   e. Clean up any spilled fuel.

9. Remove the intake air duct and air cleaner. If equipped, disconnect the intake air temperature (IAT) sensor connector from the air cleaner case.

10. Disconnect the fuel feed hose from the fuel filter. Disconnect the fuel return hose from the fuel rail.

11. Label and disconnect the following vacuum lines:
   • Intake manifold/throttle body vacuum hoses
   • Brake booster vacuum hose
   • EVAP canister vacuum hose

12. Disconnect the power steering pressure switch (PSP) and detach its clamp from the bracket below the brake booster.

13. Disconnect the transaxle ground cable. Remove the radiator hose bracket.

14. Loosen the throttle cable's locknut, and then disconnect the cable from the throttle body linkage. Don't kink the cable: move it out of the work area.

15. Loosen the power steering pump mounting bolts. Slip power steering belt off its pulleys. Unbolt the steering pump and move it out of the work area. Don't disconnect the hydraulic hoses.

16. Label and disconnect the engine wire harness connectors at the left side of the engine compartment.

17. Drain the coolant from the radiator and engine block.

18. Disconnect the upper and lower radiator hoses, and then remove them. Disconnect the heater hoses from the cylinder head.

19. If equipped with a CVT transaxle, loosen the shift cable locknut. Remove the spring clip and washers and disconnect the shift cable from its linkage. Be careful not to kink the cable or damage its boot.

20. If equipped with a manual transaxle, unbolt the hydraulic line brackets from the top of the transaxle case.

21. Attach a chain hoist to the engine lifting brackets. Don't raise the hoist to lift the engine yet.

22. Raise the vehicle and support it safely. Remove the front wheels.

23. Remove the engine splash shield.

24. Drain the engine oil.

25. Drain the fluid from the transaxle.

26. If equipped with A/C, unbolt the left front engine mount bracket from the shock tower.

27. Loosen the compressor idler pulley and adjusting bolt. Slip the belt around the engine mount stud to remove it.

28. Unbolt the compressor mounting bolts to separate the compressor from its mounting plate. Move the compressor out of the work area. Do not disconnect the air conditioning refrigerant lines.

29. If equipped with an automatic transaxle, disconnect the ATF cooler lines. Plug the cooler lines to prevent fluid leakage and contamination.

30. If equipped with a manual transaxle, unbolt the slave cylinder from the transaxle case without disconnecting its hydraulic line.

31. Separate the front exhaust pipe from the exhaust manifold and catalytic converter. Unbolt its hanger bracket and remove the exhaust pipe.

32. If equipped with a automatic transaxle, disconnect the shift cable from the transaxle control shaft.

33. If equipped with a manual transaxle, disconnect the shift rod and extension rod from the transaxle.

34. Unbolt and remove the strut damper fork. Disconnect the steering knuckle ball joint from the lower control arm using a ball joint separator.

35. Pry the inboard CV–joints from the transaxle. Then, move the half-shafts away from the transaxle and wire them to the undercarriage of the vehicle. Tie plastic bags over the inboard CV–joints to prevent damage to the boots and splined shafts.

36. Raise the hoist slightly to take up the weight of the engine and transaxle assembly.

37. Disconnect the engine mounts in the following order:

    a. Unbolt and remove the left front engine mount.

    b. Unbolt and remove the right front engine mount and bracket assembly.

    c. Remove the rear engine mount through–bolt. Then, unbolt the rear mount bracket from the engine block.

38. If necessary, lower the vehicle slightly to gain access to the side engine and transaxle mounts. Do not release the tension of the chain hoist — the engine must be securely supported.

39. Unbolt the side engine mount bracket from the engine block bracket and mount damper.

40. Unbolt the transaxle mount bracket from the transaxle case. Then, unbolt the mount from the shock tower.

41. Raise the chain hoist to lift the engine a few inches off of its mounts.

42. Verify that all electrical, vacuum, and fuel lines have been disconnected.

43. Raise the engine and transaxle assembly and remove it from the vehicle.

To install:

NOTE: Use new self-locking nuts and gaskets when installing the front exhaust pipe and when assembling the front suspension. Use new set rings on the inboard CV–joint splined shafts.

44. Lower the engine and transaxle assembly into the vehicle.

45. Install and connect the engine and transaxle mounts and brackets. Use new self–locking nuts and color–coded bolts. At this point, only tighten the mounting nuts and bolts by hand.

46. Before installing the left front engine mount, fit the A/C compressor back into place and install the compressor belt. Tighten the compressor bolts to 17 ft. lbs. (22 Nm).

NOTE: Failure to tighten the bolts in the proper sequence can cause excessive noise and vibra-tion and reduce bushing life. Be sure to check that the bushings are not twisted or offset.

47. The engine and transaxle mount and bracket fasteners must be torqued in the proper sequence with the weight of the engine resting upon them. This step is important for engine mount preloading. Torque the engine mount bolts in the following sequence:

    a. Transaxle mount bolts: 47 ft. lbs. (64 Nm); or 28 ft. lbs. (38 Nm) for CVT–equipped vehicles

    b. Side engine mount bracket nuts: 54 ft. lbs. (74 Nm)

    c. Rear mount bracket bolts: 61 ft. lbs. (83 Nm); or 43 ft. lbs. (59 Nm) for CVT–equipped vehicles

    d. Rear mount through–bolt: 43 ft. lbs. (59 Nm)

    e. Transaxle mount bracket nuts or bolts: 47 ft. lbs. (64 Nm).

    f. Transaxle mount through–bolt: 54 ft. lbs. (74 Nm).

    g. Right front mount bracket bolts: 33 ft. lbs. (44 Nm).

    h. Right front mount carrier bolts: 33 ft. lbs. (44 Nm).

    i. Left front mount: stud: 61 ft. lbs. (85 Nm); carrier bolts: 33 ft. lbs. (44 Nm); nut: 43 ft. lbs. (59 Nm).

48. Remove the chain hoist from the engine lifting hooks.

49. Install new set rings on the inboard splined shafts of each half-shaft. Check that the set ring on each inboard CV–joint clicks into place when the halfshafts are installed into the transaxle.

50. Install the damper fork and reconnect the lower ball joint. When the weight of the vehicle is resting on its suspension, tighten the pinch bolt to 32 ft. lbs. (44 Nm) and the the fork bolt to 47 ft. lbs. (65 Nm). Tighten the ball joint castle nut to 36–43 ft. lbs. (50–60 Nm). Next, tighten the castle nut only enough to install a new cotter pin.

51. If equipped, install the slave cylinder. Tighten the slave cylinder mounting bolts to 16 ft. lbs. (22 Nm). If the clutch hydraulic line was disconnected, the fluid must be bled.

52. If equipped, reconnect the transaxle shift and extension rods to the linkage at the transaxle case. Install a new 8mm spring pin into the shift rod linkage. Then, install the retainer clip and boot. Tighten the extension rod bolt to 16 ft. lbs. (22 Nm).

53. If equipped with an automatic transaxle, connect the shift cable to the control shaft. Use a new lockwasher and tighten the lockbolt to 10 ft. lbs. (14 Nm). Tighten the shift cable cover bolts to 16 ft. lbs. (22 Nm). Install the shift cable cover and tighten its bolts to 16 ft. lbs. (22 Nm).

54. Install the front exhaust pipe using new self–locking nuts.

    • If equipped with the D16Y8 engine, tighten the converter flange nuts to 16 ft. lbs. (22 Nm). Tighten the exhaust manifold nuts to 40 ft. lbs. (55 Nm).

    • If equipped with the D16Y5 or D16Y7 engine, tighten the converter flange nuts to 25 ft. lbs. (33 Nm). Tighten the exhaust flange bolts to 16 ft. lbs. (22 Nm).

55. Reconnect the ATF cooler lines. If the rubber cooler lines are cracked or stressed, they must be replaced.

56. Install the engine splash shield.

57. Refill the engine with fresh oil.

58. Refill the transaxle with the proper fluid.

59. Lower the vehicle.

60. If equipped, fit the clutch hydraulic line brackets back into place. Tighten the 8mm bolts to 17 ft. lbs. (24 Nm). Tighten the 6mm bolts to 8 ft. lbs. (11 Nm).

61. If equipped with a CVT transaxle, Reconnect the shift cable to the linkage. Use new plastic washers and a new spring clip. Tighten the locknut to 22 ft. lbs. (29 Nm).

62. Adjust the alternator and A/C compressor belt tensions.

63. Install and reconnect the upper and lower radiator hoses. Reconnect the heater hoses.

64. Install the power steering pump into its mounts. Adjust the pump belt's tension, and then tighten the mounting bolts to 17 ft. lbs. (24 Nm).

65. Reconnect the PSP switch connector and attach its harness clamp.

66. Reconnect the following vacuum lines:

    • Intake manifold/throttle body vacuum hoses

    • Brake booster vacuum hose

    • EVAP canister vacuum hose

67. Reconnect the fuel line fittings to the fuel filter and fuel rail. Use new sealing washers. Tighten the banjo fittings to 25 ft. lbs. (33 Nm), and the service bolts to 11 ft. lbs. (15 Nm). Don't overtighten the fittings.

68. Reconnect the throttle cable and adjust its deflection to 10–12mm (0.39–0.47 in.).

69. Feed the PCM harness through the hole in the bulkhead. Apply sealant to the grommet, and then install the retainer.

70. Reconnect any engine wiring harness and ground cables that were disconnected during engine removal. Make sure the grounds are free of corrosion to ensure good contact.

71. Fit the fuse and relay box back into position. Reconnect the battery and alternator cables.

72. Install the air cleaner case and air intake duct. Reconnect the IAT connector.

73. Reconnect the five PCM connectors. Install the kick panel.

74. Install the battery tray and the battery.

75. Verify that all wiring harnesses and grounds, vacuum lines, fuel lines have been reconnected.

76. Refill the radiator with fresh coolant.

77. If it was removed, install the hood. Reconnect the windshield washer tubing. After installation, check to make sure that the hood, fender, and grille panel gaps are equal.

78. Reconnect the positive and negative battery cables.

79. Turn the ignition switch to the **ON** position, but don't start the engine. Then, turn the ignition **OFF**. Repeat this procedure two or three times and check for a any indications of fuel leaks.

80. Start the engine and allow it to warm up to its normal operating temperature.

81. Bleed the air from the cooling system with the heater valve open.

82. Check the throttle cable deflection and operation.

83. Check the and adjust the ignition timing.

84. Shut the engine off and check the drive belt adjustments.

85. Check all fluid levels and top up as necessary.

86. Check the and adjust the front wheel alignment.

87. Road test the vehicle.

Prelude

1. Secure the hood as far open as possible.

2. Disconnect the negative battery cable then the positive battery cable.

3. Remove the radiator cap.

4. Raise and safely support the vehicle. Remove the front wheels and the engine splash shield.

5. Drain the engine coolant into a sealable container.

6. Drain the transaxle oil/fluid into a sealable container. Install the drain plug with a new gasket.

7. Lower the vehicle to a working level.

8. Remove the air intake duct and the air cleaner case.

9. Remove the pulsed secondary air Injection (PAIR) vacuum tank and bracket.

10. Remove the battery and the battery base. Disconnect the battery cable and starter cable harnesses from the body.

11. Relieve the pressure from the fuel system.

——— CAUTION ———
Fuel injection systems remain under pressure after the engine has been turned OFF. Properly relieve fuel pressure before disconnecting any fuel lines. Failure to do so may result in fire or personal injury.

12. Disconnect the fuel feed hose from the fuel rail and disconnect the fuel return line from the fuel pressure regulator.

13. Disconnect the injector resistor connector on the left side of the engine compartment.

14. Remove the throttle cable by loosening the locknut, then slip the cable end out of the throttle linkage. Take care not to bend the cable when removing it. Always replace any kinked cable with a new one.

15. Disconnect the engine wire harness connectors, terminal and clamps on the right side of the engine.

16. Remove the power cable from the under-hood fuse/relay box.

17. Disconnect the brake booster vacuum hose and emissions control vacuum tubes from the intake manifold.

18. Disconnect the cruise control actuator electrical connector and vacuum tube, then remove the actuator.

19. Remove the engine ground cable from the body side.

20. Remove the power steering pump drive belt then remove the pump.

21. Remove the air conditioning (A/C) condenser fan, then install a protector plate on the radiator.

22. Loosen the alternator mounting bolt, nut and adjusting nut, then remove the alternator drive belt.

23. Disconnect the A/C compressor electrical connector and loosen the compressor mounting bolts. Remove the compressor without disconnecting the A/C hoses. Support the compressor with a strong wire out of the way.

24. Remove the upper and lower radiator hoses then disconnect the heater hoses from the engine.

25. Remove the transaxle ground cable.

26. If equipped with a automatic transaxle, disconnect the cooler hoses.

27. If equipped with a manual transaxle perform the following:

    a. Disconnect the shift cable and the select cable from the transaxle. Do not bend the cables when removing them. Replace any kinked cable with a new one.

    b. Remove the clutch slave cylinder and the pipe/hose assembly. Do not operate the clutch once the slave cylinder has been removed.

    c. Remove the clutch damper assembly.

28. Remove the Vehicle Speed Sensor (VSS)/Power Steering (P/S) speed sensor assembly. Do not disconnect the hoses.

29. Remove the nuts attaching exhaust pipe A to the exhaust manifold and the catalytic converter. Remove the bolts from the exhaust pipe hanger then remove the exhaust pipe and discard the gaskets.

30. If equipped with a automatic transaxle, remove the shift cable cover then disconnect the shift cable. Do not bend the cable and replace the cable if it becomes kinked.

31. Remove the left and the right side damper forks.

32. Disconnect the lower ball joints from the lower control arms.

33. Pry the halfshafts from the transaxle. Cover the inner CV joints with plastic bags to protect them.

34. Swing the halfshafts under the fender out of the way.

35. Attach an engine hoist to the engine lifting points and raise the hoist to remove all slack from the chain.

36. Remove the rear engine mount bracket.

37. Remove the front engine mount bracket.

38. Remove the left side engine mount.

39. Remove the transaxle mount and the mount bracket.

40. Check that the engine is completely free of vacuum hoses, fuel and coolant hoses, and electrical wiring.

41. Slowly raise the engine approximately 6 in. (150mm). Check once again that all hoses and wires have been disconnected from the engine.

42. Raise the engine all the way and remove it from the vehicle.

43. Remove the transaxle.

44. If equipped with a manual transaxle, remove the clutch cover (pressure plate) and clutch disc.

45. Mount the engine on an engine stand, making sure the mounting bolts are tight. If an engine stand is not available, support the engine in an upright position with blocks.

Never leave an engine hanging from a lift or hoist.

To install:

46. Assemble the clutch disc and pressure plate to the flywheel for manual transaxle vehicles.

47. Install the transaxle.

48. Lift the engine into position and lower it into the car, aligning the mounts and bushings.

NOTE: When installing the engine mounts and vibration dampers in the following steps, they must be tightened to the correct tension in the correct order if they are to damp vibration properly.

49. Install the side engine mount and the through-bolt. Do not tighten the through-bolt at this time. Install the nut and bolt attaching the mount to the engine. Torque the nut and bolt attaching the side mount to the engine to 40 ft. lbs. (55 Nm).

50. Install the transaxle mount and through-bolt. Do not tighten the through-bolt at this time.

51. Install the rear engine mount and new bolts attaching the mount to the engine. Torque the three new bolts attaching the mount to the engine assembly to 40 ft. lbs. (55 Nm). Install a new rear engine mount through-bolt and torque the new through-bolt to 47 ft. lbs. (65 Nm).

52. Install the front mount and the three bolts attaching the mount to the engine assembly, only snug the bolts in place. Install a new through-bolt to the front mount and torque the new through-bolt to 47 ft. lbs. (65 Nm).

53. Install the nuts to the transaxle mount. Torque the nuts to 28 ft. lbs. (39 Nm).

54. Torque the side engine mount through-bolt to 47 ft. lbs. (65 Nm).

55. Torque the transaxle mount through-bolt to 47 ft. lbs. (65 Nm).

56. Torque the three bolts attaching the front mount to the engine to 28 ft. lbs. (39 Nm).

57. Remove the hoist equipment from the engine.

58. Install new spring clips to the inner CV–joints. Install the halfshafts into the transaxle, make sure that the spring clips on the inner joints click into place.

59. Connect the lower ball joints to the lower control arms. Torque the nuts to 36–43 ft. lbs. (50–60 Nm). Install a new cotter pin to the ball joint stud.

60. Install the damper forks. Torque a new self–locking bolt attaching the damper fork to the strut to 32 ft.

lbs. (44 Nm). Torque the new nut and bolt attaching the damper fork to the lower control arm to 47 ft. lbs. (65 Nm).

61. If equipped with a automatic transaxle, connect the shift cable to the transaxle. Install a new lockwasher and torque the attaching bolt to 7 ft. lbs. (10 Nm). Install the shift cable cover. Torque the shift cable cover attaching bolts to 13 ft. lbs. (18 Nm).

62. Install exhaust pipe A with new gaskets. Torque new nuts attaching the exhaust pipe to the exhaust manifold to 40 ft. lbs. (55 Nm). Torque new nuts attaching exhaust pipe A to the catalytic converter to 25 ft. lbs. (34 Nm). Install new attaching bolts to the exhaust pipe hanger and torque the bolts to 13 ft. lbs. (18 Nm).

63. Install the VSS and connect the electrical connector and torque the mounting bolt to 13 ft. lbs. (18 Nm).

64. If equipped with a manual transaxle perform the following:

a. Install the clutch damper assembly and torque the attaching bolts to 16 ft. lbs. (22 Nm).

b. Install the clutch slave cylinder and the pipe/hose assembly and torque the slave cylinder mounting bolts to 16 ft. lbs. (22 Nm).

c. Connect the shift cable and the select cable to the transaxle. Adjust the shift cable and select cable.

65. If equipped with a automatic transaxle, connect the cooler hoses.

66. Install the transaxle ground cable.

67. Install the upper and lower radiator hoses and connect the heater hoses to the engine.

68. Install the A/C compressor and connect the electrical connector. Torque the mounting bolts to 16 ft. lbs. (22 Nm).

69. Install and adjust the alternator drive belt.

70. Remove the protector plate from the radiator and install the A/C condenser fan.

71. Install and the power steering pump and drive belt. Adjust the drive belt tension then torque the attaching nuts and bolts to 16 ft. lbs. (22 Nm).

72. Attach the engine ground cable to the body.

73. Install the cruise control actuator, then connect the electrical connector and vacuum tube. Torque the mounting bolts to 7 ft. lbs. (10 Nm).

74. Connect the brake booster vacuum hose and the emissions control vacuum tubes to the intake manifold.

75. Connect the engine wiring harness connectors, terminal and clamps.

76. Install and adjust the throttle cable.

77. Connect the injector resistor connector on the left of the engine compartment.

78. Connect the fuel return hose to the regulator. Connect the fuel feed hose to the fuel rail with new washers. Torque the cap nut to 16 ft. lbs. (22 Nm).

79. Connect the battery cable and the starter cable to the body. Install the battery base and the battery. Torque the battery base attaching bolts to 16 ft. lbs. (22 Nm).

80. Install the PAIR vacuum tank and bracket. Torque the mounting bolts to 8 ft. lbs. (10 Nm).

81. Install the air cleaner duct and housing.

82. Install the engine splash shield and the front wheels.

83. Lower the vehicle.

84. Fill the engine with oil and the transaxle with oil/fluid.

85. Fill and bleed the air from the cooling system.

86. Connect the positive then the negative battery cable and enter the radio security code.

87. Switch the ignition **ON** but do not engage the starter. The fuel pump should run for approximately 2 seconds, building pressure within the lines. Switch the ignition **OFF**, then **ON** 2 or 3 more times to build full system pressure. Check for fuel leaks.

88. Disconnect the coil wire from the distributor. Insulate or protect the end of the cable so it does not arc to the engine or surrounding metal. Without touching the accelerator, turn the ignition switch to the START position and crank the engine for about 5–10 seconds; this will develop some oil pressure within the motor. Do not exceed 10 seconds cranking.

89. Switch the ignition OFF and reconnect the coil.

90. Start the engine, allowing it to idle. Check the hoses and lines carefully for any sign of leakage.

91. Check the timing and idle speed.

92. After the engine has warmed up fully and the fan(s) have come on at least once, recheck the engine for fluid leaks. Switch the engine OFF.

93. Adjust the belts and throttle cable as necessary.

94. If equipped with 4WS, start the engine and turn the steering wheel

lock–to–lock to reset the 4WS control unit.

95. Road test the vehicle then loosen and retighten the three bolts attaching the front engine mount to the engine. Torque the bolts to 28 ft. lbs. (39 Nm).

1993 Accord

1. Disconnect the battery cables, negative first. Remove the battery and battery case.
2. Raise and safely support the vehicle. Double check the security and placement of the stands.
3. Place the hood in a vertical position and safely support it in place. Do not remove the hood.
4. Remove the engine splash shield. Drain the engine oil, coolant and transaxle fluid.
5. Remove the air intake duct and the air cleaner case.
6. Relieve the fuel system pressure using the correct procedure.
7. Remove the fuel feed hose from the fuel pipe and the return hose from the pressure control valve.
8. Disconnect the 2 connectors and remove the control box from the bulkhead.

NOTE: Do not disconnect the vacuum hoses.

9. Disconnect the vacuum hose from the charcoal canister and the charcoal canister hose from the throttle body.
10. Remove the ground cable from the transaxle.
11. Remove the throttle cable by loosening the locknut, then slip the cable end out of the throttle bracket and accelerator linkage.

NOTE: Be careful not to bend the cable when removing. Do not use pliers to remove the cable from the linkage. Always replace a kinked cable with a new one.

12. Disconnect the connector and the vacuum hose, then remove the cruise control actuator.
13. Remove the brake booster vacuum hose and mount; remove the vacuum hose from the intake manifold.
14. Disconnect the 3 engine harness connectors from the main wire harness at the right side of the engine compartment. Remove the engine wire harness terminal and the starter cable terminal from the underhood relay box and clamps. Then remove the transaxle ground terminal.
15. Disconnect the 2 engine wire harness connectors from the main

harness and the resistor at the left side of the engine compartment.
16. Remove the engine ground wire from the valve cover and the power steering pump bracket.
17. Remove the mounting bolts and the power steering belt from the power steering pump, then, without disconnecting the hoses, pull the pump away from its mounting bracket. Support the pump out of the way.
18. Remove the mounting bolts and belt from the air conditioning compressor. Without disconnecting the hoses, pull the compressor away from it's mounting bracket. Support compressor with stiff wire out of the way.
19. Disconnect the heater hoses. Disconnect the radiator hoses, automatic transaxle cooler hoses and the cooling fan connectors. Remove the radiator/cooling fan assembly.
20. Remove the speed sensor without disconnecting the hoses or connector.
21. Remove the center beam.
22. Remove the nuts attaching the exhaust pipe to the exhaust manifold and the catalytic converter. Remove the exhaust pipe and discard the gaskets.
23. Remove the halfshafts as follows:
   a. Remove the front wheels.
   b. Raise the locking tab on the spindle nut and remove it.
   c. Remove the damper fork nut, damper pinch bolt and remove the damper fork.
   d. Remove the cotter pin and castle nut from the lower ball joint.
   e. Using a suitable puller, separate the lower control arm from the knuckle.
   f. Pull the knuckle outward and remove the halfshaft outboard CV-joint from the knuckle using a suitable plastic hammer.
   g. Using a suitable pry bar, pry the halfshaft out to force the set ring at the end of the halfshaft past the groove.
   h. Pull the inboard CV-joint and remove the halfshaft and CV-joint out of the differential case or intermediate shaft as an assembly.

NOTE: Do not pull on the halfshaft, as the CV-joint may come apart. Tie plastic bags over the halfshaft ends to protect them.

24. On manual transaxle equipped vehicles, remove the clutch release hose from the clutch damper on the transaxle housing. Remove the shift

cable and the select cable with the cable bracket from the transaxle.

NOTE: Be careful not to bend the cable when removing. Do not use pliers to remove the cable. Always replace a kinked cable with a new one.

25. On automatic transaxle equipped vehicles, remove the engine stiffener, then remove the torque converter cover. Remove the cable holder, then remove the shift control lever bolt and shift control cable.
26. Attach a suitable lifting device to the engine. Raise the engine to unload the engine mounts.
27. Remove the front and rear engine mounting bolts.
28. Remove the engine side mount and mounting bolt. Remove the side transaxle mount and mounting bolt.
29. Make sure the engine/transaxle assembly is completely free of vacuum hoses, fuel and coolant hoses and electrical wires.
30. Slowly raise the engine approximately 6 in. (152mm). Check again that all hoses and wires have been disconnected from the engine/transaxle assembly.
31. Raise the engine/transaxle assembly all the way and remove it from the vehicle.
32. Remove the transaxle.
33. If manual transaxle, remove the clutch cover (pressure plate) and clutch disc.
34. Mount the engine on an engine stand, making sure the mounting bolts are tight. If an engine stand is not available, support the engine in an upright position with blocks. Never leave an engine hanging from a lift or hoist.

To install:

35. Assemble the clutch disc and pressure plate to the flywheel for manual transaxle vehicles.
36. Install the transaxle.
37. Lift the engine into position and lower it into the car, aligning the mounts and bushings.

NOTE: When installing the engine mounts and vibration dampers in the following steps, the bolts running through the large rubber bushings should only be set snug or finger tight until all are in place. They must be tightened to the correct tension in the correct order if they are to damp vibration properly.

38. Install the side engine mount and mounting bolt. Install the through-bolt.
39. Install the side transaxle mount and bolt.

40. Install the front and rear engine mounting bolts.

41. At this point the engine is loosely mounted in the vehicle. Slacken the hoist chains, allowing the engine to settle into place. The vehicle must be sitting level during the next step.

42. With the vehicle sitting level, tighten the following in the order given to the proper torque. Both the order and tightness are important; the bushings play a great role in damping engine vibration.

  a. The 3 nuts holding the side transaxle mount to 28 ft. lbs. (39 Nm)

  b. The nut and bolt holding the engine side mount to 40 ft. lbs. (55 Nm)

  c. The through-bolt holding the rear engine mount to 47 ft. lbs. (65 Nm).

  d. The 3 bolts holding the rear engine mount to the rail to 40 ft. lbs. (55 Nm).

  e. The through-bolt holding the front engine mount to 47 ft. lbs. (65 Nm).

  f. The through-bolt at the transaxle side mount to 40 ft. lbs. (55 Nm).

  g. The through-bolt at the engine side mount to 40 ft. lbs. (55 Nm).

43. Remove the hoist equipment from the engine. If the car was lowered for the previous Step, re-elevate it and support it safely.

44. For automatic transaxles, install the shift control cable and cable holder. Install the torque converter cover and the engine stiffener.

45. For manual transaxles, connect the shift cable and shift select cable. Install the clutch release hose.

46. Install the driveshafts. Use a new circlip on each; install the shaft until a positive click is heard as the axle locks into place.

47. Install the ball joints to the lower arms and tie rods. Tighten the lower arm nuts to 40 ft. lbs. (55 Nm) and the tie rod nuts to 32 ft. lbs. (44 Nm). Install new cotter pins.

48. Install the exhaust pipe. Use new nuts; tighten the pipe-to-manifold nuts to 40 ft. lbs. (55 Nm).

49. Install the center beam. Tighten the bolts to 28 ft. lbs. (39 Nm).

50. Install the speed sensor. Tighten the retaining bolt to 7 ft. lbs.(10 Nm).

51. Install the radiator; connect the coolant and transaxle oil cooler lines. Connect the cooling fan connectors.

52. Connect the heater inlet and outlet hoses to the cylinder head and connecting pipe, respectively.

53. Install the A/C compressor and belt, tightening the mounting bolts to 16 ft. lbs.(22 Nm).

54. Install the power steering pump and belt. Tighten the bolt to 33 ft. lbs.(45 Nm) and the nut to 16 ft. lbs. (22 Nm).

55. Connect the engine ground wire to the valve cover and the power steering pump bracket.

56. Connect the transaxle ground strap. Connect the starter cable terminal and install the engine wire harness terminal. Connect the three connectors for the main wire harness at the right side of the engine compartment.

57. Connect the brake booster vacuum hose to the manifold and to the brake booster.

58. Connect the throttle cable and secure the locknut.

59. Connect the ground cable to the transaxle.

60. Install the control box on the bulkhead and connect the wiring connectors.

61. Using new washers, connect the fuel line and return hose to the pressure control valve.

62. Install the air cleaner and the intake ductwork.

63. Refill the transaxle fluid.

64. Refill the cooling system.

65. Refill the engine oil.

66. Install the battery base and battery.

67. Double check all installation items, paying particular attention to loose hoses or hanging wires, untightened nuts, poor routing of hoses and wires (too tight or rubbing) and tools left in the engine area.

68. Connect the battery cables, positive first.

69. Disconnect the coil wire from the distributor. Insulate or protect the end of the cable so it does not arc to the engine or surrounding metal. Without touching the accelerator, turn the ignition switch to the START position and crank the engine for about 5–10 seconds; this will develop some oil pressure within the motor. Do not exceed 10 seconds cranking.

70. Switch the ignition **OFF** and reconnect the coil.

71. Start the engine, allowing it to idle. Check the hoses and lines carefully for any sign of leakage.

72. Bleed the air from the cooling system; check the timing and idle speed.

73. After the engine has warmed up fully and the fan(s) have come on at least once, recheck the engine for fluid leaks. Switch the engine **OFF**.

74. Adjust the belts, clutch and throttle cable as necessary.

75. Install the front wheels. Lower the hood.

1994–97 Accord

2.2L Engines

1. Secure the hood as far open as possible.

2. Disconnect the negative battery cable then the positive battery cable.

3. Remove the battery and the battery base. Disconnect the engine ground cable.

4. Remove the throttle cable and the cruise control cable, by loosening the locknuts, then slip the cable ends out of the throttle linkage. Take care not to bend the cables when removing them. Always replace any kinked cable with a new one.

5. Remove the intake air duct B and intake air duct/air cleaner housing.

6. Disconnect the Intake Air Resonator (IAR) control solenoid valve connector, then remove the IAR from the vehicle.

7. Remove the battery cables from the under–hood fuse/relay box and under–hood ABS fuse/relay box.

8. Disconnect the engine wire harness connectors on the right side of the engine compartment.

9. Remove the brake booster vacuum hose then label and disconnect the other vacuum hoses from the intake manifold.

10. Relieve the pressure from the fuel system.

— **CAUTION** —
Fuel injection systems remain under pressure after the engine has been turned OFF. Properly relieve fuel pressure before disconnecting any fuel lines. Failure to do so may result in fire or personal injury.

11. Disconnect the fuel feed hose from the fuel rail and disconnect the fuel return line from the fuel pressure regulator.

12. Remove the engine wiring harness connectors, terminal and clamps on the left side of the engine compartment.

13. Disconnect the injector resistor connector on the left side of the engine compartment.

14. Remove the power steering hose clamp.

15. Remove the power steering pump mounting nuts and adjusting bolt, then remove the power steering pump drive belt and the pump.

16. Loosen the alternator mounting bolt, nut and adjusting bolt, then remove the alternator belt.

17. If equipped with a manual transaxle perform the following:

a. Disconnect the shift cable and the select cable from the transaxle. Do not bend the cables when removing them. Replace any kinked cable with a new one.

b. Disconnect the backup light switch connectors and starter motor cable.

c. Remove the clutch slave cylinder and the pipe/hose assembly. Do not operate the clutch once the slave cylinder has been removed.

18. Disconnect the vehicle speed sensor (VSS).

19. Remove the radiator cap.

20. Raise and safely support the vehicle. Remove the front wheels and the engine splash shield.

21. Drain the engine coolant into a sealable container.

22. Drain the transaxle oil/fluid into a sealable container. Install the drain plug with a new gasket.

23. Drain the engine oil into a sealable container.

24. Lower the vehicle to a working level.

25. Remove the upper and lower radiator hoses then disconnect the heater hoses from the engine.

26. If equipped with a automatic transaxle, disconnect the ATF cooler hoses.

27. Remove the radiator assembly from the vehicle.

28. Loosen the A/C mounting bolts, then remove the compressor. Do not disconnect the A/C hoses. Disconnect the compressor electrical connector and support the compressor with a strong wire.

29. Remove the center beam from under the engine.

30. Remove the nuts attaching exhaust pipe A to the exhaust manifold and the catalytic converter. Remove the nuts from the exhaust pipe hanger, then remove the exhaust pipe and discard the gaskets.

31. If equipped with a automatic transaxle, remove the shift cable cover then disconnect the shift cable. Do not bend the cable and replace the cable if it becomes kinked.

32. Remove the left and the right side damper forks.

33. Disconnect the lower ball joints from the lower control arms.

34. Pry the halfshafts from the transaxle. Cover the inner CV joints with plastic bags to protect them.

35. Swing the halfshaft under the fender out of the way.

36. Attach an engine hoist to the engine lifting points and raise the hoist to remove all slack from the chain.

37. Remove the rear engine mount bracket.

38. Remove the front engine mount bracket.

39. Remove the side engine mount.

40. Remove the transaxle mount and the mount bracket.

41. Check that the engine is completely free of vacuum hoses, fuel and coolant hoses, and electrical wiring.

42. Slowly raise the engine approximately 6 in. (150mm). Check once again that all hoses and wires have been disconnected from the engine.

43. Raise the engine all the way and remove it from the vehicle.

44. Remove the transaxle from the engine.

45. If manual transaxle, remove the clutch cover (pressure plate) and clutch disc.

46. Mount the engine on an engine stand, making sure the mounting bolts are tight. If an engine stand is not available, support the engine in an upright position with blocks. Never leave an engine hanging from a lift or hoist.

To install:
47. Assemble the clutch disc and pressure plate to the flywheel, if equipped with a manual transaxle.

48. Install the transaxle to the engine.

49. Lift the engine into position and lower it into the car, aligning the mounts and bushings.

NOTE: When installing the engine mounts and vibration dampers in the following steps, they must be tightened to the correct tension in the correct order if they are to damp vibration properly.

50. Install the side engine mount. Install a 6·100mm bolt to the mount to properly position the mount. Do not tighten the nut and bolt attaching the mount to the engine. Torque the through-bolt to 47 ft. lbs. (64 Nm) then remove the 6·100mm bolt from the mount.

51. Install the transaxle mount. Install a 6·100mm bolt to the mount to properly position the mount. Do not tighten the nuts attaching the mount to the transaxle at this time. Torque the through-bolt to 47 ft. lbs. (64 Nm)

then remove the 6·100mm bolt from the mount.

52. Install the rear engine mount bracket using new bolts. Torque the new bolts attaching the mount to the engine assembly to 40 ft. lbs. (54 Nm). Install a new rear engine mount through-bolt. Torque the new through-bolt to 47 ft. lbs. (64 Nm).

NOTE: Tighten the bolts attaching the mount to the engine assembly first then the through-bolt. If this order is not followed excessive engine vibration may be felt and mount damage may occur.

53. Install the front mount bracket. Do not tighten the nuts attaching the mount to the engine assembly, only snug the nuts in place. Install a new through-bolt to the front mount. Torque the new through-bolt to 47 ft. lbs. (64 Nm).

54. Torque the side engine mount nut and bolt to 47 ft. lbs. (64 Nm).

55. Torque the nuts attaching the transaxle mount to the transaxle to 28 ft. lbs. (38 Nm).

56. Torque the three bolts attaching the front mount bracket to the engine to 28 ft. lbs. (38 Nm).

57. Remove the hoist equipment from the engine.

58. Install new spring clips to the inner CV–joints. Install the halfshafts into the transaxle, make sure that the spring clips on the inner joints click into place.

59. Connect the lower ball joints to the lower control arms, torque the nuts to 36–43 ft. lbs. (49–59 Nm). Install a new cotter pin to the ball joint stud.

60. Install the damper forks, torque a new self–locking bolt attaching the damper fork to the strut to 32 ft. lbs. (43 Nm). Torque the new nut and bolt attaching the damper fork to the lower control arm to 47 ft. lbs. (64 Nm).

61. If equipped with a automatic transaxle, connect the shift cable to the transaxle. Install a new lockwasher and torque the attaching bolt to 10 ft. lbs. (14 Nm). Install the shift cable cover and torque the shift cable cover attaching bolts to 13 ft. lbs. (18 Nm).

62. Install exhaust pipe A with new gaskets. Torque new nuts attaching the exhaust pipe to the exhaust manifold to 40 ft. lbs. (54 Nm). Torque new nuts attaching exhaust pipe A to the catalytic converter to 16 ft. lbs. (22 Nm). Install new attaching nuts to the exhaust pipe hanger and torque the nuts to 13 ft. lbs. (18 Nm).

63. Install the center beam and torque the bolts attaching the center beam to 37 ft. lbs. (50 Nm).

64. Install the A/C compressor and connect the electrical connector and torque the mounting bolts to 16 ft. lbs. (22 Nm).

65. Install the radiator assembly.

66. If equipped with a automatic transaxle, connect the ATF cooler hoses.

67. Install the upper and lower radiator hoses and connect the heater hoses to the engine.

68. Install the engine splash shield and the front wheels.

69. Connect the VSS (vehicle speed sensor) electrical connector.

70. If equipped with a manual transaxle perform the following:

    a. Install the clutch slave cylinder and the pipe/hose assembly. Torque the slave cylinder mounting bolts to 16 ft. lbs. (22 Nm).

    b. Connect the starter motor cable and the back–up light switch connectors.

    c. Connect the shift cable and the select cable to the transaxle. Adjust the shift cable and select cable.

71. Install and adjust the alternator drive belt.

72. Install the power steering pump and drive belt. Adjust the drive belt and torque the attaching nuts and bolts to 16 ft. lbs. (22 Nm).

73. Install the power steering hose clamp.

74. Connect the injector resistor connector on the left of the engine compartment.

75. Connect the engine wiring harness connectors, terminal and clamps on the left side of the engine compartment.

76. Connect the fuel return hose to the regulator. Connect the fuel feed hose to the fuel rail with new washers and torque the banjo nut to 16 ft. lbs. (22 Nm).

77. Connect the vacuum hoses to the intake manifold.

78. Connect the engine wire harness connectors on the right side of the engine compartment.

79. Connect the battery cables to the under–hood fuse/relay box and under–hood ABS fuse/relay box.

80. Install the vacuum hose and IAR then connect the IAR control solenoid valve connector.

81. Install the intake air duct/air cleaner housing then intake air duct B.

82. Install and adjust the throttle cable.

83. Connect the engine ground cable and install the battery base. Torque the base mounting bolts to 16 ft. lbs. (22 Nm).

84. Install the battery and connect the positive then the negative battery cables. Enter the radio security code.

85. Fill the engine with oil and the transaxle with oil/fluid.

86. Fill and bleed the air from the cooling system.

87. Switch the ignition **ON** but do not engage the starter. The fuel pump should run for approximately 2 seconds, building pressure within the lines. Switch the ignition **OFF**, then **ON** 2 or 3 more times to build full system pressure. Check for fuel leaks.

88. Disconnect the coil wire from the distributor. Insulate or protect the end of the cable so it does not arc to the engine or surrounding metal. Without touching the accelerator, turn the ignition switch to the START position and crank the engine for about 5–10 seconds; this will develop some oil pressure within the motor. Do not exceed 10 seconds cranking.

89. Switch the ignition **OFF** and reconnect the coil.

90. Start the engine, allowing it to idle. Check the hoses and lines carefully for any sign of leakage.

91. Check the timing and idle speed.

92. After the engine has warmed up fully and the fan(s) have come on at least once, recheck the engine for fluid leaks. Switch the engine **OFF**.

93. Adjust the belts, clutch and throttle cable as necessary.

2.7L Engines

1. Disconnect the support struts from the engine hood, then fix the hood in a vertical position.

2. Disconnect the negative then the positive battery cables.

3. Remove the battery, battery base and bracket. Disconnect the engine ground cable, located next to the battery.

4. Remove the intake air duct.

5. Remove intake manifold cover B then disconnect the throttle cable and cruise control cables from the throttle linkage. Loosen the cable locknuts then slip the cables ends out of the accelerator linkage. Take care to not bend the cables, always replaced a kinked cable.

6. Remove the starter cable from the strut brace, then remove the strut brace.

7. Disconnect the engine wire harness connectors on the left side of the engine compartment.

8. Disconnect the injector resistor connector on the left side of the engine compartment.

9. Relieve the pressure from the fuel system.

─────── **CAUTION** ───────
Fuel injection systems remain under pressure after the engine has been turned OFF. Properly relieve fuel pressure before disconnecting any fuel lines. Failure to do so may result in fire or personal injury.
────────────────────────

10. Disconnect the fuel feed hose from the fuel filter. Disconnect the fuel return hose from the regulator.

11. Disconnect the brake booster vacuum hose and the evaporative emissions (EVAP) control canister hose.

12. Label then disconnect the vacuum hoses from the engine.

13. Remove the battery cables from the under–hood fuse/relay box and under–hood ABS fuse/relay box.

14. Disconnect the engine wire harness connectors on the right side of the engine compartment.

15. Remove the side engine mount, discard the bolts attaching the mount to the engine.

16. Loosen the air conditioning idler pulley center nut and adjusting bolt, then remove the drive belt.

17. Disconnect the ground cable from the body, located toward the front of the vehicle by the drive belts.

18. Loosen the alternator mounting nut, bolt and adjusting bolt, then remove the drive belt.

19. Loosen the power steering pump mounting and adjusting nuts, then remove the drive belt.

20. Disconnect the power steering inlet hose from the pump, plug or cap the connections.

21. Remove the power steering pump mounting nuts and adjusting bolt, then remove the pump.

22. Remove the radiator cap.

23. Raise and safely support the vehicle.

24. Remove the front wheels and the splash shield.

25. Drain the engine coolant into a sealable container.

26. Drain the transaxle fluid into a sealable container. Install the drain plug with a new gasket.

27. Drain the engine oil into a sealable container. Install the drain plug with a new gasket.

28. Remove the center beam from under the engine.

29. Disconnect the oxygen sensor electrical connector. Remove the nuts attaching exhaust pipe A to the ex-

haust manifolds. Remove the nuts attaching exhaust pipe A to the catalytic converter and remove the exhaust pipe. Discard the locknuts and gaskets.

30. Remove the crankshaft pulley bolt and remove the crankshaft pulley. Use a crank pulley holder (part # 07MAB–PY3010A) and holder handle (part # 07JAB–001020A) to hold the crankshaft pulley in place while removing the bolt.

31. Remove the oil filter.

32. Disconnect the oil pressure switch terminal, then remove the oil filter base attaching bolts.

33. Remove the oil filter base and discard the O–rings.

34. Remove the shift cable cover, then remove the shift cable attaching bolts. Discard the lockwasher.

35. Remove the left and right damper forks.

36. Disconnect the lower ball joints from the lower control arms.

37. Remove the halfshafts from the vehicle.

38. Remove the upper and lower radiator hoses, then disconnect the heater hoses from the engine.

39. Disconnect the transaxle cooler hoses, then plug the pipes and hoses.

40. Remove the radiator from the vehicle.

41. Disconnect the air conditioning (A/C) compressor electrical connector. Remove the A/C compressor mounting bolts, position the compressor out of the way and support it with a strong wire. Do not disconnect the A/C hoses from the compressor.

42. Attach an engine hoist to the engine lifting points and raise the hoist to remove all slack from the chain.

43. Remove the transaxle mount.

44. Remove the bolts attaching the front mount to the beam.

45. Disconnect the vacuum hose from the rear engine mount control solenoid valve.

46. Remove the bolts attaching the rear mount to the beam.

47. Check that the engine is completely free of vacuum hoses, fuel and coolant hoses, and electrical wiring.

48. Slowly raise the engine approximately 6 inches (150mm). Check once again that all hoses and wires have been disconnected from the engine.

49. Raise the engine all the way and remove it from the vehicle.

50. Remove the transaxle from the engine.

51. Mount the engine on an engine stand, making sure the mounting bolts are tight. If an engine stand is not available, support the engine in

an upright position with blocks. Never leave an engine hanging from a lift or hoist.

To install:

52. Install the transaxle.

53. Install the front and rear engine mounts to their mounting brackets and torque the attaching nuts to 40 ft. lbs. (54 Nm).

54. Lift the engine into position and lower it into the car, aligning the mounts and bushings.

55. Place the power steering pump drive belt over the bracket the side engine mount attaches to.

NOTE: When installing the engine mounts and vibration dampers in the following steps, they must be tightened to the correct tension in the correct order if they are to damp vibration properly.

56. Install the rear mount and torque the bolts attaching the mount to the beam to 43 ft. lbs. (59 Nm).

57. Install the front mount and torque the bolts attaching the mount to the beam to 43 ft. lbs. (59 Nm).

58. Install the side engine mount. Use three new bolts to attach the mount to the engine, torque the bolts to 40 ft. lbs. (54 Nm). Do not tighten the through-bolt at this time.

59. Install the transaxle mount and torque the three nuts attaching the mount to the transaxle to 28 ft. lbs. (38 Nm). Do not tighten the through-bolt at this time.

60. Torque the side engine mount through-bolt to 47 ft. lbs. (64 Nm).

61. Torque the transaxle mount through-bolt to 47 ft. lbs. (64 Nm).

62. Remove the engine hoist.

63. Position the A/C compressor on the engine and install the mounting bolts. Torque the mounting bolts to 16 ft. lbs. (22 Nm), then connect the A/C compressor electrical connector.

64. Install the radiator assembly.

65. Connect the transaxle cooler hoses to the transaxle cooler lines.

66. Install the upper and lower radiator hoses and connect the heater hoses to the engine.

67. Install the halfshafts with new snaprings, make sure the snaprings click into place.

68. Connect the lower ball joints to the lower control arms. Torque the ball joint nuts to 36–43 ft. lbs. (49–59 Nm) then install a new cotter pin.

69. Install the damper forks, torque the flange bolt to 32 ft. lbs. (43 Nm). Install the lower bolt and a new locknut, torque the nut to 47 ft. lbs. (64 Nm).

70. Connect the shift cable to the transaxle and install a new

lockwasher to the attaching bolt. Torque the bolt attaching the shift cable end to the transaxle to 10 ft. lbs. (14 Nm). Install the shift cable cover and torque the cover attaching bolts to 20 ft. lbs. (26 Nm). Torque the two bolts attaching the cable housing to the transaxle to 9 ft. lbs. (12 Nm).

71. Install new O–rings to the oil filter base and install the oil filter base to the engine. Torque the mounting bolts to 16 ft. lbs. (22 Nm).

72. Connect the oil pressure switch terminal and torque the attaching bolt to 1.8 ft. lbs. (2.5 Nm).

73. Install the oil filter.

74. Install the crankshaft pulley, use the pulley holder when installing and torquing the pulley bolt. Torque the bolt to 181 ft. lbs. (245 Nm).

75. Install exhaust pipe A with new gaskets and new locknuts. Torque the nuts attaching the exhaust pipe to the catalytic converter to 25 ft. lbs. (33 Nm). Torque the nuts attaching the exhaust pipe to the exhaust manifolds to 40 ft. lbs. (54 Nm). Connect the oxygen sensor electrical connector.

76. Install the center beam and torque the center beam attaching bolts to 37 ft. lbs. (50 Nm).

77. Install the splash shield and the front wheels.

78. Install the power steering pump and connect the inlet hose. Do not tighten the power steering pump mounting bolts and nuts at this time.

79. Install and adjust the power steering pump belt and torque the mounting nuts to 16 ft. lbs. (22 Nm).

80. Install and adjust the alternator belt, then torque the alternator mounting nut and bolt to 16 ft. lbs. (22 Nm).

81. Install the A/C belt and adjust the belt tension. Torque the idler center nut to 33 ft. lbs. (44 Nm).

82. Connect the engine wire harness connectors on the right side of the engine compartment.

83. Connect the battery cables to the under–hood fuse/relay box and under–hood ABS fuse/relay box.

84. Connect the vacuum hoses to the intake manifold.

85. Connect the EVAP control canister hose and the power brake booster hose to the engine assembly.

86. Connect the fuel return hose to the regulator. Connect the fuel feed hose to the fuel filter with new washers. Torque the banjo nut to 16 ft. lbs. (22 Nm) and torque the service bolt to 9 ft. lbs (12 Nm).

87. Connect the injector resistor connector on the left of the engine compartment.

88. Connect the engine wiring harness connectors on the left side of the engine compartment.

89. Install the strut brace and torque the strut brace bolts to 16 ft. lbs. (22 Nm). Install the starter cable to the strut brace.

90. Connect the throttle cable and cruise control cables to the throttle linkage, adjust the cable as necessary. Install intake manifold cover B, torque the attaching bolt to 9 ft. lbs. (12 Nm).

91. Install the intake air duct.

92. Connect the engine ground cable and install the battery base and bracket. Torque the base mounting bolts to 16 ft. lbs. (22 Nm).

93. Install the battery and connect the positive then the negative battery cables. Enter the radio security code.

94. Fill the engine with oil and the transaxle with fluid.

95. Fill and bleed the air from the cooling system.

96. Switch the ignition **ON** but do not engage the starter. The fuel pump should run for approximately 2 seconds, building pressure within the lines. Switch the ignition **OFF**, then **ON** 2 or 3 more times to build full system pressure. Check for fuel leaks.

97. Disconnect the coil wire from the distributor. Insulate or protect the end of the cable so it does not arc to the engine or surrounding metal. Without touching the accelerator, turn the ignition switch to the START position and crank the engine for about 5–10 seconds; this will develop some oil pressure within the motor. Do not exceed 10 seconds cranking.

98. Switch the ignition **OFF** and reconnect the coil.

99. Start the engine, allowing it to idle. Check the hoses and lines carefully for any sign of leakage.

100. Check the timing and idle speed.

101. After the engine has warmed up fully and the fan(s) have come on at least once, recheck the engine for fluid leaks. Switch the engine **OFF**.

Engine Mounts

REMOVAL AND INSTALLATION

Civic and del Sol

NOTE: The radio may contain a coded theft protection circuit. Always obtain the code number before disconnecting the battery. If the vehicle is equipped with 4WS, the steering control unit is shut down when the battery is disconnected. After connecting the battery, turn the steering wheel lock–to–lock to reset the steering control unit.

Side Engine Mount

1. Secure the hood as far open as possible.

2. Disconnect the negative and positive battery cables.

3. Attach an engine hoist to the engine lifting points and raise the hoist to remove all slack from the chain.

NOTE: A floor jack with a cushion or wooden plank over its pad may be substituted for the lifting chain.

4. Remove the three nuts securing the mount bracket to the engine bracket and the mount damper.

5. Unbolt the side engine mount damper from the shock tower.

To install:

6. Install the side engine mount damper. Tighten the three mount damper bolts to 33 ft. lbs. (44 Nm) each.

7. Install the mount bracket to connect the engine bracket and the mount damper. Install all three nuts, and then tighten each to 54 ft. lbs. (74 Nm).

8. Remove the hoist equipment from the engine.

9. Reconnect the positive and negative battery cables.

Left Front Engine Mount

1. Secure the hood as far open as possible.

2. Disconnect the negative and positive battery cables.

3. Raise and safely support the vehicle. Block the rear wheels.

4. Attach an engine hoist to the engine lifting points and raise the hoist to remove all slack from the chain.

NOTE: A floor jack with a cushion or wooden plank over its pad may be substituted for the lifting chain.

5. Unbolt the mount stud from the mount bracket. Don't lose the special internal–hex washer.

6. Unbolt the left front engine mount bracket from the shock tower. Remove the mount bracket from the mount stud.

7. If necessary, unscrew the mount stud from the bracket plate on the engine.

8. If equipped with air conditioning: inspect the A/C compressor drive belt and replace it if necessary, as the mount must be unbolted to change the belt.

To install:

9. If the mount stud was removed, screw it into the bracket plate and tighten it to 61 ft. lbs. (83 Nm).

10. Fit the engine mount bracket onto the stud and make sure it is fully seated. Install the two bolts that secure the bracket to the shock tower, but only snug them in place.

11. Install the internal–hex washer. Use a block of wood and a mallet to tap the washer into place so that the mount bracket is fully seated on the stud. Then, install a new self–locking nut, but don't tighten it yet.

12. Tighten the two mount bracket bolts to 33 ft. lbs. (44 Nm) each. Next, tighten the self–locking stud nut to 43 ft. lbs. (59 Nm).

13. Remove the hoist equipment from the engine.

14. Lower the vehicle.

15. Reconnect the positive and negative battery cables.

Right Front Engine Mount

1. Secure the hood as far open as possible.

2. Disconnect the negative and positive battery cables.

3. Raise and safely support the front of the vehicle. Block the rear wheels.

4. Attach an engine hoist to the engine lifting points and raise the hoist to remove all slack from the chain.

NOTE: A floor jack with a cushion or wooden plank over its pad may be substituted for the lifting chain.

5. Unbolt the right front mount bushing carrier from the shock tower.

6. Unbolt the mount bracket from the transaxle case.

To install:

7. Secure the mount bracket to the transaxle case with new self–locking bolts. Tighten the bolts to 47 ft. lbs. (64 Nm). If equipped with a CVT transaxle, tighten the three bracket bolts to 40 ft. lbs. (54 Nm).

8. Connect the mount bushing carrier to the shock tower. Snugly install the bolts.

9. Tighten the bushing carrier bolts to 33 ft. lbs. (44 Nm). If the mount stud nut was loosened, replace it with a new self–locking nut and tighten it to 43 ft. lbs. (59 Nm).

10. Remove the hoist equipment from the engine.

11. Lower the vehicle.

12. Reconnect the positive and negative battery cables.

Rear Engine Mount

1. Support the hood as far open as possible. If necessary, matchmark the hood hinges with a felt–tipped marker and remove the hood for extra hoist clearance.
2. Disconnect the negative and positive battery cables.
3. Raise and safely support the vehicle.
4. Attach an engine hoist to the engine lifting points and raise the hoist to remove all slack from the chain.
5. Remove and discard the bolts attaching the rear engine mount bracket to the engine.
6. Remove and discard the through–bolt from the rear engine mount damper on the subframe beam.
7. Remove the rear engine mount bracket.
8. Unbolt the rear mount damper form the subframe beam.

To install:

9. Install the rear mount damper to the subframe beam. Torque the bolts attaching the mount to the beam to 47 ft. lbs. (64 Nm).

─────── **WARNING** ───────
Tighten the bolts attaching the mount to the engine first. Then, tighten the through–bolt. The engine mount may be damaged by excessive engine vibration if this tightening order isn't followed.

10. Install the rear engine mount bracket using new bolts. Tighten the new self–locking bolts attaching the mount to the engine to 61 ft. lbs. (83 Nm). If equipped with a CVT transaxle, tighten the bracket bolts to 43 ft. lbs. (59 Nm).
11. Install a new rear engine mount damper through–bolt. Tighten the new through–bolt to 43 ft. lbs. (59 Nm).
12. Lower the vehicle.
13. Remove the hoist equipment from the engine.
14. Install the hood, if it was removed. After installation, check the panel gaps between the hood, fenders, headlights, and grille to make sure the hood is correctly aligned.
15. Reconnect the positive and negative battery cables.

Transaxle Mount

1. Secure the hood as far open as possible.
2. Disconnect the negative and positive battery cables.
3. Remove the air cleaner assembly to gain better access to the mount.
4. Attach an engine hoist to the engine lifting points and raise the hoist to remove all slack from the chain.

NOTE: A floor jack with a cushion or wooden plank over its pad may be substituted for the lifting chain.

5. Remove the nuts attaching the mount bracket to the transaxle.
6. Remove the through–bolt from the transaxle mount. Remove the mounting bracket.
7. Unbolt the transaxle mount from the shock tower.

To install:

8. Install the transaxle mount . Tighten the three bolts to 47 ft. lbs. (64 Nm). If equipped with a CVT transaxle, tighten the three bolts to 28 ft. lbs. (38 Nm).
9. Fit the transaxle mount bracket. Install the through–bolt, but do not tighten it yet.
10. Connect the transaxle mount bracket to the studs on the transaxle case. Install the two nuts. Then, tighten each nut to 47 ft. lbs. (64 Nm). If equipped with a CVT transaxle, tighten the two nuts and the bolt to 47 ft. lbs. (64 Nm) each.
11. Tighten mount through–bolt to 54 ft. lbs. (74 Nm).
12. Remove the hoist equipment from the engine.
13. Install the air cleaner assembly.
14. Reconnect the positive and negative battery cables.

Prelude

Side Engine Mount

1. Disconnect the negative battery cable then the positive battery cable.
2. Remove the side engine mount through-bolt, then the nut and bolt attaching the mount to the engine.
3. Remove the side engine mount.

To install:

4. Install the side engine mount and the through-bolt. Do not tighten the through-bolt at this time.
5. Install the nut and bolt attaching the mount to the engine. Torque the nut and bolt to 40 ft. lbs. (55 Nm).
6. Torque the side engine mount through-bolt to 47 ft. lbs. (65 Nm).
7. Connect the positive then the negative battery cable and enter the radio security code.
8. If equipped with 4WS, turn the steering wheel lock–to–lock to reset the 4WS control unit.

Front Engine Mount

1. Secure the hood as far open as possible.
2. Disconnect the negative battery cable then the positive battery cable.
3. Attach an engine hoist to the engine lifting points and raise the hoist to remove all slack from the chain.
4. Remove the front engine mount bracket. Discard the mount through-bolt.
5. Remove the bolts attaching the mount to the beam.

To install:

6. Install the mount to the beam. Torque the bolts attaching the mount to the beam to 69 ft. lbs. (95 Nm).
7. Position the front mount and install the bolts attaching the mount to the engine assembly, only snug the bolts in place.
8. Install a new through-bolt to the front mount. Torque the new through-bolt to 47 ft. lbs. (65 Nm).
9. Torque the bolts attaching the front mount to the engine to 28 ft. lbs. (39 Nm).
10. Remove the engine hoist.
11. Connect the positive then the negative battery cable and enter the radio security code.
12. If equipped with 4WS, turn the steering wheel lock–to–lock to reset the 4WS control unit.

Rear Engine Mount

1. Secure the hood as far open as possible.
2. Disconnect the negative battery cable then the positive battery cable.
3. Attach an engine hoist to the engine lifting points and raise the hoist to remove all slack from the chain.
4. Remove the rear engine mount bracket. Discard the mount through-bolt. If the rear bracket is equipped with a stiffener, remove the stiffener.
5. Remove the bolts attaching the mount to the beam and remove the mount.

To install:

6. Install the mount to the beam. Torque the bolts attaching the mount to the beam to 69 ft. lbs. (95 Nm).
7. Position the rear engine mount bracket and install new bolts attaching the bracket to the engine. Torque the new bolts attaching the mount to the engine assembly to 40 ft. lbs. (55 Nm). If equipped with a bracket stiffener, install the stiffener. Torque the stiffener nut and bolt to 15 ft. lbs. (21 Nm).
8. Install a new rear engine mount through-bolt. Torque the new through-bolt to 47 ft. lbs. (65 Nm).
9. Remove the engine hoist.
10. Connect the positive then the negative battery cable and enter the radio security code.

11. If equipped with 4WS, turn the steering wheel lock–to–lock to reset the 4WS control unit.

Transaxle Mount

1. Disconnect the negative battery cable then the positive battery cable.
2. Remove the transaxle mount.
3. Remove the transaxle mount from the mount bracket.
   To install:
4. Install the mount to the mount bracket. Torque the nuts to 28 ft. lbs. (39 Nm).
5. Position the transaxle mount and install through-bolt. Do not tighten the through-bolt at this time.
6. Install the nuts to the transaxle mount. Torque the nuts to 28 ft. lbs. (39 Nm).
7. Torque the transaxle mount through-bolt to 47 ft. lbs. (65 Nm).
8. Connect the positive then the negative battery cable and enter the radio security code.
9. If equipped with 4WS, turn the steering wheel lock–to–lock to reset the 4WS control unit.

1993 Accord

Side Engine Mount

1. Secure the hood as far open as possible.
2. Disconnect the negative battery cable then the positive battery cable.
3. Attach an engine hoist to the engine lifting points and raise the hoist to remove all slack from the chain.
4. Remove the nut and bolt attaching the side engine mount to the engine.
5. Remove the through-bolt from the side engine mount.
   To install:
6. Install the side engine mount. Do not tighten the mount through-bolt. Torque the side engine mount nut and bolt to 40 ft. lbs. (55 Nm).
7. Torque the through-bolt to 47 ft. lbs. (65 Nm).
8. Remove the hoist equipment from the engine.
9. Install the battery and connect the positive then the negative battery cables. Enter the radio security code.

Front Engine Mount (Torque Rod)

1. Secure the hood as far open as possible.
2. Disconnect the negative battery cable then the positive battery cable.
3. Attach an engine hoist to the engine lifting points and raise the hoist to remove all slack from the chain.

4. Remove and discard the through-bolt attaching the front engine mount to the engine.
5. Remove the through-bolt attaching the mount to the front beam.
   To install:
6. Install the mount to the front beam.
7. Install the through-bolt attaching the front mount to the engine. Torque the through-bolt to 47 ft. lbs (65 Nm).
8. Remove the hoist equipment from the engine.
9. Install the battery and connect the positive then the negative battery cables. Enter the radio security code.

Rear Engine Mount

1. Secure the hood as far open as possible.
2. Disconnect the negative battery cable then the positive battery cable.
3. Attach an engine hoist to the engine lifting points and raise the hoist to remove all slack from the chain.
4. Remove the bolts attaching the rear engine mount bracket to the engine.
5. Remove and discard the through-bolt from the rear engine mount.
6. If equipped with a automatic transaxle disconnect the vacuum hose from the rear engine mount.
7. Remove the bolts attaching the mount to the rear beam.
   To install:
8. Install the rear engine mount to the rear beam. Torque the bolts attaching the mount to the beam to 43 ft. lbs. (60 Nm).

─────── **WARNING** ───────
Tighten the bolts attaching the mount to the engine assembly first then the through-bolt. If this order is not followed excessive engine vibration may be felt and mount damage may occur.
───────────────────────

9. Install a new rear engine mount through-bolt. Torque the new through-bolt to 47 ft. lbs. (65 Nm).
10. Install the rear engine mount bracket using new bolts. Torque the new bolts attaching the mount to the engine assembly to 40 ft. lbs. (55 Nm).
11. If equipped with a automatic transaxle, connect the vacuum hose to the rear engine mount.
12. Remove the hoist equipment from the engine.
13. Install the battery and connect the positive then the negative battery cables. Enter the radio security code.

Transaxle Mount

1. Secure the hood as far open as possible.
2. Disconnect the negative battery cable then the positive battery cable.
3. Attach an engine hoist to the engine lifting points and raise the hoist to remove all slack from the chain.
4. Remove the nuts attaching the mount to the transaxle.
5. Remove the through-bolt from the transaxle mount.
6. Remove the transaxle mount from the mounting bracket.
   To install:
7. Install the transaxle mount to the mounting bracket. Torque the attaching nuts to 28 ft. lbs. (39 Nm).
8. Install the transaxle mount. Torque the nuts attaching the mount to the transaxle to 28 ft. lbs. (39 Nm).
9. Torque the transaxle mount through-bolt to 47 ft. lbs. (65 Nm).
10. Remove the hoist equipment from the engine.
11. Install the battery and connect the positive then the negative battery cables. Enter the radio security code.

1994–97 Accord with 2.2L Engines

Side Engine Mount

1. Secure the hood as far open as possible.
2. Disconnect the negative battery cable then the positive battery cable.
3. Attach an engine hoist to the engine lifting points and raise the hoist to remove all slack from the chain.
4. Remove the nut and bolt attaching the side engine mount to the engine.
5. Remove the through-bolt from the side engine mount.
   To install:
6. Install the side engine mount. Install a 6·100mm bolt to the mount to properly position the mount. Do not tighten the nut and bolt attaching the mount to the engine. Torque the through-bolt to 47 ft. lbs. (64 Nm) then remove the 6·100 mm bolt from the mount.
7. Torque the nut and bolt attaching the side engine mount to the engine to 47 ft. lbs. (64 Nm).
8. Remove the hoist equipment from the engine.
9. Connect the positive then the negative battery cables and enter the radio security code.

Front Engine Mount

1. Secure the hood as far open as possible.

2. Disconnect the negative battery cable then the positive battery cable.

3. Attach an engine hoist to the engine lifting points and raise the hoist to remove all slack from the chain.

4. Remove the bolts attaching the front engine mount bracket to the engine.

5. Remove the through-bolt from the front engine mount. Discard the bolt.

6. Remove the bolts attaching the mount to the front beam.

To install:

7. Install the front engine mount to the front beam. Torque the bolts attaching the mount to the beam to 69 ft. lbs. (93 Nm).

8. Install the front mount bracket. Do not tighten the bolts attaching the mount to the engine assembly, only snug the bolts in place. Install a new through-bolt to the front mount. Torque the new through-bolt to 47 ft. lbs. (64 Nm).

9. Torque the three bolts attaching the front mount bracket to the engine to 28 ft. lbs. (38 Nm).

10. Remove the hoist equipment from the engine.

11. Connect the positive then the negative battery cables and enter the radio security code.

Rear Engine Mount

1. Secure the hood as far open as possible.

2. Disconnect the negative battery cable then the positive battery cable.

3. Attach an engine hoist to the engine lifting points and raise the hoist to remove all slack from the chain.

4. Remove and discard the bolts attaching the rear engine mount bracket to the engine.

5. Remove and discard the through-bolt from the rear engine mount.

6. If equipped with a automatic transaxle disconnect the vacuum hose from the rear engine mount.

7. Remove the bolts attaching the mount to the rear beam.

To install:

8. Install the rear engine mount to the rear beam. Torque the bolts attaching the mount to the beam to 43 ft. lbs. (59 Nm).

─────── WARNING ───────
Tighten the bolts attaching the mount to the engine assembly first then the through-bolt. If this order is not followed excessive engine vibration may be felt and mount damage may occur.

9. Install the rear engine mount bracket using new bolts. Torque the new bolts attaching the mount to the engine assembly to 40 ft. lbs. (54 Nm).

10. Install a new rear engine mount through-bolt. Torque the new through-bolt to 47 ft. lbs. (64 Nm).

11. If equipped with a automatic transaxle, connect the vacuum hose to the rear engine mount.

12. Remove the hoist equipment from the engine.

13. Connect the positive then the negative battery cables and enter the radio security code.

Transaxle Mount

1. Secure the hood as far open as possible.

2. Disconnect the negative battery cable then the positive battery cable.

3. Attach an engine hoist to the engine lifting points and raise the hoist to remove all slack from the chain.

4. Remove the nuts attaching the mount to the transaxle.

5. Remove the through-bolt from the transaxle mount.

6. Remove the transaxle mount from the mounting bracket.

To install:

7. Install the transaxle mount to the mounting bracket. Torque the attaching nuts/bolts to 28 ft. lbs. (39 Nm).

8. Install the transaxle mount. Install a 6·100mm bolt to the mount to properly position the mount. Do not tighten the nuts attaching the mount to the transaxle at this time. Torque the through-bolt to 47 ft. lbs. (64 Nm) then remove the 6·100mm bolt from the mount.

9. Torque the nuts attaching the transaxle mount to the transaxle to 28 ft. lbs. (38 Nm).

10. Remove the hoist equipment from the engine.

11. Connect the positive then the negative battery cables and enter the radio security code.

1994–97 Accord with 2.7L Engine

Side Engine Mount

1. Disconnect the negative battery cable.

2. Remove the intake air duct.

3. Remove the side engine mount through-bolt.

4. Remove the bolts attaching the side engine mount to the engine. Discard the three bolts.

To install:

5. Install the side engine mount. Use three new bolts to attach the mount to the engine, torque the bolts to 40 ft. lbs. (54 Nm).

6. Torque the side engine mount through-bolt to 47 ft. lbs. (64 Nm).

7. Install the intake air duct.

8. Connect the negative battery cable and enter the radio security code.

Front and Rear Engine Mounts

1. Disconnect the support struts from the engine hood, then fix the hood in a vertical position.

2. Disconnect the negative then the positive battery cables.

3. Remove the intake air duct.

4. Remove the starter cable from the strut brace, then remove the strut brace.

5. Attach an engine hoist to the engine lifting points and raise the hoist to remove all slack from the chain.

6. Remove the through-bolt from the side engine mount.

7. Remove the through-bolt from the transaxle mount.

8. Remove the lower mounting bolts from the front and rear engine mounts.

9. Disconnect the rear engine mount vacuum hose from the engine mount control solenoid.

10. Remove the nut attaching the mount to the engine bracket, from the mount that is being removed.

11. Raise the engine enough to remove the engine mount.

─────── WARNING ───────
Raise the engine only enough to remove the mount, if the engine is raised too high the engine wiring and/or hoses may be damaged.

To install:

NOTE: When installing the engine mounts and vibration dampers in the following steps, they must be tightened to the correct tension in the correct order if they are to damp vibration properly.

12. Position the engine mount on the engine mount bracket. Make sure the mount is properly aligned to the bracket. Torque the nut attaching the mount to the bracket to 40 ft. lbs. (54 Nm).

13. Install the bolts attaching the rear mount to the beam. Torque the bolts to 43 ft. lbs. (59 Nm).

14. Install the bolts attaching the front mount to the beam. Torque the bolts to 43 ft. lbs. (59 Nm).

15. Install the side engine mount through-bolt and torque it to 47 ft. lbs. (64 Nm).

16. Install the transaxle mount through-bolt and torque it to 47 ft. lbs. (64 Nm).

17. Remove the engine hoist.

18. Connect the rear engine mount vacuum hose to the engine mount control solenoid valve.

19. Install the strut brace. torque the strut brace bolts to 16 ft. lbs. (22 Nm). Install the starter cable to the strut brace.

20. Install the intake air duct.

21. Connect the positive then the negative battery cables. Enter the radio security code.

22. Connect the engine hood support struts to the hood.

Transaxle Mount

1. Disconnect the negative then the positive battery cables.

2. Remove the nuts attaching the transaxle mount to the transaxle.

3. Remove the transaxle mount through-bolt and remove the mount.

To install:

4. Install the three nuts attaching the mount to the transaxle, torque the nuts to 28 ft. lbs. (38 Nm).

5. Install the transaxle mount through-bolt and torque the bolt to 47 ft. lbs. (64 Nm).

6. Loosen the side engine mount through-bolt, then torque the bolt to 47 ft. lbs. (64 Nm).

7. Connect the positive then the negative battery cables. Enter the radio security code.

Cylinder Head

REMOVAL AND INSTALLATION

NOTE: The radio may contain a coded theft protection circuit. Always obtain the code number before disconnecting the battery. If the vehicle is equipped with 4WS, the steering control unit is shut down when the battery is disconnected. After connecting the battery, turn the steering wheel lock-to-lock to reset the steering control unit.

1993–95 Civic and del Sol

1. The engine should be cold before the cylinder head is removed.

2. Disconnect the negative battery cable. Drain the cooling system.

3. Remove the brake booster vacuum hose from the brake master cylinder power booster. Remove the en-gine secondary ground cable from the valve cover.

4. Remove the air intake hose and the air chamber. Relieve the fuel pressure. Disconnect the fuel hoses and fuel return hose.

5. Remove the air intake hose and resonator hose. Disconnect the throttle cable at the throttle body. On vehicles equipped with automatic transaxles, disconnect the throttle control cable at the throttle body.

6. Disconnect the charcoal canister hose at the throttle valve.

7. Disconnect the following engine wire connectors from the cylinder head and the intake manifold:

a. 14 prong connector from the main wiring harness
b. EACV connector
c. Intake air temperature sensor connector
d. Throttle angle sensor connector
e. Injector connectors
f. Ignition coil from the distributor
g. Top dead center/crank sensor connector from the distributor.
h. Coolant temperature gauge sender connector.
i. Coolant temperature sensor connector.
j. Oxygen sensor.

8. Disconnect the vacuum hoses and the water bypass hoses from the intake manifold and throttle body.

9. Remove the upper radiator hose and the heater hoses from the cylinder head.

10. Remove the PCV hose, charcoal canister hose and vacuum hose from the intake manifold, and remove the vacuum hose from the brake master cylinder power booster.

11. Loosen the air conditioning idler pulley and remove the air conditioning belt. Remove the alternator belt. If equipped with power steering, remove the power steering belt and pump bracket.

12. Remove the intake manifold bracket and the exhaust manifold bracket.

13. Remove the exhaust manifold shroud, then remove the exhaust manifold.

14. Mark the position of the distributor in relation to the engine block, remove and tag the spark plug wires and remove the distributor assembly.

15. Remove the valve cover. Remove the timing belt cover.

16. Mark the direction of rotation on the timing belt. Loosen the timing belt adjuster bolt 180°, then remove the timing belt from the camshaft pulley. Retighten the adjuster bolt to 33 ft. lbs. (45 Nm).

NOTE: Do not crimp or bend the timing belt more than 90 degrees or less than 1 in. (25mm) in diameter (width).

17. Remove the cylinder head bolts in the reverse order of the tightening sequence. Once the bolts are all removed, remove the cylinder head along with the intake manifold from the engine. Remove the intake manifold from the cylinder head. If the head sticks to the engine block, tap it with a plastic or wooden mallet, or pry it loose with a large, flat screwdriver. Leverage slots are located at each end of the back side of the head.

To install:

NOTE: Use new O-ring, seals, and gaskets when installing the cylinder head and its components.

18. Be sure the cylinder head and the engine block surfaces are clean, level, and straight.

19. Be sure the cylinder head dowel pins and control jet are aligned. Clean the oil control orifice and reinstall it with a new O-ring.

20. Install the cylinder head onto the engine with new head gasket.

21. Be sure the UP mark on the timing belt pulley is at the top.

22. Install the intake manifold and tighten the nuts in a criss-cross pattern in 2–3 steps to 17 ft. lbs. starting with the inner nuts.

23. Install the bolts that secure the intake manifold to its bracket but do not tighten them at this point.

24. Position the cam so the TDC marks align, and install the cylinder head bolts.

25. Tighten the cylinder head bolts in 2 steps. On the first step tighten all the bolts, in sequence, to 22 ft. lbs. (30 Nm). On the final step, using the same sequence, tighten the bolts to 47 ft. lbs. (65 Nm) for D15B7 and D15B8 engines. On D16Z6 and D15Z1 engine, the final torque for the head bolts is 53 ft. lbs. (73 Nm).

26. Install the exhaust manifold and tighten the nuts in a crisscross pattern in two or three steps to 25 ft. lbs. (34 Nm) starting with the inner nuts.

27. Install the exhaust manifold to the head pipe. Tighten the bolts to the intake manifold bracket. Install the header pipe on to its bracket.

28. Install the timing belt, timing cover, and crankshaft pulley. Install the upper timing cover and valve cover. Coat the valve cover spark plug seals with oil before installation.

CYLINDER HEAD BOLTS TORQUE SEQUENCE:

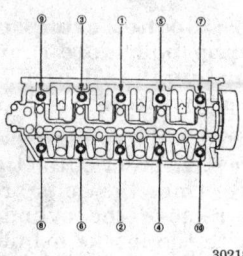

302184

Cylinder head bolt torque
sequence — 1993–95 Civic and
del Sol

29. Reconnect the throttle cable. Adjust its tension so the cable has a deflection of 10–12mm (0.39–0.47 in.).

30. Install the power steering pump bracket and torque the bolts to 33 ft. lbs. (45 Nm). Torque the power steering pump mounting bolt to 17 ft. lbs. (24 Nm).

31. Install the distributor. Install new spark plugs and reconnect the spark plug wires.

32. After the installation procedure is complete, check that all tubes, hoses, and connectors are installed correctly. Check the tension of the accessory drive belts.

33. Adjust the valve clearance and ignition timing.

34. Change the engine oil and filter.

35. Refill the cooling system with fresh coolant and bleed the system.

36. Reconnect the negative battery cable.

37. Run the the engine and road test the vehicle.

1996–97 Civic and del Sol

1.6L (D16Y5, D16Y7 and D16Y8) Engines

1. Make sure the cylinder head is cool to the touch before beginning the removal procedure. The coolant temperature must be below 100° F (38°C).

2. Disconnect the negative battery cable.

3. Drain the cooling system.

4. Label and disconnect the ignition wires.

5. Remove the air intake duct and the air cleaner assembly.

6. Relieve the fuel pressure.

7. Clean up any fuel that may have spilled on the engine or intake manifold.

8. Disconnect the upper radiator hose from the coolant inlet.

9. Disconnect the coolant bypass hoses and the heater hose from the intake manifold.

10. Loosen the power steering pump mounting bolts to release the belt tension. Remove the power steering pump belt.

11. Remove the power steering pump from its mounting bracket and lift the power steering reservoir from its mount. Move the pump and reservoir out of the work area and secure them. Don't disconnect the hydraulic lines.

12. Place a block of wood on the pad of a floor jack. Place the floor jack under the engine for support.

13. If equipped with A/C: unbolt the left–front engine mount bracket.

14. Loosen the A/C compressor idler pulley bolt. Then, loosen the adjusting bolt to release the belt's tension. Slip the A/C compressor belt around the engine mount to remove it.

15. Loosen the alternator mounts; then, remove the alternator belt.

16. Make sure the engine is supported with the padded floor jack. Loosen the nuts from left side engine mount. Remove the engine mount bracket.

17. Remove the valve cover and the upper timing belt cover.

18. Remove the crankshaft pulley and the lower timing belt cover. Separate the dipstick tube from its catches on the timing cover. Remove the timing belt.

19. With the timing belt removed, inspect the water pump and replace it if necessary.

20. Remove the distributor from the cylinder head as an assembly.

21. Unbolt and remove the camshaft sprocket.

22. Disconnect the fuel lines from the intake manifold fuel rail. Immediately plug the lines to prevent fuel leakage and contamination.

23. Disconnect the throttle cable from the linkage by first loosening its locknut and then slipping it out of its holder.

24. Label and disconnect the following engine harness connectors from the cylinder head and the intake manifold:

  a. Fuel injector wiring harness connectors

  b. VTEC solenoid valve and pressure switch connectors (D16Y5, D16Y8 engines only)

  c. Idle air control valve (IAC) connector

  d. Throttle position sensor (TPS) connector

  e. EGR valve lift sensor connectors (D16Y5 engine only)

  f. Engine coolant temperature sensor, switch, and gauge sender (ECT) connectors

  g. Manifold absolute pressure sensor (MAP) connector

  h. Primary and secondary (D16Y5, D16Y7 engines only) oxygen sensor (HO2S) connectors

25. Label and disconnect the vacuum hoses and PCV hose from the intake manifold and throttle body.

26. Disconnect the charcoal canister (EVAP) and breather hoses from the intake manifold.

27. Remove the intake manifold together with the throttle body and plenum.

28. Remove the exhaust manifold.

29. Remove the power steering pump bracket.

30. Loosen the cylinder head bolts in a three–step crisscross pattern in the reverse order of the tightening sequence. Start with the outermost bolts and work toward the middle of the cylinder head. Loosen the bolts in the reverse order of installation.

31. Remove the cylinder head. If the head sticks to the engine block, tap it with a plastic or wooden mallet.

32. Inspect the cylinder head for warpage and cracking. Repair, machine, or replace as necessary. The warpage limit is 0.002 in. (0.05mm). Standard cylinder head height is 3.659–3.663 in. (92.95–93.05mm).

33. Remove the old cylinder head gasket and thoroughly clean the mating surfaces.

34. Cover the engine block with a sheet of plastic to keep out dust and foreign objects.

To install:

NOTE: Use new O–ring, seals, and gaskets when installing the cylinder head and its components.

35. Make sure the cylinder head and the engine block surfaces are clean, level, and straight.

36. Make sure the cylinder head dowel pins and control orifice are aligned. Clean the oil control orifice and reinstall it with a new O–ring.

37. Install a new head gasket onto the engine block.

38. If the camshaft was removed, reinstall it with the keyway facing up so that the engine will remain at TDC/compression for the No. 1 cylinder. Lubricate and install a new camshaft seal.

39. Use new cylinder head bolts and washers. Used or previously–tightened bolts may be stretched, and therefore they have reduced clamping and sealing power under compression. Apply clean engine oil to the threads of each head bolt.

40. Fit the cylinder head into place. Hand–tighten all the cylinder head bolts.

41. Tighten the cylinder head bolts to their final torque specification in four steps. Use a crisscross sequence starting with the bolts at the middle of the head and working toward the outer bolts.

    a. Step 1: Tighten each bolt to 14 ft. lbs. (20 Nm).

    b. Step 2: Tighten each bolt to 36 ft. lbs. (49 Nm).

    c. Step 3: Tighten each bolt to 49 ft. lbs. (67 Nm).

    d. Step 4: Tighten only the two center bolts to an additional 49 ft. lbs. (67 Nm).

42. Apply oil to the camshaft sprocket bolt. Install the sprocket with the UP mark and the keyway pointing straight up. Tighten the sprocket bolt to 27 ft. lbs. (37 Nm).

43. Install the intake manifold with a new gasket, and tighten the nuts in a crisscross pattern in 2–3 steps to 17 ft. lbs. (24 Nm) starting with the inner nuts.

44. Install the bolts that secure the intake manifold to its bracket and tighten them to 17 ft. lbs. (24 Nm).

45. Install the power steering pump bracket and tighten its bolts to 33 ft. lbs. (44 Nm).

46. Install the exhaust manifold with a new gasket. Apply anti–seize paste to the studs, and tighten the nuts to 23 ft. lbs. (31 Nm) in a criss-cross sequence.

47. Connect the exhaust manifold to the front exhaust pipe. Tighten the self–locking nuts to 25 ft. lbs. (33 Nm). On vehicles with the D16Y8 engine, torque the nuts to 40 ft. lbs. (55 Nm).

48. Verify that the engine is at TDC/compression for the No. 1 cylinder.

49. Install the timing belt. After the timing belt has been properly tensioned, Tighten the adjusting bolt to 33 ft. lbs. (44 Nm).

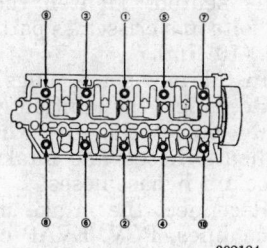

CYLINDER HEAD BOLTS TORQUE SEQUENCE:

302184

Cylinder head bolt tightening sequence — 1996–97 Civic and del Sol

50. Install the lower timing belt cover. Install the crankshaft pulley and tighten its bolt to 134 ft. lbs. (181 Nm). Fit the dipstick tube back into its catches.

51. Adjust the valves. If equipped with a VTEC engine, also check the rocker arms for free and smooth motion.

52. If equipped with a VTEC engine, remove the VTEC solenoid valve and its filter. Install a new filter, and then reinstall the VTEC solenoid valve and tighten its bolt to 9 ft. lbs. (12 Nm).

53. Install the distributor. The lugs on the distributor drive fit into the groove on the end of the camshaft. Don't fully tighten the distributor mounting bolts yet.

54. Make sure all the spark plug tube sealing gaskets are fully seated.

55. Install a new gasket onto to the valve cover. Apply liquid gasket to the corner recesses of the gasket. Don't let the sealant to cure before installing the valve cover onto the cylinder head.

56. Install the valve cover. Gently wiggle the valve cover to make sure it is fully seated. Tighten the valve cover bolts in a crisscross pattern to 7 ft. lbs. (10 Nm).

57. Install new spark plugs.

58. Reconnect the ignition wires.

59. Reconnect the upper radiator hose, heater hoses, and intake manifold coolant bypass hoses.

60. Reconnect the intake manifold vacuum lines, PCV, EVAP canister, and breather hoses.

61. Connect the fuel lines to the fuel rail. Use new sealing washers on the banjo fitting. Carefully tighten the banjo fitting to 21 ft. lbs. (28 Nm) for the D16Y5 engine, or to 16 ft. lbs. (22 Nm) for all other engines. Tighten the service bolt to 9–11 ft. lbs. (12–15 Nm).

62. Reconnect the throttle cable. Adjust its tension so the cable has a deflection of 10–12mm (0.39–0.47 in.).

63. Installation of the remaining components is the reverse of removal.

1.6L (B16A2 and B16A3) Engines

1. Before beginning the cylinder head removal procedure, make sure the engine temperature is below 100° F (38°C). To prevent warping, the cylinder head should be removed when the engine is cold.

2. Disconnect the negative battery cable.

3. Label and disconnect the ignition wires.

4. Drain the engine coolant. Remove the radiator cap to speed draining.

5. Remove the strut brace.

6. Remove the intake air duct and disconnect the breather hose.

7. Relieve the fuel pressure:

    a. Loosen the fuel filler cap.

    b. Hold the fuel filter banjo bolt with a back–up wrench. Hold the fuel filter service bolt with a box end wrench.

    c. Place a shop rag over the fuel filter to absorb fuel spray.

    d. Slowly loosen the fuel filter service bolt one complete turn.

8. Clean up any fuel that may have spilled on the engine or intake manifold.

9. Disconnect the upper radiator hose from the coolant inlet.

10. Disconnect the coolant bypass hoses and the heater hose from the intake manifold.

11. Loosen the power steering pump mounting bolts to release the belt tension. Remove the power steering pump belt.

12. Remove the power steering pump from its mounting bracket and lift the power steering reservoir from its mount. Move the pump and reservoir out of the work area and secure them. Don't disconnect the hydraulic lines.

13. Place a block of wood on the pad of a floor jack. Place the floor jack under the engine for support.

14. If equipped with A/C: unbolt the left–front engine mount bracket.

15. Loosen the A/C compressor idler pulley bolt. Then, loosen the adjusting bolt to release the belt's tension. Slip the A/C compressor belt around the engine mount to remove it.

16. Loosen the alternator mounts; then, remove the alternator belt.

17. Make sure the engine is supported with the padded floor jack. Loosen the left side engine mount nuts. Remove the engine mount bracket.

18. Remove the valve cover and the upper timing belt cover.

19. Remove the crankshaft pulley and the lower timing belt cover. Remove the timing belt.

20. With the timing belt removed, inspect the water pump and replace it if necessary.

21. Remove the distributor from the cylinder head as an assembly.

22. Disconnect the fuel lines from the intake manifold fuel rail. Immediately plug the lines to prevent fuel leakage and contamination.

23. Disconnect the throttle cable from the linkage by first loosening its

locknut and then slipping it out of its holder.

24. Label and disconnect the following engine harness connectors from the cylinder head and the intake manifold:

    a. Fuel injector wiring harness connectors

    b. VTEC solenoid valve and pressure switch connectors

    c. Idle air control valve (IAC) connector

    d. Throttle position sensor (TPS) connector

    e. Engine coolant temperature sensor, switch, and gauge sender (ECT) connectors

    f. Manifold absolute pressure sensor (MAP) connector

    g. Primary oxygen sensor (HO2S) connector

25. Label and disconnect the vacuum hoses and PCV hose from the intake manifold and throttle body.

26. Disconnect the charcoal canister (EVAP) and breather hoses from the intake manifold.

27. Loosen the intake manifold nuts in a crisscross sequence. Then, remove the intake manifold together with the throttle body and plenum.

28. Remove the exhaust manifold heat shield. Then, loosen the exhaust manifold nuts in a crisscross sequence. Remove the exhaust manifold. Be careful not to damage the oxygen sensors when removing the manifold. Cover the front exhaust pipe flange with a shop towel to keep dirt out.

29. Remove the power steering pump bracket.

30. Remove the camshaft pulleys and back cover.

31. Loosen the camshaft holder plate bolts in a crisscross sequence working toward the middle of the cylinder head.

32. Loosen the valve adjusting screws.

33. Lift the camshaft holder plates and holders from the cylinder head. The holder bolts will keep the components together. Note the positions of each camshaft holder for reassembly.

34. Lift the camshafts from the cylinder head. Mark the exhaust and intake camshafts so that they will not be confused.

35. Loosen the cylinder head bolts in a three–step crisscross pattern. Start with the outermost bolts and work toward the middle of the cylinder head.

36. Remove the cylinder head. If the head sticks to the engine block, tap it with a plastic–faced or wooden mallet.

37. Inspect the cylinder head for warpage and cracking. Repair, machine, or replace as necessary. The warpage limit is 0.002 in. (0.05mm). Standard cylinder head height is 5.589–5.593 in. (141.95–142.05mm).

To install:

NOTE: Use new O–ring, seals, and gaskets when installing the cylinder head and its components.

38. Make sure the cylinder head and the engine block surfaces are clean, level, and straight.

39. Make sure the cylinder head dowel pins and oil control orifice are aligned. Clean the oil control orifice and reinstall it with a new O–ring.

40. Install a new head gasket onto the engine block.

41. Use new cylinder head bolts and washers. Used or previously–tightened bolts may be stretched; and therefore, they have reduced clamping and sealing power under compression. Apply clean engine oil to the threads of each head bolt.

42. Fit the cylinder head into place. Hand–tighten all the cylinder head bolts.

43. Tighten the cylinder head bolts to their final torque specification in two steps. Use a crisscross sequence starting with the bolts at the middle of the head and working toward the outer bolts.

    a. Step 1: Tighten each bolt to 22 ft. lbs. (30 Nm).

    b. Step 2: Tighten each bolt to 61 ft. lbs. (85 Nm).

44. Install the dowel pin in the No. 3 cylinder head camshaft holder with a new O–ring.

45. Thoroughly clean the intake and exhaust camshaft oil control orifices. Reinstall them with new O–rings.

46. Install the camshafts.

47. Install the intake manifold with a new gasket, and tighten the nuts in

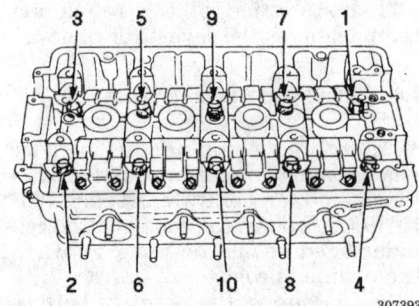

Cylinder head bolt loosening sequence — 1.6L (B16A2 and B16A3) engines

a crisscross pattern in 2–3 steps to 17 ft. lbs. (24 Nm) starting with the inner nuts.

48. Install the bolts that secure the intake manifold to its bracket and tighten them to 17 ft. lbs. (24 Nm).

49. Install the power steering pump bracket and tighten its bolts to 33 ft. lbs. (44 Nm).

50. Install the exhaust manifold with a new gasket. Apply anti–seize paste to the studs, and tighten the nuts to 23 ft. lbs. (31 Nm) in a crisscross sequence. Tighten the exhaust manifold bracket bolts to 17 ft. lbs. (24 Nm).

51. Connect the exhaust manifold to the front exhaust pipe. Tighten the self–locking nuts to 40 ft. lbs. (55 Nm).

52. Verify that the engine is at TDC/compression for the No. 1 cylinder.

53. Install the timing belt. After the timing belt has been properly tensioned, Tighten the adjusting bolt to 40 ft. lbs. (55 Nm).

54. Install the lower timing belt cover. Install the crankshaft pulley and tighten its bolt to 130 ft. lbs. (180 Nm).

55. Adjust the valves.

56. Inspect the VTEC rocker arms for free and smooth motion.

57. Remove the VTEC solenoid valve and its filter. Install a new filter, and then reinstall the VTEC solenoid valve and tighten its bolts to 9 ft. lbs. (12 Nm).

58. Install the distributor. The lugs on the distributor drive fit into the groove on the end of the intake camshaft. Don't fully–tighten the distributor mounting bolts yet.

59. Make sure all the spark plug tube sealing gaskets are fully seated.

60. Install a new gasket onto to the valve cover. Apply liquid gasket to the corners of the gasket that meet the camshaft holders. Don't let the sealant cure before installing the valve cover onto the cylinder head.

61. Install the valve cover. Gently wiggle the valve cover to make sure it is fully seated. Tighten the valve cover bolts in a crisscross pattern to 7 ft. lbs. (10 Nm).

62. Install new spark plugs.

63. Reconnect the ignition wires.

64. Reconnect the upper radiator hose, heater hoses, and intake manifold coolant bypass hoses.

65. Reconnect the intake manifold vacuum lines, PCV, EVAP canister, and breather hoses.

66. Connect the fuel lines to the fuel rail. Use new sealing washers on the banjo fitting. Carefully tighten

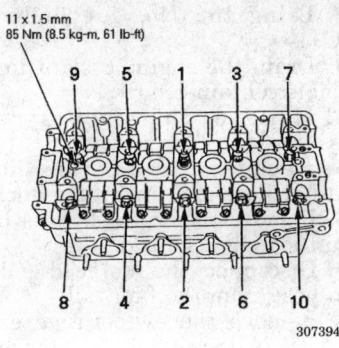

11 x 1.5 mm
85 Nm (8.5 kg-m, 61 lb-ft)

9 5 1 3 7

8 4 2 6 10

307394

**Cylinder head bolt tightening
sequence — 1.6L (B16A2 and B16A3)
engines**

the banjo fitting to 25 ft. lbs. (33 Nm).
Tighten the service bolt to 11 ft. lbs.
(15 Nm).

67. Reconnect the throttle cable.
Adjust its tension so the cable has a
deflection of 10–12mm (0.39–0.47
in.).

68. Installation of the remaining
components is the reverse of removal.

69. After the installation procedure
is complete, check that all tubes,
hoses, and connectors are installed
correctly.

Prelude and 1993 Accord

2.2L (H22A1) Engine

1. Disconnect the negative battery
cable.

2. Turn the crankshaft so the No.
1 piston is at top dead center.

**NOTE: The No. 1 piston is at
top dead center when the pointer
on the block aligns with the
white painted mark on the
flywheel.**

3. Drain the engine coolant into a
sealable container.

4. Relieve the fuel system
pressure.

5. Remove the air intake duct.

6. Remove the evaporative emissions (EVAP) control canister hose
from the intake manifold.

7. Remove the throttle cable from
the throttle body.

**NOTE: Be careful not to bend
the cable when removing. Always
replace a kinked cable with a
new one.**

8. Remove the fuel feed and return
hose.

9. Remove the brake booster vacuum hose from the intake manifold.

10. Disconnect the following engine
wire harness connectors:

    a. Fuel injector connectors

    b. Intake Air Temperature (IAT)
sensor connector

    c. Idle Air Control (IAC) valve
connector

    d. Throttle Position (TP) sensor
connector

    e. Exhaust Gas Recirculation
(EGR) valve lift sensor

    f. Ground cable terminals

    g. Engine Coolant Temperature
(ECT) switch B connector

    h. Heated oxygen sensor (HO2S)
connector

    i. ECT sensor

    j. ECT gauge sending unit
connector

    k. Ignition Control Module
(ICM) connector

    l. CKP/TDC/CYP sensor
connector

    m. Vehicle Speed Sensor (VSS)
connector

    n. Ignition coil connector

    o. VTEC solenoid valve
connector

    p. VTEC pressure switch
connector

    q. Intake air bypass solenoid
valve connector

    r. ECT switch A connector

    s. Knock sensor connector

11. Remove the engine ground
cable from the cylinder head cover.

12. Remove the connector and the
terminal from the alternator, then remove the engine wire harness from
the valve cover.

13. Remove the mounting bolts and
drive belt from the power steering
pump. Pull the pump away from the
mounting bracket, without disconnecting the hoses. Support the pump
out of the way.

14. Remove the ignition coil.

15. Tag then disconnect the emissions vacuum hoses from the intake
manifold assembly.

16. Remove the bypass hose from
the intake manifold.

17. Remove the upper radiator hose
and the heater hose from the cylinder
head.

18. Remove the lower radiator hose
and bypass hose from the thermostat
housing.

19. Remove the thermostat housing
mounting bolts. Remove the thermostat housing from the intake manifold and the connecting pipe, by pulling and twisting the housing. Discard
the O–rings.

20. Raise and safely support the
vehicle.

21. Remove the front wheel and tire
assemblies.

22. Remove the splash shield.

23. Remove the intake manifold
bracket bolts.

24. Remove the intake manifold.

25. Disconnect the exhaust pipe
from the exhaust manifold.

26. Remove the exhaust manifold
and the exhaust manifold heat
insulator.

27. Remove the plug wire cover
from the cylinder head cover. Label
then disconnect the electrical connectors from the distributor and the
spark plug wires from the spark
plugs. Mark the position of the distributor and remove it from the cylinder head.

28. Remove the Positive Crankcase
Ventilation (PCV) hose, then remove
the cylinder head cover. Replace the
rubber seals if damaged or
deteriorated.

29. Remove the timing belt middle
cover.

30. Ensure the arrows embossed on
the camshaft pulleys are aligned in
the upward position (12 o'clock).

31. Mark the rotation of the timing
belt if it is to be used again.

32. Remove the timing belt.

33. Remove the camshaft sprocket
attaching bolts then remove the
sprockets. Do not lose the sprocket
keys.

34. Loosen all of the rocker arm adjusting screws.

35. Remove the camshaft holder
bolts two turns at a time, in the
proper sequence, to prevent damaging the valves or the the rocker arm
assemblies. Remove the camshaft
holder plates and the camshaft holders, note the holders locations for
ease of installation.

36. Remove the rubber cap from the
head, located at the end of the intake
camshaft. Replace the rubber cap if
damaged or deteriorated.

37. Remove the side engine mount
bracket B, then the timing belt back
cover.

38. Remove the cylinder head bolts
in the proper sequence.

**NOTE: To prevent warpage,
unscrew the bolts in sequence ⅓
turn at a time. Repeat the sequence until all bolts are
loosened.**

39. Separate the cylinder head from
the engine block with a suitable flat
bladed pry tool.

To install:

40. Make sure all cylinder head and
block gasket surfaces are clean.
Check the cylinder head for warpage.
If warpage is less than 0.002 in.
(0.05mm), cylinder head resurfacing
is not required. Maximum resurface
limit is 0.008 in. (0.2mm) based on a
cylinder head height of 5.79 in.
(147mm).

41. Always use a new head gasket.

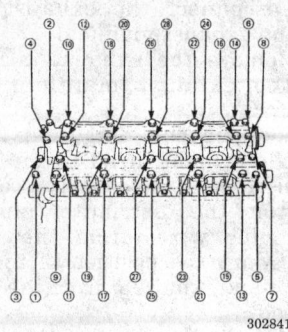

Camshaft holder bolt loosening
sequence — 2.2L (H22A1) engine

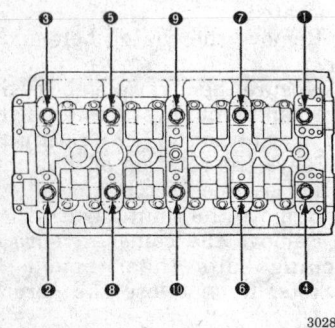

302842

Cylinder head bolt loosening
sequence — 2.2L (H22A1) engine

42. Make sure the No. 1 cylinder is at TDC.

43. Install and align the cylinder head dowel pins.

44. Install the bolts that secure the intake manifold to it's bracket but do not tighten them.

45. Apply clean oil to the cylinder head bolt threads and under the bolt threads. Install the cylinder head and bolts, then tighten the cylinder head bolts sequentially in 3 steps:
   Step 1: 29 ft. lbs. (40 Nm).
   Step 2: 51 ft. lbs. (70 Nm).
   Step 3: 72 ft. lbs. (100 Nm).

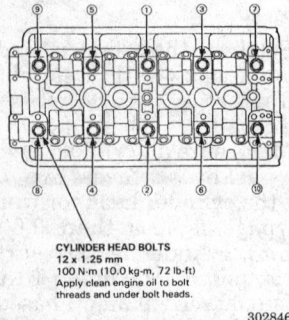

CYLINDER HEAD BOLTS
12 x 1.25 mm
100 N·m (10.0 kg·m, 72 lb·ft)
Apply clean engine oil to bolt
threads and under bolt heads.

302846

Cylinder head bolt torque
sequence — 2.2L (H22A1) engine

NOTE: A beam type torque wrench is recommended. If a bolt makes any noise while being torqued, loosen the bolt and re-tighten it.

46. Install the intake manifold with a new gasket. Beginning with the inner nuts, tighten them in a crisscross pattern using 2–3 steps. The final torque should be 16 ft. lbs. (22 Nm).

47. Connect the intake manifold bracket to the intake manifold torque the bolt to 16 ft. lbs. (22 Nm).

48. Install the exhaust manifold with a new gasket. Beginning with the inner nuts, tighten them in a crisscross pattern using 2–3 steps. Final torque should be 23 ft. lbs. (32 Nm).

49. Install the exhaust manifold brackets, then install the exhaust pipe, bracket and upper shroud.

50. Lubricate the rocker arms and camshafts with clean oil.

51. Install the camshaft seals to the end of the camshafts that the timing belt sprockets attach to. The seals open side (with spring) should be facing into the cylinder head when installed.

52. Make sure the keyways on the camshafts are facing up and install the camshafts to the cylinder head.

53. Install the rubber cap to the cylinder head at the end of the intake camshaft.

54. Clean the oil control orifice and install a new O–ring then install it into the oil passage of the No. 3 camshaft holder.

55. Install the camshafts.

56. Install the timing belt back cover and torque the attaching bolts to 9 ft. lbs. (12 Nm).

57. Install the side engine mount bracket B. Torque the bolt attaching the bracket to the cylinder head to 33 ft. lbs. (45 Nm). Torque the bolts attaching the bracket to the side engine mount to 16 ft. lbs. (22 Nm).

58. Install the remaining components in the reverse order of removal.

59. Connect the negative battery cable and enter the radio security code.

60. Start the engine, checking carefully for any leaks.

61. Check and adjust the ignition timing, then torque the distributor mounting bolts to 13 ft. lbs. (18 Nm).

62. If equipped with 4WS, start the engine and turn the steering wheel lock–to–lock to reset the 4WS control unit.

2.2L (F22A1 and F22A6) Engines

1. Disconnect the negative battery cable.

2. Bring the No. 1 cylinder to TDC.

3. Drain the engine coolant into a sealable container.

4. Relieve the fuel system pressure.

5. Remove the vacuum hose, breather hose and air intake duct.

6. Remove the water bypass hose from the cylinder head.

7. Disconnect the fuel feed and return hose from the fuel rail.

8. Remove the evaporative emissions (EVAP) control canister hose from the intake manifold.

9. Remove the brake booster vacuum hose from the intake manifold. On automatic transaxle equipped vehicles, remove the vacuum hose mount.

10. Remove the throttle cable from the throttle body. On automatic transaxle equipped vehicles, remove the throttle control cable at the throttle body.

NOTE: Be careful not to bend the cable when removing. Do not use pliers to remove the cable from the linkage. Always replace a kinked cable with a new one.

11. Remove the ignition coil.

12. Label then disconnect the electrical connectors from the distributor and the spark plug wires from the spark plugs. Mark the position of the distributor and remove it from the cylinder head. Disconnect the ignition coil wire from the distributor.

13. Remove the connector and the terminal from the alternator, then remove the engine wire harness from the valve cover.

14. Disconnect the following engine wire harness connectors:
   a. Fuel injector connectors
   b. Intake Air Temperature (IAT) sensor connector, if equipped
   c. Idle Air Control (IAC) valve connector
   d. Throttle Position (TP) sensor connector
   e. Exhaust Gas Recirculation (EGR) valve lift sensor
   f. Ground cable terminals
   g. Engine Coolant Temperature (ECT) switch B connector, if equipped
   h. Heated oxygen sensor (HO2S) connector
   i. ECT sensor
   j. ECT gauge sending unit connector
   k. CKP/TDC/CYP sensor connector, if equipped
   l. Vehicle Speed Sensor (VSS) connector
   m. ECT switch A connector

15. Remove the upper radiator hose and the heater inlet hose from the cylinder head.

16. Remove the lower radiator hose and heater outlet hose from the intake manifold.

17. Remove the bypass hose from the thermostat housing and intake manifold.

18. Remove the thermostat housing mounting bolts. Remove the thermostat housing from the intake manifold and the connecting pipe, by pulling and twisting the housing. Discard the O–rings.

19. Tag then disconnect the emissions vacuum hoses from the intake manifold assembly.

20. Disconnect the cruise control actuator electrical connector and the vacuum tube, then remove the cruise control actuator.

21. Remove the engine ground cable from the body.

22. Remove the mounting bolts and drive belt from the power steering pump. Pull the pump away from the mounting bracket, without disconnecting the hoses. Support the pump out of the way.

23. Raise and safely support the vehicle.

24. Remove the front wheel and tire assemblies.

25. Remove the splash shield.

26. Remove the intake manifold bracket bolts.

27. Remove the intake manifold.

28. Disconnect the exhaust pipe from the exhaust manifold.

29. Remove the exhaust manifold and the exhaust manifold heat insulator.

30. Remove the power steering pump mounting bracket.

31. Remove the Positive Crankcase Ventilation (PCV) hose, then remove the cylinder head cover. Replace the rubber seals if damages or deteriorated.

32. Remove the timing belt.

33. Remove the cylinder head bolts in the reverse order of installation.

NOTE: To prevent warpage, unscrew the bolts in sequence ⅓ turn at a time. Repeat the sequence until all bolts are loosened.

34. Separate the cylinder head from the engine block with a suitable flat bladed prytool.

To install:

35. Make sure all cylinder head and block gasket surfaces are clean. Check the cylinder head for warpage. If warpage is less than 0.002 in. (0.05mm), cylinder head resurfacing is not required. Maximum resurface limit is 0.008 in. (0.2mm) based on a cylinder head height of 3.94 in. (100mm).

36. Always use a new head gasket.

37. The **UP** mark on the camshaft pulley should be at the top.

38. Make sure the No. 1 cylinder is at TDC.

39. Clean the oil control orifice and install a new O–ring. Install and align the cylinder head dowel pins and oil control jet.

40. Install the bolts that secure the intake manifold to it's bracket but do not tighten them.

41. Position the camshaft correctly.

42. Install the cylinder head, then tighten the cylinder head bolts sequentially in 3 steps:
    Step 1: 29 ft. lbs. (40 Nm).
    Step 2: 51 ft. lbs. (70 Nm).
    Step 3: 72 ft. lbs. (100 Nm).

43. Install the intake manifold and tighten the nuts in a crisscross pattern, in 2–3 steps, beginning with the inner nuts. Final torque should be 16 ft. lbs. (22 Nm). Always use a new intake manifold gasket.

44. Connect the intake manifold bracket to the intake manifold. Torque the bolt to 16 ft. lbs. (22 Nm).

45. Install the heat insulator to the cylinder head and the block.

46. Install the power steering pump mounting bracket to the cylinder head. Torque the two 10·1.25mm bolts to 36 ft. lbs. (50 Nm). Torque the 8·1.25 bolt to 16 ft. lbs. (22 Nm).

47. Install the exhaust manifold and tighten the nuts in a crisscross pattern in 2–3 steps, beginning with the inner nut. Final torque should be 23 ft. lbs. (32 Nm). Always use a new exhaust manifold gasket.

48. Install the exhaust manifold bracket, then install the exhaust pipe, bracket and upper shroud.

49. Make sure the camshaft sprocket and the crankshaft pulleys are aligned to TDC. Install the timing belt.

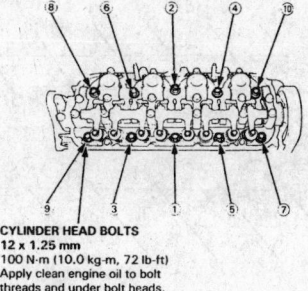

CYLINDER HEAD BOLTS
12 x 1.25 mm
100 N·m (10.0 kg-m, 72 lb-ft)
Apply clean engine oil to bolt threads and under bolt heads.

302066

Cylinder head bolt torque sequence — 2.2L (F22A1, F22A6) engines

50. Install the splash shield and the front wheels.

51. Lower the vehicle.

52. Check and adjust the valves, as necessary.

53. Torque the crankshaft pulley bolt to 181 ft. lbs. (250 Nm).

54. Installation of the remaining components is the reverse of removal.

55. Connect the negative battery cable and enter the radio security code.

56. Start the engine, checking carefully for any leaks.

57. Check the ignition timing and torque the distributor bolts to 13 ft. lbs. (18 Nm).

58. If equipped with 4WS, turn the steering wheel lock–to–lock to reset the 4WS control unit.

2.3L (H23A1) Engine

1. Disconnect the negative battery cable.

2. Turn the crankshaft so the No. 1 piston is at Top Dead Center (TDC).

NOTE: The No. 1 piston is at top dead center when the pointer on the block aligns with the white painted mark on the flywheel (manual transaxle) or driveplate (automatic transaxle).

3. Drain the engine coolant into a sealable container.

4. Relieve the fuel system pressure.

5. Remove the air intake duct.

6. Remove the evaporative emissions (EVAP) control canister hose from the intake manifold.

7. Remove the throttle cable from the throttle body. On automatic transaxle equipped vehicles, remove the throttle control cable at the throttle body.

NOTE: Be careful not to bend the cable when removing. Always replace a kinked cable with a new one.

8. Disconnect the fuel feed and return hose.

9. Remove the brake booster vacuum hose from the intake manifold.

10. Disconnect the following engine wire harness connectors:
    a. Fuel injector connectors
    b. Intake Air Temperature (IAT) sensor connector
    c. Idle Air Control (IAC) valve connector
    d. Throttle Position (TP) sensor connector
    e. Exhaust Gas Recirculation (EGR) valve lift sensor
    f. Ground cable terminals
    g. Engine Coolant Temperature (ECT) switch B connector

h. Heated oxygen sensor (HO2S) connector
i. ECT sensor
j. ECT gauge sending unit connector
k. Ignition Control Module (ICM) connector
l. CKP/TDC/CYP sensor connector
m. Vehicle Speed Sensor (VSS) connector
n. Ignition coil connector
o. Intake air bypass solenoid valve connector
p. ECT switch A connector
q. Knock sensor connector

11. Remove the engine ground cable from the cylinder head cover.

12. Remove the connector and the terminal from the alternator, then remove the engine wire harness from the valve cover.

13. Remove the mounting bolts and drive belt from the power steering pump. Pull the pump away from the mounting bracket, without disconnecting the hoses. Support the pump out of the way.

14. Remove the ignition coil.

15. Tag then disconnect the emissions vacuum hoses from the intake manifold assembly.

16. Remove the bypass hose from the intake manifold.

17. Remove the upper radiator hose and the heater hose from the cylinder head.

18. Remove the lower radiator hose and bypass hose from the thermostat housing.

19. Remove the thermostat housing mounting bolts. Remove the thermostat housing from the intake manifold and the connecting pipe, by pulling and twisting the housing. Discard the O-rings.

20. Raise and safely support the vehicle.

21. Remove the front wheel and tire assemblies.

22. Remove the splash shield.

23. Remove the intake manifold bracket bolts.

24. Remove the intake manifold.

25. Disconnect the exhaust pipe from the exhaust manifold.

26. Remove the exhaust manifold and the exhaust manifold heat insulator.

27. Label then disconnect the electrical connectors from the distributor and the spark plug wires from the spark plugs. Mark the position of the distributor and remove it from the cylinder head. Disconnect the ignition coil wire from the distributor.

28. Remove the Positive Crankcase Ventilation (PCV) hose, then remove the cylinder head cover. Replace the rubber seals if damaged or deteriorated.

29. Remove the timing belt.

30. Insert a 5.0mm pin punch in each of the camshaft caps, nearest to the sprockets, through the holes provided. Remove the camshaft sprocket attaching bolts then remove the sprockets. Do not lose the sprocket keys.

31. Loosen all of the rocker arm adjusting screws, then remove the pin punches from the camshaft caps.

32. Remove the camshaft holders, note the holders locations for ease of installation.

33. Remove the rubber cap from the head, located at the end of the intake camshaft.

34. Remove the rocker arms from the cylinder head. Note the locations of the rocker arms.

NOTE: The rocker arms have to be installed to their original locations if being reused.

35. Remove the side engine mount bracket B, then the back cover from behind the camshaft sprockets.

36. Remove the cylinder head bolts in the proper sequence.

NOTE: To prevent warpage, unscrew the bolts in sequence ⅓ turn at a time. Repeat the sequence until all bolts are loosened.

37. Separate the cylinder head from the engine block with a suitable flat bladed prytool.

To install:

38. Make sure all cylinder head and block gasket surfaces are clean. Check the cylinder head for warpage. If warpage is less than 0.002 in. (0.05mm), cylinder head resurfacing is not required. Maximum resurface limit is 0.008 in. (0.2mm) based on a cylinder head height of 5.20 in. (132.0mm).

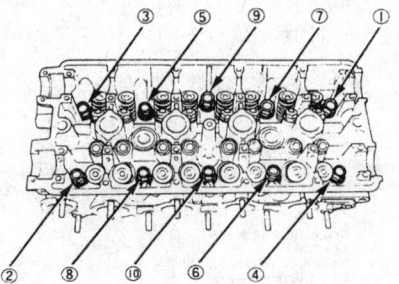

Cylinder head bolt removal sequence — 2.3L (H23A1) engine

302742

39. Always use a new head gasket.

40. Make sure the No. 1 cylinder is at TDC.

41. Clean the oil control orifice and install a new O-ring. The cylinder head dowel pins and oil control jet must be aligned.

42. Install the bolts that secure the intake manifold to it's bracket but do not tighten them.

43. Install the cylinder head, then tighten the cylinder head bolts sequentially in 3 steps:
   Step 1: 29 ft. lbs. (40 Nm).
   Step 2: 51 ft. lbs. (70 Nm).
   Step 3: 72 ft. lbs. (100 Nm).

NOTE: A beam type torque wrench is recommended. If a bolt makes any noise while being torqued, loosen the bolt and retighten it.

44. Install the intake manifold with a new gasket.

45. Install the exhaust manifold with a new gasket.

46. Install the exhaust manifold bracket, then install the exhaust pipe, bracket and upper shroud.

47. Install the camshafts and rocker arms.

48. Install the timing belt back cover.

49. Install the side engine mount bracket B. Torque the bolt attaching the bracket to the cylinder head to 33 ft. lbs. (45 Nm). Torque the bolts attaching the bracket to the side engine mount to 16 ft. lbs. (22 Nm).

50. Install the camshaft sprockets onto the camshafts.

51. Install the timing belt.

52. Adjust the valves.

53. Torque the crankshaft pulley bolt to 181 ft. lbs. (250 Nm).

54. Install the splash shield and the front wheels.

55. Lower the vehicle.

56. Install the remaining components in the reverse order of removal.

57. Drain the oil from the engine into a sealable container. Install the drain plug and refill the engine with clean oil.

58. Fill and bleed the air from the cooling system.

59. Connect the negative battery cable and enter the radio security code.

60. Start the engine, checking carefully for any leaks.

61. Check and adjust the ignition timing. Torque the distributor bolts to 13 ft. lbs. (18 Nm).

62. If equipped with 4WS, start the engine and turn the steering wheel lock-to-lock to reset the 4WS control unit.

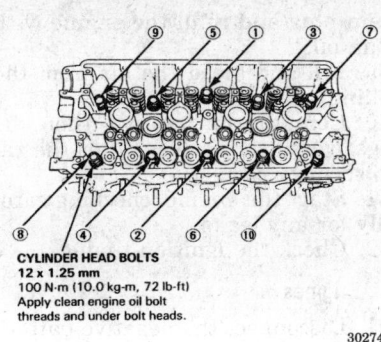

CYLINDER HEAD BOLTS
12 x 1.25 mm
100 N·m (10.0 kg-m, 72 lb-ft)
Apply clean engine oil bolt
threads and under bolt heads.

302744

Cylinder head bolt torque sequence — 2.3L (H23A1) engine

1994–97 Accord

2.2L (F22B1, F22B2) Engines

1. Disconnect the negative battery cable, then the positive battery cable.
2. Turn the engine to align the timing marks and set cylinder No.1 to TDC. The white mark on the crankshaft pulley should align with the pointer on the timing belt cover.
3. Raise and safely support the vehicle.
4. Drain the engine coolant into a sealable container.
5. Remove the front wheel and tire assemblies.
6. Remove the splash shield.
7. Disconnect the exhaust pipe from the exhaust manifold.
8. Remove the intake manifold bracket bolts.
9. Lower the vehicle to a working level without placing it on the floor.
10. Remove the throttle cable from the throttle body. On automatic transaxle equipped vehicles, remove the throttle control cable. If equipped with cruise control, remove the cruise control cable.

NOTE: Be careful not to bend the cable when removing it. Always replace a kinked cable with a new one. Do not use pliers to remove the cable from the linkage.

11. Remove the intake air duct.
12. Remove the breather hose, Positive Crankcase Ventilation (PCV) hose and evaporative emissions (EVAP) control canister hose.
13. Relieve the fuel system pressure.
14. Remove the fuel feed and return hose from the fuel rail.
15. Disconnect the vacuum hoses attached to the engine located near the fuel feed and return hoses.
16. Remove the brake booster vacuum hose from the intake manifold. Label and remove the other vacuum hoses from the intake manifold.

17. Remove the clamp holding the power steering hose to the strut tower.
18. Remove the wire harness clamp and the ground cable from the intake manifold.
19. Remove the connector and the terminal from the alternator, then remove the engine wire harness from the valve cover.
20. Remove the mounting bolts and drive belt from the power steering pump. Pull the pump away from the mounting bracket, without disconnecting the hoses. Support the pump out of the way.
21. Loosen the adjusting and mounting bolts for the alternator and remove the drive belt.
22. Remove the engine wire harness and bypass hose from the lower side of the intake manifold.
23. Disconnect the following engine wire harness connectors:
    a. Fuel injector connectors
    b. Intake Air Temperature (IAT) sensor connector
    c. Idle Air Control (IAC) valve connector
    d. Throttle Position (TP) sensor connector
    e. Manifold Absolute Pressure (MAP) sensor connector
    f. Heated oxygen sensor (HO2S) connector
    g. Engine Coolant Temperature (ECT) sensor connector
    h. ECT switch connector
    i. ECT gauge sending unit connector
    j. VTEC solenoid valve connector (VTEC engine)
    k. VTEC pressure switch connector (VTEC engine)
    l. Exhaust Gas Recirculation (EGR) valve lift sensor
    m. CKP/TDC/CYP sensor connector
    n. Ignition coil connector (Non VTEC engine)
    o. Fuel Injection Air (FIA) control solenoid valve connector (VTEC engine)
24. Label then disconnect the electrical connectors from the distributor and the spark plug wires from the spark plugs. Mark the position of the distributor and remove it from the cylinder head. Disconnect the ignition coil wire from the distributor.
25. Remove the upper radiator hose and the heater inlet hose from the cylinder head.
26. Remove the lower radiator hose from the thermostat housing.
27. Remove the coolant bypass hoses.

28. Use a jack to support the engine, make sure to place a cushion between the oil pan and the jack. Remove the through-bolt from the side engine mount and remove the mount.
29. Remove the cylinder head cover. Replace the rubber seals if damaged or deteriorated.
30. Remove the timing belt covers and the timing belt.
31. Remove the camshaft sprocket and the back cover. Do not lose the sprocket key.
32. Remove the exhaust manifold heat insulator and the exhaust manifold.
33. Remove the thermostat housing mounting bolts. Remove the thermostat housing from the intake manifold and the connecting pipe, by pulling and twisting the housing. Discard the O-rings.
34. Remove the fuel rail and fuel injectors.
35. Remove the intake manifold.
36. Remove the cylinder head bolts in the reverse order proper sequence, then remove the cylinder head.

NOTE: To prevent warpage, unscrew the bolts in sequence $1/3$ turn at a time. Repeat the sequence until all bolts are loosened.

To install:
37. Make sure all cylinder head and block gasket surfaces are clean. Check the cylinder head for warpage. If warpage is less than 0.002 in. (0.05mm), cylinder head resurfacing is not required. Maximum resurface limit is 0.008 in. (0.2mm) based on a cylinder head height of 3.94 in. (100mm).
38. Always use a new head gasket.
39. Make sure the No. 1 cylinder is at TDC.
40. Clean the oil control orifice and install a new O-ring (VTEC engine only).
41. Install the dowel pins to the engine block.
42. Install the bolts that secure the intake manifold to it's bracket but do not tighten them.
43. Position the camshaft correctly.
44. Install the cylinder head, then tighten the cylinder head bolts sequentially in 3 steps:
    Step 1: 29 ft. lbs. (39 Nm).
    Step 2: 51 ft. lbs. (69 Nm).
    Step 3: 72 ft. lbs. (98 Nm).
45. Install the intake manifold with a new gasket.
46. Connect the intake manifold bracket to the intake manifold and torque the bolt to 16 ft. lbs. (22 Nm).
47. Install the fuel rail with the fuel injectors.

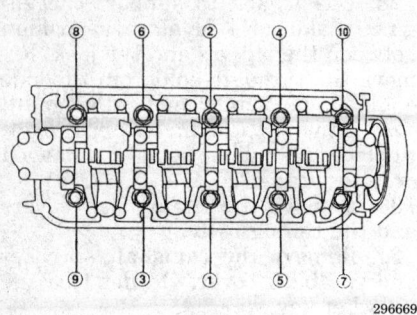

Cylinder head bolt torque sequence — 2.2L (F22B1, F22B2) engines

48. Install the exhaust manifold with a new gasket.

49. Install the exhaust manifold bracket.

50. Install the timing belt back cover to the cylinder head. Torque the cover bolt on the non VTEC engine to 9 ft. lbs. (12 Nm). On the VTEC engine torque the bolt on the intake side of the head to 9 ft. lbs. (12 Nm) and torque the bolt on the exhaust side of the head to 7 ft. lbs. (10 Nm).

51. Install the key to the camshaft, then install the camshaft sprocket. Torque the sprocket bolt to 27 ft. lbs. (37 Nm).

52. Make sure the camshaft sprocket and the crankshaft pulleys are aligned to TDC and install the timing belt.

53. Install the lower timing belt cover and torque the bolts to 9 ft. lbs. (12 Nm).

54. Install a new seal around the adjusting nut. Do not loosen the adjusting nut.

55. Install the crankshaft pulley. Coat the threads and seating face of the pulley bolt with engine oil. Install and torque the bolt to 181 ft. lbs. (250 Nm).

56. Install the side engine mount. Tighten the bolt and nut attaching the mount to the engine to 40 ft. lbs (55 Nm). Torque the through nut and bolt to 47 ft. lbs. (65 Nm), remove the jack from under the center beam.

57. Adjust the valves.

58. Install the upper timing belt cover. Torque the bolt on the intake side of the head to 9 ft. lbs. (12 Nm) and torque the bolt on the exhaust side of the head to 7 ft. lbs. (10 Nm).

59. Raise and safely support the vehicle.

60. Connect the exhaust pipe to the exhaust manifold with new gaskets. Torque the nuts 40 ft. lbs. (54 Nm).

61. Install the splash shield and the front wheels.

62. Lower the vehicle.

63. Install the cylinder head cover gasket cover to the groove of the cylinder head cover. Before installing the gasket thoroughly clean the seal and the groove. Seat the recesses for the camshaft first, then work it into the groove around the outside edges. Make sure the gasket is seated securely in the corners of the recesses.

64. Apply liquid gasket to the four corners of the recesses of the cylinder head cover gasket. Do not install the parts if 5 minutes or more have elapsed since applying liquid gasket. After assembly, wait at least 20 minutes before filling the engine with oil.

65. If equipped with a VTEC engine, install the spark plug seals on the spark plug pipes. Take care not to damage the spark plug seals when installing the cylinder head cover.

66. Clean the cylinder head cover contacting surface with a shop towel. Install the cylinder head cover, tighten the cylinder head cover bolts in two or three steps. Torque the cap nuts in the proper sequence to 7 ft. lbs. (10 Nm).

67. Install the remaining components in the reverse order of removal.

68. Drain the oil from the engine into a sealable container. Install the

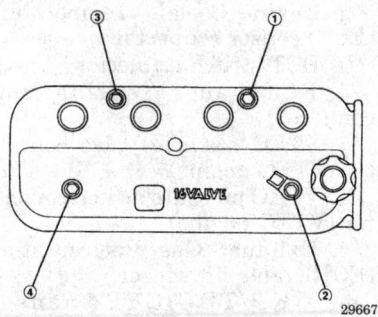

Non-VTEC engine cylinder head cover torque sequence — 2.2L (F22B1, F22B2) engines

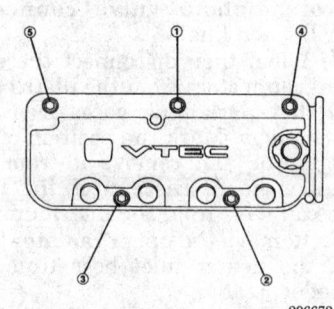

VTEC engine cylinder head cover torque sequence — 2.2L (F22B1, F22B2) engines

drain plug and refill the engine with clean oil.

69. Fill and bleed the air from the cooling system.

70. Connect the positive, then the negative battery cable. Enter the radio security code.

71. Start the engine, checking carefully for any leaks.

72. Check the ignition timing.

2.7L Engines

1. Disconnect the negative battery cable.

2. Turn the engine to align the timing marks and set cylinder No.1 to TDC. Remove the inspection caps on the upper timing belt covers to check the alignment of the timing marks. The white mark on the crankshaft pulley should align with the pointer on the timing belt cover. The pointers for the camshafts should align with the green marks on the camshaft pulleys.

3. Drain the engine coolant into a sealable container.

4. Remove the intake air duct.

5. Remove intake manifold cover B then disconnect the throttle cable and cruise control cables from the throttle linkage. Take care to not bend the cables, always replaced a kinked cable.

6. Remove the starter cable from the strut brace, then remove the strut brace.

7. Relieve the pressure from the fuel system.

── **CAUTION** ──

Fuel injection systems remain under pressure after the engine has been turned OFF. Properly relieve fuel pressure before disconnecting any fuel lines. Failure to do so may result in fire or personal injury.

8. Disconnect the fuel feed hose from the fuel filter. Disconnect the fuel return hose from the regulator.

9. Disconnect the brake booster vacuum hose and the evaporative emissions (EVAP) control canister hose.

10. Label and disconnect all vacuum hoses from the throttle body, intake manifold and cylinder head.

11. Disconnect the coolant hoses from the idle air control valve, the fast idle thermo valve and the water passage.

12. Remove the intake manifold cover.

13. Disconnect the breather hose from the cylinder head cover.

14. Remove the PCV hose from the cylinder head cover.

15. Loosen the idler pulley center nut and adjusting bolt, then remove the air conditioning compressor belt.

16. Disconnect the engine ground cable from the body, located near the drive belts.

17. Remove the vacuum pipe assembly.

18. Loosen the alternator mounting bolt, nut and adjusting bolt, then remove the alternator drive belt.

19. Support the engine with a floor jack on the oil pan (use a cushion between the jack and pan). Tension the jack so that it is just supporting the engine but not lifting it.

20. Remove the three bolts from the side engine mount, then loosen the through–bolt. Pivot the side engine mount out of the way.

21. Loosen the power steering pump mounting nuts and adjusting nut, then remove the power steering drive belt.

22. Disconnect the inlet hose from the power steering pump, then plug the hose and pump. Remove the power steering pump and place it out of the way.

23. Remove the wire harness cover and the ground cable from the water passage inlet.

24. Disconnect the following engine harness connectors from the cylinder head and the intake manifold:

  a. Six injector connectors.
  b. Intake Air Temperature (IAT) sensor connector.
  c. TDC/CYP sensor connector.
  d. Intake Air Control (IAC) valve connector.
  e. Manifold Absolute Pressure (MAP) sensor connector.
  f. Engine Coolant Temperature (ECT) sensor connector.
  g. ECT gauge sending unit connector.
  h. ECT switch connector.
  i. Exhaust Gas Recirculation (EGR) valve lift sensor connector.
  j. Throttle position sensor connector.

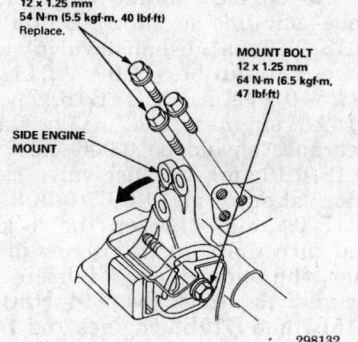

12 x 1.25 mm
54 N·m (5.5 kgf·m, 40 lbf·ft)
Replace.

MOUNT BOLT
12 x 1.25 mm
64 N·m (6.5 kgf·m, 47 lbf·ft)

SIDE ENGINE MOUNT

298132

Side engine mount — 2.7L engine

  k. Intake Air Bypass (IAB) control solenoid valve connector.
  l. Engine oil temperature sensor connector.
  m. Evaporative emissions (EVAP) purge control solenoid valve connector.
  n. Alternator connector.

25. Label then disconnect the electrical connectors from the distributor and the spark plug wires from the spark plugs. Remove the distributor from the cylinder head.

26. Remove the upper and lower radiator hoses, then disconnect the heater hoses from the engine.

27. Remove the intake air bypass vacuum tank.

28. Remove the bolts attaching the water passage to the cylinder heads then remove it from the engine. Discard the O–rings.

29. Remove the engine wire harness covers from the intake manifold.

30. Remove the nuts attaching the EGR pipe to the intake manifold and discard the gasket. Loosen the nut attaching the EGR pipe to the exhaust manifold and remove the pipe from the vehicle.

31. Remove the intake manifold.

32. Remove the exhaust manifolds from the engine.

33. Remove the cylinder head covers and the side covers.

34. Remove the timing belt covers and the timing belt.

35. Remove the camshaft sprockets and the timing belt back covers.

36. Remove the bolts attaching the camshaft holder plates and camshaft holders in the reverse order of installation.

37. Remove the camshaft holder plates, camshaft holders and the dowel pins from the cylinder head.

38. Remove the camshafts from the cylinder heads and the rubber cap from the rear cylinder head. Discard the camshaft seals.

39. Remove the intake rocker arms, exhaust inside rocker arms and the pushrods. Identify the location of the parts as they are removed, to ensure reinstallation to the original locations.

40. Remove the cylinder head bolts in the proper sequence.

NOTE: To prevent warpage, loosen the bolts in sequence ⅓ turn at a time. Repeat the sequence until all bolts are removed.

41. Remove the cylinder heads from the engine block.

42. Remove and clean the oil control orifices, then install new O–rings to the orifices.

To install:

43. Make sure all cylinder head and block gasket surfaces are clean. Check the cylinder head for warpage. If warpage is less than 0.002 in. (0.05mm), cylinder head resurfacing is not required. Maximum resurface limit is 0.008 in. (0.2mm) based on a cylinder head height of 5.24 in. (133mm).

44. Install new head gaskets.

45. Make sure the No. 1 cylinder is at TDC.

46. Install and align the cylinder head dowel pins and oil control orifices.

47. Install the cylinder heads. Apply clean oil to the threads of the cylinder head bolts and washers, then install and tighten the cylinder head bolts sequentially in 2 steps:
  Step 1: 29 ft. lbs. (39 Nm).
  Step 2: 56 ft. lbs. (76 Nm).

48. Fill the hydraulic tappet mounting hole and the oil fillers with clean engine oil.

49. Install the hydraulic tappets.

WARNING
Do not rotate the hydraulic tappets while installing them.

50. Apply clean engine oil to the rocker arms, pushrods and the camshafts.

51. Loosen the exhaust rocker arm adjusting screws and locknuts, then install the pushrods, exhaust inside rocker arms and the intake rocker arms. Install the parts to their original locations.

52. Make sure the rocker arms are properly positioned on the valve stems. Advance the crankshaft 30° from TDC to prevent interference between the pistons and valves, then install the camshafts. Position the rear camshaft on the cylinder head so the cam is not pushing on any valves.

53. Install the timing belt back covers and torque the attaching bolts to 9 ft. lbs. (12 Nm).

54. Install the camshaft sprockets and torque the attaching bolts to 23 ft. lbs. (31 Nm).

55. Set the camshaft sprockets so that the No. 1 piston is at TDC. Align the TDC marks (green mark) on the camshaft pulleys to the pointers on the back covers.

56. Turn the crankshaft counterclockwise to set it at TDC. Align the TDC mark on the tooth of the timing belt drive pulley with the pointer on the oil pump.

57. Install the timing belt and timing belt covers.

58. Set No. 1 cylinder to TDC.

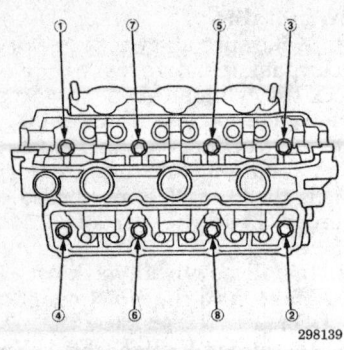

Cylinder head bolt removal sequence — 2.7L engine

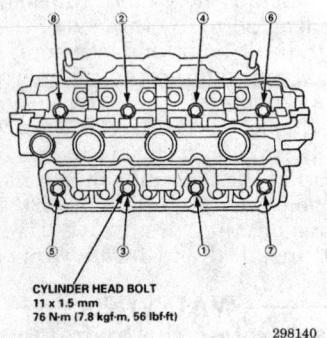

CYLINDER HEAD BOLT
11 x 1.5 mm
76 N·m (7.8 kgf·m, 56 lbf·ft)

298140

Cylinder head bolt torque sequence — 2.7L engine

59. Tighten the adjusting screws for No. 1, No. 2 and No. 4 cylinders. Tighten the screw until it contacts the valve, then tighten the screw 1 1/8 turns. Hold the screw in place and torque the locknut to 14 ft. lbs. (20 Nm).

60. Rotate the crankshaft pulley one turn clockwise, then tighten the adjusting screws for No. 3, No. 5 and No. 6 cylinders. Tighten the screw until it contacts the valve, then tighten the screw 1 1/8 turns. Hold the screw in place and torque the locknut to 14 ft. lbs. (20 Nm).

61. Install the cylinder head cover gasket into the groove of the cylinder

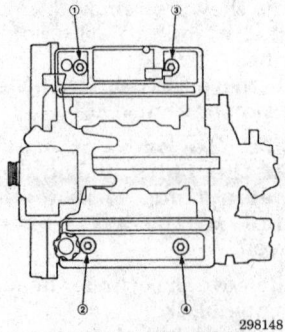

TDC MARKS

TDC MARK

298146

Timing marks for valve adjustment on cylinders 3, 5 and 6 — 2.7L engine

head cover. Seat the recesses for the camshaft first, then work it into the groove around the outside edges.

NOTE: Before installing the cylinder head cover gasket, thoroughly clean the seal groove.

62. Apply liquid gasket to the cylinder head cover gasket at the four corners of the recesses. Use a shop towel and wipe the cylinder heads where the cylinder head covers will come in contact.

63. Install the cylinder head covers, hold the gasket in the groove by placing your fingers on the camshaft contacting surfaces. With the cylinder head cover on the cylinder heads, slide the covers slightly back and forth to seat the cylinder head cover gaskets. Replace the washers if damaged or deteriorated.

64. Tighten the cylinder head cover bolts in two or three steps. In the final step, torque all the bolts, in sequence, to 11 ft. lbs. (15 Nm).

65. Install the cylinder head side covers with new O-rings and torque the bolts to 9 ft. lbs. (12 Nm).

66. Install the intake manifold.

67. Install the exhaust manifolds.

68. Install new O-rings to the water passage and install the water passage to the engine, torque the mounting bolts to 16 ft. lbs. (22 Nm).

69. Install the EGR pipe with a new gasket at the intake manifold. Torque the nuts attaching the pipe to the intake manifold to 9 ft. lbs. (12 Nm) and torque the exhaust manifold fitting to 43 ft. lbs. (59 Nm).

70. Install the side engine mount. Use three new bolts to attach the mount to the engine, torque the bolts to 40 ft. lbs. (54 Nm).

71. Torque the side engine mount through-bolt to 47 ft. lbs. (64 Nm).

72. Install the remaining components in the reverse of removal.

73. Drain the engine oil into a sealable container then refill the engine with clean oil.

Cylinder head cover torque sequence — 2.7L engine

74. Connect the negative battery cable and enter the radio security code.

75. Fill and bleed the air from the cooling system.

76. Switch the ignition **ON** but do not engage the starter. The fuel pump should run for approximately 2 seconds, building pressure within the lines. Switch the ignition **OFF**, then **ON** 2 or 3 more times to build full system pressure. Check for fuel leaks.

77. Start the engine, allowing it to idle and check for any signs of leakage.

Valve Clearance

ADJUSTMENT

NOTE: The radio may contain a coded theft protection circuit. Always obtain the code number before disconnecting the battery. If the vehicle is equipped with 4WS, the steering control unit is shut down when the battery is disconnected. After connecting the battery, turn the steering wheel lock-to-lock to reset the steering control unit.

del Sol and Civic

1. Disconnect the negative battery cable.

2. Remove the cylinder head cover and the upper timing belt cover.

3. Rotate the crankshaft to align the white TDC mark on the crankshaft pulley with the pointer on the cover for the No. 1 cylinder compression stroke. Make sure the **UP** mark on the camshaft sprocket is up and the TDC marks align with the edge of the cylinder head.

4. Hold a No. 1 cylinder rocker arm against the camshaft and use a feeler gauge to check the clearance at the valve stem. Except on B16A2 and B16A3 engines, intake valve clearance should be 0.007–0.009 in. (0.18–0.26mm), exhaust valve clearance should be 0.009–0.011 in. (0.23–0.27mm). On B16A2 and B16A3 engines, the intake valve clearance should be 0.006–0.007 in. (0.15–0.19mm), exhaust valve clearance should be 0.007–0.008 in. (0.17–0.21mm). Loosen the locknut and turn the adjusting screw to adjust the clearance. Tighten the locknut to 10 ft. lbs. (14 Nm) on D15B7 and D15B8 engines and 14 ft. lbs. (20 Nm) on all other models and recheck the clearance. Don't over-

tighten the locknut, the aluminum rockers will strip easily.

5. The adjustment order is 1–3–4–2. Rotate the crankshaft counterclockwise 180° (the camshaft sprocket will rotate 90°) to bring each cylinder to TDC/compression. Adjust each set of valves.

   a. At TDC for the No. 3 cylinder, the UP mark is pointed to the exhaust side of the cylinder head.

   b. At TDC for the No. 4 cylinder, the UP mark is pointed down, and the TDC marks align with the edge of the cylinder head.

   c. At TDC for the No. 2 cylinder, the UP mark is pointed to the intake side of the cylinder head.

6. After adjusting the valves of a VTEC engine, inspect its intake rocker arms for smooth and independent motion.

7. Apply sealant to the edges of the valve cover gasket where it meets the camshaft holders. Make sure the spark plug tube seals are properly seated.

8. Install the cylinder head and timing belt covers.

9. Torque the crankshaft pulley bolt to 134 ft. lbs. (185 Nm).

10. Reconnect the negative battery cable. Enter the radio security code.

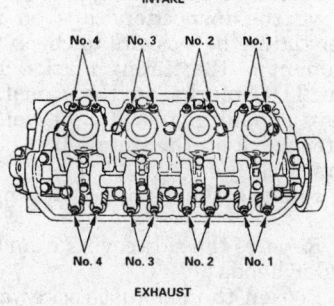

Valve adjuster locations — except 1.5L (D15B8) engine

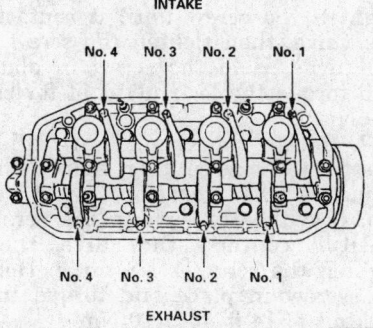

Valve adjuster locations — 1.5L (D15B8) engine

Prelude and Accord

2.2L (H22A1) Engine

NOTE: The valve clearance should be adjusted when the engine is cold. The cylinder head temperature should be less than 100°F (38°C)

1. Disconnect the negative battery cable.

2. Turn the crankshaft so the No. 1 piston is at top dead center.

NOTE: The No. 1 piston is at top dead center when the pointer on the block aligns with the white painted mark on the flywheel (manual transaxle) or driveplate (automatic transaxle).

3. Remove the air intake duct.

4. Remove the engine ground cable from the cylinder head cover.

5. Remove the connector and the terminal from the alternator, then remove the engine wire harness from the valve cover.

6. Remove the plug wire cover from the cylinder head cover. Label, then disconnect the spark plug wires from the spark plugs.

7. Remove the Positive Crankcase Ventilation (PCV) hose, then remove the cylinder head cover. Replace the rubber seals if damaged or deteriorated.

8. Ensure the arrows embossed on the camshaft pulleys are aligned in the upward position (12 o'clock).

9. Adjust the valves on cylinder No. 1:

   a. Insert a feeler gauge in between the camshaft lobe and the rocker arm.

NOTE: The intake valve clearance specification is 0.006–0.007 in (0.15–0.19mm) and the exhaust valve clearance specification is 0.007–0.008 in. (0.17–0.21mm).

   b. Loosen the locknut and turn the adjusting screw until the feeler gauge slides back and forth with a slight amount of drag.

   c. Tighten the locknut and recheck the valve clearance. Repeat the valve adjustment if necessary.

10. Rotate the crankshaft 180° counterclockwise (the camshaft pulleys will turn 90°) The arrow marks should be pointing to the exhaust side of the cylinder head.

11. Adjust the valves on cylinder No. 3:

   a. Insert a feeler gauge in between the camshaft lobe and the rocker arm.

   b. Loosen the locknut and turn the adjusting screw until the feeler gauge slides back and forth with a slight amount of drag.

   c. Tighten the locknut and recheck the valve clearance. Repeat the valve adjustment if necessary.

12. Rotate the crankshaft 180° counterclockwise (the camshaft pulleys will turn 90°) to bring No. 4 piston to TDC. The arrow marks should be pointing down, toward the crankshaft.

13. Adjust the valves on cylinder No. 4:

   a. Insert a feeler gauge in between the camshaft lobe and the rocker arm.

   b. Loosen the locknut and turn the adjusting screw until the feeler gauge slides back and forth with a slight amount of drag.

   c. Tighten the locknut and recheck the valve clearance. Repeat the valve adjustment if necessary.

14. Rotate the crankshaft 180° counterclockwise (the camshaft pulleys will turn 90°) to bring piston No. 2 to TDC. The arrow marks should be pointing to the intake side of the cylinder head.

15. Adjust the valves on cylinder No. 2:

   a. Insert a feeler gauge in between the camshaft lobe and the rocker arm.

   b. Loosen the locknut and turn the adjusting screw until the feeler gauge slides back and forth with a slight amount of drag.

   c. Tighten the locknut and recheck the valve clearance. Repeat the valve adjustment if necessary.

16. Install the cylinder head cover, torque the cap nuts to 7 ft. lbs. (10 Nm). Install the PCV hose to the cylinder head cover.

17. Connect the spark plug wires to the correct spark plugs.

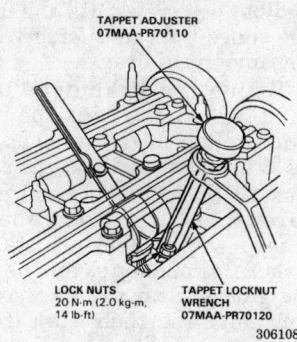

Checking and adjusting the valve clearance — 2.2L (H22A1) engine

18. Install the spark plug wire cover to the cylinder head cover and torque the cap nuts to 7 ft. lbs. (10 Nm).

19. Install the alternator wire harness to the cylinder head cover, then connect the terminal and connector to the alternator.

20. Connect the engine ground cable to the cylinder head cover.

21. Install the air intake duct.

22. Connect the positive then the negative battery cable and enter the radio security code.

23. If equipped with 4WS, Start the engine and turn the steering wheel lock-to-lock to reset the 4WS control unit.

2.2L (Except H22A1) and 2.3L (H23A1) Engines

NOTE: The valve clearance should be adjusted when the engine is cold, the cylinder head temperature should be less than 100°F (38°C).

NOTE: The radio may contain a coded theft protection circuit. Always obtain the code number before disconnecting the battery.

1. Disconnect the negative battery cable.

2. Label then disconnect the spark plug wires from the spark plugs.

3. Remove the Positive Crankcase Ventilation (PCV) hose, then remove the cylinder head cover. Replace the rubber seals if damaged or deteriorated.

4. Turn the engine to align the timing marks and set cylinder No.1 to TDC. The white mark on the crankshaft pulley should align with the pointer on the timing belt cover. The words **UP** embossed on the camshaft pulley should be aligned in the upward position. The marks on the edge of the pulley should be aligned with the cylinder head or the back cover upper edge.

5. Adjust the valves on cylinder No. 1 by performing the following:

a. Insert a feeler gauge in between the camshaft lobe and the rocker arm.

NOTE: The intake valve clearance specification is 0.009–0.011 in (0.24–0.28mm) and the exhaust valve clearance specification is 0.011–0.013 in. (0.27–0.32mm).

b. Loosen the locknut and turn the adjusting screw until the feeler gauge slides back and forth with a slight amount of drag.

c. Torque the locknut to 14 ft. lbs. (20 Nm) and recheck the valve

clearance. Repeat the valve adjustment if necessary.

6. Rotate the crankshaft 180° counterclockwise (the camshaft pulleys will turn 90°) The **UP** arrow marks should be pointing to the exhaust side of the cylinder head.

7. Adjust the valves on cylinder No. 3 by performing the following:

a. Insert a feeler gauge in between the camshaft lobe and the rocker arm.

b. Loosen the locknut and turn the adjusting screw until the feeler gauge slides back and forth with a slight amount of drag.

c. Torque the locknut to 14 ft. lbs. (20 Nm) and recheck the valve clearance. Repeat the valve adjustment if necessary.

8. Rotate the crankshaft 180° counterclockwise (the camshaft pulleys will turn 90°) to bring No. 4 piston to TDC. The **UP** arrow marks should be pointing down, toward the crankshaft.

9. Adjust the valves on cylinder No. 4 by performing the following:

a. Insert a feeler gauge in between the camshaft lobe and the rocker arm.

b. Loosen the locknut and turn the adjusting screw until the feeler gauge slides back and forth with a slight amount of drag.

c. Torque the locknut to 14 ft. lbs. (20 Nm) and recheck the valve clearance. Repeat the valve adjustment if necessary.

10. Rotate the crankshaft 180° counterclockwise (the camshaft pulleys will turn 90°) to bring piston No. 2 to TDC. The **UP** arrow marks should be pointing to the intake side of the cylinder head.

11. Adjust the valves on cylinder No. 2 by performing the following:

a. Insert a feeler gauge in between the camshaft lobe and the rocker arm.

b. Loosen the locknut and turn the adjusting screw until the feeler gauge slides back and forth with a slight amount of drag.

c. Torque the locknut to 14 ft. lbs. (20 Nm) and recheck the valve clearance. Repeat the valve adjustment if necessary.

12. Install the cylinder head cover gasket cover to the groove of the cylinder head cover. Before installing the gasket thoroughly clean the seal and the groove. Seat the recesses for the camshaft first, then work it into the groove around the outside edges. Make sure the gasket is seated securely in the corners of the recesses.

13. Apply liquid gasket to the four corners of the recesses of the cylinder head cover gasket. Do not install the parts if 5 minutes or more have elapsed since applying liquid gasket. After assembly, wait at least 20 minutes before filling the engine with oil.

14. Install the cylinder head (valve) cover. Torque the bolts attaching the cylinder head cover in the proper sequence to 7 ft. lbs. (10 Nm).

15. Connect the spark plug wires to the correct spark plugs.

16. Connect the positive then the negative battery cable and enter the radio security code.

17. If equipped with 4WS, turn the steering wheel lock-to-lock to reset the 4WS control unit.

2.7L Engine

The V-6 engine in the Accord has adjustments on the exhaust rocker arm assemblies that are located under the cylinder head side covers. The exhaust inside rocker arms and intake rocker arms operate with valve lifters (hydraulic tappets) and are not adjustable.

1. Disconnect the negative battery cable.

2. Turn the engine to align the timing marks and set cylinder No.1 to TDC. The white mark on the crankshaft pulley should align with the pointer on the timing belt cover. Remove the inspection caps on the upper timing belt covers to check the alignment of the timing marks. The green TDC marks on the camshaft pulleys should align with the yellow pointer mark on the timing belt back covers.

3. Remove the timing belt upper covers.

4. Remove the side covers from the cylinder heads.

5. Loosen the exhaust rocker arm adjusting screws and locknuts, and make sure the rocker arms are properly positioned on the valve stems.

6. Tighten the adjusting screws for No. 1, No. 2 and No. 4 cylinders. Tighten the screw until it contacts the valve, then tighten the screw 1 1/8 turns. Hold the screw in place and torque the locknut to 14 ft. lbs. (20 Nm).

7. Rotate the crankshaft pulley one turn clockwise, then tighten the adjusting screws for No. 3, No. 5 and No. 6 cylinders. Tighten the screw until it contacts the valve, then tighten the screw 1 1/8 turns. Hold the screw in place and torque the locknut to 14 ft. lbs. (20 Nm).

8. Install the cylinder head side covers with new O-rings and torque the bolts to 9 ft. lbs. (12 Nm).

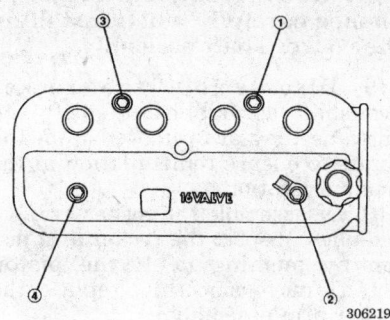

Cylinder head cover torque sequence — 2.2L (Non VTEC) engine

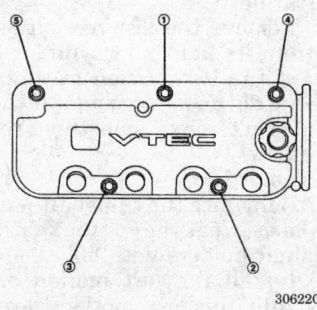

Cylinder head cover torque sequence — 2.2L (VTEC) engine

9. Install the timing belt upper covers and torque the bolts to 9 ft. lbs. (12 Nm).

10. Connect the negative battery cable.

Rocker Arm and Shaft

REMOVAL AND INSTALLATION

NOTE: The radio may contain a coded theft protection circuit. Always obtain the code number before disconnecting the battery. If the vehicle is equipped with 4WS, the steering control unit is shut down when the battery is disconnected. After connecting the battery, turn the steering wheel lock–to–lock to reset the steering control unit.

Civic and del Sol

1.5L (D15B7, D15B8) Engine

1. Disconnect the negative battery cable.

2. Remove the valve cover and bring the No. 1 cylinder to TDC for the compression stroke.

3. Loosen the valve adjusting screws.

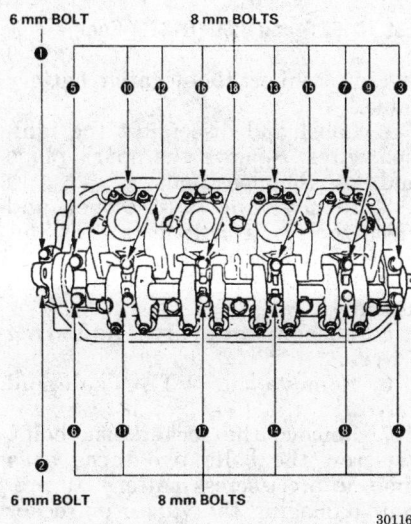

6 mm BOLT   8 mm BOLTS

6 mm BOLT   8 mm BOLTS

Rocker arm shaft bolt loosening sequence — 1.5L (D15B7) engine

4. Remove the rocker arm bolts. Unscrew the bolts two turns at a time, in a crisscross pattern, to prevent damaging the valves or rocker assembly.

NOTE: The rocker arms and shafts are an assembly; they must be removed from the engine as a unit. Always follow the torque sequence carefully when installing the rocker shaft assembly.

5. Remove the rocker arm/shaft assemblies. Do not remove the camshaft holder bolts. The bolts keep the camshaft bearing caps, springs, and rocker arms in place on the shafts.

6. If the rocker arms or shafts are to be replaced, identify the parts as they are removed from the shafts to ensure reinstallation in the original location.

To install:

7. Lubricate the camshaft journals and lobes.

8. Set the rocker arm assembly in place and loosely install the bolts. Tighten each bolt two turns at a time in the proper sequence to ensure that the rockers do not bind on the valves. Tighten the 8mm rocker arm bolts to 16 ft. lbs. (22 Nm). Tighten the 6mm bolts to 9 ft. lbs. (12 Nm).

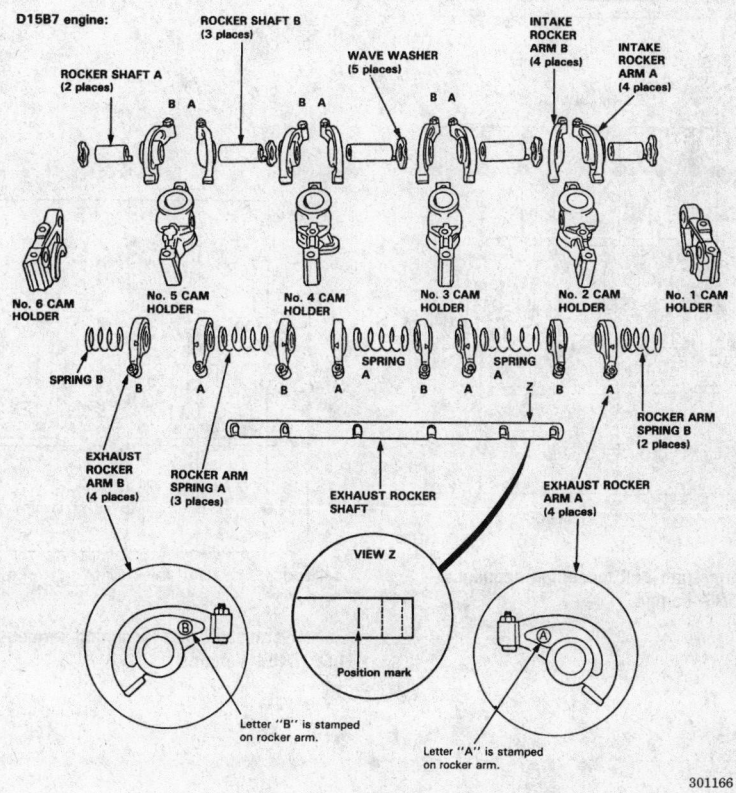

Rocker arm and shaft assembly — 1.5L (D15B7) engine

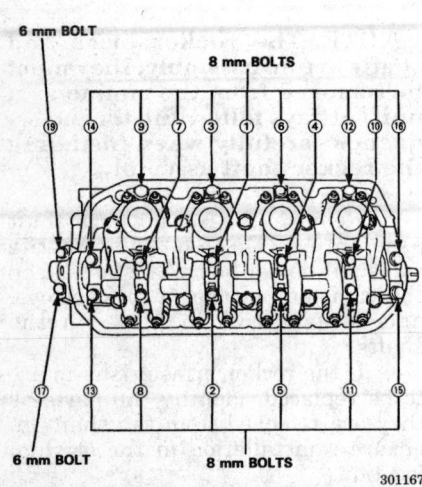

Rocker arm shaft bolt torque sequence — 1.5L (D15B7) engine

301167

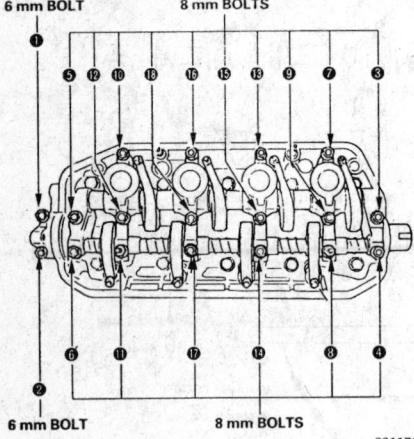

Rocker arm shaft bolt loosening sequence — 1.5L (D15B8) engine

231176

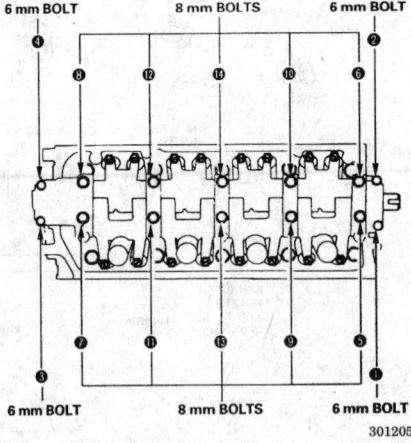

Rocker arm/shaft bolt loosening sequence — 1.5L (D15Z1) engine

301205

9. Adjust the valves and tighten the locknuts to 10 ft. lbs. (14 Nm).

10. Replace the valve cover and connect the negative battery cable.

1.5L (D15Z1) and 1.6L (D16Z6) Engine

1. Disconnect the negative battery cable.

2. Label and disconnect the ignition wires. Remove the spark plugs and note their positions.

3. Remove the valve cover and bring the No. 1 cylinder to TDC for the compression stroke.

4. Remove the distributor as an assembly.

5. Loosen the valve adjusting screws.

6. Remove the VTEC solenoid valve.

7. Remove the rocker arm bolts. Unscrew the bolts two turns at a time, in a crisscross pattern, to prevent damaging the valves or rocker assembly.

8. Remove the rocker arm/shaft assemblies. Do not remove the rocker shaft bolts yet. The bolts keep the bearing caps, springs and rocker arms in place on the shafts.

NOTE: The rocker arms and shafts are an assembly; they must be removed from the engine as a unit. Always follow the torque sequence carefully when installing the rocker shaft assembly.

9. Disassemble the rocker arm/shaft assemblies. Identify the parts as they are removed from the shafts to ensure reinstallation in the original location.

10. Disassemble the rocker arm assemblies. Inspect the rocker arm piston by pushing it. If the piston doesn't move smoothly, replace the rocker arm assembly.

11. Apply oil to the pistons and reassemble the rocker arms. Bundle the rocker arm assemblies with rubber bands to prevent the parts from separating.

12. Remove the lost motion assembly from its holder in cylinder head. Inspect the lost motion assembly by pushing down on its piston. If the piston doesn't move smoothly, replace the assembly.

To install:

13. Lubricate the camshaft journals and lobes. Coat the rocker shafts and camshaft holders with oil.

14. Install the lost motion assemblies into the lost motion assembly holder.

15. Reassemble the rocker arms assemblies onto the rocker shafts and camshaft holders. After the rocker arms and shafts are reassembled, remove the rubber bands

16. Set the rocker arm assembly in place and loosely install the bolts. Tighten each bolt two turns at a time in the proper sequence to ensure that the rockers do not bind on the valves. Tighten the 8mm rocker arm bolts to 14 ft. lbs. (20 Nm). Tighten the 6mm bolts to 9 ft. lbs. (12 Nm).

17. Install the VTEC solenoid valve with a new filter. Tighten the bolts to 9 ft. lbs. (12 Nm).

18. Install the distributor, but don't tighten the bolts yet.

19. Adjust the valves. Tighten the locknuts to 7 ft. lbs. (10 Nm).

20. Install the valve cover and connect the negative battery cable.

21. Install the spark plugs and reconnect the ignition wires.

22. Warm the engine up to normal operating temperature. Check the ignition timing, and adjust if necessary. Tighten the distributor mounting bolts to 17 ft. lbs. (24 Nm).

1.6L (D16Y5) Engine

1. Disconnect the negative battery cable.

2. Label and disconnect the ignition wires. Remove the spark plugs and note their cylinder assignments.

3. Remove the valve cover.

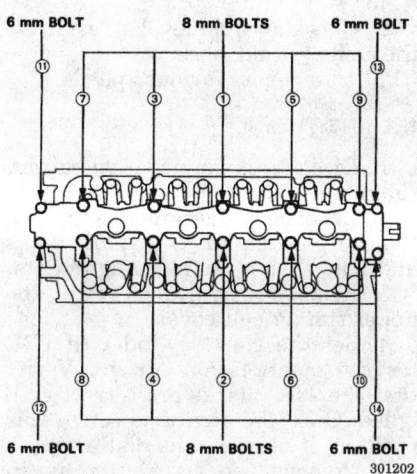

Rocker arm/shaft bolt torque sequence — 1.5L (D15Z1) engine

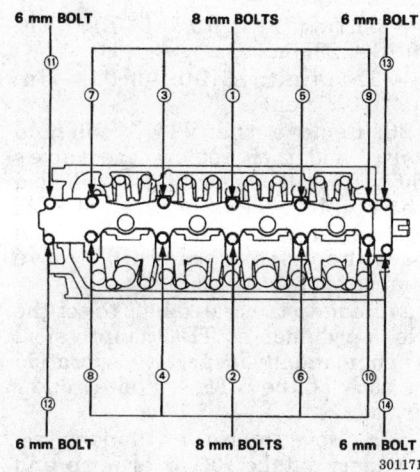

Rocker arm/shaft bolt torque sequence — 1.6L (D16Z6) engine

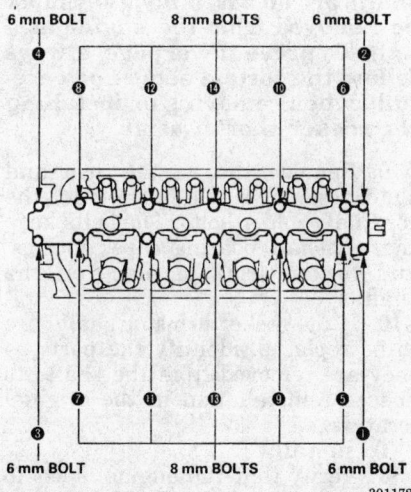

Rocker arm/shaft bolt loosening sequence — 1.6L (D16Z6) engine

4. Rotate the crankshaft to set the No. 1 cylinder to TDC for the compression stroke. The white TDC mark on the crankshaft pulley aligns with the pointers on the lower timing cover.

5. Remove the distributor.

6. Loosen the valve adjusting screws.

7. Label and disconnect the VTEC solenoid valve connector.

8. Loosen the camshaft holder bolts two turns at a time in a criss-cross pattern to prevent damaging the valves or rocker assembly.

9. Remove the rocker arm and shaft assemblies together with the camshaft holders. Do not remove the rocker shaft bolts yet. The bolts keep the bearing caps, springs, and rocker arms in place on the shafts.

NOTE: The rocker arms and shafts are an assembly; they must be removed from the engine as a unit. Always follow the torque sequence carefully when installing the rocker shaft assembly.

10. Remove the camshaft holder bolts from the rocker arm and shaft assembly.

11. Bundle the intake rocker arm assemblies with rubber bands so they don't separate when the intake rocker shaft is removed.

12. Disassemble the rocker arm and shaft assemblies. Label the parts as they are removed from the shafts to ensure reinstallation in the original location.

13. Disassemble the rocker arm assemblies taking care not to mix up any of the parts. Inspect the rocker arm synchronizing and timing pistons by pushing them with your fingers. If the pistons don't move smoothly in the rocker arm bores, replace the rocker arm assembly.

14. Apply oil to the synchronizing pistons, timing piston, and timing spring and reassemble the rocker arms. Bundle the rocker arm assemblies with rubber bands to prevent the parts from separating.

15. Inspect the timing plates and return springs which are located on the camshaft holders. Set each timing plate and return spring so that the C–shaped upper arm of the plate is position parallel to the top of the camshaft holder.

To install:

16. Verify that the engine is set at TDC/compression for the No. 1 cylinder.

17. Lubricate the camshaft journals and lobes. Coat the rocker shafts and camshaft holders with oil.

18. Remove the oil control orifice. Thoroughly clean it and reinstall it with a new O–ring.

19. Install a new camshaft seal if necessary.

20. Assemble the rocker arm and shaft assemblies. Make sure the intake shaft collars and exhaust shaft springs are in the proper locations.

21. After the rocker arms and shafts are assembled, cut the rubber bands and remove them from the intake rockers. Make sure that no rubber band fragments are left in the engine.

22. Apply fresh oil to the threads of camshaft holder bolts, and then install them.

23. Apply liquid gasket to the cylinder head mating surfaces of the No. 1 and No. 5 camshaft holders. Do not allow the sealant to cure before installation.

24. Set the rocker arm and shaft assembly in place. Install and hand–tighten the bolts. Tighten each bolt two turns at a time in the criss-cross sequence so that the rockers are evenly tightened and don't bind on the valves. Tighten the 8mm rocker arm bolts to 14 ft. lbs. (20 Nm). Tighten the 6mm bolts to 9 ft. lbs. (12 Nm).

25. Starting with the No. 1 cylinder at TDC/compression, adjust the valve

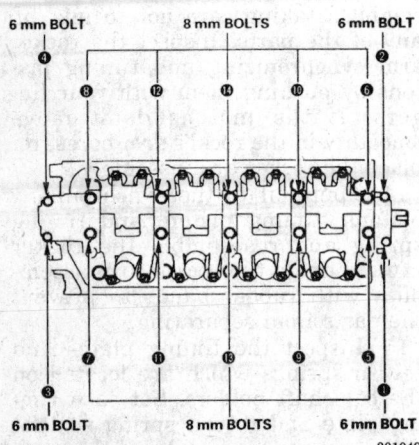

Rocker arm/shaft bolt loosening sequence — 1.6L (D16Y5) engine

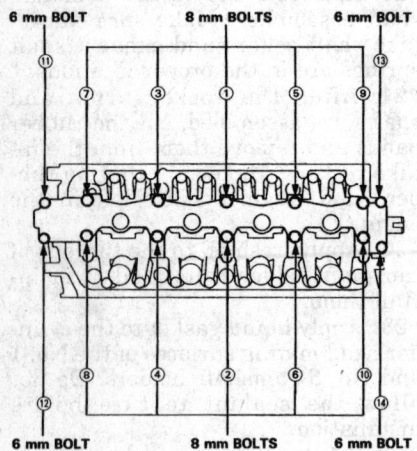

Rocker arm/shaft bolt torque sequence — 1.6L (D16Y5) engine

clearances. After the clearance has been reached, tighten the locknuts to 14 ft. lbs. (20 Nm). Set the No. 3, No. 4, and No. 2 cylinders at TDC/compression, and adjust their valve clearances.

- Intake: 0.007–0.009 in. (0.18–0.22mm)
- Exhaust: 0.009–0.011 in. (0.23–0.27mm)

26. Remove the VTEC solenoid valve, and then remove the valve's filter. Install a new VTEC solenoid valve filter. Tighten the solenoid valve bolts to 9 ft. lbs. (12 Nm) and reconnect the solenoid valve connector.

27. Rotate the crankshaft to set the No. 1 cylinder at TDC/compression. Then, manually inspect the operation of each of the VTEC intake rocker arms:

a. Move the No. 1 cylinder's secondary intake rocker arm up and down.

b. Verify that the secondary intake rocker arm moves independently of the primary intake rocker arm.

c. Repeat the rocker arm inspection for the other three cylinders with each cylinder set at TDC/compression.

28. Rotate the crankshaft back to TDC/compression for the No. 1 cylinder. Install the distributor, but do not tighten the mounting bolts yet.

29. Tighten the crankshaft pulley bolt to 134 ft. lbs. (181 Nm).

30. Install the valve cover. Make sure the gasket is in good condition, and apply sealant to the corners where the gasket meets the camshaft holders.

31. Install the spark plugs and reconnect the ignition wires.

32. Drain the engine oil and remove the oil filter. Install a new oil filter and refill the engine with fresh oil.

33. Connect the negative battery cable.

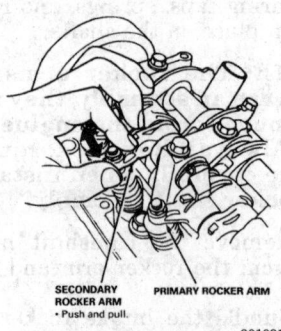

VTEC rocker arm inspection — 1.6L (D16Y5) engine

34. Warm the engine up to normal operating temperature.

35. Check the ignition timing and adjust it if necessary. Then, tighten the distributor mounting bolts to 17 ft. lbs. (24 Nm).

36. Check all fluid levels. Test drive the vehicle and observe the engine RPM changes at various speeds.

1.6L (D16Y7) Engine

1. Disconnect the negative battery cable.

2. Label and disconnect the ignition wires. Remove the spark plugs and note their cylinder assignments.

3. Remove the valve cover and the upper timing belt cover.

4. Set the No. 1 cylinder to TDC for the compression stroke. Verify that the TDC marks are correctly aligned. Once the engine is set in this position, it must not be disturbed.

5. Remove the distributor as an assembly.

6. Loosen the valve adjusting screws.

7. Cover the timing belt with a clean shop towel to protect it from engine oil. If the belt is contaminated with oil, it must be replaced.

8. Remove the camshaft holder bolts. Unscrew the bolts two turns at a time in a crisscross pattern to prevent damaging the valves. camshaft, or rocker arm assembly.

NOTE: The rocker arms and shafts are an assembly; they must be removed from the engine as a unit. To prevent warpage, always follow the torque sequence carefully when removing or installing the rocker shaft assembly.

9. Remove the rocker arm and shaft assemblies. Do not remove the camshaft holder bolts. The bolts keep the camshaft bearing caps, springs, and rocker arms in place on the shafts.

10. If the rocker arms or shafts are to be replaced, identify the parts as they are removed from the shafts to ensure reinstallation in the original location.

To install:

11. Verify that the engine is set to TDC/compression for the No. 1 cylinder. The camshaft keyway faces up when the engine is at TDC/compression.

12. Lubricate the camshaft journals and lobes with clean engine oil. Install a new camshaft seal if necessary.

13. Remove the oil control orifice. Thoroughly clean it and install it with a new O–ring.

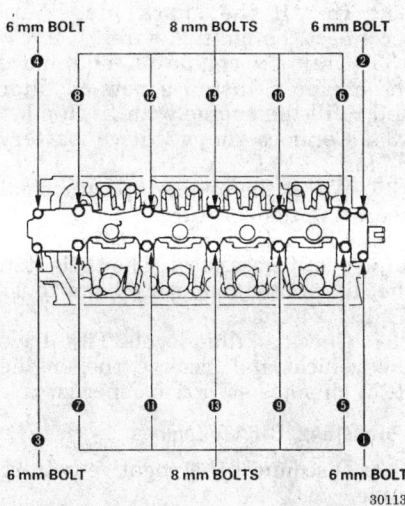

Rocker arm/shaft bolt loosening sequence — 1.6L (D16Y7) engine

301139

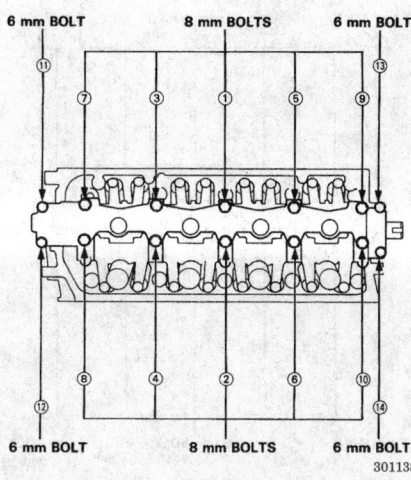

301138

Rocker arm/shaft bolt torque sequence — 1.6L (D16Y7) engine

14. Assemble the rocker arms, shafts, and camshaft bearing caps.

15. Apply sealant to the mating surfaces of the No. 1 and No. 5 camshaft bearing caps. Do not allow the sealant to cure before the rocker arm assembly is installed.

16. Set the rocker arm assembly in place. Apply engine oil to the holder bolt threads, and then loosely install the bolts. Tighten each bolt in a two–step crisscross pattern to ensure that the rockers do not bind on the valves. Tighten the 8mm bolts to 14 ft. lbs. (20 Nm). Tighten the 6mm bolts to 8.7 ft. lbs. (12 Nm).

17. Verify that the engine is at TDC/compression for the No. 1 piston, and install the distributor.

18. Adjust the valves and tighten the locknuts to 14 ft. lbs. (20 Nm).

19. Install the valve cover and upper timing belt cover.

20. Reconnect the negative battery cable.

21. Check the ignition timing and adjust if necessary. Tighten the distributor mounting bolts to 17 ft. lbs. (24 Nm).

1.6L (D16Y8) Engine

1. Disconnect the negative battery cable.

2. Label and disconnect the ignition wires. Remove the spark plugs and note their cylinder assignments.

3. Remove the valve cover.

4. Rotate the crankshaft to set the No. 1 cylinder to TDC for the compression stroke. The white TDC mark on the crankshaft pulley aligns with the TDC pointers on the lower timing belt cover.

5. Remove the distributor from the cylinder head.

6. Loosen the valve adjusting screws.

7. Label and disconnect the VTEC solenoid valve connector.

8. Loosen the camshaft holder bolts two turns at a time in a criss-cross pattern to prevent damaging the valves or rocker assembly.

9. Remove the rocker arm and shaft assemblies together with the camshaft holders and the lost motion assembly holder. Do not remove the rocker shaft bolts yet. The bolts keep the bearing caps, springs, and rocker arms in place on the shafts.

NOTE: The rocker arms and shafts are an assembly; they must be removed from the engine as a unit. Always follow the torque sequence carefully when installing the rocker shaft assembly.

10. Remove the camshaft holder bolts from the rocker arm and shaft assembly. Remove the lost motion assembly holder.

11. Bundle the intake rocker arm assemblies with rubber bands so they don't separate when the intake rocker shaft is removed.

12. Disassemble the rocker arm and shaft assemblies. Label the parts as they are removed from the shafts to ensure reinstallation in the original location.

13. Disassemble the rocker arm assemblies taking care not to mix up any of the parts. Inspect the rocker arm synchronizing pistons by pushing them with your fingers. If the pistons don't move smoothly in their rocker arm bores, replace the rocker arm assembly.

14. Apply oil to the synchronizing pistons and reassemble the rocker arms. Bundle the rocker arm assemblies with rubber bands to prevent the parts from separating.

15. Remove each lost motion assembly from its port in the lost motion assembly holder. Inspect each lost motion assembly by pushing down on its piston. If the piston doesn't move smoothly, replace the lost motion assembly. Lost motion assemblies cannot be bled like hydraulic lash adjusters.

16. Install the lost motion assemblies back into the lost motion assembly holder.

To install:

17. Verify that the engine is set at TDC/compression for the No. 1 cylinder.

18. Lubricate the camshaft journals and lobes. Coat the rocker shafts and camshaft holders with fresh oil.

19. Remove the oil control orifice. Thoroughly clean it and reinstall it with a new O–ring.

20. Install a new camshaft seal if necessary.

21. Assemble the rocker arm and shaft assemblies. Make sure the intake shaft collars and exhaust shaft springs are in the proper locations.

22. After the rocker arms and shafts are assembled, cut the rubber bands and remove them from the intake rockers. Make sure that no rubber band fragments are left in the engine.

23. Install the lost motion assembly holder onto the camshaft holder. Apply fresh oil to the threads of camshaft holder bolts, and then install them.

24. Apply liquid gasket to the cylinder head mating surfaces of the No. 1 and No. 5 camshaft holders. Don't al-

5 HONDA ACCORD/CIVIC/DEL SOL/PRELUDE

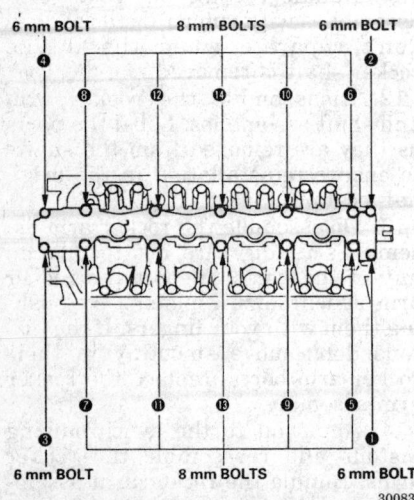

Rocker arm/shaft bolt loosening sequence — 1.6L (D16Y8) engine

300830

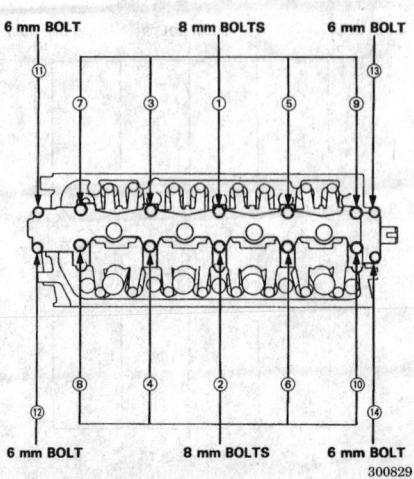

Rocker arm/shaft bolt torque sequence — 1.6L (D16Y8) engine

300829

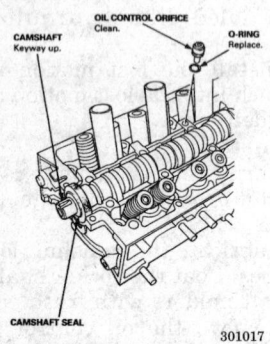

Camshaft seal and oil control orifice — 1.6L (D16Y8) engine

301017

low the sealant to cure before installation.

25. Set the rocker arm and shaft assembly in place. Install and hand–tighten the bolts. Tighten each bolt two turns at a time in the criss-cross sequence to ensure that the rockers do not bind on the valves. Tighten the 8mm rocker arm bolts to 14 ft. lbs. (20 Nm). Tighten the 6mm bolts to 9 ft. lbs. (12 Nm).

26. Starting with the No. 1 cylinder at TDC/compression, adjust the valve clearances. After the clearance has been reached, tighten the adjuster locknuts to 14 ft. lbs. (20 Nm). Set the

No. 3, No. 4, and No. 2 cylinders at TDC/compression, and then adjust their valve clearances.

- Intake: 0.007–0.009 in. (0.18–0.22mm)
- Exhaust: 0.009–0.011 in. (0.23–0.27mm)

27. Remove the VTEC solenoid valve, and then remove the valve's filter. Install a new VTEC solenoid valve filter. Tighten the solenoid valve bolts to 9 ft. lbs. (12 Nm) and reconnect the solenoid valve connectors.

28. Rotate the crankshaft to set the No. 1 cylinder at TDC/compression. Then, manually inspect the operation of each of the VTEC intake rocker arms:

   a. Push the in on the No. 1 cylinder's mid–intake rocker arm.

   b. Verify that the mid–intake rocker arm moves independently of the primary and secondary intake rocker arms.

   c. Repeat the rocker arm inspection for the other three cylinders with each cylinder set at TDC/compression.

29. Rotate the crankshaft back to TDC/compression for the No. 1 cylinder. Install the distributor, but do not tighten the mounting bolts yet.

30. Tighten the crankshaft pulley to 134 ft. lbs. (181 Nm).

31. Install the valve cover. Make sure the gasket is in good condition, and apply sealant to the corners where the gasket meets the camshaft holders.

32. Install the spark plugs and reconnect the ignition wires.

33. Drain the engine oil and remove the oil filter. Install a new oil filter and refill the engine with fresh oil.

34. Connect the negative battery cable.

35. Warm the engine up to normal operating temperature.

36. Check the ignition timing and adjust it if necessary. Then, tighten the distributor mounting bolts to 17 ft. lbs. (24 Nm).

37. Check all fluid levels. Test drive the vehicle and observe the engine RPM changes at various speeds.

1.6L (B16A2, B16A3) Engines

1. Disconnect the negative battery cable.

2. Label and disconnect the ignition wires.

3. Rotate the crankshaft to set the engine at TDC for the compression stroke of the No. 1 cylinder. The white TDC mark on the crankshaft pulley should align with the pointer on the lower timing belt cover.

4. Remove the strut brace.

5. Remove the intake air duct.

6. Loosen the power steering pump adjusting bolt to release the belt tension. Slip the belt off the pulleys. Loosen the air conditioner and alternator adjusting bolts, and slip their belts off the crankshaft pulley.

7. Use a floor jack padded with a block of wood to support the engine.

8. Remove the engine ground cable.

9. Unbolt and remove the engine side mount.

10. Remove the valve cover and the upper timing belt cover.

11. Verify that the engine is set at TDC/compression. Loosen the timing belt tensioner bolt 180°. Then, remove the crankshaft pulley, the lower timing cover, and timing belt.

--- **WARNING** ---

Inspect the timing belt for signs of cracked and broken teeth, as well as oil or coolant contamination. If the timing belt is damaged, or has been in contact with oil or coolant, it must be replaced to avoid potential failure.

---

12. Remove the distributor.

13. Disconnect and remove the VTEC solenoid valve. Remove the solenoid valve's filter and inspect it for clogging.

14. Remove the camshaft sprockets and back cover.

15. Loosen the camshaft holder plate bolts in a crisscross sequence working toward the middle of the cylinder head.

16. Loosen the valve adjusting screws.

17. Lift the camshaft holder plates and holder from the cylinder head. The holder bolts will keep the components together. Note the positions of each camshaft holder for reassembly.

18. Lift the camshafts from the cylinder head. Mark the exhaust and intake camshafts so that they will not be confused.

19. Hold each rocker arm assembly together with a rubber band to prevent them from separating.

20. Remove the intake and exhaust rocker shaft orifices from the cylinder head. The rocker shaft orifices are different and should be identified when removed. Thoroughly clean the orifices and reinstall them with new O-rings.

21. Remove the rocker arm shaft sealing bolts, discard the washers.

22. Insert 12mm bolts into the rocker arm shafts. Remove each rocker arm set while slowly pulling out the rocker arm shaft.

NOTE: Tag each rocker arm set to to assure installation in their original locations.

23. Inspect the rocker arm pistons. If they do not move smoothly, replace the rocker arm assembly.

24. Remove the two lost motion assemblies from the cylinder head. Inspect each lost motion assembly by pushing the plunger with your finger. Replace the lost motion assembly if it does not move smoothly.

To install:

25. Install the two lost motion assemblies to the cylinder head.

26. Apply engine oil to the rocker arm pistons, then bundle the rocker arms with a rubber band. Apply a light coat of clean engine oil to the rocker arms.

27. Position the rocker arms in their original locations, if they are being reused. If new assembles are being used place them in the cylinder head.

28. Lightly coat the rocker arm shafts with clean engine oil, then install the rocker arm shafts into the cylinder head. A 12mm bolt can be installed into the end of the rocker arm shafts to aid in their installation. Be sure to install the shafts in the proper positions. Remove the 12mm bolts from the rocker arm shafts, if used.

29. Clean and install the rocker arm shaft orifices with new O-rings. If the holes in the rocker arm shafts are not aligned screw a 12mm bolt into the end of the shaft to position the shaft.

30. Install the sealing bolts with new washers, torque the bolts to 47 ft. lbs. (64 Nm).

31. Lubricate the camshaft lobes and journals with clean engine oil.

32. Set the camshafts into the cylinder head. Both the intake and exhaust camshafts should be installed with their keyways pointing straight up.

33. Lubricate and install new camshaft seals. Apply liquid gasket to a new camshaft end-plug and install it. If the end-plug has is marked, the mark should be aligned with the cylinder head surface.

34. Apply liquid gasket to the cylinder head mating surfaces of the No. 1 and No. 5 camshaft holders then install them, along with No. 2, 3, and 4 holders. Be sure to pay attention to the following points:

 • Do not apply oil to the holder mating surface of camshaft seals.

 • The arrows marked on the camshaft holders should point to the timing belt.

35. Install the camshaft holder plates.

36. Lubricate the threads of the 10mm holder bolts. Then, install all the camshaft holder bolts, but don't tighten them yet.

37. Evenly hand-tighten the camshaft holders. Make sure that the rocker arms are properly positioned on the valve stems.

38. Use a two-step crisscross pattern to tighten the camshaft holder bolts. Begin tightening with the bolts in the middle of the cylinder head, and work toward the outer edges. Fi-

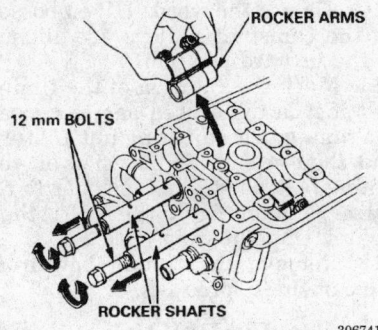

Removing the rocker arms — 1.6L (B16A2, B16A3) engines

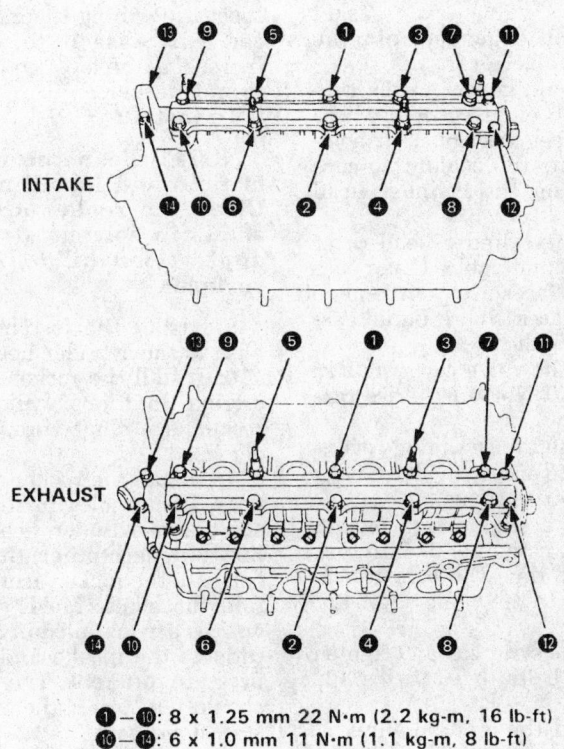

Camshaft holder bolt torque sequences — 1.6L (B16A2, B16A3) engines

nal torque specifications are as follows:

- 1996–97 B16A2 engine: 8mm bolts 20 ft. lbs. (28 Nm).
- 1994–95 B16A3 engines: 8mm bolts to 16 ft. lbs. (22 Nm)
- 6mm bolts to 7–8 ft. lbs. (10–11 Nm)

39. Verify that the camshaft keyways are pointing straight up and that the engine is at TDC/compression for the No. 1 cylinder. Fit the camshaft sprocket keys into their keyways.

40. Install the back cover and push the camshaft pulleys onto the camshafts. Then, tighten the sprocket retaining bolts to 37 ft. lbs. (51 Nm) for 1994–95 vehicles, or 41 ft. lbs. (57 Nm) for 1996–97 vehicles.

41. Install and tension the timing belt.

42. Install the lower timing cover and the crankshaft pulley. Tighten the pulley bolt to 130 ft. lbs. (180 Nm).

43. Adjust the valve clearance.

44. Inspect the VTEC rocker arms for smooth and independent movement.

45. Install the VTEC solenoid valve with a new filter. Tighten the valve mounting bolts to 9 ft. lbs. (12 Nm).

46. Install the distributor.

47. Clean the valve cover gasket surfaces. Fit the gasket into the groove of the valve cover.

48. Apply liquid gasket to the rubber seal at the eight corners where the gaskets meet the camshaft holders. Don't allow the sealant to cure before installing the cylinder head cover.

49. Install the cylinder head cover and engine ground cable. Make sure the contact surfaces are clean and do not touch surfaces where liquid gasket has been applied.

50. Tighten the valve cover nuts in to 7 ft. lbs. (10 Nm) in a crisscross sequence.

51. Install the upper timing cover.

52. Install the accessory drive belts and adjust their tensions.

53. Install the engine side mount, torque the two new nuts to 54 ft. lbs. (75 Nm) and torque the bolt attaching the mount to the vehicle to 54 ft. lbs. (74 Nm).

54. Install the strut bar and tighten the mounting bolts to 16 ft. lbs. (22 Nm).

55. Reconnect the ignition wires.

56. Drain the engine oil. Install a new oil filter and refill the engine with fresh oil.

57. Reconnect the negative battery cable.

58. Warm the engine up to its normal operating temperature. Then, check and adjust the ignition timing. Tighten the distributor mounting bolts to 17 ft. lbs. (24 Nm).

Prelude and Accord

2.2L (H22A1) Engine

1. Disconnect the negative battery cable.

2. Remove the cylinder head from the engine assembly.

3. Remove the VTEC solenoid valve and filter from the cylinder head. Discard the filter.

4. Install rubber bands to each of the rocker arm assemblies, this will hold the rocker arms together and prevent them from separating.

5. Remove the intake and exhaust rocker shaft orifices. The intake and exhaust shaft orifices are different, note their locations for installation.

6. Remove the rocker arm shaft sealing bolts from the cylinder head, discard the washers.

7. Install 12mm bolts into the rocker arm shafts. Remove each rocker arm while slowly pulling out the intake and exhaust rocker arm shafts.

8. Remove the lost motion assemblies from the cylinder head and inspect it. Pushing it gently with a finger will cause it to sink slightly. Increasing the force on it will cause it to sink deeper.

To install:

NOTE: Clean the rocker shaft orifices and install new O-rings. Clean the rocker arms and the shafts in solvent, dry them and apply clean oil to any contact surfaces.

9. Install the lost motion assemblies to the cylinder head.

10. Install the rocker arms to their original locations while passing the rocker arm shaft through the cylinder head.

11. Install the rocker arm shaft orifices. If the holes in the rocker arm shaft and cylinder head are not in line with each other, thread a 12mm bolt into the rocker arm shaft and rotate the shaft. Make sure that the orifices are installed in the correct locations, the intake and exhaust orifices are different. The rocker shafts should not turn if the orifices are installed correctly.

12. Install the rocker arm sealing bolts with new washers, torque the bolts to 43 ft. lbs. (60 Nm). Remove the rubber bands from the rocker arm assemblies.

13. Install the VTEC solenoid valve with a new filter, torque the mounting bolts to 9 ft. lbs. (12 Nm).

14. Install the cylinder head onto the engine block.

15. Adjust the valves and ignition timing.

16. Connect the negative battery cable and enter the radio security code.

17. If equipped with 4WS, start the engine and turn the steering wheel lock-to-lock to reset the 4WS control unit.

18. Run the engine and check for leaks, then road test the vehicle.

2.3L (H23A1) Engine

1. Disconnect the negative battery cable.

2. Turn the crankshaft so the No. 1 piston is at top dead center.

NOTE: The No. 1 piston is at top dead center when the pointer on the block aligns with the white painted mark on the flywheel (manual transaxle) or driveplate (automatic transaxle).

3. Remove the air intake duct.

4. Remove the engine ground cable from the cylinder head cover.

5. Remove the connector and the terminal from the alternator, then remove the engine wire harness from the valve cover.

6. Remove the ignition coil.

7. Label then disconnect the electrical connectors from the distributor and the spark plug wires from the spark plugs. Mark the position of the distributor and remove it from the cylinder head. Disconnect the ignition coil wire from the distributor.

8. Remove the Positive Crankcase Ventilation (PCV) hose, then remove the cylinder head cover. Replace the rubber seals if damaged or deteriorated.

9. Remove the timing belt middle cover.

10. Ensure the words **UP** embossed on the camshaft pulleys are aligned in the upward position.

11. Mark the rotation of the timing belt if it is to be used again. Loosen the timing belt adjusting nut 1/2 turn and then release the tension on the timing belt. Push the tensioner to release tension from the belt, then tighten the adjusting nut.

12. Remove the timing belt from the camshaft sprockets.

--- **WARNING** ---
Do not crimp or bend the timing belt more than 90° or less then 1 inch (25mm) in diameter

13. Insert a 5.0mm pin punch in each of the camshaft caps, nearest to the sprockets, through the holes provided. Remove the camshaft sprocket attaching bolts then remove the sprockets. Do not lose the sprocket keys.

14. Remove the side engine mount bracket B, then the timing belt back cover from behind the camshaft sprockets.

15. Loosen all of the rocker arm adjusting screws, then remove the pin punches from the camshaft caps.

16. Remove the camshaft holders, note the holders locations for ease of installation. Loosen the bolts in the reverse order of the holder bolts torque sequence.

17. Remove the camshafts from the cylinder head then discard the camshaft seals.

18. Remove the rubber cap from the head, located at the end of the intake camshaft.

19. Remove the rocker arms from the cylinder head. Note the locations of the rocker arms.

NOTE: The rocker arms have to be installed to their original locations if being reused.

To install:

20. Lubricate the rocker arms with clean oil, then install the rocker arms on the pivot bolts and the valve stems. If the rocker arms are being reused, install them to their original locations. The locknuts and adjustment screws should be loosened before installing the rocker arms.

21. Lubricate the camshafts with clean oil.

22. Install the camshaft seals to the end of the camshafts that the timing belt sprockets attach to. The open side (spring) should be facing into the cylinder head when installed.

23. Make sure the keyways on the camshafts are facing up and install the camshafts to the cylinder head.

24. Install the rubber plug to the cylinder head at the end of the intake camshaft.

25. Apply liquid gasket to the head mating surfaces of the No. 1 and No. 6 camshaft holders, then install them along with No. 2, 3, 4 and 5. **I** or **E** marks are stamped on the camshaft holders to identify them as Intake or Exhaust side holders. The arrows stamped on the holders should point toward the timing belt.

26. Snug the camshaft holders in place.

27. Press the camshaft seals securely into place.

28. Torque the camshaft holder bolts in two steps, following the proper sequence, to ensure that the rockers do not bind on the valves. Torque all the bolts, except the four studs, to 7 ft. lbs. (10 Nm). Torque the studs (number 5 and 7 bolts in the torque sequence) to 9 ft. lbs. (12 Nm).

29. Install the timing belt back cover.

30. Install the side engine mount bracket B. Torque the bolt attaching the bracket to the cylinder head to 33 ft. lbs. (45 Nm). Torque the bolts attaching the bracket to the side engine mount to 16 ft. lbs. (22 Nm).

31. Insert a 5.0mm pin punch in each of the camshaft caps, nearest to the pulleys, through the holes provided. Install the keys into the camshaft grooves.

32. Push the camshaft sprockets onto the camshafts, then tighten the retaining bolts to 27 ft. lbs. (38 Nm).

33. Ensure the words **UP** embossed on the camshaft pulleys are aligned in the upward position. Install the timing belt to the camshaft sprockets, then remove the two 5.0mm pin punches from the camshaft bearing caps.

34. Loosen then tighten the timing belt adjuster nut.

35. Turn the crankshaft counterclockwise until the cam pulley has moved 3 teeth; this creates tension on the timing belt. Loosen then tighten the adjusting nut and torque it to 33 ft. lbs (45 Nm).

36. Adjust the valves.

37. Torque the crankshaft pulley bolt to 181 ft. lbs. (250 Nm).

38. Install the middle timing belt cover and torque the attaching bolts to 9 ft. lbs. (12 Nm).

39. Install the cylinder head cover and torque the cap nuts to 7 ft. lbs. (10 Nm). Install the PCV hose to the cylinder head cover.

40. Install the distributor to the cylinder head, snug the attaching bolts until the timing has been checked and adjusted.

41. Connect the spark plug wires to the correct spark plugs, then connect the distributor electrical connectors. Install the ignition coil wire to the distributor.

42. Install the ignition coil.

43. Install the alternator wire harness to the cylinder head cover, then connect the terminal and connector to the alternator.

44. Connect the engine ground cable to the cylinder head cover.

45. Install the air intake duct.

46. Drain the oil from the engine into a sealable container. Install the drain plug and refill the engine with clean oil.

47. Connect the negative battery cable and enter the radio security code.

48. Start the engine, checking carefully for any leaks.

49. Check and adjust the ignition timing. Torque the distributor bolts to 13 ft. lbs. (18 Nm).

50. If equipped with 4WS, start the engine then turn the steering wheel lock–to–lock to reset the 4WS control unit.

2.2L (F22A1, F22A6, F22B1, F22B2) Engines

1. Disconnect the negative battery cable.

2. Remove the air intake duct.

3. Remove the positive crankcase ventilation (PCV) hose, then remove the cylinder head cover. Replace the rubber seals if damaged or deteriorated.

4. Remove the timing belt upper cover.

5. Bring the No. 1 cylinder to TDC. The white mark on the crankshaft pulley should align with the pointer on the timing belt cover. The words **UP** embossed on the camshaft pulley should be aligned in the upward position. The marks on the edge of the pulley should be aligned with the cylinder head or the back cover

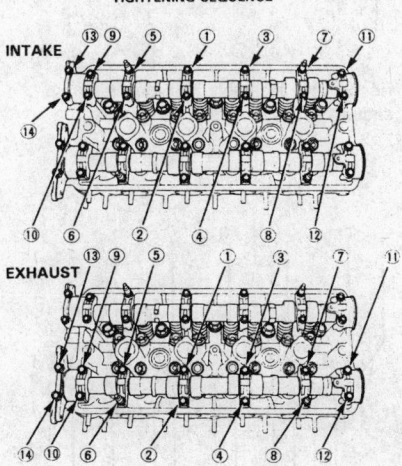

Specified torque:
Except Intake ⑤, ⑦, Exhaust ⑥, ⑧:
10 N·m (1.0 kg-m, 7 lb-ft)
Intake ⑤, ⑦, Exhaust ⑥, ⑧:
12 N·m (1.2 kg-m, 9 lb-ft)

TIGHTENING SEQUENCE

Camshaft holders torque sequence — 2.3L (H23A1) engine

upper edge. Once in this position, the engine must NOT be turned or disturbed.

6. Label then disconnect the electrical connectors from the distributor and the spark plug wires from the spark plugs. Mark the position of the distributor and remove it from the cylinder head.

7. Loosen the power steering mounting bolts and remove drive belt from the pump.

8. Mark the rotation of the timing belt if it is to be used again. Loosen the timing belt adjusting bolt 3/4 to one turn and then release the tension on the timing belt. Push the tensioner to release tension from the belt, then tighten the adjusting bolt.

9. Remove the timing belt from the camshaft sprocket.

---— WARNING ———

Do not crimp or bend the timing belt more than 90° or less then 1 inch (25mm) in diameter

---

10. Ensure the words **UP** embossed on the camshaft pulley is aligned in the upward position, then remove the camshaft sprocket bolt. Pull the sprocket from the camshaft and remove the sprocket key.

11. Remove the timing belt back cover.

12. Loosen the valve adjusting screws.

13. Loosen the camshaft holder attaching bolts two turns at a time, in the proper sequence to prevent damaging the valves or rocker arm assemblies.

NOTE: When removing the rocker arm assembly, do not remove the camshaft holder bolts. The bolts will keep the camshaft holders, springs, and the rocker arms on the shafts.

14. Carefully remove the camshaft holders and rocker arm assembly. If the rocker arm and shaft assembly needs to be disassembled for service, note the location of the components as they are removed. Install a rubber band around the VTEC rocker arm assemblies, to keep them from coming apart during disassembly of the rocker arm assembly. The rocker arms must be installed in the same position if reused.

15. Remove the camshaft from the cylinder head and discard the seal.

16. Remove the oil control orifice.

To install:

17. Wipe the camshaft and the camshaft journals clean, then lubricate both surfaces and install the camshaft.

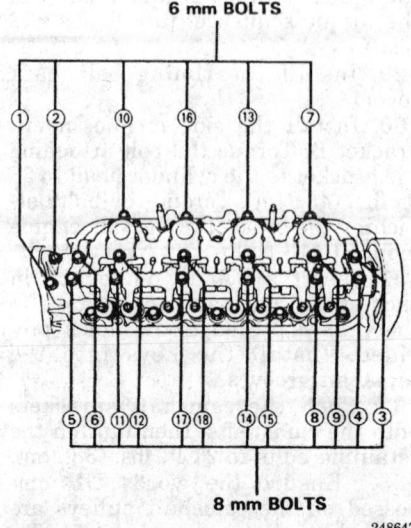

Rocker arm assembly bolt removal sequence — 2.2L (F22A1, F22A6) engines

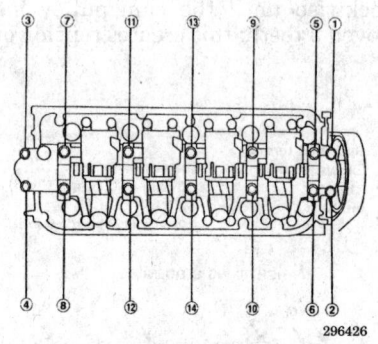

Rocker arm assembly bolt loosening sequence — 2.2L (F22B1) engine

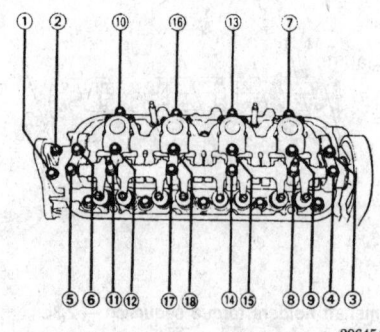

Rocker arm assembly bolt loosening sequence — 2.2L (F22B2) engine

18. Turn the camshaft so that its keyway is facing up (No. 1 cylinder will be at TDC).

19. Clean the oil control orifice and install a new O–ring, then install the oil control orifice.

20. Reassemble the rocker arm and shaft assembly, if it was disassembled. Lubricate the rocker arm and shaft assembly with clean oil, then apply liquid gasket to the head mating surfaces of the No. 1 and No. 6 camshaft holders.

21. Set the camshaft holders and rocker arm assembly in place then loosely install the attaching bolts.

22. Apply clean oil to the camshaft oil seal lip and the seal guide (part # 07NAG–PT0010A), then install the seal to the seal guide. Install the seal guide to the camshaft, then the installer cup (part # 07NAF–PT0010A) and the installer shaft (part # 07NAF–PT0020A). Tighten the nut on the installer shaft to press the seal into the cylinder head.

23. Tighten the camshaft holder bolts two turns at a time in the proper sequence. The final torque for the 8mm bolts is 16 ft. lbs. (22 Nm) and the final torque for the 6mm bolts is 9 ft. lbs. (12 Nm).

24. Install the timing belt back cover and a new gasket, if necessary. Torque the bolt toward the exhaust manifold to 7 ft. lbs. (10 Nm) and torque the bolt toward the intake manifold to 9 ft. lbs. (12 Nm).

25. Install the camshaft sprocket key to the camshaft, then install the camshaft sprocket. Install the bolt and torque it to 27 ft. lbs. (37 Nm).

26. Ensure the words **UP** embossed on the camshaft pulley is aligned in the upward position, then install the timing belt onto the camshaft sprocket. Loosen then tighten the timing belt adjusting nut.

27. Rotate the crankshaft pulley five or six turns to position the timing belt on the pulleys.

28. Set the No. 1 cylinder to TDC and loosen the timing belt adjusting nut one turn. Turn the crankshaft counterclockwise until the cam pulley has moved 3 teeth; this creates tension on the timing belt. Loosen, then tighten the adjusting nut and torque it to 33 ft. lbs (45 Nm).

29. Adjust the valves.

30. Torque the crankshaft pulley bolt to 181 ft. lbs. (245 Nm) on (F22B1) engines and 159 ft. lbs. (220 Nm) on all other engines.

31. Install the upper timing belt cover. Torque the bolt toward the exhaust manifold to 7 ft. lbs. (10 Nm)

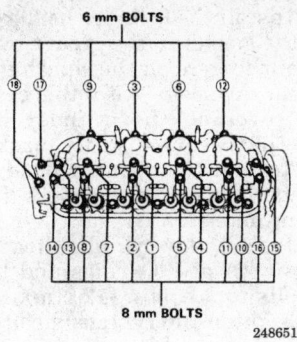

Rocker arm assembly torque sequence — 2.2L (F22A1, F22A6) engines

Specified torque:
8 mm bolts: 22 N·m (2.2 kgf·m, 16 lbf·t)
6 mm bolts: 12 N·m (1.2 kgf·m, 8.7 lbf·t)
6 mm bolts: ⑪, ⑫, ⑬, ⑭

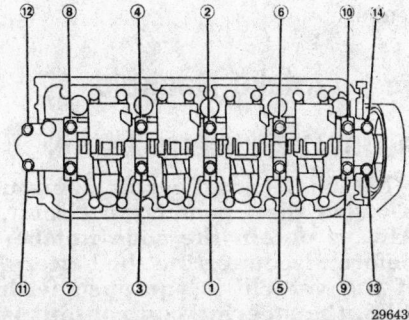

Rocker arm assembly torque sequence — 2.2L (F22B1) engine

and torque the bolt toward the intake manifold to 9 ft. lbs. (12 Nm).

32. Install the cylinder head cover gasket cover to the groove of the cylinder head cover. Before installing the gasket thoroughly clean the seal and the groove. Seat the recesses for the camshaft first, then work it into the groove around the outside edges. Make sure the gasket is seated securely in the corners of the recesses.

33. Apply liquid gasket to the four corners of the recesses of the cylinder head cover gasket. Do not install the parts if 5 minutes or more have

Specified torque:
8 mm bolts: 22 N·m (2.2 kgf·m, 16 lbf·t)
6 mm bolts: 12 N·m (1.2 kgf·m, 8.7 lbf·t)

6 mm bolts: ③, ⑥, ⑨, ⑫, ⑰, ⑱

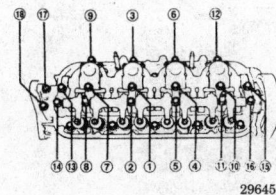

Rocker arm assembly torque sequence — 2.2L (F22B2) engine

elapsed since applying liquid gasket. After assembly, wait at least 20 minutes before filling the engine with oil.

34. Install the cylinder head (valve) cover. Torque the bolts attaching the cylinder head cover in the proper sequence to 7 ft. lbs. (10 Nm).

35. Install the PCV hose to the cylinder head cover.

36. Install and adjust the power steering belt.

37. Install the distributor to the cylinder head, snug the attaching bolts until the timing has been checked and adjusted.

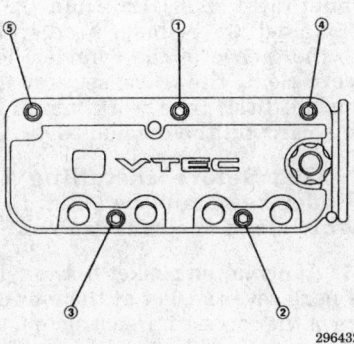

Cylinder head cover torque sequence — 2.2L (F22B1) engine

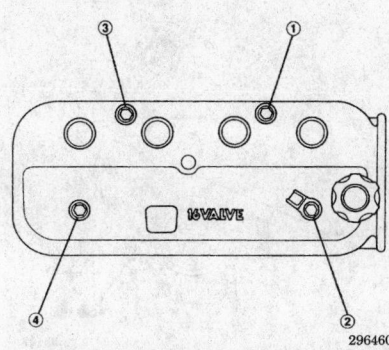

Cylinder head cover torque sequence — 2.2L (F22B2) engine

38. Connect the spark plug wires to the correct spark plugs, then connect the distributor electrical connectors.

39. Install the air intake duct.

40. Drain the oil from the engine into a sealable container. Install the drain plug and refill the engine with clean oil.

41. Connect the negative battery cable and enter the radio security code.

42. Start the engine, checking carefully for any leaks.

43. Check and adjust the the ignition timing as necessary, then torque the distributor bolts to 13 ft. lbs. (18 Nm).

2.7L Engine

1. Disconnect the negative battery cable.

2. Remove the timing belt covers and the timing belt,

3. Remove the camshafts from the cylinder heads.

4. Remove the intake rocker arms, exhaust inside rocker arms and the pushrods. Identify the location of the parts as they are removed, to ensure reinstallation to the original locations.

5. Remove the valve lifters (hydraulic tappets) from the cylinder heads.

6. Remove the intake manifold then the cylinder heads from the vehicle.

7. Remove the rocker arm shaft sealing bolt from the cylinder head and discard the washer.

8. Install a 12·1.25mm bolt into the rocker arm shaft. Slowly remove the shaft from the cylinder head and remove the exhaust rocker arms and washers. Identify the location of the parts as they are removed, to ensure reinstallation to the original locations.

To install:

9. Clean the rocker arms and rocker arm shafts in solvent, dry them, then oil the contact surfaces of the parts.

10. Install a 12·1.25mm bolt into the rocker arm shaft. Install the rocker arms to their original locations while passing the rocker arm shaft through the cylinder head.

11. Remove the bolt from the rocker arm shaft and install the sealing bolt with a new washer. Torque the sealing bolt to 33 ft. lbs. (44 Nm).

12. Install the cylinder heads and the intake manifold.

13. Fill the valve lifter (hydraulic tappets) mounting hole and the oil fillers with clean engine oil.

14. Install the valve lifters.

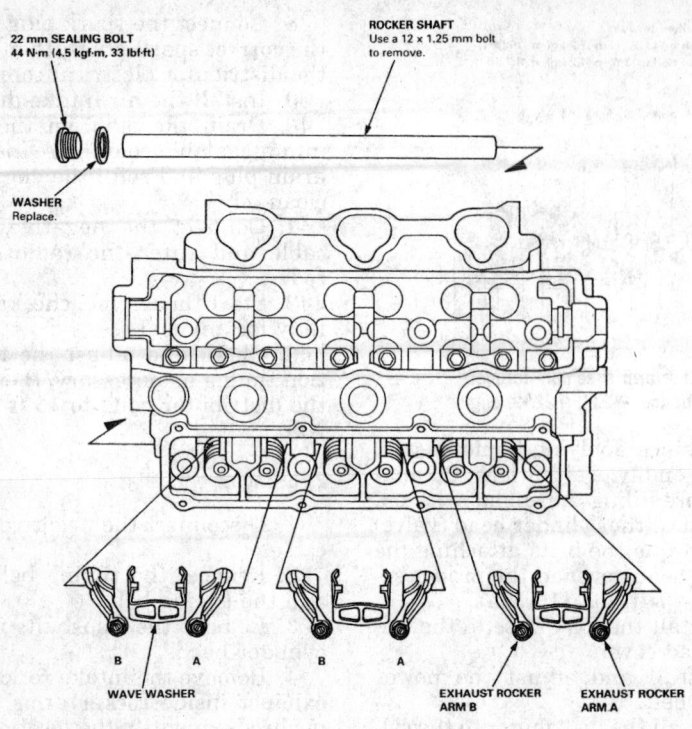

22 mm SEALING BOLT
44 N·m (4.5 kgf·m, 33 lbf·ft)

ROCKER SHAFT
Use a 12 x 1.25 mm bolt
to remove.

WASHER
Replace.

B A B A

WAVE WASHER

EXHAUST ROCKER
ARM B

EXHAUST ROCKER
ARM A

NOTE: The wave washer should be firmly fitted to the cylinder head groove.

298099

Exhaust rocker arms and rocker arm shaft component locations — 2.7L engine

WARNING
Do not rotate the valve lifters while installing them.

15. Apply clean engine oil to the rocker arms, pushrods and the camshafts.

16. Loosen the exhaust rocker arm adjusting screws and locknuts, then install the pushrods, exhaust inside rocker arms and the intake rocker arms. Install the parts to their original locations.

17. Make sure the rocker arms are properly positioned on the valve stems. Advance the crankshaft 30° from TDC to prevent interference between the pistons and valves when the camshafts are installed.

18. Install the camshafts and camshaft holders.

19. Install the timing belt and set No. 1 cylinder to TDC.

20. Tighten the adjusting screws for No. 1, No. 2 and No. 4 cylinders. Tighten the screw until it contacts the the valve, then tighten the screw 1 turn. Hold the screw in place and torque the locknut to 14 ft. lbs. (20 Nm).

21. Rotate the crankshaft pulley one turn clockwise, then tighten the adjusting screws for No. 3, No. 5 and No. 6 cylinders. Tighten the screw until it contacts the valve, then

tighten the screw 1 1/8 turns. Hold the screw in place and torque the locknut to 14 ft. lbs. (20 Nm).

22. Install the cylinder head gasket into the groove of the cylinder head cover. Seat the recesses for the camshaft first, then work it into the groove around the outside edges.

NOTE: Before installing the cylinder head cover gasket, thoroughly clean the seal groove.

23. Apply liquid gasket to the cylinder head cover gasket at the four corners of the recesses. Use a shop towel and wipe the cylinder heads where the cylinder head covers will come in contact.

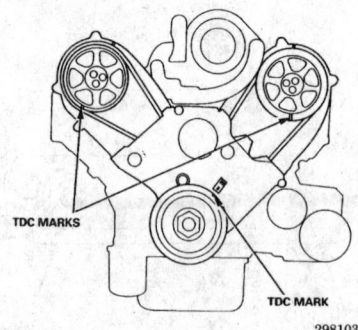

TDC MARKS

TDC MARK

298103

Timing marks for valve adjustment for cylinders 3, 5 and 6 — 2.7L engine

24. Install the cylinder head covers, hold the gasket in the groove by placing your fingers on the camshaft contacting surfaces. With the cylinder head cover on the cylinder heads, slide the covers slightly back and forth to seat the cylinder head cover gaskets. Replace the washers if damaged or deteriorated

25. Install the cylinder head side covers with new O-rings and torque the bolts to 9 ft. lbs. (12 Nm).

26. Tighten the cylinder head cover bolts in two or three steps. In the final step, torque all the bolts, in sequence, to 11 ft. lbs. (15 Nm).

27. Install the distributor to the cylinder head and torque the mounting bolt to 16 ft. lbs. (22 Nm).

28. Connect the spark plug wires to the correct spark plugs, then connect the distributor electrical connectors.

29. Drain the engine oil into a sealable container then refill the engine with clean oil.

30. Connect the negative battery cable and enter the radio security code.

31. Start the engine, allowing it to idle and check for any signs of leakage.

Intake Manifold

REMOVAL AND INSTALLATION

NOTE: The radio may contain a coded theft protection circuit. Always obtain the code number before disconnecting the battery. If the vehicle is equipped with 4WS, the steering control unit is shut down when the battery is disconnected. After connecting the battery, turn the steering wheel lock-to-lock to reset the steering control unit.

Civic and del Sol

1. Disconnect the negative battery cable.

2. Drain the cooling system to a level below the upper radiator hose.

3. Relieve the fuel system pressure by loosening the fuel filter service bolt.

CAUTION
Fuel injection systems remain under pressure even after the engine has been turned off. The fuel system pressure must be relieved before disconnecting any fuel lines. Failure to do so may result in fire and personal injury.

4. Remove the intake air duct. If equipped with the D16Y7 engine, re-

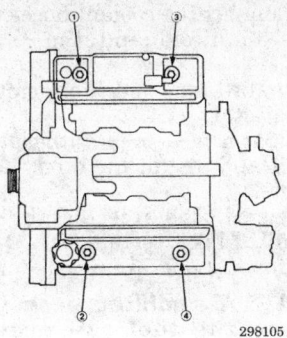

Cylinder head cover torque sequence — 2.7L engine

move the air cleaner assembly from the throttle body.

5. Cover the throttle body opening to keep dirt out.

6. Disconnect the fuel line from the fuel rail. Clean up any spilled fuel.

7. Label and disconnect the fuel injector wiring harnesses.

8. Remove the fuel rail and injectors.

9. Disconnect the throttle cable from the linkage at the throttle body.

10. Disconnect the intake manifold cooling hoses. Use a drain pan to catch any spilled coolant, also make sure no coolant spills on electrical connections.

11. Label and disconnect the engine wiring harness connectors from the intake manifold sensors.

12. Remove the intake air control (IAC) valve.

13. If equipped, remove the exhaust gas recirculation (EGR) valve.

14. Label and disconnect the throttle position sensor (TPS) and manifold absolute pressure (MAP) sensor.

15. Unbolt the manifold from its support bracket.

16. Loosen and remove the intake manifold nuts in a crisscross pattern.

17. Remove the intake manifold assembly from the vehicle.

To install:

NOTE: Use new gaskets when installing the intake manifold. Use new O-rings when installing manifold sensors and components. Use new sealing washers when reconnecting the fuel lines.

18. Clean all gasket mating surfaces and install the intake manifold assembly onto the cylinder head using new gaskets.

19. Tighten the intake manifold nuts in 2–3 steps in a crisscross pattern starting with the inside nuts. Tighten the nuts to 17 ft. lbs. (23 Nm).

20. Install the support bracket bolts and tighten them to 17 ft. lbs. (24 Nm).

21. Install the fuel rail and injectors.

22. Reconnect the fuel line using new washers.

23. If removed, install the EGR valve and tighten its nuts to 15 ft. lbs. (21 Nm).

24. Install and reconnect the IAC valve. Tighten its mounting bolts to 16 ft. lbs. (22 Nm).

25. Reconnect the fuel injector wiring harnesses.

26. Reconnect the intake manifold wiring harnesses.

27. Reconnect the intake manifold cooling hoses.

28. Reconnect the throttle cable.

29. Install the intake air duct and air cleaner assembly.

30. Refill and bleed the cooling system.

31. Connect the negative battery cable.

32. Verify that all sensors, valves, and vacuum lines are installed and connected properly. Make sure there are no loose electrical connections.

33. Turn the ignition on and off several times without starting the engine to pressurize the fuel system. Run the engine and check for proper operation. Check for vacuum leaks.

34. After the engine has warmed up, check the operation of the throttle cable and adjust it if necessary.

Prelude and Accord

2.2L and 2.3L Engines

1. Disconnect the negative battery cable.

2. Drain the engine coolant into a sealable container.

3. Disconnect the cooling hoses from the intake manifold.

4. Label and unplug the vacuum hoses and electrical connectors on the manifold and throttle body. Unplug the connector from the Exhaust Gas Recirculation (EGR) valve. Position the wiring harnesses out of the way.

5. Disconnect the throttle cable from the throttle body.

6. Relieve the fuel pressure.

7. Remove the fuel rail and fuel injectors.

8. Remove the thermostat housing mounting bolts. Remove the thermostat housing from the intake manifold and the connecting pipe, by pulling and twisting the housing. Discard the O-rings.

9. It may be necessary to remove the upper intake manifold plenum and throttle body assembly in order

to access the nuts securing the manifold to the head.

10. Remove the intake manifold support bracket bolts and the bracket. It may be necessary to access it from under the vehicle; raise and support the vehicle safely.

11. While supporting the intake manifold, remove the nuts attaching the intake manifold to the cylinder head, then remove the manifold. Remove the old gasket from the cylinder head.

12. Clean any old gasket material from the cylinder head and the intake manifold. check and clean the FIA chamber on the cylinder head.

To install:

13. Using a new gasket, place the manifold into position and support.

14. Install the support bracket to the manifold. Torque the bolt holding the bracket to the manifold to 16 ft. lbs. (22 Nm).

15. Starting with the inner or center nuts, tighten the nuts, in a criss-cross pattern, to the correct torque. The tension must be even across the entire face of the manifold if leaks are to be prevented. Correct torque is 16 ft. lbs. (22 Nm).

16. Using a new gasket, install the upper intake manifold and throttle body assembly, if removed as a separate unit. Tighten the nuts and bolts holding the chamber to 16 ft. lbs. (22 Nm).

17. Install a new O-ring to the coolant connecting pipe, and to the thermostat housing. Install the housing to the coolant pipe and the intake manifold. Torque the mounting bolts to 16 ft. lbs. (22 Nm).

18. Connect and adjust the throttle cable.

19. Install the fuel rail/injector assembly. Connect the fuel lines.

20. Properly position the wire harnesses and connect the electrical connectors.

21. Connect the vacuum hoses.

22. Fill and bleed the air from the cooling system.

23. Connect the negative battery cable and enter the radio security code.

24. If equipped with 4WS, turn the steering wheel lock-to-lock to reset the 4WS control unit.

25. Start the engine, checking carefully for any leaks of fuel, coolant or vacuum. Check the manifold gasket areas carefully for any leakage of vacuum.

2.7L Engines

1. Disconnect the negative battery cable.

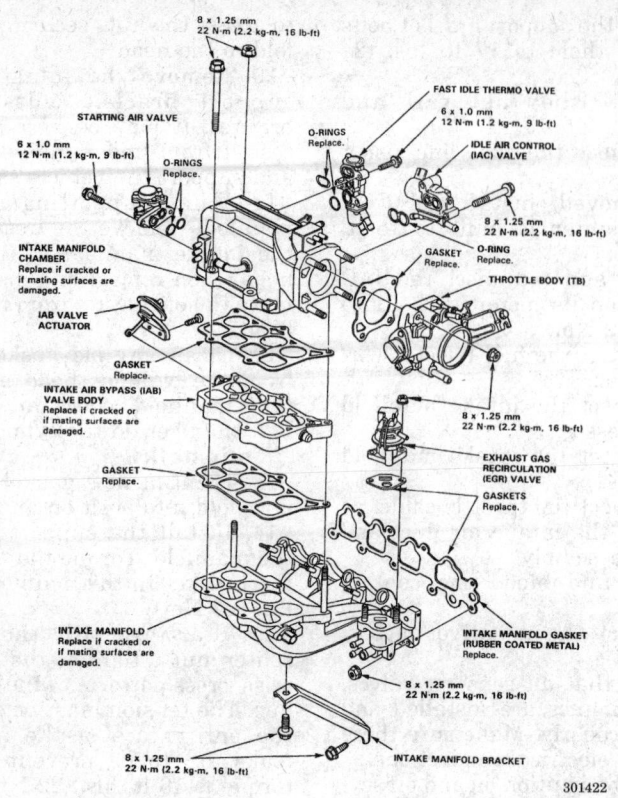

8 x 1.25 mm
22 N·m (2.2 kg-m, 16 lb-ft)

FAST IDLE THERMO VALVE

STARTING AIR VALVE

6 x 1.0 mm
12 N·m (1.2 kg-m, 9 lb-ft)

6 x 1.0 mm
12 N·m (1.2 kg-m, 9 lb-ft)

O-RINGS
Replace.

O-RINGS
Replace.

IDLE AIR CONTROL
(IAC) VALVE

INTAKE MANIFOLD
CHAMBER
Replace if cracked or
if mating surfaces are
damaged.

GASKET
Replace.

O-RING
Replace.

8 x 1.25 mm
22 N·m (2.2 kg-m, 16 lb-ft)

IAB VALVE
ACTUATOR

THROTTLE BODY (TB)

GASKET
Replace.

INTAKE AIR BYPASS (IAB)
VALVE BODY
Replace if cracked or
if mating surfaces are
damaged.

8 x 1.25 mm
22 N·m (2.2 kg-m, 16 lb-ft)

GASKET
Replace.

EXHAUST GAS
RECIRCULATION
(EGR) VALVE

GASKETS
Replace.

INTAKE MANIFOLD
Replace if cracked or
if mating surfaces are
damaged.

INTAKE MANIFOLD GASKET
(RUBBER COATED METAL)
Replace.

8 x 1.25 mm
22 N·m (2.2 kg-m, 16 lb-ft)

8 x 1.25 mm
22 N·m (2.2 kg-m, 16 lb-ft)

INTAKE MANIFOLD BRACKET

301422

Intake manifold and related components — 2.2L (F22A6) engine

2. Drain the engine coolant into a sealable container.

3. Relieve the fuel pressure.

4. Remove the feed hose from the fuel filter.

5. Remove the PCV valve from the cylinder head cover.

6. Remove the air intake duct.

7. Remove the intake manifold covers.

8. Remove the throttle cable and cruise control cable by loosening the locknut, then slip the cable end out of the accelerator linkage.

9. Label and unplug all electrical connections on the manifold and throttle body.

10. Disconnect the hoses from the brake booster and the evaporative canister.

11. Remove the wiring harness holders and position them out of the way.

12. Disconnect the vacuum hose from the fuel pressure regulator.

13. Remove the fuel rail attaching nuts.

14. Remove the fuel rail from the injectors, leaving the injectors in the manifold.

15. Remove each injector, noting its position, and remove the seal ring from each manifold port.

16. Remove the cushion ring and O-ring from each injector.

17. Label and disconnect all vacuum and coolant hoses from the intake manifold and throttle body. If necessary, remove the vacuum pipe assembly.

18. Remove the Exhaust Gas Recirculation (EGR) crossover pipe. Discard the gasket.

19. Remove the bolts and nuts securing the intake manifold to the engine. Make sure all vacuum and electrical connections are unplugged. Carefully lift the manifold from the engine. Discard the gaskets.

To install:

20. Using new gaskets, place the manifold into position. Install the nuts/bolts until just snug.

21. Starting with the inner/center bolts, tighten the nuts and bolts in a criss-cross pattern to the correct torque. The tension must be even across the entire face of the manifold if leaks are to be prevented. Correct torque is 16 ft. lbs. (22 Nm).

22. Install the EGR crossover pipe using a new gasket. Tighten the nuts (to the intake manifold) to 9 ft. lbs. (12 Nm) and the pipe (to the exhaust manifold) to 43 ft. lbs. (59 Nm).

23. If removed, install the vacuum pipe assembly. Tighten the bolts to 9 ft. lbs. (12 Nm).

24. Connect the coolant hoses to the intake manifold and the throttle body.

25. Install new cushion rings on each injector.

26. Coat new O-rings with clean engine oil and install them on the fuel injectors.

27. Install the injectors into the fuel rail. Make certain the O-rings seat properly and are not distorted.

NOTE: Assembling each injector into the fuel rail prevents damage to the O-rings. Handle the rail and injector assembly carefully when reinstalling it to the manifold. Don't drop the injectors or bang their tips.

28. Coat new seal rings with a light coat of clean, thin oil and install them into the manifold.

29. Install the fuel rail and injectors to the intake manifold.

30. With all injectors seated in the manifold, make sure each injector is positioned properly.

31. Install the fuel rail retaining nuts and tighten them evenly to 9 ft. lbs. (12 Nm).

32. Connect the vacuum hose to the regulator.

33. Install the wiring harness to the fuel rail, torque the mounting bolts to 9 ft. lbs. (12 Nm).

34. Connect the electrical harness to the injectors.

35. Connect all electrical and vacuum connections to the throttle body and manifold.

36. Connect the brake booster, evaporative canister, and fuel return hoses.

37. Connect the feed hose to the fuel filter with new gaskets. Tighten the union bolt to 16 ft. lbs. (22 Nm) and the service bolt to 9 ft. lbs. (12 Nm).

38. Install and adjust the throttle and cruise control cables.

39. Install the air intake duct.

40. Install the PCV valve to the cylinder head cover.

41. Refill and bleed the air from the cooling system.

42. Connect the negative battery cable and enter the radio security code.

43. Switch the ignition **ON** but do not engage the starter. The fuel pump should run for approximately 2 seconds, building pressure within the lines. Switch the ignition **OFF**, then **ON** 2 or 3 more times to build full system pressure. Check for fuel leaks.

44. Start the engine, checking carefully for any leaks of fuel, coolant or vacuum. Check the manifold gasket

areas carefully for any leakage of vacuum.

45. Install the intake manifold covers, torque the attaching bolts to 9 ft. lbs. (12 Nm).

Exhaust Manifold

REMOVAL AND INSTALLATION

NOTE: The radio may contain a coded theft protection circuit. Always obtain the code number before disconnecting the battery. If the vehicle is equipped with 4WS, the steering control unit is shut down when the battery is disconnected. After connecting the battery, turn the steering wheel lock–to–lock to reset the steering control unit.

─────── CAUTION ───────
The exhaust system should be serviced with the engine cold.

Civic and del Sol

1. Disconnect the negative battery cable.
2. Raise and support the front of the vehicle and block the rear wheels.
3. Unbolt the front exhaust pipe from the exhaust manifold/catalytic converter. Unbolt the exhaust manifold support brackets if their bolts are accessible from this angle. The splash shield may be removed for better access.
4. Lower the vehicle.

NOTE: Remove any rust or dirt from the exhaust manifold before removal. This will prevent dirt from entering the exhaust pipes.

5. Unbolt and remove the manifold heat shield.
6. Disconnect the oxygen sensor (HO2S) harness. Use an oxygen sensor socket or box end wrench to unscrew the sensor from the manifold. Handle the sensor carefully.
7. Unbolt the exhaust manifold brackets.
8. Unbolt the exhaust manifold and separate it from the cylinder head. Remove the exhaust manifold and its gasket.

To install:

NOTE: Use new gaskets and self-locking nuts when installing the exhaust manifold.

9. Clean the gasket mating surfaces of the manifold and cylinder head ports. Install the new gasket onto the cylinder head. Install new gaskets onto the exhaust pipe flange.

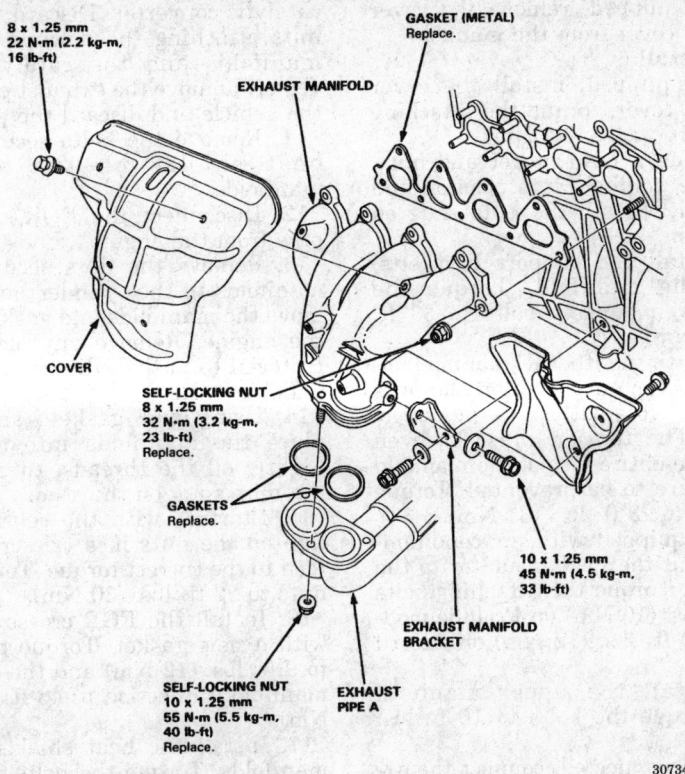

Exhaust manifold components — 1.6L (B16A3, B16A2) engines

307344

10. Install the exhaust manifold. Apply anti–seize paste to the studs. Tighten the self–locking nuts to 23 ft. lbs. (32 Nm) in a crisscross pattern starting in the center of the manifold and working outward.
11. Install the manifold brackets and torque their bolts to 17 ft. lbs. (24 Nm) for the B16A2, D15B8 and D15Z1 engines and 33 ft. lbs. (45 Nm) for all other engines.
12. Carefully coat only the threads of the oxygen sensor body with anti–seize paste. Don't get any anti–seize on the sensor probe. Install the oxygen sensor and carefully torque it to 33 ft. lbs. (45 Nm).
13. Install the heat shield and torque the bolts to 16 ft. lbs. (22 Nm).
14. Reconnect the oxygen sensor connector.
15. Raise and support the front of the vehicle and block the rear wheels.
16. Reconnect the front exhaust pipe and the exhaust manifold/catalytic converter. Torque the self–locking nuts to 40 ft. lbs. (55 Nm), if the converter is not attached to the manifold. If the converter is attached, tighten to 25 ft. lbs., (34 Nm). Install any manifold brackets and torque them to 33 ft. lbs. (45 Nm). Install the splash shield if it was removed.

17. Lower the vehicle and connect the negative battery cable.
18. Run the engine and check for exhaust leaks.

Prelude and Accord

2.2L and 2.3L Engines

1. Disconnect the negative battery cable.
2. Safely raise and support the vehicle.
3. If the oxygen sensor is located in the exhaust manifold, disconnect the oxygen sensor connector.
4. Remove the exhaust manifold upper cover.
5. If equipped with air conditioning, remove the heat insulator from the manifold.
6. Remove the nuts attaching the exhaust manifold to the front exhaust pipe. Separate the pipe from the manifold and discard the gasket. Support the pipe with wire; do not allow it to hang by itself.
7. Remove the exhaust manifold bracket(s) bolts and remove the bracket(s).
8. Using a criss–cross pattern (starting from the center), remove the exhaust manifold attaching nuts.
9. Remove the manifold and discard the gasket. Clean the manifold and cylinder head mating surfaces.

10. If equipped, remove the lower manifold cover from the manifold.

To install:

11. If equipped, install the lower manifold cover, torque the attaching bolts to 16 ft. lbs. (22 Nm).

12. Using a new gasket and nuts, place the manifold into position and support it. Install the nuts snug on the studs.

13. Install the support bracket(s) below the manifold. Torque the bracket(s) mounting bolts to 33 ft. lbs. (44 Nm).

14. Starting with the manifold inner or center nuts, tighten the nuts in a criss–cross pattern to the correct torque. The tension must be even across the entire face of the manifold if leaks are to be prevented. Torque the nuts to 23 ft. lbs. (31 Nm).

15. If equipped with air conditioning, install the heat insulator to the manifold. Torque the attaching bolts to 7 ft. lbs. (10 Nm) on Prelude models and 9 ft. lbs. (12 Nm) on Accord models.

16. Install the upper manifold cover, torque the bolts to 16 ft. lbs. (22 Nm).

17. If disconnected, connect the oxygen sensor connector.

18. Connect the front exhaust pipe, using new gaskets and nuts. Torque the exhaust pipe attaching nuts to 40 ft. lbs. (55 Nm).

19. Connect the negative battery cable and enter the radio security code.

20. Start the engine and check for exhaust leaks.

21. If equipped with 4WS, turn the steering wheel lock–to–lock to reset the 4WS control unit.

2.7L Engines

1. Disconnect the negative battery cable.

2. Remove the radiator cap and drain the cooling system into a sealable container.

3. Remove the radiator.

4. Detach the starter cable from the strut tower brace. Remove the strut tower brace.

5. If necessary for additional clearance, remove the vacuum control box on the bulkhead. Position it aside with the vacuum hoses attached.

6. Raise and safely support the vehicle.

7. Remove the front wheels, then remove the splash shield from under the engine.

8. Remove the center beam.

9. Disconnect the front oxygen sensor electrical connector.

10. Disconnect the exhaust pipe from the exhaust manifolds and the

catalytic converter. Discard the locknuts attaching the downpipe to the manifolds and the catalytic converter. Remove the exhaust pipe from the vehicle and discard the gaskets.

11. Remove the bolts securing the heat shields on the exhaust manifolds.

12. Disconnect the EGR crossover pipe from the engine.

13. Remove the nuts securing the manifolds to the cylinder heads. Remove the manifolds and gaskets from the engine. Remove any old gasket material from the cylinder heads.

To install:

14. Using new gaskets and nuts, place the manifolds into position. Lightly oil the threads, then install the nuts snug on the studs.

15. Starting with the center nuts, tighten the nuts in a crisscross pattern to the correct torque. Torque the nuts to 22 ft. lbs. (30 Nm).

16. Install the EGR crossover pipe with a new gasket. Torque the nuts to 9 ft. lbs. (12 Nm) and the exhaust manifold connection to 43 ft. lbs. (59 Nm).

17. Install the heat shields to the manifolds. Torque the bolts to 16 ft. lbs. (22 Nm).

18. Install the exhaust pipe with new nuts and gaskets. Tighten the exhaust manifold connections to 40 ft. lbs. (54 Nm) and the catalytic converter to 25 ft. lbs. (33 Nm).

19. Connect the front oxygen sensor electrical connector.

20. Install the center beam. Torque the bolts to 37 ft. lbs. (50 Nm).

21. Install the splash shield and front wheels.

22. Install the vacuum control box and the radiator.

23. Install the strut tower brace and secure the starter cable. Torque the strut tower bolts to 16 ft. lbs. (22 Nm).

24. Fill and bleed the air from the cooling system.

25. Connect the negative battery cable and enter the radio security code.

26. Start the engine and allow it to reach normal operating temperature. Check for leaks.

Front Crankshaft Seal

REMOVAL AND INSTALLATION

NOTE: The original radio may contain a coded anti–theft circuit. Obtain the security code number before disconnecting the battery cables.

1. Disconnect the negative battery cable.

2. Safely raise and support the vehicle.

3. Remove the splash shield.

4. Remove the engine accessory drive belts.

5. Turn the engine to align the timing marks and set cylinder No.1 to TDC. The white mark on the crankshaft pulley should align with the pointer on the timing belt cover. Remove the inspection caps on the upper timing belt covers to check the alignment of the timing marks. The pointers for the camshafts should align with the green marks on the camshaft sprockets.

6. Remove the upper timing belt covers and crankshaft pulley. Remove the lower timing belt cover

NOTE: Mark the direction of the timing belt's rotation if it is to be reinstalled.

7. Remove the timing belt.

8. If equipped, remove the Crankshaft Position (CKP) sensor from the oil pump, then remove the stopper plate.

9. Remove the timing belt sprocket from the crankshaft, do not lose the sprocket key.

10. Using a suitable seal removal tool, remove the seal from the front of the engine.

To install:

11. Clean the seal mounting surfaces on the engine block.

12. Apply a thin coat of grease on the crankshaft and seal lips.

13. Install the seal with the part number facing out. Use a seal driver to seat the seal against the oil pump. Clean any excess grease off the crankshaft and make sure the seal lip is not distorted.

14. Install the timing belt sprocket and key to the crankshaft.

15. Install the stopper plate and if equipped, the CKP sensor to the oil pump, torque the stopper plate and sensor mounting bolts to 9 ft. lbs. (12 Nm).

16. Verify that the engine is at TDC for the no. 1 cylinder on the compression stroke. Install and tension the timing belt.

17. Install the timing belt covers and crankshaft pulley. Torque the crankshaft pulley bolt to 181 ft. lbs. (245 Nm), with the aid of a crank pulley holder.

18. Install and adjust the accessory drive belts.

19. Verify that all engine components that may have been removed have been reinstalled correctly.

20. Install the splash shield and lower the vehicle.

21. Connect the negative battery cable.

22. Top up the engine oil if necessary.

23. Run the engine and check for leaks.

Timing Belt, Sprockets, Tensioner and Front Cover

REMOVAL AND INSTALLATION

NOTE: The radio may contain a coded theft protection circuit. Always obtain the code number before disconnecting the battery. If the vehicle is equipped with 4WS, the steering control unit is shut down when the battery is disconnected. After connecting the battery, turn the steering wheel lock–to–lock to reset the steering control unit.

Civic and del Sol

1. Disconnect the negative battery cable.

2. Remove the front splash shield.

3. Label and disconnect the ignition wires.

4. Rotate the crankshaft to set the engine at TDC/compression for the No. 1 piston. The white mark on the crankshaft pulley should align with the pointers on the timing cover. Once the engine is in this position, it must not be disturbed.

5. Loosen the power steering pump and remove the drive belt. Unbolt and remove the power steering pump, leave the inlet and outlet hoses connected. Move the pump out of the work area.

6. If equipped with A/C, unbolt the left front engine mount from the body.

7. Loosen and remove the A/C compressor idler pulley bracket. Remove the A/C compressor belt.

8. Loosen the alternator adjusting bolt and remove the alternator belt.

9. If equipped, move the cruise control actuator out of the work area.

10. Remove the valve cover. Cover the rocker arm and shaft assemblies with a towel or sheet of plastic to keep out dust and foreign objects.

11. Unbolt and remove the upper timing cover.

12. Use a floor jack with a block of wood placed on its pad to support the weight of the engine.

13. Remove the left side engine mount nuts. Loosen the mount pivot bolt. Free the mount from the studs on the engine bracket. Pivot the mount out of the way.

14. Hold the crankshaft pulley with a hex or pin type holder (depending upon engine type). Loosen the pulley bolt and remove the pulley. Remove the timing belt lower cover.

15. Remove the dipstick tube. Plug the oil pump opening to keep dirt out.

16. If equipped, unbolt the Crankshaft Speed Fluctuation (CKF) sensor from the oil pump housing. Then, disconnect and remove the sensor. Be careful not to get oil on the sensor.

17. Loosen the timing belt adjusting bolt 180°. Push the tensioner pulley down to release the belt tension. After releasing the tension, retighten the tensioner pulley bolt until snug.

NOTE: Do not remove the tensioner pulley unless it is to be replaced.

18. Remove the timing belt. Mark the direction of the belt's rotation if it is to be reinstalled.

To install:

NOTE: Inspect the water pump when replacing the timing belt — the manufacturer recommends replacing the water pump at the timing belt's service interval. Replace the timing belt if it shows any signs of wear, or if it is contaminated with oil or coolant.

19. Verify that the timing is set at TDC/compression for the No. 1 cylinder.
- The groove in the crankshaft sprocket must line up with the pointer on the oil pump.
- On 1.6L (B16A2, B16A3) engines, the TDC marks on the camshaft sprockets must line up with the pointer located between the sprockets. The TDC marks will also be in line with the upper surface of the head.
- On other engines, the TDC mark on the camshaft sprocket must line up with the pointer on the back cover.
- The UP mark on the camshaft sprocket must point up.

20. Install the timing belt onto the crankshaft sprocket, then around the adjusting pulley and water pump sprocket, and finally over the camshaft sprocket.

21. Loosen the adjusting pulley bolt 180°. Then, tighten the adjusting bolt to 40 ft. lbs. (55 Nm) 1.6L (B16A2, B16A3) engines or 33 ft. lbs. (45 Nm) on other engines.

22. Install and reconnect the CKF sensor, tighten the mounting bolt to 9 ft. lbs. (12 Nm).

23. Install the lower timing cover and the crankshaft pulley. Apply a light coat of fresh oil to the pulley bolt threads, and the tighten it to 134 ft. lbs. (181 Nm).

24. Rotate the crankshaft five to six turns counterclockwise to position the belt on the sprockets.

25. Adjust the timing belt tension:
   a. Set the No. 1 piston at TDC/compression.
   b. Loosen the adjusting pulley bolt 180°.
   c. Rotate the crankshaft counterclockwise so that the camshaft sprocket moves three teeth from the TDC/compression position.
   d. Tighten the adjusting bolt to 33 ft. lbs. (45 Nm).
   e. Tighten the crankshaft pulley to 134 ft. lbs. (181 Nm).

26. Verify that the crankshaft and camshaft sprockets will align properly at the TDC/compression position. If the camshaft pulley is not at TDC/compression, remove the timing belt, adjust the sprocket positions, and reinstall the belt.

27. Install the dipstick tube with a new O–ring.

28. Install the upper timing cover and the valve cover. Reconnect the ignition wires.

29. Reconnect the side engine mount and torque the support nuts to 54 ft. lbs. (75 Nm). Remove the jack.

30. Install the alternator belt and adjust its tension.

31. Install the A/C compressor belt and adjusting pulley. Torque the bracket bolt to 17 ft. lbs. (24 Nm). Adjust the belt tension.

32. Install the left front engine mount and tighten the bracket bolts to 33 ft. lbs. (44 Nm).

33. Install the power steering pump and belt. Torque the mounting bolts to 17 ft. lbs. (24 Nm). Adjust the belt tension.

34. If removed, install the cruise control actuator.

35. Install the splash shield.

36. Reconnect the negative battery cable. Run the engine and check its operation. Check the ignition timing.

37. Check the engine oil level and top up if necessary.

Accord and Prelude

2.2L (F22A1, F22A6) Engines

1. Disconnect the negative battery cable.

2. Turn the engine to align the timing marks and set cylinder No.1 to TDC, compression. Once in this position, the engine must NOT be turned or disturbed.

3. Remove the splash shield from below the engine.

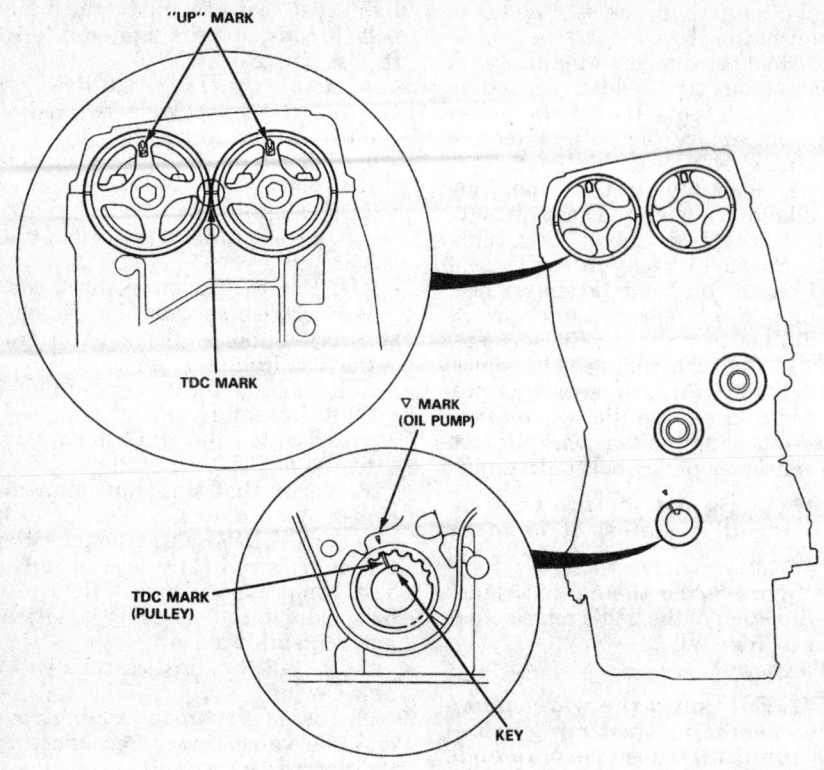

Crankshaft sprocket timing marks — 1.6L (B16A2, B16A3) engines

4. Disconnect the electrical connector at the cruise control actuator, then remove the actuator. Don't disconnect the cable; simply move the actuator out of the work area.

5. Remove the belt from the power steering pump. Remove the mounting bolts for the pump. Without disconnecting the hoses, move the pump out of the way.

6. Disconnect the alternator wiring and connectors; remove the engine wiring harness from the valve cover.

7. Loosen the adjusting and mounting bolts for the alternator and remove the drive belt.

8. Remove the cylinder head cover.

9. Remove the side engine mount support bracket, if so equipped.

10. Remove the upper timing belt cover.

11. Support the engine with a floor jack below the center of the center beam. Tension the jack so that it is just supporting the beam but not lifting it.

12. Remove the through-bolt for the side engine mount and remove the mount.

13. Remove the dipstick and dipstick tube.

14. Remove the crankshaft pulley bolt and remove the crankshaft pulley. Use a crank pulley holder (part #

07MAB–PY3010A) and holder handle (part # 07JAB–001020A) to hold the crankshaft pulley in place while removing the bolt.

15. Remove the rubber seal from the timing belt adjusting nut.

16. Remove the lower timing belt cover.

17. There are two belts in this system; the one running to the camshaft pulley is the timing belt. The other, shorter one drives the balance shafts and is referred to as the balancer belt or timing balancer belt. Lock the timing belt adjuster in position by installing one of the lower timing belt cover bolts to the adjuster arm.

18. Loosen the timing belt and balancer shafts tensioner adjuster nut, but do not loosen the nut more than one turn. Push the tensioner for the balancer belt away from the belt to relieve the tension. Hold the tensioner and tighten the adjusting nut to hold the tensioner in place.

19. Carefully remove the balancer belt. Do not crimp or bend the belt; protect it from contact with oil or coolant. Slide the belt off the pulleys.

20. Remove the balancer belt drive sprocket from the crankshaft.

21. Loosen the lockbolt installed to the timing belt adjuster and loosen the adjusting nut. Push the timing belt adjuster to remove the tension

on the timing belt, then tighten the adjuster nut.

22. Remove the timing belt. Do not crimp or bend the belt; protect it from contact with oil or coolant. Slide the belt off the pulleys.

23. Remove the bolt attaching the camshaft sprocket and remove the sprocket. Do not lose the sprocket key.

24. Remove the bolts attaching the Crankshaft Position/Top Dead Center (CKP/TDC) sensor and remove the sensor.

25. Remove the timing belt sprocket from the crankshaft, do not lose the key from the crankshaft.

26. Remove the belt tensioners by performing the following:

   a. Remove the springs from the balancer belt and the timing belt tensioners.

   b. Remove the adjusting nut.

   c. Remove the bolt from the balancer belt adjuster lever then remove the lever and the tensioner pulley.

   d. Remove the lockbolt from the timing belt tensioner lever then remove the tensioner pulley and lever from the engine.

27. This is an excellent time to check or replace the water pump. Even if the timing belt is only being replaced as part of a good maintenance schedule, consider replacing the pump at the same time.

To install:

28. If the water pump is to be replaced, install a new O-ring and make certain it is properly seated. Install the water pump and retaining bolts. Tighten the mounting bolts to 9 ft. lbs (12 Nm).

29. If the tensioners were removed perform the following to install them:

   a. Install the timing belt tensioner lever and tensioner pulley.

NOTE: The tensioner lever must be properly positioned on it's pivot pin located on the oil pump. Make sure that the timing belt lever and tensioner moves freely and does not bind.

   b. Install the lockbolt to the timing belt tensioner, do not tighten the lockbolt at this time.

   c. Install the balancer belt pulley and adjuster lever.

   d. Install the adjusting nut and the bolt to the balancer belt adjuster lever. Do not tighten the adjuster nut or bolt at this time.

NOTE: Make sure that the balancer lever and tensioner moves freely and does not bind.

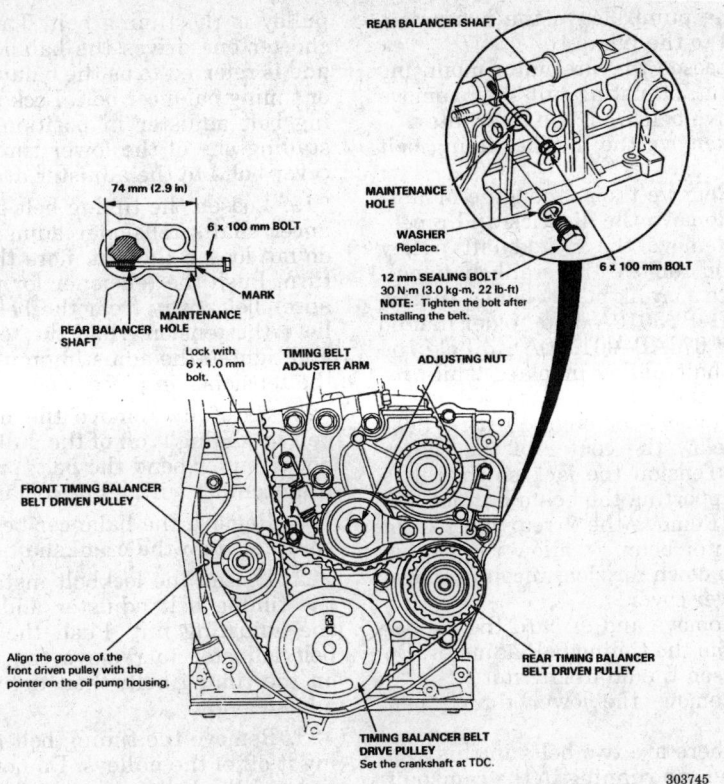

Balancer belt alignment — 2.2L (F22A1, F22A6) engines

e. Install the springs to the tensioners.

f. Move the timing belt tensioner its full deflection and tighten the lockbolt.

g. Move the balancer its full deflection and tighten the adjusting nut.

30. Install the key to the crankshaft and install the timing belt sprocket.

31. Install the CKP/TDC sensors and torque the bolts to 9 ft. lbs. (12 Nm). Connect the CKP/TDC sensors connector.

32. Install the key to the camshaft and install the camshaft timing belt sprocket. Install the attaching bolt and torque the bolt to 27 ft. lbs. (37 Nm).

33. The crankshaft timing pointer must be perfectly aligned with the white mark on the flywheel or flex–plate; the camshaft pulley must be aligned so that the word UP is at the top of the pulley and the marks on the edge of the pulley are aligned with the surfaces of the head.

34. Install the timing belt over the pulleys and tensioners.

35. Loosen the bolt used to lock the timing belt tensioner. Loosen then tighten the timing belt adjusting nut.

36. Turn the crankshaft counter-clockwise until the cam pulley has moved 3 teeth; this creates tension on the timing belt. Loosen, then tighten the adjusting nut and torque it to 33 ft. lbs (45 Nm). Tighten the bolt used to lock the timing belt tensioner.

37. Realign the timing belt timing marks then install the balancer belt drive sprocket to the crankshaft.

38. Align the front balancer pulley; the face of the front timing balancer pulley has a mark which must be aligned with the notch on the oil pump body. This pulley is the one at 10 o'clock to the crank pulley when viewed from the pulley end.

39. Align the rear timing balancer pulley (2 o'clock from the crank pulley) using a 6 x 100mm bolt or rod. Mark the bolt or rod at a point 2.9 inches (74mm) from the end. Remove the bolt from the maintenance hole on the side of the block; insert the bolt or rod into the hole. Align the 74mm mark with the face of the hole. This pin will hold the shaft in place during installation.

40. Install the balancer belt. Once the belts are in place, make sure that all the engine alignment marks are still correct. If not, remove the belts, realign the engine and reinstall the belts. Once the belts are properly installed, slowly loosen the adjusting nut, allowing the tensioner to move against the belt. Remove the pin from the maintenance hole and reinstall the bolt and washer.

41. Turn the crankshaft one turn then tighten the adjuster nut to 33 ft. lbs. (45 Nm). Remove the bolt used to lock the timing belt tensioner.

42. Install the lower cover, making certain the rubber seals are in place and correctly located. Tighten the attaching bolts to 9 ft. lbs (12 Nm). Install a new seal around the adjusting nut, DO NOT loosen the adjusting nut.

43. Install the key on the crankshaft and install the crankshaft pulley. Apply oil to the bolt threads and tighten it to 181 ft. lbs. (250 Nm).

44. Install the dipstick tube and dipstick. Torque the attaching bolt to 9 ft. lbs. (12 Nm).

45. Install the side engine mount. Tighten the bolt and nut attaching the mount to the engine to 40 ft. lbs. (55 Nm). Torque the through nut and bolt to 47 ft. lbs. (65 Nm). Remove the jack from under the center beam.

46. Adjust the valve clearances.

47. Install the upper timing belt cover and torque the attaching bolts to 9 ft. lbs. (12 Nm).

48. Install the side engine mount support bracket if it was removed.

49. Install the cylinder head cover and torque the cap nuts to 7 ft. lbs. (10 Nm). Replace the rubber seals if they are deteriorated or damaged.

50. Install the compressor and/or alternator drive belt; adjust the tension.

51. Route the wiring harness over the valve cover and connect the wiring to the alternator.

52. Install the power steering pump and drive belt.

53. Install the cruise control actuator, torque the mounting bolts to 7 ft. lbs. (10 Nm). Connect the vacuum hose and the electrical connector.

54. Install the splash shield under the engine.

55. Connect the negative battery cable and enter the radio security code.

56. Start the engine, allowing it to idle. Check for any signs of leakage or any sound of the belts rubbing or binding.

57. If equipped with 4WS, turn the steering wheel lock–to–lock to reset the 4WS control unit.

2.3L (H23A1) Engine

1. Disconnect the negative battery cable.

2. Turn the crankshaft so the No. 1 piston is at top dead center.

NOTE: The No. 1 piston is at top dead center when the pointer on the block aligns with the white painted mark on the flywheel (manual transaxle) or driveplate (automatic transaxle).

3. Disconnect the alternator terminal and connector. Then remove the engine wire harness from the cylinder head cover.

4. Remove the cylinder head (valve) cover.

5. Ensure the words **UP** embossed on the camshaft pulleys are aligned in the upward position.

6. Insert a 5.0mm pin punch in each of the camshaft caps, nearest to the pulleys, through the holes provided.

7. Remove the lower splash shield.

8. Disconnect the electrical connector from the cruise control actuator. Remove the actuator leaving the cable connected.

NOTE: Do not bend the cable, replace if kinked.

9. Remove the power steering pump drive belt. Remove the power steering pump, leave the hoses connected to the pump.

10. Loosen the alternator mounting bolt, nut, adjusting nut then remove the drive belt from the alternator.

11. Remove the middle timing belt cover.

12. Remove the side engine mount.

13. Remove the dipstick and pipe.

14. Remove the crankshaft pulley bolt and remove the crankshaft pulley. Use a crank pulley holder (part # 07MAB–PY3010A) and holder handle (part # 07JAB–001020A) to hold the crankshaft pulley in place while removing the bolt.

15. Support the engine with a floor jack below the center of the center beam. Tension the jack so that it is just supporting the beam but not lifting it. Remove the 2 rear bolts from the center beam to allow the engine to drop down for clearance to remove the lower cover.

16. Remove and discard the rubber seal from the timing belt adjuster. Do not loosen the adjusting nut.

17. Remove the lower timing belt cover.

18. There are two belts in this system; the one running to the camshaft pulley is the timing belt. The other, shorter one drives the balance shaft and is referred to as the balancer belt or timing balancer belt. Lock the timing belt adjuster in position, by installing one of the lower timing belt cover bolts to the adjuster arm.

19. Loosen the timing belt and balancer shafts tensioner adjuster nut, do not loosen the nut more than one turn. Push the tensioner for the balancer belt away from the belt to relieve the tension. Hold the tensioner and tighten the adjusting nut to hold the tensioner in place.

20. Carefully remove the balancer belt by sliding it off of the pulleys. Do not crimp or bend the belt; protect it from contact with oil or coolant.

21. Remove the balancer belt drive sprocket from the crankshaft.

22. Loosen the lockbolt installed to the timing belt adjuster and loosen the adjusting nut. Push the timing belt adjuster to remove the tension on the timing belt, then tighten the adjuster nut.

23. Remove the timing belt by sliding it off of the pulleys. Do not crimp or bend the belt; protect it from contact with oil or coolant.

24. Remove the bolts attaching the camshaft sprockets and remove the sprockets. Do not lose the sprocket keys.

25. Remove the timing belt sprocket from the crankshaft, do not lose the key from the crankshaft.

26. Remove the belt tensioners by performing the following:

a. Remove the springs from the balancer belt and the timing belt tensioners.

b. Remove the adjusting nut.

c. Remove the bolt from the balancer belt adjuster lever then remove the lever and the tensioner pulley.

d. Remove the lockbolt from the timing belt tensioner lever then remove the tensioner pulley and lever from the engine.

27. This is an excellent time to check or replace the water pump. Even if the timing belt is only being replaced as part of a good maintenance schedule, consider replacing the pump at the same time.

To install:

28. If the water pump is to be replaced, install a new O-ring and make certain it is properly seated. Install the water pump and retaining bolts. Tighten the mounting bolts to 9 ft. lbs (12 Nm).

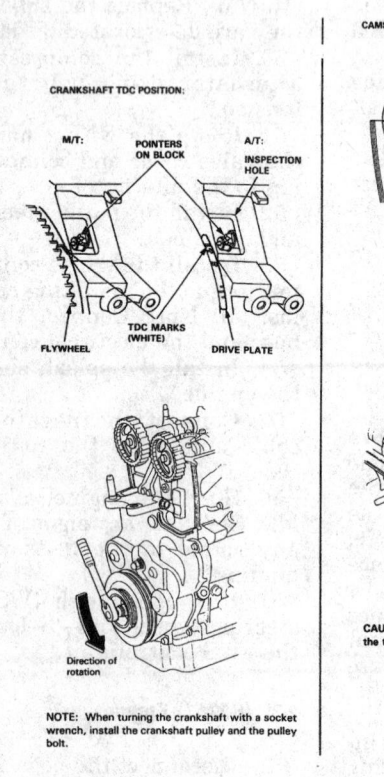
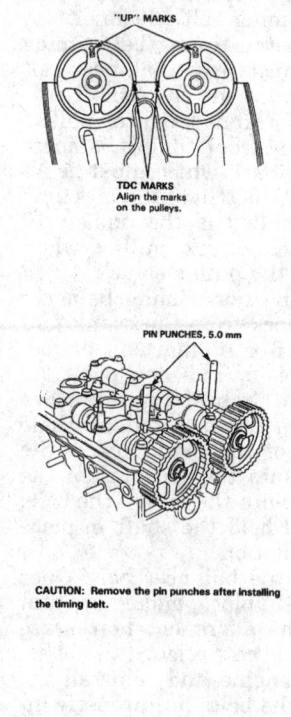

Positioning the crankshaft and camshafts to top dead center — 2.3L (H23A1) engine

29. If the tensioners were removed perform the following to install them:

    a. Install the timing belt tensioner lever and tensioner pulley.

NOTE: The tensioner lever must be properly positioned on it's pivot pin located on the oil pump. Make sure that the timing belt lever and tensioner moves freely and does not bind.

    b. Install the lockbolt to the timing belt tensioner, do not tighten the lockbolt at this time.

    c. Install the balancer belt pulley and adjuster lever.

    d. Install the adjusting nut and the bolt to the balancer belt adjuster lever. Do not tighten the adjuster nut or bolt at this time.

NOTE: Make sure that the balancer lever and tensioner moves freely and does not bind.

    e. Install the springs to the tensioners.

    f. Move the timing belt tensioner it's full deflection and tighten the lockbolt.

    g. Move the balancer it's full deflection and tighten the adjusting nut.

30. Install the key to the crankshaft and install the timing belt sprocket.

31. Install the key to the camshafts and install the camshaft timing belt sprockets. Install the attaching bolts and torque the bolts to 27 ft. lbs. (38 Nm).

32. The crankshaft timing pointer must be perfectly aligned with the white mark on the flywheel or flex–plate; the camshaft pulley must be aligned so that the word UP is at the top of the pulley and the marks on the edge of the pulley are aligned with the surfaces of the head.

33. Install the timing belt.

34. Install the balancer belt drive sprocket to the crankshaft.

35. Remove the two 5.0mm pin punches from the camshaft bearing caps.

36. Loosen the bolt used to lock the timing belt tensioner. Loosen then tighten the timing belt adjuster nut.

37. Turn the crankshaft counterclockwise until the cam pulley has moved 3 teeth; this creates tension on the timing belt. Loosen then tighten the adjusting nut and torque it to 33 ft. lbs (45 Nm). Tighten the bolt used to lock the timing belt tensioner.

38. Realign the timing belt timing marks.

39. Align the groove on the front balancer shaft pulley with the pointer on the oil pump.

40. Align the rear timing balancer pulley using a 6 x 100mm bolt or rod. Mark the bolt or rod at a point 2.913 inches (74mm) from the end. Remove the bolt from the maintenance hole on the side of the block; insert the bolt/rod into the hole and align the 74mm mark with the face of the hole. This pin will hold the shaft in place during installation.

41. Loosen the adjusting nut and ensure the timing balancer belt adjuster moves freely.

42. Install the balancer belt. Once the belts are in place, make sure that all the engine alignment marks are still correct. If not, remove the belts, realign the engine and reinstall the belts. Once the belts are properly installed, slowly loosen the adjusting nut, allowing the tensioner to move against the belt. Remove the pin from the maintenance hole and reinstall the bolt and washer.

43. Turn the crankshaft pulley 1 turn and torque the adjusting nut to 33 ft. lbs. (45 Nm).

NOTE: Both belt adjusters are spring loaded to properly tension the belts. Do not apply extra pressure to the pulleys or tensioners while performing the adjustment.

44. Remove the 6 x 10mm bolt from the timing belt adjuster arm.

45. Install the lower timing belt cover and torque the bolts to 9 ft. lbs. (12 Nm).

46. Install a new seal around the adjusting nut. Do not loosen the adjusting nut.

47. Install the crankshaft pulley. Coat the threads and seating face of the pulley bolt with engine oil, then install and torque the bolt to 181 ft. lbs. (250 Nm).

48. Position the rear center beam then, install the 2 rear center beam bolts.

49. Install a new O-ring to the dipstick tube. Install the tube and dipstick assembly and torque the mounting bolt to 9 ft. lbs. (12 Nm).

50. Install the side engine mount. Tighten the bolt and nut attaching the mount to the engine to 40 ft. lbs (55 Nm). Torque the through nut and bolt to 47 ft. lbs. (65 Nm). Remove the jack from under the center beam.

51. Install the middle timing belt cover and torque the bolts to 9 ft. lbs. (12 Nm).

52. Install and properly tension the alternator belt.

53. Install the cylinder head (valve) cover with a new gasket and seals.

54. Install the engine wire harness to the valve cover and connect the terminals.

55. Install the power steering pump.

56. Install and properly tension the power steering pump belt.

57. Install the cruise control actuator and torque the mounting bolts to 7 ft. lbs. (10 Nm). Connect the electrical connector.

58. Install the splash shield.

59. Connect the negative battery cable and enter the radio security code.

60. Start the engine, allowing it to idle. Check for any signs of leakage or any sound of the belts rubbing or binding.

61. If equipped with 4WS, turn the steering wheel lock–to–lock to reset the 4WS control unit.

2.2L (H22A1) Engine

1. Disconnect the negative battery cable.

2. Turn the crankshaft so the No. 1 piston is at top dead center. The No. 1 piston is at top dead center when the pointer on the block aligns with the white painted mark on the driveplate.

3. Disconnect the alternator terminal and connector. Then remove the engine wire harness from the cylinder head cover.

4. Remove the cylinder head (valve) cover.

5. Ensure the words UP embossed on the camshaft pulleys are aligned in the upward position.

6. Remove the lower splash shield.

7. Disconnect the electrical connector from the cruise control actuator. Remove the actuator leaving the cable connected.

NOTE: Do not bend the cable, replace if kinked.

8. Remove the power steering pump drive belt. Remove the power steering pump, leave the hoses connected to the pump.

9. Loosen the alternator mounting bolt, nut, adjusting nut then remove the drive belt from the alternator.

10. Remove the middle timing belt cover.

11. Remove the side engine mount.

12. Remove the dipstick and pipe.

13. Remove the crankshaft pulley bolt and remove the crankshaft pulley. Use a crank pulley holder (part # 07MAB–PY3010A) and holder handle (part # 07JAB–001020A) to hold the crankshaft pulley in place while removing the bolt.

14. Support the engine with a floor jack below the center of the center

beam. Tension the jack so that it is just supporting the beam but not lifting it. Remove the 2 rear bolts from the center beam to allow the engine to drop down for clearance to remove the lower cover.

15. Remove and discard the rubber seal from the timing belt adjuster. Do not loosen the adjusting nut.

16. Remove the lock pin from the maintenance bolt.

17. Remove the lower timing belt cover.

18. There are two belts in this system; the one running to the camshaft pulley is the timing belt. The other, shorter one drives the balance shafts and is referred to as the balancer belt or timing balancer belt.

19. Loosen the balancer shafts tensioner adjusting nut, do not loosen the nut more than one turn. Push the tensioner for the balancer belt away from the belt to relieve the tension. Hold the tensioner and tighten the adjusting nut to hold the tensioner in place.

20. Carefully remove the balancer belt by sliding it off of the pulleys. Do not crimp or bend the belt; protect it from contact with oil or coolant.

21. Remove the balancer belt drive sprocket from the crankshaft.

22. Remove the bolts attaching the Crankshaft Position/Top Dead Center (CKP/TDC) sensor and remove the sensor.

23. Remove the timing belt by sliding it off of the pulleys. Do not crimp or bend the belt; protect it from contact with oil or coolant.

24. Remove the two bolts mounting the timing belt auto–tensioner and remove the tensioner from the vehicle.

25. Remove the bolts attaching the camshaft sprockets and remove the sprockets. Do not lose the sprocket keys.

26. Remove the timing belt sprocket from the crankshaft, do not lose the key from the crankshaft.

27. Remove the balancer belt tensioner by performing the following:

a. Remove the spring from the balancer belt tensioner.

b. Remove the adjusting nut.

c. Remove the bolt from the balancer belt adjuster lever then remove the lever and the tensioner pulley.

28. This is an excellent time to check or replace the water pump. Even if the timing belt is only being replaced as part of a good maintenance schedule, consider replacing the pump at the same time.

To install:

29. If the water pump is to be replaced, install a new O-ring and make certain it is properly seated. Install the water pump and retaining bolts. Tighten the mounting bolts to 9 ft. lbs (12 Nm).

30. If the balancer tensioner was removed perform the following to install it:

a. Install the balancer belt pulley and adjuster lever.

b. Install the adjusting nut and the bolt to the balancer belt adjuster lever. Do not tighten the adjuster nut or bolt at this time.

NOTE: Make sure that the balancer lever and tensioner moves freely and does not bind.

c. Install the spring to the tensioner.

d. Move the balancer its full deflection and tighten the adjusting nut.

31. Install the key to the crankshaft and install the timing belt sprocket.

32. Install the key to the camshafts and install the camshaft timing belt sprockets. Install the attaching bolts and torque the bolts to 37 ft. lbs. (51 Nm).

33. Hold the auto–tensioner with the maintenance bolt pointing up. Remove the maintenance bolt and discard the gasket.

NOTE: Handle the carefully so the oil inside does not spill or leak. Replenish the auto tensioner with oil if any spills or leaks out. The auto — tensioner total capacity is 1/4 ounce (8 ml).

34. Clamp the mounting boss of the auto–tensioner in a vise. Use pieces of wood or a cloth to protect the mounting boss.

--- **WARNING** ---
Do not clamp the housing of the auto–tensioner, component damage may occur.

35. Insert a flat bladed screwdriver into the maintenance hole. Place the stopper (part # 14540–P13–003) on the auto–tensioner while turning the screw driver clockwise to compress the tensioner. Take care not to damage the threads or the gasket contact surface with the screw driver.

36. Remove the screw driver and install the maintenance bolt with a new gasket. Torque the maintenance bolt to 6 ft. lbs. (8 Nm).

37. Make sure no oil is leaking from the maintenance bolt and install the auto–tensioner to the engine. Torque

the auto–tensioner mounting bolts to 16 ft. lbs. (22 Nm).

38. The pointer on the crankshaft pulley should be aligned with the pointer on the oil pump; the camshaft pulley must be aligned so that the word UP is at the top of the pulley and the marks on the edge of the pulley are aligned with the surfaces of the head.

39. Install the timing belt.

40. Remove the stopper from the timing belt adjuster.

41. Install the CKP/TDC sensors and torque the bolts to 9 ft. lbs. (12 Nm). Connect the CKP/TDC sensors connector.

42. Install the balancer belt drive sprocket to the crankshaft.

43. Align the groove on the front balancer shaft pulley with the pointer on the oil pump.

44. Align the rear timing balancer pulley using a 6 x 100mm bolt or rod. Mark the bolt or rod at a point 2.913 inches (74mm) from the end. Remove the bolt from the maintenance hole on the side of the block; insert the bolt/rod into the hole and align the 74mm mark with the face of the hole. This pin will hold the shaft in place during installation.

45. Ensure the timing balancer belt adjuster moves freely.

46. Install the balancer belt. Once the belts are in place, make sure that all the engine alignment marks are still correct. If not, remove the belts, realign the engine and reinstall the belts. Once the belts are properly installed, slowly loosen the adjusting nut, allowing the tensioner to move against the belt. Remove the pin from the maintenance hole and reinstall the bolt and washer.

47. Turn the crankshaft pulley 1 turn, then torque the adjusting nut to 33 ft. lbs. (45 Nm).

--- **WARNING** ---
Do not apply extra pressure to the pulleys or tensioners while performing the adjustment.

48. Install the lower timing belt cover and torque the bolts to 9 ft. lbs. (12 Nm).

49. Install a new seal around the adjusting nut. Do not loosen the adjusting nut.

50. Install the lock pin to the maintenance bolt.

51. Install the crankshaft pulley. Coat the threads and seating face of the pulley bolt with engine oil. Install and torque the bolt to 181 ft. lbs. (250 Nm).

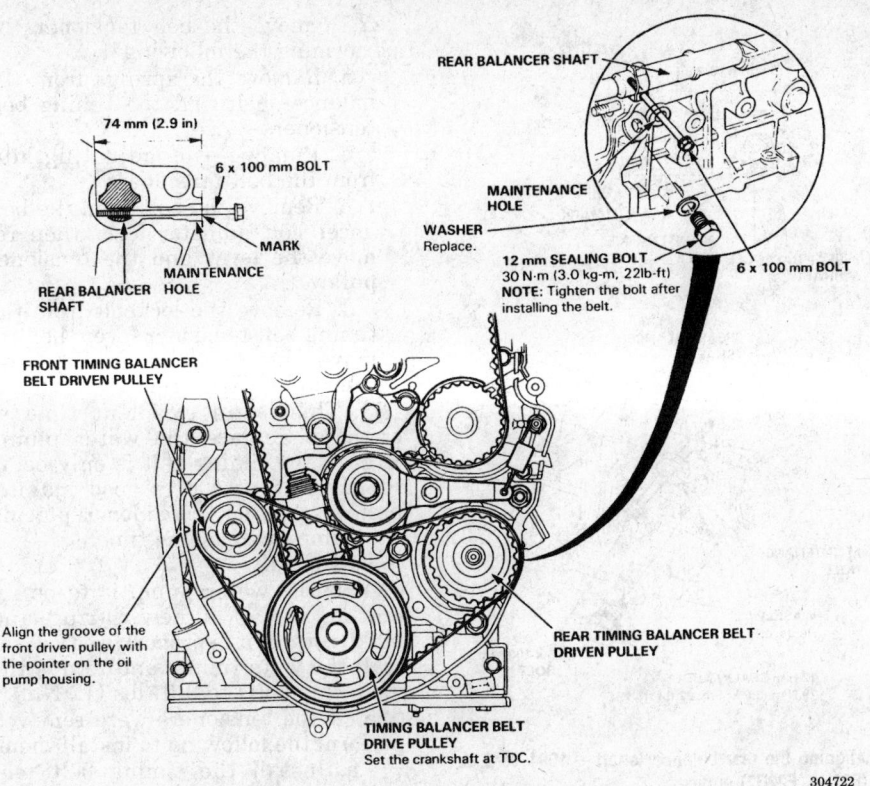

Timing belts installed and aligned — 1996–97 with 2.2L (H22A1) engine

304722

52. Position the rear center beam then, install the 2 rear center beam bolts.

53. Install a new O-ring to the dipstick tube. Install the tube and dipstick assembly and torque the mounting bolt to 9 ft. lbs. (12 Nm).

54. Install the side engine mount. Tighten the bolt and nut attaching the mount to the engine to 40 ft. lbs (55 Nm). Torque the through nut and bolt to 47 ft. lbs. (65 Nm). Remove the jack from under the center beam.

55. Install the middle timing belt cover and torque the bolts to 9 ft. lbs. (12 Nm).

56. Install and properly tension the alternator belt.

57. Install the cylinder head (valve) cover with a new gasket and seals.

58. Install the engine wire harness to the valve cover and connect the terminals.

59. Install the power steering pump.

60. Install and properly tension the power steering pump belt.

61. Install the cruise control actuator and torque the mounting bolts to 7 ft. lbs. (10 Nm). Connect the electrical connector.

62. Install the splash shield.

63. Connect the negative battery cable and enter the radio security code.

64. Start the engine, allowing it to idle. Check for any signs of leakage or any sound of the belts rubbing or binding.

65. If equipped with 4WS, turn the steering wheel lock–to–lock to reset the 4WS control unit.

2.2L (F22B1, F22B2) Engines

1. Disconnect the negative battery cable, then the positive battery cable.

2. Remove the cylinder head (valve) cover.

3. Remove the upper timing belt cover.

4. Turn the engine to align the timing marks and set cylinder No.1 to TDC. The white mark on the crankshaft sprocket should align with the pointer on the timing belt cover. The words **UP** embossed on the camshaft sprocket should be aligned in the upward position. The marks on the edge of the sprocket should be aligned with the cylinder head or the back cover upper edge. Once in this position, the engine must NOT be turned or disturbed.

5. Remove the splash shield from below the engine.

6. Remove the belt from the power steering pump.

7. Loosen the adjusting and mounting bolts for the alternator and remove the drive belt.

8. Remove the terminal and connector from the alternator.

9. Remove the through–bolt for the side engine mount and remove the mount. Use a jack to support the engine before the mount is removed. Place a cushion between the oil pan and the jack.

10. Remove the dipstick and dipstick tube.

11. Remove the crankshaft sprocket bolt and remove the crankshaft sprocket. Use a crank sprocket holder (part # 07MAB–PY3010A) and holder handle (part # 07JAB–001020A) to hold the crankshaft sprocket in place while removing the bolt.

12. Remove the rubber seal from the timing belt adjusting nut.

13. Remove the lower timing belt cover.

14. There are two belts in this system; the one running to the camshaft sprocket is the timing belt. The other, shorter one drives the balance shaft and is referred to as the balancer shaft belt or timing balancer belt. Lock the timing belt adjuster in position, by installing one of the lower timing belt cover bolts to the adjuster arm.

15. Loosen the timing belt and balancer shafts tensioner adjuster nut, do not loosen the nut more than one turn. Push the tensioner for the balancer belt away from the belt to relieve the tension. Hold the tensioner and tighten the adjusting nut to hold the tensioner in place.

16. Carefully remove the balancer belt. Do not crimp or bend the belt; protect it from contact with oil or coolant.

17. Remove the balancer belt sprocket from the crankshaft.

18. Loosen the lockbolt installed to the timing belt adjuster and loosen the adjusting nut. Push the timing belt adjuster to remove the tension on the timing belt, then tighten the adjuster nut.

19. Remove the timing belt by sliding it off the sprockets. Do not crimp or bend the belt; protect it from contact with oil or coolant.

20. Remove bolt attaching the camshaft sprocket then remove the sprocket from the camshaft. Do not loose the camshaft sprocket key.

21. Remove the Crankshaft Position/Top Dead Center (CKP/TDC) sensors then remove the timing belt sprocket from the crankshaft. Do not loose the crankshaft sprocket key.

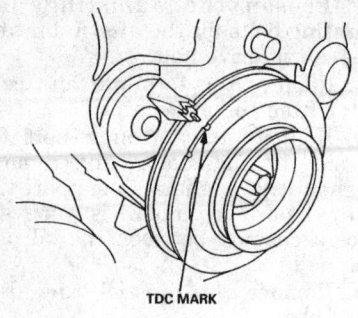

TDC MARK

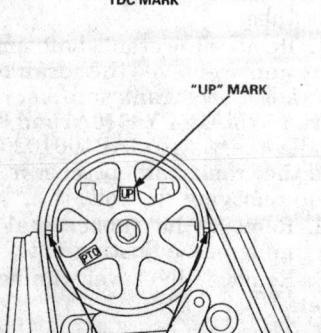
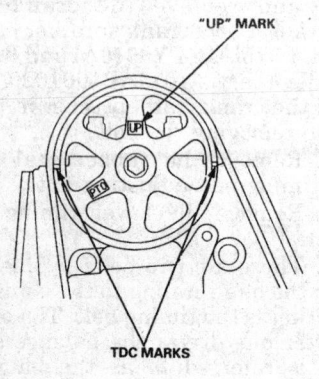

"UP" MARK

TDC MARKS

304158

Alignment of the timing marks — 1994–95 2.2L (F22B1, F22B2) engine

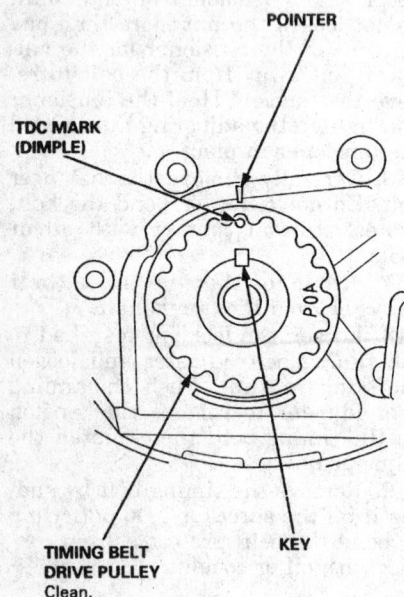

POINTER

TDC MARK (DIMPLE)

TIMING BELT DRIVE PULLEY Clean.

KEY

304159

Crankshaft timing belt pulley alignment — 1994–95 2.2L (F22B1, F22B2) engine

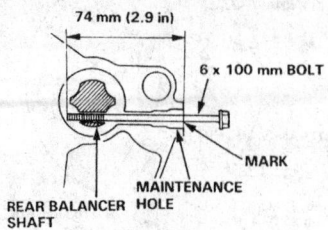

74 mm (2.9 in)

6 x 100 mm BOLT

MARK

REAR BALANCER SHAFT

MAINTENANCE HOLE

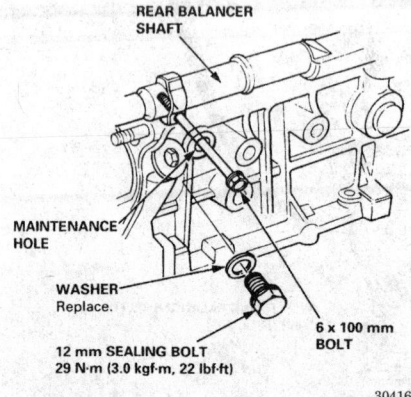

REAR BALANCER SHAFT

MAINTENANCE HOLE

WASHER Replace.

12 mm SEALING BOLT 29 N·m (3.0 kgf·m, 22 lbf·ft)

6 x 100 mm BOLT

304162

Aligning the rear balancer shaft — 1994–95 2.2L (F22B1, F22B2) engine

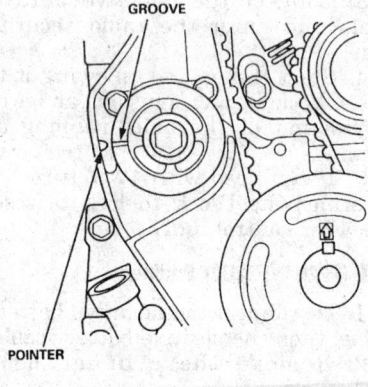

GROOVE

POINTER

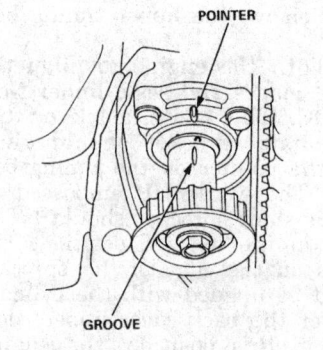

POINTER

GROOVE

304160

Front balancer shaft alignment — 1994–95 2.2L (F22B1, F22B2) engine

22. Remove the belt tensioners by performing the following:

a. Remove the springs from the balancer belt and the timing belt tensioners.

b. Remove the adjusting nut from the belt tensioners.

c. Remove the bolt from the balancer belt adjuster lever then remove the lever and the tensioner pulley.

d. Remove the lockbolt from the timing belt tensioner lever then remove the tensioner pulley and lever from the engine.

23. This is an excellent time to check or replace the water pump. Even if the timing belt is only being replaced as part of a good maintenance schedule, consider replacing the pump at the same time.

To install:

24. If the water pump is to be replaced, install a new O-ring and make certain it is properly seated. Install the water pump and torque the mounting bolts to 9 ft. lbs (12 Nm).

25. If the tensioners were removed perform the following to install them:

a. Install the timing belt tensioner lever and the tensioner pulley.

b. Install the balancer belt pulley and adjuster lever.

c. Install the adjusting nut and the bolt to the balancer belt adjuster lever.

d. Install the springs to the tensioners.

e. Install the lockbolt to the timing belt tensioner then move it it's full deflection and tighten the lockbolt.

f. Move the balancer it's full deflection and tighten the adjusting nut to hold it's position.

26. Install the key to the crankshaft and install the timing belt sprocket.

27. Install the CKP/TDC sensors and torque the mounting bolts to 9 ft. lbs. (12 Nm).

28. Install the key to the camshaft and install the camshaft timing belt sprocket. Install the attaching bolt and torque the bolt to 27 ft. lbs. (37 Nm).

29. The pointer on the crankshaft sprocket should be aligned with the pointer on the oil pump; the camshaft sprocket must be aligned so that the word UP is at the top of the sprocket and the marks on the edge of the sprocket are aligned with the surfaces of the head or the back cover upper edge.

30. Install the timing belt in the following sequence:

a. To the crankshaft.

b. The adjusting sprocket.

c. The water pump sprocket.

d. The camshaft sprocket.

31. Check the timing marks to make sure that they did not move.

32. Loosen then retighten the timing belt adjusting nut.

33. Install the timing balancer belt drive sprocket and the lower timing belt cover.

34. Install the crankshaft pulley and bolt, torque the bolt to 181 ft. lbs. (245 Nm). Rotate the crankshaft sprocket five or six turns to position the timing belt on the sprockets.

35. Set the No. 1 cylinder to TDC and loosen the timing belt adjusting nut one turn. Turn the crankshaft counterclockwise until the cam sprocket has moved 3 teeth; this creates tension on the timing belt.

36. Tighten the timing belt adjusting nut.

37. Set the crankshaft sprocket and the camshaft sprocket to TDC. If the sprockets do not align, remove the belt to realign the marks then install the belt.

38. Remove the crankshaft pulley and the lower cover.

39. With the timing marks aligned, lock the timing belt adjuster in place with one of the lower cover mounting bolts.

40. Loosen the adjusting nut and ensure the timing balancer belt adjuster moves freely.

41. Align the rear timing balancer sprocket using a 6·100mm bolt or rod. Mark the bolt or rod at a point 2.9 inches (74mm) from the end. Remove the bolt from the maintenance hole on the side of the block; insert the bolt/rod into the hole and align the 74mm mark with the face of the hole. This will hold the shaft in place during installation.

42. Align the groove on the front balancer shaft sprocket with the pointer on the oil pump.

43. Install the balancer belt. Once the belts are in place, make sure that all the engine alignment marks are still correct. If not, remove the belts, realign the engine and reinstall the belts. Once the belts are properly installed, slowly loosen the adjusting nut, allowing the tensioner to move against the belt. Remove the bolt from the maintenance hole and reinstall the bolt and washer.

44. Install the crankshaft pulley, then turn the crankshaft sprocket 1 turn counterclockwise and torque the timing belt adjusting nut to 33 ft. lbs. (45 Nm).

45. Remove the crankshaft pulley and the bolt locking the timing belt adjuster in place.

46. Install the lower timing belt cover and torque the bolts to 9 ft. lbs. (12 Nm).

47. Install a new seal around the adjusting nut. Do not loosen the adjusting nut.

48. Install the crankshaft pulley. Coat the threads and seating face of the pulley bolt with engine oil, then install and torque the bolt to 181 ft. lbs. (250 Nm).

49. Install a new O-ring to the dipstick tube. Install the tube and dipstick assembly and torque the mounting bolt to 9 ft. lbs. (12 Nm).

50. Install the side engine mount. Torque the bolt and nut attaching the mount to the engine to 40 ft. lbs (55 Nm). Torque the through nut and bolt to 47 ft. lbs. (65 Nm), then remove the jack from under the engine.

51. Install the upper timing belt cover. Torque the bolt toward the exhaust manifold to 7 ft. lbs. (10 Nm) and torque the bolt toward the intake manifold to 9 ft. lbs. (12 Nm).

52. Install and properly tension the alternator belt.

53. Install the cylinder head cover gasket cover to the groove of the cylinder head cover. Before installing the gasket thoroughly clean the seal and the groove. Seat the recesses for the camshaft first, then work it into the groove around the outside edges. Make sure the gasket is seated securely in the corners of the recesses.

54. Apply liquid gasket to the four corners of the recesses of the cylinder head cover gasket. Do not install the parts if 5 minutes or more have elapsed since applying liquid gasket. After assembly, wait at least 20 minutes before filling the engine with oil.

55. Install the cylinder head (valve) cover. Torque the bolts attaching the cylinder head cover in two or three steps to the proper sequence of 7 ft. lbs. (10 Nm).

56. Install and properly tension the power steering pump belt.

57. Connect the negative battery cable and enter the radio security code.

58. Start the engine, allowing it to idle. Check for any signs of leakage or any sound of the belts rubbing or binding.

2.7L Engines

1. Disconnect the negative battery cable, then the positive battery cable.

2. Turn the engine to align the timing marks and set cylinder No.1 to TDC. The white mark on the crankshaft pulley should align with the pointer on the timing belt cover. Remove the inspection caps on the upper timing belt covers to check the alignment of the timing marks. The pointers for the camshafts should align with the green marks on the camshaft sprockets.

3. Raise and safely support the vehicle.

4. Remove the front wheels and the engine splash shield.

5. Use a jack to support the engine. Be sure to place a cushion between the oil pan and the jack. Disconnect the side engine mount from the engine and rotate the mount out of the way.

6. Loosen the idler pulley center nut and adjusting bolt, then remove the air conditioning compressor belt.

7. Loosen the alternator mounting bolt, nut and adjusting bolt, then remove the alternator drive belt.

8. Loosen the power steering mounting nuts and adjusting nut, then remove the power steering drive belt.

9. Remove the oil dipstick and dipstick tube.

10. Remove the crankshaft pulley bolt and remove the crankshaft pulley. Use a crank pulley holder (part # 07MAB–PY3010A) and holder handle (part # 07JAB–001020A) to hold the crankshaft pulley in place while removing the bolt.

11. Remove the upper and lower timing belt covers. Do not use the covers to store removed items.

12. Loosen the timing belt adjuster bolt 180°. Push the tensioner to remove the tension from the timing belt, then retighten the adjusting bolt.

13. Remove the timing belt. Do not crimp or bend the belt; protect it from contact with oil or coolant. Slide the belt off the sprockets.

14. Remove the bolts attaching the camshaft sprockets to the camshafts, then remove the sprockets.

15. Remove the spring from the timing belt tensioner. Remove the tensioner pulley adjusting bolt and the adjuster assembly from the engine.

16. Remove the Crankshaft Position (CKP) sensor from the oil pump, then remove the stopper plate. Remove the timing belt sprocket from the crankshaft, do not lose the sprocket key.

17. This is an excellent time to check or replace the water pump. Even if the timing belt is only being replaced as part of a good maintenance schedule, consider replacing the pump at the same time.

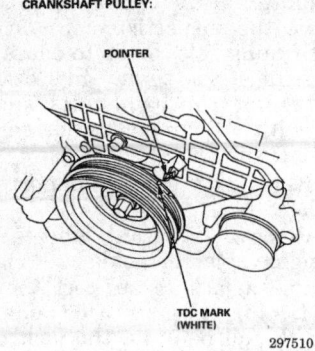

Crankshaft pulley timing mark — 2.7L engines

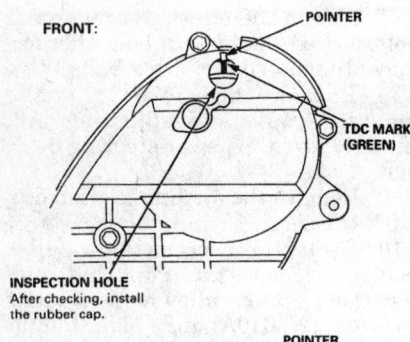

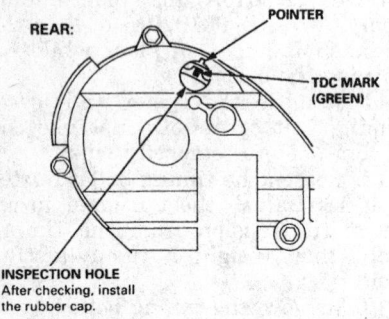

Camshaft pulleys timing marks — 2.7L engines

To install:

18. If the water pump is to be replaced, install a new O-ring and make certain it is properly seated. Install the water pump and retaining bolts. Tighten the mounting bolts to 16 ft. lbs (22 Nm).

19. Install the timing belt sprocket and key to the crankshaft. Install the stopper plate and the CKP sensor to the oil pump, torque the stopper plate and sensor mounting bolts to 9 ft. lbs. (12 Nm).

20. Install the tensioner pulley and the adjusting bolt, make sure the ten-

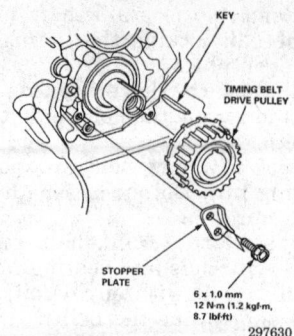

Timing belt drive pulley and stopper plate — 2.7L engines

sioner is properly positioned on its pivot pin. Install the spring to the tensioner then push the tensioner to it's full deflection and tighten the adjusting bolt.

21. Install the camshaft sprockets and attaching bolts. Torque the attaching bolts to 23 ft. lbs. (31 Nm).

22. Set the timing belt drive sprocket so that the No. 1 piston is at top dead center (TDC). Align the TDC mark on the tooth of the timing belt drive sprocket with the pointer on the oil pump.

23. Set the camshaft sprockets so that the No. 1 piston is at TDC. Align the TDC marks (green mark) on the

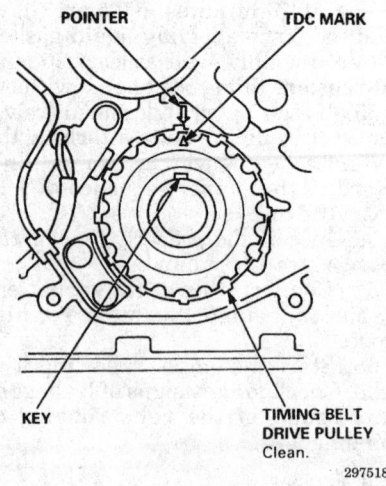

Timing belt drive pulley aligned — 2.7L engines

camshaft sprockets to the pointers on the back covers.

24. Install the timing belt in the following sequence:
   a. To the timing belt drive sprocket (crankshaft).
   b. To the adjusting pulley.
   c. To the front camshaft sprocket (toward the front of the vehicle).
   d. To the water pump pulley.
   e. To the rear camshaft sprocket (toward the bulkhead).

25. Loosen then retighten the timing belt adjuster bolt to tension the timing belt.

26. Install the lower then the upper timing belt covers. Torque the timing belt covers attaching bolts to 9 ft. lbs. (12 Nm).

27. Install the crankshaft sprocket and the crankshaft pulley bolt. Torque the bolt to 181 ft. lbs. (245 Nm) with the aid of the crank pulley holder.

28. Rotate the crankshaft five or six turns clockwise so that the timing belt positions on the sprockets.

29. Set cylinder No. 1 to TDC by aligning the timing marks. If the timing marks do not align, remove the timing belt then adjust the components and reinstall the timing belt.

30. Loosen the timing belt adjusting bolt 180° and retighten the adjusting bolt. Torque the adjusting bolt to 31 ft. lbs. (42 Nm).

31. Install the dipstick pipe and the dipstick. Install a new O-ring and torque the attaching bolts top 9 ft. lbs. (12 Nm).

32. Install and adjust the power steering drive belt. Torque the mounting nuts to 16 ft. lbs. (22 Nm).

33. Install and adjust the alternator drive belt. Torque the mounting bolt and nut to 16 ft. lbs. (22 Nm).

34. Install and adjust the A/C drive belt. Torque the idler pulley center nut to 33 ft. lbs. (44 Nm).

35. Position the side engine mount to the engine and install three new attaching bolts. Torque the new bolts to 40 ft. lbs. (54 Nm). Remove the jack from under the engine.

36. Install the engine splash shield. Torque the attaching bolts to 7 ft. lbs. (10 Nm).

37. Install the front wheels. Torque the lug nuts to 80 ft. lbs. (108 Nm).

38. Connect the positive then the negative battery cable and enter the radio security code.

39. Start the engine, allowing it to idle. Check for any sound of the belts rubbing or binding.

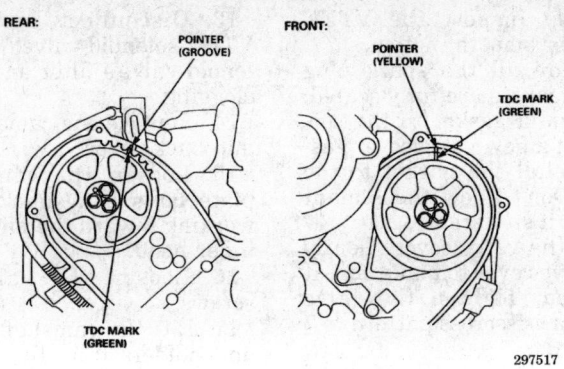

REAR: POINTER (GROOVE)

FRONT: POINTER (YELLOW)

TDC MARK (GREEN)

TDC MARK (GREEN)

297517

Camshaft pulleys aligned — 2.7L engines

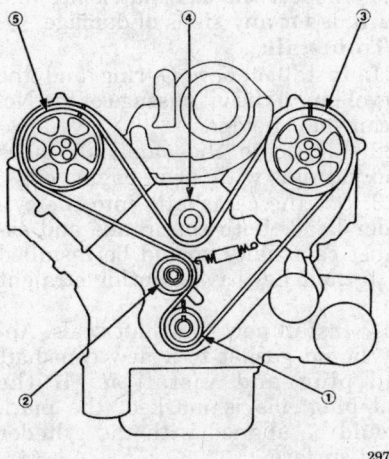

① Timing belt drive pulley (crankshaft) → ②
Adjusting pulley → ③ Front camshaft pulley → ④
Water pump pulley → ⑤ Rear camshaft pulley.

NOTE: Make sure the timing belt drive pulley and camshaft pulleys are at TDC.

297519

Timing belt installation sequence — 2.7L engines

Camshaft

NOTE: The radio may contain a coded theft protection circuit. Always obtain the code number before disconnecting the battery. If the vehicle is equipped with 4WS, the steering control unit is shut down when the battery is disconnected. After connecting the battery, turn the steering wheel lock-to-lock to reset the steering control unit.

REMOVAL AND INSTALLATION

Civic and del Sol

1.5L (D15B7, D15B8 and D15Z1) and 1.6L (D16Z6, D16Y5, D16Y7, D16Y8) Engines

1. Disconnect the negative battery cable.
2. Label and disconnect the ignition wires.
3. Remove the valve cover and the upper timing belt cover.
4. Rotate the crankshaft to set the No. 1 cylinder at TDC for the compression stroke. Once the engine is in this position, it shouldn't be disturbed.
5. Remove the timing belt. If the timing belt is contaminated with oil or coolant, it must be replaced. If the timing belt is to be reused, mark its direction of rotation.
6. Remove the distributor as an assembly.
7. Unbolt and remove the camshaft sprocket and its key. Remove the upper back cover timing cover.
8. Loosen the rocker arm locknuts and back off the valve adjusting screws.
9. Loosen the camshaft holder bolts in a two-step crisscross sequence, starting at the edges and working toward the center of the cylinder head.
10. Remove the rocker arm and shaft assembly. Leave the camshaft holder bolts in the camshaft holders to hold the rocker arm and shaft assembly together.
11. Wrap rubber bands around the VTEC rocker arm assemblies so that they do not separate.
12. Store the rocker arm and shaft assembly away from your work area. Cover the assembly with shop towels or a sheet of plastic to protect it from dust.

13. Lift the camshaft from the cylinder head. Remove the camshaft seal.
14. Inspect the camshaft journals and lobes for signs of scoring or other damage.

To install:

15. Remove the oil control orifice. Thoroughly clean it and reinstall it with a new O-ring.
16. Clean and inspect the camshaft bearing caps in the cylinder head.
17. Lubricate the lobes and journals of the camshaft prior to installation. Install the camshaft with the keyway facing up so that the camshaft will be at TDC/compression for the No. 1 cylinder.
18. Lightly lubricate a new camshaft seal with engine oil and install it.
19. Install the rocker arm and shaft assembly as follows:

　a. Remove the rubber bands from the VTEC rocker arms.

　b. Lubricate the rocker arm contact surfaces.

　c. Apply liquid gasket to the head mating surfaces of the No. 1 and No. 5 camshaft holders. Don't allow the sealant to cure before installing the rocker arm assembly.

　d. Set the rocker arm and shaft assembly in place. If equipped, install the lost motion assembly holder.

　e. Coat the threads of the camshaft holder bolts with clean oil and loosely install them.

　f. Tighten each bolt two turns at a time in the crisscross sequence to ensure that the rockers and camshaft holder do not bind on the camshaft journals.

　g. Tighten the 8mm camshaft holder bolts to 14 ft. lbs. (20 Nm), and the 6mm camshaft holder bolts to 9 ft. lbs. (12 Nm).

20. Install the camshaft sprocket and key. Tighten the retaining bolt to 27 ft. lbs. (38 Nm).
21. Verify that the engine remains at TDC/compression for the No. 1 cylinder.
22. Install the distributor. The lugs on the distributor drive fit into the groove on the end of the camshaft. Don't fully tighten the distributor mounting bolts yet.
23. Install the timing belt. Tighten the tensioner bolt to 33 ft. lbs. (44 Nm) once the belt has been properly tensioned.
24. Install the lower timing cover. Tighten the crankshaft pulley bolt to 134 ft. lbs. (181 Nm).
25. Adjust the valves.

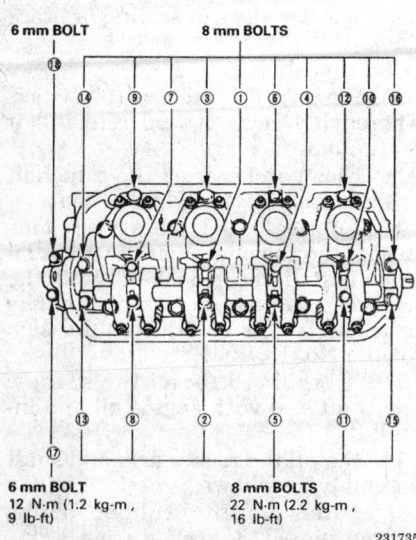

6 mm BOLT   8 mm BOLTS

6 mm BOLT
12 N·m (1.2 kg-m,
9 lb-ft)

8 mm BOLTS
22 N·m (2.2 kg-m,
16 lb-ft)

231735

**Camshaft holder bolt tightening sequence —
1.5L (D15B7, D15B8) engines**

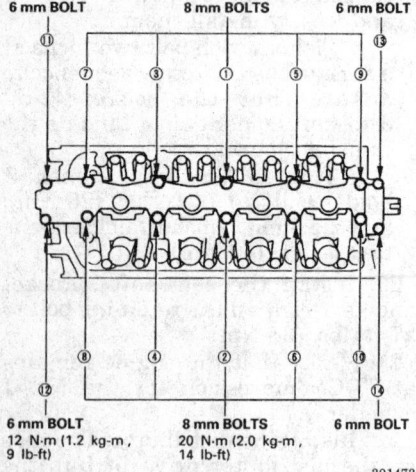

6 mm BOLT   8 mm BOLTS   6 mm BOLT

6 mm BOLT
12 N·m (1.2 kg-m,
9 lb-ft)

8 mm BOLTS
20 N·m (2.0 kg-m,
14 lb-ft)

6 mm BOLT

301473

**Camshaft holder bolt torque sequence — 1.5L
(D15Z1) and 1.6L (D16Z6, D16Y5, D16Y7, D16Y8)
engines**

26. Manually inspect the VTEC rocker arms for smooth motion.

27. Make sure all the spark plug tube sealing gaskets are fully seated.

28. Apply liquid gasket to the corner recesses of a new valve cover gasket. Then, install the gasket to the valve cover. Don't allow the sealant to cure before installation.

29. Install the valve cover. Gently wiggle the valve cover to make sure it is fully seated. Tighten the valve cover bolts in a crisscross pattern to 7 ft. lbs. (10 Nm).

30. Reconnect the ignition wires.

31. Refill the engine with fresh oil and install a new filter.

32. Reconnect the battery cable.

33. Warm the engine up to normal operating temperature. Check for oil leaks.

34. Check the ignition timing and adjust it if necessary. Then, tighten the distributor mounting bolts to 17 ft. lbs. (24 Nm).

1.6L (B16A2, B16A3) Engines

1. Disconnect the negative battery cable.

2. Label and disconnect the ignition wires.

3. Rotate the crankshaft to set the engine at TDC for the compression stroke of the No. 1 cylinder. The white TDC mark on the crankshaft pulley should align with the pointer on the lower timing belt cover.

4. Remove the strut brace.

5. Remove the intake air duct.

6. Loosen the power steering pump adjusting bolt to release the belt tension. Slip the belt off the pulleys. Loosen the air conditioner and alternator adjusting bolts, and slip their belts off the crankshaft pulley.

7. Use a floor jack padded with a block of wood to support the engine.

8. Remove the engine ground cable.

9. Unbolt and remove the engine side mount.

10. Remove the valve cover and the upper timing belt cover.

11. Verify that the engine is set at TDC/compression. Loosen the timing belt tensioner bolt 180°. Then, remove the crankshaft pulley, the lower timing cover, and timing belt.

—————— WARNING ——————
Inspect the timing belt for signs of cracked and broken teeth, as well as oil or coolant contamination. If the timing belt is damaged, or has been in contact with oil or coolant, it must be replaced to avoid potential failure.

12. Remove the distributor.

13. Disconnect and remove the VTEC solenoid valve. Remove the solenoid valve's filter and inspect it for clogging.

14. Remove the camshaft sprockets and back cover.

15. Loosen the camshaft holder plate bolts in a crisscross sequence working toward the middle of the cylinder head.

16. Loosen the valve adjusting screws.

17. Lift the camshaft holder plates and holders from the cylinder head. The holder bolts will keep the components together. Note the positions of each camshaft holder for reassembly.

18. Lift the camshafts from the cylinder head. Mark the exhaust and intake camshafts so that they will not be confused.

19. Remove the intake and exhaust oil control orifices. Thoroughly clean each, and reinstall them with new O-rings.

20. Inspect the camshaft lobes and journals for any signs of damage.

To install:

21. Install a new O-ring and the dowel pin to the oil passage of the No. 3 camshaft holder.

22. Lubricate the camshaft lobes and journals with clean engine oil.

23. Set the camshafts into the cylinder head. Both the intake and exhaust camshafts should be installed with their keyways pointing straight up.

24. Install new camshaft seals. Apply liquid gasket to a new camshaft end–plug and install it. If the end–plug has is marked, the mark should be aligned with the cylinder head surface.

25. Apply liquid gasket to the cylinder head mating surfaces of the No. 1 and No. 5 camshaft holders then install them, along with No. 2, 3, and 4 holders. Be sure to pay attention to the following points:

• Do not apply oil to the holder mating surface of camshaft seals.

• The arrows marked on the camshaft holders should point to the timing belt.

26. Install the camshaft holder plates.

27. Lubricate the threads of the 8mm holder bolts. Then, install all the camshaft holder bolts, but don't tighten them yet.

28. Evenly hand–tighten the camshaft holders. Make sure that the rocker arms are properly positioned on the valve stems.

29. Use a two–step crisscross pattern to tighten the camshaft holder bolts. Begin tightening with the bolts

in the middle of the cylinder head, and work toward the outer edges. Final torque specifications are as follows:

- 1996–97 B16A2 engine: 8mm bolts 20 ft. lbs. (28 Nm).
- 1994–95 B16A3 engines: 8mm bolts to 16 ft. lbs. (22 Nm)
- 1994–97: 6mm bolts to 7–8 ft. lbs. (10–11 Nm)

30. Verify that the camshaft keyways are pointing straight up and that the engine is at TDC/compression for the No. 1 cylinder. Fit the camshaft sprocket keys into their keyways.

31. Install the back cover and push the camshaft pulleys onto the camshafts. Then, tighten the sprocket retaining bolts to 37 ft. lbs. (51 Nm) for 1994–95 vehicles, or 41 ft. lbs. (57 Nm) for 1996 vehicles.

32. Install and tension the timing belt.

33. Install the lower timing cover and the crankshaft pulley. Tighten the pulley bolt to 130 ft. lbs. (180 Nm).

34. Adjust the valve clearance.

35. Inspect the VTEC rocker arms for smooth and independent movement.

36. Install the VTEC solenoid valve with a new filter. Tighten the valve mounting bolts to 9 ft. lbs. (12 Nm).

● — ⑫: 8 x 1.25 mm 22 N·m (2.2 kg-m, 16 lb-ft)
❶ — ⑫: 6 x 1.0 mm 11 N·m (1.1 kg-m, 8 lb-ft)
306353

Camshaft holder bolt torque sequences — 1.6L (B16A2, B16A3) engines

37. Install the distributor.

38. Clean the valve cover gasket surfaces. Fit the gasket into the groove of the valve cover.

39. Apply liquid gasket to the rubber seal at the eight corners where the gasket meets the camshaft holders. Don't allow the sealant to cure before installing the cylinder head cover.

40. Install the cylinder head cover and engine ground cable. Make sure the contact surfaces are clean and do not touch surfaces where liquid gasket has been applied.

41. Tighten the valve cover nuts in to 7 ft. lbs. (10 Nm) in a crisscross sequence.

42. Install the upper timing cover.

43. Install the accessory drive belts and adjust their tensions.

44. Install the engine side mount, torque the two new nuts to 54 ft. lbs. (75 Nm) and torque the bolt attaching the mount to the vehicle to 54 ft. lbs. (74 Nm).

45. Install the strut bar and tighten the mounting bolts to 16 ft. lbs. (22 Nm).

46. Reconnect the ignition wires.

47. Drain the engine oil. Install a new oil filter and refill the engine with fresh oil.

48. Reconnect the negative battery cable.

49. Warm the engine up to its normal operating temperature. Then, check and adjust the ignition timing. Tighten the distributor mounting bolts to 17 ft. lbs. (24 Nm).

Accord and Prelude

2.2L (H22A1) Engine

1. Disconnect the negative battery cable.

2. Turn the crankshaft so the No. 1 piston is at top dead center.

NOTE: The No. 1 piston is at top dead center when the pointer on the block aligns with the white painted mark on the flywheel (manual transaxle) or driveplate (automatic transaxle).

3. Remove the air intake duct.

4. Remove the engine ground cable from the cylinder head cover.

5. Remove the connector and the terminal from the alternator, then remove the engine wire harness from the valve cover.

6. Loosen the mounting bolts and remove drive belt from the power steering pump.

7. Remove the ignition coil.

8. Remove the plug wire cover from the cylinder head cover. Label then disconnect the electrical connec-

tors from the distributor and the spark plug wires from the spark plugs. Mark the position of the distributor and remove it from the cylinder head.

9. Remove the Positive Crankcase Ventilation (PCV) hose, then remove the cylinder head cover. Replace the rubber seals if damaged or deteriorated.

10. Remove the timing belt middle cover.

11. Ensure the arrows embossed on the camshaft pulleys are aligned in the upward position (12 o'clock).

12. Mark the rotation of the timing belt if it is to be used again.

13. Loosen the timing belt adjusting nut, do not loosen the nut more than one turn.

14. Use a open end wrench to loosen the maintenance bolt for the timing belt tensioner. If the maintenance bolt cannot be loosen with an open end wrench, a box wrench can be used after removing the lock pin.

NOTE: The use of a wrench to loosen the maintenance bolt should be limited to the bolts initial loosening only.

15. Loosen the maintenance bolt by hand until it stops. The auto–tensioner bracket is now fixed.

— **WARNING** —
Never use a tool to loosen the maintenance bolt after the initial loosening.

16. Tighten the timing belt adjusting nut.

17. Remove the timing belt from the camshaft sprockets.

— **WARNING** —
Do not crimp or bend the timing belt more than 90° or less then 1 inch (25mm) in diameter

18. Remove the camshaft sprocket attaching bolts then remove the sprockets. Do not lose the sprocket keys.

19. Remove the side engine mount bracket B, then the timing belt back cover.

20. Loosen all of the rocker arm adjusting screws.

21. Remove the camshaft holder bolts two turns at a time, in the reverse of the tightening sequence, to prevent damaging the valves or the the rocker arm assemblies. Remove the camshaft holder plates and the camshaft holders, note the holders locations for ease of installation.

22. Remove the rubber cap from the head, located at the end of the intake

camshaft. Replace the rubber cap if damaged or deteriorated.

23. Remove the camshafts and discard the camshaft seals.

To install:

24. Lubricate the rocker arms and camshafts with clean oil.

25. Install the camshaft seals to the end of the camshafts that the timing belt sprockets attach to. The open side (spring) should be facing into the cylinder head when installed.

26. Make sure the keyways on the camshafts are facing up and install the camshafts to the cylinder head.

27. Install the rubber cap to the cylinder head at the end of the intake camshaft.

28. Clean the oil control orifice and install a new O–ring. Install the oil control orifice into the oil passage of the No. 3 camshaft holder.

29. Apply liquid gasket to the head mating surfaces of the No. 1 and No. 5 camshaft holders, then install them along with No. 2, 3 and 4 camshaft holders. The arrows stamped on the holders should point toward the timing belt.

30. Install the camshaft holder plates and attaching bolts, then snug the camshaft holders in place.

31. Press the camshaft seals in securely against the base of the camshaft holders.

32. Torque the camshaft holder bolts in two steps, following the proper sequence, to ensure that the rockers do not bind on the valves. Torque the 8·1.23mm bolts to 19 ft. lbs. (26 Nm). Torque the 6·1.0mm bolts to 9 ft. lbs. (12 Nm).

33. Install the timing belt back cover and torque the attaching bolts to 9 ft. lbs. (12 Nm).

34. Install the side engine mount bracket B. Torque the bolt attaching the bracket to the cylinder head to 33 ft. lbs. (45 Nm) and torque the bolts attaching the bracket to the side engine mount to 16 ft. lbs. (22 Nm).

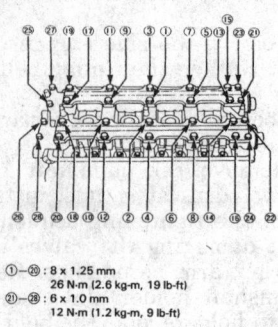

①—⑳ : 8 x 1.25 mm
26 N-m (2.6 kg-m, 19 lb-ft)
⑳—㉒ : 6 x 1.0 mm
12 N-m (1.2 kg-m, 9 lb-ft)

301998

Camshaft holders torque sequence — 2.2L (H22A1) engine

35. Install the keys into the camshaft grooves. Push the camshaft sprockets onto the camshafts, then torque the retaining bolts to 37 ft. lbs. (51 Nm).

36. Ensure the arrows embossed on the camshaft pulleys are aligned in the upward position (12 o'clock). Ensure that the TDC mark on the flywheel is aligned with the pointer on the engine block.

37. Install the timing belt slider (part # 07NAG–P130100) to the intake camshaft sprocket, then install the timing belt to the camshaft sprockets.

NOTE: If the auto — tensioner has been extended and the timing belt cannot be installed, remove the auto tensioner, compress it and reinstall it.

38. Torque the auto–tensioner maintenance bolt to 16 ft. lbs. (22 Nm), this will make the auto–tensioner functional.

39. Loosen the timing belt adjusting nut.

40. Turn the crankshaft pulley 1 turn, then torque the adjusting nut to 33 ft. lbs. (45 Nm).

41. Adjust the valves.

42. Torque the crankshaft pulley bolt to 181 ft. lbs. (250 Nm).

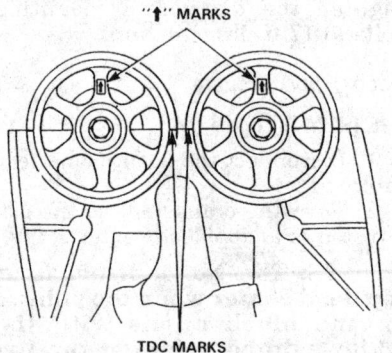

"↑" MARKS

TDC MARKS

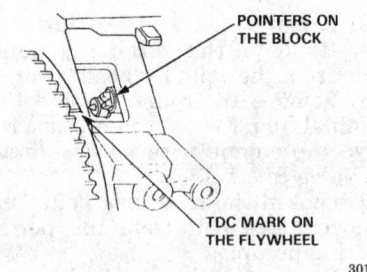

CRANKSHAFT TDC POSITION:

POINTERS ON THE BLOCK

TDC MARK ON THE FLYWHEEL

301999

Timing marks — 2.2L (H22A1) engine

43. Install the middle timing belt cover and torque the cover attaching bolts to 9 ft. lbs. (12 Nm).

44. Install the cylinder head cover, torque the cap nuts to 7 ft. lbs. (10 Nm). Install the PCV hose to the cylinder head cover.

45. Install the distributor to the cylinder head, snug the attaching bolts until the timing has been checked and adjusted.

46. Connect the spark plug wires to the correct spark plugs, then connect the distributor electrical connectors.

47. Install the spark plug wire cover to the cylinder head cover and torque the cap nuts to 7 ft. lbs. (10 Nm).

48. Install the ignition coil.

49. Install and adjust the power steering belt.

50. Install the alternator wire harness to the cylinder head cover, then connect the terminal and connector to the alternator.

51. Connect the engine ground cable to the cylinder head cover.

52. Install the air intake duct.

53. Drain the oil from the engine into a sealable container. Install the drain plug and refill the engine with clean oil.

54. Connect the negative battery cable and enter the radio security code.

55. Start the engine, checking carefully for any leaks.

56. Check and adjust the ignition timing, then torque the distributor mounting bolts to 13 ft. lbs. (18 Nm).

57. If equipped with 4WS, turn the steering wheel lock–to–lock to reset the 4WS control unit.

2.3L (H23A1) Engine

1. Disconnect the negative battery cable.

2. Turn the crankshaft so the No. 1 piston is at top dead center.

NOTE: The No. 1 piston is at top dead center when the pointer on the block aligns with the white painted mark on the flywheel (manual transaxle) or driveplate (automatic transaxle).

3. Remove the air intake duct.

4. Remove the engine ground cable from the cylinder head cover.

5. Remove the connector and the terminal from the alternator, then remove the engine wire harness from the valve cover.

6. Remove the ignition coil.

7. Label then disconnect the electrical connectors from the distributor and the spark plug wires from the spark plugs. Mark the position of the distributor and remove it from the

cylinder head. Disconnect the ignition coil wire from the distributor.

8. Remove the Positive Crankcase Ventilation (PCV) hose, then remove the cylinder head cover. Replace the rubber seals if damaged or deteriorated.

9. Remove the timing belt middle cover.

10. Ensure the words **UP** embossed on the camshaft pulleys are aligned in the upward position.

11. Mark the rotation of the timing belt if it is to be used again. Loosen the timing belt adjusting nut 1/2 turn and then release the tension on the timing belt. Push the tensioner to release tension from the belt, then tighten the adjusting nut.

12. Remove the timing belt from the camshaft sprockets.

WARNING

Do not crimp or bend the timing belt more than 90° or less then 1 inch (25mm) in diameter

---

13. Insert a 5.0mm pin punch in each of the camshaft caps, nearest to the sprockets, through the holes provided. Remove the camshaft sprocket attaching bolts then remove the sprockets. Do not lose the sprocket keys.

14. Remove the side engine mount bracket B, then the timing belt back cover from behind the camshaft sprockets.

15. Loosen all of the rocker arm adjusting screws, then remove the pin punches from the camshaft caps.

16. Remove the camshaft holders, note the holders locations for ease of installation. Loosen the bolts in the reverse order of the holder bolts torque sequence.

17. Remove the camshafts from the cylinder head then discard the camshaft seals.

18. Remove the rubber cap from the head, located at the end of the intake camshaft.

19. Remove the rocker arms from the cylinder head. Note the locations of the rocker arms.

NOTE: The rocker arms have to be installed to their original locations if being reused.

To install:

20. Lubricate the rocker arms with clean oil, then install the rocker arms on the pivot bolts and the valve stems. If the rocker arms are being reused, install them to their original locations. The locknuts and adjustment screws should be loosened before installing the rocker arms.

21. Lubricate the camshafts with clean oil.

22. Install the camshaft seals to the end of the camshafts that the timing belt sprockets attach to. The open side (spring) should be facing into the cylinder head when installed.

23. Make sure the keyways on the camshafts are facing up and install the camshafts to the cylinder head.

24. Install the rubber plug to the cylinder head at the end of the intake camshaft.

25. Apply liquid gasket to the head mating surfaces of the No. 1 and No. 6 camshaft holders, then install them along with No. 2, 3, 4 and 5. **I** or **E** marks are stamped on the camshaft holders to identify them as Intake or Exhaust side holders. The arrows stamped on the holders should point toward the timing belt.

26. Snug the camshaft holders in place.

27. Press the camshaft seals securely into place.

28. Torque the camshaft holder bolts in two steps, following the proper sequence, to ensure that the rockers do not bind on the valves. Torque all the bolts, except the four studs, to 7 ft. lbs. (10 Nm). Torque the studs (number 5 and 7 bolts in the torque sequence) to 9 ft. lbs. (12 Nm).

29. Install the timing belt back cover.

30. Install the side engine mount bracket B. Torque the bolt attaching the bracket to the cylinder head to 33 ft. lbs. (45 Nm). Torque the bolts attaching the bracket to the side engine mount to 16 ft. lbs. (22 Nm).

31. Insert a 5.0mm pin punch in each of the camshaft caps, nearest to the pulleys, through the holes provided. Install the keys into the camshaft grooves.

32. Push the camshaft sprockets onto the camshafts, then tighten the retaining bolts to 27 ft. lbs. (38 Nm).

33. Ensure the words **UP** embossed on the camshaft pulleys are aligned in the upward position. Install the timing belt to the camshaft sprockets, then remove the two 5.0mm pin punches from the camshaft bearing caps.

34. Loosen then tighten the timing belt adjuster nut.

35. Turn the crankshaft counterclockwise until the cam pulley has moved 3 teeth; this creates tension on the timing belt. Loosen then tighten the adjusting nut and torque it to 33 ft. lbs (45 Nm).

36. Adjust the valves.

37. Torque the crankshaft pulley bolt to 181 ft. lbs. (250 Nm).

38. Install the middle timing belt cover and torque the attaching bolts to 9 ft. lbs. (12 Nm).

39. Install the cylinder head cover and torque the cap nuts to 7 ft. lbs. (10 Nm). Install the PCV hose to the cylinder head cover.

40. Install the distributor to the cylinder head, snug the attaching bolts until the timing has been checked and adjusted.

41. Connect the spark plug wires to the correct spark plugs, then connect the distributor electrical connectors. Install the ignition coil wire to the distributor.

42. Install the ignition coil.

43. Install the alternator wire harness to the cylinder head cover, then connect the terminal and connector to the alternator.

44. Connect the engine ground cable to the cylinder head cover.

45. Install the air intake duct.

46. Drain the oil from the engine into a sealable container. Install the drain plug and refill the engine with clean oil.

47. Connect the negative battery cable and enter the radio security code.

48. Start the engine, checking carefully for any leaks.

49. Check and adjust the ignition timing. Torque the distributor bolts to 13 ft. lbs. (18 Nm).

50. If equipped with 4WS, start the engine then turn the steering wheel lock–to–lock to reset the 4WS control unit.

2.2L (F22A1, F22A6, F22B1 and F22B2) Engines

1. Disconnect the negative battery cable.

2. Remove the air intake duct.

3. Remove the cylinder head cover and replace the rubber seals if damaged or deteriorated.

4. Remove the timing belt upper cover.

5. Turn the engine to align the timing marks and set cylinder No.1 to TDC, compression. Once in this position, the engine must NOT be turned or disturbed.

6. Label then disconnect the electrical connectors from the distributor and the spark plug wires from the spark plugs. Mark the position of the distributor and remove it from the cylinder head.

7. Loosen the mounting bolts and remove drive belt from the power steering pump.

Specified torque:
Except Intake ⑤, ⑦, Exhaust ⑥, ⑧:
    10 N·m (1.0 kg-m, 7 lb-ft)
Intake ⑤, ⑦, Exhaust ⑥, ⑧:
    12 N·m (1.2 kg-m, 9 lb-ft)

TIGHTENING SEQUENCE

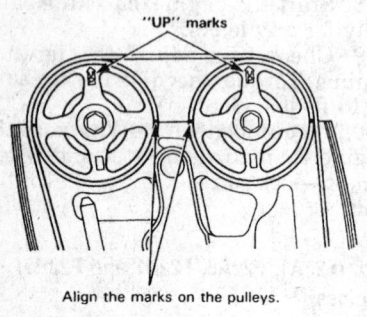

Camshaft holders torque sequence — 2.3L (H23A1) engine

"UP" marks

Align the marks on the pulleys.

301958

Camshaft sprockets alignment — 2.3L (H23A1) engine

8. Mark the rotation of the timing belt if it is to be used again. Loosen the timing belt adjusting bolt 3/4 to one turn and then release the tension on the timing belt. Push the tensioner to release tension from the belt, then tighten the adjusting bolt.

9. Remove the timing belt from the camshaft sprocket.

————— WARNING —————
Do not crimp or bend the timing belt more than 90° or less then 1 inch (25mm) in diameter

10. Ensure the words **UP** embossed on the camshaft pulley is aligned in

the upward position, then remove the camshaft sprocket bolt. Pull the sprocket from the camshaft and remove the sprocket key.

11. Remove the timing belt back cover.

12. Loosen the valve adjusting screws.

13. Loosen the camshaft holder attaching bolts two turns at a time, in the proper sequence to prevent damaging the valves or rocker arm assemblies.

NOTE: When removing the rocker arm assembly, do not remove the camshaft holder bolts.

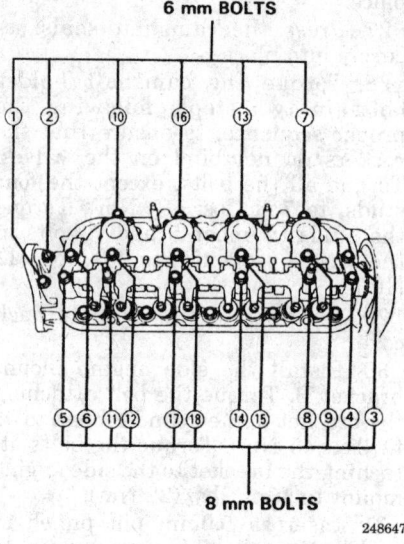

6 mm BOLTS

8 mm BOLTS

248647

Rocker arm assembly bolt removal sequence — 2.2L (F22A1, F22A6) engines

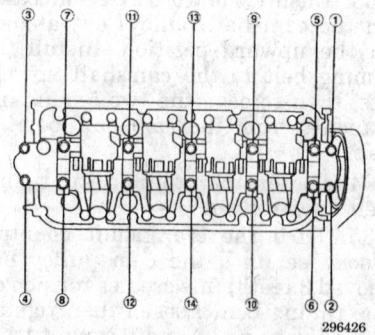

296426

Rocker arm assembly bolt loosening sequence — 2.2L (F22B1) engine

The bolts will keep the camshaft holders, springs, and the rocker arms on the shafts.

14. Carefully remove the camshaft holders and rocker arm assembly. If the rocker arm and shaft assembly needs to be disassembled for service, note the location of the components as they are removed. The rocker arms must be installed in the same position if reused.

15. Remove the camshaft from the cylinder head and discard the seal.

To install:

16. Wipe the camshaft and the camshaft journals clean, then lubricate both surfaces and install the camshaft.

17. Turn the camshaft so that its keyway is facing up (No. 1 cylinder will be at TDC).

18. Reassemble the rocker arm and shaft assembly, if it was disassembled. Lubricate the rocker arm and shaft assembly with clean oil, then apply liquid gasket to the head mating surfaces of the No. 1 and No. 6 camshaft holders.

19. Set the camshaft holders and rocker arm assembly in place then loosely install the attaching bolts.

20. Apply clean oil to the camshaft oil seal lip and the seal guide (part # 07NAG–PT0010A), then install the seal to the seal guide. Install the seal guide to the camshaft, then the installer cup (part # 07NAF–PT0010A) and the installer shaft (part # 07NAF–PT0020A). Tighten the nut on the installer shaft to press the seal into the cylinder head.

21. Tighten the camshaft holder bolts two turns at a time in the proper sequence. The final torque for the 8mm bolts is 16 ft. lbs. (22 Nm) and the final torque for the 6mm bolts is 9 ft. lbs. (12 Nm).

22. Install the timing belt back cover, torque the attaching bolt to 9 ft. lbs. (12 Nm).

23. Install the camshaft sprocket key to the camshaft, then install the camshaft sprocket. Install the bolt and torque it to 27 ft. lbs. (37 Nm).

24. Ensure the words **UP** embossed on the camshaft pulley is aligned in the upward position, then install the timing belt onto the camshaft sprocket. Loosen then tighten the timing belt adjusting nut.

25. Rotate the crankshaft pulley five or six turns to position the timing belt on the pulleys.

26. Set the No. 1 cylinder to TDC and loosen the timing belt adjusting nut one turn. Turn the crankshaft counterclockwise until the cam pulley has moved 3 teeth; this creates ten-

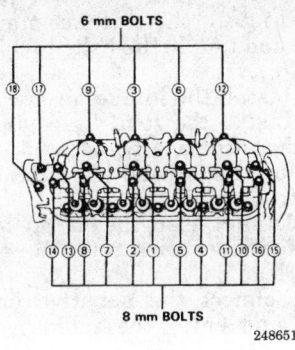

6 mm BOLTS

8 mm BOLTS

248651

Rocker arm assembly torque sequence — 2.2L (F22A1, F22A6) engines

Specified torque:
8 mm bolts: 22 N·m (2.2 kgf·m, 16 lbf·t)
6 mm bolts: 12 N·m (1.2 kgf·m, 8.7 lbf·t)
6 mm bolts: ⑪, ⑫, ⑬, ⑭

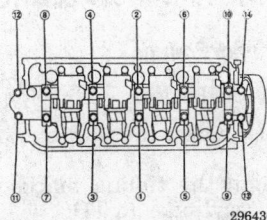

296430

Rocker arm assembly torque sequence — 2.2L (F22B1) engine

Specified torque:
8 mm bolts: 22 N·m (2.2 kgf·m, 16 lbf·t)
6 mm bolts: 12 N·m (1.2 kgf·m, 8.7 lbf·t)

6 mm bolts; ③, ⑥, ⑨, ⑫, ⑰, ⑱

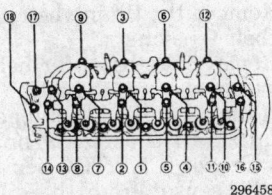

296458

Rocker arm assembly torque sequence — 2.2L (F22B2) engine

sion on the timing belt. Loosen, then tighten the timing belt adjusting nut and torque it to 33 ft. lbs (45 Nm).

27. Adjust the valves.

28. Torque the crankshaft pulley bolt to 181 ft. lbs. (245 Nm) on F22B1 and F22B2 models and 159 ft. lbs. (220 Nm) on all other models.

29. Install the upper timing belt cover and torque the bolt to 9 ft. lbs. (12 Nm).

30. Install the cylinder head cover gasket cover to the groove of the cylinder head cover. Before installing the gasket thoroughly clean the seal and the groove. Seat the recesses for the camshaft first, then work it into the groove around the outside edges. Make sure the gasket is seated securely in the corners of the recesses.

31. Install the cylinder head (valve) cover and torque the cap nuts to 7 ft. lbs. (10 Nm).

32. Install and adjust the power steering belt.

33. Install the distributor to the cylinder head, snug the attaching bolts until the timing has been checked and adjusted.

34. Connect the spark plug wires to the correct spark plugs, then connect the distributor electrical connectors. Install the ignition coil wire to the distributor.

35. Install the air intake duct.

36. Drain the oil from the engine into a sealable container. Install the drain plug and refill the engine with clean oil.

37. Connect the negative battery cable and enter the radio security code.

38. Start the engine, checking carefully for any leaks.

39. Check and adjust the the ignition timing as necessary, then torque the distributor bolts to 16 ft. lbs. (22 Nm).

2.7L Engine

1. Disconnect the negative battery cable.

2. Turn the engine to align the timing marks and set cylinder No.1 to TDC. The white mark on the crankshaft pulley should align with the pointer on the timing belt cover. Remove the inspection caps on the upper timing belt covers to check the alignment of the timing marks. The pointers for the camshafts should align with the green marks on the camshaft pulleys.

3. Remove the intake air duct.

4. Remove the starter cable from the strut brace, then remove the strut brace.

5. Remove intake manifold covers.

6. Disconnect the breather hose from the cylinder head cover.

7. Remove the PCV hose from the cylinder head cover.

8. Loosen the idler pulley center nut and adjusting bolt, then remove the air conditioning compressor belt.

9. Loosen the alternator mounting bolt, nut and adjusting bolt, then remove the alternator drive belt.

10. Loosen the power steering mounting nuts and adjusting nut, then remove the power steering drive belt.

11. Label then disconnect the electrical connectors from the distributor and the spark plug wires from the spark plugs. Remove the distributor from the cylinder head.

12. Remove the cylinder head covers and the side covers.

13. Remove the timing belt covers and the timing belt.

14. Remove the camshaft sprockets and the timing belt back covers.

15. Remove the bolts attaching the camshaft holder plates and camshaft holders in the opposite order of the installation sequence.

16. Remove the camshaft holder plates, camshaft holders and the dowel pins. Discard the O—rings.

17. Remove the camshafts from the cylinder heads and the rubber cap from the rear cylinder head. Discard the camshaft seals.

To install:

18. Apply clean engine oil to the rocker arms and the camshafts.

19. Loosen the exhaust rocker arm adjusting screws and locknuts.

20. Make sure the rocker arms are properly positioned on the valve stems. Advance the crankshaft 30° from TDC to prevent interference between the pistons and valves, then install the camshafts. Position the rear camshaft on the cylinder head so the cam is not pushing on any valves.

21. Apply liquid gasket around the rubber cap, then install it to the cylinder head.

22. Install the camshaft seals to the camshafts with the open side (spring) facing in.

23. Apply liquid gasket the the cylinder head and camshaft holder mating surfaces, then install the camshaft holders and the camshaft plates with the dowel pins. Install new O—rings to the camshaft holder plates

24. Apply clean oil to the camshaft holder bolts, then install the bolts and tighten them in the proper sequence. Torque the 8mm bolts to 20 ft. lbs. (27 Nm), and the 6mm bolts to 8.7 ft. lbs. (12 Nm).

25. Install the timing belt back covers, torque the attaching bolts to 9 ft. lbs. (12 Nm).

26. Install the camshaft sprockets and torque the attaching bolts to 23 ft. lbs. (31 Nm).

27. Set the camshaft sprockets so that the No. 1 piston is at TDC. Align the TDC marks (green mark) on the camshaft pulleys to the pointers on the back covers.

28. Turn the crankshaft counterclockwise to set it at TDC. Align the TDC mark on the tooth of the timing

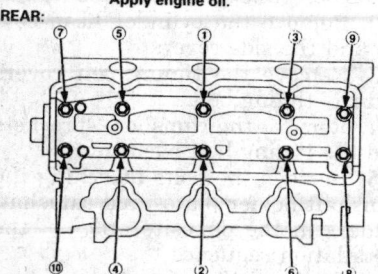

8 mm bolt: 27 N·m (2.8 kgf·m, 20 lbf·ft)
Apply engine oil.
6 mm bolt: 12 N·m (1.2 kgf·m, 8.7 lbf·ft)
Apply engine oil.

REAR:

FRONT:
6 mm bolts: ⑪, ⑫, ⑬, ⑭

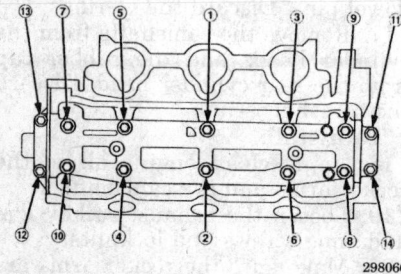

298066

Camshaft holders torque sequence — 2.7L engine

belt drive pulley with the pointer on the oil pump.

29. Install the timing belt and timing belt covers.

30. Set No. 1 cylinder to TDC.

31. Tighten the valve adjusting screws for No. 1, No. 2 and No. 4 cylinders. Tighten the screw until it contacts the valve, then tighten the screw 1 1/8 turns. Hold the screw in place and torque the locknut to 14 ft. lbs. (20 Nm).

32. Rotate the crankshaft pulley one turn clockwise, then tighten the adjusting screws for No. 3, No. 5 and No. 6 cylinders. Tighten the screw until it contacts the valve, then tighten the screw 1 1/8 turns. Hold

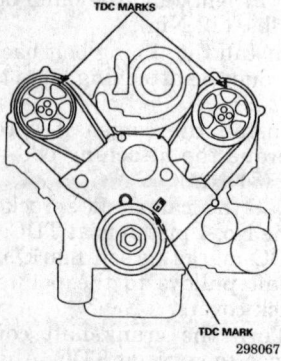

TDC MARKS

TDC MARK

298067

Timing marks — 2.7L engine

the screw in place and torque the locknut to 14 ft. lbs. (20 Nm).

33. Install the cylinder head cover gasket into the groove of the cylinder head cover. Seat the recesses for the camshaft first, then work it into the groove around the outside edges.

NOTE: Before installing the cylinder head cover gasket, thoroughly clean the seal groove.

34. Apply liquid gasket to the cylinder head cover gasket at the four corners of the recesses. Use a shop towel and wipe the cylinder heads where the cylinder head covers will come in contact.

35. Install the cylinder head covers, hold the gasket in the groove by placing your fingers on the camshaft contacting surfaces. With the cylinder head cover on the cylinder heads, slide the covers slightly back and forth to seat the cylinder head cover gaskets. Replace the washers if damaged or deteriorated

36. Tighten the cylinder head cover bolts in two or three steps. In the final step, torque all the bolts, in sequence, to 11 ft. lbs. (15 Nm).

37. Install the cylinder head side covers with new O–rings and torque the bolts to 9 ft. lbs. (12 Nm).

38. Install the distributor to the cylinder head and torque the mounting bolt to 16 ft. lbs. (22 Nm).

39. Connect the spark plug wires to the correct spark plugs, then connect the distributor electrical connectors.

40. Install and adjust the power steering belt.

41. Install and adjust the alternator belt. Torque the alternator mounting nut and bolt to 16 ft. lbs. (22 Nm).

42. Install the A/C belt and adjust the belt tension. Tighten the idler center nut to 33 ft. lbs. (44 Nm).

43. Install the PCV hose to the cylinder head cover.

44. Connect the breather hose to the cylinder head cover.

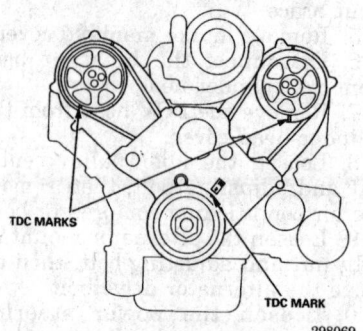

TDC MARKS

TDC MARK

298069

Timing marks for valve adjustment on cylinders 3, 5 and 6 — 2.7L engine

45. Install the intake manifold cover and torque the bolts to 9 ft. lbs. (12 Nm).

46. Install the intake air duct.

47. Install the strut brace and torque the mounting bolts to 16 ft. lbs. (22 Nm). Install the starter cable to the strut brace.

48. Drain the engine oil into a sealable container then refill the engine with clean oil.

49. Connect the negative battery cable and enter the radio security code.

50. Start the engine, allowing it to idle and check for any signs of leakage.

Balance Shaft

REMOVAL AND INSTALLATION

Accord and Prelude

2.2L and 2.3L Engines

1. Disconnect the negative battery cable.

2. Align the timing marks bringing No. 1 cylinder to TDC.

3. Remove the engine assembly from the vehicle.

4. Mount the engine on an engine stand, making sure the mounting bolts are tight.

5. Remove the timing belt covers and the timing belts.

6. Drain the engine oil into a sealable container, then remove the oil pan and the oil screen.

7. Remove the timing belt and balancer belt tensioners.

8. Remove the balancer belt drive pulley from the crankshaft.

9. If the oil pump is equipped with a Crankshaft Position/Top Dead Center Sensor (CKP/TDC sensor), remove the sensor from the oil pump and disconnect the electrical connector.

10. Remove the timing belt drive pulley from the crankshaft.

11. Insert a suitable tool into the maintenance hole in the front balancer shaft and remove the balancer driven pulley.

12. Align the rear timing balancer pulley using a 6 x 100mm bolt or rod. Mark the bolt or rod at a point 2.9 inches (74mm) from the end. Remove the bolt from the maintenance hole on the side of the block; insert the bolt/rod into the hole. Align the 74mm mark with the face of the hole. This pin will hold the shaft in place.

13. Remove the balancer gear case and the dowel pins. Discard the O–ring.

14. Remove the balancer driven gear attaching bolt and the balancer driven gear, from the rear balancer shaft.

15. Remove the bolt or rod used to align the rear balancer shaft.

16. Remove the oil pump mounting bolts and remove the oil pump assembly. Remove the dowel pins from the engine and clean the oil pump mating surfaces of old gasket material and oil. Discard the O–rings.

17. Carefully pull the rear balancer shaft from the engine block.

18. Remove the bolts from the front balancer shaft retainer and remove the retainer.

19. Carefully pull the front balancer shaft from the engine block.

To install:

20. Lubricate the front and rear balancer shafts with clean motor oil or other suitable lubricant.

21. Carefully install the balancer shafts into the engine block. Install the retainer to the front shaft, torque the retainer bolts to 14 ft. lbs. (20 Nm).

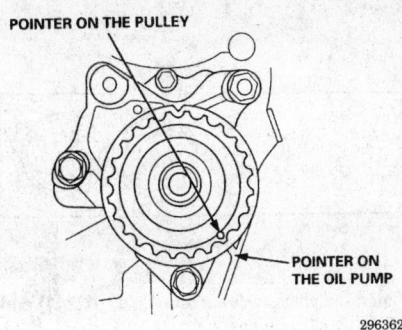

POINTER ON THE PULLEY

POINTER ON THE OIL PUMP

296362

Installed balancer pulley alignment — 2.2L and 2.3L engines

22. Install the two oil pump dowel pins and new O-rings to the cylinder block.

23. Be sure that the oil pump and engine block mating surfaces are clean and dry. Apply a liquid gasket evenly in a narrow bead, centered on the mating surface. Once the sealant is applied, do not wait longer than 20 minutes to install the parts; the sealant will become ineffective. After final assembly, wait at least 30 minutes before adding oil to the engine,

giving the sealant time to set. To prevent leakage of oil, apply a suitable thread sealer to the inner threads of the bolt holes.

24. Install the oil pump to the engine block and torque the mounting bolts to 9 ft. lbs. (12 Nm).

25. Install the oil screen and torque the mounting bolts and nuts to 9 ft. lbs. (12 Nm).

26. Install the oil pan.

27. Install the balancer driven pulley to the front balancer belt, hold the balancer shaft in place with a suitable tool. Torque the attaching bolt to 22 ft. lbs. (29 Nm).

28. Align the rear timing balancer shaft using a 6 x 100mm bolt or rod, inserted into the maintenance hole on the side of the block.

29. Install the balancer driven gear to the rear balancer shaft and torque the bolt to 18 ft. lbs. (25 Nm).

30. Before installing the balancer driven gear and the gear case, apply molybdenum disulfide (lithium grease) to the thrust surfaces of the balancer gears.

31. Align the groove on the pulley edge to the pointer on the balancer gear case.

32. Install the balancer gear case to the engine and install the mounting bolts and nut. The rear balancer shaft should still be held in place with a 6·100mm bolt/rod. Torque the mounting bolts and nut to 18 ft. lbs. (25 Nm).

33. Check the alignment of the pointer on the balancer pulley to the pointer on the oil pump.

34. Install the timing belt drive pulley to the crankshaft.

35. Install the timing belt tensioners.

36. If a CKP/TDC sensor was removed from the oil pump connect the electrical connector then install the sensor. Torque the sensor mounting bolts to 9 ft. lbs. (12 Nm).

37. Install the balancer belt drive pulley.

38. Install the timing belt and the balancer belt.

39. Install the timing belt covers and the crankshaft pulley.

40. Install the engine assembly into the vehicle.

41. Refill the engine with with clean, fresh oil.

42. Connect the negative battery cable and enter the radio security code.

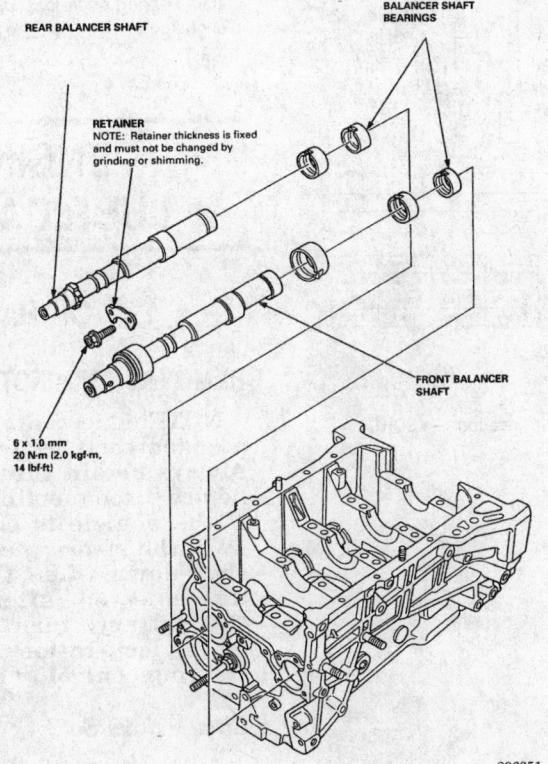

REAR BALANCER SHAFT

BALANCER SHAFT BEARINGS

RETAINER
NOTE: Retainer thickness is fixed and must not be changed by grinding or shimming.

6 x 1.0 mm
20 N·m (2.0 kgf·m, 14 lbf·ft)

FRONT BALANCER SHAFT

296351

Balancer shafts — 2.2L and 2.3L engines

Piston and Connecting Rod

POSITIONING

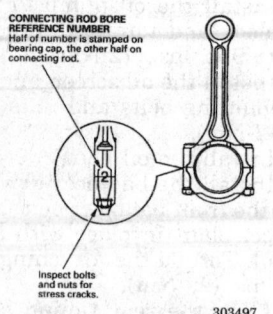

CONNECTING ROD BORE REFERENCE NUMBER
Half of number is stamped on bearing cap, the other half on connecting rod.

Inspect bolts and nuts for stress cracks.

303497

Rod bearing size code number on connecting rod — 1.5L and 1.6L engines

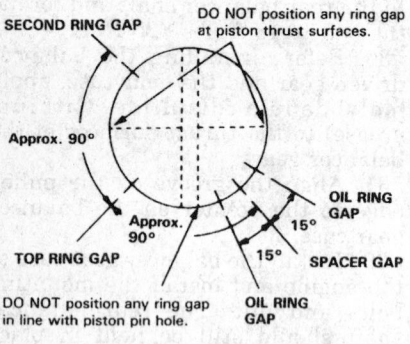

SECOND RING GAP

DO NOT position any ring gap at piston thrust surfaces.

Approx. 90°

Approx. 90°

OIL RING GAP

15°

TOP RING GAP

15°

SPACER GAP

DO NOT position any ring gap in line with piston pin hole.

OIL RING GAP

303500

Piston ring gap positions

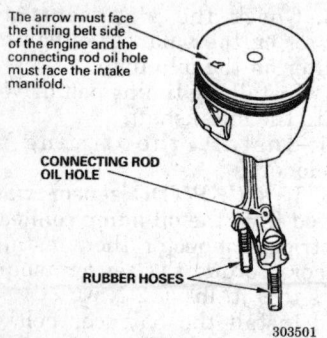

The arrow must face the timing belt side of the engine and the connecting rod oil hole must face the intake manifold.

CONNECTING ROD OIL HOLE

RUBBER HOSES

303501

Piston installation direction marks — 1.5L and 1.6L engines

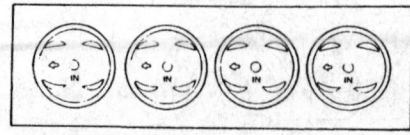

EXHAUST

INTAKE

320678

Piston installation direction — 2.2L (H22A1) and 2.3L (H23A1) engines

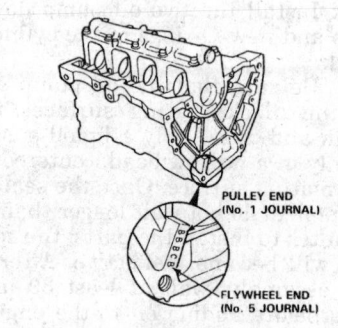

PULLEY END (No. 1 JOURNAL)

FLYWHEEL END (No. 5 JOURNAL)

Main Journal Code Location (Numbers)

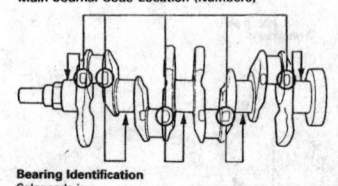

Bearing Identification
Color code is on the edge of the bearing.

Larger crank bore

A	B	C	D

Smaller bearing (thicker)

	Red	Pink	Yellow	Green
1				
2	Pink	Yellow	Green	Brown
3	Yellow	Green	Brown	Black
4	Green	Brown	Black	Blue

Smaller main journal

Smaller bearing (thicker)

303509

Main bearing code locations — SOHC engines

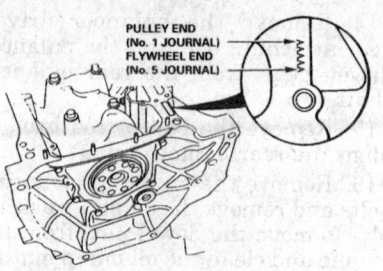

PULLEY END (No. 1 JOURNAL)
FLYWHEEL END (No. 5 JOURNAL)

Main Journal Code Locations (Numbers or Bars)

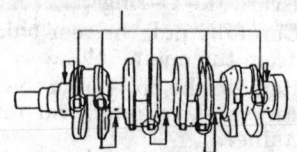

Bearing Identification

Color code is on the edge of the bearing.

Larger crank bore

	A	B	C	D

Smaller bearing (Thicker)

	Red	Pink	Yellow	Green
1 or I				
2 or II	Pink	Yellow	Green	Brown
3 or III	Yellow	Green	Brown	Black
4 or IIII	Green	Brown	Black	Blue

Smaller main journal

Smaller bearing (thicker)

303510

Main bearing code locations — DOHC engines

ENGINE LUBRICATION

Oil Pan

REMOVAL AND INSTALLATION

NOTE: The radio may contain a coded theft protection circuit. Always obtain the code number before disconnecting the battery. If the vehicle is equipped with 4WS, the steering control unit is shut down when the battery is disconnected. After connecting the battery, turn the steering wheel lock-to-lock to reset the steering control unit.

Civic and del Sol

1. Disconnect the negative battery cable.
2. Raise and safely support the vehicle.
3. Drain the oil and remove the lower splash panel.
4. Remove the nuts and bolts connecting exhaust pipe A to the cata-

lytic converter. Discard the gasket and the locknuts.

5. Remove the nuts attaching exhaust pipe A to the exhaust hanger.

6. Remove and discard the locknuts attaching exhaust pipe A to the exhaust manifold, then remove exhaust pipe A from the vehicle. Discard the exhaust gaskets.

7. Loosen the oil pan bolts in a crisscross pattern. To remove the oil pan, lightly tap the corners of the oil pan with a rubber or plastic faced mallet. Clean off all the old gasket material.

8. Inspect the oil screen and pickup tube for damaged and clogging. If the screen and tube are clogged with oil residue, they should be thoroughly cleaned or replaced.

To install:

9. If removed, install the oil screen and tube with a new gasket. Tighten the mounting nuts and bolts to 8 ft. lbs. (11 Nm).

10. Apply liquid gasket to the oil pan mating surface where the oil pump and the right side cover meet the engine block.

11. Install the oil pan gasket to the oil pan.

12. Install the oil pan. Then, install all the center and end mounting nuts and bolts. Evenly hand–tighten the oil pan nuts and bolts.

13. Tighten the oil pan mounting nuts and bolts in a three–step clockwise pattern starting with the center bolt next to the oil drain plug. The final torque value for the nuts and bolts is 9–10 ft. lbs. (12–14 Nm).

NOTE: Excessive tightening can cause distortion of the oil pan gasket and oil leakage.

14. Install the oil drain plug with a new crush washer, torque the plug to 33 ft. lbs. (44 Nm).

15. Install exhaust pipe A using new gaskets and locknuts. Torque

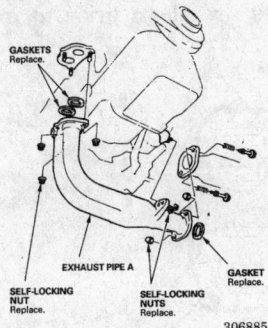

Exhaust pipe A — 1.5L (D15B7 D15B8, D15Z1) and 1.6L (D16Z6, B16A2, B16A3 engines

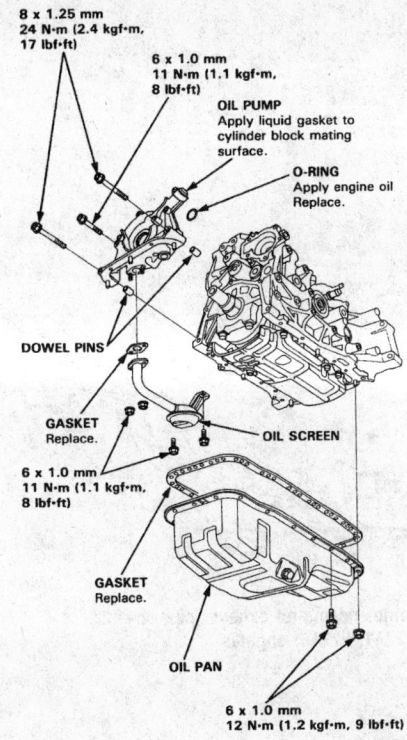

8 x 1.25 mm
24 N·m (2.4 kgf·m, 17 lbf·ft)

6 x 1.0 mm
11 N·m (1.1 kgf·m, 8 lbf·ft)

OIL PUMP
Apply liquid gasket to cylinder block mating surface.

O-RING
Apply engine oil
Replace.

DOWEL PINS

GASKET
Replace.

OIL SCREEN

6 x 1.0 mm
11 N·m (1.1 kgf·m, 8 lbf·ft)

GASKET
Replace.

OIL PAN

6 x 1.0 mm
12 N·m (1.2 kgf·m, 9 lbf·ft)

306881

Oil pan and oil screen — 1.6L (B16A2, B16A3) engines

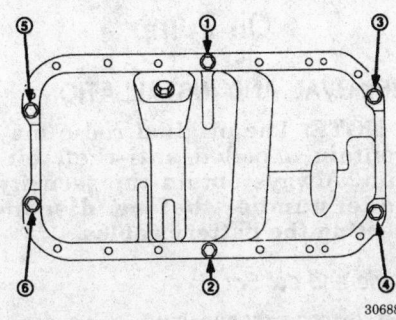

306883

Oil pan bolt tightening sequence — 1.5L (D15B7 D15B8, D15Z1) and 1.6L (D16Z6, B16A2, B16A3 engines

the nuts attaching the exhaust pipe to the exhaust manifold to 40 ft. lbs. (54 Nm). Torque the nuts attaching the exhaust pipe to the catalytic converter and the exhaust pipe hanger to 16 ft. lbs. (22 Nm).

16. Install the lower splash panel. Then, lower the vehicle.

17. Refill the engine with clean oil.

18. Connect the negative battery cable and enter the radio security code.

19. Run the engine and check for leaks.

20. Turn off the engine and check the oil level. Top off the oil level if necessary.

Prelude and Accord

2.2L, 2.3L and 2.7L Engines

1. Disconnect the negative battery cable.

2. Raise and safely support the vehicle.

3. Drain the engine oil into a sealable container.

4. Install the drain bolt with a new gasket, torque the bolt to 33 ft. lbs. (44 Nm).

5. Remove the front wheels and the splash shield.

6. Remove the center beam.

7. Disconnect the oxygen sensor electrical connector.

8. Remove the bolts from the support bracket on exhaust pipe A.

9. Remove the nuts attaching exhaust pipe A to the exhaust manifold and the catalytic converter. Remove exhaust pipe A and discard the gaskets.

10. If equipped with a automatic transaxle, remove the torque converter cover.

11. If equipped with a manual transaxle, remove the clutch cover.

12. Remove the oil pan nuts and bolts (in a criss–cross pattern) and the oil pan; if necessary, use a mallet to tap the corners of the oil pan. Do NOT pry on the pan to get it loose.

13. Clean the oil pan mounting surface of old gasket material and engine oil.

To install:

14. Install a new oil pan gasket to the oil pan. Apply liquid gasket to the corners of the curved section of the gasket.

15. Install the oil pan to the engine.

16. Install the oil pan nuts and bolts, tighten the nuts and bolts in sequence. Torque the nuts and bolts in two steps to 10 ft. lbs. (14 Nm).

17. If equipped with a automatic transaxle, install the torque converter cover. Torque the bolts to 9 ft. lbs. (12 Nm).

18. If equipped with a manual transaxle, install the clutch cover. Torque the bolts to 9 ft. lbs. (12 Nm).

19. Install exhaust pipe A with new gaskets and new locknuts. Torque the nuts attaching exhaust pipe A to the manifold to 40 ft. lbs. (54 Nm), torque the nuts attaching exhaust pipe A to the catalytic converter to 25 ft. lbs. (33 Nm). Install the bolts to the exhaust pipe support bracket and torque the bolts to 13 ft. lbs. (18 Nm).

20. Connect the oxygen sensor electrical connector.

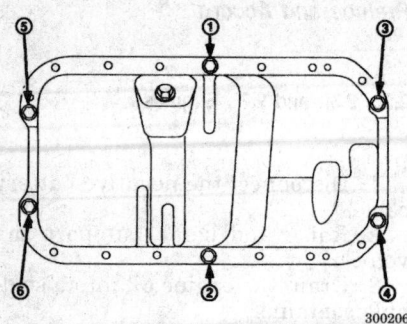

Oil pan bolt tightening sequence — 1.6L (D16Y5, D16Y7, D16Y8 engines

300206

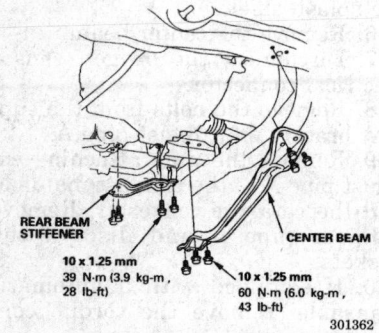

REAR BEAM STIFFENER

CENTER BEAM

10 x 1.25 mm
39 N·m (3.9 kg-m,
28 lb-ft)

10 x 1.25 mm
60 N·m (6.0 kg-m,
43 lb-ft)

301362

Center beam — 2.2L (F22A1, H22A1) and 2.3L (H23A1) engines

21. Install the center beam, torque the mounting bolts as follows;
• Prelude: 43 ft. lbs. (60 Nm)
• 1993 Accord: 28 ft. lbs. (39 Nm)
• 1994–97 Accord and 2.7L engines: 37 ft. lbs. (50 Nm)
22. Install the splash shield, torque the mounting bolts to 7 ft. lbs. (10 Nm).
23. Install the front wheels.
24. Lower the vehicle and fill the engine with oil.
25. Connect the negative battery cable and enter the radio security code.
26. Start the engine and check for leaks.
27. If equipped with 4WS, turn the steering wheel lock–to–lock to reset the 4WS control unit.
28. If the oxygen sensor is located in the mid exhaust pipe, disconnect the oxygen sensor electrical connector.

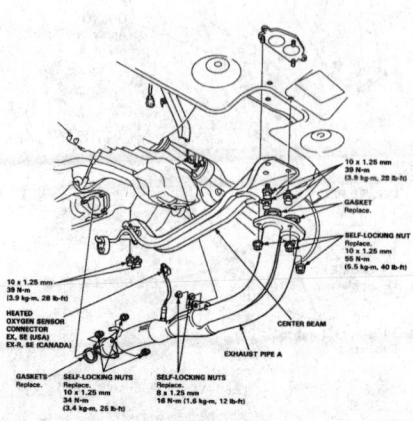

Center beam and exhaust pipe A — 2.2L (F22A1, F22A6) engines

236210

Oil Pump

REMOVAL AND INSTALLATION

NOTE: The original radio may contain a coded anti–theft circuit. Always obtain the security code number before disconnecting the battery cables.

Civic and del Sol

1.5L (D15B7 D15B8, D15Z1) and 1.6L (D16Z6) Engine

1. Disconnect the negative battery cable.
2. Raise and safely support the vehicle.

3. Drain the engine oil.
4. Set the No. 1 cylinder to TDC for the compression stroke. The mark on the crankshaft pulley should align with the index mark on the timing cover.

NOTE: Mark the direction of the timing belt's rotation if it is to be reinstalled.

5. Remove the accessory drive belts and the crankshaft pulley.
6. Remove the valve cover and the timing belt covers.
7. Remove the following components:
   a. Timing belt tensioner
   b. Timing belt
   c. Timing belt crankshaft sprocket
8. Remove the oil pan and oil screen.
9. Remove the oil pump mount bolts and the oil pump assembly.
10. Disassemble the oil pump and inspect the rotors:
   a. Inner–to–outer router clearance is 0.02–0.14mm (0.001–0.006 in.). The service limit is 0.20mm (0.008 in.).
   b. Rotor–to–pump housing axial clearance is 0.03–0.08mm (0.001–0.003 in.). The service limit is 0.15mm (0.006 in.). Check the clearance using a steel bar and feeler gauge.
   c. Outer rotor–to–pump housing clearance is 0.10–0.18mm (0.004–0.007 in.). The service limit is 0.20mm (0.008 in.).
To install:

NOTE: Replace the rotors if they are worn or damaged. Use new O–rings when assembling and installing the oil pump.

11. Assemble the oil pump and tighten the rotor cover bolts to 33 ft. lbs. (45 Nm).
12. Make sure all gasket mating surfaces are clean prior to installa-

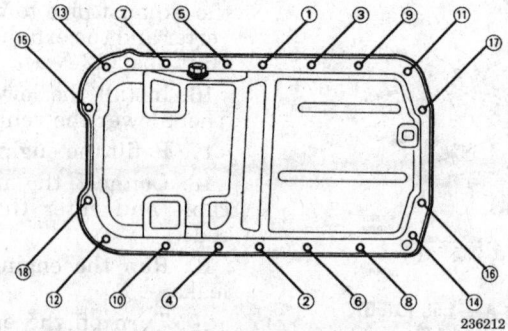

236212

Oil pan torque sequence — 2.2L (F22A1, F22B1, F22B2, H22A1 and F22A6) and 2.3L H23A1 engines

tion. Replace the crankshaft oil seal prior to installing the oil pump.

13. Apply liquid gasket to the cylinder block mating surface of the block. Apply a light coat of oil to the crankshaft seal lip. Install a new O-ring on the cylinder block and install the oil pump. Apply liquid gasket to the threads of the oil pump mounting bolts and tighten them to 8–9 ft. lbs. (11–12 Nm).

14. Install the oil screen.

15. Install the oil pan.

16. Tighten the crankshaft pulley bolt to specification.

17. Install and tension the timing belt components after installation. Install the valve cover.

18. Install and adjust the accessory drive belts.

19. Refill the engine with oil.

20. Connect the negative battery cable.

21. Run the engine and check for proper oil pressure.

22. Check for leaks. Top up the engine oil if necessary.

1.6L (D16Y5, D16Y7, D16Y8) Engines

1. Disconnect the negative battery cable.

2. Raise and safely support the vehicle.

3. Drain the engine oil.

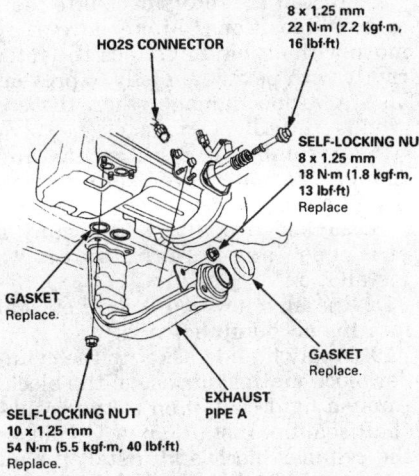

- Except sedan LX without ABS model of Engine Serial Number: F22B2 – 21XXXXX.

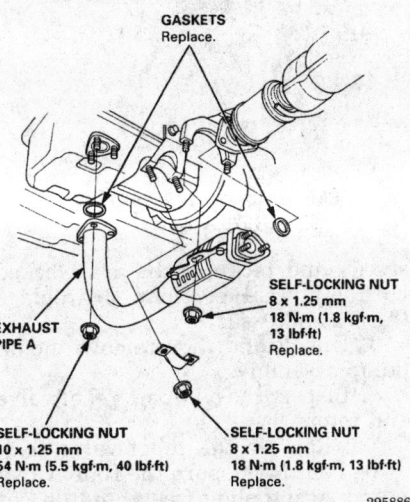

- Sedan LX without ABS model of Engine Serial Number: F22B2 – 21XXXXX.

Exhaust pipe A — 2.2L (F22B1, F22B2) engines

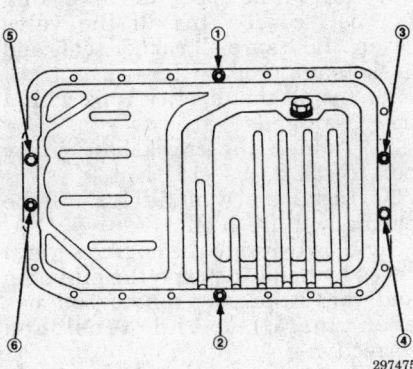

Oil pan torque sequence — 2.2L (F22B1, F22B2) engines

Oil pan tightening sequence — 2.7L engines

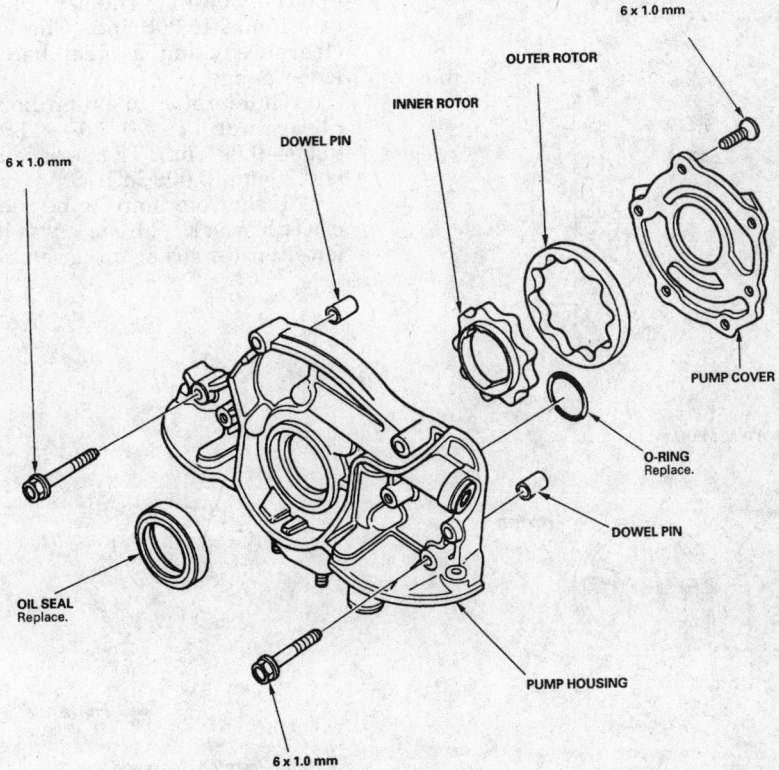

Oil pump exploded view — 1.5L (D15B7 D15B8, D15Z1) and 1.6L (D16Z6 engines

4. Rotate the crankshaft to set the No. 1 cylinder to TDC for the compression stroke. The white TDC mark on the crankshaft pulley should align with the TDC pointers on the lower timing cover.

NOTE: Mark the direction of the timing belt's rotation if it is to be reinstalled.

5. Remove the accessory drive belts and the crankshaft pulley.

6. Remove the valve cover and the upper and lower timing belt covers.

NOTE: Cover the rocker arm and shaft assemblies with a towel or sheet of plastic to keep out dirt and foreign objects.

7. Remove the dipstick and its tube from the oil pump housing.

8. Release the timing belt's tension. Then, remove the timing belt.

9. Unbolt the Crankshaft Speed Fluctuation (CKF) sensor from the oil pump cover. Disconnect the sensor and remove it so that it will not come in contact with oil or become damaged.

10. Remove the crankshaft sprocket.

11. Remove the oil pan.

12. Unbolt the oil screen and pickup tube from the oil pump housing and crankshaft buttress. If the

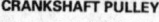

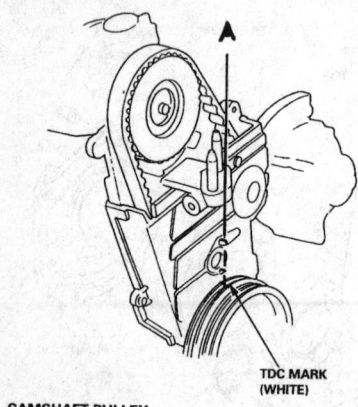

A

TDC MARK (WHITE)

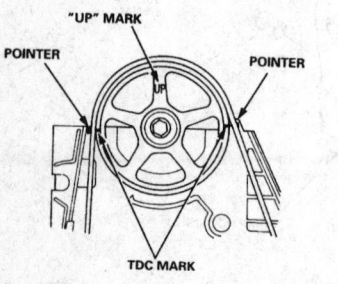

CAMSHAFT PULLEY:

"UP" MARK

POINTER   POINTER

TDC MARK

300297

Crankshaft and camshaft TDC marks — 1.6L (D16Y5, D16Y7, D16Y8) engines

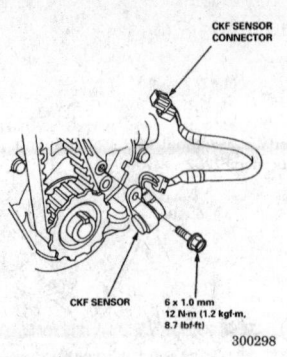

CKF SENSOR CONNECTOR

CKF SENSOR   6 x 1.0 mm
                  12 N·m (1.2 kgf·m,
                  8.7 lbf·ft)
                  300298

CKF sensor — 1.6L (D16Y5, D16Y7, D16Y8) engines

screen and pickup tube are blocked with oil residue, clean or replace them as necessary.

13. Unbolt and then remove the oil pump assembly.

14. Inspect the oil pump relief valve and rotors:

a. Remove the relief valve sealing bolt. Make sure the relief valve and spring slide freely and do not bind or stick. If the relief valve piston sticks in its bore, or is scored, replace the oil pump.

b. Inner–to–outer router clearance is 0.02–0.14mm (0.001–0.006 in.). The service limit is 0.20mm (0.008 in.).

c. Rotor–to–pump housing axial clearance is 0.03–0.08mm (0.001–0.003 in.). The service limit is 0.15mm (0.006 in.). Check the clearance using a steel bar and feeler gauge.

d. Outer rotor–to–pump housing clearance is 0.10–0.18mm (0.004–0.007 in.). The service limit is 0.20mm (0.008 in.).

e. If the rotors are to be reused, matchmark them with a felt–tipped marker for assembly.

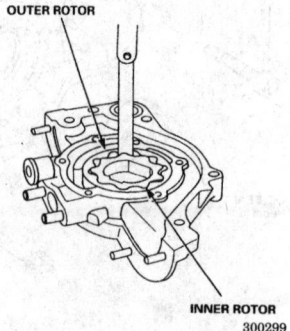

OUTER ROTOR

INNER ROTOR
300299

Rotor-to-rotor clearance inspection — 1.6L (D16Y5, D16Y7, D16Y8) engines

To install:

NOTE: Replace the rotors if they are worn or damaged. Use new O-rings when assembling and installing the oil pump.

15. Install the rotors back into their original positions. Make sure they move without binding. Pack the rotor cavity with petroleum jelly to prevent oil starvation damage when the engine is initially started.

16. Assemble the oil pump and tighten the rotor cover bolts to 5 ft. lbs. (7 Nm).

17. Make sure all gasket mating surfaces are clean prior to installation.

18. Install a new crankshaft oil seal into the oil pump housing.

19. Apply liquid gasket to the cylinder block mating surface of the block. Apply a light coat of oil to the crankshaft seal lip. Install a new O-ring on the cylinder block and install the oil pump. Apply liquid gasket to the threads of the oil pump mounting bolts and tighten them to 8 ft. lbs. (11 Nm).

20. Lightly lubricate the relief valve piston and spring, then install them. Install the sealing bolt with a new crush washer, and tighten it to 29 ft. lbs. (39 Nm).

21. Install the oil screen. Tighten the fastening nuts and bolts to 8 ft. lbs. (11 Nm).

22. Install the oil pan. Tighten the oil pan nuts and bolts to 9 ft. lbs. (12 Nm).

23. Install the crankshaft sprocket. The concave surface of the spacer must face the engine block.

24. Verify that the engine is at TDC/compression for the No. 1 cylinder.

25. Install and tension the timing belt. Tighten the tensioner adjusting bolt to 33 ft. lbs. (44 Nm).

26. Install and reconnect the CKF sensor. Tighten the sensor mounting bolt to 9 ft. lbs. (12 Nm).

27. Install the upper and lower timing belt covers. Install the valve cover. Make sure all rubber seals and gaskets are properly seated.

28. Install the dipstick tube with a new O–ring.

29. Tighten the crankshaft pulley bolt to 134 ft. lbs. (181 Nm).

30. Install a new oil filter. Refill the engine with fresh oil.

31. Slowly rotate the engine several times by hand to prime the oil pump and verify that the timing belt has been installed and tensioned correctly.

32. Install and adjust the accessory drive belts.

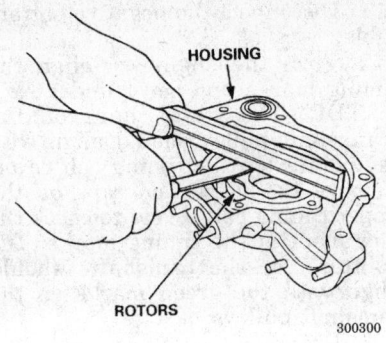

Axial clearance inspection — 1.6L (D16Y5, D16Y7, D16Y8) engines

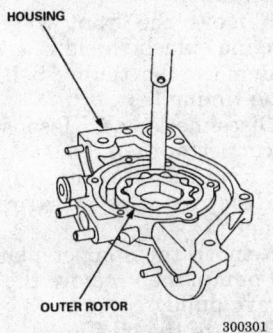

Rotor-to-housing clearance inspection — 1.6L (D16Y5, D16Y7, D16Y8) engines

33. Connect the negative battery cable.

34. Run the engine and check for proper oil pressure.

35. Check for leaks. Top up the engine oil if necessary.

1.6L (B16A2, B16A3) Engines

1. Disconnect the negative battery cable.

2. Raise and safely support the vehicle.

3. Drain the engine oil.

4. Label and disconnect the ignition wires.

5. Set the No. 1 cylinder to TDC for the compression stroke. The mark on the crankshaft pulley should align with the index mark on the timing cover.

NOTE: Mark the direction of the timing belt's rotation if it is to be reinstalled.

6. Release the tensions of the accessory drive belts and slip them off their pulleys. Remove the crankshaft pulley.

7. Remove the valve cover and the upper and lower timing belt covers.

NOTE: Cover the rocker arm and shaft assemblies with a towel or sheet of plastic to keep out dirt and foreign objects.

8. Release the timing belt's tension. Then, remove the timing belt.

9. If equipped with a Crankshaft Speed Fluctuation (CKF) sensor at the crankshaft sprocket, unbolt the sensor's bracket. Then, disconnect the sensor and remove it.

10. Remove the crankshaft sprocket.

11. Remove the oil pan and oil screen.

12. Unbolt and then remove the oil pump assembly.

13. Disassemble and inspect the oil pump:

a. Remove the relief valve sealing bolt. Make sure the relief valve and spring each slide freely and that neither binds or sticks. If the relief valve piston sticks in its bore, or is scored, replace the oil pump.

b. Inner–to–outer router clearance is 0.04–0.16mm (0.002–0.016 in.). The service limit is 0.20mm (0.008 in.).

c. Rotor–to–pump housing axial clearance is 0.02–0.07mm (0.001–0.003 in.). The service limit is 0.15mm (0.006 in.).

d. Outer rotor–to–pump housing clearance is 0.10–0.19mm (0.004–0.007 in.). The service limit is 0.20mm (0.008 in.).

e. If the rotors are to be reused, matchmark them with a felt–tipped marker for assembly.

To install:

NOTE: Replace the rotors if they are worn or damaged. Use new O-rings when assembling and installing the oil pump.

14. Install the rotors back into their original positions. Make sure they rotate without binding. Pack the rotor cavity with petroleum jelly to prevent oil starvation damage when the engine is initially started.

15. Assemble the oil pump and tighten the rotor cover bolts to 5 ft. lbs. (7 Nm).

16. Make sure all gasket mating surfaces are clean prior to installation. Replace the crankshaft oil seal prior to installing the oil pump.

17. Install a new crankshaft oil seal into the oil pump housing.

18. Apply liquid gasket to the oil pump mating surface of the cylinder block. Apply a light coat of oil to the crankshaft seal lip. Install a new oil passage O-ring on the cylinder block

and then install the oil pump. Apply liquid gasket to the threads of the oil pump mounting bolts and tighten them the 6mm bolts to 8 ft. lbs. (11 Nm). Tighten the 8mm bolts to 17 ft. lbs. (24 Nm).

19. Lightly lubricate the relief valve piston and spring, then install them. Install the sealing bolt with a new crush washer and tighten it to 29 ft. lbs. (39 Nm).

20. Install the oil screen.

21. Install the oil pan. Wait for the sealant to cure before refilling the engine with oil.

22. Install the CKF sensor. Tighten the sensor mounting bolts to 8 ft. lbs. (11 Nm).

23. Install the crankshaft sprocket. The concave surface of the spacer faces out.

24. Install and tension the timing belt. Tighten the tensioner adjusting bolt to 40 ft. lbs. (55 Nm).

25. Install and reconnect the CKF sensor. Tighten the sensor mounting bolt to 9 ft. lbs. (12 Nm).

26. Install the upper and lower timing belt covers. Install the valve cover. Make sure all rubber seals and gaskets are properly seated.

27. Tighten the crankshaft pulley bolt to 130 ft. lbs. (180 Nm).

28. Install a new oil filter. Refill the engine with fresh oil.

29. Slowly rotate the engine several times by hand to prime the oil pump and to verify that the timing belt has been installed and tensioned correctly.

30. Install and adjust the accessory drive belts.

31. Connect the negative battery cable.

32. Run the engine and check for proper oil pressure.

33. Check for leaks. Top up the engine oil if necessary.

Accord and Prelude

2.2L and 2.3L Engines

1. Disconnect the negative battery cable.

2. Drain the engine oil into a sealable container.

3. Turn the engine to align the timing marks and set cylinder No.1 to TDC. The white mark on the crankshaft pulley should align with the pointer on the timing belt cover.

4. Remove the valve cover and upper timing belt cover.

5. Remove the power steering pump belt and the alternator belt, also the air conditioning belt if so equipped.

6. Remove the crankshaft pulley and the lower timing belt cover.

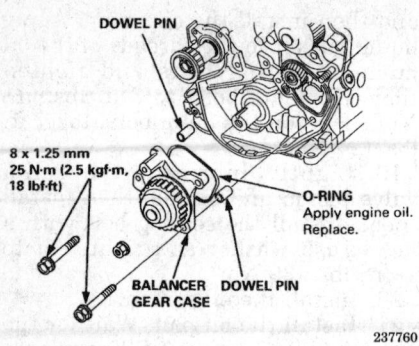

DOWEL PIN

8 x 1.25 mm
25 N·m (2.5 kgf·m,
18 lbf·ft)

O-RING
Apply engine oil.
Replace.

BALANCER
GEAR CASE

DOWEL PIN

237760

Front balancer pulley gear case — 1993–95 2.2L and 2.3L engines

7. Remove the balancer belt and the timing belt, be sure to mark the rotation of the timing belt if it is going to be reused.

8. Remove the timing belt and balancer belt tensioners.

9. If equipped, remove the bolts mounting the CKP/TDC sensor and carefully remove the CKP/TDC sensor from the oil pump. Disconnect the CKP/TDC sensor connector and remove it from the vehicle.

10. Remove the timing belt drive pulley and key from the crankshaft.

11. Insert a suitable tool into the maintenance hole in the front balancer shaft and remove the balancer driven pulley.

12. Align the rear timing balancer pulley using a 6·100mm bolt or rod. Mark the bolt or rod at a point 2.9 inches (74mm) from the end. Remove the bolt from the maintenance hole on the side of the block; insert the bolt/rod into the hole. Align the 74mm mark with the face of the hole. This pin will hold the shaft in place.

13. Remove the balancer gear case and the dowel pins. Discard the O-ring.

14. Remove the balancer driven gear attaching bolt and the balancer driven gear.

15. Remove the oil pan and the oil screen. Discard the screen gasket.

16. Remove the oil pump mounting bolts and remove the oil pump assembly. Remove the dowel pins from the engine and clean the oil pump mating surfaces of old gasket material and oil. Discard the O-rings.

To install:

17. Install the two dowel pins and new O-rings to the cylinder block.

18. Be sure that the mating surfaces are clean and dry. Apply a liquid gasket evenly in a narrow bead, centered on the mating surface. Once the sealant is applied, do not wait longer than 20 minutes to install the parts; the sealant will become ineffective. After final assembly, wait at least 30 minutes before adding oil to the engine, giving the sealant time to set. To prevent leakage of oil, apply a suitable thread sealer to the inner threads of the bolt holes.

19. Install the oil pump to the engine block. Torque the mounting bolts to 9 ft. lbs. (12 Nm).

20. Install the oil screen. Torque the screen mounting bolts and nuts to 9 ft. lbs. (12 Nm).

21. Install the oil pan.

22. Install the balancer driven pulley to the front balancer belt, hold the balancer shaft in place with a suitable tool. Torque the attaching bolt to 22 ft. lbs. (29 Nm).

23. Install the balancer driven gear to the rear balancer shaft. Torque the bolt to 18 ft. lbs. (25 Nm).

24. Before installing the balancer driven gear and the gear case, apply molybdenum disulfide (lithium grease) to the thrust surfaces of the balancer gears.

25. Align the groove on the pulley edge to the pointer on the balancer gear case.

26. Install the balancer gear case to the engine and install the mounting bolts and nut. The rear balancer shaft is being held in place with a 6·100mm bolt. Torque the mounting bolts and nut to 18 ft. lbs. (25 Nm).

27. Check the alignment of the pointer on the balancer pulley to the pointer on the oil pump.

28. Install the drive pulley to the crankshaft.

29. If equipped, connect the CKP/TDC sensor connector to the wire harness on the engine. Install the CKP/TDC sensor to the oil pump and install the mounting bolts. Torque the mounting bolts to 9 ft. lbs. (12 Nm).

30. Install the timing belt tensioners.

31. Install the timing belt and the balancer belt.

32. Install the crankshaft pulley and the lower timing belt cover.

33. Install the drive belts for the alternator, power steering and A/C compressor; adjust the tension.

34. Install the valve cover and upper timing belt cover.

35. Refill the engine with clean, fresh oil.

36. Connect the negative battery cable and enter the radio security code.

2.7L Engines

NOTE: The radio may contain a coded theft protection circuit. Always obtain the code before disconnecting the battery.

1. Disconnect the negative battery cable.

2. Turn the engine to align the timing marks and set cylinder No.1 to TDC. The white mark on the crankshaft pulley should align with the pointer on the timing belt cover. Remove the inspection caps on the upper timing belt covers to check the alignment of the timing marks. The pointers for the camshafts should align with the green marks on the camshaft pulleys.

3. Raise and safely support the vehicle.

4. Drain the engine oil into a sealable container.

5. Remove the front wheels and the engine splash shield.

6. Remove the timing belt covers and the timing belt.

7. Disconnect the CKP sensor electrical connector then remove the CKP sensor from the oil pump.

8. Remove the timing belt tensioner.

9. Remove the stopper plate from the oil pump, then remove the timing belt drive pulley.

10. Remove the oil filter.

11. Disconnect the oil pressure switch terminal, then remove the oil filter base attaching bolts.

12. Remove the oil filter base and discard the O-rings.

13. Remove the oil pan.

14. Remove the oil screen and baffle plate.

15. Remove the oil pass pipe attaching bolts then remove the pipe from the oil pump. Discard the O-rings.

16. Remove the bolts from the oil pump, note the location of the bolts.

17. Remove the oil pump and dowel pins from the engine.

To install:

18. Install the two dowel pins to the cylinder block.

19. Be sure that the mating surfaces are clean and dry. Apply a liquid gasket evenly in a narrow bead, centered on the mating surface. Once the sealant is applied, do not wait longer than 20 minutes to install the parts; the sealant will become ineffective. After final assembly, wait at least 30 minutes before adding oil to the engine, giving the sealant time to set. To prevent leakage of oil, apply a suitable thread sealer to the inner threads of the bolt holes.

20. Apply grease to the lips of the crankshaft oil seal. Install the oil pump to the engine block. Torque the 6·1.0mm mounting bolts to 9 ft. lbs. (12 Nm) and torque the 8·1.25mm mounting bolts to 16 ft. lbs. (22 Nm). Clean any excess grease from the

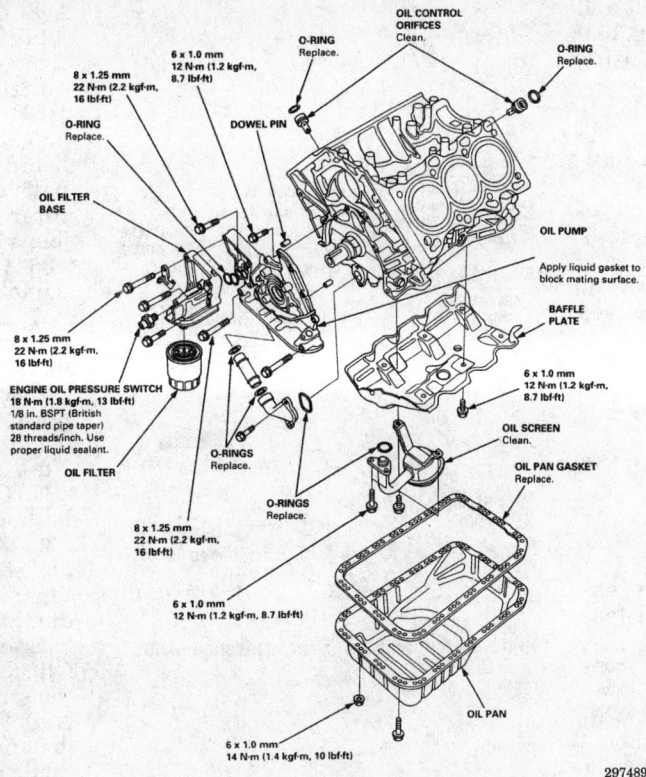

8 x 1.25 mm
22 N·m (2.2 kgf·m,
16 lbf·ft)

O-RING
Replace.

6 x 1.0 mm
12 N·m (1.2 kgf·m,
8.7 lbf·ft)

O-RING
Replace.

OIL CONTROL
ORIFICES
Clean.

O-RING
Replace.

DOWEL PIN

OIL FILTER
BASE

OIL PUMP

Apply liquid gasket to
block mating surface.

BAFFLE
PLATE

8 x 1.25 mm
22 N·m (2.2 kgf·m,
16 lbf·ft)

ENGINE OIL PRESSURE SWITCH
18 N·m (1.8 kgf·m, 13 lbf·ft)
1/8 in. BSPT (British
standard pipe taper)
28 threads/inch. Use
proper liquid sealant.

OIL FILTER

O-RINGS
Replace.

O-RINGS
Replace.

OIL SCREEN
Clean.

6 x 1.0 mm
12 N·m (1.2 kgf·m,
8.7 lbf·ft)

OIL PAN GASKET
Replace.

8 x 1.25 mm
22 N·m (2.2 kgf·m,
16 lbf·ft)

6 x 1.0 mm
12 N·m (1.2 kgf·m, 8.7 lbf·ft)

OIL PAN

6 x 1.0 mm
14 N·m (1.4 kgf·m, 10 lbf·ft)

297489

Oil pump and related components — 2.7L engines

crankshaft, then make sure that the seal lips are not distorted.

21. Install new O–rings to the oil pass pipe then install the pass pipe to the oil pump and the crankshaft bridge. torque the attaching bolts to 9 ft. lbs. (12 Nm).

22. Install the baffle plate. Torque the baffle plate mounting bolts to 9 ft. lbs. (12 Nm).

23. Install the oil screen with a new O–ring. Torque the screen mounting bolts to 9 ft. lbs. (12 Nm).

24. Install the oil pan.

25. Install new O–rings to the oil filter base and install the oil filter base to the engine. Torque the mounting bolts to 16 ft. lbs. (22 Nm).

26. Connect the oil pressure switch terminal. Torque the attaching bolt to 1.8 ft. lbs. (2.5 Nm).

27. Install the oil filter.

28. Install the timing belt drive pulley and the stopper plate. Torque the stopper plate mounting bolts to 9 ft. lbs. (12 Nm).

29. Install the timing belt tensioner.

30. Install the CKP sensor to the oil pump and connect the electrical connector. torque the mounting bolts to 9 ft. lbs. (12 Nm).

31. Install the timing belt and the timing belt covers.

32. Install the engine splash shield and the front wheels.

33. Lower the vehicle and fill the engine with oil.

34. Connect the negative battery cable and enter the radio security code.

35. Run the engine and check for leaks.

TRANSAXLE

Manual Transaxle Assembly

REMOVAL AND INSTALLATION

NOTE: The radio may contain a coded theft protection circuit. Always obtain the code number before disconnecting the battery. If the vehicle is equipped with 4WS, the steering control unit is shut down when the battery is disconnected. After connecting the battery, turn the steering wheel lock–to–lock to reset the steering control unit.

Civic and del Sol

— WARNING —

Use only genuine Honda manual transaxle fluid (MTF) — it is specially formulated for use in Honda transaxles. If Honda MTF is not available, API SG/SH 10W–30 or 10W–40 motor oil may be used as a temporary lubricant. However, motor oil will cause increased transaxle wear and shifting effort. Refill the transaxle with Honda MTF as soon as possible.

1. Disconnect the negative and positive battery cables.

2. Drain the transaxle fluid.

3. Remove the resonator, the air cleaner box, and the air intake duct.

4. Disconnect the starter cables and the transaxle ground cable.

5. Disconnect the back–up light switch connection.

6. Move the upper radiator hose out of its clamp.

7. Disconnect the vehicle speed sensor (VSS) connector.

8. Unbolt the clutch fluid line bracket. Unbolt and remove the slave cylinder. It isn't necessary to disconnect the clutch fluid line.

9. Raise and safely support the vehicle. Remove the front wheels.

10. Remove the strut pinch bolt and fork bolt. Disconnect the lower ball joint from the steering knuckle using a ball joint remover.

11. Pry the halfshaft inboard joints out of the transaxle case. Swing the steering knuckles out to free the halfshafts from the transaxle.

12. Tie the halfshafts up and out of the way with wire so that the joints will not be stressed. Tie plastic bags over the inboard joints to prevent damage to the CV–boots and splined shafts.

13. Disconnect the shift rod and extension rod from the transaxle case. Drive the shift rod retaining pin out with a pin punch.

14. Disconnect and remove the front exhaust pipe.

15. Remove the engine–to–transaxle stiffener brackets and the clutch cover plate.

16. Attach a lifting chain to the engine and lift slightly to ease the tension on the mounts.

17. Remove the splash shield from underneath the vehicle.

18. Unbolt and remove the right–front mount/bracket assembly.

19. Place a jack under the transaxle to support its weight.

20. Remove the transaxle side mount and its bracket.

21. Remove the starter's lower mounting bolt. Remove the upper three transaxle case bolts.

22. Remove the three rear transaxle mount bracket bolts. Next, remove the lower three transaxle case bolts.

23. Pull the transaxle away from the engine until it clears the mainshaft. Lower the transaxle out of the vehicle Be careful not to bend the clutch hydraulic line.

To install:

NOTE: Use new self–locking nuts and color–coded bolts when installing the transaxle and suspension components.

24. Apply high–temperature grease to the mainshaft splines, release fork contact points, and throw–out bearing. The manufacturer recommends part No. 08798–9002, Honda Super High–temp Urea Grease.

25. Place the transaxle on a transaxle jack and raise it to the level of the engine.

26. Align the transaxle and engine. Make sure the transaxle case dowel pins are securely seated, and fit the transaxle onto the engine. Install the upper and lower transaxle case bolts and the 14mm rear mount bolts and washers, only hand–tighten them at this time.

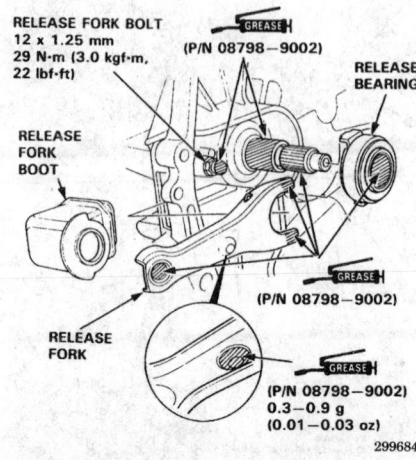

RELEASE FORK BOLT
12 x 1.25 mm
29 N·m (3.0 kgf·m,
22 lbf·ft)

GREASE (P/N 08798–9002)

RELEASE BEARING

RELEASE FORK BOOT

RELEASE FORK

GREASE (P/N 08798–9002)

GREASE (P/N 08798–9002)
0.3–0.9 g
(0.01–0.03 oz)

299684

Release fork and mainshaft lubrication points

27. Raise the transaxle and install the side mount. Tighten the upper and lower transaxle case bolts to 47 ft. lbs. (64 Nm). Tighten the 14mm rear mount bracket bolts to 61 ft. lbs. (84 Nm).

28. First, tighten the transaxle side mount bracket nuts and bolt to 47 ft. lbs. (64 Nm). each. Next, tighten the mount bushing bolts to 47 ft. lbs. (64 Nm). Finally, tighten the through–bolt to 54 ft. lbs. (74 Nm).

29. Install the right–front mount/bracket assembly. Use three new 12mm bolt and washers, and tighten them to 47 ft. lbs. (64 Nm). Tighten the two 10mm bolts to 33 ft. lbs. (45 Nm).

30. Install the clutch cover.

31. Install the engine–to–transaxle stiffener brackets and torque the 8mm bolts to 17 ft. lbs. (24 Nm). Tighten the 10mm bolts to 33 ft. lbs. (44 Nm).

32. Once the transaxle is bolted to the engine, and the transaxle mounts are securely tightened, the engine lifting chain may be removed.

33. Connect the shift rod with a new spring pin and clip. Then, fit the shift rod boot back into place. Connect the torque rod and tighten its bolt to 16 ft. lbs. (22 Nm).

34. Install the front exhaust pipe. Use new self–locking nuts and gas-

kets. Tighten the rear flange nuts to 16 ft. lbs. (22 Nm). Tighten the front flange nuts to 25 ft. lbs. (33 Nm), or 40 ft. lbs. (54 Nm) for D16Y8 and D16Y7 engines.

35. Install new set rings on the inboard CV–joint splines. Install the halfshafts into the transaxle case and intermediate shaft. The inboard joints must snap into place.

36. Connect the lower ball joint and damper fork. Install the front wheels.

37. Install the slave cylinder and clutch pipe stay. Coat the slave cylinder's tip with high temperature grease. Make sure it snaps into the release fork. Tighten the slave cylinder mounting bolts to 16 ft. lbs. (22 Nm).

38. Connect the VSS connector and back–up light switch connectors.

39. Install the wire harness clamps and starter cables. Install the resonator, air cleaner box and air intake duct. Fit the upper radiator hose back into its bracket.

40. Lower the vehicle and tighten the strut pinch bolts to 32 ft. lbs. (44 Nm). Tighten the fork bolts to 47 ft. lbs. (65 Nm). Tighten the ball joint castle nuts to 40 ft. lbs. (55 Nm), and then tighten them only enough to install new cotter pins.

41. Turn the breather cap so that the arrow with the **F** mark points toward the front of the vehicle.

42. Refill the transaxle with the Honda MTF fluid.

43. Reconnect the positive and negative battery cables.

44. Bleed the clutch hydraulic system.

45. Check the clutch and transaxle for smooth operation.

46. Check and adjust the front wheel alignment.

Prelude

1. Shift the transaxle to **R**.

2. Disconnect the negative and positive battery cables. Remove the battery.

3. Remove the intake duct and air cleaner case. Remove the battery base.

4. Remove the vacuum tank and bracket. Do not disconnect the hoses.

5. Disconnect the starter wires and remove the starter.

6. Loosen, but do not remove the two upper transaxle mounting bolts.

7. Disconnect the transaxle ground cable and the backup light switch wire. Unbolt the engine harness clamp.

8. Leave both shift cables attached their bracket. Remove the shift cables from the transaxle case and

wire them safely out of the work area.

9. Disconnect the vehicle speed sensor connector. Leave the sensor hoses connected and remove the sensor from the transaxle case.

10. Remove the slave cylinder mounting bolts. Leave the hydraulic line connected to the slave cylinder. Remove the slave cylinder from the release fork, and move it out of the work area.

NOTE: Do not operate the clutch pedal once the slave cylinder has been removed. Be careful not to kink the metal hydraulic line.

11. Raise and safely support the vehicle. Drain the transaxle fluid.

12. Remove the clutch damper mounting bolts and raise the clutch damper.

13. If equipped, remove the rear engine mount bracket stay.

14. Remove the front wheels.

15. Remove the cotter pins and lower arm ball joint nuts. Separate the ball joints and lower arms using a press–type ball joint tool.

16. Remove the damper fork bolt and the radius rod on the right side of the vehicle only.

17. Use a suitable tool to pry the right and left halfshafts out of the differential and the intermediate shaft. Pull on the inboard joint and remove the right and left halfshafts.

18. Unbolt and remove the intermediate shaft from the differential. Tie plastic bags over the halfshaft inboard joints to prevent damage to the boots and splines. Wire the halfshafts to the underbody of the vehicle so that their weight doesn't hang on their outboard joints.

19. Remove the center beam and remove the clutch cover. On vehicles with the H22A1 engine, remove the front engine stiffener plate.

20. Remove the rear beam stiffener and the intake manifold stay.

21. Remove the three rear engine mount bracket bolts.

22. Place a transaxle jack under the transaxle and raise the transaxle just enough to take its weight off the of mounts.

23. Remove the transaxle mount and mount bracket.

24. Remove the two upper transaxle housing mounting bolts and the three lower transaxle housing bolts.

25. Pull the transaxle away from the engine to clear the mainshaft.

26. Lower the transaxle from the vehicle.

To install:

NOTE: Use new self–locking nuts and set rings when assembling the front suspension components and halfshafts. Use new self–locking bolts when installing the center beam and rear engine mount bracket. These fasteners can be purchased from a Honda dealer.

27. Make sure the dowel pins are installed into the transaxle case.

28. Apply heavy duty high temperature grease (use Honda part number 08798–9002) to the mainshaft splines, release fork bolt and paws, and the throw–out bearing. Install the bearing and release fork. Make sure the release fork snaps into place.

29. Raise the transaxle into position with a transaxle jack.

30. Install the three lower and two upper transaxle mounting bolts and evenly tighten them to 47 ft. lbs. (65 Nm).

31. Install the transaxle mount and mount bracket. Install the through-bolt and tighten temporarily. Make sure the engine is level. First tighten the three bracket–to–mount nuts and two bolts to 28 ft. lbs. (39 Nm). Then, tighten the through-bolt to 47 ft. lbs. (65 Nm).

32. Install the three new rear engine mount bracket bolts on the engine side and tighten them to 40 ft. lbs. (55 Nm).

33. Install the rear beam stiffener and tighten the bolts to 28 ft. lbs. (39 Nm).

34. Install the intake manifold stay and tighten the bolts to 16 ft. lbs. (22 Nm).

35. Install the clutch cover and tighten the bolts to 9 ft. lbs. (12 Nm).

36. On Preludes equipped with the H22A1 engine, install the front engine stiffener bolts. Tighten these bolts to 28 ft. lbs. (39 Nm). First tighten the one bolt threaded into the transaxle case, and then tighten the two bolts to the engine block.

37. Install the center beam and tighten the bolts to 43 ft. lbs. (60 Nm).

38. Install the intermediate shaft. Tighten its mounting bolts to 28 ft. lbs. (39 Nm).

39. Install new set rings onto the halfshaft inboard joint splines. Install the halfshafts, making sure that they lock into place.

40. Install the radius rod and damper fork. Only hand–tighten their fasteners at this time.

41. Install the ball joint to the lower arm. Tighten the castle nut to 36–43 ft. lbs. (50–60 Nm). Then, only

tighten the nut enough to install a new cotter pin.

42. If equipped, install the rear engine mount bracket stay. Tighten the nut to 15 ft. lbs. (21 Nm) and the bolt to 28 ft. lbs. (39 Nm).

43. Install the clutch damper and tighten the bolts to 16 ft. lbs. (22 Nm).

44. Install the front wheels.

45. Lower the vehicle.

46. Use a floor jack placed under the right front control arm to raise the vehicle enough so that its weight is supported by the jack. Tighten the radius rod mounting bolts to 76 ft. lbs. (105 Nm) and the radius rod nut to 32 ft. lbs. (44 Nm). Tighten the damper pinch bolt to 32 ft. lbs. (44 Nm). Tighten the damper fork bolt to 47 ft. lbs. (65 Nm). After preloading the suspension, lower the vehicle and remove the floor jack.

47. Coat the tip of the slave cylinder with heavy duty high temperature grease. Install the clutch hose pipe and clutch slave cylinder to the transaxle housing. Make sure the slave cylinder snaps into the release fork. Tighten the slave cylinder mounting bolts to 16 ft. lbs. (22 Nm).

48. Install the speed sensor. Tighten the mounting bolt to 14 ft. lbs. (19 Nm).

49. Install the shift cable and select cable to the shift arm lever. Install the shift cable assembly onto the transaxle case. Tighten the cable bracket mounting bolts to 16 ft. lbs. (22 Nm). Install new cotter pins.

50. Connect the backup light switch coupler and the transaxle ground cable. Install the harness clamp.

51. Install the starter. Tighten the 10 x 1.25mm bolt to 32 ft. lbs. (45 Nm) and the 12 x 1.25mm bolt to 54 ft. lbs. (75 Nm). Connect the starter wires.

52. Loosen the three front engine mount bracket bolts. Torque them to 28 ft. lbs. (39 Nm).

53. Install the vacuum tank and its bracket. Install the air cleaner case and intake duct.

54. Fill the transaxle with the proper type and quantity of oil.

55. Install the battery base stay and the battery base. Tighten the battery base bolts to 16 ft. lbs. (22 Nm). Install the battery and connect the battery cables.

56. Check the clutch pedal freeplay.

57. Start the vehicle and check the transaxle and clutch for smooth operation.

58. On Preludes equipped with 4WS, start the engine and turn the

steering wheel lock–to–lock to reset the steering control unit.

59. Check and adjust the front wheel alignment.

60. Enter the radio security code.

Accord

1. Shift the transaxle into **R**.

2. Disconnect the negative and positive battery cables. Remove the battery.

3. Disconnect the IAC solenoid connector. Remove the intake duct, resonator, and air cleaner case, and battery base.

4. Disconnect the starter wires and remove the starter.

5. Disconnect the transaxle ground cable and the backup light switch wire.

6. Remove the cable stay and then disconnect the cables from the top housing of the transaxle. Remove both cables and the stay together.

7. Disconnect the vehicle speed sensor connector and remove the speed sensor. Leave the speed sensor hoses connected.

8. Remove the shift cable bracket. Then, disconnect the shift and select cables from the top of the transaxle case. Leave the cables and bracket together, and wire them out of the work area.

9. Remove the mounting bolts and clutch slave cylinder with the clutch pipe and pushrod.

10. Remove the mounting bolt and clutch hose joint with the clutch pipe and clutch hose.

NOTE: Do not operate the clutch pedal once the slave cylinder has been removed. Be careful not to kink the metal hydraulic lines.

11. Remove the two upper transaxle case bolts.

12. Raise and safely support the vehicle.

13. Remove the front wheels.

14. Remove the engine splash shield.

15. Drain the transaxle fluid.

16. Remove the clutch damper bracket and raise it out of the way.

17. Remove the subframe center beam.

18. Remove the cotter pins and lower arm ball joint nuts. Separate the ball joints and lower arms using a press–type ball joint tool.

19. Remove the right damper fork bolt. Remove the right damper pinch bolt, then separate the damper fork and damper. Remove the radius rod bolts and nut; then, remove the right radius rod.

20. Use a suitable pry tool to separate the right and left halfshafts from the differential and the intermediate shaft. Remove the left halfshaft.

21. Remove the intermediate shaft from the differential by removing its three bearing shaft mounting bolts.

22. Swing the right halfshaft out and wire it up inside the right fender well. Tie plastic bags over the inboard CV–joints to protect the boots and splines from damage.

23. Remove the engine stiffener and the clutch cover.

24. Remove the intake manifold bracket.

25. Remove the rear engine mount bracket. Remove and discard the three rear engine mount bracket mounting bolts.

26. Place a transaxle jack under the transaxle. Raise the transaxle just enough to take the weight off the its mounts.

NOTE: A chain hoist may be attached the transaxle lifting hooks to steady it and aid in lowering it from the vehicle.

27. Remove the transaxle housing mounting bolt on the engine side.

28. Remove the transaxle mount bolt and loosen the mount bracket nuts.

29. Remove the three transaxle housing mounting bolts.

30. Remove the transaxle from the vehicle.

To install:

NOTE: Use new self–locking nuts when assembling the front suspension. Install new set rings onto the inboard CV–joints. Use new self–locking bolts when installing transaxle rear mount bracket (the bolts are color coded by type). New fasteners are available from a Honda dealer.

31. Make sure the two dowel pins are installed into the transaxle case.

32. Apply heavy duty high temperature grease (use Honda part No. 08798–9002) to the release bearing, mainshaft splines, and the release fork pawls. Install the release fork and release bearing.

33. Raise the transaxle into position.

34. Install the three lower transaxle case bolts and tighten to 47 ft. lbs. (65 Nm).

35. Install the transaxle mount and mount bracket. Install the through-bolt and tighten temporarily. Make sure the engine is level and tighten the three mount bracket nuts to 40 ft. lbs. (55 Nm), or 28 ft. lbs. (38 Nm) for

1994–95 Accords. Tighten the through-bolt to 47 ft. lbs. (65 Nm).

36. Install the upper transaxle case bolts on the engine side and tighten to 47 ft. lbs. (65 Nm).

37. Install the three new rear engine bracket mounting bolts and tighten to 40 ft. lbs. (55 Nm).

38. If equipped (1993 Accords only), install the rear engine mount bracket stay. Tighten the mounting bolt to 28 ft. lbs. (39 Nm) and then tighten the mounting nut to 15 ft. lbs. (21 Nm).

39. Install the intake manifold bracket and tighten the bolts to 16 ft. lbs. (22 Nm).

40. Install the clutch cover and tighten the bolts to 9 ft. lbs. (12 Nm).

41. Install the subframe center beam with new self–locking bolts. Evenly tighten the bolts to 37 ft. lbs. (50 Nm).

42. If equipped, install the engine stiffener plate and loosely install the mounting bolts. Tighten the stiffener–to–transaxle case mounting bolt to 28 ft. lbs. (39 Nm), then tighten the two stiffener–to–engine block mounting bolts to 28 ft. lbs. (39 Nm) beginning with the bolt closest to the transaxle.

43. Install the radius rod and the damper fork. Install all the fasteners, but only hand–tighten them at this time.

44. Install the intermediate shaft, tighten its mounting bolts to 28 ft. lbs. (39 Nm).

45. Install a new set ring on the end of each halfshaft. Install the right and left halfshafts. Turn the right and left steering knuckle fully outward and slide the axle into the differential, until the set ring is felt engaging the differential side gear.

46. Reconnect the lower control arm ball joints. Tighten the castle nuts to 40 ft. lbs. (50 Nm). Then, tighten them only enough to install a new cotter pin.

47. Install the clutch damper and torque its mounting bolts to 16 ft. lbs. (22 Nm).

48. Install the front wheels. Lower the vehicle.

49. Place a floor jack under the right front knuckle, and raise the jack until it is supporting the vehicle's weight.

50. Tighten the radius rod mounting bolts to 76 ft. lbs. (105 Nm) and the radius rod nut to 32 ft. lbs. (44 Nm). Tighten the damper fork nut while holding the damper fork bolt to 40 ft. lbs. (55 Nm). Tighten the damper pinch bolt to 32 ft. lbs. (44 Nm).

51. Coat the tip of the slave cylinder with high temperature grease. Install the clutch hose joint and clutch slave cylinder to the transaxle housing. Make sure the slave cylinder's tip snaps into the release fork. Tighten the slave cylinder mounting bolts to 16 ft. lbs. (22 Nm).

52. Install the speed sensor. Tighten the mounting bolt to 13 ft. lbs. (18 Nm).

53. Install the shift cable and select cable to the shift arm lever. Tighten the cable bracket mounting bolts to 20 ft. lbs. (27 Nm). Install new cotter pins.

54. Connect the backup light switch.

55. Install the starter. Tighten the 10 x 1.25mm bolt to 32 ft. lbs. (45 Nm) and the 12 x 1.25mm bolt to 54 ft. lbs. (75 Nm). Connect the starter wires.

56. Install the transaxle ground cable.

57. Fill the transaxle with the proper type and quantity of oil.

58. Install the air cleaner case and the resonator, then the intake duct.

59. Install the battery tray bracket and battery tray, torque the bolts to 16 ft. lbs. (22 Nm).

60. Install the battery and connect the battery cables.

61. Check the clutch pedal freeplay.

62. Check and adjust the front wheel alignment.

63. Road test the vehicle and check the transaxle for smooth operation.

64. Loosen the three front engine mount bracket mounting bolts, and then retorque them to 28 ft. lbs. (38 Nm).

65. Enter the radio security code.

Clutch Assembly

REMOVAL AND INSTALLATION

NOTE: The original radio contains a coded anti-theft circuit. Obtain the security code number before disconnecting the battery cables. On vehicles equipped with 4WS, the steering control unit is shut down when the battery is disconnected. After reconnecting the battery cables, turn the steering wheel lock-to-lock to reset the control unit.

1. Disconnect the negative battery cable.

2. Raise and safely support the vehicle.

3. Remove the transaxle from the vehicle. Matchmark the flywheel and clutch for reassembly.

4. Use a flywheel ring-gear holder to lock the flywheel in position.

5. Loosen the pressure plate bolts two turns at a time working in a crisscross pattern to prevent warping the pressure plate. Remove the pressure plate and clutch disc.

6. Inspect the flywheel, disc, and pressure plate for wear, cracks, and warpage. Light scoring of the flywheel may be polished out; gouges, warpage, burn marks, cracks, or chipped teeth require replacement of the flywheel.

NOTE: If the flywheel is to be removed, but is going to be reused, matchmark it to the engine block prior to removal. Aligning the matchmarks upon reassembly will preserve driveline balance.

7. Inspect the flywheel's ball bearing: turn the inner race of the bearing with your finger, and make sure it turns smoothly and quietly. If the bearing is loose or noisy, or exhibits rough motion, replace it.

8. Remove the release fork boot, Squeeze the release fork retaining spring to disengage the fork from its pivot. Remove the release fork from the clutch housing.

9. Remove the release bearing. Spin the bearing by hand to check its degree of play. Replace the release bearing if it has excessive play or is leaking grease.

10. Inspect the rear main bearing oil seal for signs of leakage. If necessary, replace the seal to prevent oil leakage onto the clutch's friction surfaces.

To install:

11. If necessary, drive out the flywheel bearing and then use a suitably-sized bearing driver to install a new one. Use a crisscross pattern to tighten the flywheel mounting bolts in several steps to 87 ft. lbs. (118 Nm) for vehicles with SOHC engines. If equipped with the B16A2 or B16A3 engine, tighten the flywheel bolts to 76 ft. lbs. (105 Nm).

12. Install the clutch disc and pressure plate by aligning the dowels on the flywheel with the dowel holes in the pressure plate. If a new pressure plate is not being installed, align the matchmarks that were made during removal. Install and hand-tighten the pressure plate bolts.

13. Insert a suitable clutch disc alignment tool into the splined hole in the clutch disc. Align the clutch and pressure plate.

14. Tighten the pressure plate bolts in a crisscross pattern two turns at a time to prevent warping the pressure plate. The final torque is 19 ft. lbs. (26 Nm).

15. Remove the alignment tool and ring gear holder.

16. Coat the mainshaft with heavy-duty high-temperature grease. The manufacturer recommends part No. 08798–9002, Honda super high-temp urea grease.

17. Coat the release fork pawls and the inner race of the release bearing with high temperature grease and install them into the clutch housing. Make sure the release fork retainer spring snaps into place on the pivot. The bearing and fork must fit together properly and slide back and forth smoothly.

18. Coat the tip of the slave cylinder with grease. Install the release fork boot.

19. Install the transaxle, making sure the mainshaft is properly aligned with the clutch disc splines and the transaxle case dowels are properly aligned with the engine block.

20. Install the transaxle case bolts and sequentially tighten them to 47 ft. lbs. (65 Nm).

21. Bleed the clutch hydraulic system.

22. Adjust the clutch pedal freeplay.

23. Verify that all engine and transaxle components are installed and connected properly.

24. Reconnect the negative battery cable.

25. Road test the vehicle.

Clutch Master Cylinder

REMOVAL AND INSTALLATION

NOTE: The original radio may contain a coded anti-theft circuit. Obtain the security code number before disconnecting the negative battery cable. On vehicles equipped with 4WS, the steering control module is shut down when the battery is disconnected. After reconnecting the battery, turn the steering wheel lock-to-lock to reset the control unit.

1. Disconnect the negative battery cable.

2. Use a suction pump to remove the brake fluid from the clutch master cylinder reservoir.

3. Remove the reservoir bracket bolts and remove the reservoir.

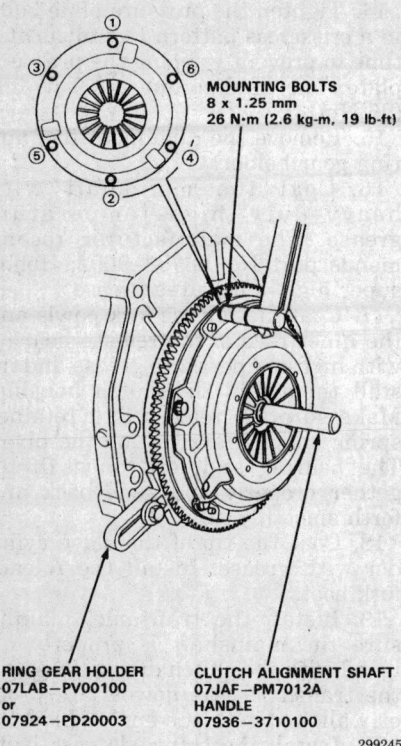

MOUNTING BOLTS
8 x 1.25 mm
26 N·m (2.6 kg-m, 19 lb-ft)

RING GEAR HOLDER
07LAB—PV00100
or
07924—PD20003

CLUTCH ALIGNMENT SHAFT
07JAF—PM7012A
HANDLE
07936—3710100

299245

Clutch alignment tools and pressure plate torque sequence

4. Disconnect the reservoir hose from the clutch master cylinder. Plug the hose to keep the fluid from draining from the reservoir.

5. Use a flare wrench to disconnect the fluid line from the clutch master cylinder body.

6. Working under the dashboard, pull out the cotter pin and remove the clevis pin from the clutch pedal linkage. Separate the clevis from the clutch pedal.

7. Remove the two clutch master cylinder mounting nuts.

8. Pull the clutch master cylinder off of the bulkhead to remove it.

To install:

9. Install the clutch master cylinder onto the bulkhead. Replace the master cylinder–to–bulkhead gasket if it's damaged.

10. Tighten the clutch master cylinder mounting nuts to 9 ft. lbs. (13 Nm).

11. Reconnect the clutch pedal to the linkage clevis. Install the clevis pin with a new cotter pin.

12. Reconnect the fluid line and carefully torque the fitting to 14 ft. lbs. (19 Nm).

13. Install the clutch master cylinder reservoir and reconnect the reservoir hose.

14. Fill the clutch master cylinder reservoir with clean brake fluid.

15. Bleed the hydraulic clutch system.

16. Refill the clutch master cylinder reservoir.

17. Reconnect the negative battery cable.

18. Check the clutch pedal adjustment and freeplay.

19. Start the engine and check the operation of the clutch starter interlock switch.

20. On vehicles with 4WS, turn the steering wheel lock–to–lock to reset the control unit.

21. Enter the radio security code.

Clutch Slave Cylinder

REMOVAL AND INSTALLATION

NOTE: The original radio contains a coded anti–theft circuit. Obtain the security code before disconnecting the battery cables. On vehicles equipped with 4WS, the steering control unit is shut down when the battery is disconnected. After reconnecting the battery cable, turn the steering wheel lock–to–lock to reset the control unit.

1. Disconnect the negative battery cable.

2. Use a flare wrench to disconnect the fluid line from the slave cylinder.

3. Unbolt the slave cylinder and remove it from the clutch housing.

To install:

4. Coat the tip of the slave cylinder

5. Install the slave cylinder onto the clutch housing. Make sure that the slave cylinder's tip is seated in the release fork. Tighten the mounting bolts to 16 ft. lbs. (22 Nm).

6. Reconnect the fluid line and carefully torque the fitting to 11 ft. lbs. (15 Nm).

7. Fill the clutch master cylinder reservoir with clean DOT 3 or 4 brake fluid.

8. Bleed the hydraulic clutch system.

9. Refill the clutch master cylinder reservoir.

10. Reconnect the negative battery cable.

11. Check the clutch operation.

12. On vehicles equipped with 4WS, turn the steering wheel lock–to–lock to reset the steering control module.

13. Enter the radio security code.

Hydraulic Clutch System Bleeding

PROCEDURE

1. Fill the clutch master cylinder reservoir with clean DOT 3 or 4 brake fluid.

2. Attach a rubber tube to the clutch slave cylinder bleed screw. Route the tube into a container of clean brake fluid.

3. Loosen the bleed screw.

4. Slowly pump the clutch pedal until the fluid draining from the slave cylinder is free of air bubbles.

5. Tighten the bleed screw to 6–7 ft. lbs. (8–10 Nm).

6. Refill the clutch master cylinder reservoir with brake fluid.

Automatic Transaxle Assembly

NOTE: The radio may contain a coded theft protection circuit. Always obtain the code number before disconnecting the battery. If the vehicle is equipped with 4WS, the steering control unit is shut down when the battery is disconnected. After connecting the battery, turn the steering wheel lock–to–lock to reset the steering control unit.

REMOVAL AND INSTALLATION

Civic and del Sol

1. Disconnect the negative and positive battery cables.

2. Remove the resonator, the air cleaner box, and the air intake duct.

3. Disconnect the starter cables and the transaxle ground cable. Remove the engine wire harness clip. Label and disconnect the lock–up control solenoid connector.

4. Label and disconnect the vehicle speed sensor (VSS) and countershaft speed sensor connectors.

5. Loosen the upper transaxle case bolts and the rear engine mounting bolt.

6. Raise and safely support the vehicle. Remove the front wheels.

7. Drain the automatic transaxle fluid. Then, install the drain plug with a new crush washer. Note the color, consistency, and odor of the drained fluid.

8. Remove the front splash shield.

9. Label and disconnect the shift control and linear solenoid connec-

tors. Disconnect the mainshaft speed sensor connector.

10. Remove the strut pinch bolt and fork bolt. Disconnect the lower ball joint using a ball joint remover.

11. Pry the halfshaft inboard joints out of the transaxle case and intermediate shaft. Swing the steering knuckles out to free the halfshafts from the transaxle.

12. Tie the halfshafts up and out of the way with wire. Tie plastic bags over the inboard joints to prevent damage to the CV–boots and splined shafts.

13. Disconnect and remove the front exhaust pipe.

14. Remove the shift cable cover. Disconnect the shift cable from the transaxle control shaft. Move the shift cable out of the way, and tie it up with wire.

15. Disconnect the ATF cooler hoses from the cooler lines. Cap the lines hoses to prevent fluid lose and contamination.

16. Remove the right–front mount and bracket assembly.

17. Remove the engine stiffener and the torque converter cover plate.

18. Remove the eight torque converter–to–driveplate bolts one at a time by rotating the crankshaft pulley. There are no gear teeth on the driveplate; the starter motor engages a ring gear on the inner edge of the torque converter.

19. After unbolting the torque converter from the driveplate, Rotate the crankshaft to set the engine at TDC/compression for the No. 1 cylinder.

20. Label and disconnect the ignition wires. Unbolt and remove the distributor assembly so that an engine lifting chain hook can be bolted to the distributor mount.

21. Attach a lifting chain to the engine and lift slightly to ease the tension on the mounts.

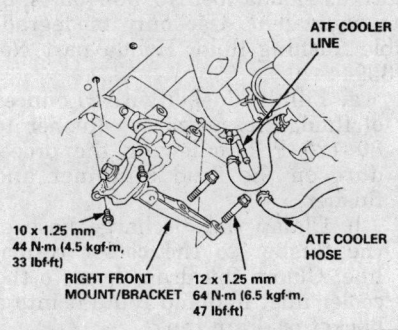

10 x 1.25 mm
44 N·m (4.5 kgf·m,
33 lbf·ft)

RIGHT FRONT
MOUNT/BRACKET

ATF COOLER
LINE

ATF COOLER
HOSE

12 x 1.25 mm
64 N·m (6.5 kgf·m,
47 lbf·ft)

299779

Right-front mount and bracket assembly — 1996–97 Civic

22. Place a transaxle jack under the transaxle and remove the transaxle side mount and bracket.

23. With the transaxle supported, remove the transaxle rear mount bracket bolts and transaxle case bolts.

24. Pull the transaxle away from the engine until it clears the locating dowel pins. Carefully lower the transaxle from the vehicle with the torque converter angled upward so it doesn't drop out of the transaxle.

25. Remove the torque converter from the transaxle. Inspect the ring gear teeth for breakage and inspect the converter's hub for burrs and scoring. Check the condition of the converter's fluid. Replace the torque converter if necessary.

26. Inspect the transaxle's front oil pump bearing and seal for signs of leakage and scoring. Inspect the mainshaft for burrs, scoring, and roughness.

27. With the transaxle removed, carefully inspect the driveplate for stress cracks, enlarged bolt holes, and other defects. Replace it if necessary.

To install:

NOTE: Use new self-locking nuts and color–coded bolts when installing the transaxle and suspension components.

28. Flush the transaxle cooler lines to remove any contaminated fluid and residual clutch material:

a. Use a pressurized flusher (Honda J38405–A or equivalent). Use only Honda flushing fluid (Honda J35944–20); other fluids may damage the system.

b. Fill the flusher with 21 ounces of fluid. Pressurize the flusher to 80–120 PSI, following the procedure on the fluid container and flusher.

c. Clamp the discharge hose of the flusher to the cooler return line. Clamp the drain hose to the cooler inlet line and route it into a bucket or drain tank.

d. Connect the flusher to air and water lines. The air line use a water trap to keep excess moisture out.

e. Open the flusher water valve and flush the cooler for ten seconds.

f. Depress the flusher trigger to mix flushing fluid with the water. Flush for two minutes, turning the air valve on and off for five seconds every 15–20 seconds to create a surging action.

g. After finishing one flushing cycle, reverse the hose and flush in the opposite direction.

h. Dry the cooler lines with compressed air for two minutes or longer to remove all excess moisture from the system.

29. If removed, install the starter motor onto the transaxle case and tighten its mounting bolts to 33 ft. lbs. (45 Nm). Install the torque converter with a new hub O–ring.

30. Place the transaxle on a transaxle jack and raise to the level of the engine.

31. Align the transaxle and engine. Install the transaxle case bolts. Install new 14mm rear mount bolts and washers.

32. Raise the transaxle and install the side mount. Tighten the case bolts to 47 ft. lbs. (64 Nm). Torque all of the 14mm rear mount bolts to 61 ft. lbs. (85 Nm).

33. Install the transaxle side mount and bracket. Torque the bracket nuts to 47 ft. lbs. (64 Nm). Torque the mount through–bolt to 54 ft. lbs. (74 Nm).

34. Remove the transaxle jack.

35. Rotate the crankshaft and install the torque converter–to–driveplate bolts. Torque the bolts to 9 ft. lbs. (12 Nm) is a crisscross pattern. Tighten the bolts to the torque specification in two steps.

36. Rotate the crankshaft to reset the engine at TDC/compression for the No. 1 cylinder. After the engine is set at TDC, it must not be disturbed until the distributor has been reinstalled.

37. Install the torque converter cover and torque the bolts to 9 ft. lbs. (12 Nm).

38. Install the engine stiffener and torque the 8mm bolts to 17 ft. lbs. (24 Nm). Torque the 10mm bolts to 33 ft. lbs. (45 Nm).

39. Install the right–front mount and bracket assembly. Tighten the 10mm bolt to 33 ft. lbs. (44 Nm), and the 12mm bolts to 47 ft. lbs. (64 Nm).

40. Remove the lifting chain and chain hooks.

41. Verify that the engine is at TDC/compression for the No. 1 cylinder. Align the tabs on the distributor drive with the grooves on the end of the camshaft. Install the distributor and hand–tighten the mounting bolts. Reconnect the ignition wires.

42. Tighten the crankshaft pulley to 134 ft. lbs. (181 Nm).

43. Reconnect the transaxle cooler lines.

44. Install new set rings on the inboard CV–joint splines. Install the halfshafts into the transaxle case and intermediate shaft. The inboard joints must snap into place.

45. Connect the lower ball joint and damper fork. Install the front wheels.

46. Connect the shift cable linkage to the transaxle control shaft. Install a new lockwasher and tighten the linkage bolt to 10 ft. lbs. (14 Nm). Install the shift cable cover and tighten its bolt to 16 ft. lbs. (22 Nm).

47. Install the front exhaust pipe. Use new self–locking nuts and gaskets. Tighten the rear flange nuts to 16 ft. lbs. (22 Nm), and the front flange nuts to 47 ft. lbs. (64 Nm).

48. Connect the vehicle speed sensor (VSS) and countershaft speed sensor connectors. Connect the lock–up control solenoid connector.

49. Connect the shift control and linear solenoid connectors. Connect the mainshaft speed sensor connector.

50. Install the wire harness clamps and starter cables. Install the resonator, air cleaner box and air intake duct.

51. Install the front splash shield.

52. Lower the vehicle and tighten the strut pinch bolts to 32 ft. lbs. (44 Nm). Tighten the fork bolts to 47 ft. lbs. (65 Nm). Tighten the ball joint castle nuts to 40 ft. lbs. (55 Nm), and then tighten them only enough to install new cotter pins.

53. Refill the transaxle with fresh ATF. Use only Honda Premium ATF or an equivalent DEXRON®II or III ATF. Reconnect the positive and negative battery cables.

  a. Leave the flusher drain hose attached to the cooler return line.

  b. With the transaxle in park, run the engine for 30 seconds, or until approximately one quart of fluid is discharged. Immediately shut the engine off. This completes the cooler flushing process.

  c. Remove the drain hose and reconnect the cooler return line.

  d. Refill the transaxle to the proper level.

54. Check shift cable and throttle cable adjustments.

55. Check the ignition timing. Rotate the distributor counterclockwise to advance the timing, or clockwise to retard the timing. When the timing has been set, tighten the distributor mounting bolts to 13 ft. lbs. (18 Nm).

56. Start the engine and shift through all the gears three times.

57. Let the engine warm up to operating temperature and check the fluid level with the transaxle in the **P** or **N** position.

58. Check and adjust the front wheel alignment.

59. Road test the vehicle. Recheck the transaxle fluid level.

Prelude

1. Disconnect both cables from the battery.

2. Shift the transaxle into **N**.

3. Remove the battery hold-down and remove the battery.

4. Drain the transaxle fluid and reinstall the drain plug with a new crush washer.

5. Remove the air intake duct, air cleaner case, and resonator.

6. Disconnect the connector from the vacuum tank and remove the vacuum tank and tank bracket. Do not remove the vacuum tube from the vacuum tank.

7. Disconnect the transaxle–to–body ground cable.

8. Remove the battery base with the ground cable and remove the battery base stay.

9. Disconnect the lock-up control solenoid valve and shift control solenoid valve connectors.

10. Disconnect the throttle control cable from the throttle control lever.

11. Disconnect the countershaft speed sensor connector.

12. Disconnect the vehicle speed sensor connector.

13. Remove the rear stiffener, then remove the vehicle speed sensor and power steering speed sensor.

NOTE: Do not disconnect the power steering pressure hoses from the vehicle speed sensor and power steering speed sensor.

14. Disconnect the ATF cooler hoses at the joint pipes. Turn the ends of the cooler hoses upward to prevent fluid loss. Plug the joint pipes.

15. Remove the starter motor.

16. Remove the upper transaxle housing mounting bolts.

17. Loosen the front engine mount bracket bolts.

18. Remove the transaxle mount.

19. Raise and support the vehicle safely. Remove the front wheels.

20. Remove the splash shield and remove the subframe center beam and rear beam stiffener.

21. Remove the cotter pins and castle nuts from the lower ball joints. Use a press–type ball joint tool to separate the ball joints from the lower arm.

22. Remove the damper fork bolts and separate the damper fork and the damper.

23. Use a suitable pry tool to separate the right and left halfshafts from the differential.

24. Pull on the inboard joint and remove the right and left halfshafts. Tie plastic bags over the halfshaft ends to protect the boots and splined shafts from damage.

25. Remove the right damper pinch bolt and separate the right damper fork from the strut.

26. Remove the right radius rod bolts and nut. Remove the radius rod.

27. Remove the torque converter cover and the shift cable cover.

28. Remove the control lever lockbolt and remove the shift cable with the lever. Do not bend the shift control cable during removal. Wire the cable to the underbody of the car our of the work area.

29. Remove the driveplate bolts while rotating the crankshaft.

30. Place a transaxle jack below the transaxle and raise it enough to take the weight off the mounts.

31. Remove the intake manifold bracket.

32. Remove the lower transaxle housing mounting bolts and lower rear engine mounting bolts.

33. Pull the transaxle away from the engine until it clears the dowel pins. Lower the transaxle out of the vehicle.

  To install:

  NOTE: Use new self–locking nuts when assembling the front suspension components. Use new set rings on the halfshaft inboard joints. Use new self–locking bolts for the subframe beams. These fasteners are available from a Honda dealer.

34. Flush the transaxle cooling lines before installing the transaxle. Use a pressurized flushing canister, such as Honda tool No. J38405–A, or its equivalent. Use only biodegradable flushing fluid, Honda part No. J35944–20.

  a. Fill the flusher with 21 ounces of fluid. Pressurize the flusher to 80–120 PSI, following the procedure on the fluid container and flusher.

  b. Clamp the discharge hose of the flusher to the cooler return line. Clamp the drain hose to the cooler inlet line and route it into a bucket or drain tank.

  c. Connect the flusher to air and water lines.Open the flusher water valve and flush the cooler for ten seconds.

d. Depress the flusher trigger to mix flushing fluid with the water. Flush for two minutes, turning the air valve on and off for five seconds every 15–20 seconds.

e. After finishing one flushing cycle, reverse the hose and flush in the opposite direction.

f. Dry the cooler lines with compressed air so that no moisture remains in the cooler lines.

35. Install the starter motor onto the transaxle case. Install the torque converter with a new hub O–ring. Torque the starter bolts to 33 ft. lbs. (45 Nm).

36. Place the transaxle on a transaxle jack and raise it to the level of the engine.

37. Align the transaxle to the engine and install the transaxle housing mounting bolts and lower rear engine mounting bolts. Tighten the rear engine mounting bolts to 40 ft. lbs. (55 Nm) and the transaxle mounting bolts to 47 ft. lbs. (65 Nm). Install the intake manifold bracket and tighten the bolts to 16 ft. lbs. (22 Nm).

38. Tighten the front engine mount bracket bolts to 28 ft. lbs. (39 Nm).

39. Install the transaxle mount. Tighten the bolt to 47 ft. lbs. (65 Nm) and the nuts to 28 ft. lbs. (39 Nm).

40. Remove the transaxle jack.

41. Attach the torque converter to the driveplate and install the mounting bolts. Turn the crankshaft to rotate the driveplate. Tighten the bolts in 2 steps, first to 4.5 ft. lbs. (6 Nm) in a crisscross pattern and finally to 9 ft. lbs. (12 Nm) for 1994–95. For 1993, tighten the driveplate bolts in a crisscross pattern to 54 ft. lbs. (75 Nm). Check for free rotation after tightening the last bolt.

42. Install the shift cable onto the control shaft and tighten the lockbolt to 10 ft. lbs. (14 Nm).

43. Install the torque converter cover and the shift cable cover.

44. Install a new set ring onto the inboard joint of each halfshaft.

45. Install the damper fork bolts and ball joint nuts to the lower arms. Tighten the ball joint nut to 47 ft. lbs. (65 Nm) and install a new cotter pin. Install the radius rod and connect the damper fork. Only hand–tighten the radius rod and damper fork fasteners at this point.

46. Turn the right steering knuckle fully outward and slide the axle into the differential until the spring clip is felt engaging the differential side gear. Repeat the procedure on the left side.

47. Install the subframe rear beam stiffener and the center beam.

Tighten the stiffener bolts to 28 ft. lbs. (39 Nm). Tighten the subframe center beam bolts to 43 ft. lbs. (60 Nm).

48. Install the front wheels and lower the vehicle.

49. Use a floor jack to place the weight of the vehicle onto the right front knuckle. Tighten the radius rod bolts to 76 ft. lbs. (105 Nm) and the nut to 40 ft. lbs. (55 Nm). Tighten the damper pinch bolt to 32 ft. lbs. (44 Nm). Tighten the nut to 47 ft. lbs. (65 Nm) while holding the damper fork bolt.

50. Install the speedometer sensor. Tighten the sensor bolt to 9 ft. lbs. (12 Nm). On 1993 Preludes only, install the rear stiffener and tighten the bolt to 28 ft. lbs. (39 Nm), tighten the nut to 15 ft. lbs. (21 Nm).

51. Connect the ATF cooler hoses to the joint pipes.

52. Connect the lock-up control solenoid and shift control solenoid valve connectors.

53. Connect the vehicle speed sensor and power steering speed sensor connectors.

54. Connect the starter motor cables and install the battery base and base stay.

55. Connect the ground cables on the body and on the transaxle.

56. Install the vacuum tank, tank bracket, and connect the tank connector.

57. Install the resonator, air cleaner case, and air intake duct.

58. Refill the transaxle with ATF. Use only Honda Premium ATF or an equivalent DEXRON®II ATF. Connect the negative and positive battery cables.

a. Leave the flusher drain hose attached to the cooler return line.

b. With the transaxle in park, run the engine for 30 seconds, or until approximately one quart of fluid is discharged. This completes the cooler flushing process.

c. Remove the drain hose and reconnect the cooler return line.

d. Refill the transaxle to the proper level with ATF.

59. Start the engine, set the parking brake and shift the transaxle through all gears 3 times. Check for proper control cable adjustment.

60. On Preludes equipped with 4WS, start the engine and turn the steering wheel lock–to–lock to reset the steering control unit.

61. Check and adjust the front wheel alignment.

62. Let the engine reach operating temperature with the transaxle in **N**

or **P** then turn the engine OFF and check the fluid level

63. After road testing the vehicle, loosen the front engine mount bolts, and torque them to 28 ft. lbs. (39 Nm).

64. Enter the radio security code.

Accord

2.2L Engines

1. Disconnect the negative then the negative battery cables.

2. Remove the battery from the vehicle.

3. Shift the transaxle into **N**.

4. Remove the air intake hose, air cleaner housing and the resonator assembly.

5. Remove the battery base and the base stay.

6. Disconnect the throttle cable from the throttle control lever.

7. Disconnect the transaxle ground cable and the speed sensor connectors. Disconnect the solenoid valve connectors.

8. Disconnect the lock-up control solenoid valve and shift control solenoid valve connectors.

9. Disconnect the transaxle cooler hoses from the joint pipes and plug the hoses.

10. Disconnect the starter cables and remove the starter.

11. Disconnect the countershaft speed sensor connector.

12. Disconnect the vehicle speed sensor connector.

13. Install a hoist to the engine.

14. Remove the four upper bolts attaching the transaxle to the engine block.

15. Loosen the three bolts attaching the front engine mount bracket to the engine.

16. Remove the transaxle mount.

17. Raise and safely support the vehicle. Remove the front wheels.

18. Drain the transaxle fluid and reinstall the drain plug with a new washer.

19. Remove the splash shield.

20. Remove the subframe center beam.

21. Remove the cotter pins and lower arm ball joint nuts, then separate the ball joints from the lower arms using a suitable tool.

22. Remove the right damper pinch bolt, then separate the damper fork and damper.

23. Remove the bolts and nut, then remove the right radius rod.

24. Using a small prying device, carefully pry the right and left halfshafts out of the differential. Remove the right and left halfshafts. Tie plastic bags over the halfshaft ends

to prevent damage to the CV boots and splines.

25. Remove the bolts mounting the intermediate shaft then remove the intermediate shaft from the differential.

26. Remove the torque converter cover and shift cable cover.

27. Remove the shift control cable by removing the lockbolt. Remove the shift cable lever from the control shaft. Don't disconnect the control lever from the shift cable. Wire the shift cable out of the work area and be careful not to kink it.

28. Remove the eight drive plate bolts one at a time while rotating the crankshaft pulley.

29. Place a suitable jack under the transaxle and raise the jack just enough to take weight off of the mounts.

30. Remove the intake manifold bracket.

31. Remove the transaxle housing mounting bolts.

32. Remove the mounting bolts from the rear engine mount bracket.

33. Remove the four transaxle housing mounting bolts and three mount bracket nuts.

34. Pull the transaxle away from the engine until it clears the 14mm dowel pins, then lower it using the jack.

To install:

NOTE: Use new self-locking nuts when assembling the front suspension components. Install new set rings onto the halfshaft inboard joint splines. Replace any color-coded self-locking bolts.

35. Flush the transaxle cooler lines before installing the transaxle. Use a pressurized flushing unit such as Honda J38405–A or equivalent. Use only Honda biodegradable flushing fluid, Honda J35944–20. Other fluids will damage the A/T cooling system.

 a. Fill the flusher with 21 ounces of fluid. Pressurize the flusher to 80–120 PSI, following the procedure on the fluid container and flusher.

 b. Clamp the discharge hose of the flusher to the cooler return line. Clamp the drain hose to the cooler inlet line and route it into a bucket or drain tank.

 c. Connect the flusher to air and water lines. Open the flusher water valve and flush the cooler for ten seconds. The air line should be equipped with a water trap to keep the system dry.

 d. Depress the flusher trigger to mix flushing fluid with the water.

Flush for two minutes, turning the air valve on and off for five seconds every 15–20 seconds to create a surging action.

 e. After finishing one flushing cycle, reverse the hose and flush in the opposite direction following the same steps.

 f. Dry the cooler lines with compressed air so that no moisture is left in the cooler system.

36. Make sure the two 14mm dowel pins are installed into the torque converter housing.

37. Install the torque converter onto the transaxle mainshaft with a new hub O–ring. Install the starter motor onto the transaxle case and tighten the mounting bolts to 33 ft. lbs. (44 Nm).

38. Raise the transaxle into position and install the transaxle housing mounting bolts. Tighten the bolts to 47 ft. lbs. (65 Nm).

39. Install the rear engine mounting bolts and tighten to 40 ft. lbs. (54 Nm).

40. Install the intake manifold bracket and tighten the bolts to 16 ft. lbs. (22 Nm).

41. Install the upper bolts attaching the transaxle to the engine and tighten the bolts to 47 ft. lbs. (64 Nm).

42. Tighten the front engine mount bracket bolts to 28 ft. lbs. (38 Nm).

43. Install the transaxle mount and loosely install the nuts and bolt that attach the mount. Tighten the nuts first to 28 ft. lbs. (38 Nm), then tighten the bolt to 47 ft. lbs. (64 Nm).

44. Remove the jack from the transaxle.

45. Attach the torque converter to the drive plate with the eight bolts. Tighten the bolts in two steps in a crisscross pattern: first to 4.5 ft. lbs. (6 Nm), and finally to 9 ft. lbs. (12 Nm). Check for free rotation after tightening the last bolt.

46. Install the shift control cable and control cable holder. Tighten the shift cable lockbolt to 10 ft. lbs. (14 Nm). Tighten the shift cable cover bolts to 13 ft. lbs. (18 Nm).

47. Install the torque converter cover and tighten the bolts to 9 ft. lbs. (12 Nm).

48. Remove the engine hoist.

49. Install the radius rod and damper fork.

50. Install the intermediate shaft into the differential and tighten the mounting bolts to 28 ft. lbs. (38 Nm).

51. Install a new set ring on the end of each halfshaft.

52. Turn the right steering knuckle fully outward and slide the axle into

the differential until the set ring snaps into the differential side gear. Repeat the procedure on the left side.

53. Install the damper fork bolts and ball joint nuts to the lower arms. Tighten the ball joint nut to 40 ft. lbs. (55 Nm) and install a new cotter pin.

54. Install the subframe center beam and tighten the center beam bolts to 28 ft. lbs. (39 Nm).

55. Install the splash shield.

56. Install the front wheels and lower the vehicle.

57. Reconnect the speed sensor connector.

58. Support the right front knuckle with a floor jack until the weight of the car is held by the jack. Tighten the damper fork pinch bolt to 32 ft. lbs. (44 Nm). Tighten the radius rod bolts to 76 ft. lbs. (105 Nm), and the radius rod nut to 32 ft. lbs. (44 Nm). Hold the damper fork bolt with a wrench, and tighten the nut to 40 ft. lbs. (55 Nm).

59. Connect the cables to the starter.

60. Reconnect the throttle control cable.

61. Connect the lock-up control solenoid valve and shift control solenoid valve connectors.

62. Connect the speed sensor connectors and the transaxle ground cable.

63. Connect the transaxle cooler inlet hose to the joint pipe. Attach a drain hose to the return line.

64. Install the battery base stay and the battery base.

65. Install the resonator assembly, the air cleaner assembly and the air intake hose.

66. Install the battery and connect the positive then the negative battery cables to the battery.

67. Refill the transaxle with ATF. Use only Honda Premium ATF or an equivalent DEXRON®II ATF.

 a. With the flusher drain hose attached to the cooler return line.

 b. Place the transaxle in **P**, run the engine for 30 seconds, or until approximately one quart of fluid is discharged. Immediately shut off the engine. This completes the cooler flushing process.

 c. Remove the drain hose and reconnect the cooler return line.

 d. Refill the transaxle to the proper level with ATF.

68. Start the engine, set the parking brake and shift the transaxle through all gears 3 times. Check for proper shift cable adjustment.

69. Let the engine reach operating temperature with the transaxle in **P**

or **N**. Then, shut off the engine and check the fluid level.

70. Road test the vehicle.

71. After road testing the vehicle, loosen the front engine mount bracket bolts, and then retighten them to 28 ft. lbs. (39 Nm).

72. Check and adjust the vehicle's front end alignment.

73. Enter the radio security code.

2.7L Engine

NOTE: Several of the engine mounts must be removed during the transaxle removal procedure. The objective of this procedure is to allow the engine to tilt so that the transaxle will clear the left shock tower. A chain host is necessary for this operation.

1. Disconnect the negative and positive battery cables.

2. Prop the hood open and remove the support struts. Secure the hood open in a vertical position.

3. Remove the battery. Remove the battery tray and bracket. Unbolt the cable holder from the battery tray.

4. Remove the intake air duct.

5. Detach the clips securing the starter cable to the strut brace. Remove the strut brace.

6. Drain the transaxle fluid. Install the drain plug with a new crush washer.

7. Uncouple the ATF cooler lines. Plug the lines, and turn them upward to lessen fluid spillage.

8. Remove the starter cables. Unbolt the starter cable clamp and the transaxle ground cable. Unbolt the engine sub–harness clamp from the transaxle case, and move the harness holder out of the way.

9. Disconnect the following electrical couplings:
- Shift control solenoid valve
- Mainshaft speed sensor connector
- Lock–up control solenoid valve
- Vehicle speed sensor connector
- Linear solenoid connector
- Counter shaft speed sensor

10. Remove the intake air bypass (IAB) vacuum tank from the transaxle. Do not disconnect its vacuum hoses.

11. Loosen the top two upper transaxle case bolts, they may not be easily accessible when the vehicle is raised.

12. Raise and safely support the vehicle. Remove the front wheels.

13. Remove the splash shield.

14. Remove the subframe center beam.

15. Remove the lower ball joint castle nuts. Separate the ball joints from the lower control arms using a suitable press–type tool.

16. Remove the damper fork bolts, and separate the damper forks from the lower control arm.

17. Use a flat-bladed tool to carefully pry the halfshafts out of the differential and intermediate shaft.

18. Pull the inboard joints away from the differential and intermediate shaft. Tie plastic bags over the inboard joints to protect the boots and splines.

19. Remove the left damper pinch bolt; then, separate the damper fork from the strut. Unbolt and remove the left radius rod. Use wire to support the left control arm and halfshaft out of the work area.

20. Remove the intermediate shaft.

21. Remove the shift cable holder and cover. Remove the shift cable lockbolt, and slide the cable lever off of the control rod. Be careful not to kink the shift cable.

22. Remove the torque convertor cover. Remove the eight driveplate bolts one at a time while rotating the crankshaft pulley

23. Attach a chain hoist to the transaxle. Place a jack under the engine for support.

24. Remove the rear mount stiffener; and then, unbolt and remove the rear mount.

25. Remove the front mount bracket; and then, remove the front mount.

26. Remove the four upper transaxle case bolts.

27. Remove the side transaxle mount.

28. Remove the two lower transaxle case bolts.

----------- CAUTION -----------
Make sure the transaxle is securely supported before removing the rear mounting bracket bolts. Place a transaxle jack under the transaxle for additional support.

29. Remove the rear mount bolts. Raise the engine and transaxle slightly; and then, remove the rear mount bracket from the rear mount.

30. Carefully lower the transaxle from the vehicle. Tilt the engine just enough for the transaxle to clear the left shock tower.

NOTE: Do not allow the A/C compressor to hit the right shock tower when the engine is being tilted.

31. Pull the transaxle away from the engine until it clears the dowel pins, then lower it from the vehicle.

To install:

NOTE: Use new self–locking nuts when assembling the front suspension components. Install a new set ring onto the inboard halfshafts. Replace any color–coded self–locking nuts and bolts when installing the engine and transaxle mounts. These fasteners are available from a Honda dealer.

32. Flush the transaxle cooler lines before installing the transaxle. Use a pressurized flushing unit such as Honda J38405–A or equivalent. Use only Honda biodegradable flushing fluid, Honda J35944–20. Other fluids will damage the A/T cooling system.

a. Fill the flusher with 21 ounces of fluid. Pressurize the flusher to 80–120 PSI, following the procedure on the fluid container and flusher.

b. Clamp the discharge hose of the flusher to the cooler return line. Clamp the drain hose to the cooler inlet line and route it into a bucket or drain tank.

c. Connect the flusher to air and water lines. Open the flusher water valve and flush the cooler for ten seconds. The air line should be equipped with a water trap to keep the system dry.

d. Depress the flusher trigger to mix flushing fluid with the water. Flush for two minutes, turning the air valve on and off for five seconds every 15–20 seconds to create a surging action.

e. After finishing one flushing cycle, reverse the hose and flush in the opposite direction following the same steps.

f. Dry the cooler lines with compressed air so that no moisture is left in the cooler system.

33. Make sure the two 14mm dowel pins are installed into the torque converter housing.

34. Install the torque convertor onto the transaxle mainshaft with a new O–ring. Install the starter motor onto the transaxle case and tighten the mounting bolts to 33 ft. lbs. (44 Nm).

35. Position the transaxle under the vehicle and attach the chain hoist to it.

36. Jack or hoist the engine into place; then, lift the transaxle into place.

37. Mate the transaxle to the engine. Install the transaxle case bolts, and tighten them to 47 ft. lbs. (65

Nm). The one long bolt fits behind the starter.

38. Install the rear mount bracket onto the rear mount and hand–tighten the nut and upper bolt. Install two new rear mount bracket–to–transaxle case bolts, and tighten them to 40 ft. lbs. (55 Nm).

39. Verify that all of the transaxle case bolts have been installed.

40. Install the front mount, tighten the three bolts to 43 ft. lbs. (59 Nm). Install the front mount bracket onto the transaxle and tighten the bolts to 28 ft. lbs. (39 Nm). Install and hand–tighten the mount nut.

41. Install the side transaxle mount. Install and hand–tighten the three nuts and the through-bolt. Tighten the nuts to 28 ft. lbs. (39 Nm), and then tighten the bolt to 47 ft. lbs. (65 Nm).

42. With all the engine and transaxle mounts installed, tighten the mount nuts to 40 ft. lbs. (55 Nm). Install the rear mount stiffener. Tighten the rear stiffener bolt and upper rear mount bracket bolt to 28 ft. lbs. (39 Nm).

43. Remove the chain hoist and jacks from the engine and transaxle.

44. Install the eight driveplate bolts and tighten them to their final torque specification in two steps in a crisscross pattern. The first tightening step is 4.3–4.5 ft. lbs. (6 Nm). The second tightening step is 8.7 ft. lbs. (12 Nm). After tightening the last bolt, check the crankshaft for free rotation.

45. Install the torque convertor cover and tighten its bolts to 8.7 ft. lbs. (12 Nm).

46. Use a holder tool and a torque wrench to retighten the crankshaft pulley to 181 ft. lbs. (245 Nm).

47. Reconnect the shift cable lever to the control shaft with a new lockwasher, and tighten the bolt to 10 ft. lbs. (14 Nm). Install the shift cable holder and cover. Torque the cover bolts to 20 ft. lbs. (26 Nm).

48. Install the intermediate shaft and tighten the bolts to 16 ft. lbs. (22 Nm).

49. Install the left radius rod and damper fork. Hand–tighten the bolts.

50. Install new set rings on the halfshaft inboard CV–joints. Install the halfshafts into the differential and intermediate shaft. Make sure the set rings snap into place.

51. Reassemble the damper forks and lower control arm ball joints. Tighten the ball joint castle nuts to 36–43 ft. lbs. (49–59 Nm), and then tighten them only enough to install new cotter pins.

52. Install the center beam and tighten the bolts to 37 ft. lbs. (50 Nm). Install the splash shield.

53. Install the front wheels and lower the vehicle.

54. Use a floor jack to raise the left front knuckle until it is supporting the weight of the vehicle. Tighten the damper pinch bolt to 32 ft. lbs. (43 Nm). Tighten the radius rod bolts to 76 ft. lbs. (101 Nm), and the nuts to 32 ft. lbs. (43 Nm). Tighten the damper fork nuts to 47 ft. lbs. (65 Nm).

55. Install the IAB vacuum tank and the engine wiring harness clamp.

56. Reconnect the following electrical couplings:
- Shift control solenoid valve
- Mainshaft speed sensor connector
- Lock–up control solenoid valve
- Vehicle speed sensor connector
- Linear solenoid connector
- Counter shaft speed sensor

57. Install and connect the starter cables and the transaxle ground cable.

58. Reconnect the ATF cooler hoses.

59. Install the strut brace and tighten its mounting bolts to 16 ft. lbs. (22 Nm). Attach the starter cable clips to the strut brace.

60. Install the intake air duct. Install the battery tray and tighten its mounting bolts to 16 ft. lbs. (22 Nm). Reconnect the cable holder and ground terminal.

61. Install the battery and reconnect the battery cables.

62. Prop the hood and install the support struts.

63. Refill the transaxle with ATF. Use only Honda Premium ATF or an equivalent DEXRON®II ATF. Connect the battery cables.

   a. Leave the flusher drain hose attached to the cooler return line.

   b. With the transaxle in **P**, run the engine for 30 seconds, or until approximately one quart of fluid is discharged. Immediately shut off the engine. This completes the cooler flushing process.

   c. Remove the drain hose and reconnect the cooler return line.

   d. Refill the transaxle to the proper level with ATF.

64. Start the engine, set the parking brake and shift the transaxle through all gears three times. Check for proper shift cable adjustment.

65. Let the engine reach operating temperature with the transaxle in **P** or **N**. Then, shut off the engine and check the fluid level.

66. Road test the vehicle.

67. After road testing the vehicle, loosen the front and rear engine mount nuts, and then retighten them to 40 ft. lbs. (54 Nm).

68. Check and adjust the vehicle's front end alignment.

69. Enter the radio security code.

Throttle Cable

ADJUSTMENT

Accord and Prelude

NOTE: The radio may contain a coded theft protection circuit. Always obtain the code number before disconnecting the battery.

1. Warm the engine up to operating temperature.

2. Make sure the idle speed is correct.

3. There must be 0.39–0.47 in. (10–12mm) of play in the throttle cable at the throttle linkage. If the cable deflection is not within these specs, loosen the locknut, and tighten the adjusting nut until the defection is correct. Then, retighten the locknut.

4. Verify that the throttle control cable is securely clamped and that the throttle linkage is in the fully closed position.

5. Disconnect the negative battery cable.

6. Loosen the locknut on the throttle control cable at the throttle control lever.

7. Turn the locknut to remove the throttle control cable freeplay. Do this while pushing down on the lever to hold it in the fully closed position.

8. Tighten the locknuts.

9. Reconnect the negative battery cable.

10. Inspect the movement of the throttle control lever. Check the cable deflection.

11. Test drive the vehicle and check the transaxle's shift points.

DRIVE AXLE

Halfshaft

REMOVAL AND INSTALLATION

1. Loosen the front spindle nut.

2. Raise and safely support the vehicle.

3. Remove the front wheels and the spindle nut.

4. Drain the transaxle fluid and install the drain plug with a new crush washer. If the halfshaft to be removed is installed into the intermediate shaft, the transaxle fluid does not need to be drained.

5. Remove the damper fork nut and damper pinch bolt.

6. Remove the damper fork.

7. Remove the cotter pin and castle nut from the lower arm ball joint. Install a hex nut flush onto the ball joint stud to prevent the ball joint tool from damaging the stud threads.

8. Using a ball joint tool, separate the lower arm from the knuckle.

9. Pull the knuckle outward. Remove the halfshaft outboard joint from the hub by tapping it with a plastic hammer.

10. Carefully pry the inner CV-joint away from the transaxle case to force the halfshaft set ring out of the groove.

11. Pull on the inboard CV-joint and remove the halfshaft from the differential case or intermediate shaft.

NOTE: Do not pull on the halfshaft as the CV-joint may come apart. Use care when prying out the assembly and pull it straight to avoid damaging the differential oil seal or intermediate shaft oil or dust seals.

To install:

12. Replace the differential oil seal or intermediate shaft seal if either were damaged during removal.

13. Install new set rings on the ends of the halfshafts.

14. Install the halfshafts and make sure the set ring locks in the differential gear groove and the halfshaft bottoms in the differential or intermediate shaft.

15. Install the outboard joint into the hub. Make sure the splines mesh together and the joint is fully seated into the hub.

16. Fit the ball joint stud into the lower control arm. Install the damper fork into position. Torque the upper damper pinch bolt to 32 ft. lbs. (44 Nm) and the fork nut to 47 ft. lbs. (65 Nm).

17. Torque the ball joint castle nut to 40 ft. lbs. (55 Nm); then, tighten the nut just enough to install a new cotter pin.

18. Install the front wheels. Install a new spindle nut, but don't tighten it yet.

19. Lower the vehicle.

20. Torque the spindle nut to 181 ft. lbs. (245 Nm) and stake its tab.

Tighten the wheel nuts to 80 ft. lbs. (110 Nm).

21. Fill the transaxle with the proper type and quantity of fluid.

22. Warm the engine up, check the transaxle fluid level, and road test the vehicle.

CV-Joint Boot

REPLACEMENT

1. Raise and safely support the vehicle.

2. Remove the halfshaft.

3. Place the halfshaft in a vise with padded jaws.

4. Remove the boot bands. Welded bands must be cut to be removed. After removing the bands, push the CV-boot away from the inboard joint to reveal the spider and roller assemblies.

5. Remove the inboard CV-joint. Mark the locations of the rollers on the spider during disassembly to ensure proper positioning during reassembly.

6. Remove the rollers and the circlip. Mark the position of spider on the shaft.

7. Use a bearing puller to remove the spider from the shaft. Remove the stopper ring.

8. Remove the inboard CV-boot.

9. Remove the dynamic damper.

10. Remove the boot bands and the outboard CV-boot.

11. Inspect the outboard joint for excess play and rough movement. The outboard joint can't be disassembled. If it is damaged or worn out, it must be replaced.

To install:

12. Check the CV-joint components and replace any worn parts. Replace the boots if they show any signs of cracks or rips.

13. Wrap the inboard halfshaft splines with tape to prevent the new boots from being ripped.

14. Install the outboard boot onto the shaft and pack it with the grease included in the boot replacement kit. Install the boot onto the outboard joint with a new band.

15. Install the dynamic damper with a new band.

16. Install the inboard boot onto the shaft and remove the tape from the splines.

17. Install a new stopper ring onto the halfshaft. Align the matchmarks on the spider and the shaft, install the spider onto the halfshaft. Install a new circlip.

18. Install the rollers onto the spider in their original positions. The

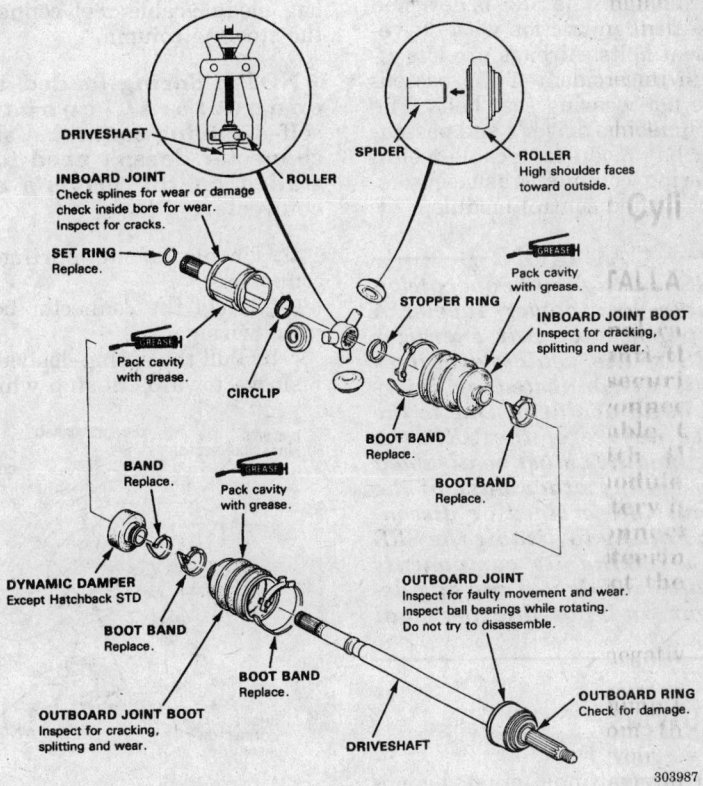

Halfshaft components exploded view

303987

high shoulders of the rollers must face out.

19. Pack the inboard joint and inboard boot with the grease included in the boot kit.

20. Install the inboard joint onto the halfshaft.

21. Adjust the halfshafts so that the boots are partially extended.

22. Install a new boot band on the inboard halfshaft.

23. Install a new set ring onto the inboard shaft groove.

24. Install the halfshafts into the vehicle.

25. Install the front wheels and lower the vehicle.

STEERING

Air Bag

DESCRIPTION

The Supplemental Restraint System (SRS/air bag) is a passive safety device designed to reduce the risk and severity of injuries to the front seat passengers of a vehicle involved in a frontal collision. The SRS is designed to be used in conjunction with the vehicle's seat belts. Airbags are less effective in an accident if the passengers are not wearing seat belts. The SRS includes the driver's and passenger's air bag modules, a cable reel in the steering column, crash sensors, and a dedicated control module.

— **CAUTION** —

The SRS is designed to operate on extremely low power levels. A back–up power circuit energizes the SRS if the vehicle's battery power is disconnected or interrupted in an accident. Due to the sensitive nature of the SRS, the air bag modules must be disabled if they, or any other part of the SRS, must be serviced or disconnected. Failing to disable the SRS before servicing its components may cause accidental air bag deployment and possible personal injury.

PRECAUTIONS

Several precautions must be observed when handling the inflator module to avoid accidental deployment and possible personal injury.

• Never carry the inflator module by the wires or connector on the underside of the module.

• When carrying a live inflator module, hold securely with both hands, and ensure that the bag and trim cover are pointed away.

• Place the inflator module on a bench or other surface with the bag and trim cover facing up.

• With the inflator module on the bench, never place anything on or close to the module which may be thrown in the event of an accidental deployment.

DISARMING

NOTE: The radio may contain a coded theft protection circuit. Always obtain the code number before disconnecting the battery.

Driver's Side

1. Disconnect the negative and positive battery cables.

2. Always wait at least three minutes after disconnecting the battery before working around the air bag.

3. Remove the steering wheel lower access cover.

4. Remove the clip securing the air bag module/cable reel connection to the steering column.

NOTE: Spring–loaded air bag connectors contain a self-disabling contact. A shorting connector doesn't need to be installed on the driver's air bag connector.

5. Uncouple the spring–loaded connectors:

   a. Hold the connector body, not the wiring.

   b. Pull the spring–loaded locking sleeve toward its stop while holding the opposite half of the connector.

   c. After releasing the locking sleeve, uncouple the connectors.

6. After servicing has been completed, couple the air bag and cable reel connectors. Press the sleeve side of the connector into the pawl side until the sleeve locks the connectors together.

7. Install the clip securing the air bag/cable reel connection to the steering column.

8. Install the access cover.

9. Reconnect the positive and negative battery cables.

10. Turn the ignition switch to the **ON** position, but don't start the engine. The air bag indicator light should turn on for six seconds and then turn off. If the air bag indicator light doesn't come on, or stays on longer than six seconds, the system fault must be diagnosed.

11. Enter the radio security code.

Passenger's Side

1. Disconnect the negative and positive battery cables.

2. Always wait at least three minutes after disconnecting the battery before working around the air bag.

3. Remove the glove box door and frame.

4. If equipped, remove any lower mounting brackets that may cover the air bag connection.

5. Disconnect the passenger's air bag connector. Pull the spring loaded sleeve toward the stop while holding the opposite half of the connector and pull the connector apart.

6. After servicing has been completed, immediately couple the air bag and cable reel connectors.

7. If equipped, install any lower mounting brackets that may have been removed.

8. Install the glove box frame and glove box door.

9. Reconnect the positive and negative battery cables.

10. Turn the ignition switch to the **ON** position, but don't start the engine. The air bag indicator light should turn on for six seconds and then turn off. If the air bag indicator light doesn't come on, or stays on longer than six seconds, the system fault must be diagnosed.

11. Enter the radio security code.

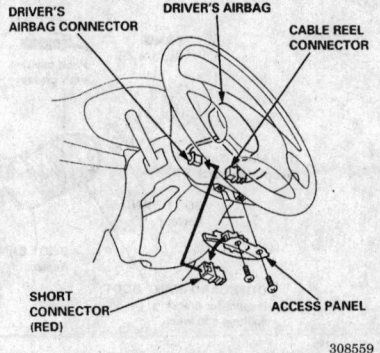

DRIVER'S AIRBAG CONNECTOR   DRIVER'S AIRBAG   CABLE REEL CONNECTOR

SHORT CONNECTOR (RED)   ACCESS PANEL

308559

Driver's air bag

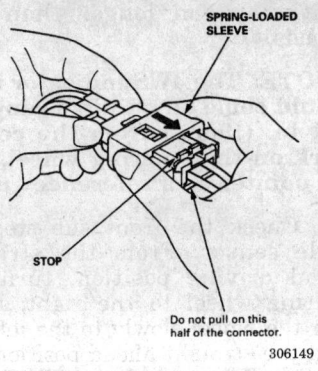

Spring-loaded connectors

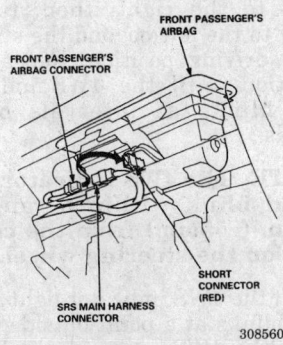

Passenger's air bag — 1993–95 Accord, Civic, del Sol and 1993–97 Prelude

Steering Wheel

REMOVAL AND INSTALLATION

NOTE: The original radio contains a coded anti–theft circuit. Obtain the security code number before disconnecting the battery cables.

CAUTION

The supplemental restraint system (air bag) must be disabled before removing the steering wheel. Failure to disarm the air bag system may cause accidental air bag deployment, resulting in unnecessary air bag system repairs and the risk of personal injury.

1. Disconnect the negative and positive battery cables and disable the Supplemental Restraint System (air bag).
2. Remove the covers on either side of the steering wheel.
3. Remove the TORX® bolts from either side of the steering wheel. Carefully lift the air bag module from the steering wheel and remove it from the vehicle.

CAUTION

The air bag module is live. Carry the air bag module with the cushion facing away from your body. Store the air bag module with the cushion facing up on a work bench. Following these precautions will lessen the chance of personal injury if the air bag module accidentally deploys.

4. Disconnect the horn and cruise control switch connectors.
5. Remove the steering wheel nut or bolt. Pull the steering wheel off of the shaft by rocking it slightly side–to–side.

WARNING

If a steering wheel puller is used to remove the steering wheel, do not thread the puller bolts more than five threads into the steering wheel. Install a nut five threads up on each puller bolt to act as a stop. If the puller bolts are threaded into the steering wheel more than five threads, the cable reel will be damaged.

To install:

NOTE: Use new TORX®bolts when reinstalling the air bag module. They are available at a Honda dealer.

6. The cable reel must be centered before installing the steering wheel.
• Rotate the cable reel clockwise until it stops.
• Rotate the cable reel counterclockwise two turns.
• The yellow gear tooth must line up with the alignment mark on the cover. The arrow on the cable reel label points straight up.
7. If equipped with 4WS:
  a. Temporarily install the steering wheel. Verify that the slot on the steering wheel shaft engages with the tabs on the turn signal cancelling sleeve.

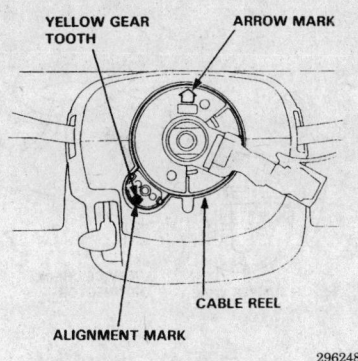

Cable reel alignment marks

  b. Turn the steering wheel from lock–to–lock right to left. Return the steering wheel to the straight ahead position to center the steering rack.
  c. Remove the wheel and install it with the spokes in a horizontal position.
  d. Turn the steering wheel until the mark is facing down.
  e. Return the steering wheel spokes to the horizontal angle. Be careful not to push in on the steering wheel.
  f. Remove the steering wheel.
8. Install the steering wheel. Verify that the slot on the steering wheel shaft engages with the tabs on the turn signal cancelling sleeve. Tighten the steering wheel nut to 36 ft. lbs. (50 Nm).
9. Reconnect the horn and cruise control switch connectors.
10. Install the air bag module with new TORX®bolts. Install the side covers.
11. Remove the shorting connector from the air bag. Reconnect the air bag and cable reel connectors.
12. Install the shorting connector back into its holder on the lower access cover. Install the access cover.
13. Reconnect the positive and negative battery cables.
14. Turn the ignition switch to the **ON** position. The air bag indicator light should come on for six seconds and then turn off. This light sequence indicates that the air bag system is enabled and functioning normally. If the air bag light stays on longer, the system must be diagnosed.
15. Check the operation of the horn and cruise control switches.
16. Turn the steering wheel counter clockwise and verify that the cable reel is aligned.
17. Turn the steering wheel lock–to–lock to reset the 4WS control unit.
18. Check the front wheel alignment and the steering wheel spoke angle. Adjust the tie rod ends if the alignment is out of specification.

Tie Rod Ends

REMOVAL AND INSTALLATION

Two–Wheel Steering

1. Raise and support the vehicle.
2. Remove the front wheels.
3. Use a ball joint separator to separate the tie rod end from the steering knuckle. Be careful not to tear the tie rod end boots.
4. Loosen the tie rod locknut.

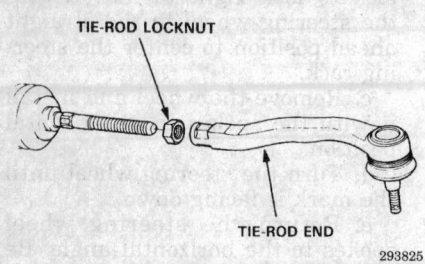

Tie rod end components

5. Remove the tie rod end. Leave the locknut as a guide for installing and adjusting the new tie rod end.

To install:

6. Install the tie rod end onto the steering rack ends.

7. Install the tie rod end onto the steering knuckle and tighten the ball joint castle nut to 29–35 ft. lbs. (40–48 Nm). Install a new cotter pin.

8. Install the front wheels and lower the vehicle

9. Check the vehicle's front wheel alignment and adjust it by turning the tie rods (shaft extending from the bellows at either end of the rack) equally until the alignment is within the proper specification.

10. After the alignment has been set, tighten the tie rod locknuts to 33 ft. lbs. (45 Nm).

11. Road test the vehicle.

Four–Wheel Steering

1. Raise and support the vehicle.
2. Remove the front or rear wheels.
3. Use a ball joint separator to separate the tie rod end from the steering knuckle. Be careful not to tear the tie rod end boots.
4. Loosen the tie rod locknut.
5. Remove the tie rod end. Leave the locknut as a guide for installing and adjusting the new tie rod end.

To install:

6. Install the tie rod end onto the steering rack ends. The front and rear tie rod ends are not interchangeable.

7. Install the tie rod end onto the steering knuckle and tighten the ball joint castle nut to 29–35 ft. lbs. (40–48 Nm). Install a new cotter pin.

8. Install the front wheels or rear wheels.

9. If the battery was disconnected, reset the steering PCM by turning the steering wheel lock–to–lock after

reconnecting the battery and starting the engine.

NOTE: The 4WS system electronic neutral check must be performed before the front and rear wheel alignment is inspected or adjusted.

10. Raise and support the vehicle and place each of the four wheels on the center of turning radius gauge turn tables. Computerized alignment equipment must be used to service the 4WS system.

11. Measure the outer circumference of the steering wheel and mark the center with tape. Make a pointer from a piece coat hanger wire. Align the pointer with the center mark and tape the pointer to the dashboard.

12. Make sure the steering wheel is in the neutral position (straight ahead). The center mark and the pointer align.

13. Use a jumper wire to bridge the two terminal of the service check connector, Blue 2P, located behind the left leading edge of the center console.

14. Pull the parking brake lever up and turn the ignition switch **ON**, but do not start the engine. This procedure sets the front 4WS sensors to the inspection mode.

15. From the straight ahead driving position, turn the steering wheel to the left; then, turn the steering wheel slowly to the right beyond the straight ahead position. Do not turn the steering wheel in the opposite direction. Turn the steering wheel to the left; then, turn it slowly to the right beyond the straight ahead driving position. Repeat this operation until the 4WS indicator light starts to come on steady. Steady means the

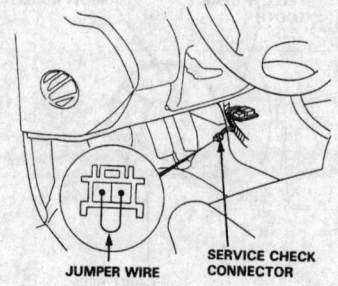

Service check connector — Prelude with 4–wheel steering

light stays on longer than two seconds.

NOTE: The 4WS indicator light should come on within a range of ±0.4 in. (±9mm) from the center mark on the steering wheel. Use the pointer as a reference guide.

16. Check the front sub–steering angle sensor. From the straight ahead driving position, turn the steering wheel to the right; then, turn the wheel slowly to the left beyond the straight ahead position. Do not turn the steering wheel in the opposite direction. Turn the steering wheel to the right; then, turn it slowly to the left beyond the straight ahead driving position. Repeat this operation until the 4WS indicator light blinks at intervals of 0.2 seconds.

NOTE: The 4WS indicator light should blink within a range of ±2.2 in. (±55mm) from the center mark on the steering wheel.

17. If the 4WS indicator light comes on or blinks at a point outside of the range, the 4WS system must be adjusted. The adjustment process should be performed by a Honda dealer.

18. Fully release the parking brake to turn the parking brake indicator off. This procedure sets the rear 4WS sensors to the inspection mode.

19. Turn the ignition switch **OFF**. Remove the cap bolt and sealing washer from the rear steering actuator. Install special tool No. 07NAJ–SS0020A, the rear steering center lock pin.

20. Set the steering wheel in the straight ahead position.

21. Turn the ignition switch **ON** engine **OFF**. To check the rear sub–steering angle sensor, push the left rear wheel fully to the right by hand; then, push it slowly to the left. Take note of the narrow range in which the 4WS indicator light blinks at 0.2 second intervals.

22. To check the rear main angle sensor, push the left rear wheel fully to the left by hand; then, push it slowly to the right. Take note of the narrow range in which the 4WS indicator light stays on for longer than two seconds.

23. Turn the ignition switch **OFF**.

24. Remove the special tool from the rear steering actuator. Then, install the cap bolt with a new crush washer and tighten it to 16 ft. lbs. (22 Nm). Remove the jumper wire from the service check connector and tuck it back under the console edge.

25. Adjust the front wheel alignment by turning the tie rod adjuster in or out. When the front wheel alignment is within the specifications, tighten the locknuts to 33 ft. lbs. (45 Nm).

26. Adjust the rear wheel alignment by turning the rear tie rod adjusters in or out. When the rear wheel alignment is within the specifications, tighten the locknuts to 33 ft. lbs. (45 Nm).

27. Lower the vehicle and tighten the wheel nuts to 80 ft. lbs. (110 Nm).

28. Road test the vehicle and check the operation of the four wheel steering system.

Manual Rack and Pinion

REMOVAL AND INSTALLATION

Civic and del Sol

----- **CAUTION** -----

The supplemental restraint system (air bag) must be disabled before removing the steering wheel to center the cable reel. Failure to disarm the air bag system may cause accidental air bag deployment, resulting in unnecessary air bag system repairs and the risk of personal injury.

1. Position the front wheels straight ahead. Lock the steering column and remove the ignition key.

2. Disconnect the negative battery cable and positive battery cables.

3. Disable the Supplemental Restraint System (air bag).

4. Remove the steering joint cover. Remove the upper and lower steering joint bolts.

5. Raise and support the vehicle safely.

6. Remove the front wheels.

7. Remove the tie–rod end cotter pins and castle nuts. Using a ball

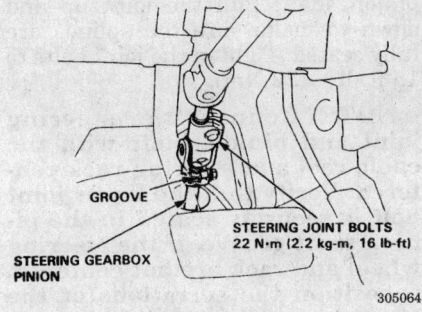

GROOVE

STEERING JOINT BOLTS
22 N•m (2.2 kg-m, 16 lb-ft)

STEERING GEARBOX
PINION

305064

Steering joint bolts — 1993–95 Civic and del Sol

joint tool, disconnect the tie rod ends from the steering knuckles.

8. Remove the left tie–rod end and slide the rack all the way to the right.

9. Remove the self–locking nuts and separate the catalytic converter or front exhaust pipe from the rear exhaust pipes. Remove the catalytic converter or front exhaust pipe.

10. Manual transaxle: Disconnect the shift lever extension rod from the clutch housing. Slide the pin retainer out of the way, drive out the spring pin, and disconnect the shift rod.

11. Automatic transaxle: Unbolt the shift cable bracket and holder. Disconnect the shift cable from the control shaft. Suspend the cable from the underbody with a piece of wire.

12. Unbolt and remove the steering rack stiffener plate.

13. Remove the steering rack mounting bracket.

14. Pull the steering rack down to release it from the pinion shaft.

15. Drop the steering rack far enough to permit the end of the pinion shaft and the grommet to come out of the hole in the bulkhead.

16. Slide the gearbox to the right until the left tie rod clears the subframe, then drop it down and out of the vehicle to the left.

To install:

NOTE: Use new self–locking nuts and gaskets when installing the catalytic converter.

17. Install the steering rack into position. Install the pinion shaft grommet and insert the pinion through the hole in the bulkhead.

18. Install the steering rack mounting cushion, bracket, and bolts. The arrow on the bracket faces the front of the vehicle. Tighten the bracket bolts to 28 ft. lbs. (39 Nm).

19. Install the steering rack stiffener plate. Tighten the steering rack mounting bolts to 43 ft. lbs. (59 Nm). Tighten the stiffener plate bolts to 28 ft. lbs. (39 Nm).

20. Center the rack ends within their steering strokes.

21. Install the tie rod ends onto the rack ends. Connect the tie rod ends to the steering knuckles and install the castle nuts. Install the front wheels.

22. Install the catalytic converter using new gaskets and self–locking nuts. Tighten the front nuts to 16 ft. lbs. (22 Nm), and the rear nuts to 25 ft. lbs. (34 Nm).

23. Manual transaxle: Reconnect the shift linkage by installing a new spring pin and clip. Install the extension rod and tighten its bolt to 16 ft. lbs. (22 Nm).

24. Automatic transaxle: Reconnect the shift cable and brackets. Tighten the bracket bolts to 9 ft. lbs. (12 Nm). Tighten the cable lockbolt to 10 ft. lbs. (14 Nm). Tighten the cable holder bolts to 16 ft. lbs. (22 Nm).

25. Verify that the rack is centered within its strokes. Lower the vehicle.

26. Center the air bag cable reel:
- Remove the steering wheel.
- Turn the cable reel clockwise until it stops.
- Turn the steering wheel counterclockwise (approximately two turns) until the arrow on the label points straight up.
- Install the steering wheel.

27. During steering wheel installation, verify that the slot on the steering wheel shaft engages with the tabs on the turn signal cancelling sleeve. The pins on the cable reel fit into the holes on the steering wheel body. Install a new steering wheel nut and tighten it to 36 ft. lbs. (50 Nm).

28. Line up the bolt hole in the steering joint with the groove in the pinion shaft. Slip the joint onto the pinion shaft. Pull the joint up and down to make sure the splines are fully seated. Tighten the joint bolts to 16 ft. lbs. (22 Nm).

NOTE: Connect the steering joint and pinion shaft with the cable reel and steering rack centered. Verify that the lower joint bolt is securely seated in the pinion shaft groove. If the steering wheel and rack are not centered, reposition the serrations at the lower end of the steering joint.

29. Install the steering joint cover.

30. Torque the ball joint castle nuts to 29–35 ft. lbs. (40–48 Nm). Then tighten them only enough to install new cotter pins.

31. Enable the Supplemental Restraint System (air bag).

32. Install the steering wheel's lower access cover.

33. Reconnect the negative and positive battery cables.

34. Turn the ignition switch to the **ON** position. The air bag indicator light should come on for six seconds and then turn off. This light sequence indicates that the air bag system is enabled and functioning normally. If the air bag light stays on longer, or doesn't turn on, the system must be diagnosed.

35. Check the front wheel alignment and steering wheel spoke angle. Make adjustments by turning the left and right tie–rod ends equally.

36. Road test the vehicle.

Power Rack and Pinion

REMOVAL AND INSTALLATION

NOTE: The radio may contain a coded theft protection circuit. Always obtain the code number before disconnecting the battery. If the vehicle is equipped with 4WS, the steering control unit is shut down when the battery is disconnected. After connecting the battery, turn the steering wheel lock–to–lock to reset the steering control unit.

Civic and del Sol

--------- CAUTION ---------
The supplemental restraint system (air bag) must be disabled before removing the steering wheel to center the cable reel. Failure to disarm the air bag system may cause accidental air bag deployment, resulting in unnecessary air bag system repairs and the risk of personal injury.

1. Lift the power steering reservoir off of its mount and disconnect the inlet hose.
2. Insert a length of tubing into the inlet hose and route the tubing into a drain container.
3. With the engine running at idle, turn the steering wheel lock–to–lock several times until fluid stops running out of the hose.
4. Position the front wheels straight ahead. Shut off the engine and lock the steering column and remove the ignition key. Reconnect the reservoir inlet hose.
5. Disconnect the negative battery cable and positive battery cables. Wait three minutes before working around the air bags.
6. Remove the steering wheel's lower access cover.

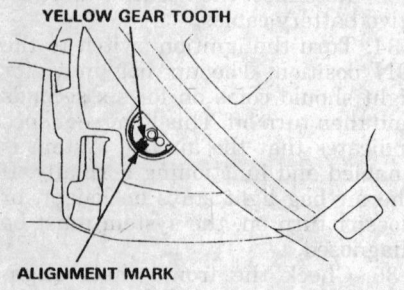

YELLOW GEAR TOOTH

ALIGNMENT MARK

305083

Cable reel alignment — 1993–95 Civic and del Sol

7. Uncouple the air bag connector from the cable reel connector:
   a. Hold the cable reel connector. With your other hand, slide the spring–loaded sleeve toward the stop tab on the air bag connector.
   b. Separate the two connectors. There is no need to install a shorting connector, as the connectors are automatically grounded when they are uncoupled.
8. Remove the steering joint cover and remove the upper and lower steering joint bolts.
9. Raise and support the vehicle safely.
10. Remove the front wheels.
11. Remove the tie rod end cotter pins and castle nuts. Using a ball joint tool, disconnect the tie rod ends from the steering knuckles.
12. Manual transaxle: Disconnect the shift lever torque rod from the clutch housing. Slide the pin retainer out of the way, drive out the spring pin and disconnect the shift rod.
13. Automatic transaxle: Unbolt the shift cable bracket and holder. Disconnect the shift cable from the control shaft. Suspend the cable from the underbody with a piece of wire.
14. Remove the self–locking nuts and separate the catalytic converter from the exhaust pipes. Remove the catalytic converter.
15. Use a flare wrench to disconnect the hydraulic line and hose from the rack valve body.
16. Remove the left tie rod end and slide the rack all the way to the right.
17. Remove the steering rack mounting bolts.
18. Pull the steering rack down to release it from the pinion shaft.
19. Drop the gearbox far enough to permit the end of the pinion shaft to come out of the hole in the frame channel.
20. Slide the gearbox to the right until the left tie rod clears the subframe, then drop it down and out of the vehicle to the left.

To install:

NOTE: Use new self–locking nuts when installing the catalytic converter.

--------- WARNING ---------
Use only genuine Honda power steering fluid. Any other type or brand of fluid will damage the power steering pump.

21. Install the steering rack into position. Install the pinion shaft grommet and insert the pinion through the hole in the bulkhead.

22. Install the rack mounting bolts. Tighten the bracket bolts to 28 ft. lbs. (39 Nm). Tighten the mounting bolt under the valve body to 43 ft. lbs. (59 Nm).
23. Reconnect the two hydraulic lines to the rack valve body. Carefully tighten the hydraulic line fitting to 28 ft. lbs. (38 Nm). Securely tighten the return hose clamp.
24. Center the rack ends within their steering strokes.
25. Install the tie rod ends onto the rack ends. Connect the tie rod ends to the steering knuckles and install the castle nuts. Install the front wheels.
26. Install the catalytic converter using new gaskets and self–locking nuts. Tighten the front nuts to 16 ft. lbs. (22 Nm), and the rear nuts to 25 ft. lbs. (34 Nm).
27. Manual transaxle: Reconnect the shift linkage by installing a new spring pin and clip. Install the extension rod and tighten its bolt to 16 ft. lbs. (22 Nm).
28. Automatic transaxle: Reconnect the shift cable and brackets. Tighten the bracket bolts to 9 ft. lbs. (12 Nm). Tighten the cable lockbolt to 10 ft. lbs. (14 Nm). Tighten the cable holder bolts to 16 ft. lbs. (22 Nm).
29. Verify that the rack is centered within its strokes. Lower the vehicle.
30. Center the air bag cable reel:
   a. Remove the steering wheel.
   b. Turn the cable reel clockwise until it stops.
   c. Turn the steering wheel counterclockwise (approximately two turns) until the arrow on the label points straight up.
   d. Install the steering wheel.
31. During steering wheel installation, verify that the slot on the steering wheel shaft engages with the tabs on the turn signal cancelling sleeve. The pins on the cable reel fit into the holes on the steering wheel body. Install a new steering wheel nut and tighten it to 36 ft. lbs. (50 Nm).
32. Line up the bolt hole in the steering joint with the groove in the pinion shaft. Slip the joint onto the pinion shaft. Pull the joint up and down to make sure the splines are fully seated. Tighten the joint bolts to 16 ft. lbs. (22 Nm).

NOTE: Connect the steering joint and pinion shaft with the cable reel and steering rack centered. Verify that the lower joint bolt is securely seated in the pinion shaft groove. If the steering wheel and rack are not centered, reposition the serrations at the lower end of the steering joint.

33. Install the steering joint cover.

34. Torque the ball joint castle nuts to 29–35 ft. lbs. (40–48 Nm). Then tighten them only enough to install new cotter pins.

35. Reconnect the air bag and cable reel connectors: Make sure the connectors fit squarely together. Then, press the connectors to couple them. The spring–loaded sleeve will lock into place as the two connectors are coupled.

36. Install the steering wheel's lower access cover.

37. Reconnect the negative and positive battery cables.

38. Turn the ignition switch to the **ON** position. The air bag indicator light should come on for six seconds and then turn off. This light sequence indicates that the air bag system is enabled and functioning normally. If the air bag light stays on longer, or doesn't turn on, the system must be diagnosed.

39. Make sure the reservoir inlet line has been reconnected. Fill the reservoir to the upper line with Honda power steering fluid. Run the engine at idle and turn the steering wheel lock–to–lock several times to bleed air from the system and fill the rack valve body. Recheck the fluid level and add more if necessary.

40. Check the power steering system for leaks.

41. Check the front wheel alignment and steering wheel spoke angle. Make adjustments by turning the left and right tie rod ends equally.

42. Road test the vehicle.

Prelude

NOTE: The electronic neutral check must be performed on 4WS equipped Preludes any time the steering rack, steering wheel, or steering column is removed, and before the wheels are aligned.

1. Lift the power steering reservoir off of its mount and disconnect the inlet hose.

2. Insert a length of tubing into the inlet hose and route the tubing into a drain container.

3. With the engine running at idle, turn the steering wheel lock–to–lock several times until fluid stops running out of the hose. Shut off the engine.

4. Position the front wheels straight ahead. Lock the steering column with the ignition key. Reconnect the reservoir inlet hose.

5. Disconnect the negative battery cable.

6. Remove the steering joint cover and remove the upper and lower steering joint bolts.

7. Raise and support the vehicle safely.

8. Remove the front wheels.

9. Remove the tie rod end cotter pins and castle nuts. Install a 12mm nut onto the end of the ball joint stud to protect the threads from damage. Using a ball joint tool, disconnect the tie rod ends from the steering knuckles.

10. Disconnect the heated oxygen sensor.

11. Remove the self–locking nuts and separate the catalytic converter from exhaust pipe A. Unbolt exhaust pipe A from the intake manifold, and remove it from the vehicle.

12. On vehicles equipped with automatic transaxles, remove the shift cable cover, disconnect the shift cable, and wire it up and out of the way.

13. Clean any oil or dirt off of the valve body with solvent.

14. Remove the center beam from the subframe.

15. Remove the valve body shield.

16. Use a flare wrench to disconnect the four hydraulic lines from the rack valve body. Plug the lines to keep dirt and moisture out.

17. On models with 4WS, carefully cut the wire tie securing the cover to the front sub–steering angle sensor. Remove the cover.

18. Remove the sensor wire harness from the two securing clamps; then, disconnect sensor connector from the 4WS steering main wiring harness.

19. Remove the steering joint bolt and slide the pinion shaft out of the joint.

20. Remove the left mounting bracket; then remove the right mounting brackets.

21. Remove the left tie rod end and slide the rack all the way to the right.

22. Pull the steering rack down to release it from the pinion shaft.

23. Slide the steering rack to the right until the left tie rod clears the

subframe, then drop it down and out of the vehicle to the left.

To install:

NOTE: Use new gaskets and self-locking nuts when installing exhaust pipe A.

----- **WARNING** -----
Use only genuine Honda power steering fluid. Any other type or brand of fluid will damage the power steering pump.

24. Install the steering rack into position. Install the pinion shaft grommet and insert the pinion through the hole in the firewall.

25. Install the right and left mounting brackets. Tighten the short bolts to 28 ft. lbs. (39 Nm), and the long bolts to 32 ft. lbs. (44 Nm).

26. Center the rack ends within their steering strokes.

27. Center the air bag cable reel:
• Turn the steering wheel clockwise until it stops.
• Turn the steering wheel counterclockwise until the yellow gear tooth lines up with the alignment mark on the lower column cover.

28. Line up the bolt hole in the steering joint with the groove in the pinion shaft. Slip the joint onto the pinion shaft. Pull the joint up and down to make sure the splines are fully seated. Tighten the joint bolts to 16 ft. lbs. (22 Nm).

NOTE: Connect the steering joint and pinion shaft with the cable reel and steering rack centered. Verify that the lower joint bolt is securely seated in the pinion shaft groove. If the steering wheel and rack are not centered, reposition the serrations at the lower end of the steering joint.

29. Reconnect the four hydraulic lines to the rack valve body. Carefully tighten the 12mm fittings to 9 ft. lbs. (13 Nm), the 14mm inlet fitting to 28 ft. lbs. (37 Nm), and the 17mm oil cooler fitting to 21 ft. lbs. (29 Nm).

30. Connect the front sub–steering angle sensor to the 4WS harness. Place the wire back into its clamps, making sure that it doesn't interfere with the stabilizer bar. Install the sensor cover with a new wire tie.

31. Install the valve body shield.

32. Install the center beam. Use new self–locking bolts and tighten them to 43 ft. lbs. (60 Nm).

33. On automatic transaxle equipped vehicles, reconnect the shift cable and tighten the locknut to 10 ft. lbs. (14 Nm). Install the cable holder

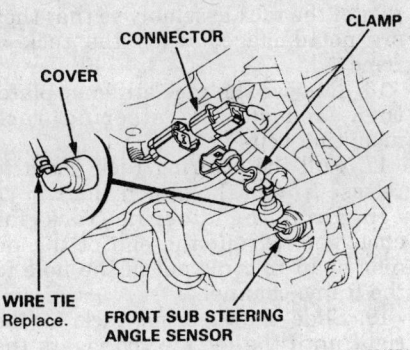

304954

4WS front sub-steering angle sensor — Prelude

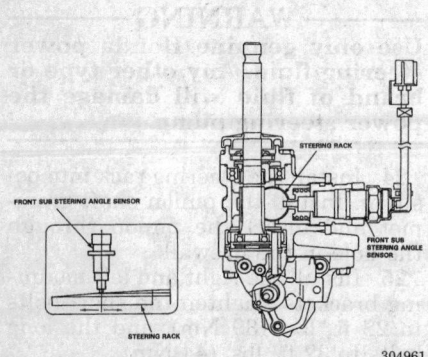

Front sub-steering angle sensor, harness, and steering rack — Prelude

304961

and tighten its bolts to 13 ft. lbs. (18 Nm).

34. Install the catalytic convertor using new gaskets and self–locking nuts. Tighten the exhaust manifold nuts to 40 ft. lbs. (55 Nm), and the rear nuts to 25 ft. lbs. (34 Nm).

35. Reconnect the heated oxygen sensor.

36. Install the tie rod ends onto the rack ends. Connect the tie rod ends to the steering knuckles and install the castle nuts. Install the front wheels.

37. Verify that the rack is centered within its strokes. Lower the vehicle.

38. Install the steering joint cover.

39. Torque the ball joint castle nuts to 36–43 ft. lbs. (50–60 Nm). Then, tighten them only enough to install new cotter pins.

40. Reconnect the negative battery cable.

41. Make sure the reservoir inlet line has been reconnected. Fill the reservoir to the upper line with Honda power steering fluid. Run the engine at idle and turn the steering wheel lock–to–lock several times to bleed air from the system and fill the rack valve body. Recheck the fluid level and add more if necessary.

42. Check the power steering system for leaks.

43. On Preludes without 4WS, check and adjust the front wheel

alignment. On Preludes with 4WS, the electronic neutral check must be performed on the 4WS system.

Accord

1. Lift the power steering reservoir off of its mount and disconnect the inlet hose.

2. Insert a length of tubing into the inlet hose and route the tubing into a drain container.

3. With the engine running at idle, turn the steering wheel lock–to–lock several times until fluid stops running out of the hose. Immediately shut off the engine.

4. Position the front wheels straight ahead. Lock the steering column with the ignition key. Reconnect the reservoir inlet hose.

5. Disconnect the negative battery cable.

6. Remove the steering joint cover and remove the upper and lower steering joint bolts.

7. Raise and support the vehicle safely.

8. Remove the front wheels.

9. Remove the tie rod end cotter pins and castle nuts. Using a ball joint tool, disconnect the tie rod ends from the steering knuckles.

10. Remove the left tie rod end and slide the rack all the way to the right.

11. Disconnect the heated oxygen sensor.

12. Remove the self–locking nuts and separate the catalytic converter from exhaust pipe A. Remove the catalytic converter.

13. On vehicles equipped with manual transaxles, disconnect the shift linkage from the transaxle case.

14. On vehicles equipped with automatic transaxles, remove the shift cable cover, disconnect the cable, and wire it up and out of the way.

15. Use a flare wrench to disconnect the two hydraulic lines from the rack valve body. Plug the lines to keep dirt and moisture out. Carefully move the disconnected lines to the rear of the rack assembly so that they are not damaged when the rack is removed.

16. Remove the rack stiffener plate; then, remove the steering rack mounting bolts.

17. Pull the steering rack down to release it from the pinion shaft.

18. Drop the steering rack far enough to permit the end of the pinion shaft to come out of the hole in the frame channel.

19. Slide the steering rack to the right until the left tie rod clears the subframe, then drop it down and out of the vehicle to the left.

To install:

NOTE: Use new gaskets and self–locking nuts when installing the catalytic converter.

------- **WARNING** -------
Use only genuine Honda power steering fluid. Any other type or brand of fluid will damage the power steering pump.

20. Before installing the rack & pinion, slide the ends all the way to the right. Install the pinion shaft grommet. The lug on the pinion shaft grommet aligns with the slot on the valve body.

21. Install the steering rack into position. Install the pinion shaft grommet and insert the pinion through the hole in the bulkhead.

22. Install the rack mounting bolts. Tighten the bracket bolts to 28 ft. lbs. (39 Nm). Tighten the stiffener plate mounting bolts to 32 ft. lbs. (43 Nm).

23. Center the rack ends within their steering strokes.

24. Center the air bag cable reel:
 • Turn the steering wheel left approximately 150°, to check the cable reel position with the indicator.
 • If the cable reel is centered, the yellow gear tooth lines up with the alignment mark on the cover.
 • Return the steering wheel right approximately 150° to position the steering wheel in the straight ahead position.

25. Line up the bolt hole in the steering joint with the groove in the pinion shaft. Slip the joint onto the pinion shaft. Pull the joint up and down to make sure the splines are fully seated. Tighten the joint bolts to 16 ft. lbs. (22 Nm).

NOTE: Connect the steering joint and pinion shaft with the cable reel and steering rack centered. Verify that the lower joint bolt is securely seated in the pinion shaft groove. If the steering wheel and rack are not centered, reposition the serrations at the lower end of the steering joint.

26. Install the steering joint cover and the rack & pinion cover.

27. On 1993 models, reconnect the four hydraulic lines to the rack valve body. Carefully tighten the hydraulic line fittings: reservoir line 17mm fitting to 21 ft. lbs. (29 Nm), oil cooler line 12mm fitting to 9 ft. lbs. (13 Nm), pump outlet line 14mm fitting to 28 ft. lbs. (38 Nm), vehicle speed sensor 12mm fitting to 9 ft. lbs. (13 Nm).

28. On 1994–97 models, reconnect the two hydraulic lines to the rack

valve body. Carefully tighten the 14mm inlet fitting to 27 ft. lbs. (37 Nm) and the 16mm outlet fitting to 21 ft. lbs. (28 Nm).

29. If equipped with a manual transaxle, connect the shift cable and the select cable to the transaxle with new cotter pins.

30. If equipped with a automatic transaxle, connect the shift cable to the transaxle using a new lockwasher. Torque the lockbolt to 10 ft. lbs. (14 Nm).

31. Install the catalytic convertor using new gaskets and self–locking nuts. Tighten the front nuts to 16 ft. lbs. (22 Nm), and the rear nuts to 25 ft. lbs. (34 Nm).

32. Reconnect the heated oxygen sensor.

33. Install the tie rod ends onto the rack ends. Connect the tie rod ends to the steering knuckles and install the castle nuts.

34. Torque the ball joint castle nuts to 29–35 ft. lbs. (40–48 Nm). Then tighten them only enough to install new cotter pins.

35. Install the front wheels.

36. Lower the vehicle.

37. Reconnect the negative battery cable.

38. Make sure the reservoir inlet line has been reconnected. Fill the reservoir to the upper line with Honda power steering fluid. Run the engine at idle and turn the steering wheel lock–to–lock several times to bleed air from the system and fill the rack valve body. Recheck the fluid level and add more if necessary.

39. Check the power steering system for leaks.

40. Check the front wheel alignment and steering wheel spoke angle. Make adjustments by turning the left and right tie rod ends equally.

41. Road test the vehicle.

Power Steering Pump

REMOVAL AND INSTALLATION

NOTE: The radio may contain a coded theft protection circuit. Always obtain the code number before disconnecting the battery. If the vehicle is equipped with 4WS, the steering control unit is shut down when the battery is disconnected. After connecting the battery, start the engine and turn the steering wheel lock–to–lock to reset the steering control unit.

1. Disconnect the negative battery cable.

2. Drain the power steering fluid from the reservoir.

3. Cover the alternator with some shop rags to protect it from spilled fluid.

4. Loosen the the pump adjusting bolt and mounting bolts. Slip the pump drive belt off of its pulley.

5. Disconnect and plug the outlet and inlet hoses.

6. Remove the pump mounting bolts. Remove the pump.

To install:

— CAUTION —
Use only genuine Honda power steering fluid. Any other brand and type of fluid will damage the power steering pump.

7. Install the power steering pump into its bracket. Loosely install the mounting and adjusting bolts.

8. Install a new O–ring onto the outlet hose fitting. Connect the outlet and inlet hoses to the power steering pump. Tighten the outlet hose bolts to 8 ft. lbs. (11 Nm).

9. Fit the pump belt onto the pulley.

10. Adjust the pump belt tension to the proper specification for the vehicle.

11. Tighten the mounting nuts and bolts to the following specifications:
• 1993 Accord: mounting nut 16 ft. lbs. (22 Nm), bolt 33 ft. lbs. (45 Nm)
• 1993–97 Civic and del Sol and 1994–97 Accord: mounting nuts 17 ft. lbs. (24 Nm)
• Prelude: mounting nut and bolt 16 ft. lbs. (22 Nm)

12. Fill the fluid reservoir to the upper level line with Honda power steering fluid.

13. Connect the negative battery cable.

14. On vehicles equipped with 4WS, start the engine and turn the steer-

ing wheel lock–to–lock to reset the steering control unit.

15. Bleed the power steering system by running the engine at idle and turning the steering wheel lock–to–lock several times. Add more fluid if necessary.

16. Check the hose connections for leaks.

17. Enter the radio security code and reset the clock.

BRAKES

Anti-Lock Brake System Service

PRECAUTIONS

• Certain components within the Anti-Lock Brake System (ABS) are not intended to be serviced or repaired individually. Only those components with removal and installation procedures should be serviced.

• Do not use rubber hoses or other parts not specifically specified for and ABS system. When using repair kits, replace all parts included in the kit. Partial or incorrect repair may lead to functional problems and require the replacement of components.

• Lubricate rubber parts with clean, fresh brake fluid to ease assembly. Do not use lubricated shop air to clean parts; damage to rubber components may result.

• Use only specified brake fluid from an unopened container.

• If any hydraulic component or line is removed or replaced, it may be necessary to bleed the entire system.

• A clean repair area is essential. Always clean the reservoir and cap thoroughly before removing the cap. The slightest amount of dirt in the fluid may plug an orifice and impair the system function. Perform repairs after components have been thoroughly cleaned; use only denatured alcohol to clean components. Do not allow ABS components to come into contact with any substance containing mineral oil; this includes used shop rags.

• The Anti-Lock control unit is a microprocessor similar to other computer units in the vehicle. Ensure that the ignition switch is **OFF** before removing or installing controller harnesses. Avoid static electricity discharge at or near the controller.

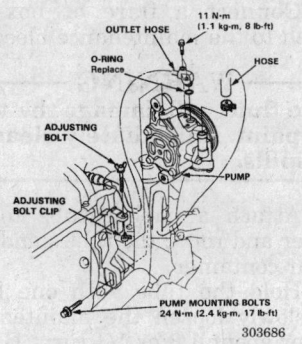

Power steering pump components — Civic and del Sol

• If any arc welding is to be done on the vehicle, the control unit should be unplugged before welding operations begin.

DEPRESSURIZING

1993 Accord, 1993–95 Civic and 1993–97 del Sol and Prelude

———— CAUTION ————

The hydraulic accumulator contains brake fluid and nitrogen gas at extremely high pressures. Certain portions of the hydraulic system also contain brake fluid at high pressure. The system must be depressurized before disconnecting any hoses, lines, or fittings; or personal injury may result.

1. Disconnect the negative battery cable.
2. Clean any dirt or grime off of the modulator reservoir and bleeder screw.
3. Remove the red service cap from the bleeder screw on the body of the ABS modulator.
4. Fit a bleeder T–wrench, tool No. 07HAA–SG00101 or equivalent, squarely onto the bleeder screw.
5. Slowly turn the T–wrench 90° to release the high–pressure fluid

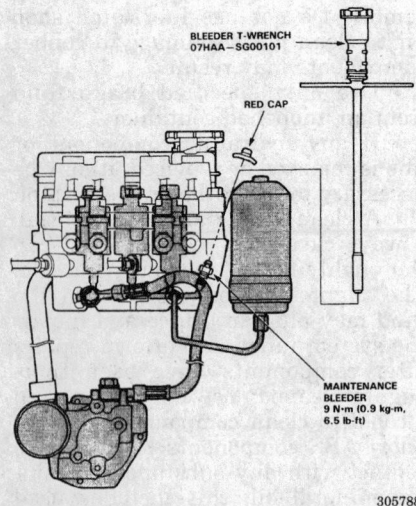

BLEEDER T–WRENCH
07HAA–SG00101

RED CAP

MAINTENANCE BLEEDER
9 N·m (0.9 kg-m, 6.5 lb-ft)

305788

Bleeder T-wrench and ABS components, shaded areas are pressurized — 1993–95 Civic and 1993–97 del Sol

from the modulator. Then, continue turning the T–wrench until it has been rotated one complete turn. This will completely drain the brake fluid from the modulator.
6. If necessary, a suction pump or syringe may be used to suck the brake fluid from the modulator reservoir.
7. After servicing, use the T–wrench to tighten the bleeder screw to 6.5 ft. lbs. (9 Nm). Reinstall the red service cap.
8. Refill and bleed the brake system. Use only clean DOT 3 or 4 fluid. Don't mix brands or types of brake fluid.

1996–97 Civic

The Honda anti–lock braking system used on the 1996 Civic does not require any special pressure–relief procedure. This system does not use an nitrogen gas–filled accumulator. The brake fluid replacement and bleeding procedures are the same as those used for conventional (non–ABS) braking systems. Remember to plug the brake lines when they are disconnected to prevent brake fluid loss and system contamination.

1994–97 Accord

———— CAUTION ————

The hydraulic accumulator contains brake fluid and nitrogen gas at extremely high pressures. Certain portions of the hydraulic system also contain brake fluid at high pressure. Do not loosen the relief plug on the accumulator. The system must be depressurized before disconnecting any hoses, lines, or fittings, or personal injury may result.

Relieving Brake System Pressure

1. Remove the cap from the modulator maintenance bleeder.
2. Connect a flare or box–end wrench to the maintenance bleeder.

———— WARNING ————

Brake fluid will damage the vehicle's paint. Immediately clean up any spills.

3. Attach a rubber tube to the bleeder and route the tube's end into a clear container.
4. Hold the tube with one hand and slowly loosen the maintenance bleeder about 1/8 or 1/4 turn. Do not loosen the maintenance bleeder too much, as the highly–pressurized fluid may burst out.

5. After relieving the pressure, tighten the bleeder to 8 ft. lbs. (11 Nm).
6. After servicing, start the engine and make sure that the ABS indicator light turns off. Check the performance of the brake system.

Draining the ABS Modulator

1. Relieve the brake system pressure. Then, retighten the maintenance bleeder.
2. Start the engine and allow it to idle for one minute. Then, shut the engine off.
3. Check to see that the brake fluid in the modulator reservoir is below the MAX line.
4. To drain the remaining brake fluid from the modulator, repeat steps 1, 2, and 3. The total fluid capacity of the modulator is 5 fl. oz. (150 ml); about 1.3–1.5 fl. oz. (40–45 ml) of fluid is drained during each cycle.

NOTE: The modulator should be bled if air enters it during the draining and brake fluid replacement processes.

5. Remove the modulator reservoir cap and fill the reservoir to the MAX line with clean brake fluid. Recap the reservoir.
6. Repeat steps 1, 2, and 3 two more times. Then, refill the reservoir with clean brake fluid.
7. Tighten the bleeder to 8 ft. lbs. (11 Nm).
8. After servicing, start the engine and make sure that the ABS indicator light turns off.

Bleeding the ABS Modulator

1. Fill the modulator reservoir to the MAX line.
2. Attach a snug–fitting rubber tube to the maintenance bleeder and route the end of the tube into a clean container.
3. Loosen the maintenance bleeder. Then start the engine to active the modulator's pump motor.
4. Tighten the maintenance bleeder when fluid starts to flow out of the bleeder.
5. Shut the engine off after the modulator pump motor stops.

NOTE: If the ABS indicator light turns on and the pump motor stops, restart the engine and repeat steps 3, 4, and 5.

6. Make sure the brake fluid level is at the reservoir's MAX line.
7. After servicing, start the engine and make sure that the ABS indicator light turns off.

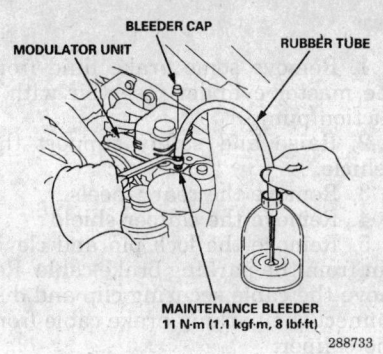

Modulator maintenance bleeder and bleeding equipment — 1994–97 Accord

Master Cylinder

REMOVAL AND INSTALLATION

NOTE: The original radio contains a coded anti–theft circuit. Obtain the security code number before disconnecting the battery.

1. Disconnect the negative battery cable.
2. Disconnect the fluid level sensor wires.

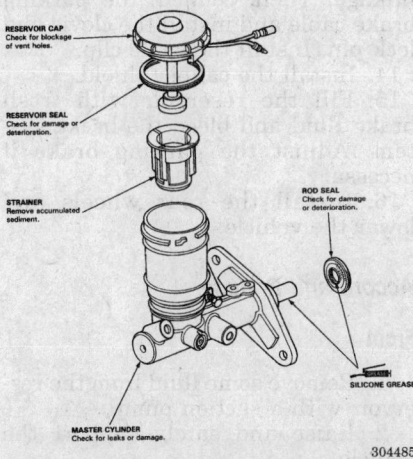

Master cylinder components — 1993–95 Civic and 1993–97 del Sol

3. Remove the fluid from the reservoir using a suction pump.
4. Use a flare wrench to disconnect the brake lines from the master cylinder. Plug the lines to prevent dirt and dust contamination.
5. Remove the mounting nuts and remove the master cylinder from the power booster.

To install:

NOTE: The master cylinder piston–to–booster pushrod clearance must be checked and adjusted before the master cylinder is installed.

6. Install a pushrod adjustment gauge, Honda tool No. 07JAG–SD40100, or equivalent, onto the master cylinder body. Turn the tool's adjusting nut until its center shaft touches the master cylinder piston.
7. Reverse the adjustment tool and install it onto the power booster. Install the master cylinder nuts and tighten them to 11 ft. lbs. (15 Nm).
8. Connect a vacuum gauge in–line with the booster vacuum line. Connect the battery cable and run the engine at a steady idle speed to deliver 20 in. (500mm) Hg of vacuum.
9. Measure the clearance with a feeler gauge. The clearance must be within the specification. If the clearance is not correct, it must be adjusted by tightening or loosening the power booster pushrod adjuster (located inside the vehicle).
• Civic and del Sol: 0–0.02 in. (0–0.4mm)
• Accord: 0–0.02 in. (0–0.4mm)
• Prelude with ABS: 0–0.01 in. (0–0.2mm)
• Prelude without ABS: 0–0.02 in. (0–0.04mm)
10. Coat a new rod seal with silicon grease and install it onto the master cylinder. Remove the adjustment gauge and install the master cylinder onto the booster. Tighten the nuts to 11 ft. lbs. (15 Nm).
11. Connect the brake lines and carefully tighten the fittings to 14 ft. lbs. (19 Nm).
12. Fill the reservoir with fluid and install the cap. Reconnect the fluid level sensor wire.
13. Bleed the brake system.
14. Pump the pedal several times to build pedal pressure and test the feel of the pedal action.
15. Check brake pedal height and freeplay.
16. Road test the vehicle.

Brake Caliper

REMOVAL AND INSTALLATION

Civic and del Sol

Front

NOTE: Two distinct types of front calipers are used on these vehicles. The caliper body will be marked either "5410" or "2056" depending on type. Servicing procedures are similar, but the different calipers use different pads.

1. Remove some brake fluid from the master cylinder reservoir with a suction pump.
2. Raise and safely support the vehicle.
3. Remove the front wheels.
4. Remove the banjo bolt and disconnect the brake hose from the caliper. Plug the hose to prevent fluid loss and contamination.
5. Remove the mounting bolts, and then remove the caliper from its mounting bracket.
6. If necessary for servicing, remove the caliper mounting bracket from the steering knuckle.
To install:
7. If the caliper mounting bracket was removed, install it. Apply brake seal grease to the caliper pins and install them with new pin boots. Apply anti–seize paste to the caliper mounting bolts and tighten them to 80 ft. lbs. (108 Nm).
8. Compress the piston into the caliper. Use a large C–clamp or a piston compressor tool. The piston should fit squarely into its bore, and must not bind on any surface.
9. Install the brake pads.
10. Fit the caliper over the pads and onto its mounting bracket.
11. On vehicles equipped with type 2056 calipers, torque the top caliper bolt to 25 ft. lbs. (35 Nm). Torque the lower bolt to 20 ft. lbs. (27 Nm).
12. On vehicles equipped with type 5410 calipers, torque both caliper bolts to 24 ft. lbs. (33 Nm).
13. Reconnect the brake hose to the caliper using new sealing washers. Carefully torque the banjo bolt to 25 ft. lbs. (35 Nm).
14. Fill the reservoir with fresh brake fluid and bleed the brake system.
15. Install the front wheels and lower the vehicle.

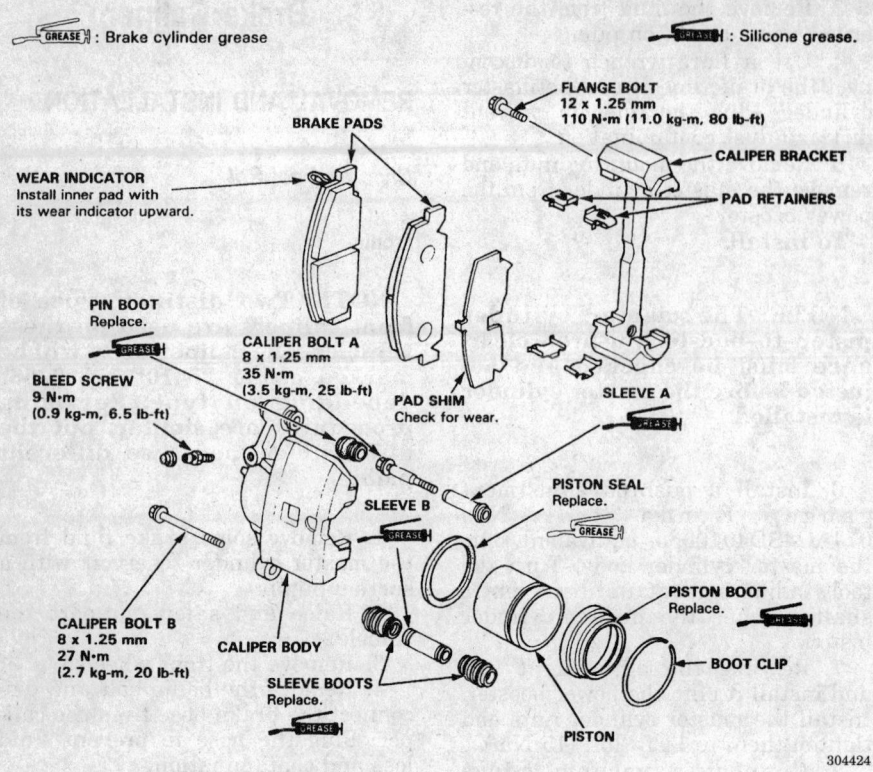

GREASE : Brake cylinder grease

GREASE : Silicone grease.

BRAKE PADS

FLANGE BOLT
12 x 1.25 mm
110 N·m (11.0 kg-m, 80 lb-ft)

CALIPER BRACKET

WEAR INDICATOR
Install inner pad with
its wear indicator upward.

PAD RETAINERS

PIN BOOT
Replace.
GREASE

CALIPER BOLT A
8 x 1.25 mm
35 N·m
(3.5 kg-m, 25 lb-ft)

BLEED SCREW
9 N·m
(0.9 kg-m, 6.5 lb-ft)

PAD SHIM
Check for wear.

SLEEVE A
GREASE

PISTON SEAL
Replace.
GREASE

SLEEVE B
GREASE

PISTON BOOT
Replace.
GREASE

CALIPER BOLT B
8 x 1.25 mm
27 N·m
(2.7 kg-m, 20 lb-ft)

CALIPER BODY

SLEEVE BOOTS
Replace.
GREASE

BOOT CLIP

PISTON

304424

Akebono front caliper — Civic and del Sol

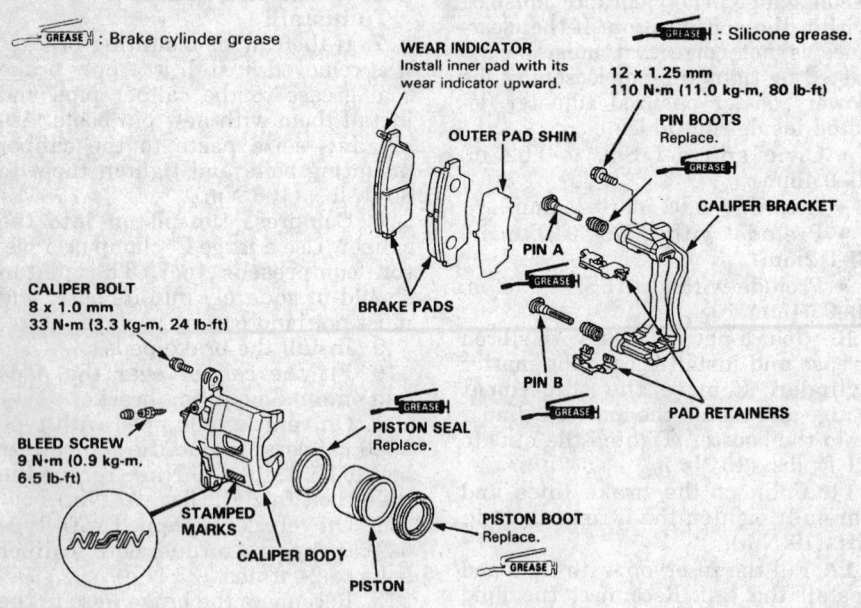

GREASE : Brake cylinder grease

GREASE : Silicone grease.

WEAR INDICATOR
Install inner pad with its
wear indicator upward.

12 x 1.25 mm
110 N·m (11.0 kg-m, 80 lb-ft)

OUTER PAD SHIM

PIN BOOTS
Replace.
GREASE

CALIPER BRACKET

PIN A
GREASE

CALIPER BOLT
8 x 1.0 mm
33 N·m (3.3 kg-m, 24 lb-ft)

BRAKE PADS

PIN B
GREASE

PAD RETAINERS

BLEED SCREW
9 N·m (0.9 kg-m,
6.5 lb-ft)

STAMPED
MARKS

NISIN

PISTON SEAL
Replace.

CALIPER BODY

PISTON

PISTON BOOT
Replace.
GREASE

304423

Nissin front caliper — Civic and del Sol

Rear

1. Remove some brake fluid from the master cylinder reservoir with a suction pump.

2. Raise and safely support the vehicle.

3. Remove the rear wheels.

4. Remove the caliper shield.

5. Remove the lock pin and clevis pin from the parking brake cable. Remove the cable securing clip and disconnect the parking brake cable from the caliper.

6. Remove the banjo bolt and disconnect the brake hose from the caliper. Plug the hose to prevent fluid loss and contamination.

7. Remove the two caliper mounting bolts. Remove the caliper from its mounting bracket.

8. If necessary for servicing, remove the caliper mounting bracket from the trailing arm.

To install:

9. If the caliper mounting bracket was removed, install it. Apply brake seal grease to the caliper pins and install them with new pin boots. Apply anti–seize paste to the caliper bracket mounting bolts and tighten them to 28 ft. lbs. (39 Nm).

10. Install the brake pads.

11. Rotate the caliper piston clockwise into place in the cylinder and then align the groove in the piston with the tab on inner pad. Fit the caliper over the pads and onto its mounting bracket.

12. Tighten the caliper bolts to 17 ft. lbs. (23 Nm).

13. Grease the parking brake linkage. Then, connect the parking brake cable and install the clevis and lock pin. Install the cable clip.

14. Install the caliper shield.

15. Fill the reservoir with fresh brake fluid and bleed the brake system. Adjust the parking brake if necessary.

16. Install the rear wheels and lower the vehicle.

Accord and Prelude

Front

1. Remove some fluid from the reservoir with a suction pump.

2. Raise and safely support the vehicle.

3. Remove the front wheels.

4. Remove the banjo bolt and disconnect the brake hose from the caliper. Plug the hose to prevent fluid loss and contamination.

5. Remove the mounting bolts and remove the caliper from its mounting bracket.

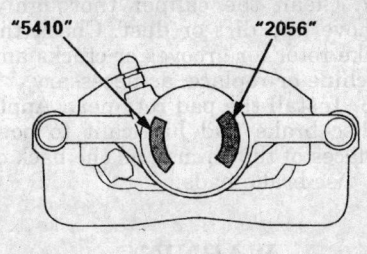

"5410"   "2056"

295526

Front caliper identification marks — 1996–97 Civic

To install:

6. Fit the caliper over the pads and onto its mounting bracket.

7. On vehicles equipped with long caliper pins, torque the top caliper bolts to 54 ft. lbs. (74 Nm).

8. On vehicles equipped with short caliper bolts, torque the caliper bolts to 36 ft. lbs. (50 Nm).

9. Reconnect the brake hose to the caliper using new sealing washers. Carefully torque the banjo bolt to 25 ft. lbs. (35 Nm).

10. Fill the reservoir with fluid and bleed the brakes.

11. Install the front wheels and lower the vehicle.

Rear

1. Remove some fluid from the reservoir with a suction pump.

2. Raise and safely support the vehicle.

3. Remove the rear wheels.

4. Remove the caliper shield.

5. Remove the lock pin and clevis pin from the parking brake cable. Remove the cable securing clip and disconnect the parking brake cable from the caliper.

6. Remove the banjo bolt and disconnect the brake hose from the caliper. Plug the hose to prevent fluid loss and contamination.

7. Remove the two caliper mounting bolts. Remove the caliper from its mounting bracket.

To install:

8. Fit the caliper over the pads and onto the mounting bracket. Rotate the piston clockwise into place in the cylinder and then align the groove in the piston with the tab on inner pad.

9. Tighten the caliper bolts to 17 ft. lbs. (23 Nm) .

10. Reconnect the brake hose to the caliper using new sealing washers, then torque the banjo bolt to 25 ft. lbs. (35 Nm).

11. Connect the parking brake cable and install the clevis and lock pin. Install the cable clip.

12. Install the caliper shield.

13. Fill the reservoir with fluid and bleed the brake system. Adjust the parking brake if necessary.

14. Install the rear wheels and lower the vehicle.

Disc Brake Pads

REMOVAL AND INSTALLATION

Civic and del Sol

Front

NOTE: Two distinct types of front caliper are used on these vehicles. The caliper body will be marked either "5410" or "2056", according to type. Servicing procedures are similar, but the different calipers use different pads.

1. Use a suction pump to remove some brake fluid from the master cylinder reservoir.

2. Raise and support the vehicle safely. Remove the front wheels.

3. Unbolt the brake hose clamp from the steering knuckle.

4. Remove the lower caliper retaining bolt and pivot the caliper up.

5. Remove the disc brake pads, shims, and pad retainers from the caliper.

To install:

6. Clean the caliper thoroughly; remove any rust. Check the brake rotor for grooves or cracks and machine or replace if necessary.

7. Install the pad retainers. Apply the lubricant included with the pad kit, or molybdenum brake grease to the inner side of the shims and the back of the disc brake pads. Do not get any lubricant on the contact surface of the pad.

8. Install the pads, shims, and pad retainers. Make sure the wear indicator on the inner pad is facing up.

9. Push in the caliper piston with a suitable tool so that the caliper will fit over the pads.

10. Pivot the caliper down into position. Install the caliper bolts and tighten them to 24 ft. lbs. (33 Nm) for 5410 calipers; or to 25 ft. lbs. (35 Nm) (top) and 20 ft. lbs. (27 Nm) (bottom) for 2056 calipers.

11. Fill the reservoir with clean brake fluid. Slowly press the brake pedal several times to seat the pads and test the system. Bleed the brakes

if the pedal action feels spongy or weak.

12. Install the front wheels and lower the vehicle.

Rear

1. Use a suction pump to remove some brake fluid from the master cylinder reservoir.

2. Raise and safely support the vehicle. Remove the rear wheels.

3. Remove the caliper dust shield.

4. Remove the two caliper mounting bolts. Remove the caliper from the bracket and hang it out of the way with a piece of wire. The parking brake cable doesn't need to be disconnected.

5. Remove the pads, shims, and pad retainers from the caliper.

To install:

6. Clean the caliper thoroughly; remove any rust. Check the brake rotor for grooves or cracks and machine or replace if necessary.

7. Install the pad retainers. Apply the lubricant included with the pad kit, or molybdenum brake grease to the inner side of the shims and to the back of the disc brake pads. Do not get any lubricant on the contact surface of the pad.

8. Install the pads, shims, and pad retainers.

9. Rotate the caliper piston clockwise into the caliper bore enough to allow the caliper to fit over the brake pads. Lubricate the piston boot with silicon grease. Avoid twisting the piston boot.

10. Install the brake caliper. Align the groove in the piston with the tab on the inner pad. Tighten the mounting bolts to 17 ft. lbs. (23 Nm).

11. Fill the master cylinder reservoir with clean brake fluid. Slowly press the brake pedal several times to seat the pads and test the system. Bleed the brakes if the pedal action feels spongy or weak.

12. Install the rear wheels and lower the vehicle.

Accord and Prelude

Front

----- **CAUTION** -----

Brake pads and shoes contain asbestos, which has been determined to be a cancer causing agent. Never clean the brake surfaces with compressed air. Avoid inhaling any dust from brake surfaces. When cleaning brakes, use commercially available brake cleaning fluids.

NOTE: If the car was recently driven, the brake components (particularly the rotor) will be hot. Wear gloves.

1. Raise and support the vehicle safely.
2. Remove the wheel.
3. Remove a small amount of brake fluid from the reservoir using a suction pump.
4. Unbolt the brake hose clamp from the strut or knuckle by removing the retaining bolts.
5. Remove the lower caliper retaining bolt and pivot the caliper upward, off of the pads.
6. Remove the pad shim and pad retainers. Remove the disc brake pads from the caliper.

To install:

7. Clean the caliper thoroughly; remove any rust from the lip of the disc or rotor. Check the brake rotor for grooves or cracks. If any heavy scoring is present, the rotor must be replaced.
8. Install the pad retainers. Apply a disc brake pad lubricant to both surfaces of the shims and the back of the disc brake pads.

—— WARNING ——
DO NOT get any lubricant on the braking surface of the pad.

9. Install the pads and shims. The pad with the wear indicator goes in the inboard position.
10. Push in the caliper piston so the caliper will fit over the pads. This is most easily accomplished with a large C-clamp.

—— WARNING ——
As the piston is forced back into the caliper, fluid will be forced back into the master cylinder reservoir. It may be necessary to siphon some fluid out to prevent overflowing.

11. Pivot the caliper down into position and tighten the mounting bolt to 33 ft. lbs (45 Nm), or 54 ft. lbs. (75 Nm) for long caliper bolts (integral bolt and pin).
12. Connect the brake hose to the strut or knuckle, if removed. Install the wheel and lower the vehicle to the ground.
13. Add brake fluid to the master cylinder reservoir and install the cap.
14. Depress the brake pedal several times and make sure that the movement feels normal. The first brake pedal application may result in a very long pedal action due to the pis-

tons being retracted. Always make several brake applications before starting the vehicle. Bleed the system if necessary.

NOTE: Braking should be moderate for the first 5 miles or so until the new pads seat correctly. The new pads will seat correctly if put through several moderate heating and cooling cycles. Avoid hard braking until the brakes have experienced several long, slow stops, with time to cool in between. Taking the time to properly seat the brakes will yield quieter operation, more efficient stopping, extended brake life.

Rear

1. Raise and safely support the vehicle.
2. Remove a small amount of brake fluid from the reservoir using a suction pump.
3. Remove the rear wheel.
4. Remove the dust shield.
5. Remove the two caliper mounting bolts and remove the caliper from the bracket.
6. Remove the pads, shims and pad retainers.

To install:

7. Clean the caliper thoroughly; remove any dirt or dust. Check the brake rotor for grooves or cracks and machine or replace, as necessary.
8. Install the pad retainers. Apply a disc brake pad lubricant to both surfaces of the shims and the back of the disc brake pads.

—— WARNING ——
DO NOT get any lubricant on the braking surface of the pad.

9. Install the pads and shims. The wear retainer on the inboard pad faces down.
10. Use a suitable tool to rotate the caliper piston clockwise into the caliper bore, enough to enable the caliper to fit over the pads. Lubricate the piston boot with silicon grease, and avoid twisting the boot.
11. Install the brake caliper, aligning the cutout in the piston with the tab on the inner pad. Tighten the mounting bolts to 17 ft. lbs. (23 Nm).
12. Install the wheel. Lower the vehicle.
13. Add brake fluid to the master cylinder reservoir. Depress the brake pedal several times to seat the pads. Bleed the brakes if necessary.

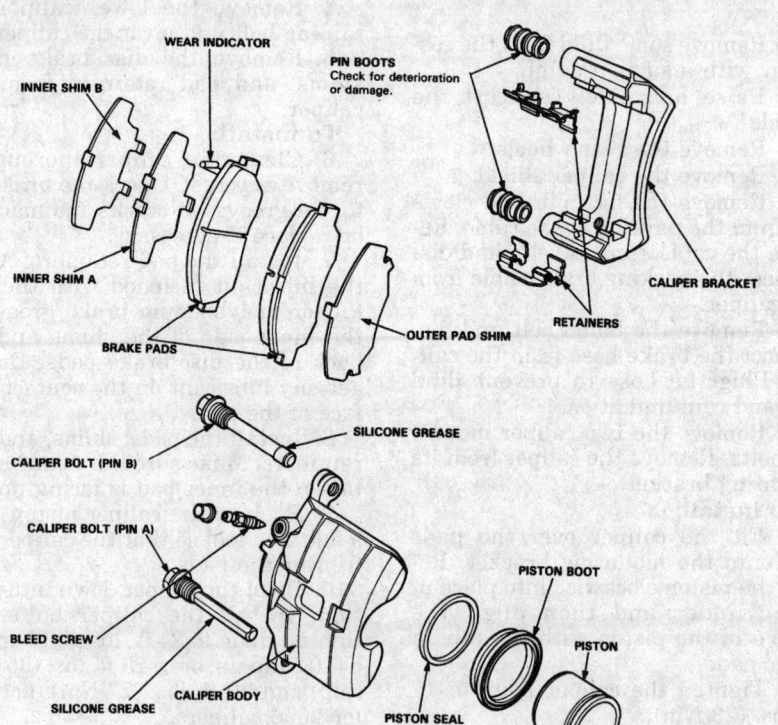

Front disc brake pads and shims — Accord and Prelude

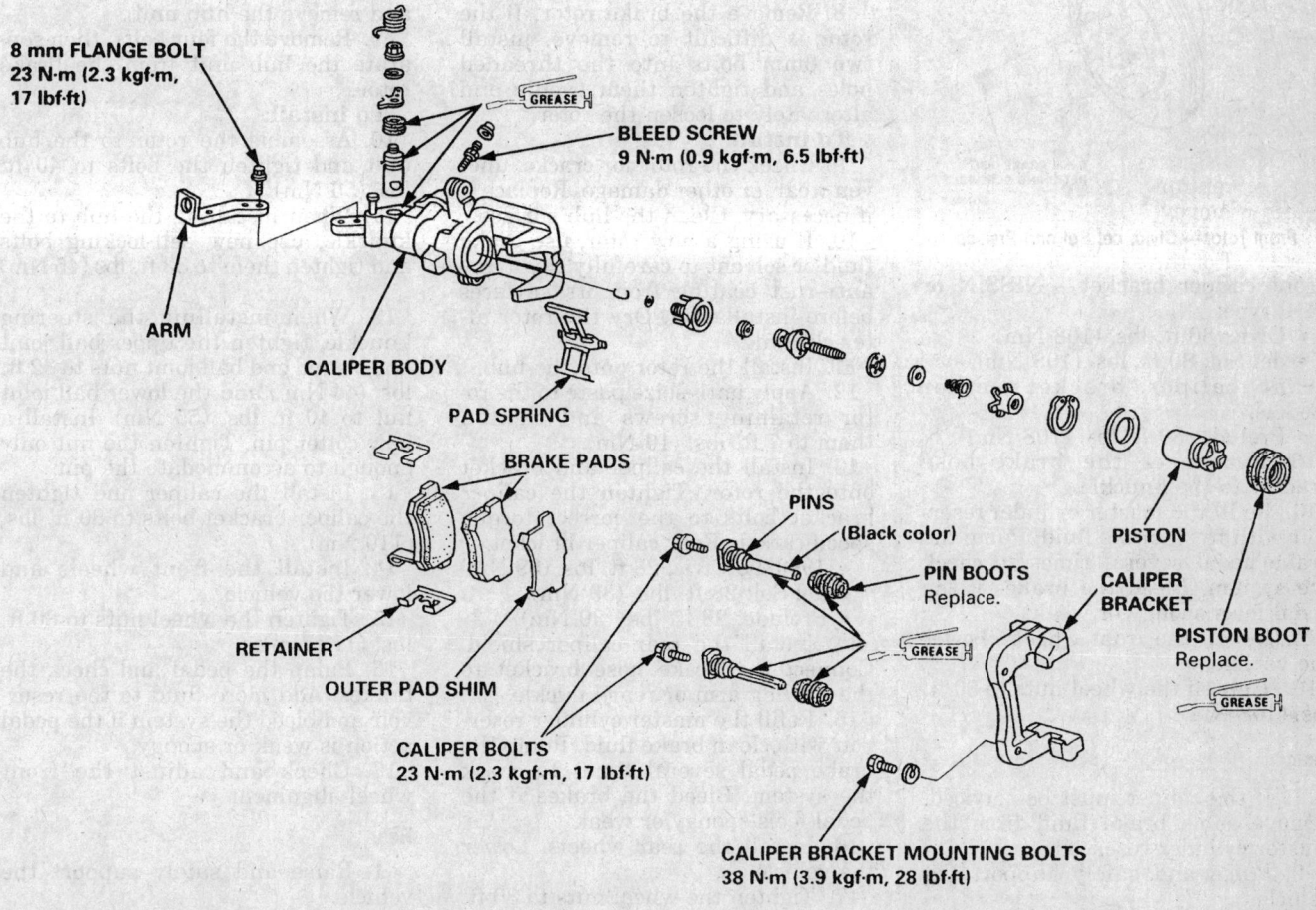

8 mm FLANGE BOLT
23 N·m (2.3 kgf·m,
17 lbf·ft)

GREASE

BLEED SCREW
9 N·m (0.9 kgf·m, 6.5 lbf·ft)

ARM

CALIPER BODY

PAD SPRING

BRAKE PADS

PINS
(Black color)

PIN BOOTS
Replace.

GREASE

PISTON

CALIPER
BRACKET

PISTON BOOT
Replace.

GREASE

RETAINER

OUTER PAD SHIM

CALIPER BOLTS
23 N·m (2.3 kgf·m, 17 lbf·ft)

GREASE

CALIPER BRACKET MOUNTING BOLTS
38 N·m (3.9 kgf·m, 28 lbf·ft)

289830

Rear disc brake pad and shim components — Accord and Prelude

Brake Rotor

REMOVAL AND INSTALLATION

Civic, del Sol and Prelude

Front

1. If the caliper must be serviced, remove some brake fluid from the master cylinder reservoir.
2. Raise and safely support the vehicle.
3. Remove the front wheels.
4. Unbolt the brake hose bracket from the knuckle.
5. Unbolt the caliper bracket from the knuckle. To avoid stressing the brake hose, hang the caliper out of the way with a length of wire.
6. Remove the two 6mm retaining screws. If the screws are seized, tap them with a mallet or use a screw extractor to remove them. Discard the retaining screws if they are rusted or damaged.
7. Remove the brake rotor. If the rotor is difficult to remove, install two 8mm bolts into the threaded holes and tighten them evenly and alternately to loosen the rotor.

To install:

8. Check the rotor for cracks, uneven wear, or other damage. Replace it if necessary. Clean the surface of the hub.
9. If using a new rotor, use brake fluid or solvent to carefully clean any anti–rust coating from its surfaces before installation. Dry the rotor after cleaning.
10. Install the rotor onto the hub.
11. Apply anti–seize paste to the rotor retaining screws and tighten them to 7 ft. lbs. (10 Nm).
12. Install the caliper and bracket onto the rotor. Tighten the caliper bracket bolts to the correct torque specifications:Front caliper bracket — AKEBONO or 2056 type:
- Civic: 80 ft. lbs. (108 Nm)
- del Sol: 80 ft. lbs. (108 Nm)

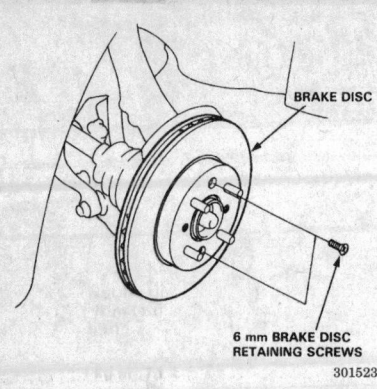

6 mm **BRAKE DISC RETAINING SCREWS**

301523

Front rotor — Civic, del Sol and Prelude

Front caliper bracket — NISSIN or 5410 type:
- Civic: 80 ft. lbs. (108 Nm)
- del Sol: 80 ft. lbs. (108 Nm)

Front caliper bracket — other models:
- Prelude: 80 ft. lbs. (108 Nm)

13. Reconnect the brake hose bracket to the knuckle.

14. Refill the master cylinder reservoir with fresh brake fluid. Pump the brake pedal several times to check the system. Bleed the brakes if the pedal feels spongy or weak.

15. Install the front wheels. Lower the vehicle.

16. Tighten the wheel nuts to 80 ft. lbs. (108 Nm).

Rear

1. If the caliper must be serviced, remove some brake fluid from the master cylinder reservoir.

2. Raise and safely support the vehicle.

3. Remove the rear wheels.

4. Remove the rear caliper shield.

5. If equipped, unbolt the brake hose bracket from the trailing arm or rear knuckle.

6. Unbolt the caliper bracket from the trailing arm or rear knuckle. To avoid stressing the brake hose, hang the caliper out of the way with a length of wire.

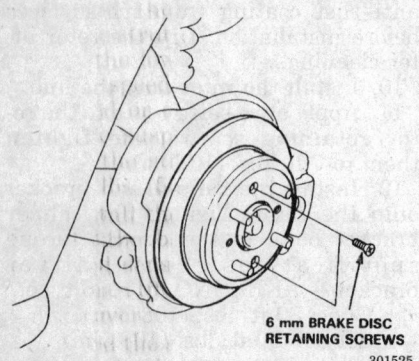

6 mm **BRAKE DISC RETAINING SCREWS**

301525

Rear brake rotor — Civic, del Sol and Prelude

7. Remove the two 6mm retaining screws. If the screws are seized, tap them with a mallet or use a screw extractor to remove them. Discard the retaining screws if they are rusted or damaged.

8. Remove the brake rotor. If the rotor is difficult to remove, install two 8mm bolts into the threaded holes and tighten them evenly and alternately to loosen the rotor.

To install:

9. Check the rotor for cracks, uneven wear, or other damage. Replace it if necessary. Clean the hub surface.

10. If using a new rotor, use brake fluid or solvent to carefully clean any anti–rust coating from its surfaces before installation. Dry the rotor after cleaning.

11. Install the rotor onto the hub.

12. Apply anti–seize paste to the rotor retaining screws and tighten them to 7 ft. lbs. (10 Nm).

13. Install the caliper and bracket onto the rotor. Tighten the caliper bracket bolts to the correct torque specifications:Rear caliper bracket:
- 1993–95 Civic: 28 ft. lbs. (39 Nm)
- del Sol: 28 ft. lbs. (39 Nm)
- Prelude: 28 ft. lbs. (39 Nm)

14. Install the rear caliper shield. Connect the brake hose bracket to the trailing arm or rear knuckle.

15. Refill the master cylinder reservoir with clean brake fluid. Pump the brake pedal several times to check the system. Bleed the brakes if the pedal feels spongy or weak.

16. Install the rear wheels. Lower the vehicle.

17. Tighten the wheel nuts to 80 ft. lbs. (108 Nm).

Accord

Front

1. Pry the spindle nut stake away from the spindle nut, then loosen the nut.

2. Raise and safely support the vehicle.

3. Remove the wheel. Remove the spindle nut.

4. Remove the caliper and support it out of the way with a length of wire. Do not allow the caliper to hang from the brake hose. Remove the caliper bracket.

5. Remove the cotter pin and tie rod ball joint nut. Separate the tie rod from the steering knuckle using a tie rod separator tool.

6. Remove the cotter pin and loosen the lower arm ball joint nut half the length of the joint threads. Separate the ball joint and lower arm using the proper puller. Remove the lower ball joint nut.

7. Pull the steering knuckle outward and remove the halfshaft outboard CV-joint from the knuckle, using a plastic hammer.

8. Remove the four bolts retaining the hub unit to the steering knuckle and remove the hub unit.

9. Remove the four bolts, then separate the hub unit from the brake rotor.

To install:

10. Assemble the rotor to the hub unit and tighten the bolts to 40 ft. lbs. (55 Nm)

11. When installing the hub to the knuckle, use new self-locking bolts and tighten them to 33 ft. lbs (45 Nm)

12. When installing the steering knuckle, tighten the upper ball joint and tie rod end ball joint nuts to 32 ft. lbs. (44 Nm) and the lower ball joint nut to 40 ft. lbs. (55 Nm). Install a new cotter pin. Tighten the nut only enough to accommodate the pin.

13. Install the caliper and tighten the caliper bracket bolts to 80 ft. lbs. (110 Nm).

14. Install the front wheels and lower the vehicle.

15. Tighten the wheel nuts to 80 ft. lbs. (110 Nm).

16. Pump the pedal and check the brakes. Add more fluid to the reservoir and bleed the system if the pedal action is weak or spongy.

17. Check and adjust the front wheel alignment.

Rear

1. Raise and safely support the vehicle.

2. Remove the wheel.

3. Remove the caliper and support it out of the way with a length of wire. Do not allow the caliper to hang from the brake hose. Remove the caliper bracket.

4. Remove the brake hose mounting bolts.

5. Remove the brake caliper bracket mounting bolts and support it out of the way with a length of wire. Do not allow the caliper to hang from the brake hose.

6. Remove the 6mm brake rotor retaining screws.

7. Screw two 8 x 1.25mm bolts into the rotor to push it away from the hub.

NOTE: Turn each bolt two turns at a time to prevent cocking the rotor excessively.

To install:

8. Install the brake rotor to the hub.

9. Install the brake caliper with the caliper bracket mounting bolts.

6. Remove the brake shoe assembly from the backing plate.

7. Disconnect the parking brake cable from the brake shoe lever.

8. Remove the upper return spring, self-adjuster lever, and self-adjuster spring. Separate the brake shoes.

9. Separate the wave washer, parking brake lever, and pivot pin from the brake shoe by removing the U-clip.

To install:

10. Apply brake cylinder grease to the sliding surface of the pivot pin and insert the pin into the brake shoe.

11. Install the parking brake lever and wave washer on the pivot pin and pinch the U-clip with a pair of pliers to secure the pivot pin to the shoe. Make sure the pivot pin is securely locked by the U-clip.

12. Connect the parking brake cable to the parking brake lever.

13. Apply high-temperature molybdenum grease to each sliding surface of the brake backing plate. Also apply grease to the edges of the brake shoes and to areas where the metal body of the shoe contacts the backing plate.

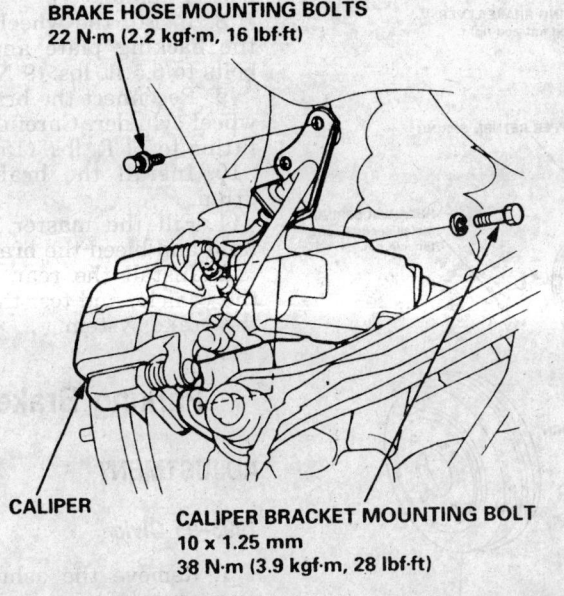

BRAKE HOSE MOUNTING BOLTS
22 N·m (2.2 kgf·m, 16 lbf·ft)

CALIPER

CALIPER BRACKET MOUNTING BOLT
10 x 1.25 mm
38 N·m (3.9 kgf·m, 28 lbf·ft)

289899

Rear brake caliper and hose mounting bolts — Accord

NOTE: Do not allow grease to come in contact with the brake linings. Grease will contaminate the linings and reduce braking power.

Torque the caliper bracket mounting bolts to 28 ft. lbs. (38 Nm).

10. Install the brake hose with the brake hose mounting bolts. Torque the mounting bolts to 16 ft. lbs. (22 Nm).

11. Install the brake rotor mounting screws and torque to 7 ft. lbs. (9.8 Nm).

12. Install the wheel and lower the vehicle.

13. Test the operation of the brakes. Fill the reservoir with fluid and bleed the brakes if the pedal action feels spongy or weak.

Brake Drums

REMOVAL AND INSTALLATION

Accord, Civic and del Sol

1. Block the front wheels securely. Raise and support the rear of the vehicle on jackstands. Remove the rear wheels. Make sure that the parking brake is OFF.

2. Pull off the rear brake drum. If the drum is difficult to remove, use a brake drum puller or tap it a few times with a rubber mallet. A drum that is stuck can be removed by screwing in two 8mm bolts into the two threaded holes and tightening them alternately until the drum breaks loose. Make certain to remove it squarely; if it is cocked to one side, it will jam.

To install:

3. Make certain the brake shoes are adjusted to allow the drum clearance during installation. Fit the drum into position.

4. Install the rear wheels and lower the vehicle.

Brake Shoes

REMOVAL AND INSTALLATION

Accord, Civic and del Sol

1. Raise and safely support the vehicle.

2. Remove the rear wheels and brake drums.

3. Disconnect the upper return spring from the brake shoes.

4. Push the retainer springs and turn the tension pins to release the shoes from the backing plate.

5. Lower the brake shoe assembly and remove the lower return spring.

14. Clean the threaded portions of the clevises of the adjuster bolt. Coat the threads with grease. Turn the adjuster bolt to shorten the clevises.

15. Hook the adjuster spring to the adjuster lever first, then to the brake shoe.

16. Install the adjuster bolt/clevis assembly and the upper return spring.

17. Install the brake shoes to the backing plate.

18. Install the lower return spring, the tension pins and retaining springs.

19. Connect the upper return spring.

20. Turn the adjuster bolt to force the brake shoes out until the brake drum will not easily go on. Back off the adjuster bolt just enough that the brake drum will go on and turn easily.

21. Install the wheels; lower the vehicle to the ground.

22. Depress the brake pedal several times to set the self-adjusting brake. Adjust the parking brake.

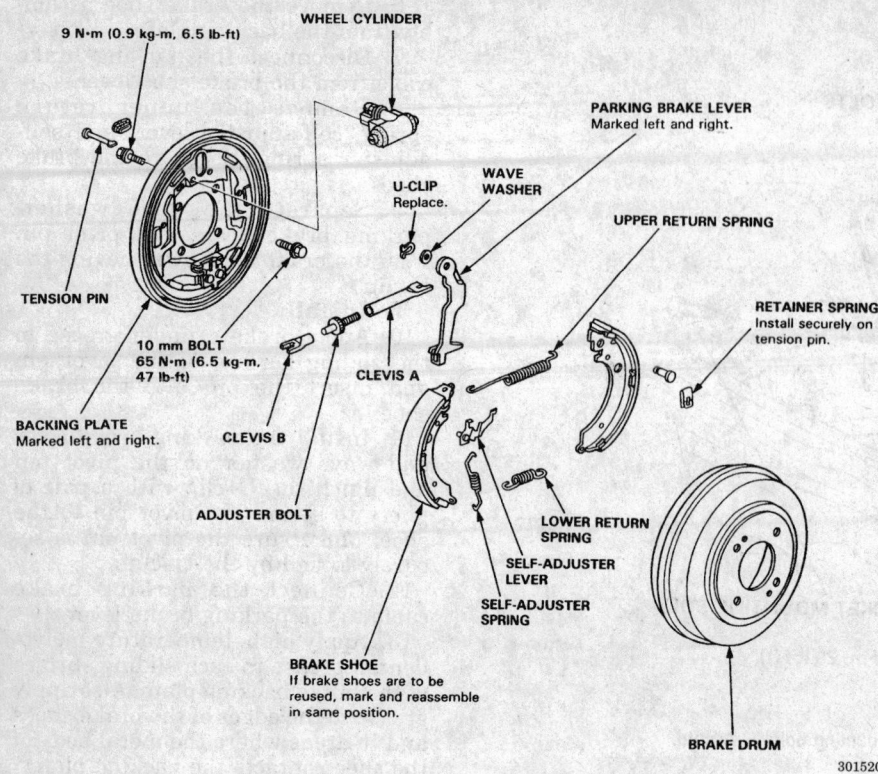

9 N·m (0.9 kg-m, 6.5 lb-ft)

WHEEL CYLINDER

PARKING BRAKE LEVER
Marked left and right.

U-CLIP
Replace.

WAVE WASHER

UPPER RETURN SPRING

RETAINER SPRING
Install securely on tension pin.

TENSION PIN

10 mm BOLT
65 N·m (6.5 kg-m, 47 lb-ft)

CLEVIS A

BACKING PLATE
Marked left and right.

CLEVIS B

ADJUSTER BOLT

LOWER RETURN SPRING

SELF-ADJUSTER LEVER

SELF-ADJUSTER SPRING

BRAKE SHOE
If brake shoes are to be reused, mark and reassemble in same position.

BRAKE DRUM

301520

Drum brake components — Accord, Civic and del Sol

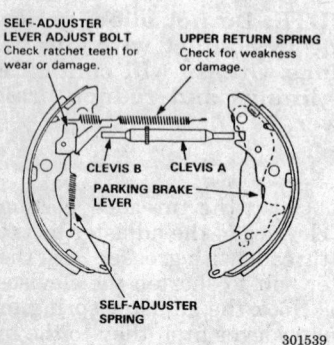

SELF-ADJUSTER LEVER ADJUST BOLT
Check ratchet teeth for wear or damage.

UPPER RETURN SPRING
Check for weakness or damage.

CLEVIS B   CLEVIS A

PARKING BRAKE LEVER

SELF-ADJUSTER SPRING

301539

Brake shoe springs and clevis — Accord, Civic and del Sol

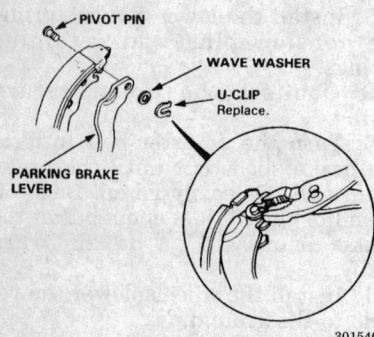

PIVOT PIN

WAVE WASHER

U-CLIP
Replace.

PARKING BRAKE LEVER

301540

Pivot pin and U-clip — Accord, Civic and del Sol

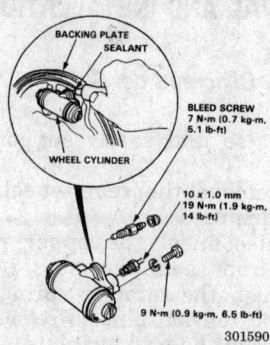

BACKING PLATE SEALANT

BLEED SCREW
7 N·m (0.7 kg-m, 5.1 lb-ft)

WHEEL CYLINDER

10 x 1.0 mm
19 N·m (1.9 kg-m, 14 lb-ft)

9 N·m (0.9 kg-m, 6.5 lb-ft)

301590

Wheel cylinder components — Accord, Civic and del Sol

Wheel Cylinder

REMOVAL AND INSTALLATION

Accord, Civic and del Sol

1. Raise and support the vehicle safely.

2. Remove the rear wheels.

3. Remove the brake drum and shoes.

4. Disconnect the brake line from the backing plate; use a drip pan to catch the brake fluid.

5. Remove the two wheel cylinder-to-backing plate bolts. Remove the wheel cylinder.

6. Clean the backing plate.

To install:

7. Apply sealant to the back of the wheel cylinder where it contacts the backing plate.

8. Install the wheel cylinder onto the backing plate and tighten the bolts to 6.5 ft. lbs. (9 Nm).

9. Reconnect the brake line to the wheel cylinder. Carefully tighten the fitting to 11 ft. lbs. (15 Nm).

10. Install the brake shoes and drum.

11. Fill the master cylinder with fluid and bleed the brakes.

12. Install the rear wheels, lower the vehicle, and test the operation of the brake system.

Parking Brake Cable

ADJUSTMENT

1993–95 Civic

1. Remove the ashtray from the rear console to reveal the parking brake cable adjusting nut.

2. Pull the parking brake lever up one click only.

3. Raise the rear wheels off of the ground. Support the vehicle safely and block the front wheels.

4. On vehicles equipped with disc brakes, make sure that the parking brake arm on the caliper contacts the caliper pin. On vehicles equipped with drum brakes, make sure the brake shoes do not drag when the wheel is turned by hand.

5. Turn the parking brake cable adjusting nut clockwise until the rear wheels drag slightly when you turn them by hand.

6. Release the parking brake lever. Check that the rear wheels do not drag when you turn them by hand.

7. With the parking brake cable adjusted properly, the parking brakes should be fully applied when the lever is pulled up six to ten clicks.

8. Install the ashtray back into the rear console.

1996–97 Civic

1. Remove rear section of the console to reveal the parking brake cable adjusting nut.

2. Raise the rear wheels off of the ground. Support the vehicle safely and block the front wheels.

3. Pull the parking brake lever up one click only.

4. Make sure the brake shoes do not drag when the wheel is turned by hand.

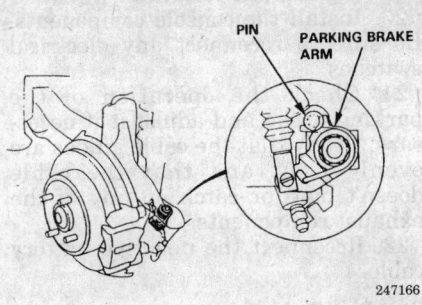

247166

Parking brake lever rear disc brakes — 1993–95 Civic

5. Turn the parking brake cable adjusting nut clockwise until the rear wheels drag slightly when you turn them by hand.

6. Release the parking brake lever. Check that the rear wheels do not drag when you turn them by hand.

7. With the parking brake cable adjusted properly, the parking brakes should be fully applied when the lever is pulled up six to nine clicks.

8. Install the rear section of the console.

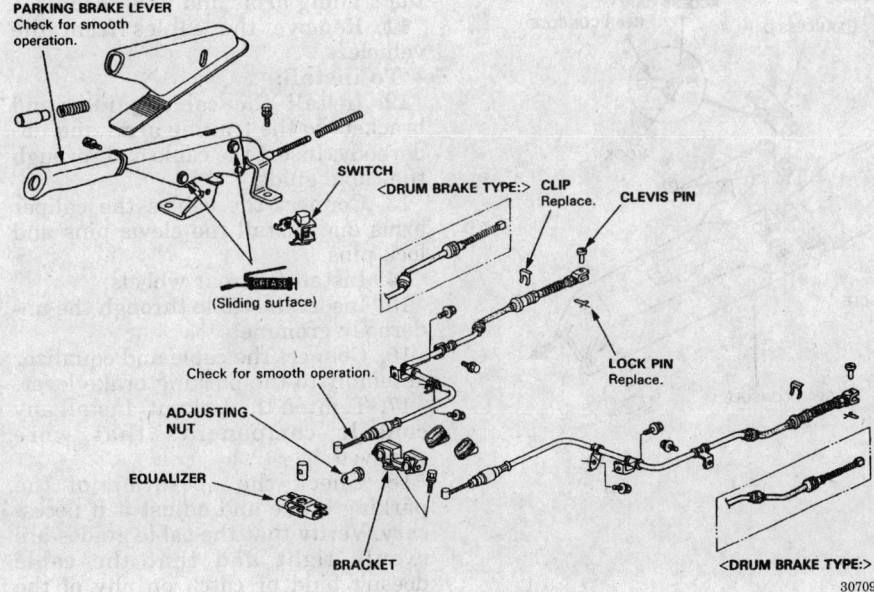

307099

Parking brake cable, guides, and brackets

del Sol

1. Carefully pull up the parking brake lever boot. Remove the plastic cap from the cable adjuster's end.

2. Loosen the parking brake adjusting nut. Then, start the engine and pump the brake pedal several time to set the self–adjuster. Then, shut the engine off.

3. Raise the rear wheels off of the ground. Support the vehicle safely and block the front wheels.

4. On vehicles equipped with disc brakes, make sure that the parking brake arm on the caliper contacts the

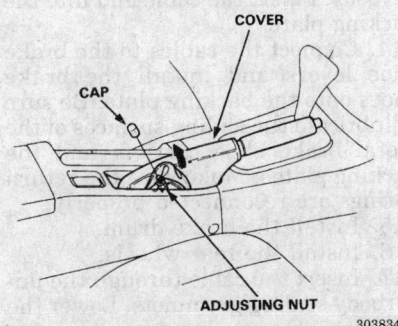

303834

Parking brake lever boot and adjusting nut — del Sol

caliper pin. On vehicles equipped with drum brakes, make sure the brake shoes do not drag when the wheel is turned by hand.

5. Pull the parking brake lever up one click only.

6. Turn the parking brake cable adjusting nut clockwise until the rear wheels drag slightly when you turn them by hand.

7. Release the parking brake lever. Check that the rear wheels do not drag when you turn them by hand.

8. With the parking brake cable adjusted properly, the parking brakes should be fully applied when the lever is pulled up six to ten clicks.

9. Place the cap back on the cable adjuster's end. Fit the parking brake lever boot back into place.

1993 Accord and 1993–97 Prelude

NOTE: Inspect the cables for excess wear, signs of fraying or any damage which may hamper proper operation. The cables should not bind through their travel. Replace any cable which show signs of damage. Check the ratchet mechanism for signs of wear, and replace any components which are excessively worn.

1. Raise and safely support the vehicle.

2. On rear disc brake equipped vehicles, make sure the lever of the rear brake caliper contacts the brake caliper pin.

3. On drum brake equipped vehicles, make sure the rear brakes are properly adjusted.

4. Pull the parking brake lever up 1 notch.

5. Remove the access cover at the rear of the console and tighten the adjusting nut until the rear wheels drag slightly when turned.

6. Release the parking brake lever and check that the rear wheels do not drag when turned. Readjust if necessary.

7. With the equalizer properly adjusted, the parking brake should be fully applied when the parking brake lever is pulled up 6–10 clicks.

REMOVAL AND INSTALLATION

Civic, del Sol and Prelude

NOTE: The original radio contains a coded anti–theft circuit. Obtain the security code before disconnecting the negative battery cable.

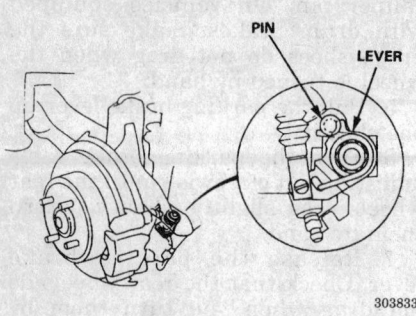

Parking brake lever — del Sol with rear disc brakes

Rear Drum Brakes

1. Disconnect the negative battery cable.

2. Remove the console screw access cap. Remove the four screws and remove the rear part of the console. If necessary, disconnect any electrical switches.

3. Release the parking brake lever.

4. Loosen the adjusting nut and remove the cables from the equalizer at the rear of the parking brake lever.

5. Raise and safely support the vehicle.

6. Remove the rear wheels.

7. Carefully pull the cables through the sealing grommets on the underbody of the vehicle. Be careful not to bend or twist the cables when moving them over and around the exhaust system.

8. Remove the brake drum and brake shoes.

9. Disconnect the brake cable from the brake shoe lever .

10. Use a 12mm box–end wrench to disconnect the cable from the backing plate.

11. Unbolt or disengage the cable guides from the trailing arms and underbody.

12. Remove the parking brake cables from the vehicle.

To install:

13. Install the cable guides and brackets to the trailing arms and underbody. Insert the cable end into the backing plate.

14. Connect the cables to the brake shoe levers and install the brake shoes onto the backing plate. Be sure to lubricate the sliding surfaces of the metal brake shoe carriers and the backing plates. Make sure the return springs are reconnected properly.

15. Install the brake drum.

16. Install the rear wheels.

17. Insert the cable through the underbody sealing grommets. Lower the vehicle.

18. Connect the cable and equalizer assembly to the parking brake lever.

19. Tighten the locknut.

20. Install the console components. Be sure to reconnect any electrical switches.

21. Check the operation of the parking brake and adjust it if necessary. Verify that the cable guides are evenly tight and that the cable doesn't bind or catch on any of the exhaust components.

22. Reconnect the negative battery cable.

Rear Disc Brakes

1. Disconnect the negative battery cable.

2. Remove the six rear console screws. Lift the console up and disconnect the power mirror and hazard flasher switches.

3. Release the parking brake lever.

4. Loosen the adjusting nut and remove the cables from the equalizer at the rear of the parking brake lever.

5. Raise and safely support the vehicle.

6. Remove the rear wheels.

7. Carefully pull the cables through the grommets on the underbody of the vehicle. Be careful not to bend or twist the cables when moving them over and around the exhaust system.

8. Remove the caliper dust shields.

9. Pull out the lock pin and remove the clevis pin from the arm on the caliper. Remove the cable clip.

10. Unbolt the cable guides from the trailing arms and underbody.

11. Remove the cables from the vehicle.

To install:

12. Install the cable guides and brackets to the trailing arms and underbody. Insert the cable end through the cable guide.

13. Connect the cables the caliper arms and install the clevis pins and lock pins.

14. Install the rear wheels.

15. Insert the cable through the underbody grommets.

16. Connect the cable and equalizer assembly to the parking brake lever.

17. Tighten the locknut. Install any console components that were removed.

18. Check the operation of the parking brake and adjust it if necessary. Verify that the cable guides are evenly tight and that the cable doesn't bind or catch on any of the exhaust components.

19. Reconnect the negative battery cable.

20. On vehicles equipped with 4WS, start the engine and turn the steer-

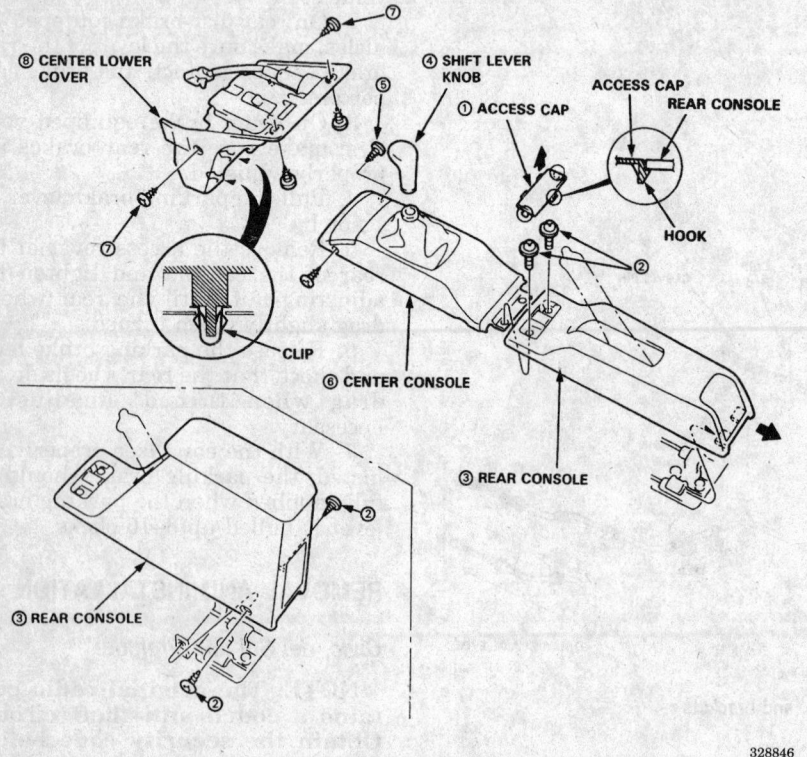

Console components — 1993–95 Civic with rear drum brakes

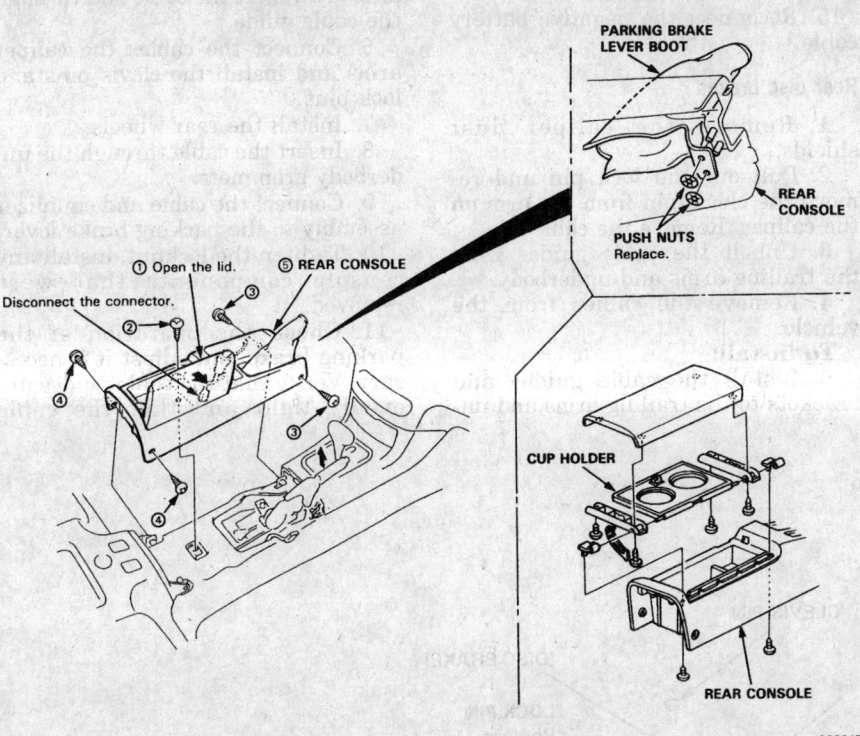

Console components — del Sol with rear drum brakes

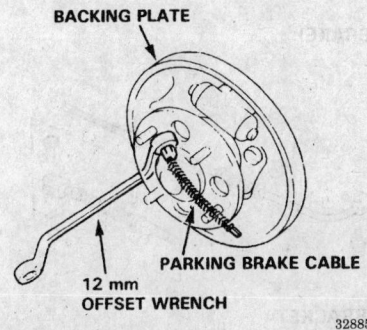

Parking brake cable — Civic and del Sol with rear drum brakes

ing wheel lock–to–lock to reset the control unit.

21. Enter the radio security code.

1993 Accord

With Drum Brakes

1. Remove the access cover at the rear of the console. Loosen the adjusting nut until the cable ends can be disconnected from the equalizer.

2. Raise and safely support the vehicle.

3. Remove the rear wheels.

4. Remove the brake drum and brake shoes.

5. Disconnect the cable from the parking brake lever.

6. Detach the cable from the backing plate by sliding a box wrench on the cable to compress the anchor clips.

7. Detach the cables from the cable guides and remove the cables from the vehicle.

To install:

8. Connect the cables to the cable guides. Connect the cables at the parking brake levers and install the brake shoes.

9. Install the brake drum.

10. Install the rear wheels. Connect the cable at the equalizer assembly.

11. Adjust the parking brake.

With Rear Disc Brakes

1. Remove the access cover at the rear of the console. Loosen the adjusting nut until the cable ends can be disconnected from the equalizer.

2. Raise and safely support the vehicle.

3. Remove the rear wheels.

4. Pull out the lock pin, remove the clevis pin and remove the clip.

5. Detach the cables from the cable guides and remove the cables from the vehicle.

To install:

6. Connect the cables to the cable guides. connect the cable to the cali-

per lever and install the lock pin and clevis pin.

7. Install the rear wheels. Connect the cable at the equalizer assembly.

8. Adjust the parking brake.

1994–97 Accord

NOTE: The original radio may contain a coded anti–theft circuit. Always obtain the security code before disconnecting the negative battery cable.

1. Disconnect the negative battery cable.

2. Lift the armrest and remove the console access lid. The rear console may be removed for more access, it is secured by three screws.

3. Release the parking brake lever.

4. Loosen the adjusting nut and remove the cables from the equalizer at the rear of the parking brake lever.

5. Raise and safely support the vehicle.

6. Remove the rear wheels.

7. Carefully pull the cables through the grommets on the underbody of the vehicle. Be careful not to bend or twist the cables when moving them over and around the exhaust system.

Rear drum brakes

1. Remove the brake drum and brake shoes.

2. Disconnect the brake cable from the brake shoe lever .

3. Use a 12mm box–end wrench to disconnect the cable from the backing plate.

4. Unbolt the cable guides from the trailing arms and underbody.

5. Remove the cables from the vehicle.

To install:

6. Install the cable guides and brackets to the trailing arms and underbody. Insert the cable end into the backing plate.

7. Connect the cables to the brake shoe levers and install the brake shoes onto the backing plate.

8. Install the brake drum.

9. Install the rear wheels.

10. Insert the cable through the underbody grommets.

11. Connect the cable and equalizer assembly to the parking brake lever.

12. Tighten the locknut. Install any console components that were removed.

13. Check the operation of the parking brake and adjust it if necessary. Verify that the cable guides are evenly tight and that the cable doesn't bind or catch on any of the exhaust components.

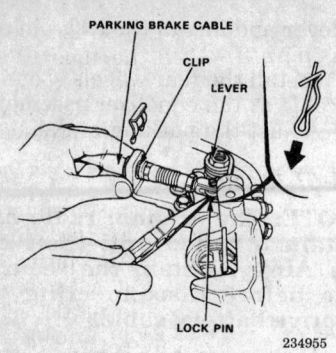

Disconnecting the parking brake cable from the rear caliper — 1993 Accord wtih drum brakes

14. Lower the vehicle.
15. Reconnect the negative battery cable.

Rear disc brakes

1. Remove the caliper dust shields.
2. Pull out the lock pin and remove the clevis pin from the arm on the caliper. Remove the cable clip.
3. Unbolt the cable guides from the trailing arms and underbody.
4. Remove the cables from the vehicle.

To install:

5. Install the cable guides and brackets to the trailing arms and un-

derbody. Insert the cable end through the cable guide.
6. Connect the cables the caliper arms and install the clevis pins and lock pins.
7. Install the rear wheels.
8. Insert the cable through the underbody grommets.
9. Connect the cable and equalizer assembly to the parking brake lever.
10. Tighten the locknut. install any console components that were removed.
11. Check the operation of the parking brake and adjust it if necessary. Verify that the cable guides are evenly tight and that the cable

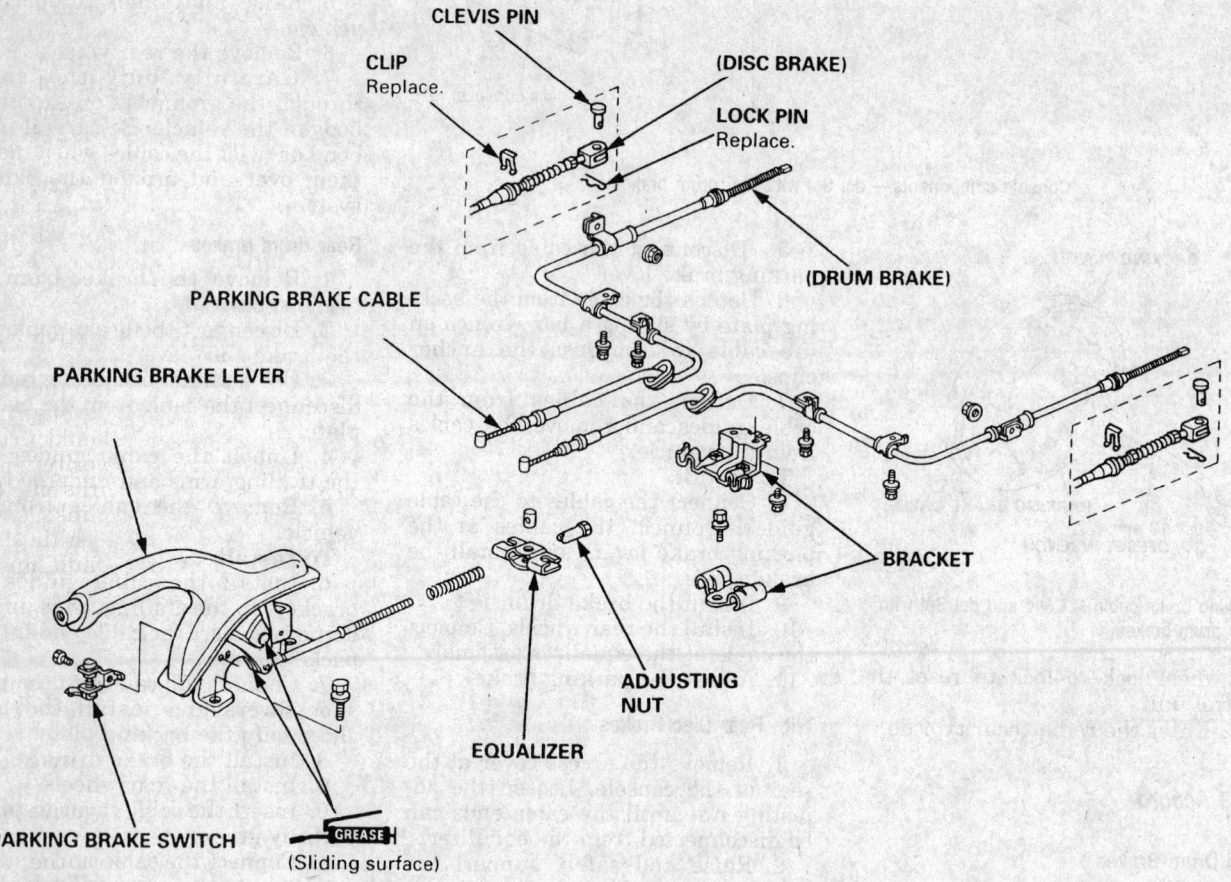

Parking brake cable components — 1994–97 Accord and Prelude

doesn't bind or catch on any of the exhaust components.

12. Lower the vehicle.

13. Reconnect the negative battery cable.

Brake System Bleeding

PROCEDURE

————— **CAUTION** —————

The Honda anti-lock brake system contains brake fluid under extremely high pressure within the pump, accumulator and modulator assembly. Do not disconnect or loosen any lines, hoses, fittings or components without properly relieving the system pressure. If a tool is required to relieve system pressure, use only a bleeder T-wrench 07HAA-SG00100 or equivalent to relieve pressure. For modulators with maintenance bleeders, follow the proper depressurizing procedure. Improper procedures or failure to discharge the system pressure may result in severe personal injury and/or property damage.

NOTE: The master cylinder must be full at the start of the bleeding procedure and checked after bleeding the brake at each wheel. Add fluid as required. Use only DOT 3 or 4 brake fluid. If a pressure bleeder is not available, an assistant will be necessary to perform this brake bleeding operation.

1. Fill the master cylinder.

2. Have an assistant slowly pump the brake pedal several times and then apply a steady pressure to the brake pedal.

3. Attach a bleed hose to the bleed screw and place it into a clear container. Loosen the brake bleed screw at the brake caliper and allow the fluid to flow. Close the bleeder.

4. Repeat this procedure for each wheel, until no air bubbles appear in the brake fluid. Torque the bleeders to 7 ft. lbs. (9 Nm). Use the following sequence in order to bleed the brake system properly:
- 1993–96 Civic: RR, LF, LR, RF
- 1993 Accord: LF, RR, RF, LR
- 1994–95 Accord: RR, LF, LR, RF
- 1993–95 Prelude: RR, LF, LR, RF
- del Sol: RR, LF, LR, RF

NOTE: LF (Left Front), RF (Right Front), LR (Left Rear), RR (Right Rear).

5. Check the fluid level in the master cylinder and add fluid, if necessary. Road test the vehicle and check the brake performance.

Wheel Speed Sensor

REMOVAL AND INSTALLATION

Civic and del Sol

NOTE: The radio may contain a coded theft protection circuit. Always obtain the code number before disconnecting the battery.

Front

1. Make sure the ignition switch is turned **OFF**. Remove the key from the ignition.

2. Disconnect the negative battery cable.

3. Detach the wheel sensor cable from the ABS harness at its connector located on the vehicle's shock tower. Carefully detach the plastic clip from the shock tower.

4. Raise and support the vehicle safely. Remove the front wheels.

5. Unbolt the speed sensor cable brackets from the inside of the shock tower and the steering knuckle.

6. Unbolt the wheel speed sensor assembly from the steering knuckle.

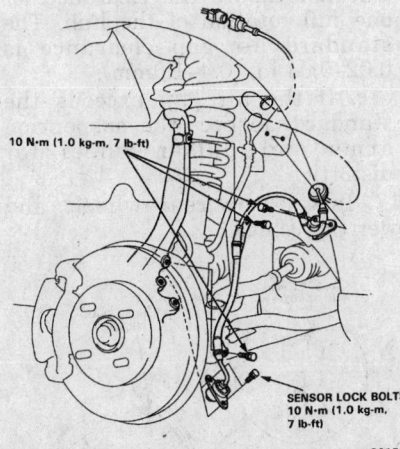

10 N·m (1.0 kg-m, 7 lb-ft)

SENSOR LOCK BOLTS
10 N·m (1.0 kg-m, 7 lb-ft)

301504

Front wheel sensor — 1993–95 Civic and 1993–97 del Sol

7. Pull the speed sensor cable through the hole in the shock tower. Be careful not to bend or kink the cable if it is to be reused. Remove the speed sensor and cable from the vehicle as an assembly.

To install:

8. Feed the speed sensor connector through the hole in the shock tower and connect it to the ABS harness.

9. Install the speed sensor cable brackets to the inside of the shock tower and the steering knuckle. Do not bend or twist the speed sensor cable.

10. Install the wheel speed sensor assembly into its cavity on the steering knuckle. Torque the speed sensor cable bracket bolts and mounting bolts to 7 ft. lbs. (10 Nm).

11. Check the wheel sensor air gap:

  a. Inspect the pulser wheel for damaged teeth.

  b. Rotate the driveshaft by hand and measure the clearance between the wheel speed sensor and the pulser wheel. Measure the clearance for one full rotation of the driveshaft. The standard air gap clearance is 0.02–0.04 in. (0.4–1.0mm).

  c. If the air gap exceeds the standard, inspect the steering knuckle and pulser wheel for distortion.

12. Install the front wheels and lower the vehicle.

13. Reconnect the negative battery cable.

14. Turn the ignition to the **ON** position, but don't start the engine. The ABS indicator light should turn on. Start the engine and verify that the ABS indicator light turns off. If the ABS indicator light blinks, doesn't come on, or stays on with the engine running, the system fault must be diagnosed.

Rear

1. Make sure the ignition switch is turned **OFF**. Remove the key from the ignition.

2. Remove the spare tire compartment cover.

3. Disconnect the negative battery cable.

4. Raise and support the vehicle safely. Remove the rear wheels.

5. Uncouple the wheel speed sensor cable connector from its junction on the trunk floor located under the trim flap at the forward edge of the spare tire compartment.

6. Detach the plastic cable clips from the trailing arm.

7. Unbolt the speed sensor cable brackets from the trailing arm and

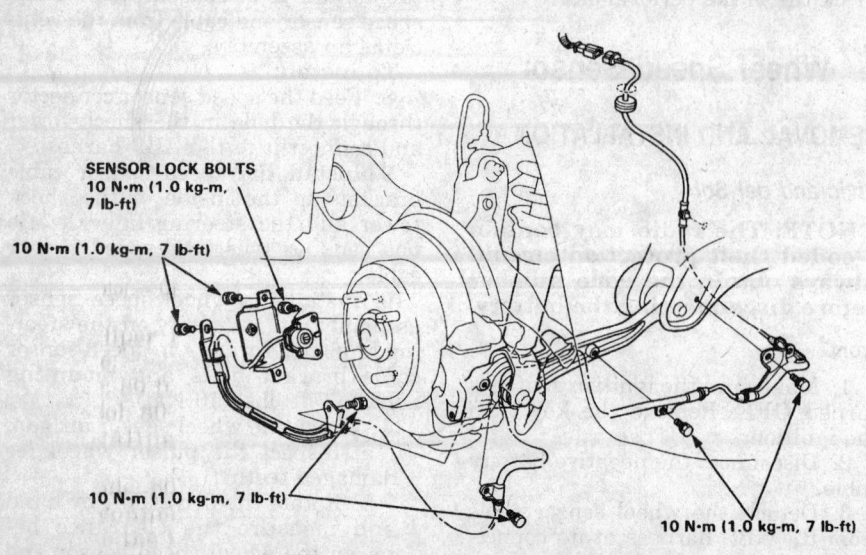

SENSOR LOCK BOLTS
10 N·m (1.0 kg-m, 7 lb-ft)

10 N·m (1.0 kg-m, 7 lb-ft)

10 N·m (1.0 kg-m, 7 lb-ft)

10 N·m (1.0 kg-m, 7 lb-ft)

301506

Rear wheel sensor — 1993–95 Civic and 1993–97 del Sol

the underbody of the vehicle. Remove the speed sensor cover.

8. Unbolt the speed sensor from the brake backing plate.

9. Support the lower control arm with a floor jack.

10. Remove the strut mount flange bolt from the lower control arm. Then, move the strut far enough out of the way to remove the speed sensor cable.

11. Remove the speed sensor and cable. Be careful not to bend or kink the cable if it is to be reused.

To install:

12. Install the speed sensor cable brackets to the trailing arm and underbody of the vehicle.

13. Feed the speed sensor cable under the strut and secure in position against the lower control arm. Make sure the speed sensor cable will not be rubbed or contacted by the strut when the suspension is under load. Tighten the strut mount flange bolt to 40 ft. lbs. (55 Nm).

14. Install the speed sensor assembly into the backing plate. Tighten the speed sensor cable bracket bolts to 7 ft. lbs. (10 Nm). Install the speed sensor cover.

15. Feed the speed sensor connector through the hole in the underbody and reconnect the connector to its junction on the rear suspension beam. Fasten the plastic cable clips.

16. Check the wheel sensor air gap:

a. Inspect the pulser wheel for damaged teeth.

b. Rotate the hub by hand and measure the clearance between the wheel speed sensor and the pulser wheel. Measure the clearance for one full rotation of the hub. The standard air gap clearance is 0.02–0.04 in. (0.4–1.0mm).

c. If the air gap exceeds the standard, inspect the suspension arms and pulser wheel for distortion.

17. Install the rear wheels and lower the vehicle.

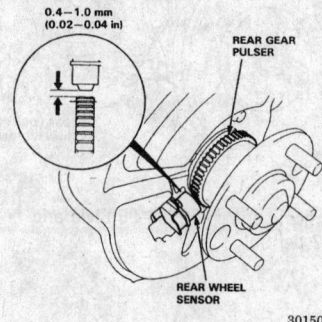

0.4–1.0 mm
(0.02–0.04 in)

REAR GEAR
PULSER

REAR WHEEL
SENSOR

301507

Rear wheel sensor air gap — 1993–95 Civic and 1993–97 del Sol

18. Tighten the wheel nuts to 80 ft. lbs. (110 Nm).

19. Reconnect the negative battery cable.

20. Turn the ignition to the **ON** position, but don't start the engine. The ABS indicator light should turn on. Start the engine and verify that the ABS indicator light turns off. If the ABS indicator light blinks, doesn't come on, or stays on with the engine running, the system fault must be diagnosed.

21. Reinstall the spare tire compartment cover.

Accord and Prelude

NOTE: The radio may contain a coded theft protection circuit. Always obtain the code number before disconnecting the battery. If the vehicle is equipped with 4WS, the steering control unit is shut down when the battery is disconnected. After connecting the battery, turn the steering wheel lock-to-lock to reset the steering control unit.

Front

1. Disconnect the negative battery cable.

2. Disconnect the front wheel speed sensor connector located on the strut tower.

3. Safely raise and support the vehicle.

4. Remove the front wheel.

5. Remove the two bolts mounting the speed sensor to the steering knuckle.

6. Remove the bolts securing the speed sensor wiring to the steering knuckle and the vehicle's body, then remove the sensor from the vehicle.

To install:

7. Position the speed sensor wiring on the body and steering knuckle, make sure that the wiring is not twisted. Feed the connector through the hole in the strut tower then install the wiring mounting bolts and torque the bolts to 7 ft. lbs. (10 Nm).

8. Install the speed sensor to the steering knuckle, make sure that the sensor does not contact the pulser wheel.

9. Install the speed sensor mounting bolts and torque the bolts to 16 ft. lbs. (22 Nm).

10. Check the wheel sensor air gap:

a. Inspect the pulser wheel for damaged teeth.

b. Rotate the halfshaft by hand and measure the clearance between the wheel speed sensor and the pulser wheel. Measure the clearance for one full rotation of

NOTE:
- Be careful when installing the sensors to avoid twisting the wires.
- After sensor replacement, confirm proper operation

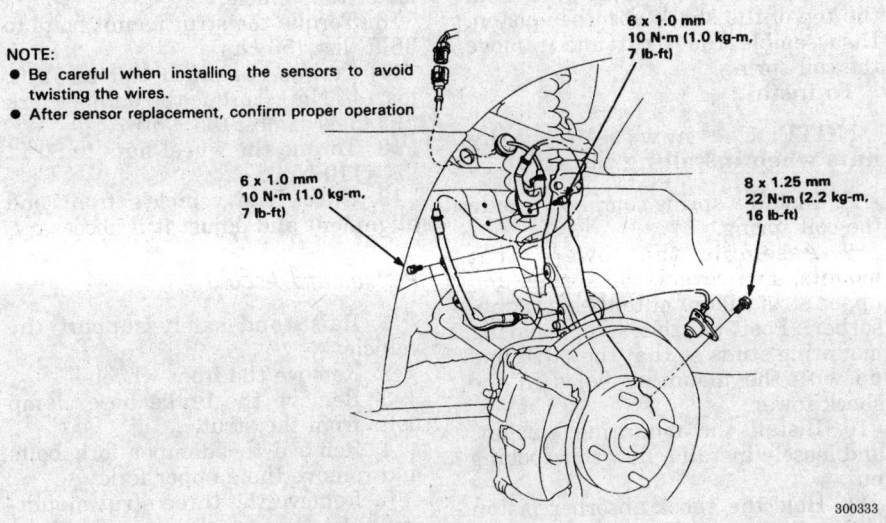

6 x 1.0 mm
10 N·m (1.0 kg-m, 7 lb-ft)

6 x 1.0 mm
10 N·m (1.0 kg-m, 7 lb-ft)

8 x 1.25 mm
22 N·m (2.2 kg-m, 16 lb-ft)

300333

Front wheel speed sensor — Prelude

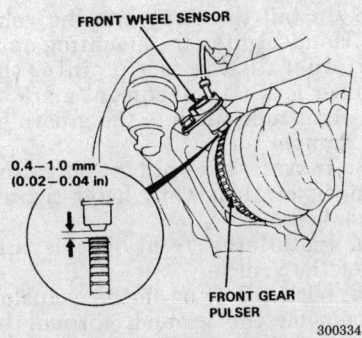

FRONT WHEEL SENSOR

0.4—1.0 mm
(0.02—0.04 in)

FRONT GEAR PULSER

300334

Measuring the front wheel speed sensor gap — Prelude

the halfshaft. The air gap clearance is 0.02–0.04 in. (0.4–1.0mm).

c. If the air gap exceeds the standard, inspect the steering knuckle and pulser wheel for distortion.

11. Install the front wheel and lower the vehicle.

12. Connect the speed sensor connector and secure it to the strut tower.

13. Connect the positive then the negative battery cable and enter the radio security code.

14. Turn the ignition to the **ON** position, but don't start the engine. The ABS indicator light should turn on.

Start the engine and verify that the ABS indicator light turns off. If the ABS indicator light blinks, doesn't come on, or stays on with the engine running, the system fault must be diagnosed.

15. If equipped with 4WS, start the engine and turn the steering wheel lock–to–lock to reset the 4WS control unit.

Rear

1. Disconnect the negative battery cable.

2. Safely raise and support the vehicle.

3. Remove the rear wheel.

4. Remove the brake rotor.

5. Disconnect the speed sensor connector from the wiring harness connector.

6. Remove the two bolts mounting the speed sensor to the knuckle and remove the sensor from the knuckle.

7. Remove the bolts securing the speed sensor wiring to the lower control arm and the vehicle's body, then remove the sensor from the vehicle.

To install:

8. Position the speed sensor wiring on the body and lower control arm, make sure that the wiring is not twisted. Install the wiring mounting bolts and torque the bolts to 7 ft. lbs. (10 Nm).

9. Install the speed sensor to the knuckle, make sure that the sensor does not contact the pulser wheel.

10. Install the speed sensor mounting bolts and torque the bolts to 16 ft. lbs. (22 Nm).

11. Check the wheel sensor air gap:

a. Inspect the pulser wheel for damaged teeth.

b. Rotate the hub by hand and measure the clearance between the wheel speed sensor and the pulser wheel. Measure the clearance for one full rotation of the hub. The air gap clearance is 0.02–0.04 in. (0.4–1.0mm).

c. If the air gap exceeds the standard, inspect the knuckle and pulser wheel for distortion.

12. Install the brake rotor.

13. Install the wheel and lower the vehicle.

14. Connect the positive then the negative battery cable and enter the radio security code.

15. Turn the ignition to the **ON** position, but don't start the engine. The ABS indicator light should turn on. Start the engine and verify that the ABS indicator light turns off. If the ABS indicator light blinks, doesn't come on, or stays on with the engine running, the system fault must be diagnosed.

16. If equipped with 4WS, start the engine and turn the steering wheel lock–to–lock to reset the 4WS control unit.

FRONT SUSPENSION

Strut

REMOVAL AND INSTALLATION

Civic and del Sol

1. Raise and safely support the vehicle. Remove the front wheels.

2. Unbolt the brake hose brackets from the bottom of the strut tube. Do not disconnect the brake hoses.

NOTE: Some Civic models may not have brake hose brackets on their struts. In these cases, there is no need to unbolt the brackets.

3. Remove the damper pinch bolt.

4. Remove the damper fork nut and bolt. Remove the damper fork.

5. Remove the two strut mounting bolts from the shock tower and remove the strut from the vehicle.

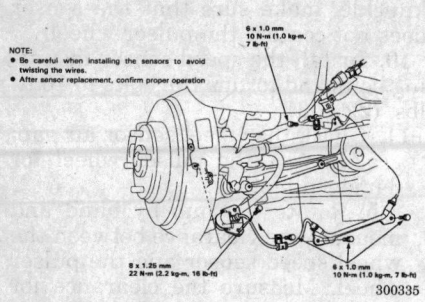

NOTE:
• Be careful when installing the sensors to avoid twisting the wires.
• After sensor replacement, confirm proper operation

Rear wheel speed sensor — Prelude

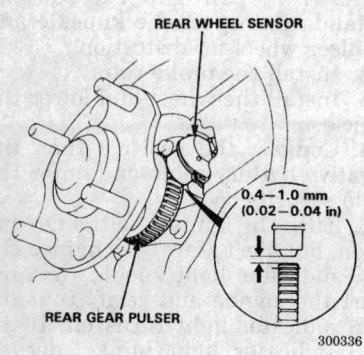

REAR WHEEL SENSOR

0.4–1.0 mm
(0.02–0.04 in)

REAR GEAR PULSER

300336

Measuring the rear wheel speed sensor gap — Prelude

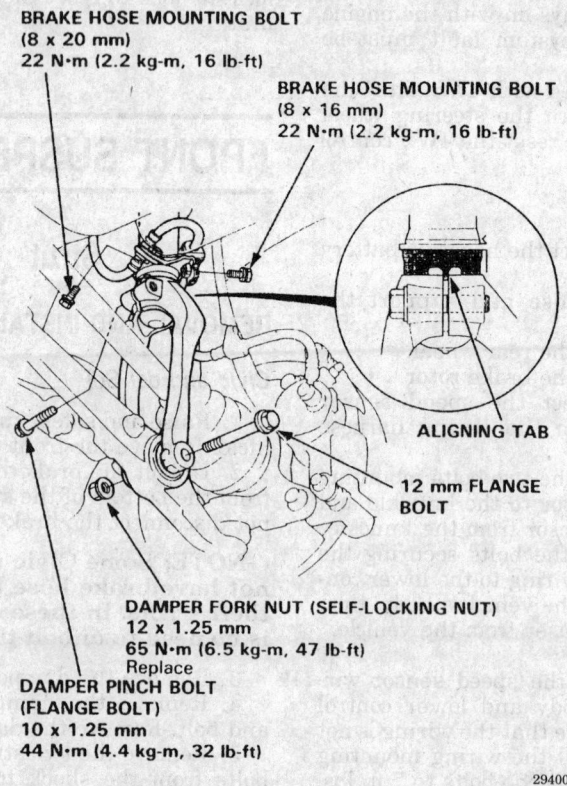

BRAKE HOSE MOUNTING BOLT
(8 x 20 mm)
22 N·m (2.2 kg-m, 16 lb-ft)

BRAKE HOSE MOUNTING BOLT
(8 x 16 mm)
22 N·m (2.2 kg-m, 16 lb-ft)

ALIGNING TAB

12 mm FLANGE BOLT

DAMPER FORK NUT (SELF-LOCKING NUT)
12 x 1.25 mm
65 N·m (6.5 kg-m, 47 lb-ft)
Replace

DAMPER PINCH BOLT
(FLANGE BOLT)
10 x 1.25 mm
44 N·m (4.4 kg-m, 32 lb-ft)

294005

Damper fork components — Civic and del Sol

6. Install a spring compressor onto the strut assembly and tighten the compressor according to the manufacturer's instructions.

7. Remove the locking nut from the top of the shock absorber piston. Disassemble the strut and remove the coil spring.

To install:

NOTE: Use new self-locking nuts when installing the strut.

8. Install a spring compressor onto the coil spring.

9. Assemble the lower strut mounts, dust covers, coil spring, and upper strut mount onto the shock absorber. Position the strut bearing mounting studs so that they will line up with the mounting holes in the shock tower.

10. Install the mounting washer, and loosely install a new self–locking nut.

11. Hold the shock absorber piston with a hex wrench and tighten the self–locking nut. Torque the self–locking nut to 22 ft. lbs. (30 Nm).

12. Install the strut into the vehicle. Hand-tighten the strut mounting bolts. The alignment mark on the strut tube faces away from the wheel.

13. Install the damper fork onto the strut and lower control arm. Install the pinch and fork bolts.

14. Connect the brake hose brackets to the strut tube and torque them to 16 ft. lbs. (22 Nm).

15. Install the front wheels and lower the vehicle.

16. Torque the strut mount bolts to 36 ft. lbs. (50 Nm).

17. Torque the pinch bolt to 32 ft. lbs. (44 Nm). Torque the damper fork nut to 47 ft. lbs. (65 Nm).

18. Torque the wheel nuts to 80 ft. lbs. (110 Nm).

19. Check the vehicle's front end alignment and adjust it if necessary.

Prelude and Accord

1. Raise and safely support the vehicle.

2. Remove the front wheels.

3. Remove the brake hose clamp bolts from the strut.

4. Remove the damper fork bolts and remove the damper fork.

5. Remove the three strut mounting nuts. Remove the strut from the vehicle.

To install:

NOTE: Use new self–locking bolts when installing the struts and assembling the damper forks.

6. Install the strut into the vehicle. Hand–tighten the mounting nuts.

7. Install the strut into the damper fork. The alignment mark on the strut tube fits into the groove on the damper fork.

8. Install the pinch bolt and damper fork bolt. Only hand–tighten these bolts.

9. Install the front wheels and lower the vehicle.

10. With all four of the vehicle's wheels on the ground, torque the damper fork nut to 47 ft. lbs. (65 Nm) while holding the damper fork bolt. Torque the damper fork pinch bolt to 32 ft. lbs. (44 Nm). Tighten the strut mounting nuts to 28 ft. lbs. (39 Nm).

11. Tighten the wheel nuts to 80 ft. lbs. (110 Nm).

12. Check and adjust the vehicle's front end alignment. On Preludes equipped with 4WS, the electronic neutral check must be performed before aligning all four wheels.

Coil Spring

REMOVAL AND INSTALLATION

Civic and del Sol

1. Raise and safely support the vehicle. Remove the front wheels.

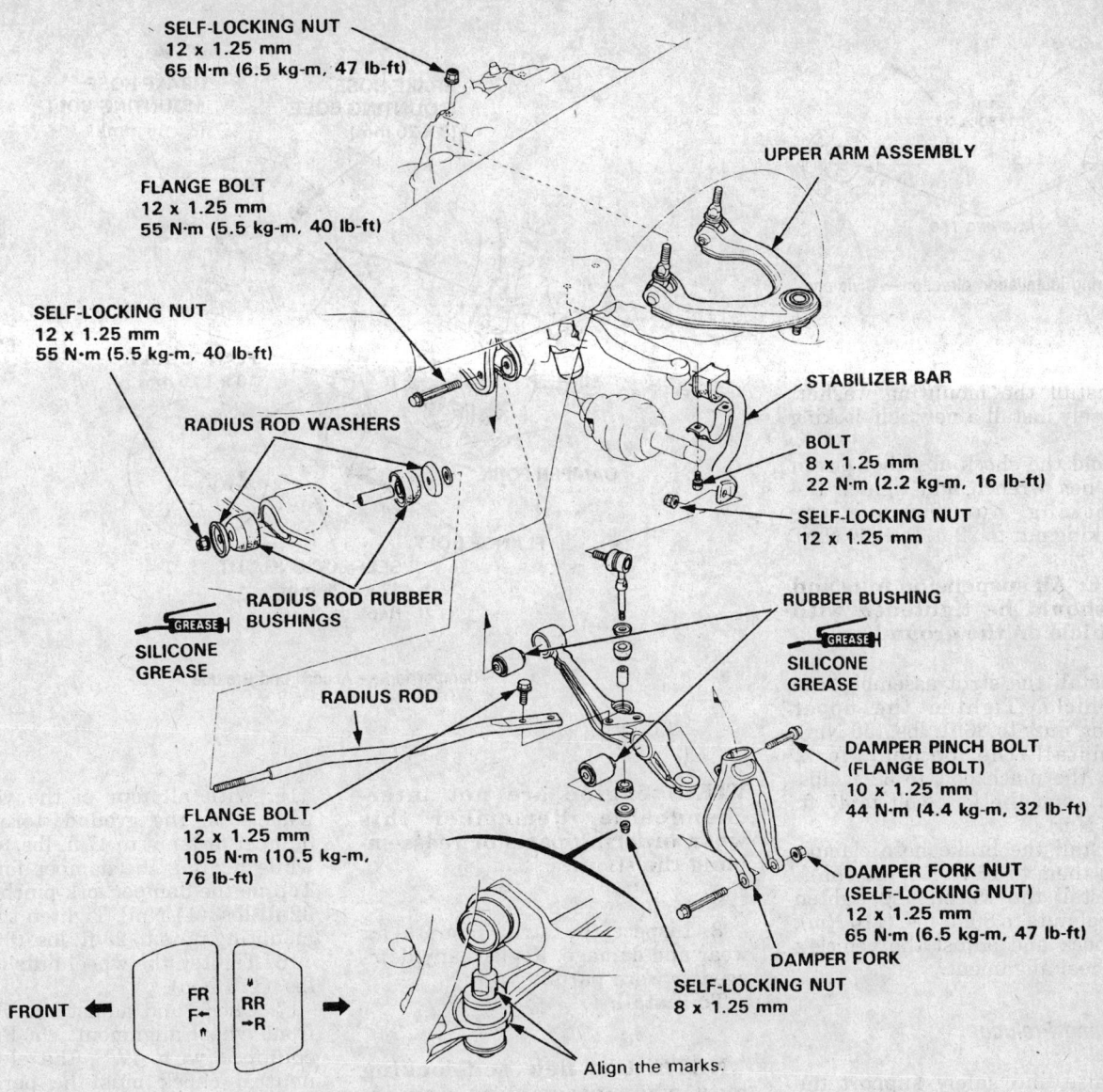

SELF-LOCKING NUT
12 x 1.25 mm
65 N·m (6.5 kg-m, 47 lb-ft)

UPPER ARM ASSEMBLY

FLANGE BOLT
12 x 1.25 mm
55 N·m (5.5 kg-m, 40 lb-ft)

SELF-LOCKING NUT
12 x 1.25 mm
55 N·m (5.5 kg-m, 40 lb-ft)

STABILIZER BAR

BOLT
8 x 1.25 mm
22 N·m (2.2 kg-m, 16 lb-ft)

RADIUS ROD WASHERS

SELF-LOCKING NUT
12 x 1.25 mm

RADIUS ROD RUBBER BUSHINGS

RUBBER BUSHING

GREASE
SILICONE GREASE

GREASE
SILICONE GREASE

RADIUS ROD

DAMPER PINCH BOLT (FLANGE BOLT)
10 x 1.25 mm
44 N·m (4.4 kg-m, 32 lb-ft)

FLANGE BOLT
12 x 1.25 mm
105 N·m (10.5 kg-m, 76 lb-ft)

DAMPER FORK NUT (SELF-LOCKING NUT)
12 x 1.25 mm
65 N·m (6.5 kg-m, 47 lb-ft)

DAMPER FORK

SELF-LOCKING NUT
8 x 1.25 mm

FRONT ← | FR F← ↑ | RR →R | →

Align the marks.

289162

Front suspension components — Prelude and Accord

NUTS
10 x 1.25 mm

DAMPER ASSEMBLY

289163

Front strut and strut mount — Prelude and Accord

2. Unbolt the brake hose brackets from the bottom of the strut tube. Do not disconnect the brake hoses.

NOTE: Some Civic models may not have brake hose brackets on their struts.

3. Remove the damper fork pinch bolt and flange bolt; then, remove the damper fork.
4. Remove the strut's upper mounting nuts. Remove the strut assembly from the vehicle.
5. Install a spring compressor onto the strut assembly and tighten the compressor according to the manufacturer's instructions.

6. Remove the locking nut from the top of the shock absorber piston. Disassemble the strut and remove the coil spring.
To install:

NOTE: Use new self-locking nuts when assembling the strut.

7. Install a spring compressor onto the coil spring.
8. Assemble the lower strut mounts, dust covers, coil spring, and upper strut mount onto the shock absorber. Position the strut bearing mounting studs so that they will line up with the mounting holes in the shock tower.

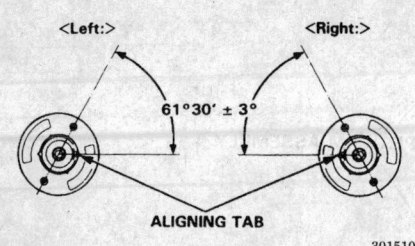

Strut bearing installation direction — Civic and del Sol

301510

9. Install the mounting washer, and loosely install a new self–locking nut.

10. Hold the shock absorber piston with a hex wrench and tighten the self–locking nut. Torque the self–locking nut to 22 ft. lbs. (30 Nm).

NOTE: All suspension nuts and bolts should be tightened with the vehicle on the ground.

11. Install the strut assembly into the vehicle. Tighten the upper mounting nuts to 36 ft. lbs. (50 Nm).

12. Install the damper fork. Tighten the pinch bolt to 32 ft. lbs. (44 Nm), and the fork bolt to 47 ft. lbs. (64 Nm).

13. Install the brake hose clamps. Tighten them to 16 ft. lbs. (22 Nm).

14. Install the wheel, and tighten the wheel nuts to 80 ft. lbs. (110 Nm).

15. Check and adjust the vehicle's front wheel alignment.

Accord and Prelude

1. Raise and safely support the vehicle.

2. Remove the front wheels.

3. Unbolt the brake hose clamp from the strut.

4. Remove the damper fork bolts and remove the damper fork.

5. Remove the three strut mounting nuts. Remove the strut from the vehicle.

6. Place the strut in vice and install a spring compressor onto the coil spring. Follow the spring compressor manufacturer's instructions.

7. Compress the spring and remove the self–locking nut from the top of the strut. Disassemble the strut mounts and remove the coil spring.

NOTE: The left and right front coil springs on 1994–97 Accords equipped with the F22B1 (2.2L

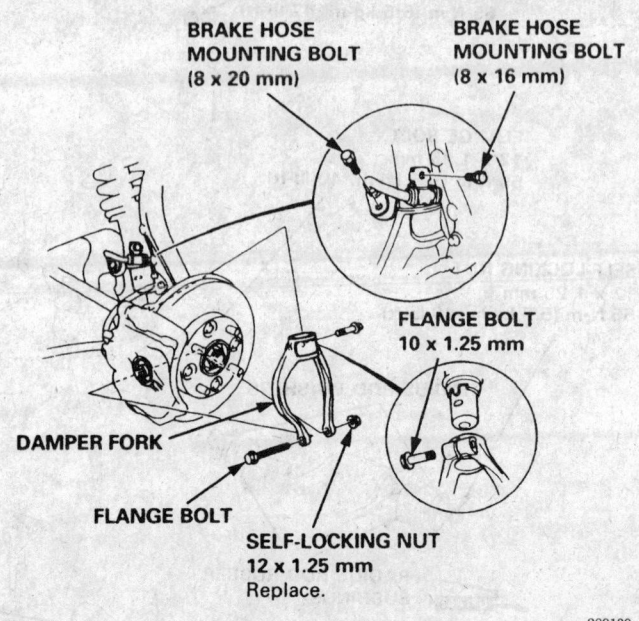

Damper fork — Accord and Prelude

289139

VTEC) engine are not interchangeable. Remember this when ordering parts or reassembling the strut.

8. Inspect the strut mounts for wear and damage. Replace any damaged or worn parts.
To install:

NOTE: Use new self–locking nuts when assembling and installing the struts.

9. Install the spring compressor onto the coil spring. Set the spring onto the strut cartridge. The flat part of the coil spring is its top.

10. Assemble the strut mount and its washer onto the strut. Tighten the self–locking nut to 22 ft. lbs. (29 Nm). Remove the spring compressor.

11. Install the strut into the vehicle. Hand–tighten the mounting nuts.

12. Install the strut into the damper fork. The alignment mark on the strut tube fits into the groove on the damper fork.

13. Install the pinch bolt and damper fork bolt. Only hand–tighten these bolts.

14. Install the front wheels and lower the vehicle.

15. With all four of the vehicle's wheels on the ground, torque the damper fork nut to 47 ft. lbs. (65 Nm) while holding the damper fork bolt. Torque the damper fork pinch bolt to 32 ft. lbs. (44 Nm). Tighten the strut mounting nuts to 28 ft. lbs. (39 Nm).

16. Tighten the wheel nuts to 80 ft. lbs. (110 Nm).

17. Check and adjust the vehicle's front wheel alignment. On Preludes equipped with 4WS, the electronic neutral check must be performed before all four wheels are aligned.

Upper Ball Joints

REMOVAL AND INSTALLATION

All Models

The upper ball joint cannot be removed from the upper control arm. If the ball joint is faulty or worn, the entire control arm must be replaced. If the upper ball joint boot is damaged and the ball joint itself is still usable, the boot can be replaced.

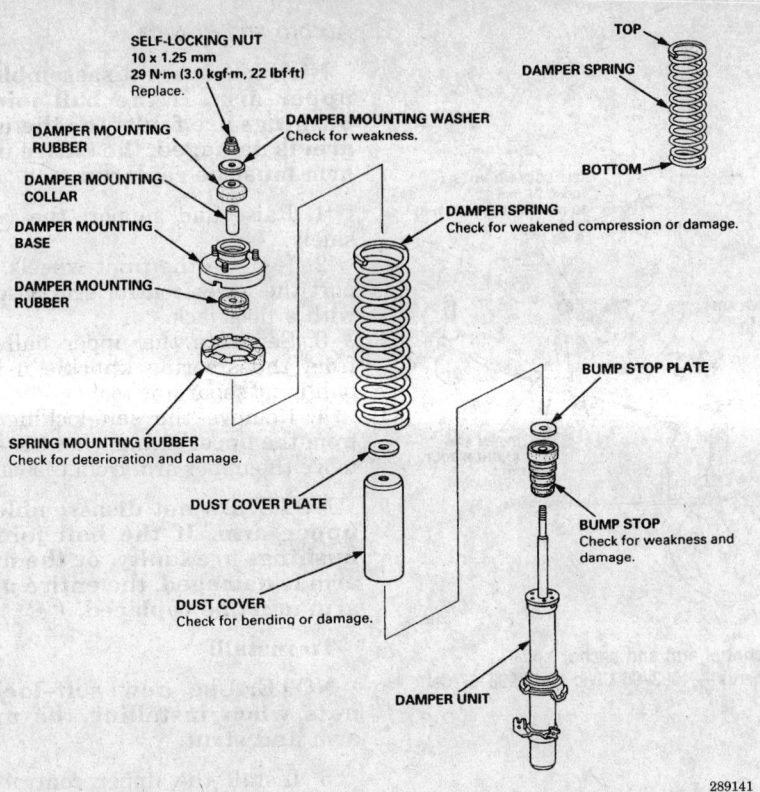

SELF-LOCKING NUT
10 x 1.25 mm
29 N·m (3.0 kgf·m, 22 lbf·ft)
Replace.

DAMPER MOUNTING RUBBER

DAMPER MOUNTING WASHER
Check for weakness.

DAMPER MOUNTING COLLAR

DAMPER MOUNTING BASE

DAMPER MOUNTING RUBBER

SPRING MOUNTING RUBBER
Check for deterioration and damage.

DUST COVER PLATE

DUST COVER
Check for bending or damage.

DAMPER UNIT

TOP

DAMPER SPRING

BOTTOM

DAMPER SPRING
Check for weakened compression or damage.

BUMP STOP PLATE

BUMP STOP
Check for weakness and damage.

289141

Coil spring, strut cartridge, and strut mount components — Accord and Prelude

Lower Ball Joints

REMOVAL AND INSTALLATION

Civic and del Sol

NOTE: The steering knuckle must be removed from the vehicle for the ball joint to be replaced. The following special tools or their equivalents are needed to press the ball joint in and out of the knuckle: ball joint installer base tool 07965–SB00200, ball joint installer/remover tool 07965–SB00100, and ball joint remover base tool 07965–SH20200. A large vise will be required to hold the knuckle and the press tools. A ball joint clip guide tool 07974–SA50700 or 07GAG–SD40700 is used to install the retaining clip on the joint boot.

1. Remove the steering knuckle assembly from the vehicle. Remove the ball joint boot snapring and the boot.
2. Pry the snapring out of the groove in the ball joint body.
3. Install the ball joint removal tool onto the ball joint with the large

end facing out. Install the ball joint nut to attach the tool to the joint.
4. Position the removal base tool on the ball joint and set the assembly in a large vise. Press the ball joint out of the steering knuckle.

To install:
5. Position the new ball joint into the hole of the steering knuckle.
6. Install the ball joint installer tool over the ball joint with the small end facing out.
7. Position the installation base tool on the ball joint and set the assembly in a large vise. Press the ball joint into the steering knuckle.
8. Seat the snapring in the groove of the ball joint.
9. Adjust the boot clip tool with the adjusting bolt until the end of the tool aligns with the groove on the boot. Slide the clip over the tool and into position.

Upper Control Arms

REMOVAL AND INSTALLATION

Civic and del Sol

1. Raise and support the vehicle safely.
2. Remove the front wheels.

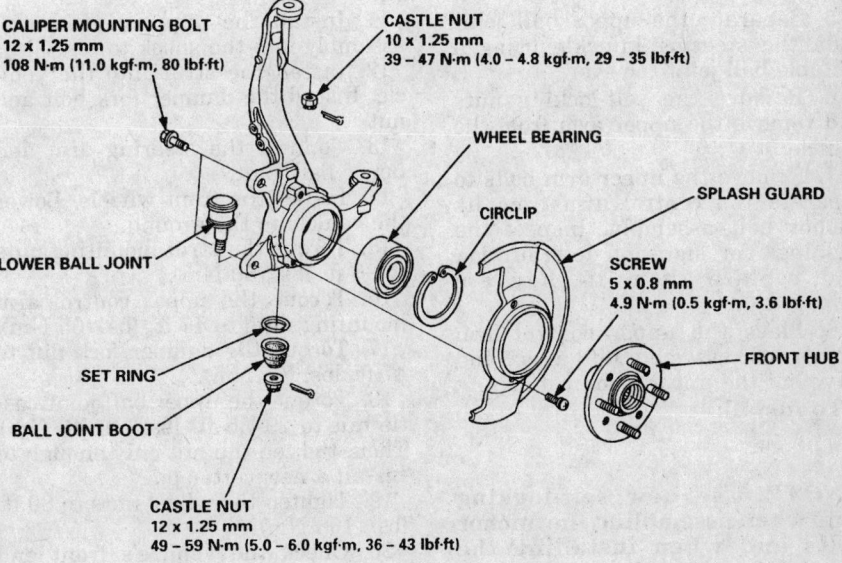

CALIPER MOUNTING BOLT
12 x 1.25 mm
108 N·m (11.0 kgf·m, 80 lbf·ft)

CASTLE NUT
10 x 1.25 mm
39 – 47 N·m (4.0 – 4.8 kgf·m, 29 – 35 lbf·ft)

WHEEL BEARING

CIRCLIP

SPLASH GUARD

LOWER BALL JOINT

SCREW
5 x 0.8 mm
4.9 N·m (0.5 kgf·m, 3.6 lbf·ft)

FRONT HUB

SET RING

BALL JOINT BOOT

CASTLE NUT
12 x 1.25 mm
49 – 59 N·m (5.0 – 6.0 kgf·m, 36 – 43 lbf·ft)

294265

Knuckle components — Civic and del Sol

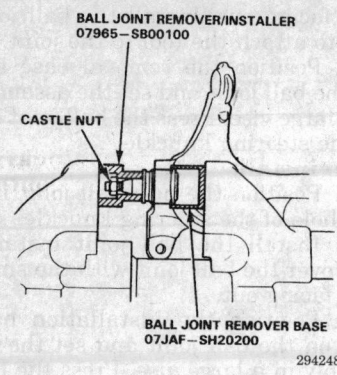

Ball joint removal tools — Civic and del Sol

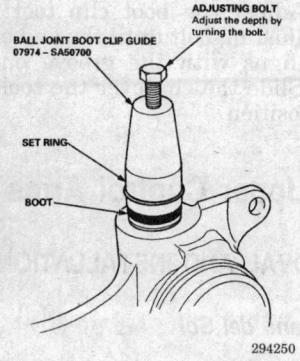

Ball joint boot clip guide — Civic and del Sol

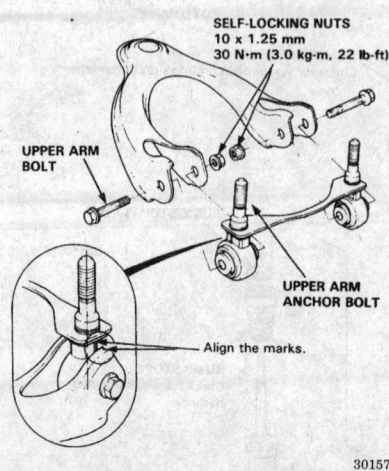

Upper control arm and anchor bolt components — 1993–95 Civic and 1993–97 del Sol

3. Unbolt the damper fork from the lower control arm.

4. Unbolt the strut mounting nuts and remove the strut from the vehicle.

5. Separate the upper ball joint from the steering knuckle using a suitable ball joint remover.

6. Remove the self-locking nuts and remove the upper arm from the vehicle.

7. Remove the upper arm bolts to separate the control arm from its anchor bolt assembly. Inspect the bushings for signs of deterioration and replace them if they are damaged.

8. Place the upper control arm anchor bolt assembly into a vice and drive out the upper arm bushings.

To install:

NOTE: Use new self-locking nuts when assembling the anchor bolts and when installing the control arm into the vehicle.

9. Drive the new upper arm bushings into the upper arm anchor bolts. Center the bushing in the anchor bolt

so that equal amounts of the bushing sleeve protrude on either side.

10. Install the anchor bolt assembly onto the control arm. Align the marks on the arm and anchor assembly. Tighten the nuts to 22 ft. lbs. (30 Nm).

11. Install the upper control arm assembly into the shock tower.

12. Install the strut into the vehicle. Install the damper fork bolt and nut.

13. Connect the steering arm and upper ball joint.

14. Install the front wheels. Lower the vehicle to the ground.

15. Torque the strut mounting nuts to 36 ft. lbs. (50 Nm).

16. Torque the upper control arm mounting nuts to 47 ft. lbs. (65 Nm).

17. Torque the damper fork nut to 47 ft. lbs. (65 Nm).

18. Torque the upper ball joint castle nut to 29–35 ft. lbs. (40–48 Nm). Then, tighten the nut only enough to Install a new cotter pin.

19. Tighten the wheel nuts to 80 ft. lbs. (108 Nm).

20. Check the vehicle's front end alignment and adjust it if necessary. Road test the vehicle.

Accord and Prelude

NOTE: Do not disassemble the upper arm. If the ball joint or bushings are faulty, or the upper arm is damaged, the entire upper arm must be replaced.

1. Raise and support the vehicle safely.

2. Remove the front wheels. Support the lower control arm assembly with a floor jack.

3. Separate the upper ball joint from the steering knuckle using a ball joint separator tool.

4. Remove the self–locking nuts from the upper arm anchor bolts. Remove the upper arm from the vehicle.

NOTE: Do not disassemble the upper arm. If the ball joint or bushings are faulty, or the upper arm is damaged, the entire upper arm must be replaced.

To install:

NOTE: Use new self–locking nuts when installing the upper arm and strut.

5. Install the upper control arm assembly into the strut tower.

6. Connect the upper ball joint.

7. Install the front wheels and lower the vehicle.

8. With all four of the vehicle's wheels on the ground, torque the upper control arm nuts to 47 ft. lbs. (65 Nm). Torque the castle nut to 32 ft. lbs. (44 Nm); then, only tighten it only enough to install a new cotter pin.

9. Tighten the wheel nuts to 80 ft. lbs. (110 Nm).

10. Check and adjust the vehicle's front end alignment. On Preludes equipped with 4WS, the electronic neutral check must be performed before all four wheels are aligned.

Lower Control Arms

REMOVAL AND INSTALLATION

Civic and del Sol

1. Raise and safely support the vehicle.

2. Remove the front wheels.

3. Separate the ball joint from the lower control arm with a ball joint remover. Do not damage the joint or tear its boot.

4. Unbolt the damper fork and the stabilizer bar linkage from the lower control arm.

5. Unbolt the control arm from its front and rear subframe brackets.

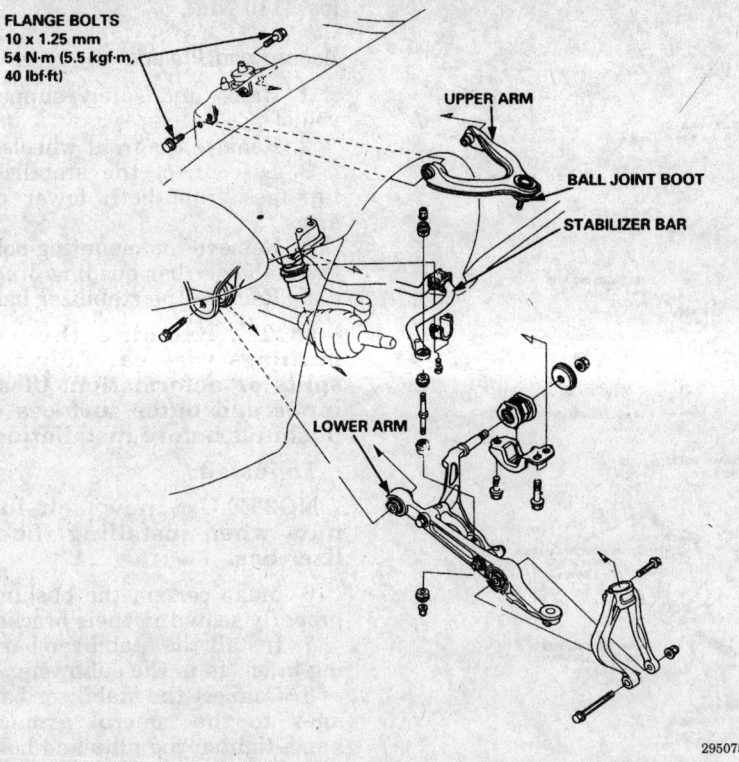

FLANGE BOLTS
10 x 1.25 mm
54 N·m (5.5 kgf·m,
40 lbf·ft)

UPPER ARM

BALL JOINT BOOT

STABILIZER BAR

LOWER ARM

295075

Front suspension components — 1996–97 Civic

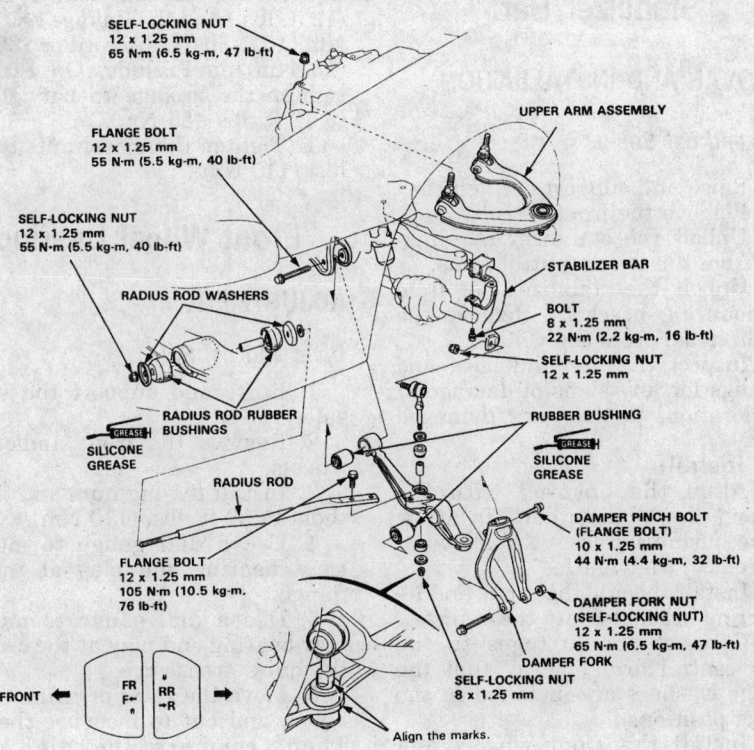

SELF-LOCKING NUT
12 x 1.25 mm
65 N·m (6.5 kg·m, 47 lb·ft)

FLANGE BOLT
12 x 1.25 mm
55 N·m (5.5 kg·m, 40 lb·ft)

UPPER ARM ASSEMBLY

SELF-LOCKING NUT
12 x 1.25 mm
55 N·m (5.5 kg·m, 40 lb·ft)

RADIUS ROD WASHERS

STABILIZER BAR

BOLT
8 x 1.25 mm
22 N·m (2.2 kg·m, 16 lb·ft)

SELF-LOCKING NUT
12 x 1.25 mm

RADIUS ROD RUBBER BUSHINGS

RUBBER BUSHING

SILICONE GREASE

SILICONE GREASE

RADIUS ROD

DAMPER PINCH BOLT
(FLANGE BOLT)
10 x 1.25 mm
44 N·m (4.4 kg·m, 32 lb·ft)

FLANGE BOLT
12 x 1.25 mm
105 N·m (10.5 kg·m,
76 lb·ft)

DAMPER FORK NUT
(SELF-LOCKING NUT)
12 x 1.25 mm
65 N·m (6.5 kg·m, 47 lb·ft)

DAMPER FORK

SELF-LOCKING NUT
8 x 1.25 mm

FRONT ←

FR
F←
↑

#
RR
→R
↑

→

Align the marks.

300763

Front suspension components — Prelude and Accord

6. Remove the control arm from the vehicle. Inspect the arm and bushings for any signs of damage or deterioration.

To install:

NOTE: Use new self-locking nuts when installing the control arm.

7. Install the control arm into the rear and then the front subframe brackets. Verify that the rear bracket washers have been reinstalled in their original positions. Install the mounting bolts and nuts.

8. Connect the stabilizer bar linkage to the control arm.

9. Connect the damper fork to the control arm and install the through-bolt with a new nut.

10. Connect the ball joint to the lower control arm.

11. Install the front wheels and lower the vehicle to the ground.

12. Torque the control arm subframe bolt to 47 ft. lbs. (65 Nm), and the nut to 61 ft. lbs. (85 Nm). Tighten the rear bushing bracket bolt to 66 ft. lbs. (91 Nm).

13. Torque the damper fork nut to 47 ft. lbs. (65 Nm). Torque the stabilizer bar linkage nuts to 16 ft. lbs. (22 Nm).

14. Torque the ball joint castle nut to 36–43 ft. lbs. (50–60 Nm). Then, tighten it only enough to install a new cotter pin.

15. Tighten the wheel nuts to 80 ft. lbs. (108 Nm).

16. Check the vehicle's front wheel alignment and adjust it if necessary.

Accord and Prelude

1. Raise and safely support the vehicle.

2. Remove the front wheels.

3. Disconnect the sway bar link from the lower control arm.

4. Remove the cotter pin and nut from the lower ball joint. Disconnect the ball joint from the lower control arm a using ball joint separator, tool 07MAC - SL00200, or equivalent.

5. Remove the nut from the radius rod front beam bushing.

6. Unbolt the radius rod from the lower control arm.

7. Support the lower control arm with a floor jack and remove the damper fork bolt.

8. Unbolt the lower control arm from the subframe and remove it from the vehicle.

To install:

NOTE: Use new self-locking nuts when installing the lower control arm.

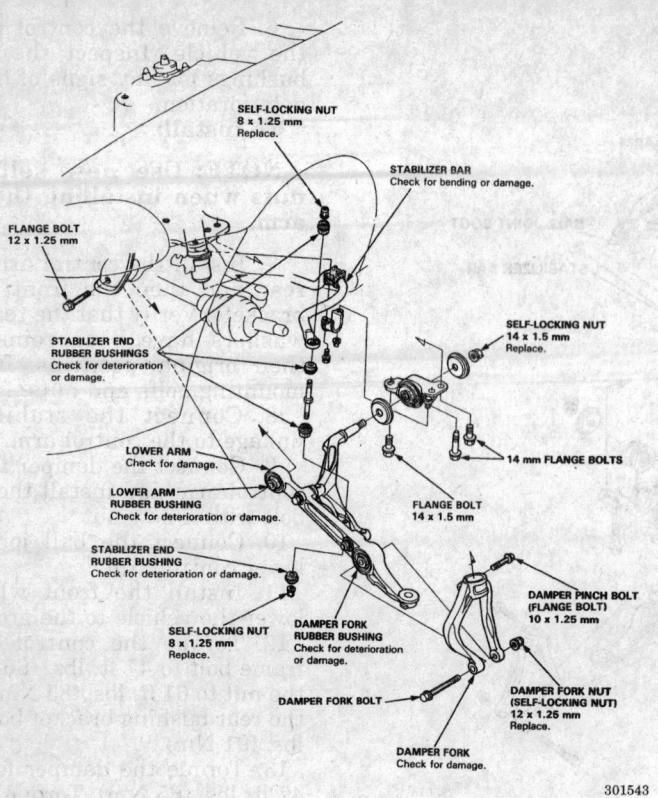

FLANGE BOLT
12 x 1.25 mm

SELF-LOCKING NUT
8 x 1.25 mm
Replace.

STABILIZER BAR
Check for bending or damage.

STABILIZER END
RUBBER BUSHINGS
Check for deterioration
or damage.

SELF-LOCKING NUT
14 x 1.5 mm
Replace.

LOWER ARM
Check for damage.

LOWER ARM
RUBBER BUSHING
Check for deterioration or damage.

14 mm FLANGE BOLTS

STABILIZER END
RUBBER BUSHING
Check for deterioration or damage.

FLANGE BOLT
14 x 1.5 mm

SELF-LOCKING NUT
8 x 1.25 mm
Replace.

DAMPER FORK
RUBBER BUSHING
Check for deterioration
or damage.

DAMPER PINCH BOLT
(FLANGE BOLT)
10 x 1.25 mm

DAMPER FORK BOLT

DAMPER FORK NUT
(SELF-LOCKING NUT)
12 x 1.25 mm
Replace.

DAMPER FORK
Check for damage.

301543

Lower control arm components — 1993–95 Civic and 1993–97 del Sol

WARNING

All suspension fasteners must be tightened to torque specifications with all four of the vehicle's wheels on the ground. This step is important for the preloading of the suspension.

9. Install the lower control arm and install the radius rod bolts.

10. Install the lower control subframe flange bolt and tighten to 40 ft. lbs. (55 Nm). Tighten the radius rod bolts to 76 ft. lbs. (105 Nm). Compress the strut with a floor jack and install the damper fork bolt and tighten its flange bolt to 47 ft. lbs. (64 Nm).

11. Connect the lower ball joint and the lower control arm and install the nut. Tighten the nut to 40 ft. lbs. (55 Nm) and install a new cotter pin.

NOTE: If the hole in the ball joint stud does not align with the nut castellation, further tighten the nut until the cotter pin can be installed; never loosen the nut to allow cotter pin installation.

12. Attach the sway bar link and tighten the nut to 14 ft. lbs. (19 Nm).

13. Install the front wheels and lower the vehicle.

Stabilizer Bar

REMOVAL AND INSTALLATION

Civic and del Sol

1. Raise and support the vehicle.
2. Remove the front wheels.
3. Unbolt the stabilizer bar linkages from the lower control arms.
4. Unbolt the stabilizer bar from its mounting brackets. Remove the stabilizer bar from the vehicle.
5. Inspect the bar, linkages, and bushings for any signs of damage or deterioration. Replace any damaged parts.

To install:

6. Align the bushing with the mark on the stabilizer bar. The arrow on the bushing bracket faces toward the front of the vehicle.
7. Install the stabilizer bar and its mounting brackets onto the vehicle.
8. Connect the linkages to the lower control arms. Verify that the linkage washers are installed in the correct positions.
9. Install the front wheels and lower the vehicle to the ground.
10. Torque the mounting bracket nuts and bolts to 16 ft. lbs. (22 Nm). Torque the linkage nuts to 16 ft. lbs. (22 Nm).

11. Tighten the wheel nuts to 80 ft. lbs. (110 Nm).

Accord and Prelude

1. Raise and safely support the vehicle.
2. Remove the front wheels.
3. Disconnect the stabilizer bar linkages from both lower control arms.
4. Remove the mounting bolts and the stabilizer bar bushing brackets.
5. Remove the stabilizer bar.

NOTE: Examine the rubber bushings very carefully for any splits or deformation. Clean the inner and outer surfaces of the bushings before installation.

To install:

NOTE: Use new self–locking nuts when installing the stabilizer bar.

6. Make certain the bushings are properly seated in their brackets.
7. Install the stabilizer bar bushing brackets to the subframe.
8. Connect the stabilizer bar linkages to the control arms. Only hand–tighten the nuts and bolts.
9. Install the front wheels.
10. With the vehicle on the ground, tighten the brackets bolts to 16 Nm (12 ft. lbs.) and the linkage bolts to 22 Nm (16 ft. lbs) on Accord or 13 ft. lbs. (18 Nm) on Prelude. On Preludes, tighten the linkage–to–bar joint nut to 40 ft. lbs. (55 Nm).
11. Tighten the wheel nuts to 80 ft. lbs. (110 Nm).

Front Wheel Bearings

ADJUSTMENT

Civic and del Sol

1. Raise and support the vehicle safely.
2. Remove the front and/or rear wheels.
3. Install the lug nuts and tighten them to 80 ft. lbs. (110 Nm).
4. Use a dial gauge to measure front bearing end play at the hub flange.
5. Use a dial gauge to measure rear bearing end play at the center of the hub's grease cap.
6. Move the rotor or drum assembly in and out to measure the play. Then, compare the dial gauge readings.
7. The standard bearing end play for both front and rear wheels is 0–0.002 in. (0–0.05mm). If the end play measurement exceeds the stan-

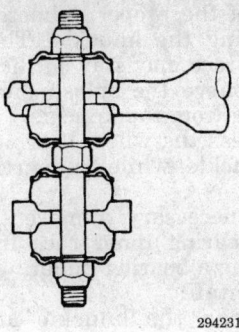

Sway bar linkages — Civic and del Sol

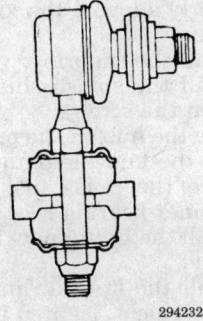

Sway bar linkages — del Sol with VTEC engine

dard, the wheel bearings must be replaced. The wheel bearings cannot be adjusted.

Prelude and Accord

The wheel bearings are not adjustable or repairable and should be replaced if found defective.

REMOVAL AND INSTALLATION

Civic and del Sol

NOTE: A hydraulic press and several bearing drivers and attachments are needed to remove and install the hub and bearing.

1. Pry the spindle nut stake away from the spindle, then loosen the nut.
2. Raise and safely support the vehicle.
3. Remove the front wheel and the spindle nut.
4. Remove the wheel sensor wire bracket from the knuckle, but don't disconnect it.
5. Remove the caliper mounting bolts and the caliper. Support the caliper out of the way with a length of wire. Do not let the caliper hang from the brake hose.

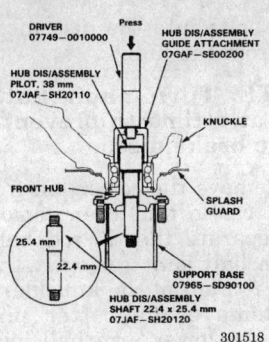

Hub installation driver, guide, and base — 1993–95 Civic and 1993–97 del Sol

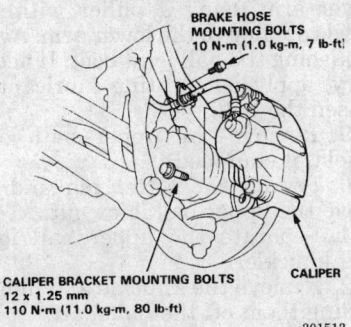

Caliper bracket mounting bolts — 1993–95 Civic and 1993–97 del Sol

6. Remove the 6mm brake disc retaining screws. Screw two 12mm bolts into the disc to push it away from the hub.
7. Remove the tie rod castle nut. Disconnect the tie rod ball joint using a suitable ball joint remover.
8. Remove the cotter pin and loosen the lower arm ball joint nut half the length of the joint threads.
9. Separate the ball joint and lower arm using a suitable puller with the pawls applied to the lower arm.

NOTE: Avoid damaging the ball joint boot. If necessary, apply penetrating type lubricant to loosen the ball joint.

10. Remove the ball joint nut cover. Remove the cotter pin and remove the upper ball joint nut.
11. Separate the upper ball joint and knuckle using a ball joint remover.
12. Use a plastic mallet to free the halfshaft from the knuckle. Pull the knuckle out to remove it.

NOTE: A new wheel bearing must be used when the hub is removed.

13. Place the knuckle in a press and use a base and pilot to press the hub assembly out of the wheel bearing.
14. Remove the knuckle ring seal and circlip. Remove the splash guard from the knuckle.
15. Press the wheel bearing out of the knuckle using a driving attachment.
    To install:
16. Clean the knuckle and hub assembly and inspect them for damage.
17. Press a new wheel bearing into the hub using a driving tool.
18. Install the circlip in the outer groove of the knuckle.
19. Install the splash guard.
20. Press the hub assembly into the steering knuckle using a base and a driving and guide tool.
21. Install the knuckle ring seal.
22. Install the knuckle onto the spindle.
23. Install the knuckle onto the upper and lower ball joints and tighten the castle nuts. Install the tie rod ball joint onto the steering knuckle.
24. Torque the upper ball joint nut and tie rod nut to 29–35 ft. lbs. (40–48 Nm) and the lower ball joint castle nut to 36–43 ft. lbs. (50–60 Nm).
25. Install the ABS wheel sensor wire brackets onto the knuckle. Torque the mounting bolts to 7 ft. lbs. (10 Nm).
26. Install the brake disc and use two lug nuts to evenly draw the disc onto the hub. Install the retainer screws and torque them to 7 ft. lbs. (10 Nm). Install the spindle washer and nut. Don't torque the nut until the vehicle is on the ground.
27. Install the brake caliper and torque the bolts to 80 ft. lbs. (110 Nm).
28. Install the front wheels and lower the vehicle.
29. Torque the spindle nut to 134 ft. lbs. (185 Nm), stake the nut, and install the grease cap.
30. Check and adjust the vehicle's front wheel alignment.

NOTE: Avoid damaging the ball joint boot. If necessary, apply penetrating–type lubricant to loosen the ball joint.

Prelude and Accord

NOTE: Once the hub has been removed, the wheel bearings must be replaced. A hydraulic press and bearing drivers must be used to remove and install the bearing.

1. Pry the spindle nut stake away from the spindle and loosen the nut.

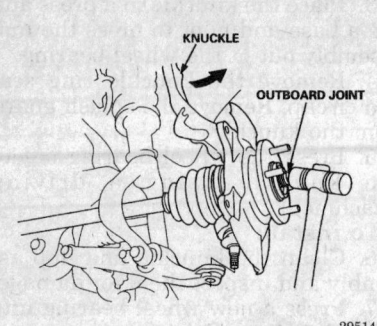

Separating the halfshaft from the hub/knuckle assembly — 1996–97 Civic

Do not tighten or loosen a spindle nut unless the vehicle is sitting on all four wheels. The torque required is high enough to cause the vehicle to fall off the stands even when properly supported.

2. Raise and safely support the vehicle.

3. Remove the wheel and the spindle nut.

4. Remove the caliper mounting bolts and the caliper. Support the caliper out of the way with a length of wire. Do not let the caliper hang from the brake hose.

5. Remove the 6mm brake disc retaining screws. Screw two 8·1.25

12mm bolts into the disc to push it away from the hub.

NOTE: Turn each bolt two turns at a time to prevent cocking the brake disc.

6. Remove the cotter pin from the tie rod castle nut, then remove the nut. Separate the tie rod ball joint using a ball joint remover, then lift the tie rod out of the knuckle.

7. Remove the cotter pin and loosen the lower arm ball joint nut half the length of the joint threads. The nut will retain the arm when the joint comes loose.

8. Separate the ball joint and lower arm using a puller with the pawls applied to the lower arm. Avoid damaging the ball joint boot. If necessary, apply penetrating lubricant to loosen the ball joint.

9. Remove the upper ball joint shield, if equipped.

10. Pry off the cotter pin and remove the upper ball joint nut.

11. Separate the upper ball joint and knuckle.

12. Remove the knuckle and hub by sliding them off the halfshaft.

13. Remove the splash guard screws from the knuckle.

14. Position the knuckle/hub assembly in a hydraulic press. Press the hub from the knuckle using a

driver of the proper diameter while supporting the knuckle. The inner bearing race may stay on the hub.

15. Remove the splash guard and snapring from the knuckle.

16. Press the wheel bearing out of the knuckle while supporting the knuckle.

17. If necessary, remove the outboard bearing inner race from the hub using a bearing puller.

To install:

18. Clean the knuckle and hub thoroughly.

19. Press a new wheel bearing into the knuckle. Make sure the press tool contacts only the outer bearing race and properly support the knuckle so it is stable.

20. Install the snapring.

21. Install the splash shield. Don't overtighten the screws.

22. Place the hub on the press table and press the knuckle onto the hub. Make sure the press tool contacts only the inner bearing race.

23. Install the front knuckle ring on the knuckle.

24. Install the knuckle/hub assembly on the vehicle. Tighten the upper ball joint nut and tie rod end nut to 32 ft. lbs. (44 Nm). Install new cotter pins. Tighten the lower ball joint nut to 40 ft. lbs. (55 Nm) and install a new cotter pin.

25. Install the brake disc and caliper. Tighten the caliper bracket bolts to 80 ft. lbs. (110 Nm).

26. Install the front wheels and lower the vehicle.

27. Torque the spindle nut to 180 ft. lbs. (250 Nm). Torque the wheel nuts to 80 ft. lbs. (110 Nm).

28. Check and adjust the vehicle's front wheel alignment.

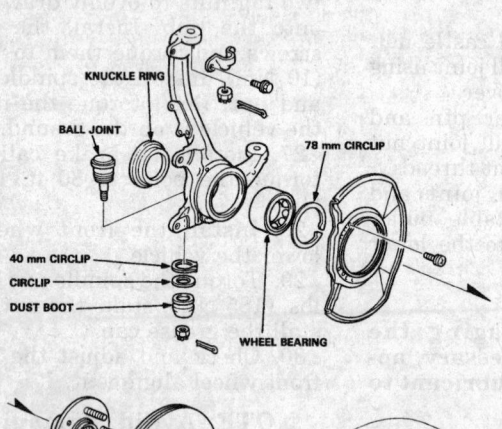
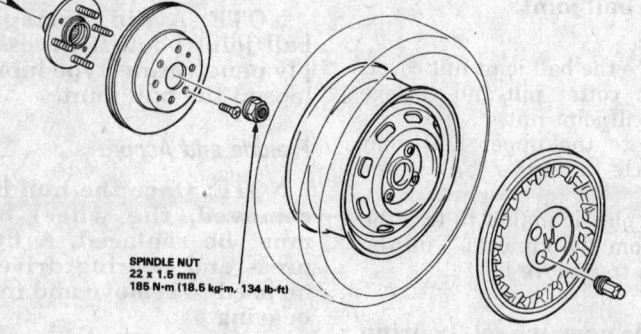

Hub and steering knuckle components — Prelude

REAR SUSPENSION

Strut

REMOVAL AND INSTALLATION

Civic and del Sol

CAUTION
Removing rear suspension components may make the vehicle front-heavy and cause it to tip forward when raised on a hoist. Use under-lift support stands, or place additional weight in the trunk of the vehicle before hoisting it.

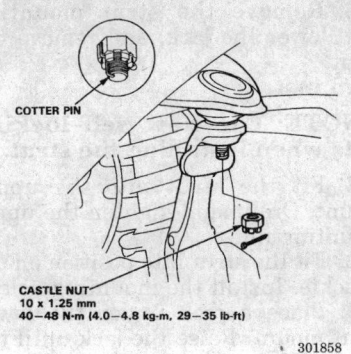

Upper control arm ball joint — Prelude

COTTER PIN

CASTLE NUT
10 x 1.25 mm
40—48 N·m (4.0—4.8 kg-m, 29—35 lb-ft)

301858

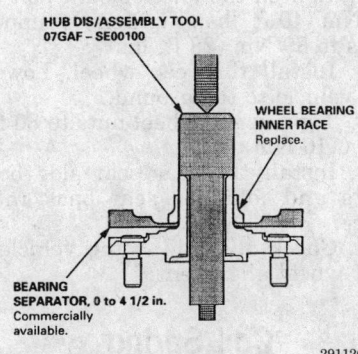

Wheel bearing inner race — Accord

HUB DIS/ASSEMBLY TOOL
07GAF – SE00100

WHEEL BEARING
INNER RACE
Replace.

BEARING
SEPARATOR, 0 to 4 1/2 in.
Commercially
available.

291126

1. Remove the interior or trunk trim pieces that cover the strut mount:

a. **Sedan and coupe models:** Fold down the upper rear seat cushion. Carefully pry out the clips that secure the trunk and shock tower trim to the body. Remove the trunk trim to expose the strut mounts.

b. **Hatchback models:** Fold down the rear seat. Unbolt and remove the rear side shelf/speaker grille assemblies. Disconnect and remove the speaker. Carefully pry out the clips and remove the screws to remove shock tower trim panel.

c. **del Sol models:** Support the trunk lid in a fully–open position. Remove the trunk lid support struts. Lift the roof storage frame up to get it out of the work area. It is not necessary to remove the roof storage frame. Carefully loosen and remove the screw clips. Then, position the trunk side trim panels away from the strut mounts.

2. Raise and support the vehicle. Remove the rear wheels.

3. Unbolt the two upper mounting nuts.

4. Unbolt the wheel sensor bracket from the lower control arm.

5. Remove the lower strut bolt and the knuckle flange bolt.

6. Remove the strut from the vehicle.

7. Install a spring compressor onto the strut assembly and tighten the compressor according to the manufacturer's instructions.

8. Remove the locking nut from the top of the shock absorber. Disassemble the strut and remove the coil spring.

To install:

NOTE: Use new self-locking nuts when installing the strut.

9. Install a spring compressor onto the coil spring.

10. Assemble the upper and lower strut mounts, dust covers, and coil spring onto the shock absorber.

11. Install the mounting washer, and loosely install a new self–locking nut.

12. Hold the shock absorber piston with a hex wrench and tighten the self–locking nut. Torque the self–locking nut to 22 ft. lbs. (30 Nm).

NOTE: All suspension nuts and bolts should be tightened with the vehicle on the ground. Alternatively, raise the lower control arm with a floor jack until the jack is supporting the weight of the vehicle. This method preloads the suspension and allows room to work.

13. Install the strut into the vehicle with the locknut facing the front of the vehicle. Hand-tighten the upper mounting nuts.

14. Install the wheel sensor bracket onto the lower control arm. Tighten the bolts to 7ft. lbs. (10 Nm).

15. Install the knuckle flange bolt and the lower strut bolt. Hand-tighten the bolts.

16. Install the wheels and lower the vehicle.

17. Torque the upper mounting nuts to 36 ft. lbs. (50 Nm). Torque the knuckle flange bolt and strut bolts to 40 ft. lbs. (55 Nm). Torque the wheel nuts to 80 ft. lbs. (110 Nm).

18. Install the trunk side trim panels.

19. Check and adjust the vehicle's rear wheel alignment.

Prelude

1. Raise and safely support the vehicle.

2. Remove the trunk side trim and remove the 2 top strut nuts.

3. Remove the upper ball joint cover.

4. Remove the cotter pin and upper ball joint nut.

UPPER ARM

OUTBOARD
JOINT

LOWER ARM

CASTLE NUT
12 x 1.25 mm
49 – 59 N·m
(5.0 – 6.0 kgf·m,
36 – 43 lbf·ft)

COTTER PIN
Replace.

TIE-ROD
END

COTTER PIN
Replace.
On reassembly,
bend the cotter pin
as shown.

CASTLE NUT
10 x 1.25 mm
39 – 47 N·m
(4.0 – 4.8 kgf·m,
29 – 35 lbf·ft)

COTTER PINS
Replace.

CASTLE NUT
10 x 1.25 mm
39 – 47 N·m
(4.0 – 4.8 kgf·m,
29 – 35 lbf·ft)

291131

Steering knuckle components — Accord

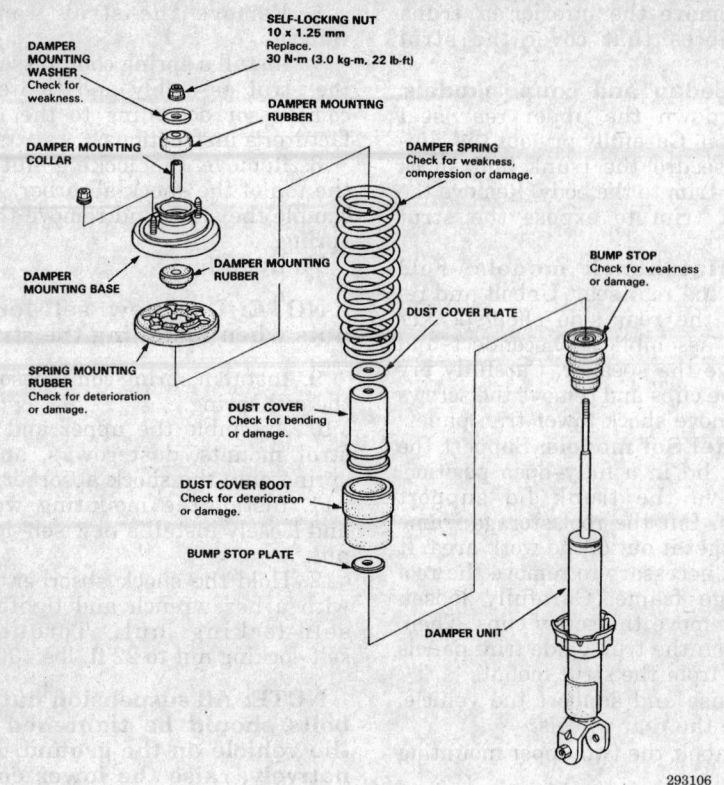

DAMPER MOUNTING WASHER
Check for weakness.

SELF-LOCKING NUT
10 x 1.25 mm
Replace.
30 N·m (3.0 kg-m, 22 lb-ft)

DAMPER MOUNTING RUBBER

DAMPER MOUNTING COLLAR

DAMPER SPRING
Check for weakness, compression or damage.

DAMPER MOUNTING BASE

DAMPER MOUNTING RUBBER

BUMP STOP
Check for weakness or damage.

SPRING MOUNTING RUBBER
Check for deterioration or damage.

DUST COVER PLATE

DUST COVER
Check for bending or damage.

DUST COVER BOOT
Check for deterioration or damage.

BUMP STOP PLATE

DAMPER UNIT

293106

Strut components — Civic and del Sol

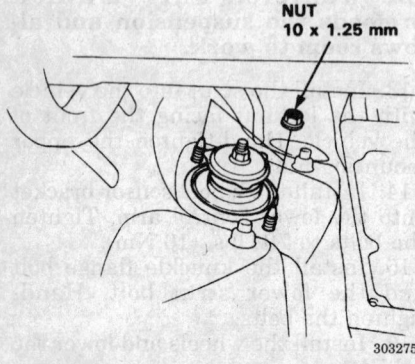

NUT
10 x 1.25 mm

303275

Upper mounting of rear strut — Prelude

5. Fit a 10mm nut on the ball joint and separate the ball joint and the knuckle by using a ball joint removal tool.

6. Remove the lower strut mounting bolt and lower the suspension.

7. Remove the strut from the vehicle.

To install:

NOTE: Use new self-locking nuts when installing the rear struts.

8. Install the strut and loosely install the lower mounting bolt. Do not tighten.

9. Install the upper strut mounting bolts. Tighten the bolts to 28 ft. lbs. (39 Nm).

10. Connect the upper arm and knuckle and tighten the castle nut to 29–35 ft. lbs. (40–48 Nm).

11. Install the upper ball joint cover.

12. Raise the rear suspension with a floor jack until the weight is on the strut.

13. Tighten the lower strut mounting bolt to 47 ft. lbs. (65 Nm).

14. Install the rear wheels and lower the vehicle.

15. Tighten the rear wheel nuts to 80 ft. lbs. (110 Nm).

16. Check and adjust the vehicle's rear wheel alignment.

Accord

1. Fold the rear seat forward and remove the side bolster cushions. The side bolster cushions are secured by a screw at the bottom and two clips at the top.

2. Remove the strut mount cap. Remove the upper strut mounting nuts.

3. Raise and safely support the vehicle.

4. Remove the rear wheels.

5. Support the knuckle with a floor jack.

6. Remove the strut mounting bolt, lower the jack, and remove the strut.

To install:

NOTE: Use new self-locking nuts when installing the strut.

7. Fit the strut into the upper mount. Only hand-tighten the upper mounting nuts.

8. Fit the strut into position on the knuckle. Install the mounting bolt.

9. Place a jack under the lower strut mount. Raise the jack until the weight of the car is on the jack.

10. With the suspension under load, tighten the lower mount bolt to 55 Nm (40 ft. lbs). Tighten the upper nuts to 39 Nm (28 ft. lbs).

11. Install the rear wheel. Lower the vehicle to the ground.

12. Tighten the wheel nuts to 80 ft. lbs. (110 Nm).

13. Install the rear seat side bolsters and fold the seat back into place.

14. Check and adjust the vehicle's rear wheel alignment.

Coil Spring

REMOVAL AND INSTALLATION

Civic and del Sol

———— CAUTION ————
Removing rear suspension components may make the vehicle front-heavy and cause it to tip forward when raised on a hoist. Use under-lift support stands, or place additional weight in the trunk of the vehicle before hoisting it.

1. Remove the interior or trunk trim pieces that cover the strut mount:

a. **Sedan and coupe models:** Fold down the upper rear seat cushion. Carefully pry out the clips that secure the trunk and shock tower trim to the body. Remove the trunk trim to expose the strut mounts.

b. **Hatchback models:** Fold down the rear seat. Unbolt and remove the rear side shelf/speaker grille assemblies. Disconnect and remove the speaker. Carefully pry out the clips and remove the screws to remove shock tower trim panel.

c. **del Sol models:** Support the trunk lid in a fully-open position. Remove the trunk lid support struts. Lift the roof storage frame up to get it out of the work area. It is not necessary to remove the roof storage frame. Carefully loosen and remove the screw clips. Then, position the trunk side trim panels away from the strut mounts.

2. Raise and safely support the vehicle.

3. Remove the two strut mounting bolts.

4. Unbolt the wheel sensor brackets from the lower control arm. Do not disconnect the sensor.

5. Support the lower control arm with a floor jack.

6. Remove the strut mounting flange bolt and the knuckle flange bolt.

7. Lower the floor jack and remove the strut from the vehicle.

8. Install a spring compressor onto the strut assembly and tighten the compressor according to the manufacturer's instructions.

9. Remove the locking nut from the top of the shock absorber. Disassemble the strut and remove the coil spring.

To install:

NOTE: Use new self-locking nuts when assembling the strut.

10. Install a spring compressor onto the coil spring.

11. Assemble the upper and lower strut mounts, dust covers, and coil spring onto the shock absorber.

12. Install the mounting washer, and loosely install a new self-locking nut.

13. Hold the shock absorber piston with a hex wrench and tighten the self-locking nut. Torque the self-locking nut to 22 ft. lbs. (30 Nm).

NOTE: All suspension nuts and bolts should be tightened with the vehicle on the ground. Alternatively, raise the lower control arm with a floor jack until the jack is supporting the weight of the vehicle. This method preloads the suspension and allows room to work.

14. Install the strut assembly into the vehicle. Tighten the upper mounting nuts to 36 ft. lbs. (50 Nm).

15. Install the shock mounting bolt at the knuckle and tighten to 40 ft. lbs. (55 Nm).

16. Install the knuckle flange bolt and tighten it to 40 ft. lbs. (55 Nm).

17. Install the wheel sensor brackets.

18. Install the wheel, and tighten the wheel nuts to 80 ft. lbs. (110 Nm).

19. Install the trunk side trim.

20. Check and adjust the rear wheel alignment.

Prelude

1. Raise and safely support vehicle.

2. Remove the trunk side trim and remove the two strut mounting nuts.

3. Remove the upper ball joint cover.

4. Remove the cotter pin and upper ball joint nut.

5. Fit a 10mm nut on the ball joint and separate the ball joint and the knuckle by using a ball joint removal tool.

6. Remove the lower strut mounting bolt and lower the suspension.

7. Remove the strut from the vehicle.

8. Place the strut in vice and install a spring compressor onto the coil spring. Follow the spring compressor manufacturer's instructions.

9. Compress the spring and remove the self-locking nut from the strut. Disassemble the strut mounts and remove the coil spring.

10. Inspect the strut mounts for wear and damage. Replace any damaged or worn parts.

To install:

NOTE: Use new self-locking nuts when installing the rear struts.

11. Install the spring compressor onto the coil spring. Set the spring

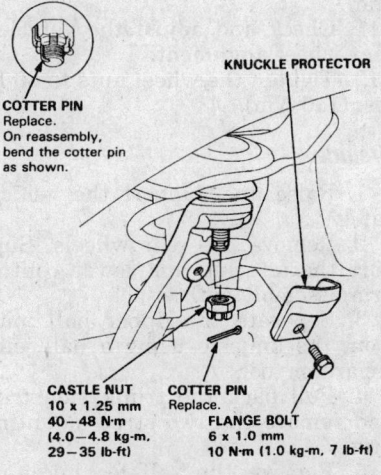

COTTER PIN
Replace.
On reassembly,
bend the cotter pin
as shown.

KNUCKLE PROTECTOR

CASTLE NUT
10 x 1.25 mm
40–48 N·m
(4.0–4.8 kg-m,
29–35 lb-ft)

COTTER PIN
Replace.

FLANGE BOLT
6 x 1.0 mm
10 N·m (1.0 kg-m, 7 lb-ft)

300814

Knuckle cover — Prelude

onto the strut cartridge. The flat part of the coil spring is its top.

12. Assemble the strut mount and its washer onto the strut. Tighten the self-locking nut to 22 ft. lbs. (29 Nm). Remove the spring compressor.

13. Install the strut to the vehicle and loosely install the lower mounting bolt. Do not tighten.

14. Install the upper strut mounting bolts. Tighten the bolts to 28 ft. lbs. (39 Nm).

15. Connect the upper arm and knuckle and tighten the castle nut to 29–35 ft. lbs. (40–48 Nm).

16. Install the upper ball joint cover.

17. Raise the rear suspension with a floor jack until the weight is on the strut.

18. Tighten the lower strut mounting bolt to 47 ft. lbs. (65 Nm).

19. Install the rear wheels and lower the vehicle.

20. Tighten the rear wheel nuts to 80 ft. lbs. (110 Nm).

21. Install the trunk trim.

22. Check and adjust the vehicle's rear wheel alignment. On Preludes equipped with 4WS, the electronic neutral check must be performed before all four wheels are aligned.

Accord

1. Fold the rear seat forward and remove the side bolster cushions. The side bolster cushions are secured by a screw at the bottom and two clips at the top.

2. Remove the upper strut mounting nuts.

3. Raise and safely support the vehicle.

4. Remove the rear wheels.

5. Support the knuckle with a floor jack.

6. Remove the strut mounting bolt, lower the jack, and remove the strut.

7. Place the strut in vice and install a spring compressor onto the coil spring. Follow the spring compressor manufacturer's instructions.

8. Compress the spring and remove the self-locking nut from the strut. Disassemble the strut mounts and remove the coil spring.

9. Inspect the strut mounts for wear and damage. Replace any damaged or worn parts.

To install:

NOTE: Use new self-locking nuts when assembling and installing the struts.

10. Install the spring compressor onto the coil spring. Set the spring

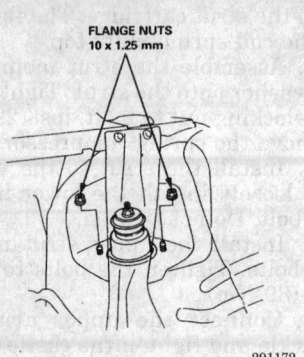

FLANGE NUTS
10 x 1.25 mm

291172

Strut mounting nuts — Accord

onto the strut cartridge. The flat part of the coil spring is its top.

11. Assemble the strut mount and its washer onto the strut. Tighten the self–locking nut to 22 ft. lbs. (29 Nm). Remove the spring compressor.

12. Install the strut into the vehicle. Hand–tighten the mounting nuts.

13. Fit the strut into position on the knuckle. Install the mounting bolt.

14. Place a jack under the lower strut mount. Raise the jack until the weight of the car is on the jack.

15. With the suspension under load, tighten the lower mount bolt to 55 Nm (40 ft. lbs). Tighten the upper nuts to 39 Nm (28 ft. lbs).

16. Install the rear wheel. Lower the vehicle to the ground.

17. Tighten the wheel nuts to 80 ft. lbs. (110 Nm).

18. Install the rear seat side bolsters and fold the seat back into place.

19. Check and adjust the vehicle's rear wheel alignment.

Upper Control Arms

REMOVAL AND INSTALLATION

Civic and del Sol

— CAUTION —

Removing rear suspension components may make the vehicle front-heavy and cause it to tip forward when raised on a hoist. Use under–lift support stands, or place additional weight in the trunk of the vehicle before hoisting it.

1. Raise and safely support the vehicle.

2. Remove the rear wheels.

3. Support the lower control arm with a floor jack.

4. Unbolt the upper control arm from the trailing arm.

5. Unbolt the upper control arm flange bar from its vehicle body mount. Remove the upper control arm.

6. Inspect the upper control and its bushings for signs of wear and distortion. The bushings are replaceable:

a. Press the bushings out of the upper control arm using suitably–sized press fixtures.

b. Matchmark the bolt flange bar to the body of the upper control arm.

c. Lubricate the new bushings with silicon grease before installation.

d. Press the new bushings into the control arm. Make sure the bolt flange bar matchmarks align. The leading edges of the control arm bushings must be flush with the edges of the control arm body.

To install:

NOTE: Use new self–locking nuts and color–coded bolts when assembling suspension components.

7. Install the control arm to its body mount. Hand–tighten the flange bolts.

8. Install the control arm to the trailing arm. Hand–tighten the flange bolt.

9. Install the rear wheel and lower the vehicle.

10. Torque the bolts with the vehicle on the ground. Tighten the control arm bolts-to-body to 29 ft. lbs. (40 Nm). Tighten the control arm-to-trailing arm bolt to 40 ft. lbs. (55 Nm).

11. Check and adjust the vehicle's rear wheel alignment.

12. Tighten the wheel nuts to 80 ft. lbs. (110 Nm).

Prelude

1. Raise and support the vehicle safely.

2. Remove the rear wheels. Support the knuckle and lower control arm assembly with a jack.

3. Separate the upper ball joint from the knuckle using a ball joint separator tool.

4. Pull back the trunk side trim and remove the two strut mounting nuts.

5. Remove the self–locking nuts from the upper arm anchor bolts. Remove the upper arm from the vehicle.

NOTE: Do not disassemble the upper arm. If the ball joint or bushings are faulty, or the upper arm is damaged, the entire upper arm must be replaced.

To install:

NOTE: Use new self–locking nuts when installing the upper arm and strut.

6. Install the upper control arm assembly into the strut tower.

7. Connect the upper ball joint.

8. Install the rear wheels and lower the vehicle.

9. With all four of the vehicle's wheels on the ground, torque the upper control arm nuts to 47 ft. lbs. (65 Nm). Torque the castle nut to 32 ft. lbs. (44 Nm); then, only tighten it only enough to install a new cotter pin.

10. Tighten the wheel nuts to 80 ft. lbs. (110 Nm).

11. Put the trunk side trim back into position.

12. Check and adjust the vehicle's rear end wheel alignment. On Preludes equipped with 4WS, the electronic neutral check must be performed before all four wheels are aligned.

Accord

1. Raise and safely support the vehicle.

2. Remove the rear wheels.

3. Support the knuckle and lower control arm with a floor jack to compress the strut.

4. Remove the castle nut cap, cotter pin and castle nut from the upper ball joint. Use a ball joint separator tool to separate the ball joint from the knuckle.

5. Unbolt and remove the upper control arm.

6. Check upper arm and its bushing for signs of wear and damage. Replace the upper arm if the ball joint is faulty.

To install:

NOTE: Use new self–locking nuts when assembling suspension components.

7. Install the upper arm into the vehicle. Install the mounting bolts and only hand–tighten them. Reconnect the upper arm to the knuckle.

8. Tighten the castle nut at the ball joint to 32 ft. lbs. (44 Nm). Tighten the castle nut only enough to install a new cotter pin. Install the castle nut cap.

9. Install the rear wheels and lower the vehicle.

10. Tighten the upper mounting bolts to 28 ft. lbs. (39 Nm).

11. Tighten the wheel nuts to 80 ft. lbs. (110 Nm).

12. Check and adjust the vehicle's rear wheel alignment.

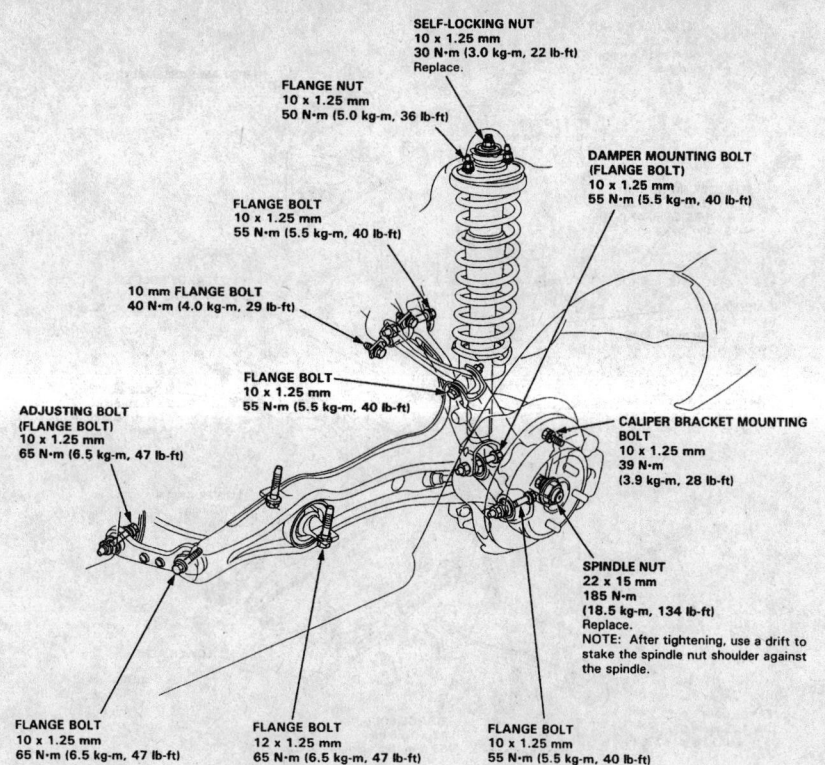

SELF-LOCKING NUT
10 x 1.25 mm
30 N·m (3.0 kg-m, 22 lb-ft)
Replace.

FLANGE NUT
10 x 1.25 mm
50 N·m (5.0 kg-m, 36 lb-ft)

DAMPER MOUNTING BOLT
(FLANGE BOLT)
10 x 1.25 mm
55 N·m (5.5 kg-m, 40 lb-ft)

FLANGE BOLT
10 x 1.25 mm
55 N·m (5.5 kg-m, 40 lb-ft)

10 mm FLANGE BOLT
40 N·m (4.0 kg-m, 29 lb-ft)

FLANGE BOLT
10 x 1.25 mm
55 N·m (5.5 kg-m, 40 lb-ft)

ADJUSTING BOLT
(FLANGE BOLT)
10 x 1.25 mm
65 N·m (6.5 kg-m, 47 lb-ft)

CALIPER BRACKET MOUNTING
BOLT
10 x 1.25 mm
39 N·m
(3.9 kg-m, 28 lb-ft)

SPINDLE NUT
22 x 15 mm
185 N·m
(18.5 kg-m, 134 lb-ft)
Replace.
NOTE: After tightening, use a drift to
stake the spindle nut shoulder against
the spindle.

FLANGE BOLT
10 x 1.25 mm
65 N·m (6.5 kg-m, 47 lb-ft)

FLANGE BOLT
12 x 1.25 mm
65 N·m (6.5 kg-m, 47 lb-ft)

FLANGE BOLT
10 x 1.25 mm
55 N·m (5.5 kg-m, 40 lb-ft)

293856

Rear suspension components — Civic and del Sol

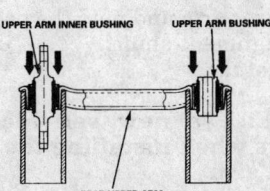

UPPER ARM INNER BUSHING   UPPER ARM BUSHING

REAR UPPER ARM

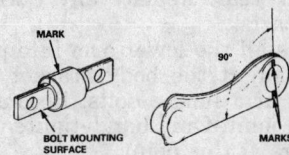

MARK

90°

BOLT MOUNTING
SURFACE

MARKS

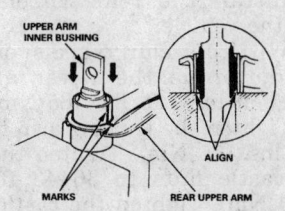

UPPER ARM
INNER BUSHING

ALIGN

MARKS   REAR UPPER ARM

293928

**Details of control arm bushing
replacement — Civic and del Sol**

Lower Control Arms

REMOVAL AND INSTALLATION

1993–95 Civic and 1993–97 del Sol

—————— **CAUTION** ——————
*Removing rear suspension compo-
nents may make the vehicle
front–heavy and cause it to tip
forward when raised on a hoist.
Use under–lift support stands, or
place additional weight in the
trunk of the vehicle before hoisting it.*

1. Raise and safely support the
vehicle.
2. Remove the rear wheels.
3. Unbolt the wheel sensor wire
brackets from the lower control arm.
Don't disconnect the sensor wire.
4. Use a floor jack to slightly compress the strut.
5. Unbolt the strut mounting bolt
and the knuckle–to–control arm
flange bolt.
6. Unbolt the control arm from its
mounting bracket. Remove the control arm from the vehicle.
To install:
**NOTE: Use new self–locking
nuts when assembling suspension components.**

7. Install the control arm into its
mounting bracket and position it on
the knuckle. Install both flange bolts.
8. Install the strut mounting bolt
and the control arm flange bolt. Install the wheel sensor wire brackets
and torque them to 7 ft. lbs. (10 Nm).
Be careful not to pinch or damaged
the wheel sensor wire when connecting the strut bracket to the control
arm.
9. Install the rear wheel and lower
the vehicle.
10. Torque the control arm flange
bolt and the strut mounting bolt to 40
ft. lbs. (55 Nm) each.
11. Torque the wheel nuts to 80 ft.
lbs. (110 Nm).
12. Check and adjust the vehicle's
rear wheel alignment.

Prelude

Standard Rear Suspension

1. Raise and safely support the
vehicle.
2. Remove the rear wheels.
3. Remove the cotter pin and
loosen the castle nut from the lower
control arm B ball joint. Separate the
ball joint from the knuckle using a
ball joint separator. Remove the castle nut.
4. Remove the lower control arm
upper mounting bolt. Note the posi-

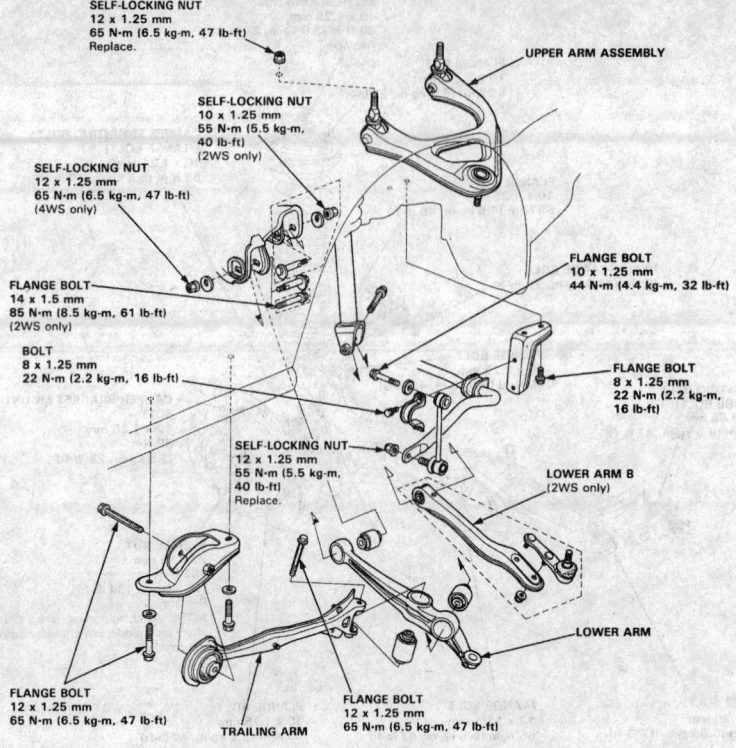

SELF-LOCKING NUT
12 x 1.25 mm
65 N·m (6.5 kg-m, 47 lb-ft)
Replace.

UPPER ARM ASSEMBLY

SELF-LOCKING NUT
10 x 1.25 mm
55 N·m (5.5 kg-m, 40 lb-ft)
(2WS only)

SELF-LOCKING NUT
12 x 1.25 mm
65 N·m (6.5 kg-m, 47 lb-ft)
(4WS only)

FLANGE BOLT
14 x 1.5 mm
85 N·m (8.5 kg-m, 61 lb-ft)
(2WS only)

FLANGE BOLT
10 x 1.25 mm
44 N·m (4.4 kg-m, 32 lb-ft)

BOLT
8 x 1.25 mm
22 N·m (2.2 kg-m, 16 lb-ft)

FLANGE BOLT
8 x 1.25 mm
22 N·m (2.2 kg-m, 16 lb-ft)

SELF-LOCKING NUT
12 x 1.25 mm
55 N·m (5.5 kg-m, 40 lb-ft)
Replace.

LOWER ARM B
(2WS only)

LOWER ARM

FLANGE BOLT
12 x 1.25 mm
65 N·m (6.5 kg-m, 47 lb-ft)

TRAILING ARM

FLANGE BOLT
12 x 1.25 mm
65 N·m (6.5 kg-m, 47 lb-ft)

300682

Rear suspension components — Prelude

tion of the alignment shims for reassembly.

5. Remove the lower control arm B.

6. Support the knuckle with a floor jack. Remove lower control arm A flange bolts at the trailing arm and the underbody bracket.

7. Remove the cotter pin and loosen the lower arm A castle nut. Use a ball joint separator to disconnect the lower arm from the knuckle. Remove lower arm A and note the position of its alignment shims.

To install:

NOTE: Use new self–locking nuts when installing the lower control arms.

8. Inspect the lower control arms and ball joints for any signs of damage and wear. Replace any worn or faulty parts.

9. Install lower arm A into position. Connect the body bracket and trailing arm flange bolts. Reconnect the ball joint. Only hand–tighten the fasteners at this point.

10. Install lower arm B into position. Connect the upper mounting bolt and the ball joint. Hand–tighten the fasteners.

11. Install the rear wheels and lower the vehicle.

12. With all four wheels on the ground, torque the lower arm B nut to 40 ft. lbs. (55 Nm). Torque the lower arm A flange bolt to 61 ft. lbs. (85 Nm).

13. Tighten the trailing arm bolt to 47 ft. lbs. (65 Nm). Tighten the ball joint castle nuts to 36–43 ft. lbs. (50–60 Nm). Tighten the castle nuts only enough to install new cotter pins.

14. Tighten the wheel nuts to 80 ft. lbs. (110 Nm).

15. Check and adjust the vehicle's rear wheel alignment.

Four–Wheel Steering

1. Raise and safely support the vehicle.

2. Remove the rear wheels.

3. Remove the cotter pin and loosen the castle nut from the rear tie rod end ball joint. Separate the ball joint from the knuckle using a ball joint separator. Remove the castle nut.

4. Support the knuckle with a floor jack. Remove lower control arm flange bolts at the trailing arm and the underbody bracket.

5. Remove the cotter pin and loosen the lower arm castle nut. Use a ball joint separator to disconnect the lower arm from the knuckle. Re-

move lower arm and note the position of its alignment shims.

To install:

NOTE: Use new self–locking washers when installing the rear control arm.

6. Inspect the lower control arms and ball joints for any signs of damage and wear. Replace any worn or faulty parts.

7. Install the lower arm into position. Connect the body bracket and trailing arm flange bolts. Reconnect the ball joint. Only hand–tighten the fasteners at this point.

8. Connect the tie rod end ball joint to the steering knuckle. Hand–tighten the fasteners.

9. Install the rear wheels and lower the vehicle.

10. With all four wheels on the ground, torque the lower arm A flange bolt to 61 ft. lbs. (85 Nm).

11. Tighten the trailing arm bolt to 47 ft. lbs. (65 Nm). Tighten the ball joint castle nuts to 36–43 ft. lbs. (50–60 Nm). Tighten the castle nuts only enough to install new cotter pins.

12. Tighten the wheel nuts to 80 ft. lbs. (110 Nm).

13. Check and adjust the vehicle's rear wheel alignment.

Accord

1. Raise and safely support the vehicle.
2. Remove the rear wheels.
3. Support the knuckle assembly with a floor jack. Remove the through–bolt securing lower control arms A and B to the knuckle.
4. Remove the body bracket flange bolts and remove both lower control arms. Note the positions of the washers and alignment shims so that they can be reinstalled correctly. Remove the control arms from the vehicle.
5. Inspect the control arms and their bushings for signs of wear and damage. Replace any damaged parts.

To install:

NOTE: Use new self–locking nuts when installing the lower control arms.

6. Install lower control arms A and B onto the vehicle. Install the body bracket flange bolts, washers, and alignment shims. Only hand–tighten the bolts and nuts at this point.
7. Install the knuckle through–bolt and washers. Hand–tighten the self–locking nut.
8. Install the rear wheels. Lower the vehicle.
9. With all four wheels sitting on the ground, torque the body bracket flange bolt to 47 ft. lbs. (64 Nm). Torque the body bracket self–locking nut to 40 ft. lbs. (54 Nm). Torque the knuckle through-bolt to 47 ft. lbs. (64 Nm).
10. Torque the wheel nuts to 80 ft. lbs. (110 Nm).
11. Check and adjust the vehicle's rear wheel alignment.

Sway Bar

REMOVAL AND INSTALLATION

Civic and del Sol

1. Raise and safely support the rear of the vehicle. Block the front wheels.
2. Remove the rear wheels.
3. Unbolt the sway bar linkages from the trailing arms.
4. Unbolt the sway bar brackets from their underbody mounts.
5. Remove the sway bar from the vehicle. Inspect the sway bar, its bushings, and linkages for wear and damage. Replace any parts that are damaged.

To install:

NOTE: Use new self–locking nuts when installing the sway bar linkages.

6. Assemble the bushings and linkages onto the sway bar.
7. Install the sway bar brackets onto their underbody mounts. Hand–tighten the mounting bolts.
8. Connect the linkages to the trailing arms. Hand–tighten the mounting nuts and bolts.
9. Install the rear wheels and lower the vehicle.
10. With all four wheels on the ground, torque the sway bar mount bolts to 16 ft. lbs. (22 Nm). Torque the linkage–to–trailing arm self–locking nuts to 9 ft. lbs. (13 Nm).
11. Tighten the wheel nuts to 80 ft. lbs. (110 Nm)

Prelude and Accord

1. Raise and support the rear of the vehicle.
2. Remove the rear wheels.
3. Unbolt the sway bar linkages from the upper part of the knuckles.
4. Unbolt the sway bar mounts from their underbody brackets.
5. Remove the sway bar from the vehicle. Inspect the sway bar, its mounts, and linkages for wear and damage. Replace any parts that are damaged.

To install:

NOTE: Use new self–locking nuts when installing the sway bar linkages.

6. Assemble the mounts and linkages onto the sway bar.
7. Install the sway bar mounts onto their underbody brackets. Hand–tighten the mounting bolts.
8. Connect the linkages to the knuckles. Hand–tighten the mounting nuts and bolts.
9. Install the rear wheels and lower the vehicle.
10. With all four wheels on the ground, torque the sway bar mount bolts to 16 ft. lbs. (22 Nm). Torque the linkage–to–knuckle flange bolt to 32 ft. lbs. (44 Nm). Tighten the linkage–to–bar nuts to 40 ft. lbs. (55 Nm).

Wheel Bearings

ADJUSTMENT

Civic and del Sol

1. Raise and support the vehicle safely.

2. Remove the front and/or rear wheels.
3. Install the lug nuts and tighten them to 80 ft. lbs. (110 Nm).
4. Use a dial gauge to measure front bearing end play at the hub flange.
5. Use a dial gauge to measure rear bearing end play at the center of the hub's grease cap.
6. Move the rotor or drum assembly in and out to measure the play. Then, compare the dial gauge readings.
7. The standard bearing end play for both front and rear wheels is 0–0.002 in. (0–0.05mm). If the end play measurement exceeds the standard, the wheel bearings must be replaced. The wheel bearings cannot be adjusted.

Prelude and Accord

The wheel bearings are not adjustable or repairable and should be replaced if found defective.

REMOVAL AND INSTALLATION

Civic and del Sol

1. Remove the hub dust cap and loosen the spindle nut.
2. Raise and safely support the vehicle. Remove the rear wheels.
3. Engage the parking brake for added leverage and remove the two brake rotor or drum retaining screws. Then, release the parking brake. If the retaining screws are stuck or stripped, drill them out, or use an extractor.
4. Remove the caliper shield. Unbolt the brake hose bracket.
5. If equipped with drum brakes, remove the brake drum.
6. If equipped with disc brakes, unbolt the caliper bracket and hang the caliper out of the way with a piece of wire.
7. Remove the brake rotor.
8. Remove the hub assembly from the spindle.
9. Clean the hub assembly in solvent.
10. Inspect the hub assembly for any signs of wear or damage. If the wheel bearings are damaged, the hub assembly must be replaced.

To install:
11. Clean the spindle and the brake rotor/drum mounting surfaces.
12. Install the hub assembly onto the spindle. Install the spindle washer.
13. Install the brake rotor or brake drum. Apply anti–seize paste to the retaining screws and tighten them to

7 ft. lbs. (10 Nm) — Don't overtighten the retaining screws.

14. Install the brake caliper and tighten the mounting bolts to 28 ft. lbs. (39 Nm). Install the brake hose bracket onto its mount. Install the caliper dust shield and tighten the bolts to 7 ft. lbs. (10 Nm).

15. Install a new spindle nut. Install the wheel and lower the vehicle.

16. Torque the spindle nut to 134 ft. lbs. (185 Nm). Tighten the wheel nuts to 80 ft. lbs. (110 Nm). Stake the spindle nut with a punch. If the dust cap was bent during removal — install a new one.

Prelude and Accord

NOTE: The rear wheel bearing and hub unit are replaced as a unit.

1. Set the parking brake then loosen the rear wheel nuts and the spindle nut.

2. Raise the vehicle and support it safely.

3. Remove the rear wheels.

4. Remove the brake disc retaining screws.

5. Release the parking brake.

6. Unbolt the brake hose brackets from the knuckle.

7. Remove the caliper bracket mounting bolts and hang the caliper out of the way with a piece of wire.

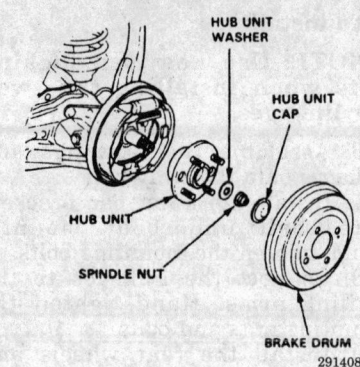

Hub unit, drum brakes — Accord

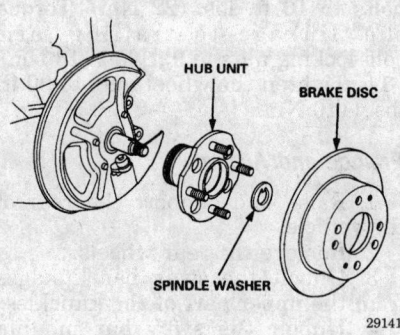

Hub unit, disc brakes — Accord and Prelude

8. Remove the brake disc. If the disc is frozen on the hub, screw two 8 x 1.25mm bolts evenly into the disc to push it away from the hub.

9. Remove the spindle nut and pull the hub unit off of the spindle.

NOTE: Clean the backing plate and the mating surfaces of the brake disc and hub with brake cleaner. Clean the spindle, washer, and hub with solvent.

To install:

10. Inspect the hub unit for signs of damage or wear. If the bearings are worn, the entire unit must be replaced.

11. Install the hub unit and spindle washer onto the spindle. Install the spindle nut but do not tighten it.

12. Install the brake disc and torque the retaining screws to 7 ft. lbs. (10 Nm).

13. Install the brake caliper and tighten the mounting bolts to 28 ft. lbs. (39 Nm). Install the brake hose brackets onto the knuckle and tighten the bolts to 16 ft. lbs. (22 Nm).

14. Install the rear wheels and lower the vehicle.

15. With the vehicle on the ground, torque the new spindle nut to 185 Nm (134 ft. lbs.), then stake the nut with a punch.

16. Tighten the wheel nuts to 80 ft. lbs. (110 Nm).

17. Test the operation of the brakes.

FIRING ORDERS

NOTE: To avoid confusion, always replace spark plug wires one at a time.

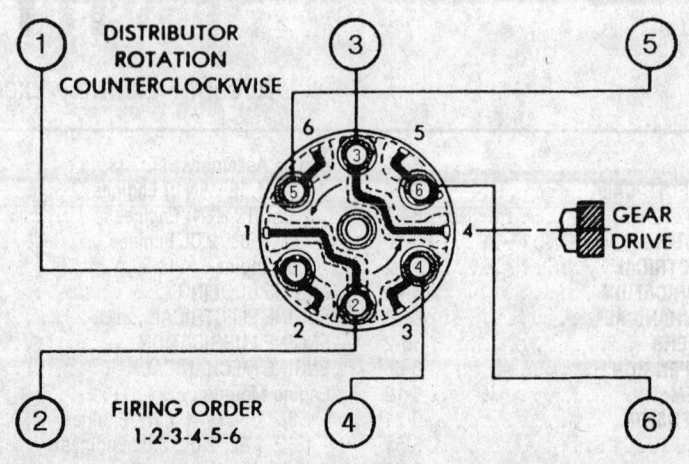

3.0L (VIN T) engine
Firing Order: 1–2–3–4–5–6
Distributor Rotation: Counterclockwise

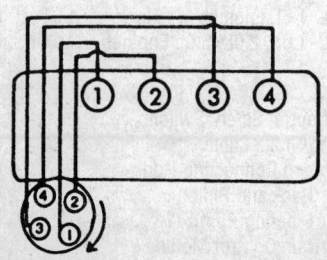

1.3L (VIN M) and 1993–94 1.5L (VINs J and N) engines
Engine Firing Order: 1–3–4–2
Distributor Rotation: Clockwise

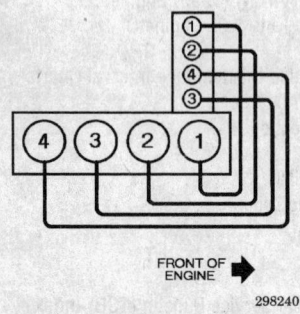

1.5L (VIN K) engine
Engine Firing Order: 1–3–4–2
Distributorless Ignition System

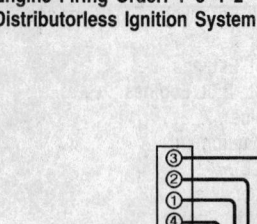

1995 1.5L (VIN N) and All 1.6L, 1.8L and 2.0L engines
Firing Order 1–3–4–2
Distributorless Ignition

ENGINE ELECTRICAL

NOTE: Disconnecting the negative battery cable on some vehicles may interfere with the functions of the on board computer systems and may require the computer to undergo a relearning process, once the negative battery cable is reconnected.

Distributor

REMOVAL AND INSTALLATION

1.3L (VIN M), 1.5L (VINs J and N), 3.0L (VIN T) Engines

1. Rotate the engine and bring up the No. 1 cylinder to top dead center of its compression stroke.
2. Disconnect the negative battery cable.
3. Disconnect and tag (if necessary) the electrical connectors along with the spark plug wires from the distributor.
4. Disconnect the wiring harness from the distributor lead wire.

5. If applicable, disconnect the vacuum hoses from the vacuum controller.
6. Using a scribe tool or a marker, scribe or mark an alignment mark on the base of the distributor and the cylinder head. This can be used for an easier installation procedure.
7. Remove the distributor cap and mark the direction that the rotor tip is pointing to.
8. Remove the distributor mounting nut and remove the distributor assembly from the engine cylinder head.
To install:
9. Install the distributor assembly into the cylinder block, being sure to align the scribe marks made previously. Lightly tighten the distributor mounting bolt.
10. Connect the wiring harness to the distributor lead wire. If applicable, connect the vacuum hoses to the vacuum controller.
11. Connect all the electrical connectors along with the spark plug wires to the distributor.
12. Connect the negative battery cable. Start the engine and set the timing. Then tighten the distributor mounting bolt.

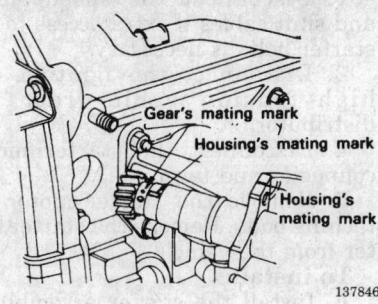

Distributor housing mating marks — 1.5L (VINs J and N) engines, others similar

Ignition Timing

ADJUSTMENT

1.3L (VIN M) Engine

1. Place the vehicle in **P** or **N**, the emergency brake applied and the drive wheels blocked. Start the engine and let it reach normal operating temperature.
2. Connect a suitable tachometer and timing light to the engine. Turn **OFF** all accessories. Check the curb idle and adjust as necessary. The curb idle should be as follows:
   a. Manual Transmissions — 600–800 rpm
   b. Automatic Transmission — 650–850 rpm
3. If the timing marks are difficult to see, use chalk or a dab of paint to make them more visible.
4. If equipped with vacuum advance, disconnect the vacuum line from the distributor vacuum advance and plug the line.
5. Aim the timing light onto the timing marks and check the engine timing. If the timing is out of specifications, loosen the distributor mounting bolt and advance or retard the timing as necessary.
6. Once the proper timing has been reached, tighten the distributor mounting bolt and then recheck the timing again. The correct timing specifications are as follows:
   a. California — 1–5 degrees BTDC
   b. Federal — 3–7 degrees BTDC
   c. Canada — 2–6 degrees BTDC
7. When the timing has been set properly, recheck the curb idle speed and adjust as necessary.
8. When all final adjustments have been made, shut **OFF** the engine, reconnect all vacuum lines and remove all test equipment.

All Except 1.3L (VIN M) Engine

This engine's ignition timing is controlled by the Powertrain Control Module and cannot be adjusted. If the ignition timing is faulty or out of adjustment, the engine management system should be investigated for possible faults.

Alternator

PRECAUTIONS

Several precautions must be observed with alternator equipped vehicles to avoid damage to the unit.
- If the battery is removed for any reason, make sure it is reconnected with the correct polarity. Reversing the battery connections may result in damage to the 1-way rectifiers.
- When utilizing a booster battery as a starting aid, always connect the positive to positive terminals and the negative terminal from the booster battery to a good engine ground on the vehicle being started.
- Never use a fast charger as a booster to start vehicles.
- Disconnect the battery cables when charging the battery with a fast charger.
- Never attempt to polarize the alternator.
- Do not use test lights of more than 12 volts when checking diode continuity.
- Do not short across or ground any of the alternator terminals.
- The polarity of the battery, alternator and regulator must be matched and considered before making any electrical connections within the system.
- Never separate the alternator on an open circuit. Make sure all connections within the circuit are clean and tight.
- Disconnect the battery ground terminal when performing any service on electrical components.
- Disconnect the battery if arc welding is to be done on the vehicle.

REMOVAL AND INSTALLATION

All Engines

1. Disconnect the negative battery cable.
2. For 1.3L (VIN M), 1.5L (VIN J), 2.0L (VIN F) and 1993–94 1.6L (VIN R) and 1.8L (VIN M) engines, if equipped with power steering, remove the oil pump and position the pump above the bracket without disconnecting the fluid lines.
3. For 3.0L (VIN T) engines, remove the distributor cap and power steering pressure hose nut.
4. Remove or disconnect any component that interferes with access to the alternator mounting hardware.
5. Loosen the alternator mounting bolts to relieve the belt tension and remove the drive belt from the alternator pulley.
6. Raise the front of the vehicle and support safely.
7. Remove any mud guards or splash shield that obstruct the alternator.
8. Disconnect the B+ terminal wire from the alternator.
9. Support the alternator by hand, finish removing the mounting bolts and lift the unit up and out of the engine compartment.
   To install:
10. Position the alternator onto its mounting and install the mounting bolts.
11. Connect the B+ wire to the alternator terminal. Make sure the wire nut is tight and the protective cap is firmly placed over the wire connection.
12. Install the left side mud guard.
13. Lower the vehicle.
14. Install the steering pump, if removed.
15. Install and adjust the belt to specifications. Torque the adjusting bolt to 14–18 ft. lbs. (20–25 Nm). Torque the support nut to 11–16 ft. lbs. (15–22 Nm)
16. Connect the negative battery cable.

Drive Belt

REMOVAL AND INSTALLATION

All Except 3.0L (VIN T) Engine

Alternator Belt

1. Loosen the alternator support nut.
2. Loosen the adjuster lock bolt.
3. Rotate the adjuster bolt counter clockwise to release the tension on the belt.
4. Remove the belt.
   To install:
5. Install the belt on the pulleys.
6. Rotate the adjuster bolt clockwise until the proper tension is reached.
7. Tighten the adjuster lock bolt and the alternator support nut.

A/C Compressor Belt

1. Loosen the tension pulley and remove the belt.

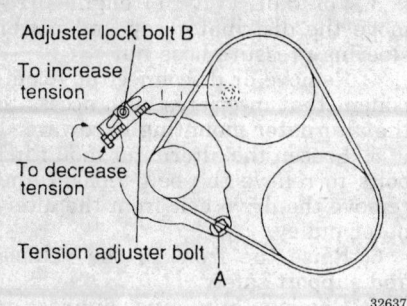

Alternator drive belt — 1995–97 2.0L (VIN F) engines

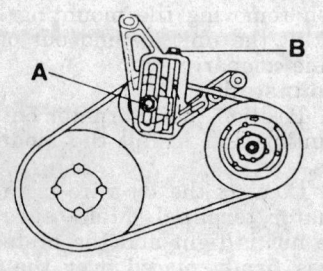

A/C compressor belt — 1995–97 2.0L (VIN F) engines

2. Reverse the procedure to install the belt.

3.0L (VIN T) Engine

Alternator and Power Steering Belt

1. Disconnect the negative battery cable.
2. Loosen the tension pulley locking nut.
3. Rotate the adjuster bolt clockwise to release the tension on the belt.
4. Remove the belt.
To install:
5. Install the belt on the pulleys.
6. Rotate the adjuster bolt counter-clockwise until the proper tension is reached.
7. Tighten the alternator tension pulley nut to 28–43 ft. lbs. (39–60 Nm).
8. Connect the negative battery cable.

A/C Compressor Belt

1. Disconnect the negative battery cable.
2. Loosen the two tension pulley adjustment bolts.
3. Remove the belt.
To install:
4. Install the belt.
5. Adjust the belt.

6. Tighten the adjustment bolts.
7. Connect the negative battery cable.

Starter

REMOVAL AND INSTALLATION

All Except 1.5L (VINs K and N) Engines

1. Disconnect the negative battery cable.
2. Remove the battery and tray to gain additional clearance.
3. Raise and support the vehicle safely.
4. Remove the engine splash shield.
5. Disconnect the electrical harness from the starter solenoid, noting position of wires for correct installation.
6. Remove the starter mounting bolts and the starter from the vehicle.
7. Clean the surfaces of the starter motor flange and the flywheel housing where the starter attaches.
To install:
8. Install the starter motor and secure with the retainer bolts. Tighten the bolts to 20–25 ft. lbs. (27–34 Nm).
9. Connect the electrical harness to the starter solenoid. Tighten the mounting nut to 7.3–11.7 ft. lbs. (10–16 Nm).
10. Install the engine splash shield.
11. Install the battery and tray, if removed.
12. Reconnect the negative battery cable.

1.5L (VIN K) Engine

1. Disconnect the negative battery cable.
2. Disconnect the electrical harness from the starter solenoid, noting position of wires for correct installation.
3. Remove the starter mounting bolts and the starter from the vehicle.
4. Clean the surfaces of the starter motor flange and the flywheel housing where the starter attaches.
To install:
5. Install the starter motor and secure with the mounting bolts. Torque the bolts to 23 ft. lbs. (30 Nm).
6. Connect the electrical harness to the starter solenoid.
7. Reconnect the negative battery cable.

1.5L (VIN N) Engine

1. Disconnect the negative battery cable.

2. Disconnect the speedometer and shift cables if extra access to the starter bolts is necessary.
3. Disconnect the ignition coil high–tension cable from the distributor.
4. Disconnect the starter motor connector and terminal.
5. Remove the starter motor attaching bolts. Remove the starter motor from the vehicle.
To install:
6. Install the starter assembly to the engine and secure using the attaching bolts, tightening to 20 to 25 ft. lbs. (26 to 33 Nm).
7. Connect the starter motor electrical connector and terminal.
8. Install the speedometer cable and shifter cable.
9. Connect the ignition coil high–tension cable to the distributor.
10. Connect the negative battery cable and check the starting system for proper operation.

CHASSIS ELECTRICAL

Blower Motor

REMOVAL AND INSTALLATION

1993–94 Excel, Scoupe and Elantra

1. Disconnect the negative battery cable.
2. Remove the glove box housing cover.
3. Remove the lower crash pad, crash pad center fascia and lower crash pad center skin.
4. Disconnect the blower motor and blower resistor connectors.
5. Disconnect the blower motor cooling tube.
6. Remove the screws that attach the blower motor to the lower case and lower the blower motor far enough so that the FRESH/RECIRC vacuum connector can be disconnected. Remove the blower motor from the car.
To install:
7. Inspect the blower wheel for damage and replace it as necessary. Replace the blower motor mounting seal.
8. Raise the blower motor and seal up onto the lower case half and connect the FRESH/RECIRC vacuum connector. Mount the blower onto the

lower case half and install the three retaining screws.

9. Connect the blower motor cooling tube.

10. Connect the blower motor and resistor wire connectors.

11. Install the lower crash pad center skin, crash pad center fascia and lower crash pad.

12. Install the glove box housing cover.

13. Connect the negative battery cable.

14. Check the blower for proper operation at all speeds.

Accent

1. Remove the instrument panel undercover on the passenger side of the vehicle.

2. Remove the blower motor cooling tube.

3. Disconnect the blower motor harness connector.

4. Remove the three mounting screws and remove the blower motor from the housing.

To install:

5. Install the blower motor in the housing.

6. Connect the harness connector.

7. Install the cooling tube.

8. Install the instrument panel undercover.

1995 Scoupe

1. Disconnect the negative battery cable.

2. Remove the glove box housing cover.

3. Remove the radio and storage pocket trim assembly.

4. Remove the passenger's side dashboard lower trim panel.

5. Disconnect the blower motor and blower resistor connectors.

6. Disconnect the blower motor cooling tube.

7. Remove the three screws that secure the blower motor to the blower case. Remove the blower motor.

To install:

8. Inspect the blower wheel for damage and replace it as necessary. Replace the blower motor mounting seal.

9. Mount the blower onto the case and install the three retaining screws.

10. Connect the blower motor cooling tube.

11. Connect the blower motor and resistor wire connectors.

12. Install the passenger's side dashboard lower trim panel.

13. Install the radio and storage pocket trim assembly.

14. Install the glove box housing cover.

15. Connect the negative battery cable.

16. Check the blower for proper operation at all speeds.

Sonata

——————— **CAUTION** ———————
The air bag system (SRS) must be disarmed before removing the heater assembly. Failure to do so may cause accidental deployment, property damage or personal injury.

NOTE: The A/C lines use spring-lock couplings that require the uses of special coupling tool 09977-33600 (A/B) to disconnect and connect.

1. Disconnect the negative battery cable and drain the coolant from the radiator.

2. Discharge the refrigerant into an approved container and disconnect the A/C lines from the evaporator. Cap the open lines tightly with plastic to prevent contamination.

3. From inside the engine compartment, disconnect the heater hoses and evaporator drain hose from the heater core tubes. Disconnect the drain hose. These three hoses are located just above the fuel filter.

4. Remove the front and rear console assemblies.

5. Remove the left and right side covers by removing the two retaining screws on each side.

6. Remove the glove box, center crash pad cover, center crash pad and cassette assembly.

7. Remove the lower crash pad.

8. Remove the console and center support mounting brackets.

9. Disconnect the rear passenger heating duct assembly and rear heating duct Y-joint from the heater case.

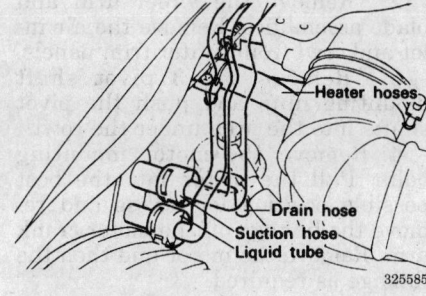

Heater and A/C connections — Sonata models

10. Remove the heater control assembly from the dash by removing the four retaining screws.

11. Disconnect the blower motor wiring connector.

12. Remove the four retaining screws and pull the heater case down and out of the dash.

13. Separate the case halves and remove the blower motor, blower wheel and gasket from the lower case half. Discard the gasket. Inspect the blower wheel for damage and replace it as necessary.

To install:

14. Install the blower assembly into the lower heater case half with a new gasket.

15. Assemble the upper and lower case halves together using new gaskets and sealant as required.

16. Position the heater case under the dash and secure it with the four retaining screws.

17. Connect the blower motor wiring.

18. Install the heater control assembly into the dash and secure it with the four retaining screws.

19. Connect the rear heating Y-joint and left and right rear passenger heating ducts to the heater case openings.

20. Install the center and console support mounting brackets.

21. Install the lower crash pad.

22. Install the cassette assembly, center crash pad and cover and the glove box.

23. Install the left and right side covers.

24. Install the front and rear console assembly.

25. Connect the drain hose. Connect the hoses to the heater core tubes.

26. Recharge the A/C system using the proper equipment.

27. Fill the cooling system to the proper level with a 50mix of antifreeze and connect the negative battery cable. Start the engine and check for coolant leaks.

Windshield Wiper Motor

REMOVAL AND INSTALLATION

All Except Sonata

Front

1. Disconnect the negative battery cable.

2. Remove the wiper arm and blade assemblies. Take care not to scratch the paint on the cowl.

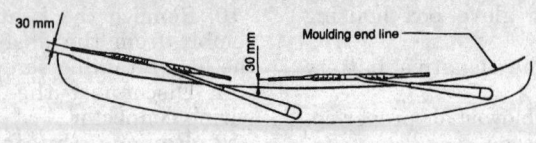

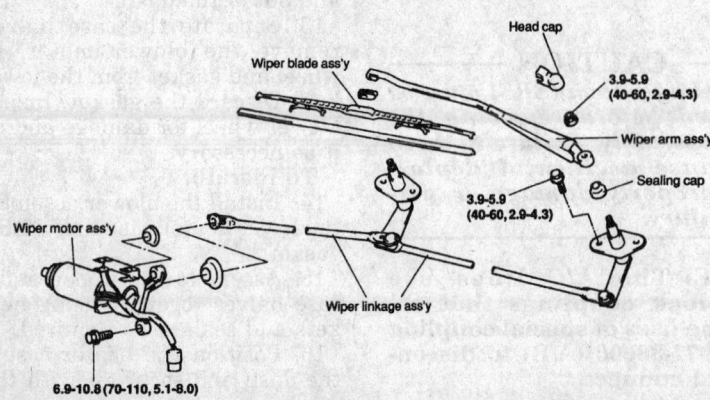

TORQUE : Nm (kg.cm, lb.ft)

136639

Front windshield wiper assembly — 1993–94 Excel, Scoupe, Accent and Elantra models

3. Remove the air inlet and cowl front center trim panels. Remove the 3 pivot shaft mounting nuts and push the pivot shafts into the area under the cowl.

4. Remove the motor mounting bolts. Pull the motor into the best possible position for access and remove the linkage off the motor crank arm. Remove the motor and then the linkage as required.

To install:

5. Connect the wiper motor to the link securely then install the motor attaching bolts.

6. Install the cowl top cover.

7. Install the cowl top sealing caps.

8. Install the wiper arm and blade assemblies to the pivot shafts. Torque the pivot shaft nuts to 2.9 to 4.3 ft. lbs. (4 to 6 Nm).

9. Position the wiper arms so the blades are about 30mm above the lower windshield molding.

10. Connect the negative battery cable.

Rear

1. Remove the wiper blade and arm by lifting the wiper blade locknut cover and removing the locknut. Then, pull the arm from the shaft.

2. Remove the lift gate trim panel and disconnect the wiring harness connector.

3. Remove the inside and outside motor mounting nuts and remove the motor.

To install:

4. Install the motor assembly to the tailgate.

5. Connect the electrical connector to the wiper motor.

6. Install the tail gate trim.

7. Install the wiper arm.

Sonata

1993–94 Models

1. Disconnect the negative battery cable.

2. Remove the wiper arm and blade assemblies. Remove the air inlet and cowl front center trim panels.

3. Remove the 3 pivot shaft mounting nuts and push the pivot shafts into the area under the cowl.

4. Remove the motor mounting bolts. Pull the motor into the best possible position for access and remove the linkage off the motor crank arm. Remove the motor and then the linkage as required.

5. If the motor is being replaced, matchmark the position of the crank arm of the motor shaft of the new

motor and then remove the nut and crank arm, transferring both to the new motor.

To install:

6. Torque the pivot shaft nuts to 4.3 to 5.8 ft. lbs. (5 to 8 Nm).

7. Position the wiper arms so each blade tip is about 30mm above the molding. Torque the nut to 12 to 15 ft. lbs. (17 to 21 Nm).

8. Make sure the wiper motor is securely grounded.

9. Connect the negative battery cable.

1995–97 Models

1. Disconnect the negative battery cable.

2. Remove the wiper arm and blade assemblies.

3. Remove the cowl top sealing cap and cover.

4. Disconnect the harness connector from the wiper motor.

5. Remove the wiper motor mounting bolts. Pull the motor into the best possible position for access and remove the linkage from the motor crank arm. Remove the motor and then the linkage as required.

6. If the motor is being replaced, transfer the crank arm to the new motor. Make sure it is in the proper position.

To install:

7. Install the wiper motor, tighten the mounting bolts to 5–8 ft. lbs. (7–11 Nm).

8. Connect the harness connector to the wiper motor.

9. Install the cowl top sealing cap. Install the cowl top cover.

10. Position the wiper arms so each blade tip is about 30mm above the molding. Torque the nut to 12–15 ft. lbs. (17–21 Nm).

11. Make sure the wiper motor is securely grounded.

12. Connect the negative battery cable.

Combination Switch

REMOVAL AND INSTALLATION

Excel, Scoupe and 1993–94 Sonata

NOTE: The headlights, turn signals, dimmer switch, horn switch, windshield wiper/washer, intermittent wiper switch and the cruise control function are all built into 1 multi-function combination switch that is mounted on the steering column.

1. Disconnect the negative battery cable.

2. Matchmark the steering shaft and remove the steering wheel.

3. Remove the steering column covers.

4. Unplug the electrical connectors and remove the electrical harness retainers.

5. Remove the retaining screws and slide the switch off the steering column.

To install:

6. Position the switch on the steering column and install the 4 mounting screws.

7. Connect the electrical connectors and install the harness retainers.

8. Install the steering column covers.

9. Align the matchmark and install the steering wheel in the original position.

10. Connect the negative battery cable.

1995–97 Sonata

NOTE: The headlights, turn signals, dimmer switch, horn switch, windshield wiper/washer, intermittent wiper switch and the cruise control function are all built into one multi-function combination switch that is mounted on the steering column.

───── CAUTION ─────

The air bag system must be disarmed before removing the steering wheel. Failure to do so may cause accidental deployment, property damage or personal injury.

1. Make sure the front wheels are in the straight ahead position and turn the ignition to the lock position.

2. Disconnect the negative battery cable and wait at least 30 seconds before doing any work.

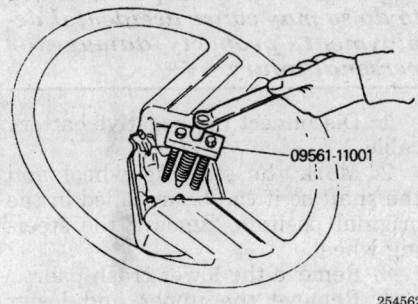

09561-11001

254562

Steering wheel removal — Excel, Scoupe and 1993–94 Sonata models

3. Remove the air bag module mounting nut, disengage the electrical connector and remove the air bag. To disconnect the clock spring connector, press the air bag's lock toward the outer side to spread it open. Gently pry the connector from the module.

4. Matchmark the steering wheel to the shaft and remove the steering wheel.

5. Remove the steering column covers.

6. Unplug the electrical connectors, and remove the electrical harness retainers.

7. Remove the retaining screws and slide the switch off the steering column.

To install:

8. Align the mating mark and neutral position indicator on the clock spring.

9. Install the column switch to the clock spring. Install the three retaining screws.

10. Connect the switch electrical connectors.

11. Install the steering column covers.

12. Align the matchmark and install the steering wheel.

13. Engage the air bag module connector. Install the air bag module and tighten the mounting nut.

14. Connect the negative battery cable.

Accent

───── CAUTION ─────

The air bag system (SRS) must be disarmed before removing the horn pad. Failure to do so may cause accidental deployment, property damage or personal injury.

1. Turn the front wheels to the straight ahead position.

2. Turn the ignition switch to the **LOCK** position and disconnect the negative battery cable. Wait at least 30 seconds before starting work.

3. Remove the horn pad or air bag module. Spread the connector locking tabs outward and gently pry the connector up.

4. Remove the steering wheel using the proper puller. Do not hammer on the steering wheel or related components to remove it.

5. Remove the steering column covers.

6. If equipped with an air bag, remove the clock spring from the combination switch.

7. Disconnect the harness connectors for the combination switch.

8. Remove the three straps.

9. Remove the three mounting screws and the combination switch.

To install:

10. Position the switch on the steering column and install the four mounting screws.

11. Connect the harness connectors and install the three harness straps.

12. Align the mating mark and the neutral position indicator on the clock spring. With the front wheels in the straight ahead position install the clock spring on the combination switch.

13. Install the steering column covers.

14. Install the steering wheel.

15. Install the horn pad/air bag module.

16. Connect the negative battery cable. Turn the ignition switch to the on position and check that the SRS warning lamp in the instrument panel turns on for about six seconds and then goes out.

1993–95 Elantra

NOTE: The headlights, turn signals, dimmer switch, horn switch, windshield wiper/washer, intermittent wiper switch and the cruise control function are all built into 1 multi-function combination switch that is mounted on the steering column.

Without Air Bag

1. Disconnect the negative battery cable.

2. Remove the horn cover.

3. Remove the steering wheel and the steering column covers.

4. Remove the lower crash pad and the rheostat connector.

5. Unplug the electrical connectors and remove the electrical harness retainers.

6. Remove the retaining screws and slide the switch off the steering column.

To install:

7. Slide the switch on the steering shaft and install the 3 retaining screws.

8. Connect the switch electrical connectors.

9. Install the steering column covers and the steering wheel.

10. Install the horn cover.

11. Install the rheostat connector and the lower crash pad.

12. Connect the negative battery cable.

With Air Bag

———— **CAUTION** ————

The air bag system must be disarmed before removing the steering wheel. Failure to do so may cause accidental deployment, property damage or personal injury.

1. Disconnect the negative battery cable.
2. Remove the air bag module mounting nut, and remove the air bag.
3. To disconnect the clock spring connector from the air bag module, press the air bags lock toward the outer side to spread it open. Gently pry the connector from the module.
4. Remove the steering wheel and the steering column covers.
5. Remove the lower crash pad.
6. Unplug the electrical connectors and remove the electrical harness retainers.
7. Remove the retaining screws and slide the switch off the steering column.
To install:
8. Align the mating mark and **NEUTRAL** position indicator on the clock spring.
9. After turning the front wheels to the straight-ahead position, install the column switch to the clock spring. Install the three retaining screws.
10. Connect the switch electrical connectors.
11. Install the lower crash pad.
12. Install the steering column covers and the steering wheel.
13. Install the air bag module and mounting nut.
14. Connect the negative battery cable.

1996–97 Elantra

NOTE: The headlights, turn signals, dimmer switch, horn switch, windshield wiper/washer, intermittent wiper switch and the cruise control function are all built into one multi-function combination switch that is mounted on the steering column.

Without Air Bag

1. Disconnect the negative battery cable.
2. Remove the horn pad.
3. Remove the steering wheel and the steering column covers.
4. Unplug the electrical connectors and remove the electrical harness retainers.

5. Remove the retaining screws and slide the switch off the steering column.
To install:
6. Slide the switch on the steering shaft and install the three retaining screws.
7. Connect the switch electrical connectors.
8. Install the steering column covers and the steering wheel.
9. Install the horn pad.
10. Connect the negative battery cable.

With Air Bag

———— **CAUTION** ————

The air bag system must be disarmed before removing the steering wheel. Failure to do so may cause accidental deployment, property damage and personal injury.

1. Turn the front wheels to the straight ahead position.
2. Disconnect the negative battery cable. Wait at least 30 seconds before doing any work to allow the air bag to disarm.
3. Remove the four air bag module mounting nuts, and remove the air bag.
4. To disconnect the clock spring connector from the air bag module, press the air bags lock toward the outer side to spread it open. Gently pry the connector from the module.
5. Remove the steering wheel and the steering column covers.
6. Unplug the electrical connectors and remove the electrical harness retainers.
7. Remove the retaining screws and slide the switch off the steering column.
To install:
8. Align the mating mark and **NEUTRAL** position indicator on the clock spring.
9. Install the combination switch and the three retaining screws.
10. Connect the switch electrical connectors.
11. Install the steering column covers and the steering wheel.
12. Connect the air bag connector and install the air bag module and mounting nuts.
13. Connect the negative battery cable.

Ignition Lock Cylinder

REMOVAL AND INSTALLATION

———— **CAUTION** ————

The air bag system (SRS) must be disarmed before removing the steering column covers. Failure to do so may cause accidental deployment, property damage or personal injury.

1. Disconnect the negative battery cable.
2. Carefully pry off the upper steering column cover.
3. Remove the screws attaching the lower steering column cover to the column and remove the lower cover.
4. Insert the key in the ignition lock and turn the key to the **ACC** position. Use a small punch or equivalent to depress the small button on the side of the cylinder housing and withdraw the lock cylinder.
To install:
5. Install the lock cylinder with key into the cylinder housing. Make sure the key is in the **ACC** position. The small button on the lock cylinder should engage the hole in the cylinder housing.
6. Turn the key to the **LOCK** position and remove the key.
7. Install the steering column covers.
8. Connect the negative battery cable and check the operation of the switch.

Ignition Switch

REMOVAL AND INSTALLATION

All 1993–94 Models and 1995 Scoupe

———— **CAUTION** ————

The air bag system (SRS) must be disarmed before removing the air bag module or horn pad. Failure to do so may cause accidental deployment, property damage or personal injury.

1. Disconnect the negative battery cable.
2. Mark the steering wheel and the shaft so it can be installed in the original position. Remove the steering wheel.
3. Remove the lower crash pad.
4. Remove the upper and lower steering column covers.
5. Disconnect the ignition switch electrical connector.

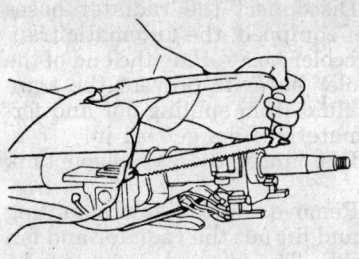

255468

Removing the ignition switch — 1993–94 models and 1995 Scoupe

6. Remove the combination switch mounting screws and remove the combination switch.

7. Using a hacksaw, cut a slot in the head of the screw and back it out using a flat blade screwdriver.

8. Remove the ignition switch assembly.

To install:

9. Align the steering lock with the column boss and insert the key to confirm the steering lock operation before tightening the switch assembly.

NOTE: Special screws should be used to secure the ignition switch assembly.

10. Install the combination switch.
11. Connect the electrical connectors.
12. Install the steering column covers.
13. Install the lower crash pad.
14. Install the steering wheel in the original position.
15. Connect the negative battery cable.

1995 Accent, Elantra and Sonata

——— CAUTION ———
The air bag system (SRS) must be disarmed before removing the air bag module or horn pad. Failure to do so may cause accidental deployment, property damage or personal injury.

1. Disarm the air bag assembly.
2. Disconnect the negative battery cable.
3. Remove the air bag.
4. Mark the steering wheel and the shaft so it can be installed in the original position. Remove the steering wheel.
5. Remove the lower crash pad.
6. Remove the upper and lower steering column covers.

7. Disconnect the ignition switch electrical connector.

8. Remove the combination switch mounting screws and remove the combination switch.

9. Using a hacksaw, cut a slot in the head of the screw and back it out using a flat blade screwdriver.

10. Remove the ignition switch assembly.

To install:

11. Align the steering lock with the column boss and insert the key to confirm the steering lock operation before tightening the switch assembly.

NOTE: Special screws should be used to secure the ignition switch assembly.

12. Install the combination switch.
13. Connect the electrical connectors.
14. Install the steering column covers.
15. Install the lower crash pad.
16. Install the steering wheel in the original position.
17. Install the air bag.
18. Connect the negative battery cable.

All 1996–97 Models

——— CAUTION ———
The air bag system (SRS) must be disarmed before removing the steering column covers. Failure to do so may cause accidental deployment, property damage or personal injury.

1. Disconnect the negative battery cable.
2. Carefully pry off the upper steering column cover.
3. Remove the screws attaching the lower steering column cover to the column and remove the lower cover.
4. Remove the ignition switch from the back of the lock cylinder and disconnect the harness connector.

To install:

5. Align the slot in the switch with the blade on the back of the lock cylinder and install the switch on the cylinder.
6. Connect the harness connector to the switch.
7. Install the steering column covers.
8. Connect the negative battery cable and check the operation of the switch.

Park/Neutral Safety Switch

REMOVAL AND INSTALLATION

All Except Accent

1. Disconnect the negative battery cable.
2. Apply the parking brake and place the selector lever in the neutral position.
3. Disconnect the selector control lever from the park/neutral switch.
4. Disconnect the connector from the park/neutral switch.
5. Remove the 2 mounting screws from the switch and remove the switch.

To install:

6. Position the switch on the transaxle and install the two mounting screws finger–tight.
7. Reconnect the selector control lever, making sure the transaxle is still in neutral.
8. Turn the switch body so the flat on the control lever aligns with the flange on the switch body.
9. Connect the electrical connector and the negative battery cable.
10. Check the operation of the switch.

Accent

1. Apply the parking brake and place the selector lever in the neutral position.
2. Remove the selector cable from the manual control lever.
3. Disconnect the connector from the park/neutral switch.
4. Remove the nut from the manual shaft.
5. Remove the mounting screws from the switch and remove the switch.

To install:

6. With the transaxle still in neutral, position the switch on the transaxle and install the mounting screws finger tight.
7. Turn the switch body so the flat on the control lever aligns with the flange on the switch body. Torque the mounting bolts to 7–9 ft. lbs. (10–12 Nm).
8. Install the washer and nut on the manual shaft.
9. Connect the selector cable to the manual control lever.
10. Connect the electrical connector and the negative battery cable.

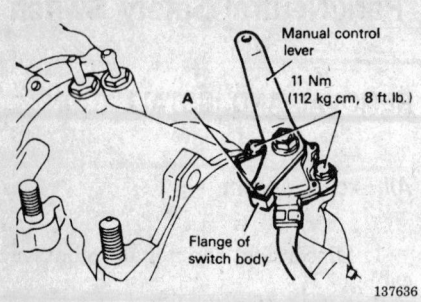

Park/Neutral switch mounting and adjustment — All except Accent

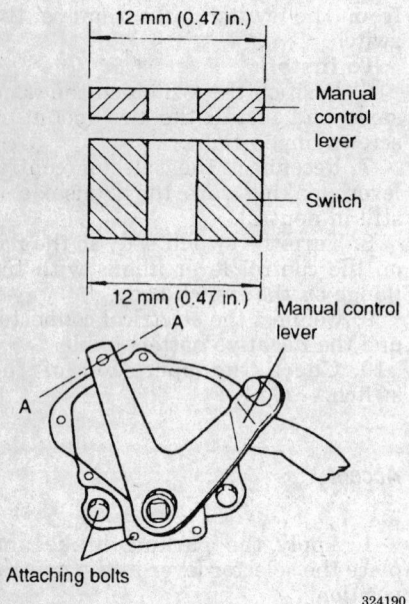

Park/Neutral switch mounting and adjustment — Accent

Powertrain Control Module

REMOVAL AND INSTALLATION

All Models

1. Disconnect the negative battery cable.
2. Remove the any trim panel that interferes.
3. Disconnect the electrical connectors at the control module.
4. Remove the control module from the vehicle.

To install:
5. Install the control module into the mounting.
6. Connect the electrical connectors to the control module.
7. Install the lower trim panel.
8. Connect the negative battery cable.

ENGINE COOLING

Radiator

REMOVAL AND INSTALLATION

All Models

1. Disconnect the negative battery cable.
2. Set the warm water flow control lever of the heater control to the **HOT** position.
3. Drain the radiator. Raise the vehicle and support it safely. Remove the splash shield from under the vehicle.
4. Remove the fan shroud and disconnect the fan motor wiring harness.

5. Disconnect the radiator hoses and, if equipped, the automatic transaxle cooler hoses. Plug the end of the oil cooler hoses to prevent the transaxle fluid from spilling out and foreign material from getting in.
6. Disconnect the expansion tank hose.
7. Remove the radiator mounting bolts and lift out the radiator and fan assembly. The fan and motor may be left attached to the radiator and removed with the radiator as one unit.

To install:
8. Install the radiator. Tighten the retaining bolts gradually in a criss-cross pattern.
9. Connect the expansion tank hose, the radiator hoses and the automatic transaxle oil cooler lines.
10. Install the fan shroud and connect the fan wiring.
11. Install the splash shield and refill the engine with coolant.
12. Connect the negative battery cable, run the vehicle until the thermostat opens, fill the radiator completely and check the automatic transaxle fluid level, if equipped.
13. Once the vehicle has cooled, recheck the coolant level.

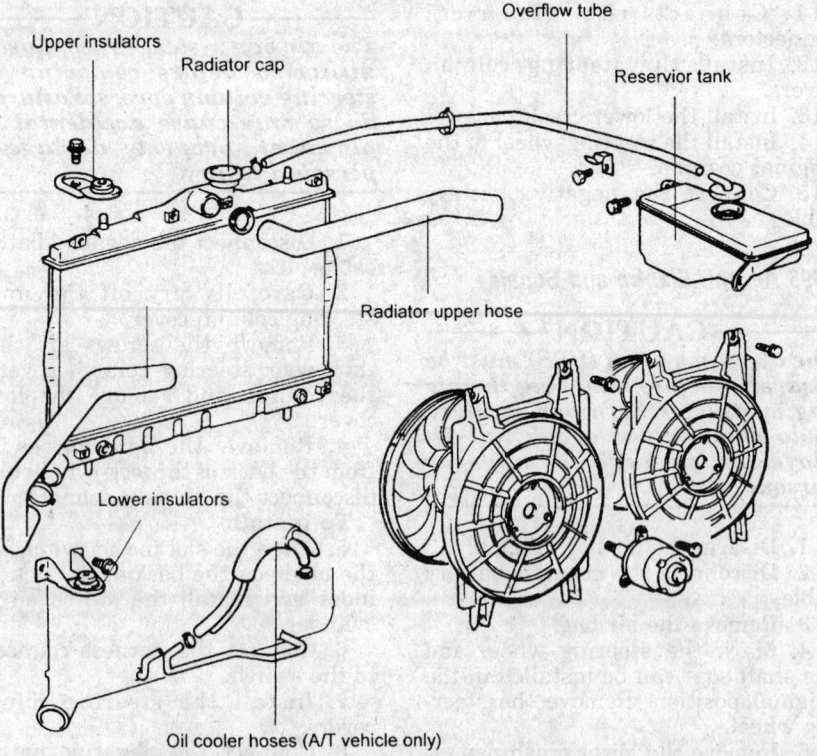

Common radiator and cooling fan assembly

Water Pump

REMOVAL AND INSTALLATION

Except 3.0L (VIN T) Engine

1. Disconnect the negative battery cable.

2. Remove the radiator cap, open the draincock at the bottom of the radiator and drain the coolant from the radiator into a clean container.

3. Loosen the four bolts attaching the water pump pulley to the pulley flange. Loosen the alternator mounting bolts and remove the alternator belt.

4. Remove the water pump pulley and any other drive belts as necessary.

5. Rotate the crankshaft clockwise and align the timing marks so that the No. 1 piston will be at TDC of the compression stroke. Remove the timing belt covers.

6. Remove the timing belt and tensioner. Mark the timing belt's direction of rotation if it is to be reinstalled. Do not disturb the engine after the timing marks have been aligned.

7. Remove the water pump mounting bolts, noting the three different lengths and locations. Remove the pump and gasket, disconnecting the water pipe from the coolant outlet pipe.

To install:

8. Clean the gasket surfaces and coat a new gasket with sealant. Then, position the gasket on the front of the block with all bolt holes aligned. Install a new O-ring in the coolant outlet pipe port.

9. Install the pump and connect the coolant outlet pipe. Install the bolts with the shortest at the bottom; two just slightly longer at the 1 and 4 o'clock positions on the right side of the pump; next-to-longest bolt at the 8 o'clock position, just under the outlet; and the longest bolt at the 11

o'clock position and also attaching the alternator brace. Torque the bolts with a head mark 4, to 9–11 ft. lbs. (12–15 Nm); those with a head mark 7, to 14–20 ft. lbs. (20–26 Nm).

10. Install the timing belt and tensioner.

11. Install the upper and lower timing covers.

12. Install the water pump pulley.

13. Install and adjust the engine accessory drive belts.

14. Tighten the water pump pulley bolts to 6–7 ft. lbs. (8–10 Nm) after the drive belts have been installed and tensioned. Recheck belt tension after the pulley bolts are tightened.

15. Close the radiator drain and refill the system. Run the engine until the thermostat opens and then add coolant until the level stabilizes before replacing the radiator cap. Check for leaks.

16. Connect the negative battery cable.

3.0L (VIN T) Engine

1. Disconnect the negative battery cable.

2. Drain the cooling system.

3. Remove the timing cover. If the same timing belt will be reused, mark the direction of the timing belt's rotation, for installation in the same direction. Make sure the engine is positioned so the No. 1 cylinder is at the TDC of its compression stroke and the sprocket timing marks are aligned with the engine's timing mark indicators.

4. Loosen the timing belt tensioner bolt and remove the belt.

5. Position the tensioner as far away from the center of the engine as possible and tighten the bolt.

6. Remove the water pump mounting bolts, separate the pump from the water inlet pipe and remove the pump from the engine.

To install:

7. Install the pump with a new gasket to the engine. Torque the water pump mounting bolts to 14–20 ft. lbs. (20–27 Nm).

8. If not already done, position both camshafts so the marks align with those on the alternator bracket (rear bank) and inner timing cover (front bank). Rotate the crankshaft so the timing mark aligns with the mark on the oil pump.

9. Install the timing belt on the crankshaft sprocket and while keeping the belt tight on the tension side (right side), install the belt on the front camshaft sprocket.

10. Install the belt on the water pump pulley, then the rear camshaft sprocket and the tensioner.

11. Rotate the front camshaft counterclockwise to tension the belt between the front camshaft and the crankshaft. If the timing marks became misaligned, repeat the procedure.

12. Install the crankshaft sprocket flange.

13. Loosen the tensioner bolt and allow the spring to tension the belt.

14. Turn the crankshaft two full turns in the clockwise direction only until the timing marks align again. Now that the belt is properly tensioned, torque the tensioner lock bolt to 21 ft. lbs. (29 Nm).

15. Refill the cooling system. Connect the negative battery cable and road test the vehicle.

Thermostat

REMOVAL AND INSTALLATION

All Engines

1. Disconnect the negative battery cable. Remove the air cleaner.

2. Drain the cooling system to a point below the level of the thermostat or below.

3. Disconnect the hose from the thermostat housing.

4. Unbolt and remove the coolant outlet fitting.

5. Remove the gasket and lift out the thermostat. Discard the gasket.

To install:

6. Clean the mating surfaces of the outlet fitting and the thermostat housing.

7. Install the thermostat with the spring facing downward and position a new gasket on the thermostat housing.

8. Install the coolant outlet fitting. Torque the bolts to 11–15 ft. lbs. (15–20 Nm).

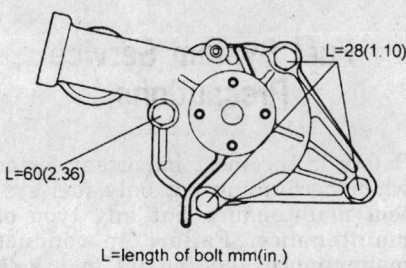

L=28(1.10)

L=60(2.36)

L=length of bolt mm(in.)

324246

Water pump bolt locations — 1.5L (VIN K) engine

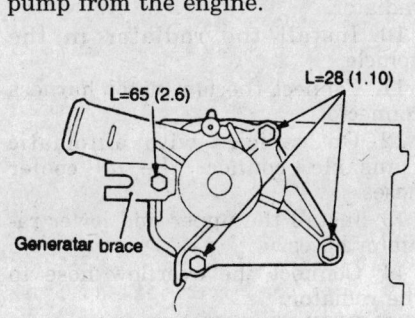

L=65 (2.6)   L=28 (1.10)

Generatar brace

L=length of bolt mm (in.)

257128

Water pump bolt locations — 1.3L (VIN M), 1.5L (VINs J and N)

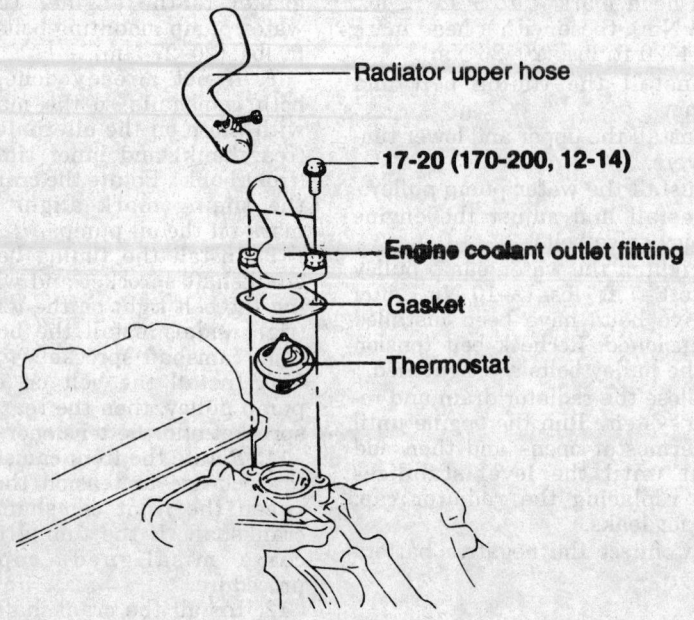

Radiator upper hose

17-20 (170-200, 12-14)

Engine coolant outlet fiitting

Gasket

Thermostat

334090

Thermostat mounting — 1.3L (VIN M) and 1.5L (VIN J) engines shown, others similar

9. Connect the upper radiator hose to the coolant outlet fitting.
10. Refill the cooling system.

Electric Cooling Fan

REMOVAL AND INSTALLATION

All Except 1.5L (VIN K) Engines

1. Disconnect the negative battery cable.
2. Remove the upper cooling fan shroud-to-radiator support bolts.
3. Disconnect the electrical leads from the fan motor. Remove the fan and shroud as an assembly.
To install:
4. Position the fan and shroud assembly onto the radiator support and install the mounting bolts.
5. Connect the electrical leads to the fan motor.
6. Connect the negative battery cable. Start the engine and check for proper fan operation.

1.5L (VIN K) Engine

1. Disconnect the harness connector at the fan motor.
2. Drain the coolant from the radiator.

3. Remove the upper and lower radiator hoses.
4. Disconnect the overflow hose from the radiator.
5. If equipped with automatic transaxle, disconnect the oil cooler hoses from the radiator. Plug the hoses to prevent fluid from leaking out and dirt from entering.
6. Remove the radiator mounting bolts.
7. Remove the radiator with the fan still attached.
8. Separate the fan from the radiator.
To install:
9. Install the fan assembly on the radiator.
10. Install the radiator in the vehicle.
11. Connect the fan motor harness connector.
12. On vehicles with automatic transaxle, connect the oil cooler hoses.
13. Install the upper and lower radiator hoses.
14. Connect the overflow hose to the radiator.
15. Refill the radiator with coolant.
16. On vehicles with automatic transaxle, start the engine and check the level of the transaxle fluid, add if necessary.

Cooling System Bleeding

PROCEDURE

Except 1995 1.5L (VIN K) Engine

1. Set the heater control knob to **HOT**.
2. Slowly remove the radiator cap.
3. Loosen the drain plug to drain the coolant.
4. Drain the coolant from the reservoir.
5. Securely tighten the drain plug.
6. Fill the radiator up to the filler neck with coolant.
7. Fill the reservoir with coolant.
8. Install the radiator cap.
9. Warm the engine until the cooling fan comes on and the thermostat opens.
10. Turn the engine **OFF** and allow it to cool.
11. Remove the radiator cap and fill the radiator up to the filler neck with coolant.
12. Install the radiator cap.
13. Fill the reservoir to the **FULL** line with coolant.

1995 1.5L (VIN K) Engine

1. Set the heater control knob to **HOT**.
2. Fill the radiator and reservoir tank with clean coolant mixture.
3. Install the radiator cap and run the engine until the thermostat opens and turn off the engine.
4. Allow the engine to cool and remove the radiator cap. Fill the radiator if necessary and install the cap.
5. Fill the reservoir to the **FULL** line with coolant.

FUEL SYSTEM

Fuel System Service Precautions

Safety is the most important factor when performing not only fuel system maintenance but any type of maintenance. Failure to conduct maintenance and repairs in a safe manner may result in serious personal injury or death. Maintenance and testing of the vehicle's fuel system components can be accomplished safely and effectively by adhering to the following rules and guidelines.

• To avoid the possibility of fire and personal injury, always disconnect the negative battery cable unless the repair or test procedure requires that battery voltage be applied.

• Always relieve the fuel system pressure prior to disconnecting any fuel system component (injector, fuel rail, pressure regulator, etc.), fitting or fuel line connection. Exercise extreme caution whenever relieving fuel system pressure to avoid exposing skin, face and eyes to fuel spray. Please be advised that fuel under pressure may penetrate the skin or any part of the body that it contacts.

• Always place a shop towel or cloth around the fitting or connection prior to loosening to absorb any excess fuel due to spillage. Ensure that all fuel spillage (should it occur) is quickly removed from engine surfaces. Ensure that all fuel soaked cloths or towels are deposited into a suitable waste container.

• Always keep a dry chemical (Class B) fire extinguisher near the work area.

• Do not allow fuel spray or fuel vapors to come into contact with a spark or open flame.

• Always use a backup wrench when loosening and tightening fuel line connection fittings. This will prevent unnecessary stress and torsion to fuel line piping. Always follow the proper torque specifications.

• Always replace worn fuel fitting O-rings with new. Do not substitute fuel hose or equivalent, where fuel pipe is installed.

Fuel System Pressure

RELIEVING

All Except 1.5L (VIN K) Engines

CAUTION

Fuel injection systems remain under pressure after the engine has been turned OFF. Properly relieve fuel pressure before disconnecting any fuel lines. Failure to do so may result in fire or personal injury. Do not allow fuel spray or fuel vapors to come in contact with a spark or open flame. Keep a dry chemical fire extinguisher nearby. Never store fuel in an open container due to risk of fire or explosion.

1. Reduce pressure in the fuel lines as follows:
   a. Turn the ignition to the **OFF** position.
   b. Loosen the fuel filler cap to release fuel tank pressure.
   c. Disconnect the fuel pump harness connector located in the area of the fuel tank.
   d. Start the vehicle and allow it to run until it stalls from lack of fuel. Turn the key to the **OFF** position.
   e. Disconnect the negative battery cable, then reconnect the fuel pump connector.

1.5L (VIN K) Engine

1. Remove the rear seat cushion.
2. Remove the fuel pump service cover and disconnect the fuel pump connector.
3. Start the engine and let it run until it stalls from lack of fuel. Crank the engine for a couple of seconds to further reduce the pressure in the lines.
4. Disconnect the negative battery cable.
5. Connect the fuel pump connector, replace the service cover and install the rear seat.

Idle Speed

ADJUSTMENT

1993–94 1.3L (VIN M) and 1.5L (VINs J and N) Engines

1. Warm the engine to operating temperature, leave lights, electric cooling fan and accessories **OFF**. The transaxle should be in **N**. The steering wheel in a neutral position for vehicles with power steering.
2. Loosen the accelerator cable.
3. Turn the ignition switch **ON** but do not start the engine. Leave the key in this position for at least 15 seconds. Check to see that the ISC servo is fully retracted to the curb idle position.

NOTE: When the ignition switch is turned to the ON position, the ISC plunger extends to the fast idle position opening. After 15 seconds, it retracts to the fully closed (curb idle) position.

4. Turn the ignition switch **OFF**.
5. Disconnect the ISC motor connector and secure the ISC motor at the fully retracted position.
6. In order to prevent the throttle valve from sticking, open it 2 or 3 times, then allow it to click shut and loosen the fixed SAS sufficiently.

7. Start the engine and allow it to run at idle speed. Ensure that the engine is running at idle speed of 700 rpm ±100. Adjust as necessary by turning the ISC adjust screw.

8. Tighten the fixed SAS until the engine speed starts to increase. Then, loosen the screw until the engine speed ceases to drop (touch point) and loosen an additional ½ turn.

9. Turn the ignition switch to the **OFF** position.

10. Adjust the accelerator cable to 0 to 0.04 in. (0 to 1mm) for a manual transaxle and 0.08 to 0.12 in. (2 to 3mm) for an automatic transaxle.

11. Connect the ISC motor connector.

12. Start the engine and check to be sure that the idle speed is correct.

13. Turn the ignition switch **OFF** and disconnect the negative battery cable for 15 seconds and re-connect. This will erase the data stored in memory during ISC adjustment.

1.6L (VIN R), 1.8L (VIN M), 2.0L (VIN F) and 1993–95 3.0L (VIN T) Engines

1. Warm the engine to operating temperature, leave lights, electric cooling fan and accessories **OFF**. The transaxle should be in **N**. The steering wheel in a neutral position for vehicles with power steering.

2. Loosen the accelerator cable.

3. Connect a multi-use tester to the diagnostic connector in the fuse box. If a multi-use tester is not being used, connect a tachometer to the engine and ground the self-diagnostic terminal.

4. Ground the ignition timing adjustment terminal in the engine compartment.

5. Run the engine for more than 5 seconds at an engine speed of 2000–3000 rpm. Allow the engine to return to idle for a minimum of 2 minutes.

6. Check that the engine is idling at 600–800 rpm. If the multi-use tester is being used, press code No. 22 and read the idle speed.

NOTE: The engine speed on a new vehicle driven less than 300 miles may be 100 rpm lower than the desired rpm, in which case it would not require adjustment. Break-in should take approximately 300 miles. If the vehicle stalls or has a very low idle speed, suspect a deposit buildup on the throttle valve which must be cleaned.

7. If the basic engine idle speed is out of specification, adjust by turning the Speed Adjusting Screw (SAS), until the desired reading is obtained.

8. Turn the ignition switch **OFF** and stop the engine. Disconnect the jumper wire from the diagnosis connector, disconnect the jumper wire from the ignition timing connector and reinstall the waterproof connector cover. Disconnect the tachometer.

9. Disconnect the negative battery cable for at least 15 seconds.

10. Restart the engine and allow it to run for at least 5 minutes; check for good idle quality.

1996–97 3.0L (VIN T) Engine

NOTE: **Idle speed adjustment is usually not necessary, the idle speed control motor adjusts the idle speed as needed under all conditions.**

1. Warm the engine to operating temperature, leave lights, electric cooling fan and accessories **OFF**. The transaxle should be in **N**. The steering wheel in a neutral position for vehicles with power steering.

2. Loosen the accelerator cable.

3. Connect a multi-use tester to the diagnostic connector in the fuse box. If a multi-use tester is not being used, connect a tachometer to the engine and ground the self-diagnostic terminal.

4. Ground the ignition timing adjustment terminal in the engine compartment.

5. Run the engine for more than 5 seconds at an engine speed of 2000–3000 rpm. Allow the engine to return to idle for a minimum of 2 minutes.

6. Check that the engine is idling at 600–800 rpm. If the multi-use tester is being used, press code No. 22 and read the idle speed.

NOTE: **The engine speed on a new vehicle driven less than 300 miles may be 100 rpm lower than the desired rpm, in which case it would not require adjustment. Break-in should take approximately 300 miles. If the vehicle stalls or has a very low idle speed, suspect a deposit buildup on the throttle valve which must be cleaned.**

7. If the basic engine idle speed is out of specification, check the idle speed control system.

Mixture

ADJUSTMENT

All Engines

The air/fuel mixture is automatically adjusted by the engine control system. If the mixture is too rich or too lean, use the appropriate diagnostic procedure to locate the problem.

Fuel Filter

REMOVAL AND INSTALLATION

All Engines

——————— CAUTION ———————
Fuel injection systems remain under pressure after the engine has been turned OFF. Properly relieve fuel pressure before disconnecting any fuel lines. Failure to do so may result in fire or personal injury. Do not allow fuel spray or fuel vapors to come in contact with a spark or open flame. Keep a dry chemical fire extinguisher nearby. Never store fuel in an open container due to risk of fire or explosion.

1. Relieve the fuel pressure.

2. Disconnect the negative battery cable.

3. Remove the fuel line inlet and outlet union bolts while holding the fuel filter stationary with a backup wrench.

4. Remove the fuel filter mounting bolts.

5. Pull the old filter from the mounting bracket.
To install:

6. Place the new fuel filter into the mounting bracket.

7. Install and tighten the fuel filter mounting bolts to 18–25 ft. lbs. (25–35 Nm).

8. Using new gaskets, connect the fuel inlet and outlet lines to the filter. Carefully tighten the union bolts to avoid stripping the threads.

9. Connect the negative battery cable.

10. Start the engine and check the filter connections for leaks by running the tip of your finger around each union bolt connection.

Fuel Pump

REMOVAL AND INSTALLATION

Except 1.5L (VIN K) Engine

——————— CAUTION ———————
Fuel injection systems remain under pressure after the engine has been turned OFF. Properly relieve fuel pressure before disconnecting any fuel lines. Failure to do so may result in fire or personal injury.

1. Relieve the fuel system pressure.

2. Raise the vehicle and support it safely.

3. Remove the fuel tank from the vehicle. Disconnect the hoses at the pump.

4. Unbolt and remove the pump from the tank.
To install:

5. Install the new pump with a new gasket in place into the fuel tank. Evenly tighten the fuel pump bolts in a crisscross pattern.

6. Install the fuel tank into the vehicle.

7. Connect the fuel lines and the electrical connectors to the fuel pump and sending unit. Tighten the high–pressure fittings to 22–29 ft. lbs. (30–40 Nm).

8. Connect the negative battery cable.

9. Turn the ignition switch to the **ON** position to pressurize the fuel system. Carefully check for leaks at the fuel line connections and around the fuel pump sealing surface.

1.5L (VIN K) Engine

——————— CAUTION ———————
Fuel injection systems remain under pressure after the engine has been turned OFF. Properly relieve fuel pressure before disconnecting any fuel lines. Failure to do so may result in fire or personal injury.

1. Remove the rear seat cushion and the fuel pump service cover.

2. Relieve the fuel system pressure.

3. Disconnect the negative battery cable.

4. Disconnect the fuel pump and gauge harness connector.

5. Disconnect the high pressure hose and the return hose from the fuel pump.

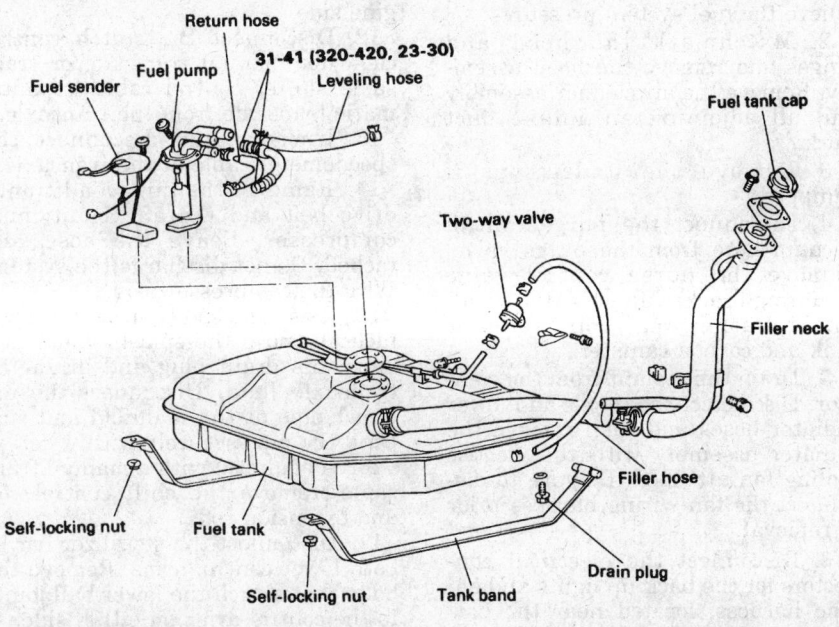

Fuel pump and tank assembly — 1995 1.5L (VIN N) engine shown, others similar

257327

6. Remove the mounting screws holding the fuel pump/sender in the fuel tank.

7. Remove the fuel pump from the tank.

To install:

8. Use a new gasket and install the fuel pump/sender in the fuel tank.

9. Connect the high pressure hose and the return hose to the fuel pump. Torque the high pressure flare nut to 22.1–29.5 ft. lbs. (30–40 Nm).

10. Connect the wiring connector to the fuel pump/sender.

11. Install the service cover and the rear seat cushion.

12. Connect the negative battery cable.

Fuel Injector

REMOVAL AND INSTALLATION

Except 1.5L (VIN K) Engine

1. Relieve the fuel system pressure.

2. Disconnect the negative battery cable.

3. Disconnect and remove the air intake hoses as required.

4. Wrap the fuel line connection with a shop towel to contain leaks.

Disconnect the high–pressure fuel line at the fuel rail.

5. Disconnect the fuel return hose.

6. Disconnect the accelerator cable connection from the throttle body and position it aside.

7. Disconnect the vacuum connection from the fuel pressure regulator.

8. Disconnect the electrical harness connector from each fuel injector.

9. Remove the injector rail retaining bolts.

10. Lift the rail assembly up and away from engine. Be very careful not to drop any of the injectors.

11. Gently pull the injectors from the rail. Remove and discard the O-rings.

12. Check the resistance through the injector. The specification is 13–16 ohms at 70°F (20°C).

To install:

13. Install a new grommets and O-rings to the injectors. Coat the O-rings with lightweight oil.

14. Install the injector to the fuel rail port. Turn the injector body gently side–to–side to seat it into the port.

15. Install the fuel rail and injectors to the manifold.

16. Tighten the retaining bolts to 9 ft. lbs. (12 Nm).

17. Connect the electrical connectors to the injectors.

18. Replace the O-ring on the fuel pressure regulator, lightly lubricate and install on the delivery pipe. Connect the vacuum hose to the fuel pressure regulator.

19. Connect the fuel return hose.

20. Replace the O-ring on the high pressure fuel line, lightly lubricate it and connect to delivery pipe.

21. Reconnect the accelerator cable to the throttle body and adjust to specifications.

22. Connect the negative battery cable and check the entire system for proper operation and leaks.

1.5L (VIN K) Engine

— **CAUTION** —

Fuel injection systems remain under pressure after the engine has been turned OFF. Properly relieve fuel pressure before disconnecting any fuel lines. Failure to do so may result in fire or personal injury.

1. Relieve the fuel system pressure.

2. Remove the air hose for the idle speed control.

3. Disconnect the fuel injector connectors.

4. Disconnect the fuel supply and return hoses from the fuel delivery pipe.

5. Remove the fuel delivery pipe mounting bolts.

6. Carefully pull the delivery pipe with the injectors out of the intake manifold.

NOTE: Fuel may flow out of the delivery pipe when the injector is removed from the pipe.

7. Remove the injector retaining clip. Rotate and pull the injector out of the delivery pipe.

To install:

8. Install new O-rings on the injector. Lubricate the O-rings with gasoline to ease installation.

9. Install the injector in the delivery pipe while turning it left and right.

10. Install the retaining clip.

11. Install the delivery pipe assembly and injectors on the intake manifold. Torque the mounting bolts to 7–11 ft. lbs. (10–15 Nm).

12. Use a new gasket and connect the fuel supply hose to the delivery pipe.

13. Connect the fuel return hose to the pressure regulator.

14. Connect the fuel injector connectors.

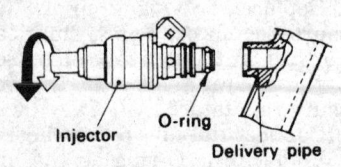

Install the injector to the fuel rail — 1.3L (VIN M) and 1.5L (VIN J) engines shown, others similar

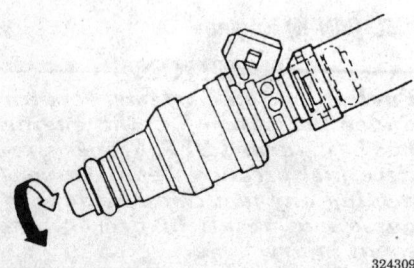

Turn the injector left and right to install or remove from delivery pipe — 1.5L (VIN K) engine shown, others similar

15. Install the idle control air hose.
16. Connect the negative battery cable.
17. Check for fuel leaks and repair if necessary.

ENGINE MECHANICAL

Engine Assembly

REMOVAL AND INSTALLATION

1.3L (VIN M), 1.5L (VIN J), 1.6L (VIN R), 1.8L (VIN M) and 2.0L (VIN F) Engines

The manufacturer recommends that the engine and transaxle be removed as a unit on all models.

Tag all hoses, wires, tubes, cables, etc., to ease reinstallation. If there is an instant camera handy, a couple of pre-removal photographs of the en-gine compartment may also be useful.

1. If equipped with fuel injection, relieve the fuel system pressure.
2. Matchmark the hood and hinges and remove the hood assembly. Remove the air cleaner assembly and all adjoining air intake duct work.
3. Remove the undercover if equipped.
4. Disconnect the purge control vacuum hose from the purge valve. Remove the purge control valve mounting bracket. Remove the windshield washer reservoir, radiator tank and carbon canister.
5. Drain the coolant from the radiator. Disconnect the upper and lower radiator hoses and then remove the radiator assembly with the electric cooling fan attached. Be sure to disconnect the fan wiring harness prior to removal.
6. Disconnect the electrical connectors for the back-up lights and engine harness, located near the battery tray. If equipped with a 5 speed transaxle, disconnect the select control valve connector. Disconnect the alternator harness connectors and the oil pressure sending unit.
7. Label and disconnect the automatic transaxle oil cooler hoses. Avoid spilling oil and cap the openings.
8. Label and disconnect all low tension wires and the one high tension wire going to the coil from the distributor. Disconnect the engine ground.
9. Disconnect the brake booster vacuum hose at the intake manifold.
10. Disconnect the fuel supply, return and vapor hoses at the side of the engine. Before disconnecting the fuel supply and return lines, wrap shop towels around the fuel fitting that is being disconnected to absorb any fuel spray caused by residual pressure in the lines.

Front roll stopper mounting — 1.6L (VIN R), 1.8L (VIN M) and 2.0L (VIN F) engines

11. Disconnect the heater hoses from the side of the engine. Disconnect the accelerator cable at the engine side.
12. Disconnect the clutch control cable for manual transaxle or transaxle shifter control cable for automatic transaxle from the transaxle.
13. Unscrew and disconnect the speedometer cable at the transaxle.
14. Remove the air conditioner drive belt and the air conditioning compressor. Leave the hoses attached. Do not discharge the system. Wire the compressor aside.
15. Raise and safely support the vehicle. Remove the splash shield. Remove the drain plug and drain the transaxle fluid. Disconnect the exhaust pipe at the manifold and suspend the pipe securely with wire.
16. If equipped with a manual transaxle, remove the shift control rod and extension rod.
17. Disconnect the stabilizer bar at both lower control arms. Remove the bolts that attach the lower ball joints to the control arms on either side.
18. Remove the front halfshafts. Then, seal off the openings in the transaxle to prevent damage caused by the introduction of foreign substances into the transaxle. Be sure to replace the circlips holding the half-shafts in the transaxle during assembly.
19. Attach an engine lift, via chains or cables, to both the engine lifting hooks. Put just a little tension on the cables. Then, remove the nut and bolt from the front roll stopper; unbolt the brace from the top of the engine damper.
20. Separate the rear roll stopper from the No. 2 crossmember. Remove the attaching nut from the left mount insulator bolt, but do not remove the bolt.
21. Raise the engine just enough that the lifting device is supporting its weight. Check that everything is disconnected from the engine.
22. Remove the blind cover from the inside of the right fender inner shield. Remove the transaxle mounting bracket bolts.
23. Remove the left mount insulator bolt. Then, press downward on the transaxle while lifting the engine/transaxle assembly to guide it up and out of the vehicle.

NOTE: Make sure the transaxle does not hit the battery bracket during engine and transaxle removal.

To install:
24. Using a lifting device, lower the engine and transaxle carefully into

position and loosely install the mounting bolts. Temporarily tighten the front and rear roll control rods mounting bolts. Lower the full weight of the engine and transaxle onto the mounts and tighten the nuts and bolts. Loosen and retighten the roll control rods.

25. Install the transaxle mounting bracket bolts. Install the blind cover to the inside of the right fender inner shield.

26. Assemble the rear roll stopper to the No. 2 crossmember. Install retaining nut and bolt.

27. Install new circlips on the halfshafts and install in position.

28. Attach the lower ball joints to the control arms on either side. Connect the stabilizer bar to both lower control arms.

29. If equipped with a manual transaxle, install the shift control rod and extension rod.

30. Raise the vehicle and support it safely. Connect the exhaust pipe at the manifold. Install the splash shield.

31. The balance of the installation is the reverse of the removal procedure.

32. Fill all fluids to the proper levels. Adjust the transaxle control cables, accessory drive belts and accelerator linkages as required. Reconnect the negative battery cable.

33. Start the engine and check for leaks as well as proper gauge operation. Allow the vehicle to reach normal operating temperature and recheck all fluid levels.

34. Replace the hood making sure to align the matchmarks made during removal. Allow the engine to cool and recheck the coolant level.

1.5L (VIN K) Engine

─── CAUTION ───
Fuel injection systems remain under pressure after the engine has been turned OFF. Properly relieve fuel pressure before disconnecting any fuel lines. Failure to do so may result in fire or personal injury.

NOTE: The manufacturer recommends that the engine and transaxle be removed as a unit on all models. Tag all hoses, wires, tubes, cables, etc., so you'll remember where they go when it comes time for installation.

1. Relieve the fuel system pressure and remove the battery.
2. Remove the air intake duct from the throttle body.

3. Disconnect the back-up light and the engine harness connectors.
4. Disconnect the alternator harness and the oil pressure gauge.
5. Drain the engine coolant.
6. On vehicles with automatic transaxles, disconnect the oil cooler hoses and plug the hoses.
7. Disconnect the upper and lower radiator hoses from the engine and remove the radiator assembly.
8. Disconnect the engine ground.
9. Disconnect the brake booster vacuum hose.
10. Disconnect the fuel pressure, return and vapor hoses from the engine.
11. Disconnect the heater hoses from the engine.
12. Disconnect the accelerator cable from the throttle body.
13. On vehicles with manual transaxle, disconnect the clutch fluid hose from the transaxle and disconnect the shift select and control cables. On vehicles with automatic transaxle, disconnect the selector cable.
14. Disconnect the speedometer cable from the transaxle.
15. Remove the A/C compressor from the bracket and position it out of the way. Do not disconnect the hoses.
16. Drain the transaxle fluid or oil.
17. Disconnect the front exhaust pipe from the manifold and use wire to support it.
18. Remove the right and left halfshafts.
19. Attach an engine hoist to the engine. Raise the engine assembly slightly to take the weight off of the mounts.
20. Remove the front and rear roll stoppers.
21. Remove the engine mount insulator bolts and remove the engine mount bracket from the engine.
22. Remove the caps from inside the fender and remove the transaxle mount bracket bolts.
23. Check that all cables, hoses, harnesses and connectors are disconnected from the engine and transaxle assembly.
24. Direct the transaxle side downward and lift the engine assembly out of the vehicle.

To install:
25. Carefully lower the engine assembly into the vehicle.
26. Install the transaxle mount, front and rear roll stoppers.
27. Install the engine mount.
28. Install the right and left halfshafts.
29. Use a new gasket and connect the exhaust pipe to the manifold.

30. The balance of the installation is the reverse of the removal procedure.
31. Check that all cables, hoses, harnesses and connectors are connected to the engine and transaxle assembly.
32. Install the battery.
33. Refill the engine with coolant and fill the transaxle with fluid (A/T) or oil (M/T).
34. Start the engine and check for leaks and proper operation of gauges and instruments.

1.5L (VIN N) Engine

NOTE: The manufacturer recommends that the engine and transaxle be removed as a unit.

1. Relieve the fuel system pressure:
   a. Turn the ignition to the OFF position.
   b. Loosen the fuel filler cap to release fuel tank pressure.
   c. Disconnect the fuel pump harness connector located in the area of the fuel tank.
   d. Start the vehicle and allow it to run until it stalls from lack of fuel. Turn the key to the OFF position.
   e. Disconnect the negative battery cable, then reconnect the fuel pump connector.
2. Use a magic marker to match-mark the hood and hinges. Remove the hood.
3. Remove the air cleaner assembly and all adjoining air intake duct work.
4. Remove the splash cover.
5. Disconnect the purge control vacuum hose from the purge valve. Remove the purge control valve mounting bracket. Remove the windshield washer reservoir, radiator tank, and carbon canister.
6. Drain the engine oil.
7. Drain the coolant from the radiator.
8. Disconnect the upper and lower radiator hoses. Disconnect the fan wiring harnesses. Remove the radiator with the electric cooling fan attached.
9. Disconnect the electrical connectors for the back-up lights and engine harness, located near the battery tray. Disconnect the alternator harness connectors and the oil pressure sending unit.
10. Label and disconnect the automatic transaxle oil cooler hoses. Avoid spilling oil and cap the openings.
11. Label and disconnect all low tension wires and the one high ten-

sion wire going to the coil from the distributor. Disconnect the engine ground.

12. Disconnect the brake booster vacuum hose at the intake manifold.

13. Disconnect the fuel supply, return, and vapor hoses at the side of the engine. Immediately clean up any spilled fuel.

14. Disconnect the heater hoses from the side of the engine. Disconnect the throttle cable from the throttle body.

15. If equipped with a manual transaxle, disconnect the shift and select cables.

16. If equipped with and automatic transaxle, disconnect the shifter control cable.

17. Unscrew and disconnect the speedometer cable at the transaxle.

18. Remove the air conditioner drive belt and the air conditioning compressor. Leave the hoses attached. Do not discharge the system. Wire the compressor aside.

19. Raise and safely support the vehicle. Remove the splash shield. Remove the drain plug and drain the transaxle fluid. Disconnect the exhaust pipe at the manifold and suspend the pipe securely with wire.

20. If equipped with a manual transaxle, unbolt the slave cylinder from the transaxle case and suspend it out of the way with wire.

21. Disconnect the stabilizer bar at both lower control arms. Remove the bolts that attach the lower control arms to the body on either side. Support the arms from the body.

22. Remove the front halfshafts. Then, seal off the openings in the transaxle to prevent damage caused by the introduction of foreign objects into the transaxle. Be sure to replace the circlips holding the halfshafts in the transaxle during assembly.

23. Attach an engine lifting chain to both the engine lifting hooks. Put just a little tension on the chain. Then, remove the nut and bolt from the front roll stopper; unbolt the brace from the top of the engine damper.

24. Separate the rear roll stopper from the No. 2 crossmember. Remove the attaching nut from the right mount insulator bolt, but do not remove the bolt.

25. Raise the engine just enough that the lifting chain is supporting its weight. Check that everything is disconnected from the engine.

26. Remove the bolt cover from the inside of the left fender inner shield.

Front roll stopper — 1.5L (VIN N) engine

Remove the transaxle mounting bracket bolts.

27. Remove the right mount insulator bolt. Then, press downward on the transaxle while lifting the engine/transaxle assembly to guide it up and out of the vehicle.

NOTE: Make sure the transaxle does not hit the battery bracket during engine and transaxle removal.

To install:

28. Using a lifting device, lower the engine and transaxle carefully into position and loosely install the mounting bolts. Temporarily tighten the front and rear roll control rods mounting bolts. Lower the full weight of the engine and transaxle onto the mounts and tighten the nuts and bolts. Loosen and retighten the roll control rods.

29. Install the transaxle mounting bracket bolts. Install the blind cover to the inside of the right fender inner shield.

30. Assemble the rear roll stopper to the No. 2 crossmember. Install retaining nut and bolt.

31. With the weight of the engine on the mounts, torque the fasteners to the proper specification:
- Side engine mount: bracket nuts and bolts 36–47 ft. lbs. (50–65 Nm); through-bolt nut 65–80 ft. lbs. (90–110 Nm); small nut 32–43 ft. lbs. (50–60 Nm).
- Rear roll stopper: through-bolt and bracket bolts 33–43 ft. lbs. (45–60 Nm)
- Front roll stopper: through-bolt 33–43 ft. lbs. (45–60 Nm); bracket bolts 22–29 ft. lbs. (30–40 Nm)
- Transaxle mount: through-bolt 65–85 ft. lbs. (90–110 Nm); bracket bolts 22–29 ft. lbs. (30–40 Nm)
- Center beam: bolts 43–58 ft. lbs. (60–80 Nm)

32. Install new circlips on the half-shafts and install them into the transaxle.

33. Attach the lower control arms to the body on either side. Connect the stabilizer bar to both lower control arms.

34. If equipped with a manual transaxle, connect the shift and select cables.

35. If equipped with an automatic transaxle, connect the selector control cable.

36. Raise the vehicle and support it safely. Connect the exhaust pipe at the manifold. Install the splash shield.

37. Install the clutch master cylinder.

38. Connect the speedometer cable at the transaxle. Connect the air conditioning compressor to the mounting bracket.

39. Lower the vehicle. Connect the heater hoses to the engine. Connect the throttle cable to the throttle body.

40. Connect the fuel supply, return, and vapor hoses using new O-rings.

41. Connect the brake booster vacuum hose at the intake manifold.

42. Connect all low tension wires and the high tension wire going to the coil from the distributor. Connect the engine ground and the fuel injectors.

43. Connect the automatic transaxle oil cooler hoses.

44. Connect the electrical connectors for the back-up lights and engine harness, located near the battery tray. Connect the alternator harness connectors and the oil pressure sending unit.

45. Install the radiator and electric cooling fan assembly. Connect the upper and lower radiator hoses.

46. Fill all fluids to the proper levels. Adjust the transaxle control cables, accessory drive belts and throttle linkages as required. Reconnect the negative battery cable.

47. Start the engine and check for leaks as well as proper gauge operation. Allow the vehicle to reach normal operating temperature and recheck all fluid levels.

48. Install the hood making sure to align the matchmarks made during removal.

49. Allow the engine to cool and recheck the coolant level.

3.0L (VIN T) Engine

NOTE: The manufacturer recommends that the engine and transaxle be removed as a unit.

--- CAUTION ---

Fuel injection systems remain under pressure after the engine has been turned OFF. Properly relieve fuel pressure before disconnecting any fuel lines. Failure to do so may result in fire or personal injury. Do not allow fuel spray or fuel vapors to come in contact with a spark or open flame. Keep a dry chemical fire extinguisher nearby. Never store fuel in an open container due to risk of fire or explosion.

1. Relieve the fuel system pressure.
2. Remove the battery
3. Remove the undercover, if equipped.
4. Matchmark the hood and hinges and remove the hood assembly. Remove the air cleaner assembly and all adjoining air intake duct work.
5. Disconnect the fuel supply line, return line and vent hoses. Prior to disconnecting the fuel supply or return lines, wrap shop towels around the fitting that is being disconnected to absorb any fuel spray caused by residual pressure in the lines.
6. Disconnect the backup light, engine, alternator and oil pressure harnesses.
7. Drain the engine coolant.
8. Label and disconnect the transaxle oil cooler lines, radiator hoses and remove the radiator assembly.
9. Disconnect the brake booster, fuel, evaporative canister and heater hoses.
10. Disconnect the accelerator, transaxle, cruise control and speedometer cables.
11. Detach the air conditioning compressor from the mounting bracket and hang out of the way with a piece of wire. Do not disconnect the refrigerant lines.
12. Remove the power steering pump and wire aside.

13. Raise the vehicle and support safely.
14. Remove the oil pan shield and drain the transaxle.
15. Disconnect the front exhaust pipe.
16. Remove the lower arm ball joint and stabilizer bar at the point where it is mounted to the lower arms.
17. Remove the halfshaft from the housing by prying against the transaxle housing with a prybar.
18. Suspend the lower arm and driveshafts aside using wire attached to the vehicle underbody.
19. Attach an engine lifting device to the engine. Raise the engine just enough to take the tension off the engine mounts.
20. Remove the front roll stopper, engine damper and rear roll stopper.
21. Remove the engine mount bolts.
22. Remove the blind plugs from the inside of the right fender shield and remove the transaxle mounting bracket bolts.
23. Remove the left mount insulator bolt.
24. Raise the engine and transaxle slightly and inspect to make sure all cables, hoses and harness connectors are disconnected.
25. While directing the transaxle side downward, lift the engine and transaxle assembly up and out of the vehicle.

To install:
26. While directing the transaxle side downward, direct the engine and transaxle assembly into the vehicle. Install all mounting hardware and control bracket retainers.
27. Tighten the center crossmember-to-body bolts to 43–58 ft. lbs. (58–77 Nm).
28. Once the engine is securely in place, remove the engine lifting device.
29. Connect the lower arm and the remaining suspension components disassembled during engine removal.

30. Install the halfshafts to the transaxle housing. Make sure the new C-clips are fully engaged into the differential assembly.
31. Connect the front exhaust pipe. Refill the transaxle with fluid.
32. Install the oil pan shield and lower the vehicle.
33. Install the power steering pump.
34. Reconnect the fuel lines using new O-rings where required.
35. Install the air conditioning compressor to the mounting bracket.
36. Connect the accelerator, transaxle, cruise control and speedometer cables.
37. Connect the brake booster, fuel, evaporative canister and heater hoses.
38. Install the radiator.
39. Connect the transaxle oil cooler lines and radiator hoses.
40. Refill the engine coolant.
41. Connect the backup light, engine, alternator and oil pressure harnesses.
42. Connect the negative battery cable.
43. Install the air cleaner assembly.
44. Start the engine and check for leaks. Allow the engine to run until normal operating temperature is reached and recheck all fluid levels.
45. Replace the hood making sure to align the matchmarks made during removal.
46. Allow the engine to cool and recheck the coolant level.

Engine Mounts

REMOVAL AND INSTALLATION

1.3L (VIN M) and 1.5L (VIN J) Engines

1. Disconnect the negative battery cable.
2. Attach an engine hoist to the engine hooks and raise the engine just enough to take the pressure off of the engine mounts.
3. The mounts can now be removed and serviced as needed.
4. If working under the vehicle, an adjustable stand may be used to raise the engine and take the pressure off of the engine mounts.
5. Connect the negative battery cable.

1.5L (VIN N) Engine

1. Disconnect the negative battery cable.
2. Attach an engine hoist to the engine hooks, or use a floor jack with a block of wood on its paw for pad-

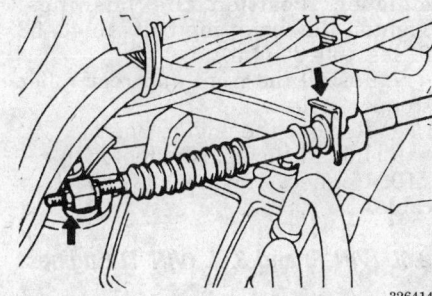

Transaxle control cable — 3.0L (VIN T) Engine

326414

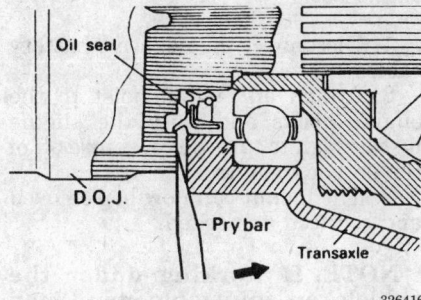

Remove the halfshafts from the transaxle — 3.0L (VIN T) Engine

326416

ding. Raise the engine just enough to take the pressure off of the engine mounts.

3. Raise and safely support the vehicle if necessary.

4. The mounts can now be removed and serviced as needed.

5. If working under the vehicle, an adjustable stand may be used to raise the engine and take the pressure off of the engine mounts.

To install:

6. Install the new engine mount. Only hand-tighten the mounting nuts and bolts.

7. Remove the engine lifting equipment.

8. With the weight of the engine on the mounts, torque the fasteners to the proper specification:

• Side engine mount: bracket nuts and bolts 36–47 ft. lbs. (50–65 Nm); through-bolt nut 65–80 ft. lbs. (90–110 Nm); small nut 32–43 ft. lbs. (50–60 Nm)

• Rear roll stopper: through-bolt and bracket bolts 33–43 ft. lbs. (45–60 Nm)

• Front roll stopper: through-bolt 33–43 ft. lbs. (45–60 Nm); bracket bolts 22–29 ft. lbs. (30–40 Nm)

• Transaxle mount: through-bolt 65–85 ft. lbs. (90–110 Nm); bracket bolts 22–29 ft. lbs. (30–40 Nm)

• Center beam: bolts 43–58 ft. lbs. (60–80 Nm)

9. Lower the vehicle.

10. Connect the negative battery cable.

1.6L (VIN R) and 1.8L (VIN M) Engine

Transaxle Mount

1. Disconnect the negative battery cable.

2. If equipped with a manual transaxle, remove the select control valve.

3. Attach an engine hoist to the engine hooks and raise the engine just enough to take the pressure off of the engine mounts.

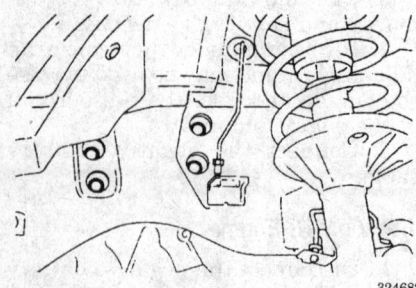

324682

Transaxle bracket mounting bolts — 1.5L (VIN K) engine

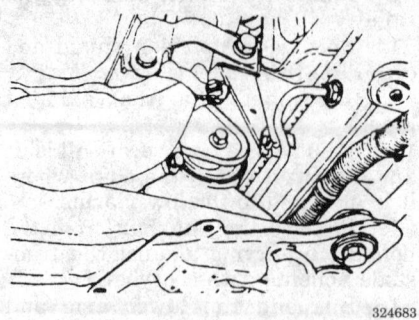

324683

Front roll stopper — 1.5L (VIN K) engine

324684

Rear roll stopper — 1.5L (VIN K) engine

4. The mount can now be removed and serviced as needed.

NOTE: If working under the vehicle, an adjustable stand may be used to raise the engine and take the pressure off of the engine mounts.

To install:

5. Position and install the mount. Tighten the through-bolt to 65–80 ft. lbs. (90–110 Nm). Tighten the mounting bolts to 22–29 ft. lbs. (30–40 Nm).

6. Remove the engine hoist.

7. Install the select control valve.

8. Connect the negative battery cable.

Engine Mount

1. Disconnect the negative battery cable.

2. Attach an engine hoist to the engine hooks and raise the engine just enough to take the pressure off of the engine mounts.

3. The mount can now be removed and serviced as needed.

NOTE: If working under the vehicle, an adjustable stand may be used to raise the engine and take the pressure off of the engine mounts.

To install:

4. Position and install the mount.

5. Remove the engine hoist.

6. Connect the negative battery cable.

Front and Rear Roll Stoppers

1. Disconnect the negative battery cable.

2. Attach an engine hoist to the engine hooks and raise the engine just enough to take the pressure off of the engine mounts.

3. The roll stoppers can now be removed and serviced as needed.

NOTE: If working under the vehicle, an adjustable stand may be used to raise the engine and take the pressure off of the engine mounts.

To install:

4. Position and install the roll stopper. Tighten the through-bolt to 33–43 ft. lbs. (45–60 Nm). Tighten the front roll stopper mounting bolts to 22–29 ft. lbs. (30–40 Nm), and the rear roll stopper to 33–43 ft. lbs. (45–60 Nm).

5. Remove the engine hoist.

6. Connect the negative battery cable.

Center Member

1. Disconnect the negative battery cable.

2. Remove the splash shield.

3. Attach an engine hoist to the engine hooks and raise the engine just enough to take the pressure off of the engine mounts.

4. Remove the front and rear roller stopper mounts.

5. The center member can now be removed and serviced as needed.

NOTE: If working under the vehicle, an adjustable stand may be used to raise the engine and take the pressure off of the engine mounts.

To install:

6. Position and install the center member. Position the bushings. Tighten the mounting bolts to 43–58 ft. lbs. (60–80 Nm).

7. Install the front and rear roller stopper mounts.

8. Remove the engine hoist.

9. Install the splash shield.

10. Connect the negative battery cable.

2.0L (VIN F) and 3.0L (VIN T) Engines

Transaxle Mount

1. Disconnect the negative battery cable.

2. If equipped with a manual transaxle, remove the select control valve.

3. Attach an engine hoist to the engine hooks and raise the engine just enough to take the pressure off of the engine mounts.

4. The mount can now be removed and serviced as needed.

NOTE: If working under the vehicle, an adjustable stand may be used to raise the engine and take the pressure off of the engine mounts.

To install:

5. Position and install the mount. Tighten the through-bolt to 65–80 ft. lbs. (90–110 Nm). Tighten the mounting bolts to 29–36 ft. lbs. (40–50 Nm).

6. Remove the engine hoist.

7. Install the select control valve.

8. Connect the negative battery cable.

Engine Mount

1. Disconnect the negative battery cable.

2. Attach an engine hoist to the engine hooks and raise the engine just enough to take the pressure off of the engine mounts.

3. The mount can now be removed and serviced as needed.

NOTE: If working under the vehicle, an adjustable stand may be used to raise the engine and take the pressure off of the engine mounts.

To install:

4. Position and install the mount.

5. Remove the engine hoist.

6. Connect the negative battery cable.

Front and Rear Roll Stoppers

1. Disconnect the negative battery cable.

2. Attach an engine hoist to the engine hooks and raise the engine just enough to take the pressure off of the engine mounts.

3. The roll stoppers can now be removed and serviced as needed.

NOTE: If working under the vehicle, an adjustable stand may be used to raise the engine and take the pressure off of the engine mounts.

To install:

4. Position and install the roll stopper. Tighten the front roll stopper through-bolt to 36–47 ft. lbs. (50–65 Nm), and the rear roll stopper to 22–29 ft. lbs. (30–40 Nm). Tighten the front roll stopper mounting bolts to 29–36 ft. lbs. (40–50 Nm), and the

rear roll stopper to 29–36 ft. lbs. (40–50 Nm).

5. Remove the engine hoist.

6. Connect the negative battery cable.

Cylinder Head

REMOVAL AND INSTALLATION

1.3L (VIN M) and 1.5L (VINs J and N) Engines

— **WARNING** —

Do not remove the cylinder head unless the engine is cold, a hot cylinder head will warp.

1. Release the fuel system pressure on fuel injected engines and disconnect the negative battery cable.

2. Drain the cooling system and then disconnect the upper radiator hose.

3. Remove the PCV hose that runs between the air cleaner and the rocker cover.

4. Remove the air cleaner. Label and disconnect any vacuum lines running to the cylinder head, manifold or carburetor from other parts of the engine compartment. Disconnect the intake pipe on turbocharged engines.

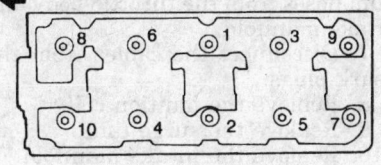

Crankshaft pulley side

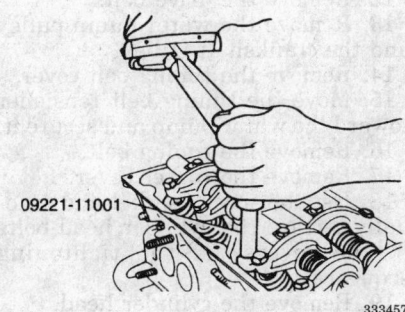

09221-11001

333457

Cylinder head bolt tightening sequence — 1.3L (VIN M) and 1.5L (VINs J and N) engines

5. Disconnect the heater hoses going to the head.

6. Disconnect the fuel supply line, return line and vent hoses. Prior to disconnecting the fuel supply or return lines, wrap shop towels around the fitting that is being disconnected to absorb any fuel spray caused by residual pressure in the lines.

7. Label and disconnect the spark plug wires, injectors and any other electronic components that will be removed with the cylinder head. Remove the rocker cover.

8. Turn the crankshaft until the TDC timing marks align and both No. 1 cylinder valves are closed, both rockers are off the cams. The engine will now be at top dead center with the No. 1 piston on its compression stroke.

9. Remove the distributor cap and matchmark the rotor tip to the distributor housing and the housing to the engine. Remove the distributor or the ignition coil assembly.

NOTE: With the distributor removed, do not crank the engine. If rotated with the distributor removed, the engine will have to be positioned with the No. 1 cylinder at top dead center of its compression stroke prior to installing the distributor.

10. Remove the carburetor, if equipped.

11. Remove the intake and the exhaust manifolds.

12. Remove the timing belt cover. Note the location of the camshaft sprocket timing mark. Loosen both timing belt tensioner mounting bolts and move tensioner toward the water pump as far as it will go. Retighten the adjusting bolt to hold the tensioner in this position. Remove the timing belt.

13. Loosen the head bolts in the reverse of the proper installation sequence. When all have been loosened, remove them. Then, pull the head off the engine block, rocking it slightly to break it loose.

14. Inspect the head with a straightedge and a flat feeler gauge of 0.002 in. (0.05mm) thickness. The tolerance for warping of a used head is 0.002 in. (0.05mm). The block deck must be flat within the same tolerance. The height of the head should be 3.5 in. (89mm) with a maximum machining limit of 0.012 in. (0.3mm).

To install:

15. Clean the combustion chambers of carbon with a scraper that is not excessively sharp, being careful not to damage the aluminum surface. Sharp edges in the combustion cham-

bers can cause detonation. Clean the gasket mating surfaces with a scraper and solvent.

16. The oil and water passages should be cleaned thoroughly. Also, blow compressed air through all the small oil passages to ensure that they are clear. Check that the EGR and air pump passages are also clear. Both gasket surfaces must be completely free of dirt and oil.

17. Install a new head gasket, without sealant, and position the head on the cylinder block. Install all the bolts finger tight. Torque the bolts in sequence. First step to 15 ft. lbs. (20 Nm), 2nd step to 25 ft. lbs. (35 Nm). Finally, to 51–54 ft. lbs. (69–74 Nm).

18. Align the matchmarks and install the distributor to the engine.

19. Install the timing belt on the crankshaft and camshaft sprockets and rotate the camshaft sprocket backward so the belt is tight on what is normally the tension side. Make sure all the timing marks are now aligned. That is, timing marks on the crankshaft sprocket and front case must align; and the marks on the camshaft sprocket and the tab on the cylinder head must be simultaneously aligned with the side of the belt away from the tensioner under tension. Now, loosen the timing belt tensioner adjusting bolt and allow spring tension to tension the belt. Make sure all timing marks are still aligned. If not, the belt is out of time and must be shifted with the tensioner shifted back, toward the water pump and locked there. Now, torque the adjusting bolt on the right side, working through a slot, to 15–18 ft. lbs. (20–25 Nm). After the tensioner adjusting bolt is torqued, torque the hinged mounting bolt located on the opposite side. Don't torque the mounting bolt first or the tension on the belt will be too great.

20. Turn the crankshaft one full turn in the normal direction of rotation. Loosen first the tensioner pivot bolt and then the adjusting bolt. Now torque them exactly as before. This extra step is necessary to ensure the timing belt is properly seated before final tension is adjusted.

21. Install the cylinder head cover and tighten the bolts to 13–16 inch lbs. (1.5–2.0 Nm).

22. Install the timing belt cover.

23. Install the intake manifold using a new gasket and tighten the bolts and nut to 12–14 ft. lbs. (16–19 Nm).

24. Install the exhaust manifold using a new gasket and tighten the nuts to 12–14 ft. lbs. (16–19 Nm).

25. Install the carburetor or throttle body assembly.

26. Install the distributor. Connect all hoses, lines, and the air cleaner. Reconnect the intake pipe on turbocharged engines.

27. Refill and bleed the cooling system. Operate the engine and check for leaks. After the engine has reached normal operating temperature, turn it off and remove the air cleaner and rocker cover. Retighten the cylinder head bolts to 58–61 ft. lbs. (78–83 Nm), in the proper sequence.

28. Reinstall the rocker cover and the air cleaner.

1.5L (VIN K) Engine

— **CAUTION** —
Fuel injection systems remain under pressure after the engine has been turned OFF. Properly relieve fuel pressure before disconnecting any fuel lines. Failure to do so may result in fire or personal injury.

1. Disconnect the negative battery cable.
2. Drain the engine coolant and remove the upper radiator hose.
3. Remove the breather hose between the rocker cover and the air cleaner.
4. Remove the air intake hose.
5. Relieve the fuel system pressure and disconnect the fuel hose from the delivery pipe.
6. Label and disconnect the vacuum hoses from the throttle body and intake manifold.
7. Disconnect the cables from the spark plugs.
8. Remove the ignition coil.
9. Remove the surge tank.
10. Remove the intake manifold.
11. Remove the heat shield and the exhaust manifold.
12. Remove the drive belts.
13. Remove the water pump pulley and the crankshaft pulley.
14. Remove the timing belt cover.
15. Move the timing belt tensioner toward the water pump and secure it.
16. Remove the timing belt.
17. Remove the rocker cover.
18. Use the special socket and gradually remove the cylinder head bolts in the reverse of the tightening sequence.
19. Remove the cylinder head.

To install:
20. Clean all gasket surfaces on the cylinder block and head.
21. Align the mark on the crankshaft sprocket with the timing mark on the front case.

22. Turn the camshaft sprocket so the timing mark is in the straight up position.
23. Position a new gasket on the cylinder block.
24. Carefully place the cylinder head on the engine block.
25. Install the bolts and torque them using two stages in the proper sequence to 51–54 ft. lbs. (70–75 Nm) on a cold engine.
26. Install the timing belt. Make sure that when the camshaft sprocket is turned slightly in reverse to tighten the tension side that the timing marks are aligned.
27. Install the rocker cover. Torque the bolts to 11–14 ft. lbs. (15–20 Nm).
28. Install the timing belt cover.
29. Install the water pump pulley and the crankshaft pulley.
30. Install the exhaust manifold and heat shield.
31. Use a new gasket and install the intake manifold.
32. Use a new gasket and install the surge tank.
33. Connect the fuel hose to the delivery pipe and connect all vacuum hoses.
34. Install the ignition coil and connect the spark plug cables.
35. Install the air intake hose and the breather hose.
36. Install the drive belts.
37. Install the upper radiator hose and refill the engine with coolant.
38. Connect the negative battery cable.

1.6L (VIN R), 1.8L (VIN M) and 2.0L (VIN F) Engines

— **CAUTION** —
Fuel injection systems remain under pressure after the engine has been turned OFF. Properly relieve fuel pressure before disconnecting any fuel lines. Failure to do so may result in fire or personal injury. Do not allow fuel spray or fuel vapors to come in contact with a spark or open flame. Keep a dry chemical fire extinguisher nearby. Never store fuel in an open container due to risk of fire or explosion.

1. Release the fuel system pressure.
2. Disconnect the negative battery cable.
3. Drain the cooling system.
4. Disconnect the accelerator cable, PCV hoses, breather hoses, spark plug cables and remove the valve cover.
5. Disconnect the fuel lines.

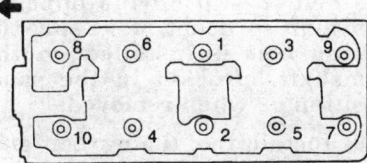

Crankshaft pulley side

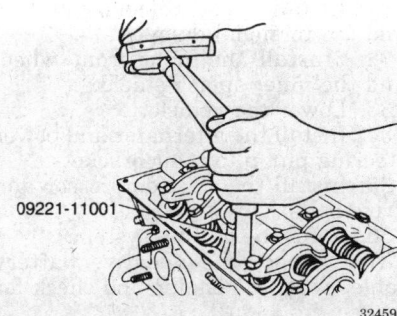

09221-11001

324594

Cylinder head bolt tightening sequence — 1.5L (VIN K) engine

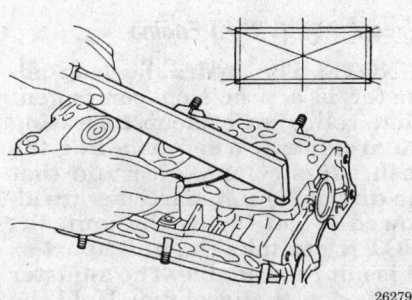

262792

Typical method of checking the cylinder head for warpage

6. Remove the intake and exhaust manifolds.

7. Rotate the crankshaft clockwise and align the timing marks so No. 1 piston will be at TDC of the compression stroke. At this time the timing marks on the camshaft sprocket and the upper surface of the cylinder head should coincide, and the dowel pin of the camshaft sprocket should be at the upper side.

NOTE: Always rotate the crankshaft in a clockwise direction. Make a mark on the back of the timing belt indicating the di-

rection of rotation so it may be reassembled in the same direction if it is to be reused.

8. Remove the timing belt upper and lower covers.

9. Remove the timing belt.

10. Remove the crank angle sensor.

11. Loosen the cylinder head mounting bolts in 3 steps in the reverse order of the tightening sequence. Lift off the cylinder head assembly and remove the head gasket.

To install:

12. Thoroughly clean and dry the mating surfaces of the head and block. Check the cylinder head for cracks, damage or engine coolant leakage. Remove scale, sealing compound and carbon. Clean oil passages thoroughly. Check the head for flatness. End to end, the head should be within 0.002 in. normally, with 0.008 in. the maximum allowed out of true. The total thickness allowed to be removed from the head and block is 0.008 in. maximum.

13. Place a new head gasket on the cylinder block with the identification marks at the front top (upward) position. Make sure the gasket has the proper identification mark for the engine. Do not use sealer on the gasket.

14. Carefully install the cylinder head on the block. Using 3 even steps, torque the head bolts in sequence to 76–83 ft. lbs. (105–115 Nm). These torques apply to a cold engine.

15. Align the punch mark on the crank angle sensor housing with the notch on the plate. Install the crank angle sensor into the cylinder head.

NOTE: The crank angle sensor can be installed even when the punch mark is positioned opposite the notch; however, the position results in incorrect fuel injection and ignition timing.

16. Install the timing belt and covers.

17. Install the intake and exhaust manifolds.

18. Apply sealer to the perimeter of the half-round seal and to the lower edges of the half-round portions of the belt-side of the new gasket. Install the valve cover.

19. Connect the accelerator cable.

20. Install the spark plug wires and the PCV valve.

21. Install the spark plug cable center cover.

22. Replace the O-rings and connect the fuel lines.

23. Install the air cleaner and intake hose. Connect the breather hose.

24. Change the engine oil and oil filter.

25. Fill the system with coolant.

26. Connect the negative battery cable, run the vehicle until the thermostat opens and fill the radiator completely.

27. Adjust the accelerator cable.

28. Check and adjust the idle speed and ignition timing.

29. Once the vehicle has cooled, recheck the coolant level.

3.0L (VIN T) Engine

———— CAUTION ————

Fuel injection systems remain under pressure after the engine has been turned OFF. Properly relieve fuel pressure before disconnecting any fuel lines. Failure to do so may result in fire or personal injury. Do not allow fuel spray or fuel vapors to come in contact with a spark or open flame. Keep a dry chemical fire extinguisher nearby. Never store fuel in an open container due to risk of fire or explosion.

1. Relieve the fuel pressure.

2. Disconnect the negative battery cable.

3. Drain the cooling system.

4. Remove the compressor drive belt and the air conditioning compressor from its mount and support it aside. Using a ½ in. drive breaker bar, insert it into the square hole of the serpentine drive belt tensioner, rotate it counterclockwise to reduce the belt tension and remove the belt. Remove the alternator and power steering pump from the brackets and move them aside.

5. Raise the vehicle and support safely. Remove the right front wheel and the inner splash shield.

6. Remove the crankshaft pulleys and the torsional damper.

7. Lower the vehicle. Using a floor jack and a block of wood positioned under the oil pan, raise the engine slightly. Remove the engine mount bracket from the timing cover end of the engine and the timing belt covers.

8. To remove the timing belt, perform the following procedures:

a. Rotate the crankshaft to position the No. 1 cylinder on TDC of its compression stroke; the crankshaft sprocket timing mark should align with the oil pan timing indicator and the camshaft sprocket timing marks (triangles) should align with the rear timing belt cover timing marks.

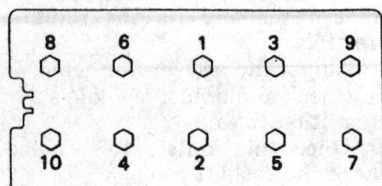

← Camshaft sprocket side

Cylinder head bolt tightening sequence — 1.6L (VIN R), 1.8L (VIN M) and 2.0L (VIN F) engines

09221-32001

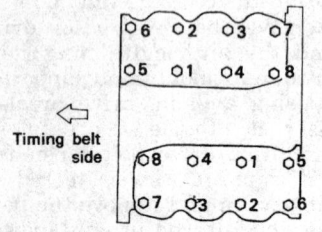

← Timing belt side

Cylinder head bolt tightening sequence — 3.0L (VIN T) Engine

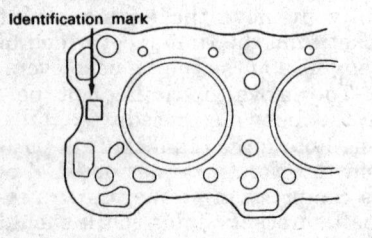

Identification mark

Install the head gasket with the identification mark facing up — 3.0L (VIN T) Engine

b. Mark the timing belt in the direction of rotation for reinstallation purposes.

c. Loosen the timing belt tensioner and remove the timing belt.

NOTE: When removing the timing belt from the camshaft sprocket, make sure the belt does not slip off of the other camshaft sprocket. Support the belt so it cannot slip off of the crankshaft sprocket and opposite side camshaft sprocket.

9. Remove the air cleaner assembly.

10. Label and disconnect the spark plug wires and the vacuum hoses.

11. Remove the valve cover.

12. Install auto lash adjuster retainer tools 09246-32000 (MD998443) or equivalent, on the rocker arms.

13. If removing the front cylinder head, matchmark the distributor rotor to the distributor housing and the housing to distributor extension locations. Remove the distributor and the distributor extension.

14. Remove the camshaft bearing assembly to cylinder head bolts but do not remove the bolts from the assembly. Remove the rocker arms, rocker shafts and bearing caps as an assembly, as required. Remove the camshafts from the cylinder head and inspect them for damage.

15. Remove the intake manifold assembly.

16. Remove the exhaust manifold.

17. Remove the cylinder head bolts in the reverse of the tightening sequence, (starting from the outside and working inward). Remove the cylinder head from the engine.

18. Clean the gasket mounting surfaces and check the heads for warpage; maximum warpage is 0.008 in. (0.20mm).

To install:

19. Install the new cylinder head gasket over the dowels on the engine block.

20. Install the cylinder head(s) on the engine and torque the cylinder head bolts, in sequence, using 3 even steps, to 70 ft. lbs. (95 Nm).

21. Install the exhaust manifold.

22. Install the intake manifold assembly.

23. Install the camshafts.

24. Install the rocker arms, rocker shafts and bearing caps as an assembly.

25. Install, if removed, the distributor and the distributor extension.

26. Install the valve cover.

27. Connect the spark plug wires and the vacuum hoses.

28. Install the air cleaner assembly.

NOTE: When installing the timing belt on the camshaft sprocket, use care not to allow the belt to slip off the opposite camshaft sprocket. Make sure the timing belt is installed on the camshaft sprocket in the same position as when removed.

29. Install the timing belt and covers.

30. Install the engine mount bracket.

31. Install the crankshaft pulleys and the torsional damper.

32. Install the right front wheel and the inner splash shield.

33. Lower the vehicle.

34. Install the alternator and power steering pump to their brackets.

35. Install the A/C compressor and drive belts.

36. Refill the cooling system.

37. Connect the negative battery cable. Start the engine and check for leaks.

38. Adjust the timing, as required.

Lash Adjusters

BLEEDING

Except 1.5L (VIN K) Engine

NOTE: The hydraulic lash adjuster is a precision component that relies on a clean operating environment. When handling the lash adjuster, make certain that no dirt or foreign particles are allowed to get inside the unit. DO NOT try to take the unit apart as it is not rebuildable. The adjuster is filled with diesel fuel. Hold the adjuster in the upright position so that the diesel fuel does not spill out. When cleaning the unit, only use clean diesel fuel.

To bleed the hydraulic lash adjusters when disassembling the rocker arms or as part of an engine overhaul, perform the following:

1. Mount the adjuster into special bleeding tool 09246-32100 or equivalent and immerse the tool and adjuster in a container of clean diesel fuel.

2. With air bleed wire 09246-32200 or equivalent inserted in the adjuster, lightly press down on the steel ball and compress the plunger four or five times.

3. Remove the air bleed wire from the adjuster and push down firmly on the plunger. If the plunger moves even slightly, repeat Steps 1 and 2

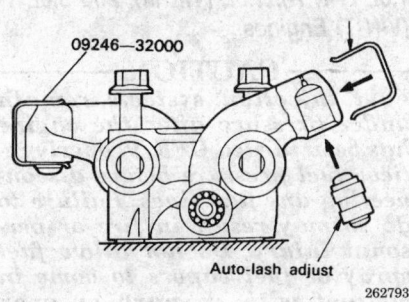

09246–32000

Auto-lash adjust

262793

Auto lash adjuster holding tool — 3.0L (VIN T) Engine

until the plunger stops moving. If the plunger continues to move, replace it.

1.5L (VIN K) Engine

1. Place the hydraulic lash adjuster (HLA) in a container of diesel fuel.

2. Insert a wire into the hole to open the check valve and pump the lash adjuster up and down about five times. Remove the wire and try to compress the lash adjuster. If it moves even slightly, replace it.

Valve Lash

ADJUSTMENT

1.3L (VIN M) and 1.5L (VIN J) Engines

1. Run the engine until the coolant temperature reaches 176°–205°F (80°–95°C).

2. Make sure the piston in the No. 1 cylinder is TDC on the compression stroke. Adjust cylinder numbers 1 and 2 intake valves and cylinder numbers 1 and 3 exhaust valves.

3. Loosen the locknut on the adjusting screw for valve to be adjusted.

4. Turn the adjusting screw while measuring the clearance with a feeler gauge. Clearance should be 0.006

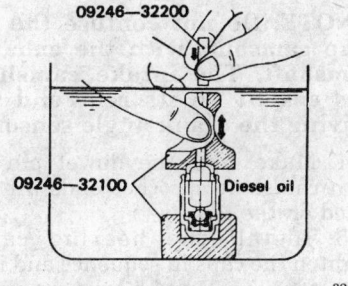

09246–32200

09246–32100   Diesel oil

320613

Bleed the lash adjuster in clean diesel fuel — except 1.5L (VIN K) Engine

inch (0.15 mm) for the intake valve and 0.010 inch (0.25 mm) for the exhaust valve. Tighten the locknut securely while holding the adjusting screw with a screwdriver to prevent it from turning.

5. Turn the crankshaft so the piston in the No. 4 cylinder is TDC on the compression stroke. Adjust cylinder numbers 3 and 4 intake valves and cylinder numbers 2 and 4 exhaust valves using the same specifications. Tighten the locknut securely while holding the adjusting screw with a screwdriver to prevent it from turning.

6. Check the clearance between the adjusting screw and the valve stem. Repeat the procedure if necessary.

1.5L (VIN K), 2.0L (VIN F) and 3.0L (VIN T) Engines

Valve clearance is not adjustable on these vehicles. If the valve makes noise, look for a leaky lash adjuster, low oil pressure or excessive camshaft wear.

Rocker Arms and Shafts

REMOVAL AND INSTALLATION

1.3L (VIN M) and 1.5L (VIN J) Engines

1. Disconnect the negative battery cable.

2. Remove the PCV hose running from the rocker cover and the air cleaner. Remove the air cleaner.

3. Remove the upper timing belt cover. Remove the rocker cover.

4. Loosen the rocker shaft mounting bolts but do not remove them. Remove each rocker shaft, rocker arms and springs as an assembly. Disassemble the whole assembly by progressively removing each bolt and then the associated springs and rockers, keeping all parts in the exact order of disassembly. The left and right springs have different tension ratings and free length. Observe the location of the rocker arms as they are removed. Exhaust and intake, right and left are different. Do not mix them up.

5. Check the rocker arm face contacting the cam lobe and the adjusting screw that contacts the valve stem for excess wear. Inspect the fit of the rockers on the shaft. Replace adjusting screws, rockers, and/or shafts that show excessive wear. Pay special attention to the contact pad ends of the rocker arms and the ball surface of the adjusting studs. Check

the diameter of the shaft at the rocker mounting points and subtract that number from the measured inside diameter of the corresponding rocker arm. Clearance should be 0.0005–0.0017 in. (0.013–0.043mm). The service limit is 0.004 in. (0.1mm). Check the rocker shaft bend. Total rocker shaft bend should be 0.002 in. (0.05mm). Check the spring free length. Maximum free length should be 2.1 in. (53.3mm) for the exhaust side springs; 2.6 in. (66mm) for intake side springs.

To install:

6. Assemble all the parts, noting the differences between intake and exhaust parts. The intake rocker shaft is much longer; the intake rocker shaft springs are over 3 in. long, while those for the exhaust side are less than 2 in. long; rockers are labeled 1-3 and 2-4 for the cylinder with which they are associated. Torque the rocker shaft mounting bolts to 15–19 ft. lbs. (20–26 Nm).

7. Adjust the valve clearances. This step may be omitted only if all parts are being reused.

8. Install the rocker cover with a new gasket, torquing the bolts to 12–18 inch lbs. (1.5–2.0 Nm).

9. Install the air cleaner and PCV valve. Connect the battery cable.

10. Run the engine at idle speed until it is hot. Then, unless valves did not require adjustment, remove the valve cover again and adjust the valve clearances with the engine hot.

11. Replace the rocker cover and timing belt cover, air cleaner and PCV valve.

1.5L (VIN K) Engine

1. Disconnect the negative battery cable.

2. Remove the breather hose and PCV valve from the cylinder head cover.

3. Remove the timing belt cover from the cylinder head cover.

4. Remove the cylinder head cover.

NOTE: Rocker arms must be installed in the same position they were removed from.

5. Gradually loosen the bolts securing the rocker arm shaft to the cylinder head.

6. Remove the shaft with the rocker arms as an assembly. Do not drop the lash adjusters into the cylinder head.

7. Remove the bolts from the shaft and slide the rocker arms off as needed. Keep them in order for reassembly.

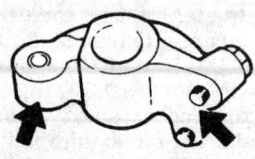

333748

Inspect the rocker arm for wear at the points shown — 1.3L (VIN M) and 1.5L (VIN J) engines

To install:

8. Inspect the rocker arms and shafts for wear and replace any parts that show excessive wear.

9. Check the lash adjusters by placing them is diesel oil and bleeding the air from them. This is done by inserting a small wire in the lash adjuster and pumping the lash adjuster up and down about five times. Remove the wire and try to compress the lash adjuster. If it moves even slightly, replace it.

10. Position the rocker shaft assemble on the head and tighten the bolts gradually to 15–18 ft. lbs. (20–24 Nm).

11. Install the cylinder head cover.

12. Install the timing belt cover to the cylinder head cover.

13. Connect the breather hose and the PCV valve.

14. Connect the negative battery cable.

1995 1.5L (VIN N) Engine

1. Disconnect the negative battery cable.

2. If equipped with a turbocharger, remove the air intake pipe and cover the throttle body and turbocharger ports to keep out dust and foreign objects.

3. Disconnect the PCV valve and breather hose from the valve cover.

4. Remove the valve cover and the upper timing belt cover.

5. Bring the No. 1 cylinder to TDC for the compression stroke.

6. Loosen the rocker assembly retaining bolts in crisscross pattern working from the edges of the assembly toward the center. Loosen each bolt one or two turns at a time. Do not loosen the No. 1 and No. 5 camshaft holders.

7. Lift the rocker assembly from the cylinder head. Leave the retain-

ing bolts in the holders to keep the rockers and shafts together.

NOTE: Cover the timing belt with a shop towel to protect it from spilled oil.

8. If the rocker and shaft assembly is to be disassembled, label each part for reassembly. The intake and exhaust springs and shafts are not interchangeable.

9. Check the rocker arm faces and cam lobes for wear and damage. Replace any damaged parts.

10. Remove the hydraulic lifter assemblies from their ports in the rocker arms. Bleed and inspect the lifters:

   a. Submerge the lifter into a cup of clean diesel oil.

   b. Insert a length of wire into the lifter's bleed hole.

   c. Use the wire to move the plunger up and down until no more air bubbles appear in the diesel oil.

   d. Remove the wire and press on the plunger with your finger. If the plunger moves, repeat the bleeding procedure. If the plunger continues to move under light pressure, even after repeated bleeding, replace the lifter.

NOTE: Store the lifters upright in a cup of diesel oil until they are installed.

To install:

11. Assemble the hydraulic lifters onto the rocker arms.

12. Assemble the rocker arm and shaft assemblies and lubricate them with clean engine oil.

13. Apply clean engine oil to the camshaft lobes and journals.

14. Install the rocker arm and shaft assembly. Install and only hand-tighten the retaining bolts.

15. Tighten the rocker assembly retaining bolts in a crisscross pattern working from the center of the head toward the edges to 14–18 ft. lbs. (20–24 Nm).

16. Apply a 0.4 in. (10mm) bead of sealant to the cylinder head surfaces of the No. 1 and No. 5 camshaft holders. Then, install the valve cover with a new gasket. Tighten the valve cover retaining bolts to 6–7.5 ft. lbs. (8–10 Nm), or 2.2–3.0 ft. lbs. (3–4 Nm) for turbocharged vehicles.

17. Install the upper timing belt cover.

18. Reconnect the PCV valve and breather hose.

19. Install the turbocharger air intake pipe.

20. Change the engine oil and filter.

21. Reconnect the negative battery cable.

1.6L (VIN R), 1.8L (VIN M), and 3.0L (VIN T) Engines

——— CAUTION ———

Fuel injection systems remain under pressure after the engine has been turned OFF. Properly relieve fuel pressure before disconnecting any fuel lines. Failure to do so may result in fire or personal injury. Do not allow fuel spray or fuel vapors to come in contact with a spark or open flame. Keep a dry chemical fire extinguisher nearby. Never store fuel in an open container due to risk of fire or explosion.

1. Relieve the fuel system pressure.

2. Disconnect battery negative cable.

3. Disconnect the accelerator cable.

4. Remove the timing belt cover and timing belt.

5. Remove the center cover, breather and PCV hoses, and spark plug cables.

6. Remove the rocker cover, semi-circular packing, throttle body stay, crankshaft angle sensor, both camshaft sprockets, and oil seals.

7. Loosen the bearing cap bolts in 2–3 steps. Label and remove all camshaft bearing caps.

NOTE: If the bearing caps are difficult to remove, use a plastic hammer to gently tap the rear part of the camshaft.

8. Remove the intake and exhaust camshafts.

9. Remove the rocker arms.

10. Remove the lash adjuster (valve lifter).

To install:

11. Bleed the lash adjuster.

12. Lubricate the lash adjuster and rocker arms with clean engine oil and install them in the cylinder head.

13. Lubricate the camshafts with heavy engine oil and position the camshafts on the cylinder head.

NOTE: Do not confuse the intake camshaft with the exhaust camshaft. The intake camshaft has a split on its rear end for driving the crank angle sensor.

14. Make sure the dowel pin on both camshaft sprocket ends are located on the top.

15. Install the bearing caps. Tighten the caps in sequence and in 2 or 3 steps. No. 2 and 5 caps are of the same shape. Check the markings on the caps to identify the cap number and intake/exhaust symbol. Only **L**

(intake) or **R** (exhaust) is stamped on No. 1 bearing cap. Also, make sure the rocker arm is correctly mounted on the lash adjuster and the valve stem end. Torque the retaining bolts to 15 ft. lbs. (20 Nm).

16. Apply a coating of engine oil to the oil seal. Using tool 09221-21000 or equivalent, press-fit the seal into the cylinder head.

17. Align the punch mark on the crank angle sensor housing with the notch in the plate. With the dowel pin on the sprocket side of the intake camshaft at top, install the crank angle sensor on the cylinder head.

NOTE: Do not position the crank angle sensor with the punch mark positioned opposite the notch; this position will result in incorrect fuel injection and ignition timing.

18. Install the valve cover.
19. Install the throttle body bracket and the semi-circular packing.
20. Install the center cover, breather and PCV hoses, and spark plug cables.
21. Install the timing belt and covers.
22. Connect the negative battery cable, and check for leaks.

Intake Manifold

REMOVAL AND INSTALLATION

1.3L (VIN M) Engine

1. Disconnect the negative battery cable.
2. Remove the air cleaner assembly.
3. Disconnect the fuel line and the EGR lines, if equipped with EGR. Tag and disconnect all vacuum hoses.
4. Disconnect the throttle position sensor and fuel cut-off solenoid wires.
5. Disconnect the throttle linkage.
6. If equipped with an automatic transaxle, disconnect the shift cable linkage.
7. Disconnect the power brake booster vacuum line.
8. Drain the cooling system.
9. Disconnect the choke water hose at the manifold.
10. Remove the heater and water outlet hoses, disconnect the water temperature sending unit.
11. Remove the mounting nuts that hold the manifold to the cylinder head. Remove the intake manifold.
   To install:
12. Clean all mounting surfaces. Before installing the manifold, coat

both sides of a new gasket with a gasket sealer.

NOTE: If equipped with jet air system, take care not to get any sealer into the jet air intake passage.

13. Install the intake manifold assembly to the engine block. Torque the mounting bolts/nuts to 12–14 ft. lbs. (16–19 Nm).
14. Reconnect the heater and water outlet hoses. Connect the water temperature sending unit.
15. Connect the brake booster vacuum line. Connect the choke hose at the manifold.
16. Connect the throttle and shift cable linkages, the fuel lines and all vacuum hoses. Install the air cleaner.
17. Refill the engine with coolant. Connect the negative battery cable.

1.5L (VIN J) Engine

1. Disconnect the air intake hose from the throttle body. Position the hose off to the side and out of the way.
2. Disconnect the accelerator cable from the throttle lever. To do this loosen the two cable adjusting bracket bolts so that there is enough play in the cable to disengage the cable end from the throttle lever. Move the cable out of the way.
3. Disconnect the water hose from the water outlet fitting.
4. Remove the throttle body and gasket.
5. Disconnect the PCV hose from the rocker cover and disconnect the brake vacuum hoses.
6. Disconnect all vacuum hoses from their respective connections on the intake manifold. Make sure that you label them to avoid confusion during installation. Disconnect the air intake pipe.
7. Relieve the fuel system pressure, then disconnect the high pressure fuel hose connection from the fuel delivery pipe. Cover the connection joint with a rag to absorb the excess fuel.
8. Unbolt and remove the air intake surge tank (and gasket) from the intake manifold.
9. Disconnect the fuel injector harness connectors.
10. Unbolt and remove the fuel delivery pipe (with pressure regulator) from the intake manifold. Take care not to drop the injectors when removing the delivery pipe.
11. Disconnect the wiring harness that runs between the water temperature gauge and the water temperature sensor assembly.

12. Remove the water outlet fitting and gasket from the intake manifold. Remove the thermostat.
13. Disconnect the spark plug wires from the distributor cap. Make sure that you label them first.
14. Remove the distributor and the ignition coil.
15. Remove the intake manifold stay.
16. Remove the intake manifold retaining nuts. Rock the intake manifold back and forth to separate it from the cylinder head. Remove the intake manifold gasket.
17. Clean all the intake manifold and cylinder head gasket surfaces making sure that all existing gasket material is removed. Do the same for the surge tank and all other gasket mating surfaces. Inspect the intake manifold and surge tank for cracks and check the coolant passages for restrictions. Replace all gaskets.
   To install:
18. Install the intake manifold gasket. Position the intake manifold to the cylinder head studs and lower it onto the gasket. Install the intake manifold retaining nuts and torque them to 11–14 ft. lbs., starting from the center and working outwards.
19. Install the intake manifold stay and torque the retaining bolts to 13–18 ft. lbs.
20. Install the ignition coil, distributor, and spark plug wires.
21. Install the thermostat into the intake manifold.
22. Install the water outlet fitting with a new gasket. Torque the retaining bolts to 12–14 ft. lbs.
23. Connect the wiring harness that runs between the temperature sensor and temperature gauge.
24. Install the fuel delivery pipe onto the intake manifold. Torque the retaining bolts to 7 ft. lbs.
25. Connect the fuel injector harness connectors.
26. Place the surge tank and gasket onto the intake manifold. Install the retaining bolts and torque them to 11–14 ft. lbs.
27. Connect the high pressure hose to the fuel delivery pipe.
28. Connect the intake manifold, brake and PCV vacuum hoses.
29. Install the throttle body with a new gasket. Torque the throttle body bolts to 7–9 ft. lbs.
30. Connect the water hose to the outlet fitting.
31. Connect the accelerator cable to the throttle lever and tighten the cable bracket adjusting bolts.
32. Connect the air intake hose to the throttle body.

33. Start the engine and check/adjust the ignition timing and the idle speed. Check for fuel and coolant leaks.

1.5L (VIN N and K), 1.6L (VIN R), 1.8L (VIN M) and 2.0L (VIN F) Engines

1. Relieve the fuel system pressure.
2. Disconnect the negative battery cable.
3. Remove the idle speed actuator.
4. Remove the intake air hose from the throttle body. If equipped with a turbocharger, disconnect the air intake pipe and cover the throttle body and turbocharger ports to keep out dirt and foreign objects.
5. Disconnect the accelerator cable.
6. Disconnect the PCV hose and the brake booster vacuum hose.
7. Disconnect the vacuum hose connections from the intake manifold and throttle body and label them if necessary for installation.
8. Disconnect the high pressure fuel hose from the fuel delivery pipe.
9. Remove the intake manifold bracket.
10. Remove the surge tank assembly and gasket.
11. Disconnect the fuel injector connectors.
12. Remove the fuel delivery pipe with the fuel injectors and pressure regulator attached.
13. Remove the insulator from the intake manifold and disconnect the heater hose.
14. Unbolt the intake manifold bracket.
15. Loosen the intake manifold bolts in a crisscross pattern. Remove the intake manifold and gasket from the cylinder head.
   To install:
16. Clean the sealing surface of the cylinder head and the intake manifold. Use a new gasket and install the manifold on the cylinder head. Torque the nuts and bolts in a crisscross pattern to 11–15 ft. lbs. (15–20 Nm).
17. Install new O-rings onto the fuel injectors. Install the fuel delivery pipe and fuel injector assembly on the intake manifold.
18. Connect the fuel injector connectors.
19. Use a new gasket and install the surge tank to the intake manifold. Torque the nuts and bolts in a crisscross pattern to 11–15 ft. lbs. (15–20 Nm).
20. Install the intake manifold bracket and tighten the bolts 13–18 ft. lbs. (18–25 Nm).

21. Connect the high pressure fuel hose to the delivery pipe using new sealing washers.
22. Connect the vacuum hoses, PCV, and brake booster hoses.
23. Connect the accelerator cable to the throttle body.
24. Connect the intake air hose to the throttle body. On turbocharged vehicles, reconnect the intake pipe to the turbocharger.
25. Install the idle speed actuator.
26. Connect the negative battery cable and check for fuel leaks.

3.0L (VIN T) Engine

——————— **CAUTION** ———————
Fuel injection systems remain under pressure after the engine has been turned OFF. Properly relieve fuel pressure before disconnecting any fuel lines. Failure to do so may result in fire or personal injury. Do not allow fuel spray or fuel vapors to come in contact with a spark or open flame. Keep a dry chemical fire extinguisher nearby. Never store fuel in an open container due to risk of fire or explosion.

1. Relieve the fuel system pressure.
2. Disconnect the negative battery cable.
3. Drain the cooling system.
4. Remove the throttle body to air cleaner hose.
5. Remove the throttle body and transaxle kickdown linkage.
6. Remove the AIS motor and TPS wiring connectors from the throttle body.
7. Remove and label the vacuum hose harness from the throttle body.
8. From the air intake plenum, remove the PCV and brake booster hoses and the EGR tube flange.
9. Disconnect and label the charge and temperature sensor wiring at the intake manifold.
10. Remove the vacuum connections from the air intake plenum vacuum connector.
11. Remove the fuel hoses from the fuel rail.
12. Remove the air intake plenum mounting bolts and the plenum.
13. Remove the vacuum hoses from the fuel rail and pressure regulator.
14. Disconnect the fuel injector wiring harness from the engine wiring harness.
15. Remove the fuel pressure regulator mounting bolts and the regulator from the fuel rail.

16. Remove the fuel rail mounting bolts and the fuel rail from the intake manifold.
17. Separate the radiator hose from the thermostat housing and heater hoses from the heater pipe.
18. Remove the intake manifold mounting bolts and the manifold from the engine.
19. Clean the gasket mounting surfaces on the engine and intake manifold.
   To install:
20. Using new gaskets, position the intake manifold on the engine and install the mounting nuts and washers.

NOTE: The adhesive side of the gasket should face upward toward the intake manifold.

21. Torque the mounting nuts gradually and evenly, in sequence, to 15 ft. lbs. (20 Nm).
22. Make sure the injector holes are clean. Lubricate the injector O-rings with a drop of clean engine oil and install the injector assembly onto the engine.
23. Install and torque the fuel rail mounting bolts to 10 ft. lbs. (14 Nm).
24. Install the fuel pressure regulator onto the fuel rail.
25. Install the fuel supply and return tube and the vacuum crossover hold-down bolt.
26. Connect the fuel injection wiring harness to the engine wiring harness.
27. Connect the vacuum harness to the fuel pressure regulator and fuel rail assembly.
28. Remove the cover from the lower intake manifold and clean the mating surface.
29. Place the intake plenum gasket with the beaded sealant side up, on the intake manifold. Install the air intake plenum and torque the mounting bolts gradually and evenly, in sequence, to 10 ft. lbs. (14 Nm).
30. Install the fuel hoses to the fuel rail.
31. Install the vacuum connections to the air intake plenum vacuum connector.
32. Connect the charge and temperature sensor wiring at the intake manifold.
33. Install the PCV and brake booster hoses and the EGR tube flange.
34. Install the vacuum hose harness to the throttle body.
35. Install he AIS motor and TPS wiring connectors to the throttle body.
36. Install the throttle body and transaxle kickdown linkage.

37. Install the throttle body to air cleaner hose.

38. Refill the cooling system. Connect the negative battery cable and check for leaks using the Multi use tester to activate the fuel pump.

Exhaust Manifold

REMOVAL AND INSTALLATION

1.3L (VIN M), 1.5L (VINs J and N), 1.6L (VIN R), 1.8L (VIN M) and 2.0L (VIN F) Engines

NOTE: A hot engine should be allowed to cool before the manifold is removed.

1. Disconnect the negative battery cable.
2. Remove the air cleaner.
3. Remove the exhaust manifold heat shield. Soak all manifold nuts and studs with a liquid penetrate.

NOTE: Only 6-point sockets should be used on the exhaust manifold bolts to prevent the bolt heads from being rounded off.

4. Raise and safely support the vehicle.
5. Disconnect the front exhaust pipe from the exhaust manifold.
6. Disconnect and remove the oxygen sensor. If there is a secondary air line connected to the exhaust manifold, disconnect it.
7. Loosen the manifold nuts in a crisscross pattern. Support the manifold and remove all attaching nuts and washers. Slide the manifold from the cylinder head.
8. If the manifold is equipped with a catalytic converter, unbolt it to separate it from the converter. If necessary, rock the converter to break it loose.
9. Thoroughly clean the sealing surfaces on the cylinder head and manifold. Replace any nuts, washers or studs that are excessively rusted or may have been damaged during removal. Use a straightedge to check the manifold and cylinder head sealing surfaces for flatness. Correct problems by replacing the manifold or machining the cylinder head surface.
   To install:
10. If equipped, assemble the catalytic converter onto the manifold using a new gasket.
11. Install a new exhaust manifold gasket. Install the exhaust manifold.

Install and hand-tighten all the manifold fasteners.

12. For 1.3L (VIN M) and 1.5L (VINs J and N) engines, torque the exhaust manifold nuts to 14 ft. lbs. (20 Nm) in a crisscross pattern. If the manifold has an integral catalytic converter, torque the mounting nuts to 22–26 ft. lbs. (30–35 Nm).

13. For 1.6L (VIN R), 1.8L (VIN M) and 2.0L (VIN F) engines, torque the exhaust manifold nuts to 18–22 ft. lbs. (25–30 Nm), alternately and in several stages.

14. Install the oxygen sensor and secondary air line.

15. Connect the front exhaust pipe to the exhaust manifold.

16. Lower the vehicle.

17. Install the exhaust manifold heat shield.

18. Install the air cleaner.

19. Connect the negative battery cable. Operate the engine and check for air leaks.

1.5L (VIN K) Engines

1. Disconnect the negative battery cable. Remove the heat shield on the exhaust manifold. With the manifold cool, soak all manifold nuts and studs with a liquid penetrate.

NOTE: Only 6-point sockets should be used on the exhaust manifold nuts to prevent the nuts from being rounded off.

2. Disconnect the oxygen sensor connector to prevent wire damage when the exhaust pipe is disconnected from the manifold.

3. Disconnect the exhaust pipe at the exhaust manifold.

4. Remove all attaching nuts and washers. Slide the manifold from the cylinder head.

5. Remove the exhaust manifold gasket.

6. Thoroughly clean the sealing surfaces on the cylinder head and manifold. Replace any nuts, washers or studs that are excessively rusted or may have been damaged during removal. Use a straightedge to check the manifold and cylinder head sealing surfaces for flatness. Correct problems by replacing the manifold or machining the cylinder head surface.
   To install:
7. Install a new gasket on the cylinder head.
8. Make sure all the nuts turn freely, oiling them lightly, if necessary. Also, make sure all the studs

are screwed all the way into the block.

9. Place the manifold in position and install all washers and nuts hand tight.

10. Torque the exhaust manifold-to-cylinder head nuts to 11–15 ft. lbs. (15–20 Nm) starting from the center and working outward.

11. Use a new gasket and connect the exhaust pipe to the manifold.

12. Install the heat shield.

13. Connect the oxygen sensor.

14. Connect the negative battery cable. Operate the engine and check for leaks.

3.0L (VIN T) Engine

1. Disconnect the negative battery cable.

2. Raise the vehicle and support safely.

3. Disconnect the exhaust pipe from the rear exhaust manifold at the articulated joint.

4. Disconnect the EGR tube from the rear manifold, and unplug the oxygen sensor wire.

5. Remove the crossover pipe to manifold bolts.

6. Remove the rear manifold to cylinder head nuts and the manifold.

7. Lower the vehicle and remove the heat shield from the manifold.

8. Remove the front manifold nuts from the cylinder head. Remove the manifold.

9. Clean the gasket mounting surfaces. Inspect the manifolds for cracks, flatness and/or damage.
   To install:
10. When installing, the numbers 1–3–5 on the gaskets are used with the rear cylinders and 2–4–6 are on the gasket for the front cylinders. Torque the front manifold mounting nuts to 14 ft. lbs. (20 Nm).

11. Install the heat shield.

12. Raise and safely support the vehicle.

13. Torque the rear manifold mounting nuts to 14 ft. lbs. (20 Nm).

14. Install the crossover pipe to the manifold.

15. Connect the EGR tube and oxygen sensor wire.

16. Connect the exhaust pipe to the rear exhaust manifold, at the articulated joint.

17. Lower the vehicle.

18. Connect the negative battery cable, and check the manifolds for leaks.

Turbocharger

REMOVAL AND INSTALLATION

1.5L (VIN N) Engine

NOTE: Clean and fresh engine oil and coolant are essential to the proper operation of turbocharged engines. Due to the high operating temperatures of turbocharged engines, contaminated oil and coolant can cause extensive and costly turbocharger and engine damage. Use only engine oil approved for turbocharged vehicles.

1. Disconnect the negative battery cable.
2. Drain the engine oil into a sealable container.
3. Drain the coolant to a level below the upper radiator hose.
4. Remove the turbocharger air intake pipe and the air intake hose. Cover both the throttle body and turbocharger ports to keep out any dust or foreign objects.
5. Raise and safely support the vehicle.
6. Remove the splash shield.
7. Remove the exhaust manifold heat shield.
8. Disconnect the oil and cooler lines from the turbo housing. Catch any fluid in a drain pan.
9. Disconnect the front exhaust pipe from the turbocharger discharge pipe.
10. Remove the turbocharger discharge pipe and bracket from the outlet (exhaust) side of the turbo.
11. Unbolt the turbocharger from the exhaust manifold mounting bolts and remove it. The exhaust manifold may be removed with the turbocharger attached to it if so desired.

NOTE: The turbocharger cannot be disassembled and rebuilt.

To install:

12. Install the turbo in position on the manifold, using a new gasket. Tighten the turbo-to-manifold bolts to 18 to 25 ft. lbs. (25 to 35 Nm).
13. Connect the discharge pipe and bracket to the turbo, tighten the bolts to 18 to 25 ft. lbs. (25 to 35 Nm).
14. Reconnect the exhaust pipe to the discharge pipe and tighten the bolts to 22–29 ft. lbs. (30–40 Nm).
15. Prior to connecting the turbocharger oil lines, pour clean engine oil into the turbocharger through the inlet port. Follow any preparation instructions that may come with the new turbocharger.

16. Connect the oil and cooler lines to the turbo housing.
17. Install the exhaust manifold heat shield.
18. Install the air intake pipe and hose.
19. Change the engine oil filter. Refill the engine with clean oil.
20. Install the splash shield.
21. Lower the vehicle.
22. Refill and bleed the cooling system.
23. Connect the negative battery cable, run the engine and check for exhaust, oil, and coolant leaks.

Front Crankshaft Seal

REMOVAL AND INSTALLATION

1.5L (VIN K) Engine

1. Disconnect the negative battery cable.
2. Loosen the water pump pulley bolts.
3. Remove the alternator belt.
4. Remove the water pump pulley.
5. Remove the crankshaft pulley.
6. Remove the timing belt cover.
7. Torn the crankshaft clockwise and align the timing marks.
8. Move the timing belt tensioner pulley toward the water pump and remove the timing belt.
9. Remove the crankshaft sprocket.
10. Carefully pry the oil seal out of the front case assembly. Do not damage the crankshaft sealing surface or the bore of the seal.

To install:

11. Apply clean engine oil to the oil seal and install the seal using 09221-21000 seal installer or equivalent.
12. Install the crankshaft sprocket. Torque the bolt to 110–118 ft. lbs. (150–160 Nm).
13. Install the timing belt.
14. Install the front cover.

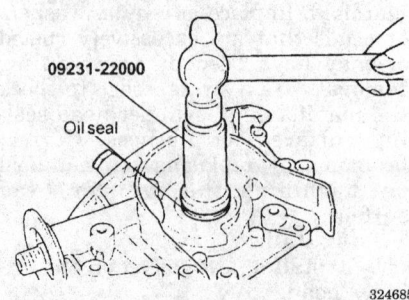

Front crankshaft oil seal installation — 1.5L (VIN K) engine

15. Install the water pump and the crankshaft pulleys. Torque the crankshaft pulley bolt to 7.2–8.7 ft. lbs. (10–12 Nm).
16. Install the alternator belt.
17. Connect the negative battery cable.

1.3L (VIN M), 1.5L (VINs J and N), 1.6L (VIN R), 1.8L (VIN M) and 2.0L (VIN F) Engines

1. Disconnect the negative battery cable.
2. Remove the valve cover.
3. Remove the crankshaft pulley retainer bolts and remove the pulley.
4. Remove the timing belt covers and timing belt(s).

NOTE: When reusing the timing belt(s), be sure to mark the direction of rotation on the belt. This will extend belt life, ensuring the same direction of rotation.

5. Remove the crankshaft sprocket retainer bolt and washer from the sprocket, if used, and remove sprocket. If sprocket is difficult to remove, the appropriate puller may be used.
6. Pry the seal from the bore. Take care not to damage the crankshaft.

To install:

7. Using proper size driver, install a new seal.
8. Install the crankshaft sprocket and torque the retaining bolt to 80–94 ft. lbs. (110–130 Nm).
9. Install the timing belt(s) and timing belt covers.
10. Install the crankshaft pulley. Tighten the mounting bolt to 14–22 ft. lbs. (20–30 Nm).
11. Install the valve cover.
12. Connect the negative battery cable.
13. Start the engine and check for leaks.

3.0L (VIN T) Engine

1. Disconnect the negative battery cable.
2. Remove the accessory drive belts.
3. Remove the crankshaft pulley.
4. Remove the timing belt covers and the timing belt.
5. Remove the crankshaft sprocket.
6. Pry out the oil seal with a flat prying tool, being careful not to damage the crankshaft.

To install:

7. Coat the lip of the new seal with oil and install the seal using the proper seal driver.

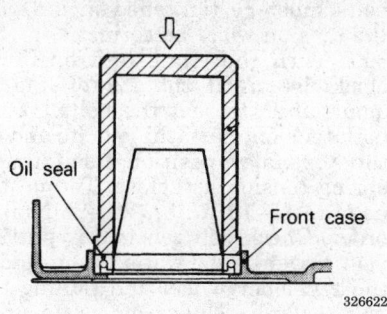

Front crankshaft oil seal installation — 1.6L (VIN R), 1.8L (VIN M) and 2.0L (VIN F) engines

8. Install the crankshaft sprocket and the timing belt.

9. Install the timing belt covers.

10. Install the crankshaft pulley.

11. Install the accessory drive belts.

12. Connect the negative battery cable.

13. Start the engine and check for proper operation.

Timing Belt, Sprockets, Tensioner and Front Cover

ADJUSTMENT

1.3L (VIN M) and 1.5L (VINs J and N) Engines

1. Rotate the crankshaft clockwise and align the timing marks so No. 1 piston will be at TDC of the compression stroke. Disconnect the negative battery cable.

2. Remove the timing belt covers.

3. Loosen the tensioner lower mounting bolt first, then the uppermost bolt.

4. Check to insure that the timing marks are in correct alignment and secure, tightening the uppermost bolt first.

5. Rotate the crankshaft one revolution in operating direction (clockwise), and realign the timing marks. Loosen the tensioner lower mounting bolt first, then the uppermost bolt.

6. Retighten the attaching bolts, uppermost first, to 14–20 ft. lbs. (20–27 Nm).

7. Apply a moderate pressure to the tension side of the timing belt and measure the belt deflection. The inner (cog) side of the belt should be depressed to the center of the tensioner mounting bolt head. If the deflection point is correct, the tension adjustment of the timing belt is correct.

8. If a tension gauge is used, measure the tension in the middle of the tension side span. The desired reading is 32–47 lbs.

9. Install the timing belt covers and all related components.

10. Reconnect the negative battery cable.

1.5L (VIN K) Engine

1. Disconnect the negative battery cable.

2. Remove the timing belt cover.

3. Align the crankshaft and camshaft timing marks.

4. Loosen the timing belt tensioner bolts 1 and 2 in that order. This will apply spring tension to the timing belt.

5. Tighten the bolts 1 and 2 in the same order.

6. Turn the crankshaft clockwise in the normal direction of rotation until the timing mark on the crankshaft sprocket is aligned again.

7. Loosen the timing belt tensioner bolts 1 and 2 in that order. This will apply spring tension to the timing belt.

8. Tighten the bolts 1 and 2 in the same order. Torque the bolts to 14–20 ft. lbs. (20–27 Nm)

9. Verify that the belt tension is correct by squeezing the timing belt and tensioner pulley together with a force of about 11 lbs. (49 N). The tip of the timing belt cog should deflect across ½ of the bolt head radius (across the flats).

1.6L (VIN R), 1.8L (VIN M) and 2.0L (VIN F) Engines

1. Disconnect the negative battery cable.

2. Remove the timing belt covers.

3. Adjust the silent shaft (inner) belt tension first. Loosen the idler pulley center bolt so the pulley can be moved.

4. Move the pulley by hand so the long side of the belt deflects about ¼ inch.

5. Hold the pulley tightly so the pulley cannot rotate when the bolt is tightened. Tighten the bolt to 15 ft. lbs. (20 Nm) and recheck the amount of deflection.

6. To adjust the timing (outer) belt, turn the crankshaft ¼ turn counterclockwise, then turn it clockwise to move No. 1 cylinder to TDC.

7. Loosen the center bolt. Using tool 09244-28100 or equivalent and a torque wrench, apply a torque of 1.88–2.03 ft. lbs. (2.6–2.8 Nm). If the body of the vehicle interferes with the special tool and the torque wrench, use a jack and slightly raise the engine assembly. Holding the tensioner pulley, tighten the center bolt.

8. Screw special tool 09244-28000 or exact equivalent into the engine left support bracket until its end makes contact with the tensioner arm. At this point, screw the special tool in some more and remove the set wire attached to the auto tensioner, if wire was not previously removed. Then remove the special tool.

9. Rotate the crankshaft 2 complete turns clockwise and let it sit for approximately 15 minutes. Then, measure the auto tensioner protrusion (the distance between the tensioner arm and auto tensioner body) to ensure that it is within 0.15–0.18 in. (3.8–4.5mm). If out of specification, repeat Steps 1 to 4 until the specified value is obtained.

10. If the timing belt tension adjustment is being performed with the engine mounted in the vehicle, and clearance between the tensioner arm and the auto tensioner body cannot be measured, the following alternative method can be used:

　a. Screw in special tool 09244-28000 or equivalent, until its end makes contact with the tensioner arm.

　b. After the special tool makes contact with the arm, screw it in some more to retract the auto tensioner pushrod while counting the number of turns the tool makes until the tensioner arm is brought into contact with the auto tensioner body. Make sure the number of turns the special tool makes conforms with the standard value of 2½ to 3 turns.

　c. Install the rubber plug to the timing belt rear cover.

11. Install the timing belt covers and all related items.

12. Connect the negative battery cable.

3.0L (VIN T) Engine

NOTE: The timing belt does not normally require adjustment, however if noise results from the belt hitting the cover, make the following adjustment.

1. Disconnect the negative battery cable.

2. If equipped with air conditioning, remove the compressor drive belt.

3. Remove the access cover located in the lower timing belt cover.

4. Loosen the timing belt tensioner mounting bolt 1–2 turns.

5. Rotate the crankshaft clockwise two revolutions.

6. Tighten the timing belt tensioner mounting bolt.

7. Install the access cover and the air compressor drive belt.

8. Reconnect the negative battery cable.

REMOVAL AND INSTALLATION

1.3L (VIN M) and 1.5L (VINs J and N) Engines

1. Disconnect the negative battery cable.

2. Remove the engine accessory drive belts.

3. Remove the water pump pulley.

4. Remove the timing belt cover.

5. Rotate the crankshaft clockwise and align the timing marks so the No. 1 piston will be at TDC of the compression stroke. Loosen the tensioning bolt in the slotted portion of the tensioner, and the pivot bolt on the timing belt tensioner. Move the tensioner as far as it will go toward the water pump. Tighten the adjusting bolt. If the belt is to be reused, mark its direction of rotation with an arrow.

6. Remove the timing belt.

7. Remove the camshaft sprocket as required.

8. Remove the crankshaft sprocket bolts and remove the crankshaft sprocket and flange, noting the direction of installation for each. Remove the timing belt tensioner.

9. Inspect the belt thoroughly. The back surface must be pliable and rough. Replace the belt if it is hard and glossy. Replace the belt if it shows any cracks or missing teeth. The canvas cover should be intact on all the teeth. If rubber is exposed anywhere, the belt should be replaced.

10. Inspect the tensioner for grease leaking from the grease seal and any roughness in rotation. Replace a tensioner if either condition is present.

11. The sprockets should be inspected and replaced if there is any sign of damaged teeth or cracking. Do not immerse the sprockets in solvent, as solvent that has soaked into the metal may cause deterioration of the timing belt later. Do not clean the tensioner in solvent either, as this may wash the grease out of the bearing.

To install:

12. Install the flange and crankshaft sprocket. The flange must go on first with the chamfered area outward. The sprocket is installed with the boss forward and the studs for the fan belt pulley outward. Install and torque the crankshaft sprocket bolt to 51–72 ft. lbs. (69–98 Nm), or 140–147 ft. lbs. (190–200 Nm) for 1993–95 Scoupe vehicles. Install the camshaft sprocket and bolt, torquing it to 47 to 54 ft. lbs. (64 to 74 Nm), or 58–72 ft. lbs. (80–100 Nm) for 1993–95 Scoupe vehicles.

13. Align the camshaft sprocket timing marks. Check that the crankshaft timing marks are still in alignment (the locating pin on the front of the crankshaft sprocket is aligned with a mark on the front case).

14. Mount the spring and spacer onto the tensioner. Then, install the bolts and tighten the adjusting bolt slightly with the tensioner moved as far as possible away from the water pump. Install the free end of the spring into the locating tang on the front case. Position the belt over the crankshaft sprocket and then over the camshaft sprocket. Slip the back of the belt over the tensioner wheel. Turn the camshaft sprocket in the opposite of its normal direction of rotation until the straight side of the belt is tight and the timing marks align.

15. Loosen the tensioner mounting bolts so the tensioner works under spring pressure without the interference of any friction.

16. Torque the tensioner adjusting bolt to 15–18 ft. lbs. (20–26 Nm). Then, torque the tensioner pivot bolt to 15–18 ft. lbs. (20–26 Nm). The bolts must be tightened in order or the tension won't be correct.

17. Turn the crankshaft one turn clockwise until the timing marks again align to seat the belt. Loosen both tensioner attaching bolts and let the tensioner position itself under spring tension as before. Torque the bolts to 15–18 ft. lbs. (20–26 Nm) in order. Check belt tension by putting your finger on the water pump side of the tensioner wheel and pulling the belt toward . Under moderate pressure, the belt should move toward the pump until the teeth are about $4\frac{1}{2}$ of the way across the head of the tensioner adjusting bolt.

18. Install the timing belt covers.

19. Install the crankshaft pulley, making sure the pin on the crankshaft sprocket fits through the hole in the rear surface of the pulley. Install the bolts and torque to 7.5–8.5 ft. lbs. (9–12 Nm).

20. Install the water pump pulley.

21. Install and tension the accessory drive belts.

22. Connect the negative battery cable.

1.5L (VIN K) Engine

1. Disconnect the negative battery cable. Remove the timing belt cover.

2. Rotate the crankshaft clockwise and align the timing marks so No. 1 piston will be at TDC of the compression stroke. Loosen the tensioning bolt (in the slotted portion of the tensioner) and the pivot bolt on the timing belt tensioner. Move the tensioner as far as it will go toward the water pump. Tighten the adjusting bolt. Mark the timing belt with an arrow showing direction of rotation if the belt is to be reused.

3. Remove the timing belt.

4. Remove the camshaft sprocket as required.

5. Remove the crankshaft sprocket bolts and remove the crankshaft sprocket and flange, noting the direction of installation for each. Remove the timing belt tensioner.

6. Inspect the belt thoroughly. The back surface must be pliable and rough. If it is hard and glossy, the belt should be replaced. Any cracks in the belt backing or teeth or missing teeth mean the belt must be replaced. The canvas cover should be intact on all the teeth. If rubber is exposed anywhere, the belt should be replaced.

7. Inspect the tensioner for grease leaking from the grease seal and any roughness in rotation. Replace a tensioner for either defect.

8. The sprockets should be inspected and replaced if there is any

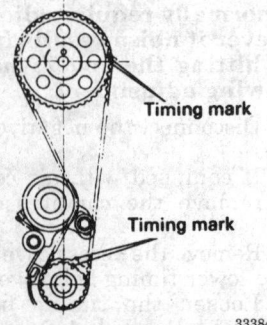

Timing mark identification — 1.3L (VIN M) and 1.5L (VINs J and N) engines

333847

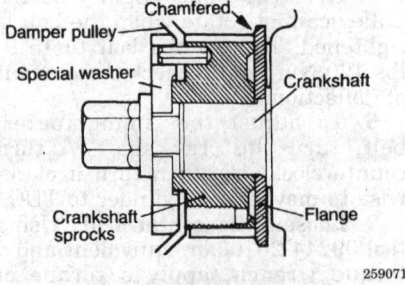

Crankshaft sprocket installation — 1.3L (VIN M) and 1.5L (VINs J and N) engines

259071

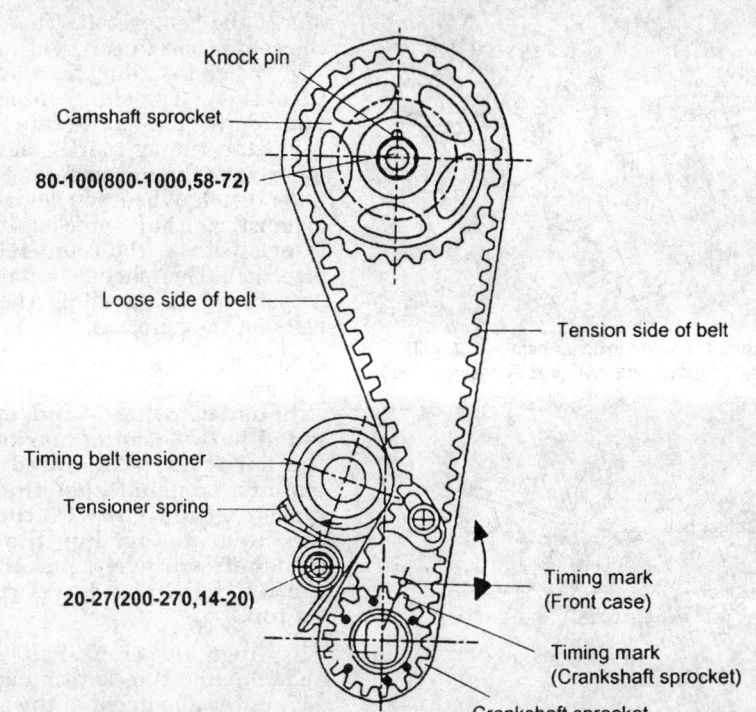

Knock pin

Camshaft sprocket

80-100(800-1000,58-72)

Loose side of belt

Tension side of belt

Timing belt tensioner

Tensioner spring

20-27(200-270,14-20)

Timing mark
(Front case)

Timing mark
(Crankshaft sprocket)

Crankshaft sprocket

TORQUE : Nm (kg.cm, lb.ft)

324589

Timing mark locations — 1.5L (VIN K) Engine

sign of damaged teeth or cracking. Do not immerse sprockets in solvent, as solvent that has soaked into the metal may cause deterioration of the timing belt later. Do not clean the tensioner in solvent either, as this may wash the grease out of the bearing.

To install:

9. Install the flange and crankshaft sprocket. The flange must go on first with the chamfered area outward. The sprocket is installed with the boss forward and the studs for the fan belt pulley outward. Install and torque the crankshaft sprocket bolt to 110–118 ft. lbs. (150–160 Nm). If removed, install the camshaft sprocket. Torque the bolt to 59–73 ft. lbs. (80–100 Nm).

10. Align the camshaft sprocket timing mark. Check that the crankshaft timing marks are still in alignment (the locating pin on the front of the crankshaft sprocket is aligned with a mark on the front case).

11. Mount the tensioner, spring and spacer with the bottom end of the spring free. Then, install the bolts and tighten the adjusting bolt slightly with the tensioner moved as far as possible away from the water pump. Install the free end of the spring into the locating tang on the front case. Position the belt over the

crankshaft sprocket and then over the camshaft sprocket. Slip the back of the belt over the tensioner wheel. Turn the camshaft sprocket in the opposite of its normal direction of rotation until the straight side of the belt is tight and make sure the timing marks align. If not, shift the belt one tooth at a time in the appropriate direction until this occurs.

12. Loosen the tensioner mounting bolts so the tensioner works, without the interference of any friction, under spring pressure. Make sure the belt follows the curve of the camshaft pulley so the teeth are engaged all the way around.

13. Correct the path of the belt, if necessary. Torque the tensioner adjusting bolt to 15–18 ft. lbs. (20–26 Nm). Then, torque the tensioner pivot bolt to the same figure. Bolts must be torqued, in order, or tension won't be correct.

14. Turn the crankshaft one turn clockwise until timing marks again align to seat the belt. Loosen both tensioner attaching bolts and let the tensioner position itself under spring tension as before. Torque the bolts in order. Check belt tension by putting a finger on the water pump side of the tensioner wheel and pull the belt toward it. The belt should move toward the pump until the teeth are about 1/4

of the way across the head of the tensioner adjusting bolt. Retension the belt, if necessary.

15. Install the timing belt covers.

16. Install the crankshaft pulley, making sure the pin on the crankshaft sprocket fits through the hole in the rear surface of the pulley. Install the bolts and torque to 7.5–8.5 ft. lbs. (9–12 Nm).

17. Connect the negative battery cable.

1.6L (VIN R), 1.8L (VIN M) and 2.0L (VIN F) Engines

1. Disconnect the negative battery cable.

2. Remove the timing belt upper and lower covers.

3. Rotate the crankshaft clockwise and align the timing marks so No. 1 piston will be at TDC of the compression stroke. At this time, the timing marks on the camshaft sprocket and the upper surface of the cylinder head should coincide, and the dowel pin of the camshaft sprocket should be at the upper side.

NOTE: Always rotate the crankshaft in a clockwise direction. Make a mark on the back of the timing belt indicating the direction of rotation so it may be reassembled in the same direction if it is to be reused.

4. Remove the auto tensioner and remove the outermost timing belt.

5. Remove the timing belt tensioner pulley, tensioner arm, idler pulley, oil pump sprocket, special washer, flange and spacer.

6. Remove the silent shaft (inner) belt tensioner and remove the inner belt.

To install:

7. Align the timing marks on the crankshaft sprocket and the silent shaft sprocket. Fit the inner timing belt over the crankshaft and silent shaft sprocket. Ensure that there is no slack in the belt.

8. While holding the inner timing belt tensioner with your fingers, adjust the timing belt tension by applying a force towards the center of the belt, until the tension side of the belt is taut. Tighten the tensioner bolt.

NOTE: When tightening the bolt of the tensioner, ensure that the tensioner pulley shaft does not rotate with the bolt. Allowing it to rotate with the bolt can cause excessive tension on the belt.

9. Check belt for proper tension by depressing the belt on its long side

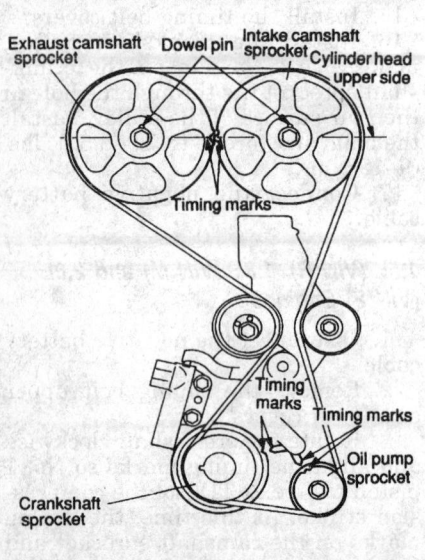

Timing marks when No. 1 cylinder is at TDC on compression — 1.6L (VIN R), 1.8L (VIN M) and 2.0L (VIN F) engines

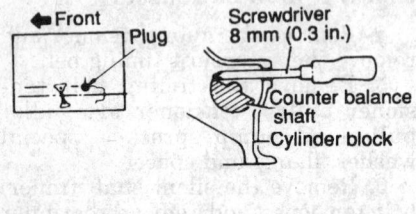

Use a screwdriver to hold the oil pump shaft — 1.6L (VIN R), 1.8L (VIN M) and 2.0L (VIN F) engines

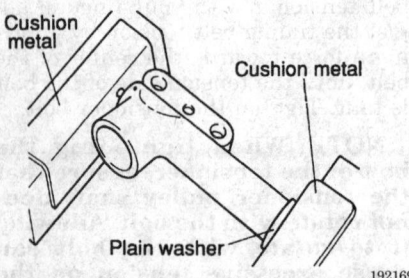

Compress the auto tensioner in a vise — 1.6L (VIN R), 1.8L (VIN M) and 2.0L (VIN F) engines

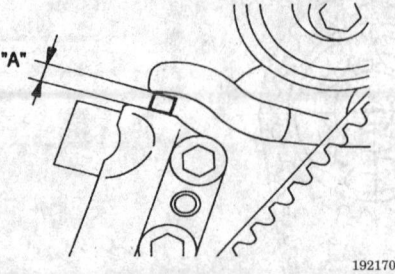

Measure the auto tensioner here — 1.6L (VIN R), 1.8L (VIN M) and 2.0L (VIN F) engines

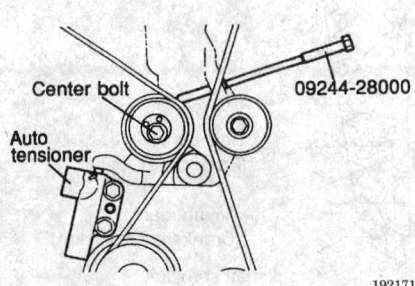

Using the special tool to hold the tensioner arm — 1.6L (VIN R), 1.8L (VIN M) and 2.0L (VIN F) engines

with your finger and noting the belt deflection. The desired reading is 0.20–0.28 in. (5–7mm). If tension is not correct, readjust and check belt deflection.

10. Install the flange, crankshaft and washer to the crankshaft. The flange on the crankshaft sprocket must be installed towards the inner timing belt sprocket. Tighten bolt to 80 to 94 ft. lbs. (110 to 130 Nm).

11. To install the oil pump sprocket, insert a Phillips screwdriver with a shaft 0.31 in. (8mm) in diameter into the plug hole in the left side of the cylinder block to hold the left silent shaft. Tighten the nut to 36 to 43 ft. lbs. (50 to 60 Nm).

12. Using a wrench, hold the camshaft at its hexagon between journal No. 2 and 3 and tighten bolt to 58 to 72 ft. lbs. (80 to 100 Nm). If no hexagon is present between journal No. 2 and 3, hold the sprocket stationary with a wrench while tightening the retainer bolt.

13. Carefully push the auto tensioner rod in until the set hole in the rod is aligned with the hole in the cylinder. Place a wire into the hole to retain the rod.

14. Install the tensioner pulley onto the tensioner arm. Locate the pinhole in the tensioner pulley shaft to the left of the center bolt. Then tighten the center bolt finger-tight.

15. When installing the timing belt, turn the 2 camshaft sprockets so their dowel pins are located on top. Align the timing marks facing each other with the top surface of the cylinder head. When you let go of the exhaust camshaft sprocket, it will rotate 1 tooth in the counterclockwise direction. This should be taken into account when installing the timing belts on the sprocket.

NOTE: Both camshaft sprockets are used for the intake and exhaust camshafts and are provided with 2 timing marks. When the sprocket is mounted on the exhaust camshaft, use the timing mark on the right with the dowel pin hole on top. For the intake camshaft sprocket, use the 1 on the left with the dowel pin hole on top.

16. Align the crankshaft sprocket and oil pump sprocket timing marks.

17. After alignment of the oil pump sprocket timing marks, remove the plug on the cylinder block and insert a Phillips screw driver with a shaft diameter of 0.31 in. (8mm) through the hole. If the shaft can be inserted 2.4 in. deep, the silent shaft is in the correct position. If the shaft of the tool can only be inserted 0.8 to 1.0 in. (20 to 25mm) deep, turn the oil pump sprocket 1 turn and realign the marks. Reinsert the tool making sure it is inserted 2.4 in. deep. Keep the tool inserted in hole for the remainder of this procedure.

NOTE: The above step assures that the oil pump socket is in correct orientation to the silent shafts. This step must not be skipped or a vibration may develop during engine operation.

18. Install the timing belt as follows:

a. Install the timing belt around the intake camshaft sprocket and retain it with 2 spring clips or binder clips.

b. Install the timing belt around the exhaust sprocket, aligning the timing marks with the cylinder head top surface using 2 wrenches. Retain the belt with 2 spring clips.

c. Install the timing belt around the idler pulley, oil pump sprocket, crankshaft sprocket and the tensioner pulley. Remove the 2 spring clips.

d. Lift upward on the tensioner pulley in a clockwise direction and tighten the center bolt. Make sure all timing marks are aligned.

e. Rotate the crankshaft ¼ turn counterclockwise. Then, turn in clockwise until the timing marks are aligned again.

19. To adjust the timing (outer) belt, turn the crankshaft ¼ turn counterclockwise, then turn it clockwise to move No. 1 cylinder to TDC.

20. Loosen the center bolt. Using tool 09244-28100 or equivalent and a torque wrench, apply a torque of 1.88 to 2.03 ft. lbs. (2.6 to 2.8 Nm). Tighten the center bolt.

21. Screw the special tool into the engine left support bracket until its end makes contact with the tensioner arm. At this point, screw the special tool in some more and remove the set wire attached to the auto tensioner, if the wire was not previously removed. Then remove the special tool.

22. Rotate the crankshaft 2 complete turns clockwise and let it sit for approximately 15 minutes. Then, measure the auto tensioner protrusion (the distance between the tensioner arm and auto tensioner body) to ensure that it is within 0.15 to 0.18 in. (3.8 to 4.5mm). If out of specification, repeat Steps 1 to 4 until the specified value is obtained.

23. If the timing belt tension adjustment is being performed with the engine mounted in the vehicle, and clearance between the tensioner arm and the auto tensioner body cannot be measured, the following alternative method can be used:

a. Screw in special tool 09244-28000 or equivalent, until its end makes contact with the tensioner arm.

b. After the special tool makes contact with the arm, screw it in some more to retract the auto tensioner pushrod while counting the number of turns the tool makes until the tensioner arm is brought into contact with the auto tensioner body. Make sure the number of turns the special tool makes conforms with the standard value of 2½ to 3 turns.

c. Install the rubber plug to the timing belt rear cover.

24. Install the timing belt covers and all related items.

25. Connect the negative battery cable.

3.0L (VIN T) Engine

1. Disconnect the negative battery cable.

2. To remove the air conditioning compressor belt, loosen the adjustment pulley locknut, turn the screw counterclockwise to reduce the drive belt tension and remove the belt.

3. To remove the serpentine drive belt, insert a ½ inch breaker bar into the square hole of the tensioner pulley, rotate it counterclockwise to reduce the drive belt tension and remove the belt.

4. Remove the air conditioning compressor and the air compressor bracket, power steering pump and alternator from the mounts and support them to the side. Remove power steering pump/alternator automatic belt tensioner bolt and the tensioner.

5. Raise and safely support the vehicle. Remove the right inner fender splash shield.

6. Remove the crankshaft pulley bolt and the pulley/damper assembly from the crankshaft.

7. Lower the vehicle and place a floor jack under the engine to support it.

8. Separate the front engine mount insulator from the bracket. Raise the engine slightly and remove the mount bracket.

9. Remove the upper and lower timing belt covers from the engine.

10. Turn the crankshaft until the timing marks on the camshaft sprocket and cylinder head are aligned.

11. Loosen the tensioning bolt, it runs in the slotted portion of the tensioner, and the pivot bolt on the timing belt tensioner.

12. Move the tensioner counterclockwise as far as it will go. Tighten the adjusting bolt.

13. Mark the timing belt with an arrow showing direction of rotation.

14. Remove the timing belt from the camshaft sprocket.

15. Remove the crankshaft pulley. Then, remove the timing belt. Remove the timing belt tensioner. Remove the retainer bolts from the timing sprockets and remove as required.

16. Inspect the belt thoroughly. The back surface must be pliable and rough. If it is hard and glossy, the belt should be replaced. Any cracks in the belt backing or teeth, or missing teeth mean the belt must be replaced. The canvas cover should be intact on all the teeth. If rubber is exposed anywhere, the belt should be replaced.

17. Inspect the tensioner for grease leaking from the grease seal and check for any roughness in rotation. Replace a tensioner for either defect.

18. The sprockets should be inspected and replaced if there is any sign of damaged teeth or cracking anywhere.

19. Do not immerse sprockets in solvent, as solvent that has soaked into the metal may cause deterioration of the timing belt later.

20. Do not clean the tensioner in solvent either, as this may wash the grease out of the bearing.

To install:

21. Align the timing marks of the camshaft sprocket. Check that the crankshaft timing marks are still in alignment, the locating pin on the front of the crankshaft sprocket is aligned with a mark on the front case.

22. Mount the tensioner, spring and spacer with the bottom end of the spring free. Then, install the bolts and tighten the adjusting bolt slightly with the tensioner moved as far as possible away from the water pump. Install the free end of the spring into the locating tang on the front case. Position the belt over the crankshaft sprocket and then over the camshaft sprocket. Slip the back of the belt over the tensioner wheel. Turn the camshaft sprocket in the opposite of its normal direction of rotation until the straight side of the belt is tight and make sure the timing marks align. If not, shift the belt one tooth at a time in the appropriate direction until this occurs.

23. Loosen the tensioner mounting bolts so the tensioner works, without the interference of any friction, under spring pressure. Make sure the belt follows the curve of the camshaft pulley so the teeth are engaged all the way around.

24. Correct the path of the belt, if necessary. Torque the tensioner adjusting bolt to 16–21 ft. lbs. (22–29 Nm). Then, torque the tensioner pivot bolt to the same figure. Bolts must be torqued in that order, or tension won't be correct.

25. Turn the crankshaft one turn clockwise until timing marks again align to seat the belt. Then loosen both tensioner attaching bolts and let the tensioner position itself under spring tension as before. Finally, torque the bolts in the proper order exactly as before. Check belt tension by putting a finger on the water pump side of the tensioner wheel and pull the belt toward it. The belt should move toward the pump until the teeth are about ¼ of the way across the head of the tensioner adjusting bolt. Retension the belt, if necessary.

26. Install the timing belt covers.

27. Install the crankshaft pulley, making sure the pin on the crankshaft sprocket fits through the hole in the rear surface of the pulley. Install the retaining bolt and torque to 108–116 ft. lbs. (147–157 Nm).

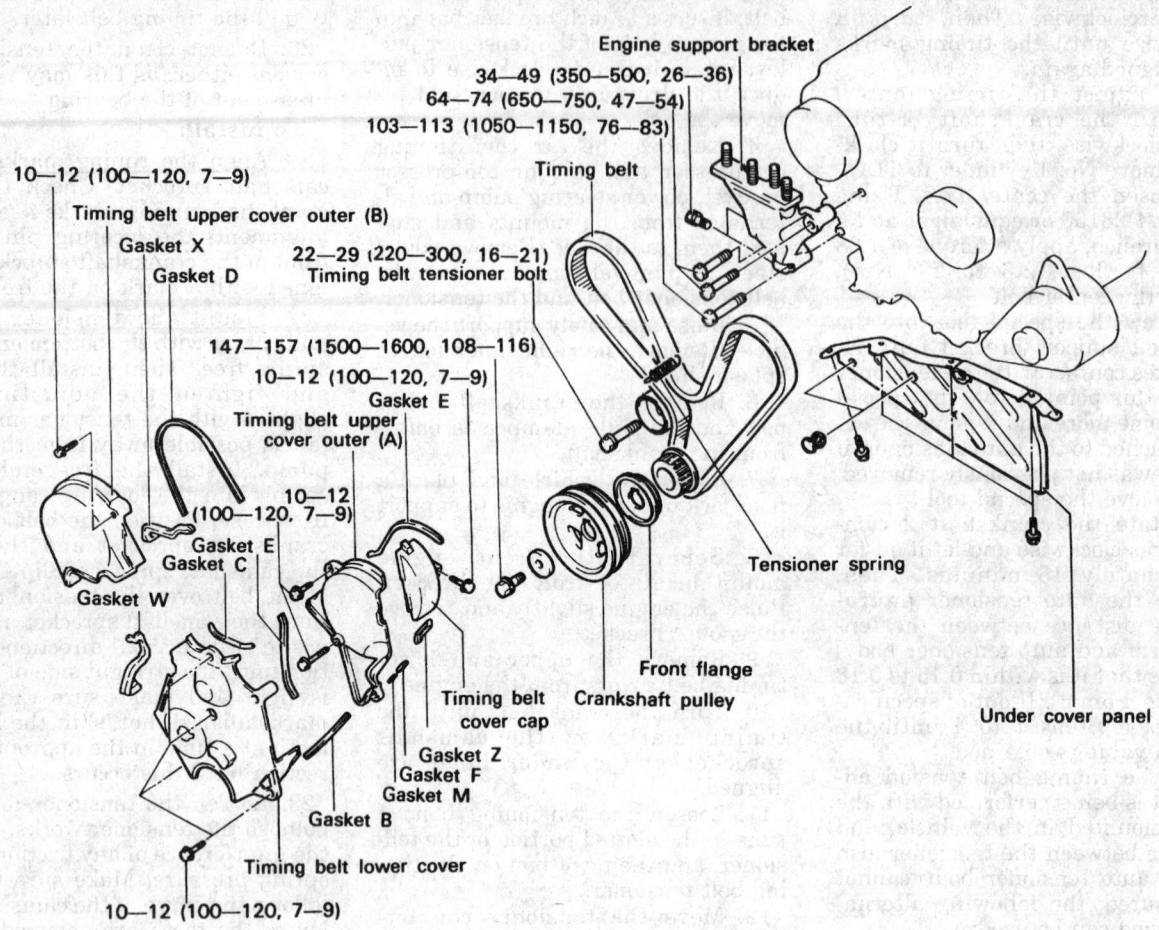

Engine support bracket
34—49 (350—500, 26—36)
64—74 (650—750, 47—54)
103—113 (1050—1150, 76—83)

10—12 (100—120, 7—9)
Timing belt upper cover outer (B)
Gasket X
Gasket D

Timing belt

22—29 (220—300, 16—21)
Timing belt tensioner bolt

147—157 (1500—1600, 108—116)
10—12 (100—120, 7—9)
Gasket E
Timing belt upper cover outer (A)

10—12
(100—120, 7—9)
Gasket E
Gasket C

Gasket W

10—12
(100—120, 7—9)

Tensioner spring

Timing belt
cover cap

Front flange
Crankshaft pulley

Under cover panel

Gasket Z
Gasket F
Gasket M

Gasket B

Timing belt lower cover

10—12 (100—120, 7—9)

TORQUE : Nm (kg.cm, lb.ft)

327696

Exploded view of the timing belt and related components — 3.0L (VIN T) engine

28. Install the engine mount bracket and secure with the mounting hardware.

29. Install the pulley damper assembly to the crankshaft. Torque the bolt to 110 ft. lbs. (149 Nm). Install the splash shield.

30. Install the power steering pump/alternator automatic belt tensioner.

31. Install the air conditioning compressor bracket, compressor, power steering pump and alternator.

32. Install the accessory drive belt.

33. Connect the negative battery cable and check all disturbed components for proper operation.

Camshaft

REMOVAL AND INSTALLATION

1.3L (VIN M) and 1.5L (VIN J) Engines

1. Disconnect the negative battery cable.

2. Disconnect the breather hose and the secondary air hose.

3. Remove the air cleaner.

4. Remove the rocker cover.

5. Remove the timing belt cover.

6. Remove the distributor.

7. Loosen the two bolts and move the timing belt tensioner toward the water pump as far as it will go, then retighten the timing belt tensioner

adjusting bolt. Disengage the timing belt from the camshaft sprocket and unbolt and remove the sprocket. The timing belt may be left engaged with the crankshaft sprocket and tensioner.

8. Remove the rocker shaft assembly. Remove the small, square cover that sits directly behind the camshaft on the transaxle side of the head. Remove the camshaft thrust case tightening bolt that sits on the top of the head right near that cover.

9. Carefully, slide the camshaft out of the head through the hole in the transaxle side of the head, being careful that the cam lobes do not strike the bearing bores in the head.

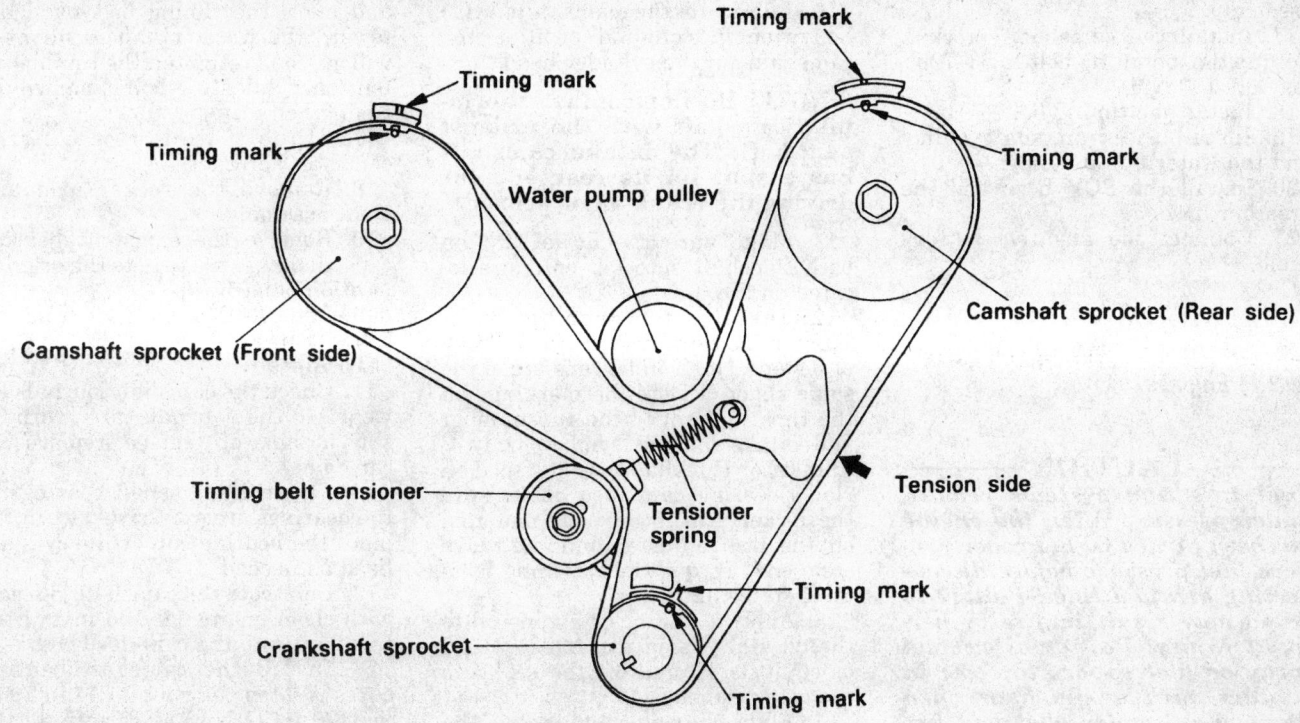

327697

Timing mark locations — 3.0L (VIN T) engine

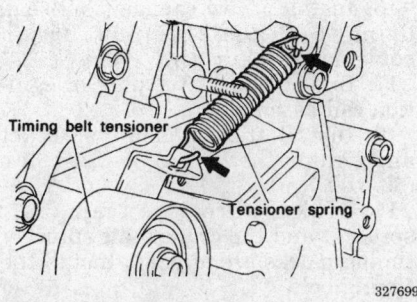

327699

Correct timing belt tensioner and spring installation — 3.0L (VIN T) engine

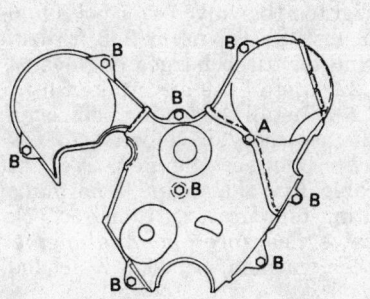

327698

Install the long bolt in the location marked "A" — 3.0L (VIN T) engine

To install:

10. Lubricate all journal and thrust surfaces with clean engine oil.

11. Carefully insert the camshaft into the engine. Make sure the camshaft goes in with the threaded hole in the top of the thrust case straight upward.

12. Align the threaded hole with the bolt hole in the cylinder head surface.

13. Install the thrust case bolt and tighten firmly.

14. Install the rear cover with a new gasket and install and tighten the bolts to 5.8–7.2 ft. lbs. (8–10 Nm).

15. Coat the external surface of the front oil seal with engine oil.

16. Using special installer tool MD 998306-01 or equivalent, drive a new front camshaft oil seal into the clearance between the cam and head at the forward end, making sure the seal seats fully.

17. Install the camshaft sprocket and torque the bolt to 47–54 ft. lbs. (64–74 Nm).

18. Reconnect the timing belt, check the timing and adjust the belt tension.

19. Reinstall the rocker shaft assembly. Adjust the valves and install the rocker and timing belt covers.

1.5L (VIN K) Engine

1. Disconnect the negative battery cable.

2. Disconnect the PCV hose and the breather hose.

3. Remove the water pump pulley and crankshaft pulley.

4. Remove the timing belt cover.

5. Move the timing belt tensioner toward the water pump and secure it.

6. Remove the timing belt from the camshaft sprocket.

7. Remove the ignition coil assembly.

8. Remove the cylinder head cover.

9. Remove the rocker arm shaft assembly.

10. Remove the camshaft bearing caps. Keep them in order so they can be installed in the same positions.

11. Remove the camshaft.

To install:

12. Lubricate the camshaft and journals with clean engine oil and position the camshaft on the journals.

13. Install the bearing caps in their original positions. Torque the bolts to 14–20 ft. lbs. (20–27 Nm).

14. Install the rocker arm shaft assembly.

15. Apply sealant to the cylinder head and cam cap area and install the cylinder head cover. Torque the bolts to 5.9–7.4 ft. lbs. (10–12 Nm).

16. Install the ignition coil assembly.

17. Install the camshaft sprocket. Torque the mounting bolt to 58–72 ft. lbs. (80–100 Nm).

18. Install the timing belt.

19. Install the crankshaft pulley and the water pump pulley.

20. Install the PCV hose and the breather hose.

21. Connect the negative battery cable.

1.6L (VIN R), 1.8L (VIN M) and 2.0L (VIN F) Engines

——————— CAUTION ———————

Fuel injection systems remain under pressure after the engine has been turned OFF. Properly relieve fuel pressure before disconnecting any fuel lines. Failure to do so may result in fire or personal injury. Do not allow fuel spray or fuel vapors to come in contact with a spark or open flame. Keep a dry chemical fire extinguisher nearby. Never store fuel in an open container due to risk of fire or explosion.

1. Relieve the fuel system pressure.

2. Disconnect battery negative cable.

3. Disconnect the accelerator cable.

4. Remove the timing belt cover and timing belt.

5. Remove the center cover, breather and PCV hoses, and spark plug cables.

6. Remove the rocker cover, semi-circular packing, throttle body bracket, crankshaft angle sensor, both camshaft sprockets, and oil seals.

7. Loosen the bearing cap bolts in 2–3 steps. Label and remove all camshaft bearing caps.

NOTE: If the bearing caps are difficult to remove, use a plastic hammer to gently tap the rear part of the camshaft.

8. Remove the intake and exhaust camshafts.

9. Check the camshaft journals for wear or damage. Check the cam lobes for damage. Also, check the cylinder head oil holes for clogging.

To install:

10. Lubricate the camshafts with heavy engine oil and position the camshafts on the cylinder head.

NOTE: Do not confuse the intake camshaft with the exhaust camshaft. The intake camshaft has a split on its rear end for driving the crank angle sensor.

11. Make sure the dowel pin on both camshaft sprocket ends are located on the top.

12. Install the bearing caps. Tighten the caps in sequence and in 2 or 3 steps. No. 2 and 5 caps are of the same shape. Check the markings on the caps to identify the cap number and intake/exhaust symbol. Only **L** (intake) or **R** (exhaust) is stamped on No. 1 bearing cap. Also, make sure the rocker arm is correctly mounted on the lash adjuster and the valve stem end. Torque the retaining bolts to 15 ft. lbs. (20 Nm).

13. Apply a coating of engine oil to the oil seal. Using tool 09221-21000 or equivalent, press-fit the seal into the cylinder head.

14. Align the punch mark on the crank angle sensor housing with the notch in the plate. With the dowel pin on the sprocket side of the intake camshaft at top, install the crank angle sensor on the cylinder head.

NOTE: Do not position the crank angle sensor with the punch mark positioned opposite the notch; this position will result in incorrect fuel injection and ignition timing.

15. Install the valve cover.

16. Install the throttle body bracket and the semi-circular packing.

17. Install the center cover, breather and PCV hoses, and spark plug cables.

18. Install the timing belt and covers.

19. Connect the negative battery cable, and check for leaks.

1.5L (VIN N) Engine

1. Disconnect the negative battery cable.

2. If equipped with a turbocharger, remove the air intake duct and cover the throttle body and turbocharger ports to keep out dirt.

3. Remove the valve cover. Remove the upper timing belt cover.

4. Remove the distributor or ignition coil assembly.

5. Set the No. 1 cylinder at TDC for the compression stroke. Once the engine is in this position, it must not be disturbed.

6. Loosen the two tensioner bolts and move the timing belt tensioner toward the water pump as far as it will go, then retighten the timing belt tensioner adjusting bolt. Remove the timing belt.

7. Unbolt and remove the camshaft sprocket.

8. Remove the rocker arm and shaft assembly.

9. Remove the camshaft bearing caps, making sure to note the original location of each cap.

10. Remove the camshaft from the cylinder head.

To install:

11. Check the camshaft journals for wear. If the journals are worn or shown signs of damage, replace the camshaft.

12. Check the camshaft bearings. If the bearings are excessively worn, replace the bearing cap or the cylinder head as needed.

13. Lubricate the camshaft journals with clean engine oil and install the camshaft into the cylinder head.

14. Install the camshaft bearing caps. Tighten the bolts to 14 to 20 ft. lbs. (20 to 27 Nm) starting from the center and working outward to the end caps.

15. Install the rocker arm and shaft assembly.

16. Install a new camshaft oil seal, using tool 09221-21000 or an equivalent seal driver.

17. Install the distributor or ignition coil assembly.

18. Install the camshaft sprocket and tighten the bolt to 58–72 ft. lbs. (80–100 Nm).

19. Make sure the camshaft sprocket and the crankshaft sprocket timing marks are aligned. Install the timing belt.

20. Adjust the valve clearance on vehicles not equipped with hydraulic lifters.

21. Apply 0.4 in. (10mm) beads of sealant on either side of the No. 1 and No. 5 camshaft holders. Install the rocker cover using a new gasket. Tighten the valve cover bolts to 6–7.5 ft. lbs. (8–10 Nm) or 2–3 ft. lbs. (3–4 Nm) for turbocharged engines.

22. Install the air intake tube.

23. Install the timing belt cover and the accessory drive belts.

24. Connect the negative battery cable. Run the engine to normal operating temperature, then recheck the valve clearance once the engine has been warmed, adjust it if needed.

3.0L (VIN T) Engine

1. Disconnect the negative battery cable.

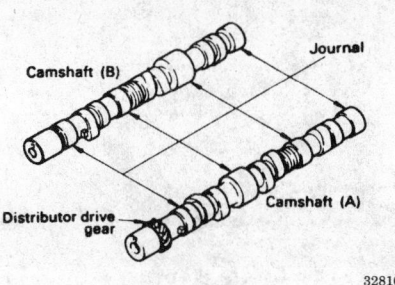

Camshaft identification — 3.0L (VIN T) engine

328103

2. Remove the air cleaner assembly and valve covers.

3. Turn the engine with a wrench until No. 1 cylinder is at TDC on the compression stroke. Check the timing marks for proper alignment.

4. Remove the timing belt covers and the timing belt.

5. Install auto lash adjuster retainer tools 09246-32000 (MD998443) or equivalent on the rocker arms.

6. If removing the right side (front) camshaft, remove the distributor extension.

7. Loosen the camshaft bearing cap bolts but do not remove the bolts from the caps.

8. Remove the rocker arms, rocker shafts and bearing caps, as an assembly.

9. Remove the camshaft from the cylinder head.

10. Inspect the bearing journals on the camshaft, cylinder head and bearing caps.

To install:

11. Lubricate the camshaft journals and camshaft with clean engine oil and install the camshaft in the cylinder head.

12. Align the camshaft bearing caps with the arrow mark depending on cylinder numbers and install in numerical order.

13. Apply sealer at the ends of the bearing caps and install the assembly.

14. Torque the rocker arm and shaft assembly bolts to 15 ft. lbs. (21 Nm).

15. Install the distributor extension, if removed.

16. Install the timing belt and covers.

17. Install the valve cover. Torque the valve cover retaining bolts to 7 ft. lbs. (10 Nm).

18. Install the air cleaner assembly.

19. Connect the negative battery cable and road test the vehicle.

Balance Shaft

REMOVAL AND INSTALLATION

1.8L (VIN M) and 2.0L (VIN F) Engines

NOTE: A number of special tools are required to perform this operation. They are listed as follows:

• Bearing puller 09212–32000 or equivalent — to remove right countershaft front bearing.
• Bearing puller 09212–32100 and holding fixture 09212–32300 or equivalents — to remove left counter shaft rear bearing.
• Bearing installer 09212–32200 or equivalent — to install left and right countershaft rear bearings.
• Seal installer 09214–32100 or equivalent — to install the crankshaft front seal

1. Disconnect the negative battery cable.

2. Remove the oil filter, oil pressure switch, oil gauge sending unit, oil filter mounting bracket and gasket.

3. Raise and safely support the vehicle. Drain engine oil. Remove engine oil pan.

4. Lower the vehicle. Remove the timing belts.

5. Remove the crankshaft sprocket (inner) and counterbalance shaft sprocket.

6. Remove the front engine cover which is also the oil pump cover. Different length bolts are used. Take note of their locations. Discard the shaft seal and gasket.

7. Remove the oil pump driven gear flange bolt. When loosening this bolt, first remove the plug at the bottom of the left side of the cylinder block and insert a tool approximately 3/8 inch in diameter into the hole. The tool will hold the silent shaft in position. The tool must be inserted at least 2.4 inches into the hole. If depth

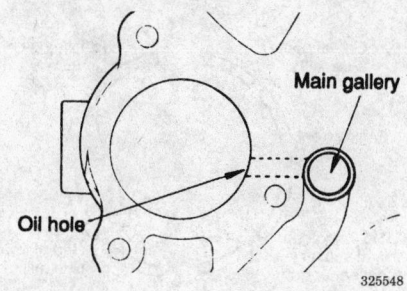

Align the bearing oil hole with the hole in the block — 1.8L (VIN M) and 2.0L (VIN F) engines

325548

of insertion is not correct, rotate the oil pump sprocket 1 revolution, and align the timing marks. Insert the tool shaft again, and watch the amount of insertion, which should be at least 2.4 inches.

8. Remove the oil pump gears and remove the front case assembly. Remove the threaded plug, the oil pressure relief spring and plunger.

9. Remove the shaft alignment tool, front cover and oil pump as a unit, with the left countershaft attached.

10. Remove the oil pump gear and left counterbalance shaft.

NOTE: To aid in removal of the front cover, a driver groove is provided on the cover, above the oil pump housing. Avoid prying on the thinner parts of the housing flange or hammering on it to remove the case.

11. Remove the right counterbalance shaft from the engine block.

To replace the counterbalance shaft bearings:

12. Using special bearing puller tool 09212-32000 or equivalent, remove the right counterbalance shaft bearing from the cylinder block.

13. Using special bearing puller tool 09212-32100 and holding fixture 09212–32300 (to hold the puller) or equivalents, remove the left countershaft rear bearing from the cylinder block.

14. Coat the inner surfaces of the new bearings and the block bores with clean engine oil.

15. Install the left rear countershaft bearing into the cylinder block bore and install it using special bearing installer tool 09212–32200 and the special holding fixture 09212–32300 used before to remove the bearing. The fixture serves as a guide for the bearing installer tool.

16. Using special bearing installer tool 09212-32200 or equivalent, install the right countershaft front bearing into the block.

To install:

17. Install a new front seal in the cover. Install the oil pump drive and driven gears in the front case, aligning the timing marks on the pump gears.

18. Install the left counterbalance shaft in the driven gear and temporarily tighten the bolt.

19. Install the right counterbalance shaft into the cylinder block.

20. Install an oil seal guide on the end of the crankshaft and install a new gasket on the front of the engine block for the front cover.

21. Install a new front case packing.

22. Insert the left counterbalance shaft into the engine block and at the same time, guide the front cover into place on the front of the engine block.

23. Install an O-ring on the oil pump cover and install it on the front cover.

24. Tighten the oil pump cover bolts and the front cover bolts to 11–13 ft. lbs. (15–18 Nm).

25. Install the upper and lower undercovers.

26. Install the spacer on the end of the right counterbalance shaft, with the chamfered edge toward the rear of the engine.

27. Install the counterbalance shaft sprocket and temporarily tighten the bolt.

28. Install the inner crankshaft sprocket and align the timing marks on the sprockets with those on the front case.

29. Install the inner tensioner (B) with the center of the pulley on the left side of the mounting bolt and with the pulley flange toward the front of the engine.

30. Lift the tensioner by hand, clockwise, to apply tension to the belt. Tighten the bolt to secure the tensioner.

31. Check that all alignment marks are in their proper places and the belt deflection is approximately ¼–½ inch on the tension side.

NOTE: When the tensioner bolt is tightened, make sure the shaft of the tensioner does not turn with the bolt. If the belt is too tight there will be noise and if the belt is too loose, the belt and sprocket may come out of mesh.

32. Tighten the counterbalance shaft sprocket bolt to 22.0–28.5 ft. lbs. (29–40 Nm).

33. Install the flange and crankshaft sprocket. Tighten the bolt to 43–50 ft. lbs. (58–67 Nm).

34. Install the camshaft spacer and sprocket. Tighten the bolt to 44–57 ft. lbs. (61–75 Nm).

35. Align the camshaft sprocket timing mark with the timing mark on the upper inner cover.

36. Install the oil pump sprocket, tightening the nut to 25–28 ft. lbs. (34–39 Nm). Align the timing mark on the sprocket with the mark on the case.

NOTE: To be assured that the phasing of the oil pump sprocket and the left counterbalance shaft is correct, a metal rod should be inserted in the plugged hole on the left side of the cylinder block. If it can be inserted more than 2.4 inches, the phasing is correct. If the tool can only be inserted approximately 1.0 inch, turn the oil pump sprocket through one turn and realign the timing marks. Keep the metal rod inserted until the installation of the timing belt is completed. Remove the tool from the hole and install the plug, before starting the engine.

37. Install the tensioner spring and tensioner. Temporarily tighten the nut. Install the front end of the tensioner spring (bent at right angles) on the projection of the tensioner and the other end (straight) on the water pump body.

38. If the timing belt is correctly tensioned, there should be about 12mm clearance between the outside of the belt and the edge of the belt cover. This is measured about halfway down the side of the belt opposite the tensioner.

39. Complete the assembly by installing the oil screen, gasket and oil pan.

40. Install the crankshaft pulley, alternator and accessory belts and adjust to specifications.

41. Install the radiator.

42. Fill the cooling system with antifreeze and the crankcase with clean engine oil.

43. Connect the negative battery cable and start the engine.

Piston and Connecting Rod

POSITIONING

All Engines

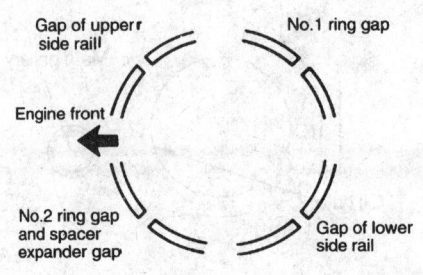

Piston ring gap locations

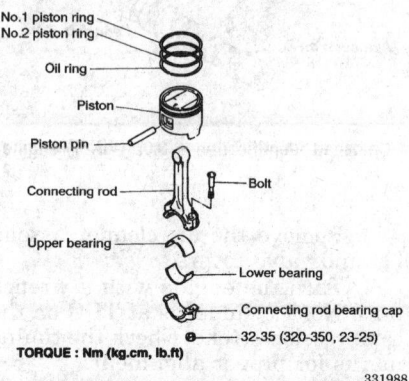

TORQUE : Nm (kg.cm, lb.ft)

331988

Piston and connecting rod assembly

ENGINE LUBRICATION

Oil Pan

REMOVAL AND INSTALLATION

Except 3.0L (VIN T) Engine

1. Disconnect the negative battery cable.

2. Raise the vehicle and support it safely.

3. Drain the oil.

4. Remove the underbody splash shield.

5. Remove the oil pan bolts.

6. Knock in the special tool between the oil pan and the block. Then tap it sideways to break the oil pan loose from the engine block.

7. Remove the oil pan.

To install:

8. Clean the mating surfaces of the oil pan and the engine block.

9. Apply a ⅛ in. (3mm) bead of RTV sealer along the groove in the oil pan.

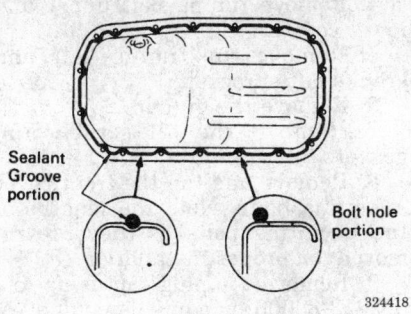

Apply sealant to the oil pan flange — 1.5L (VIN K) engines

10. Install the oil pan. Hand tighten the retaining bolts.

11. Starting at the middle of the pan, gradually tighten the retaining bolts to 4–6 ft. lbs. (6–8 Nm) in a crisscross pattern.

12. Install the oil pan drain plug.

13. Install the splash shield and lower the vehicle.

14. Refill the crankcase with oil. Start the engine and check for leaks.

3.0L (VIN T) Engine

1. Disconnect the negative battery cable.

2. Raise the vehicle and support safely.

3. On automatic transaxles only, remove the torque converter bolt access cover.

4. Drain the engine oil.

5. Remove the oil pan retaining screws and remove the oil pan and gasket.

To install:

6. Thoroughly clean and dry all sealing surfaces, bolts and bolt holes.

7. Apply silicone sealer to the chain cover to block mating seam and the rear main seal retainer to block seam, if equipped.

8. Install a new pan gasket or apply silicone sealer to the sealing surface of the pan and install to the engine.

9. Install the retaining screws and torque to 3.7–5.0 ft. lbs. (5–7 Nm).

10. Install the torque converter bolt access cover. Lower the vehicle.

11. Install the dipstick. Fill the engine with the proper amount of oil.

12. Connect the negative battery cable and check for leaks.

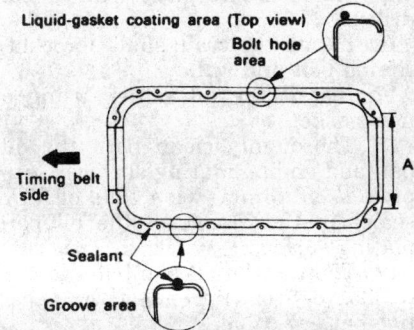

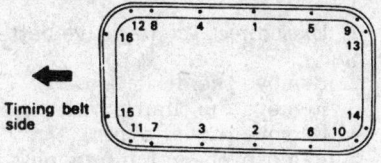

Oil pan bolt tightening sequence and sealant application — 3.0L (VIN T) engine

Oil Pump

REMOVAL AND INSTALLATION

1.3L (VIN M) and 1.5L (VINs J and N) Engines

NOTE: Whenever the oil pump is disassembled or the cover removed, the gear cavity must be filled with petroleum jelly for priming purposes. Do not use grease.

1. Disconnect the negative battery cable. Remove the timing belt.

2. Remove the oil pan.

3. Remove the oil screen.

4. Unbolt and remove the front case assembly.

5. Remove the oil pump cover.

6. Remove the inner and outer gears from the front case.

NOTE: The outer gear has no identifying marks to indicate direction of rotation. Clean the gear and mark it with an indelible marker.

7. Remove the plug, relief valve spring and relief valve from the case.

To install:

8. Check the front case for damage or cracks. Replace the front seal. Re-

place the oil screen O-ring. Clean all parts thoroughly with a safe solvent.

9. Check the pump gears for wear or damage. Clean the gears thoroughly and place them in their original position in the case to check the clearances. There is a crescent-shaped piece between the 2 gears.

10. Check that the relief valve can slide freely in the case.

11. Check the relief valve spring for damage. The relief valve free length should be 1.8 in. (47mm). Load length should be 13.4 lbs. at 1.6 in. (40mm).

12. Thoroughly coat both oil pump gears with clean engine oil and install them in the correct direction of rotation.

13. Install the pump cover and torque the bolts to 6–8 ft. lbs. (8–12 Nm).

14. Coat the relief valve and spring with clean engine oil, install them and tighten the plug to 30–36 ft. lbs. (39–49 Nm).

15. Position a new front case gasket, coated with sealer, on the engine and install the front case. Torque the bolts to 8–11 ft. lbs. (12–15 Nm). Note that the bolts have different shank lengths. Use the following guide to determine which bolts go where. Bolts marked:

A: 0.8 in. (20mm)
B: 1.2 in. (30mm)
C: 2.4 in. (60mm)

16. Coat the lips of a new seal with clean engine oil and slide it along the crankshaft until it touches the front case. Drive it into place with a seal driver.

17. Install the sprocket, timing belt and pulley.

18. Install the oil screen.

19. Thoroughly clean both the oil pan and engine mating surfaces. Apply a 1/8in. (3mm) wide bead of RTV sealer in the groove of the oil pan mating surface.

20. Tighten the oil pan bolts to 60–72 inch lbs. Connect the negative battery cable.

1.5L (VIN K) Engines

SOHC Engines

1. Disconnect the negative battery cable.

2. Remove the crankshaft pulley.

3. Remove the timing belt.

4. Remove the oil pan.

5. Remove the oil screen.

6. Unbolt and remove the front case assembly.

7. Remove the oil pump cover.

8. Remove the inner and outer gears from the front case.

To install:

9. Check the front case for damage or cracks.

10. Clean the gears thoroughly. Align the mating marks and place the gears in their original position in the case to check the clearances.

• Clearance between the outer gear and front case; 0.005–0.007 in. (0.12–0.18mm)

• Tip clearance on the pump rotor; 0.001–0.003 in. (0.025–0.069mm)

• Clearance between the outer rotor and cover; 0.0016–0.0034 in. (0.04–0.087mm)

11. Check the relief valve spring for damage. The relief valve free length should be 1.835 in. (46.6mm). Load length should be 13.4 lbs. at 1.579 in. (6.1kg/40.1mm).

12. Pack the gear cavity with petroleum jelly.

13. Install the pump cover and torque the bolts to 6–8.8 ft. lbs. (8–12 Nm).

14. Position a new front case gasket on the engine and install the front case. Torque the bolts to 8.7–11 ft. lbs. (12 to 15 Nm). Note that the bolts have different shank lengths. Use the following guide to determine which bolts go where. Bolts marked:

A: 0.98 in. (25mm)
B: 1.18 in. (30mm)
C: 1.77 in. (45mm)
D: 2.36 in (60mm)

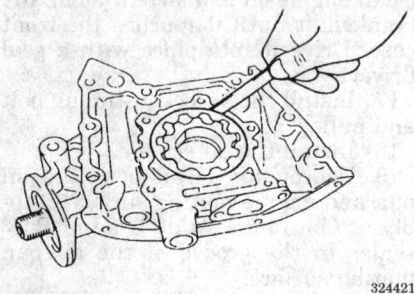

324421

Check the clearance between the outer rotor and front case — 1.5L (VIN K) engines (SOHC) shown, others similar

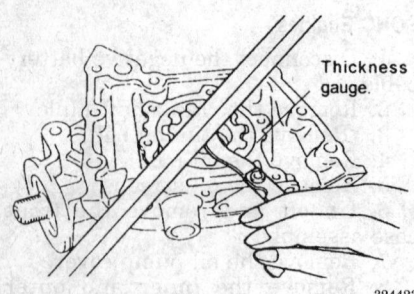

Thickness gauge.

324423

Check the clearance between the rotor and the pump cover — 1.5L (VIN K) engines (SOHC) shown, others similar

15. Coat the lips of a new seal with clean engine oil and slide it along the crankshaft until it touches the front case. Drive it into place with a seal driver.

16. Install the crankshaft sprocket, timing belt and pulley.

17. Install the oil screen using a new gasket.

18. Thoroughly clean both the oil pan and engine mating surfaces. Apply a ⅛ in. (3mm) wide bead of RTV sealer in the groove of the oil pan mating surface.

19. Tighten the oil pan bolts to 4–6 ft. lbs. (6–8 Nm). Connect the negative battery cable.

DOHC Engines

1. Disconnect the negative battery cable.
2. Remove the drive belts.
3. Remove the timing belt.
4. Remove the oil pan.
5. Remove the oil pump pick-up and screen.
6. Remove the front case assembly.

NOTE: The bolts attaching the front case to the engine block are different lengths. Take note of the locations of the bolts.

7. Remove the oil pump cover. Take note of the mating marks on the oil pump gears. These marks indicate the direction of gear installation. Oil pump gears must be installed in the same direction they were before removal.
8. Remove the oil pump gears if necessary.

To install:
9. If removed, install the oil pump gears. Make sure the mating marks are on the same side.
10. Install the oil pump cover. Torque the bolts to 6–9 ft. lbs. (8–12 Nm).
11. Use a new gasket and install the front case assembly. Torque the bolts to 15–20 ft. lbs. (20–27 Nm).
12. Install the oil pump pickup and screen.
13. Install the oil pan.
14. Install the timing belt.
15. Install the drive belts
16. Connect the negative battery cable.

1.6L (VIN R), 1.8L (VIN M) and 2.0L (VIN F) Engines

1. Disconnect the negative battery cable.
2. Remove the front engine mount bracket.
3. Remove the accessory drive belts.

4. Remove timing belt upper and lower covers.
5. Remove the timing belt and crankshaft sprocket.
6. Remove the oil pan.
7. Remove the oil screen and gasket.
8. Remove and tag the front cover mounting bolts. Note the lengths of the mounting bolts as they are removed for proper installation.
9. Remove the plug cap using tool 092213-33000 or equivalent, and remove the oil pressure switch.
10. Remove the front case cover and oil pump assembly. If necessary, the silent shaft can come out with the assembly. Disassemble as required.

To install:
11. Thoroughly clean all gasket material from all mounting surfaces.
12. Apply engine oil to the entire surface of the gears or rotors.
13. Install the drive/driven gears with the 2 timing marks aligned.
14. Assemble the front case cover and oil pump assembly to the engine block using a new gasket. Tighten the upper bolt to 20–25 ft. lbs. (27–34 Nm), the lower case bolts to 14–20 ft. lbs. (20–27 Nm), and the driven gear bolt to 25–29 ft. lbs. (34–40 Nm) .
15. Install the oil pressure switch.
16. Install the oil screen with new gasket.
17. Install the oil pan.
18. Install the crankshaft sprocket and timing belt.
19. Install the timing belt covers.
20. Install the accessory drive belts.
21. Install the front engine mount bracket.
22. Connect the negative battery cable and check for adequate oil pressure.

3.0L (VIN T) Engines

1. Disconnect the negative battery cable.
2. Remove the dipstick.
3. Raise the vehicle and support safely.
4. Remove the timing belt covers and timing belt.
5. Drain the engine oil and remove the oil pan from the engine.
6. Remove the oil pickup.
7. Remove the oil pump mounting bolts and remove the pump from the front of the engine. Note the different length bolts and their position in the pump for installation.

To install:
8. Clean the gasket mounting surfaces of the pump and engine block.
9. Prime the pump by pouring fresh oil into the pump and turning the rotors. Using a new gasket, in-

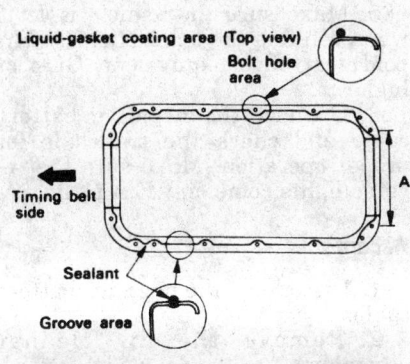

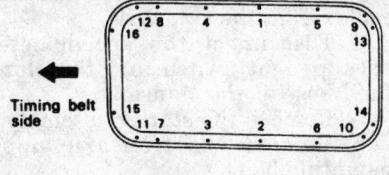

327200

Oil pan bolt tightening sequence and sealant application — 3.0L (VIN T) engines

stall the oil pump on the engine and torque all bolts to 9–10 ft. lbs. (12–15 Nm).

10. Install the crankshaft sprocket to the end of the crankshaft.

11. Clean out the oil pickup or replace, if necessary. Replace the oil pickup gasket ring and install the pickup to the pump.

12. Install the oil pan.

13. Install the timing belt and covers.

14. Lower the vehicle.

15. Install the dipstick.

16. Fill the engine with the proper amount of oil.

17. Connect the negative battery cable and check the oil pressure.

TRANSAXLE

Manual Transaxle Assembly

REMOVAL AND INSTALLATION

1993–94 Excel, Scoupe and Elantra

NOTE: If the vehicle is going to be rolled while the halfshafts are out of the vehicle, obtain 2 outer CV-joints or proper equivalent tools and install to the hubs. If the vehicle is rolled without the proper torque applied to the front wheel bearings, the bearings will no longer be usable.

1. Disconnect the negative battery cable. Remove the air cleaner assembly, battery and battery tray as required.

2. On 5-speed transaxle, disconnect the electrical connector for the selector control valve.

NOTE: The actuator-to-shaft coupling pin collar is not reusable and must be replaced.

3. Disconnect and remove the speedometer cable.

4. If equipped with a cable operated clutch, disconnect the clutch cable from the transaxle assembly.

5. If equipped with a hydraulically operated clutch, remove the clevis pin connecting the slave cylinder to the release fork shaft and remove the slave cylinder mounting bolts. Remove the bolts attaching the hydraulic line support bracket to the transaxle. Remove and support the slave cylinder assembly out of the way with a length of mechanics wire.

6. Disconnect the back-up lamp electrical connector. Remove the starter motor electrical harness.

7. Remove the transaxle mounting bolts accessible from the top side of the transaxle.

8. Unbolt and remove the starter motor.

9. Raise the vehicle and support it safely. Then, remove the splash shield from under the engine. Drain the transaxle fluid.

10. Disconnect the extension rod and the shift rod at the transaxle end and lower them.

11. Disconnect the stabilizer bar at the lower control arm.

12. Remove the halfshafts from the transaxle assembly.

13. Support the transaxle from below with a floor jack. Make sure the support is widely enough spread that the transaxle pan will not be damaged. Then, remove the attaching bolts and the bell housing cover.

14. Remove the lower bolts attaching the transaxle to the engine.

15. Remove the transaxle insulator mount bolt. Remove the cover from inside the right fender shield and remove the transaxle support bracket.

16. Remove the transaxle mount bracket.

17. Pull the assembly away from the engine and lower it from the vehicle.

To install:

18. Install the transaxle to the engine and install the mounting bolts. Tighten the mounting bolts as follows:

M10–7T engine-to-transaxle bolts — 35 ft. lbs. (48 Nm)

M8–10T engine-to-transaxle bolts — 25 ft. lbs. (34 Nm)

19. The balance of the installation is the reverse of the removal procedure.

20. Make sure the vehicle is level when refilling the transaxle. Use Hypoid gear oil or equivalent, GL-4 or higher.

21. Connect the negative battery cable and check the transaxle for proper operation. Make sure the reverse lights come on when in R.

1995 Scoupe

1. Shift the transaxle into neutral.

2. Disconnect the negative and positive battery cables.

3. Remove the air cleaner assembly, battery, and battery tray.

4. Disconnect and speedometer cable.

5. Remove the clevis pin connecting the slave cylinder to the release fork shaft and remove the slave cylinder mounting bolts. Remove the bolts attaching the hydraulic line support bracket to the transaxle. Unbolt the slave cylinder and support it out of the way with a length of mechanics wire. Do not disconnect the hydraulic line.

6. Disconnect the back-up light electrical connector.

7. Disconnect the shift and select cables from the transaxle linkage.

8. Disconnect the starter motor electrical harness.

9. Unbolt and remove the starter motor.

10. Remove the transaxle mounting bolts accessible from the top side of the transaxle.

11. Raise the vehicle and support it safely. Then, remove the splash shield from under the engine.

12. Drain the transaxle fluid.

13. Remove the front wheels.

14. Disconnect the stabilizer bar from the lower control arm.

15. Remove the lower ball joint castle nuts. Use a ball joint separator to separate the lower control arms from the steering knuckles.

16. Remove the halfshafts from the transaxle assembly.

17. If equipped with a turbocharged engine, remove the intermediate shaft and bearing assembly from the transaxle case.

18. Support the transaxle with a transmission jack. Make sure the jack doesn't damage the transaxle oil pan. Then, remove the attaching bolts and the bell housing cover.

19. Remove the lower bolts attaching the transaxle to the engine.

20. Remove the transaxle insulator mount bolt. Remove the cover from inside the left fender shield and remove the transaxle support bracket.

21. Remove the transaxle mount bracket.

22. Pull the transaxle assembly away from the engine and lower it from the vehicle.

To install:

23. Apply high-temperature grease to the mainshaft splines.

24. Install the transaxle to the engine, making sure any dowel pins fit securely into place. Install the mounting bolts and hand-tighten them. Then, tighten the mounting bolts to 32–41 ft. lbs. (43–55 Nm).

25. Remove the transmission jack.

26. Install the intermediate shaft and bearing assembly, if removed. Tighten the bolts to 27–34 ft. lbs. (36–46 Nm).

27. The balance of the installation is the reverse of the removal procedure. When installing the transmission mount and brackets, hand-tighten the mount bolts. Torque the mounting bracket bolt to 44–59 ft. lbs. (60–80 Nm). Torque the mounting bracket-to-body bolts to 66–81 ft. lbs. (90–110 Nm).

28. Refill the transaxle with the oil.

29. Connect the positive and negative battery cables.

1995–97 Elantra

NOTE: If the vehicle is going to be rolled while the halfshafts are out of the vehicle, obtain 2 outer CV-joints or proper equivalent tools and install to the hubs. If the vehicle is rolled without the proper torque applied to the front wheel bearings, the bearings will no longer be usable.

1. Disconnect the negative battery cable.

2. If applicable, remove the clevis pin connecting the slave cylinder to the release fork shaft and remove the slave cylinder mounting bolts. Remove the bolts attaching the hydraulic line support bracket to the transaxle. Remove and support the slave cylinder assembly out of the way with a length of mechanics wire.

3. Remove the air cleaner assembly.

4. Remove the select and shift cables.

5. Disconnect the back-up lamp electrical connector.

6. Disconnect and remove the speedometer cable.

7. Remove the starter motor electrical harness.

8. Unbolt and remove the starter motor.

9. Remove the transaxle mounting bolts accessible from the top side of the transaxle.

10. Raise the vehicle and support it safely.

11. Remove the splash shield from under the engine.

12. Drain the transaxle fluid.

13. Disconnect the extension rod and the shift rod at the transaxle end and lower them.

14. Disconnect the tie rod end and lower arm ball joint.

15. Remove the halfshafts from the transaxle assembly.

16. Support the transaxle from below with a floor jack. Make sure the support is widely enough spread that the transaxle pan will not be damaged. Then, remove the attaching bolts and the bell housing cover.

17. Remove the lower bolts attaching the transaxle to the engine.

18. Remove the transaxle insulator mount bolt. Remove the cover from inside the right fender shield and remove the transaxle support bracket.

19. Remove the transaxle mount bracket.

20. Pull the assembly away from the engine and lower it from the vehicle.

To install:

21. Install the transaxle to the engine and install the mounting bolts. Tighten the mounting bolts to 32–39 ft. lbs. (43–55 Nm).

22. Install the mounting brackets and torque the mounting bracket bolt to 65 ft. lbs. (90 Nm).

23. Install the bell housing cover.

24. The balance of the installation is the reverse of the removal procedure.

25. Make sure the vehicle is level when refilling the transaxle. Use Hypoid gear oil or equivalent, GL-4 or higher.

26. Connect the negative battery cable and check the transaxle for proper operation. Make sure the reverse lights come on when in **R**.

Accent

1. Disconnect the negative battery cable.

2. Remove the air cleaner assembly.

3. Disconnect the select and shift cables from the transaxle.

4. Disconnect the speedometer, backup light switch and the clutch fluid hose at the transaxle.

5. Remove the starter assembly.

6. Remove the transaxle upper mounting bolts.

7. Remove the transaxle mounting bracket bolts.

8. Safely raise and support the vehicle.

9. Drain the gear oil from the transaxle.

10. Remove the clutch release cylinder.

11. Remove the left wheel and splash shield.

12. Disconnect the tie rod end and the ball joint from the left knuckle.

13. Remove the left halfshaft assembly.

14. Remove the bell housing cover.

15. Support the transaxle with a jack and remove the lower mounting bolts.

16. Remove the transaxle while removing the right halfshaft from the differential case.

To install:

17. Install a new circlip on the right halfshaft.

18. Raise the transaxle into position while guiding the right halfshaft into the differential. Make sure the transaxle is flush against the engine before installing the lower bolts. Torque the lower mounting bolts to 32–39 ft. lbs. (43–55 Nm).

19. The balance of the installation is the reverse of the removal procedure. When installing the transaxle mounting bracket bolts, torque the bolts on the transaxle to 43–58 ft. lbs. (60–80 Nm) and the bolts on the body to 65–80 ft. lbs. (90–110 Nm). When installing the transaxle upper mounting bolts, torque the bolts to 32–39 ft. lbs. (43–55 Nm).

20. Refill the transaxle with SAE 75W/85W gear oil to the proper level.

21. Refill the clutch master cylinder reservoir and bleed the system.

22. Connect the negative battery cable.

Sonata

1. Disconnect the negative battery cable.

2. Remove the clutch release cylinder from the transaxle case as described later in this section.

3. Drain the transaxle oil into a suitable waste container.

4. Remove the air cleaner assembly.

5. Remove the pin clips and cotter pins and disconnect the select and shift cables from the control levers.

6. Disconnect the back-up lamp switch connector and route the wiring harness off to the side and out of the way.

7. Disconnect the speedometer cable from the transaxle.

8. Disconnect and label the starter wiring. Remove the starter.

9. Raise the vehicle and remove the front wheels. Remove the splash shield.

10. Remove the sway bar, tie rod end and lower arm ball joint.

11. Remove the left and right driveshafts.

12. Unbolt and remove the bell housing cover.

13. Connect a chain hoist or equivalent to the engine lifting bracket. Tension the hoist to support the weight of the engine.

14. Support the bottom of the transaxle with a transmission jack.

15. Remove the 4 plastic mounting bracket bolt access caps and remove the 4 transaxle mounting bracket bolts.

16. Remove the upper and lower transaxle-to-engine and engine-to-transaxle mounting bolts.

17. Lower the transaxle away from the engine.

To install:

18. Raise the transaxle up and onto the engine. Align the bolt holes and install the upper and lower mounting bolts. Install the four transaxle bracket bolts. Make sure the weight of the engine is supported with the chain hoist. Torque the bolts to 32–39 ft. lbs. (43–55 Nm). Install the four plastic caps.

19. Install the bell housing cover.

20. The balance of the installation is the reverse of the removal procedure.

21. Install the clutch release cylinder. Make sure the hose fittings are tight.

22. Connect the negative battery cable.

23. Fill the transaxle and clutch master cylinder to the proper level with the specified fluid.

24. Move the shift lever through all the gears to make sure they engage and disengage properly.

Clutch Assembly

REMOVAL AND INSTALLATION

Excel, Accent, Scoupe, 1993–95 Elantra and Sonata

1. Remove the transaxle. Insert the forward end of an old transaxle input shaft or a clutch disc guide tool through the splined center of the clutch disc, pressure plate and the pilot bearing in the crankshaft. This will keep the disc from dropping when the pressure plate is removed from the flywheel.

2. Loosen the clutch cover mounting bolts alternately and diagonally in very small increments, no more than two turns at a time, so as to avoid warping the cover flange.

3. Remove the pressure plate and disc.

4. Remove the return clip and the clutch release bearing.

5. Insert tool 09414–24000 in the spring pin and attach the round nut to the end of the tool. While holding the shaft of the special tool, rotate the sleeve with a wrench to force the spring pin out.

6. Remove the clutch release shaft, packings, return spring and the release fork.

To install:

7. Apply a light coating of multipurpose grease to the release fork shaft and the throw out bearing contact surfaces.

8. Align the lock pin holes of the release fork and shaft and drive 2 new spring pins into the holes. Make sure the spring pin slot is at right angles to the centerline of the control shaft.

9. Apply grease into the groove in the release bearing and install bearing into the front bearing retainer in the transaxle. Install the return clip to the release bearing and fork.

10. Make sure the surfaces of the pressure plate and flywheel are wiped clean of grease and lightly sand them with crocus cloth. Lightly grease the clutch disc and transaxle input shaft splines making sure not to allow any grease to contact the clutch disc material or clutch slip may result.

11. Locate the clutch disc on the flywheel with the stamped mark facing outward. Use a clutch disc guide or old input shaft to center the disc on the flywheel and then install the pressure plate over it. Install the bolts and tighten them evenly. Tighten them in increments of 2 turns or less to avoid warping the pressure plate. Torque to 11 to 15 ft. lbs. (15 to 21 Nm).

12. Remove the clutch disc centering tool. Install the transaxle. Adjust the clutch free-play.

1996–97 Elantra

NOTE: The release lever must be removed from the transaxle before the transaxle can be removed.

1. Disconnect the negative battery cable.

2. Remove the air intake duct.

3. Remove the nut attaching the lever to the release fork shaft.

4. Remove the clevis pin attaching the lever to the release cylinder. Remove the release lever.

5. Remove the release cylinder. It is not necessary to disconnect the fluid line.

6. Remove the transaxle assembly.

7. Using a snapring pliers, spread the snapring and remove the release bearing from the pressure plate.

8. Install the clutch alignment tool through the clutch disc and remove the pressure plate.

To install:

9. Lightly grease the splines of the clutch disc. Place the disc on the flywheel and hold it in place with the clutch alignment tool.

10. Position the pressure plate on the disc and install the bolts. Torque the bolts in a star pattern to 11–16 ft. lbs. (15–22 Nm).

11. Align the ears of the release bearing assembly with the release fork and install the bearing on the transaxle. Install the release lever to the release fork.

12. Remove the alignment tool and install the transaxle assembly to the engine.

13. Push the release lever toward the release cylinder until the bearing assembly snaps into place on the pressure plate.

NOTE: The operating range of the release lever should be 3° or less after the bearing assembly has engaged the pressure plate. If the range is over 3°, the bearing has not engaged the pressure plate properly.

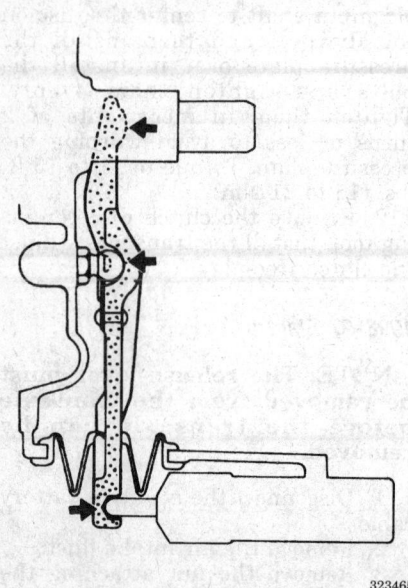

Apply grease to the pivot points of the release fork — 1993–95 Elantra, Scoupe and Sonata

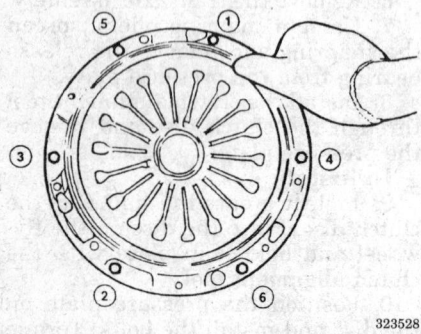

Pressure plate bolt torque sequence — 1996–97 Elantra

14. Install the air intake duct and connect the negative battery cable.

Clutch Master Cylinder

REMOVAL AND INSTALLATION

All Models

1. Disconnect the negative battery cable.
2. Unscrew the reservoir cap. Loosen the bleeder plug and drain the clutch fluid into a container. Then, retighten the bleeder plug.

3. From inside the vehicle; remove the cotter pin, clevis pin and washer from the clutch pedal to release the master cylinder pushrod.
4. Unbolt the clutch reservoir bracket, if equipped.
5. Loosen the clutch reservoir hose clamps to permit movement of the clutch hose. Then, disconnect the clutch hose from the master cylinder.
6. Remove the two nuts and pull the master cylinder and gasket from the firewall.
7. Inspect the gasket for damage and replace as required.
To install:
8. Position the master cylinder assembly with gasket onto the firewall and loosely install the fluid reservoir mounting bracket bolts and the two master cylinder mounting nuts. Tighten the nuts to 7–10 ft. lbs. (9–14 Nm).
9. Connect the clutch tube to the master cylinder and tighten the tube clamps. Tighten the clutch reservoir bracket bolts.
10. Connect the pushrod to the clutch pedal and insert the clevis pin. Lubricate the clevis pin with multi-purpose grease and secure it with a new split pin.
11. Refill and bleed the clutch hydraulic system.
12. Check and adjust the clutch pedal height and free-play.

Clutch Slave Cylinder

REMOVAL AND INSTALLATION

All Models

1. Disconnect the negative battery cable.
2. Drain the fluid from the clutch master cylinder reservoir.
3. Disconnect the clutch hose from the slave cylinder.
4. Unbolt the cylinder from the clutch housing and remove it.

5. Inspect the cylinder for leakage or torn boots and repair or replace as required.
To install:
6. Apply a thin coating of grease to the contact points of the release fork and the cylinder and install the slave cylinder to the clutch housing. Tighten the cylinder retainer bolts to 11–16 ft. lbs. (15–22 Nm).
7. Connect the fluid line to the cylinder and tighten the fitting to 7–10 ft. lbs. (13–17 Nm).
8. Refill and bleed the clutch hydraulic system.
9. Reconnect the negative battery cable.

Hydraulic Clutch System Bleeding

PROCEDURE

All Models

Whenever a clutch system component is removed; a clutch hydraulic line disconnected; or the system is opened for any reason; the system must be bled to remove any air that may be trapped inside the system.
1. Unscrew the clutch fluid reservoir cap and fill the reservoir with new brake fluid.
2. Fill a plastic container halfway with brake fluid.
3. Connect a hose to the slave cylinder bleeder screw and place the other end of the hose into the container of brake fluid.
4. Loosen the bleeder screw enough so fluid starts to leak out of the bleeder hole. Have an assistant push the clutch pedal slowly to the floor. You will notice air bubbles rising to the top of the fluid container. Close the bleeder screw when the pedal is at the floor. Keep the reservoir full at all times.

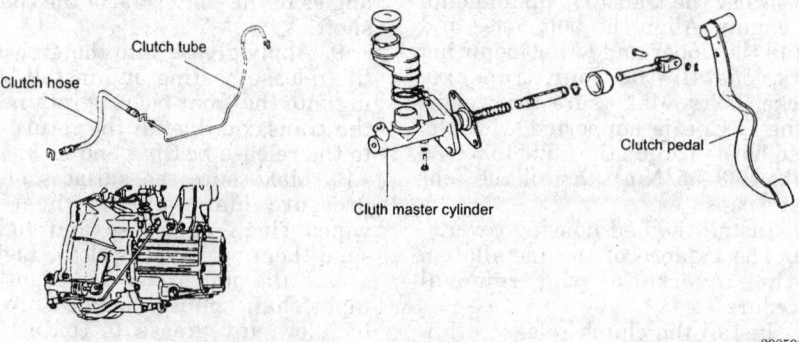

Clutch master cylinder and related components — Accent model shown, others similar

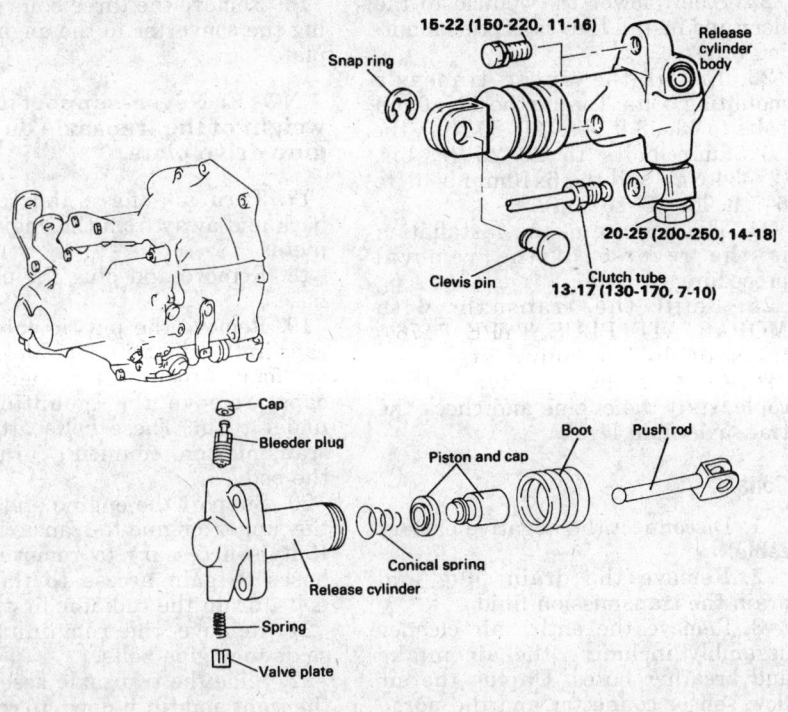

Snap ring

15-22 (150-220, 11-16)

Release cylinder body

20-25 (200-250, 14-18)

Clevis pin

Clutch tube
13-17 (130-170, 7-10)

Cap

Bleeder plug

Piston and cap

Boot

Push rod

Conical spring

Release cylinder

Spring

Valve plate

TORQUE : Nm (kg.cm, lb.ft)

258063

Clutch slave cylinder — Excel and Scoupe

5. Repeat Step 4 until all the air bubbles are gone.

6. Tighten the bleeder screw and check the fluid level. Add fluid as necessary until the proper level is reached.

7. Install the reservoir cap.

Automatic Transaxle Assembly

REMOVAL AND INSTALLATION

Excel, Scoupe and Elantra

1. Shift the transaxle into the **N** position.

2. Disconnect the negative battery cable.

3. Drain the fluid from the transaxle.

4. Remove the entire air cleaner assembly including the air intake and breather hoses. Disconnect the air flow sensor connector and the purge control solenoid valve connector and hoses. Label all the connections to avoid confusion during assembly.

5. Disconnect the cooler supply and return lines from the transaxle. Plug the line openings to prevent the entry of dirt and foreign matter.

6. Disconnect the control cable from the shift lever.

7. Disconnect the speedometer cable from the transaxle. Route the cable off to the side and out of the way.

8. Unplug the connectors to the solenoid valves, inhibitor switch, pulse generators, kickdown servo switch and oil temperature sensor.

9. Working from the top of the transaxle, remove the upper transaxle mounting bolts.

10. Remove the transaxle mounting bracket.

11. Support the transaxle with a jack and remove the center member mounting bolts.

NOTE: To avoid placing excessive pressure on the oil pan, support the transaxle in a wide area.

12. Raise and safely support the vehicle.

13. Remove both front wheels and both axle-end cotter pins, nuts and washers.

14. Remove the ball joint castle nuts and separate the ball joints from the knuckles.

15. Remove the halfshafts from the transaxle.

NOTE: If desired, the axles may be left attached to the vehicle. Some prefer to have them out

of the way. If left attached, be sure to suspend them with a strong rope or wire; if they hang free, the CV-joints or boots could get damaged.

16. Unbolt and remove the bell housing cover.

17. Remove the three special bolts attaching the torque converter to the drive plate. To remove these bolts, engage the crankshaft pulley nut with a socket and turn the crankshaft until each bolt comes into view. Remove each bolt with the proper size box end wrench by working through the bell housing cover access hole.

18. After the drive plate bolts are removed, reinstall the center member temporarily.

19. Remove the lower transaxle mounting bolts and lower the transaxle from the engine.

To install:

20. Place the transaxle on the transaxle jack securely and install onto the engine, using the dowels as guides. Make sure all heater tube clamps, wires, etc. are out of the way or it will not be possible to install the transaxle flush with the block.

21. Install the lower transaxle mounting bolts.

22. For models except 1996–97 Elantra, torque the 2.4 in. (60mm) bolt to 22–26 ft. lbs. (30–35 Nm). Torque the 2.6 in. (65mm) bolt to 31–40 ft. lbs. (43–55 Nm).

23. For 1996–97 Elantra, torque the 2.2 inch (55 mm) bolt to 20–25 ft. lbs. (27–34 Nm). Torque the 2.7 inch (65 mm) bolt to 31–40 ft. lbs. (43–55 Nm).

24. Remove the center member.

25. Install the torque converter-to-drive plate bolts and torque them to 34–39 ft. lbs. (46–53 Nm).

26. Install the bell housing cover and torque the bolts to 7–9 ft. lbs. (10–12 Nm).

27. Install the halfshafts.

28. Connect the ball joints to the steering knuckles. Tighten the castle nuts to 43–53 ft. lbs. (60–72 Nm), and then only tighten them enough to install new cotter pins.

29. Install the front wheels.

30. Lower the vehicle.

31. Install the center member.

32. Install the transaxle mounting bracket and torque the bolts to 43–58 ft. lbs. (60–80 Nm).

33. Install the upper transaxle mounting bolts. Torque the upper two bolts to 31–40 ft. lbs. (43–55 Nm). Torque the two bolts at the starter motor housing to 20–25 ft. lbs. (27–34 Nm).

34. The balance of the installation is the reverse of the removal procedure.

35. Reconnect the negative battery cable.

36. Fill the transaxle to the proper level.

Accent

1. Disconnect the negative battery cable.

2. Remove the air cleaner assembly.

3. Disconnect the fluid cooler hoses from the transaxle.

4. Disconnect the control cable and the speedometer cable at the transaxle.

5. Disconnect the following connectors;
Pulse generator
Range switch
Kickdown servo switch
Solenoid valve
Oil temperature sensor

6. Remove the transaxle mounting bracket.

7. Remove the upper bolts attaching the transaxle to the engine.

8. Raise and safely support the vehicle.

9. Drain the transaxle fluid.

10. Support the transaxle with a jack and remove the center member.

11. Remove the bell housing cover.

12. Remove the six bolts attaching the torque converter to the flexplate. Turn the crankshaft with a wrench to access the bolts.

13. After removing the torque converter bolts, reinstall the center member.

14. Disconnect the left ball joint from the knuckle and remove the halfshaft.

15. Remove the remaining transaxle mounting bolts and remove the transaxle. Be careful that the torque converter does not fall from the transaxle.

To install:

16. Position the transaxle on the engine while guiding the right halfshaft into the differential. Install the lower mounting bolts. Torque the 10x70mm bolts to 31–40 ft. lbs. (43–55 Nm) and the 6x10mm bolts to 6–7 ft. lbs. (8–10 Nm).

17. Install the left halfshaft and connect the ball joint to the knuckle.

18. Remove the center member.

19. Install the torque converter bolts. Torque the bolts to 33–38 ft. lbs. (46–53 Nm).

20. Install the bell housing cover.

21. Install the center member.

22. Safely lower the vehicle to the floor and install the transaxle mounting bracket.

23. Install the upper transaxle mounting bolts. Torque the 12x40mm bolts to 43–58 ft. lbs. (60–80 Nm); the 10x55mm bolts to 20–25 ft. lbs. (27–34 Nm) and the 6x10mm bolts to 6–7 ft. lbs. (8–10 Nm).

24. The balance of the installation is the reverse of the removal procedure.

25. Refill the transaxle with MOPAR ATF PLUS TYPE 7176® transaxle fluid or equivalent.

26. Connect the negative battery cable, start the engine and check the transaxle fluid level.

Sonata

1. Disconnect the negative battery cable.

2. Remove the drain plug and drain the transmission fluid.

3. Remove the entire air cleaner assembly including the air intake and breather hoses. Unplug the air flow sensor connector and the purge control solenoid valve connector and hoses.

4. Disconnect the manual control cable from the control lever.

5. Disconnect the speedometer cable from the transaxle.

6. Unplug the connectors to the solenoid valves, inhibitor switch, pulse generators, kickdown servo switch, oil temperature sensor and unplug the engine wiring harness.

7. Remove the starter motor.

8. Raise the vehicle and support it safely.

9. Remove the left side undercover.

10. Remove the stabilizer bar.

11. Remove both wheels and both axle-end cotter pins, nuts and washers.

12. Remove the ball joint nuts and separate the ball joints from the knuckles.

13. Remove the driveshaft from the transaxle.

NOTE: If desired, the axles may be left attached to the vehicle. Some prefer to have them out of the way. If left attached, be sure to suspend them with a strong rope or wire; if they hang free, the CV-joints or boots could get damaged.

14. Unbolt and remove the bell housing inspection cover.

15. Support the transaxle with a transmission jack and remove the center member from the crossmember.

16. Remove the three bolts connecting the converter to the engine drive plate.

NOTE: Never support the full weight of the transaxle on the engine drive plate.

17. Turn and force the converter back and away from the engine drive plate.

18. Remove and plug the oil cooler hoses.

19. Remove the plastic outer cover caps in the right fender shield. There are four of them. After removing the caps, remove the mounting bolts under them. These bolts attach the transmission mounting bracket to the body.

20. Support the engine and remove the upper engine-to-transaxle bolts. If it is necessary to remove heater hoses to gain access to the upper bolts, drain the radiator first.

21. Remove the remaining transaxle-to-engine bolts.

22. Slide the transaxle assembly to the right and tilt it down to remove it from the vehicle.

To install:

23. Install the torque converter to the transaxle. The converter is correctly installed when the distance from the ring gear to the bell housing surface is 12mm.

24. Place the transaxle on the transaxle jack securely and install the transaxle to the engine using the dowels as guides. Make sure all heater tube clamps, wires, etc. are out of the way or it will not be possible to install the transaxle flush with the block.

25. Install the lower transaxle-to-engine bolts. Torque the bolts to 11–16 ft. lbs. (15–22 Nm).

26. Install the four transaxle-to-body mounting bolts and torque the to 43–57 ft. lbs. (58–77 Nm). Install the four plastic caps.

27. Connect the oil cooler lines.

28. Turn and push the torque converter towards the drive plate.

29. Apply Loctite to the threads and install the three torque converter bolts. Torque the bolts to 94–101 ft. lbs. (130–140 Nm).

30. Install the torque converter inspection plate.

31. Install the transaxle center member onto the crossmember. Torque the bolts to 58–71 ft. lbs. (78–96 Nm).

32. Install the bell housing inspection cover and torque the retaining bolts to 7–9 ft. lbs. (9–12 Nm).

33. Install the driveshafts. Make sure the shafts are completely seated

in place and an circlips and snaprings are securely in place.

34. Install the ball joints to the knuckles and torque the locknuts to 43–52 ft. lbs. (58–70 Nm).

35. If removed, install the axle-end washers, nuts and cotter pins. Torque the nuts to 145–188 ft. lbs. (200–260 Nm).

36. The balance of the installation is the reverse of the removal procedure.

37. Fill the transaxle with the specified fluid to the proper level. Road test the vehicle.

DRIVE AXLE

Halfshaft

REMOVAL AND INSTALLATION

Excel, Scoupe, Accent and Elantra

1. Disconnect the negative battery cable.
2. Remove the hub center cap and loosen the driveshaft (axle) nut.
3. Raise and support the front of the vehicle safely.
4. Remove the front wheels.
5. Remove the engine splash shield.
6. Remove the lower ball joint bolts from the lower control arm.

NOTE: Place the lower arm ball joint on the lower arm to prevent damage to the ball joint dust boot.

7. Detach the tie rod end from the steering knuckle.
8. Remove the sway bar self locking nut.
9. Drain the transaxle fluid into a suitable waste container.

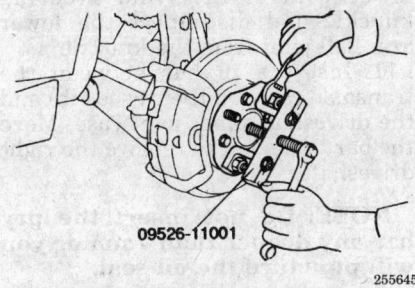

09526-11001

255645

Removing the halfshaft from the hub — Scoupe and 1993–94 Elantra

10. Insert a prybar between the transaxle case (on the raised rib) and the driveshaft inner joint case. Move the bar to the right to withdraw the left driveshaft; left, to remove the right driveshaft.

NOTE: Do not insert the pry bar too deeply (more than 7mm), or you will damage the oil seal.

11. Plug the transaxle case with a clean rag to prevent dirt from entering the case.
12. Use a puller/driver mounted on the wheel studs to push the driveshaft from the front hub. Take care to prevent the spacer shims from falling out of place.

To install:
13. Insert the driveshaft into the hub first, then install the transaxle end.
14. Install the lower ball joint mounting bolts and the sway bar self locking nut to the lower control arm. Tighten the ball joint mounting bolts and to 44–53 ft. lbs. (60–72 Nm).
15. Install the tie rod end onto the steering knuckle. Tighten the mounting nut to 11–25 ft. lbs. (15–34 Nm).
16. Install the wheel and tire assembly.
17. Install the splash shield.
18. Lower the vehicle to the floor.
19. Install the hub nut washer. Torque the axle shaft hub nut to 148–192 ft. lbs. (200–260 Nm).

NOTE: Always use a new inner joint retaining ring every time you remove the driveshaft.

20. Replace the hub center cap.
21. Connect the negative battery cable.
22. Check and add transaxle fluid if necessary.

Sonata

All Models with 4 Cylinder Engines

1. Disconnect the negative battery cable.
2. Remove the hub center cap and loosen the driveshaft (axle) nut.
3. Loosen the wheel lug nuts.
4. Raise and support the front end on jackstands.
5. Remove the front wheels.
6. Remove the engine splash shield, and drain the transaxle fluid.
7. Remove the cotter pin from the tie rod end and loosen, do not remove, the tie rod end nut.
8. Using special puller tool 09568-3100 or equivalent, disconnect the tie rod end from the steering knuckle. Tie the tool off to a suspension member component before using it. Remove the tie rod end nut.

9. Position the tool between the lower control arm and steering knuckle, and disconnect the lower arm ball joint from the knuckle.
10. Insert a prybar between the transaxle case (on the raised rib) and the driveshaft inner joint case. Move the bar to the right to withdraw the left driveshaft; to the left to remove the right driveshaft.

NOTE: Do not insert the pry bar too deeply (more than 7mm) or you will puncture the oil seal.

11. Plug the transaxle case with a clean rag to prevent dirt from entering the case.
12. Use a puller/driver mounted on the wheel studs to push the driveshaft from the front hub. Take care to prevent the spacer shims from falling out of place.

To install:

NOTE: Always use a new inner joint retaining ring every time you remove the driveshaft.

13. Insert the driveshaft into the hub, then install the transaxle end.
14. Connect the lower ball joint to the steering knuckle and torque the nut to 44–53 ft. lbs. (60–72 Nm).
15. Install the tie rod end onto the steering knuckle. Tighten the mounting nut to 18–25 ft. lbs. (24–34 Nm). Install a new cotter pin.
16. Install the splash shield.
17. Mount the front wheels, tighten the lug nuts, and lower the vehicle to the ground.
18. Install the axle washer and nut. Make sure the washer is installed properly. Torque the axle nut to 148–192 ft. lbs. (200–260 Nm), and secure the nut with a new cotter pin.
19. Fill the transaxle to the proper level with the specified fluid.
20. Install the hub center cap.
21. Connect the negative battery cable.

6 Cylinder Engines — Left Halfshaft

1. Disconnect the negative battery cable.
2. Remove the hub center cap and remove the split pin, driveshaft (axle) nut and washer.

NOTE: Make a mental note of how the washer is installed.

3. Loosen the wheel lug nuts.
4. Raise and support the front end on jackstands.
5. Remove the front wheels.
6. Remove the engine splash shield and drain the transaxle fluid.
7. Remove the cotter pin from the tie rod end and loosen, do not remove, the tie rod end nut.

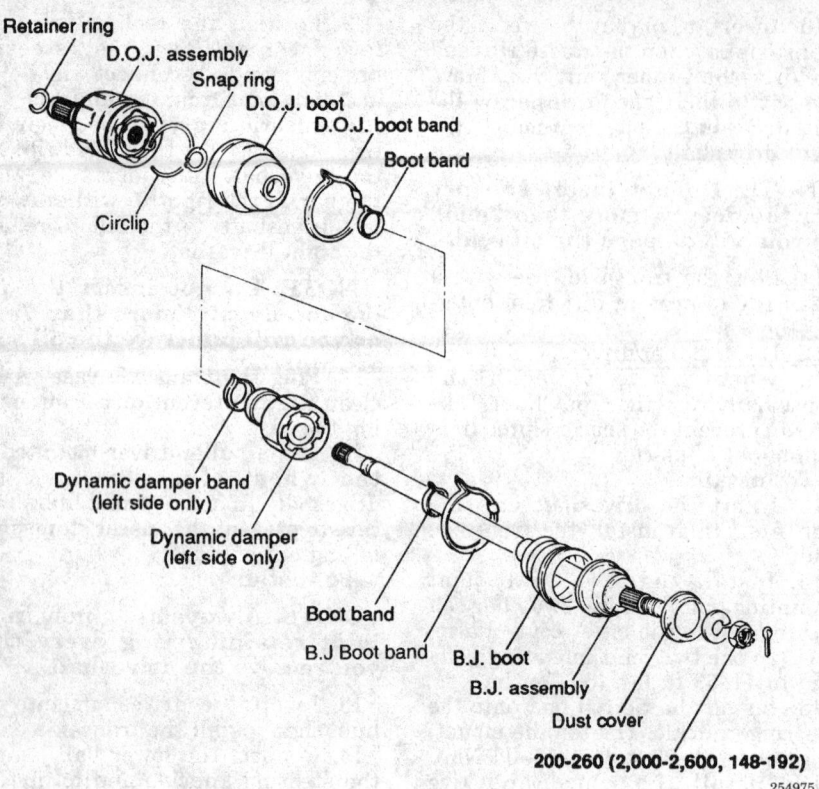

Halfshaft assembly — Accent shown, others similar

200-260 (2,000-2,600, 148-192)

254975

8. Using special puller tool 09568-3100 or equivalent, disconnect the tie rod end from the steering knuckle. Tie the tool off to a suspension member component before using it. Remove the tie rod end nut.

9. Position the tool between the lower control arm and steering knuckle and disconnect the lower arm ball joint from the knuckle.

10. Using puller tool 09526-11001 or equivalent, pull the left driveshaft from the wheel hub.

11. Insert a prybar between the center bearing bracket and the driveshaft. Separate the driveshaft from the center bracket as shown.

NOTE: Do not insert the pry bar any deeper than 7mm or you will puncture the oil seal and also damage the joint. When separating the driveshaft, do not allow the full weight of the vehicle to be placed on the wheel bearing. If the weight of the vehicle must be applied for any reason, support the wheel bearing with holding tool 09517-21500 or equivalent.

12. Remove the oxygen sensor connector from the center bearing bracket. One screw holds the connector to the bracket.

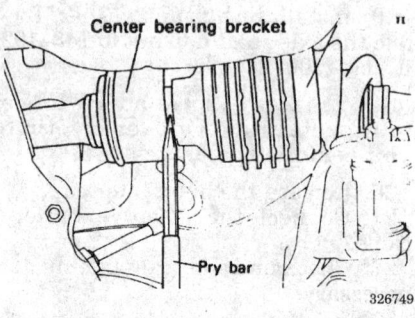

Disconnect the halfshaft from the center bearing — Sonata with 6-cylinder engine

326749

13. Remove the two center bracket mounting bolts. Insert a pry bar between the center bearing bracket, inner shaft and cylinder block. Then, pull the center bracket and inner shaft assembly from the transaxle case. Plug the transaxle case with a clean rag to prevent dirt from entering the case.

To install:

NOTE: Always use a new inner joint retaining ring every time you remove the driveshaft.

14. Insert the inner shaft and bracket assembly into the transaxle case and install the center bracket

mounting bolts. Torque the bolts to 26-33 ft. lbs. (35-45 Nm).

15. Connect the oxygen sensor connector to the center mounting bracket with the mounting screw.

16. Insert the driveshaft into the center bearing, then into the wheel hub.

17. Connect the lower ball joint to the steering knuckle and torque the nut to 44-53 ft. lbs. (60-72 Nm).

18. Connect the tie rod end to the steering knuckle and torque the tie rod end nut to 18-25 ft. lbs. (24-34 Nm). Secure the nut with a new cotter pin.

19. Install the splash shield.

20. Mount the front wheels, tighten the lug nuts, and lower the vehicle to the ground.

21. Install the axle washer and nut. Make sure the washer is installed properly. Torque the axle nut to 148-192 ft. lbs. (200-260 Nm), and secure the nut with a new cotter pin.

22. Install the hub center cap.

23. Connect the negative battery cable.

24. Fill the transaxle to the proper level with the specified fluid.

6 Cylinder Engines — Right Halfshaft

1. Disconnect the negative battery cable.

2. Remove the hub center cap and loosen the driveshaft (axle) nut.

3. Loosen the wheel lug nuts.

4. Raise and support the front end on jackstands.

5. Remove the front wheels.

6. Remove the engine splash shield and drain the transaxle fluid.

7. Remove the split pin from the tie rod end and loosen, do not remove, the tie rod end nut.

8. Using special puller tool 09568-3100 or equivalent, disconnect the tie rod end from the steering knuckle. Tie the tool off to a suspension member component before using it.

9. Remove the tie rod end nut.

10. Position the tool between the lower control arm and steering knuckle and disconnect the lower arm ball joint from the knuckle.

11. Insert a prybar between the transaxle case (on the raised rib) and the driveshaft inner joint case, Move the bar to the left to remove the right driveshaft.

NOTE: Do not insert the pry bar any deeper than 7mm or you will puncture the oil seal.

12. Plug the transaxle case with a clean rag to prevent dirt from entering the case.

13. Use a puller/driver mounted on the wheel studs to push the driveshaft from the front hub. Take care to prevent the spacer shims from falling out of place.

To install:

NOTE: Always use a new inner joint retaining ring every time you remove the driveshaft.

14. Insert the driveshaft into the hub, then install the transaxle end.

15. Connect the lower ball joint to the steering knuckle and torque the nut to 44–53 ft. lbs. (60–72 Nm).

16. Install the tie rod end onto the steering knuckle. Tighten the mounting nut to 18–25 ft. lbs. (24–34 Nm). Install a new cotter pin.

17. Install the splash shield.

18. Mount the front wheels, tighten the lug nuts, and lower the vehicle to the ground.

19. Install the axle washer and nut. Make sure the washer is installed properly. Torque the axle nut to 148–192 ft. lbs. (200–260 Nm), and secure the nut with a new cotter pin.

20. Fill the transaxle to the proper level with the specified fluid.

21. Install the hub center cap.

22. Connect the negative battery cable.

CV-Joint Boot

REPLACEMENT

All Models

——— **WARNING** ———
Do not disassemble a Birfield joint. Service with a new joint or clean and repack using a new boot kit. The driveshaft joints use special grease, do not add any grease other than that supplied with the kit.

NOTE: The Double Offset Joint (D.O.J.) is bigger than other joints and in these applications, is normally used as an inboard joint.

1. Remove the halfshaft from the vehicle.

2. Side cutter pliers can be used to cut the metal retaining bands. Remove the boot from the joint outer race.

3. Locate and remove the large circlip at the base of the joint. Remove the outer race (the body of the joint).

4. Remove the small snapring and take off the inner race, cage and balls as an assembly. Clean the inner race, cage and balls without disassembling.

5. If the boot is to be reused, wipe the grease from the splines and wrap the splines in vinyl tape before sliding the boot from the shaft.

6. Remove the inner (D.O.J.) boot from the shaft.

7. If the outer (B.J.) boot is to be replaced, remove the boot retainer rings and slide the boot down and off of the shaft at this time.

To install:

8. Be sure to tape the shaft splines before installing the boots. Fill the inside of the boot with the specified grease. Often the grease supplied in the replacement parts kit is meant to be divided in half, with half being used to lubricate the joint and half being used inside the boot.

9. Install the cage onto the halfshaft so the small diameter side of the cage is installed first. With a brass drift pin, tap lightly and evenly around the inner race to install the race until it comes into contact with the rib of the shaft. Apply the specified grease to the inner race and cage and fit them together. Insert the balls into the cage.

10. Install the outer race (the body of the joint) after filling with the specified grease. The outer race should be filled with this grease.

11. Tighten the boot bands securely. Make sure the distance between the boot bands is correct.

12. Install the halfshaft to the vehicle.

STEERING

Air Bag

——— **CAUTION** ———
Some vehicles are equipped with an air bag system, also known as the Supplemental Restraint System (SRS). The system must be disabled before performing service on or around system components, steering column, instrument panel components, wiring and sensors. Failure to follow safety and disabling procedures could result in accidental air bag deployment, possible personal injury and unnecessary system repairs.

PRECAUTIONS

Several precautions must be observed when handling the inflator module to avoid accidental deployment and possible personal injury.

• Never carry the inflator module by the wires or connector on the underside of the module.

• When carrying a live inflator module, hold securely with both hands, and ensure that the bag and trim cover are pointed away.

• Place the inflator module on a bench or other surface with the bag and trim cover facing up.

• With the inflator module on the bench, never place anything on or close to the module which may be thrown in the event of an accidental deployment.

DISARMING

Accent, Elantra and Sonata

——— **CAUTION** ———
The air bag system must be disarmed before removing the steering wheel or air bag module. Failure to do so may cause accidental deployment, property damage or personal injury.

1. Disconnect the negative battery cable.

2. Wait 30 seconds.

3. Remove the air bag module mounting nuts and lift the module from the steering wheel.

4. Locate and unplug the connector. Protect this connector from accidental electrical contact, including static electricity.

5. The air bag is now disarmed and the battery can be connected to perform electrical testing on other systems.

6. When ready to reconnect the air bag; disconnect the battery, connect the air bag connector, and install the cover or panel.

7. Install the mounting nuts. Make sure no one is in the vehicle when connecting the battery.

Steering Wheel

REMOVAL AND INSTALLATION

All 1993–94 Models

Without Air-Bag

1. Disconnect the negative battery cable.

2. Remove the horn pad. Then, disconnect the horn wire connector.

3. Remove the steering wheel retaining nut. Matchmark the relationship between the wheel and shaft.

4. Remove the steering wheel dynamic dampener.

5. Screw the two bolts of a steering wheel puller into the wheel. Then, turn the bolt at the center of the puller to force the wheel off the steering shaft. Do not pound on the wheel to remove it or the collapsible steering shaft may be damaged.

To install:

6. The steering wheel can be pushed onto the shaft splines by hand far enough to start the retaining nut. Do not hammer on the steering wheel. Install the retaining nut and torque it to 26–32 ft. lbs. (34–44 Nm).

With Air-Bag

1. Disconnect and insulate the negative battery cable and wait at least 30 seconds before working on the vehicle.

2. Remove the air-bag mounting nut from the back of the steering wheel.

3. Carefully remove the connector by spreading the lock toward the outer edge. Use a screw driver and gently pry off the connector.

4. Matchmark the steering shaft and the steering wheel so it can be installed in the original position.

5. Using a steering wheel puller, remove the steering wheel.

To install:

6. Install the steering wheel. Torque the nut to 26–32 ft. lbs. (34–44 Nm).

7. Reconnect and install the air-bag module.

8. Connect the negative battery cable.

9. Turn the ignition switch on and check for proper operation of the SRS warning light.

Accent

Without Air-Bag

1. Disconnect the negative battery cable.

2. Remove the horn pad. Then, disconnect the horn wire connector.

3. Remove the steering wheel retaining nut. Matchmark the relationship between the wheel and shaft.

4. Remove the steering wheel dynamic dampener.

5. Screw the two bolts of a steering wheel puller into the wheel. Then, turn the bolt at the center of the puller to force the wheel off the steering shaft. Do not pound on the wheel to remove it or the collapsible steering shaft may be damaged.

To install:

6. The steering wheel can be pushed onto the shaft splines by hand far enough to start the retaining nut. Do not hammer on the steering wheel. Install the retaining nut and torque it to 26–32 ft. lbs. (34–44 Nm).

With Air-Bag

————— CAUTION —————
The air bag system (SRS) must be disarmed before removing the steering wheel. Failure to do so may cause accidental deployment, property damage or personal injury.
—————————————————

1. Turn the front wheels to the straight ahead position and turn the ignition switch to the **LOCK** position.

2. Disconnect and insulate the negative battery cable and wait at least 30 seconds before working on the vehicle.

3. Remove the air-bag mounting nut from the back of the steering wheel.

4. Carefully remove the connector by spreading the lock toward the outer edge. Use a screw driver and gently pry off the connector.

5. Remove the nut from the steering shaft.

6. Matchmark the steering shaft and the steering wheel so it can be installed in the original position.

7. Using a steering wheel puller, remove the steering wheel.

To install:

8. Align the matchmark and install the steering wheel on the shaft. Torque the retaining nut to 26–32 ft. lbs. (34–44 Nm).

9. Reconnect and install the air-bag module.

10. Connect the negative battery cable.

11. Turn the ignition switch on and check for proper operation of the SRS warning light.

1995–97 Elantra and Sonata

————— CAUTION —————
The air bag system must be disarmed before removing the steering wheel. Failure to do so may cause accidental deployment, property damage or personal injury.
—————————————————

1. Disconnect and insulate the negative battery cable and wait at least 30 seconds before working on the vehicle.

2. Remove the air bag mounting nuts from the back of the steering wheel.

3. Carefully remove the connector by spreading the lock toward the outer edge. Use a small pry tool and gently pry off the connector.

4. Matchmark the steering shaft and the steering wheel so it can be installed in the original position.

5. Using a steering wheel puller, remove the steering wheel.

To install:

6. Install the steering wheel. Tighten the mounting nuts to 26–33 ft. lbs. (35–45 Nm).

7. Reconnect and install the air-bag module.

8. Connect the negative battery cable.

9. Turn the ignition switch on and check for proper operation of the SRS warning light.

Tie Rod Ends

REMOVAL AND INSTALLATION

Excel and Scoupe

1. Raise and safely support the vehicle.

2. Remove the front wheels.

3. Loosen the locknut on the tie rod end to be removed.

4. Remove the cotter pin and the castellated nut.

5. Use a ball joint separator to separate the tie rod end from the knuckle.

6. Remove the tie rod end counting the number of turns it takes to remove it from the tie rod.

To install:

7. Install the new tie rod end on the shaft the same amount of turns that as it took to remove the old one.

8. Reconnect the tie rod end to the steering knuckle. Install the castle nut and tighten it to 11–25 ft. lbs. (15–34 Nm). Then, tighten the castle nut only enough to install a new cotter pin.

9. Tighten the tie rod end locknut making sure the tie rod end is perpendicular to the steering knuckle mounting surface.

10. Install the front wheels.

11. Lower the vehicle to the floor.

12. Check and adjust the front wheel alignment.

Accent

1. Raise and support the vehicle.

2. Remove the wheel and tire assembly.

3. Loosen the locknut on the tie rod end to be removed.

4. Remove the cotter pin and the castellated nut.

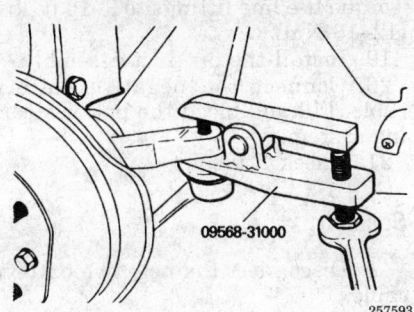

09568-31000

257593

Proper method of disconnecting the tie rod
ends — Excel and Scoupe

5. Using the special tool, separate
the ball stud from the knuckle.

6. Remove the tie rod end counting
the number of turns it takes to remove it.

To install:

7. Install the new tie rod end on
the shaft the same amount of turns
that it took to remove the old one.

8. Install the ball stud to the
steering knuckle. Torque the nut to
11–25 ft. lbs. (15–34 Nm) and install
the cotter pin.

9. Tighten the locknut to 37–41 ft.
lbs. (50–55 Nm) making sure the tie
rod end is perpendicular to the steering knuckle mounting surface.

10. Install the wheel and tire
assembly.

11. Lower the vehicle to the floor.

12. Check and adjust the toe if
necessary.

Elantra and Sonata

1. Disconnect the negative battery
cable.

2. Raise and support the vehicle
under the frame.

3. Remove the front wheels.

4. Disconnect the tie rod ends with
separator tool 09568-31000.

5. The tie rod ends can now be removed. Prior to removal, count the
exact number of exposed threads on
the tie rod ends, then loosen the
locknut and unscrew the tie rod end.

To install:

6. When installing new tie rod
ends, oil the threads and screw them
into place so that the previously
noted number of threads is visible
with the locknut tight. Check the
length of the tie rod.

7. For Elantra models, the tie-rod
length should be 181mm (7.1 inches).
Remember that the left and right tie
rod ends should be equal.

8. For all 4-cylinder Sonata models and on V6 Sonata models with a
Mando power steering box, the length
should be 187.4mm. On V6 Sonata

models with TRW power steering, the
length should be 176.1mm.

9. For Elantra models, connect the
tie rods to the knuckle arms using
new cotter pins. Torque the tie rod
nuts to 11–15 ft. lbs. (15–34 Nm).

10. For Sonata models, connect the
tie rods to the knuckle arms using
new cotter pins. Torque the tie rod
nuts to 17–25 ft. lbs. (24–34 Nm).

11. Mount the front wheels and
lower the vehicle.

12. Check the front wheel
alignment.

13. Connect the negative battery
cable.

Manual Rack and Pinion

REMOVAL AND INSTALLATION

Excel and Scoupe

1. Disconnect the negative battery
cable.

2. Position the front wheels in the
straight-ahead position.

3. Slide the steering joint dust
cover up to reveal the lower U-joint
bolt.

4. Raise the vehicle and support it
safely.

5. Remove the front wheels.

6. Use a ball joint separator to disconnect the tie rod ends.

7. Remove the steering shaft lower
U-joint bolt.

8. Remove the clamps securing the
rack to the crossmember.

9. Slide the rack away from the exhaust pipe and remove it from the
vehicle.

To install:

10. Install the rubber mount for the
gear box with the slit on the
downside.

11. Fit the rack into the vehicle and
install the mounting bolts. Tighten
the bolts to 44–59 ft. lbs. (60–80 Nm).

12. Center the rack ends within
their strokes.

13. Install the steering shaft coupling bolt and tighten it to 11–15 ft.
lbs. (15–20 Nm).

14. Connect the tie rod ends to the
steering knuckles. Tighten the castle
nuts to 11–25 ft. lbs. (15–34 Nm), and
then tighten them only enough to install new cotter pins.

15. Install the front wheels.

16. Lower the vehicle to the floor.

17. Fit the steering joint dust cover
back into place.

18. Reconnect the negative battery
cable.

19. Check and adjust the front
wheel alignment and the steering
wheel spoke angle.

Accent

1. Safely raise and support the
vehicle.

2. Remove both wheel assemblies.

3. Remove the coupling bolt.

4. Using special tool 09568-31000
or equivalent, disconnect the tie rod
ends from the knuckles.

5. Remove the rear roll stopper
bracket.

6. Loosen the rear mounting bolts
for the center member halfway.

7. Remove the steering gear
mounting clamps.

8. Move the steering gear first to
the right and then down through the
left side of the center member.

To install:

9. Install the rack assembly from
the left side of the center member.
Then move it towards the left into the
proper position and install the
mounting clamps. Torque the clamp
bolts to 44–59 ft. lbs. (45–60 Nm).

10. Install the center member rear
mounting bolts. Torque the bolts to
44–59 ft. lbs. (45–60 Nm).

11. Install the rear roll stopper
bracket.

12. Connect the tie rod ends to the
knuckles. Torque the nut to 11–25 ft.
lbs. (15–34 Nm) and install a new cotter pin.

13. Install the coupling bolt. Torque
the bolt to 11–15 ft. lbs. (15–20 Nm).

14. Install both wheel assemblies.

15. Safely lower the vehicle to the
floor.

Power Rack and Pinion

REMOVAL AND INSTALLATION

Excel and Scoupe

1. Disconnect the negative battery
cable.

2. Position the front wheels in the
straight-ahead position.

3. Drain the fluid from the power
steering reservoir.

4. Slide the steering joint dust
cover up to reveal the lower U-joint
bolt.

5. Raise the vehicle and support it
safely.

6. Remove the front wheels.

7. Use a ball joint separator to disconnect the tie rod ends.

8. Remove the steering shaft lower
U-joint bolt.

9. Disconnect the fluid lines from
the rack valve body. Plug the lines to
prevent fluid loss and contamination.

10. If equipped, unbolt and remove
the stabilizer bar.

11. If necessary, remove the rear roll stopper from the center member and move the rear roll stopper forward.

12. Remove the rack unit mounting clamp bolts. Move the rack to the side to clear the exhaust pipe and take it out the right side of the vehicle.

To install:

13. Install the rack with the projection on the rubber mount aligned with the hole in the clamp. Apply adhesive to the cylinder side of the mount so that the slit part will not open. Tighten the rack mounting bolts to 44–59 ft. lbs. (60–80 Nm).

14. Install the steering shaft lower U-joint bolt.

15. Connect the fluid pressure and return lines.

16. Connect the tie rod ends to the steering knuckles. Tighten the castle nuts to 11–25 ft. lbs. (15–34 Nm), and then tighten them only enough to install new cotter pins.

17. Install the stabilizer bar if it was removed. Reconnect the rear roll stopper if it was disconnected.

18. Install the front wheels.

19. Lower the vehicle to the floor.

20. Fit the steering joint dust cover back into place.

21. Refill and bleed the power steering system.

22. Check and adjust the front wheel alignment and the steering wheel spoke angle.

Accent

1. Safely raise and support the vehicle.

2. Remove both wheel assemblies.

3. Disconnect the tie rod ends from the knuckles.

4. Remove the coupling bolt from the universal joint.

5. Drain the power steering fluid.

6. Disconnect the pressure and return lines from the rack assembly.

7. Remove the band from the steering joint cover.

8. Remove the steering gear mounting bolts.

9. Remove the center member and the stabilizer bar assemblies.

10. Remove the steering rack through the side of the vehicle. Be careful not to damage the boots.

To install:

11. Position the steering rack on the vehicle. Apply adhesive to the slit part of the mounting rubber so that it will not open.

12. Install the mounting bolts. Torque the bolts to 44–59 ft. lbs. (60–80 Nm).

13. Install the stabilizer bar and the center member.

14. Install the band on the steering joint cover.

15. Connect the pressure and return tubes to the rack assembly. Torque the nuts to 9–13 ft. lbs. (12–18 Nm).

16. Connect the tie rod ends to the knuckles. Torque the nuts to 11–25 ft. lbs. (15–34 Nm) and install a new cotter pin.

17. Install both wheel assemblies.

18. Safely lower the vehicle to the floor.

19. Refill and bleed the power steering system.

Elantra

1. Disconnect the negative battery cable.

2. Remove the air intake hose assembly.

3. Place a drain pan under the fluid connections and disconnect the pressure hose and the return hose from the steering rack.

4. From inside the vehicle, disconnect the steering shaft from the rack assembly.

5. Remove the strap, push out the dust cover and remove the dust cover mounting plate.

6. Safely raise and support the vehicle.

7. Remove both front wheels and disconnect the tie rod ends from the knuckles using special tool 09568-31000 or equivalent.

8. Remove the right side stabilizer mounting bracket.

9. Remove the front exhaust pipe assembly.

10. Remove the power steering rack mounting clamps and the clamp holding the fluid tubes.

11. Remove the rack assembly through the right side of the vehicle. Do not damage the boots while removing the rack assembly.

To install:

12. Carefully position the rack assembly in the vehicle and install the mounting clamps. Torque the clamp bolts to 44–59 ft. lbs. (60–80 Nm).

13. Install the clamp holding the fluid tubes.

14. Install the exhaust pipe and the right side stabilizer bracket.

15. Connect the tie rods to the knuckles. Torque the nuts to 17.7–25 ft. lbs. (24–34 Nm).

16. Install both front wheels and lower the vehicle to the floor.

17. From inside the car, connect the steering shaft to the rack assembly and install the dust boot.

18. Connect the pressure hose and return hose to the rack assembly.

Torque the line fittings to 9–13 ft. lbs. (12–18 Nm).

19. Install the air duct assembly.

20. Connect the negative battery cable. Fill and bleed the power steering system.

21. Check for leaks.

Sonata

1. Disconnect the negative battery cable.

2. Raise the vehicle and support it safely.

3. Remove the wheels.

4. Drain the fluid from the power steering system.

5. Remove the steering shaft-to-pinion coupling bolt.

6. Remove the retainer nut and disconnect both tie rod ends using an appropriate puller.

7. Disconnect the left lower arm from the steering knuckle. This will increase clearance as the gear assembly will be removed from the left side of the vehicle.

8. Disconnect the fluid hoses from the gear box.

9. Remove the center member and temporarily loosen the front muffler.

10. Unbolt and remove the stabilizer bar.

11. Remove the rack unit mounting clamp bolts and move the rack towards the right and then take the unit out on the left side of the vehicle. The tie rod ends can now be removed. Prior to removal, count the exact number of exposed threads on the tie rod ends, then loosen the locknut and unscrew the tie rod end.

To install:

12. When installing new tie rod ends, oil the threads and screw them into place so the previously noted number of threads is visible with the locknut tight. As a further reference, the distance between the end of the tie rod boot and the point at which the locknut touches the tie rod ball socket body should be 187.4mm. Torque the locknut to 37–41 ft. lbs. (50–55 Nm).

13. When installing the power steering rack, make sure the rubber isolators have their nubs aligned with the holes in the clamps.

14. Apply rubber cement to the slits in the gear mounting grommet. Tighten the clamp bolt to 43–58 ft. lbs. (60–80 Nm).

15. Install the stabilizer bar.

16. Tighten the front muffler.

17. Install the center member.

18. Connect the fluid hoses to the gear box. Tighten the fittings to 9–13 ft. lbs. (12–18 Nm).

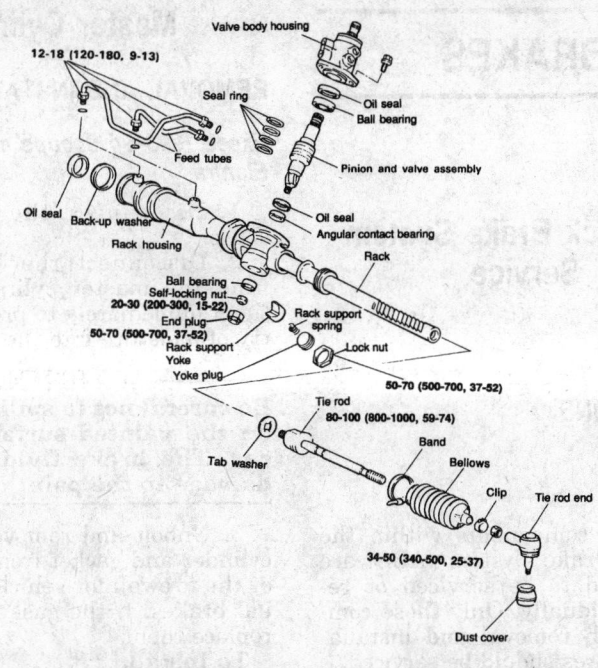

12-18 (120-180, 9-13)
Valve body housing
Seal ring
Oil seal
Ball bearing
Feed tubes
Pinion and valve assembly
Oil seal
Oil seal
Back-up washer
Angular contact bearing
Rack housing
Rack
Ball bearing
Self-locking nut
20-30 (200-300, 15-22)
End plug
50-70 (500-700, 37-52)
Rack support spring
Rack support
Lock nut
Yoke
Yoke plug
50-70 (500-700, 37-52)
Tie rod
80-100 (800-1000, 59-74)
Band
Bellows
Tab washer
Clip
Tie rod end
34-50 (340-500, 25-37)
Dust cover

TORQUE : Nm (kg.cm, lb.ft)

257035

Power rack and pinion assembly — 1993–95 Elantra shown, later models similar

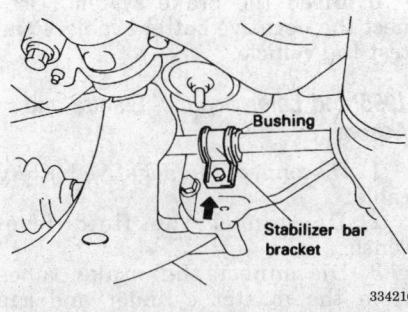

Bushing

Stabilizer bar bracket

334216

Stabilizer bar mounting — 1993–94 Sonata shown, later models similar

19. Connect the lower arm to the steering knuckle.
20. Connect the tie rod ends to the knuckles. Tighten the mounting nuts to 17–25 ft. lbs. (24–34 Nm).
21. Connect the steering shaft to the rack. Tighten the coupling bolt to 11–14 ft. lbs. (15–20 Nm).
22. Fill the system with DEXRON II ATF.
23. Install the wheels and lower the vehicle.
24. Connect the negative battery cable.

Power Steering Pump

BLEEDING

All Models

1. Ensure that the reservoir is full of DEXRON II automatic transmission fluid.
2. Raise and safely support the front wheels of the vehicle.
3. Turn the steering wheel from lock to lock 5 or 6 times.
4. Disconnect the coil wire and connect to a solid ground. Operate the starter motor intermittently for 15 to 20 seconds and turn the steering wheel from lock to lock 5 or 6 times.

NOTE: Ensure that the reservoir is full during air bleeding to prevent the fluid level from falling below the lower position of the filter.

5. Connect the coil wire and start the engine.
6. Turn the steering wheel from lock to lock until no more air bubbles are visible in the reservoir.
7. Confirm that the oil is not milky and that the fluid level is correct.
8. Confirm that there is little change in the fluid level when the steering wheel is turned to the left and right.

NOTE: An abrupt rise in the fluid level after stopping the engine is a sign of incomplete bleeding. If this occurs, repeat the bleeding procedure.

REMOVAL AND INSTALLATION

Excel and Scoupe

1. Disconnect the negative battery cable.
2. Drain the fluid from the power steering pump reservoir.
3. Loosen the pump mounting bolts and remove the drive belt.
4. Disconnect the suction hose from the pump and plug it.
5. Disconnect the pressure hose from the pump and the return hose from the reservoir. Plug both hoses to prevent fluid loss and contamination
6. Remove the pump mounting bracket bolts and lift out the pump. If necessary, unbolt and remove the pump reservoir.
To install:
7. Install the pump reservoir, if removed. Position the pump onto the mounting bracket and torque the mounting bolts to 15–20 ft. lbs. (20–27 Nm).
8. Connect the suction hose to the pump. Install the hose so that the painted portion is toward the pump body.
9. Connect the pressure hose to the pump and the return hose to the fluid reservoir. Use new crush washers and tighten the pressure hose banjo bolt to 41–44 ft. lbs. (55–60 Nm). When installing the return line, make sure you push it at least 25–30mm onto the return tube. The hoses should not be twisted or allowed to come in contact with an other component in the engine compartment.
10. Install the drive belt and adjust the tension. Tighten the tension adjusting bolt to 18–24 ft. lbs. (25–33 Nm).
11. Connect the negative battery cable.
12. Fill the system with DEXRON II ATF and bleed the system. Check for leaks and the hydraulic fittings.

Accent

——— **WARNING** ———
Place a rag over the alternator when working on the steering pump.

1. Remove the pressure hose from the steering pump.

2. Disconnect the suction hose from the suction connector and drain the fluid.

3. Loosen the pump mounting bolts and remove the belt.

4. Disconnect the suction hose from the pump.

5. Disconnect the pressure switch connector at the pump.

6. Remove the mounting bolts and remove the pump.

To install:

7. Install the pump on the pump bracket. Torque the adjusting bolt to 18–24 ft. lbs. (25–33 Nm) and the bracket bolt to 15–20 ft. lbs. (20–27 Nm).

8. Connect the suction hose to the pump.

9. Install and adjust the belt.

10. Connect the pressure hose to the pump and the suction hose to the reservoir.

11. Refill the reservoir and bleed the system.

Elantra and Sonata

1. Disconnect the negative battery cable.

2. Place a drain pan under the pump.

3. Remove the pressure hose from the oil pump.

4. Disconnect the suction hose and drain the fluid in the container.

5. Loosen the pump mounting bolts and remove the drive belt.

6. Disconnect the pressure switch connector.

7. Remove the pump-to-mounting bracket bolts and lift out the pump.

To install:

8. Position the pump onto the mounting bracket and torque the mounting bolts to 33–41 ft. lbs. (45–55 Nm).

9. Connect the suction hose to the pump.

10. Install the drive belt and adjust the tension.

11. Connect the pressure switch.

12. Connect the pressure hose to the pump and the return hose to the oil reservoir. When installing the return line, make sure you push it at least 30mm onto the return tube. The hoses should not be twisted or allowed to come in contact with an other component in the engine compartment.

13. Fill the system with DEXRON II ATF and bleed the system.

14. Connect the negative battery cable.

BRAKES

Anti-Lock Brake System Service

PRECAUTIONS

- Certain components within the Anti-Lock Brake System (ABS) are not intended to be serviced or repaired individually. Only those components with removal and installation procedures should be serviced.

- Do not use rubber hoses or other parts not specifically specified for and ABS system. When using repair kits, replace all parts included in the kit. Partial or incorrect repair may lead to functional problems and require the replacement of components.

- Lubricate rubber parts with clean, fresh brake fluid to ease assembly. Do not use lubricated shop air to clean parts; damage to rubber components may result.

- Use only specified brake fluid from an unopened container.

- If any hydraulic component or line is removed or replaced, it may be necessary to bleed the entire system.

- A clean repair area is essential. Always clean the reservoir and cap thoroughly before removing the cap. The slightest amount of dirt in the fluid may plug an orifice and impair the system function. Perform repairs after components have been thoroughly cleaned; use only denatured alcohol to clean components. Do not allow ABS components to come into contact with any substance containing mineral oil; this includes used shop rags.

- The Anti-Lock control unit is a microprocessor similar to other computer units in the vehicle. Ensure that the ignition switch is **OFF** before removing or installing controller harnesses. Avoid static electricity discharge at or near the controller.

- If any arc welding is to be done on the vehicle, the control unit should be unplugged before welding operations begin.

Master Cylinder

REMOVAL AND INSTALLATION

Excel, 1993–94 Scoupe and 1993–94 Elantra

1. Disconnect the fluid level sensor.

2. Disconnect the brake tubes from the master cylinder and cap them immediately to prevent the entry of moisture into the system.

——— WARNING ———
Be careful not to spill brake fluid on the painted surfaces of your car. The brake fluid will cause damage to the paint.

3. Unbolt and remove the master cylinder and gasket from the booster or the firewall for vehicles with manual brakes. If the gaskets are worn, replace them.

To install:

4. Position the master cylinder to the booster and install the mounting bolts. Torque the mounting bolts to 6 to 9 ft. lbs. (8 to 12 Nm).

5. Connect the brake tubes to the master cylinder. Torque the tubes to 9–12 ft. lbs. (13–17 Nm).

6. Bleed the brake system. Connect the negative battery cable. Road test the vehicle.

1995 and Later Scoupe, Elantra and Accent

1. Disconnect the negative battery cable.

2. Disconnect the fluid level sensor.

3. Disconnect the brake tubes from the master cylinder and cap them immediately to prevent the entry of moisture into the system.

——— WARNING ———
Be careful not to spill brake fluid on the painted surfaces of the car. The brake fluid will cause damage to the paint.

4. Remove the proportioning valve mounting bracket from the master cylinder.

5. Unbolt the master cylinder and gasket from the booster. If the gaskets are worn, replace them.

To install:

6. Position the master cylinder to the booster and install the mounting bolts. Torque the mounting bolts to 6–9 ft. lbs. (8–12 Nm).

7. Install the proportioning valve mounting bracket to the master cylinder.

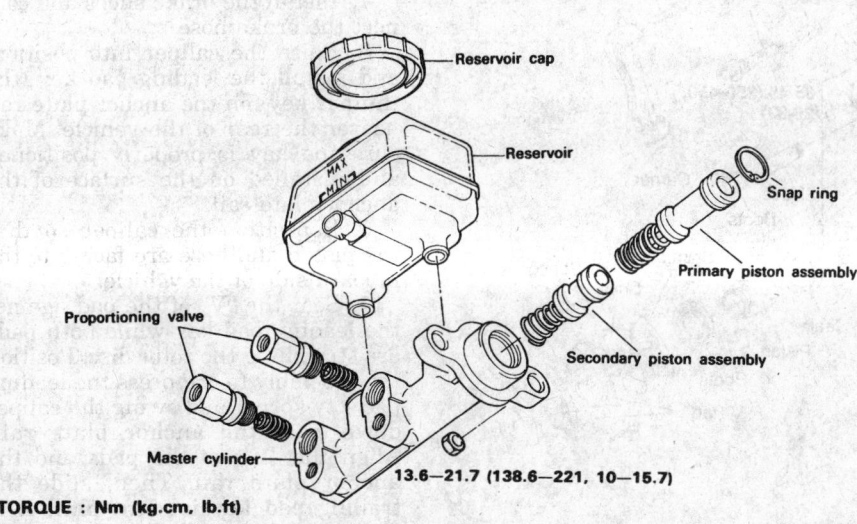

TORQUE : Nm (kg.cm, lb.ft)

13.6—21.7 (138.6—221, 10—15.7)

140639

Master cylinder assembly — 1993–94 Scoupe

8. Connect the brake tubes to the master cylinder. Torque the tubes to 9–12 ft. lbs. (13–17 Nm).

9. Connect the fluid level sensor.

10. Bleed the brake system.

11. Connect the negative battery cable. Road test the vehicle.

Sonata

1. Disconnect the fluid level sensor.

2. Disconnect the brake tubes from the master cylinder and cap

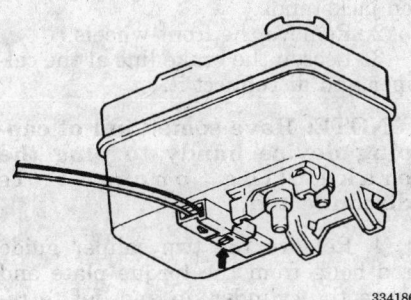

334186

Fluid level sensor connector — 1993–94 Sonata

them immediately to prevent the entry of moisture into the system. On V6 Sonata with rear disc brakes, disconnect the brake tubes without removing the proportioning valves.

3. Unbolt and remove the master cylinder and gasket from the booster. If the gaskets are worn, replace them. On V6 Sonata with rear disc brakes, there is a rubber O-ring located on the master cylinder cartridge that seals the unit to the brake booster. If this seal is worn or damaged in any way, replace it.

To install:

4. Before installing the master cylinder, check the pushrod-to-master cylinder clearance as follows:

a. On Sonata with rear drum brakes, use a Vernier caliper to measure the distance from the booster front shell surface to the booster pushrod end (dimension "A"). It should be 37.35–37.65mm.

NOTE: Master cylinder pushrod clearance adjustment is not required on Sonata with rear disc brakes.

5. If the pushrod-to-master cylinder clearance is not as specified, ad-

just by changing the pushrod length by turning the adjusting screw on the pushrod.

6. Position the master cylinder to the booster and install the mounting nuts.

7. Connect the brake tubes to the master cylinder. Torque the tubes to 9–12 ft. lbs. (13–17 Nm).

8. Bleed the brake system. Connect the negative battery cable. Road test the vehicle.

Brake Caliper

REMOVAL AND INSTALLATION

Excel, Accent and Scoupe

—— **CAUTION** ——
Brake linings may contain asbestos. Asbestos is a known cancer-causing agent. When working on brakes, remember that the dust which accumulates on the brake parts, contains asbestos. Always wear a protective face covering, such as a painter's mask, when working on the brakes. NEVER blow the dust from the brakes or drum.

1. Raise and support the front end on jackstands.

2. Remove the front wheels.

3. Loosen the brake line at the caliper and disconnect it.

NOTE: Have some kind of capping device handy to plug the brake line once it is disconnected.

4. Remove the brake pads.

5. Remove the pin and sleeve boots.

6. Remove the lower caliper bolt and raise the caliper up and out to remove it.

To install:

7. Position the caliper onto its mounting and install the lower mounting bolt. Torque the bolt to 16 to 24 ft. lbs. (22 to 32 Nm).

8. Install the pin boots, sleeve boots and brake pads.

9. Connect the brake line to the caliper with two new metal gaskets. Torque the brake line union bolt to 18–22 ft. lbs. (25–30 Nm).

10. Bleed the system.

11. Mount the front wheels and lower the vehicle.

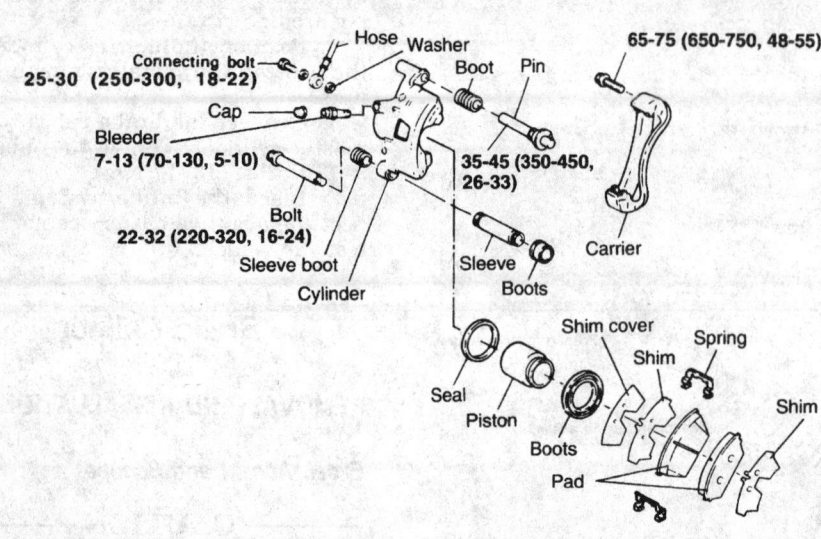

Connecting bolt
25-30 (250-300, 18-22)

Hose Washer
Boot Pin

65-75 (650-750, 48-55)

Cap

Bleeder
7-13 (70-130, 5-10)

Bolt
22-32 (220-320, 16-24)

Sleeve boot

Cylinder

35-45 (350-450, 26-33)

Sleeve Carrier

Boots

Seal Piston

Shim cover
Shim Spring

Shim

Boots

Pad

TORQUE : Nm (kg.cm, lb.ft)

298343

Front brake caliper assembly — Accent shown, others similar

Elantra

> **— CAUTION —**
> *Brake linings may contain asbestos. Asbestos is a known cancer-causing agent. When working on brakes, remember that the dust which accumulates on the brake parts, contains asbestos. Always wear a protective face covering, such as a painter's mask, when working on the brakes. NEVER blow the dust from the brakes or drum.*

Front

1. Raise and support the front end on jackstands.
2. Remove the front wheels.
3. Loosen the brake line at the caliper and disconnect it.

NOTE: Use a capping device to plug the brake line once it is disconnected.

4. Remove the brake pads.
5. Remove the pin and sleeve boots.
6. Remove the lower caliper bolt and raise the caliper up and out to remove it.
To install:
7. Position the caliper onto its mounting and install the lower mounting bolt. Torque the bolt to 16–24 ft. lbs. (22–32 Nm).
8. Install the pin boots, sleeve boots and brake pads.
9. Connect the brake line to the caliper with two new metal gaskets. Torque the brake line union bolt to 18–22 ft. lbs. (24–30 Nm).
10. Bleed the system.
11. Mount the front wheels and lower the vehicle.

Rear

1. Loosen the rear wheel lug nuts.
2. Raise and support the rear end safely.
3. Remove the screw that attaches the trailing pad key to the anchor plate.
4. Press the caliper assembly against the leading key (this is the thinner of the two keys) and slide the trailing shoe key outward. To remove the caliper, swing the front end of the caliper up and away from the anchor plate rail. After the keys are removed, the caliper should be free.
5. Once the caliper is free, disconnect the brake hose. Plug the end of the hose to prevent excess loss of fluid. The hose need not be disconnected if you are replacing the brakes. Support the caliper assembly with a piece of wire. Support in a manner that will allow you to work on the brake pads.
6. Remove the brake pads.
To install:
7. Install the brake shoes and connect the brake hose.
8. Raise the caliper into position, and install the leading pad key (the thinner key) on the anchor plate rail nearer the rear of the vehicle. Make sure the key is properly positioned and installed on the surface of the anchor plate rail.
9. Reposition the caliper so that the piston and hose are facing to the inboard side of the vehicle.
10. Seat the "V" of the pad against the leading pad key while both pads are straddling the rotor disc. Position the assembly to compress the leading pad key spring and swing the caliper down past the anchor plate rail. Align the "V"s of the pads and the anchor plate rail. Then, slide the trailing pad key inboard until it is fully seated. Install the trailing pad key screw to hold it in place. Tighten the caliper mounting bolts to 25 ft. lbs. (34 Nm).
11. Mount the rear wheels and lower the vehicle.
12. Refill the master cylinder reservoir. Bleed the system. Pump the brake pedal several times to set the pads.

Sonata

NOTE: The 1993–94 Sonata is equipped with two types of front disc brakes: Mando and Bendix. Mando front disc brakes are used on 4 and 6 cylinder Sonatas (1993–94 models) equipped with rear drum brakes. Bendix front disc brakes are used on V6 Sonatas equipped with rear disc brakes.

Mando Front Caliper

1. Raise and support the front end on jackstands.
2. Remove the front wheels.
3. Loosen the brake line at the caliper and disconnect it.

NOTE: Have some kind of capping device handy to plug the brake line once it is disconnected.

4. Remove the two caliper guide rod bolts from the torque plate and raise the cylinder up and out to remove it.
To install:
5. Mount the caliper and install the two guide bolts. Torque the bolts to 16–23 ft. lbs.

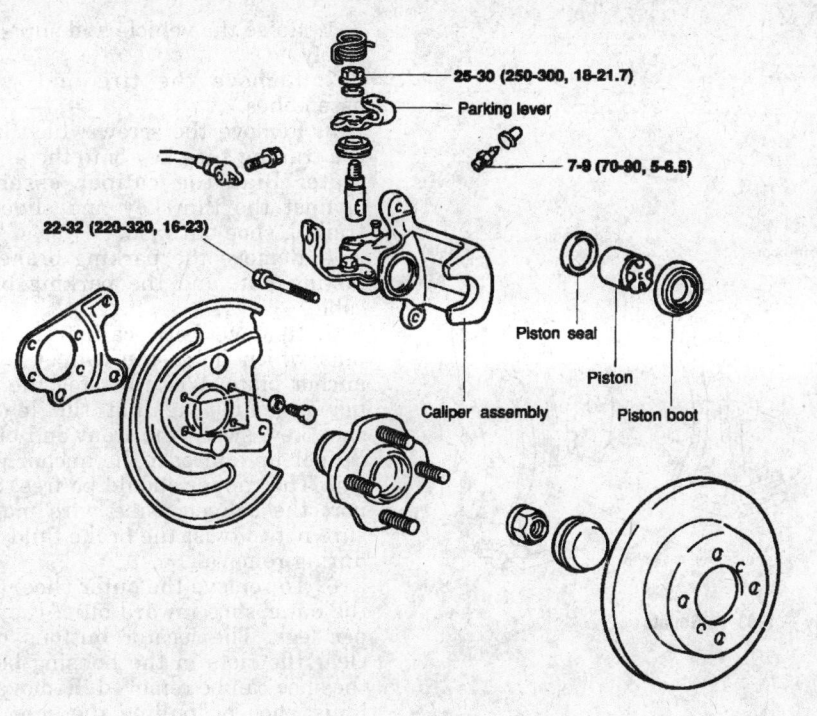

25-30 (250-300, 18-21.7)

Parking lever

7-9 (70-90, 5-6.5)

22-32 (220-320, 16-23)

Piston seal

Piston

Caliper assembly

Piston boot

314200

Rear disc brake assembly — Elantra

6. Connect the brake line to the caliper and tighten the fitting to 9–12 ft. lbs.

7. Bleed the system.

8. Mount the front wheels and lower the vehicle.

Bendix Front Caliper

1. Raise and support the front end on jackstands.

2. Remove the front wheels.

3. Disconnect the brake tube from the brake hose. Release the brake hose clip and remove the brake hose from the strut.

NOTE: Have some kind of capping device handy to plug the brake line once it is disconnected.

4. Loosen and disconnect the brake line from the caliper.

5. Remove the small retaining pin from the lower part of the caliper.

6. Swing the caliper up until it clears the rotor and pads.

7. Slide the caliper inboard until the locating pin disengages from it's groove in the caliper. Pull the caliper from the locating pin.

8. Inspect the locating pin for wear and damage. Replace the pin as required. Protect the pin from exposure to dirt and dust, If the pin is contaminated it will wear prema-

turely and eventually fail. Inspect the anchor plate for wear and damage.

To install:

9. Lubricate the locating pin bore with white silicone compound and mount the caliper onto the locating pin. Push the caliper firmly onto the pin making sure that the pin snaps into the caliper groove.

10. Lower the caliper until the small retaining pin holes are aligned. Install a new retaining pin into the lower part of the caliper. Tighten the pin.

11. Connect the brake line to the caliper and bleed the brakes.

12. Install the wheel assembly.

13. Lower the vehicle to the floor.

Rear Brake Caliper

1. Raise and support the rear end safely.

2. Remove the wheel(s).

3. Remove the screw that attaches the trailing shoe key to the anchor plate.

4. Press the caliper assembly against the leading key (this is the thinner of the two keys) and slide the trailing shoe key outward. To remove the caliper, swing the front end of the caliper up and away from the anchor plate rail. After the keys are removed, the caliper should be free.

5. Once the caliper is free, disconnect the brake hose. Plug the end of the hose to prevent excess loss of fluid. The hose need not be disconnected if you are replacing the brakes. Support the caliper assembly with a piece of wire. Support in a manner that will allow you to work on the brake pads.

6. Remove the brake shoes.

To install:

7. Install the brake shoes and connect the brake hose.

8. Raise the caliper into position, and install the leading shoe key (the thinner key) on the anchor plate rail nearer the rear of the vehicle. Make sure the key is properly positioned and installed on the surface of the anchor plate rail.

9. Reposition the caliper so that the piston and hose are facing to the inboard side of the vehicle.

10. Seat the V of the shoes against the leading shoe key while both shoes are straddling the rotor disc. Position the assembly to compress the leading shoe key spring and swing the caliper down past the anchor plate rail. Align the V's of the shoes and the anchor plate rail. Then, slide the trailing shoe key inboard until it is fully seated. Install the trailing shoe key screw to hold it in place. Tighten the caliper mounting bolts to 25 ft. lbs. (34 Nm).

11. Mount the rear wheels and lower the vehicle.

12. Refill the master cylinder reservoir. Bleed the system. Pump the brake pedal several times to set the pads.

Disc Brake Pads

REMOVAL AND INSTALLATION

Excel, Scoupe, Accent and 1993 Elantra

— **CAUTION** —

Brake linings may contain asbestos. Asbestos is a known cancer-causing agent. When working on brakes, remember that the dust which accumulates on the brake parts, contains asbestos. Always wear a protective face covering, such as a painter's mask, when working on the brakes. NEVER blow the dust from the brakes or drum! There are solvents made for the purpose of cleaning brake parts.

1. Raise the vehicle and support it safely.

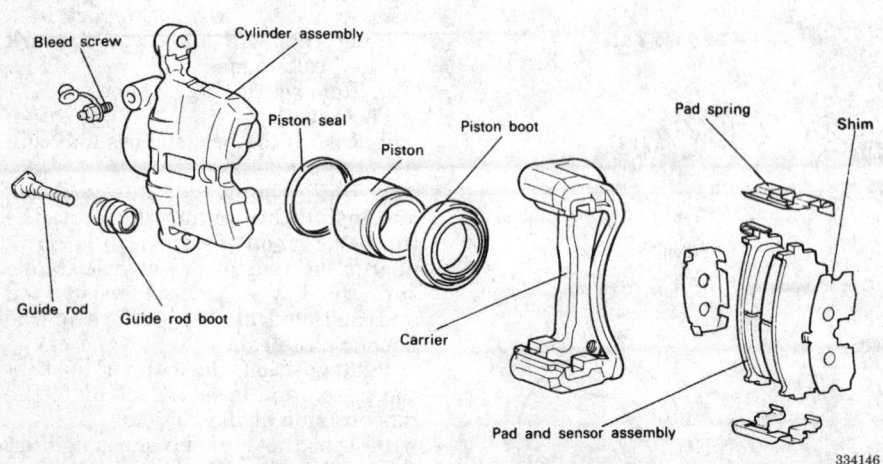

Mando brake caliper assembly — 1993-94 Sonata

2. Remove the front wheels.

3. Remove the lower caliper mounting bolt and rotate the caliper upward. Secure the caliper with a wire or heavy string.

4. Remove the pads from the caliper.

To install:

5. Install the pad clips.

6. Install the pads onto the pad clips.

7. Compress the caliper piston using a C-clamp.

8. Rotate the caliper downward and install the mounting bolt.

NOTE: Replace all brake pads at the same time. Never replace the pads on 1 wheel only.

9. Install the wheels and lower the vehicle.

10. Depress the brake pedal a few times. The first couple of strokes on the pedal will feel overly long. However, the pads will set themselves and the stroke will return to normal.

1994-97 Elantra

— **CAUTION** —

Brake linings may contain asbestos. Asbestos is a known cancer-causing agent. When working on brakes, remember that the dust which accumulates on the brake parts, contains asbestos. Always wear a protective face covering, such as a painter's mask, when working on the brakes. NEVER blow the dust from the brakes or drum! There are solvents made for the purpose of cleaning brake parts.

Front Brakes

1. Raise the vehicle and support it safely.

2. Remove the front wheels.

3. Remove the lower caliper mounting bolt and rotate the caliper upward. Secure the caliper with a wire or heavy string.

4. Remove the pads from the caliper.

To install:

5. Install the pad clips.

6. Install the pads onto the pad clips.

7. Compress the caliper piston using a C-clamp.

8. Rotate the caliper downward and install the mounting bolt.

9. Install the wheels and lower the vehicle.

10. Depress the brake pedal a few times. The first couple of strokes on the pedal will feel overly long. However, the pads will set themselves and the stroke will return to normal.

Rear Brakes

1. Raise the vehicle and support it safely.

2. Remove the tire and wheel assemblies.

3. Remove the screw which holds the trailing shoe key onto the anchor plate. Bias the caliper assembly against the thin key and slide the trailing shoe outward.

4. Remove the parking brake adjusting nut, and the parking brake cable.

5. Remove both caliper support pins which holds the caliper to the anchor plate. While biasing the caliper assembly against the leading shoe key, swing the front end of the caliper up and past the anchor plate rail. The caliper should be free. Support the caliper using wire making sure not to twist the brake fluid hose during removal.

6. To remove the outer shoe, push the outer shoe inward off of the caliper legs. The locator buttons must clear the slots in the housing before the shoe can be removed. Remove the inner shoe by pulling the shoe outward until the shoe clears the piston.

To install:

7. Remove the cap from the master cylinder reservoir and, using a clean suction gun, remove about 6mm of fluid.

8. Using a large C-clamp, press the piston all the way back into the caliper.

9. Install the new pads and clips. Position the pad with the wear sensor on the piston side and upwards.

10. Position the caliper and install the bolts. Torque the bolts to 25 ft. lbs. (34 Nm).

11. Position the parking brake cable and tighten the adjustment nut.

12. Refill the master cylinder, install the tire and wheel assemblies and pump the brake pedal until firm.

Sonata

NOTE: The 1993-94 Sonata is equipped with two types of front disc brakes: Mando and Bendix. Mando front disc brakes are used on 4 and 6 cylinder Sonatas (1993-94 models) equipped with rear drum brakes. Bendix front disc brakes are used on V6 Sonatas equipped with rear disc brakes.

Mando Front Disc Brakes

1. Raise and support the front end on jackstands.

2. Remove the front wheels.

3. Remove the 2 caliper guide rod bolts and raise the caliper up and out of the way. Tie the caliper assembly off to a suspension member with a piece of string.

4. Lift the pads and anti-squeal shims from the caliper.

5. Depress the center of the outboard spring clip and remove the clip by slipping the ends from its anchors.

6. Remove the inboard spring clip with pliers.

7. Clean all caliper parts, especially the torque plate shafts, with a solvent made for brake parts.

— CAUTION —
Replace all brake pads at the same time. Never replace the pads on one wheel only.

8. If the spring clips are weak, damaged or deformed, replace them.

9. Remove the cap from the master cylinder reservoir and, using a clean suction gun, remove about 6mm of fluid.

10. Install the pad clips.

11. Install the pads onto each clip.

NOTE: Position the inner pad so that the pad wear indicator is facing toward the piston side and up.

12. Using a C-clamp or a piece of wood, force the caliper piston back into the caliper as far as it will go.

13. Untie the caliper and install the guide bolts. Torque the bolts to 16-23 ft. lbs.

14. Install the wheels and lower the car. Get in the car and depress the brake pedal a few times. The first couple of strokes on the pedal will feel overly long. However, the pads will set themselves and the stroke will return to normal.

Bendix Front Disc Brakes

1. Raise and support the front end on jackstands.

2. Remove the front wheels.

3. Remove the small retaining pin from the bottom of the caliper. This pin must be replaced whenever the front pads are replaced.

4. Rotate the caliper on the large locating pin until the brake pads are accessible.

— CAUTION —
DO NOT remove the locating pin from the caliper. This pin is sealed for life and exposure to dust or dirt will cause it to fail prematurely.

5. Lift the pads and anti-squeal spring from the caliper.

6. Clean all caliper parts, especially the torque plate shafts, with a solvent made for brake parts.

7. Remove the cap from the master cylinder reservoir and, using a clean suction gun, remove about 6mm of fluid.

8. Using a C-clamp, a piece of wood, or special tool 09581-11000, force the caliper piston back into the caliper as far as it will go.

9. Install the pads and anti-rattle spring.

NOTE: Position the inner pad so that the pad wear indicator is facing toward the large caliper pin.

10. Swing the caliper upward and install a new small retaining pin.

11. Add brake fluid to the fluid reservoir until the full line is reached.

12. Install the wheels and lower the car. Get in the car and depress the brake pedal a few times. The first couple of strokes on the pedal will feel overly long. However, the pads will set themselves and the stroke will return to normal. Replenish the master cylinder fluid reservoir as necessary.

Rear Disc Brakes

1. Raise the vehicle and support it safely.

2. Remove the tire and wheel assemblies.

3. Remove the screw which holds the trailing shoe key onto the anchor plate. Bias the caliper assembly against the thin key and slide the trailing shoe outward.

4. Remove both caliper support pins which holds the caliper to the anchor plate. While biasing the caliper assembly against the leading shoe key, swing the front end of the caliper up and past the anchor plate rail. The caliper should be free. Support the caliper using wire making sure not to twist the brake fluid hose during removal.

5. To remove the outer shoe, push the outer shoe inward off of the caliper legs. The locator buttons must clear the slots in the housing before the shoe can be removed. Remove the inner shoe by pulling the shoe outward until the shoe clears the piston.

To install:

6. Remove the cap from the master cylinder reservoir and, using a clean suction gun, remove about 6mm of fluid.

7. Using a large C-clamp, press the piston all the way back into the caliper.

8. Install the new pads and clips. Position the pad with the wear sensor on the piston side and upwards.

9. Position the caliper and install the bolts. Torque the bolts to 25 ft. lbs. (34 Nm).

10. Refill the master cylinder, install the tire and wheel assemblies and pump the brake pedal until firm.

Brake Rotor

REMOVAL AND INSTALLATION

Except 1995–97 Sonata

NOTE: Not all models utilize rear disc brakes. Refer to the rear disc brake procedure only if applicable.

— CAUTION —
Brake linings may contain asbestos. Asbestos is a known cancer-causing agent. When working on brakes, remember that the dust which accumulates on the brake parts, contains asbestos. Always wear a protective face covering, such as a painter's mask, when working on the brakes. NEVER blow the dust from the brakes or drum! There are solvents made for the purpose of cleaning brake parts.

Front Brake Rotor

1. Remove the center hub cap and halfshaft nut. Then raise the vehicle and support it safely. Allow the wheels to hang freely. Then remove the front wheel.

2. Remove the brake caliper without disconnecting the hydraulic line and suspend it out of the way with a piece of wire.

3. Disconnect the stabilizer bar and strut bar from the lower control arm.

4. Disconnect the ball joint stud from the knuckle.

5. Remove the halfshaft from the transaxle and press the halfshaft out of the hub using tool 09526-11001 or equivalent.

6. Unbolt and remove the hub and knuckle from the bottom of the strut and remove the hub and knuckle assembly from the vehicle.

7. Several special tools are required to press the hub and disc from the steering knuckle and to remount them. Use 09517–21600 or equivalent. Do not attempt to hammer the parts apart, or the bearing will be damaged. Install the arm of the special tool then the body onto the knuckle and tighten the nut man-

ually. Using special tool 09517–21500, separate the hub from the knuckle. Pull the bearings out, noting their positions and direction of installation (smaller diameter inward).

8. Matchmark the relationship between the brake disc and hub. Then place the knuckle in a vise and separate the rotor from the hub by removing the attaching bolts.

To install:

9. Install the hub to the rotor.

10. Install the hub and rotor to the knuckle.

11. Install the hub and knuckle to the bottom of the strut.

12. Install the halfshaft to the transaxle.

13. Connect the stabilizer bar and strut bar to the lower control arm, if disconnected.

14. Install the lower ball joint stud to the knuckle.

15. Install the brake caliper.

16. Install the halfshaft nut and center hub cap and halfshaft nut.

17. Install the front wheel and lower the vehicle.

18. Torque the hub nut to 145 to 188 ft. lbs. (200 to 260 Nm).

Rear Brake Rotor

1. Raise and safely support the vehicle.

2. Remove the wheel and tire assembly.

3. Remove the brake caliper and support it with and wire to prevent damage to the brake hose.

4. Remove the retaining screw and pull the rotor from the hub.

To install:

5. Position the rotor on the hub and install the retaining screw.

6. Install the brake caliper on the rotor assembly.

7. Install the wheel and tire assembly.

8. Lower the vehicle to the floor and apply the brakes a few times to insure good brake operation before moving the vehicle.

1995–97 Sonata

— **CAUTION** —
Brake linings may contain asbestos. Asbestos is a known cancer-causing agent. When working on brakes, remember that the dust which accumulates on the brake parts and/or in the drum contains asbestos. Always wear a protective face covering, such as a painter's mask, when working on the brakes. NEVER blow the dust

from the brakes or drum! There are solvents made for the purpose of cleaning brake parts.

Front Rotor

1. Raise and safely support the vehicle.

2. Remove the front wheel(s).

3. Remove the brake caliper and support it with and wire to prevent damage to the brake hose.

4. Remove the retaining screw and pull the rotor from the hub.

To install:

5. Position the rotor on the hub and install the retaining screw.

6. Install the brake caliper on the rotor assembly.

7. Install the front wheel(s).

8. Lower the vehicle to the floor and apply the brakes a few times to insure good brake operation before moving the vehicle.

Rear Rotor

1. Raise and safely support the vehicle.

2. Remove the rear wheel(s).

3. Remove the brake caliper and support it with wire to prevent damage to the brake hose.

4. Remove the retaining screw and pull the rotor from the hub.

To install:

5. Position the rotor on the hub and install the retaining screw.

6. Install the brake caliper on the rotor assembly.

7. Install the rear wheel(s).

8. Lower the vehicle to the floor and apply the brakes a few times to insure good brake operation before moving the vehicle.

Brake Drums

REMOVAL AND INSTALLATION

Except Accent

— **CAUTION** —
Brake linings may contain asbestos. Asbestos is a known cancer-causing agent. When working on brakes, remember that the dust which accumulates on the brake parts and/or in the drum contains asbestos. Always wear a protective face covering, such as a painter's mask, when working on the brakes. NEVER blow the dust from the brakes or drum.

1. Raise the vehicle and support safely.

2. Remove the wheel and tire assembly.

3. Remove the dust cap, cotter pin, nut lock, wheel bearing nut and washer from the spindle. Remove the outer wheel bearing. Remove the drum with the inner wheel bearing from the spindle. If the drum is difficult to remove, remove the plug from the rear of the backing plate and push the self adjuster lever away from the star wheel. Rotate the star wheel to retract the shoes.

4. Remove the brake drum and the grease seal.

To install:

5. Lubricate and install the inner wheel bearing. Install a new grease seal. Install the drum to the spindle. Lubricate and install the outer wheel bearing, washer and nut. Adjust the bearing preload as required. When the bearing preload is properly set, install the nut lock and a new cotter pin. Install the grease cap.

6. Install the wheel and tire assembly. Adjust the rear brakes as required.

7. Apply the brakes until a firm pedal is obtained, prior to moving the vehicle.

Accent

1. Raise and safely support the vehicle.

2. Remove the rear wheel.

3. Matchmark the drum to the hub assembly.

4. Remove the brake drum from the hub assembly. Tap the drum with a plastic hammer if needed.

To install:

5. Align the matchmark and install the drum on the hub assembly.

6. Install the rear wheel.

7. Safely lower the vehicle to the floor.

Brake Shoes

REMOVAL AND INSTALLATION

All Models

1. Raise and support the vehicle.

2. Remove the wheel and tire assembly.

3. Remove the brake drum.

4. Remove the self adjuster spring and the adjuster lever.

5. Spread the shoes and remove the adjuster strut.

6. Remove the shoe to shoe spring and the hold- down springs.

7. Remove the primary brake shoe.

8. Remove the horseshoe clip and the parking brake lever from the secondary brake shoe.

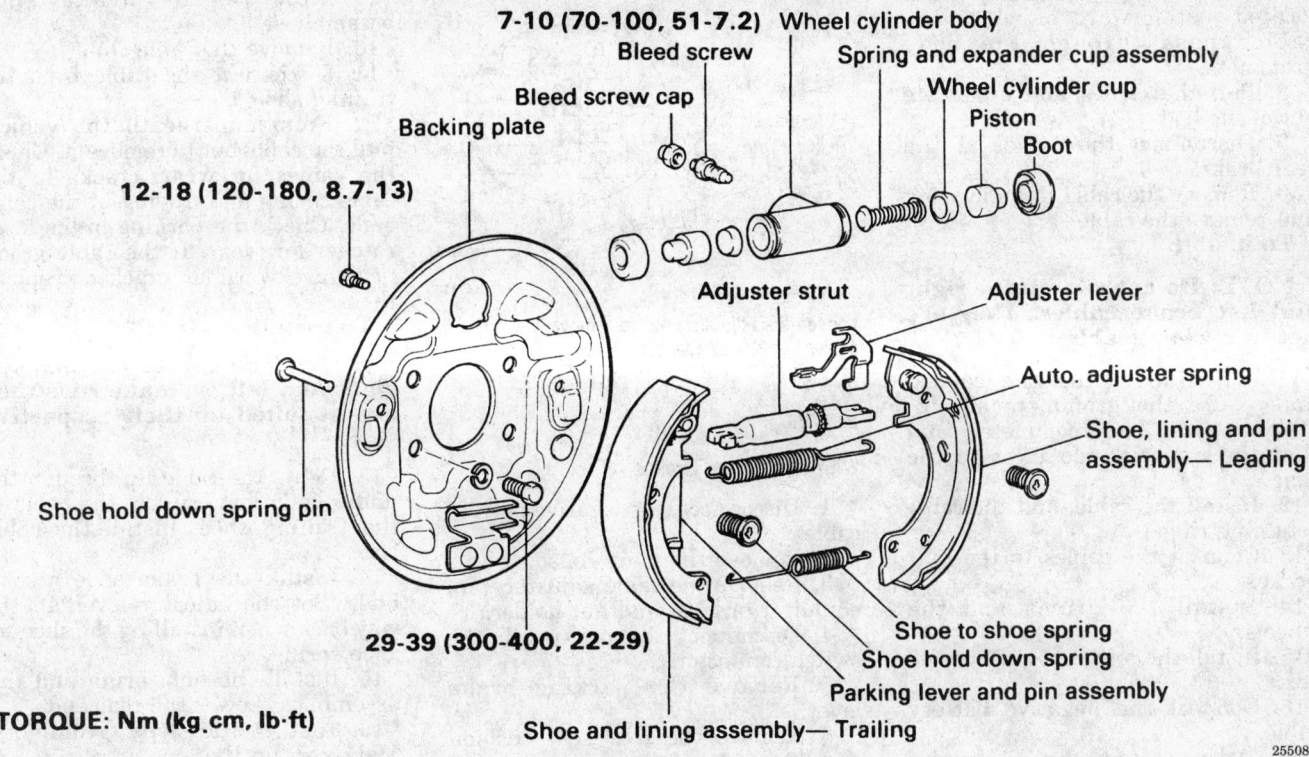

7-10 (70-100, 51-7.2) Wheel cylinder body
Bleed screw   Spring and expander cup assembly
Bleed screw cap   Wheel cylinder cup
Backing plate   Piston
Boot

12-18 (120-180, 8.7-13)

Adjuster strut   Adjuster lever

Auto. adjuster spring

Shoe, lining and pin
assembly— Leading

Shoe hold down spring pin

Shoe to shoe spring
Shoe hold down spring
Parking lever and pin assembly

29-39 (300-400, 22-29)

Shoe and lining assembly— Trailing

TORQUE: Nm (kg.cm, lb·ft)

255088

Rear brake assembly — 1995 Scoupe

To install:

9. Clean the backing plate with brake cleaning solvent.

10. Apply a light coating of lithium grease to the friction points on the backing plate.

11. Position the primary shoe on the backing plate and install the hold down spring and pin.

12. Install the parking brake lever to the secondary shoe and install the shoe to the backing plate.

13. Install the adjuster strut assembly and the adjuster lever and spring.

14. Install the lower shoe to shoe spring.

15. Install the brake drum and the wheel and tire assembly.

16. Adjust the brake shoes and lower the vehicle to the floor.

Wheel Cylinder

REMOVAL AND INSTALLATION

All Models

1. Raise and support the vehicle.
2. Remove the wheel and tire assembly.
3. Remove the brake drum.
4. Remove the brake shoes.

5. Place a container or some old rags under the brake backing plate to catch the brake fluid that will run out of the wheel cylinder.

6. Disconnect the brake line(s) and remove the cylinder mounting bolts. Remove the cylinder from the backing plate.

To install:

7. Install the wheel cylinder to the brake backing plate. Torque the wheel cylinder mounting bolts to 4–9 ft. lbs. (5.5–12 Nm).

8. Connect the brake fluid line to the wheel cylinder.

9. Install the brake shoes and the drum.

10. Install the wheel and tire assembly.

11. Adjust the brakes.

12. Refill and bleed the brake system.

Parking Brake Cable

ADJUSTMENT

All Models

1. If equipped with rear drum brakes, adjust the shoe to drum clearance before making the parking brake cable adjustment.

2. Remove the center console to expose the adjusting nut on the parking brake equalizer.

3. Pull the parking brake lever up with a force of 44 lbs. (20 kg) and count the number of clicks until the lever stops.

4. If needed, adjust the nut at the equalizer to obtain a stroke of 6–7 clicks.

5. Install the center console assembly.

REMOVAL AND INSTALLATION

Except Sonata

1. Disconnect the negative battery cable.

2. Excel and Scoupe, if applicable, remove the console box and rear seat.

3. Elantra models, remove the rear seat cushion, and roll up the carpet.

4. Release the hand brake and then disconnect the cable connectors at the equalizer. It may be necessary to loosen the cable adjusting nuts to do this.

5. Raise the vehicle and support it safely.

6. Remove the rear wheels and brake drums.

7. Disconnect all cable clamps from the body. Remove the mounting

bolts for the large mounting clamp located just forward of where the cables pass through the body grommets.

8. Pull the cables and grommets out of the body.

9. Disconnect the cables at the rear brakes.

10. Remove the cable retaining ring and remove the cable.

To install:

NOTE: Do not mix up the right and left brake cables. They are not interchangeable.

11. When installing the cables, make sure the grommets are installed in the body completely and that the concave side faces to the rear.

12. Install the cable, and the cable retaining ring.

13. Connect the cables to the rear brakes.

14. Install the drums and the wheels.

15. Install the rear seat and console box.

16. Connect the negative battery cable.

17. Adjust the hand brake mechanism.

18. Adjust the switch for the indicator light so the light comes on when the lever is pulled one notch.

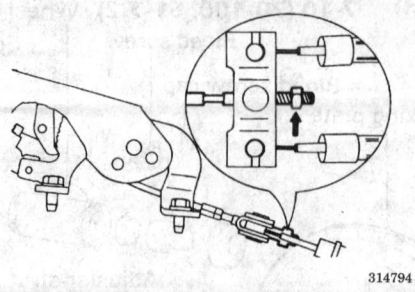

314794

Loosen the adjusting nut to remove the cables — 1995–97 Elantra

Sonata

With Rear Drum Brakes

1. Disconnect the negative battery cable.

2. Remove the main console.

3. Remove the cable adjuster, pin, equalizer bracket and nut holder.

4. Disconnect the parking brake switch connector.

5. Remove the parking brake lever.

6. Remove the rear seat cushion and lift up the carpet.

7. Remove the parking brake cable clamp and grommet.

8. Raise the rear end and remove the tires, drums and hub assemblies.

9. Remove the brake shoe assemblies.

10. Remove the cable clip.

11. Disconnect the cable from the trailing shoe.

12. From underneath the vehicle pull the cable out to remove it. Check the cables for wear, cracks in the cable casing and fraying in the cable ends. Check the parking brake lever ratchet for wear. If the cable grommets are worn or cracked, replace them.

To install:

NOTE: The cables are marked right and left, so make sure they are installed on their respective sides.

13. Route the cable up through the under body and connect the cable to the trailing shoe. Install the cable clip.

14. Install the brake shoe assemblies. Set the adjuster lever all the way back when installing the shoe-to-shoe spring.

15. Install the hub, drum and tire assemblies. Lower the rear end.

16. Install the body grommets. Make certain that the grommets are facing in the right direction.

17. Replace the carpet and the rear seat cushion.

18. Install the parking brake lever and connect the parking brake switch.

19. Install the nut holder, equalizer bracket, pin and cable adjuster. Lubricate the ratchet plate and ratchet pawl sliding surfaces with multi-purpose grease.

20. Adjust the parking brake lever stroke.

21. Install the main console. Check the parking brake and parking brake switch for proper operation.

22. Connect the negative battery cable.

With Rear Disc Brakes

1. Disconnect the negative battery cable.

2. Raise the rear and support safely. Remove the rear wheels.

3. Remove the rear caliper, rotor disc, hub and bearing assemblies.

4. Remove the lower shoe-to-shoe spring.

5. Remove the parking brake lever adjusters.

6. Remove the upper shoe-to-shoe return spring.

7. Disconnect the shoe hold down pin and spring.

8. Disconnect the cable clevis from the adjusting lever. Be careful not to tear cable dust boot. Pull the lever and slider assembly out towards the

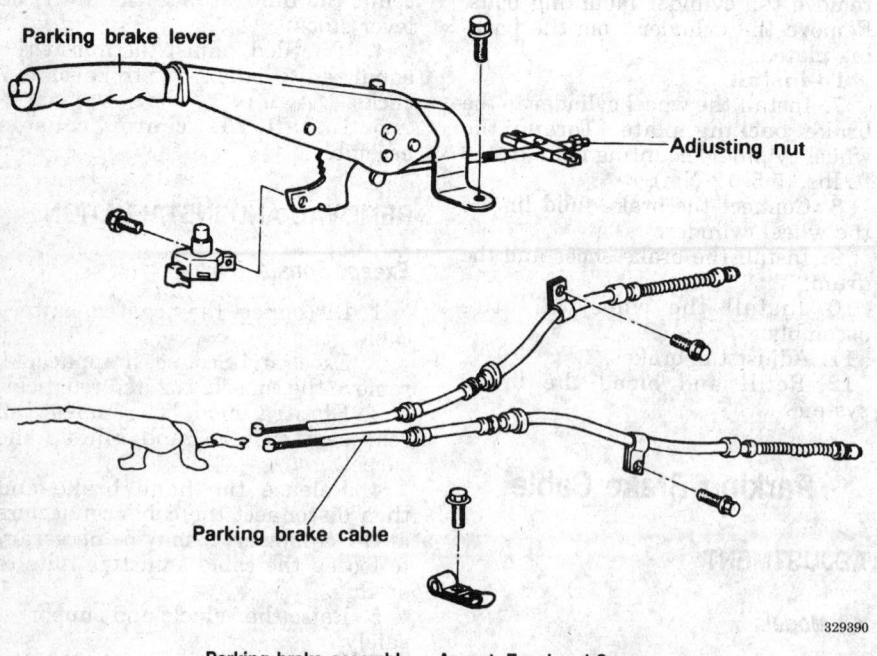

Parking brake lever

Adjusting nut

Parking brake cable

329390

Parking brake assembly — Accent, Excel and Scoupe

outboard side of the brake and through the access window in the backing plate.

9. Remove the cable clips.

10. Unbolt the backing plate from the suspension arm.

11. Loosen the cable adjusting nut and remove the cables from the adjuster bracket.

12. Remove the cables. Check the cables for wear, cracks in the cable casing and fraying in the cable ends. Check the parking brake lever ratchet for wear. If the cable grommets are worn or cracked, replace them. Check the dust boots for rips and tears. Check the brake shoe linings for wear and oil contamination. Minimum brake lining thickness is 2mm. Check the shoe webs and adjusters for bending. Check the return springs for bent hooks, over-extension and breaks. Check the lever and slider assemblies for cracks, bending and wear. Check the drum surface for scoring, wear and oil contamination. Replace any worn or damaged component as necessary.

To install:

13. Mount the backing plate to the suspension flange and torque the bolts in a diagonal pattern to 45–60 ft. lbs. (61–81 Nm).

14. Lubricate the area between the lever and slider and where the slider contacts the backing plate. Use multi-purpose grease.

15. Install the lever and slider assembly through the window in the backing plate. Be careful not damage the dust boot. The lever should be positioned with the long portion into the backing plate first and then the indent for the cable clevis facing toward the rear of the vehicle last. The slider should be positioned on top of the lever. Lubricate the slider with multipurpose grease for easy rotation. The long arm of the slider should be positioned on the rear of the backing plate. The assembly should slide and rotate freely. Connect the cable clevis to the lever indent.

16. Install the brake shoes and shoe holddown pin.

17. Shorten the adjuster assembly and install between the slots in the bottom portion of the shoes.

18. Install the upper and lower shoe-to-shoe springs. The upper portion of the shoe web should contact the anchor block. Adjust the cable as necessary to make this contact.

19. Check the lever function by pushing the end of the lever by the cable clevis and observe the movement of the brake shoes. The tab end

of the lever should move against the shoe and the slider tab end should contact fully against the outer shoe.

20. Install the parking brake adjusters.

21. Install the rear hub and bearing, rotor disc and caliper assemblies.

22. Mount the rear wheels and lower the vehicle.

23. Adjust the parking brake lever stroke. Check the parking brake and parking brake switch for proper operation. Lubricate the ratchet plate and ratchet pawl sliding surfaces with multi-purpose grease.

Brake System Bleeding

PROCEDURE

Except ABS-Equipped Models

The brakes should be bled whenever a brake line, caliper, wheel cylinder, or master cylinder has been removed or when the brake pedal is low or soft.

NOTE: If using a pressure bleeder, follow the instructions furnished with the unit and choose the correct adapter for the application. Do not substitute an adapter that "almost fits" as it will not work and could be dangerous.

1. Fill the master cylinder with fresh brake fluid. Check the level often during the procedure.

2. Starting with the left rear wheel, remove the protective cap from the bleeder, if equipped, and place where it will not be lost. Clean the bleed screw.

─── **CAUTION** ───
When bleeding the brakes, keep face away from the brake area. Spewing fluid may cause facial and/or visual injury. Do not allow brake fluid to spill on the vehicle's finish; it will remove the paint.

3. If the system is empty, the most efficient way to get fluid down to the wheel is to loosen the bleeder about ½ to ¾ turn, place a finger firmly over the bleeder and have a helper pump the brakes slowly until fluid comes out the bleeder. Once fluid is at the bleeder, close it before the pedal is released inside the vehicle.

NOTE: If the pedal is pumped rapidly, the fluid will churn and create small air bubbles, which are almost impossible to remove

from the system. These air bubbles will eventually congregate and a spongy pedal will result.

4. Once fluid has been pumped to the caliper or wheel cylinder, open the bleed screw again, have an assistant press the brake pedal to the floor, lock the bleeder and have an assistant slowly release the pedal. Wait 15 seconds and repeat the procedure, including the 15 second wait, until no more air comes out of the bleeder upon application of the brake pedal. Remember to close the bleeder before the pedal is released inside the vehicle each time the bleeder is opened. If not, air will be induced into the system.

5. If a helper is not available, connect a small hose to the bleeder, place the end in a container of brake fluid and proceed to pump the pedal from inside the vehicle until no more air comes out the bleeder. The hose will prevent air from entering the system.

6. Repeat the procedure on remaining wheel cylinders in order:
   a. Right front caliper
   b. Right rear wheel cylinder or caliper
   c. Left front caliper

7. Hydraulic brake systems must be totally flushed, if the fluid becomes contaminated with water, dirt or other corrosive chemicals. To flush, bleed the entire system until all fluid has been replaced with the correct type of new fluid.

8. Install the bleeder cap(s) on the bleeder to keep dirt out. Always road test the vehicle.

ABS-Equipped Models

The brake system must be bled any time a line, hose or component is loosened or removed. Any air trapped within the system can affect pedal feel and system function. Bleeding the complete system including the modulator assembly requires the use of the Hyundai Multi-Use Tester (MUT) to cycle all of the build/decay and isolation valves of the hydraulic modulator. Make certain the fluid level in the reservoir is maintained at or near correct levels during bleeding operations. Use an inline filter when adding new brake fluid to the master cylinder reservoir. To bleed the system proceed as follows:

1. Bleed the normal brake system using the standard procedure. Bleed the wheel circuits in the following order: left rear, right front, right rear and left front.

2. Connect the MUT to the diagnostic connector located under the dash near the dash.

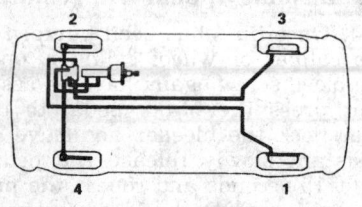

Brake bleeding sequence

136998

3. Turn the ignition switch **ON**, but do not start the engine.

4. Connect a clear vinyl tube to the bleeder screw to be opened and place the other end in a clear container partially filled with new brake fluid.

5. Depress the brake pedal lightly and use the MUT to actuate the appropriate valves. When each valve cycles, fluid from the respective fitting will emerge. Proceed with bleeding each of the 4 modulator circuits in the order shown.

------ **CAUTION** ------

Fluid emerging from the modulator bleed ports will be at high pressure, Insure the vinyl tube is firmly connected to each port prior to opening the corresponding bleeder to avoid personal injury and/or damage to the vehicle.

6. The pedal will drop slightly as the valve is cycled. If pedal drops completely, close the bleed valve and re-apply light pressure to the pedal.

7. When air bubbles no longer emerge from the bleed port, close the bleeder and proceed to the next port.

8. Check and add brake fluid to the reservoir as needed.

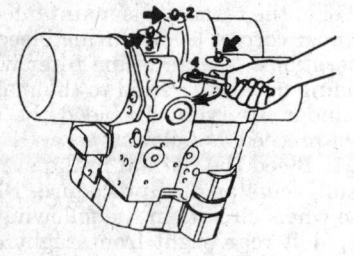

ABS Modulator bleeding sequence

325564

Wheel Speed Sensor

REMOVAL AND INSTALLATION

Accent, Elantra and Sonata

1. Disconnect the negative battery cable.

2. Raise and safely support the vehicle.

3. Remove the speed sensor mounting bolt and the harness retainers. Disconnect the remove the sensor.

To install:

4. Install the wheel speed sensor and torque the mounting bolt to 5.8–6.9 ft. lbs. (8–9.5 Nm) for the front speed sensor and 11.2–11.9 ft. lbs. (15.5–16.5 Nm) for the rear wheel speed sensor.

5. Route the wiring in the original position and install the retainers.

6. Connect the harness connector and lower the vehicle to the floor.

7. Connect the negative battery cable.

FRONT SUSPENSION

Strut

REMOVAL AND INSTALLATION

Excel, Scoupe and Elantra

1. Remove the front wheels.

2. Raise and safely support the vehicle.

3. Detach the brake hose from the clip on the strut.

4. Unbolt the strut from the knuckle.

5. Remove the four strut-to-fender apron nuts.

6. Pull the strut away from the steering knuckle and wheelhousing and out from the car.

To install:

NOTE: Before installing the strut, make sure the surface where the strut attaches to the knuckle is clean. This ensures a good connection.

7. For Excel and Scoupe models, install the strut to the inner fender. Tighten the mounting nuts to 11–14 ft. lbs. (15–20 Nm).

8. For Elantra models, install the strut to the inner fender. Tighten the mounting nuts to 29–36 ft. lbs. (40–50 Nm).

9. For Excel and Scoupe models, install the lower strut to the steering knuckle. Tighten the mounting nuts to 65–76 ft. lbs. (90–105 Nm).

10. For Elantra models, install the lower strut to the steering knuckle. Tighten the mounting nuts to 80–94 ft. lbs. (110–130 Nm).

11. Attach the brake line the bracket on the strut body.

12. Install the wheel and tire assembly.

13. Install the wheel and tire assembly, and lower the vehicle to the floor.

14. Check and adjust the front wheel alignment as needed.

Sonata

1. Remove the front wheels.

2. Raise and safely support the vehicle.

3. Detach the brake hose from the clip on the strut.

4. Unbolt the strut from the knuckle.

5. Jack up the lower control arm to prevent damage to the brake line, wheel speed sensor and to keep the driveshaft from being pulled out of the transaxle.

6. Remove the four strut-to-fender apron nuts.

7. Pull the strut away from the steering knuckle and wheelhousing and out from the car.

To install:

8. Install the strut to the inner fender. Tighten the mounting nuts to 18–25 ft. lbs. (25–35 Nm).

9. Install the lower strut to the steering knuckle. Tighten the mounting nuts to 65–76 ft. lbs. (90–105 Nm).

10. Attach the brake hose into the clip on the strut body.

11. Install the wheel and tire assembly and lower the vehicle to the floor.

NOTE: Before installing the strut, make sure the surface where the strut attaches to the knuckle is clean. This ensures a good connection.

Lower Ball Joints

REMOVAL AND INSTALLATION

Excel and Scoupe

1. Raise the vehicle and support it safely.

2. Remove the wheel and tire assembly.

3. Unbolt the ball joint from the control arm.

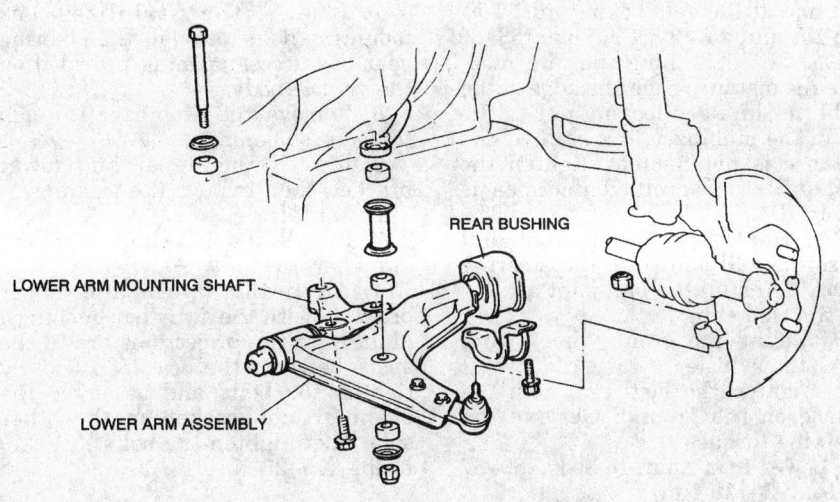

Lower control arm — Excel and Scoupe

255970

4. Remove the stud retaining nut.

5. Use a ball joint removing tool and separate the ball joint from the steering knuckle.

To install:

6. Replace the ball joint and tighten the ball joint to control arm nut to 69–87 ft. lbs. (99–118 Nm). Torque the ball joint stud nut 43–52 ft. lbs. (59–71 Nm).

7. Install the wheel and tire assembly.

8. Lower the vehicle to the floor.

Accent

1. Raise the vehicle and support it safely.

2. Remove the front wheel and tire assembly.

3. Remove the undercover.

4. Remove the ball joint stud nut and the ball joint stud from the lower control arm.

5. Disconnect the stabilizer bar from the lower arm. Remove the nut from under the control arm and take off the washer and spacer.

6. Remove the lower control arm bracket and the through-bolt from the body, and remove the lower control arm from the vehicle.

7. Remove the dust cover and snap-ring from the ball joint.

8. Using the proper tools, press the ball joint from the lower control arm.

To install:

9. Using the proper tools, press the ball joint into the lower control arm.

10. Install the snap-ring and the dust cover.

11. Install the lower control arm to the body. Tighten the mounting bracket bolts to 43–58 ft. lbs. (60–80 Nm), and the lower arm mounting through-bolt to 69–87 ft. lbs. (95–120 Nm).

12. Install the stabilizer bar to the control arm. The nut on the stabilizer bar bolt must be torqued until the link shows 1.02–1.1 inch (26–28mm) of threads below the bottom of the nut.

13. Install the undercover.

14. Install the wheel and tire assembly.

15. Lower the vehicle.

Elantra

1. Raise the vehicle and support it safely.

2. Remove the tire and wheel assembly.

3. Remove the nut from the ball joint at the steering knuckle and using the special tool 09568-31000 or

equivalent, disconnect the lower arm ball joint from the steering knuckle.

4. Remove the stabilizer bar link self-locking nut and detach the stabilizer bar from the lower control arm.

5. Remove the ball joint mounting bolts and remove the joint from the arm.

6. Remove the dust cover from the ball joint.

To install:

7. Pack the specified grease in the new dust cover and press the cover to the ball joint using cup driver tool 09545-21100 or equivalent.

8. Install the joint to the lower arm. Tighten the ball joint mounting bolts.

9. Connect the ball joint to the steering knuckle and tighten the nut to 52 ft. lbs. (72 Nm).

10. Connect the stabilizer bar to the lower control arm and tighten the bar link self-locking nut to 40 ft. lbs. (55 Nm).

11. Install the tire and wheel assembly to the vehicle.

Sonata

1. Raise the vehicle and support it safely.

2. Remove the tire and wheel assemblies.

3. Remove the lower control arm from the vehicle.

4. Remove the ball joint dust cover. Press the ball joint from the control arm.

To install:

5. Apply grease to the lip of the control arm and to the ball joint contact surfaces.

6. Place the ball joint in the control arm. Using a special tool, press the ball joint into the control arm. The ball joint must be pressed evenly into the control arm.

7. Install a new dust cover on the ball joint. Install the control arm assembly into the vehicle. Torque the ball joint retaining nut to 52 ft. lbs. (71 Nm).

8. Install the wheel and tire assembly. Lower the vehicle.

Lower Control Arms

REMOVAL AND INSTALLATION

Except Sonata

1. Raise the vehicle and support it safely.

2. Remove the front wheel and tire assembly.

3. Remove the undercover.

4. Remove the ball joint stud nut and the ball joint stud from the lower control arm.

5. Disconnect the stabilizer bar from the lower arm. Remove the nut from under the control arm and take off the washer and spacer.

6. Remove the lower control arm brackets from the body, and remove the lower control arm from the vehicle.

To install:

7. Install the lower control arm to the body. For 1995–97 Elantra models, do not torque the hardware until the weight of the vehicle is on the ground.

8. For Excel, Scoupe and Accent models, tighten the mounting bracket bolts to 43–58 ft. lbs. (60–80 Nm), and the lower arm mounting bolts to 116–137 ft. lbs. (160–190 Nm).

9. For 1993–94 Elantra models, tighten the front mounting nut to 69–87 ft. lbs. (95–120 Nm), and the rear mounting bolts to 43–58 ft. lbs. (60–80 Nm).

10. Install the stabilizer bar to the control arm. The nut on the stabilizer bar bolt must be torqued until the link shows 0.9–1.0 inch (24–26mm) of threads below the bottom of the nut.

11. Install the ball joint to the steering knuckle. Tighten the nut to 43–52 ft. lbs. (60–72 Nm).

12. Install the undercover.

13. Install the wheel and tire assembly.

14. Lower the vehicle.

15. For 1995–97 Elantra models, torque the bolts attaching the lower control arm to the body. Torque the through-bolt to 72–87 ft. lbs. (100–120 Nm). Torque the nuts on the front bracket to 25–33 ft. lbs. (35–45 Nm) and the bolt to 58–72 ft. lbs. (80–100 Nm).

Sonata

1. Remove the front wheels.

2. Loosen the lower arm ball joint nut, but do not remove it at this time.

3. Using the special tool, disconnect the ball joint from the steering knuckle. Tie the tool off to a chassis member during use.

4. Disconnect the stabilizer link ball joint from the stabilizer bar.

5. Remove the lower arm shaft bolts and rod bushing clamp fasteners. Remove the lower control arm.

6. Inspect all parts and replace any cracked, dry or deformed parts.

To install:

7. Bolt the lower control arm in place and snug all the fasteners. Final fastener torque will be applied

with the full weight of the vehicle on the ground.

8. Connect the stabilizer bar to the stabilizer links. Engage the stabilizer link with a 5/8 inch open end wrench and install the self-locking nut. Torque the nut to 43–51 ft. lbs. (58–69 Nm). After tightening the nut, measure the distance from the edge of the stabilizer link self-locking nut to the top of the stabilizer link stud. If the distance is not 5–8mm, tighten the nut until the specified distance is obtained.

9. Reverse the special tool and press the ball joint into the steering knuckle. Install the ball joint attaching nut and snug it.

10. Mount the front wheels and lower the vehicle.

11. Tighten the fasteners with the vehicle on the ground. Observe the following torques:
• Lower arm shaft-to-body: 69–87 ft. lbs. (95–110 Nm)
• Ball joint-to-knuckle: 43–52 ft. lbs. (60–72 Nm)
• Rod bushing clamp nuts: 25–34 ft. lbs. (35–47 Nm)
• Rod bushing clamp center bolt: 72–87 ft. lbs. (100–118 Nm)
• Rod bushing clamp outside bolts: 58–72 ft. lbs. (80–100 Nm)

Sway Bar

REMOVAL AND INSTALLATION

Excel and Scoupe

1. Raise and safely support the vehicle.

2. Remove the tire and wheel assembly.

3. Disconnect the tie rod end ball joint from the knuckle using special tool 09568–31000 or equivalent.

4. Remove the rear roll stopper mounting bolt and rear roll bracket assembly mounting bolt.

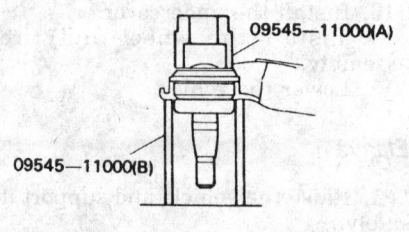

Install the ball joint into the control arm — Sonata

5. Pull the rear roll bracket assembly forward.

6. Loosen the stabilizer link bolt and nut, then separate the stabilizer bar from the lower arm.

7. Loosen the stabilizer bar mounting bolts through the steering gear box access opening provided on the vehicle body.

8. Remove the stabilizer through the access opening.

9. Detach the upper and lower bracket, then remove the bushing.

To install:

10. Install the bushings onto the sway bar.

11. Align the upper and lower brackets with the sway bar bushings. Make sure the projections are in the space between the brackets. Loosely tighten the bolts and assemble the bushing and bracket on the other side. Then tighten the bolts to 12–19 ft. lbs. (19–26 Nm).

NOTE: Bushing brackets are marked for left and right applications. They are not interchangeable.

12. Install the sway bar link.

13. Install the rear roll stopper assembly.

14. Connect the tie rod end to the steering knuckle.

15. Install the wheel and tire assembly.

16. Lower the vehicle to the floor.

Accent

1. Raise and safely support the vehicle.

2. Remove the front wheels.

3. Disconnect the tie rod end from the knuckle using the proper tool.

4. Remove the stabilizer link from the lower control arm.

5. Remove the upper mounting brackets.

6. Remove the stabilizer through the access opening in the wheel well.

To install:

7. Position the bar with the bushings on the vehicle.

8. Install the bracket bolts but do not tighten them at this time.

9. Install the stabilizer bar to the control arm. The nut on the stabilizer bar bolt must be torqued until the link shows 1.02–1.1 inch (26–28mm) of threads below the bottom of the nut.

10. Connect the tie rod ends to the knuckles. Torque the nuts to 11–15 ft. lbs. (15–34 Nm) and install a new cotter pin.

11. Install the wheels and lower the vehicle to the floor.

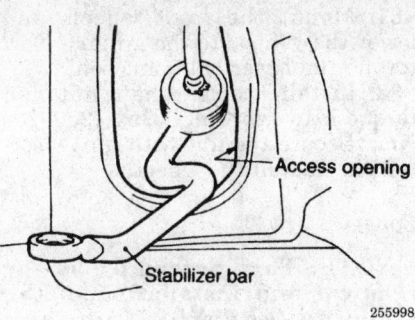

Steering gear access opening — Excel and Scoupe

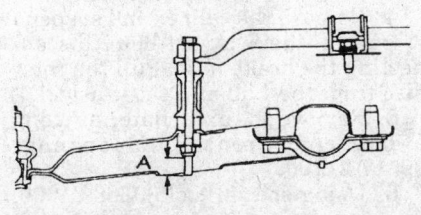

Correct stabilizer link installation — Accent

12. Torque the bushing bracket bolts to 12–19 ft. lbs. (17–26 Nm).

1993–95 Elantra

1. Raise the vehicle and support it safely.
2. Disconnect and separate the tie rod end and the ball joint from the lower control arm and the steering knuckle.
3. Remove the stabilizer link self locking nut using a 14mm spanner wrench.
4. Remove the stabilizer bar through the access opening. Detach the upper and the lower brackets, and remove the bushings.
To install:
5. Install the bushings onto the bar. Align the upper and the lower brackets with the bushings making sure the projections are securely in the space between the brackets.

NOTE: Distinguish the side the fixtures to be installed by locating the identification marks stamped on each; R will denote the right side and L will denote the left side fixture. They are not the same and should be installed as labeled.

6. Using the access opening, install the rod to the vehicle. Temporarily tighten the bushing brackets.
7. Tighten the stabilizer bar link with a spanner wrench. Install the self-locking nut tightening it to 51 ft. lbs. (70 Nm).
8. Connect the tie rod end and the ball joint to the steering knuckle and control arm tightening the ball joint stud nut to 52 ft. lbs. (72 Nm).
9. Install the tire and wheel assembly, and lower the vehicle.
10. Tighten the upper and lower bracket mounting bolts to 12–19 ft. lbs. (17–26 Nm).

1996–97 Elantra

1. Safely raise and support the vehicle.
2. Remove the front wheels.
3. Disconnect the tie rod end from the steering knuckle.
4. Remove the sway bar link mounting nuts from the lower control arm and the sway bar.
5. Remove the sway bar mounting brackets on both sides.
6. Remove the rack and pinion assembly.
7. Remove the sway bar through the access opening.
To install:
8. Position the sway bar on the vehicle and install the mounting brackets. Make sure the bushings are in the proper positions. Torque the mounting bolts to 25–33 ft. lbs. (35–45 Nm).
9. Install the rack and pinion assembly.
10. Install the sway bar links to the lower control arms. Torque the mounting nuts to 25–33 ft. lbs. (35–45 Nm).
11. Connect the tie rod ends to the steering knuckles.
12. Install the front wheels.
13. Safely lower the vehicle to the floor.

Sonata

1. Raise the vehicle and support it safely.
2. Remove the stabilizer bar brackets from the crossmember.
3. Lower the rear of the center member and lower the stabilizer bar.
4. Disconnect the end links and remove the stabilizer.
To install:
5. Hold the sway bar in position and connect the end links to the control arms. Torque the end link nuts to 43–51 ft. lbs. (60–70 Nm).

6. Install the sway bar brackets to the crossmember. Torque the bracket bolts to 22–30 ft. lbs. (30–41 Nm).
7. Position the center member, and install the rear mounting bolt with bushing, and tighten to 58–72 ft. lbs. (80–100 Nm).
8. Lower the vehicle to the floor.

Front Wheel Bearings

ADJUSTMENT

Excel and Sonata

The front wheel bearings are not adjustable. If the bearings make noise or turn roughly, replace them. Torque the axle nut to the proper specification.

REMOVAL AND INSTALLATION

Except Sonata

1. Remove wheel ornaments then, remove the axle shaft nut.
2. Raise and support the car on a lift positioned so that the front wheels hang freely.
3. Remove the front wheels.
4. Remove the caliper and suspend it out of the way without disconnecting the brake hose.
5. Disconnect the lower ball joint from the knuckle.
6. Disconnect the tie rod end from the knuckle using the proper tool.
7. Using a two-jawed puller, press the axle shaft from the hub.
8. Unbolt the strut from the knuckle. Remove the hub and knuckle assembly from the car.
9. Install first the arm, then the body of special tool 09517-21600 on the knuckle and tighten the nut.
10. Using special tool 09517-21500, separate the hub from the knuckle.

NOTE: Prying or hammering will damage the bearing. Use these special tools, or their equivalent to separate the hub and knuckle.

11. Place the knuckle in a protected jaw vise and separate the rotor from the hub by removing the four attaching bolts.
12. Using special tools 09532-11000, 0953211301 and 09517-21100, remove the outer bearing inner race.
13. Drive the oil seal and inner bearing inner race from the knuckle with a brass drift .

14. Drive out both outer races in a similar fashion.

NOTE: Always replace bearings and races as a set. Never replace just an inner or outer bearing. If either is in need of replacement, both sets must be replaced.

15. Thoroughly clean and inspect all parts. Any suspect part should be replaced.

———— CAUTION ————
When drying the wheel bearings with compressed air, do not spin the bearings. They may come apart causing serious injury.

To install:
16. Pack the wheel bearings with lithium based wheel bearing grease. Coat the inside of the knuckle with similar grease and pack the cavities in the knuckle.

NOTE: Apply a thin coating of grease to the outer surface of the race before installation.

17. Using special tools 09500-21000, 09517- 21300, and 09517-21200, install the outer races.
18. Install the rotor on the hub and torque the bolts to 36–43 ft. lbs. (50–60 Nm).
19. Drive the outer bearing inner race into position.
20. Coat the out ring and lip of the oil seal and drive the hub side oil seal into place, using a seal driver.
21. Place the inner bearing in the knuckle.
22. Mount the knuckle in a vise. Position the hub and knuckle together. Install tool 0951721500 and tighten the tool to 145–188 ft. lbs. (200–260 Nm). Rotate the hub to seat the bearing.
23. With the knuckle still in the vise, measure the hub starting torque with an inch lbs. torque wrench and tool 09517-215000. Starting torque should be 11.5 inch lbs. If the starting torque is 0, measure the hub bearing axial play with a dial indicator. If axial play exceeds 0.11mm, while the nut is tightened to specification, the assembly has not been done correctly. Disassemble the knuckle and hub and start again.
24. Remove the special tool.
25. Place the outer bearing in the hub and drive the seal into place.
26. Position the knuckle and hub assembly onto the strut and install the attaching bolts. Tighten the bolt for Excel, Scoupe or Accent to 65–76 ft. lbs. (90–105 Nm). Tighten the bolt

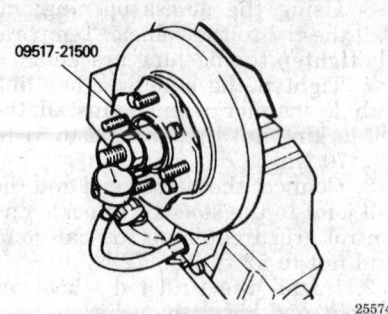

Measuring the axial play with a dial indicator — Excel, 1993–94 Elantra and Scoupe

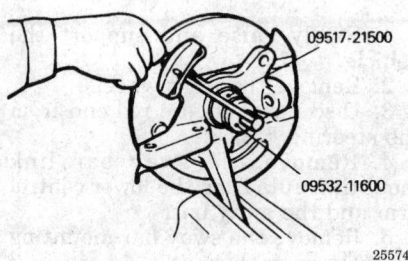

Measuring the starting torque — Excel, 1993–94 Elantra and Scoupe

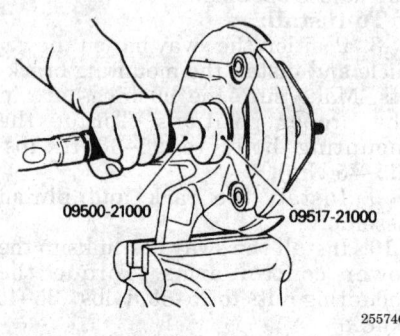

Installing the inner oil seal — Excel, 1993–94 Elantra and Scoupe

for Elantra to 81-96 ft. lbs. (110-130 Nm).
27. Using the proper tool, press the axle shaft into the hub.
28. Connect the tie rod end to the steering knuckle. Install the tie rod end nut and torque it to 11–25 ft. lbs. (15–34 Nm).
29. Connect the lower ball joint to the steering knuckle. Install the attaching bolt and snug it. The bolt will be torqued properly when the wheels are on the ground.
30. Mount the brake caliper assembly and brake hose.

31. Mount the front wheels and lower the vehicle to the ground. Now torque the lower ball joint bolt.
32. Install the axleshaft nut and torque it to 145–188 ft. lbs. (200–260 Nm). Secure the nut with a new cotter pin. Install the wheel.

Sonata

NOTE: This procedure is for removal and installation of the steering knuckle, hub and bearing.

1. Remove the hub cap.
2. Raise and support the car with jackstands positioned so that the front wheels hang freely.
3. Remove the front wheels.
4. Remove the caliper and suspend it out of the way without disconnecting the brake hose. Pull the rotor disc from the hub and set it aside.
5. Remove the axle shaft nut using a box end wrench and spanner 09517-21700.
6. Using special tool 09568-31000, disconnect the tie rod end from the knuckle. Tie the tool off to a chassis member during use.
7. Loosen, but do not remove, the lower arm ball joint nut. Using special tool 09568-31000, disconnect the ball joint from the steering knuckle, then remove the nut. Tie the tool off to a chassis member during use.
8. Press the driveshaft from the hub using special tool 09526-11001.
9. Unbolt and remove the knuckle/hub assembly from the strut.
10. Examine the hub and knuckle for cracks and check the spline for wear. Check the oil seal for deterioration. Replace any damaged component as required.
11. Install first the arm, then the body of special tool 09517-21600 on the knuckle and tighten the nut.
12. Mount the knuckle in a protected jaw vise.
13. Using special tool 09517-21500, separate the hub from the knuckle.

NOTE: Prying or hammering will damage the bearing. Use special tools or their equivalents to separate the hub and knuckle.

14. Using a drift punch, crush the hub oil seal in two places. This is done in preparation to install the bearing race removal tool. Then, install special puller tool 09455-21000 to separate the outside bearing inner race from the hub.
15. Pry the oil seal from the knuckle and discard it.
16. Remove the snapring from the knuckle and press the bearing out

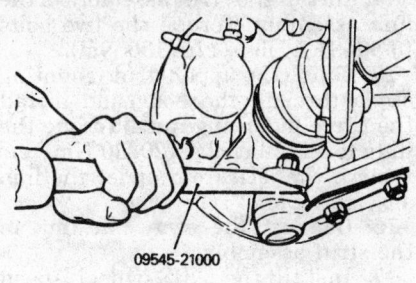

09545-21000

324903

Remove the ball joint stud from the knuckle — Accent

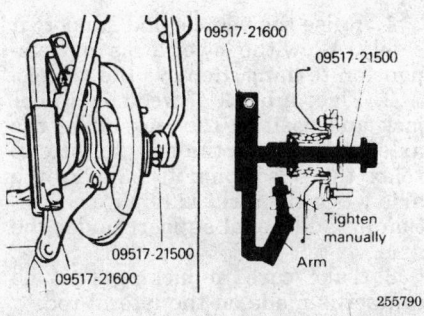

09517-21600
09517-21500
09517-21500
09517-21600
Tighten manually
Arm

255790

Removing the hub from the knuckle — 1993–94 Sonata

from the knuckle using the proper tools.

17. Examine the hub and brake disc for galling or any other surface imperfections. Check the bearing seating surface for cracks and galling. Check the bearing for damage. Replace any damaged component as required.

To install:

18. Pack the bearing and the cavities of the knuckle with multi-purpose grease. Press the bearing into the knuckle using the proper tools. Install the wheel bearing inner race.

19. Drive the new outer oil seal into the knuckle using the proper tools. The oil seal should be flush with the surface of the knuckle. After the oil seal is installed lubricate the seal lip and the areas where the seal contacts the hub with multi-purpose grease.

20. Mount the knuckle in a vise. Position the hub and knuckle together. Install tool 0951721500 and tighten the tool to 145–188 ft. lbs. (200–260 Nm). Rotate the hub to seat the bearing.

21. With the knuckle still in the vise, measure the hub starting torque with an inch lbs. torque wrench and tool 09532-11600. Starting torque should be 16.0 inch lbs. If the start-

ing torque is 0.0, measure the hub bearing axial play with a dial indicator. If axial play exceeds 0.10 mm, while the nut is tightened, the assembly has not been done correctly. Disassemble the knuckle and hub and start again.

22. Remove the special tool.

23. Place the outer bearing in the hub and drive the seal into place. After the oil seal is installed lubricate the seal lip and the areas where the seal contacts the hub with multi-purpose grease.

24. Position the knuckle and hub assembly onto the strut and install the attaching bolts. Torque the bolts to 65–76 ft. lbs. (90–105 Nm).

25. Press the driveshaft into the hub using the proper tool.

26. Connect the ball joint to the steering knuckle, and torque the nut to 43–52 ft. lbs. (60–72 Nm).

27. Connect the tie rod end to the steering knuckle and torque the nut to 17–25 ft. lbs. (24–34 Nm). Secure the nut with a new cotter pin.

28. Mount the rotor disc onto the hub and install the brake caliper assembly. Torque the caliper bolts to 58–72 ft. lbs. (80–100 Nm).

29. Mount the front wheels and lower the vehicle.

30. Install the hub cap.

REAR SUSPENSION

Shock Absorber

REMOVAL AND INSTALLATION

Excel and Scoupe

1. Raise the vehicle and support it safely.

2. Remove the wheel.

3. Jack up the control arm slightly, and remove the upper mounting bolt and nut.

4. Remove the lower mounting bolt and nut.

5. Remove the shock absorber.

6. Check the shock for:

a. Excessive oil leakage, some minor weeping is permissible;

b. Bent center rod, damaged outer case, or other defects.

c. Pump the shock absorber several times, if it offers even resistance on full strokes it may be considered serviceable.

To install:

7. Install the upper shock mounting nut and bolt. Hand-tighten the nut.

8. Install the bottom eye of the shock in the mounting bracket.

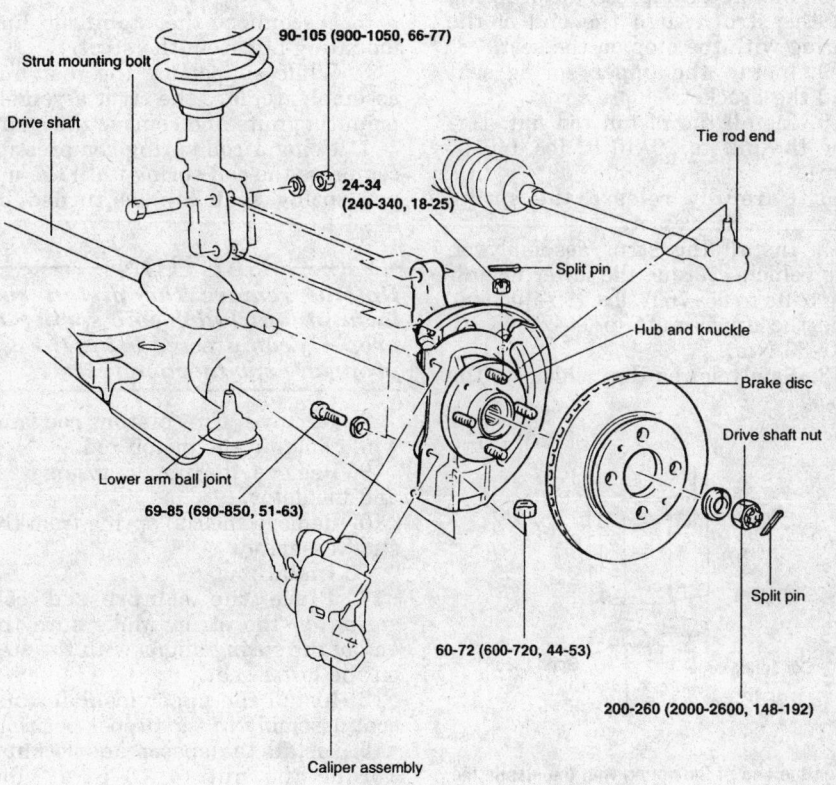

Strut mounting bolt
90-105 (900-1050, 66-77)
Drive shaft
24-34 (240-340, 18-25)
Tie rod end
Split pin
Hub and knuckle
Brake disc
Drive shaft nut
Lower arm ball joint
69-85 (690-850, 51-63)
Split pin
60-72 (600-720, 44-53)
200-260 (2000-2600, 148-192)
Caliper assembly

328551

Hub and knuckle assembly — 1995–97 Sonata

Tighten the nut to 47–58 ft. lbs. (64–78 Nm).

9. Tighten the upper fasteners to 47–58 ft. lbs. (64–78 Nm).

10. Remove the jack from under the control arm.

11. Install the wheel and tire assemblies.

12. Lower the vehicle to the floor.

Accent

1. Raise and safely support the vehicle.

2. Remove the wheel assembly.

3. Remove the three upper mounting nuts.

CAUTION

Do not remove the piston rod nut until the coil spring is compressed.

---

4. Remove the two bolts securing the strut to the knuckle and remove the strut assembly.

5. Using a spring compressor, compress the coil spring.

6. While holding the piston rod, remove the piston rod nut.

7. Remove the bracket, upper spring seat and coil spring.

To install:

8. Install a compressed coil spring on the strut. Align the end of the spring with the step on the seat.

9. Install the upper spring seat and the bracket.

10. Install the piston rod nut. Torque the nut to 29–40 ft. lbs. (40–55 Nm).

11. Carefully release the spring from the compressor.

12. Install the strut assembly on the vehicle. Torque the lower mounting bolts to 65–76 ft. lbs. (90–105 Nm) and the upper nuts to 14–22 ft. lbs. (20–30 Nm).

13. Safely lower the vehicle to the floor.

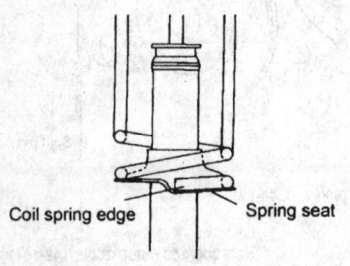

Align the end of the spring with the step in the seat — Accent

324935

1993–95 Elantra

1. Raise the vehicle and support it safely. Remove the wheel.

2. Remove the trim cover inside the rear compartment for access to the top mounting nuts.

3. Support the lower arm with a jack and compress the coil spring. Remove the lower mounting nut.

4. Remove the cap from the upper end of the shock.

5. Remove the upper mounting nut and the shock from the vehicle.

To install:

6. Install the shock absorber to the lower arm, install the lower nut and tighten to 58–72 ft. lbs. (80–100 Nm).

7. Install the upper nut and torque to 29–36 ft. lbs. (40–50 Nm).

8. Install the cap and cover.

9. Lower the arm and remove the jack.

10. Install the trim cover.

11. Install the wheel.

12. Lower the vehicle.

1996–97 Elantra

1. Remove the three rear strut upper mounting nuts.

2. Safely raise and support the vehicle.

3. Remove the rear wheels.

4. If equipped, remove the ABS wheel speed sensor wiring from the strut.

5. Disconnect the stabilizer link mounting nut from the strut.

6. While supporting the rear hub assembly, remove the strut assembly mounting nuts and remove the strut.

7. Using a coil spring compressor, compress the coil spring until the upper spring seat can be turned by hand.

CAUTION

Do not remove the piston rod locknut until the coil spring is properly compressed using the appropriate spring compressor.

---

8. Remove the piston rod nut while holding the piston rod.

9. Remove the upper spring seat and insulator.

10. Remove the coil spring from the strut assembly.

To install:

11. Place the compressed coil spring on the strut. Make sure the end of the spring aligns with the step on the lower seat.

12. Install the upper insulator and seat assembly in the proper position.

13. Install the spacer and locknut. Torque the nut to 43–51 ft. lbs. (60–70 Nm). Carefully release the compressor from the spring.

14. Install the strut assembly to the hub assembly. Torque the two bolts to 80–94 ft. lbs. (110–130 Nm).

15. Guide the upper strut mounting studs through the body and install the three mounting nuts. Torque the nuts to 14–22 ft. lbs. (20–30 Nm) and remove the jack from under the hub assembly.

16. Connect the sway bar link to the strut assembly.

17. Install the ABS wheel speed sensor wiring to the strut assembly.

18. Install the rear wheels and lower the vehicle to the floor.

1993–94 Sonata

1. Raise the vehicle and support it safely. Allow the lower arms and suspension to hang. Remove the wheels.

2. Place a block of wood on a floor jack and position the jack under the axle beam. Raise the axle slightly to relax the shock and to support the axle when the strut is removed. Position an additional support under the axle.

3. Take care in jacking that no contact is made on the lateral rod.

4. Remove the upper dust cover cap from the strut assembly.

5. Remove the upper mounting nuts. Remove the lower mounting bolt and nut.

6. Remove the shock absorber and spring assembly.

CAUTION

Do not remove the self-locking nut on the piston rod unless the coil spring is properly compressed with a coil spring compressor.

To install:

7. Install the shock upper mounting nuts. Torque the upper mounting nuts to 18–25 ft. lbs. (25–34 Nm).

8. Install the lower mounting bolt and nut. Torque the lower mounting bolt and nut to 58–72 ft. lbs. (79–96 Nm).

9. Remove the floor jack from under the rear axle.

10. Install the wheels, and lower the vehicle to the floor.

1995–97 Sonata

1. Remove the upper dust cover cap from the strut assembly.

2. Remove the upper mounting nuts.

3. Raise the vehicle and support it safely.

4. Allow the suspension to hang. Remove the wheels.

5. Remove the sway bar link mounting nut.

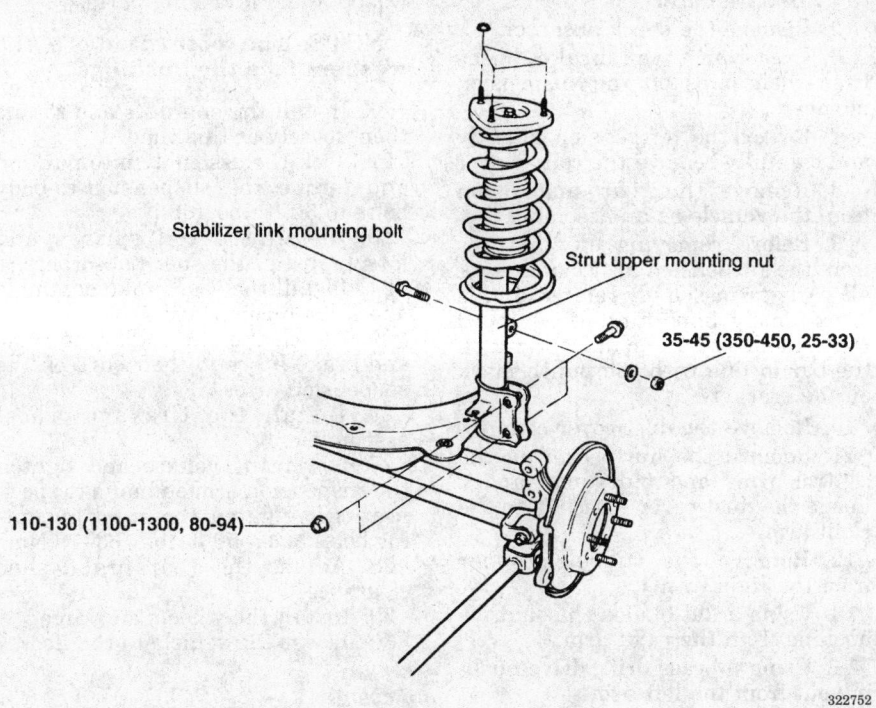

Stabilizer link mounting bolt

Strut upper mounting nut

35-45 (350-450, 25-33)

110-130 (1100-1300, 80-94)

322752

Rear strut mounting — 1996–97 Elantra

6. Remove the lower mounting bolt and nut.

7. Support the crossmember. Loosen, do not remove, the opposite side crossmember mounting nut. Remove the crossmember mounting nut from the side the strut is being removed from.

8. While lowering the upper arm, remove the strut and spring assembly.

--- **CAUTION** ---

Do not remove the self-locking nut on the piston rod unless the coil spring is properly compressed with a coil spring compressor.

To install:

9. Install the lower mounting bolt and nut. Torque the lower mounting bolt and nut to 58–72 ft. lbs. (79–96 Nm).

10. Install the crossmember mounting nut. Tighten both sides mounting nuts to 58–72 ft. lbs. (79–96 Nm).

11. Remove the floor jack from under the crossmember.

12. Install the sway bar link mounting nut. Torque the upper mounting nuts to 25–32 ft. lbs. (35–45 Nm).

13. Install the wheels, and lower the vehicle to the floor.

14. Install the shock upper mounting nuts. Torque the upper mounting nuts to 29–36 ft. lbs. (40–50 Nm).

15. Install the upper dust cover cap to the strut assembly.

Coil Spring

REMOVAL AND INSTALLATION

Excel and Scoupe

1. Raise the vehicle and support it safely.

2. Remove the rear wheels.

3. Support the rear suspension arm with a floor jack. Then, remove the lower shock absorber attaching bolt, nut and lock washer.

4. Slowly, lower the jack just to the point where the spring can be removed and remove the spring.

To install:

5. If the spring is being replaced, transfer the spring seat to the new spring.

6. When installing the coil spring, make sure the smaller diameter is upward. Make sure the spring identification and load markings match up.

7. Install and torque the lower shock mounting nut/bolt to 47–58 ft. lbs. (64–78 Nm).

8. Install the rear wheels.

9. Lower the vehicle.

1993–95 Elantra

1. Raise the vehicle and support it safely. Remove the wheel.

2. Remove the trim cover inside the rear compartment for access to the top mounting nuts.

3. Support the lower arm with a jack and remove the lower mounting nut.

4. Remove the cap from the upper end of the shock.

5. Remove the upper mounting nuts and the shock from the vehicle.

6. Compress the spring on the shock using spring compressor 09546-11000 or equivalent. Remove the piston rod tightening nut at the top of the strut while holding the piston rod with a wrench.

7. Disassemble the upped dust cover, bushing, pad, bracket and cup assembly taking note of positioning to assure correct installation. Remove the spring from the shock absorber.

To install:

8. Install the compressed spring to the shock absorber assembly and install the bump rubber, cup assembly, upper bushing, upper spring pad, collar, bracket assembly, upper bushing, washer and self-locking nut to the shock absorber in that order. Make sure the components are in the same orientation as was prior to removal.

9. Tighten the upper rod nut to 14–22 ft. lbs. (20–30 Nm).

10. Remove the spring compressor tool.

11. Install the shock absorber to the lower arm, install the lower mounting nut and tighten to 72 ft. lbs. (100 Nm).

12. Install the upper nut and torque to 29–36 ft. lbs. (40–50 Nm).

13. Install the cap and cover.

14. Lower the arm and remove the jack.

15. Install the wheel.

16. Lower the vehicle.

Sonata

1. Raise the vehicle and support it safely. Remove the wheel.

2. Remove the rear strut from the vehicle.

3. Compress the spring on the strut using spring compressor 09546-11000 or equivalent. Remove the piston rod tightening nut at the top of the strut while holding the piston rod with a wrench.

4. Disassemble the upper dust cover, bushing, pad, bracket and cup assembly taking note of positioning

to assure correct installation. Remove the spring from the strut.

To install:

5. Install the compressed spring to the strut assembly and install the bump rubber, cup assembly, upper bushing, upper spring pad, collar, bracket assembly, upper bushing, washer and self-locking nut to the strut in that order. Make sure the components are in the same orientation as was prior to removal.

6. Tighten the upper rod nut to 14–18 ft. lbs. (20–25 Nm).

7. Remove the spring compressor tool.

8. Install the strut.

9. Install the wheel.

10. Lower the vehicle to the floor.

Control Arms

REMOVAL AND INSTALLATION

1995–97 Sonata

1. Raise and safely support the vehicle.

2. Remove the wheels.

3. Remove the strut assembly.

4. Remove the brake line clamp bolt.

5. Disconnect the ball joint from the knuckle using special tool 09568-34000, or equivalent.

6. Remove the upper arm.

7. Remove the lower arm.

To install:

8. Position the lower arm, and install the ball joint to the knuckle, tighten the mounting nut to 54–64 ft. lbs. (75–89 Nm).

9. Install the lower arm mounting bolt to the crossmember, and tighten the mounting bolt to 102–116 ft. lbs. (140–160 Nm).

10. Install the upper arm. Tighten the mounting nut to the hub to 54–64 ft. lbs. (75–89 Nm).

11. Install the upper arm mounting bolt to the crossmember, and tighten the mounting bolt to 102–116 ft. lbs. (140–160 Nm).

12. Install the brake line clamp bolt.

13. Install the strut assembly.

14. Install the wheels, and lower the vehicle.

Excel and Scoupe

1. Raise the vehicle and support it safely.

2. Remove the muffler assembly.

3. Raise the suspension arm assembly slightly and keep in this position.

4. Disconnect the parking brake cable from the arm.

5. Remove the shock absorber.

6. Disconnect the brake hoses from their clips on the suspension members.

7. Lower the suspension slightly and carefully remove the coil spring.

8. Remove the rear suspension from the vehicle as an assembly.

9. Before removing any fixtures from the suspension arm matchmark all parts for assembly reference; this is extremely important! If equipped with stabilizer bars, make a mark on the bar in line with the punch mark on the bracket.

10. Remove the dust cover clamp.

11. Remove the nuts securing the control arms and pull them apart. Leave the dust cover attached to the right arm.

12. Remove the rubber stopper from the right arm.

13. Using a flat bladed chisel, drive bushing from the right arm.

14. Using a brass drift, drive bushing out from the left arm.

To install:

15. Coat the inside of the left arm and the outside of the bushing with chassis lube and drive it into place with driver tools 09555-21100 and 09555-21000 or equivalent. Drive the bushing in until the notch on 09555-21000 or equivalent, reaches the end of the arm.

16. Coat the inside of the arm and the outside of the bushing with chassis lube and drive it into the arm until it is fully seated.

17. Install the dust cover to the center position of the right arm, about 400mm.

18. Apply chassis lube to the surface of the right arm and install the rubber stopper.

19. Align all matchmarks, including the stabilizer bar and slowly push the suspension halves together.

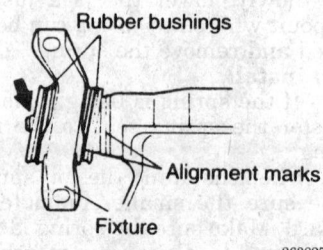

Bushing alignment marks — Excel and Scoupe

20. Install all remaining bushing, washers and attaching parts.

NOTE: The toothed sides of the washers face the bushings.

21. Install the end nuts and torque them loosely at this time.

22. Jack the assembly into position and torque the suspension-to-body bolts to 50 ft. lbs. (68 Nm).

23. Install the coil springs and loosely install the shock absorbers.

24. Install the rear brake assembly and related components.

25. Attach the parking brake cable and brake hoses to their clips on the suspension arm.

26. Install the tire and wheel assemblies.

27. Lower the vehicle and tighten the suspension arm end nuts to specifications. Tighten the shock mounting bolts to 47–58 ft. lbs. (63–79 Nm).

28. Adjust the rear brake shoe clearance.

29. Install the wheels and tires.

30. Lower the vehicle to the floor.

Accent

1. Raise and safely support the vehicle.

2. Remove the wheel assembly.

3. Disconnect the stabilizer link from the control arm.

4. Remove the mounting bolts and remove the control arm(s).

To install:

5. Install the control arms. Do not tighten the nuts and bolts until the vehicle is on the floor at normal riding height.

6. Connect the stabilizer link to the rear control arm.

7. Install the wheel assembly.

8. Lower the vehicle to the floor.

9. Torque the nuts at both ends of the front control arm to 72–87 ft. lbs. (100–120 Nm).

10. Torque the nut at the knuckle end of the rear control arm to 72–87 ft. lbs. (100–120 Nm). Torque the nut at the body end to 58–72 ft. lbs. (80–100 Nm).

Sway Bar

REMOVAL AND INSTALLATION

All Models

1. Raise and safely support the vehicle.

2. Remove the rear wheels.

3. Remove the rear axle assembly.

4. Mark the sway bar for proper installation.

5. Pull apart the suspension arms slightly and remove the sway bar from the axle assembly.

To install:

6. Install the sway bar on the axle assembly.

7. Install the axle assembly into the vehicle.

8. Install the rear wheels, and lower the vehicle to the floor.

Wheel Bearings

ADJUSTMENT

Excel, Scoupe and Elantra

With Rear Drum Brakes

1. Safely raise and support the rear of the vehicle.

2. Remove the rear wheels.

3. Remove the grease cap and loosen the axle shaft nut.

4. For Excel, Scoupe and 1993–95 Elantra, tighten the nut to 108–145 ft. lbs. (150–200 Nm). Check for correct bearing end-play by placing a dial indicator on the hub surface and moving the hub outward. Note the movement of the gauge and compare to the desired reading of 0.008 in. or less (0.2mm or less). If end-play exceeds the desired reading, retighten the rear hub bearing nut and recheck end-play. If reading is still excessive, replace the hub unit.

5. For 1996–97 Elantra, the spindle nut should be tightened to 130–159 ft. lbs. (180–220 Nm). Using a dial indicator, measure the axial play of the bearing assembly. If the axial play exceeds 0.001 inch (0.025 mm), replace the bearing assembly.

6. If end-play is correct, check the starting torque by attaching a spring balance to the hub lug bolts and pulling at a 90 degree angle while noting the required force to turn the hub. If the torque required is above the desired reading of 4.9 lbs. or less (22 N

or less), loosen the nut and again tighten to the desired torque. Recheck the starting torque. If torque is still above the desired reading, replace the rear bearings.

7. Install the rear wheels and lower the vehicle.

8. Prior to moving the vehicle, pump the brakes until a firm pedal is obtained.

With Rear Disc Brakes

On vehicles equipped with rear disc brakes, the spindle nut should be tightened to 11–15 ft. lbs. (15–20 Nm).

Sonata

1. Safely raise and support the rear of the vehicle.

2. Remove the tire and wheel assembly.

3. Remove the grease cap and loosen the axle shaft nut.

4. For 1993–95 Sonata models, tighten the new rear wheel bearing nut to 174–217 ft. lbs. (240–300 Nm).

5. For 1996–97 Sonata, models, tighten a new rear wheel bearing nut to 146–189 ft. lbs. (200–260 Nm).

6. Check for correct bearing end-play by placing a dial indicator on the hub surface and moving the hub outward. Note the movement of the gauge and compare to the desired reading of 0.004 inch (0.01mm) or less . If end-play exceeds the desired reading, retighten the rear hub bearing nut and recheck end-play. If reading is still excessive, replace the hub unit.

7. If end-play is correct, check the starting torque by attaching a spring balance to the hub lug bolts and pulling at a 90 degree angle while noting the required force to turn the hub. If the torque required is above the desired reading of 7 lbs. (9 Nm) or less, loosen the nut and again tighten to the desired torque. Recheck the starting torque. If torque is still above the desired reading, replace the rear hub bearing unit.

8. After final tightening of the wheel bearing nut, crimp the edge of the nut to meet the spindle's indentation.

9. Install the grease cap.

10. Install the tire and wheel assembly and lower the vehicle.

11. Prior to moving the vehicle, pump the brakes until a firm pedal is obtained.

REMOVAL AND INSTALLATION

Excel, Scoupe and 1993–95 Elantra

1. Safely raise and support the rear of the vehicle.

2. Remove the tire and wheel assembly.

3. Remove the grease cap, cotter pin, serrated nut cap, axle shaft nut and washer from the spindle.

4. Pull outward on the brake drum slightly to remove the outer wheel bearing.

5. Slide the drum down the spindle and remove assembly from the vehicle.

6. Pry the inner grease seal from the rear hub of the drum and discard.

7. Remove the inner wheel bearing. If the bearings are being replaced, drive the bearing races from the hub taking care not to damage the inner surface of the drum.

To install:

8. Coat the new races with wheel bearing grease and drive them into the hub, making sure they are fully and squarely seated.

9. Pack the hub cavity with wheel bearing grease.

10. Install the inner bearing and drive a new grease seal into place. Make sure to pack the bearings completely with grease prior to installing into the drum.

11. Install the brake drum onto the spindle. Install the outer bearing, washer and shaft nut onto spindle.

12. Install and tighten bearing locknut as follows:

   a. Prior to installation, inspect the rear bearing nut by threading the nut onto the spindle until the distance between the shoulder of the spindle and the inner flat on the nut is 0.07–0.11 inch (2.0–3.0mm).

   b. Install the brake drum and outer bearing onto the spindle. Install and torque the nut to 108–145 ft. lbs. (150–200 Nm).

   c. Check for correct bearing end-play by placing a dial indicator on the hub surface and moving the drum outward. Note the movement of the gauge and compare to the desired reading of 0.0043 in. or less (0.11mm or less). If end-play exceeds the desired reading, retighten the rear hub bearing nut and recheck end-play. If reading is still excessive, replace the hub assembly.

13. If end-play is correct, check the starting torque by attaching a spring balance to the hub lug bolts and pulling at a 90 degree angle while noting the required force to turn the hub. If

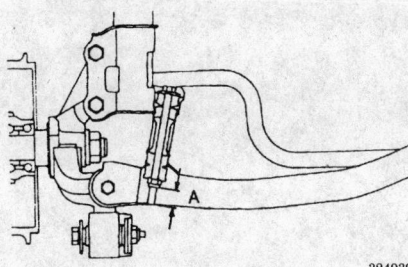

324939

Correct link installation — Accent

the torque required is above the desired reading of 4.9 lbs. or less (22 N or less), loosen the nut and again tighten to the desired torque. Recheck the starting torque. If torque is still above the desired reading, replace the rear bearing.

14. After final tightening the wheel bearing nut, align with the spindle's indentation and crimp the edge of the nut to wedge in position.

15. Install the tire and wheel assembly, and lower the vehicle.

16. Prior to moving the vehicle, pump the brakes until a firm pedal is obtained.

1996–97 Elantra

1. Raise and safely support the vehicle.

2. Remove the rear wheel.

3. If equipped with ABS, remove the wheel speed sensor.

4. Remove the caliper assembly and suspend it with a wire to prevent damage to the brake hose.

5. Remove the brake rotor.

6. Remove the hub cap, wheel bearing nut and tongued washer.

7. Remove the rear hub and bearing assembly.

────── WARNING ──────
Do not disassemble the hub/bearing assembly. If the bearing is defective, replace the complete assembly.

To install:

8. Install the hub and bearing assembly on the spindle. Torque the wheel bearing nut to 130–159 ft. lbs. (180–220 Nm) and stake the nut on the concave portion of the spindle. Install the hub cap.

9. Install the brake rotor.

10. Install the brake caliper assembly.

11. If equipped with ABS, install the wheel speed sensor.

12. Install the rear wheel and lower the vehicle to the floor.

Accent

1. Safely raise and support the vehicle.

2. Remove the rear wheel.

3. Remove the brake drum.

4. Remove the dust cap from the center of the hub.

5. Remove the wheel bearing nut and washer.

NOTE: Do not attempt to disassemble the bearing assembly. If the bearing is bad, replace it.

To install:

6. Install the hub assembly on the spindle. Torque the nut to 130–159 ft. lbs. (180–220 Nm).

7. Stake the nut on the slotted part of the spindle.

8. Install the brake drum.

9. Install the wheel assembly.

10. Safely lower the vehicle to the floor.

Sonata

NOTE: The rear hub bearing unit can not be disassembled. If the hub shows signs of wear or damage, replacement of the unit is required.

1. Safely raise and support rear of the vehicle.

2. Remove the tire and wheel assembly.

3. If equipped with ABS, remove the rear speed sensor.

4. If equipped with rear disc brakes, remove the brake caliper and support out of the way using wire. Do not disconnect the brake hose from the caliper. Remove the brake rotor. If equipped with drum brakes, remove the brake drum from the hub assembly.

5. Remove the grease cap, wheel bearing nut and washer from the center of the hub bearing unit. Remove the rear hub unit from the vehicle.

To install:

6. Install the rear bearing unit onto the spindle. Install the outer bearing and the tongued washer into the rear hub unit.

NOTE: Press the inner race further until the inner race contacts with the spindle end.

7. For 1993–95 Sonata models, install and tighten the new rear wheel bearing nut to 174–217 ft. lbs. (240–300 Nm).

8. For 1996–97 Sonata models, install and tighten the new rear wheel bearing nut to 146–189 ft. lbs. (200–260 Nm).

9. Check for correct bearing end-play by placing a dial indicator on the hub surface and moving the hub outward. Note the movement of the gauge and compare to the desired reading of 0.004 inch (0.01 mm) or less. If end-play exceeds the desired reading, retighten the rear hub bearing nut and recheck end-play. If reading is still excessive, replace the hub unit.

10. If end-play is correct, check the starting torque by attaching a spring balance to the hub lug bolts and pulling at a 90 degree angle while noting the required force to turn the hub. If the torque required is above the desired reading of 7 lbs. (9 Nm) or less, loosen the nut and again tighten to the desired torque. Recheck the starting torque. If torque is still above the desired reading, replace the rear hub bearing unit.

11. After final tightening the wheel bearing nut, align with the spindle's indentation and crimp the edge of the nut to wedge in position.

12. If equipped with rear disc brakes, install the brake disc and caliper to the vehicle. If equipped with drum brakes, install drum to hub assembly.

13. Install the rear speed sensor, if equipped.

14. Install the grease cap.

15. Install the tire and wheel assembly and lower the vehicle.

16. Pump the brake pedal to assure correct brake operation, prior to moving the vehicle.

FIRING ORDERS

NOTE: To avoid confusion, always replace spark plug wires one at a time.

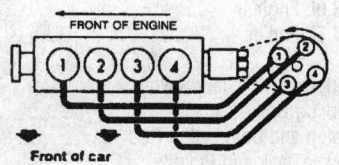

285581

2.0L Engine
Engine Firing Order: 1–3–4–2
Distributor Rotation: Counterclockwise

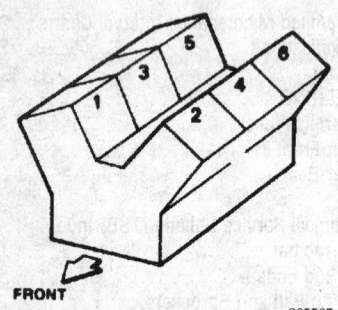

285587

3.0L Engine
Engine Firing Order: 1–2–3–4–5–6
Distributorless Ignition System

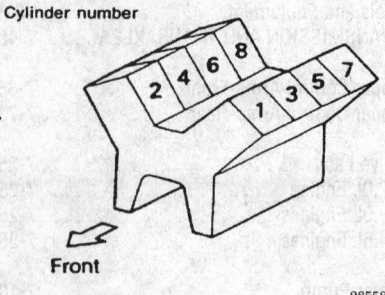

285582

4.5L Engine
Engine Firing Order: 1–8–7–3–6–5–4–2
Distributorless Ignition System

ENGINE ELECTRICAL

NOTE: Disconnecting the negative battery cable on some vehicles may interfere with the functions of the on board computer systems and may require the computer to undergo a relearning process, once the negative battery cable is reconnected.

Distributor

REMOVAL AND INSTALLATION

2.0L (SR20DE) Engine

1. Disconnect the negative battery cable.
2. Remove the splash shield, if equipped. Unplug the distributor connections but leave the ignition wires in place.
3. Unscrew the distributor cap hold-down screws and lift off the distributor cap with all ignition wires still connected.
4. Matchmark the rotor to the distributor housing and the distributor housing to the engine.

NOTE: Do not crank the engine during this procedure. If the engine is cranked, the matchmark must be disregarded.

5. Remove the hold-down bolt.
6. Remove the distributor from the engine.

Engine Not Disturbed

1. If the engine was not disturbed, proceed as follows:
   a. Install a new distributor housing O-ring.
   b. Install the distributor in the engine so the rotor is aligned with the matchmark on the housing and the housing is aligned with the matchmark on the engine. Make sure the distributor is fully seated and the distributor gear is fully engaged.
   c. Install and snug the hold-down bolt.
   d. Connect the distributor pickup lead wires.
   e. Install the distributor cap and tighten the screws. Install the splash shield.
   f. Connect the negative battery cable.
   g. Check and/or adjust the ignition timing and tighten the hold-down bolt.

Engine Disturbed

1. If the engine was disturbed (cranked or turned over with the distributor removed), proceed as follows:
   a. Install a new distributor housing O-ring.
   b. Position the engine so the No. 1 piston is at TDC of its compression stroke and the mark on the vibration damper is aligned with 0 on the timing indicator.
   c. Install the distributor in the engine so the rotor is aligned with the position of the No. 1 ignition wire on the distributor cap. Make sure the distributor is fully seated and that the distributor shaft is fully engaged.

NOTE: There are distributor cap runners inside the cap on 2.0L engine. Make sure the rotor is pointing to where the No. 1 runner originates inside the cap.

   d. Install and snug the hold-down bolt.
   e. Connect the distributor pickup lead wires.
   f. Install the distributor cap and tighten the screws. Install the splash shield, if equipped.
   g. Connect the negative battery cable.
   h. Check and/or adjust the ignition timing and tighten the hold-down bolt to 9–12 ft. lbs. (13–16 Nm).

3.0L (VG30DE and VQ30DE) and 4.5L (VH45DE) Engines

The J30 and I30 3.0L engine is equipped with a distributorless ignition. As a result, there is no distributor to remove.

Ignition Timing

ADJUSTMENT

2.0L (SR20DE) Engines

NOTE: The engine should be in good mechanical condition and all electrical connectors and vacuum hoses connected before making this adjustment.

1. Start the engine and let it warm up to normal operating temperature.
2. Open the hood and run the engine under no load at about 2,000 rpm for about two minutes.
3. Perform Diagnostic Test Mode II and repair any causes of trouble codes as needed.
4. Run the engine under no load at 2,000 rpm for about two minutes. Rev

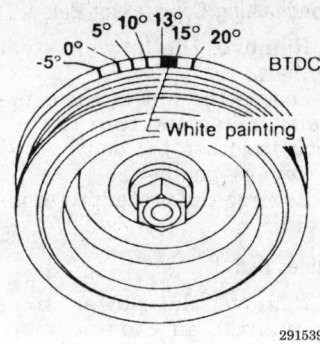

Crankshaft pulley and timing marks — 2.0L engine

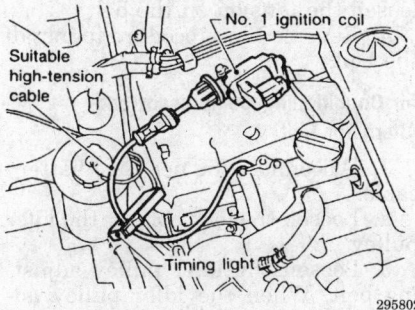

Timing light connection to spark plug cable — 3.0L (VG30DE) engine shown

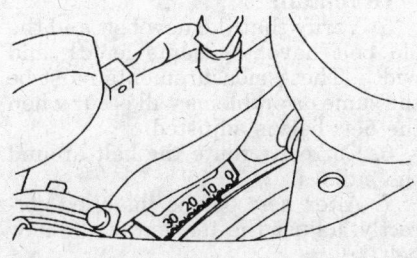

Location of timing marks — 3.0L (VQ30DE) engine

the engine two or three times and let it idle for one minute.

5. Turn **OFF** the engine and disconnect the throttle position sensor connector. Connect a timing light to the No. 1 spark plug wire. Start the engine.

6. Adjust the timing to 15°±2° BTDC by loosening the distributor mounting bolts and turning the distributor. When the timing is correct, tighten the mounting bolts and turn the engine **OFF**.

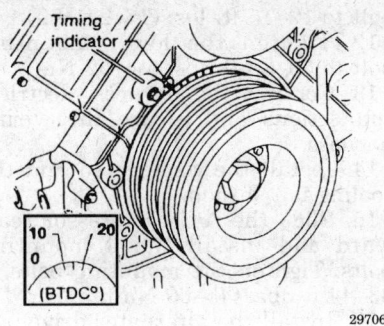

Location of timing marks — 4.5L engine

7. Reconnect the throttle position sensor connector. Start the engine and check the ignition timing again.

3.0L (VG30DE and VQ30DE) and 4.5 (VH45DE) Engines

NOTE: The engine should be in good mechanical condition and all electrical connectors and vacuum hoses connected before making this adjustment.

1. Start the engine and let it warm up to normal operating temperature.

2. Open the hood and run the engine under no load at about 2,000 rpm for about two minutes.

3. Perform Diagnostic Test Mode II and repair any causes of trouble codes as needed.

4. Run the engine under no load at 2,000 rpm for about two minutes. Rev the engine two or three times and let it idle for one minute.

5. Turn **OFF** the engine and disconnect the throttle position sensor connector. Remove the No. 1 (or 6 ignition coil.) Connect the coil to the spark plug using a spare piece of high-tension wire so you have a place to connect your timing light. Start the engine.

6. Run the engine under no load at 2,000 rpm for about two minutes. Rev the engine two or three times and let it idle.

7. Check the ignition timing and adjust if needed.

• Correct ignition timing for 3.0L engines is 10°±2° BTDC.

• Correct ignition timing for 4.5L is 15°±2° BTDC.

8. Adjustment is made by loosening the screws and turning the camshaft position sensor until the mark on the crankshaft pulley is pointing at 10°BTDC. Tighten the mounting screws and confirm ignition timing has not changed.

9. Turn the engine **OFF** and connect the throttle position sensor connector.

Alternator

PRECAUTIONS

Several precautions must be observed with alternator equipped vehicles to avoid damage to the unit.

• If the battery is removed for any reason, make sure it is reconnected with the correct polarity. Reversing the battery connections may result in damage to the 1-way rectifiers.

• When utilizing a booster battery as a starting aid, always connect the positive to positive terminals and the negative terminal from the booster battery to a good engine ground on the vehicle being started.

• Never use a fast charger as a booster to start vehicles.

• Disconnect the battery cables when charging the battery with a fast charger.

• Never attempt to polarize the alternator.

• Do not use test lights of more than 12 volts when checking diode continuity.

• Do not short across or ground any of the alternator terminals.

• The polarity of the battery, alternator and regulator must be matched and considered before making any electrical connections within the system.

• Never separate the alternator on an open circuit. Make sure all connections within the circuit are clean and tight.

• Disconnect the battery ground terminal when performing any service on electrical components.

• Disconnect the battery if arc welding is to be done on the vehicle.

REMOVAL AND INSTALLATION

2.0L (SR20DE) Engine

1. Disconnect the negative battery cable.

2. Loosen the alternator belt and remove from the pulley.

3. Disconnect the harness connector and cable from the rear of the alternator.

4. Remove the adjusting bracket.

NOTE: On some models, the front mounting bolt cannot be removed separately because of insufficient clearance between the alternator and engine coolant inlet tube.

5. Remove the rear mounting bolt and loosen the front mounting bolt.

6. Remove the alternator with the front mounting bolt.

To install:

7. Install the alternator and connect the wiring. Torque the mounting bolts to 15 ft. lbs. (20 Nm).

8. Install and adjust all the belts as required.

9. Connect the negative battery cable and check the alternator for proper operation.

3.0L (VG30DE and VQ30DE) Engines

1. Disconnect the negative battery cable.

2. Loosen the alternator belt and remove from the pulley.

3. Remove the alternator adjusting bar.

4. Disconnect the harness connector and cable from the rear of the alternator.

NOTE: On some models, the front mounting bolt cannot be removed separately because of insufficient clearance between the alternator and engine coolant inlet tube.

5. Remove the rear mounting bolt and loosen the front mounting bolt.

6. Remove the alternator with the front mounting bolt.

To install:

7. Install the alternator and connect the wiring. Torque the mounting bolts to 15 ft. lbs. (20 Nm).

8. Install and adjust the alternator belt.

9. Connect the negative battery cable and check the alternator for proper operation.

4.5L (VH45DE) Engine

1. Disconnect the negative battery cable.

2. Remove the engine RH undercover and remove the RH side inspection cover.

3. Loosen the nut that secures the alternator belt idler pulley and turn the adjusting bolt to loosen the alternator belt.

4. Remove the alternator belt from the alternator pulley.

5. Remove the 4 air conditioner compressor mounting bolts.

6. Remove the cooling fan and the fan shroud.

7. Slide the A/C compressor forward and disconnect the alternator connectors. Be sure to note electrical connections for installation purposes.

8. Remove the alternator upper and lower mounting bolts.

9. Carefully remove the alternator from the vehicle.

To install:

10. Install the alternator to the vehicle and install the mounting bolts.

11. Tighten the upper mounting bolt to 12–15 ft. lbs. (16–21 Nm).

12. Tighten the lower mounting bolt to 33–38 ft. lbs. (44–52 Nm).

13. Connect the harness electrical connections and secure the wiring harness.

14. Install the cooling fan and the cooling fan shroud.

15. Slide the A/C compressor rearward and install the 4 mounting bolts. Tighten the mounting bolts to 33–44 ft. lbs. (45–60 Nm).

16. Install the alternator drive belt and tighten the idler pulley adjusting screw to 35–61 inch lbs. (4–7 Nm).

17. Tighten the idler pulley mounting nut to 19–24 ft. lbs. (25–32 Nm).

18. Install the engine RH inspection cover and install the RH undercover.

19. Connect the negative battery cable.

Drive Belts

REMOVAL AND INSTALLATION

2.0L (SR20DE) Engine

Alternator and A/C Compressor Belt

1. Loosen the alternator pivot bolts.

2. Loosen the lock bolt and turn the adjusting bolt to loosen the tension on the belt.

3. Remove the belt from the pulleys.

4. Reverse the procedure to install the belt. Torque the lock bolt to 12–16 ft. lbs. (16–22 Nm).

Power Steering and Water Pump Belt

1. Remove the alternator belt.

2. Loosen the power steering pump pivot bolts.

3. Loosen the lock bolt and turn the adjusting bolt to loosen the tension on the belt.

4. Remove the belt from the pulleys.

5. Reverse the procedure to install the belt. Torque the lock bolt to 12–16 ft. lbs. (16–22 Nm)

3.0L (VG30DE) Engines

Power Steering Pump Belt

1. Loosen the lock bolt on the power steering pump bracket.

2. Loosen the power steering pump mounting bolts and turn the adjusting bolt to loosen the tension on the belt.

3. Remove the belt.

4. Reverse the procedure to install the belt.

Air Conditioning Compressor Belt

1. Remove the power steering pump belt.

2. Loosen the nut in the center of the idler pulley and turn the adjusting bolt to loosen the belt.

3. Remove the belt.

4. Reverse the procedure to install the belt.

Alternator Belt

1. Remove the power steering pump and the air conditioning compressor belts.

2. Loosen the pivot bolt and the locknut on the alternator mounting.

3. Turn the adjusting bolt to loosen the tension on the belt.

4. Reverse the procedure to install the belt.

Air Conditioning Compressor and Alternator Belt

1. Disconnect the negative battery cable.

2. Loosen the lock bolt for the idler pulley.

3. Loosen the idler pulley adjusting bolt. When the idler pulley adjustment bolt is loosened, the drive belt tension will slowly be released.

4. When there is enough slack in the belt, remove the belt from the pulleys.

To install:

5. Verify that the new belt and the old belt have the same length and width. These measurements must be the same or problems will occur when the new belt is adjusted.

6. Correctly route the belt around the pulleys.

7. After new belt is installed correctly, adjust the tension of the new belt.

8. Tighten the mounting nut to 19–24 ft. lbs. (25–32 Nm).

9. Connect the negative battery cable.

Power Steering Pump Belt

1. Disconnect the negative battery cable.

2. Loosen the power steering oil pump mounting and pivot bolts.

3. Loosen the drive belt adjustment locking bolt. The drive belt adjustment bolt is located on the power steering oil pump. The drive belt adjustment bolt will move the power steering oil pump and increase or decrease belt tension.

4. Turn the power steering oil pump adjusting bolt until there is enough slack in the drive belt to remove it.

5. Remove the drive belt from around the pulleys.

To install:

6. Verify that the new belt and the old belt have the same length and width. These measurements must be the same or problems will occur when the new belt is adjusted.

7. Correctly route the belt around the pulleys.

8. After the drive belt is installed correctly, adjust the drive belt tension.

9. Tighten the mounting nut to 12–15 ft. lbs. (16–15 Nm). Also tighten the pivot bolts.

10. Connect the negative battery cable.

4.5L (VH45DE) Engine

To remove a drive belt, loosen the belt by loosening the center bolt in the appropriate idler pulley and turning the adjusting bolt.

Inspect the belt for cracks, fraying, wear or oil contamination. Replace the belt with a new one if necessary.

To install:

1. Install the belt on the proper pulleys and turn the adjuster to tighten the belt. To check for proper belt tension, push on the belt midway between the pulleys with a force of 22 lbs. (98 N) and measure the deflection as follows:

   a. Alternator belt; new — 0.295–0.335 inch (7.5–8.5 mm), used — 0.35–0.39 inch (9–10 mm)

   b. A/C compressor belt; new — 0.295–0.335 inch (7.5–8.5 mm), used — 0.335–0.374 inch (8.5–9.5 mm)

   c. Power steering belt; With active suspension; new — 0.217–0.256 inch (5.5–6.5 mm), used — 0.28–0.31 inch (7–8 mm). Without active suspension; new — 0.31–0.35 inch (8–9 mm), used — 0.35–0.39 inch (9–10 mm)

2. After the belt is tightened to the correct specification, tighten the bolt in the center of the tensioner to 22–25 ft. lbs. (29–33 Nm).

Starter

REMOVAL AND INSTALLATION

2.0L (SR20DE) Engine

1. Disconnect the negative battery cable.
2. Remove the air intake duct.
3. Raise the vehicle and support safely.
4. Remove the engine undercover.
5. Remove the starter motor mounting bolts.

6. Disconnect the electrical connections from the starter.

To install:

7. Connect the electrical connections to the starter.

8. Install the starter motor. Torque the mounting bolts to 25 ft. lbs. (34 Nm).

9. Install the engine under cover.

10. Lower the vehicle and install the air duct.

11. Connect the negative battery cable and check the starter for proper operation.

3.0L (VG30DE) and 4.5L (VH45DE) Engines

1. Disconnect the negative battery cable.
2. Raise the vehicle and support safely.
3. Remove the engine undercover.
4. Disconnect the electrical connections on the starter.
5. Remove the starter mounting bolts and remove the starter.

To install:

6. Install the starter motor. Torque the mounting bolts to 25 ft. lbs. (34 Nm).

7. Connect the electrical connections to the starter.

8. Install the engine under cover.

9. Connect the negative battery cable and check the starter for proper operation.

3.0L (VQ30DE) Engine

The gear reduction starter has a set of ratio reduction gears; the brushes on the gear reduction starter are located on a plate behind the starter drive housing. The extra gears make the starter pinion gear turn at about 1/2 the speed of the starter, giving the starter twice the turning power of a conventional starter.

1. Disconnect the negative battery cable from the battery.
2. Remove the air duct assembly.
3. Disconnect the wiring harness connector at the starter.
4. Remove the starter-to-engine bolts and remove the starter from the vehicle.

To install:

5. Install the starter to the bell housing.

6. Tighten the starter mounting bolts as follows:

   a. Tighten the long starter bolt to 58–72 ft. lbs. (78–98 Nm).

   b. Tighten the short starter bolt to 23–30 ft. lbs. (30–41 Nm).

NOTE: Tighten the starter bolts evenly to prevent damage to the starter nose housing.

7. Connect the starter wiring harness connector.

8. Install the air duct assembly.

9. Connect the negative battery cable.

10. Start the engine and verify proper starter operation.

CHASSIS ELECTRICAL

Blower Motor

REMOVAL AND INSTALLATION

G20, J30 and Q45 Models

1. Disconnect the negative battery cable.
2. Remove the lower right side instrument panel cover.
3. Disconnect the harness connector to the blower motor.
4. Remove the screws that attach the blower housing to the intake unit.
5. Remove the blower motor from the housing.

To install:

6. Install the blower motor into the housing.

7. Connect the harness connector to the blower motor.

8. Install the lower instrument panel.

9. Connect the negative battery cable and check the climate control system for proper operation.

I30 Models

NOTE: It may be necessary to remove the glove box assembly to gain clearance for the blower motor removal and installation. The blower motor is located behind the glove box, facing the floor.

1. Disconnect the negative battery cable.
2. Remove all panels and ducting necessary to gain access to the blower motor.
3. Disconnect the blower motor harness wiring connectors.
4. Remove the blower motor retaining screws and lower the motor/wheel from the intake housing.

NOTE: On some models, release the clips that attach the blower casing to the intake housing to remove the blower motor.

To install:

5. Transfer the old blower wheel to the shaft of the new motor.

6. Raise the blower/wheel assembly up and onto the intake housing. Use a new gasket, if required.

7. Install the blower motor retaining screws or lock the clips.

8. Connect the blower motor wiring.

9. Install all ducting and panels.

10. Connect the negative battery cable.

11. Check the blower for proper operation at all speeds.

Windshield Wiper Motor

REMOVAL AND INSTALLATION

G20, J30 and Q45 Models

1. Turn the wiper switch to the **OFF** position and disconnect the motor leads at the connector.

2. Disconnect the wiring at the motor.

3. Remove the motor mounting bolts. Pull the motor out and remove the wiper motor linkage attaching nut.

To install:

4. Position the wiper near the dash panel and connect the motor to the linkage.

5. Install the wiper motor mounting bolts and connect the wiring.

6. Connect the motor leads and check all windshield wiper and washer functions for proper operation.

I30 Models

1. Disconnect the negative battery cable.

2. Disconnect the harness connector from the wiper motor.

3. Remove the four wiper motor mounting bolts.

4. Disconnect the wiper motor from the linkage at the ball joint.

To install:

5. Grease the ball joint and attach the wiper motor to the linkage.

6. Mount the wiper motor on the bulkhead and install the four mounting bolts. Torque the bolts to 34–45 in. lbs. (3.8–5.1 Nm).

7. Connect the harness connector to the wiper motor and connect the negative battery cable.

Combination Switch

REMOVAL AND INSTALLATION

G20, J30 and I30 Models

— **CAUTION** —

The Air Bag system must be disarmed before removing the steering wheel or combination switch. Failure to do so may cause accidental deployment, property damage or personal injury.

The combination switch assembly is made up of three units; the windshield wiper/washer switch is located to the right side of the steering column, the combination switch base is attached to the steering column and the headlight, dimmer and turn signal switch is located to the left of steering column. The two switches can be removed without removing the switch base.

Combination Switch

1. Disconnect the negative battery cable and wait at least 10 minutes for the Air Bag power supply to discharge.

2. Remove the steering column covers. Disconnect the connector.

3. Remove the combination switch mounting screws and remove the switch(s) from the switch base.

To install:

4. Install the switch to the switch base.

5. Install the steering column covers.

6. Connect the negative battery cable and check all functions of the combination switch for proper operation.

Combination Switch Base

1. Make sure the wheels are pointing straight ahead. Disconnect the

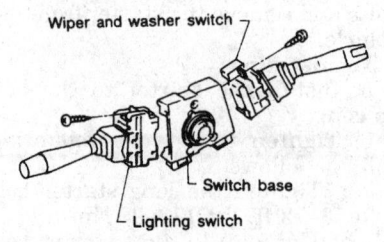

Combination switch component identification — G20, J30 and I30 models

negative battery cable and wait at least 10 minutes for the Air Bag power supply to discharge.

2. Remove the steering wheel.

— **CAUTION** —

Always carry an Air Bag assembly with the bag and cover facing away from your body. Store the assembly with the bag and cover facing up; never place the assembly face down on the floor or workbench.

3. Disconnect all combination and windshield wiper switch connectors and remove the switches from the switch base.

4. Remove the base attaching screw, push and turn the base clockwise to remove the base.

To install:

5. Install the switch base and the switches on the steering column.

6. Install the steering column covers and the steering wheel.

7. Connect the negative battery cable and check all functions of the combination and windshield wiper switches for proper operation.

Q45 Models

The combination switch cannot be disassembled and refers to all column-mounted stalk switches as a single unit. The wiring harness includes wiring for the rear wheel steering angle sensor. Before removing the switch, unplug the existing switch and connect each replacement one at a time to make sure all functions operate properly.

— **CAUTION** —

The Air Bag system must be disarmed before removing the steering wheel. Failure to do so may cause accidental deployment, property damage or personal injury.

1. Make sure the wheels are pointing straight-ahead. Disconnect the negative battery cable and wait at least 10 minutes for the Air Bag power supply to discharge.

2. Remove the steering wheel pad and the steering wheel.

— **CAUTION** —

Always carry an Air Bag assembly with the bag and trim cover away from your body. Store the assembly with the bag and trim cover facing up; never place the assembly face down on the floor or workbench.

3. Disconnect all combination switch connectors from underneath the steering column.

4. Remove the screws that fasten the combination switch to the steering column and remove it from the vehicle.

5. The installation is the reverse of the removal procedure.

6. Connect the negative battery cable and check all functions of the combination switch for proper operation.

Ignition Lock

REMOVAL AND INSTALLATION

The steering lock/ignition switch assembly is attached to the steering column by special screws or bolts whose heads shear off upon installation. The screws must be drilled out to remove the assembly or removed with an appropriate tool. The bolts may also be removed with a hammer and chisel by notching the bolts and tapping them counterclockwise with the hammer and chisel.

NOTE: The ignition switch or warning switch can be replaced without removing the steering lock assembly. The ignition switch is on the back of the assembly and the warning switch on the side.

1. Disconnect the negative battery cable and insulate the terminal end.
2. Disarm the Air Bag system.

────── **CAUTION** ──────
The Air Bag system must be disarmed before removing the steering wheel. Failure to do so may cause accidental deployment, property damage or personal injury.

3. Remove the steering wheel and the steering column upper and lower covers.

4. Remove the combination switch from the steering column.

5. Remove steering column support nuts and lower the steering column.

6. Disconnect the ignition switch wiring.

7. Remove the set screw that holds the ignition switch to the steering lock and remove the remove the switch from the lock assembly.

To install:

8. Install the ignition switch to the steering lock and secure with set screw.

9. Install the ignition switch assembly and secure with new shear type bolts.

10. Connect the switch wiring.

11. Raise the steering column and secure with mounting nuts. Tighten the mounting nuts to 11–14 ft. lbs. (15–19 Nm).

12. Install the combination switch, steering column covers, and steering wheel.

13. Connect the negative battery cable and enable the Air Bag system.

Ignition Switch

REMOVAL AND INSTALLATION

G20 and J30 Models

────── **CAUTION** ──────
The Air Bag system must be disarmed before removing the steering wheel. Failure to do so may cause accidental deployment, property damage or personal injury.

1. Make sure the wheels are pointing straight-ahead. Disconnect the negative battery cable and wait at least 10 minutes for the Air Bag power supply to discharge.

2. Remove the steering wheel and combination switch.

────── **CAUTION** ──────
Always carry an Air Bag assembly with the bag and trim cover away from your body. Store the assembly with the bag and trim cover facing up; never place the assembly face down on the floor or workbench.

3. Disconnect the ignition switch wiring.

4. Lower the steering column.

5. Use a hacksaw blade to cut a groove into the heads of the self-

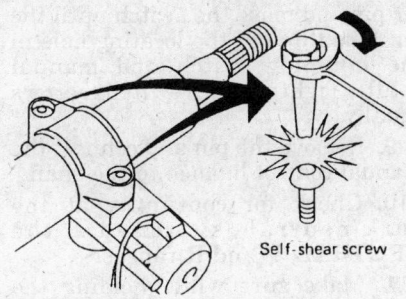

Self-shear screw

288109

Tighten the new ignition lock bolts until the heads break off

shearing screws and remove the screws.

6. Remove the lock/switch assembly from the column.

To install:

7. With the key inserted in the switch, install the assembly onto the column with new self-shearing screws. Tighten the screws gradually, testing the key for binding often. Tighten the screws until the heads shear off.

8. Connect the ignition switch connector.

9. Raise and secure the steering column.

10. Install the combination switch, column covers and the steering wheel.

11. Connect the negative battery cable and check the ignition switch for proper operation in all positions.

I30 and Q45 Models

1. Disconnect the negative battery cable and insulate the terminal end.
2. Disarm the Air Bag system.

────── **CAUTION** ──────
The Air Bag system must be disarmed before removing the steering wheel. Failure to do so may cause accidental deployment, property damage or personal injury.

3. Remove the steering wheel and the steering column upper and lower covers.

4. Remove the combination switch from the steering column.

5. Remove steering column support nuts and lower the steering column.

6. Disconnect the ignition switch wiring.

7. Remove the set screw that holds the ignition switch to the steering lock and remove the remove the switch from the lock assembly.

To install:

8. Install the ignition switch to the steering lock and secure with set screw.

9. Install the ignition switch assembly and secure with new shear bolts.

10. Connect the switch wiring.

11. Raise the steering column and secure with mounting nuts. Tighten the mounting nuts to 11–14 ft. lbs. (15–19 Nm).

12. Install the combination switch, steering column covers, and the steering wheel.

13. Connect the negative battery cable and enable the Air Bag system.

Park/Neutral Safety Switch

REMOVAL AND INSTALLATION

G20 Models

1. Disconnect the negative battery cable.
2. Disconnect the manual control linkage from the manual shift shaft.
3. Set the manual shaft on the transmission to the **NEUTRAL** detent.
4. Disconnect the switch harness connector.
5. Remove the neutral safety switch mounting bolts and remove the switch.
To install:
6. Install the switch on the transaxle but do not tighten the mounting screws.
7. Connect the harness connector to the switch.
8. Align the switch by inserting a suitable drill bit through the alignment holes in the shift lever on the transmission and the switch.
9. Tighten the switch mounting bolts and connect the control linkage.
10. Make sure the vehicle does not start in any gear except **P** or **NEUTRAL** and does start in both **P** and **NEUTRAL**.

J30 and Q45 Models

1. Disconnect the negative battery cable.
2. Raise the vehicle and support safely.
3. The switch is mounted on the transmission at the shift shaft. Disconnect the wires to the switch.
4. Remove the switch mounting screws and remove the switch.
To install:
5. Install and adjust the switch on the transmission.

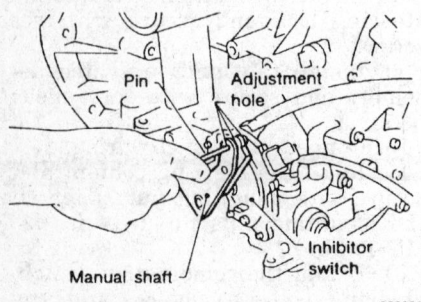

Park/neutral switch adjustment — G20 models

289932

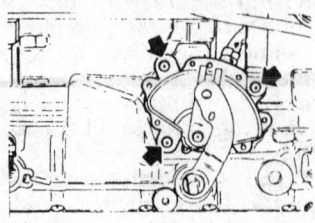

293693

Inhibitor switch mounting screws — J30 and Q45 models

6. After adjusting the switch, make sure the vehicle does not start in any gear except **P** or **NEUTRAL** and that it does start in both **P** and **NEUTRAL**.

I30 Models

The switch unit is bolted to the left side of the transmission shift lever.

1. Disconnect the manual control linkage from the manual shaft.
2. Disconnect the harness connectors from the park/neutral switch.
3. Remove the bolts that secure the switch to the side of the transfer case and remove the switch.
To install:
4. Install the switch and secure with mounting bolts.
5. Connect the harness connectors to the park/neutral switch.
6. Set the manual shaft on the side of the transaxle to the **NEUTRAL** position.
7. Loosen the park/neutral switch mounting screws enough to allow for movement of the switch.
8. Insert a 0.16 in. (4 mm) diameter pin and move the switch until the pin falls through the locating hole in the inhibitor switch and manual shaft. Tighten the switch screws equally.
9. Remove the pin and connect the manual control linkage to the shaft.
10. Check for continuity at the park/neutral switch in the **NEUTRAL**, **P** and **R** ranges.
11. Make sure while holding the brakes on, that the engine will start only in **P** or **NEUTRAL** and that the backup lights only illuminate in reverse.

Powertrain Control Module

REMOVAL AND INSTALLATION

G20 Models

NOTE: This unit should be removed before placing the vehicle in a paint oven or when using electric welding equipment on the vehicle.

1. Disconnect the negative battery cable.
2. The control unit is mounted below the center console next to the accelerator pedal. To access the diagnostic mode selector, remove the panel on the right side of the console. To remove the control unit, remove the panel on both sides.
3. Unplug the connector, remove the screws and slide the control unit out towards the right side of the vehicle.
To install:
4. Install the control unit and connect the harness connector.
5. Install the trim panel and connect the negative battery cable.

J30 and Q45 Models

NOTE: This unit should be removed before placing the vehicle in a paint oven or when using electric welding equipment on the vehicle.

1. Disconnect the negative battery cable.
2. Remove the passenger side kick panel below the dashboard.
3. Unplug the connector and remove the screws to remove the module control unit.
To install:
4. Mount the control unit to the body and connect the harness connector.
5. Replace the kick panel and connect the negative battery cable.

I30 Models

NOTE: Whenever handling the powertrain control module it is recommended that the technician wear a grounding strap. A grounding strap will prevent a static electrical discharge. Static electricity can cause severe damage to electrical components, such as a powertrain control module.

1. Disconnect the negative battery cable.
2. Remove the powertrain control module bolts.

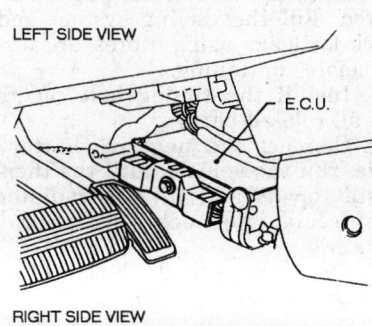

LEFT SIDE VIEW

E.C.U.

RIGHT SIDE VIEW

E.C.U.

DIAGNOSTIC MODE SELECTOR

RED L.E.D.

289652

Powertrain control module location — G20 models

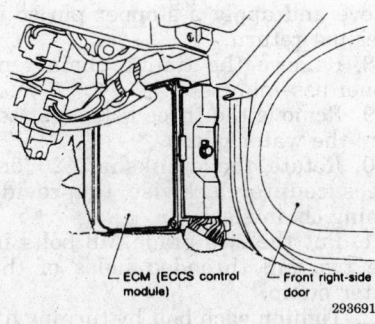

ECM (ECCS control module)

Front right-side door

293691

Powertrain control module location — J30 and Q45 models

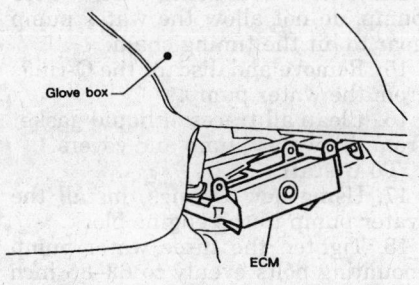

Glove box

ECM

287631

Powertrain control module location — I30 models

3. Disconnect the wiring harness connector.

4. Remove the module.

To install:

───── **WARNING** ─────
Before installing a new powertrain control module, perform input/output signal inspection at the powertrain control module connector terminals to prevent damage to the new module.

5. Install the powertrain control module in the proper position.

6. Connect the harness connector and tighten the securing bolt until the gap between the orange indicators disappears.

ENGINE COOLING

Radiator

REMOVAL AND INSTALLATION

G20 Models

1. Disconnect the negative battery cable.

2. Drain the coolant system, remove the upper radiator hose and reservoir tank.

3. Remove the lower radiator hose and transmission cooler lines.

4. Unplug the radiator fan motor connector and remove the radiator fan.

5. Remove all radiator attaching bolts and remove the radiator.

To install:

6. Lower the radiator into position. Take care not to damage the radiator fins as this will affect cooling efficiency.

7. Install all attaching bolts and tighten securely.

8. Install the radiator fan and reconnect the radiator fan motor connector.

9. Install the radiator upper and lower hoses, and the reservoir tank.

10. Fill the cooling system, start the engine and allow it to reach normal operating temperature. Bleed the cooling system and check for leaks.

J30 Models

1. Disconnect the negative battery cable.

2. Drain the coolant.

3. Remove the engine undercover.

4. Disconnect the radiator upper and lower hoses. Disconnect the overflow tank hose.

5. Disconnect and plug the automatic transmission oil cooler hoses.

6. Remove the radiator lower shroud.

7. Remove the radiator.

To install:

8. Install the radiator and tighten the mounting brackets to 4.6–6.1 ft. lbs. (6.3–8.3 Nm).

9. Connect the automatic transmission oil cooler hoses.

10. Connect the radiator upper and lower hoses.

11. Connect the overflow tank hose.

12. Fill the system with coolant.

13. Install the engine undercover. Start the engine and allow it to reach operating temperature. Check for leaks.

I30 Models

1. Disconnect the negative battery cable.

2. Remove the cap and drain the radiator via the drain plug in the bottom tank.

3. Disconnect upper and lower radiator hoses.

4. Disconnect the fan electrical connectors. Remove the fan/shroud assembly from the radiator.

5. If equipped with an automatic transaxle disconnect and plug the transaxle cooling lines at the radiator.

6. Disconnect the overflow hose from the radiator.

7. Remove the nuts that secure the radiator support brackets.

8. Remove the radiator from the vehicle.

To install:

NOTE: Be sure the rubber mounting bushings are in position before the radiator is installed.

9. Install the radiator into the vehicle.

10. Position the support brackets and tighten the mounting nuts to 33–44 inch lbs. (3.7–4.7 Nm).

11. Install and position the fan/shroud assembly in the vehicle and connect it to the radiator. Tighten the mounting bolts to 33–44 inch lbs. (3.7–4.7 Nm).

12. Connect the electrical connectors to the cooling fans.

13. Using new clamps, connect the upper and lower radiator hoses. Connect the overflow hose to the radiator.

14. If equipped with an automatic transaxle, connect the oil cooler lines

to the radiator. Check and/or refill the transaxle.

15. Refill the cooling system with 50/50 antifreeze/water mix and bleed the system.

16. Start the engine and check the cooling system for leaks.

Q45 Models

1. Disconnect the negative battery cable.

2. Remove the engine undercover. Drain the coolant.

3. Remove the upper hose and coolant reserve tank hose from the radiator.

4. Unbolt the shroud and move it backward in order to remove the fan and coupling. Remove the fan to water pump bolts and remove the fan, coupling, water pump pulley and shroud.

5. Raise the vehicle and support safely. Remove the lower hose from the radiator.

6. Disconnect and plug the automatic transmission cooler hoses. Disconnect the coolant thermo switch. Lower the vehicle.

7. Remove the mounting bolts from the support and carefully lift out of the engine compartment.

8. Remove the cooling fans from the radiator.

To install:

9. Lower the radiator into position.

10. Install the mounting bolts.

11. Raise the vehicle and support safely. Connect the automatic transmission cooler lines and the thermo switch connector.

12. Connect the lower hose. Lower the vehicle.

13. Install the shroud, pulley, coupling and fan. Torque the water pump pulley nuts to 7 ft. lbs. (10 Nm). Adjust the belt.

14. Connect the upper hose and coolant reserve tank hose.

15. Connect the negative battery cable, run the vehicle until the thermostat opens, fill the radiator completely and check the automatic transmission fluid level. Recheck for coolant leaks.

16. Once the vehicle has cooled, recheck the coolant level.

Water Pump

REMOVAL AND INSTALLATION

2.0L (SR20DE) Engine

1. Disconnect the negative battery cable.

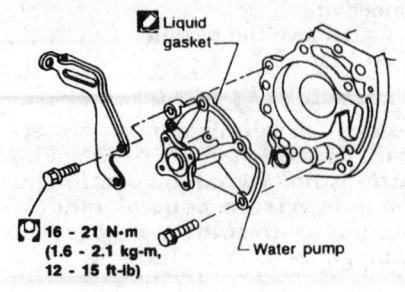

Liquid gasket

16 - 21 N·m
(1.6 - 2.1 kg-m,
12 - 15 ft-lb)

Water pump

291119

Water pump mounting — 2.0L engine

2. Drain the coolant from the radiator and engine block. The drain plug in the engine block is located at the left front of the cylinder block.

3. Remove the right wheel and the engine side cover.

4. Remove the drive belts.

5. Remove the front engine mount.

6. Loosen the water pump attaching bolts and remove the water pump. Take care not to drip coolant on the drive belts.

To install:

7. Clean all mating surfaces and place a 2–3mm bead of liquid gasket on the water pump mating surface.

8. Install the water pump and tighten the bolts to 12–15 ft. lbs. (16–21 Nm).

9. Install and tighten drive belts.

10. Install the front engine mount.

11. Install the engine side cover and the right wheel.

12. Using a radiator tester or equivalent, check the system for leaks.

13. Connect the negative battery cable.

14. Refill with coolant and bleed the system of air.

3.0L (VG30DE) Engine

1. Disconnect the negative battery cable.

2. Drain the coolant from the radiator and from the drain plugs on both sides of the cylinder block.

3. Remove the cooling fan assembly. Remove the timing belt covers.

NOTE: Use the proper precautions to avoid getting coolant on the timing belt.

4. Remove the water pump mounting bolts and remove the pump from the engine.

To install:

5. Thoroughly clean and dry the mating surfaces, bolts and bolt holes.

6. Apply liquid gasket to the water pump and install to the engine. Tor-

que the bolts to 12–15 ft. lbs. (16–21 Nm).

7. Open the air release plug, as required. Fill the cooling system and check for leaks using a pressure tester before continuing.

8. Install the timing belt covers and all related parts.

9. Connect the negative battery cable, run the vehicle until the thermostat opens and fill the radiator completely. Recheck for coolant leaks.

10. Once the vehicle has cooled, recheck the coolant level.

3.0L (VQ30DE) Engine

1. Disconnect the negative battery cable.

2. Drain the coolant from the plugs on the radiator and both sides of the engine block.

3. Position a jack under the oil pan for support. Be sure to place a block of wood on the jack for protection to the engine parts.

4. Remove the right side engine mount and engine mounting bracket.

5. Remove the drive belts and the idler pulley bracket.

6. Remove the chain tensioner cover and the water pump cover.

7. Push the timing chain tensioner sleeve and apply a stopper pin so it does not return.

8. Remove the timing chain tensioner assembly.

9. Remove the three bolts that secure the water pump.

10. Rotate the crankshaft 20 degrees counterclockwise to provide timing chain slack.

11. Put the two grade M8 bolts in the two M8 threaded holes of the water pump.

12. Tighten each bolt by turning alternately ½ turn until they reach the timing chain rear case. Be sure to turn each bolt ½ turn at a time to prevent damage.

13. Lift up the water pump and remove it.

14. When removing the water pump, do not allow the water pump gear to hit the timing chain.

15. Remove and discard the O-rings from the water pump.

16. Clean all traces of liquid gasket from the water pump and covers.

To install:

17. Using new O-rings, install the water pump to the engine block.

18. Tighten the three water pump mounting bolts evenly to 62–86 inch lbs. (7–10 Nm).

19. Rotate the crankshaft pulley to its original position by turning it 20 degrees clockwise.

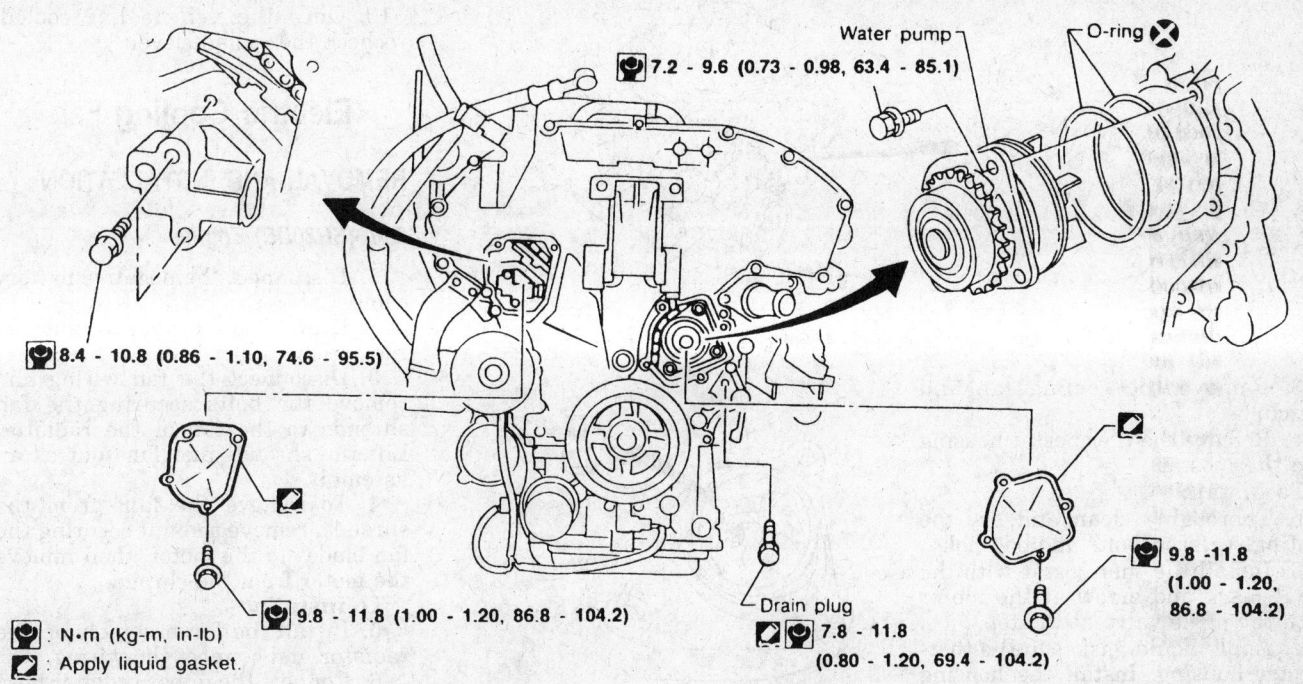

Water pump and timing cover assembly — 3.0L (VQ30DE) engine

20. Install the timing chain tensioner and tighten the mounting bolts to 75–96 inch lbs. (9–10 Nm).
21. Remove the stopper pin from the timing chain tensioner.
22. Apply a continuous 0.091–0.130 inch (2.3–3.3mm) bead of liquid sealant to the mating surfaces of the timing chain tensioner and water pump covers.
23. Install the timing chain tensioner and water pump covers to the engine block. Tighten the cover mounting bolts to 84–108 inch lbs. (10–13 Nm).
24. Install the drive belts and the idler pulley bracket.
25. Install the right side engine mounting bracket and the engine mount.
26. Remove the jack from under the engine and install the drain plugs to the cylinder block.
27. Connect the negative battery cable and refill the cooling system.
28. Start the engine, bleed the cooling system, and check for leaks.

4.5L (VH45DE) Engine

1. Disconnect the negative battery cable.
2. Drain the coolant from the radiator and from the drain cocks on both sides of the cylinder block.

3. Remove the drive belts.
4. Unbolt the fan shroud and move it backward in order to remove the fan and coupling. Remove the fan to water pump bolts and remove the fan, coupling, water pump pulley and shroud.
5. Remove all necessary accessories to gain access to the water pump.
6. Note the positioning of the clamp and disconnect the hose from the water pump.
7. Remove the water pump mounting bolts and remove the pump from the engine.
To install:
8. Thoroughly clean and dry the mating surfaces, bolts and bolt holes.
9. Apply liquid gasket to the water pump and install it to the engine. Torque the bolts to 12–15 ft. lbs. (16–21 Nm).
10. Connect the hose and install the clamp in the same position as when it was removed. Fill the cooling system and check for leaks using a pressure tester before continuing.
11. Install all removed accessories.
12. Install the shroud, pulley, coupling and fan. Torque the water pump pulley nuts to 7 ft. lbs. (10 Nm). Install and adjust all the belts.
13. Connect the negative battery cable, run the vehicle until the thermostat opens and fill the radiator

completely. Recheck for coolant leaks.
14. Once the vehicle has cooled, recheck the coolant level.

Thermostat

REMOVAL AND INSTALLATION

2.0L (SR20DE) Engine

1. Drain the engine coolant.
2. Remove the lower radiator hose.
3. Remove the water inlet, then remove the thermostat.
To install:
4. Install the new thermostat with the air bleeder or jiggle valve facing upward.
5. Clean all mating surfaces and apply a 2–3mm bead of liquid gasket to the water inlet.
6. Install the water pump inlet and tighten bolts to 5–6 ft. lbs. (6–8 Nm).
7. Install the lower radiator hose, refill and bleed the coolant system and check for leaks.

3.0L (VG30DE) Engine

1. Disconnect the negative battery cable. Drain the cooling system.
2. Disconnect the upper radiator hose from the thermostat housing.

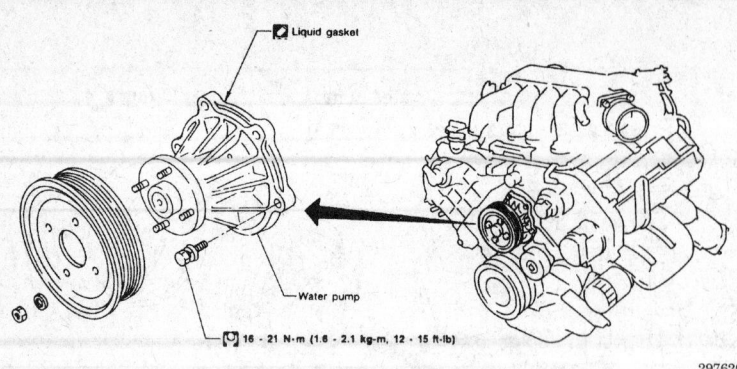

16 - 21 N·m (1.6 - 2.1 kg-m, 12 - 15 ft-lb)

297620

Water pump assembly — 4.5L engine

3. Remove the cooling fan and clutch.

4. Remove the thermostat housing and thermostat.

To install:

5. Thoroughly clean and dry the mating surfaces, bolts and bolt holes.

6. Install the thermostat with the **UPR** mark and arrow at the top or with the jiggle valve at the top.

7. Apply liquid gasket to the thermostat housing. Install the housing and torque the bolts to 12–15 ft. lbs. (16–21 Nm). Install cooling fan assemble, if necessary.

8. Open the air release plug, as required. Fill the cooling system.

9. Connect the negative battery cable, run the vehicle until the thermostat opens and fill the radiator completely. Recheck or coolant leaks.

10. Once the vehicle has cooled, recheck the coolant level.

3.0L (VQ30DE) Engine

1. Disconnect the negative battery cable.

2. Drain the cooling system from the radiator and both sides of the engine block.

3. Disconnect the radiator hose from the water inlet housing.

4. Remove the water inlet housing mounting nuts. Separate the water inlet housing from the engine.

5. Remove the thermostat from the thermostat housing.

To install:

6. Clean the gasket mounting surfaces.

7. Install the thermostat into the thermostat housing with the thermostat spring facing toward the engine. Be sure to position the jiggle valve facing upward.

8. Using a new gasket, install the water inlet housing and torque the water inlet housing mounting nuts to 74–99 inch lbs. (9–11 Nm).

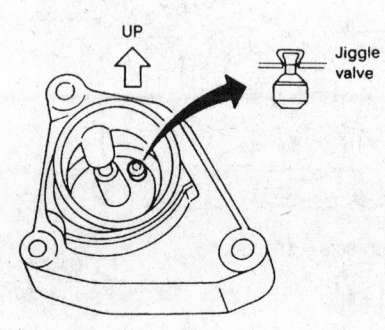

UP

Jiggle valve

297214

Positioning of the thermostat jiggle valve — 3.0L (VQ30DE) engine

9. Connect the radiator hose to the water inlet housing. Be sure to use a new clamp.

10. Refill the cooling system.

11. Connect the negative battery cable.

12. Start the engine, bleed the cooling system, and allow it to reach normal operating temperatures. After cooling, check for leaks and recheck the coolant level.

4.5L (VQ45DE) Engine

1. Disconnect the negative battery cable. Drain the cooling system.

2. Remove the front ornament cover.

3. Disconnect the upper hose from the coolant inlet.

4. Remove the water inlet and thermostat.

To install:

5. Thoroughly clean and dry the mating surfaces, bolts and bolt holes.

6. Install the thermostat with the jiggle valve at the top.

7. Apply liquid gasket to the inlet and to the bolts. Install and torque the bolts to 14 ft. lbs. (19 Nm).

8. Fill the cooling system.

9. Connect the negative battery cable, run the vehicle until the thermostat opens and fill the radiator

completely. Recheck for coolant leaks.

10. Once the vehicle has cooled, recheck the coolant level.

Electric Cooling Fan

REMOVAL AND INSTALLATION

2.0L (SR20DE) Engine

1. Disconnect the negative battery cable.

2. Drain about 1 qt. of coolant and remove the upper radiator hose.

3. Disconnect the fan wiring and remove the bolts securing the fan shrouds to the top of the radiator. Lift the shrouds and fans out as an assembly.

4. To remove the fans from the shrouds, remove the nut securing the fan blades to the motor, then remove the motor from the shroud.

To install:

5. Install the fan assembly on the radiator and connect the wiring.

6. Connect the upper radiator hose and refill the engine with coolant.

7. Connect the negative battery cable.

3.0L (VG30DE) and 4.5L (VH45DE) Engines

1. While the belts are tight, loosen the four fan clutch mounting nuts.

2. Remove the belts on the water pump pulley.

3. Remove the fan clutch mounting nuts and remove the cooling fan with the clutch.

4. Remove the fan from the clutch as needed.

To install:

5. Install the fan on the clutch.

6. Install the clutch on the water pump studs and torque the nuts to 52–87 in lbs. (6–10 Nm).

7. Install and adjust the drive belts.

3.0L (VQ30DE) Engine

1. Disconnect the negative battery cable.

2. Remove the radiator cap and drain the radiator via the drain plug in the bottom tank.

3. Remove the upper radiator hose.

4. Disconnect the fan electrical connectors. Remove the fan/shroud assembly from the radiator.

To install:

5. Install and position the fan/shroud assembly in the vehicle and connect it to the radiator.

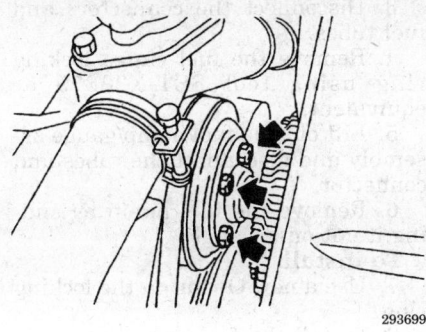

Fan clutch mounting studs — 3.0L (VG30DE) and 4.5L engines

Tighten the mounting bolts to 3–4 ft. lbs. (3.7–4.7 Nm).

6. Connect the electrical connectors to the cooling fans.

7. Using new clamps, install the upper hose.

8. Refill the cooling system with 50/50 antifreeze/water mix and bleed the system.

9. Connect the negative battery cable.

10. Run engine and check the cooling system for leaks and verify the operation of the cooling fans.

Cooling System Bleeding

PROCEDURE

2.0L (SR20DE) Engine

1. Set the heater temperature control lever to MAX hot position. Remove the radiator cap, air relief plug (located at the thermostat housing) and the air bleeder cap (located near the heater core).

2. Refill the reservoir bottle to the MAX line. Reinstall the air relief plug when coolant spills from the hole. Reinstall the air bleeder cap.

3. Install a steel wire between the negative pressure valve and the seat of the radiator cap. Install the cap and warm the engine to normal operating temperature.

4. Run the engine at 2500 rpm for 10 seconds and return to idle. Repeat this 2–3 times. Turn the engine off and allow the engine to cool.

5. Remove the radiator cap and check the coolant level. If necessary refill the radiator with coolant up to the filler neck.

6. Remove the radiator cap and remove the steel wire. Install the cap and warm the engine and check for the sound of coolant flow with engine running from idle to 4000 rpm. If a

sound is heard, bleed air from the cooling system as follows:

a. Cool engine and remove the air bleeder cap on the heater inlet hose.

b. Attach a suitable transparent hose at the air bleeder pipe and put the opposite end of the hose into the coolant reservoir.

c. Install the radiator cap with the steel wire inserted and check for proper connection of all coolant related hoses.

d. Start the engine and check for bubbles in the reservoir tank.

e. Set the heater control lever to MAX cool and run the engine up to 2300 rpm until the bubbles disappear in the hose.

f. After bubbles disappear, set the heater control lever to MAX hot and listen for coolant system sound. If sound is heard, perform the previous steps.

g. After all air has been bled from the system, remove the steel wire from the radiator cap, remove the transparent hose, install the air bleeder cap and check the coolant reservoir to ensure it is full.

3.0L (VG30DE) and 4.5L (VH45DE) Engines

The Q45 and J30 use a thermostat which is equipped with a jiggle valve. This valve automatically bleeds air as the system is being filled. The cooling system requires no further bleeding.

NOTE: Pour coolant into the filler neck slowly to allow air in system to escape.

3.0L (VQ30DE) Engine

1. Disconnect the negative battery cable.

2. With the coolant drained, close the radiator drain plug and install the engine drain plugs.

3. Slowly fill the radiator with the proper mixture of coolant and water.

4. Fill the reservoir and install the radiator cap.

5. Connect the negative battery cable.

6. Start the engine and allow it to reach normal operating temperature, watch the coolant temperature gauge for signs of overheating. Race the engine two or three times with no-load.

7. Shut the engine **OFF** and let it cool down completely.

8. After the engine has cooled down remove the radiator cap and fill the radiator to the filler opening. Fill the reservoir to the **H** level.

9. Check the radiator drain plug and the engine drain plugs for leakage.

10. Install the radiator cap.

FUEL SYSTEM

Fuel System Service Precautions

Safety is the most important factor when performing not only fuel system maintenance but any type of maintenance. Failure to conduct maintenance and repairs in a safe manner may result in serious personal injury or death. Maintenance and testing of the vehicle's fuel system components can be accomplished safely and effectively by adhering to the following rules and guidelines.

• To avoid the possibility of fire and personal injury, always disconnect the negative battery cable unless the repair or test procedure requires that battery voltage be applied.

• Always relieve the fuel system pressure prior to disconnecting any fuel system component (injector, fuel rail, pressure regulator, etc.), fitting or fuel line connection. Exercise extreme caution whenever relieving fuel system pressure to avoid exposing skin, face and eyes to fuel spray. Please be advised that fuel under pressure may penetrate the skin or any part of the body that it contacts.

• Always place a shop towel or cloth around the fitting or connection prior to loosening to absorb any excess fuel due to spillage. Ensure that all fuel spillage (should it occur) is quickly removed from engine surfaces. Ensure that all fuel soaked cloths or towels are deposited into a suitable waste container.

• Always keep a dry chemical (Class B) fire extinguisher near the work area.

• Do not allow fuel spray or fuel vapors to come into contact with a spark or open flame.

• Always use a backup wrench when loosening and tightening fuel line connection fittings. This will prevent unnecessary stress and torsion to fuel line piping. Always follow the proper torque specifications.

• Always replace worn fuel fitting O-rings with new. Do not substitute fuel hose or equivalent, where fuel pipe is installed.

Fuel System Pressure

RELIEVING

1. Remove the fuel pump fuse.
2. Start the engine.
3. Allow the engine to run until it stalls.
4. After the engine stalls, crank the engine two or three times to release the remaining fuel pressure.
5. Turn the ignition switch **OFF**. Reinstall the fuel pump fuse into the fuse block.

NOTE: Do not crank the engine or turn the ignition switch ON after the fuel pump fuse has been reinstalled, or the fuel pressure will be reestablished.

Idle Speed

ADJUSTMENT

On all engines, the idle speed mixture is computer controlled and is not adjustable. If the idle speed is incorrect, other problems with the engine/engine control system may exist.

Air/Fuel Mixture

ADJUSTMENT

On all engines, the air/fuel mixture is computer controlled according to the needs of the engine and is not adjustable. If the air/fuel mixture is not correct, other problems with the engine/engine control system exist.

Fuel Filter

REMOVAL AND INSTALLATION

All Engines

———— **CAUTION** ————
Do not use conventional fuel filters, hoses or clamps when servicing this fuel system. They are not compatible with the high pressures of the injection system and could fail, causing personal injury. Use only components specifically designed for fuel injection.

1. Relieve the fuel system pressure.
2. Disconnect the negative battery cable.
3. Disconnect the fuel hoses from the fuel filter, located at the right side of the engine compartment.

4. Remove the filter mounting screws and remove from the vehicle.
To install:
5. Inspect all hoses and clamps for damage of any type. Replace parts, as required.
6. The fuel filters are directional and should be installed with the arrow facing the direction of fuel flow.
7. Install a new filter in the bracket and install new hose clamps
8. Connect the negative battery cable.
9. Start vehicle and check for fuel leaks.

NOTE: On some vehicles, a code will be set and/or the check engine light will remain on after starting the vehicle. This is because a code was set for an open fuel pump circuit when the fuel pressure was released. If you did not disconnect the negative battery cable during this procedure, do it now so the code will be erased. The negative battery cable should be disconnected for at least 1 minute. Also, remember to reset the clock and radio stations when finished.

Fuel Pump

REMOVAL AND INSTALLATION

2.0L (SR20DE) Engine

———— **CAUTION** ————
Fuel injection systems remain under pressure after the engine has been turned OFF. Properly relieve fuel pressure before disconnecting any fuel lines. Failure to do so may result in fire or personal injury.

1. Release the fuel system pressure.
2. Remove the inspection hole cover located beneath the rear seat.

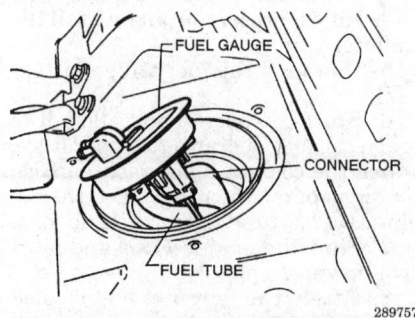

Fuel pump — 2.0L engine

3. Disconnect the connectors and fuel tubes.
4. Remove the fuel gauge locking ring using tool SST-X38879 or equivalent.
5. Lift out the fuel pump/gauge assembly and disconnect the tubes and connector.
6. Remove the fuel pump by sliding it out on an angle.
To install:
7. Use a new O-ring on the locking ring.
8. Install the fuel pump/gauge assembly and attach all fuel lines and connectors.
9. Using tool SST-X38879 or equivalent, tighten the locking ring to 22–26 ft. lbs. (30–35 Nm).
10. Install the inspection cover and test fuel system pressure at the injectors.

3.0L (VG30DE) and 4.5L (VH45DE) Engines

1. Relieve the fuel system pressure.
2. Disconnect the negative battery cable.
3. Remove the trunk front finish panel.
4. Disconnect the wiring harness connector and fuel tubes. Remove the fuel tank sender unit attaching bolts. Remove the fuel tank sender and discard the O-ring.
5. Remove the fuel pump from the sender unit.
To install:
6. Install the new fuel pump on the sender unit assembly.
7. Using a new O-ring, install the sender unit in the fuel tank. Tighten the bolts to 1.4–1.9 ft. lbs. (2–2.5 Nm).
8. Connect the wiring harness connectors and fuel tubes.
9. Install the trunk room finish panel.
10. Connect the negative battery cable, start the engine and check for leaks.

3.0L (VQ30DE) Engine

1. Relieve the fuel system pressure

———— **CAUTION** ————
Fuel injection systems remain under pressure after the engine has been turned OFF. Properly relieve fuel pressure before disconnecting any fuel lines. Failure to do so may result in fire or personal injury.

2. Disconnect the negative battery cable.

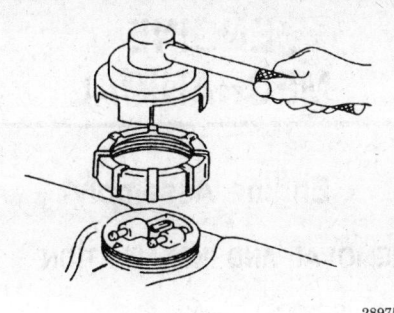

Locking ring removal and installation using special tool SST-X38879 — 2.0L engine

3. Remove the access panel under the rear seat.

NOTE: If the vehicle has no fuel pump access cover, the fuel tank must be lowered or removed to gain access to the in-tank fuel pump.

4. Disconnect the fuel gauge electrical connector and pump electrical connector.

5. Disconnect the fuel outlet and the return hoses. If necessary, remove the fuel tank.

6. Remove the fuel pump assembly-to-fuel tank bolts and lift the fuel pump assembly from the fuel tank. Discard the O-ring. Plug the fuel tank opening with a clean rag to prevent dirt from entering the system.

NOTE: When removing or installing the fuel pump assembly, be careful not to damage or deform it and always install a new O-ring.

To install:

7. Remove the rag; using a new O-ring, install fuel pump assembly into the fuel tank. Torque the bolts to 1.4–1.9 ft. lbs. (2.0–2.5 Nm). If removed, install the fuel tank assembly

8. Connect the fuel lines and the electrical connectors. Always use new clamps when reconnecting fuel line hoses

9. Install the fuel pump access cover.

10. Connect the negative battery cable.

11. Start the engine and check for fuel leaks.

NOTE: On some models, the Check Engine Light will stay ON after installation is completed. The memory code in the control unit must be erased. This code is stored for an open fuel pump circuit, this is caused when the fuel pressure is released. To erase the code, disconnect the battery cable for 10 seconds then reconnect after installation of fuel pump.

Fuel Injector

REMOVAL AND INSTALLATION

2.0L (SR20DE) Engine

————— CAUTION —————
Fuel injection systems remain under pressure after the engine is turned OFF. Properly relieve fuel pressure before disconnecting any fuel lines. Failure to do so may result in fire or personal injury.

1. Relieve fuel system pressure.
2. Disconnect the negative battery cable.

NOTE: Do not remove injector by pinching connector.

3. Disconnect injector harness connectors.
4. Disconnect vacuum hose from pressure regulator.
5. Disconnect fuel hoses from fuel tube assembly.
6. Remove injectors with fuel tube assembly.
7. To remove injector, push out of the fuel tube assembly.

NOTE: Do not remove injector by pinching connector.

To install:

8. Replace or clean injector as necessary.
9. Install injector on fuel tube assembly using a new O-ring and insulator. Lubricate O-rings with clean engine oil.
10. Install injectors with fuel tube assembly onto intake manifold. Torque fuel tube assembly bolts to 7–8 ft. lbs. (9–10 Nm), and then to 15–20 ft. lbs. (21–26 Nm).
11. Install fuel hoses, lubricating them with oil.
12. Connect the injector harness connector.
13. Connect the negative battery cable, start the engine and check for leaks.

3.0L (VG30DE) Engine

————— CAUTION —————
Fuel injection systems remain under pressure after the engine is turned OFF. Properly relieve fuel pressure before disconnecting any fuel lines. Failure to do so may result in fire or personal injury.

1. Relieve fuel system pressure.
2. Disconnect the negative battery cable.

NOTE: Do not remove injector by pinching connector.

3. Disconnect injector harness connectors.
4. Disconnect vacuum hose from pressure regulator.
5. Disconnect fuel hoses from fuel tube assembly.
6. Remove injectors with fuel tube assembly.
7. To remove injector, push out of the fuel tube assembly.

To install:

8. Replace or clean injector as necessary.
9. Install injector on fuel tube assembly using a new O-ring and insulator. Lubricate O-rings with oil.
10. Install injectors with fuel tube assembly onto intake manifold. Torque fuel tube assembly bolts to 7–8 ft. lbs. (9–10 Nm), and then to 15–20 ft. lbs. (21–26 Nm).
11. Install fuel hoses, lubricating them with oil.
12. Connect the injector harness connector.
13. Connect the negative battery cable, start the engine and check for leaks.

3.0L (VQ30DE) Engine

1. Release the fuel system pressure.

————— CAUTION —————
The fuel injection system remains under pressure after the engine has been turned OFF. Properly relieve the fuel pressure before disconnecting any fuel lines. Failure to do so may result in fire or personal injury.

2. Disconnect the negative battery cable and drain the cooling system.
3. Remove the throttle body coolant hoses.
4. Label and disconnect the electrical connectors from the throttle position sensor.
5. Label and disconnect the hoses from the throttle body, the EGR valve, intake manifold collector, IAC valve, and the fuel pressure regulator.
6. Disconnect the canister purge hose and blow-by hose.
7. Disconnect the EGR guide tube.
8. Disconnect the accelerator cable from the throttle body.
9. Remove the intake manifold collector support brackets.
10. Disconnect the right side electrical connectors from the ignition coils.

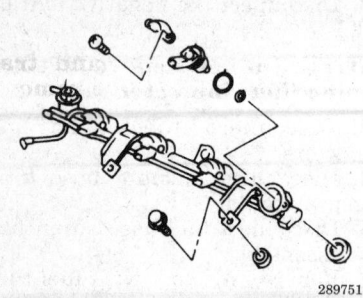

289751

**Fuel injectors and tube assembly —
2.0L engine**

11. If necessary, disconnect the electrical connector from the crank angle sensor and the power transistor.

12. Remove the intake manifold collector-to-intake manifold bolts/nuts and remove the intake manifold collector.

13. Remove the fuel injector assembly by performing the following procedures:

a. Disconnect the electrical connectors from the fuel injectors.

b. Disconnect the fuel lines from the fuel injector assembly.

c. Remove the fuel rail-to-cylinder head bolts.

d. Remove the fuel rail assembly from the engine.

14. To remove the fuel injector from the fuel rail, remove the fuel injector-to-fuel rail bolts and remove the fuel injector from the fuel rail; discard the O-rings.

To install:

15. To install the fuel injector to the fuel rail, perform the following procedures:

a. Install new O-rings onto the fuel injector.

b. Wet the new O-rings with fuel and press the injector into the fuel rail.

c. Install the bolts and tighten the fuel injector retainer to 27–33 inch lbs. (2.9–3.8 Nm).

16. Clean the gasket mounting surfaces.

17. Install the fuel injector assembly by performing the following procedures:

a. Install the fuel rail assembly to the engine.

b. Install the fuel rail-to-cylinder head bolts and torque the bolts to 15–20 ft. lbs. (21–26 Nm) in two progressive steps.

c. Connect the fuel lines to the fuel injector assembly.

d. Connect the electrical connectors to the fuel injectors.

18. Using a new gasket, install the intake manifold collector and torque the intake manifold collector-to-intake manifold bolts/nuts to 13–16 ft. lbs. (18–22 Nm).

19. Install the intake manifold collector supports and torque the bolts to 14–18 ft. lbs. (20–25 Nm).

20. Connect the electrical connectors to the ignition coils and tighten the mounting bolts to 27–33 inch lbs. (2.9–3.8 Nm)

21. Connect the accelerator cable to the throttle body.

22. Connect the EGR guide tube and tighten the bolts to 15–20 ft. lbs. (21–26 Nm) in two progressive steps.

23. Connect the canister purge hose and blow-by hose.

24. Connect the hoses to the throttle body, EGR valve, intake manifold collector, IAC valve, and the fuel pressure regulator.

25. If disconnected, connect the electrical connectors to the crank angle sensor and the power transistor.

26. Connect the electrical connectors to the throttle position sensor.

27. Connect the throttle body coolant hoses.

28. Refill the cooling system and connect the negative battery cable.

29. Start the engine, bleed cooling system, and check for leaks.

4.5L (VH45DE) Engine

1. Relieve the fuel system pressure. Disconnect the negative battery cable.

2. Drain the coolant.

3. Remove the EGR control valve.

4. Remove the intake manifold collector.

5. The injectors can be removed separately. Disconnect the harness connector(s) from the fuel injector(s).

6. Remove the injector(s) from the injector tube assembly. Do not reuse the O-rings.

To install:

7. Using new O-rings, install the injector(s) to the injector tube.

8. Connect the harness connector(s).

9. Install the intake manifold collector.

10. Install the EGR control valve.

11. Fill the cooling system.

12. Connect the negative battery cable and check for leaks.

ENGINE MECHANICAL

Engine Assembly

REMOVAL AND INSTALLATION

2.0L (SR20DE) Engine

1. Disconnect the negative battery cable.

2. Raise and support the vehicle safely.

3. Remove the engine undercover.

4. Matchmark the hood with the hood hinges and remove.

5. Drain the coolant from both the cylinder block and radiator.

6. Drain the engine and transaxle oil.

—————— **CAUTION** ——————
Fuel injection systems remain under pressure after the engine has been turned OFF. Properly relieve fuel pressure before disconnecting any fuel lines. Failure to do so may result in fire or personal injury.
————————————————————————

7. Release fuel system pressure and remove fuel line.

8. Label and remove all vacuum lines and wiring harness connectors.

9. Remove exhaust tubes, ball joints, and driveshafts.

10. Remove the radiator and fans.

11. Remove the drive belts.

12. Remove the alternator, A/C compressor, and the power steering pump from the engine and lay them aside. Do not disconnect the compressor or power steering pump lines.

13. Support the engine with a hoist and the transaxle with a suitable jack. Raise the engine and transaxle slightly and remove the center member.

14. Remove the engine mounting bolts from both sides and slowly lower the hoist and transaxle jack.

15. Remove the engine and transaxle from beneath the vehicle.

To install:

16. Install the center member bracket (manual transmission) on the engine, if removed. Ensure that all insulators are correctly positioned on the brackets. Torque insulator through bolts to 32–41 ft. lbs. (43–55 Nm).

17. If equipped with manual transaxle, ensure that the distance between the center of the insulator

through bolt and the center member is 2.28–2.36 in. (58–60mm). Torque through bolt to 46–58 ft. lbs. (62–78 Nm).

18. Carefully install the engine and torque the center member-to-frame bolts to 57–72 ft. lbs. (77–98 Nm).

19. Install the alternator, A/C compressor, and power steering pump.

20. Connect all vacuum hoses and wiring harness connectors. Connect the fuel line.

21. Install the exhaust tubes, ball joints, driveshafts, the radiator and fans and drive belts.

22. Fill the cooling system and fill the crankcase and transaxle with the proper oil.

23. Install the engine undercover and hood.

24. Connect the negative battery cable and road test the vehicle for proper operation.

3.0L (VG30DE) Engine

––––––– **CAUTION** –––––––
Fuel injection systems remain under pressure after the engine has been turned OFF. Properly relieve fuel pressure before disconnecting any fuel lines. Failure to do so may result in fire or personal injury.

1. Relieve the fuel system pressure.

2. Disconnect the negative battery cable.

3. Remove the engine undercover. Remove the hood after matchmarking it to the hinges.

4. Drain the cooling system. Use the cylinder block drain plugs to drain the engine.

5. Label and disconnect all the vacuum hoses, fuel tubes, wires, harnesses and connectors from the engine.

6. Remove the driveshaft.

7. Remove the radiator.

8. Remove the drive belts, cooling fan, and the coupling.

9. Raise and support the vehicle safely.

10. Remove the power steering pump, alternator, air conditioner compressor, and the starter motor from the engine.

11. Remove the exhaust tube front nuts. Remove the front tubes after removing the exhaust tube bracket.

12. Remove the fluid filler pipe from the transmission.

13. Remove the cooler pipes from the transmission.

14. Remove the control linkage from the selector lever.

15. Disconnect the inhibitor switch and solenoid harness connectors.

16. Remove the gusset securing the transmission to the engine.

17. Remove the crankshaft position sensor to prevent it from being damaged.

18. Remove the bolts securing torque converter to the driveplate.

19. Support the engine and separate engine from transmission.

20. Lift the engine slightly and remove the engine mounting bolts from both sides.

21. Carefully remove the engine from the top side of the vehicle.

To install:

22. Lower the engine into the vehicle and torque the engine mounting bolts to 32–41 ft. lbs. (43–55 Nm).

23. Install the transmission. Torque the converter bolts to 33–43 ft. lbs. (44–59 Nm). Torque the engine-to-transmission bolts as follows:

a. Torque the upper six bolts to 29–36 ft. lbs. (39–49 Nm).

b. Torque the lower four bolts to 22–29 ft. lbs. (29–39 Nm).

24. Install the crankshaft position sensor.

25. Connect the inhibitor switch and solenoid harness connectors.

26. Install the control linkage on the selector lever.

27. Install the fluid cooler pipes on the transmission.

28. Install the fluid filler pipe on the transmission.

29. Install the front exhaust tubes and exhaust tube bracket. Torque the nuts to 33–44 ft. lbs. (45–60 Nm).

30. Install the power steering pump, alternator, air conditioner compressor, and the starter motor.

31. Install the drive belts, cooling fan, and the coupling. Adjust the accessory drive belts.

32. Install the radiator.

33. Install the driveshaft.

34. Connect all the vacuum hoses, fuel tubes, wires, harnesses and connectors to the engine.

35. Check and fill the cooling system, engine oil, transmission fluid and power steering fluid.

36. Install the engine hood and undercover.

37. Connect the negative battery cable.

38. Start the engine and allow it to reach normal operating temperature. Make any necessary adjustments and check for leaks. Road test the vehicle for proper operation.

3.0L (VQ30DE) Engine

It is recommended the engine and transaxle be removed as a single unit. If necessary, the units may be separated after removal.

NOTE: The engine and transaxle assembly must be removed from the under side of the vehicle.

1. Matchmark the hood hinge relationship and remove the hood.

2. Release the fuel system pressure and disconnect the negative battery cable.

3. Raise and safely support the vehicle.

4. Drain the coolant from the cylinder block and the radiator. Drain the crankcase and the automatic transaxle, if equipped.

5. Remove the air cleaner, the air intake tube, the air flow meter and disconnect the throttle linkage.

6. Disconnect and/or remove the following:

• Drive belts
• Engine ground cable
• Electrical connector from the crank angle sensor
• Engine electrical harness connectors

––––––– **CAUTION** –––––––
The fuel injection system remains under pressure after the engine has been turned OFF. Properly relieve the fuel pressure before disconnecting any fuel lines. Failure to do so may result in fire or personal injury.

• Fuel feed and fuel return hoses
• Upper and lower radiator hoses
• Heater inlet and outlet hoses
• Engine vacuum hoses
• Carbon canister hoses
• Any interfering engine accessory: power steering pump, air conditioning compressor, and the alternator

7. Remove the carbon canister.

8. Remove the auxiliary fan, washer tank, and the radiator (with the fan assembly).

9. If equipped with a manual transaxle, remove the clutch release cylinder from the clutch housing.

10. If equipped with a manual transaxle, disconnect the shift control rod and disconnect the shift support rod.

11. If equipped with a automatic transaxle, disconnect the control cable from the transaxle.

12. Install engine slingers to the block and connect a suitable lifting device to the slingers. Do not tension the lifting device at this point.

13. Disconnect the exhaust pipe at both the manifold connections and remove the front exhaust pipe from the vehicle.

14. If equipped with a manual transaxle, drain the transaxle gear oil.

15. Support the engine and transaxle assembly with proper jack.

16. Disconnect the right and left side halfshafts from their side flanges and remove the bolt holding the radius link support.

17. Lower the shifter and selector rods and remove the bolts from the motor mount brackets. Remove the nuts holding the front and rear motor mounts to the frame.

18. Remove the center crossmember assembly from the vehicle.

19. Lower the engine/transaxle assembly down and onto an engine stand.

To install:

20. Raise the engine/transaxle assembly into the vehicle. When raising the engine onto the mounts, make sure to keep it as level as possible.

21. Check the clearance between the frame and clutch housing and make sure the engine mount bolts are seated in the groove of the mounting bracket.

22. Remove the transaxle and engine jack assembly.

23. Install the center crossmember and secure the engine mounting bolts.

24. If removed, raise the shifter and selector rods to their normal operating positions and secure them with the mounting bolts.

25. Install the halfshafts.

26. Connect the exhaust pipe assembly.

27. Disconnect and remove the engine slingers.

28. If equipped with a automatic transaxle, connect the control cable to the transaxle.

29. If equipped with a manual transaxle, install the clutch release cylinder to the clutch housing.

30. Install the auxiliary fan, washer tank, and the radiator (with the fan assembly).

31. Install the carbon canister.

32. Connect and/or install the following:

• Any interfering engine accessory: power steering pump, air conditioning compressor, and the alternator
• Carbon canister hoses
• Engine vacuum hoses
• Heater inlet and outlet hoses
• Upper and lower radiator hoses
• Fuel feed and fuel return hoses
• Engine electrical harness connectors

• Electrical connector from the crank angle sensor
• Engine ground cable
• Drive belts

33. Connect the throttle linkage, then install the air cleaner, the air flow meter, and the air intake duct.

34. Fill the transaxle, the engine, and the cooling system to the proper levels with the appropriate fluids.

35. Install the hood and connect the negative battery cable.

36. Make all the necessary engine adjustments. Charge the air conditioning system, if discharged. Road test the vehicle for proper operation.

4.5L (VH45DE) Engine

────── **CAUTION** ──────
Fuel injection systems remain under pressure after the engine has been turned OFF. Properly relieve fuel pressure before disconnecting any fuel lines. Failure to do so may result in fire or personal injury.

1. Disconnect the negative battery cable.

2. Relieve the pressure from the fuel system.

3. Mark the relation of the hood to the hinge brackets and remove the hood.

4. Raise and safely support the vehicle.

5. Remove the engine splash shield.

6. Drain the coolant and the engine oil.

7. Disconnect the transmission cooler lines from the radiator.

8. Remove the radiator hoses and remove the radiator and shroud.

9. Tag and disconnect all the vacuum hoses, fuel lines, and electrical connectors.

10. Disconnect the exhaust pipes from the exhaust manifolds.

11. Mark the position of the driveshaft on the flanges and remove the driveshaft.

12. Remove the accessory drive belts.

13. Remove the alternator, air conditioning compressor, and the power steering pump. Do not disconnect the coolant or hydraulic lines from the compressor and pump.

14. Remove the lower steering joint.

15. Remove the sway bar, transverse link, and the tension rod with bracket.

16. Place a suitable jack under the transmission and attach a hoist to the engine.

17. Disconnect the transmission rear mount.

18. Remove the suspension member attaching bolts.

19. Remove the engine mounting bolts.

20. Attach a suitable hoist to the engine. Lower the transmission jack and the hoist and lower the engine and transmission from under the vehicle.

To install:

21. Install the engine and transmission into position. Check the clearance between the frame and transaxle and make sure the engine mount bolts are seated in the groove of the mounting bracket.

22. Torque the engine mounts to the following specifications:
   a. Front engine mount bracket-to-engine bolts to 32–41 ft. lbs. (43–55 Nm).
   b. Front engine mount-to-frame bolts to 41–49 ft. lbs. (55–67 Nm).
   c. Rear engine mount-to-crossmember bolts to 16–21 ft. lbs. (22–28 Nm).
   d. Rear crossmember-to-frame bolts to 32–41 ft. lbs. (43–55 Nm).

23. Install the suspension member bolts.

24. Remove the hoist and the transmission jack.

25. Install the sway bar, transverse link, and the tension rod with bracket.

26. Install the lower steering joint.

27. Install the alternator, air conditioning compressor, and the power steering pump.

28. Install and adjust the accessory drive belts.

29. Install the radiator and shroud. Install the radiator hoses.

30. Install the driveshaft, aligning the marks, that were made during the removal procedure.

31. Connect the exhaust pipes to the exhaust manifolds.

32. Connect all the electrical connectors, fuel lines, and vacuum hoses.

33. Install the transmission cooling lines and install the engine splash shield.

34. Fill the crankcase with the proper type of engine oil to the required level. Fill the cooling system with the proper type and quantity of coolant.

35. Install the hood, aligning the marks that were made during the removal procedure.

36. Connect the negative battery cable, start the engine and check for leaks. Road test the vehicle for proper operation.

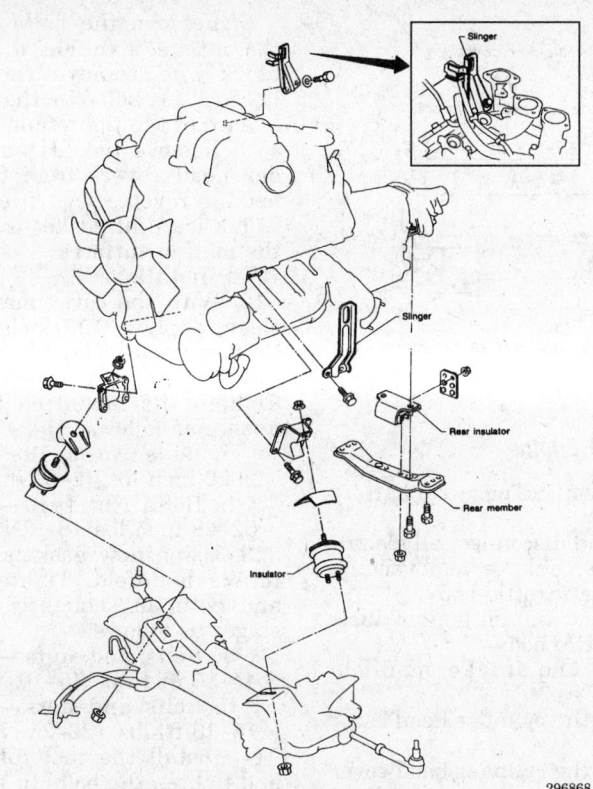

Engine mounts and brackets — 4.5L engine

Engine Mounts

REMOVAL AND INSTALLATION

2.0L (SR20DE) Engine

1. Disconnect the negative battery cable.
2. Matchmark the engine mount to its frame mounting location.
3. Raise the vehicle and support safely. Using the proper equipment, support the weight of the engine.
4. Inspect all mounts to determine which is defective. A defective mount will have the rubber portion of the mount separated from the metal backing or stud.
5. Remove all bolts and nuts that attach the mount to the engine, transaxle or frame and remove the mount assembly from the vehicle.
6. Remove the through bolt and separate the insulator from the bracket, as required.
To install:
7. Position the mount between the engine and frame and install the mounting bolts.
8. If equipped with manual transaxle, ensure that the distance between the center of the front mounting bracket through bolt and the center member is 2.28–2.36 in.

(58–60mm). Torque through bolt to 46–58 ft. lbs. (62–78 Nm).
9. Torque center member bolts to 57–72 ft. lbs. (77–98 Nm); front and rear engine mount to 32–41 ft. lbs. (43–55 Nm).

3.0L (VG30DE) Engines

1. Disconnect the negative battery cable.
2. Raise and support the vehicle safely.
3. Attach a hoist to the engine and lift until the slack in the chain is taken up.
4. Remove the nuts from the engine mounts.

NOTE: Inspect the engine compartment for components that may bind when the engine is raised. Disconnect these components.

5. Lift the engine only as needed to remove the engine mount. Do not lift any higher.
6. Remove the engine mounts.
To install:
7. Install the engine mounts.
8. Lower the engine and torque the engine mount-to-engine bolts to 30–38 ft. lbs. (41–52 Nm). Torque the engine mount-to-frame nuts to 32–41 ft. lbs. (43–55 Nm).

9. Remove the engine hoist and lower the vehicle.
10. Connect the negative battery cable.

3.0L (VQ30DE) Engine

1. Disconnect negative battery cable.
2. Raise and support the vehicle.

— **WARNING** —
Make certain vehicle is properly supported when raising engine.

3. Raise the engine slightly to take the tension off the motor engine mounts.

NOTE: When raising the engine to replace the engine mounts, DO NOT place the jack directly under the oil pan or the crankshaft torsional damper. Use a block of wood between engine and jack to prevent damage at jacking point.

4. Remove the through bolt nut(s) and through bolt(s).
5. Remove the engine mount-to-engine block bolts.
6. Raise the engine enough to remove the engine mount(s) and remove the engine mount(s).

NOTE: Only raise the engine enough to remove the mounts. Raising the engine too far could result in damage to some engine components.

To install:
7. Install the engine mount(s) and engine mount-to-engine block bolts. Tighten the bolts to correct specification.
8. Lower the engine enough to install the engine mount through bolt(s) and nut(s). Tighten the through bolts to correct specification.
9. Lower the vehicle and connect the negative battery cable.

4.5L (VH45DE) Engine

1. Disconnect the negative battery cable.
2. Matchmark the engine mount to its frame mounting location.
3. Raise the vehicle and support safely, if necessary. Using the proper equipment, support the weight of the engine.
4. Inspect all mounts to determine which is defective. A defective mount will have the rubber portion of the mount separated from the metal backing or stud.
5. Remove all bolts and nuts that attach the mount to the engine,

transmission or frame and remove the mount assembly from the vehicle.

6. Remove the through bolt and separate the insulator from the bracket, as required.

To install:

7. Position the insulator to the bracket and install the through bolt.

8. Attach the mount to the engine and install the nuts or bolts and tighten.

9. Torque bolts as follows:

Engine mount-to-frame bolt to 41–49 ft. lbs. (55–67 Nm)

Engine mount-to-engine bolt to 32–41 ft. lbs. (43–55 Nm)

Transmission mount-to-crossmember bolt to 16–21 ft. lbs. (22–28 Nm)

Crossmember-to-frame bolt to 32–41 ft. lbs. (43–55 Nm).

Cylinder Head Cover

REMOVAL AND INSTALLATION

2.0L (SR20DE) Engine

1. Disconnect the negative battery cable.

2. Remove the air duct to the intake manifold.

3. Label and remove the vacuum hoses, fuel hoses and wire harness connectors around the cylinder head cover.

4. Remove all the spark plugs, the AIV valve and resonator.

5. Remove the cylinder head cover by loosening the cover bolts in 2 to 3 steps, in the opposite sequence of tightening.

6. Remove the cylinder head cover and oil separator. Clean all oil and gasket material from both mating surfaces.

To install:

7. With the cylinder head cover clean, apply a continuous bead of liquid gasket to the mating surface.

8. Install the cover and oil separator. Tighten the cylinder cover bolts in sequence:

a. Tighten nuts 1, 10, 11 and 8 in that order to 3 ft. lbs. (4 Nm).

b. Tighten nuts 1 through 13 in proper sequence to 6–7 ft. lbs. (8–10 Nm).

9. Install the spark plugs, AIV valve and the resonator.

10. Connect the lines, vacuum hoses and wiring connectors.

11. Install the intake manifold and air duct.

12. Connect the negative battery cable.

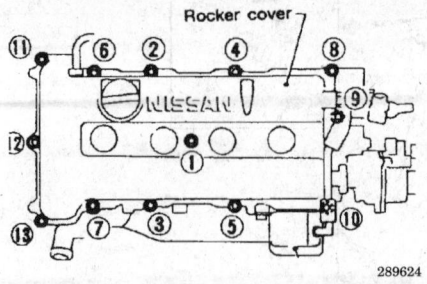

Cylinder head cover bolt torque sequence — 2.0L engine

3.0L (VG30DE) Engine

1. Disconnect the negative battery cable.

2. Label and disconnect all electrical connectors, linkage and vacuum hoses from the throttle body.

3. Disconnect the intake air ducts from the throttle body.

4. Remove the intake manifold collector.

5. Remove the cylinder head cover ornament.

6. Remove the cylinder head cover bolts, and lift the valve cover off.

7. Clean all gasket material from the mating surfaces.

To install:

8. Install the cylinder head cover bolts.

9. Install the intake manifold collector and air duct.

10. Connect all electrical connectors and vacuum hoses.

11. Connect the negative battery cable.

3.0L (VQ30DE) Engine

1. Disconnect the negative battery cable.

2. Remove the left side cylinder head cover ornament.

NOTE: Before disconnecting any hoses or connectors, note the locations for reassembly.

3. Remove the air duct to intake manifold hose, collector hose, blow-by hose, and vacuum hoses.

4. Remove the fuel hoses and disconnect the harness connections.

5. Disconnect the canister purge hoses.

6. Disconnect and remove all six ignition coils from the spark plugs. Remove the spark plugs.

7. Remove the bolts that secure the EGR tube and remove the tube.

8. Remove the intake manifold collector supports and remove the collector.

9. Remove the bolts that secure the intake manifold to the engine block and remove the manifold. Loosen the bolts in the reverse sequence of the tightening procedure.

10. Remove the LH and RH cylinder head covers from the cylinder head in reverse sequence.

11. Clean all gasket material from the mating surfaces.

To install:

12. With the cover mating surface clean, apply a 0.12 inch (3mm) continuous bead of liquid gasket to the rocker covers and install the covers. Tighten the mounting bolts in sequence as follows:

a. Bolts No. 1–10 — Tighten to 9–26 inch lbs. (1–3 Nm)

b. Bolts No. 1–10 — Tighten to 52–69 inch lbs. (6–8 Nm)

13. Using new gaskets, install the intake manifold. Tighten the nuts and bolts in sequence and in two stages as follows:

a. Bolts and nuts — Tighten to 44–86 inch lbs. (5–10 Nm)

b. Bolts and nuts — Tighten to 16–18 ft. lbs. (22–25 Nm)

14. Install the fuel tube assembly and tighten the bolts to 15–20 ft. lbs. (21–26 Nm).

15. Install the intake manifold collector assembly and tighten the mounting bolts to 16–18 ft. lbs. (22–25 Nm). Install the intake manifold collector support brackets.

16. Install the spark plugs and the ignition coils.

17. Install the rocker cover ornament.

18. Install the water hoses to the cylinder head and intake manifold.

19. Connect the canister purge hoses.

20. Connect the fuel hoses and wiring harness connections to the fuel rail.

21. Connect the negative battery cable.

4.5L (VH45DE) Engine

1. Disconnect the negative battery cable.

2. Remove the cylinder head cover ornament.

3. Remove the fuel hoses and disconnect the harness connections.

4. Disconnect the canister purge hoses.

5. Disconnect and remove all ignition coils from the spark plugs. Remove the spark plugs.

6. Remove the bolts that secure the EGR tube and remove the tube.

7. Remove the cylinder head covers from the cylinder head in reverse sequence.

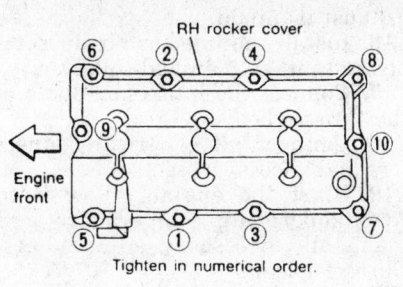

Right cylinder head cover torque sequence — 3.0L (VQ30DE) engine

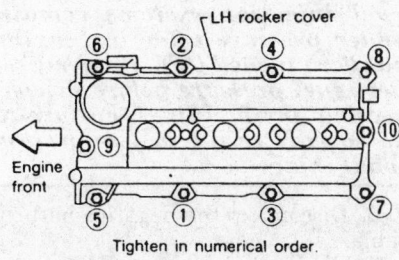

Left cylinder head cover torque sequence — 3.0L (VQ30DE) engine

Cylinder head cover bolt torque sequence — 4.5L engine

8. Clean all gasket material from the mating surfaces.

To install:

9. Make sure all mating surfaces are clean before installation.

10. Install the cylinder head covers, tightening the bolts in sequence.

11. Instal the EGR tube.

12. Position and secure each spark plug and ignition coil.

13. Install the rocker cover ornament.

14. Connect the fuel hoses and wiring harness connections to the fuel rail.

15. Connect the negative battery cable.

Cylinder Head

REMOVAL AND INSTALLATION

2.0L (SR20DE) Engine

CAUTION

Fuel injection systems remain under pressure after the engine has been turned OFF. Properly relieve fuel pressure before disconnecting any fuel lines. Failure to do so may result in fire or personal injury.

1. Relieve the fuel system pressure and disconnect the negative battery cable.

2. Drain the coolant and remove the radiator.

3. Remove the right front wheel and engine side cover.

4. Remove the drive belts, water pump pulley, alternator and power steering pump and bracket and oil filter bracket.

5. Label and remove the vacuum hoses, fuel hoses and wire harness connectors.

6. Remove the cylinder head cover.

7. Remove the intake manifold supports.

8. Set No. 1 piston at TDC on the compression stroke. Rotate the camshaft until the alignment marks are at the 11 o'clock and 1 o'clock positions.

NOTE: Ensure that the left camshaft key is at 12 o'clock and the right camshaft key is at 10 o'clock.

9. Remove the chain tensioner and distributor, timing chain guide, camshaft sprockets, brackets, oil tubes and baffle plate.

10. Remove the starter motor and water pipe bolt.

11. Loosen the cylinder head bolts in 2–3 steps.

12. Remove the cylinder head with the intake and exhaust manifolds attached.

To install:

13. Apply liquid gasket to the top of the chain cover where it meets the cylinder block before installing the head gasket.

14. Install the gasket and cylinder head on the block.

NOTE: Cylinder head bolts may be reused providing the dimension from the bottom of the head to the end of the bolt does not exceed 6.228 in. (158.2mm). If the dimension exceeds the specification, replace the cylinder head bolts.

15. Torque cylinder head bolts as follows:

a. Torque all bolts to 29 ft. lbs. (39 Nm) using the proper sequence.

b. Torque all bolts to 58 ft. lbs. (78 Nm) using the proper sequence.

c. Loosen all bolts completely.

d. Torque all bolts to 25–33 ft. lbs. (34–44 Nm) using the proper sequence.

e. Torque all bolts 90–100 degrees.

f. Torque all bolts an additional 90–100 degrees.

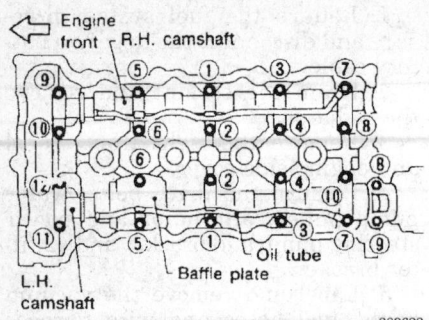

Camshaft bracket (bearing cap) bolt torque sequence — 2.0L engine

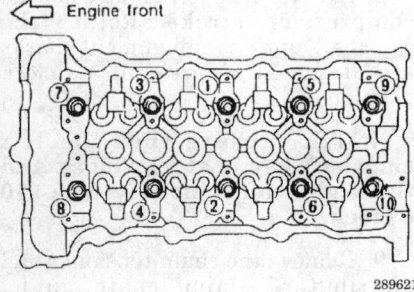

Cylinder head bolt torque sequence — 2.0L engine

16. Install the water pipe bolt and starter motor.

17. Install the camshafts, camshaft brackets, oil tubes and baffle plate. Ensure that the camshaft keys are at 12 o'clock and 10 o'clock.

18. The procedure for tightening camshaft bolts must be followed exactly to prevent camshaft damage. Torque bolts as follows:

a. Torque the right camshaft bolts No. 9 and No.10 to 18 inch lbs. (2 Nm). Torque bolts 1 through 8 to the same amount.

b. Torque the left camshaft bolts No. 11 and No. 12 to 18 inch lbs. (2 Nm). Torque bolts 1 through 10 to the same amount.

c. Torque all bolts in sequence to 54 inch lbs. (6 Nm).

d. Torque all bolts in sequence again. Torque type A, B, and C bolts to 6.5–8.5 ft. lbs. (9–12 Nm) and type D bolts to 13–19 ft. lbs. (18–25 Nm).

19. Line up the mating marks on the timing chain and camshaft sprockets and install the sprockets. Torque the sprocket bolts to 101–116 ft. lbs. (137–157 Nm).

20. Install the timing chain guide, distributor, chain tensioner, oil filter bracket and power steering oil pump bracket.

21. Install the intake manifold supports.

22. Install the cylinder head cover and oil separator.

23. Install the AIV, spark plugs, power steering pump, alternator water pump pulley and drive belts, air duct to the intake manifold and the radiator.

24. Install all vacuum and fuel hoses and reconnect all electrical connections.

25. Install the engine side cover, right front wheel and engine undercover.

26. Refill the cooling system.

27. Connect the negative battery cable, start the engine, and check for leaks.

3.0L (VG30DE) Engine

CAUTION
Fuel injection systems remain under pressure after the engine has been turned OFF. Properly relieve fuel pressure before disconnecting any fuel lines. Failure to do so may result in fire or personal injury.

1. Relieve the fuel system pressure and disconnect the negative battery cable.

2. Drain the cooling system.

3. Label and disconnect all electrical connectors, linkage and vacuum hoses from the throttle body.

4. Disconnect the intake air ducts from the throttle body.

5. Disconnect the fuel injector connectors and remove the injector pipe assembly.

6. Remove the cylinder head covers, timing belt, idler pulley and the mounting bolt.

7. Remove the intake manifold.

8. Disconnect the exhaust pipe from the exhaust manifold.

9. Loosen the cylinder head bolts in several steps.

10. Remove the cylinder head with the exhaust manifold attached.

To install:

11. Set the No. 1 piston at TDC on its compression stroke. Ensure that the crankshaft sprocket alignment mark is aligned with the one on the oil pump body and the camshaft sprocket alignment mark is aligned with the one on the timing belt rear cover.

12. Fit the cylinder head into place with a new gasket. Install the cylinder head bolts with washers.

13. Torque the cylinder head bolts in sequence.

14. Install the cylinder head covers, timing belt and intake manifold.

15. Install the exhaust pipe on the exhaust manifold.

16. Install the injector pipe assembly and intake manifold collector.

17. Connect the intake air ducts to the throttle body.

18. Connect all electrical connectors and vacuum hoses.

19. Start the engine, allow it to reach operating temperature and make any necessary adjustments. Check for leaks.

3.0L (VQ30DE) Engine

1. Relieve the fuel system pressure.

CAUTION
Fuel injection systems remain under pressure after the engine has been turned OFF. Properly relieve fuel pressure before disconnecting any fuel lines. Failure to do so may result in fire or personal injury.

2. Disconnect the negative battery cable.

3. Drain the engine oil and the cooling system.

NOTE: Before disconnecting any hoses or connectors, note the locations for reassembly.

4. Remove the air duct to intake manifold hose, collector hose, blow-by hose, and vacuum hoses.

5. Remove the fuel hoses and disconnect the harness connections.

6. Disconnect the canister purge hoses.

7. Disconnect and remove all six ignition coils from the spark plugs.

8. Remove the bolts that secure the EGR tube and remove the tube.

9. Remove the intake manifold collector supports and remove the collector. Remove the manifold from the cylinder head.

10. Remove the cylinder head covers from the cylinder head.

11. Remove the right front wheel and engine side covers.

12. Remove the drive belts and idler pulley followed by the power steering oil pump belt and remove the power steering oil pump assembly.

13. Remove the camshaft position sensor (PHASE) and crankshaft position sensors (REF)/(POS).

14. Set the No. 1 piston to TDC of compression stroke by rotating the crankshaft.

15. Remove the ring gear cover access plate and remove the crankshaft pulley bolt.

16. Remove the timing chain tensioner and slack side chain guide.

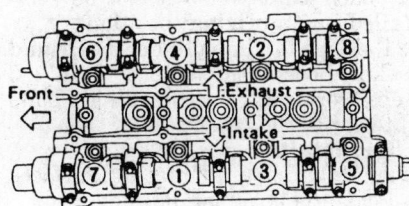

Right cylinder head

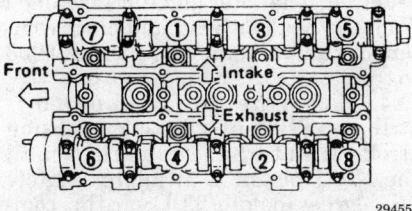

Left cylinder head

Cylinder head bolt torque sequence — 3.0L (VG30DE) engine

17. Remove the engine oil pan.
18. Remove the camshaft sprockets first. Be sure to hold the flats of the camshafts while removing the sprocket bolts.
19. Loosen the camshaft bearing caps in several steps. The bearing caps MUST be loosened in sequence.

NOTE: Keep all bearing caps and camshafts in proper order for reinstallation.

20. Remove the cylinder head bolts in sequence.
To install:
21. Turn the crankshaft until the No. 1 piston is set 240 degrees before TDC on compression stroke.

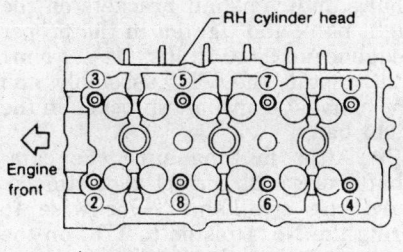

Right cylinder head loosening sequence — 3.0L (VQ30DE) engine

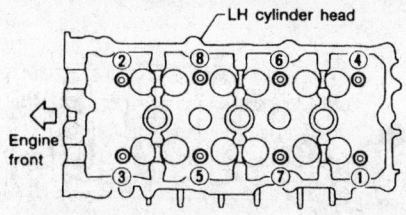

Left cylinder head loosening sequence — 3.0L (VQ30DE) engine

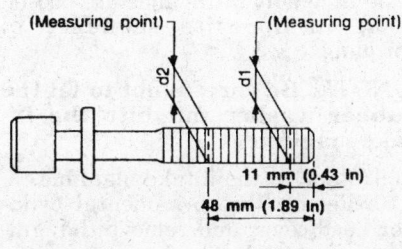

Measuring the cylinder head bolts — 3.0L (VQ30DE) engine

22. Using new head gaskets, install the cylinder heads.

NOTE: If possible, replacement of the head bolts is suggested.

23. If replacement of the head bolts is not possible, perform the following bolt measurement:
   a. Measure the diameter of the head bolt 0.43 inches (11mm) from the bottom of the bolt.
   b. Measure the diameter of the head bolt 1.89 inches (48mm) from the bottom of the bolt.
   c. Whenever the size difference between the two measurements exceeds 0.0043 inches (0.11mm) the head bolts must be replaced.
24. Install the cylinder head bolts and tighten in sequence.
25. Install the camshaft tensioners. Tighten the tensioner mounting bolts to 75–96 inch lbs. (8.4–10.8 Nm).

NOTE: The camshafts can be identified by the paint marks on the camshaft. The left cylinder head camshafts have a YELLOW paint mark and the right cylinder head camshafts have a WHITE paint mark.

NOTE: When installing the camshafts, position the camshaft keys at the 12 o'clock position in respect to the cylinder head angle.

26. Install the camshafts and install the bearing caps.
27. Install new O-rings to the front of the engine block.
28. Install the crankshaft sprocket with the mating mark facing out.
29. Rotate the crankshaft clockwise and position the crankshaft to TDC of compression stroke and align the dowels of the camshaft sprockets to the 12 o'clock position.
30. Install the lower chain guide on the dowel pin with the front mark on the guide facing upward.
31. Install the timing chains and sprockets to the intake camshafts. Be sure to align the timing chain and sprocket mating marks.
32. Remove the left and right camshaft tensioner stopper pins.
33. Align the mating mark on the crankshaft with the matchmark (gold link) on the lower timing chain.
34. Attach the lower timing chain to the water pump sprocket.
35. Working counterclockwise, install the lower timing chain camshaft sprockets. Be sure to align the sprocket marks with the blue links of the timing chain during installation.
36. Install the intake sprocket bolts and tighten to 88–95 ft. lbs. (119–128 Nm). Be sure to secure the camshafts while tightening the bolts.
37. Install the timing chain guide, upper timing chain guide, lower timing chain tensioner and slack side timing chain guide.
38. Install the timing cover evenly and gently. Be sure to align the dowel pin holes. Tighten the mounting bolts in sequence.
39. Install the front exhaust pipe and its support.
40. Install the A/C compressor and bracket.
41. Install the crankshaft pulley to the crankshaft and install the mounting bolt.
42. Tighten the mounting bolt to 14–22 ft. lbs. (20–29 Nm). Tighten the crankshaft bolt an additional 60–66 degrees clockwise. This is about the angle from one hexagon bolt head corner to another.
43. Install the ring gear cover plate.
44. Install the camshaft position sensor (PHASE) and crankshaft position sensors (REF)/(POS).
45. Install the power steering pump assembly.
46. Install the drive belts and the idler pulley.

47. Install the right front inner wheel cover and install the right front wheel.

48. Install the engine undercovers.

49. Using new gaskets, install the intake manifold. Tighten the nuts and bolts in sequence.

50. Install the intake manifold collector gasket with the arrow facing forward.

51. Install the intake manifold collector assembly and tighten the mounting bolts to 16–18 ft. lbs. (22–25 Nm).

52. Install the intake manifold collector support brackets.

53. Using new gaskets, install the EGR tube and tighten the mounting bolts to 15–20 ft. lbs. (21–26 Nm) in two progressive steps.

54. Install the spark plugs.

55. Install the ignition coils and tighten the mounting bolts to 27–33 inch lbs. (2.9–3.8 Nm).

56. Install the cylinder head cover ornament on the left side.

57. Install the water hoses to the cylinder head and intake manifold.

58. Connect the canister purge hoses.

59. Connect the fuel hoses and wiring harness connections to the fuel rail.

60. Install and connect the air duct to intake manifold hose, collector hose, blow-by hose, and vacuum hoses.

61. Refill the engine crankcase and cooling system with the proper type and amount of fluid.

62. Connect the negative battery cable.

63. Start the engine and run at 3000 RPM under no load to purge the air from the high pressure chamber. The engine may produce a rattling noise. This indicates that air still remains in the chamber and is not a matter of concern.

64. Verify that there are no leaks.

4.5L (VH45DE) Engine

1. Disconnect the negative battery cable.

2. Remove the engine and transmission assembly from the vehicle.

3. Remove the suspension member and engine mounts from the engine.

4. Remove the air compressor bracket and the exhaust manifolds.

5. Remove the cooling fan with coupling and the engine gusset.

6. Separate the engine from the transmission and mount the engine on a suitable workstand.

7. Remove the oil pan. Remove the intake collector.

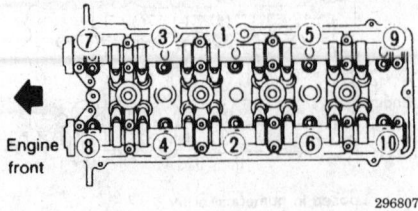

Cylinder head torque sequence — 4.5L engine

8. Disconnect the injector harness connector and remove the injector tube assembly with injector. Loosen bolts in opposite sequence of torquing..

NOTE: Be careful not to let the rubber washer fall into the intake manifold.

9. Remove the intake manifold.

10. Remove the ornamental cylinder head cover and remove the ignition coils and spark plugs.

11. Bring the No. 1 piston to TDC on the compression stroke.

12. Use a suitable puller to remove the crankshaft pulley.

13. Remove the cylinder heads cover.

14. Remove the crank angle sensor and the Valve Timing Control (VTC) solenoid.

15. Remove the chain tensioners and the upper front covers.

16. Remove the front timing chain cover.

NOTE: The timing chain will not be disengaged or dislocated from the crankshaft sprocket unless the front cover is removed. The cast portion of the front cover is located on the lower side of the crankshaft sprocket so the timing chain is not disengaged from the sprocket.

17. Remove the VTC assembly and the camshaft sprocket.

18. Remove the oil pump chain and the timing chains.

NOTE: Do not attempt to disassemble the VTC assembly since they are difficult to reassemble accurately in the field. If it should be disassembled, the VTC assembly must be replaced with a new one.

19. Remove the camshaft brackets in the reverse order of torquing sequence. Use 2–3 steps. Remove the camshafts. Mark the parts so they

can be reinstalled in their original positions.

20. Remove the rocker arm and hydraulic lash adjuster. Be sure to identify each adjuster so it can be reinstalled in it's original position.

21. Remove the cylinder head and gasket. Loosen the head bolts in 2–3 steps working from the outside bolts in towards the center bolts.

To install:

22. Make sure all mating surfaces are clean before installation.

23. Check the cylinder head surface for warpage using a feeler gauge and a suitable straightedge. If the cylinder head is warped more than 0.004 in. (0.1mm), it must be resurfaced or replaced. The total amount machined from the head or head and block combined, cannot total more than 0.008 in. (0.2mm).

24. Make sure the No. 1 piston is still at TDC of the compression stroke, then turn the crankshaft until the No. 1 piston is at approximately 45 degrees before TDC on the compression stroke. At this point, the No. 3 piston will be at the same height as the No. 1 piston to prevent interference of the valves and pistons.

25. Install the cylinder heads with new gaskets. Temporarily tighten the cylinder head bolts to avoid damaging the cylinder head gaskets. Be sure to install washers between the bolts and the cylinder heads. Do not rotate the crankshaft or camshaft separately or the valves will hit the pistons.

26. Install the hydraulic lash adjusters and check them as follows:

   a. When the rocker arm can be moved at least 0.04 in. (1.0mm) by pushing at the hydraulic lash adjuster location, it indicates that there is air in the high pressure chamber The adjuster will have to be bled.

NOTE: Air cannot be bled from the lash adjusters by running the engine.

27. Install the rocker arms, camshafts and camshaft brackets on the right bank and tighten in the proper sequence to 9–10 ft. lbs. (12–14 Nm).

28. Install the VTC assembly and the exhaust camshaft sprocket on the right bank.

29. After making sure the camshafts are still correctly positioned, turn the crankshaft clockwise to bring the No. 1 piston to TDC on the compression stroke.

30. Install the timing chain on the right bank, aligning the mating marks on the chain with those on the crankshaft and camshaft sprockets.

31. Install the chain tensioner on the right bank.

32. Turn the crankshaft approximately 120 degrees clockwise from the point where the No. 1 piston is at TDC on the compression stroke. At this point, the valves on the left bank still remain closed.

33. Correctly position the camshafts and tighten brackets in the proper sequence to 9–10 ft. lbs. (12–14 Nm). Install the VTC assembly and the exhaust cam sprocket.

34. Install the timing chain on the left bank, aligning the mating marks on the chain with those on the crankshaft and camshaft sprockets.

35. Install the oil pump chain and sprockets.

36. Install the oil pump chain guides. Place a 0.04 in. (1.0mm) feeler gauge between the upper chain guide and chain before assembling the chain guides. The force applied to the chain is equivalent to the upper chain guide weight.

37. Apply suitable sealer and install the front covers.

38. Install the chain tensioner for the left bank.

39. Apply suitable sealer to the rubber plugs and install them on the cylinder head.

40. Install the crank angle sensor, VTC solenoid, rocker cover and crank pulley.

41. Bring the piston in No. 1 cylinder to TDC on the compression stroke.

42. Tighten the cylinder head bolts in the proper torque sequence.

43. Install the intake manifold bolts in their proper positions on the cylinder head and lightly tighten the mounting bolts.

44. Connect the injector tube assemblies, including the fuel injectors, to the intake manifolds and lightly tighten the mounting bolts.

NOTE: Be careful not to let the rubber washer fall into the intake manifold.

45. Install the intake collector and lightly tighten the mounting bolts.

46. Tighten the intake manifold mounting bolts at the cylinder head, remove the intake collectors and tighten the intake manifolds to 12–15 ft. lbs. (16–21 Nm).

47. Tighten the sub-fuel tubes, in sequence, first to 3.1–4.3 ft. lbs. (4.2–5.9 Nm) and then to 6.2–8.0 ft. lbs. (8.4–10.8 Nm).

48. Tighten the injector tube assemblies, in sequence, first to 6.9–8.0 ft. lbs. (9.3–10.8 Nm) and then to 15–20 ft. lbs. (21–26 Nm).

49. Install the intake collectors and tighten to 9–11 ft. lbs. (12–15 Nm).

50. Install the exhaust manifolds.

51. Install the cylinder head covers and tighten in the proper sequence to 5–7 ft. lbs. (7–10 Nm).

52. Install all other remaining components. Join the engine and transmission and install the assembly in the vehicle.

Lash Adjusters

BLEEDING

2.0L (SR20DE) and 4.5L (VH45DE) Engines

NOTE: The hydraulic lash adjuster is a precision component that relies on a clean operating environment. When handling the lash adjuster, make certain that no dirt or foreign particles are allowed to get inside the unit. DO NOT try to take the unit apart as it is not rebuildable. The adjuster is filled with engine oil. Hold the adjuster in the upright position so that the oil does not spill out. When cleaning the unit, only use clean engine oil.

To bleed the hydraulic lash adjusters when disassembling the rocker arms or as part of an engine overhaul, perform the following:

1. Place the adjuster in a container of clean engine oil. While pressing down on the plunger insert a thin rod in the lash adjuster to move the check ball off of the seat and allow the air to escape. Bleeding is completed when the plunger no longer moves.

2. Remove the air bleed rod wire from the adjuster and push down firmly on the plunger. If the plunger moves even slightly, repeat the bleeding procedure until the plunger stops

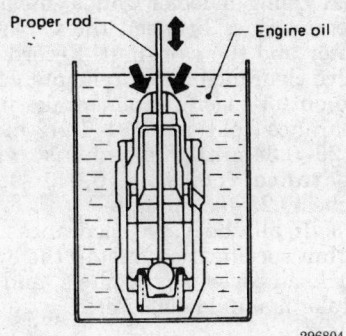

Bleeding air from the hydraulic lash adjusters

moving. If the plunger continues to move, replace the lash adjuster.

NOTE: Air can't be bled from this type of adjuster by running the engine.

3.0L (VG30DE and VQ30DE) Engines

When removed, the hydraulic lash adjusters should be immersed in a container of clean engine oil in the upright position.

Valve Lash

ADJUSTMENT

2.0L (SR20DE) Engine

NOTE: A special gauge plate and collar will be needed to complete this procedure.

1. Remove the camshafts.

2. Install the J38957-1 gauge plate to the cylinder head. Use the bolts supplied in the kit to secure the plate to the cam bearing journals.

3. Install the collar J38957-2 on the dial indicator. Make sure the dished side of the collar is toward the gauge and tighten the set screw.

4. Place the gauge on the No. 1 intake valve (shim side). Make sure the shim has been removed. Place the tip of the dial gauge on the top of the valve stem and the collar on the gauge plate. Zero the dial gauge.

5. Move the dial gauge to the other intake valve (rocker guide side). Place the tip of the dial gauge on the rocker guide and the collar of the gauge plate. Record the measurement.

6. Select the correct size shim using the chart. Shims are available in 17 different sizes ranging from 0.1102 in. (2.800 mm) to 0.1260 in. (3.200 mm) in increments of 0.001 in. (0.025 mm).

3.0L (VG30DE) and 4.5L (VH45DE) Engines

No valve lash adjustment is possible on this engine. If valve noise occurs, look for excessive wear on the camshaft, rocker arm and auto lash adjuster.

3.0L (VQ30DE) Engine

NOTE: Check and adjust the valve clearances while the engine is cold and not running.

Checking

1. Remove the intake manifold collector.

Available shim

Thickness mm (in)	Identification mark
2.800 (0.1102)	28 00
2.825 (0.1112)	28 25
2.850 (0.1122)	28 50
2.875 (0.1132)	28 75
2.900 (0.1142)	29 00
2.925 (0.1152)	29 25
2.950 (0.1161)	29 50
2.975 (0.1171)	29 75
3.000 (0.1181)	30 00
3.025 (0.1191)	30 25
3.050 (0.1201)	30 50
3.075 (0.1211)	30 75
3.100 (0.1220)	31 00
3.125 (0.1230)	31 25
3.150 (0.1240)	31 50
3.175 (0.1250)	31 75
3.200 (0.1260)	32 00

291639

Select the correct valve lash adjusting shim using the chart — 2.0L engine

2. Remove the left and right rocker covers.

3. Remove the spark plugs.

4. Set the No. 1 cylinder at TDC on its compression stroke. Align the pointer with the TDC mark on the crankshaft pulley. Check that the valve adjusters on the No. 1 cylinder are loose and valve adjusters on the No. 4 cylinder are tight. If not, turn the crankshaft one revolution (360 degrees) and align the pointer with the TDC mark on the crankshaft pulley.

5. Check the following valves:
   a. Both No. 1 intake valves.
   b. Both No. 2 exhaust valves.
   c. Both No. 3 exhaust valves.
   d. Both No. 6 intake valves.

6. Using a feeler gauge, measure the clearance between the valve adjuster and the camshaft. Record any valve clearance measurements which are out of specification. Intake valve clearance (cold) is 0.010–0.013 inches (0.26–0.34mm) and exhaust valve clearance (cold) is 0.011–0.015 inches 0.29–0.37mm).

7. Turn the crankshaft 240 degrees and set the No. 3 cylinder to TDC of its compression stroke.

8. Check the following valves:
   a. Both No. 2 intake valves.
   b. Both No. 3 intake valves.

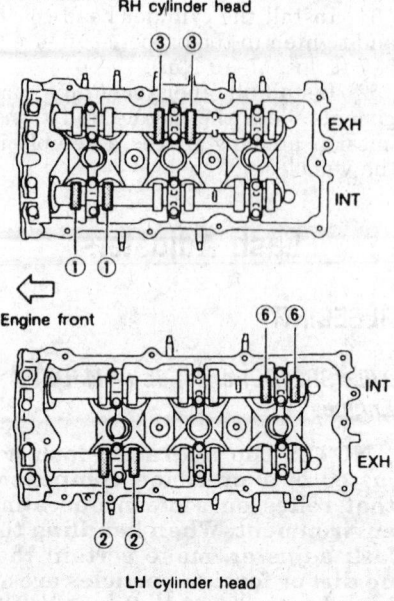

RH cylinder head

EXH

INT

Engine front

LH cylinder head

INT

EXH

289216

Valve lash checking sequence at TDC of cylinder No. 1 — 3.0L (VQ30DE) engine

   c. Both No. 4 exhaust valves.
   d. Both No. 5 exhaust valves.

9. Using a feeler gauge, measure the clearance between the valve adjuster and the camshaft. Record any valve clearance measurements which are out of specification. Intake valve clearance (cold) is 0.010–0.013 inches (0.26–0.34mm) and exhaust valve clearance (cold) is 0.011–0.015 inches (0.29–0.37mm).

10. Turn the crankshaft 240 degrees and set the No. 5 cylinder to TDC of its compression stroke.

11. Check the following valves:
    a. Both No. 1 exhaust valves.
    b. Both No. 4 intake valves.
    c. Both No. 5 intake valves.
    d. Both No. 6 exhaust valves.

12. Using a feeler gauge, measure the clearance between the valve adjuster and the camshaft. Record any valve clearance measurements which are out of specification. Intake valve clearance (cold) is 0.010–0.013 inches (0.26–0.34mm) and exhaust valve clearance (cold) is 0.011–0.015 inches(0.29–0.37mm).

13. If all the valve clearances are within specification, install the cylinder head cover, spark plugs, and the intake manifold collector.

Adjusting

1. If an adjustment is necessary, adjust the valve clearance while engine is cold by removing the adjusting shim. The adjusting shim can be removed by using the following procedures:
   a. Turn the crankshaft so the camshaft lobe of the valve to be adjusted is pointed straight up.
   b. Turn the adjuster so the notch is pointed towards the center of the cylinder head; this will facilitate the shim removal process.
   c. Using a depressor tool No. KV10115110 or equivalent, push down on the adjuster and insert a keeper tool on the edge of the adjuster to keep the adjuster in the depressed position.
   d. Remove the depressor tool and remove the shim with a magnet.

NOTE: Compressed air can be blown into the hole of the adjuster to separate the adjusting shim from the adjuster.

2. Determine the replacement adjusting shim size by using the following procedures and formula:
   a. Using a micrometer determine thickness of the removed shim.
   b. Calculate the thickness of a new adjusting shim so valve clearance is within the specified values.
   • R= thickness of the removed shim
   • N= thickness of the new shim
   • M= measured valve clearance
   • Calculate the Intake Shim as follows: N = R + M -0.0118 in. (0.30mm)]
   • Calculate the Exhaust Shim as follows: N = R + M -0.0130 in. (0.33mm)]

3. Shims are available in 64 sizes from 0.0913–0.1161 inches (2.32–2.95mm) in steps of 0.004 inches (0.01mm). The thickness is stamped on the shim; this side is always installed facing down. Select new shims with thickness as close as possible to calculated valve and install it in the adjuster.

4. Install the new shim onto the adjuster.

5. Depress the adjuster and remove the keeper tool. Remove the depressor tool and recheck the valve clearance. Repeat this procedure for any other valves requiring adjustment.

6. When all valve adjustments are finished, install the cylinder head cover, spark plugs, and the intake manifold collector.

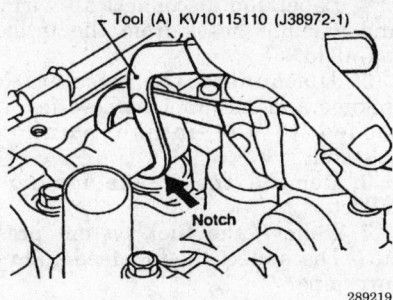

Install the depressor tool around the camshaft being careful not to damage the surfaces — 3.0L (VQ30DE) engine

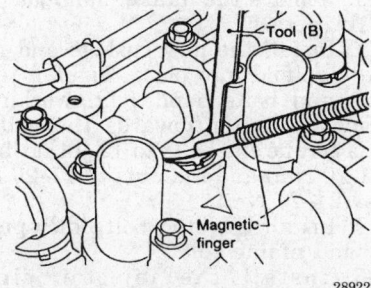

Use a magnet to remove the shim from the adjuster. Sometimes a shot of compressed air can help lift the shim up — 3.0L (VQ30DE) engine

Rocker Arms

REMOVAL AND INSTALLATION

2.0L (SR20DE) Engine

1. Relieve the fuel system pressure and disconnect the negative battery cable.

2. Drain the coolant from the radiator and engine block. Remove the radiator.

3. Raise and support the vehicle safely. Remove the right front wheel and engine side cover.

4. Remove the air duct to the intake manifold.

5. Remove the drive belts, water pump pulley, alternator and power steering pump.

6. Label and remove the vacuum hoses, fuel hoses and wire harness connectors.

7. Remove all the spark plugs, the AIV valve and resonator.

8. Remove the rocker cover and oil separator. Loosen rocker cover bolts, using 2 to 3 steps, in the opposite sequence of tightening

9. Remove the intake manifold supports.

10. Remove the oil filter bracket and power steering oil pump bracket.

11. Set No. 1 piston at TDC on the compression stroke by rotating the crankshaft.

12. Remove the chain tensioner.

13. Remove the distributor. Do not turn the rotor with the distributor removed.

14. Remove the timing chain guide, camshaft sprockets, camshafts, brackets, oil tubes and baffle plate. The camshaft bracket bolts must be loosened in sequence to prevent damage to the camshafts or the head.

15. Remove rocker arm assembly.

To install:

16. Check the hydraulic lash adjusters to ensure they did not bleed down during disassembly. If bleed down has occurred, remove the lash adjuster and prime.

NOTE: Air cannot be bled from the lash adjusters by running the engine.

17. Clean the camshaft end bracket and coat with liquid gasket. Install the camshafts, camshaft brackets, oil tubes and baffle plate. Ensure the left camshaft key is at 12 o'clock and the right camshaft key is at 10 o'clock.

NOTE: The procedure for tightening camshaft bracket bolts must be followed exactly to prevent camshaft damage.

18. Line up the mating marks on the timing chain and camshaft sprockets and install the sprockets. Tighten sprocket bolts to 101–116 ft. lbs. (137–157 Nm).

19. Install the timing chain guide, distributor (ensure that rotor head is at 5 o'clock position) and chain tensioner.

20. Install intake manifold supports. Clean the rocker cover and mating surfaces and apply a continuous bead of liquid gasket to the mating surface.

21. Install the rocker cover and oil separator. Tighten the rocker cover bolts in sequence.

22. Install the oil filter bracket and the power steering pump bracket.

23. Install the spark plugs, AIV valve and the resonator.

24. Connect the fuel lines, vacuum hoses and wiring connectors.

25. Install the water pump pulley, alternator and power steering pump.

26. Install the adjust the drive belts.

27. Install the intake manifold air duct, engine side cover and right wheel.

28. Install the radiator and refill the cooling system.

29. Connect the negative battery cable.

4.5L (VH45DE) Engine

CAUTION
Fuel injection systems remain under pressure after the engine has been turned OFF. Properly relieve fuel pressure before disconnecting any fuel lines. Failure to do so may result in fire or personal injury.

1. Disconnect the negative battery cable.

2. Remove the engine and transmission assembly from the vehicle.

3. Remove the suspension member and engine mounts from the engine.

4. Remove the air compressor bracket.

5. Remove the cooling fan with coupling and the engine gusset.

6. Separate the engine from the transmission and mount the engine on a suitable workstand.

7. Remove the oil pan.

8. Remove the crank angle sensor and the Valve Timing Control (VTC) solenoid.

9. Remove the chain tensioners and the upper front covers.

10. Remove the front timing chain cover.

NOTE: The timing chain will not be disengaged or dislocated from the crankshaft sprocket unless the front cover is removed. The cast portion of the front cover is located on the lower side of the crankshaft sprocket so the timing chain is not disengaged from the sprocket.

11. Remove the VTC assembly and the camshaft sprocket.

12. Remove the oil pump chain and the timing chains.

NOTE: Do not attempt to disassemble the VTC assembly since they are difficult to reassemble accurately in the field. If it should be disassembled, the VTC assembly must be replaced with a new one.

13. Remove the camshaft brackets and the camshafts. Mark the parts so they can be reinstalled in their original positions.

14. Remove the rocker arms. Be sure to identify each rocker arm so it can be reinstalled in it's original position.

To install:

15. Make sure all mating surfaces are clean before installation.

16. Install the rocker arms, camshafts and camshaft brackets on the right bank. Properly lubricate the rocker arms and camshafts prior to installation.

17. Install the VTC assembly and the exhaust cam sprocket on the right bank.

18. Make sure the camshafts are still correctly positioned and the piston in the No. 1 cylinder is still at TDC.

19. Install the timing chain on the right bank, aligning the mating marks on the chain with those on the crankshaft and camshaft sprockets.

20. Install the chain tensioner on the right bank.

21. Turn the crankshaft approximately 120 degrees clockwise from the point where the No. 1 piston is at TDC on the compression stroke. At this point, the valves on the left bank still remain closed.

22. Correctly position the camshafts and rocker arms for the left cylinder head. Properly lubricate the rocker arms and camshafts prior to installation. Install the VTC assembly and the exhaust cam sprocket.

23. Install the timing chain on the left bank, aligning the mating marks on the chain with those on the crankshaft and camshaft sprockets.

24. Install the oil pump chain and sprockets.

25. Install the oil pump chain guides. Place a 0.04 in. (1.0mm) feeler gauge between the upper chain guide and chain before assembling the chain guides. The force applied to the chain is equivalent to the upper chain guide weight.

26. Apply suitable sealer and install the front covers.

27. Install the chain tensioner for the left bank.

28. Apply suitable sealer to the rubber plugs and install them on the cylinder head.

29. Install the crank angle sensor, VTC solenoid, rocker cover and crank pulley.

30. Install the transmission on the engine and install the engine assembly in the vehicle.

Intake Manifold

REMOVAL AND INSTALLATION

2.0L (SR20DE) Engine

1. Disconnect the negative battery cable.

─── **CAUTION** ───

Fuel injection systems remain under pressure after the engine has been turned OFF. Properly relieve fuel pressure before disconnecting any fuel lines. Failure to do so may result in fire or personal injury.

2. Properly relieve the fuel system pressure.

3. Drain the cooling system.

4. Tag and disconnect the fuel lines, vacuum hoses and electrical connectors. Disconnect the throttle linkage.

5. Remove the intake manifold collector support.

6. Remove the intake manifold collector.

7. Remove the injector tube assembly.

8. Remove the intake manifold-to-cylinder head bolts, working from the ends towards the center.

9. Remove the manifold support bolts and remove the manifold.

To install:

10. Make sure all mating surfaces are clean prior to installation.

11. Fit a new gasket and the manifold into place. Start the support bolts to hold the manifold in place.

12. Install the intake manifold bolts. Torque the bolts to 13–15 ft. lbs. (18–21 Nm). Tighten the bolts in two steps, starting at the center and working towards the ends.

13. Install the injector tube assembly. Torque the bolts first to 6.9–8.0 ft. lbs. (9.3–10.8 Nm) and then to 15–20 ft. lbs. (21–26 Nm).

14. Use a new gasket and install the intake manifold collector. Torque the mounting bolts to 13–15 ft. lbs. (18–21 Nm).

15. Reconnect the fuel lines, vacuum hoses, electrical connectors and the throttle linkage.

16. Refill the cooling system, connect the negative battery cable, start engine and test for leaks.

3.0L (VG30DE) Engine

─── **CAUTION** ───

Fuel injection systems remain under pressure, after the engine has been turned OFF. Properly relieve fuel pressure before disconnecting any fuel lines. Failure to do so may result in fire or personal injury.

1. Disconnect the negative battery cable.

2. Remove the timing belt.

3. Remove the air ducts.

4. Label and disconnect all wiring and vacuum hoses from the intake manifold.

5. Disconnect the accelerator linkage and all other accessories attached to the intake manifold collector.

6. Remove the intake manifold collector.

7. Relieve the fuel system pressure. Disconnect the fuel feed and return pipe.

8. Remove the injector pipe assembly.

9. Remove the timing belt idler pulley.

10. Remove the intake manifold.

To install:

11. Install the new gaskets and fit the manifold into place. Tighten the nuts and bolts hand-tight, working from the center to the ends, then torque the bolts to 12–14 ft. lbs. (16–20 Nm) and the nuts to 17–20 ft. lbs. (24–27 Nm).

12. Install the stud bolt, idler pulley and timing belt.

13. Install the injector pipe assembly.

14. Connect the fuel feed and return pipes.

15. Install the intake manifold collector and torque the bolts to 12–15 ft. lbs. (16–21 Nm).

16. Connect the accelerator linkage and all other accessories attached to the intake manifold collector.

17. Connect all electrical harnesses and vacuum hoses.

18. Connect the air ducts.

19. Fill the cooling system.

20. Make any necessary adjustments. Start the engine and allow it to reach operating temperature. Check for leaks.

3.0L (VQ30DE) Engine

1. Disconnect the negative battery cable and drain the cooling system.

2. Release the fuel system pressure.

─── **CAUTION** ───

The fuel injection system remains under pressure after the engine has been turned OFF. Properly relieve the fuel pressure before disconnecting any fuel lines. Failure to do so may result in fire or personal injury.

3. Remove the throttle body coolant hoses.

4. Label and disconnect the electrical connectors from the throttle position sensor.

5. Label and disconnect the hoses from the throttle body, the EGR

valve, intake manifold collector, IAC valve, and the fuel pressure regulator.

6. Disconnect the canister purge hose and blow-by hose.

7. Disconnect the EGR guide tube.

8. Disconnect the accelerator cable from the throttle body.

9. Remove the intake manifold collector support brackets.

10. Disconnect the right side electrical connectors from the ignition coils.

11. If necessary, disconnect the electrical connector from the crank angle sensor and the power transistor.

12. Remove the intake manifold collector-to-intake manifold bolts/nuts and remove the intake manifold collector.

13. Remove the fuel injector assembly by performing the following procedures:

    a. Disconnect the electrical connectors from the fuel injectors.

    b. Disconnect the fuel lines from the fuel injector assembly.

    c. Remove the fuel rail-to-cylinder head bolts.

    d. Remove the fuel rail assembly from the engine.

14. Remove the intake manifold bolts/nuts in the reverse sequence of the torque procedure.

15. Remove the intake manifold from the engine and discard the gaskets.

16. Clean all gasket mounting surfaces.

To install:

17. Using new gaskets, install the intake manifold to the engine.

18. Tighten the bolts/nuts in sequence as follows:

    a. Tighten nuts and bolts to 44–86 inch lbs. (5–10 Nm).

    b. Tighten nuts and bolts to 20–23 ft. lbs. (26–31 Nm).

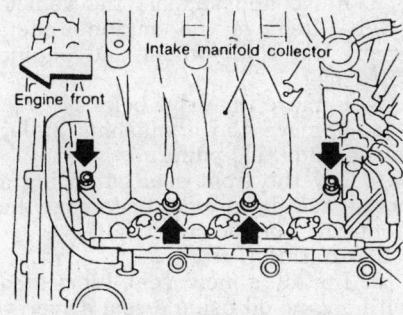

Intake manifold collector mounting bolts — 3.0L (VQ30DE) engine

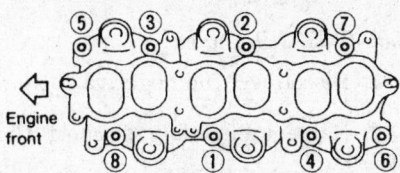

Tighten in numerical order.

287893

Intake manifold torque procedure-Loosen bolts in the reverse sequence — 3.0L (VQ30DE) engine

19. Install the fuel injector assembly by performing the following procedures:

    a. Install the fuel rail assembly to the engine.

    b. Install the fuel rail-to-cylinder head bolts and torque the bolts to 15–20 ft. lbs. (21–26 Nm) in two progressive steps.

    c. Connect the fuel lines to the fuel injector assembly.

    d. Connect the electrical connectors to the fuel injectors.

20. Using a new gasket, install the intake manifold collector and torque the intake manifold collector-to-intake manifold bolts/nuts to 13–16 ft. lbs. (18–22 Nm).

21. Install the intake manifold collector supports and torque the bolts to 14–18 ft. lbs. (20–25 Nm).

22. If disconnected, connect the electrical connector to the crank angle sensor and the power transistor.

23. Connect the electrical connectors to the ignition coils and tighten the mounting bolts to 27–33 inch lbs. (2.9–3.8 Nm)

24. Connect the accelerator cable to the throttle body.

25. Connect the EGR guide tube and tighten the bolts to 15–20 ft. lbs. (21–26 Nm) in two progressive steps.

26. Connect the canister purge hose and blow-by hose.

27. Connect the hoses to the throttle body, EGR valve, intake manifold collector, IAC valve, and the fuel pressure regulator.

28. Connect the electrical connectors to the throttle position sensor.

29. Connect the throttle body coolant hoses.

30. Refill the cooling system and connect the negative battery cable.

31. Start the engine, bleed cooling system, and check for leaks.

4.5L (VH45DE) Engine

CAUTION

Fuel injection systems remain under pressure after the engine has been turned OFF. Properly relieve fuel pressure before disconnecting any fuel lines. Failure to do so may result in fire or personal injury.

---

1. Properly relieve the fuel system pressure.

2. Disconnect the negative battery cable.

3. Drain the cooling system.

4. Tag and disconnect the fuel lines, vacuum hoses and electrical connectors. Disconnect the throttle linkage.

5. Remove the intake manifold collector.

6. Remove the injector tube assembly and remove the intake manifold.

To install:

7. Make sure all mating surfaces are clean prior to installation. Fit new gaskets and install the intake manifolds. Make the bolts only finger tight at this time.

8. Install the injectors and tube assembly. Make the bolts only finger tight.

NOTE: Be careful not to let the rubber washer fall into the intake manifold.

9. Install the intake collector without the gaskets. Start the bolts to hold the manifolds in place.

10. Remove the intake collector, then tighten the intake manifold-to-cylinder head bolts to 12–15 ft. lbs. (16–21 Nm).

11. Torque the sub-fuel tube assemblies in sequence, first to 37–52 inch lbs. (4.2–5.9 Nm) and then to 74–96 inch lbs. (8.4–10.8 Nm).

12. Torque the injector tube assemblies in sequence, first to 83–96 inch lbs. (9.3–10.8 Nm) and then to 15–20 ft. lbs. (21–26 Nm).

13. Install the intake collector and torque to 9–11 ft. lbs. (12–15 Nm).

14. Install the remaining components in the reverse order of their removal.

Exhaust Manifold

REMOVAL AND INSTALLATION

2.0L (SR20DE) Engine

1. Disconnect the negative battery cable. Raise and support the vehicle safely.

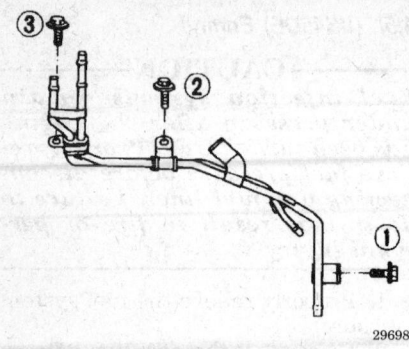

296981

Sub-fuel tube bolt torque sequence — 4.5L engine

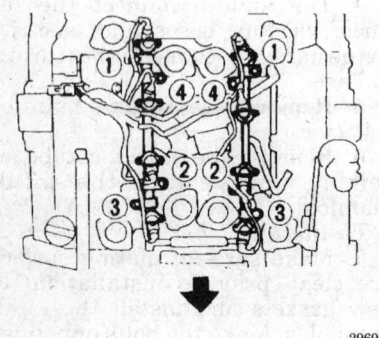

296982

Fuel tube/injector assembly bolt torque sequence — 4.5L engine

2. If equipped, remove the undercover and dust covers. Disconnect the exhaust pipe at the manifold flange.

3. Remove the AIV, AIV tube, and the attaching bracket.

4. Disconnect the exhaust gas sensor electrical connection and remove the sensor.

5. Remove the exhaust manifold cover.

6. Remove the exhaust manifold nuts, starting at the outside and working towards the middle.

7. Remove the exhaust manifold and gasket.

To install:

8. Clean the gasket mating surface and install a new exhaust manifold gasket.

9. Install the exhaust manifold and tighten the manifold nuts, in two steps, starting at the middle and alternating towards the outside of each end. Torque the nuts to 27–35 ft. lbs. (37–48 Nm).

10. Install the exhaust manifold cover and exhaust gas sensor. Reconnect the sensor electrical connection.

11. Install the AIV, AIV tube, and the attaching bracket.

12. Install the exhaust pipe to the manifold flange and tighten the nuts to 30–35 ft. lbs. (41–48 Nm).

13. Lower the vehicle, start the engine, and check for leaks.

3.0L (VG30DE) Engine

1. Disconnect the negative battery cable.

2. Raise and safely support the vehicle.

NOTE: If necessary, soak the exhaust pipe retaining nuts with penetrating oil to loosen them.

3. Disconnect the exhaust manifolds from the exhaust pipes.

4. Remove the protective covers from the manifolds.

5. Remove the exhaust manifold-to-engine mounting nuts. Remove the manifolds from the engine an discard the gaskets.

To install:

6. Clean all gasket mounting surfaces. Install new gaskets.

7. Install the exhaust manifold to the engine and tighten the mounting nuts in two progressive steps to 22–24 ft. lbs. (30–32 Nm).

8. Install the protective shields and tighten the mounting bolts in two progressive steps to 46–57 inch lbs. (5–7 Nm).

9. Install the exhaust manifolds to the exhaust pipes and tighten the mounting nuts to 32–37 ft. lbs. (43–50 Nm).

10. Connect the negative battery cable, start the engine, and check for exhaust leaks.

3.0L (VQ30DE) Engine

1. Disconnect the negative battery cable.

2. Raise and safely support the vehicle.

NOTE: If necessary, soak the exhaust pipe retaining nuts with penetrating oil to loosen them.

3. Disconnect the exhaust manifolds from the exhaust pipes.

4. Remove the protective covers from the manifolds.

5. Remove the exhaust manifold-to-engine mounting nuts. Remove the manifolds from the engine an discard the gaskets.

To install:

6. Clean all gasket mounting surfaces. Install new gaskets.

7. Install the exhaust manifold to the engine and tighten the mounting nuts in two progressive steps to 22–24 ft. lbs. (30–32 Nm).

8. Install the protective shields and tighten the mounting bolts in two progressive steps to 46–57 inch lbs. (5.1–6.5 Nm).

9. Install the exhaust manifolds to the exhaust pipes and tighten the mounting nuts to 32–37 ft. lbs. (43–50 Nm).

10. Connect the negative battery cable, start the engine, and check for exhaust leaks.

4.5L (VH45DE) Engine

1. Disconnect the negative battery cable. Raise and support the vehicle safely.

2. Remove the engine undercovers. Disconnect the exhaust pipe at the manifold flange.

3. Disconnect the exhaust gas sensor electrical connection and if necessary, remove the sensor.

4. Remove the exhaust manifold nuts, starting at the ends and working towards the center.

5. Remove the exhaust manifold and gasket.

To install:

6. Clean the gasket mating surface and install a new exhaust manifold gasket.

7. Install the exhaust manifold and tighten the nuts in two steps and in sequence. Torque the nuts to 20–23 ft. lbs. (27–31 Nm).

8. Install exhaust gas sensor and tighten to 30–37 ft. lbs. (40–50 Nm). Reconnect the sensor electrical connection.

9. Install the exhaust pipe to the manifold flange and tighten the nuts to 33–44 ft. lbs. (45–60 Nm).

10. Lower the vehicle and connect the negative battery cable.

11. Start the engine and check for leaks.

Front Cover Seal

REMOVAL AND INSTALLATION

2.0 (SR20DE) Engine

1. Disconnect the negative battery cable.

2. Raise and support the vehicle safely. Remove the engine undercover, the right wheel and engine side cover.

3. Remove the drive belts.

4. Remove the crankshaft pulley using a suitable puller.

5. Pry the front seal out using a prybar taking care not to damage the front cover or the crankshaft.

To install:

6. Install a new seal lubricated with engine oil using a seal driver.

7. Install the crankshaft pulley. Torque the pulley bolt to 105–112 ft. lbs. (142–152 Nm).

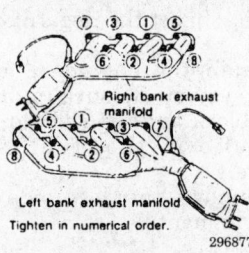

Exhaust manifold torque sequence — 4.5L engine

8. Install the drive belts, engine covers and right wheel assembly.

9. Connect the negative battery cable.

10. Start engine and check for leaks.

3.0L (VQ30DE) Engine

1. Disconnect the negative battery cable.

2. Remove the right front wheel assembly.

3. Remove the engine undercovers and remove the inner cover from the right front wheel.

4. Remove the engine drive belts.

5. Remove the bolt that secures the crankshaft pulley and remove the pulley.

6. Using a suitable tool, pry the oil seal from the front cover.

NOTE: When removing the oil seal, be careful not the gouge or scratch the seal bore or crankshaft surfaces.

7. Wipe the seal bore with a clean rag.

To install:

8. Lubricate the lip of the new seal with clean engine oil.

9. Install the seal into the front cover with a suitable seal installer.

10. Install the crankshaft pulley and tighten the mounting bolt to 14–22 ft. lbs. (20–29 Nm). Turn the crankshaft bolt an additional 60–66 degrees clockwise.

11. Install and adjust the engine drive belts.

12. Install the engine covers and install the right front wheel.

13. Connect the negative battery cable.

14. Start the engine and check for leaks.

4.5L (VH45DE) Engine

1. Disconnect the negative battery cable.

2. Raise and safely support the vehicle.

3. Remove the engine splash shield.

4. Remove the cooling fan and the engine gusset.

5. Remove the necessary accessory drive belts.

6. Remove the lower rear plate in order to remove the crankshaft pulley bolt.

7. Remove the crankshaft pulley bolt and the crankshaft pulley.

8. Use a suitable tool to remove the front cover oil seal.

To install:

9. Lubricate the seal lip prior to installation. Position the oil seal in the front cover. Install with special tool NT049, or equivalent.

10. Install the crankshaft pulley, and tighten the crank pulley bolt to 260–275 ft. lbs. (353–373 Nm). Install the lower rear plate.

11. Install the removed drive belts.

12. Install the engine gusset.

13. Install the cooling fan. Tighten the mounting bolts to 4–7 ft. lbs. (6–10 Nm).

14. Install the engine splash shield.

15. Lower the vehicle. Adjust the drive belts.

16. Connect the negative battery cable.

Front Crankshaft Seal

REMOVAL AND INSTALLATION

3.0 (VG30DE) Engine

1. Disconnect the negative battery cable.

2. Remove the timing belt.

3. Remove the crankshaft sprocket.

4. Remove the oil pan and oil pump assembly.

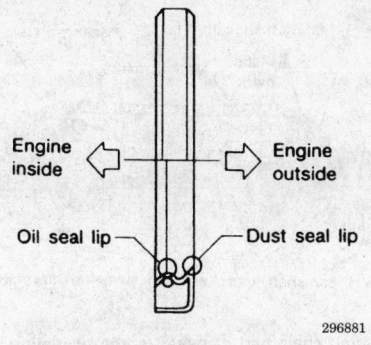

Oil seal positioning (cross-sectional view) — 4.5L engine

5. Using a suitable tool, pry the oil seal from the oil pump assembly.

NOTE: When removing the oil seal, be careful not the gouge or scratch the seal bore or crankshaft surface.

To install:

6. Wipe the seal bore with a clean rag.

7. Lubricate the lip of the new seal with clean engine oil.

8. Install the seal into the oil pump assembly using a seal installer.

9. Install the oil pump and oil pan.

10. Install the crankshaft sprocket.

11. Install the timing belt.

12. Connect the negative battery cable.

Timing Chain and Sprockets

REMOVAL AND INSTALLATION

2.0L (SR20DE) Engine

——— **CAUTION** ———

Fuel injection systems remain under pressure after the engine has been turned OFF. Properly relieve fuel pressure before disconnecting any fuel lines. Failure to do so may result in fire or personal injury.

1. Relieve the fuel system pressure and disconnect the negative battery cable.

2. Raise and support the vehicle safely. Remove the engine under covers. Remove the right front wheel and engine side cover, then lower the vehicle.

3. Drain the cooling system and remove the radiator.

4. Remove the intake manifold air duct.

5. Remove the drive belts, water pump pulley, alternator and power steering pump.

6. Label and remove the vacuum hoses, fuel hoses and wire harness connectors.

7. Remove the spark plugs.

8. Remove the cylinder head cover and oil separator.

9. Remove the intake manifold supports.

10. Remove the oil filter bracket and the power steering oil pump bracket.

11. Place the No. 1 piston at TDC on the compression stroke.

12. Remove the chain tensioner.

13. Remove the distributor. Do not turn the rotor while the distributor is removed.

14. Remove the timing chain guide.

15. Remove the camshaft sprockets.

16. Remove the camshafts, camshaft brackets, oil tubes and baffle plate.

17. Remove the starter.

18. Disconnect the heater hoses and the water hoses from the cylinder head.

19. Disconnect the knock sensor harness connector.

20. Remove the cylinder head outside bolts.

21. Use two or three steps and gradually remove the cylinder head bolts.

22. Remove the cylinder head with the intake and exhaust manifolds.

23. Raise and support the vehicle safely.

24. Remove the oil pans.

25. Remove the oil strainer and baffle plate.

26. Remove the crankshaft pulley.

27. Place a transmission jack under the main bearing beam and raise the engine slightly to take the weight off of the front engine mount.

28. Remove the front engine mount.

29. Remove the timing chain cover.

30. Loosen and remove the timing chain sprocket bolts.

31. Remove the timing chain guides, then timing chain and sprockets.

To install:

32. Make sure all sealing surfaces are clean and prepared for assembly.

33. Install the crankshaft sprocket. Position the crankshaft so No. 1 piston is set at TDC (keyway at 12 o'clock, mating mark at 4 o'clock).

34. Fit the timing chain to crankshaft sprocket with the gold mating mark on the chain aligned with the mark on the sprocket. (The mating marks for the camshaft sprockets are silver.)

35. Install the timing chain guides and hang the chain off the left (front) guide. If necessary, secure the chain so it does not disengage from the crankshaft sprocket during assembly.

36. Apply a bead of liquid gasket to the front cover. Install the oil pump drive spacer and front cover. Torque the bolts evenly to 60 inch lbs. (6.7 Nm) and wipe away any excess liquid gasket.

37. Install the front engine mount.

38. Install the crankshaft pulley and temporarily tighten the bolt to hold the sprocket in place. The timing mark should align with the TDC mark.

39. Install the oil strainer, baffle plate and oil pan.

40. Install the cylinder head, camshafts, oil tubes and baffles. Position the left camshaft key at 12 o'clock and the right camshaft key at 10 o'clock.

41. Install the camshaft sprockets by lining up the mating marks on the timing chain with the mating marks on the camshaft sprockets. Torque the camshaft bolts to 101–116 ft. lbs. (137–157 Nm). Torque the crankshaft pulley bolt to 105–112 ft. lbs. (142–152 Nm).

42. Install the upper timing chain guide and distributor. Ensure that the rotor is at the 5 o'clock position.

43. Before installing the chain tensioner, press the cam stopper down and the push in the sleeve until the hook can be engaged on the pin. When tensioner is bolted in position, the hook will release automatically. Ensure the arrow on the outside faces the front of the engine.

44. Install the oil filter bracket and the power steering pump bracket.

45. Install the intake manifold supports.

46. Install the oil separator and the cylinder head cover.

47. Install the spark plugs.

48. Connect the vacuum hoses, fuel hoses, and wire harness connectors.

49. Install the alternator and power steering pump.

50. Install the water pump pulley. Install and adjust the drive belts.

51. Install the radiator and refill the cooling system.

52. Connect the negative battery cable.

53. Bleed the power steering hydraulic system and the cooling system.

54. Inspect the engine for any fluid leaks.

55. Replace the engine under covers, side cover and right front wheel.

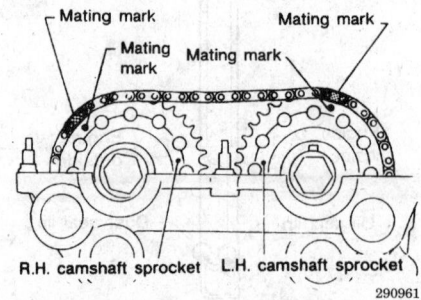

Timing chain and camshaft sprocket mating marks — 2.0L engine

3.0L (VQ30DE) Engine

1. Disconnect the negative battery cable.

2. Drain the engine oil and the cooling system. Be sure to drain the engine block and the radiator.

3. Relieve the fuel system pressure.

4. Remove the left side rocker cover ornament.

NOTE: Before disconnecting any hoses or connectors, note the locations for reassembly.

5. Remove the air duct to intake manifold hose, collector hose, blow-by hose, and vacuum hoses.

———— CAUTION ————
Fuel injection systems remain under pressure after the engine has been turned OFF. Properly relieve fuel pressure before disconnecting any fuel lines. Failure to do so may result in fire or personal injury.

6. Remove the fuel hoses and disconnect the harness connections.

7. Disconnect the canister purge hoses.

8. Remove the water hoses from the cylinder head and intake manifold.

9. Disconnect and remove all six ignition coils from the spark plugs.

10. Remove the spark plugs.

11. Remove the bolts that secure the EGR tube and remove the tube.

12. Remove the intake manifold collector supports and remove the collector.

13. Remove the bolts that secure the fuel tube and remove the fuel tube from the vehicle.

14. Remove the bolts that secure the intake manifold to the engine block and remove the manifold. Loosen the bolts in the reverse sequence of the tightening procedure.

15. Remove the LH and RH rocker covers from the cylinder head.

16. Remove the engine undercovers.

17. Remove the right front wheel and the engine side covers.

18. Remove the drive belts and the idler pulley.

19. Remove the power steering oil pump belt and remove the power steering oil pump assembly.

20. Remove the camshaft position sensor (PHASE) and crankshaft position sensors (REF)/(POS).

21. Set the No. 1 piston to TDC of compression stroke by rotating the crankshaft.

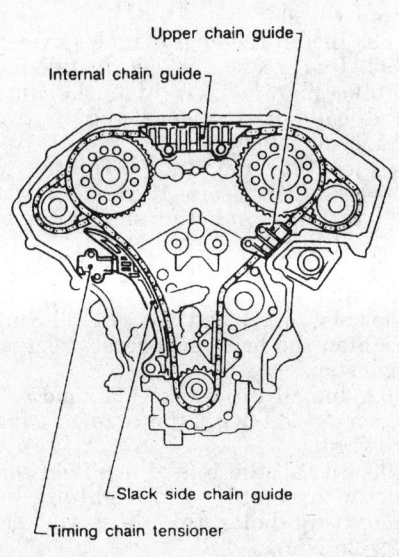

Timing chain tensioner and guides — 3.0L (VQ30DE) engine

22. Remove the ring gear cover access plate.
23. Loosen the crankshaft pulley bolt while securing the ring gear so the crankshaft cannot rotate.

NOTE: Use care not to damage the ring gear teeth.

24. Using a suitable puller, remove the crankshaft pulley.
25. Remove the A/C compressor and bracket.
26. Remove the front exhaust pipe and its support.
27. Hang the engine at the right and left side engine slingers with a suitable hoist.
28. Support the transaxle with jack.
29. Remove the right side engine mounting, mounting bracket, and nuts.
30. Remove the center crossmember assembly.
31. Remove the steel (lower) oil pan bolts in the reverse sequence of the torque sequence.
32. Insert a seal cutter between the steel and aluminum oil pan.
33. Tapping the cutter with a hammer, slide it around the entire edge of the oil pan. Be careful not to damage the aluminum mating surface of the upper oil pan.

34. Remove the steel oil pan and the oil strainer.
35. Remove the aluminum (upper) oil pan bolts in the reverse sequence of the torque sequence.
36. Remove the transaxle bolts that secure the oil pan.
37. Insert a seal cutter between the aluminum oil pan and the engine block.
38. Tapping the cutter with a hammer, slide it around the entire edge of the oil pan. Be careful not to damage the mating surfaces of the oil pan or engine block.
39. Remove the oil pan from the vehicle.
40. Remove the water pump cover and remove the bolts that secure the front timing chain case.
41. Using the seal cutter, remove the timing chain case cover.
42. Remove the internal timing chain guide and the upper chain guide.
43. Remove the timing chain tensioner and slack side chain guide.
44. Remove the left and right intake camshaft sprockets first. Be sure to hold the flats of the camshafts while removing the sprocket bolts.
45. Remove the lower timing chain assembly. Be sure to note the aligning marks of the chain before removal.
46. Insert a suitable stopper pin for the left and right camshaft tensioners.
47. Remove the left and right exhaust camshaft sprocket bolts. Be sure to hold the flats of the camshafts while removing the sprocket bolts.
48. Remove the upper timing chain assembly. Be sure to note the aligning marks of the chain before removal.
49. Remove the lower timing chain guide.
50. Remove the crankshaft sprocket.
51. Remove all traces of liquid gasket from the front timing chain case and from the water pump.
52. Inspect the timing chain for excessive wear or damage and replace as necessary.

To install:
53. Install the crankshaft sprocket with the mating mark facing out.
54. Position the crankshaft to TDC of compression stroke and align the dowels of the camshaft sprockets to the 12 o'clock position in respect to the cylinder head.
55. Install the lower timing chain guide. The front mark on the guide should face upwards.

56. On a work bench, align the marks on the intake and exhaust camshaft sprockets with the marks of the chain.
57. Put the exhaust camshaft sprockets onto the dowel pin and tighten the mounting bolts to 88–95 ft. lbs. (119–128 Nm). Be sure to secure the camshafts while tightening the bolts.
58. Install the timing chains and sprockets to the intake camshafts. Be sure to align the timing chain and sprocket mating marks.
59. Remove the left and right camshaft tensioner stopper pins.
60. Align the mating mark on the crankshaft with the matchmark (gold link) on the lower timing chain.
61. Attach the lower timing chain to the water pump sprocket.
62. Working counterclockwise, install the lower timing chain camshaft sprockets. Be sure to align the sprocket marks with the blue links of the timing chain during installation.
63. Install the intake sprocket bolts and tighten to 88–95 ft. lbs. (119–128 Nm). Be sure to secure the camshafts while tightening the bolts.
64. Install the internal timing chain guide, upper timing chain guide, lower timing chain tensioner and slack side timing chain guide.
65. Tighten the tensioner mounting bolt to 75–96 inch lbs. (8.4–10.8 Nm) and tighten the guide bolts to 108–168 inch lbs. (13–19 Nm).
66. Apply a 0.102–0.142 inch (2.6–3.6mm) continuous bead of liquid gasket to all necessary areas as shown on the front timing cover.
67. Install the timing cover evenly and gently. Be sure to align the dowel pin holes.
68. Tighten the mounting bolts in sequence as follows:
   a. Bolts No. 1 and 2 — Tighten to 19–23 ft. lbs. (26–31 Nm)
   b. Bolts No. 3 to 20 — Tighten to 105–121 inch lbs. (11.8–13.7 Nm)

NOTE: Leave the bolts unattended for 30 minutes or more after tightening. This will allow the liquid gasket to cure sufficiently.

69. Apply a 0.091–0.130 inch (2.3–3.3mm) continuous bead of liquid gasket to the water pump cover and install the cover. Tighten the bolts to 84–108 inch lbs. (10–13 Nm).
70. Apply a 0.12 inch (3mm) continuous bead of liquid gasket to the rocker covers and install the covers. Tighten the mounting bolts in sequence as follows:
   a. Bolts No. 1 to 10 — Tighten to 9–26 inch lbs. (1–3 Nm)

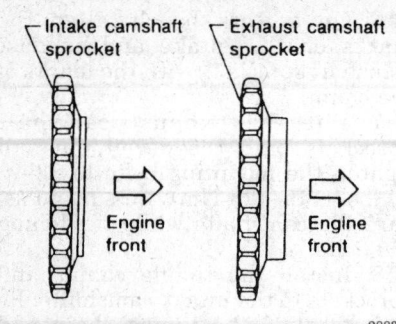

288351

Identification of the intake and exhaust camshaft sprockets — 3.0L (VQ30DE) engine

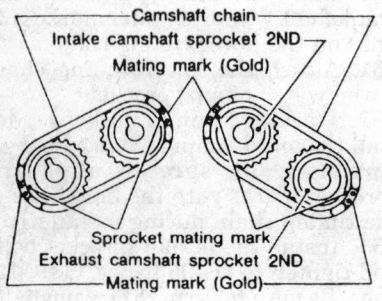

288346

Upper timing chains alignment marks — 3.0L (VQ30DE) engine

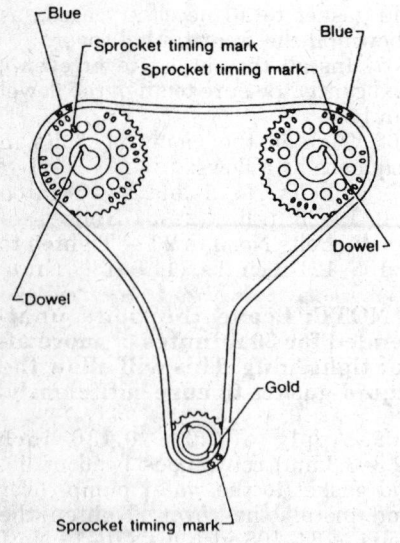

288344

Lower timing chain alignment marks — 3.0L (VQ30DE) engine

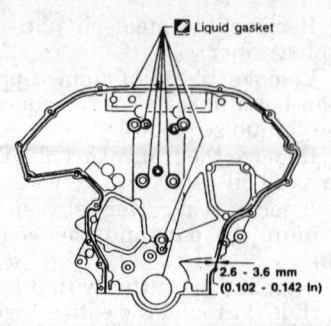

288352

Application of liquid gasket to the front timing case — 3.0L (VQ30DE) engine

b. Bolts No. 1 to 10 — Tighten to 52–69 inch lbs. (6–8 Nm).

71. Apply sealant to the front and rear seal of the oil pan.

72. Apply a 0.177–0.217 inch (4.5–5.5mm) continuous bead of liquid gasket to the upper oil pan mating surface and install the oil pan. Tighten the mounting bolts in sequence to 12–14 ft. lbs. (16–19 Nm).

73. Install the transaxle bolts that secure the oil pan.

74. Install the oil pan strainer and tighten the mounting bolts to 12–14 ft. lbs. (16–19 Nm).

75. Apply a 0.177–0.217 inch (4.5–5.5mm) continuous bead of liquid gasket to the lower oil pan mating surface and install the oil pan. Tighten the mounting bolts in sequence to 57–66 inch lbs. (6.4–7.5 Nm).

76. Tighten the oil pan drain plug to 22–29 ft. lbs. (29–39 Nm).

77. Install the center crossmember assembly.

78. Install the right side engine mounting bracket and mount assembly.

79. Remove the engine slinger assembly.

80. Install the front exhaust pipe and its support.

81. Install the A/C compressor and bracket.

82. Install the crankshaft pulley to the crankshaft and install the mounting bolt.

83. Tighten the mounting bolt to 14–22 ft. lbs. (20–29 Nm). Tighten the crankshaft bolt an additional 60–66 degrees clockwise. This is about the angle from one hexagon bolt head corner to another.

84. Install the ring gear cover plate.

85. Install the camshaft position sensor (PHASE) and crankshaft position sensors (REF)/(POS).

86. Install the power steering pump assembly.

87. Install the drive belts and the idler pulley.

88. Install the right front inner wheel cover and install the right front wheel.

89. Install the engine undercovers.

90. Using new gaskets, install the intake manifold. Tighten the nuts and bolts in sequence and in two stages as follows:
    a. Bolts and nuts — Tighten to 44–86 inch lbs. (5–10 Nm).
    b. Bolts and nuts — Tighten to 16–18 ft. lbs. (22–25 Nm).

91. Using new insulators, install the fuel tube assembly and tighten the bolts to 15–20 ft. lbs. (21–26 Nm). Tighten the bolts in several progressive steps.

92. Install the intake manifold collector gasket with the arrow facing forward.

93. Install the intake manifold collector assembly and tighten the mounting bolts to 16–18 ft. lbs. (22–25 Nm).

94. Install the intake manifold collector support brackets.

95. Using new gaskets, install the EGR tube and tighten the mounting bolts to 15–20 ft. lbs. (21–26 Nm) in two progressive steps.

96. Install the spark plugs.

97. Install the ignition coils and tighten the mounting bolts to 27–33 inch lbs. (2.9–3.8 Nm).

98. Install the rocker cover ornament on the left side.

99. Install the water hoses to the cylinder head and intake manifold.

100. Connect the canister purge hoses.

101. Connect the fuel hoses and wiring harness connections to the fuel rail.

102. Install and connect the air duct to intake manifold hose, collector hose, blow-by hose, and vacuum hoses.

103. Refill the engine oil and coolant with the proper type and amount of fluid.

104. Connect the negative battery cable.

105. Start the engine and run at 3000 RPM under no load to purge the air from the high pressure chamber. The engine may produce a rattling noise. This indicates that air still remains in the chamber and is not a matter of concern.

4.5L (VH45DE) Engine

— **CAUTION** —

Fuel injection systems remain under pressure, after the engine has been turned OFF. Properly relieve fuel pressure before disconnecting any fuel lines. Failure to do so may result in fire or personal injury.

1. Disconnect the negative battery cable.
2. Remove the engine and transmission assembly from the vehicle.
3. Remove the suspension member and engine mounts from the engine.

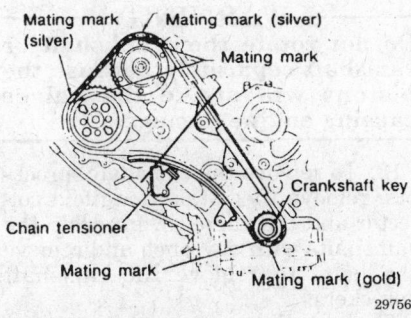

Timing chain mating mark alignment for right bank — 4.5L engine

4. Remove the air compressor bracket.
5. Remove the cooling fan with coupling and the engine gusset.
6. Separate the engine from the transmission and mount the engine on a suitable workstand.
7. Remove the oil pan.
8. Remove the ornamental rocker cover and remove the ignition coils and spark plugs.
9. Bring the No. 1 piston to TDC on the compression stroke.
10. Use a suitable puller to remove the crankshaft pulley.
11. Remove the rocker covers.
12. Remove the crank angle sensor or camshaft position sensor on 1996 and later models.
13. Remove the left bank chain tensioner from the upper cover and remove both upper front covers.
14. Remove the lower front timing chain cover. Note how the dowels on the camshafts are aligned away from the crankshaft.
15. Loosen and remove the sprocket bolts.

— **WARNING** —

Do not turn the crankshaft or the camshafts with the timing chains removed. The valves will contact the pistons and severe engine damage can occur.

16. Remove the right bank chain tensioner, then carefully pry the sprocket off.
17. Remove the oil pump chain and the timing chain. Keep the chains separated or tag them.
18. If necessary, carefully pry off the crankshaft sprockets.

To install:

19. Make sure all mating surfaces are clean before installation.
20. Install the any removed sprocket, making sure it is properly aligned.
21. Make sure the camshafts are still correctly positioned and the piston in the No. 1 cylinder is still at TDC. If removed, install the crankshaft chain sprockets with the collar towards the engine block. The mating mark for the right bank (inner) sprocket will be at the bottom when the crankshaft is at TDC.
22. Install the timing chain on the right bank, aligning the mating marks on the chain with those on the crankshaft and camshaft sprockets.
23. Install the timing chain on the left bank, aligning the mating marks on the chain with those on the crankshaft and camshaft sprockets.
24. Install the oil pump chain and sprocket.
25. Install the oil pump chain guides. Place a 0.040 in. (1.0mm) feeler gauge between the upper chain guide and chain before assembling the chain guides. The force applied to the chain is equivalent to the upper chain guide weight.
26. Apply a liquid gasket sealer and install the front covers. Be sure to seal around the bolt holes on the inside of the cover. Torque the bolts in sequence to 84 inch lbs. (9.5 Nm).
27. Install the chain tensioner for the left bank.
28. Apply suitable sealer to the rubber plugs and install them on the cylinder head.
29. Install the crank angle sensor or camshaft position sensor, VTC solenoid, rocker cover and crankshaft pulley. Torque the pulley bolt to 260–275 ft. lbs. (353–373 Nm).
30. Install the oil pan. Torque the bolts in the proper sequence to 4.6–6.1 ft. lbs. (6.3–8.3 Nm).
31. Install the transmission to the engine.
32. Install the engine mounts and the suspension member.
33. Install the engine and transmission assembly into the vehicle.

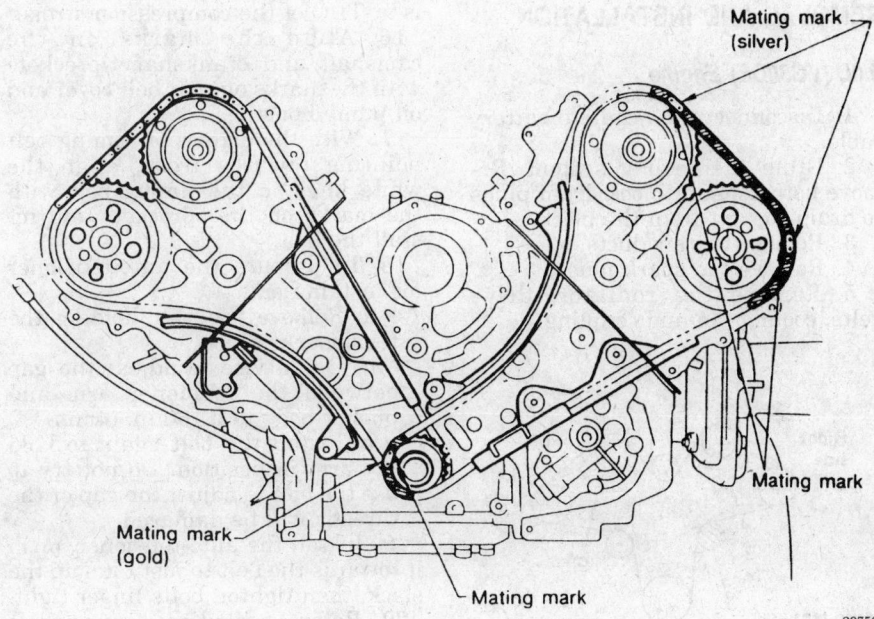

Timing chain mating mark alignment for left bank — 4.5L engine

Timing Belt and Sprockets

ADJUSTMENT

3.0L (VG30DE) Engine

Auto-Tensioner Removed

1. Prepare the auto-tensioner for installation:
   a. Remove the bolt holding the tensioner position.
   b. Use a vise to adjust the gap between the tensioner arm and pusher body to 0.160 in. (4mm).
   c. Install the bolt again to hold the arm in position. Do not try to use the bolt to adjust the gap or the threads will be damaged.
2. Install the auto-tensioner, push it towards the belt to just take up the slack, then tighten bolts finger tight.
3. Before adjusting timing belt tension, the slack must be properly distributed:
   a. Turn the crankshaft 10 degrees clockwise and tighten the tensioner bolts and nut to 12–15 ft. lbs. (16–21 Nm). Do not push the auto-tensioner hard or the belt will be adjusted too tight.
   b. Turn the crankshaft 120 degrees counterclockwise.
   c. Loosen the tensioner bolts and nut 1/2 turn and move the tensioner body away from the timing belt as far as it will move.
   d. Turn the crankshaft clockwise to TDC again.
   e. Push the tensioner against the belt with a force of 13 lbs. (59 N) using a spring scale or similar tool and tighten the bolts again to 12–15 ft. lbs. (16–21 Nm). The pressure specification is important and a special spring scale tool, J–38387, is available to measure the tensioner force.

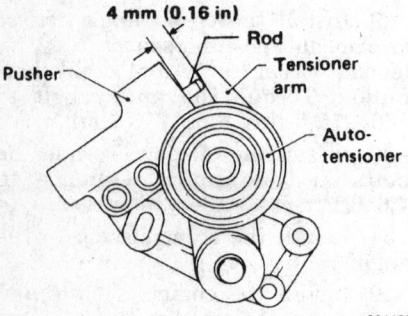

Adjust the tensioner dimension and install the stopper bolt — 3.0L (VG30DE) engine

Auto-Tensioner Installed

1. To check timing belt tension:
   a. Turn the crankshaft 120 degrees clockwise, then turn counterclockwise and return to TDC.
   b. Prepare a steel plate that is about 3/8 in. (10mm) wide and longer than the width of the belt.
   c. Set the plate on the timing belt between two camshaft sprockets and push against the plate with a force of 11 lbs. (49 N). Note the belt deflection.
   d. Repeat the procedure between the other camshaft sprockets and between the exhaust sprockets and idler/tensioner pulleys. There will be a total of four measurements.
   e. Add the deflection measurements and divide by 4. The average deflection must be 0.240–0.280 in. (6–7mm). If belt tension is not correct, start the entire adjustment procedure again.
2. Confirm the auto-tensioner mounting nuts are torqued to 12–15 ft. lbs. (16–21 Nm) and remove the auto-tensioner stopper bolt.
3. After 5 minutes, measure the clearance between the tensioner arm and the pusher. It should be 0.138–0.205 in. (3.5–5.2mm).
4. Make sure all the sprocket timing marks are correctly aligned. Install the timing belt covers and torque the bolts to 24–38 inch lbs. (3–5 Nm).

REMOVAL AND INSTALLATION

3.0L (VG30DE) Engine

1. Disconnect the negative battery cable.
2. Drain the cooling system. Remove both cylinder block drain plugs to drain coolant from the block.
3. Remove the air ducts.
4. Remove the spark plugs.
5. Remove the radiator, drive belts, cooling fan and coupling.

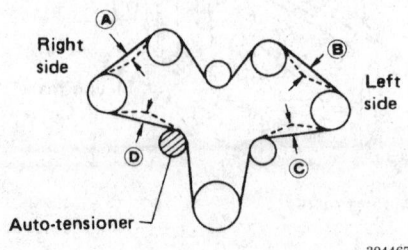

Check belt tension at the indicated points — 3.0L (VG30DE) engine

6. Remove the camshaft position sensor.
7. Remove the starter and use a suitable tool to prevent the flywheel from turning. Remove the crankshaft pulley bolt. Remove the crankshaft pulley using a puller.
8. Remove the water inlet and outlet.
9. Remove the front timing covers.
10. Set the No. 1 cylinder on TDC of the compression stroke.
11. The automatic belt tensioner is oil damped and spring operated. Install a 6mm bolt to hold the tensioner back against the spring and release tension on the belt.
12. Remove the auto-tensioner and timing belt.

--------- **WARNING** ---------
Do not rotate the crankshaft or camshaft separately because the pistons will strike the valves causing engine damage.

13. To remove the camshaft sprockets, remove the intake manifold collector and the valve covers. Hold the camshafts with a wrench and remove the bolts to remove the camshaft sprockets.
To install:
14. If removed, install the camshaft sprockets. Torque the intake sprocket bolts to 90–98 ft. lbs. (123–132 Nm) and use a new O-ring when installing the cover plate.
15. Confirm that the No. 1 cylinder is at TDC of the compression stroke.
16. Align the marks on the camshaft and crankshaft sprockets with the marks on rear belt cover and oil pump housing.
17. With the arrows on timing belt pointing towards front, align the white lines on the timing belt with the marks on the sprockets and install the belt.
18. To prepare the auto-tensioner for installation:
   a. Remove the bolt holding the tensioner position.
   b. Use a vise to adjust the gap between the tensioner arm and pusher body to 0.160 in. (4mm).
   c. Install the bolt again to hold the arm in position. Do not try to use the bolt to adjust the gap or the threads will be damaged.
19. Install the auto-tensioner, push it towards the belt to just take up the slack, then tighten bolts finger tight.
20. Before adjusting timing belt tension, the slack must be properly distributed:
   a. Turn the crankshaft 10 degrees clockwise and tighten the tensioner bolts and nut to 12–15 ft.

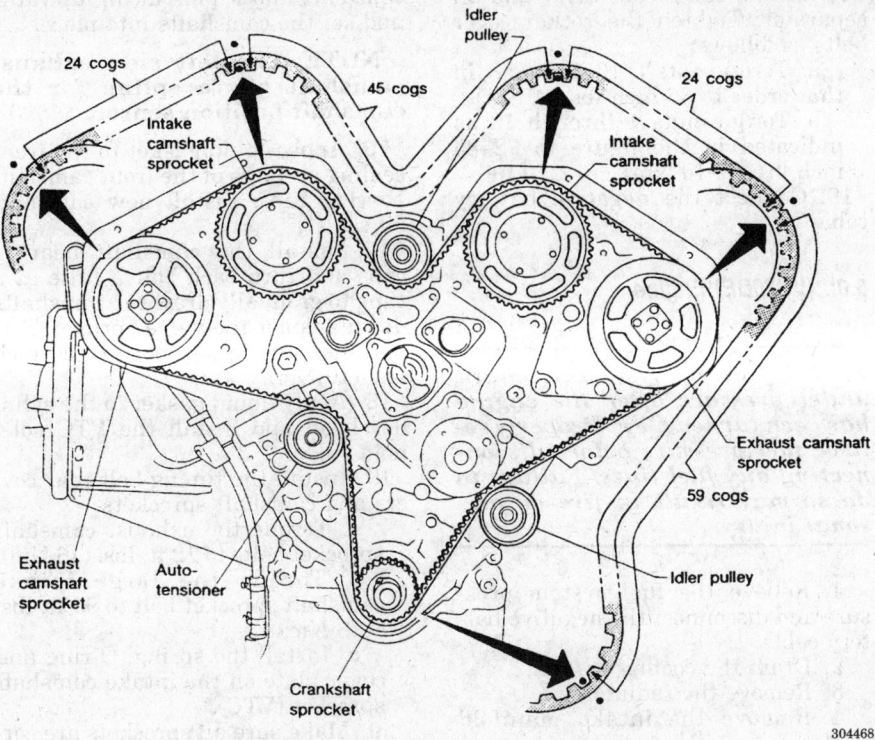

Timing belt alignment marks — 3.0L (VG30DE) engine

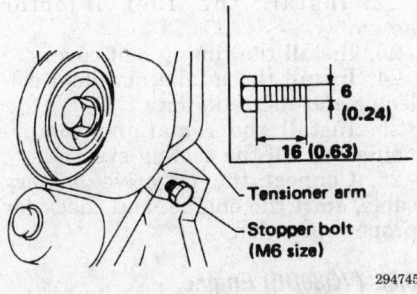

Install a 6mm bolt to push the tensioner away from the belt — 3.0L (VG30DE) engine

lbs. (16–21 Nm). Do not push the auto-tensioner hard or the belt will be adjusted too tight.

b. Turn the crankshaft 120 degrees counterclockwise.

c. Loosen the tensioner bolts and nut ½ turn and move the tensioner body away from the timing belt as far as it will move.

d. Turn the crankshaft clockwise to TDC again.

e. Push the tensioner against the belt with a force of 13 lbs. (59 N) using a spring scale or similar tool and tighten the bolts again to 12–15 ft. lbs. (16–21 Nm). The pressure specification is important and

a special spring scale tool, J–38387, is available to measure the tensioner force.

21. To check timing belt tension:

a. Turn the crankshaft 120 degrees clockwise, then turn counterclockwise and return to TDC.

b. Prepare a steel plate that is about ³/₈ in. (10mm) wide and longer than the width of the belt.

c. Set the plate on the timing belt between two camshaft sprockets and push against the plate with a force of 11 lbs. (49 N). Note the belt deflection.

d. Repeat the procedure between the other camshaft sprockets and between the exhaust sprockets and idler/tensioner pulleys. There will be a total of four measurements.

e. Add the deflection measurements and divide by 4. The average deflection must be 0.240–0.280 in. (6–7mm). If belt tension is not correct, start the entire adjustment procedure again.

22. Confirm the auto-tensioner mounting nuts are torqued to 12–15 ft. lbs. (16–21 Nm) and remove the auto-tensioner stopper bolt.

23. After 5 minutes, measure the clearance between the tensioner arm and the pusher. It should be 0.138–0.205 in. (3.5–5.2mm).

24. Make sure all the sprocket timing marks are correctly aligned. Install the timing belt covers and torque the bolts to 24–38 inch lbs. (3–5 Nm).

25. Install the water inlet and outlet.

26. Install the crankshaft pulley and bolt. Torque the crankshaft pulley bolt to 159–174 ft. lbs. (216–235 Nm).

27. Install the starter assembly.

28. Install the camshaft position sensor.

29. If removed, install the valve covers and the intake manifold collector.

30. Install the cooling fan and fan coupling, drive belts and radiator.

31. Install the air ducts.

32. Connect the negative battery cable.

33. Fill the cooling system and install the engine undercover.

Camshaft

REMOVAL AND INSTALLATION

2.0L (SR20DE) Engine

CAUTION

Fuel injection systems remain under pressure, after the engine has been turned OFF. Properly relieve fuel pressure before disconnecting any fuel lines. Failure to do so may result in fire or personal injury.

1. Relieve the fuel system pressure.

2. Disconnect the negative battery cable. Remove the rocker cover and oil separator.

3. Rotate the crankshaft until the No. 1 piston is at TDC on the compression stroke and the mating marks on the camshaft sprockets line up with the mating marks on the timing chain.

4. Remove the timing chain tensioner.

5. Remove the distributor.

6. Remove the timing chain guide.

7. Remove the camshaft sprockets. Use a wrench to hold the camshaft while loosening the sprocket bolt.

8. Loosen the camshaft bearing cap bolts in the opposite order of the tightening sequence.

9. Remove the camshaft from the cylinder head.

10. When removing the rocker arm, be careful not to drop the valve shims into the cylinder head. After removing the adjuster, set them upright or lay them down in a pan of clean en-

gine oil. Do not lay them down on the bench or the oil will drain out and the adjuster will become air bound. Keep all these parts in order so they can be installed in the same locations.

To install:

11. Install the adjusters, shims and rockers into their original locations.

12. Clean the left hand camshaft end bearing cap and coat the mating surface with liquid gasket. Install the camshafts, bearing caps, oil tubes and baffle plate. Ensure the left camshaft key is at 12 o'clock and the right camshaft key is at 10 o'clock.

13. The procedure for tightening bearing cap bolts must be followed exactly to prevent camshaft damage. Tighten bolts as follows:

   a. Torque right camshaft bolts 9 and 10 (in that order) to 18 inch lbs. (2 Nm) then tighten bolts 1 through 8 (in that order) to the same specification.

   b. Torque left camshaft bolts 11 and 12 (in that order) to 18 inch lbs. (2 Nm) then tighten bolts 1 through 10 (in that order) to the same specification.

   c. Torque all bolts in sequence to 54 inch lbs. (6 Nm).

   d. Torque all bolts again in sequence. Torque type A, B and C bolts to 6.5–8.5 ft. lbs. (9–12 Nm) and type D bolts to 13–19 ft. lbs. (18–25 Nm).

14. Line up the mating marks on the timing chain and camshaft sprockets and install the sprockets. Torque sprocket bolts to 101–116 ft. lbs. (137–157 Nm).

15. Install the timing chain guide and chain tensioner.

16. Install the distributor making sure that rotor head is at 5 o'clock position.

17. Clean the rocker cover and mating surfaces and apply a continuous bead of liquid gasket to the mating surface.

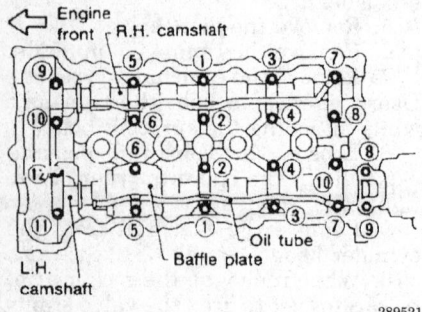

Camshaft bearing cap bolt torque sequence — 2.0L engine

18. Install the rocker cover and oil separator. Tighten the rocker cover bolts as follows:

   a. Torque nuts 1, 10, 11 and 8, in that order to 36 inch lbs. (4 Nm).

   b. Torque nuts 1 through 13 as indicated in the figure to 72–84 inch lbs. (8–10 Nm).

19. Connect the negative battery cable.

3.0L (VG30DE) Engine

———— **CAUTION** ————
Fuel injection systems remain under pressure after the engine has been turned OFF. Properly relieve fuel pressure before disconnecting any fuel lines. Failure to do so may result in fire or personal injury.

1. Relieve the fuel system pressure and disconnect the negative battery cable.

2. Drain the cooling system.

3. Remove the radiator.

4. Remove the intake manifold collector.

5. Remove the fuel injector assembly.

6. Remove the valve covers.

7. Remove the timing belt cover and the timing belt.

8. Remove the camshaft sprockets and the timing belt rear cover.

9. Remove the VTC solenoid valve.

10. Measure the camshaft end-play for installation reference.

11. Loosen the camshaft bearing cap bolts in several steps. Remove the camshaft bearing caps.

12. Remove the oil seals and camshafts.

NOTE: Keep the hydraulic lash adjusters upright and immersed in engine oil to prevent air from entering them. Keep the lash adjusters in order and install them in their original positions.

13. If necessary, remove the hydraulic lash adjusters, keeping them in order for installation into the same location. Make sure the lash adjusters are set upright to prevent them from becoming air bound.

To install:

14. Install the hydraulic lash adjusters into their original locations.

NOTE: Apply new engine oil to the camshafts before installation.

15. Make sure the crankshaft is set at TDC of No. 1 cylinder. Place the

camshaft knock pins facing upwards and set the camshafts into place.

NOTE: The left side exhaust camshaft has a spline for the camshaft position sensor.

16. Apply liquid gasket to the front sealing surfaces of the front camshaft bearing caps. Install new camshaft oil seals.

17. Install the camshaft bearing caps and turn each bolt a little at a time to gradually draw the camshafts down against the valve springs. Torque the bolts in sequence to 90 inch lbs. (10 Nm).

18. Apply liquid gasket to the cylinder head and install the VTC solenoid valve.

19. Install the timing belt rear covers and camshaft sprockets.

   a. Torque the exhaust camshaft sprocket bolts to 12 ft. lbs. (16 Nm).

   b. Torque the large intake camshaft sprocket bolt to 95 ft. lbs. (128 Nm).

   c. Install the spring, O-ring and cover plate on the intake camshaft sprocket (VTC).

20. Make sure all sprockets are correctly aligned and install the timing belt.

21. Install the valve covers with new gaskets.

22. Install the fuel injector assembly.

23. Install the timing belt covers.

24. Install the intake manifold collector and the radiator.

25. Install the remaining components and fill the cooling system.

26. Connect the negative battery cable, start the engine, and check for proper operation.

3.0L (VQ30DE) Engine

1. Relieve the fuel system pressure.

———— **CAUTION** ————
Fuel injection systems remain under pressure after the engine has been turned OFF. Properly relieve fuel pressure before disconnecting any fuel lines. Failure to do so may result in fire or personal injury.

2. Disconnect the negative battery cable.

3. Drain the engine oil and the cooling system. Be sure to drain the engine block and the radiator.

4. Remove the left side rocker cover ornament.

NOTE: Before disconnecting any hoses or connectors, note the locations for reassembly.

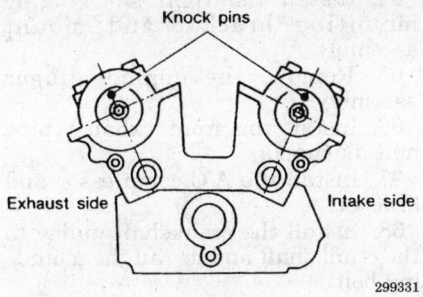

Knock pins

Exhaust side   Intake side

299331

Positioning of the camshaft keys during
installation — 3.0L (VG30DE) engine

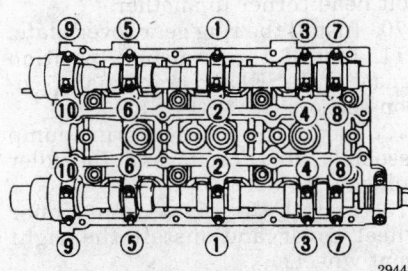

294481

Camshaft bearing cap bolt torque sequence —
3.0L (VG30DE) engine

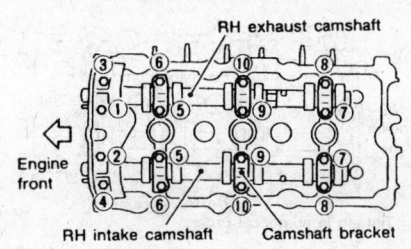

RH exhaust camshaft

Engine
front

RH intake camshaft   Camshaft bracket

Loosen in numerical order.

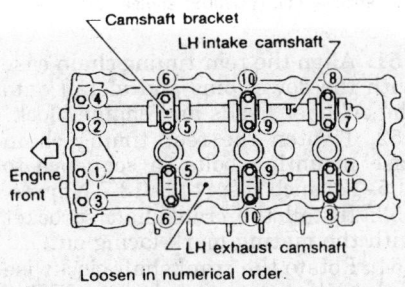

Camshaft bracket
LH intake camshaft

Engine
front

LH exhaust camshaft

Loosen in numerical order.

296148

Camshaft bearing cap loosening sequence —
3.0L (VQ30DE) engine

5. Remove the air duct to intake manifold hose, collector hose, blow-by hose, and vacuum hoses.

6. Remove the fuel hoses and disconnect the harness connections.

7. Disconnect the canister purge hoses.

8. Remove the water hoses from the cylinder head and intake manifold.

9. Disconnect and remove all six ignition coils from the spark plugs.

10. Remove the spark plugs.

11. Remove the bolts that secure the EGR tube and remove the tube.

12. Remove the intake manifold collector supports and remove the collector.

13. Remove the bolts that secure the fuel tube and remove the fuel tube from the vehicle.

14. Remove the bolts that secure the intake manifold to the engine block and remove the manifold. Loosen the bolts in the reverse sequence of the tightening procedure.

15. Remove the LH and RH rocker covers from the cylinder head.

16. Remove the engine undercovers.

17. Remove the right front wheel and engine side covers.

18. Remove the drive belts and idler pulley.

19. Remove the power steering oil pump belt and remove the power steering oil pump assembly.

20. Remove the camshaft position sensor (PHASE) and crankshaft position sensors (REF)/(POS).

21. Set the No. 1 piston to TDC of compression stroke by rotating the crankshaft.

22. Remove the ring gear cover access plate.

23. Loosen the crankshaft pulley bolt while securing the ring gear so the crankshaft cannot rotate.

NOTE: Use care not to damage the ring gear teeth.

24. Using a suitable puller, remove the crankshaft pulley.

25. Remove the A/C compressor and bracket.

26. Remove the front exhaust pipe and its support.

27. Hang the engine at the right and left side engine slingers with a suitable hoist.

28. Support the transaxle with jack.

29. Remove the right side engine mounting, mounting bracket and nuts.

30. Remove the center crossmember assembly.

31. Remove the oil pan bolts and oil pans.

32. Remove the timing chain.

NOTE: Remove the O-rings from the front of the engine block.

33. Loosen the camshaft bearing caps in several steps. The bearing caps MUST be loosened in sequence.

NOTE: Keep all bearing caps and camshafts in proper order for reinstallation.

34. Remove the LH and RH camshaft tensioners from the cylinder head.

35. Remove the camshafts from the cylinder heads.

NOTE: The valve adjusters have a replaceable shim on the top of the adjuster. Note the proper locations of each shim to adjuster and remove the shims from the adjusters.

36. Using a magnet, remove the valve adjusting shim from the adjuster.

37. Remove the adjuster assembly from the bore. Be sure to note the locations from where each adjuster came.

38. Check the diameter of the valve adjuster and the valve adjuster guide bore.

39. The diameter of the adjuster should be 1.3764–1.3770 inches (34.960–34.975mm) and the diameter of the bore should be 1.3780–1.3788 inches (35.000–35.021mm).

40. Remove all traces of liquid gasket from the timing chain case and from the water pump covers.

41. Remove all traces of liquid gasket from the engine block.

42. Inspect the camshafts for excessive wear or damage and replace as necessary.

To install:

43. Lubricate the valve adjusters with clean engine oil and install the adjusters into the bore from which they were removed.

44. Lubricate the valve adjuster shims with clean engine oil and install the shims into the adjuster from which they were removed.

45. Turn the crankshaft clockwise until the No. 1 piston is set 240 degrees before TDC on compression stroke.

46. Install the camshaft tensioners on both sides of the cylinder heads.

Tighten the tensioner mounting bolts to 75–96 inch lbs. (8.4–10.8 Nm).

NOTE: The camshafts can be identified by the paint marks on the camshaft. The left cylinder head camshafts have a YELLOW paint mark and the right cylinder head camshafts have a WHITE paint mark. When installing the camshafts, position the camshaft keys at the 12 o'clock position in respect to the cylinder head angle.

47. Install the exhaust and intake camshafts and install the bearing caps. Before installing the No. 1 bearing cap, apply liquid gasket to the corners of the cap.
48. Tighten the camshaft bearing caps as follows:
   a. Tighten bolts No. 7–10 to 17 inch lbs. (2 Nm).
   b. Tighten bolts No. 1–6 to 17 inch lbs. (2 Nm).
   c. Tighten bolts No. 1–10 to 52 inch lbs. (6 Nm).
   d. Tighten bolts No. 1–10 to 81–104 inch lbs. (9–11 Nm).
49. Install new O-rings to the front of the engine block.
50. Apply sealant to the hatched portion of the of the rear timing chain case.

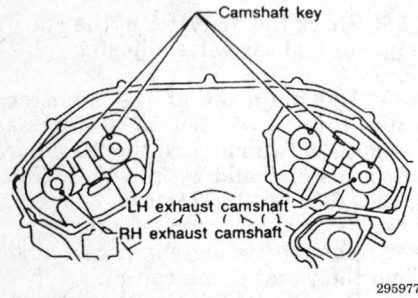

Positioning of the camshaft keys during installation — 3.0L (VQ30DE) engine

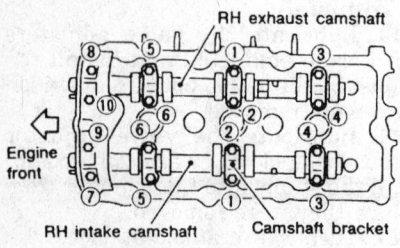

Right cylinder head camshaft bearing cap torque sequence — 3.0L (VQ30DE) engine

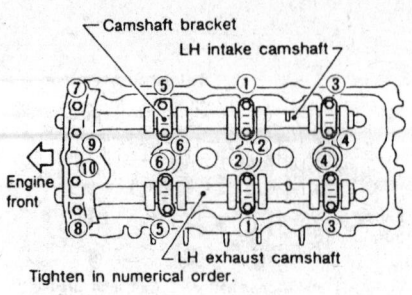

Left cylinder head camshaft bearing cap torque sequence — 3.0L (VQ30DE) engine

51. Align the rear timing chain case with the dowel pins and install onto the cylinder heads and engine block.
52. Tighten the rear timing chain case mounting bolts in sequence to 105–121 inch lbs. (11.8–13.7 Nm).
53. Install the crankshaft sprocket with the mating mark facing out.
54. Rotate the crankshaft clockwise and position the crankshaft to TDC of compression stroke and align the dowels of the camshaft sprockets to the 12 o'clock position in respect to the cylinder head.
55. Install the lower chain guide on the dowel pin with the front mark on the guide facing upward.
56. On a work bench, align the marks on the intake and exhaust camshaft sprockets with the marks of the chain.
57. Put the exhaust camshaft sprockets onto the dowel pin and tighten the mounting bolts to 88–95 ft. lbs. (119–128 Nm). Be sure to secure the camshafts while tightening the bolts.
58. Align and install the timing chains and sprockets to the camshafts.
59. Install the timing cover evenly and gently. Be sure to align the dowel pin holes. Tighten the bolts in sequence

NOTE: Leave the bolts unattended for 30 minutes or more after tightening.

60. Apply a 0.091–0.130 inch (2.3–3.3mm) continuous bead of liquid gasket to the water pump cover and install the cover. Tighten the bolts to 84–108 inch lbs. (10–13 Nm).
61. Install the rocker covers. Tighten the mounting bolts in sequence.
62. Apply sealant to the front and rear seal of the oil pan. Install the bolts and tighten the bolts.
63. Install the center crossmember assembly.

64. Install the right side engine mounting bracket and mount assembly.
65. Remove the engine slinger assembly.
66. Install the front exhaust pipe and its support.
67. Install the A/C compressor and bracket.
68. Install the crankshaft pulley to the crankshaft and install the mounting bolt.
69. Tighten the mounting bolt to 14–22 ft. lbs. (20–29 Nm). Tighten the crankshaft bolt an additional 60–66 degrees clockwise. This is about the angle from one hexagon bolt head corner to another.
70. Install the ring gear cover plate.
71. Install the camshaft position sensor (PHASE) and crankshaft position sensors (REF)/(POS).
72. Install the power steering pump assembly, drive belts and the idler pulley.
73. Install the right front inner wheel cover and install the right front wheel.
74. Install the engine undercovers.
75. Using new gaskets, install the intake manifold. Tighten the nuts and bolts in sequence and in two stages as follows:
   a. Bolts and nuts — Tighten to 44–86 inch lbs. (5–10 Nm)
   b. Bolts and nuts — Tighten to 16–18 ft. lbs. (22–25 Nm)
76. Using new insulators, install the fuel tube assembly and tighten the bolts to 15–20 ft. lbs. (21–26 Nm). Tighten the bolts in several progressive steps.
77. Install the intake manifold collector gasket with the arrow facing forward.
78. Install the intake manifold collector assembly and support bracket. Tighten the mounting bolts to 16–18 ft. lbs. (22–25 Nm).
79. Using new gaskets, install the EGR tube and tighten the mounting bolts to 15–20 ft. lbs. (21–26 Nm) in two progressive steps.
80. Install the spark plugs, ignition coils and tighten the mounting bolts to 27–33 inch lbs. (2.9–3.8 Nm).
81. Install the water hoses to the cylinder head and intake manifold.
82. Connect the fuel hoses and wiring harness connections to the fuel rail.
83. Install and connect the air duct to intake manifold hose, collector hose, blow-by hose, and vacuum hoses.
84. Refill the engine oil and coolant with the proper type and amount of fluid.

85. Connect the negative battery cable.

86. Start the engine and run at 3000 RPM under no load to purge the air from the high pressure chamber. The engine may produce a rattling noise. This indicates that air still remains in the chamber and is not a matter of concern.

4.5L (VH45DE) Engine

1. Disconnect the negative battery cable.
2. Remove the engine and transmission assembly from the vehicle.
3. Remove the suspension member and engine mounts from the engine.
4. Remove the air compressor bracket.
5. Remove the cooling fan with coupling and the engine gusset.
6. Separate the engine from the transmission and mount the engine on a suitable workstand.
7. Remove the oil pan.
8. Remove the ornamental rocker cover and remove the ignition coils and spark plugs.
9. Bring the No. 1 piston to TDC on the compression stroke.
10. Use a suitable puller to remove the crankshaft pulley.
11. Remove the rocker cover.
12. Remove the crank angle sensor and the Valve Timing Control (VTC) solenoid.
13. Remove the chain tensioners and the upper front covers.
14. Remove the front timing chain cover.

NOTE: The timing chain will not be disengaged or dislocated from the crankshaft sprocket unless the front cover is removed. The cast portion of the front cover is located on the lower side

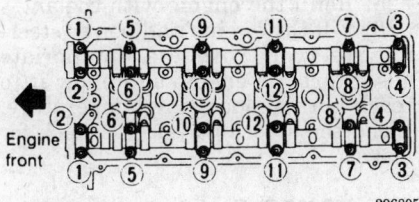

Camshaft bearing cap bolt torque sequence — 4.5L engine

of the crankshaft sprocket so the timing chain is not disengaged from the sprocket.

15. Remove the VTC assembly and the camshaft sprocket.
16. Remove the oil pump chain and the timing chains.

NOTE: Do not attempt to disassemble the VTC assembly since they are difficult to reassemble accurately in the field. If it should be disassembled, the VTC assembly must be replaced with a new one.

17. Remove the camshaft brackets and the camshafts. Mark the parts so they can be reinstalled in their original positions.
18. Remove the rocker arms. Be sure to identify each rocker arm so it can be reinstalled in it's original position.

To install:

19. Make sure all mating surfaces are clean before installation.
20. Install the rocker arms, camshafts and camshaft brackets on the right bank. Properly lubricate the rocker arms and camshafts prior to installation. Tighten the camshaft bracket bolts to 9–10 ft. lbs. (12–14 Nm) in the proper sequence.
21. Install the VTC assembly and the exhaust cam sprocket on the right bank.
22. Make sure the camshafts are still correctly positioned and the piston in the No. 1 cylinder is still at TDC.
23. Install the timing chain on the right bank, aligning the mating marks on the chain with those on the crankshaft and camshaft sprockets.
24. Install the chain tensioner on the right bank.
25. Turn the crankshaft approximately 120 degrees clockwise from the point where the No. 1 piston is at TDC on the compression stroke. At this point, the valves on the left bank still remain closed.
26. Correctly position the camshafts and rocker arms for the left cylinder head. Properly lubricate the rocker arms and camshafts prior to installation. Tighten the camshaft bracket bolts to 9–10 ft. lbs. (12–14 Nm) in the proper sequence. Install the VTC assembly and the exhaust cam sprocket.
27. Install the timing chain on the left bank, aligning the mating marks

on the chain with those on the crankshaft and camshaft sprockets.

28. Install the oil pump chain and sprockets.
29. Install the oil pump chain guides. Place a 0.04 in. (1.0mm) feeler gauge between the upper chain guide and chain before assembling the chain guides. The force applied to the chain is equivalent to the upper chain guide weight.
30. Apply suitable sealer and install the front covers.
31. Install the chain tensioner for the left bank.
32. Apply suitable sealer to the rubber plugs and install them on the cylinder head.
33. Install the crank angle sensor, VTC solenoid, rocker cover and crank pulley.
34. Installation of the remaining components is the reverse of the removal procedure.

Piston and Connecting Rod

POSITIONING

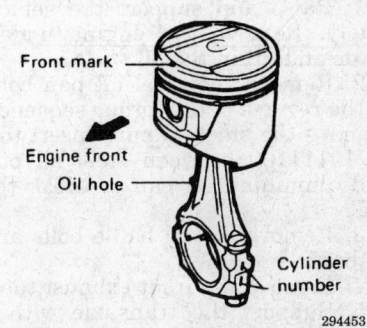

Correct piston and connecting rod alignment — all except 3.0L (VQ30DE) engines

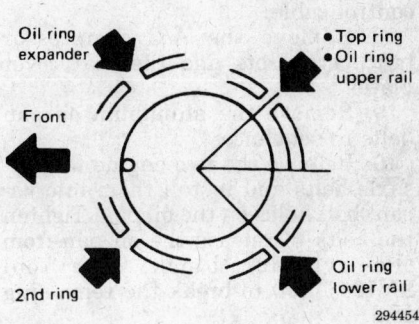

Correct piston ring gap placement — all engines

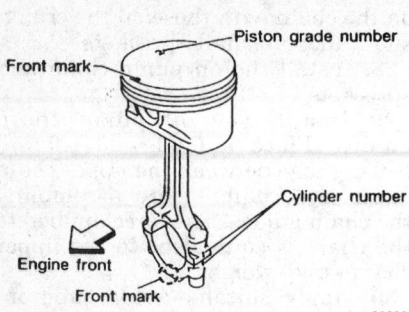

Correct piston and connecting rod alignment — 3.0L (VQ30DE) engine

ENGINE LUBRICATION

Oil Pan

REMOVAL AND INSTALLATION

2.0L (SR20DE) Engine

1. Raise and support the vehicle safely. Remove the engine undercover and drain the oil.
2. Remove the steel oil pan bolts in the reverse of the torque sequence. Remove the steel oil pan. Insert tool KV10111100 between steel oil pan and aluminum oil pan to break the seal.
3. Remove the oil baffle bolts and oil baffle.
4. Remove the front exhaust tube.
5. Support the transaxle with a suitable jack and raise the engine with an engine hoist.
6. Remove the center crossmember.
7. If equipped with an automatic transaxle, remove the transaxle shift control cable.
8. Remove the A/C compressor bracket gussets and the rear cover plate.
9. Remove the aluminum oil pan bolts in sequence.
10. Remove the two engine to transaxle bolts and install them into vacant bolt holes on the oil pan. Tighten the bolts to release the oil pan from the cylinder block. Use tool KV10111100 to break the remaining seal.
  To install:
11. Remove the two bolts previously installed in the oil pan.

12. Clean the oil pan rail of all liquid gasket and apply a new bead of 1/8 inch thickness to the oil pan rail.
13. Install the aluminum oil pan and torque them in the numbered sequence. Torque bolts 1 through 16 to 12–14 ft. lbs. (16–19 Nm) and bolts 17 through 18 to 60–72 inch lbs. (6–8 Nm).
14. Install the two engine-to-transaxle bolts, rear cover plate, compressor bracket gussets, automatic transmission shift control cable (if equipped), center member, front exhaust tube and baffle plate.
15. Clean the oil pan rail of all liquid gasket and apply a new bead of 1/8 inch thickness to the oil pan rail.
16. Install the steel oil pan. Torque the bolts in numbered sequence to 56–66 inch lbs. (6.4–7.5 Nm). Wait 30 minutes before refilling crankcase with oil.

3.0L (VG30DE) Engine

1. Disconnect the negative battery cable. Raise and support the vehicle safely.
2. Remove the engine undercover and drain the engine oil and coolant.
3. Remove the radiator from the engine compartment.
4. Remove the air ducts.

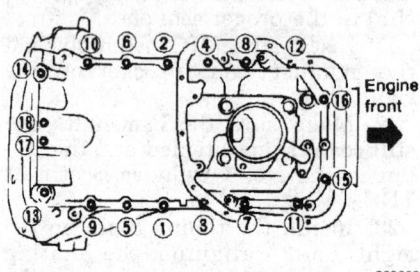

Aluminum oil pan bolt installation torque sequence — 2.0L engine

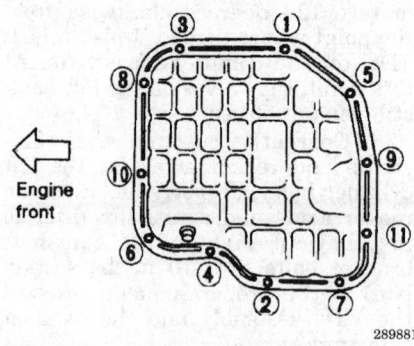

Steel oil pan bolt installation torque sequence — 2.0L engine

5. Remove the upper and lower radiator shrouds. Disconnect the oil cooler lines on vehicles with automatic transmissions.
6. Remove the fan coupling.
7. Disconnect the power steering oil hoses.
8. Remove the power steering oil pump.
9. Remove the stabilizer bar.
10. Place a suitable support under the transmission.
11. Remove the engine mounting insulator lower attaching nuts from both sides of the engine. Hoist the engine so enough clearance is provided for oil pan removal.
12. Remove the oil pan bolts in the reverse order of installation. Tap the oil pan removal tool J–37228 or equivalent, between the cylinder block and oil pan. Tap the tool around oil pan to break the gasket seal.
13. Remove the oil strainer and lay it in the oil pan. Remove the oil pan.
  To install:
14. Clean all gasket mating surfaces thoroughly.
15. Apply sealant to oil pump gasket and rear oil seal retainer gasket.
16. Apply a continuous bead of liquid gasket to the oil pan mating surface. Be sure the bead is 3/16 inch wide.
17. Place oil pan under engine and install oil strainer.
18. Install oil pan and tighten bolts in the proper sequence to 60–72 inch lbs. (7–8 Nm).

  NOTE: Wait at least 30 minutes before refilling the engine with oil

19. Lower the engine and install the engine mount insulator bolts.
20. Install the stabilizer bar.
21. Install the power steering oil pump and hoses.
22. Install the radiator, radiator hoses, shroud and transmission oil cooler lines.
23. Install the fan coupling.
24. Install the air ducts and engine undercover.
25. Refill the engine with coolant.
26. Fill the engine and power steering reservoir with the appropriate oils. Start the engine and allow it to reach normal operating temperature. Bleed the steering system. Check for leaks.

3.0L (VQ30DE) Engine

1. Disconnect the negative battery cable.
2. Drain the engine oil and remove the engine undercovers.

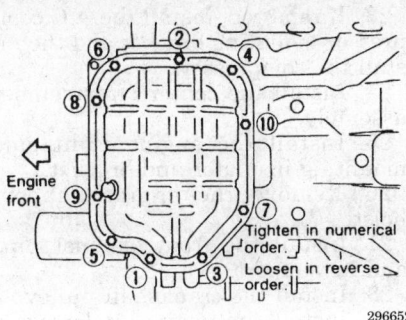

Steel oil pan torque sequence–loosen in
reverse sequence — 3.0L (VG30DE) engine

3. Remove the steel (lower) oil pan bolts in the reverse sequence of the torque sequence.

4. Insert a seal cutter between the steel and aluminum oil pan.

5. Tapping the cutter with a hammer, slide it around the entire edge of the oil pan. Be careful not to damage the aluminum mating surface of the upper oil pan.

6. Remove the steel oil pan and the oil strainer.

7. Remove the front exhaust pipe and its support.

8. Hang the engine at the right and left side engine slingers with a suitable hoist.

9. Position a suitable jack under the transaxle.

10. Remove the crankshaft position sensors (REFERENCE and POSITION) from the oil pan.

11. Remove the front and rear engine mounting nuts and bolts.

12. Remove the center crossmember assembly.

13. Remove the engine drive belts.

14. Remove the air conditioner compressor and the compressor mounting bracket.

15. Remove the rear cover plate and the lower transaxle bolts.

16. Remove the aluminum (upper) oil pan bolts in the reverse sequence of the torque sequence.

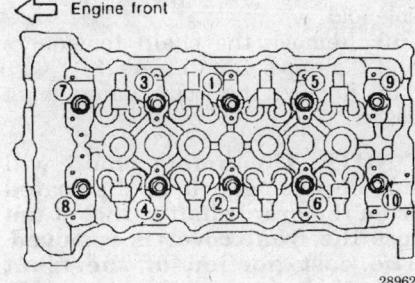

Aluminum oil pan torque sequence–loosen in
reverse sequence — 3.0L (VQ30DE) engine

17. Insert a seal cutter between the aluminum oil pan and the engine block.

18. Tapping the cutter with a hammer, slide it around the entire edge of the oil pan. Be careful not to damage the mating surfaces of the oil pan or engine block.

19. Remove the oil pan assembly.

20. Remove the bolts that secure the baffle plate and remove the baffle plate.

21. Remove the O-rings from the cylinder block and oil pump body.

To install:

22. Install the baffle plate to the oil pan and tighten the mounting bolts to 22–27 inch lbs. (2.5–3.1 Nm).

23. Apply sealant to the front and rear seal of the oil pan.

24. Install new O-rings to the cylinder block and the oil pump body.

25. Apply a 0.177–0.217 inch (4.5–5.5mm) continuous bead of liquid gasket to the upper oil pan mating surface and install the oil pan. Tighten the mounting bolts in sequence to 12–14 ft. lbs. (16–19 Nm).

26. Install the oil pan strainer and tighten the mounting bolts to 12–14 ft. lbs. (16–19 Nm).

27. Install the rear cover plate and the lower transaxle bolts.

28. Install the A/C compressor mounting bracket and compressor.

29. Install and adjust the engine drive belts.

30. Install the center crossmember assembly.

31. Install the front and rear engine mounting nuts and bolts.

32. Remove the support jack and the engine hoist.

33. Install the crankshaft position sensors (REFERENCE and POSITION) to the oil pan. Tighten the sensor mounting bolts to 75–96 inch lbs. (9–10 Nm).

34. Install the front exhaust pipe and its support.

35. Install the oil strainer.

36. Apply a 0.177–0.217 inch (4.5–5.5mm) continuous bead of liquid gasket to the lower oil pan mating surface and install the oil pan. Tighten the mounting bolts in sequence to 57–66 inch lbs. (6.4–7.5 Nm).

NOTE: Wait at least 30 minutes before refilling the engine oil.

37. Tighten the oil pan drain plug to 22–29 ft. lbs. (29–39 Nm).

38. Install the engine undercovers.

39. Refill the engine oil with the proper type and amount of fluid.

40. Start the engine and check for leaks.

4.5L (VH45DE) Engine

1. Disconnect the negative battery cable. Raise and support the vehicle safely.

2. Remove the engine undercover and drain the engine oil.

3. Remove the fan coupling with the fan.

4. Remove the drive belts, alternator, air conditioning compressor and engine gusset.

5. Matchmark and remove the steering lower joint.

6. Place a suitable support under the transmission. Hoist the engine with engine sling (support fixture).

7. Remove the suspension member assembly.

8. Remove the oil pan bolts. Insert oil pan removal tool J–37228 or equivalent, between the cylinder block and oil pan. Tap the tool with a hammer around the oil pan to break the seal.

9. Remove the oil pan.

To install:

10. Clean all gasket mating surfaces thoroughly.

11. Apply a continuous bead of liquid gasket to the oil pan mating surface. Be sure the bead is 1/8 inch wide.

12. Install oil pan and torque the bolts in the proper sequence 60–72 inch lbs. (7–8 Nm). Wait at least 30 minutes for the sealant to cure before filling the engine with oil.

13. Install all remaining components in the reverse order of removal.

14. Fill the engine with oil. Start the engine and allow it to reach normal operating temperature. Check for leaks.

Oil Pump

REMOVAL AND INSTALLATION

2.0L (SR20DE) Engine

── **CAUTION** ──
Fuel injection systems remain under pressure after the engine has been turned OFF. Properly relieve fuel pressure before disconnecting any fuel lines. Failure to do so may result in fire or personal injury.
────────────

1. Relive the fuel system pressure and disconnect the negative battery cable.

2. Remove the drive belts.

3. Remove the cylinder head with the intake and exhaust manifolds attached.

4. Remove the oil pans.

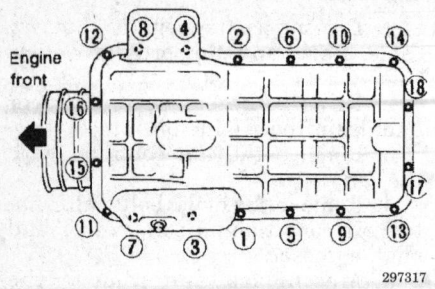

Engine front

Oil pan bolt installation torque sequence — 4.5L engine

297317

5. Remove the oil strainer and baffle plate.

6. Remove the crankshaft pulley and the front cover assembly.

7. Remove the screws to remove the oil pump from the inside of the front cover.

To install:

8. Coat the oil pump gears with oil and fit the pump to the cover, using a new oil seal and O-ring.

9. Clean the mating surfaces of liquid gasket and apply a fresh bead of ⅛ inch sealer to the sealing surface of the front cover. Install the front cover assembly.

10. Install the crankshaft pulley.

11. Install the oil strainer, baffle plate, oil pans, cylinder head and drive belts.

12. Connect the negative battery cable.

3.0L (VG30DE) Engine

1. Raise and safely support the vehicle.

2. Drain the engine oil. Remove the oil level gauge (dipstick).

3. Remove the timing belt and crankshaft sprocket. Remove the oil pan.

4. Remove the oil pump mounting bolts and lift out the oil pump.

To install:

5. Always replace with a new oil seal and gasket. Apply oil to the inner and outer gears when installing.

6. Install the oil pump and tighten the long mounting bolt to 9–12 ft. lbs. (12–16 Nm) and the short bolts to 4.3–5.1 ft. lbs. (6–7 Nm).

7. Install the oil pan and oil level gauge.

8. Install the crankshaft sprocket and the timing belt.

9. Fill the engine with oil. Start the engine and check for leaks.

3.0L (VQ30DE) Engine

NOTE: The oil pump bolts to the front of the engine block and is driven by the crankshaft. Removal of the timing cover and chains are necessary for oil pump service.

1. Disconnect the negative battery cable and drain the engine oil.

2. Rotate the engine and position it to TDC compression stroke of cylinder No. 1.

3. Remove the drive belts.

4. Remove the camshaft position sensor (PHASE) and the crankshaft position sensor (REF/POS).

5. Remove the right front wheel and inner fender cover.

6. Remove the engine undercovers.

7. Remove the bolt that secures the crankshaft pulley and remove the pulley.

8. Remove the front exhaust pipe and its support.

9. Support the engine at the left and right side slingers with a suitable hoist.

10. Remove the engine right side mounting insulator and bracket nuts and bolts.

11. Remove the center crossmember assembly.

12. If equipped, remove the A/C compressor and the mounting bracket.

13. Remove the lower and upper oil pans.

14. Remove the oil strainer from the oil pump.

15. Remove the water pump cover and remove the front cover assembly.

16. Remove the lower timing chain assembly.

17. Remove the bolts that secure the oil pump to the engine block and remove the oil pump.

To install:

NOTE: When installing the oil pump, be sure to apply engine oil to the gears.

18. Install the oil pump to the engine block. Tighten the mounting bolts to 57 inch lbs. (6.5 Nm) and tighten the mounting screws to 33–44 inch lbs. (4–5 Nm).

19. Install the lower timing chain assembly.

20. Install the front timing cover and water pump covers.

21. Using a new gasket, install the oil strainer and tighten the mounting bolts to 12–14 ft. lbs. (16–19 Nm).

22. Install the upper and lower oil pans. Be sure to use new O-rings at the oil pump to upper oil pan mating surface.

23. If removed, install the A/C compressor mounting bracket and the install the compressor.

24. Install the center crossmember assembly.

25. Install the engine right side mounting insulator and bracket.

26. Remove the engine support hoist.

27. Install the front exhaust pipe and its support.

28. Install the crankshaft pulley.

29. Install the engine undercovers and install the right side inner fender cover.

30. Install the right front wheel.

31. Install the camshaft position sensor (PHASE) and the crankshaft position sensor (REF/POS).

32. Install and adjust the engine drive belts.

33. Refill the engine oil with the proper type and amount.

34. Start the engine, check the oil pressure, and check for oil leaks.

4.5L (VH45DE) Engine

1. The oil pump is mounted in the cylinder block below the left bank and behind the left timing chain. Disconnect the negative battery cable.

2. Remove the engine and transmission assembly from the vehicle.

3. Remove the suspension member and engine mounts from the engine.

4. Remove the air compressor bracket.

5. Remove the cooling fan with coupling and the engine gusset.

6. Separate the engine from the transmission and mount the engine on a suitable workstand.

7. Remove the oil pan.

8. Remove the ornamental rocker cover and remove the ignition coils and spark plugs.

9. Bring the No. 1 piston to TDC on the compression stroke.

10. Remove the rocker cover.

11. Use a suitable puller to remove the crankshaft pulley.

12. Remove the crank angle sensor and the Valve Timing Control (VTC) solenoid.

13. Remove the chain tensioners and the upper front covers.

14. Remove the front timing chain cover.

NOTE: The timing chain will not be disengaged or dislocated from the crankshaft sprocket unless the front cover is removed. The cast portion of the front cover is located on the lower side of the crankshaft sprocket so the timing chain is not disengaged from the sprocket.

15. Remove the VTC assembly and the camshaft sprocket.

16. Remove the oil pump chain and the timing chains.

17. Remove the mounting bolts and lift out the oil pump.

To install:

18. Thoroughly clean the mounting surfaces. Apply engine oil to the gears.

19. Install the oil pump with a new seal and gasket. Torque the long bolts to 12–15 ft. lbs. (16–20 Nm) and the short bolts to 3.3–4.3 ft. lbs. (4–6 Nm).

20. Make sure all mating surfaces are clean before installation.

21. Install the VTC assembly and the exhaust cam sprocket on the right bank.

22. Make sure the camshafts are still correctly positioned and the piston in the No. 1 cylinder is still at TDC.

23. Install the timing chain on the right bank, aligning the mating marks on the chain with those on the crankshaft and camshaft sprockets.

24. Install the chain tensioner on the right bank.

25. Turn the crankshaft approximately 120 degrees clockwise from the point where the No. 1 piston is at TDC on the compression stroke. At this point, the valves on the left bank still remain closed.

26. Correctly position the camshafts and rocker arms for the left cylinder head. Properly lubricate the rocker arms and camshafts prior to installation. Install the VTC assembly and the exhaust cam sprocket.

27. Install the timing chain on the left bank, aligning the mating marks on the chain with those on the crankshaft and camshaft sprockets.

28. Install the oil pump chain and sprockets.

29. Install the oil pump chain guides. Place a 0.040 in. (1.0mm) feeler gauge between the upper chain guide and chain before assembling the chain guides. The force applied to the chain is equivalent to the upper chain guide weight.

30. Apply suitable sealer and install the front covers.

31. Install the chain tensioner for the left bank.

32. Apply suitable sealer to the rubber plugs and install them on the cylinder head.

33. Install the crank angle sensor, VTC solenoid, rocker cover and crank pulley.

34. Install the drive belts.

35. Install the engine mounts and the suspension crossmember.

36. Install the engine assembly.

37. Connect all vacuum hoses, cables and electrical connections.

38. Refill the engine with coolant and engine oil.

39. Start the engine and check for leaks.

TRANSMISSION AND TRANSAXLE

Manual Transaxle Assembly

REMOVAL AND INSTALLATION

G20 Models

1. Disconnect the negative battery cable and remove the air duct.

2. Raise and safely support the vehicle.

3. Disconnect the clutch control cable and speedometer cable from the transaxle.

4. Disconnect the backup light switch, neutral switch and ground harness connectors.

5. Remove the starter, shift control rod and support rod from the transaxle.

6. Drain the gear oil from the transaxle and remove the exhaust front tube.

7. Remove the halfshafts.

8. Support the engine with a suitable jack under the oil pan.

9. Remove the rear and left engine mounts

10. Raise the jack and remove the lower transaxle housing bolts. Lower jack and remove the upper housing bolts. Keep the bolts in order as they are different lengths and must be returned to the same position.

11. Lower the transaxle.

To install:

12. Raise the transaxle into place and install the attaching bolts. Torque the two shortest bolts to 22–30 ft. lbs. (30–40 Nm) and the remaining bolts to 51–59 ft. lbs. (70–79 Nm)

13. Install the rear and left engine mounts.

14. Install the driveshafts.

15. Install the shift control rods, support rod and starter on the transaxle.

16. Connect the backup light switch, neutral switch and ground harness connectors.

17. Connect the clutch control cable and speedometer cable from the transaxle.

18. Refill the transaxle with oil and lower the vehicle to the floor.

19. Install the air duct and connect the negative battery cable. Road test the vehicle.

I30 Models

1. Disconnect the negative and the positive battery cables.

2. Remove the battery, battery bracket and tray.

3. Remove the air cleaner assembly with the mass air flow sensor.

4. Raise and safely support the vehicle so there is clearance to remove the transaxle from underneath. Securely support the engine via the oil pan using a cushioning wooden block and a floor jack.

5. Remove the clutch operating cylinder; do not disconnect the hydraulic line from the cylinder.

6. Remove the clutch hose clamp.

7. Disconnect the speedometer pinion and the neutral position switch connectors and the ground harness connectors.

8. Remove the starter motor assembly from the transaxle.

9. Disconnect the back-up lamp switch and the neutral position switch.

10. Remove the crankshaft position sensor (POS) from the transaxle front side.

11. Remove the shifter control rod and the support rod bracket from the transaxle.

12. Drain the fluid from the transaxle.

13. Remove both driveshafts from the transaxle assembly. Securely support the transaxle with another jack.

14. Support the engine of the transaxle by placing a jack under the oil pan. Be sure to use a block of wood between the oil pan and jack.

15. Remove the bolts that secure the center crossmember.

16. Remove the LH engine mounts.

NOTE: The transaxle bolts are of different lengths, be sure to note the location of the bolts for reassembly.

17. Remove the transaxle bolts. Remove the transaxle from the vehicle by sliding the transaxle input shaft out of the clutch, lowering the rear of the transaxle and then lowering the transaxle from of the vehicle.

To install:

18. Install the transaxle assembly to the bell housing while aligning the

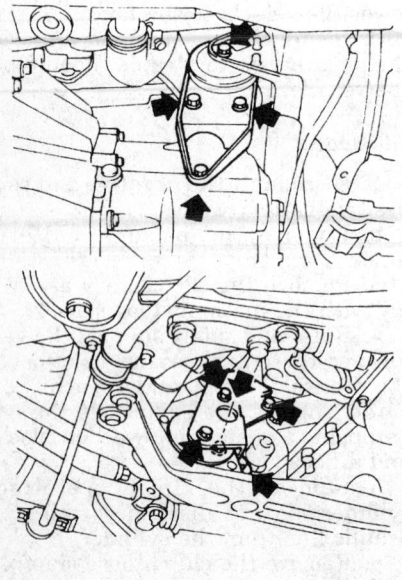

Transaxle mount locations — G20 models

output shaft of the transaxle with the clutch disc. Torque the transaxle bolts to specifications.

19. Install the LH engine mount and torque the through bolt to 32–41 ft. lbs. (43–55 Nm)

20. Install the center crossmember assembly and tighten the mounting bolts to 57–72 ft. lbs. (77–98 Nm).

21. Install both driveshafts to the transaxle assembly.

22. Install the shifter control rod and the support rod bracket to the transaxle.

23. Install the crankshaft position sensor (POS) to the transaxle front side.

24. Connect the back-up lamp switch and the neutral position switch.

25. Install the starter motor assembly to the transaxle.

26. Connect the speedometer pinion and the ground harness connectors.

27. Install the clutch hose clamp.

28. Install the clutch operating cylinder and tighten the bolts to 22–30 ft. lbs. (30–40 Nm).

29. Install the air cleaner assembly with the mass air flow sensor.

30. Install the battery tray , bracket and battery.

31. Connect the positive and the negative battery cables.

32. Refill the transaxle with proper amount and type of fluid.

33. Road test the vehicle for proper shift operation.

Clutch Assembly

REMOVAL AND INSTALLATION

G20 Models

1. Disconnect the negative battery cable.

2. Raise and support the vehicle safely.

3. Remove the transaxle.

4. Insert tool KV30101000 or equivalent alignment tool into the clutch disc hub and loosen the pressure plate bolts in small increments using a star-type pattern.

5. Remove the pressure plate and clutch disc as an assembly.

6. Remove the release bearing by pulling the bearing retainers outward from the transaxle case.

7. Inspect the clutch disc for surface wear. Measure from the friction surface to the top of the rivets. Wear limit is 0.012 in. (0.3mm). Replace clutch disc as necessary.

8. Inspect the contact surface of the flywheel for burns or discoloration. Check flywheel run-out. Maximum run-out is 0.0059 in. (0.15mm).

9. Using tools ST20050100 and ST20050010 or equivalent, check pressure plate diaphragm springs. Measure from the pressure plate/flywheel mating surface to the top of the diaphragm spring. Height should be 1.201–1.280 in. (30.5–32.5mm). Replace pressure plate as necessary.

10. Inspect the release bearing for damage. Spin the bearing to see that it rolls freely.

To install:

11. Lightly lubricate the transaxle input shaft, input shaft collar, clutch lever assembly and the clutch release bearing with a lithium based grease.

NOTE: Keep clutch disc and all clutch components clean during installation. Do not allow grease to contact the clutch disc.

12. Insert tool KV30101000 or equivalent alignment tool into the clutch disc hub. Install the clutch disc and pressure plate on the tool and torque the pressure plate bolts to 16–22 ft. lbs. (22–29 Nm) in 2–3 steps using a crisscross pattern. Remove the tool.

13. Install release bearing in the transaxle. Ensure that the bearing retainer clips are fully engaged.

14. Install the transaxle.

15. Adjust clutch pedal height and free-play.

16. Lower vehicle, connect negative battery cable and road test vehicle.

I30 Models

1. Remove the battery and battery bracket.

2. Remove the air cleaner and the air flow meter.

3. Raise and safely support the vehicle so there is clearance to remove the transaxle from underneath. Securely support the engine via the oil pan using a cushioning wooden block and jack.

4. Drain the fluid from the transaxle.

5. Remove the transaxle from the engine and lower to the floor.

6. Insert a clutch aligning bar or similar tool all the way into the clutch disc hub. This must be done so as to support the weight of the clutch disc during removal. Mark the clutch assembly-to-flywheel relationship with paint or a center punch so the clutch assembly can be assembled in the same position from which it is removed.

7. Loosen the bolts in reverse order of tightening sequence, a turn at a time. Remove the bolts.

8. Remove the pressure plate and clutch disc.

9. Remove the release mechanism from the transaxle housing.

10. Inspect the pressure plate for wear, scoring, etc., and resurface or replace, as necessary.

11. Measure the thickness of the clutch plate lining to the rivet heads; if the it is worn to a minimum of 0.012 in. (0.3mm), replace the clutch plate.

12. Inspect the release bearing and replace as necessary.

13. Using a dial indicator, mount it to the engine and inspect the flywheel run-out; if the run-out exceeds 0.0059 in. (0.15mm), replace it.

To install:

14. Apply a small amount of grease to the transaxle input shaft splines. Install the disc on the splines and slide back and forth a few times. Remove the disc and remove excess grease on hub. Be sure no grease contacts the disc or pressure plate.

15. Apply lithium based molybdenum disulfide grease to the bearing sleeve inside groove, the contact point of the withdrawal lever and bearing sleeve, the contact surface of the lever ball pin and lever.

16. Install the release mechanism and release bearing.

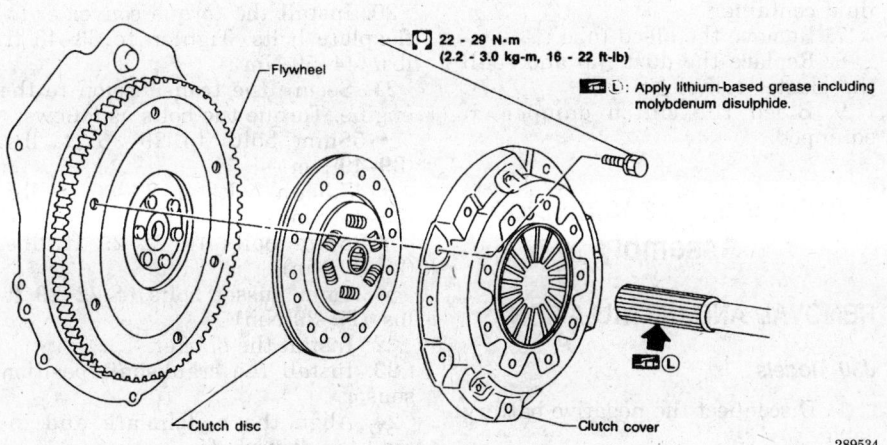

Clutch disc and pressure plate — G20 models

289534

17. Install the pressure plate and clutch disc, aligning it with a splined dummy shaft tool KV301010000 or equivalent.

18. Torque the pressure plate bolts in sequence to 25–33 ft. lbs. (34–44 Nm).

19. Remove the dummy shaft.

20. Install the transaxle in the correct position. Torque the transaxle-to-engine bolts.

21. Connect the rear and LH mounts.

22. Connect the speedometer cable.

23. Connect the electrical harness connector.

24. Install the clutch release cylinder.

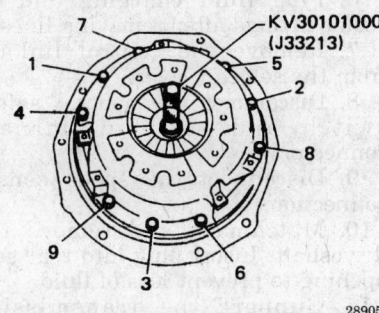

289058

Pressure plate bolt tightening sequence — I30 models

25. Install the starter assembly.

26. Securely support the transaxle and install the driveshafts.

27. Refill the transaxle with the required amount of approved fluid.

28. Install the air flow meter and the air cleaner.

29. Install the battery and battery bracket.

30. Road test the vehicle for proper shift operation.

Clutch Cable

ADJUSTMENT

G20 Models

NOTE: Clutch pedal height must be correct before the cable can be adjusted.

1. Adjust pedal height by loosening the locknut on the pedal stopper or cruise control cancel switch. Pedal height should be 6.28–6.67 in. (159.5–169.5mm) from the top of the pedal to the floor, measured at a 90 degree angle to the top of the pedal.

2. To adjust the clutch cable:

  a. Push the withdrawal lever until resistance is felt.

  b. Tighten the adjusting the nut.

  c. Turn the adjusting nut 2.5–3.5 turns back and then

tighten the locknut. Free-play at the withdrawal lever should be 0.098–0.138 in. (2.5–3.5mm).

3. As a final check, measure pedal free travel at the center of the pedal pad. Pedal free travel should be 0.425–0.594 in. (10.8–15.1mm).

4. Make sure the starting switch on the pedal bracket still functions correctly. The starter should operate with the ignition switch when the pedal is depressed fully.

REMOVAL AND INSTALLATION

G20 Models

The transaxle available with the 2.0L engine is the only transaxle which utilizes a cable driven clutch.

1. Raise and support the vehicle safely.

2. Loosen the locknut and adjusting nut on the clutch cable at the transaxle end and disconnect the cable.

3. Disconnect the cable from the clutch pedal under the dash.

4. Remove any clips or ties holding the cable to the chassis and remove the cable.

To install:

5. Install the cable using the original routing.

6. Connect the cable at the clutch pedal and the transaxle.

7. Adjust the pedal height and free-play.

8. Lower the vehicle and road test.

Clutch Master Cylinder

REMOVAL AND INSTALLATION

I30 Models

1. Disconnect the clutch pedal arm from the pushrod.

2. Disconnect the clutch hydraulic line from the master cylinder.

――――― **WARNING** ―――――
Take precautions to keep brake fluid from coming in contact with any painted surfaces.

3. Remove the nuts attaching the master cylinder and remove the master cylinder and pushrod toward the engine compartment side.

4. If equipped, remove the dust boot from the pushrod. The boot will not fit through the cowl without tearing.

To install:

5. Install the master cylinder assembly into the firewall.

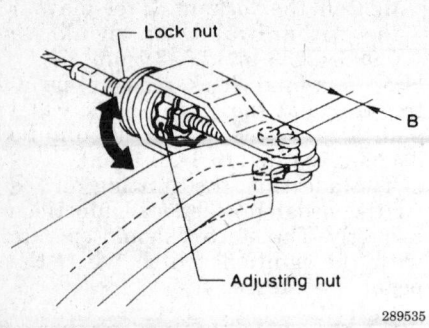

Clutch cable adjusting nut at the withdrawal lever — G20 models

6. Install the attaching nuts and torque to 5.8–8.0 ft. lbs. (8–11 Nm).
7. Connect the hydraulic line.
8. Connect clutch pedal.
9. Bleed system and check for leaks.

Clutch Slave Cylinder

REMOVAL AND INSTALLATION

I30 Models

1. Disconnect the flexible fluid hose from the slave cylinder.
2. Remove the slave cylinder attaching bolts.
3. Remove the slave cylinder and the pushrod.
To install:
4. Install the slave cylinder and pushrod.
5. Install attaching bolts and torque to 22–30 ft. lbs. (30–40 Nm).
6. Connect the flexible fluid hose.
7. Bleed system and check for leaks.

Bleeding

PROCEDURE

I30 Models

Bleeding is required to remove air trapped in the hydraulic system. The bleed screw is located on the clutch slave (release) cylinder.
1. Remove the bleed screw dust cap.
2. Attach a transparent vinyl tube to the bleed screw, immersing the free end in a clean container of clean brake fluid.
3. Fill the clutch master cylinder with the proper fluid.
4. Slowly depress the clutch pedal all the way several times and hold it down.

5. Have an assistant open the bleeder valve about ¾ turn to release the air. Then close the bleeder valve while the pedal is still depressed.
6. Repeat the above procedure until no more air bubbles are seen in the fluid container.
7. Remove the bleed tube.
8. Replace the dust cap and refill the master cylinder.
9. Bleed the clutch damper, if equipped.

Automatic Transmission Assembly

REMOVAL AND INSTALLATION

J30 Models

1. Disconnect the negative battery cable.
2. Raise the vehicle and support safely.
3. Remove the exhaust tube.
4. Drain the fluid from the transmission pan.
5. Remove the dipstick tube.
6. Remove the oil cooler lines.
7. Plug dipstick tube hole and oil cooler fittings after removing lines.
8. Remove the control linkage from the selector lever.
9. Disconnect the neutral safety switch and solenoid harness connectors.
10. Disconnect the speedometer cable.
11. Matchmark and remove the driveshaft. Insert plug into rear seal opening to prevent loss of fluid.
12. Support the transmission safely.
13. Remove the bolts securing the torque converter to the flexplate.
14. Remove the starter.
15. Remove the gussets securing the transmission to the engine. Remove the bolts attaching the transmission to the engine.

NOTE: The bolts securing the transmission to the engine are of different lengths. Note the length of the bolts as they are removed.

16. Support the engine safely. Avoid jacking directly under the oil pan drain plug.
17. Remove the transmission from the vehicle.
To install:
18. Install the torque converter in the transmission. Make sure the torque converter is fully seated in the front pump assembly. The distance from the front edge of the transmission to the bolt hole of the torque con-

verter should be 1.02 inches (26.0 mm) or more.
19. Position the transmission to the engine and install a few bolts to hold the transmission in place. Do not fully tighten the bolts at this time.
20. Install the torque converter-to-flexplate bolts. Tighten to 33–43 ft. lbs. (44–59 Nm).
21. Secure the transmission to the engine. Torque the bolts as follows:
• 58mm bolts to 29–36 ft. lbs. (39–49 Nm)
• 47.5mm bolts to 29–36 ft. lbs. (39–49 Nm)
• 25mm bolts to 22–29 ft. lbs. (29–39 Nm)
• 20mm gusset bolts to 22–29 ft. lbs. (29–39 Nm)
22. Install the starter.
23. Install the crankshaft position sensor.
24. Align the matchmark and install the driveshaft.
25. Connect the speedometer cable.
26. Connect the neutral safety switch and solenoid harness connectors.
27. Install the control linkage to the selector lever.
28. Install the dipstick tube and oil cooler lines.
29. Connect the exhaust tube.
30. Lower the vehicle.
31. Fill the transmission with new fluid. Use the same amount of fluid that was drained before removal.
32. Connect negative battery cable and start the engine. Allow the engine to reach normal operating temperature and check the transmission fluid level. Add fluid as needed.

Q45 Models

1. Disconnect the negative battery cable.
2. Raise the vehicle and support safely.
3. Remove the exhaust tubes.
4. Remove the fluid charging pipe.
5. Remove the oil cooler lines.
6. Plug fluid charging and oil cooler fittings after removing lines.
7. Remove the control linkage from the selector lever.
8. Disconnect the neutral safety switch and solenoid harness connectors.
9. Disconnect the speed sensor connection.
10. Matchmark and remove the driveshaft. Insert plug into rear seal opening to prevent loss of fluid.
11. Support the transmission safely.
12. Remove the bolts securing the torque converter to the flexplate.

13. Remove the gussets securing the transmission to the engine. Remove the bolts attaching the transmission to the engine.

NOTE: The bolts securing the transmission to the engine are of different lengths. Note the length of the bolts as they are removed.

14. Support the engine safely. Avoid jacking directly under the oil pan drain plug.
15. Remove the transmission from the vehicle.
To install:
16. Position the transmission in the vehicle and install the torque converter-to-flexplate bolts. Tighten to 33–43 ft. lbs. (44–59 Nm).
17. Secure the transmission to the engine. Torque the bolts as follows:
• 70mm bolts to 80–87 ft. lbs. (108–118 Nm)
• 30mm bolts to 51–58 ft. lbs. (69–78 Nm)
• 30mm (gusset) bolts to 51–58 ft. lbs. (69–78 Nm)
18. Install the torque converter-to-drive plate bolts and torque in two steps to 33–43 ft. lbs. (44–59 Nm).
19. Align the matchmark and install the driveshaft.
20. Connect the speed sensor connection.
21. Connect the neutral safety switch and solenoid harness connectors.
22. Install the control linkage to the selector lever.
23. Install the fluid charging and oil cooler lines.
24. Connect the exhaust tubes.
25. Lower the vehicle.
26. Connect negative battery cable.

Automatic Transaxle Assembly

REMOVAL AND INSTALLATION

G20 Models

1. Disconnect the negative battery cable and the air duct.
2. Raise and support the vehicle safely. Disconnect the transaxle solenoid harness and inhibitor switch harness connector. Disconnect the throttle wire at the engine side.
3. Drain the transaxle fluid.
4. Disconnect the control cable and transaxle coolant lines.
5. Remove the halfshafts, the front exhaust tube, and the starter.
6. Remove the rear cover plate and the bolts securing the torque con-

verter to the driveplate. Rotate the crankshaft to gain access to the bolts.
7. Support the engine with a suitable stand and use a suitable jack to support the transaxle.

NOTE: Bolts are of different lengths, note the locations that the bolts are removed from.

8. Remove the transaxle mounts.
9. Remove the transaxle mounting bolts and lower the transaxle.
To install:
10. Place a straightedge across the bell housing of the transaxle and measure the distance to the mounting bosses on the torque converter. The distance should be 0.626 in. (15.9mm). If not, the torque converter is not installed correctly.
11. Check the driveplate run-out with a dial indicator. Maximum allowable run-out is 0.008 in. (0.2mm).
12. Raise the transaxle into position and install the transaxle mounting bolts. Torque the 50, 55, and 65 mm long bolts to 51–59 ft. lbs. (70–79 Nm). Torque the 35 and 45 mm long bolts to 12–15 ft. lbs. (16–21 Nm).
13. Install the torque converter bolts. Torque the bolts to 33–43 ft. lbs. (44–59 Nm). Rotate the crankshaft to gain access to the bolts.
14. Install the transaxle mounts.
15. Install the halfshafts, the front exhaust tube, and the starter.
16. Connect the control cable and transaxle coolant lines.
17. Connect the transaxle solenoid harness and inhibitor switch harness connector. Connect the throttle wire at the engine side.
18. Fill the transaxle with fluid, connect the negative battery cable and road test the vehicle.

I30 Models

NOTE: The radio may contain a coded theft protection circuit. Always obtain the code number from the customer before disconnecting the battery.

1. Disconnect the negative battery cable.
2. Raise the vehicle and support safely.
3. Remove the exhaust tube.
4. Drain the fluid from the transmission pan.
5. Remove the dipstick tube.
6. Remove the oil cooler lines.
7. Plug dipstick tube hole and oil cooler fittings after removing lines.
8. Remove the control linkage from the selector lever.
9. Disconnect the neutral safety switch and solenoid harness connectors.

10. Disconnect the speedometer cable.
11. Matchmark and remove the driveshaft. Insert plug into rear seal opening to prevent loss of fluid.
12. Support the transmission safely.
13. Remove the bolts securing the torque converter to the flexplate.
14. Remove the starter.
15. Remove the crankshaft position sensor.
16. Remove the gussets securing the transmission to the engine. Remove the bolts attaching the transmission to the engine.

NOTE: The bolts securing the transmission to the engine are of different lengths. Note the length of the bolts as they are removed.

17. Support the engine safely. Avoid jacking directly under the oil pan drain plug.
18. Remove the transmission from the vehicle.
To install:
19. Install the torque converter in the transmission. Make sure the torque converter is fully seated in the front pump assembly. The distance from the front edge of the transmission to the bolt hole of the torque converter should be 1.02 inches (26.0 mm) or more.
20. Position the transmission to the engine and install a few bolts to hold the transmission in place. Do not fully tighten the bolts at this time.
21. Install the torque converter-to-flexplate bolts. Tighten to 33–43 ft. lbs. (44–59 Nm).
22. Secure the transmission to the engine. Torque the bolts as follows:
• 58mm bolts to 29–36 ft. lbs. (39–49 Nm)
• 47.5mm bolts to 29–36 ft. lbs. (39–49 Nm)
• 25mm bolts to 22–29 ft. lbs. (29–39 Nm)
• 20mm gusset bolts to 22–29 ft. lbs. (29–39 Nm)
23. Install the starter.
24. Install the crankshaft position sensor.
25. Align the matchmark and install the driveshaft.
26. Connect the speedometer cable.
27. Connect the neutral safety switch and solenoid harness connectors.
28. Install the control linkage to the selector lever.
29. Install the dipstick tube and oil cooler lines.
30. Connect the exhaust tube.
31. Lower the vehicle.

32. Fill the transmission with new fluid. Use the same amount of fluid that was drained before removal.

33. Connect negative battery cable and start the engine. Allow the engine to reach normal operating temperature and check the transmission fluid level. Add fluid as needed.

DRIVE AXLE

Driveshaft

REMOVAL AND INSTALLATION

J30 and Q45 Models

NOTE: The universal joint is not serviceable. If the U-joint is defective, replace the driveshaft assembly.

1. Raise the vehicle and support safely.

2. Matchmark the final drive flange, driveshaft flanges, center bearing flanges, transmission yoke and transmission.

3. Remove the attaching bolts and separate the driveshaft from the differential carrier.

4. Remove the bolts attaching the driveshaft to the center bearing flange and remove the rear driveshaft.

5. Remove the nuts attaching the center bearing. Slide the front driveshaft rearward to remove it from the transmission. Plug the rear opening of the transmission extension housing to prevent oil spills.

6. Inspect the rear driveshaft run-out. Run-out should not exceed 0.024 in. (0.6mm).

------ **WARNING** ------
Do not disassembly the rubber coupling.

To install:
7. Apply a coat of multi-purpose lithium grease containing molybdenum disulfide to the end face of the center bearing and both sides of the washer.

8. Remove the plug from the transmission. Align and install the front driveshaft. Tighten the center bearing attaching nuts to 18–22 ft. lbs. (25–29 Nm).

9. Align and install the rear driveshaft. Tighten the differential carrier-to-driveshaft bolts to 41–48 ft. lbs. (55–65 Nm) on J30 models or

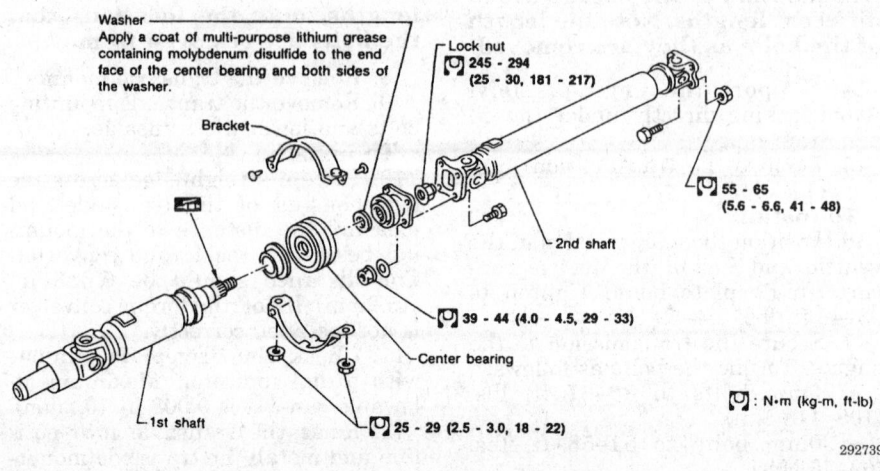

Driveshaft assembly — J30 models

65–72 ft. lbs. (88–98 Nm) on the Q45 model. Torque the center bearing flange-to-driveshaft bolts to 29–33 ft. lbs. (39–44 Nm) on J30 models or 40–47 ft. lbs. (54–65 Nm) on Q45 models.

10. Lower the vehicle.

Halfshaft

REMOVAL AND INSTALLATION

G20 Models

1. Raise and support the vehicle safely. Remove the wheel bearing locknut.

2. Remove the brake caliper assembly and rotor. Using a piece of wire, position the caliper so it is not supported by the brake line.

3. Separate the tie-rod from the ball joint.

4. Separate the kingpin from the knuckle.

5. Remove the halfshaft from the wheel hub/knuckle by lightly tapping it with a wood drift. Take care not to damage the CV-boots.

6. Remove the halfshaft from the transaxle by prying outward with a suitable tool at the transaxle case.

7. On automatic transaxle models, remove the left halfshaft by tapping

it out with a drift from the right side of the transaxle case. Take care not to damage the pinion mate shaft and side gear.

To install:
8. Drive a new oil seal into the transaxle. For the right side use tool KV38106800 or equivalent, along the inner circumference of the oil seal. For the left side use tool KV38106700

9. Insert the halfshaft into the transaxle. Ensure that the serration's are aligned. Remove the tool.

10. Push the halfshaft inward and install the circular clip in the groove of the side gear. After inserting the clip, pull outward on the flange of the slide joint to ensure the clip is properly meshed with the side gear. If it pulls out, the clip was not installed properly.

11. Install the halfshaft into the wheel hub/knuckle. Tighten the upper knuckle nut to 72–87 ft. lbs. (98–118 Nm) and wheel bearing locknut to 174–231 ft. lbs. (235–314 Nm).

12. Using a dial indicator, check wheel bearing axial end-play. Specification calls for 0.0020 in. (0.05mm) or less.

13. Install the rotor and brake caliper.

14. Install the wheel and lower the vehicle to the floor.

J30 and Q45 Models

— CAUTION —

The torque on the rear wheel bearing nut is high enough to cause the vehicle to fall off the jack stands while loosing it. Loosen and tighten this nut with the vehicle sitting on the ground.

1. Remove the rear wheel cotter pin, adjusting cap and insulator. Loosen the wheel bearing nut with the brakes applied and the vehicle sitting on the ground.
2. Raise the vehicle and support safely.
3. Remove the rear wheel.
4. Remove the differential side flange bolts and nuts and separate shaft from the differential.
5. Remove the wheel bearing locknut and washer from halfshaft.
6. Remove the halfshaft by lightly tapping it with a copper hammer.
7. Remove the halfshaft assembly from the vehicle.

To install:

8. Insert halfshaft into wheel hub and install washer and wheel bearing locknut. Temporarily tighten the locknut.
9. Connect the halfshaft with the differential side flange. Install the nuts and bolts and torque to 61–69 ft. lbs. (83–93 Nm).
10. Install the wheels and lower the vehicle to the ground.
11. Tighten the wheel bearing locknut with the brakes applied to 152–203 ft. lbs. (206–275 Nm). Install the insulator, adjusting cap and a new cotter pin.

I30 Models

Right Halfshaft

1. Raise and support the front of the vehicle safely and remove the wheels.
2. Remove the ABS wheel sensor and move it out of the way.
3. Remove the brake hose from the strut.
4. Remove wheel bearing locknut.
5. Matchmark and remove the bolts attaching the steering knuckle to the strut.

NOTE: Cover axle boots with waste cloth or equivalent so as not to damage them when removing halfshaft.

6. Separate the halfshaft from the knuckle by slightly tapping it.
7. Pry the halfshaft from the transaxle with a flat bladed tool. Remove and discard the circlip on the end of the halfshaft.

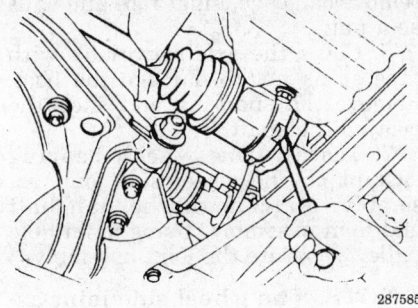

287585

Separating the right halfshaft from the transaxle — I30 models

8. Remove the seal from the transaxle.

To install:

9. Install a new seal into the transaxle and install a halfshaft alignment tool KV38106800 or equivalent into the transaxle seal.
10. Install a new circlip to the halfshaft, then insert the halfshaft into the transaxle.
11. With the serration's aligned remove the alignment tool.
12. Push the halfshaft fully into the transaxle to seat the circlip. Try to pull the halfshaft from the transaxle by hand to verify that the circlip is properly seated.
13. Insert the halfshaft into the steering knuckle and install the hub locknut, do not tighten the hub nut at this time.
14. Connect the steering knuckle to the strut.
15. Install the strut mounting bolts and align the matchmarks. Torque the bolts to 103–117 ft. lbs. (140–159 Nm).
16. Install the brake hose to the strut.
17. Install the ABS wheel sensor and torque the attaching bolt to 13–17 ft. lbs. (18–24 Nm).
18. Install the front wheels, lower the vehicle and torque hub locknut to 174–231 ft. lbs. (235–314 Nm).

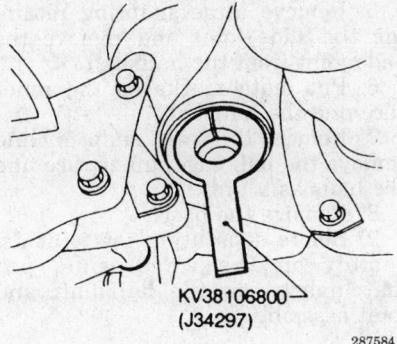

KV38106800
(J34297)
287584

Right halfshaft alignment tool — I30 models

19. Check and/or adjust the wheel alignment as necessary.

Left Halfshaft

1. Raise and support the front of the vehicle safely and remove the wheels.
2. Remove the ABS wheel sensor and move it out of the way.
3. Remove the brake hose from the strut.
4. Remove wheel bearing locknut.
5. Matchmark and remove the bolts attaching the steering knuckle to the strut.

NOTE: Cover axle boots with waste cloth or equivalent so as not to damage them when removing halfshaft.

6. Separate the halfshaft from the knuckle by slightly tapping it.
7. Loosen the bolts attaching the support bearing to the support bearing bracket.
8. If equipped with a manual transaxle, pry the halfshaft from the transaxle with a flat bladed tool.
9. If equipped with a automatic transaxle perform the following:
   a. Remove the right halfshaft from the vehicle.
   b. Insert a flat bladed tool into the transaxle where the right halfshaft was, place the end of the tool on the halfshaft, then drive the left shaft from the pinion side gear.
10. Remove the support bearing bolts and remove the halfshaft from the vehicle.
11. Remove and discard the circlip on the end of the halfshaft. Remove the seal from the transaxle.

To install:

12. Install a new seal into the transaxle and install a halfshaft alignment tool KV38106700 or equivalent into the transaxle seal.
13. Install a new circlip to the halfshaft, then insert the halfshaft into the transaxle.
14. With the serration's aligned remove the alignment tool.
15. Push the halfshaft fully into the transaxle to seat the circlip. Try to pull the halfshaft from the transaxle by hand to verify that the circlip is properly seated.
16. Install the support bearing bolts and torque the bolts to 9–14 ft. lbs. (13–19 Nm).
17. Insert the halfshaft into the steering knuckle and install the hub locknut, do not tighten the hub nut at this time.
18. Connect the steering knuckle to the strut.
19. Install the strut mounting bolts and align the matchmarks. Torque

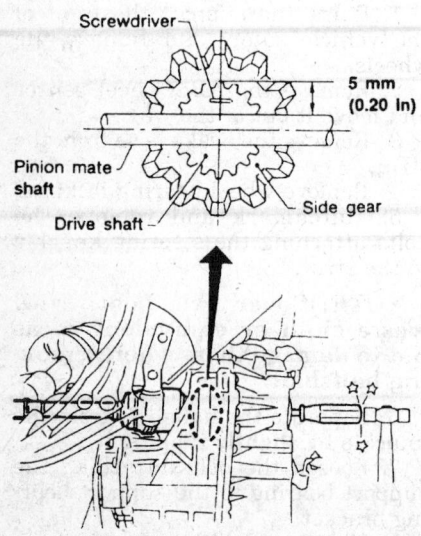

Separating the left halfshaft from an automatic transaxle — I30 models

the bolts to 103–117 ft. lbs. (140–159 Nm).

20. Install the brake hose to the strut.

21. Install the ABS wheel sensor and torque the attaching bolt to 13–17 ft. lbs. (18–24 Nm).

22. Install the front wheels, lower the vehicle and torque hub locknut to 174–231 ft. lbs. (235–314 Nm).

23. Check and/or adjust the wheel alignment as necessary.

CV-Joint Boot

REPLACEMENT

G20 Models

1. Raise and support the vehicle safely.

2. Remove halfshaft assembly from vehicle and place in a suitable working fixture.

3. Remove the boot bands. Matchmark the transaxle side slide joint housing and the inner race before separating the joint assembly.

4. Remove the snapring and disassemble the slide joint housing.

5. Make matching marks on inner race and drive shaft, pry off snapring,

remove ball cage, inner race and balls as a unit.

6. Cover the axle serration's with tape so as not to damage the boot. Remove the snapring and slide the boot off the shaft.

7. Install the wheel bearing locknut on the wheel side joint assembly. Matchmark the halfshaft and joint assembly. Using a suitable puller, separate the joint assembly.

NOTE: The wheel side joint assembly cannot be disassembled.

8. Cover the axle serration's with tape so as not to damage the boot. Remove the snapring and slide the boot off the shaft.
To install:
9. Clean the joint of old grease and any debris which may be present.

10. Install the transaxle side boot and joint assembly. Ensure that all snaprings are secure. If a snapring is loose or damaged, replace it. Ensure that the matchmarks made during assembly are mated.

11. Install the wheel side boot and joint by setting the joint assembly on the halfshaft. Lightly tap the joint to seat it on the shaft. Ensure that the matchmarks made during disassembly are mated.

12. Pack the joint assemblies with 3.6–4.2 oz. (105–125ml) of grease. Install the boots so the length is 3.86 in. (98.5mm) for the wheel side and 3.96 in. (100.5mm) for the transaxle side. Lock the boot bands securely in place.

13. Install the halfshaft assembly and lower the vehicle.

J30 and Q45 Models

1. Raise the vehicle and support safely.

2. Remove the halfshaft assembly from the vehicle and place it in a vise.

3. Remove the boot bands on both inner and outer joints.

4. Put matchmarks on the slide joint housing and inner race before separating the joint assembly.

5. Remove large snapring retaining the slide joint and remove the slide joint from the halfshaft.

6. Put matchmarks on the inner race and the halfshaft.

7. Remove the small snapring and remove the ball cage, inner race and the balls as a unit.

8. Remove the boot.

9. Before separating the joint assembly on the wheel side, put matchmarks on the halfshaft and joint assembly.

NOTE: The joint on the wheel side cannot be disassembled.

10. Separate the joint assembly from the halfshaft using a slide hammer or equivalent.

11. Remove the boot.
To install:
12. Clean the joint of old grease and any debris which may be present.

13. Apply tape to the halfshaft splines to prevent damage to the boots.

14. Install a new small boot band and a new boot on the wheel side of the halfshaft.

15. Position the joint assembly onto the halfshaft and seat the joint by lightly tapping it. Ensure that the matchmarks are aligned when assembling.

16. Pack the CV-joint and boot with 6.00–6.70 oz. (170–190g) of the grease supplied in the repair kit.

17. Position the boot so it does not swell or deform when installed in the vehicle.

18. Lock the new large and small boot bands securely with a suitable tool.

19. Install a new small boot band and a new boot on the differential side of the halfshaft.

20. Install the ball cage, inner race and balls as a unit. Ensure that the matchmarks are aligned when assembling.

21. Install a new large snapring.

22. Pack the CV-joint and boot with 6.35–7.05 oz. (180–200g) of the grease supplied in the repair kit.

23. Install slide joint housing and install a new small snapring.

24. Set the boot so it does not swell or deform when installed in the vehicle.

25. Lock the new larger and smaller boot bands securely with a suitable tool.

26. Install the halfshaft assembly in the vehicle.

27. Install the rear wheel.

28. Lower the vehicle.

I30 Models

Transaxle Side Joint

1. Remove the halfshaft assembly from the vehicle.

2. Remove the boot bands and pull the boot back to expose the CV-joint components.

3. Before separating the joint assembly from the halfshaft, matchmark the slide joint housing and inner race.

4. Pry off the outer snapring and pull out the slide joint housing.

5. Place matchmarks on the inner race and the halfshaft.

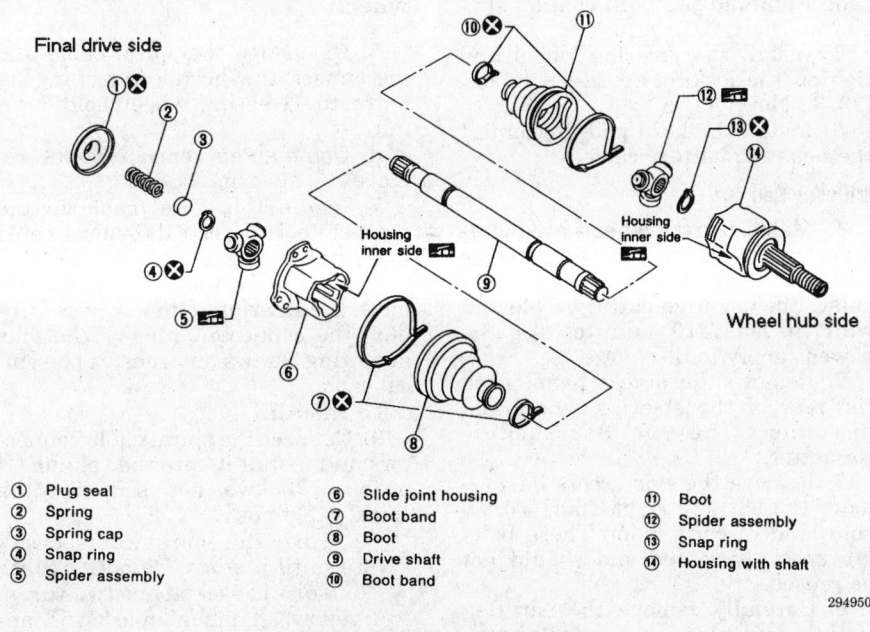

Final drive side

Housing inner side

Housing inner side

Wheel hub side

① Plug seal
② Spring
③ Spring cap
④ Snap ring
⑤ Spider assembly

⑥ Slide joint housing
⑦ Boot band
⑧ Boot
⑨ Drive shaft
⑩ Boot band

⑪ Boot
⑫ Spider assembly
⑬ Snap ring
⑭ Housing with shaft

294950

Halfshaft assembly — J30 models

6. Pry off the snapring, then remove the ball cage, the inner race, and the balls as a unit.
7. Remove the CV-joint boot.

NOTE: Cover the halfshaft serration with tape so not to damage the boot.

To install:
8. Clean the joint of old grease and any debris which may be present.
9. Install the boot and a small boot band.
10. Install the ball cage, the inner race, and the balls as a unit. Be sure that the matchmarks on the inner race and halfshaft are aligned; install a new snapring.
11. Pack the CV-joint housing and boot with 5.87–6.17 oz. (165–175 g) of Nissan genuine grease or equivalent.
12. Install the slide joint housing and align the matchmarks made during disassembly.
13. Install a new outer snapring.
14. Install the boot and check the boot length, the boot length should be 3.82–3.90 inches (97–99 mm). The boot should not be deformed or swell at the specified length.
15. Secure the boot with the boot bands.
16. Install the halfshaft in the vehicle.

Wheel Side Joint

NOTE: The joint on the wheel side cannot be disassembled.

1. Remove the halfshaft from the vehicle.
2. Matchmark the halfshaft and the joint assembly.
3. Separate the joint assembly from the halfshaft by placing the halfshaft securely in a vise and place the locknut on the end of the joint. Attach a slide hammer to the locknut and pull the joint from the halfshaft.
4. Remove the boot bands and remove the outer joint assembly and the boot.
To install:
5. Clean the joint of old grease and any debris which may be present.

NOTE: Cover halfshaft serration with tape so not to damage the boot.

6. Install the boot with new boot bands.
7. Align the matchmarks and lightly tap the joint assembly onto the shaft.
8. Pack CV-joint and boot with 6.70–7.05 oz. (190–200 g) of Moly grease or equivalent.
9. Install the boot and check the boot length, the boot length should be

3.78–3.86 inches (96–98 mm). The boot should not be deformed or swell at the specified length.
10. Secure the boot with the boot bands.
11. Install the halfshaft in the vehicle.

STEERING

Air Bag

─────── **CAUTION** ───────
Some vehicles are equipped with an Air Bag system, also known as the Supplemental Inflatable Restraint (SIR) or Supplemental Restraint System (SRS). The system must be disabled before performing service on or around system components, steering column, instrument panel components, wiring and sensors. Failure to follow safety and disabling procedures could result in accidental Air Bag deployment, possible personal injury and unnecessary system repairs.

PRECAUTIONS

Several precautions must be observed when handling the inflator module to avoid accidental deployment and possible personal injury.
• Never carry the inflator module by the wires or connector on the underside of the module.
• When carrying a live inflator module, hold securely with both hands, and ensure that the bag and trim cover are pointed away.
• Place the inflator module on a bench or other surface with the bag and trim cover facing up.
• With the inflator module on the bench, never place anything on or close to the module which may be thrown in the event of an accidental deployment.

DISARMING

─────── **CAUTION** ───────
To avoid rendering the Air Bag inoperative, which could lead to personal injury or death in the event of a severe frontal collision, extreme caution must be taken when servicing the electrical related systems.

NOTE: All Air Bag electrical wiring harnesses and connectors are covered with YELLOW outer insulation. Do not use electrical test equipment on any circuit related to the Air Bag sensors. When installing Air Bag components, always install with the arrow marks facing the front of the vehicle.

Disarming

To disarm the Air Bag system turn the ignition switch to the **OFF** position. Then disconnect both battery cables starting with the negative cable first and wait at least 10 minutes after the cables are disconnected. Be sure to insulate the battery terminal ends.

Arming

To arm the Air Bag system, turn the ignition switch to the **OFF** position. Connect both battery cables starting with the positive cable first.

NOTE: The Air Bag or Air Bag system is equipped with a self-diagnostic operation. After turning the ignition key to the ON or START position, the AIR BAG warning lamp will illuminate for 7 seconds. After 7 seconds, the AIR BAG lamp will extinguish if no malfunction is detected. If the AIR BAG lamp does not extinguish after 7 seconds, check the Air Bag self diagnostic system for a malfunction.

Steering Wheel

REMOVAL AND INSTALLATION

G20, J30 and Q45 Models

─── CAUTION ───
The Air Bag system must be disarmed before removing the steering wheel. Failure to do so may cause accidental deployment, property damage or personal injury.

Without Air Bag

1. Disconnect the negative battery cable.
2. Ensure that the steering wheel and front tires are positioned in the straight-ahead position.
3. Using an appropriate tool, pry the horn pad off the steering wheel.
4. Remove the steering wheel locknut.
5. Using an appropriate puller, remove the steering wheel.

To install:

6. Apply multi-purpose grease to the entire surface of the turn signal cancel pin and the horn contact slip-ring.
7. Install the steering wheel and tighten the locknut to 22–29 ft. lbs. (29–39 Nm).
8. Install the horn pad. Reconnect the negative battery cable.

With Air Bag

1. Make sure the wheels are pointing straight-ahead and the steering wheel is centered. Disconnect and insulate the negative battery cable and wait at least 10 minutes for the power supply to discharge.
2. Remove the access panel from the rear of the steering wheel and disconnect the Air Bag module connector.
3. Remove the side access lids, remove the left and right T50H Torx® bolts and discard them. These bolts are specially coated and should not be reused.
4. Carefully remove the Air Bag module and place in a safe location with the pad side facing upward.

NOTE: The Air Bag module is a fragile component. Always place it with the pad side facing upward. Do not allow oil, grease or water to come in contact with the module. Do not drop the module; if it is damaged in any way, do not reinstall it to the steering wheel.

5. Disengage the spiral cable and disconnect the horn connector. Remove the steering wheel hold-down nut.
6. Using an appropriate puller, remove the steering wheel.
7. Install the spiral cable to the stopper tool to keep the spiral cable in its correct alignment.
8. To remove the spiral cable, remove the steering column covers. Unplug the connector, remove the four mounting screws and remove the spiral cable.

To install:

9. Connect the spiral cable connectors and install it onto the column. If a stopper tool was not used, align the spiral cable now.
   a. Turn the spiral cable clockwise until it stops. Do not force it.
   b. Turn it back about two turns. A yellow alignment mark will appear on the gear to the left.
   c. Align the arrow with the yellow alignment mark.
10. Pull the spiral cable through the steering wheel opening and install the steering wheel, engaging the spiral cable pin guides.

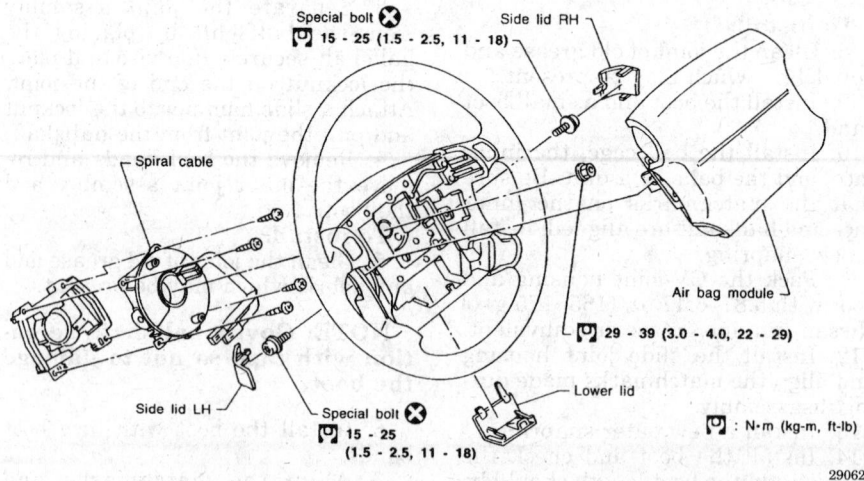

Steering wheel assembly with Air Bag — G20, J30 and Q45

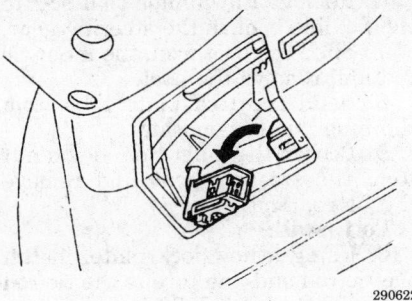

Air Bag module connector location — G20, J30 and Q45 models

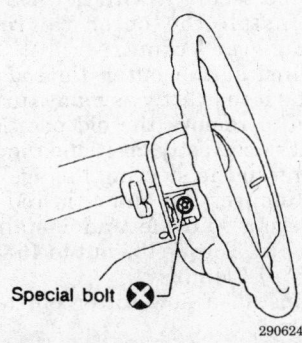

Air Bag mounting bolts — All models

Special bolt ⊗

11. Connect the horn connector and engage the spiral cable with the pawls in the steering wheel.

12. Install the hold-down nut and torque to 22–29 ft. lbs. (29–39 Nm).

13. Carefully position the Air Bag module. Install new Torx® bolts and torque to 15 ft. lbs. (20 Nm). Connect the Air Bag module connector.

14. Install the three access lids and the column covers.

15. Make sure no one is in the vehicle and connect the negative battery cable.

16. If the Nissan Consult tool is not available, perform the following:

  a. From the passenger seat, turn the ignition switch to the **ON** position.

  b. Observe the **AIR BAG** warning light on the instrument cluster. The warning light should illuminate for about seven seconds, then go out.

  c. If the warning light illuminates in any sequence except the above, the control unit has detected a fault in the Air Bag system and it will not operate.

17. The Nissan Consult or equivalent scan tool is required to check the fault diagnosis memory and turn the **AIR BAG** warning light off.

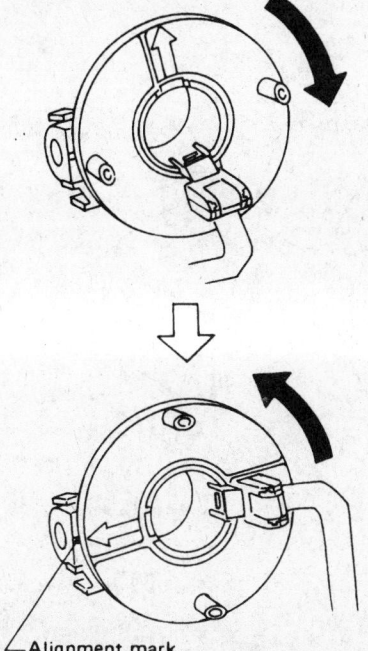

Alignment mark

Align the arrow with the yellow alignment mark on the spiral cable gear — G20, J30 and Q45 models

I30 Models

1. Position the wheels in the straight-ahead direction. The steering wheel should be right side up and level.

2. Disconnect both battery cables.

—— CAUTION ——
Wait 10 minutes after the battery cable has been disconnected, before attempting to work on the Air Bag unit. The Air Bag unit is still armed and can inflate, during the 10 minute period, and possibly causing bodily injury.

3. Remove the lower lid from the steering wheel and disconnect the Air Bag unit electrical connector.

4. Remove both side lids from the steering wheel.

5. Using a T50H Torx® bit, remove the Air Bag-to-steering wheel bolts from both sides of the steering wheel; discard the bolts.

6. Lift the Air Bag unit upward and place it in a safe, clean, dry place with the pad side facing upward; be sure the temperature in the area will not exceed 212°F (100°C).

7. Disconnect the horn connector and remove the nuts.

8. Matchmark the top of the steering column shaft and the steering wheel flange.

9. Remove the steering wheel-to-steering column nut. Using the steering wheel puller tool ST27180001 or equivalent, pull the steering wheel from the steering column.

—— WARNING ——
Do not strike the shaft with a hammer, which may cause the column to collapse.

10. Apply multi-purpose grease to the entire surface of the turn signal cancel pin (both portions) and the horn contact slip ring.

11. Install the steering wheel by aligning the matchmarks. Do not drive or hammer the steering wheel into place or you may cause the collapsible steering column to collapse, in which case you'll have to buy a whole new steering column unit.

12. Torque the steering wheel-to-steering column nut to 25–29 ft. lbs. (33–39 Nm).

13. Connect the horn electrical connector and install the nuts.

14. Align the spiral cable correctly with the alignment mark and install the Air Bag unit into the steering wheel.

15. Using new Air Bag-to-steering wheel Torx® bolts, install them into both sides of the steering wheel and torque to 11–18 ft. lbs. (15–25 Nm).

16. Install both side lids to the steering wheel.

17. Connect the Air Bag unit electrical connector and install the lower lid to the steering wheel.

18. Connect the negative battery cable and make sure the Air Bag **red** indicator light turns **ON**. The AIR BAG light should extinguish after about seven seconds. If the light does not extinguish, check the Air Bag self diagnostic system for a malfunction.

Tie Rod Ends

REMOVAL AND INSTALLATION

G20, J30 and Q45 Models

1. Raise the vehicle and support safely.

2. Remove the front wheel.

3. Matchmark the position of tie rod end locknut on the threaded section of the tie rod.

4. Loosen the tie rod end locknut.

5. Remove the cotter pin and tie rod end nut.

6. Separate the tie rod end from the steering knuckle using a suitable tool.

7. Remove the tie rod end from the tie rod.

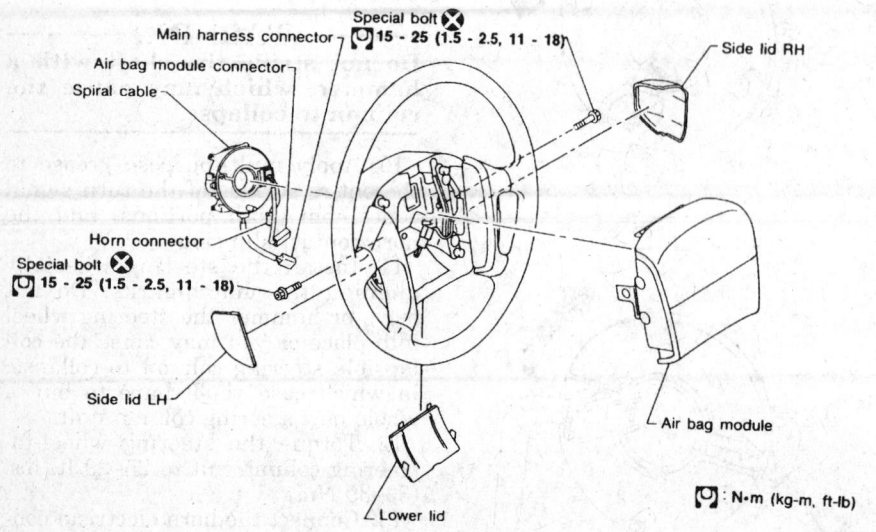

Special bolt ⊗ 🔧 15 - 25 (1.5 - 2.5, 11 - 18)

Main harness connector

Air bag module connector

Spiral cable

Side lid RH

Horn connector

Special bolt ⊗
🔧 15 - 25 (1.5 - 2.5, 11 - 18)

Side lid LH

Air bag module

Lower lid

🔧 : N·m (kg-m, ft-lb)

288860

Steering wheel and related parts — I30 model

6. Remove the clamps that secure the flexible boot to the steering gear.

7. Slide the boot from the inner tie rod and remove the boot.

8. Bend the lock plate tabs from the inner tie rod end nut.

9. Loosen the inner tie rod end nut from the steering gear and remove the tie rod end.

To install:

10. Using a new lock plate, install the tie rod end and torque the tie rod to 58–72 ft. lbs. (78–98 Nm).

11. Bend the tabs of the new lock plate to secure the inner tie rod end.

12. Slide the boot onto the steering gear and secure it with new clamps.

13. Install the outer tie rod-to-steering gear locknut.

14. Install the outer tie rod end, turning it in exactly as many turns as it was to remove the old one. Make sure it is correctly positioned in relationship to the steering linkage.

15. Connect the outer tie rod end-to-steering knuckle and install the castle nut. Torque the nut to 46–54 ft. lbs. (63–73 Nm.).

16. Install a new cotter pin to the castle nut.

17. Tighten the tie rod end locking nut to 58–72 ft. lbs. (78–98 Nm).

18. Install the wheel and tire assembly.

19. Lower the vehicle and perform a front end alignment.

Power Rack and Pinion

REMOVAL AND INSTALLATION

G20 Models

— **CAUTION** —

Some vehicles are equipped with an Air Bag system. The system must be disabled before performing service on or around system components, steering column, instrument panel components, wiring and sensors. Failure to follow safety and disabling procedures could result in accidental Air Bag deployment, possible personal injury and unnecessary system repairs.

1. Raise and safely support the vehicle and remove the front wheels.

2. Remove the cotter pins and use a ball joint press to disconnect the tie rod ends from the steering knuckles.

3. Disconnect the hydraulic lines from the steering rack and plug the lines to keep dirt out.

4. Remove the front exhaust tube.

5. Remove the clamp bolt from the pinion shaft universal joint. Match-

To install:

8. Install the new tie rod end on the tie rod.

9. Install the tie rod end on the steering knuckle. Initially, tighten the tie rod-to-steering knuckle nuts to 22–29 ft. lbs. (29–39 Nm). Tighten the nut further to expose first pin hole and install a new cotter pin.

10. Adjust the toe-in to the matchmark made on the threaded section of the tie rod. Tighten the locknut.

11. Install the front wheel.

12. Lower the vehicle.

13. Check alignment to verify proper toe-in setting.

I30 Models

Outer

1. Raise the front of the vehicle and support it on jackstands. Remove the wheel.

2. Remove the cotter pin and the tie rod ball joint stud nut. Note the position of the steering linkage.

3. Using a suitable ball joint separator tool, remove the tie rod ball joint from the steering knuckle.

4. Loosen the locknut and remove the tie rod end from the tie rod. Count the number of complete turns it takes to completely remove it.

To install:

5. Install the new tie rod end, turning it in exactly as many turns as it was to remove the old one. Make sure it is correctly positioned in relationship to the steering linkage.

6. Connect the outer tie rod end-to-steering knuckle and install the castle nut. Torque the nut to 46–54 ft. lbs. (63–73 Nm.).

7. Install a new cotter pin to the castle nut.

8. Tighten the tie rod end locking nut to 58–72 ft. lbs. (78–98 Nm).

9. Install the wheel and tire assembly.

10. Lower the vehicle and perform a front end alignment.

Inner

1. Raise the front of the vehicle and support it on jackstands. Remove the wheel.

2. Remove the cotter pin and the tie rod ball joint stud nut. Note the position of the steering linkage.

3. Using a suitable ball joint separator tool, remove the tie rod ball joint from the steering knuckle.

4. Loosen the locknut and remove the tie rod end from the tie rod. Count the number of complete turns it takes to completely remove it.

5. Remove the tie rod-to-steering gear locknut.

mark the joint to the pinion shaft. Secure the steering wheel so that it cannot rotate.

——— WARNING ———
Do not turn the steering wheel with the steering rack removed. The spiral cable for the Air Bag will be damaged if turned beyond full travel.

6. Remove the rack bracket bolts and lay the rack on the frame rails.
7. Support the transaxle with a jack and remove the center engine support member with the engine mount.
8. Tilt the rack and remove it towards the right side of the vehicle.
To install:
9. Fit the rack into place with the universal joint properly aligned. Torque the bracket bolts to 54–72 ft. lbs. (73–97 Nm).
10. Install the engine mount support. Torque the support bolts to 57–72 ft. lbs. (77–98 Nm) and the engine mount bolts to 36–43 ft. lbs. (49–59 Nm).
11. Connect the hydraulic lines. Torque the low pressure fitting to 20–29 ft. lbs. (27–39 Nm) and the high pressure fitting to 11–18 ft. lbs. (15–25 Nm). Do not over tighten these fittings or the seals will split and leak.
12. Connect the tie rod ends and torque the nuts to 22–29 ft. lbs. (29–39 Nm). Tighten as required to install a new cotter pin but do not exceed 36 ft. lbs. (49 Nm).
13. Install the universal joint clamp bolt and torque to 17–22 ft. lbs. (22–29 Nm).
14. Install the exhaust pipe.
15. Fill the power steering reservoir with fluid and bleed the air from the power steering system.
16. Check the vehicle front end alignment and adjust as necessary.

J30 Models

——— CAUTION ———
Some vehicles are equipped with an Air Bag system. The system must be disabled before performing service on or around system components, steering column, instrument panel components, wiring and sensors. Failure to follow safety and disabling procedures could result in accidental Air Bag deployment, possible personal injury and unnecessary system repairs.

1. Raise the vehicle and support safely.

2. Remove the front wheels.
3. Disconnect the outer tie rods from the steering knuckle.
4. Remove the clamping bolt from the lower steering column universal joint. Matchmark the joint to the pinion shaft. Secure the steering wheel so that it cannot rotate.

——— WARNING ———
Do not turn the steering wheel with the steering rack removed. The spiral cable for the Air Bag will be damaged if turned beyond full travel.

5. Disconnect the hydraulic lines from the rack assembly.
6. Remove the bolts and remove the rack assembly from the vehicle.
To install:
7. Make sure the matchmarks on the pinion shaft and universal joint are aligned and fit the rack into place.
8. Install the mounting brackets and torque the bolts to 65–80 ft. lbs. (88–108 Nm).
9. Connect the hydraulic lines. Torque the low pressure fitting to 20–29 ft. lbs. (27–39 Nm) and the high pressure fitting to 11–18 ft. lbs. (15–25 Nm). Do not over tighten the fittings or the seals will split and leak.
10. Connect the tie rod ends and torque the nut to 22–29 ft. lbs. (29–39 Nm). Tighten as required to install a new cotter pin. Do not exceed 72 ft. lbs. (98 m).
11. Fill the power steering reservoir with fluid and bleed the air from the power steering system.
12. Check the vehicle front end alignment and adjust as necessary.

I30 Models

——— CAUTION ———
Some vehicles are equipped with an Air Bag system. The system must be disabled before performing service on or around system components, steering column, instrument panel components, wiring and sensors. Failure to follow safety and disabling procedures could result in accidental Air Bag deployment, possible personal injury and unnecessary system repairs.

1. Disconnect both battery cables and wait at least 10 minutes after the battery cables are disconnected. This will disarm the Air Bag system so the steering wheel can be removed.

2. Point the front tires straight ahead and lock the steering in this position.

——— WARNING ———
Do not turn the steering wheel or column with the lower joint removed from the steering column or the spiral cable may be damaged.

3. Remove the steering wheel.

NOTE: The steering wheel must be removed before disconnecting the steering column lower joint to avoid damaging the Air Bag spiral cable.

4. Raise and support the vehicle safely and remove the front wheels.
5. Disconnect the tie rod ends from the steering knuckles.
6. Remove the carbon canister from the vehicle.
7. Support the engine then remove the bolts attaching the engine mounts to the engine mounting center member. Remove the engine mounting center member.
8. Remove the front stabilizer bar from the vehicle.
9. Remove the nuts attaching the hole cover to the bulkhead.
10. Move the hole cover aside and disconnect the lower joint from the rack and pinion. Matchmark the pinion shaft and the pinion housing to record the steering neutral position.
11. Disconnect the power steering fluid pipes from the rack and pinion.
12. Remove the bolts attaching the mounting brackets and remove the rack and pinion from the vehicle.
To install:
13. Position the rack and pinion in the vehicle and install the mounting brackets. Torque the mounting nuts and bolts in the proper sequence to 54–72 ft. lbs. (73–97 Nm).
14. Install new O-rings to the power steering fluid pipes and connect them to the rack and pinion. Torque the low pressure line 20–29 ft. lbs. (27–39 Nm). Torque the high pressure line to 11–18 ft. lbs. (15–25 Nm).
15. Align the lower steering joint to the pinion shaft and install the joint onto the pinion shaft. Install the bolt and torque the bolt to 17–22 ft. lbs. (24–29 Nm).
16. Properly position the hole cover and install the attaching nuts, torque the nuts to 2.9–3.6 ft. lbs. (4–5 Nm).
17. Install the front stabilizer.
18. Install the engine mounting center member and torque the attaching bolts to 57–72 ft. lbs. (77–98 Nm). Attach the engine mounts to

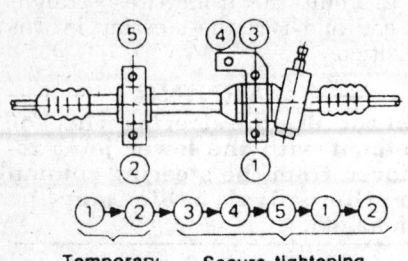

Temporary
tightening

Secure tightening

288833

Rack and pinion bolt tightening sequence — I30 models

the center member and torque the bolts to 57–72 ft. lbs. (77–98 Nm). Remove the support from the engine.

19. Install the carbon canister into the vehicle.

20. Connect the tie rod ends to the steering knuckles, torque the nuts to 46–61 ft. lbs. (63–82 Nm) and install a new cotter pin.

21. Install the front wheels and lower the vehicle.

22. Install the steering wheel.

23. Connect the positive then the negative battery cable.

24. Fill the power steering reservoir with fluid and bleed the air from the power steering system.

25. Check the vehicle front end alignment and adjust as necessary.

Q45 Models

Front Steering Rack

—— **CAUTION** ——
Some vehicles are equipped with an Air Bag system. The system must be disabled before performing service on or around system components, steering column, instrument panel components, wiring and sensors. Failure to follow safety and disabling procedures could result in accidental Air Bag deployment, possible personal injury and unnecessary system repairs.

1. Make sure the front wheels are pointed straight ahead. Disconnect the negative battery cable.

2. Raise the vehicle and support safely. Remove the front wheels.

3. Disconnect the outer tie rod ends from the steering knuckle.

4. Remove the lower clamping bolt from the steering column universal joint. Matchmark the joint and the pinion shaft for assembly. Secure the steering wheel so it cannot rotate.

—— **WARNING** ——
Do not turn the steering wheel with the steering rack removed. The spiral cable for the Air Bag will be damaged if turned beyond full travel.

5. Disconnect the hydraulic lines from the rack assembly and drain the fluid into a pan.

6. Remove the mounting bolts and remove the steering rack assembly.

To install:

7. Make sure the rack is centered and fit it into place with the universal joint matchmarks aligned.

8. Start all of the mounting bolts. Make sure the assembly is properly positioned and torque the bolts to 65–80 ft. lbs. (88–108 Nm).

9. Connect the hydraulic lines. Torque the low pressure connection to 27–30 ft. lbs. (36–40 Nm) and the high pressure connection to 22–26 ft. lbs. (30–35 Nm). Do not over torque these connections or the O-rings will split and leak.

NOTE: The low pressure side has the larger O-ring.

10. Install the universal joint clamping bolt and torque to 17–22 ft. lbs. (24–29 Nm).

11. Connect the tie rod ends to the steering knuckles. Torque the nuts to 22–29 ft. lbs. (29–39 Nm), then tighten as required to install a new cotter pin. Do not exceed 36 ft. lbs. (49 Nm).

12. Connect the battery and refill the steering fluid reservoir.

13. Bleed the air from the power steering system.

14. Check the vehicle front end alignment and adjust as necessary.

Rear Steering Power Cylinder

1. Raise and safely support the vehicle.

2. Remove the cotter pins from the rear steering links and press the joints apart.

3. Disconnect the hydraulic fittings and plug the lines to keep out dirt and prevent draining the whole system.

4. Remove the bolts and remove the power cylinder.

To install:

5. Make sure the mounting surfaces on the chassis and power cylinder are clean and dry.

6. Fit the cylinder into place and start both mounting bolts. Torque the bolts to 62–80 ft. lbs. (84–108 Nm).

7. Connect the hydraulic fittings and torque to 36–51 ft. lbs. (49–69 Nm).

8. Connect the steering links and torque the nut to 33–44 ft. lbs. (45–60 Nm). Tighten as required to install a new cotter pin.

9. There are two bleeder valves on top of the cylinder. To bleed the system:

a. Top off the fluid level. Run the engine at more than 2000 rpm and turn the steering wheel 180 degrees to the right.

b. Connect a tube to the right side bleeder valve and loosen the valve to bleed the air. Tighten the valve again.

c. Check the fluid level. Run the engine at more than 2000 rpm and turn the steering wheel 180 degrees to the left.

d. Connect the tube to the left bleeder valve and open the valve to let the air out.

e. Repeat this procedure alternating left and right until there is no more air in the system.

10. Check and adjust rear wheel alignment.

Power Steering Pump

BLEEDING

All Except Q45 with 4-Wheel Steering

1. Raise and safely support the front of the vehicle so the front wheels are off of the floor.

2. Add power steering fluid to the reservoir to the FULL COLD level.

3. Turn the steering wheel quickly to the right and left several times. Repeat this until the fluid level in the reservoir no longer decreases.

—— **WARNING** ——
Do not let the reservoir run dry while the engine is running or damage to the pump can occur.

4. Start the engine, turn the steering wheel quickly to the right and left several times. Add fluid as needed. Repeat this until the fluid level in the reservoir no longer decreases.

5. If there are still air bubbles in the reservoir, clicking or buzzing noise from the pump, repeat the procedure.

Q45 with 4-Wheel Steering

1. To bleed the system, run the engine and turn the steering wheel lock-to-lock with the front wheels off the ground. When the fluid level stops decreasing, the system is properly bled.

2. With four wheel steering, the rear steering power cylinder may

need to be bled. There are two bleeder valves on top of the cylinder. To bleed the system:

a. Top off the fluid level. Run the engine at more than 2000 rpm and turn the steering wheel 180 degrees to the right.

b. Connect a tube to the right side bleeder valve and loosen the valve to bleed the air. Tighten the valve again.

c. Check the fluid level. Run the engine at more than 2000 rpm and turn the steering wheel 180 degrees to the left.

d. Connect the tube to the left bleeder valve and open the valve to let the air out.

e. Repeat this procedure alternating left and right until there is no more air in the system.

REMOVAL AND INSTALLATION

G20, J30 and Q45 Models

1. Disconnect the negative battery cable.
2. Remove the power steering pump drive belt from the pump pulley.
3. Place a catch pan under the pump and disconnect the power steering fluid lines.
4. Remove the power steering pump mounting bolts and remove the pump.

To install:

5. Install the power steering pump.
6. Use new gaskets and connect the pressure line. Torque the union bolt to 36–51 ft. lbs. (49–69 Nm).
7. Install and adjust the belt.
8. Connect the negative battery cable and bleed the power steering system.

I30 Models

1. Safely raise and support the vehicle.
2. Remove the right wheel and splash shield.
3. Loosen and remove the belt from the power steering pump pulley.
4. Place a drain pan under the power steering pump and disconnect the hoses from the pump.
5. Remove the power steering pump mounting bolts.
6. Remove the pump from the vehicle.

To install:

7. Position the pump on the mounting bracket and install the bolts. Do not tighten the bolts until the belt has been adjusted.

8. Reconnect the hoses using new gaskets. Torque the pressure hose union bolt to 36–51 ft. lbs. (49–69 Nm).
9. Place the belt on the pulley and adjust the belt tension. Torque the mounting bolts to 15–20 ft. lbs. (20–26 Nm).
10. Install the splash shield and wheel.
11. Safely lower the vehicle to the floor.
12. Refill the reservoir and bleed the system.

BRAKES

Anti-Lock Brake System Service

PRECAUTIONS

• Certain components within the Anti-Lock Brake System (ABS) are not intended to be serviced or repaired individually. Only those components with removal and installation procedures should be serviced.

• On vehicles equpped with ABS, it is important to depressurize the brake system, prior to performing repairs on the hydraulic system.

• Do not use rubber hoses or other parts not specifically specified for and ABS system. When using repair kits, replace all parts included in the kit. Partial or incorrect repair may lead to functional problems and require the replacement of components.

• Lubricate rubber parts with clean, fresh brake fluid to ease assembly. Do not use lubricated shop air to clean parts; damage to rubber components may result.

• Use only specified brake fluid from an unopened container.

• If any hydraulic component or line is removed or replaced, it may be necessary to bleed the entire system.

• A clean repair area is essential. Always clean the reservoir and cap thoroughly before removing the cap. The slightest amount of dirt in the fluid may plug an orifice and impair the system function. Perform repairs after components have been thoroughly cleaned; use only denatured alcohol to clean components. Do not allow ABS components to come into contact with any substance containing mineral oil; this includes used shop rags.

• The Anti-Lock control unit is a microprocessor similar to other computer units in the vehicle. Ensure that the ignition switch is OFF before removing or installing controller harnesses. Avoid static electricity discharge at or near the controller.

• If any arc welding is to be done on the vehicle, the control unit should be unplugged before welding operations begin.

Depressurizing

——— **WARNING** ———
Prior to opening hydraulic connections on vehicles equipped with ABS, it is important to depressurize the brake system.

1. Turn the ignition key to the OFF position.
2. Connect a vinyl tube to each air bleeder valve.
3. Drain the brake fluid from each bleeder valve by depressing the brake pedal.
4. After repairs, refill the reservoir with new DOT 3 brake fluid until brake fluid comes out of each bleeder valve and close the valve. Bleed the brake system.

Master Cylinder

REMOVAL AND INSTALLATION

Non ABS Equipped Models

G20, J30 and Q45 Models

NOTE: Prevent brake fluid from coming in contact with painted surfaces. Clean up any spills immediately.

1. Disconnect the negative battery terminal and the electrical connector from the reservoir.
2. Clean the outside of the master cylinder thoroughly, particularly around the cap and fluid lines.
3. Disconnect the brake fluid tubes, then plug the openings to prevent dirt from entering the system.
4. Remove the master cylinder mounting nuts and remove the master cylinder from the vehicle.

To install:

5. Bench bleed the master cylinder assembly prior to installation.
6. Install the master cylinder and torque the master cylinder mounting nuts to 9–11 ft. lbs. (12–15 Nm).
7. Connect the brake lines.
8. Refill the reservoir with brake fluid and bleed the system. Adjust the brake system if necessary.

9. Connect the negative battery cable and the electrical to the bottom of the reservoir.

NOTE: Ordinary brake fluid will boil and cause brake failure under the high temperatures developed in disc brake systems; use DOT 3 brake fluid in the brake systems. The adjustable pushrod is used to adjust the brake pedal free height.

10. Check for fluid leaks and verify proper brake system operation.

ABS Equipped Models

G20, J30 and Q45 Models

NOTE: Prevent brake fluid from coming in contact with painted surfaces. Clean up any spills immediately.

1. Loosen the brake line flarenuts and remove brake lines from master cylinder fittings.
2. Remove the master cylinder mounting nuts.
3. Remove the master cylinder.
To install:
4. Bench bleed the master cylinder.
5. Install the master cylinder in the vehicle. Tighten bolts to 9–11 ft. lbs. (12–15 Nm).
6. Connect the brake lines to the master cylinder and torque the flarenuts to 11–13 ft. lbs. (15–18 Nm).
7. Bleed the air from the master cylinder and brake lines.

I30 Models

1. Disconnect the negative battery terminal and the electrical connector from the reservoir.
2. Clean the outside of the master cylinder thoroughly, particularly around the cap and fluid lines.
3. Disconnect the brake fluid tubes, then plug the openings to prevent dirt from entering the system.
4. Remove the master cylinder mounting nuts and remove the master cylinder from the vehicle.
To install:
5. Bench bleed the master cylinder assembly prior to installation.
6. Install the master cylinder and torque the master cylinder mounting nuts to 9–11 ft. lbs. (12–15 Nm).
7. Connect the brake lines.
8. Refill the reservoir with brake fluid and bleed the system. Adjust the brake system if necessary.

9. Connect the negative battery cable and the electrical to the bottom of the reservoir.

NOTE: Ordinary brake fluid will boil and cause brake failure under the high temperatures developed in disc brake systems; use DOT 3 brake fluid in the brake systems. The adjustable pushrod is used to adjust the brake pedal free height.

10. Check for fluid leaks and verify proper brake system operation.

Bench Bleeding

1. Place the master cylinder in a vise.
2. Connect 2 lines to the fluid outlet orifices and place them into the reservoir.
3. Fill the reservoir with brake fluid.
4. Using a wooden dowel, depress the pushrod slowly, allowing the pistons to return. Do this several times until the air bubbles in the brake fluid are gone.
5. Remove the bleeding tubes from the master cylinder, plug the outlets, and install the caps.

Brake Caliper

REMOVAL AND INSTALLATION

G20 Models

Front Caliper

NOTE: Prevent brake fluid from coming in contact with painted surfaces. Clean up any spills immediately.

1. Raise the vehicle and support it safely.
2. Remove the wheel.
3. Loosen the brake hose connecting bolt.
4. Remove the bolts connecting the caliper to the torque member.
5. Slide the caliper out from the rotor and remove the pads, shims and shim covers.
6. Remove the brake hose connecting bolt from the caliper.
7. Remove the caliper from the vehicle.
To install:
8. Fit the pads and shims into place.
9. Fit the caliper onto the torque member and torque the bolts to 16–23 ft. lbs. (22–31 Nm).
10. Use new copper washers and connect the hydraulic hose to the cali-

per. Torque the union bolt to 12–14 ft. lbs. (17–19 Nm).
11. Bleed the air from the system.

Rear Caliper

1. Raise and safely support the vehicle.
2. Remove the rear wheel(s).
3. Remove the brake cable mounting bolt and lock spring.
4. Release the parking brake control lever, then disconnect the cable from the caliper.
5. Place a drain pan under the caliper and disconnect the brake fluid hose.
6. Remove the torque member mounting bolts and remove the caliper assembly.
To install:
7. Fit the pads and shims into place.
8. Fit the caliper onto the torque member and torque the bolts to 16–23 ft. lbs. (22–31 Nm).
9. Use new copper washers and connect the hydraulic hose to the caliper. Torque the union bolt to 12–14 ft. lbs. (17–19 Nm).
10. Connect the parking brake cable to the rear caliper.
11. Bleed the brake system.
12. Install the rear wheel and lower the vehicle to the floor.

J30 Models

Front Caliper

NOTE: Prevent brake fluid from coming in contact with painted surfaces. Clean up any spills immediately.

1. Raise the vehicle and support it safely.
2. Remove the wheel.
3. Place a drain pan under the caliper and loosen the brake hose connecting bolt.
4. Remove the torque member mounting bolts and disconnect the brake fluid hose from the caliper.
5. Slide the caliper off of the rotor and remove the pads, shims and shim covers.
6. Remove the caliper from the vehicle.
To install:
7. Fit the pads and shims into place.
8. Position the torque member on the knuckle assembly and install the mounting bolts. Torque the bolts to 53–72 ft. lbs. (72–97 Nm).
9. Use new copper washers and connect the hydraulic hose to the caliper. Torque the union bolt to 12–14 ft. lbs. (17–19 Nm).
10. Bleed the air from the system.

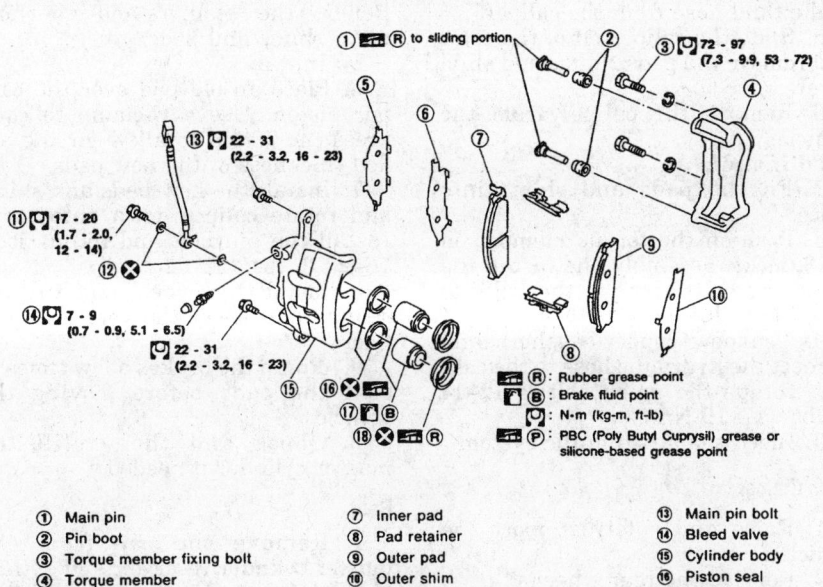

① Main pin
② Pin boot
③ Torque member fixing bolt
④ Torque member
⑤ Shim cover
⑥ Inner shim
⑦ Inner pad
⑧ Pad retainer
⑨ Outer pad
⑩ Outer shim
⑪ Connecting bolt
⑫ Copper washer
⑬ Main pin bolt
⑭ Bleed valve
⑮ Cylinder body
⑯ Piston seal
⑰ Piston
⑱ Piston boot

291938

Front brake caliper assembly — J30 models

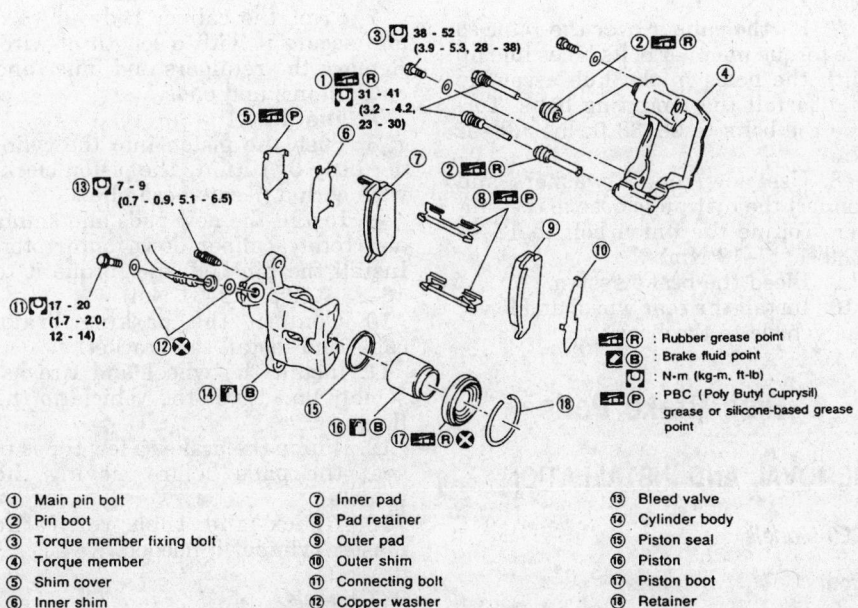

① Main pin bolt
② Pin boot
③ Torque member fixing bolt
④ Torque member
⑤ Shim cover
⑥ Inner shim
⑦ Inner pad
⑧ Pad retainer
⑨ Outer pad
⑩ Outer shim
⑪ Connecting bolt
⑫ Copper washer
⑬ Bleed valve
⑭ Cylinder body
⑮ Piston seal
⑯ Piston
⑰ Piston boot
⑱ Retainer

291939

Rear brake caliper assembly — J30 models

Rear Caliper

1. Raise and safely support the vehicle.

2. Remove the rear wheel(s).

3. Place a drain pan under the caliper and disconnect the brake fluid hose.

4. Remove the torque member mounting bolts and remove the caliper assembly.

5. Remove the pads and shims from the caliper.

To install:

6. Fit the pads and shims into place.

7. Fit the caliper over the rotor so the torque member bolts holes line up with the holes in the hub assembly and install the mounting bolts. Torque the bolts to 28–38 ft. lbs. (38–52 Nm).

8. Use new copper washers and connect the hydraulic hose to the caliper. Torque the union bolt to 12–14 ft. lbs. (17–19 Nm).

9. Bleed the brake system.

10. Install the rear wheel and lower the vehicle to the floor.

I30 Models

Front Caliper

NOTE: All brake pads must always be replaced as a complete set.

1. Raise and safely support vehicle.

2. Remove the front wheels.

3. Remove both guide pin bolts securing the caliper to the steering knuckle.

4. Loosen and remove the brake hose connector from the caliper. Be sure to plug the openings to prevent dirt from entering the system.

5. Remove the caliper assembly from the vehicle.

To install:

6. Using new copper washers, install the brake line to the brake caliper and torque the connecting bolt to 12–14 ft. lbs. (17–20 Nm).

7. Install the caliper to the steering knuckle using the guide pins bolts.

8. Install the wheels and tighten the lug nuts to the proper specification.

9. Bleed the brake system and top off the master cylinder as necessary.

NOTE: Make sure you pump the brakes and get a hard pedal before moving the vehicle!

Rear Caliper

NOTE: All rear brake pads must always be replaced as a complete set.

1. Raise and safely support the vehicle. Remove the rear wheels.
2. Remove the parking brake cable stay fixing bolt and the lock spring.
3. Remove the brake fluid hose from the caliper.
4. Remove the lower caliper-to-torque member pin bolt and raise the caliper.
5. Remove the pad retainers, the pads and the shims.
6. Remove the upper caliper-to-torque member pin bolt and remove the caliper.

To install:

7. Clean the piston end of the caliper body and the area around the pin holes. Be careful not to get oil on the rotor.
8. Using a pair of needle nose pliers, carefully turn the piston clockwise back into the caliper body; remove some brake fluid from the master cylinder, if necessary. Take care not to damage the piston boot.
9. Coat the pad contact area on the mounting support with grease.
10. Install the new pads, shims and the pad springs.

NOTE: Always use new shims.

11. Install the caliper body into position and torque the caliper-to-torque member pin bolts to 16–23 ft. lbs. (22–31 Nm).
12. Reconnect the brake fluid hose and tighten the flare nut to 12–14 ft. lbs. (17–20 Nm).
13. Install the lock spring and the parking brake stay fixing bolt.
14. Bleed the brake system and top off the master cylinder as necessary.
15. Replace the wheel, lower the vehicle.

NOTE: Make sure you pump the brakes and get a hard pedal before moving the vehicle!

Q45 Models

Front Caliper

NOTE: Prevent brake fluid from coming in contact with painted surfaces. Clean up any spills immediately.

1. Raise the vehicle and support it safely.
2. Remove the wheel.
3. Place a drain pan under the caliper and loosen the brake hose connecting bolt.

4. Remove the torque member mounting bolts and disconnect the brake fluid hose from the caliper.
5. Slide the caliper off of the rotor and remove the pads, shims and shim covers.
6. Remove the caliper from the vehicle.

To install:

7. Fit the pads and shims into place.
8. Position the torque member on the knuckle assembly and install the mounting bolts. Torque the bolts to 118–137 ft. lbs. (87–101 Nm).
9. Use new copper washers and connect the hydraulic hose to the caliper. Torque the union bolt to 12–14 ft. lbs. (17–19 Nm).
10. Bleed the air from the system.

Rear Caliper

1. Raise and safely support the vehicle.
2. Remove the rear wheel(s).
3. Place a drain pan under the caliper and disconnect the brake fluid hose.
4. Remove the torque member mounting bolts and remove the caliper assembly.
5. Remove the pads and shims from the caliper.

To install:

6. Fit the pads and shims into place.
7. Fit the caliper over the rotor so the torque member bolts holes line up with the holes in the hub assembly and install the mounting bolts. Torque the bolts to 28–38 ft. lbs. (38–52 Nm).
8. Use new copper washers and connect the hydraulic hose to the caliper. Torque the union bolt to 12–14 ft. lbs. (17–19 Nm).
9. Bleed the brake system.
10. Install the rear wheel and lower the vehicle to the floor.

Disc Brake Pads

REMOVAL AND INSTALLATION

G20 Models

Front

1. Remove the cap from the master cylinder reservoir and extract about 1/3 of the brake fluid from the reservoir to prevent overflow when the caliper piston is compressed.
2. Raise the vehicle and support safely.
3. Remove the wheel.
4. Remove the lower pin bolt.

5. Pivot the caliper body upward and secure it with a length of wire. Remove the retainers and inner and outer shims and pads.

To install:

6. Place an old pad over the caliper piston. Use a C-clamp to compress the piston to allow for the added thickness of the new pads.
7. Install the new pads and shims and rotate caliper down onto rotor. Install the pin bolt and torque it to 16–23 ft. lbs. (22–31 Nm).
8. Install the wheel and tire assembly and lower the vehicle to the floor.
9. Pump the brakes a few times to seat the pads before moving the vehicle.
10. Check and then refill the master cylinder if needed.

Rear

1. Remove the cap from the master cylinder reservoir and extract about 1/3 of the brake fluid from the reservoir to prevent overflow when the caliper piston is compressed.
2. Raise the vehicle and support safely.
3. Remove the wheel.
4. Remove the brake cable mounting bracket bolt and lock spring.
5. Disconnect the parking brake cable.
6. Remove the lower pin bolt.
7. Pivot the caliper body upward and secure it with a length of wire. Remove the retainers and inner and outer shims and pads.

To install:

8. Push the piston into the cylinder body by turning the piston clockwise with a needle nose pliers.
9. Install the new pads and shims and rotate caliper down onto rotor. Install the pin bolt and torque it to 16–23 ft. lbs. (22–31 Nm).
10. Connect the parking brake cable and install the bracket.
11. Install the wheel and tire assembly and lower the vehicle to the floor.
12. Pump the brakes a few times to seat the pads before moving the vehicle.
13. Check and then refill the master cylinder if needed.

J30 Models

Front and Rear Disc Brake Pads

1. Remove the cap from the master cylinder reservoir and extract about 1/3 of the brake fluid from the reservoir to prevent overflow when the caliper piston is compressed.
2. Raise the vehicle and support safely.

3. Remove the wheel.

4. Remove the lower pin bolt.

5. Pivot the caliper body upward and secure it with a length of wire. Remove the retainers and inner and outer shims and pads.

To install:

6. Place an old pad over the caliper piston. Use a C-clamp to compress the piston to allow for the added thickness of the new pads.

7. Install the new pads and shims and rotate caliper down onto rotor. Install the pin bolt and torque it to 16–23 ft. lbs. (22–31 Nm) for the front or 28–38 ft. lbs. (38–52 Nm) for the rear.

8. Install the wheel and tire assembly and lower the vehicle to the floor.

9. Pump the brakes a few times to seat the pads before moving the vehicle.

10. Check and then refill the master cylinder if needed.

I30 Models

----- CAUTION -----

Brake shoes contain asbestos, which has been determined to be a cancer causing agent. Never clean the brake surfaces with compressed air. Avoid inhaling any dust from any brake surface. When cleaning brake surfaces, use a commercially available brake cleaning fluid.

Front

1. Raise and support the front of the vehicle, then remove the wheels.

2. Remove the bottom guide pin from the caliper and swing the caliper cylinder body upward; support the caliper with a wire.

3. Remove the brake pad retainers and the pads.

To install:

4. Compress the piston of the disc brake caliper.

5. Install the brake pads and caliper assembly. Torque the guide pin to 16–23 ft. lbs. (22–31 Nm).

6. Install the wheels.

7. Apply the brakes a few times to seat the pads. Check the master cylinder and add fluid if necessary. Bleed the brakes, if necessary.

Rear

----- WARNING -----

Do not press the piston into the bore as performed on the front disc brakes. Due to the parking brake mechanism, the caliper piston must be turned into the bore using a special tool.

1. Raise and support the vehicle safely.

2. Remove the rear wheels.

3. Remove the parking brake cable mounting bolt and lock spring.

4. Release the parking brake control lever and disconnect the cable from the caliper.

5. Remove the upper pin bolt.

6. Pivot the caliper body downward.

7. Pull out the pad springs and then remove the pads and shims.

To install:

8. Clean the piston end of the caliper body and the area around the pin holes. Be careful not to get oil on the rotor.

9. Using the proper tool, carefully turn the piston clockwise back into the caliper body. Take care not to damage the piston boot.

10. Coat the pad contact area on the mounting support with grease.

11. Install the pads, shims, and the pad springs. Always use new shims.

12. Position the caliper body in the mounting support and tighten the pin bolts to 16–23 ft. lbs. (22–31 Nm).

13. Install the wheels.

14. Apply the brakes a few times to seat the pads. Check the master cylinder and add fluid if necessary. Bleed the brakes, if necessary.

Q45 Models

1. Remove the cap from the master cylinder reservoir and extract about 1/3 of the brake fluid from the reservoir to prevent overflow when the caliper piston is compressed.

2. Raise the vehicle and support safely.

3. Remove the wheel.

4. On the right side, disconnect the sensor harness by pushing the connector pin and pulling the connector. To remove the rear wire connector, push it toward the pad and turn it counterclockwise. Then pull it out. Remove the bracket from the cylinder body.

5. Remove the lower pin bolt.

6. Pivot the caliper body upward and secure it with a length of wire. Remove the retainers and inner and outer shims and pads.

To install:

7. Place an old pad over the caliper piston. Use a C-clamp to compress the piston to allow for the added thickness of the new pads.

8. Install the new pads and shims and rotate caliper down onto rotor. Install the pin bolt and torque it to 61–69 ft. lbs. (83–93 Nm) for the front or 23–30 ft. lbs. (31–41 Nm) for the rear.

9. Install and connect the sensor.

10. Install the wheel and tire assembly and lower the vehicle to the floor.

11. Pump brakes a few times to seat pads before moving the vehicle.

12. Check and then refill master cylinder if needed.

Brake Rotor

REMOVAL AND INSTALLATION

G20, J30 and Q45 Models

1. Raise the vehicle and support it safely.

2. Remove the wheel.

3. Remove the caliper from the torque member and support using a length of mechanics wire.

4. Remove the pads and shims.

5. Remove the bolts attaching the torque member and remove.

6. Remove the rotor from the hub assembly.

To install:

7. Install the rotor on the front wheel hub.

8. Fit the torque member into place and torque the bolts as follows:

- G20 front: 53–72 ft. lbs. (72–97 Nm), rear: 25–31 ft. lbs. (34–42 Nm)
- J30 front: 53–72 ft. lbs. (72–97 Nm), rear: 28–38 ft. lbs. (38–52 Nm)
- Q45 front: 53–72 ft. lbs. (72–97 Nm), rear: 28–38 ft. lbs. (38–52 Nm)

9. Install the brake pads, shims and calipers. Torque the sliding pin bolts as follows:

- G20 front: 16–23 ft. lbs. (22–31 Nm), rear: 16–23 ft. lbs. (22–31 Nm)
- J30 front: 16–23 ft. lbs. (22–31 Nm), rear: 28–38 ft. lbs. (38–52 Nm)
- Q45 front: 61–69 ft. lbs. (83–93 Nm), rear: 23–30 ft. lbs. (31–41 Nm)

10. Install the wheel and lower the vehicle to the floor.

11. Apply the brakes a few times before moving the vehicle.

I30 Models

1. Raise and support the vehicle safely.

2. Remove the wheel assembly.

3. Remove the disc brake caliper assembly. Using a wire, support the caliper assembly; do not disconnect the brake line from the caliper.

4. Remove the caliper support mounting bolts and remove the support.

5. Remove the brake disc (rotor) from the wheel hub.

6. Inspect the disc for wear, cracks, run-out and thickness; if necessary, replace the disc.

To install:

7. Install disc onto the wheel hub.

8. Install the caliper support on the knuckle and torque the mounting bolts to the following specifications:
- Front wheel — torque the mounting bolts to: 53–72 ft. lbs.
- Rear wheel — torque the mounting bolts to: 16–23 ft. lbs. (22–31 Nm)

9. Install the caliper assembly to the caliper support bracket. Tighten the brake caliper mounting pin bolts to 23 ft. lbs. (31 Nm).

10. Check the level of fluid in the master cylinder. Top off fluid as necessary.

11. Install the wheels and torque the wheel lug nuts to the proper specification.

12. If the brake system was opened, bleed the brake system.

NOTE: Make sure to pump the brakes and get a hard pedal before driving the vehicle!

Parking Brake Cable

ADJUSTMENT

G20 Models

1. Remove the center console assembly.

2. Loosen the parking brake cable with the adjusting nut.

3. Depress the brake pedal fully more than five times.

4. Adjust the nut so the lever moves 6–8 notches when pulled with a force of 44 lb. (196 N).

5. Make sure the warning lamp comes on when the lever is pulled one notch. Bend the switchplate to adjust. The light should go out when the lever is fully released.

6. Install the center console assembly.

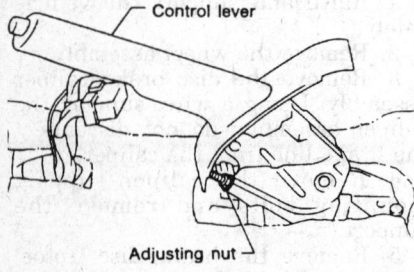

Parking brake adjusting nut location — G20 models

289940

J30 and I30 Models

NOTE: Make sure the rear brakes are properly adjusted prior to adjusting the parking brake.

1. Pull the parking brake lever with 44 lbs. of force and note the number of notches.

2. The parking brake lever should raise:
   a. 7–8 notches for G20
   b. 8–11 notches for J30
   c. 10–11 notches for I30

3. Locate the parking brake adjuster at the base of the hand lever and rotate the adjuster nut on the threaded rod to obtain the proper parking brake lever adjustment.

4. Bend the parking brake warning lamp switch plate so that brake warning light comes on when parking brake lever is pulled up 1 notch. The brake light should turn off when the lever is fully released.

Q45 Models

1. With the wheel removed, remove the adjuster plug.

2. Turn the adjuster wheel down until the rotor cannot be turned by hand. Back off the adjustment 5–6 notches.

3. Install the wheel and turn it to make sure the brake is not dragging.

4. Adjust the cable adjuster for pedal travel of 3.54–4.13 inches when a force of 44 lbs. is applied to the parking brake pedal.

NOTE: To bed in new shoes or drums, drive the vehicle on a dry road at about 20 mph. Carefully apply the parking brake to drag the brake for about 300 feet. Release the brake and continue driving to cool the drum. Repeat the process two more times.

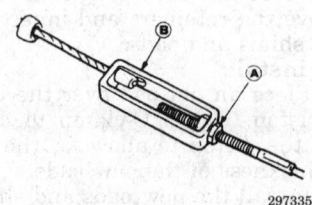

Parking brake cable adjuster — Q45 models

297335

REMOVAL AND INSTALLATION

G20 Models

1. Remove the center console assembly.

2. Disconnect the brake warning lamp connector.

3. Remove the adjusting nut.

4. Raise and safely support the vehicle.

5. Remove the rear wheel.

6. Remove the cable mounting bracket and lock plate.

7. Remove the cable retainers and disconnect the cable from the brake caliper.

To install:

8. Position the cable in the proper position and install the cable retainers.

9. Connect the cable to the brake caliper and install the mounting bracket and lock plate.

10. Connect the cable to the equalizer. Make sure the front cable is installed in the lever assembly.

11. Safely lower the vehicle to the floor and install the adjusting nut.

12. Connect the brake warning switch and install the center console assembly.

J30 Models

Front Cable

NOTE: It is possible to remove the front parking brake cable without removing the lever assembly.

1. Raise the vehicle and support safely.

2. Disconnect the front cable from the equalizer.

3. Lower the vehicle.

4. Remove the clip retaining the cable end to the pedal assembly.

5. Remove the center console.

6. Remove the cable brackets and remove the cable from the floor grommet.

To install:

7. Place the cable in the proper position and install the brackets.

8. Install the center console and attach the cable to the pedal assembly.

9. Raise the vehicle and connect the cable to the equalizer.

10. Adjust the parking brake cable and safely lower the vehicle to the floor.

Rear Cable

1. Raise the vehicle and support safely.

2. Remove rear wheels.

3. Remove the bolts attaching the rear disc brake caliper mounting

bracket. Remove bracket with caliper attached. Support using a length of mechanics wire.

4. Remove 2 bolts attaching rear brake rotor and remove rotor.

5. Disconnect parking brake cable end from toggle lever. Remove cable from backing plate mounting.

6. Remove cable brackets.

7. Disconnect rear cable from equalizer.

8. Complete the installation of the rear parking brake cable by reversing the removal procedure. Pay close attention to the following:

  a. Install the rear parking brake cable into the backing plate by tapping the flanged section of the cable cover with a hammer and punch.

  b. Check the shoe clearance adjustment before adjusting the cable.

  c. Adjust parking brake cable.

I30 Models

Front Cable

1. Remove the center console assembly.

2. Loosen and remove the parking brake cable adjusting nut at the base of the parking brake lever.

3. Under the vehicle, disconnect the parking brake cables at the equalizer.

4. Unbolt the parking brake lever from the center console.

5. Unbolt and remove the front parking brake cable from the vehicle.

To install:

6. Install the front parking brake cable into the parking brake lever and start the adjusting nut on the front cable.

7. Install the lever and cable assembly into the vehicle. Tighten the mounting bolts.

8. Connect the parking brake cables at the equalizer.

9. Install the center console assembly.

Rear Cable

1. Remove the rear wheel and the brake drum or disc assembly.

2. At the cable adjuster, loosen the adjusting nut, then separate the rear cable from the equalizer.

3. On rear drum brakes, remove the brake shoes from the backing plate, then separate the rear cable from the toggle lever.

4. On rear disc brakes, remove the cable retainer and the cable end from the toggle lever.

5. Remove the bolts that secure the cable to the vehicle under body.

6. Pull the cable through the backing plate and remove it from the vehicle.

To install:

7. On rear drum brakes, install the brake cable to the vehicle through the backing plate.

8. On rear drum brakes, install the brake shoes with cable attached and the brake drum.

9. On rear disc brakes, connect the cable to the toggle lever and tighten the cable retainer.

10. Install the bolts that secure the cable to the vehicle under body.

11. At the cable equalizer, connect the parking brake cable.

12. Install the rear wheels.

13. Adjust the parking brake cable.

Q45 Models

Front Cable

NOTE: It is possible to remove the front parking brake cable without removing the lever assembly.

1. Raise the vehicle and support safely.

2. Disconnect the front cable from the equalizer.

3. Lower the vehicle.

4. Remove the clip retaining the cable end to the pedal assembly.

5. Remove the center console.

6. Remove the cable brackets and remove the cable from the floor grommet.

To install:

7. Place the cable in the proper position and install the brackets.

8. Install the center console and attach the cable to the pedal assembly.

9. Raise the vehicle and connect the cable to the equalizer.

10. Safely lower the vehicle to the floor.

Rear Cable

1. Raise the vehicle and support safely.

2. Remove rear wheels.

3. Remove the bolts attaching the rear disc brake caliper mounting bracket. Remove bracket with caliper attached. Support using a length of mechanics wire.

4. Remove 2 bolts attaching rear brake rotor and remove rotor.

5. Disconnect parking brake cable end from toggle lever. Remove cable from backing plate mounting.

6. Remove cable brackets.

7. Disconnect rear cable from equalizer.

8. Complete the installation of the rear parking brake cable by reversing the removal procedure. Pay close attention to the following:

  a. Install the rear parking brake cable into the backing plate by tapping the flanged section of the cable cover with a hammer and punch.

  b. Check the shoe clearance adjustment before adjusting the cable.

  c. Adjust parking brake cable.

Brake System Bleeding

The purpose of bleeding the brakes is to expel air trapped in the hydraulic system. The system must be bled whenever the pedal feels spongy, indicating that compressible air has entered the system. It must also be bled whenever the system has been opened or repaired. You will need a helper for this job.

—— **WARNING** ——

Be careful! Brake fluid is extremely harmful to painted surfaces. Brake fluid picks up moisture from the air. Don't leave the master cylinder or the fluid container uncovered any longer than necessary. Never reuse brake fluid which has been bled from the brake system.

1. The sequence for bleeding is as follows:

• Models not equipped with ABS: left rear wheel cylinder, right rear wheel cylinder, left front caliper, right front caliper.

• Model equipped with ABS: left rear caliper or wheel cylinder, right rear caliper or wheel cylinder, left front caliper, right front caliper, ABS actuator.

NOTE: On models with ABS, be sure to turn the ignition OFF and disconnect the actuator connector.

2. Clean all the bleeder screws. You may want to give each one a shot of a penetrating lubricant to loosen it up; seizure is a common problem with bleeder screws, which then break off, usually requiring replacement of the part to which they are attached.

3. Fill the master cylinder with DOT 3 brake fluid.

NOTE: Check the level of the fluid often when bleeding, and refill the reservoirs as necessary. Don't let them run dry, or you will have to repeat the process.

4. Attach a length of clear vinyl tubing to the bleeder screw on the

wheel cylinder (or master cylinder). Insert the other and of the tube into a clear, clean jar half-filled with brake fluid.

5. Have your helper slowly depress the brake pedal. As this is done, open the bleeder screw ⅓–½ of a turn, and allow the fluid to run through the tube. Then close the bleeder screw before the pedal reaches the end of its travel. Have your assistant slowly release the pedal. Repeat this process until no air bubbles appear in the expelled fluid.

NOTE: If the brake pedal is depressed too fast, small air bubbles will form in the brake fluid.

6. Repeat the procedure on the other bleeder screws, checking the level of fluid in the cylinder reservoirs often.

7. When all the air has been bleed from the system, perform the following steps:

   a. If disconnected, reconnect the ABS actuator.

   b. Pressurize the system and check for leaks.

   c. Check and fill the fluid reservoir.

Wheel Speed Sensor

REMOVAL AND INSTALLATION

—————— WARNING ——————
Always remove the wheel speed sensor before removing the front hub assembly or the final drive assembly to prevent damage to the wheel speed sensor.

1. Raise and safely support the vehicle.

2. Remove the wheel sensor mounting bolt.

3. Disconnect the sensor harness connector and remove the sensor.

To install:

4. Position the sensor in the hub or differential assembly and install the mounting bolt. Torque the bolt to the following amounts:

• For G20 models, torque the mounting bolt to 8–11 ft. lbs. (11–15 Nm)

• For J30 models, torque the bolt on the front speed sensor to 13–17 ft. lbs. (18–24 Nm) and the bolt on the rear speed sensor to 18–25 ft. lbs. (25–33 Nm)

• For I30 models, torque the bolt on the front speed sensor to 13–17 ft. lbs. (18–24 Nm) and the bolt on the

rear speed sensor to 18–25 ft. lbs. (25–33 Nm)

• For Q45 models, torque the bolts to 8–12 ft. lbs. (11–16 Nm)

5. Connect the sensor harness connector.

6. Safely lower the vehicle to the floor.

FRONT SUSPENSION

Strut

REMOVAL AND INSTALLATION

G20 Models

1. Raise and support the vehicle safely. Remove the strut mounting bolt at the lower suspension member and the three nuts inside the engine compartment. Do not remove the piston rod locknut.

2. Remove the strut assembly and place in a suitable holding device.

3. Using a prybar to hold the upper spring mount, loosen but do not remove the piston rod locknut.

4. Compress the spring with a spring compressor so the strut mounting insulator can be turned by hand.

5. Remove the piston rod locknut. Remove the coil spring from strut assembly.

To install:

6. Inspect all components carefully for damage or wear. Replace as necessary.

7. Install the coil spring on the strut assembly and tighten the locknut to 13–17 ft. lbs. (18–24 Nm).

8. Install the strut assembly in the vehicle. Ensure the bend in the lower shock bracket faces rearward on the left side and forward on the right side of the vehicle.

9. Install the upper spring seat with the cutout facing the inside of the vehicle.

10. Tighten the upper strut mounting bolts to 31–40 ft. lbs. (42–54 Nm) and the lower through bolt to 82–93 ft. lbs. (112–126 Nm). Final tightening must take place with the suspension loaded (vehicle at normal ride height).

J30 Models

1. Raise and support the vehicle safely.

2. Remove the front wheel.

3. Remove the brake caliper and hang it from the body with wire. Do not let the caliper hang by the brake hose.

4. Remove the nut and separate the stabilizer bar linkage from the strut.

5. Remove the lower ball joint nut and separate the ball joint from the strut.

6. Remove the nut and separate the tie rod end from the strut.

————— CAUTION —————
Do not remove the center nut from the strut piston rod unless the coil spring is compressed using the proper spring compressor.

7. Remove the nuts from the upper strut mount insulator and lower the strut assembly from the vehicle.

8. To disassemble the strut, secure the strut assembly in a vise using special tool ST-35652000, or equivalent, strut holding fixture.

9. Loosen, but do not remove the piston rod locknut.

10. Compress the spring with a spring compressor so the strut mounting insulator can be turned by hand.

11. Remove the piston rod locknut and coil spring.

12. To remove the internal shock absorber components, clean the top of the shock tube and remove the gland packing nut. Push the rod all the way down, then slowly pull it up again to withdraw the piston rod and guide.

To install:

13. Inspect the rubber parts for deterioration. If there is oil on the spring seat, the gland packing nut and O-ring should replaced. Some seepage on the rod or upper surface of the nut is normal.

14. Lubricate the sealing lip of the gland packing nut. Cover the piston rod with tape to prevent damage to the oil seal and install the nut.

15. Tighten the gland packing to 43–80 ft. lbs. (49–108 Nm) total torque. If a crows foot or other special tool is used to torque the nut, its length must be calculated into the torque.

16. Set the spring into the lower spring seat. The flat portion of the spring goes on top.

17. Install the spring seat with its cutout facing the outer side of the vehicle. Install the upper plate and strut mount insulator and the piston rod nut. Torque the nut to 43–58 ft. lbs. (58–79 Nm).

18. Remove the spring compressor and fit the strut into the vehicle. In-

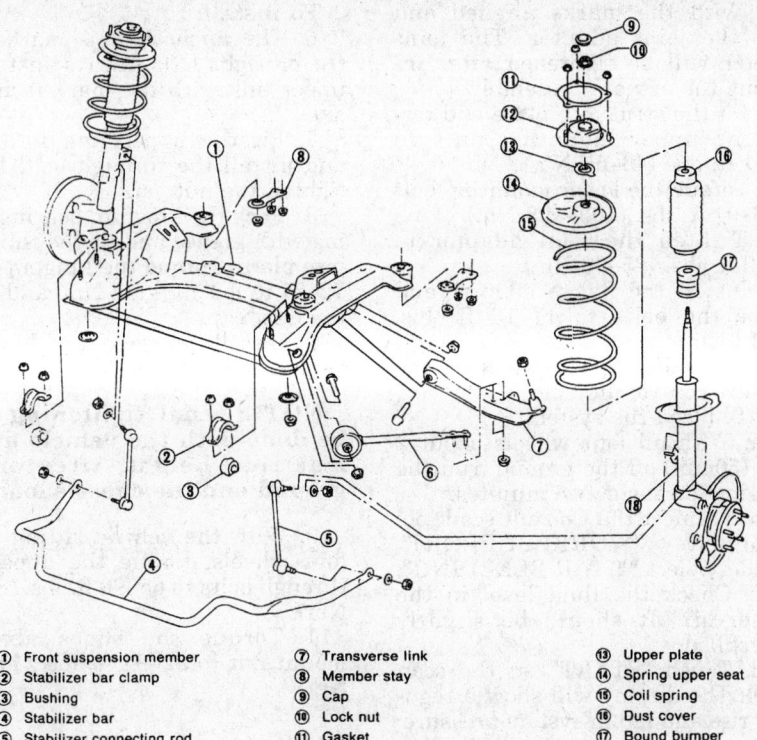

① Front suspension member
② Stabilizer bar clamp
③ Bushing
④ Stabilizer bar
⑤ Stabilizer connecting rod
⑥ Tension rod
⑦ Transverse link
⑧ Member stay
⑨ Cap
⑩ Lock nut
⑪ Gasket
⑫ Strut mounting insulator
⑬ Upper plate
⑭ Spring upper seat
⑮ Coil spring
⑯ Dust cover
⑰ Bound bumper
⑱ Strut assembly

295007

Front strut and spring assembly — J30 models

stall the upper strut mount nuts and torque to 30–35 ft. lbs. (41–48 Nm).

19. Fit the lower ball joint into place and torque the nut to 71–88 ft. lbs. (96–120 Nm).

20. Connect the tie rod end and torque the nut.

21. Connect the stabilizer bar linkage and torque the nut to 54–67 ft. lbs. (74–90 Nm).

22. Install the brake caliper and wheel and check front wheel alignment.

I30 Models

1. Raise and safely support the vehicle.

2. Remove the wheel. Matchmark the position of the strut-to-steering knuckle location.

3. Disconnect the brake hose from the strut.

4. Remove the ABS wheel sensor and move it out of the way.

5. Matchmark and remove the bolts attaching the steering knuckle to the strut.

6. Open the hood and remove the strut attaching nuts while holding the strut.

— **CAUTION** —
Do not remove the center locknut from the strut assembly until the strut is safely compressed.

7. Remove the strut from the vehicle.

8. Place the strut assembly in a vise with the special holding tool ST35652000 or in a spring compressor.

9. Loosen the piston rod locknut.

— **CAUTION** —
Do not remove the piston rod locknut, the spring is under tension and can cause serious personal injury.

10. Compress the spring with the spring compressor then remove the piston rod locknut.

NOTE: Before removing the strut from the coil spring, note the positioning of the strut in relationship to the coil spring for reassembly.

11. Remove the strut mounting insulator bracket, strut mounting bearing, upper spring seat, and the upper spring rubber seat.

12. Remove the strut, leaving the coil spring compressed.

13. Remove the piston boot and rebound bumper from the strut.

To install:

14. Install the rebound bumper and the boot to the strut piston.

15. Install the strut into the coil spring, make sure the strut and spring are properly positioned.

16. Install the upper spring rubber seat, upper spring seat, strut mounting bearing, and the strut mounting insulator bracket. Make sure that the cutout on the upper spring seat is facing the outside of the vehicle.

17. Install the piston rod locknut then remove the spring compressor.

18. Torque the piston rod locknut to 43–58 ft. lbs. (59–76 Nm).

19. Install the strut into the strut tower and install new attaching nuts. Torque the nuts to 29–40 ft. lbs. (39–54 Nm).

20. Install the bolts attaching the steering knuckle to the strut and align the matchmarks. Torque the bolts to 103–117 ft. lbs. (140–159 Nm).

21. Install the ABS wheel sensor and torque the attaching bolt to 13–17 ft. lbs. (18–24 Nm).

22. Install the brake hose to the strut.

23. Install the front wheels and lower the vehicle.

24. Check and/or adjust the wheel alignment as necessary.

Q45 Models

Standard Suspension

1. Remove the upper mounting insulator bolts.

2. Raise and safely support the vehicle.

3. Remove the lower strut mounting bolt and lift out the strut assembly.

4. Secure the strut in a suitable holding fixture.

5. Loosen the piston rod locknut. Do not remove the locknut.

6. Compress the spring with the proper tool so the strut assembly mounting insulator can be turned by hand.

7. Remove the piston rod locknut. Remove the spring assembly, dust cover and rubber seat. Remove the strut insert.

To install:

8. Inspect the rubber parts for deterioration. If the rubber is pulling away from the metal, the mounting insulator should be replaced.

9. Fit the spring into the lower seat, install the dust cover/bumper and upper seat and mounting insulator.

10. Install the piston rod locknut and torque to 13–17 ft. lbs. (17–23 Nm).

11. Fit the strut into place and torque the upper mounting nuts to 30–35 ft. lbs. (40–47 Nm).

12. Torque the lower mounting bolt to 80–94 ft. lbs. (108–128 Nm).

Active Suspension

NOTE: The Nissan Consult or an equivalent scan tool that can issue commands to the control unit is required for bleeding the hydraulics in the Full Active Suspension system.

1. Relieve the hydraulic pressure:
   a. Raise all four wheels off the ground and wait at least 3 minutes for the system to stabilize.
   b. Remove both front inner fenders and the rear pressure control unit cover.
   c. Loosen the locknut and slowly open the bypass valve on each pressure control unit. Open the valves all the way and leave them open until the job is finished.
2. Remove the flange joint from the top of the actuator.
3. Install two 15mm bolts into the actuator in the flange joint mounting bolt holes.
4. Insert a bar between the bolts and loosen the joint adapter. Do not remove it yet.
5. Remove the upper mount insulator nuts.
6. Raise and safely support the vehicle.
7. Disconnect the hydraulic lines. Cap the lines to keep the system clean.
8. Remove the lower actuator mounting nut and remove the assembly.
9. Secure the actuator/spring assembly in a suitable holding fixture. Scribe alignment marks on the spring, upper mount insulator and actuator unit.
10. Compress the spring with the proper tool so the joint adapter can be turned by hand. Remove the joint adapter and lift off the mount insulator, spring, and any other components necessary.

To install:
11. If the actuator is being replaced, the rubber bumper should also be replaced. Fit the bumper, dust cover and rubber seat onto the actuator.
12. Fit the spring into the lower seat with the matchmarks aligned. Install the upper seat/mounting insulator with the marks aligned and start the joint adapter. The joint adapter will be tightened after installing the actuator assembly.

13. Fit the strut into place and torque the upper mounting nuts to 30–41 ft. lbs. (40–55 Nm).
14. Torque the lower mounting bolt to 76–94 ft. lbs. (103–128 Nm).
15. Tighten the joint adapter to 63–72 ft. lbs. (85–98 Nm).
16. Install the flange adapter and torque the bolts to 11–13 ft. lbs. (15–18 Nm).
17. Close the bypass valves on the pressure control units.
18. To bleed the system:
   a. With all four wheels about 2 in. (50mm) off the ground, run the engine for about two minutes.
   b. Connect the Consult scan tool and enter "WORK SUPPORT" mode. Select "4. AIR BLEEDING".
   c. Check the fluid level in the reservoir. It should be slightly overfilled.
   d. Touch "START" on the scan tool. The display will show a regular rise and fall in system pressure. When the pressure stabilizes, stop the engine.
   e. Connect a clear tube to the air bleeder at the actuator and place the other end in a container.

NOTE: Do not allow the fluid to contact the body or the paint will be damaged.

   f. Open the bleeder and watch the fluid move through the tube. If there are still air bubbles in the fluid when the flow stops, check the fluid level, pressurize the system again and repeat the process.

Upper Control Arms

REMOVAL AND INSTALLATION

G20 Models

1. Raise and safely support the vehicle. Support the hub assembly with a suitable jack.
2. Remove the cap and the upper kingpin mounting nut.
3. Remove the shock absorber mounting nut and the upper link through bolts.
4. Remove the third link and the upper link.
5. The bushings cannot be removed from the upper link. The king pin bearings can be removed with a hammer and drift pin. If removed, they must be replaced with new ones.

To install:
6. The upper link is marked left (L) or right (R). Always install the upper link with the mark facing the axle.
7. Fit the upper link into place and install the through bolt. Do not tighten the nut yet.
8. Pack the king pin bearing housing with grease and fit the third link into place. Torque the kingpin nut to 72–87 ft. lbs. (98–118 Nm) and install the dust cap.
9. Install the wheel and lower the vehicle to the ground.

NOTE: Final tightening must be done with the vehicle at normal ride height, tires on the ground and the chassis loaded.

10. With the vehicle sitting on all four wheels, torque the upper link through bolts to 65–90 ft. lbs. (88–123 Nm).
11. Torque the shock absorber mount nut to 82–94 ft. lbs. (112–127 Nm).

Q45 Models

1. Raise and safely support the vehicle. Support the hub assembly with a suitable jack.
2. Remove the cap and the upper kingpin mounting nut. Do not remove the lower nut.
3. Remove the shock absorber mounting nut and the upper link through bolts.
4. Remove the third link and the upper link.
5. The bushings cannot be removed from the upper link. The king pin bearings can be removed with a hammer and drift pin. If removed, they must be replaced with new ones.

To install:
6. The upper link is marked left (L) or right (R). Always install the upper link with the (A) facing the axle.
7. Fit the upper link into place and install the through bolt. Do not tighten the nut yet.
8. Pack the king pin bearing housing with grease and fit the third link into place. Torque the kingpin nut to 72–87 ft. lbs. (98–118 Nm) and install the dust cap.
9. Install the wheel and lower the vehicle to the ground.
10. With the vehicle sitting on all four wheels, torque the upper link through bolts to 65–80 ft. lbs. (88–108 Nm).
11. Torque the shock absorber mount nut to 76–94 ft. lbs. (103–127 Nm).

Lower Control Arms

REMOVAL AND INSTALLATION

G20 Models

1. Raise and support the vehicle safely.
2. Remove the stabilizer bar.

NOTE: Take note of paint mark and clamp position when removing stabilizer bar for correct reinstallation.

3. Support the steering knuckle with a suitable jack and remove the lower ball joint nut. Separate the ball joint from the knuckle.
4. Remove the bolts attaching the lower control arm to the chassis. Remove the lower control arm.

To install:

5. If the lower ball joint is worn or damaged, the lower control arm must be replaced. The ball joint is not serviceable separately.
6. Reattach lower control arm to the chassis with the attaching bolts and nut.
7. Reinstall the ball joint stud in the knuckle and torque the nut to 52–64 ft. lbs. (71–86 Nm).
8. Install the stabilizer bar and wheel. Safely lower the vehicle.

NOTE: Final tightening must be done with the vehicle at normal ride height, tires on the ground and the chassis loaded.

9. Torque front control arm bolts to 87–108 ft. lbs. (118–147 Nm) and rear gusset nut to 69–87 ft. lbs. (93–118 Nm).

J30 Models

1. Raise and safely support the vehicle.
2. Remove the wheel and tire assembly.
3. Remove the nut and disconnect the transverse arm ball joint from the knuckle assembly, using the proper tool.
4. Remove the nuts and disconnect the tension rod from the transverse arm.
5. Remove the transverse arm from the front suspension member.

To install:

6. Install the transverse arm to the suspension member. Do not tighten the bolt and nut until the vehicle is on the floor.

NOTE: Final tightening must be done with the vehicle at normal ride height with the weight of the vehicle on the tires.

7. Connect the transverse arm ball joint to the knuckle. Torque the nut to 71–88 ft. lbs. (96–120 Nm).
8. Install the tension rod to the transverse arm. Torque the nuts to 72–87 ft. lbs. (98–118 Nm).
9. Install the wheel and tire assembly and lower the vehicle to the floor.
10. Torque the transverse arm mounting nut to 72–87 ft. lbs. (98–118 Nm).

I30 Models

1. Raise and safely support the vehicle.
2. Remove the front wheels.
3. Remove the ABS wheel sensor and move it out of the way.
4. Remove the wheel bearing locknut.
5. Disconnect the tie rod from steering knuckle.
6. Matchmark then remove the bolts attaching the strut to the steering knuckle.
7. Separate the halfshaft from steering knuckle by lightly tapping the end of the shaft.
8. Separate the steering knuckle and the lower ball joint.
9. Disconnect the stabilizer bar from the lower control arm.
10. Remove the bolts attaching the link bushing pin to the chassis. If necessary, remove the nut attaching the link to the control arm and remove the link.
11. Remove the bolts attaching the compression rod bushing clamp and remove the lower control arm/traverse link.

To install:

12. Install the lower control arm and the compression rod bushing clamp into the vehicle.
13. Install the link bushing pin, if removed from the control arm.
14. Tighten all bolts and nuts until they are snug enough to support the weight of the vehicle but not fully tight, the bolts should be torqued to specification with the vehicle on the floor.
15. Install the steering knuckle to the lower control arm and connect the ball joint. Torque the ball joint nut to 46–56 ft. lbs. (62–76 Nm).

NOTE: Always use a new nut when installing the ball joint to the control arm.

16. Connect the steering knuckle to the strut and to the halfshaft.
17. Install the strut mounting bolts and align the matchmarks. Torque the bolts to 103–117 ft. lbs. (140–159 Nm).

18. Install the tie rod ball joint and torque the nut to 46–54 ft. lbs. (63–73 Nm).
19. Install the wheel bearing locknut
20. Install the ABS wheel sensor and torque the attaching bolt to 13–17 ft. lbs. (18–24 Nm).
21. Install the front wheels, lower the vehicle and torque hub locknut to 174–231 ft. lbs. (235–314 Nm).
22. Torque the bolts attaching the compression rod bushing clamp and the link bushing pin, in the proper sequence to 87–108 ft. lbs. (118–147 Nm).
23. If the link bushing pin was removed from the control arm torque the attaching nut to 87–108 ft. lbs. (118–147 Nm).
24. Torque the sway bar attaching nut to 30–35 ft. lbs. (41–47 Nm).
25. Check the vehicle alignment.

Ball Joint

REMOVAL AND INSTALLATION

J30 and I30 Models

The lower ball joint assembly is part of the lower control arm/transverse link. If replacement of the ball joint is required, the lower control arm needs to be replaced.

Sway Bar

REMOVAL AND INSTALLATION

G20, J30 and Q45 Models

1. Raise and support the vehicle safely.
2. Using a backup wrench to support the connecting rod (link), remove the stabilizer to connecting rod bolt.
3. Remove the stabilizer bracket bolts. Remove the stabilizer.

To install:

4. Install the stabilizer with the paint mark to the right of the bracket when viewed from the front of the vehicle. Install the bracket with the elongated hole toward the rear of the vehicle.

NOTE: Final tightening must be done with the vehicle at normal ride height, tires on the ground and the chassis loaded.

5. Ensure that the ball socket on the connecting rod is straight, then attach the connecting rod and tighten the bolt to 30–38 ft. lbs. (41–51 Nm). Use a backup wrench to keep the connecting rod straight.

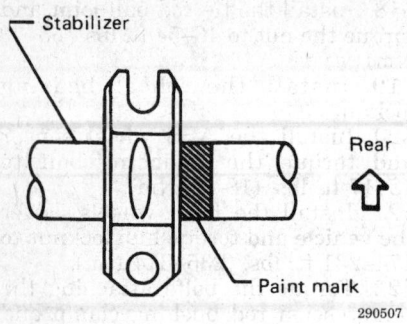

Stabilizer

Rear

Paint mark

290507

Make sure the paint mark is properly positioned.

I30 Models

1. Raise and safely support the vehicle.
2. Remove the ball joint socket nuts connecting the stabilizer bar to the lower control arm.
3. Remove the 4 stabilizer bar bracket bolts and then pull the bar from the vehicle.

To install:
4. Make sure the stabilizer ball joint socket is positioned properly.
5. Install the stabilizer bar and mounting brackets. Never fully tighten the mounting bolts unless the vehicle is resting on the ground with normal weight upon the wheels. Be sure the stabilizer bar ball joint socket is properly positioned.

NOTE: When installing the stabilizer bar, make sure the paint mark and the bushings are aligned. Make sure the clamp is facing in the right direction.

6. Lower the vehicle.
7. Bounce the vehicle to stabilize the suspension. Torque the stabilizer bar bracket bolts to 36–43 ft. lbs. (49–59 Nm).
8. Torque the nuts connecting the stabilizer bar to the control arm to 30–35 ft. lbs. (40–47 Nm)

Front Wheel Bearings

ADJUSTMENT

The front or rear wheel bearing assemblies on all models are pressed in and are not adjustable. If the bearing assembly does not turn smoothly or has more than 0.002 in. (0.05 mm) of axial play, replace the bearing assembly.

REMOVAL AND INSTALLATION

G20 Models

1. The axle nut torque is very high and should be loosened and tightened with the vehicle still on the ground. Remove the cotter pin, adjusting cap and insulator and loosen the front axle nut.
2. Raise and safely support the vehicle.
3. Remove the brake caliper, carrier, and the rotor. Hang the caliper from the body with wire; do not let it hang by the brake hose.
4. Remove the cotter pin and nut and use a ball joint press to disconnect the tie rod end.
5. Remove the cap and the upper kingpin mounting nut and separate the kingpin from the third link.
6. Hold a block of wood against the axle stub and strike it with a hammer to release it from the hub. Withdraw the axle from the hub and fold the steering knuckle down on the ball joint.
7. Remove the cotter pin and nut and use a ball joint press to disconnect the ball joint. Remove the steering knuckle.

NOTE: Wheel bearings must be replaced any time the hub is removed.

8. Pry the grease seals out of the steering knuckle.
9. Support the steering knuckle and press the hub out of the bearing.
10. Remove the snaprings and press the bearing out towards the inside of the knuckle.

To install:
11. Make sure all parts are clean and dry. The hub and steering knuckle should be inspected for cracks using dye or a magnetic crack detection process.
12. Install the inner snapring and carefully press the new bearing into the steering knuckle. Make sure the press tool contacts only the outer bearing race or the bearing will be damaged.
13. Install the outer snapring. Pack the new grease seals with clean grease and install them. If removed, install the splash guard.
14. Support the inner race on the press table and carefully press the hub into the bearing. Make sure the hub turns smoothly in both directions.
15. Fit the steering knuckle onto the lower ball joint and start the nut. Fit the axle shaft through the hub and start the nut.

16. Pack the king pin bearing housing with grease and fit the third link into place. Torque the kingpin nut to 72–87 ft. lbs. (98–118 Nm) and install the dust cap.
17. Torque the lower ball joint nut to 52–64 ft. lbs. (71–86 Nm). Install a new cotter pin.
18. Connect the tie rod end and torque the nut to 22–29 ft. lbs. (29–39 Nm). Tighten as needed to install a new cotter pin but do not exceed 36 ft. lbs. (49 Nm).
19. Install the brake caliper, carrier, rotor, and the wheel and lower the vehicle to the ground.
20. Torque the front axle nut to 174–231 ft. lbs. (235–314 Nm). Install the insulator, adjusting cap and cotter pin.

J30 Models

1. With the vehicle still on the ground, remove the hub cap and loosen the front wheel bearing nut.
2. Raise and safely support the vehicle. Remove the brake caliper, torque member and brake rotor. Support the caliper with wire, do not let it hang by the brake line.
3. Remove the knuckle assembly from the vehicle by separating the ball joint and tie rod end from the assembly and removing the hardware securing the spindle to the strut assembly.
4. Remove the nut and wheel hub from the spindle by forcing the hub out of the spindle assembly with a piece of wood and hammer.
5. Pry the grease seal out of the hub and remove the snapring.

NOTE: If the bearing is removed from the hub, it must be replaced.

6. Using a suitable shop press, press the bearing out towards the inside of the hub.
To install:
7. Press a new wheel bearing assembly into the wheel hub. Be sure the press tool contacts only the outer bearing race or the bearing will be damaged.
8. Install the snapring.
9. Pack grease seal lip with multipurpose grease.
10. Install the grease seal.
11. Install the knuckle assembly on to the strut. Connect the ball joint and tie rod end.
12. Install the hub and washer and start the nut.
13. Install the brake caliper and wheel and lower the vehicle to the ground.

14. Torque the wheel bearing nut to 152–210 ft. lbs. (206–284 Nm).

I30 Models

NOTE: Whenever the hub or bearing assembly is removed, the wheel bearing assembly must be replaced. Never reuse the old bearing assembly.

1. Remove the knuckle assembly from the vehicle by separating the ball joint and tie rod end, then removing the retaining hardware securing the knuckle to the strut.

2. Using a shop press and a suitable tool, press the hub with the inner race from the steering knuckle.

3. Using a shop press and a suitable tool, press the bearing inner race from the hub and remove the outer grease seal.

4. Use snapring pliers to remove the snaprings from the steering knuckle.

5. Inspect the hub, steering knuckle and snaprings for cracks and/or wear; if necessary, replace the damaged part(s).

To install:

6. Install the inner snapring in the steering knuckle groove.

7. Using a shop press and a suitable tool, press the new wheel bearing assembly into the steering knuckle,

until it seats, using a maximum pressure of 3 tons.

8. Install the outer snapring.

9. Pack the new grease seal lips with multi-purpose grease.

10. Using a shop press and a suitable tool, press the new outer grease seal into the steering knuckle.

11. Using a shop press and a suitable tool, press the new inner grease seal into the steering knuckle.

12. Using a shop press and a suitable tool, press the hub into the steering knuckle, until it seats, using a maximum pressure of 5.5 tons; be careful not to damage the grease seal.

13. To check the bearing operation, perform the following procedures:

   a. Increase the press pressure to 3.5–5.0 tons.

   b. Spin the steering knuckle, several turns, in both directions.

   c. Make sure the wheel bearings operate smoothly.

14. If the wheel bearings do not operate smoothly, replace the wheel bearing assembly.

15. Install the knuckle assembly.

16. Install the halfshaft into the hub and tighten the locknut to 174–231 ft. lbs. (235–314 Nm).

17. Install the wheel assembly and lower the vehicle.

18. Road test the vehicle and verify proper operation.

Q45 Models

1. Raise and safely support the vehicle. Support the hub assembly with a suitable jack.

2. Remove the brake caliper, carrier and rotor. Hang the caliper from the body with wire, do not let it hang by the brake hose.

3. Remove the cotter pins and nuts and use a ball joint press to disconnect the lower ball joint and tie rod end.

4. Remove the kingpin lower mounting nut to remove the steering knuckle assembly.

NOTE: Wheel bearings must be replaced any time the hub is removed.

5. Use a vise or a wheel to hold the hub and remove the hub cap and nut from the back of the hub. Remove the wheel speed sensor rotor.

6. Use a press or large drift pin to press the hub out of the steering knuckle.

7. Remove the snapring and press the bearings and grease seal out of the steering knuckle.

To install:

8. Make sure all parts are clean. Carefully press the new bearing into the steering knuckle. Make sure the press tool contacts only the outer bearing race or the bearing will be damaged.

9. Install a new grease seal and the snapring. If removed, install the splash guard.

10. Lightly lubricate the lips of the seal with clean grease. Be careful not to grease the bearing or hub mating surfaces.

11. Carefully press the hub into the bearing. Support the inner race on the press table or the bearing will be damaged. Do not exceed 3.9 tons (34.3 kN) pressure.

12. Install the speed sensor rotor and nut on the hub and torque to 152–210 ft. lbs. (206–284 Nm). Stake the nut into place.

13. Lightly tap the cap into place and install the bolts.

14. Install the steering knuckle to the king pin. Torque the nut to 108–137 ft. lbs. (88–108 Nm).

15. Connect the lower ball joint and torque the nut to 65–80 ft. lbs. (88–108 Nm). Install a new cotter pin.

16. Connect the tie rod end and torque the nut to 22–29 ft. lbs. (29–39 Nm). Tighten as needed to install a new cotter pin but do not exceed 36 ft. lbs. (49Nm).

17. Install the rotor and the brake caliper.

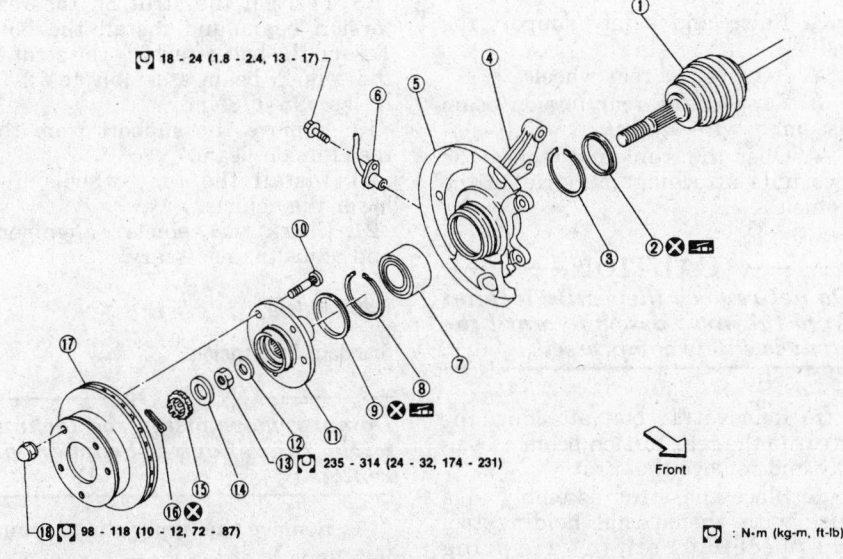

① Drive shaft
② Inner grease seal
③ Snap ring
④ Knuckle
⑤ Baffle plate
⑥ ABS sensor
⑦ Wheel bearing assembly
⑧ Snap ring
⑨ Outer grease seal
⑩ Hub bolt
⑪ Wheel hub
⑫ Plain washer
⑬ Wheel bearing lock nut
⑭ Insulator
⑮ Adjusting cap
⑯ Cotter pin
⑰ Disc rotor
⑱ Wheel nut

296419

Exploded view of steering knuckle components — I30 models

18. Install the wheel and tire assembly.

REAR SUSPENSION

Strut

REMOVAL AND INSTALLATION

G20 Models

1. Raise and support the vehicle safely.
2. Remove the rear seat to gain access to the top strut assembly bolts. Remove the three top mount bolts.
3. Remove the rear stabilizer connecting rod where it attaches the knuckle assembly.
4. Remove the strut through bolts at the knuckle assembly and remove the shock absorber assembly.
5. Set the strut assembly in a vise using attachment ST-25652000 or equivalent.

——— WARNING ———
Loosen the piston rod locknut but do not remove.

6. Compress the spring with a suitable tool so that the strut mounting insulator can be turned by hand.
7. Remove the piston rod locknut and spring with compressor attached.
To install:
8. Replace the bound rubber bumpers. Install the coil spring on the strut and tighten the piston rod locknut to 43–58 ft. lbs. (59–78 Nm). Gradually release the spring compressor. When the coil spring is located correctly, there should be two identification color codes on the lower side.
9. Tighten the strut assembly upper attaching bolts to 31–40 ft. lbs. (42–54 Nm); lower attaching bolts to 72–87 ft. lbs. (98–118 Nm) and the stabilizer bar connecting rod bolts to 30–35 ft. lbs. (41–47 Nm)
10. Check and adjust the wheel alignment if needed.

J30 Models

1. Raise and support the vehicle safely.
2. Remove the rear parcel shelf to gain access to shock upper mounting nuts.

——— CAUTION ———
Do not remove the piston rod locknut until the coil spring is compressed with a spring compressor.

3. Remove the strut upper mounting nuts.
4. Remove the lower strut mounting nut and bolt.
5. Remove the strut assembly.
6. Place the assembly into a suitable holding fixture.
7. Compress the spring until the upper spring seat can be turned by hand.
8. Remove the locknut, spring seat components, spring, bushings and bumper.
To install:
9. Fit the bumper, spring, seat and other components onto the strut and install the locknut. Torque the nut to 13–17 ft. lbs. (18–24 Nm) and remove the spring compressor.
10. Install the strut assembly.
11. Torque the upper mounting nuts to 12–14 ft. lbs. (16–19 Nm) and the lower mounting bolt to 72–87 ft. lbs. (98–118 Nm).
12. Install the parcel shelf and lower the vehicle.

I30 Models

1. Raise and safely support the vehicle.
2. Remove the rear wheels.
3. Support the rear torsion beam assembly with a jack.
4. Open the trunk and remove the two nuts attaching the strut to the vehicle.

——— CAUTION ———
Do not remove the center locknut from the strut assembly until the strut is safely compressed.

5. Remove the bolt attaching the strut to the rear torsion beam assembly and remove the strut.
6. Place the strut assembly in a vise with the special holding tool HT71780000 or in a spring compressor.
7. Loosen the piston rod locknut.

——— CAUTION ———
Do not remove the piston rod locknut, the spring is under tension and can cause serious personal injury.

8. Compress the spring with the spring compressor then remove the piston rod locknut.

NOTE: Before removing the strut from the coil spring, note the positioning of the strut in relationship to the coil spring for reassembly.

9. Remove the bushing, strut mounting bracket, and the upper spring seat rubber.
10. Remove the strut, leaving the coil spring compressed.
11. Remove the bushing, bound bumper cover, and the bound bumper.
To install:
12. Install the bound bumper, bound bumper cover, and the bushing.
13. Install the strut into the coil spring, make sure the strut and spring are properly positioned.
14. Install the upper spring seat rubber, strut mounting bracket, and the bushing. Make sure that the mounting bracket is properly positioned.
15. Install the piston rod locknut then remove the spring compressor.
16. Torque the piston rod locknut to 13–17 ft. lbs. (18–24 Nm).
17. Install the strut into the vehicle and install new attaching nuts. Torque the nuts to 12–14 ft. lbs. (16–19 Nm).
18. Position the strut on the rear torsion beam and install the bolt. Torque the bolt attaching the strut to the torsion beam assembly to 72–87 ft. lbs. (98–118Nm).
19. Remove the support from the rear torsion beam.
20. Install the rear wheels and lower the vehicle.
21. Check the vehicle's alignment and adjust as necessary.

Q45 Models

Standard Suspension

——— CAUTION ———
Do not remove piston rod locknut with the shock absorber on vehicle.

1. Remove the upper strut mounting nuts.
2. Raise and safely support the vehicle and remove the lower mounting bolt. Remove coil spring/strut absorber assembly.
3. Place the assembly into a suitable holding fixture and matchmark the spring, strut and upper seat. Loosen but do not remove the piston rod locknut.

4. Install a spring compressor and compress the spring until the upper spring seat can be turned by hand.

5. Remove the locknut, spring seat components, spring, bushings and bumper.

To install:

6. Fit the bumper, spring, upper seat and other components onto the strut with the matchmarks aligned. The top of the spring is flat.

7. Install the locknut and torque it to 13–17 ft. lbs. (18–24 Nm) and remove the spring compressor.

8. Install strut assembly. Torque the upper shock mounting nuts to 12–14 ft. lbs. (16–19 Nm) and the lower shock mounting bolt to 57–72 ft. lbs. (77–98 Nm).

Active Suspension

NOTE: The Nissan Consult or an equivalent scan tool that can issue commands to the control unit is required for bleeding the hydraulics in the Full Active Suspension system.

1. Relieve the hydraulic pressure:

a. Raise and safely support the vehicle with all four wheels off the ground and wait at least three minutes for the system to stabilize.

b. Remove both front inner fenders and the rear pressure control unit cover.

c. Loosen the locknut and slowly open the bypass valve on each pressure control unit. Do not open the bleeder valves.

d. Open the bypass valves all the way and leave them open until the job is finished.

2. Remove the upper mount insulator nuts.

3. Disconnect the hydraulic lines. Cap the lines to keep the system clean.

4. Remove the lower actuator mounting bolt and remove the actuator/spring assembly.

5. Secure the actuator/spring assembly in a suitable holding fixture. Scribe alignment marks on the spring, upper mount insulator and actuator unit.

6. Compress the spring with the proper tool. Remove the piston rod locknut lift off the mount insulator, hose joint adapter, spring, and any other components necessary.

To install:

7. Fit the bumper and dust cover onto the actuator.

8. Fit the spring into the lower seat with the matchmarks aligned. Install the upper seat, mounting insulator and other components with the marks aligned.

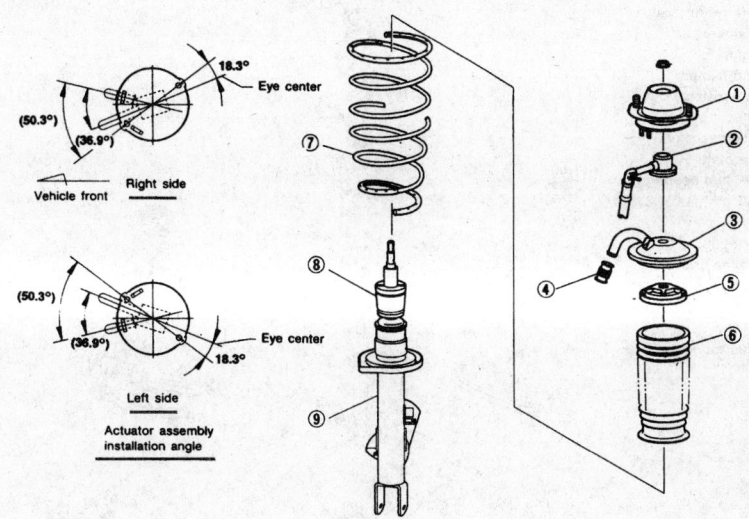

Figure shows rear left actuator.

① Mount insulator   ④ Air tube connector   ⑦ Coil spring
② Rear joint hose   ⑤ Bound bumper cover   ⑧ Bound bumper
③ Spring upper seat   ⑥ Rear actuator dust cover   ⑨ Rear actuator

297524

Rear actuator removal with Active Suspension — Q45 models

9. Install the locknut and torque to 43–54 ft. lbs. (59–74 Nm).

10. Fit the assembly onto the vehicle and torque the upper mounting nuts to 12–24 ft. lbs. (16–19 Nm).

11. Torque the lower mounting bolt to 58–72 ft. lbs. (78–98 Nm).

12. To bleed the system:

a. With all four wheels off the ground, run the engine for about two minutes.

b. Connect the Consult or equivalent scan tool and enter "WORK SUPPORT" mode. Select "4. AIR BLEEDING".

c. Check the fluid level in the reservoir and make it slightly overfilled.

d. Touch "START" on the scan tool. The display will show a regular rise and fall in system pressure that may last for several minutes. When the pressure stabilizes, stop the engine.

e. Connect a clear tube to the air bleeder at the actuator and place the other end in a container. Do not allow fluid to contact the body or the paint will be damaged.

f. Open the bleeder and watch the fluid move through the tube. If there are still air bubbles in the fluid when the flow stops, close the bleeder and check the fluid level.

Pressurize the system again and repeat the bleeding process.

Upper Control Arms

REMOVAL AND INSTALLATION

J30 Models

1. Raise and safely support the vehicle.

2. Each upper control link can be removed separately by removing the through bolts. The rear upper link has an eccentric used for rear wheel alignment. Matchmark the position of the eccentric and remove both bolts to remove the link.

3. To remove the lower control arm:

a. Remove the stabilizer bar linkage.

b. Remove the cotter pin and remove the ball joint nut. Use a ball joint press to separate the joint from the axle housing.

c. Remove the inner bushing bolts and remove the control arm.

4. Suspension bushings cannot be replaced. If the bushings are faulty, replace the suspension link.

To install:

5. Fit the link into place and install the bushing bolts. Start the nuts

① Gasket
② Strut mounting insulator
③ Upper spring seat
④ Dust cover
⑤ Coil spring
⑥ Bound bumper
⑦ Strut assembly
⑧ Connecting rod
⑨ Knuckle assembly
⑩ Baffle plate
⑪ Wheel hub bearing
⑫ Cotter pin
⑬ Cap
⑭ Parallel link
⑮ Mounting bracket
⑯ Bushing
⑰ Clamp
⑱ Stabilizer bar
⑲ Radius rod

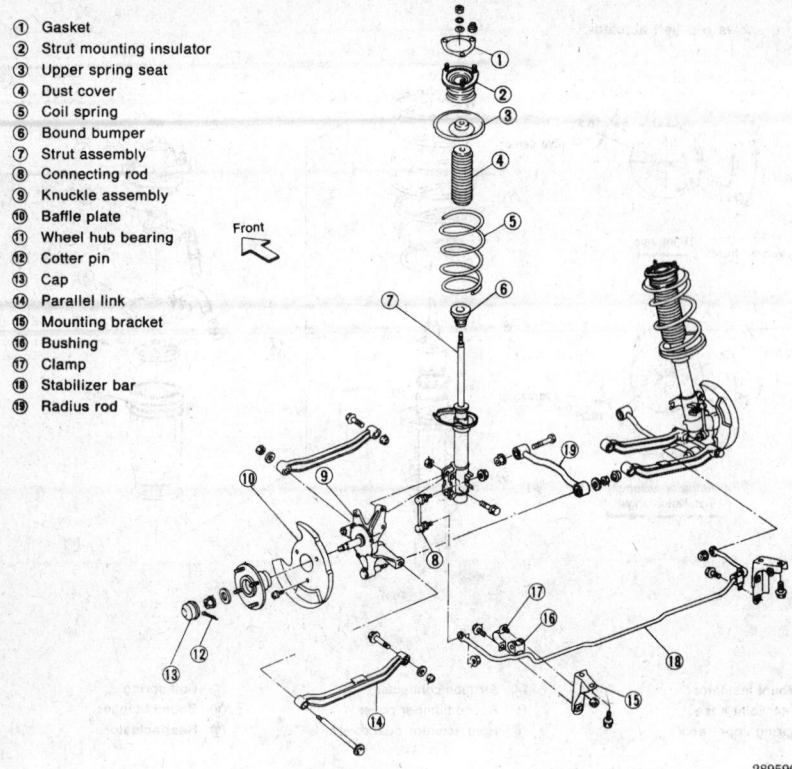

Front

289590

Rear suspension assembly — J30 models

but do not tighten them until the vehicle is sitting on all four wheels.

6. To install the lower control arm:

a. Fit the arm into place and install the bushing bolts. Start the nuts but do not tighten them until the vehicle is sitting on all four wheels.

b. Connect the ball joint. Torque the ball joint nut to 58–69 ft. lbs. (78–93 Nm) and install a new cotter pin.

c. Connect the stabilizer bar linkage and torque the nut to 78–104 inch. lbs. (9–12 Nm).

7. Install the wheel and lower the vehicle to the ground. Roll the vehicle forward and back several times to settle the bushings.

8. Align the matchmarks on the rear upper link eccentric and torque that bolt to 51–65 ft. lbs. (68–88 Nm).

9. Torque the remaining link and control arm bolts to 57–72 ft. lbs. (77–98 Nm).

10. If any links were replaced, the rear wheels should be aligned after the vehicle has been driven.

Lower Control Arms

REMOVAL AND INSTALLATION

G20, J30 and I30 Models

1. Raise and safely support the vehicle.

2. Each link or rod can be removed individually as required by removing the nuts and bolts.

3. To remove the rear parallel link, matchmark the toe adjustment eccentric to the link before removing the adjusting bolt.

To install:

4. The bushings cannot be removed from the links. If the bushings are faulty, the link must be replaced.

5. When installing the rear parallel link, make sure the matchmarks on the toe adjusting bolt are aligned. Start the nut but do not tighten it yet.

NOTE: Final tightening must be done with the vehicle at normal ride height, tires on the ground and the chassis loaded.

6. When all link bolts are installed, lower the vehicle so it is resting on all four wheels. Roll it forward and back several times to settle the bushings.

7. Torque the parallel link nuts and bolts to 80–94 ft. lbs. (108–127 Nm). Torque the radius rod nuts and bolts to 65–80 ft. lbs. (88–108 Nm).

Q45 Models

—————— **WARNING** ——————
On vehicles equipped with full active suspension, wait at least three minutes after turning the ignition to the OFF position before raising the vehicle on the lift.
——————————————————

NOTE: The ball joint is an integral part of the lower control arm and cannot be removed separately.

1. On vehicles with full active suspension, wait at least three minutes after turning the ignition to the **OFF** position before raising the vehicle on the lift.

2. Raise and safely support the vehicle.

3. Remove the wheel and tire assembly.

4. Disconnect the stabilizer bar link from the lower control arm.

5. Remove the cotter pin from the ball joint stud and loosen the nut.

6. Using the proper tool, disconnect the ball joint from the knuckle and then remove the nut.

7. Remove the lower control arm from the suspension crossmember.

To install:

8. Install the lower control arm to the suspension crossmember. Do not tighten the bolts until the vehicle is on the floor at normal riding height.

9. Connect the ball joint to the knuckle assembly and install a new cotter pin.

10. Connect the stabilizer link to the lower control arm.

11. Install the wheel and tire assembly and lower the vehicle to the floor.

12. Torque the lower control arm mounting bolts to specifications.

Sway Bar

REMOVAL AND INSTALLATION

1. Raise and safely support the vehicle.

2. Remove the nuts and bolts and disconnect the stabilizer bar linkage from the suspension.

3. Remove the bracket bolts and lower the bar from the vehicle.

To install:

4. Position the bar on the vehicle and install the brackets.

5. Connect the bar to the lower control arms.

6. Tighten the linkage nuts and bolts with the vehicle sitting on the ground.

Wheel Bearings

ADJUSTMENT

The rear wheel bearing assemblies are pressed in and are not adjustable. If the bearing assembly does not turn smoothly or has more than 0.002 in. (0.05 mm) of axial play, replace the bearing assembly.

REMOVAL AND INSTALLATION

G20

1. Raise and support the vehicle safely.

2. Remove the rear caliper and rotor. Hang the caliper from the body with wire, do not let hang by the brake hose.

3. Remove the rear wheel hub cap, cotter pin and locknut.

4. Slide the hub off the stub axle.

NOTE: The wheel bearing is integral with the hub and cannot be serviced separately.

To install:

5. Install the new hub assembly onto the axle stub.

6. Replace the washer and wheel bearing locknut and torque to 137–188 ft lbs. (186–255 Nm). Install a new cotter pin.

7. Reinstall the brake rotor, caliper and wheel. Lower the vehicle to the ground.

J30 Models

1. Raise and safely support the vehicle.

2. Remove the cotter pin and adjusting cap and loosen the wheel bearing nut. Carefully tap the end of the axle shaft or use a puller to loosen the shaft from the hub.

3. Remove the brake caliper and rotor. Do not let the caliper hang by the brake hose, support it with wire.

4. Remove the parking brake assembly.

5. Remove the nuts and through bolts to remove the axle housing from the suspension.

6. Secure the axle housing in a vise and use a pull hammer to remove the hub from the bearing. The bearing race may stay on the hub. Remove it with a standard bearing puller.

7. Remove the grease seal and snaprings and press the bearing out of the axle housing.

To install:

8. Install the inner snapring into the axle housing and press a new bearing into place. Make sure the press tool contacts only the outer bearing race or the bearing will be damaged. Install the outer snapring.

9. Lightly lubricate the lip of the new grease seals and press the seals into the axle housing.

10. Support the inner bearing race on the press table and carefully press the hub into the bearing.

11. Fit the axle housing onto the lower ball joint, torque the nut to 58–69 ft. lbs. (78–93 Nm) and install a new cotter pin.

12. Fit the axle shaft into the hub and install the bolts through the suspension bushings. Tighten the bolts temporarily, they will be torqued with the vehicle resting on the wheels.

13. Install the brake components and apply the brake to hold the hub from turning.

14. Install the wheel bearing locknut and torque to 152–203 ft. lbs. (206–275 Nm). Install the insulator and adjusting cap and a new cotter pin.

15. Install the wheel and lower the vehicle to the ground. Torque the suspension bushing bolts to 57–72 ft. lbs. (77–98 Nm) and the shock absorber mounting bolt to 72–87 ft lbs. (98–113 Nm).

I30 Models

NOTE: If the vehicle is equipped with ABS, the sensor must be removed to protect the sensor and its wiring.

1. Raise and safely support the vehicle. Remove the rear wheel(s).

2. Remove the wheel speed sensor.

3. Perform the following procedures:

   a. Remove the brake caliper and hang it by a piece of wire.

   b. Remove the brake caliper support.

   c. Remove the disc brake pads.

   d. Remove the brake disc.

4. Remove the grease cap.

5. Remove the cotter pin, wheel bearing locknut, washer, and the wheel hub bearing assembly. A slide hammer may be needed to remove the hub bearing assembly.

NOTE: The wheel hub bearing assembly is not repairable; it must be replaced when defective.

To install:

6. Install the wheel hub bearing assembly, the washer and the wheel bearing locknut. Torque the wheel bearing locknut to 137–188 ft. lbs. (186–255 Nm).

7. Verify that the wheel bearings operate smoothly.

8. Install a new cotter pin into the spindle to hold the wheel bearing locknut.

9. Install a dial micrometer to the rear wheel hub bearing assembly and check the axial end-play; it should be less than 0.0020 in. (0.05 mm).

10. Install the grease cap.

11. Install the ABS wheel sensor and its wiring.

12. Install the brake assembly and the wheels.

Q45 Models

1. Raise and safely support the vehicle.

2. Remove the cotter pin and adjusting cap and loosen the wheel bearing nut. Carefully tap the end of the axle shaft or use a puller to loosen the shaft from the hub.

3. Remove the brake caliper and rotor. Do not let the caliper hang by the brake hose, support it with wire.

4. Remove the parking brake assembly.

5. Remove the nuts and through bolts to remove the axle housing from the suspension. If equipped with rear wheel steering, use a ball joint press to separate the tie rod end.

6. Remove the four bolts at the back and remove the bearing flange and hub from the bearing housing.

7. Press the hub out of the bearing flange and use a puller to remove the bearing from the hub. If it is not damaged, the hub can be used again but the bearing and flange are supplied as a single unit.

To install:

NOTE: The wheel bearing and flange are supplied as an assembly.

8. Place the hub on a press table and press the new bearing and flange onto the hub. Make sure the press tool contacts only the inner bearing race and take care not to damage the seal.

9. Assemble the bearing flange onto the axle housing and torque the bolts to 58–72 ft. lbs. (78–98 Nm).

10. Fit the axle housing onto the lower ball joint, torque the nut to 58–69 ft. lbs. (78–93 Nm) and install a new cotter pin.

11. If equipped with rear wheel steering, torque the tie rod end nut to

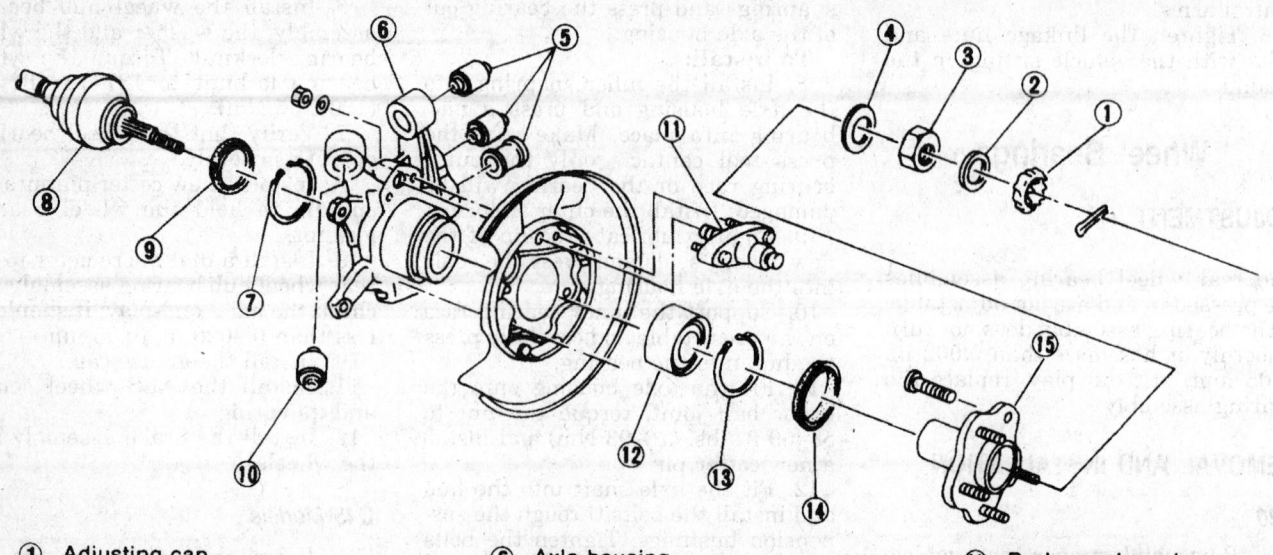

① Adjusting cap
② Insulator
③ Wheel bearing lock nut
④ Washer
⑤ Bushing

⑥ Axle housing
⑦ Snap ring
⑧ Drive shaft
⑨ Grease seal
⑩ Bushing

⑪ Brake anchor pin
⑫ Wheel bearing
⑬ Snap ring
⑭ Grease seal
⑮ Wheel hub

295313

The wheel bearing and flange are supplied as an assembly — J30 models

33–44 ft. lbs. (45–60 Nm) and install a new cotter pin.

12. Fit the axle shaft into the hub and install the bolts through the suspension bushings. Tighten the bolts temporarily, they will be torqued

with the vehicle resting on the wheels.

13. Install the brake components and apply the brake to hold the hub from turning.

14. Install the wheel bearing locknut and torque to 152–203 ft. lbs.

(206–275 Nm). Install the insulator and adjusting cap and a new cotter pin.

15. Install the wheel and lower the vehicle to the ground. Torque the suspension bushing bolts to 57–72 ft. lbs. (77–98 Nm).

FIRING ORDERS

NOTE: To avoid confusion, always replace spark plug wires one at a time.

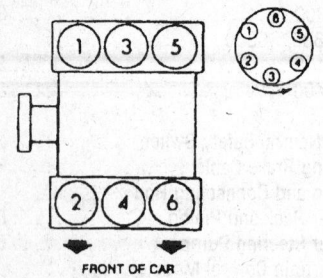

50948

3.0L (3VZ-FE) Engine
Engine Firing Order: 1-2-3-4-5-6
Distributor Rotation: Counterclockwise

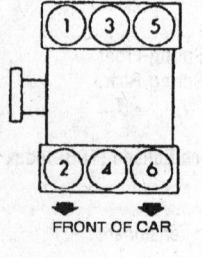

285579

3.0L (1MZ-FE) Engine
Engine Firing Order: 1-2-3-4-5-6
Direct Ignition System:
One Coil Per Cylinder

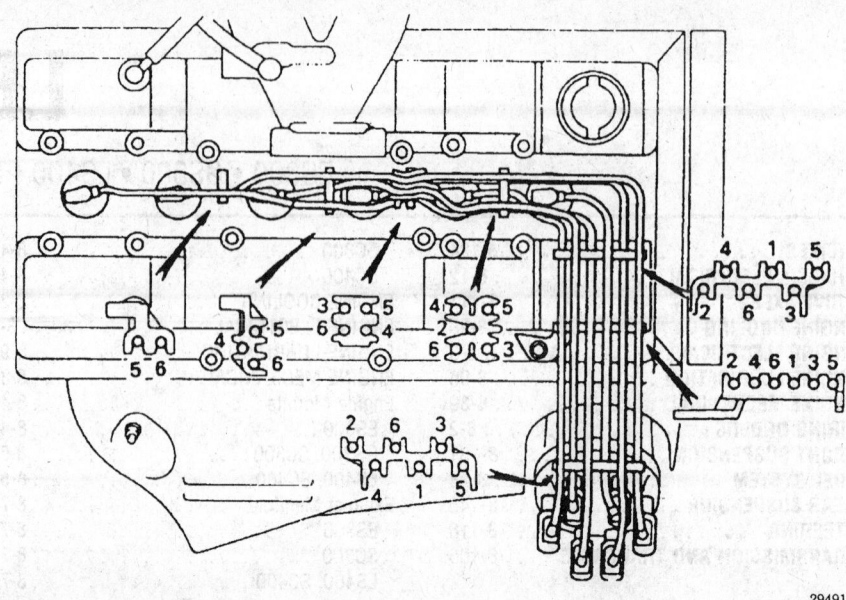

294911

3.0L (2JZ-GE) Engine
Engine Firing Order: 1-5-3-6-2-4
Distributor Rotation: Clockwise

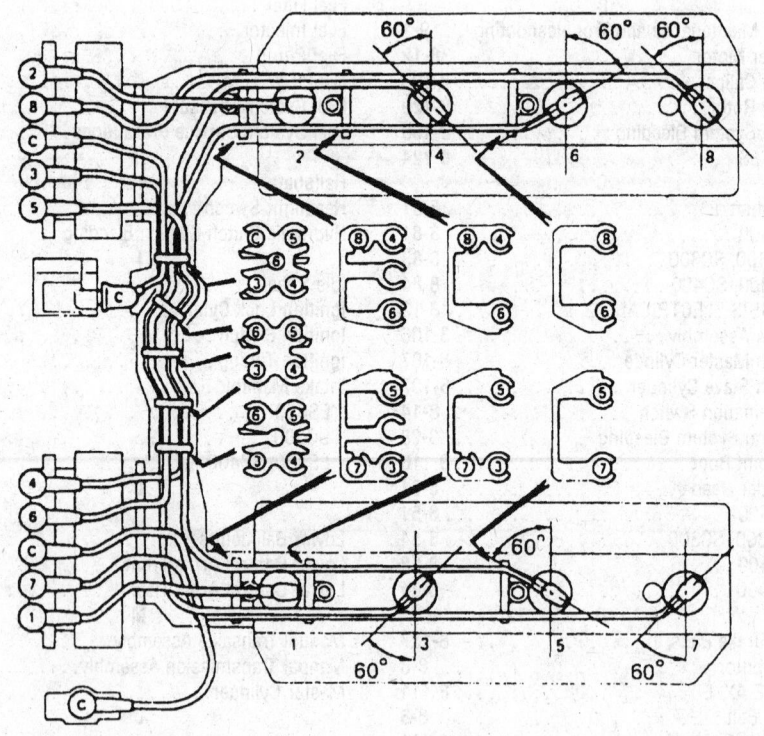

285580

4.0L (1UZ-FE) Engine
Engine Firing Order: 1-8-4-3-6-5-7-2
Distributor Rotation: Clockwise

ENGINE ELECTRICAL

NOTE: Disconnecting the negative battery cable on some vehicles may interfere with the functions of the on board computer systems and may require the computer to undergo a relearning process, once the negative battery cable is reconnected.

Distributor

REMOVAL AND INSTALLATION

3VZ-FE Engine

1. Disconnect the negative battery cable.

—————— CAUTION ——————

If the vehicle is equipped with an air bag, wait at least 90 seconds from the time the ignition switch is turned to the "LOCK" position and the negative battery cable is disconnected before starting work.

2. Remove the air cleaner cover, air flow meter and air duct.
3. Remove the wires from the distributor cap and disconnect the distributor wire connector.
4. Remove the two hold-down bolts and pull out the distributor. Remove the O-ring.
5. Remove the distributor.
To install:
6. Turn the crankshaft pulley and align the groove on the pulley with the **0** brand on the No. 1 timing belt cover. Make sure the No. 1 cylinder is on compression and not exhaust.
7. Position the slit of the intake camshaft (right side cylinder head) in the proper position for distributor installation.
8. Install a new O-ring in the housing. Lubricate the O-ring with engine oil.
9. Align the cutout marks of the coupling and the housing.
10. Insert the distributor, aligning the line of the distributor housing with the cutout of the distributor attachment bearing cap. Tighten the hold-down bolts. Correct torque is 13 ft. lbs. (18 Nm).
11. Install the rotor and the distributor cap. Connect the cables to the distributor cap. Align the spline of the distributor cap with the spline groove of the holder.

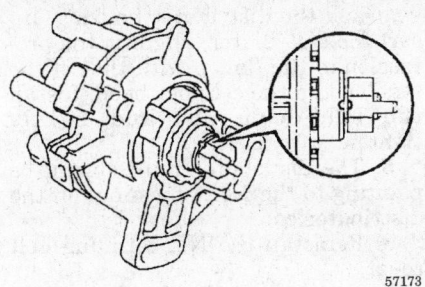

Aligning the marks on the distributor — 3VZ-FE engine

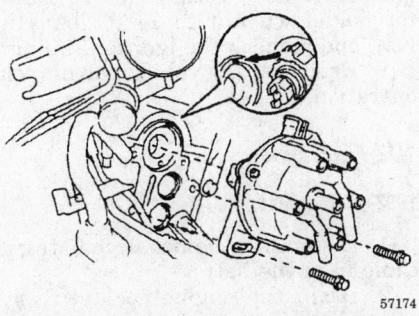

Installing the distributor — 3VZ-FE engine

12. Connect the electrical connections to the distributor. Replace the air cleaner cover, air flow meter and air duct.
13. Connect the negative battery cable.
14. Adjust the timing.

1MZ-FE Engine

The 1MZ-FE engine is not equipped with a distributor. The spark is controlled by the Engine Control Module (ECM) and spark is sent to the plugs via 6 separate ignition coils.

2JZ-GE Engine — 1993–95 GS300 and 1993–97 SC300

1. Disconnect the negative battery cable.
2. Set the No. 1 piston is at TDC of its compression stroke.
3. Tag and disconnect the spark plug wires at the distributor. Disconnect the distributor connector.
4. Loosen the hold-down bolt and remove the distributor.

Engine Undisturbed

To install:
1. Lubricate a new O-ring with engine oil and install it.

2. Align the groove on the distributor housing with the protrusion on the drive gear. Insert the distributor so the center of the flange is aligned with that of the bolt hole on the cylinder head.
3. The rotor should be pointing towards the No. 1 tower on the distributor cap.
4. Lightly tighten the hold-down bolt.
5. Connect the distributor connector and the spark plug wires.
6. Connect the battery cable and check the ignition timing. When the timing is properly adjusted, tighten the hold-down bolt to 10 ft. lbs. (14 Nm).

Engine Disturbed

1. Lubricate the new O-ring with engine oil, then install it.
2. Set the No. 1 cylinder to TDC of the compression stroke. Remove the oil filler cap. Rotate the crankshaft clockwise until the small end of the camshaft lobe can be seen through the hole. Turn the crankshaft counter-clockwise approximately 120°. Turn it as additional 10–40° clockwise until the timing marks on the crankshaft pulley and timing cover are aligned.
3. Align the groove on the distributor housing with the protrusion on the drive gear. Insert the distributor so the center of the flange is aligned with that of the bolt hole on the cylinder head.
4. Lightly tighten the hold-down bolt.
5. Connect the distributor connector and the spark plug wires.
6. Connect the battery cable and check the ignition timing. When the timing is properly adjusted, tighten the hold-down bolt to 10 ft. lbs. (14 Nm).

2JZ-GE — 1996–97 GS300

1. Disconnect the negative battery cable. Wait at least 90 seconds before performing any work.
2. Set the No. 1 piston is at TDC of its compression stroke.
3. Disconnect the distributor connector.
4. Disconnect the 7 high tension wires from the distributor.

NOTE: Pulling or bending the wires may damage the conductor inside.

5. Remove the distributor hold-down nut, remove the distributor and

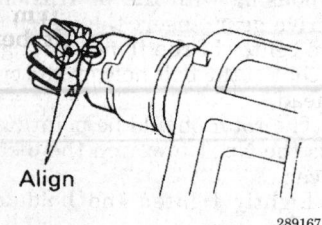

Aligning the marks on the distributor — 1993–95 GS300 and 1993–97 SC300 2JZ-GE engine

the O-ring from the distributor housing.

NOTE: The marks on the distributor drive gear and distributor housing should be aligned. If the marks are not aligned, mark the distributor housing and rotor position.

To install:

Engine Undisturbed

1. Apply a light coat of oil to a new distributor O-ring and install the O-ring to the distributor housing.
2. Align the marks of the drive gear and the distributor housing. Insert the distributor, aligning the protrusion of the flange with that of the stud bolt on the cylinder head. Install the hold-down nut and lightly tighten.
3. The distributor rotor should be pointing to the No. 1 terminal on the distributor cap.
4. Reconnect the high tension wires and reconnect the distributor connector.
5. Reconnect the negative battery cable. Start the engine and adjust the ignition timing. Torque the distributor hold-down nut to 14 ft. lbs. (19 Nm) and recheck the ignition timing.
6. Road test the vehicle for proper operation.

Engine Disturbed

1. Remove the No. 3 timing belt cover.
2. Turn the crankshaft pulley and align its groove with timing mark **0** of the No. 1 timing belt cover. Check that the timing marks of the camshaft timing pulleys are aligned. If not, turn the crankshaft 1 revolution (360°). This will set No. 1 cylinder to TDC compression.
3. Apply a light coat of oil to a new distributor O-ring and install the O-ring to the distributor housing.

4. Align the marks of the drive gear and the distributor housing. Insert the distributor, aligning the protrusion of the flange with that of the stud bolt on the cylinder head. Install the hold-down nut and lightly tighten.
5. The distributor rotor should be pointing to the No. 1 terminal on the distributor cap.
6. Reinstall the No. 3 timing belt cover.
7. Reconnect the high tension wires and reconnect the distributor connector.
8. Reconnect the negative battery cable. Start the engine and adjust the ignition timing. Torque the distributor hold-down nut to 14 ft. lbs. (19 Nm) and recheck the ignition timing.
9. Road test the vehicle for proper operation.

1UZ-FE Engine

1993–95 Models

1. Disconnect the negative battery cable from the battery.
2. Drain the engine coolant from the radiator.
3. Remove the intake air connector.
4. Remove the high tension cord cover by removing the two bolts.
5. Remove the right hand engine wire cover by removing the bolt.
6. Remove the left hand engine wire cover by removing the two bolts.
7. Remove the right hand No. 3 timing belt cover by removing the four bolts.
8. Remove the left hand No. 3 timing belt cover by removing the four bolts.
9. Disconnect the radiator hose from the water inlet pipe.
10. Remove the drive belt idler pulley by removing the bolt and cover plate.
11. Remove the right and left hand No. 2 timing belt covers.
12. Disconnect the high tension wires from the distributor caps. Mark the position of the high tension wires for installation.
13. Loosen the three bolts and remove the distributor cap. Remove both caps. Make sure to replace each side to its original position.
14. Mark the position of the rotor to the engine. Also place matchmarks on the distributor and engine for proper alignment when installing the distributor.
15. Loosen the two bolts and remove the distributor rotor. Remove both rotors. Make sure to replace each side to its original position.

16. Remove the distributor housings as follows:
   a. For the right side, disconnect the camshaft position sensor connector.
   b. Remove the three bolts and the distributor housing. Remove both distributors. Make sure to replace each side to its original position.

To install:

17. Install the right and left hand distributor housings by installing the three bolts for each distributor. When installing the distributor, align the marks on the distributor and the engine. When installing the distributor, make sure to install the distributor rotor as follows:
   a. Align the protrusion of the distributor rotor with the groove the camshaft timing pulley.
   b. Install the distributor rotor with the two bolts. Install the two distributor rotors.
18. Install the distributor cap with the three bolts. Install the two distributor caps. Torque the bolts to 34 inch lbs. (4 Nm).
19. Connect the high tension cord to the distributor caps.
20. Install the right hand No. 2 timing belt cover, then tighten the bolts to 13 ft. lbs. (18 Nm).
21. Install the left hand No. 2 timing belt cover.
22. Install the idler pulley and cover plate with the bolt. Torque the bolt to 27 ft. lbs. (37 Nm).
23. Connect the radiator hose to the water inlet pipe.
24. Install the right hand No. 3 timing belt cover by installing the gaskets and four bolts.
25. Install the left hand No. 3 timing belt cover by installing the gaskets and four bolts.
26. Install the right hand engine wire cover.
27. Install the left hand engine wire cover.
28. Install the intake air connector.
29. Connect the negative battery cable to the battery.
30. Fill the engine coolant.
31. Warm up the engine and check the timing.

1996–97 Models

1. Disconnect the negative battery cable.

CAUTION
Work must be started after 90 seconds from the time the ignition switch is turned to the LOCK position and the negative battery cable is disconnected.
---

2. Remove the No. 2 timing belt covers.

3. Disconnect the spark plug wires from the distributor caps.

4. Loosen the three bolts and remove the distributor cap. Remove both caps.

5. Remove the two rubber caps from each distributor cap.

6. Loosen the bolt and remove the distributor rotor. Remove both distributor rotors.

7. Remove the No. 2 distributor housing by removing the camshaft position sensor connector and removing the three bolts.

8. Remove the bolt, screw and camshaft position sensor.

To install:

Engine Not Disturbed

1. Install the camshaft position sensor with the bolt and screw. Install the both sensors and torque the bolt to 34 inch lbs. (4 Nm).

NOTE: The No. 1 distributor housing is marked with an L and the No. 2 with an R.

2. Install the left distributor housing with the three bolts. Torque the bolts to 13 ft. lbs. (18 Nm). The two inner bolts are 3.15 inches and the outer bolt is 1.50 inches.

3. Install the right distributor housing with the three bolts. Torque the bolts to 13 ft. lbs. (18 Nm). The two inner bolts are 3.78 inches and the outer bolt is 1.50 inches.

4. Align the protrusion of the distributor rotor with the groove of the camshaft timing pulley. Install both rotors and torque the bolts to 34 inch lbs. (4 Nm).

5. Connect the camshaft position sensor connectors.

6. Install the distributor cap and torque the three bolts to 34 inch lbs. (4 Nm). Install both caps. Install the two rubber caps to each distributor cap.

7. Connect the spark plug wires to the distributors.

8. Install the No. 2 timing belt covers.

9. Connect the negative battery cable. Start the engine and allow normal operating temperature to be reached.

10. Check the ignition timing.

Engine Disturbed

1. Make sure that the No. 1 cylinder is in TDC of the compression stroke.

   a. Turn the crankshaft pulley and align its groove with the tim-

ing mark **0** of the timing chain cover.

   b. Check that the timing marks on the camshafts align with the timing marks.

   c. If not, turn the crankshaft 1 revolution (360 degrees) and align the crankshaft pulley groove with the timing mark **0** of the timing chain cover.

2. Install the camshaft position sensor with the bolt and screw. Install the both sensors and torque the bolt to 34 inch lbs. (4 Nm).

NOTE: The No. 1 distributor housing is marked with an L and the No. 2 with an R.

3. Install the left distributor housing with the three bolts. Torque the bolts to 13 ft. lbs. (18 Nm). The two inner bolts are 3.15 inches and the outer bolt is 1.50 inches.

4. Install the right distributor housing with the three bolts. Torque the bolts to 13 ft. lbs. (18 Nm). The two inner bolts are 3.78 inches and the outer bolt is 1.50 inches.

5. Align the protrusion of the distributor rotor with the groove of the camshaft timing pulley. Install both rotors and torque the bolts to 34 inch lbs. (4 Nm).

6. Connect the camshaft position sensor connectors.

7. Install the distributor cap and torque the three bolts to 34 inch lbs. (4 Nm). Install both caps. Install the two rubber caps to each distributor cap.

8. Connect the spark plug wires to the distributors.

9. Install the No. 2 timing belt covers.

10. Connect the negative battery cable. Start the engine and allow normal operating temperature to be reached.

11. Check the ignition timing.

Ignition Timing

ADJUSTMENT

3VZ-FE Engine

1. Allow the engine to reach normal operating temperature.

2. Connect a tachometer to terminal IG (−) of the check connector.

—— WARNING ——
Never allow the tachometer test probe to touch ground; it could result in damage to the igniter and or ignition coil. Some ta-

chometers are not compatible with this ignition system; it is recommended to confirm the compatibility of the unit before use.

3. Check the idle speed.

4. Connect the proper jumper wire to terminals **TE₁** and **E₁** of the check connector in the engine compartment.

5. Connect the timing light to spark plug wire for the No. 1 cylinder.

6. Start the engine and check the timing with the transmission in **N** position.

7. Check the ignition timing. If not within specifications, loosen the hold-down bolts and adjust the timing by turning the distributor.

8. Tighten the hold-down bolts and recheck the timing and idle speed, adjust as necessary.

NOTE: With the jumper wire removed, the timing mark will move between 10–20 degrees. This is controlled by the ECM.

9. Remove the jumper wire and test equipment.

1MZ-FE Engine

The ignition timing may be checked to verify the setting, but cannot be adjusted.

1. Allow the engine reach normal operating temperature.

2. Connect a Lexus Hand-Held Tester or OBD II scan tool to the DLC3.

3. Connect a timing light to the No. 1 high tension cord for No. 4 cylinder.

4. Race the engine speed at 2,500 rpm for at least 90 seconds and check that idle speed is 700 ±50 rpm.

5. Using SST 09843–18020, connect terminals TE1 and E1 of the DLC 1. With the timing light check the ignition timing. It should be: 8–12° BTDC at idle with the transmission in the neutral position. Remove the SST from the DLC1.

6. Further checking with the transmission in the neutral position and the engine at idle will be in the range of 7–24° BTDC.

7. Stop the engine and disconnect the timing light and the scan tool.

2JZ-GE Engine

GS300 and SC300

1. Allow the engine to reach normal operating temperature.

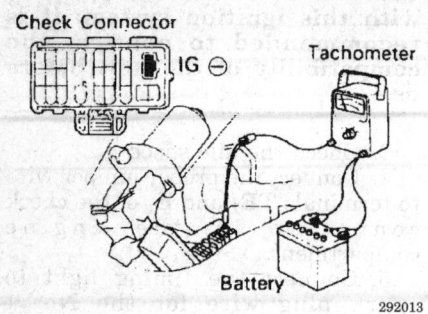

Connecting the tachometer — 3VZ-FE engine

292013

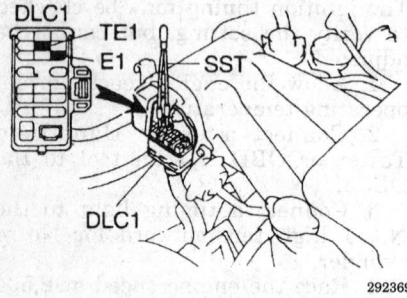

Connecting the scan tool — 1MZ-FE engine

292367

Using the SST tool — 1MZ-FE engine

292369

2. Connect a tachometer to terminal IG (-) of the check connector.

NOTE: Never allow the tachometer test probe to touch ground as it could result in damage to the igniter and or ignition coil. Some tachometers are not compatible with this ignition system, always confirm the compatibility of the unit before use.

3. With the transmission in **P** or **N**, raise the engine idle to 2500 rpm for 90 seconds, then release the throttle.

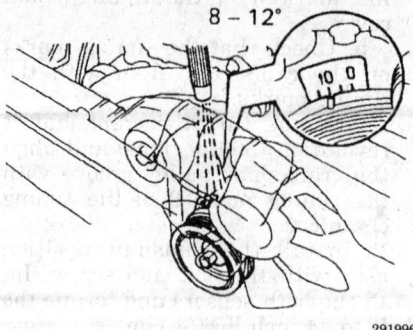

Checking the ignition timing — 1UZ-FE engine

291996

4. Check the idle speed.
5. Connect the proper jumper wire to terminals TE_1 and E_1 of the check connector in the engine compartment.
6. Connect the timing light to spark plug wire for the No. 1 cylinder.
7. Start the engine and check the timing with the transmission in **N** position.
8. Check the ignition timing; it should be 10° BTDC. If not within specifications, loosen the hold-down bolt(s) and adjust the timing by turning the distributor.
9. Tighten the hold-down bolts and recheck the timing and idle speed, adjust as necessary.
10. Remove the jumper wires. Recheck the timing; it should fluctuate between 7–19°. The change in timing is normal and is controlled by the ECU when the jumper wire is removed.
11. Remove the remaining test equipment.

1UZ-FE Engine

The ignition timing can be checked using this procedure. The timing, however, can not be adjusted.
1. Remove the upper spark plug wire cover.
2. Start the engine and allow it to reach normal operating temperature.
3. Connect a tachometer to terminal IG (–) of the check connector.
4. Set the tachometer to the 4 cylinder range.

NOTE: Never allow the tachometer test probe to touch ground as it could damage the igniter and or ignition coil. Some tachometers are not compatible with this ignition system; always confirm the compatibility of the unit before use.

5. Check the idle speed; it should be 650 ±50 rpm.
6. Connect a proper jumper wire to terminals TE_1 and E_1 of the check connector in the engine compartment.
7. Connect the timing light to spark plug wire for the No. 6 cylinder.
8. Check the ignition timing. Specification is 8–12° BTDC at idle in NEUTRAL.

NOTE: No adjustment is possible.

9. If the timing is not within specification, check these items:
   a. The throttle valve must be fully closed.
   b. Continuity must exist between throttle position sensor terminals IDL1 (or IDL) and E2.
   c. The valve timing must be correct.
10. Remove the jumper from the data link connector. Disconnect and remove the test equipment.

Alternator

PRECAUTIONS

Several precautions must be observed with alternator equipped vehicles to avoid damage to the unit.
• If the battery is removed for any reason, make sure it is reconnected with the correct polarity. Reversing the battery connections may result in damage to the 1-way rectifiers.
• When utilizing a booster battery as a starting aid, always connect the positive to positive terminals and the negative terminal from the booster battery to a good engine ground on the vehicle being started.
• Never use a fast charger as a booster to start vehicles.
• Disconnect the battery cables when charging the battery with a fast charger.
• Never attempt to polarize the alternator.
• Do not use test lights of more than 12 volts when checking diode continuity.
• Do not short across or ground any of the alternator terminals.
• The polarity of the battery, alternator and regulator must be matched and considered before making any electrical connections within the system.
• Never separate the alternator on an open circuit. Make sure all connections within the circuit are clean and tight.

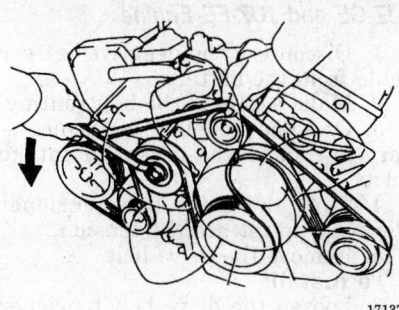

171378

Loosen the belt tension before removing the belt — 1993-94 LS400 1UZ-FE engine

• Disconnect the battery ground terminal when performing any service on electrical components.
• Disconnect the battery if arc welding is to be done on the vehicle.

REMOVAL AND INSTALLATION

3VZ-FE Engine

1. Disconnect the negative battery cable. Wait at least 90 seconds before performing any other work.
2. Remove the 2 bolts and No. 3 right hand engine mounting stay.
3. Remove the bolt and nut and then remove the No. 2 right hand engine mounting stay.
4. Disconnect the electrical connector and wire (and nut) from the alternator.
5. Unfasten the bolts which attach the adjusting link to the alternator.
6. Remove the alternator drive belt from the pulley.
7. Unfasten the alternator attaching bolt and then withdraw the alternator from its bracket.
   To install:
8. Mount the alternator on the alternator brackets with the pivot and adjusting bolts. Do not tighten the bolts at this time.
9. Install the No. 2 stay and tighten the bolts to 55 ft. lbs. (75 Nm), the nut to 46 ft. lbs. (62 Nm). Install the No. 3 stay and tighten the 2 bolts to 54 ft. lbs. (73 Nm).
10. Install the drive belt onto the pulley, making sure the grooves on the belt and the grooves on the pulley are properly aligned.
11. Adjust the drive belt tension and properly tighten the pivot and adjusting bolts.
12. Connect the wire (nut) and connector to the alternator.
13. Connect the negative battery cable.
14. Start the engine.

15. Visually inspect the drive belt and listen for any abnormal vibration.
16. Check the charging system for proper function.
17. Turn the engine off and recheck the belt tension.

1MZ-FE Engine

1. Disconnect the negative battery cable.
2. Remove the alternator drive belt by loosening the pivot bolt, adjusting lock bolt and the adjusting bolt.
3. Disconnect the electrical connector to the alternator.
4. Remove the nut and disconnect the electrical wire from the alternator.
5. Remove the pivot bolt, plate washer and adjusting lock bolt from the alternator.
6. Remove the alternator from its bracket.
   To install:
7. Mount the alternator on the alternator brackets with the pivot and adjusting lock bolt. Do not tighten the bolts at this time.
8. Install the drive belt onto the pulley, making sure the grooves on the belt and the grooves on the pulley are properly aligned.
9. Adjust the drive belt tension by tightening the adjusting bolt. Torque the pivot bolt to 41 ft. lbs. (46 Nm) and the adjusting lock bolt to 13 ft. lbs. (18 Nm).
10. Connect the electrical wire and nut to the alternator.
11. Connect the electrical connector to the alternator.
12. Connect the negative battery cable to the battery.
13. Start the engine and allow it warm.
14. Visually inspect the drive belt and listen for any abnormal vibration.
15. Check the charging system function.
16. Turn the engine off and recheck the belt tension.

2JZ-GE Engine

1. Disconnect the negative battery cable. Wait at least 90 seconds before performing any other work.
2. Remove the engine undercover.
3. Turn the drive belt tensioner clockwise and remove the drive belt from the engine.
4. Disconnect the generator electrical connector.
5. Remove the rubber cap and nut and then disconnect the alternator wire.

6. Disconnect the alternator wire clamp from the wire clip on the alternator.
7. Disconnect the A/T coolant hoses and bracket from the alternator by removing the bolt and pipe clamp.
8. Remove the alternator bolt, nut, pipe bracket and then remove the alternator from the engine.
   To install:
9. Install the alternator and pipe bracket. Torque the bolt and nut to 27 ft. lbs. (37 Nm).
10. Install the A/T oil cooler pipes to the alternator and install the bolt and pipe clamp.
11. Connect the alternator wire clamp to the wire clip on the alternator.
12. Connect the electrical wire to the alternator and install the nut.
13. Install the alternator electrical connector.
14. Install the drive belt.
15. Install the lower engine cover.
16. Connect the negative battery cable.
17. Start the engine and check the charging system for proper operation.

1UZ-FE Engine

1. Turn the ignition switch to the **LOCK** position. Disconnect the negative battery cable.

— **CAUTION** —
Work must begin after 90 seconds from the time the ignition switch is turned to LOCK and the negative battery cable is disconnected.

2. Remove the clip holding the batter clamp cover. disconnect the claws of the cover from the upper radiator support and the end of the No. 2 air cleaner hose.
3. If necessary, remove the air cleaner cover. Then, remove the No. 2 air cleaner hose.
4. Loosen the belt tension by tuning the belt tensioner counter-clockwise. Remove the drive belt.

NOTE: The pulley bolt for the tensioner has a left-hand thread.

5. Remove the engine undercover and remove the right rear engine side undercover.
6. Remove the three bolts holding the power steering oil cooler line to the engine oil pan. Remove the line from the oil pan.
7. Remove the connector and battery wire from the alternator. Disconnect the wire clamp from the alternator.

8. Remove the bolt and clamp. Remove the oil cooler lines from the bracket.

9. Remove the bolt, nut and bracket. Remove the alternator.

To install:

10. Install the alternator and bracket with the bolt and nut. Tighten to 27 ft. lbs. (37 Nm).

11. Connect the wiring connector and install the wire to the terminal post. Install the rubber cap. Install the wire clamp.

12. Install the oil cooler lines with the clamp and bolt. Install the oil cooler line to the engine oil pan with the three bolts.

13. Install the drive belt.

14. Install the No. 2 air cleaner hose. Install the air cleaner cover. Install the battery clamp cover.

15. Connect the negative battery cable to the battery.

16. Start the engine and test the charging system.

17. Stop the engine. Reinstall the engine undercovers.

Drive Belt

REMOVAL AND INSTALLATION

1MZ-FE and 3VZ-FE

Alternator and A/C

1. Disconnect the negative battery cable from the battery.

2. Loosen the alternator pivot bolt, adjusting lock bolt and the adjusting bolt.

3. Remove the alternator drive belt from the engine.

To install:

4. Route the belt on the pulleys.

5. Tighten the adjusting bolt until the belt meets one of the following tensions:

- New belt — 170–180 ft. lbs. (230–243 Nm)
- Used belt — 95–135 ft. lbs. (129–183 Nm)

6. Torque the alternator pivot bolt to 41 ft. lbs. (56 Nm) and the lock bolt to 13 ft. lbs. (18 Nm).

7. Connect the negative battery cable to the battery.

Power Steering

1. Remove the alternator belt.

2. Raise and support the vehicle safely.

3. Remove the right front wheel.

4. Remove the fender apron seal.

5. Loosen the power steering pump mounting bolts and remove the belt.

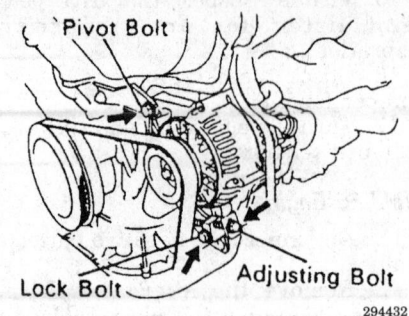

Alternator mounting bolts — 1MZ-FE and 3VZ-FE engines

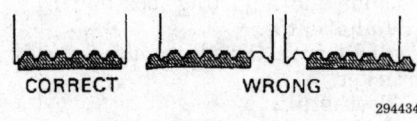

Correct drive belt installation

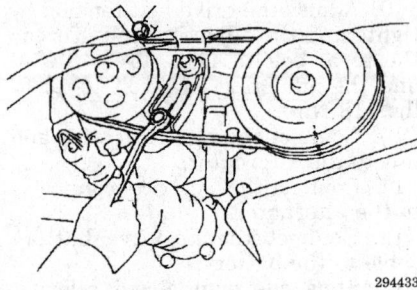

Power steering pump mounting bolts — 1MZ-FE and 3VZ-FE engines

To install:

6. Route the belt on the pulleys.

7. Adjust the belt tension to the following tensions:

- New belt — 150–185 ft. lbs. (203–251 Nm)
- Used belt — 95–135 ft. lbs. (122–183 Nm)

8. Torque the power steering mounting bolts to 32 ft. lbs. (43 Nm).

9. Install the fender apron seal.

10. Install the wheel.

11. Lower the vehicle.

12. Install the alternator belt.

2JZ-GE and 1UZ-FE Engine

1. Disconnect the negative battery cable from the battery.

2. Make a note of the belt routing.

3. On some models it will be necessary to remove the upper air intake duct.

4. Turn the drive belt tensioner clockwise to release the tension.

5. Remove the drive belt.

To install:

6. Loosen the drive belt tensioner and position the belt over the pulleys.

7. If installing a used belt, ensure the arrow mark on the tensioner falls within the area **A** of the scale.

8. If installing a new belt, ensure the arrow mark on the tensioner falls within the area **B** of the scale.

9. If not as specified, replace the belt.

10. Install the upper intake duct.

11. Connect the negative battery cable to the battery.

Starter

REMOVAL AND INSTALLATION

1MZ-FE and 3VZ-FE Engines

1. Disconnect the negative and positive battery cables.

—— **CAUTION** ——

On models with an air bag, wait at least 90 seconds from the time that the ignition switch is turned to the LOCK position and the battery is disconnected before performing any further work.

2. Remove the battery from the vehicle by removing the battery clamp.

3. On models with cruise control, remove the actuator cover and disconnect the electrical connector.

4. Remove the three bolts and then lift out the cruise control actuator.

5. Disconnect the starter connector from the starter.

6. Remove the nut and disconnect the starter wire.

7. Support the starter by hand and remove the two mounting bolts.

8. Remove the starter from the transaxle.

To install:

9. Place the starter motor in the transaxle and support it by hand.

10. Install the two mounting bolts and torque them to 30 ft. lbs. (41 Nm).

11. Place the starter wire into position and tighten the nut.

12. Connect the electrical connector to the starter.

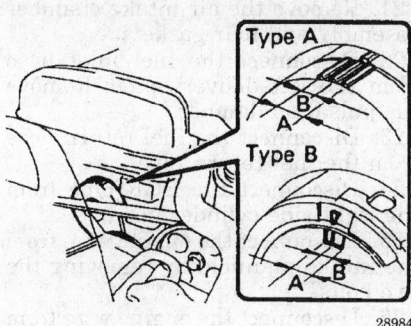

Aligning the tensioner arrow mark to the scale
— 2JZ-GE engine

13. If removed, install the cruise control actuator and the three mounting bolts. Connect the electrical connector to the actuator and install the actuator cover.

14. Install the battery and tighten down the battery clamp.

15. Connect the negative and positive battery cables.

16. Verify proper starter operation.

2JZ-GE Engine

GS300 and SC300

1. Disconnect the negative battery cable. Wait at least 90 seconds before performing any other work.

2. If necessary, disconnect the fuel hose and tubes from the starter body by removing the nut and bolt.

3. Remove the rubber cap and nut at the starter and then remove the starter wire.

4. Disconnect the electrical connector from the starter.

5. Remove the engine wire clamp with the two bolts and then remove the starter.

To install:

6. Place the engine wire clamp in place and install the starter. Torque the mounting bolts to 29 ft. lbs. (39 Nm).

7. Connect the electrical connectors to the starter.

8. Install the starter wire, nut and rubber cap.

9. Connect the fuel line and tubes to the starter body.

10. Connect the negative battery cable.

11. Check the starter for proper operation.

1UZ-FE Engine

LS400

NOTE: The starter is located in the V of the block, under the intake manifold.

1. Disconnect the negative battery cable.

2. Raise and safely support the vehicle.

3. Remove the oil pan protector.

4. Remove the engine undercover.

5. Relieve the fuel pressure from the fuel lines.

—————— **CAUTION** ——————

Fuel injection systems remain under pressure after the engine has been turned OFF. Properly relieve fuel pressure before disconnecting any fuel lines. Failure to do so may result in fire or personal injury.

6. Drain the engine coolant from the radiator.

7. Remove the battery clamp cover.

8. Remove the air cleaner inlet.

9. Remove the V-bank cover by removing the bolt and two cap nuts.

10. Remove the air cleaner and intake air connector assembly.

11. Remove the left hand No. 3 timing belt cover.

12. Remove the air intake chamber. Disconnect the following connectors:
- TPS connector
- With TRAC system, sub throttle position sensor connector
- With TRAC system, sub throttle actuator connector
- IAC valve connector
- EGR valve connector
- VSV connector for fuel pressure control
- VSV connector for EVAP system
- EGR gas temperature sensor connector

13. Disconnect the following hoses:
- Brake booster vacuum hose from the union on the air intake chamber
- PCV hose from the PCV valve on the left hand cylinder head
- Water bypass hose (from the EGR valve) from the rear water bypass joint
- Water bypass hose (from the throttle body) from the rear water bypass joint
- Vacuum hose (from the VSV for fuel pressure control) from the fuel pressure regulator
- EVAP hose (from charcoal canister) from the VSV for EVAP

14. Disconnect the fuel inlet hose from the delivery pipe.

15. Disconnect the fuel return hose from the fuel return pipe.

16. Disconnect the engine wire from the delivery pipes and rear water bypass joint.

17. Disconnect the fuel hose from the fuel pressure regulator.

18. Remove the two bolts and fuel return pipe from the intake manifold.

19. Disconnect the eight injector connectors.

20. Remove the six bolts, four nuts, the intake manifold assembly and two gaskets.

21. Remove the right hand front catalytic converter.

22. Remove the rear water bypass joint and No. 1 EGR pipe assembly as follows:

 a. Remove the two nuts holding the EGR pipe to the right hand exhaust manifold.

 b. Disconnect the heater water hose from the water bypass pipe.

 c. Disconnect the heater water hose from the water bypass joint.

 d. Remove the rear water bypass joint from the cylinder heads by removing the four nuts.

 e. Remove the water bypass joint, EGR pipe assembly and three gaskets.

23. Remove the water bypass pipe by removing the bolt and then pulling out the water bypass pipe from the water pump.

24. Remove the two bolts holding the starter to the cylinder block.

25. Disconnect the starter connector.

26. Remove the nut and disconnect the starter wire.

27. Disconnect the wire clamp from the wire bracket and remove the starter from the engine.

To install:

28. Install the starter to the engine and connect the wire clamp to the wire bracket.

29. Connect the starter wire with the nut.

30. Connect the starter connector.

31. Install the starter to the cylinder block with the two bolts. Torque the two bolts to 29 ft. lbs. (39 Nm).

32. Install the water bypass pipe.

33. Install the rear water bypass joint and No. 1 EGR pipe.

34. Install the front right hand catalytic converter.

35. Install the delivery pipe and intake manifold.

36. Install the return pipe with two new gaskets. Torque the union bolt to 26 ft. lbs. (35 Nm).

37. Install the engine wire to the delivery pipes and rear water bypass joint.

38. Connect the fuel return hose to the fuel return pipe.

39. Connect the fuel inlet hose to the left hand delivery pipe.

40. Connect the fuel hose to the fuel pressure regulator.

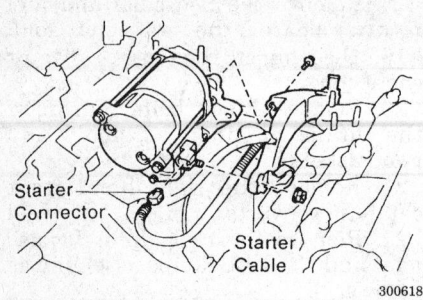

Starter electrical connectors — 1995–97 LS400 1UZ-FE engine

41. Install the air intake chamber assembly.

42. Attach the hoses previously disconnected during removal.

43. Attach all of the connectors removed previously.

44. Install the accelerator bracket with the two bolts.

45. Connect the accelerator cable, A/T throttle control cable and the cruise control actuator cable.

46. Install the left hand No. 3 timing belt cover.

47. Install the air cleaner and intake air connector assembly.

48. Install the V bank cover.

49. Fill the engine with coolant.

50. Connect the negative battery cable to the battery.

51. Start the engine and check for leaks.

52. Recheck engine coolant level.

53. Install the air cleaner inlet.

54. Install the battery clamp cover.

55. Install the engine undercover.

56. Install the oil pan protector.

1993–95 SC400

The starter is located below the intake manifold. The manifold must be removed in order to access the starter.

1. Disconnect the negative battery cable. Wait at least 90 seconds before proceeding with any other work.

— **CAUTION** —

Fuel injection systems remain under pressure after the engine has been turned OFF. Properly relieve fuel pressure before disconnecting any fuel lines. Failure to do so may result in fire or personal injury.

2. Relieve the fuel pressure in the fuel lines.

— **CAUTION** —

Do not allow fuel spray or fuel vapors to come in contact with a spark or open flame. Keep a dry chemical fire extinguisher nearby. Never store fuel in an open container due to risk of fire or explosion.

3. Drain the engine coolant.

4. Disconnect the throttle body.

5. Remove the accelerator, transmission and cruise control cables from the throttle body.

6. Remove the intake air connector.

7. Remove the upper spark plug wire cover.

8. Remove the right hand engine wire cover by removing the bolt.

9. Remove the left hand engine wire cover by removing the two bolts.

10. Remove the right hand No. 3 timing belt cover by removing the four bolts.

11. Remove the left hand No. 3 timing belt cover by removing the four bolts.

12. Remove the lower spark plug wire cover by removing the bolt. The spark plug wires will have to be disconnected from the upper spark plug cover and the right side ignition coil.

13. Remove the throttle body.

14. Disconnect the following connectors from the engine:

    a. IAC valve connector

    b. EGR valve connector

    c. VSV connector for fuel pressure control

    d. VSV connector for EVAP system

    e. EGR gas temperature sensor connector

15. Disconnect the following hoses:

    a. Water bypass hose from the IAC valve

    b. AIR hose from the air intake chamber

    c. Vacuum sensing hose from the fuel pressure regulator

    d. Water bypass hose from the EGR valve

    e. Vacuum hose from the air intake chamber

    f. PCV hose from the left hand cylinder head cover

16. Remove the accelerator bracket from the air intake chamber by removing the bolt and stud bolt.

17. Disconnect the EGR pipe from the air intake chamber by removing the two bolts and the gasket.

18. Remove the transmission throttle cable bracket from the intake chamber by removing the four bolts and eight nuts.

19. Disconnect the DLC1 bracket from the air intake chamber.

20. Lift the air intake chamber assembly and disconnect the EGR gas temperature sensor connector.

21. Remove the air intake chamber assembly and four gaskets.

22. Disconnect the fuel inlet hose from the left delivery pipe. Remove the pulsation damper.

23. Disconnect the fuel return hose from the fuel return pipe.

24. Disconnect the EGR pipe from the right side cylinder head.

25. Disconnect the engine wire from the intake manifold by removing the two bolts.

26. Disconnect the engine wire from the delivery pipes, rear water bypass joint and the right hand cylinder head.

27. Remove the fuel return pipe and rear fuel pipe.

28. Remove the delivery pipes and intake manifold assembly.

29. Disconnect the heater water hoses from the water bypass pipe and the rear water bypass joint.

30. Remove the water bypass pipe by removing the bolt and then pulling out the water bypass pipe from the water pump.

31. Remove the rear water bypass joint by removing the four nuts.

32. Remove the two bolts holding the starter to the cylinder block.

33. Disconnect the starter connector.

34. Remove the nut and disconnect the starter wire.

35. Disconnect the wire clamp from the wire bracket and remove the starter from the engine.

To install:

36. Install the starter to the engine and connect the wire clamp to the wire bracket.

37. Connect the starter wire with the nut.

38. Connect the starter connector.

39. Install the starter to the cylinder block with the two bolts. Torque the two bolts to 29 ft. lbs. (39 Nm).

40. Install the rear water bypass joint by installing the four gaskets and four nuts. Torque the nuts to 13 ft. lbs. (18 Nm). Install the bolt holding the water bypass pipe to the left hand engine hanger.

41. Install the water bypass pipe.

42. Connect the heater water hose to the water bypass pipe and rear water bypass joint.

43. Install the delivery pipe and intake manifold. Torque the bolts and nuts to 13 ft. lbs. (18 Nm).

44. Install the return pipe with two new gaskets. Torque the union bolt to 26 ft. lbs. (35 Nm).

45. Install the engine wire to the delivery pipes, rear water bypass joint and the right hand cylinder head.

46. Install the engine wire to the intake manifold with the two bolts.

47. Temporarily install the EGR pipe to the right hand cylinder head by installing the bolt. Do not torque the bolt at this time.

48. Connect the fuel return hose to the fuel return pipe.

49. Connect the fuel inlet hose to the left hand delivery pipe.

50. Install the air intake chamber assembly and the following parts with the four bolts and eight nuts:
- DLC1 connector
- Transmission throttle cable bracket
- VSV for the fuel pressure control
- VSV for EVAP

51. Torque the eight nuts and four bolts to the intake chamber to 13 ft. lbs. (18 Nm).

52. Torque the bolts holding the EGR pipe to the air intake chamber to 13 ft. lbs. (18 Nm).

53. Torque the bolt holding the EGR pipe to the right cylinder head to 13 ft. lbs. (18 Nm).

54. Install the accelerator bracket with the bolt and stud bolt.

55. Connect the hoses removed previously.

56. Attach the connectors detached prior.

57. Install the throttle body with a new gasket. Tighten the bolts and nuts to 13 ft. lbs. (18 Nm).

58. Install the remaining components. Fill the engine coolant.

59. Connect the negative battery cable to the battery.

60. Start the engine, bleed the cooling system and check for leaks.

61. Verify that the starter is operating correctly.

1996–97 SC400

1. Disconnect the negative battery cable to the battery.

2. Raise and safely support the vehicle.

3. Remove the oil pan protector.

4. Remove the engine undercover.

5. Relieve the fuel pressure from the fuel lines.

————— CAUTION —————

Fuel injection systems remain under pressure after the engine has been turned OFF. Properly relieve fuel pressure before disconnecting any fuel lines. Failure to do so may result in fire or personal injury.

---

6. Drain the engine coolant from the radiator.

7. Remove the battery clamp cover.

8. Remove the air cleaner inlet.

9. Remove the V bank cover by removing the bolt and two cap nuts.

10. Remove the air cleaner and intake air connector assembly.

11. Remove the left hand No. 3 timing belt cover.

12. Remove the air intake chamber assembly.
    a. Tag and disconnect the following connectors:
    - TPS connector
    - With TRAC system, sub throttle position sensor connector
    - With TRAC system, sub throttle actuator connector
    - IAC valve connector
    - EGR valve connector
    - VSV connector for fuel pressure control
    - VSV connector for EVAP system
    - EGR gas temperature sensor connector
    b. Disconnect the following hoses:
    - Brake booster vacuum hose from the union on the air intake chamber
    - PCV hose from the PCV valve on the left hand cylinder head
    - Water bypass hose (from the EGR valve) from the rear water bypass joint
    - Water bypass hose (from the throttle body) from the rear water bypass joint
    - Vacuum hose (from the VSV for fuel pressure control) from the fuel pressure regulator
    - EVAP hose (from charcoal canister) from the VSV for EVAP

13. Disconnect the fuel inlet hose from the delivery pipe.

14. Disconnect the fuel return hose from the fuel return pipe.

15. Disconnect the engine wire from the delivery pipes and rear water bypass joint.

16. Disconnect the fuel hose from the fuel pressure regulator.

17. Remove the two bolts and fuel return pipe from the intake manifold.

18. Disconnect the eight injector connectors.

19. Remove the six bolts, four nuts, the intake manifold assembly and two gaskets.

20. Remove the right hand front catalytic converter.

21. Remove the rear water bypass joint and No. 1 EGR pipe assembly.

22. Remove the water bypass pipe by removing the bolt and then pulling out the water bypass pipe from the water pump.

23. Remove the two bolts holding the starter to the cylinder block.

24. Disconnect the starter connector.

25. Remove the nut and disconnect the starter wire.

26. Disconnect the wire clamp from the wire bracket and remove the starter from the engine.

To install:

27. Install the starter to the engine and connect the wire clamp to the wire bracket.

28. Connect the starter wire with the nut.

29. Connect the starter connector.

30. Install the starter to the cylinder block with the two bolts. Torque the two bolts to 29 ft. lbs. (39 Nm).

31. Install the water bypass pipe using a new O-ring. Torque the bolts to 13 ft. lbs. (18 Nm).

32. Install the rear water bypass joint and No. 1 EGR pipe. Torque the EGR nuts to 17 ft. lbs. (23 Nm).

33. Install the front right hand catalytic converter.

34. Install the delivery pipe and intake manifold. On the intake manifold, install the six bolts and four nuts and torque to 13 ft. lbs. (18 Nm).

35. Install the return pipe with two new gaskets. Torque the union bolt to 26 ft. lbs. (35 Nm).

36. Install the engine wire to the delivery pipes and rear water bypass joint.

37. Connect the fuel return hose to the fuel return pipe.

38. Connect the fuel inlet hose to the left hand delivery pipe.

39. Connect the fuel hose to the fuel pressure regulator.

40. Install the air intake chamber assembly using new gaskets. Uniformly tighten the bolts and nuts in several passes. Torque them to 13 ft. lbs. (18 Nm).

41. Install the No. 2 EGR pipe with the two bolts and two nuts. Torque the two bolts and two nuts to 17 ft. lbs. (23 Nm). Connect the heater hose to the water bypass pipe.

42. Connect the hoses removed earlier.

43. Attach the connectors detached in removal. Install the remaining components.

44. Fill the engine with coolant.

45. Connect the negative battery cable to the battery.

46. Start the engine and check for leaks.

47. Recheck engine coolant level.

48. Install the air cleaner inlet.

49. Install the battery clamp cover.

50. Install the engine undercover.

51. Install the oil pan protector.

CHASSIS ELECTRICAL

Blower Motor

REMOVAL AND INSTALLATION

ES300

1. Disconnect the negative battery cable from the battery.
2. Remove the lower finish panel and the No. 2 undercover from the dashboard.
3. Remove the two screws to the connector bracket and remove the connector bracket.
4. Disconnect the electrical connector from the blower motor.
5. Remove the three screws from the blower motor and remove the blower motor.
   To install:
6. Install the blower motor and install the three screws.
7. Connect the electrical connector to the blower motor.
8. Install the connector bracket and install the two screws.
9. Install the lower finish panel and the No. 2 undercover to the dashboard.
10. Connect the negative battery cable to the battery and check the operation of the blower motor.

GS300

1. Disconnect the negative battery cable.
2. Set the air inlet mode to FRESH.
3. For the 1993–94 vehicles, pry the passenger side lower instrument panel out and remove it. For the 1995–97 vehicles, remove the two screws and remove the panel from the vehicle.

4. Remove the passenger scuff plate.
5. Pull back the passenger side portion of the carpet.
6. Disconnect the blower motor relay connectors.
7. Remove the two screws and the relay.
8. Remove the shaft cover from the blower motor cover.
9. Disconnect the shaft from the control lever of the blower motor cover.
10. Remove the blower motor cover.
11. Disconnect the blower motor electrical connector.
12. Remove the screws and blower motor.
    To install:
13. Install the blower motor, tighten the screws and connect the electrical connector.
14. Install the blower motor cover.
15. Install the control shaft and cover.
16. Install the blower motor relay.
17. Position the carpet and install the scuff plate.
18. Install the lower instrument panels.
19. Connect the negative battery cable.
20. Check the blower motor for proper operation.

LS400

1993–94 Models

1. Disconnect the negative battery cable.
2. Remove the right side instrument panel cover.
3. Remove the lower instrument panel and the No. 2 undercover.
4. Loosen the 2 screws and remove the connector bracket.
5. Remove the front passenger's door scuff plate. Pull back cowl side portion of the floor carpet.
6. Disconnect the shaft from the blower motor case.

7. Remove the blower lower case.
8. Disconnect the electrical lead, remove the 3 screws and remove the blower motor.
   To install:
9. Install the blower motor and tighten the screws.
10. Install the blower motor case.
11. Install the blower lower case control lever on the shaft by turning it counterclockwise.
12. Position the carpet and install the scuff plate.
13. Install the connector bracket.
14. Install the No. 2 undercover and lower instrument panel.
15. Install the right side instrument panel cover.
16. Connect the negative battery cable.

1995–97 Models

1. Disconnect the negative battery cable from the battery.
2. Remove the glove compartment from the vehicle.
3. Remove the electrical connector from the blower motor.
4. Remove the three screws and the blower motor.
   To install:
5. Install the blower motor to the vehicle and install the three screws.
6. Connect the electrical connector to the blower motor.
7. Install the glove compartment.
8. Connect the negative battery cable to the battery.
9. Verify that the blower motor is operating correctly.

SC300 and SC400

1. Disconnect the negative battery cable from the battery.
2. Remove the glove box.
3. Lift up the front edge of the carpet on the passenger side, remove the two screws and remove the Engine Control Module (ECM) cover.
4. Remove the connector bracket by removing the two screws.
5. Disconnect the electrical connector at the blower motor.
6. Remove the three screws and lift out the motor.
   To install:
7. Install the motor and tighten the three screws.
8. Connect the electrical connector to the motor.
9. Install the connector bracket by installing the two screws.
10. Install the ECM cover.
11. Install the glove box.
12. Connect the negative battery cable.
13. Check the blower motor for proper operation.

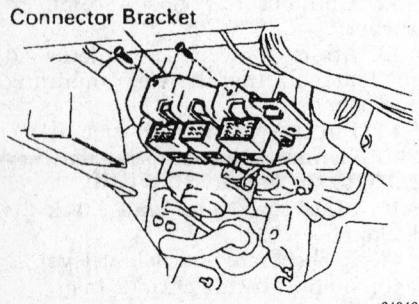

Connector Bracket

64046

Connector bracket mounting screws — ES300

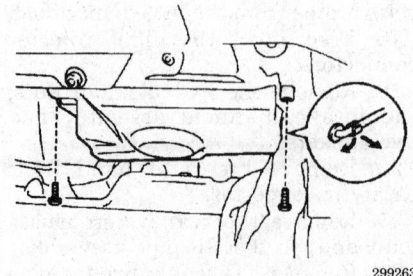

299263

Blower motor cover — GS300

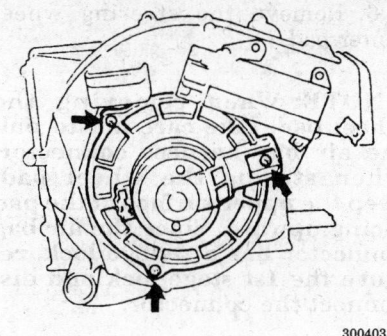

300403

Blower motor screws mount the unit to the vehicle

Windshield Wiper Motor

REMOVAL AND INSTALLATION

ES300, SC300 and SC400

1. Disconnect the negative battery cable.
2. Unscrew the nuts and remove the wiper blades and arms.
3. Remove the wiper arms.
4. Using a clip remover tool, remove the eleven cowl louver retaining clips and the weather stripping.
5. Remove the remaining clips and pull the cowl out and forward.
6. Remove the wiper link retaining bolts.
7. Disconnect the wiper motor electrical lead and then remove the four motor mounting bolts.
8. Connect the claw on the link to the panel, disconnect the motor from the link and remove it. Remove the link through the service hole.
To install:
9. Slide the link through the service hole and install the bolts.
10. Slide on the wire protector, hook the link claw on the panel and reconnect the motor. Install the bolts.
11. Position the cowl with the two clips, install the weather-stripping and install the remaining clips.
12. Install the wiper arm and blade assembly and tighten the retaining nuts to 18 ft. lbs. (25 Nm).
13. Connect the negative battery cable.
14. Check the wiper motor for proper operation.

GS300 and LS400

1. Disconnect the negative battery cable.
2. Remove the wiper arm and blade assembly.
3. Remove the cowl seal and louver.

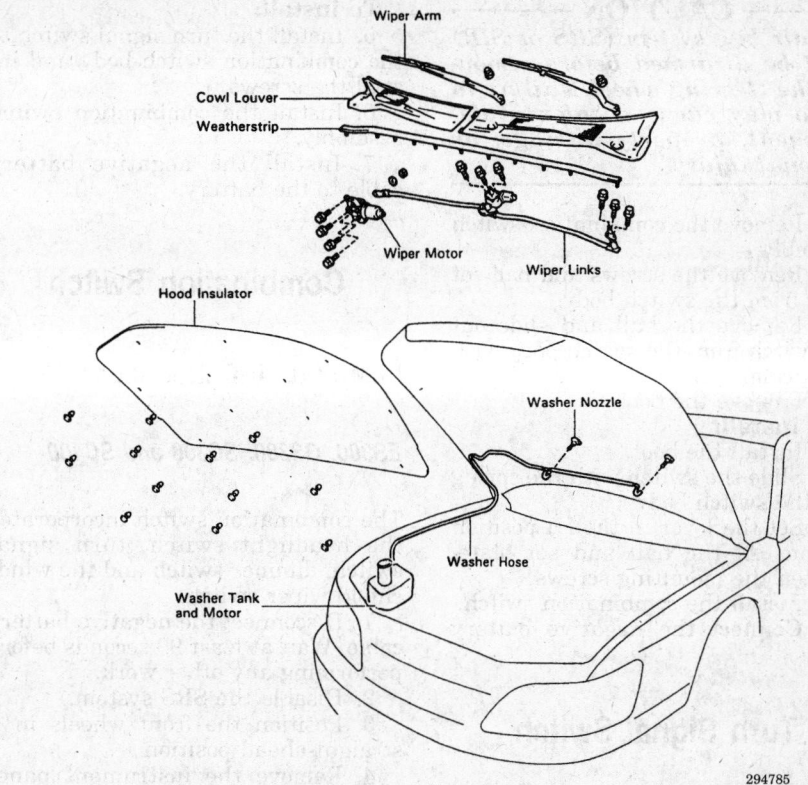

294785

Exploded view of the wiper and washer system — ES300

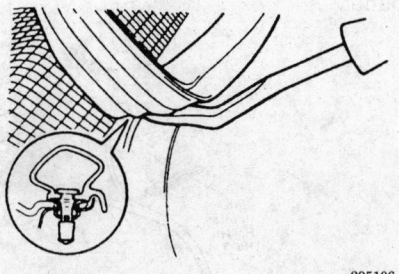

295106

Cowl louver seal removal — SC300 and SC400

4. Remove the wiper motor cover.
5. Remove the four wiper motor mounting bolts.
6. Disconnect the electrical connection.

NOTE: The wiper link rod must be kept at a 90° angle to the shaft to remove it on some models.

7. Remove the wiper motor and link assembly.
8. Using a clip remover, remove the clamp.
9. Disconnect the link rod.

10. Matchmark the position of the crank arm on the wiper frame. Remove the nut and crank arm.
11. Remove the three bolts and wiper motor.
To install:
12. Install the wiper motor and torque the bolts to 48 inch lbs. (5 Nm).
13. Install the crank arm at the mark on the wiper frame and torque the nut to 13 ft. lbs. (17 Nm).
14. Connect the link rod.
15. Install the clamp.
16. Install the motor and link assembly, tighten to 48 inch lbs. (5 Nm).
17. Connect the electrical connector.
18. Connect the wiper motor cover.
19. Install the cowl louver and seal.
20. Connect the negative battery cable.
21. Check that the wiper motor is operating correctly.

Headlight Switch

REMOVAL AND INSTALLATION

1. Disconnect the negative battery cable.

---

2. Remove the combination switch assembly.

3. Remove the screws and ball set plate from the switch body.

4. Remove the ball and slide out the switch from the switch body with the spring.

5. Remove the boot.

To install:

6. Install the boot.

7. Slide the switch with the spring into the switch body.

8. Set the lever in the HI position and install the ball and set plate. Tighten the mounting screws.

9. Install the combination switch.

10. Connect the negative battery plate.

Turn Signal Switch

REMOVAL AND INSTALLATION

1. Disconnect the negative battery cable. Wait at least 90 seconds before performing any other work.

2. Disable the SRS system.

3. Remove the combination switch assembly.

4. Remove the mounting screws and separate the turn signal switch from the combination switch body.

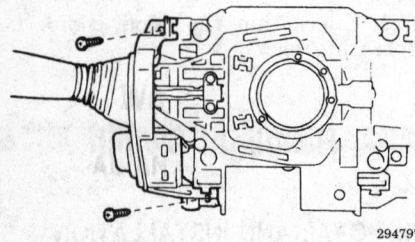

The headlight switch is mounted by two screws

294797

To install:

5. Install the turn signal switch to the combination switch body and install the screws.

6. Install the combination switch assembly.

7. Install the negative battery cable to the battery.

Combination Switch

REMOVAL AND INSTALLATION

ES300, GS300, SC300 and SC400

The combination switch incorporates the headlight switch, turn signal switch, dimmer switch and the windshield wiper switch.

1. Disconnect the negative battery cable. Wait at least 90 seconds before performing any other work.

2. Disable the SRS system.

3. Position the front wheels in a straight-ahead position.

4. Remove the instrument panel No. 1 undercover sub-assembly.

5. Remove the instrument panel lower finish panel and cluster finish panel.

6. Remove the steering wheel center pad.

NOTE: When removing the wheel pad, take care not to pull the air bag harness connector. When storing the wheel pad, keep the upper surface of the pad facing upward. Since the air bag connector has a 2-stage lock, remove the 1st stage lock and disconnect the connector.

7. Remove the steering wheel assembly.

8. Remove the steering column covers. Remove the combination switch retaining screws. Disconnect the connectors and remove the combination switch assembly from the steering column.

9. Loosen the mounting screws and the turn signal switch from the switch body.

10. Separate the bracket from the wiper switch body. Remove the wiper switch from the switch body.

To install:

11. Install the wiper switch to the switch body and connect the mounting bracket.

12. Install the turn signal switch to the switch body and tighten the mounting screws.

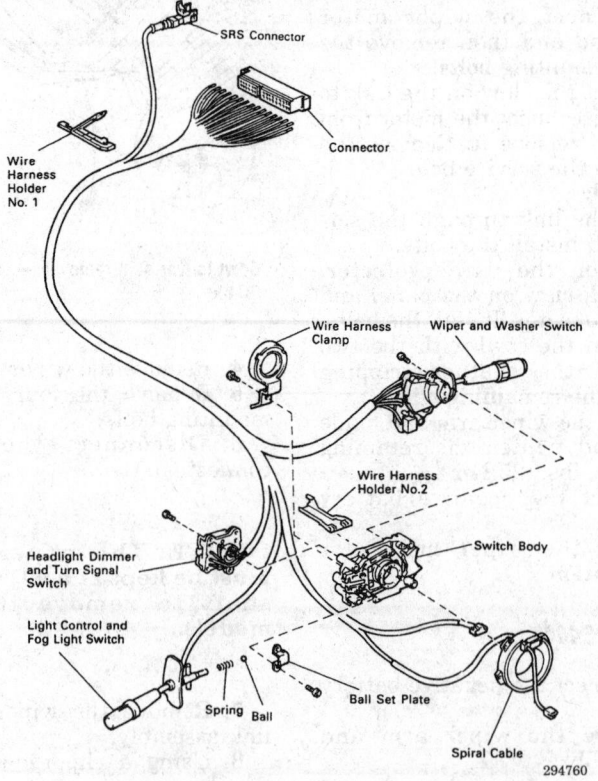

Combination switch components — ES300

294760

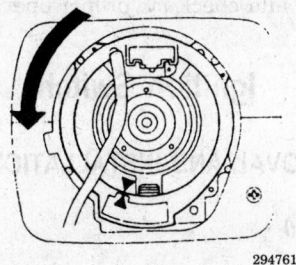

Step 1: turn the spiral cable all the way to the left until it stops

294761

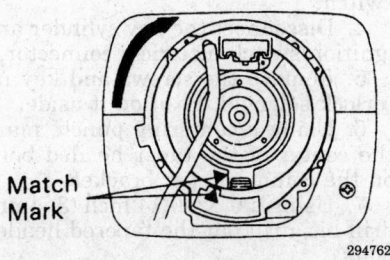

Match Mark

294762

Step 2: turn the spiral cable back to the right 3 turns and check that the matchmarks align

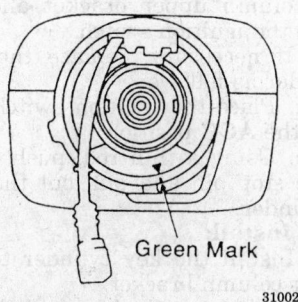

Green Mark

310020

Aligning the spiral cable — 1995–97 LS400

13. Install the combination switch assembly to the steering column and tighten the mounting screws.

14. Connect the electrical connectors. Push in the terminals until they are securely locked in the connector lug.

NOTE: The spiral cable matchmarks must be aligned for correct installation with the front wheels in the straight head position.

15. Install both steering column covers.

16. Center the spiral cable as follows:

   a. Check that the front wheels are facing straight ahead.

   b. Turn the cable counterclockwise by hand until it becomes harder to turn the cable.

   c. Rotate the cable clockwise about three turns to align the red mark.

17. Install the steering wheel, using the proper tool.

18. Connect the air bag connector and replace the 1st stage lock.

19. Install the cluster finish panel. Install the steering wheel center pad and connect the wire connector.

20. Install the lower instrument trim panel and cover assembly.

21. Enable the SRS system.

22. Connect the negative battery cable.

LS400

1993–94 Models

The combination switch incorporates the headlight switch, turn signal switch, dimmer switch and the windshield wiper switch.

1. Disconnect the negative battery cable. Wait at least 90 seconds before performing any other work.

2. Disable the SRS system.

3. Position the front wheels in a straight-ahead position.

4. Remove the undercover, lower pad, key cylinder pad. The No. 3 finish panel mounting bracket. The No. 2 heater to register duct.

5. Remove the steering wheel center pad.

6. Remove the steering wheel assembly.

7. Remove the steering column cover. Remove the 4 combination switch retaining screws. Disconnect the connectors and remove the combination switch assembly from the steering column.

NOTE: Since the air bag connector has a 2 stage lock, remove the 1st stage lock and then disconnect the connector.

8. Loosen the mounting screws and remove the dimmer switch and the turn signal switch from the switch body.

9. Separate the bracket from the wiper switch body. Remove the wiper switch from the switch body.

To install:

10. Install the wiper switch to the switch body and connect the mounting bracket.

11. Install the dimmer and turn signal switch to the switch body and tighten the mounting screws.

12. Install the combination switch assembly to the steering column and tighten the mounting screws.

13. Connect the electrical connectors. Push in the terminals until they are securely locked in the connector lug.

NOTE: The spiral cable matchmarks must be aligned for correct installation with the front wheels in the straight head position.

14. Replace both steering column covers. Install the steering wheel, using the proper tool.

15. Connect the air bag connector and replace the 1st stage lock.

16. Replace the cluster finish panel. Install the steering wheel center pad and connect the wire connector.

17. Replace the key cylinder trim pad assembly and heater duct register.

18. Replace the key cylinder trim pad assembly and heater duct register.

19. Install the lower instrument trim panel and cover assembly.

20. Connect the negative battery cable.

1995–97 Models

The combination switch incorporates the headlight switch, turn signal switch, dimmer switch and the windshield wiper switch.

1. Disconnect the negative battery cable. Wait at least 90 seconds before performing any other work.

2. Disable the SRS system.

3. Position the front wheels in a straight-ahead position.

CAUTION
The air bag system must be disarmed before removing the steering wheel. Failure to do so may cause accidental deployment, property damage or personal injury.

4. Remove the steering wheel center pad.

5. Remove the steering wheel assembly.

6. Remove the No. 1 instrument panel undercover as follows:

   a. Disconnect the two connectors and DLC2.

   b. Remove the two screws with the DLC3.

   c. Remove the two cover set screws.

7. Remove the instrument panel end pad and instrument panel finish plate.

8. Remove the No. 1 instrument panel safety pad as follows:

a. Remove the two screws and disconnect the engine hood release lever from the panel.

b. Remove the four panel set bolts.

9. Remove the No. 2 heater to register duct.

10. Remove the steering column cover.

11. Remove the combination switch as follows:

a. Disconnect the two connectors.

b. Disconnect the air bag connector.

c. Remove the three screws.

12. Loosen the mounting screws and remove the turn signal switch from the switch body.

13. Separate the bracket from the wiper switch body. Remove the wiper switch from the switch body.

To install:

14. Install the wiper switch to the switch body and connect the mounting bracket.

15. Install the turn signal switch to the switch body and tighten the mounting screws.

16. Install the combination switch assembly to the steering column and tighten the mounting screws.

17. Connect the electrical connectors. Push in the terminals until they are securely locked in the connector lug.

18. Replace both steering column covers.

19. Install the No. 2 heater to register duct.

20. Install the No. 1 instrument panel safety pad as follows:

a. Tighten the four panel set bolts.

b. Connect the engine hood release lever and tighten the two screws to the panel.

21. Install the instrument panel end pad and finish plate.

22. Install the No. 1 instrument panel undercover.

23. Check that the front wheels are facing straight-ahead. Center the spiral cable. When centering the spiral cable be sure to use the following procedure:

a. Check that the front wheels are pointing straight-ahead.

b. Turn the spiral cable counter-clockwise by hand until it becomes harder to turn the cable. The spiral cable will rotate approximately 3 turns to either the left or right of the center.

c. Rotate the spiral cable clockwise approximately 3 turns to align the green mark.

NOTE: The spiral cable matchmarks must be aligned for correct installation with the front wheels in the straight head position.

24. Install the steering wheel and pad, using the proper tool.

25. Connect the negative battery cable.

26. Enable the air bag system.

Ignition Lock Cylinder

REMOVAL AND INSTALLATION

────────── CAUTION ──────────
The Supplemental Inflatable Restraint (SIR) system must be disarmed before removing the steering column. Failure to do so may result in accidental deployment of the air bag, resulting in unnecessary SIR repairs and/or personal injury.

1. Disconnect the negative battery cable. Wait at least 90 seconds from the time the negative battery cable is disconnected to start any work.

2. Remove the upper and lower column covers by removing the five screws.

3. Remove the two nuts securing the upper steering column to the dash panel support and carefully lower the steering column assembly.

4. Place the ignition key in the **ACC** position. push down the stop pin with a small, flat tool and pull out the key cylinder.

To install:

5. Install the key cylinder making sure that it is in the **ACC** position. The key cylinder should snap into place as it passes the locking mechanism. Move the key through all the positions to ensure smooth and proper operation.

6. Reattach the upper steering column to the dash panel support and secure with the two nuts.

7. Reinstall the upper and lower column covers with the 5 screws.

8. Reconnect the negative battery cable and check for proper operation.

Ignition Switch

REMOVAL AND INSTALLATION

GS300

1. Disconnect the negative battery cable. Wait at least 90 seconds before performing any other work.

2. Disarm the air bag and remove the steering wheel.

3. Remove the combination switch.

4. Disconnect the key cylinder and ignition switch electrical connector.

5. Remove the screws and key interlock solenoid. Position it aside.

6. Using a centering punch, mark the center of the taper headed bolts on the column upper bracket.

7. Using a 0.12–0.14 inch (3–4mm) drill bit, drill out the tapered headed bolts.

8. Using a screw extractor, remove the tapered headed bolts.

9. Remove the column upper bracket assembly.

10. Remove the five screws from the column upper bracket and remove the ignition switch.

11. If necessary, replace the key cylinder as follows:

a. Place the ignition switch key at the **ACC** position.

b. Using a thin rod, push down the stop pin and pull out the key cylinder.

To install:

12. Install the key cylinder to the upper column bracket.

13. Position the ignition switch and install the five bolts.

14. Install the upper column bracket on the steering column. Tighten 2 new shelf-shear bolts until the heads break off.

15. Install the key interlock solenoid and tighten the two screws.

16. Connect the ignition switch electrical connector.

17. Install the combination switch.

18. Install the steering wheel.

19. Connect the negative battery cable to the battery.

20. Enable the air bag.

21. Verify that the key lock cylinder and the ignition switch are operating correctly.

ES300

1993–94 Models

────── CAUTION ──────

The Supplemental Inflatable Restraint (SIR) system must be disarmed before removing the ignition switch. Failure to do so may cause accidental deployment of the air bag, resulting in unnecessary SIR system repairs and/or personal injury.

1. With the ignition switch in the **LOCK** position, disconnect the negative battery terminal. If equipped with an air bag system, wait at least 90 seconds or longer before performing any other work.

2. Remove the steering wheel.

3. Remove the combination switch.

4. Remove the screw and key cylinder illumination. Position it aside.

5. Remove the screws and key interlock solenoid. Position it aside.

6. Using a centering punch, mark the center of the taper headed ignition switch bolts.

7. Using a 0.12–0.14 inch (3–4mm) drill bit, drill out the tapered-headed bolts.

8. Using a screw extractor, remove the tapered-headed bolts.

9. Disconnect the electrical connector.

10. Remove the ignition switch and lock assembly.

11. Place the ignition switch in the **ACC** position. Push down the stop key with a thin rod and pull out the key cylinder.

To install:

12. Turn the ignition key plate to the **ACC** position and install the new key cylinder.

13. Connect the electrical connector.

14. Position the ignition switch and lock assembly on the steering col-umn. Tighten two new retaining bolts until the heads break off.

15. Install the key interlock solenoid and illumination ring.

16. Install the combination switch.

17. Install the steering wheel.

18. Connect the negative battery cable.

1995–97 Models

1. Disconnect the negative battery cable to the battery.

2. Remove the instrument lower finish panel from the drivers side of the vehicle.

3. Loosen the four steering column assembly set nuts. Lower the steering column enough to remove the column covers.

4. Remove the steering column covers by removing the screws in the lower cover.

5. Remove the ignition switch electrical connector.

6. Remove the column upper bracket and column upper clamp as follows:

   a. Using a punch, mark the center of the two tapered head bolts.

   b. Using a 0.12–0.16 inch (3–4mm) drill, drill into the two bolts.

   c. Using a screw extractor, remove the two bolts.

7. If replacing the key cylinder, follow the following steps:

   a. Place the ignition key at the **ACC** position.

   b. Push down the stop pin with a thin rod and pull out the key cylinder.

8. Remove the ignition switch as follows:

   a. Remove the one screw at the bottom of the column upper bracket.

   b. Remove the two screws holding the key unlock warning switch wire to the column upper bracket.

   c. Disconnect the key unlock warning switch connector.

   d. Remove the ignition switch from the column upper bracket.

To install:

9. Install the ignition key to the column upper bracket by installing the one screw.

10. Connect the key unlock warning switch connector.

11. Place the key unlock warning switch wire in place and install the two screws.

12. With the ignition switch is the **ACC** position, install the ignition key to the column upper bracket.

13. Place the upper and lower bracket in place and install the two tapered head bolts. Tighten the bolts until the bolt heads break off.

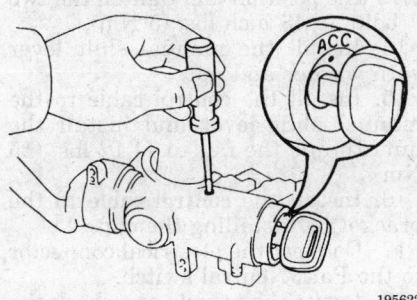

Press down the stop key to remove the lock cylinder

195622

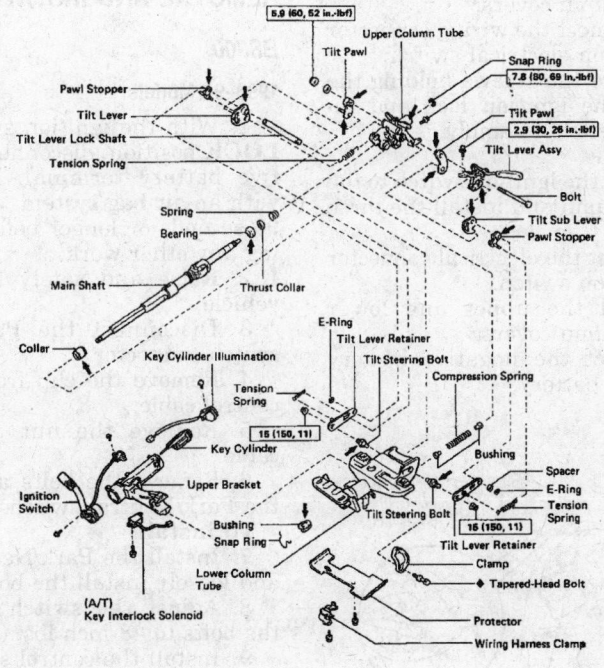

N·m (kgf·cm, ft·lbf) : Specified torque
◆ Non-reusable part
◀ Apply molybdenum disulphide lithium base grease

57612

Exploded view of the steering column — 1993–94 ES300

14. Install the ignition switch electrical connector to the ignition switch.

15. Install the upper and lower column cover and then install the screws.

16. Tighten and torque the steering column assembly nuts to 19 ft. lbs. (25 Nm).

17. Install the instrument lower finish panel to the drivers side and install the screws.

18. Connect the negative battery cable to the battery.

LS400

1. With the ignition switch **OFF**, disconnect the negative battery terminal first. Allow 90 seconds to elapse before performing any other work.

---- **CAUTION** ----

The Supplemental Inflatable Restraint (SIR) system must be disarmed before removing the ignition switch. Failure to do so may cause accidental deployment of the air bag, resulting in unnecessary SIR system repairs and/or personal injury.

2. Remove the upper and lower steering column covers.

3. Disconnect the wiring connector to the ignition electrical switch.

4. Remove the screws holding the switch to the ignition lock and remove the switch assembly.

To install:

5. Install the ignition switch to the steering column and install the three screws.

6. Connect the electrical connector to the ignition switch.

7. Install the upper and lower steering column covers.

8. Connect the negative battery cable to the battery.

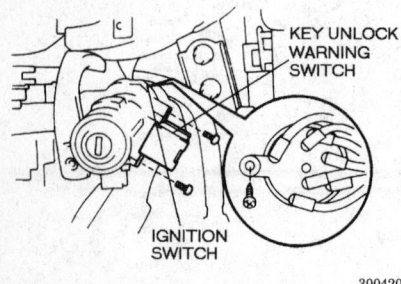

300420

Ignition switch mounting screws — SC300, SC400 and LS400

SC300 and SC400

1. With the ignition switch **OFF**, disconnect the battery cables, negative terminal first. Allow 90 seconds to elapse before performing any other work.

2. Remove the following components:
• No. 1 (left) undercover
• Lower the finish panel, then detach the connectors and hose from the pad
• Console box assembly
• Lower center instrument panel

3. Disconnect the wiring connector to the ignition electrical switch.

4. Remove the three screws holding the switch to the ignition lock and remove the switch assembly.

To install:

5. Install the ignition switch to the ignition lock and install the three (3) screws.

6. Connect the electrical connectors for the ignition switch.

7. Install the components removed earlier.

8. Connect the negative battery cable to the battery.

9. Verify that the ignition switch is operating correctly.

Park/Neutral Safety Switch

REMOVAL AND INSTALLATION

ES300

1993–94 Models

1. With the ignition switch in the **LOCK** position, disconnect the negative battery terminal. If equipped with an air bag system, wait at least 90 seconds or longer before performing any other work.

2. Raise and safely support the vehicle.

3. Disconnect the Park/Neutral switch connector.

4. Remove the clip from the shift control cable.

5. Remove the nut and control cable.

6. Remove the bolts and pull out the Park/Neutral switch.

To install:

7. Install the Park/Neutral switch and loosely install the bolts.

8. Adjust the switch and torque the bolts to 48 inch lbs. (5 Nm).

9. Install the control shaft lever.

10. Install the control shaft cable and torque the nut to 9 ft. lbs. (13 Nm).

11. Install the clip to the shift control cable.

12. Lower the vehicle.

13. Connect the negative battery cable.

14. Make sure the vehicle will start only in the **PARK** or **NEUTRAL** position.

1995–97 Models

1. With the ignition switch **OFF**, disconnect the battery cables, negative terminal first. Allow 90 seconds to elapse before performing any other work.

2. Apply the parking brake fully and block the rear wheels.

3. Disconnect the electrical connector from the Park/Neutral position switch.

4. Remove the locknut and disconnect the control cable from the manual shift lever.

5. Remove the clip from the holding the control cable to the bracket.

6. Remove the nut, washer and the manual shift lever from the manual valve shaft.

7. Using a flat bladed tool, bend the lock washer back away from the nut.

8. Remove the nut and lock washer from the Park/Neutral switch.

9. Remove the two bolts and the Park/Neutral position switch.

To install:

10. Install the Park/Neutral position switch to the manual valve shaft. Leave the two bolts loose at this time.

11. Install the lock washer and nut. Torque the nut to 61 inch lbs. (7 Nm).

12. Using a flat bladed tool, bend the lock washer over the nut to secure the nut in place.

13. Adjust the Park/Neutral as follows:

   a. Install the manual shift lever to the Park/Neutral switch and set the shift lever to the neutral (**N**) position.

   b. Align the groove and the neutral basic line on the Park/Neutral switch.

   c. Hold the Park/Neutral switch in this position and tighten the two bolts to 48 inch lbs. (5 Nm).

14. Install the manual shift lever with washer and nut.

15. Install the control cable to the manual shift lever and install the nut. Torque the nut to 11 ft. lbs. (15 Nm).

16. Install the control cable to the bracket by installing the clip.

17. Connect the electrical connector to the Park/Neutral switch.

18. Lower the vehicle and check the vehicles shifting positions.

19. Connect the negative battery cable to the battery.

Powertrain Control Module

REMOVAL AND INSTALLATION

NOTE: It is recommended that the technician wear a grounding strap whenever servicing the ECM. This will prevent any shocks resulting from static electricity. A static electrical shock can severely damage the ECM.

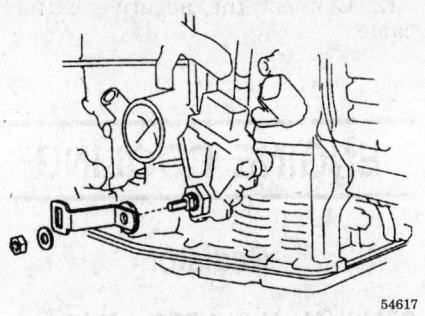

The transaxle control shaft lever has a nut and washer to retain it — 1993–94 ES300

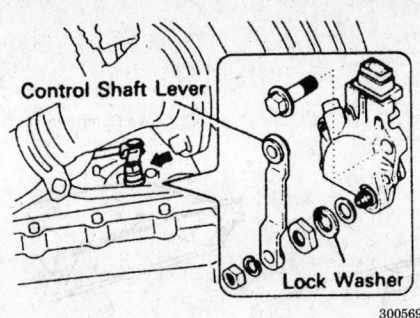

Park/Neutral switch removal and installation — GS300, LS400, SC300 and SC400

Removing the Park/Neutral switch — 1993–94 ES300

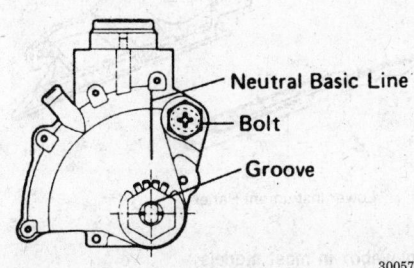

Park/Neutral switch adjustment — GS300, LS400, SC300 and SC400

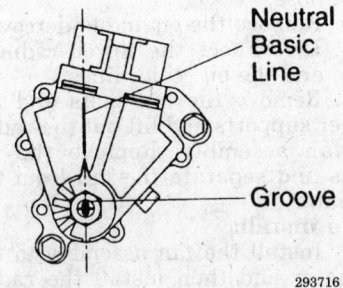

Aligning the neutral base line and groove — 1995–97 ES300

20. Verify that the vehicle only starts in **PARK** or **NEUTRAL**.

GS300, LS400, SC300 and SC400

1. With the ignition switch **OFF**, disconnect the battery cables, negative terminal first. Allow 90 seconds to elapse before performing any other work.
2. Raise and safely support the vehicle.
3. If necessary, remove the front exhaust pipe.
4. Disconnect the neutral start switch connector.

5. Remove the control shaft shift lever.
6. Pry off the washer and remove the nut.
7. Remove the bolt and pull out the neutral start switch.
To install:
8. Install the Park/Neutral switch and loosely install the bolt.
9. Adjust the switch as follows:
   a. Loosen the neutral start switch bolt and place the shift lever in the **N** position.
   b. Align the groove and the neutral basic line.
   c. Hold in position and tighten the bolt to 9 ft. lbs. (13 Nm).
10. Install the nut and washer.
11. Install the control shaft shift lever.
12. Connect the electrical connector.
13. Install the front exhaust pipe. Torque the front nuts to 46 ft. lbs. (62 Nm). Torque the pipe support bracket to bolts to 32 ft. lbs. (43 Nm). Torque the No. 2 front exhaust pipe to the front exhaust pipe bolts to 32 ft. lbs. (43 Nm).
14. Lower the vehicle.
15. Connect the negative battery cable to the battery.
16. Verify that the vehicle will only start in the **N** or **P** positions.

All Except SC300 and SC400

1. With the ignition switch **OFF**, disconnect the battery cables, negative terminal first. Allow 90 seconds to elapse before performing any other work.
2. Remove the instrument panel undercover.
3. On some models you may need to remove the air heater guide.
4. If necessary, remove the glove compartment door by removing the nuts at the bottom of the door. Remove the glove compartment. If applicable, disconnect the sir bag harness from the glove compartment.
5. On the 1995–97 LS400, remove the power shoulder belt anchor relay from the body by removing the nut.
6. Disconnect the ECM from its bracket by removing the nut and screws.
7. Disconnect the electrical connectors from the ECM.
8. Remove the ECM from the vehicle.
To install:
9. Install the ECM to the vehicle.
10. Connect the electrical connector to the ECM.
11. Connect the ECM to its bracket and install the screws and the nut.
12. Connect the power shoulder belt anchor relay to the vehicle by installing the nut.
13. Install the glove compartment and glove compartment door.
14. Install the instrument panel undercover.
15. Install the instrument lower panel and install the screws.
16. Connect the negative battery cable to the battery.

SC300 and SC400

1. Disconnect the negative battery cable. Wait at least 90 seconds before performing any other work.
2. Remove the passenger side lower instrument panel undercover.
3. On the SC300 and SC400, pull back the front of the floor carpet.
4. Remove the two nuts and ECM protective cover.

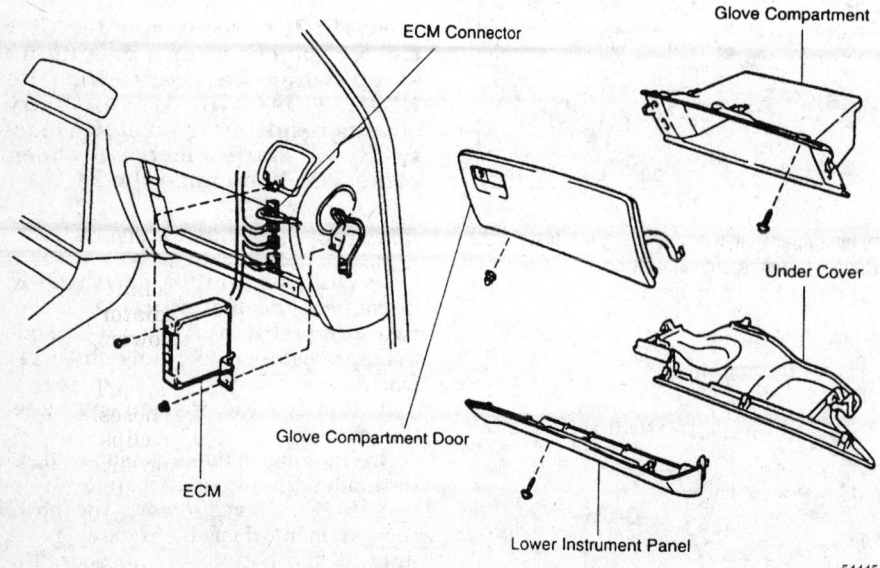

The ECM is located behind the glovebox in most models

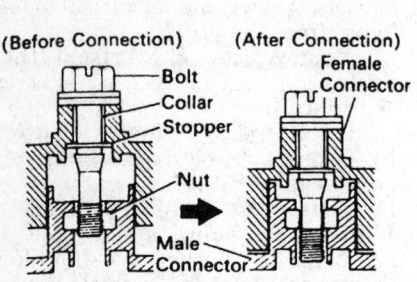

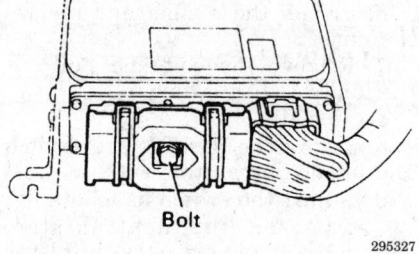

ECM connector securing bolt — SC300 and SC400

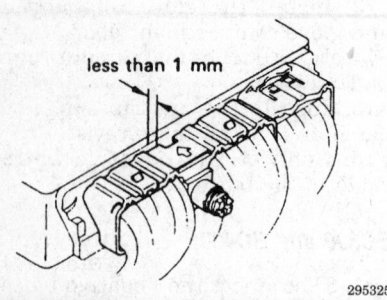

Harness installation — SC300 and SC400

5. Remove the nut and remove the ECM from the floor panel.

6. Fully loosen the bolt and disconnect the electrical connectors.

7. Remove the ECM.

To install:

8. Connect the electrical connectors to the ECM and tighten the bolt until there is less than 0.04 inch (1mm) clearance between the bottom of the male connector and end of the female connector.

9. Position the ECM and install the mounting nut.

10. Install the ECM protector.

11. Install the lower instrument panel.

12. Connect the negative battery cable.

ENGINE COOLING

Radiator

REMOVAL AND INSTALLATION

ES300

1. With the ignition switch in the **LOCK** position, disconnect the negative battery terminal. If equipped with an air bag system, wait at least 90 seconds or longer before performing any other work.

2. Drain the engine coolant into a container.

3. Remove the cruise control actuator cover.

4. Remove the union bolt and gasket and then disconnect the pressure line from the hydraulic fan motor. You may lose some hydraulic fluid; so have a container ready.

5. Disconnect the upper radiator hose and the coolant reservoir hose. Disconnect the hydraulic motor return hose.

6. Remove the engine undercover.

7. Disconnect the lower radiator hose and the oil cooler lines.

8. Remove the two bolts and the upper supports and lift out the radiator/fan assembly. Remove the six bolts and separate the fan from the radiator.

To install:

9. Install the fan assembly to the radiator and then install the radiator/fan assembly in the vehicle. Install the upper radiator supports and tighten the two support bolts to 9 ft. lbs. (13 Nm).

NOTE: Be sure the rubber support cushions are not pinched.

10. Connect the oil cooler hoses and the lower radiator hose.

11. Install the engine undercover.

12. Connect the hydraulic motor return line.

13. Connect the reservoir and upper radiator hoses.

14. Connect the pressure line to the hydraulic motor with a new gasket. Tighten the union bolt to 47 ft. lbs. (64 Nm).

15. Install the cruise control actuator cover.

16. Fill the engine with coolant, connect the battery, start the vehicle and bleed the cooling system.

17. Check for leaks and check the automatic transmission fluid. Add transmission fluid as needed.

GS300

1. Disconnect the negative battery cable. Wait at least 90 seconds before performing any other work.
2. Remove the lower engine cover.
3. Drain the cooling system.
4. Remove the air cleaner duct.
5. Remove the lower (No. 2) fan shroud.
6. Disconnect the upper and lower radiator hoses from the radiator.
7. Disconnect and plug the automatic transmission lines.
8. Disconnect the reservoir tank inlet hose.
9. Detach the cooling fan motor connector and the coolant temperature switch wiring.
10. Remove the upper radiator supports and radiator.
11. Remove both lower supports.
12. Remove the electric fan and bracket, then remove the No. 1 fan shroud. If the radiator is to be serviced or replaced, remove the coolant temperature switch from the radiator.

To install:

13. Install the coolant temperature switch if it was removed. Tighten it to 65 inch lbs. (8 Nm).
14. Install the No. 1 fan shroud to the radiator.
15. Install the electric fan and bracket.
16. Install the two lower supports to the radiator. Install the radiator with the two upper supports and torque the bolts to 9 ft. lbs. (12 Nm).

NOTE: Insure that the rubber cushion of the upper support is not compressed.

17. Connect the cooling fan connector.
18. Connect the engine coolant temperature switch connector.
19. Connect the upper and lower radiator hoses to the radiator.
20. Connect the reservoir tank hose and transmission lines to the radiator.
21. Install the lower (No. 2) fan shroud.
22. Install the air intake duct.
23. Fill the system to the correct level with coolant. Check the level of the automatic transmission fluid.
24. Connect the negative battery cable.
25. Start the engine and check for leaks. Bleed the cooling system.
26. Install the lower engine cover.

LS400

1. Disconnect the negative battery cable. Wait at least 90 seconds before performing any other work.
2. Remove the lower engine cover.
3. Drain the cooling system.
4. Remove the battery clamp cover.
5. Remove the air cleaner cover.
6. Remove the air intake duct.
7. Disconnect the hoses from the radiator.
8. Disconnect the reservoir tank inlet hose.
9. If equipped, disconnect and plug the automatic transmission lines.
10. Remove the No. 2 (lower) fan shroud. Use care to disconnect all the clips holding the shrouds together.
11. Disconnect the coolant temperature switch connector for the A/C cooling fan.
12. Disconnect the upper sides of the right and left cool air ducts.
13. Remove the two upper radiator supports.
14. Lift out the radiator assembly; remove the two lower supports from the radiator.
15. Remove the No. 1 fan shroud from the radiator.

To install:

16. Double check the installation of the coolant temperature switch, the drain-cock and the reservoir tank inlet pipe if they were removed.
17. Install the No. 1 fan shroud by installing the four bolts. Torque the bolts to 44 ft. lbs. (5 Nm).
18. Install the lower supports on the radiator and place the radiator into position. Install the two upper radiator supports. Tighten the bolts to 9 ft. lbs. (12 Nm).

NOTE: Make sure the rubber cushions of the upper supports are not depressed.

19. Install the upper sides of the right and left cool air ducts.
20. Connect the coolant temperature switch connector for the A/C cooling fan.
21. Install the No. 2 fan shroud; make certain it is properly placed and that all the clips are engaged.
22. Connect the radiator hoses and tank inlet hose to the radiator.
23. If equipped with A/T, connect the transmission lines to the radiator.
24. Install the air intake duct.
25. Install the air cleaner cover.
26. Install the battery clamp cover.
27. Connect the negative battery cable.

28. Fill the system to the correct level with coolant.
29. Start the engine, bleed the cooling system and check for leaks. Recheck the coolant level.
30. Install the lower engine cover.

SC300

1. Disconnect the battery cables. Wait 90 seconds before performing any other work. Remove the battery and battery tray.
2. Remove the engine undercover and drain the coolant.
3. Disconnect the coolant reservoir hose and the upper radiator hose.
4. Disconnect the lower radiator hose.
5. If equipped with an A/T, disconnect the two oil cooler hoses.
6. Remove the two clips and remove the No. 2 radiator shroud.
7. Remove the two bolts and one screw; remove the upper radiator support.
8. Remove the two radiator upper supports.
9. Lift out the radiator and remove the two lower supports. If the radiator is to replaced or worked on, remove the fan shroud and cushions.

To install:

10. Install the fan shroud if it was removed. Install the lower supports and install the radiator in the vehicle.
11. Install the upper radiator supports with the two bolts. Tighten the bolts and the screw to 9 ft. lbs. (12 Nm).
12. Install the No. 2 fan shroud.
13. Connect the upper and lower radiator hoses and the reservoir hose.
14. If equipped with an automatic transmission, connect the two oil cooler lines.
15. Install the battery, battery tray and then connect the cables.
16. Fill the engine with coolant.
17. Check the level of the automatic transmission fluid.
18. Start the engine, bleed the cooling system and check for leaks.
19. Install the undercover.

SC400

1. Disconnect the battery cables and remove the battery. After disconnecting the battery cables, wait at least 90 seconds before doing anything else.
2. Remove the engine undercovers and drain the coolant.
3. Remove the coolant reservoir tank.
4. Loosen the clamps and remove the two radiator hoses. If equipped

with an A/T, remove the two A/T oil cooler hoses.

5. Disconnect the pressure and return lines from the cooling fan hydraulic motor.

6. Remove the two bolts, the brackets and bushings and then disconnect the cooling fan inlet pipe from the shroud.

7. Disconnect the cooling fan hydraulic reservoir tank from the fan shroud.

8. Disconnect the water temperature sensor connector and then disconnect the ECT wiring from the fan shroud.

9. Remove the two bolts, the screw and the upper radiator supports.

10. Lift the radiator slightly and disconnect the two oil cooler hoses (for the cooling fan) from the clamp on the fan shroud. Remove the radiator.

11. Remove the fan shroud from the radiator by removing the four bolts.

To install:

12. Attach the fan shroud to the radiator by installing the four bolts.

13. Press the lower radiator supports onto the radiator and then slide it into position. Lift the radiator slightly and reconnect the two oil cooler hoses for the cooling fan.

14. Install the upper supports and torque the bolts to 9 ft. lbs. (12 Nm).

15. Connect the engine temperature sensor.

16. Install the reservoir tank and the inlet pipe to the fan shroud and torque the four bolts to 43 inch lbs. (5 Nm).

17. Connect the two hydraulic lines for the fan motor and torque the bolts to 47 ft. lbs. (64 Nm).

18. Connect the upper and lower radiator hoses to the radiator.

19. If equipped with A/T, install the two oil cooler hoses for the transmission to the radiator.

20. Install the coolant tank.

21. Install the battery to the vehicle. Fill the engine with coolant. Check the level of the automatic transmission fluid.

22. Start the engine and check for leaks.

23. Install the engine undercover.

Water Pump

REMOVAL AND INSTALLATION

ES300

1993 Models

1. With the ignition switch in the **LOCK** position, disconnect the nega-

tive battery terminal. If equipped with an air bag system, wait at least 90 seconds or longer before performing any other work.

2. Drain the cooling system.

3. Disconnect the lower radiator hose from the inlet pipe.

4. Remove the timing belt from the water pump pulley.

5. Remove the water inlet pipe.

6. Remove the water inlet and the thermostat.

7. Remove the water pump mounting bolts.

8. Lift out the water pump by carefully prying between the pump and the cylinder head.

9. Remove all the old sealant and clean the mounting surfaces.

To install:

10. Install new seal packing (sealant) to the water pump groove.

11. Install the water pump and tighten the mounting bolts to 14 ft. lbs. (20 Nm).

12. Install the thermostat and the water inlet.

13. Use a new O-ring and install the water inlet pipe.

14. Install the timing belt.

15. Connect the lower radiator hose to the inlet pipe.

16. Refill the cooling system.

17. Connect the negative battery cable.

18. Run the engine and check for leaks.

1994-97 Models

1. With the ignition switch in the **LOCK** position, disconnect the negative battery terminal. If equipped with an air bag system, wait at least 90 seconds or longer before performing any other work.

2. Drain the engine coolant.

3. Remove the timing belt.

4. Mark the left and right camshaft pulleys with a touch of paint. Using SST tools 09249-63010 and 09960-1000 or equivalents, remove the bolts to the right and left camshaft pulleys. Remove the pulleys from the engine. Be sure not to mix up the pulleys.

5. Remove the No. 2 idler pulley by removing the bolt.

6. Disconnect the three clamps and engine wire from the rear timing belt cover.

7. Remove the six bolts holding the rear timing belt cover to the engine block.

8. Remove the four bolts and two nuts to the water pump.

9. Remove the water pump.

10. Remove all the old packing (sealant) and gasket material from the water pump and clean the mounting surfaces.

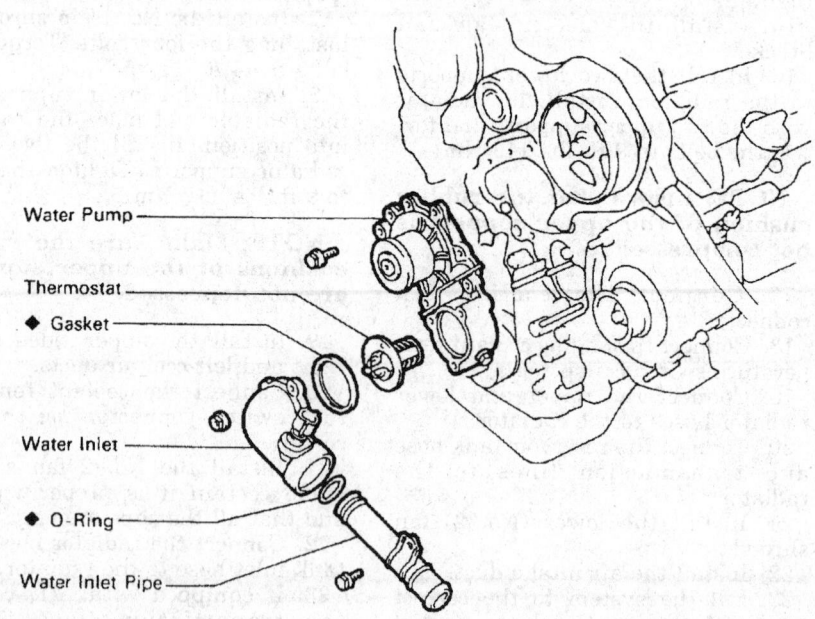

◆ Non-reusable part

56009

Water pump and related components — 1993 ES300 shown, others similar

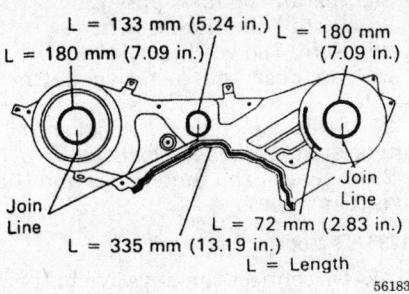

L = 133 mm (5.24 in.) L = 180 mm (7.09 in.)
L = 180 mm (7.09 in.)
Join Line
Join Line
L = 72 mm (2.83 in.)
L = 335 mm (13.19 in.)
L = Length
56183

Upper inner timing belt cover gasket installation positions — 1994 ES300

11. Scrape and clean all gasket material from the upper inner timing belt cover.

To install:

12. Check that the water pump turns smoothly. Also check the air hole for coolant leakage.

13. Using a new gasket, apply liquid sealer to the gasket, water pump and engine block.

14. Install the gasket and pump to the engine and install the four bolts and two nuts. Torque the nuts and bolts to 53 inch lbs. (6 Nm).

15. Install the rear timing belt cover and torque the six bolts to 74 inch lbs. (9 Nm).

16. Connect the engine wire with the three clamps to the rear timing belt cover.

17. Install the No. 2 idler pulley with the bolt. Torque the bolt to 32 ft. lbs. (43 Nm). After torquing the bolt, make sure the idler pulley moves smoothly.

18. With the flange side **outward**, install the right hand camshaft pulley to the engine. Make sure to align the knock pin hole on the camshaft pulley with the knock pin on the camshaft. Using the same tools as removal, torque the camshaft bolt to 65 ft. lbs. (88 Nm).

19. With the flange side **inward**, install the left hand camshaft pulley to the engine. Make sure to align the knock pin hole on the camshaft pulley with the knock pin on the camshaft. Using the same tools as removal, torque the camshaft bolt to 94 ft. lbs. (125 Nm).

20. Install the timing belt to the engine.

21. Fill the engine coolant.

22. Connect the negative battery cable to the battery and start the engine.

23. Top off the engine coolant and check for leaks.

GS300

1. Disconnect the negative battery cable. Wait at least 90 seconds before performing any other work.

2. Drain the cooling system.

3. Remove the lower engine cover.

4. Remove the nut and then remove the air cleaner duct.

5. Disconnect the high tension lead from the ignition coil.

6. Disconnect the high tension lead from the clamp on the VAF meter.

7. Disconnect the VAF meter electrical connector and disconnect the harness from the meter.

8. Disconnect the power steering air hose from the timing belt cover.

9. Disconnect the PCV hose from the cylinder head cover.

10. Loosen the hose clamp bolt securing the intake air connector pipe to the throttle body.

11. Remove the three (3) bolts, air cleaner, VAF meter and intake air connector pipe assembly.

12. Remove the drive belt.

13. Place matchmarks on the cooling fan clutch and the cooling fan pulley. Remove the cooling fan by removing the four (4) nuts.

14. Remove the water pump pulley.

15. Remove the radiator.

16. Remove the two (2) nuts and disconnect the water inlet from the water pump.

17. Remove the thermostat.

18. Remove the distributor.

19. On all except California vehicles, remove the four(4) nuts and exhaust manifold heat insulator.

20. Remove the two (2) bolts, the No. 1 water bypass outlet and pipe. Discard the O-rings.

21. Remove the timing belt.

22. Remove the idler pulley.

23. Remove the mounting bolt and disconnect the engine wire bracket.

24. Remove the alternator mounting bolt and disconnect the alternator from the water pump.

25. Remove the two (2) nuts and disconnect the No. 2 water bypass pipe from the water pump.

26. Remove the water pump mounting bolts. Note the position and type of each bolt for correct installation.

27. Lift out the water pump. If prying is necessary, use a protected blade and take great care not to damage the mating surfaces.

28. Remove all the old sealant and gasket material; clean the mounting surfaces.

To install:

29. Install a new O-ring to the cylinder block and a new gasket to the water pump.

30. Connect the water bypass pipe to the water pump. Do not install the nuts yet.

31. Install the water pump. Fit the bolts and hand tighten them in the correct order; then torque the mounting bolts to 15 ft. lbs. (21 Nm).

NOTE: Hand tighten the A bolts prior to hand tightening the B bolts.

32. Install the nuts to the No. 2 water bypass pipe and torque them to 15 ft. lbs. (21 Nm).

33. Install the alternator mounting bolt and nut and torque them to 27 ft. lbs. (37 Nm).

34. Connect the engine wire harness. Install and tighten the bolts.

35. Install the idler pulley.

36. Install the timing belt.

37. Install new O-rings to the No. 1 water bypass pipe and outlet. Install the bolts and torque them to 78 inch lbs. (9 Nm).

38. Install the exhaust manifold heat insulator if it was removed and torque the nuts to 13 ft. lbs. (18 Nm).

39. Install the distributor.

40. Install the thermostat and align the jiggle valve with the protrusion on the water inlet housing.

41. Install the water inlet housing; tighten the bolts to 78 inch lbs. (9 Nm).

42. Install the radiator.

43. Align the matchmarks and install the water pump pulley and fan assembly. Do not torque the nuts at this time.

44. Install the drive belt and torque the fan assembly nuts to 12 ft. lbs. (16 Nm).

45. Connect the intake air connector pipe to the throttle body.

46. Install the air cleaner, VAF meter and intake air connector pipe assembly with the three (3) bolts.

47. Install the hose clamp.

48. Connect the power steering air hose to the timing belt cover.

49. Connect the PCV hose to the cylinder head cover.

50. Connect the VAF meter harness and electrical connector.

51. Connect the high tension lead to the VAF meter and ignition coil.

52. Connect the air cleaner duct to the air cleaner.

53. Install the air cleaner duct.

54. Refill the cooling system.

55. Connect the negative battery cable.

56. Start the engine and check for leaks.

57. Check the fluid level on vehicles with A/T.

58. Inspect the engine ignition timing.

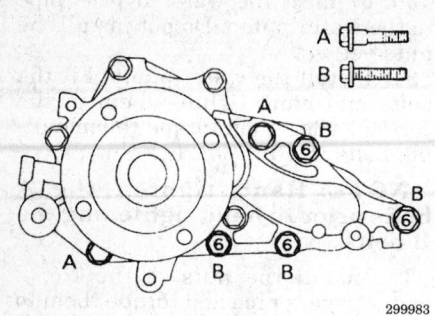

Water pump mounting bolt installation sequence — GS300 and 1996–97 SC300

59. Install the lower engine cover.
60. Road test the vehicle.

LS400

1. Disconnect the negative battery cable.
2. Drain the cooling system.
3. Remove the timing belt and the No. 2 idler pulley.
4. Remove the right side ignition coil.
5. Remove the two water inlet housing to water pump bolts.
6. Disconnect the IAC valve bypass hose from the water inlet housing.
7. Remove the water inlet housing and discard the O-ring.
8. Remove the mounting bolts, studs and nut. Lift out the water pump by carefully prying between the pump and the cylinder head.
9. Remove all the old packing and clean all mounting surfaces. Remove the O-ring from the water bypass pipe.
   To install:
10. Install new seal packing to the water pump groove and a new O-ring to the water bypass pipe.
11. Connect the water pump to the water bypass pipe end.
12. Install the water pump and tighten the mounting bolts to 13 ft. lbs. (18 Nm).
13. Apply sealant to the groove of the water inlet housing.
14. Install a new O-ring to the water inlet housing.
15. Push the water inlet housing end into the water pump hole.
16. Connect the IAC valve bypass hose to the water inlet housing.
17. Install the water inlet and housing assembly with the two bolts. Alternately torque the bolts to 13 ft. lbs. (18 Nm).
18. Install the ignition coil.
19. Install the No. 2 idler pulley for the timing belt.
20. Install the timing belt.

21. Refill the cooling system.
22. Connect the negative battery cable.
23. Start the engine and check for leaks.

SC300

1993–95 Models

1. Disconnect the negative battery cable. Wait at least 90 seconds before performing any other work.
2. Drain the cooling system.
3. Remove the timing belt.
4. Remove the idler pulley.
5. Remove the thermostat.
6. Remove the two bolts, water bypass outlet and the No. 1 water bypass pipe. Discard the three O-rings.
7. Remove the mounting bolt and disconnect the engine wire harness bracket above the alternator.
8. Remove the alternator mounting nut.
9. Remove the alternator mounting bolt and disconnect the alternator from the water pump.
10. Remove the two nuts and disconnect the No. 2 water bypass pipe from the water pump.
11. Remove the six water pump mounting bolts. The bolts are of different lengths and styles; note the correct position of each bolt during removal.
12. Lift out the water pump by carefully prying between the pump and the cylinder head.
13. Remove all the old packing and clean the mounting surfaces.
14. Remove the O-ring from the cylinder block.
   To install:
15. Install a new O-ring to the cylinder block.
16. Apply a thin layer of liquid sealant to the water pump and install a new gasket.
17. Connect the No. 2 water bypass pipe to the water pump. Do not install the nuts yet.
18. Install the water pump. Install the bolts in the correct positions and tighten them finger tight. Tighten the mounting bolts to 15 ft. lbs. (21 Nm).
19. Install the nuts to the No. 2 water bypass pipe and torque them to 15 ft. lbs. (21 Nm).
20. Install the alternator mounting bolt and nut and torque them to 27 ft. lbs. (37 Nm).
21. Connect the engine wire harness. Install and tighten the bolts.
22. Install new O-rings to the No. 1 water bypass pipe and outlet. Install the bolts and torque them to 78 inch lbs. (9 Nm).
23. Install the thermostat.

24. Install the idler pulley.
25. Install the timing belt.
26. Refill the cooling system.
27. Connect the negative battery cable.
28. Start the engine, bleed the cooling system and check for leaks.
29. Recheck the fluid levels and the ignition timing.

1996–97 Models

1. Disconnect the negative battery cable. Wait at least 90 seconds before performing any work.
2. Drain the engine coolant and remove the radiator assembly.
3. Remove the air cleaner, MAF meter and the intake air connector pipe assembly.
4. Remove the timing belt.
5. Remove the idler pulley.
6. Remove the water inlet and the thermostat.
7. Remove the 2 bolts, the water bypass outlet and the No. 1 water bypass pipe. Remove the three O-rings from the water bypass outlet and the No. 1 water bypass pipe.
8. Remove the generator.
9. Remove the bolt and disconnect the engine wire bracket. Remove the bolt and disconnect the clamp bracket for the crankshaft position sensor connector. Remove the nuts and disconnect the No. 2 water bypass pipe from the water pump. Remove the six bolts and the water pump and gasket.
10. Remove the drain hose and the O-ring from the cylinder block.
   To install:
11. Install a new O-ring to the cylinder block.
12. Install the drain hose.
13. Install a new gasket to the water pump. Connect the water pump to the water bypass pipe. Do not install the nut yet. Install the water pump with the two bolts (A) and the four bolts (B).

   NOTE: Hand tighten the (A) bolts first. Torque all six bolts to 15 ft. lbs. (21 Nm).

14. Install the two nuts holding the No. 2 water bypass pipe to the water pump. Torque the nuts to 15 ft. lbs. (21 Nm). Install the clamp bracket for the crankshaft position sensor connector.
15. Install the engine wire bracket.
16. Install the generator.
17. Install new O-rings to the No. 1 water bypass pipe. Install a new O-ring and the water bypass outlet with the two bolts and torque them to 78 inch lbs. (9 Nm).
18. Install the thermostat and the water inlet.

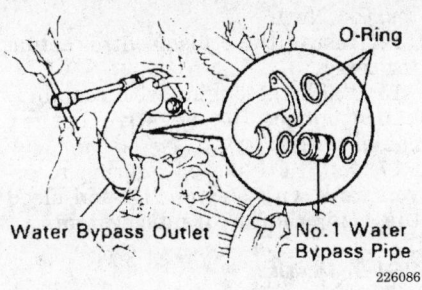

Water bypass pipe removal and installation — SC300

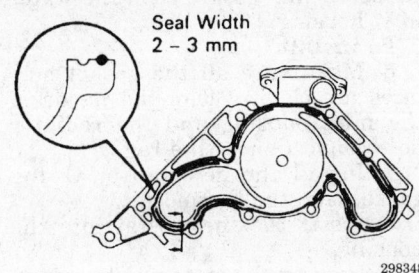

Water pump sealant application — SC400

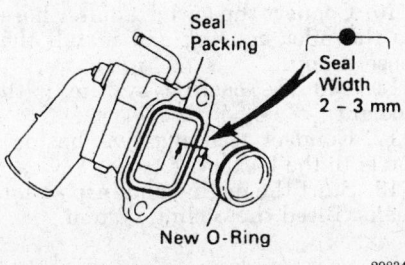

Water inlet housing sealant application — SC400

19. Install the idler pulley.

20. Install the timing belt.

21. Install the air cleaner, the MAF meter and the intake air connector pipe assembly.

22. Install the radiator assembly.

23. Reconnect the negative battery cable. Refill the cooling system. Start the engine, check for leaks and bleed the cooling system.

24. Road test for proper operation.

SC400

1. Disconnect the negative battery cable.

2. Drain the cooling system.

3. Remove the timing belt.

4. Remove the No. 2 idler pulley.

5. Remove the right side ignition coil by removing the ignition coil connector and two bolts.

6. Disconnect the bypass hose(s) from the water inlet housing.

7. Remove the two bolts holding the water inlet housing to water pump.

8. Remove the water inlet housing and discard the gasket.

9. Remove the mounting bolts, studs and the nut to the water pump. Remove the water pump by carefully prying between the pump and the cylinder head.

10. Remove all the old gasket and clean all mounting surfaces.

To install:

11. Install new seal packing to the water pump groove and a new O-ring to the water bypass pipe end.

12. Connect the water pump to the water bypass pipe end.

13. Install the water pump and tighten the mounting bolts and nut to 13 ft. lbs. (18 Nm).

14. Apply sealant to the groove of the water inlet housing.

15. Install a new O-ring to the water inlet housing.

16. Push the water inlet housing end into the water pump hole.

17. Install the water inlet and housing assembly with the two bolts. Alternately torque the bolts to 13 ft. lbs. (18 Nm).

18. Connect the bypass hose(s) to the water inlet housing.

19. Replace the ignition coil by installing the two bolts and coil connector.

20. Install the No. 2 idler pulley.

21. Install the timing belt.

22. Refill the cooling system.

23. Connect the negative battery cable.

24. Start the engine, bleed the cooling system and check for leaks.

Thermostat

REMOVAL AND INSTALLATION

ES300

1993 Models

1. With the ignition switch in the **LOCK** position, disconnect the negative battery terminal. If equipped with an air bag system, wait at least 90 seconds or longer before performing any other work.

2. Position a suitable drain pan under the radiator drain cock and drain the cooling system.

3. Disconnect the lower radiator hose from the water inlet pipe.

4. Remove the water inlet pipe.

5. Disconnect the water temperature sensor connector from the water inlet housing.

6. Remove the three nuts from the water inlet housing and remove the housing from the water pump studs.

7. Remove the thermostat and rubber O-ring gasket from the water inlet housing.

To install:

8. Make sure all the gasket surfaces are clean. Clean the inside of the inlet housing and the radiator hose connection with a rag.

9. Install the new rubber O-ring gasket onto the thermostat.

10. Align the jiggle valve with the upper stud bolt.

11. Position the water inlet housing with the thermostat over the studs on the water pump and install the three nuts. Torque the nuts to 14 ft. lbs. (20 Nm).

12. Connect the water temperature sensor connector.

13. Use a new O-ring and install the water inlet pipe.

14. Connect the lower radiator hose to the water inlet pipe.

15. Connect the negative battery cable.

16. Refill the cooling system.

17. Start the engine and inspect for leaks.

1994–97 Models

1. Disconnect the negative battery cable from the battery.

2. Drain the cooling system.

3. Remove the air cleaner cap assembly.

4. Disconnect the heater hose.

5. Remove the bolt and disconnect the hydraulic cooling motor pressure hose from the water inlet.

6. Disconnect the engine wire from the thermostat housing and cylinder head by removing the two nuts.

7. Disconnect the two ECT switch connectors from the thermostat housing.

8. Remove the bolt holding the water inlet pipe to the cylinder head.

9. Disconnect the water inlet pipe from the thermostat housing and remove the O-ring.

10. Remove the three nuts and thermostat housing from the engine.

11. Remove the thermostat and gasket.

To install:

12. Clean the gasket mating surfaces.

13. Be sure to use a new thermostat and gasket. Install the thermostat with the spring inside the engine block. Align the jiggle valve on the thermostat with the stud bolt on the thermostat housing.

14. Install the thermostat housing and the three retaining nuts. Torque the nuts to 69 inch lbs. (8 Nm).

15. Using a new O-ring, connect the water inlet pipe to the thermostat housing.

16. Install the bolt holding the water inlet pipe to the cylinder head. Torque the bolt to 14 ft. lbs. (19 Nm).

17. Connect the ECT switch connectors.

18. Connect the engine wire to the thermostat housing and cylinder head with the two nuts.

19. Install the hydraulic cooling motor pressure hose to the water inlet and torque the bolt to 69 inch lbs. (8 Nm).

20. Connect the heater hose.

21. Install the air cleaner cap assembly.

22. Connect all electrical connectors and all hoses.

23. Fill the engine coolant.

24. Connect the negative battery cable to the battery and start the engine.

25. Top off the engine coolant and check for leaks. Make sure the thermostat works properly. Bleed the cooling system.

All Except ES300

1. Disconnect the negative battery cable from the battery.

2. With the engine cool, position a drain pan under the radiator drain cock and drain the cooling system.

3. Loosen the hose clamp and disconnect the lower radiator hose from the water inlet housing.

4. Remove the two nuts from the water inlet housing and remove the housing from the water pump studs.

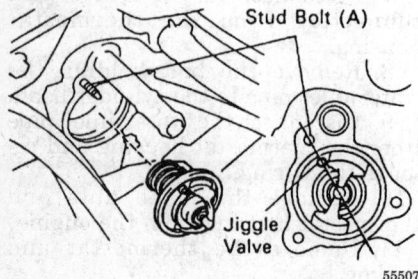

Stud Bolt (A)

Jiggle Valve

55507

Aligning the thermostat jiggle valve

5. Remove the thermostat and rubber O-ring gasket from the water inlet housing.

To install:

6. Make sure all the gasket surfaces are clean. Clean the inside of the inlet housing and the radiator hose connection with a rag.

7. Install the new rubber O-ring gasket onto the thermostat.

8. Insert the thermostat into the housing.

9. Position the water inlet housing with the jiggle valve of the thermostat aligned with the protrusion of the water inlet; install the two nuts. Torque the two nuts to 8–13 ft. lbs. (9–18 Nm).

10. Connect the lower radiator hose to the inlet housing and install the hose clamp.

11. Fill the cooling system with coolant.

12. Connect the negative battery cable to the battery.

13. Start the engine and inspect for leaks. Bleed the cooling system.

Electric Cooling Fan

REMOVAL AND INSTALLATION

ES300

1993 Models

1. Disconnect the negative battery cable. Wait at least 90 seconds to perform any work on the vehicle once the negative battery cable has been disconnected.

2. Drain the engine coolant.

3. Remove the cruise control actuator cover.

4. Place a drain pan under the hydraulic motor and remove the pressure hose from the motor.

5. Disconnect the radiator upper hose.

6. Disconnect the coolant reservoir hose.

7. Disconnect the hydraulic motor return hose.

8. Remove the six mounting bolts and the cooling fan.

To install:

9. Position the cooling fan assembly into place and torque the mounting bolts to 43 inch lbs. (5 Nm).

10. Connect the hydraulic motor return hose.

11. Connect the coolant reservoir hose.

12. Connect the upper radiator hose.

13. Use new gaskets and connect the pressure hose to the hydraulic

motor. Torque the union bolt to 47 ft. lbs. (64 Nm).

14. Install the cruise control actuator cover.

15. Refill the engine with coolant.

16. Connect the negative battery cable.

17. Refill the power steering reservoir tank with Dexron II® and bleed the cooling fan hydraulic system.

1994–97 Models

1. Disconnect the negative battery cable. Wait at least 90 seconds to perform any work on the vehicle once the negative battery cable has been disconnected.

2. Drain the engine coolant.

3. Place a drain pan under the hydraulic motor and remove the pressure hose from the motor.

4. Disconnect the radiator upper hose.

5. Disconnect the hydraulic motor return hose.

6. Disconnect the lower radiator hose from the water inlet pipe.

7. Disconnect the two oil hoses from the oil cooler pipes.

8. On Canadian models, remove the No. 7 relay block.

9. Remove the cruise control actuator wire from the clamp.

10. Remove the 2 upper radiator mounting brackets.

11. Remove the radiator and cooling fan as an assembly.

12. Remove the six mounting bolts and the cooling fan.

To install:

13. Position the cooling fan assembly into place and torque the 6 mounting bolts to 43 inch lbs. (5 Nm).

14. If removed, install the lower radiator hose and the oil cooler hoses to the radiator.

15. Install the radiator and cooling fan as an assembly. Make sure the insulators are in place on the bottom of the radiator.

16. Install the 2 radiator upper mounting brackets. Torque the bolts to 9 ft. lbs. (13 Nm).

17. Install the cruise control actuator wire to the clamp.

18. If removed, install the No. 7 relay block.

19. Connect the oil cooler and lower radiator hoses to the water inlet pipe and oil cooler pipes.

20. Connect the hydraulic motor return hose and the upper radiator hose.

21. Use new gaskets and connect the pressure hose to the hydraulic motor. Torque the union bolt to 47 ft. lbs. (64 Nm).

22. Refill the engine with coolant.

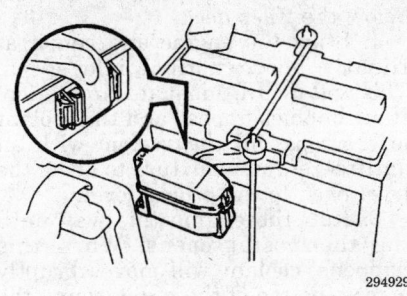

No. 7 relay block (Canada only) — 1994–97 ES300

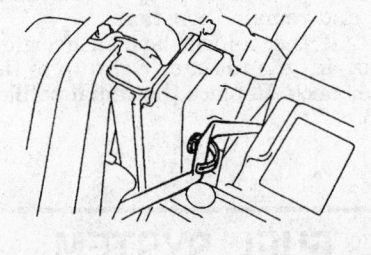

Cruise control actuator wire — 1994–97 ES300

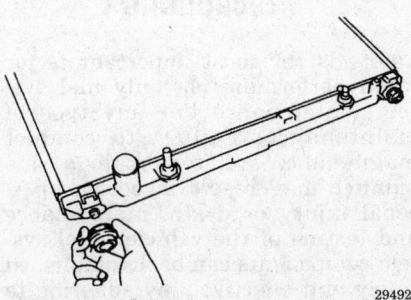

Lower radiator insulators look like a rubber grommet

23. Connect the negative battery cable.

24. Refill the power steering reservoir with Dexron II®. Bleed the cooling fan hydraulic system and the cooling system.

25. Check for leaks.

GS300

1. Disconnect the negative battery cable. Wait at least 90 seconds before performing any other work.

2. Remove the engine undercover.

3. Remove the air cleaner duct.

4. Disconnect the upper radiator hose from the radiator.

5. Disconnect the electrical connector from the fan motor and free the sensor wire from the clamp.

6. Remove the three bolts holding the fan frame and then lift the fan assembly from the engine compartment.

7. Remove the fan from the motor by removing the screws.

8. Disconnect the wire and connector from the fan shroud.

9. Remove the three screws and the fan motor.

To install:

10. Install the motor to the frame with the three screws.

11. Connect the wire and connector to the fan shroud.

12. Connect the fan with the three screws.

13. Position the fan and install the mounting bolts.

14. Connect the electrical connector and secure the sensor wire in the clamp.

15. Connect the upper radiator hose.

16. Install the air cleaner duct.

17. Install the undercover

18. Connect the negative battery cable.

19. Check coolant level and add coolant as necessary.

20. Start the engine and verify proper cooling fan operation.

LS400

1993–94 Models

1. Disconnect the negative battery cable.

2. Raise and safely support the vehicle.

3. Remove the lower engine cover.

4. Remove the lower wind guide.

5. Lower the vehicle to a level where the bumper and headlights can be removed.

6. Remove the clearance light mounting bolts. Carefully pry and pull the lights until free. Disconnect the electrical connectors and remove the lights.

7. Remove the headlight/foglight nut and 3 bolts and pull the assemblies out. Disconnect the electrical connectors and remove the assemblies.

8. Remove the 8 upper clips and bumper retainer.

9. Disconnect the ambient temperature sensor and remove the wire harness from the bumper.

10. Disconnect the 4 turn signal/marker lamp connectors.

11. Remove the 4 clips holding the front bumper to the reinforcement.

12. Remove the 4 bolts, 4 nuts and 2 retainers securing the bumper to the front fenders.

13. Remove the 12 screws securing the bumper to the fender liners.

14. Remove the 4 bolts holding the bumper to the body.

15. Remove the bumper.

16. Remove the 10 bolts and bumper reinforcement.

17. Remove the right side horn and disconnect the electrical connector.

18. Disconnect the 2 cooling fan electrical connectors.

19. Disconnect the cooling fan wires from the clamps.

20. Remove the locknut cap. Remove the 3 nuts, then 2 nuts and hood lock support.

21. Remove the mounting bolts and cooling fans.

To install:

22. Install the cooling fans and loosely install the bolts.

23. Install the hold lock support, tighten the 3 nuts and install the locknut cap.

24. Install the 2 bolts holding the hood lock support to the cooling fans and tighten the shroud bolts.

25. Connect the cooling fan electrical connectors and secure the clamps.

26. Install the right side horn.

27. Install the bumper reinforcement.

28. Install the bumper and torque the bumper-to-body bolts to 67 ft. lbs. (91 Nm).

29. Connect the turn signal/marker light electrical connectors.

30. Install the bumper retainer and clips.

31. Connect the ambient temperature sensor and route the electrical lead.

32. Install the headlight/foglight assembly.

33. Install the clearance lights.

34. Install the lower wind guide.

35. Install the lower engine cover.

36. Connect the negative battery cable.

1995–97 Models

1. Disconnect the negative battery cable to the battery.

2. Remove the air cleaner inlet.

3. Loosen the four nuts holding the fluid coupling to the fan bracket.

4. Remove the alternator drive belt by turning the tensioner counterclockwise.

5. Remove the four nuts and remove the fan.

To install:

6. Install the fan to the engine and install the four nuts.

7. Install the drive belt.

8. Torque the fan bolts to 16 ft. lbs. (21 Nm).

9. Install the air cleaner inlet.

10. Connect the negative battery cable to the battery.

SC300

1. Disconnect the negative battery cable to the battery.

2. Loosen the four (4) nuts holding the fluid coupling to the water pump.

3. Loosen the drive belt tension by turning the tensioner clockwise; remove the drive belt.

4. Remove the four (4) cooling fan pulley nuts. Remove the fan assembly.

5. Remove the clutch from the fan.

To install:

6. Install the clutch to the fan.

7. Install the fan assembly to the engine and install the four (4) bolts. Do not torque the nuts at this time.

8. Install the drive belt by turning the drive tensioner clockwise.

9. Tighten the four (4) fan nuts to 12 ft. lbs. (16 Nm).

10. Connect the negative battery cable to the battery.

SC400

1. Disconnect the negative battery cable. Wait at least 90 seconds before performing any work.

2. Remove the engine undercover and drain the engine coolant.

3. Remove the radiator reservoir.

4. Disconnect the upper radiator hose.

5. Disconnect the pressure and return hoses from the hydraulic motor.

6. Remove the two bolts, brackets and bushings and disconnect the cooling fan inlet pipe from the fan shroud.

7. Remove the cooling fan reservoir from the fan shroud.

8. Disconnect the ECT sensor connector for the cooling fan.

9. Remove the four bolts attaching the fan shroud to the radiator.

Slightly lift the shroud and disconnect the two oil cooler hoses (for the cooling fan) from the hose clamp on the fan shroud. Remove the fan shroud.

10. Loosen the pulley nut clockwise, remove the nut, plate washer and cooling fan from the hydraulic motor.

To install:

11. Install the cooling fan to the hydraulic motor with the plate washer and nut. Tighten the nut by turning it counterclockwise. Torque the nut to 43 inch lbs. (15 Nm).

12. Place the fan shroud on the radiator. Slightly lift the shroud and connect the two oil cooler hoses for the cooling fan to the hose clamp on the fan shroud. Install the fan shroud with the four bolts and torque them to 43 inch lbs. (5 Nm).

13. Connect the ECT sensor connector for the cooling fan.

14. Install the reservoir tank with the four bolts. Torque the bolts to 43 inch lbs. (5 Nm). Install the suction hose to the clamp on the fan shroud.

15. Install the cooling fan inlet pipe to the fan shroud.

16. Connect the upper radiator hose and install the hydraulic pressure hose with a new gasket and union bolt to the hydraulic motor. Torque the union bolt to 47 ft. lbs. (64 Nm).

17. Install the radiator reservoir, fill the cooling system with coolant and fill with cooling fan fluid. Connect the negative battery cable and start the engine to check for leaks.

18. Install the engine undercover.

Cooling System Bleeding

After working on the cooling system, even to replace the thermostat, it must be bled. Air trapped in the system will prevent proper filling and leave the radiator coolant level low, causing a risk of overheating.

To bleed the system, start with the system cool, the radiator cap off and

the radiator filled to about an inch below the filler neck.

1. Start the engine and run it at slightly above normal idle speed. This will insure adequate circulation. If air bubbles appear and the coolant level drops, fill the system with an antifreeze/water mixture to bring the level back to the proper level.

2. Run the engine this way until the thermostat opens. When this happens, coolant will move abruptly across the top of the radiator and the temperature of the radiator will suddenly rise.

3. At this point, air is often expelled and the level may drop quite a bit. Keep refilling the system until the level is near the top of the radiator and remains constant.

4. If the vehicle has an overflow tank, fill the radiator right up to the filler neck. Replace the radiator filler cap.

FUEL SYSTEM

Fuel System Service Precautions

Safety is the most important factor when performing not only fuel system maintenance but any type of maintenance. Failure to conduct maintenance and repairs in a safe manner may result in serious personal injury or death. Maintenance and testing of the vehicle's fuel system components can be accomplished safely and effectively by adhering to the following rules and guidelines.

• To avoid the possibility of fire and personal injury, always disconnect the negative battery cable unless the repair or test procedure requires that battery voltage be applied.

• Always relieve the fuel system pressure prior to disconnecting any fuel system component (injector, fuel rail, pressure regulator, etc.), fitting or fuel line connection. Exercise extreme caution whenever relieving fuel system pressure to avoid exposing skin, face and eyes to fuel spray. Please be advised that fuel under pressure may penetrate the skin or any part of the body that it contacts.

• Always place a shop towel or cloth around the fitting or connection prior to loosening to absorb any excess fuel due to spillage. Ensure that all fuel spillage (should it occur) is

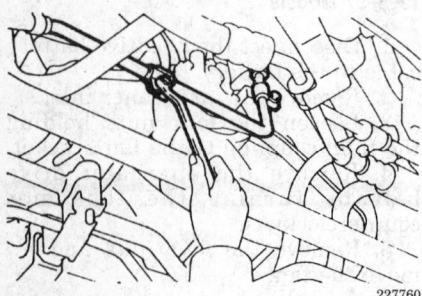

227760

Removing the fan inlet pipe from the shroud — 1993-95 SC400

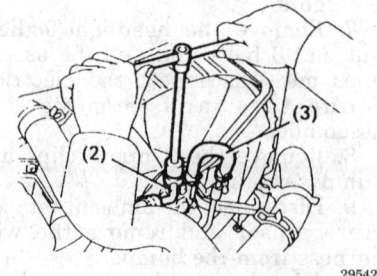

295420

Connecting the hydraulic pressure hose for the cooling fan motor — 1996-97 SC400

quickly removed from engine surfaces. Ensure that all fuel soaked cloths or towels are deposited into a suitable waste container.

• Always keep a dry chemical (Class B) fire extinguisher near the work area.

• Do not allow fuel spray or fuel vapors to come into contact with a spark or open flame.

• Always use a backup wrench when loosening and tightening fuel line connection fittings. This will prevent unnecessary stress and torsion to fuel line piping. Always follow the proper torque specifications.

• Always replace worn fuel fitting O-rings with new. Do not substitute fuel hose or equivalent, where fuel pipe is installed.

Fuel System Pressure

RELIEVING

-------- CAUTION --------
Failure to relieve fuel pressure before repairs or disassembly can cause serious personal injury and/or property damage. Fuel pressure is maintained within the fuel lines, even if the engine is OFF or has not been run in a period of time. This pressure must be safely relieved before any fuel-bearing line or component is loosened or removed. On vehicles equipped with inflatable restraints or air bag systems, wait at least 90 seconds after disconnecting the battery cable before performing any other work. The backup power will keep the restraint system energized for a period of time after the battery is disconnected.

1. Remove the fuse for the electronic fuel pump.
2. Start the engine until the engine stalls.
3. Disconnect the negative battery terminal.
4. Place a catch-pan under the joint to be disconnected. A large quantity of fuel may be released when the joint is opened.
5. Wear eye or full face protection.
6. Place a shop towel over the area and slowly release the joint using a wrench of the correct size.
7. Allow the any fuel left in the line to bleed off slowly before fully disconnecting the joint.
8. Plug the opened lines immediately to prevent fuel spillage or the entry of dirt.

9. Dispose of the released fuel properly.
10. After connecting fuel lines, install the fuse for the fuel pump and start the engine.
11. Check for leaks and repair as needed.

Idle Speed

ADJUSTMENT

Idle speed is controlled by the ECU and is not adjustable.

Fuel Filter

REMOVAL AND INSTALLATION

The fuel filter on the ES300 is located under the hood, on the driver's side, by the fenderwell. The fuel filter SC300 is located under the vehicle, on the driver's side, in front of the rear axle.

The fuel filter for the LS400 and SC400 is located under the vehicle on the left side before the rear axle. The fuel filter on the GS300 is located under the vehicle, next to the left rear exhaust resonator.

1. Disconnect the negative battery cable. Wait at least 90 seconds before performing any other work.
2. Raise and safely support the vehicle if necessary.
3. On the GS300, remove the rear body protector.
4. Place a drain pan or plastic container under the fuel filter.
5. Slowly loosen the lower flare nut fitting until all the pressure is relieved and all the fuel is collected.
6. Loosen the union bolt on the upper portion of the filter and remove the banjo fitting and two metal gaskets. Discard the gaskets.

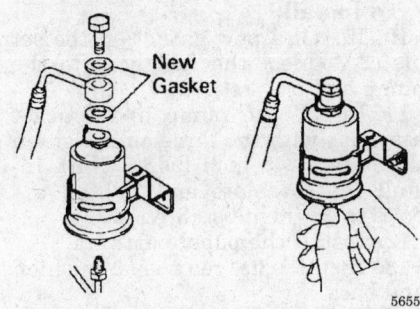

Fuel filter removal and installation — 1993–94 ES300

7. Loosen the fuel filter bracket bolt, remove the fuel line with the flared nut from the filter and pull the filter from the mounting bracket.

To install:
8. Install a new fuel filter to the vehicle and tighten the bracket bolt.
9. Install the banjo fitting with a new metal gasket on each side and install the union bolt. Torque the union bolt to 22 ft. lbs. (30 Nm).
10. Connect the flare nut to the lower connection. Torque the flare nut to 22 ft. lbs. (30 Nm).
11. On the GS300, install the body protector.
12. Lower the vehicle if raised.
13. Remove the drain pan and/or rags and connect the negative battery cable.
14. Start the engine and visually inspect the upper and lower connections for leaks.

Fuel Pump

REMOVAL AND INSTALLATION

ES300

-------- CAUTION --------
Fuel injection systems remain under pressure even after the engine has been turned OFF. The fuel system pressure must be relieved before disconnecting any fuel lines. Failure to do so may result in fire and/or personal injury.

1. Relieve the fuel system pressure.
2. With the ignition switch in the **LOCK** position, disconnect the negative battery terminal. If equipped with an air bag system, wait at least 90 seconds or longer before performing any other work.
3. Remove the rear seat cushion.
4. Disconnect the fuel pump connector.

-------- CAUTION --------
The Supplemental Inflatable Restraint (SIR) system must be disarmed before removing the fuel pump. Failure to do so may cause accidental deployment of the air bag, resulting in unnecessary SIR system repairs and/or personal injury.

5. Remove the floor service hole cover.

NOTE: Do not lift the fuel pump assembly up using the wire harness.

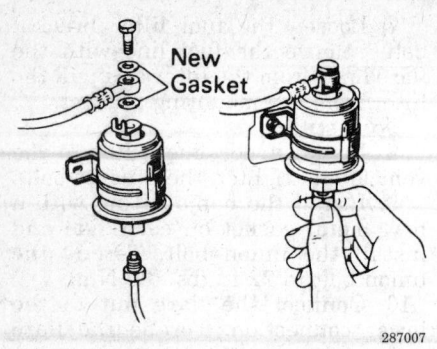

287007

Fuel filter components — 1995–97 ES300 and 1996–97 SC300 shown, others similar

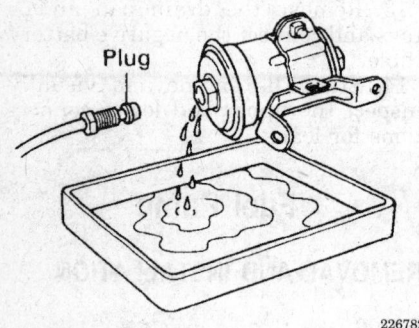

226782

Be sure to have a pan ready for fuel leakage!

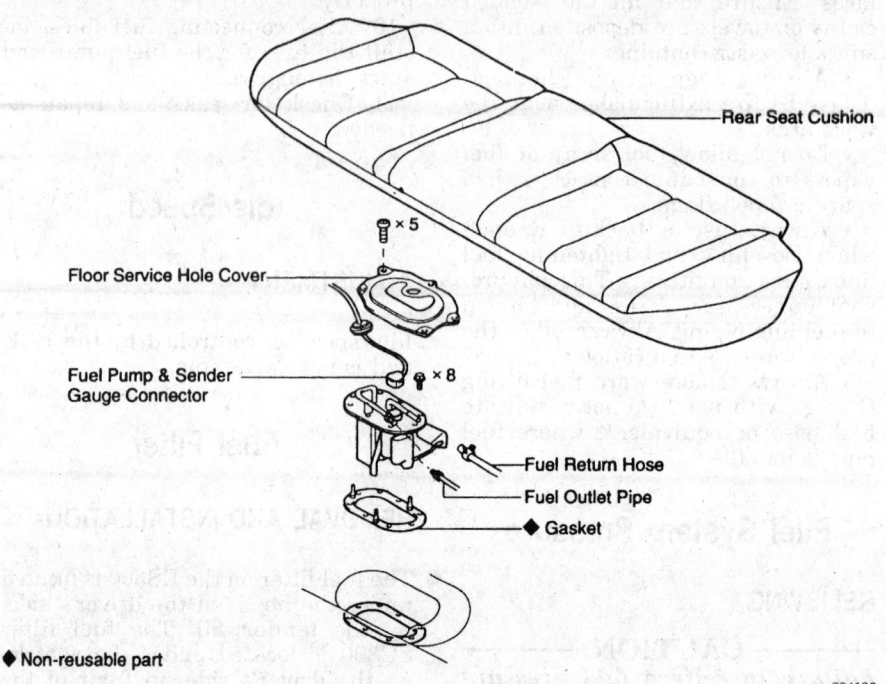

◆ Non-reusable part

294126

Fuel pump assembly — ES300

6. Remove the fuel filler cap. Disconnect the fuel outlet pipe and the return hose from the pump bracket.

7. Remove the 8 screws and lift out the pump/bracket assembly with gasket.

8. Remove the fuel pump lead wire.

9. Pull the lower end of the pump off of the bracket.

10. Disconnect the fuel hose from the pump and remove the pump.

11. Remove the rubber cushion from the pump.

To install:

12. Install the filter and rubber cushion on the new pump and install the pump on the bracket.

13. Connect the fuel hose and the wire connector on the pump.

14. Using a new gasket, install the pump and tighten the eight screws to 35 inch lbs. (4 Nm).

15. Connect the fuel pipe and return hose to the pump and tighten the bolts to 22 ft. lbs. (29 Nm).

16. Connect the wire, install the service cover and replace the rear seat.

17. Connect the negative battery cable.

18. Start the engine and check for leaks.

GS300, LS400, SC300 and SC400

1. Disconnect the negative battery cable. Wait at least 90 seconds before performing any other work.

2. Remove the trunk floor mat.

3. Remove the trunk trim cover.

4. Disconnect the fuel pump electrical connector.

5. Remove the rear seat bottom and seat back.

6. Remove the partition cover.

7. Remove the mounting bolts and remove the fuel pump set plate.

8. Remove the three nuts and disconnect the fuel pump bracket from the tank.

9. Disconnect the fuel hose from the bracket. Remove the pump, bracket and set plate as an assembly.

To install:

10. Install a new gasket on the set plate. Connect the fuel hose to the pump and bracket,

11. Install the pump and bracket assembly with the three nuts; tighten the nuts to 48 inch lbs. (5 Nm). Install the set plate and tighten the bolts to 26 inch lbs. (3 Nm).

12. Install the panel partition

13. Install the rear seat cushion and back.

14. Connect the fuel pump electrical connector.

15. Install the trim panel.

16. Install the spare tire and the trunk floor mat.

17. Connect the negative battery cable.

18. Start the engine; check the fuel system for leaks

19. Road test the vehicle for proper operation.

Fuel Injector

REMOVAL AND INSTALLATION

ES300

1993 Models

1. Relieve the fuel system pressure.

2. With the ignition switch in the **LOCK** position, disconnect the negative battery terminal. If equipped with an air bag system, wait at least 90 seconds or longer before performing any other work.

3. Drain the engine coolant.

4. Disconnect the accelerator cable from the throttle linkage.

5. If equipped with automatic transmission, disconnect the throttle cable from the throttle linkage.

6. Remove the air cleaner cap, air flow meter and the air cleaner hose as a unit.

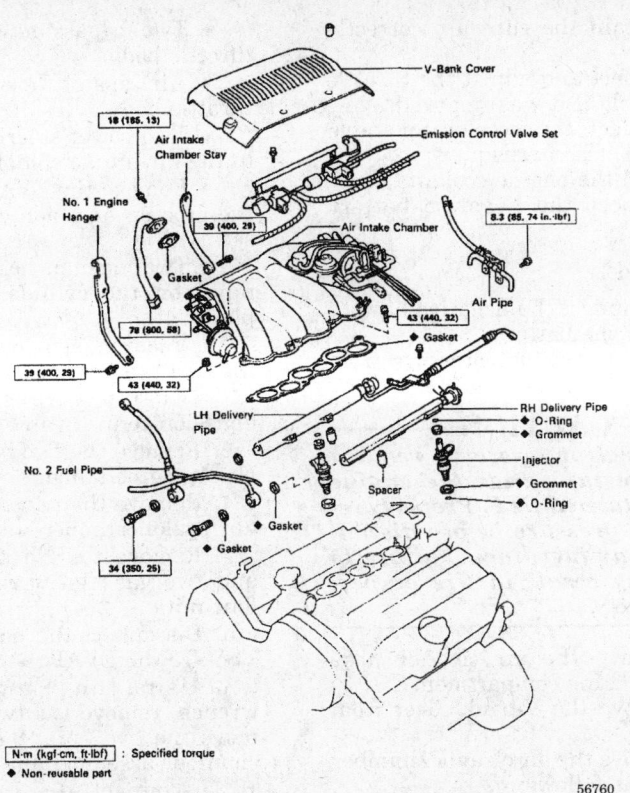

N·m (kgf-cm, ft-lbf) : Specified torque
◆ Non-reusable part

56760

Injector related components — 1993 ES300

7. Remove the two 5mm bolts holding the V-cover; remove the cover.

8. Disconnect the EGR temperature sensor connector clamp from the set of emission control valves.

9. Label and remove the two hoses from the fuel pressure control VSV. Disconnect the hoses from the IACV, disconnect the two VSV wiring connectors and remove the emission control valve set.

10. Label and disconnect the brake booster vacuum hose, PS air hose, PCV hose and IACV vacuum hose.

11. Disconnect the two ground straps.

12. Remove the wiring connector from the cold start injector. Disconnect the fuel line from the cold start injector.

13. Remove the No. 1 engine hanger and the air intake chamber support.

14. Remove the EGR pipe.

15. Remove the bolt and disconnect the hydraulic pressure pipe bracket from the air intake chamber.

16. Disconnect the three hoses at the air intake chamber, disconnect the two coolant bypass hoses and disconnect the EGR temperature sensor connector, (Calif. only).

17. Disconnect the throttle position sensor connector. Detach the connector for the IAC valve and remove the air hoses from the IAC valve. Remove the PS air hose.

18. Remove the bolts and nuts holding the air intake chamber, remove the air intake chamber and gasket.

19. Disconnect the fuel return hoses from the No. 1 fuel pipe; then disconnect the fuel inlet hose from the filter.

20. Disconnect the wiring connector from each injector.

21. Remove the No. 2 fuel pipe.

22. Remove the left delivery pipe or fuel rail; be careful not to drop the injectors during removal.

23. Remove the three injectors from the delivery pipe. Remove the rail spacers from the intake manifold.

24. Disconnect the two air hoses; remove the air pipe with the hoses attached.

25. Remove the right fuel rail and injectors. Take care not to drop the injectors during removal. Remove the injectors from the rail.

To install:

26. Install new grommets on each injector.

27. Apply a light coat of clean gasoline to new O-rings and install two on each injector.

28. Install each injector into the fuel rail while turning the injector left and right. Once installed, the injector should turn freely in the rail. If not, remove the injector and inspect the O-ring for damage or dislocation.

29. Place the rail spacers on the manifold. Clean the injector ports and install the right rail and injector assembly. Again check that the injectors turn freely in place.

30. Position the injector wiring connector upward. Install the bolts holding the delivery pipe and tighten them to 9 ft. lbs. (13 Nm).

31. Install the air pipe and hoses; tighten the retaining bolts only to 74 inch lbs. (8 Nm).

32. Place the two spacers on the intake manifold and install the left side delivery pipe and injectors. Torque the mounting bolts to 9 ft. lbs. (13 Nm).

33. Install the No. 2 fuel pipe connecting the two fuel rails. Use new gaskets at each union bolt. Tighten the union bolts to 22 ft. lbs. (29 Nm).

34. Connect the IACV vacuum hose.

35. Attach the wiring connectors to their proper injectors.

36. Install the inlet hose to the fuel filter using new gaskets; tighten the bolt to 22 ft. lbs. (29 Nm). Connect the return hose to the No. 1 fuel pipe.

37. Using a new gasket, install the air intake chamber. Tighten the mounting nuts to 32 ft. lbs. (43 Nm).

38. Connect the throttle position sensor harness, ISC valve wiring connector, ISC valve air hose and PS air hose.

39. Connect the EGR temperature sensor wiring (Calif. only).

40. Install the coolant bypass hose to the throttle body. Install the coolant bypass hose to the EGR cooler.

41. Connect the vacuum hose (from the EGR) to the BVSV (for EGR).

42. Connect the vacuum hose (from the EGR vacuum modulator) to the BVSV (for EGR).

43. Connect the vacuum hose (from throttle body) to BVSV (for EVAP).

44. Attach the hydraulic pressure pipe bracket to the air intake chamber.

45. Install the EGR pipe with a new gasket and new sleeve ball. Tighten the bolts to 13 ft. lbs. (18 Nm) and the union nut to 58 ft. lbs. (78 Nm).

46. Install the No. 1 engine hanger and the air intake chamber stay (support). Tighten the bolts to 29 ft. lbs. (39 Nm).

47. Connect the injector pipe with new gaskets to the cold start injector. Tighten the bolts to 11 ft. lbs. (15 Nm). Attach the cold start injector wiring connector.

48. Connect the two ground straps.

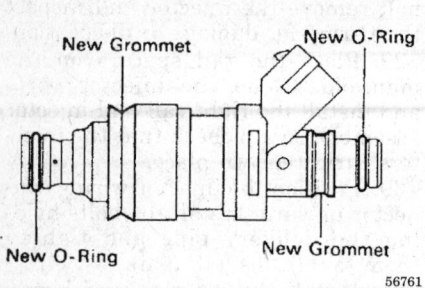

Injector O-ring installation — 1993 ES300

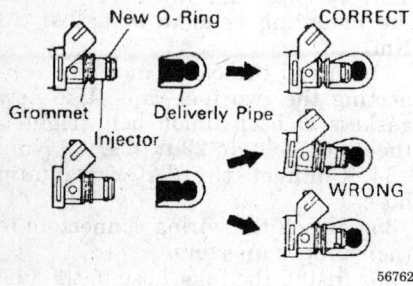

Correct injector installation to fuel rail — 1993 ES300

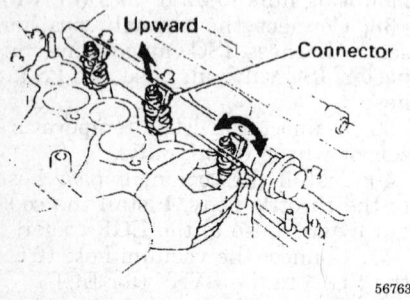

Injector connector positioning — 1993 ES300

49. Connect the brake booster vacuum hose, PS air hose, PCV hose and the IACV vacuum hose.

50. Install the emission valve set and tighten the bolts. Connect the VSV connectors and attach the two vacuum hoses to the IACV VSV. Install the two hoses to the fuel pressure control VSV. Connect the EGR temperature sensor connector clamp to the valve set.

51. Install the V-bank cover on the engine.

52. Install the air cleaner cover, air flow meter and air hose as a unit.

Make certain the clips are correctly engaged.

53. Connect and adjust the throttle control cable if it was removed.

54. Connect the accelerator cable and adjust it as needed.

55. Refill the engine coolant.

56. Connect the negative battery cable.

1994–97 Models

1. Disconnect the negative battery cable from the battery.

2. Relieve the fuel pressure from the fuel lines.

— **CAUTION** —

Fuel injection systems remain under pressure after the engine has been turned OFF. Properly relieve fuel pressure before disconnecting any fuel lines. Failure to do so may result in fire or personal injury.

3. Remove the air cleaner hose from the engine compartment.

4. Remove the V-bank cover from the engine.

5. Remove the air cleaner chamber assembly as follows:

a. Drain the engine coolant from the vehicle.

b. Disconnect the following connectors and cables:
- Accelerator cable
- A/T throttle cable
- TPS connector
- IAC valve connector
- EGR gas temperature sensor connector
- A/C idle up valve connector
- VSV connector for the ACIS
- VSV connector for the fuel pressure control
- Disconnect the VSV for the EVAP
- VSV connector for the EGR
- DLC1 from the bracket on the intake air control valve

c. Remove the power steering pressure tube from the No. 1 engine hanger by removing the two bolts.

d. Disconnect the following hoses, clamps and cables:
- Brake booster vacuum hose from the intake air control valve for the ACIS
- PCV hose from the PCV valve on the right hand cylinder head
- Ground strap and cable from the air intake air control valve from the ACIS
- Ground cable from the air intake chamber
- Vacuum hose clamp from fuel pipe
- Two bypass hoses from the throttle body
- Air assist hose from the throttle body
- Two power steering air hoses to the air intake chamber.
- Remove the EVAP hose from the pipe on emission control valve set
- Two vacuum hoses from the pipes on the cylinder head rear plate
- Vacuum sensing hose from the fuel pressure regulator
- Engine wire clamp from emission control valve set

e. Remove the two bolts and the No. 1 engine hanger.

f. Remove the two bolts and the air intake chamber stay.

g. Remove the No. 2 EGR pipe and two gaskets by removing the four nuts.

h. Disconnect the hose from the VSV fro the EVAP.

i. Using an 8mm hexagon wrench, remove the two bolts. Remove the two nuts, the air intake chamber assembly and the gasket.

6. Disconnect the fuel injector connectors.

7. Remove the air assist hoses and pipe.

8. Disconnect the fuel return hose from the No. 1 fuel pipe.

9. Disconnect the fuel inlet hose from the fuel filter. Catch any fuel leaking from the filter in a shop rag. Dispose of the rag properly.

10. Remove the delivery pipes and injectors from the engine as follows:

a. Loosen the two union bolts holding the No. 2 fuel pipe to the delivery pipes.

b. Disconnect the fuel return hose from the fuel pressure regulator.

c. Remove the union bolt and two gaskets for the right hand delivery pipe.

d. Remove the two bolts to the left hand delivery pipe and then remove the left hand delivery pipe, three injectors and the No. 2 fuel pipe as an assembly.

e. Remove the union bolt and two gaskets from the left hand delivery pipe. Disconnect the No. 2 fuel pipe from the left hand delivery pipe.

f. Remove the right hand delivery pipe by removing the three bolts. Remove the delivery pipe, injectors and the fuel inlet hose as an assembly.

g. Remove the four spacers from the intake manifold.

h. Pull out the six injectors from the delivery pipes.

i. Remove the two O-rings and two grommets from each injector.

To install:

11. Install the injectors as follows:

a. Install two new grommets to each injector.

b. Apply a light coat of spindle oil or gasoline to the two O-rings and install them to each injector.

c. Install the injector into the delivery pipe by turning the injector back and forth. Install all the injectors into the delivery pipes. Make sure to position the injector electrical connector outward.

d. Place the four spacers in position on the intake manifold.

e. Place the right hand delivery pipe and the No. 1 fuel pipe together with the three injectors in position on the intake manifold.

f. Temporarily install the two bolts holding the right hand delivery pipe to the intake manifold.

g. Temporarily install the bolt holding the No. 1 fuel pipe to the intake manifold.

h. Place the left hand delivery pipe and the No. 2 fuel pipe together with the three injectors in position.

i. Connect the fuel return hose to the fuel pressure regulator.

j. Temporarily install the two bolts holding the left hand delivery pipe to the intake manifold.

k. Temporarily install the No. 2 fuel pipe to the left hand delivery pipe with the union bolts and two new gaskets.

l. Check that the injectors rotate smoothly. If the injectors do not rotate smoothly, the probable cause is incorrect installation of the O-rings. Replace the O-rings.

m. Tighten the four bolts holding the delivery pipes to the intake manifold. Torque the bolts to 7 ft. lbs. (10 Nm).

n. Tighten the bolt holding the No. 1 fuel pipe to the intake manifold. Torque the bolt to 14 ft. lbs. (20 Nm).

o. Tighten the two union bolts holding the No. 2 fuel pipe to the delivery pipes. Torque the bolts to 24 ft. lbs. (33 Nm).

12. Connect the fuel inlet hose to the fuel filter by installing the union bolt. Use two new gaskets when installing the union bolt.

13. Connect the fuel return hose to the No. 1 fuel pipe. When routing the fuel return hose, pass the hose under the heater hoses.

14. Connect the air assist hoses to the intake manifold and then install the air assist pipe to the bracket on the No. 1 fuel pipe.

15. Connect the injector connectors.

16. Install the air intake chamber assembly as follows:

a. Using an 8mm hexagon wrench, install the air intake chamber with a new gasket. Install the two bolts and two nuts. Uniformly tighten the bolts and nuts in several passes and then torque the bolts and nuts to 32 ft. lbs. (43 Nm).

b. Connect the hose to the VSV for the EVAP system.

c. Install the two new gaskets and No. 2 EGR pipe with the four nuts. Torque the nuts to 9 ft. lbs. (12 Nm).

d. Install the No. 1 engine hanger with the two bolts. Torque the bolts to 19 ft. lbs. (39 Nm).

e. Install the air intake chamber stay with the two bolts. Torque the bolts to 14 ft. lbs. (20 Nm).

f. Connect the hoses, clamp and the cables as follows:

• Brake booster vacuum hose to the intake air control valve for the ACIS.

• PCV hose to the PCV valve on the right hand cylinder head.

• Ground strap and cable to the intake air control valve for the ACIS.

• Connect the ground cable and strap with the nut. Torque the nut to 10 ft. lbs. (14.5 Nm)

• Ground cable to air intake chamber

• Vacuum hose clamp to fuel pipe.

• Two water bypass hoses to the throttle body.

• Air assist hose to the throttle body.

• Two power steering air hoses to the air intake chamber.

• Connect the EVAP hose to the pipe on the emission control valve set

• Two vacuum hoses to the pipes on the cylinder head rear plate.

• Vacuum sensing hose to the fuel pressure regulator.

• Engine wire clamp to the emission control valve set.

17. Install the power steering pressure tube with the two nuts.

18. Connect the following connectors and cable:

• TPS sensor connector
• IAC valve connector
• EGR gas temperature sensor connector

• A/C idle up valve connector
• VSV connector for the ACIS
• VSV connector for the fuel pressure control
• For California vehicles, install the VSV connector for the EVAP
• VSV connector for the EGR
• DLC1 to the bracket on the intake air control valve
• Accelerator cable
• A/T throttle cable

19. Install the V-bank cover.

20. Install the air cleaner hose to the engine.

21. Fill the engine coolant.

22. Connect the negative battery cable to the battery and start the engine.

23. Check the engine coolant level and check for leaks.

GS300 and SC300

1. Disconnect the negative battery cable. Wait at least 90 seconds before performing any other work.

2. Drain the engine coolant.

3. Disconnect the control cables and intake air connector pipe.

4. Remove the EGR pipe by removing the union nut and two bolts.

5. Disconnect the EGR gas temperature sensor wiring connector.

6. Disconnect the throttle body bracket from the throttle body and cylinder head.

7. To remove the throttle body and intake air connector assembly disconnect these hoses and connectors:

a. The throttle position sensor connector.

b. The IAC valve connector.

c. The VSV connector for the EGR.

d. The PCV hose from the intake air connector.

e. The water bypass hose (from the No. 2 water bypass pipe) from the throttle body.

f. W/TRAC: the sub-throttle position sensor connector.

g. W/TRAC: the sub-throttle actuator connector.

8. Remove the nut holding the VSV for the EGR to the air intake chamber.

9. Remove the four bolts and two nuts holding the intake air connector to the air intake chamber.

10. Disconnect these hoses and remove the throttle body, intake air connector assembly and gasket:

a. The three vacuum hoses (from the No. 2 vacuum pipe) from the No. 1 vacuum pipe.

b. The water bypass hose (from the water outlet) from the throttle body.

c. The vacuum hose (from the actuator for ACIS) from the No. 1 vacuum pipe.

11. Relieve the fuel system pressure.

------ CAUTION ------

Fuel injection systems remain under pressure after the engine has been turned OFF. Properly relieve fuel pressure before disconnecting any fuel lines. Failure to do so may result in fire or personal injury.

12. Remove the two nuts and disconnect the No. 2 vacuum pipe and VSV assembly from the air intake chamber and intake manifold.

13. Remove the union bolt and two gaskets and then disconnect the No. 1 fuel pipe at the fuel delivery pipe.

14. Remove the union bolt and two gaskets and then disconnect the No. 2 fuel pipe at the fuel pressure regulator.

15. Remove the air intake chamber stays.

16. Tag and disconnect the vacuum sensing hose at the regulator and disconnect the injector connectors.

17. Remove the two bolts and turn the control valve actuator sideways to disconnect it. Wrap the actuator with tape and attach it to the intake chamber.

NOTE: Do not apply any force on the actuator rod. Do not allow the rod clip to become disconnected.

18. Remove the three bolts and then lift out the delivery pipe along with the injectors. Take care not to drop the injectors.

19. Pull the injectors out of the pipe.

20. Remove the O-ring and grommet from each injector.

21. Remove the six insulators and three spacers from the cylinder head.

To install:

22. Install the grommet to each injector.

23. Coat new O-rings with gasoline and slide them onto each injector.

24. While swiveling the injector back and forth, press it onto the delivery pipe so the connector is facing outward.

25. Position the insulators and spacers on the head and press the delivery pipe with the injectors installed into place. Hand-tighten the three mounting bolts and then check that the injectors rotate smoothly. Position the injector connector so it is now pointing upward and then

tighten the mounting bolts to 15 ft. lbs. (21 Nm).

NOTE: If the injectors do not rotate smoothly, remove them and install new O-rings.

26. Install the control valve actuator and tighten the two bolts to 61 inch lbs. (7 Nm).

27. Install the injector connectors. The No. 1, 3 and 5 connectors are gray, while the No. 2, 4 and 6 connectors are dark gray.

28. Connect the vacuum sensing hose. Install the air intake chamber stays and tighten to 13 ft. lbs. (18 Nm).

29. Connect the No. 2 fuel pipe to the regulator and tighten the union bolt to 20 ft. lbs. (27 Nm). Don't forget to use new gaskets.

30. Install the throttle body.

31. Connect the No. 1 fuel pipe to the delivery pipe and tighten the union bolt to 30 ft. lbs. (42 Nm). Don't forget to use new gaskets.

32. Install the throttle body bracket. Torque the nuts to 15 ft. lbs. (21 Nm).

33. Install the EGR pipe by tightening the union nut, gasket and two bolts. Torque the two bolts to 20 ft. lbs. (27 Nm) and the union nut to 47 ft. lbs. (64 Nm).

34. Connect the EGR temperature sensor.

35. Install the intake air connector pipe and connect the control cables.

36. Fill the engine with coolant.

37. Connect the battery cable.

38. Start the engine, bleed the cooling system and check for leaks.

LS400

1993–94 Models

1. Disconnect the negative battery cable. Wait at least 90 seconds before proceeding with any other work.

2. Drain the engine coolant.

3. Remove the battery clamp cover.

4. Remove the air cleaner cover. Remove the air cleaner hose or duct.

5. Remove the battery.

6. Remove the cover from the engine compartment relay box.

7. Remove the air cleaner.

8. Remove the throttle body cover.

9. At the throttle body, disconnect the accelerator cable, the transmission throttle control cable and the cruise control actuator cable.

10. Disconnect the air hose from the IAC valve and the hose from the PS air control valve at the intake air connector. Remove the clamp bolt holding the intake air connector to

the throttle body. Remove the intake air connector.

11. Remove the upper ignition wire cover. Remove the right engine wire harness cover.

12. Remove the left engine wiring cover.

13. Remove the right No. 3 (upper) timing belt cover.

14. Disconnect the EVAP hose from the clamp on the left timing belt No. 3 cover. Remove the 4 bolts. Disconnect the grommet from the timing belt cover and remove the cover. Remove the cord grommet.

15. Remove the lower ignition wire cover. It will be necessary to disconnect the high-tension wire from the right ignition coil.

16. Disconnect the radiator hoses from the water inlet housing and from the front water bypass joint.

17. Remove the throttle body.

18. Disconnect the following:
   a. IAC valve connector
   b. EGR valve connector
   c. VSV connectors for the fuel pressure control system and the EVAP system
   d. EGR gas temperature sensor connector
   e. VSV for the AIR system, if so equipped.

19. Disconnect the following:
   a. Water bypass hose from the IAC valve.
   b. PS air control hose from the air intake chamber.
   c. Fuel pressure regulator vacuum sensing hose from the vacuum pipe.
   d. Coolant bypass hose from the EGR valve.
   e. Brake booster vacuum hose from the air intake chamber.
   f. AIR system VSV hose from the AIR pump outlet pipe.
   g. PCV hose from the left cylinder head cover.

20. Remove the VSVs. Disconnect the DIAGNOSIS data link connector from the air intake chamber. Remove the air intake chamber.

21. Disconnect the fuel inlet pipe from the delivery pipe. Remove the pulsation damper.

22. Disconnect the fuel return hose from the return pipe.

23. Disconnect the wiring harness from the clips and brackets on the right cylinder head. Remove the wire brackets from the cylinder head. Disconnect the engine wiring harness from the intake manifold.

24. Disconnect the engine wiring harness from the delivery pipes. Disconnect the injector wiring connectors.

25. Disconnect the wire harness from the rear water bypass joint.

26. Remove the fuel return pipe. Remove the fuel delivery pipes and the remove the injectors. Take care not to drop or damage the injectors.

27. Remove the spacers and insulators. Pull out the injectors from the delivery pipes. Remove the O-ring and grommet from each injector.

To install:

28. Install a new grommet to the injector. Apply a light coat of gasoline to each new O-ring and carefully install them on each injector.

29. Install the injectors to the delivery pipes; push them into place while turning them to the left and right. Make sure the injectors rotate smoothly in the delivery pipe. Position the wiring connector outward.

30. Place the insulators and the spacers in the proper position on the intake manifold.

31. Install the injectors together with the delivery pipe in the proper position on the intake manifold.

NOTE: Make sure the injectors rotate smoothly. If not, the probable cause is the incorrect installation of the O-rings. Disassemble the injectors and replace the O-rings.

32. Install the fuel return pipe. Tighten the union bolt at the pressure regulator to 26 ft. lbs. (35 Nm).

33. Tighten the locknut on the fuel pressure regulator to 22 ft. lbs. (29 Nm).

34. Connect the engine wiring harness to the rear water bypass joint. On California vehicles, connect the harness connectors.

35. Connect the injector connectors to the injectors. Connect the wiring connectors to the left delivery pipe. Install the harness protector.

36. Install the engine wiring to the intake manifold.

37. Install the wire harness brackets onto the right cylinder head. Secure the harness to the brackets and install the protector.

38. Connect the fuel return hose to the return pipe; connect the fuel inlet hose to the left delivery pipe. Always use new gaskets. Using a torque wrench with a fulcrum length of 30mm (11.81 in), tighten the union bolt at the pulsation damper to 29 ft. lbs. (39 Nm).

39. Install the air intake chamber. Tighten the bolts holding the EGR pipe to the chamber and to the right cylinder head to 13 ft. lbs. (18 Nm). Install the accelerator bracket. Install new gaskets and connect the

union with the union bolt. Tighten the bolt to 22 ft. lbs. (29 Nm).

40. Connect the brake booster vacuum hose to the union on the intake chamber. Connect the vacuum hose from the AIR VSV to the intake chamber. Connect the other vacuum hose to the AIR outlet pipe.

41. Install the following:
  a. Water hose to the EGR
  b. Brake booster hose to the air intake chamber
  c. AIR VSV hose to the AIR pump outlet pipe, if so equipped.
  d. PCV hose to the left cylinder head cover.
  e. Water bypass hose to the IAC valve.
  f. PS air control valve to the air intake chamber.
  g. Fuel pressure regulator vacuum hose to the vacuum pipe.

42. Connect the following:
  a. IAC valve connector
  b. EGR valve connector
  c. VSV connectors for the fuel pressure control system and the EVAP system
  d. EGR gas temperature sensor connector
  e. VSV for the AIR system, if so equipped.

43. Before installing the throttle body, place a new gasket in position on the air intake chamber. Install the throttle body. Use the 40mm (1.57 in) bolts and tighten them to 13 ft. lbs. (18 Nm). Connect the water bypass pipe to the clamp on the wire cover. Connect the electrical connectors to the throttle body.

44. Connect the radiator hoses to the front water bypass joint and the water inlet housing.

45. Install the lower high-tension wire cover. Connect the coil lead to the right coil.

46. Install the right timing belt cover.

47. Install the 3 gaskets on the left timing belt cover. Install the cord grommet to the high tension cord. Install the grommet to the timing belt cover and install the cover with the 4 bolts. Connect the vacuum hose to the EVAP BVSV.

48. Install the upper ignition wire cover and the right engine wire harness cover.

49. Install the left engine wire cover. Make certain all lines and hoses are properly retained within the clips.

50. Install the intake air connector pipe. Connect the air connector pipe to the throttle body. Connect the air hoses to the IAC valve and to the PS air control valve.

51. Reconnect the accelerator, transmission throttle control and cruise control cables.

52. Reinstall the throttle body cover.

53. Install the air cleaner. Connect the VAF meter connector.

54. Install the relay box cover.

55. Reinstall the battery; do not connect the negative terminal.

56. Install the air hose and air cleaner cover.

57. Install the battery clamp cover. Connect the negative battery cable.

58. Refill the engine with coolant.

59. Start the engine and check for leaks of fuel or coolant. Recheck the coolant level.

1995–97 Models

1. Disconnect the negative battery cable. Wait at least 90 seconds before proceeding with any other work.

2. Drain the engine coolant.

3. Relieve the fuel pressure from the fuel lines.

4. Remove the V bank cover by removing the bolt and two cap nuts.

5. Remove the air cleaner and intake air connector assembly.

6. Remove the left hand No. 3 timing belt cover.

7. Remove the air intake chamber assembly.

8. Tag and disconnect the following connectors:
 • TPS connector
 • With TRAC system, sub throttle position sensor connector
 • With TRAC system, sub throttle actuator connector
 • IAC valve connector
 • EGR valve connector
 • VSV connector for fuel pressure control
 • VSV connector for EVAP system
 • EGR gas temperature sensor connector

9. Disconnect the following hoses:
 • Brake booster vacuum hose from the union on the air intake chamber
 • PCV hose from the PCV valve on the left hand cylinder head
 • Water bypass hose (from the EGR valve) from the rear water bypass joint
 • Water bypass hose (from the throttle body) from the rear water bypass joint
 • Vacuum hose (from the VSV for fuel pressure control) from the fuel pressure regulator
 • EVAP hose (from charcoal canister) from the VSV for EVAP

10. Disconnect the heater hose from the water bypass pipe.

11. Disconnect the fuel inlet hose from the delivery pipe.

12. Disconnect the vacuum sensing hose and fuel return hose from the pressure regulator.

13. Disconnect the engine wire from the delivery pipes.

14. Remove the four nuts holding the delivery pipe to the intake manifold.

15. Remove the two connector brackets, the two delivery pipes and the fuel injector assembly. Remove the four spacers and eight insulators.

16. Pull out the eight injectors from the delivery pipes.

17. Remove the O-ring and grommet from each injector.

To install:

18. Install a new grommet to each injector.

19. Apply a light coat of gasoline to a new O-ring and install it to each injector.

20. While turning the injector clockwise and counterclockwise, push it onto the delivery pipes. Install all eight injectors in the same manner.

21. Position the injector connector outward.

22. Place eight new insulators and the four spacers on the intake manifold.

23. Place the eight injectors and two delivery pipes assembly in position on the intake manifold.

24. Temporarily install the two connector brackets and four mounting bolts.

25. Check that the injectors rotate smoothly.

26. Position the injector connector outward.

27. Tighten the four nuts holding the delivery pipes to the intake manifold. Torque the nuts to 13 ft. lbs. (18 Nm).

28. Connect the engine wire to the delivery pipes.

29. Connect the fuel return hose and vacuum sensing hose to the pressure regulator.

30. Connect the fuel inlet hose to the left hand delivery pipe.

31. Install the air intake chamber assembly, VSV for fuel pressure control and the VSV for EVAP by installing the four bolts and eight nuts. Uniformly tighten the bolts and nuts in several passes. Torque the bolts and nuts to 13 ft. lbs. (18 Nm).

32. Install the No. 2 EGR pipe with the two bolts and two nuts. Torque the two bolts and two nuts to 17 ft. lbs. (23 Nm).

33. Connect the heater hose to the water bypass pipe.

34. Connect the following hoses:
- Brake booster vacuum hose to the union on the air intake chamber
- PCV hose to the PCV valve on the left hand cylinder head
- Water bypass hose (from EGR valve) to the rear water bypass joint
- Water bypass hose (from throttle body) to the rear water bypass joint
- Vacuum hose (from VSV for fuel pressure control) to the fuel pressure regulator
- EVAP hose (from charcoal canister) from the VSV for EVAP

35. Connect the following connectors:
- TPS connector
- With TRAC system, sub TPS connector
- With TRAC system, sub throttle actuator connector
- IAC valve connector
- EGR valve connector
- EGR gas temperature sensor connector
- VSV connector for fuel pressure control
- VSV connector for EVAP

36. Install the accelerator bracket with the two bolts.

37. Connect the accelerator cable, A/T throttle control cable and the cruise control actuator cable.

38. Install the left hand No. 3 timing belt cover.

39. Install the air cleaner and intake air connector assembly.

40. Install the V bank cover.

41. Fill the engine with coolant.

42. Connect the negative battery cable to the battery.

43. Start the engine and check for leaks.

44. Recheck engine coolant level.

SC400

1993–95 Models

--- **CAUTION** ---

Fuel injection systems remain under pressure, even after the engine has been turned OFF. The fuel pressure in the system must be released prior to disconnecting any fuel lines. Failure to do so may result in fire or personal injury.

1. Relieve the fuel system pressure.

2. Disconnect the negative battery cable. Wait at least 90 seconds before proceeding with any other work.

3. Drain the engine coolant.

4. Remove the throttle body cover.

5. At the throttle body, disconnect the accelerator cable, the transmission throttle control cable and the cruise control actuator cable.

6. Disconnect the air hose from the IAC valve and the hose from the Power Steering (P/S) air control valve at the intake air connector. Remove the clamp bolt holding the intake air connector to the throttle body.

7. Remove the intake air connector.

8. Remove the upper ignition wire cover by removing the two bolts.

9. Remove the right hand side engine wire cover by removing the bolt.

10. Remove the left hand side engine wire cover by removing the two bolts.

11. Remove the EVAP VSV from the head and valve cover.

12. Remove the right hand side No. 3 timing belt cover by removing the three bolts, timing belt cover and the timing belt cover gaskets.

13. Remove the left hand side No. 3 timing belt cover by removing the four bolts and disconnect the spark plug wire holding grommet.

14. Remove the lower ignition wire cover. It will be necessary to disconnect the spark plug wire from the right ignition coil.

15. Remove the throttle body assembly.

16. Disconnect the following:
   a. Cold start injector connector.
   b. IAC valve connector
   c. EGR valve connector
   d. VSV connectors for the fuel pressure control system and the EVAP system
   e. EGR gas temperature sensor connector

17. Disconnect the following:
   a. Water bypass hose from the IAC valve.
   b. P/S air control hose from the air intake chamber.
   c. Fuel pressure regulator vacuum sensing hose from the vacuum pipe.
   d. Coolant bypass hose from the EGR valve.
   e. Brake booster vacuum hose from the air intake chamber.
   f. EVAP system VSV vacuum hoses from the vacuum pipe.
   g. PCV hose from the left hand side cylinder head cover.
   h. Vacuum hose (from the VSV for the heater water valve) from the air intake chamber.
   i. Vacuum hose (from the charcoal canister) from the vacuum pipe.

18. Remove the intake chamber assembly.

19. Disconnect the fuel inlet pipe from the left hand side delivery pipe and remove the pulsation damper.

20. Disconnect the fuel return hose from the return pipe.

21. Disconnect the EGR pipe from the right hand side cylinder head.

22. Disconnect the engine wiring harness from the intake manifold.

23. Disconnect the engine wiring harness from the delivery pipes, rear water bypass joint and right hand cylinder head.

24. Remove the fuel return pipe and the rear fuel pipe.

25. Remove the fuel delivery pipes and injectors as follows:

   a. Remove the four nuts holding the delivery pipes to the intake manifold.

   b. Remove the connector bracket, the two delivery pipes and the eight injectors as an assembly.

   c. Pull out the eight insulators and four spacers from the cylinder head. Take care not to drop or damage the injectors.

   d. Remove the fuel injectors from the rail.

   e. Remove the O-ring and grommet from each injector.

To install:

26. Install a new grommet to the injector. Apply a light coat of gasoline to each new O-ring and carefully install them on each injector.

27. Install the injectors to the delivery pipes; push them into place while turning them to the left and right. Make sure the injectors rotate smoothly in the delivery pipe. Position the wiring connector outward.

28. Place the insulators and the spacers in the proper position on the intake manifold.

29. Install the injectors together with the delivery pipe in the proper position on the intake manifold.

NOTE: Make sure the injectors rotate smoothly. If not, the probable cause is the incorrect installation of the O-rings. Disassemble the injectors and replace the O-rings.

30. Torque the four nuts holding the delivery pipes to the intake manifold to 13 ft. lbs. (18 Nm).

31. Install the fuel return pipe. Tighten the union bolt at the pressure regulator to 26 ft. lbs. (35 Nm).

32. Tighten the locknut on the fuel pressure regulator to 22 ft. lbs. (29 Nm).

33. Connect the injector connectors to the injectors. Install the engine harness to the left hand side delivery pipe bracket. Make sure all wiring is properly retained within the clips.

34. Connect the engine wiring harness to the rear water bypass joint. Using the two bolts, attach the wire harness to the right hand side cylinder head.

35. Install the engine wiring harness to the delivery pipes.

36. Install the engine wiring to the intake manifold.

37. Loosely install the EGR pipe to the right hand side cylinder head.

38. Connect the fuel return hose to the return pipe; connect the fuel inlet hose to the left delivery pipe. Always use new gaskets. Using a torque wrench with a fulcrum length of 30mm (11.81 in), tighten the union bolt at the pulsation damper to 29 ft. lbs. (39 Nm).

39. Install the cold start injector with the two bolts. Torque the bolts to 69 inch lbs. (8 Nm).

40. Install the cold start injector tube with a new gasket and the union bolt. Torque the union bolt to 11 ft. lbs. (15 Nm).

41. Connect the cold start injector connector.

42. Install the air intake chamber, place four new gaskets on the manifold. Place the air intake chamber on the manifold and install a new gasket for the EGR pipe. Connect the cold start injector tube to the right hand side delivery pipe with two new gaskets and the union bolt. Torque the union bolt to 11 ft. lbs. (15 Nm).

43. Install the following:

   a. Water bypass hose to the EGR

   b. Brake booster hose to the air intake chamber

   c. VSV vacuum hose for the heater water valve to the air intake chamber.

   d. PCV hose to the left hand side cylinder head cover.

   e. Water bypass hose to the IAC valve.

   f. P/S air control valve hose to the air intake chamber.

   g. Fuel pressure regulator vacuum hose to the vacuum pipe.

   h. EVAP system VSV vacuum hoses to the vacuum pipe.

   i. Vacuum hose (from the charcoal canister) to the vacuum pipe.

44. Connect the following:

   a. IAC valve connector

   b. EGR valve connector

   c. VSV connectors for the fuel pressure control system and the EVAP system

   d. EGR gas temperature sensor connector

   e. Cold start injector connector.

45. Install a new gasket in position on the air intake chamber and install the throttle body using two 40mm (1.57 in) bolts. Torque the two bolts and two nuts on the throttle body to 13 ft. lbs. (18 Nm). Connect the water bypass and PCV valve hoses to the throttle body.

46. Install the water bypass pipe (from the rear water bypass joint) to the clamp on the engine wire cover. Connect the electrical connectors to the throttle body.

47. Connect the water hose to the heater water valve.

48. Install the lower spark plug wire cover. Clamp the spark plug wires to the wire cover and connect the coil lead to the right coil.

49. Install the right hand side No. 3 timing belt cover.

NOTE: Be sure to match the timing belt cover opening with the lower spark plug wire cover.

50. Install the EVAP VSV to the head and the valve cover.

51. Install the left hand side No. 3 timing belt cover.

52. Install the left and right engine wire harness cover.

53. Install the upper spark plug wire cover.

54. Install the intake air connector pipe.

55. Connect the accelerator, transmission throttle control and cruise control cables.

56. Install the throttle body cover.

57. Connect the negative battery terminal.

58. Fill the engine with coolant.

59. Start the engine and check for leaks of fuel or coolant. Recheck the coolant level.

60. Road test the vehicle for proper operation.

1996-97 Models

1. Disconnect the negative battery cable. Wait at least 90 seconds before proceeding with any other work.

2. Drain the engine coolant.

3. Relieve the fuel pressure from the fuel lines.

4. Remove the V bank cover by removing the bolt and two cap nuts.

5. Remove the air cleaner and intake air connector assembly.

6. Remove the left hand No. 3 timing belt cover.

7. Remove the air intake chamber assembly.

8. Tag and disconnect the following connectors:

• TPS connector

• With TRAC system, sub throttle position sensor connector

• With TRAC system, sub throttle actuator connector

• IAC valve connector

• EGR valve connector

26. Check that the injectors rotate smoothly.

27. Position the injector connector outward.

28. Tighten the four nuts holding the delivery pipes to the intake manifold. Torque the nuts to 13 ft. lbs. (18 Nm).

29. Connect the engine wire to the delivery pipes.

30. Connect the fuel return hose and vacuum sensing hose to the pressure regulator.

31. Connect the fuel inlet hose to the left hand delivery pipe.

32. Install the air intake chamber assembly. Uniformly tighten the bolts and nuts in several passes. Torque the bolts and nuts to 13 ft. lbs. (18 Nm).

33. Connect the following hoses:
- Brake booster vacuum hose to the union on the air intake chamber
- PCV hose to the PCV valve on the left hand cylinder head
- Water bypass hose (from EGR valve) to the rear water bypass joint
- Water bypass hose (from throttle body) to the rear water bypass joint
- Vacuum hose (from VSV for fuel pressure control) to the fuel pressure regulator
- EVAP hose (from charcoal canister) from the VSV for EVAP

34. Connect the following connectors:
- TPS connector
- With TRAC system, sub TPS connector
- With TRAC system, sub throttle actuator connector
- IAC valve connector
- EGR valve connector
- EGR gas temperature sensor connector
- VSV connector for fuel pressure control
- VSV connector for EVAP

35. Install the accelerator bracket with the two bolts.

36. Connect the accelerator cable, A/T throttle control cable and the cruise control actuator cable.

37. Install the left hand No. 3 timing belt cover.

38. Install the air cleaner and intake air connector assembly.

39. Install the V bank cover.

40. Fill the engine with coolant.

41. Connect the negative battery cable to the battery.

42. Start the engine and check for leaks.

43. Recheck engine coolant level.

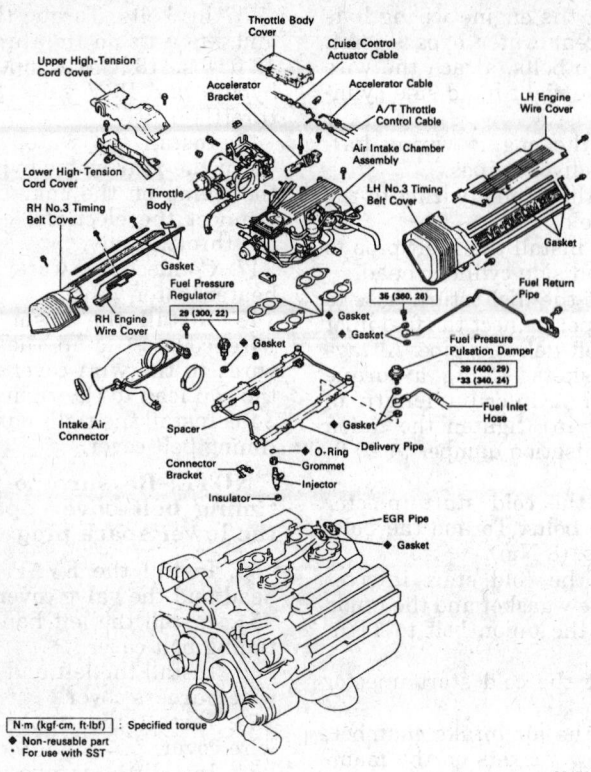

N·m (kgf·cm, ft-lbf) : Specified torque
♦ Non-reusable part
* For use with SST

234511

Fuel injector removal related components — 1995 SC400

- VSV connector for fuel pressure control
- VSV connector for EVAP system
- EGR gas temperature sensor connector

9. Disconnect the following hoses:
- Brake booster vacuum hose from the union on the air intake chamber
- PCV hose from the PCV valve on the left hand cylinder head
- Water bypass hose (from the EGR valve) from the rear water bypass joint
- Water bypass hose (from the throttle body) from the rear water bypass joint
- Vacuum hose (from the VSV for fuel pressure control) from the fuel pressure regulator
- EVAP hose (from charcoal canister) from the VSV for EVAP

10. Disconnect the heater hose from the water bypass pipe.

11. Remove the air intake chamber assembly.

12. Disconnect the fuel inlet hose from the delivery pipe.

13. Disconnect the vacuum sensing hose and fuel return hose from the pressure regulator.

14. Disconnect the engine wire from the delivery pipes.

15. Remove the four nuts holding the delivery pipe to the intake manifold.

16. Remove the two connector brackets, the two delivery pipes and the fuel injector assembly. Remove the four spacers and eight insulators.

17. Pull out the eight injectors from the delivery pipes.

18. Remove the O-ring and grommet from each injector.

To install:

19. Install a new grommet to each injector.

20. Apply a light coat of gasoline to a new O-ring and install it to each injector.

21. While turning the injector clockwise and counterclockwise, push it onto the delivery pipes. Install all eight injectors in the same manner.

22. Position the injector connector outward.

23. Place eight new insulators and the four spacers on the intake manifold.

24. Place the eight injectors and two delivery pipes assembly in position on the intake manifold.

25. Temporarily install the two connector brackets and four mounting bolts.

ENGINE MECHANICAL

Engine Assembly

REMOVAL AND INSTALLATION

ES300

1993 Models

— CAUTION —
The Supplemental Inflatable Restraint (SIR) system must be disarmed before removing the engine assembly. Failure to do so may cause accidental deployment of the air bag, resulting in unnecessary SIR system repairs and/or personal injury.

1. With the ignition switch in the **LOCK** position, disconnect the negative battery terminal. If equipped with an air bag system, wait at least 90 seconds or longer before performing any other work.
2. Remove the battery and its tray.
3. Matchmark the hood to the hinges and remove the hood.
4. Remove the engine undercover and then drain the engine coolant and oil.
5. Disconnect the accelerator cable from the throttle body. On models with automatic transaxle, disconnect the throttle cable also.
6. Remove the air cleaner assembly, air flow meter and the air intake duct.
7. On models with cruise control, remove the actuator cover, unplug the connector, remove the 3 bolts and then disconnect the actuator with the bracket.
8. Disconnect the ground strap at the battery carrier.
9. Remove the radiator and then disconnect the coolant reservoir hose.
10. Remove the 3 washer tank mounting bolts, disconnect the connector and hose and then lift out the tank.
11. Tag and disconnect:
    a. Three connectors to the engine relay box; remove the relay box
    b. Two connectors from the left side fender apron
    c. Igniter connector
    d. Ignition coil connector
    e. Coil wire from the ignition coil
    f. Noise filter connector
    g. Connector at the fender apron
    h. Check connector
    i. Ground strap at the right fender apron
    j. Backup light switch and speed sensor models with manual transaxle

— CAUTION —
Fuel injection systems remain under pressure even after the engine has been turned OFF. The fuel system pressure must be relieved before disconnecting any fuel lines. Failure to do so may result in fire and/or personal injury.

12. Disconnect the heater hoses, fuel return hose and the fuel inlet hose.
13. On models with manual transaxle, remove the starter and then remove the clutch release cylinder. Don't disconnect the hydraulic line, simply hang the cylinder out of the way.
14. Disconnect the transaxle control cables at the transaxle.
15. Disconnect the following:
    a. Brake booster vacuum hose from the intake chamber
    b. Charcoal canister vacuum hose
    c. IACV vacuum hose
16. Remove the undercover beneath the glove box. Remove the lower instrument panel, the glove box door and the box itself. Tag and disconnect the 3 ECU connectors, the 5 cowl wire connectors and the cooling fan ECU connector. Remove the 2 nuts and then pull the engine harness into the engine compartment.
17. Without disconnecting the refrigerant lines, remove the A/C compressor and hang it carefully out of the way.
18. Loosen the 2 bolts and disconnect the front exhaust pipe bracket. Remove the 3 nuts attaching the front pipe to the manifold. Disconnect the pipe.
19. Remove the halfshafts.
20. Without disconnecting the hydraulic lines, remove the power steering pump and hang it aside. Disconnect the hydraulic cooling fan pressure hose.
21. Remove the 3 bolts on manual transaxle or 4 bolts on automatic transaxle and then disconnect the left engine mounting insulator.
22. Pop out the plugs, remove the 4 nuts and then remove the rear engine mounting insulator.
23. Remove the 4 bolts and remove the mount absorber.

24. Remove the 3 bolts and disconnect the front engine mounting insulator.
25. Attach an engine lifting device to the lift hooks. Remove the 3 bolts and disconnect the control rod. Slowly and carefully, lift the engine/transaxle assembly out of the engine compartment.

To install:

26. Install the front and rear engine mounts on the engine. Torque the mounting bolts to 57 ft. lbs.(77 Nm).
27. Carefully lower the engine into the engine compartment. Align the right and left engine mounts with the body brackets.

— WARNING —
Do not damage the power steering gear housing or the Park/Neutral switch when installing the engine and transaxle assembly.

28. Install the No. 2 engine mounting bracket and the engine moving control rod. Tighten the 3 bolts in sequence to 47 ft. lbs. (64 Nm).
29. Install the right side mounting stays and tighten the small bolt to 23 ft. lbs. (31 Nm) and the larger bolts to 46 ft. lbs. (62 Nm).
30. Install the front engine mount to the body. Torque the bolts to 59 ft. lbs. (80 Nm).
31. Install the engine mount absorber and tighten the bolts to 35 ft. lbs. (48 Nm).
32. Install the rear mount and tighten the nuts to 48 ft. lbs. (66 Nm). Don't forget the plugs.
33. Install the left mount and tighten the bolts to 47 ft. lbs. (64 Nm).
34. Remove the hoist from the engine.
35. Install the power steering pump and tighten the bolts to 31 ft. lbs. (43 Nm).
36. Install the halfshafts. Connect the cooling fan pressure hose.
37. Install the front pipe to the manifold and tighten the new nuts to 46 ft. lbs. (62 Nm), tighten the converter nuts to 32 ft. lbs. (43 Nm). Don't forget to install the bracket.
38. Install the A/C compressor and tighten the cylinder block bolts to 20 ft. lbs. (27 Nm); tighten the alternator bracket bolts to 14 ft. lbs. (20 Nm). Install the belt and connect the electrical connector.
39. Feed the engine harness through the cowl and reconnect it. Install the glove box and the lower instrument panel cover.

40. Connect the vacuum hoses and the transaxle control cables.

41. Install the release cylinder and the starter.

42. Connect the fuel inlet hose to the filter and tighten it to 22 ft. lbs. (29 Nm). Connect the return hose and the 2 heater hoses.

43. Reconnect all wires disconnected previously.

44. Install the washer tank and connect the electrical lead and hose.

45. Install the coolant reservoir hose and the radiator.

46. Connect the ground strap to the battery carrier and then install the cruise control actuator. Install the air cleaner assembly and air flow meter.

47. Connect the throttle/accelerator cable to the throttle body and adjust it.

48. Fill the engine with oil and coolant. Connect the battery cable, start the engine and check for any leaks.

1994–97 Models

1. Release the fuel pressure.

2. With the ignition switch in the **LOCK** position, disconnect the negative battery terminal. If equipped with an air bag system, wait at least 90 seconds or longer before performing any other work.

3. Remove the battery and tray.

4. Matchmark the hood to the hinges and remove the hood.

5. Drain the engine coolant.

6. Drain the engine oil.

7. Disconnect the accelerator cable.

8. Disconnect the throttle cable.

9. Remove the air cleaner cover, volume air flow meter and air cleaner duct as an assembly.

10. If equipped, remove the cruise control actuator.

11. Remove the radiator.

12. Remove the two bolts and disconnect the engine relay box. Disconnect the following connectors:
   • 5 Connections from the relay box
   • 2 Igniter connectors
   • Left fender apron connector
   • Noise filter connector
   • 2 Ground straps

13. Disconnect the engine wire harness from the engine.

14. Disconnect the vacuum hoses from the following connections:
   • Intake air control valve vacuum tank
   • Charcoal canister
   • Brake booster vacuum hose from the intake chamber

15. Disconnect the two heater hoses from the bulkhead.

16. Disconnect the fuel feed and return lines.

17. Disconnect the control cable from the transaxle.

18. Disconnect the wire harness from the Engine Control Module (ECM) and route it through the bulkhead.

19. Remove the A/C compressor from the engine without disconnecting the lines and position it out of the way.

20. Remove the front exhaust pipe.

21. Remove the halfshafts.

22. Disconnect the two power steering air hoses from the engine.

23. Disconnect the hydraulic cooling fan pressure hose.

24. Remove the power steering pump without disconnecting the lines and position it out of the way.

25. Disconnect the right and left lower engine mounts from the body.

26. Remove the engine mounting shock absorber.

27. Remove the three front engine mounting bolts from the body.

28. Attach a lifting device to the engine.

29. Remove the coolant reservoir tank.

30. Remove the right engine mounting bracket.

31. Remove the engine moving control rod and right No. 2 engine mounting bracket.

32. Remove the engine and transaxle as an assembly.

─────── **WARNING** ───────
Be careful not to hit the power steering gear housing or Park/Neutral switch.

33. Remove the engine mounting insulator below the oil filter.

34. Remove the right rear engine mounting insulator.

35. Remove the front exhaust pipe stay.

36. Disconnect the following connectors:
   • O/D solenoid
   • PNP switch speedometer
   • Starter terminal
   • Speed sensor

37. Disconnect the two wire clamps from the transaxle.

38. Remove the oil dipstick and guide.

39. Remove the starter.

40. Remove the flywheel housing cover.

41. Turn the crankshaft pulley to gain access to the eight torque converter bolts. Secure the crankshaft and remove them as they become accessible.

42. Remove the two exhaust manifold stays and plate.

43. Remove the two bolts attaching the transaxle to the oil pan.

44. Remove the six transaxle mounting bolts.

45. Remove the transaxle.

To install:

46. Position the transaxle to the engine. Torque the transaxle mounting bolts to 47 ft. lbs. (64 Nm).

47. Install the bolts that attach the transaxle to the oil pan bolts and torque them to 34 ft. lbs. (46 Nm).

48. Install the exhaust manifold support. Torque the bolts to 14 ft. lbs. (20 Nm).

49. Install the bolts that attach the flywheel to the torque converter. Coat the threads with a locking compound. Rotate the engine and torque the bolts alternately to 30 ft. lbs. (41 Nm).

50. Install the starter to the engine.

51. Install the flywheel cover and torque the bolts to 13 ft. lbs. (18 Nm).

52. Install the dipstick and tube with a new O-ring.

53. Connect the clamps and following connectors:
   • O/D solenoid
   • PNP switch speedometer
   • Starter terminal
   • Speed sensor

54. Install the exhaust pipe stay and torque the bolts to 15 ft. lbs. (21 Nm).

55. Install the right rear insulator and torque the bolts to 47 ft. lbs. (64 Nm).

56. Install the front engine mounting insulator and torque the bolts to 47 ft. lbs. (64 Nm).

57. Lower the engine and transaxle into the engine compartment. Tilt the transaxle downward and clear the left mount.

58. Keep the engine level and align the right and left engine mounts.

59. Install the engine mounting bracket and moving control rod. Torque the bolts to 47 ft. lbs. (64 Nm).

60. Install the right engine stay and torque the bolts to 23 ft. lbs. (32 Nm).

61. Connect the ground straps.

62. Install the coolant reservoir.

63. Connect the front engine mounting insulator to the body and torque the bolts to 59 ft. lbs. (81 Nm).

64. Install the engine mounting shock absorber and torque the bolts to 35 ft. lbs. (48 Nm).

65. Connect the right engine mount and torque the bolts to 48 ft. lbs. (66 Nm).

66. Connect the left engine mount and torque the bolts to 47 ft. lbs. (64 Nm).

67. Remove the engine lifting device.

68. Install the power steering pump and the belt.

69. Connect the hydraulic cooling fan pressure hose and torque the fitting to 33 ft. lbs. (44 Nm).

70. Connect the power steering air tube and hoses.

71. Install the halfshafts.

72. Install the front exhaust pipe with new gaskets.

73. Install the A/C compressor and torque the bolts to 18 ft. lbs. (25 Nm).

74. Connect the harness to the ECM and assemble the instrument panel.

75. Connect the control cable to the transaxle.

76. Connect the fuel lines and torque the fittings to 22 ft. lbs. (30 Nm).

77. Connect the heater hoses.

78. Connect the vacuum hoses to the following connections:
- Intake air control valve vacuum tank
- Charcoal canister
- Air intake chamber from the brake booster

79. Connect the engine wire harness to the engine.

80. Connect the engine relay box. Install the two bolts and connect the following connectors:
- 5 Connections from the relay box
- 2 Ignitor connectors
- Left fender apron connector
- Noise filter connector
- 2 Ground straps

81. Install the radiator.

82. If equipped, install the cruise control actuator.

83. Install the air cleaner cover, volume airflow meter and air cleaner duct assembly.

84. Connect the throttle cable.

85. Connect the accelerator cable.

86. Fill the engine to the proper level with the recommended grade of oil.

87. Align the matchmarks and install the hood.

88. Fill the engine to the proper level with coolant.

89. Bleed the cooling system.

90. Install the battery and tray.

91. Check and/or adjust the ignition timing.

92. Start the engine and check for leaks.

93. Road test the vehicle.

94. Recheck the engine oil and coolant levels.

GS300

1. Release the fuel pressure.

CAUTION

The Supplemental Inflatable Restraint (SIR) system must be disarmed before removing the engine assembly. Failure to do so may cause accidental deployment of the air bag, resulting in unnecessary SIR system repairs and/or personal injury.

2. Disconnect the negative battery cable. Wait at least 90 seconds before performing any other work.

3. Remove the hood insulator pad and remove the hood.

4. Remove the engine undercover and then drain the engine coolant and oil.

5. Drain the fuel from the tank.

6. Disconnect the accelerator cable, cruise control actuator cable and the A/T throttle control cable from the throttle body.

7. Remove the air cleaner assembly, volume air flow meter and the air intake hose. Remove the air cleaner duct.

8. Remove the drive belt.

9. Remove the radiator.

10. Disconnect the following wires and electrical connectors:
    a. Igniter
    b. Ignition coil
    c. Wire harness from the wire clamp and coolant tank
    d. Alternator
    e. Ground strap from the left engine mount
    f. Starter

11. Disconnect the fuel lines from the intake and return lines.

12. Remove the power steering pump without disconnecting the lines and position it aside.

13. Remove the A/C compressor without disconnecting the A/C lines and position it aside.

14. Disconnect the brake booster vacuum hose.

15. Disconnect the EVAP hose.

16. Disconnect the heater hoses from the firewall.

17. Disconnect the heater valve and engine wire from the firewall.

18. Disconnect the electrical harness from the Engine Control Module (ECM) and route it through the firewall.

19. Disconnect and remove the sub heated oxygen sensor (if so equipped) from the front exhaust pipe.

20. Remove the front exhaust pipes and heat insulator.

21. Remove the rear center floor crossmember brace.

22. Disconnect the transmission control rod.

23. Remove the driveshaft.

24. Support the transmission with a jack. Use a piece of wood to prevent damage to the transmission oil pan.

25. Remove the rear transmission crossmember.

26. Attach a lifting device to the engine.

27. Remove the two hole plugs in the front crossmember. Remove the two nuts holding the engine insulators to the front crossmember.

28. Slowly and carefully, remove the engine and transmission from the engine compartment as an assembly. Take great care not to damage the A/C compressor or the cooling fan. Once removed, place the engine and transmission assembly onto a proper stand.

To install:

29. Lower the engine and transmission into the engine compartment.

30. Insert the stud bolts of the front engine mount into their bores in the front engine crossmember. Temporarily install the two nuts.

31. Remove the engine hoist.

32. Temporarily install rear engine support with the four nuts. Install the four support bolts and torque them to 19 ft. lbs. (25 Nm). Torque the nuts to 10 ft. lbs. (13 Nm).

33. Torque the front engine crossmember to mount nuts to 54 ft. lbs. (74 Nm) and install the hole plugs.

34. Install the driveshaft.

35. Shift the transmission control shift rod into **N** by shifting the lever all the way back and returning it two notches. Connect the shift rod to the lever and torque it to 9 ft. lbs. (13 Nm).

36. Install the rear center floor crossmember brace and torque the bolts to 9 ft. lbs. (13 Nm).

37. Install the exhaust pipe heat insulator.

38. Install the front exhaust pipes.

39. Install the sub heated oxygen sensor, if equipped.

40. Connect the engine wire harness to the ECM. Install the ECM and its cover. Reassemble the lower portion of the passenger side instrument panel, the vent, the carpet and the scuff panel.

41. Connect the heater water valve and engine wire to the cowl panel.

42. Connect the heater hoses.

43. Connect the EVAP hose.

44. Connect the brake booster hose.

45. Install the A/C compressor and torque the Torx® bolt to 19 ft. lbs. (26 Nm). Torque the nut and bolts to 38 ft. lbs. (52 Nm).

46. Install the power steering pump.

47. Connect the fuel lines with new gaskets and torque the union bolts to 22 ft. lbs. (29 Nm).
48. Connect the following wires and electrical connectors:
    a. Igniter
    b. Ignition coil
    c. Wire harness from the wire clamp and coolant tank
    d. Alternator
    e. Ground strap from the left engine mount
    f. Starter
49. Install the radiator.
50. Install the drive belt.
51. Install the air cleaner, volume air flow meter and intake air connector pipe as an assembly.
52. Install the air cleaner duct.
53. Connect the accelerator cable, cruise control cable and the A/T throttle control cable.
54. Fill the tank with fuel.
55. Fill the engine oil to the proper level.
56. Fill the engine coolant to the proper level
57. Connect the negative battery cable.
58. Start the engine and check for leaks. Bleed the cooling system.
59. Check the automatic transmission fluid level.
60. Check and/or adjust the ignition timing.
61. Install the hood and the hood insulator pad
62. Road test the vehicle.
63. Recheck the fluid levels.

LS400

1993–94 Models

1. Disconnect the negative battery cable and the positive battery cable. Remove the hood assembly.
2. Remove the dust covers and the air duct above the radiator assembly. Drain the cooling system.
3. Remove the battery from the vehicle. Raise and safely support the vehicle.
4. Remove the engine undercover and drain the engine oil. Lower the vehicle.
5. Disconnect the radiator upper hose from the water inlet. Loosen the nuts holding the fluid coupling to the fan bracket.
6. Loosen the drive belt tension by turning the belt tensioner counterclockwise. Remove the drive belt.
7. Remove the radiator assembly.
8. Disconnect the air flow meter connector, the mounting bolts and the air cleaner hose. Remove the air cleaner, the air flow meter and hose assembly.

9. Remove the igniter cover and disconnect the igniter connectors.
10. Remove the bolts, nut and the throttle body cover. Disconnect the accelerator and cruise control actuator cables.
11. Disconnect the air hose from the ISC valve and the power steering air control valve.
12. Disconnect the air connector pipe from the throttle body and remove the air connector pipe. Remove the bolt and connector pipe bracket.
13. Disconnect the air hose from the air intake chamber. Remove the power steering pump mounting bolts and nut. Position the pump aside.
14. Disconnect the coolant level sensor connector and remove the radiator reservoir tank. Remove the mounting bolts and reservoir tank bracket.
15. Disconnect the following hoses:
    a. Heater and bypass hoses
    b. Fuel hoses (plug the open end and catch the fuel in a suitable container)
    c. Vacuum hose from the brake booster on the air intake chamber
    d. Air conditioning control valve vacuum hoses
    e. EVAP and BVSV vacuum hoses
16. Remove the relay box cover. Disconnect the connector and ground cables from the engine compartment relay box. Remove the ground straps from under the fender aprons.
17. Remove the cruise control actuator cover.
18. Remove the instrument panel undercover and the lower the trim panel the ECU for the engine and transmission.
19. Disconnect the glove box door, the glove box light and remove the glove box assembly.
20. Disconnect the ABS ECU and the heater air duct.
21. Disconnect the following connectors:
    a. The 3 engine and Electronic Controlled Transmission (ECT) ECU connectors
    b. Circuit opening relay connector
    c. Cowl wire connector
    d. Instrument panel wire connector
22. Remove the mounting bolts and pull out the engine wire from the cowl panel.
23. Raise and safely support the vehicle.
24. Remove the mounting bolts and disconnect the power steering oil cooler pipe from the oil pan.

25. Remove the mounting bolts and the steering damper.
26. Disconnect the grommet from the floor and the sub-oxygen sensor from the exhaust pipe. Disconnect the 2 sub-oxygen sensors.
27. Remove the sub-oxygen sensor covers and the exhaust pipe. Remove the exhaust pipe support brackets.
28. Remove the catalytic converters and the exhaust pipe heat insulator.
29. Remove the center floor crossmember braces. Remove the driveshaft.
30. Disconnect the shift control rod from the shift lever. Lower the vehicle.
31. Attach a suitable engine hoist to the engine hangers and support the engine.
32. Remove the nuts holding the engine mounting insulators to the front suspension crossmember.
33. Remove the rear engine mounting member. Disconnect the ground strap.
34. Lift out the engine with the transmission attached. Place the engine assembly on a suitable holding fixture. Separate the engine from the transmission.

To install:
35. Connect the engine to the transmission. Attach a suitable engine hoist to the engine hangers.
36. Lower the engine assembly into the vehicle. Insert the stud bolts of the front engine mounting brackets into the stud bolt holes of the front suspension crossmember.
37. Install the rear engine mounting member and tighten the bolts to 19 ft. lbs. (26 Nm) and the nuts to 10 ft. lbs. (14 Nm). Install the ground strap.
38. Remove the engine hoist. Raise and safely support the vehicle.
39. Install the nuts holding the engine mounting brackets to the front suspension crossmember. Tighten the nuts to 43 ft. lbs. (59 Nm).
40. Connect the transmission control rod to the shift lever. Install the propeller shaft.
41. Install the center floor crossmember braces. Tighten the bolts to 9 ft. lbs. (11 Nm).
42. Install the exhaust pipe heat insulator. Replace the catalytic converters and tighten the bolts to 46 ft. lbs. (62 Nm).
43. Install the front exhaust pipe and the sub-oxygen sensor covers. Tighten the bolts to 32 ft. lbs. (43 Nm). Install the sub-oxygen sensors to the exhaust pipe and tighten to 33 ft. lbs. (44 Nm).

44. Install the steering damper and tighten the mounting bolts to 20 ft. lbs. (26 Nm). Connect the engine wire to the wire bracket on the front suspension crossmember.

45. Install the power steering oil cooler pipe to the engine oil pan. Lower the vehicle.

46. Push in the engine wire through the cowl panel and install the wire retainer.

47. Connect the following connectors:

   a. Three engine and ECT ECU connectors
   b. Circuit opening relay connector
   c. Cowl wire connector
   d. Instrument panel wire connector

48. Install the heater duct and the glove compartment.

49. Install the right side lower instrument panel pad and the engine and ECT electronic control units. Replace the right side instrument panel undercover.

50. Install the cruise control actuator. Connect the connectors and ground cables to the relay box.

51. Install the upper cover to the relay box. Connect the 2 ground straps to the underside of the fender aprons.

52. Connect the following hoses:

   a. The heater bypass water hoses
   b. Fuel hoses
   c. Vacuum hose to the brake booster on the air intake chamber
   d. Vacuum hose to the EVAP BVSV

53. Install the air conditioning compressor. Tighten the bolts to 36 ft. lbs. (49 Nm) and the nut to 22 ft. lbs. (29 Nm). Connect the electrical connectors.

54. Install the radiator reservoir tank bracket, the reservoir and connect the coolant level sensor connector.

55. Install the power steering pump and tighten the mounting bolts to 29 ft. lbs. (39 Nm) and the nuts to 32 ft. lbs. (43 Nm). Connect the air hose to the air intake manifold.

56. Connect the air connector pipe to the throttle body. Connect the air hose to the ISC valve and the power steering control valve.

57. Connect the accelerator cable and the cruise control actuator cable to the throttle body. Install the throttle body cover.

58. Connect the igniter connectors and install the igniter cover.

59. Connect the air cleaner hose to the intake air connector pipe.

60. Install the air cleaner, the air flow meter and hose assembly. Connect the air flow meter connector.

61. Install the radiator.

62. Temporarily install the fan pulley, the fan and the fluid coupling assembly. Install the drive belt by turning the belt tensioner counterclockwise.

63. Tighten the bolts holding the fluid coupling to the fan bracket to 16 ft. lbs. (21 Nm).

64. Install the battery. Replace the air ducts and dust covers.

65. Refill the cooling system and the crankcase to the proper levels.

66. Install the engine undercover and hood assembly.

67. Connect the battery cables. Start the engine and check for leaks. Check the timing.

68. Recheck the fluid levels.

1995–97 Models

1. Remove the battery clamp cover.

2. Relieve the fuel pressure from the fuel system.

3. Disconnect the battery cables, negative cable first.

4. Remove the battery from the engine compartment.

5. Remove the hood from the vehicle.

6. Raise and safely support the vehicle.

7. Remove the oil pan protector from the engine.

8. Drain the engine coolant from the engine.

9. Drain the engine oil from the engine.

10. Remove the air cleaner inlet.

11. Remove the air cleaner and intake air connector assembly.

12. Remove the drive belt, fan clutch and the fan pulley.

13. Disconnect the accelerator cable, cruise control actuator and automatic transmission throttle cable from the throttle body.

14. Remove the radiator assembly from the engine compartment.

15. Disconnect the following connectors, wires, straps, clamps and hoses:

• Engine oil level sensor connector
• Alternator connector and wire
• Engine wire clamp from the bracket on the alternator
• Two ignitor connectors
• Engine wire clamp from the igniter bracket
• Ground strap from the right hand engine mounting bracket
• Ground strap from under the left hand fender apron
• Engine wire clamp from the cowl panel
• Radiator reservoir hose from the water bypass pipe
• Brake booster vacuum hose from the air intake chamber
• Heater hose from the heater water valve and water bypass pipe
• Fuel inlet hose from the fuel inlet pipe
• Fuel return hose to the return pipe
• Power steering air hose from the air intake chamber
• Two power steering air hoses from the clamp on the right hand No. 3 timing belt cover
• For California vehicles, remove the two EVAP hoses from the pipes (from charcoal canister)
• For all vehicles except California, disconnect the EVAP hose from the pipe (from the charcoal canister)

16. Disconnect the engine wire from the cabin as follows:

   a. Remove the undercover from under the glove compartment.
   b. Remove the glove compartment.
   c. Disconnect the three ECM connectors.
   d. Disconnect the two cowl wire connectors from the connector on the bracket.
   e. Disconnect the wire clamp from the bracket.
   f. Disconnect the grommet from the cowl panel and pull the engine wire out.

17. Disconnect the power steering oil cooler pipe from the oil pan.

18. Disconnect the heated oxygen sensors from the front exhaust pipe.

19. Remove the front exhaust pipe.

20. Remove the two catalytic converters by removing the three nuts at each converter.

21. Remove the center exhaust pipe.

22. Remove the heat insulator from the rear side of the front exhaust pipe.

23. Remove the front center floor crossmember brace.

24. Remove the rear center floor crossmember brace.

25. Remove the driveshaft from the vehicle.

26. Remove the A/C compressor from the engine without disconnecting the A/C lines.

27. Disconnect the power steering pump from the engine by removing the nut and three bolts. Do not disconnect the power steering lines from the power steering pump. Set the power steering pump aside and support the pump.

28. Remove the heat insulators for the front side of the front exhaust pipe.

29. Remove the engine and transmission assembly as follows:

a. Disconnect the heater water valve from the cowl panel by removing the two nuts.

b. Attach the engine chain hoist to the engine hangers.

c. Remove the engine mounting insulators from the engine suspension crossmember by removing the two nuts.

d. Disconnect the transmission control rod from the shift lever by removing the nut.

e. Remove the rear engine mounting member by removing the four nuts and four bolts.

f. Lift the engine and transmission assembly out of the vehicle slowly and carefully.

30. Disconnect the engine from the transmission as follows:

a. Disconnect the following connectors

• Vehicle speed sensor connector

• Park/Neutral switch connector

• Solenoid connector

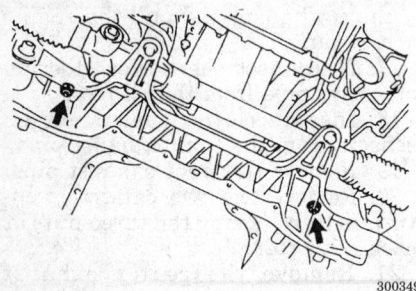

Engine mounting insulators — 1995–97 LS400

Engine mounting member bolts — 1995–97 LS400

• Direct clutch speed sensor connector

• Four engine wire clamps from the brackets

b. Remove the oil dipstick and guide from the transmission.

c. Remove the oil cooler pipes from the transmission and clamps.

d. Remove the flywheel housing undercover by removing the two bolts.

e. Turn the crankshaft pulley bolt to gain access to each torque converter bolt.

f. Hold the crankshaft pulley bolt with a wrench and remove the six torque converter bolts.

g. Remove the ten bolts holding the transmission to the engine.

h. Remove the transmission together with the torque converter clutch.

To install:

31. Install the transmission to the engine and install the 10 bolts. Torque the bolts as follows:

• 14mm head bolt to 27 ft. lbs. (37 Nm)

• 17mm head bolt to 53 ft. lbs. (72 Nm)

32. Install the torque converter clutch bolts as follows:

a. Apply adhesive to two or three threads of the bolt end.

b. Hold the crankshaft pulley bolt with a wrench and install the six bolts evenly. Torque the bolts to 30 ft. lbs. (41 Nm).

c. Install the flywheel housing undercover with the two bolts. Torque the bolts to 14 ft. lbs. (19 Nm).

33. Install the oil cooler pipe for the transmission.

34. Install the dipstick guide and dipstick for the transmission.

35. Connect the engine wire to the transmission and connect the following connectors:

• Vehicle speed sensor connector

• Park/Neutral switch connector

• Solenoid connector

• Direct clutch speed sensor connector

• Four wire clamps to the brackets

36. Install the engine and transmission assembly to the vehicle.

37. Install the rear engine mounting member to the vehicle and install the four bolts and four nuts. Torque the bolts to 19 ft. lbs. (25 Nm) and the nuts to 10 ft. lbs. (14 Nm).

38. Connect the transmission control rod to the shift lever with the nut. Torque the nut to 9 ft. lbs. (13 Nm).

39. Install the two nuts holding the engine mounting brackets to the front suspension crossmember. Torque the two nuts to 52 ft. lbs. (70 Nm).

40. Install the heater water valve to the cowl panel with the two nuts.

41. Remove the engine hoist.

42. Install the heat insulators for the front side of the front exhaust pipe.

43. Install the power steering pump with the nut and three bolts. Torque the nut to 32 ft. lbs. (43 Nm) and the bolts to 29 ft. lbs. (39 Nm).

44. Install the A/C compressor, torque the bolts to 36 ft. lbs. (49 Nm) and the nut to 22 ft. lbs. (29 Nm).

45. Install the driveshaft to the vehicle.

46. Install the front center floor crossmember brace and torque the bolts to 9 ft. lbs. (13 Nm).

47. Install the rear center floor crossmember brace and torque the bolts to 9 ft. lbs. (13 Nm).

48. Install the heat insulator for the rear side of the front exhaust pipe.

49. Install the center exhaust pipe.

50. Install the two front catalytic converters with three new nuts each. Torque the nuts to 46 ft. lbs. (62 Nm).

51. Install the front exhaust pipe.

52. Tighten the four bolts holding the pipe support bracket to the transmission. Torque the bolts to 32 ft. lbs. (44 Nm).

53. Install the heated oxygen sensors and torque the sensors to 33 ft. lbs. (44 Nm).

54. Install the power steering oil cooler pipe.

55. Connect the engine wire to the cabin as follows:

a. Push in the engine wire through the cowl panel. Install the grommet.

b. Connect the following connectors:

• Three ECM connectors

• Two engine wire connectors to the connector on the bracket

• Engine wire clamp to bracket

c. Install the glove compartment and the dash undercover to the vehicle.

56. Connect the following connectors, wires, straps, clamps and hoses:

• Engine oil level sensor connector

• Alternator connector

• Alternator wire

• Engine wire clamp to the bracket on the alternator

• Two igniter connectors

• Engine wire clamp to the igniter bracket

• Ground strap to the right hand engine mounting bracket

- Ground strap under the left hand fender apron
- Engine wire clamp to the bracket on the cowl panel
- Radiator reservoir hose to the water bypass pipe
- Brake booster vacuum hose to the air intake chamber
- Heater hose to the heater water valve
- Heater hose to the water by-pass pipe
- Fuel inlet hose to the fuel inlet pipe
- Fuel return hose to the return pipe
- Power steering air hose to the air intake chamber
- Two power steering air hoses to the clamp on the right hand No. 3 timing belt cover
- For California vehicles, connect the two EVAP hoses to the pipe (from the charcoal canister)
- Except for California vehicles, connect the EVAP hose to the pipe (from the charcoal canister)

57. Install the radiator assembly.
58. Connect the accelerator cable, cruise control cable to the throttle body. If equipped with A/T, connect the throttle control cable to the throttle body.
59. Install the fan pulley, fan, fan clutch and the drive belt. Torque the four nuts for the fan to 16 ft. lbs. (21 Nm).
60. Install the air cleaner and intake air connector assembly.
61. Install the air cleaner inlet.
62. Fill the engine coolant.
63. Install the battery to the vehicle.
64. Fill the engine with oil.
65. Connect the battery cables, positive cable first.
66. Install the battery cover.
67. Start the engine and check for leaks.
68. Install the engine undercover.
69. Install the oil pan protector.
70. Lower the vehicle.
71. Install the hood.
72. Recheck the all fluids and make all necessary engine adjustments.

SC300

1. Release the fuel system pressure.
2. Disconnect the negative battery cable. Wait at least 90 seconds before performing any other work.
3. Remove the battery and tray.
4. Remove the hood.
5. Remove the engine undercover and then drain the engine coolant and oil.

6. Drain the fuel from the fuel tank.

------ CAUTION ------

Do not allow fuel spray or fuel vapors to come in contact with a spark or open flame. Keep a dry chemical fire extinguisher nearby. Never store fuel in an open container due to risk of fire or explosion.

7. Disconnect the accelerator cable, the cruise control cable and the throttle control cable (A/T only) from the throttle body.
8. Remove the air cleaner assembly, resonator and the air intake hose.
9. Remove the drive belt, fan (with fluid coupling attached) and the water pump pulley.
10. Remove the radiator.
11. Disconnect the EVAP hoses(vacuum hose and air hose) from the charcoal canister and remove the charcoal canister.
12. Without disconnecting the hydraulic or refrigerant lines, remove the power steering pump and the A/C compressor and position them out of the way.
13. Tag and disconnect all wires, electrical leads and vacuum hoses from the block. Disconnect any wiring clips or brackets.
14. On models with manual transmission, remove the shift lever assembly.
15. Remove the undercover beneath the glove box. Remove the lower instrument panel, the trim panel and the glove box door.
16. Remove the right door sill trim (scuff plate). Lift the front edge of the carpet and remove the protective cover from the ECM.
17. Tag and disconnect the ECM connectors, the cowl wire connectors and the control unit connectors behind the glove box.

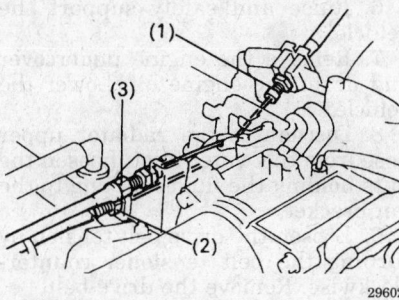

(1): Accelerator cable, (2): Throttle control cable, (3): Cruise control cable — 1996–97 SC300

18. Remove the two nuts holding the harness to the firewall and carefully pull the engine harness into the engine compartment.

NOTE: This is best done with an assistant guiding the harness inside the cabin.

19. Remove the two clamp bolts and disconnect the power steering pipe from the engine block.
20. Remove the union bolt and two gaskets and disconnect the fuel inlet hose. Contain any dripping fuel; clean up spills immediately.
21. Disconnect the starter wiring and disconnect the starter wiring from the clip.
22. For manual transmissions, unbolt and remove the clutch master cylinder; move it aside without disconnecting the hydraulic lines.
23. Disconnect the front exhaust pipe and then remove the heat shield.
24. For automatic transmissions, disconnect the transmission control rod at the shift lever.
25. Remove the intermediate shaft. Place matchmarks on the flanges for proper reassembly. Some vehicles are not equipped with adjusting washers.
26. Attach an engine lift to the lift hooks.
27. Remove the two nuts holding the engine to the front suspension crossmember. Remove the four bolts and four nuts holding the engine to the rear crossmember; lift out the rear engine mount.
28. Slowly and carefully, lift the engine/transaxle assembly out of the engine compartment. Take great care to avoid damaging the A/C compressor or power steering solenoid.
29. Place the engine and transmission assembly onto a stand.
30. For vehicle with A/T, remove the oil dipstick guide and dipstick for the transmission.
31. For vehicle with A/T, remove the oil cooler tubes.
32. Disconnect the engine wire from the transmission.
33. Remove the starter from the engine by removing the bolts..
34. Separate the engine and transmission.
35. For vehicles with manual transmission, remove the clutch cover and disc.

To install:

36. For vehicles with manual transmission, install the clutch cover and disc.
37. Connect the engine and transmission.
38. Install the starter to the engine by installing the bolts..

39. Connect the engine wire to the transmission.

40. For vehicle with A/T, install the oil cooler tubes.

41. For vehicle with A/T, install the oil dipstick guide and dipstick to the transmission.

42. Carefully lower the engine into the engine compartment. With the engine level and all the mounts aligned with their brackets, install the rear mount. Tighten the four nuts to 10 ft. lbs. (13 Nm). Tighten the four bolts to 19 ft. lbs. (25 Nm).

43. Connect the intermediate shaft to the rear differential and tighten the bolts and nuts to 54 ft. lbs. (74 Nm). Install the center support bearing set bolts with the adjusting washers. Tighten the bolts to 36 ft. lbs. (49 Nm).

44. Connect the transmission control rod to the shift lever (automatic transmission only) by installing the nut.

45. Install the exhaust heat insulator by installing the four nuts.

46. Attach the No. 2 front exhaust pipe, torque the nuts to 46 ft. lbs. (62 Nm). Install the pipe support bracket to the transmission with the two bolts. Tighten the bolts to 32 ft. lbs. (43 Nm). Install a new gasket and the No. 2 front exhaust pipe to the front exhaust pipe with the two bolts and nuts. Torque the bolts to 32 ft. lbs. (43 Nm).

47. For manual transmissions, install the clutch release cylinder and tighten the bolts to 9 ft. lbs. (12 Nm).

48. Connect the starter wiring and secure the harness in the clips.

49. Connect the fuel inlet hose with two new gaskets and tighten the union bolt to 22 ft. lbs. (29 Nm).

50. Install the power steering pipe below the engine.

51. Carefully push the engine harness back into the cabin. Secure the bolts holding the harness.

52. Connect each connector to the proper ECM, controller, or relay. Make certain each connector is square and secure.

53. Install the ECM and cover; connect the wiring harnesses. Refit the carpet and install the scuff plate.

54. Install the lower instrument panel trim, the glove box door and the undercover.

55. Install the shift lever on models with manual transmission. Install the upper and lower center console pieces.

56. Secure the engine wire harness in the clips and retainers. Reconnect all wires, electrical leads and vacuum hoses. Make certain each wiring lead is firmly connected and is properly routed; double check the security of clamps and wiring retainers.

57. Install the A/C compressor and tighten the through-bolt to 19 ft. lbs. (26 Nm). Tighten the other bolt and nut to 38 ft. lbs. (52 Nm).

58. Install the power steering pump. Tighten the long bottom bolt to 43 ft. lbs. (58 Nm); tighten the others to 29 ft. lbs. (39 Nm). Connect the power steering air hoses.

59. Install the charcoal canister and connect the hoses.

60. Install the radiator, coolant hoses and the transmission lines.

61. Install the water pump pulley, the fan and the drive belt. Tighten the four pulley nuts to 12 ft. lbs. (16 Nm).

62. Install the air cleaner assembly, resonator and the air intake hose.

63. Connect the accelerator cable, throttle control cable (A/T only) and the cruise control cable to the throttle body.

64. Install battery tray and battery. Connect the battery cables.

65. Refill all fluids, including fuel.

66. Start the engine and check for leaks. Allow the engine to warm to normal operating temperature.

67. Check the automatic transmission fluid level.

68. Check the ignition timing.

69. Shut the engine off and install the engine undercovers.

70. Install the hood.

71. Road test the vehicle for proper operation. Recheck all fluid levels.

SC400

1993–94 Models

1. Disconnect the negative battery cable and the positive battery cable.

2. Remove the hood assembly.

3. Remove the dust covers and the air duct above the radiator assembly.

4. Drain the cooling system.

5. Remove the battery from the vehicle.

6. Raise and safely support the vehicle.

7. Remove the engine undercover and drain the engine oil. Lower the vehicle.

8. Disconnect the radiator upper hose from the water inlet. Loosen the nuts holding the fluid coupling to the fan bracket.

9. Loosen the drive belt tension by turning the belt tensioner counterclockwise. Remove the drive belt.

10. Remove the radiator assembly.

11. Disconnect the air flow meter connector, the mounting bolts and the air cleaner hose.

12. Remove the air cleaner, the air flow meter and hose assembly.

13. Remove the igniter cover and disconnect the igniter connectors.

14. Remove the bolts, nut and the throttle body cover. Disconnect the accelerator and cruise control actuator cables.

15. Disconnect the air hose from the ISC valve and the power steering air control valve.

16. Disconnect the air connector pipe from the throttle body and remove the air connector pipe. Remove the bolt and connector pipe bracket.

17. Disconnect the air hose from the air intake chamber. Remove the power steering pump mounting bolts and nut. Position the pump aside.

18. Disconnect the coolant level sensor connector and remove the radiator reservoir tank. Remove the mounting bolts and reservoir tank bracket.

19. Disconnect the following hoses:
   a. Heater and bypass hoses
   b. Fuel hoses (plug the open end and catch the fuel in a suitable container)
   c. Vacuum hose from the brake booster on the air intake chamber
   d. Air conditioning control valve vacuum hoses
   e. EVAP and BVSV vacuum hoses

20. Remove the relay box cover. Disconnect the connector and ground cables from the engine compartment relay box. Remove the ground straps from under the fender aprons.

21. Remove the cruise control actuator cover.

22. Remove the instrument panel undercover and the lower the trim panel the ECU for the engine and transmission.

23. Disconnect the glove box door, the glove box light and remove the glove box assembly.

24. Disconnect the ABS ECU and the heater air duct.

25. Disconnect the following connectors:
   a. The 3 engine and Electronic Controlled Transmission (ECT) ECU connectors
   b. Circuit opening relay connector
   c. Cowl wire connector
   d. Instrument panel wire connector

26. Remove the mounting bolts and pull out the engine wire from the cowl panel.

27. Raise and safely support the vehicle.

28. Remove the mounting bolts and disconnect the power steering oil cooler pipe from the oil pan.

29. Remove the mounting bolts and the steering damper.

30. Disconnect the grommet from the floor and the sub-oxygen sensor from the exhaust pipe. Disconnect the 2 sub-oxygen sensors.

31. Remove the sub-oxygen sensor covers and the exhaust pipe. Remove the exhaust pipe support brackets.

32. Remove the catalytic converters and the exhaust pipe heat insulator.

33. Remove the center floor crossmember braces. Remove the driveshaft.

34. Disconnect the shift control rod from the shift lever. Lower the vehicle.

35. Attach a suitable engine hoist to the engine hangers and support the engine.

36. Remove the nuts holding the engine mounting insulators to the front suspension crossmember.

37. Remove the rear engine mounting member. Disconnect the ground strap.

38. Lift out the engine with the transmission attached. Place the engine assembly on a suitable holding fixture. Separate the engine from the transmission.

To install:

39. Connect the engine to the transmission. Attach a suitable engine hoist to the engine hangers.

40. Lower the engine assembly into the vehicle. Insert the stud bolts of the front engine mounting brackets into the stud bolt holes of the front suspension crossmember.

41. Install the rear engine mounting member and tighten the bolts to 19 ft. lbs. (26 Nm) and the nuts to 10 ft. lbs. (14 Nm). Install the ground strap.

42. Remove the engine hoist. Raise and safely support the vehicle.

43. Install the nuts holding the engine mounting brackets to the front suspension crossmember. Tighten the nuts to 43 ft. lbs. (59 Nm).

44. Connect the transmission control rod to the shift lever. Install the propeller shaft.

45. Install the center floor crossmember braces. Tighten the bolts to 9 ft. lbs. (11 Nm).

46. Install the exhaust pipe heat insulator. Replace the catalytic converters and tighten the bolts to 46 ft. lbs. (62 Nm).

47. Install the front exhaust pipe and the sub-oxygen sensor covers. Tighten the bolts to 32 ft. lbs. (43 Nm). Install the sub-oxygen sensors

to the exhaust pipe and tighten to 33 ft. lbs. (44 Nm).

48. Install the steering damper and tighten the mounting bolts to 20 ft. lbs. (26 Nm). Connect the engine wire to the wire bracket on the front suspension crossmember.

49. Install the power steering oil cooler pipe to the engine oil pan. Lower the vehicle.

50. Push in the engine wire through the cowl panel and install the wire retainer.

51. Connect the following connectors:
   a. Three engine and ECT ECU connectors
   b. Circuit opening relay connector
   c. Cowl wire connector
   d. Instrument panel wire connector

52. Install the heater duct and the glove compartment.

53. Install the right side lower instrument panel pad and the engine and ECT electronic control units. Replace the right side instrument panel undercover.

54. Install the cruise control actuator. Connect the connectors and ground cables to the relay box.

55. Install the upper cover to the relay box. Connect the 2 ground straps to the underside of the fender aprons.

56. Connect the following hoses:
   a. The heater bypass water hoses
   b. Fuel hoses
   c. Vacuum hose to the brake booster on the air intake chamber
   d. Vacuum hose to the EVAP BVSV

57. Install the air conditioning compressor. Tighten the bolts to 36 ft. lbs. (49 Nm) and the nut to 22 ft. lbs. (29 Nm). Connect the electrical connectors.

58. Install the radiator reservoir tank bracket, the reservoir and connect the coolant level sensor connector.

59. Install the power steering pump and tighten the mounting bolts to 29 ft. lbs. (39 Nm) and the nuts to 32 ft. lbs. (43 Nm). Connect the air hose to the air intake manifold.

60. Connect the air connector pipe to the throttle body. Connect the air hose to the ISC valve and the power steering control valve.

61. Connect the accelerator cable and the cruise control actuator cable to the throttle body. Install the throttle body cover.

62. Connect the igniter connectors and install the igniter cover.

63. Connect the air cleaner hose to the intake air connector pipe.

64. Install the air cleaner, the air flow meter and hose assembly. Connect the air flow meter connector.

65. Install the radiator.

66. Temporarily install the fan pulley, the fan and the fluid coupling assembly. Install the drive belt by turning the belt tensioner counterclockwise.

67. Tighten the bolts holding the fluid coupling to the fan bracket to 16 ft. lbs. (21 Nm).

68. Install the battery. Replace the air ducts and dust covers.

69. Refill the cooling system and the crankcase to the proper levels.

70. Install the engine undercover and hood assembly.

71. Connect the battery cables. Start the engine and check for leaks. Check the timing.

72. Recheck the fluid levels.

1995–97 Models

1. Disconnect the battery cables and remove the battery. Wait at least 90 seconds before proceeding with any other work.

────────── **CAUTION** ──────────

Fuel injection systems remain under pressure after the engine has been turned off. Properly relieve fuel pressure before disconnecting any fuel lines. Failure to do so may result in fire or personal injury.

─────────────────────────────

2. Relieve the fuel pressure from the fuel lines.

3. Remove the hood.

4. Drain the engine coolant from the cooling system.

5. Remove the V-bank cover, if equipped.

6. Raise and safely support the vehicle.

7. Remove the engine undercover and drain the engine oil. Lower the vehicle.

8. Loosen the drive belt tension by turning the belt tensioner counterclockwise. Remove the drive belt.

9. Disconnect the throttle body.

10. Remove the accelerator, transmission and cruise control cables from the throttle body.

11. Remove the air cleaner assembly.

12. Disconnect the vacuum hose (from the power steering air control valve) from the air intake chamber.

13. Remove the intake air connector.

14. Remove the coolant reservoir tank.

15. Remove the radiator.

16. Disconnect the igniter connectors and then remove the wire clamp from the body.

17. Disconnect the engine wires located next to the relay box. The relay box is located next to the left strut tower.

18. Disconnect the engine ground cable.

19. Disconnect the power steering solenoid valve connector.

20. Remove the alternator.

21. Disconnect the power steering tubes from the suspension crossmember.

22. Disconnect the power steering reservoir tank and bracket from the body by removing the three bolts.

23. Remove the power steering pump by removing the pump mounting bolts and nut. Do not disconnect the power steering lines and place the pump off to the side.

24. Disconnect the air conditioning compressor from the engine. Do not remove the compressor pressure lines.

25. Disconnect the following hoses and ground straps:
   • Heater water hose from the water bypass hose.
   • Heater water hose from the heater water valve.
   • Brake booster hose from the union on the air intake chamber.
   • Vacuum hose (from VSV for heater water valve) from the air intake chamber.
   • Ground strap from the bracket on the body.
   • Fuel inlet hose from fuel tube

26. Remove the charcoal canister from the engine.

27. Disconnect the engine wire from the cabin as follows:
   a. Remove the passenger side lower instrument panel undercover.
   b. Remove the four screws to the lower instrument panel finish panel and glove compartment door assembly. Remove the glove compartment and finish panel.
   c. Pull out the right scuff plate.
   d. Take out the front side of the floor carpet.
   e. Remove the two nuts and ECM protector.
   f. Remove the nut and disconnect the ECM from the floor panel.
   g. Disconnect the following connectors:
   • Two connectors from the ECM
   • Connector from the ABS and TRAC ECU
   • Two connectors from the TRAC ECU

   • Four connectors from connector cassette
   • Connector from A/C control assembly
   h. Remove the bolt holding the engine wire clamp to the heater water valve bracket.
   i. Remove the two bolts holding the engine wire clamp to the body.
   j. Pull the engine wiring (through the cowl panel) from the vehicle cabin.

28. Disconnect the oxygen sensors from the front exhaust pipe.

29. Remove the front exhaust pipe.

30. Remove the front catalytic converter by removing the three nuts and gasket.

31. Remove the tailpipes.

32. Remove the center exhaust pipe by disconnecting the two hooks.

33. Remove the heat insulator by removing the four nuts.

34. Remove the center floor crossmember brace by removing the four bolts.

35. Remove the driveshaft from the vehicle using the proper tools (two of tool SST 09922–10010), loosen the adjusting nut on the driveshaft. Place matchmarks on the transmission flange and the flexible coupling.

36. Disconnect the transmission control rod from the shift lever by removing the nut.

37. Attach the engine chain hoist to the engine hangers.

38. Remove the two nuts holding the engine mounting insulators to the front suspension crossmember.

39. Remove the four bolts, four nuts and the rear engine mounting member. Disconnect the ground strap to the rear mounting member.

40. Lift the engine out of the vehicle slowly and carefully.

41. Separate the engine from the transmission as follows:
   a. Remove the oil dipstick guide and dipstick for transmission.
   b. Remove the oil cooler pipes for the transmission.
   c. Disconnect all the engine wiring.
   d. Remove the engine bolts holding the transmission to the engine.
   e. Disconnect the engine from the transmission.

To install:

42. Attach the transmission to the engine as follows:
   a. Install the transmission to the engine and install the bolts. Torque the bolts to 42 ft. lbs. (57 Nm).
   b. Connect the engine wiring.
   c. Install the oil cooler pipe for the transmission. Torque the un-

ions on the pipes to 25 ft. lbs. (34 Nm).
   d. Install the engine oil dipstick guide and the dipstick for the transmission.

43. Install the engine and transmission to the vehicle.

44. Install the rear engine mounting member with the four bolts and four nuts. Torque the bolts to 19 ft. lbs. (25 Nm) and the nuts to 10 ft. lbs. (13 Nm).

45. Install the two nuts holding the engine mounting brackets to the front suspension crossmember. Torque the nuts to 43 ft. lbs. (59 Nm).

46. Remove the engine chain hoist.

47. Connect the transmission control rod to the shift lever by installing the nut.

48. Install the driveshaft.

49. Install the center floor crossmember brace by installing the four bolts. Torque the bolts to 9 ft. lbs. (13 Nm).

50. Install the heat insulator for the front exhaust pipe by installing the four bolts.

51. Install the center exhaust pipe by installing the two hooks.

52. Install the tailpipe and torque the two bolts to 14 ft. lbs. (19 Nm).

53. Install the front catalytic converter and torque the nuts to 46 ft. lbs. (62 Nm).

54. Install the front exhaust. Tighten the four bolts and nuts holding the catalytic converter to the front exhaust pipe to 32 ft. lbs. (43 Nm). Tighten the two bolts and nuts holding the front exhaust pipe to the center exhaust pipe. Torque the bolts to 32 ft. lbs. (43 Nm). Tighten the four bolts holding the pipe support bracket to the transmission. Torque the bolt to 32 ft. lbs. (43 Nm).

55. Install the oxygen sensors to the front exhaust and torque the sensors to 33 ft. lbs. (44 Nm).

56. Connect the engine wire to the cabin as follows:
   a. Push in the engine wire through the cowl panel.
   b. Install the engine wire retainer with the three bolts.
   c. Connect the following connectors under the dash panel;
   • Two connectors to the ECM
   • Connector to the ABS and TRAC ECM
   • Two connectors to the TRAC ECM
   • Four connectors to the connector cassette
   • Connector to the A/C control assembly
   d. Install the ECM with the nut.

e. Install the ECM protector with the two nuts.

f. Install the floor carpet.

g. Install the scuff plate.

h. Connect the connectors.

i. Install the lower instrument panel finish panel and glove compartment door assembly with the four screws.

j. Install the instrument panel undercover with the two screws.

57. Install the charcoal canister.

58. Connect the following hoses and grounds:

• Heater water hose to the water bypass hose

• Heater water hose to the water valve

• Brake booster hose to the union on the air intake chamber

• Vacuum hose (from VSV for heater water valve) to the air intake chamber

• Ground strap to the bracket on the body

• Fuel inlet hose to the fuel tube. Torque to 22 ft. lbs. (30 Nm)

• Fuel return hose to the return pipe

59. Install the air conditioning compressor with the nut and three bolts. Torque the bolts to 36 ft. lbs. (49 Nm) and the nut to 22 ft. lbs. (29 Nm).

60. Install the power steering pump with the nut and three bolts. Torque the bolts to 29 ft. lbs. (39 Nm) and the nut to 32 ft. lbs. (43 Nm).

61. Install the power steering reservoir tank and bracket with the three bolts.

62. Install the power steering tubes with the clamp and bolt.

63. Install the alternator, tighten the nut and bolt to 27 ft. lbs. (37 Nm).

64. Install the power steering solenoid valve connector.

65. Connect the engine wire connectors.

66. Connect the theft deterrent horn connector.

67. Install the ground cable to the body from the engine.

68. Connect the igniter connectors. Connect the yellow taped connector to the igniter on the rear side.

69. Install the radiator assembly. Install the reservoir tank and the inlet pipe to the fan shroud and torque the four bolts to 43 inch lbs. (5 Nm). Connect the two hydraulic lines for the fan motor and torque the bolts to 47 ft. lbs. (64 Nm).

70. Connect the upper and lower radiator hoses to the radiator. Install the two oil cooler hoses for the transmission to the radiator.

71. Install the coolant tank.

72. Install the intake air connector.

73. Connect the vacuum hose (from the power steering air control valve) to the air intake chamber.

74. Install the air cleaner.

75. Connect the accelerator cable, transmission throttle control cable and the cruise control actuator cable to the engine.

76. Install the throttle cover and hose clamp with the cap nut and two bolts. Also install the EVAP hose to the hose clamp.

77. Install the drive belt to the engine.

78. Install the battery to the engine compartment and connect the electrical connectors.

79. Fill the engine coolant.

80. Install the V-bank cover if it was removed.

81. Fill the engine oil and check the transmission oil.

82. Start the engine, bleed the cooling system and check for leaks.

83. Install the engine undercover. Install the hood.

84. Perform a road test and recheck all fluids.

Engine Mounts

REMOVAL AND INSTALLATION

1993 ES300

Front

1. Raise and safely support the vehicle.

2. Support the engine with a suitable jacking device.

3. Remove the three bolts on the frame and the three bolts on the engine block.

4. Remove the engine mount.

To install:

5. Install the engine mount to the block and the frame. Torque the bolts to 57 ft. lbs. (77 Nm).

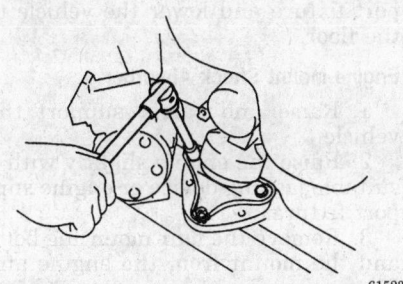

Front engine mount — 1993 ES300

61523

6. Remove the jack or the support fixture and lower the vehicle to the floor.

Rear

1. Raise and safely support the vehicle.

2. Support the engine with a suitable jacking device.

3. Remove the hole plugs and four nuts.

4. Remove the four bolts attaching the mount to the engine and remove the mount.

To install:

5. Position the engine mount on the engine and install the four mounting bolts. Torque the bolts to 57 ft. lbs. (77 Nm).

6. Install the four nuts attaching the mount to the engine. Torque the nuts to 48 ft. lbs. (66 Nm).

7. Remove the jack or the engine support fixture and lower the vehicle to the floor.

Left Engine Mount

1. Raise and safely support the vehicle.

2. Support the engine slightly with a suitable jacking device near the transmission or install an engine support fixture.

3. Remove the two nuts attaching the mount to the crossmember.

4. Remove the four bolts attaching the mount to the transaxle (three bolts for a manual transaxle), raise the engine enough to remove the mount and remove the engine mount.

To install:

5. Position the mount on the crossmember and install the two mounting nuts. Torque the nuts to 48 ft. lbs. (66 Nm).

6. Lower the engine and install the four mounting bolts (three for a manual transaxle). Torque the nuts to 47 ft. lbs. (64 Nm).

7. Remove the jack or the engine support fixture and lower the vehicle to the floor.

Engine Mounting Damper

1. Raise and safely support the vehicle.

2. Support the engine slightly with a suitable jacking device or install an engine support fixture.

3. Remove the four mounting bolts and the damper.

To install:

4. Position the damper on the engine and frame and install the mounting bolts. Torque the bolts to 35 ft. lbs. (48 Nm).

5. Remove the jack or the engine support fixture and lower the vehicle to the floor.

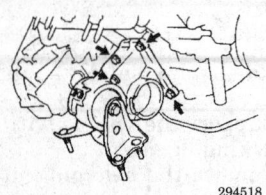

Rear engine mount — 1994–97 ES300

Left engine mount — 1994–97 ES300

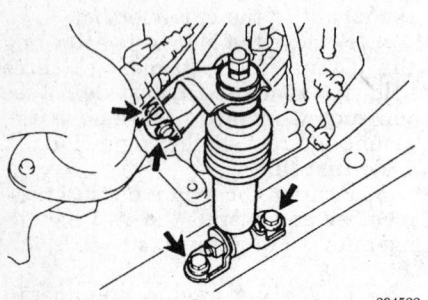

Engine mount shock absorber — 1994–97 ES300

1994–97 ES300

Front

1. Raise and safely support the vehicle.
2. Raise the engine slightly with a suitable jacking device or engine support fixture.
3. Remove the four bolts attaching the mount to the engine.
4. Remove the bolts attaching the mount to the crossmember.
5. Remove the engine mount.

To install:

6. Position the mount next to the engine and install the mounting bolts. Tighten the bolts to 47 ft. lbs. (64 Nm).
7. Install the bolts that attach the mount to the crossmember. Torque the bolts to 59 ft. lbs. (80 Nm).
8. Remove the jack or engine support fixture and lower the vehicle to the floor.

Rear

1. Raise and safely support the vehicle.
2. Raise the engine slightly with a suitable jacking device or engine support fixture.
3. Remove the four nuts attaching the mount to the frame.
4. Remove the four bolts attaching the mount to the engine.
5. Remove the engine mount.

To install:

6. Position the mount next to the engine and install the four mounting bolts. Torque the bolts to 47 ft. lbs. (64 Nm).
7. Install the four nuts attaching the mount to the frame. Torque the nuts to 48 ft. lbs. (65 Nm).
8. Remove the jack or engine support fixture and lower the vehicle to the floor.

Left Engine Mount

1. Raise and safely support the vehicle.
2. Raise the engine slightly with a suitable jacking device or engine support fixture.
3. Remove the two bolts that attach the mount to the frame.
4. Remove the four bolts that attach the mount to the transaxle.
5. Remove the engine mount.

To install:

6. Position the mount on the frame and install the four bolts that attach the mount to the transaxle. Torque the bolts to 47 ft. lbs. (64 Nm).
7. Lower the engine and install the two bolts that attach the mount to the frame. Torque the bolts to 47 ft. lbs. (64 Nm).
8. Remove the jack or engine support fixture and lower the vehicle to the floor.

Engine Mount Shock Absorber

1. Raise and safely support the vehicle.
2. Raise the engine slightly with a suitable jacking device or engine support fixture.
3. Remove the four mounting bolts and the mount from the engine and frame.

To install:

4. Position the mount on the engine and frame, install the mounting bolts. Torque the bolts to 35 ft. lbs. (47 Nm).
5. Remove the jack or engine support fixture and lower the vehicle to the floor.

GS300 and SC300

Left

1. Disconnect the negative battery cable to the battery.
2. Raise and safely support the vehicle.
3. Support the engine with a suitable jacking device.
4. Remove the 3 bolts and nut connecting the mount to the engine.
5. Remove the stud mount-to-crossmember nut.
6. Remove the mounting bracket.

To install

7. Install the mounting bracket.
8. Tighten the mount to engine bolts and nut to 43 ft. lbs. (58 Nm).
9. Remove the jacking device and lower the vehicle.
10. Connect the negative battery cable to the battery.

Right

1. Raise and safely support the vehicle.
2. Support the engine with a suitable jacking device.
3. Remove the 3 bolts and nut connecting the mount to the engine.
4. Remove the stud mount to crossmember nut.
5. Remove the mounting bracket.

To install

6. Install the mounting bracket.
7. Tighten the mount to engine bolts and nut to 43 ft. lbs. (58 Nm).
8. Remove the jacking device and lower the vehicle.
9. Connect the negative battery cable to the battery.

LS400 and SC400

Right

1. Raise and safely support the vehicle.
2. Support, but do not lift, the engine with a suitable hoist.
3. Remove the four engine bolts holding the engine mount to the engine.
4. Remove the two nuts holding the engine mount to the suspension crossmember.
5. Remove the mounting insulator from the vehicle.

To install:

6. Install the engine mount to the vehicle.
7. Install and tighten the mount to engine bolts to 27 ft. lbs. (37 Nm).

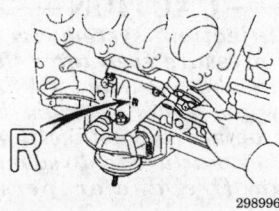

298996

Right engine mount has the letter R stamped for identification

8. Install and tighten the nuts holding the engine mount to the front crossmember to 43 ft. lbs. (59 Nm).

9. Remove the jacking device and lower the vehicle.

Left

1. Raise and safely support the vehicle.

2. Support, but do not lift, the engine with a suitable hoist.

3. Remove the four bolts holding the engine mount to the engine.

4. Remove the two nuts holding the engine mount to the crossmember.

5. Remove the mounting insulator.

To install:

6. Install the engine mount to the vehicle.

7. Install and tighten the mount to engine bolts to 27 ft. lbs. (37 Nm).

8. Install and tighten the nuts holding the engine mount to the front crossmember to 43 ft. lbs. (59 Nm).

9. Remove the jacking device and lower the vehicle.

Cylinder Head

REMOVAL AND INSTALLATION

ES300

1993 Models

1. Disconnect the negative battery cable. Wait at least 90 seconds from the time the ignition switch is turned to the LOCK position and the negative battery cable is disconnected before starting work.

2. Drain the cooling system.

3. If equipped with automatic transmission, disconnect the throttle cable and bracket from the throttle body.

4. Disconnect the accelerator cable and bracket from the throttle body and intake chamber.

5. Remove the air cleaner cover, air flow meter and air duct.

6. Remove the alternator.

7. Remove the oil pressure gauge, engine hangers and alternator upper bracket.

8. Loosen the lug nuts on the right wheel and raise and support the vehicle safely.

9. Remove the right tire and wheel assembly.

10. Remove the right undercover.

11. Remove the front exhaust pipe.

12. Remove the V-bank cover.

13. Remove the IAC valve.

14. Remove the throttle body.

15. Remove the EGR valve and vacuum modulator assembly.

16. Remove the distributor.

17. Remove the EGR pipe.

18. Disconnect the two vacuum hoses of the fuel pressure control VSV.

19. Disconnect the two vacuum hoses of the IAC VSV.

20. Disconnect the two VSV connectors.

21. Remove the two bolts and the emission control valve set.

22. Disconnect the knock sensor connector, cold start injector time switch connector, ECT sensor connector, oxygen sensor connector and the 3 fuel injector connectors.

23. Remove the left engine wire harness from the engine.

24. Remove the cylinder head rear plate.

25. Unbolt and remove the water outlet and gasket.

26. Disconnect the brake booster vacuum hose, PS air hose, PCV hose, EGR water bypass hose.

27. Disconnect the cold start injector connector.

28. Remove the No. 2 fuel pipe.

29. Remove the air intake chamber.

30. Remove the EGR cooler.

31. Disconnect and remove the right side wire harness.

32. Remove the left fuel delivery pipe and the injectors.

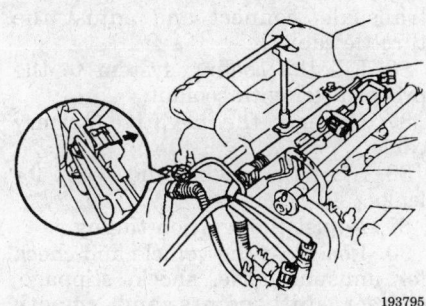

193795

Remove the engine wire harness and position it out of the way — 1993 ES300

33. Remove the air pipe.

34. Remove the right fuel delivery pipe and the injectors.

35. Remove the water bypass pipe with O-rings and gasket.

36. Remove the intake manifold.

37. Remove the exhaust manifolds.

38. Remove the spark plugs.

39. Remove the oil dipstick.

40. Remove the timing belt, all camshaft timing pulleys and the No. 2 idler pulley.

41. Remove the No. 3 timing belt cover. Support the belt carefully so the belt and pulley mesh does not shift.

42. Remove the cylinder head covers.

43. Remove the intake and exhaust camshafts from each cylinder head..

44. Remove the power steering pump bracket and the left side engine hanger.

45. Insure that the cylinder head is near ambient temperature and remove the two (one on each head) 8mm hex bolts. Loosen and remove the eight head bolts evenly, in the reverse order of tightening sequence, in three passes. Carefully lift the head from the engine and place it on wood blocks in a clean work area.

NOTE: If the cylinder head bolts are loosened out of sequence, warpage or cracking could result.

46. Remove the cylinder head gasket. With a gasket scraper, carefully remove all the old gasket material from the cylinder head and engine block surfaces.

To install:

47. Place the new cylinder head gasket onto the cylinder block. Place the cylinder head onto the gasket.

48. Coat the threads of the eight cylinder head bolts (12-sided) with clean engine oil and install the bolts into the cylinder head. Uniformly tighten the bolts in sequence in three passes to 25 ft. lbs. (34 Nm). If any of the bolts does not meet the torque, replace it.

49. Mark the forward edge of each bolt with paint and then retighten each bolt an additional 90 degrees, in sequence. Now repeat the process once more, for an additional 90 degrees. Check that each painted mark is now at a 180 degrees angle to the front, facing the rear.

50. Coat the threads of the two remaining 8mm bolts with engine oil and install them. Tighten to 13 ft. lbs. (18 Nm).

51. Install the left engine hanger and tighten it to 27 ft. lbs. (37 Nm).

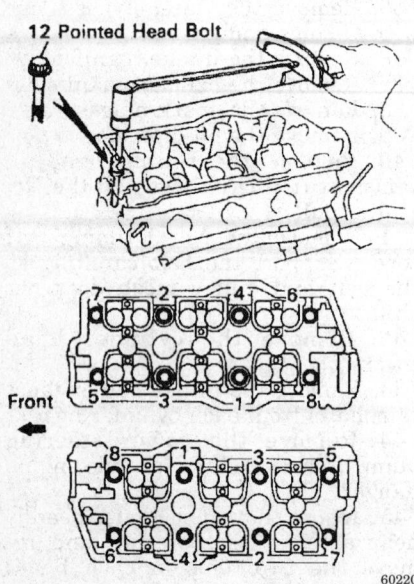

12 Pointed Head Bolt

Front ◀—

Cylinder head bolt tightening sequence — 1993 ES300

Install the power steering pump bracket.

52. Install the camshafts.

53. Install the cylinder head covers and tighten the bolts to 52 inch lbs. (6 Nm).

54. Install the No. 3 timing belt cover and tighten the six bolts to 65 inch lbs. (7 Nm). Install the No. 2 idler pulley, the camshaft timing pulleys and the timing belt.

55. Install the spark plugs.

56. Install the dipstick assembly.

57. Install the right and left side exhaust manifolds and tighten them to 29 ft. lbs. (39 Nm).

58. Install the intake manifold and the No. 2 idler pulley bracket. Tighten all bolts to 13 ft. lbs. (18 Nm).

59. Install the right fuel delivery pipe and fuel injectors. Tighten the bolts to 9 ft. lbs. (13 Nm).

60. Install the air pipe. Tighten the pipe to 73 inch lbs. (8.3 Nm) and the cooler to 13 ft. lbs. (18 Nm).

61. Install the left fuel delivery pipe and fuel injectors. Tighten the bolts to 9 ft. lbs. (13 Nm).

62. Install and connect the right engine wire harness.

63. Install the EGR cooler assembly.

64. Install the air intake chamber. Tighten the mounting bolts to 29 ft. lbs. (39 Nm).

65. Use new gaskets and install the No. 2 fuel pipe. Torque the union bolts to 25 ft. lbs. (34 Nm).

66. Connect the cold start injector connector.

67. Connect the brake booster vacuum hose, PS air hose, PCV hose and the EGR water bypass hose.

68. Install the water outlet and tighten the bolts to 74 inch lbs. (8.3 Nm).

69. Install the cylinder head rear plate.

70. Install and connect the left engine wire harness.

71. Install the emission control valve set.

72. Connect the two VSV connectors.

73. Connect the two vacuum hoses to the IAC VSV.

74. Connect the two vacuum hoses to the fuel pressure control VSV.

75. Install the distributor.

76. Install the EGR pipe and tighten the bolt to 13 ft. lbs. (18 Nm) and the union nut to 58 ft. lbs. (78 Nm).

77. Use a new gasket and install the EGR valve and vacuum modulator. Tighten the EGR bolts to 13 ft. lbs. (18 Nm).

78. Install the throttle body and the IAC valve. Tighten both sets of bolts to 9 ft. lbs. (13 Nm).

79. Install the V-bank cover.

80. Install the front exhaust pipe and tighten the manifold nuts to 46 ft. lbs. (62 Nm), tighten the torque converter nuts to 32 ft. lbs. (43 Nm).

81. Install the alternator and adjust the drive belt tension.

82. Install the air cleaner cover, air flow meter and air duct.

83. If equipped, install the cruise control actuator and bracket.

84. Install and adjust the accelerator cable.

85. If equipped with automatic transaxle, connect and adjust the throttle cable.

86. Fill the cooling system to the proper level with coolant.

87. Connect the negative battery cable.

88. Start the engine and check for leaks.

89. Adjust the ignition timing.

90. Road test the vehicle and check for unusual noise, shock, slippage, correct shift points and smooth operation.

91. Recheck the coolant and engine oil levels.

1994–97 Models

— CAUTION —

Fuel injection systems remain under pressure even after the engine has been turned OFF. The fuel system pressure must be relieved before disconnecting any fuel lines. Failure to do so may result in fire and/or personal injury.

1. Relieve the fuel pressure.

2. Disconnect the negative battery cable. If the vehicle is equipped with an air bag, wait at least 90 seconds from the time the ignition switch is turned to the "LOCK" position and the negative battery cable is disconnected before starting work.

3. Drain the cooling system.

4. Disconnect the accelerator cable and the throttle cable on vehicles equipped with an automatic transaxle.

5. Remove the air cleaner cover, air flow meter and the air duct.

6. Remove the cruise control actuator and bracket, if equipped.

7. Disconnect the two engine ground straps.

8. Remove the right engine mounting support.

9. Disconnect the radiator hoses.

10. Disconnect the two heater hoses.

11. Disconnect and plug the fuel feed and return lines from the fuel rail assembly.

12. Disconnect and plug the pressure hose from the hydraulic motor.

13. Remove the V-bank cover.

14. Disconnect the following vacuum hoses:
   a. Fuel pressure control VSV
   b. Fuel pressure regulator
   c. Cylinder head rear plate
   d. Intake air control valve VSV
   e. EGR vacuum modulator
   f. EGR valve

15. Disconnect the following connectors:
   a. Intake air control valve
   b. Fuel pressure regulator
   c. EGR VSV

16. Remove the two nuts and the emission control valve set.

17. Disconnect the following hoses;
   a. Brake booster vacuum hose
   b. PCV hose
   c. Intake air control valve vacuum hose

18. Remove the data link connector from the mounting bracket.

19. Remove the two ground straps from the intake chamber.

20. Remove the hydraulic motor pressure hose from the intake chamber.

21. Remove the right oxygen sensor connector from the P/S pressure tube.

22. Remove the two nuts and the P/S pressure tube from the intake chamber.

23. Disconnect the two P/S air hoses.

24. Remove the engine hanger and the intake chamber support.

25. Remove the EGR pipe and gaskets.

26. Disconnect the following connectors;

   a. Throttle position sensor connector

   b. IAC valve connector

   c. EGR gas temperature connector

   d. A/C idle up connector

27. Disconnect the following vacuum hoses:

   a. Two vacuum hoses from the TVV

   b. Vacuum hose from the cylinder head rear plate

   c. Vacuum hose from the charcoal canister

28. Disconnect the air assist hose and the two water bypass hoses.

29. Remove the air intake chamber.

30. Disconnect the left engine wire harness and position it out of the way.

31. Remove the wire harness from the rear of the engine.

32. Disconnect the right engine wire harness and position it out of the way.

33. Remove the ignition coils and the spark plugs.

34. Remove the timing belt.

35. Remove the camshaft pulleys and the timing belt rear cover.

36. Remove the cylinder head rear plate.

37. Remove the water inlet pipe.

38. Remove the air assist hose and vacuum hose.

39. Remove the intake manifold and fuel rail assembly.

40. Remove the water outlet.

41. Remove the EGR pipe from the right exhaust manifold.

42. Remove the exhaust manifolds.

43. Remove the dipstick assembly and the P/S pump bracket.

44. Remove the valve covers and the camshaft position sensor.

45. Remove the camshafts.

46. Make sure the engine is at or near ambient temperature and remove the two (one on each head) 8mm recessed hex bolts. Loosen and remove the 8 head bolts evenly, in 3 passes, in the reverse order of the tightening sequence. Carefully lift the head from the engine; if it is necessary to pry the head loose, take great care not to damage the mating surfaces. Place the head on wood blocks in a clean work area.

NOTE: If the cylinder head bolts are loosened out of sequence, warpage or cracking could result.

47. Remove the cylinder head gasket. With a gasket scraper, carefully remove all the old gasket material from the cylinder head and engine block surfaces.

To install:

48. Place the new cylinder head gasket onto the cylinder block. Place the cylinder head onto the gasket.

49. Coat the threads of the 8 cylinder head bolts (12-sided) with clean engine oil and install the bolts into the cylinder head. Uniformly tighten the bolts in sequence in three steps to an ultimate torque of 40 ft. lbs. (54 Nm). If any of the bolts does not meet the torque, replace it.

50. Mark the forward edge of each bolt with paint and then retighten each bolt, in proper sequence, an additional 90 degrees. Check that each painted mark is now at a 90 degrees angle to the front. The paint mark should have been applied to the bolt in the 9 o'clock position and should now be in the 12 o'clock position.

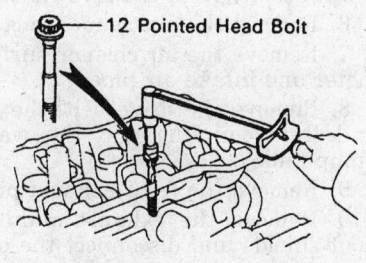

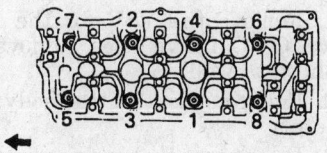

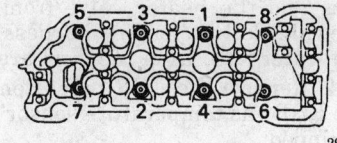

294418

Cylinder head bolt tightening sequence — 1994-97 ES300

51. Coat the threads of the two remaining 8mm bolts with engine oil and install them. Tighten to 13 ft. lbs. (18 Nm).

52. Install the camshafts and adjust the valves.

53. Apply sealant to the cylinder heads where the camshaft supports meet the cylinder heads.

54. Use new gaskets and install the cylinder head covers.

55. Install the dipstick and power steering pump bracket.

56. Install the exhaust manifolds. Torque the nuts to 36 ft. lbs. (49 Nm).

57. Install the EGR pipe to the right exhaust manifold.

58. Install the water outlet.

59. Install the intake manifold and the fuel rail assembly. Torque the intake manifold nuts and bolts to 11 ft. lbs. (15 Nm).

60. Install the air assist hose and the two water bypass hoses.

61. Install the water inlet pipe and the cylinder head rear plate.

62. Install the timing belt rear cover and the camshaft pulleys.

63. Install the timing belt.

64. Install the spark plugs and the ignition coils.

65. Install the right engine wire harness.

66. Install the wire harness to the rear of the engine.

67. Install the left engine wire harness.

68. Install the air intake chamber.

69. Use new gaskets and install the EGR pipe.

70. Connect the following vacuum hoses:

   a. The two TVV vacuum hoses.

   b. The vacuum hose to the rear cylinder head plate.

   c. Charcoal canister vacuum hose.

71. Connect the following electrical connectors:

   a. Throttle position sensor connector.

   b. IAC valve connector.

   c. EGR gas temperature connector.

   d. A/C idle up connector.

72. Install the engine hanger and the intake chamber support.

73. Connect the two P/S air hoses.

74. Install the P/S pressure tube to the intake chamber.

75. Install the oxygen sensor connector to the pressure tube.

76. Install the two ground straps to the intake chamber.

77. Install the data link connector to the bracket.

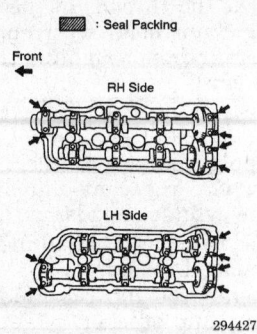

:::::: : Seal Packing

Front ←

RH Side

LH Side

294427

Apply sealant to the shaded areas on the cylinder head — 1994–97 ES300

78. Connect the following hoses:
   a. Power brake booster vacuum hose.
   b. PCV hose.
   c. IAC valve vacuum hose.

79. Install the emission control valve set and related vacuum hoses and connectors.

80. Install the V-bank cover.

81. Connect the pressure hose to the hydraulic motor.

82. Connect the fuel lines to the fuel rail assembly.

83. Connect the heater and radiator hoses.

84. Install the right engine mounting support.

85. Connect the two engine ground straps.

86. Install the cruise control actuator and bracket.

87. Install the air cleaner, air flow meter and air duct assembly.

88. Connect the accelerator cable and the throttle cable on vehicles equipped with an automatic transaxle.

89. Fill the cooling system to the proper level with coolant.

90. Connect the negative battery cable.

91. Start the engine and check for leaks. Bleed the air from the cooling system.

92. Adjust the ignition timing.

93. Road test the vehicle and check for unusual noise, shock, slippage, correct shift points and smooth operation.

94. Recheck the coolant and engine oil levels.

GS300 and SC300

1993–95 Models

1. Relieve the fuel system pressure.

2. Disconnect the negative battery cable. Wait at least 90 seconds before performing any other work.

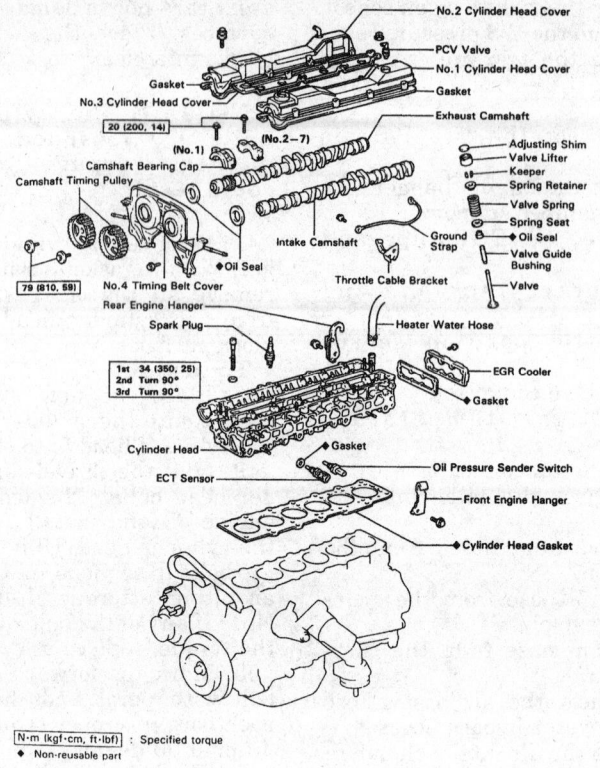

N·m (kgf·cm, ft·lbf) : Specified torque
♦ Non-reusable part

60291

Cylinder head and related components — 1993–94 GS300

3. Drain the engine coolant.

4. Remove the undercovers.

5. Disconnect the accelerator, throttle control (A/T only) and cruise control cables from the throttle body.

6. Remove the air cleaner duct.

7. Remove the air cleaner, airflow meter and intake air pipe.

8. Remove the drive belt, the fan and fluid coupling and the water pump pulley.

9. Remove the front exhaust pipe.

10. Remove the exhaust manifold heat shields and disconnect the oxygen sensor connector(s). Remove the exhaust manifolds.

11. Remove the VSV for the EGR valve. On Calif. vehicles, remove the air hose as well.

12. Remove the throttle body and bracket.

13. Disconnect the engine wire harness and the heater valve from the firewall. Pull the engine harness and the water hose from the cowl area.

14. Remove the EGR pipe. Remove the EGR gas temperature sensor if so equipped.

15. Remove the EGR valve and vacuum modulator.

16. Remove the PCV hose; disconnect the brake booster vacuum hose. Label and disconnect the hoses at the

No. 2 vacuum pipe. Remove the No. 2 vacuum pipe assembly.

17. Remove the intake air connector.

18. Remove the No. 3 (upper) timing belt cover.

19. Disconnect the spark plug cables and free the wires from the clips. Remove the spark plugs.

20. Remove the distributor with the spark plug wires attached.

21. Remove the water bypass outlet and the No. 1 bypass pipe.

22. Remove the No. 2 (middle) timing belt cover.

23. Remove the drive belt tensioner.

24. Set the engine to TDC/compression for cylinder No. 1. Turn the crankshaft pulley clockwise to align the pulley's groove with the timing mark **0** on the lower cover. Check that the timing marks of the camshaft pulleys are aligned with the marks on the rear belt cover. If the cam pulley marks do not align. rotate the crankshaft an additional full turn.

25. Alternately loosen the two bolts holding the timing belt tensioner; remove the tensioner and dust boot. Remove the timing belt from the camshaft pulleys. Support the belt so

that it remains in contact with the crankshaft pulley.

NOTE: Protect the belt from fluids and grease. Take great care not to drop any objects into the lower timing belt cover.

26. Disconnect the DLC (data link) connector from the bracket; remove the bracket. Disconnect the power steering air hose, the fuel pressure sensor vacuum sensing hose and the vacuum hose from the air intake chamber.
27. Remove the air intake chamber.
28. Disconnect the fuel return hose.
29. Remove the power steering pump rear support; remove the intake manifold support.
30. Remove the dipstick tubes for the engine and transmission oil.
31. Remove the vacuum control set under the intake manifold.
32. Disconnect the engine wiring harness from the bracket at the water pump. Free the harness from the intake manifold clamp and disconnect the 2 ground straps from the intake manifold.
33. Disconnect the wiring to the sensors, switches and injectors. These include the ECT sensors, knock sensors, oil pressure and oil level sensors. Disconnect the wiring to the A/C compressor.
34. Remove the water outlet and the bypass hose.
35. Remove the fuel pressure pulsation damper. Remove the fuel pipes from the injector rail.
36. Remove the delivery pipe and the injectors.
37. Remove the intake manifold.
38. Remove the cylinder head covers (valve covers).
39. Remove the camshaft timing pulleys. Remove the rear (No. 4) timing belt cover.
40. Remove the camshafts. Uniformly loosen the bearing cap bolts in several passes in the correct sequence. See camshaft removal and installation.
41. Uniformly loosen the cylinder head bolts in several passes and in the reverse order of tightening sequence. The cylinder head may crack or warp if the correct order is not followed.
42. Remove the head from the engine. If prying is necessary, use a protected blade; take great care not to damage the mating surfaces of the head and block.
43. Clean the head and block of all gasket material. Take great care not to gouge or scratch the mating surfaces.

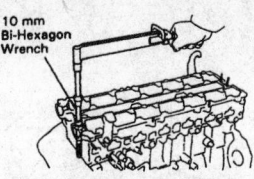

60293

Cylinder head bolt tightening sequence — 1993–97 GS300 and SC300

To install:

44. Place a new gasket on the cylinder block. Make sure it is positioned correctly. Install the cylinder head in position.
45. Lightly coat the head bolt threads and plate washers with engine oil. Install the bolts in the head.
46. Uniformly tighten the head bolts in several passes in sequence, to 25 ft. lbs. (34 Nm).
47. Mark the front of each bolt with a dot of paint. Following the correct order, tighten each bolt an additional 90 degrees. When complete, all the paint marks should face to the side of the engine.
48. Again following the correct order, tighten the bolts another 90 degrees of rotation. When complete, the paint marks should face the rear of the engine, exactly 180 degrees away from the original starting point. Correct bolt torque is expressed as 25 ft. lbs. (34 Nm) + 90° + 90°.

NOTE: Correct bolt torque must be achieved in 3 steps; do not attempt to shorten the procedure by combining the two 90 degree steps.

49. Coat the thrust portions of each camshaft with engine oil and then position them in the cylinder head with the cam lobes and the knock pins in the correct position.
50. Position the No. 3 and No. 7 bearing caps in place, coat the bolt threads with oil and then tighten them temporarily.
51. Coat new oil seals with multipurpose grease and then slide them over the camshafts.
52. Clean the mating surfaces of the two No. 1 bearing caps and then squeeze on some sealant. Install the bolts.
53. Install all remaining bearing caps, coat the threads of each bolt with clean oil and then tighten them,

in several passes, in the reverse order of the loosening sequence, to 14 ft. lbs. (20 Nm). Note that there are separate sequences for the intake and exhaust sides. See the camshaft removal and installation procedure.
54. Press the oil seal in as far as it will go.
55. Rotate each camshaft until the forward straight (knock) pin is straight up. Loosen exhaust Nos. 1, 2 and 6 bearing cap bolts until they can be turned by hand; re-tighten them to 14 ft. lbs. (20 Nm). Loosen intake Nos. 1 and 2 and re-tighten to 14 ft. lbs. (20 Nm). See the camshaft procedure.
56. Turn each camshaft ⅓ of a revolution (120 degrees). Loosen exhaust Nos. 4 and 7 bearing cap bolts; retighten them to 14 ft. lbs. (20 Nm). Loosen intake Nos. 4 and 6 bearing cap bolts; retighten them to 14 ft. lbs. (20 Nm). See the camshaft procedure.
57. Turn each camshaft an additional ⅓ of a revolution, loosen exhaust bearing cap bolts Nos. 3 and 5 and then retighten them to 14 ft. lbs. (20 Nm). Loosen intake bearing cap bolts Nos. 3 and 7 and then retighten them to 14 ft. lbs. (20 Nm). See the camshaft procedure.
58. Check and adjust the valve clearance.
59. Install the rear (No. 4) timing belt cover.
60. Install the camshaft pulleys. Align the shaft pin with the pulley groove and slide the pulley on. Install the bolt temporarily. Hold the hex portion of the camshaft with a wrench; tighten the pulley bolt to 59 ft. lbs. (79 Nm).
61. Install the cylinder head covers.
62. Install the intake manifold with a new gasket. Tighten the bolts to 15 ft. lbs. (21 Nm).
63. Install the injectors and the delivery pipe. Tighten the bolts holding the pipe to the manifold to 15 ft. lbs. (21 Nm).
64. Install the fuel pipes to the fuel rail. Tighten the No. 1 pipe union bolt to 30 ft. lbs. (42 Nm) and the No. 2 pipe union bolt to 20 ft. lbs. (27 Nm). Install the 2 clamp bolts.
65. Install the fuel pressure pulsation damper.
66. Install the water outlet and the bypass hose.
67. Install the engine wire harness. Secure the wiring in all clamps and retainers. Connect each wiring lead to the proper sender, sensor or switch. Connect the injector leads.
68. Install the vacuum control set.
69. Install the dipstick tubes. Always use a new O-ring on each tube.

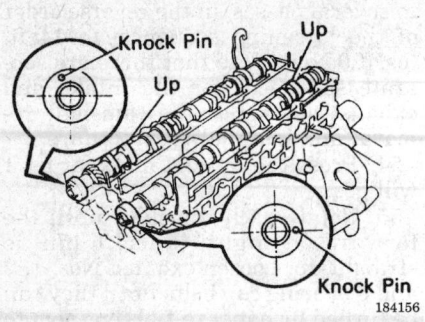

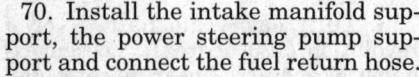

Camshaft installation positions — 1993–97 GS300 and SC300

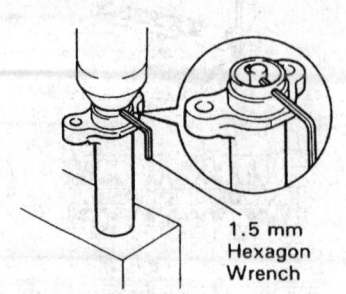

1.5 mm Hexagon Wrench

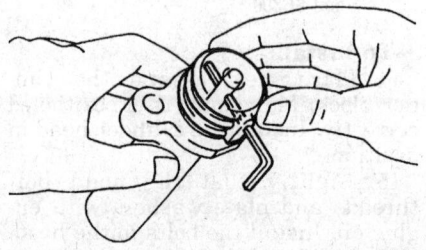

184152

Setting the timing belt tensioner — 1993–97 SC300 and GS300

70. Install the intake manifold support, the power steering pump support and connect the fuel return hose.

71. Install the air intake chamber with a new gasket. Tighten the bolts to 15 ft. lbs. (21 Nm).

72. Install the brake booster union. Install the intake chamber supports and tighten the bolts to 13 ft. lbs. (18 Nm). The supports are marked F and R for the front and rear positions.

73. Install the DLC connector and bracket. Connect the power steering air hose and the vacuum hoses to the fuel pressure regulator and air intake chamber.

74. Double check that the engine is still set to TDC/compression for cylinder No. 1. Check the alignment of both the crank and camshaft timing marks. Install the timing belt.

75. Compress the timing belt tensioner in a vise and retain the pin with a 1.5mm hex wrench. Install the dust boot onto the tensioner.

76. Install the tensioner. Alternately tighten the bolts to 20 ft. lbs. (26 Nm). Remove the hex wrench with a pair of pliers, allowing the tensioner to act on the timing belt.

77. Turn the crankshaft 2 full revolutions clockwise. Check that all timing marks align as before. If the marks (cam and crankshaft) do not align, remove the timing belt and reinstall it.

78. Install the accessory drive belt tensioner. Take great care not to drop the bolts inside the lower timing cover. Tighten the bolts to 15 ft. lbs. (21 Nm).

79. Install the No. 2 timing belt cover.

80. Install the water bypass outlet and the bypass pipe. Always use new O-rings.

81. Install the spark plugs.

82. Install the distributor, making sure all reference marks are aligned. Connect the wiring to the spark plugs.

83. Install the No. 3 timing belt cover.

84. Install the air intake connector. Note that the protrusion on the gasket must face the air intake chamber. Attach the vacuum hoses and tighten the bolts to 15 ft. lbs. (21 Nm).

85. Install the No. 2 vacuum pipe assembly and connect the hoses.

86. Connect the brake booster vacuum hose and install the PCV hose.

87. Install the EGR valve and vacuum modulator.

88. Install the EGR gas temperature sensor if it was removed. Tighten the bolts to 14 ft. lbs. (20 Nm). Connect the vacuum hoses.

89. Install the EGR pipe with new gaskets.

90. Install the engine wire harness and heater valve to the firewall.

91. Install the throttle body and bracket. Install the VSV for the EGR system. On Calif. vehicles, reinstall the air hose.

92. Install the exhaust manifolds with new gaskets. Tighten the bolts to 29 ft. lbs. (39 Nm). Connect the oxygen sensor leads.

93. Install the front exhaust pipe. Tighten the bolts to 46 ft. lbs. (62 Nm).

94. Install the support brackets for the exhaust pipe.

95. Install the water pump pulley, the fan and coupling and the drive belt.

96. Install the air cleaner, airflow meter and the intake air connector pipe.

97. Install the air cleaner duct.

98. Connect the control cables to the throttle body.

99. Refill the engine coolant

100. Connect the negative battery cable. Start the engine and check for leaks.

101. Check the ignition timing.

102. Install the engine undercovers. Road test the vehicle. Recheck the engine coolant level.

1996–97 Models

1. Disconnect the negative battery cable. Wait at least 90 seconds before performing any other work.

2. Drain the engine coolant.

3. Relieve the fuel pressure from the fuel lines.

───── **CAUTION** ─────
Fuel injection systems remain under pressure after the engine has been turned OFF. Properly relieve fuel pressure before disconnecting any fuel lines. Failure to do so may result in fire or personal injury.

4. Remove the undercovers.

5. Disconnect the accelerator, throttle control (A/T only) and cruise control cables from the throttle body.

6. Remove the air cleaner duct.

7. Remove the air cleaner, airflow meter and the intake air pipe.

8. Remove the drive belt, the fan and fluid coupling and the water pump pulley.

9. Remove the No. 2 front exhaust pipe.

10. If equipped with an exhaust manifold cover, remove the four nuts and remove the cover from the manifold.

11. Disconnect the 2 heated oxygen sensor connector(s). Remove the exhaust manifolds and gaskets by removing the eight bolts.

12. Remove the water bypass outlet and the No. 1 water bypass pipe.

13. Remove the power steering air hose from the No. 4 timing belt cover. Remove the power steering hose from the air intake chamber. Remove the two bolts and disconnect the vane pump from the pump bracket.

NOTE: Put aside the vane pump and suspend it. Remove the two bolts and the pump rear stay.

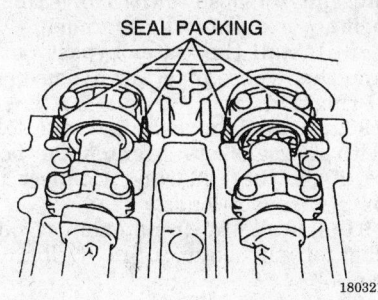

SEAL PACKING

180321

Apply sealant to the head — 1993–97 GS300 and SC300

14. Disconnect the fuel return hose from the fuel return pipe. Plug the hose end. Disconnect the fuel return hose from the oil dipstick guide.

15. Remove the bolt and bracket and disconnect the engine wire from the intake manifold stay.

16. Remove the throttle body and intake air connector assembly.

17. Remove the bolt, pull out the oil dipstick guide with the dipstick and remove the O-ring from the dipstick guide. If equipped with A/T, remove that transmission dipstick and guide.

18. Disconnect the connector from the No. 2 vacuum pipe. Disconnect the EGR gas temperature sensor connector from the wiring connector.

19. Remove the 2 nuts and disconnect the vacuum pipe from the air intake chamber and intake manifold. Remove the No. 2 vacuum pipe and VSV assembly.

20. Remove the nuts and disconnect the vacuum tank from the intake manifold. Disconnect the VSV connector and hoses, the vacuum hose(from the air intake chamber) from port B of the vacuum tank and the vacuum hose(from actuator) from the VSV. Remove the vacuum control valve set.

21. Remove the DLC1 bracket and VSV assembly.

22. Disconnect the vacuum hose from the brake booster union and the EVAP hose from the No. 2 vacuum pipe. Remove the bolt holding the engine wire protector to the air intake chamber. Remove the five bolts, nut, air intake chamber and gasket.

23. Remove the No. 3 (top) timing belt cover by removing the oil filler cap and the 6 bolts using a 5mm hexagon wrench.

24. Using a 5mm hexagon wrench, remove the four bolts and the rear cylinder head cover.

25. Disconnect the spark plug cables and free the wires from the clips. Remove the spark plugs.

26. Remove the distributor with the spark plug wires attached.

27. Remove the spark plugs.

28. Remove the drive belt tensioner by removing the three bolts.

29. Set the engine to TDC/compression for cylinder No. 1. Turn the crankshaft pulley clockwise to align the pulley's groove with the timing mark **0** on the lower cover. Check that the timing marks of the camshaft pulleys are aligned with the marks on the rear belt cover. If the cam pulley marks do not align. rotate the crankshaft an additional full turn.

30. Alternately loosen the two bolts holding the timing belt tensioner; remove the tensioner and dust boot. Remove the timing belt from the camshaft pulleys. Support the belt so that it remains in contact with the crankshaft pulley.

NOTE: Protect the belt from fluids and grease. Take great care not to drop any objects into the lower timing belt cover.

31. To disconnect the engine wire, remove and disconnect the following:
a. The wire clamp from the bracket.
b. The heated oxygen sensor connectors and the crankshaft position sensor connector.
c. Remove the two bolts and disconnect the two ground straps from the intake manifold.
d. Disconnect the ECT connector, the ECT sender gauge connector, the knock sensor connectors, the oil pressure switch connector, the oil level sensor connector, the AC compressor connector and the six injector connectors.
e. Remove the three nuts and disconnect the engine wire protector from the intake manifold.

32. Disconnect the water bypass hose from the clamp on the oil filter bracket. Remove the two nuts, bolt and water outlet with the water bypass hose.

33. Remove the two bolts and the intake manifold stay.

34. Remove the fuel pressure pulsation damper.

35. Remove the clamp bolt from the intake manifold, remove the union bolt and gaskets and disconnect the fuel inlet pipe.

36. Remove the six bolts, two nuts, the intake manifold and the delivery pipe assembly and gasket.

37. Remove the cylinder head covers (valve covers).

38. Remove the camshaft timing pulleys.

39. Remove the rear (No. 4) timing belt cover.

40. Remove the camshafts.

41. Uniformly loosen the cylinder head bolts in several passes and in the reverse order of the tightening sequence. The cylinder head may crack or warp if the correct order is not followed.

42. Remove the head from the engine. If prying is necessary, use a protected blade; take great care not to damage the mating surfaces of the head and block.

43. Clean the head and block of all gasket material. Take great care not to gouge or scratch the mating surfaces.

To install:

44. Place a new gasket on the cylinder block. Make sure it is positioned correctly. Install the cylinder head in position.

45. Lightly coat the head bolt threads and plate washers with engine oil. Install plate washers and bolts to the head.

46. Uniformly tighten the head bolts in several passes in sequence to 25 ft. lbs. (34 Nm).

47. Mark the front (towards the front of the engine) of each bolt with a dot of paint. Following the correct order and tighten each bolt an additional 90 degrees. When complete, all the paint marks should face to the side of the engine.

48. Again following the correct order, tighten the bolts another 90 degrees of rotation. When complete, the paint marks should face the rear of the engine, exactly 180 degrees away from the original starting point. Correct bolt torque is expressed as 25 ft. lbs. (34 Nm) + 90° + 90°.

NOTE: Correct bolt torque must be achieved in 3 steps; do not attempt to shorten the procedure by combining the two 90 degree steps.

49. Coat the thrust portions of each camshaft with engine oil and then position them in the cylinder head with the cam lobes and the knock pins in the correct position.

50. Position the No. 3 and No. 7 bearing caps in place, coat the bolt threads with oil and then uniformly and alternately tighten them temporarily.

51. Coat new oil seals with multi-purpose grease and then slide them over the camshafts.

52. Clean the surfaces of the No. 1 bearing cap and cylinder head with cleaner. Apply seal packing to the No. 1 bearing cap.

53. Install the remaining bearing caps in their proper locations.

54. Coat the threads of each bolt with clean oil and then tighten them, in several passes, in the correct sequence, to 14 ft. lbs. (20 Nm).

55. Using SST tool 09316-60010 or equivalent, press the two oil seals in as far as it will go.

56. Rotate each camshaft until the forward straight (knock) pin is straight up. Loosen the exhaust Nos. 1, 2 and 6 bearing cap bolts until they can be turned by hand; retighten the bolts, in several passes, to 14 ft. lbs. (20 Nm). Loosen the intake Nos. 1, 2 and 5 bearing cap bolts and retighten the bolts, in several passes, to 14 ft. lbs. (20 Nm).

57. Turn each camshaft ⅓ of a revolution (120 degrees). Loosen the exhaust Nos. 4 and 7 bearing cap bolts; retighten the bolts, in several passes, to 14 ft. lbs. (20 Nm). Loosen the intake Nos. 4 and 6 bearing cap bolts; retighten the bolts, in several passes, to 14 ft. lbs. (20 Nm).

58. Turn each camshaft an additional ⅓ of a revolution, loosen the exhaust bearing cap bolts Nos. 3 and 5 and then retighten the bolts, in several passes, to 14 ft. lbs. (20 Nm). Loosen the intake bearing cap bolts Nos. 3 and 7 and then retighten the bolts, in several passes, to 14 ft. lbs. (20 Nm).

59. Check and adjust the valve clearance.

60. Install the rear (No. 4) timing belt cover. Torque the bolts to 78 inch lbs. (9 Nm).

61. Install the camshaft timing pulleys. Align the shaft pin with the pulley groove and slide the pulley on. Install the bolt temporarily. Hold the hex portion of the camshaft with a wrench and tighten the pulley bolt to 59 ft. lbs. (79 Nm).

62. Install the cylinder head covers.

63. Install the intake manifold and delivery pipe with a new gasket. Tighten the 6 bolts and 2 nuts to 20 ft. lbs. (27 Nm).

64. Install the fuel inlet pipe to the fuel rail. Tighten the union bolt to 30 ft. lbs. (42 Nm). Install the clamp bolt to the intake manifold.

65. Install the fuel pressure pulsation damper.

66. Install the intake manifold stay and torque the bolts to 29 ft. lbs. (39 Nm)

67. Install the water outlet and the bypass hose. Torque the bolts to 15 ft. lbs. (21 Nm).

68. Install the engine wire harness. Secure the wiring in all clamps and retainers. Connect each wiring lead

to the proper sender, sensor or switch. Connect the injector leads.

69. Compress the timing belt tensioner in a vise and retain the pin with a 1.5mm hex wrench. Install the dust boot onto the tensioner.

70. Install the tensioner. Alternately tighten the bolts to 20 ft. lbs. (26 Nm). Remove the hex wrench with a pair of pliers, allowing the tensioner to be applied to the timing belt.

71. Turn the crankshaft two full revolutions clockwise. Check that all timing marks align as before. If the marks (cam and crankshaft) do not align, remove the timing belt and re-install it.

72. Install the accessory drive belt tensioner. Take great care not to drop the bolts inside the lower timing cover. Tighten the bolts to 15 ft. lbs. (21 Nm).

73. Double check that the engine is still set to TDC/compression for cylinder No. 1. Check the alignment of both the crank and camshaft timing marks. Install the timing belt.

74. Install the spark plugs.

75. Install the distributor, making sure all reference marks are aligned. Connect the wiring to the spark plugs.

76. Install the No. 3 timing belt cover.

77. Install the cylinder head rear cover.

78. Install the air intake chamber with a new gasket. Tighten the bolts to 20 ft. lbs. (27 Nm). Install the bolt to hold the engine wire protector to the air intake chamber. Connect the vacuum hose to the brake booster union and the EVAP hose to the No. 2 vacuum pipe.

79. Install the DLC connector and bracket and VSV connector.

80. Install the vacuum control set.

81. Install the No. 2 vacuum pipe assembly and connect the hoses. Torque the nuts to 20 ft. lbs. (27 Nm).

82. Install the EGR gas temperature sensor. Torque it to 14 ft. lbs. (20 Nm). Connect the vacuum hoses.

83. Install the dipstick tubes. Always use a new O-ring on each tube.

84. Install the intake chamber supports and tighten the bolts to 13 ft. lbs. (18 Nm). The supports are marked F and R for the front and rear positions.

85. Install the throttle body and intake air connector assembly.

86. Install the engine wire bracket.

87. Connect the fuel return hose.

88. Connect the vane pump to the pump bracket. Install the power

steering air hose to the No. 4 timing belt cover and intake chamber.

89. Install the water bypass outlet and the bypass pipe. Always use new O-rings.

90. Install the exhaust manifolds with new gaskets. Tighten the bolts to 29 ft. lbs. (39 Nm). Connect the oxygen sensor leads.

91. Install the front exhaust pipe. Tighten the bolts to 46 ft. lbs. (62 Nm).

92. If equipped with an exhaust manifold cover, install the cover and four nuts.

93. Install the water pump pulley, the fan and coupling and the drive belt. Torque the four nuts to 12 ft. lbs. (16 Nm).

94. Install the air cleaner, airflow meter and the intake air connector pipe.

95. Install the air cleaner duct.

96. Connect the control and accelerator cables to the throttle body.

97. Refill and bleed the engine coolant system.

98. Connect the negative battery cable. Start the engine and check for leaks.

99. Check the ignition timing.

100. Install the engine undercovers. Road test the vehicle. Recheck the engine coolant level.

LS400

1993-94 Models

1. Disconnect the negative battery cable. Wait at least 90 seconds before proceeding with any other work.

2. Drain the engine coolant.

3. Disconnect the timing belt from the camshaft timing pulleys.

4. Remove the camshaft timing pulleys. Do not drop any parts into the timing belt cover. Protect the belt from grease, fluids, etc.

5. Remove the fan bracket.

6. Disconnect the air hose from the air intake chamber. Remove the power steering pump pulley. Remove the power steering pump from the engine.

7. Disconnect the front exhaust pipe from the front catalytic converters.

8. Remove the front catalytic converters.

9. Remove the throttle body cover.

10. Disconnect the accelerator cable, the automatic transmission throttle cable and the cruise control actuator cable.

11. Remove the right ignition coil.

12. Remove the inlet housing from the water pump. Disconnect the coolant bypass hose from the IAC valve. Remove the coolant inlet and housing

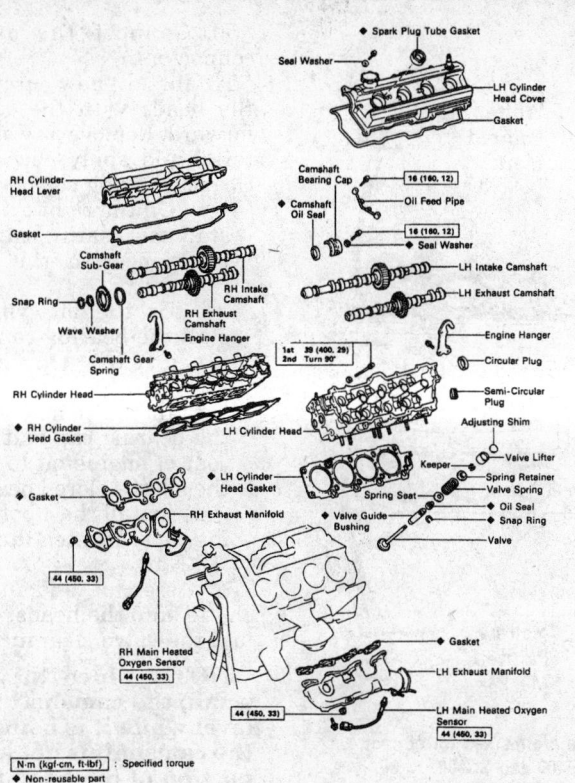

Spark Plug Tube Gasket
Seal Washer
LH Cylinder Head Cover
Gasket
RH Cylinder Head Lever
Camshaft Bearing Cap
16 (160, 12)
Oil Feed Pipe
Gasket
Camshaft Oil Seal
16 (160, 12)
Seal Washer
Camshaft Sub-Gear
Snap Ring
RH Intake Camshaft
LH Intake Camshaft
Wave Washer
RH Exhaust Camshaft
LH Exhaust Camshaft
Camshaft Gear Spring
Engine Hanger
Engine Hanger
RH Cylinder Head
1st 39 (400, 29) 2nd Turn 90°
Circular Plug
Semi-Circular Plug
RH Cylinder Head Gasket
LH Cylinder Head
Adjusting Shim
Valve Lifter
Keeper
Spring Retainer
LH Cylinder Head Gasket
Spring Seat
Valve Spring
Gasket
Oil Seal
RH Exhaust Manifold
Snap Ring
Valve Guide Bushing
Valve
44 (450, 33)
Gasket
RH Main Heated Oxygen Sensor
44 (450, 33)
LH Exhaust Manifold
44 (450, 33)
LH Main Heated Oxygen Sensor
44 (450, 33)

N·m (kgf-cm, ft-lbf) : Specified torque
◆ Non-reusable part

172290

Cylinder head removal component view — LS400

assembly; remove the O-ring from the water inlet housing.

13. Remove the timing belt rear plates.

14. Disconnect the hoses from the throttle body and the fuel pressure regulator. Disconnect the vacuum hose from the air intake chamber (from fuel pressure VSV) and disconnect the charcoal canister hose line from the vacuum pipe. Disconnect the vacuum hoses from the VSVs for fuel pressure regulator, EVAP and the charcoal canister. Remove the 2 bolts and remove the vacuum pipe and hose assembly.

15. To remove the EGR, disconnect the coolant bypass pipe from the hose (from the IAC valve). Also disconnect the coolant pipe from the hose at the rear bypass joint. Remove the EGR valve and gasket.

16. Disconnect the EGR temperature sensor connector. Disconnect the PCV hose from the cylinder head. Remove the EGR valve adapter.

17. Remove the IAC valve.

18. Disconnect the wires and hoses; remove the throttle body assembly.

19. Disconnect the brake booster vacuum hose at the union on the air intake chamber. Disconnect the vacuum hoses from the air intake chamber.

20. Remove the accelerator bracket. Remove the union bolt and the union. Disconnect the VSV connectors for the fuel pressure, EVAP and AIR. Disconnect the EGR pipe from the air intake chamber.

21. Remove the VSVs. Disconnect the DIAGNOSIS data link connector from the air intake chamber. Remove the air intake chamber.

22. Disconnect the heater hoses from the rear bypass joint and pipes.

23. Disconnect the fuel inlet pipe from the delivery pipe. Remove the pulsation damper.

24. Disconnect the wiring harness from the clips and brackets on the right cylinder head. Remove the wire brackets from the cylinder head. Disconnect the engine wiring harness from the intake manifold.

25. Disconnect the engine wiring harness from the delivery pipes. Disconnect the injector wiring connectors.

26. Disconnect the wire harness from the rear water bypass joint. for California vehicles, disconnect the harness connector.

27. Remove the fuel return pipe. Remove the fuel delivery pipes and the remove the injectors. Take care not to drop or damage the injectors.

28. Remove the intake manifold.

29. Remove the front coolant bypass joint, then remove the rear bypass joint.

30. For California vehicles, remove the water bypass pipe.

31. Remove the oil dipsticks and tubes for both the engine and transmission.

32. For California vehicles, remove the secondary air injection (AIR) manifolds.

33. Remove the rear water bypass joint.

34. For non-Calif. vehicles, remove the water bypass pipe.

35. Remove the EGR pipe.

36. Disconnect the engine wiring harness and the ground straps from the cylinder heads.

37. Remove the engine hangers (hoist hooks).

38. Remove the cylinder head covers (valve covers). If necessary, remove the semi-circular plugs.

39. Remove the camshafts.

NOTE: Since the thrust clearance of the camshaft is small, the camshaft must be held level while it is being removed. If the camshaft is not kept level, the portion of the head receiving the thrust may crack or be damaged, causing the camshaft to seize or break during operation.

40. Uniformly loosen the 10 bolts on 1 cylinder head in the reverse order of tightening. Make several repetitive passes in sequence, loosening each bolt slightly each time. Repeat the loosening procedure for the other head.

NOTE: The head(s) may crack or warp if correct bolt loosening procedure is not followed.

41. Remove the 20 cylinder head bolts and plate washers. Lift the heads, with the exhaust manifold attached, from the dowels on the block. Place the heads on blocks of wood on the workbench.

NOTE: If necessary to pry the head loose, take great care not to damage the contact surfaces of the head or block.

42. For each exhaust manifold, remove the main heated oxygen sensor, then remove the manifold and gasket.**To install:**

43. Install the right exhaust manifold. The new gasket must be installed with the yellow marks facing the manifold side. Tighten the bolts to 33 ft. lbs. (44 Nm). Connect the right oxygen sensor connector.

44. Install the left exhaust manifold. The new gasket must be in-

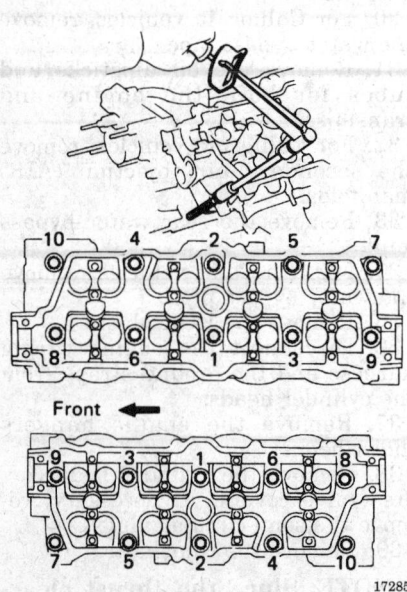

Cylinder head bolt tightening sequence — LS400 and SC400

172858

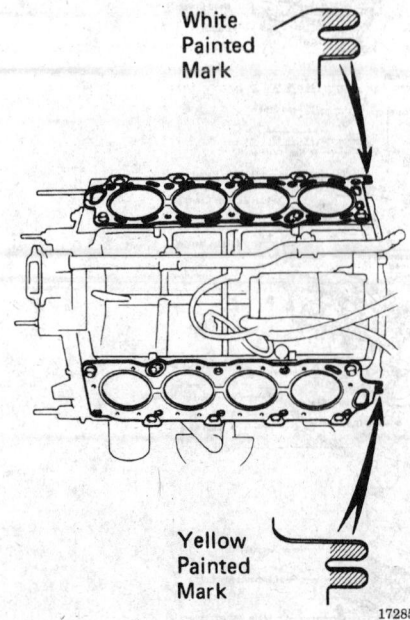

The head gaskets are marked for correct placement — LS400 and SC400

172857

stalled with the yellow marks facing the manifold side. Tighten the bolts to 33 ft. lbs. (44 Nm). Connect the left oxygen sensor connector.

45. Place the two new cylinder head gaskets in position on the engine block. Each gasket has a painted mark denoting the rear of the gasket. The gasket for the right bank has a white mark and the gasket for the left bank has a yellow mark. Double check the gasket position and placement.

46. Place the two cylinder heads in position on the block.

47. Apply a light coat of engine oil to the threads and under the head of each bolt. Temporarily install the 20 plate washers and bolts.

48. Uniformly tighten the 10 bolts on 1 cylinder head in the correct tightening sequence. Make several repetitive passes in this order, tightening each bolt slightly each time. Final torque for each bolt is 29 ft. lbs. (39 Nm). Repeat the tightening procedure for the other head.

49. Carefully mark the front of each bolt with a small dot of paint. Following the tightening sequence used earlier, retighten each bolt by turning it exactly 90 degrees. Check that each paint mark is now at a 90 degree angle to the front of the engine.

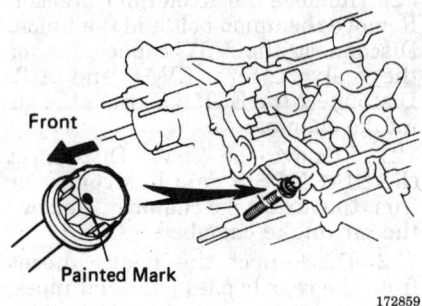

Apply a dot of paint at the front of each bolt — LS400

172859

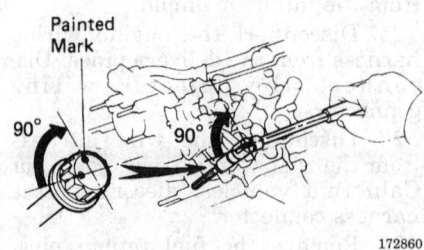

The paint mark must be 90 degrees from the starting point — LS400

172860

50. Connect the oxygen sensor connectors.

51. Install new circular plugs on the heads with the cup side facing forward. Remove any old packing material and apply new sealant to the correct bearing caps.

   a. On the right cylinder head, install the bearing cap marked **I1** in position with the arrow facing rearward.

   b. On the left cylinder head, install the bearing cap marked **I6** in place with the arrow facing forward.

   c. Install a new seal washer to the bearing cap bolt. Apply a light coat of engine oil to the threads of the silver-colored bearing cap bolts.

   d. Install the 4 bolts for the bearing caps. Tighten them to 12 ft. lbs. (16 Nm).

52. Assemble and install the camshafts onto the heads. Check and adjust the valve clearance.

NOTE: Since the camshaft is small, the camshaft must be held level while it is being installed. If the camshaft is not kept level, the portion of the head receiving the thrust may crack or be damaged, causing the camshaft to seize or break during operation.

53. Install the semi-circular plugs. Clean the plug of old sealant and apply new sealant to the plug groove.

54. Clean the cylinder head covers. Apply new sealant in the correct locations and install the gaskets. Install the cylinder head covers and bolts. Tighten the bolts to 52 inch lbs. (6 Nm).

55. Install the engine hangers. Tighten the hanger bolts to 27 ft. lbs. (37 Nm).

56. Install the engine harness to the cylinder heads; install the protectors and connect the ground strap.

57. Install the EGR pipe with new gaskets. Use the 25mm (0.98 in) bolts and tighten them to 13 ft. lbs. (18 Nm).

58. On non-California vehicles, install the water bypass pipe.

59. Install the rear water bypass joint.

60. For California vehicles, install the secondary air injection manifold on each side. Tighten the 4 nuts to 25 ft. lbs. (34 Nm). Tighten the bolt to 13 ft. lbs. (18 Nm). Connect a new three-way air hose to the left AIR manifold.

61. Install the oil dipsticks and tubes.

62. Install the front water bypass joint and the water bypass pipe.. Use new gaskets and connect the hoses. Tighten the bolts to 13 ft. lbs. (18

Nm). On California vehicles, install the AIR inlet pipe, then connect the ECT sensor and gauge sender connectors.

63. Before installing the intake manifold, place two new gaskets on the cylinder heads with the white painted mark facing upward. Double check the positioning and alignment of the gasket. Place the manifold on the heads with the arrow facing forward. Install the nuts and 30mm (1.18 in) bolts. Tighten to 13 ft. lbs. (18 Nm).

64. Install the fuel injectors and the delivery pipes.

65. Install the fuel return pipe. Tighten the union bolt at the pressure regulator to 26 ft. lbs. (35 Nm).

66. Connect the engine wiring harness to the rear water bypass joint. On California vehicles, connect the harness connectors.

67. Connect the injector connectors to the injectors. Connect the wiring connectors to the left delivery pipe. Install the harness protector.

68. Install the engine wiring to the intake manifold.

69. Install the wire harness brackets onto the right cylinder head. Secure the harness to the brackets and install the protector.

70. Connect the fuel return hose to the return pipe; connect the fuel inlet hose to the left delivery pipe. Always use new gaskets. Using a torque wrench with a fulcrum length of 30mm (11.81 in), tighten the union bolt at the pulsation damper to 29 ft. lbs. (39 Nm).

71. Connect the heater hoses.

72. Install the air intake chamber.

73. Before installing the throttle body, place a new gasket in position on the air intake chamber. Install the throttle body. Use the 40mm (1.57 in) bolts and tighten them to 13 ft. lbs. (18 Nm). Connect the water bypass pipe to the clamp on the wire cover. Connect the electrical connectors to the throttle body.

74. Install the IAC valve. Tighten the bolts to 13 ft. lbs. (18 Nm). Install the hoses and attach the connectors.

75. Install the EGR valve adapter with a new gasket. Do not touch the contact faces of the gasket with your fingers. Tighten the bolts to 13 ft. lbs. (18 Nm). Connect the PCV hose and attach the wiring connector.

76. Install the EGR valve with a new gasket. Do not touch the contact faces of the gasket with your fingers. Double check the alignment of the gasket. Tighten the bolts to 13 ft. lbs. (18 Nm). Connect the coolant hoses to the IAC valve and to the bypass pipe.

77. Install the vacuum pipe and hose assembly and install the EGR water bypass pipe. Connect the vacuum hoses removed prior.

78. Install the timing belt rear plates.

79. Install the water inlet and inlet housing. Make sure the contact faces are clean and that new sealant is applied to the sealing groove of the inlet housing. Alternately and gradually, tighten the bolts to 13 ft. lbs. (18 Nm). Connect the coolant hoses.

80. Install the right ignition coil.

81. Connect the accelerator cable, the transmission throttle control cable and the cruise control cable to the throttle body.

82. Install the throttle body cover.

83. Install the front catalytic converters.

84. Install the front exhaust pipe.

85. Install the power steering pump. Tighten the bolts to 29 ft. lbs. (39 Nm) and the nuts to 32 ft. lbs. (43 Nm). Install the power steering pump pulley; tighten the nut to 32 ft. lbs. (43 Nm). Connect the air hose to the air intake chamber.

86. Install the fan bracket.

87. Install the right camshaft timing pulley. The **R** mark must face forward. Install the pulley bolt; tighten it to 80 ft. lbs. (108 Nm).

88. Install the left camshaft timing pulley. The **L** mark must face forward. Install the pulley bolt; tighten it to 80 ft. lbs. (108 Nm).

89. Install the timing belt.

90. Connect the negative battery terminal.

91. Refill the engine with coolant.

92. Start the engine and check for leaks. Check the ignition timing and other engine settings.

93. Road test the vehicle.

94. Install the engine undercovers.

1995–97 Models

1. Relieve the fuel system pressure.

2. Disconnect the negative battery cable. Wait at least 90 seconds before performing any other work.

3. Raise and safely support the vehicle.

4. Remove the oil pan protector.

5. Remove the engine undercover.

6. Drain the engine coolant from the radiator.

7. Remove the battery clamp cover.

8. Remove the air cleaner inlet.

9. Remove the V bank cover by removing the bolt and two cap nuts.

10. Remove the air cleaner and intake air connector assembly.

11. Remove the drive belt, fluid coupling and the fan pulley. The

drive belt tension may be slackened by turning the tensioner counterclockwise. The pulley bolt for the drive belt tensioner has a left handed thread.

12. Remove the radiator.

13. Remove the right hand No. 3 timing belt cover.

14. Remove the left hand No. 3 timing belt cover.

15. Remove the drive belt idler pulley by removing the pulley bolt and cover plate.

16. Remove the right hand No. 2 timing belt cover.

17. Remove the left hand No. 2 timing belt cover.

18. Remove the distributor housings.

19. Remove the No. 1 ignition coil.

20. Disconnect the A/C compressor from the engine.

21. Remove the fan bracket by removing the two bolts and two nuts.

22. Set the engine to TDC on cylinder No. 1. Turn the crankshaft pulley and align its groove with the timing mark **0** of the No. 1 timing cover. Check that the timing marks of the camshaft timing pulleys and timing belt rear plates are aligned. If not, turn the crankshaft 1 full revolution (360 degrees).

NOTE: Since the thrust clearance of the camshaft is small, the camshaft must be kept level while it is being removed. If the camshaft is not kept level, the portion of the cylinder head receiving the shaft thrust may crack or be damaged, causing the camshaft to seize or break.

23. Turn the crankshaft pulley approximately 50° clockwise and put the timing mark of the crankshaft pulley in line with the centers of the crankshaft pulley bolt and the idler pulley bolt.

— **WARNING** —
If the timing belt is disengaged, having the crankshaft pulley at the wrong angle can cause the piston head and valve head to come into contact with each other when you remove the camshaft timing pulley. Always set the crankshaft pulley at the correct angle before removing the timing belt.

24. If the timing belt is to be reused, turn the crank pulley slowly; check that the 3 installation marks are present on the belt. If the marks are not present, make new installation marks before removing the belt. The marks should align with the tim-

ing marks on each camshaft pulley and the crank pulley.

25. Remove the timing belt tensioner. Alternately loosen the two bolts; remove the bolts, the tensioner and the dust protector.

26. Using the proper tool, 09278-54012 or equivalent, loosen the tension between the left side and the right side timing pulleys by slightly turning the left side camshaft clockwise.

27. Disconnect the timing belt from the camshaft timing pulleys. Using the proper tool, remove the bolt and the camshaft timing pulleys.

28. Disconnect the power steering pump from the engine. Do not disconnect the hoses or lines from the power steering pump. Support the power steering pump with a piece of wire. Do not allow the pump to hang.

29. Remove the front catalytic converter.

30. Remove the high tension spark plug wires, wire clamps and the wire cover assembly.

31. Remove the No. 2 ignition coil by removing the connector and the two bolts.

32. Remove the two bolts and the rear timing belt plate. Remove both plates.

33. Remove the intake chamber assembly.

34. Tag and disconnect the following connectors:
• TPS connector
• With TRAC system, sub throttle position sensor connector
• With TRAC system, sub throttle actuator connector
• IAC valve connector
• EGR valve connector
• VSV connector for fuel pressure control
• VSV connector for EVAP system
• EGR gas temperature sensor connector

35. Disconnect the following hoses:
• Brake booster vacuum hose from the union on the air intake chamber
• PCV hose from the PCV valve on the left hand cylinder head
• Water bypass hose (from the EGR valve) from the rear water bypass joint
• Water bypass hose (from the throttle body) from the rear water bypass joint
• Vacuum hose (from the VSV for fuel pressure control) from the fuel pressure regulator
• EVAP hose (from charcoal canister) from the VSV for EVAP

36. Disconnect the heater hose from the water bypass pipe.

CAUTION

Fuel injection systems remain under pressure after the engine has been turned OFF. Properly relieve fuel pressure before disconnecting any fuel lines. Failure to do so may result in fire or personal injury.

---

37. Disconnect the fuel inlet hose from the delivery pipe.

38. Disconnect the fuel return hose from the fuel return pipe.

39. Disconnect the engine wire from the delivery pipes and rear water bypass joint.

40. Disconnect the fuel hose from the fuel pressure regulator.

41. Remove the two bolts and fuel return pipe from the intake manifold.

42. Disconnect the eight injector connectors.

43. Remove the six bolts, four nuts, the intake manifold assembly and the two gaskets.

44. Remove the water inlet and inlet housing.

45. Remove the front water bypass joint.

46. Remove the rear water bypass joint and No. 1 EGR pipe assembly.

47. Remove the oil dipstick and guide for the automatic transmission.

48. Remove the oil dipstick and guide for the engine.

49. Remove the engine hangers.

50. Remove the right and left cylinder head covers by removing the eight bolts, seal washers and gaskets.

51. If necessary, remove the semicircular plugs.

52. Remove the exhaust camshaft from the right side cylinder head. See the camshaft procedure for tightening sequence.

53. Remove the intake camshaft from the right side cylinder head. See the camshaft procedure for tightening sequence.

54. Remove the exhaust camshaft of the left side cylinder head. See the

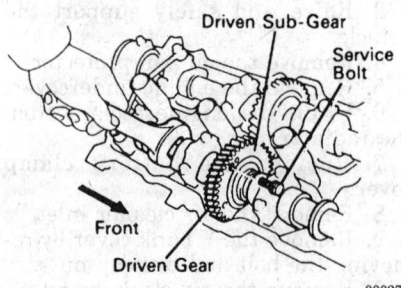

Securing the exhaust camshaft on the right cylinder head — LS400

camshaft procedure for tightening sequence.

NOTE: When removing the camshaft, make sure the torsional spring force of the subgear has been eliminated.

55. Remove the intake camshaft from the left side cylinder head. See the camshaft procedure for tightening sequence.

56. Disconnect the main oxygen sensor connectors.

57. Remove the bolt and disconnect the ground cable from the right cylinder head.

58. Remove the bolt and disconnect the ground strap from the left cylinder head.

59. Remove the bolt and disconnect the engine wire protector from the left hand cylinder head.

60. Remove the two bolts, seal washers, bearing cap and the camshaft housing plug from the right hand cylinder head.

61. Remove the 10 cylinder head bolts and plate washers to each cylinder head. Loosen the bolts in the reverse order of the tightening sequence. Lift the heads from the dowels on the block with the exhaust manifolds attached. Place the heads on blocks of wood on the workbench.

NOTE: Do not drop anything in the opening in the front of the right side cylinder head. The opening leads through the block and into the oil pan. If anything falls into the opening the oil pan will have to be removed in order to retrieve it.

WARNING

If necessary to pry the head loose, take great care not to damage the contact surfaces of the head or block.

---

62. Remove the two bolts, seal washers, bearing cap and camshaft housing plug from the right hand cylinder head.

63. Remove the right exhaust manifold from the cylinder head by removing the heat insulator, eight nuts and the gasket.

64. Remove the left exhaust manifold from the cylinder head by removing the heat insulator, eight nuts and the gasket.

To install:

65. Install the right exhaust manifold. The new gasket must be installed with the white marks facing the manifold side. Tighten the bolts to 33 ft. lbs. (44 Nm). Install the right oxygen sensor.

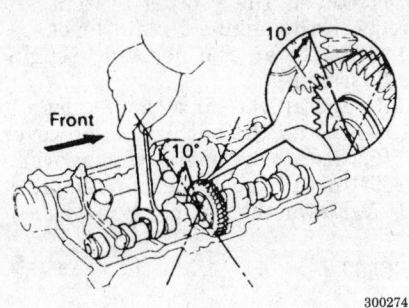

300274

Turning the exhaust camshaft 10 degrees on the right cylinder head — LS400 and SC400

66. Install the left exhaust manifold. The new gasket must be installed with the white marks facing the manifold side. Tighten the bolts to 33 ft. lbs. (44 Nm). Install the left oxygen sensor.

67. Place the two new cylinder head gaskets in position on the engine block. Each gasket has a painted mark denoting the rear of the gasket. The gasket for the right bank has a white mark and the gasket for the left bank has a yellow mark. Double check the gasket position and placement.

68. Place the two cylinder heads in position on the block.

69. Apply a light coat of engine oil to the threads and under the head of each bolt. Temporarily install the plate washers and bolts.

70. Uniformly tighten the 10 bolts on 1 cylinder head in the order shown. Make several repetitive passes in the correct sequence, tightening each bolt slightly each time. Final torque for each bolt is 29 ft. lbs. (39 Nm). Repeat the tightening procedure for the other head.

71. Carefully mark the front of each bolt with a small dot of paint. Following the tightening sequence used earlier, retighten each bolt by turning it exactly 90 degrees. Check that each paint mark is now at a 90 degree angle to the front of the engine.

72. Connect the oxygen sensor connectors.

73. Install the engine wire to the right hand cylinder head with the two bolts.

74. Install the ground cable to the right hand cylinder head with the bolt.

75. Install the engine wire protector to the left hand cylinder head with the bolt.

76. Install the ground cable to the left hand cylinder head with the bolt.

77. Remove any old packing and apply new seal packing to the bearing caps.

78. Install the bearing cap on the right side cylinder head, marked **I1**, in position with the arrow mark facing the rear. Install the bearing cap on the left side cylinder head, marked **I6**, in position with the arrow mark facing the front.

79. Apply a light coat of oil on the threads of the cap bolts. Install the bearing cap bolts with new washers and alternately tighten each bolt to 12 ft. lbs. (16 Nm).

NOTE: Use silver colored bolts 1.50 inches (38mm) in length.

80. Install new camshaft housing plugs on the cylinder heads. Be sure to face the cupped side forward.

81. Turn the crankshaft pulley clockwise or counterclockwise and put the timing mark of the crankshaft pulley in line with the centers of the crankshaft pulley bolt and the idler pulley bolt

NOTE: Since the thrust clearance of the camshaft is small, the camshaft must be kept level while it is being installed. If the camshaft is not kept level, the portion of the cylinder head receiving the shaft thrust may crack or be damaged, causing the camshaft to seize or break.

82. Install the right side cylinder head intake camshaft. Tighten the bracket bolt in the reverse order of the loosening sequence.

83. Install the right side cylinder head exhaust camshaft. Tighten the bracket bolt in the reverse order of the loosening sequence.

84. Install the left side cylinder head intake camshaft. Tighten the bracket bolt in the reverse order of the loosening sequence.

85. Install the left side cylinder head exhaust camshaft. Tighten the bracket bolt in the reverse order of the loosening sequence.

86. Check and adjust the valve clearance.

87. Install the camshaft oil seals with the proper tool (SST 09223–46011). Make sure to apply MP grease to the new oil seal lip.

88. Install the semi-circular plugs.

89. Clean the cylinder head covers. Apply new sealant in the correct locations and install the gaskets.

90. Install the right cylinder head cover and bolts. Tighten the bolts to 52 inch lbs. (6 Nm).

91. Install the left cylinder head cover and bolts. Tighten the bolts to 52 inch lbs. (6 Nm).

92. Install the engine hanger with the two bolts. Install both engine

hangers. Torque the bolts to 27 ft. lbs. (37 Nm).

93. Install the oil dipstick guide for the engine.

94. Install the oil dipstick for the transmission.

95. Install the rear water bypass joint and No. 1 EGR pipe.

96. Install the front water bypass joints. Install two gaskets and alternately tighten the nuts to 13 ft. lbs. (18 Nm).

97. Install the water inlet and inlet housing, alternately torque the bolts to 13 ft. lbs. (18 Nm).

98. Install the delivery pipe and intake manifold.

99. Install the return pipe with two new gaskets. Torque the union bolt to 26 ft. lbs. (35 Nm).

100. Install the engine wire to the delivery pipes and rear water bypass joint.

101. Connect the fuel return hose to the fuel return pipe.

102. Connect the fuel inlet hose to the left hand delivery pipe.

103. Connect the fuel hose to the fuel pressure regulator.

104. Install the air intake chamber assembly.

105. Connect the following hoses:
- Brake booster vacuum hose to the union on the air intake chamber
- PCV hose to the PCV valve on the left hand cylinder head
- Water bypass hose (from EGR valve) to the rear water bypass joint
- Water bypass hose (from throttle body) to the rear water bypass joint
- Vacuum hose (from VSV for fuel pressure control) to the fuel pressure regulator
- EVAP hose (from charcoal canister) from the VSV for EVAP

106. Connect the following connectors:
- TPS connector
- With TRAC system, sub TPS connector
- With TRAC system, sub throttle actuator connector
- IAC valve connector
- EGR valve connector
- EGR gas temperature sensor connector
- VSV connector for fuel pressure control
- VSV connector for EVAP

107. Install the accelerator bracket with the two bolts.

108. Connect the accelerator cable, A/T throttle control cable and the cruise control actuator cable.

109. Connect the spark plug wires and clamps to the right and left cylinder head cover.

110. Install the timing belt rear plates by installing the bolts. Torque the bolts to 66 inch lbs. (8 Nm).

111. Install the No. 2 ignition coil.

112. Install a new gasket to the exhaust manifold and install the catalytic converters. Torque the three nuts to each converter to 46 ft. lbs. (62 Nm).

113. Install the front exhaust pipe, torque the bolts and nuts to 32 ft. lbs. (44 Nm). Tighten the four bolts holding the pipe support bracket to the transmission. Torque the bolts to 32 ft. lbs. (44 Nm).

114. Install the power steering pump with the nut and three bolts. Torque the nut to 32 ft. lbs. (43 Nm) and the bolts to 29 ft. lbs. (39 Nm).

115. Align the knock pin on the right side camshaft with the knock pin of the timing pulley. Slide on the timing pulley with the right side mark facing forward. Tighten the bolt to 80 ft. lbs. (108 Nm).

116. Align the knock pin on the left side camshaft with the knock pin of the timing pulley. Slide on the timing pulley with the left side mark facing forward. Tighten the bolt to 80 ft. lbs. (108 Nm).

117. Install the timing belt to the left side camshaft timing pulley:

a. Using the proper tool, slightly turn the left side timing pulley clockwise. Align the installation mark of the timing belt with the timing mark of the camshaft timing pulley and hang the timing belt on the left side camshaft pulley.

b. Align the timing marks of the left side camshaft pulley and the timing belt rear plate.

c. Check that the timing belt has tension between crankshaft timing pulley and the left side camshaft pulley.

118. Install the timing belt to the right side camshaft timing pulley by:

a. Using the proper tool, slightly turn the right side timing pulley clockwise. Align the installation mark of the timing belt with the timing mark of the camshaft timing pulley and hang the timing belt on the right side camshaft pulley.

b. Align the timing marks of the right side camshaft pulley and the timing belt rear plate.

c. Check that the timing belt has tension between crankshaft timing pulley and the right side camshaft pulley.

119. The timing belt tensioner must be set prior to installation. The tensioner can be set by:

a. Place a plate washer between the tensioner and a block. Using a press, press in the pushrod using 220–2205 lbs. of pressure.

b. Align the holes of the pushrod and housing, pass a 1.27mm Allen wrench through the holes to keep the setting position of the pushrod.

c. Release the press and install the dust boot to the tensioner.

120. Loosely install the tensioner. Evenly and alternately tighten the bolts to 20 ft. lbs. (26 Nm). Remove the tool from the tensioner.

121. Turn the crankshaft pulley two complete revolutions from TDC to TDC. Always turn the crankshaft clockwise. Check that all belt and pulley marks align with their reference marks. If any mark is out of perfect alignment, the timing belt must be removed and reinstalled.

122. Install the drive belt tensioner and tighten the bolt and nuts to 12 ft. lbs. (16 Nm).

123. Install both distributor housings and tighten the mounting bolts to 13 ft. lbs. (18 Nm). The distributors are marked L or R for correct installation.

124. Replace the distributor rotors and caps.

125. Install the fan bracket by installing the two bolts and two nuts. Torque as follows:

- 12mm head to 12 ft. lbs. (16 Nm)
- 14mm head to 24 ft. lbs. (32 Nm)

126. Install the A/C compressor. Torque the bolts to 36 ft. lbs. (49 Nm) and the nut to 22 ft. lbs. (29 Nm).

127. Install the No. 1 ignition coil.

128. Install the right side No. 2 timing belt cover.

129. Install the left side No. 2 timing belt cover.

130. Install the drive belt idler pulley and cover plate. Tighten the bolt to 27 ft. lbs. (37 Nm).

131. Install and secure the ignition wires. Make certain that all clips and retainers are securely engaged and that the wires are properly routed.

132. Install the right side No. 3 timing belt.

133. Install the left hand No. 3 timing belt cover.

134. Install the radiator assembly.

135. Install the fan pulley, fan, fluid coupling and the drive belt.

136. Install the air cleaner and intake air connector assembly.

137. Install the V bank cover.

138. Fill the radiator with engine coolant.

139. Connect the negative battery cable to the battery.

140. Start the engine and check for leaks.

141. Bleed the cooling system and recheck the engine coolant level.

142. Make all the necessary engine adjustments.

143. Install the air cleaner inlet.

144. Install the battery clamp cover.

145. Install the engine undercover.

146. Install the oil pan protector.

147. Lower the vehicle.

SC400

1993–94 Models

1. Disconnect the negative battery cable. Wait at least 90 seconds before proceeding with any other work.

2. Drain the engine coolant.

3. Disconnect the timing belt from the camshaft timing pulleys.

4. Remove the camshaft timing pulleys. Do not drop any parts into the timing belt cover. Protect the belt from grease, fluids, etc.

5. Remove the fan bracket.

6. Disconnect the air hose from the air intake chamber. Remove the power steering pump pulley. Remove the power steering pump from the engine.

7. Remove the front catalytic converter.

8. Remove the throttle body cover.

9. Disconnect the accelerator cable, the automatic transmission throttle cable and the cruise control actuator cable.

10. Remove the right ignition coil.

11. Remove the inlet housing from the water pump. Disconnect the coolant bypass hose from the IAC valve. Remove the coolant inlet and housing assembly; remove the O-ring from the water inlet housing.

12. Remove the timing belt rear plates.

13. Disconnect the hoses from the throttle body and the fuel pressure regulator and fuel pressure control. Disconnect the vacuum hose from the air intake chamber (from fuel pressure VSV) and disconnect the charcoal canister hose line from the vacuum pipe. Disconnect the vacuum hoses from the VSVs for fuel pressure regulator, EVAP and the charcoal canister. Remove the 2 bolts and remove the vacuum pipe and hose assembly.

14. Remove the EGR valve (California) or EGR valve and vacuum modulator (non-Calif.) disconnect the coolant bypass pipe from the hose (from the IAC valve). Also disconnect the coolant pipe from the hose at the rear bypass joint.

15. Remove the EGR valve adapter.

16. Remove the IAC/ISC valve.

17. Disconnect the wires and hoses; remove the heater water valve from the body.

18. Disconnect the wiring harnesses and disconnect the water and PCV hoses from the throttle body. Remove the throttle body.

19. Disconnect the brake booster vacuum hose at the union on the air intake chamber. Disconnect the vacuum hoses from the air intake chamber.

20. Remove the accelerator bracket. remove the union bolt and union. Disconnect the VSV connectors for the fuel pressure control and disconnect the cold start injector wiring. Disconnect the EGR pipe from the air intake chamber.

21. Remove the VSVs. Remove the transmission throttle cable bracket. Disconnect the DIAGNOSIS data link connector (DLC1) from the air intake chamber. Remove the air intake chamber.

22. Remove the cold start injector, tube and wiring assembly.

23. Disconnect the heater hoses from the rear bypass joint and pipes.

24. Disconnect the fuel inlet pipe from the delivery pipe. Remove the pulsation damper. Disconnect the fuel return hose from the pipe.

25. Disconnect the wiring harness from the clips and brackets on the intake manifold. Disconnect the wiring harness from the right cylinder head, the rear coolant bypass joint and the fuel delivery pipes.

26. Disconnect the injector wiring connectors.

27. Remove the fuel return pipe. Remove the fuel delivery pipes and the remove the injectors. Take care not to drop or damage the injectors.

28. Remove the intake manifold.

29. Remove the front coolant bypass joint, then remove the rear bypass joint.

30. Remove the oil dipsticks and tubes for both the engine and transmission.

31. Remove the EGR pipe.

32. Disconnect the engine wiring harness and the ground straps from the cylinder heads.

33. Remove the engine hangers (hoist hooks).

34. Remove the cylinder head covers (valve covers). If necessary, remove the semi-circular plugs.

35. Remove the camshafts. After the camshafts are removed, remove the bearing caps and circular plugs from each head.

NOTE: Since the thrust clearance of the camshaft is small, the camshaft must be held level while it is being removed. If the camshaft is not kept level, the portion of the head receiving the thrust may crack or be damaged, causing the camshaft to seize or break during operation.

36. Disconnect the wiring to the heated exhaust sensors on both sides.

37. Uniformly loosen the 10 bolts on 1 cylinder head in the reverse order of tightening sequence. Make several repetitive passes in sequence, loosening each bolt slightly each time. Repeat the loosening procedure for the other head.

NOTE: The head(s) may crack or warp if correct bolt loosening procedure is not followed.

38. Remove the 20 cylinder head bolts and plate washers. Lift the heads — with the exhaust manifold attached — from the dowels on the block. Place the heads on blocks of wood on the workbench.

NOTE: If necessary to pry the head loose, take great care not to damage the contact surfaces of the head or block.

39. For each exhaust manifold, remove the main heated oxygen sensor, then remove the manifold and gasket. **To install:**

40. Install the right exhaust manifold. The new gasket must be installed with the white marks facing the manifold side. Tighten the bolts to 29–33 ft. lbs. (39–44 Nm). Install the right oxygen sensor.

41. Install the left exhaust manifold. The new gasket must be installed with the white marks facing the manifold side. Tighten the bolts to 29–33 ft. lbs. (39–44 Nm). Install the left oxygen sensor.

42. Place the two new cylinder head gaskets in position on the engine block. Each gasket has a painted mark denoting the rear of the gasket. The gasket for the right bank has a white mark and the gasket for the left bank has a yellow mark. Double check the gasket position and placement.

43. Place the two cylinder heads in position on the block.

44. Apply a light coat of engine oil to the threads and under the head of each bolt. Temporarily install the 20 plate washers and bolts.

45. Uniformly tighten the 10 bolts on 1 cylinder head in the order shown. Make several repetitive passes in the correct sequence, tightening each bolt slightly each time. Final torque for each bolt is 29 ft. lbs.

(39 Nm). Repeat the tightening procedure for the other head.

46. Carefully mark the front of each bolt with a small dot of paint. Following the tightening sequence used earlier, retighten each bolt by turning it exactly 90 degrees. Check that each paint mark is now at a 90 degree angle to the front of the engine.

47. Connect the oxygen sensor connectors.

48. Install new circular plugs on the heads with the cup side facing forward. Remove any old packing material and apply new sealant to the correct bearing caps.

   a. On the right cylinder head, install the bearing cap marked **I1** in position with the arrow facing rearward.

   b. On the left cylinder head, install the bearing cap marked **I6** in place with the arrow facing forward.

   c. Install a new seal washer to the bearing cap bolt. Apply a light coat of engine oil to the threads of the silver-colored bearing cap bolts.

   d. Install the 4 bolts for the bearing caps. Tighten them to 12 ft. lbs. (16 Nm).

49. Assemble and install the camshafts onto the heads. Check and adjust the valve clearance.

NOTE: Since the thrust clearance of the camshaft is small, the camshaft must be held level while it is being installed. If the camshaft is not kept level, the portion of the head receiving the thrust may crack or be damaged, causing the camshaft to seize or break during operation.

50. Install the semi-circular plugs. Clean the plug of old sealant and apply new sealant to the plug groove.

51. Clean the cylinder head covers. Apply new sealant in the correct locations and install the gaskets. Install the cylinder head covers and bolts. Tighten the bolts to 52 inch lbs. (6 Nm).

52. Install the engine hangers. Tighten the hanger bolts to 27 ft. lbs. (37 Nm).

53. Install the engine harness to the cylinder heads; install the protectors and connect the ground strap.

54. Install the EGR pipe with new gaskets.

55. Install the oil dipsticks and tubes.

56. Install the front water bypass joint. Attach the connectors for the ECT sensor and the ECT gauge sender.

57. Before installing the intake manifold, place two new gaskets on

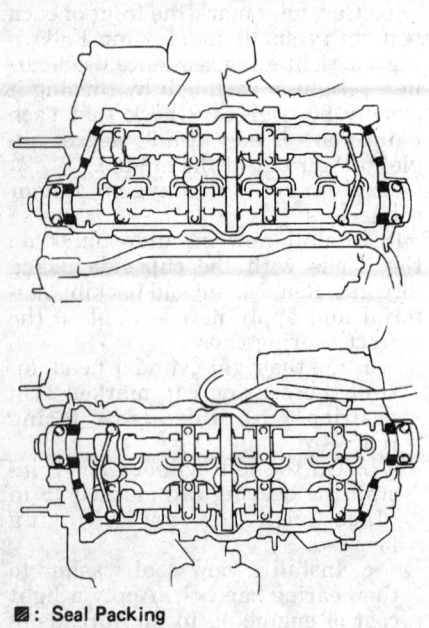

☑ : Seal Packing

172870

Apply sealant to the cylinder heads — 1994 SC400

the cylinder heads with the white painted mark facing upward. Double check the positioning and alignment of the gasket. Place the manifold on the heads with the arrow facing forward. Install the nuts and 30mm (1.18 in) bolts. Tighten to 13 ft. lbs. (18 Nm).

58. Install the fuel injectors and the delivery pipes.

59. Install the fuel return pipe. Tighten the union bolt at the pressure regulator to 26 ft. lbs. (35 Nm).

60. Connect the engine wiring harness to the rear water bypass joint, the delivery pipes and the right cylinder head.

61. Install the engine wiring to the intake manifold.

62. Loosely install the EGR pipe to the right cylinder head.

63. Connect the fuel return hose to the return pipe; connect the fuel inlet hose to the left delivery pipe. Always use new gaskets. Using a torque wrench with a fulcrum length of 30mm (11.81 in), tighten the union bolt at the pulsation damper to 29 ft. lbs. (39 Nm).

64. Connect the heater hoses.

65. To install the air intake chamber:

a. Place 4 new gaskets on the manifold. Place the air intake chamber on the manifold and in-

stall a new gasket for the EGR pipe.

b. Install the bolts holding the EGR pipe to the air intake chamber. Loosely install the air intake chamber and install the VSV for fuel pressure and EGR. Install the transmission throttle cable bracket. Install the data link connector.

c. Tighten the bolts holding the EGR pipe to the chamber and to the right cylinder head to 13 ft. lbs. (18 Nm).

d. Install the cold start injector tube to the right delivery pipe.

e. Attach the electrical connectors to the VSVs.

f. Install the accelerator bracket.

g. Install 2 new gaskets and connect the union with the union bolt. Tighten the bolt to 22 ft. lbs. (29 Nm).

h. Connect the brake booster vacuum hose to the union on the intake chamber. Connect the vacuum hose from the AIR VSV to the intake chamber. Connect the other vacuum hose to the AIR outlet pipe.

66. Before installing the throttle body, place a new gasket in position on the air intake chamber. Install the throttle body. Use the 40mm (1.57 in) bolts and tighten them to 13 ft. lbs. (18 Nm). Connect the water bypass pipe to the clamp on the wire cover. Connect the electrical connectors to the throttle body.

67. Install the heater water valve; connect the wiring and hoses.

68. Install the IAC valve. Tighten the bolts to 13 ft. lbs. (18 Nm). Install the hoses and attach the connectors.

69. Install the EGR valve adapter with a new gasket. Do not touch the contact faces of the gasket with your fingers. Tighten the bolts to 13 ft. lbs. (18 Nm). Connect the PCV hose and attach the wiring connector.

70. Install the EGR valve (California) or EGR valve and vacuum modulator with new gaskets. Do not touch the contact faces of the gasket with your fingers. Double check the alignment of the gasket. Tighten the bolts to 13 ft. lbs. (18 Nm). Connect the coolant hoses to the IAC valve and to the bypass pipe.

71. Install the vacuum pipe and hose assembly and install the EGR water bypass pipe. Connect the vacuum hoses removed during removal.

72. Install the timing belt rear plates.

73. Install the water inlet and inlet housing. Make sure the contact faces

are clean and that new sealant is applied to the sealing groove of the inlet housing. Alternately and gradually, tighten the bolts to 13 ft. lbs. (18 Nm). Connect the coolant hoses.

74. Install the right ignition coil.

75. Connect the accelerator cable, the transmission throttle control cable and the cruise control cable to the throttle body.

76. Install the throttle body cover.

77. Install the power steering pump. Tighten the bolts to 29 ft. lbs. (39 Nm) and the nuts to 32 ft. lbs. (43 Nm). Install the power steering pump pulley; tighten the nut to 32 ft. lbs. (43 Nm). Connect the air hose to the air intake chamber.

78. Install the front catalytic converter.

79. Install the hydraulic pump. Note that the pump is retained by bolts of 2 different lengths and head sizes. The longer, 114mm (4.49 in) bolt with the 14mm head is installed in the lower position and should be tightened to 22 ft. lbs. (30 Nm). The 106mm (4.17) bolt with the 12mm head installs in the upper position and should be tightened to 12 ft. lbs. (16 Nm.).

80. Install the right camshaft timing pulley. The **R** mark must face forward. Install the pulley bolt; tighten it to 80 ft. lbs. (108 Nm).

81. Install the left camshaft timing pulley. The **L** mark must face forward. Install the pulley bolt; tighten it to 80 ft. lbs. (108 Nm).

82. Install the timing belt.

83. Connect the negative battery terminal.

84. Refill the engine with coolant.

85. Start the engine and check for leaks. Check the ignition timing and other engine settings.

86. Road test the vehicle.

87. Install the engine undercovers.

1995–97 Models

1. Relieve the fuel system pressure.

2. For 1996–97 vehicles, remove the V-bank cover.

3. Disconnect the negative battery cable. Wait at least 90 seconds before performing any other work. Disconnect the positive battery cable and remove the battery.

4. Remove the engine undercover.

5. Drain the cooling system.

6. Remove the accessory drive belt by turning the tensioner counterclockwise.

7. Remove the radiator.

8. Remove the engine coolant reservoir tank.

9. Disconnect the air conditioning compressor from the engine. Do not

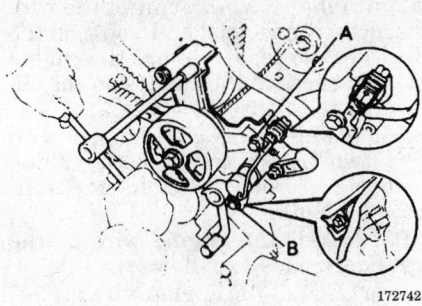

Install the hydraulic pump bolts in the proper locations — 1994 SC400

remove the compressor pressure lines.

10. Remove the intake air connector.

11. Remove the left ignition coil.

12. Remove the upper high tension cord cover by removing the two bolts.

13. Remove the right side engine wire cover by removing the bolt.

14. Remove the left side engine wire cover by removing the two bolts.

15. Remove the right side No. 3 timing cover by removing the two air control valve hoses and four bolts.

16. Remove the left No. 3 timing belt cover by removing the four bolts.

17. Disconnect and tag the vacuum hoses and remove the left side engine wire cover. Disconnect the spark plug wires.

18. Remove the pulley bolt, cover plate and idler pulley.

19. Remove the right hand No. 2 timing belt cover.

20. Remove the left hand No. 2 timing belt cover.

21. Remove the distributor caps and rotors. Mark both caps and rotors (left or right) for correct installation. Disconnect and remove both distributor housings.

22. Disconnect and remove the alternator from the engine.

23. Remove the drive belt tensioner.

24. Remove the spark plugs.

25. Set the engine to TDC on cylinder No. 1. Turn the crankshaft pulley and align its groove with the timing mark **0** of the No. 1 timing cover. Check that the timing marks of the camshaft timing pulleys and timing belt rear plates are aligned. If not, turn the crankshaft 1 full revolution (360 degrees).

26. If the timing belt is to be reused, turn the crank pulley slowly; check that the four installation marks are present on the belt. If the marks are not present, make new installation marks before removing the belt. The marks should align with the

timing marks on each camshaft pulley, the crank pulley and a matchmark on the belt at the end of the fan bracket.

27. Remove the timing belt tensioner. Alternately loosen the two bolts; remove the bolts, the tensioner and the dust protector.

28. Using the proper tool, 09278-54012 or equivalent, loosen the tension between the left side and right side timing pulleys by slightly turning the left side camshaft clockwise.

29. Disconnect the timing belt from the camshaft timing pulleys. Using the proper tool, remove the bolt and the camshaft timing pulleys.

30. Remove the fan bracket by removing the two bolts and two nuts.

31. Remove the power steering pump.

32. Remove the front catalytic converter.

33. Disconnect the throttle body cover.

34. Remove the accelerator, transmission and cruise control cables from the throttle body.

35. Remove the right hand ignition coil.

36. Remove the water inlet and inlet housing.

37. Remove the two bolts and rear timing belt plate. Remove both plates.

38. Remove the following vacuum hoses:
• Vacuum hose from the throttle body and VSV for the EVAP system.
• Vacuum hose from the fuel pressure regulator and VSV for the fuel pressure regulator.
• Vacuum hose from the VSV for fuel pressure and air intake chamber.
• Vacuum hose from the charcoal canister and VSV for the EVAP.

39. Remove the EGR valve and adapter.

40. Remove the idle air control valve from the engine.

41. Remove the heater water valve from the body.

42. Remove the throttle body assembly.

43. Remove the air intake chamber assembly.

44. Remove the transmission throttle cable bracket from the intake chamber by removing the four bolts and eight nuts.

45. Disconnect the DLC1 bracket from the air intake chamber.

46. Lift the air intake chamber assembly and disconnect the EGR gas temperature sensor connector.

47. Remove the air intake chamber assembly and four gaskets.

48. Disconnect the heater water hoses from the bypass pipes.

----- **CAUTION** -----

Fuel injection systems remain under pressure after the engine has been turned OFF. Properly relieve fuel pressure before disconnecting any fuel lines. Failure to do so may result in fire or personal injury.

49. Disconnect the fuel inlet hose from the left delivery pipe. Remove the pulsation damper.

50. Disconnect the fuel return hose from the fuel return pipe.

51. Disconnect the EGR pipe from the right hand cylinder head by removing the bolt.

52. Disconnect the engine wire from the intake manifold by removing the two bolts.

53. Disconnect the engine wire from the delivery pipes, rear water bypass joint and the right hand cylinder head.

54. Remove the fuel return pipe and rear fuel pipe.

55. Remove the delivery pipes and intake manifold assembly.

56. Remove the front water bypass joint.

57. Remove the oil dipstick guide and dipstick guide for the transmission.

58. Remove the bolt holding the bypass joint to the left hand rear engine hanger. Remove the water bypass joint by removing the four bolts.

59. Remove the two bolts holding the EGR pipe to the right hand exhaust pipe. Remove the EGR pipe and gasket.

60. Remove the engine hangers.

61. On the right cylinder head cover, release the clips and retainers holding the spark plug wires to the cylinder head cover. Remove the cylinder head cover.

62. On the left cylinder head cover:
  a. Release the clips and retainers holding the spark plug wires.
  b. Remove the eight bolts, seal washers, cylinder head cover and gaskets.

63. If necessary, remove the semicircular plugs.

64. Remove the exhaust camshaft from the right side cylinder head.

65. Remove the intake camshaft from the right side cylinder head.

66. Remove the exhaust camshaft of the left side cylinder head.

NOTE: When removing the camshaft, make sure the torsional spring force of the subgear has been eliminated.

67. Remove the intake camshaft from the left side cylinder head.

68. Remove the circular plugs by removing the two bolts, seal washers and the bearing caps. Remove both circular plugs. Arrange the bearing caps in order.

69. Disconnect the main oxygen sensor connectors.

70. Remove the 10 cylinder head bolts and plate washers to each cylinder head. Loosen the bolts in the reverse order of tightening. Lift the heads from the dowels on the block with the exhaust manifolds attached. Place the heads on blocks of wood on the workbench.

NOTE: Do not drop anything in the opening in the front of the right side cylinder head. The opening leads through the block and into the oil pan. If anything falls into the opening the oil pan will have to be removed in order to retrieve it.

——— WARNING ———
If necessary to pry the head loose, take great care not to damage the contact surfaces of the head or block.

71. Remove the main heated oxygen sensor from each exhaust manifold. Remove the manifolds and gaskets from the cylinder head.

To install:

72. Install the right exhaust manifold. The new gasket must be installed with the white marks facing the manifold side. Tighten the bolts to 33 ft. lbs. (44 Nm). Install the right oxygen sensor.

73. Install the left exhaust manifold. The new gasket must be installed with the white marks facing the manifold side. Tighten the bolts to 33 ft. lbs. (44 Nm). Install the left oxygen sensor.

74. Place the two new cylinder head gaskets in position on the engine block. Each gasket has a painted mark denoting the rear of the gasket. The gasket for the right bank has a white mark and the gasket for the left bank has a yellow mark. Double check the gasket position and placement.

75. Place the two cylinder heads in position on the block.

76. Apply a light coat of engine oil to the threads and under the head of each bolt. Temporarily install the 20 plate washers and bolts.

77. Uniformly tighten the 10 bolts on 1 cylinder head in the order shown. Make several repetitive passes in the correct sequence, tight-

ening each bolt slightly each time. Final torque for each bolt is 29 ft. lbs. (39 Nm). Repeat the tightening procedure for the other head.

78. Carefully mark the front of each bolt with a small dot of paint. Following the tightening sequence used earlier, retighten each bolt by turning it exactly 90 degrees. Check that each paint mark is now at a 90 degree angle to the front of the engine.

79. Connect the oxygen sensor connectors.

80. Install new circular plugs on the heads with the cup side facing forward.

81. Remove any old packing and apply new seal packing to the bearing caps.

82. Install the bearing cap on the right side cylinder head, marked **I1**, in position with the arrow mark facing the rear. Install the bearing cap on the left side cylinder head, marked **I6**, in position with the arrow mark facing the front.

83. Apply a light coat of oil on the threads of the cap bolts. Install the bearing cap bolts with new washers and alternately tighten each bolt to 12 ft. lbs. (16 Nm).

NOTE: Use silver colored bolts 1.50 inches (38mm) in length.

84. Install the right side cylinder head intake camshaft. See the camshaft procedure for tightening sequence.

85. Install the right side cylinder head exhaust camshaft. See the camshaft procedure for tightening sequence.

86. Install the left side cylinder head intake camshaft. See the camshaft procedure for tightening sequence.

87. Install the left side cylinder head exhaust camshaft. See the camshaft procedure for tightening sequence.

88. Check and adjust the valve clearance.

89. Install the camshaft oil seals with the proper tool (SST 09223–46011). Make sure to apply MP grease to the new oil seal lip.

90. Install the semi-circular plugs, remove any old packing material. Apply seal packing to the semi-circular plug grooves. Install the four semi-circular plugs to the cylinder heads.

91. Clean the cylinder head covers. Apply new sealant in the correct locations and install the gaskets.

92. Install the right cylinder head cover and bolts. Tighten the bolts to 52 inch lbs. (6 Nm).

93. Install the left cylinder head cover and bolts. Tighten the bolts to

52 inch lbs. (6 Nm). Connect the wire harness connectors and connect the fuel injector leads. Secure the engine wire harness in the clamps on the delivery pipe.

94. Install the engine hanger with the two bolts. Install both engine hangers. Torque the bolts to 27 ft. lbs. (37 Nm).

95. Install the engine wire to the cylinder heads as follows:

   a. Install the engine wire protector to the right hand cylinder head with the two bolts.

   b. Install the ground strap to the right hand cylinder head with the bolt.

   c. Install the engine wire protector to the left hand cylinder head with the five bolts.

96. Install a new gasket and EGR pipe to the right hand exhaust manifold with the two nuts.

97. Install the rear water bypass joint by installing the gasket and four nuts. Torque the nuts to 13 ft. lbs. (18 Nm).

98. Install the oil dipstick guide for the engine.

99. Install the front water bypass joint by installing the gaskets and four nuts. Torque the nuts to 13 ft. lbs. (18 Nm).

100. Install the engine wire protector to the front water bypass joint. Connect the ECT sensor connector and the ECT sender gauge connector.

101. Install the delivery pipe and intake manifold.

102. Install the return pipe with two new gaskets. Torque the union bolt to 26 ft. lbs. (35 Nm).

103. Install the engine wire to the delivery pipes, rear water bypass joint and the right hand cylinder head.

104. Install the engine wire to the intake manifold with the two bolts.

105. Temporarily install the EGR pipe to the right hand cylinder head by installing the bolt. Do not torque the bolt at this time.

106. Connect the fuel return hose to the fuel return pipe.

107. Connect the fuel inlet hose to the left hand delivery pipe. Install the pulsation damper

108. Connect the water hose to the water bypass pipe and the water hose to the rear water bypass joint.

109. Install the air intake chamber assembly. Torque the eight nuts and four bolts to the intake chamber to 13 ft. lbs. (18 Nm). Torque the bolts holding the EGR pipe to the air intake chamber to 13 ft. lbs. (18 Nm). Torque the bolt holding the EGR pipe

to the right cylinder head to 13 ft. lbs. (18 Nm).

110. Install the accelerator bracket with the bolt and stud bolt.

111. Install the brake booster union by installing the gaskets and the union bolt. Torque the bolt to 22 ft. lbs. (29 Nm).

112. Connect the following connectors and hoses:

    a. VSV connector for fuel pressure control.

    b. VSV connector for EVAP.

    c. Brake booster vacuum hose to the union on the air intake chamber.

    d. Vacuum hose (from the VSV for heater water valve) to the air intake chamber.

113. Install the throttle body.

114. Install the heater water valve.

115. Install the IAC valve with the two nuts. Torque the nuts to 13 ft. lbs. (18 Nm).

116. Connect the water bypass hose (from the throttle body) to the IAC valve. Connect the IAC valve connector.

117. Install the EGR valve adapter. Torque the bolts and nuts to 13 ft. lbs. (18 Nm).

118. Connect the PCV hose to the cylinder head.

119. Connect the EGR gas temperature sensor connector.

120. Install the EGR valve with the gasket and two nuts. Torque the nuts to 13 ft. lbs. (18 Nm).

121. Connect water bypass hose to the IAC valve and then connect the water bypass hose to the water bypass pipe.

122. Connect the EGR valve connector.

123. Connect the following hoses:

    a. Vacuum hose to the throttle body and VSV for EVAP.

    b. Vacuum hose to the fuel pressure regulator and VSV for fuel pressure.

    c. Vacuum hose to the VSV for fuel pressure and air intake chamber.

    d. Vacuum hose to the charcoal canister and VSV for EVAP.

124. Install the rear timing belt plate with the two bolts. Install both sides. Torque the bolts to 69 inch lbs. (8 Nm).

125. Install the water inlet and inlet housing. Torque the bolts to 13 ft. lbs. (18 Nm).

126. Connect the water bypass hose (from the IAC valve) to the water inlet housing.

127. Connect the water hose (from the reservoir tank) to the water inlet housing.

128. Install the right ignition coil with the two bolts. Connect the ignition coil connector.

129. Connect the accelerator, transmission and cruise control cables to the throttle body.

130. Install the throttle body cover.

131. Install the power steering pump with the nut and three bolts. Torque the bolts to 29 ft. lbs. (39 Nm) and the nut to 32 ft. lbs. (43 Nm).

132. Install the power steering reservoir tank and bracket with the three bolts.

133. Install the power steering tubes with the clamp and bolt.

134. Install the front catalytic converter.

135. Install the hydraulic pump. Tighten the 12mm fasteners to 12 ft. lbs. (16 Nm) and the other fasteners to 22 ft. lbs. (30 Nm).

136. Install the fan bracket with the two bolts and two nuts.

137. Align the knock pin on the right side camshaft with the knock pin of the timing pulley. Slide on the timing pulley with the right side mark facing forward. Tighten the bolt to 80 ft. lbs. (108 Nm).

138. Align the knock pin on the left side camshaft with the knock pin of the timing pulley. Slide on the timing pulley with the left side mark facing forward. Tighten the bolt to 80 ft. lbs. (108 Nm).

139. Turn the crankshaft pulley and align its groove with the **0** timing mark on the timing belt cover.

140. Turn each camshaft timing pulley and align the timing marks of the pulley with the timing belt rear plate.

141. Install the timing belt to the left side camshaft timing pulley.

142. Install the timing belt to the right side camshaft timing pulley.

143. The timing belt tensioner must be set prior to installation. The tensioner can be set by:

    a. Place a plate washer between the tensioner and a block. Using a

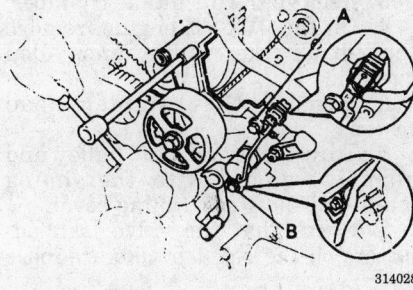

Install the hydraulic pump bolts in the proper locations — 1995–97 SC400

press, press in the pushrod using 220–2205 lbs. of pressure.

    b. Align the holes of the pushrod and housing, pass a 1.27mm Allen wrench through the holes to keep the setting position of the pushrod.

    c. Release the press and install the dust boot to the tensioner.

144. Loosely install the tensioner. Tighten the bolts evenly and alternately to 20 ft. lbs. (26 Nm). Remove the tool from the tensioner.

145. Turn the crankshaft pulley two complete revolutions from TDC to TDC. Always turn the crankshaft clockwise. Check that all belt and pulley marks align with their reference marks. If any mark is out of perfect alignment, the timing belt must be removed and reinstalled.

146. Install the spark plugs and tighten to 13 ft. lbs. (18 Nm).

147. Install the drive belt tensioner and tighten the bolt to 12 ft. lbs. (16 Nm).

148. Install the alternator and engine wire bracket. Tighten the nut and bolt to 27 ft. lbs. (37 Nm). Connect the electrical connections at the alternator.

149. Install both distributor housings and tighten the mounting bolts to 13 ft. lbs. (18 Nm). The distributors are marked L or R for correct installation.

150. Install the distributor rotors and caps.

151. Install the right side No. 2 timing belt cover. Except California vehicles, install the camshaft position sensor connector to the ignition coil bracket.

152. Install the left side No. 2 timing belt cover.

153. Install the drive belt idler pulley and cover plate. Tighten the bolt to 27 ft. lbs. (37 Nm).

154. Install and secure the ignition wires. Make certain that all clips and retainers are securely engaged and that the wires are properly routed.

155. Install the right side No. 3 timing belt cover by install the gaskets, four bolts and the two air control valve hoses.

156. Install the left side No. 3 timing belt cover, connect the vacuum hose and the connectors.

157. Install the right side engine wire cover.

158. Install the left side engine wire cover and connect the vacuum hoses.

159. Install the upper high tension cord covers. Fit the front side claw groove of the upper cover to claw of the lower cover.

160. Install the left side ignition coil and connect the coil connector.

161. Install the intake air connector and connect the air hoses.
162. Install the A/C compressor to the engine.
163. Install the radiator.
164. Install the radiator reservoir tank.
165. Install the accessory drive belt.
166. Install the battery and connect the cables.
167. For 1996–97 vehicles, install the V-bank cover.
168. Fill the engine coolant and check the transmission fluid level.
169. Start the engine and check carefully for fluid leaks. Check the ignition timing.
170. Install the engine under-covers.
171. Road test the vehicle and verify proper operation.

Valve Lash Adjuster

REMOVAL AND INSTALLATION

1. Disconnect the negative battery cable.
2. Remove the camshaft from the cylinder head assembly.
3. Remove the lash adjuster and the adjusting shims, using the proper tool.

To install:

NOTE: Be sure to store the lash adjusters and shims in the proper order.

4. Install the valve lash adjusters and shims.
5. Check that the valve lash adjuster rotates smoothly.
6. Install the camshaft in the cylinder head assembly.
7. Connect the negative battery cable.

Valve Lash

ADJUSTMENT

——— CAUTION ———

On models with an air bag, wait at least 90 seconds from the time that the ignition switch is turned to the LOCK position and the battery is disconnected before performing any further work.

ES300

NOTE: Adjust the valve clearance when the engine is cold.

1. Disconnect the negative battery cable.

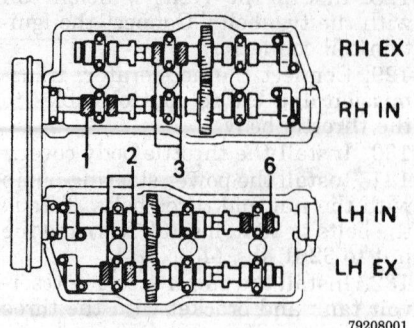

Adjust these valves FIRST — ES300

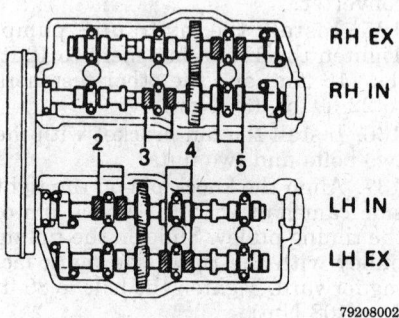

Adjust these valves SECOND — ES300

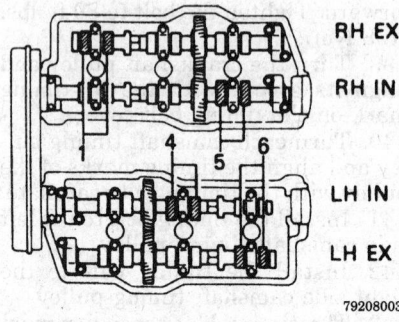

Adjust these valves THIRD — ES300

2. Disconnect the accelerator/throttle cable from the throttle linkage.
3. Remove the air intake chamber.
4. On the 3VZ-FE engine, remove the V-bank cover with a 5mm Allen wrench.
5. Remove the cylinder head covers.
6. Turn the crankshaft pulley and align it's groove with the timing mark **0** of the No. 1 timing cover.
7. Check that the valve lash adjusters on the No. 1 intake are loose

and the exhaust are tight. If not, turn the crankshaft on complete revolution (360 degrees).
8. Measure the clearance between the valve lash adjuster and the camshaft. Record the measurements on valves No. 1, 2, 3 and 6.
   a. The intake valve clearance cold is 0.005–0.009 in. (0.13–0.23mm).
   b. The exhaust valve clearance cold is 0.011–0.015 in. (0.27–0.37mm).
9. Turn the crankshaft ⅔ of a revolution (240 degrees) and check the clearance on valves No. 2, 3, 4 and 5 and record.
10. Turn the crankshaft another ⅔ of a revolution and check valves; No. 1, 4, 5 and 6 and record.
11. Remove the adjusting shim and turn the crankshaft to position the cam lobe of the camshaft on the adjusting valve upward. Press down the valve lash adjuster with the proper tool and place the proper tool between the camshaft and the valve lash adjuster. Remove the tool.
12. Remove the adjusting shim with the proper tool.
13. Use the accompanying charts to determine the correct size replacement shim. Install the specified valve shim on the valve lash adjuster with the proper tool.
14. Recheck the valve clearance.
15. Install the cylinder head covers and intake chamber.
16. Connect the negative battery cable.

SC300

NOTE: Adjust the valve clearance when the engine is cold.

1. Disconnect the negative battery cable.
2. Disconnect the accelerator/throttle cable from the throttle linkage.
3. Remove the cylinder head covers.
4. Turn the crankshaft pulley and align it's groove with the timing mark **0** of the No. 1 timing cover.
5. Check that the timing marks on the camshaft sprockets are in alignment with the marks on the No. 4 timing cover. If not, turn the crankshaft 1 complete revolution (360 degrees).
6. Uniformly tighten the camshaft bearing cap bolts in several passes, in the sequence, to 14 ft. lbs. (20 Nm).
7. Measure the clearance between the valve lash adjuster and the

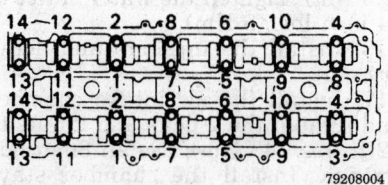

Camshaft bearing cap bolt tightening sequence — SC300

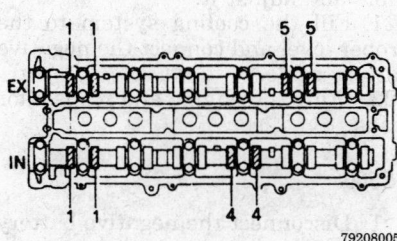

Adjust these valves FIRST — SC300

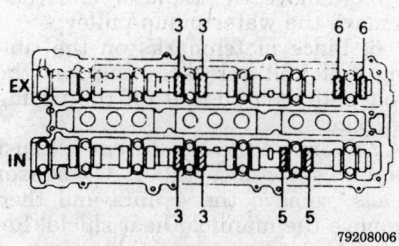

Adjust these valves SECOND — SC300

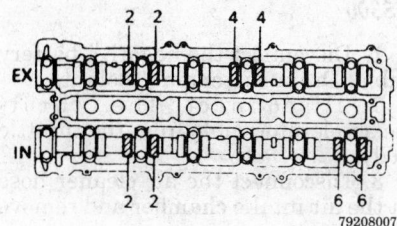

Adjust these valves THIRD — SC300

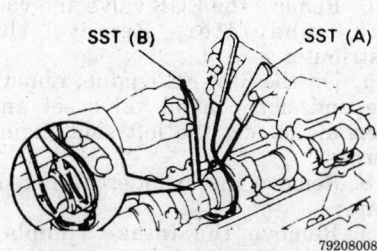

Press down the valve lash adjuster with a special tool — SC300 shown — others similar

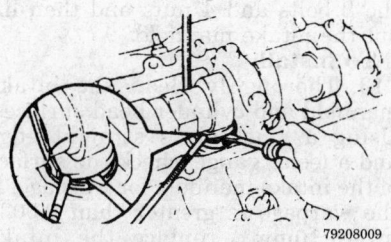

Removing the adjusting shim — SC300 shown — others similar

camshaft. Record the measurements on valves No. 1, 4 and 5.

    a. The intake valve clearance cold is 0.006–0.010 in. (0.15–0.25mm).

    b. The exhaust valve clearance cold is 0.010–0.014 in. (0.25–0.35mm).

    8. Turn the crankshaft 2/3 of a revolution (240 degrees) and check the clearance on valves No. 3, 5 and 6 and record.

    9. Turn the crankshaft another 2/3 of a revolution and check valves; No. 2, 4 and 6 and record.

    10. Remove the adjusting shim and turn the crankshaft to position the cam lobe of the camshaft on the adjusting valve upward. The notches should be perpendicular to the camshaft. Press down the valve lash adjuster with the proper tool and place the proper tool between the camshaft and the valve lash adjuster. Remove the tool.

    11. Remove the adjusting shim with the proper tool (a magnetic finger).

    12. Use the accompanying charts to determine the correct size replace-

ment shim. Install the specified valve shim on the valve lash adjuster with the proper tool.

    13. Recheck the valve clearance.

    14. Install the cylinder head covers and intake chamber.

    15. Connect the negative battery cable.

LS400 and SC400

    1. Disconnect the negative battery cable.

    2. Remove the No. 3 timing belt covers.

    3. Disconnect the spark plug wires and remove the cylinder head covers.

    4. Turn the crankshaft pulley and align it's groove with the timing mark **0** of the No. 1 timing cover. Check that the timing marks of the camshaft timing pulleys and timing belt rear plates are aligned. If not, turn the crankshaft 1 revolution (360 degrees) and align the mark.

    5. Measure the clearance between the valve lash adjuster and the camshaft on the valves in the first sequence and record.

    a. The intake valve clearance cold is 0.006–0.010 in. (0.15–0.25mm).

    b. The exhaust valve clearance cold is 0.010–0.014 in. (0.25–0.35mm).

    6. Turn the crankshaft 1 full revolution (360 degrees) and align the mark.

    7. Measure the clearance between the valve lash adjuster and the camshaft on the valves in the second sequence and record.

    8. Remove the adjusting shim and turn the crankshaft to position the cam lobe of the camshaft on the adjusting valve upward. Position the hole in the shim toward the outside of the cylinder head. Press down the valve lash adjuster with the proper tool and place the proper tool between the camshaft and the valve lash adjuster. Remove the tool.

    9. Remove the adjusting shim with the proper tool.

    10. Use the accompanying charts to determine the correct replacement shim. Install the specified valve shim on the valve lash adjuster with the proper tool.

    11. Recheck the valve clearance. Install the cylinder head covers.

    12. Connect the spark plug wires and install the No. 3 timing belt covers.

    13. Connect the negative battery cable.

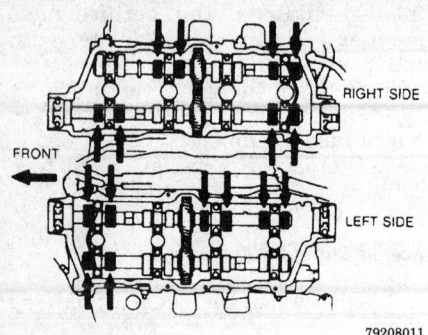

Adjust these valves FIRST — LS400 and SC400

79208011

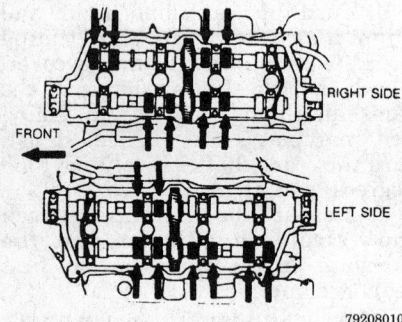

Adjust these valves SECOND — LS400 and SC300

79208010

ADJUSTING SHIM SELECTION CHART

New shim thickness   mm (in.)

Shim No.	Thickness	Shim No.	Thickness
01	2.50 (0.0984)	38	2.95 (0.1161)
63	2.55 (0.1004)	43	3.00 (0.1181)
06	2.60 (0.1024)	48	3.05 (0.1201)
66	2.65 (0.1043)	51	3.10 (0.1220)
13	2.70 (0.1063)	77	3.15 (0.1240)
18	2.75 (0.1083)	56	3.20 (0.1260)
23	2.80 (0.1102)	80	3.25 (0.1280)
28	2.85 (0.1122)	61	3.30 (0.1299)
33	2.90 (0.1142)		

79208014

Valve shim selection chart

Intake Manifold

CAUTION

On models with an air bag, wait at least 90 seconds from the time that the ignition switch is turned to the LOCK position and the battery is disconnected before performing any further work.

REMOVAL AND INSTALLATION

ES300

1. Disconnect the negative battery cable. Drain the engine coolant.
2. Disconnect the throttle/accelerator cable from the throttle body.
3. Disconnect the air cleaner hose at the air intake chamber and remove it.
4. Remove the V-bank cover on the 3VZ-FE engine.
5. Tag and disconnect all lines and hoses and then remove both the ISC valve and the throttle body.
6. Remove the EGR valve and vacuum modulator. Remove the distributor.
7. On the 3VZ-FE engine, remove the emission control valve set and then disconnect the left side engine harness.
8. Remove the cylinder head rear plate.
9. Remove the intake chamber stays, any wires and then remove the air intake chamber.
10. Remove the fuel injection delivery pipe and the injectors.
11. Remove the water outlet and the bypass outlet.
12. Remove the 2 bolts and the No. 2 idler pulley bracket stay. Remove the 8 bolts and 4 nuts and then lift out the intake manifold.

To install:

13. Thoroughly clean the intake manifold and cylinder head surfaces. Using a machinist's straight edge and a feeler gauge, check the surface of the intake manifold for warpage. If the warpage is greater than 0.0039 in. (0.10mm), replace the intake manifold.
14. Place new gaskets onto the intake manifold and position the intake manifold between the cylinder heads. Tighten the nuts and bolts to 13 ft. lbs. (18 Nm). Tighten the No. 2 pulley bracket bolts to 13 ft. lbs. (18 Nm).

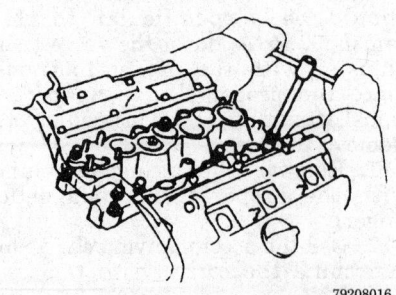

79208016

Removing the intake manifold — ES300

15. Install the water bypass outlet and tighten the bolts to 74 inch lbs. (8.3 Nm). Tighten the water outlet to 74 inch lbs. (8 Nm).
16. Install the injectors and delivery pipe.
17. Install the air intake chamber and tighten the 2 bolts and 2 nuts to 32 ft. lbs. (43 Nm); use an 8mm hex wrench. Install the chamber stays and tighten the mounting bolts to 29 ft. lbs. (39 Nm).
18. Install the remaining components. Tighten the emission control valve set to, 73 inch lbs. (8 Nm).
19. Unplug and connect all hoses.
20. If equipped with automatic transaxle, connect the accelerator cable and adjust it.
21. Fill the cooling system to the proper level and connect the negative battery cable.
22. Start the engine and inspect for leaks.

SC300

1. Disconnect the negative battery cable.
2. Drain the cooling system.
3. Tag and disconnect the spark plug wires at the spark plugs.
4. Remove the spark plugs. Remove the distributor with the spark plug leads attached.
5. Remove the radiator and then remove the water pump pulley.
6. Place matchmarks on the timing belt and sprockets, support the belt and then slide it off the timing sprockets.
7. Remove the No. 2 front exhaust pipe. Disconnect the 2 O_2 sensor leads, remove the 4 nuts and then remove the manifold heat shield. Remove the exhaust manifolds.
8. Loosen the 2 bolts and remove the water bypass outlet and the No. 1 bypass pipe. Remove the 3 O-rings from the outlet and the pipe.
9. Loosen the 2 bolts and nut and remove the water outlet. Loosen the clamp and remove the No. 1 bypass hose.
10. Remove the vacuum control valve set and the No. 2 vacuum pipe.
11. Disconnect the fuel return hose from the oil dipstick guide, remove the mounting bolt and pull the guide and dipstick from the pan. Plug the hole.
12. Remove the air intake chamber. Remove the fuel delivery pipe and then pull out the injectors. Remove the No. 1 and 2 fuel pipes.
13. Disconnect the engine harness from the intake manifold.
14. Loosen the 2 bolts and remove the intake manifold stay. Loosen the

6 bolts and 2 nuts and then lift out the intake manifold.

To install:

15. Using a new gasket, install the intake manifold and tighten the bolts and nuts to 15 ft. lbs. (21 Nm).

16. Install the mounting stay and tighten the bolts to 29 ft. lbs. (39 Nm).

17. Connect the engine harness to the manifold.

18. Install the 2 fuel pipes and tighten the bolts to 78 inch lbs. (9 Nm). Install the delivery pipe and injectors. Tighten the pipe bolts to 15 ft. lbs. (21 Nm).

19. Install the air intake chamber and tighten it to 15 ft. lbs. (21 Nm). Install the 2 stays and tighten them to 13 ft. lbs. (18 Nm); The No. 1 stay is marked with an **F** and the No. 2 stay is marked with an **R**.

20. Use a new O-ring and install the oil dipstick and guide.

21. Install the VCV set and the vacuum pipe. Tighten the set mounting bolts to 15 ft. lbs. (21 Nm).

22. Install the water bypass outlet and the pipe, tighten the bolts to 78 inch lbs. (9 Nm).

23. Using a new gasket, install the exhaust manifolds. Tighten the bolts to 29 ft. lbs. (39 Nm). Install the heat shield and tighten it to 13 ft. lbs. (18 Nm). Install the No. 2 front pipe.

24. Install the timing belt.

25. Install the radiator and water pump pulley.

26. Install the distributor and spark plugs. Connect the plug wires to the plugs.

27. Fill the cooling system to the proper level with coolant.

28. Connect the negative battery cable. Start the engine and check for leaks.

29. Road test the vehicle and check for unusual noise, shock, slippage, correct shift points and smooth operation.

30. Recheck the coolant and engine oil levels.

LS400 and SC400

1. Disconnect the negative battery cable. Drain the cooling system.

2. Remove the camshaft timing pulleys. Remove the cooling fan hydraulic pump on the SC400.

3. Disconnect the accelerator cable, the throttle control cable, if equipped with automatic transmission and the cruise control actuator cable.

4. Remove the high tension cord cover and the right side ignition coil.

5. Remove the water inlet housing mounting bolts and disconnect the

water bypass hose from the ISC valve.

6. Remove the water inlet and inlet housing assemblies. Remove the O-ring from the water inlet housing.

7. Remove the EGR pipe.

8. Disconnect the following:

   a. VSV connector

   b. Vacuum pipe hose

   c. EGR water bypass pipe

   d. Fuel pressure VSV

9. Disconnect the EGR vacuum hoses and remove the EGR VSV.

10. Disconnect the following hoses:

   a. Water bypass pipe hose from the ISC valve.

   b. Water bypass joint hose.

   c. Vacuum pipe hoses.

11. Disconnect the EGR gas temperature sensor, California only. Remove the EGR valve adapter.

12. Disconnect the following:

   a. Fuel pressure regulator vacuum hose.

   b. Air intake chamber vacuum hose.

   c. Vacuum hose from the EVAP BVSV.

13. Remove the mounting bolts, hoses and the vacuum pipe.

14. Remove the ISC valve.

15. Remove the throttle body sensor connectors and the water bypass pipe from the rear water bypass joint.

16. Disconnect the PCV valve hose. Remove the throttle body and gasket.

17. Disconnect the accelerator cable bracket and the brake booster vacuum union and hose.

18. Disconnect the cold start injector connector and the cold start injector tube from the right side delivery pipe, if equipped.

19. Disconnect the check connector from the intake chamber and remove the mounting nuts and bolts.

20. Remove the air intake chamber and the cold start injector, if equipped, tube and wire assembly.

21. Disconnect the engine wire from the intake manifold and from the

right side cylinder head. Disconnect the heater hoses.

22. Remove the delivery pipes and the fuel injectors. Remove the mounting bolts and nuts. Lift up the intake manifold.

To install:

23. Install the intake manifold, using new gaskets. Tighten the mounting nuts and bolts to 13 ft. lbs. (18 Nm).

NOTE: Align the port holes of the gasket and cylinder head. Be careful of the installation direction.

24. Install the delivery pipes and fuel injectors. Install the fuel return pipe with new gaskets. Tighten the union bolt to 26 ft. lbs. (35 Nm).

25. Connect the fuel hoses and the injector connectors. Connect the engine wire to the delivery pipes.

26. Connect the connectors on the left side delivery pipe, the water temperature sensor connector, cold start injector time switch connector and the water temperature sender gauge connector.

27. Connect the heater hoses and engine wire bracket. Install the engine wire to the bracket.

28. Install the cold start injector, tube and wire assembly if equipped. Tighten the mounting bolts to 69 inch lbs. (8 Nm).

29. Install the air intake chamber with new gaskets and tighten the mounting bolts to 13 ft. lbs. (18 Nm).

30. Connect the cold start injector tube to the right side delivery pipe and tighten the union bolt to 11 ft. lbs. (15 Nm), if equipped.

31. Connect the cold start injector connector as necessary. Install the accelerator cable bracket.

32. Install the brake booster union and connect the vacuum hose. Tighten the union bolt to 22 ft. lbs. (29 Nm).

33. Connect the water bypass hose to the throttle body and the PCV hose to the cylinder head cover.

34. Install the throttle body, using a new gasket. Tighten the mounting bolts to 13 ft. lbs. (18 Nm).

35. Install the water bypass pipe and connect the sensor connectors. Install the ISC valve and tighten the mounting bolts to 13 ft. lbs. (18 Nm). Connect the water bypass hose.

36. Install the vacuum pipe and the following hoses:

   a. Fuel pressure regulator vacuum hose.

   b. Vacuum hose to the upper port of the EVAP BVSV.

   c. Air intake chamber vacuum hose.

79208017

Removing the intake manifold — LS400 and SC400

d. Throttle body vacuum hoses.

37. Install the EGR valve adapter with a new gasket. Tighten the mounting bolts to 13 ft. lbs. (18 Nm).

38. Install the remaining components. Install the timing belt rear plates and tighten the bolts to 69 inch lbs. (8 Nm). Install the water inlet and inlet hosing and tighten the bolts to 13 ft. lbs. (18 Nm).

39. Fill the cooling system and connect the negative battery cable. Start the engine and check for leaks.

40. Recheck all the fluid levels. Roadtest the vehicle for proper operation.

Exhaust Manifold

CAUTION

On models with an air bag, wait at least 90 seconds from the time that the ignition switch is turned to the LOCK position and the battery is disconnected before performing any further work.

REMOVAL AND INSTALLATION

ES300

1. Disconnect the negative battery cable.

2. Raise the vehicle, support it on safety stands and then remove the engine undercovers.

3. Remove the 2 front exhaust pipe stay bolts. Disconnect the front pipe from the center pipe and remove the gasket. Loosen the 3 nuts and then remove the front pipe.

4. Disconnect the O_2 sensor at the right side manifold. Remove the 3 mounting nuts and lift off the outside heat insulator.

5. Remove the 6 nuts and lift off the right side manifold and gasket.

6. Loosen the 2 nuts and bolt and lift off the left side heat insulator. Remove the 6 nuts and lift off the left side manifold and gaskets.

To install:

7. Scrape the mating surfaces of all old gasket material.

8. Install the right manifold with a new gasket. Tighten the nuts to 29 ft. lbs. (39 Nm). Install the outer insulator.

9. Use a new gasket and install the left manifold. Tighten the nuts to 29 ft. lbs. (39 Nm). Install the outer insulator.

10. Install the front exhaust pipe and tighten the manifold-to-pipe nuts to 46 ft. lbs. (62 Nm). Tighten the pipe-to-converter nuts to 32 ft. lbs. (43 Nm).

11. Connect the O_2 sensor, install the undercovers and then lower the vehicle. Connect the battery cable.

SC300

1. Disconnect the negative battery cable.

2. Raise the vehicle, support it on safety stands and then remove the engine undercovers.

3. Remove the No. 2 front exhaust pipe bolts and disconnect it from the front exhaust pipe. Loosen the 4 nuts and then remove the front pipe.

4. Disconnect the two O_2 sensors at the manifold. Remove the 4 mounting nuts and lift off the outside heat insulator.

5. Remove the 4 nuts and disconnect the manifolds from the pipe. Loosen the mounting bolts and remove the two manifolds and the gasket.

To install:

6. Scrape the mating surfaces of all old gasket material.

7. Install the manifolds with a new gasket. Tighten the nuts to 29 ft. lbs. (39 Nm). Install the outer insulator and tighten the nuts to 13 ft. lbs. (18 Nm).

8. Use a new gasket and install the No. 2 front pipe. Tighten the nuts to 46 ft. lbs. (62 Nm).

9. Connect the front exhaust pipe and tighten the bolts and nuts to 32 ft. lbs. (43 Nm).

10. Connect the O_2 sensors, install the undercovers and then lower the vehicle. Connect the battery cable.

LS400 and SC400

1. Disconnect the negative battery cable. Drain the cooling system.

2. Remove the camshaft timing pulleys. Remove the cooling fan hydraulic pump on the SC400.

3. Disconnect the accelerator cable, the throttle control cable, if equipped with automatic transaxle and the cruise control actuator cable.

Removing the exhaust manifolds — SC300

79208018

4. Remove the high tension cord cover and the right side ignition coil.

5. Remove the water inlet housing mounting bolts and disconnect the water bypass hose from the ISC valve.

6. Remove the water inlet and inlet housing assemblies. Remove the O-ring from the water inlet housing.

7. Remove the EGR pipe.

8. Disconnect the following:
   a. VSV connector
   b. Vacuum pipe hose
   c. EGR water bypass pipe
   d. Fuel pressure VSV

9. Disconnect the EGR vacuum hoses and remove the EGR VSV.

10. Disconnect the following hoses:
   a. Water bypass pipe hose from the ISC valve.
   b. Water bypass joint hose.
   c. Vacuum pipe hoses.

11. Disconnect the EGR gas temperature sensor, California only. Remove the EGR valve adapter.

12. Disconnect the following:
   a. Fuel pressure regulator vacuum hose.
   b. Air intake chamber vacuum hose.
   c. Vacuum hose from the EVAP BVSV.

13. Remove the mounting bolts, hoses and the vacuum pipe.

14. Remove the ISC valve.

15. Remove the throttle body sensor connectors and the water bypass pipe from the rear water bypass joint.

16. Disconnect the PCV valve hose. Remove the throttle body and gasket.

17. Disconnect the accelerator cable bracket and the brake booster vacuum union and hose.

18. Disconnect the cold start injector connector and the cold start injector tube from the right side delivery pipe, if equipped.

19. Disconnect the check connector from the intake chamber and remove the mounting nuts and bolts.

20. Remove the air intake chamber and the cold start injector, tube and wire assembly, if equipped.

21. Disconnect the engine wire from the intake manifold and from the right side cylinder head. Disconnect the heater hoses.

22. Remove the delivery pipes and the fuel injectors. Remove the mounting bolts and nuts. Lift up the intake manifold.

23. Remove the front and rear water bypass joint.

24. Raise and safely support the vehicle. Remove the front exhaust pipe and the main catalytic converters. Lower the vehicle.

25. Disconnect the right side oxygen sensor. Remove the mounting bolts and nuts and remove the right side exhaust manifold.

26. Remove the oil dipstick and guide. Disconnect the left side oxygen sensor.

27. Remove the mounting bolts and nuts and remove the left side exhaust manifold.

To install:

28. Install the right side exhaust manifold with a new gasket (the painted marks should face the manifold) and tighten the mounting bolts to 29 ft. lbs. (39 Nm). Connect the right side oxygen sensor connector.

29. Install the left side exhaust manifold with a new gasket (the painted marks should face the manifold) and tighten the mounting bolts to 29 ft. lbs. (39 Nm). Connect the left side oxygen sensor connector.

30. Install the oil dipstick and guide. Raise and safely support the vehicle.

31. Install the catalytic converters and front exhaust pipe. Lower the vehicle.

32. Install the front and rear water bypass joints. Tighten the mounting bolts to 13 ft. lbs. (18 Nm).

33. Install the intake manifold, using new gaskets. Tighten the mounting nuts and bolts to 13 ft. lbs. (18 Nm).

34. Install the delivery pipes and fuel injectors. Install the fuel return pipe with new gaskets. Tighten the union bolt to 26 ft. lbs. (35 Nm).

35. Connect the fuel hoses and the injector connectors. Connect the engine wire to the delivery pipes.

36. Connect the connectors on the left side delivery pipe, the water temperature sensor connector, cold start injector time switch connector and the water temperature sender gauge connector.

37. Connect the heater hoses and engine wire bracket. Install the engine wire to the bracket.

38. Install the cold start injector, tube and wire assembly. Tighten the mounting bolts to 69 inch lbs. (8 Nm), if equipped.

39. Install the air intake chamber with new gaskets and tighten the mounting bolts to 13 ft. lbs. (18 Nm).

40. Connect the cold start injector tube to the right side delivery pipe and tighten the union bolt to 11 ft. lbs. (15 Nm), if equipped.

41. Connect the cold start injector connector, if necessary. Install the accelerator cable bracket.

42. Install the brake booster union and connect the vacuum hose.

Tighten the union bolt to 22 ft. lbs. (29 Nm).

43. Connect the water bypass hose to the throttle body and the PCV hose to the cylinder head cover.

44. Install the throttle body, using a new gasket. Tighten the mounting bolts to 13 ft. lbs. (18 Nm).

45. Install the water bypass pipe and connect the sensor connectors. Install the ISC valve and tighten the mounting bolts to 13 ft. lbs. (18 Nm). Connect the water bypass hose.

46. Install the vacuum pipe and the assorted hoses. Install the remaining components.

47. Connect and adjust the accelerator cable, the automatic transmission throttle cable and the cruise control actuator cable. Install the cooling fan hydraulic pump on the SC400. Install the camshaft timing pulleys.

48. Fill the cooling system and connect the negative battery cable. Start the engine and check for leaks.

49. Recheck all the fluid levels. Roadtest for proper operation.

Timing Belt Front Cover

─────── **CAUTION** ───────
On models with an air bag, wait at least 90 seconds from the time that the ignition switch is turned to the LOCK position and the battery is disconnected before performing any further work.

REMOVAL AND INSTALLATION

ES300

1. Disconnect the cable from the negative battery terminal.

2. Remove the power steering pump reservoir and position it out of the way. Remove the right fender apron seal and then remove the alternator and power steering belts.

3. Remove the coolant reservoir hose, the washer tank and then the coolant overflow tank.

4. Remove the right side engine mount stays.

5. Position a piece of wood on a floor jack and then slide the jack under the oil pan. Raise the jack slightly until the pressure is off the engine mounts.

6. Remove the engine control rod.

7. Remove the spark plugs.

8. Remove the right side engine mounting bracket.

9. Remove the 8 bolts and lift off the upper (No. 2) cover.

10. Paint matchmarks on the timing belt at all points where it meshes

with the pulleys and the lower timing cover.

11. Set the No. 1 cylinder to TDC of the compression stroke and check that the timing marks on the camshaft timing pulleys are aligned with those on the No. 3 timing cover. If not, turn the engine 1 complete revolution (360 degrees) and check again.

12. Remove the timing belt tensioner and the dust boot.

13. Turn the right camshaft pulley clockwise slightly to release tension and then remove the timing belt from the pulleys.

14. Use a spanner wrench to hold the pulley, loosen the set bolt and then remove the camshaft timing pulleys along with the knock pin. Be sure to keep track of which is which.

15. Remove the No. 2 idler pulley.

16. Remove the crankshaft pulley and then pull off the lower (No. 1) timing belt cover.

To install:

17. Install the lower (No. 1) timing cover and tighten the bolts.

18. Align the crankshaft pulley set key with the key groove on the pulley and slide the pulley on. Tighten the bolt to 181 ft. lbs. (245 Nm).

19. Install the No. 2 idler pulley and tighten the bolt to 29 ft. lbs. (39 Nm). Check that the pulley moves smoothly.

20. Install the left camshaft pulley with the flange side outward. Align the knock pin hole in the camshaft with the knock pin groove on the pulley and then install the pin. Tighten the bolt to 80 ft. lbs. (108 Nm).

21. Set the No. 1 cylinder to TDC again. Turn the right camshaft until the knock pin hole is aligned with the timing mark on the No. 3 belt cover. Turn the left pulley until the marks on the pulley are aligned with the mark on the No. 3 timing cover.

22. Check that the mark on the belt matches with the edge of the lower cover. If not, shift it on the crank pulley until it does. Turn the left pulley clockwise a bit and align the mark on the timing belt with the timing mark on the pulley. Slide the belt over the left pulley. Now move the pulley until the marks on it align with the 1 on the No. 3 cover. There should be tension on the belt between the crankshaft pulley and the left camshaft pulley.

23. Align the installation mark on the timing belt with the mark on the right side camshaft pulley. Hang the belt over the pulley with the flange facing inward. Align the timing marks on the right pulley with the 1

on the No. 3 cover and slide the pulley onto the end of the camshaft. Move the pulley until the camshaft knock pin hole is aligned with the groove in the pulley and then install the knock pin. Tighten the bolt to 55 ft. lbs. (75 Nm).

24. Position a plate washer between the timing belt tensioner and the a block and then press in the pushrod until the holes are aligned between it and the housing. Slide a 1.27mm (3VZ-FE engine — 1.5mm) Allen wrench through the hole to keep the pushrod set. Install the dust boot and then install the tensioner. Tighten the bolts to 20 ft. lbs. (26 Nm). Don't forget to pull out the Allen wrench.

25. Turn the crankshaft clockwise 2 complete revolutions and check that all marks are still in alignment. If they aren't, remove the timing belt and start over again.

26. Install the remaining components. Install and adjust the drive belts.

27. Install the fender apron seal and the wheel.

28. Install the No. 2 stay and tighten the bolt to 55 ft. lbs. (75 Nm), the nut to 46 ft. lbs. (62 Nm). Install the No. 3 stay and tighten it to 54 ft. lbs. (73 Nm).

29. Install the coolant overflow tank and the washer tank.

30. Install the power steering reservoir tank and the cruise control actuator.

31. Connect the battery cable, start the vehicle and check for any leaks.

SC300

1. Disconnect the negative battery cable.

2. Drain the engine coolant. Remove the water pump pulley. Remove the radiator.

3. Remove the oil filler cap.

4. Using a 5mm Allen wrench, remove the 9 bolts and lift off the No. 2 and No. 3 timing covers, the top 2.

5. Rotate the crankshaft pulley clockwise so its groove is aligned with the **0** mark in the No. 1 (lower) timing cover. Check that the timing marks on the camshaft timing sprockets are aligned with the marks on the No. 4 (inner) cover, if not, rotate the crankshaft 1 complete revolution (360 degrees).

6. Alternately loosen the 2 tensioner mounting bolts and remove them, the tensioner and the dust boot. Slide the timing belt off of the 2 camshaft sprockets. Its a good idea to matchmark the belt to the pulleys.

7. Making sure the timing belt is securely supported, hold the crankshaft pulley with a spanner wrench and loosen the mounting bolt. Remove the bolt and the pulley.

8. Remove the 5 bolts and then lift off the lower No. 1 timing cover.

To install:

9. Position 2 new gaskets on the lower cover and then install the cover.

10. Align the crankshaft pulley set key with the groove in the pulley and slide it onto the shaft. Secure the pulley with a spanner and tighten the bolt to 239 ft. lbs. (324 Nm).

11. Make sure the crankshaft timing marks and the camshaft sprocket marks are still in alignment and carefully slide the timing belt back over the sprockets. The marks you made previously on the belt and sprockets should still be in alignment.

12. Press the tensioner pushrod into the housing and slide an 1.5mm hex key through the holes to keep it retracted. Install the dust boot and tensioner and tighten the bolts to 20 ft. lbs. (26 Nm). Remove the hex key.

13. Rotate the crankshaft 2 complete revolutions and check that all timing marks are still in alignment. If they aren't, reinstall the belt and try it again.

14. Install the 2 upper covers.

15. Install the radiator and water pump pulley. Refill the engine with coolant. Connect the battery and road test the vehicle.

LS400 and SC400

1. Disconnect the negative battery cable and the positive battery cable. Remove the battery.

2. Remove the air duct and dust covers. Remove the engine undercover.

3. Drain the cooling system. Remove the drive belt, fan, fluid coupling and fan pulley.

4. Remove the radiator, air cleaner and throttle body cover. Remove the air intake connector pipe.

5. Remove the air conditioning compressor and power steering pump. Do not disconnect the hoses.

6. Remove the upper high tension cord cover and the right side engine wire cover.

7. Disconnect the PCV hose and remove the left side engine wire cover. Remove the right side No. 3 timing cover.

8. Disconnect and tag the vacuum hoses and remove the left side engine wire cover. Disconnect the spark plug wires.

9. Remove the bolt, cover plate and idler pulley.

10. Disconnect the camshaft position sensor connector and remove the right side No. 2 timing belt cover.

11. Disconnect and remove the ignition coil. Disconnect the hoses and wires from the water bypass pipe. Remove the water bypass pipe.

12. Disconnect the camshaft position sensor connector and remove the left No. 2 timing belt cover.

13. Remove the distributor caps and rotors. Disconnect and remove both distributor housings.

14. Disconnect and remove the alternator. Remove the drive belt tensioner and the spark plugs.

15. Turn the crankshaft pulley and align it's groove with the timing mark **0** of the No. 1 timing cover. Check that the timing marks of the camshaft timing pulleys and timing belt rear plates are aligned. If not, turn the crankshaft 1 full revolution (360 degrees).

16. Remove the timing belt tensioner. Using the proper tool, loosen the tension between the left side and right side timing pulleys by slightly turning the left side camshaft clockwise.

17. Disconnect the timing belt from the camshaft timing pulleys. Using the proper tool, remove the bolt and the timing pulleys.

18. Remove the bolt and the crankshaft pulley with the proper tool. Remove the fan bracket. Remove the hydraulic pump on the SC400.

19. Remove the mounting bolts and the No. 1 timing belt cover.

To install:

20. Install the timing belt guide (No. 1 crank angle sensor plate) with the cup side facing forward. Replace the timing belt cover spacer.

21. Install the No. 1 timing belt cover and tighten the mounting bolts. Install the hydraulic pump on the SC400. Install the fan bracket.

22. Align the pulley set key on the crankshaft with the key groove of the pulley. Install the pulley, using the proper tool to tap in the pulley. Tighten the pulley bolt to 181 ft. lbs. (245 Nm).

23. Align the knock pin on the right side camshaft with the knock pin of the timing pulley. Slide on the timing pulley with the right side mark facing forward. Tighten the bolt to 80 ft. lbs. (108 Nm).

24. Align the knock pin on the left side camshaft with the knock pin of the timing pulley. Slide on the timing pulley with the left side mark facing

forward. Tighten the bolt to 80 ft. lbs. (108 Nm).

25. Turn the crankshaft pulley and align it's groove with the **0** timing mark on the No. 1 timing belt cover. Using the proper tool, turn the crankshaft timing pulley and align the timing marks of the camshaft timing pulley and the timing belt rear plate.

26. Install the timing belt to the left side camshaft timing pulley by:

 a. Using the proper tool, slightly turn the left side timing pulley clockwise. Align the installation mark of the timing belt with the timing mark of the camshaft timing pulley and hang the timing belt on the left side camshaft pulley.

 b. Using the proper tool, align the timing marks of the left side camshaft pulley and the timing belt rear plate.

 c. Check that the timing belt has tension between crankshaft timing pulley and the left side camshaft pulley.

27. Install the timing belt to the right side camshaft timing pulley by:

 a. Using the proper tool, slightly turn the right side timing pulley clockwise. Align the installation mark of the timing belt with the timing mark of the camshaft timing pulley and hang the timing belt on the right side camshaft pulley.

 b. Using the proper tool, align the timing marks of the right side camshaft pulley and the timing belt rear plate.

 c. Check that the timing belt has tension between crankshaft timing pulley and the right side camshaft pulley.

28. The timing belt tensioner must be set prior to installation. The tensioner can be set by:

 a. Place a plate washer between the tensioner and a block. Using a suitable press, press in the pushrod using 220–2205 lbs. of pressure.

 b. Align the holes of the pushrod and housing, pass the proper tool (an 1.27mm Allen wrench) through the holes to keep the setting position of the pushrod.

 c. Release the press and install the dust boot to the tensioner.

29. Install the tensioner and tighten the bolts to 20 ft. lbs. (26 Nm). Remove the tool from the tensioner.

30. Turn the crankshaft pulley 2 complete revolutions from TDC to TDC. Always turn the crankshaft clockwise. Check that each pulley aligns with the timing marks.

31. Install the spark plugs and tighten to 13 ft. lbs. (18 Nm). Install

the drive belt tensioner and tighten the bolt to 12 ft. lbs. (16 Nm).

32. Install the alternator and engine wire bracket. Tighten the nut and bolt to 26 ft. lbs. (35 Nm) on the LS400 or 27 ft. lbs. (37 Nm) on the SC400. Connect the electrical connections at the alternator.

33. Install both distributor housings and tighten the mounting bolts to 13 ft. lbs. (18 Nm). Replace the distributor rotors and caps.

34. Install the right side No. 2 timing belt cover and tighten the 10mm bolts to 69 inch lbs. (8 Nm) and the 12mm bolts to 12 ft. lbs. (16 Nm). Connect the camshaft position sensor connector.

35. Install the left side No. 2 timing belt cover and connect the camshaft position sensor connector.

36. Install the water bypass pipe and connect the hoses and connectors.

37. Replace the left side ignition coil and connect the coil connector. Install the idler pulley and cover plate. Tighten the bolt to 27 ft. lbs. (37 Nm).

38. Install and secure the ignition wires. Install the right side No. 3 timing belt cover.

39. Install the left side No. 3 timing belt cover and connect the vacuum hose and connectors. Install the right side engine wire cover.

40. Install the left side engine wire cover and connect the vacuum hoses.

41. Install the upper high tension cord covers. Fit the front side claw groove of the upper cover to claw of the lower cover.

42. Install the power steering pump and the air conditioning compressor.

43. Install the throttle body cover and the air cleaner.

44. Install the radiator, fan pulley, fan coupling, fan and drive belt.

45. Install the engine undercover and replace the battery.

46. Install the air ducts and dust covers. Connect the battery cables.

47. Refill the cooling system. Check the ignition timing.

Timing Belt and Tensioner

REMOVAL AND INSTALLATION

ES300

1. Remove the upper and lower timing belt covers as previously detailed.

2. Remove the timing belt guide.

3. Remove the timing belt.

NOTE: If the timing belt is to be reused, draw a directional arrow on the timing belt in the direction of engine rotation (clockwise) and place matchmarks on the timing belt and crankshaft gear to match the drilled mark on the pulley.

4. With a 10mm hex wrench, remove the setbolt, plate washer and the No. 1 idler pulley.

 To install:

5. Turn the crankshaft until the key groove in the crankshaft timing pulley is facing upward. Slide the timing pulley on so the flange side faces inward.

6. Apply bolt adhesive to the first few threads of the No. 1 idler pulley setbolt, install the plate washer and pulley and then tighten the bolt to 25 ft. lbs. (34 Nm).

7. Install the timing belt on the crankshaft timing, No. 1 idler and water pump pulleys.

NOTE: If the old timing belt is being reinstalled, make sure the directional arrow is facing in the original direction and that the belt and crankshaft gear matchmarks are properly aligned.

8. Install the lower (No. 1) timing cover and tighten the bolts.

9. Align the crankshaft pulley set key with the key groove on the pulley and slide the pulley on. Tighten the bolt to 181 ft. lbs. (245 Nm).

10. Install the No. 2 idler pulley and tighten the bolt to 29 ft. lbs. (39 Nm). Check that the pulley moves smoothly.

11. Install the left camshaft pulley with the flange side outward. Align the knock pin hole in the camshaft with the knock pin groove on the pulley and then install the pin. Tighten the bolt to 80 ft. lbs. (108 Nm).

12. Set the No. 1 cylinder to TDC again. Turn the right camshaft until the knock pin hole is aligned with the timing mark on the No. 3 belt cover. Turn the left pulley until the marks on the pulley are aligned with the mark on the No. 3 timing cover.

13. Check that the mark on the belt matches with the edge of the lower cover. If not, shift it on the crank pulley until it does. Turn the left pulley clockwise a bit and align the mark on the timing belt with the timing mark on the pulley. Slide the belt over the left pulley. Now move the pulley until the marks on it align with the 1 on the No. 3 cover. There should be tension on the belt between the crank-

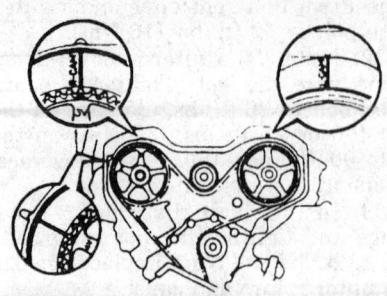

Before removing the timing belt, make sure all marks are there — ES300

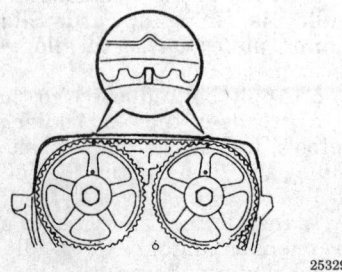

Aligning the timing marks — 1993–97 GS300 and SC300

253296

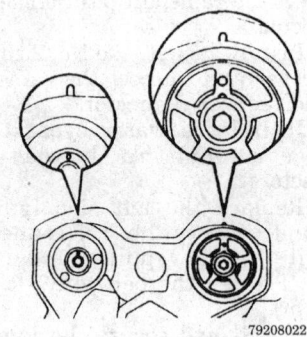

Camshaft sprocket alignment — ES300

79208022

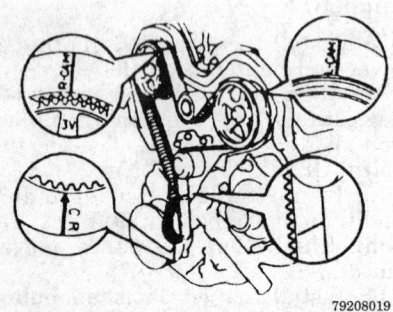

Before removing the timing belt, make sure all marks are there — ES300

79208019

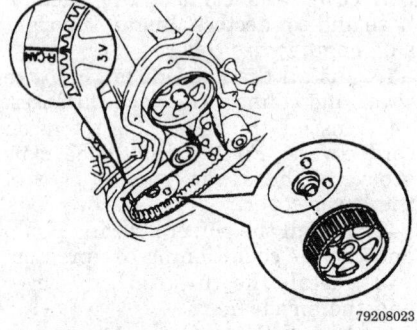

Align the installation mark on the belt with the mark on the right side sprocket — ES300

79208023

Check that the marks on the camshaft sprocket and the No. 3 cover are there — ES300

79208020

15. Position a plate washer between the timing belt tensioner and the a block and then press in the pushrod until the holes are aligned between it and the housing. Slide a 1.27mm, except 3VZ-FE engine or 1.5mm for 3VZ-FE Allen wrench through the hole to keep the pushrod set. Install the dust boot and then install the tensioner. Tighten the bolts to 20 ft. lbs. (26 Nm). Don't forget to pull out the Allen wrench.

16. Turn the crankshaft clockwise 2 complete revolutions and check that all marks are still in alignment. If they aren't, remove the timing belt and start over again.

17. Install the remaining components. Install and adjust the drive belts.

18. Install the fender apron seal and the wheel.

19. Install the No. 2 stay and tighten the bolt to 55 ft. lbs. (75 Nm), the nut to 46 ft. lbs. (62 Nm). Install the No. 3 stay and tighten it to 54 ft. lbs. (73 Nm).

20. Install the coolant overflow tank and the washer tank.

21. Install the power steering reservoir tank and the cruise control actuator.

22. Connect the battery cable, start the vehicle and check for any leaks.

SC300

1. Remove the 2 upper and lower timing belt covers.
2. Remove the timing belt guide.
3. Remove the timing belt.

NOTE: If the timing belt is to be reused, draw a directional arrow on the timing belt in the direction of engine rotation (clockwise) and place matchmarks on the timing belt and crankshaft gear to match the drilled mark on the pulley.

4. With a 10mm hex wrench, remove the pivot bolt, plate washer and the idler pulley.
To install:
5. Turn the crankshaft until the key groove in the crankshaft timing pulley is facing upward. Slide the timing pulley on so the flange side faces inward.
6. Apply bolt adhesive to the first few threads of the idler pulley pivot bolt, install the plate washer and pulley and then tighten the bolt to 25 ft. lbs. (34 Nm).

shaft pulley and the left camshaft pulley.
14. Align the installation mark on the timing belt with the mark on the right side camshaft pulley. Hang the belt over the pulley with the flange facing inward. Align the timing marks on the right pulley with the 1 on the No. 3 cover and slide the pulley onto the end of the camshaft. Move the pulley until the camshaft knock pin hole is aligned with the groove in the pulley and then install the knock pin. Tighten the bolt to 55 ft. lbs. (75 Nm).

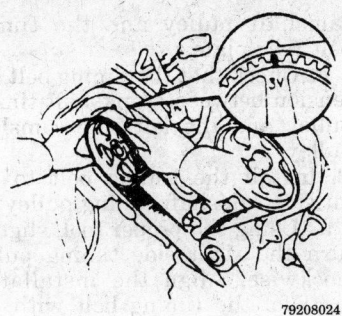

Align the marks on the right sprocket and the No. 3 cover and then slide the belt on — ES300

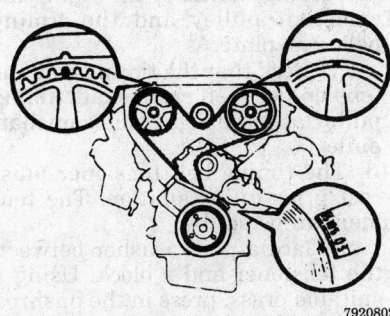

Check that the sprocket aligns with the timing marks — ES300

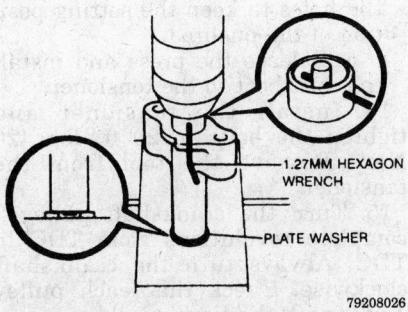

1.27MM HEXAGON WRENCH

PLATE WASHER

Align the holes on the pushrod and the housing and then insert an Allen wrench — ES300

7. Install the timing belt on the crankshaft timing pulley and the idler pulleys.

NOTE: If the old timing belt is being reinstalled, make sure the directional arrow is facing in the original direction and that the belt and crankshaft gear matchmarks are properly aligned.

8. Install the timing belt guide. Install the lower (No. 1) timing cover and tighten the bolts.
9. Align the crankshaft pulley set key with the key groove on the pulley

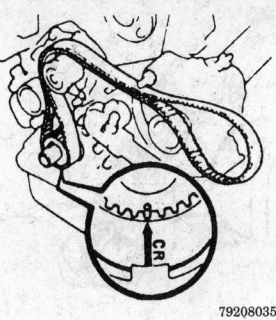

Matchmark the belt to the drilled mark on the crankshaft sprocket — ES300

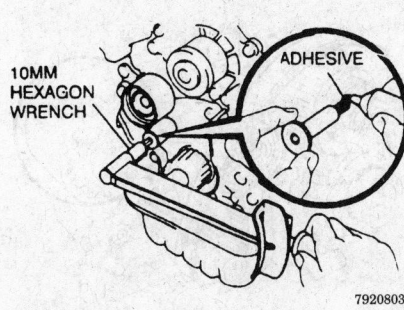

10MM HEXAGON WRENCH

ADHESIVE

Coat the end of the idler pulley set bolt with adhesive — ES300

Check that the timing marks on the camshaft sprocket are aligned with the marks on the No. 4 cover — SC300

and slide the pulley on. Tighten the bolt to 239 ft. lbs. (324 Nm).
10. Install the camshaft pulleys. Align the knock pin on the camshaft with the key groove on the pulley and then install the pulley. Tighten the bolt to 59 ft. lbs. (79 Nm).
11. Set the No. 1 cylinder to TDC again. Turn the camshaft until the sprocket timing marks are aligned with the timing marks on the No. 4 belt cover.
12. Check that the marks on the belt matches with those on the sprockets and then slide it over the

sprockets. If not, shift it on the crank pulley until it does.
13. Position a plate washer between the timing belt tensioner and the a block and then press in the pushrod until the holes are aligned between it and the housing. Slide a 1.5mm Allen wrench through the hole to keep the pushrod set. Install the dust boot and then install the tensioner. Tighten the bolts to 20 ft. lbs. (26 Nm). Don't forget to pull out the Allen wrench.
14. Turn the crankshaft clockwise 2 complete revolutions and check that all marks are still in alignment. If they aren't, remove the timing belt and start over again.
15. Position new gaskets and then install the upper (No. 2 and No. 3) timing covers.
16. Install and adjust the drive belts.
17. Connect the battery cable, start the vehicle and check for any leaks.

LS400 and SC400

1. Remove the 2 upper and lower timing belt covers.
2. Remove the timing belt guide (No. 1 crank position sensor plate).
3. Remove the timing belt.

NOTE: If the timing belt is to be reused, draw a directional arrow on the timing belt in the direction of engine rotation (clockwise) and place matchmarks on the timing belt and crankshaft gear to match the drilled mark on the pulley.

To install:
4. Align the installation mark on the timing belt with the drilled mark of the crankshaft timing pulley. Install the timing belt on the crankshaft timing pulley, No. 1 idler pulley and the No. 2 idler pulley.

NOTE: If the old timing belt is being reinstalled, make sure the directional arrow is facing in the original direction and that the belt and crankshaft gear matchmarks are properly aligned.

5. Install the timing belt guide (No. 1 crank angle sensor plate) with the cup side facing forward. Replace the timing belt cover spacer.
6. Install the No. 1 timing belt cover and tighten the mounting bolts. Install the hydraulic pump on the SC400. Install the fan bracket.
7. Align the pulley set key on the crankshaft with the key groove of the pulley. Install the pulley, using the proper tool to tap in the pulley.

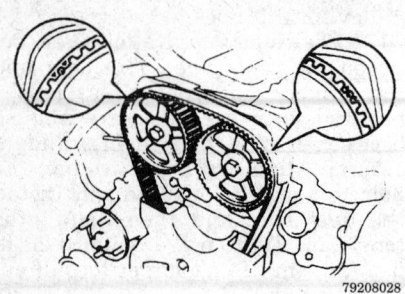

Place matchmarks on the belt and sprockets — SC300

79208028

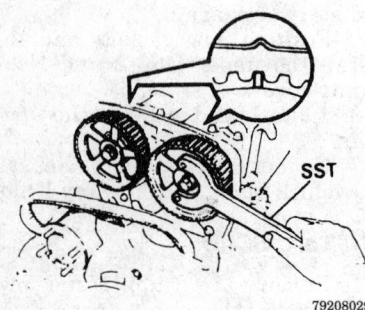

SST

79208029

Rotate the camshaft until the timing marks on the sprockets and No. 4 timing cover are aligned — SC300

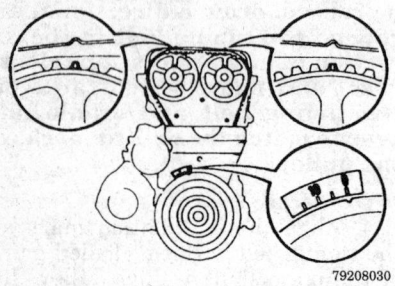

79208030

Rotate the crankshaft and check that the timing marks are still in alignment — SC300

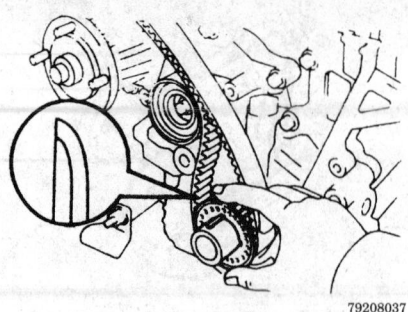

Install the timing belt guide with the cup side out — SC300

79208037

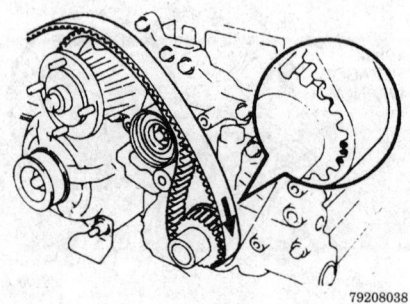

Draw a directional arrow on the belt — SC300

79208038

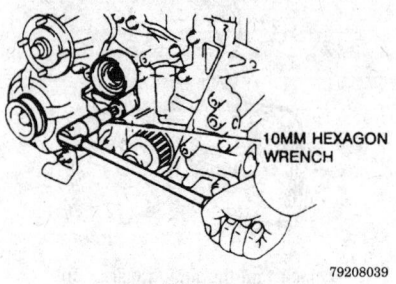

10MM HEXAGON WRENCH

79208039

Removing the idler pulley — SC300

Tighten the pulley bolt to 181 ft. lbs. (245 Nm).

8. Align the knock pin on the right side camshaft with the knock pin of the timing pulley. Slide on the timing pulley with the right side mark facing forward. Tighten the bolt to 80 ft. lbs. (108 Nm).

9. Align the knock pin on the left side camshaft with the knock pin of the timing pulley. Slide on the timing pulley with the left side mark facing forward. Tighten the bolt to 80 ft. lbs. (108 Nm).

10. Turn the crankshaft pulley and align it's groove with the **0** timing

mark on the No. 1 timing belt cover. Using the proper tool, turn the crankshaft timing pulley and align the timing marks of the camshaft timing pulley and the timing belt rear plate.

11. Install the timing belt to the left side camshaft timing pulley by:

a. Using the proper tool, slightly turn the left side timing pulley clockwise. Align the installation mark of the timing belt with the timing mark of the camshaft timing pulley and hang the timing belt on the left side camshaft pulley.

b. Using the proper tool, align the timing marks of the left side

camshaft pulley and the timing belt rear plate.

c. Check that the timing belt has tension between crankshaft timing pulley and the left side camshaft pulley.

12. Install the timing belt to the right side camshaft timing pulley by:

a. Using the proper tool, slightly turn the right side timing pulley clockwise. Align the installation mark of the timing belt with the timing mark of the camshaft timing pulley and hang the timing belt on the right side camshaft pulley.

b. Using the proper tool, align the timing marks of the right side camshaft pulley and the timing belt rear plate.

c. Check that the timing belt has tension between crankshaft timing pulley and the right side camshaft pulley.

13. The timing belt tensioner must be set prior to installation. The tensioner can be set by:

a. Place a plate washer between the tensioner and a block. Using a suitable press, press in the pushrod using 220–2205 lbs. of pressure.

b. Align the holes of the pushrod and housing, pass the proper tool (1.27mm Allen wrench) through the holes to keep the setting position of the pushrod.

c. Release the press and install the dust boot to the tensioner.

14. Install the tensioner and tighten the bolts to 20 ft. lbs. (26 Nm). Remove the tool from the tensioner.

15. Turn the crankshaft pulley 2 complete revolutions from TDC to TDC. Always turn the crankshaft clockwise. Check that each pulley aligns with the timing marks.

16. Install the spark plugs and tighten to 13 ft. lbs. (18 Nm). Install the drive belt tensioner and tighten the bolt to 12 ft. lbs. (16 Nm).

17. Install the alternator and engine wire bracket. Tighten the nut and bolt to 26 ft. lbs. (35 Nm) on the LS400 or 27 ft. lbs. (37 Nm) on the SC400. Connect the electrical connections at the alternator.

18. Install both distributor housings and tighten the mounting bolts to 13 ft. lbs. (18 Nm). Replace the distributor rotors and caps.

19. Install the right side No. 2 timing belt cover and tighten the 10mm bolts to 69 inch lbs. (8 Nm) and the 12mm bolts to 12 ft. lbs. (16 Nm). Connect the camshaft position sensor connector.

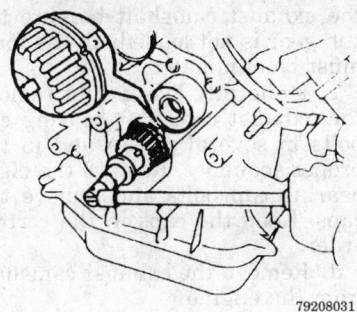

Aligning the timing mark of the crankshaft timing pulley and oil pump body — LS400

Aligning the timing mark of the camshaft timing pulley and timing belt rear cover — LS400

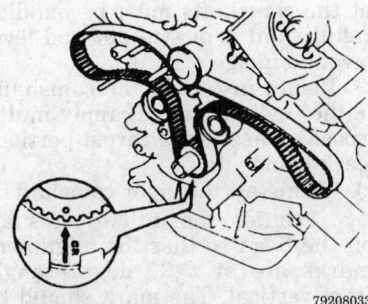

Aligning the installation mark on the timing belt with the drilled mark of the crankshaft timing pulley — LS400

20. Install the left side No. 2 timing belt cover and connect the cam position sensor connector.
21. Install the water bypass pipe and connect the hoses and connectors.
22. Replace the left side ignition coil and connect the coil connector. Install the idler pulley and cover plate. Tighten the bolt to 27 ft. lbs. (37 Nm).
23. Install and secure the ignition wires. Install the right side No. 3 timing belt cover.
24. Install the left side No. 3 timing belt cover and connect the vacuum

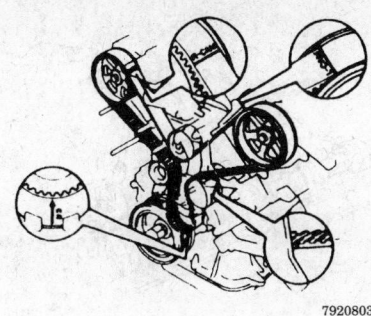

Checking the timing belt installation marks on a reinstalled used timing belt — LS400

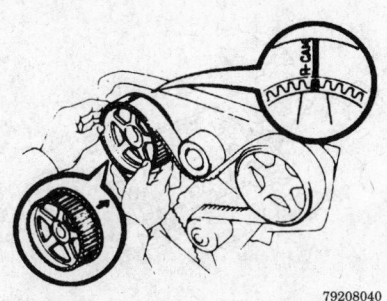

Installing the right camshaft sprocket — ES300

Installing the right and left camshaft sprockets — LS400 and SC400

hose and connectors. Install the right side engine wire cover.
25. Install the remaining components. Install the air ducts and dust covers. Connect the battery cables.
26. Refill the cooling system. Check the ignition timing.

Timing Sprockets

REMOVAL AND INSTALLATION

1. Remove the timing belt.

2. Using a spanner wrench to hold the camshaft pulley, remove the setbolt.
3. Install a 2-armed puller and remove the camshaft pulley. Be careful, the pulley may spring off so don't drop it.
To install:
4. Align the camshaft knockpin with the groove in the camshaft timing pulley and slide the pulley onto the shaft (flange side outward). Install the setbolt and plate washer and tighten it to 80 ft. lbs. (108 Nm) on the ES300, LS400 and SC400 or 59 ft. lbs. (79 Nm) on the SC300.
5. Install the timing belt and covers.

Camshaft

REMOVAL AND INSTALLATION

The following procedures have the valve lash adjuster removal and installation incorporated in.

ES300

1. Remove the timing belt and idler pulley.
2. Remove the camshaft timing pulleys.
3. Remove the cylinder head covers.

NOTE: The thrust clearance on both the intake and exhaust camshafts is very small; the camshafts must be kept level during removal. If the camshafts are removed without being kept level, the camshaft may be caught in the cylinder head causing the head to break or the camshaft to seize.

4. To remove the exhaust and intake camshafts from the right side cylinder head:
a. Turn the camshaft with a wrench until the 2 pointed marks drive and driven gears are aligned. (The right camshaft gears have 2 marks apiece; the left side camshaft gears have one mark each.)
b. Secure the exhaust camshaft sub-gear to the main gear using a service bolt. A bolt 0.63–0.79 in. (16–20mm) long with a 6mm thread diameter and a 1mm pitch is recommended. When removing the exhaust camshaft be sure the sub-gear is not loaded; all the force must be eliminated.
c. Uniformly loosen and remove the exhaust camshaft bearing cap bolts in several passes and in the

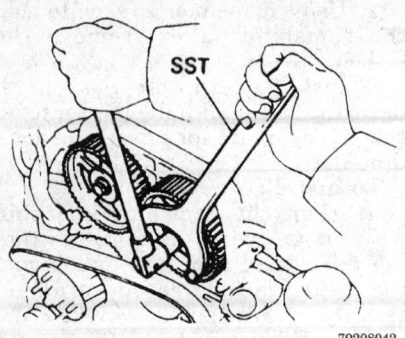

79208042

Removing the camshaft sprockets — SC300

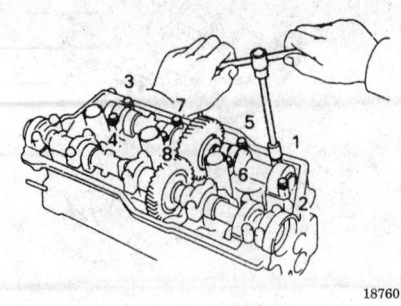

187601

Right exhaust camshaft bearing loosening sequence — 1993 ES300

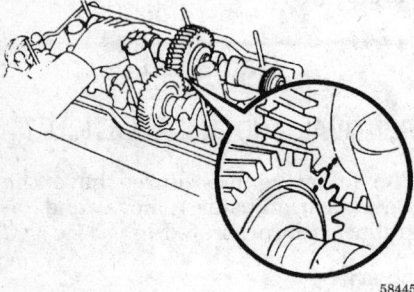

58445

Aligning the right side camshaft timing marks — 1993 ES300

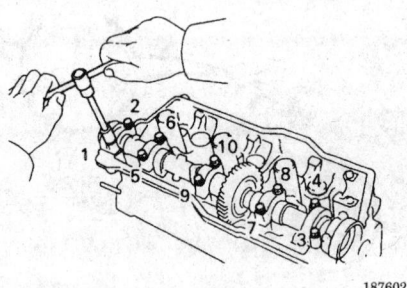

187602

Right intake camshaft bearing loosening sequence — 1993 ES300

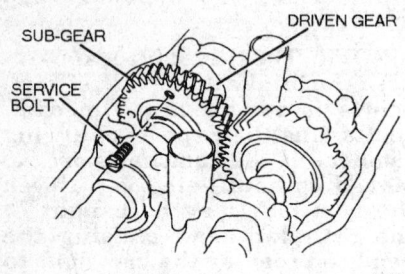

187600

Securing the sub-gear and driven gear, right side — 1993 ES300

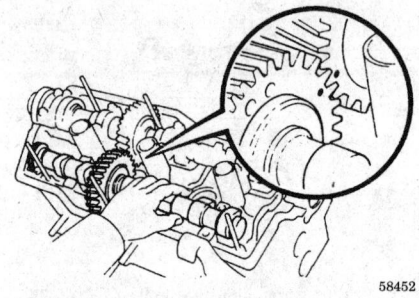

58452

Aligning the left side camshaft timing marks — 1993 ES300

proper sequence. Remove the eight bearing cap bolts and remove the caps, keeping them in the correct order.

d. Remove the exhaust camshaft from the engine.

e. Uniformly loosen and remove the 10 bearing cap bolts in several passes, in the proper sequence. Remove the bearing caps, keeping them in order, remove the oil seal and then lift out the intake camshaft.

5. To remove the exhaust and intake camshafts from the left side cylinder head:

a. Turn the camshaft with a wrench until the pointed marks on the drive and driven gears are aligned. (The right camshaft gears have 2 marks apiece; the left side camshaft gears have one mark each.)

b. Secure the exhaust camshaft sub-gear to the main gear using a service bolt. A bolt 0.63–0.79 in. (16–20mm) long with a 6mm thread diameter and a 1mm pitch is recommended. When removing

the exhaust camshaft be sure the sub-gear is not loaded; all the force must be eliminated.

c. Uniformly loosen and remove the exhaust camshaft bearing cap bolts in several passes and in the proper sequence. Remove the eight bearing cap bolts and remove the caps. Keep the caps in the correct order.

d. Remove the exhaust camshaft from the engine.

e. Uniformly loosen and remove the 10 bearing cap bolts in several passes, in the proper sequence. Remove the bearing caps, keeping them in order, remove the oil seal and then lift out the intake camshaft.

6. Remove the valve lash adjuster shims and hydraulic lash adjusters. Identify each lash adjuster and shim as it is removed so it can be reinstalled in the same position. If the lash adjusters are to be reused, store them upside down in a sealed container.

To install:

7. Install the valve lash adjusters into their original positions and install the shims. Check valve clearance and replace the shims as necessary.

8. When reinstalling, remember that the camshafts must be handled carefully and kept straight and level to avoid damage.

9. Before installing the camshafts in either cylinder head, apply multipurpose grease to the thrust portions of each camshaft.

10. To install the right camshafts:

a. Position the intake camshaft on the head so that the alignment marks are at a 90 degree angle from vertical. The mark should be at the 3 o'clock position.

b. Apply sealant to the No. 1 bearing cap.

c. Apply a light coat of clean engine oil to the bolt threads and under the bolt head. Install the bearing caps to their proper position. Tighten the bolts evenly and in several passes in the reverse order of loosening to 12 ft. lbs. (16 Nm) in the proper sequence.

d. Position the exhaust camshaft on the head so that the alignment marks are at a 90 degree angle from vertical. The mark should be at the 9 o'clock position and must align with the marks on the other gear.

e. Apply a light coat of clean engine oil to the bolt threads and under the bolt head. Install the bearing caps to their proper position. Tighten the bolts evenly and

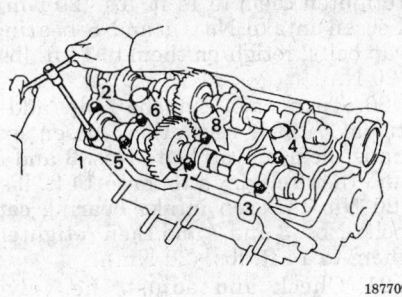

Left exhaust camshaft bearing loosening sequence — 1993 ES300

187709

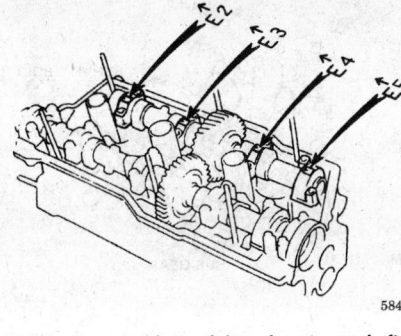

Bearing cap positions, right exhaust camshaft — 1993 ES300

58447

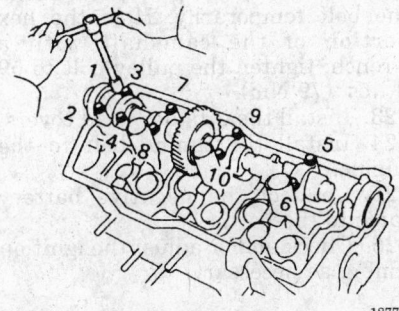

Left intake camshaft bearing loosening sequence — 1993 ES300

187710

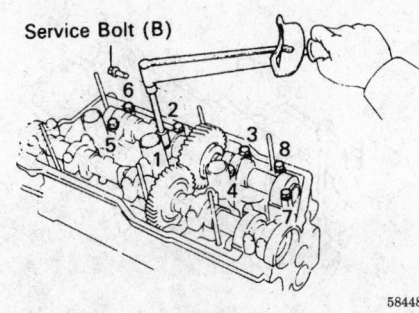

Right exhaust camshaft bearing tightening sequence — 1993 ES300

58448

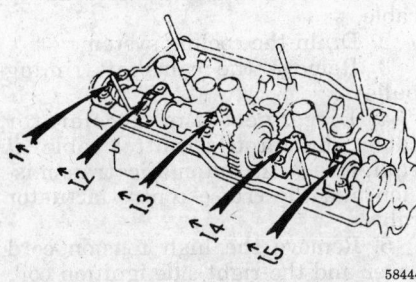

Bearing cap positions, right intake camshaft — 1993 ES300

58444

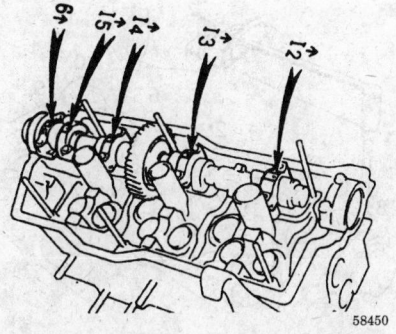

Bearing cap positions, left intake camshaft — 1993 ES300

58450

in several passes in the reverse order of loosening to 12 ft. lbs. (16 Nm) in the proper sequence.

f. Remove the service bolt.

11. To install the left camshafts:

a. Position the intake camshaft on the head so that the alignment mark is at a 90 degree angle from vertical. The mark should be at the 9 o'clock position.

b. Apply sealant to the No. 1 bearing cap.

c. Apply a light coat of clean engine oil to the bolt threads and under the bolt head. Install the bearing caps to their proper posi-

tion. Tighten the bolts evenly and in several passes to 12 ft. lbs. (16 Nm) in the proper sequence.

d. Position the exhaust camshaft on the head so that the alignment marks are at a 90 degree angle from vertical. The mark should be at the 3 o'clock position and must align with the marks on the other gear.

e. Apply a light coat of clean engine oil to the bolt threads and under the bolt head. Install the bearing caps to their proper position. Tighten the bolts evenly and

in several passes to 12 ft. lbs. (16 Nm) in the proper sequence.

f. Remove the service bolt.

12. Apply multi-purpose grease to new camshaft oil seals. Install the seals.

13. Install the No. 3 (rear) timing belt cover.

14. Install the camshaft timing gears.

15. Install the idler pulley, timing belt and covers.

16. Check and adjust the valve clearance.

17. Install the cylinder head (valve) covers.

18. Start the engine. Check the ignition timing.

19. Test drive the vehicle.

20. Check all fluid levels.

GS300 and SC300

1. Disconnect the negative battery cable from the battery.

2. Remove the timing belt from the engine.

3. Remove the cylinder head covers.

4. While holding each camshaft with a wrench, loosen the camshaft sprocket bolt and remove the sprocket.

5. Remove the four bolts and lift out the No. 4 (inner) timing belt cover.

6. Uniformly loosen and then remove the four No. 1 camshaft bearing cap bolts. These are the bolts directly behind the sprockets. Remove the bearing caps.

7. Uniformly and in the correct sequence, loosen and remove the remaining bearing cap bolts. Note that there are separate sequences for the exhaust and intake camshafts. Lift off all 12 bearing caps.

8. Lift out the exhaust and intake camshafts.

9. Remove the valve lash adjuster shims and hydraulic lash adjusters. Identify each lash adjuster and shim as it is removed so it can be reinstalled in the same position. If the lash adjusters are to be reused, store them upside down in a sealed container.

To install:

10. Install the valve lash adjusters into their original positions and install the shims. Check valve clearance and replace the shims as necessary.

11. When reinstalling, remember that the camshafts must be handled carefully and kept straight and level to avoid damage.

12. Coat the thrust portions of each camshaft with engine oil and then position them in the cylinder head with

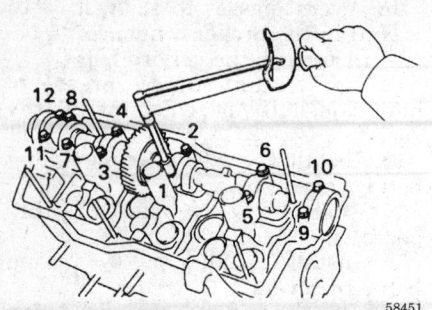

Left intake camshaft bearing tightening sequence — 1993 ES300

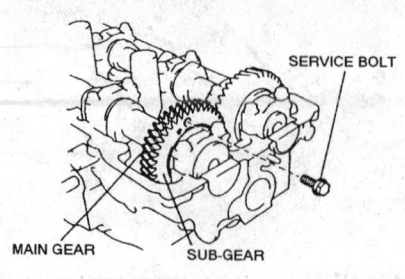

Left intake camshaft bearing removal sequence — 1994–97 ES300

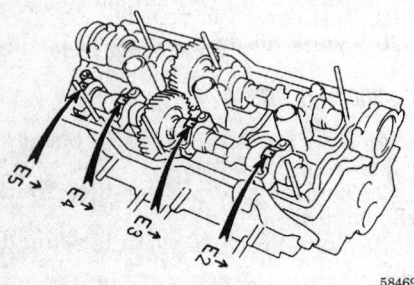

Bearing cap positions, left exhaust camshaft — 1993 ES300

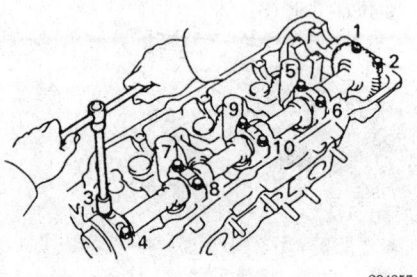

Left exhaust camshaft bearing removal sequence — 1994–97 ES300

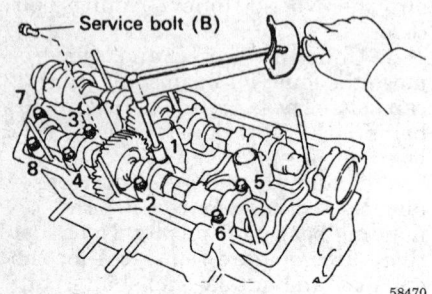

Left exhaust camshaft bearing tightening sequence — 1993 ES300

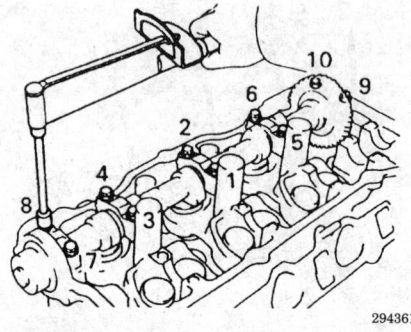

Bearing cap bolt tightening sequence right exhaust camshaft — 1994–97 ES300

the cam lobes and the knock pins in the correct position.

13. Position the No. 3 and No. 7 bearing caps in place, coat the bolt threads with oil and then tighten them temporarily.

14. Coat new oil seals with multi-purpose grease and then slide them over the camshafts.

15. Clean the mating surfaces of the two No. 1 bearing caps and then apply some sealant. Install the bolts.

16. Install all remaining bearing caps, coat the threads of each bolt with clean oil and then tighten them, in several passes, in sequence, to 14

ft. lbs. (20 Nm). Note that there are separate sequences for the intake and exhaust sides.

17. Press the oil seal in as far as it will go.

18. Rotate each camshaft until the forward straight (knock) pin is straight up. Loosen exhaust Nos. 1, 2 and 6 bearing cap bolts until they can be turned by hand; re-tighten them to 14 ft. lbs. (20 Nm). Loosen intake Nos. 1 and 2 and re-tighten to 14 ft. lbs. (20 Nm).

19. Turn each camshaft 1/3 of a revolution (120 degrees). Loosen exhaust Nos. 4 and 7 bearing cap bolts;

retighten them to 14 ft. lbs. (20 Nm). Loosen intake Nos. 4 and 6 bearing cap bolts; retighten them to 14 ft. lbs. (20 Nm).

20. Turn each camshaft an additional 1/3 of a revolution, loosen exhaust bearing cap bolts Nos. 3 and 5 and then retighten them to 14 ft. lbs. (20 Nm). Loosen intake bearing cap bolts Nos. 3 and 7 and then retighten them to 14 ft. lbs. (20 Nm).

21. Check and adjust the valve clearance.

22. Install the No 4. inside timing belt cover and the camshaft pulleys. Align the shaft pin with the pulley groove and slide the pulley on. Install the bolt temporarily. Hold the hex portion of the camshaft with a wrench; tighten the pulley bolt to 59 ft. lbs. (79 Nm).

23. Install the cylinder head covers.

24. Install the timing belt to the engine.

25. Connect the negative battery cable to the battery.

26. Check and/or adjust the ignition timing as necessary.

LS400 and SC400

1993–94 Models

1. Disconnect the negative battery cable.

2. Drain the cooling system.

3. Remove the camshaft timing pulleys.

4. Disconnect the accelerator cable, the throttle control cable, if equipped with automatic transmission and the cruise control actuator cable.

5. Remove the high tension cord cover and the right side ignition coil.

6. Remove the water inlet housing mounting bolts and disconnect the water bypass hose from the ISC valve.

7. Remove the water inlet and inlet housing assemblies. Remove the O-ring from the water inlet housing.

8. Remove the EGR pipe.

9. Disconnect the following:
   a. VSV connector
   b. Vacuum pipe hose
   c. EGR water bypass pipe
   d. Fuel pressure VSV

10. Disconnect the EGR vacuum hoses and remove the EGR VSV.

11. Disconnect these hoses:
   a. Water bypass pipe hose from the ISC valve.
   b. Water bypass joint hose.
   c. Vacuum pipe hoses.

12. Disconnect the EGR gas temperature sensor, California only. Remove the EGR valve adapter.

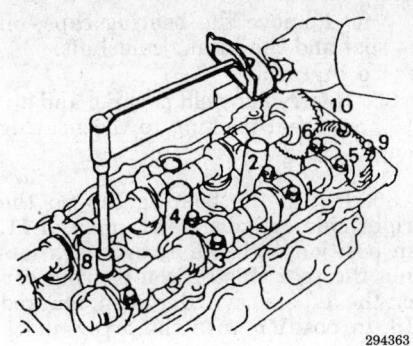

Bearing cap bolt tightening sequence right intake camshaft — 1994–97 ES300

294363

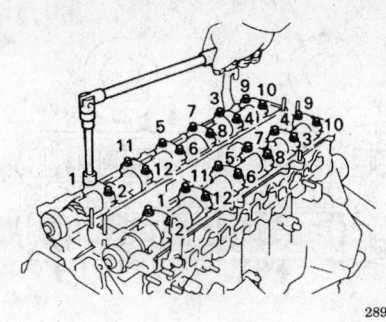

Camshaft bearing cap bolt removal sequence — GS300 and SC300

289577

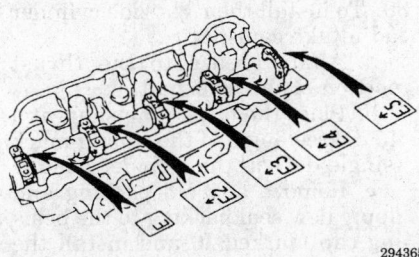

Bearing cap positions left exhaust camshaft — 1994–97 ES300

294365

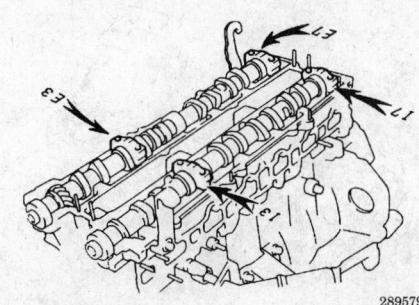

Installing No. 3 and 7 bearing caps — GS300 and SC300

289579

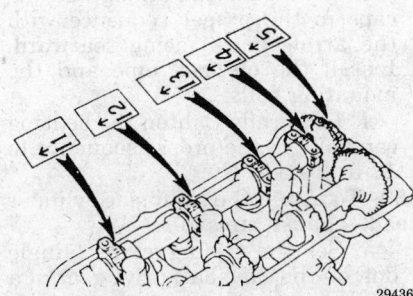

Bearing cap positions left intake camshaft — 1994–97 ES300

294367

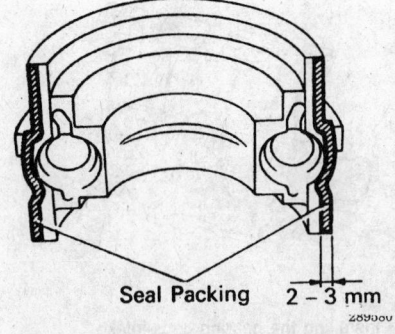

Seal Packing   2 – 3 mm

289580

Applying sealant to the No. 1 bearing cap — GS300 and SC300

13. Disconnect the following:

 a. Fuel pressure regulator vacuum hose.

 b. Air intake chamber vacuum hose.

 c. Vacuum hose from the EVAP BVSV.

14. Remove the mounting bolts, hoses and the vacuum pipe.

15. Remove the ISC valve.

16. Remove the throttle body sensor connectors and the water bypass pipe from the rear water bypass joint.

17. Disconnect the PCV valve hose. Remove the throttle body and gasket.

18. Disconnect the accelerator cable bracket and the brake booster vacuum union and hose.

19. Disconnect the cold start injector connector and the cold start injector tube from the right side delivery pipe, if equipped.

20. Disconnect the check connector from the intake chamber and remove the mounting nuts and bolts.

21. Remove the air intake chamber and the cold start injector, tube and wire assembly, if equipped.

22. Disconnect the engine wire from the intake manifold and from the

right side cylinder head. Disconnect the heater hoses.

23. Remove the delivery pipes and the fuel injectors. Remove the mounting bolts and nuts. Lift up the intake manifold.

24. Remove the front and rear water bypass joint.

25. Raise and safely support the vehicle. Remove the front exhaust pipe and the main catalytic converters. Lower the vehicle.

26. Disconnect the right side oxygen sensor. Remove the mounting bolts and nuts and remove the right side exhaust manifold.

27. Remove the oil dipstick and guide. Disconnect the left side oxygen sensor.

28. Remove the mounting bolts and nuts and remove the left side exhaust manifold.

29. Remove the 2 engine hangers and the wire brackets from the right side cylinder head.

30. Remove the bolts, washers and the cylinder head cover. Remove the semi-circular plugs, if necessary.

31. To remove the exhaust camshaft from the right side cylinder head:

 a. Position the service bolt hole of the drive sub-gear to the upright position. Secure the camshaft sub-gear to drive gear with a service bolt.

 b. Set the timing mark (single dot) on the camshaft drive gear at approximately 10 degrees. Turn the camshaft with a wrench on the hexagonal flats.

 c. Alternately loosen and remove the bearing cap bolts holding the intake camshaft side of the oil feed pipe to the cylinder head.

 d. Uniformly loosen (in several passes) and remove the bearing cap bolts, in sequence.

 e. Remove the oil feed pipe and the bearing caps. Remove the camshaft.

32. To remove the intake camshaft from the right side cylinder head:

 a. Set the timing mark (single dot) on the camshaft drive gear at approximately 45 degrees. Turn the camshaft with a wrench on the hexagonal flats.

 b. Uniformly loosen (in several passes) and remove the bearing cap bolts in the proper sequence.

 c. Remove the bearing caps, oil seal and the intake camshaft.

33. To remove the exhaust camshaft of the left side cylinder head:

 a. Position the service bolt hole of the drive sub-gear to the upright

Camshaft bearing cap bolt torquing sequence — GS300 and SC300

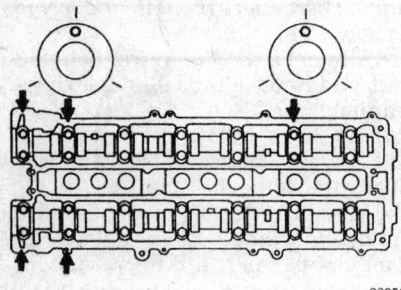

Retorquing the camshafts (Step 1) — GS300 and SC300

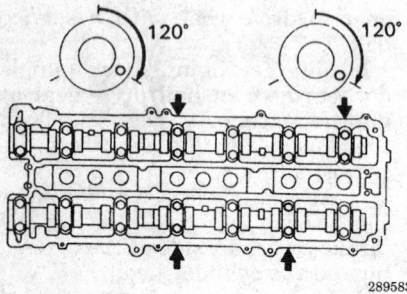

Retorquing the camshafts (Step 2) — GS300 and SC300

position. Secure the camshaft sub-gear to drive gear with a service bolt.

NOTE: When removing the camshaft, make sure the torsional spring force of the sub-gear has been eliminated.

b. Set the timing mark (2 dots) on the camshaft drive gear at approximately 15 degrees, by turning the camshaft with the proper tool.

c. Alternately loosen and remove the bearing cap bolts holding the intake camshaft side of the oil feed pipe to the cylinder head.

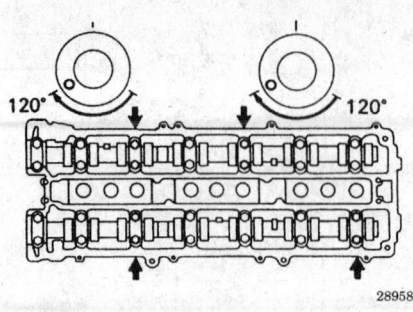

Retorquing the camshafts (Step 3) — GS300 and SC300

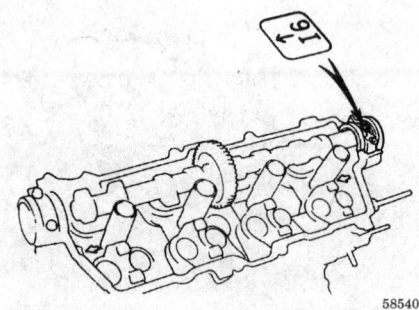

Installing the I6 bearing cap-intake camshaft, right cylinder head — 1993–94 LS400 and SC400

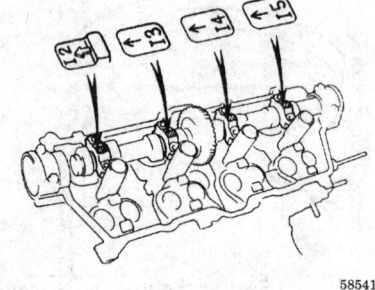

Installing the bearing caps-intake camshaft, right cylinder head — 1993–94 LS400 and SC400

d. Uniformly loosen (in several passes) and remove the bearing cap bolts in the proper sequence.

e. Remove the oil feed pipe and the bearing caps. Remove the camshaft.

34. To remove the intake camshaft from the left side cylinder head:

a. Set the timing mark (single dot) of the camshaft drive gear at approximately 60 degrees, by turning the camshaft with the proper tool.

b. Uniformly loosen (in several passes) and remove the bearing cap bolts, in sequence.

c. Remove the bearing caps, oil seal and the intake camshaft.

To install:

35. Remove any old packing and apply new seal packing to the bearing caps.

36. Install the bearing cap on the right side cylinder head, marked **I1**, in position with the arrow mark facing the rear. Install the bearing cap on the left side cylinder head, marked **I6**, in position with the arrow mark facing the front.

37. Apply a light coat of oil on the threads of the cap bolts. Install the bearing cap bolts with new washers and tighten to 12 ft. lbs. (16 Nm).

38. To install the right side cylinder head intake camshaft:

a. Apply grease to the thrust portion of the camshaft.

b. Place the intake camshaft at a 45 degree angle of the timing mark (single dot) on the cylinder head.

c. Remove any old packing and apply new seal packing to the bearing cap marked **I6** and install the front bearing cap, marked **I6** with the arrow facing rearward.

d. Align the arrows at the front and rear of the cylinder head with the bearing cap.

e. Install the remaining bearing caps in the proper sequence with the arrow mark facing rearward. Install the oil feed pipe and the mounting bolts.

f. Uniformly tighten the bearing cap bolts in the proper sequence to 12 ft. lbs. (16 Nm).

39. To install the right side cylinder head exhaust camshaft:

a. Set the timing mark (single dot) on the camshaft drive gear at a 10 degree angle by turning the intake camshaft with the proper tool.

b. Apply grease to the thrust portion of the camshaft.

c. Align the timing marks (single dots) on the camshaft drive and driven gears.

d. Place the exhaust camshaft in the cylinder head. Install the rear bearing cap with the arrow mark facing rearward.

e. Align the arrow marks at the front and rear of the cylinder head with the mark on the bearing cap. Apply a light coat of oil on the threads of the bearing cap bolts.

f. Uniformly tighten the bearing cap bolts in the proper sequence to 12 ft. lbs. (16 Nm).

g. Bring the service bolt installed upward by turning the camshaft with the proper tool. Remove the service bolt.

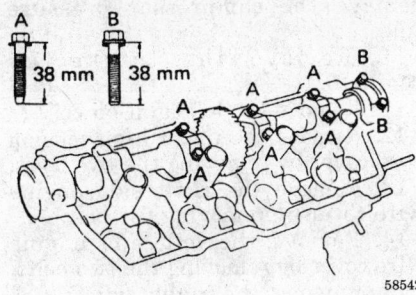

Installing the bearing cap bolts-intake camshaft, right cylinder head — 1993–94 LS400 and SC400

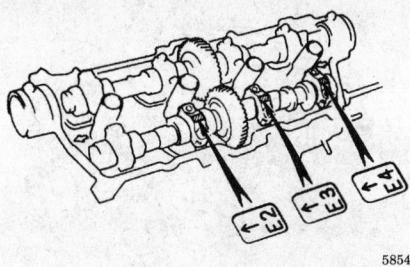

Installing the bearing caps-exhaust camshaft, right cylinder head — 1993–94 LS400 and SC400

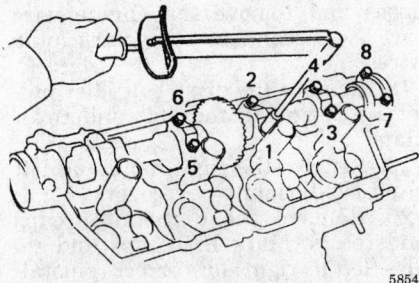

Camshaft torquing sequence-intake camshaft, right cylinder head — 1993–94 LS400 and SC400

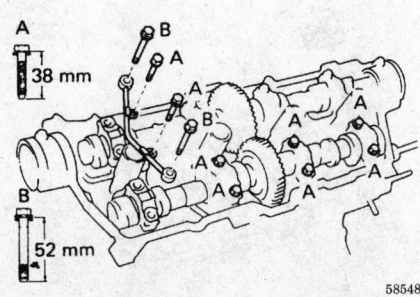

Installing the bearing cap bolts-exhaust camshaft, right cylinder head — 1993–94 LS400 and SC400

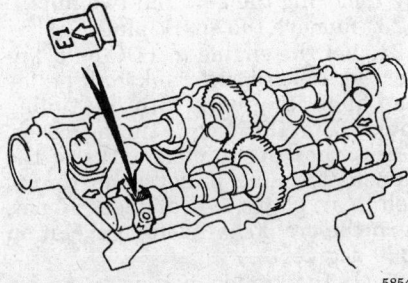

Installing the E1 bearing cap-exhaust camshaft, right cylinder head — 1993–94 LS400 and SC400

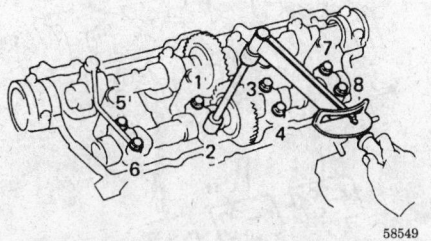

Camshaft bearing cap torquing sequence-exhaust camshaft, right cylinder head — 1993–94 LS400 and SC400

40. To install the left side cylinder head intake camshaft:

a. Apply grease to the thrust portion of the camshaft.

b. Place the intake camshaft with the timing mark (single dot) at a 60 degree angle on the cylinder head.

c. Remove any old packing and apply new seal packing to the bearing cap marked **I6** and install the front bearing cap, marked **I1** with the arrow facing rearward.

d. Align the arrows at the front and rear of the cylinder head with the bearing cap. Apply a light coat

of oil on the threads of the bearing cap bolts.

e. Install the remaining bearing caps in the proper sequence with the arrow mark facing rearward. Install the oil feed pipe and the mounting bolts.

f. Uniformly tighten the bearing cap bolts in the proper sequence to 12 ft. lbs. (16 Nm).

41. Install the left side cylinder head exhaust camshaft:

a. Set the timing mark (2 dots) on the camshaft drive gear at a 15 degree angle by turning the intake camshaft with the proper tool.

b. Apply grease to the thrust portion of the camshaft.

c. Align the timing marks (2 dots each) on the camshaft drive and driven gears.

d. Place the exhaust camshaft ion the cylinder head. Install the rear bearing cap with the arrow mark facing rearward.

e. Align the arrow marks at the front and rear of the cylinder head with the mark on the bearing cap. Apply a light coat of oil on the threads of the bearing cap bolts.

f. Uniformly tighten the bearing cap bolts in the proper sequence to 12 ft. lbs. (16 Nm).

g. Bring the service bolt installed upward by turning the camshaft with the proper tool. Remove the service bolt.

42. Install the camshaft oil seals with the proper tool. Install the semicircular plugs with the proper seal packing.

43. Install the cylinder head covers with the proper seal packing and gasket. Tighten the mounting bolts to 52 inch lbs. (6 Nm).

44. Install the engine wire bracket and hangers. Tighten the hanger bolts to 27 ft. lbs. (37 Nm).

45. Install the right side exhaust manifold with a new gasket and tighten the mounting bolts to 29 ft. lbs. (40 Nm). Connect the right side oxygen sensor connector.

46. Install the right side exhaust manifold with a new gasket and tighten the mounting bolts to 29 ft. lbs. (40 Nm). Connect the right side oxygen sensor connector.

47. Install the left side exhaust manifold with a new gasket and tighten the mounting bolts to 29 ft. lbs. (40 Nm). Connect the left side oxygen sensor connector.

48. Install the remaining components. Tighten the intake manifold nuts and bolts to 13 ft. lbs. (18 Nm).

49. Tighten the timing belt rear plates to 69 inch lbs. (7.7 Nm).

50. Connect and adjust the accelerator cable, the automatic transmission throttle cable and the cruise control actuator cable. Install the camshaft timing pulley.

51. Fill the cooling system.

52. Connect the negative battery cable.

53. Start the engine and check for leaks.

54. Recheck all the fluid levels and check the ignition timing.

1995 SC400

1. Disconnect the negative battery cable. Wait at least 90 seconds before performing any other work. Discon-

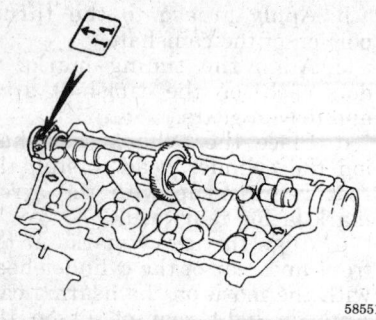

Installing the I1 bearing cap-intake camshaft, left cylinder head — 1993-94 LS400 and SC400

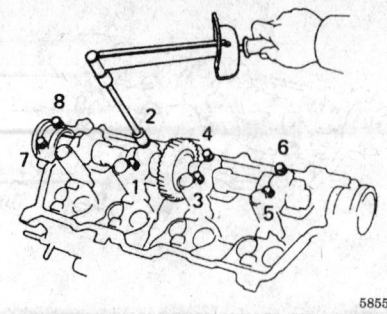

Camshaft bearing cap torquing sequence-intake camshaft, left cylinder head — 1993-94 LS400 and SC400

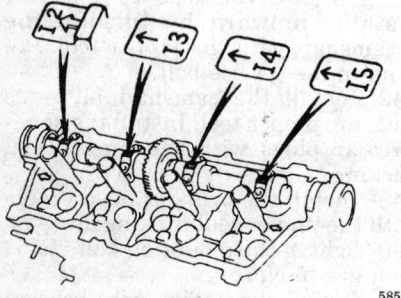

Installing the bearing caps-intake camshaft, left cylinder — 1993-94 LS400 and SC400

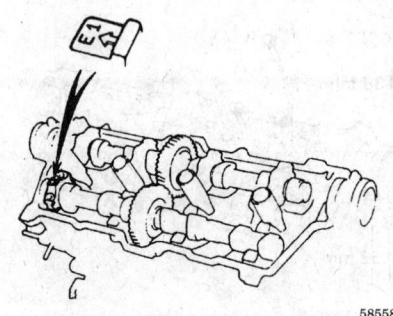

Installing the E1 bearing cap-exhaust camshaft, left cylinder head — 1993-94 LS400 and SC400

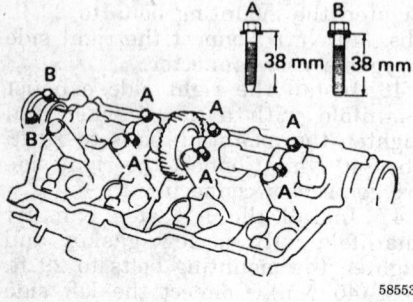

Installing the bearing cap bolts-intake manifold, left cylinder head — 1993-94 LS400 and SC400

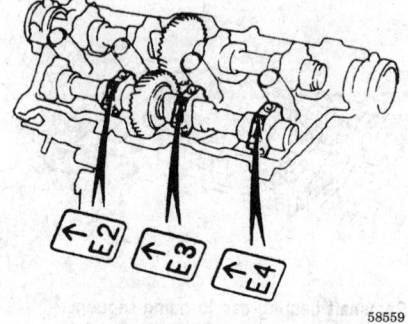

Exhaust bearing caps-exhaust camshaft, left cylinder head — 1993-94 LS400 and SC400

nect the positive battery cable. Remove the battery.

2. Release the fuel pressure from the fuel lines.

———— CAUTION ————

Fuel injection systems remain under pressure after the engine has been turned OFF. Properly relieve fuel pressure before disconnecting any fuel lines. Failure to do so may result in fire or personal injury.

3. Remove the engine undercovers.

4. Drain the engine coolant from the radiator.

5. Remove the drive belt from the engine by turning the tensioner counterclockwise.

6. Remove the engine coolant reservoir tank.

7. Remove the radiator. Disconnect the two transmission oil cooler hoses from the radiator. Plug the hose ends.

8. Disconnect the suction and pressure hoses from the hydraulic pump.

9. Disconnect the air conditioning compressor from the engine. Do not

remove the compressor pressure lines.

10. Remove the intake air connector.

11. Remove the left ignition coil.

12. Remove the upper high tension cord cover by removing the two bolts.

13. Remove the right side engine wire cover by removing the bolt.

14. Remove the left side engine wire cover by removing the two bolts.

15. Remove the right side No. 3 timing cover by removing the two air control valve hoses and four bolts.

16. Remove the left No. 3 timing belt cover by removing the four bolts.

17. Disconnect and tag the vacuum hoses and remove the engine wire covers. Disconnect the spark plug wires.

18. Remove the drive belt idler pulley by removing the bolt and cover plate.

19. Remove the right hand and left hand No. 2 timing belt covers.

20. Remove the distributor caps and rotors. Mark both caps and rotors (left or right) for correct reinstallation. Disconnect and remove both distributor housings.

21. Disconnect and remove the alternator from the engine.

22. Remove the drive belt tensioner by removing the bolt and two nuts.

23. Remove the spark plugs.

24. Set the engine to TDC on cylinder No. 1. Turn the crankshaft pulley and align its groove with the timing mark **0** of the No. 1 timing cover. Check that the timing marks of the camshaft timing pulleys and timing belt rear plates are aligned. If not, turn the crankshaft 1 full revolution (360 degrees).

25. If the timing belt is to be reused, turn the crank pulley slowly; check that the installation marks are present on the belt. If the marks are not present, make new installation marks before removing the belt. The marks should align with the timing marks on each camshaft pulley.

26. Remove the timing belt tensioner. Alternately loosen the two (2) bolts and then remove the bolts, the tensioner and the dust protector.

27. Using the proper tool, 09960-10010 or equivalent, loosen the tension spring between the left side and right side camshaft timing pulleys by slightly turning the left side camshaft clockwise.

28. Disconnect the timing belt from the camshaft timing pulleys.

29. Using the proper tool (SST 09960-10010 or equivalent), remove the bolt and the camshaft timing pulleys.

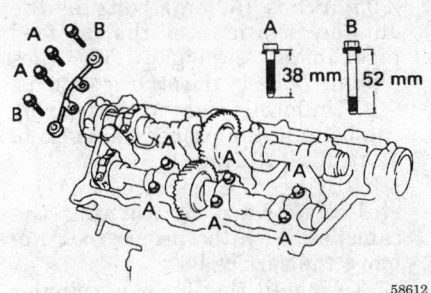

Installing the bearing cap bolts-exhaust
camshaft, left cylinder head — 1993–94 LS400
and SC400

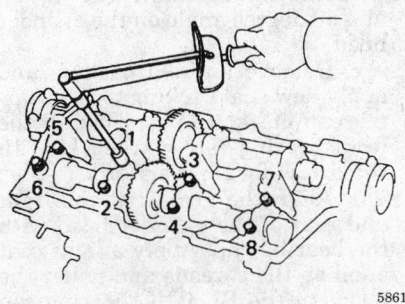

Camshaft bearing cap torquing sequence-
exhaust camshaft, left cylinder head —
1993–94 LS400 and SC400

30. Remove the right hand ignition coil.

31. Remove the timing belt rear plates by removing the two bolts for each plate.

32. Remove the throttle body cover.

33. Disconnect the accelerator cable, automatic transmission throttle control cable and the cruise control cable from the throttle body.

34. Disconnect the fuel return hose from the fuel return pipe.

35. Disconnect the PCV hose from the left cylinder head.

36. Remove the heater water valve from the firewall.

37. Remove the throttle body from the engine.

38. On the right cylinder head cover, release the clips and retainers holding the spark plug wires to the cylinder head cover. Remove the cylinder head cover by removing the eight bolts and eight washers.

39. On the left cylinder head cover:

a. Release the clips and retainers holding the spark plug wires.

b. Remove the bolt holding the fuel inlet hose to the delivery pipe.

c. Remove the pulsation damper and two gaskets and disconnect the fuel inlet hose from the delivery pipe.

d. Free the engine wiring harness from the clamps on the delivery pipe and disconnect the four injector connectors.

e. Remove the eight bolts, seal washers, cylinder head cover and gaskets.

40. Remove the semi-circular plugs, if necessary.

41. To remove the exhaust camshaft from the right side cylinder head :

a. Position the service bolt hole of the drive sub-gear to the upright position. Secure the camshaft sub-gear to drive gear with a service bolt.

b. Set the timing mark (single dot) on the camshaft drive gear at approximately 10 degrees. Turn the camshaft with a wrench on the hexagonal flats.

c. Alternately loosen and remove the bearing cap bolts holding the intake camshaft side of the oil feed pipe to the cylinder head.

d. Uniformly loosen (in several passes) and remove the bearing cap bolts, in sequence.

e. Remove the oil feed pipe and the bearing caps. Remove the camshaft.

42. To remove the intake camshaft from the right side cylinder head:

a. Set the timing mark (single dot) on the camshaft drive gear at approximately 45 degrees. Turn the camshaft with a wrench on the hexagonal flats.

b. Uniformly loosen (in several passes) and remove the bearing cap bolts in the proper sequence.

c. Remove the bearing caps, oil seal and the intake camshaft.

43. To remove the exhaust camshaft of the left side cylinder head:

a. Position the service bolt hole of the drive sub-gear to the upright position. Secure the camshaft sub-

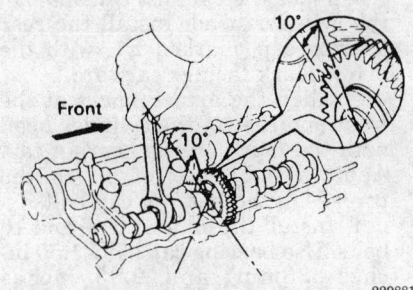

Turning the exhaust camshaft 10 degrees on
the right cylinder head — 1995 SC400

gear to drive gear with a service bolt.

NOTE: When removing the camshaft, make sure the torsional spring force of the sub-gear has been eliminated.

b. Set the timing mark (2 dots) on the camshaft drive gear at approximately 15 degrees, by turning the camshaft with the proper tool.

c. Alternately loosen and remove the bearing cap bolts holding the intake camshaft side of the oil feed pipe to the cylinder head.

d. Uniformly loosen (in several passes) and remove the bearing cap bolts in the proper sequence.

e. Remove the oil feed pipe and the bearing caps. Remove the camshaft.

44. To remove the intake camshaft from the left side cylinder head:

a. Set the timing mark (single dot) of the camshaft drive gear at approximately 60 degrees, by turning the camshaft with the proper tool.

b. Uniformly loosen (in several passes) and remove the bearing cap bolts, in sequence.

c. Remove the bearing caps, oil seal and the intake camshaft.

45. Remove the valve lash adjuster shims and hydraulic lash adjusters. Identify each lash adjuster and shim as it is removed so it can be reinstalled in the same position. If the lash adjusters are to be reused, store them upside down in a sealed container.

To install:

46. Install the valve lash adjusters into their original positions and install the shims. Check valve clearance and replace the shims as necessary.

47. When reinstalling, remember that the camshafts must be handled carefully and kept straight and level to avoid damage.

48. To install the right side cylinder head intake camshaft:

a. Apply grease to the thrust portion of the camshaft.

b. Place the intake camshaft at a 45 degree angle of the timing mark (single dot) on the cylinder head.

c. Remove any old packing and apply new seal packing to the bearing cap marked **I6** and install the bearing cap with the arrow facing rearward.

d. Align the arrows at the front and rear of the cylinder head with the bearing cap.

e. Install the remaining bearing caps in the proper sequence with the arrow mark facing rearward. Apply oil to the threads and under

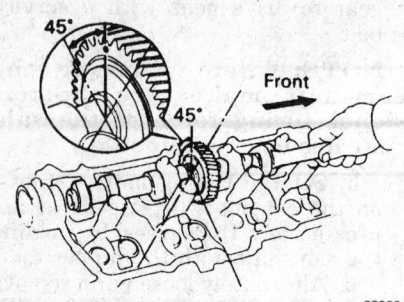

Turning the intake camshaft on the right cylinder head 45 degrees — 1995 SC400

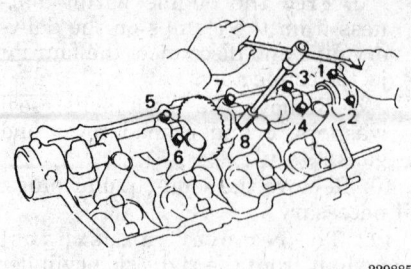

Bolt sequence removal for the intake camshaft on the right cylinder head — LS400 and SC400

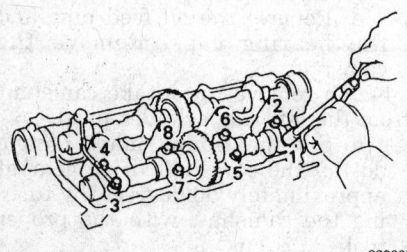

Bolt sequence removal for the exhaust camshaft on the right cylinder head — LS400 and SC400

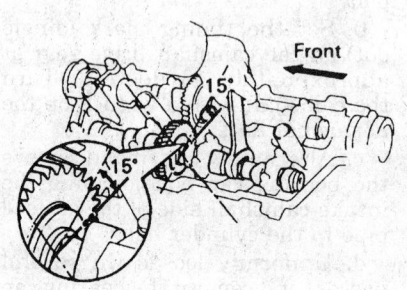

Turning the exhaust camshaft on the left cylinder head 15 degrees — SC400

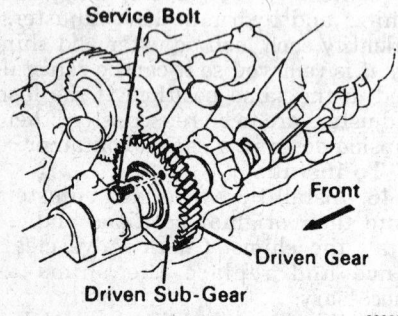

Securing the exhaust camshaft on the left cylinder head — 1995 SC400

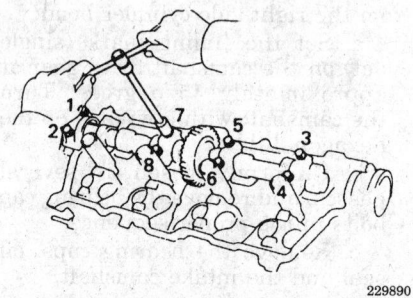

Bolt sequence removal for the intake camshaft on the left cylinder head — LS400 and SC400

the heads of the **BLACK** colored bolts. Only apply oil to the threads of the **SILVER** colored bolts.

f. Install a new seal washer to the silver bearing cap bolts.

g. Uniformly tighten the bearing cap bolts in the proper sequence to 12 ft. lbs. (16 Nm).

49. To install the right side cylinder head exhaust camshaft:

a. Set the timing mark (single dot) on the camshaft drive gear at a 10 degree angle by turning the intake camshaft with the proper tool.

b. Apply grease to the thrust portion of the camshaft.

c. Align the timing marks (single dots) on the camshaft drive and driven gears.

d. Place the exhaust camshaft in the cylinder head. Install the rear bearing cap, marked **E1**, with the arrow mark facing rearward.

e. Align the arrow marks at the front and rear of the cylinder head with the mark on the bearing cap. Apply a light coat of oil on the threads of the bearing cap bolts.

f. Install the oil feed pipe and 10 bolts. Use bearing cap bolts 1.50 inches (38mm) and 2.05 inches (52mm) in length. Install the two 2.05 inches (52mm) bolts in the outside positions of the oil feed pipe. Install the eight 1.50 inches (38mm) bolts in the other positions.

g. Uniformly tighten the bearing cap bolts in the proper sequence to 12 ft. lbs. (16 Nm).

h. Bring the service bolt installed upward by turning the camshaft with the proper tool. Remove the service bolt.

50. To install the left side cylinder head intake camshaft:

a. Apply grease to the thrust portion of the camshaft.

b. Place the intake camshaft with the timing mark (single dot) at a 60 degree angle on the cylinder head.

c. Remove any old packing and apply new seal packing to the bearing cap marked **I1** and install the front bearing cap, marked **I1** with the arrow facing rearward.

d. Align the arrows at the front and rear of the cylinder head with the bearing cap. Apply a light coat of oil on the threads and under the heads of the **BLACK** bearing cap bolts. Only apply oil to the threads on the **SILVER** bearing cap bolts.

e. Install the remaining bearing caps in the proper sequence with the arrow mark facing rearward.

f. Install a new seal washer to the silver bearing cap bolts.

g. Uniformly tighten the bearing cap bolts in the proper sequence to 12 ft. lbs. (16 Nm).

51. Install the left side cylinder head exhaust camshaft by:

a. Set the timing mark (2 dots) on the camshaft drive gear at a 15 degree angle by turning the intake camshaft with the proper tool.

b. Apply grease to the thrust portion of the camshaft.

c. Align the timing marks (2 dots each) on the camshaft drive and driven gears.

d. Place the exhaust camshaft ion the cylinder head. Install the rear bearing cap, marked **E1** with the arrow mark facing forward.

e. Align the arrow marks at the front and rear of the cylinder head with the mark on the bearing cap. Apply a light coat of oil on the threads and under the heads of the **BLACK** bearing cap bolts. Only apply oil to the threads on the **SILVER** bearing cap bolts.

f. Install the oil feed pipe and 10 bolts. Use bearing cap bolts 1.50 inches (38mm) and 2.05 inches (52mm) in length. Install the two 2.05 inches (52mm) bolts in the outside positions of the oil feed

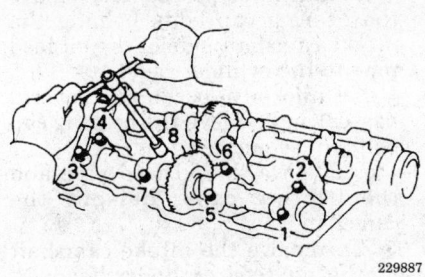

Bolt sequence removal for the exhaust camshaft on the left cylinder head — LS400 and SC400

pipe. Install the eight 1.50 inches (38mm) bolts in the other positions.

g. Uniformly tighten the bearing cap bolts in the proper sequence to 12 ft. lbs. (16 Nm).

h. Bring the service bolt installed upward by turning the camshaft with the proper tool. Remove the service bolt.

52. Install the camshaft oil seals with the proper tool (SST 09223–46011). Make sure to apply MP grease to the new oil seal lip.

53. Install the semi-circular plugs as follows:

a. Remove any old packing material.

b. Apply seal packing to the semi-circular plug grooves.

c. Install the four semi-circular plugs to the cylinder heads.

54. Clean the cylinder head covers. Apply new sealant in the correct locations and install the gaskets.

55. Install the left cylinder head cover, torque the bolts to 52 inch lbs. (6 Nm).

56. Install the right cylinder head cover and bolts. Tighten the bolts to 52 inch lbs. (6 Nm).

57. On the right cylinder head cover, install the spark plug wires and connect the clips and retainers to the cylinder head cover.

58. Install the throttle body. Before installing the throttle body, place a new gasket in position on the air intake chamber. Connect the water bypass hose and the PCV hose to the throttle body. Torque the bolts and nuts to 13 ft. lbs. (18 Nm).

a. Connect the water bypass pipe to the clamp on the engine wire cover.

b. Connect the water bypass hose to the IAC valve.

c. Connect the vacuum hose to the throttle body.

d. Connect the sensor wiring to the throttle body.

59. Install the heater water valve and connect the hoses and wiring.

60. Reconnect the PCV hose to the left cylinder head.

61. Connect the fuel return line to the pipe and connect the fuel inlet hose with the pulsation damper to the left delivery pipe. Tighten the bolts to 29 ft. lbs. (39 Nm).

62. Install the upper high tension cord cover.

63. Install the intake air connector. Connect the air connector from the throttle body and connect the air cleaner hose. Connect the hoses to the IAC valve and to the PS air control valve.

64. Connect the control cables to the throttle body.

65. Install the throttle body cover.

66. Install the two rear timing plates with the four bolts.

67. Install the right hand ignition coil.

68. Align the knock pin on the right side camshaft with the knock pin of the timing pulley. Slide on the timing pulley with the right side mark facing forward. Tighten the bolt to 80 ft. lbs. (108 Nm).

69. Align the knock pin on the left side camshaft with the knock pin of the timing pulley. Slide on the timing pulley with the left side mark facing forward. Tighten the bolt to 80 ft. lbs. (108 Nm).

70. Turn the crankshaft pulley and align its groove with the **0** timing mark on the timing belt cover.

71. Turn each camshaft timing pulley and align the timing marks of the pulley with the timing belt rear plate.

72. Install the timing belt to the left side camshaft timing pulley:

a. Using the proper tool, slightly turn the left side timing pulley clockwise. Align the installation mark of the timing belt with the timing mark of the camshaft timing pulley and hang the timing belt on the left side camshaft pulley.

b. Align the timing marks of the left side camshaft pulley and the timing belt rear plate.

c. Check that the timing belt has tension between crankshaft timing pulley and the left side camshaft pulley.

73. Install the timing belt to the right side camshaft timing pulley:

a. Using the proper tool, slightly turn the right side timing pulley clockwise. Align the installation mark of the timing belt with the timing mark of the camshaft timing pulley and hang the timing belt on the right side camshaft pulley.

b. Align the timing marks of the right side camshaft pulley and timing belt rear plate.

c. Check that the timing belt has tension between crankshaft timing pulley and the right side camshaft pulley.

74. The timing belt tensioner must be set prior to installation. The tensioner can be set by:

a. Place a plate washer between the tensioner and a block. Using a press, press in the pushrod using 220–2205 lbs. of pressure.

b. Align the holes of the pushrod and housing, pass a 1.27mm Allen wrench through the holes to keep the setting position of the pushrod.

c. Release the press and install the dust boot to the tensioner.

75. Loosely install the timing belt tensioner. Evenly and alternately tighten the bolts to 20 ft. lbs. (26 Nm). Remove the tool from the tensioner.

76. Turn the crankshaft pulley two complete revolutions from TDC to TDC. Always turn the crankshaft clockwise. Check that all belt and pulley marks align with their reference marks. If any mark is out of perfect alignment, the timing belt must be removed and reinstalled.

77. Install the remaining components. Install the accessory drive belt.

78. Install the battery and connect the cables.

79. Fill the engine coolant and check the transmission fluid level.

80. Start the engine. Check carefully for fluid leaks. Check the ignition timing.

81. Install the engine under-covers.

1995–97 LS400 and 1996–97 SC400

1. Disconnect the negative battery cable. Wait at least 90 seconds before performing any other work. Disconnect the positive battery cable and remove the battery.

2. Remove the air cleaner inlet.

3. Remove the V bank cover by removing the bolt and two cap nuts.

4. Relieve the fuel pressure from the fuel lines.

— **CAUTION** —
Fuel injection systems remain under pressure after the engine has been turned OFF. Properly relieve fuel pressure before disconnecting any fuel lines. Failure to do so may result in fire or personal injury.

5. Remove the air cleaner and intake air connector assembly.

6. Remove the drive belt, fluid coupling and the fan pulley. The drive belt tension may be slackened by turning the tensioner counter-

clockwise. The pulley bolt for the drive belt tensioner has a left handed thread.

7. Remove the radiator.

8. Remove the right hand No. 3 timing belt cover.

9. Remove the left hand No. 3 timing belt cover, for 1995 California vehicles and all 1996–97 vehicles, disconnect the EVAP hose clamp from the timing belt cover. For all other vehicles, disconnect the EVAP hose from the hose clamp on the timing belt cover. Remove the four bolts to the left hand timing belt cover. Disconnect the cord grommet from the timing belt cover and remove the timing belt cover.

10. Remove the drive belt idler pulley by removing the pulley bolt and cover plate.

11. Remove the right hand and left No. 2 timing belt covers.

12. Remove the distributor housings.

13. Remove the No. 1 ignition coil.

14. Disconnect the A/C compressor from the engine. Do not disconnect the A/C pressure lines.

15. Remove the fan bracket by removing the two bolts and two nuts.

16. Disconnect and remove the alternator from the engine.

17. Remove the drive belt tensioner.

18. Set the engine to TDC on cylinder No. 1. Turn the crankshaft pulley and align its groove with the timing mark **0** of the No. 1 timing cover. Check that the timing marks of the camshaft timing pulleys and timing belt rear plates are aligned. If not, turn the crankshaft 1 full revolution (360 degrees).

19. Turn the crankshaft pulley approximately 50° clockwise and put the timing mark of the crankshaft pulley in line with the centers of the crankshaft pulley bolt and the idler pulley bolt.

——— WARNING ———

If the timing belt is disengaged, having the crankshaft pulley at the wrong angle can cause the piston head and valve head to come into contact with each other when you remove the camshaft timing pulley. Always set the crankshaft pulley at the correct angle before removing the timing belt.

20. If the timing belt is to be reused, turn the crank pulley slowly; check that the three installation marks are present on the belt. If the marks are not present, make new installation marks before removing the belt. The marks should align with the timing marks on each camshaft pulley and the crank pulley.

21. Remove the timing belt tensioner. Alternately loosen the two bolts; remove the bolts, the tensioner and the dust protector.

22. Using the proper tool, 09278–54012, or equivalent, loosen the tension between the left side and right side timing pulleys by slightly turning the left side camshaft clockwise.

23. Disconnect the timing belt from the camshaft timing pulleys.

24. Using the proper tool (SST 09960–10010 or equivalent), remove the bolt and the camshaft timing pulleys.

25. Remove the No. 2 ignition coil.

26. Remove the rear timing belt plates by removing the two bolts to each plate.

27. Remove the throttle body.

28. Remove the spark plug wires, wire clamps and the wire cover assembly from the right cylinder head.

29. Remove the right cylinder head cover by removing the eight bolts and eight washers.

30. Remove the transmission oil dipstick.

31. Disconnect the EVAP hose (from the charcoal canister) from the vacuum switching valve.

32. Disconnect the engine wire clamp from the wire bracket on the delivery pipe.

33. Disconnect the spark plug wires and clamps from the left hand cylinder head cover.

34. Remove the left cylinder head cover by removing the eight bolts and eight seal washers.

35. Remove the semi-circular plugs, if necessary.

NOTE: Since the thrust clearance of the camshaft is small, the camshaft must be kept level while it is being removed. If the camshaft is not kept level, the portion of the cylinder head receiving the shaft thrust may crack or be damaged, causing the camshaft to seize or break.

36. To remove the exhaust camshaft from the right side cylinder head:

   a. Position the service bolt hole of the drive sub-gear to the upright position. Secure the camshaft sub-gear to drive gear with a service bolt.

   b. Set the timing mark (single dot) on the camshaft drive gear at approximately 10 degrees. Turn the camshaft with a wrench on the hexagonal flats.

   c. Alternately loosen and remove the bearing cap bolts holding the intake camshaft side of the oil feed pipe to the cylinder head.

   d. Uniformly loosen (in several passes) and remove the bearing cap bolts, in sequence.

   e. Remove the oil feed pipe and the bearing caps. Remove the camshaft.

37. To remove the intake camshaft from the right side cylinder head:

   a. Set the timing mark (single dot) on the camshaft drive gear at approximately 45 degrees. Turn the camshaft with a wrench on the hexagonal flats.

   b. Uniformly loosen (in several passes) and remove the bearing cap bolts in the proper sequence.

   c. Remove the bearing caps, oil seal and the intake camshaft.

38. To remove the exhaust camshaft of the left side cylinder head:

   a. Position the service bolt hole of the drive sub-gear to the upright position. Secure the camshaft sub-gear to drive gear with a service bolt.

NOTE: When removing the camshaft, make sure the torsional spring force of the sub-gear has been eliminated.

   b. Set the timing mark (2 dots) on the camshaft drive gear at approximately 15 degrees, by turning the camshaft with the proper tool.

   c. Alternately loosen and remove the bearing cap bolts holding the intake camshaft side of the oil feed pipe to the cylinder head.

   d. Uniformly loosen (in several passes) and remove the bearing cap bolts in the proper sequence.

   e. Remove the oil feed pipe and the bearing caps. Remove the camshaft.

39. To remove the intake camshaft from the left side cylinder head:

   a. Set the timing mark (single dot) of the camshaft drive gear at approximately 60 degrees, by turning the camshaft with the proper tool.

   b. Uniformly loosen (in several passes) and remove the bearing cap bolts, in sequence.

   c. Remove the bearing caps, oil seal and the intake camshaft.

40. Remove the valve lash adjuster shims and hydraulic lash adjusters. Identify each lash adjuster and shim as it is removed so it can be reinstalled in the same position. If the lash adjusters are to be reused, store them upside down in a sealed container.

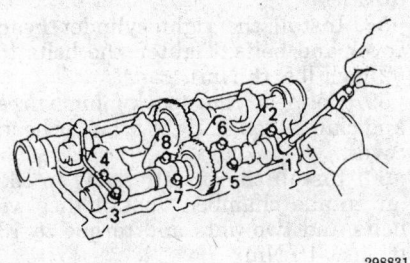

Right side exhaust camshaft bracket bolts removal sequence — 1995–97 LS400

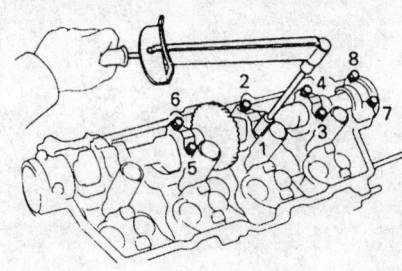

Right side intake camshaft bracket bolt tightening sequence — 1995–97 LS400

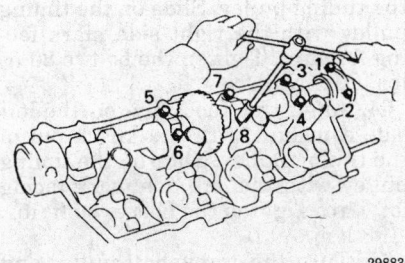

Right side intake camshaft bracket bolts removal sequence — 1995–97 LS400

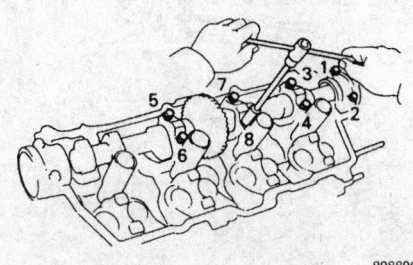

Right side intake camshaft bracket bolts removal sequence — 1996–97 SC400

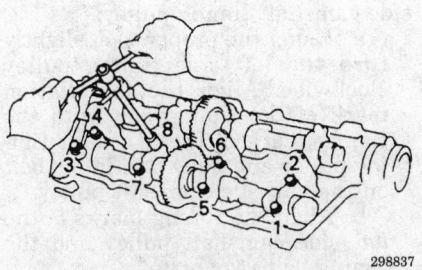

Left side exhaust camshaft bracket bolts removal sequence — 1995–97 LS400

To install:

41. Install the valve lash adjusters into their original positions and install the shims. Check valve clearance and replace the shims as necessary.

42. When reinstalling, remember that the camshafts must be handled carefully and kept straight and level to avoid damage.

43. Remove any old packing and apply new seal packing to the bearing caps.

44. Install the bearing cap on the right side cylinder head, marked **I1** ,

in position with the arrow mark facing the rear. Install the bearing cap on the left side cylinder head, marked **I6**, in position with the arrow mark facing the front.

45. Apply a light coat of oil on the threads of the cap bolts. Install the bearing cap bolts with new washers and tighten to 12 ft. lbs. (16 Nm).

46. To install the right side cylinder head intake camshaft:

   a. Apply grease to the thrust portion of the camshaft.

   b. Place the intake camshaft at a 45 degree angle of the timing mark (single dot) on the cylinder head.

   c. Remove any old packing and apply new seal packing to the bearing cap marked **I6** and install the front bearing cap, marked **I6** with the arrow facing rearward.

   d. Align the arrows at the front and rear of the cylinder head with the bearing cap.

   e. Install the remaining bearing caps in the proper sequence with the arrow mark facing rearward. Install the oil feed pipe and the mounting bolts.

   f. Uniformly tighten the bearing cap bolts in the proper sequence to 12 ft. lbs. (16 Nm).

47. To install the right side cylinder head exhaust camshaft:

   a. Set the timing mark (single dot) on the camshaft drive gear at a 10 degree angle by turning the intake camshaft with the proper tool.

   b. Apply grease to the thrust portion of the camshaft.

   c. Align the timing marks (single dots) on the camshaft drive and driven gears.

   d. Place the exhaust camshaft in the cylinder head. Install the rear bearing cap with the arrow mark facing rearward.

   e. Align the arrow marks at the front and rear of the cylinder head with the mark on the bearing cap. Apply a light coat of oil on the threads of the bearing cap bolts.

   f. Uniformly tighten the bearing cap bolts in the proper sequence to 12 ft. lbs. (16 Nm).

   g. Bring the service bolt installed upward by turning the camshaft with the proper tool. Remove the service bolt.

48. To install the left side cylinder head intake camshaft:

   a. Apply grease to the thrust portion of the camshaft.

   b. Place the intake camshaft with the timing mark (single dot) at a 60 degree angle on the cylinder head.

   c. Remove any old packing and apply new seal packing to the bearing cap marked **I6** and install the front bearing cap, marked **I1** with the arrow facing rearward.

   d. Align the arrows at the front and rear of the cylinder head with the bearing cap. Apply a light coat of oil on the threads of the bearing cap bolts.

   e. Install the remaining bearing caps in the proper sequence with the arrow mark facing rearward. Install the oil feed pipe and the mounting bolts.

   f. Uniformly tighten the bearing cap bolts in the proper sequence to 12 ft. lbs. (16 Nm).

49. Install the left side cylinder head exhaust camshaft by:

   a. Set the timing mark (2 dots) on the camshaft drive gear at a 15 degree angle by turning the intake camshaft with the proper tool.

   b. Apply grease to the thrust portion of the camshaft.

   c. Align the timing marks (2 dots each) on the camshaft drive and driven gears.

   d. Place the exhaust camshaft on the cylinder head. Install the rear bearing cap with the arrow mark facing rearward.

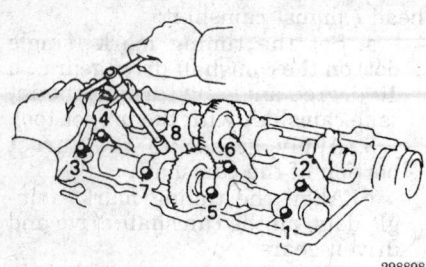

Left side exhaust camshaft bracket bolts
removal sequence — 1996–97 SC400

298898

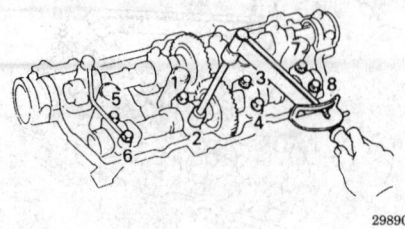

Right side exhaust camshaft bracket bolt
tightening sequence — 1996–97 SC400

298903

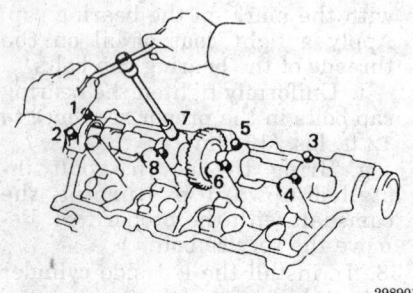

Left side intake camshaft bracket bolts removal
sequence — 1996–97 SC400

298901

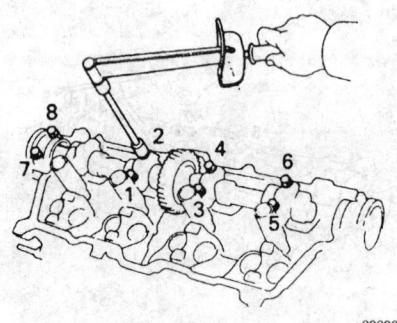

Left side intake camshaft bracket bolt
tightening sequence — LS400 and SC400

298904

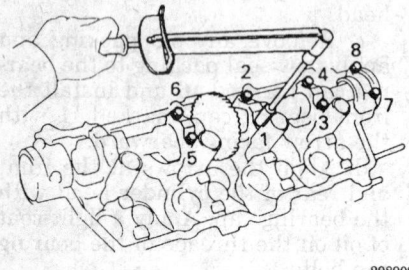

Right side intake camshaft bracket bolt
tightening sequence — 1996–97 SC400

298902

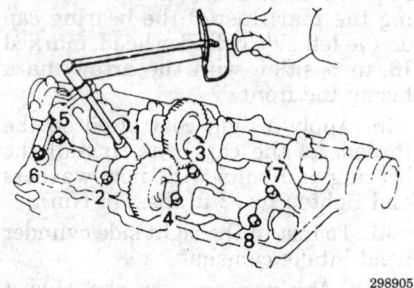

Left side exhaust camshaft bracket bolt
tightening sequence — LS400 and SC400

298905

e. Align the arrow marks at the front and rear of the cylinder head with the mark on the bearing cap. Apply a light coat of oil on the threads of the bearing cap bolts.

f. Uniformly tighten the bearing cap bolts in the proper sequence to 12 ft. lbs. (16 Nm).

g. Bring the service bolt installed upward by turning the camshaft with the proper tool. Remove the service bolt.

50. Install the camshaft oil seals with the proper tool (SST 09223–46011). Make sure to apply MP grease to the new oil seal lip.

51. Install the semi-circular plugs to the cylinder heads.

52. Clean the cylinder head covers. Apply new sealant in the correct locations and install the gaskets.

53. Install the left cylinder head cover and bolts. Tighten the bolts to 52 inch lbs. (6 Nm).

54. Connect the spark plug wires and clamps to the left cylinder head cover.

55. Connect the engine wire clamp to the wire bracket on the delivery pipe.

56. Connect the EVAP hose to the VSV.

57. Install the transmission oil dipstick.

58. Install the right cylinder head cover and bolts. Tighten the bolts to 52 inch lbs. (6 Nm).

59. Connect the spark plug wires and clamps to the right cylinder head cover.

60. Install the throttle body to the air intake chamber. Install the two bolts and two nuts and torque to 13 ft. lbs. (18 Nm).

61. Install the timing belt rear plates by installing the bolts. Torque the bolts to 66 inch lbs. (8 Nm).

62. Install the No. 2 ignition coil.

63. Align the knock pin on the right side camshaft with the knock pin of the timing pulley. Slide on the timing pulley with the right side mark facing forward. Tighten the bolt to 80 ft. lbs. (108 Nm).

64. Align the knock pin on the left side camshaft with the knock pin of the timing pulley. Slide on the timing pulley with the left side mark facing forward. Tighten the bolt to 80 ft. lbs. (108 Nm).

65. Turn the crankshaft pulley and align its groove with the 0 timing mark on the timing belt cover.

66. Turn each camshaft timing pulley and align the timing marks of the pulley with the timing belt rear plate.

67. Install the timing belt to the left side camshaft timing pulley:

a. Using the proper tool, slightly turn the left side timing pulley clockwise. Align the installation mark of the timing belt with the timing mark of the camshaft timing pulley and hang the timing belt on the left side camshaft pulley.

b. Align the timing marks of the left side camshaft pulley and the timing belt rear plate.

c. Check that the timing belt has tension between crankshaft timing pulley and the left side camshaft pulley.

68. Install the timing belt to the right side camshaft timing pulley:

a. Using the proper tool, slightly turn the right side timing pulley clockwise. Align the installation mark of the timing belt with the timing mark of the camshaft timing pulley and hang the timing belt on the right side camshaft pulley.

b. Align the timing marks of the right side camshaft pulley and the timing belt rear plate.

c. Check that the timing belt has tension between crankshaft timing pulley and the right side camshaft pulley.

81. Lower the vehicle.

Piston and Connecting Rod

POSITIONING

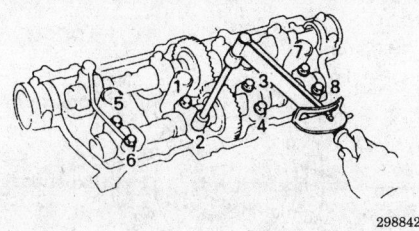

Right side exhaust camshaft bracket bolt tightening sequence — 1995–97 LS400

69. The timing belt tensioner must be set prior to installation. The tensioner can be set by:

a. Place a plate washer between the tensioner and a block. Using a press, press in the pushrod using 220–2205 lbs. of pressure.

b. Align the holes of the pushrod and housing, pass a 1.27mm Allen wrench through the holes to keep the setting position of the pushrod.

c. Release the press and install the dust boot to the tensioner.

70. Loosely install the tensioner. Evenly and alternately tighten the bolts to 20 ft. lbs. (26 Nm). Remove the tool from the tensioner.

71. Turn the crankshaft pulley two complete revolutions from TDC to TDC. Always turn the crankshaft clockwise. Check that all belt and pulley marks align with their reference marks. If any mark is out of perfect alignment, the timing belt must be removed and reinstalled.

72. Install the remaining components. Install the V bank cover.

73. Fill the radiator with engine coolant.

74. Install the battery and battery tray. Connect the battery cables, positive cable first.

75. Start the engine and check for leaks.

76. Recheck the engine coolant level.

77. Install the air cleaner inlet.

78. Install the battery clamp cover.

79. Install the engine undercover.

80. Install the oil pan protector.

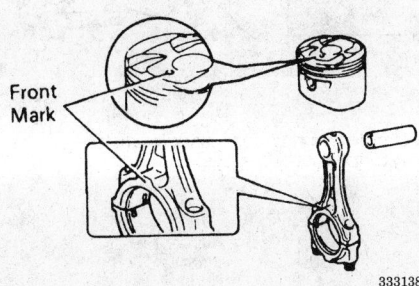

The position of the piston and connecting rod assemblies — GS300 and SC300

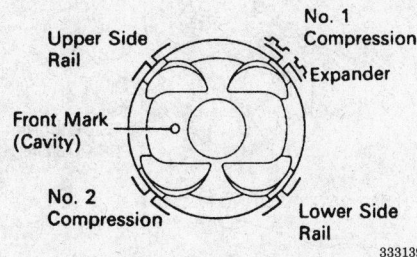

Piston ring end gap positioning — GS300 and SC300

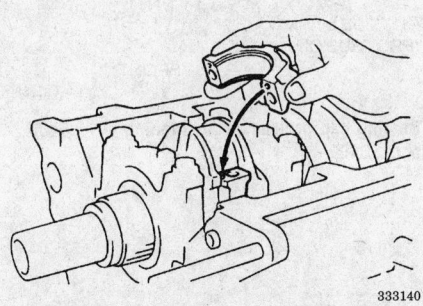

Connecting rod installation — GS300 and SC300

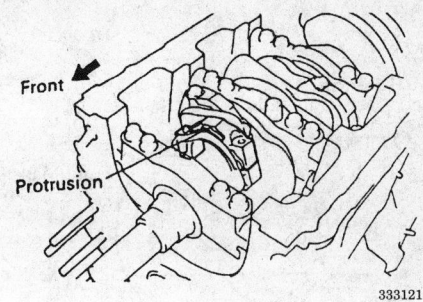

Connecting rod cap protrusion — ES300

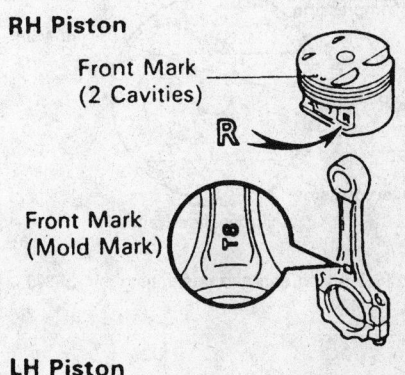

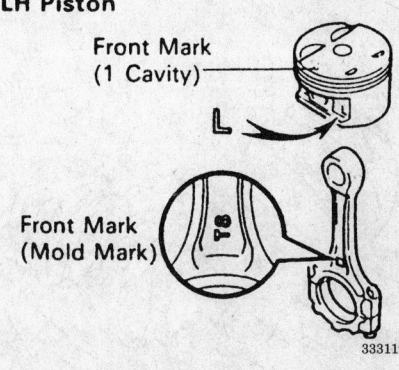

Matching the pistons to the connecting rods — ES300

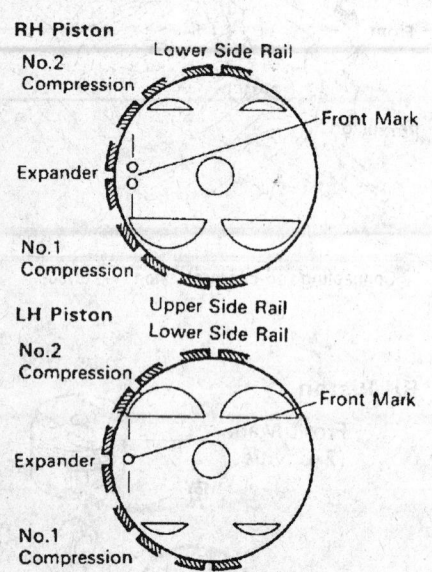

Piston ring end gap positioning — ES300

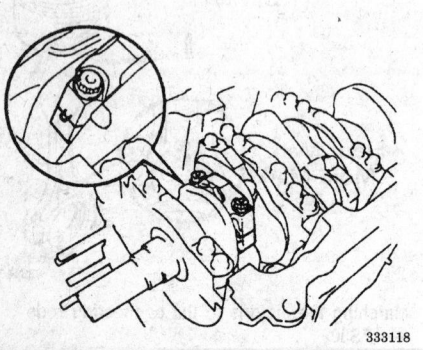

Connecting rod and cap matchmarks — ES300

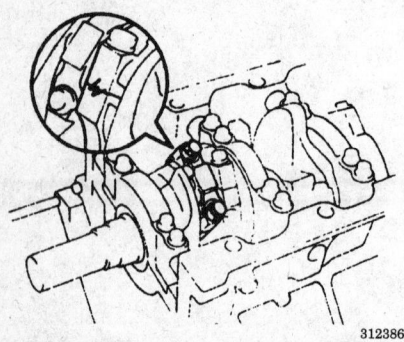

Connecting rod-to-cap matchmark — 1993–95 LS400 and SC400

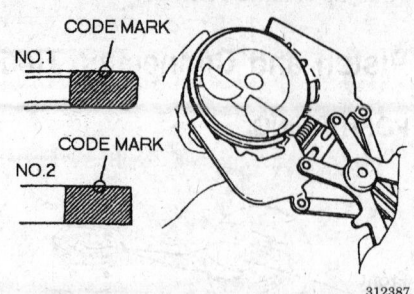

Compression ring code mark location — 1993–95 LS400 and SC400

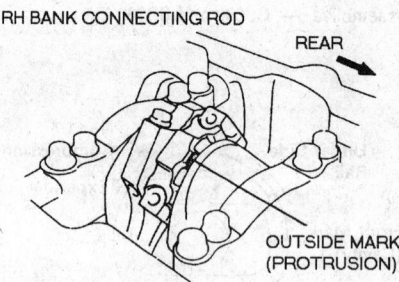

Bearing cap protrusion positioning — LS400 and SC400

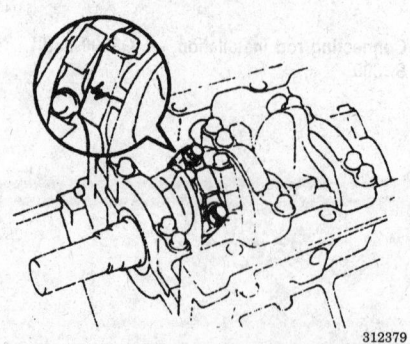

Connecting rod-to-cap matchmark — 1996–97 LS400 and SC400

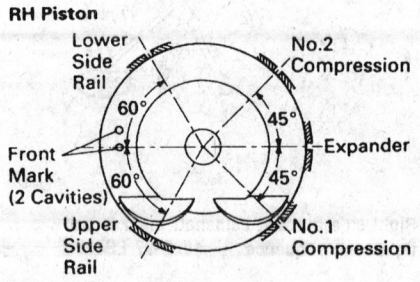

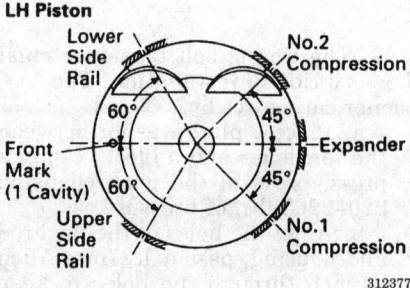

Arranging the piston rings — LS400 and SC400

ENGINE LUBRICATION

Oil Pan

REMOVAL AND INSTALLATION

ES300

1993 Models

— CAUTION —

The Supplemental Inflatable Restraint (SIR) system must be disarmed before removing the oil pan or pump. Failure to do so may cause accidental deployment of the air bag, resulting in unnecessary SIR system repairs and/or personal injury.

1. Raise and support the vehicle safely.
2. Drain the oil.
3. Remove the front exhaust pipe.
4. Remove the stiffener plate.
5. Remove the bolts and the nuts attaching the oil pan to the engine block.

6. Using SST 09032-00100 or equivalent, break the seal between the oil pan and the engine block.

7. Lower the oil pan to the ground being careful not to damage the oil pan flange.

To install:

8. Using a gasket scraper and a wire brush, remove all the old packing material (sealant) from the oil pan and cylinder block gasket surfaces. Wipe the oil pan interior with a rag. Clean the contact surfaces with a non-residue type solvent.

9. Apply a thin 1/8 inch (3–4mm) bead of No. 08826-00080 seal packing or equivalent to the oil pan.

NOTE: Avoid applying too much sealant to the oil pan. The oil pan must be assembled to the block within 5 minutes after the sealant is applied. If not, the sealant must be removed and re-applied.

10. Place the oil pan against the block and install the bolts and nuts. Torque the nuts and bolts to 65 inch lbs. (6 Nm).

11. Install the stiffener plate and torque the mounting bolts to 27 ft. lbs. (37 Nm) for 14mm bolts or 13 ft. lbs. (18 Nm) for 12mm bolts.

12. Install the front exhaust pipe and lower the vehicle to the ground.

13. Fill the engine with oil to the proper level.

14. Start the engine and check for leaks. Recheck the engine oil level.

1994 Models

-------- **CAUTION** --------

The Supplemental Inflatable Restraint (SIR) system must be disarmed before removing the oil pan or pump. Failure to do so may cause accidental deployment of the air bag, resulting in unnecessary SIR system repairs and/or personal injury.

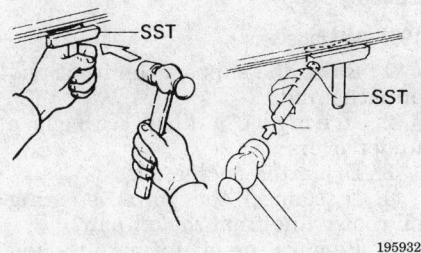

195932

Oil pan removal — 1993 ES300

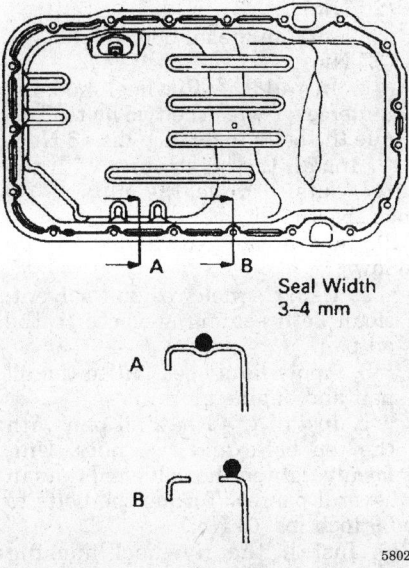

Seal Width
3–4 mm

58025

Apply sealant to the oil pan flange — 1993 ES300

1. With the ignition switch in the **LOCK** position, disconnect the negative battery terminal. If equipped with an air bag system, wait at least 90 seconds or longer before performing any other work.

2. Raise and support the vehicle safely.

3. Drain the oil from the engine.

4. Remove the dipstick.

5. Remove the front exhaust pipe.

6. Remove the flywheel housing cover.

7. Remove the oil level sensor.

8. Remove the two bolts attaching the lower oil pan to the transaxle.

9. Remove the ten bolts, two nuts attaching the lower oil pan to the upper oil pan.

10. Using SST 09032-00100 or equivalent, break the seal between the lower oil pan and the upper oil pan.

11. Remove the 17 bolts attaching the upper oil pan to the engine block. Pry the upper oil pan from the engine block.

To install:

12. Remove all gasket material from the oil pans.

13. Clean the gasket sealing surfaces with a non-residue type solvent.

14. Apply sealant to the upper oil pan and install it to the engine block.

NOTE: Parts must be assembled within three minutes after sealant application. Otherwise, remove the sealant and reapply.

15. Torque the bolts to:
• 10mm bolts — 69 inch lbs. (8 Nm)
• 12mm bolts — 14 ft. lbs. (15 Nm)

16. Apply sealant to the lower oil pan.

17. Install the lower oil pan and torque the nuts and bolts to 69 inch lbs. (8 Nm).

18. Install the bolts attaching the lower oil pan to the transaxle. Torque the bolts to 27 ft. lbs. (37 Nm).

19. Install the oil level sensor with a new gasket.

20. Install the flywheel housing cover and torque the bolts to 69 inch lbs. (8 Nm).

21. Install the front exhaust pipe support. Torque the two bolts to 15 ft. lbs. (21 Nm).

22. Install the front exhaust pipe with new gaskets.

23. Install the drain plug and a new gasket.

24. Lower the vehicle.

25. Fill the engine to the proper level with the recommended grade of oil.

26. Connect the negative battery cable, start the engine and check for leaks.

1995–97 Models

1. Disconnect the negative battery cable from the battery.

2. Raise and safely support the front of the vehicle.

3. Remove the right front wheel.

4. Remove the fender apron seal.

5. Remove the engine undercover.

6. Drain the engine oil from the engine.

7. Remove the front exhaust pipe.

8. Remove the front exhaust pipe bracket from the No. 1 oil pan.

9. Remove the flywheel housing undercover.

10. Remove the ten bolts and two nuts to the No. 2 oil pan.

11. Insert the blade of SST tool 09032–00100 or equivalent between the No. 1 and No. 2 oil pans. Clean the surfaces of the oil pans.

12. Remove the oil strainer and gasket from the engine by removing the three nuts.

13. Remove the No. 1 oil pan as follows:

a. Remove the two bolts to the flywheel housing undercover. Remove the flywheel undercover.

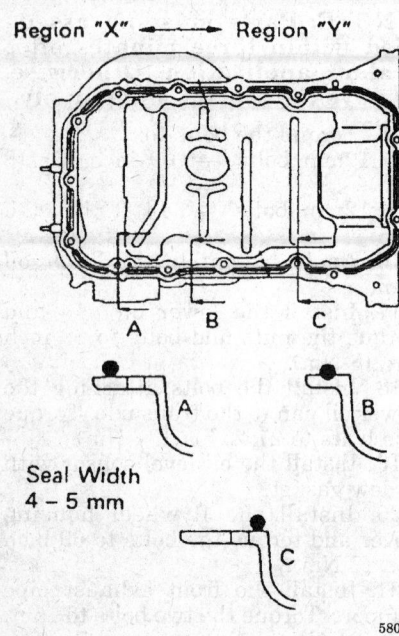

Region "X" ←——→ Region "Y"

Seal Width
4 – 5 mm

58078

Lower oil pan sealant application — 1994 ES300

b. Remove the 17 bolts and 2 nuts to the No. 1 oil pan. Make a note of the position of the each bolt. When replacing the bolts into the oil pan, place each bolt in the position from which it was removed.

c. Remove the oil pan by prying the portions between the cylinder block and the oil pan. Be careful not to damage the contact surfaces.

14. Remove the baffle plate from the No. 1 oil pan.

To install:

15. Clean all mating surfaces of the oil pans.

16. Install the baffle plate to the No. 1 oil pan and torque to 69 inch lbs. (8 Nm).

17. Install the No. 1 oil pan as follows:

a. Using a non residue solvent, clean both sealing surfaces to the oil pan.

b. Apply liquid sealant to the oil pan and engine block.

c. Install the oil pan with the 17 bolts and 2 nuts. Uniformly tighten the bolts and nuts in several passes.

d. Torque the bolts as follows:
• 10mm head bolt–69 inch lbs. (8 Nm)
• 12mm head bolt–14 ft. lbs. (20 Nm)
• 14mm head bolt–27 ft. lbs. (37 Nm)

e. Install the flywheel housing undercover with the two bolts. Torque the bolts to 69 inch lbs. (8 Nm).

18. Install the oil strainer with the three nuts. Torque the nuts to 69 inch lbs. (8 Nm).

19. Install the No. 2 oil pan as follows:

a. Using a non residue solvent, clean both sealing surfaces to the oil pan.

b. Apply liquid sealant to the oil pan and engine block.

c. Install the No. 2 oil pan with the ten bolts and two nuts. Uniformly tighten the bolts and nuts in several passes. Torque the bolts to 69 inch lbs. (8 Nm).

20. Install the flywheel housing undercover.

21. Install the front exhaust pipe bracket to the No. 1 oil pan. Torque the bolts to 15 ft. lbs. (21 Nm).

22. Install the front exhaust pipe, Tighten the four nuts holding the exhaust manifolds to the front exhaust pipe. Torque the four nuts to 46 ft. lbs. (62 Nm). Tighten the two bolts and two nuts holding the front exhaust pipe to the center exhaust pipe. Torque the bolts and nuts to 41 ft. lbs. (56 Nm). Install the bracket with the two bolts and torque to 14 ft. lbs. (19 Nm). Install the support stay with the two bolts and torque to 22 ft. lbs. (30 Nm).

23. Install the engine undercover.

24. Install the right fender apron seal.

25. Install the right front wheel and lower the vehicle.

26. Fill the engine with oil.

27. Start the engine and check for leaks.

28. Recheck the engine oil.

GS300 and SC300

NOTE: The No. 1 oil pan can not be removed with the engine in the vehicle. The engine/transmission assembly must be removed. The manufacturer does not provide any on vehicle information for the No. 2 oil pan removal and installation. If only the No. 2 oil pan is being serviced, the engine/transmission assembly can remain in the vehicle.

1. Remove the engine/trans assembly and then separate the transmission from the engine.

2. With the engine on a stand, remove the timing belt, the idler pulley and the crankshaft timing pulley.

3. Remove the oil dipstick and guide.

4. Disconnect the oil sensor lead, remove the four attaching bolts and lift off the oil level sensor. Be careful not to drop this sensor.

5. Remove the 14 bolts (16 bolts for GS300) and two nuts and pry off the lower (No. 2) oil pan. Be careful not to damage the No. 1 pan while performing this procedure.

6. Remove the bolt and two nuts and drop down the oil strainer and gasket.

7. Remove the five bolts and two nuts and drop down the baffle plate.

8. Remove the 22 bolts and the carefully pry off the upper (No. 1) oil pan. Remove the O-ring from the cylinder block.

To install:

9. Position a new O-ring in the block and scrape off any old sealant. Apply sealant to the pan mating surface with a 1/8 inch (3–4mm) bead. Install the upper pan and tighten the 12mm bolts to 15 ft. lbs. (21 Nm) and the 14mm bolts to 29 ft. lbs. (39 Nm).

10. Install the baffle plate and oil strainer. Tighten them both to 78 inch lbs. (9 Nm).

11. Install the lower pan in the same manner as the upper pan and tighten the bolts to 78 inch lbs. (9 Nm).

12. Using a new gasket, install the oil level sensor and tighten it to 48 inch lbs. (5 Nm).

13. Install the oil dipstick and guide, the timing pulleys and belt and reconnect the transmission to the engine.

14. Install the engine and transmission.

15. Refill all fluids.

16. Start the engine and check for leaks.

17. Road test the vehicle.

LS400

1993–94 Models

1. Raise the support the front end on jackstands.

2. Remove the engine undercovers.

3. Drain the engine oil.

4. Disconnect the power steering oil cooler line from the oil pan.

5. Remove the mounting nuts and bolts. Use a gasket cutter (tool no. 09032-00100 or equivalent) to cut the sealant holding the lower oil pan to

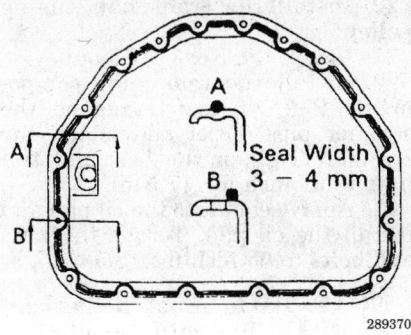

Lower oil pan sealant application — GS300 and SC300

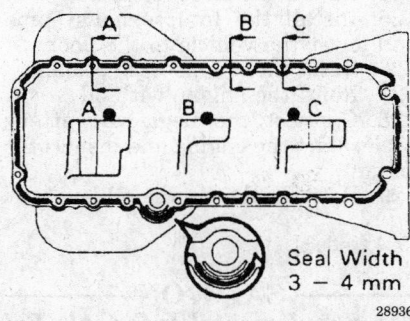

Upper oil pan sealant application — GS300 and SC300

the upper oil pan. Remove the lower oil pan.

NOTE: Use great care to avoid damaging the contact surfaces or flanges with the gasket cutter.

To install:
6. Using a gasket scraper and a wire brush, remove all the old packing material from the lower and upper oil pan surfaces. Wipe the oil pan interior with a rag. Clean the contact surfaces with a non-residue solvent. Do not use a solvent which will affect the painted surface.
7. Apply a thin, 1/8 inch (3–4mm) bead of seal packing or form-in-place gasket (FIPG) or equivalent to the oil pan. To ensure the proper size bead, cut the nozzle on the tube a bit smaller than the desired bead.

NOTE: Avoid applying too much sealant to the oil pan. The oil pan must be assembled to the block 3-5 minutes after the sealant is applied. If not, the sealant must be removed and re-applied.

8. Hold the oil pan in place and install the bolts and nuts. Torque the nuts and bolts to 69 inch lbs. (8 Nm).
9. Reattach the oil cooler lines to the pan.

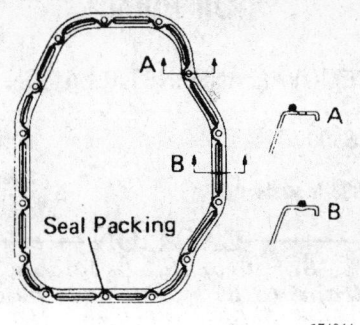

Sealant application — 1993–94 LS400

10. Fill the engine with oil to the proper level.
11. Start the engine and check for leaks.
12. Install the engine cover or covers.
13. Recheck the engine oil level.

1995–97 Models

1. Remove the engine/trans assembly. Separate the transmission from the engine.
2. With the engine on a stand, remove the timing belt, the idler pulleys and the crankshaft timing pulley.
3. Remove the oil dipstick and guide.
4. Disconnect the oil level sensor lead.
5. Remove the four bolts and lift off the oil level sensor. Be careful not to drop this sensor.
6. Remove the oil filter and the bracket assembly by removing the stud bolt and two nuts.
7. Disconnect the engine crankshaft position sensor connector. Remove the sensor by removing the bolt.
8. Remove the 12 bolts and 2 nuts to the No. 2 oil pan. Use a gasket cutting tool to separate the No. 2 (lower) oil pan. Be careful not to damage the No. 1 pan while performing this procedure.
9. Remove the two bolts and three nuts and drop down the baffle plate.
10. Remove the oil strainer by removing the bolts and nuts.
11. Remove the bolts and then carefully pry off the No. 1 oil pan. There are slots for inserting the prybar.
To install:
12. Scrape off any old sealant and then install the No. 1 pan. Apply sealant to the pan mating surface with a 1/8 inch (3–4mm) bead.

13. Torque the bolts for the No. 1 oil pan as follows:
• 10mm head to 66 inch lbs. (8 Nm)
• 12mm head to 21 ft. lbs. (28 Nm)
14. Install the oil strainer by installing the bolts and nuts. Tighten the bolts to 66 inch lbs. (8 Nm).
15. Install the baffle plate and tighten the bolts and nuts to 66 inch lbs. (8 Nm).
16. Install the No. 2 pan in the same manner as the No. 1 oil pan and tighten the bolts to 66 inch lbs. (8 Nm). Make sure the bolts are 14mm in length.
17. Install the engine crankshaft position sensor. Tighten the bolt to 56 inch lbs. (6 Nm). Connect the crankshaft position sensor connector.
18. Place a new O-ring in position on the oil filter bracket. Install the bracket and tighten the bolt and nuts to 13 ft. lbs. (18 Nm). Connect the wiring to the pressure switch.
19. Using a new gasket, install the oil level sensor and tighten the four bolts to 48 inch lbs. (5 Nm).
20. Install the dipstick and guide, the timing belt pulleys and the timing belt components.
21. Connect the transaxle to the engine.
22. Install the engine and transaxle, refill all fluids and road test the vehicle.

SC400

NOTE: The No. 1 oil pan cannot be removed with the engine in the vehicle. The engine and transmission must be removed as a unit and then separated. It may be possible to remove the No. 2 oil pan from the vehicle while the engine is still in the vehicle.

1. Remove the engine/trans assembly. Separate the transmission from the engine.
2. Remove the oil dipstick and guide.
3. Remove the 12 bolts and 2 nuts. Use a gasket cutting tool to separate the No. 2 (lower) oil pan. Be careful not to damage the No. 1 pan while performing this procedure.
4. Remove the six bolts and two nuts; remove the baffle plate.
5. Remove the 16 bolts and then carefully pry off the No. 1 oil pan.

NOTE: There are slots for inserting the prybar.

To install:
6. Scrape off any old sealant and then install the No. 1 pan. Apply sealant to the pan mating surface with a 1/8 inch (3–4mm) bead.

7. Torque the bolts as follows:
a. For 1993:
- 12mm bolts — 13 ft. lbs. (18 Nm)
- 10mm bolts — 69 inch lbs. (8 Nm).
b. For 1994–97:
- 12mm bolts — 69 inch lbs. (8mm)
- 14mm bolts — 20 ft. lbs. (28 Nm)

8. Install the baffle plate and tighten the bolts and nuts to 69 inch lbs. (8 Nm).

9. Scrape off any old sealant, apply sealant to the pan mating surface with a 1/8 inch (3–4mm) diameter bead and install the No. 2 oil pan. Tighten the bolts to 69 inch lbs. (8 Nm). Make sure the bolts are 14mm in length.

10. Install the dipstick and guide.

11. Install the engine to the transaxle and install the assembly to the vehicle.

12. Refill all fluids, check for leaks and road test the vehicle.

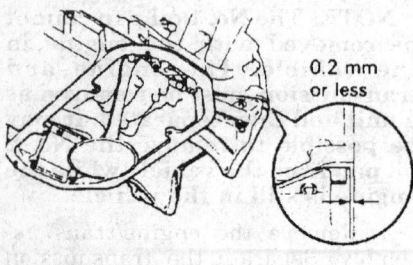

0.2 mm or less

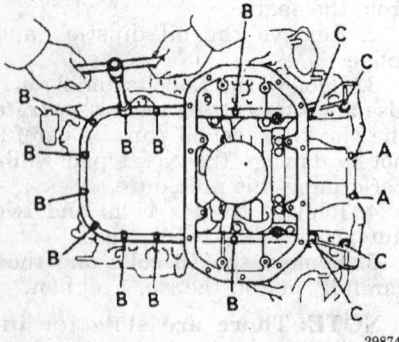

298746

Oil pan bolt locations: A–0.78 in (20mm), B–1.38 in. (35mm), C–0.78 in (20mm) w/12mm head — SC400

Oil Pump

REMOVAL AND INSTALLATION

ES300

1993 Models

— CAUTION —

The Supplemental Inflatable Restraint (SIR) system must be disarmed before removing the oil pan or pump. Failure to do so may cause accidental deployment of the air bag, resulting in unnecessary SIR system repairs and/or personal injury.

1. With the ignition switch in the **LOCK** position, disconnect the negative battery terminal. If equipped with an air bag system, wait at least 90 seconds or longer before performing any other work.

2. Raise and support the vehicle safely.

3. Remove the oil pan.

4. Remove the oil pan baffle plate.

5. Remove the oil strainer and O-ring.

6. Remove the windshield washer reservoir.

7. Remove the alternator.

8. Disconnect the radiator lower hose.

9. Remove the A/C compressor, it is not necessary to disconnect the refrigerant lines. Also, remove the compressor bracket.

10. Remove the water inlet pipe.

11. Support the engine with a hoist and remove the timing belt.

12. Remove the No. 1 idler pulley and pry off the crankshaft timing pulley.

13. Remove the nine oil pump retaining bolts.

14. With a soft-faced hammer or rubber mallet, tap the oil pump loose from the block.

15. Remove the discard the O-ring from the engine block.

To install:

16. Clean the cylinder block and oil pump gasket sealing surfaces.

17. Draw a 1/8 inch (3–4mm) bead of seal packing.

18. Insert a new O-ring and then engage the spline teeth on the drive gear with the large teeth on the end of the crankshaft.

19. Tighten the oil pump mounting bolts as follows:
- 12mm bolts (C and D) — 14 ft. lbs. (20 Nm)
- 14mm bolts (A and B) — 30 ft. lbs. (41 Nm)

20. Install the crankshaft timing pulley.

21. Install the No. 1 idler pulley.

22. Install the remaining components. Place a new O-ring on the strainer pipe outlet and install the strainer. Tighten the bolt and two nuts to 61 inch lbs. (7 Nm).

23. Apply sealant to the oil pan and install the oil pan. Torque the nuts and bolts to 65 inch lbs. (5.5 Nm).

NOTE: Parts must be assembled within five minutes of sealant application or the sealant must be removed and reapplied.

24. Install the stiffener plate.

25. Install the front exhaust pipe and lower the vehicle to the floor.

26. Refill the engine with coolant.

27. Refill the engine with oil.

28. Connect the negative battery cable, start the engine and inspect for leaks.

29. Recheck the engine oil level.

1994 Models

— CAUTION —

The Supplemental Inflatable Restraint (SIR) system must be disarmed before removing the oil pan or pump. Failure to do so may cause accidental deployment of the air bag, resulting in unnecessary SIR system repairs and/or personal injury.

1. With the ignition switch in the **LOCK** position, disconnect the negative battery terminal. If equipped with an air bag system, wait at least 90 seconds or longer before performing any other work.

2. Drain the engine oil.

3. Remove the dipstick and tube.

4. Remove the timing belt.

5. Remove the timing pulleys.

6. Disconnect the engine wire harness from the front of the engine.

7. Remove the upper inner timing belt cover.

8. Remove the alternator.

9. Remove the crankshaft position sensor.

10. Remove the oil lever sensor.

11. Remove the A/C compressor leaving the lines attached.

12. Remove the A/C compressor housing bracket.

13. Remove the front exhaust pipe.

14. Remove the front exhaust pipe stay.

15. Remove the flywheel cover.

16. Remove the oil pans.

17. Remove the oil strainer (pick-up).

18. Remove the oil plate baffle.

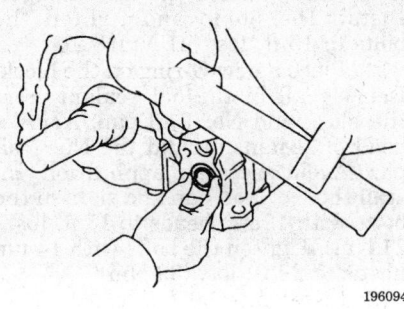

Remove the O-ring from the engine block —
1993 ES300

196094

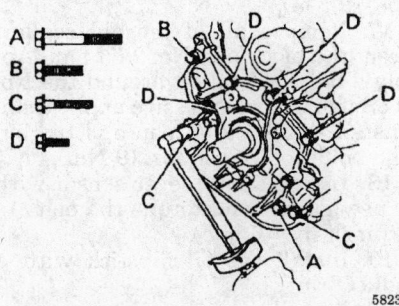

Oil pump mounting bolt installation locations
— 1993 ES300

58234

19. Remove the nine bolts and pry the oil pump from the block. Discard the O-ring.

To install:

20. Clean all gaskets mating surfaces.

21. Apply sealant to the oil pump.

22. Place a new O-ring on the cylinder block.

23. Engage the spline teeth of the oil pump drive gear with the crankshaft teeth. Slide the oil pump onto the crankshaft. Install the bolts and torque them as follows:

• 10mm bolts — 69 inch lbs. (8 Nm)

• 12mm bolts — 14 ft. lbs. (20 Nm)

24. Install the oil baffle and torque the bolts to 69 inch lbs. (8 Nm).

25. Install the oil strainer.

26. Install the oil pans.

27. Install the flywheel cover.

28. Install the exhaust pipe stay.

29. Install the front exhaust pipe with new gaskets.

30. Install the A/C compressor housing bracket and torque the bolts to 18 ft. lbs. (25 Nm).

31. Install the A/C compressor and torque the bolts to 18 ft. lbs. (25 Nm).

32. Install the oil level sensor with a new gasket. Torque the bolts to 69 inch lbs. (8 Nm).

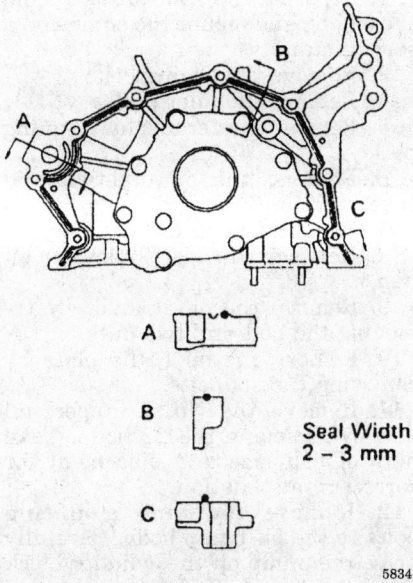

Seal Width
2 – 3 mm

Oil pump sealant application — 1994 ES300

58341

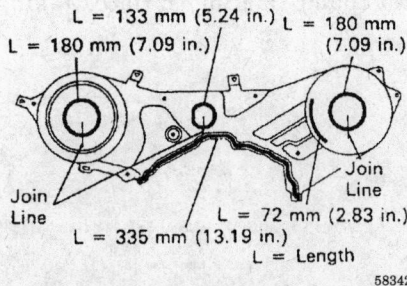

L = 133 mm (5.24 in.) L = 180 mm (7.09 in.)
L = 180 mm (7.09 in.)
Join Line
Join Line
Join Line
L = 72 mm (2.83 in.)
L = 335 mm (13.19 in.)
L = Length

Upper inner timing belt cover gasket locations
— 1994 ES300

58342

33. Install the crankshaft position sensor and torque the bolts to 69 inch lbs. (8 Nm).

34. Install the remaining components. Install the engine oil dipstick and housing with a new O-ring.

35. Fill the engine to the correct level with the recommended oil.

1995–97 Models

1. Disconnect the negative battery cable from the battery.

2. Raise and safely support the front of the vehicle.

3. Remove the right front wheel.

4. Remove the fender apron seal.

5. Remove the engine undercover.

6. Drain the engine oil from the engine.

7. Remove the front exhaust pipe.

8. Remove the front exhaust pipe bracket from the No. 1 oil pan.

9. Remove the alternator drive belt from the engine.

10. Disconnect the A/C compressor from the engine, without disconnecting the compressor lines.

11. Remove the power steering pump drive belt and adjusting strut.

12. Remove the timing belt from the engine.

13. Remove the timing belt pulleys.

14. Remove the rear timing belt cover from the engine by removing the wire clamps and six bolts.

15. Remove the A/C compressor housing bracket by removing the three bolts.

16. Remove the ten bolts and two nuts to the No. 2 oil pan.

17. Insert the blade of SST tool 09032–00100 or equivalent between the No. 1 and No. 2 oil pans. Remove the No. 2 oil pan from the engine. Clean the surfaces of the oil pans.

18. Remove the oil strainer and gasket from the engine by removing the three nuts.

19. Remove the No. 1 oil pan.

20. Remove the baffle plate from the No. 1 oil pan.

21. Remove the crankshaft position sensor by removing the connector and bolt.

22. Remove the oil pump as follows:

a. Remove the nine bolts. Make a note of the position of the each bolt. When replacing the bolts into the oil pump body, place each bolt in the position from which it was removed.

b. Remove the oil pump body by prying between the oil pump and main bearing cap.

c. Remove the O-ring from the cylinder block.

d. Remove the plug, gasket, spring and relief valve from the oil pump body.

e. Remove the nine screws, pump body cover, drive and driven rotors.

To install:

23. To install the oil pump:

a. Install the driven rotors, drive, pump body cover and then install the nine screws.

b. Install the oil pump relief valve, spring, gasket and the plug to the oil pump body.

c. Place a new O-ring on the cylinder block.

d. Using a non residue solvent, clean both sealing surfaces to the oil pump.

e. Apply liquid sealant to the oil pump and engine block.

f. Install the oil pump to the engine block. Make sure to engage the spline teeth of the oil pump drive gear with the large teeth of the crankshaft.

g. Install the nine bolts to the oil pump and uniformly tighten the bolts in several passes. Torque the bolts as follows:

- 10mm head–69 inch lbs. (8 Nm)
- 12mm head–14 ft. lbs. (20 Nm)

24. Install the crankshaft position sensor and install the bolt. Torque the bolt to 69 inch lbs. (8 Nm).

25. Install the baffle plate to the No. oil pan and torque to 69 inch lbs. (8 Nm).

26. Install the No. 1 oil pan as follows:

a. Using a non residue solvent, clean both sealing surfaces to the oil pan.

b. Apply liquid sealant to the oil pan and engine block.

c. Install the oil pan with the seventeen bolts and two nuts. Uniformly tighten the bolts and nuts in several passes.

d. Torque the bolts as follows:

- 10mm head bolt–69 inch lbs. (8 Nm)
- 12mm head bolt–14 ft. lbs. (20 Nm)
- 14mm head bolt–27 ft. lbs. (37 Nm)

e. Install the flywheel housing undercover with the two bolts. Torque the bolts to 69 inch lbs. (8 Nm).

27. Install the oil strainer with the three nuts. Torque the nuts to 69 inch lbs. (8 Nm).

28. Install the No. 2 oil pan as follows:

a. Using a non residue solvent, clean both sealing surfaces to the oil pan.

b. Apply liquid sealant to the oil pan and engine block.

c. Install the No. 2 oil pan with the ten bolts and two nuts. Uniformly tighten the bolts and nuts in several passes. Torque the bolts to 69 inch lbs. (8 Nm).

29. Install the remaining components. Install the right front wheel and lower the vehicle.

30. Fill the engine with oil.

31. Connect the negative battery cable to the battery.

32. Start the engine and check for leaks.

33. Recheck the engine oil.

GS300 and SC300

1. Remove the engine and transmission.

2. Separate the transmission from the engine and mount the engine on a service stand.

3. Remove the timing belt.

4. Remove the idler pulley.

5. Remove the crankshaft timing pulley.

6. Remove the oil dipstick and tube.

7. Remove the oil level sensor.

8. Remove the No. 2 (lower) oil pan.

9. Remove the oil strainer by removing the bolt and two nuts.

10. Remove the oil baffle plate by removing the six bolts.

11. Remove the No. 1 (upper) oil pan by removing the 22 bolts. Take note of bolt size and placement for correct re-installation.

12. Remove the nine mounting bolts to the oil pump body. Carefully drive the pump off the cylinder block using a brass drift. Remove the two O-rings.

To install:

13. Position two new O-rings in the cylinder block. Scrape any old sealant from the mating surfaces. Draw a ⅛ inch (3–4mm) bead of sealant around the pump mating surface, taking great care around the oil passages. Install the pump and tighten the bolts to 15 ft. lbs. (21 Nm).

14. Place a new O-ring on the block. Remove all of the old sealant from the block and No. 1 oil pan. Apply a bead of sealant around the No. 1 oil pan. Avoid excessive application. Install the No. 1 oil pan and tighten the bolts with 12mm heads to 15 ft. lbs. (21 Nm). Tighten the bolts with 14mm heads to 29 ft. lbs. (39 Nm).

15. Install the oil baffle plate and tighten the nuts and bolts to 78 inch lbs. (9 Nm).

16. Install the oil strainer and tighten the nuts and bolts to 78 inch lbs. (9 Nm).

17. Remove all of the old sealant from the block and No. 2 oil pan. Apply a bead of sealant around the No. 2 oil pan. Avoid excessive application. Install the No. 2 oil pan and tighten the bolts to 78 inch lbs. (9 Nm).

18. Install the oil level sensor with a new gasket and torque the bolts to 48 inch lbs. (6 Nm).

19. Install the oil dipstick with a new O-ring.

20. Install the remaining components. Fill all fluids.

21. Connect the negative battery cable.

22. Start the engine and check for leaks.

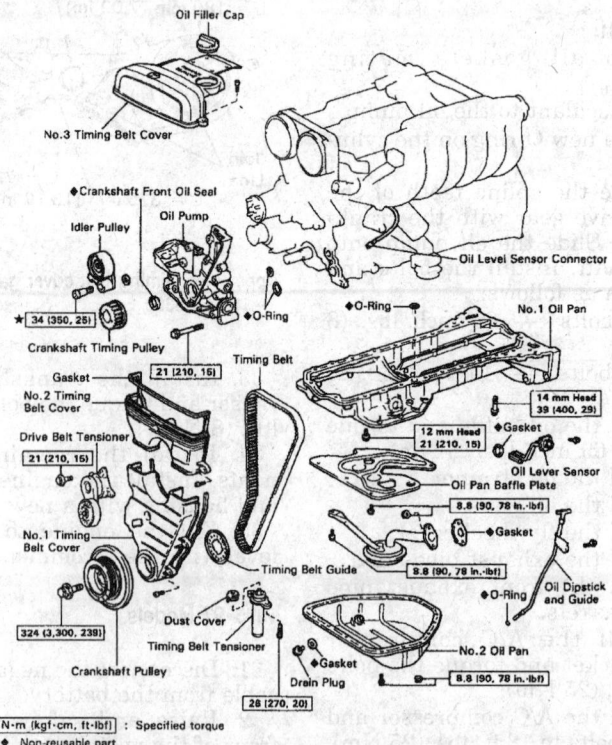

Oil pump and related components — GS300 and SC300

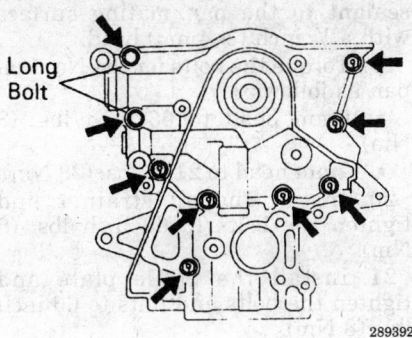

Oil pump mounting bolt installation locations — GS300 and SC300

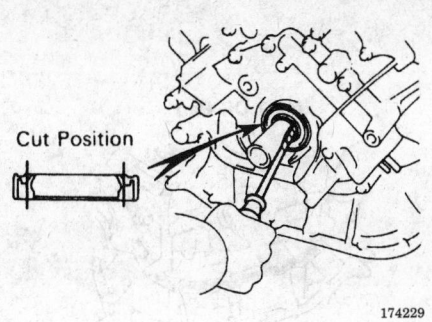

Front oil seal removal — 1993–94 LS400

LS400

1993–94 Models

1. Disconnect the negative battery cable.
2. Raise and safely support the vehicle.
3. Remove the engine undercover.
4. Drain the oil.
5. Disconnect the power steering oil cooler lines from the No. 1 (upper) oil pan.
6. Remove the timing belt.
7. Remove the No. 2 (lower) idler pulley.
8. Remove the No. 1 (upper) idler pulley.
9. Remove the crankshaft timing pulley.
10. Remove the mounting nuts and bolts. Use a gasket cutter (tool no. 09032-00100 or equivalent) to cut the sealant holding the lower oil pan to the upper oil pan. Remove the lower oil pan.

NOTE: Use great care to avoid damaging the contact surfaces or flanges with the gasket cutter.

11. Remove the oil baffle plate.
12. Remove the oil strainer.
13. Disconnect the oil pressure switch connector. Remove the oil filter and bracket assembly. Remove the gasket from the bracket.
14. Disconnect the crankshaft position sensor wiring connector. Remove the sensor.
15. Use a small sharp knife to cut the lip off the front crankshaft oil seal. Use a small, flat tool with tape over the edge to pry out the seal. Take care not to damage the crankshaft surface.
16. Using two nuts, remove the left stud bolt. Remove the 3 bolts and the stud bolt holding the upper oil pan to the oil pump.

17. Remove the bolts holding the oil pump to the block.

NOTE: The bolts are of different lengths and must be reinstalled in the proper holes. Take note of bolt placement during removal.

18. Gently pry the oil pump away from the block and remove it. Remove the O-ring from the cylinder block.

To install:

19. Using a gasket scraper and a razor blade, remove all the old packing material from the lower and upper oil pans, the oil pump and the cylinder block. Wipe the oil pan interior with a rag. Clean the contact surfaces with a non-residue solvent. Do not use a solvent which will affect the painted surfaces.
20. Apply a thin, 1/8 inch (3–4mm) bead of seal packing or form-in-place gasket (FIPG) or equivalent to the No. 1 oil pan and the oil pump. To ensure the proper size bead, cut the nozzle on the tube a bit smaller than the desired bead.

NOTE: Avoid applying too much sealant. The components must be assembled to the block 3-5 minutes after the sealant is applied. If not, the sealant must be removed and re-applied.

21. Install a new O-ring to the block.
22. Install the pump to the crankshaft with the spline teeth of the drive gear engaged with the large teeth of the crankshaft. Install the mounting bolts in the proper locations. Uniformly tighten the mounting bolts, in several passes, to:
- 12mm bolts — 12 ft. lbs. (16 Nm)
- 14mm bolts — 22 ft. lbs. (30 Nm)
23. Fit the No. 1 oil pan and install the bolts. Uniformly tighten the bolts in several passes to 69 inch lbs. (8

Nm). Install the stud bolt for the oil strainer; tighten it to 35 in lbs. (3.9 Nm).
24. Install a new crankshaft seal.
25. Install the crankshaft position sensor with the bolt and tighten to 56 inch lbs. (6 Nm). Connect the wiring to the sensor.
26. Install the remaining components. Refill the engine oil.
27. Start the engine and check for leaks. Recheck all fluid levels.
28. Install the undercovers.

1995–97 Models

NOTE: The oil pump cannot be removed with the engine in the vehicle. The engine and transmission must be removed as a unit and then separated.

1. Remove the engine/transmission assembly. Separate the transmission from the engine.
2. With the engine on a stand, remove the timing belt, the idler pulleys and the crankshaft timing pulley.
3. Remove the oil dipstick and guide.
4. Disconnect the oil level sensor lead.
5. Remove the four bolts and lift off the oil level sensor. Be careful not to drop this sensor.
6. Remove the oil filter and filter bracket assembly by removing the stud bolt and two nuts.
7. Disconnect the engine crankshaft position sensor connector. Remove the sensor by removing the bolt.
8. Remove the 12 bolts and 2 nuts to the No. 2 oil pan. Use a gasket cutting tool to separate the No. 2 (lower) oil pan. Be careful not to damage the No. 1 pan while performing this procedure.
9. Remove the two bolts and three nuts and drop down the baffle plate.
10. Remove the oil strainer by removing the bolts and nuts.
11. Remove the bolts and then carefully pry off the No. 1 oil pan. There are slots for inserting the prybar.
12. Remove the eight bolts holding the oil pump to the engine.

NOTE: Make certain to observe bolt position during removal. The bolts are different lengths and sizes. Record their position for proper reassembly.

13. Carefully pry the oil pump away from the engine block. Use great care not to damage the contact faces of the block or pump. Remove the O-ring from the block.

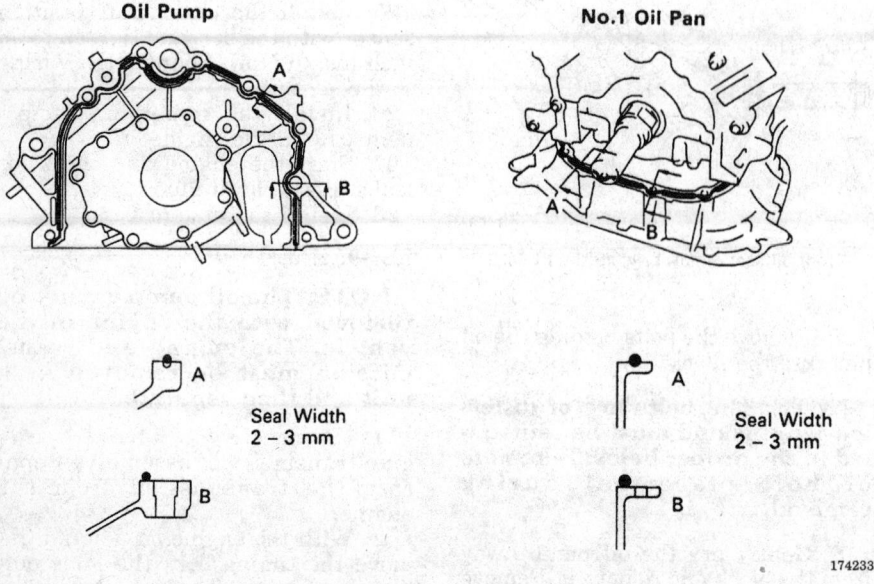

Sealant application for oil pump and No. 1 pan — 1993-94 LS400

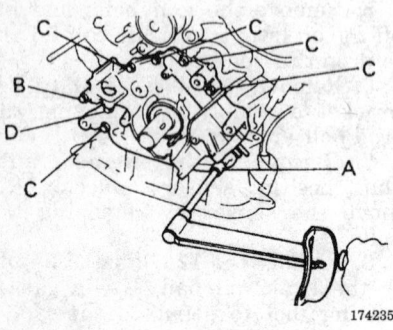

Oil pump bolts: A=1.97 in. (50mm), B=4.17 in (106mm), C=1.18 in (30mm), D=1.57 in. (40mm) — 1993-94 LS400

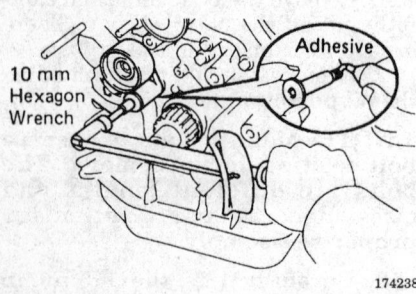

Installing the No. 1 idler pulley — 1993-94 LS400

To install:

14. Remove any old gasket material from the oil pump face and the engine block. Make certain the sealing groove is clean. Use a solvent which is free of residue to clean the surfaces.

NOTE: Prior to installing the oil pump, lubricate the gears with clean engine oil.

15. Apply a bead of new sealant to the oil pump. The gasket bead should be 2–3mm wide (0.08–0.12 in). Avoid excessive application and be careful around oil passages.

NOTE: Parts must be assembled within 5 minutes after the sealant is applied. If not, the sealant must be cleaned off and the parts re-coated.

16. Place a new O-ring in position on the block. Mount the oil pump to the engine by engaging the spline teeth of the oil pump drive with the large teeth of the crankshaft. Slide the oil pump into position.

17. Install the eight bolts in their correct locations. Tighten the bolts with 12mm heads to 12 ft. lbs. (16 Nm) and the bolts with 14mm heads to 22 ft. lbs. (30 Nm).

18. Scrape off any old sealant and then install the No. 1 pan. Apply sealant to the pan mating surface with a ⅛ inch (3–4mm) bead.

19. Torque the bolts for the No. 1 oil pan as follows:
- 10mm head to 66 inch lbs. (8 Nm)
- 12mm head to 21 ft. lbs. (28 Nm).

20. Install the oil strainer and tighten the bolts to 66 inch lbs. (8 Nm).

21. Install the baffle plate and tighten the bolts and nuts to 66 inch lbs. (8 Nm).

22. Install the remaining components. Connect the transaxle to the engine.

23. Install the engine and transaxle, refill all fluids and road test the vehicle.

SC400

NOTE: The oil pump cannot be removed with the engine in the vehicle. The engine and transmission must be removed as a unit and then separated.

1. Remove the engine/transmission assembly. Separate the transmission from the engine.

2. With the engine on a stand, remove the timing belt, the idler pulleys and the crankshaft timing pulley.

3. Remove the oil dipstick and guide.

4. Remove the main O_2 sensor bracket.

5. Disconnect the oil level sensor lead.

6. Remove the four bolts and lift off the oil level sensor. Be careful not to drop this sensor.

7. Remove the 12 bolts and 2 nuts. Use a gasket cutting tool to separate the No. 2 (lower) oil pan. Be careful not to damage the No. 1 pan while performing this procedure.

8. Remove the six bolts and two nuts and drop down the baffle plate.

9. Remove the 16 bolts and then carefully pry off the No. 1 oil pan. There are slots for inserting the prybar.

10. Remove the oil strainer by removing the two bolts and two nuts.

11. Disconnect the oil pressure switch connector. Remove the bolts and stud, then remove the filter bracket and filter.

12. Disconnect the engine crankshaft position sensor connector. Remove the sensor by removing the bolt.

13. Remove the eight bolts holding the oil pump to the engine.

NOTE: Make certain to observe bolt position during removal. The bolts are different lengths and sizes. Record their position for proper reassembly.

14. Carefully pry the oil pump away from the engine block. Use great care not to damage the contact faces of the block or pump. Remove the O-ring from the block.

To install:

15. Remove any old gasket material from the oil pump face and the engine block. Make certain the sealing groove is clean. Use a solvent which is free of residue to clean the surfaces.

16. Apply a bead of new sealant to the oil pump. The gasket bead should be 2–3mm wide (0.08–0.12 in). Avoid excessive application and be careful around oil passages.

NOTE: Parts must be assembled within 5 minutes after the sealant is applied. If not, the sealant must be cleaned off and the parts re-coated.

17. Place a new O-ring in position on the block. Mount the oil pump to the engine by engaging the spline teeth of the oil pump drive with the large teeth of the crankshaft. Slide the oil pump into position.

18. Install the eight bolts in their correct locations. Tighten the bolts with 12mm heads to 12 ft. lbs. (16 Nm) and the bolts with 14mm heads to 22 ft. lbs. (30 Nm).

19. Install the engine crankshaft position sensor. Tighten the bolt to 56 inch lbs. (6 Nm). Connect the crankshaft position sensor connector.

20. Place a new O-ring in position on the oil filter bracket. Install the bracket and tighten the bolts to 13 ft. lbs. (18 Nm). Connect the wiring to the pressure switch.

21. Install the oil strainer by installing the two bolts and two nuts. Tighten the bolts to 69 inch lbs. (8 Nm).

22. Scrape off any old sealant and then install the No. 1 pan. Apply sealant to the pan mating surface with a ⅛ inch (3–4mm) bead.

23. Torque the bolts as follows:
1993 Models
• 12mm bolts — 13 ft. lbs. (18 Nm)
• 10mm bolts — 69 inch lbs. (8 Nm)

1994–97 Models
• 12mm bolts — 69 inch lbs. (8 Nm)
• 14mm bolts — 20 ft. lbs. (28 Nm)

24. Install the baffle plate and tighten the bolts and nuts to 69 inch lbs. (8 Nm).

25. Install the No. 2 pan in the same manner as the No. 1 oil pan and tighten the bolts to 69 inch lbs. (8 Nm). Make sure the bolts are 14mm in length.

26. Using a new gasket, install the oil level sensor and tighten the four bolts to 48 inch lbs. (5 Nm).

27. Install the bracket to the oxygen sensor and install the bolt.

28. Install the dipstick and guide, the timing belt pulleys and the belt.

29. Connect the transaxle to the engine.

30. Install the engine and transaxle, refill all fluids and road test the vehicle.

TRANSMISSION AND TRANSAXLE

Manual Transmission Assembly

REMOVAL AND INSTALLATION

SC300

1. Disconnect the negative battery cable. Wait at least 90 seconds before performing any other work.

2. Pry out the rear side of the cup holder and remove the cup holder.

3. Unscrew the shift lever knob.

4. Pry up the upper rear console panel.

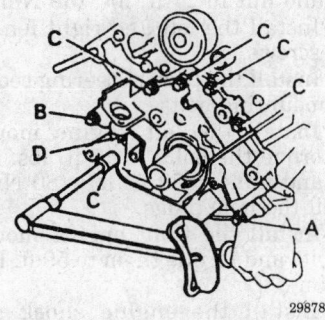

Oil pump bolts: A=1.97 in (50mm), B=4.17 in. (106mm), C=1.18 in (30mm), D=1.57 in. (40mm) w/14mm head — SC400

5. Remove the six mounting screws and then pry out the upper console panel.

6. Remove the four bolts and lift out the shift lever boots.

7. Remove the four bolts holding the shift lever to the transmission.

8. Raise the vehicle and support it on safety stands.

9. Remove the undercover and drain the transmission fluid.

10. Remove the engine undercover.

11. Remove the front exhaust pipe and support bracket.

12. Remove the center exhaust pipe.

13. Remove the heat insulator by removing the four bolts.

14. Remove the four bolts to the crossmember brace. Remove the crossmember brace.

15. Remove the driveshaft.

16. Disconnect the shift lever linkage from the transmission by removing the bolt and nut. Remove the shift linkage from inside the vehicle.

17. Remove the clutch release cylinder by removing the two bolts.

18. Disconnect the ground cable.

19. Remove the starter by removing the electrical wires and the two bolts.

20. Disconnect the backup light switch connector and the speed sensor connector.

21. Remove the two lower transmission mounting bolts.

22. Raise the transmission slightly until its weight is off the rear support.

23. Remove the four nuts and bolts and then remove the rear mounting member.

24. Remove the five mounting bolts, lower the rear of the engine and remove the transmission.

To install:

25. Align the input spline with the clutch disc and install the transmission to the engine. Tighten the five mounting bolts to 53 ft. lbs. (72 Nm).

26. Install the rear engine mount and tighten the nuts to 10 ft. lbs. (13 Nm) and the bolts to 19 ft. lbs. (25 Nm).

27. Connect the speed sensor.

28. Connect the backup light switch.

29. Install the two lower transmission mounting bolts and tighten it to 27 ft. lbs. (37 Nm).

30. Install the starter by installing the two bolts. Torque the bolts to 29 ft. lbs. (39 Nm). Connect the electrical connectors.

31. Install the release cylinder and tighten the two bolts to 9 ft. lbs. (12 Nm). Connect and tighten the ground wire to 27 ft. lbs. (37 Nm).

32. Install the transmission shift lever inside the vehicle. From below the vehicle, connect the shift lever by installing the bolt and nut. Torque the bolt and nut to 14 ft. lbs. (19 Nm).

33. Install the driveshaft.

34. Install the crossmember brace and tighten the bolts to 9 ft. lbs. (13 Nm).

35. Install the heat insulator and exhaust pipes. Tighten the center pipe to 14 ft. lbs. (19 Nm), the front pipe to 32 ft. lbs. (43 Nm) and the bracket to 27 ft. lbs. (37 Nm).

36. Fill the transmission with oil.

37. Install the undercover(s).

38. Lower the vehicle.

39. Install the shifter lever and the shifter boot.

40. Install the shift console and install the shifter knob.

41. Connect the negative battery cable.

42. Perform a road test and check all fluids.

Transaxle Assembly

REMOVAL AND INSTALLATION

ES300

― CAUTION ―

The Supplemental Inflatable Restraint (SIR) system must be disarmed before removing the transaxle assembly. Failure to do so may cause accidental deployment of the air bag, resulting in unnecessary SIR system repairs and/or personal injury.

1. Turn the ignition switch to the "LOCK" position and disconnect the negative battery cable. Wait at least 20 seconds or longer before doing any work on the vehicle.

2. Remove the air cleaner case and hose.

3. If equipped, remove the cruise control actuator and bracket.

4. Remove the clutch slave cylinder.

5. Remove the starter.

6. Remove the clutch accumulator and tube clamp.

7. Disconnect the backup light switch electrical connector.

8. Disconnect the wire clamps from the transaxle.

9. Remove the clutch release cylinder bracket.

10. Disconnect the ground cables from the transaxle.

11. Disconnect the control cables from the transaxle.

12. Remove the three upper transaxle mounting bolts.

13. Disconnect the speed sensor electrical connector.

14. Raise and support the vehicle safely.

15. Remove the front wheels.

16. Remove the undercovers and side covers.

17. Drain the transaxle oil.

18. Remove the front exhaust pipe.

19. Remove the halfshafts.

20. Disconnect the sway bar bushing brackets.

21. Disconnect the steering gear from the front suspension crossmember and suspend it.

22. Remove the stiffener plate.

23. Remove the engine mounting absorber.

24. Remove the clutch inspection plate.

25. Remove the front engine mounting set bolts and nut.

26. Remove the rear engine mounting set nuts.

27. Remove the left engine mounting bolts and nuts.

28. Remove the power steering cooler pipe mounting bolts.

29. Remove the left and right fender liner set bolts.

30. Remove the front suspension member.

31. Support the engine with a suitable jack.

32. Remove the transaxle mounting bolts.

33. Lower the left side of the engine and remove the transaxle.

To install:

34. Install the transaxle and torque the bolts to the following specifications:
• 10mm bolts — 34 ft. lbs. (46 Nm)
• 12mm bolts — 47 ft. lbs. (64 Nm)

35. Install the front suspension member and torque the four main bolts to 134 ft. lbs. (181 Nm). Torque the outer four bolts to 24 ft. lbs. (32 Nm) and nut to 27 ft. lbs. (36 Nm).

36. Install the left and right fender liner screws.

37. Install the power steering cooler pipe mounting bolts.

38. Install the left engine mount and torque the bolts to 47 ft. lbs. (64 Nm) and nuts to 59 ft. lbs. (80 Nm). Install the hole plugs.

39. Install the front engine mount set bolts and torque them to 59 ft. lbs. (80 Nm).

40. Install the engine shock absorber and torque the bolts to 35 ft. lbs. (48 Nm).

41. Install the clutch inspection plate. Torque the four outer bolts to

13 ft. lbs. (18 Nm) and two inner bolts to 27 ft. lbs. (37 Nm).

42. Connect the steering gear to the front suspension crossmember and torque the bolts and nuts to 134 ft. lbs. (181 Nm).

43. Install the sway bar brackets and torque the bolts to 14 ft. lbs. (19 Nm).

44. Install the halfshafts. Install the remaining components.

45. Fill the transaxle to the proper level with the recommended oil.

46. Install the front wheels.

47. Lower the vehicle.

48. Connect the negative battery cable.

49. Check the front wheel alignment.

Clutch Assembly

REMOVAL AND INSTALLATION

1. Disconnect the negative battery cable. Wait at least 90 seconds before proceeding with any other work.

2. Remove the transaxle/transmission assembly from the vehicle.

3. Place matchmarks on the flywheel and clutch cover.

4. Remove the clutch pressure plate retaining bolts. Loosen the bolts 1 turn at a time, in a criss-cross pattern, until all the spring tension is released.

5. Remove the pressure plate and clutch disc. Do not drop the clutch disc or plate.

6. Remove the release bearing, fork and boot from the transmission.

To install:

7. Apply grease to the following;
• Release fork and hub contact points
• Release fork and pushrod contact point
• Release fork pivot point
• Clutch disc spline

8. Install a new release bearing on the fork and install the fork to the transaxle/transmission.

9. Install the clutch disc and pressure plate on the flywheel, aligning them with 09301–17010 or equivalent.

10. Tighten the pressure plate mounting bolts to 14 ft. lbs. (19 Nm) in X-type pattern.

11. Install the transaxle/transmission.

12. Connect the negative battery cable.

13. Adjust the clutch pedal as necessary.

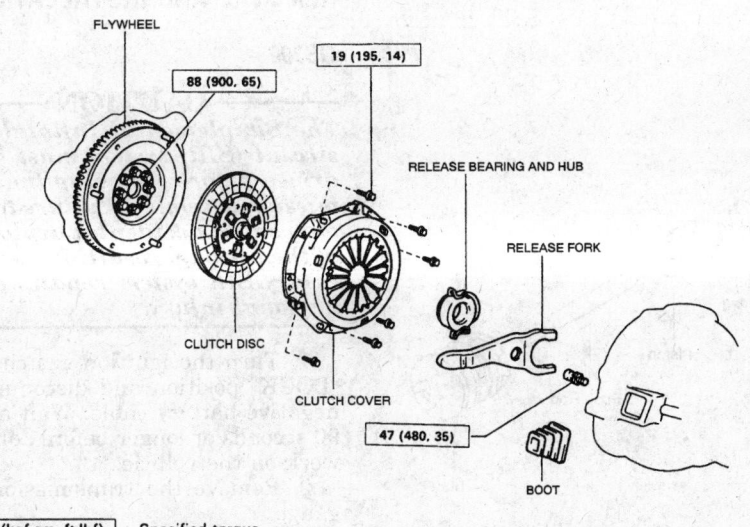

N·m (kgf·cm, ft·lbf) : Specified torque

197111

Clutch disc and pressure plate assembly — ES300

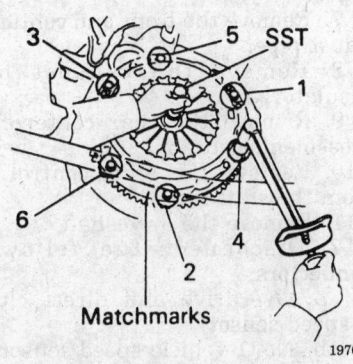

197090

Pressure plate bolt torque sequence — ES300

14. Road test the vehicle and verify proper clutch operation.

Clutch Master Cylinder

REMOVAL AND INSTALLATION

ES300

1. Disconnect the negative battery cable.
2. Disconnect the fluid line.
3. Remove the clutch master cylinder retaining nuts.
4. Remove the master cylinder from the vehicle.

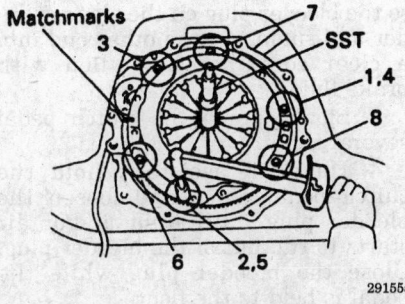

291555

Pressure plate bolt tightening sequence — SC300

To install:

5. Install the master cylinder and tighten the nuts to 58 inch lbs. (8 Nm).
6. Connect the hydraulic line and tighten the flare nut to 11 ft. lbs. (15 Nm).
7. Connect the pushrod and install the clevis pin.
8. Fill the master cylinder with fluid and bleed the system.

SC300

1. Disconnect the negative battery cable to the battery.
2. Remove the shifter finish panel.

3. Remove the upper console panel.
4. Remove the console box assembly.
5. Remove the lower instrument cluster pad.
6. Remove the lower instrument panel pad end.
7. Remove the lower instrument finish panel.
8. Remove the clutch pedal retaining clip and clevis pin with the spring washer.
9. Under the hood, disconnect the fluid line.
10. Remove the clutch master cylinder retaining bolts.
11. Remove the master cylinder from the vehicle.

To install:

12. Install the master cylinder and tighten to 9 ft. lbs. (12 Nm).
13. Connect the hydraulic line and tighten the union bolt to 11 ft. lbs. (15 Nm).
14. Fill the clutch master cylinder with fluid.
15. Under the dashboard, connect the pushrod and install the clevis pin.
16. Check for leaks and adjust the clutch pedal as necessary.
17. Bleed the clutch system.
18. Install the remaining components. Connect the negative battery cable to the battery.

Clutch Slave Cylinder

REMOVAL AND INSTALLATION

1. Raise and support the vehicle safely.
2. Disconnect the fluid line from the assembly. Use a container to catch any fluid. Dispose of the fluid properly.
3. Remove the slave cylinder retaining bolts.
4. On some models, it may be necessary to remove the bolts from the LH clutch housing cover.
5. Remove the clutch slave cylinder from the vehicle.

To install:

6. Install the slave cylinder and tighten the bolts to the following specifications:
   - A — 9 ft. lbs. (12 Nm)
   - B — 43 inch lbs. (5 Nm)
7. Connect the hydraulic line and tighten the union bolt to 11 ft. lbs. (15 Nm).
8. Fill the master cylinder with fluid.
9. Bleed the system and lower the vehicle to the floor.

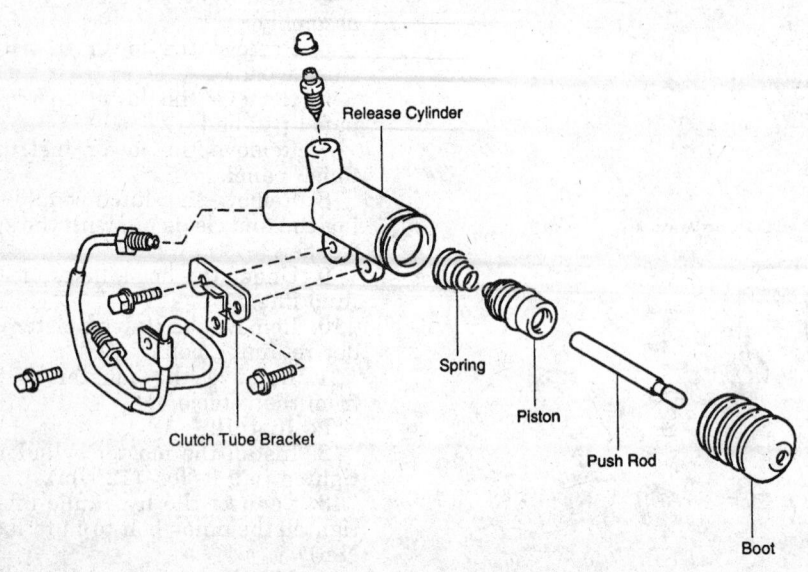

Release Cylinder

Spring

Piston

Push Rod

Clutch Tube Bracket

Boot

◆ Non-reusable part

291627

Slave cylinder with the mounting bolts

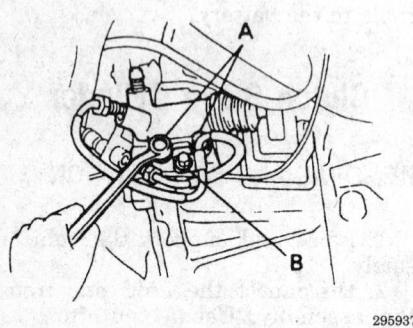

A

B

295937

Slave cylinder mounting bolt identification

Hydraulic Clutch System Bleeding

NOTE: If any maintenance on the clutch system was performed or the system is suspected of containing air, bleed the system. Use care; brake fluid will remove the paint from any surface. If the brake fluid spills onto any painted surface, wash it off immediately with soap and water.

1. Fill the clutch reservoir with brake fluid. Check the reservoir level frequently and add fluid as needed.

2. Connect one end of a vinyl tube to the bleeder plug on the slave cylinder and submerge the other end into a clear container half-filled with brake fluid.

3. Slowly pump the clutch pedal several times.

4. Have an assistant hold the clutch pedal down and loosen the bleeder plug until fluid and/or air starts to run out of the bleeder plug. Close the bleeder plug while the pedal is held to the floor.

NOTE: Do not allow the pedal to rise backup while the bleeder is still open. If this happens, it will allow air to re-enter the slave cylinder and cause the clutch system not to work properly.

5. Repeat Steps 2 and 3 until all the air bubbles are removed from the system.

6. Tighten the bleeder plug when all the air is gone.

7. Refill the master cylinder to the proper level as required.

8. Check the system for leaks.

Automatic Transmission Assembly

REMOVAL AND INSTALLATION

GS300

——— CAUTION ———

The Supplemental Inflatable Restraint (SIR) system must be disarmed before removing the transmission assembly. Failure to do so may cause accidental deployment of the air bag, resulting in unnecessary SIR system repairs and/or personal injury.

1. Turn the ignition switch to the "LOCK" position and disconnect the negative battery cable. Wait at least 90 seconds or longer before doing any work on the vehicle.

2. Remove the transmission level gauge.

3. Remove the transmission dipstick and tube.

4. Disconnect the throttle cable from the throttle body.

5. Disconnect the oxygen sensor from the exhaust system.

6. Remove the left and right tail pipes.

7. Remove the front and center exhaust pipe.

8. Remove the exhaust heat insulator.

9. Remove the rear center floor crossmember brace.

10. Remove the shift control rod from the shift lever.

11. Remove the driveshaft.

12. Disconnect the following connectors;

    a. Overdrive and direct clutch speed sensor

    b. No. 1 vehicle speed sensor

    c. No. 2 vehicle speed sensor

    d. Solenoid wire

    e. Park/Neutral position switch

13. Disconnect the wiring from the starter.

14. Remove the oil cooler pipes as follows:

    a. Disconnect the two oil cooler union nuts.

    b. Disconnect the oil cooler hoses from the oil cooler pipes.

    c. Remove the front oil cooler pipe bracket.

    d. Remove the center and rear oil cooler pipe brackets.

    e. Remove the two oil cooler pipes.

15. Remove the torque converter inspection plate.

16. Turn the crankshaft to gain access to the torque converter bolts and

remove them as they become accessible.

17. Support the transmission with a suitable jack.

18. Support the engine with a jack and a block of wood.

19. Remove the rear transmission mount.

20. Remove the wire harness clamps.

21. Remove the starter.

22. Remove the nine transmission mounting bolts and transmission.

To install:

23. Install the transmission and torque the bolts to 53 ft. lbs. (52 Nm)

24. Install the starter and torque the bolts to 27 ft. lbs. (37 Nm).

25. Install the rear transmission mount and torque the bolts to 19 ft. lbs. (25 Nm).

26. Install and torque the torque converter bolts to 30 ft. lbs. (41 Nm) while rotating the crankshaft.

27. Install the converter inspection plate.

28. Connect the oil cooler lines as follows:

   a. Install the two oil cooler pipes.

   b. Install the center and rear oil cooler pipe brackets.

   c. Install the front oil cooler pipe bracket and torque to 49 inch lbs. (5.5 Nm).

   d. Connect the oil cooler hoses to the oil cooler pipes.

   e. Install the two oil cooler union nuts and torque to 32 ft. lbs. (44 Nm).

29. Connect the wiring to the starter.

30. Connect the transmission electrical connectors.

31. Install the driveshaft.

32. Insztall the remaining components. Connect the negative battery cable.

33. Install the transmission level gauge.

34. Fill the transmission to the proper level with Dexron II® or equivalent.

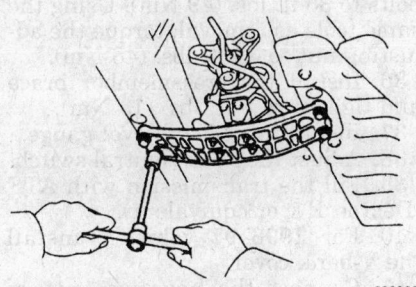

299908

Transmission rear mount — GS300

1993–94 LS400

1. Disconnect the negative battery cable. Wait at least 90 seconds before performing any other work.

2. Remove the transmission dipstick and tube.

3. Disconnect the throttle cable.

4. Remove the exhaust pipe and converters.

5. Remove the exhaust heat insulator.

6. Remove the rear center floor crossmember brace.

7. Remove the shift control rod.

8. Remove the driveshaft.

9. Disconnect the electrical harness from the transmission.

10. Remove the oil cooler tube clamps and disconnect the tubes.

11. Remove the torque converter inspection plate.

12. Turn the crankshaft to gain access to the torque converter bolts and remove them as they become accessible.

13. Support the transmission with a suitable jack.

14. Remove the rear transmission mount.

15. Remove the 10 transmission mounting bolts and transmission.

To install:

16. Install the transmission and torque the bolts as follows:

   • 14mm bolts — 29 ft. lbs. (39 Nm)

   • 17mm bolts — 42 ft. lbs. (57 Nm)

17. Install the rear transmission mount and torque the bolts to 19 ft. lbs. (20 Nm).

18. Install and torque the torque converter bolts to 25 ft. lbs. (33 Nm) while rotating the crankshaft.

19. Install the converter inspection plate.

20. Install the lower engine cover.

21. Connect the oil cooler lines and tighten the clamps.

22. Connect the transmission electrical connectors.

23. Install the driveshaft. Install the remaining components.

24. Connect the negative battery cable.

25. Fill the transmission to the proper level with Dexron II® or equivalent.

1995–1997 LS400

1. Disconnect the negative battery cable. Wait at least 90 seconds before performing any other work.

2. Remove the transmission dipstick and tube.

3. Disconnect the throttle cable.

4. Remove the driveshaft.

5. Remove the engine undercover.

6. Remove the shift control rod.

7. Remove the exhaust pipe support bracket by removing the two bolts.

8. Remove the catalytic converters by removing the six nuts.

9. Remove both side heat insulators.

10. Remove the oil cooler tube clamps and disconnect the tubes.

11. Remove the torque converter inspection plate by removing the two bolts.

12. Turn the crankshaft to gain access to the torque converter bolts and remove them as they become accessible.

13. Support the transmission with a suitable jack.

14. Remove the rear transmission mount by removing the four bolts and four nuts.

15. Tilt down the transmission and disconnect the following connectors:

   a. O/D direct clutch speed sensor connector.

   b. Vehicle speed sensor connector.

   c. Park/Neutral position switch connector.

   d. Solenoid connector.

16. Disconnect the three wire harness clamp from the bracket on the transmission.

17. Remove the 10 transmission mounting bolts and the transmission.

To install:

18. Install the transmission and torque the bolts as follows:

   • 14mm bolts — 27 ft. lbs. (37 Nm)

   • 17mm bolts — 53 ft. lbs. (72 Nm)

19. Install the three wire harness clamp to the bracket on the transmission.

20. Connect the following connectors:

   a. Solenoid connector

   b. Park/Neutral position switch connector

   c. Vehicle speed sensor connector

   d. O/D direct clutch speed sensor connector

21. Install the rear transmission mount and torque the bolts to 19 ft. lbs. (20 Nm) and the nuts to 10 ft. lbs. (13 Nm).

22. Install and torque the torque converter bolts to 30 ft. lbs. (41 Nm).

23. Install the converter inspection plate.

24. Remove the support from the transmission.

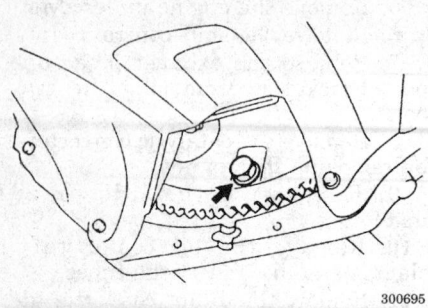

Torque converter clutch mounting bolts — 1995–97 LS400 A340E

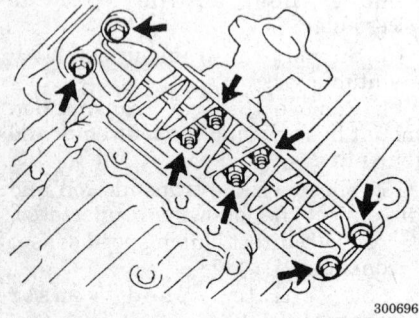

Engine rear mounting bolts and nuts — 1995–97 LS400 A340E

25. Install the oil cooler pipes and torque the union nuts to 32 ft. lbs. (44 Nm).

26. Install the side heat insulators.

27. Install the catalytic converters with new gaskets and new nuts. Torque the nuts to 46 ft. lbs. (62 Nm).

28. Install the exhaust pipe support bracket with the two bolts and torque the bolts to 32 ft. lbs. (44 Nm).

29. Install the shift control rod.

30. Install the engine undercover.

31. Install the driveshaft.

32. Connect and adjust the throttle control cable.

33. Install the transmission tube and dipstick.

34. Connect the negative battery cable.

35. Fill the transmission to the proper level with Dexron II® or equivalent.

SC300 and SC400

1. Disconnect the negative battery cable. Wait at least 90 seconds before proceeding with any other work.

2. For the 1996–97 vehicle, remove the V-bank cover.

3. Remove the A/T oil level gauge if equipped.

4. Remove the transmission dipstick and tube.

5. Disconnect the throttle cable and clamps.

6. Remove the exhaust pipe and converters.

7. Remove the exhaust heat insulator.

8. Remove the rear center floor crossmember brace.

9. Remove the shift control rod.

10. Remove the driveshaft using two of tool SST 09922–10010 or equivalent, loosen the adjusting nut on the driveshaft. Place matchmarks on the transmission on the flanges prior to removal.

NOTE: The bolts inserted from the driveshaft side should not be removed.

11. Disconnect the electrical harness from the transmission.

12. Remove the oil cooler tube clamp and disconnect the tubes.

13. Remove the lower engine cover.

14. Remove the torque converter inspection plate.

15. Turn the crankshaft to gain access to the torque converter bolts and remove them as they become accessible.

16. Support the transmission with a suitable jack.

17. If necessary, remove the starter.

18. Remove the rear transmission mount by removing the bolts.

19. Remove the transmission mounting bolts and the transmission.

To install:

20. Before installing the transmission, use calipers and a straightedge to check the distance between the installed surface of the torque converter and the front edge of the transmission case. Correct distance is 0.673 in. (17.1mm). If this distance is not correct, check the torque converter installation.

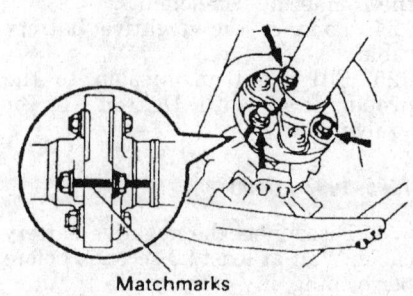

Matchmarks

Matchmarking the driveshaft to the transmission — SC400 A340E

21. Install the transmission and torque the bolts as follows:
- SC300 — 14mm bolts 27 ft. lbs. (37 Nm)
- SC300 — 17mm bolts 53 ft. lbs. (52 Nm)
- SC400 — 14mm bolts 29 ft. lbs. (39 Nm)
- SC400 — 17mm bolts 42 ft. lbs. (57 Nm)

22. If removed, install the starter. Torque the bolts to 27 ft. lbs. (37 Nm). Connect the electrical connectors to the starter.

23. Install the rear transmission mount and torque the bolts to 19 ft. lbs. (20 Nm).

24. Install and torque the torque converter bolts to 25 ft. lbs. (33 Nm) while rotating the crankshaft.

25. Install the converter inspection plate.

26. Install the lower engine cover.

27. Connect the oil cooler lines and torque the lines to 25 ft. lbs. (34 Nm). Install the oil cooler pipe bracket and tighten the bolt.

28. Connect the transmission electrical connectors.

29. Connect the shift control rod and adjust the shift linkage. Torque the nut to 12 ft. lbs. (16 Nm).

30. Install the rear center floor crossmember brace by installing the four bolts. Torque the bolts to 9 ft. lbs. (13 Nm).

31. Install the heat insulator.

32. Install the transmission filler tube and dipstick.

33. Install the front exhaust pipe and converters with new gaskets.

34. Connect and adjust the throttle control cable.

35. Install the driveshaft. Align the matchmarks and connect the driveshaft to the transmission. Insert the bolts from the transmission side and tighten to 58 ft. lbs. (79 Nm). Align the matchmarks and install the driveshaft to the differential. Insert the bolts from the differential side and tighten to 58 ft. lbs. (79 Nm). Tighten the center bearing support bolts to 36 ft. lbs. (49 Nm). Using the same tools as removal, torque the adjusting nut to 35 ft. lbs. (48 Nm).

36. Install the crossmember brace and tighten to 8 ft. lbs. (13 Nm).

37. Install the A/T oil level gauge.

38. Adjust the Park/Neutral switch.

39. Fill the transmission with ATF (Dexron II® or equivalent).

40. For 1996–97 vehicles, install the V-bank cover.

41. Connect the negative battery cable.

42. Start the engine and check the ATF fluid level. If necessary, add

ATF (Dexron II® or equivalent) to the transmission to obtain the proper fluid level.

Automatic Transaxle Assembly

REMOVAL AND INSTALLATION

ES300

—————— **CAUTION** ——————
The Supplemental Inflatable Restraint (SIR) system must be disarmed before removing the transaxle assembly. Failure to do so may cause accidental deployment of the air bag, resulting in unnecessary SIR system repairs and/or personal injury.

1. Turn the ignition switch to the LOCK position and disconnect the negative battery cable. Wait at least 90 seconds or longer before doing any work on the vehicle.
2. Remove the battery.
3. Remove the air cleaner assembly.
4. Disconnect the throttle cable from the throttle body.
5. Remove the cruise control actuator cover and disconnect the connector, if equipped.
6. Remove the ground wire.
7. Remove the starter.
8. Disconnect speed sensor connectors, direct clutch speed sensor and the Park/Neutral position switch connector on the transaxle.
9. Disconnect the solenoid connector on the transaxle.
10. Disconnect shift control cable.
11. Disconnect oil cooler hoses.
12. Remove the two front side transaxle mounting bolts.
13. Remove the two front engine mounting bolts.
14. Remove the oil cooler line mounting bolts from the front frame.
15. Remove the three upper transaxle to engine mounting bolts.
16. Install a engine support fixture. Tie steering gear housing to engine support fixture.
17. Raise and safely support the vehicle.
18. Drain the transaxle/differential fluid.
19. Remove the front wheels.
20. Remove the front exhaust pipe.
21. Remove the engine side covers and undercovers.
22. Disconnect both halfshafts.
23. Remove the front side engine mounting nut.

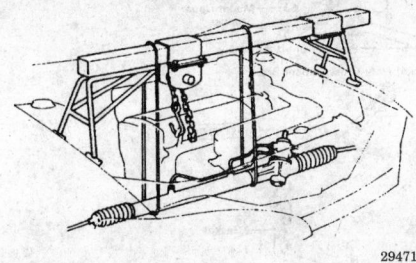

294714

Tie the steering rack to the engine support fixture — ES300

24. Remove the rear side engine mounting bolts (remove hole plugs).
25. Remove the four left side transaxle mounting bolts.
26. Remove the steering gear housing.
27. Remove the front frame assembly.
28. Properly support the transaxle assembly.
29. Remove the rear end plate mounting bolts.
30. Remove the torque converter cover.
31. Remove the torque converter retaining bolts.
32. Remove the remaining transaxle mounting bolts.
33. Carefully remove the transaxle assembly from the vehicle.

To install:

34. Install the transaxle aligning the two dowel pins on the block with the converter housing. Torque the bolts as follows:
- 10mm bolts — 34 ft. lbs. (46 Nm)
- 12mm bolts — 47 ft. lbs. (64 Nm)
35. Coat the threads of the torque converter bolts with sealer. Install the bolts starting with the green bolt followed by the rest and torque the bolts evenly to 20 ft. lbs. (27 Nm).
36. Install the rear end plate and torque the bolts to 27 ft. lbs. (37 Nm).
37. Install the front frame assembly and torque the fasteners as follows:
- 12mm bolts — 24 ft. lbs. (32 Nm)
- 19mm bolts — 134 ft. lbs. (181 Nm)
- Nut — 27 ft. lbs. (36 Nm)
38. Install the two fender liner set screws.
39. Connect the steering gear to the frame and torque the bolts and nuts to 134 ft. lbs. (181 Nm).
40. Connect the sway bar brackets and toque the bolts to 14 ft. lbs. (19 Nm).

41. Install the left transaxle mounting bolts and torque them to 38 ft. lbs. (52 Nm).
42. Install the rear side mounting bolts and nuts and torque them to 48 ft. lbs. (66 Nm). Install the plugs.
43. Install the front engine mounting nut and torque it to 59 ft. lbs. (80 Nm).
44. Install the halfshafts.
45. Install the right and left engine side covers.
46. Install the lower engine cover.
47. Fill the transaxle/differential to the proper level with Dexron II® or equivalent.
48. Install the exhaust pipe to the engine with new gaskets and torque the nuts to 46 ft. lbs. (62 Nm). Connect the exhaust pipe to the converter with a new gasket and torque the nuts and bolts to 32 ft. lbs. (43 Nm).
49. Install the wheel.
50. Lower the vehicle.
51. Remove the engine support.
52. Install the four upper transaxle mounting bolts and torque them to 47 ft. lbs. (64 Nm).
53. Install the oil cooler clamping bolts to the front frame.
54. Install the two front side engine mounting bolts and torque them to 59 ft. lbs. (80 Nm).
55. Install the two front side transaxle mounting bolts and torque them to 59 ft. lbs. (80 Nm).
56. Install the remaining components. Install the battery and connect the battery cables.
57. Check the transaxle/differential fluid level.
58. Check the front wheel alignment.

DRIVE AXLE

Driveshaft

REMOVAL AND INSTALLATION

SC300

1. Raise and safely support the vehicle.
2. Remove the front exhaust pipe and the heat insulator.
3. Remove the center exhaust pipe and the heat insulator.
4. Remove the four (4) bolts and lift out the center crossmember brace.
5. Place matchmarks on the differential companion flange and the flex-

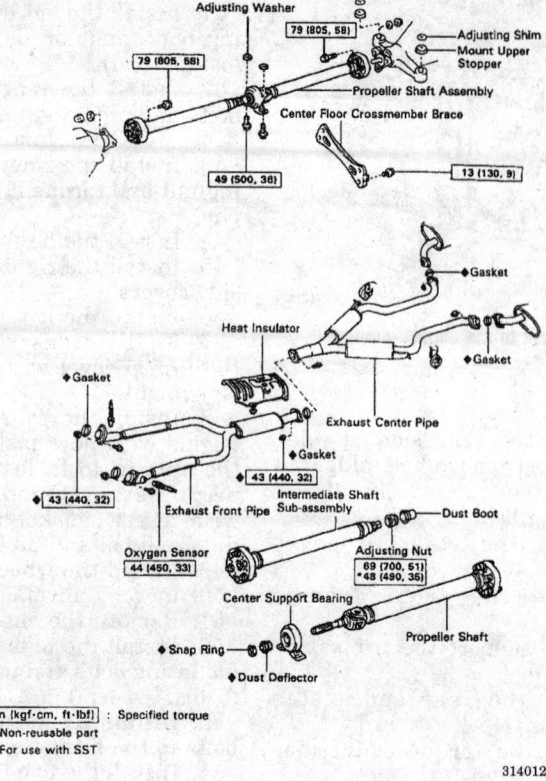

Driveshaft removal and installation — 1995–97 SC400

8. Remove the bolts inserted from the transmission side.

NOTE: The bolts inserted from the driveshaft side should not be removed.

9. Place matchmarks on the differential flange and the flexible coupling.
10. Remove the bolts inserted from the differential side. Separate the flexible couplings from the transmission and the differential.
11. Remove the center support bearing set bolts and the adjusting washers. Some vehicles are not equipped with the adjusting washers.

NOTE: When removing the set bolts, support the center support bearing so the transmission and intermediate shaft and the driveshaft and differential remain in straight line.

12. Push the rear driveshaft forward to compress the driveshaft and pull out the driveshaft from the centering pin of the differential.
13. Remove the driveshaft by pulling out toward the rear of the vehicle.
To install:
14. Apply a suitable grease to the flexible coupling centering bushings.
15. Insert the driveshaft from the rear of the vehicle and connect the driveshaft to the transmission and differential.
16. Temporarily, install the center support bearing set bolts with the adjusting washers. Use the adjusting washers that were removed.
17. Align the matchmarks and connect the driveshaft to the transmission. Insert the bolts from the transmission side and tighten to 58 ft. lbs. (79 Nm).
18. Align the matchmarks and install the driveshaft to the differential. Insert the bolts from the differential side and tighten to 58 ft. lbs. (79 Nm).
19. Tighten the center bearing support bolts to 36 ft. lbs. (49 Nm).
20. Using the same tools as removal, torque the adjusting nut to 35 ft. lbs. (48 Nm).
21. Install the crossmember brace and tighten to 8 ft. lbs. (13 Nm).
22. Install the heat insulator.
23. Install the center exhaust pipe and torque the four bolts to 14 ft. lbs. (19 Nm).
24. Install the front exhaust pipe and torque the six bolts and nuts to 32 ft. lbs. (43 Nm).
25. Connect the two oxygen sensors to the front exhaust pipe. Torque the sensors to 33 ft. lbs. (44 Nm).

ible coupling. Remove the three (3) bolts inserted in the companion flange. Do not remove the three (3) bolts in the driveshaft flange. Separate the flexible couplings from the differential side. It may be necessary to insert a small prybar into a bolt hole and carefully pry them apart.
6. Remove the two (2) center bearing bolts and the adjustment washers (some vehicles may not have the washers).

NOTE: When removing these bolts, support the center bearing so all shafts remain in a straight line.

7. Pull the yoke from the transmission and remove the driveshaft.
To install:
8. Insert the yoke into the transmission. Coat the center bushing on the flexible coupling with grease and slide it into position.
9. Coat the threads of the coupling bolts with clean engine oil, align the matchmarks and insert the bolts from the driveshaft side. Torque as follows:
• 1993 to 69 ft. lbs. (93 Nm)
• 1994–95 to 58 ft. lbs. (79 Nm)
10. Position the center bearing and tighten the bolts to 36 ft. lbs. (49 Nm).

11. Install the crossmember brace and tighten it to 9 ft. lbs. (13 Nm).
12. Install the heat insulator and torque the four (4) bolts to 48 inch lbs. (5 Nm).
13. Install the exhaust center pipe with the four (4) nuts and tighten them to 14 ft. lbs. (19 Nm).
14. Install the front pipe with the four (4) nuts and tighten them to 32 ft. lbs. (43 Nm). Tighten the O₂ sensors to 33 ft. lbs. (44 Nm).
15. Lower the vehicle.

GS300, SC400 and LS400

1. Raise and safely support the vehicle.
2. Remove the front exhaust pipe.
3. Remove the center exhaust pipe.
4. Remove the heat insulator by removing the four nuts.
5. Remove the center floor crossmember brace by removing the bolts.
6. Using the proper tools (two of tool SST 09922–10010 or equivalent), loosen the adjusting nut on the driveshaft.
7. Place matchmarks on the transmission flange and the flexible coupling.

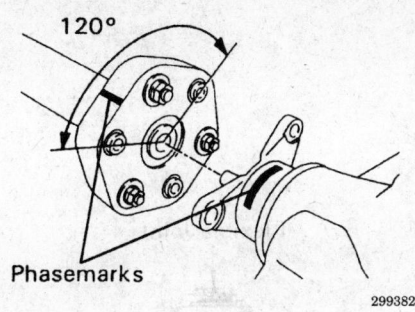

Align the phasemark on the driveshaft and flange — GS300

299382

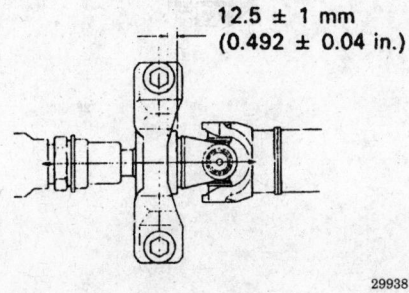

299381

Correct center bearing installation — GS300

26. Connect the oxygen sensor grommets to the floor.

27. Lower the vehicle.

U-Joints

REMOVAL AND INSTALLATION

The driveshaft does not use a U-style joint. Instead, flexible rubber couplings are used. These couplings are incorporated into the driveshaft and are non-serviceable.

Halfshaft

REMOVAL AND INSTALLATION

ES300

1993–94 Models

—————— **WARNING** ——————
The hub bearing could be damaged if it is subjected to the vehicle weight, such as when moving the vehicle with halfshaft removed.

1. Disconnect the negative battery cable.

2. Raise and safely support the vehicle.

3. Remove the front fender apron seal.

4. Remove the wheel assembly.

5. Remove the cotter pin, locknut cap and locknut. Have a helper apply the brakes when removing the locknut.

6. Drain the oil from the differential.

7. Disconnect the steering knuckle from the lower ball joint.

8. Disconnect the tie rod end from the steering knuckle.

9. Disconnect the stabilizer bar link from the lower control arm.

10. Place matchmarks on the halfshaft and center halfshaft or the side gear shaft. Loosen but do not remove the mounting bolts.

NOTE: Do not remove the bolts. Finger-tighten them so the halfshaft does not fall.

11. Pull the left hub toward the outside of the vehicle and disconnect the halfshaft from the axle hub.

12. Remove the six mounting bolts and the halfshaft.

13. Push the side gear shaft to the differential. Measure and note the distance between the transaxle case and the side gear shaft so it can be installed in the same position.

14. Pry the left sidegear shaft out of the differential.

15. Pull the right hub toward the outside of the vehicle and disconnect the halfshaft from the axle hub.

16. Remove the right side bearing lock bolt and the snapring. Pull out the right side halfshaft with the center shaft.

NOTE: If the halfshaft cannot be pulled out, tap out the driveshaft with a plastic hammer.

17. If necessary, replace the side gear shaft oil seal.

To install:

18. Using a new snapring, install the left side gear shaft using a suitable tool to tap in the driveshaft until it makes contact with the pinion shaft. Ensure the new snapring is positioned securely in the groove of the side gear shaft.

19. Check that the side gear shaft will not come out by hand. Push the side gear shaft to the differential and measure the distance between side gear shaft and the transaxle case. Make sure the distance is the same measurement taken before removing the side gear shaft.

20. Pack the side gear shaft with a suitable grease.

21. Align the matchmarks on the side gear shaft and the halfshaft. Install the left side halfshaft and finger-tighten the bolts.

22. Slide the axle shaft through the hub and install the hub nut but do not tighten it.

23. Install the right side halfshaft with the center halfshaft to the transaxle through the bearing bracket. Install the snapring and a new lock bolt. Torque the lockbolt to 24 ft. lbs. (32 Nm).

24. Install the outboard joint side of the halfshaft to the axle hub. Temporarily, connect the steering knuckle to the lower ball joint.

25. Connect the tie rod ends to the steering knuckles and tighten the bolts to 36 ft. lbs. (49 Nm). Tighten the lower ball joint mounting bolts to 94 ft. lbs. (127 Nm).

26. Connect the stabilizer bar link to the lower arms and tighten it to 47 ft. lbs. (64 Nm).

27. Have a helper apply the brakes and tighten the six hexagon bolts to 48 ft. lbs. (65 Nm). Torque the locknuts to 217 ft. lbs. (294 Nm).

28. Install the locknut caps and new cotter pins.

29. Install the front fender apron seals.

30. Replace the front wheels assembly.

31. Lower the vehicle to the floor.

32. Check the front wheel alignment.

1995–97 Models

1. Disconnect the negative battery cable to the battery.

2. Raise and support the vehicle safely.

3. Remove the front wheel(s).

4. Remove the front fender apron seal.

5. Drain the transaxle.

6. Disconnect the tie rod end from the steering knuckle by removing the cotter pin and nut. Using tool SST 09628–62011 or equivalent, separate the tie rod from the steering knuckle.

7. Disconnect the stabilizer bar link from the lower control arm. Make note of the washers and cushions positions.

8. Disconnect the lower ball joint from the steering knuckle by removing the bolt and two nuts. Push down on the lower control arm and separate the steering knuckle from the ball joint.

9. Remove the cotter pin, lock cap and locknut holding the halfshaft to the steering knuckle.

10. Using a plastic hammer, disconnect the halfshaft from the steering knuckle.

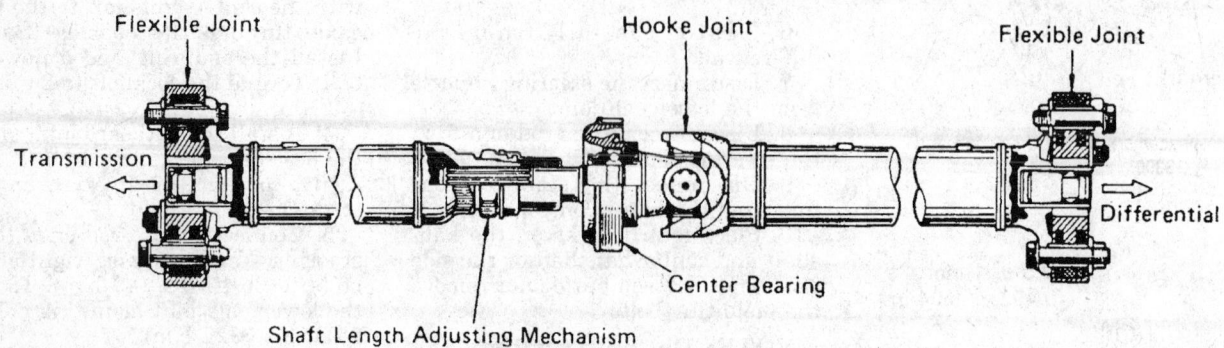

Driveshaft component identification

287502

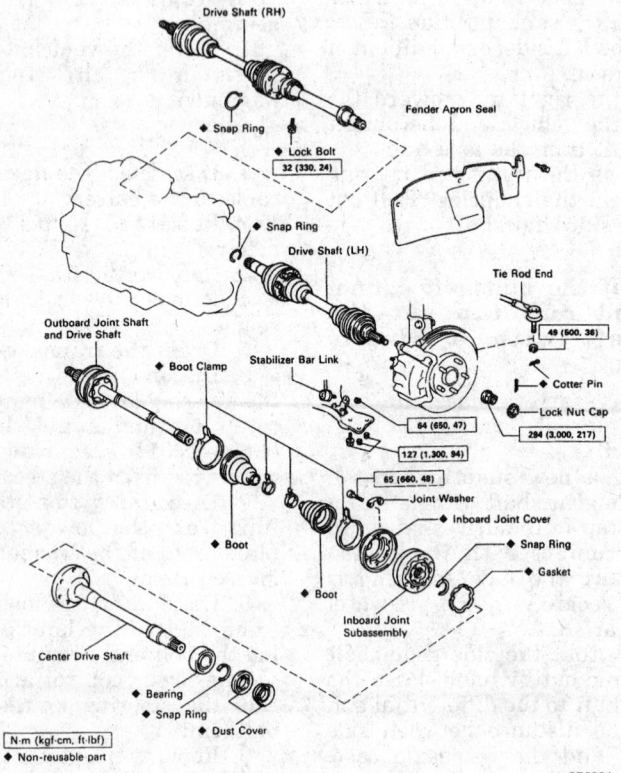

Front axle shaft components — 1993–94 ES300

276291

11. Remove the left halfshaft from the transaxle as follows:

a. Use a brass bar and hammer to tap the inner joint out of the transaxle.

b. Remove the halfshaft.

c. Once the halfshaft is removed from the vehicle, remove the snapring from the halfshaft.

12. Remove the right halfshaft from the transaxle as follows:

a. Remove the bearing lockbolt. The lockbolt is located in the center of the halfshaft, near the dampener.

b. Using snapring pliers, remove the snapring and pull the halfshaft from the transaxle.

To install:

13. To install the right halfshaft to the transaxle:

a. Coat the side gear shaft and differential case sliding surface with gear oil.

b. Using snapring pliers, install the snapring to the halfshaft.

c. Install the halfshaft and the bearing lockbolt. Torque the lockbolt to 24 ft. lbs. (32 Nm).

14. To install the left halfshaft to the transaxle:

a. Install a new snapring to the inner spline of the halfshaft.

b. Coat the side gear shaft and differential case sliding surface with gear oil.

c. Install the halfshaft to the transaxle with the snapring opening facing down. The halfshaft should click into place when installing.

d. After installation of the halfshaft, check that the halfshaft cannot be removed by hand.

15. Connect the halfshaft to the steering knuckle and then install the locknut. Torque the locknut to 217 ft. lbs. (294 Nm).

16. Install the lock cap and a new cotter pin to the halfshaft.

17. Connect the steering knuckle to the lower ball joint. Install the two nuts and bolt. Torque the nuts and bolt to 94 ft. lbs. (127 Nm).

18. Connect the stabilizer bar link to the lower control arm. Torque the nut to 29 ft. lbs. (39 Nm).

19. Connect the tie rod to the steering knuckle and torque the nut to 36 ft. lbs. (49 Nm). Install a new cotter pin to the tie rod end.

20. Install the front fender apron seal.

21. Install the wheel(s) and lower the vehicle. Torque the lug nuts to 76 ft. lbs. (103 Nm).

22. Refill the transaxle and check for leaks.

23. Connect the negative battery cable to the battery.

GS300

—————— **CAUTION** ——————

The air bag system (SRS or SIR) must be disarmed before removing the halfshafts. Failure to do so may cause accidental deployment, property damage or personal injury.

———————————————————————

1. Disconnect the negative battery cable from the battery.

2. Raise and safely support the vehicle.

3. Remove the rear tire and wheel assembly.

4. Remove the cotter pin, locknut cap and locknut while having someone pressing down on the brake pedal.

5. Secure the rear exhaust assembly with mechanics wire or equivalent.

6. Remove the two exhaust pipe support brackets.

7. Have someone press down on the brake pedal. Place matchmarks on the halfshaft and the side gear shaft. Remove the six hex bolts and two washers.

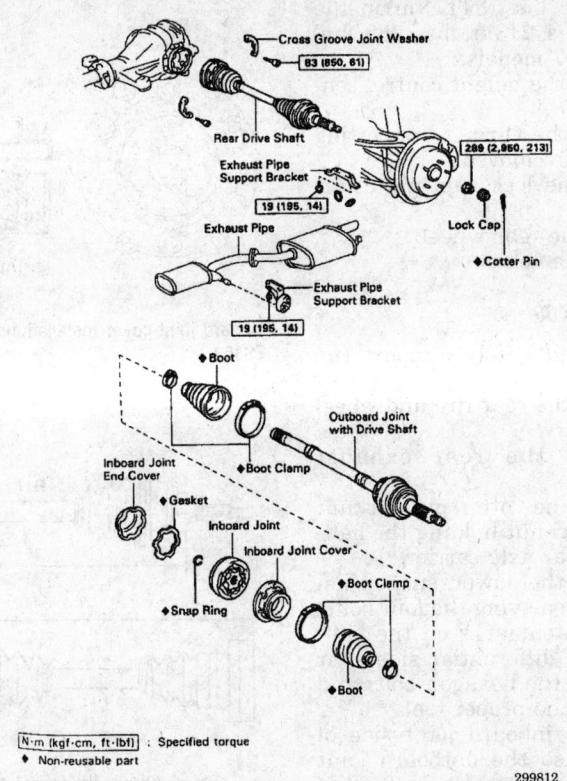

N·m (kgf·cm, ft-lbf) : Specified torque
◆ Non-reusable part

299812

Rear halfshaft and related components — GS300

8. Hold the inboard joint side of the halfshaft so the outboard joint side does not bend too much. Tap the end of the halfshaft with a rubber mallet to loosen it from the axle hub and remove the halfshaft.

To install:

9. Insert the outboard joint side of the halfshaft through the axle hub. Align the matchmarks on the side gear shaft and the halfshaft.

10. Coat the threads with clean oil and install the hex bolts. Tighten the bolts to 61 ft. lbs. (83 Nm).

11. Install the exhaust pipe support brackets and torque to 14 ft. lbs. (19 Nm).

12. Install the bearing locknut and have a helper apply the brakes. Torque the locknut to 213 ft. lbs. (289 Nm).

13. Install the lock cap and a new cotter pin.

14. Replace the rear tire and wheel assembly.

15. Lower the vehicle and connect the negative battery cable.

LS400

1. Raise and safely support the rear of the vehicle.

2. Remove the rear wheel.

3. Remove the cotter pin, lock cap and the nut holding the halfshaft to the rear knuckle.

4. On some models it will be necessary to, remove the tail pipe O-rings and suspend the tail pipe, using a piece of wire.

5. Disconnect the height control sensor, if equipped.

6. Remove the suspension member brace by removing the two bolts.

7. Place matchmarks on the halfshaft and the side gear shaft. Remove the hexagon bolts and washers with the proper tool.

8. Hold the inboard joint side of the halfshaft so the outboard joint side does not bend too much. Tap the end of the halfshaft with a rubber mallet and disengage the halfshaft from the knuckle.

9. Remove the halfshaft.

To install:

10. Insert the outboard joint side of the halfshaft and align the matchmarks on the side gear shaft and the halfshaft.

11. Coat the threads with clean oil and install the hexagon bolts. Torque bolts to 61 ft. lbs. (83 Nm).

12. Install the suspension member brace with the two bolts. Torque the two bolts to 37 ft. lbs. (50 Nm).

13. Install the nut to hold the halfshaft to the rear knuckle. Torque the

nut to 253 ft. lbs. (344 Nm) on the 1993–94 models, 213 ft. lbs. (289 Nm) on the 1995–97 models.

14. Connect the height control sensor, if equipped.

15. Replace the O-rings supporting the tail pipe if removed.

16. Install the lock cap and cotter pin.

17. Install the rear wheel.

18. Lower the vehicle.

SC300 and SC400

1. Raise and safely support the vehicle.

2. Remove the rear tire and wheel assembly.

3. Remove the rear exhaust assembly.

4. Remove the cotter pin, locknut cap and the locknut holding the half-shaft to the rear axle carrier.

5. Remove the lower suspension arm brace by removing the four bolts.

6. Place matchmarks on the half-shaft and the differential side gear shaft. Remove the hexagon bolts and washers with the proper tool.

7. Hold the inboard joint side of the halfshaft so the outboard joint side does not bend too much. Tap the end of the halfshaft with a rubber mallet and disengage the halfshaft from the axle carrier.

8. Remove the halfshaft.

To install:

9. Insert the outboard joint side of the halfshaft and align the matchmarks on the side gear shaft and the halfshaft.

10. Coat the threads with clean oil and install the hexagon bolts. Torque the bolts to 61 ft. lbs. (83 Nm).

11. Install the lower suspension arm brace and torque the four bolts to 13 ft. lbs. (18 Nm).

12. Install the bearing locknut and torque the locknut to 213 ft. lbs. (289 Nm).

13. Install the locknut cap and install a new cotter pin.

14. Install the rear exhaust assembly.

15. Replace the rear tire and wheel assembly.

16. Lower the vehicle.

CV-Joint Boot

REPLACEMENT

ES300

1. Raise and safely support the vehicle.

2. Remove the front tire and wheel assembly.

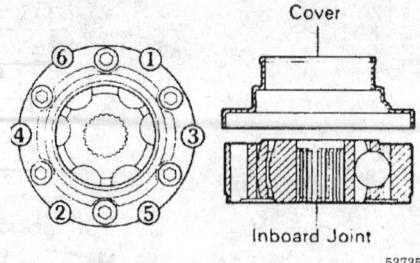

Inboard joint cover installation sequence — ES300

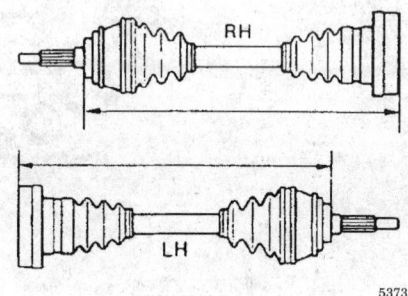

Axle shaft measuring points — ES300

3. Remove the front halfshaft.

4. Remove the six bolts and three washers and disconnect the center axle shaft from the right axle shaft. Remove the joint end cover gasket from the axle shaft.

NOTE: Do not compress the inboard boot.

5. Install nuts, bolts and washers finger-tight to keep the inboard joint together.

6. Remove the inboard and outboard CV-boot clamps.

7. Matchmark the inboard joint and axle shaft.

8. Remove the snapring.

9. Remove the inboard joint from the axle shaft, using the proper tool to press out the joint.

10. Remove the inboard joint from the inboard axle shaft.

11. Remove the nuts, bolts and the washers.

12. Using a prybar and hammer, pry around the whole perimeter of the inboard joint cover.

13. Secure the inner and outer races and remove the inboard joint from the cover.

14. Remove the inner and outer boots.

To install:

15. Wrap vinyl tape around the spline of the shaft to prevent damage to the boot.

16. Pack the outer CV-boot with the grease provided in the kit.

17. Install the outboard CV-boot.

18. Install the inboard CV-boot.

19. Apply seal packing 08826–00801 or Three Bond 1121® or equivalent, to the inboard joint cover.

NOTE: Avoid applying an excess amount to the surface.

20. Align the bolt holes of the cover with those of the inboard joint, then insert the hexagon bolts.

21. Using a plastic faced hammer, tap the rim of the inboard joint cover into place. Repeat several times until seated.

22. Install the bolts, nuts and the washers finger-tight to keep the joint assembled.

23. Align the matchmarks placed before removal.

24. Using a brass bar and hammer, tap the inboard joint onto the axle shaft.

WARNING

Ensure the brass bar contacts the inner race, not the cage.

25. Remove the nuts, bolts and the washers.

26. Install a new snapring.

27. Pack grease into the inboard boot and joint.

28. Install new clamps onto the CV-boots. Ensure the boots are on the shaft groove and are not stretched or contracted when the driveshaft is at the standard length of:

- Right side — 17.949 inches (455.0mm)
- Left side — 17.953 inches (455.9mm)

29. Pack grease into the center axle shaft or side gear shaft.

30. Ensure there is not play in the inboard and outboard joint.

31. Ensure the inboard joint slides smoothly in the thrust direction.

32. Install the axle shaft.

33. Install the wheel.

34. Lower the vehicle.

All Except ES300

If the Outboard joint CV-boot is being replaced, the Inboard CV-joint and boot must be removed first.

1. Raise and safely support the vehicle.

2. Remove the rear tire and wheel assembly.

3. Remove the halfshaft.

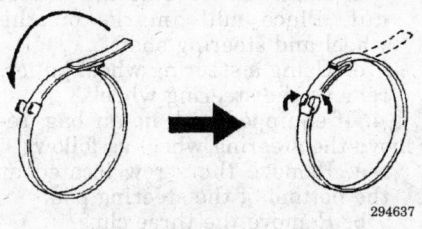

Clamp installation — ES300

294637

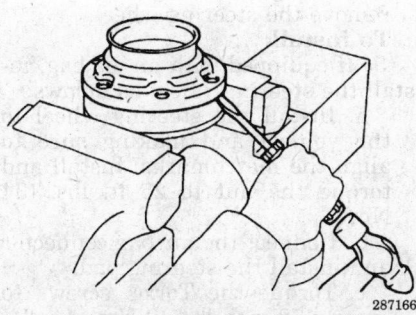

Removing the inboard joint cover from the joint — excluding ES300

287166

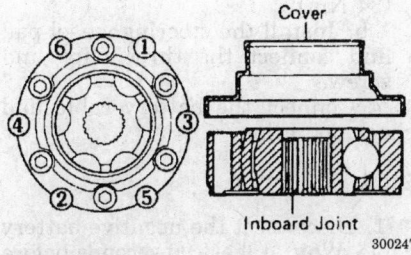

Inboard joint cover installation sequence — excluding ES300

300247

4. Secure the halfshaft in a suitable holding fixture.

5. Wrap a shop towel around the speed sensor rotor to prevent damage.

6. Pry out the end cover.

7. Install nuts, bolts and washers finger-tight to keep the inboard joint together.

8. Remove the boot clamps from the inboard and outboard joint boots.

9. Matchmark the inboard joint and the halfshaft.

NOTE: Do not use punchmarks as the matchmarks.

10. Remove the snapring, using the proper pliers.

11. Press out the inboard joint from the halfshaft with the proper tools. Secure the inboard joint in a suitable holding fixture.

12. Tap out the inboard joint cover.

NOTE: Ensure the cage and inner race are not positioned too much to one side of the outer race.

13. Wrap the shaft splines with tape.

14. Remove both the inboard and outboard joint boots.

To install:

15. If the joint has been disassembled:

a. Align the inboard joint matchmarks made prior to disassembly.

b. Install the inner race to the cage so that the indented beveled part of the inner race is on the opposite side to the beveled top of the cage.

c. Install the outer race so that the indented side of the outer race is facing the same side as the beveled surface of the cage.

d. Match the narrow projections of the inner race with the wide projections of the outer race.

e. Tilt the cage and the inner race to the side and insert the balls one by one.

NOTE: When the cage and inner race are tilted over, support the joint with your hand to prevent the balls from falling out.

16. Temporarily install the four CV-boot clamps. Install the boots and pack them with grease.

a. Outboard and inboard CV-joint grease capacity: 100–105g (3.5–3.7 oz). Use all the grease supplied in the boot kit and thoroughly pack the ball contact surface inside the joint.

17. Apply seal packing 08826–00801, Three Bond 1121®, or equivalent to the inboard joint cover as shown in the illustration.

NOTE: Avoid applying an excessive amount to the surface.

18. Remove the grease from the surface of the inboard joint facing the cover.

19. Align the bolt holes of the cover with those of the inboard joint, then insert the hexagon bolts.

20. Using a plastic tipped hammer, tap the inboard joint cover onto the joint. Strike the cover in the order shown and repeat several times.

21. Align the matchmarks for the joint and the halfshaft.

22. Using a brass bar and a hammer, tap the inboard joint onto the halfshaft.

NOTE: Ensure the brass bar is contacting the inner race and not the cage.

23. Install a new snapring.

24. Remove the hexagon bolts, nuts and washers.

25. Position the boots and tighten the clamps.

26. Ensure the boots are not stretched with the halfshaft at normal length of:

• Left halfshaft — 21.791 inches (553.5mm)

• Right halfshaft — 23.602 inches (598.5mm)

• 1993–94 LS400 Left halfshaft — 22.855 inches (580.5mm)

• 1993–94 LS400 Right halfshaft — 24.666 inches (626.5mm)

• 1995 LS400 Left halfshaft — 22.579 inches (573.5mm)

• 1995 LS400 Right halfshaft — 24.390 inches (619.5mm)

27. Pack grease into the end cover.

a. End cover grease capacity: 50–55g (1.8–1.9 oz).

28. Wipe off any excess grease and install a new gasket on the end cover.

29. Align the bolt holes of the end cover with those of the inboard joint.

30. Install the 6 hexagon bolts and washers from the end cover side and 6 nuts to the boot side. Tighten the bolts in the following sequence several times.

31. Ensure the claw of the end cover touches the inboard joint.

32. Ensure the operation of the joint is smooth within the sliding region in the axial direction.

NOTE: If a large angle is used for the cross-groove type joint, the joint will feel like it is catching, but this does not indicate an abnormality.

33. Install the halfshaft.

34. Install the wheel.

35. Lower the vehicle.

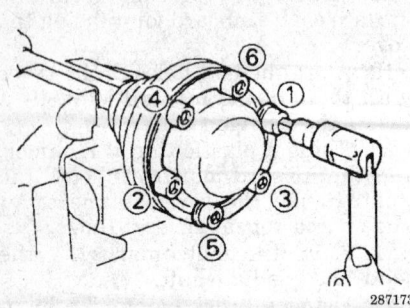

Cover torquing sequence — excluding ES300

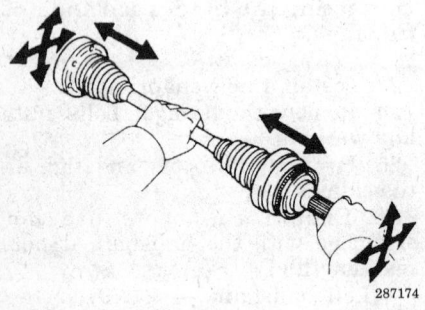

Checking the CV-joint operation — excluding ES300

STEERING

Air Bag

——— CAUTION ———

Some vehicles are equipped with an air bag system, also known as the Supplemental Inflatable Restraint (SIR) or Supplemental Restraint System (SRS). The system must be disabled before performing service on or around system components, steering column, instrument panel components, wiring and sensors. Failure to follow safety and disabling procedures could result in accidental air bag deployment, possible personal injury and unnecessary system repairs.

PRECAUTIONS

Several precautions must be observed when handling the inflator module to avoid accidental deployment and possible personal injury.

• Never carry the inflator module by the wires or connector on the underside of the module.

• When carrying a live inflator module, hold securely with both hands and ensure that the bag and trim cover are pointed away.

• Place the inflator module on a bench or other surface with the bag and trim cover facing up.

• With the inflator module on the bench, never place anything on or close to the module which may be thrown in the event of an accidental deployment.

DISARMING

To avoid personal injury when working on vehicles equipped with an air bag, the negative battery cable must be disconnected and at least 90 seconds must elapse before working on the system. Failure to do so may result in deployment of the air bag.

Steering Wheel

REMOVAL AND INSTALLATION

ES300

NOTE: Do not attempt to remove or install the steering wheel by hammering on it. Damage to the energy-absorbing steering column could result.

1. Disconnect the negative battery cable. Wait at least 90 seconds before performing any work on the steering wheel with an air bag.

——— CAUTION ———

The Supplemental Inflatable Restraint (SIR) system must be disarmed before removing the steering wheel. Failure to do so may cause accidental deployment of the air bag, resulting in unnecessary SIR system repairs and/or personal injury.

2. Place the front wheels facing straight ahead.

3. If equipped with an air bag, remove the steering wheel as follows;

a. Remove the steering wheel screw covers.

b. Using a Torx®, loosen the screws until the groove trailing the screw circumference catches on the screw case.

c. Pull the wheel pad out from the steering wheel and disconnect the air bag connector. Store the wheel pad with the upper surface of the pad facing upward.

d. Remove the steering wheel nut. Place matchmarks on the wheel and steering shaft.

e. Using a steering wheel puller, remove the steering wheel.

4. If equipped with no air bag, remove the steering wheel as follows:

a. Remove the screw located at the bottom of the steering pad.

b. Remove the three clips.

c. Remove the steering pad from the steering wheel.

d. Remove the steering wheel nut.

e. Using a steering wheel puller, remove the steering wheel.

To install:

5. If equipped with an air bag, install the steering wheel as follows:

a. Install the steering wheel to the vehicle and making sure to align the matchmarks. Install and torque the nut to 25 ft. lbs. (34 Nm).

b. Connect the air bag connector and install the steering pad.

c. Torque the Torx® screws to exactly 78 inch lbs. (9 Nm).

d. Install the screw covers.

6. If equipped with no air bag, install the steering wheel as follows:

a. Install the steering wheel and nut. Torque the nut to 25 ft. lbs. (34 Nm).

b. Install the steering wheel pad and connect the three clips and screw.

7. Connect the battery cable and check operation.

All Except ES300

1. Disconnect the negative battery cable. Wait at least 90 seconds before performing any work on the steering wheel.

2. Position the front wheels in a straight-ahead position.

3. If necessary, remove the steering wheel lower covers.

4. Loosen the Torx® screws until the groove along the screw circumference catches on the screw case.

5. Pull the wheel pad away from the steering wheel and disconnect the air bag connector.

NOTE: When removing the wheel pad, take care not to pull the air bag harness connector. When storing the wheel pad, keep the upper surface of the pad facing upward.

6. Disconnect the wire connector and remove the set nut.

7. Place matchmarks on the steering wheel and the main shaft.

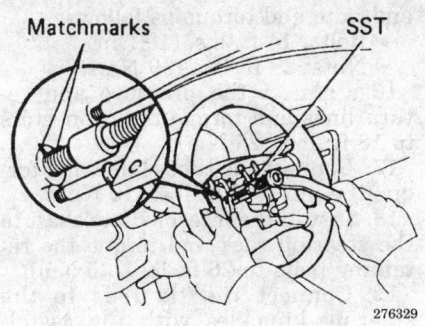

Removing the steering wheel — ES300

276329

8. Remove the steering with a suitable steering wheel puller.

To install:

9. Check that the front wheels are facing straight-ahead. Center the spiral cable. When centering the spiral cable be sure to use the following procedure:

a. Check that the front wheels are pointing straight-ahead.

b. Turn the spiral cable counterclockwise by hand until it becomes harder to turn the cable. The spiral cable will rotate approximately 3 turns to either the left or right of the center.

c. Rotate the spiral cable clockwise approximately 3 turns to align the green mark.

10. Align the matchmarks and install the steering wheel. Tighten the set nut to 26 ft. lbs. (35 Nm).

11. Connect the connector.

12. Install the steering wheel pad after confirming that the circumference groove of the screws is caught on the screw case.

NOTE: Make sure the wheel pad is installed to the specified torque. If the wheel pad has been dropped, or there are cracks, dents or other defects in the case or connector, replace the wheel pad with a new one. When install-

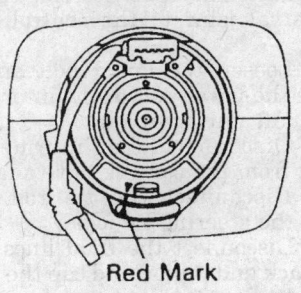

Red Mark

291207

Aligning the spiral cable red mark — SC300 and SC400

ing the wheel pad, be sure the wires and connectors do not interfere with other parts and are not pinched between other parts.

13. Tighten the screws to 65 inch lbs. (7 Nm) on 1993–94 LS400 78 inch lbs. (9 Nm) on others.

14. Install the steering wheel lower No. 2 and No. 3 covers.

15. Check the steering wheel center point.

16. Connect the negative battery cable to the battery.

Tie Rod Ends

REMOVAL AND INSTALLATION

1. Raise and safely support the vehicle. Remove the front wheels.

2. Working at the steering knuckle arm, pull out the cotter pin and then remove the nut to the tie rod.

3. Using a tie rod end puller, disconnect the tie rod from the steering knuckle arm.

4. Loosen the locknut for the tie rod. Turn the tie rod in the opposite direction of the locknut until it is removed from steering rack.

NOTE: When removing the tie rods, count the amount of turns needed to remove the tie rod. This will help with getting the steering alignment close when installing the tie rod.

To install:

5. Install the tie rod end.

6. Install the tie rod the same amount of turns that were needed to remove the tie rod.

7. Tighten the tie rod end to steering knuckle nut to 36–41 ft. lbs. (49–56 Nm). Install a new cotter pin.

NOTE: If the hole on the tie rod does not line up with the nut, always tighten the nut until the hole lines up.

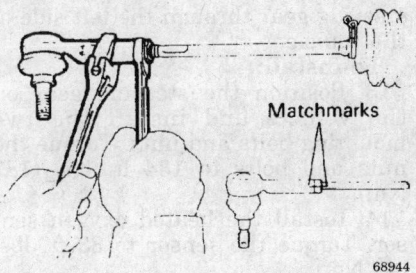

Matchmarks

68944

Place matchmarks, then using two wrenches, detach the tie rod

8. Install the front wheels and lower the vehicle. Check the front end alignment.

9. Torque the tie rod locknut to 54 ft. lbs. (74 Nm).

10. Check steering wheel center point.

Power Rack and Pinion

REMOVAL AND INSTALLATION

ES300

1993–94 Models

― **CAUTION** ―
The air bag system (SRS or SIR) must be disarmed before removing the rack and pinion. Failure to do so may cause accidental deployment, property damage or personal injury.

1. Disconnect the negative battery cable.

2. Place the front wheels in a straight-ahead position, secure the steering wheel with a suitable device to prevent the wheel from turning.

3. Place matchmarks in the universal joint and control valve shaft. Disconnect the joint.

4. Disconnect and plug the hydraulic lines to rack assembly.

5. Raise and safely support the vehicle.

6. Disconnect and remove the stabilizer bar brackets.

7. Disconnect the tie rod ends from the steering knuckle with the proper tool.

8. Disconnect the solenoid connector.

9. Remove the mounting brackets and slide the gear housing out through the right side of the vehicle.

― **WARNING** ―
Do not damage the turn pressure tube or the transmission control cables.

To install:

10. Center the rack and pinion.

11. Connect the hydraulic tubes to the rack assembly and tighten to 38 ft. lbs. (51 Nm). Insert the rack assembly into position.

12. Replace the mounting brackets and torque the mounting bolts to 134 ft. lbs. (182 Nm).

13. Connect the solenoid connector.

14. Connect the tie rod ends to the steering knuckle. Tighten to 36 ft. lbs. (49 Nm).

15. Lower the vehicle.

15. Install the stabilizer bar bolts and nuts and torque as follows:
- Bolts: 14 ft. lbs. (19 Nm)
- Nuts: 29 ft. lbs. (39 Nm)

16. Connect the pressure and return lines and torque the connectors to 18 ft. lbs. (25 Nm).

17. Connect the tube clamp and torque the nut to 7 ft. lbs. (10 Nm).

18. Install the intermediate shaft to the steering rack and torque the retaining bolts to 26 ft. lbs. (35 Nm).

19. Connect the tie rods to the steering knuckles with the castellated nuts. Torque the nut to 36 ft. lbs. and install a new cotter pin. The prongs of the cotter pin should be firmly wrapped around the flats of the nut.

20. Install the front fender apron seals by installing the two bolts.

21. Install the front wheels and lower the vehicle.

22. Fill the power steering reservoir tank to the proper level with power steering fluid.

23. Connect the negative battery cable to the battery.

24. Release the steering wheel.

25. Bleed the system.

26. Check for leaks, adjust the toe-in and check the steering wheel center point.

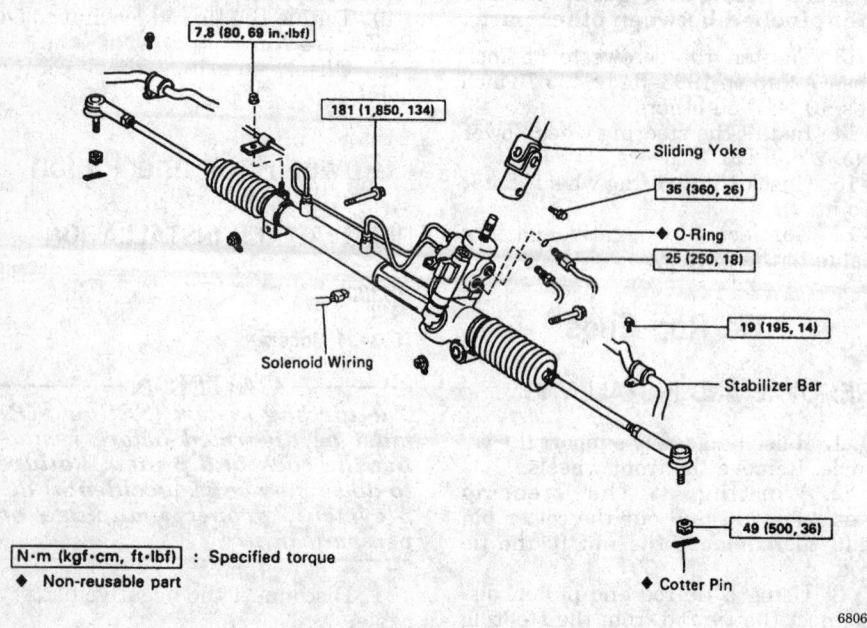

7,8 (80, 69 in.·lbf)

181 (1,850, 134)

Sliding Yoke

35 (360, 26)

◆ O-Ring

25 (250, 18)

Solenoid Wiring

19 (195, 14)

Stabilizer Bar

49 (500, 36)

N·m (kgf·cm, ft·lbf) : Specified torque
◆ Non-reusable part

◆ Cotter Pin

68068

Rack and pinion and related components — 1993–94 ES300

16. Connect the hydraulic lines to the rack assembly.

17. Align the matchmarks on the universal joint and the control valve shaft and connect. Tighten the connecting bolt to 26 ft. lbs. (35 Nm).

18. Check the steering wheel center point and the toe setting.

19. Connect the negative battery cable.

20. Refill the power steering reservoir.

1995–97 Models

1. Disconnect the negative battery cable and wait at least 90 seconds before working on the vehicle to disarm the air bag.

2. Secure the steering wheel in a straight forward position.

3. Raise and support the vehicle safely. Remove the front wheels.

4. Remove the left and right front fender apron seals by removing the two bolts.

5. Remove the cotter pin and nut holding the steering knuckle to the tie rod end. Using a tie rod puller, disconnect the tie rod end from the steering knuckle.

6. Place matchmarks on the intermediate shaft and the control valve shaft.

7. Loosen the upper bolt and remove the lower bolt holding the control valve shaft to the intermediate shaft. Disconnect the intermediate shaft from steering rack housing.

8. Remove the nut to the tube clamp. Remove the clamp from the vehicle.

9. Using SST No. 09631–22020 or equivalent, disconnect the return line and the pressure line from the control valve housing. Use a small plastic container to catch the fluid.

10. Remove the four stabilizer bar bolts and two nuts. Position the stabilizer bar out of the way. Do not remove the sway bar from the vehicle.

11. Remove the heated oxygen sensor (bank 1 sensor 1).

12. Remove the two steering gear mounting bolts and nuts. Remove the steering gear through the left side of the vehicle.

To install:

13. Position the steering gear on the vehicle and install the two mounting bolts and nuts. Torque the nuts and bolts to 134 ft. lbs. (181 Nm).

14. Install the heated oxygen sensor. Torque the sensor to 33 ft. lbs. (44 Nm).

GS300

CAUTION

The air bag system (SRS or SIR) must be disarmed before removing the rack and pinion. Failure to do so may cause accidental deployment, property damage or personal injury.

1. Disconnect the negative battery cable from the battery.

2. Position the wheels in the straight ahead position and secure the steering wheel.

3. Raise and support the vehicle safely.

4. Remove the front wheels.

5. Matchmark the steering column universal joint to the control valve shaft.

6. Loosen the upper bolt and remove the lower bolt to the intermediate shaft universal joint.

7. Disconnect the intermediate shaft from the control valve shaft.

8. Disconnect the tie rod ends from the steering knuckle.

9. Disconnect the fluid lines from the rack and pinion and cap the lines.

10. Disconnect the two tube clamps by removing the bolt.

11. Remove the mounting bolts and nuts. Remove the rack and pinion.

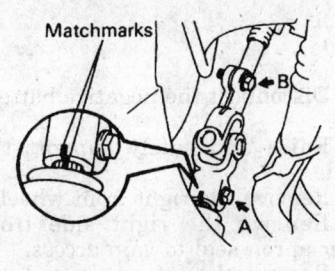
Matchmarks

299791

Matchmarking the intermediate shaft to the control valve shaft — GS300

To install:

12. Center the rack and pinion to the following dimensions.
- Dimension **A** — 1.14 inch (28.9mm)
- Dimension **B** — 23.54 inches (589mm)

13. Install the rack and torque the bolts to 72 ft. lbs. (98 Nm).

14. Connect the two tube clamps and torque the bolt to 12 ft. lbs. (17 Nm).

15. Align the matchmarks on the intermediate shaft and control valve shaft. Torque the intermediate shaft bolts to 26 ft. lbs. (35 Nm).

16. Connect the fluid lines to the rack and pinion with new washers. Torque the union bolts to 36 ft. lbs. (49 Nm).

17. Connect the ties rod ends.

18. Install the wheels

19. Lower the vehicle.

20. Check the steering wheel center point.

21. Connect the negative battery cable.

22. Check the front wheel alignment.

LS400

1993–94 Models

1. Disconnect the negative battery cable.

2. Place the front wheels in a straight-ahead position, secure the steering wheel with a suitable device to prevent the wheel from turning.

3. Place matchmarks in the universal joint and control valve shaft. Disconnect the joint.

4. Disconnect and plug the hydraulic lines to rack assembly.

5. Raise and safely support the vehicle.

6. Disconnect or raise the brake calipers and support them with a piece of wire.

7. Disconnect the tie rod end from the steering knuckle with the proper tool.

8. Remove the steering damper and rack boot protector.

9. Disconnect the solenoid wiring.

10. Remove the mounting grommets and brackets. Remove the rack assembly.

To install:

11. Center the rack correctly.

12. Install the rack. Torque mounting bracket bolts to 56 ft. lbs. (76 Nm).

13. Connect the solenoid wiring.

14. Install the rack boot protector and steering damper. Torque the bolts to 20 ft. lbs. (27 Nm).

15. Connect the tie rod ends.

16. Install the calipers.

17. Connect the fluid lines to the rack and pinion with new washers. Torque the union bolts to 36 ft. lbs. (49 Nm).

18. Align the universal joint to the control valve shaft and torque the bolt to 26 ft. lbs. (35 Nm).

1995–97 Models

1. Raise and safely support the vehicle.

2. Remove the wheel(s).

3. Remove the engine undercover by removing the eight bolts and five screws.

4. Remove the cotter pin and nut holding each tie rod to the steering knuckle.

5. Disconnect the tie rod end from the steering knuckle with a tie rod end puller.

6. Place matchmarks on the sliding yoke and control valve shaft.

7. Loosen the top bolt holding the sliding yoke to the intermediate shaft. Remove the bottom bolt holding the sliding yoke to the steering rack.

8. Disconnect the pressure feed and return lines to the rack and pinion.

9. Disconnect the power steering connector.

10. Remove the four mount bolts and nuts to the power steering rack.

11. Remove the two brackets and grommets.

12. Remove the power steering rack from the vehicle.

To install:

13. Install the power steering rack to the vehicle.

14. Install the two brackets and grommets to the power steering rack.

15. Install the four bolts and torque the bolts to 56 ft. lbs. (76 Nm).

16. Connect the power steering solenoid connector.

17. Connect the pressure feed and return tubes. Torque the union bolt to 36 ft. lbs. (49 Nm).

18. Align the matchmarks on the sliding yoke and control valve shaft.

19. Torque the bolt holding the sliding yoke to the steering rack to 26 ft. lbs. (35 Nm).

20. Torque the bolt holding the sliding yoke to the intermediate shaft to 26 ft. lbs. (35 Nm).

21. Connect the tie rod end to the steering knuckle. Tighten the nut to 48 ft. lbs. (65 Nm). Install a new cotter pin.

22. Install the engine undercover.

23. Bleed the power steering system.

24. Install the wheel(s) and check the front end alignment.

SC300 and SC400

1993–95 Models

1. Disconnect the negative battery cable.

2. Place the front wheels in a straight ahead position, secure the steering wheel with a suitable device to prevent the wheel from turning.

3. Place matchmarks on the universal joint and the control valve shaft.

4. Loosen the bolt on the upper side of the intermediate shaft.

5. Remove the bolt on the lower side and disconnect the intermediate shaft from the steering rack.

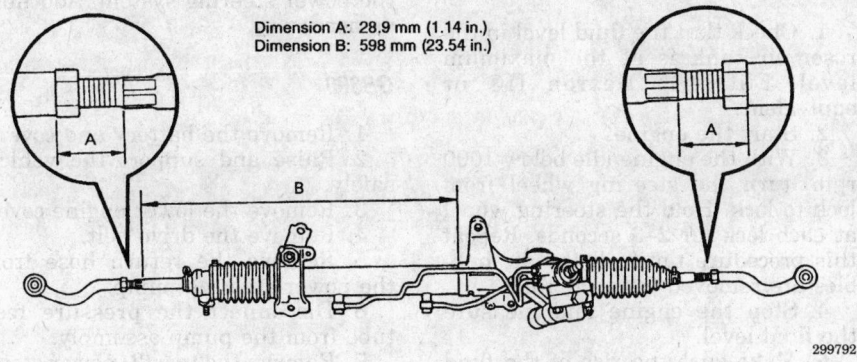

Dimension A: 28.9 mm (1.14 in.)
Dimension B: 598 mm (23.54 in.)

299792

Centering the rack and pinion — GS300

6. Disconnect and plug the hydraulic lines to the rack assembly.

7. Raise and safely support the vehicle. Remove the wheels.

8. Disconnect the tie rod ends from the steering knuckles.

9. Disconnect the solenoid wiring to the rack and pinion.

10. Remove the bolts and brackets holding the steering rack to the frame.

11. Remove the rack assembly.

To install:

12. Install the rack. Torque mounting bracket bolts to 56 ft. lbs. (76 Nm).

13. Connect the solenoid wiring.

14. Connect the tie rod ends to the steering knuckles. Torque the nuts to 36 ft. lbs. (49 Nm).

15. Connect the fluid lines to the rack and pinion with new washers. Torque the union bolts to 36 ft. lbs. (49 Nm).

16. Align the universal joint to the control valve shaft and torque the upper and lower bolts to 26 ft. lbs. (35 Nm).

17. Install the wheels to the vehicle and lower the vehicle to the ground.

18. Connect the negative battery cable to the battery.

19. Check and adjust the vehicle's alignment as necessary.

1996–97 Models

1. Disconnect the negative battery cable. Wait at least 90 seconds before performing any work.

2. Place the front wheels facing straight ahead.

3. Remove the steering wheel pad.

———— **WARNING** ————
Keep the upper surface of the wheel pad pointed away from you at all times. Store the pad with the upper surface facing upward.

4. Disconnect the intermediate shaft.

5. Raise and safely support the vehicle.

6. Disconnect the right and left tie rod ends.

7. Remove the union bolt and gasket and remove the pressure tube.

8. Remove the union bolt and two gaskets; remove the return tube.

9. Disconnect the PPS solenoid connector.

10. On SC400 models, remove the bolt and disconnect the tube clamp.

11. Remove the two bolts and nuts and remove the bracket and grommet.

12. Remove the two bolts and nuts; remove the rack and pinion assembly.

To install:

13. Install the rack and pinion assembly with the two set bolts and nuts. Torque the bolts to 56 ft. lbs. (76 Nm).

14. Install the bracket and grommet with the two bolts and nuts. Torque the bolts to 56 ft. lbs. (76 Nm).

15. On SC400 models, reconnect the tube clamp with the bolt.

16. Reconnect the PPS solenoid.

17. Reconnect the return tube with the bolt and new gaskets. Torque the union bolt to 36 ft. lbs. (49 Nm).

18. Connect the pressure tube with the union bolt and a new gasket. Torque the union bolt to 36 ft. lbs. (49 Nm).

19. Connect the right and left tie rod ends.

20. Reconnect the intermediate shaft.

21. Position the front wheels facing straight ahead and safely lower the vehicle.

22. Align the matchmarks and install the steering wheel. Temporarily tighten the wheel set nut and connect the connector.

23. Reconnect the negative battery cable, refill the steering fluid and bleed the steering system.

24. Check the steering wheel center point and torque the steering wheel set nut to 26 ft. lbs. (35 Nm).

25. Disconnect the negative battery cable. Wait at least 90 seconds before performing the next step.

26. Install the steering wheel pad.

27. Reinstall the negative battery cable and check the front wheel alignment.

28. Road test the vehicle for proper operation.

Power Steering Pump

BLEEDING

1. Check that the fluid level in the reservoir tank is at the maximum level. Fill with Dexron II® or equivalent.

2. Start the engine.

3. With the engine idle below 1000 rpm, turn the steering wheel from lock to lock. Hold the steering wheel at each lock for 2–3 seconds. Repeat this procedure until all the air bubbles are removed from the fluid.

4. Stop the engine and measure the fluid level.

5. Make sure the rise of the fluid is not over 0.020 inch (5mm).

REMOVAL AND INSTALLATION

ES300

1. Disconnect the negative battery cable.

2. Raise and safely support the vehicle.

3. Remove the right front wheel.

4. Remove the right side front fender apron seal to gain access.

5. Remove the cotter pin and nut from the right side tie rod end. Using the proper tool, disconnect the right side tie rod end from the knuckle assembly.

6. Remove the two hydraulic return lines from the pump assembly.

7. Remove the two pressure feed hoses from the pump.

8. Disconnect the solenoid valve electrical connector.

9. Loosen the adjusting and through-bolts on the power steering pump. Slide the power steering pump forward and remove the drive belt.

10. Remove the bolts and remove the power steering pump.

To install:

11. Install the power steering pump and temporarily tighten the adjusting and through-bolts.

12. Install the drive belt. Adjust the drive belt tension and tighten the power steering pump bolts to 32 ft. lbs. (43 Nm).

13. Connect the solenoid valve electrical connector.

14. Connect the two pressure feed hoses and torque the fittings to 26 ft. lbs. (36 Nm).

15. Connect the two return hoses.

16. Install the right side tie rod end and torque the nut to 36 ft. lbs. (49 Nm). Install a new cotter pin.

17. Install the right side fender apron seal.

18. Connect the negative battery cable.

19. Start the engine and check for leaks.

20. Check the fluid level and bleed the power steering system. Add fluid as necessary.

GS300

1. Remove the battery and cover.

2. Raise and support the vehicle safely.

3. Remove the lower engine cover.

4. Remove the drive belt.

5. Remove the return hose from the power steering pump.

6. Disconnect the pressure feed tube from the pump assembly.

7. Remove the two (2) power steering pump mounting bolts.

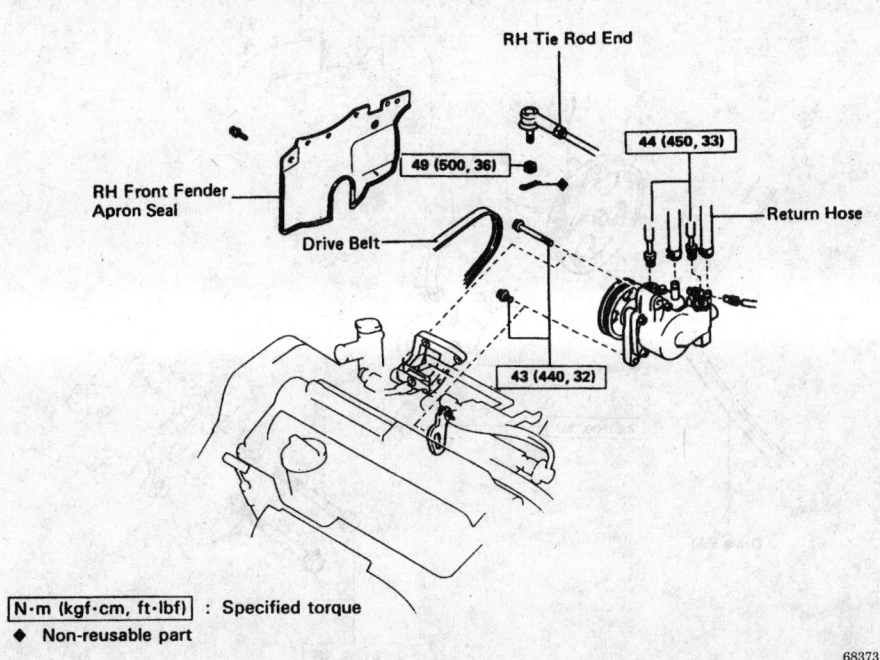

RH Tie Rod End

RH Front Fender Apron Seal

Drive Belt

49 (500, 36)

44 (450, 33)

Return Hose

43 (440, 32)

N·m (kgf·cm, ft·lbf) : Specified torque

◆ Non-reusable part

68373

Steering pump and related components — 1993 ES300

8. Remove the power steering pump assembly.

To install:

9. Install the pump and loosely install the bolts.

10. Install and tension the drive belts. Torque the steering pump mounting bolts to 43 ft. lbs. (58 Nm).

11. Connect the return hose to the power steering pump.

12. Connect the hydraulic pressure tube to the pump with new sealing washers. Torque the union bolt to 36 ft. lbs. (49 Nm).

13. Install the lower engine cover.

14. Lower the vehicle.

15. Install the battery and cover.

16. Refill and bleed the power steering system.

LS400

1993–94 Models

1. Disconnect the negative battery cable.

2. Remove the battery cover, air intake cover and air cleaner cover.

3. Raise and support the vehicle safely.

4. Remove the lower engine cover and right engine splash.

5. Remove the drive belt.

6. Disconnect the hydraulic lines from the pump assembly.

7. Remove the pump mounting bolts and pump assembly.

To install:

8. Install the pump and loosely install the bolts.

9. Install and tension the drive belts. Torque the steering pump mounting bolts to 29 ft. lbs. (39 Nm) and nut to 32 ft. lbs. (43 Nm).

10. Connect the hydraulic hoses to the pump.

11. Connect the hydraulic pressure tube to the pump with new sealing washers. Torque the union bolt to 36 ft. lbs. (49 Nm).

12. Install the lower engine cover and right engine splash.

13. Lower the vehicle.

14. Install the battery cover, air intake duct and air cleaner cover.

15. Connect the negative battery cable.

16. Bleed the steering system.

1995–97 Models

1. Disconnect the negative battery cable from the battery.

2. Remove the engine undercover by removing the eight bolts and five screws.

3. Remove the right hand rear engine undercover by removing the three bolts.

4. Remove the air cleaner inlet and battery clamp cover by removing the bolt.

5. Remove the air cleaner assembly with the air cleaner hose.

6. If equipped with TRAC, disconnect the TRAC actuator as follows:

a. Using SST 09023–00100 or equivalent, disconnect the four brake lines from the TRAC actuator.

b. Disconnect the brake hose from the brake line of the actuator.

c. Remove the nut and two bolts from the body and ABS actuator bracket.

d. Disconnect the brake line clamp from the TRAC actuator bracket.

e. Remove the TRAC actuator assembly from the engine compartment.

7. Remove the ABS actuator as follows:

a. Using SST 09023–00100 or equivalent, disconnect the five (w/o TRAC) or seven (w/TRAC) brake lines from the ABS actuator.

b. Disconnect the two connectors from the ABS actuator.

c. If equipped without TRAC system, remove the nut, two bolts and the ABS actuator.

d. If equipped with TRAC system, remove the nuts, bolt and the ABS actuator.

8. Loosen the drive belt tension by turning the drive belt tensioner counterclockwise and remove the drive belt.

9. Disconnect the power steering return hose.

10. Disconnect the two vacuum hoses to the power steering pump.

11. Disconnect the pressure feed line to the power steering pump by removing the union bolt and gasket.

12. Remove the nut and three bolts from the power steering pump and remove the pump.

To install:

13. Install the power steering pump to the engine and install the nut and three bolts. Torque the bolts to 29 ft. lbs. (39 Nm) and the nut to 32 ft. lbs. (43 Nm).

14. Install the pressure feed tube and torque the union bolt over a new gasket to 36 ft. lbs. (49 Nm).

15. Install the drive belt.

16. Connect the two vacuum hoses to the power steering pump.

17. Connect the return hose to the power steering pump.

18. Install the ABS actuator as follows:

a. If equipped with TRAC, install the ABS actuator assembly

with the bolt and nuts. Torque the bolts and nuts to 14 ft. lbs. (19 Nm).

b. If equipped without TRAC, install the ABS actuator assembly with the bolts and nut. Torque the bolts and nuts to 14 ft. lbs. (19 Nm).

c. Connect the two connectors to the ABS actuator.

d. Using the same tool as installation, connect the brake lines to the ABS actuator. Torque the lines to 11 ft. lbs. (15 Nm).

19. Install the TRAC actuator as follows:

a. Install the TRAC actuator and install the brake line clamp.

b. Install the nut and two bolts to the body and ABS actuator bracket. Torque the two bolts to 14 ft. lbs. (19 Nm).

c. Connect the connector to the actuator.

d. Connect the brake hose to the brake line of the actuator.

e. Using the same tool as installation, connect the four brake lines to the TRAC actuator. Torque the lines to 11 ft. lbs. (15 Nm).

f. Bleed the brake system.

20. Install the air cleaner assembly with the air cleaner hose.

21. Install the right hand rear engine undercover with the three bolts.

22. Install the engine undercover by installing the eight bolts and five screws.

23. Bleed the power steering system.

24. Connect the negative battery cable to the battery.

SC300

1993–95 Models

1. Disconnect the negative battery cable.

2. Remove the drive belt by loosening the drive belt tensioner. Turn the tensioner clockwise to relieve tension on the drive belt.

3. Drain the fluid from the power steering pump.

4. Disconnect the hydraulic pressure line and the return line from the power steering pump assembly.

5. Remove the pump mounting bolts and remove the pump assembly.

To install:

6. Install the pump and loosely install the bolts.

7. Install the drive belt. Torque the steering pump mounting bolts to 43 ft. lbs. (58 Nm).

8. Connect the hydraulic return line to the pump.

9. Connect the hydraulic pressure tube to the pump with new sealing

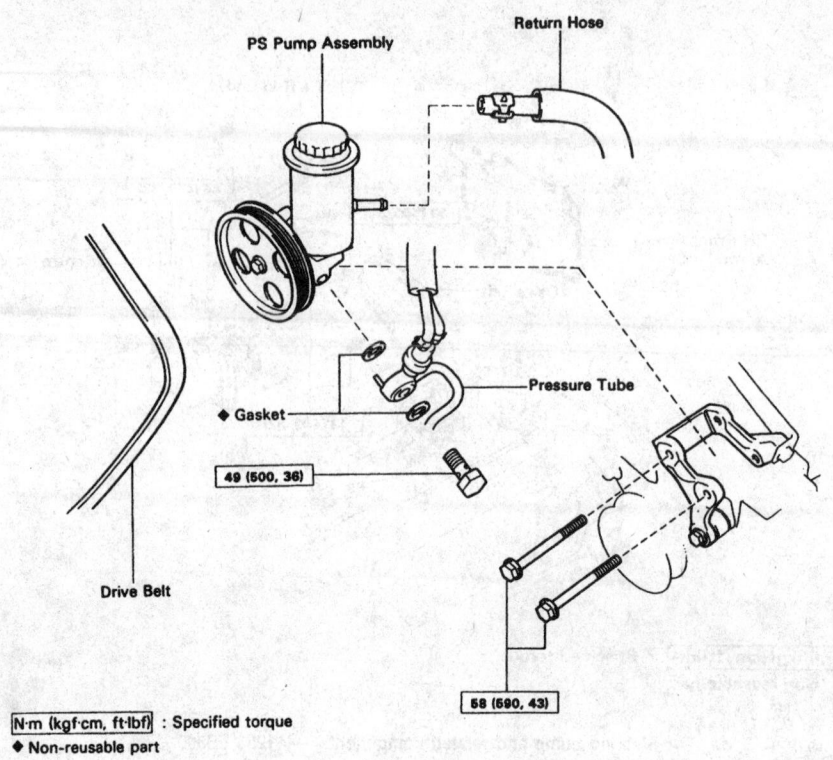

PS Pump Assembly **Return Hose** **Pressure Tube** **Gasket** **Drive Belt**

49 (500, 36)

58 (590, 43)

N·m (kgf·cm, ft·lbf) : Specified torque
◆ Non-reusable part

226692

Steering pump removal and installation — 1993–95 SC300

washers. Torque the union bolt to 36 ft. lbs. (49 Nm).

10. Connect the negative battery cable to the battery.

SC400

1. Disconnect the negative battery cable from the battery.

2. Raise and support the vehicle safely.

3. Remove the lower engine cover.

4. Remove the drive belt by turning the tensioner counterclockwise.

5. Disconnect the hydraulic and vacuum lines from the pump assembly.

6. Remove the pump mounting bolts and nut.

7. Remove the pump assembly from the engine.

To install:

8. Install the pump and loosely install the bolts and nut.

9. Install the drive belt.

10. Torque the steering pump mounting bolts to 29 ft. lbs. (39 Nm) and the nut to 32 ft. lbs. (43 Nm).

11. Connect the hydraulic and vacuum hoses to the pump.

12. Connect the hydraulic pressure tube to the pump with new sealing washers. Torque the union bolt to 36 ft. lbs. (49 Nm).

13. Install the lower engine cover.

14. Lower the vehicle.

15. Connect the negative battery cable.

16. Refill and bleed the steering system. Check for leaks and proper operation.

BRAKES

Anti-Lock Brake System Service

PRECAUTIONS

• Certain components within the Anti-Lock Brake System (ABS) are not intended to be serviced or repaired individually. Only those components with removal and installation procedures should be serviced.

• Do not use rubber hoses or other parts not specifically specified for and ABS system. When using repair kits, replace all parts included in the kit. Partial or incorrect repair may lead to functional problems and require the replacement of components.

• Lubricate rubber parts with clean, fresh brake fluid to ease assembly. Do not use lubricated shop air to clean parts; damage to rubber components may result.

• Use only specified brake fluid from an unopened container.

• If any hydraulic component or line is removed or replaced, it may be necessary to bleed the entire system.

• A clean repair area is essential. Always clean the reservoir and cap thoroughly before removing the cap. The slightest amount of dirt in the fluid may plug an orifice and impair the system function. Perform repairs after components have been thoroughly cleaned; use only denatured alcohol to clean components. Do not allow ABS components to come into contact with any substance containing mineral oil; this includes used shop rags.

• The Anti-Lock control unit is a microprocessor similar to other computer units in the vehicle. Ensure that the ignition switch is **OFF** before removing or installing controller harnesses. Avoid static electricity discharge at or near the controller.

• If any arc welding is to be done on the vehicle, the control unit should be unplugged before welding operations begin.

Master Cylinder

REMOVAL AND INSTALLATION

1. Disconnect the negative battery cable. Wait at least 90 seconds before performing any other work.

2. If equipped with a Traction Control System (TRAC), perform the following procedure:

   a. Remove the air cleaner.

   b. Connect a vinyl tube from a container to the bleeder plug of the TRAC actuator, then loosen the bleeder plug with the ignition in the **OFF** position.

   c. Tighten the plug when the fluid stops flowing out.

─────── WARNING ───────
The fluid is under high pressure and could spray out with great force. Use caution when opening the bleeder plug.

3. Disconnect the level warning switch connector.

4. Remove the fluid from the reservoir, using a suitable syringe.

5. If equipped with TRAC, remove the hoses from the master cylinder.

6. Disconnect the brake lines from the master cylinder.

7. Remove the mounting nuts and 2-way union. Remove the master cyl-

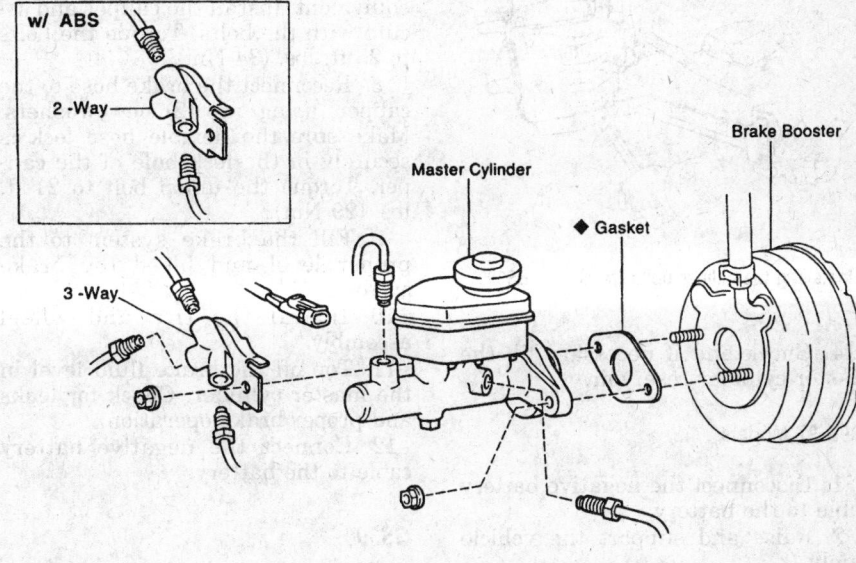

◆ **Non-reusable part**

293108

Master cylinder and related components

inder and gasket from the brake booster.

To install:

8. Adjust the length of the booster pushrod before installation.

9. Install the master cylinder and replace the 2-way union. Tighten the mounting nuts to 9 ft. lbs. (13 Nm).

10. Connect the brake lines to the master cylinder and the 2-way union. Tighten the nuts to 11 ft. lbs. (15 Nm).

11. If equipped with TRAC, connect the hoses to the master cylinder.

12. Connect the level warning switch connector to the master cylinder.

13. Reconnect the battery cable.

14. Bleed the brake system and, if equipped, bleed the TRAC system.

15. Check for leaks. Check and adjust the brake pedal, if needed.

Brake Caliper

REMOVAL AND INSTALLATION

ES300

1993–94 Models

1. Raise and safely support the vehicle. Remove the tire and wheel assembly.

2. Disconnect and plug the brake line at the caliper.

3. Hold the sliding pin with a wrench and remove the mounting bolts.

4. Remove the caliper assembly.

To install:

5. Remove a small amount of fluid from the master cylinder to prevent overflow when the caliper piston is compressed.

6. Press the caliper piston in slightly with a hammer handle or equivalent, to ease installation.

7. On the rear, install the main pin and tighten it to 19 ft. lbs. (27 Nm).

8. Install the caliper. Hold the sliding pin and torque the mounting bolts to:

• Front — 25 ft. lbs. (34 Nm)

• Rear — 14 ft. lbs. (20 Nm)

9. Connect the brake line with two new gaskets and tighten the union to 21 ft. lbs. (29 Nm).

NOTE: Insert the brake line securely into the lock hole in the caliper.

10. Bleed the brake system.

11. Install the wheel.

12. Lower the vehicle.

13. Depress the brake pedal several times until full braking power is restored.

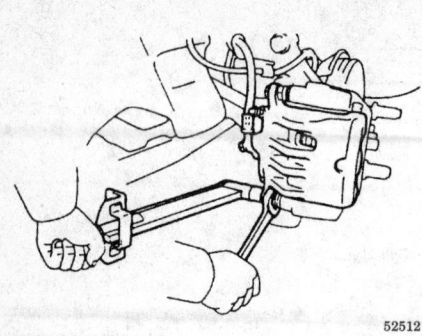

Installing the caliper bolts (front) — ES300

14. Check and if necessary fill the master cylinder reservoir.

1995–97 Models

1. Disconnect the negative battery cable to the battery.
2. Raise and support the vehicle safely.
3. Remove the front wheels.
4. Disconnect the brake hose from the caliper by removing the union bolt and two gaskets. Plug the end of the hose to prevent loss of fluid.
5. Remove the bolts that attach the caliper to the torque plate.
6. Lift the bottom of the caliper up and remove the caliper assembly.

To install:

7. Grease the caliper slides and bolts with lithium grease or equivalent. Install the caliper and secure with the bolts. Torque the bolts to 25 ft. lbs. (34 Nm).
8. Reconnect the brake hose to the caliper, using two (2) new washers. Make sure the flexible hose lock is securely in the lock hole of the caliper. Torque the union bolt to 21 ft. lbs. (29 Nm).
9. Fill the brake system to the proper level and bleed the brake system.
10. Install the tire and wheel assembly.
11. Top off the brake fluid level in the master cylinder. Check for leaks and proper brake operation.
12. Connect the negative battery cable to the battery.

GS300

1. Raise and safely support the vehicle. Remove the tire and wheel assembly.
2. Disconnect and plug the brake line at the caliper.
3. Hold the sliding pin with a wrench and remove the mounting bolts.
4. Remove the caliper assembly.

To install:

5. Press the caliper piston in slightly with a hammer handle or equivalent, to ease installation.
6. Install the caliper. Hold the sliding pin and tighten the mounting bolts to 25 ft. lbs. (34 Nm).
7. Connect the brake line with two new gaskets and tighten the union bolt to 22 ft. lbs. (30 Nm).

NOTE: Insert the brake line securely into the lock hole in the caliper.

8. Bleed the brake system.
9. Install the wheel.
10. Lower the vehicle.
11. Depress the brake pedal several times until full braking power is restored.
12. Check and if necessary fill the master cylinder reservoir.

1993–94 LS400

1. Raise and safely support the vehicle.
2. Remove the tire and wheel assembly.
3. Disconnect and plug the brake line at the caliper.
4. Hold the sliding pin with a wrench and then remove the mounting bolts.
5. Remove the caliper.

To install:

6. Install the caliper. Hold the sliding pin and tighten the mounting bolts to 25 ft. lbs. (34 Nm) on the others.
7. Connect the brake line with two new gaskets and tighten the union to 22 ft. lbs. (30 Nm) on the others.

NOTE: Insert the brake line securely into the lock hole in the caliper.

8. Refill the reservoir as necessary and bleed the brake system.

1995–97 LS400

Front

1. Raise and safely support the vehicle.
2. Remove the front tire and wheel assembly.
3. Disconnect and plug the brake line at the caliper.
4. Remove the two bolts to the holding the caliper to the steering knuckle and remove the caliper assembly.

To install:

5. Install the caliper. Install the two bolts and torque the bolts to 87 ft. lbs. (118 Nm).

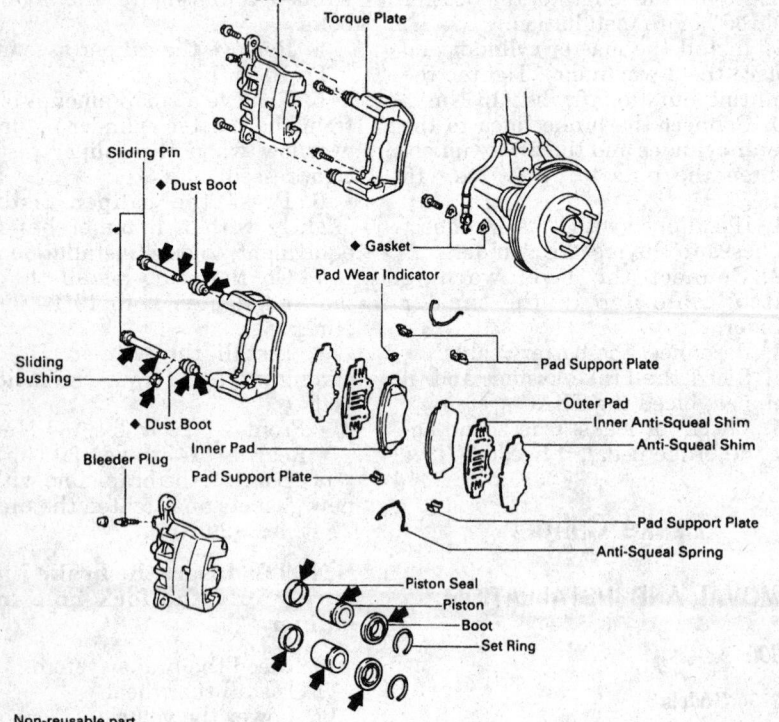

Caliper
Torque Plate
Sliding Pin
◆ Dust Boot
◆ Gasket
Pad Wear Indicator
Sliding Bushing
◆ Dust Boot
Inner Pad
Pad Support Plate
Bleeder Plug
Pad Support Plate
Outer Pad
Inner Anti-Squeal Shim
Anti-Squeal Shim
Pad Support Plate
Anti-Squeal Spring
Piston Seal
Piston
Boot
Set Ring

◆ Non-reusable part
➤ Lithium Soap Base Glycol Grease

Front brake assembly (two piston type) — 1995–97 ES300

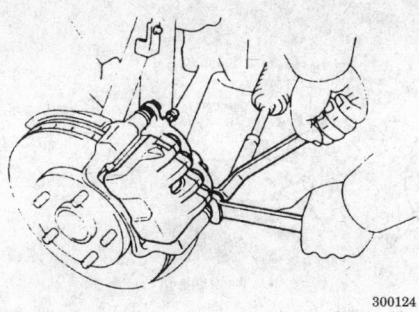

Removing the mounting bolts on the front caliper

6. Connect the brake line with two new gaskets and tighten the union to 29 ft. lbs. (39 Nm) on the others.

NOTE: Insert the brake line securely into the lock hole in the caliper.

7. Refill the reservoir as necessary and bleed the brake system.

Rear

1. Raise and safely support the vehicle. Remove the rear tire and wheel assembly.
2. Disconnect and plug the brake line at the caliper.
3. Remove the mounting bolts and the caliper assembly.
To install:
4. Temporarily install the caliper on the torque plate with the two installation bolts.
5. Hold the sliding pin and tighten the mounting bolts to 25 ft. lbs. (34 Nm).
6. Connect the brake line with two new gaskets and tighten the union to 29 ft. lbs. (39 Nm).

NOTE: Insert the brake line securely into the lock hole in the caliper.

7. Refill the reservoir as necessary and bleed the brake system.

SC300 and SC400

1. Raise and safely support the vehicle. Remove the tire and wheel assembly.
2. Disconnect and plug the brake line at the caliper by removing the union bolt and two (2) gaskets. Use a suitable container to catch the brake fluid.
3. Hold the sliding pin with a wrench and then remove the mounting bolts.
4. Remove the caliper from the caliper support.

To install:
5. Press the caliper piston in slightly with a hammer handle or equivalent.
6. Install the caliper to the caliper support.
7. Install the caliper bolts. Hold the sliding pin and tighten the mounting bolts to 25 ft. lbs. (34 Nm).
8. Connect the brake line with (2) new gaskets and tighten the union to 22 ft. lbs. (30 Nm).

NOTE: Insert the brake line securely into the lock hole in the caliper.

9. Bleed the brake system.
10. Install the wheel.
11. Lower the vehicle.
12. Depress the brake pedal several times until full braking power is restored.
13. Check and fill the master cylinder reservoir, if needed.

Disc Brake Pads

REMOVAL AND INSTALLATION

1993–94 ES300

1. Raise and safely support vehicle.
2. Remove the tire and wheel assembly.
3. Temporarily install two lug nuts on the wheel studs to hold the rotor in place.
4. Hold the sliding pin on the lower mounting bolt and remove the bolt. Swivel the caliper upward and out of the way. Support the caliper. Do not disconnect the brake line.
5. Remove the brake pads, shims, springs and indicators from the caliper.
To install:
6. Remove a small amount of fluid from the master cylinder to prevent overflow when the caliper piston is compressed.
7. Install the pad support plates on the torque plate.
8. Install the pad wear indicator clip on the inside pad.
9. Apply disc brake grease to both sides of the inner anti-squeal shims.
10. Install both anti-squeal shims to each pad.
11. Install the inner pad with the wear indicator clip facing up. Install the outer pad.
12. Install the anti-squeal springs.
13. Press in the caliper pistons with large pliers or a C-clamp and install the caliper over the pads. Hold the sliding pin and torque the bolt to 25 ft. lbs. (34 Nm) for the front and 14 ft. lbs. (20 Nm) on the rear.
14. Remove the two lug nuts and install the front tire and wheel assembly.
15. Lower the vehicle.
16. Depress the brake pedal several times until full braking power is restored before moving the vehicle.

1995–97 ES300

─────── **CAUTION** ───────
Brake shoes contain asbestos, which has been determined to be a cancer causing agent. Never clean the brake surfaces with compressed air! Avoid inhaling any dust from any brake surfaces. When cleaning brake surfaces, use a commercially available brake cleaning fluid.
────────────────────────

1. Raise and safely support the front of the vehicle.
2. Remove the front wheels and temporarily fasten the rotor disc with the hub nuts.

NOTE: Always replace the front disc pads as a set.

3. Hold the sliding pin on the bottom and loosen the installation bolt.
4. Remove the lower installation bolt.
5. Lift up the caliper and suspend it securely. Do not remove the upper installation bolt.
6. Remove the following parts:
   a. The 2 anti-squeal springs.
   b. The 2 brake pads.
   c. The 4 anti-squeal shims.
   d. The 1 pad wear indicator.
   e. The 4 pad support plates.
To install:
7. Install the pad support plates.
8. Install a pad wear indicator plate to the pad. Install the anti-squeal shims and support plates to each pad.

NOTE: It recommended that a suitable anti-squeal compound (available at your local parts house) be applied to both sides of the inner anti-squeal shim.

9. Install the two pads so that the wear indicator plate is facing upward. Do not allow oil or grease to get in the rubbing face.
10. Draw out a small amount of brake fluid from the brake reservoir. Press in the caliper piston with a suitable tool.
11. Press the brake piston in carefully so the boot will not become wedged.

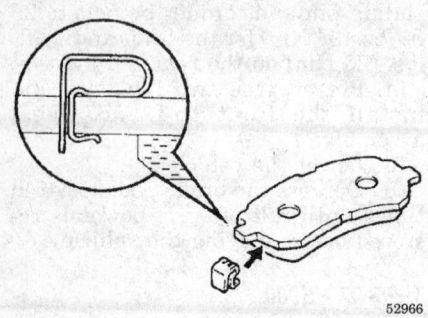

Front wear indicator clip installation — 1993–94 ES300

52966

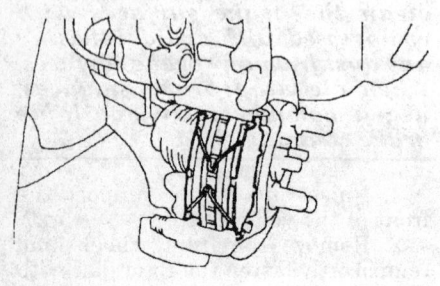

Anti-squeal spring installation — 1993–94 ES300

52967

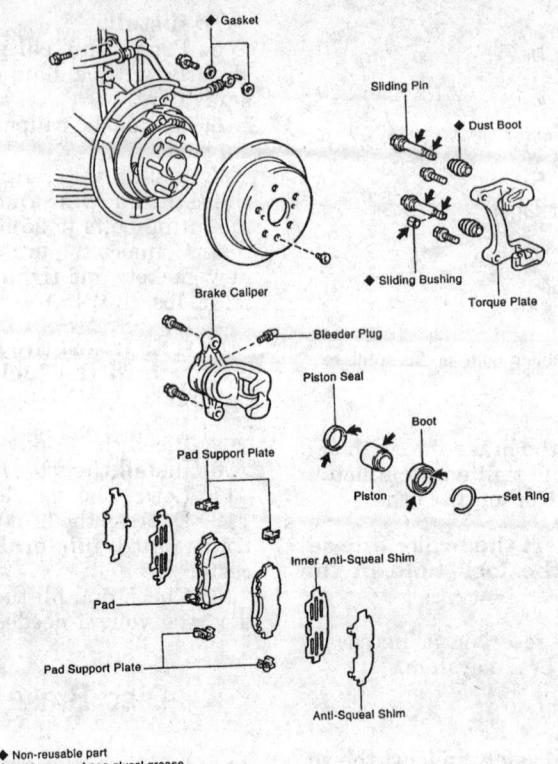

◆ Non-reusable part
→ Lithium soap base glycol grease

293064

Rear brake assembly — 1995–97 ES300

12. Lower and install the caliper. Torque the sliding main pin to 25 ft. lbs. (34 Nm).

NOTE: When installing the sliding main pin, be careful that the plug installed in the torque plate does not come loose.

13. Install the wheels and lower the vehicle.
14. Check the fluid level in the master cylinder and add as necessary. Be sure to pump the brake pedal a few times before road testing the vehicle.

1993–97 GS300, SC300 and SC400 1993–94 LS400

1. Raise and safely support the vehicle.
2. Remove the tire and wheel assembly.
3. Hold the sliding pin on the lower mounting bolt and remove the bolt. Swivel the caliper upward and out of the way. Support the caliper. Do not disconnect the brake line.
4. Remove the two anti-squeal springs, the brake pads, the four anti-squeal shims, the wear indicators and the four pad support plates.

To install:
5. Install the pad support plates and the pad wear indicator plate on the inside pad.
6. Apply disc brake grease to both sides of the anti-squeal shims. Install the shims to the pads. Do not allow any grease to contact the friction surface of the pad.
7. Install both pads with the wear indicator plates facing downward.
8. Install the anti-squeal springs. Press in the caliper pistons with a C-clamp and the old inner brake pad. Install the caliper over the pads.
9. Hold the sliding pin and tighten the mounting bolts to 25 ft. lbs. (34 Nm).
10. Install the tire and wheel assembly.
11. Lower the vehicle and check the brake fluid level in the reservoir.

NOTE: Be sure to pump the brakes several times to attain a firm brake pedal prior to moving the vehicle.

12. Road test the vehicle for proper operation.

1995–97 LS400

Front

1. Raise and safely support the vehicle.

2. Remove the front tire and wheel assembly.
3. Remove the two bolts holding the caliper to the steering knuckle. Remove the caliper upward and out of the way. Support the caliper. Do not disconnect the brake line.
4. Remove the anti-squeal springs, the brake pads, the anti-squeal shims, the wear indicators and the four pad support plates.

To install:
5. Install the pad support plates and the pad wear indicator plate on the inside pad.
6. Apply disc brake grease to both sides of the anti-squeal shims. Install the shims to the pads. Do not allow any grease to contact the friction surface of the pad.
7. Install both pads with the wear indicator plates.
8. Install the anti-squeal springs. Press in the caliper pistons with a C-clamp and the old inner brake pad. Install the caliper over the pads.
9. Install the two bolts to hold the caliper to the steering knuckle. Torque the bolts to 87 ft. lbs. (118 Nm).
10. Install the front tire and wheel assembly.

11. Lower the vehicle and check the fluid level in the reservoir.

NOTE: Be sure to pump the brakes several times to attain a firm brake pedal prior to moving the vehicle.

12. Road test the vehicle for proper operation.

Rear

1. Raise and safely support the vehicle.
2. Remove the rear tire and wheel assembly.
3. Hold the sliding pin on the lower mounting bolt and remove the bolt. Swivel the caliper upward and out of the way. Support the caliper. Do not disconnect the brake line.
4. Remove the two anti-squeal springs, the brake pads, the four anti-squeal shims, the wear indicators and the four pad support plates.

To install:
5. Install the pad support plates and the pad wear indicator plate on the inside pad.
6. Apply disc brake grease to both sides of the anti-squeal shims. Install the shims to the pads. Do not allow any grease to contact the friction surface of the pad.
7. Install both pads with the wear indicator plates facing downward.
8. Install the anti-squeal springs. Press in the caliper pistons with a C-clamp and the old inner brake pad. Install the caliper over the pads.
9. Hold the sliding pin and tighten the mounting bolts to 25 ft. lbs. (34 Nm).
10. Install the rear tire and wheel assembly.
11. Lower the vehicle and check the brake fluid level in the reservoir.

NOTE: Be sure to pump the brakes several times to attain a firm brake pedal prior to moving the vehicle.

12. Road test the vehicle for proper operation.

Brake Rotor

REMOVAL AND INSTALLATION

ES300
Front

1. Disconnect the negative battery cable to the battery.
2. Loosen the wheel lugs slightly, then raise and safely support the vehicle.

3. Remove the wheel(s) and temporarily install two of the wheel lug nuts.
4. Hold the sliding pin on the bottom with a wrench and loosen the installation bolt.
5. Remove the two installation bolts holding the caliper to the torque plate.
6. Remove the caliper from the torque plate and hang the caliper from a piece of wire. Do not disconnect the brake hose.

------ **WARNING** ------
Do not allow the caliper to hang freely from the vehicle. Always support the caliper with a wire from the vehicle.

7. Unbolt and remove the torque plate.
8. Remove the two wheel nuts and pull the disc from the axle hub.

To install:
9. Position the new rotor disc onto the hub and reinstall the two wheel nuts temporarily.
10. Install the torque plate onto the vehicle. Torque the two torque plate bolts to 79 ft. lbs. (107 Nm). Make sure the brake pads are seated correctly within the torque plate.
11. Install the caliper to the torque plate. Torque the installation bolts to 25 ft. lbs. (34 Nm).
12. Remove the wheel lug nuts and install the wheels. Secure the wheel lugs.
13. Lower the vehicle and tighten the lug nuts. Before moving the vehicle, make sure to pump the brake pedal to seat the brake pads against the rotors.
14. Connect the negative battery cable to the battery.

Rear

1. Raise and safely support the vehicle.
2. Remove the rear and wheel assembly.
3. Remove the two retaining bolts from the rear disc brake caliper. Suspend the caliper with a suitable piece of wire so as not to stretch the brake hose.
4. Matchmark the rotor disc to the rear hub.
5. Remove the two rotor retaining screws from the rotor and remove the rotor.

NOTE: If the rotor disc cannot be removed easily, turn the shoe adjuster until the wheel turns freely.

To install:
6. Align the matchmark and install the rotor.
7. Install the torque plate and torque the bolts to 34 ft. lbs. (47 Nm).
8. Install the caliper.
9. Install the wheel.
10. Lower the vehicle.

All Except ES300
Front

1. Raise and safely support the vehicle.
2. Remove the front tire and wheel assembly.
3. Remove the mounting bolts and lift off the caliper assembly. Support the caliper. Do not disconnect the brake line.
4. Remove the torque plate from the steering knuckle by removing the two bolts.
5. Matchmark the rotor to the hub.
6. Remove the mounting bolts and pull the brake rotor off the hub.

To install:
7. Install the brake rotor and torque the mounting screws to 48 inch lbs. (5 Nm).
8. Install the torque plate and torque the mounting bolts to 79 ft. lbs. (107 Nm).
9. Install the caliper.
10. Install the wheel.
11. Lower the vehicle.

Rear

1. Raise and safely support the vehicle.
2. Remove the rear wheel assembly.
3. Remove the two retaining bolts from the rear disc brake caliper. Suspend the caliper with a suitable piece of wire so as not to stretch the brake hose.
4. Matchmark the rotor disc and rear axle shaft.
5. Remove the two rotor retaining screws from the rotor and remove the rotor.

NOTE: If the rotor disc cannot be removed easily, loosen the parking brake shoe adjuster until the wheel turns freely.

To install:
6. Install the rotor and torque the mounting screws to 48 inch lbs. (5 Nm).
7. Install the torque plate and torque the bolts to 77 ft. lbs. (104 Nm).
8. Install the caliper. Adjust the parking brake shoes if they were loosened during rotor removal.
9. Install the wheel.
10. Lower the vehicle.

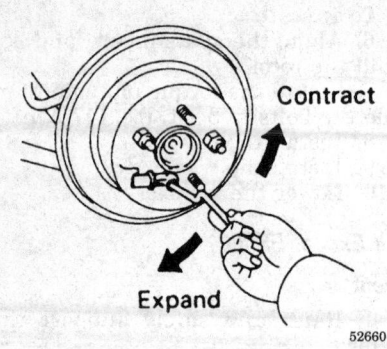

Parking brake shoe adjustment

52660

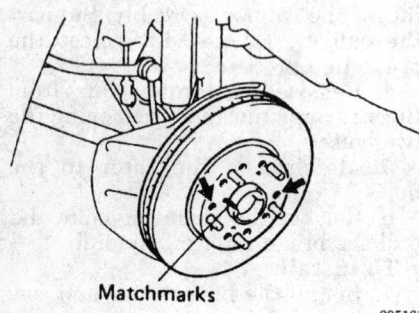

295161

Matchmarking the rotor front — All except ES300

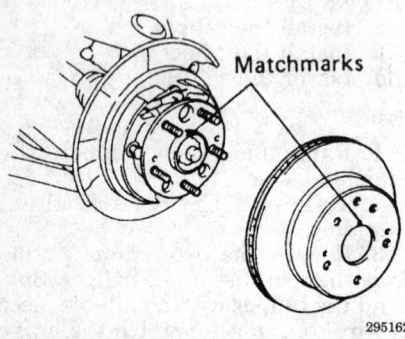

295162

Matchmarking the rotor rear — All except ES300

Parking Brake Cable

ADJUSTMENT

1. Raise and support the vehicle safely.
2. Remove the rear wheels.
3. Temporarily install the hub nuts to hold the rotor on the vehicle.
4. Remove the brake rotor adjustment hole plug.
5. Turn the adjuster to expand the shoes until the rotor disc locks.
6. Return the adjuster 8 notches.
7. Install the plug.

8. Remove the lug nuts and then install the wheels. Install the lug nuts and torque to 77 ft. lbs. (104 Nm).
9. Lower the vehicle.

Brake System Bleeding

Start the bleeding procedure at the caliper or wheel cylinder the furthest from the master cylinder.
1. Connect a vinyl tube to the bleeder screw on the brake cylinder and submerge the other end of the tube in a transparent container half filled with clean brake fluid.
2. Pump the brake pedal several times and loosen the bleeder screw with the pedal held down.
3. When brake fluid stops coming out of the tube with the brake pedal held to the floor, tighten the bleeder screw and release the brake pedal.
4. Repeat Steps 2 and 3 until no air bubbles can be seen in the container.
5. Repeat the procedure for each wheel.
6. Check the level in the master cylinder. Add fluid as necessary.

Wheel Speed Sensor

REMOVAL AND INSTALLATION

Front

1. Disconnect the negative battery cable.
2. Raise and safely support the vehicle.
3. Remove the fender liner from the front fender.
4. Disconnect the speed sensor connector.
5. Remove the three clamp bolts holding the sensor harness to the body and strut.
6. Remove the bolt holding the speed sensor to the steering knuckle. Remove the speed sensor from the vehicle.
   To install:
7. Install the speed sensor to the steering knuckle and install the bolt. Torque the bolt to 69 inch lbs. (8 Nm).
8. Torque the clamp bolts holding the sensor harness to the body and strut to 48 inch lbs. (5 Nm).
9. Connect the speed sensor connector.
10. Install the fender liner and install the wheels.
11. Lower the vehicle.
12. Connect the negative battery cable.

13. After installation, check speed sensor signal.

Rear

ES300

1. Disconnect the negative battery cable.
2. Remove the back seat cushion and seat back.
3. Disconnect the sensor connector and pull out the sensor wire harness with the grommet.
4. Raise and safely support the vehicle.
5. Remove the two clamp bolts holding the sensor wire harness to the body and strut.
6. Remove the lock bolt from the axle carrier and remove the speed sensor.
   To install:
7. Install the speed sensor and tighten the lock bolt to 69 inch lbs. (8 Nm).
8. Install the two clamp bolts to hold the sensor wire harness to the body and strut. Torque the bolts to 44 inch lbs. (5 Nm).
9. Lower the vehicle.
10. Connect the speed sensor connector and install the grommet.
11. Install the seat cushion and seat back.
12. Connect the negative battery cable.
13. After installation, check speed sensor signal.

GS300 and LS400

1. Disconnect the negative battery cable.
2. Raise and safely support the vehicle. Remove the rear wheel.
3. Remove the front cover of the luggage compartment.
4. Disconnect the speed sensor connector and pull out the sensor wire harness with the grommet.
5. Remove the 2 clamp bolts holding the sensor wire harness to the body and the upper suspension arm.
6. Remove the bolt and the speed sensor from the axle carrier.
   To install:
7. Install the speed sensor to the axle carrier with the bolt and torque to 69 inch lbs. (8 Nm).
8. Install the sensor wire harness clamp to the upper suspension arm and the body with the 2 clamp bolts and torque to 48 inch lbs. (5 Nm).
9. Install the sensor wire harness and grommet and reconnect the speed sensor connector.
10. Install the front cover of the luggage compartment.

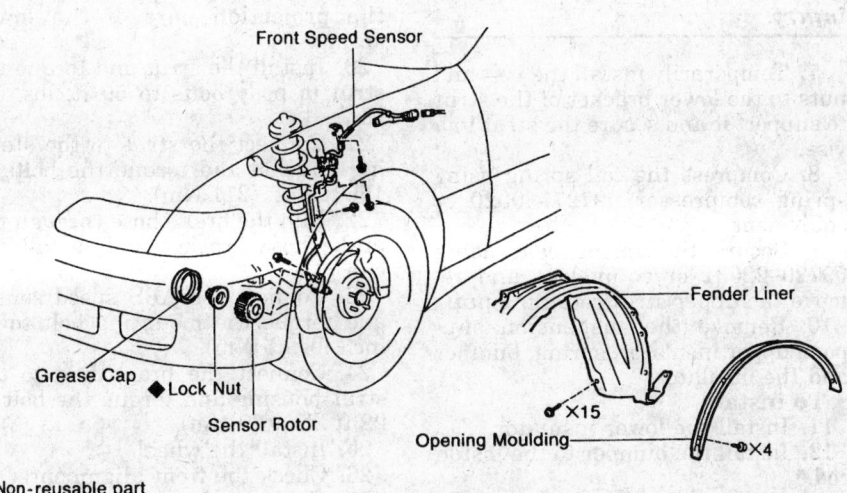

Front speed sensor components for removal and installation — GS300 and LS400

◆ **Non-reusable part**

333217

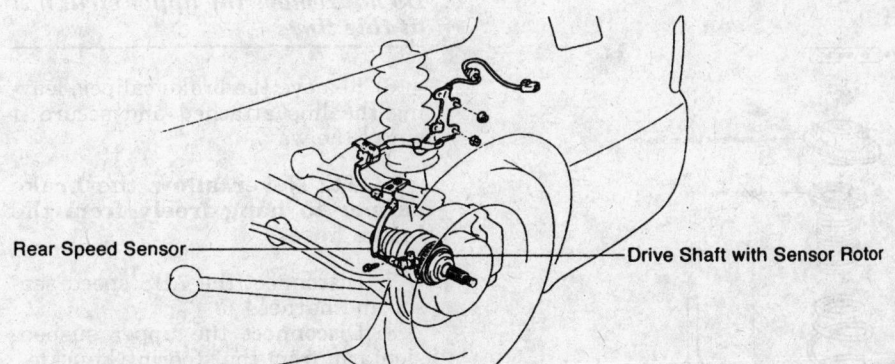

Rear speed sensor components for removal and installation — GS300 and LS400

333218

11. Install the wheel, safely lower the vehicle and reconnect the negative battery cable.

SC300 and SC400

1. Disconnect the negative battery cable.

2. Raise and safely support the vehicle. Remove the rear wheel.

3. Remove the side cover of the luggage compartment.

4. Disconnect the speed sensor connector and pull out the sensor wire harness with the grommet.

5. Remove the 2 clamp bolts holding the sensor wire harness to the body and the upper suspension arm.

6. Remove the bolt and the speed sensor from the axle carrier.

To install:

7. Install the speed sensor to the axle carrier with the bolt and torque to 69 inch lbs. (8 Nm).

8. Install the sensor wire harness clamp to the upper suspension arm and the body with the 2 clamp bolts and torque to 48 inch lbs. (5 Nm).

9. Reinstall the sensor wire harness and grommet and reconnect the speed sensor connector.

10. Reinstall the side cover of the luggage compartment.

11. Reinstall the wheel, safely lower the vehicle and reconnect the negative battery cable.

FRONT SUSPENSION

Strut and Coil Spring

REMOVAL AND INSTALLATION

ES300

1. Disconnect the negative battery cable from the battery.

2. Raise and safely support the vehicle.

3. Remove the tire and wheel assembly.

4. If equipped with ABS, disconnect the ABS speed sensor connector and brake line from the strut housing. Do not remove the brake line from the brake caliper.

5. Disconnect the strut assembly from the steering knuckle by removing the two nuts and bolts.

6. Remove the three upper mounting nuts from the strut tower and remove the strut assembly.

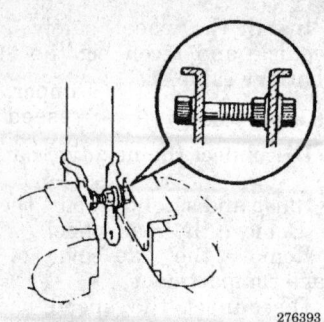

Temporarily install the support nuts and bolt to the strut — 1993–94 ES300

276393

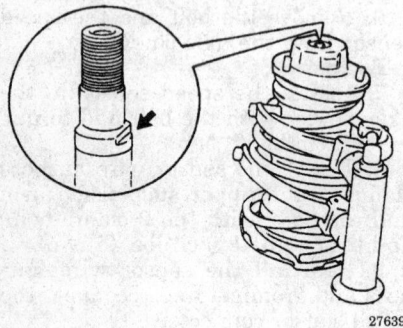

276391

Aligning the upper support to the strut rod — 1993–94 ES300 and LS400

CAUTION

Do not remove the center nut to the strut at this time. The spring on the strut is under high pressure and can cause serious injury.

7. Temporarily install the bolt and nuts to the lower bracket of the strut to support it and secure the strut in a vise.

8. Compress the coil spring using spring compressor 09727–30020 or equivalent.

9. Secure the spring seat using 09729–22031 or equivalent and remove the upper strut retaining nut.

10. Remove the suspension support, upper insulator, spring, bumper and the insulator.

To install:

11. Install the lower insulator.

12. Install the bumper to the piston rod.

13. Align the (compressed) coil spring end into the gap of the lower seat.

14. Install the upper insulator.

15. Install the upper support to the piston rod, aligning it with the groove in the strut rod.

16. Secure the spring seat and torque a new upper strut retaining nut to 36 ft. lbs. (49 Nm).

17. Remove the spring compressor.

18. Remove the strut from the vise and disassemble the securing nuts and bolt.

19. Rotate the upper support so the lowest bolt on the support aligns with the projection part of the lower spring.

20. Install the strut and torque the strut to body bolts to 59 ft. lbs. (80 Nm).

21. Connect the strut to the steering knuckle and torque the bolts to 156 ft. lbs. (211 Nm).

22. Run the brake hose through the brake hose bracket and install the clip.

23. Connect the ABS speed sensor and torque the mounting bolt to 48 inch lbs. (5 Nm).

24. Connect the brake line to the strut housing and torque the bolt to 22 ft. lbs. (29 Nm).

25. Install the wheel.

26. Check the front alignment.

27. Connect the negative battery cable.

GS300

1. Disconnect the negative battery cable.

2. Raise and support the vehicle safely.

3. Remove the front wheel.

4. Loosen the three upper strut mounting nuts.

5. Loosen, but do not remove, the upper strut rod nut.

CAUTION

Do not remove the upper strut nut at this time.

6. Remove the brake caliper, leaving the line attached and secure it out of the way.

NOTE: Never allow the brake caliper to hang freely from the brake hose.

7. Disconnect the ABS speed sensor and harness.

8. Disconnect the upper suspension arm from the steering knuckle.

9. Disconnect the stabilizer bar from the link and remove the bracket.

10. Disconnect the strut from the lower suspension arm.

11. Remove the three upper strut mounting nuts and remove the strut.

12. Using a spring compressor, 09727–30020 or equivalent, compress the coil spring.

13. Remove the piston rod locknut.

14. Remove the suspension support, coil spring and bumper.

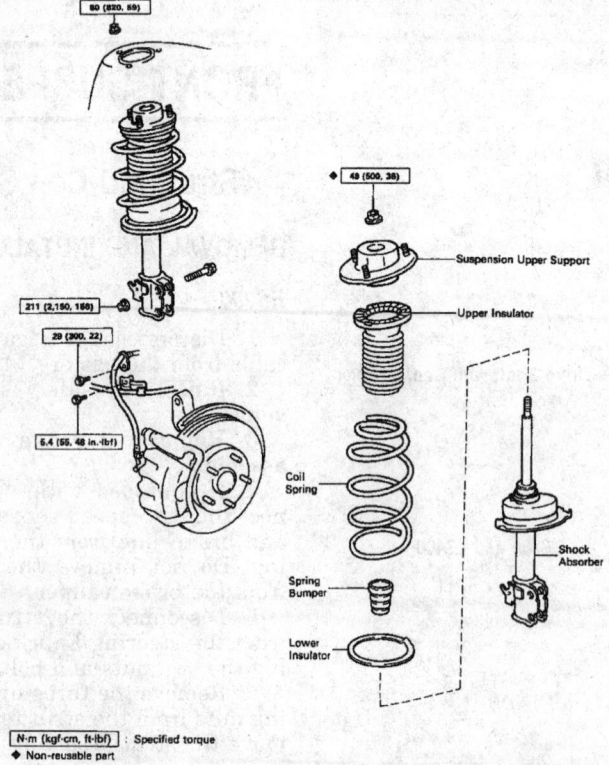

80 (820, 59)

49 (500, 36)

211 (2,150, 156)

29 (300, 22)

5.4 (55, 48 in.·lbf)

Suspension Upper Support

Upper Insulator

Coil Spring

Spring Bumper

Lower Insulator

Shock Absorber

N·m (kgf·cm, ft·lbf) : Specified torque
◆ Non-reusable part

276390

Exploded view of the strut and coil spring — 1993–94 ES300

To install:

15. If disposing the strut, perform the following procedure:
    a. Fully extend the strut rod.
    b. Drill a hole in the body of the shock to remove the gas inside.

NOTE: The gas is harmless, but be careful of chips which may fly up when the gas is released.

16. Match the bolt of the suspension support with the cut out portion of the insulator.
17. Install the spring bumper.
18. Install the compressed coil spring. Match the end of the coil into the recess of the strut spring seat.
19. Install the suspension support to the rod and temporarily install a new nut.
20. Turn the suspension support so one of the bolts on the support faces the same direction as shown in the illustration.

NOTE: Align the bolt so a line drawn between the rod and bolt would be at 90° to the direction of the lower bushing.

21. Remove the spring compressor.
22. Install the strut and torque the upper retaining nuts to 41 ft. lbs. (56 Nm).

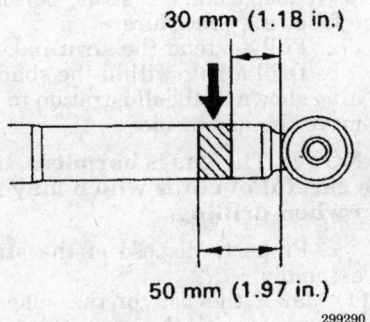

30 mm (1.18 in.)

50 mm (1.97 in.)

299290

Drill a hole to release the gas from the strut

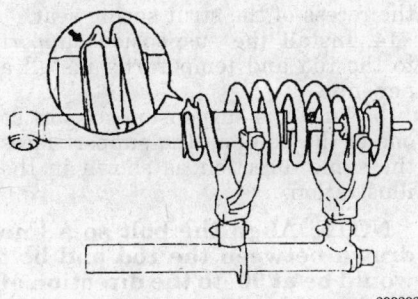

299292

Matching the spring to the seat

23. Torque the new upper strut rod nut to 20 ft. lbs. (27 Nm).
24. Connect the strut to the lower arm and temporarily tighten the nut and bolt.
25. Install the stabilizer bar bracket and torque the bolts to 21 ft. lbs. (28 Nm).
26. Connect the stabilizer bar to the link and torque the bolts to 29 ft. lbs. (39 Nm).
27. Connect the upper suspension arm to the steering knuckle. Torque the nut to 64 ft. lbs. (87 Nm) and install a new cotter pin.
28. Install the ABS speed sensor and torque the bolt to 69 inch lbs. (8 Nm).
29. Install the caliper.
30. Install the wheel.
31. Lower the vehicle.
32. Bounce the vehicle several times to stabilize the suspension.
33. Torque the lower strut bolt and nut to 116 ft. lbs. (157 Nm).
34. Check the front wheel alignment.

1993–94 LS400

Except Air Suspension

1. Raise and safely support the vehicle.
2. Remove the tire and wheel assembly.
3. Remove the steering knuckle from the upper ball joint with the proper tool. Support the steering knuckle using a piece of wire.
4. Disconnect the strut assembly from the lower strut bracket. Remove the plug from the upper strut mount.
5. Loosen the nut on the middle of the strut mount support. Do not remove it.
6. Remove the other 3 mounting nuts and remove the strut assembly with the coil spring from the vehicle.
7. Using compressor 09727–30020 or equivalent, compress the coil spring.
8. Remove the piston rod locknut.
9. Remove the suspension support, coil spring and bumper.

To install:
10. If disposing the strut, perform the following procedure:
    a. Fully extend the strut rod.
    b. Drill a hole within the shaded area shown in the illustration to remove the gas inside.

NOTE: The gas is harmless, but be careful of chips which may fly up when drilling.

    c. Properly dispose of the strut assembly.

11. Match the bolt of the suspension support with the cut out portion of the insulator.
12. Install the spring bumper.
13. Install the compressed coil spring. Match the end of the coil into the recess of the strut spring seat.
14. Install the suspension support to the rod and temporarily install a new nut.
15. Turn the suspension support so one of the bolts on the support faces the same direction as shown in the illustration.

NOTE: Align the bolt so a line drawn between the rod and bolt would be at 90° to the direction of the lower bushing.

16. Remove the spring compressor.
17. Install the strut and torque the upper retaining nuts to 27 ft. lbs. (36 Nm).
18. Torque a new strut rod nut to 20 ft. lbs. (27 Nm) and install the cap.
19. Connect the strut to the lower bracket and temporarily install the nut and bolt.
20. Connect the upper suspension arm to the steering knuckle. Torque the nut to 48 ft. lbs. (65 Nm) and install a new cotter pin.
21. Install the wheel.
22. Lower the vehicle.
23. Bounce the vehicle several times to stabilize the suspension.
24. Torque the lower strut bolt and nut to 106 ft. lbs. (143 Nm).
25. Check the front wheel alignment.

With Air Suspension

1. Move the height control switch to OFF.
2. Raise and support the vehicle safely.
3. Remove the wheel.
4. Disconnect the upper ball joint from the steering knuckle. Suspend the knuckle with wire to prevent excessive force on the brake line and ABS speed sensor.
5. Remove the height control sensor link from the lower strut bracket.
6. Remove the strut from the lower bracket.
7. Disconnect the air tube from the strut.
8. Remove the nuts and suspension control actuator.
9. Remove the suspension control actuator and position it aside.
10. Remove the 3 upper mounting nuts and strut from the vehicle.

To install:
11. If disposing the strut perform the following procedure:
    a. Using a screwdriver, remove the air from inside the cylinder.

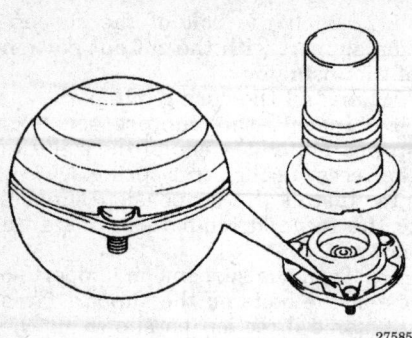

Aligning the insulator to the support-except air suspension

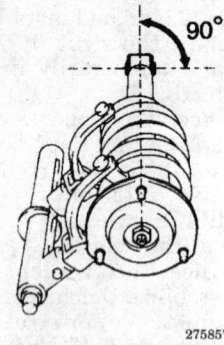

Aligning the suspension support to the strut housing-except air suspension — 1993-94 LS400

b. Fully extend the cylinder.

c. Drill a hole in the cylinder in the area shown to remove the gas inside.

NOTE: The gas coming out is harmless, but be careful of chips which may fly up while drilling.

12. Install the strut and torque the upper mounting nuts to 27 ft. lbs. (36 Nm).

13. Match the rods of the strut with the holes in the suspension control actuator.

14. Install the suspension control actuator and tighten the bolts.

15. Install the suspension control actuator cover and torque the nuts to 27 ft. lbs. (36 Nm).

16. Install 2 new O-rings to the air tube. Install the tube and torque it to 13 ft. lbs. (17 Nm). Install the grommet.

17. Install the strut to the lower strut bracket and temporarily install the nut and bolt.

18. Connect the height control sensor link and torque a new nut to 48 inch lbs. (5 Nm).

19. Connect the steering knuckle to the upper ball joint. Torque the nut to 48 ft. lbs. (65 Nm) and install a new cotter pin.

20. Install the wheel.

21. Lower the vehicle.

22. Turn the height control switch ON.

23. Start the engine to fill the strut with air.

24. Bounce the vehicle several times to normalize the suspension.

25. Support the lower control arm with a jack.

26. Remove the front wheel.

27. Torque the lower strut mounting nut and bolt to 76 ft. lbs. (106 Nm).

28. Install the wheel.

29. Lower the vehicle.

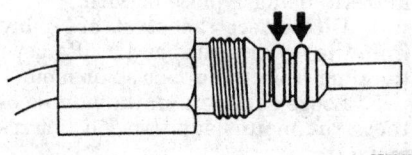

Air tube O-ring installation on air suspension vehicles

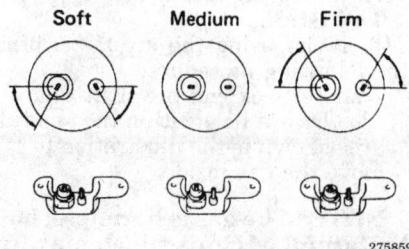

Matching the strut rods to the suspension control actuator

30. Check the front wheel alignment.

1995-97 LS400

Except Air Suspension

1. Raise and safely support the vehicle.

2. Remove the tire and wheel assembly.

3. Remove the steering knuckle from the upper ball joint with the proper tool. Support the steering knuckle using a piece of wire.

4. Disconnect the strut assembly from the lower strut bracket.

5. Remove the strut cover from the upper strut mount.

6. Remove the three mounting nuts and remove the strut assembly with the coil spring from the vehicle.

— **CAUTION** —
Do not remove the center nut to the strut at this time. The spring on the strut is under high pressure and can cause serious injury.

7. Using compressor 09727-30020 or equivalent, compress the coil spring.

8. Remove the piston rod locknut.

9. Remove the suspension support, coil spring and the bumper.

To install:

10. If disposing the strut, perform the following procedure:

a. Fully extend the strut rod.

b. Drill a hole within the shaded area shown in the illustration to remove the gas inside.

NOTE: The gas is harmless, but be careful of chips which may fly up when drilling.

c. Properly dispose of the strut assembly.

11. Match the bolt of the suspension support with the cut out portion of the insulator.

12. Install the spring bumper.

13. Install the compressed coil spring. Match the end of the coil into the recess of the strut spring seat.

14. Install the suspension support to the rod and temporarily install a new nut.

15. Turn the suspension support so one of the bolts on the support faces the same direction as shown in the illustration.

NOTE: Align the bolt so a line drawn between the rod and bolt would be at 90° to the direction of the lower bushing.

16. Torque the strut rod nut to 20 ft. lbs. (27 Nm) and install the cap.

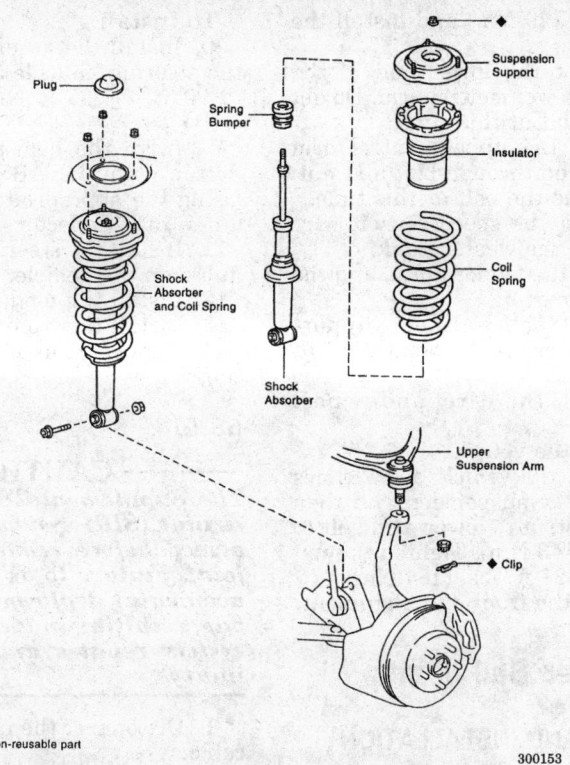

◆ Non-reusable part

300153

Exploded view of the strut and spring removal and installation procedure — 1995–97 LS400-except air suspension

17. Remove the spring compressor.

18. Install the strut and torque the upper retaining nuts to 43 ft. lbs. (58 Nm).

19. Connect the strut to the lower bracket and temporarily install the nut and bolt.

20. Connect the upper control arm to the steering knuckle. Torque the nut to 48 ft. lbs. (65 Nm) and install a new cotter pin.

21. Install the wheel.

22. Lower the vehicle.

23. Bounce the vehicle several times to stabilize the suspension.

24. Torque the lower strut bolt and nut to 116 ft. lbs. (157 Nm).

25. Check the front wheel alignment.

With Air Suspension

1. Move the height control switch to OFF.

2. Raise and support the vehicle safely.

3. Remove the wheel.

4. Bleed the air in the suspension.

5. Remove the height control sensor link from the lower strut bracket.

6. Remove the cotter pin and nut holding the upper control arm to the steering knuckle.

7. Disconnect the upper ball joint from the steering knuckle. Suspend the knuckle with wire to prevent excessive force on the brake line and ABS speed sensor.

8. Disconnect the pneumatic cylinder from the lower bracket by removing the through-bolt.

9. Disconnect the air tube from the strut.

10. Remove the three nuts holding the actuator cover to the strut tower. Remove the actuator cover.

CAUTION
Do not remove the center nut from the pneumatic cylinder.

11. Disconnect the actuator electrical connector.

12. Remove the two bolts to the suspension control actuator and position the actuator aside.

13. Remove the three upper mounting nuts and the strut from the vehicle.

To install:

14. If disposing the strut perform the following procedure:

   a. Using a screwdriver, remove the air from inside the cylinder.

   b. Fully extend the cylinder.

   c. Drill a hole in the cylinder at a point above 1.57 in. (40mm) from the bottom of the strut assembly. This will release the gas charge in the strut. Do not puncture the pneumatic cylinder.

NOTE: The gas coming out is harmless, but be careful of chips which may fly up while drilling.

15. Install the strut and torque the upper mounting nuts to 43 ft. lbs. (58 Nm).

16. Match the rods of the strut with the holes in the suspension control actuator.

17. Install the suspension control actuator and tighten the bolts. Torque the two nuts to 13 ft. lbs. (17 Nm).

18. Install the suspension control actuator cover and torque the nuts to 43 ft. lbs. (58 Nm).

19. Install two new O-rings to the air tube. Install the tube and torque it to 13 ft. lbs. (17 Nm). Install the grommet.

20. Install the strut to the lower strut bracket and temporarily install the nut and bolt.

21. Connect the steering knuckle to the upper ball joint. Torque the nut to 48 ft. lbs. (65 Nm) and install a new cotter pin.

22. Connect the height control sensor link and torque a new nut to 48 inch lbs. (5 Nm).

23. Install the wheel.

24. Lower the vehicle.

25. Turn the height control switch ON.

26. Start the engine to fill the strut with air.

27. Bounce the vehicle several times to normalize the suspension.

28. Support the lower control arm with a jack.

29. Remove the front wheel.

30. Torque the lower strut mounting nut and bolt to 76 ft. lbs. (106 Nm).

31. Install the wheel.

32. Lower the vehicle.

33. Check the front end alignment.

SC300 and SC400

1. Raise and safely support the vehicle.

2. Remove the tire and wheel assembly.

3. Remove the brake caliper support bracket by removing the two bolts. Suspend it with a piece of wire.

4. Remove the fender apron, engine undercover and the front fender wheel opening molding.

5. If removing the left side strut, disconnect the windshield washer tank.

6. Remove the bolt and disconnect the ABS speed sensor at the steering knuckle. Remove the three bolts and

then disconnect the wire harness clamp in order to prevent the harness from being damaged when removing the through-bolt.

7. Remove the plug from the upper strut mount. Do not remove the center bolt.

CAUTION

Do not remove the center bolt to the strut at this time. The spring on the strut is under high pressure and can cause serious injury or vehicle damage.

8. Disconnect the upper control arm through-bolt from the subframe. Disconnect the upper control arm and turn the control arm completely around. It is not necessary to remove the upper ball joint.

9. Disconnect the strut at the lower control arm by removing the nut and bolt.

10. Remove the three upper mounting nuts and remove the strut assembly with the coil spring from the vehicle.

11. Using compressor 09727–30020 or equivalent, compress the coil spring.

12. Remove the piston rod locknut.

13. Remove the suspension support, coil spring and bumper.

To install:

14. If disposing the strut, perform the following procedure:
   a. Fully extend the strut rod.
   b. Drill a hole within the shaded area shown in the illustration to remove the gas inside and dispose the old strut.

NOTE: The gas is harmless, but be careful of chips which may fly up when drilling.

15. Match the bolt of the suspension support with the cut out portion of the insulator.

16. Install the spring bumper.

17. Install the compressed coil spring. Match the end of the coil into the recess of the strut spring seat.

18. Install the suspension support to the rod and temporarily install a new nut.

19. Turn the suspension support so one of the bolts on the support faces the same direction as shown in the illustration.

NOTE: Align the bolt so a line drawn between the rod and bolt would be at 90° to the direction of the lower bushing.

20. Remove the spring compressor.

21. Install the strut and tighten the three upper strut mount nuts to 26 ft. lbs. (35 Nm). Tighten the middle nut

to 22 ft. lbs. (29 Nm) and install the plug.

22. Connect the lower end of the strut to the lower control arm. Do not torque the bolt at this time.

23. Install the upper control arm and install the through-bolt and nut. Do not torque the bolt at this time.

24. Connect the speed sensor, wire harness and the washer tank.

25. Install the fender apron and the engine undercover.

26. Install the caliper support bracket and torque the bolts to 87 ft. lbs. (118 Nm).

27. Install the tire and wheel assembly.

28. Lower the vehicle.

29. Bounce the vehicle a few times to stabilize the suspension and then tighten the strut to lower arm bolt to 106 ft. lbs. (143 Nm). Tighten the upper arm to 121 ft. lbs. (164 Nm).

30. Check the front end alignment.

Upper Ball Joints

REMOVAL AND INSTALLATION

The upper ball joint is an integral part of the upper arm and is not replaced separately. The upper ball joint replacement is accomplished by replacing the upper arm.

Lower Ball Joints

REMOVAL AND INSTALLATION

ES300

CAUTION

The Supplemental Inflatable Restraint (SIR) system must be disarmed before removing the ball joint. Failure to do so may cause accidental deployment of the air bag, resulting in unnecessary SIR system repairs and/or personal injury.

1. Disconnect the negative battery cable.

2. Raise the front of the vehicle and support it safely.

3. Remove the front wheel(s).

4. Remove side fender apron seal.

5. Remove the steering knuckle with the axle hub, from the vehicle.

6. Pry the dust deflector from the knuckle.

7. Remove the cotter pin and the nut from the ball joint stud.

8. Using SST 09628–62011 or equivalent, remove the lower ball joint from the steering knuckle.

To install:

9. Install the lower ball joint onto the steering knuckle and tighten nut to 90 ft. lbs. (123 Nm). Install new cotter pin.

10. Align the hole in the dust deflector with the ABS speed sensor. Using the appropriate driver, install a new dust deflector.

11. Install the steering knuckle and hub onto the vehicle.

12. Install the fender apron seal.

13. Install the front wheel(s).

14. Connect the negative battery cable.

GS300

CAUTION

The Supplemental Inflatable Restraint (SIR) system must be disarmed before removing the ball joint. Failure to do so may cause accidental deployment of the air bag, resulting in unnecessary SIR system repairs and/or personal injury.

1. Disconnect the negative battery cable.

2. Raise and support the vehicle safely.

3. Remove the wheel(s).

4. Remove the caliper, leaving the brake line connected and suspend it out of the way.

NOTE: Never allow the brake caliper to hang freely from the brake hose.

5. Remove the rotor.

6. Remove the ABS speed sensor and harness.

7. Disconnect the tie rod end from the arm on the lower ball joint.

8. Remove the cotter pin and nut. Using SST 09610–20012 or equivalent, disconnect the upper control arm from the steering knuckle.

9. Remove the cotter pin and nut. Using SST 09628–62011 or equivalent, disconnect the steering knuckle from the lower control arm.

10. Remove the steering knuckle and ball joint assembly from the vehicle.

11. Remove the two ball joint mounting bolts and then remove the ball joint from the steering knuckle.

To install:

12. Install the ball joint and torque the bolts to 83 ft. lbs. (113 Nm).

13. Connect the steering knuckle to the lower and upper suspension arms. Torque the lower control arm nut to 95 ft. lbs. (127 Nm) and install a new cotter pin. Torque the upper control arm to 64 ft. lbs. (87 Nm) and install a new cotter pin.

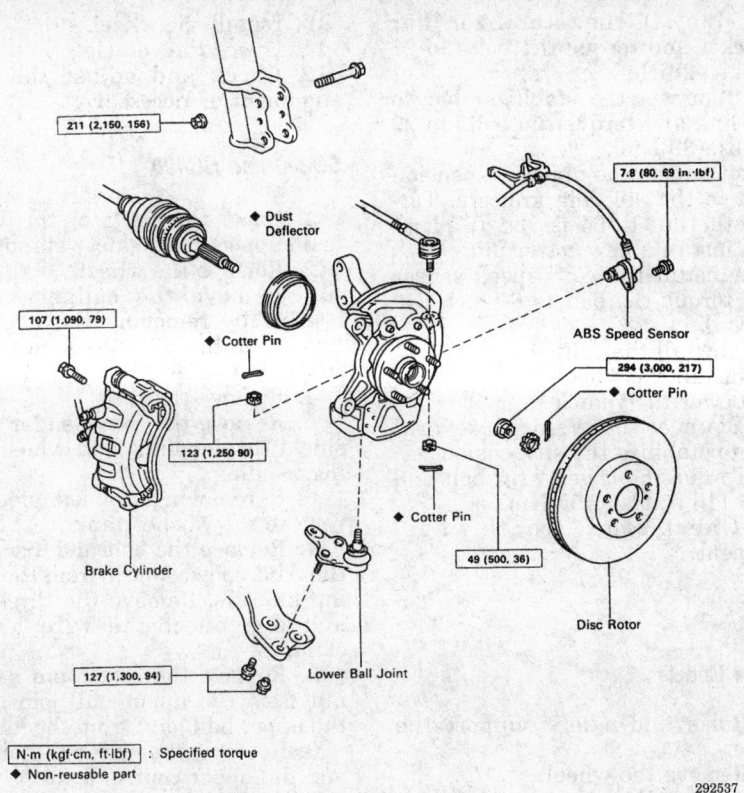

211 (2,150, 156)

♦ Dust Deflector

7.8 (80, 69 in.·lbf)

107 (1,090, 79)

♦ Cotter Pin

ABS Speed Sensor

294 (3,000, 217)

♦ Cotter Pin

123 (1,250 90)

Brake Cylinder

♦ Cotter Pin

49 (500, 36)

127 (1,300, 94)

Lower Ball Joint

Disc Rotor

N·m (kgf·cm, ft·lbf) : Specified torque
♦ Non-reusable part

292537

Exploded view of the lower suspension — ES300

14. Connect the tie rod end to the ball joint arm. Torque the nut to 64 ft. lbs. (87 Nm) and install a new cotter pin.
15. Install the rotor.
16. Install the caliper.
17. Install the ABS speed sensor and harness. Torque the sensor retaining bolt to 69 inch lbs. (8 Nm).
18. Install the wheel(s).
19. Lower the vehicle and connect the negative battery cable.
20. Connect the negative battery cable.
21. Check the front wheel alignment.

LS400

1. If equipped with air suspension, move the height control switch (located in the trunk) to the OFF position.
2. Raise and safely support the vehicle.
3. Remove the tire and wheel assembly.
4. Remove the ABS speed sensor and wire harness from the steering knuckle.
5. Disconnect the brake caliper support bracket by removing the two bolts. Leave the brake line connected. Support the caliper aside by using a piece of wire.

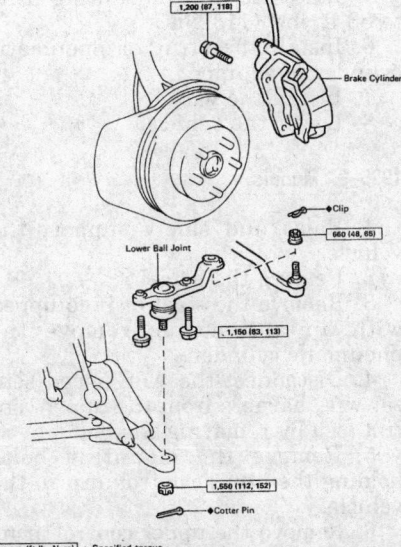

1,200 (87, 118)

Brake Cylinder

♦ Clip

660 (48, 65)

Lower Ball Joint

1,150 (83, 113)

1,550 (112, 152)

♦ Cotter Pin

kg-cm (ft-lb, N·m) : Specified torque
♦ Non-reusable part

300459

Exploded view of the lower ball joint removal and installation procedure — LS400

6. Loosen the two lower ball joint mounting bolts.

NOTE: Do not remove the bolts.

7. Remove the clip and nut from the tie rod end.
8. Disconnect the tie rod end from the steering arm with the proper tool.
9. Remove the lower ball joint mounting bolts from the steering knuckle.
10. Remove the cotter pin and nut from the lower ball joint.
11. Using SST 09628–62011 or equivalent, disconnect the lower ball joint from the lower control arm.
To install:
12. Install the ball joint to the lower control arm. Torque the nut to 112 ft. lbs. (152 Nm) and install a new cotter pin.
13. Temporarily tighten the mounting bolts holding the ball joint to the steering knuckle.
14. Connect the tie rod end to the steering knuckle. Torque the nut to 48 ft. lbs. (65 Nm) and install a new cotter pin.
15. Torque the lower ball joint bolts to 83 ft. lbs. (113 Nm).
16. Install the brake caliper support bracket and torque the two bolts to 87 ft. lbs. (118 Nm).
17. Install the ABS speed sensor and wire harness to the steering knuckle.
18. Install the wheel.
19. Lower the vehicle.
20. Turn the height control switch ON.

SC300 and SC400

The lower ball joint is not replaceable. If the lower ball joint is defective, replace the lower arm and ball joint as an assembly.

Upper Control Arms

REMOVAL AND INSTALLATION

GS300

1. Disconnect the negative battery cable from the battery.
2. Raise and safely support the vehicle.
3. Remove the wheel.
4. Remove the strut and coil spring assembly as follows:
   a. Loosen the three upper strut mounting nuts.
   b. Loosen, but do not remove, the upper strut rod nut.

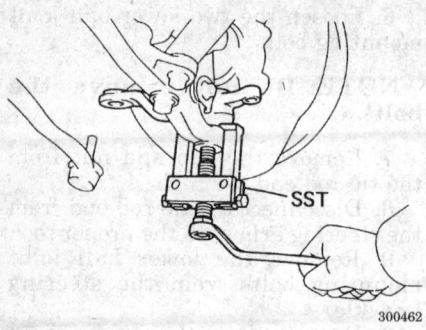

Disconnecting the ball joint from the lower suspension arm — LS400

300462

—————— CAUTION ——————
DO NOT completely remove the upper strut nut at this time.

c. Remove the brake caliper, leaving the line attached and secure it out of the way.

NOTE: Never allow the brake caliper to hang freely from the brake hose.

d. Disconnect the ABS speed sensor and harness.
e. Remove the cotter pin and nut from the upper control arm.
f. Using SST 09610–20010 or equivalent, disconnect the upper control arm from the steering knuckle.
g. Disconnect the stabilizer bar from the link and remove the bracket.
h. Remove the cotter pin and nut from the lower control arm.
i. Disconnect the strut from the lower suspension arm.
j. Remove the three upper strut mounting nuts and remove the strut.
5. Remove the mounting bolts holding the upper control arm to the frame.
6. Remove the upper control arm from the vehicle.
To install:
7. Install the upper suspension arm and tighten the mounting bolts to 39 ft. lbs. (53 Nm).
8. Install the strut and spring assembly as follows:
a. Install the strut and torque the upper retaining nuts to 41 ft. lbs. (56 Nm).
b. Torque the new upper strut rod nut to 20 ft. lbs. (27 Nm).
c. Connect the strut to the lower arm and temporarily tighten the nut and bolt.

d. Install the stabilizer bar bracket and torque the bolts to 21 ft. lbs. (28 Nm).
e. Connect the stabilizer bar to the link and torque the bolts to 29 ft. lbs. (39 Nm).
f. Connect the upper suspension arm to the steering knuckle. Torque the nut to 64 ft. lbs. (87 Nm) and install a new cotter pin.
g. Install the ABS speed sensor and torque the bolt to 69 inch lbs. (8 Nm).
h. Install the caliper.
9. Install the wheel.
10. Lower the vehicle.
11. Bounce the vehicle several times to stabilize the suspension.
12. Torque the lower strut bolt and nut to 116 ft. lbs. (157 Nm).
13. Check the front wheel alignment.

LS400

1993–94 Models

1. Raise and safely support the vehicle.
2. Remove the wheel.
3. Remove the strut or pneumatic cylinder, if equipped.
4. Remove the mounting bolts and the upper suspension arm.
To install:
5. Install the upper suspension arm and tighten the mounting bolts to 83 ft. lbs. (113 Nm).
6. Install the strut or pneumatic cylinder, if equipped.
7. Install the wheel.
8. Lower the vehicle.

1995–97 Models

1. Raise and safely support the vehicle.
2. Remove the wheel.
3. Remove the strut or if equipped with air suspension, remove the pneumatic cylinder.
4. Disconnect the ABS speed sensor wire harness from the upper control arm by removing the bolt.
5. Remove the mounting bolts holding the upper control arm to the vehicle.
6. Remove the upper control arm.
To install:
7. Install the upper control arm and torque the two mounting bolts to 83 ft. lbs. (113 Nm).
8. Connect the ABS speed sensor wire harness to the upper control arm with the attaching bolt.
9. Install the strut or if equipped with air suspension, install the pneumatic cylinder.

10. Install the wheel.
11. Lower the vehicle.
12. Check and adjust the wheel alignment as necessary.

SC300 and SC400

1. Raise the front of the vehicle and support it on safety stands.
2. Remove the wheel.
3. Remove the caliper support bracket by removing the two bolts. Leave the brake line connected and suspend it aside.
4. Remove the rotor.
5. Remove the front fender splash shield, fender liner and wheel opening molding.
6. If removing the left side arm, remove the washer tank.
7. Remove the bolt and disconnect the ABS speed sensor from the steering knuckle. Remove the three bolts and disconnect the wire harness clamp.
8. Remove the cotter pin and the nut from the upper ball joint; press the upper ball joint from the knuckle.
9. Remove the through-bolt, nut and the upper control arm.
To install:
10. Install the upper control arm. Connect the upper control arm to the subframe and install the through-bolt. Do not torque the bolt at this time.

NOTE: The upper control arm mounting bolts are not torqued until the suspension has been assembled and vehicle is on the ground.

11. Install the ball joint to the knuckle and tighten the nut to 76 ft. lbs. (103 Nm). Install a new cotter pin.
12. Connect the wire harness and ABS speed sensor. Torque the speed sensor to knuckle bolt to 69 inch lbs. (8 Nm).
13. Install the washer tank, the fender liner, splash shield and molding.
14. Install the rotor.
15. Install the caliper support bracket and torque the bolts to 87 ft. lbs. (118 Nm).
16. Install the wheel.
17. Lower the vehicle.
18. Bounce the suspension several times to set the suspension.
19. Support the lower arm and tighten the upper control arm through-bolt and nut to 121 ft. lbs. (164 Nm).
20. Check the front wheel alignment and adjust as necessary.

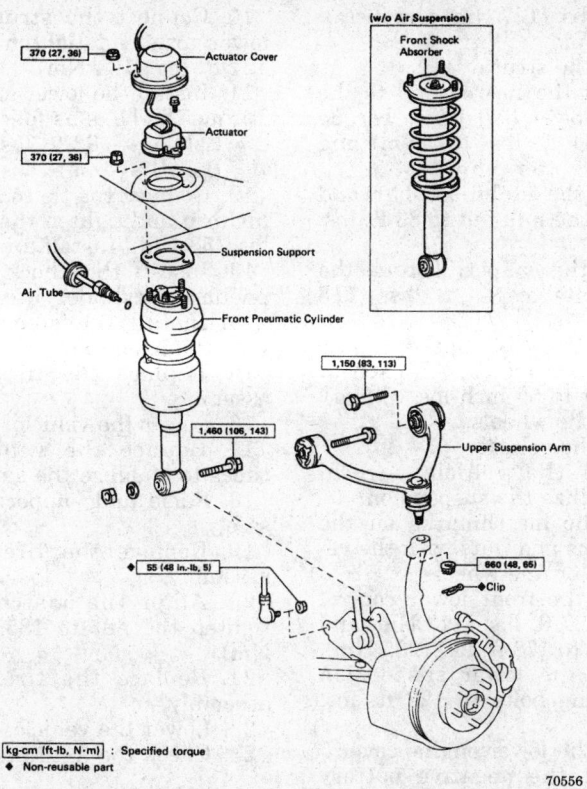

370 (27, 36) — Actuator Cover
Actuator
370 (27, 36)
Suspension Support
Air Tube
Front Pneumatic Cylinder
1,150 (83, 113)
1,460 (106, 143)
Upper Suspension Arm
55 (48 in.-lb, 5)
660 (48, 65)
Clip

(w/o Air Suspension)
Front Shock Absorber

kg-cm (ft-lb, N·m) : Specified torque
◆ Non-reusable part

70556

Upper control arm and related components — 1993–94 LS400

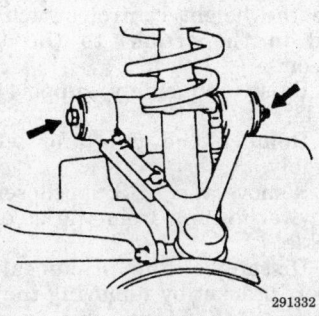

291332

Torque the upper control arm bolts — SC300 and SC400

Lower Control Arms

REMOVAL AND INSTALLATION

ES300

— **CAUTION** —
The air bag system (SRS or SIR) must be disarmed before removing the lower control arm. Failure to do so may cause accidental deployment, property damage or personal injury.

1. Disconnect the negative battery cable.
2. Raise the vehicle and support it safely.
3. Remove the front wheel(s).
4. Remove the fender apron seal.
5. Remove the cotter pin and lock cap from the halfshaft.
6. Have an assistant apply the brakes firmly; with the front brakes locked, remove the halfshaft locknut.
7. Disconnect and separate the tie rod end from the steering knuckle.
8. Remove the left and right stabilizer bar end brackets from the lower control arms.
9. Remove the bolt and two nuts and disconnect the lower arm from the ball joint.
10. Remove the halfshaft from the axle hub. Secure the shaft out of the way using wire.

— **WARNING** —
Be careful not to damage the shaft joint boot or ABS sensor rotor.

11. Remove the bolts from the front side of the lower control arm.
12. Remove the bolts and nuts from the rear side of the lower control arm and remove control arm from the vehicle.

To install:
13. Place the lower control arm onto the vehicle and temporarily install the rear mounting nut and bolt.
14. Install the lower control arm bushing stopper to the lower control arm shaft. Install the bolts on the front side of the control arm and tighten to 152 ft. lbs. (206 Nm).
15. Tighten the bolts on the rear side of the lower control arm to 152 ft. lbs. (206 Nm).
16. Install the halfshaft to the axle hub.
17. Connect the lower control arm to the lower ball joint and tighten the fasteners to 94 ft. lbs. (127 Nm).
18. Install both side stabilizer end brackets to the lower control arm and tighten to 41 ft. lbs. (56 Nm).
19. Connect the tie rod end to the steering knuckle and tighten nut to 36 ft. lbs. (49 Nm). Install new cotter pin.
20. Have a helper apply the brakes and install the halfshaft locknut and tighten it to 217 ft. lbs. (294 Nm). Install the lock cap and a new cotter pin.
21. Install front fender apron seal.
22. Install the front wheel(s).
23. Lower the vehicle to the floor and connect the negative battery cable.

GS300

— **CAUTION** —
The air bag system (SRS or SIR) must be disarmed before removing the lower control arms. Failure to do so may cause accidental deployment, property damage, or personal injury.

1. Disconnect the negative battery cable.
2. Raise and support the vehicle safely.
3. Remove the lower engine cover.
4. Remove the wheels.
5. Remove the caliper without disconnecting the brake line and suspend it with wire out of the way.
6. Remove the ABS speed sensor and wire harness.
7. Remove the stabilizer bar and link.
8. Disconnect the tie rod end from the arm on the lower ball joint.
9. Remove the strut.
10. Remove the ball joint cotter pin and nut and disconnect the lower control arm from the steering knuckle.
11. Remove the steering rack.
12. Matchmark the adjusting cams to the front suspension crossmember.

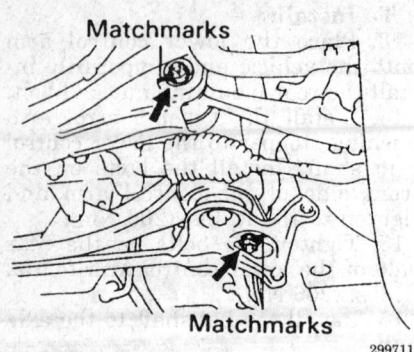

Matchmarking the adjusting cams — GS300

13. Loosen the front suspension crossmember nuts.

14. Disconnect the wire harness from the front suspension crossmember.

15. Remove the front lower suspension crossmember brace bolts.

16. Remove the rear lower suspension crossmember brace bolts.

17. Using a suitable engine brace, suspend the engine.

18. Remove the grommets and nuts. Disconnect the engine mounting from the front suspension crossmember.

19. Supporting the crossmember with a jack, remove the four bolts and lower the front suspension crossmember with the front and rear braces.

20. Remove the two nuts and disconnect the lower control arms from the front suspension crossmember.

21. Remove the two bolts and separate the lower suspension arms.

22. Remove the strut bracket from the lower control arm.

To install:

23. Install the strut bracket to the lower control arm and torque the bolt to 43 ft. lbs. (59 Nm).

24. Connect the two lower control arms with the two bolts.

25. Connect the lower control arms to the suspension crossmember.

26. Install the front suspension crossmember and torque the nuts to 133 ft. lbs. (180 Nm).

27. Connect the engine mounting to the suspension crossmember and torque the bolts to 54 ft. lbs. (74 Nm). Install the grommets.

28. Install the rear lower brace and torque the bolts to 22 ft. lbs. (29 Nm).

29. Install the front lower brace and torque the bolts to 22 ft. lbs. (29 Nm).

30. Connect the wire harness to the front suspension crossmember.

31. Install the steering rack.

32. Connect the lower control arm to the steering knuckle. Torque the

nut to 95 ft. lbs. (127 Nm) and install a new cotter pin.

33. Install the strut.

34. Connect the tie rod end to the arm on the lower ball joint. Torque the nut to 64 ft. lbs. (87 Nm) and install a new cotter pin.

35. Install the stabilizer bar and link. Torque the link nut to 83 ft. lbs. (113 Nm).

36. Install the caliper. Torque the mounting bolts to 87 ft. lbs. (118 Nm).

37. Install the ABS speed sensor and harness. Torque the ABS sensor retaining bolt to 69 inch lbs. (8 Nm).

38. Install the wheels.

39. Lower the vehicle.

40. Bounce the vehicle several times to stabilize the suspension.

41. Align the matchmarks on the adjusting cams and nuts with the vehicles weight on the wheels.

42. Torque the front lower control arm bolt to 127 ft. lbs. (172 Nm) and rear arm bolt to 173 ft. lbs. (235 Nm).

43. Torque the lower suspension arm connecting bolts to 122 ft. lbs. (166 Nm).

44. Install the lower engine cover.

45. Connect the negative battery cable.

46. Check the front wheel alignment.

LS400

1993–94 Models

1. Raise and safely support the vehicle.

2. Remove the tire and wheel assembly.

3. Remove the shock absorber or pneumatic cylinder, if equipped.

4. Disconnect the tie rod end from the steering knuckle with the proper tool.

5. Remove the lower shock bracket.

6. Remove the nuts and disconnect the lower strut bar from the lower arm.

7. Place matchmarks on the camber adjusting cam.

8. Remove the nut, the adjusting cam and lower arm with the lower ball joint.

NOTE: To help remove the adjusting cam, fully pull the steering wheel toward the lower arm being removed.

To install:

9. Insert the camber adjusting cam from the rear side of the vehicle and temporarily tighten the nut. Put 2 strut bar bolts into the holes of the lower arm beforehand.

10. Connect the strut bar to the lower arm and tighten the bolts to 121 ft. lbs. (164 Nm).

11. Install the lower shock bracket. Torque the bolts as follows:
• Bolt A — 83 ft. lbs. (113 Nm)
• Bolt B — 43 ft. lbs. (59 Nm)

12. Connect the tie rod to the steering arm and tighten the nut to 43 ft. lbs. (58 Nm). Install a new clip.

13. Install the shock absorber or pneumatic cylinder, if equipped.

14. Install the steering knuckle with the axle hub.

15. Replace the tire and wheel assembly.

16. Lower the vehicle.

17. Bounce the vehicle several times to stabilize the suspension.

18. Raise and support the vehicle safely.

19. Remove the tire and wheel assembly.

20. Align the matchmarks and tighten the nut to 185 ft. lbs. (250 Nm).

21. Replace the tire and wheel assembly.

22. Lower the vehicle.

23. Check the front end alignment.

1995–97 Models

1. If equipped with air suspension, move the height control switch (located in the trunk) to the OFF position.

2. Raise and safely support the vehicle.

3. Remove the tire and wheel assembly.

4. Remove the ABS speed sensor and wire harness from the steering knuckle.

5. Disconnect the brake caliper support bracket by removing the two bolts. Leave the brake line connected. Support the caliper aside using a piece of wire.

6. Remove the clip and nut from the tie rod end.

7. Disconnect the tie rod end from the steering arm with the proper tool.

8. Remove the cotter pin and nut from the lower ball joint.

9. Using SST 09628–62011 or equivalent, disconnect the lower ball joint from the lower control arm.

10. Remove the shock absorber or if equipped with air suspension, remove the pneumatic cylinder.

11. Remove the lower shock bracket by removing the bolt and the through-bolt.

12. Remove the nuts and disconnect the lower strut bar from the lower control arm.

13. Remove the two nuts and remove the suspension member brace.

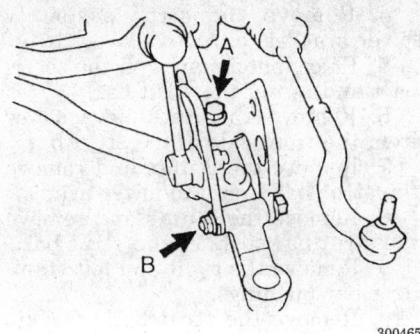

Lower control arm bracket bolts — 1995–97 LS400

14. Place matchmarks on the camber adjusting cam.

15. Remove the nut, the adjusting cam and the lower control arm.

NOTE: To help remove the adjusting cam, fully pull the steering wheel toward the lower arm being removed.

To install:

16. Insert the camber adjusting cam from the front side of the vehicle and install the lower control arm. Temporarily tighten the cam nut.

17. Install the suspension member brace to the vehicle and install the two nuts. Torque the nuts to 29 ft. lbs. (39 Nm).

18. Connect the strut bar to the lower arm and tighten the nuts to 121 ft. lbs. (164 Nm).

19. Install the lower shock bracket. Torque the through-bolt to 83 ft. lbs. (113 Nm) and the top bolt to 43 ft. lbs. (59 Nm).

20. Install the shock absorber or if equipped with air suspension, install pneumatic cylinder.

21. Install the steering knuckle with the ball joint.

22. Connect the lower ball joint to the lower control arm and torque the nut to 112 ft. lbs. (152 Nm). Install a new cotter pin to the ball joint.

23. Connect the tie rod to the steering arm and tighten the nut to 48 ft. lbs. (65 Nm). Install a new clip.

24. Install the brake caliper support bracket and torque the two bolts to 87 ft. lbs. (118 Nm).

25. Install the ABS speed sensor and wire harness to the steering knuckle.

26. Replace the tire and wheel assembly.

27. Lower the vehicle.

28. Turn the height control switch ON.

29. Bounce the vehicle several times to stabilize the suspension.

30. Turn OFF the height control switch.

31. Raise and support the vehicle safely.

32. Remove the tire and wheel assembly.

33. Align the matchmarks on the adjusting cam and tighten the nut to 185 ft. lbs. (250 Nm).

34. Replace the tire and wheel assembly.

35. Lower the vehicle.

36. Turn the height control switch ON.

37. Check the front end alignment.

SC300 and SC400

1. Raise the front of the vehicle and support it on safety stands.

2. Remove the wheel and the engine undercover.

3. Remove the caliper support bracket from the vehicle by removing the two bolts. Support the caliper and bracket with a wire. Do not let the assembly hang from the brake line.

4. Remove the nut and disconnect the stabilizer bar from the lower control arm.

5. Remove the cotter pin and nut from the lower ball joint. Press the lower ball joint out of the steering knuckle.

6. Disconnect the lower end of the strut by removing the nut and bolt.

7. Remove the nut, two bolts and the front lower arm bracket stay.

8. Matchmark the front and rear adjustment cams to the body and then remove the nuts and adjusting cams.

9. Lift out the lower control arm.

10. Remove the bracket from the control arm by removing the two bolts.

To install:

11. Install the bracket to the lower control arm by installing the two bolts. Torque the bolts to 38 ft. lbs. (52 Nm).

12. Install the lower control arm to the body and temporarily install the adjusting cams and nuts. Do not torque the nuts at this time

13. Connect the lower control arm to the knuckle and tighten the ball joint nut to 92 ft. lbs. (125 Nm). Install a new cotter pin.

14. Connect the strut to the arm and tighten the bolt and nut to 106 ft. lbs. (143 Nm).

15. Connect the stabilizer bar link and tighten the nut to 54 ft. lbs. (74 Nm).

16. Install the brake caliper support bracket to the vehicle and torque the bolts to 87 ft. lbs. (118 Nm.)

17. Install the wheel.

18. Lower the vehicle.

19. Bounce it several times to set the suspension.

20. Support the lower arm, align the matchmarks on the adjusting cams and tighten the nuts to 166 ft. lbs. (226 Nm).

21. Check the front end wheel alignment.

Sway Bar

REMOVAL AND INSTALLATION

ES300

1. Raise the front of the vehicle and support it safely. Remove the front wheels.

2. Remove both right and left side fender apron seals.

3. Disconnect the right and left tie rod ends from the steering knuckles.

4. Remove stabilizer bar links (nuts) from each control arm.

5. Remove the right and left sway bar bushing retainers and the bushings.

6. Remove the front exhaust pipe.

7. Remove the rack and pinion mounting bolts and nuts.

8. Lift up the rack and pinion and remove the stabilizer bar from the vehicle.

To install:

9. Lift the rack and pinion and install bar into position.

10. Install the rack and pinion mounting bolts and nuts and torque to 134 ft. lbs. (181 Nm).

11. Install the front exhaust pipe with new gaskets.

12. Install the left and the right stabilizer bar retainers and bushings. Secure the retainers with the retaining bolts. Torque the retainer bolts to 14 ft. lbs. (19 Nm).

13. Install both side stabilizer bar links and torque the nuts to 47 ft. lbs. (64 Nm).

14. Connect both side tie rod ends to the steering knuckles and tighten the nuts to 36 ft. lbs. (49 Nm). Install new cotter pins.

15. Install the left and the right fender apron seals.

16. Install the front wheels.

GS300

1. Disconnect the negative battery cable.

2. Raise and support the vehicle safely.

3. Remove the lower engine cover.

4. Disconnect the sway bar from the sway bar links by removing the two (2) nuts and two (2) bolts.

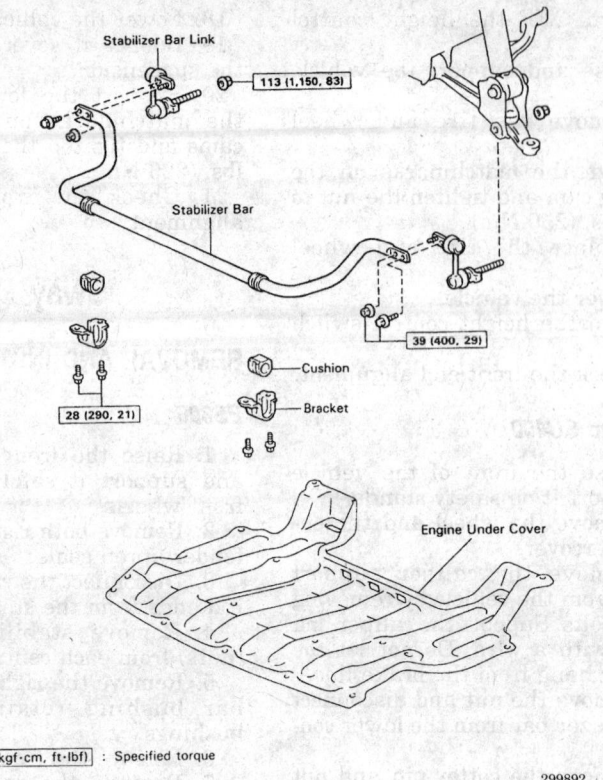

N·m (kgf·cm, ft·lbf) : Specified torque

299892

Sway bar and related components — GS300

5. Disconnect the sway bar brackets from the vehicle body by removing the two (2) bolts at each bracket.

6. Remove the sway bar brackets and cushions from the sway bar.

7. Remove the sway bar from the vehicle.

8. If necessary, remove the nuts connecting the sway bar links to the lower control arm. Remove the links from the lower control arms.

To install:

9. Install the sway bar links to the lower control arm and torque the nuts to 83 ft. lbs. (113 Nm).

10. Install the sway bar and torque the brackets to 21 ft. lbs. (28 Nm).

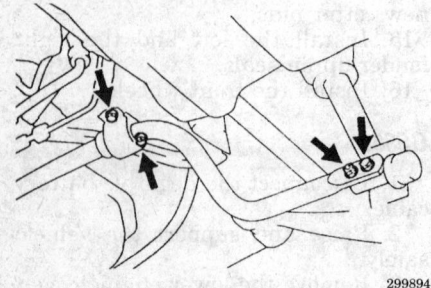

299894

Sway bar mounting bolts — GS300

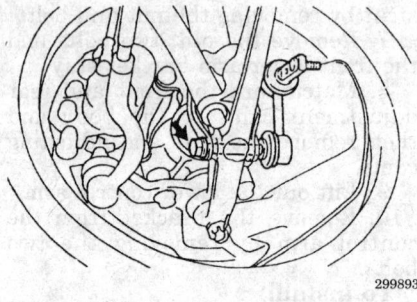

299893

Sway bar link — GS300

11. Connect the sway bar to the links and torque the bolts to 29 ft. lbs. (39 Nm).

12. Install the lower engine cover.

13. Lower the vehicle and connect the negative battery cable.

LS400

1993–94 Models

1. Disconnect the negative battery cable. Raise and support the vehicle safely.

2. Remove the steering knuckle with the axle hub.

3. Remove the strut assembly or the pneumatic cylinder.

4. Remove the strut assembly lower bracket.

5. Place matchmarks on the screw part and nut of the strut bar.

6. Remove the nut and washer from the front side of the strut bar.

7. Remove the 2 nuts and remove the strut bar from the lower arm.

8. Remove the 3 nuts and remove the strut bar cushion and strut bar.

9. Remove the right and left stabilizer bar bushings.

10. Remove the strut bar brackets with the stabilizer bar.

11. Remove the stabilizer bar and inspect the stabilizer bar link as follows:

 a. Flip the ball joint stud back and forth 5 times.

 b. Using a torque gauge, turn the stud continuously 1 turn per 2–4 seconds and take the torque reading on the 5th turn.

 c. Turning torque should be 0.4–13 inch lbs. (0.05–1.0 Nm).

To install:

12. Install the strut bar bushings and brackets with the stabilizer bar to lower arm. Torque the bolts to 53 ft. lbs. (73 Nm).

13. Install the strut assembly lower bracket. Torque the nuts to 59 ft. lbs. (80 Nm).

14. Install the strut assembly or the pneumatic cylinder. Torque the nuts to 121 ft. lbs. (164 Nm).

15. Install the strut lower bracket and torque the bolts as follows:

 • Bolts A — 83 ft. lbs. (113 Nm)
 • Bolts B — 43 ft. lbs. (59 Nm)

16. Install the stabilizer bar links and torque the nuts to 70 ft. lbs. (95 Nm).

NOTE: The left and right stabilizer bars are different.

17. Install the stabilizer bar bushings and torque the bolts to 21 ft. lbs. (28 Nm).

18. Install the struts.

19. Install the steering knuckle and hub.

1995–97 Models

1. Disconnect the negative battery cable. Raise and support the vehicle safely.

2. Remove the steering knuckle with the axle hub.

3. Remove the strut assembly or if equipped with air suspension, remove the pneumatic cylinder.

4. Remove the right and left sway bar links by removing the nut from the strut lower bracket and the nut from the sway bar.

5. Remove the strut assembly lower bracket from the lower control arm.

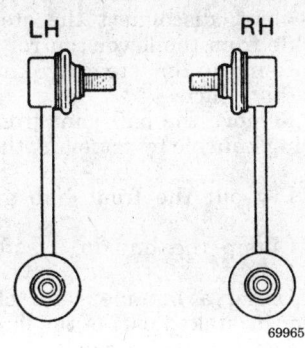

Left and right sway bar links — 1993–94 LS400

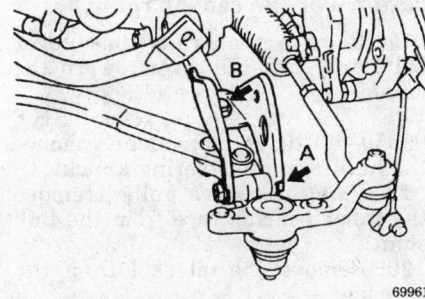

Strut lower bracket bolts — 1993–94 LS400

6. Remove the brake strut bar as follows:

a. Place matchmarks on the caster adjust cam and bracket.

b. Remove the nut, caster adjust cam and disconnect the strut bar.

c. Remove the two nuts and remove the strut bar from the lower control arm.

7. Remove the right and left sway bar brackets and bushings by removing the bolts.

8. Remove the four brake strut bar bracket bolts and then pull out the strut bar bracket from the sway bar.

9. Pull out the sway bar from the other brake strut bar bracket.

10. Remove the sway bar from the vehicle.

11. Inspect the sway bar link as follows.

a. Flip the ball joint stud back and forth 5 times.

b. Using a torque gauge, turn the stud continuously 1 turn per 2–4 seconds and take the torque reading on the 5th turn.

c. Turning torque should be 0.4–13 inch lbs. (0.05–1.0 Nm).

To install:

12. Install the sway bar to the vehicle and the brake strut bar bracket.

13. Install the other brake strut bar bracket to the sway bar.

14. Install the four strut bar bracket bolts and torque the bolts to 53 ft. lbs. (72 Nm).

15. Install the sway bar bushings and brackets to the sway bar. Torque the bolts to 21 ft. lbs. (28 Nm).

16. Install the strut bar and two bolts to the lower control arm. Torque the bolts to 121 ft. lbs. (164 Nm).

17. Install bracket strut to the brake strut bracket and install the caster adjust cam and nut. Line up the matchmarks for the caster adjust cam and torque the nut to 121 ft. lbs. (164 Nm).

18. Install the strut assembly lower bracket to the lower control arm. Torque the through-bolt to 83 ft. lbs. (113 Nm) and the inside bolt to 43 ft. lbs. (59 Nm).

19. Install the sway bar links and torque the nuts to 70 ft. lbs. (95 Nm).

NOTE: The left and right sway links are different.

20. Install the strut assembly or if equipped with air suspension, install the pneumatic cylinder.

21. Install the steering knuckle and hub.

22. Connect the negative battery cable to the battery.

23. Check the vehicle's front end alignment.

SC300 and SC400

1. Raise the front of the vehicle and support it on safety stands.

2. Remove the wheels.

3. Remove the engine undercovers.

4. Remove the stabilizer bar links from the stabilizer bar and the lower control arm by removing the nuts.

5. Remove the four bracket mounting bolts and remove the bar and cushions.

To install:

6. Install the cushions so they touch the inside of the line painted on the stabilizer bar. Install the bar and tighten the bracket bolts to 13 ft. lbs. (18 Nm).

7. Install the links and torque the nuts as follows:

• 1993 to 47 ft. lbs. (64 Nm).
• 1994–95 to 54 ft. lbs. (103 Nm).

8. Install the undercovers.

9. Install the wheels.

10. Lower the vehicle.

11. Check the front wheel alignment.

Front Wheel Bearings

ADJUSTMENT

Check the backlash in bearing shaft direction and the axle hub deviation. Maximum for backlash should be 0.0020 in. (0.05mm) and for axle hub deviation 0.020 inch (0.05mm).

NOTE: The wheel bearing is non-adjustable. If the wheel bearing is out of specifications, replace the wheel bearing.

REMOVAL AND INSTALLATION

ES300

1. Disconnect the negative battery cable.

2. Raise the vehicle and support safely.

3. Remove the front wheels.

4. Remove the fender apron seal.

5. Remove the cotter pin and lock cap from the end of the halfshaft.

6. While applying the front brakes, remove the halfshaft locknut.

7. Remove the brake caliper and use a wire to support it out of the way.

NOTE: Never allow the caliper to hang freely from the brake hose.

8. Matchmark the rotor to the hub and remove the rotor.

9. If equipped with ABS brakes, remove the ABS speed sensor from the steering knuckle.

10. Loosen the nuts on the lower end of the strut.

11. Disconnect and separate the tie rod end from the steering knuckle.

12. Disconnect the lower control arm from the ball joint by removing the three bolts.

13. Remove the driveshaft from the axle hub. Secure the shaft out of the way using a wire. Be careful not to damage the shaft boot or ABS sensor rotor.

14. Remove the two nuts on the lower end of the strut and remove the steering knuckle.

15. Clamp the steering knuckle in a vise with soft jaws to protect the knuckle.

16. Carefully pry the dust deflector from the hub.

17. Remove the ball joint from the steering knuckle.

18. Using SST 09520–00031 or equivalent, remove the hub from the knuckle.

19. Using SST 09950–00020 or equivalent, remove the inner race from the hub.

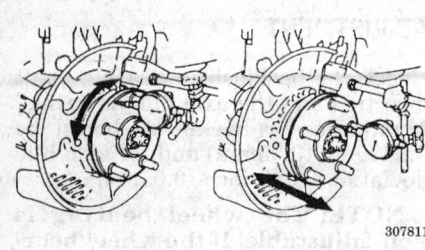

Checking wheel bearings

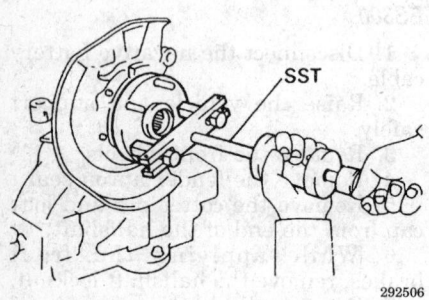

Removing the axle hub from the steering knuckle

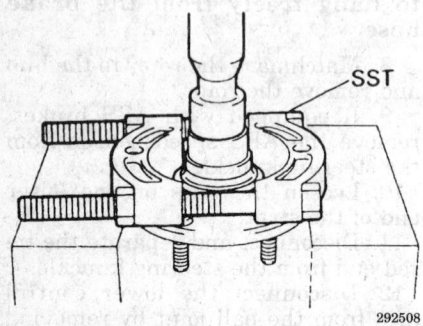

Using SST 09950–00020 or equivalent, remove the inner race from the hub

20. Remove the four bolts to the dust cover and then remove dust cover.

21. Using snapring pliers, remove the snapring.

22. Take the inner race (removed from the hub) and install it on the outside of the bearing.

23. Using a bearing driver, drive the bearing from the steering knuckle.

To install:

24. Clean bearing seating surfaces with a clean, dry rag.

---

25. Using a press and SST 09608–32010 or equivalent, install the bearing into the knuckle.

26. Install the snapring.

27. Install the dust cover. Torque the four bolts to 74 inch lbs. (8.3 Nm).

28. Press the hub into the steering knuckle.

29. Install the lower ball joint to the steering knuckle. Torque the nut to 90 ft. lbs. (123 Nm) and install a new cotter pin.

30. Align the hole in the dust deflector and the hole for the ABS speed sensor and install the dust deflector.

31. Position the knuckle to the lower strut and install the bolts.

32. Install the lower ball joint to the lower arm. Torque the bolts to 94 ft. lbs. (127 Nm).

33. Connect the tie rod end to the steering knuckle. Torque the nut to 36 ft. lbs. (49 Nm).

34. Install and torque the nuts on the lower strut to 156 ft. lbs. (211 Nm).

35. Install the ABS speed sensor. Torque the mounting bolt to 69 inch lbs. (8 Nm).

36. Align the matchmark and install the rotor on the hub. Install the brake caliper. Torque the mounting bolts to 79 ft. lbs. (107 Nm).

37. Have a helper apply the brakes and install the axle locknut. Torque the nut to 217 ft. lbs. (294 Nm). Install the lock cap and a new cotter pin.

38. Install front fender apron seal.

39. Install the wheel.

40. Turn the wheel by hand, verify that the wheel turns without noise and without binding.

41. Lower the vehicle.

GS300

1. Disconnect the negative battery cable.

2. Raise and support the vehicle safely.

3. Remove the front wheel.

4. Remove the caliper, leaving the brake line connected and suspend it out of the way.

NOTE: Never allow the brake caliper to hang freely from the brake hose.

5. Remove the rotor.

6. Remove the ABS speed sensor and harness.

7. Disconnect the tie rod from the arm on the lower ball joint.

8. Remove the cotter pin and nut. Disconnect the upper suspension arm from the steering knuckle.

9. Remove the cotter pin and nut. Using SST 09628–62011 or

---

equivalent, disconnect the steering knuckle from the lower control arm.

10. Remove the steering knuckle from the vehicle.

11. Remove the ball joint from the steering knuckle by removing the two bolts.

12. Pry out the front hub grease cap.

13. Clamp the hub in a soft jaw vise.

14. Using a hammer and chisel, loosen the staked part of the locknut.

15. Remove the locknut.

16. Remove the ABS speed sensor rotor.

NOTE: Do not scratch the serration's of the sensor rotor.

17. Remove the brake dust cover bolts and shift the cover toward the outside.

18. Using a puller SST 09950–40010 or equivalent, remove the hub from the steering knuckle.

19. Using the same puller, remove the inner bearing race from the hub shaft.

20. Remove the oil seal from the knuckle.

21. Remove the bearing snapring from the steering knuckle.

22. Using a press and a bearing driver, press the bearing from the steering knuckle.

To install:

23. Using a press and a bearing driver, press a new bearing into the steering knuckle.

NOTE: If the inner race and balls come loose from the bearing outer race, be sure to install them on the same side as before.

24. Install the snapring.

25. Install a new outside inner race and tap in the new seal using SST 09608–32010 or equivalent. Tap the seal until it is flush with the end surface of the steering knuckle.

26. Install the brake dust cover to the knuckle and torque the bolts to 74 inch lbs. (8 Nm).

27. Press the hub into the steering knuckle.

28. Install the ABS speed sensor rotor.

29. Install the axle hub locknut. Torque the nut to 147 ft. lbs. (199 Nm) and stake it.

30. Install the grease cap to the steering knuckle by tapping lightly around the circumference of the cap with a hammer.

31. Install the ball joint to the steering knuckle. Torque the two bolts to 83 ft. lbs. (113 Nm).

32. Connect the steering knuckle to the upper and lower suspension

arms. Torque the upper nut to 64 ft. lbs. (87 Nm) and the lower nut to 95 ft. lbs. (127 Nm). Install a new cotter pin on the lower nut. Install the clip on the upper suspension arm nut.

33. Connect the tie rod end to the steering knuckle. Torque the nut to 64 ft. lbs. (87 Nm) and install a new cotter pin.

34. Install the rotor, disc brake pads and the brake caliper.

35. Install the ABS speed sensor and harness. Torque the sensor retaining bolt to 69 inch lbs. (8 Nm).

36. Install the wheel.

37. Lower the vehicle and connect the negative battery cable.

38. Check the front wheel alignment.

LS400

1993–94 Models

1. If equipped with air suspension, move the height control switch in the trunk area to the OFF position.

2. Raise and safely support the vehicle.

3. Remove the front tire and wheel assembly.

4. Disconnect the brake caliper from the steering knuckle, leaving the brake line connected and support with a piece of wire.

5. Remove the brake rotor.

6. Remove the ABS speed sensor from the steering knuckle.

7. Remove the grease cap from the steering knuckle, chisel off the nut caulking.

8. Loosen the axle shaft nut.

9. Remove the steering knuckle from the lower ball joint.

10. Remove the steering knuckle from the upper ball joint.

11. Remove the steering knuckle with the axle hub.

12. Remove the nut and the speed sensor rotor.

13. Using the proper tool, remove the axle hub from the steering knuckle.

14. Remove the outside inner race from the axle and oil seal from the steering knuckle, using the proper tools.

15. Remove the snapring and remove the bearing from the steering knuckle.

To install:

16. Using the proper tool, install the bearing to the steering knuckle. Replace the snapring.

17. Install the inner race (outside) and press in a new oil seal until it is flush with the end surface of the steering knuckle.

18. Install the brake dust cover to the steering knuckle and using the

proper tools, press the axle hub to the steering knuckle.

19. Install the speed sensor. Install the steering knuckle to the lower ball joint and temporarily, tighten the bolts.

20. Install the steering knuckle to the upper arm and tighten the nut to 48 ft. lbs. (65 Nm).

21. Install brake rotor.

22. Install the brake caliper.

23. Tighten the axle shaft nut to 147 ft. lbs. (199 Nm).

24. Install the speed sensor to the steering knuckle.

25. Install the front tire and wheel assembly.

26. Lower the vehicle.

27. If equipped with air suspension, turn the height control switch to the ON position.

1995–97 Models

1. If equipped with air suspension, move the height control switch in the trunk area to the OFF position.

2. Raise and safely support the vehicle.

3. Remove the front tire and wheel assembly.

4. Disconnect the brake caliper bracket from the steering knuckle, leaving the brake line connected. Support the caliper with a piece of wire.

5. Remove the brake rotor.

6. Remove the ABS speed sensor from the steering knuckle.

7. Remove the steering knuckle from the lower ball joint by removing the two bolts.

8. Remove the steering knuckle from the upper ball joint.

9. Remove the steering knuckle with the axle hub from the vehicle.

10. Using a prytool, remove the grease cap from the hub.

11. Remove the nut and the speed sensor rotor.

12. Remove the four bolts and shift the brake dust cover towards the hub side.

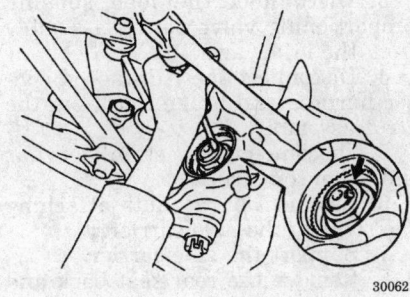

300621

Axle hub nut — 1995–97 LS400

13. Using SST 09950–00020 or equivalent, press out the axle hub from the steering knuckle.

14. Using SST 09950–00020 or equivalent, press out the outside inner race from the axle.

15. Using SST 09308–00010 or equivalent, remove the oil seal from the steering knuckle.

16. Remove the snapring and using SST 09950–60010 or equivalent, remove the bearing from the steering knuckle.

To install:

17. Using the SST 09608–35014 or equivalent, install the bearing to the steering knuckle. Install the snapring.

18. Install the inner race (outside) and press in a new oil seal until it is flush with the end surface of the steering knuckle.

19. Install the brake dust cover to the steering knuckle and torque the bolts to 74 inch lbs. (8.4 Nm).

20. Using SST 09608–32010 and 09608–35014 or equivalent, press the axle hub to the steering knuckle.

21. Install the ABS speed sensor.

22. Install and torque a new nut to the axle shaft. Torque the nut to 147 ft. lbs. (199 Nm). Stake the nut and install the grease cap.

23. Install the steering knuckle to the lower ball joint and torque the bolts to 83 ft. lbs. (113 Nm).

24. Install the steering knuckle to the upper ball joint and tighten the nut to 48 ft. lbs. (65 Nm).

25. Install brake rotor.

26. Install the brake caliper and torque the two bolts to 87 ft. lbs. (118 Nm).

27. Install the speed sensor to the steering knuckle.

28. Install the front tire and wheel assembly.

29. Lower the vehicle.

30. If equipped with air suspension, turn the height control switch to the ON position.

SC300 and SC400

1. Raise and safely support the vehicle.

2. Remove the front tire and wheel assembly.

3. Remove the brake caliper support bracket, leaving the brake line connected and support it using a piece of wire.

4. Remove the rotor by removing the two screws.

5. Disconnect the ABS speed sensor.

6. Remove the cotter pin and nut and disconnect the tie rod from the steering knuckle.

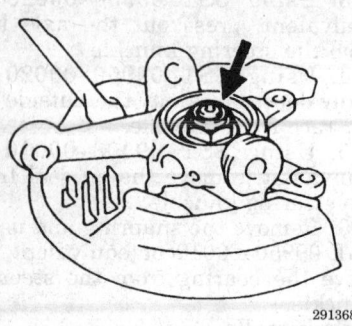

291368

Axle hub locknut — SC300 and SC400

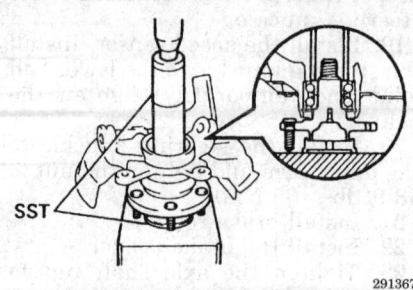

SST

291367

Pressing the hub into the knuckle — SC300 and SC400

7. Remove the cotter pin and nut and disconnect the steering knuckle from the upper control arm.

8. Remove the clip and nut and press the knuckle off the lower control arm.

9. Remove the steering knuckle from the vehicle.

10. Pry the hub bearing cap from the steering knuckle. Using a hammer and chisel, loosen the staked part of the hub nut and remove it.

11. Remove the ABS sensor rotor.

12. Remove the four bolts and shift the brake dust shield toward the hub (outside).

13. Using a two-arm puller, remove the axle hub from the knuckle.

14. With a puller, remove the inner bearing race from the axle hub. Pry out the oil seal.

15. Remove the bearing snapring and then position the inner race above the bearing on the inner side. Press the bearing out.

To install:

16. Press the bearing into the knuckle. If the inner race and balls come loose from the outer race, be sure to install them on the same side as before.

17. Install the snapring and inner race and then tap in a new oil seal until it is flush with the end surface of the knuckle.

18. Install the brake dust cover and tighten the bolts to 74 inch lbs. (8.3 Nm).

19. Press the hub into the knuckle and install the speed sensor.

20. Install a new locknut and tighten it to 147 ft. lbs. (199 Nm). Stake the nut with a chisel. Tap the bearing cap into place.

21. Connect the knuckle to the upper control arm and tighten the nut to 76 ft. lbs. (103 Nm). Install a new cotter pin.

22. Connect the knuckle to the lower control arm and tighten the nut to 92 ft. lbs. (125 Nm). Install a new clip.

23. Connect the tie rod end to the steering knuckle with the nut. Torque the nut to 36 ft. lbs. (49 Nm). Install a new cotter pin.

24. Install the rotor by installing the two screws.

25. Install the caliper support bracket and torque the bolt to 87 ft. lbs. (118 Nm).

26. Connect the speed sensor to the knuckle and tighten the bolt to 69 inch lbs. (8 Nm).

27. Install the front wheel and tighten the lug nuts to 76 ft. lbs. (103 Nm).

28. Lower the vehicle.

29. Check the front end alignment and ABS speed sensor signal.

REAR SUSPENSION

Strut and Coil Spring

REMOVAL AND INSTALLATION

ES300

1. Raise and safely support the vehicle.

2. Remove the tire and wheel assembly.

3. Disconnect the load sensing proportioning valve spring assembly from the lower arm.

4. Disconnect the ABS speed sensor harness and brake line from the strut assembly.

5. Disconnect the stabilizer bar link from the strut.

6. Loosen the two nuts attaching the strut to the axle carrier.

7. Support the axle carrier.

8. Remove the rear seat back and package tray trim.

9. Remove the upper mounting nuts.

10. Remove the two lower mounting bolts and remove the strut assembly.

11. Using spring compressor SST 09727–30020 or equivalent, compress the coil spring.

12. Temporarily install a bolt and two nuts on the bracket at the lower end of the strut and secure it in a vice.

13. Secure the upper support with SST 09729–22031 or equivalent and remove the strut rod retaining nut.

14. Remove the upper suspension support, upper insulator, coil spring, spring bumper and lower insulator.

To install:

15. If discarding the strut, perform the following:

a. Fully extend the strut rod.

b. Drill a hole in the side of the strut to release the gas.

— **WARNING** —

The gas coming out is harmless, but be careful of chips which may fly up while drilling.

16. Install the lower insulator to the strut.

17. Install the spring bumper to the strut piston rod.

18. Install the (compressed) coil spring.

19. Position the coil spring with the end butted against the gap in the lower seat.

20. Install the upper insulator and support matching the bolt of the support with the cut-off part of the insulator.

21. Install the upper suspension support.

22. Secure the upper suspension support and torque a new strut piston rod nut to 36 ft. lbs. (49 Nm).

23. Remove the spring compressor.

24. Install the strut rod piston nut cap.

25. Install the strut and torque the three nuts to 29 ft. lbs. (39 Nm).

26. Connect the strut to the axle carrier. Coat the nuts with engine oil and torque the nuts and bolts to 188 ft. lbs. (255 Nm).

27. Connect the ABS harness to the strut and torque the bolt to 48 inch lbs. (6 Nm).

28. Connect the brake line to the strut and torque the retaining nut to 22 ft. lbs. (29 Nm).

29. Connect the spring to the lower arm and torque the nut to 10 ft. lbs. (13 Nm).

30. Connect the LSPV to the lower arm. Torque the nut to 9 ft. lbs. (12 Nm).

31. Install the rear wheel.

32. Lower the vehicle.

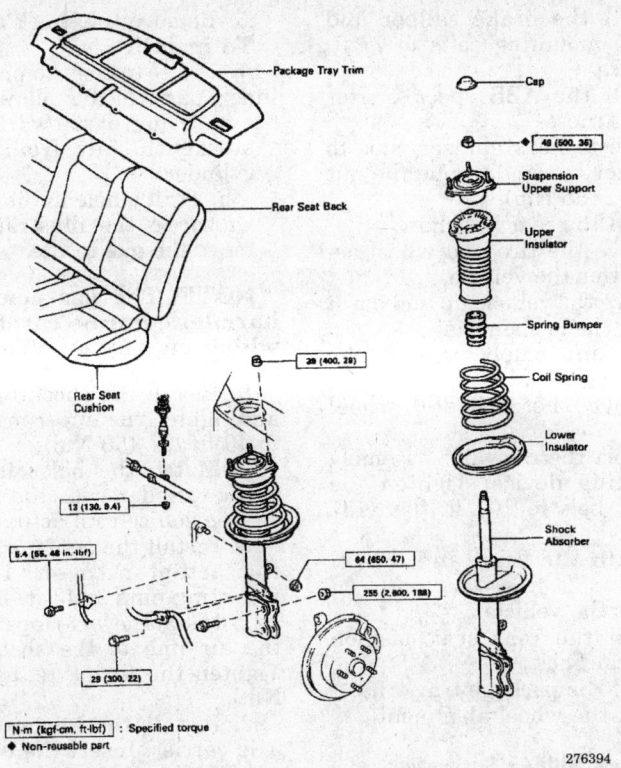

Exploded view of the strut and coil spring removal and installation procedure — 1993-94 ES300

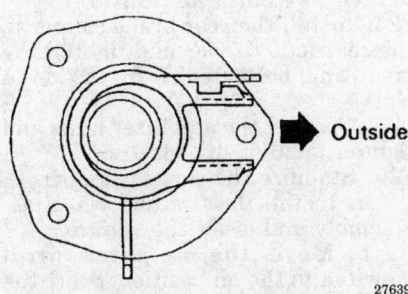

Positioning the upper suspension support — 1993-94 ES300

33. Install the rear seat and package tray.

GS300

1. Remove the front trunk compartment trim cover.
2. Raise and support the vehicle safely.
3. Remove the wheel(s).
4. Remove the brake caliper support bracket from the rear axle carrier by removing the two (2) bolts.

Leave the brake line connected and position it out of the way.

NOTE: Never allow the brake caliper to hang freely from the brake hose.

5. Remove the nut and disconnect the sway bar link from the lower control arm.
6. Remove the nut and bolt on the lower end of the strut.
7. Remove the three upper nuts and lift out the strut. Do not remove the center nut to the strut.
8. Using spring compressor SST 09727-30020 or equivalent, compress the coil spring.
9. Secure the upper support with SST 09729-22031 or equivalent and remove the strut rod retaining nut.
10. Remove the upper suspension support, upper insulator, coil spring, spring bumper and lower insulator.
   To install:
11. If discarding the strut, perform the following:
   a. Fully extend the strut rod.
   b. Drill a hole in the side of the strut drain the gas inside

─── CAUTION ───
The gas coming out is harmless, but be careful of chips which may fly up while drilling.

12. Install the lower insulator to the strut.
13. Install the spring bumper to the strut piston rod.
14. Install the (compressed) coil spring. Position the coil spring with the end butted against the gap in the lower seat.
15. Install the upper insulator and suspension support.
16. Secure the upper suspension support and torque a new strut piston rod nut to 20 ft. lbs. (27 Nm).
17. Remove the spring compressor.
18. Install the strut to the vehicle and tighten the three upper mounting nuts to 14 ft. lbs. (20 Nm). Install the cap.
19. Install and torque the lower strut bolt and nut to 101 ft. lbs. (137 Nm).
20. Connect the sway bar link to the lower control arm. Torque the nut to 33 ft. lbs. (44 Nm).
21. Install the brake caliper to the rear axle carrier by installing the two (2) bolts. Torque the bolts to 77 ft. lbs. (104 Nm).
22. Install the wheel(s).
23. Lower the vehicle.
24. Install the trunk compartment cover trim.
25. Check and adjust the vehicle alignment as necessary.

LS400

Except Air Suspension

1. Remove the rear seat cushion and seat back.
2. Remove the tray trim.
3. Raise and safely support the vehicle.
4. Remove the tire and wheel assembly.
5. Remove the rear halfshaft.
6. Disconnect the stabilizer bar link from the stabilizer bar.
7. Disconnect the ABS speed sensor and wire harness.
8. Disconnect the brake caliper bracket from the axle carrier, leaving the brake line connected. Suspend the brake caliper aside with a piece of wire.
9. Remove the nut on the lower side of the strut. Do not remove the bolt.
10. Support the rear axle assembly with a lifting device.
11. Remove the strut cap by removing the three nuts.
12. Remove the three mounting nuts holding the strut assembly to the strut tower. Do not remove the center bolt.

CAUTION

Do not remove the center nut to the strut at this time. The spring on the strut is under high pressure and can cause serious injury or vehicle damage.

---

13. Lower the rear axle assembly and remove the bolt on the lower side of the strut assembly.

14. Remove the strut assembly with the coil spring.

15. Using compressor 09727–30020 or equivalent, compress the coil spring.

16. Secure the strut housing in a vice.

17. Remove the strut rod retaining nut.

18. Remove the upper suspension support, upper insulator, coil spring, spring bumper and the lower insulator.

To install:

19. If discarding the strut, perform the following:

  a. Fully extend the strut rod.

  b. Drill a hole in the strut (about 1 inch above the strut lower mount) and drain the gas inside

NOTE: The gas coming out is harmless, but be careful of chips which may fly up while drilling.

20. Install the lower insulator to the strut.

21. Install the spring bumper to the strut piston rod.

22. Install the (compressed) coil spring.

23. Position the coil spring with the end butted against the gap in the lower seat.

24. Install the upper insulator and support. Match the bolt of the support with the cut off part of the insulator.

25. Install the upper suspension support.

26. Temporarily install the upper strut rod retaining nut.

27. Rotate the suspension support so that the rod and one of the bolts on the suspension support are aligned with the lower bushing.

28. Remove the spring compressor.

29. Install the strut assembly to the vehicle and tighten the three nuts to 47 ft. lbs. (64 Nm).

30. Tighten the strut rod retaining nut to 20 ft. lbs. (27 Nm).

31. Install the strut assembly cap and install the three nuts.

32. Install the strut to the rear axle carrier. Install the bolt from the rear of the vehicle and temporarily tighten the nut.

33. Install the brake caliper and tighten the mounting bolts to 77 ft. lbs. (104 Nm).

34. Install the ABS speed sensor and wire harness.

35. Connect the stabilizer link to the stabilizer bar and torque the nut to 48 ft. lbs. (65 Nm).

36. Install the rear halfshaft.

37. Replace the tire and wheel assembly. Lower the vehicle.

38. Bounce the vehicle up and down to stabilize the suspension.

39. Raise and safely support the vehicle.

40. Remove the tire and wheel assembly.

41. Support the rear axle assembly with a lifting device. Tighten the lower strut bolt to 101 ft. lbs. (137 Nm).

42. Install the tire and wheel assembly.

43. Lower the vehicle.

44. Install the rear seat cushion and rear seat back.

45. Install the package tray trim.

46. Check the wheel alignment.

1993–94 Models With Air Suspension

1. Remove the rear seat cushion and seat back.

2. Remove the rear scuff plates, roof side inner trim panel and speaker panel.

3. Remove the trunk trim panel. Move the height control switch, located in the trunk area to the OFF position.

4. Raise and safely support the vehicle.

5. Remove the tire and wheel assembly.

6. Disconnect the stabilizer links from the stabilizer bar.

7. Disconnect and support the brake caliper, using a piece of wire. Do not disconnect the brake line.

8. Disconnect the height control sensor link from the suspension arm.

9. Remove the nut on the lower side of the shock absorber. Do not remove the bolt.

10. Support the rear axle assembly with a lifting device.

11. Remove the grommet and disconnect the air tube from the shock absorber.

12. Remove the mounting bolts and actuator cover.

13. Remove the mounting bolts and actuator.

14. Remove the upper mounting nuts and lower the rear axle assembly.

15. Remove the bolt on the lower side of the shock absorber.

16. Remove the shock absorber.

To install:

17. If discarding the pneumatic cylinder, perform the following:

  a. Using a screwdriver, depressurize the air from inside the cylinder.

  b. Drill a hole in the shaded area shown in the illustration and remove the gas inside.

NOTE: The gas coming out is harmless, but be careful of chips which may fly up when drilling.

18. Install the shock to the vehicle and tighten the upper mounting nuts to 43 ft. lbs. (58 Nm).

19. Match the holes in the pneumatic cylinder with the holes in the suspension control actuator.

20. Install the actuator and replace the actuator cover. Tighten the mounting nuts to 13 ft. lbs. (18 Nm).

21. Install new O-rings and connect the air line to the shock absorber. Tighten the fitting to 13 ft. lbs. (18 Nm).

22. Install the shock to the rear axle carrier. Insert the bolt from the vehicle's rear and temporarily tighten the nut.

23. Connect the height control sensor link to suspension arm and tighten to 48 inch lbs. (5 Nm).

24. Install the rear brake caliper to the rear axle carrier and tighten the mounting bolts to 77 ft. lbs. (104 Nm).

25. Connect the stabilizer links and tighten to 26 ft. lbs. (35 Nm).

26. Stabilize the suspension by:

  a. Install the tire and wheel assembly and lower the vehicle.

  b. Move the height control switch to the on position. Start the engine an fill the pneumatic cylinder with air.

  c. Bounce the vehicle up and down several times to stabilize the suspension.

27. Raise and safely support the vehicle.

28. Remove the tire and wheel assembly.

29. Support the rear axle carrier with a lifting device. Tighten the lower shock bolt to 101 ft. lbs. (137 Nm).

30. Replace the tire and wheel assembly.

31. Lower the vehicle.

32. Install the rear seat cushion and seat back.

33. Replace the rear scuff plates, the roof side inner trim panel and speaker panel.

34. Install the trunk trim panel.

35. Check the wheel alignment.

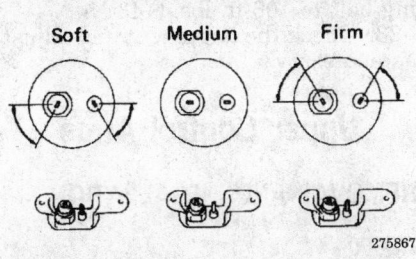

Matching the rods on the strut to the holes in the suspension control actuator-air suspension — 1993–94 LS400

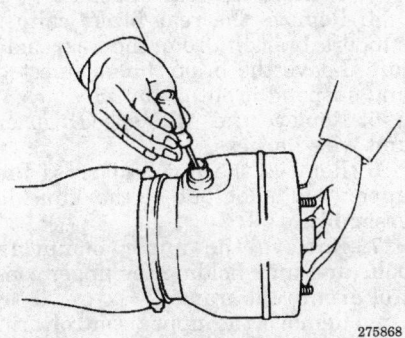

Releasing the air from the strut-air suspension — 1993–94 LS400

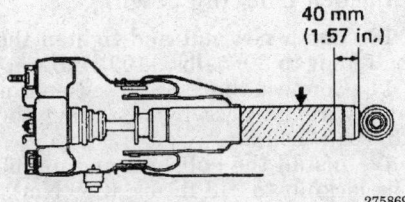

Strut disposal drilling area-air suspension

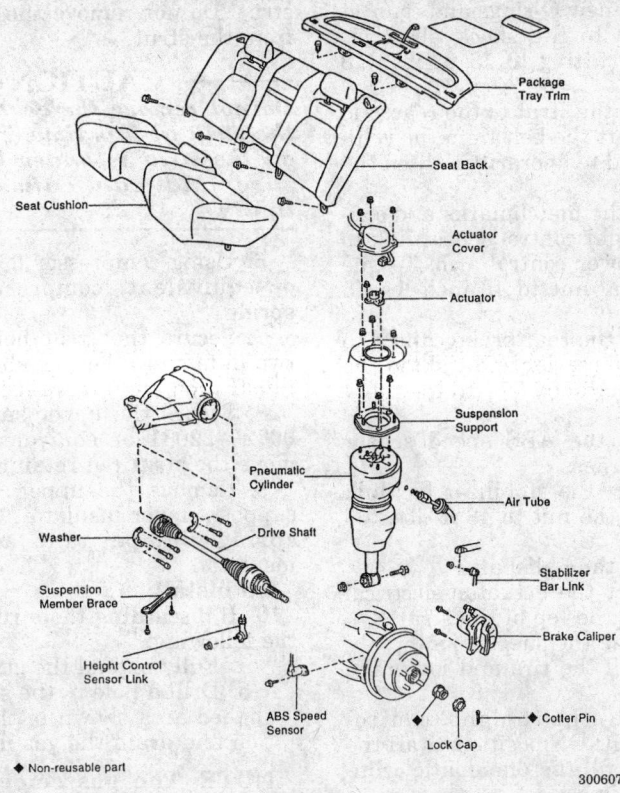

◆ Non-reusable part

Pneumatic cylinder (strut) component overview (air suspension)

1995–97 Models With Air Suspension

1. Remove the rear seat cushion and seat back.
2. Remove the package tray trim.
3. Remove the trunk trim panel. Move the height control switch, located in the trunk area, to the OFF position.
4. Raise and safely support the vehicle.
5. Remove the tire and wheel assembly.
6. Bleed the air system from the suspension.
7. Remove the rear halfshaft.

8. Disconnect the stabilizer links from the stabilizer bar.
9. Disconnect the ABS speed sensor and wire harness.
10. Disconnect the brake caliper bracket from the rear axle carrier. Support the caliper using a piece of wire. Do not disconnect the brake line.
11. Place matchmarks on the height control sensor link and bracket. Disconnect the height control sensor link from the No. 1 lower control arm.
12. Remove the nut on the lower side of the shock absorber. Do not remove the bolt.
13. Support the rear axle assembly with a lifting device.
14. Remove the grommet and disconnect the air tube from the shock absorber.
15. Remove the actuator cover from the strut tower by removing the three nuts.
16. Disconnect the actuator electrical connector from the top of the strut.
17. Remove the actuator by removing the two nuts.
18. Remove the three upper mounting nuts holding the strut to the strut tower.
19. Lower the rear axle assembly.

20. Remove the bolt on the lower side of the shock absorber.
21. Remove the pneumatic cylinder strut assembly from the vehicle.
22. Remove the suspension support from the strut assembly by removing the three nuts.

To install:

23. If discarding the pneumatic cylinder, perform the following:

   a. Using a screwdriver, depressurize the air from inside the cylinder.

   b. Drill a hole in the shaded area shown in the illustration and remove the gas inside.

NOTE: The gas coming out is harmless, but be careful of chips which may fly up when drilling.

24. Install the suspension support to the pneumatic cylinder (strut) and torque the nuts to 27 ft. lbs. (36 Nm).
25. Install the strut assembly to the vehicle and tighten the upper mounting nuts to 47 ft. lbs. (64 Nm).
26. Match the holes in the pneumatic cylinder with the holes in the suspension control actuator.
27. Install the actuator and torque the mounting nuts to 69 inch lbs. (8 Nm).
28. Install the actuator cover and torque the three nuts to 18 ft. lbs. (25 Nm).

29. Install new O-rings and connect the air line to the shock absorber. Tighten the fitting to 13 ft. lbs. (18 Nm).

30. Install the strut to the rear axle carrier. Insert the bolt from the vehicle's rear and temporarily tighten the nut.

31. Align the matchmarks and connect the height control sensor link to the No. 1 lower control arm. Torque the mounting nut to 48 inch lbs. (5 Nm).

32. Install the rear brake caliper to the rear axle carrier and tighten the mounting bolts to 77 ft. lbs. (104 Nm).

33. Install the ABS speed sensor and wire harness.

34. Connect the stabilizer bar link and tighten the nut to 48 ft. lbs. (65 Nm).

35. Install the halfshaft.

36. Connect the actuator electrical connector to the top of the strut.

37. Stabilize the suspension by:
   a. Install the tire and lower the vehicle.
   b. Move the height control switch to the ON position. Start the engine an fill the pneumatic cylinder with air.
   c. Bounce the vehicle up and down several times to stabilize the suspension.

38. Turn the suspension height control to the OFF position.

39. Raise and safely support the vehicle.

40. Remove the tire and wheel assembly.

41. Support the rear axle carrier with a lifting device. Tighten the lower strut bolt to 101 ft. lbs. (137 Nm).

42. Replace the tire and wheel assembly.

43. Lower the vehicle.

44. Install the package tray trim.

45. Install the rear seat cushion and seat back.

46. Turn the suspension control switch to the ON position.

47. Check the wheel alignment.

SC300 and SC400

1. Raise the rear of the vehicle and support it with safety stands.

2. Remove the wheel(s).

3. Remove the brake caliper support bracket by removing the two bolts. Leave the brake line connected and position it aside.

4. Remove the nut and bolt on the lower end of the strut.

5. Remove the cap nut on the upper end of the strut. Remove the three upper nuts and lift out the

strut. Do not remove the center nut from the strut.

---- CAUTION ----
Do not remove the center nut on the strut at this time. The spring on the strut is under high pressure and can cause serious injury.

6. Using compressor 09727–30020 or equivalent, compress the coil spring.

7. Secure the strut housing with two nuts and a bolt as shown in the illustration and secure it in a vice.

8. Secure the upper support with 09729–22031 or equivalent and remove the strut rod retaining nut.

9. Remove the upper suspension support, upper insulator, coil spring, spring bumper and the lower insulator.

To install:

10. If discarding the strut, perform the following:
   a. Fully extend the strut rod.
   b. Drill a hole in the strut in the shaded area shown in the illustration and drain the gas inside

NOTE: The gas coming out is harmless, but be careful of chips which may fly up while drilling.

11. Install the lower insulator to the strut.

12. Install the spring bumper to the strut piston rod.

13. Install the (compressed) coil spring.

14. Position the coil spring with the end butted against the gap in the lower seat.

15. Install the upper insulator and support matching the bolt of the support with the cut off part of the insulator.

16. Install the upper suspension support.

17. Secure the upper suspension support and torque a new strut piston rod nut to 20 ft. lbs. (27 Nm).

18. Remove the spring compressor.

19. Install the strut rod piston nut cap.

20. Install the strut and the three nuts. Torque the nuts to 19 ft. lbs. (25 Nm). Install the cap.

21. Install the lower bolt to hold the strut to the lower control arm. Do not torque the bolt at this time.

22. Install the caliper support bracket and torque the bolts to 77 ft. lbs. (104 Nm).

23. Install the wheel(s).

24. Lower the vehicle.

25. Bounce the vehicle several times to normalize the suspension.

26. Support the lower arm.

27. Torque the lower strut mounting bolt to 106 ft. lbs. (143 Nm).

28. Check the alignment and adjust as necessary.

Upper Control Arms

REMOVAL AND INSTALLATION

GS300

1. Raise and safely support the rear of the vehicle.

2. Remove the tire and wheel assembly.

3. Remove the rear halfshaft.

4. Remove the rear brake caliper support bracket from the rear axle hub. Leave the brake line connected and suspend it out of the way.

5. Remove the ABS speed sensor and wire harness.

6. Remove the nut and press the upper ball joint out of the knuckle assembly.

7. Remove the upper mounting bolts and nuts holding the upper control arm to the frame.

8. Remove the upper control arm from the vehicle.

To install:

9. Position the upper control arm on the vehicle and temporarily install the upper bolts and nuts.

NOTE: Ensure the tip of the nut lock is facing down.

10. Use a new nut and tighten the ball joint to 80 ft. lbs. (109 Nm).

11. Connect the ABS speed sensor and wire harness by installing the three (3) bolts.

12. Install the halfshaft and torque the locknut to 213 ft. lbs. (289 Nm).

13. Install the caliper and torque the two bolts to 77 ft. lbs.

14. Install the wheel to the vehicle.

15. Lower the vehicle.

16. Bounce the vehicle several times to set the suspension.

17. Torque the upper control arm bolts and nuts to 121 ft. lbs. (164 Nm).

18. Check the vehicle's alignment.

LS400

1993–94 Models

1. Raise and safely support the vehicle.

2. Remove the tire and wheel assembly.

3. Remove the rear axle carrier with the upper arm assembly.

4. Install the axle carrier in a suitable holding fixture.

5. Disconnect the upper control arm from the axle carrier.

To install:

6. Install the upper control arm and torque the bolts to 80 ft. lbs. (108 Nm).

7. Install the axle carrier.

8. Install the wheel.

9. Lower the vehicle.

1995–97 Models

1. If equipped with air suspension, move the height control switch in the trunk area to the **OFF** position.

2. Disconnect the negative battery cable from the battery.

3. Raise and safely support the vehicle.

4. Remove the rear tire and wheel assembly.

5. If equipped with height control suspension, matchmark and disconnect the height control sensor link from the lower control arm by removing the nut.

6. Remove the ABS speed sensor and wire harness.

7. Disconnect the brake caliper bracket from the rear axle carrier by removing the two bolts. Support the caliper with a piece of wire. Do not allow the caliper to hang freely from the brake hose.

8. Place matchmarks on the disc brake rotor and the axle hub. Remove the brake rotor.

9. Remove the parking brake shoes and the cable.

10. Remove the halfshaft as follows:

   a. Remove the cotter pin, lock cap and the nut holding the halfshaft to the rear axle.

   b. Remove the suspension member brace by removing the two bolts.

   c. Place matchmarks on the halfshaft and the side gear shaft. Remove the hexagon bolts and washers with the proper tool.

   d. Hold the inboard joint side of the halfshaft so the outboard joint side does not bend too much. Tap the end of the halfshaft with a rubber mallet and disengage the axle hub.

   e. Remove the halfshaft from the vehicle.

11. Remove the strut rod as follows:

   a. Remove the nut and bolt and disconnect the strut rod from the rear axle carrier.

   b. Remove the nut, bolt and the strut rod from the body.

   c. Remove the strut rod.

12. Remove the lower control arms as follows:

   a. Place matchmarks on the adjusting cam and body for the No. 1

control arm. Remove the nut and adjusting cam.

   b. Remove the nut on the axle carrier side of the No. 1 lower control arm.

   c. Using a tie rod removal tool, remove the No. 1 lower control arm.

   d. Disconnect the stabilizer bar link from the No. 2 lower control arm.

   e. Place matchmarks on the adjusting cam and body. Remove the nut and adjusting cam from the No. 2 lower control arm.

   f. Remove the nut and bolt holding the No. 2 lower control arm to the axle carrier.

   g. Remove the No. 2 control arm from the vehicle.

13. Remove the nut and bolt on the lower side of the strut assembly.

14. Remove the two upper control arm set nuts and bolts.

15. Remove the axle carrier with the upper control arm.

16. Secure the axle carrier in a vise.

17. Remove the nut holding the upper control arm to the axle carrier and remove the control arm.

To install:

18. Install the upper control arm to the axle carrier with the nut. Torque the nut to 80 ft. lbs. (108 Nm),

19. Install the axle carrier and upper control arm to the vehicle as an assembly.

20. Install the two upper control arm set bolts and torque the bolts to 121 ft. lbs. (164 Nm).

21. Install the bolt and nut holding the strut to the axle carrier. Torque to 101 ft. lbs. (137 Nm).

22. Install the lower control arms as follows:

   a. Install the bolt and nut connecting the No. 2 lower control arm to the axle carrier. Torque the bolt to 60 ft. lbs. (81 Nm).

   b. Install the nut and adjusting cam to hold the No. 2 lower control arm to the body. Align the adjusting cam matchmarks and torque the nut to 57 ft. lbs. (78 Nm).

   c. Install the stabilizer bar link to the No. 2 lower control arm and torque the nut to 48 ft. lbs. (65 Nm).

   d. Install the No. 1 lower control arm to the axle carrier and body. Install the nut to hold the No. 1 lower control arm to the axle carrier. Torque the nut to 43 ft. lbs. (59 Nm).

   e. Install the nut and adjusting cam to hold the No. 1 lower control

arm to the body. Align the matchmarks and torque the nut to 57 ft. lbs. (78 Nm).

23. Install the strut rod as follows:

   a. Install the strut rod to the axle carrier and body. Install the bolt and nut to hold the strut rod to the body. Torque to 57 ft. lbs. (78 Nm).

   b. With the strut connected to the axle carrier, install the bolt and nut to hold the strut rod to the axle carrier. Torque to 136 ft. lbs. (184 Nm)

24. Install the parking brake shoes and the cable.

25. Install the halfshaft to the vehicle as follows:

   a. Insert the outboard joint side of the halfshaft and align the matchmarks on the side gear shaft and the halfshaft.

   b. Coat the threads with clean oil and install the hexagon bolts. Torque bolts to 61 ft. lbs. (83 Nm).

   c. Install the suspension member brace with the two bolts. Torque the two bolts to 37 ft. lbs. (50 Nm).

   d. Install the nut to hold the halfshaft to the rear axle. Torque the nut to 213 ft. lbs. (289 Nm).

   e. Install the lock cap and cotter pin.

26. Install the brake rotor to the axle hub with the matchmarks aligned. Install the two screws and torque the screws to 48 inch lbs. (5 Nm).

27. Install the brake caliper to the vehicle and install the two bolts. Torque the bolts to 77 ft. lbs. (104 Nm).

28. Install the ABS speed sensor and wire harness.

29. Install the height control sensor link with the matchmarks aligned. Torque the nut to 48 inch lbs. (5 Nm).

30. Install the rear wheel.

31. Lower the vehicle and turn ON the air suspension switch.

SC300 and SC400

1. Raise and safely support the vehicle. Remove the tire and wheel assembly.

2. Remove the rear halfshaft.

3. Remove the brake caliper support bracket from the vehicle by removing the two bolts.

4. Remove the two bolts and the ABS wire harness clamp.

5. Remove the nut from the upper ball joint; press the upper ball joint out of the axle carrier.

6. Remove the upper mounting bolts and nuts. Remove the control arm from the vehicle.

To install:

7. Position the arm and tighten the upper bolts and nuts to 121 ft. lbs. (164 Nm).

NOTE: Ensure the tip of the nut lock is facing down.

8. Use a new nut and tighten the ball joint to 80 ft. lbs. (109 Nm).
9. Install the halfshaft.
10. Install the ABS clamp by installing the two bolts.
11. Install the brake caliper support bracket to the vehicle by installing the two bolts. Torque the bolts to 77 ft. lbs. (104 Nm).
12. Install the wheel and lower the vehicle.
13. Check the vehicle alignment and adjust as necessary.

Lower Control Arms

REMOVAL AND INSTALLATION

ES300

NOTE: There are two lower control arm bars on each side of the vehicle. The bar to the rear of the vehicle is referred to as No. 2 and the one towards the front of the vehicle is referred to as No. 1.

1. Loosen the lug nuts to the rear wheel.
2. Raise and safely support the vehicle and then remove the rear wheel. Support the vehicle under the frame.
3. Remove the two bolts and nuts from the strut rod and remove the strut rod from the vehicle.
4. If the vehicle is equipped with ABS brakes, disconnect the load sensing proportioning valve (LSPV) from the No. 2 lower control arm.
5. Remove the two nuts and washers at each end of the No. 2 lower control arm.
6. Remove the No. 2 lower control arm from the vehicle.

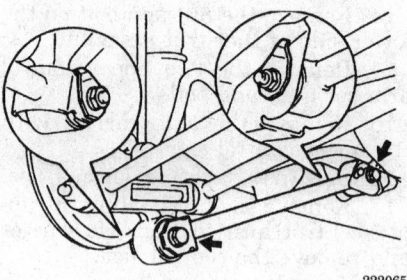

Strut rod removal — ES300

222065

7. Remove the left and right stabilizer bushing retainers.
8. Remove the exhaust center pipe and tailpipe from the vehicle.
9. Support the rear suspension crossmember with a jack.
10. Remove the six nuts to the suspension crossmember. Remove the left and right suspension crossmember lower supports.
11. Lower the suspension crossmember enough to gain access to the No. 1 lower control arm bolts.
12. Remove the two bolts and the washers from the No. 1 lower control arm and then remove the control arm from the vehicle.
13. To disassemble the No. 2 lower control arm:

    a. loosen the two locknuts and turn the adjusting tube.

    b. remove the locknuts from the No. 2 control arm.

To install:

14. To assemble the No. 2 lower control arm:

    a. install the locknuts to the control arm.

    b. turn the adjusting tube and assemble the No. 2 lower control arm.

    c. adjust the No. 2 lower control arm length by turning the adjusting tube. The arm length should be on the 1993–95 models 23.0 inches (584.2mm) and 23.366 inches (593.5mm) on 1996–97 models.

NOTE: When adjusting the control arm length, try to adjust the arm so that the splines on each side of the locknuts are the same length. There should be a maximum difference on 1993–95 models (between A and B) of 0.12mm (3mm) and on 1996–97 models (between A and B) 0.118mm (3.0mm).

    d. Temporarily tighten the two locknuts. Make sure to tighten the nuts after the rear wheels are aligned.

15. Install the No. 1 lower control arm with the paint mark facing the rear of the vehicle.
16. Install the washers and through-bolts to the lower control arm.
17. Once the control arm is in place, lift the suspension crossmember and install the suspension crossmember support brackets and six nuts. Torque the four suspension crossmember to body nuts to 83 ft. lbs. (113 Nm) on 1993–954 models and 38 ft. lbs. (51 Nm). on the 1996–97 models. Torque the two front bracket bolt to 28 ft. lbs. (38 Nm).

164 (1,670, 121)

Upper Suspension Arm

ABS Speed Sensor

7.8 (80, 69 in.lbf)

Brake Caliper

109 (1,100, 80)

104 (1,065, 77)

83 (850, 64)

Drive Shaft

289 (2,950, 213)

Lock Cap

◆ Cotter Pin

Exhaust Pipe

N·m (kgf·cm, ft·lbf) : Specified torque
◆ Non-reusable part

291412

Upper control arm and related components — SC300 and SC400

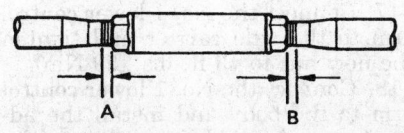

No. 2 control arm spline difference — ES300

222070

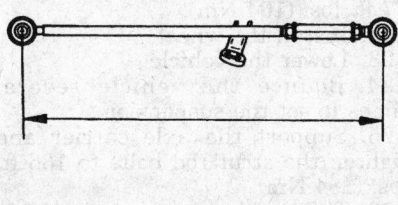

222071

Measuring the No. 2 control arm length — ES300

18. Install the left and right stabilizer bushing retainers and torque the bolts to 14 ft. lbs. (19 Nm).

19. Install the exhaust center pipe and tailpipe.

20. Install the No. 2 lower suspension arm with the paint mark facing rearward. Install the washers and nuts but do not tighten completely at this point.

21. If the vehicle is equipped with ABS brakes, install the load sensing proportioning valve to the No. 2 lower control arm. Torque the nut to 9.4 ft. lbs. (13 Nm).

22. Install the strut rod and temporarily install the two bolts and nuts.

23. Torque the nut on the outside of the lower control arm to 134 ft. lbs. (181 Nm).

24. Install the rear wheel and lower the vehicle.

25. Bounce the vehicle up and down several times to stabilize the suspension.

26. Jack up the vehicle and support the body with stands.

27. Support the rear suspension member with a jack.

28. Torque the nut on the inside of the lower control arm to 134 ft. lbs. (181 Nm).

29. Torque the strut rod set bolts to 83 ft. lbs. (113 Nm).

30. Install the rear wheel and lower the vehicle.

31. Inspect and adjust the rear wheel alignment.

32. Torque the No. 2 lower suspension arm locknuts to 41 ft. lbs. (56 Nm).

GS300

— CAUTION —
The air bag system (SRS or SIR) must be disarmed before removing the lower control arms. Failure to do so may cause accidental deployment, property damage or personal injury.

1. Disconnect the negative battery cable.

2. Raise and support the vehicle safely.

3. Remove the rear driveshaft.

4. Remove the wheel.

5. Remove the caliper, leaving the brake line connected and suspend it with wire out of the way.

6. Remove the strut rod.

7. Remove the No. 1 lower control arm as follows:

a. Remove the parking brake cable bracket.

b. Remove the exhaust support bracket.

c. Remove the nut and using SST 09628–10011 or equivalent, press the No. 1 arm out of the axle carrier.

d. Place matchmarks on the adjusting cam, remove the cam and nut and lift out the No. 1 arm.

8. Disconnect the strut from the No. 2 suspension arm by removing the bolt and nut.

9. Disconnect the stabilizer bar link from the No. 2 suspension arm.

10. Remove the nut and press the lower arm out of the axle carrier.

11. Place matchmarks on the adjusting cam, remove the cam and nut and lift out the No. 2 arm.

To install:

12. Install the No. 2 arm to the axle carrier with a new nut and tighten it to 110 ft. lbs. (150 Nm).

13. Connect the No. 2 arm to the body and install the adjusting cam and nut, align the matchmarks. Do not tighten the adjusting bolt at this time.

14. Connect the strut to the No. 2 arm. Torque the bolt to 101 ft. lbs. (137 Nm).

15. Connect the stabilizer link and tighten the nut to 33 ft. lbs. (44 Nm).

16. Connect the No. 1 lower control arm to the axle carrier and tighten the new nut to 43 ft. lbs. (59 Nm).

17. Connect the No. 1 lower control arm to the body and install the adjusting cam and nut, align the matchmarks. Do not tighten the adjusting bolt at this time.

18. Install the exhaust support bracket.

19. Connect the parking brake cable.

20. Install the strut rod.

21. Install the caliper.

22. Install the wheel.

23. Lower the vehicle.

24. Bounce the vehicle several times to set the suspension.

25. Support the axle carrier and tighten the strut rod to 136 ft. lbs. (184 Nm).

26. Tighten the strut bolt to 136 ft. lbs. (184 Nm).

27. Align the adjusting cam matchmarks and tighten the nuts to 136 ft. lbs. (184 Nm).

28. Connect the negative battery cable.

29. Check the vehicle alignment.

LS400

1. Disconnect the negative battery cable. If equipped with air suspension. move the height control ON/OFF switch (located in the luggage compartment) to the OFF position.

2. Raise and support the vehicle safely.

3. Remove the rear wheel assembly.

4. Remove the strut rod:

a. Remove the nut and bolt and disconnect the strut rod from the rear axle carrier.

b. Remove the nut, bolt and the strut rod from the body.

5. If equipped with air suspension, place matchmarks on the height control link and bracket. Disconnect the height control sensor link from the No. 1 lower control arm.

6. Place matchmarks on the adjusting cam and body. Remove the nut and adjusting cam.

7. Remove the nut on the axle carrier side of the No. 1 lower control arm.

8. Using a tie rod removal tool, remove the No. 1 lower control arm.

9. Disconnect the stabilizer bar link from the No. 2 lower control arm.

10. Place matchmarks on the adjusting cam and body. Remove the nut and adjusting cam from the No. 2 lower control arm.

11. Remove the nut and bolt holding the No. 2 lower control arm to the axle carrier.

12. Inspect the No. 1 lower suspension arm ball joint as follows.

    a. Flip the ball joint stud back and forth 5 times.

    b. Using a torque gauge, turn the stud continuously 1 turn per 2–4 seconds and take the torque reading on the 5th turn.

    c. Turning torque should be 7–30 inch lbs. (0.8–3.4 Nm).

    d. If not within specifications, replace the No. 1 suspension arm.

To install:

13. Install the No. 2 lower control arm to the body and axle carrier as follows:

    a. Install the bolt and nut connecting the No. 2 lower control arm to the axle carrier. Torque the bolt to 60 ft. lbs. (81 Nm).

    b. Install the nut and adjusting cam to hold the No. 2 lower control arm to the body. Align the adjusting cam and torque the nut to 57 ft. lbs. (78 Nm).

    c. Install the stabilizer bar link to the No. 2 lower control arm and torque the nut to 48 ft. lbs. (65 Nm).

14. Install the No. 1 lower control arm to the axle carrier and body. Install the nut to hold the No. 1 lower control arm to the axle carrier. Torque the nut to 43 ft. lbs. (59 Nm).

15. Install the nut and adjusting cam to hold the No. 1 lower control arm to the body. Align the matchmarks and torque the nut to 57 ft. lbs. (78 Nm).

16. Connect the height control sensor link to the No. 1 lower control arm with a new nut. Align the matchmarks on the height control sensor and torque the nut to 48 inch lbs. (5 Nm).

17. Install the strut rod to the axle carrier and body. Install the bolt and nut to hold the strut rod to the body. Torque to 57 ft. lbs. (78 Nm).

18. With the strut connected to the axle carrier, install the bolt and nut to hold the strut rod to the axle carrier. Torque the bolt and nut to 136 ft. lbs. (184 Nm)

19. Install the rear wheel assemblies, lower the vehicle and torque the rear wheel retaining nuts to 76 ft. lbs. (103 Nm).

20. Connect the negative battery cable to the battery.

21. If equipped with air suspension, turn the height control switch back to the ON position. Bounce the vehicle up and down several times to stabilize the suspension.

22. Check the rear wheel alignment.

SC300 and SC400

1. Raise the rear of the vehicle and support it on safety stands.

2. Remove the wheel.

3. Remove the caliper support bracket by removing the two bolts. Leave the brake line connected and suspend it aside.

4. Remove the brake rotor.

5. Remove the bolt and nut and disconnect the strut rod from the axle carrier.

6. Remove the bolt and nut holding the strut rod to the body. Remove the strut rod from the vehicle.

7. Remove the nut and press the No. 1 control arm out of the axle carrier.

8. Place matchmarks on the adjusting cam and frame. Remove the cam and nut and lift out the No. 1 control arm.

9. Disconnect the strut from the No. 2 lower control arm by removing the bolt and nut.

10. Disconnect the stabilizer bar link from the No. 2 lower control arm by removing the nut.

11. Remove the nut and press the No. 2 lower control arm out of the axle carrier.

12. Place matchmarks on the adjusting cam and frame for the No. 2 lower control arm. Remove the cam and nut and lift out the No. 2 arm.

  To install:

13. Install the No. 2 arm to the axle carrier with a new nut and tighten it to 110 ft. lbs. (150 Nm).

14. Connect the No. 2 arm to the body and install the adjusting cam and nut, align the matchmarks. Do not torque the bolts and nuts at this time.

15. Connect the strut to the lower arm by installing the bolt and nut. Do not torque the bolt at this time

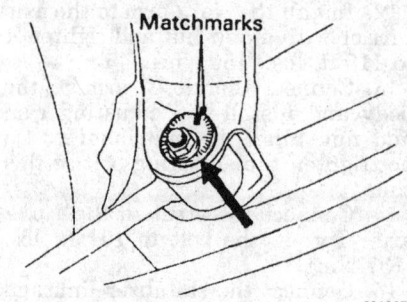

Matchmarks

291387

Matchmarking the No. 1 suspension arm adjusting cam to the body — SC300 and SC400

16. Connect the stabilizer link and torque the nut to the following specifications:
- 1993 to 26 ft. lbs. (35 Nm)
- 1994–95 to 54 ft. lbs. (74 Nm)

17. Connect the No. 1 lower control arm to the axle carrier and tighten the new nut to 43 ft. lbs. (59 Nm).

18. Connect the No. 1 lower control arm to the body and install the adjusting cam and nut. Align the matchmarks. Do not torque the bolts and nuts at this time.

19. Install the strut rod to the body and axle carrier. Do not tighten the bolts at this time.

20. Install the brake rotor.

21. Install the caliper support bracket and torque the two bolts to 77 ft. lbs. (104 Nm).

22. Install the wheel.

23. Lower the vehicle.

24. Bounce the vehicle several times to set the suspension.

25. Support the axle carrier and tighten the strut rod bolts to 136 ft. lbs. (184 Nm).

26. Tighten the strut bolt to 106 ft. lbs. (143 Nm).

27. Align the adjusting cam matchmarks and tighten the nuts to 136 ft. lbs. (184 Nm).

Sway Bar

REMOVAL AND INSTALLATION

ES300

1. Raise and support the vehicle safely.

2. Remove the stabilizer bar links. If the ball joint stud turns together with the nut, use a hexagon wrench to hold the stud.

3. Remove the four bolts, stabilizer bar brackets and the cushions.

4. Remove the stabilizer bar from the vehicle.

5. Rotate the ball joint stud in all directions. If the movement is not smooth and free, replace the stabilizer link.

  To install:

6. Install the stabilizer bar links to the stabilizer bar. Torque the nuts to 47 ft. lbs. (64 Nm).

7. Install the stabilizer bar assembly. Torque the stabilizer bar bracket retaining bolts to 14 ft. lbs. (19 Nm).

8. Connect the stabilizer bar links to the struts and torque the stabilizer bar link nuts to 47 ft. lbs. (64 Nm).

9. Install the rear wheels.

10. Lower the vehicle.

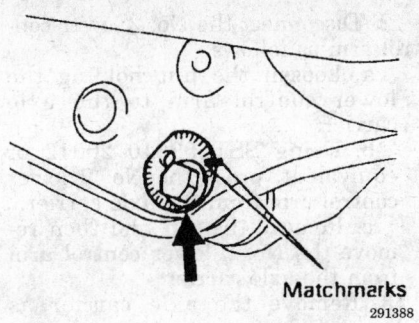

Matchmarking the No. 2 suspension arm adjusting cam to the body — SC300 and SC400

LS400

1. Raise and support the vehicle safely.

2. Remove the stabilizer bar links. If the ball joint stud turns together with the nut, use a hexagon wrench to hold the stud.

3. Remove the four bolts, stabilizer bar brackets and the cushions.

4. Remove the stabilizer bar from the vehicle.

5. Rotate the ball joint stud in all directions. If the movement is not smooth and free, replace the stabilizer link.

To install:

6. Install the stabilizer bar assembly. Torque the stabilizer bar cushion retaining bolts to 13 ft. lbs. (18 Nm).

7. Torque the stabilizer bar link to 26 ft. lbs. (35 Nm) on 1993–94 models and 48 ft. lbs. (65 Nm) on 1995–97 models.

8. Lower the vehicle.

GS300

1. Safely raise and support the rear of the vehicle.

2. Disconnect the rear exhaust pipe support brackets from the body.

3. Remove the sway bar link mounting nuts and then remove the sway bar links. If the ball joint stud turns together with the nut, use a hexagon wrench to hold the stud.

4. Remove the bolts and disconnect sway bar brackets from the body.

5. Remove the brackets and cushions from the sway bar.

6. Remove the sway bar from the vehicle.

To install:

7. Install the sway bar to the vehicle.

8. Install the cushions and brackets to the sway bar.

9. Install the four (4) bolts to the brackets and attach the brackets to

the vehicle body. Torque the bolts to 21 ft. lbs. (28 Nm).

10. Install the sway bar links and torque the nuts to 33 ft. lbs. (44 Nm).

11. Connect the exhaust brackets and torque the bolts to 14 ft. lbs. (19 Nm).

12. Lower the vehicle.

SC300 and SC400

1. Safely raise and support the rear of the vehicle.

2. Disconnect the rear exhaust pipe support brackets from the body.

3. Remove the sway bar link mounting nuts and then remove the sway bar links. If the ball joint stud turns together with the nut, use a hexagon wrench to hold the stud.

4. Remove the bolts and disconnect sway bar brackets from the body.

5. Remove the brackets and cushions from the sway bar.

6. Remove the sway bar from the vehicle.

To install:

7. Install the sway bar to the vehicle.

8. Install the stabilizer bar brackets and cushions. Install the four bolts and torque the bolts to 21 ft. lbs. (28 Nm).

9. Connect the stabilizer bar links and torque the nuts to 54 ft. lbs. (74 Nm)

10. Connect the exhaust to the brackets.

11. Lower the vehicle.

Wheel Bearings

ADJUSTMENT

Check the backlash in bearing shaft direction and the axle hub deviation. Maximum for backlash should be 0.0020 in. (0.05mm) and for axle hub deviation 0.0028 inch (0.07mm).

NOTE: The wheel bearing is non-adjustable. If the wheel bearing is out of specifications, replace the wheel bearing.

REMOVAL AND INSTALLATION

ES300

1. Raise and safely support the vehicle.

2. Remove the rear tire and wheel assembly.

3. If equipped with rear disc brakes, remove the caliper mounting bolts. Leave the brake line connected and suspend the assembly out of the way.

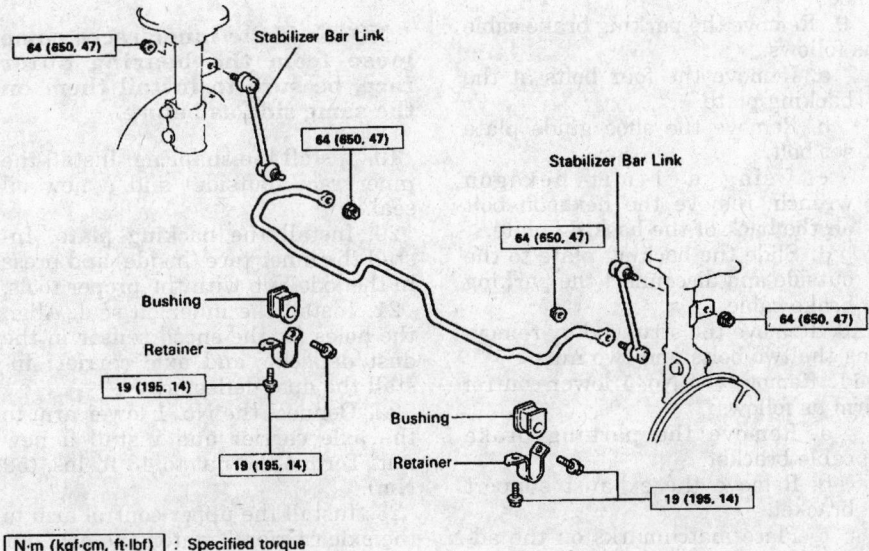

N·m (kgf·cm, ft·lbf) : Specified torque
◆ Non-reusable part

Exploded view of the sway bar removal and installation procedure — ES300

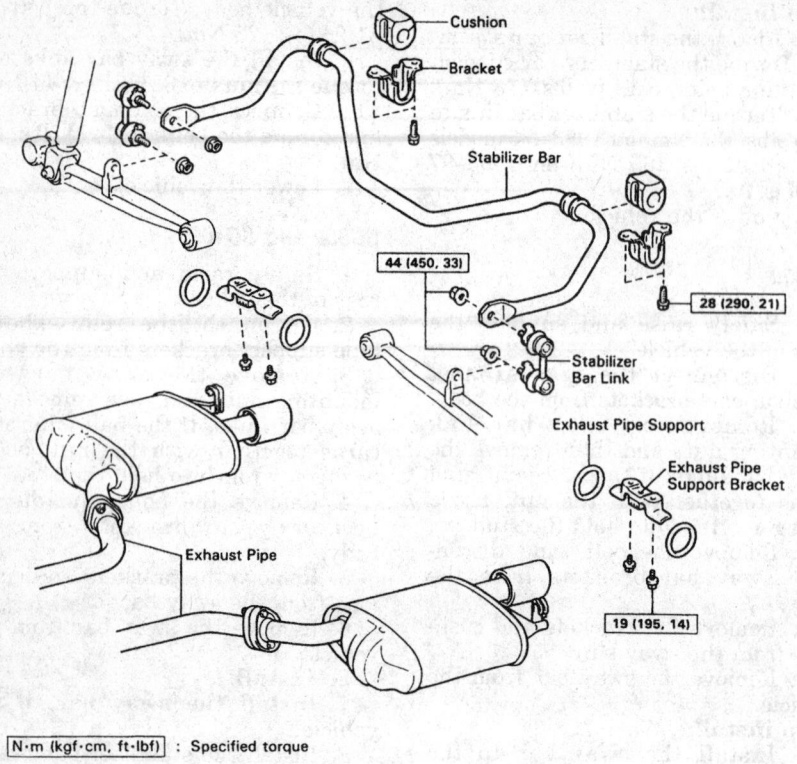

N·m (kgf·cm, ft·lbf) : Specified torque

299879

Exploded view of the sway bar removal and installation procedure — GS300

4. Remove the brake rotor or drum.

5. Remove the four bolts and pull off the rear axle hub. Remove the O-ring.

NOTE: If it is necessary to replace the hub or bearing, replace the components as an assembly.

To install:

6. Position the hub on the carrier and torque the bolts to 59 ft. lbs. (80 Nm).

7. Install the rotor or drum.

8. If equipped with rear disc brakes, install the caliper and torque the bolts to 34 ft. lbs. (64 Nm).

9. Install the wheel.

10. Lower the vehicle to the ground.

11. Road test the vehicle for proper operation.

GS300

1. Disconnect the negative battery cable.

2. Raise and safely support the vehicle.

3. Remove the rear tire and wheel assembly.

4. Disconnect the brake caliper support from the rear axle carrier and support it with a piece of wire.

5. Place matchmarks on the disc brake rotor and the axle hub. Remove the brake rotor.

6. Remove the speed sensor.

7. Remove the rear halfshaft.

8. Remove the parking brake shoes.

9. Remove the parking brake cable as follows:

a. Remove the four bolts at the backing plate.

b. Remove the shoe guide plate set bolt.

c. Using a 14mm hexagon wrench, remove the hexagon bolt on the back of the backing plate.

d. Slide the backing plate to the outside and disconnect the parking brake cable.

10. Remove the strut rod by removing the two bolts and two nuts.

11. Remove the No. 1 lower control arm as follows:

a. Remove the parking brake cable bracket.

b. Remove the exhaust support bracket.

c. Place matchmarks on the adjusting cam and rear control crossmember.

d. Remove the nut, adjusting cam and the washer to the No. 1 control arm.

e. Disconnect the No. 1 lower control arm from the crossmember.

12. Disconnect the No. 2 lower control arm as follows:

a. Loosen the nut holding the lower control arm to the axle carrier.

b. Using SST 09610–20012 or equivalent, press the No. 2 lower control arm from the axle carrier.

c. Remove the nut and then remove the No. 2 lower control arm from the axle carrier.

13. Remove the axle carrier as follows:

a. Remove the nut holding the upper control arm to the axle carrier.

b. Using SST 09628–62011 or equivalent, disconnect the upper control arm and remove the axle carrier.

c. With the axle carrier out of the vehicle, remove the nut holding the No. 1 control arm to the axle carrier.

d. Using SST 09628–10011 or equivalent, disconnect the No. 1 lower control arm from the axle carrier.

14. Remove the dust deflector.

15. Using a two-arm puller, remove the axle hub from the carrier. Remove the backing plate.

16. Pull out the inner race (outside) and then remove the oil seal. Remove the snapring.

17. Install the inner race over the bearing and press the bearing out.

To install:

18. Install the bearing to the axle carrier.

NOTE: If the inner races come loose from the bearing outer race, be sure to install them on the same side as before.

19. Install the snapring. Install the inner race (outside) and a new oil seal.

20. Install the backing plate. Install the inner race (inside) and press in the axle hub with the proper tools.

21. Install the inner oil seal. Align the holes for the speed sensor in the dust deflector and axle carrier. Install the dust deflector.

22. Connect the No. 1 lower arm to the axle carrier and install a new nut. Torque the nut to 43 ft. lbs. (59 Nm).

23. Install the upper control arm to the axle carrier. Tighten the new nut and bolt to 80 ft. lbs. (109 Nm).

24. Connect the No. 2 lower control arm to the axle carrier and tighten a new nut to 110 ft. lbs. (150 Nm).

25. Connect the No. 1 lower control arm to the rear crossmember. Torque the nut to 136 ft. lbs. (184 Nm).

To install:

13. Install the bearing to the axle carrier.

NOTE: If the inner races come loose from the bearing outer race, be sure to install them on the same side as before.

14. Install the snapring. Replace the backing plate to the axle carrier and tighten the mounting bolts 43 ft. lbs. (58 Nm).

15. Install the inner race (outside) and a new oil seal.

16. Install the inner race (inside) and press in the axle hub with the proper tools.

17. Install the inner oil seal. Align the holes for the speed sensor in the dust deflector and axle carrier. Install the dust deflector.

18. Install the upper arm to the axle carrier. Tighten the nut and bolt to 80 ft. lbs. (108 Nm).

19. Replace the nut on the lower side of the shock absorber.

20. Install the speed sensor. Replace the strut rod and lower suspension rods.

21. Install the brake rotor.

22. Connect the brake caliper to the rear axle carrier.

23. Replace the rear tire and wheel assembly.

24. Lower the vehicle.

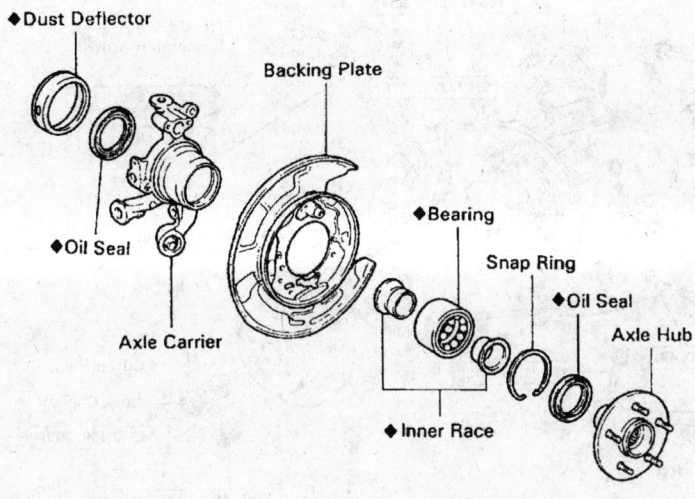

◆ Non-reusable part

Exploded view of the axle carrier — GS300

26. Connect the strut rod to the axle carrier. Torque the nuts and bolts to 134 ft. lbs. (184 Nm).

27. Connect the parking brake cable and slide the backing plate to the inside. Install the hex bolt and tighten it to 132 ft. lbs. (180 Nm).

28. Install the shoe guide plate set bolt. Torque the bolt to 13 ft. lbs. (18 Nm).

29. Install the four hub bolts and tighten them to 19 ft. lbs. (26 Nm).

30. Install the bolts at the speed sensor and tighten them to 69 inch lbs. (8 Nm).

31. Install the parking brake shoes.

32. Install the halfshafts. Apply the brakes and tighten the locknut to 213 ft. lbs. (289 Nm).

33. Install the brake rotor.

34. Connect the brake caliper support to the rear axle carrier. Torque the bolts to 77 ft. lbs. (104 Nm).

35. Install the rear tire and wheel assembly.

36. Lower the vehicle and bounce it a few times to stabilize the suspension.

37. Connect the negative battery cable.

LS400

1993–94 Models

1. If equipped with air suspension, move the height control switch in the trunk area to the **OFF** position.

2. Raise and safely support the vehicle.

3. Remove the rear tire and wheel assembly.

4. Disconnect the brake caliper from the rear axle carrier and support with a piece of wire.

5. Place matchmarks on the disc brake rotor and the axle hub. Remove the brake rotor.

6. Remove the speed sensor.

7. Remove the strut rod and lower suspension rods.

8. Remove the nut on the lower side of the shock absorber. Do not remove the bolt.

9. Remove the upper arm set bolts and the bolt on the lower side of the shock absorber. Remove the axle with the arm.

10. Remove the upper arm and the dust deflector, using the proper tools. Remove the inner oil seal.

11. Remove the axle hub and the backing plate. Remove the inner race (outside) from the axle hub.

12. Pry out the outer oil seal. Remove the snapring and the bearing, using the proper tools.

1995–97 Models

1. If equipped with air suspension, move the height control switch in the trunk area to the **OFF** position.

2. Disconnect the negative battery cable from the battery.

3. Raise and safely support the vehicle.

4. Remove the rear wheel(s).

5. If equipped with height control suspension, disconnect the height control sensor link from the lower control arm by removing the nut.

6. Remove the ABS speed sensor and wire harness.

7. Disconnect the brake caliper bracket from the rear axle carrier by removing the two bolts. Support the caliper with a piece of wire.

8. Place matchmarks on the disc brake rotor and the axle hub. Remove the brake rotor.

9. Remove the parking brake shoes and cable.

10. Remove the halfshaft as follows:

  a. Remove the cotter pin, lock cap and the nut holding the halfshaft to the rear axle.

  b. Remove the suspension member brace by removing the two bolts.

  c. Place matchmarks on the halfshaft and the side gear shaft.

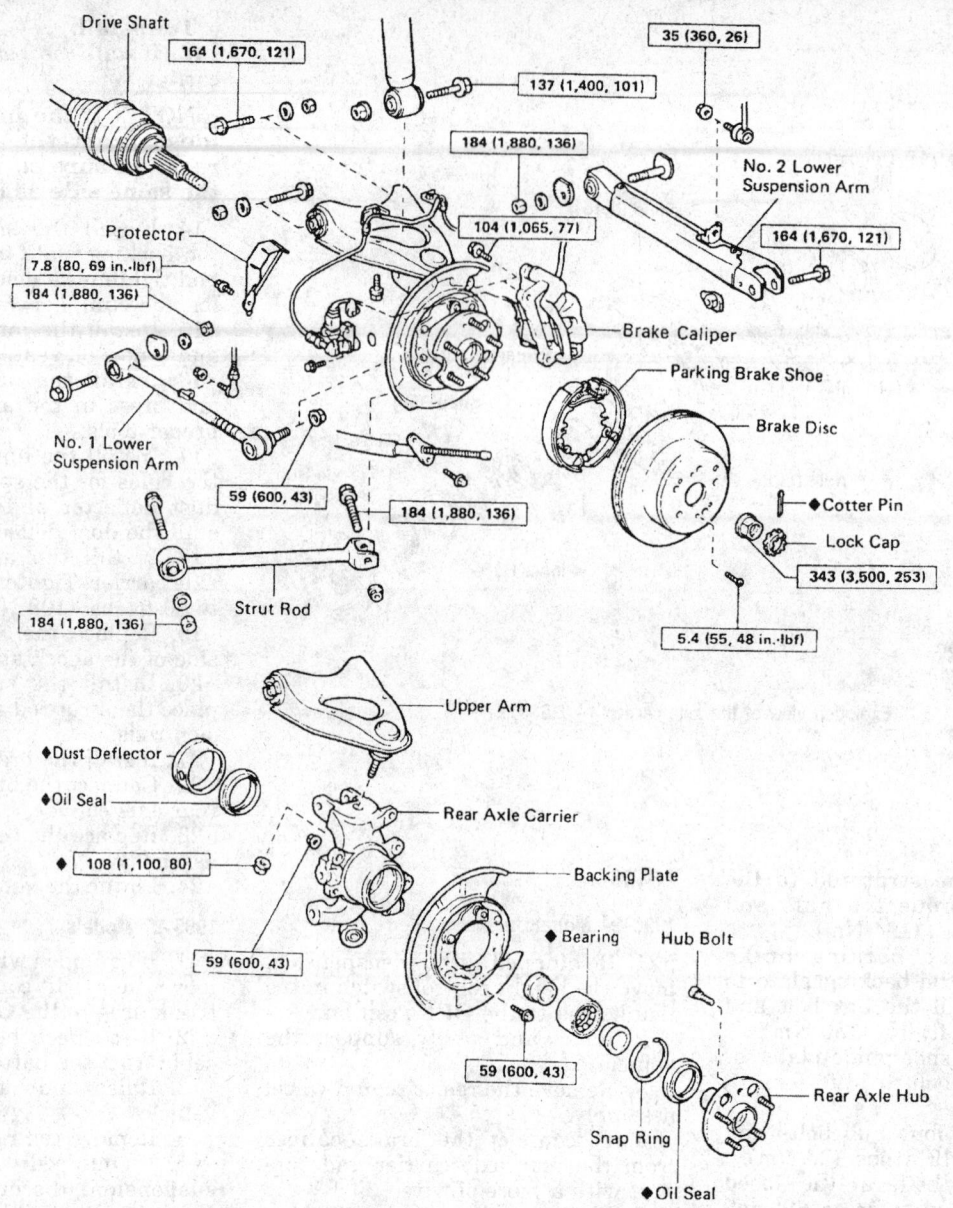

Drive Shaft

164 (1,670, 121)

35 (360, 26)

137 (1,400, 101)

184 (1,880, 136)

No. 2 Lower
Suspension Arm

104 (1,065, 77)

164 (1,670, 121)

Protector

7.8 (80, 69 in.-lbf)

184 (1,880, 136)

Brake Caliper

Parking Brake Shoe

No. 1 Lower
Suspension Arm

Brake Disc

59 (600, 43)

184 (1,880, 136)

Cotter Pin

Lock Cap

343 (3,500, 253)

184 (1,880, 136)

Strut Rod

5.4 (55, 48 in.-lbf)

Upper Arm

◆Dust Deflector

◆Oil Seal

Rear Axle Carrier

◆ 108 (1,100, 80)

Backing Plate

◆ Bearing

Hub Bolt

59 (600, 43)

59 (600, 43)

Rear Axle Hub

Snap Ring

◆Oil Seal

275896

Exploded view of the rear suspension — 1993–94 LS400

Remove the hexagon bolts and washers with the proper tool.

d. Hold the inboard joint side of the halfshaft so the outboard joint side does not bend too much. Tap the end of the halfshaft with a rubber mallet and disengage the axle hub.

e. Remove the halfshaft from the vehicle.

11. Remove the strut rod as follows:

a. Remove the nut and bolt and disconnect the strut rod from the rear axle carrier.

b. Remove the nut, bolt ,and the strut rod from the body.

c. Remove the strut rod.

12. Remove the lower control arms as follows:

a. Place matchmarks on the adjusting cam and body for the No. 1 control arm. Remove the nut and adjusting cam.

b. Remove the nut on the axle carrier side of the No. 1 lower control arm.

c. Using a tie rod removal tool separate the control arm from the axle carrier. Remove the No. 1 lower control arm.

d. Disconnect the stabilizer bar link from the No. 2 lower control arm.

e. Place matchmarks on the adjusting cam and body. Remove the

nut and adjusting cam from the No. 2 lower control arm.

f. Remove the nut and bolt holding the No. 2 lower control arm to the axle carrier.

g. Remove the No. 2 control arm from the vehicle.

13. Remove the nut and bolt on the lower side of the strut assembly.

14. Remove the two upper control arm set nuts and bolts.

15. Remove the axle carrier with the upper control arm.

16. Secure the axle carrier in a vise.

17. Remove the nut holding the upper control arm to the axle carrier and remove the control arm.

18. Using a suitable prytool, remove the dust deflector.

19. Using SST 09308–00010 or equivalent, remove the oil seal.

20. Remove the two bolts and nuts and shift the backing the plate towards the hub side (outside).

21. Using SST 09950–40010 and SST 09950–60010 or equivalent, press the axle hub out.

22. Remove the backing plate.

23. Using a press, remove the inner race (outside) from the axle hub.

24. Using SST 09308–00010 or equivalent, remove the oil seal (outer) from the axle.

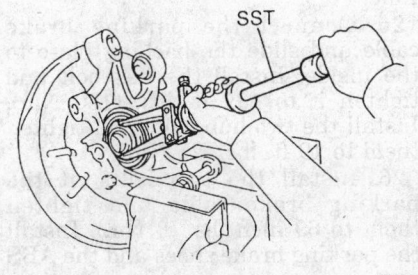

Removing the oil seal (inner) — 1995–97 LS400

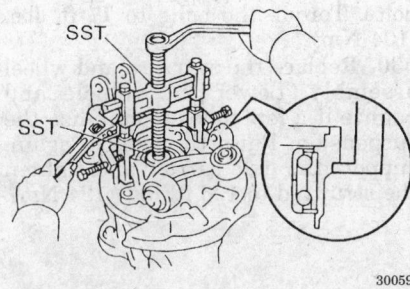

Removing the axle hub from the axle carrier — 1995–97 LS400

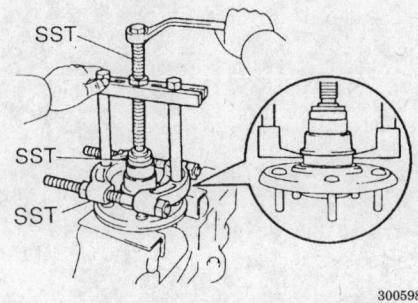

Removing the inner race (outside) from the axle hub — 1995–97 LS400

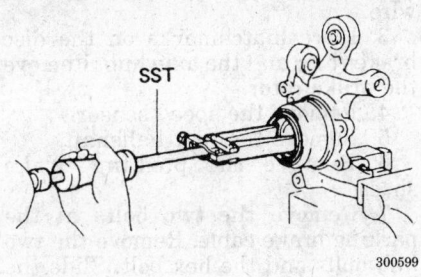

Removing the oil seal (outer) — 1995–97 LS400

25. Using snapring pliers, remove the snapring from inside the axle housing.

26. Using SST 09950–60010 and 09950–70010 or equivalent, remove the bearing from the axle housing.

To install:

27. Using SST 09527–17011 and 09608–32010 or equivalent, install a new bearing to the axle housing.

28. Using snapring pliers, install the snapring to the axle carrier.

29. Install a new outer oil seal. Coat the oil seal lip with multipurpose grease.

30. Install the backing plate to the axle housing. Do not install the bolts or nuts at this time.

31. Install the inner race (inside) to the axle housing.

32. Using a press, install the axle hub to the axle housing.

33. Place the backing plate in position install the bolts and nuts. Torque the bolts and nuts to 43 ft. lbs. (59 Nm).

34. Install a new oil seal (inner) to the axle housing. Coat the oil seal lip with multipurpose grease.

35. Using a press, install a new dust deflector. Make sure to align the hose for the ABS speed sensor in the dust deflector and axle carrier.

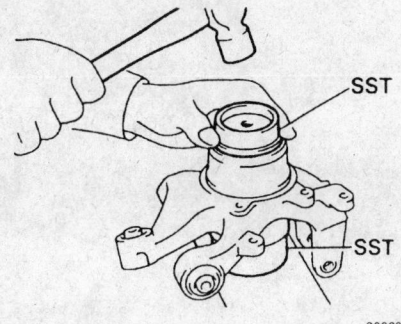

Installing the oil seal (outer) — 1995–97 LS400

36. Install the upper control arm to the axle carrier by installing the nut. Torque the nut to 80 ft. lbs. (108 Nm).

37. Install the axle carrier and upper control arm to the vehicle as an assembly.

38. Install the two upper control arm set bolts and torque the bolts to 121 ft. lbs. (164 Nm).

39. Install the bolt and nut holding the strut to the axle carrier. Torque to 101 ft. lbs. (137 Nm).

40. Install the lower control arms as follows:

    a. Install the bolt and nut connecting the No. 2 lower control arm to the axle carrier. Torque the bolt to 60 ft. lbs. (81 Nm).

    b. Install the nut and adjusting cam to hold the No. 2 lower control arm to the body. Align the adjusting cam marks and torque the nut to 57 ft. lbs. (78 Nm).

    c. Install the stabilizer bar link to the No. 2 lower control arm and torque the nut to 48 ft. lbs. (65 Nm).

    d. Install the No. 1 lower control arm to the axle carrier and body. Install the nut to hold the No. 1 lower control arm to the axle carrier. Torque the nut to 43 ft. lbs. (59 Nm).

    e. Install the nut and adjusting cam to hold the No. 1 lower control arm to the body. Align the matchmarks and torque the nut to 57 ft. lbs. (78 Nm).

41. Install the strut rod as follows:

    a. Install the strut rod to the axle carrier and body. Install the bolt and nut to hold the strut rod to the body. Torque to 57 ft. lbs. (78 Nm).

    b. With the strut connected to the axle carrier, install the bolt and nut to hold the strut rod to the axle carrier. Torque to 136 ft. lbs. (184 Nm).

42. Install the parking brake shoes and cable.

43. Install the axle shaft to the vehicle as follows:

    a. Insert the outboard joint side of the halfshaft and align the matchmarks on the side gear shaft and the halfshaft.

    b. Coat the threads with clean oil and install the hexagon bolts. Torque bolts to 61 ft. lbs. (83 Nm).

    c. Install the suspension member brace with the two bolts. Torque the two bolts to 37 ft. lbs. (50 Nm).

    d. Install the nut to hold the halfshaft to the rear axle. Torque the nut to 213 ft. lbs. (289 Nm).

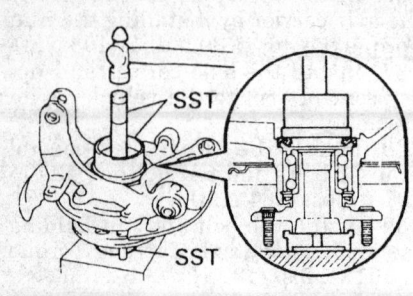

Installing the oil seal (inner) — 1995–97 LS400

e. Install the lock cap and cotter pin.

44. Install the brake disc to the axle hub with the matchmarks aligned. Install the two screws and torque the screws to 48 inch lbs. (5 Nm).

45. Install the brake caliper to the vehicle and install the two bolts. Torque the bolts to 77 ft. lbs. (104 Nm).

46. Install the ABS speed sensor and wire harness.

47. Install the height control sensor link with the matchmarks aligned. Torque the nut to 48 inch lbs. (5 Nm).

48. Install the rear wheel(s).

49. Connect the negative battery cable.

50. Lower the vehicle and turn **ON** the air suspension switch.

SC300 and SC400

1. Raise and safely support the vehicle. Remove the rear tire and wheel assembly.

2. Disconnect the brake caliper support bracket from the rear axle carrier and support it with a piece of wire.

3. Place matchmarks on the disc brake rotor and the axle hub. Remove the brake rotor.

4. Remove the speed sensor.

5. Remove the rear halfshaft.

6. Remove the parking brake shoes.

7. Remove the two bolts at the parking brake cable. Remove the two hub bolts and the hex bolt. Slide the backing plate to the outside and disconnect the parking brake cable.

8. Disconnect the strut rod at the axle carrier.

9. Remove the nut and then press out the No. 1 lower suspension arm.

10. Remove the nut and then press out the No. 2 lower suspension arm.

11. Remove the nut and then press out the upper suspension arm. Remove the axle carrier.

12. Remove the dust deflector and pull out the oil seal.

13. Using a two arm puller, remove the axle hub from the carrier.

14. Remove the backing plate.

15. Press the inner race (outside) from the hub. Then, remove the oil seal and the snapring.

16. Place the inner race (outside) over the bearing and tap out the bearing and inner race (inside).

To install:

17. Install the bearing to the axle carrier.

NOTE: If the inner races come loose from the bearing outer race, be sure to install them on the same side as before.

18. Install the snapring, the inner race (outside) and a new oil seal.

19. Install the backing plate. Install the inner race (inside) and press in the axle hub with the proper tools.

20. Install the inner oil seal. Align the holes for the speed sensor in the dust deflector and axle carrier. Install the dust deflector.

21. Install the upper arm to the axle carrier. Tighten the nut and bolt to 80 ft. lbs. (109 Nm).

22. Connect the No. 2 lower arm to the carrier and tighten a new nut to 110 ft. lbs. (150 Nm).

23. Connect the No. 1 lower arm to the carrier and tighten a new nut to 43 ft. lbs. (59 Nm).

24. Connect the strut rod to the carrier. Do not torque the bolt at this time.

25. Connect the parking brake cable and slide the backing plate to the inside. Install the hex bolt and tighten it to 132 ft. lbs. (180 Nm). Install the two hub bolts and tighten them to 19 ft. lbs. (26 Nm).

26. Install the two bolts at the parking brake cable and tighten them to 69 inch lbs. (8 Nm). Install the parking brake shoes and the ABS sensor.

27. Install the halfshafts. Tighten the locknut to 213 ft. lbs. (289 Nm).

28. Install the brake rotor.

29. Connect the brake caliper to the rear axle carrier by installing the two bolts. Torque the bolts to 77 ft. lbs. (104 Nm).

30. Replace the rear tire and wheel assembly. Lower the vehicle and bounce it a few times to stabilize the suspension. Raise the vehicle again, support the axle carrier and tighten the strut rod to 136 ft. lbs. (184 Nm).

FIRING ORDERS

NOTE: To avoid confusion, always replace spark plug wires one at a time.

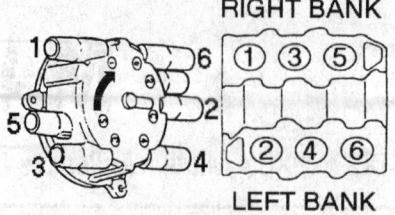

RIGHT BANK / LEFT BANK

253849

1993–94 929 3.0L, (JE-ZE) Engine
Engine Firing Order: 1–2–3–4–5–6
Distributor Rotation: Clockwise

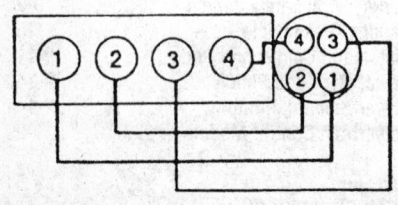

287316

4-cylinder except 626, MX6 2.0L (FS) and Miata 1.6L (B6E) and 1.8L (BPD) Engine
Firing Order: 1–3–4–2
Distributor Rotation: Counterclockwise

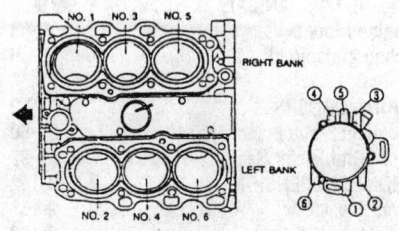

287321

1993 MX3, 1.8L (K8), 1995–97 626/MX6 2.5L (KL), and 1995–97 Millenia, 2.5L (KLD) Engine
Engine Firing Order: 1–2–3–4–5–6
Distributor Rotation: Counterclockwise

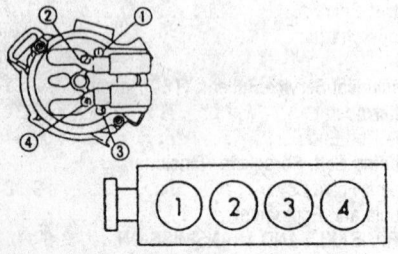

287320

1993–97 626 and MX6, 2.0L (FS) Engine
Engine Firing Order: 1–3–4–2
Distributor Rotation: Clockwise

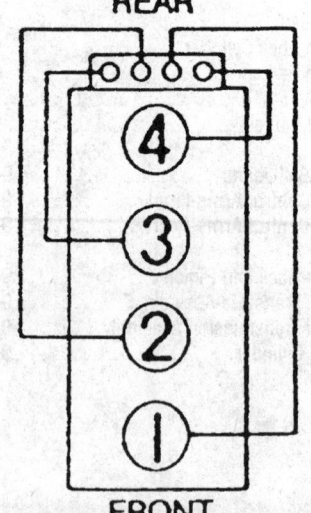

REAR / FRONT

287324

Miata 1.6L (B6E) and 1.8L (BPD)
Firing Order: 1–3–4–2
Distributorless Ignition

287328

1993–95 RX7 1.2L and 1.3L (13B) Rotary Engine

ENGINE ELECTRICAL

NOTE: Disconnecting the negative battery cable on some vehicles may interfere with the functions of the on board computer systems and may require the computer to undergo a relearning process, once the negative battery cable is reconnected.

Distributor

REMOVAL AND INSTALLATION

1993–94 323 1.6L (B6E) and Protege 1.8L (BPD and BPE) Engines

1. Disconnect the negative battery cable.

2. Remove the distributor cap and position it aside, leaving the ignition wires connected.

3. On SOHC engines, remove the air intake hose from it's position next to the distributor.

4. Disconnect the distributor electrical connector(s) from the side of the distributor.

5. Using a wrench on the crankshaft pulley, rotate the crankshaft to position the No. 1 piston on Top Dead Center (TDC) of the compression stroke; the crankshaft pulley mark should align with the timing indicator.

6. Using chalk or paint, mark the position of the distributor housing on the cylinder head. Also mark the position of the distributor rotor in relation to the distributor housing.

7. Remove the distributor hold-down bolt(s) and remove the distributor.

8. Inspect the O-ring on the distributor housing and replace it, if it is damaged or worn.

To install:

Timing Not Disturbed

1. Apply clean engine oil to a new O-ring and install it on the distributor.

2. Install the distributor in the cylinder head. Make sure the distributor rotor is pointing toward the No. 1 spark plug tower position on the distributor cap.

3. Install the distributor mounting bolts. Align the marks made on the distributor housing and cylinder head during removal and loosely tighten the bolts.

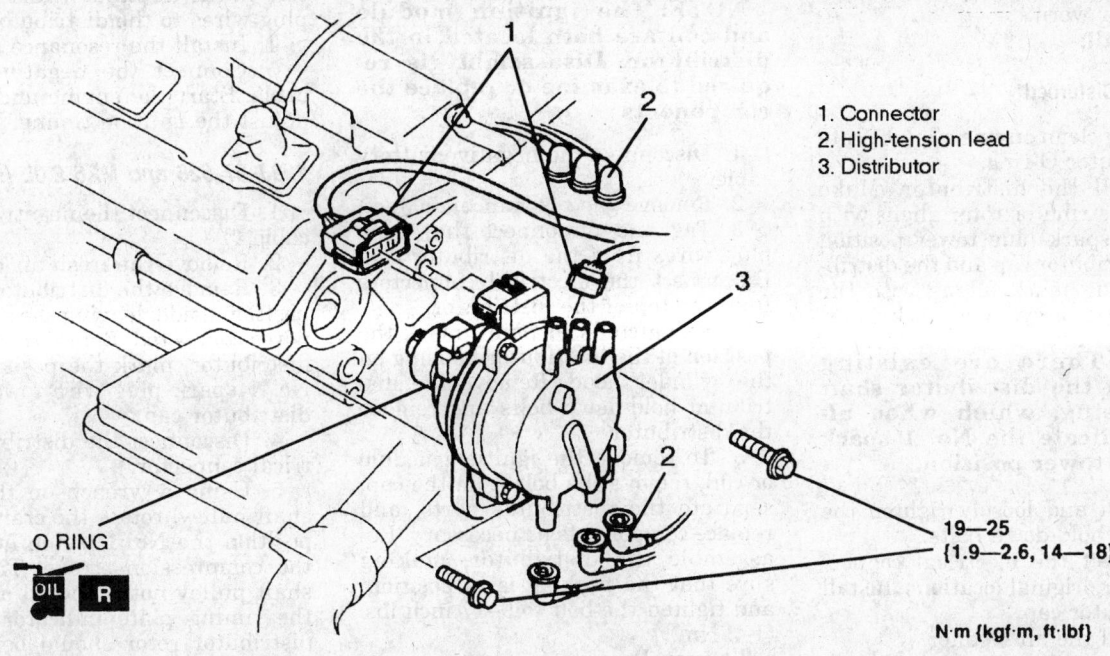

1. Connector
2. High-tension lead
3. Distributor

19—25
{1.9—2.6, 14—18}

N·m {kgf·m, ft·lbf}

308261

Distributor assembly — Millenia 2.5L (KLD) engines

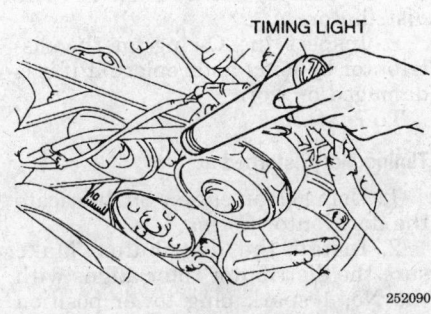

TIMING LIGHT

252090

Aim the timing light in the appropriate position

4. Connect the distributor wiring and, if equipped, vacuum hoses.

5. Install the distributor cap and secure the screws.

6. Connect the spark plug wires to the distributor cap in their original locations.

7. Connect the negative battery cable. Check and adjust the ignition timing.

Timing Disturbed

1. Disconnect the spark plug wire and remove the spark plug from the No. 1 cylinder. Place a finger over the spark plug hole.

2. Turn the crankshaft pulley in the normal direction of rotation until compression is felt; the piston is approaching TDC on the compression stroke. Continue rotating the crankshaft pulley until the pulley mark aligns with the mark on the timing belt cover.

3. Apply clean engine oil to a new O-ring and install it on the distributor.

4. Install the distributor in the cylinder head. Make sure the distributor rotor is pointing toward the No. 1 spark plug tower position on the distributor cap.

5. Install the distributor mounting bolts. Align the marks made on the distributor housing and cylinder head during removal and loosely tighten the bolts.

6. Connect the distributor wiring and, if equipped, vacuum hoses.

7. Install the distributor cap and secure the screws.

8. Install the spark plug in the No. 1 cylinder and connect the spark plug wire.

9. Connect the spark plug wires to the distributor cap in their original locations.

10. Connect the negative battery cable. Check and adjust the ignition timing.

1995–97 Protege 1.5L (Z5D), 1.8L (BPD) and 1995 MX3 1.6L (B6-ZE) Engines

NOTE: The ignition coil and module are integral to the distributor. The assembly must be replaced as a whole.

1. Disconnect the negative battery cable.

2. Remove the air cleaner.

3. Remove the distributor cap and position aside, leaving the spark plug wires connected. Before removing the distributor, mark the position of the No. 1 spark plug wire tower on the distributor cap.

4. Disconnect the distributor electrical connectors.

5. Using a wrench on the crankshaft pulley, rotate the crankshaft to position the No. 1 piston at TDC on the compression stroke. The crankshaft pulley notch should align with the timing plate indicator and the distributor rotor should be pointing to the No. 1 spark plug tower position on the distributor cap.

6. Using chalk or paint, mark the position of the distributor housing on the cylinder head.

7. Remove the distributor hold-down bolts and remove the distributor.

8. Inspect the O-ring on the distributor housing and replace it if it is damaged or worn.

To install:

Timing Not Disturbed

1. Using clean engine oil, lubricate the distributor O-ring.
2. Install the distributor. Make sure the distributor rotor aligns with the No. 1 spark plug tower position on the distributor cap and the distributor housing mark aligns with the cylinder head or cylinder block mark.

NOTE: There are existing marks on the distributor shaft and housing, which when aligned, indicate the No. 1 spark plug wire tower position.

3. Install and loosely tighten the distributor hold-down bolts.
4. Connect the electrical connectors to their original locations. Install the distributor cap.
5. Install the air cleaner.
6. Connect the negative battery cable. Start the engine, and check or adjust the ignition timing.

Timing Disturbed

1. Using clean engine oil, lubricate the distributor O-ring.
2. Disconnect the spark plug wire from the No. 1 cylinder spark plug. Remove the spark plug from the No. 1 cylinder and press a thumb over the spark plug hole.
3. Using a wrench on the crankshaft pulley, rotate the crankshaft until pressure is felt at the spark plug hole, indicating the piston is approaching TDC on the compression stroke. Continue rotating the crankshaft until the crankshaft pulley mark aligns with the timing cover indicator.
4. Position the distributor rotor so it aligns with the No. 1 spark plug wire tower on the distributor cap.
5. Install the distributor. Align the mark that was made on the distributor housing with the mark that was made on the cylinder block. Loosely tighten the distributor hold-down bolts.
6. Connect the electrical connectors to their original locations. Install the distributor cap.
7. Install the spark plug in the No. 1 cylinder and connect the spark plug wire.
8. Install the air cleaner.
9. Connect the negative battery cable. Start the engine, and check or adjust the ignition timing.

1995 MX3 1.8L (K8) Engines

NOTE: The ignition module and coil are both located in the distributor. Disassembly is required to examine or replace the components.

1. Disconnect the negative battery cable.
2. Remove the resonance air duct.
3. Tag and disconnect the spark plug wires from the distributor cap. Disconnect the electrical connectors from the top of the distributor.
4. Using chalk or paint, mark the position of the distributor housing on the cylinder head. Remove the distributor hold-down bolts and remove the distributor.
5. To remove the ignition module or coil, remove the bolt from the cap, separate the distributor parts, and replace components as necessary. Reassemble the distributor, making sure that the packing is in position, and tighten the bolt to 9–17 inch lbs. (1–2 Nm).

To install:

Timing Not Disturbed

1. Align the distributor shaft with the camshaft end and install the distributor.

NOTE: The tangs on the distributor shaft are different sizes, allowing the distributor to be installed in only one position.

2. Install the distributor hold-down bolts. Align the mark that was made on the distributor housing with the mark that was made on the cylinder head and loosely tighten the bolts.
3. Connect the electrical connectors to the distributor and the spark plug wires to the distributor cap.
4. Install the resonance air duct.
5. Connect the negative battery cable. Start the engine, and check or adjust the ignition timing.

Timing Disturbed

1. Align the distributor shaft with the camshaft end and install the distributor.

NOTE: The tangs on the distributor shaft are different sizes, allowing the distributor to be installed in only 1 position.

2. Install the distributor hold-down bolts. Align the mark that was made on the distributor housing with the mark that was made on the cylinder head and loosely tighten the bolts.

3. Connect the electrical connectors to the distributor and the spark plug wires to the distributor cap.
4. Install the resonance air duct.
5. Connect the negative battery cable. Start the engine and check or adjust the ignition timing.

1993–97 626 and MX6 2.0L (FS)

1. Disconnect the negative battery cable.
2. Remove the fresh air duct.
3. Remove the distributor cap and position aside, leaving the spark plug wires connected. Before removing the distributor, mark the position of the No. 1 spark plug wire tower on the distributor cap.
4. Disconnect the distributor electrical connector.
5. Using a wrench on the crankshaft pulley, rotate the crankshaft to position the No. 1 piston at TDC on the compression stroke. The crankshaft pulley notch should align with the timing plate indicator and the distributor rotor should be pointing to the No. 1 spark plug tower position on the distributor cap.
6. Using chalk or paint, mark the position of the distributor housing on the cylinder head.
7. Remove the distributor hold-down bolt(s) and remove the distributor.
8. Inspect the O-ring on the distributor housing and replace it if it is damaged or worn.

To install:

Timing Not Disturbed

1. Using clean engine oil, lubricate the distributor O-ring.
2. Install the distributor. Make sure the distributor rotor aligns with the No. 1 spark plug tower position on the distributor cap and the distributor housing mark aligns with the cylinder head or cylinder block mark.

NOTE: There are existing marks on the distributor shaft and housing, which when aligned, indicate the No. 1 spark plug wire tower position.

3. Install and loosely tighten the distributor hold-down bolt(s).
4. Connect the electrical connectors to their original locations. Install the distributor cap.
5. Install the fresh air duct.
6. Connect the negative battery cable. Start the engine and check or adjust the ignition timing.

Timing Disturbed

1. Using clean engine oil, lubricate the distributor O-ring.

2. Disconnect the spark plug wire from the No. 1 cylinder spark plug. Remove the spark plug from the No. 1 cylinder and press a thumb over the spark plug hole.

3. Using a wrench on the crankshaft pulley, rotate the crankshaft until pressure is felt at the spark plug hole, indicating the piston is approaching TDC on the compression stroke. Continue rotating the crankshaft until the crankshaft pulley mark aligns with the timing cover indicator.

4. Position the distributor rotor so it aligns with the No. 1 spark plug wire tower on the distributor cap.

5. Install the distributor. Align the mark that was made on the distributor housing with the mark that was made on the cylinder block. Loosely tighten the distributor hold-down bolts.

6. Connect the electrical connectors to their original locations. Install the distributor cap.

7. Install the spark plug in the No. 1 cylinder and connect the spark plug wire.

8. Install fresh air duct.

9. Connect the negative battery cable. Start the engine and check or adjust the ignition timing.

1993–97 626 and MX6 2.5L (KL) Engines

NOTE: The ignition module and coil are both located in the distributor. Disassembly is required to examine or replace the components.

1. Disconnect the negative battery cable.

2. Remove the 2 fresh air duct nuts and 3 bolts. Loosen the spring clamp at the front of the air cleaner assembly and slide it forward. Remove the fresh air duct.

3. Loosen the clamp on the front of the air flow meter and disconnect the air duct. Disconnect the air flow meter electrical connector at the left side of the air cleaner.

4. Disconnect the evaporative canister hose from the routing clip on the front of the air cleaner. Remove the fuel pressure regulator control solenoid from the air cleaner and position aside.

5. Remove the nuts and bolt and remove the air cleaner assembly.

6. Tag and disconnect the spark plug wires from the distributor cap. Disconnect the electrical connectors from the top of the distributor.

7. Using chalk or paint, mark the position of the distributor housing on

the cylinder head. Remove the distributor hold-down bolts and remove the distributor.

8. To remove the ignition module or coil, remove the bolt from the cap, separate the distributor parts, and replace components as necessary. Reassemble the distributor, making sure that the packing is in position, and tighten the bolt to 27–46 inch lbs. (3–5 Nm).

To install:

Timing Not Disturbed

1. Align the distributor shaft with the camshaft end and install the distributor.

NOTE: The tangs on the distributor shaft are different sizes, allowing the distributor to be installed in only one position.

2. Install the distributor hold-down bolts. Align the mark that was made on the distributor housing with the mark that was made on the cylinder head and loosely tighten the bolts.

3. Connect the electrical connectors to the distributor and the spark plug wires to the distributor cap.

4. Install the air cleaner assembly and tighten the nuts and bolt to 18 ft. lbs. (25 Nm).

5. Install the fuel pressure regulator solenoid and connect the evaporative canister hose into the routing clip.

6. Connect the air flow meter electrical connector. Connect the air duct and tighten the clamp.

7. Align the fresh air duct and install the hose to the air cleaner assembly. Loosen the spring clamp and slide it into position. Install the fresh air duct nuts and bolts and tighten to 71–88 inch lbs. (8–10 Nm).

8. Connect the negative battery cable. Start the engine, and check or adjust the ignition timing.

Timing Disturbed

1. Align the distributor shaft with the camshaft end and install the distributor.

NOTE: The tangs on the distributor shaft are different sizes, allowing the distributor to be installed in only 1 position.

2. Install the distributor hold-down bolts. Align the mark that was made on the distributor housing with the mark that was made on the cylinder head and loosely tighten the bolts.

3. Connect the electrical connectors to the distributor and the spark plug wires to the distributor cap.

4. Install the air cleaner assembly and tighten the nuts and bolt to 18 ft. lbs. (25 Nm).

5. Install the fuel pressure regulator solenoid and connect the evaporative canister hose into the routing clip.

6. Connect the air flow meter electrical connector. Connect the air duct and tighten the clamp.

7. Align the fresh air duct and install the hose to the air cleaner assembly. Loosen the spring clamp and slide it into position. Install the fresh air duct nuts and bolts and tighten to 71–88 inch lbs. (8–10 Nm).

8. Connect the negative battery cable. Start the engine and check or adjust the ignition timing.

1993–95 929 3.0L (JE-ZE)

1. Disconnect the negative battery cable.

2. Turn the crankshaft pulley, in the normal direction of rotation, until the No. 1 cylinder piston is at Top Dead Center (TDC) on the compression stroke. The mark on the crankshaft pulley should be aligned with the **T** on the timing belt cover.

3. Label and remove the spark plug wires from the distributor cap. Remove the distributor cap.

4. Disconnect the wiring harness from the distributor.

5. Mark the position of the distributor housing on the cylinder head.

6. Remove the distributor mounting bolt and remove the distributor. Remove and discard the distributor O-ring.

To install:

Timing Not Disturbed

1. Apply clean engine oil to a new O-ring and install it on the distributor.

2. Install the distributor in the cylinder head. Make sure the distributor rotor is pointing toward the No. 1 spark plug tower position on the distributor cap.

3. Install the distributor mounting bolt. Align the marks made on the distributor housing and cylinder head during removal and loosely tighten the bolts.

4. Connect the distributor wiring.

5. Install the distributor cap and secure the screws.

6. Connect the spark plug wires to the distributor cap in their original locations.

7. Connect the negative battery cable. Check and adjust the ignition timing.

Timing Disturbed

1. Disconnect the spark plug wire and remove the spark plug from the No. 1 cylinder. Place a finger over the spark plug hole.

2. Turn the crankshaft pulley in the normal direction of rotation until compression is felt; the piston is approaching TDC on the compression stroke. Continue rotating the crankshaft pulley until the pulley mark aligns with the **T** mark on the timing belt cover.

3. Apply clean engine oil to a new O-ring and install it on the distributor.

4. Install the distributor in the cylinder head. Make sure the distributor rotor is pointing toward the No. 1 spark plug tower position on the distributor cap.

5. Install the distributor mounting bolt. Align the marks made on the distributor housing and cylinder head during removal and loosely tighten the bolt.

6. Connect the distributor wiring.

7. Install the distributor cap and secure the screws.

8. Install the spark plug in the No. 1 cylinder and connect the spark plug wire.

9. Connect the spark plug wires to the distributor cap in their original locations.

10. Connect the negative battery cable. Check and adjust the ignition timing.

Millenia

NOTE: The ignition module and coil are both located in the distributor. Disassembly is required to examine or replace the components.

1. Disconnect the negative battery cable.

2. Tag and disconnect the spark plug wires from the distributor cap. Disconnect the electrical connectors from the top of the distributor.

3. Using chalk or paint, mark the position of the distributor housing on the cylinder head. Remove the distributor hold-down bolts and remove the distributor.

4. To remove the ignition module or coil, remove the bolt from the cap, separate the distributor parts, and replace components as necessary. Reassemble the distributor, making sure that the packing is in position, and tighten the bolt to 27–46 inch lbs. (3–5 Nm).

To install:

Timing Not Disturbed

1. Align the distributor shaft with the camshaft end and install the distributor.

NOTE: The tangs on the distributor shaft are different sizes, allowing the distributor to be installed in only one position.

2. Install the distributor hold-down bolts. Align the mark that was made on the distributor housing with the mark that was made on the cylinder head and tighten the bolts to 14–18 ft. lbs. (19–25 Nm).

3. Connect the electrical connectors to the distributor and the spark plug wires to the distributor cap.

4. Connect the negative battery cable. Start the engine, and check or adjust the ignition timing.

Timing Disturbed

1. Align the distributor shaft with the camshaft end and install the distributor.

NOTE: The tangs on the distributor shaft are different sizes, allowing the distributor to be installed in only 1 position.

2. Install the distributor hold-down bolts. Align the mark that was made on the distributor housing with the mark that was made on the cylinder head and tighten the bolts to 14–18 ft. lbs. (19–25 Nm).

3. Connect the electrical connectors to the distributor and the spark plug wires to the distributor cap.

4. Connect the negative battery cable. Start the engine and check or adjust the ignition timing.

Ignition Timing

ADJUSTMENT

1993 MX3, 1.8L (K8), 1993–95 626, MX6 2.5L (KL), Millenia 2.5L (KLD), 929 3.0L (JE-ZE)1995–97 Protege 1.5L (Z5D), 1995 Protege 1.8L (BPD) and MX3 1.6L (B6-ZE) Engines

1. Apply the parking brake. Place the shift lever in NEUTRAL on manual transaxles, and **P** on automatic transaxles.

2. Locate the timing marks on the crankshaft pulley and the timing indicator scale on the engine front cover. If the marks are hard to see, clean them with degreaser and a stiff brush.

3. Start the engine and bring to normal operating temperature. Make sure all accessories are OFF.

4. Connect a tachometer and timing light to the engine according to the manufacturer's instructions.

5. Verify that the idle speed is for the 1995–97 Protege 1.5L (Z5D) engine, between 650–750 rpms on manual, and 700–800 rpms on automatic transaxles. Idle speed on the 1993 MX3 1.6L (B6-ZE) and Protege 1.8L (BPD) engines should be between 700–800 rpms. Adjust the idle speed as necessary.

6. On the 1993–95 MX3, 1.8L (K8), 1993–95 626, MX6 2.5L (KL) and Millenia 2.5L (KLD) engines, check the idle speed and adjust, if necessary; it should be 600–700 rpm.

7. On the 1993–94 929 3.0L (JE and ZE) engines, the idle speed should be 680–720 rpm.

8. Connect a jumper wire between the **TEN** terminal and the **GND** terminal of the data link connector.

9. Verify that the timing mark (white) on the crankshaft pulley, and the T mark on the timing belt cover are aligned. Aim the timing light at the timing marks; the timing should be on the 1995–97 Protege 1.5L (Z5D), 1995 Protege 1.8L (BPD) 0–1 degrees.

10. On the 1995 MX3 1.6L (B6-ZE), 1993–95 MX3, 1.8L (K8), 1993–95 626, MX6 2.5L (KL) and Millenia 2.5L (KLD) engines, the timing should be 9–11 degrees.

11. On the 1993–94 929 3.0L (JE and ZE) engines, the timing should be 11–13 degrees BTDC.

12. If the timing marks are not aligned, loosen the distributor hold-down bolt and turn the distributor housing to adjust. When the marks align, tighten the distributor hold-down bolts to 14–19 ft. lbs. (19–25 Nm) and recheck the timing.

13. Remove the jumper wire.

14. Increase the engine speed and verify that the timing advances.

15. Remove all test equipment.

1993 626 and MX6 2.0L (FS) Engines

1. Apply the parking brake. If equipped with manual transaxle, place the shift lever in neutral. If equipped with automatic transaxle, place the shift lever in **P**.

2. Locate the timing marks on the crankshaft pulley and the timing indicator scale on the engine front cover. If the marks are hard to see, clean them with degreaser and a wire brush.

3. Start the engine and bring to normal operating temperature. Make sure all accessories are **OFF**.

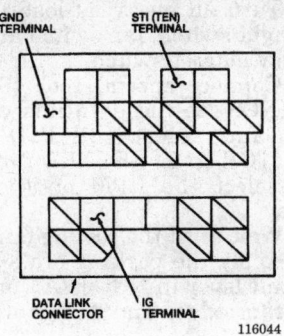

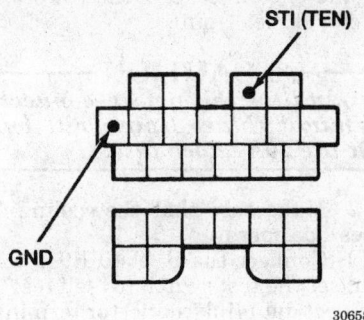

Connect a jumper wire to the Data Link Connector

306585

Data link connector (ATX only) — 1993 626 and MX6 2.0L (FS) engines

116044

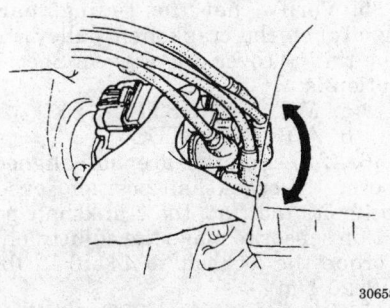

306584

Adjust the distributor position — 1995–97 Protege 1.5L (Z5D) and 1.8L (BPD) shown

4. Connect a tachometer and timing light to the engine according to the manufacturer's instructions.

5. If equipped with manual transaxle, remove the shorting bar from the double wire SPOUT connector. If equipped with automatic transaxle, connect a jumper wire between the STI **TEN** terminal and the **GND** terminal on the data link connector.

6. Check the idle speed and adjust, if necessary; it should be 700 ± 50 rpm.

7. Aim the timing light at the timing marks; the timing should be 10 ± 1 degrees if equipped with manual transaxle or 12 ± 1 degrees if equipped with automatic transaxle.

8. If the timing marks are not aligned, loosen the distributor hold-down bolt and turn the distributor housing to adjust. When the marks align, tighten the hold-down bolt to 19 ft. lbs. (25 Nm). Recheck the timing after the bolt has been tightened.

9. If equipped with manual transaxle, install the shorting bar to the double wire SPOUT connector. If equipped with automatic transaxle, remove the jumper wire from the data link connector.

10. Remove all test equipment.

1994–97 626 and MX6 2.0L (FS) Engines

Manual Transaxle

1. Apply the parking brake. Place the shift lever in neutral.

2. Locate the timing marks on the crankshaft pulley and the timing indicator scale on the engine front cover. If the marks are hard to see, clean them with degreaser and a stiff brush.

3. Start the engine and bring to normal operating temperature. Make sure all accessories are off.

4. Connect a tachometer and timing light to the engine according to the manufacturer's instructions.

5. Verify that the idle speed is between 650–750 rpm; adjust if necessary.

6. Connect a jumper wire between the **TEN** terminal and the **GND** terminal of the data link connector.

7. Verify that the idle speed is now 500–800 rpm.

8. Aim the timing light at the timing marks; the timing should be 12 ± 1 degrees,

9. If the timing marks are not aligned, loosen the distributor hold-down bolt and turn the distributor housing to adjust. When the marks align, tighten the hold-down bolt to 19 ft. lbs. (25 Nm). Recheck the timing after the bolt has been tightened.

10. Remove the jumper wire.

11. Remove all test equipment.

Automatic Transaxle

1. Apply the parking brake. Place the shift lever in **P**.

2. Locate the timing marks on the crankshaft pulley and the timing indicator scale on the engine front cover. If the marks are hard to see, clean them with degreaser and a stiff brush.

3. Start the engine and bring to normal operating temperature. Make sure all accessories are off.

4. Check the idle speed and adjust, if necessary; it should be 700 ± 50 rpm.

5. Connect a tachometer and timing light to the engine according to the manufacturer's instructions.

6. Remove the shorting bar from the double wire SPOUT connector.

7. Check the idle speed, it should be 500–800 rpm.

8. Aim the timing light at the timing marks; the timing should be 10 ± 1 degrees,

9. If the timing marks are not aligned, loosen the distributor hold-down bolt and turn the distributor housing to adjust. When the marks align, tighten the hold-down bolt to 19 ft. lbs. (25 Nm). Recheck the timing after the bolt has been tightened.

10. Install the shorting bar to the double wire SPOUT connector.

11. Remove all test equipment.

1995 Millenia 2.3L (KJS) Engines

NOTE: The ignition timing can NOT be adjusted on this engine. Follow this procedure for inspection only.

1. Apply the parking brake. Place the shift lever in **P**.

2. Locate the timing marks on the crankshaft pulley and timing belt cover. If the marks are hard to see, clean them off with some degreasing cleaner and a wire brush.

3. Start the engine and allow it to come to normal operating temperature. Make sure all accessories are **OFF**.

4. Connect a tachometer and inductive timing light according to the manufacturer's instructions.

5. Check the idle speed and adjust, if necessary; it should be 650 ± 50 rpm.

6. Connect terminals **TEN** and **GND** on the data link connector with a jumper wire.

7. Check the idle speed, it should be 550–750 rpm.

8. Connect the diagnostic tool to the ignition timing checking harness with the arrow facing the connector.

9. Aim the timing light at the timing marks. The timing should be 6–8° BTDC.

10. If it isn't in the correct range, inspect the on-board diagnostic system, the camshaft position sensor and/or the crankshaft position sensor.

11. Stop the engine. Remove the jumper wire and test equipment.

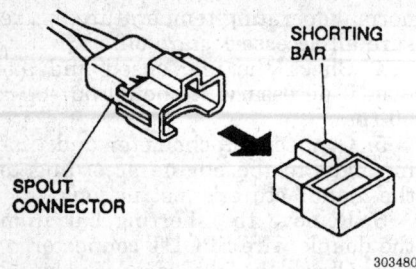

SPOUT CONNECTOR

SHORTING BAR

303480

Removing the shorting bar — 1994–97 626 and MX6 2.0L (FS) engines with an automatic transmission

1994–95 Miata 1.8L Engine

1. Apply the parking brake. Place the shift lever in NEUTRAL on manual transaxles, and **P** on automatic transaxles.

2. Locate the timing marks on the crankshaft pulley and the timing indicator scale on the engine front cover. If the marks are hard to see, clean them with degreaser and a stiff brush.

3. Start the engine and bring to normal operating temperature. Make sure all accessories are OFF.

4. Connect a tachometer and timing light to the engine according to the manufacturer's instructions.

5. Verify that the idle speed is between 800–900 rpm on manual, and 750–850 rpm on automatic transaxles. Adjust the idle speed as necessary.

6. Connect a jumper wire between the **TEN** terminal and the **GND** terminal of the data link connector.

7. Verify that the timing mark (white) on the crankshaft pulley, and the indicator pin are aligned. Aim the timing light at the timing marks; the timing should be 0 ± 1 degrees,

8. If the timing marks are not aligned:

   a. Remove the ignition coil bracket, and move aside for clearance.

   b. Loosen the camshaft position sensor lockbolts, and turn the camshaft position sensor to adjust.

   c. When the marks align, tighten the lockbolts to 14–18 ft. lbs. (19–25 Nm). Recheck the timing after the bolts have been tightened.

9. Remove the jumper wire.

10. Increase the engine speed and verify that the timing advances.

11. Remove all test equipment.

1996–97 Miata 1.8L Engine

1. Warm up the engine to normal operating temperature.

2. Turn all electrical loads OFF - headlight switch, fan switch and rear window defrost switch.

3. Connect special tool 49 T088 0A0 (NGS) to the data link connector-2 and select the PID/DATA MONITOR AND RECORD function.

4. Select the RPM on the NGS display.

5. Wait until the cooling fan stops.

6. Apply the parking brake. Place the shift lever in NEUTRAL on manual transaxles, and **P** on automatic transaxles.

7. Locate the timing marks on the crankshaft pulley and the timing indicator scale on the engine front cover. If the marks are hard to see, clean them with degreaser and a stiff brush.

8. Connect a timing light to the high-tension lead No. 1.

9. Connect special tool 49 B019 9A0 (system selector) to the data link connector.

10. Set switch A to position 1.

11. Set the test switch to SELF TEST.

12. Verify that the idle speed is within the specification of 650–975 rpm.

13. Verify that the timing mark (white) on the crankshaft pulley, and the T mark on the timing belt cover are aligned. Aim the timing light at the timing marks; the timing should be 9°–11° BTDC (yellow timing mark).

14. If the timing marks are not aligned:

   a. Remove the ignition coil bracket, and move aside for clearance.

   b. Loosen the camshaft position sensor lockbolts, and turn the camshaft position sensor to adjust.

   c. When the marks align, tighten the lockbolts to 14–18 ft. lbs. (19–25 Nm). Recheck the timing after the bolts have been tightened.

15. Disconnect the special tool.

16. Increase the engine speed and verify that the timing advances.

17. Remove all remaining test equipment.

1993 Miata 1.6L Engine

Before making any adjustment, make sure of the condition of the engine (spark plugs, leaks in hoses, etc.). Make sure all accessories are OFF and warm the engine to normal operating temperature.

1. Connect a timing light and a tachometer to the data link connector terminal IG-. When using an externally powered timing light and/or ta-chometer, connect it to the power connector (blue: 1-pin).

—————— CAUTION ——————
Grounding the power connector terminal (blue: 1-pin) will burn out the 20A wiper fuse.
—————————————————————

2. Make sure that the cooling fan does not operate.

3. Connect the SST 49 B019 9AO and set the test switch to "self test" or connect data link connector terminals TEN and GRD with a jumper wire.

4. Check the idle speed and set it to specification:

   a. A/T - park 800±50 rpm.

5. Verify that the timing mark (white) on the crankshaft pulley and the timing cover are aligned. Specification is:

   a. M/T - 9–11° BTDC.

   b. A/T - 7–9° BTDC.

6. If the marks are not aligned, loosen the crankshaft position sensor lockbolt, and turn the crankshaft position sensor to make the adjustment. Torque the lockbolt to 14–18 ft. lbs. (19–25 Nm).

7. Disconnect the SST 49 B019 9AO.

8. Verify that the ignition timing is within specification.

9. Disconnect all other test equipment and check for proper operation.

1993–94 Protege 1.8L Engine

1. Apply the parking brake. Place the shift lever in NEUTRAL on manual transaxles, and **P** on automatic transaxles.

2. Locate the timing marks on the crankshaft pulley and the timing indicator scale on the engine front cover. If the marks are hard to see, clean them with degreaser and a stiff brush.

3. Start the engine and bring to normal operating temperature. Make sure all accessories are OFF.

4. Connect a tachometer and timing light to the engine according to the manufacturer's instructions.

5. Verify that the idle speed is between 700–800 rpm. Adjust the idle speed as necessary.

6. Connect a jumper wire between the **TEN** terminal and the **GND** terminal of the data link connector.

7. Verify that the timing mark (yellow) on the crankshaft pulley, and the T mark on the timing belt cover are aligned. Aim the timing light at the timing marks; the timing should be 10 ± 1 degrees for the DOHC engine and 5 ± 1 degree for the SOHC engine.

8. If the timing marks are not aligned, loosen the distributor hold-down bolts and turn the distributor housing to adjust. When the marks align, tighten the hold-down bolt to 18 ft. lbs. (25 Nm). Recheck the timing after the bolt has been tightened.

9. Remove the jumper wire.

10. Increase the engine speed and verify that the timing advances.

11. Remove all test equipment.

Alternator

PRECAUTIONS

Several precautions must be observed with alternator equipped vehicles to avoid damage to the unit.

• If the battery is removed for any reason, make sure it is reconnected with the correct polarity. Reversing the battery connections may result in damage to the 1-way rectifiers.

• When utilizing a booster battery as a starting aid, always connect the positive to positive terminals and the negative terminal from the booster battery to a good engine ground on the vehicle being started.

• Never use a fast charger as a booster to start vehicles.

• Disconnect the battery cables when charging the battery with a fast charger.

• Never attempt to polarize the alternator.

• Do not use test lights of more than 12 volts when checking diode continuity.

• Do not short across or ground any of the alternator terminals.

• The polarity of the battery, alternator and regulator must be matched and considered before making any electrical connections within the system.

• Never separate the alternator on an open circuit. Make sure all connections within the circuit are clean and tight.

• Disconnect the battery ground terminal when performing any service on electrical components.

• Before disconnecting the battery, obtain the radio theft protection code.

• Do not reverse the battery connections or the alternator and various control computers will be damaged.

• Do not use high voltage testers when testing the rectifiers.

• Do not disconnect the battery or the alternator while the engine is running. Do not start the engine with the alternator disconnected.

• Remember that terminal **B** on the alternator is always the battery connection.

• Disconnect the battery and control computers if electric welding is to be done anywhere on the vehicle.

REMOVAL AND INSTALLATION

1993–94 323 and MX3, 1.6L (B6E), 1993–95 Protege 1.8L (BPE, BPD), 1995 MX3 1.6L (B6-ZE) and 1995–97 Protege 1.5L (Z5D) and 1.8L (BPD) Engines

1. Disconnect the negative battery cable.

2. On 323/Protege, disconnect the vacuum hose and remove the solenoid bracket, if equipped.

3. On 1995–97 1.8L engines, remove the pressure pipe bracket and the EGR solenoid valve bracket.

4. Remove the alternator drive belt.

5. Label and disconnect the electrical connectors from the alternator.

6. On the 1995–97 1.8L engines, remove the alternator bracket.

7. Remove the alternator pivot and adjusting bar bolts and remove the alternator.

To install:

8. Install the alternator with the through–bolt.

9. On the 1995–97 1.8L engines, install the alternator bracket.

10. Connect the alternator electrical connectors.

11. Install the drive belt and upper mounting bolt. Adjust the belt tension. Tighten the lower through–bolt to 27–38 ft. lbs. (37–52 Nm) and the upper mounting bolt to 12–19 ft. lbs. (16–26 Nm).

12. On the 1995–97 1.8L engines, install the pressure pipe bracket and the EGR solenoid valve bracket.

13. Connect the negative battery cable.

1993–95 MX3 1.8L (K8) and 1993–97 626 and MX6 2.5L (KL) Engines

1. Disconnect the negative battery cable.

2. Remove the fresh air duct and radiator upper bracket.

3. If equipped, remove the condenser fan.

4. Disconnect the electrical connectors from the alternator.

5. Loosen the belt tensioner locknut and tension adjusting bolt. Remove the alternator upper mounting bolt.

6. Raise and safely support the vehicle.

7. Remove the right splash shield.

8. Remove the drive belt from the alternator pulley.

9. If necessary, remove the A/C compressor mounting bolts and and support the compressor aside, leaving the refrigerant lines connected.

10. Remove the alternator through-bolt and the alternator.

To install:

11. Position the alternator and install the through-bolt. Tighten the alternator through-bolt to 24–33 ft. lbs. (32–46 Nm) on the 1.8L, and to 38 ft. lbs. (51 Nm) on the 2.5L engine.

12. If removed, install the mounting bolts and tighten the A/C compressor mounting bolts to 26 ft. lbs. (35 Nm).

13. Install the right splash shield, and lower the vehicle.

14. Install the drive belt and adjust the tension.

15. Install the alternator upper mounting bolt and tighten to 18 ft. lbs. (25 Nm).

16. Tighten the belt tensioner locknut and tension adjusting bolt.

17. Connect the electrical connectors from the alternator.

18. If removed, install the condenser fan.

19. Install the radiator upper bracket and fresh air duct.

20. Connect the negative battery cable.

626 and MX6 2.0L (FS) Engines

1. Disconnect the negative battery cable.

2. Remove the alternator upper mounting bolt.

3. Loosen the alternator adjusting bolt and remove the drive belt from the alternator pulley.

4. Raise and safely support the vehicle.

5. Remove the 6 bolts and remove the transverse member.

6. Disconnect the electrical connectors from the alternator.

7. Remove the front exhaust pipe as follows:

   a. Support the exhaust system at the catalytic converter with a jack.

   b. Disconnect the oxygen sensor electrical connector and remove the sensor using a sensor wrench.

   c. Remove the 3 exhaust manifold flange nuts and remove the clamp from the hold-down bracket.

   d. Remove the exhaust pipe-to-converter nuts and pry the rubber hangers from the mounting hooks. Remove the pipe.

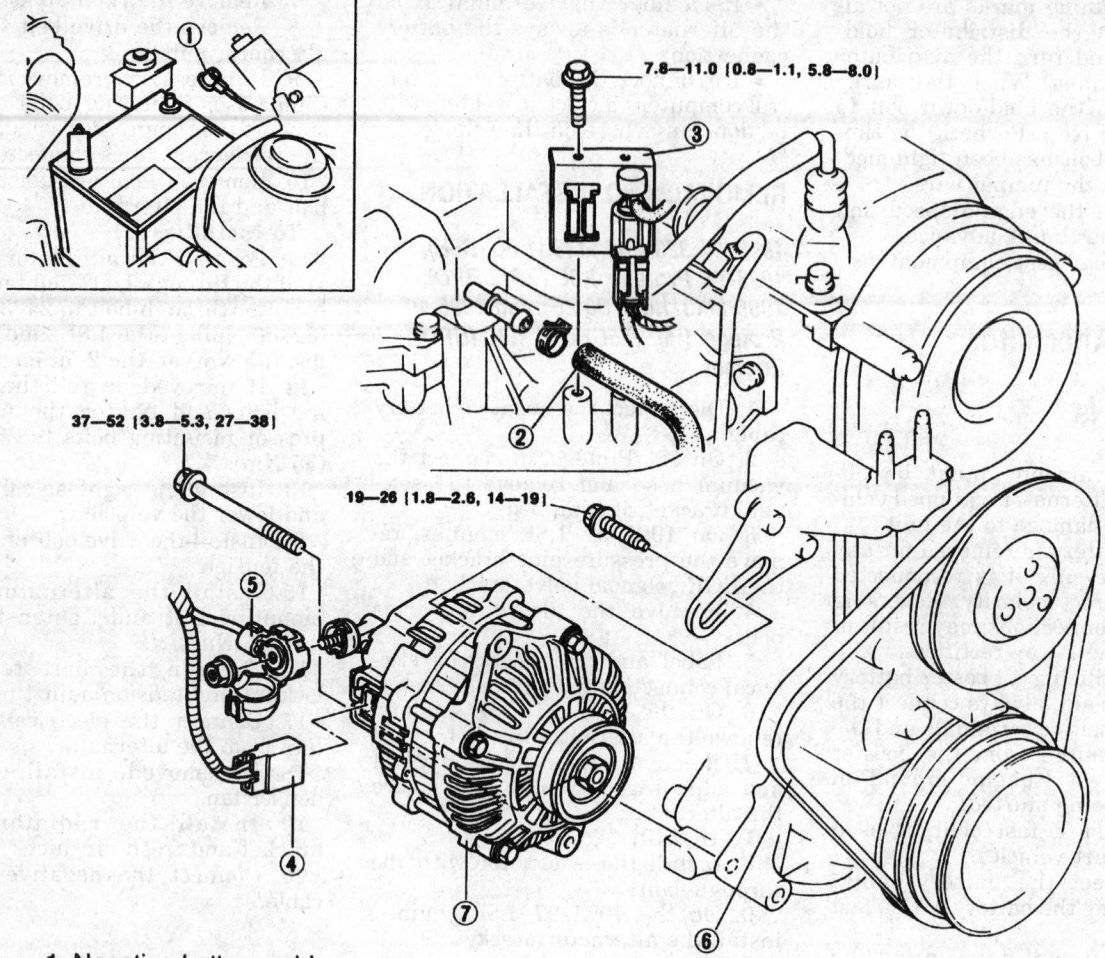

7.8—11.0 (0.8—1.1, 5.8—8.0)

37—52 (3.8—5.3, 27—38)

19—26 (1.8—2.6, 14—19)

N·m (kgf·m, ft·lbf)

1. Negative battery cable
2. Vacuum hose
3. Solenoid bracket (If equipped)
4. Connector
5. B terminal wire
   Inspect for damage and corrosion
6. Drive belt
7. Alternator

125424

Alternator removal and installation — 1993–94 323 and MX3, 1.6L (B6E) and Protege 1.8L (BPE, BPD) engines shown

8. Remove the alternator lower through-bolt and remove the alternator.

To install:

9. Install the alternator with the through-bolt.

10. Install the exhaust pipe, using new gaskets. Tighten the pipe-to-converter nuts to 28–38 ft. lbs. (38–51 Nm) and the exhaust manifold flange nuts to 28–38 ft. lbs. (38–51 Nm). Tighten the exhaust clamp nuts to 14–18 ft. lbs. (19–25 Nm).

11. Install the oxygen sensor, using a sensor wrench, and tighten to 36 ft. lbs. (49 Nm). Connect the oxygen sensor electrical connector.

12. Connect the alternator electrical connectors.

13. Install the transverse member and tighten the bolts to 68–96 ft. lbs. (94–131 Nm). Lower the vehicle.

14. Install the drive belt and upper mounting bolt. Adjust the belt tension. Tighten the lower through-bolt to 24–33 ft. lbs. (32–46 Nm) and the upper mounting bolt to 12–16 ft. lbs. (16–22 Nm).

15. Connect the negative battery cable.

1993–95 929 3.0L (JE-ZE) Engines

1. Disconnect the negative battery cable. Disconnect the positive battery cable and remove the battery and battery tray.

2. Remove the B terminal wire from the alternator.

3. Disconnect the electrical connector from the alternator.

4. Remove the alternator drive belt.

5. Remove the alternator through and adjusting bar bolts, and remove the alternator.

To install:

6. Position the alternator and install the through-bolt. Tighten the alternator through-bolt to 28–38 ft. lbs. (38–51 Nm).

7. Install the adjusting bar, and tighten the mounting bolt to 14–18 ft. lbs. (19–25 Nm).

8. Install the drive belt. Adjust the belt tension.

9. Connect the electrical connector to the alternator.

10. Install the B terminal wire from the alternator.

11. Install the battery tray and the battery. Connect the positive and negative battery cables.

Millenia

1. Disconnect the negative battery cable.

2. On some models it may be necessary to remove the front charge air cooler, radiator upper seal board and the condenser fan assembly.

3. Disconnect the electrical connectors from the alternator.

4. Raise and safely support the vehicle.

5. Remove the right splash shield.

6. Remove the drive belt from the alternator pulley.

7. Remove the A/C compressor mounting bolts and and support the compressor aside, leaving the refrigerant lines connected.

8. Remove the upper and lower alternator mounting bolts and the alternator.

To install:

9. Position the alternator and install the through-bolt. Tighten the alternator lower bolt to 24–33 ft. lbs. (32–46 Nm). Install the alternator upper mounting bolt and tighten to 12–16 ft. lbs. (16–22 Nm).

10. Install the mounting bolts and tighten the A/C compressor mounting bolts to 12–16 ft. lbs. (16–22 Nm).

11. Install the right splash shield, and lower the vehicle.

12. Install the drive belt.

13. Connect the electrical connectors to the alternator.

14. Install the condenser fan assembly.

15. Install the radiator upper seal board.

16. Install the front charge air cooler using new O-rings. Tighten the mounting bolts to 12–16 ft. lbs. (16–22 Nm).

17. Connect the negative battery cable.

1993 Miata 1.6L (B6-ZE) Engines

NOTE: Voltage regulator is of the internal type.

1. Disconnect the negative battery cable.

2. Disconnect the vacuum hose and remove the solenoid bracket, if equipped.

3. Remove the alternator drive belt.

4. Label and disconnect the electrical connectors from the alternator.

5. Remove the alternator pivot and adjusting bar bolts and remove the alternator.

6. Installation is the reverse of the removal procedure. Adjust the belt tension.

1994–97 Miata 1.8L (BPD) Engines

NOTE: The voltage regulator is of the internal type.

1. Disconnect the negative battery cable.

2. Label and disengage the power steering pressure switch, water thermoswitch and idle air control valve electrical connectors.

3. Remove the intake air pipe.

4. Disconnect the electrical connectors from the alternator.

5. Remove the alternator upper mounting bolt.

6. Loosen the alternator adjusting bolt and remove the drive belt from the alternator pulley.

7. Remove the alternator lower through–bolt, and remove the alternator.

To install:

8. Install the alternator with the through–bolt.

9. Connect the alternator electrical connectors.

10. Install the drive belt and upper mounting bolt. Adjust the belt tension. Tighten the lower through–bolt to 28–38 ft. lbs. (38–51 Nm) and the upper mounting bolt to 14–18 ft. lbs. (19–25 Nm).

11. Install the intake air pipe.

12. Engage the power steering pressure switch, water thermoswitch and idle air control valve electrical connectors.

13. Connect the negative battery cable.

1993–95 RX7 1.3L (13B) Engines

1. Disconnect the negative battery cable.

2. Partially drain the cooling system.

3. Remove the air intake hose and air relief hose.

4. Disconnect the accelerator cable.

5. Disconnect the vacuum hoses and remove the pressure chamber.

6. Remove the turbocharger-to-intercooler pipe and bracket.

7. Label and disconnect the electrical connectors from the alternator.

8. Disconnect the air pump hoses, and remove the air pump.

9. Loosen the tension and remove the alternator drive belt.

10. Remove the coolant hose at the rear of the water pump.

11. Remove the alternator mounting bolt and nut, and remove the alternator.

To install:

12. Position the alternator, and install the mounting bolt and nut. Tighten the mounting bolt to 28–38 ft. lbs. (38–51 Nm), and the mounting nut to 14–18 ft. lbs. (19–25 Nm).

13. Install the coolant hose at the rear of the water pump.

14. Install the drive belt. Adjust the belt tension.

15. Install the air pump and connect the hoses.

16. Connect the electrical connectors to the alternator.

17. Install the turbocharger-to-intercooler bracket and pipe.

18. Install the pressure chamber and connect the vacuum hoses.

19. Connect and adjust the accelerator cable.

20. Install the air relief and intake hoses.

21. Fill and bleed the cooling system.

22. Connect the negative battery cable.

Drive Belt

REMOVAL AND INSTALLATION

1993–94 323 1.6L (B6E), Miata 1.6L (B6-ZE), 1.8L (BPD), MX3 1.6L (B6-ZE) and 1993–95 Protege 1.8L (BPE, BPD) Engines

Alternator

1. Position a ruler perpendicular to the drive belt midway between the pulleys on the longest accessible belt span. Press firmly on the belt with your thumb to test the belt tension. The belt should deflect 0.31–0.35 in. (8–9mm) if it is new or 0.35–0.39 in. (9–10mm) for a used belt.

2. If the belt tension is not as specified in Step 1, loosen the alternator adjustment bolt and the through-bolt. Turn the alternator adjustment screw to adjust the belt tension.

3. After adjustment, tighten the through-bolt to 27–38 ft. lbs. (37–52 Nm) and the adjusting bolt to 14–19 ft. lbs. (19–25 Nm).

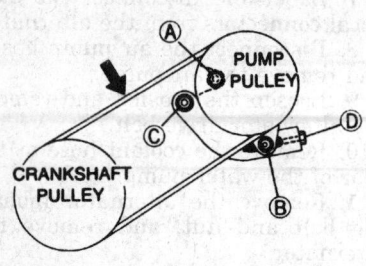

Power steering belt tensioning — 1993–94 323 1.6L (B6E) and 1994 Protege 1.8L (BPD) engines

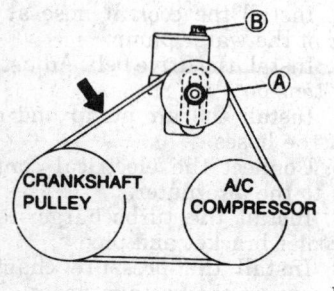

Air conditioner compressor belt tensioning — 1993–94 323 1.6L (B6E) and 1994 Protege 1.8L (BPD) engines

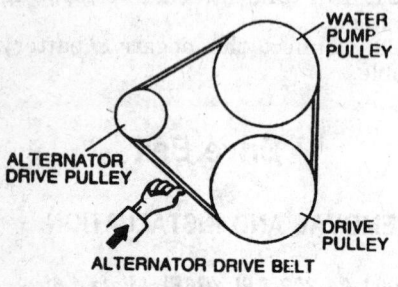

Alternator belt — 1993–94 Miata, 1.6L (B6-ZE) 1.8L (BPD), 1994 MX3 1.6L (B6-ZE) and 1993–94 Protege 1.8L (BPD, BPE) engines

Power Steering/Air Conditioning

1. Position a ruler perpendicular to the drive belt midway between the pulleys on the longest accessible belt span. Press firmly on the belt with your thumb to test the belt tension. The belt should deflect 0.31–0.35 in. (8–9mm) if it is new or 0.35–0.39 in. (9–10mm) if it is used.

2. If the belt tension is not as specified in Step 1, loosen the upper and lower air conditioning compressor through-bolts.

3. Using a suitable prybar against the compressor body, move the com-

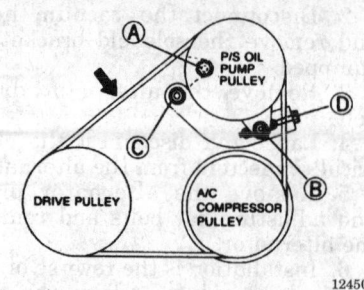

Power steering belt tensioning — 1993–94 Miata, 1.6L (B6-ZE) 1.8L (BPD), 1994 MX3 1.6L (B6-ZE) and 1993–94 Protege 1.8L (BPD, BPE) engines

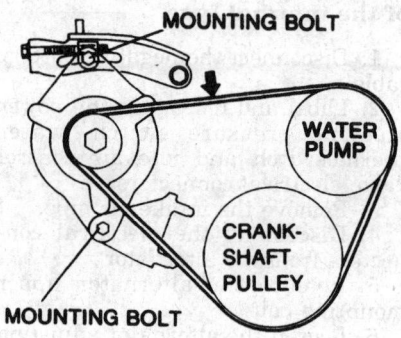

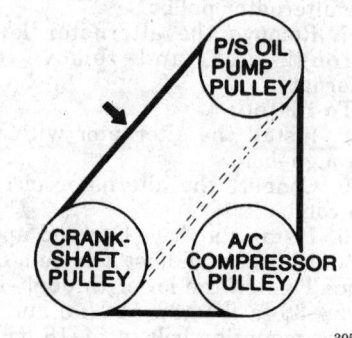

Drive belt routing — 1995–97 Protege 1.5L (Z5D), 1.8L (BPD), 1995 MX3 1.6L (B6-ZE) and 1993–97 626 and MX6 2.0L (FS) engines

pressor until the belt tension is as specified in Step 1.

4. After adjustment, tighten the upper and lower air conditioning compressor through-bolts.

Air Conditioning

1. Position a ruler perpendicular to the drive belt midway between the pulleys on the longest accessible belt span. Press firmly on the belt with your thumb to test the belt tension. The belt should deflect 0.31–0.35 in. (8–9mm) if it is new or 0.35–0.39 in. (9–10mm) if it is used.

2. If the belt tension is not as specified in Step 1, loosen the upper and

lower air conditioning compressor through-bolts.

3. Using a suitable prybar against the compressor body, move the compressor until the belt tension is as specified in Step 1.

4. After adjustment, tighten the upper and lower air conditioning compressor through-bolts.

1995–97 Protege 1.5L (Z5D) and 1.8L (BPD) Engines

Power Steering

NOTE: This belt includes the air conditioner compressor, if equipped.

1. Turn the ignition **OFF** and remove the key. Allow the engine to cool.

2. Remove the power steering pump belt shield.

3. Loosen the adjusting bolt, lockbolts, and through-bolt.

4. Remove the power steering belt.

To install:

5. Install the power steering belt, and make sure it is correctly lined up on the pulley.

6. Adjust the power steering belt tension/deflection by turning the adjusting bolt. A new belt should deflect 0.32–0.35 inch (8–9mm), and a used belt should deflect 0.36–0.39 inch (9–10mm).

7. Tighten the lockbolt to 32–38 ft. lbs. (44–51 Nm). Tighten the through-bolt to 28–38 ft. lbs. (38–51 Nm).

8. Install the belt shield and tighten the attaching bolts to 61–86 inch lbs. (7–9 Nm).

Alternator

1. Turn the ignition **OFF** and remove the key. Allow the engine to cool.

2. Remove the power steering belt.

3. Loosen the alternator adjusting bolt and upper mounting bolt.

4. Raise and safely support the vehicle.

5. Remove the right splash shield.

6. Loosen the lower through-bolt.

7. Lower the vehicle, and remove the alternator belt.

To install:

8. Install the alternator belt and make sure it is correctly lined up on the pulley.

9. Adjust the alternator belt deflection by turning the adjusting bolt. A new belt should deflect 0.22–0.27 inch (6–7mm), and a used belt should deflect 0.24–0.29 inch (6–8mm).

10. Tighten the upper mounting bolt to 14–18 ft. lbs. (19–25 Nm).

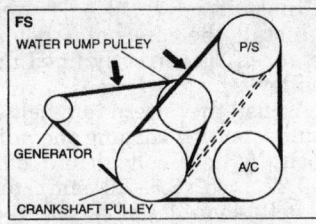

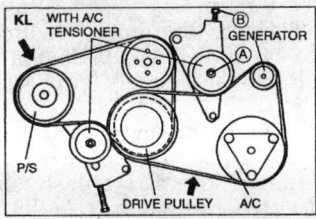

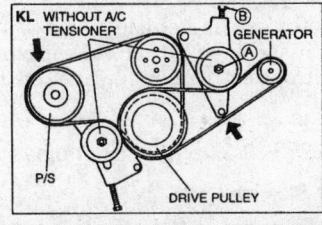

Drive belt routing — 1993–95 MX3 1.8L (K8), 1993–97 626, MX6 (KL), 1993–97 626 and MX6 2.0L (FS), 1995–97 Millenia 2.3L (KJS) and 2.5L (KLD) engines

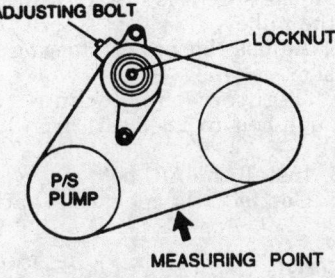

Power steering belt — 1993–94 929 3.0L (JE-ZE) engines

11. Raise and safely support the vehicle.
12. Tighten the lower through-bolt to 28–38 ft. lbs. (38–51 Nm).
13. Install the right splash shield and tighten the bolts to 71–88 inch lbs. (8–10 Nm).
14. Lower the vehicle.
15. Install the power steering belt.

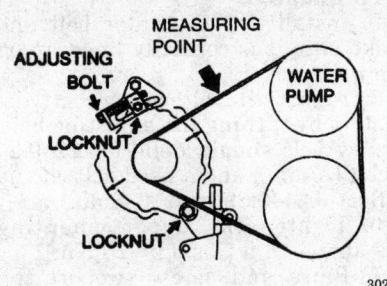

Alternator belt — 1993–94 929 3.0L (JE-ZE) engines

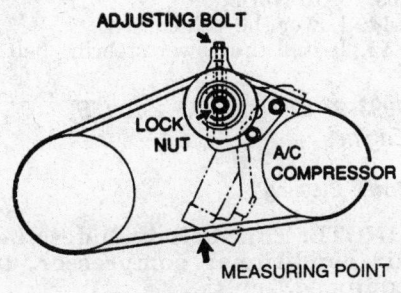

Air conditioner belt — 1993–94 929 3.0L (JE-ZE) engines

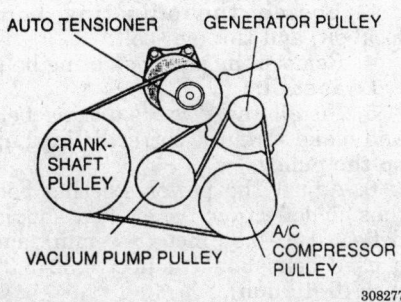

Alternator drive belt routing — 1995–97 Millenia 2.3L (KJS) engines

1993–95 MX3 1.8L (K8), 1993–97 626, MX6 (KL) and 1995–97 Millenia 2.3L (KJS) and 2.5L (KLD) Engines

Alternator

NOTE: If the belt is to be re-used, mark the direction of normal belt rotation.

1. Switch the ignition **OFF** and remove the key. Allow the engine to cool.
2. Loosen the tensioner locknut on the alternator belt tensioner.

3. Loosen the tensioner adjusting bolt until there is enough slack to remove the belt.
4. Remove the belt from each of the pulleys, remove from the vehicle and inspect the belt.

To install:

NOTE: Make sure that the belt is properly seated on all of the pulleys before adjusting the tensioner.

5. Route the alternator belt around the alternator and crankshaft pulleys and, if equipped, the A/C pulley.
6. Adjust the alternator to the proper deflection, and tighten the tensioner locknut to 24–34 ft. lbs. (32–46 Nm).

Power Steering and Water Pump

NOTE: If the belt is to be re-used, mark the direction of normal belt rotation.

1. Switch the ignition **OFF** and remove the key. Allow the engine to cool.
2. Loosen the tensioner locknut on the power steering and water pump belt tensioner.
3. Loosen the tensioner adjusting bolt until there is enough slack to remove the belt.
4. Remove the belt from each of the pulleys, remove from the vehicle and inspect the belt.

To install:

NOTE: Make sure that the belt is properly seated on all of the pulleys before adjusting the tensioner.

5. Route the power steering and water pump drive belt around the three pulleys.
6. Adjust the belt to the proper tension and tighten the tensioner locknut to 24–34 ft. lbs. (32–46 Nm).

1995 MX3 1.6L (B6-ZE) Engines

Power Steering

NOTE: This belt includes the air conditioner compressor, if equipped.

1. Turn the ignition **OFF** and remove the key. Allow the engine to cool.
2. Remove the power steering pump belt shield.
3. Without A/C, loosen the adjusting bolt, lockbolt, and through-bolt. With A/C, loosen the adjusting bolt and the locknut.
4. Remove the power steering belt.

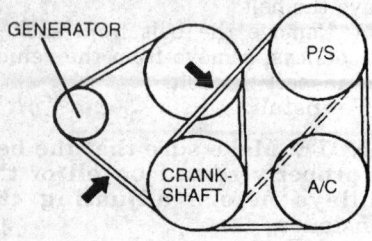

P/S or P/S and A/C equipped drive belt routing — 1995–97 Miata 1.8L (BPD) engine

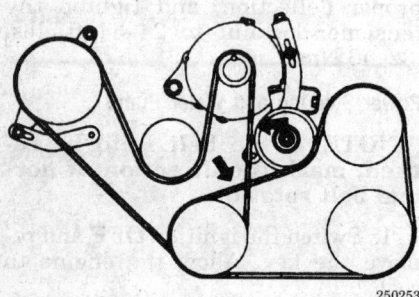

Drive belt routing — 1993–95 RX7 1.3L (13B) engines

To install:

5. Install the power steering belt and make sure it is correctly lined up on the pulley.

6. Adjust the power steering belt tension/deflection by turning the adjusting bolt. A new belt should deflect 0.32–0.35 inch (8–9mm), and a used belt should deflect 0.36–0.39 inch (9–10mm).

7. Without A/C, tighten the lockbolt to 24–33 ft. lbs. (32–46 Nm). Tighten the through-bolt to 14–18 ft. lbs. (19–25 Nm). With A/C, tighten the locknut to 24–25 ft. lbs. (32–34 Nm).

8. Install the belt shield and tighten the attaching bolts to 61–86 inch lbs. (7–9 Nm).

Alternator

1. Turn the ignition **OFF** and remove the key. Allow the engine to cool.

2. Remove the power steering belt.

3. Loosen the alternator adjusting bolt and upper mounting bolt.

4. Raise and safely support the vehicle.

5. Remove the right splash shield.

6. Loosen the lower through-bolt.

7. Lower the vehicle and remove the alternator belt.

To install:

8. Install the alternator belt and make sure it is correctly lined up on the pulley.

9. Adjust the alternator belt deflection by turning the adjusting bolt. A new belt should deflect 0.22–0.27 inch (6–7mm), and a used belt should deflect 0.24–0.29 inch (6–8mm).

10. Tighten the upper mounting bolt to 14–18 ft. lbs. (19–25 Nm).

11. Raise and safely support the vehicle.

12. Tighten the lower through-bolt to 28–38 ft. lbs. (38–51 Nm).

13. Install the right splash shield and tighten the bolts to 71–88 inch lbs. (8–10 Nm).

14. Lower the vehicle.

15. Install the power steering belt.

1993–97 626 and MX6 2.0L (FS) Engines

Power Steering

NOTE: This belt includes the air conditioner compressor, if equipped.

1. Disconnect the negative battery cable.

2. Remove the power steering pump belt shield.

3. Loosen the adjusting bolt, lockbolt, and through-bolt.

4. Remove the power steering belt.

To install:

5. Install the power steering belt and make sure it is correctly lined up on the pulley.

6. Adjust the power steering belt tension/deflection. A new belt should deflect 0.30–0.35 inch (8–9 mm), and a used belt should deflect 0.32–0.37 inch (8–10 mm).

7. Tighten the lockbolt to 24–34 ft. lbs. (32–46 Nm).

8. Tighten the through-bolt to 32–44 ft. lbs. (44–60 Nm).

9. Install the belt shield and tighten the attaching bolts to 61–86 inch lbs. (7–9 Nm).

10. Connect the negative battery cable.

Alternator

1. Disconnect the negative battery cable.

2. Remove the power steering belt.

3. Loosen the alternator adjusting bolt and upper mounting bolt.

4. Raise and safely support the vehicle.

5. Remove the RH splash shield.

6. Loosen the lower through-bolt.

7. Lower the vehicle and remove the alternator belt.

To install:

8. Install the alternator belt and make sure it is correctly lined up on the pulley.

9. Adjust the alternator belt tension/deflection by turning the adjusting bolt. A new belt should deflect 0.26–0.27 inch (6.5–7.0 mm), and a used belt should deflect 0.27–0.35 inch (7–9 mm).

10. Tighten the upper mounting bolt to 14–18 ft. lbs. (19–25 Nm).

11. Raise and safely support the vehicle.

12. Tighten the lower through-bolt to 28–38 ft. lbs. (38–51 Nm).

13. Install the RH splash shield and tighten the bolts to 71–88 inch lbs. (8–10 Nm).

14. Lower the vehicle.

15. Install the power steering belt.

16. Connect the negative battery cable.

1993–95 929 3.0L (JE-ZE) Engines

Power Steering

1. Disconnect the negative battery cable.

2. Remove the A/C belt.

3. Loosen the adjusting bolt, locknut, and through-bolt.

4. Remove the power steering belt.

To install:

5. Install the power steering belt and make sure it is correctly lined up on the pulley.

6. Adjust the power steering belt tension.

7. Tighten the locknut and through-bolt to 28–38 ft. lbs. (38–51 Nm).

8. Install the A/C belt.

9. Connect the negative battery cable.

Alternator

1. Disconnect the negative and positive battery cables and remove the battery.

2. Remove the power steering and A/C belts.

3. Loosen the alternator adjusting bolt and upper mounting bolt.

4. Raise and safely support the vehicle.

5. Loosen the lower through-bolt.

6. Lower the vehicle and remove the alternator belt.

To install:

7. Install the alternator belt and make sure it is correctly lined up on the pulley.

8. Adjust the alternator belt tension.

9. Tighten the upper mounting bolt to 14–18 ft. lbs. (19–25 Nm).

10. Raise and safely support the vehicle.

11. Tighten the lower through-bolt to 28–38 ft. lbs. (38–51 Nm).

12. Lower the vehicle.

13. Install the power steering and A/C belts.

14. Install the battery. Connect the positive and negative battery cables.

Air Conditioner

1. Disconnect the negative battery cable.

2. Loosen the through-bolt.

3. Remove the A/C belt.

To install:

4. Install the A/C and make sure it is correctly lined up on the pulley.

5. Adjust the belt tension.

6. Tighten the through-bolt to 28–38 ft. lbs. (38–51 Nm).

7. Connect the negative battery cable.

1995–97 Millenia 2.3L (KJS) Engines

Alternator, A/C, and Vacuum Pump

NOTE: If the belt is to be re-used, mark the direction of normal belt rotation.

1. Switch the ignition **OFF** and remove the key. Allow the engine to cool.

2. Disconnect the negative battery cable.

3. Remove the right-hand splash shield and dust cover.

4. Using a wrench, turn the tensioner pulley locknut clockwise to remove tension on the belt.

5. Remove the belt from each of the pulleys, remove from the vehicle and inspect the belt.

To install:

6. Route the alternator belt around the pulleys.

7. Install the right-hand splash shield and dust cover.

8. Connect the negative battery cable.

Power Steering, Supercharger, and Water Pump

NOTE: If the belt is to be re-used, mark the direction of normal belt rotation.

1. Disconnect the negative battery cable.

2. Switch the ignition **OFF** and remove the key. Allow the engine to cool.

3. Remove the right-hand splash shield and dust cover.

4. Remove the alternator belt.

5. Using a wrench, turn the tensioner pulley locknut clockwise to remove tension on the belt.

6. Remove the belt from each of the pulleys, remove from the vehicle and inspect the belt.

To install:

7. Route the power steering drive belt around the pulleys.

8. Install the alternator belt.

9. Install the right-hand splash shield and dust cover.

10. Connect the negative battery cable.

1994–97 Miata 1.8L (BPD) Engine

Power Steering or Power Steering and A/C

NOTE: This belt includes the air conditioner compressor, if equipped.

1. Turn the ignition **OFF** and remove the key. Allow the engine to cool.

2. Loosen the adjusting bolt, oil pump bolt and both nuts.

3. Remove the power steering belt.

To install:

4. Install the power steering belt and make sure it is correctly lined up on the pulley.

5. Adjust the power steering belt tension/deflection by turning the adjusting bolt. A new belt should deflect 0.32–0.35 inch (8–9mm), and a used belt should deflect 0.36–0.39 inch (9–10mm).

6. Tighten the lockbolt to 24–33 ft. lbs. (32–46 Nm). Tighten the upper nut to 14–18 ft. lbs. (19–25 Nm), and tighten the lower locknut to 27–39 ft. lbs. (37–53 Nm).

A/C

1. Turn the ignition **OFF** and remove the key. Allow the engine to cool.

2. Loosen the adjusting bolt and the locknut.

3. Remove the A/C belt.

To install:

4. Install the A/C belt, and make sure it is correctly lined up on the pulley.

5. Adjust the A/C belt tension/deflection by turning the adjusting bolt. A new belt should deflect 0.32–0.35 inch (8–9mm), and a used belt should deflect 0.36–0.39 inch (9–10mm).

6. Tighten the locknut to 28–38 ft. lbs. (38–51 Nm).

Alternator

1. Turn the ignition **OFF** and remove the key. Allow the engine to cool.

2. Remove the power steering or A/C belt.

3. Loosen the alternator adjusting bolt and upper mounting bolt.

4. Raise and safely support the vehicle.

5. Loosen the lower through-bolt.

6. Lower the vehicle and remove the alternator belt.

To install:

7. Install the alternator belt and make sure it is correctly lined up on the pulley.

8. Adjust the alternator belt deflection by turning the adjusting bolt. A new belt should deflect 0.22–0.27 inch (6–7mm), and a used belt should deflect 0.24–0.29 inch (6–8mm).

9. Tighten the upper mounting bolt to 14–18 ft. lbs. (19–25 Nm).

10. Raise and safely support the vehicle.

11. Tighten the lower through-bolt to 28–38 ft. lbs. (38–51 Nm).

12. Lower the vehicle.

13. Install the power steering belt.

1993–95 RX7 1.3L (13B) Engine

Power Steering and Air Conditioning

1. Disconnect the negative battery cable.

2. Disconnect the air hoses.

3. Loosen the idler pulley locknut **A** and the adjusting bolt **B**.

4. Remove the belt.

To install:

5. Install the belt and make sure it is correctly lined up on the pulleys.

6. Adjust the belt deflection by turning the adjusting bolt **B**. The deflection should be 0.14–0.15 inches (3.5–4.0 mm)

7. Tighten the locknut **A** to 27–39 ft. lbs. (37–53 Nm).

8. Connect the air hoses and the negative battery cable.

Alternator

1. Disconnect the negative battery cable.

2. Disconnect the air hoses.

3. Remove the power steering and A/C belt.

4. Loosen the air pump mounting bolts **A** and **B**.

5. Loosen the alternator mounting bolt **A** and locknut **B**.

6. Loosen adjusting bolt **C**.

7. Remove the alternator belt.

To install:

8. Install the alternator belt and make sure it is correctly lined up on the pulleys.

9. Tighten the air pump mounting bolts **A** and **B** while applying pressure to the drive belt. Tighten the bolts to 14–18 ft. lbs. (19–25 Nm).

10. Adjust the alternator belt deflection by turning adjusting bolt **C**. The deflection should be 0.24–0.27 inches (6–7 mm).

11. Tighten the mounting bolt **A** to 28–38 ft. lbs. (38–51 Nm).

12. Tighten the locknut **B** to 14–18 ft. lbs. (19–25 Nm).

13. Connect the air hoses.

14. Install the power steering belt.

15. Connect the negative battery cable.

Starter

REMOVAL AND INSTALLATION

——— **CAUTION** ———

Fuel injection systems remain under pressure after the engine has been turned OFF. Properly relieve fuel pressure before disconnecting any fuel lines. Failure to do so may result in fire or personal injury. Do not allow fuel spray or fuel vapors to come in contact with a spark or open flame. Keep a dry chemical fire extinguisher nearby. Never store fuel in an open container due to risk of fire or explosion.

1993–94 323 1.6L (B6E), 1993–95 Protege 1.8L (BPE, BPD) and MX3 1.6L (B6-ZE) Engines

NOTE: Always disconnect the negative battery cable first!

1. Disconnect the negative then positive battery cables. Remove the battery and battery tray.

2. Raise and safely support the vehicle. Remove the engine undercover.

3. Remove the intake manifold bracket.

4. Remove the starter bracket, if equipped.

5. Label and disconnect the wiring from the starter.

6. Remove the starter mounting bolts and remove the starter.

To install:

7. Position the starter and loosely tighten the lower mounting bolt.

8. Connect the electrical wiring to the starter.

9. Install the intake manifold support bracket bolts and the bracket. Tighten the bolts to 28–38 ft. lbs. (38–51 Nm).

10. Install the upper starter mounting bolts. Tighten all starter mounting bolts to 28–38 ft. lbs. (38–51 Nm). The upper mounting bolts must be tightened first.

11. Install the battery and tray.

12. Connect the negative battery cable.

1995–97 Protege 1.5L (Z5D) and 1.8L (BPE, BPD) Engines

1. Disconnect the negative battery cable.

2. Remove the air cleaner.

3. Raise and safely support the vehicle.

4. Disconnect the catalytic converter pipe from the front pipe.

5. Remove the intake manifold support bracket bolts and the bracket.

6. Disconnect the electrical connectors from the starter and solenoid.

7. Remove the starter mounting bolts, and remove the starter.

To install:

8. Position the starter and loosely tighten the lower starter mounting bolt.

9. Connect the electrical connectors to the starter solenoid and the starter.

10. Install the intake manifold support bracket bolts and the bracket. Tighten the bolts to 28–38 ft. lbs. (38–51 Nm).

11. Install the upper starter mounting bolts. Tighten all starter mounting bolts to 24–33 ft. lbs. (32–46 Nm). The upper mounting bolts must be tightened first.

12. Connect the catalytic converter pipe to the front pipe. Tighten the bolts to 28–38 ft. lbs. (38–51 Nm).

13. Lower the vehicle.

14. Install the air cleaner.

15. Connect the negative battery cable.

1993–95 MX3 1.8L (K8) Engines

1. On automatic transaxles, relieve the fuel system pressure.

2. Disconnect the negative battery cable.

3. On manual transaxles, remove the strut bar.

4. Remove the intake air hose.

5. Remove the S-terminal wire from the starter solenoid. Remove the nut and the B-terminal wire from the starter solenoid.

6. Remove the fuel filter.

7. Remove the starter mounting bolts and remove the starter.

To install:

8. Position the starter and tighten the starter mounting bolts to 28–38 ft. lbs. (38–51 Nm).

9. Install the fuel filter if removed.

10. Install the B-terminal wire and nut. Tighten the nut to 7–9 ft. lbs. (10–12 Nm).

11. Install the S-terminal wire to the solenoid.

12. Install the intake air hose.

13. On manual transaxles, install the strut bar.

14. Connect the negative battery cable.

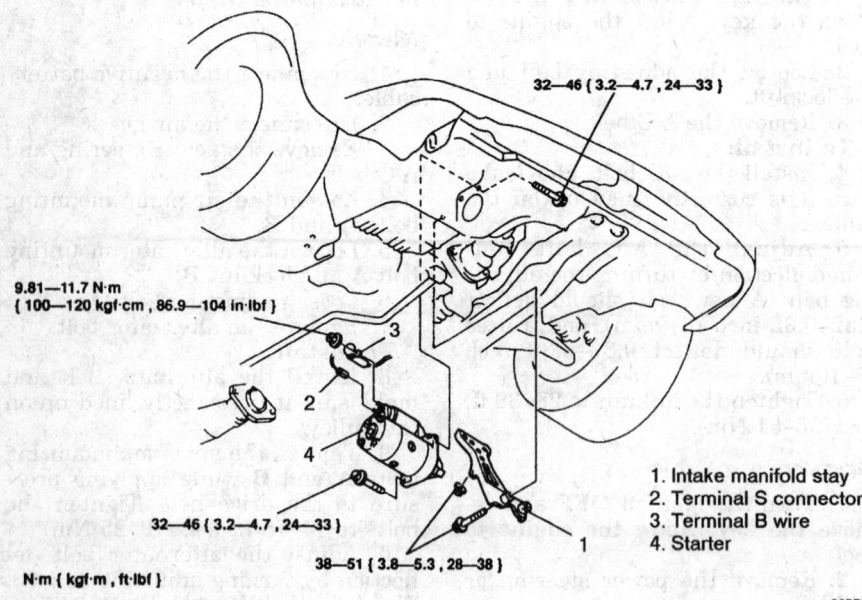

32—46 { 3.2—4.7 , 24—33 }

9.81—11.7 N·m
{ 100—120 kgf·cm , 86.9—104 in·lbf }

3

2

4

32—46 { 3.2—4.7 , 24—33 }

38—51 { 3.8—5.3 , 28—38 }

1

1. Intake manifold stay
2. Terminal S connector
3. Terminal B wire
4. Starter

N·m { kgf·m , ft·lbf }

305794

Starter motor removal and installation — Protege 1.5L (Z5D) and 1.8L (BPD) engines

1993–97 626 and MX6 2.0L (FS) Engines

1. Disconnect the negative battery cable.
2. Remove the air duct and air cleaner assembly.
3. Remove the upper starter mounting bolts.
4. Raise and safely support the vehicle.
5. Remove the intake manifold support bracket bolts and the bracket.
6. Disconnect the electrical connectors from the starter and solenoid.
7. Remove the lower starter mounting bolt and remove the starter.

To install:

8. Position the starter and loosely tighten the lower starter mounting bolt.
9. Connect the electrical connectors to the starter solenoid and the starter.
10. Install the intake manifold support bracket bolts and the bracket. Tighten the bolts to 28–38 ft. lbs. (38–51 Nm).
11. Lower the vehicle and install the upper starter mounting bolts. Tighten all starter mounting bolts to 24–33 ft. lbs. (32–46 Nm). The upper mounting bolts must be tightened first.
12. Install the air duct and air cleaner assembly.
13. Connect the negative battery cable.

1993–97 626 and MX6 2.5L (KL) Engines

Manual Transaxle

1. Disconnect the negative battery cable.
2. Remove the air duct and air cleaner assembly.
3. Remove the S-terminal wire from the starter solenoid.
4. Remove the nut and the B-terminal wire from the starter solenoid.
5. Remove the starter mounting bolts and remove the starter.
6. Installation is the reverse of removal. Torque the mounting bolts to 24–43 ft. lbs. (32–46 Nm).

Automatic Transaxle

1. Properly relieve the fuel system pressure.
2. Disconnect the negative battery cable.
3. Remove the air duct and air cleaner assembly.
4. Remove the shift cable from the selector lever using a screwdriver or suitable tool.

5. Squeeze the lock tabs on the shift cable and remove the cable from the bracket.
6. Disconnect the coolant hose and the two fuel lines.
7. Disconnect the electrical connectors from the starter solenoid.
8. Position the wiring harness out of the way.
9. Remove the two selector cable bracket mounting bolts and the bracket.
10. Remove the two nuts and the bolt from the starter bracket and remove the bracket.
11. Disconnect the S-terminal wire from the starter solenoid.
12. Remove the nut and the B-terminal wire from the starter solenoid.
13. Remove the starter mounting bolts and remove the starter.

To install:

14. Position the starter and tighten the starter mounting bolts to 24–33 ft. lbs. (32–46 Nm).
15. Install the B-terminal wire and nut. Tighten the nut to 12–16 ft. lbs. (16–22 Nm).
16. Connect the S-terminal wire to the solenoid.
17. Position the starter bracket and install the bolt and nuts.
18. Install the selector cable bracket and tighten the bolts to 60.8–86.8 inch lbs. (7–10 Nm).
19. Connect the electrical connectors to the starter solenoid.
20. Connect the fuel lines and coolant hose.
21. Install the shift cable into the cable bracket and into the selector lever.
22. Install the air duct and air cleaner assembly.
23. Connect the negative battery cable.

1993–95 929 3.0L (JE-ZE) Engines

1. Disconnect the negative battery cable.
2. Raise and safely support the vehicle. Remove the engine undercover.
3. Label and disconnect the wiring from the starter.
4. Remove the starter mounting bolts, and remove the starter.
5. Installation is the reverse of the removal procedure. Tighten the starter mounting bolts to 24–33 ft. lbs. (32–46 Nm). Tighten the B-terminal nut to 87–104 inch lbs. (10–11 Nm).

1995–97 Millenia 2.3L (KJS) Engine

1. Disconnect the negative battery cable.
2. Remove the charge air cooler duct.

3. Remove the battery clamp, box, and battery. Remove the battery tray.
4. Remove the rear charge air cooler.
5. Remove the bolt from the pipe bracket and remove the bracket.
6. Remove the S-terminal wire from the starter solenoid.
7. Remove the nut and the B-terminal wire from the starter solenoid.
8. Remove the starter mounting bolts and remove the starter.

To install:

9. Position the starter and tighten the starter mounting bolts to 24–33 ft. lbs. (32–46 Nm).
10. Install the B-terminal wire and nut. Tighten the nut to 12–16 ft. lbs. (16–22 Nm).
11. Install the S-terminal wire to the solenoid.
12. Position the pipe bracket and install the bolt. Tighten the bolt to 14–18 ft. lbs. (19–25 Nm).
13. Install the rear charge air cooler, using new O-rings. Tighten the nuts to 14–18 ft. lbs. (19–25 Nm).
14. Install the battery tray. Install the battery, box, and clamp.
15. Install the charge air cooler duct.
16. Connect the negative battery cable.

1995–97 Millenia 2.5L (KLD) Engine

1. Disconnect the negative battery cable.
2. Remove the battery clamp, box, and battery. Remove the battery tray.
3. Remove the shift cable from the selector lever using a screwdriver or suitable tool.
4. Squeeze the lock tabs on the shift cable and remove the cable from the bracket.
5. Disconnect the electrical connectors from the starter solenoid.
6. Position the wiring harness out of the way.
7. Remove the two selector cable bracket mounting bolts and the bracket.
8. Remove the two nuts and the bolt from the starter bracket and remove the bracket.
9. Remove the S-terminal wire from the starter solenoid.
10. Remove the nut and the B-terminal wire from the starter solenoid.
11. Remove the starter mounting bolts and remove the starter.

To install:

12. Position the starter and tighten the starter mounting bolts to 24–33 ft. lbs. (32–46 Nm).

13. Install the B-terminal wire and nut. Tighten the nut to 12–16 ft. lbs. (16–22 Nm).

14. Install the S-terminal wire to the solenoid.

15. Position the starter bracket and install the bolt and nuts.

16. Install the selector cable bracket and tighten the bolts to 5–7 ft. lbs. (7–9 Nm).

17. Connect the electrical connectors to the starter solenoid.

18. Install the shift cable into the cable bracket and into the selector lever.

19. Install the battery tray. Install the battery, box, and clamp.

20. Connect the negative battery cable.

1993–97 Miata 1.6L (B6E) and 1.8L (BPD) Engine

1. Disconnect the negative battery cable.

2. Disconnect the electrical connectors from the starter and solenoid.

3. For (M/T), remove the starter bracket.

4. Safely raise and support the vehicle.

5. Remove the starter mounting bolts, and remove the starter.

To install:

6. Position the starter and tighten the starter mounting bolts to 28–38 ft. lbs. (38–51 Nm).

7. Lower the vehicle.

8. Install the starter bracket (M/T) and tighten the lower bolt to 28–38 ft. lbs. (38–51 Nm), and the upper bolt to 12–16 ft. lbs. (16–22 Nm).

9. Connect the electrical connectors to the starter solenoid and the starter.

10. Connect the negative battery cable.

1993–95 RX7 1.3L (13B) Engines

1. Disconnect the negative battery cable.

2. Raise and safely support the vehicle.

3. Remove the engine undercover.

4. Label and disconnect the wiring from the starter.

5. Remove the starter mounting bolts and remove the starter.

6. Installation is the reverse of the removal procedure.

CHASSIS ELECTRICAL

Blower Motor

REMOVAL AND INSTALLATION

1993–94 323, Protege and 1993–97 MX3

1. Disconnect the negative battery cable.

2. Open the glove box and remove the glove box retaining screws. Remove the glove box assembly.

3. Remove the inner glove box assembly screws and remove the inner glove box.

4. Unclip and remove the heater blower seal plate.

5. Remove the 3 heater blower mounting nuts and remove the blower unit case.

6. Remove the screws and separate the two halves of the blower unit case.

7. Remove the blower fan from the blower motor. Remove the mounting screws and separate the blower motor from the case.

To install:

8. Install the blower motor to the case and install the screws.

9. Install the blower fan to the blower motor.

10. Connect the 2 halves of the blower unit case and install the screws.

11. Install the heater blower unit case to the vehicle and tighten the mounting nuts.

12. Install the heater blower seal plate.

13. Install the inner glove box assembly.

14. Install the glove box door.

15. Connect the negative battery cable.

1995–97 Millenia and Protege

NOTE: The blower motor is located on the right side of the vehicle underneath the dashboard.

1. Disconnect the negative battery cable.

2. Disconnect the air hose from the unit.

3. Remove the mounting screws and remove the blower motor.

4. Disconnect the electrical connections.

5. Installation is the reverse of removal.

1993–94 626 and MX6

1. Disconnect the negative battery cable.

2. Remove the two instrument panel insulator screws.

3. Disconnect the courtesy lamp electrical connector.

4. Remove the instrument panel insulator.

5. Disconnect the blower motor electrical connector.

6. Remove the four attaching screws from the housing and remove the blower motor and wheel assembly from the vehicle.

To install:

7. Position the blower motor and wheel assembly in the housing and install the four attaching screws and the electrical connector.

8. Install the instrument panel insulator, courtesy light and panel screws.

9. Connect the negative battery cable.

1995–97 626 and MX6

——————— **CAUTION** ———————
The air bag system must be disarmed before removing the steering wheel. Failure to do so may cause accidental deployment, property damage or personal injury.

1. Disconnect the negative battery cable.

2. Properly recover the refrigerant from the A/C system.

3. Remove the A-pillar, and front side trim.

4. Disarm and remove the driver-side air bag module.

——————— **CAUTION** ———————
Always carry an air bag assembly, with the bag and trim cover, away from your body. Store the assembly facing upward; never place the assembly face down on any surface.

5. Remove the steering wheel.

6. Remove the steering column covers and the combination switch.

7. Remove the steering column.

8. Remove the side panel by pulling it forward to disengage the clips.

9. Remove the hood release knob.

10. Pull the lap duct from the rear of the lower panel. Remove the mounting screws, and pull the panel to disengage the clips.

11. Remove the front console by pulling it forward to disengage the clips.

12. Remove the rear console mounting screws, and remove the console.

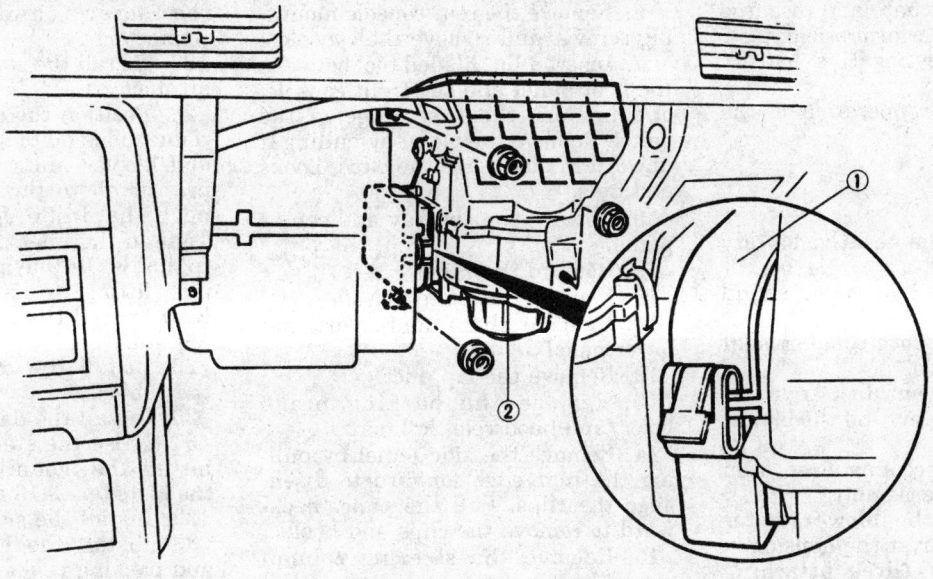

1. Seal plate
2. Blower unit case

126019

Blower motor removal and installation — 1993–94 323, MX3 and Protege

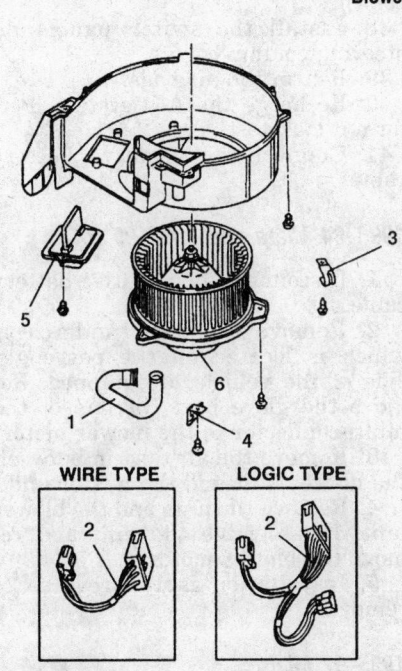

WIRE TYPE   **LOGIC TYPE**

1. Air hose   4. Connector bracket
2. Blower harness   5. Resistor
3. Harness clip   6. Blower motor

305973

Blower motor, exploded view — 1995–97 Protege

13. Remove the side wall mounting screws and fasteners, and remove the panel.

14. Remove the undercover mounting fasteners, and remove the cover.

15. Remove the glove compartment mounting screws, and remove the compartment.

16. Remove the switch panel mounting screws. Remove the panel by pulling it forward to disengage the clips.

17. Remove the passenger-side air bag module.

18. Remove the meter hood mounting screws. Remove the hood by pulling it forward to disengage the clips.

19. Pull the upper garnish upwards to disengage the clips, and remove the garnish.

20. Remove the instrument cluster.

21. Remove the dashboard.

22. Remove the A/C duct, and disconnect the A/C amplifier connector.

23. Disconnect and plug the inlet and outlet pipes. Remove the cooling unit.

24. Disconnect the blower motor electrical connector.

25. Remove the three attaching nuts from the housing, and remove the blower motor and wheel assembly from the vehicle.

To install:

26. Position the blower motor and wheel assembly in the housing and install the three attaching nuts and the electrical connector.

27. Position the cooling unit so the connections match those of the heater and blower units. Apply clean compressor oil to the O-rings, and connect the inlet and outlet pipes. Tighten the inlet pipe fitting to 8–14 ft. lbs. (10–19 Nm), and the outlet pipe fitting to 15–21 ft. lbs. (20–29 Nm). Install the cooling unit mounting nuts.

28. Connect the A/C amplifier, and the A/C duct.

29. Install the dashboard and instrument cluster.

30. Install the upper garnish and meter hood.

31. Install the passenger-side air bag.

32. Install the switch panel, glove compartment, undercover and side wall.

33. Install the rear and front consoles.

34. Install the lower panel, hood release knob and side panel.

35. Install the steering column, combination switch, column cover and steering wheel.

36. Install the driver-side air bag.

37. Install the front side and A-pillar trim.

38. Properly evacuate and recharge the A/C system. Perform a leak test.

39. Connect the negative battery cable.

40. Verify proper operation of the blower motors.

1993-94 929

1. Disconnect the negative battery cable.

2. Remove the passenger sound deadening panel.

3. Remove the glove compartment assembly and brace.

4. Remove the air duct from between the heater unit and the blower unit.

5. Remove the cooling hose from the blower motor assembly.

6. Disconnect the blower motor electrical connector at the housing.

7. Remove the three attaching screws from the housing and remove the blower motor from the vehicle.

8. Remove the retaining clip from the blower wheel assembly.

9. Remove the blower wheel by pulling it straight off the blower motor shaft.

To install:

10. Install the blower wheel and retaining clip.

11. Position the blower motor in the housing and install the three attaching screws and the electrical connector.

12. Install the cooling hose, air duct, glove compartment and brace.

13. Install the passenger sound deadening panel.

14. Connect the negative battery cable.

1995 929

——————— CAUTION ———————

The air bag system must be disarmed before removing the dashboard. Failure to do so may cause accidental deployment, property damage or personal injury.

————————————————————

1. Disconnect the negative battery cable.

2. Disarm the driver and passenger air bags.

3. Properly recover the refrigerant from the A/C system.

4. Remove the steering column covers.

5. Insert a flat-bladed tool into the switch panel holes, and pry out the front edge of the panel. Remove the panel by pulling it upward to disengage the clips. Disconnect the solar

ventilation switch electrical harness, if equipped.

6. Remove the rear console mounting screws, and remove the console.

7. Insert a flat bladed tool between the boot panel and the front console, and pry out the front edge of the panel. Remove the panel by pulling it upward to disengage the clips, hooks and pins.

8. Remove the ashtray and center panel.

9. Remove the radio.

10. Remove the front console.

11. Remove the undercover, and lower panel.

12. Remove the lap duct.

13. Remove the parking brake lever, and hood release knob.

14. Remove the side panel by pulling the rear edge forward to disengage the clips. Pull the panel backward to remove the clips and hooks.

15. Remove the steering column mounting bolts, and lower the column.

16. Remove the dashboard.

17. Remove the main air duct from the vehicle.

18. Disconnect and plug the cooler hose and cooler pipe. Remove the cooling unit mounting nuts, and remove the cooling unit.

19. Remove the air duct from between the heater unit and the blower unit.

20. Remove the cooling hose from the blower motor assembly.

21. Disconnect the blower motor electrical connector at the housing.

22. Remove the three attaching screws from the housing and remove the blower motor from the vehicle.

23. Remove the retaining clip from the blower wheel assembly.

24. Remove the blower wheel by pulling it straight off the blower motor shaft.

To install:

25. Install the blower wheel and retaining clip.

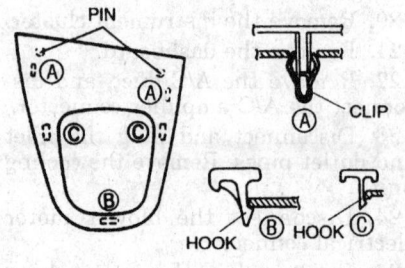

Boot panel clip locations — 1995-97 929

237135

26. Position the blower motor in the housing and install the three attaching screws and the electrical connector.

27. Install the cooling hose, and the air duct.

28. Position the cooling unit so the connections match those of the heater and blower units. Apply clean compressor oil to the O-rings, and connect the inlet and outlet pipes. Tighten the inlet pipe fitting to 8–14 ft. lbs. (10–19 Nm), and the outlet pipe fitting to 15–21 ft. lbs. (20–29 Nm). Install the cooling unit mounting nuts.

29. Install the main air duct into the vehicle.

30. Install the dashboard.

31. Raise the steering column, and install the mounting bolts. Tighten the bolts to 12–16 ft. lbs. (16–22 Nm).

32. Install the side panel.

33. Install the hood release knob and parking brake lever.

34. Install the lap duct, lower panel and undercover.

35. Install the front console and radio.

36. Install the center panel and the ashtray.

37. Install the boot panel and rear console.

38. Install the switch panel and steering column covers.

39. Rearm the air bags.

40. Recharge the A/C system. Perform a leak test.

41. Connect the negative battery cable.

1993-94 Miata and 1993-95 RX7

1. Disconnect the negative battery cable.

2. Remove the dash undercover which is located on the passenger side of the vehicle, if equipped. Remove the glove box. Disconnect the multi-connector to the blower motor.

3. Remove the air duct in between the blower unit and the heater unit.

4. Remove the nuts and the blower unit. Disassemble the unit and remove the blower motor.

5. Installation is the reverse of removal.

1995-97 Miata

1. Disconnect the negative battery cable.

2. Open the glove box and remove the glove box retaining screws. Remove the glove box assembly.

3. Remove the inner glove box assembly screws and remove the inner glove box.

4. Remove the air duct.

5. Unclip and remove the heater blower seal plate.

6. Remove the three heater blower mounting nuts and remove the blower unit case.

7. Remove the screws and separate the two halves of the blower unit case.

8. Remove the blower fan from the blower motor. Remove the mounting screws, and separate the blower motor from the case.

To install:

9. Install the blower motor into the case and install the screws.

10. Install the blower fan to the blower motor.

11. Connect the two halves of the blower unit case and install the screws.

12. Install the heater blower unit case to the vehicle and tighten the mounting nuts.

13. Install the heater blower seal plate.

14. Install the air duct.

15. Install the inner glove box assembly.

16. Install the glove box door.

17. Connect the negative battery cable.

Windshield Wiper Motor

REMOVAL AND INSTALLATION

1993–94 323 and Protege

Front Wiper Motor

1. Disconnect the negative battery cable.

2. Remove the wiper arm nut covers, the wiper arm nuts and the wiper arms.

3. Remove the cowl grille nut caps, the nuts, and the cowl grille assembly.

4. Remove the baffle.

5. Disconnect the electrical connector and remove the three wiper motor mounting bolts. Remove the wiper link from the wiper motor arm and remove the wiper motor from the vehicle.

6. Make an alignment mark on the wiper motor bracket before removing the arm from the motor. Remove the motor arm.

To install:

7. Install the wiper arm to the wiper motor with the nut and aligning the mark made earlier. Torque the nut to 95–156 in. lbs. (11–18 Nm).

8. Reinstall the wiper link to the wiper motor and mount the wiper motor with the three bolts. Torque the mounting bolts to 61–87 in. lbs.

Wiper
1. Wiper arm cover
2. Wiper arm and blade
3. Cowl grille
4. Baffle
5. Wiper motor and bracket
6. Wiper link

Washer
7. Washer tank assembly
   a. Washer tank
   b. Washer motor
   c. Grommet
8. Washer pipe
9. Washer nozzle

327835

Exploded view of typical wiper system — 1993–95 323 and Protege shown, others similar

(6.9–9.8 Nm). Reconnect the electrical connector.

9. Install the baffle.

10. Reinstall the cowl grille assembly with the mounting screws and install the screw caps.

11. Reconnect the negative battery cable.

12. Turn the wiper motor on to operate the motor. Turn the wiper switch off to set the automatic park position. Install the wiper arms and set to the correct height. Install and torque the wiper arm nuts to 87–122 in. lbs. (9.8–14 Nm). Reinstall wiper arm nut caps.

Rear Wiper Motor

1. Disconnect the negative battery cable.

2. Remove the wiper arm cover, remove the nut, and remove the wiper arm.

3. Remove the seal cap and bushing.

4. Remove the rear hatch lower trim.

5. Disconnect the wiper motor electrical connector and remove the wiper motor mounting bolts and the motor.

To install:

6. Install the wiper motor with its mounting bolts. Torque the mounting

bolts to 61–87 in. lbs. (6.9–9.8 Nm). Reconnect the electrical connector.

7. Install the outer bushing and the seal cap.

8. Reconnect the negative battery cable. Set the wiper motor shaft to the park position by turning the rear wiper switch from on to off. Install the rear wiper arm and adjust to the correct height.

9. Install the rear wiper arm nut and torque the nut to 52–87 in. lbs. (5.9–9.8 Nm). Replace the wiper arm cover.

10. Reinstall the rear hatch lower trim panel.

1993–95 MX3

Front Wiper Motor

1. Disconnect the negative battery cable.

2. Remove the wiper arm covers, the nuts, and remove the wiper blade and arm assemblies.

3. Remove the screw covers, remove the attaching screws and remove the cowl cover. Once the cowl cover is removed, remove the lower baffle from the left side of the vehicle.

4. Disconnect the wiper link from the wiper motor shaft. DO NOT disconnect the motor arm unless it absolutely necessary. The relationship of the motor arm to the motor shaft de-

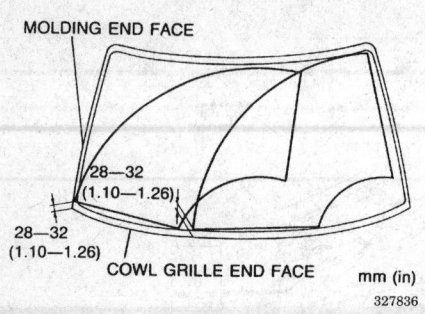

MOLDING END FACE

28—32
(1.10—1.26)

28—32
(1.10—1.26)

COWL GRILLE END FACE   mm (in)

327836

Adjusting the front wiper arms — 1993–95 323 and Protege

termines the position of the automatic stop angle. If this relationship is disturbed, wiper arm positioning will be affected.

5. Disconnect the wiper motor connector. Remove the attaching bolts and remove the wiper motor.

To install:

6. Position the motor onto its mounting and install the attaching bolts. Torque the mounting bolts to 61–86.7 in. lbs. (6.9–9.80 Nm). Connect the wiring to the motor.

7. Connect the wiper link to the motor shaft and tighten the nut to 7.9–13 ft. lbs. (10–17 Nm).

8. Install the left lower baffle, install the cowl panel with the screws, and replace the screw caps.

9. Install the wiper arm/blade assemblies, adjust to the correct position, and install the wiper arm nuts and torque them to 86.8–121 in. lbs. (9.81–13.7 Nm). Replace the wiper arm nut caps.

10. Connect the negative battery cable. Check the wiper system for proper operation.

Rear Wiper Motor

1. Disconnect the negative battery cable.

2. Remove the nut and remove the rear wiper arm.

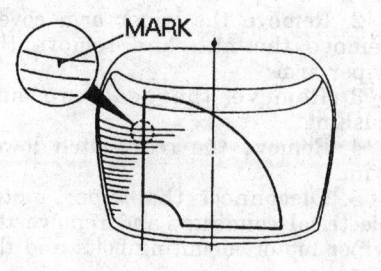

mm (in)   30 ± 2
(1.2 ±
0.08)

30 ± 2
(1.2 ± 0.08)

328147

Adjusting the front wipers — 1993–95 MX3

3. Remove the seal cap, the nut, and the outer bushing.

4. Raise the rear hatch and remove the rear wiper motor cover.

5. Disconnect the wiper motor wiring and remove the mounting bolts and the wiper motor.

To install:

6. Position the motor onto its mounting and install the attaching bolts. Torque the mounting bolts to 61–86.7 in. lbs. (6.9–9.80 Nm). Reconnect the motor wiring.

7. Install the wiper motor cover.

8. Lower the hatch and install the outer bushing. Install the nut and torque it to 27–52 in. lbs. (3.0–5.8 Nm). Replace the seal cap.

9. Connect the negative battery cable. Set the wiper motor shaft to park by turning the rear wiper switch to on and off. Set the wiper height to the correct position. Torque the wiper arm nut to 53–86.7 in. lbs. (5.9–9.80 Nm). Test the wiper system for proper operation.

1993–97 626, MX6 and Protege

1. Disconnect the negative battery cable.

2. If necessary remove the cowl grille.

3. Disconnect the electrical harness.

4. Remove the mounting bolts and remove the motor assembly.

5. Installation is the reverse of removal. Torque the mounting bolts to 61–87 ft. lbs. (7–10 Nm).

1993–95 929

1. Disconnect the negative battery cable.

2. Remove the wiper arm cover. Remove the wiper arm and blade.

3. Remove the cowl plate weatherstrip.

4. Remove the cowl plate screws, and remove the cowl plate.

MARK

328149

Adjusting the rear wiper arm — 1993–94 MX3

5. Disconnect the wiring and linkage from the wiper motor.

6. Unbolt and remove the motor.

To install:

7. Position the wiper motor, and install the mounting bolts.

8. Connect the wiring and linkage to the wiper motor.

9. Position the cowl plate and install the mounting screws. Install the weatherstrip.

10. Install the wiper arm and blade, and replace the cover.

11. Connect the negative battery cable.

12. Check the system for proper operation.

Millenia

1. Disconnect the negative battery cable.

2. Remove the cowl grille and remove the wiper arm mounting nuts and the wiper arm assemblies.

3. Disconnect the wiper motor electrical connector.

4. Remove the wiper frame mounting bolts and wiper frame and link.

5. Remove the nut and the wiper arm bell crank.

6. Remove the mounting bolts and remove the motor.

To install:

7. Install the wiper motor to the wiper frame and link assembly. Torque the mounting bolts to 61–88 inch lbs. (7–10 Nm).

8. Install the wiper motor bell crank with the lock washer and the nut.

9. Install the wiper frame and link assembly to the cowl with the mounting bolts and torque the bolts to 61–88 inch lbs. (7–10 Nm).

10. Reconnect the wiper motor connector.

11. Reinstall the wiper arms to the wiper pivots with the nuts. Torque the nuts to 12–14 ft. lbs. (16–20 Nm).

12. Replace the cowl grille.

13. Reconnect the negative battery cable and check for proper operation.

Miata

Without ABS

1. Disconnect the negative battery cable.

2. On non-ABS equipped vehicles only:
   a. Remove the wiper arms.
   b. Remove the cowl plate screws and remove the cowl plate.
   c. Remove the baffle cover.

3. Remove the main fuse block.

4. Disconnect the wiring and linkage from the wiper motor. Unbolt and remove the motor.

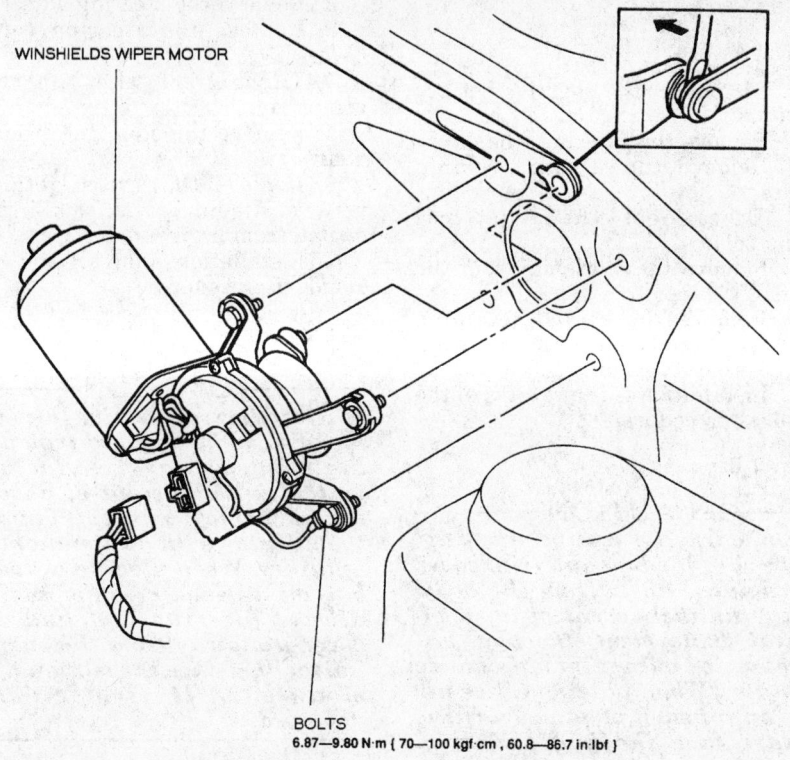

WINSHIELDS WIPER MOTOR

BOLTS
6.87—9.80 N·m (70—100 kgf·cm , 60.8—86.7 in·lbf)

304250

Front wiper assembly (exploded view) — 1995–97 Protege

5. Installation is the reverse of removal. Check the system for proper operation.

With ABS

1. Disconnect the negative battery cable.
2. Remove the main fuse block.
3. Disconnect the wiring and linkage from the wiper motor. Unbolt and remove the motor.
4. Installation is the reverse of removal. Check the system for proper operation.

RX7

Front

1. Disconnect the negative battery cable. Remove the wiper arms.
2. Remove the cowl grille screws and remove the cowl grille.
3. Disconnect the wiring and linkage from the wiper motor.
4. Unbolt and remove the motor.
5. Installation is the reverse of removal. Check the system for proper operation.

Rear

1. Disconnect the negative battery cable.
2. Remove the wiper arm and outer bushing.

3. Remove the liftgate lower trim panel.
4. Disconnect the electrical connector and remove the wiper motor.
5. Installation is the reverse of the removal procedure.

Concealed Headlights

MANUAL OPERATION

1993–97 Miata

Retractor Operational

1. Push the independent headlight retractor switch to raise or retract the headlights. The ignition switch may be at any position.
2. The headlights should be raised at all times in cold weather.

Retractor Doesn't Operate

1. Turn **OFF** the ignition switch.
2. Lift the hood and open the main fuse panel.
3. Pull the RETRACTOR (30 A) fuse straight out.
4. Inspect the removed fuse. If it is blown, replace it with a new one of the same rating. If it isn't blown or a new one doesn't solve the problem, remove the fuse and manually operate the headlight motor.

5. To manually operate the headlight motor:
   a. Remove the cap from the top of the headlight motor. Then turn the knob clockwise until the headlight is fully raised or retracted.
   b. Replace the cap.
   c. Reinstall the fuse and the fuse panel cover.

Headlight Switch

REMOVAL AND INSTALLATION

1993–94 323 and Protege

1. Disconnect the negative battery cable.

------ **CAUTION** ------
The supplemental inflatable restraint (SIR) system must be disarmed before removing the steering wheel. Failure to do so may cause accidental deployment of the air bag, resulting in unnecessary SIR repairs and/or personal injury.

2. After the air bag is disarmed; the battery can be connected to perform electrical testing on other systems.
3. On models so equipped, remove the air bag mounting screws and remove the air bag from the steering wheel.

------ **CAUTION** ------
Always carry an air bag assembly with the bag and trim cover away from your body. Store the assembly with the bag and trim cover facing up; never place the assembly face down on the floor or workbench.

4. Remove the steering wheel.
5. Remove the 2 column cover screws and remove the cover.
6. Remove the meter hood assembly and remove the instrument cluster screws.
7. Carefully pull the cluster module outward and disconnect the electrical connectors and the speedometer cable connector from the instrument cluster.
8. Remove the instrument cluster.
9. Individual instruments may be removed as necessary.
   To install:
10. Reconnect the instrument connectors and speedometer cable; carefully install the instrument cluster module and secure it with the screws.
11. Reattach column covers with screws.

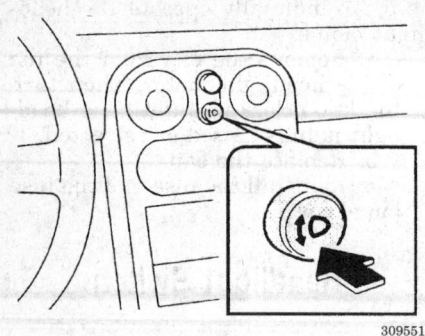

Headlight retractor switch — 1993–97 Miata

309551

12. Reinstall the steering wheel.

13. When ready to reconnect the air bag: disconnect the battery (if not already done), connect the air bag connector, install the cover or panel and make sure no one is in the vehicle when connecting the battery.

14. Check that all gauges are working properly, and road test for speedometer operation.

Combination Switch

REMOVAL AND INSTALLATION

1995 MX3

------ CAUTION ------
The air bag system must be disarmed before removing the steering wheel. Failure to do so may cause accidental deployment, property damage or personal injury.

1. Remove the driver-side side and lower panels.
2. Disconnect the negative battery cable, and disarm the air bag.
3. Remove the steering wheel.

------ CAUTION ------
Always carry an air bag assembly, with the bag and trim cover, away from your body. Store the assembly facing upward; never place the assembly face down on any surface.

4. Remove the 2 column cover screws and remove the cover.
5. Disconnect the electrical connections from the combination switch.
6. Remove the combination switch as an assembly.
7. Installation is the reverse of the removal procedure. Check the switch operation.

626 and MX6

Without Air Bag

1. Disconnect the negative battery cable.
2. Remove the steering wheel.
3. Remove the steering column covers.
4. Disconnect the electrical connectors.
5. Remove the stop ring from the shaft.
6. Remove the switch retaining screws. Remove the combination switch from its mounting.
7. Installation is the reverse of the removal procedure.

With Air Bag

------ CAUTION ------
When carrying a live air bag, make sure the bag and trim cover are pointed away from the body. In the unlikely event of an accidental deployment, the bag will then deploy with minimal chance of injury. When placing a live air bag on a bench or other surface, always face the bag and trim cover up, away from the surface. This will reduce the motion of the module if it is accidently deployed.

1. Disconnect the negative battery cable and disarm the air bag system.
2. Remove the driver-side side panel and lower panel. Disconnect the clock spring connector.
3. Remove the steering wheel and the column covers.
4. Disconnect the wiring and remove the screws for the combination switch. Remove the combination switch.
5. Installation of the switch is the reverse of removal. To install the steering wheel, the clock-spring must be reset.
   a. Make sure the front wheels are straight ahead.
   b. Turn the clock-spring connector clockwise until it stops. Don't force it.
   c. Turn the clock-spring back about 2.75 turns and align the marks on the clock spring connector to the marks on the outer housing.
   d. Connect the wiring and install the steering wheel.

1993–97 929, Miata and RX7

Without Air Bag

1. Disconnect the negative battery cable.

2. Remove the steering wheel.
3. Remove the steering column covers.
4. Disconnect the electrical connectors.
5. Remove the stop ring from the shaft.
6. Remove the switch retaining screws. Remove the combination switch from its mounting.
7. Installation is the reverse of the removal procedure.

With Air Bag

------ CAUTION ------
When carrying a live air bag, make sure the bag and trim cover are pointed away from the body. In the unlikely event of an accidental deployment, the bag will then deploy with minimal chance of injury. When placing a live air bag on a bench or other surface, always face the bag and trim cover up, away from the surface. This will reduce the motion of the module if it is accidently deployed.

1. Disconnect the negative battery cable, and disarm the air bag system.
2. Remove the steering wheel and the column covers.
3. To remove the combination switch, disconnect the wiring, and remove the screws.
4. Installation of the switch is the reverse of removal. To install the steering wheel, the clock-spring must be reset.
   a. Make sure the front wheels are straight ahead.
   b. Turn the clock-spring all the way to the right. Don't force it.
   c. Turn the clock-spring back about 2 ¾ turns and align the marks.
   d. Connect the wiring and install the steering wheel.
5. Reconnect the battery cable and re-arm the air bag system.

Millenia

------ CAUTION ------
The air bag system must be disarmed before removing the steering wheel. Failure to do so may cause accidental deployment, property damage or personal injury.

1. Disconnect the negative battery cable and disarm the air bag.
2. Remove the air bag module and the steering wheel.

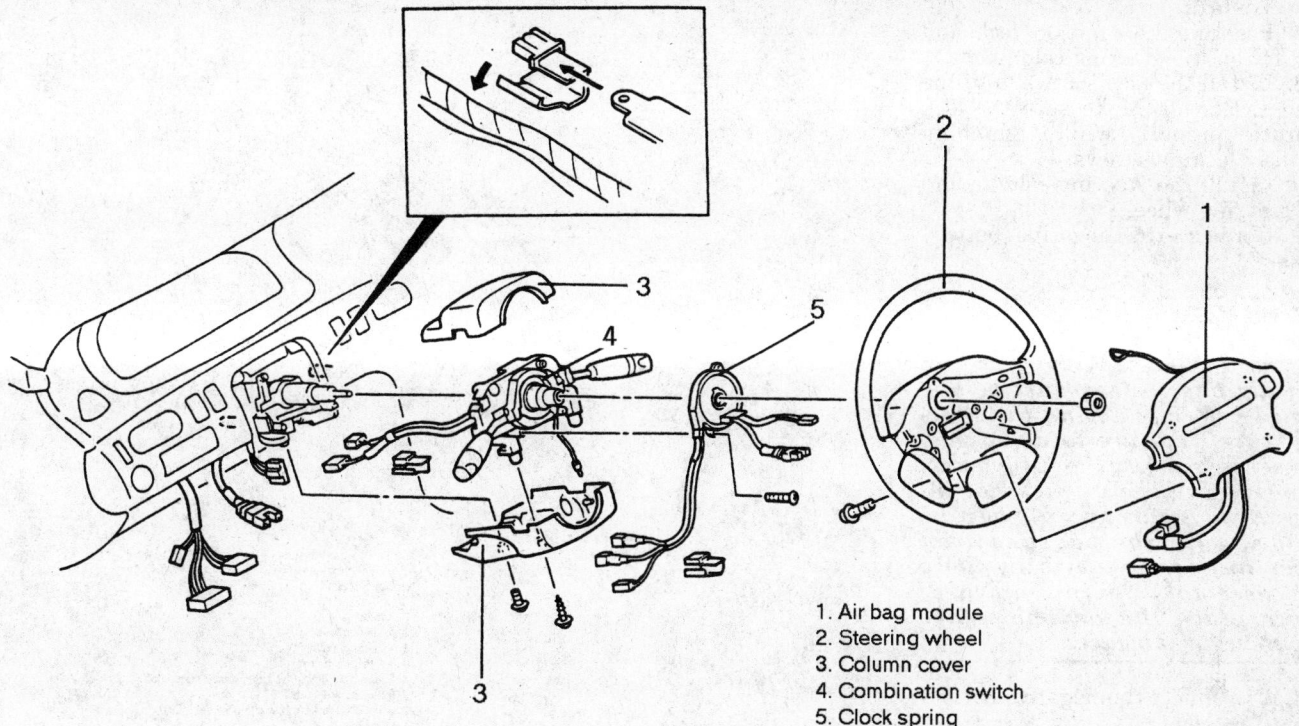

1. Air bag module
2. Steering wheel
3. Column cover
4. Combination switch
5. Clock spring

294136

Combination switch — 1995–97 626 and MX6

TURN CLOCKWISE

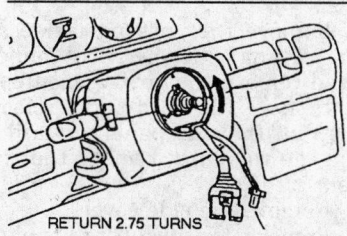

RETURN 2.75 TURNS

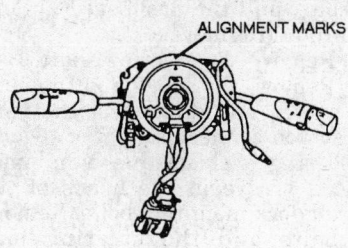

ALIGNMENT MARKS

294147

Clock spring alignment marks — 1995–97 626 and MX6

— **CAUTION** —

Always carry an air bag assembly, with the bag and trim cover, away from your body. Store the assembly facing upward; never place the assembly face down on any surface.

3. Remove the steering column cover.

4. Remove the electrical connectors and mounting screws, then remove the combination switch.

5. Installation is the reverse of removal.

1993–94 323, MX3 and 1993–97 Protege

— **CAUTION** —

Before removing the steering wheel or horn pad, the air bag system connectors, Orange and Blue, must be disconnected under the steering wheel.

1. Disconnect the negative battery cable.

2. Remove the horn cap.

3. Remove the steering wheel.

4. Remove the 2 column cover screws and remove the cover.

5. Disconnect the electrical connections from the combination switch.

6. Remove the combination switch as an assembly.

7. Installation is the reverse of the removal procedure. Check the switch operation.

Ignition Lock Cylinder

REMOVAL AND INSTALLATION

1993–94 323, 1993–95 MX3 and 1993–97 Protege and Miata

— **CAUTION** —

The air bag system must be disarmed before removing the steering wheel. Failure to do so may cause accidental deployment, property damage or personal injury.

1. Disconnect thew negative battery cable.

2. If equipped, disarm and remove the air bag.

3. Remove the steering wheel.

4. Remove the steering column.

5. Place the steering column securely on a workbench. Use a chisel and hammer to make slots in the head of the lock screws. Remove the screws.

6. Remove the ignition lock assembly.

To install:

7. Position the ignition lock and bracket on the steering column.

8. Install the new screws until the head twists off. Make sure the lock operates properly while tightening the new locking screws.

9. Install the steering column and the steering wheel.

10. Connect the negative battery cable.

626 and MX6

------- **CAUTION** -------

The air bag system must be disarmed before removing the steering wheel. Failure to do so may cause accidental deployment, property damage or personal injury. Always carry an air bag assembly, with the bag and trim cover, away from your body. Store the assembly facing upward; never place the assembly face down on any surface.

1. Disconnect the negative battery cable.

2. Disarm and remove the air bag.

3. Remove the steering wheel.

4. Remove the steering column.

5. Place the steering column securely on a workbench. Use a chisel and hammer to make slots in the head of the lock screws. Remove the screws.

6. Remove the ignition lock assembly.

To install:

7. Position the ignition lock and bracket on the steering column.

8. Install the new screws until the head twists off. Make sure the lock operates properly while tightening the new locking screws.

9. Install the steering column, wheel and air bag.

10. Connect the negative battery cable.

Millenia

The ignition lock is an integral component of the steering lock mechanism. If the ignition lock must be replaced, perform the following procedure.

------- **CAUTION** -------

The air bag system must be disarmed before removing the steering wheel. Failure to do so may cause accidental deployment, property damage or personal injury. Always carry an air bag assembly, with the bag and trim

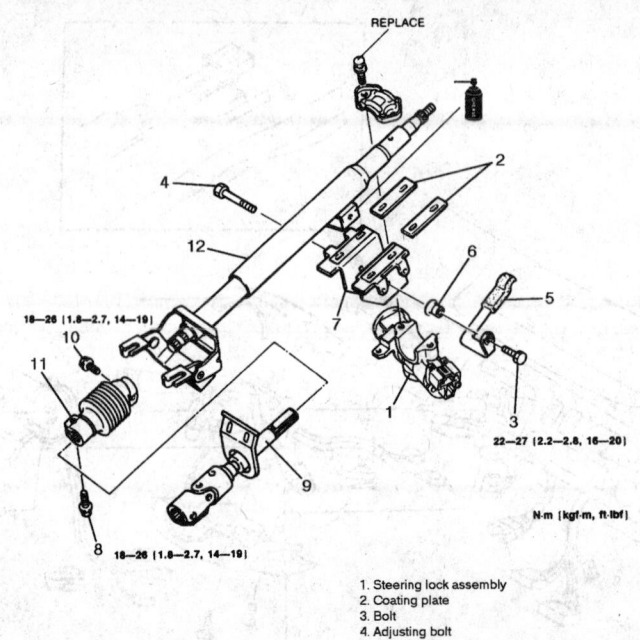

REPLACE

18—26 [1.8—2.7, 14—19]

22—27 [2.2—2.8, 16—20]

18—26 [1.8—2.7, 14—19]

N·m [kgf·m, ft-lbf]

1. Steering lock assembly
2. Coating plate
3. Bolt
4. Adjusting bolt
5. Adjusting lever
6. Adjusting nut
7. Tilt lever bracket
8. Fixing bolt (universal joint/intermediate shaft)
9. Intermediate shaft
10. Fixing bolt (steering shaft/universal joint)
11. Universal joint
12. Steering shaft assembly

332464

Ignition switch and steering column — 1993–94 323 and MX3

cover, away from your body. Store the assembly facing upward; never place the assembly face down on any surface.

1. Disconnect the negative battery cable.

2. Properly disarm the air bag system.

3. Remove the rear console box, brake boot and center panel (disconnect the harness connectors).

4. Remove the rear console.

5. Remove the undercover, glove compartment and compartment cover.

6. Remove the upper and lower steering column covers.

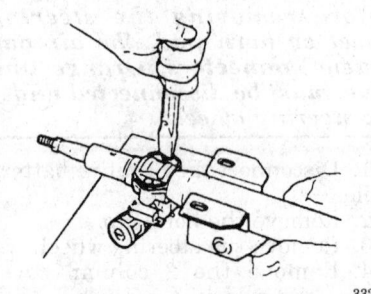

332465

Common method of removing the steering lock mounting bolts

7. Tilt the steering shaft down. Remove the cap and panel light control switch. Remove the meter hood mounting screws. Pull the meter hood forward to disengage the clips.

8. Remove the instrument cluster,

9. Remove the drivers side air bag and the steering wheel.

10. Remove the combination switch.

11. Remove the passenger side air bag mounting bolts, electrical connector and remove the air bag module.

12. Remove the hood panel release lever. Pull the side panel forward to disengage the clips, remove the side panel.

13. Remove the side covers.

14. Remove the dashboard mounting bolts, pull the dashboard up and forward and remove.

15. Remove the steering shaft.

16. Remove the screw and the ignition switch and disconnect the ignition switch connector.

17. Using a chisel and a hammer make a groove in the heads of the steering lock mounting bolts. Remove the bolts and the steering lock assembly.

To install:

18. Install the steering lock mechanism to the steering column jacket. Make sure that the lock operates correctly. Install new mounting bolts

and tighten the bolts till the heads break off.

19. Reconnect the ignition switch connector to the ignition switch and install the ignition switch to the steering lock mechanism with the screw.

20. Replace the dashboard close to its mounts and connect all harness connectors. Install all mounting bolts. Torque to 14–18 ft. lbs. (19–25 Nm).

21. Install the side covers.

22. Instal the side panels.

23. Install the steering shaft and torque the mounting bolts to 12–16 ft. lbs. (16–22 Nm).

24. Connect the passenger side air bag module electrical connector and replace the module to the dash and install the mounting bolts.

25. Install the combination switch on the steering column.

26. Install the steering wheel and drivers side air bag module.

27. Install the instrument cluster.

28. Install the meter hood.

29. Install the upper and lower steering column covers.

30. Install the glove compartment cover, compartment and the undercover.

31. Install the rear console, bracket, center panel, brake lever boot and the rear console box.

32. Reconnect the negative battery cable,

33. Place the steering column securely on a workbench. Use a chisel and hammer to make slots in the head of the lock screws. Remove the screws.

Ignition Switch

REMOVAL AND INSTALLATION

1993–94 323 1.6L B6E), 626/MX6 2.0L (FS), 2.5L (KL), 929 3.0L (JE-ZE) MX3 K8), Protege 1.8L (BPD, BPE), Miata 1.6L (B6-ZE and 1.8L BPD) and RX7 1.3L (13B) Engines

1. Disconnect the negative battery cable.

2. Remove the steering wheel.

3. Remove the steering column covers. Remove the lower panel or air duct, if necessary.

4. Disconnect the electrical connectors.

5. Remove the switch screws and the combination switch.

6. Disconnect the ignition switch wires.

7. Use a chisel to make slots in the lock screws. Remove the screws.

8. Installation is the reverse of the removal procedure. Install the new screws until the head twists off. Make sure the lock operates properly while tightening the new locking screws.

1995–97 Protege 1.5L (Z5D) and 1.8L (BPD) engines

1. Disconnect the negative battery cable.

2. Remove the steering column cover.

3. Remove the key reminder switch connector from the ignition switch.

4. Disconnect the ignition switch connector.

5. Remove the mounting screws and remove the switch.

NOTE: If the screw is difficult to remove, remove the steering shaft mounting bolts and lower panel to remove the ignition switch.

6. Installation is the reverse of removal.

1995 MX3 1.6L (B6-ZE) and 1.8L (K8) Engines

———— **CAUTION** ————

The air bag system must be disarmed before removing the steering wheel. Failure to do so may cause accidental deployment, property damage or personal injury. Always carry an air bag assembly, with the bag and trim cover, away from your body. Store the assembly facing upward; never place the assembly face down on any surface.

1. Disconnect the negative battery cable.

2. Disarm and remove the air bag.

3. Remove the steering wheel.

4. Remove the steering column.

5. Place the steering column securely on a workbench. Use a chisel and hammer to make slots in the head of the lock screws. Remove the screws.

6. Remove the ignition lock assembly.

To install:

7. Position the ignition lock and bracket on the steering column.

8. Install the new screws until the head twists off. Make sure the lock operates properly while tightening the new locking screws.

9. Install the steering column, wheel and air bag.

10. Connect the negative battery cable.

1995–97 626 and MX6 2.0L (FS) and 2.5L (KL) Engines

1. Disconnect the negative battery cable.

2. Remove the steering column cover.

3. Disconnect the ignition switch connector.

4. Remove the screw and the ignition switch.

5. Installation is the reverse of removal.

1995 929 3.0L (JE-ZE) Engine

———— **CAUTION** ————

The air bag system must be disarmed before removing the steering wheel. Failure to do so may cause accidental deployment, property damage or personal injury. Always carry an air bag assembly with the bag and trim cover away from your body. Store the assembly facing upward; never place the assembly face down on any surface.

1. Disconnect the negative battery cable.

2. Disarm and remove the air bag.

3. Remove the steering wheel.

4. Remove the steering column.

5. Place the steering column securely on a workbench. Use a chisel and hammer to make slots in the head of the lock screws. Remove the screws.

6. Remove the ignition lock assembly.

To install:

7. Position the ignition lock and bracket on the steering column.

8. Install the new screws until the head twists off. Make sure the lock operates properly while tightening the new locking screws.

9. Install the steering column, wheel and air bag.

10. Connect the negative battery cable.

1995–97 Miata 1.8L (BPD) Engines

1. If equipped, deactivate the audio antitheft system.

2. Disconnect the negative battery cable.

3. Remove the steering column cover.

4. Disconnect the connector and remove the ignition switch.

5. Installation is the reverse of removal.

1995 RX7 1.3L engine

─────────── CAUTION ───────────

The air bag system must be disarmed before removing the steering wheel. Failure to do so may cause accidental deployment, property damage or personal injury. Always carry an air bag assembly with the bag and trim cover away from your body. Store the assembly facing upward; never place the assembly face down on any surface.

1. Disconnect the negative battery cable.
2. Remove the steering wheel.
3. Remove the steering column covers. Remove the lower panel or air duct, if necessary.
4. Disconnect the electrical connectors.
5. Remove the switch screws and the combination switch.
6. Disconnect the ignition switch wires.
7. Remove the steering column to dash panel structure mounting bolts and carefully lower the steering column.
8. Using a chisel, make slots in the steering lock mounting screws. Remove the screws.

To install:

9. Install the steering lock to the steering column with new bolts. Tighten the bolts until the heads break off.
10. Reconnect the ignition switch wiring harness.
11. Carefully reposition the steering column to the dash panel structure and secure the steering column with the mounting bolts.
12. Reinstall the combination switch with its mounting screws and reconnect the electrical connectors.
13. Reinstall the steering column covers and the steering wheel.
14. Install the steering wheel cover or air bag assembly.
15. Reconnect the negative battery cable and check for proper operation.

1995–97 Millenia 2.3L (KJS) and 2.5L (KLS) engines

1. Disconnect the negative battery cable. Wait at least 90 seconds before performing any work. This will allow time for the SRS backup power supply system to deplete its stored energy.
2. Remove the lap louver duct.
3. Remove the screw and withdrawal the ignition switch from the steering lock assembly.

4. Disconnect the ignition switch electrical connector and remove the switch.
5. Installation is the reverse of the removal procedure.

Park/Neutral Safety Switch

REMOVAL AND INSTALLATION

1993 MX3 1.6L (B6-ZE), 1993–94 323 1.6L B6E), Miata 1.6L (B6-ZE), 929 3.0L (JE-ZE) and 1993–95 Protege 1.8L(BPE, BPD) and RX7 1.3L (13B) Engines

1. Remove the shift cable or selector rod from the manual lever shaft on the transmission.
2. Remove the electrical connector from the switch.
3. Remove the mounting bolts and remove the switch.

To Install And Adjust:

4. Install the park/neutral switch to the transmission manual shift lever with the mounting bolts, but do not tighten the bolts until the adjustment is made.
5. Make sure that the transmission manual shaft lever is in the N position.
6. Align the holes of the park/neutral switch and the manual shaft by inserting a 0.16 in. (4.0mm) outer diameter pin.
7. Tighten the park/neutral switch mounting bolts to 22–34 in. lbs. (2.5–3.9 Nm). Remove the aligning pin.
8. Reconnect the electrical connector and reinstall the shift cable or selector rod to the manual lever shaft on the transmission.
9. Check for proper operation.

1995–97 Protege 1.5L (Z5D), 1.8L (BPD), 1993–95 MX3 1.6L (B6E), 1.8L (K8), 1993–97 626 and MX6 2.0L (FS) and 2.5L (KL) Engines

1. Disconnect the negative battery cable.
2. Remove the resonance chamber, fresh air duct and air cleaner assembly.
3. Remove the spring pin and clip and disconnect the shift cable at the transmission.
4. Disconnect the neutral safety switch connector.
5. Remove the manual shaft nut, lockwasher and lever.
6. Remove the switch mounting bolts and the switch.

To install:

7. Rotate the manual shaft to **N**.

8. Turn the neutral safety switch so the neutral mark is in line with the flat, straight surfaces on either side of the manual shaft.
9. Loosely tighten the switch bolts and adjust the switch. Tighten the bolts to 70–95 inch lbs. (8–11 Nm).
10. Install the manual shaft nut, lockwasher and lever. Torque to 24–33 ft. lbs. (32–46 Nm).
11. Connect the neutral safety switch electrical connector.
12. Connect the shift cable at the transmission. Install the spring pin and clip.
13. Install the resonance chamber, fresh air duct and air cleaner assembly.
14. Connect the negative battery cable.

1995 929 3.0L (JE-ZE) Engine

NOTE: The switch is called a P position switch, and the vehicle can only be started with the transmission in park.

1. Disconnect the negative battery cable.
2. Remove the power window switch.
3. Remove the rear console.
4. Remove the boot panel.
5. Remove the ashtray.
6. Remove the radio.
7. Remove the front console.
8. Shift the selector lever to the **N** position. Remove the spring pin and washers, and disconnect the selector rod. Lift the selector lever.
9. Disconnect the P position switch connector.
10. Remove the switch.

To install:

11. Install the P position switch.
12. Connect the P position switch connector.
13. Lower the selector lever, position the selector rod, and install a new spring pin and washers.
14. Install the front console, radio, ashtray and boot panel.
15. Install the rear console and power window switch.
16. Connect the negative battery cable.

1995–97 Millenia 2.3L (KJS) and 2.5L (KLD) Engines

GF4A-EL Transaxle

1. Disconnect the negative battery cable. Wait at least 90 seconds before performing any work.
2. Remove the air cleaner assembly.
3. Disconnect the transaxle range switch connector.

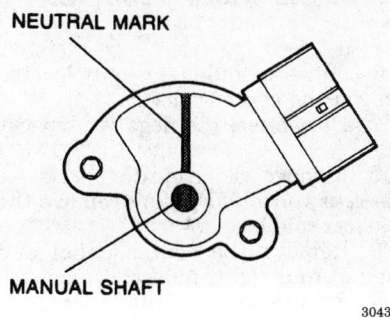

Park/Neutral switch

304323

4. Remove the nut and the clip, and disconnect the selector cable.

5. Remove the transaxle manual shaft nut, lock washer and nut.

6. Remove the mounting screws and the range switch.

To install:

7. Install the transaxle range switch over the manual shaft lever.

8. Rotate the manual shaft to the neutral position.

9. Turn the transaxle range switch so that the neutral mark is in line with the flat, straight surfaces on either side of the manual shaft.

10. Hand tighten the range switch mounting screws.

11. Verify that there is continuity between terminals A and H of the range switch connector.

12. Torque the transaxle range switch mounting screws to 70–95 inch lbs. (8–11 Nm).

13. Install the manual shaft lever, lock washer and nut. Torque the nut to 24–33 ft. lbs. (32–46 Nm).

14. Verify that the selector lever range position and transaxle range switch are aligned, then connect with the nut and clip.

15. Reconnect the range switch connector and install the air cleaner assembly.

16. Reconnect the negative battery cable and check for proper operation.

LJ4A-EL Transaxle

1. Disconnect the negative battery cable. Wait at least 90 seconds before performing any work.

2. Remove the selector rod from the manual shaft lever.

3. Move the manual shaft to the N position.

4. Disconnect the transaxle range switch connector, remove the mounting bolts and remove the switch.

To install:

5. Install the transaxle range switch and hand tighten the mounting bolts.

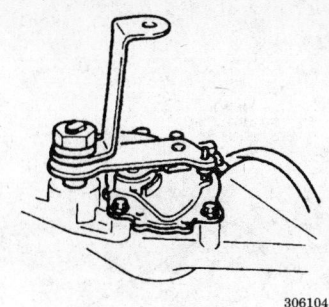

306104

Aligning the range switch with a pin — 1995–97 Millenia 2.3L (KJS) and 2.5L (KLD) engines

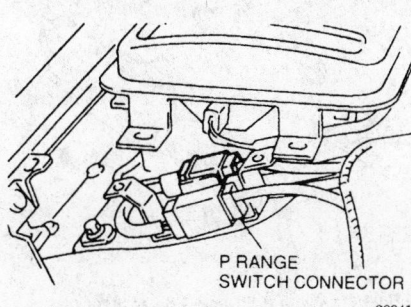

P RANGE
SWITCH CONNECTOR

289416

Park switch — 1995–97 Miata 1.8L (BPD)

6. Align the holes of the range switch and the manual shaft lever by inserting a pin with an outside diameter of 0.157 in. (4 mm).

7. Tighten the range switch mounting bolts to 22–34 inch lbs. (3–4 Nm).

8. Reconnect the range switch connector and install the selector rod to the manual shaft lever.

9. Reconnect the negative battery cable and check for proper operation.

1995–97 Miata 1.8L (BPD) Engine

NOTE: The switch is called a P position switch, and the vehicle can only be started with the transmission in park.

1. Disconnect the negative battery cable.

2. Remove the rear console.

3. Remove the screws and lift the indicator panel.

4. Disconnect the P position switch connector.

5. Remove the switch.

To install:

6. Install the P position switch.

7. Connect the P position switch connector.

8. Lower the indicator panel, and install the mounting screws.

9. Install the rear console.

10. Connect the negative battery cable.

Powertrain Control Module

REMOVAL AND INSTALLATION

1995–97 Protege 1.5L (Z5D) and 1.8L (BPD) Engines

The control module is usually located under the front console.

1. Disconnect the negative battery cable.

2. Remove the rear console.

3. Remove the ashtray.

4. On manual transaxles only, remove the shift knob.

5. Remove the front console to gain access to the module.

6. Remove the ECM mounting bolts/nuts, and disconnect the module electrical connectors.

To install:

7. Position the module, and install the mounting nuts and bolts.

8. Install the front console.

9. If equipped, install the shift knob.

10. Install the ashtray.

11. Install the rear console.

12. Connect the negative battery cable.

1995 MX3 1.6L (B6E) and 1.8L (K8) Engines

The control module is usually located behind the center console side trim panels.

1. Disconnect the negative battery cable.

2. Remove both center console side trim panels to gain access to the module.

3. Remove the module mounting bolts/nuts, and disconnect the module electrical connectors.

4. Installation is the reverse of the removal procedure.

626 and MX6 2.0L (FS) and 2.5L (KL) Engines

The control module is usually located behind the center console.

1. Disconnect the negative battery cable.

2. Remove the center console assembly, by pulling it forward to disengage the six clips, to gain access to the module.

3. Remove the module mounting bolts/nuts and disconnect the module electrical connectors.

4. Installation is the reverse of the removal procedure.

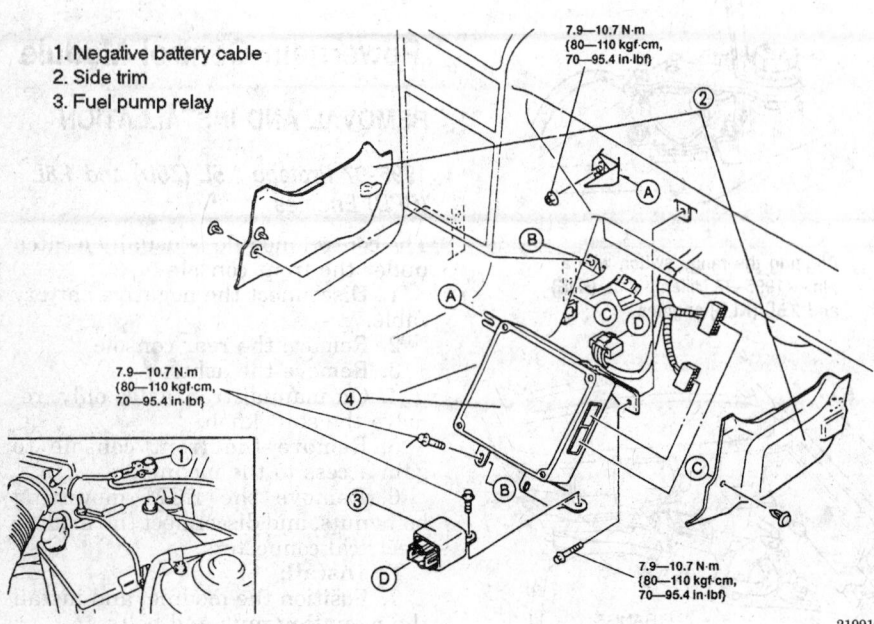

1. Negative battery cable
2. Side trim
3. Fuel pump relay

7.9—10.7 N·m
{80—110 kgf·cm,
70—95.4 in·lbf}

7.9—10.7 N·m
{80—110 kgf·cm,
70—95.4 in·lbf}

7.9—10.7 N·m
{80—110 kgf·cm,
70—95.4 in·lbf}

219919

Powertrain control module mounting location — 1995 MX3 with 1.6L and 1.8L engines

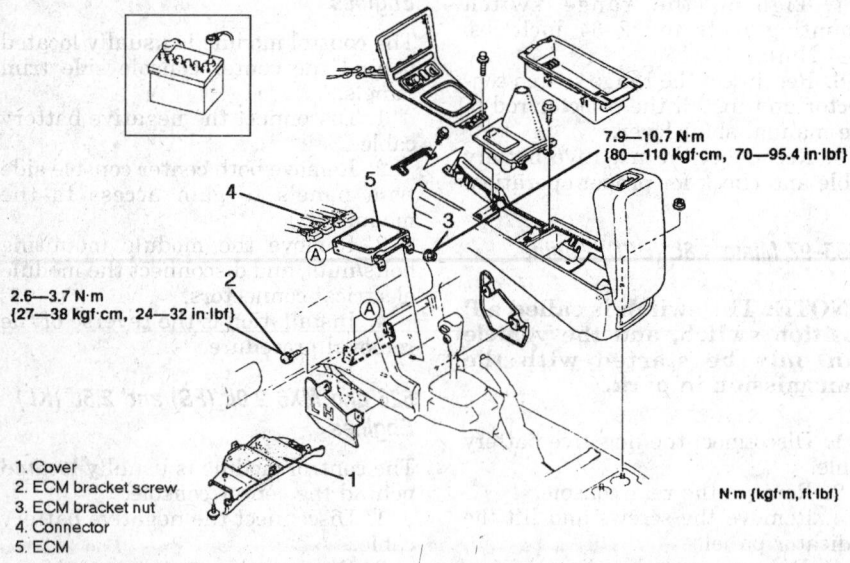

2.6—3.7 N·m
{27—38 kgf·cm, 24—32 in·lbf}

7.9—10.7 N·m
{80—110 kgf·cm, 70—95.4 in·lbf}

1. Cover
2. ECM bracket screw
3. ECM bracket nut
4. Connector
5. ECM

N·m {kgf·m, ft·lbf}

308604

Powertrain control module mounting location — Millenia with 2.3L and 2.5L engines

Millenia 2.3L (KJS) and 2.5L (KLD) Engines

The control module is usually located behind the center console.

1. Disconnect the negative battery cable.

2. Remove the mounting nuts for the rear console box, and remove the rear console box.

3. Remove the center panel and the left and right undercovers.

4. Remove the module cover.

5. Remove the module bracket screw and nut.

6. Remove the rear heater duct fasteners.

7. Separate the rear heater duct to both sides.

8. Disconnect the module connector, and remove the module from the vehicle.

To install:

9. Position the module, and engage the connector.

10. Assemble the rear heater duct, and attach the fasteners.

11. Install the module bracket nut and screw.

12. Install the left and right undercovers.

13. Install the center panel.

14. Install the rear console box.

15. Connect the negative battery cable.

1993–95 RX7 1.3L (13B) and 1995 929 3.0L (JE-ZE) Engines

The control module is usually located behind the passenger side scuff plate.

1. Disconnect the negative battery cable.

2. Remove the passenger side scuff plate and front side trim panels to gain access to the module.

3. Remove the module mounting bolts/nuts, and disconnect the module electrical connectors.

4. Installation is the reverse of the removal procedure.

1993–94 323 1.6L (B6E), MX3 1.8L (K8, BPD) and Protege 1.8L (BPD) Engines

The control is usually located behind the center console driver and passenger side panels.

1. Disconnect the negative battery cable.

2. Remove the passenger and driver side cover walls from the center console unit.

3. Remove the module mounting bolts/nuts and disconnect the module electrical connectors.

4. Installation is the reverse of the removal procedure.

Miata 1.6L (B6-ZE) and 1.8L (BPD) Engines

The control module is usually located beneath the passenger seat or floor mat.

1. Disconnect the negative battery cable.

2. On the 1993 models, move the passenger seat forward.

3. On the 1994–97 models, lift the passenger side floor mat.

4. Remove the module protective cover.

5. Remove the module mounting bolts/nuts and disconnect the electrical connectors.

6. Installation is the reverse of the removal procedure.

ENGINE COOLING

Radiator

REMOVAL AND INSTALLATION

1993–94 323 1.6 (B6E), 626/MX6 2.0L (FS), 2.5L (KL) and Protege 1.8L (BPE, BPD) engines

1. Disconnect the negative battery cable.

2. Remove the engine undercover, if equipped and drain the cooling system.

3. Remove the necessary air ducts.

4. Disconnect the electric cooling fan connector and, if equipped, temperature sensor connector.

5. Disconnect the coolant reservoir and upper and lower radiator hoses. If equipped with automatic transaxle or transmission, disconnect the oil cooler lines and plug the hoses.

6. Remove the upper radiator mounting brackets.

7. Lift the radiator/cooling fan(s) assembly from the vehicle. Remove the radiator/cooling fans assembly.

8. If necessary, remove the cooling fan(s)/shroud assembly from the radiator.

To install:

9. If removed, install the fan and shroud assembly. Tighten the mounting bolts to 61–87 inch lbs. (7–10 Nm).

10. Install the radiator, making sure the lower tank engages the insulators.

11. Install the upper radiator insulators and tighten the retaining bolts to 69–95 inch lbs. (8–11 Nm).

12. Unplug and connect the cooler lines, if required.

13. Reattach the wiring harness to the routing clips and install the upper and lower radiator hoses to the radiator.

14. Connect the overflow tube to the radiator and connect the cooling fan wiring connectors.

15. Close the radiator drain valve and fill the system with coolant. Install the pressure cap to the first stop only.

16. Connect the negative battery cable. Start the engine and run it at fast idle until the upper radiator hose feels warm, indicating the thermostat has opened and coolant is flowing throughout the system.

17. Stop the engine. Carefully remove the radiator cap and top off the radiator with coolant, if required.

18. Install the radiator cap securely and fill the coolant reservoir to the **FULL** mark.

19. Run the engine and check for leaks.

1995 MX3 1.6L (B6-ZE), 1995–97 Protege 1.5L (Z5D) and 1.8L (BPE, BPD) Engines

1. Disconnect the negative battery cable.

2. Drain the cooling system.

3. Remove the upper seal board and radiator grille.

4. Remove the hood safety lever.

5. Disconnect the coolant reservoir and upper and lower radiator hoses. If equipped with automatic transaxle, disconnect the oil cooler lines and plug the hoses.

6. Disconnect the electric cooling fan connector.

7. Remove the cooling fan/shroud assembly from the radiator. Move the assembly to the side of the engine to gain clearance space.

8. Remove the radiator brackets.

9. Lift and remove the radiator from the vehicle.

To install:

10. Install the radiator, making sure the lower tank engages the insulators.

11. Install the radiator brackets. Tighten the mounting bolts to 70–95 inch lbs. (8–11 Nm).

12. Install the fan and shroud assembly. Tighten the mounting bolts to 70–95 inch lbs. (8–11 Nm).

13. Unplug and connect the cooler lines, if required.

14. Reattach the wiring harness to the routing clips, and install the upper and lower radiator hoses to the radiator.

15. Connect the overflow tube to the radiator, and connect the cooling fan wiring connector.

16. Install the hood safety lever. Tighten the mounting bolts to 70–95 inch lbs. (8–11 Nm).

17. Install the radiator grille and upper seal board.

18. Close the radiator drain valve and fill the system with coolant. Install the pressure cap to the first stop only.

19. Connect the negative battery cable. Start the engine and run it at fast idle until the upper radiator hose feels warm, indicating the thermostat has opened and coolant is flowing throughout the system.

20. Stop the engine. Carefully remove the radiator cap and top off the radiator with coolant, if required.

21. Install the radiator cap securely and fill the coolant reservoir to the **FULL** mark.

22. Run the engine and check for leaks.

1994–95 MX3 1.8L (K8) Engine

1. Disconnect the negative battery cable.

2. Drain the cooling system.

3. Remove the fresh air duct.

4. Disconnect the electric cooling fan connector.

5. Disconnect the coolant reservoir and upper and lower radiator hoses. If equipped with automatic transaxle, disconnect the oil cooler lines and plug the hoses.

6. Remove the shroud upper panel.

7. Lift the radiator/cooling fan assembly from the vehicle. Remove the radiator/cooling fan assembly.

8. If necessary, remove the cooling fan/shroud assembly from the radiator.

To install:

9. If removed, install the fan and shroud assembly. Tighten the mounting bolts to 70–95 inch lbs. (8–11 Nm).

10. Install the radiator, making sure the lower tank engages the insulators.

11. Install the shroud upper panel.

12. Unplug and connect the cooler lines, if required.

13. Reattach the wiring harness to the routing clips, and install the upper and lower radiator hoses to the radiator.

14. Connect the overflow tube to the radiator and connect the cooling fan wiring connector.

15. Install the air duct. Tighten the mounting bolts to 70–95 inch lbs. (8–11 Nm).

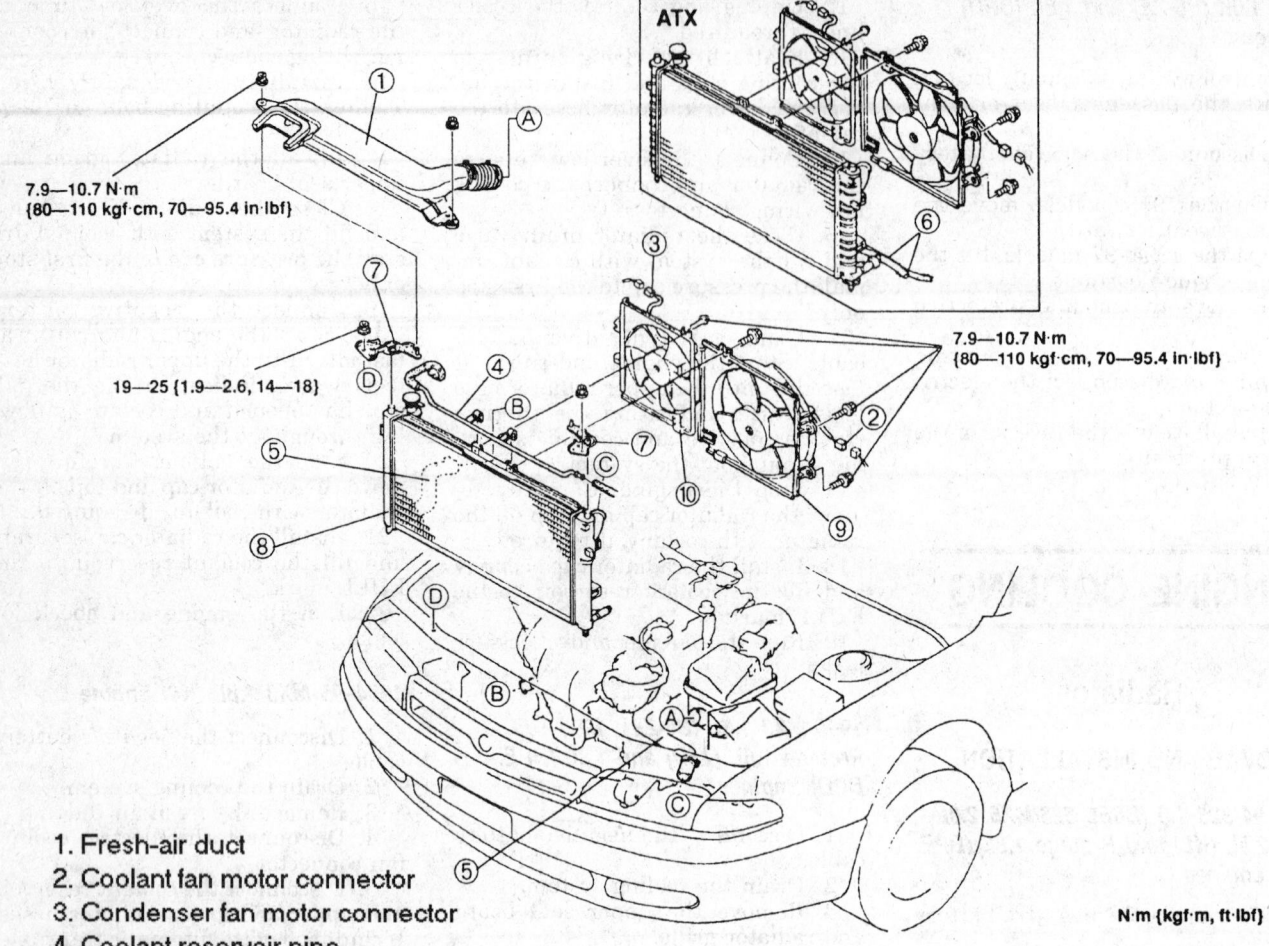

7.9—10.7 N·m
{80—110 kgf·cm, 70—95.4 in·lbf}

19—25 {1.9—2.6, 14—18}

ATX

7.9—10.7 N·m
{80—110 kgf·cm, 70—95.4 in·lbf}

1. Fresh-air duct
2. Coolant fan motor connector
3. Condenser fan motor connector
4. Coolant reservoir pipe
5. Radiator hose (upper and lower)
6. Oil cooler hose (ATX)
7. Radiator bracket
8. Radiator
9. Coolant fan and radiator cowling assy
10. Condenser fan and radiator cowling assy

N·m {kgf·m, ft·lbf}

294971

Exploded view of the radiator and related components — 626 and MX6 with 2.0L (FS) engine

16. Close the radiator drain valve and fill the system with coolant. Install the pressure cap to the first stop only.

17. Connect the negative battery cable. Start the engine and run it at fast idle until the upper radiator hose feels warm, indicating the thermostat has opened and coolant is flowing throughout the system.

18. Stop the engine. Carefully remove the radiator cap and top off the radiator with coolant, if required.

19. Install the radiator cap securely and fill the coolant reservoir to the FULL mark.

20. Run the engine and check for leaks.

1995–97 626 and MX6 Engines

1. Disconnect the negative battery cable.
2. Drain the cooling system.
3. Remove the fresh air duct.
4. Disconnect the electric cooling fan and condenser connectors.
5. Disconnect the coolant reservoir and upper and lower radiator hoses.
6. Remove the upper radiator mounting brackets.
7. Lift the radiator/cooling fans assembly from the vehicle.
8. If necessary, remove the cooling fan(s)/shroud assembly from the radiator.

To install:

9. If removed, install the fan and shroud assembly. Tighten the mounting bolts to 70–95 inch lbs. (8–11 Nm).

10. Install the radiator/fan assembly, making sure the lower tank engages the insulators.

11. Install the upper radiator mounting brackets, and tighten the retaining bolts to 14–18 ft. lbs. (19–25 Nm).

12. Reattach the wiring harness to the routing clips, and install the upper and lower radiator hoses to the radiator.

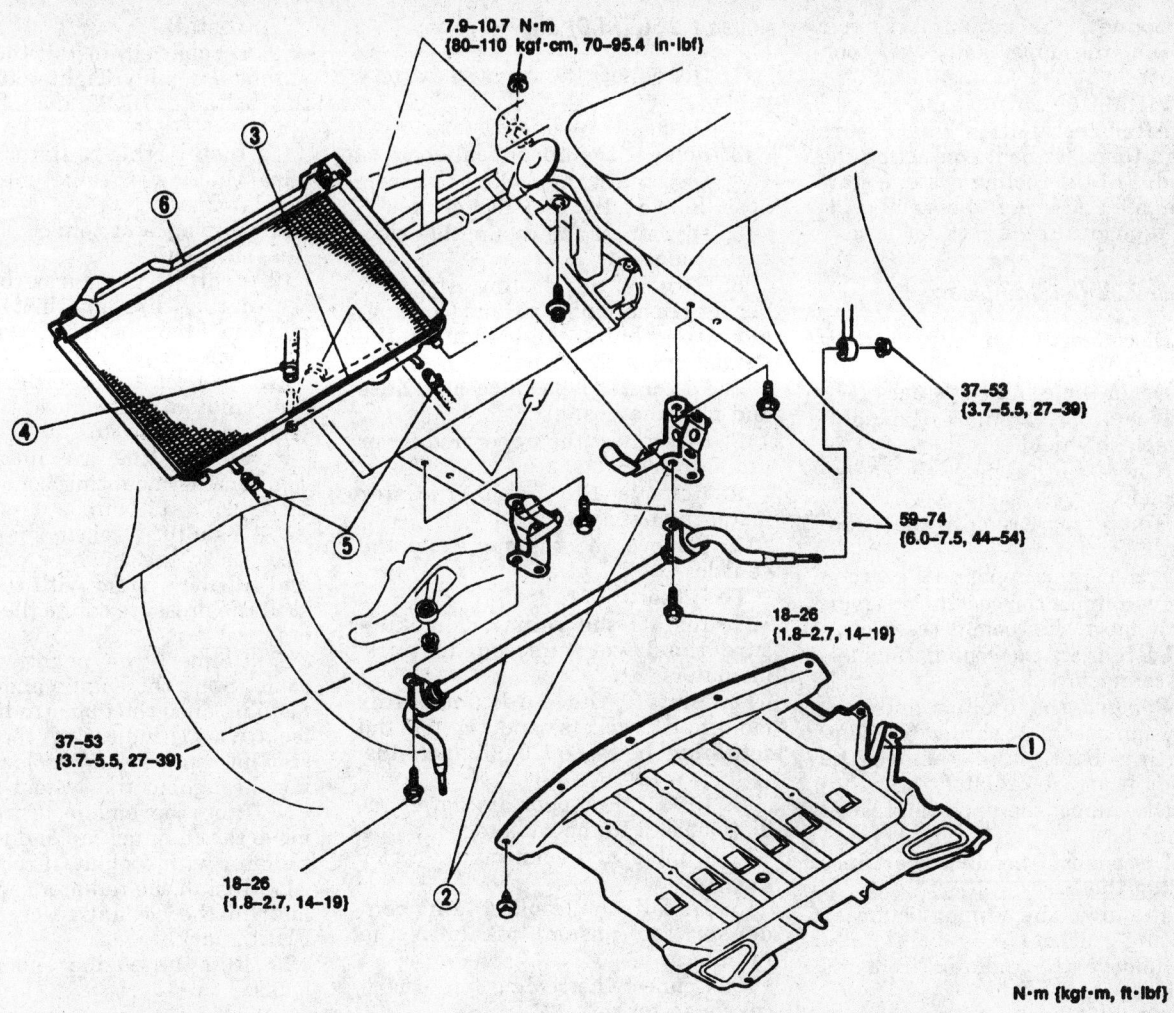

7.9–10.7 N·m
{80–110 kgf·cm, 70–95.4 in·lbf}

37–53
{3.7–5.5, 27–39}

59–74
{6.0–7.5, 44–54}

18–26
{1.8–2.7, 14–19}

37–53
{3.7–5.5, 27–39}

18–26
{1.8–2.7, 14–19}

N·m {kgf·m, ft·lbf}

1. Undercover
2. Stabilizer and bracket
3. Radiator hose (lower)
4. Air separation hose

5. Oil cooler hose (AT)
6. Radiator and coolant fan

249340

Exploded view of the radiator and related components — RX7 with 1.3L (BPD) engine

13. Connect the overflow tube to the radiator and connect the cooling fan wiring connectors.

14. Install the air duct, and tighten the mounting nuts to 70–95 inch lbs. (8–11 Nm).

15. Close the radiator drain valve and fill the system with coolant. Install the pressure cap to the first stop only.

16. Connect the negative battery cable. Start the engine and run it at fast idle until the upper radiator hose feels warm, indicating the thermostat has opened and coolant is flowing throughout the system.

17. Stop the engine. Carefully remove the radiator cap and top off the radiator with coolant, if required.

18. Install the radiator cap securely and fill the coolant reservoir to the FULL mark.

19. Run the engine and check for leaks.

1993–95 929 3.0L (JE-ZE) Engine

1. Disconnect the negative battery cable and drain the coolant.

2. Remove the air inlet duct.

3. Remove the upper and lower radiator hoses and the coolant reservoir hose. If equipped with an automatic transmission, disconnect and plug the cooling hoses.

4. Remove the shroud mounting nuts, and move it back towards the engine to access the nuts holding the fan/clutch assembly.

5. Remove the fasteners for the fan and shroud, and lift them out together.

6. Remove the bolts and lift the radiator out of the vehicle.

To install:

7. Position the radiator. Install the cooling fan and shroud. Tighten the mounting bolts to 70–95 inch lbs. (8–11 Nm), and the nuts to 14–18 ft. lbs. (19–25 Nm).

8. On automatic transmissions, connect the oil cooler lines.

9. Connect the coolant reservoir hose, and the upper and lower coolant hoses.

10. Install the air inlet duct.

11. After installation, make sure the fan turns without contacting the shroud. Fill the cooling system, start the engine and bring to normal operating temperature. Check for leaks.

Millenia 2.3L (KJS) Engine

1. Disconnect the negative battery cable.

2. Drain the cooling system.

3. If equipped, remove the right-hand splash shield.

4. If necessary, remove the charge air cooler.

5. If necessary, remove the radiator grill.

6. Remove the upper seal board.

7. Disconnect the coolant reservoir hose. Remove the coolant reservoir.

8. Disconnect the cooling fan electrical connector.

9. Remove the cooling and condenser fan assembly shroud (cowling) mounting bolts, and remove the shrouds from the radiator.

10. Disconnect the upper and lower radiator hoses.

11. Disconnect the oil cooler lines and plug the hoses.

12. Remove the upper radiator mounting brackets.

13. Remove the radiator from the vehicle.

To install:

14. Install the radiator, making sure the lower tank engages the insulators.

15. Install the upper radiator mounting brackets, and tighten the retaining bolts to 70–95 inch lbs. (8–11 Nm).

16. Unplug and connect the cooler lines.

17. Install the upper and lower radiator hoses to the radiator.

18. Install the coolant and condenser fan assemblies onto the radiator.

19. Connect the cooling fan electrical connector.

20. Install the coolant reservoir. Connect the coolant reservoir hose.

21. If removed, install the radiator grill.

22. If removed, install the charge air cooler.

23. Install the right-hand splash shield, if equipped.

24. Install the upper seal board.

25. Properly fill the cooling system.

26. Connect the negative battery cable.

Millenia 2.5L (KLD) Engine

1. Disconnect the negative battery cable.

2. Drain the cooling system.

3. Remove the upper seal board.

4. Disconnect the coolant reservoir hose. Remove the coolant reservoir.

5. Disconnect the cooling fan electrical connector.

6. Remove the cooling and condenser fan assembly shroud (cowling) mounting bolts, and remove the shrouds from the radiator.

7. Disconnect the oil cooler lines and plug the hoses.

8. Disconnect the upper and lower radiator hoses.

9. Remove the upper radiator mounting brackets.

10. Remove the radiator from the vehicle.

To install:

11. Install the radiator, making sure the lower tank engages the insulators.

12. Install the upper radiator mounting brackets, and tighten the retaining bolts to 70–95 inch lbs. (8–11 Nm).

13. Install the upper and lower radiator hoses to the radiator.

14. Unplug and connect the cooler lines.

15. Install the coolant and condenser fan assemblies onto the radiator.

16. Connect the cooling fan electrical connector.

17. Install the coolant reservoir. Connect the coolant reservoir hose.

18. Install the upper seal board.

19. Properly fill the cooling system.

20. Connect the negative battery cable.

1993–97 Miata 1.6L (B6E) and 1.8L (BPD) Engine

1. Disconnect the negative battery cable.

2. Drain the cooling system.

3. Remove the splash shield.

4. Remove the air intake pipe.

5. Disconnect the coolant reservoir and upper and lower radiator hoses. If equipped with automatic transaxle, disconnect the oil cooler lines and plug the hoses.

6. Disconnect the electric cooling fan connector.

7. Lift the radiator/cooling fan assembly from the vehicle. Remove the radiator/cooling fan assembly.

8. If necessary, remove the cooling fan/shroud assembly from the radiator.

To install:

9. If removed, install the fan and shroud assembly. Tighten the mounting bolts to 70–95 inch lbs. (8–11 Nm).

10. Install the radiator, making sure the lower tank engages the insulators.

11. Unplug and connect the cooler lines, if required.

12. Reattach the wiring harness to the routing clips, and install the upper and lower radiator hoses to the radiator.

13. Connect the overflow tube to the radiator, and connect the cooling fan wiring connector.

14. Install the air intake pipe. Tighten the mounting bolts to 70–95 inch lbs. (8–11 Nm).

15. Install the splash shield.

16. Close the radiator drain valve and fill the system with coolant. Install the pressure cap to the first stop only.

17. Connect the negative battery cable. Start the engine and run it at fast idle until the upper radiator hose feels warm, indicating the thermostat has opened and coolant is flowing throughout the system.

18. Stop the engine. Carefully remove the radiator cap and top off the radiator with coolant, if required.

19. Install the radiator cap securely and fill the coolant reservoir to the FULL mark.

20. Run the engine and check for leaks.

1993–95 RX7 1.3L (13B) Engine

1. Disconnect the negative battery cable.

2. Raise and safely support the vehicle.

3. Drain the cooling system.

4. Remove the fresh air duct and the air cleaner assembly.

5. Remove the battery box, battery and battery tray.

6. Remove the upper radiator hose.

7. Remove the relay box and position aside, leaving the wiring connected.

8. Disconnect the cooling fan electrical connectors.

9. Remove the engine undercover.

10. Remove the stabilizer bar and brackets assembly.

11. Remove the lower radiator hose and the air separation hose.

12. If equipped with automatic transmission, disconnect and plug the oil cooler hoses.

13. Remove the A/C condenser mounting bolts, position the condenser away from the radiator and

secure it with wire. Do not disconnect the A/C lines.

14. Remove the power steering oil line bracket and the A/C high pressure line bracket.

15. Remove the radiator brackets and radiator mounting bolts and remove the radiator/cooling fans assembly. Be careful not to damage the condenser during radiator removal.

16. If necessary, remove the cooling fans/shroud assembly from the radiator.

To install:

17. Install the cooling fans/shroud assembly to the radiator, if removed.

18. Install the radiator/cooling fans assembly into the vehicle and tighten the radiator bracket mounting bolts to 95 inch lbs. (11 Nm).

19. Install the power steering and A/C line brackets.

20. Install the A/C condenser and secure with the mounting bolts.

21. If equipped with automatic transmission, unplug and connect the oil cooler hoses.

22. Install the air separation and lower radiator hoses.

23. Install the stabilizer bar/brackets assembly. Tighten the stabilizer bar end nuts to 39 ft. lbs. (53 Nm) and the bracket bolts to 19 ft. lbs. (25 Nm).

24. Install the engine undercover.

25. Connect the cooling fan electrical connectors.

26. Install the relay box and tighten the mounting bolt to 95 inch lbs. (11 Nm).

27. Install the upper radiator hose.

28. Install the battery tray, battery and battery box.

29. Install the air cleaner assembly and connect the ducts.

30. Connect the battery cables.

31. Fill the cooling system, start the engine and bring to normal operating temperature. Check for leaks.

Water Pump

REMOVAL AND INSTALLATION

1993–94 323 1.6L (B6E), Miata 1.6L (B6E) 1.8L (BPD), Protege 1.8L (BPE, BPD), 1993–95 MX3 1.6L (B6E, B6-ZE) and 1995–97 Protege 1.5L (Z5D) and 1.8L (BPE, BPD) Engines

------ **CAUTION** ------
When draining the coolant, keep in mind that cats and dogs are attracted by the ethylene glycol antifreeze, and are quite likely to drink any that is left in an uncovered container or in puddles on

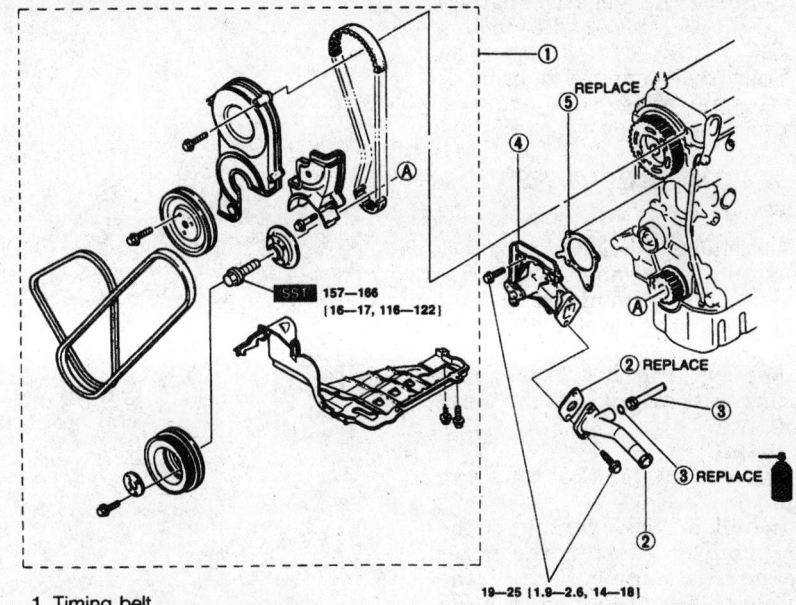

1. Timing belt
2. Water inlet pipe and gasket
3. Water bypass pipe and O-ring
4. Water pump assembly
5. Water pump gasket

SST 157—166 {16—17, 116—122}

19—25 {1.9—2.6, 14—18}

N·m {kgf·m, ft·lbf}

122739

Water pump removal and installation — 1993–94 323 1.6L (B6E), Miata 1.6L (B6E) 1.8L (BPD), Protege 1.8L (BPE, BPD) 1993–95 MX3 1.6L (B6E, B6-ZE) and 1995–97 Protege1.5L (Z5D) and 1.8L (BPD) engines

the ground. This will prove fatal in sufficient quantity. Always drain the coolant into a sealable container. Coolant should be reused unless it is contaminated or several years old.

1. Disconnect the negative battery cable. Drain the cooling system.

2. Remove the timing belt covers, and remove the timing belt.

3. Disconnect the coolant inlet pipe and gasket.

4. Remove the timing belt idler pulleys still attached to the water pump.

5. Remove the water pump mounting bolts, and remove the water pump.

To install:

6. Clean all gasket mating surfaces.

7. Install a new rubber seal on the water pump.

8. Using a new gasket, install the water pump on the engine. Tighten the mounting bolts to 14–18 ft. lbs. (19–25 Nm). Tighten the bolt from the water pump to the alternator bracket to 28–38 ft. lbs. (38–51 Nm).

9. Install the timing belt idler pulleys that were removed.

10. Install the coolant inlet pipe, using a new gasket. Tighten the bolts to 14–18 ft. lbs. (19–25 Nm).

11. Install the timing belt and the timing belt covers.

12. Fill and bleed the cooling system. Connect the negative battery cable, start the engine and bring to normal operating temperature. Check for leaks.

1993–95 MX3 1.8L (K8) and 1993–97 626/MX6 2.5L (KL) Engines

1. Disconnect the negative battery cable. Drain the cooling system.

2. Remove the timing belt.

3. Remove the No.3 engine mount bracket.

4. Position a drain pan under the water pump.

5. Remove the 5 water pump mounting bolts, and remove the water pump.

To install:

6. Clean the mating surfaces of the water pump and the engine block.

7. Install a NEW rubber seal onto the water pump.

8. Install the water pump and torque the bolts 14–18 ft. lbs. (19–25 Nm).

9. Install the engine mount bracket, and tighten the mounting bolt to 32–44 ft. lbs. (44–60 Nm).

10. Install the timing belt.

11. Connect the negative battery cable. Fill and bleed the cooling system.

12. Start the engine and bring to normal operating temperature. Check for leaks.

1993–97 626 and MX6 2.0L (FS) Engines

1. Disconnect the negative battery cable. Drain the cooling system.

2. Remove the timing belt.

3. Remove the power steering oil pump adjuster.

4. Remove the 5 water pump mounting bolts and remove the water pump.

To install:

5. Clean all gasket mating surfaces.

6. Install a NEW gasket on the water pump and install the water pump on the engine. Install the mounting bolts and tighten to 14–18 ft. lbs. (19–25 Nm).

7. Install the power steering oil pump adjuster, torque the mounting bolts to 12–16 ft. lbs. (16–22 Nm).

8. Install the timing belt.

9. Connect the negative battery cable. Fill and bleed the cooling system.

10. Start the engine and bring to normal operating temperature. Check for leaks.

1993–95 929 3.0L (JE-ZE) Engines

1. Disconnect the negative battery cable and drain the cooling system.

2. Turn the crankshaft to place the No. 1 piston at TDC of the compression stoke.

3. Remove the timing belt.

4. Remove the retaining bolts and nuts and separate the water pump from the cylinder block. Remove the gasket and purchase a new one.

5. Remove all the old gasket material from the water pump and cylinder block contact surfaces.

To install:

6. Install the gasket onto the water pump and position the water pump onto the block. Install the retaining nuts and bolts. Torque the nuts and bolts to 14–19 ft. lbs.

7. Install the timing belt.

8. Fill the cooling system to the proper level and connect the negative battery cable. Start the engine and check for leaks.

1995–97 Millenia

1. Disconnect the negative battery cable. Drain the cooling system.

2. Remove the timing belt covers and the timing belt.

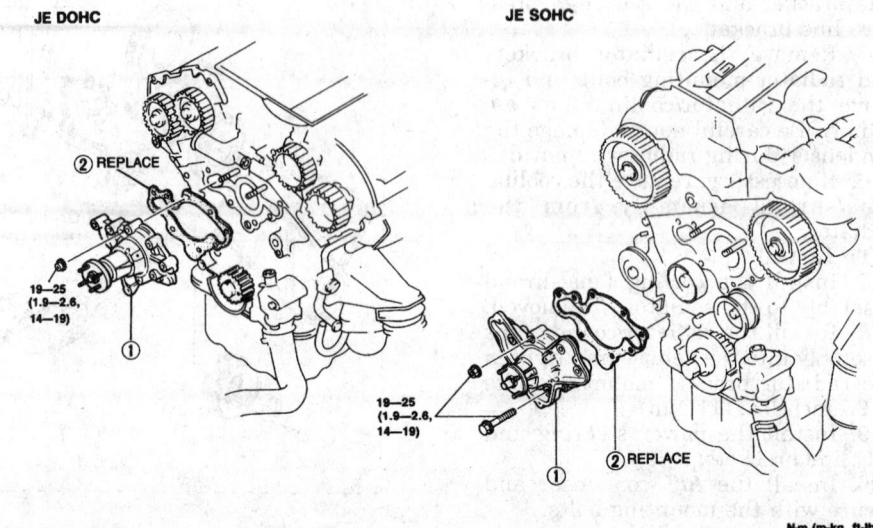

JE DOHC   JE SOHC

② REPLACE

19–25
(1.9–2.6,
14–19)

①

19–25
(1.9–2.6,
14–19)

① ② REPLACE

N·m (m-kg, ft-lb)

1. Water pump assembly
Inspect for cracks, damaged mounting surface, bearing condition, and leakage

2. Water pump gasket

124835

Water pump assembly — 1993–95 929 3.0L (JE-ZE) engines

3. On the 2.5L engine, remove the engine mount bracket.

4. Use a pulley removal tool to hold the water pump pulley and remove the bolts. Remove the water pump pulley.

5. Position a drain pan under the water pump.

6. Remove the water pump mounting bolts, and remove the water pump.

To install:

7. Clean the mating surfaces of the water pump and the engine block.

8. Install a new O-ring onto the water pump.

9. Install the water pump and torque the bolts 18 ft. lbs. (25 Nm).

10. Install the water pump pulley with the bolts. Hold the pulley with the tool and tighten the bolts to 88 inch lbs. (10 Nm).

11. On the 2.5L engine, install the engine mount bracket, and tighten the mounting bolt to 32–44 ft. lbs. (44–60 Nm).

12. Install the timing belt and timing covers.

13. Connect the negative battery cable. Fill and bleed the cooling system.

14. Start the engine and bring to normal operating temperature. Check for leaks.

1995–97 Miata 1.8L (BPD) Engine

1. Disconnect the negative battery cable. Drain the cooling system.

2. Remove the timing belt covers, and remove the timing belt.

3. Remove the power steering pump with the hoses still connected, and position it out of the way.

4. Disconnect the coolant inlet pipe and gasket.

5. Remove the timing belt idler pulleys still attached to the water pump.

6. Remove the water pump mounting bolts, and remove the water pump.

To install:

7. Clean all gasket mating surfaces.

8. Install a new rubber seal on the water pump.

9. Using a new gasket, install the water pump on the engine. Tighten the mounting bolts to 14–18 ft. lbs. (19–25 Nm).

10. Install the timing belt idler pulleys that were removed.

11. Install the coolant inlet pipe, using a new gasket. Tighten the bolts to 14–18 ft. lbs. (19–25 Nm).

12. Install the power steering pump.

13. Install the timing belt and the timing belt covers.

14. Fill and properly bleed the cooling system. Connect the negative battery cable, start the engine and bring to normal operating temperature. Check for leaks.

1993–95 RX7 1.3L (13B) Engines

1. Disconnect the battery cables and remove the battery box, battery and battery tray.
2. Drain the cooling system.
3. Remove the fresh air duct and the air cleaner assembly.
4. Remove the intercooler-to-throttle body hose and funnel.
5. Disconnect the accelerator cable.
6. Disconnect the filler port coolant hose from the water pump.
7. Remove the turbocharger-to-intercooler pipe and hose.
8. Remove the water pump pulley and drive belt.
9. Remove the water pump body coolant hose from the rear of the water pump.
10. Label and disconnect the electrical connectors and remove the alternator and alternator bracket.
11. Remove the air pump and air pump bracket.
12. Remove the upper radiator hose.
13. Remove the intercooler and air separation tank, but do not remove the air duct from the body.
14. Remove the subframe at the front of the vehicle.
15. Remove the lower radiator hose.
16. Remove the remaining coolant hoses from the water pump. Disconnect the metering oil lines.
17. Remove the metering oil pump connector from the engine hanger. Remove bolt **A**, then position the metering oil line and metering oil pump harness under the lower radiator hose.
18. Remove nuts **B** and remove the water pump.

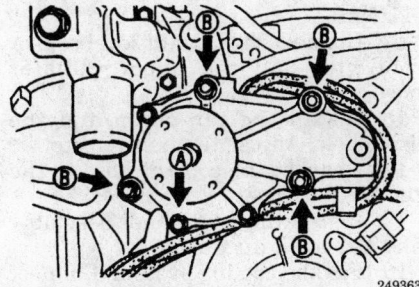

Water pump attaching bolt locations — 1993–97 RX7 1.3L (13B) engine

To install:
19. Clean all gasket mating surfaces.
20. Install the water pump, using a new gasket. Tighten the nuts to 19 ft. lbs. (25 Nm).
21. Reposition the metering oil pump harness and line and install the bolt in the water pump. Tighten to 19 ft. lbs. (25 Nm).
22. Install the metering oil pump connector to the engine hanger. Connect the metering oil lines, using new gaskets, and tighten the bolts to 10 ft. lbs. (13 Nm).
23. Install the coolant hose at the rear of the water pump and connect the heater hose. Install the lower radiator hose.
24. Install the subframe at the front of the vehicle and tighten the bolts to 95 inch lbs. (11 Nm).
25. Install the intercooler and air separation tank. Tighten the bolts to 95 inch lbs. (11 Nm).
26. Install the upper radiator hose.
27. Install the air pump and alternator with the brackets.
28. Connect the remaining coolant hose at the rear of the water pump. Install the water pump pulley and drive belt. Tighten the pulley bolts to 95 inch lbs. (11 Nm).
29. Install the turbocharger-to-intercooler hose and pipe.
30. Connect the filler port coolant hose at the water pump. Install the air cleaner assembly and fresh air duct.
31. Connect and adjust the accelerator cable.
32. Install the intercooler-to-throttle body hose and funnel.
33. Install the battery tray, battery and battery box. Connect the battery cable.
34. Fill and bleed the cooling system.
35. Start the engine and bring to normal operating temperature. Check for leaks and check the coolant level.

Thermostat

REMOVAL AND INSTALLATION

1993–94 323 1.6L (B6E), MX3 1.6L (B6-ZE), Miata 1.6L (B6-ZE), 1.8L (BPD) and Protege 1.8L (BPE, BPD) Engines

1. Disconnect the negative battery cable. Drain the radiator to below the level of the thermostat.

2. Disconnect the coolant temperature switch at the thermostat housing.
3. Remove the upper radiator hose.
4. Remove the mounting nuts, thermostat housing, thermostat and gasket.

NOTE: Do not pry the housing off.

To install:
5. Clean the thermostat housing and the cylinder head mating surfaces.
6. Insert the thermostat into the rear cylinder head housing with the jiggle pin at the top. The spring side of the thermostat should face the housing.
7. Position a new gasket onto the studs with the seal print side facing the rear cylinder housing.
8. Install the thermostat housing and 2 nuts. Tighten the nuts to 14–22 ft. lbs. (19–30 Nm).
9. Connect the coolant temperature switch and install the upper radiator hose.
10. Fill the cooling system. Connect the negative battery cable, start the engine and check for leaks. Check the coolant level and add coolant, as necessary.

1993–97 626, MX6 2.0L (FS), 1995 MX3 1.6L (B6-ZE) and 1995–97 Miata 1.8L (BPD), Protege 1.5L (Z5D) and 1.8L (BPD) Engines

1. Disconnect the negative battery cable. Drain the radiator to below the level of the thermostat.
2. Remove the air cleaner.
3. Remove the upper radiator hose.
4. Remove the housing nut, thermostat housing, thermostat and gasket.

NOTE: Do not pry the housing off.

To install:
5. Clean the thermostat housing and the cylinder head mating surfaces.
6. Make sure that the thermostat jiggle pin is aligned with the gasket projection.
7. Install the thermostat to the housing. Align the gasket projection with the opening in the housing. The spring side of the thermostat should face into the housing.
8. Install the thermostat housing cover. Tighten the mounting bolt to 14–18 ft. lbs. (19–25 Nm).
9. Install the upper radiator hose.
10. Install the air cleaner.

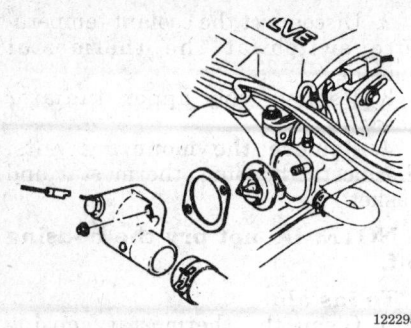

122295

Thermostat — 1993–94 323 1.6L (B6E), MX3 1.6L (B6-ZE), Miata 1.6L (B6-ZE), 1.8L (BPD) and Protege 1.8L (BPE, BPD) engines

11. Fill, and bleed the cooling system.

12. Connect the negative battery cable, start the engine and check for leaks. Check the coolant level and add coolant, as necessary.

1993–95 MX3 1.8L (K8) and 1993–97 2.5L 626 and MX6 (KL) Engines

1. Disconnect the negative battery cable. Drain the radiator to below the level of the thermostat.

2. Remove the water inlet pipe mounting bolt and remove the pipe.

3. Remove the engine harness bracket bolt and position the harness aside.

4. Remove the housing bolt, thermostat housing, thermostat and gasket.

NOTE: Do not pry the housing off.

To install:

5. Clean the thermostat housing and the cylinder head mating surfaces.

6. Insert the thermostat into the housing with the jiggle pin at the top.

7. Position a new gasket with the projection facing the same direction as the thermostat jiggle pin.

8. Install the thermostat housing cover and loosely install the lower bolt. Install the engine harness bracket and loosely install the bolt. Tighten both bolts to 14–18 ft. lbs. (19–25 Nm).

9. Fill the cooling system. Connect the negative battery cable, start the engine and check for leaks. Check coolant level and add coolant, as necessary.

1993–95 929 3.0L (JE-ZE) Engines

1. Disconnect the negative battery cable. Raise and safely support the vehicle and remove the engine undercover.

2. Drain the cooling system.

3. Remove the fresh air duct and the air cleaner assembly. Remove the upper radiator hose.

4. Remove the cooling fan and shroud. Remove the alternator belt and the water pump pulley.

5. Remove the water bypass hose and the lower radiator hose.

6. Remove the thermostat housing from the engine. Remove the housing cover and thermostat.

To install:

7. Make sure the cover, housing and cylinder block contact surfaces are clean. If the integral thermostat gasket is damaged, replace the thermostat.

8. Install the thermostat in the cover with the spring facing the thermostat housing. Install the cover on the housing and tighten the bolts to 19 ft. lbs. (25 Nm).

9. Using a new gasket, install the thermostat housing on the engine. Tighten the bolts to 19 ft. lbs. (25 Nm).

10. Install the lower radiator hose and water bypass hose to the thermostat housing.

11. Install the water pump pulley and alternator belt. Adjust the belt tension.

12. Install the cooling fan and shroud. Tighten the fan nuts to 95 inch lbs. (11 Nm).

13. Install the air cleaner assembly and fresh air duct. Install the engine undercover and lower the vehicle.

14. Fill the cooling system. Connect the negative battery cable, start the engine and bring to normal operating temperature. Check for leaks and coolant level.

1995–97 Millenia 2.3L (KJS) and 2.5L (KLD) Engines

1. Disconnect the negative battery cable.

2. Drain the radiator to below the level of the thermostat.

3. Remove the air cleaner housing assembly.

4. Remove the water inlet pipe mounting bolt and remove the pipe.

5. Remove the engine harness bracket bolt and position the harness aside.

6. Remove the housing bolt, thermostat housing, thermostat and gasket.

NOTE: Do not pry the housing off.

To install:

7. Clean the thermostat housing and the cylinder head mating surfaces.

8. Insert the thermostat into the housing with the jiggle pin at the top. The spring side of the thermostat should face the housing.

9. Position a new gasket with the projection facing the same direction as the thermostat jiggle pin.

10. Install the thermostat housing cover and loosely install the lower bolt. Install the engine harness bracket and loosely install the bolt. Tighten both bolts to 14–19 ft. lbs. (19–25 Nm).

11. Install the air cleaner housing assembly.

12. Fill the cooling system.

13. Connect the negative battery cable, start the engine and check for leaks.

14. Check the coolant level and add coolant, as necessary.

1993–95 RX7 1.3L (13B) Engines

1. Disconnect the negative battery cable.

2. Drain the cooling system.

3. Remove the fresh air duct and the air cleaner assembly.

4. Remove the coolant hose from the thermostat cover.

5. Remove the drive belt and remove the air pump.

6. Remove the upper radiator hose.

7. Disconnect the coolant level sensor connector.

8. Remove the thermostat cover and remove the thermostat.

To install:

9. Make sure the thermostat cover and engine contact surfaces are clean. Replace the thermostat if the gasket is damaged.

10. Using a new gasket, install the thermostat in the engine with the spring toward the engine and the jiggle pin at the top.

11. Install the thermostat cover and tighten the bolts to 95 inch lbs. (11 Nm).

12. Connect the coolant level sensor connector and the upper radiator hose.

13. Install the air pump and the drive belt. Adjust the belt tension.

14. Connect the coolant hose to the thermostat cover.

15. Install the air cleaner assembly and the fresh air duct.

16. Fill the cooling system. Connect the negative battery cable, start the engine and bring to normal operating temperature. Check for leaks and coolant level.

Electric Cooling Fan

REMOVAL AND INSTALLATION

1993–94 323 1.6L (BPE), Protege 1.8L (BPE, BPD) and 1993–95 626/MX6 2.0L (FS), 2.5L (KL), 1994–95 MX3 1.6L (B6E, B6-ZE) and 1.8L (K8) Engines

1. Disconnect the negative battery cable, and drain the cooling system.
2. Remove the fresh air duct, and the shroud upper panel.
3. Disconnect the cooling fan electrical connector.
4. Remove the shroud (cowling) mounting bolts and remove the shroud from the radiator.
5. Remove the mounting nut, and remove the fan from the fan motor.
6. Remove the mounting bolts, and remove the fan motor from the shroud.

To install:
7. Position the fan motor on the shroud and install the mounting bolts.
8. Install the fan onto the fan motor. Apply a locking compound on the mounting nut, and install the mounting nut.
9. Install the shroud onto the radiator.
10. Connect the cooling fan electrical connector.
11. Install the shroud upper panel, and the fresh air duct.
12. Properly fill the cooling system.
13. Connect the negative battery cable.

1995–97 Protege 1.5L (Z5D) and 1.8L (BPD) Engines

1. Disconnect the negative battery cable.
2. Drain the cooling system.
3. Remove the upper seal board and radiator grille.
4. Remove the hood safety lever.
5. Disconnect the coolant reservoir and upper and lower radiator hoses. If equipped with automatic transaxle, disconnect the oil cooler lines and plug the hoses.
6. Disconnect the electric cooling fan connector.
7. Remove the shroud (cowling) mounting bolts, and remove the shroud from the radiator. Move the assembly to the side of the engine to gain clearance space.
8. Remove the radiator brackets.
9. Lift and remove the radiator from the vehicle.
10. Remove the cooling fan from the vehicle. Remove the mounting nut,

and remove the fan blade from the fan motor.
11. Remove the mounting bolts, and remove the fan motor from the shroud.

To install:
12. Position the fan motor on the shroud and install the mounting bolts.
13. Install the fan blade onto the fan motor. Apply a locking compound on the mounting nut, and install the mounting nut.
14. Install the radiator, making sure the lower tank engages the insulators.
15. Install the radiator brackets. Tighten the mounting bolts to 70–95 inch lbs. (8–11 Nm).
16. Install the fan and shroud assembly. Tighten the mounting bolts to 70–95 inch lbs. (8–11 Nm).
17. Unplug and connect the cooler lines, if required.
18. Reattach the wiring harness to the routing clips, and install the upper and lower radiator hoses to the radiator.
19. Connect the overflow tube to the radiator, and connect the cooling fan wiring connector.
20. Install the hood safety lever. Tighten the mounting bolts to 70–95 inch lbs. (8–11 Nm).
21. Install the radiator grille and upper seal board.
22. Close the radiator drain valve and fill the system with coolant. Install the pressure cap to the first stop only.
23. Connect the negative battery cable. Start the engine and run it at fast idle until the upper radiator hose feels warm, indicating the thermostat has opened and coolant is flowing throughout the system.
24. Stop the engine. Carefully remove the radiator cap and top off the radiator with coolant, if required.
25. Install the radiator cap securely and fill the coolant reservoir to the FULL mark.
26. Run the engine and check for leaks.

1996–97 626 and MX6 2.0L (FS) Engine

1. Disconnect the negative battery cable.
2. Remove the fresh air duct.
3. Disconnect the cooling fan electrical connectors.
4. Remove the shroud (cowling) mounting bolts and remove the shroud from the radiator.
5. Remove the mounting nuts and remove the cooling fan.

6. Remove the mounting bolts and remove the fan motor.

To install:
7. Position the fan motor on the shroud and install the mounting bolts. Torque the cooling fan motor mounting bolts to 16–19 in lbs. (1.8–2.1 Nm).
8. Install the fan onto the motor. Apply a locking compound on the mounting nut, and install the mounting nut. Torque the cooling fan nut 24–32 in. lbs. (2.7–3.7 Nm).
9. Install the shroud onto the radiator.
10. Connect the cooling fan electrical connector.
11. Install the fresh air duct.
12. Connect the negative battery cable.

1996–97 626 and MX6 2.5L (KL) Engine

1. Disconnect the negative battery cable and drain the cooling system.
2. Remove the fresh air duct.
3. Remove the radiator.
4. Disconnect the cooling fan and condenser fan electrical connectors.
5. Remove the shroud (cowling) mounting bolts and remove the shroud from the radiator.
6. Remove the mounting nuts and remove the condenser and cooling fans.
7. Remove the mounting bolts and remove the fan motors.

To install:
8. Position the fan motor on the shroud and install the mounting bolts. Torque the condenser fan mounting bolts to 35–43 in. lbs. (4.0–4.9 Nm). Torque the cooling fan motor mounting bolts to 16–19 in lbs. (1.8–2.1 Nm).
9. Install the fans onto each motor. Apply a locking compound on the mounting nut, and install the mounting nut. Torque the condenser fan nut to 14–21 in. lbs. (1.5–2.4 Nm). Torque the cooling fan nut 38.46 in. lbs. (4.3–5.1 Nm).
10. Install the shroud onto the radiator.
11. Connect the cooling fan electrical connector.
12. Install the radiator and the fresh air duct.
13. Properly fill the cooling system.
14. Connect the negative battery cable.

1995–97 Millenia

1. Disconnect the negative battery cable.
2. Drain the cooling system.

3. If equipped, remove the right-hand splash shield.

4. If necessary, remove the charge air cooler.

5. If necessary, remove the radiator grill.

6. Remove the upper seal board.

7. Disconnect the coolant reservoir hose. Remove the coolant reservoir.

8. Disconnect the cooling fan electrical connector.

9. Remove the shroud (cowling) mounting bolts and remove the shroud from the radiator.

10. Remove the mounting nut and remove the fan from the fan motor.

11. Remove the mounting bolts and remove the fan motor from the shroud.

To install:

12. Position the fan motor on the shroud and install the mounting bolts.

13. Install the fan onto the fan motor. Apply a locking compound on the mounting nut, and install the mounting nut.

14. Install the shroud onto the radiator.

15. Connect the cooling fan electrical connector.

16. Install the coolant reservoir. Connect the coolant reservoir hose.

17. Install the upper seal board.

18. If removed, install the radiator grill.

19. If removed, install the charge air cooler.

20. If equipped, install the right-hand splash shield.

21. Properly fill the cooling system.

22. Connect the negative battery cable.

1995–97 Miata 1.8L (BPD) Engine

1. Disconnect the negative battery cable, and drain the cooling system.

2. Remove the air intake pipe.

3. Disconnect the cooling fan electrical connector.

4. Remove the shroud (cowling) mounting bolts and remove the shroud from the radiator.

5. Remove the mounting nut, and remove the fan from the fan motor.

6. Remove the mounting bolts, and remove the fan motor from the shroud.

To install:

7. Position the fan motor on the shroud and install the mounting bolts, torque to 35–43 in. lbs. (4.0–4.9 Nm).

8. Install the fan onto the fan motor. Apply a locking compound on the mounting nut, and install the mounting nut. Torque the nut to 14–21 in. lbs. (1.5–2.4 Nm).

9. Install the shroud onto the radiator.

10. Connect the cooling fan electrical connector.

11. Install the air intake pipe.

12. Properly fill the cooling system.

13. Connect the negative battery cable.

1993–95 RX7 1.3L (13B) Engines

1. Disconnect the negative battery cable.

2. Remove the attaching screws from the radiator cowling.

3. Remove the center attaching screw from each coolant fan and remove the fans.

4. Remove the three attaching bolts from both of the coolant fan motors.

5. Installation is the reverse of the removal procedure.

Cooling System

BLEEDING

1993–94 626, MX6, 323, MX3 Miata, Protege and 1993–95 929 and RX7

1. Slowly refill the coolant, at an even pace of 1.1 qt (1.0 L) per minute, up to the coolant filler port.

2. Fill the coolant reservoir up to the F mark.

3. Fully install the radiator cap.

4. Start the engine and let it idle until it warms up. If the temperature rises above normal, there is air in the system; repeat the last three steps.

5. Run the engine at 2200–2800 rpm for five minutes.

6. Stop the engine and allow it to cool.

── **CAUTION** ──
Allow the engine to cool completely. Never remove the cap from a hot radiator.

7. Properly remove the radiator cap and verify that the engine coolant level is near the filler neck. If the coolant level is low, repeat each of the previous steps.

8. Fill the reservoir up to the F mark.

1995 MX3 1.6L (K8, B6-ZE) and 1995–97 626/MX6 2.0L (FS), Miata 1.8L (BPD), Millenia 2.3L (KJS), 2.5L (KLD) and Protege 1.5L (Z5D), 1.8L (BPD) Engines

1. Slowly refill the coolant, at an even pace of 1.1 qt (1.0 L) per minute, up to the coolant filler port.

2. Fill the coolant reservoir up to the F mark.

3. Fully install the radiator cap.

4. Start the engine and let it idle until it warms up. If the temperature rises above normal, there is air in the system; repeat the last three steps.

5. Run the engine at 2200–2800 rpm for five minutes.

6. Stop the engine and allow it to cool.

── **CAUTION** ──
Allow the engine to cool completely. Never remove the cap from a hot radiator.

7. Properly remove the radiator cap and verify that the engine coolant level is near the filler neck. If the coolant level is low, repeat each of the previous steps.

8. Fill the reservoir up to the F mark.

GASOLINE FUEL SYSTEM

Fuel System Service Precautions

Safety is the most important factor when performing not only fuel system maintenance but any type of maintenance. Failure to conduct maintenance and repairs in a safe manner may result in serious personal injury or death. Maintenance and testing of the vehicle's fuel system components can be accomplished safely and effectively by adhering to the following rules and guidelines.

• To avoid the possibility of fire and personal injury, always disconnect the negative battery cable unless the repair or test procedure requires that battery voltage be applied.

• Always relieve the fuel system pressure prior to disconnecting any fuel system component (injector, fuel rail, pressure regulator, etc.), fitting or fuel line connection. Exercise extreme caution whenever relieving fuel system pressure to avoid exposing skin, face and eyes to fuel spray. Please be advised that fuel under pressure may penetrate the skin or any part of the body that it contacts.

• Always place a shop towel or cloth around the fitting or connection prior to loosening to absorb any excess fuel due to spillage. Ensure that all fuel spillage (should it occur) is quickly removed from engine surfaces. Ensure that all fuel soaked cloths or towels are deposited into a suitable waste container.

• Always keep a dry chemical (Class B) fire extinguisher near the work area.

• Do not allow fuel spray or fuel vapors to come into contact with a spark or open flame.

• Always use a backup wrench when loosening and tightening fuel line connection fittings. This will prevent unnecessary stress and torsion to fuel line piping. Always follow the proper torque specifications.

• Always replace worn fuel fitting O-rings with new. Do not substitute fuel hose or equivalent, where fuel pipe is installed.

Fuel System Pressure

RELIEVING

— **CAUTION** —

Fuel injection systems remain under pressure after the engine has been turned OFF. Properly relieve fuel pressure before disconnecting any fuel lines. Failure to do so may result in fire or personal injury. Do not allow fuel spray or fuel vapors to come in contact with a spark or open flame. Keep a dry chemical fire extinguisher nearby. Never store fuel in an open container due to risk of fire or explosion.

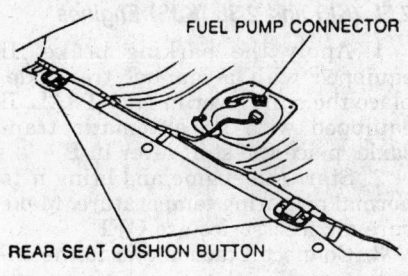

Fuel pump connector — 1993–94 323 1.6L (B6E) and 1993–97 MX3 1.6L (B6-ZE), 1.8L (K8) and Protege 1.5L (Z5D), 1.8L (BPD, BPE) engines

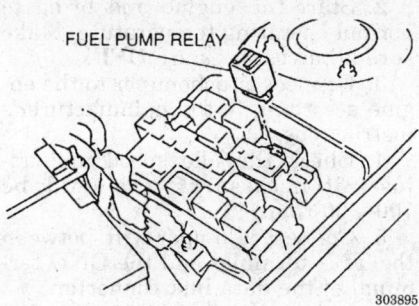

Fuel pump relay location — 1993–97 626 and MX6 2.0L (FS) and 2.5L (KL) engines

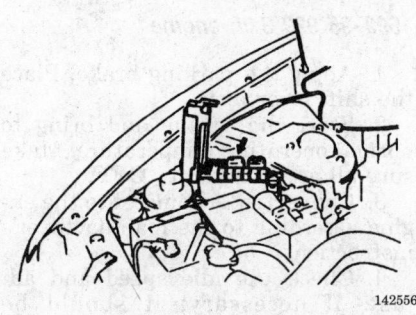

Circuit opening relay connector — 1993–94 Miata 1.6L (B6-ZE) and 1.8L (BPD) Engine

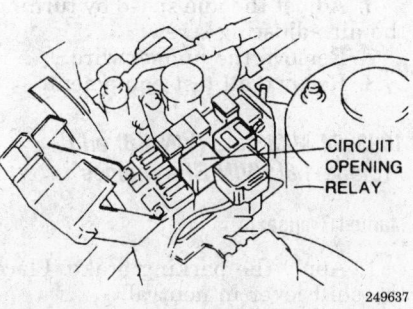

Circuit opening relay connector location–1993–95 RX7 1.3L (13B) engines

1993–94 323 1.6L (B6E) and 1993–97 MX3 1.6L (B6-ZE), 1.8L (K8) and Protege 1.5L (Z5D), 1.8L (BPD, BPE) Engines

1. Release the rear seat retainers (clips or catches) and remove rear seat cushion.
2. Remove the fuel pump cover.
3. Disconnect the fuel pump electrical connector.
4. Start the engine, allowing it to idle until it runs out of fuel.

5. After the engine stalls, reconnect the fuel pump connector and turn the ignition switch **OFF**.

1993–97 626 and MX6 2.0L (FS) and 2.5L (KL) engines

1. Start the engine.
2. Remove the fuel pump relay from the relay box, located in the right side of the engine compartment.
3. After the engine stalls, turn the ignition switch to **OFF** and reinstall the relay connector.

1993–95 929 3.0L (JE-ZE) engines

1. Start the engine.
2. Remove the circuit opening relay connector from the relay box, located in the right side of the engine compartment.
3. After the engine stalls, reinstall the relay connector and turn the ignition switch **OFF**.

1995–97 Millenia 2.3L (KJS) Engine

1. Remove the cruise actuator mounting nuts, and move the cruise actuator to the side.
2. Start the engine.
3. Remove the fuel pump relay from the relay box, located in the right side of the engine compartment.
4. After the engine stalls, turn the ignition switch to **OFF** and reinstall the relay connector.
5. Position and install the cruise actuator.

1995–97 Millenia 2.5L (KLD) Engine

1. Start the engine.
2. Remove the fuel pump relay from the relay box, located in the right side of the engine compartment.
3. After the engine stalls, turn the ignition switch to **OFF** and reinstall the relay connector.

1993–97 Miata 1.6L (B6-ZE) and 1.8L (BPD) Engine

1. Start the engine.
2. Disconnect the circuit opening relay connector.
3. After the engine stalls, reinstall the relay connector and turn the ignition switch **OFF**.

1993–95 RX7 1.3L (13B) Engines

1. Start the engine.
2. Remove the circuit opening relay connector from the relay box, located in the right side of the engine compartment.
3. After the engine stalls, turn the ignition switch **OFF** and reinstall the relay connector.

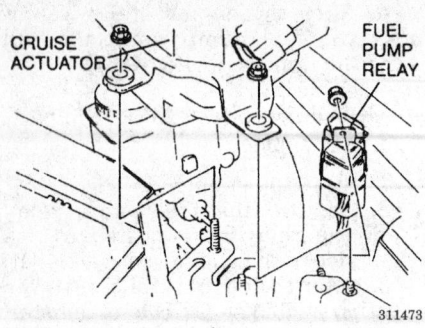

Fuel pump relay location — 1995–97 Millenia 2.3L (KJS) engines

Idle Speed

ADJUSTMENT

1993–94 323 and 1993–95 MX3 1.6L, 1995 MX3 1.6L (B6-ZE), 1993–97 626/MX6 2.0L (FS), 1995–97 Protege 1.5L (Z5D) Engine

Manual Transaxle

1. Apply the parking brake. Place the shift lever in neutral.
2. Start the engine and bring to normal operating temperature. Make sure all accessories are **OFF**.
3. Connect a tachometer to the engine according to the manufacturer's instructions.
4. Verify that the idle speed is between 650–750 rpm, adjust if necessary.
5. Connect a jumper wire between the **TEN** terminal and the **GND** terminal of the data link connector.
6. Adjust the idle speed by turning the air adjusting screw.
7. Remove the jumper wire.
8. Remove all test equipment.

Automatic Transaxle

1. Apply the parking brake. Place the shift lever in **P**.

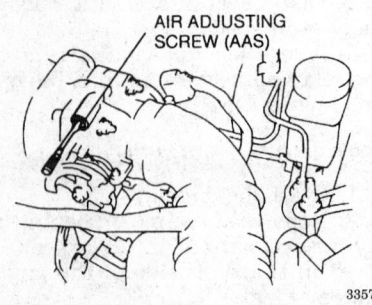

Turn the air adjusting screw — 1993–94 323 and 1993–95 MX3 1.6L engine

2. Start the engine and bring to normal operating temperature. Make sure all accessories are **OFF**.
3. Connect a tachometer to the engine according to the manufacturer's instructions.
4. Check the idle speed and adjust, if necessary; it should be 700–800 rpm.
5. Connect a jumper wire between the **TEN** terminal and the **GND** terminal of the data link connector.
6. Adjust the idle speed by turning the air adjusting screw.
7. Remove the jumper wire.
8. Remove all test equipment.

1993–95 929 3.0L engine

1. Apply the parking brake. Place the shift lever in **P**.
2. Start the engine and bring to normal operating temperature. Make sure all accessories are **OFF**.
3. Connect a tachometer to the engine according to the manufacturer's instructions.
4. Check the idle speed and adjust, if necessary; it should be 680–720 rpm.
5. Connect a jumper wire between the **TEN** terminal and the **GND** terminal of the data link connector.
6. Adjust the idle speed by turning the air adjusting screw.
7. Remove the jumper wire.
8. Remove all test equipment.

1993–94 MX3 1.8L (VIN K8) and Protege 1.8L (VIN BP) engines

Manual Transaxle

1. Apply the parking brake. Place the shift lever in neutral.
2. Start the engine and bring to normal operating temperature. Make sure all accessories are **OFF**.
3. Connect a tachometer to the engine according to the manufacturer's instructions.

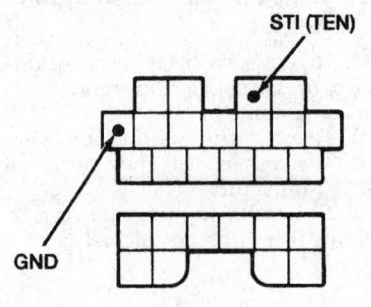

Connect a jumper wire to the data link connector when adjusting the idle speed

4. Verify that the idle speed is between 700–800 rpm, adjust if necessary.
5. Connect a jumper wire between the **TEN** terminal and the **GND** terminal of the data link connector.
6. Adjust the idle speed by turning the air adjusting screw.
7. Remove the jumper wire.
8. Remove all test equipment.

Automatic Transaxle

1. Apply the parking brake. Place the shift lever in **P**.
2. Start the engine and bring to normal operating temperature. Make sure all accessories are **OFF**.
3. Connect a tachometer to the engine according to the manufacturer's instructions.
4. Check the idle speed and adjust, if necessary; it should be 700–800 rpm.
5. Connect a jumper wire between the **TEN** terminal and the **GND** terminal of the data link connector.
6. Adjust the idle speed by turning the air adjusting screw.
7. Remove the jumper wire.
8. Remove all test equipment.
9. Adjust the idle speed by turning the air adjusting screw.

1995 MX3 1.8L (K8) Engine

1. Apply the parking brake. If equipped with a manual transaxle, place the shift lever in Neutral. If equipped with an automatic transaxle, place the shift lever in **P**.
2. Start the engine and bring it to normal operating temperature. Make sure all accessories are off.
3. Connect a tachometer to the engine according to the manufacturer's instructions.
4. Verify that the idle speed is between 640–700 rpm; adjust if necessary.
5. Adjust the idle speed by turning the air adjusting screw.
6. Remove all test equipment.

1993–94 626/MX6 and 1995–97 Millenia 2.5L (KL) and 2.3L (KJS) Engines

1. Apply the parking brake. If equipped with a manual transaxle, place the shift lever in NEUTRAL. If equipped with an automatic transaxle, place the shift lever in **P**.
2. Start the engine and bring it to normal operating temperature. Make sure all accessories are **OFF**.
3. Connect a tachometer to the engine according to the manufacturer's instructions.
4. Verify that the idle speed is between 600–700 rpm; adjust if necessary.

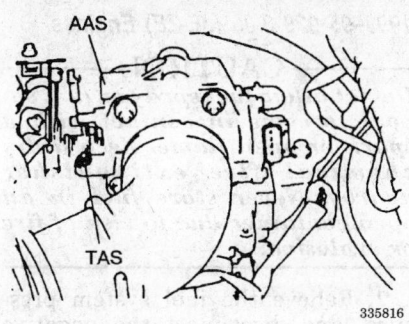

Turn the air adjusting screw — 1993–94 MX3 1.8L (VIN K8) and Protege 1.8L (VIN BP) engine

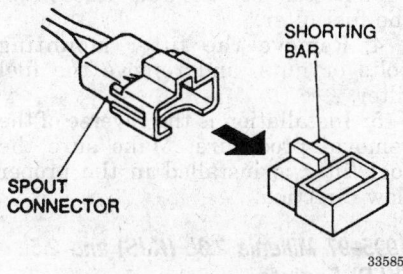

Remove the shorting bar

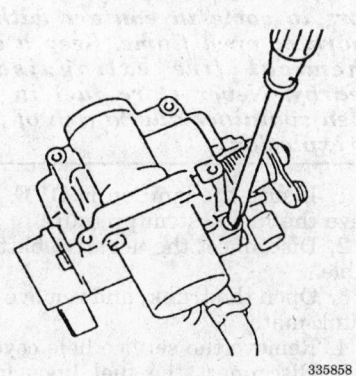

Turn the air adjusting screw

5. Connect a jumper wire between the data link terminals **TEN** and **GND**.

6. Adjust the idle speed by turning the air adjusting screw.

7. Remove the jumper wire.

8. Remove all test equipment.

1995–97 626 and MX6 2.0L (FS) Engine

Manual Transaxle

1. Apply the parking brake. Place the shift lever in NEUTRAL.

2. Start the engine and bring to normal operating temperature. Make sure all accessories are OFF.

3. Connect a tachometer to the engine according to the manufacturer's instructions.

4. Verify that the idle speed is between 650–750 rpm, adjust if necessary.

5. Connect a jumper wire between the **TEN** terminal and the **GND** terminal of the data link connector.

6. Remove the jumper wire.

7. Remove all test equipment.

Automatic Transaxle

1. Apply the parking brake. Place the shift lever in **P**.

2. Start the engine and bring to normal operating temperature. Make sure all accessories are OFF.

3. Connect a tachometer to the engine according to the manufacturer's instructions.

4. Check the idle speed and adjust, if necessary; it should be 700 ± 50 rpm.

5. Remove the shorting bar from the double wire SPOUT connector.

6. Adjust the idle speed by turning the air adjusting screw.

7. Install the shorting bar to the double wire SPOUT connector.

8. Remove all test equipment.

1995–97 626, MX6 2.5L (KL) Engines

1. Apply the parking brake. If equipped with a manual transaxle, place the shift lever in NEUTRAL. If equipped with an automatic transaxle, place the shift lever in **P**.

2. Start the engine and bring it to normal operating temperature. Make sure all accessories are OFF .

3. Connect a tachometer to the engine according to the manufacturer's instructions.

4. Verify that the idle speed is between 600–700 rpm; adjust if necessary.

5. Connect a jumper wire between the data link terminals **TEN** and **GND** .

6. Adjust the idle speed by turning the air adjusting screw.

7. Remove the jumper wire.

8. Remove all test equipment.

1995–97 Miata 1.8L (BPD) Engines

Manual Transaxle

1. Apply the parking brake. Place the shift lever in neutral.

2. Start the engine and bring to normal operating temperature. Make sure all accessories are **OFF**.

3. Connect a tachometer to the engine according to the manufacturer's instructions.

4. Verify that the idle speed is between 800–900 rpm, adjust if necessary.

5. Connect a jumper wire between the **TEN** terminal and the **GND** terminal of the data link connector.

6. Adjust the idle speed by turning the air adjusting screw.

7. Remove the jumper wire.

8. Remove all test equipment.

Automatic Transaxle

1. Apply the parking brake. Place the shift lever in **P**.

2. Start the engine and bring to normal operating temperature. Make sure all accessories are **OFF**.

3. Connect a tachometer to the engine according to the manufacturer's instructions.

4. Check the idle speed and adjust, if necessary; it should be 750–850 rpm.

5. Connect a jumper wire between the **TEN** terminal and the **GND** terminal of the data link connector.

6. Adjust the idle speed by turning the air adjusting screw.

7. Remove the jumper wire.

8. Remove all test equipment.

1993–95 RX7 1.3L (13B)

The idle speed is controlled automatically by the idle speed control solenoid valve (ISC) and adjustment is not necessary.

Mixture

ADJUSTMENT

On all Mazda vehicles with electronic fuel injection, fuel-air mixture is constantly adjusted electronically by the engine control unit and cannot be adjusted manually.

Fuel Filter

REMOVAL AND INSTALLATION

— CAUTION —

Fuel injection systems remain under pressure after the engine has been turned OFF. Properly relieve fuel pressure before disconnecting any fuel lines. Failure to do so may result in fire or personal injury. Do not allow fuel spray or fuel vapors to come in contact with a spark or open flame. Keep a dry chemical fire extinguisher nearby. Never store fuel in an open container due to risk of fire or explosion.

1993–94 323 1.6L (B6E) and Protege 1.8L (BPD, BPE) and 1994 MX3 1.6L (B6-ZE) Engines

1. On fuel injected engines, relieve the fuel system pressure.
2. Disconnect the negative battery cable.
3. If equipped, remove the fuel line clamps.
4. Disconnect the fuel lines from the filter and plug the ends to prevent leakage.
5. Loosen the bolt and nut and remove the fuel filter from its mounting bracket. Note the direction of the flow arrow on the filter so the replacement filter can be installed in the correct position.

To install:

6. Install the fuel filter in its mounting bracket, making sure the flow arrow is pointing in the proper direction. Tighten the bracket bolt and nut.
7. Unplug the fuel lines and connect them to the fuel filter.
8. If equipped, install the fuel line clamps.
9. Connect the negative battery cable.

1995–97 Protege 1.5L (Z5D), 1.8L (BPD) and 1993–97 626 and MX6 2.0L (FS) and 1995 MX3 1.6L (B6-ZE) Engines

1. Relieve the fuel system pressure.
2. Disconnect the negative battery cable.
3. Remove the air intake hose.
4. Remove the two nuts an the fuel filter bracket, and remove the fuel tube clamps.
5. Disconnect the fuel lines from both ends of the fuel filter. Plug the lines to prevent leakage.

6. Remove the filter from the mounting bracket.

To install:

7. Position the filter in the mounting bracket.
8. Unplug the fuel lines and connect them to the filter.
9. Install the fuel tube clamps, and install and tighten the bracket nuts to 70–95 inch lbs. (8–11 Nm).
10. Install the air intake hose.
11. Connect the negative battery cable.
12. Run the engine and check for any fuel leaks.

1994–95 MX3 1.8L (K8) and 1993–97 626 and MX6 2.5L (KL) Engines

— CAUTION —

Do not allow fuel spray or fuel vapors to come in contact with a spark or open flame. Keep a dry chemical fire extinguisher nearby. Never store fuel in an open container due to risk of fire or explosion.

1. Insure the ignition is **OFF**. Relieve the fuel system pressure.
2. Remove the nuts on the fuel filter bracket, and remove the fuel tube clamps.
3. Disconnect the fuel lines from both ends of the fuel filter. Plug the lines to prevent leakage.
4. Remove the filter from the mounting bracket.

To install:

5. Position the filter in the mounting bracket.
6. Unplug the fuel lines and connect them to the filter.
7. Install the fuel tube clamps, and install and tighten the bracket nuts to 79–104 inch lbs. (9–12 Nm) on MX3 1.8L engines and 70–95.4 inch. lbs. (7.9–10.7 Nm) on 626/MX6 2.5L engines.
8. Run the engine and check for any fuel leaks.

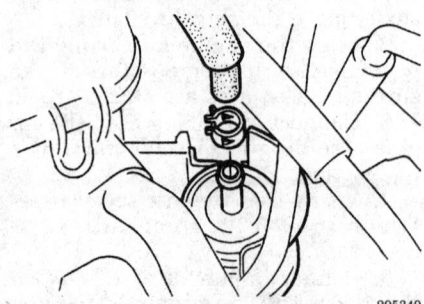

295349

Fuel filter assembly — 1993–97 626 and MX6 2.0L (FS) and 1995 MX3 1.6L (B6-ZE) engines

1993–95 929 3.0L (JE-ZE) Engines

— CAUTION —

Do not allow fuel spray or fuel vapors to come in contact with a spark or open flame. Keep a dry chemical fire extinguisher nearby. Never store fuel in an open container due to risk of fire or explosion.

1. Relieve the fuel system pressure, and disconnect the negative battery cable.
2. Raise and safely support the vehicle. The fuel filter is located at the rear of the vehicle, next to the fuel tank.
3. Disconnect the fuel lines from the fuel filter.
4. Remove the filter mounting bolts or nuts, and remove the fuel filter.
5. Installation is the reverse of the removal procedure. Make sure the fuel filter is installed in the proper flow direction.

1995–97 Millenia 2.3L (KJS) and 2.5L (KLD) Engines

— CAUTION —

Do not allow fuel spray or fuel vapors to come in contact with a spark or open flame. Keep a dry chemical fire extinguisher nearby. Never store fuel in an open container due to risk of fire or explosion.

1. Insure the ignition is **OFF**. Relieve the fuel system pressure.
2. Disconnect the negative battery cable.
3. Open the trunk, and remove the trunk mat.
4. Remove the service hole cover.
5. Disconnect the fuel lines from both ends of the fuel filter. Plug the lines to prevent leakage.
6. Remove the nut from the fuel filter bracket, and remove the filter and bracket from the vehicle.
7. Remove the filter from the mounting bracket.

To install:

8. Position the filter in the mounting bracket.
9. Install and tighten the bracket nut to 70–95 inch lbs. (8–11 Nm).
10. Unplug the fuel lines and connect them to the filter.
11. Install the service hole cover.
12. Replace the trunk mat, and close the trunk.
13. Connect the negative battery cable.
14. Run the engine and check for any fuel leaks.

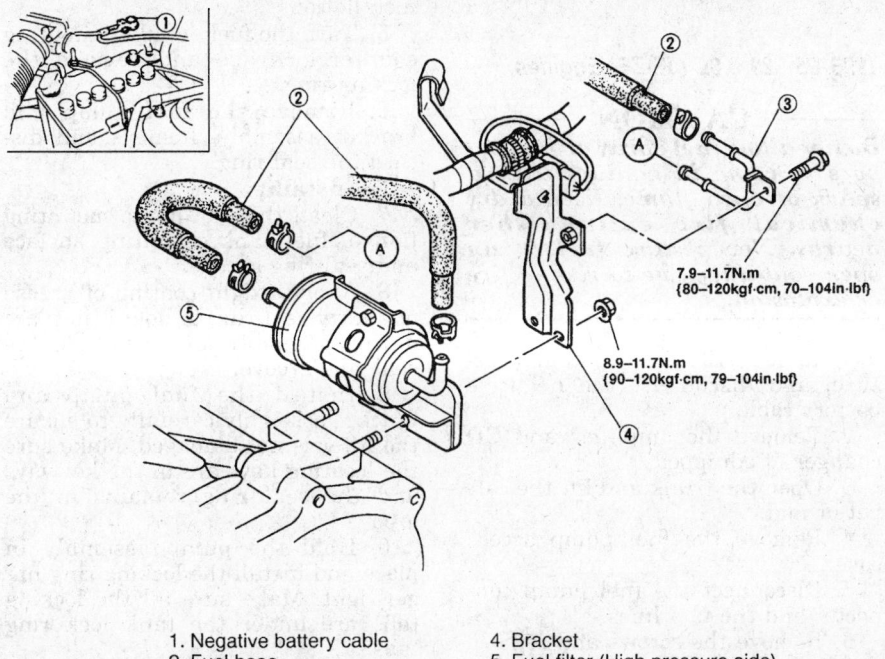

1. Negative battery cable
2. Fuel hose
3. Fuel pipe
4. Bracket
5. Fuel filter (High pressure side)

7.9–11.7N.m {80–120kgf·cm, 70–104in·lbf}

8.9–11.7N.m {90–120kgf·cm, 79–104in·lbf}

220391

Fuel filter assembly (exploded view) — 1994–95 MX3 1.8L (K8) engine

Miata 1.6L (B6-ZE) and 1.8L (BPD) Engines

——— **CAUTION** ———

Do not allow fuel spray or fuel vapors to come in contact with a spark or open flame. Keep a dry chemical fire extinguisher nearby. Never store fuel in an open container due to risk of fire or explosion.

1. Relieve the fuel system pressure and disconnect the negative battery cable.
2. Raise and safely support the vehicle. The fuel filter is located at the

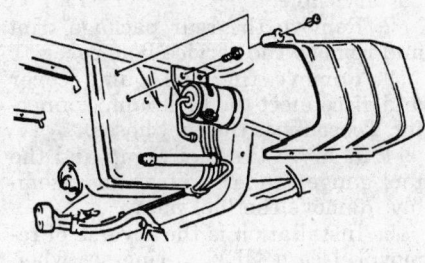

289646

Fuel filter cover plate identification — Miata with 1.6L and 1.8L engines

rear of the vehicle, next to the fuel tank.
3. Remove the fuel filter cover.
4. Disconnect and plug the fuel lines from the fuel filter.
5. Remove the filter mounting bolts or nuts and remove the fuel filter.
6. Installation is the reverse of the removal procedure. Make sure the fuel filter is installed in the proper flow direction. Torque the fuel filter bracket mounting bolts to 70–95.4 in lbs. (7.9–10.7 Nm).

1993–95 RX7 1.3L (13B) Engines

——— **CAUTION** ———

Do not allow fuel spray or fuel vapors to come in contact with a spark or open flame. Keep a dry chemical fire extinguisher nearby. Never store fuel in an open container due to risk of fire or explosion.

NOTE: The fuel filter is located at the rear of the vehicle, next to the fuel tank.

1. Relieve the fuel system pressure.
2. Disconnect the negative battery cable.
3. Raise and safely support the vehicle. Remove the splash shield.

4. Disconnect the fuel lines from the fuel filter.
5. Remove the filter mounting bolts and the fuel filter.
6. Installation is the reverse of the removal procedure. Make sure the fuel filter is installed in the proper flow direction.

Fuel Pump

REMOVAL AND INSTALLATION

——— **CAUTION** ———

Fuel injection systems remain under pressure after the engine has been turned OFF. Properly relieve fuel pressure before disconnecting any fuel lines. Failure to do so may result in fire or personal injury. Do not allow fuel spray or fuel vapors to come in contact with a spark or open flame. Keep a dry chemical fire extinguisher nearby. Never store fuel in an open container due to risk of fire or explosion.

1993–94 323 1.6L (B6E) and 1993–97 Protege 1.8L (BPD, BPE), 1.5L (Z5D) and 1994 MX3 1.8L (B6-ZE) Engines

1. Relieve the fuel pressure and disconnect the negative battery cable.
2. Depress the clips on each end of the rear seat cushion and remove the cushion.
3. Disconnect the electrical connector from the fuel pump/sending unit.
4. Remove the attaching screws from the fuel pump/sending unit access cover and remove the cover.
5. Disconnect the fuel supply and return hoses from the fuel pump/sending unit.
6. Remove the attaching screws and the fuel pump/sending unit from the fuel tank.
7. Disconnect the sending unit electrical connector, remove the sending unit attaching nuts and remove the sending unit from the fuel pump assembly.
To install:
8. Attach the sending unit to the fuel pump assembly and install the nuts. Connect the sending unit electrical connector.
9. Install the fuel pump/sending unit into the fuel tank with a new gasket and install the mounting screws.
10. Connect the fuel supply and return lines.
11. Install the access cover and the mounting screws.

12. Connect the sending unit electrical connector.

13. Position the rear seat cushion over the floor, making sure to align the retaining pins with the clips. Push down firmly until the 2 retaining pins are locked into the rear seat retaining clips.

14. Connect the negative battery cable, start the engine and check for proper system operation and for fuel leaks.

1995 MX3 1.6L (B6-ZE) and 1.8L (K8) Engines

— CAUTION —

Do not allow fuel spray or fuel vapors to come in contact with a spark or open flame. Keep a dry chemical fire extinguisher nearby. Never store fuel in an open container due to risk of fire or explosion.

1. Relieve the fuel system pressure.
2. Disconnect the negative battery cable.
3. Remove the rear seat cushion from the vehicle.
4. Remove any dirt that has accumulated around the fuel pump cover so it will not enter the tank during pump removal and installation.
5. Remove the fuel pump cover.
6. Disconnect the fuel gauge connector, hoses, and the gauge.
7. Disconnect the fuel pump electrical connector.
8. Remove the fuel pump from the bracket assembly. Remove and discard the seal ring.

To install:
9. Clean the fuel pump mounting flange, fuel tank mounting surface and seal ring groove.
10. Apply a light coating of grease on a new seal ring to hold it in place during assembly and install in the seal ring groove.
11. Install the fuel pump to the bracket assembly carefully to ensure the filter is not damaged. Make sure the seal ring remains in the groove.
12. Hold the pump assembly in place, and pull the fuel pump down so that it is tight against the bracket.
13. Connect the fuel pump electrical connector.
14. Install the fuel gauge, hoses, and gauge connector.
15. Install the fuel pump cover.
16. Install the rear seat cushion.
17. Connect the negative battery cable, start the engine and check for

proper system operation and for fuel leaks.

1993–95 929 3.0L (JE-ZE) Engines

— CAUTION —

Do not allow fuel spray or fuel vapors to come in contact with a spark or open flame. Keep a dry chemical fire extinguisher nearby. Never store fuel in an open container due to risk of fire or explosion.

1. Relieve the fuel system pressure, and disconnect the negative battery cable.
2. Remove the amplifier and CD changer, if equipped.
3. Open the trunk and lift the carpet or mat.
4. Remove the fuel pump access cover.
5. Disconnect the fuel pump connector and the fuel lines.
6. Remove the screws and remove the fuel pump/sending unit assembly from the fuel tank.
7. If necessary, remove the fuel pump from the tank gauge sending unit.

To install:
8. Using a new rubber seal, install the fuel pump into the fuel tank.
9. Connect the fuel pump connector and the fuel lines.
10. Install the access cover, and replace the mat or carpet.
11. Install the amplifier and CD changer, if removed.
12. Connect the negative battery cable.

1995–97 Millenia 2.3L (KJS), 2.5L (KLD) and 1993 MX3 (K8) and 1993–97 626, MX6, 2.0L (FS) and 2.5L (KL) Engines

— CAUTION —

Do not allow fuel spray or fuel vapors to come in contact with a spark or open flame. Keep a dry chemical fire extinguisher nearby. Never store fuel in an open container due to risk of fire or explosion.

1. Relieve the fuel system pressure.
2. Disconnect the negative battery cable.
3. Remove the fuel tank and place it on a bench.
4. Remove any dirt that has accumulated around the fuel pump retaining flange so it will not enter the

tank during pump removal and installation.
5. Turn the fuel pump locking ring counterclockwise and remove the locking ring.
6. Remove the fuel pump and bracket assembly. Remove and discard the seal ring.

To install:
7. Clean the fuel pump mounting flange, fuel tank mounting surface and seal ring groove.
8. Apply a light coating of grease on a new seal ring to hold it in place during assembly and install in the seal ring groove.
9. Install the fuel pump and bracket assembly carefully to ensure the filter is not damaged. Make sure the locating keys are in the keyways and the seal ring remains in the groove.
10. Hold the pump assembly in place and install the locking ring finger-tight. Make sure all the locking tabs are under the tank lock ring tabs.
11. Rotate the locking ring clockwise until the ring is against the stops.
12. Install the fuel tank in the vehicle. Add a minimum of 10 gallons of fuel to the tank and check for leaks.
13. Connect the negative battery cable. Start the engine and check for proper system operation and for fuel leaks.

1993–97 Miata 1.6L (B6-ZE) and 1.8L (BPD) Engines

— CAUTION —

Do not allow fuel spray or fuel vapors to come in contact with a spark or open flame. Keep a dry chemical fire extinguisher nearby. Never store fuel in an open container due to risk of fire or explosion.

1. Properly relieve the fuel pressure, and disconnect the negative battery cable.
2. Remove the rear package trim and remove the service hole cover.
3. Remove the fuel pump cover and disconnect the fuel pump connector. Disconnect the fuel hoses.
4. Remove the fuel pump and the fuel gauge sender unit as an assembly. Remove the fuel pump.
5. Installation is the reverse of removal. Use a NEW O-ring set when installing. After install the fuel pump to the bracket, pull the fuel pump down so that it is tight against the bracket.

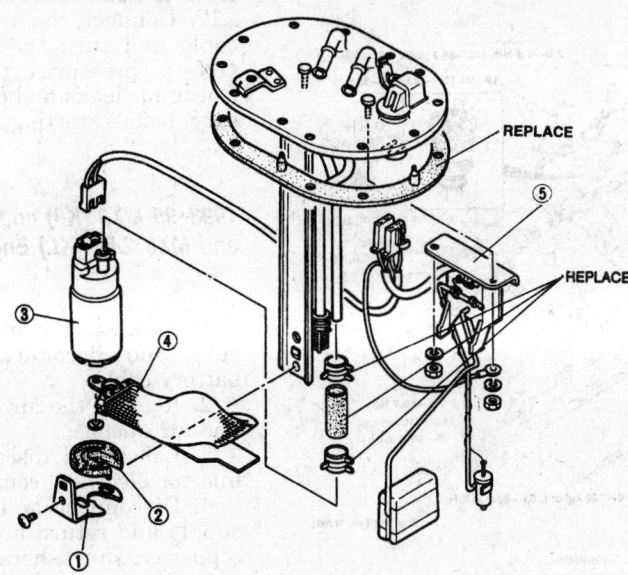

REPLACE

⑤

REPLACE

1. Bracket
2. Rubber mount
3. Fuel pump
4. Fuel filter
5. Fuel gauge sender unit

234381

Fuel pump assembly — 1993–94 929 3.0L (JE-ZE) engines

1993–95 RX7 1.3L (13B) Engines

----- **CAUTION** -----

Do not allow fuel spray or fuel vapors to come in contact with a spark or open flame. Keep a dry chemical fire extinguisher nearby. Never store fuel in an open container due to risk of fire or explosion.

1. Relieve the fuel system pressure and disconnect the negative battery cable.
2. Open the trunk or hatch and lift the carpet or mat.
3. Remove the fuel pump access cover.
4. Disconnect the fuel pump connector and the fuel lines.
5. Remove the screws and remove the fuel pump/sending unit assembly from the fuel tank. If necessary, remove the fuel pump from the tank gauge sending unit.
6. Installation is the reverse of the removal procedure. Use a new rubber seal and tighten the mounting bolts to 70–95 inch lbs. (8–11 Nm).

Fuel Injector

REMOVAL AND INSTALLATION

----- **CAUTION** -----

Fuel injection systems remain under pressure after the engine has been turned OFF. Properly relieve fuel pressure before disconnecting any fuel lines. Failure to do so may result in fire or personal injury. Do not allow fuel spray or fuel vapors to come in contact with a spark or open flame. Keep a dry chemical fire extinguisher nearby. Never store fuel in an open container due to risk of fire or explosion.

1993–94 323 (1.6L (B6E), Miata 1.6L (B6-ZE), 1.8L (BPD) and Protege 1.8L (BPE, BPD) Engines

1. Relieve the fuel system pressure.
2. Disconnect the negative battery cable.
3. Disconnect the electrical connectors at the injectors.
4. Remove the fuel injector harness from the fuel delivery pipe.
5. Remove the delivery pipe bolt(s) and remove the delivery pipe with the injectors and the pressure regulator.
6. Separate the injectors, grommets and the insulators from the delivery pipe assembly.

   To install:
7. Replace the fuel injector O-rings and apply a small amount of clean engine oil to the O-rings before installation.
8. Replace the fuel injector insulator seals.
9. Install the injectors to the delivery pipe and install the assembly to the engine. Tighten the delivery pipe bolt to 14–18 ft. lbs. (19–25 Nm).
10. Install the injector harness to the fuel delivery pipe and connect the injector wiring.
11. Connect the negative battery cable.

1995–97 Protege 1.5L (Z5D) and 1.8L (BPD, BPE) Engines

1. Relieve the fuel system pressure, and disconnect the negative battery cable.
2. Disconnect the throttle and accelerator cables. Remove the cable bracket.
3. Label and disconnect the fuel injector wiring harness.
4. Disconnect and plug the fuel lines at the fuel rail.
5. Disconnect the vacuum hose from the fuel pressure regulator.
6. Remove the fuel line mounting bracket bolt.
7. Remove the fuel rail mounting bolts, spacers, insulators and the fuel rail, with the injectors attached.
8. Remove the fuel injectors, grommets and O-rings from the fuel rail. Remove the O-rings from the fuel injectors.

   To install:
9. Apply a small amount of clean engine oil to new O-rings and install them and the grommets on the fuel injectors.
10. Install the insulators and injectors on the intake manifold.
11. Install the grommets and the fuel rail onto the injectors.
12. Install the fuel rail attaching bolts and tighten to 14–18 ft. lbs. (19–25 Nm).
13. Connect the vacuum hose to the fuel pressure regulator and the fuel lines to the fuel rail.
14. Install the fuel line mounting bracket, and tighten the bolt to 70–95 inch lbs. (8–11 Nm).
15. Connect the fuel injector wiring harness.
16. Install the cable bracket, and tighten the bolt to 70–95 inch lbs.

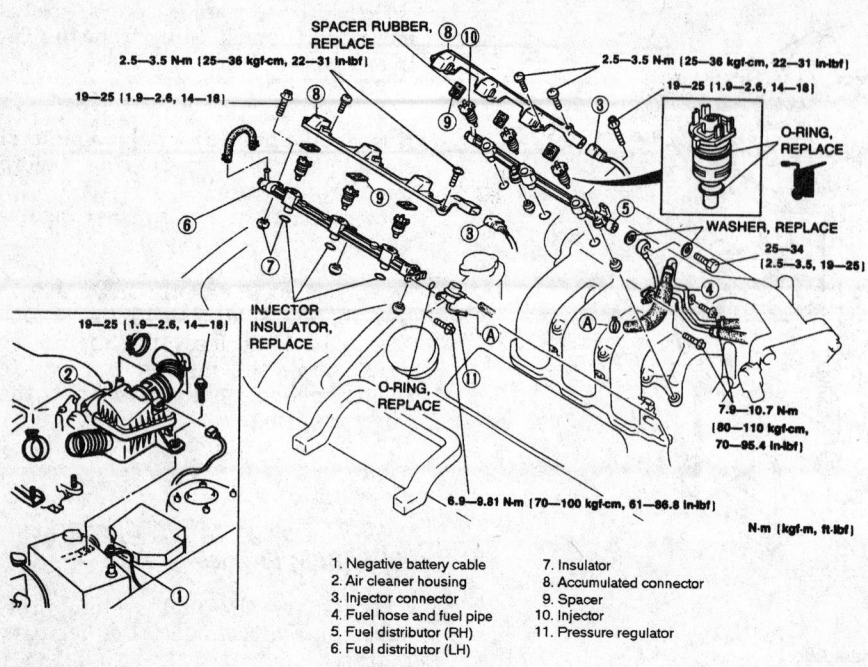

Exploded view of the fuel rail assembly — 626 and MX6 with 2.5L (KL) and 1993–95 MX3 1.8L engines

1. Negative battery cable
2. Air cleaner housing
3. Injector connector
4. Fuel hose and fuel pipe
5. Fuel distributor (RH)
6. Fuel distributor (LH)
7. Insulator
8. Accumulated connector
9. Spacer
10. Injector
11. Pressure regulator

295845

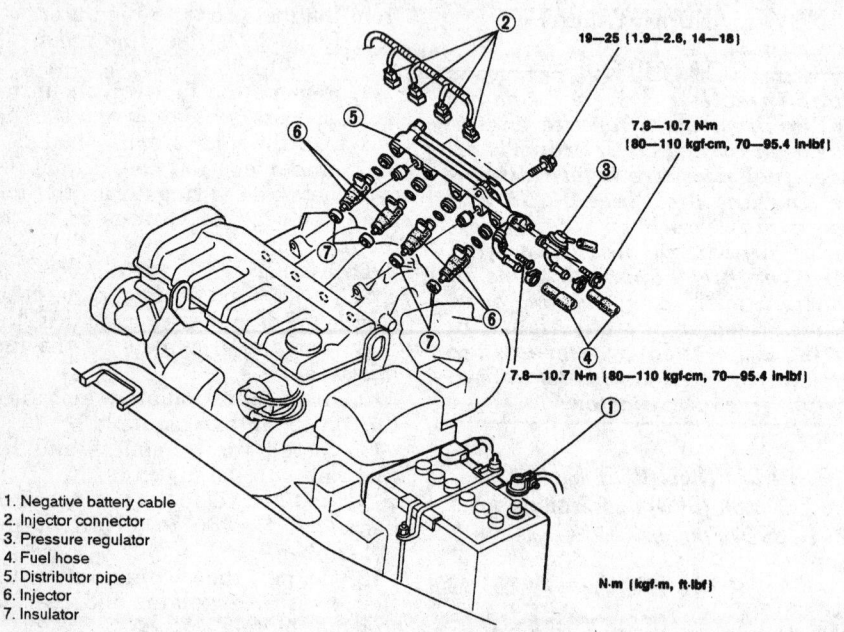

Fuel injectors and related components — 626 and MX6 2.0L (FS) engines

1. Negative battery cable
2. Injector connector
3. Pressure regulator
4. Fuel hose
5. Distributor pipe
6. Injector
7. Insulator

295805

(8–11 Nm). Connect and adjust the throttle and accelerator cables.

17. Connect the negative battery cable and turn the ignition switch **ON** to pressurize the fuel system. Check for leaks and correct as necessary, before starting the engine.

1993–95 MX3 (K8) and 1993–97 626 and MX6 2.5L (KL) Engines

1. Relieve the fuel system pressure, and disconnect the negative battery cable.
2. Remove the air cleaner housing and air ducts.
3. Label and disconnect the fuel injector electrical connectors.
4. Disconnect and plug the fuel supply and return lines. Discard the copper crush washers.
5. Remove the fuel pressure regulator mounting bolts and the fuel pressure regulator.
6. Remove the fuel rail mounting bolts and the fuel rails.
7. Remove the 6 insulators.
8. Remove the distribution harness (accumulated connector) attaching screws, and remove the distribution harness from the fuel rails.
9. Remove and discard the spacer from the top of each fuel injector. Remove the fuel injectors from the fuel rails by rotating back and forth.

To install:

10. Apply clean engine oil to new O-rings and install them on the injectors. Install the injectors into the fuel rails.
11. Install new spacers on the injectors, then install the distribution harness with the screws. Tighten the screws to 22–31 in. lbs. (2.5–3.5 Nm).
12. Install the 6 insulators and the fuel rails. Install the fuel rail mounting bolts and tighten to 14–18 ft. lbs. (19–25 Nm).
13. Install the fuel pressure regulator and tighten the bolts to 61–86 in. lbs. (7–10 Nm).
14. Using new copper crush washers, connect the fuel supply and return lines.
15. Connect the fuel injector electrical connectors.
16. Install the air ducts and air cleaner housing.
17. Connect the negative battery cable. Turn the ignition switch **ON** to pressurize the fuel system. Check for fuel leaks and correct as necessary before starting the engine.

1993–97 626 and MX6 2.0L (FS) Engines

1. Relieve the fuel system pressure and disconnect the negative battery cable.
2. Label and disconnect the fuel injector wiring harness.
3. Disconnect and plug the fuel lines at the fuel rail.
4. Disconnect the vacuum hose from the fuel pressure regulator.
5. Remove the fuel line mounting bracket bolt.
6. Remove the fuel rail mounting bolts, spacers, insulators and the fuel rail, with the injectors attached.
7. Remove the fuel injectors, grommets and O-rings from the fuel rail. Remove the O-rings from the fuel injectors.

To install:

8. Apply a small amount of clean engine oil to new O-rings and install them and the grommets on the fuel injectors.
9. Install the insulators and injectors on the intake manifold.
10. Install the grommets and the fuel rail onto the injectors.
11. Install the fuel rail attaching bolts and tighten to 14–18 ft. lbs. (19–25 Nm).
12. Connect the vacuum hose to the fuel pressure regulator and the fuel lines to the fuel rail.
13. Install the fuel line mounting bracket and tighten the bolt to 70–95 inch lbs. (8–11 Nm).
14. Connect the fuel injector wiring harness.
15. Connect the negative battery cable and turn the ignition switch **ON** to pressurize the fuel system. Check for leaks and correct as necessary, before starting the engine.

1993–95 929 3.0L (JE-ZE) Engines

1. Properly relieve the fuel system pressure.
2. Disconnect the negative battery cable, and drain the cooling system.
3. Disconnect the air intake hose from the throttle body. If necessary, disconnect the air flow meter electrical connector and remove the air cleaner assembly and air ducts.
4. Disconnect the accelerator cable. Disconnect and plug the fuel lines.
5. Label and disconnect the necessary hoses and electrical connectors from the throttle body. Remove the throttle body.
6. Disconnect the electrical connector, water hoses and air hose from the air bypass valve. Remove the re-

taining bolts/nuts and remove the air bypass valve. Remove the air intake pipe from the upper intake manifold.
7. Label and disconnect the necessary hoses and electrical connectors from the upper intake manifold. Remove the upper intake manifold retaining bolts and the upper intake manifold.
8. Label and disconnect the electrical connectors from the fuel injectors.
9. Remove the retaining bolts and cap from the top of the injector. Remove the injector by grasping the plastic part of the injector and twisting out. Disconnect the vacuum hose from the fuel pressure regulator and remove the fuel rail.

To install:

10. Install the fuel rail. Apply a small amount of clean engine oil to new injector O-rings and install on the injectors. Install the injectors in the fuel rail using a turning motion, to prevent damaging the O-rings. Install the injector retainer caps and tighten the screws to 30 inch lbs. (3.4 Nm).
11. Connect the injector electrical connectors and connect the vacuum hose to the fuel pressure regulator. Connect the fuel lines.
12. Using a new gasket, install the upper intake manifold on the lower intake manifold. Tighten the mounting bolts to 19 ft. lbs. (25 Nm).
13. Connect the electrical connectors and hoses to the upper intake manifold.
14. Install the air intake pipe on the upper intake manifold using a new gasket, and tighten the mounting bolts to 17 ft. lbs. (23 Nm). Install the air bypass valve and tighten the nuts to 19 ft. lbs. (25 Nm). Connect the coolant hoses, air hose and electrical connector to the air bypass valve.
15. Install the throttle body, using a new gasket. Tighten the nuts to 19 ft. lbs. (25 Nm). Connect the hoses and electrical connectors to the throttle body.
16. If removed, install the air cleaner and air ducts. Connect the air intake to the throttle body.
17. Connect the negative battery cable and turn the ignition **ON** to pressurize the fuel system. Check for leaks.
18. Fill and bleed the cooling system.
19. Start the engine and bring to normal operating temperature. Check for leaks. Check the idle speed.

1995–97 Millenia 2.3L (KJS) and 2.5L (KLD) Engines

1. Relieve the fuel system pressure, and disconnect the negative battery cable.
2. Remove the charge air cooler air duct.
3. Remove the air cleaner assembly.
4. Remove the resonator.
5. Remove the left and right-hand charge air coolers.
6. Disconnect the accelerator cable.
7. Remove the air intake pipe assembly.
8. Remove the vacuum hose assembly.
9. Label and disconnect the fuel injector electrical connectors.
10. Disconnect and plug the fuel supply and return lines. Discard the copper crush washers.
11. Remove the fuel rail mounting bolts and the fuel rails.
12. Remove the six insulators.
13. Remove the distribution harness (accumulated connector) attaching screws, and remove the distribution harness from the fuel rails.
14. Remove and discard the spacer from the top of each fuel injector. Remove the fuel injectors from the fuel rails by rotating back and forth.
15. Remove the fuel pressure regulator mounting bolts and the fuel pressure regulator.

To install:

16. Install the fuel pressure regulator and tighten the bolts to 61–86 inch lbs. (7–10 Nm).
17. Apply clean engine oil to new O-rings and install them on the injectors. Install the injectors into the fuel rails.
18. Install new spacers on the injectors, then install the distribution harness with the screws. Tighten the screws to 22–31 inch lbs. (2.5–3.5 Nm).
19. Install the 6 insulators and the fuel rails. Install the fuel rail mounting bolts and tighten to 14–18 ft. lbs. (19–25 Nm).
20. Using new copper crush washers, connect the fuel supply and return lines.
21. Connect the fuel injector electrical connectors.
22. Install the vacuum hose assembly. Tighten the mounting nuts to 70–95 inch lbs. (8–11 Nm).
23. Install the air intake pipe assembly. Tighten the mounting nuts to 70–95 inch lbs. (8–11 Nm), and the mounting bolts to 44–78 inch lbs. (5–9 Nm).

24. Connect and adjust the accelerator cable. Tighten the mounting bolt to 70–95 inch lbs. (8–11 Nm).

25. Install the left and right-hand charge air coolers. Install new O-rings, and tighten the bolts to 14–18 ft. lbs. (19–25 Nm).

26. Install the resonator. Tighten the mounting nuts to 70–95 inch lbs. (8–11 Nm).

27. Install the air cleaner assembly. Tighten the mounting nuts to 70–95 inch lbs. (8–11 Nm).

28. Install the charge air cooler air duct. Tighten the mounting nuts to 70–95 inch lbs. (8–11 Nm).

29. Connect the negative battery cable. Turn the ignition switch **ON** to pressurize the fuel system. Check for fuel leaks and correct as necessary before starting the engine.

1995 MX3 1.6L (B6-ZE) and 1995–97 Miata 1.8L (BPD) Engines

1. Properly relieve the fuel system pressure, and disconnect the negative battery cable.

2. Remove the PCV hose.

3. Disconnect the vacuum hose from the fuel pressure regulator.

4. Label and disconnect the fuel injector connectors.

5. Remove the fuel rail mounting bolts, spacers, insulators and the fuel rail, with the injectors attached.

6. Remove the fuel injectors, grommets and O-rings from the fuel rail. Remove the O-rings from the fuel injectors.

To install:

7. Apply a small amount of clean engine oil to NEW O-rings and install them and the grommets on the fuel injectors.

8. Install the insulators and injectors on the intake manifold.

9. Install the grommets and the fuel rail onto the injectors.

10. Install the fuel rail attaching bolts and tighten to 14–18 ft. lbs. (19–25 Nm).

11. Connect the vacuum hose to the fuel pressure regulator and the fuel lines to the fuel rail.

12. Connect the fuel injector connectors.

13. Install the PCV hose.

14. Connect the negative battery cable and turn the ignition switch **ON** to pressurize the fuel system. Check for leaks and correct as necessary, before starting the engine.

1993–95 RX7 1.3L (13B) Engines

1. Relieve the fuel system pressure.

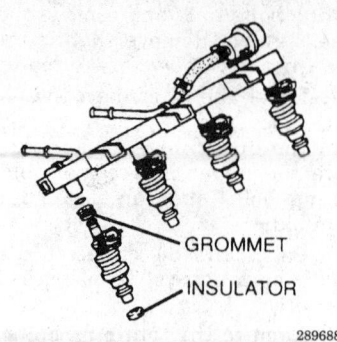

Fuel injector assembly — 1995–97 Miata 1.8L (BPD) engines

2. Disconnect the negative battery cable.

3. Drain the cooling system.

4. Remove the fresh air duct, and disconnect the accelerator cable.

5. Disconnect the hoses and remove the air cleaner assembly. Remove the air cleaner-to-turbocharger hoses.

6. Disconnect the vacuum hoses and remove the pressure chamber.

7. Remove the intercooler hoses and the turbocharger-to-intercooler pipe and intercooler-to-throttle body pipe.

8. Label and disconnect the electrical connectors, vacuum and coolant hoses from the throttle body and extension manifold.

9. Remove the mounting bolts and nuts, and the throttle body/extension manifold assembly.

10. Label and disconnect the fuel injector electrical connectors. Disconnect and plug the fuel lines.

11. Remove the primary and secondary fuel distributor assemblies.

12. Remove the screws and remove the injector covers. Remove the injectors, insulators and air bleed sockets from the fuel distributors.

To install:

13. Apply clean engine oil to new O-rings and install on the injectors. Install the injectors into the fuel distributors with a twisting motion. Install the injector caps and tighten the screws to 30 inch lbs. (3.5 Nm).

14. Align the tabs of the air bleed socket with the notches in the intermediate housing. Install the air bleed socket, insulators and fuel distributor assemblies. Tighten the fuel distributor mounting bolts to 19 ft. lbs. (25 Nm).

15. Connect the fuel lines and the injector connectors.

16. Install the throttle body/extension manifold assembly, using a new gasket, and tighten the nuts and bolts to 19 ft. lbs. (25 Nm).

17. Connect the coolant and vacuum hoses and the electrical connectors to the throttle body and extension manifold.

18. Install the intercooler-to-throttle body, turbocharger-to-intercooler pipes and the intercooler connecting hoses.

19. Install the pressure chamber and connect the vacuum hoses.

20. Connect the accelerator cable.

21. Install the air cleaner assembly and the air hoses.

22. Connect the negative battery cable. Turn the ignition key **ON** to pressurize the system and check for leaks. Correct as necessary.

23. Fill and bleed the cooling system. Start the engine and check for leaks.

ENGINE MECHANICAL

Engine Assembly

REMOVAL AND INSTALLATION

—— **CAUTION** ——

When draining the coolant, keep in mind that cats and dogs are attracted by the ethylene glycol antifreeze, and are quite likely to drink any that is left in an uncovered container or in puddles on the ground. This will prove fatal in sufficient quantity. Always drain the coolant into a sealable container. Coolant should be re-used unless it is contaminated or several years old.

1993–94 323 and 1994 MX3 1.6L (B6E), 1993–95 Protege 1.8L (BPD, BPE) and 1994 MX3 1.6L (B6-ZE) Engines

1. Properly relieve the fuel system pressure. Raise and safely support the vehicle, as necessary.

2. Disconnect the battery cables and remove the battery and the battery tray. Raise and safely support the vehicle.

3. Remove the splash shield(s) from under the vehicle and drain the engine and transaxle oil and the coolant.

4. Remove the air cleaner assembly and resonance chamber, including the air flow meter and all of the ducting. Remove the oil dipstick.

5. Remove the radiator hoses. If equipped with automatic transaxle, disconnect the oil cooler lines from the radiator. Disconnect the cooling fan and, if equipped, radiator switch electrical connectors and remove the radiator/cooling fan assembly.

6. Disconnect the throttle and the speedometer cable.

7. Label and disconnect the vacuum hoses and wiring.

8. Disconnect the fuel supply and return hoses and the heater hoses.

9. Disconnect the exhaust pipe from the manifold.

10. Without disconnecting the hydraulic hoses, remove the power steering pump and hang it from the body with wire.

11. Without disconnecting the refrigerant lines, remove the air conditioning compressor and hang it from the body with wire.

12. If equipped with manual transaxle, disconnect the clutch cable and shift control rod. If equipped with hydraulic clutch, remove the slave cylinder from the transaxle without disconnecting the hydraulic line.

13. If equipped with automatic transaxle, disconnect the shift control cable.

14. Remove the nuts and disconnect the tie rod ends from the steering knuckles. Disconnect the stabilizer bar from the lower control arms.

15. Attach an engine lifting chain to the engine lifting eyes. Attach the chain to a suitable engine hoist and raise the hoist until there is tension on the chain.

16. Remove the engine mount nuts and the engine mount member bolts and nuts and remove the engine mount member.

NOTE: Be careful so the engine does not fall when removing the engine mount member.

17. Remove the pinch bolts from the steering knuckle and pry the control arm down to slip lower ball joint out of the knuckle.

18. If equipped, remove the bolts from the right side intermediate shaft support and, using a suitable prybar, pry the intermediate shaft from the transaxle. Insert a suitable prybar between the inner CV-joint and transaxle case and carefully pry the inner CV-joints out of the transaxle. Suspend the halfshafts with wire.

19. Remove the dynamic damper from the right side engine mount, if equipped. Remove the engine/transaxle mount nuts/bolts and right engine and, if equipped, left

transaxle mounts. Carefully lift the engine/transaxle assembly from the vehicle.

20. Properly support the engine/transaxle assembly. Remove the intake manifold bracket, starter, torque converter nuts, stiffener, if equipped and No. 2 engine mount. Disconnect the throttle cable.

21. Remove the transaxle mounting bolts and separate the transaxle from the engine.

To install:

22. Attach the transaxle to the engine and install the transaxle-to-engine bolts. If equipped with automatic transaxle, install the torque converter nuts and tighten to 25–36 ft. lbs. (34–49 Nm).

23. Connect the throttle cable. Install the No. 2 engine mount, stiffener, if equipped, starter and intake manifold bracket.

24. On 2WD vehicles, proceed as follows:

 a. Install the engine mount member and tighten the bolts/nuts to 47–66 ft. lbs. (64–89 Nm).

 b. Carefully lower the engine/transaxle assembly into the engine compartment and align the engine mount bolts with the engine mount member mounting holes. Install the mount-to-mount member nuts and tighten to 27–38 ft. lbs. (37–52 Nm).

 c. Install the right side engine mount. Tighten the mount-to-engine nut(s) to 54–76 ft. lbs. (74–103 Nm) on 1993 vehicles. Tighten the mount through-bolt to 49–69 ft. lbs. (67–93 Nm) on 1993 vehicles.

 d. On 1993 vehicles, install the dynamic damper, to the right side mount and tighten to 41–59 ft. lbs. (55–80 Nm).

 e. If equipped, install the left side transaxle mount. On 1993 vehicles, loosely install the mount-to-transaxle nuts and align the mount bracket holes with the body holes. Install the mount-to-body bolts and tighten, in sequence, to 32–45 ft. lbs. (43–61 Nm). Tighten the mount-to-transaxle nuts to 49–69 ft. lbs. (67–93 Nm).

25. Install new circlips on the inner CV-joint stub shafts and, if equipped, intermediate shaft. Grease the shaft splines and install the half-shaft/intermediate shaft into the transaxle.

26. If equipped, install the right intermediate shaft support bolts and tighten, in sequence, to 31–46 ft. lbs. (42–62 Nm).

27. Install the lower ball joint and torque the clamping bolt to 43 ft. lbs.

(59 Nm). Install the tie rod ends and torque the nut to 42 ft. lbs. (57 Nm), then tighten as required to install a new cotter pin.

28. Attach the stabilizer bar and tighten the nuts so there is ³⁄₄in. (19mm) of thread showing above the nut.

29. If equipped with manual transaxle, connect the extension bar and shift control rod. Connect the clutch cable or install the hydraulic slave cylinder, as necessary.

30. If equipped with automatic transaxle, connect the shift control cable.

31. Install a new gasket and connect the exhaust pipe to the manifold. Use new self-locking nuts and torque to 34 ft. lbs. (46 Nm).

32. Connect the wiring and all heater, fuel and vacuum hoses.

33. Install the air conditioner compressor and power steering pump, if equipped.

34. Install the radiator/cooling fan assembly and connect the radiator hoses and the necessary electrical connectors.

35. Connect the accelerator and speedometer cables.

36. Install the battery tray assembly and battery. Install the air cleaner and air flow meter assembly and all the ducting. Connect the air flow sensor connector.

37. Install the splash shield(s).

38. Fill the engine and the transaxle with the proper types and quantities of oil. Fill the cooling system.

39. Connect the negative battery cable, start the engine and check for leaks. Check the ignition timing and the idle speed. Check all fluid levels.

1995–97 Protege 1.5L (Z5D) and 1.8L (BPD) Engines

1. Relieve the fuel system pressure, and disconnect the battery cables.

2. Mark the position of the hood on its hinges, and remove the hood.

3. Drain the cooling system, engine oil and transaxle oil.

4. Slightly raise the vehicle and remove the front wheels.

5. Remove the upper and lower radiator hose and, on automatic transaxles (ATX) only, the oil cooler hose.

6. Disconnect the cooling fan motor connector, and remove the coolant reservoir pipe.

7. Remove the radiator bracket. Remove the radiator and cooling fan assembly.

8. Disconnect the oxygen sensor connector. Remove and discard the

exhaust pipe-to-catalytic converter nuts.

9. Remove the exhaust support bolts. Remove and discard the exhaust pipe-to-exhaust manifold nuts and remove the exhaust pipe. Support the remaining exhaust system with mechanics wire.

10. Remove the exhaust manifold.

11. Remove the air cleaner assembly.

12. Remove the battery clamp, battery and battery tray.

13. Remove the coolant reservoir, and the battery duct bracket.

14. Remove the resonance chamber.

15. Disconnect the accelerator cable.

16. If equipped with an automatic transaxle, remove the throttle cable.

17. Remove the splash shield.

18. Remove the power steering belt shield and the power steering belt. Remove the power steering hose brackets from the cylinder head cover.

19. Remove the power steering belt adjuster and disconnect the power steering pressure switch connector. Remove the power steering pump and position aside, leaving the hoses connected.

20. If equipped, remove the A/C compressor and position aside, leaving the refrigerant lines attached. Support the compressor with suitable wire.

21. Label and disconnect the electrical connectors from the distributor, mass airflow sensor, intake air temperature sensor, engine coolant temperature sensor, cooling fan temperature sensor, coolant temperature gauge sensor, throttle position sensor, air bypass valve, idle switch, fuel injectors, EGR solenoids and alternator. Disconnect and label any remaining engine harness connectors.

22. Label, disconnect and plug the fuel lines at the fuel rail.

23. Disconnect the heater hoses.

24. Raise and safely support the vehicle. Attach suitable engine lifting equipment to the lifting eyes on the engine.

25. On manual transaxles (MTX), remove the bolts and clips from the clutch slave cylinder. Remove the slave cylinder with the hose still connected, and position it away from the transaxle.

26. On MTX, remove the extension bar nut and washer, then disengage the bar from the transaxle. Remove the transaxle control rod through-bolt and nut, then disengage the rod from the transaxle.

27. On ATX, disconnect the shift control cable from the transaxle.

28. Remove the stabilizer control link and the tie-rod end ball joint.

29. Remove the bolts and nuts from the lower arm ball joints, and pry the lower arms from the knuckles.

30. Separate the driveshafts and the joint shaft from the transaxle. Install transaxle plug tools T88C–7025–AH or equivalent, into the differential side gears.

NOTE: If the plugs are not installed, the differential side gears may become mispositioned. If the gears are. mispositioned, the differential may have to be removed to reposition them.

31. Remove the number one engine mount bracket.

32. Remove the transverse (engine mount) member, and the number 2 engine mount bracket nuts.

33. Remove the number three and four engine mount brackets.

34. Lower the vehicle. Slowly lift the engine and transaxle assembly as a unit out of the vehicle. Keep the engine from swinging or bumping into components in the engine compartment.

35. On ATX, remove the torque converter-to-flexplate nuts.

36. Remove the engine-to-transaxle bolts and the transaxle-to-engine mounting bolts.

37. Separate the engine from transaxle.

38. On MTX, remove the clutch assembly, flywheel and crankshaft rear cover plate. On ATX, remove the flexplate and throttle cable.

39. Place the engine on a workstand.

To install:

40. Remove the engine from the workstand.

41. On ATX, remove the old sealant from the flexplate mounting bolts and bolt holes. If reusing the flexplate bolts, apply silicone sealant to the bolt threads. Install the flexplate and loosely install the bolts. Tighten the flexplate bolts in 2–3 steps to 75 ft. lbs. (103 Nm) in a crisscross pattern.

NOTE: New flexplate mounting bolts come with sealant already on them.

42. On MTX, install the crankshaft rear cover plate and tighten the bolt to 88 inch lbs. (10 Nm). Install the flywheel and clutch assembly.

43. On ATX:

a. Install the transaxle-to-engine bolts and tighten the mounting bolts to 41–59 ft. lbs. (55–80 Nm).

b. Install the torque converter to the flexplate and tighten the nuts to 26–36 ft. lbs. (35–49 Nm). Rotate the flexplate, as necessary, to gain access to all of the nuts.

44. On MTX, install the transaxle mounting bolts and tighten the bolts to 48–65 ft. lbs. (64–89 Nm).

45. Hoist the engine and transaxle assembly. Slowly lower and position the assembly into the engine compartment. Align the engine mounts with the mount holes. Install the engine mounts.

46. Install the transverse member. Tighten the mounting bolts to 50–65 ft. lbs. (67–89 Nm), and the nuts (including engine mount number two) to 28–38 ft. lbs. (38–51 Nm).

47. Apply grease to the end of the joint shaft and the end of the driveshaft.

48. Remove the plugs and install the driveshaft. Install a new clip with the end gap facing upward. Push the driveshaft into the joint shaft. Pull the hub outward to confirm that the driveshaft is held securely.

49. Install the lower arm ball joint to the knuckle, and tighten the through-bolt to 32–43 ft. lbs. (44–58 Nm).

50. Install the tie-rod end to the knuckle, and tighten to 32–41 ft. lbs. (43–56 Nm). Install a new cotter pin.

51. Install the stabilizer control link, and tighten to 32–44 ft. lbs. (44–60 Nm).

52. On MTX, install the clutch slave cylinder. and tighten the bolts to 14–18 ft. lbs. (19–25 Nm).

53. On MTX, connect the extension bar to the transaxle with the washer and nut. Tighten the nut to 38 ft. lbs. (51 Nm). Connect the control rod to the transaxle with the through-bolt and nut. Tighten the through-bolt to 12–16 ft. lbs. (16–22 Nm).

54. On ATX, install the selector cable, and tighten the mounting nut to 12–16 ft. lbs. (16–22 Nm).

55. Connect the heater hoses.

56. Connect the fuel lines to the fuel rail.

57. Connect the engine wiring harness connectors including: the electrical connectors from the distributor, mass airflow sensor, intake air temperature sensor, engine coolant temperature sensor, cooling fan temperature sensor, coolant temperature gauge sensor, throttle position sensor, air bypass valve, idle switch, fuel injectors, EGR solenoids and alternator.

58. If equipped, install the A/C compressor on the mounting bracket and

tighten the bolts to 26 ft. lbs. (35 Nm).

59. Install the alternator belt and adjust the tension. Tighten the alternator upper mounting bolt to 18 ft. lbs. (25 Nm).

60. Raise and safely support the vehicle. Tighten the alternator through-bolt to 38 ft. lbs. (52 Nm).

61. Install the splash shields and lower the vehicle.

62. Loosely install the power steering pump through-bolt and lockbolt. Connect the power steering switch connector and install the power steering belt.

63. Adjust the power steering belt tension, then tighten the through-bolt to 38 ft. lbs. (51 Nm) and the lockbolt to 34 ft. lbs. (46 Nm).

64. Install the power steering pump belt shield and tighten the bolts to 86 inch lbs. (9 Nm). Install the power steering hose brackets to the cylinder head cover and tighten the bolts to 88 inch lbs. (10 Nm).

65. Install and adjust the accelerator and throttle cables.

66. Install the resonance chamber.

67. Install the battery duct bracket and coolant reservoir.

68. Install the battery tray, clamp and battery.

69. Install the air cleaner assembly.

70. Install the exhaust manifold.

71. Install the exhaust pipe to the catalytic converter and tighten the new nuts to 24–33 ft. lbs. (32–46 Nm). Attach the exhaust pipe support bracket to the engine and tighten the bolts to 24–33 ft. lbs. (32–46 Nm).

72. Install the new exhaust pipe-to-exhaust manifold gasket and nuts, and tighten to 24–33 ft. lbs. (32–46 Nm). Connect the oxygen sensor connector and lower the vehicle.

73. Install the radiator and coolant fan.

74. Connect the radiator hoses, coolant reservoir hose and the cooling fan motor connector. On ATX, connect the oil cooler lines.

75. Install the shroud upper panel.

76. Install the front wheels.

77. Install the hood, aligning the marks that were made during removal.

78. Connect the battery cables.

79. Fill the engine with the proper type and quantity of oil. Fill and bleed the cooling system.

80. Start the engine and bring to normal operating temperature. Check for leaks and proper engine operation.

81. Inspect the ignition timing and idle speed.

1995 MX3 1.6L (B6-ZE), 1994–95 MX3 1.8L (K8) and 1993–97 626 and MX6 2.5L (KL) Engines

1. Relieve the fuel system pressure and disconnect the battery cables. Remove the battery and battery tray.

2. Mark the position of the hood on its hinges and remove the hood.

3. Drain the cooling system, engine oil and transaxle oil.

4. Slightly raise the vehicle and remove the front wheels.

5. Remove the splash shield.

6. Remove the fresh air dust, air cleaner assembly and coolant reservoir.

7. Disconnect the accelerator cable.

8. Remove the upper and lower radiator hose and, on automatic transaxles (ATX) only, the oil cooler hose.

9. Disconnect the cooling fan motor connector, and remove the coolant reservoir pipe.

10. Remove the radiator bracket. Remove the radiator and cooling fan assembly.

11. Loosen the locknut and adjuster bolt on the power steering belt tensioner and remove the belt. Remove the 3 power steering pump mounting bolts working through the pulley holes.

12. Remove the rear bracket bolt from the power steering pump and secure the pump aside with mechanics wire.

13. If equipped, remove the 4 A/C compressor mounting bolts and secure the compressor aside with mechanics wire, leaving the refrigerant lines attached. Do not let the compressor hang by the refrigerant lines.

14. Remove the power steering hose bracket from the pump. Loosen the alternator belt tensioner locknut and adjuster bolt and remove the belt.

15. If equipped, disconnect the cruise control actuator electrical connector. Remove the 2 nuts from the actuator bracket and position the actuator and bracket aside.

16. On ATX, disconnect the shift linkage from the transaxle.

17. Label and disconnect the electrical connectors from the distributor, mass airflow sensor, intake air temperature sensor, engine coolant temperature sensor, cooling fan temperature sensor, coolant temperature gauge sensor, throttle position sensor, air bypass valve, idle switch, fuel injectors, EGR solenoids and alternator. Label and disconnect any remaining engine harness connectors.

18. Label, disconnect and plug the fuel lines at the fuel rail.

19. If equipped, disconnect the cruise control vacuum hose from the back right-hand side of the intake manifold.

20. Disconnect the vacuum line connecting the evaporative canister and the metal EGR vacuum line.

21. Disconnect the power booster vacuum line from the back left-hand side of the intake manifold.

22. Disconnect the heater hoses and remove the upper starter mounting bolts.

23. Raise and safely support the vehicle. Attach suitable engine lifting equipment to the lifting eyes on the engine.

24. On the 2.5L engine, remove the fuel filter and bracket.

25. On manual transaxles (MTX), remove the bolts and clips from the clutch slave cylinder. Remove the slave cylinder with the hose still connected, and position it away from the transaxle.

26. Remove the transverse member.

27. Disconnect the oxygen sensor connector. Remove and discard the exhaust pipe-to-catalytic converter nuts.

28. Remove the exhaust support bolts. Remove and discard the exhaust pipe-to-exhaust manifold nuts and remove the exhaust pipe. Support the remaining exhaust system with mechanics wire.

29. On MTX, remove the extension bar nut and washer, then disengage the bar from the transaxle. Remove the transaxle shift linkage through-bolt and nut, then disengage the linkage from the transaxle.

30. Remove the stabilizer control link and the tie-rod end ball joint.

31. Remove the bolts and nuts from the lower arm ball joints, and pry the lower arms from the knuckles.

32. Separate the driveshafts and the joint shaft from the transaxle. Install transaxle plug tools T88C–7025–AH or equivalent, into the differential side gears.

NOTE: If the plugs are not installed, the differential side gears may become mispositioned. If the gears are is positioned, the differential may have to be removed to reposition them.

33. On the 1.6L B6E–ZE) engine remove the transverse (engine mount) member, and the number 2 engine mount nuts. Remove the number one, three and four engine mounts.

34. On the 1.8L (K8) engine, remove the number three and four en-

gine mounts. Remove the cross-member. Remove the nuts from the engine mount number two.

35. On the 2.5L engine, remove the fuse box, engine mounts Nos. 1, 3 and 4. Remove the crossmember, then remove the nuts from the engine mount No. 2, and remove engine mount No. 5.

36. Lower the vehicle. Slowly lift the engine and transaxle assembly as a unit out of the vehicle. Keep the engine from swinging or bumping into components in the engine compartment.

37. Remove the starter. Remove the engine mount number two.

38. On ATX, remove the throttle cable and torque converter-to-flexplate nuts.

39. Remove the engine-to-transaxle bolts and the transaxle-to-engine mounting bolts.

40. Separate the engine from the transaxle.

41. On MTX, remove the clutch assembly, flywheel and crankshaft rear cover plate. On ATX, remove the flexplate.

42. Place the engine on a workstand.

To install:

1.6L (B6-ZE) Engine

1. Remove the engine from the workstand.

2. On ATX, remove the old sealant from the flexplate mounting bolts and bolt holes. If reusing the flexplate bolts, apply silicone sealant to the bolt threads. Install the flexplate and loosely install the bolts. Tighten the flexplate bolts in 2–3 steps to 75 ft. lbs. (103 Nm) in a crisscross pattern.

NOTE: New flexplate mounting bolts come with sealant already on them.

3. On MTX, install the crankshaft rear cover plate and tighten the bolt to 88 inch lbs. (10 Nm). Install the flywheel and clutch assembly.

4. On ATX:
   a. Install the transaxle-to-engine bolts and tighten mounting bolts **A** to 41–59t. lbs. (55–80m).
   b. Install the torque converter to the flexplate and tighten the nuts to 26–36 ft. lbs. (35–49 Nm). Rotate the flexplate, as necessary, to gain access to all of the nuts.
   c. Install the remaining transaxle-to-engine mounting bolts. Tighten mounting bolt **B** to 28–38 ft. lbs. (38–51 Nm).

5. On MTX, install the transaxle mounting bolts and tighten bolts **A** to 48–65 ft. lbs. (64–89 Nm), and bolts **B** to 28–38 ft. lbs. (38–51 Nm).

6. Install engine mount number two onto the transaxle and tighten to 28–38 ft. lbs. (38–51 Nm).

7. Install the starter and tighten the bolts to 28–38 ft. lbs. (38–51 Nm). Install the intake manifold support bracket and tighten the bolts to 28–38 ft. lbs. (38–51 Nm).

8. Suspend the engine and transaxle assembly. Slowly lower and position the assembly into the engine compartment. Align the engine mounts with the mount holes. Install the engine mounts.

9. Install a new clip on the joint shaft and install the shaft. Tighten the three mounting bolts to 32–45 ft. lbs. (43–61 Nm). Apply grease to the end of the joint shaft and the end of the driveshaft.

10. Install the transverse member. Tighten the mounting bolts to 48–65 ft. lbs. (64–89 Nm), and the nuts (including engine mount number two) to 28–38 ft. lbs. (38–51 Nm).

11. Remove the plugs and install the driveshaft. Install a new clip with the end gap facing upward. Push the driveshaft into the joint shaft. Pull the hub outward to confirm that the driveshaft is held securely.

12. Install the lower arm ball joint to the knuckle, and tighten the through-bolt to 32–43 ft. lbs. (44–58 Nm).

13. Install the tie-rod end to the knuckle, and tighten to 32–41 ft. lbs. (43–56 Nm). Install a new cotter pin.

14. Install the stabilizer control link, and tighten to 32–44 ft. lbs. (44–60 Nm).

15. On MTX, connect the extension bar to the transaxle with the washer and nut. Tighten the nut to 38 ft. lbs. (51 Nm). Connect the shift linkage to the transaxle with the through-bolt and nut. Tighten the through-bolt to 18 ft. lbs. (25 Nm).

16. Install the exhaust pipe to the catalytic converter and tighten the new nuts to 24–33 ft. lbs. (32–46 Nm). Attach the exhaust pipe support bracket to the engine and tighten the bolts to 24–33 ft. lbs. (32–46 Nm).

17. Install the new exhaust pipe-to-exhaust manifold gasket and nuts, and tighten to 24–33 ft. lbs. (32–46 Nm). Connect the oxygen sensor connector and lower the vehicle.

18. On MTX, install the clutch slave cylinder. and tighten the bolts to 12–16 ft. lbs. (16–22 Nm).

19. Connect the heater hoses. If equipped, connect the cruise control vacuum line to the back right-hand side of the intake manifold.

20. Connect the vacuum line connecting the evaporative canister to the metal EGR vacuum line. Connect the power brake booster vacuum line to the back left-hand side of the intake manifold.

21. Connect the fuel lines to the fuel rail.

22. Connect the engine wiring harness connectors including: the electrical connectors from the distributor, mass airflow sensor, intake air temperature sensor, engine coolant temperature sensor, cooling fan temperature sensor, coolant temperature gauge sensor, throttle position sensor, air bypass valve, idle switch, fuel injectors, EGR solenoids and alternator.

23. If equipped, install the A/C compressor on the mounting bracket and tighten the bolts to 26 ft. lbs. (35 Nm).

24. Install the alternator belt and adjust the tension. Tighten the alternator upper mounting bolt to 18 ft. lbs. (25 Nm).

25. Raise and safely support the vehicle. Tighten the alternator through-bolt to 38 ft. lbs. (52 Nm).

26. Install the splash shields and lower the vehicle.

27. Loosely install the power steering pump through-bolt and lockbolt. Connect the power steering switch connector and install the power steering belt.

28. Adjust the power steering belt tension, then tighten the through-bolt to 38 ft. lbs. (51 Nm) and the lockbolt to 34 ft. lbs. (46 Nm).

29. Install the power steering pump belt shield and tighten the bolts to 86 inch lbs. (9 Nm). Install the power steering hose brackets to the cylinder head cover and tighten the bolts to 88 inch lbs. (10 Nm).

30. Install the radiator and coolant fan.

31. Connect the radiator hoses, coolant reservoir hose and the cooling fan motor connector. On ATX, connect the oil cooler lines.

32. Install and adjust the accelerator cable. Install the air cleaner assembly.

33. Install the shroud upper panel and the fresh air duct.

34. Install the battery tray and battery. Connect the battery cables.

35. Install the front wheels.

36. Install the hood, aligning the marks that were made during removal.

37. Fill the engine with the proper type and quantity of oil. Fill and bleed the cooling system.

38. Start the engine and bring to normal operating temperature. Check for leaks and proper engine operation.

1.8L (K8) Engine

1. Remove the engine from the workstand.

2. On ATX, remove the old sealant from the flexplate mounting bolts and bolt holes. If reusing the flexplate bolts, apply silicone sealant to the bolt threads. Install the flexplate and loosely install the bolts. Tighten the flexplate bolts in 2–3 steps to 75 ft. lbs. (103 Nm) in a crisscross pattern.

NOTE: New flexplate mounting bolts come with sealant already on them.

3. On MTX, install the crankshaft rear cover plate and tighten the bolt to 88 inch lbs. (10 Nm). Install the flywheel and clutch assembly.

4. On ATX:

a. Install the transaxle-to-engine bolts and tighten the mounting bolts to 50–73 ft. lbs. (68–99 Nm).

b. Install the torque converter to the flexplate and tighten the nuts to 28–45 ft. lbs. (38–60 Nm). Rotate the flexplate, as necessary, to gain access to all of the nuts.

5. On MTX, install the transaxle mounting bolts and tighten to 50–73 ft. lbs. (68–99 Nm).

6. Install engine mount number two onto the transaxle and tighten to 28–38 ft. lbs. (38–51 Nm).

7. On ATX, install the throttle cable.

8. Install the starter and tighten the bolts to 28–38 ft. lbs. (38–51 Nm).

9. Suspend the engine and transaxle assembly. Slowly lower and position the assembly into the engine compartment. Align the engine mounts with the mount holes.

10. Install the crossmember, and tighten the four mounting bolts to 48–65 ft. lbs. (64–89 Nm). Install the engine mounts.

11. Install a new clip on the joint shaft and install the shaft. Tighten the three mounting bolts to 32–45 ft. lbs. (43–61 Nm). Apply grease to the end of the joint shaft and the end of the driveshaft.

12. Remove the plugs and install the driveshaft. Install a new clip with the end gap facing upward. Push the driveshaft into the joint shaft. Pull the hub outward to confirm that the driveshaft is held securely.

13. Install the lower arm ball joints to the knuckles, and tighten the through-bolt to 32–43 ft. lbs. (44–58 Nm).

14. Install the tie-rod end to the knuckle, and tighten to 32–41 ft. lbs. (43–56 Nm). Install a new cotter pin.

15. Install the stabilizer control link, and tighten to 32–44 ft. lbs. (44–60 Nm).

16. On MTX, connect the extension bar to the transaxle with the washer and nut. Tighten the nut to 24–33 ft. lbs. (32–46 Nm). Connect the shift linkage to the transaxle with the through-bolt and nut. Tighten the through-bolt to 12–16 ft. lbs. (16–22 Nm).

17. Install the exhaust pipe to the catalytic converter and tighten the new nuts to 48–65 ft. lbs. (64–89 Nm).

18. Install the new exhaust pipe-to-exhaust manifold gasket and nuts and tighten to 30–40 ft. lbs. (41–54 Nm). Connect the oxygen sensor connector and lower the vehicle.

19. Install the transverse member and tighten the four mounting bolts to 69–93 ft. lbs. (94–126 Nm).

20. On MTX, install the clutch slave cylinder. and tighten the bolts to 12–16 ft. lbs. (16–22 Nm).

21. If equipped, align the cruise control actuator and install the nuts. Connect the actuator electrical connector.

22. Connect the heater hoses. If equipped, connect the cruise control vacuum line to the back right-hand side of the intake manifold.

23. Connect the vacuum line connecting the evaporative canister to the metal EGR vacuum line. Connect the power brake booster vacuum line to the back left-hand side of the intake manifold.

24. Connect the fuel lines to the fuel rail.

25. Connect the engine wiring harness connectors including: the electrical connectors from the distributor, mass airflow sensor, intake air temperature sensor, engine coolant temperature sensor, cooling fan temperature sensor, coolant temperature gauge sensor, throttle position sensor, air bypass valve, idle switch, fuel injectors, EGR solenoids and alternator.

26. If equipped, install the A/C compressor on the mounting bracket and tighten the bolts to 18–26 ft. lbs. (24–35 Nm).

27. Install the alternator belt and adjust the tension. Tighten the alternator upper mounting bolt to 18 ft. lbs. (25 Nm).

28. Raise and safely support the vehicle. Tighten the alternator through-bolt to 38 ft. lbs. (52 Nm).

29. Install the splash shields and lower the vehicle.

30. Align the power steering pump and install the rear bracket bolt. Tighten to 34 ft. lbs. (46 Nm). Install the power steering hose bracket bolt and tighten to 34 ft. lbs. (46 Nm).

31. Install the 3 power steering pump bolts through the pulley and tighten to 34 ft. lbs. (46 Nm). Install the power steering belt and adjust the tension.

32. Install the radiator and coolant fan.

33. Connect the radiator hoses, coolant reservoir hose. Connect the cooling fan motor connector. On ATX, connect the oil cooler lines.

34. Install and adjust the accelerator cable. Install the air cleaner, fresh air ducts and coolant reservoir.

35. Install the battery tray and battery. Connect the battery cables.

36. Install the front wheels.

37. Install the hood, aligning the marks that were made during removal.

38. Fill the engine with the proper type and quantity of oil. Fill and bleed the cooling system.

39. Start the engine and bring to normal operating temperature. Check for leaks and proper engine operation.

1993–97 626 and MX6 2.5L (KL) Engine

1. Remove the engine from the workstand.

2. On ATX, remove the old sealant from the flexplate mounting bolts and bolt holes. If reusing the flexplate bolts, apply silicone sealant to the bolt threads. Install the flexplate and loosely install the bolts. Tighten the flexplate bolts in 2–3 steps to 75 ft. lbs. (103 Nm) in a crisscross pattern.

NOTE: New flexplate mounting bolts come with sealant already on them.

3. On MTX, install the crankshaft rear cover plate and tighten the bolt to 88 inch lbs. (10 Nm). Install the flywheel and clutch assembly.

4. On ATX:

a. Install the transaxle-to-engine bolts and tighten the mounting bolts to 50–73 ft. lbs. (68–99 Nm).

b. Install the torque converter to the flexplate and tighten the nuts to 28–45 ft. lbs. (38–60 Nm). Rotate the flexplate, as necessary, to gain access to all of the nuts.

5. On MTX, install the transaxle mounting bolts and tighten to 50–73 ft. lbs. (68–99 Nm).

6. Install engine mount number two onto the transaxle and tighten to 28–38 ft. lbs. (38–51 Nm).

7. Install the starter and tighten the bolts to 28–38 ft. lbs. (38–51 Nm).

8. Suspend the engine and transaxle assembly. Slowly lower and position the assembly into the engine compartment. Align the engine mounts with the mount holes.

9. Install the crossmember, and tighten the four mounting bolts to 50–68 ft. lbs. (67–93 Nm). Install the engine mounts.

10. Install the fuse box.

11. Install a new clip on the joint shaft and install the shaft. Tighten the three mounting bolts to 32–45 ft. lbs. (43–61 Nm). Apply grease to the end of the joint shaft and the end of the driveshaft.

12. Remove the plugs and install the driveshaft. Install a new clip with the end gap facing upward. Push the driveshaft into the joint shaft. Pull the hub outward to confirm that the driveshaft is held securely.

13. Install the lower arm ball joints to the knuckles, and tighten the through-bolt to 32–43 ft. lbs. (44–58 Nm).

14. Install the tie-rod end to the knuckle, and tighten to 32–41 ft. lbs. (43–56 Nm). Install a new cotter pin.

15. Install the stabilizer control link, and tighten to 32–44 ft. lbs. (44–60 Nm).

16. On MTX, connect the extension bar to the transaxle with the washer and nut. Tighten the nut to 24–33 ft. lbs. (32–46 Nm). Connect the shift linkage to the transaxle with the through-bolt and nut. Tighten the through-bolt to 12–16 ft. lbs. (16–22 Nm).

17. Install the exhaust pipe to the catalytic converter and tighten the new nuts to 48–65 ft. lbs. (64–89 Nm).

18. Install the new exhaust pipe-to-exhaust manifold gasket and nuts and tighten to 28–38 ft. lbs. (38–51 Nm). Connect the oxygen sensor connector and lower the vehicle.

19. Install the transverse member and tighten the six mounting bolts to 69–93 ft. lbs. (94–126 Nm).

20. On MTX, install the clutch slave cylinder. and tighten the bolts to 12–16 ft. lbs. (16–22 Nm).

21. If equipped, align the cruise control actuator and install the nuts. Connect the actuator electrical connector.

22. Install the fuel filter.

23. Connect the heater hoses. If equipped, connect the cruise control vacuum line to the back right-hand side of the intake manifold.

24. Connect the vacuum line connecting the evaporative canister to the metal EGR vacuum line. Connect the power brake booster vacuum line to the back left-hand side of the intake manifold.

25. Connect the fuel lines to the fuel rail.

26. Connect the engine wiring harness connectors including: the electrical connectors from the distributor, mass airflow sensor, intake air temperature sensor, engine coolant temperature sensor, cooling fan temperature sensor, coolant temperature gauge sensor, throttle position sensor, air bypass valve, idle switch, fuel injectors, EGR solenoids and alternator.

27. If equipped, install the A/C compressor on the mounting bracket and tighten the bolts to 18–26 ft. lbs. (24–35 Nm).

28. Install the alternator belt and adjust the tension. Tighten the alternator upper mounting bolt to 18 ft. lbs. (25 Nm).

29. Raise and safely support the vehicle. Tighten the alternator through-bolt to 38 ft. lbs. (52 Nm).

30. Install the splash shields and lower the vehicle.

31. Align the power steering pump and install the rear bracket bolt. Tighten to 34 ft. lbs. (46 Nm). Install the power steering hose bracket bolt and tighten to 34 ft. lbs. (46 Nm).

32. Install the three power steering pump bolts through the pulley and tighten to 34 ft. lbs. (46 Nm). Install the power steering belt and adjust the tension.

33. Install the radiator and coolant fan.

34. Connect the radiator hoses, coolant reservoir hose. Connect the condenser and cooling fan motor connectors. On ATX, connect the oil cooler lines.

35. Install and adjust the accelerator cable. Install the air cleaner, fresh air ducts and air hose.

36. Install the battery tray and battery. Connect the battery cables.

37. Install the front wheels.

38. Install the hood, aligning the marks that were made during removal.

39. Fill the engine with the proper type and quantity of oil. Fill and bleed the cooling system.

40. Start the engine and bring to normal operating temperature. Check for leaks and proper engine operation.

626 and MX6 2.0L (FS) Engines

1. Relieve the fuel system pressure and disconnect the battery cables. Remove the battery and battery tray.

2. Mark the position of the hood on its hinges and remove the hood.

3. Drain the cooling system, engine oil and transaxle oil.

4. Slightly raise the vehicle and remove the front wheels.

5. Remove the splash shield.

6. Remove the fresh air dust, air cleaner assembly, air hose and resonance chamber.

7. Label and disconnect the electrical connectors from the distributor, mass airflow sensor, intake air temperature sensor, engine coolant temperature sensor, cooling fan temperature sensor, coolant temperature gauge sensor, throttle position sensor, air bypass valve, idle switch, fuel injectors, EGR solenoids and alternator.

8. Disconnect the accelerator cable.

9. Remove the fuel filter.

10. Remove the upper and lower radiator hose and, on automatic transaxles (ATX) only, the oil cooler hose.

11. Disconnect the cooling fan motor connector, and remove the coolant reservoir pipe.

12. Remove the radiator bracket. Remove the radiator and cooling fan assembly.

13. Remove the power steering belt shield and the power steering belt. Remove the power steering hose brackets from the cylinder head cover.

14. Remove the power steering belt adjuster and disconnect the power steering pressure switch connector. Remove the power steering pump and position aside, leaving the hoses connected.

15. If equipped, remove the A/C compressor and position aside, leaving the refrigerant lines attached. Support the compressor with suitable wire.

16. Loosen the alternator adjusting bolt, remove the upper mounting bolt and remove the alternator belt.

17. Disconnect and label any remaining engine harness connectors.

18. Label, disconnect and plug the fuel lines at the fuel rail.

19. If equipped, disconnect the cruise control vacuum hose from the back right-hand side of the intake manifold.

20. Disconnect the vacuum line connecting the evaporative canister and the metal EGR vacuum line.

21. Disconnect the power booster vacuum line from the back left-hand side of the intake manifold.

22. Disconnect the heater hoses and remove the upper starter mounting bolts. Raise and safely support the

vehicle. Attach suitable engine lifting equipment to the lifting eyes on the engine.

23. On manual transaxles (MTX), remove the bolts and clips from the clutch slave cylinder. Remove the slave cylinder with the hose still connected, and position it away from the transaxle.

24. On ATX, disconnect the shift control cable from the transaxle.

25. Remove the transverse member, the No. 5 engine mount rubber and the No. 2 engine mount nuts.

26. Remove the crossmember.

27. Disconnect the oxygen sensor connector. Remove and discard the exhaust pipe-to-catalytic converter nuts.

28. Remove the exhaust support bolts. Remove and discard the exhaust pipe-to-exhaust manifold nuts and remove the exhaust pipe. Support the remaining exhaust system with mechanics wire.

29. On MTX, remove the extension bar nut and washer, then disengage the bar from the transaxle. Remove the transaxle shift linkage through-bolt and nut, then disengage the linkage from the transaxle.

30. Remove the stabilizer control link and the tie-rod end ball joint.

31. Remove the bolts and nuts from the lower arm ball joints, and pry the lower arms from the knuckles.

32. Separate the driveshafts and the joint shaft from the transaxle. Install transaxle plug tools T88C–7025–AH or equivalent, into the differential side gears.

NOTE: If the plugs are not installed, the differential side gears may become mispositioned. If the gears are mispositioned, the differential may have to be removed to reposition them.

33. Remove the number one, three and four engine mounts.

34. Lower the vehicle. Slowly lift the engine and transaxle assembly as a unit out of the vehicle. Keep the engine from swinging or bumping into components in the engine compartment.

35. Remove the starter and the intake manifold support bracket. Remove the engine mount number two.

36. On ATX, remove the torque converter-to-flexplate nuts.

37. Remove the engine-to-transaxle bolts and the transaxle-to-engine mounting bolts.

38. Separate the engine from the transaxle.

39. On MTX, remove the clutch assembly, flywheel and crankshaft rear

cover plate. On ATX, remove the flexplate.

40. Place the engine on a workstand.

To install:

41. Remove the engine from the workstand.

42. On ATX, remove the old sealant from the flexplate mounting bolts and bolt holes. If reusing the flexplate bolts, apply silicone sealant to the bolt threads. Install the flexplate and loosely install the bolts. Tighten the flexplate bolts in 2–3 steps to 75 ft. lbs. (103 Nm) in a crisscross pattern.

NOTE: New flexplate mounting bolts come with sealant already on them.

43. On MTX, install the crankshaft rear cover plate and tighten the bolt to 88 inch lbs. (10 Nm). Install the flywheel and clutch assembly.

44. On ATX:

  a. Install the transaxle-to-engine bolts and tighten mounting bolts **A** to 66–86 ft. lbs. (90–116 Nm).

  b. Install the torque converter to the flexplate and tighten the nuts to 21–33 ft. lbs. (28–46 Nm). Rotate the flexplate, as necessary, to gain access to all of the nuts.

  c. Install the remaining transaxle-to-engine mounting bolts. Tighten mounting bolt **B** to 28–38 ft. lbs. (38–51 Nm) and mounting bolt **C** to 14–18 ft. lbs. (19–25 Nm).

45. On MTX, install the transaxle mounting bolts and tighten to 66–86 ft. lbs. (90–116 Nm).

46. Install engine mount number two onto the transaxle and tighten to 32–44 ft. lbs. (44–60 Nm).

47. Install the starter and tighten the bolts to 28–38 ft. lbs. (38–51 Nm). Install the intake manifold support bracket and tighten the bolts to 28–38 ft. lbs. (38–51 Nm).

48. Suspend the engine and transaxle assembly. Slowly lower and position the assembly into the engine compartment. Align the engine mounts with the mount holes. Install the engine mounts.

49. Install a new clip on the joint shaft and install the shaft. Tighten the three mounting bolts to 32–45 ft. lbs. (43–61 Nm). Apply grease to the end of the joint shaft and the end of the driveshaft.

50. Remove the plugs and install the driveshaft. Install a new clip with the end gap facing upward. Push the driveshaft into the joint shaft. Pull the hub outward to confirm that the driveshaft is held securely.

51. Install the lower arm ball joint to the knuckle, and tighten the

through-bolt to 28–41 ft. lbs. (35–56 Nm).

52. Install the tie-rod end to the knuckle, and tighten to 24–32 ft. lbs. (32–44 Nm). Install a new cotter pin.

53. Install the stabilizer control link, and tighten to 27–39 ft. lbs. (37–53 Nm).

54. On MTX, connect the extension bar to the transaxle with the washer and nut. Tighten the nut to 38 ft. lbs. (51 Nm). Connect the shift linkage to the transaxle with the through-bolt and nut. Tighten the through-bolt to 18 ft. lbs. (25 Nm).

55. Install the exhaust pipe to the catalytic converter and tighten the new nuts to 28–38 ft. lbs. (38–51 Nm). Attach the exhaust pipe support bracket to the engine and tighten the bolts to 28–38 ft. lbs. (38–51 Nm).

56. Install the new exhaust pipe-to-exhaust manifold gasket and nuts and tighten to 28–38 ft. lbs. (38–51 Nm). Connect the oxygen sensor connector and lower the vehicle.

57. Install the crossmember, and tighten the six mounting bolts to 69–97 ft. lbs. (94–131 Nm).

58. On MTX, install the clutch slave cylinder. and tighten the bolts to 12–16 ft. lbs. (16–22 Nm).

59. Connect the heater hoses. If equipped, connect the cruise control vacuum line to the back right-hand side of the intake manifold.

60. Connect the vacuum line connecting the evaporative canister to the metal EGR vacuum line. Connect the power brake booster vacuum line to the back left-hand side of the intake manifold.

61. Connect the fuel lines to the fuel rail.

62. Connect the engine wiring harness connectors including: the electrical connectors from the distributor, mass airflow sensor, intake air temperature sensor, engine coolant temperature sensor, cooling fan temperature sensor, coolant temperature gauge sensor, throttle position sensor, air bypass valve, idle switch, fuel injectors, EGR solenoids and alternator.

63. If equipped, install the A/C compressor on the mounting bracket and tighten the bolts to 26 ft. lbs. (35 Nm).

64. Install the alternator belt and adjust the tension. Tighten the alternator upper mounting bolt to 18 ft. lbs. (25 Nm).

65. Raise and safely support the vehicle. Tighten the alternator through-bolt to 38 ft. lbs. (52 Nm).

66. Install the splash shields and lower the vehicle.

67. Loosely install the power steering pump through-bolt and lockbolt. Connect the power steering switch connector and install the power steering belt.

68. Adjust the power steering belt tension, then tighten the through-bolt to 45 ft. lbs. (61 Nm) and the lockbolt to 34 ft. lbs. (46 Nm).

69. Install the power steering pump belt shield and tighten the bolts to 86 inch lbs. (9 Nm). Install the power steering hose brackets to the cylinder head cover and tighten the bolts to 88 inch lbs. (10 Nm).

70. Install the radiator and coolant fan.

71. Connect the radiator hoses, coolant reservoir hose and the cooling fan motor connector. On ATX, connect the oil cooler lines.

72. Install the fuel filter.

73. Install and adjust the accelerator cable. Install the air cleaner, air ducts and resonance chamber.

74. Install the battery tray and battery. Connect the battery cables.

75. Install the front wheels.

76. Install the hood, aligning the marks that were made during removal.

77. Fill the engine with the proper type and quantity of oil. Fill and bleed the cooling system.

78. Start the engine and bring to normal operating temperature. Check for leaks and proper engine operation.

1993–95 929 3.0L (JE-ZE) Engines

1. Properly relieve the fuel system pressure. Disconnect the negative battery cable.

2. Mark the position of the hood on the hinges and remove the hood.

3. Remove the air cleaner assembly and the air ducts.

4. Remove the battery and battery tray.

5. Raise and safely support the vehicle. Remove the front wheels and splash shield.

6. Drain the cooling system, and the engine and transmission oil.

7. Remove the cooling fan and fan shroud.

8. Disconnect the radiator hoses and transmission cooler lines, and remove the radiator.

9. Remove the accessory drive belts.

10. Remove the A/C compressor and position aside, leaving the refrigerant lines attached.

11. Remove the alternator.

12. Remove the power steering pump pulley, using holder tool

49–W023–585A or equivalent, to hold the pulley while the nut is removed. Remove the power steering pump and position aside, leaving the hoses attached.

13. Disconnect the accelerator cable.

14. Label and disconnect the engine electrical harness, vacuum hoses, heater hoses and fuel lines.

15. Remove the dynamic chamber (upper intake manifold) and throttle body assembly.

16. Disconnect the exhaust pipes from the exhaust manifolds.

17. Remove the engine mount nuts.

18. Remove the starter, and disconnect the transmission from the engine.

19. Attach suitable lifting equipment to the engine. Carefully remove the engine from the vehicle and position on a workstand.

To install:

20. Carefully lower the engine into the engine compartment. Loosely install the engine mount nuts.

21. Connect the transmission. After the transmission and engine are bolted together, tighten the engine mount nuts to 36 ft. lbs. (49 Nm).

22. Install the starter.

23. Using new gaskets, connect the exhaust pipes to the exhaust manifolds. Tighten the nuts to 34 ft. lbs. (46 Nm).

24. Connect the electrical connectors, vacuum hoses, heater hoses and fuel lines.

25. Use a new gasket and install the dynamic chamber (upper intake manifold) and throttle body assembly. Tighten the bolts to 19 ft. lbs. (25 Nm).

26. Connect the accelerator cable, and the electrical connectors and hoses to the upper intake manifold and throttle body assembly.

27. Install the power steering pump, and tighten the mounting bolts to 34 ft. lbs. (46 Nm). Install the pump pulley, and tighten the nut to 43 ft. lbs. (59 Nm).

28. Install the alternator, and tighten the mounting bolts to 38 ft. lbs. (52 Nm).

29. Install the A/C compressor, and tighten the mounting bolts to 38 ft. lbs. (52 Nm).

30. Install the accessory drive belts and adjust the belt tension.

31. Install the radiator and tighten the mounting bolts to 19 ft. lbs. (26 Nm). Connect the radiator hoses and the transmission oil cooler lines.

32. Install the cooling fan and shroud.

33. Install the air cleaner assembly and ducts. Install the battery tray and battery.

34. Install the splash shield and wheels, and lower the vehicle.

35. Install the hood, aligning the marks that were made during removal.

36. Connect the battery cables. Fill the engine and transmission with the proper type and quantity of oil. Fill and bleed the cooling system.

37. Start the engine and bring to normal operating temperature. Check for leaks.

38. Check the ignition timing and idle speed. Check all fluid levels and road test the vehicle for proper operation.

1995–97 Millenia 2.3L (KJS) Engines

1. Relieve the fuel system pressure, and disconnect the battery cables.

2. Mark the position of the hood on its hinges and remove the hood.

3. Drain the cooling system, engine oil and transaxle oil.

4. Slightly raise the vehicle and remove the front wheels.

5. Remove the dynamic chamber cover.

6. Remove the splash shields.

7. Remove the charge air cooler air duct.

8. Remove the air cleaner assembly and resonator.

9. Remove the battery clamp, battery and battery tray.

10. Remove the battery carrier and duct.

11. Remove the air cleaner air duct.

12. Disconnect the accelerator cable.

13. Remove the charge air cooler from the back of the engine.

14. Remove the dust cover from the drive belt.

15. Remove the drive belts.

16. Remove the crankshaft pulley.

17. Remove the vacuum pump pulley using a hexagon wrench to hold the pulley. Remove the left-handed thread mounting nut.

18. Remove the four A/C compressor mounting bolts, and secure the compressor aside with mechanics wire, leaving the refrigerant lines attached. Do not let the compressor hang by the refrigerant lines.

19. Remove the power steering pump pulley. Remove the power steering pump and secure the pump aside with mechanics wire.

20. Remove the front charge air cooler.

21. Remove the front radiator grill.

22. Remove the upper seal board (radiator bracket).
23. Remove the coolant reservoir and hose.
24. Disconnect the condenser and cooling fan motor connectors.
25. Remove the condenser and cooling fan motor assemblies.
26. Remove and plug the the oil cooler hose.
27. Remove the upper and lower radiator hoses.
28. Remove the radiator brackets.
29. Remove the radiator.
30. Disconnect the shift linkage from the transaxle.
31. Label and disconnect the electrical connectors from the distributor, mass airflow sensor, intake air temperature sensor, engine coolant temperature sensor, cooling fan temperature sensor, coolant temperature gauge sensor, throttle position sensor, air bypass valve, idle switch, fuel injectors, EGR solenoids and alternator. Label and disconnect any remaining engine harness connectors.
32. Label, disconnect and plug the fuel lines at the fuel rail.
33. Disconnect the heater hoses.
34. Raise and safely support the vehicle. Attach suitable engine lifting equipment to the lifting eyes on the engine.
35. Disconnect the oxygen sensor connector. Remove and discard the exhaust pipe-to-catalytic converter nuts.
36. Remove the exhaust support bolts. Remove and discard the exhaust pipe-to-exhaust manifold nuts and remove the exhaust pipe. Support the remaining exhaust system with mechanics wire.
37. Remove the upper lateral link link ball joint.
38. Remove the bolts and nuts from the lower arm ball joints, and pry the lower arms from the knuckles.
39. Separate the driveshafts and the joint shaft from the transaxle. Install transaxle plug tools T88C-7025-AH or equivalent, into the differential side gears.

NOTE: If the plugs are not installed, the differential side gears may become mispositioned. If the gears are mispositioned, the differential may have to be removed to reposition them.

40. Remove the number one engine mount bracket.
41. Remove the crossmember. Remove the nuts from the engine mount number two.
42. Remove engine mount number four bracket.
43. Remove engine mount number three sub-bracket.
44. Lower the vehicle. Slowly lift the engine and transaxle assembly as a unit out of the vehicle. Keep the engine from swinging or bumping into components in the engine compartment.
45. Remove the starter. Remove the engine mount number two.
46. Remove the torque converter-to-flexplate nuts.
47. Remove the engine-to-transaxle bolts and the transaxle-to-engine mounting bolts.
48. Separate the engine from the transaxle.
49. Remove the flexplate.
50. Place the engine on a workstand.

To install:
51. Remove the engine from the workstand.
52. Remove the old sealant from the flexplate mounting bolts and bolt holes. If reusing the flexplate bolts, apply silicone sealant to the bolt threads. Install the flexplate and loosely install the bolts. Tighten the flexplate bolts in 2–3 steps to 75 ft. lbs. (103 Nm) in a crisscross pattern.

NOTE: New flexplate mounting bolts come with sealant already on them.

53. Install the transaxle-to-engine bolts and tighten the mounting bolts to 50–73 ft. lbs. (68–99 Nm).
54. Install the torque converter to the flexplate and tighten the nuts to 28–45 ft. lbs. (38–60 Nm). Rotate the flexplate, as necessary, to gain access to all of the nuts.
55. Install engine mount number two onto the transaxle and tighten to 28–38 ft. lbs. (38–51 Nm).
56. Install the starter and tighten the bolts to 28–38 ft. lbs. (38–51 Nm).
57. Suspend the engine and transaxle assembly. Slowly lower and position the assembly into the engine compartment. Align the engine mounts with the mount holes.
58. Install the crossmember, and tighten the three mounting bolts to 50–68 ft. lbs. (67–93 Nm).
59. Install the engine mounts.
60. Install a new clip on the joint shaft and install the shaft. Tighten the three mounting bolts to 32–45 ft. lbs. (43–61 Nm). Apply grease to the end of the joint shaft and the end of the driveshaft.
61. Remove the plugs and install the driveshaft. Install a new clip with the end gap facing upward. Push the driveshaft into the joint shaft. Pull the hub outward to confirm that the driveshaft is held securely.

62. Install the lower arm ball joint to the knuckle, and tighten the through-bolt to 58–86 ft. lbs. (79–116 Nm).
63. Install the upper lateral link ball joint using a new cotter pin, and tighten to 41–59 ft. lbs. (55–80 Nm).
64. Install the exhaust pipe to the catalytic converter and tighten the new nuts to 28–38 ft. lbs. (38–51 Nm).
65. Install the new exhaust pipe-to-exhaust manifold gasket and nuts and tighten to 28–38 ft. lbs. (38–51 Nm). Connect the oxygen sensor connector and lower the vehicle.
66. Connect the heater hoses.
67. Connect the fuel lines to the fuel rail.
68. Connect the engine wiring harness connectors including: the electrical connectors from the distributor, mass airflow sensor, intake air temperature sensor, engine coolant temperature sensor, cooling fan temperature sensor, coolant temperature gauge sensor, throttle position sensor, air bypass valve, idle switch, fuel injectors, EGR solenoids and alternator.
69. Connect the shift linkage to the transaxle.
70. Install the radiator.
71. Install the radiator brackets. Tighten the mounting bolts to 70–95 inch lbs. (8–11 Nm).
72. Connect the radiator hoses.
73. Unplug and connect the oil cooler lines.
74. Install the condenser and cooling fan motor assemblies.
75. Connect the condenser and cooling fan motor connectors.
76. Install the coolant reservoir and hose.
77. Install the upper seal board.
78. Install the front radiator grill.
79. Install the front charge air cooler, using new O-rings. Tighten the mounting bolts to 14–18 ft. lbs. (19–25 Nm).
80. Position the power steering pump and install the mounting bolts. Install the pulley and tighten the mounting nut to 37–43 ft. lbs. (50–58 Nm).
81. Position the A/C compressor and install the mounting bolts.
82. Position the vacuum pump pulley. Tighten the left-hand thread mounting nut to 73–101 ft. lbs. (99–137 Nm).
83. Install the crankshaft pulley. Tighten the center mounting bolt to 116–122 ft. lbs. (157–166 Nm), and the surrounding mounting bolts to 19–22 ft. lbs. (26–30 Nm).
84. Install the drive belts.

85. Install the dust cover for the drive belt.

86. Install the rear charge air cooler, using new O-rings. Tighten the mounting bolts to 14–18 ft. lbs. (19–25 Nm).

87. Install and adjust the accelerator cable.

88. Install the air cleaner air duct.

89. Install the battery duct and carrier.

90. Install the battery tray, battery and battery clamp.

91. Install the resonator and air cleaner assembly.

92. Install the charge air cooler air duct.

93. Install the dynamic chamber cover.

94. Install the splash shields and front wheels.

95. Install the hood, aligning the marks that were made during removal.

96. Connect the battery cables.

97. Fill the engine with the proper type and quantity of oil. Fill and bleed the cooling system.

98. Start the engine and bring to normal operating temperature. Check for leaks and proper engine operation.

1995–97 Millenia 2.5L (KLD) Engine

1. Relieve the fuel system pressure, and disconnect the battery cables.

2. Remove the battery clamp, battery and battery tray.

3. Remove the battery carrier and duct.

4. Mark the position of the hood on its hinges and remove the hood.

5. Drain the cooling system, engine oil and transaxle oil.

6. Slightly raise the vehicle and remove the front wheels.

7. Remove the splash shields.

8. Remove the fresh air duct, air cleaner assembly and air hose.

9. Remove the upper seal board (radiator bracket).

10. Remove the coolant reservoir and hose.

11. Disconnect the condenser and cooling fan motor connectors.

12. Remove the condenser and cooling fan motor assemblies.

13. Remove and plug the the oil cooler hose.

14. Remove the upper and lower radiator hoses.

15. Remove the radiator.

16. Disconnect the accelerator cable.

17. Loosen the locknut and adjuster bolt on the power steering belt tensioner and remove the belt. Remove

the three power steering pump mounting bolts working through the pulley holes.

18. Remove the rear bracket bolt from the power steering pump and secure the pump aside with mechanics wire.

19. Remove the four A/C compressor mounting bolts, and secure the compressor aside with mechanics wire, leaving the refrigerant lines attached. Do not let the compressor hang by the refrigerant lines.

20. Remove the power steering hose bracket from the pump. Loosen the alternator belt tensioner locknut and adjuster bolt and remove the belt.

21. Disconnect the shift linkage from the transaxle.

22. Label and disconnect the electrical connectors from the distributor, mass airflow sensor, intake air temperature sensor, engine coolant temperature sensor, cooling fan temperature sensor, coolant temperature gauge sensor, throttle position sensor, air bypass valve, idle switch, fuel injectors, EGR solenoids and alternator. Label and disconnect any remaining engine harness connectors.

23. Label, disconnect and plug the fuel lines at the fuel rail.

24. Disconnect the heater hoses.

25. Raise and safely support the vehicle. Attach suitable engine lifting equipment to the lifting eyes on the engine.

26. Disconnect the oxygen sensor connector. Remove and discard the exhaust pipe-to-catalytic converter nuts.

27. Remove the exhaust support bolts. Remove and discard the exhaust pipe-to-exhaust manifold nuts and remove the exhaust pipe. Support the remaining exhaust system with mechanics wire.

28. Remove the upper lateral link link ball joint.

29. Remove the bolts and nuts from the lower arm ball joints, and pry the lower arms from the knuckles.

30. Separate the driveshafts and the joint shaft from the transaxle. Install transaxle plug tools T88C–7025–AH or equivalent, into the differential side gears.

NOTE: If the plugs are not installed, the differential side gears may become mispositioned. If the gears are mispositioned, the differential may have to be removed to reposition them.

31. Remove the number one engine mount stay (holder) and bracket.

32. Remove the crossmember. Remove the nuts from the engine mount number two.

33. Remove engine mount number four bracket.

34. Remove engine mount number three sub-bracket.

35. Lower the vehicle. Slowly lift the engine and transaxle assembly as a unit out of the vehicle. Keep the engine from swinging or bumping into components in the engine compartment.

36. Remove the starter. Remove the engine mount number two.

37. Remove the torque converter-to-flexplate nuts.

38. Remove the engine-to-transaxle bolts and the transaxle-to-engine mounting bolts.

39. Separate the engine from the transaxle.

40. Remove the flexplate.

41. Place the engine on a workstand.

To install:

42. Remove the engine from the workstand.

43. Remove the old sealant from the flexplate mounting bolts and bolt holes. If reusing the flexplate bolts, apply silicone sealant to the bolt threads. Install the flexplate and loosely install the bolts. Tighten the flexplate bolts in 2–3 steps to 75 ft. lbs. (103 Nm) in a crisscross pattern.

NOTE: New flexplate mounting bolts come with sealant already on them.

44. Install the transaxle-to-engine bolts and tighten the mounting bolts to 50–73 ft. lbs. (68–99 Nm).

45. Install the torque converter to the flexplate and tighten the nuts to 28–45 ft. lbs. (38–60 Nm). Rotate the flexplate, as necessary, to gain access to all of the nuts.

46. Install engine mount number two onto the transaxle and tighten to 28–38 ft. lbs. (38–51 Nm).

47. Install the starter and tighten the bolts to 28–38 ft. lbs. (38–51 Nm).

48. Suspend the engine and transaxle assembly. Slowly lower and position the assembly into the engine compartment. Align the engine mounts with the mount holes.

49. Install the crossmember, and tighten the three mounting bolts to 50–68 ft. lbs. (67–93 Nm).

50. Install the engine mounts.

51. Install a new clip on the joint shaft and install the shaft. Tighten the three mounting bolts to 32–45 ft. lbs. (43–61 Nm). Apply grease to the end of the joint shaft and the end of the driveshaft.

52. Remove the plugs and install the driveshaft. Install a new clip with the end gap facing upward. Push the driveshaft into the joint shaft. Pull

the hub outward to confirm that the driveshaft is held securely.

53. Install the lower arm ball joint to the knuckle, and tighten the through-bolt to 58–86 ft. lbs. (79–116 Nm).

54. Install the upper lateral link ball joint using a new cotter pin, and tighten to 41–59 ft. lbs. (55–80 Nm).

55. Install the exhaust pipe to the catalytic converter and tighten the new nuts to 48–65 ft. lbs. (64–89 Nm).

56. Install the new exhaust pipe-to-exhaust manifold gasket and nuts and tighten to 28–38 ft. lbs. (38–51 Nm). Connect the oxygen sensor connector and lower the vehicle.

57. Connect the heater hoses.

58. Connect the fuel lines to the fuel rail.

59. Connect the engine wiring harness connectors including: the electrical connectors from the distributor, mass airflow sensor, intake air temperature sensor, engine coolant temperature sensor, cooling fan temperature sensor, coolant temperature gauge sensor, throttle position sensor, air bypass valve, idle switch, fuel injectors, EGR solenoids and alternator.

60. Connect the shift linkage to the transaxle.

61. Install the alternator belt and adjust the tension. Tighten the alternator upper mounting bolt to 18 ft. lbs. (25 Nm).

62. Raise and safely support the vehicle. Tighten the alternator through-bolt to 38 ft. lbs. (52 Nm).

63. Lower the vehicle.

64. Install the A/C compressor on the mounting bracket and tighten the bolts to 12–16 ft. lbs. (16–22 Nm).

65. Align the power steering pump and install the rear bracket bolt. Tighten to 34 ft. lbs. (46 Nm). Install the power steering hose bracket bolt and tighten to 34 ft. lbs. (46 Nm).

66. Install the three power steering pump bolts through the pulley and tighten to 34 ft. lbs. (46 Nm). Install the power steering belt and adjust the tension.

67. Install and adjust the accelerator cable.

68. Install the radiator.

69. Connect the radiator hoses.

70. Unplug and connect the oil cooler lines.

71. Install the condenser and cooling fan motor assemblies.

72. Connect the condenser and cooling fan motor connectors.

73. Install the coolant reservoir and hose.

74. Install the upper seal board.

75. Install the air cleaner, fresh air ducts and air hose.

76. Install the battery duct and carrier.

77. Install the battery tray, battery and battery clamp. Connect the battery cables.

78. Install the splash shields and front wheels.

79. Install the hood, aligning the marks that were made during removal.

80. Fill the engine with the proper type and quantity of oil. Fill and bleed the cooling system.

81. Start the engine and bring to normal operating temperature. Check for leaks and proper engine operation.

1993–94 Miata 1.6L (B6-ZE) and 1.8L (BPD) Engines

1. Mark the position of the hood on its hinges and remove the hood.

2. Properly relieve the fuel system pressure.

3. Raise the trunk lid and disconnect the negative battery cable. Remove the fresh air duct and the air cleaner/air flow meter assembly.

4. Disconnect the accelerator cable from the throttle body.

5. Raise and safely support the vehicle and remove the undercover. Drain the engine and transmission oil and the coolant.

6. Disconnect the radiator hoses and the cooling fan electrical connector. Remove the radiator/cooling fans assembly.

7. Remove the accessory drive belts. Without disconnecting the hydraulic hoses, remove the power steering pump and secure it aside.

8. Without disconnecting the refrigerant, remove the air conditioner compressor and secure it aside.

9. Label and disconnect the wiring and all vacuum, fuel and coolant hoses.

10. Disconnect the exhaust pipe from the exhaust manifold.

11. Without disconnecting the hydraulic line, remove the clutch slave cylinder.

12. Remove the center console and the shifter assembly.

13. Disconnect the transmission electrical connectors and the speedometer cable.

14. Mark the position of the driveshaft on the differential flange and remove the bolts. Slide the driveshaft yoke from the transmission and remove the driveshaft.

15. The frame member between the transmission and differential must be removed. With the transmission

properly supported, remove the bolts at both ends and remove the frame.

NOTE: Do not remove the upper frame-to-differential spacers from the frame. If they are removed, the entire frame must be replaced as a unit.

16. Install suitable lifting equipment onto the engine and make sure all hoses, wires and cables are disconnected.

17. Remove the engine mount nuts and lift the engine and transmission as an assembly from the vehicle.

18. Remove the starter.

19. If equipped with automatic transmission, remove the torque converter-to-flywheel bolts.

20. Remove the bolts to separate the transmission from the engine.

To install:

21. Assemble the engine to the transmission and torque the bolts to 66 ft. lbs. (89 Nm). If equipped with automatic transmission, install the torque converter-to-flywheel bolts and torque to 40 ft. lbs. (54 Nm).

22. Install the starter and torque the bolts to 38 ft. lbs. (52 Nm).

23. Carefully install the engine and transmission assembly into the vehicle. Start but do not tighten the mount nuts.

24. Install the transmission-to-differential frame and torque the bolts to 91 ft. lbs. (124 Nm). Torque the engine mount nuts to 58 ft. lbs. (78 Nm).

25. Connect the transmission wiring and speedometer cable and install the driveshaft. Torque the driveshaft bolts to 22 ft. lbs. (30 Nm).

26. Use a new gasket and attach the exhaust pipe to the manifold. Torque the nuts to 34 ft. lbs. (46 Nm).

27. Install the clutch slave cylinder and torque bolts to 19 ft. lbs. (25 Nm).

28. Connect and adjust the shift linkage as required.

29. Install the air conditioner compressor and power steering pump. Install and adjust the drive belts.

30. Install the radiator and fans and connect all cooling system hoses.

31. Connect all wiring and hoses.

32. Connect and adjust the accelerator cable as required.

33. Install the air cleaner and air flow meter assembly.

34. Check to make sure all wiring and hoses are properly connected. Fill and bleed the cooling system. Fill the engine and transmission with the proper type and quantity of oil.

35. Connect the negative battery cable, start the engine and bring to

normal operating temperature. Check for leaks.

36. Check the ignition timing and idle speed. Check all fluid levels.

1995–97 Miata 1.8L (BPD) Engine

1. Properly relieve the fuel system pressure.
2. Raise the trunk lid and disconnect the negative battery cable. Remove the fresh air duct and the air cleaner/air flow meter assembly.
3. Remove the transmission.
4. Disconnect the accelerator cable from the throttle body.
5. Raise and safely support the vehicle and remove the undercover. Drain the engine oil and the coolant.
6. Disconnect the radiator hoses, including the coolant reservoir hose and the cooling fan electrical connector.
7. On automatic transmission vehicles only, remove the oil cooler hose.
8. Remove the radiator/cooling fan assembly.
9. Remove the accessory drive belts. Without disconnecting the hydraulic hoses, remove the power steering pump and secure it aside.
10. Without disconnecting the refrigerant, remove the air conditioner compressor and secure it aside.
11. Label and disconnect the following electrical connectors: Steering pressure sensor, throttle position sensor, idle air control valve, heated oxygen sensor, ignition coil, crankshaft position sensor, ground, fuel injector, alternator, oil pressure sensor and the starter.
12. Label and disconnect the following hoses: Brake vacuum, fuel, purge control vacuum, cruise control vacuum, water inlet and heater.
13. Disconnect the exhaust pipe from the exhaust manifold.
14. Install suitable lifting equipment onto the engine and make sure all hoses, wires and cables are disconnected.
15. Remove the engine mount nuts, and lift the engine from the vehicle.

To install:

16. Carefully and slowly install the engine assembly into the vehicle. Tilt the engine downward, and align the engine mounts with the crossmember mounting holes. Tighten the mount nuts to 42–57 ft. lbs. (57–78 Nm).
17. Use a new gasket and attach the exhaust pipe to the manifold. Torque the nuts to 34 ft. lbs. (46 Nm).
18. Install the following hoses: Brake vacuum, fuel, purge control vacuum, cruise control vacuum, water inlet and heater.

19. Install the following electrical connectors: Steering pressure sensor, throttle position sensor, idle air control valve, heated oxygen sensor, ignition coil, crankshaft position sensor, ground, fuel injector, alternator, oil pressure sensor and the starter.
20. Install the air conditioner compressor and power steering pump. Install and adjust the drive belts.
21. Install the radiator and fans, and connect all cooling system hoses.
22. Connect and adjust the accelerator cable as required.
23. Install the air cleaner and air flow meter assembly.
24. Install the transmission.
25. Check to make sure all wiring and hoses are properly connected. Fill and bleed the cooling system. Fill the engine and transmission with the proper type and quantity of oil.
26. Connect the negative battery cable, start the engine and bring to normal operating temperature. Check for leaks.
27. Check the ignition timing and idle speed. Check all fluid levels.

1993–95 RX7 1.3L (13B)

1. Mark the position of the hood on its hinges and remove the hood.
2. Properly relieve the fuel system pressure and disconnect the negative battery cable.
3. Remove the engine undercover and drain the engine oil and cooling system.
4. Remove the transmission.
5. Disconnect the Engine Control Module (ECM) as follows:
   a. Open the passenger side door and remove the sill plate.
   b. Remove the passenger side kick panel.
   c. Disconnect the electrical connectors from the ECM.
6. Remove the fresh air duct, air cleaner intake hose and air cleaner.
7. Remove the battery, battery box and carrier.
8. Remove the strut bar, and temporarily tighten the locknut to the stud bolt.
9. Disconnect the accelerator cable, and remove the throttle body air intake hose.
10. Remove the upper and lower radiator hoses, and disconnect the heater hose.
11. Remove the (engine mounted) fuse box and position aside, leaving the wiring harness connected.
12. Remove the accessory drive belts.
13. Hold the power steering pump pulley using a suitable tool and remove the pulley nut. Remove the pul-

ley. Remove the power steering pump and position aside, leaving the hoses connected.
14. Remove the A/C compressor and position aside, leaving the hoses connected.
15. Label and disconnect the engine wiring harness electrical connectors, and vacuum hoses. Disconnect and plug the fuel lines.
16. Position a drain pan under the engine oil cooler lines. Remove the clips and disconnect the hoses from the engine oil cooler.
17. Remove the insulators from the exhaust pipe and turbocharger. Do not let oil get on the insulators.
18. Disconnect the oxygen sensor connector and remove the front exhaust pipe. Remove the split air pipe.
19. If equipped with automatic transmission, disconnect the oil cooler lines at the radiator. Remove the bolt and nut from the line support brackets and remove the lines.
20. Attach suitable lifting equipment to the engine and remove the engine mount nuts. Carefully remove the engine from the vehicle and position on a workstand.

To install:

21. Carefully lower the engine into the vehicle, aligning the engine mounts with the crossmember mounting holes.
22. Install the engine mount nuts and tighten to 34–49 ft. lbs. (46–67 Nm). Remove the engine lifting equipment.
23. If equipped with automatic transmission, install and connect the transmission oil cooler lines. Tighten the support bracket nut and bolt to 95 inch lbs. (10.7 Nm).
24. Install the front exhaust pipe, using new gaskets. Tighten the exhaust pipe-to-turbocharger nuts to 38 ft. lbs. (51 Nm) and the exhaust pipe-to-main converter nuts to 65 ft. lbs. (89 Nm). Connect the split air pipe.
25. Install the front exhaust pipe insulator, turbocharger insulator and center insulator, in that order. Tighten the bolts, after all three insulators have been installed, to 95 inch lbs. (10.7 Nm).
26. Install the engine mount insulator and tighten the bolts to 95 inch lbs. (10.7 Nm).
27. Connect the engine oil cooler hoses to the engine oil cooler and install the clips. Make sure the clips are securely locked.
28. Connect the fuel lines, hoses and electrical connectors.
29. Install the A/C compressor and tighten the bolts to 18 ft. lbs. (25 Nm).

30. Install the power steering pump and tighten the bolts to 33 ft. lbs. (46 Nm). Connect the electrical connector to the pump.

31. Install the power steering pump pulley and loosely tighten the nut. Hold the pulley with a suitable tool and tighten the nut to 43 ft. lbs. (58 Nm).

32. Install the accessory drive belts, and adjust the belt tension.

33. Install the (engine mounted) fuse box and tighten the nut to 95 inch lbs. (10.7 Nm).

34. Install the radiator and heater hoses.

35. Install the throttle body air intake hose using a new gasket. Tighten the nuts to 95 inch lbs. (10.7 Nm). Connect the accelerator cable.

36. Remove the upper nuts and install the strut bar. Tighten the nuts to 26 ft. lbs. (36 Nm).

37. Install the air cleaner assembly and ducts. Tighten the bolts to 95 inch lbs. (10.7 Nm).

38. Install the carrier and battery box, and tighten the bolts to 95 inch lbs. (10.7 Nm). Install the battery.

39. Install the transmission.

40. Install the engine undercover.

41. Connect the ECM and install the kick panel and sill plate.

42. Fill the engine with the proper type and quantity of oil. Fill and bleed the cooling system.

43. Install the hood, aligning the marks that were made during removal.

44. Connect the battery cables, start the engine and bring to normal operating temperature. Check for leaks.

45. Check all fluid levels and road test.

Engine Mounts

REMOVAL AND INSTALLATION

1993–94 323 and Protege 1.6L (B6E), 1994 MX3 1.6L (B6-ZE) and Protege 1.8L (BPE, BPD) Engines

1. Raise and support the vehicle safely.

2. Attach a hoist to the engine and lift until the slack in the chain is taken up.

3. Remove the nuts from the engine mounts.

NOTE: Inspect the engine compartment for components that may bind when the engine is raised. Disconnect these components.

4. Lift the engine the exact amount needed to remove the engine mount. Do not lift any higher.

5. Remove the engine mounts.

To install:

6. Install the engine mounts.

7. Lower the engine and tighten the engine mount-to-engine nuts to 32–41 ft. lbs. (43–55 Nm). Tighten the engine mount-to-frame nuts to 41–49 ft. lbs. (55–67 Nm).

8. Remove the engine hoist and lower the vehicle.

1993–94 MX3 1.8L (K8) Engine

Number 1 and 2 Engine Mounts

NOTE: The number 1 and 2 engine mounts are located on the crossmember.

1. Disconnect the negative battery cable.

2. Raise and support the vehicle safely.

3. Using an engine lifting device, attach it to the engine and support it's weight.

4. Remove the four crossmember mounting bolts, and remove the crossmember.

5. Remove the two engine mount-to-engine nuts and the through-bolt. Remove the mount.

To install:

6. Install the engine mount on the crossmember, and hand tighten the nuts.

7. Install the crossmember on the chassis, and tighten the bolts to 48–65 ft. lbs. (64–89 Nm).

8. Tighten the engine mount nuts to 27–38 ft. lbs. (38–52 Nm).

9. Install the engine mount through-bolt and tighten to 49–68 ft. lbs. (67–93 Nm).

10. Remove the lifting device and lower the vehicle.

11. Connect the negative battery cable.

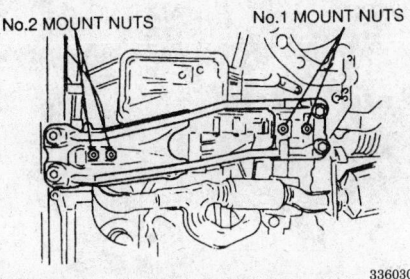

Crossmember and engine mounts No, 1 and 2 — 1993–94 MX3 1.8L engine

Number 3 Engine Mount

1. Disconnect the negative battery cable.

2. Raise and support the vehicle safely.

3. Using an engine lifting device, attach it to the engine and support it's weight.

4. Remove the two engine mount-to-engine nuts and through-bolt. Remove the mount.

To install:

5. Position the engine mount making sure to align the holes.

6. Install the mounting nuts and tighten to 54–76 ft. lbs. (74–103 Nm). Install the through-bolt and tighten to 49–68 ft. lbs. (67–93 Nm).

7. Remove the lifting device and lower the vehicle.

8. Connect the negative battery cable.

Number 4 Engine Mount

1. Disconnect the negative battery cable.

2. Raise and support the vehicle safely.

3. Using an engine lifting device, attach it to the engine and support it's weight.

4. Remove the mounting bolts for the engine mount rubber, and remove the engine mount rubber.

5. Remove the mounting bolts for the engine mount bracket, and remove the bracket.

To install:

6. Position the bracket and engine mount rubber. Hand tighten the "A" nuts.

7. Tighten the mounting bolts ("B"), in two to three steps, to 32–44 ft. lbs. (44–60 Nm) using a crossing pattern. Tighten the mounting nuts ("A") to 49–68 ft. lbs. (67–93 Nm).

8. Remove the lifting device and lower the vehicle.

9. Connect the negative battery cable.

1995–97 Protege 1.5L (Z5D) and 1.8L (BPD) Engines

Number 1 Engine Mount

1. Disconnect the negative battery cable.

2. Raise and support the vehicle safely.

3. Using an engine lifting device, attach it to the engine and support it's weight.

4. Remove the three engine mount bracket bolts from the engine, and remove the through-bolt. Remove the mount bracket.

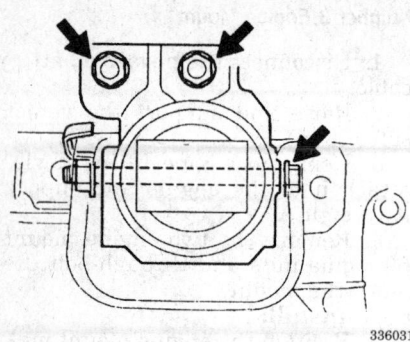

Engine mount No. 3 — 1993–94 MX3 1.8L engine

336031

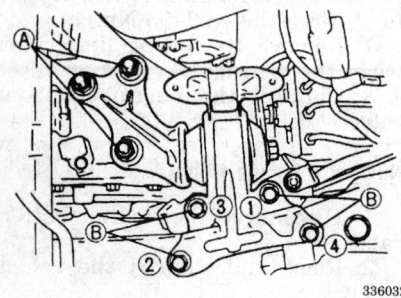

Engine mount No.4 bracket — 1993–94 MX3 1.8L engine

336032

5. Remove the two mounting bolts, and remove the engine mount from the body.

To install:

6. Position the engine mount making sure to align the holes. Install the mounting bolts.

7. Position the bracket. Install the mounting bolts and tighten to 50–68 ft. lbs. (67–93 Nm). Install the through-bolt and tighten to 48–65 ft. lbs. (64–89 Nm).

8. Remove the lifting device and lower the vehicle.

9. Connect the negative battery cable.

Number 2 Engine Mount

NOTE: The number 2 engine mount is located on the crossmember.

1. Disconnect the negative battery cable.

2. Raise and support the vehicle safely.

3. Using an engine lifting device, attach it to the engine and support its weight.

4. Remove the four crossmember mounting bolts, and remove the crossmember.

5. Remove the two engine mount-to-engine nuts and the through-bolt. Remove the mount.

6. Remove the engine mount bracket bolts and remove the engine mount bracket.

To install:

7. Install the engine mount bracket, and tighten the mounting bolts to 28–38 ft. lbs. (38–51 Nm).

8. Install the engine mount on the crossmember, and hand tighten the nuts.

9. Install the crossmember on the chassis, and tighten the bolts to 50–65 ft. lbs. (67–89 Nm).

10. Tighten the engine mount nuts to 28–38 ft. lbs. (38–51 Nm).

11. Install the engine mount through-bolt and tighten to 63–86 ft. lbs. (86–116 Nm).

12. Remove the lifting device and lower the vehicle.

13. Connect the negative battery cable.

Number 3 Engine Mount

1. Disconnect the negative battery cable.

2. Raise and support the vehicle safely.

3. Using an engine lifting device, attach it to the engine and support it's weight.

4. Remove the engine mount-to-engine nuts and through-bolt. Remove the mount.

To install:

5. Position the engine mount making sure to align the holes.

6. Install the mounting nuts and tighten to 55–76 ft. lbs. (74–102 Nm). Install the through-bolt and tighten to 63–86 ft. lbs. (86–116 Nm).

7. Remove the lifting device and lower the vehicle.

8. Connect the negative battery cable.

Number 4 Engine Mount

1. Disconnect the negative battery cable.

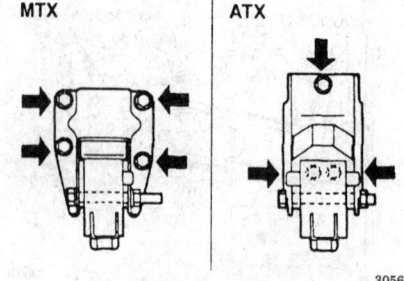

MTX   ATX

305669

Engine mount number 2 bracket — 1995–97 Protege 1.5L (Z5D) and 1.8L (BPD) engines

2. Raise and support the vehicle safely.

3. Using an engine lifting device, attach it to the engine and support its weight.

4. Remove the through-bolt. Remove the mounting bolts for the engine mount rubber, and remove the engine mount rubber.

5. Remove the mounting bolts for the engine mount bracket, and remove the bracket.

To install:

6. Position the bracket and engine mount rubber. Hand tighten the nuts. Tighten the bolts, in two to three steps, to a torque of 32–44 ft. lbs. (44–60 Nm) in a crossing pattern. Tighten the nuts to 50–68 ft. lbs. (67–93 Nm).

7. Remove the lifting device and lower the vehicle.

8. Connect the negative battery cable.

1995 MX3 1.6L (B6-ZE) and 1.8L (K8) Engines

Number 1 and 2 Engine Mounts

NOTE: The number 1 and 2 engine mounts are located on the crossmember.

1. Disconnect the negative battery cable.

2. Raise and support the vehicle safely.

3. Using an engine lifting device, attach it to the engine and support its weight.

4. Remove the four crossmember mounting bolts, and remove the crossmember.

5. Remove the two engine mount-to-engine nuts and the through-bolt. Remove the mount(s).

To install:

6. Install the engine mount(s) on the crossmember, and hand tighten the nuts.

7. Install the crossmember on the chassis, and tighten the bolts to 48–65 ft. lbs. (64–89 Nm).

8. On the 1.6L engine, tighten the engine mount nuts to 28–38 ft. lbs. (38–51 Nm).

9. On the 1.8L engine, install the engine mount through-bolt and tighten to 49–68 ft. lbs. (67–93 Nm).

10. Install the engine mount through-bolt and tighten to 63–86 ft. lbs. (86–116 Nm).

11. Remove the lifting device and lower the vehicle.

12. Connect the negative battery cable.

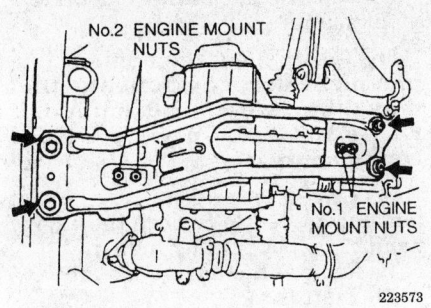

No.2 ENGINE MOUNT NUTS

No.1 ENGINE MOUNT NUTS

223573

Crossmember and engine mount numbers 1 and 2 — 1995 MX3 1.6L (B6-ZE) engine

Number 3 Engine Mount

1. Disconnect the negative battery cable.
2. Raise and support the vehicle safely.
3. Using an engine lifting device, attach it to the engine and support it's weight.
4. Remove the two engine mount-to-engine nuts and through-bolt. Remove the mount.
To install:
5. Position the engine mount making sure to align the holes.
6. Install the mounting nuts and tighten to 55–76 ft. lbs. (74–102 Nm). Install the through-bolt and tighten on the 1.6L engine to 63–86 ft. lbs. (86–116 Nm) and the 1.8L engine, 49–68 ft. lbs. (67–93 Nm).
7. Remove the lifting device and lower the vehicle.
8. Connect the negative battery cable.

Number 4 Engine Mount

1. Disconnect the negative battery cable.
2. Raise and support the vehicle safely.
3. Using an engine lifting device, attach it to the engine and support its weight.
4. Remove the through-bolt. Remove the mounting bolts for the engine mount rubber, and remove the engine mount rubber.
5. Remove the mounting bolts for the engine mount bracket, and remove the bracket.
To install:
6. Position the bracket and engine mount rubber. Hand tighten the A nuts. Tighten the B bolts, in two to three steps, to a torque of 32–44 ft. lbs. (44–60 Nm) in a crossing pattern. Tighten nuts A to 50–68 ft. lbs. (67–93 Nm).
7. Remove the lifting device and lower the vehicle.

8. Connect the negative battery cable.

626 and MX6 2.0L (FS) and 2.5L (KL) Engines

Number 1 Engine Mount

1. Disconnect the negative battery cable.
2. Raise and support the vehicle safely.
3. Using an engine lifting device, attach it to the engine and support it's weight.
4. Remove the engine mount-to-engine bolts, through-bolt and the engine mount-to-chassis bolts. Remove the mount.
To install:
5. Position the engine mount making sure to align the holes.
6. On manual transaxles, torque the bolts to 50–68 ft. lbs. (67–93 Nm). On automatic transaxles, tighten the three bolts "A" to 50–77 ft. lbs. (67–104 Nm), through-bolt "B" to 63–86 ft. lbs. (86–116 Nm) and the three bolts "C" to 50–68 ft. lbs. (67–93 Nm).
7. Remove the lifting device and lower the vehicle.
8. Connect the negative battery cable.

Number 2 Engine Mount

NOTE: The number 2 engine mount is located on the crossmember.

1. Disconnect the negative battery cable.
2. Raise and support the vehicle safely.
3. Using an engine lifting device, attach it to the engine and support it's weight.
4. Remove the four crossmember mounting bolts, and remove the crossmember.
5. Remove the two engine mount-to-engine nuts and the through-bolt. Remove the mount.
To install:
6. Install the engine mount on the crossmember, and hand tighten the nuts ("B").
7. Install the crossmember on the chassis, and tighten the bolts ("A") to 50–68 ft. lbs. (67–93 Nm).
8. Tighten the engine mount nuts ("B") to 55–77 ft. lbs. (75–104 Nm).
9. Install the engine mount through-bolt and tighten to 63–86 ft. lbs. (86–116 Nm).
10. Remove the lifting device and lower the vehicle.

11. Connect the negative battery cable.

Number 3 Engine Mount

1. Disconnect the negative battery cable.
2. Raise and support the vehicle safely.
3. Using an engine lifting device, attach it to the engine and support it's weight.
4. Remove the two engine mount-to-engine nuts and through-bolt. Remove the mount.
To install:
5. Position the engine mount making sure to align the holes.
6. Install the mounting nuts and tighten to 55–77 ft. lbs. (75–104 Nm). Install the through-bolt and tighten to 63–86 ft. lbs. (86–116 Nm).
7. Remove the lifting device and lower the vehicle.
8. Connect the negative battery cable.

Number 4 Engine Mount

1. Disconnect the negative battery cable.
2. Raise and support the vehicle safely.
3. Using an engine lifting device, attach it to the engine and support it's weight.
4. Remove the through-bolt. Remove the mounting bolts for the engine mount rubber, and remove the engine mount rubber.
5. Remove the mounting bolts for the engine mount bracket, and remove the bracket.
To install:
6. Position the bracket, and tighten the "B" bolts first and then the "A" bolts to the same torque of 44–59 ft. lbs. (59–80 Nm).
7. Install the engine mount rubber. On manual transaxles, tighten the mounting bolts ("A") to 50–68 ft. lbs. (67–93 Nm), and the through-bolt ("B") to 63–86 ft. lbs. (86–116 Nm). On automatic transaxles, tighten the through-bolt ("B") to 63–86 ft. lbs. (86–116 Nm) and the mounting bolts ("C") to 28–38 ft. lbs. (38–51 Nm).
8. Remove the lifting device and lower the vehicle.
9. Connect the negative battery cable.

1993–95 929 3.0L (JE-ZE) Engines

1. Raise and support the vehicle safely.
2. Attach a hoist to the engine and lift until the slack in the chain is taken up.

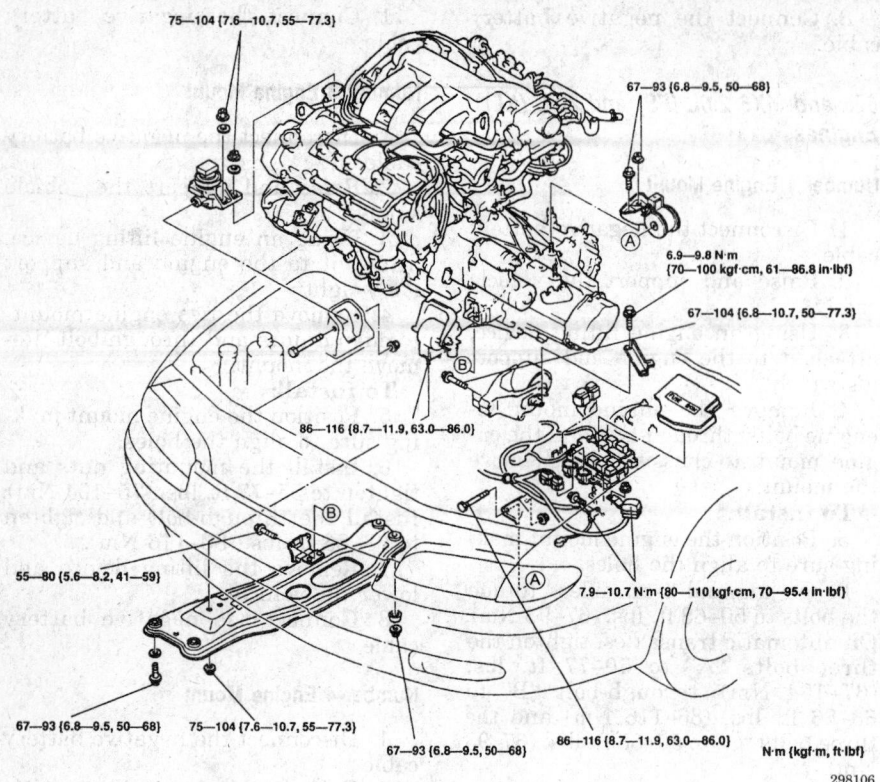

75—104 {7.6—10.7, 55—77.3}

67—93 {6.8—9.5, 50—68}

6.9—9.8 N·m {70—100 kgf·cm, 61—86.8 in·lbf}

67—104 {6.8—10.7, 50—77.3}

86—116 {8.7—11.9, 63.0—86.0}

55—80 {5.6—8.2, 41—59}

7.9—10.7 N·m {80—110 kgf·cm, 70—95.4 in·lbf}

67—93 {6.8—9.5, 50—68}   75—104 {7.6—10.7, 55—77.3}

67—93 {6.8—9.5, 50—68}   86—116 {8.7—11.9, 63.0—86.0}

N·m {kgf·m, ft·lbf}

298106

Engine mount torque values — 2.0L (FS) and 2.5L (KL) engines

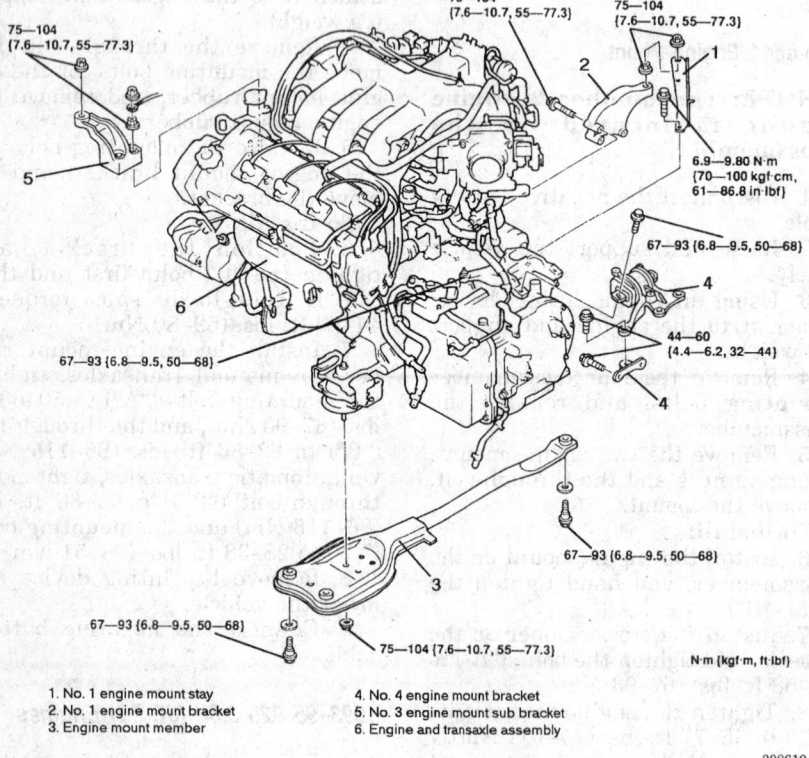

75—104 {7.6—10.7, 55—77.3}   75—104 {7.6—10.7, 55—77.3}

75—104 {7.6—10.7, 55—77.3}

6.9—9.80 N·m {70—100 kgf·cm, 61—86.8 in·lbf}

67—93 {6.8—9.5, 50—68}

44—60 {4.4—6.2, 32—44}

67—93 {6.8—9.5, 50—68}

67—93 {6.8—9.5, 50—68}

75—104 {7.6—10.7, 55—77.3}

N·m {kgf·m, ft·lbf}

1. No. 1 engine mount stay
2. No. 1 engine mount bracket
3. Engine mount member
4. No. 4 engine mount bracket
5. No. 3 engine mount sub bracket
6. Engine and transaxle assembly

308610

Engine mount locations — 1995-97 Millenia 2.5L (KLD) engine

3. Remove the nuts from the engine mounts.

NOTE: Inspect the engine compartment for components that may bind when the engine is raised. Disconnect these components.

4. Lift the engine the exact amount needed to remove the engine mount. Do not lift any higher.

5. Remove the engine mounts.

To install:

6. Install the engine mounts.

7. Lower the engine and tighten the engine mount-to-engine nuts to 29–42 ft. lbs. (39–57 Nm). Tighten the engine mount-to-frame nuts to 41–49 ft. lbs. (55–67 Nm).

8. Remove the engine hoist and lower the vehicle.

1995-97 Millenia

Number 1 Engine Mount

1. Disconnect the negative battery cable.

2. Raise and support the vehicle safely.

3. Using an engine lifting device, attach it to the engine and support it's weight.

4. Remove the engine mount bracket.

5. Remove the engine mount bolts and the through-bolt. Remove the mount.

To install:

6. Position the engine mount making sure to align the holes. Tighten the four bolts to 55–77 ft. lbs. (75–104 Nm).

7. Install the engine mount bracket. Tighten the upper bolt to 55–77 ft. lbs. (75–104 Nm), and the lower bolt to 61–87 inch lbs. (7–10 Nm).

8. Remove the lifting device and lower the vehicle.

9. Connect the negative battery cable.

Number 2 Engine Mount

NOTE: The number 2 engine mount is located on the crossmember.

1. Disconnect the negative battery cable.

2. Raise and support the vehicle safely.

3. Using an engine lifting device, attach it to the engine and support it's weight.

4. Remove the three crossmember mounting bolts, and remove the crossmember.

5. Remove the two engine mount-to-engine nuts and the through-bolt. Remove the mount.

To install:

6. Install the engine mount on the crossmember, and hand tighten the **A** nuts.

7. Install the crossmember on the chassis, and tighten the **B** bolts to 50–68 ft. lbs. (67–93 Nm).

8. Tighten the engine mount **A** nuts to 55–77 ft. lbs. (75–104 Nm).

9. Install the engine mount through-bolt and tighten to 63–86 ft. lbs. (86–116 Nm).

10. Remove the lifting device and lower the vehicle.

11. Connect the negative battery cable.

Number 3 Engine Mount

1. Disconnect the negative battery cable.

2. Raise and support the vehicle safely.

3. Using an engine lifting device, attach it to the engine and support it's weight.

4. Remove the three mounting nuts. Remove the sub bracket mount.

To install:

5. Position the engine mount sub bracket making sure to align the holes.

6. Install the mounting nuts and tighten to 55–77 ft. lbs. (75–104 Nm).

7. Remove the lifting device and lower the vehicle.

8. Connect the negative battery cable.

Number 4 Engine Mount

1. Disconnect the negative battery cable.

2. Raise and support the vehicle safely.

3. Using an engine lifting device, attach it to the engine and support it's weight.

4. Remove the mounting bolts for the engine mount, and remove the engine mount.

5. Remove the mounting bolts for the engine mount bracket, and remove the bracket.

To install:

6. Position the bracket, and tighten the **E** bolts to 32–44 ft. lbs. (44–60 Nm), and hand tighten the **F** bolts.

7. Install the engine mount. Tighten the **C** and **F** mounting bolts to 50–68 ft. lbs. (67–93 Nm)

8. Remove the lifting device and lower the vehicle.

9. Connect the negative battery cable.

Miata 1.6L (B6-ZE) and 1.8L (BPD) Engines

1. Raise and support the vehicle safely.

2. Attach a hoist to the engine and lift until the slack in the chain is taken up.

3. Remove the nuts from the engine mounts.

NOTE: Inspect the engine compartment for components that may bind when the engine is raised. Disconnect these components.

4. Lift the engine the exact amount needed to remove the engine mount. Do not lift any higher.

5. Remove the engine mounts.

To install:

6. Install the engine mounts.

7. Lower the engine and tighten the engine mount-to-engine nuts to 27–41 ft. lbs. (37–55 Nm). Tighten the engine mount-to-frame nuts to 41–57 ft. lbs. (55–78 Nm).

8. Remove the engine hoist and lower the vehicle.

Cylinder Head

REMOVAL AND INSTALLATION

1993–94 323 1.6L (B6E) and Protege 1.8L (BPD, BPE) Engines

――――― CAUTION ―――――

When draining the coolant, keep in mind that cats and dogs are attracted by the ethylene glycol antifreeze, and are quite likely to drink any that is left in an uncovered container or in puddles on the ground. This will prove fatal in sufficient quantity. Always drain the coolant into a sealable container. Coolant should be reused unless it is contaminated or several years old.

―――――――――――――

1. Disconnect the negative battery cable and remove the engine undercover.

2. Remove the air ducts from the air cleaner and throttle body.

3. Tag and disconnect the spark plug wires from the spark plugs. Remove the spark plugs and the distributor cap and wires assembly. Remove the distributor.

4. Drain the cooling system and disconnect the radiator and heater hoses.

5. Disconnect the exhaust pipe and remove the exhaust manifold.

6. On DOHC engines, remove the coolant bypass pipe.

7. Disconnect the accelerator cable.

8. Label and disconnect all necessary electrical connections and vacuum hoses. Disconnect the fuel lines.

9. Remove the intake manifold bracket and the intake manifold.

10. Remove the cylinder head cover bolts and the cylinder head cover.

11. Remove the timing belt cover(s). Rotate the crankshaft, in the normal direction of rotation, until the No. 1 cylinder piston is at TDC on the compression stroke. Make sure the timing marks on the crankshaft and camshaft sprocket(s) are properly aligned and mark the direction of rotation of the belt.

12. Loosen the timing belt tensioner and remove the belt. Do not rotate the crankshaft until the timing belt is reinstalled.

13. When everything is disconnected, loosen the cylinder head bolts in the reverse of the tightening sequence. Remove the bolts and lift the head off the engine.

To install:

14. Thoroughly, clean the cylinder head and the block contact surfaces. Examine the head gasket and check the cylinder head for cracks. Check the cylinder head for warpage using a feeler gauge and straightedge. The maximum allowable distortion is 0.004 in. (0.10mm) on 1.8L engine.

15. Clean the cylinder head bolts and the threads in the block. Make sure the bolts turn freely in the block.

16. Install a new head gasket on the engine block. Make sure the camshaft sprocket timing marks are still aligned, as set during the removal procedure. Install the cylinder head.

17. Lubricate the bolt threads and seat surfaces with clean engine oil and install them. Torque the bolts in 2–3 steps to 56–60 ft. lbs. (75–81 Nm) in the proper sequence.

18. Make sure the crankshaft and camshaft sprocket timing marks are aligned, install the timing belt and set the tension. Carefully rotate the crankshaft 2 turns to make sure the timing marks still line up.

19. Apply a thin bead of sealant to the cylinder head cover and install the new gasket. Install the cover and torque the cover bolts to 78 inch lbs. (9 Nm).

20. Install the timing belt cover(s) and tighten the bolts to 95 inch lbs. (11 Nm).

21. Use new gaskets and install the manifolds. Torque the intake manifold bolts/nuts to 19 ft. lbs. (25 Nm) and install the intake manifold

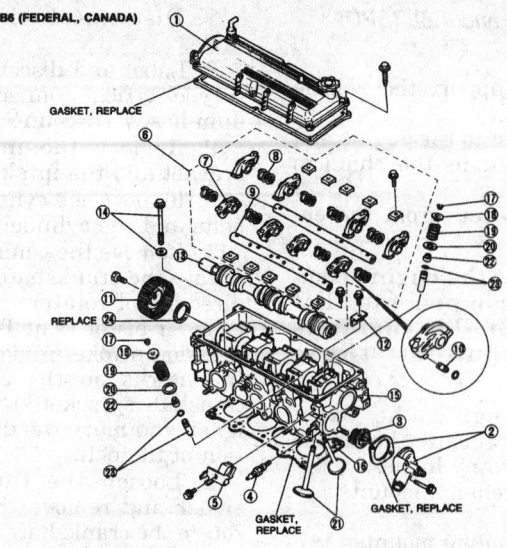

B6 (FEDERAL, CANADA)

GASKET, REPLACE

REPLACE

GASKET, REPLACE

GASKET, REPLACE

1. Cylinder head cover and gasket
2. Thermostat cover and gasket
3. Thermostat
4. Heat gauge unit
5. Bracket
6. Rocker arm and shaft assembly
7. Rocker arm
8. Rocker arm spring
9. Rocker shaft
10. Hydraulic lash adjuster (HLA)
11. Camshaft pulley
12. Thrust plate
13. Camshaft
14. Cylinder head bolts
15. Cylinder head
16. Cylinder head gasket
17. Valve keeper
18. Valve spring seat, upper
19. Valve spring
20. Valve spring seat, lower
21. Valve
22. Valve seal
23. Valve guide
24. Camshaft oil seal

142028

Cylinder head components — 1993–94 323 1.6L (B6E) engines

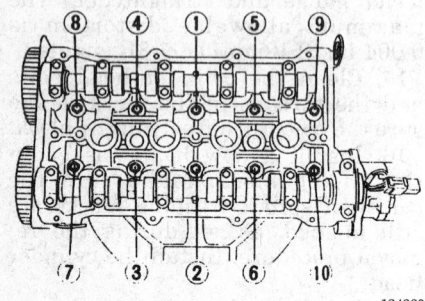

124389

Cylinder head torque sequence — 1993–94 323 and Protege

bracket. Torque the exhaust manifold nuts to 34 ft. lbs. (46 Nm).

22. Use a new gasket to connect the exhaust pipe and torque the nuts to 34 ft. lbs. (46 Nm).

23. If removed, install the radiator and connect all cooling system hoses. On DOHC engines, install the coolant bypass pipe.

24. Install the distributor, spark plugs, distributor cap and wires.

25. Connect all vacuum and fuel system hoses and connect all wiring.

26. Connect the accelerator cable and install the air ducts and engine undercover.

27. Connect the negative battery cable. Fill and bleed the cooling system. Change the engine oil.

28. Start the engine and bring to normal operating temperature. Check for leaks. Check the ignition timing and idle speed.

1995–97 Protege 1.5L (Z5D) and 1.8L (BPD) Engines

1. Relieve the fuel system pressure and disconnect the negative battery cable. Drain the cooling system.

2. Raise and safely support the vehicle.

3. Remove the right front wheel and splash shield.

4. Remove the power steering belt shield. Loosen the power steering adjusting bolt, lockbolt and through-bolt and remove the power steering belt.

5. Loosen the alternator adjusting bolt and upper mounting bolt. Remove the alternator belt.

6. Remove the water pump pulley.

7. Using a holder tool, hold the crankshaft pulley and remove the pulley bolt. Use a suitable puller to remove the pulley, then remove the guide plate.

8. Remove the power steering hose brackets from the cylinder head

cover. Label and disconnect the spark plug wires and wire clips.

9. Disconnect the breather tube and PCV valve from the cylinder head cover. Remove the bolts, in 2 steps, in the reverse order of the tightening sequence. Remove the cylinder head cover.

10. Remove the oil dipstick and bracket.

11. Remove the timing belt upper cover.

12. Use a suitable engine support tool, and remove the number three engine mount bracket.

13. Remove the timing belt middle and lower covers.

14. Rotate the crankshaft, in the normal direction of rotation, until the No. 1 cylinder piston is at TDC on the compression stroke. Make sure the timing marks on the crankshaft and camshaft sprocket(s) are properly aligned and mark the direction of rotation of the belt.

15. Loosen the timing belt tensioner and remove the belt. Do not rotate the crankshaft until the timing belt is reinstalled.

16. Remove air cleaner assembly and front pipe.

17. Remove the exhaust manifold.

18. Disconnect the accelerator and throttle cables.

19. Tag and disconnect the spark plug wires from the spark plugs. Remove the spark plugs and the distributor cap and wires assembly. Remove the distributor.

20. Disconnect the following hoses: Heater, brake vacuum, purge, fuel, water and upper radiator.

21. Label and disconnect the distributor/coil connectors, engine coolant temperature sensor connector, cooling fan coolant temperature sensor connector, and temperature gauge sensor connector.

22. Remove the camshaft sprockets and the camshaft.

23. On the 1.5L engine, remove the tappets. Identify each tappet as it is removed so it can be reinstalled in the same position.

24. On the 1.8L engine, remove the hydraulic lifters. Identify each lifter as it is removed so it can be reinstalled in the same position. If the lifters are to be reused, store them upside down in an oil-filled sealed container.

25. Remove the intake manifold bracket and the intake manifold.

26. Loosen the cylinder head bolts, in 2–3 steps, in sequence. Remove the bolts and the cylinder head.

To install:

27. Thoroughly, clean the cylinder head and the block contact surfaces.

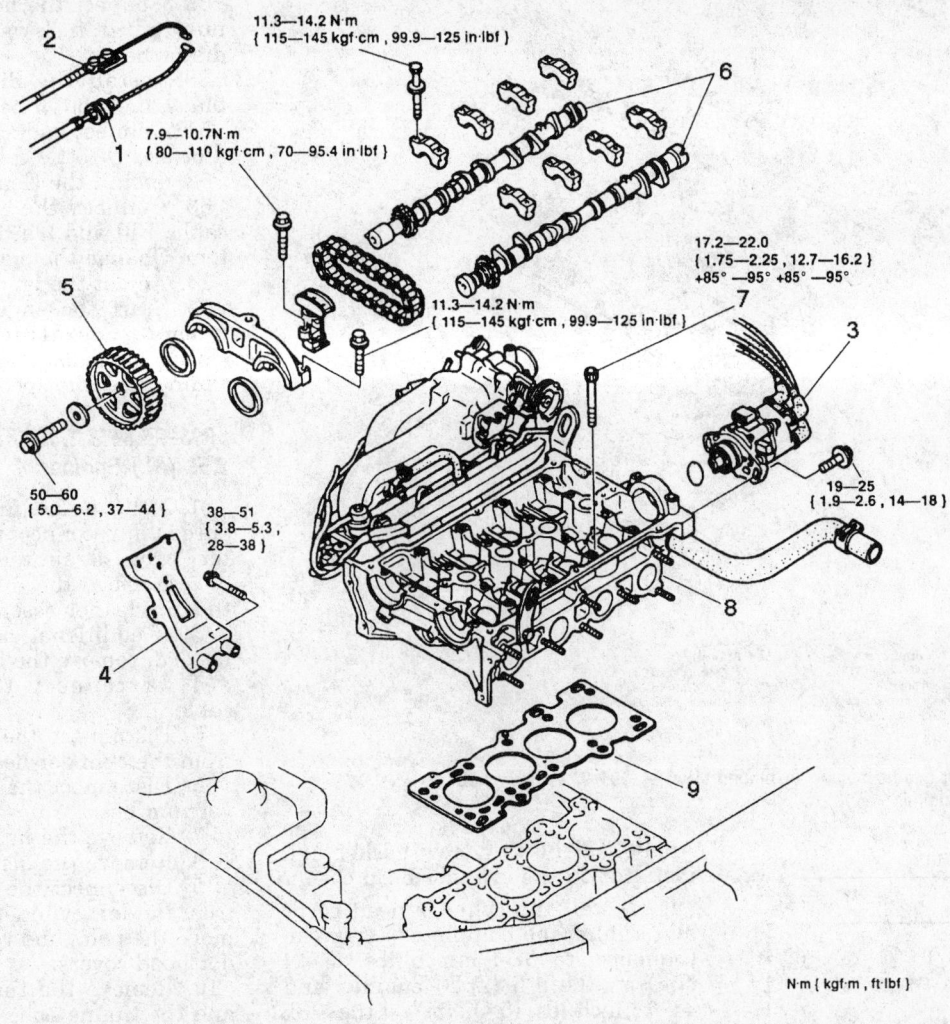

11.3—14.2 N·m
{ 115—145 kgf·cm , 99.9—125 in·lbf }

7.9—10.7 N·m
{ 80—110 kgf·cm , 70—95.4 in·lbf }

17.2—22.0
{ 1.75—2.25 , 12.7—16.2 }
+85°—95° +85°—95°

11.3—14.2 N·m
{ 115—145 kgf·cm , 99.9—125 in·lbf }

50—60
{ 5.0—6.2 , 37—44 }

38—51
{ 3.8—5.3 ,
28—38 }

19—25
{ 1.9—2.6 , 14—18 }

N·m { kgf·m , ft·lbf }

1. Accelerator cable
2. Throttle cable (ATX)
3. Distributor
4. Intake manifold stay
5. Camshaft pulley
6. Camshaft
7. Cylinder head bolt
8. Cylinder head assembly
9. Cylinder head gasket

305438

Cylinder head assembly (exploded view) — 1995–97 Protege 1.5L (Z5D) engines

Examine the head gasket and check the cylinder head for cracks. Check the cylinder head for warpage using a feeler gauge and straightedge. The maximum allowable distortion is 0.004 in. (0.10mm).

28. Clean the cylinder head bolts and the threads in the block. Make sure the bolts turn freely in the block.

29. Install a new head gasket on the engine block. Make sure the camshaft sprocket timing marks are still aligned, as set during the removal procedure. Install the cylinder head.

30. Lubricate the bolt threads and seat surfaces with clean engine oil and install them as follows;

  a. On the 1.5L (Z5D) engines, torque the bolts in 2–3 steps to 13–16 ft. lbs. (17–22 Nm) in the proper sequence. Paint a reference mark on each bolt head and turn the bolts, in sequence, 90°, and then an additional 90°.

  b. On the 1.8L (BPD) engines, torque the bolts in 2–3 steps to 56–60 ft. lbs. (75–81 Nm) in the proper sequence.

31. Use new gaskets and install the manifolds. Torque the intake manifold bolts/nuts to 19 ft. lbs. (25 Nm) and install the intake manifold bracket. Torque the exhaust manifold nuts to 34 ft. lbs. (46 Nm).

32. Use a new gasket to connect the exhaust pipe and torque the nuts to 34 ft. lbs. (46 Nm).

33. Apply clean engine oil to the tappets and install them in their original positions.

34. Install the camshaft and sprockets.

35. Make sure the crankshaft and camshaft sprocket timing marks are aligned, install the timing belt and set the tension. Carefully rotate the crankshaft 2 turns to make sure the timing marks still line up.

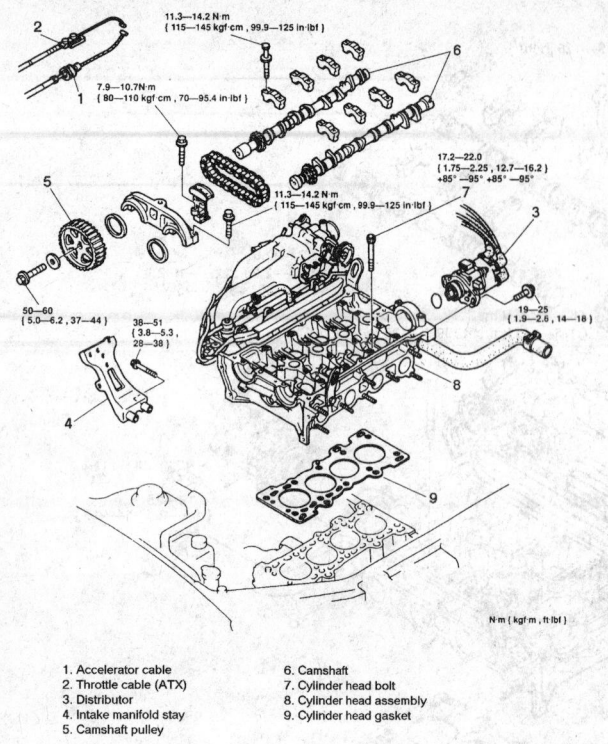

1. Accelerator cable
2. Throttle cable (ATX)
3. Distributor
4. Intake manifold stay
5. Camshaft pulley
6. Camshaft
7. Cylinder head bolt
8. Cylinder head assembly
9. Cylinder head gasket

305438

Cylinder head assembly (exploded view) — 1995–97 Protege 1.5L (Z5D) engines

```
CRANKSHAFT      ③  ⑦  ⑩  ⑥  ②
PULLEY
SIDE            ④  ⑧  ⑨  ⑤  ①
```

305447

Cylinder head bolt (loosening sequence; tighten in reverse order) — 1995–97 Protege 1.8L (BPD) and 1.5L (Z5D) engines

36. Install the timing belt middle and lower covers, and tighten the bolts to 70–95 inch lbs. (8–11 Nm).

37. Install the number three engine mount bracket. Tighten the nut to 70–95 inch lbs. (8–11 Nm), and the bolt to 14–16 ft. lbs. (19–22 Nm). Remove the engine support tool.

38. Install the upper timing belt cover and tighten the bolts to 70–95 inch lbs. (8–11 Nm).

39. Install the oil dipstick and bracket.

40. Apply silicone sealant to the cylinder surface in the area adjacent to the front camshaft bearing caps.

Apply sealant to a new gasket and install it on the cylinder head cover.

41. Install the cylinder head cover and tighten the bolts in 5–6 steps, in sequence, to 61–95 inch lbs. (7–11 Nm) on the 1.5 (Z5D) engines and 44–78 inch lbs. (5–9 Nm) on the 1.8L (BPD) engines.

42. Install the power steering hose brackets and tighten the bolts to 70–95 inch lbs. (7–11 Nm). Connect the spark plug wires and wire clips. Connect the breather tube and PCV valve.

43. Install the guide plate, crankshaft pulley and pulley bolt. Hold the pulley with the holder tool and tighten the bolt to 116–122 ft. lbs. (157–166 Nm).

44. Install the water pump pulley.

45. Install the alternator belt and adjust the tension.

46. Install the power steering belt and adjust the tension. Tighten the through-bolt to 32–44 ft. lbs. (44–60 Nm) and the lockbolt to 24–33 ft. lbs. (32–46 Nm). Install the power steering belt shield and tighten the bolts to 86 inch lbs. (9 Nm).

47. Install the splash shield and wheel. Lower the vehicle.

48. Connect the electrical engine harness connectors.

49. Connect the heater, brake vacuum, purge, fuel, water and upper radiator hoses.

50. Install the distributor, spark plugs, distributor cap and wires.

51. Connect and adjust the accelerator and throttle cables.

52. Install the cleaner.

53. Connect the negative battery cable. Fill and bleed the cooling system. Change the engine oil.

54. Adjust the ignition timing.

55. Start the engine and bring to normal operating temperature. Check for leaks. Check the ignition timing and idle speed.

1993–95 MX3 1.8L (K8), 626 and MX6 2.5L (KL) Engines

1. Relieve the fuel system pressure and disconnect the negative battery cable. Drain the cooling system.

2. Remove the fresh air duct and the air cleaner assembly.

3. If additional clearance space is needed, remove the battery.

4. Disconnect the accelerator cable.

5. Disconnect the wiring harness from the cylinder heads.

6. Disconnect the fuel, heater and vacuum hoses.

7. Remove the intake manifold.

8. Remove the distributor.

9. Disconnect the ventilation pipe from the left cylinder head cover, remove the bolts and remove the cylinder head covers.

10. Remove the timing belt covers and the timing belt.

11. Remove the camshafts. Remove the 3 bolts and the seal plate from the front of the engine.

12. Remove the upper radiator hose. Raise and safely support the vehicle.

13. Disconnect the oxygen sensor connectors. Remove the exhaust pipe-to-manifold nuts and lower the exhaust pipes. Lower the vehicle.

14. Remove the hydraulic lifters. Identify each lifter as it is removed so it can be reinstalled in the same position. If the lifters are to be reused, store them upside down in an oil-filled, sealed container.

15. Loosen the cylinder head bolts, in 2–3 steps, in the reverse order of the torque sequence. Remove the bolts and remove the cylinder heads.

16. Clean all gasket mating surfaces. Inspect the cylinder head for damage, cracks, and water and oil leakage. Check the head gasket surface for distortion using a straightedge and feeler gauge. Maximum allowable distortion is 0.004 in. (0.10mm).

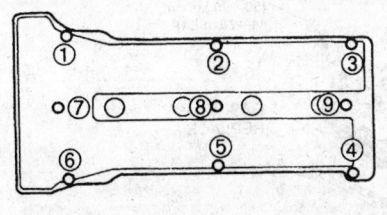

Cylinder head cover (bolt tightening sequence) — 1995–97 Protege 1.5L (Z5D) engines

305440

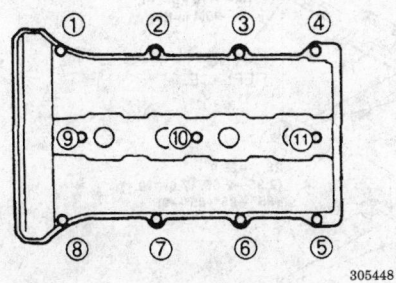

Cylinder head cover (bolt tightening sequence) — 1995–97 Protege 1.8L (BPD) engines

305448

To install:

17. Position NEW head gaskets on the cylinder block. The gaskets cannot be interchanged between sides and are marked R and L for right and left side.

18. Install the cylinder heads. Apply clean engine oil to the threads of new cylinder head bolts and install. Tighten the cylinder head bolts in 2–3 steps, in sequence, to 17–19 ft. lbs. (23–26 Nm).

19. Paint a mark on the edge of each cylinder head bolt to use as a reference. Turn each bolt, in sequence, 90 degrees. Again, turn each bolt, in sequence, an additional 90 degrees.

20. Apply clean engine oil to the hydraulic lifters and install them in their original positions. Make sure they move freely in the bores.

21. Install the camshafts. Raise and safely support the vehicle.

22. Connect the exhaust pipes to the manifolds and tighten the nuts to 41 ft. lbs. (55 Nm). Connect the oxygen sensor connectors. Lower the vehicle.

23. Install the timing belt and timing belt covers.

24. Apply sealant to the cylinder head surface in the area of the front

and rear camshaft caps. Install new gaskets and install the cylinder head covers. Tighten the bolts in 5–6 steps, in sequence, to 44–78 inch lbs. (5–9 Nm).

25. Install the intake manifold using new gaskets. Tighten the mounting bolts to 14–18 ft. lbs. (19–25 Nm).

26. Install the distributor:
a. Apply clean engine oil to a new O-ring, and position it on the distributor.
b. Apply clean engine oil to the drive blade. Install the distributor with the blade fit into the camshaft groove.
c. Hand tighten the mounting bolts.

27. Connect the vacuum, heater and fuel hoses.

28. Connect the wiring harness to the cylinder heads.

29. Connect and adjust the accelerator cable.

30. If removed, install the battery.

31. Install the air cleaner assembly and the fresh air duct.

32. Connect the negative battery cable. Fill and bleed the cooling system. Adjust the ignition timing and idle speed. Run the engine and check for proper operation.

1994–95 MX3 1.6L (B6-ZE) Engines

1. Relieve the fuel system pressure and disconnect the negative battery cable. Drain the cooling system.

2. Remove the splash shield, fresh air duct and air cleaner assembly.

3. Disconnect the accelerator and throttle cables.

4. Tag and disconnect the spark plug wires from the spark plugs. Remove the spark plugs and the distributor cap and wires assembly. Remove the distributor.

5. Disconnect the following hoses: Heater, brake vacuum, purge, fuel, water and upper radiator.

6. Label and disconnect the distributor/coil connectors, engine coolant temperature sensor connector, cooling fan coolant temperature sensor connector and temperature gauge sensor connector.

7. Remove the coolant bypass pipe.

8. Remove the accessory drive belts. Remove the power steering pump bolts and secure the pump aside with mechanics wire, leaving the hoses attached.

9. Remove the water pump pulley.

10. Remove the alternator bracket nut and bolt and position the bracket aside.

11. Disconnect the hoses from the cylinder head cover and loosen the

cover bolts in 5–6 step sequences. Remove the cylinder head cover.

12. Remove the timing belt cover(s). Rotate the crankshaft, in the normal direction of rotation, until the No. 1 cylinder piston is at TDC on the compression stroke. Make sure the timing marks on the crankshaft and camshaft sprocket(s) are properly aligned and mark the direction of rotation of the belt.

13. Loosen the timing belt tensioner and remove the belt. Do not rotate the crankshaft until the timing belt is reinstalled.

14. Remove the camshaft sprockets and the camshaft.

15. Remove the hydraulic lifters. Identify each lifter as it is removed so it can be reinstalled in the same position. If the lifters are to be reused, store them upside down in an oil-filled sealed container.

16. Disconnect the exhaust pipe and remove the exhaust manifold.

17. Remove the intake manifold bracket and the intake manifold.

18. Loosen the cylinder head bolts, in 2–3 steps, in sequence. Remove the bolts and the cylinder head.

To install:

19. Thoroughly, clean the cylinder head and the block contact surfaces. Examine the head gasket and check the cylinder head for cracks. Check the cylinder head for warpage using a feeler gauge and straightedge. The maximum allowable distortion is 0.004 in. (0.10mm).

20. Clean the cylinder head bolts and the threads in the block. Make sure the bolts turn freely in the block.

21. Install a new head gasket on the engine block. Make sure the camshaft sprocket timing marks are still aligned, as set during the removal procedure. Install the cylinder head.

22. Lubricate the bolt threads and seat surfaces with clean engine oil and install them. Torque the bolts in 2–3 steps to 56–60 ft. lbs. (75–81 Nm) in the proper sequence.

23. Use new gaskets and install the manifolds. Torque the intake manifold bolts/nuts to 19 ft. lbs. (25 Nm) and install the intake manifold bracket. Torque the exhaust manifold nuts to 34 ft. lbs. (46 Nm).

24. Use a new gasket to connect the exhaust pipe and torque the nuts to 34 ft. lbs. (46 Nm).

25. Apply clean engine oil to the hydraulic lifters and install them in their original positions. Make sure they move freely in the bores.

26. Install the camshaft and sprockets.

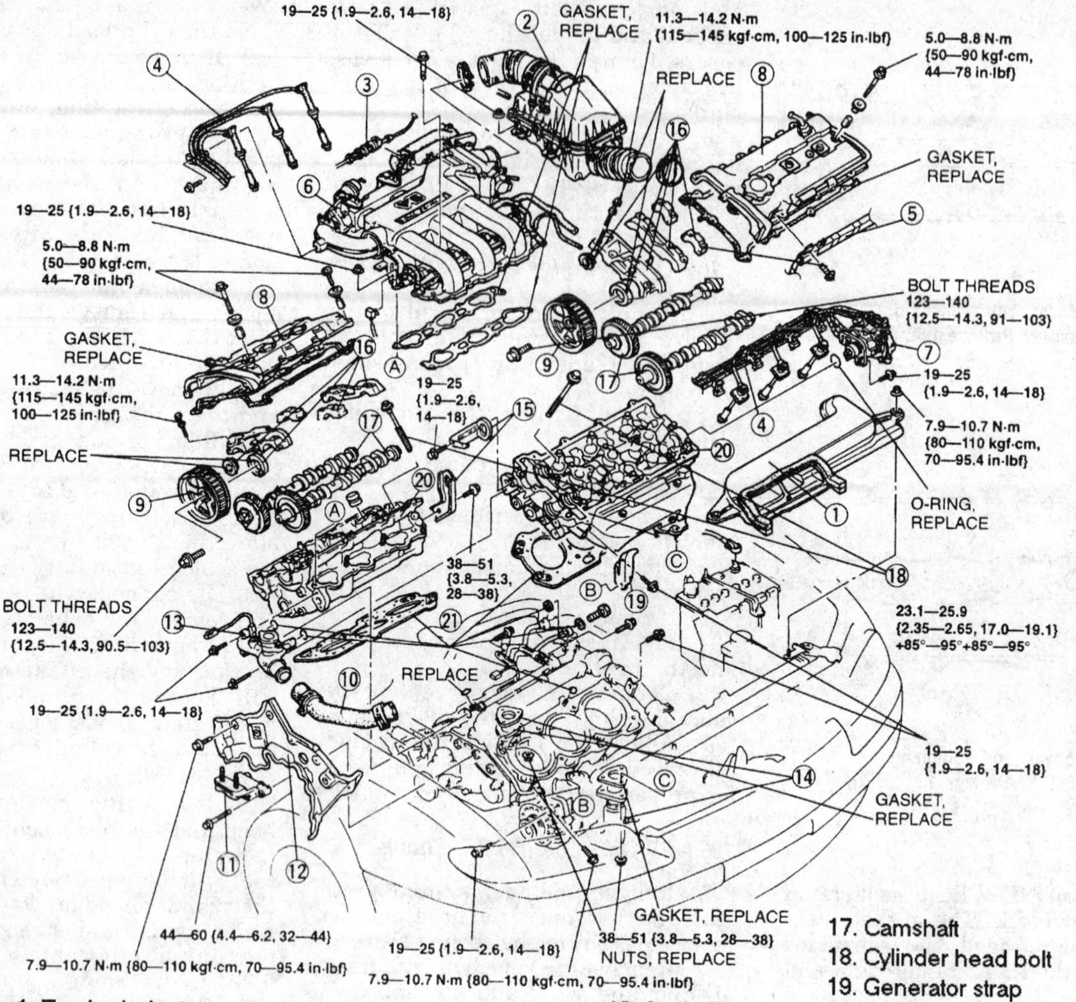

GASKET, REPLACE

19—25 {1.9—2.6, 14—18}

11.3—14.2 N·m {115—145 kgf·cm, 100—125 in·lbf}

REPLACE

5.0—8.8 N·m {50—90 kgf·cm, 44—78 in·lbf}

GASKET, REPLACE

19—25 {1.9—2.6, 14—18}

5.0—8.8 N·m {50—90 kgf·cm, 44—78 in·lbf}

BOLT THREADS 123—140 {12.5—14.3, 91—103}

GASKET, REPLACE

11.3—14.2 N·m {115—145 kgf·cm, 100—125 in·lbf}

19—25 {1.9—2.6, 14—18}

REPLACE

7.9—10.7 N·m {80—110 kgf·cm, 70—95.4 in·lbf}

O-RING, REPLACE

BOLT THREADS 123—140 {12.5—14.3, 90.5—103}

38—51 {3.8—5.3, 28—38}

19—25 {1.9—2.6, 14—18}

23.1—25.9 {2.35—2.65, 17.0—19.1} +85°—95°+85°—95°

REPLACE

19—25 {1.9—2.6, 14—18}

GASKET, REPLACE

44—60 {4.4—6.2, 32—44}

7.9—10.7 N·m {80—110 kgf·cm, 70—95.4 in·lbf}

19—25 {1.9—2.6, 14—18}

GASKET, REPLACE

38—51 {3.8—5.3, 28—38} NUTS, REPLACE

7.9—10.7 N·m {80—110 kgf·cm, 70—95.4 in·lbf}

1. Fresh-air duct
2. Air cleaner assy
3. Accelerator cable
4. High-tension leads
5. Harness
6. Intake manifold assy
7. Distributor
8. Cylinder head cover
9. Camshaft pulley
10. Radiator hose, upper
11. No. 3 engine mount bracket
12. Seal plate
13. Water outlet
14. Exhaust pipe
15. Engine hanger
16. Camshaft cap, blind cap and oil seal
17. Camshaft
18. Cylinder head bolt
19. Generator strap
20. Cylinder head
21. Cylinder head gasket and o-ring

N·m {kgf·m, ft·lbf}

297485

Exploded view of the cylinder head assembly — 1993–95 MX3 1.8L (K8), 626 and MX6 2.5L (KL) engines

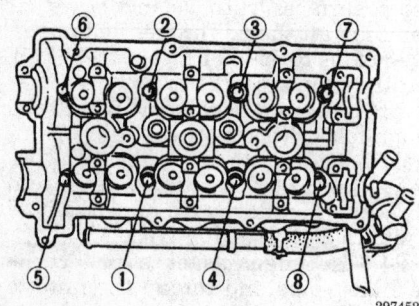

297452

Cylinder head bolt (tightening sequence) — 1993–95 MX3 1.8L (K8), 1993–97 626 and MX6 2.5L (KL) engines

27. Make sure the crankshaft and camshaft sprocket timing marks are aligned, install the timing belt and set the tension. Carefully rotate the crankshaft 2 turns to make sure the timing marks still line up.

28. Apply a thin bead of sealant to the cylinder head cover and install the new gasket. Install the cover and torque the cover bolts to 78 inch lbs. (9 Nm).

29. Install the timing belt cover(s) and tighten the bolts to 95 inch lbs. (11 Nm).

30. Install the alternator bracket. Tighten the bracket nut and bolt to 19 ft. lbs. (25 Nm).

31. Install the water pump pulley.

32. Install the alternator belt and adjust the tension.

33. Loosely install the power steering pump through and lockbolts. Connect the pump pressure switch connector.

34. Install the power steering pump belt and adjust the tension. Tighten the pump through-bolt to 45 ft. lbs. (61 Nm) and the lockbolt to 34 ft. lbs. (46 Nm).

35. Install the power steering pump belt shield and tighten the bolts to 86 inch lbs. (9 Nm). Install the power steering hose brackets to the cylinder head cover, and tighten the bolts to 88 inch lbs. (10 Nm).

36. Install the coolant bypass pipe.

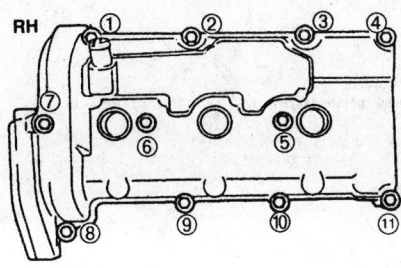

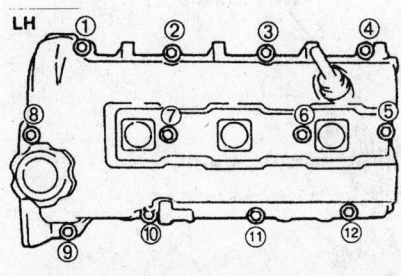

297455

Cylinder head cover (tightening sequence) — 1993–95 MX3 1.8L (K8), 1993–97 626 and MX6 2.5L (KL) engines

37. Connect the electrical engine harness connectors.

38. Connect the heater, brake vacuum, purge, fuel, water and upper radiator hoses.

39. Install the distributor, spark plugs, distributor cap and wires.

40. Connect and adjust the accelerator and throttle cables.

41. Install the air ducts, cleaner and splash shield.

42. Connect the negative battery cable. Fill and bleed the cooling system. Change the engine oil.

43. Adjust the ignition timing.

44. Start the engine and bring to normal operating temperature. Check for leaks. Check the ignition timing and idle speed.

626 and MX6 2.0L (FS) Engines

1. Relieve the fuel system pressure and disconnect the negative battery cable. Drain the cooling system.

2. Remove the splash shield, fresh air duct and air cleaner assembly.

3. Disconnect the accelerator cable.

4. Disconnect the following hoses: heater, brake vacuum, purge, fuel, water and upper radiator.

5. Remove the accessory drive belts. Remove the power steering pump bolts and secure the pump

aside with mechanics wire, leaving the hoses attached.

6. Remove the alternator bracket nut and bolt and position the bracket aside. Remove the exhaust manifold.

7. Label and disconnect the spark plug wires. Remove the power steering hose brackets from the cylinder head cover.

8. Label and disconnect the distributor/coil connectors, engine coolant temperature sensor connector, cooling fan coolant temperature sensor connector and temperature gauge sensor connector.

9. Disconnect the hoses from the cylinder head cover and loosen the cover bolts in 5–6 steps, in the reverse order of the tightening sequence. Remove the cylinder head cover.

10. Remove the timing belt cover and the timing belt.

11. Remove the coolant temperature sensor housing from the cylinder head. Remove the distributor.

12. Remove the camshaft sprockets and the camshaft.

13. Remove the hydraulic lifters. Identify each lifter as it is removed so it can be reinstalled in the same position. If the lifters are to be reused, store them upside down in an oil-filled sealed container.

14. Loosen the cylinder head bolts, in 2–3 steps, in sequence. Remove the bolts and the cylinder head.

15. Clean all gasket mating surfaces. Inspect the cylinder head for damage, cracks, and water and oil leakage. Check the head gasket surface for distortion using a straight-edge and feeler gauge. Maximum allowable distortion is 0.004 inch (0.10mm).

To install:

16. Position a new cylinder head gasket on the cylinder block, and install the cylinder head.

17. Apply clean engine oil to the bolt threads and seating faces. Install new cylinder head bolts and tighten in 2–3 steps, in sequence, to 13–16 ft. lbs. (17–22 Nm).

18. Paint a mark on the edge of each cylinder head bolt to use as a reference. Turn each bolt, in sequence, 90 degrees. Again, turn each bolt, in sequence, an additional 90 degrees.

19. Apply clean engine oil to the hydraulic lifters and install them in their original positions. Make sure they move freely in the bores.

20. Install the camshafts and sprockets. Install the distributor and connect the distributor/coil connectors.

21. Install the timing belt and cover.

22. Install a new cylinder head cover gasket on the cylinder head cover. Apply sealant to the cylinder head surface in the area adjacent to the front camshaft caps, then install the cover. Tighten the bolts in 5–6 steps, in sequence, to 61–95 inch lbs. (7–11 Nm).

23. Connect the hoses to the cylinder head cover. Connect the spark plug wires.

24. Install the exhaust manifold and the alternator bracket. Tighten the bracket nut and bolt to 19 ft. lbs. (25 Nm).

25. Install the alternator belt and adjust the tension.

26. Loosely install the power steering pump through and lockbolts. Connect the pump pressure switch connector.

27. Install the power steering pump belt and adjust the tension. Tighten the pump through-bolt to 45 ft. lbs. (61 Nm) and the lockbolt to 34 ft. lbs. (46 Nm).

28. Install the power steering pump belt shield and tighten the bolts to 86 inch lbs. (9 Nm). Install the power steering hose brackets to the cylinder head cover, and tighten the bolts to 88 inch lbs. (10 Nm).

29. Install the coolant temperature sensor housing with a new gasket. Tighten the bolts to 19 ft. lbs. (25 Nm). Connect the electrical connectors at the housing.

30. Connect and adjust the accelerator cable.

31. Connect the heater, brake vacuum, purge, fuel, water and upper radiator hoses.

32. Install the air cleaner assembly, fresh air duct and splash shield.

33. Connect the negative battery cable. Fill and bleed the cooling system. Run the engine and check for proper operation.

1993–95 929 3.0L (JE-ZE) Engines

1. Properly relieve the fuel system pressure.

2. Disconnect the negative battery cable, and drain the cooling system.

3. Remove the air cleaner and air intake pipe.

4. Remove the cooling fan and fan shroud.

5. Disconnect and mark the spark plug wires and remove the spark plugs. Remove the idler pulleys and the accessory drive belts.

6. Remove the coolant bypass hose and the upper radiator hose. Disconnect the electrical connector and remove the distributor.

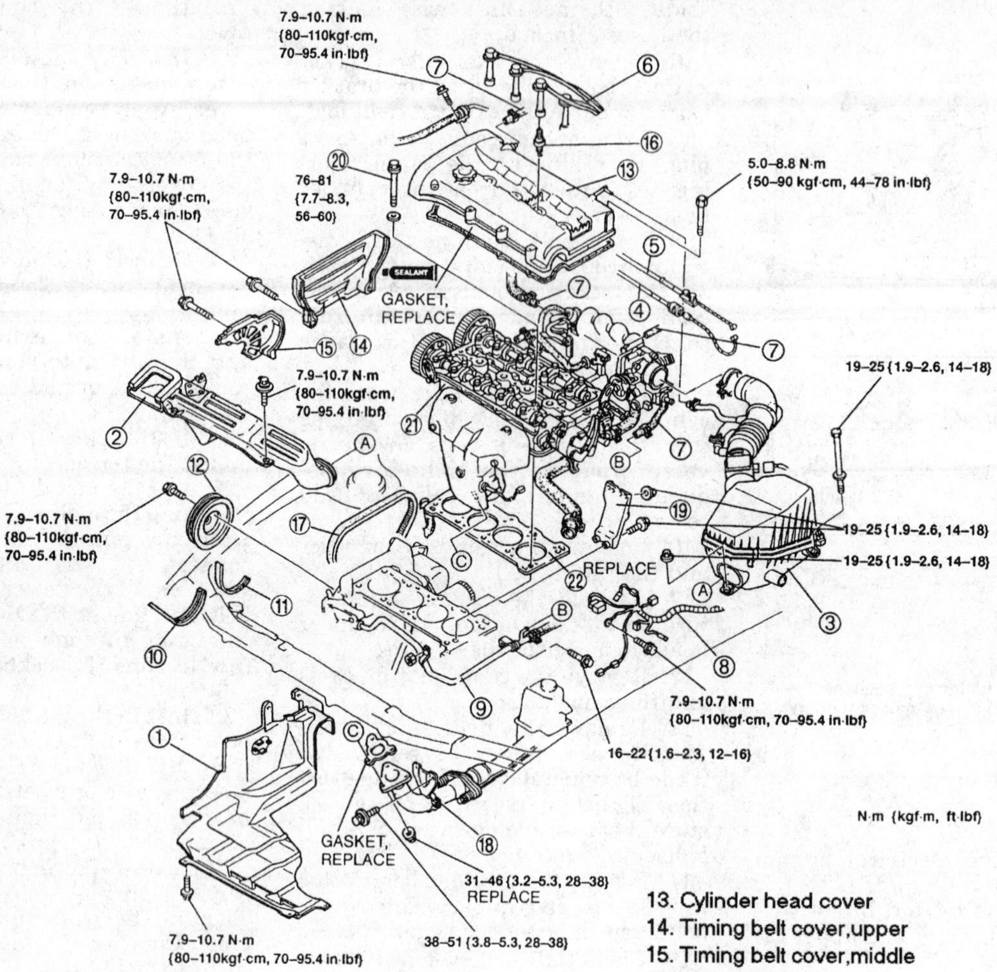

7.9–10.7 N·m
{80–110kgf·cm,
70–95.4 in·lbf}

7.9–10.7 N·m
{80–110kgf·cm,
70–95.4 in·lbf}

76–81
{7.7–8.3,
56–60}

SEALANT

GASKET,
REPLACE

5.0–8.8 N·m
{50–90 kgf·cm, 44–78 in·lbf}

19–25 {1.9–2.6, 14–18}

7.9–10.7 N·m
{80–110kgf·cm,
70–95.4 in·lbf}

19–25 {1.9–2.6, 14–18}

19–25 {1.9–2.6, 14–18}

7.9–10.7 N·m
{80–110kgf·cm,
70–95.4 in·lbf}

REPLACE

7.9–10.7 N·m
{80–110kgf·cm, 70–95.4 in·lbf}

16–22 {1.6–2.3, 12–16}

N·m {kgf·m, ft·lbf}

GASKET,
REPLACE

31–46 {3.2–5.3, 28–38}
REPLACE

38–51 {3.8–5.3, 28–38}

7.9–10.7 N·m
{80–110kgf·cm, 70–95.4 in·lbf}

1. Splash shield
2. Fresh-air duct
3. Air cleaner assembly
4. Accelerator cable
5. Throttle cable
6. High-tension lead
7. Hoses
8. Harness
9. Water bypass pipe
10. P/S and/or A/C drive belt
11. Generator drive belt
12. Water pump pulley
13. Cylinder head cover
14. Timing belt cover, upper
15. Timing belt cover, middle
16. Spark plug
17. Timing belt
18. Front exhaust pipe
19. Intake manifold bracket
20. Cylinder head bolt
21. Cylinder head
22. Cylinder head gasket

223464

Exploded view of the cylinder head assembly — 1994–95 MX3 1.6L (B6-ZE) engines

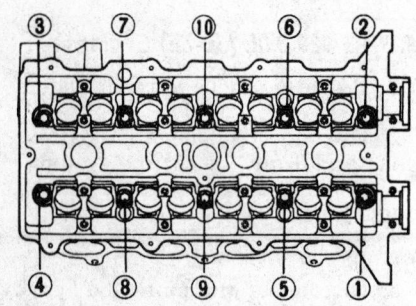

223465

Cylinder head bolt (loosening sequence; tighten in reverse order) — 1994–95 MX3 1.6L (B6-ZE) engines

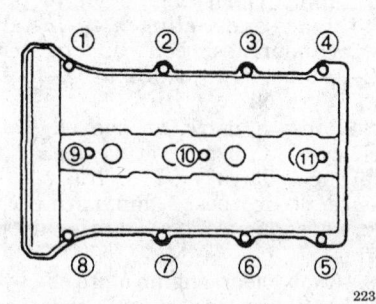

223466

Cylinder head cover (bolt tightening sequence) — 1994–95 MX3 1.6L (B6-ZE) engines

7. Hold the crankshaft pulley with a suitable tool and remove the bolt. Remove the crankshaft pulley. Be careful not to damage the sensor rotor.

8. Remove the timing belt covers. Rotate the crankshaft until the camshaft and crankshaft sprocket timing marks are aligned.

9. Remove the upper idler pulley and the tensioner and pulley. Mark the direction of rotation on the timing belt and remove the timing belt.

10. Disconnect the accelerator cable.

11. Disconnect the harness connectors and remove the mounting bolts

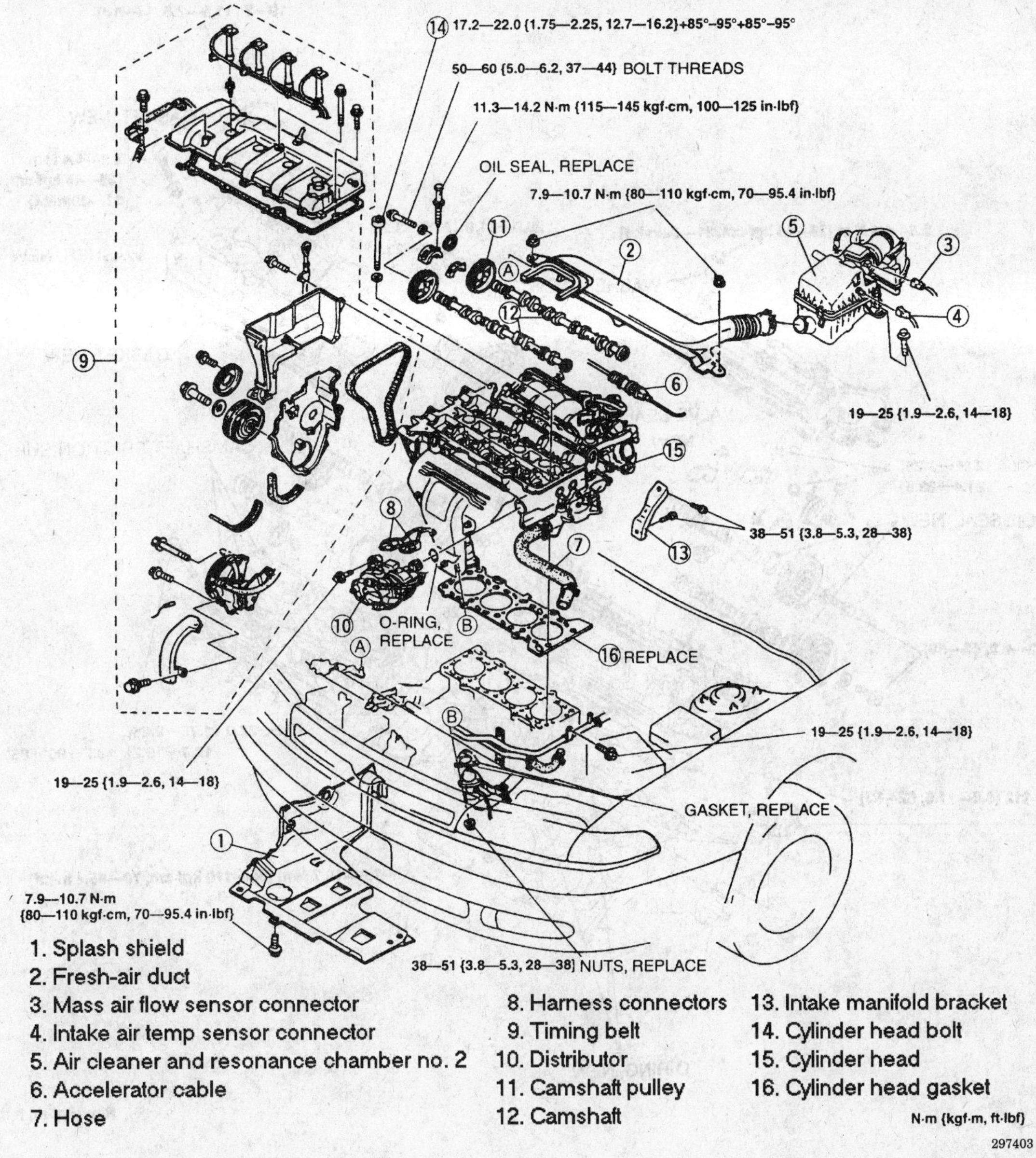

17.2—22.0 {1.75—2.25, 12.7—16.2}+85°—95°+85°—95°

50—60 {5.0—6.2, 37—44} BOLT THREADS

11.3—14.2 N·m {115—145 kgf-cm, 100—125 in-lbf}

OIL SEAL, REPLACE

7.9—10.7 N·m {80—110 kgf-cm, 70—95.4 in-lbf}

19—25 {1.9—2.6, 14—18}

38—51 {3.8—5.3, 28—38}

19—25 {1.9—2.6, 14—18}

O-RING, REPLACE

REPLACE

19—25 {1.9—2.6, 14—18}

GASKET, REPLACE

7.9—10.7 N·m {80—110 kgf-cm, 70—95.4 in-lbf}

38—51 {3.8—5.3, 28—38} NUTS, REPLACE

1. Splash shield
2. Fresh-air duct
3. Mass air flow sensor connector
4. Intake air temp sensor connector
5. Air cleaner and resonance chamber no. 2
6. Accelerator cable
7. Hose
8. Harness connectors
9. Timing belt
10. Distributor
11. Camshaft pulley
12. Camshaft
13. Intake manifold bracket
14. Cylinder head bolt
15. Cylinder head
16. Cylinder head gasket

N·m {kgf-m, ft-lbf}

297403

Exploded view of the cylinder head assembly — 626 and MX6 2.0L (FS) engine

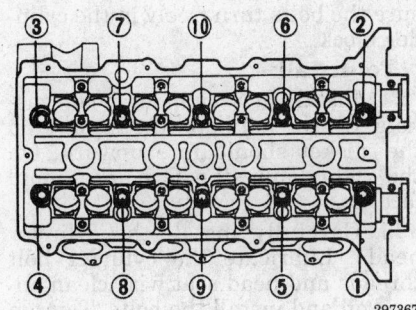

297367

Cylinder head bolt (loosening sequence; tighten in reverse order) — 626 and MX6 2.0L (FS) engines

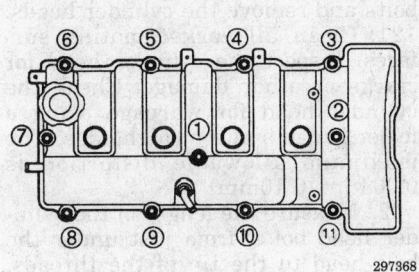

297368

Cylinder head cover (bolt tightening sequence) — 626 and MX6 2.0L (FS) engines

from the dynamic chamber (upper intake manifold plenum).

12. Disconnect and plug the fuel lines. Label and disconnect the engine electrical connectors.

13. Loosen the lower intake manifold bolts in 2–3 steps. Remove the lower intake manifold.

14. Remove the bolts from the transmission dipstick tube bracket and the coolant bypass pipe.

15. Disengage the heated oxygen sensor connector. Disconnect the exhaust pipe from the exhaust manifolds. Remove the exhaust manifold insulators and the manifolds.

16. Remove the cylinder head covers.

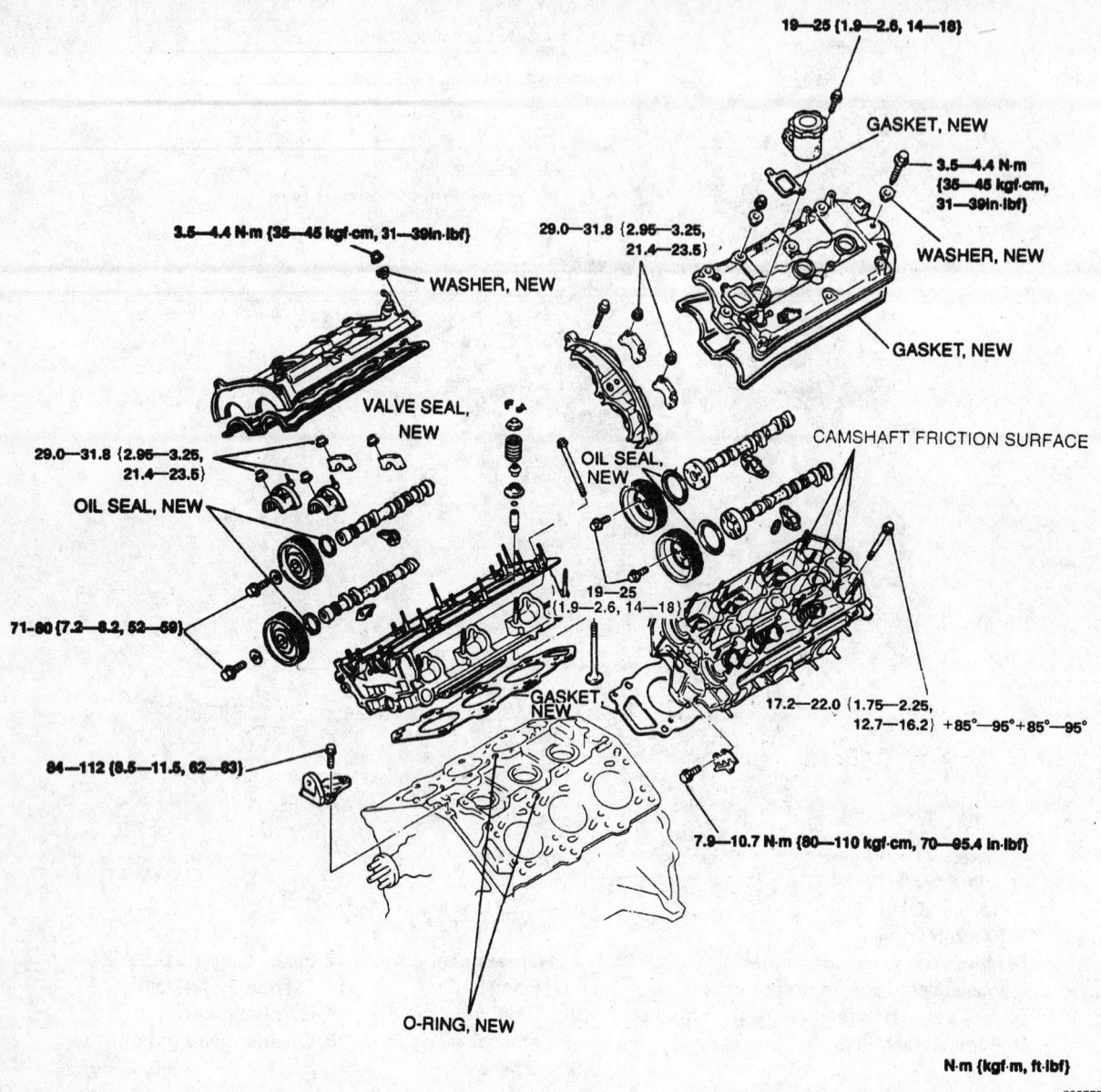

19—25 {1.9—2.6, 14—18}

GASKET, NEW

3.5—4.4 N·m {35—45 kgf·cm, 31—39In·lbf}

WASHER, NEW

GASKET, NEW

29.0—31.8 {2.95—3.25, 21.4—23.5}

3.5—4.4 N·m {35—45 kgf·cm, 31—39In·lbf}

WASHER, NEW

VALVE SEAL, NEW

CAMSHAFT FRICTION SURFACE

29.0—31.8 {2.95—3.25, 21.4—23.5}

OIL SEAL, NEW

OIL SEAL, NEW

71—80 {7.2—8.2, 52—59}

19—25 {1.9—2.6, 14—18}

17.2—22.0 {1.75—2.25, 12.7—16.2} +85°—95° +85°—95°

84—112 {8.5—11.5, 62—83}

GASKET NEW

7.9—10.7 N·m {80—110 kgf·cm, 70—95.4 in·lbf}

O-RING, NEW

N·m {kgf·m, ft·lbf}

235777

Exploded view of the cylinder head assembly — 929 with 3.0L (JE-ZE) engines

17. Hold the camshaft with a wrench on the hexagon cast into the camshaft and remove the camshaft sprocket bolt. Remove the camshaft sprockets.

18. Remove the bolts from the front camshaft cap on the left cylinder head. Loosen the remaining camshaft cap bolts in 2–3 steps in the proper sequence. Mark the position of the caps so they can be reinstalled in their original positions. Remove the caps and remove the camshafts. Label the camshafts so they can be reinstalled in their proper locations.

19. Remove the rocker arms.

20. Loosen the cylinder head bolts in 2–3 steps, in the reverse of the tightening sequence. Remove the bolts and remove the cylinder heads.

21. Clean all gasket mating surfaces. Inspect the cylinder head for cracks or other damage. Check the cylinder head for warpage using a feeler gauge and a straightedge. The maximum allowable distortion is 0.004 in. (0.10mm).

22. Measure the length of the cylinder head bolts, from just under the bolt head to the tip of the threads. The bolts should not exceed 4.29 inches (109mm). Replace any bolts that exceed specification.

23. Clean the bolt threads and the threads in the cylinder block. Make sure the bolts turn freely in the cylinder block.

To install:

24. Install new cylinder head gaskets. On the left bank the **L** mark on the gasket should face upward. On the right bank, the **R** mark should face up.

25. Carefully install the cylinder heads. Lubricate the cylinder bolt threads and head seat with clean engine oil and install the bolts. Tighten the bolts as follows:

a. Tighten the bolts, in sequence, to 12.7–16.2 ft. lbs. (17–22 Nm).

b. Make a paint mark on each bolt head.

c. Tighten each bolt, in sequence, 90 degrees.

d. Turn each bolt, in sequence, an additional 90 degrees.

26. Lubricate the rocker arms with clean engine oil and install them over the valve stems and hydraulic lash adjusters.

27. Lubricate the camshaft lobes and journals with clean engine oil and install the camshafts on the cylinder heads. Apply clean engine oil to the lips of the new camshaft seals and install them on the camshafts.

28. Apply a small amount of silicone sealant to the cylinder head on the front camshaft cap mating surfaces. Do not allow sealant to get on the camshaft journal, oil seal face or camshaft thrust face.

29. Install the camshaft caps in their original locations. Gradually tighten the cap bolts, in sequence, to 24 ft. lbs. (32 Nm). Tighten the front cap bolts on the left cylinder head to 95 inch lbs. (11 Nm).

30. Hold the camshafts using a wrench on the hexagon cast into the camshaft. Install the camshaft sprockets with the retaining bolts. Tighten the sprocket bolts on the right cylinder head to 59 ft. lbs. (80 Nm) and the sprocket bolts on the left cylinder head to 19 ft. lbs. (25 Nm).

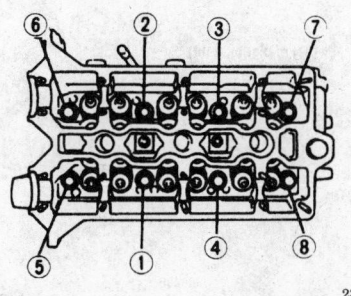

235778

Cylinder head bolt tightening sequence (loosen in reverse order) — 1993–95 929 3.0L (JE-ZE) engines

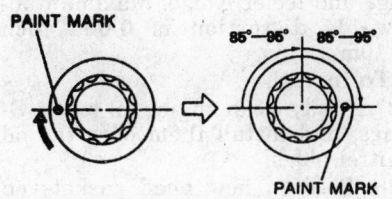

235779

Cylinder head bolt torque procedure — 1993–95 929 3.0L (JE-ZE) engines

31. Apply sealant to the cylinder head in the areas adjacent to the front camshaft caps. Install the cylinder head covers and tighten the nuts to 39 inch lbs. (4 Nm).

32. Using new gaskets, install the exhaust manifolds. Tighten the nuts to 21 ft. lbs. (28 Nm). Install the exhaust manifold insulators and tighten the bolts to 19 ft. lbs. (25 Nm).

33. Connect the exhaust pipes to the exhaust manifolds with new gaskets. Tighten the flange nuts to 38 ft. lbs. (52 Nm). Connect the heated oxygen sensor.

34. Install the bolts attaching the coolant bypass pipe and the transmission dipstick tube.

35. Install the lower intake manifold using new gaskets. Tighten the mounting nuts, in 2–3 steps, to 19 ft. lbs. (25 Nm). Connect the fuel lines.

36. Connect the engine electrical connectors.

37. Install the upper intake manifold plenum (dynamic chamber) to the lower intake manifold, using a new gasket. Tighten the bolts to 19 ft. lbs. (25 Nm), and engage the harness connectors.

38. Connect the accelerator cable.

39. Position the timing belt tensioner on a press. Place a flat washer at the bottom of the tensioner body to prevent damage to the body plug. Slowly press in the tensioner rod, but do not exceed 2200 lbs. Insert a pin in the tensioner body to hold the rod in place, then install the tensioner and tighten the mounting bolts to 19 ft. lbs. (25 Nm).

40. Make sure the crankshaft and camshaft sprocket timing marks are properly aligned. Install the timing belt over the sprockets and pulleys in the following order: crankshaft sprocket, lower idler pulley, left cylinder head exhaust cam sprocket, left cylinder head intake cam sprocket, tensioner pulley, right cylinder head exhaust cam sprocket and right cylinder head intake cam sprocket.

41. Push the belt down and install the upper idler pulley. Tighten the upper idler pulley bolt to 38 ft. lbs. (52 Nm). Make sure the camshaft and crankshaft sprocket timing marks are still aligned after installing the upper idler pulley.

42. Turn the crankshaft 2 revolutions in the normal direction of rotation and realign the timing marks. Make sure all timing marks are correctly aligned.

43. Remove the pin from the automatic tensioner and again rotate the

crankshaft 2 turns. Confirm that the timing marks are aligned.

44. Install the timing belt covers and tighten the bolts to 95 inch lbs. (11 Nm).

45. Install the crankshaft pulley. Hold the pulley with a suitable tool and tighten the pulley bolt to 123 ft. lbs. (167 Nm). Be careful not to damage the sensor rotor.

46. Install the spark plugs and the distributor assembly. Connect the spark plug wires and the distributor electrical connector.

47. Install the upper radiator hose and water bypass hose. Install the idler pulleys and accessory drive belts. Adjust the belt tension.

48. Install the shroud and cooling fan. Install the air cleaner and ducts.

49. Connect the negative battery cable. Fill and bleed the cooling system. Change the engine oil.

50. Start the engine and bring to normal operating temperature. Check for leaks. Check the ignition timing and idle speed.

1995–97 Millenia 2.3L (KJS) Engines

1. Relieve the fuel system pressure. Disconnect the negative battery cable.

2. Drain the engine coolant.

3. Raise and safely support the vehicle.

4. Disconnect the oxygen sensor connectors. Remove the exhaust pipe-to-manifold nuts and lower the exhaust pipes.

5. Remove the right-hand three-way catalytic converter. Lower the vehicle.

6. Remove the Lysholm compressor (supercharger).

7. Remove the intake manifold.

8. Remove the timing belt covers and timing belt.

9. Remove the spacer and O-ring from the front of the camshaft.

10. Remove the ignition coils.

11. Remove the cylinder head cover mounting bolts, in 5–6 steps, using the reverse of the tightening sequence. Remove the cylinder head cover.

12. Remove the camshaft sprockets.

13. Turn the camshafts so the knock pins are aligned with the marks on the camshaft caps. This will reduce the pressure on the adjustment shims.

14. Note the markings on the camshaft caps prior to removal, so they can be reinstalled in the same positions. The right hand (rear) caps are marked with numbers and the left hand (front) caps are marked with letters.

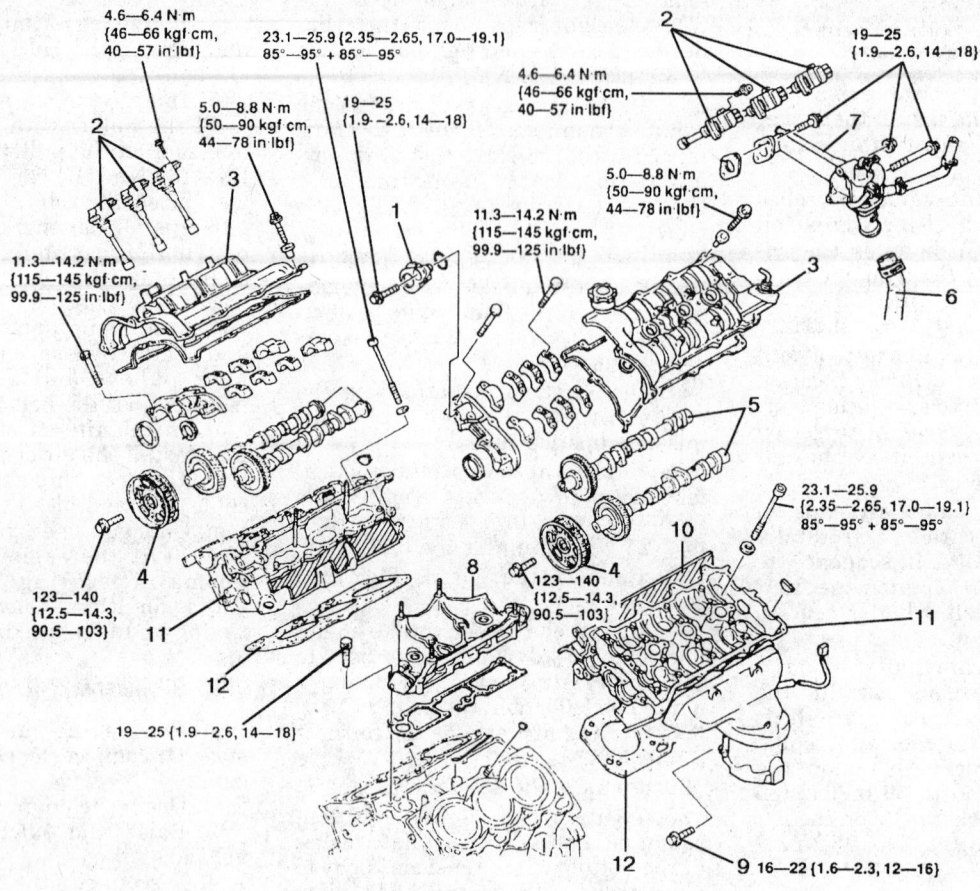

1. Spacer
2. Ignition coil
3. Cylinder head cover
4. Camshaft pulley
5. Camshaft
6. Lower radiator hose
7. Water inlet pipe
8. Lysholm compressor bracket
9. Generator bolt
10. Rubber insulator (LH)
11. Cylinder head
12. Cylinder head gasket

308238

Cylinder head assembly (exploded view) — 1995–97 Millenia 2.3L (KJS) engines

15. Loosen the front camshaft cap bolts in sequence, in 5–6 steps. Remove the front camshaft caps.

16. Remove the remaining camshaft cap bolts in the proper sequence. Remove the caps, being sure to remove the thrust caps last. Do not damage the cylinder head thrust bearing support.

17. Remove the camshafts and oil seals.

18. Remove the lifters and adjustment shims. Identify and mark each lifter as it is removed so it can be reinstalled in the same position.

19. Remove the lower radiator hose and water inlet pipe.

20. Remove the Lysholm compressor bracket.

21. Remove the alternator bracket bolt to gain additional clearance.

22. Remove the rubber insulator from the left-hand cylinder head.

23. Temporarily install the number three engine mount, which was removed with the timing belt, to support the engine. Remove the engine support device.

24. Loosen the cylinder head bolts, in 2–3 steps, in the reverse order of the torque sequence. Remove the bolts and remove the cylinder heads.

25. Remove the oil control plug O-rings.

26. Clean all gasket mating surfaces. Inspect the cylinder head for damage, cracks, and water and oil leakage. Check the head gasket surface for distortion using a straight-edge and feeler gauge. Maximum allowable distortion is 0.004 inch (0.10mm).

To install:

27. Apply clean engine oil to the O-rings, and install them onto the oil control plugs.

28. Position new head gaskets on the cylinder block. The gaskets cannot be interchanged between sides and are marked **R** and **L** for right and left side.

29. Install the cylinder heads. Apply clean engine oil to the threads of new cylinder head bolts and install. Tighten the cylinder head bolts in 2–3 steps, in sequence, to 17–19 ft. lbs. (23–26 Nm).

30. Paint a mark on the edge of each cylinder head bolt to use as a reference. Turn each bolt, in sequence, 90 degrees. Again, turn each bolt, in sequence, an additional 90 degrees.

31. Install the rubber insulator onto the left-hand cylinder head.

32. Fit the knock sensor harness into the drill hole on the cylinder block. Pass the harness under the rubber insulator.

33. Install an engine support device, and remove the number three engine mount.

34. Install the alternator bracket bolt. Tighten the mounting bolt to 12–16 ft. lbs. (16–22 Nm).

35. Install the Lysholm compressor bracket. Tighten the mounting bolts to 14–18 ft. lbs. (19–25 Nm).

36. Install the water inlet pipe. Tighten the mounting bolts to 14–18 ft. lbs. (19–25 Nm). Install the lower radiator hose.

37. Apply clean engine oil to the lifters, then install them in their original positions. Verify that they move smoothly in their bore.

RH

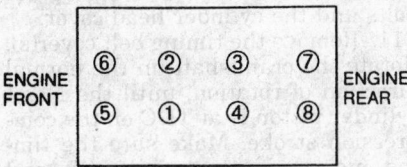

LH

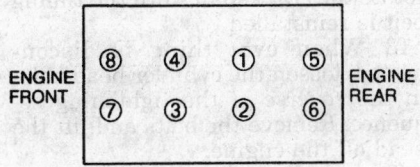

308240

Cylinder head bolt (tightening sequence) — 1995–97 Millenia 2.3L (KJS) engines

38. Install new oil seals on the camshafts. Apply clean engine oil to the camshaft lobes, journals and supports.

39. Install the camshafts so the gear marks align.

40. Remove all oil and dirt from the mating surfaces between the front camshaft cap and the cylinder head.

41. Install the thrust caps. Tighten the thrust cap bolts, in 5–6 steps, until the caps are fully seated on the cylinder head.

42. Apply silicone sealant, at a thickness of 0.06–0.09 inch (1.5–2.5mm), to the cylinder head surface in the area forward of the camshaft gear cavity.

43. Install the remaining camshaft caps in their original positions. Tighten the caps, in sequence, in five equal steps, with the final step being 100–125 inch lbs. (11–14 Nm).

44. Apply clean engine oil to the lip of the new camshaft oil seal. Push the seal in lightly by hand. Tap the seal in evenly with a seal installer (49 F401 337A or equivalent) with a final protrusion of 0–0.02 inch (0–0.5mm). Tap in a new blind cap.

45. Install the camshaft sprockets. Tighten the mounting bolts to 91–103 ft. lbs. (123–140 Nm).

46. Measure and adjust valve clearances.

47. Remove any sealant and gasket material from the cylinder head cover contact surfaces.

48. Apply silicone sealant to the cylinder head in the area adjacent to the front and rear camshaft caps. Install a new gasket on the cylinder head.

49. Install the cylinder head cover. Tighten the bolts in 5–6 steps, in sequence, to 44–78 inch lbs. (5–9 Nm).

50. Using a new O-ring, install the distributor.

51. Install the ignition coils.

52. Install the spacer, using a new O-ring. Tighten the mounting bolt to 14–18 ft. lbs. (19–25 Nm).

53. Install the timing belt and timing belt cover.

54. Install the intake manifold.

55. Install the Lysholm compressor (supercharger).

56. Raise and safely support the vehicle.

57. Install the right-hand three-way catalytic converter.

58. Connect the exhaust pipes to the manifolds and tighten the nuts to 28–38 ft. lbs. (38–51 Nm). Connect the oxygen sensor connectors. Lower the vehicle.

59. Connect the negative battery cable.

60. Fill and bleed the coolant system.

61. Run the engine and check for leaks.

1995–97 Millenia 2.5L (KLD) Engine

1. Relieve the fuel system pressure, and disconnect the negative battery cable.

2. Drain the cooling system.

3. Remove the fresh air duct and the air cleaner assembly.

4. Install an engine support device. Remove the timing belt covers and timing belt.

5. Remove the spark plug wires and distributor.

6. Remove the intake manifold.

7. Remove the upper radiator hose and water outlet.

8. Temporarily install the number three engine mount bracket, which was removed with the timing belt, to support the engine. Remove the engine support device.

9. Remove the seal plate from the front of the engine.

10. Raise and safely support the vehicle.

11. Disconnect the oxygen sensor connectors. Remove the exhaust pipe-to-manifold nuts and lower the exhaust pipes. Lower the vehicle.

12. Remove the alternator bracket.

13. Disconnect the ventilation pipe from the left cylinder head cover, remove the bolts and remove the cylinder head covers.

14. Remove the camshaft pulleys.

15. Remove the camshafts.

16. Remove the hydraulic lifters. Identify each lifter as it is removed so it can be reinstalled in the same position. If the lifters are to be reused, store them upside down in an oil-filled sealed container.

17. Loosen the cylinder head bolts, in 2–3 steps, in the reverse order of the torque sequence. Remove the bolts and remove the cylinder heads.

18. Clean all gasket mating surfaces. Inspect the cylinder head for damage, cracks, and water and oil leakage. Check the head gasket surface for distortion using a straightedge and feeler gauge. Maximum allowable distortion is 0.004 in. (0.10mm).

To install:

19. Position new head gaskets on the cylinder block. The gaskets cannot be interchanged between sides and are marked **R** and **L** for right and left side.

20. Install the cylinder heads. Apply clean engine oil to the threads of new cylinder head bolts and install. Tighten the cylinder head bolts in

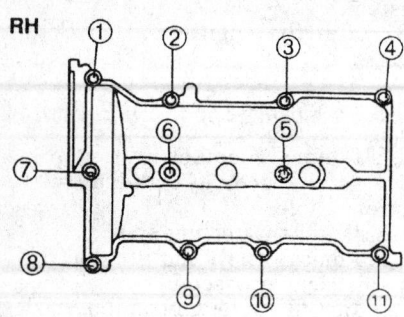

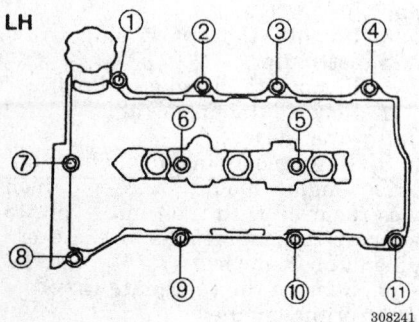

Cylinder head cover (tightening sequence) — 1995–97 Millenia 2.3L (KJS) engines

2–3 steps, in sequence, to 17–19 ft. lbs. (23–26 Nm).

21. Paint a mark on the edge of each cylinder head bolt to use as a reference. Turn each bolt, in sequence, 90 degrees. Again, turn each bolt, in sequence, an additional 90 degrees.

22. Apply clean engine oil to the hydraulic lifters and install them in their original positions. Make sure they move freely in the bores.

23. Install the camshafts.

24. Install the camshaft pulleys.

25. Apply sealant to the cylinder head surface in the area of the front and rear camshaft caps. Install new gaskets and install the cylinder head

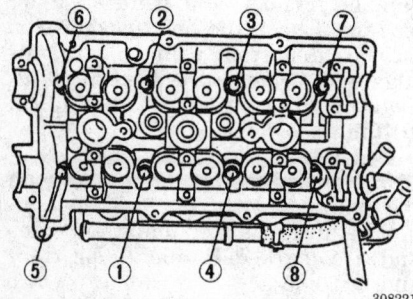

Cylinder head bolt (tightening sequence) — 1995–97 Millenia 2.5L (KLD) engine

covers. Tighten the bolts in 5–6 steps, in sequence, to 44–78 inch lbs. (5–9 Nm).

26. Install the alternator bracket. Tighten the mounting bolts to 14–18 ft. lbs. (19–25 Nm).

27. Raise and safely support the vehicle.

28. Connect the exhaust pipes to the manifolds and tighten the nuts to 28–38 ft. lbs. (38–51 Nm). Connect the oxygen sensor connectors. Lower the vehicle.

29. Install the seal plate, and tighten the mounting bolts to 70–95 inch lbs. (8–11 Nm).

30. Install an engine support device, and remove the number three engine mount bracket.

31. Install the water outlet. Tighten the mounting bolts to 14–18 ft. lbs. (19–25 Nm). Install the upper radiator hose.

32. Install the intake manifold using new gaskets. Tighten the mounting bolts to 14–18 ft. lbs. (19–25 Nm).

33. Install the distributor:

a. Apply clean engine oil to a new O-ring, and position it on the distributor.

b. Apply clean engine oil to the drive blade. Install the distributor with the blade fit into the camshaft groove.

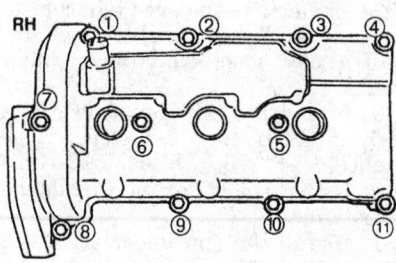

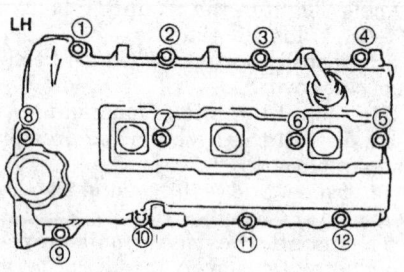

Cylinder head cover (tightening sequence) — 1995–97 Millenia 2.5L (VIN KLD) engine

c. Hand tighten the mounting bolts.

34. Install the spark plug wires.

35. Install the timing belt and timing belt covers.

36. Install the air cleaner assembly and the fresh air duct.

37. Connect the negative battery cable.

38. Fill and bleed the cooling system.

39. Adjust the ignition timing and idle speed. Run the engine and check for proper operation.

1993–94 Miata 1.6L (B6-ZE) and 1.8L (BPD) Engines

1. Disconnect the negative battery cable and remove the engine undercover.

2. Remove the air ducts from the air cleaner and throttle body.

3. Tag and disconnect the spark plug wires from the spark plugs. Remove the spark plugs and the distributor cap and wires assembly. Remove the distributor.

4. Drain the cooling system and disconnect the radiator and heater hoses.

5. Disconnect the exhaust pipe and remove the exhaust manifold.

6. On DOHC engines, remove the coolant bypass pipe.

7. Disconnect the accelerator cable.

8. Label and disconnect all necessary electrical connections and vacuum hoses. Disconnect the fuel lines.

9. Remove the intake manifold bracket and the intake manifold.

10. Remove the cylinder head cover bolts and the cylinder head cover.

11. Remove the timing belt cover(s). Rotate the crankshaft, in the normal direction of rotation, until the No. 1 cylinder piston is at TDC on the compression stroke. Make sure the timing marks on the crankshaft and camshaft sprocket(s) are properly aligned and mark the direction of rotation of the belt.

12. Loosen the timing belt tensioner and remove the belt. Do not rotate the crankshaft until the timing belt is reinstalled.

13. When everything is disconnected, loosen the cylinder head bolts in the reverse of the tightening sequence. Remove the bolts and lift the head off the engine.

To install:

14. Thoroughly, clean the cylinder head and the block contact surfaces. Examine the head gasket and check the cylinder head for cracks. Check the cylinder head for warpage using a feeler gauge and straightedge. The

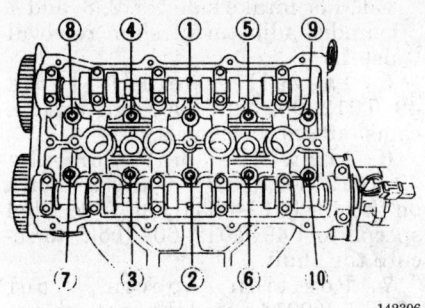

Cylinder head torque sequence — 1993 Miata 1.6L (B6-ZE) engine

maximum allowable distortion is 0.006 in. (0.15mm) on 1.6L engine. The maximum allowable distortion is 0.004 in. (0.10mm) on 1.8L engine.

15. Clean the cylinder head bolts and the threads in the block. Make sure the bolts turn freely in the block.

16. Install a new head gasket on the engine block. Make sure the camshaft sprocket timing marks are still aligned, as set during the removal procedure. Install the cylinder head.

17. Lubricate the bolt threads and seat surfaces with clean engine oil and install them. Torque the bolts in 2–3 steps to 56–60 ft. lbs. (75–81 Nm) in the proper sequence.

18. Make sure the crankshaft and camshaft sprocket timing marks are aligned, install the timing belt and set the tension. Carefully rotate the crankshaft 2 turns to make sure the timing marks still line up.

19. Apply a thin bead of sealant to the cylinder head cover and install the new gasket. Install the cover and torque the cover bolts to 78 inch lbs. (9 Nm).

20. Install the timing belt cover(s) and tighten the bolts to 95 inch lbs. (11 Nm).

21. Use new gaskets and install the manifolds. Torque the intake manifold bolts/nuts to 19 ft. lbs. (25 Nm) and install the intake manifold bracket. Torque the exhaust manifold nuts to 34 ft. lbs. (46 Nm).

22. Use a new gasket to connect the exhaust pipe and torque the nuts to 34 ft. lbs. (46 Nm).

23. If removed, install the radiator and connect all cooling system hoses. On DOHC engines, install the coolant bypass pipe.

24. Install the distributor, spark plugs, distributor cap and wires.

25. Connect all vacuum and fuel system hoses and connect all wiring.

26. Connect the accelerator cable and install the air ducts and engine undercover.

27. Connect the negative battery cable. Fill and bleed the cooling system. Change the engine oil.

28. Start the engine and bring to normal operating temperature. Check for leaks. Check the ignition timing and idle speed.

1995–97 Miata 1.8L (BPD) Engines

1. Relieve the fuel system pressure, and disconnect the negative battery cable. Drain the cooling system.

2. Remove the air cleaner assembly.

3. Disconnect the accelerator cable.

4. Tag and disconnect the spark plug wires from the spark plugs. Remove the spark plugs.

5. Disconnect the following hoses: heater, brake vacuum, purge, fuel, water and cruise control.

6. Label and disconnect the camshaft position sensor and coil, heated oxygen sensor, steering pressure sensor, idle air control valve, throttle position sensor, fuel injector and ground connectors.

7. Remove the exhaust manifold heat shield.

8. Remove the coolant bypass pipe.

9. Remove the accessory drive belts. Remove the power steering pump bolts and secure the pump aside with mechanics wire, leaving the hoses attached.

10. Remove the water pump pulley.

11. Remove the alternator bracket nut and bolt, and position the bracket aside.

12. Disconnect the hoses from the cylinder head cover, and loosen the cover bolts in 5–6 step sequences. Remove the cylinder head cover.

13. Remove the timing belt covers. Rotate the crankshaft, in the normal direction of rotation, until the No. 1 cylinder piston is at TDC on the compression stroke. Make sure the timing marks on the crankshaft and camshaft sprocket(s) are properly aligned and mark the direction of rotation of the belt.

14. Loosen the timing belt tensioner and remove the belt. Do not rotate the crankshaft until the timing belt is reinstalled.

15. Remove the camshaft sprockets, and the camshaft.

16. Remove the hydraulic lifters. Identify each lifter as it is removed so it can be reinstalled in the same position. If the lifters are to be reused, store them upside down in an oil-filled sealed container.

17. Disconnect the exhaust pipe and remove the exhaust manifold.

18. Loosen the cylinder head bolts, in 2–3 steps, in the correct sequence. Remove the bolts and the cylinder head.

To install:

19. Thoroughly clean the cylinder head and the block contact surfaces. Examine the head gasket and check the cylinder head for cracks. Check the cylinder head for warpage using a feeler gauge and straightedge. The maximum allowable distortion is 0.004 in. (0.10mm).

20. Clean the cylinder head bolts and the threads in the block. Make sure the bolts turn freely in the block.

21. Install a new head gasket on the engine block. Make sure the camshaft sprocket timing marks are still aligned, as set during the removal procedure. Install the cylinder head.

22. Lubricate the bolt threads and seat surfaces with clean engine oil and install them. Torque the bolts in 2–3 steps to 56–60 ft. lbs. (75–81 Nm) in the proper sequence.

23. Use new gaskets and install the manifold. Torque the exhaust manifold nuts to 34 ft. lbs. (46 Nm).

24. Use a new gasket to connect the exhaust pipe and torque the nuts to 34 ft. lbs. (46 Nm).

25. Apply clean engine oil to the hydraulic lifters and install them in their original positions. Make sure they move freely in the bores.

26. Install the camshaft and sprockets.

27. Make sure the crankshaft and camshaft sprocket timing marks are aligned, install the timing belt and set the tension. Carefully rotate the crankshaft 2 turns to make sure the timing marks still line up.

28. Apply a thin bead of sealant to the cylinder head cover and install the new gasket. Install the cover and torque the cover bolts to 44–78 in. lbs. (5–9 Nm).

29. Install the timing belt cover(s) and tighten the bolts to 95 inch lbs. (11 Nm).

30. Install the alternator bracket. Tighten the bracket nut and bolt to 19 ft. lbs. (25 Nm).

31. Install the water pump pulley.

32. Install the alternator belt and adjust the tension.

33. Loosely install the power steering pump through and lockbolts. Connect the pump pressure switch connector.

34. Install the power steering pump belt and adjust the tension. Tighten the pump through-bolt to 45 ft. lbs.

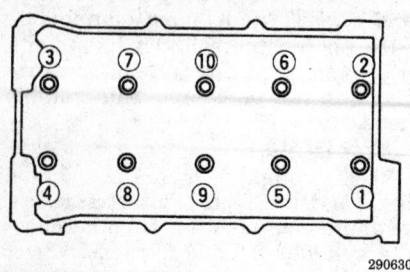

Cylinder head bolt (loosening sequence; tighten in reverse order) — 1995–97 Miata 1.8L (BPD) engines

(61 Nm) and the lockbolt to 34 ft. lbs. (46 Nm).

35. Install the power steering pump belt shield and tighten the bolts to 86 inch lbs. (9 Nm). Install the power steering hose brackets to the cylinder head cover, and tighten the bolts to 88 inch lbs. (10 Nm).

36. Install the coolant bypass pipe.

37. Install the exhaust manifold heat shield.

38. Connect the electrical engine harness connectors.

39. Connect the heater, brake vacuum, purge, fuel, water and cruise control hoses.

40. Install the spark plugs and wires.

41. Connect and adjust the accelerator cable.

42. Install the air cleaner assembly.

43. Connect the negative battery cable. Fill and bleed the cooling system. Change the engine oil.

44. Check the ignition timing and properly adjust if necessary.

45. Start the engine and bring to normal operating temperature. Check for leaks. Check the ignition timing and idle speed.

Valve Lash

ADJUSTMENT

Many of these procedures require the use of special service tools. Read the procedure first prior to disassembly.

1996–97 626, MX6 2.0L (FS), 2.5L (KL), Miata 1.8L (BPD), Protege 1.8L (BPE), and 1995–97 Millenia 2.5L (KLD) Engines

There is no valve adjustment on these engines. Measure the valve clearance with a feeler gauge. If it is more than 0.0059 in. (0.15 mm), then you must replace the hydraulic lash adjuster (HLA).

Press down HLA by hand.
If it moves, replace HLA.
If it does not move, HLA normal.

Measure valve clearance.

FEELER GAUGE

If more than 0.15 mm (0.0059 in), replace HLA.

312383

HLA clearance measurement

1995–97 Protege 1.8L (BPD) Engines

NOTE: With the engine cold, standard valve clearance is 0.010–0.012 inch (0.25–0.31mm) on intake and exhaust sides.

1. Measure the valve clearance by turning the crankshaft clockwise until the number one piston is at TDC.

2. Measure the valve clearance at **A**. Turn the crankshaft clockwise 360° until the number four piston is at TDC. Measure the valve clearance at **B**.

NOTE: If the valve clearance exceeds the standard, replace the adjustment shim.

3. Turn the crankshaft clockwise until the cam, on the camshaft requiring the adjustment shim replacement, is positioned straight up.

4. Remove the camshaft cap bolts as follows:

   a. For exhaust side No. 1, 2, and 3 cylinder adjustment shim removal use **A**.

   b. For intake side No. 1, 2, and 3 cylinder adjustment shim removal use **B**.

   c. For exhaust side No. 2, 3, and 4 cylinder adjustment shim removal use **C**.

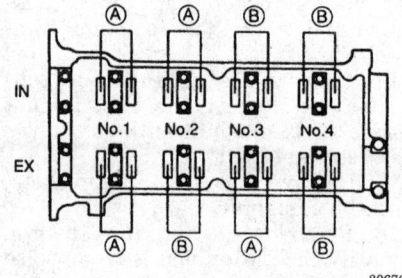

Measuring the valve clearance — 1995–97 Protege 1.8L (BPD) engines

d. For intake side No. 2, 3, and 4 cylinder adjustment shim removal use **D**.

5. Install special tools 49–T012–002 and 003, using the camshaft cap bolt holes.

6. Align the mark on the 49–T012–002 (shaft) with the mark on the 49–T012–003 (clamp). Tighten special tool 49–T012–004 (bolt) to secure the shaft.

7. Position special tool 49–T012–001 toward the center of the cylinder head, and mount it on the shaft where the adjustment shim needs replacement.

8. Position the notch of the tappet to allow a small pry tool to be inserted.

9. Set the special tool on the tappet by its notch. Tighten the mounting bolt **B** securing it on the shaft.

10. Tighten bolt **C**, and press down the tappet.

11. Using a small pry tool, pry the adjustment shim upwards through the notch on the tappet. Remove the shim with a magnet.

12. Select and install the proper adjustment shim. Loosen bolt **C** to allow the tappet to move up, and loosen bolt **B** to remove special tool 49–T012–001.

13. Remove special tools 49–T012–002, 003 and 004, and tighten the camshaft cap bolts to 100–125 inch lbs. (11–14 Nm).

14. Repeat the procedure for all necessary adjustment shims. Check the valve clearance.

1993 MX3 1.8L (K8), 1993–94 626 and MX6 2.0L FS) and 2.5L (KL) Engines

The hydraulic lifters are not adjustable. When the lifters are removed from the engine, check the friction surfaces for wear or damage. Hold the lifter and try to press the plunger by hand. If the lifter is worn or damaged, or the plunger can be moved by hand, replace the lifter.

1995–97 Millenia 2.3L (KJS) Engine

NOTE: With the engine cold, standard valve clearance is 0.011–0.012 inch (0.27–0.31mm) on intake and exhaust sides.

1. Measure the valve clearance by turning the crankshaft clockwise until the number one piston is at TDC.

2. Measure the valve clearance at **A**. Turn the crankshaft clockwise 240° until the number three piston is at TDC. Measure the valve clearance at **B**. Turn the crankshaft clockwise 240° until the number five piston is

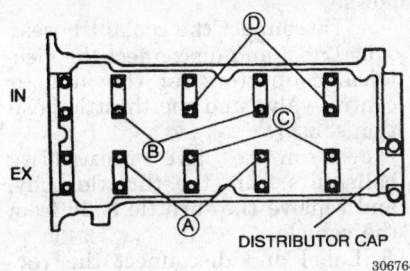

Camshaft cap bolt removal — 1995–97 Protege 1.8L (BPD) engines

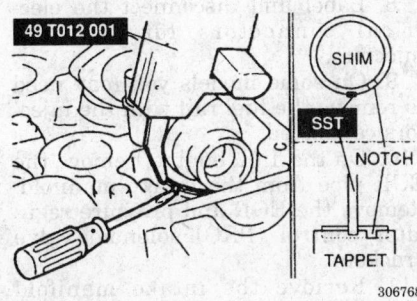

Inserting a pry tool — 1995–97 Protege 1.8L (BPD) engines

at TDC. Measure the valve clearance at **C**.

NOTE: If the valve clearance exceeds the standard, replace the adjustment shim.

3. Turn the crankshaft clockwise until the cam, on the camshaft requiring the adjustment shim replacement, is positioned straight up.

4. Remove the camshaft cap bolts as follows:

    a. For right-hand (RH) exhaust side adjustment shim removal use **1**.

    b. For RH intake side adjustment shim removal use **2**.

    c. For left-hand (LH) exhaust side adjustment shim removal use **3**.

    d. For LH intake side adjustment shim removal use **4**.

5. Install special tools 49–T012–002 and 003, using the camshaft cap bolt holes.

6. Align the mark on the 49–T012–002 (shaft) with the mark on the 49–T012–003 (clamp).

7. Position special tool 49–T012–001 toward the center of the cylinder head, and mount it on the shaft where the adjustment shim needs replacement.

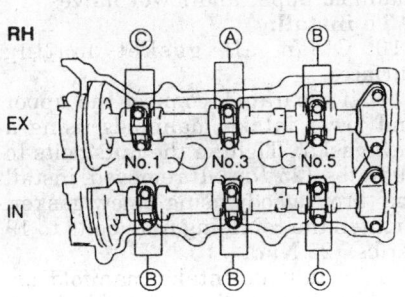

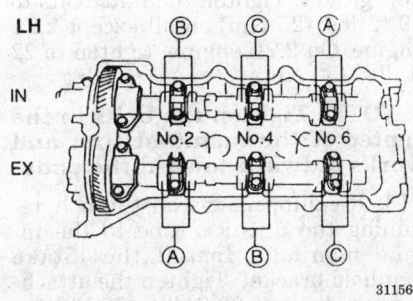

Measuring the valve clearance — 1995–97 Millenia 2.3L (KJS) engine

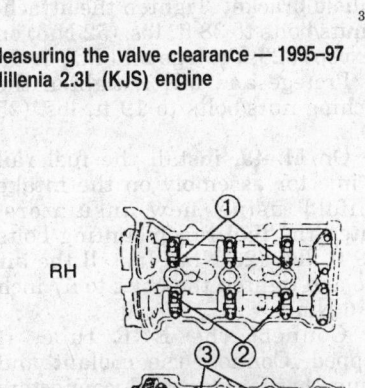

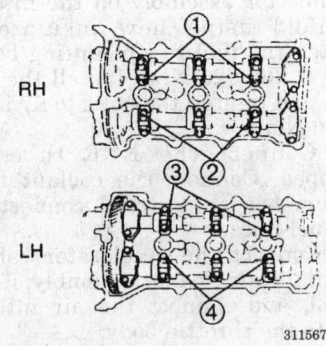

Camshaft cap bolt removal — 1995–97 Millenia 2.3L (KJS) engine

8. Position the notch of the tappet to allow a small pry tool to be inserted.

9. Set the special tool on the tappet by its notch. Tighten the mounting bolt **B** securing it on the shaft.

10. Tighten bolt **C**, and press down the tappet.

11. Using a small pry tool, pry the adjustment shim upwards through the notch on the tappet. Remove the shim with a magnet.

12. Select and install the proper adjustment shim. Loosen bolt **C** to allow the tappet to move up, and loosen bolt **B** to remove special tool 49–T012–001.

13. Remove special tools 49–T012–002, 003 and 004, and tighten the camshaft cap bolts to 100–125 inch lbs. (11–14 Nm).

14. Repeat the procedure for all necessary adjustment shims. Check the valve clearance.

Rocker Arm Shaft

REMOVAL AND INSTALLATION

1993–94 323 1.6L (B6E), 1994 Protege 1.8L (BPD) Engine

1. Disconnect the negative battery cable.

2. Disconnect and mark the spark plug wires and the breather hoses and remove the cylinder head cover.

3. Loosen the rocker arm shaft bolts in 2–3 steps, in the reverse order of the torque sequence. Remove the rocker arms/shafts assembly with the bolts.

4. If necessary, disassemble the shaft assemblies. Note the position of the rocker arms and shafts so they can be reassembled in the same locations.

5. Inspect the rocker arms and shafts for wear or damage and replace parts, as necessary.

To install:

6. If disassembled, lubricate the rocker arms and shafts with clean engine oil and assemble with the springs and bolts. If reusing the original parts, make sure they are installed in the same locations.

NOTE: The installation bolt holes are different for the exhaust side and intake side shafts. On the 1.6L 8-valve engine, the shaft oil holes must face downward. On the 1.6L and 1.8L 16-valve engines, the identification marks at the end of the shafts must face upward.

7. Apply clean engine oil to the valve stems and camshaft lobes. Install the rocker arms/shafts assembly.

8. Tighten the rocker arm shaft bolts, in 2–3 steps, to 21 ft. lbs. (28 Nm) in the proper sequence.

9. Install the cylinder head cover. Tighten the cylinder head cover-to-cylinder head bolts to 78 inch lbs. (8.8 Nm) and the timing cover-to-cylinder head cover bolts to 95 inch lbs. (11 Nm).

10. Connect the breather hoses and spark plug wires. Connect the negative battery cable, start the engine and check for leaks.

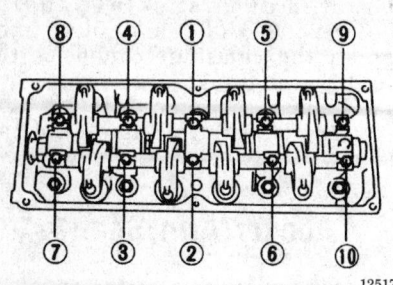

Rocker arm shaft bolt tightening sequence — 1993-94 323 1.6L (B6E), 1994 Protege 1.8L (BPD) engine

Intake Manifold

REMOVAL AND INSTALLATION

1993–94 323 1.6L (B6E), Protege 1.8L (BPD, BPE), 1994 Miata 1.8L (BPD) and MX3 1.8L (K8) Engines

---
CAUTION

When draining the coolant, keep in mind that cats and dogs are attracted by the ethylene glycol antifreeze, and are quite likely to drink any that is left in an uncovered container or in puddles on the ground. This will prove fatal in sufficient quantity. Always drain the coolant into a sealable container. Coolant should be re-used unless it is contaminated or several years old.

1. Properly relieve the fuel system pressure. Disconnect the negative battery cable and drain the cooling system.
2. Disconnect the air intake hose from the throttle body. Remove the hose and air cleaner assembly if necessary.
3. Disconnect the accelerator cable. Disconnect and plug the fuel lines.
4. Label and disconnect all necessary vacuum hoses and electrical connectors. Disconnect the coolant hoses.
5. Disconnect the EGR tube, if equipped.
6. If equipped, remove the air valve and remove the fuel rail attaching bolts. Remove the fuel rail and injectors as an assembly.
7. Remove the intake manifold support bracket. If necessary, remove the bolt retaining the dipstick tube bracket to the intake manifold.
8. Remove the intake manifold-to-cylinder bolts/nuts and remove the intake manifold assembly.

9. If necessary, remove the throttle body and separate the intake manifold upper and lower halves.
To install:
10. Clean all gasket mating surfaces.
11. If separated, connect the upper and lower intake manifolds using a new gasket. Tighten the nuts/bolts to 19 ft. lbs. (25 Nm). If removed, install the throttle body using a new gasket. Tighten the retaining nuts/bolts to 19 ft. lbs. (25 Nm).
12. Install the intake manifold assembly to the cylinder head using a new gasket. Tighten the nuts/bolts to 19 ft. lbs. (25 Nm) on all except 2.2L engine. On 2.2L engine, tighten to 22 ft. lbs. (30 Nm).

NOTE: Tighten the bolts in the center of the manifold first and works outward toward the ends.

13. If equipped, install the bolt retaining the dipstick tube to the intake manifold. Install the intake manifold bracket. Tighten the attaching nuts/bolts to 38 ft. lbs. (52 Nm) on all except 323, Protege and MX-3. On 323, Protege and MX-3, tighten the attaching nuts/bolts to 19 ft. lbs. (25 Nm).
14. On Miata, install the fuel rail and injector assembly on the intake manifold using new insulators. Tighten the fuel rail mounting bolts to 19 ft. lbs. (25 Nm). Install the air valve and tighten the bolts to 57 inch lbs. (6 Nm).
15. Connect the EGR tube, if equipped. Connect the coolant and vacuum hoses, electrical connectors and fuel lines.
16. Connect the accelerator cable. Install the air cleaner assembly, if removed, and connect the air intake tube to the throttle body.
17. Connect the negative battery cable. Fill and bleed the cooling system.
18. Start the engine and bring to normal operating temperature. Check for leaks. Check the idle speed.

1995–97 Protege 1.5L (Z5D) and 1.8L (BPD) Engines

1. Relieve the fuel system pressure, and disconnect the negative battery cable. Drain the cooling system.
2. Disconnect the mass air flow sensor electrical connector. Remove the air ducts, air cleaner assembly, mass air flow sensor and resonance chamber.
3. Disconnect the throttle and accelerator cables. Disconnect and plug the fuel lines.

4. Remove the throttle body as follows:
   a. Disconnect the coolant hoses.
   b. Label and disconnect the electrical connectors for the idle air control valve and the throttle position sensor.
   c. Remove the mounting bolts/nuts from the throttle body, and remove the throttle body from the vehicle.
5. Label and disconnect the vacuum lines at the intake manifold.
6. Remove the dynamic chamber (upper intake manifold).
7. Disconnect and plug the fuel hoses from the fuel rail.
8. Label and disconnect the electrical connectors for the fuel injectors.
9. On some models you may need to remove the fuel rail with the injectors connected.
10. On the 1.5L engine, remove the EGR pipe from the intake manifold. Remove the EGR and pressure regulator control (PRC) solenoid valve brackets.
11. Remove the intake manifold support bracket.
12. Remove the bolts and nuts, and remove the intake manifold.
To install:
13. Clean all gasket mating surfaces.
14. Install the intake manifold, using a new gasket. Tighten the nuts and bolts to 14–18 ft. lbs. (19–25 Nm).
15. Attach the EGR pipe to the manifold and install the intake manifold support bracket. Tighten the support bracket bolts to 38 ft. lbs. (51 Nm).
16. If removed, install the EGR and pressure regulator control (PRC) solenoid valve brackets.
17. Install the fuel rail and injector assembly. Connect the electrical connectors to the injectors, and the fuel lines to the rail.
18. Install the upper intake manifold to the intake manifold using new gaskets. Tighten the nuts to 14–18 ft. lbs. (19–25 Nm).
19. Install the throttle body, using a new mounting gasket. Tighten the mounting bolts to 14–18 ft. lbs. (19–25 Nm).
20. Connect the electrical connectors for the idle air control valve and the throttle position sensor.
21. Connect the vacuum and coolant lines.
22. Connect and adjust the throttle and accelerator cables.
23. Connect the fuel lines.
24. Install the resonance chamber. Install the air cleaner assembly,

mass air flow sensor and ducts. Connect the mass air flow sensor connector.

25. Connect the negative battery cable. Fill and bleed the cooling system. Run the engine and check for leaks.

1993 MX3 1.8L (K8), 1993–94 626 and MX6 2.5L (KL) Engines

1. Relieve the fuel system pressure and disconnect the negative battery cable. Drain the cooling system.

2. Disconnect the vacuum hoses and electrical connectors from the air cleaner housing. Remove the air cleaner assembly.

3. Disconnect the knock sensor connector and remove the knock sensor bracket from the intake manifold. Remove the crankshaft position sensor bracket from the right side of the intake manifold.

4. Remove the right bank (rear) spark plug wires from the spark plugs and the routing clips. Remove the Variable Resource Induction System (VRIS) solenoid connector bracket from the rear of the intake manifold.

5. Label and disconnect the necessary vacuum hoses from the rear of the intake manifold and EGR valve. Disconnect the PCV valve hose from the intake manifold, near the throttle body.

6. Label and disconnect the throttle position sensor and fuel rail electrical connectors. Disconnect the throttle cable from the throttle body and the vacuum hose from the evaporative canister.

7. Disconnect and plug the fuel supply line at the fuel rails and discard the copper crush washers. Disconnect the fuel and vacuum lines from the fuel pressure regulator.

8. Disconnect the EGR breather tube. Remove the intake manifold mounting nuts and bolts in 2–3 steps, then remove the intake manifold.

To install:

9. Clean all gasket mating surfaces.

10. Position new gaskets and install the intake manifold. Tighten the nuts and bolts in 2–3 steps to 18 ft. lbs. (25 Nm).

11. Connect the EGR breather tube and connect the fuel and vacuum lines to the fuel pressure regulator.

12. Connect the fuel supply line to the fuel rail, using new copper crush washers.

13. Connect the vacuum hoses to the evaporative canister, intake manifold, throttle body and EGR valve.

14. Connect the throttle position sensor electrical connector. Install the VRIS solenoid connector bracket.

15. Connect the spark plug wires to the spark plugs and routing clips. Install the crankshaft position sensor bracket.

16. Install the knock sensor bracket and connect the knock sensor electrical connector.

17. Install the air cleaner assembly and connect the vacuum hoses and electrical connectors to the air cleaner housing.

18. Connect the negative battery cable. Fill and bleed the cooling system. Run the engine and check for leaks.

1995 MX3 1.6L (B6-ZE) Engine

1. Relieve the fuel system pressure and disconnect the negative battery cable. Drain the cooling system.

2. Disconnect the mass air flow sensor electrical connector. Remove the air ducts, air cleaner assembly and resonance chamber.

3. Remove the fuel line mounting bracket and disconnect the throttle cable. Disconnect and plug the fuel lines.

4. Remove the throttle body as follows:

   a. Disconnect the coolant hoses.

   b. Label and disconnect the electrical connectors for the idle air control valve and the throttle position sensor.

   c. Remove the mounting bolts/nuts from the throttle body, and remove the throttle body from the vehicle.

5. Label and disconnect the vacuum lines at the intake manifold.

6. Remove the bypass air control (BAC) valve, variable inertia charging system (VICS) solenoid valve, pressure regulator control (PRC) solenoid valves, and the EGR solenoid vent and vacuum valves from the intake manifold.

7. Disconnect and plug the fuel hoses from the fuel rail.

8. Label and disconnect the electrical connectors for the fuel injectors.

9. Remove the fuel rail with the injectors connected.

10. Remove the intake manifold support bracket and remove the EGR pipe from the intake manifold.

11. Lower the vehicle. Remove the bolts and nuts, and remove the intake manifold.

To install:

12. Clean all gasket mating surfaces.

13. Install the intake manifold, using a new gasket. Tighten the nuts and bolts to 16–22 ft. lbs. (22–30 Nm).

14. Raise and safely support the vehicle.

15. Attach the EGR pipe to the manifold and install the intake manifold support bracket. Tighten the support bracket bolts to 38 ft. lbs. (51 Nm).

16. Install the fuel rail and injector assembly. Connect the electrical connectors to the injectors, and the fuel lines to the rail.

17. Install the bypass air control (BAC) valve, variable inertia charging system (VICS) solenoid valve, pressure regulator control (PRC) solenoid valves, and the EGR solenoid vent and vacuum valves to the intake manifold.

18. Install the throttle body, using a new mounting gasket. Tighten the mounting bolts to 14–18 ft. lbs. (19–25 Nm).

19. Connect the electrical connectors, vacuum lines and coolant lines.

20. Connect the throttle cable and the fuel lines. Install the fuel line mounting bracket and tighten the bolt to 97 inch lbs. (11 Nm).

21. Install the resonance chamber. Install the air cleaner assembly and ducts. Connect the mass air flow sensor connector.

22. Connect the negative battery cable. Fill and bleed the cooling system. Run the engine and check for leaks.

1995 MX3 1.8L (K8) Engine

1. Relieve the fuel system pressure and disconnect the negative battery cable. Drain the cooling system.

2. Remove the top engine compartment strut bar from the vehicle.

3. Remove the fresh air duct. Label and disconnect the vacuum hoses and electrical connectors from the air cleaner housing. Remove the air cleaner assembly.

4. Remove the resonance duct.

5. Remove the volume air flow sensor and the air intake hose from the throttle body.

6. Disconnect the knock sensor connector and remove the knock sensor bracket from the intake manifold. Remove the crankshaft position sensor bracket from the right side of the intake manifold.

7. Remove the Variable Resource Induction System (VRIS) solenoid connector bracket from the rear of the intake manifold.

8. Label and disconnect the necessary vacuum hoses from the rear of the intake manifold and EGR valve.

Disconnect the PCV valve hose from the intake manifold, near the throttle body.

9. Label and disconnect the throttle position sensor and fuel rail electrical connectors. Disconnect the throttle cable from the throttle body and the vacuum hose from the evaporative canister.

10. Disconnect and plug the fuel supply line at the fuel rails and discard the copper crush washers. Disconnect the fuel and vacuum lines from the fuel pressure regulator.

11. Disconnect the EGR breather tube. Remove the intake manifold mounting nuts and bolts in 2–3 steps, then remove the intake manifold.

12. If the intake manifold is being replaced, remove the throttle body, air bypass (BAC) valve and the air intake pipe from the manifold. Remove the fuel rail and injectors from the manifold.

To install:

13. Clean all gasket mating surfaces.

14. If installing a new intake manifold:

a. Position the fuel rail, with injectors, onto the manifold. Tighten the mounting bolts to 14–18 ft. lbs. (19–25 Nm).

b. Lubricate new O-rings with clean engine oil, and install the air intake pipe.

c. Install the BAC valve and the throttle body onto the manifold.

15. Position new gaskets and install the intake manifold. Tighten the nuts and bolts in 2–3 steps, from the center to the ends, to 14–18 ft. lbs. (19–25 Nm).

16. Connect the EGR breather tube and connect the fuel and vacuum lines to the fuel pressure regulator.

17. Connect the fuel supply line to the fuel rail, using new copper crush washers.

18. Connect and adjust the throttle cable.

19. Connect the vacuum hoses to the evaporative canister, intake manifold, throttle body, PCV and EGR valve.

20. Connect the throttle position sensor electrical connector. Install the VRIS solenoid connector bracket.

21. Install the crankshaft position sensor bracket.

22. Install the knock sensor bracket and connect the knock sensor electrical connector.

23. Install the air intake hose onto the throttle body. Install the volume air flow sensor.

24. Install the air cleaner assembly and connect the vacuum hoses and electrical connectors to the air cleaner housing. Install the fresh air duct, and tighten the mounting nut to 70–95 inch lbs. (8–11 Nm).

25. Install the strut bar, and tighten the bolts to 34–46 ft. lbs. (47–62 Nm).

26. Connect the negative battery cable. Fill and bleed the cooling system. Run the engine and check for leaks.

1995–97 626 and MX6 2.5L (KL) Engine

1. Relieve the fuel system pressure and disconnect the negative battery cable.

2. Remove the fresh air duct. Label and disconnect the vacuum hoses and electrical connectors from the air cleaner housing. Remove the air cleaner assembly.

3. Remove the volume air flow sensor and the air intake hose from the throttle body.

4. Disconnect the accelerator cable.

5. Disconnect the fuel pipe from the fuel distributor.

6. Remove the intake manifold.

7. Remove the following from the manifold if necessary: the throttle body, air bypass (BAC) valve and the air intake pipe from the manifold. Remove the fuel rail and injectors from the manifold.

To install:

8. Clean all gasket mating surfaces.

9. If installing a new intake manifold:

a. Position the fuel rail, with injectors, onto the manifold. Tighten the mounting bolts to 12–16 ft. lbs. (16–22 Nm).

b. Lubricate new O-rings with clean engine oil, and install the air intake pipe.

c. Install the BAC valve and the throttle body onto the manifold.

10. Position new gaskets and install the intake manifold. Tighten the nuts and bolts in 2–3 steps, from the center to the ends, to 14–18 ft. lbs. (19–25 Nm).

11. Connect the fuel supply line to the fuel rail, using new copper crush washers.

12. Connect and adjust the accelerator cable.

13. Install the air intake hose onto the throttle body. Install the volume air flow sensor.

14. Install the air cleaner assembly and connect the vacuum hoses and electrical connectors to the air cleaner housing. Install the fresh air duct, and tighten the mounting nut to 70–95 inch lbs. (8–11 Nm).

15. Connect the negative battery cable.

1993–97 626 and MX6 2.0L (FS) Engine

1. Relieve the fuel system pressure and disconnect the negative battery cable.

2. Remove the fresh air duct and the air cleaner assembly.

3. Disconnect the Mass Air Flow Sensor and remove the air hose.

4. Remove the resonance chamber.

5. Disconnect the accelerator cable.

6. Remove the throttle body and the BAC valve.

7. Remove the fuel distributor.

8. Remove the intake manifold bracket and remove the intake manifold.

To install:

9. Clean all gasket mating surfaces.

10. Install the intake manifold, using a new gasket. Tighten the nuts and bolts, in sequence, to 14–18 ft. lbs. (19–25 Nm).

11. Install the manifold bracket. Torque the bolt to 28–38 ft. lbs. (38–51 Nm).

12. Install the fuel distributor and torque the bolts to 14–18 ft. lbs. (19–25 Nm).

13. Install the BAC valve, torque the mounting bolt to 22–31 in. lbs. (2.5–3.5 Nm).

14. Install the throttle body, torque the mounting nuts and bolts to 14–18 ft. lbs. (19–25 Nm).

15. Install the accelerator cable.

16. Install the resonance chamber, air hose and connect the MAF sensor.

17. Install the air cleaner assembly and the fresh air duct.

18. Connect the negative battery cable.

1993–95 929 3.0L (JE-ZE) Engines

1. Properly relieve the fuel system pressure.

2. Disconnect the negative battery cable, and drain the cooling system.

3. Disconnect the air intake hose from the throttle body. Disconnect the air flow meter electrical connector and remove the air cleaner assembly and air ducts.

4. Disconnect the accelerator cable. Disconnect and plug the fuel lines.

5. Label and disconnect the necessary hoses and electrical connectors from the throttle body, and remove the throttle body.

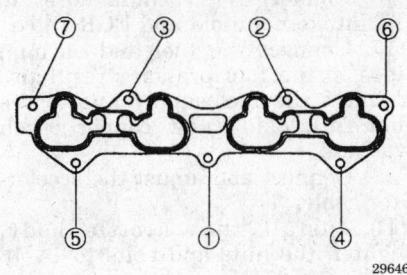

Intake manifold mounting bolt (torque sequence) — 1993–97 626 and MX6 2.0L (FS) engine

296464

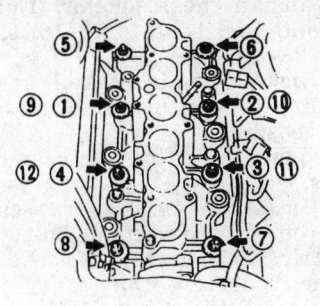

Intake manifold nut torque sequence — 1993–95 929 3.0L (JE-ZE) engines

234759

6. Disconnect the electrical connector, water hoses and air hose from the air bypass valve. Remove the retaining bolts/nuts and remove the air bypass valve. Remove the air intake pipe from the upper intake manifold.

7. Label and disconnect the necessary hoses and electrical connectors from the upper intake manifold. Remove the upper intake manifold retaining bolts and the upper intake manifold.

8. Label and disconnect the electrical connectors from the fuel injectors.

9. Remove the retaining bolts and cap from the top of the injector. Remove the injector by grasping the plastic part of the injector and twisting out. Disconnect the vacuum hose from the fuel pressure regulator and remove the fuel rail.

10. Label and disconnect the remaining hoses and electrical connectors from the lower intake manifold. Loosen the retaining nuts in 2–3 steps working gradually inward from the ends of the manifold. Remove the lower intake manifold.

11. Clean all gasket mating surfaces. Inspect the manifolds for cracks or other damage.

To install:

12. Position new gaskets and install the lower intake manifold. Install the retaining nut washers with the white paint mark facing upward. Tighten the retaining nuts in 2–3 steps to 19 ft. lbs. (25 Nm), in the proper sequence.

13. Connect the hoses and electrical connectors to the lower intake manifold.

14. Install the fuel rail. Apply a small amount of clean engine oil to new injector O-rings and install on the injectors. Install the injectors in the fuel rail using a turning motion, to prevent damaging the O-rings. Install the injector retainer caps and tighten the screws to 30 inch lbs. (3.4 Nm).

15. Connect the injector electrical connectors and connect the vacuum hose to the fuel pressure regulator. Connect the fuel lines.

16. Using a new gasket, install the upper intake manifold on the lower intake manifold. Tighten the mounting bolts to 19 ft. lbs. (25 Nm).

17. Connect the electrical connectors and hoses to the upper intake manifold.

18. Install the air intake pipe on the upper intake manifold using a new gasket, and tighten the mounting bolts to 17 ft. lbs. (23 Nm). Install the air bypass valve and tighten the nuts to 19 ft. lbs. (25 Nm). Connect the coolant hoses, air hose and electrical connector to the air bypass valve.

19. Install the throttle body, using a new gasket. Tighten the nuts to 19 ft. lbs. (25 Nm).

20. If removed, install the air cleaner and air ducts. Connect the air intake to the throttle body.

21. Connect the negative battery cable. Fill and bleed the cooling system.

22. Start the engine and bring to normal operating temperature. Check for leaks. Check the idle speed.

1995–97 Millenia 2.3L (KJS) Engine

1. Relieve the fuel system pressure, and disconnect the negative battery cable.

2. Drain the cooling system.

3. Remove the dynamic chamber cover.

4. Remove the charge air cooler air duct.

5. Label and disconnect the vacuum hoses and electrical connectors from the air cleaner housing. Remove the air cleaner assembly.

6. Remove the air and fresh air ducts.

7. Remove the mass air flow sensor and the air intake hose from the throttle body.

8. Remove the resonator.

9. Remove the right-hand charge air cooler.

10. Remove the left-hand charge air cooler.

11. Disconnect the accelerator cable.

12. Label and disconnect the necessary vacuum hoses from the rear of the intake manifold and EGR valve.

13. Remove the EGR valve.

14. Remove the air intake pipe assembly.

15. Remove the charge air cooler pipe.

16. Disconnect and plug the fuel supply line at the fuel rails and discard the copper crush washers. Disconnect the fuel and vacuum lines from the fuel pressure regulator.

17. Disconnect and plug the coolant hoses.

18. Remove the harness from the intake manifold.

19. Remove the intake manifold mounting nuts and bolts in 2–3 steps, then remove the intake manifold.

20. Label and disconnect the fuel hoses and electrical connectors from the throttle body. Remove the throttle body.

To install:

21. Clean all gasket mating surfaces.

22. Install the throttle body. Tighten the nuts and bolts to 14–18 ft. lbs. (19–25 Nm), and connect the fuel hoses and electrical connectors.

23. Position new gaskets and install the intake manifold. Tighten the nuts and bolts in 2–3 steps, from the center to the ends, to 14–18 ft. lbs. (19–25 Nm).

24. Install the harness onto the intake manifold.

25. Unplug and and connect the coolant hoses.

26. Connect the fuel and vacuum lines to the fuel pressure regulator. Connect the fuel supply line to the fuel rail, using new copper crush washers.

27. Install the charge air cooler pipe.

28. Position the air intake pipe assembly using new gaskets. Hand tighten the nuts and bolts in the order shown in the graphic until the air intake pipe contacts the intake manifold. Verify that the rubber gaskets are not twisted or distorted. Tighten the bolts marked **A** to 70–95 inch lbs. (8–11 Nm), and all others, in sequence, to 14–18 ft. lbs. (19–25 Nm).

29. Install the EGR valve using a new gasket.

30. Connect the vacuum hoses to the intake manifold and EGR valve.

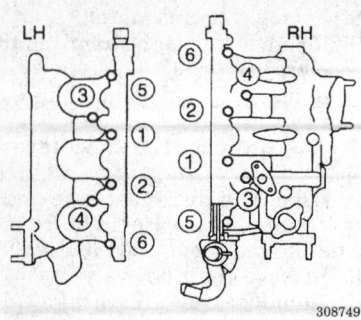

Intake manifold mounting bolt torque sequence — 1995–97 Millenia 2.3L (KJS) engine

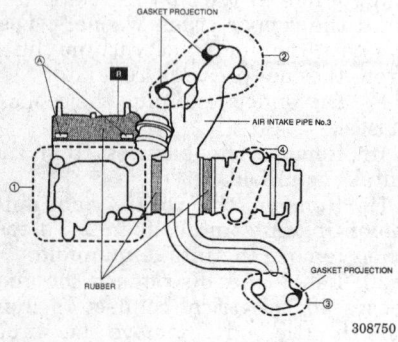

Air intake pipe mounting bolt locations — 1995–97 Millenia 2.3L (KJS) Engine

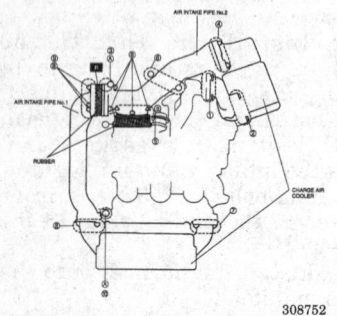

Charge air cooler mounting bolt locations — 1995–97 Millenia 2.3L (KJS) engine

31. Connect and adjust the accelerator cable.

32. Using new gaskets, position the left and right-hand charge air coolers. Hand tighten the nuts and bolts in the order shown in the graphic until the air intake pipes and charge air coolers contact the intake manifold. Verify that the rubber gaskets are not twisted or distorted. Tighten the bolts marked **A** to 44–78 inch lbs. (5–9 Nm). Tighten the bolts marked **B** to 70–95 inch lbs. (8–11 Nm), and all others, in sequence, to 14–18 ft. lbs. (19–25 Nm).

33. Install the resonator. Tighten the nuts and bolts to 12–16 ft. lbs. (16–22 Nm)

34. Install the air intake hose onto the throttle body. Install the mass air flow sensor.

35. Install the fresh air and air ducts.

36. Install the air cleaner assembly and connect the vacuum hoses and electrical connectors to the air cleaner housing.

37. Install the charge air cooler air duct. Tighten the mounting bolts to 70–95 inch lbs. (8–11 Nm).

38. Install the dynamic chamber cover.

39. Connect the negative battery cable.

40. Fill and bleed the cooling system. Run the engine and check for leaks.

1995–97 Millenia 2.5L (KLD) Engine

1. Relieve the fuel system pressure and disconnect the negative battery cable.

2. Drain the cooling system.

3. Label and disconnect the vacuum hoses and electrical connectors from the air cleaner housing. Remove the air cleaner assembly.

4. Remove the mass air flow sensor and the air intake hose from the throttle body.

5. Label and disconnect the fuel hoses and electrical connectors from the throttle body. Remove the throttle body.

6. Disconnect the accelerator cable.

7. Disconnect and plug the fuel supply line at the fuel rails and discard the copper crush washers. Disconnect the fuel and vacuum lines from the fuel pressure regulator.

8. Label and disconnect the necessary vacuum hoses from the rear of the intake manifold and EGR valve.

9. Remove the harness from the intake manifold.

10. Remove the EGR valve. Disconnect the EGR breather tube.

11. Remove the intake manifold mounting nuts and bolts in 2–3 steps, then remove the intake manifold.

To install:

12. Clean all gasket mating surfaces.

13. Position new gaskets and install the intake manifold. Tighten the nuts and bolts in 2–3 steps, from the center to the ends, to 14–18 ft. lbs. (19–25 Nm).

14. Connect the EGR breather tube and install the EGR valve.

15. Install the harness onto the intake manifold.

16. Connect the vacuum hoses to the intake manifold and EGR valve.

17. Connect the fuel and vacuum lines to the fuel pressure regulator. Connect the fuel supply line to the fuel rail, using new copper crush washers.

18. Connect and adjust the accelerator cable.

19. Install the throttle body. Tighten the nuts and bolts to 14–18 ft. lbs.(19–25 Nm), and connect the fuel hoses and electrical connectors.

20. Install the air intake hose onto the throttle body. Install the mass air flow sensor.

21. Install the air cleaner assembly and connect the vacuum hoses and electrical connectors to the air cleaner housing.

22. Connect the negative battery cable.

23. Fill and bleed the cooling system. Run the engine and check for leaks.

1995–97 Miata 1.8L (BPD) Engines

1. Relieve the fuel system pressure, and disconnect the negative battery cable. Drain the cooling system.

2. Remove the air pipe.

3. Remove the throttle cable. Disconnect and plug the fuel lines.

4. Remove the throttle body as follows:

   a. Disconnect the coolant hoses.

   b. Label and disconnect the electrical connectors for the idle air control valve and the throttle position sensor.

   c. Remove the mounting bolts/nuts from the throttle body, and remove the throttle body from the vehicle.

5. Label and disconnect the vacuum lines at the intake manifold.

6. Disconnect and plug the fuel hoses from the fuel rail.

7. Label and disconnect the electrical connectors for the fuel injectors.

8. Remove the fuel rail with the injectors connected.

9. Remove the fuel injector harness.

10. Remove the intake manifold support bracket.

11. Lower the vehicle. Remove the nuts, and remove the intake manifold.

To install:

12. Clean all gasket mating surfaces.

13. Install the intake manifold, using a NEW gasket. Tighten the nuts to 14–18 ft. lbs. (19–25 Nm).

14. Raise and safely support the vehicle.

15. Install the intake manifold support bracket. Tighten the support bracket bolts to 38 ft. lbs. (51 Nm).

16. Install the fuel injector harness.

17. Install the fuel rail and injector assembly. Connect the electrical connectors to the injectors, and the fuel lines to the rail.

18. Install the throttle body, using a new mounting gasket. Tighten the mounting bolts to 14–18 ft. lbs. (19–25 Nm).

19. Connect the electrical connectors, vacuum lines and coolant lines.

20. Connect the throttle cable and the fuel lines.

21. Install the air pipe.

22. Connect the negative battery cable. Fill and bleed the cooling system. Run the engine and check for leaks.

1993–95 RX7 1.3L (13B) Engines

1. Relieve the fuel system pressure.

2. Disconnect the negative battery cable.

3. Drain the cooling system.

4. Remove the resonance chamber.

5. Disconnect the accelerator cable.

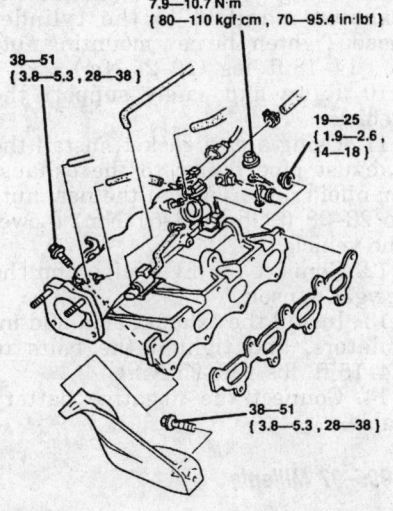

7.9–10.7 N·m { 80–110 kgf·cm , 70–95.4 in·lbf }

38–51 { 3.8–5.3 , 28–38 }

19–25 { 1.9–2.6 , 14–18 }

38–51 { 3.8–5.3 , 28–38 }

N·m { kgf·m , ft·lbf }

289779

Intake manifold and related components (exploded view) — 1995–97 Miata 1.8L (BPD) engines

6. Remove the air intake hose, air cleaner and air bypass valve.

7. Remove the pressure chamber and air intake pipe.

8. Remove the water hose from the extension manifold and the throttle body.

9. Label and disconnect the electrical connectors, vacuum and coolant hoses from the throttle body and extension manifold.

10. Remove the mounting bolts and nuts, and the throttle body/extension manifold assembly.

11. Label and disconnect the electrical connectors and hoses at the intake manifold. Disconnect and plug the fuel lines.

12. Remove the intake manifold retaining bolts and remove the intake manifold. Cover the intake ports while the manifold is removed to keep dirt out of the engine.

To install:

13. Make sure all gasket mating surfaces are clean.

14. Install a new manifold-to-engine gasket and install the intake manifold assembly with the retaining bolts.

15. Tighten the intake manifold mounting bolts to 14–18 ft. lbs. (19–25 Nm). Connect the fuel lines and all hoses and wiring to the manifold.

16. Install the throttle body/extension manifold assembly, using a new gasket, and tighten the nuts and bolts to 14–18. lbs. (19–25 Nm).

17. Connect the coolant and vacuum hoses and the electrical connectors to the throttle body and extension manifold.

18. Install the water hose to the extension manifold and the throttle body.

19. Install the air intake pipe.

20. Install the pressure chamber and connect the vacuum hoses.

21. Install the air bypass valve, air cleaner assembly and the air intake hose.

22. Connect the accelerator cable.

23. Install the resonance chamber.

24. Connect the negative battery cable. Turn the ignition key **ON** to pressurize the system and check for leaks. Correct as necessary.

25. Fill and bleed the cooling system.

26. Start the engine and bring to normal operating temperature. Check for leaks. Check the idle speed.

Exhaust Manifold

REMOVAL AND INSTALLATION

1993–94 323 1.6L (B6E)

1. Disconnect the negative battery cable.

2. Remove the retaining bolts and remove the exhaust manifold insulator.

3. Disconnect the oxygen sensor electrical connector. Remove the oxygen sensor, if necessary, if it is installed in the manifold.

4. Disconnect the EGR pipe, if equipped.

5. Raise and safely support the vehicle. Remove the nuts from the exhaust pipe flange and disconnect the exhaust pipe from the manifold or turbocharger, if equipped.

6. Lower the vehicle.

7. Remove the mounting nuts/bolts and remove the exhaust manifold.

8. Installation is the reverse of the removal procedure. Make sure all gasket mating surfaces are clean prior to assembly.

9. Use new gaskets and tighten the exhaust manifold-to-cylinder head nuts/bolts to specifications.

10. Use a new gasket and tighten the exhaust pipe flange-to-exhaust manifold nuts to 34 ft. lbs. (46 Nm).

1993–97 Protege 1.5L (Z5D) and 1.8L (BPE,BPD) Engines

1. Disconnect the negative battery cable.

2. Remove the air cleaner, and disconnect the air hose.

3. Remove the water bypass pipe bolt.

4. Remove the exhaust manifold heat shield bolts and the heat shield.

5. Disconnect the oxygen sensor electrical connector.

6. Raise and safely support the vehicle.

7. Remove and discard the exhaust pipe-to-exhaust manifold nuts. Suspend the exhaust system with wire.

8. Disconnect the EGR pipe from the exhaust manifold and lower the vehicle.

9. Remove the nuts and bolts and remove the exhaust manifold. Discard the nuts.

To install:

10. Clean all gasket mating surfaces.

11. Position a new exhaust manifold gasket over the studs and install the exhaust manifold. Tighten the mounting nuts and bolts to 14–16 ft.

lbs. (19–22 Nm) on the 1.5L, and for the 1.8L to 29–34 ft. lbs. (39–47 Nm).

12. Raise and safely support the vehicle.

13. Connect the exhaust pipe to the manifold. Install new nuts and tighten to 38 ft. lbs. (52 Nm). Connect the oxygen sensor connector.

14. Connect the EGR pipe to the back of the exhaust manifold and tighten to 34 ft. lbs. (47 Nm). Lower the vehicle.

15. Install the heat shield and tighten the bolts to 88 inch lbs. (10 Nm).

16. Install the water bypass pipe bolt, and tighten to 48–65 ft. lbs. (64–89 Nm).

17. Connect the air hose, and install the air cleaner.

18. Connect the negative battery cable.

1993–95 MX3 1.8L (K8) and 1993–97 626 and MX6 2.5L (KL) Engines

1. Disconnect the negative battery cable. Raise and safely support the vehicle.

2. Disconnect the oxygen sensor connectors.

3. Remove the nuts from the front and rear exhaust pipes and lower the exhaust system. Both pipes must be disconnected, even if only one manifold is to be removed.

4. If removing the rear (right side) manifold, disconnect the EGR pipe.

5. Remove the 3 heatshield bolts and remove the heat shield.

6. Remove the 2 nuts and 5 bolts and remove the exhaust manifold.

To install:

7. Clean all gasket mating surfaces.

8. Install the exhaust manifold, using a new gasket, and tighten the nuts to 15–20 ft. lbs. (20–28 Nm), and the bolts to 12–16 ft. lbs. (16–22 Nm).

9. Install the heat shield and tighten the bolts to 88 inch lbs. (10 Nm).

10. If installing the rear (right side) manifold, connect the EGR pipe.

11. Connect the exhaust pipes to the manifolds, using new gaskets and nuts, and tighten the nuts to 38 ft. lbs. (51 Nm).

12. Connect the oxygen sensor connectors and lower the vehicle.

13. Connect the negative battery cable.

1993–95 MX3 1.6L (B6-ZE) Engine

1. Disconnect the negative battery cable.

2. Remove the 4 exhaust manifold heat shield bolts and the heat shield.

3. Disconnect the oxygen sensor electrical connector.

4. Raise and safely support the vehicle. Remove the nuts from the exhaust pipe flange and disconnect the exhaust pipe from the manifold. Suspend the exhaust system with wire.

5. Disconnect the EGR pipe.

6. Lower the vehicle.

7. Remove the mounting nuts/bolts and remove the exhaust manifold.

To install:

8. Clean all gasket mating surfaces.

9. Position a new exhaust manifold gasket over the studs and install the exhaust manifold. Tighten the 8 mounting bolts/nuts to 29–33 ft. lbs. (39–46 Nm).

10. Raise and safely support the vehicle.

11. Connect the exhaust pipe to the manifold. Install new nuts and tighten to 38 ft. lbs. (52 Nm). Connect the oxygen sensor connector.

12. Install the two inlet pipe bracket bolts.

13. Connect the EGR pipe to the back of the exhaust manifold. Lower the vehicle.

14. Install the heat shield and tighten the bolts to 70–95 inch lbs. (8–11 Nm). Connect the negative battery cable.

1993–97 626 and MX6 2.0L (FS) Engines

1. Disconnect the negative battery cable.

2. Raise and safely support the vehicle.

3. Disconnect the front exhaust pipe from the exhaust manifold.

4. Lower the vehicle.

5. Remove the power steering oil pump and the mounting bracket.

6. Remove the exhaust manifold heat shield bolts and the heat shield.

7. Disconnect the EGR pipe from the manifold.

8. Disconnect the oxygen sensor electrical connector from the manifold.

9. Remove the manifold mounting nuts and bolts, remove the manifold.

To install:

10. Clean all gasket mating surfaces.

11. Position a NEW exhaust manifold gasket over the studs and install the exhaust manifold. Tighten the mounting bolts to 12–16 ft. lbs. (16–22 Nm).

12. Install NEW manifold mount nuts and tighten to 15–20 ft. lbs. (20–28 Nm). Raise and safely support the vehicle.

13. Connect the oxygen sensor electrical connector and the EGR pipe to the manifold.

14. Install the heat shield and tighten the bolts to 70–95.4 in. lbs. (7.9–10.7 Nm).

15. Install the power steering oil pump bracket and the oil pump.

16. Raise and safely support the vehicle.

17. Connect the front exhaust pipe to the manifold, use a NEW gasket. Install NEW nuts and tighten to 28–38 ft. lbs. (38–51 Nm).

18. Connect the negative battery cable.

1993–95 929 3.0L (JE-ZE) Engines

1. Disconnect the negative battery cable.

2. Remove the retaining bolts, and remove the exhaust manifold insulator.

3. Disconnect the oxygen sensor electrical connector.

4. Disconnect the EGR pipe.

5. Raise and safely support the vehicle. Remove and discard the nuts from the exhaust pipe flange, and disconnect the exhaust pipe from the manifold.

6. Lower the vehicle.

7. Remove and discard the mounting nuts, and remove the exhaust manifold.

To install:

8. Make sure all gasket mating surfaces are clean prior to assembly.

9. Using a new gasket, install the exhaust manifold to the cylinder head. Tighten the new mounting nuts to 14–18 ft. lbs. (19–25 Nm).

10. Raise and safely support the vehicle.

11. Using a new gasket, install the exhaust pipe flange to the exhaust manifold nuts. Tighten the new nuts to 28–38 ft. lbs. (38–51 Nm). Lower the vehicle.

12. Connect the EGR pipe and the oxygen sensor.

13. Install the exhaust manifold insulators, and tighten the bolts to 14–18 ft. lbs. (19–25 Nm).

14. Connect the negative battery cable.

1995–97 Millenia

1. Disconnect the negative battery cable.

2. Raise and safely support the vehicle.

3. Remove the nuts from the front and rear exhaust pipes and lower the exhaust system. Both pipes must be disconnected, even if only one manifold is to be removed.

4. If removing the rear (right side) manifold, disconnect the EGR pipe.

5. If removing the front (left side) manifold, remove the charge air cooler and the coolant/condenser fans.

6. Disconnect the front and rear oxygen sensor connectors.

7. Remove the three heat shield bolts and remove the heat shield.

8. Remove the 2 nuts and 5 bolts and remove the exhaust manifold.

To install:

9. Clean all gasket mating surfaces.

10. Install the exhaust manifold, using a new gasket, and tighten the nuts and bolts to 12–16 ft. lbs. (16–22 Nm).

11. Install the heat shield and tighten the bolts to 70–95 inch lbs. (8–11 Nm).

12. Connect the oxygen sensor connectors.

13. If installing the front (left side) manifold, install the the coolant/condenser fans and the charge air cooler. Tighten the charge air cooler mounting bolts to 14–18 ft. lbs. (19–25 Nm)

14. If installing the rear (right side) manifold, connect the EGR pipe.

15. Connect the exhaust pipes to the manifolds, using new gaskets and nuts, and tighten the nuts to 28–38 ft. lbs. (38–51 Nm).

16. Lower the vehicle.

17. Connect the negative battery cable.

Miata 1.6L (B6-ZE) and 1.8L (BPD) Engines

1. Disconnect the negative battery cable.

2. Remove the retaining bolt, and remove the exhaust manifold insulator.

3. Disconnect the oxygen sensor electrical connector. Remove the oxygen sensor.

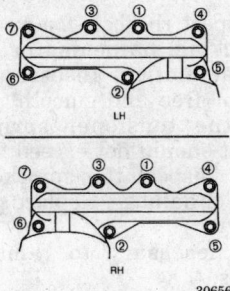

Exhaust manifold mounting bolt torque sequence — 1995–97 Millenia 2.3L (KJS) engines

4. Raise and safely support the vehicle. Remove the nuts from the exhaust pipe flange and disconnect the exhaust pipe from the manifold.

5. Lower the vehicle.

6. Remove the mounting nuts/bolts and remove the exhaust manifold.

To install:

7. Make sure all gasket mating surfaces are clean prior to assembly.

8. Use new gaskets and tighten the exhaust manifold nuts/bolts to the cylinder head. Torque to 29–33 ft. lbs. (39–46 Nm).

9. Raise and safely support the vehicle.

10. Use a new gasket and tighten the exhaust pipe flange, to the exhaust manifold nuts, to 34 ft. lbs. (46 Nm).

11. Lower the vehicle. Install the oxygen sensor, and connect the electrical connector.

12. Install the exhaust manifold insulator, and tighten the mounting bolt to 70–95 inch lbs. (8–11 Nm).

13. Connect the negative battery cable.

1993–95 RX7 1.3L (13B)

1. Disconnect the negative battery cable.

2. Remove the intake manifold assembly.

3. Remove the exhaust shields.

4. Disconnect the oxygen sensor connector.

5. Disconnect the exhaust pipe from the turbocharger.

6. Remove the turbocharger and wastegate assembly and cover the air, water and oil openings to keep them clean.

7. Remove the nuts to remove the exhaust manifold from the engine. Cover the exhaust ports to keep dirt out of the engine.

To install:

8. Clean all gasket mating surfaces.

9. Using a new gasket, install the exhaust manifold to the engine.

10. Install new nuts, and tighten to 57 ft. lbs. (78 Nm).

11. Install the turbocharger and wastegate assembly.

12. Connect the exhaust pipe to the turbocharger. Tighten the nuts to 28–38 ft. lbs. (38–51 Nm).

13. Connect the oxygen sensor connector, and install the exhaust shields.

14. Install the intake manifold assembly.

15. Connect the negative battery cable.

16. Start the engine and bring to normal operating temperature. Check for leaks.

Supercharger

REMOVAL AND INSTALLATION

1995–97 Millenia 2.3L (KJS) Engine

——— CAUTION ———

Fuel injection systems remain under pressure after the engine has been turned OFF. Properly relieve fuel pressure before disconnecting any fuel lines. Failure to do so may result in fire or personal injury.

1. Relieve the fuel system pressure, and disconnect the negative battery cable.

2. Drain the cooling system.

3. Remove the dynamic chamber cover.

4. Remove the charge air cooler air duct.

5. Label and disconnect the vacuum hoses and electrical connectors from the air cleaner housing. Remove the air cleaner assembly.

6. Remove the air and fresh air ducts.

7. Remove the mass air flow sensor and the air intake hose from the throttle body.

8. Remove the resonator.

9. Remove the right-hand charge air cooler.

10. Remove the left-hand charge air cooler.

11. Disconnect the accelerator cable.

12. Label and disconnect the necessary vacuum hoses from the rear of the intake manifold and EGR valve.

13. Remove the EGR valve.

14. Remove the air intake pipe assembly.

15. Remove the charge air cooler pipe.

16. Disconnect and plug the fuel supply line at the fuel rails and discard the copper crush washers. Disconnect the fuel and vacuum lines from the fuel pressure regulator.

——— CAUTION ———

Do not allow fuel spray or fuel vapors to come in contact with a spark or open flame. Keep a dry chemical fire extinguisher nearby. Never store fuel in an open container due to risk of fire or explosion.

17. Disconnect and plug the coolant hoses.

18. Remove the harness from the intake manifold.

19. Remove the intake manifold mounting nuts and bolts in 2–3 steps, then remove the intake manifold.

20. Label and disconnect the fuel hoses and electrical connectors from the throttle body. Remove the throttle body.

21. Remove the drive belt from the Lysholm compressor (supercharger).

22. Remove the mounting bolts from the Lysholm compressor, and remove the compressor from the vehicle.

To install:

23. Clean all gasket mating surfaces.

24. Position the rubber shield for the Lysholm compressor onto the compressor using double sided adhesive tape. Place the compressor onto the mounting studs and tighten the mounting nuts to 14–18 ft. lbs. (19–25 Nm).

25. Install and adjust the drive belt.

26. Install the throttle body. Tighten the nuts and bolts to 14–18 ft. lbs. (19–25 Nm), and connect the fuel hoses and electrical connectors.

27. Position new gaskets and install the intake manifold. Tighten the nuts and bolts in 2–3 steps, from the center to the ends, to 14–18 ft. lbs. (19–25 Nm).

28. Install the harness onto the intake manifold.

29. Unplug and and connect the coolant hoses.

30. Connect the fuel and vacuum lines to the fuel pressure regulator. Connect the fuel supply line to the fuel rail, using new copper crush washers.

31. Install the charge air cooler pipe.

32. Position the air intake pipe assembly using new gaskets. Hand tighten the nuts and bolts in the order shown in the graphic until the air intake pipe contacts the intake manifold. Verify that the rubber gaskets are not twisted or distorted. Tighten the bolts marked **A** to 70–95 inch lbs. (8–11 Nm), and all others, in sequence, to 14–18 ft. lbs. (19–25 Nm).

33. Install the EGR valve using a new gasket.

34. Connect the vacuum hoses to the intake manifold and EGR valve.

35. Connect and adjust the accelerator cable.

36. Using new gaskets, position the left and right-hand charge air coolers. Hand tighten the nuts and bolts in the order shown in the graphic until the air intake pipes and charge air coolers contact the intake manifold. Verify that the rubber gaskets are not twisted or distorted. Tighten the bolts marked **A** to 44–78 inch lbs. (5–9 Nm). Tighten the bolts marked **B** to 70–95 inch lbs. (8–11 Nm), and all others, in sequence, to 14–18 ft. lbs. (19–25 Nm).

37. Install the resonator. Tighten the nuts and bolts to 12–16 ft. lbs. (16–22 Nm)

38. Install the air intake hose onto the throttle body. Install the mass air flow sensor.

39. Install the fresh air and air ducts.

40. Install the air cleaner assembly and connect the vacuum hoses and electrical connectors to the air cleaner housing.

41. Install the charge air cooler air duct. Tighten the mounting bolts to 70–95 inch lbs. (8–11 Nm).

42. Install the dynamic chamber cover.

43. Connect the negative battery cable.

44. Fill and bleed the cooling system. Run the engine and check for leaks.

Front Crankshaft Seal

REMOVAL AND INSTALLATION

1. Disconnect the negative battery cable.

2. Raise and safely support the vehicle. On some models it may be necessary to remove the right front tire and splash shield.

3. Remove the crankshaft pulley.

4. Remove the timing belt covers and belt.

5. Remove the crankshaft pulley boss and sprocket. It may be necessary to use a puller to remove the sprocket. Remove the key from the sprocket.

6. Protect the oil pump housing with a rag. Cut the oil seal lip with a knife. Using a small prybar, pry the oil seal from the engine block; be careful not to score the crankshaft or the seal seat.

To install:

7. Lubricate the seal lip with clean engine oil and push the seal slightly in by hand.

8. Tap the seal in evenly using a seal installer. Install the seal until it is flush with the oil pump body on 1993–94 323 1.6L (B6E) and 1994 Protege and MX3 1.8L (BPD), 1993–94 Miata 1.6L (B6-ZE), 1.8L (BPD) and Protege 1.8L (BPE) and 1993–95 MX3 1.6L (B6-ZE) engines.

On other engines, install the seal until it protrudes 0.020 in. (0.5mm)

9. Install the crankshaft sprocket. Install the sprocket key with the tapered side toward the oil pump body.

10. Install the remaining components in the reverse order of removal.

Timing Belt, Sprockets, Tensioner and Front Cover

REMOVAL AND INSTALLATION

1993–94 323 1.6L (B6E) and Protege 1.8L (BPE) Engines

1. Disconnect the negative battery cable. Remove the engine undercover.

2. Remove the accessory drive belts.

3. Remove the water pump pulley.

4. Remove the crankshaft pulley bolts and remove the crankshaft pulley and baffle plate. Using a suitable tool to hold the crankshaft pulley and remove the pulley lockbolt. Remove the crankshaft pulley boss.

5. Remove the upper and lower timing belt covers.

6. Tag and disconnect the spark plug wires. Remove the spark plugs.

NOTE: Spark plugs are removed to make it easier to rotate the engine.

7. Temporarily reinstall the crankshaft pulley boss and lockbolt.

8. Turn the crankshaft, using the bolt, until the camshaft sprocket and crankshaft sprocket timing marks are aligned. Mark the direction of rotation on the timing belt.

9. Remove the belt tensioner lockbolt, the tensioner wheel and the spring. Remove the timing belt.

NOTE: Do not rotate the engine after the timing belt has been removed.

10. Inspect the belt for wear, peeling, cracking, hardening or signs of oil contamination. Inspect the tensioner for free and smooth rotation. Check the tensioner spring free length; it should not exceed 2.520 in. (64mm). Inspect the sprocket teeth for wear or damage. Replace parts, as necessary.

11. If necessary to remove the sprockets:

   a. Insert a small prybar through one of the camshaft sprocket holes to keep it from turning.

   b. Remove the sprocket bolt and the sprocket from the camshaft.

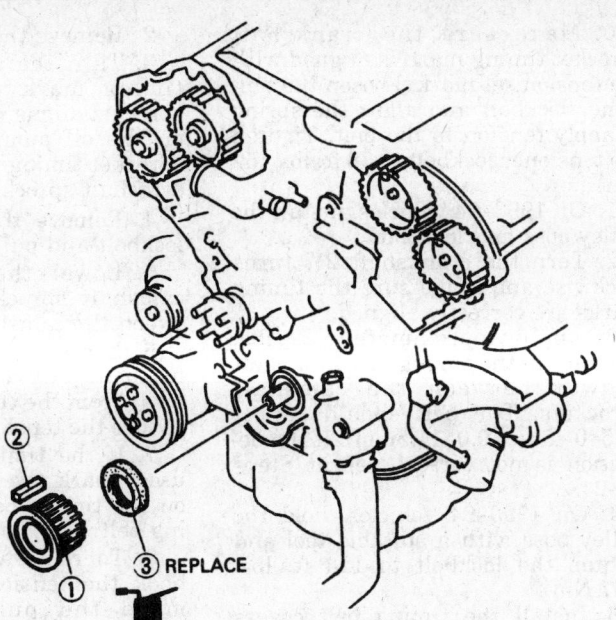

1. Timing belt pulley
2. Crankshaft key
3. Oil seal

245610

Front crankshaft seal — 1993–95 929 3.0L (JE-ZE) engines

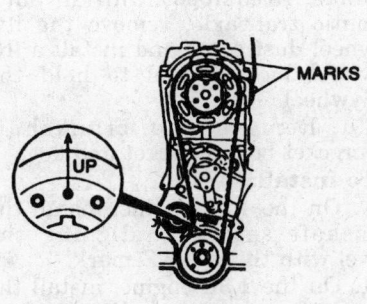

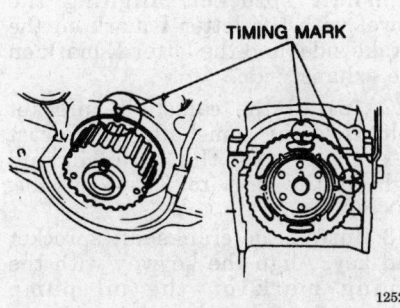

125221

Timing marks — 1993–94 323 1.6L (B6E) and Protege 1.8L (BPE) engines

To install:

12. Align the dowel on the camshaft with the dowel pin facing straight up. The dowel pin on the camshaft should also be facing upward.

13. Install the camshaft sprocket bolt. Hold the sprocket with the prybar and tighten the bolt(s) to 36–45 ft. lbs. (49–61 Nm).

14. Make sure the timing marks on the sprockets are properly aligned.

15. Install the timing belt tensioner and spring. Temporarily tighten the bolt with the spring fully extended.

16. Install the timing belt so there is no looseness on the tension side. If reusing the old timing belt, make sure it is reinstalled in the same direction of rotation.

17. Turn the crankshaft 2 turns clockwise and check the timing mark alignment. If the marks are not aligned, repeat Steps 11–14.

18. Loosen the tensioner lockbolt to set the tension, then torque the bolt to 19 ft. lbs. (25 Nm).

19. Turn the crankshaft 2 turns clockwise and check the alignment of the timing marks. If they are not aligned, repeat Steps 11–16.

20. Apply approximately 22 lbs. pressure to the timing belt on the side opposite the tensioner, at a point midway between the sprockets. The belt should deflect 0.43–0.51 in.

(11–13mm). If the tension is not as specified, repeat Steps 14–17 or, if necessary, replace the tensioner spring.

21. Install the spark plugs and connect the spark plug wires.

22. Install the upper and lower timing belt covers. Tighten the bolts to 95 inch lbs. (11 Nm).

23. Install the crankshaft pulley boss and tighten the lockbolt to 123 ft. lbs. (167 Nm), while holding the pulley boss with a suitable tool.

24. Install the crankshaft pulley and baffle plate.

25. Install the undercover or side cover. Connect the negative battery cable.

26. Start the engine and check for proper operation. Check the ignition timing.

1993–94 MX3 1.6L and Protege 1.8L (BPD) Engines

1. Disconnect the negative battery cable. Remove the engine undercover.

2. Remove the accessory drive belts.

3. Remove the crankshaft pulley bolts and remove the crankshaft pulley.

4. Remove the outer timing belt guide plate. Remove the inner timing belt guide plate if so equipped.

5. Tag and disconnect the spark plug wires. Remove the spark plugs.

NOTE: Spark plugs are removed to make it easier to rotate the engine.

6. Remove the engine oil dipstick.

7. Remove the upper, middle and lower timing belt covers.

8. Turn the crankshaft until the timing marks on the crankshaft and camshaft sprockets are aligned. On 1993–94 vehicles, the pin on the pulley boss must face upward.

9. On 1993–94 vehicles, hold the crankshaft pulley boss with a suitable tool and remove the pulley lockbolt, being careful not to rotate the crankshaft. Remove the crankshaft pulley boss.

10. Mark the direction of rotation on the timing belt. Loosen the tensioner lockbolt and pry the tensioner outward. Tighten the lockbolt with the tensioner spring fully extended. Remove the timing belt.

NOTE: Protect the tensioner with a shop towel before prying on it. Do not rotate the crankshaft after the timing belt has been removed.

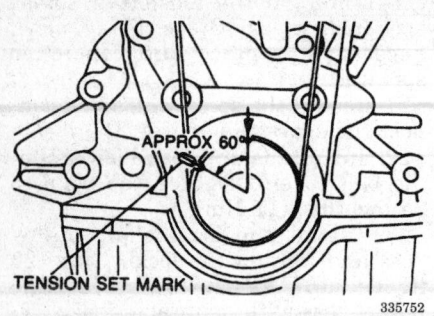

Tensioner set mark — 1993-94 MX3 1.6L and Protege 1.8L (BPD) engines

335752

11. Remove the tensioner and spring. If necessary, remove the idler pulley.

12. Inspect the belt for wear, peeling, cracking, hardening or signs of oil contamination. Inspect the tensioner for free and smooth rotation. Check the tensioner spring free length; it should not exceed 2.315 in. (58.8mm). Inspect the sprocket teeth for wear or damage. Replace parts, as necessary.

To install:

13. If removed, install the idler pulley and tighten the bolt to 38 ft. lbs. (52 Nm).

14. Install the tensioner and tensioner spring. Pry the tensioner outward and temporarily tighten the tensioner lockbolt with the tensioner spring fully extended.

15. Make sure the crankshaft sprocket timing mark is aligned with the mark on the oil pump housing and the camshaft sprocket timing marks are aligned with the marks on the seal plate.

16. Install the timing belt so there is no looseness at the idler pulley side or between the camshaft sprockets. If reusing the old belt, make sure it is installed in the same direction of rotation.

17. On 1993-94 vehicles, temporarily install the pulley boss and lockbolt.

18. Turn the crankshaft 2 turns clockwise and align the crankshaft sprocket timing mark. On 1993-94 vehicles, face the pin on the pulley boss upright. Make sure the camshaft sprocket timing marks are aligned. If they are not, repeat Steps 15-19.

19. Turn the crankshaft 1⅚ turns clockwise and align the crankshaft sprocket timing mark with the tension set mark for proper belt tension adjustment. On 1992-94 vehicles, remove the lockbolt and pulley boss.

20. Make sure the crankshaft sprocket timing mark is aligned with the tension set mark. Loosen the tensioner lockbolt and allow the spring to apply tension to the belt. Tighten the tensioner lockbolt to 38 ft. lbs. (52 Nm).

21. On 1993-94 vehicles, install the pulley boss and lockbolt.

22. Turn the crankshaft 2⅙ turns clockwise and make sure the timing marks are correctly aligned.

23. Apply approximately 22 lbs. pressure to the timing belt at a point midway between the camshaft sprockets. The belt should deflect 0.35-0.45 in. (9.0-11.5mm). If the deflection is not correct, repeat Steps 21-24.

24. On 1993-94 vehicles, hold the pulley boss with a suitable tool and tighten the lockbolt to 123 ft. lbs. (167 Nm).

25. Install the timing belt covers and tighten the bolts to 95 inch lbs. (11 Nm). Install the engine oil dipstick.

26. Install the spark plugs and connect the spark plug wires.

27. Install the timing belt inner guide plate, if equipped. Make sure the dished side of the plate faces away from the timing belt. Install the outer guide plate, if equipped.

28. Install the crankshaft pulley and tighten the bolts to 13 ft. lbs. (17 Nm).

29. Install the water pump pulley and the accessory drive belts. Adjust the belt tension.

30. Install the engine side or undercover, as necessary. Connect the negative battery cable.

31. Start the engine and check for proper operation. Check the ignition timing.

1995-97 Protege 1.5L (Z5D) and 1.8L (BPD) Engines

1. Disconnect the negative battery cable.

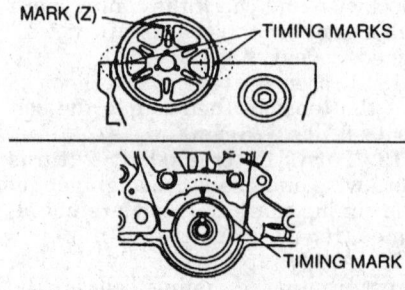

Timing mark locations — 1995-97 Protege (Z5D) engines

305810

2. Remove the timing belt covers.

3. Turn the crankshaft until the timing mark on the crankshaft sprocket aligns with the timing mark on the oil pump and the camshaft sprocket timing marks line up on the camshaft sprockets.

4. Remove the crankshaft pulley lockbolt and pulley boss.

5. Lower the vehicle. Insert a camshaft sprocket holding tool between the camshaft sprockets.

6. Loosen the tensioner pulley lockbolt. Pull the tensioner pulley away from the center of the engine to reduce the tension on the timing belt.

7. If the timing belt is to be reused, mark the direction of rotation on the timing belt. Remove the timing belt.

8. To remove the tensioner, unhook the tensioner spring, and remove the pulley lockbolt and tensioner.

9. If necessary to remove the sprockets:

a. Hold the camshaft by using a wrench on the cast hexagon, and loosen the sprocket mounting bolt.

b. Remove the sprocket bolt and the sprocket from the camshaft.

c. If equipped with a manual transaxle, place the shift lever in **4th** gear and apply the parking brake. If equipped with an automatic transaxle, remove the flywheel dust cover and install a flywheel locking tool to hold the flywheel.

d. Remove the crankshaft sprocket bolt, sprocket and key.

To install:

10. On the 1.5L engine, install the camshaft sprocket, aligning the dowel with the letter **Z** mark.

11. On the 1.8L engine, install the camshaft sprocket, aligning the dowel with the letter **I** mark on the intake side, and the letter **E** mark on the exhaust side.

12. Install the camshaft sprocket bolt. Hold the camshaft, on the cast hexagon, with the wrench and tighten the bolt to 37-44 ft. lbs. (50-60 Nm).

13. Install the crankshaft sprocket and key. Align the keyway with the timing mark on the oil pump housing.

14. Install the crankshaft sprocket bolt. Install the flywheel locking tool, if equipped with automatic transaxle, or place the shift lever in **4th** gear and apply the parking brake, if equipped with manual transaxle. Tighten the bolt to 116-122 ft. lbs. (157-166 Nm).

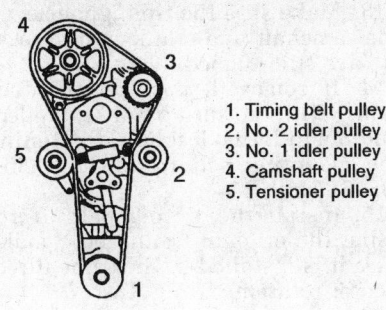

Timing belt installation pulley order — 1995–97 Protege (Z5D) engines

1. Timing belt pulley
2. No. 2 idler pulley
3. No. 1 idler pulley
4. Camshaft pulley
5. Tensioner pulley

305812

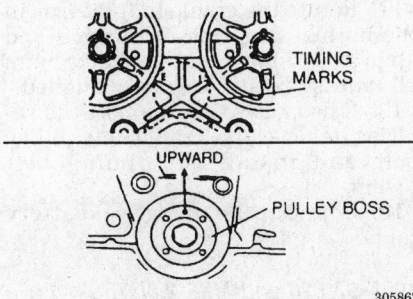

Timing mark locations — 1995–97 Protege 1.8L (BPD) engines

305867

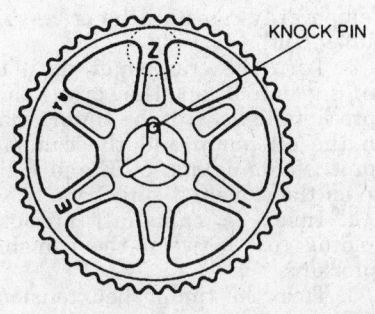

Camshaft sprocket dowel pins — 1995–97 Protege 1.5L (Z5D) engines

305882

15. Make sure the timing marks on the camshaft and crankshaft sprockets are still aligned.

16. If removed, position the tensioner with the spring fully extended, and install the lockbolt tightening the mounting bolt to 28–38 ft. lbs. (38–51 Nm).

17. Install the timing belt. If reusing the original timing belt, make sure it is installed in the same direction of rotation.

18. Rotate the crankshaft clockwise 1 5/6 turns and align the timing marks. Make sure all marks are still correctly aligned.

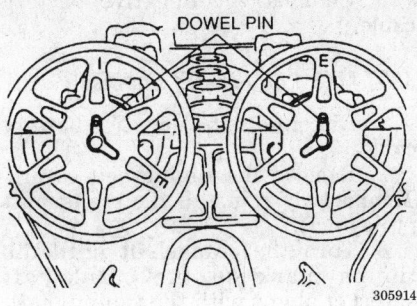

Camshaft sprocket dowel pins — 1995–97 Protege 1.8L (BPD) engines

DOWEL PIN

305914

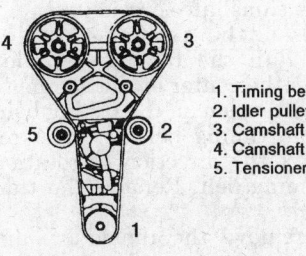

Timing belt installation pulley order — 1995–97 Protege 1.8L (BPD) engines

1. Timing belt pulley
2. Idler pulley
3. Camshaft pulley (LH)
4. Camshaft pulley (RH)
5. Tensioner pulley

305869

19. Loosen the tensioner lockbolt to apply tension to the timing belt. Tighten the tensioner lockbolt to 28–38 ft. lbs. (38–51 Nm). Remove the holding tool from between the camshaft sprockets.

20. Rotate the crankshaft clockwise 2 1/6 turns and make sure all marks are still correctly aligned.

21. Raise and safely support the vehicle. Install the crankshaft pulley lockbolt and boss. Tighten the bolt to 116–122 ft. lbs. (157–166 Nm).

22. Install the timing belt covers.

23. Connect the negative battery cable.

1993–95 MX3 1.8L (K8) and 1993–97 626 and MX6 2.5L (KL) Engines

1. Disconnect the negative battery cable.

2. Support the engine, and remove the nuts and through-bolt from the right side engine mount Remove the mount.

3. Remove the timing belt covers. Temporarily reinstall the crankshaft pulley bolt.

4. Turn the crankshaft until the timing mark on the crankshaft sprocket aligns with the timing mark on the oil pump and the camshaft sprocket timing marks align with the

marks on the cylinder head. The number one piston should be at TDC of the compression stroke.

5. Remove the 2 bolts from the automatic tensioner, removing the lower one first. Keep the bolt holes aligned by holding the tensioner to reduce the chance of stripping the threads on the bolts.

6. If the timing belt is to be reused, mark the direction of rotation on the timing belt.

7. Remove the number one idler pulley. Remove the timing belt.

8. If necessary to remove the sprockets:

   a. Insert a proper tool through one of the camshaft sprocket holes to keep it from turning.

   b. Remove the sprocket bolt and the sprocket from the camshaft.

   c. If equipped with a manual transaxle, place the shift lever in **4th** gear and apply the parking brake. If equipped with an automatic transaxle, remove the flywheel dust cover and install a flywheel locking tool to hold the flywheel.

   d. Remove the crankshaft sprocket bolt, sprocket and key.

To install:

9. Install the camshaft sprocket, aligning the dowel with the number **1** mark.

10. Install the camshaft sprocket bolt. Hold the sprocket with a suitable tool and tighten the bolt to 35–48 ft. lbs. (47–65 Nm).

11. Install the crankshaft sprocket and key. Align the keyway with the timing mark on the oil pump housing.

12. Install the crankshaft sprocket bolt. Install the flywheel locking tool, if equipped with automatic transaxle, or place the shift lever in **4th** gear and apply the parking brake, if equipped with manual transaxle. Tighten the bolt to 116–122 ft. lbs. (157–166 Nm).

13. Position the automatic tensioner in a suitable press. Set a flat washer under the tensioner body to prevent damage to the body plug.

14. Compress the tensioner until the hole in the piston is aligned with the 2nd hole in the tensioner case. Insert a 0.063 in. (1.6mm) diameter wire or pin through the 2nd hole to keep the piston compressed.

15. Make sure the camshaft sprocket timing marks are still aligned. Turn the crankshaft counterclockwise until the timing sprocket is offset from TDC by 1 tooth.

16. With the number one idler pulley removed, install the timing belt.

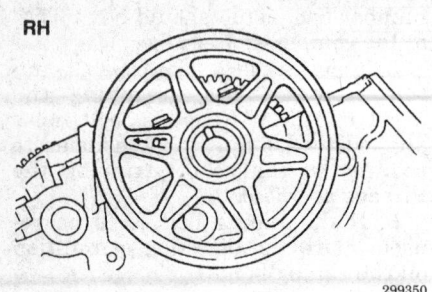

RH camshaft sprocket — 1993–95 MX3 1.8L (K8) and 1993–97 626 and MX6 2.5L (KL) engines

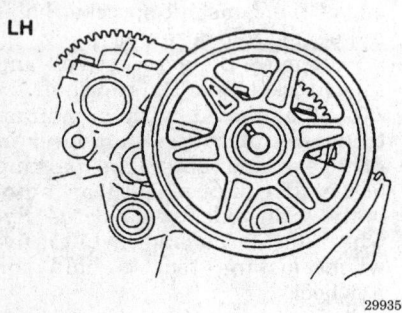

LH camshaft sprocket — 1993–95 MX3 1.8L (K8) and 1993–97 626 and MX6 2.5L (KL) engines

If the original belt is being reused, make sure it is installed in the same direction of rotation. The order of installation is: timing belt (crankshaft) sprocket, number two idler pulley, LH camshaft sprocket, tensioner pulley and RH camshaft sprocket.

17. Install the number one idler pulley while applying pressure on the timing belt. Tighten the bolt to 28–38 ft. lbs. (38–51 Nm).

18. Install the automatic belt tensioner and tighten the bolts to 14–18 ft. lbs. (19–25 Nm). Remove the wire or pin from the tensioner.

19. Rotate the crankshaft 2 turns in the normal direction of rotation and align the timing marks. Make sure all marks are still correctly aligned.

20. Inspect the timing belt deflection, 0.24–0.31 ft. lbs. (6–8mm), between the crankshaft sprocket and the tensioner pulley. If it is out of specification, replace the auto tensioner.

21. Remove the crankshaft damper bolt and install the timing belt covers.

22. Install the right side engine mount. Tighten the nuts to 55–77 ft. lbs. (75–104 Nm) and the through-bolt to 63–86 ft. lbs. (86–116 Nm). Remove the engine support.

23. Connect the negative battery cable.

1995 MX3 1.6L (B6-ZE) Engines

1. Disconnect the negative battery cable.

2. Remove the timing belt covers. Temporarily reinstall the crankshaft pulley bolt.

3. Turn the crankshaft until the timing mark on the crankshaft sprocket aligns with the timing mark on the oil pump and the camshaft sprocket timing marks, **E** and **I**, line up on the camshaft sprockets.

4. Lower the vehicle. Insert a camshaft sprocket holding tool between the camshaft sprockets.

5. Loosen the tensioner pulley lockbolt. Pull the tensioner pulley away from the center of the engine to reduce the tension on the timing belt.

6. If the timing belt is to be reused, mark the direction of rotation on the timing belt. Remove the timing belt.

7. To remove the tensioner, unhook the tensioner spring, and remove the pulley lockbolt and tensioner.

8. If necessary to remove the sprockets:

    a. Hold the camshaft by using a wrench on the cast hexagon, and loosen the sprocket mounting bolt.

    b. Remove the sprocket bolt and the sprocket from the camshaft.

    c. If equipped with a manual transaxle, place the shift lever in **4th** gear and apply the parking brake. If equipped with an automatic transaxle, remove the flywheel dust cover and install a flywheel locking tool to hold the flywheel.

    d. Remove the crankshaft sprocket bolt, sprocket and key.

To install:

9. Install the camshaft sprocket, aligning the dowel with the number **1** mark.

10. Install the camshaft sprocket bolt. Hold the camshaft, on the cast hexagon, with the wrench and tighten the bolt to 37–44 ft. lbs. (50–60 Nm).

11. Install the crankshaft sprocket and key. Align the keyway with the timing mark on the oil pump housing.

12. Install the crankshaft sprocket bolt. Install the flywheel locking tool, if equipped with automatic transaxle, or place the shift lever in **4th** gear and apply the parking brake, if equipped with manual transaxle. Tighten the bolt to 108–116 ft. lbs. (147–157 Nm).

13. Make sure the timing marks on the camshaft and crankshaft sprockets are still aligned.

14. If removed, position the tensioner with the spring fully extended, and install the lockbolt tightening the mounting bolt to 28–38 ft. lbs. (38–51 Nm).

15. Install the timing belt. If reusing the original timing belt, make sure it is installed in the same direction of rotation.

16. Loosen the tensioner lockbolt to apply tension to the timing belt. Tighten the tensioner lockbolt. Remove the holding tool from between the camshaft sprockets.

17. Rotate the crankshaft 2 turns in the normal direction of rotation and align the timing marks. Make sure all marks are still correctly aligned.

18. Raise and safely support the vehicle. Remove the crankshaft pulley bolt and install the timing belt covers.

19. Connect the negative battery cable.

1993–97 626 and MX6 2.0L (FS) Engines

1. Disconnect the negative battery cable.

2. Remove the timing belt covers. Temporarily reinstall the crankshaft pulley bolt.

3. Turn the crankshaft until the timing mark on the crankshaft sprocket aligns with the timing mark on the oil pump and the camshaft sprocket timing marks, **E** and **I**, line up on the camshaft sprockets.

4. Insert a camshaft sprocket holding tool between the camshaft sprockets.

5. Turn the timing belt tensioner with an Allen wrench and remove the tensioner spring from the hook pin.

6. If the timing belt is to be reused, mark the direction of rotation on the timing belt. Remove the timing belt.

7. If necessary to remove the sprockets:

    a. Hold the camshaft by using a wrench on the cast hexagon, and loosen the sprocket mounting bolt.

    b. Remove the sprocket bolt and the sprocket from the camshaft.

    c. If equipped with a manual transaxle, place the shift lever in **4th** gear and apply the parking brake. If equipped with an automatic transaxle, remove the flywheel dust cover and install a flywheel locking tool to hold the flywheel.

    d. Remove the crankshaft sprocket bolt, sprocket and key.

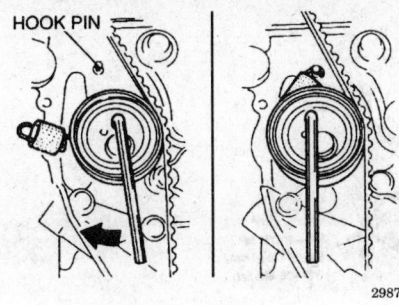

HOOK PIN

298741

Turning the tensioner spring — 1993–97 626 and MX6 2.0L (FS) engines

To install:

8. Install the camshaft sprocket, aligning the dowel with the number **1** mark.

9. Install the camshaft sprocket bolt. Hold the camshaft, on the cast hexagon, with the wrench and tighten the bolt to 35–48 ft. lbs. (47–65 Nm).

10. Install the crankshaft sprocket and key. Align the keyway with the timing mark on the oil pump housing.

11. Install the crankshaft sprocket bolt. Install the flywheel locking tool, if equipped with automatic transaxle, or place the shift lever in **4th** gear and apply the parking brake, if equipped with manual transaxle. Tighten the bolt to 108–116 ft. lbs. (147–157 Nm).

12. Make sure the timing marks on the camshaft and crankshaft sprockets are still aligned.

13. Install the timing belt. If reusing the original timing belt, make sure it is installed in the same direction of rotation.

14. Turn the tensioner clockwise with an Allen wrench and install the tensioner spring. Remove the holding tool from between the camshaft sprockets.

15. Rotate the crankshaft 2 turns in the normal direction of rotation and align the timing marks. Make sure all marks are still correctly aligned.

16. Remove the crankshaft pulley bolt and install the timing belt covers.

17. Connect the negative battery cable.

1993–94 929 3.0L (JE-ZE) Engines

1. Disconnect the negative battery cable and drain the cooling system.

2. Remove the fresh air duct.

3. Remove the cooling fan and fan shroud.

4. Remove the air intake pipe from the throttle body and air cleaner.

5. Disconnect the spark plug wires and remove the spark plugs.

6. Remove the idler pulleys and the accessory drive belts.

7. Remove the coolant bypass hose and the upper radiator hose.

8. Disconnect the electrical connector and remove the distributor.

9. Hold the crankshaft pulley with a suitable tool and remove the bolt. Remove the crankshaft pulley. Be careful not to damage the sensor rotor, if equipped.

10. Remove the timing belt covers. Rotate the crankshaft until the camshaft and crankshaft sprocket timing marks are aligned.

11. Remove the upper idler pulley and the automatic tensioner and pulley. Mark the direction of rotation on the timing belt and remove the timing belt.

NOTE: Do not rotate the engine after the timing belt has been removed.

12. Inspect the belt for wear, peeling, cracking, hardening or signs of oil contamination. Inspect the tensioner pulley for free and smooth rotation. Check the automatic tensioner for oil leakage. Check the tensioner rod projection (free length); it should be 0.47–0.55 in. (12–14mm). Inspect the sprocket teeth for wear or damage. Replace parts, as necessary.

13. If necessary to remove the sprockets:

a. Insert a proper tool through one of the camshaft sprocket holes to keep it from turning.

NOTE: The right and left camshaft sprockets are different and need to be installed on the same camshaft from which they were removed.

b. Remove the sprocket bolt and the sprocket from the camshaft.

c. Remove the flywheel dust cover and install a flywheel locking tool to hold the flywheel.

d. Remove the crankshaft sprocket bolt, sprocket and key.

To install:

14. Install the camshaft sprocket, aligning the dowel with the number **1** mark.

15. Install the camshaft sprocket bolt. Hold the sprocket with a suitable tool and tighten the right-hand bolt to 53–59 ft. lbs. (71–80 Nm), and the left-hand bolt to 14–18 ft. lbs. (19–254 Nm).

16. Install the crankshaft sprocket and key. Align the keyway with the timing mark on the oil pump housing.

17. Install the crankshaft sprocket bolt. Install the flywheel locking tool. Tighten the bolt to 116–122 ft. lbs. (157–166 Nm). Remove the locking tool.

18. Install the timing belt and cover. Connect the negative battery cable.

19. Position the automatic tensioner in a suitable press. Place a flat washer under the tensioner body to prevent damage to the body plug.

20. Slowly press in the tensioner rod, but do not exceed 2200 lbs. force. Insert a pin into the tensioner body to hold the rod in place.

21. Install the tensioner and tighten the bolts to 19 ft. lbs. (25 Nm).

22. Make sure the crankshaft and camshaft sprocket timing marks are aligned.

23. Install the timing belt over the crankshaft sprocket, lower idler pulley, left exhaust camshaft sprocket, left intake camshaft sprocket, tensioner pulley, right exhaust camshaft sprocket and right intake camshaft sprocket, in that order.

24. Push the belt down and install the upper idler pulley. Tighten the bolt to 38 ft. lbs. (52 Nm). Make sure the timing marks are still aligned after installing the upper idler pulley.

25. Turn the crankshaft 2 revolutions in the direction of normal rotation and realign the timing marks. If the timing marks do not align, repeat Steps 16–20.

26. Remove the pin from the automatic tensioner. Again rotate the crankshaft 2 turns and make sure the timing marks are aligned.

27. Apply approximately 22 lbs. (98 N) pressure to the timing belt at a point midway between the right exhaust camshaft sprocket and the tensioner pulley. The belt should deflect 0.20–0.28 in. (5–7mm). If the deflection is not as specified, replace the automatic tensioner.

28. Install the timing belt covers, and tighten the bolts to 95 inch lbs. (11 Nm).

29. Install the crankshaft pulley. Hold the pulley with a suitable tool and tighten the lockbolt to 123 ft. lbs. (167 Nm). Be careful not to damage the sensor rotor, if equipped.

30. Install the distributor and connect the electrical connector.

31. Install the upper radiator hose and coolant bypass hose.

32. Install the idler pulleys and the accessory drive belts. Adjust the belt tension.

33. Install the spark plugs and connect the spark plug wires.

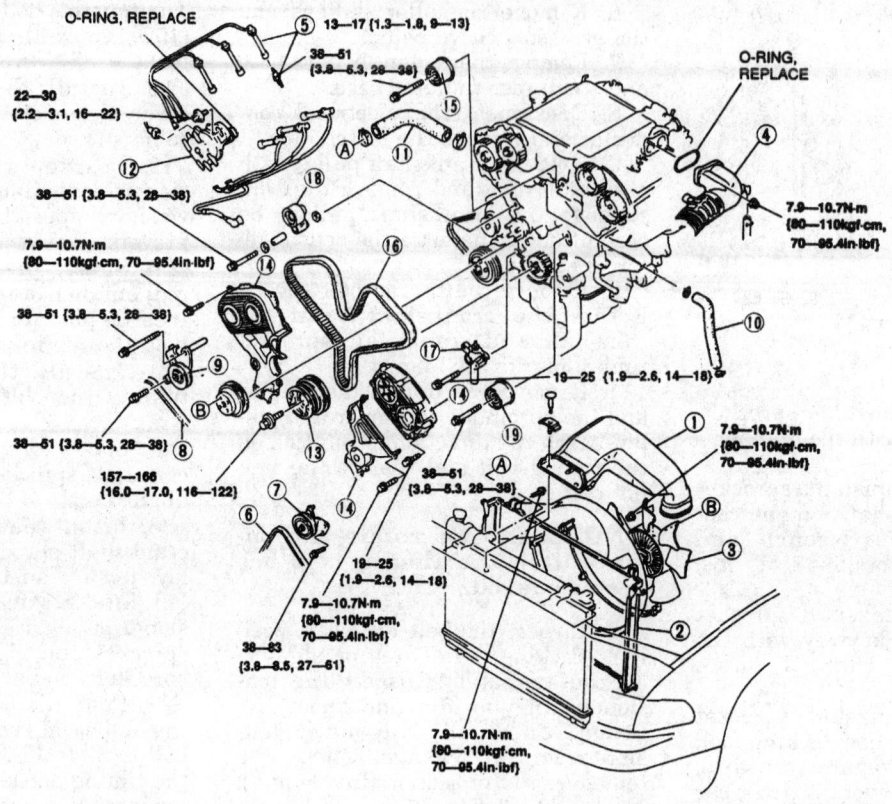

1. Air duct
2. Radiator cowling
3. Coolant fan
4. Air intake pipe
5. High-tension leads and spark plugs
6. A/C compressor belt
7. A/C compressor belt idler pulley
8. P/S oil pump belt
9. P/S oil pump belt idler pulley
10. Water bypass hose
11. Upper radiator hose
12. Distributor
13. Crankshaft pulley
14. Timing belt cover
15. No.2 Idler pulley
16. Timing belt
17. Timing belt auto tensioner
18. Tensioner pulley
19. No.1 Idler pulley

235102

Timing belt assembly (exploded view) — 929 3.0L (JE-ZE) engines

34. Install the air intake pipe to the throttle body and air cleaner.

35. Install the cooling fan and radiator shroud. Install the fresh air duct.

36. Connect the negative battery cable. Fill and bleed the cooling system.

37. Start the engine and bring to normal operating temperature. Check for leaks and for proper operation. Check the ignition timing.

1995–97 Millenia 2.3L (KJS) Engines

1. Disconnect the negative battery cable.

2. Remove the timing belt covers. Temporarily reinstall the crankshaft pulley bolt.

3. Remove the power steering auto tensioner and pulley.

4. Turn the crankshaft until the timing mark on the crankshaft sprocket aligns with the timing mark on the oil pump and the camshaft sprocket timing marks align with the marks on the cylinder head. The number one piston should be at TDC of the compression stroke.

5. Remove the two bolts from the automatic tensioner, removing the lower one first. Keep the bolt holes aligned by holding the tensioner to

reduce the chance of stripping the threads on the bolts.

6. If the timing belt is to be reused, mark the direction of rotation on the timing belt.

7. Remove the timing belt.

8. If necessary to remove the sprockets:

a. Insert a proper tool through one of the camshaft sprocket holes to keep it from turning.

NOTE: The right and left camshaft sprockets are different and need to be installed on the same camshaft from which they were removed.

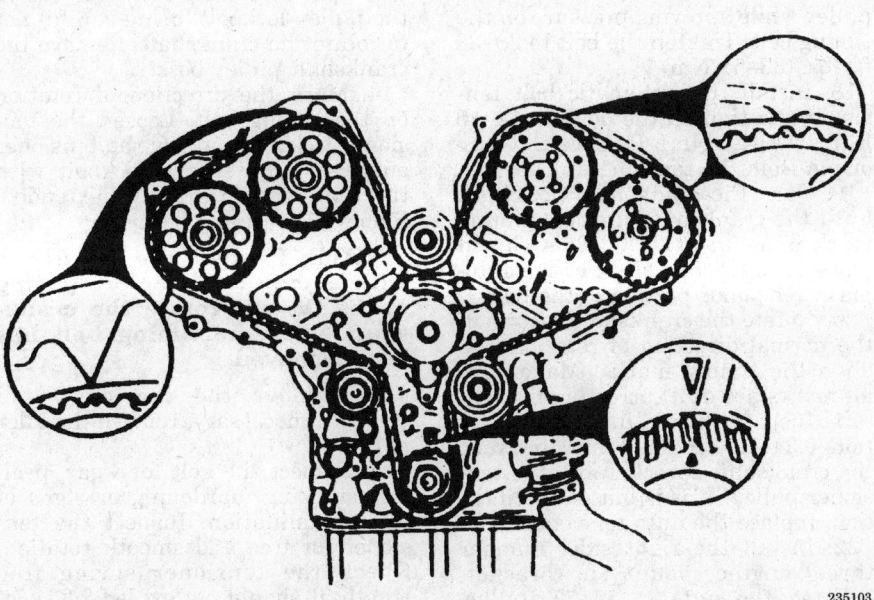

Timing marks — 1993–94 929 3.0L (JE-ZE) engines

235103

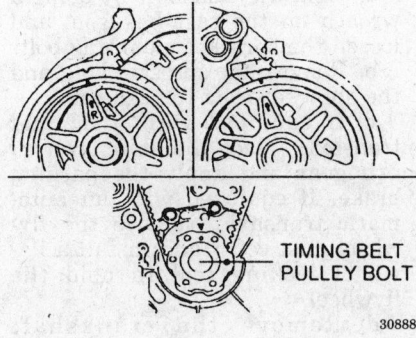

TIMING BELT
PULLEY BOLT

308885

Timing mark locations — 1995–97 Millenia 2.3L (KJS) engines

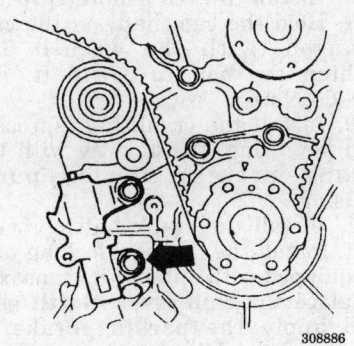

308886

Lower auto tensioner bolt location — 1995–97 Millenia 2.3L (KJS) engines

b. Remove the sprocket bolt and the sprocket from the camshaft.

c. Remove the flywheel dust cover and install a flywheel locking tool to hold the flywheel.

d. Remove the crankshaft sprocket bolt, sprocket and key.

To install:

9. Install the camshaft sprocket so that the **R** (on right-hand) and **L** (on left-hand) face out, and align the timing marks with the knock/dowel pin.

10. Apply clean engine oil to the bolt threads, and install the camshaft sprocket bolt. Hold the sprocket with a suitable tool and tighten the bolt to 91–103 ft. lbs. (123–140 Nm).

11. Install the crankshaft sprocket and key. Align the keyway with the timing mark on the oil pump housing.

12. Install the crankshaft sprocket bolt. Install the flywheel locking tool. Tighten the bolt to 116–122 ft. lbs. (157–166 Nm). remove the flywheel locking tool.

13. Position the automatic tensioner in a press. Set a flat washer under the tensioner body to prevent damage to the body plug.

14. Compress the tensioner until the hole in the piston is aligned with the 2nd hole in the tensioner case. Insert a 0.063 inch (1.6 mm) diame-

ter wire or pin through the 2nd hole to keep the piston compressed.

15. Make sure the camshaft sprocket timing marks are still aligned. Turn the crankshaft clockwise until the timing sprocket is aligned.

16. Install the timing belt. If the original belt is being reused, make sure it is installed in the same direction of rotation. The order of installation is: timing belt (crankshaft) sprocket, number two idler pulley, LH camshaft sprocket, both number one idler pulleys, RH camshaft sprocket and the tensioner pulley.

17. Install the automatic belt tensioner and tighten the bolts to 14–18 ft. lbs. (19–25 Nm). Remove the wire or pin from the tensioner.

18. Turn the crankshaft clockwise, until the crankshaft sprocket timing mark is again at TDC. This should place all of the belt slack in the automatic tensioner portion of the belt.

19. Rotate the crankshaft two turns in the normal direction of rotation and align the timing marks. Make sure all marks are still correctly aligned.

20. Inspect the timing belt deflection, 0.24–0.31 inches (6–8 mm), between the crankshaft sprocket and the tensioner pulley. If it is out of specification, replace the auto tensioner.

21. Install the power steering auto tensioner and tighten the bolts to 14–18 ft. lbs. (19–25 Nm). Install the pulley, and tighten the bolt to 29–34 ft. lbs. (40–47 Nm).

22. Remove the crankshaft damper bolt and install the timing belt covers.

23. Connect the negative battery cable. Start the engine, and check the ignition timing.

1995–97 Millenia 2.5L (KLD) Engine

1. Disconnect the negative battery cable.

2. Remove the timing belt covers. Temporarily reinstall the crankshaft pulley bolt.

3. Support the engine, and remove the nuts and through-bolt from the right side (number three) engine mount sub bracket. Remove the sub bracket.

4. Turn the crankshaft until the timing mark on the crankshaft sprocket aligns with the timing mark on the oil pump and the camshaft sprocket timing marks align with the marks on the cylinder head. The number one piston should be at TDC of the compression stroke.

5. Remove the two bolts from the automatic tensioner, removing the

lower one first. Keep the bolt holes aligned by holding the tensioner to reduce the chance of stripping the threads on the bolts.

6. If the timing belt is to be re-used, mark the direction of rotation on the timing belt.

7. Remove the number one idler pulley. Remove the timing belt.

8. If necessary to remove the sprockets:

a. Insert a proper tool through one of the camshaft sprocket holes to keep it from turning.

NOTE: The right and left camshaft sprockets are different and need to be installed on the same camshaft from which they were removed.

b. Remove the sprocket bolt and the sprocket from the camshaft.

c. Remove the flywheel dust cover and install a flywheel locking tool to hold the flywheel.

d. Remove the crankshaft sprocket bolt, sprocket and key.

To install:

9. Install the camshaft sprocket so that the **R** (on right-hand) and **L** (on left-hand) face out, and align the timing marks with the knock/dowel pin.

10. Apply clean engine oil to the bolt threads, and install the camshaft sprocket bolt. Hold the sprocket with a suitable tool and tighten the bolt to 91–103 ft. lbs. (123–140 Nm).

11. Install the crankshaft sprocket and key. Align the keyway with the timing mark on the oil pump housing.

12. Install the crankshaft sprocket bolt. Install the flywheel locking tool. Tighten the bolt to 116–122 ft. lbs. (157–166 Nm). remove the flywheel locking tool.

13. Position the automatic tensioner in a suitable press. Set a flat washer under the tensioner body to prevent damage to the body plug.

14. Compress the tensioner until the hole in the piston is aligned with the 2nd hole in the tensioner case. Insert a 0.060 in. (1.6 mm) diameter wire or pin through the 2nd hole to keep the piston compressed.

15. Make sure the camshaft sprocket timing marks are still aligned. Turn the crankshaft counter-clockwise until the timing sprocket is aligned.

16. With the number one idler pulley removed, install the timing belt. If the original belt is being reused, make sure it is installed in the same direction of rotation. The order of installation is: timing belt (crankshaft) sprocket, number two idler pulley,

LH camshaft sprocket, tensioner pulley and RH camshaft sprocket.

17. Install the number one idler pulley while applying pressure on the timing belt. Tighten the bolt to 28–38 ft. lbs. (38–51 Nm).

18. Install the automatic belt tensioner and tighten the bolts to 14–18 ft. lbs. (19–25 Nm). Remove the wire or pin from the tensioner.

19. Turn the crankshaft clockwise, until the crankshaft sprocket timing mark is again at TDC. This should place all of the belt slack in the automatic tensioner portion of the belt.

20. Rotate the crankshaft 2 turns in the normal direction of rotation and align the timing marks. Make sure all marks are still correctly aligned.

21. Inspect the timing belt deflection, 0.24–0.31 in. (6–8 mm), between the crankshaft sprocket and the tensioner pulley. If it is out of specification, replace the auto tensioner.

22. Install the right side (number three) engine mount sub bracket. Tighten the nuts to 55–77 ft. lbs. (75–104 Nm) and the through-bolt to 63–86 ft. lbs. (86–116 Nm). Remove the engine support.

23. Remove the crankshaft damper bolt and install the timing belt covers.

24. Connect the negative battery cable. Start the engine, and check the ignition timing.

1993–94 Miata 1.6L (B6-ZE) and 1993–97 1.8L (BPD) Engines

1. Disconnect the negative battery cable. Remove the engine undercover.

2. Drain the engine coolant.

3. Remove the air intake pipe.

4. Remove the upper radiator hose, then remove the water hoses connected to the thermostat housing.

5. Remove the accessory drive belts.

6. Remove the water pump pulley.

7. Remove the crankshaft pulley front plate. Remove the bolts and remove the crankshaft pulley.

8. Remove the outer timing belt guide plate.

9. Tag and disconnect the spark plug wires. Remove the coil.

10. Remove the spark plugs.

NOTE: Spark plugs are removed to make it easier to rotate the engine.

11. Remove the valve cover (cylinder head cover).

12. Remove the upper, middle and lower timing belt covers.

13. Turn the crankshaft until the timing marks on the crankshaft and camshaft sprockets are aligned. The

pin on the pulley boss must face upward. Hold the crankshaft pulley boss with a suitable tool and remove the pulley lockbolt, being careful not to rotate the crankshaft. Remove the crankshaft pulley boss.

14. Mark the direction of rotation on the timing belt. Loosen the tensioner lockbolt and pry the tensioner outward. Tighten the lockbolt with the tensioner spring fully extended. Remove the timing belt.

NOTE: Protect the tensioner with a shop towel before prying on it. Do not rotate the crankshaft after the timing belt has been removed.

15. Remove the tensioner and spring. If necessary, remove the idler pulley.

16. Inspect the belt for wear, peeling, cracking, hardening or signs of oil contamination. Inspect the tensioner for free and smooth rotation. Check the tensioner spring free length; it should not exceed 2.331 in. (59.2mm) on the 1995–97 1.8L engine and 2.315 in. (58.8mm) on all others. Inspect the sprocket teeth for wear or damage. Replace parts, as necessary.

17. If necessary to remove the sprockets:

a. Hold the camshaft by using a wrench on the cast hexagon, and loosen the sprocket mounting bolt.

b. Remove the sprocket bolt, and the sprocket from the camshaft.

c. If equipped with a manual transaxle, place the shift lever in **4th** gear and apply the parking brake. If equipped with an automatic transaxle, remove the flywheel dust cover and install a flywheel locking tool to hold the flywheel.

d. Remove the crankshaft sprocket bolt, sprocket and key.

To install:

18. Install the camshaft sprocket, aligning the dowel with the number **1** mark.

19. Install the camshaft sprocket bolt. Hold the camshaft, on the cast hexagon, with the wrench and tighten the bolt to 37–44 ft. lbs. (50–60 Nm).

20. Install the crankshaft sprocket and key. Align the keyway with the timing mark on the oil pump housing.

21. Install the crankshaft sprocket bolt. Install the flywheel locking tool, if equipped with automatic transaxle, or place the shift lever in **4th** gear and apply the parking brake, if equipped with manual transaxle. Tighten the bolt to 108–116 ft. lbs. (147–157 Nm).

22. If removed, install the idler pulley and tighten the bolt to 38 ft. lbs. (52 Nm).

23. Install the tensioner and tensioner spring. Pry the tensioner outward and temporarily tighten the tensioner lockbolt with the tensioner spring fully extended.

24. Make sure the crankshaft sprocket timing mark is aligned with the mark on the oil pump housing. Make sure the camshaft sprocket timing marks are aligned with the marks on the seal plate.

25. Install the timing belt so there is no looseness at the idler pulley side or between the camshaft sprockets. If reusing the old belt, make sure it is installed in the same direction of rotation.

26. Temporarily install the pulley boss and lockbolt.

27. Turn the crankshaft 2 turns clockwise and align the crankshaft sprocket timing mark. Face the pin on the pulley boss upright. Make sure the camshaft sprocket timing marks are aligned. If they are not, repeat the alignment steps.

28. Turn the crankshaft 1⅚ turns clockwise and align the crankshaft sprocket timing mark with the tension set mark for proper belt tension adjustment. Remove the lockbolt and pulley boss.

29. Make sure the crankshaft sprocket timing mark is aligned with the tension set mark. Loosen the tensioner lockbolt, and allow the spring to apply tension to the belt. Tighten the tensioner lockbolt to 28–38 ft. lbs. (38–52 Nm).

30. Install the pulley boss and lockbolt.

31. Turn the crankshaft 2⅙ turns clockwise and make sure the timing marks are correctly aligned.

32. Apply approximately 22 lbs. pressure to the timing belt at a point midway between the camshaft sprockets. The belt should deflect 0.35–0.45 in. (9.0–11.5mm). If the deflection is not correct, repeat the alignment and tensioning procedure.

33. Hold the pulley boss with a suitable tool, and tighten the lockbolt to 123 ft. lbs. (167 Nm).

34. Install the timing belt covers and tighten the bolts to 95 inch lbs. (11 Nm).

35. Install the valve cover.

36. Install the PCV valve.

37. Install the coil.

38. Install the spark plugs and connect the spark plug wires.

39. Install the crankshaft pulley and tighten the bolts to 13 ft. lbs. (17 Nm).

40. Install the water pump pulley and the accessory drive belts. Adjust the belt tension.

41. Connect the radiator and coolant hoses.

42. Install the air intake pipe.

43. Refill the coolant.

44. Connect the negative battery cable.

45. Start the engine and check for proper operation. Check the ignition timing and idle speed. Check carefully for any fluid leaks.

46. Install the engine side or undercover, as necessary.

Camshaft

REMOVAL AND INSTALLATION

1993 Protege 1.8L (BPD) Engine

1. Disconnect the negative battery cable. Drain the cooling system.

2. Label and disconnect the spark plug wires and remove the spark plugs.

3. Disconnect the hoses from the cylinder head cover, if equipped.

4. Remove the cylinder head cover bolts and remove the cylinder head cover.

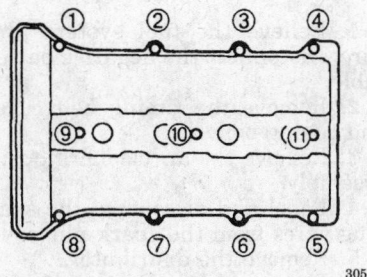

Cylinder head cover (tightening sequence) — Protege 1.8L (BPD) engine

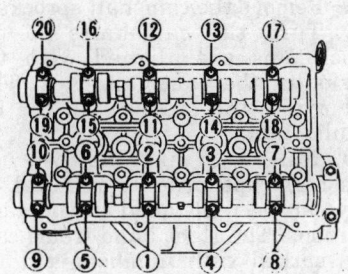

Camshaft cap bolt tightening sequence — 1993 Protege 1.8L (BPD) engine

5. Remove the timing belt. Remove the distributor.

6. Hold the camshaft with a wrench on the hexagon cast into the camshaft. Remove the sprocket bolts and remove the sprockets.

7. Label the caps so they can be reinstalled in their original positions. Loosen the camshaft cap bolts in 2–3 steps in the reverse of the torque sequence, then remove the camshaft caps.

8. Remove the camshafts. Remove the camshaft oil seals from the camshafts.

To install:

9. Lubricate the camshaft journals and lobes with clean engine oil. Install the camshafts in the cylinder head.

10. Apply silicone sealant to the cylinder head on the front camshaft cap mating surfaces. Do not allow any sealant on the camshaft journals.

11. Install the camshaft caps in their original positions. Loosely install the cap bolts.

12. Tighten the camshaft cap bolts in 2–3 steps to 125 inch lbs. (14 Nm) in the proper sequence.

13. Apply clean engine oil to the lip of a new camshaft seal. Push the seal slightly in by hand. Tap the seal into position, using a seal installer, until it is flush with the edge of the camshaft cap.

14. Turn the camshafts until the dowel pins face straight up. Install the camshaft sprockets and the sprocket bolts.

15. Hold the camshaft with the wrench on the cast hexagon and tighten the sprocket bolts to 44 ft. lbs. (60 Nm).

16. Install the remaining components in the reverse order of removal.

1995–97 Millenia 2.3L (KJS) Engine

1. Relieve the fuel system pressure. Disconnect the negative battery cable.

2. Remove the timing belt covers and timing belt.

3. Remove the spacer and O-ring from the front of the camshaft.

4. Remove the ignition coils.

5. Remove the intake manifold.

6. Remove the bolts, in 5–6 steps, using the reverse of the tightening sequence. Remove the cylinder head cover.

7. Remove the camshaft sprockets.

8. Turn the camshafts so the knock pins are aligned with the marks on the camshaft caps. This will reduce the pressure on the adjustment shims.

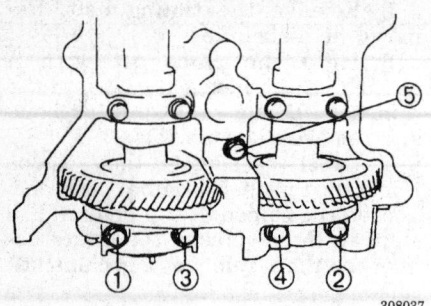

Camshaft front cap bolts (loosening sequence) — 1995–97 Millenia 2.3L (KJS) engine

308037

9. Note the markings on the camshaft caps prior to removal, so they can be reinstalled in the same positions. The right hand (rear) caps are marked with numbers and the left hand (front) caps are marked with letters.

10. Loosen the front camshaft cap bolts in the reverse of the torque sequence, in 5–6 steps. Remove the front camshaft caps.

11. Remove the remaining camshaft cap bolts in the proper sequence. Remove the caps, being sure to remove the thrust caps last. Do not damage the cylinder head thrust bearing support.

12. Remove the camshafts and oil seals.

13. If necessary, remove the lifters and adjustment shims. Identify and mark each lifter as it is removed so it can be reinstalled in the same position.

To install:

14. Apply clean engine oil to the lifters, then install them in their original positions. Verify that they move smoothly in their bore.

15. Install new oil seals on the camshafts. Apply clean engine oil to the camshaft lobes, journals and supports.

16. Install the camshafts so the gear marks align.

17. Remove all oil and dirt from the mating surfaces between the front camshaft cap and the cylinder head.

18. Install the thrust caps. Tighten the thrust cap bolts, in 5–6 steps, until the caps are fully seated on the cylinder head.

19. Apply silicone sealant, at a thickness of 0.06–0.09 inch (1.5–2.5mm), to the cylinder head surface in the area forward of the camshaft gear cavity.

20. Install the remaining camshaft caps in their original positions. Tighten the caps, in sequence, in five equal steps, with the final step being 100–125 inch lbs. (11–14 Nm).

21. Apply clean engine oil to the lip of the new camshaft oil seal. Push the seal in lightly by hand. Tap the seal in evenly with a seal installer (49 F401 337A or equivalent) with a final protrusion of 0–0.02 inch (0–0.5mm). Tap in a new blind cap.

22. Install the camshaft sprockets. Tighten the mounting bolts to 91–103 ft. lbs. (123–140 Nm).

23. Measure and adjust valve clearances.

24. Remove any sealant and gasket material from the cylinder head cover contact surfaces.

25. Apply silicone sealant to the cylinder head in the area adjacent to the front and rear camshaft caps. Install a new gasket on the cylinder head.

26. Install the cylinder head cover. Tighten the bolts in 5–6 steps, in sequence, to 44–78 inch lbs. (5–9 Nm).

27. Using a new O-ring, install the distributor.

28. Install the ignition coils.

29. Install the intake manifold.

30. Install the spacer, using a new O-ring. Tighten the mounting bolt to 14–18 ft. lbs. (19–25 Nm).

31. Install the timing belt and timing belt cover.

32. Connect the negative battery cable. Run the engine and check for leaks.

1995–97 Millenia 2.5L (KLD) Engine

1. Relieve the fuel system pressure. Disconnect the negative battery cable.

2. Remove the timing belt covers and timing belt.

3. Remove the air cleaner housing assembly.

4. Label and disconnect the spark plug wires from the spark plugs.

5. Remove the distributor.

6. Remove the intake manifold.

7. Remove the bolts, in 5–6 steps, using the reverse of the tightening sequence. Remove the cylinder head cover.

8. Remove the camshaft sprockets.

9. Turn the camshafts so the knock pins are aligned with the marks on the camshaft caps. This will reduce the pressure on the hydraulic lifters.

10. Note the markings on the camshaft caps prior to removal, so they can be reinstalled in the same positions. The right hand (rear) caps are marked with numbers and the left hand (front) caps are marked with letters.

11. Loosen the front camshaft cap bolts in sequence, in 5–6 steps. Remove the front camshaft caps.

12. Remove the remaining camshaft cap bolts in the proper sequence. Remove the caps, being sure to remove the thrust caps last. Do not damage the cylinder head thrust bearing support.

13. Remove the camshafts and oil seals.

14. If necessary, remove the hydraulic lifters. Identify and mark each lifter as it is removed so it can be reinstalled in the same position. If the lifters are to be reused, store them upside down in an oil-filled sealed container.

To install:

15. If removed, inspect each lifter by holding it in your hand and pressing the bottom up against the top of the lifter. If there is movement, replace the lifter.

16. Apply clean engine oil to the lifters, then install them in their original positions. Verify that they move smoothly in their respective bores.

17. Install new oil seals on the camshafts. Apply clean engine oil to the camshaft lobes, journals and supports.

18. Install the camshafts so the gear marks align.

19. Remove all oil and dirt from the mating surfaces between the front camshaft cap and the cylinder head.

20. Install the thrust caps. Tighten the thrust cap bolts, in 5–6 steps, until the caps are fully seated on the cylinder head.

21. Apply silicone sealant, at a thickness of 0.06–0.09 inch (1.5–2.5mm), to the cylinder head surface in the area forward of the camshaft gear cavity and to the left cylinder head on the rear exhaust camshaft cap mating surface.

22. Install the remaining camshaft caps in their original positions. Tighten the caps, in sequence, in five equal steps, with the final step being 100–125 inch lbs. (11–14 Nm).

23. Apply clean engine oil to the lip of the new camshaft oil seal. Push the seal in lightly by hand. Tap the seal in evenly with a seal installer (49 F401 337A or equivalent) with a final protrusion of 0–0.02 inch (0–0.5mm).

24. Install the camshaft sprockets. Tighten the mounting bolts to 91–103 ft. lbs. (123–140 Nm).

25. Remove any sealant and gasket material from the cylinder head cover contact surfaces.

26. Apply silicone sealant to the cylinder head in the area adjacent to the front and rear camshaft caps. Install a new gasket on the cylinder head.

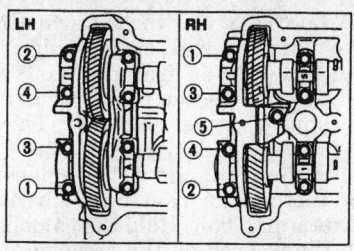
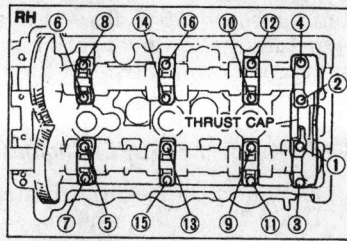
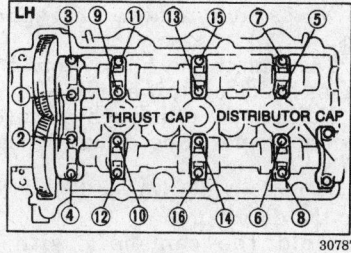

Camshaft assembly (loosening sequence) — 1995–97 Millenia 2.5L (KLD) engine

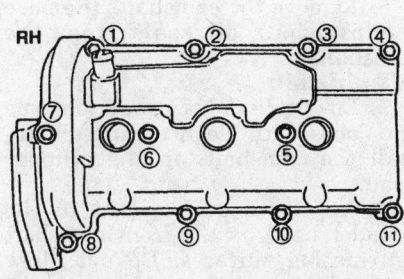

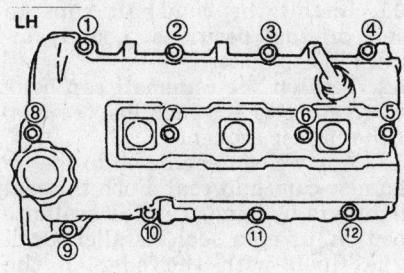

Cylinder head cover (tightening sequence) — 1995–97 Millenia 2.5L (KLD) engine

27. Install the cylinder head cover. Tighten the bolts in 5–6 steps, in sequence, to 44–78 inch lbs. (5–9 Nm).
28. Install the intake manifold.
29. Using a new O-ring, install the distributor.
30. Install the spark plug wires.
31. Install the air cleaner housing assembly.
32. Install the timing belt and timing belt cover.
33. Connect the negative battery cable. Run the engine and check for leaks.

1993–95 929 3.0L (JE-ZE) Engines

1. Properly relieve the fuel system pressure.
2. Disconnect the negative battery cable, and drain the cooling system.
3. Remove the fresh air duct and air intake pipe.
4. Remove the cooling fan and shroud.
5. Disconnect the spark plug wires and remove the spark plugs. Remove the idler pulleys and the accessory drive belts.
6. Remove the coolant bypass hose and the upper radiator hose. Disconnect the electrical connector and remove the distributor.

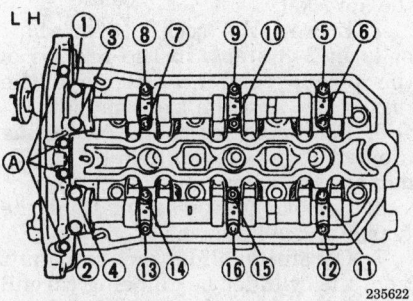

Camshaft loosening sequence (LH cylinder head); reverse for tightening — 1993–95 929 3.0L (JE-ZE) engines

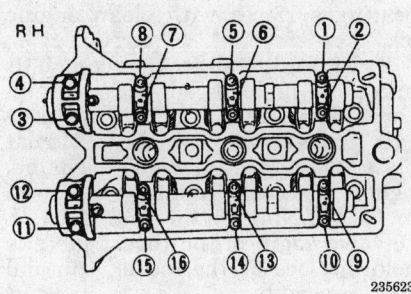

Camshaft loosening sequence (RH cylinder head); reverse for tightening — 1993–95 929 3.0L (JE-ZE) engines

7. Hold the crankshaft pulley with a suitable tool and remove the bolt. Remove the crankshaft pulley. Be careful not to damage the sensor rotor.
8. Remove the timing belt covers. Rotate the crankshaft until the camshaft and crankshaft sprocket timing marks are aligned.
9. Remove the upper idler pulley and the tensioner and pulley. Mark the direction of rotation on the timing belt and remove the timing belt.
10. Disconnect the accelerator cable. Disconnect and plug the fuel lines. Label and disconnect the necessary electrical connectors and vacuum hoses.
11. Loosen and remove the lower intake manifold bolts in 2–3 steps. Remove the lower intake manifold.
12. Remove the cylinder head covers.
13. Hold the camshaft with a wrench on the hexagon cast into the camshaft and remove the camshaft sprocket bolt. Remove the camshaft sprockets.
14. Remove bolts **A** from the front camshaft cap on the left cylinder head. Loosen the remaining camshaft cap bolts in 2–3 steps in the reverse of the torque sequence. Mark the position of the caps so they can be reinstalled in their original positions.
15. Remove the camshaft caps and remove the camshafts. Remove the camshaft oil seals. Label the camshafts so they can be reinstalled in their proper locations.
16. Lift out and label the rocker arms, if necessary.
To install:
17. If removed, position the rocker arms.
18. Lubricate the camshaft lobes and journals with clean engine oil and install the camshafts on the cylinder heads. Apply clean engine oil to the lips of the new camshaft seals, if replaced, and install them on the camshafts.
19. Apply a small amount of silicone sealant to the cylinder head on the front camshaft cap mating surfaces. Do not allow sealant to get on the camshaft journal, oil seal face or camshaft thrust face.
20. Install the camshaft caps in their original locations. Gradually tighten the cap bolts, in sequence, to 24 ft. lbs. (32 Nm). Tighten the front cap bolts, **A**, on the left cylinder head to 95 inch lbs. (11 Nm).
21. Hold the camshafts using a wrench on the hexagon cast into the camshaft. Install the camshaft sprockets with the retaining bolts.

Tighten the sprocket bolts on the right cylinder head to 59 ft. lbs. (80 Nm) and the sprocket bolts on the left cylinder head to 19 ft. lbs. (25 Nm).

22. Apply sealant to the cylinder head in the areas adjacent to the front camshaft caps. Install the cylinder head covers and tighten the nuts to 39 inch lbs. (4 Nm).

23. Install the lower intake manifold using new gaskets. Tighten the mounting nuts in 2–3 steps, in the proper sequence, to 19 ft. lbs. (25 Nm). Connect the fuel lines.

24. Install the upper intake manifold to the lower intake manifold, using a new gasket. Tighten the bolts to 19 ft. lbs. (25 Nm).

25. Connect the electrical connectors and vacuum hoses. Connect the accelerator cable.

26. Position the timing belt tensioner on a suitable press. Place a flat washer at the bottom of the tensioner body to prevent damage to the body plug. Slowly press in the tensioner rod, but do not exceed 2200 lbs. Insert a pin in the tensioner body to hold the rod in place, then install the tensioner and tighten the mounting bolts to 19 ft. lbs. (25 Nm).

27. Make sure the crankshaft and camshaft sprocket timing marks are properly aligned. Install the timing belt over the sprockets and pulleys in the following order: crankshaft sprocket, lower idler pulley, left cylinder head exhaust cam sprocket, left cylinder head intake cam sprocket, tensioner pulley, right cylinder head exhaust cam sprocket and right cylinder head intake cam sprocket.

28. Push the belt down and install the upper idler pulley. Tighten the upper idler pulley bolt to 38 ft. lbs. (52 Nm). Make sure the camshaft and crankshaft sprocket timing marks are still aligned after installing the upper idler pulley.

29. Turn the crankshaft 2 revolutions in the normal direction of rotation and realign the timing marks. Make sure all timing marks are correctly aligned.

30. Remove the pin from the automatic tensioner and again rotate the crankshaft 2 turns. Confirm that the timing marks are aligned.

31. Install the timing belt covers and tighten the bolts to 95 inch lbs. (11 Nm).

32. Install the crankshaft pulley. Hold the pulley with a suitable tool and tighten the pulley bolt to 123 ft. lbs. (167 Nm). Be careful not to damage the sensor rotor.

33. Install the spark plugs and the distributor assembly. Connect the spark plug wires and the distributor electrical connector.

34. Install the upper radiator hose and water bypass hose. Install the idler pulleys and accessory drive belts. Adjust the belt tension.

35. Install the shroud and cooling fan. Install the air intake pipe and fresh air duct.

36. Connect the negative battery cable. Fill and bleed the cooling system.

37. Start the engine and bring to normal operating temperature. Check for leaks. Check the ignition timing and idle speed.

1993–94 323 1.6L (B6E) and 1994 Protege 1.8L (BPD) Engines

NOTE: The camshaft is removed through the front of the cylinder head.

1. Remove the cylinder head from the vehicle and position in a suitable holding fixture.

NOTE: Do not lay the cylinder head flat on the head gasket surface as the valves may be damaged.

2. On 1.6L and 1.8L engines, hold the camshaft with a wrench on the hexagon cast into the front of the camshaft.

3. Remove the sprocket bolt and the sprocket.

4. Loosen the rocker arm shaft bolts in 2–3 steps, in the reverse of the torque sequence. Remove the rocker arm and shaft assemblies.

5. Pry out the camshaft seal using a small prybar, being careful not to damage the camshaft or seal bore.

6. Remove the thrust plate at the rear of the cylinder head.

7. Carefully slide the camshaft from the cylinder head, being careful not to damage the cylinder head bearing surfaces.

To install:

8. Lubricate the camshaft lobes and journals and the cylinder head bearing surfaces with clean engine oil.

9. Carefully slide the camshaft into the cylinder head, being careful not to damage the bearing surfaces.

10. Install the camshaft thrust plate. On the 1.6L 8-valve engine, tighten the thrust retaining bolt to 95 in. lbs. (11 Nm). On the 1.6L and 1.8L 16-valve engines, the thrust plate is held in place by the rocker arm and shaft assembly.

11. Lubricate the lip of a new camshaft seal with clean engine oil and install in the cylinder head, using a seal installer.

12. Lubricate the rocker arms and valve stem tips with clean engine oil. Install the rocker arm and shaft assemblies and tighten the bolts, in 2–3 steps, in the proper sequence. The final torque should be 21 ft. lbs. (28 Nm) on the 1.6L and 1.8L engines.

13. Install the camshaft sprocket and retaining bolt. Hold the camshaft with the wrench on the hexagon and tighten the bolt to 45 ft. lbs. (61 Nm).

14. Install the cylinder head and the remaining components in the reverse order of removal.

1993–94 Protege 1.8L (BPE) Engines

1. Disconnect the negative battery cable. Drain the cooling system.

2. Label and disconnect the spark plug wires and remove the spark plugs.

3. Disconnect the hoses from the cylinder head cover, if equipped.

4. Remove the cylinder head cover bolts and remove the cylinder head cover.

5. Remove the timing belt. Remove the distributor.

6. Hold the camshaft with a wrench on the hexagon cast into the camshaft. Remove the sprocket bolts and remove the sprockets.

7. Label the caps so they can be reinstalled in their original positions. Loosen the camshaft cap bolts in 2–3 steps in the reverse of the torque sequence, then remove the camshaft caps.

8. Remove the camshafts. Remove the camshaft oil seals from the camshafts.

To install:

9. Lubricate the camshaft journals and lobes with clean engine oil. Install the camshafts in the cylinder head.

10. Apply silicone sealant to the cylinder head on the front camshaft cap mating surfaces. Do not allow any sealant on the camshaft journals.

11. Install the camshaft caps in their original positions. Loosely install the cap bolts.

12. Tighten the camshaft cap bolts in 2–3 steps to 125 inch lbs. (14 Nm) in the proper sequence.

13. Apply clean engine oil to the lip of a new camshaft seal. Push the seal slightly in by hand. Tap the seal into position, using a seal installer, until it is flush with the edge of the camshaft cap.

14. Turn the camshafts until the dowel pins face straight up. Install the camshaft sprockets and the sprocket bolts.

15. Hold the camshaft with the wrench on the cast hexagon and

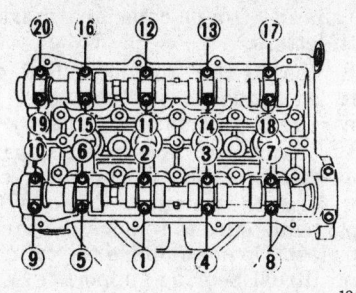

Camshaft cap bolt tightening sequence — 1993–94 Protege 1.8L (BPE) and 1994 MX3 1.6L (B6-ZE) engines

tighten the sprocket bolts to 44 ft. lbs. (60 Nm).

16. Install the remaining components in the reverse order of removal.

1995–97 Protege 1.5L (Z5D) and 1.8L (BPE, BPD) Engines

1. Disconnect the negative battery cable.

2. Remove the power steering hose brackets from the cylinder head cover.

3. Label and disconnect the spark plug wires and spark plug wire clips.

4. Disconnect the breather tube and PCV valve from the cylinder

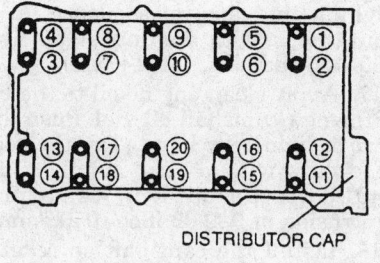

Camshaft assembly (loosening sequence) — 1995–97 Protege 1.8L (BPD) engines

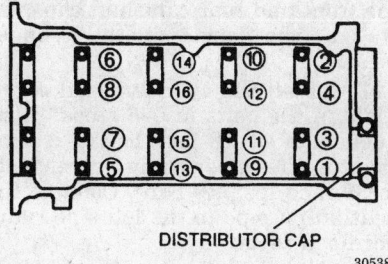

Camshaft assembly (loosening sequence) — 1995–97 Protege 1.5L (Z5D) engines

head cover. Loosen the cylinder head cover in 2–3 steps. Remove the cylinder head cover.

5. Remove the accessory drive belts, water pump pulley, timing belt covers and timing belt.

6. Remove the distributor.

7. Hold the camshaft with a wrench on the cast hexagon, and loosen the camshaft sprocket mounting bolt. Remove the camshaft sprockets.

8. Remove the seal plate.

9. Rotate the camshafts clockwise so the cams don't press on the tappets.

10. Loosen the front camshaft cap bolts in 5–6 steps, starting on the two outside bolts and finishing on the two inside bolts. Remove the front camshaft bolts and caps.

NOTE: Note the location of the numbers on top of the camshaft caps, so the caps can be reinstalled in their original positions.

11. Loosen the camshaft cap bolts in 5–6 steps, in the reverse order of removal. Remove the camshaft caps.

12. Remove the camshafts. Remove the chain and oil seals from the camshafts.

To install:

1.5L (Z5D) Engine

1. Insert the chain adjuster between the camshafts.

2. Lubricate the camshaft lobes and journals with clean engine oil and install the camshafts on the cylinder head. Make sure none of the lobes are located directly on the tappets. Align the marks on the camshaft gear and the timing chain.

3. Apply silicone sealant to the cylinder head on the front camshaft caps mating surface. Do not get sealant on the camshaft journals.

4. Install the camshaft bearing caps in their original locations. Hand tighten the camshaft cap bolts numbered: 5, 7, 2, and 4. Install all the

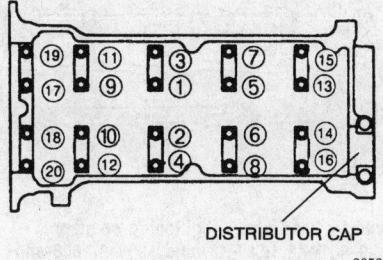

Camshaft assembly (tightening sequence) — 1995–97 Protege 1.5L (Z5D) engines

bolts and tighten, in sequence, in 5–6 steps with a final torque of 100–125 inch lbs. (11–14 Nm).

5. Apply clean engine oil to the lips of new camshafts seals. Install the seals using a suitable seal installer flush with the edge of the camshaft cap.

6. Install the seal plate.

7. Install the camshaft sprockets, timing belt and timing belt covers. Install the water pump pulley and accessory drive belts. Adjust the tension.

8. Apply silicone sealant to a new cylinder head cover gasket, and install the gasket on the cylinder head cover.

9. Apply silicone sealant to the cylinder head in the area adjacent to the front camshaft caps.

10. Install the distributor.

11. Install the cylinder head cover. Tighten the bolts in 2 steps, in reverse of the loosening sequence, to 61–95 inch lbs. (7–11 Nm).

12. Install the power steering hose brackets and tighten the bolts to 88 inch lbs. (10 Nm). Connect the spark plug wires and clips.

13. Connect the breather hose and PCV valve.

14. Adjust the valve clearance.

15. Adjust the ignition timing and idle speed.

16. Connect the negative battery cable, run the engine and check for leaks.

1.8L (BPE, BPD) Engine

1. Lubricate the camshaft lobes and journals with clean engine oil and install the camshafts on the cylinder head. Make sure none of the lobes are located directly on the HLAs.

2. Apply silicone sealant to the cylinder head on the front camshaft caps mating surface. Do not get sealant on the camshaft journals.

3. Install the camshaft bearing caps in their original locations. Install all the bolts and tighten, in sequence, in 5–6 steps with a final torque of 100–125 inch lbs. (11–14 Nm).

4. Apply clean engine oil to the lips of new camshafts seals. Install the seals using a suitable seal installer flush with the edge of the camshaft cap.

5. Install the seal plate.

6. Install the camshaft sprockets, timing belt and timing belt covers. Install the water pump pulley and accessory drive belts. Adjust the tension.

7. Apply silicone sealant to a new cylinder head cover gasket, and in-

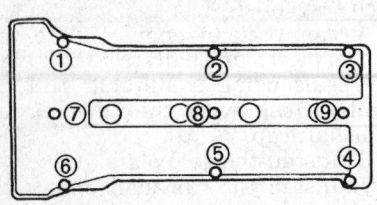

Cylinder head cover (tightening sequence) —
1995-97 Protege 1.5L (Z5D) engines

stall the gasket on the cylinder head cover.

8. Apply silicone sealant to the cylinder head in the area adjacent to the front camshaft caps.

9. Install the distributor.

10. Install the cylinder head cover. Tighten the bolts in 2 steps, in reverse of the loosening sequence, to 61-95 inch lbs. (7-11 Nm).

11. Install the power steering hose brackets and tighten the bolts to 88 inch lbs. (10 Nm). Connect the spark plug wires and clips.

12. Connect the breather hose and PCV valve.

13. Adjust the ignition timing and idle speed.

14. Connect the negative battery cable, run the engine and check for leaks.

1993-95 MX3 1.8L (K8) and 1993-97 626 and MX6 2.5L (KL) Engines

1. Relieve the fuel system pressure. Disconnect the negative battery cable.

2. Remove the intake manifold. Label and disconnect the spark plug wires from the spark plugs.

3. Remove the upper timing belt cover bolts. If removing the left cylinder head cover, disconnect the ventilation pipe from the front of the left side (front) cylinder head cover.

4. Remove the bolts, in 5-6 steps, using the reverse of the tightening sequence. Remove the cylinder head cover.

5. Remove the timing belt and the camshaft sprockets.

6. Turn the camshafts so the knock pins are aligned with the marks on the camshaft caps. This will reduce the pressure on the hydraulic lifters.

7. Note the markings on the camshaft caps prior to removal, so they can be reinstalled in the same positions. The right hand (rear) caps are marked with numbers and the

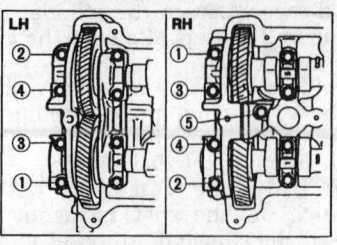

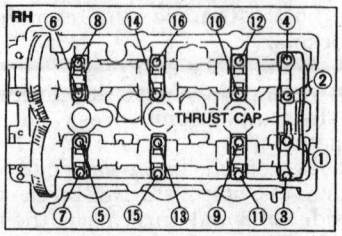

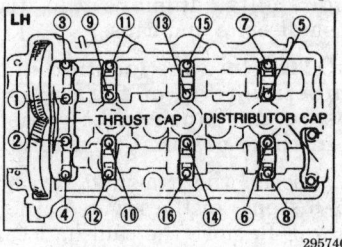

Camshaft assembly (loosening sequence) — 1993-95 MX3 1.8L (K8) and 1993-97 626 and MX6 2.5L (KL) engines

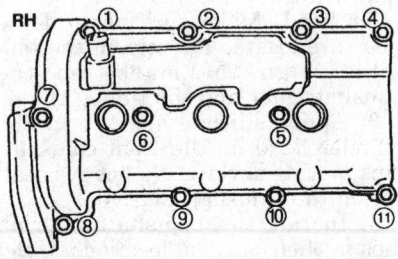

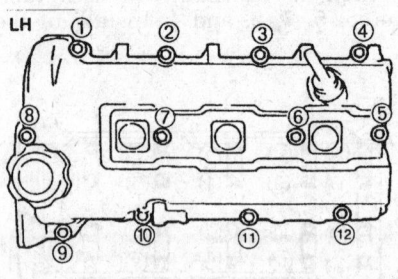

Cylinder head cover (tightening sequence) — 1993-95 MX3 1.8L (K8) and 1993-97 626 and MX6 2.5L (KL) engines

left hand (front) caps are marked with letters.

8. Loosen the front camshaft cap bolts in sequence, in 5-6 steps. Remove the front camshaft caps.

9. Remove the remaining camshaft cap bolts in the reverse of the torque sequence. Remove the caps, being sure to remove the thrust caps last. Do not damage the cylinder head thrust bearing support.

10. Remove the camshafts and oil seals.

To install:

11. Install new oil seals on the camshafts. Apply clean engine oil to the camshaft lobes, journals and supports.

12. Install the camshafts so the gear marks align.

13. Remove all oil and dirt from the mating surfaces between the front camshaft cap and the cylinder head.

14. Apply silicone sealant, at a thickness of 0.06-0.09 inch (1.5-2.5mm), to the cylinder head surface in the area forward of the camshaft gear cavity and to the left cylinder head on the rear exhaust camshaft cap mating surface.

15. Install the thrust caps. Tighten the thrust cap bolts, in 5-6 steps, until the caps are fully seated on the cylinder head.

16. Install the remaining camshaft caps in their original positions. Tighten the caps, in sequence, in 5 equal steps, with the final step being 100-125 inch lbs. (11-14 Nm).

17. Apply clean engine oil to the lip of the new camshaft oil seal. Push the seal in lightly by hand. Tap the seal in evenly with a seal installer (49 F401 337A or equivalent) with a final protrusion of 0-0.02 inch (0-0.5mm).

18. Install the camshaft sprockets, timing belt and timing belt cover.

19. Remove any sealant and gasket material from the cylinder head cover contact surfaces.

20. Apply silicone sealant to the cylinder head in the area adjacent to the front and rear camshaft caps. Install a new gasket on the cylinder head.

21. Install the cylinder head cover. Tighten the bolts in 5-6 steps, in sequence, to 44-78 inch lbs. (5-9 Nm). Tighten the upper timing cover bolts to 88 inch lbs. (10 Nm). Connect the ventilation pipe to the left side cylinder head cover.

22. Install the intake manifold.

23. Connect the negative battery cable. Run the engine and check for leaks.

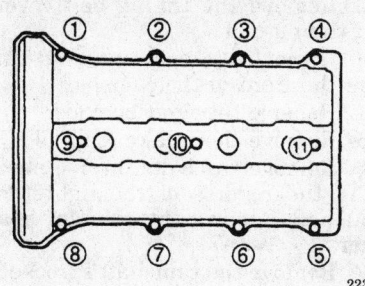

Cylinder head cover (tightening sequence) — 1995 MX3 1.6L (B6-ZE) and 1995–97 Protege 1.8L (BPD) engines

223158

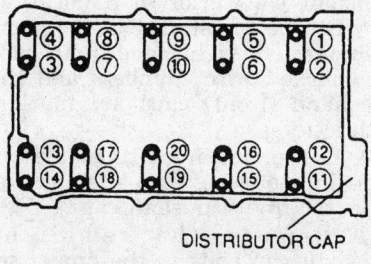

Camshaft assembly (loosening sequence) — 1995 MX3 1.6L (B6-ZE) engine

223156

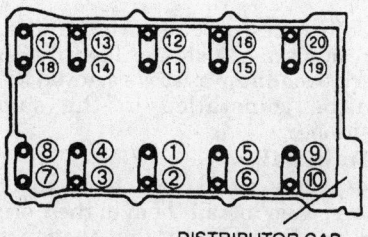

Camshaft assembly (tightening sequence) — 1995 MX3 1.6L (B6-ZE) and 1995–97 Protege 1.8L (BPD) engines

223159

1993–95 MX3 1.6L (B6-ZE), 1993–97 626 and MX6 2.0L (FS) Engines

1. Disconnect the negative battery cable.

2. Remove the power steering hose brackets from the cylinder head cover.

3. Label and disconnect the spark plug wires and spark plug wire clips.

4. Disconnect the breather tube and PCV valve from the cylinder head cover. Loosen the cylinder head cover in 2–3 steps. Remove the cylinder head cover.

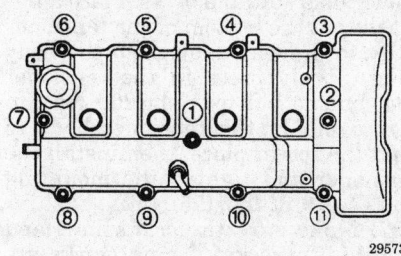

Cylinder head cover (loosening sequence) — 1993–97 626 and MX6 2.0L (FS) engines

295736

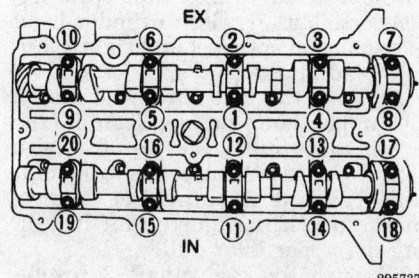

Camshaft assembly (tightening sequence) — 1993–97 626 and MX6 2.0L (FS) engines

295737

5. Remove the accessory drive belts, timing belt covers and timing belt.

6. Hold the camshaft with a wrench on the cast hexagon, and loosen the camshaft sprocket mounting bolt. Remove the camshaft sprockets.

7. Note the location of the numbers on top of the camshaft caps, so the caps can be reinstalled in their original positions.

8. Loosen the camshaft cap bolts in 5–6 steps, in the reverse order of installation. Remove the camshaft caps and the oil seals.

9. Remove the camshafts.

To install:

10. Lubricate the camshaft lobes and journals with clean engine oil and install the camshafts on the cylinder head. Make sure none of the lobes are located directly on the hydraulic lifters.

11. Apply silicone sealant to the cylinder head on the front camshaft caps mating surface. Do not get sealant on the camshaft journals.

12. Install the camshaft bearing caps in their original locations. Install the bolts and tighten, in sequence, in 5–6 steps with a final torque of 100–125 inch lbs. (11.3–14.2 Nm).

13. Apply clean engine oil to the lips of new camshafts seals. Install the seals using a suitable seal installer.

14. Install the camshaft sprockets, timing belt and timing belt covers. Install the accessory drive belts and adjust the tension.

15. Apply silicone sealant to a new cylinder head cover gasket, and install the gasket on the cylinder head cover.

16. Apply silicone sealant to the cylinder head in the area adjacent to the front camshaft caps.

17. Install the cylinder head cover. Tighten the bolts in 2 steps, in reverse of the loosening sequence, to 61–95 inch lbs. (7–11 Nm).

18. Install the power steering hose brackets and tighten the bolts to 88 inch lbs. (10 Nm). Connect the spark plug wires and clips.

19. Connect the breather hose and PCV valve.

20. Connect the negative battery cable, run the engine and check for leaks.

1993–95 929 3.0L (JE-ZE) Engines

1. Properly relieve the fuel system pressure.

2. Disconnect the negative battery cable, and drain the cooling system.

3. Remove the fresh air duct and air intake pipe.

4. Remove the cooling fan and shroud.

5. Disconnect the spark plug wires and remove the spark plugs. Remove the idler pulleys and the accessory drive belts.

6. Remove the coolant bypass hose and the upper radiator hose. Disconnect the electrical connector and remove the distributor.

7. Hold the crankshaft pulley with a suitable tool and remove the bolt. Remove the crankshaft pulley. Be careful not to damage the sensor rotor.

8. Remove the timing belt covers. Rotate the crankshaft until the camshaft and crankshaft sprocket timing marks are aligned.

9. Remove the upper idler pulley and the tensioner and pulley. Mark the direction of rotation on the timing belt and remove the timing belt.

10. Disconnect the accelerator cable. Disconnect and plug the fuel lines. Label and disconnect the necessary electrical connectors and vacuum hoses.

11. Loosen and remove the lower intake manifold bolts in 2–3 steps. Remove the lower intake manifold.

12. Remove the cylinder head covers.

13. Hold the camshaft with a wrench on the hexagon cast into the camshaft and remove the camshaft sprocket bolt. Remove the camshaft sprockets.

14. Remove bolts **A** from the front camshaft cap on the left cylinder head. Loosen the remaining camshaft cap bolts in 2–3 steps in sequence. Mark the position of the caps so they can be reinstalled in their original positions.

15. Remove the camshaft caps and remove the camshafts. Remove the camshaft oil seals. Label the camshafts so they can be reinstalled in their proper locations.

16. Lift out and label the rocker arms, if necessary.

To install:

17. If removed, position the rocker arms.

18. Lubricate the camshaft lobes and journals with clean engine oil and install the camshafts on the cylinder heads. Apply clean engine oil to the lips of the new camshaft seals, if replaced, and install them on the camshafts.

19. Apply a small amount of silicone sealant to the cylinder head on the front camshaft cap mating surfaces. Do not allow sealant to get on the camshaft journal, oil seal face or camshaft thrust face.

20. Install the camshaft caps in their original locations. Gradually tighten the cap bolts, in sequence, to 24 ft. lbs. (32 Nm). Tighten the front cap bolts, **A**, on the left cylinder head to 95 inch lbs. (11 Nm).

21. Hold the camshafts using a wrench on the hexagon cast into the camshaft. Install the camshaft sprockets with the retaining bolts. Tighten the sprocket bolts on the right cylinder head to 59 ft. lbs. (80 Nm) and the sprocket bolts on the left cylinder head to 19 ft. lbs. (25 Nm).

22. Apply sealant to the cylinder head in the areas adjacent to the front camshaft caps. Install the cylinder head covers and tighten the nuts to 39 inch lbs. (4 Nm).

23. Install the lower intake manifold using new gaskets. Tighten the mounting nuts in 2–3 steps, in the proper sequence, to 19 ft. lbs. (25 Nm). Connect the fuel lines.

24. Install the upper intake manifold to the lower intake manifold, using a new gasket. Tighten the bolts to 19 ft. lbs. (25 Nm).

25. Connect the electrical connectors and vacuum hoses. Connect the accelerator cable.

26. Position the timing belt tensioner on a suitable press. Place a flat washer at the bottom of the tensioner body to prevent damage to the body plug. Slowly press in the tensioner rod, but do not exceed 2200 lbs. Insert a pin in the tensioner body to hold the rod in place, then install the tensioner and tighten the mounting bolts to 19 ft. lbs. (25 Nm).

27. Make sure the crankshaft and camshaft sprocket timing marks are properly aligned. Install the timing belt over the sprockets and pulleys in the following order: crankshaft sprocket, lower idler pulley, left cylinder head exhaust cam sprocket, left cylinder head intake cam sprocket, tensioner pulley, right cylinder head exhaust cam sprocket and right cylinder head intake cam sprocket.

28. Push the belt down and install the upper idler pulley. Tighten the upper idler pulley bolt to 38 ft. lbs. (52 Nm). Make sure the camshaft and crankshaft sprocket timing marks are still aligned after installing the upper idler pulley.

29. Turn the crankshaft 2 revolutions in the normal direction of rotation and realign the timing marks. Make sure all timing marks are correctly aligned.

30. Remove the pin from the automatic tensioner and again rotate the crankshaft 2 turns. Confirm that the timing marks are aligned.

31. Install the timing belt covers and tighten the bolts to 95 inch lbs. (11 Nm).

32. Install the crankshaft pulley. Hold the pulley with a suitable tool and tighten the pulley bolt to 123 ft. lbs. (167 Nm). Be careful not to damage the sensor rotor.

33. Install the spark plugs and the distributor assembly. Connect the spark plug wires and the distributor electrical connector.

34. Install the upper radiator hose and water bypass hose. Install the idler pulleys and accessory drive belts. Adjust the belt tension.

35. Install the shroud and cooling fan. Install the air intake pipe and fresh air duct.

36. Connect the negative battery cable. Fill and bleed the cooling system.

37. Start the engine and bring to normal operating temperature. Check for leaks. Check the ignition timing and idle speed.

1995–97 Millenia 2.3L (KJS) Engines

1. Relieve the fuel system pressure. Disconnect the negative battery cable.

2. Remove the timing belt covers and timing belt.

3. Remove the spacer and O-ring from the front of the camshaft.

4. Remove the ignition coils.

5. Remove the intake manifold.

6. Remove the bolts, in 5–6 steps, using the reverse of the tightening sequence. Remove the cylinder head cover.

7. Remove the camshaft sprockets.

8. Turn the camshafts so the knock pins are aligned with the marks on the camshaft caps. This will reduce the pressure on the adjustment shims.

9. Note the markings on the camshaft caps prior to removal, so they can be reinstalled in the same positions. The right hand (rear) caps are marked with numbers and the left hand (front) caps are marked with letters.

10. Loosen the front camshaft cap bolts in sequence, in 5–6 steps. Remove the front camshaft caps.

11. Remove the remaining camshaft cap bolts in the proper sequence. Remove the caps, being sure to remove the thrust caps last. Do not damage the cylinder head thrust bearing support.

12. Remove the camshafts and oil seals.

13. If necessary, remove the lifters and adjustment shims. Identify and mark each lifter as it is removed so it can be reinstalled in the same position.

To install:

14. Apply clean engine oil to the lifters, then install them in their original positions. Verify that they move smoothly in their bore.

15. Install new oil seals on the camshafts. Apply clean engine oil to the camshaft lobes, journals and supports.

16. Install the camshafts so the gear marks align.

17. Remove all oil and dirt from the mating surfaces between the front camshaft cap and the cylinder head.

18. Install the thrust caps. Tighten the thrust cap bolts, in 5–6 steps, until the caps are fully seated on the cylinder head.

19. Apply silicone sealant, at a thickness of 0.06–0.09 inch (1.5–2.5mm), to the cylinder head surface in the area forward of the camshaft gear cavity.

20. Install the remaining camshaft caps in their original positions. Tighten the caps, in sequence, in five equal steps, with the final step being 100–125 inch lbs. (11–14 Nm).

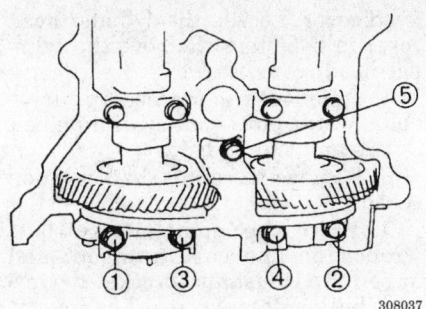

Camshaft front cap bolts (loosening sequence) — 1995–97 Millenia 2.3L (KJS) engines

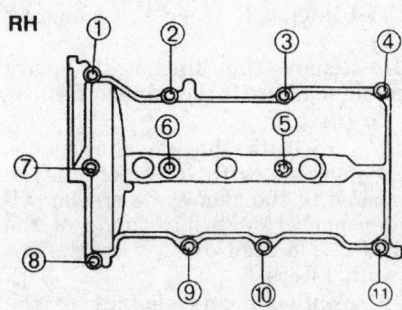

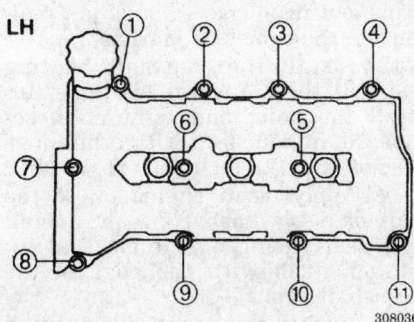

Cylinder head cover (tightening sequence) — 1995–97 Millenia 2.3L (KJS) engines

21. Apply clean engine oil to the lip of the new camshaft oil seal. Push the seal in lightly by hand. Tap the seal in evenly with a seal installer (49 F401 337A or equivalent) with a final protrusion of 0–0.02 inch (0–0.5mm). Tap in a new blind cap.

22. Install the camshaft sprockets. Tighten the mounting bolts to 91–103 ft. lbs. (123–140 Nm).

23. Measure and adjust valve clearances.

24. Remove any sealant and gasket material from the cylinder head cover contact surfaces.

25. Apply silicone sealant to the cylinder head in the area adjacent to the front and rear camshaft caps. Install a new gasket on the cylinder head.

26. Install the cylinder head cover. Tighten the bolts in 5–6 steps, in sequence, to 44–78 inch lbs. (5–9 Nm).

27. Using a new O-ring, install the distributor.

28. Install the ignition coils.

29. Install the intake manifold.

30. Install the spacer, using a new O-ring. Tighten the mounting bolt to 14–18 ft. lbs. (19–25 Nm).

31. Install the timing belt and timing belt cover.

32. Connect the negative battery cable. Run the engine and check for leaks.

1995–97 Millenia 2.5L (KLD) Engine

1. Relieve the fuel system pressure. Disconnect the negative battery cable.

2. Remove the timing belt covers and timing belt.

3. Remove the air cleaner housing assembly.

4. Label and disconnect the spark plug wires from the spark plugs.

5. Remove the distributor.

6. Remove the intake manifold.

7. Remove the bolts, in 5–6 steps, using the reverse of the tightening

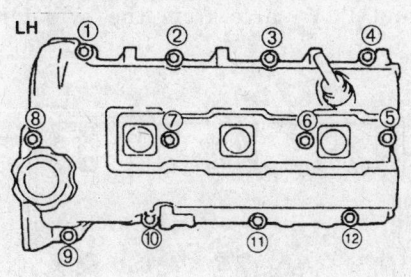

Cylinder head cover (tightening sequence) — 1995–97 Millenia 2.5L (KLD) engine

sequence. Remove the cylinder head cover.

8. Remove the camshaft sprockets.

9. Turn the camshafts so the knock pins are aligned with the marks on the camshaft caps. This will reduce the pressure on the hydraulic lifters.

10. Note the markings on the camshaft caps prior to removal, so they can be reinstalled in the same positions. The right hand (rear) caps are marked with numbers and the left hand (front) caps are marked with letters.

11. Loosen the front camshaft cap bolts in sequence, in 5–6 steps. Remove the front camshaft caps.

12. Remove the remaining camshaft cap bolts in the proper sequence. Remove the caps, being sure to remove the thrust caps last. Do not damage the cylinder head thrust bearing support.

13. Remove the camshafts and oil seals.

14. If necessary, remove the hydraulic lifters. Identify and mark each lifter as it is removed so it can be reinstalled in the same position. If the lifters are to be reused, store them upside down in an oil-filled sealed container.

To install:

15. If removed, inspect each lifter by holding it in your hand and pressing the bottom up against the top of the lifter. If there is movement, replace the lifter.

16. Apply clean engine oil to the lifters, then install them in their original positions. Verify that they move smoothly in their respective bores.

17. Install new oil seals on the camshafts. Apply clean engine oil to the camshaft lobes, journals and supports.

18. Install the camshafts so the gear marks align.

19. Remove all oil and dirt from the mating surfaces between the front camshaft cap and the cylinder head.

20. Install the thrust caps. Tighten the thrust cap bolts, in 5–6 steps, until the caps are fully seated on the cylinder head.

21. Apply silicone sealant, at a thickness of 0.06–0.09 inch (1.5–2.5mm), to the cylinder head surface in the area forward of the camshaft gear cavity and to the left cylinder head on the rear exhaust camshaft cap mating surface.

22. Install the remaining camshaft caps in their original positions. Tighten the caps, in sequence, in five equal steps, with the final step being 100–125 inch lbs. (11–14 Nm).

23. Apply clean engine oil to the lip of the new camshaft oil seal. Push the seal in lightly by hand. Tap the seal in evenly with a seal installer (49 F401 337A or equivalent) with a final protrusion of 0–0.02 inch (0–0.5mm).

24. Install the camshaft sprockets. Tighten the mounting bolts to 91–103 ft. lbs. (123–140 Nm).

25. Remove any sealant and gasket material from the cylinder head cover contact surfaces.

26. Apply silicone sealant to the cylinder head in the area adjacent to the front and rear camshaft caps. Install a new gasket on the cylinder head.

27. Install the cylinder head cover. Tighten the bolts in 5–6 steps, in sequence, to 44–78 inch lbs. (5–9 Nm).

28. Install the intake manifold.

29. Using a new O-ring, install the distributor.

30. Install the spark plug wires.

31. Install the air cleaner housing assembly.

32. Install the timing belt and timing belt cover.

33. Connect the negative battery cable. Run the engine and check for leaks.

1993–94 Miata 1.6L (B6E) and 1.8L (BPD) Engines

1. Disconnect the negative battery cable. Drain the cooling system.

2. Label and disconnect the spark plug wires and remove the spark plugs.

3. Disconnect the hoses from the cylinder head cover, if equipped.

4. Remove the cylinder head cover bolts and remove the cylinder head cover.

5. Remove the timing belt. Remove the crank angle sensor.

6. Hold the camshaft with a wrench on the hexagon cast into the camshaft. Remove the sprocket bolts and remove the sprockets.

7. Label the caps so they can be reinstalled in their original positions. Loosen the camshaft cap bolts in 2–3 steps in the reverse of the torque sequence, then remove the camshaft caps.

8. Remove the camshafts. Remove the camshaft oil seals from the camshafts.

To install:

9. Lubricate the camshaft journals and lobes with clean engine oil. Install the camshafts in the cylinder head.

10. Apply silicone sealant to the cylinder head on the front camshaft cap mating surfaces. Do not allow any sealant on the camshaft journals.

11. Install the camshaft caps in their original positions. Loosely install the cap bolts.

12. Tighten the camshaft cap bolts in 2–3 steps to 125 inch lbs. (14 Nm) in the proper sequence.

13. Apply clean engine oil to the lip of a new camshaft seal. Push the seal slightly in by hand. Tap the seal into position, using a seal installer, until it is flush with the edge of the camshaft cap.

14. Turn the camshafts until the dowel pins face straight up. Install the camshaft sprockets and the sprocket bolts.

15. Hold the camshaft with the wrench on the cast hexagon and tighten the sprocket bolts to 44 ft. lbs. (60 Nm).

16. Install the remaining components in the reverse order of removal.

1995–97 Miata 1.8L (BPD) Engine

1. Disconnect the negative battery cable.

2. Remove the power steering hose brackets from the cylinder head cover.

3. Label and disconnect the spark plug wires and spark plug wire clips.

4. Disconnect the breather tube and PCV valve from the cylinder head cover. Loosen the cylinder head cover in 2–3 steps. Remove the cylinder head cover.

5. Remove the accessory drive belts, water pump pulley, timing belt covers and timing belt.

6. Remove the crankshaft position sensor.

7. Hold the camshaft with a wrench on the cast hexagon, and loosen the camshaft sprocket mounting bolt. Remove the camshaft sprockets.

8. Note the location of the numbers on top of the camshaft caps, so the caps can be reinstalled in their original positions.

9. Loosen the camshaft cap bolts in 5–6 steps. Remove the camshaft caps.

10. Remove the camshafts. Remove the oil seals from the camshafts.

To install:

11. Lubricate the camshaft lobes and journals with clean engine oil and install the camshafts on the cylinder head. Make sure none of the lobes are located directly on the hydraulic lifters.

12. Apply silicone sealant to the cylinder head on the front camshaft caps mating surface. Do not get sealant on the camshaft journals.

13. Install the camshaft bearing caps in their original locations. Install the bolts and tighten, in sequence, in 5–6 steps with a final torque of 100–125 inch lbs. (11–14 Nm).

14. Apply clean engine oil to the lips of new camshafts seals. Install the seals using a suitable seal installer flush with the edge of the camshaft cap.

15. Install the camshaft position sensor. Tighten the mounting bolt to 14–18 ft. lbs. (19–25 Nm).

16. Install the camshaft sprockets, timing belt and timing belt covers. Install the water pump pulley and accessory drive belts. Adjust the tension.

17. Apply silicone sealant to a new cylinder head cover gasket, and install the gasket on the cylinder head cover.

18. Apply silicone sealant to the cylinder head in the area adjacent to the front camshaft caps.

19. Install the cylinder head cover. Tighten the bolts to 44–78 inch lbs. (5.0–8.8 Nm).

20. Install the power steering hose brackets and tighten the bolts to 88 inch lbs. (10 Nm). Connect the spark plug wires and clips.

21. Connect the breather hose and PCV valve.

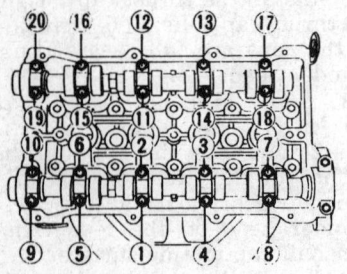

141251

Camshaft cap bolt tightening sequence — 1993–94 Miata 1.6L (B6E) and 1.8L (BPD) engines

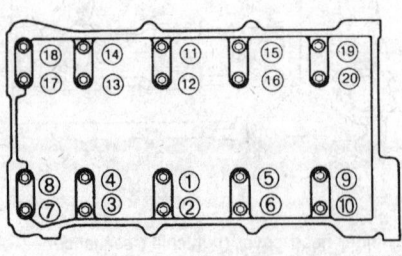

290585

Tightening sequence for the camshaft bearing caps — 1995–97 Miata 1.8L (BPD) engine

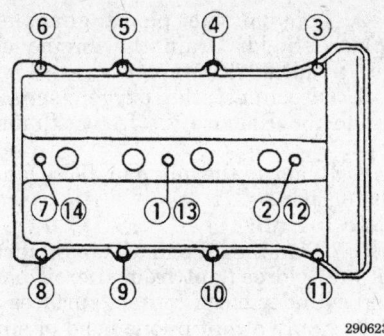

Cylinder head cover bolt torque sequence — 1995–97 Miata 1.8L (BPD) engine

22. Connect the negative battery cable, run the engine and check for leaks.

Piston and Connecting Rod

POSITIONING

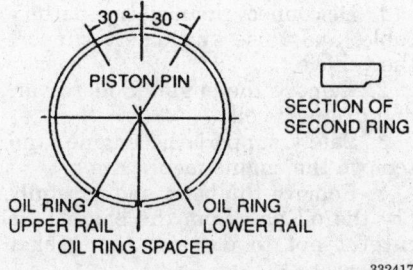

Piston ring end gap positioning — except 626, MX6 and Millenia 2.5L engines

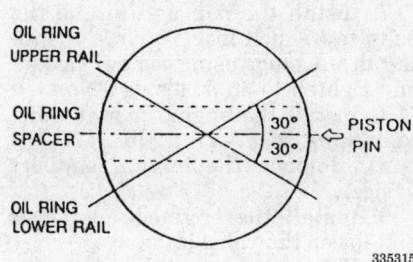

Positioning the piston rings — 626, MX6 and Millenia 2.5L engines

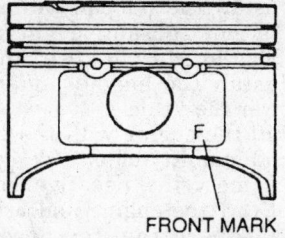

Piston position in the cylinder block — 1993–95 323 1.6L (B6E) and MX3 1.6L (B6-ZE), 1.8L (K8) and 1993–97 Protege 1.8L (Z5D, BPD, BPE) Engines

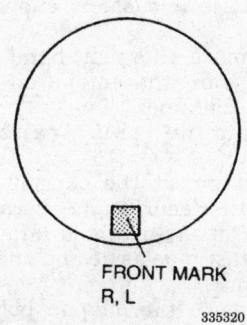

Position of the piston in the cylinder block — 1995–97 Millenia 2.3L engine

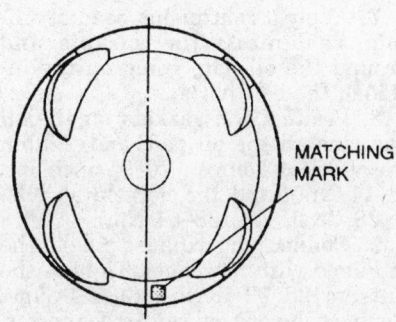

Position of the piston in the cylinder block — 626 and MX6 2.0L engine

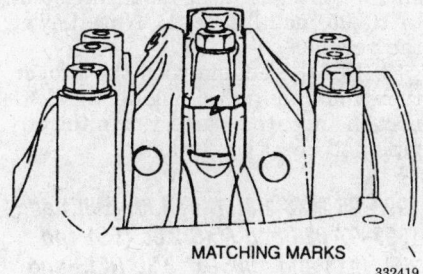

Connecting rod and connecting rod cap matchmarks — 1993–95 323 1.6L (B6E) and MX3 1.6L (B6-ZE), 1.8L (K8) and 1993–97 Protege 1.8L (Z5D, BPD, BPE) Engines

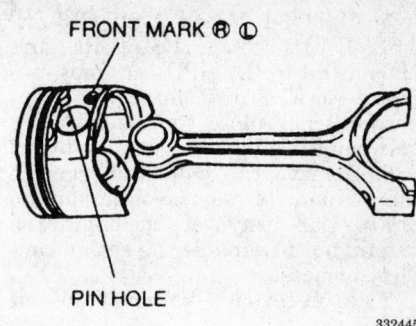

Piston position in the cylinder block — V6 engines

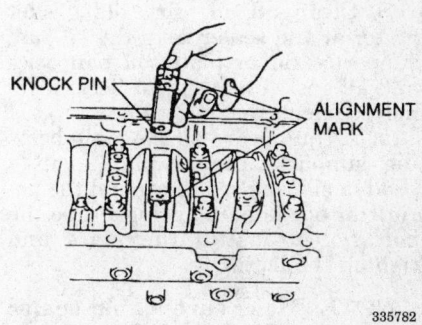

Aligning the connecting rods and rod caps

ENGINE LUBRICATION

Oil Pan

REMOVAL AND INSTALLATION

1993–94 323 1.6L (B6E), 1994 Protege 1.8L (BPE, BPD), 1993–94 Miata 1.6L (B6-ZE) 1.8L (BPD) and 1994 MX3 1.8L (BPD) Engines

1. Disconnect the negative battery cable. Raise and safely support the vehicle.
2. Remove the engine undercover, if equipped. Position a suitable container under the oil pan. Remove the drain plug and drain the oil.
3. Remove the exhaust pipe from the exhaust manifold and from the catalytic converter. If necessary, remove the exhaust pipe bracket from the engine block.
4. On 1993–94 323 1.6L (B6E) and 1994 Protege 1.8L (BPD) engines, remove the integrated stiffener from the engine block and transaxle.

5. 1993–94 Miata 1.6L (B6-ZE) 1.8L (BPD), Protege 1.8L (BPE) and 1994 MX3 1.8L (BPD) engines, remove the main bearing support/stiffener plate that is installed between the oil pan and engine block.

6. Remove the bolts and remove the oil pan. It may be necessary to pry the pan away from the engine; be careful not to damage the gasket contact surfaces.

7. If necessary remove the oil strainer.

8. Remove the main bearing support/stiffener plate that is installed between the oil pan and engine block.

To install:

9. Clean all oil, dirt, old gasket material and sealer from the oil pan, support/stiffener plate, oil pan bolts and all gasket mating surfaces. If removed, clean the oil strainer.

10. If equipped with the main bearing support/stiffener plate, run a bead of silicone sealer around the perimeter of the plate, going inside the bolt holes. Install the plate and tighten the bolts.

NOTE: Make sure all old sealer is removed from the bolts prior to installation. Installing a bolt coated with old sealer could result in cracking of the bolt holes.

11. If removed, install the oil strainer using a new gasket. Tighten the bolts.

12. If used, apply silicone sealer to new rubber end gaskets and press them into place on the engine.

13. Apply a bead of silicone to the perimeter of the oil pan, going around the inside of the bolt holes and install the pan to the engine. Install the oil pan bolts finger tight.

14. Tighten the oil pan bolts.

NOTE: Make sure all old sealer is removed from the bolts prior to installation. Installing a bolt coated with old sealer could result in cracking of the bolt holes.

15. On 1993–94 323 1.6L (B6E) and 1994 Protege 1.8L (BPD) engines, install the integrated stiffener to the engine block and transaxle. Tighten the bolts to 38 ft. lbs. (52 Nm).

16. 1993–94 Miata 1.6L (B6-ZE) 1.8L (BPD) and Protege 1.8L (BPE) and 1994 MX3 1.8L (BPD) engines, install the transverse member. Tighten the bolts to 93 ft. lbs. (126 Nm).

17. Install the front exhaust pipe bracket, if equipped. Install the front exhaust pipe, using new gaskets. Tighten the exhaust manifold flange nuts to 34 ft. lbs. (46 Nm).

18. Install the oil pan drain plug using a new gasket. Tighten the drain plug to 30 ft. lbs. (41 Nm).

19. Install the engine undercover and lower the vehicle.

20. Fill the engine with the proper type and quantity of oil.

21. Connect the negative battery cable. Start the engine and bring to normal operating temperature. Check for leaks.

1995–97 Protege 1.5L (Z5D) and 1.8L (BPD) Engines

1. Disconnect the negative battery cable. Raise and safely support the vehicle.

2. Remove the right-hand splash shield. Drain the engine oil into a suitable container.

3. Remove the transverse member.

4. Disconnect the oxygen sensor connector. Remove and discard the exhaust pipe-to-manifold nuts. Move the exhaust pipe aside and support it with a jack.

5. Remove the oil pan bolts and the oil pan.

To install:

6. Clean the oil pan. Clean all dirt, oil, gasket and old sealant from the oil pan and cylinder block contact surfaces.

7. Apply a continuous bead of silicone sealant on the gaskets and around the oil pan, going on the inside of the bolt holes.

8. Position the gaskets on the oil pan. Install the oil pan, and tighten the vertical bolts to 70–95 inch lbs. (8–11 Nm), and the horizontal bolts to 28–38 ft. lbs. (38–51 Nm).

9. Connect the exhaust pipe to the manifold with new nuts. Tighten the nuts to 28–38 ft. lbs. (38–51 Nm). Connect the oxygen sensor connector.

10. Install the transverse member, and tighten the mounting bolts to 69–97 ft. lbs. (94–131 Nm).

11. Install the right-hand splash shield and tighten the mounting bolts to 70–95 inch lbs. (8–11 Nm). Lower the vehicle.

12. Fill the engine with the proper type and quantity of engine oil. Connect the negative battery, run the engine and check for leaks.

1993–95 MX3 1.8L (K8) 1.6L (B6E) and 1993–97 626 and MX6 2.0L (FS) and 2.5L (KL) and 1995–97 2.3L (KJS and 2.5L (KLD) Engines

1. Disconnect the negative battery cable. Raise and safely support the vehicle.

2. Remove the passenger side splash shield. Drain the engine oil into a suitable container.

3. Disconnect the oxygen sensor connector. Remove the front exhaust pipe.

4. Remove the oil pan bolts and the oil pan.

To install:

5. Clean the oil pan. Clean all dirt, oil and old sealant from the oil pan and cylinder block contact surfaces.

6. Apply a continuous bead of silicone sealant around the oil pan, going on the inside of the bolt holes.

7. Install the oil pan and tighten the bolts to 14–18 ft. lbs. (19–25 Nm).

8. Install the front pipe. Tighten the nuts to 28–38 ft. lbs. (38–51 Nm). Connect the oxygen sensor connector.

9. Install the splash shield and tighten the mounting bolts to 70–95 inch lbs. (8–11 Nm). Lower the vehicle.

10. Fill the engine with the proper type and quantity of engine oil. Connect the negative battery, run the engine and check for leaks.

1993–95 929 3.0L (JE-ZE) Engines

1. Disconnect the negative battery cable, and raise and safely support the vehicle.

2. Remove the engine undercover, and drain the oil.

3. Safely support the engine, and remove the engine mount nuts.

4. Remove the bolts and carefully pry the oil pan from the engine. Be careful not to damage the gasket surfaces.

To install:

5. Clean the old sealer from the engine and pan bolts. Thoroughly clean the oil, dirt and old sealer from the oil pan.

6. Apply a bead of silicone sealer to the pan, making sure it is inside the bolt holes.

7. Install the pan and torque the bolts to 87 inch lbs. (10 Nm). Install the drain plug, using a new gasket, and tighten to 30 ft. lbs. (41 Nm).

8. Install the engine mount nuts, tightening to 57–77 ft. lbs. (77–104 Nm). Remove the engine support device.

9. Install the engine undercover and lower the vehicle.

10. Fill the engine with the proper type and quantity of engine oil. Connect the negative battery cable, start the engine and bring to normal operating temperature. Check for leaks.

1995–97 Miata 1.8L (BPD) Engine

1. Disconnect the negative battery cable. Raise and safely support the vehicle.
2. Remove the splash shield. Drain the engine oil into a suitable container.
3. Remove the dipstick and pipe.
4. Remove the intermediate steering shaft bolt.
5. Loosen the oil pan mounting bolts.
6. Safely support the engine with a suitable lifting device.
7. Remove the left and right engine mounting nuts.
8. Using a hoist lift the engine slightly.
9. Support the crossmember with a suitable transmission jack.
10. Remove the crossmember mounting bolts and nuts.
11. Separate the intermediate steering shaft from the pinion shaft.
12. Lower the crossmember until there is about four inches (100mm) of clearance between the oil pan and the steering gear housing.
13. Remove the oil pan bolts. Insert a suitable prying tool at the rear of the oil pan and remove the oil pan.

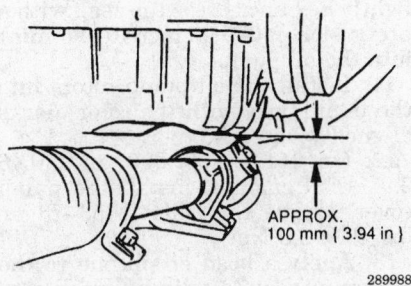

APPROX. 100 mm (3.94 in)

289988

Oil pan clearance — 1995–97 Miata 1.8L (BPD) engines

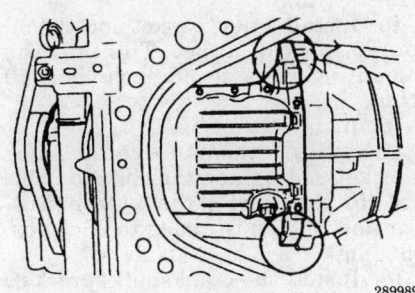

289989

Oil pan removal prybar insertion points — 1995–97 Miata 1.8L (BPD) engines

To install:

14. Clean the oil pan. Clean all dirt, oil and old sealant from the oil pan and cylinder block contact surfaces.
15. Apply silicone sealant to the oil pan where the oil pump body and rear cover are located. Install new gaskets at the same two locations.
16. Apply a continuous bead of silicone sealant around the oil pan, going on the inside of the bolt holes.
17. Install the oil pan and the mounting bolts.
18. Reposition the crossmember, and install the intermediate steering shaft to the pinion shaft.
19. Install the crossmember mounting bolts and nuts, and remove the support jack.
20. Reposition the engine, and install the engine mounting nuts. Remove the engine lifting device.
21. Install the intermediate steering shaft bolt and tighten to 14–19 ft. lbs. (18–26 Nm).
22. Install the dipstick and pipe.
23. Install the splash shield and tighten the mounting bolts to 70–95 inch lbs. (8–11 Nm). Lower the vehicle.
24. Fill the engine with the proper type and quantity of engine oil. Connect the negative battery, run the engine and check for leaks.

1993–95 RX7 1.3L (13B) Engines

1. Disconnect the negative battery cable.
2. Raise and safely support the vehicle.
3. Remove the engine undercover.
4. Remove the drain plug from the oil pan and drain the oil into a suitable container.
5. Remove the stabilizer bar.
6. Install engine support tool. Remove the engine mount nuts and take the weight from the mounts with the engine support tool.
7. Disconnect the steering gear from the crossmember.
8. Support the crossmember with a jack. Remove the power steering hose bracket and the mounting bolts and nuts from the crossmember. Lower the crossmember from the vehicle.
9. Remove the engine mount brackets from the engine. Disconnect the oil level sensor connector and remove it from the harness bracket.
10. Remove the oil pan bolts and the oil pan. If necessary, pry the oil pan from the engine, using the prying tool only at the rear of the engine and being careful not to damage the contact surfaces.

To install:

11. Thoroughly clean the oil, dirt and old sealer from the oil pan. Clean the old sealer and any foreign material from the contact surfaces on the engine and from the oil pan bolts.
12. Apply silicone sealer to the contact surfaces of the oil pan and the engine side of the new gasket.
13. Install the oil pan and tighten the bolts to 79–104 inch lbs. (9–12 Nm).
14. Connect the oil level sensor connector.
15. Install the engine mount brackets and tighten the bolts to 68 ft. lbs. (93 Nm).
16. Raise the crossmember into position. Install the bolts/nuts and tighten to 86 ft. lbs. (117 Nm).
17. Install the power steering hose bracket.
18. Install the steering gear to the crossmember and tighten the bolts to 38 ft. lbs. (51 Nm).
19. Remove the engine support tool. Install the mount nuts and tighten to 49 ft. lbs. (68 Nm).
20. Install the stabilizer bar. Tighten the stabilizer bar bracket bolts to 19 ft. lbs. (25 Nm).
21. Install the drain plug, using a new gasket, and tighten to 30 ft. lbs. (41 Nm). Install the engine undercover and lower the vehicle.
22. Fill the engine with the proper type and quantity of engine oil. Connect the negative battery cable, start the engine and bring to normal operating temperature. Check for leaks.

Oil Pump

REMOVAL AND INSTALLATION

1993–94 323 1.6L (B6E), 1994 Protege 1.8L (BPE, BPD), 1993–94 Miata 1.6L (B6-ZE) 1.8L (BPD) and 1994 MX3 1.8L (BPD) Engines

1. Disconnect the negative battery cable. Raise and safely support the vehicle.
2. Remove the timing belt. Remove the retaining bolt and remove the crankshaft sprocket. Drain the engine oil and remove the oil pan.
3. Remove the oil pump pickup tube-to-oil pump bolts, the tube and gasket.
4. Remove the oil pump-to-cylinder block bolts, the pump and gasket.
5. If necessary, pry the oil seal from the pump and clean the seal bore.

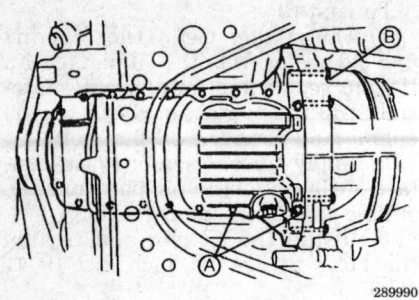

Oil pan bolt torque, A : 70–95.4 in. lbs. (7.9–10.7 Nm), B: 48–65 ft. lbs. (64–89 Nm) — 1995–97 Miata 1.8L (BPD) engines

6. Clean the gasket mounting surfaces. Inspect the pump and gears for wear.

1995–97 Protege 1.5L (Z5D) and 1.8L (BPD) Engines

1. Disconnect the negative battery cable.
2. Remove the crankshaft pulley, timing belt cover, belt and the crankshaft sprocket.
3. Remove the oil pan.
4. Remove the oil pickup tube and discard the gasket.
5. Remove the oil pump attaching bolts and remove the oil pump.
6. Remove the front crankshaft seal from the oil pump if the pump is being replaced.
To install:
7. Clean the oil, dirt and old sealant from all contact surfaces. Replace the O-rings on the oil pump.
8. If the oil seal was removed from the oil pump, apply clean engine oil to the lip of the seal. Push the seal in lightly be hand. Press the seal, with a protrusion of 0.02–0.04 inch (0.5–1.0 mm), into the oil pump with a suitable tool (49 B014 401 or equivalent).
9. Apply a bead of silicone to the oil pump at the cylinder block contact surface, going inside the bolt holes.
10. Install the oil pump and tighten the bolts to 14–18 ft. lbs. (19–25 Nm).
11. Install a new gasket and the oil pump pickup tube. Tighten the mounting bolts to 70–95 inch lbs. (8–11 Nm).
12. Install the oil pan.
13. Install the crankshaft sprocket, timing belt and cover.
14. Install the crankshaft pulley
15. Connect the negative battery cable.
16. Fill the engine with the proper type and quantity of oil. Run the engine and check for leaks.

1993–95 MX3 1.8L (K8) and 1993–97 626 and MX6 2.5L (KL) Engines

1. Disconnect the negative battery cable.
2. Remove the oil pan.
3. Properly discharge the refrigerant from the A/C system.
4. Remove the A/C compressor and the compressor bracket.
5. Remove the power steering pump and tensioner bolts from the engine block. Remove the pump and tensioner and position aside.
6. Remove the crankshaft pulley, timing belt cover, timing belt and the crankshaft sprocket.
7. Remove the oil pump mounting bolts (4 long 'A' and 5 short 'B'), and the 2 oil strainer-to-pump bolts. Remove the oil pump body.
8. Press the oil seal from the housing. Remove the pump cover mounting bolts with an impact screwdriver. Remove the cover and remove the oil pump rotors.
To install:
9. Clean the oil, dirt and old sealant from all contact surfaces.
10. Press a new oil seal into the pump housing with a protrusion of 0–0.03 inch (0–0.7mm).
11. Install the rotors into the oil pump body with the marks aligned with each other. Install the pump cover and torque the mounting bolts to 53–78 in. lbs. (5.9–8.8 Nm).
12. Apply a continuous bead of silicone sealant to the oil pump mating surface, and install the pump body.
13. Install the oil pump body mounting bolts. Tighten bolts to 14–18 ft. lbs. (19–25 Nm).
14. Replace the oil strainer-to-pump gasket, and install the mounting bolts. Tighten the bolts to 70–95 inch lbs. (8–11 Nm).
15. Install the crankshaft sprocket and key, timing belt and cover. Install the crankshaft pulley.
16. Install the power steering pump and tensioner. Tighten the 2 power steering belt tensioner upper bolts and the power steering pump rear bracket bolt to 33 ft. lbs. (46 Nm). Tighten the tensioner lower bolt to 18 ft. lbs. (25 Nm).
17. Install the A/C compressor bracket and tighten the bolts to 38 ft. lbs. (51 Nm). Install the A/C compressor and tighten the bolts to 38 ft. lbs. (51 Nm).
18. Install the oil pan. Fill the engine with the proper type and quantity of oil.
19. Connect the negative battery cable. Run the engine and check for leaks.

20. Evacuate and charge the A/C system.

1995 MX3 1.6L (B6-ZE), 1995–97 Miata 1.8L (BPD) and 1993–97 626 and MX6 2.0L (FS) Engines

1. Disconnect the negative battery cable.
2. Remove the crankshaft pulley, timing belt cover, belt and the crankshaft sprocket.
3. Remove the A/C compressor and secure it aside, leaving the refrigerant lines attached. Remove the compressor mounting bracket.
4. Remove the oil pan.
5. Remove the oil pickup tube and discard the gasket.
6. Remove the oil pump body attaching bolts and remove the oil pump body.
7. Remove the front crankshaft seal from the oil pump if the pump is being replaced.
8. On 1993—97 626 and MX6 2.0L (FS) engines, remove the pump cover mounting bolts with an impact screwdriver, remove the cover and the rotors.
To install:
9. Clean the oil, dirt and old sealant from all contact surfaces. Replace the O-rings on the oil pump.
10. If the oil seal was removed from the oil pump, apply clean engine oil to the lip of the seal. Push the seal in lightly be hand. Press the seal, with a protrusion of 0–0.02 inch (0–0.5 mm), into the oil pump.
11. Install the oil pump rotors into the pump body with the rotor marks aligned with each other.
12. On 1993—97 626 and MX6 2.0L (FS) engines, install the pump cover, torque the bolts to 53–78 in. lbs. (5.9–8.8 Nm).
13. Apply a bead of silicone to the oil pump body-to-cylinder block contact surface, going inside the bolt holes.
14. Install the oil pump body and tighten the bolts to 14–18 ft. lbs. (19–25 Nm).
15. Install a new gasket and the oil pump pickup tube. Tighten the mounting bolts to 88 inch lbs. (10 Nm).
16. Install the oil pan.
17. Install the A/C compressor bracket and tighten the bolts to 38 ft. lbs. (52 Nm). Install the A/C compressor and tighten the bolts to 26 ft. lbs. (35 Nm).
18. Install the crankshaft sprocket, timing belt assembly and cover.
19. Install the crankshaft pulley
20. Connect the negative battery cable.

21. Fill the engine with the proper type and quantity of oil. Run the engine and check for leaks.

1995–97 Millenia 2.3L (KJS) and 2.5L (KLD) engines

Due to space requirements, the engine assembly must be removed in order to replace the oil pump.

1. Disconnect the negative battery cable. Wait at least 90 seconds before performing any work. The backup power supply system for the SRS must deplete its stored energy.
2. Raise and safely support the vehicle.
3. Drain the engine oil and the engine coolant.
4. Remove and tag all electrical connections, hoses, and cables necessary to remove the engine assembly. Remove the front exhaust pipe.
5. Remove the engine assembly and secure to a suitable engine holding device.
6. Remove the oil pan, the oil strainer, and the oil pan baffle.
7. Remove the timing belt.
8. Using a suitable tool, remove the front timing belt pulley and key.
9. If equipped with a vacuum pump, remove the mounting bolts and nuts and remove the vacuum pump, O-ring, and gasket.
10. Remove the front seal, the mounting bolts, the oil pump, and the oil pump O-rings.

To install:

11. Make sure that the oil pump mating surfaces are clean. Install new O-rings coated with fresh engine oil to the oil pump cavity. Coat a new oil seal with fresh engine oil and install it to the oil pump. Apply silicone sealant to the oil pump contact surface and install the oil pump with the mounting bolts. Torque the mounting bolts to A: 16–22 ft. lbs. (22–30 Nm). All others: 14–18 ft. lbs. (19–25 Nm). Torque the bolts in sequence.

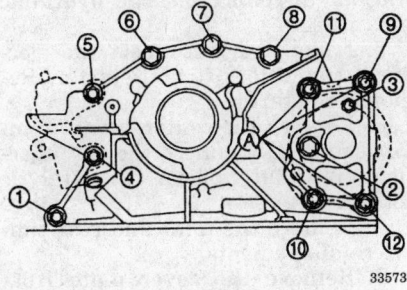

335734

Tightening the oil pump mounting bolts in sequence — 1995–97 Millenia 2.3L and 2.5L engines

12. If equipped with a vacuum pump, reinstall the pump using a new gasket and O-ring.
13. Install the key and the timing belt pulley.
14. Reinstall the timing belt.
15. Install the oil pan baffle, the oil pump strainer, and install the oil pan.
16. Remove the engine from the holding device and install the engine assembly into the vehicle.
17. Reinstall the front exhaust pipe. Reconnect all electrical connectors, cables, hoses, and install any components necessary to complete the engine installation.
18. Refill the engine with engine oil and coolant. Safely lower the vehicle.
19. Reconnect the negative battery cable. Start the engine, bleed the cooling system, make any necessary adjustments, check for leaks, and road test for proper operation.

1993–94 929 3.0L (JE-ZE) Engines

1. Disconnect the negative battery cable.
2. Remove the crankshaft pulley, timing belt covers, belt and the crankshaft sprocket.
3. Remove the A/C compressor and secure it aside, leaving the refrigerant lines attached. Remove the compressor mounting bracket.
4. Remove the oil pan, oil strainer and baffle plate.
5. Remove the oil pan block assembly.
6. Remove the thermostat housing assembly.
7. Remove the oil pump mounting bolts and remove the oil pump. If necessary, pry out the old oil seal being careful not to damage the seal housing area.

To install:

8. Clean the oil, dirt and old sealant from all contact surfaces. Replace the O-rings on the oil pump.
9. If the oil seal was removed from the oil pump, apply clean engine oil to the lip of the seal. Push the seal in lightly be hand. Press the seal flush into the oil pump with a suitable seal installer tool.
10. Apply a bead of silicone to the oil pump at the cylinder block contact surface, going inside the bolt holes.
11. Install the oil pump, using a new gasket, and tighten the bolts to 14–19 ft. lbs. (19–25 Nm).
12. Install the thermostat housing using a new gasket. Tighten the bolts to 14–19 ft. lbs. (19–25 Nm).
13. Cut away any portion of the oil pan gasket that projects toward the oil pan.

14. Install the baffle plate to the oil pan block assembly and tighten the bolts to 26 inch lbs. (3 Nm).
15. Apply a bead of silicone sealer to the perimeter of the oil pan block assembly and install it on the cylinder block. Tighten the bolts, in 2–3 steps, to 95 inch lbs. (11 Nm).
16. Install the oil strainer and oil pan.
17. Install the A/C compressor bracket and tighten the bolts to 38 ft. lbs. (52 Nm). Install the A/C compressor and tighten the bolts to 26 ft. lbs. (35 Nm).
18. Install the crankshaft sprocket, timing belt and cover.
19. Install the crankshaft pulley
20. Connect the negative battery cable.
21. Fill the engine with the proper type and quantity of oil. Run the engine and check for leaks.

1993–95 RX7 1.3L (13B) Engines

NOTE: The engine must be removed from the vehicle and disassembled to remove and install the main oil pump.

1. Disconnect the negative battery cable.
2. Remove the engine, and place on a suitable stand.
3. Remove the shaft lockbolt from the front engine cover.
4. Remove the pulley boss with the shaft bypass valve and the spring.
5. Remove the front cover.
6. Remove the oil seal by gently prying out of the front cover.
7. Remove the control valve spring, the control valve, the o-ring and the drive gear.
8. Lift the lock washer tab and remove the sprocket locknut.
9. Remove the oil pump drive gear, the sprocket wheel, and the drive chain as an assembly.
10. Remove the oil pump.

To install:

11. Install the oil pump. Torque the bolts to 5–7 ft. lbs. (7–10 Nm).
12. Install the oil pump drive gear, the sprocket wheel, and the drive chain as an assembly.
13. Install a new washer and the oil pump locknut. Torque the nut to 24–34 ft. lbs. (32–46 Nm).
14. Install the drive gear so that the chamfered surface faces the housing.
15. Apply clean oil to the new front oil seal and install into the front cover.
16. Install the oil pressure control valve in the front cover. Torque to 29–36 ft. lbs. (40–49 Nm).

17. Apply petroleum jelly and install the new o–ring and backup ring.

18. Install the front cover with a new gasket. Torque bolts to 12–17 ft. lbs. (16–22 Nm).

19. Install the shaft pulley boss.

20. Temporarily install the lockbolt, and tighten it by hand.

21. Remove the lockbolt, and measure the pulley boss protrusion. If it is over the limit, the needle bearing may be caught by the spacer. Remove and reinstall the needle bearing, if necessary. Protrusion: 2.44 mm Max.

22. Install the bypass valve, spring and the o–ring into the shaft.

23. Apply sealant to the flange face of a new lockbolt.

24. Install the lockbolt. Torque to 180–200 ft. lbs. (240–270 Nm).

25. Install the engine.

26. Connect the negative battery cable.

TRANSAXLE AND TRANSMISSION

Manual Transmission Assembly

REMOVAL AND INSTALLATION

1993–95 Miata 1.6L (B6-ZE) and 1994–97 1.8L (BPD) with M15M-D Transmission

1. Disconnect the negative battery cable. Raise the vehicle and support it safely. Drain the transmission.

2. Remove the shifter knob, and the center console. Remove the gearshift lever.

3. Remove the engine undercover and the performance rod.

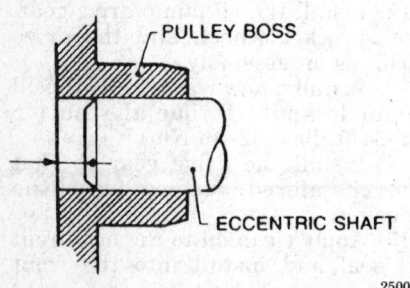

PULLEY BOSS

ECCENTRIC SHAFT

250062

Pulley shaft and boss — 1993–95 RX7 1.3L (13B) engines

4. Disconnect the exhaust pipe from the manifold. Remove the entire exhaust system as an assembly.

5. Matchmark the driveshaft flange at the rear, and remove the driveshaft.

6. Without disconnecting the hydraulic hose, remove the clutch release cylinder and set it aside. Disconnect the wiring and remove the starter.

7. Disconnect the speedometer cable, and remove the wiring from the frame member.

8. Support the transmission and differential with jacks. Remove the transmission-to-differential frame (also called the power plant frame or PPF) as follows:

   a. Remove the frame-to-transmission bracket.

   b. Remove the bolts from the underside of the frame at the differential end, noting their location. Pry out the spacer from the frame.

   c. Remove the differential mounting spacer from the underside of the differential.

   d. Insert a 14x1.5mm bolt through the frame hole, and turn it into the sleeve. Twist and pull the bolt downward.

   e. Install a 6x1mm bolt in the side hole to hold the sleeve, and remove the long bolt. Remove the short bolt.

NOTE: Do not remove the spacers from the end of the PPF. Doing so will reduce the performance of the frame. If the spacers are removed, the PPF must be replaced as an assembly.

   f. Remove the transmission side bolts, and remove the frame member.

9. Remove the bolts from the clutch housing, and slide the transmission back away from the engine. Lower the transmission from the vehicle.

To install:

10. Lightly lubricate the main shaft spline and the release bearing fork contact points with molybdenum grease and install the fork. Place a wood block on a floor jack and use it to tilt the engine up in front.

11. Carefully guide the transmission into place, making sure the main shaft spline fits properly into the clutch disc. Start all the transmission retaining bolts by hand. Tighten them alternately, in several passes, to 48–65 ft. lbs. (64–89 Nm).

12. Install the transmission-to-differential frame, and loosely install all the bolts. Torque the frame-to-transmission bolts, then the frame-to-dif-

ferential bolts to 76–91 ft. lbs. (104–124 Nm). Install the bracket, and torque the bracket-to-transmission bolts to 27–40 ft. lbs. (37–54 Nm), the bracket-to-frame bolts to 77–91 ft. lbs. (104–124 Nm).

NOTE: After the frame installation, position a straightedge between the body frame members on each side of the vehicle. Measure the distance between the bottom of the frame to the straightedge; it should be 2.403–2.797 in. (61–71mm). If the distance is not as specified, reposition the frame member at the transmission.

13. Connect the speedometer cable and wiring, and install the starter,.

14. Install the clutch release cylinder.

15. Install the driveshaft, and torque the nuts to 22 ft. lbs. (30 Nm).

16. Use a new gasket and install the exhaust system. Torque the nuts to 34 ft. lbs. (46 Nm).

17. Install the performance rod and engine undercover.

18. Lower the vehicle.

19. Fill the transmission with the proper type and quantity of oil.

20. Install the shift lever, rear console and the shift lever knob.

21. Connect the negative battery cable.

22. Verify proper operation of the transmission.

1993–95 RX7 1.3L (13B) with R15M-D Transmission

1. Disconnect the negative battery cable. Raise and safely support the vehicle. Drain the transmission oil.

2. Remove the shift lever knob and the console panel assembly. Remove the insulators and the shift lever.

3. Remove the transmission undercovers.

4. Remove the clutch slave cylinder bolts and hydraulic line bracket bolt. Secure the slave cylinder aside, without disconnecting the hydraulic line.

5. Label and disconnect the electrical connectors to the starter. Remove the starter.

6. Remove the center tunnel reinforcement. Disconnect the air injection pipe and remove the catalytic converter.

7. Remove the front and rear tunnel reinforcements.

8. Remove the cover plate from under the transmission tailshaft. Mark the position of the driveshaft on the rear axle flange and remove the driveshaft.

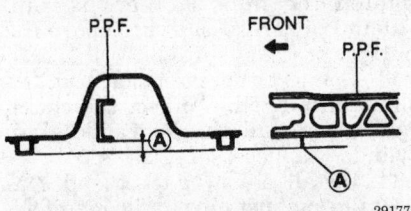

Measure distance A after installing the transmission-to-differential frame member — 1993–95 Miata 1.6L (B6-ZE) and 1.8L (BPD) with M15M-D transmission

9. Support the engine using engine support tool 49 G017 5A0 or equivalent. Support the differential with a jack.

10. Remove the transmission-to-differential frame member.

11. Label and disconnect the necessary electrical connectors. Remove the service hole covers from the transmission and remove the backup light switch.

12. Working through the service hole **A**, swing the clutch release fork so the release bearing is pushed and held toward the clutch pressure plate.

13. Insert a small prybar through service hole **B** and into the space between the wedge collar and the release bearing. Pry and separate the release bearing from the clutch pressure plate.

14. Swing the clutch release fork back and forth to make sure the release bearing and clutch pressure plate are separated.

15. If Steps 12–14 do not work, gradually loosen the 6 clutch pressure plate bolts in a crisscross pattern, working through service hole **B**. Remove the bolts and separate the pressure plate from the flywheel.

16. Support the transmission with a jack and remove the transmission-to-engine bolts. Carefully remove the transmission.

17. If the transmission was removed as per Step 15, remove the wire ring from the release bearing and separate the release bearing from the clutch pressure plate.

To install:

18. If the clutch pressure plate was removed, proceed as follows:

   a. Install a new wedge collar on the pressure plate. Apply a small amount of grease to a new wire ring and install on the wedge collar.

   b. Clean the clutch disc splines and apply a small amount of molybdenum grease to the splines.

   c. Install the clutch disc on the flywheel and install a suitable alignment tool.

   d. Install the clutch pressure plate, aligning the dowel holes with the flywheel dowels. Install the bolts and gradually tighten to 19 ft. lbs. (25 Nm), in a crisscross pattern.

   e. Remove the clutch alignment tool.

19. Coat the splines of the transmission input shaft with molybdenum grease.

20. Raise the transmission into position and install the transmission-to-engine bolts. Tighten the bolts to 38 ft. lbs. (51 Nm).

21. Working through service hole **A**, push the release cylinder end of the clutch release fork toward the transmission and connect the clutch release bearing to the clutch pressure plate.

22. Swing the clutch release fork back and forth to make sure the clutch release bearing is connected to the clutch pressure plate. Push the release cylinder end of the clutch release fork toward the engine and make sure it does not move past the engine–transmission mating line.

23. Install the service hole covers. Install the backup light switch and tighten to 25 ft. lbs. (34 Nm). Connect the electrical connectors.

24. Position the jack under the differential so the differential is perfectly level. Install the transmission-to-differential frame member and hold in place with a new bolt and 8 new nuts.

25. Tighten the frame-to-differential nuts to 130 ft. lbs. (176 Nm) and the frame-to-differential bolt to 68 ft. lbs. (93 Nm).

26. Tighten the frame to transmission nuts to 130 ft. lbs. (176 Nm), tightening the upper nuts first.

27. Remove the jack from under the differential. Lower the vehicle and remove the engine support tool.

28. Raise and safely support the vehicle. Position a straightedge between the right and left floorpans and measure the distance between the bottom of the frame member and the straightedge. The distance should be 2.91 in. (74mm) minimum. If the distance is not correct, readjust the frame member.

29. Install the driveshaft, aligning the marks made during removal. Tighten the bolts to 43 ft. lbs. (58 Nm).

30. Install the cover under the transmission tailshaft. Install the

front and rear tunnel reinforcements and tighten to 19 ft. lbs. (25 Nm).

31. Install the catalytic converter, using new gaskets, and tighten the nuts to 65 ft. lbs. (89 Nm). Connect the air injection pipe.

32. Install the center tunnel reinforcement and tighten the bolts to 19 ft. lbs. (25 Nm).

33. Install the starter and tighten the bolts to 38 ft. lbs. (51 Nm). Connect the starter wiring.

34. Install the clutch slave cylinder and tighten the bolts to 15 ft. lbs. (22 Nm).

35. Install the transmission undercovers. Fill the transmission with the proper type and quantity of oil and lower the vehicle.

36. Install the shift lever assembly and insulators. Install the console panel and the shift lever knob. Connect the negative battery cable.

Manual Transaxle Assembly

REMOVAL AND INSTALLATION

1993–94 323, Protege and 1994 MX3 with F25MR, FA4A and G25MR Transmissions

1. Properly relieve the fuel system pressure.

2. Disconnect the battery cables and remove the battery and the battery tray. Raise and safely support the vehicle.

3. Remove the splash shield(s) from under the vehicle and drain the engine and transaxle oil and the coolant.

4. Remove the air cleaner assembly and resonance chamber, including the air flow meter and all of the ducting. Remove the oil dipstick.

5. Remove the radiator hoses. If equipped with automatic transaxle, disconnect the oil cooler lines from the radiator. Disconnect the cooling fan and, if equipped, radiator switch electrical connectors and remove the radiator/cooling fan assembly.

6. Disconnect the throttle and the speedometer cable.

7. Label and disconnect the vacuum hoses and wiring.

8. Disconnect the fuel supply and return hoses and the heater hoses.

9. Disconnect the exhaust pipe from the manifold. On 4WD vehicles, remove the exhaust manifold. If equipped, remove the water inlet pipe and gasket.

10. Without disconnecting the hydraulic hoses, remove the power

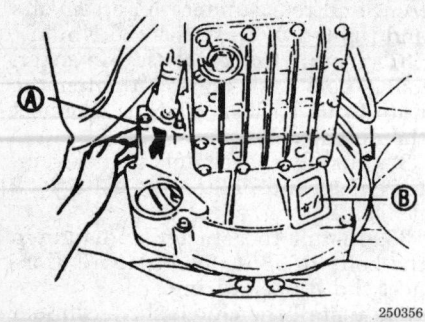

Service hole identification — 1993–95 RX7 1.3L (13B) with R15M-D transmission

steering pump and hang it from the body with wire.

11. Without disconnecting the refrigerant lines, remove the air conditioning compressor and hang it from the body with wire.

12. If equipped with manual transaxle, disconnect the clutch cable and shift control rod. If equipped with hydraulic clutch, remove the slave cylinder from the transaxle without disconnecting the hydraulic line.

13. If equipped with automatic transaxle, disconnect the shift control cable.

14. Remove the nuts and disconnect the tie rod ends from the steering knuckles. Disconnect the stabilizer bar from the lower control arms.

15. Attach an engine lifting chain to the engine lifting eyes. Attach the chain to a suitable engine hoist and raise the hoist until there is tension on the chain.

16. Remove the engine mount nuts and the engine mount member bolts and nuts and remove the engine mount member. On 4WD vehicles, remove the front transaxle mount.

NOTE: Be careful so the engine does not fall when removing the engine mount member.

17. Remove the pinch bolts from the steering knuckle and pry the control arm down to slip lower ball joint out of the knuckle.

18. If equipped, remove the bolts from the right side intermediate shaft support and, using a suitable prybar, pry the intermediate shaft from the transaxle. Insert a suitable prybar between the inner CV-joint and transaxle case and carefully pry the inner CV-joints out of the transaxle. Suspend the halfshafts with wire.

19. Remove the dynamic damper from the right side engine mount, if equipped. Remove the engine/transaxle mount nuts/bolts and right engine and, if equipped, left

transaxle mounts. Carefully lift the engine/transaxle assembly from the vehicle.

20. Properly support the engine/transaxle assembly. Remove the intake manifold bracket, starter, torque converter nuts, stiffener, if equipped and No. 2 engine mount. Disconnect the throttle cable.

21. Remove the transaxle mounting bolts and separate the transaxle from the engine.

To install:

22. Attach the transaxle to the engine and install the transaxle-to-engine bolts. Install the torque converter nuts and tighten to 25–36 ft. lbs. (34–49 Nm).

23. Connect the throttle cable. Install the No. 2 engine mount, stiffener, if equipped, starter and intake manifold bracket.

24. Install the engine mount member and tighten the bolts/nuts to 47–66 ft. lbs. (64–89 Nm).

25. Carefully lower the engine/transaxle assembly into the engine compartment and align the engine mount bolts with the engine mount member mounting holes. Install the mount-to-mount member nuts and tighten to 27–38 ft. lbs. (37–52 Nm).

26. Install the right side engine mount. Tighten the mount-to-engine nut(s) to 54–76 ft. lbs. (74–103 Nm) on 1993 vehicles. Tighten the mount through-bolt to 49–69 ft. lbs. (67–93 Nm) on 1993 vehicles.

27. On 1993 vehicles, install the dynamic damper, to the right side mount and tighten to 41–59 ft. lbs. (55–80 Nm).

28. If equipped, install the left side transaxle mount. On 1993 vehicles, loosely install the mount-to-transaxle nuts and align the mount bracket holes with the body holes. Install the mount-to-body bolts and tighten, in sequence, to 32–45 ft. lbs. (43–61 Nm). Tighten the mount-to-transaxle nuts to 49–69 ft. lbs. (67–93 Nm).

29. Install new circlips on the inner CV-joint stub shafts and, if equipped, intermediate shaft. Grease the shaft splines and install the half-shaft/intermediate shaft into the transaxle.

30. If equipped, install the right intermediate shaft support bolts and tighten, in sequence, to 31–46 ft. lbs. (42–62 Nm).

31. Install the lower ball joint and torque the clamping bolt to 43 ft. lbs. (59 Nm). Install the tie rod ends and torque the nut to 42 ft. lbs. (57 Nm), then tighten as required to install a new cotter pin.

32. Attach the stabilizer bar and tighten the nuts so there is 3/4in. (19mm) of thread showing above the nut.

33. Connect the extension bar and shift control rod. Connect the clutch cable or install the hydraulic slave cylinder, as necessary.

34. Install a new gasket and connect the exhaust pipe to the manifold. Use new self-locking nuts and torque to 34 ft. lbs. (46 Nm).

35. Connect the wiring and all heater, fuel and vacuum hoses.

36. Install the air conditioner compressor and power steering pump, if equipped.

37. Install the radiator/cooling fan assembly and connect the radiator hoses and the necessary electrical connectors.

38. Connect the accelerator and speedometer cables.

39. Install the battery tray assembly and battery. Install the air cleaner and air flow meter assembly and all the ducting. Connect the air flow sensor connector.

40. Install the splash shield(s).

41. Fill the engine and the transaxle with the proper types and quantities of oil. Fill the cooling system.

42. Connect the negative battery cable, start the engine and check for leaks. Check the ignition timing and the idle speed. Check all fluid levels.

1995 Protege with G25M-R Transaxle and 1995–97 Protege with F25M-R Transaxle

1. Drain the transaxle oil.

2. Remove the battery box and the battery.

3. Remove the air cleaner assembly.

4. Remove the battery carrier.

5. Disconnect the back-up light switch and remove the bracket.

6. Disconnect the neutral switch connector and the vehicle speedometer sensor connector.

7. Remove the harness bracket.

8. Remove the wheels.

9. Remove the splash shield.

10. Remove the transverse member.

11. Remove the extension bar and the change control rod.

12. Disconnect the tie-rod ends.

13. Remove the stabilizer control link.

14. Remove the driveshafts and the joint shaft.

15. Remove the intake manifold bracket.

16. Remove the starter.

17. Support the engine and remove the engine mounting member.

18. Remove the rear engine/trans mount.

19. Remove the front engine/trans mount.

20. Remove the clutch release cylinder.

21. Remove the side engine/trans mount.

22. Support the transaxle on a jack and remove the trans mounting bolts.

23. Remove the transaxle.

To install:

24. Place the transaxle into position and install the mounting bolts. Torque to 48–65 ft. lbs. (64–89 Nm).

25. Remove the support jack.

26. Install the side mount. Torque the body side nuts and bolts to 32–44 ft. lbs. (44–60 Nm). Torque the trans side nuts to 50–68 ft. lbs. (67–93 Nm).

27. Install the clutch release cylinder.

28. Install the front mount, loosely tighten the mount nut and bolt.

29. Install the rear mount, align and set all bolts then torque to 50–68 ft. lbs. (67–93 Nm).

30. Install the engine mounting member. Torque the four outer nuts and bolts to 50–65 ft. lbs. (67–89 Nm) and the two remaining nuts to 28–38 ft. lbs. (38–51 Nm).

31. Torque the front mount nut and bolt to 50–68 ft. lbs. (67–93 Nm).

32. Install the starter.

33. Install the manifold bracket.

34. Install the joint shaft and the driveshafts.

35. Install the stabilizer control link.

36. Connect the tie-rod ends.

37. Connect the change control rod and the extension bar.

38. Install the transverse member.

39. Install the splash shield.

40. Install the wheels.

41. Install the harness bracket.

42. Install the vehicle speedometer sensor connector and the neutral switch connector.

43. Install the back-up light switch connector bracket and the switch.

44. Install the battery carrier.

45. Install the air cleaner assembly.

46. Install the battery and battery box.

47. Fill the trans with the proper fluid.

48. Check for proper clutch operation.

1993–95 MX3 with F25M, F25MR, F2SM-R and G25MR Transaxles

1. Relieve the fuel system pressure and disconnect the negative battery cable. Raise and safely support

the vehicle and drain the fluid from the transaxle.

2. Remove the wheel and tire assemblies and the splash shield.

3. Remove the resonance duct and the air cleaner assembly.

4. Disconnect the positive battery cable and remove the battery and battery tray.

5. Remove the starter and the wiring harness bracket. Disconnect the fuel lines and remove the fuel filter.

6. Label and disconnect the connectors for the neutral switch, backup light switch and speedometer sensor.

7. Remove the mounting bolts and line clip and position the clutch slave cylinder aside without disconnecting the hydraulic line.

8. Remove the transverse member and the front exhaust pipe/converter assembly.

9. Disconnect the extension bar and shift control rod from the transaxle.

10. Remove the cotter pins and nuts and separate the tie rod ends from the knuckles. Disconnect the stabilizer links from the lower control arms.

11. Remove the halfshafts and the intermediate shaft.

12. Support the engine using engine support tool 49 G017 5A0 or equivalent. Remove the bolts/nuts and remove the engine mount member.

13. Remove the left side transaxle mount and the transaxle inspection plate.

14. Loosen the engine support tool and lean the engine toward the transaxle. Support the transaxle with a jack. Remove the transaxle mounting bolts and carefully remove the transaxle.

To install:

15. Raise the transaxle into position and install the transaxle-to-engine bolts. Tighten to 66 ft. lbs. (89 Nm) on the 1.6L engine and 73 ft. lbs. (99 Nm) on the 1.8L engine.

16. Install the transmission inspection plate and tighten the bolts to 38 ft. lbs. (52 Nm).

17. Loosely tighten the left side transaxle mount bolts.

18. Install the engine mount member, aligning the front and rear transaxle mount stud bolts. Install the mount member-to-mount nuts and tighten to 38 ft. lbs. (52 Nm) and the mount member bolts/nuts to the body, and tighten to 66 ft. lbs. (89 Nm).

19. Tighten left side transaxle mount-to-transaxle nut **A** to 69 ft.

lbs. (93 Nm) and nuts **B** to 30 ft. lbs. (40 Nm). Remove the engine support tool.

20. Install the halfshafts and joint shaft.

21. Connect the stabilizer bar link to the lower control arm and tighten the nut to 45 ft. lbs. (61 Nm). Connect the tie rod end to the steering knuckle and tighten the nut to 42 ft. lbs. (57 Nm). Install a new cotter pin.

22. Connect the extension bar to the transaxle and tighten the bolt to 38 ft. lbs. (52 Nm). Connect the shift control rod and tighten the bolt to 17 ft. lbs. (23 Nm).

23. Install the exhaust pipe, using new gaskets. Tighten the manifold flange nuts to 38 ft. lbs. (52 Nm).

24. Install the transverse member and tighten the bolts to 93 ft. lbs. (127 Nm).

25. Connect the electrical connectors. Install the fuel filter and connect the fuel lines.

26. Install the starter and the wiring harness bracket.

27. Install the battery tray and battery.

28. Install the air cleaner assembly and the resonance duct. Install the splash shield and the wheel and tire assemblies.

29. Fill the transaxle with the proper type and quantity of oil.

30. Lower the vehicle and connect the battery cables. Start the engine and check for leaks and proper transaxle operation.

1993–97 626 and MX6 with G25M-R Transaxle

1. Disconnect the battery cables and remove the battery and battery tray. Raise and safely support the vehicle. Drain the transaxle fluid.

2. Remove the ducts and the air cleaner assembly.

3. If equipped with 2.5L engine, remove the starter.

4. Label and disconnect the electrical connectors for the neutral switch, backup light switch and vehicle speed sensor. Disconnect the ground wires.

5. Remove the fuel filter mounting nuts and position the filter aside, leaving the fuel lines attached. Remove the wiring harness bracket.

6. Remove the mounting bolts and the hydraulic line clips, then position the clutch slave cylinder aside, leaving the hydraulic line attached.

7. Remove the wheel and tire assemblies and the splash shields. Remove the transverse member.

8. If equipped with 2.5L engine, disconnect the oxygen sensor connec-

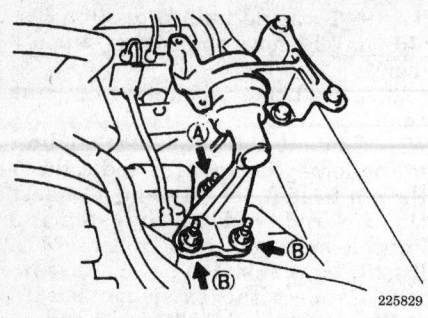

Number four engine mount bolt locations (1.8L only) — 1993-95 MX3 with F25M, F25MR, F2SM-R and G25MR transaxles

tors and remove the front exhaust pipe/catalytic converter assembly.

9. Disconnect the extension bar and shift control rod from the transaxle.

10. Remove the pinch bolts from the steering knuckles. Pry the lower control arms down to separate the ball joints from the knuckles.

11. Remove the cotter pins and nuts and separate the tie rod ends from the knuckles. Remove the nuts and disconnect the stabilizer links from the lower control arms.

12. Remove the brake hose clips from the struts, and (if equipped) remove the ABS speed sensor harness mounting nuts.

13. Separate the right side half-shaft from the intermediate shaft using a hammer and brass drift. Pry the left side shaft from the transaxle using a prybar inserted between the transaxle and the inner CV-joint. Be careful not to damage the oil seal.

14. Suspend the halfshafts with rope, in a level position. Insert plug tools 49 G030 455 or equivalent into the differential side gears to keep them from becoming mispositioned.

15. Remove the intermediate/joint shaft.

16. If equipped with 2.0L engine, remove the starter.

17. Remove the engine mount rubber at the right side of the engine mount member. Remove the rear transaxle mount.

18. Support the engine using engine support tool 49 G017 5A0 or equivalent. Remove the nuts and bolts and remove the engine mount member.

19. If equipped with 2.5L engine, remove the transaxle housing inspection plate.

20. Remove the left side transaxle mount.

21. Loosen the engine support tool and lean the engine toward the transaxle. Support the transaxle on a jack and remove the transaxle mounting

bolts. Carefully remove the transaxle from the vehicle.

22. Remove the front transaxle mount.

To install:

23. Install the front transaxle mount and tighten the mount-to-transaxle bolts to 44 ft. lbs. (60 Nm). Loosely tighten the mount through-bolt and nut.

24. Raise the transaxle into position and install the transaxle mounting bolts. If equipped with 2.5L engine, install the inspection cover. Tighten the mounting bolts to the proper torque values.

25. Loosely tighten the left side transaxle mount bolts and nuts.

26. Use the engine support tool to make sure the transaxle bolt holes and the rear transaxle mount are aligned. Install the bolts and tighten to 68 ft. lbs. (93 Nm).

27. Install the engine mount member, making sure the engine mount rubbers are properly installed. Tighten the mount member-to-body bolts/nuts to 68 ft. lbs. (93 Nm) and the front transaxle mount nuts to 77 ft. lbs. (104 Nm).

28. Tighten the front transaxle mount through-bolt and nut to 86 ft. lbs. (116 Nm).

29. Tighten the left side transaxle mount bolts and nut to 68 ft. lbs. (93 Nm) and remove the engine support tool.

30. Install the engine mount rubber on the right side of the mount member and tighten the bolts to 44 ft. lbs. (60 Nm).

31. If equipped with 2.0L engine, install the starter.

32. Remove the plug tool from the differential side gear. Install the intermediate/joint shaft into the transaxle. Install the support bearing to the engine and tighten the bolts, in sequence, to 45 ft. lbs. (61 Nm).

33. Remove the plug from the other differential side gear. Install a new circlip on the halfshaft with end gap facing upward. Install the halfshaft into the transaxle, being careful not to damage the seal. Make sure the circlip seats in the differential side gear, by pulling out on the shaft; it must not pull out.

34. Install a new circlip on the end of the other halfshaft and connect it to the intermediate shaft.

35. Connect the lower ball joints to the steering knuckles. Install the pinch bolts and tighten to 43 ft. lbs. (58 Nm).

36. Connect the stabilizer links to the lower control arms and tighten the nuts to 39 ft. lbs. (53 Nm). Con-

nect the tie rod ends to the steering knuckles and tighten the nuts to 32 ft. lbs. (44 Nm). Install new cotter pins.

37. Connect the shift control rod to the transaxle and tighten the bolt to 18 ft. lbs. (25 Nm). Connect the extension bar and tighten the nut to 38 ft. lbs. (51 Nm).

38. If equipped with 2.5L engine, install the exhaust pipe, using new gaskets. Tighten the exhaust manifold flange nuts to 38 ft. lbs. (51 Nm). Connect the oxygen sensor connectors.

39. Install the transverse member and tighten the bolts to 96 ft. lbs. (131 Nm).

40. Install the splash shields and the wheel and tire assemblies.

41. Install the clutch slave cylinder and tighten the bolts to 16 ft. lbs. (22 Nm). Install the wiring harness bracket and tighten the bolts to 130 inch lbs. (14 Nm).

42. Install the fuel filter mounting nuts and tighten to 95 inch lbs. (11 Nm). Connect the grounds and electrical connectors.

43. If equipped with 2.5L engine, install the starter.

44. Install the air cleaner assembly and air ducts. Install the battery tray and battery.

45. Fill the transaxle with the proper type and quantity of fluid and lower the vehicle.

46. Connect the battery cables and start the engine. Check for leaks and for proper transaxle operation.

Clutch Assembly

REMOVAL AND INSTALLATION

Except RX7

1. Disconnect the negative battery cable. Raise and safely support the vehicle.

2. Remove the transaxle.

3. Gradually loosen the clutch pressure plate bolts, in a crisscross pattern. Support the pressure plate and remove the bolts. Remove the pressure plate and clutch disc.

4. Inspect the pilot bearing. If it is worn or damaged and does not turn easily by hand, remove it using a puller/slide hammer.

5. Check the flywheel surface for scoring, cracks or burning and machine or replace, as necessary.

6. Install holder tool 49 E011 1A0 or equivalent, to keep the flywheel from turning. Loosen the flywheel bolts evenly and gradually in a crisscross pattern. Remove the flywheel.

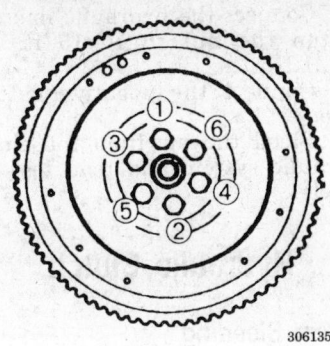

Flywheel bolt removal and
installation sequence — except RX7

306135

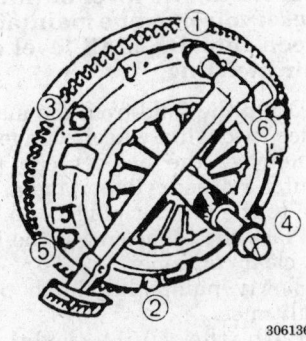

Clutch cover bolt torque
sequence — except RX7

306136

7. Inspect the clutch release bearing for wear. Replace it if it sticks or does not turn easily.

8. Inspect the release fork for wear or damage and replace as necessary.

To install:

9. Lubricate the release fork fingers and pivot with molybdenum grease and install in the release fork boot.

10. Install the clutch release bearing on the release fork.

11. If removed, install a new pilot bearing in the flywheel, using a suitable installation tool.

12. Make sure the flywheel mounting surface and the crankshaft or eccentric shaft mounting surfaces are clean. Remove any old sealant from the flywheel bolt hole threads and the flywheel bolts.

13. Apply sealant to the flywheel bolt threads and install them hand tight. Install the flywheel holding tool. Tighten the bolts, in a crisscross pattern, to 49 ft. lbs. (67 Nm) on the Miata 1.8L engine and 75 ft. lbs. (102 Nm) on all other models.

14. Apply a small amount of molybdenum grease to the clutch disc splines and install the clutch disc on the flywheel, spring side toward the

transmission or transaxle. Install a suitable alignment tool in the pilot bearing to position the clutch disc.

15. Install the clutch pressure plate, aligning the dowel holes with the flywheel dowels. Install the pressure plate bolts and gradually tighten, in a crisscross pattern to 20 ft. lbs. (26 Nm). Remove the alignment tool.

16. Install the transaxle and lower the vehicle.

RX7

1. Disconnect the negative battery cable. Raise and safely support the vehicle.

2. Remove the transmission.

3. Remove the clutch release fork assembly bolts and remove the release fork and bearing as an assembly. Inspect the release fork for wear or damage; make sure it swings freely without the return spring installed. Inspect the release bearing for wear, damage or sticking when it is turned. Replace parts as necessary.

4. Install flywheel holder tool 49 F011 101 or equivalent. Loosen the clutch pressure plate bolts gradually, in a crisscross pattern. Support the pressure plate, remove the bolts and remove the pressure plate and clutch disc.

5. Remove the wire ring from the wedge collar and remove the wedge collar from the pressure plate.

6. Inspect the pilot bearing for wear or damage, make sure it turns freely by hand. Remove the pilot bearing, with the oil seal, using a puller/slide hammer.

7. Inspect the flywheel for scoring, cracks or burning. Machine or replace as necessary. To remove, install holder tool 49 F011 101 or equivalent and remove the locknut. Remove the key from the eccentric shaft.

To install:

8. Install the key in the eccentric shaft and install the flywheel. Apply thread locking compound to the locknut threads and sealant to the locknut contact surface. Install the locknut.

9. Install the holder tool and tighten the locknut to 360 ft. lbs. (490 Nm).

10. Install a new pilot bearing using a suitable installer. The bearing must be installed to a depth of 0.453–0.482 in. (11.5–12.3mm). Install a new oil seal.

11. Apply a small amount of molybdenum grease to the clutch disc splines. Install the clutch disc on the flywheel, spring side toward the transmission. Install an alignment

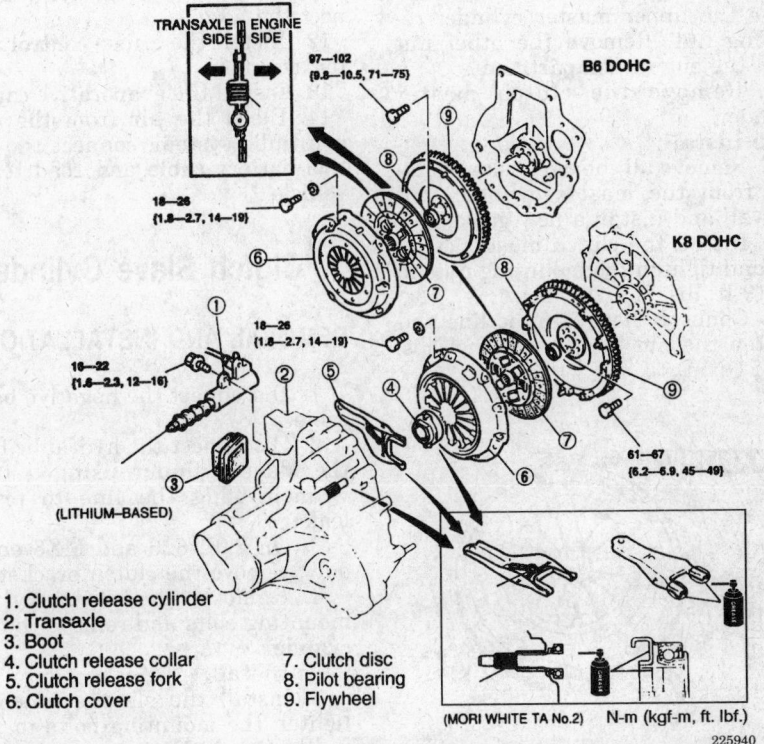

1. Clutch release cylinder
2. Transaxle
3. Boot
4. Clutch release collar
5. Clutch release fork
6. Clutch cover
7. Clutch disc
8. Pilot bearing
9. Flywheel

Clutch assembly (exploded view) — 1993–95 MX3 with F25M, F25MR, F2SM-R and G25MR transaxles

tool in the pilot bearing to position the clutch disc.

12. Install a new wedge collar on the pressure plate. Apply a small amount of grease to a new wire ring and install.

13. Align the pressure plate dowel holes with the flywheel dowels and install the flywheel. Install the pressure plate bolts and gradually tighten, in a crisscross pattern, to 19 ft. lbs. (26 Nm).

14. Lubricate the release fork and install the release fork and release bearing assembly in the transmission. Tighten the bolts to 33 ft. lbs. (46 Nm).

15. Install the transmission and lower the vehicle.

Clutch Master Cylinder

REMOVAL AND INSTALLATION

Except RX7

1. Disconnect the negative battery cable.

2. Move the charcoal canister out of the way. Disconnect and plug the hose from the brake fluid reservoir.

3. Disconnect the hydraulic line at the master cylinder, using a tubing wrench.

4. Working inside the vehicle, remove the upper master cylinder retaining nut. Remove the other nut from the engine compartment.

5. Remove the clutch master cylinder.

To install:

6. Remove all the old gasket material from the master cylinder and firewall and install a new gasket.

7. Install the clutch master cylinder and tighten the mounting nuts to 14–19 ft. lbs. (19–26 Nm).

8. Connect the hydraulic line and tighten the nut to 113–190 inch lbs. (13–21 Nm).

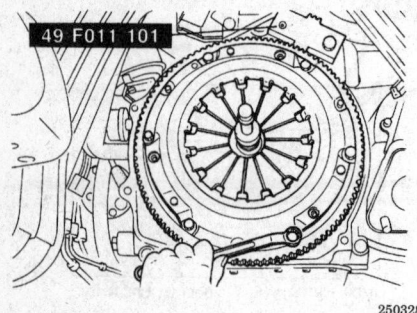

`49 F011 101`

`250328`

Ring gear brake tool (49 F011 101) — RX7

9. Unplug and connect the hose to the brake fluid reservoir.

10. Replace the charcoal canister.

11. Bleed the air from the clutch hydraulic system, connect the negative battery cable and road test the vehicle.

RX7

1. Disconnect the negative battery cable.

2. Remove the evaporative canister.

3. Remove the cruise control actuator assembly.

4. Disconnect the hose from the brake master cylinder reservoir and plug the reservoir port.

5. Disconnect the hydraulic line at the master cylinder, using a tubing wrench.

6. Working inside the vehicle, remove the master cylinder mounting nuts.

7. Remove the clutch master cylinder.

To install:

8. Remove all the old gasket material from the master cylinder and firewall and install a new gasket.

9. Install the clutch master cylinder and tighten the mounting nuts.

10. Connect the hydraulic line and tighten the nut securely.

11. Remove the plug from the brake master cylinder reservoir and connect the hose.

12. Install the cruise control actuator assembly.

13. Install the evaporative canister.

14. Bleed the air from the clutch hydraulic system, connect the negative battery cable and road test the vehicle.

Clutch Slave Cylinder

REMOVAL AND INSTALLATION

1. Disconnect the negative battery cable.

2. Disconnect the hydraulic line at the slave cylinder using a tubing wrench. Plug the line to prevent leakage.

3. On 2.0L 626 and MX6 engines only, remove the clutch bracket.

4. Remove the slave cylinder mounting bolts and remove the slave cylinder.

To install:

5. Install the slave cylinder and tighten the mounting bolts to 12–16 ft. lbs. (16–22 Nm).

6. Install the clutch bracket, if removed.

7. Connect the hydraulic line and tighten the nut to 10–15 ft. lbs. (13–21 Nm).

8. Connect the negative battery cable.

9. Bleed the air from the clutch hydraulic system and road test the vehicle.

Hydraulic Clutch

System Bleeding

NOTE: A common reservoir is used for the clutch and the brake system fluid. The level of fluid in the reservoir must be maintained between the MIN/MAX level during air bleeding.

1. Drain the fluid from the master cylinder through a wheel cylinder.

2. Remove the bleeder cap from the clutch release cylinder and attach a hose to the bleeder plug.

3. Place the other end of the hose into a clear container.

4. Slowly pump the clutch pedal several times.

5. With the clutch pedal depressed, loosen the bleeder screw to let the fluid escape. Close the bleeder valve.

6. Repeat bleeding until clean fluid with no air bubbles is seen.

7. Properly bleed the brake system.

8. Fill the reservoir to MAX with new fluid.

9. Slowly pump the clutch pedal several times to verify that there are no fluid leaks.

10. Check operation of the clutch and brake systems.

Automatic Transmission Assembly

REMOVAL AND INSTALLATION

1993–97 Miata with N4A-HL and NC4A-EL Transmissions

1. Shift the selector lever to the **N** position.

2. Disconnect the negative battery cable. Raise and safely support the vehicle. Drain the transmission fluid.

3. Remove the engine undercover, and disconnect the shift rod.

4. Remove the performance rod.

5. Remove the complete exhaust system from the exhaust manifold.

6. Mark the position of the driveshaft on the rear axle flange, and remove the driveshaft.

7. Disconnect the speedometer cable, and disconnect the vacuum hose from the vacuum diaphragm.

8. Label and disconnect the electrical connectors from the: inhibitor switch, kickdown solenoid, overdrive cancel solenoid, oil pressure switch and lockup solenoid.

9. Remove the dipstick and dipstick tube. Disconnect the oil cooler lines.

10. Support the transmission and differential with jacks. Remove the transmission-to-differential frame as follows:

   a. Disconnect the wiring harness from the frame.

   b. Remove the bolts from the underside of the frame at the differential end, noting their location. Pry out the spacer from the frame.

   c. Remove the differential mounting spacer from the underside of the differential.

   d. Insert a 14x1.5mm bolt through the frame hole, and turn it into the sleeve. Twist and pull the bolt downward.

   e. Install a 6x 1mm bolt in the side hole to hold the sleeve, and remove the long bolt. Remove the short bolt.

   f. Remove the transmission side bolts, and remove the frame member.

11. Remove the torque converter bolts, and the starter.

12. Remove the transmission mounting bolts and remove the transmission, being careful not to drop the torque converter.

   To install:

13. Make sure the torque converter is fully installed in the transmission. The distance between 1 of the bolt hole lugs and a straightedge laid across the bellhousing should be 0.89 in. (22.5mm).

14. Raise the transmission into position, and install the mounting bolts. Tighten to 48–66 ft. lbs. (64–89 Nm).

15. Install the starter. Install the torque converter bolts. Align the holes by turning the torque converter. Hold the flexplate with a small prybar, and tighten the bolts to 27–40 ft. lbs. (36–54 Nm).

16. Install the transmission-to-frame member as follows:

   a. Install the differential mounting spacer on the underside of the differential. Tighten the mounting bolts to 38 ft. lbs. (52 Nm).

   b. Position the jack under the transmission so the transmission is level.

   c. Position the frame and install the transmission side bolts. Snug the bolts.

   d. Make sure the sleeve is installed in the block. Install the spacer and bolts with the reamer bolt in the front hole. Snug the bolts.

   e. Tighten the transmission side bolts to 77–91 ft. lbs. (104–123 Nm), then tighten the differential bolts to the same specification.

   f. Connect the wiring harness to the frame member and remove the jacks.

NOTE: After the frame installation, position a straightedge between the body frame members on each side of the vehicle. Measure the distance between the bottom of the frame to the straightedge; it should be 2.023–2.417 in. (51.5–61.5mm). If the distance is not as specified, reposition the frame member at the transmission.

17. Connect the oil cooler lines, using new gaskets. Install the dipstick tube and dipstick.

18. Connect the electrical connectors and the vacuum hose. Connect the speedometer cable.

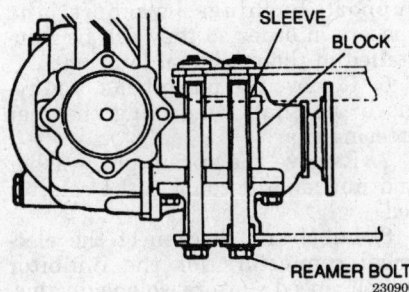

Sleeve and reamer bolt positioning — 1993–95 Miata with N4A-HL and NC4A-EL transmissions

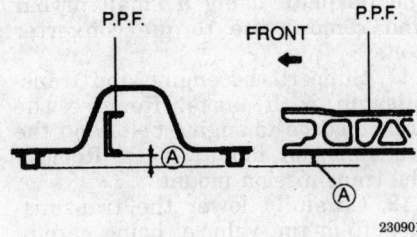

Measure distance A after installing the transmission-to-differential frame member — 1993–95 Miata with N4A-HL and NC4A-EL transmissions

19. Install the driveshaft, aligning the marks made during removal. Tighten the bolts to 22 ft. lbs. (30 Nm).

20. Install the exhaust system. Connect the front pipe to the exhaust manifold, using a new gasket, and tighten the nuts to 34 ft. lbs. (46 Nm).

21. Install the performance rod.

22. Install the engine undercover and connect the shift rod.

23. Lower the vehicle and connect the negative battery cable. Fill the transmission with the proper type and quantity of fluid. Start the engine and check for leaks and proper operation.

1993–95 RX7 with RB4A-EL Transmission

1. Disconnect the negative battery cable. Raise and safely support the vehicle. Drain the transmission fluid.

2. Remove the transmission dipstick.

3. Remove the right and left undercovers. Remove the starter.

4. Remove the center tunnel reinforcement. Disconnect the air injection pipe and remove the catalytic converter.

5. Remove the front and rear tunnel reinforcements. Remove the cover at the front of the transmission-to-differential frame member.

6. Mark the position of the driveshaft on the axle flange and remove the driveshaft. Plug the transmission tailshaft to prevent fluid leakage.

7. Support the engine with engine support tool 49 G017 5A0 or equivalent. Support the differential with a jack. Remove the transmission-to-differential frame member.

8. Disconnect the oxygen sensor and remove the front exhaust pipe.

9. Label and disconnect the electrical connectors to the inhibitor switch, speed sensors, pulse generator and solenoid valve.

10. Disconnect the transmission selector rod and remove the torque converter cover.

11. Lock the flexplate with a small prybar and remove the torque converter bolts.

12. Remove the upper and lower oil filler tubes. Disconnect the oil cooler lines.

13. Support the transmission with a jack. Remove the transmission-to-engine bolts and carefully lower the transmission from the vehicle. Be careful not to drop the torque converter.

To install:

14. Make sure the torque converter is completely installed in the transmission. Lay a straightedge across the bellhousing and measure the distance to 1 of the converter bolt hole lugs; it should be 1.1 in. (29mm). If the distance is less than specification, push the converter into the pump while rotating it, to properly engage the pump drive.

15. Raise the transmission into position and install the transmission-to-engine bolts. Tighten to 38 ft. lbs. (51 Nm).

16. Position the jack under the differential so the differential is perfectly level. Install the transmission-to-differential frame member and hold in place with a new bolt and 8 new nuts.

17. Tighten the frame-to-differential nuts to 130 ft. lbs. (176 Nm) and the frame-to-differential bolt to 68 ft. lbs. (93 Nm).

18. Tighten the frame to transmission nuts to 130 ft. lbs. (176 Nm), tightening the upper nuts first.

19. Remove the jack from under the differential. Lower the vehicle and remove the engine support tool.

20. Raise and safely support the vehicle. Position a straightedge between the right and left floor pans and measure the distance between the bottom of the frame member and the straightedge. The distance should be 2.91 in. (74mm) minimum. If the distance is not correct, readjust the frame member.

21. Turn the torque converter to align the holes. Hold the flexplate with a small prybar and tighten the bolts, evenly and gradually, to 26–36 ft. lbs. (35–49 Nm).

22. Connect the oil cooler lines, using new washers, and tighten the banjo bolts to 26 ft. lbs. (35 Nm). Install the upper and lower oil filler tubes, using a new O-ring.

23. Install the service hole cover. Connect the selector rod, using the washer and a new spring clip.

24. Connect the electrical connectors.

25. Install the front exhaust pipe, using a new gasket, and tighten the nuts to 65 ft. lbs. (89 Nm). Connect the oxygen sensor connector.

26. Install the driveshaft, aligning the marks made during removal. Tighten the bolts to 43 ft. lbs. (58 Nm).

27. Install the cover at the front of the transmission-to-differential frame member. Install the front and rear tunnel reinforcements and

tighten the bolts to 19 ft. lbs. (26 Nm).

28. Install the catalytic converter, using new gaskets. Tighten the nuts to 65 ft. lbs. (89 Nm) and connect the air injection pipe.

29. Install the center tunnel reinforcement and tighten the bolts to 19 ft. lbs. (26 Nm). Install the starter and the right and left undercovers.

30. Lower the vehicle and connect the negative battery cable. Fill the transmission with the proper type and quantity of fluid.

31. Start the engine and bring to normal operating temperature. Check for leaks and proper operation.

1993–95 929 with RA4A-EL Transmission

1. Disconnect the negative battery cable. Raise and safely support the vehicle. Drain the transmission fluid.

2. Remove the transmission dipstick.

3. Disconnect the oxygen sensors. Remove the exhaust system, except for the rear muffler.

4. Remove the exhaust pipe bracket from the transmission, and remove the heat shield.

5. Mark the position of the driveshaft on the axle flange and remove the driveshaft. Plug the tailshaft to prevent fluid leakage. Be sure to keep all the center bearing support bushings, washers and spacers in order so they can be reinstalled in their original locations.

6. Remove the pinch bolts, and remove the shaft connecting the rear steering gear.

7. Remove the spring pin, washer and nut, and disconnect the selector rod.

8. Label and disconnect the electrical connectors for the inhibitor switch, speed sensors, solenoid valve, pulse generator and knock sensor.

9. Disconnect the oil cooler lines, and remove the oil filler tube.

10. Remove the starter and the transmission service hole cover. Lock the flexplate using a small prybar and remove the torque converter bolts.

11. Support the engine and transmission with jacks. Remove the transmission-to-engine bolts and the transmission mount bolts. Remove the transmission mount.

12. Carefully lower the transmission from the vehicle, being careful not to drop the torque converter.

To install:

13. Make sure the torque converter is completely installed in the transmission. Lay a straightedge across

the bellhousing and measure the distance to one of the converter bolt hole lugs; it should be 1.161 in. (29.5mm). If the distance is less than specification, push the converter into the pump while rotating it, to properly engage the pump drive.

14. Raise the transmission into position, and install the transmission-to-engine bolts. Tighten to 38 ft. lbs. (52 Nm).

15. Install the transmission mount and tighten the mount-to-body bolts to 45 ft. lbs. (61 Nm). Remove the engine and transmission jacks, then tighten the mount-to-transmission bolts/nuts to 56 ft. lbs. (77 Nm).

16. Turn the torque converter to align the bolt holes. Lock the flexplate using a small prybar and install the torque converter bolts. Tighten the bolts gradually and evenly to 25–36 ft. lbs. (34–49 Nm).

17. Install the service hole cover and the starter.

18. Install the oil filler tube using a new O-ring. Connect the oil cooler lines using new washers.

19. Connect the electrical connectors. Connect the selector rod with the washer and a new spring clip.

20. Install the rear steering gear shaft, and tighten the pinch bolt to 36 ft. lbs. (49 Nm).

21. Install the driveshaft, aligning the marks made during removal. Tighten the driveshaft-to-axle flange bolts to 43 ft. lbs. (59 Nm). Install the center bearing support, making sure all bushings, washers and spacers are installed in their original locations. Tighten the bolts to 39 ft. lbs. (53 Nm).

22. Install the heat shield and the exhaust pipe bracket. Install the exhaust system, using new gaskets, and tighten the flange nuts to 41 ft. lbs. (55 Nm). Connect the oxygen sensors.

23. Lower the vehicle and connect the negative battery cable. Fill the transmission with the proper type and quantity of fluid.

24. Start the engine and bring to normal operating temperature. Check for leaks and proper operation.

Automatic Transaxle Assembly

REMOVAL AND INSTALLATION

1993–94 323 and Protege with FA4A Transaxle

1. Raise and safely support the vehicle and remove the front wheels.

Remove the battery and battery box and the air cleaner and ducting.

2. Remove the splash shield and drain the transaxle oil.

3. Disconnect the speedometer cable, throttle cable, shift cable and the wiring from the transaxle.

4. Disconnect the wiring and remove the starter.

5. Disconnect the exhaust pipe from the manifold and the catalytic converter and remove the pipe.

6. Disconnect the tie rod ends and lower ball joints and remove the half-shafts. Use special tool 49 G030 455 or equivalent to hold the differential side gears in place when the half-shafts are removed.

7. If equipped, remove the torque converter–to–flywheel nuts.

8. Disconnect the oil cooler hoses and plug them to prevent leakage.

9. Install the necessary lifting equipment and support the engine from above. Remove the lower mounting frame and support the transaxle from below with a jack.

10. Remove the front and left rear mounts and allow the engine/transaxle to tilt towards the left.

11. Remove the bolts and slide the transaxle away from the engine to lower it out of the vehicle. Do not let the torque converter fall out.

To install:

12. Make sure the torque converter is properly placed and carefully guide the transaxle into place. Start all the transaxle–to–engine bolts, then torque them to 59 ft. lbs. (80 Nm).

13. Install the left rear mount but do not torque the bolts yet.

14. Install the torque converter–to–flywheel nuts and torque to 25 ft. lbs. (34 Nm).

15. Install the halfshafts, making sure the inner joint is firmly seated into place. Torque the extension shaft bracket bolts to 46 ft. lbs. (62 Nm).

16. Assemble the suspension. Torque the lower ball joint pinch bolt and the tie rod end nuts to 43 ft. lbs. (59 Nm). If equipped with a stabilizer bar, adjust the link with ³/₄ in. (19mm) of thread showing above the locknut.

17. Install the front mount to the transaxle and torque the bolts to 38 ft. lbs. (52 Nm).

18. Install the lower mounting frame and torque the frame–to–body nuts and bolts **A** to 66 ft. lbs. (89 Nm). Torque the mount–to–frame nuts and bolts **B** to 38 ft. lbs. (52 Nm).

19. Connect the cooler hoses, making sure the clamp does not interfere with other parts.

20. Connect the shift cable, speedometer cable, throttle cable and the wiring.

21. Install the starter and connect the wiring.

22. Install the splash shields and wheels.

23. Install the air cleaner and battery.

1995–97 Protege with FA4A-EL Transaxle

1. Raise the vehicle on a hoist.
2. Drain the transaxle fluid.
3. Remove the battery and battery cover.
4. Remove the air cleaner assembly.
5. Remove the battery carrier.
6. Disconnect the solenoid connector and the transaxle range switch connector.
7. Disconnect the selector cable.
8. Disconnect the speedometer sensor connector.
9. Remove the harness bracket.
10. Disconnect the throttle cable.
11. Remove the front wheels.
12. Remove the splash shields.
13. Remove the transverse member, if equipped.
14. Disconnect the tie-rod ends.
15. Disconnect the stabilizer control links.
16. Disconnect the lower arm by removing the cinch bolt from the lower arm ball joints. Pry the lower arm out of the knuckle.
17. Support the engine and remove the engine mounting member.
18. Remove the left and right driveshafts and install special tool 49 G030 455 or equivalent to hold the side gears.
19. Remove the joint shaft.
20. Remove the manifold bracket.
21. Remove the starter.
22. Remove the front engine/trans mount.
23. Remove the rear engine/trans mount.

Crossmember and mount installation — 1993–94 323 and Protege with a FA4A transmission

24. Disconnect the inner and outer oil hoses.

25. Remove the side engine/trans mount.

26. Hold the drive plate and remove the converter nuts.

27. Support the transaxle on a jack and remove the mounting bolts.

28. Remove the transaxle.

To install:

29. Support the transaxle on a jack and lift it into place. Align the transaxle with the engine and install the mounting bolts. Torque to 41–59 ft. lbs. (55–80 Nm).

30. Hold the driveplate and install the torque converter mount nuts. 26–36 ft. lbs. (35–49 Nm).

31. Install the side engine/trans mount. Loosely tighten the nuts of the transaxle side. Tighten the nuts and bolts of the body side. Torque to 32–44 ft. lbs. (44–60 Nm). Tighten the nuts of the transaxle side, torque to 50–68 ft. lbs.

32. Connect the inner and outer oil hoses.

33. Install the rear engine/trans mount. Torque the bolts to 50–68 ft. lbs. (67–93 Nm).

34. Install the front engine/trans mount. Torque the mount bracket to the transaxle to 28–38 ft. lbs. (38–51 Nm). Loosely tighten the nuts and bolts of the engine mount rubber, then torque to 50–68 ft. lbs. (67–98 Nm).

35. Install the starter.

36. Install the manifold bracket.

37. Insert the joint shaft into the transaxle. Install the joint shaft to the cylinder block and tighten the bolts in sequence (counterclockwise). Torque to 32–46 ft. lbs. (42–62 Nm).

38. Install the driveshafts, make sure that the shafts are properly installed and do not pull out.

39. Replace the engine mounting member. Install the mounting nuts/bolts and torque the nuts/bolts at the far corners to 48–65 ft. lbs. (64–89 Nm), torque the remaining two nuts to 28–38 ft. lbs. (38–51 Nm).

40. Connect the lower arm to the knuckle.

41. Install the stabilizer control link.

42. Connect the tie-rod ends.

43. Install the transverse member, if removed.

44. Install the splash shields.

45. Install the wheels.

46. Connect the throttle cable.

47. Install the harness bracket.

48. Connect the vehicle speedometer sensor connector.

49. Connect the selector cable.

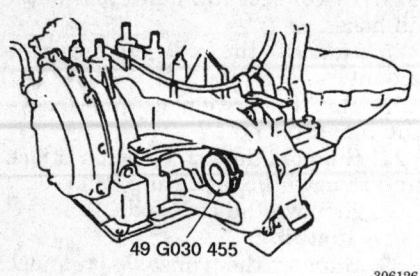

49 G030 455

306126

Special tool for differential side gear — 1995–97 Protege with a FA4A-EL transmission

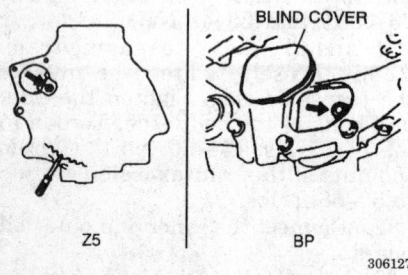

BLIND COVER

Z5 BP

306127

View of torque converter nut access for Z5D and BP engines — 1995–97 Protege with a FA4A-EL transmission

50. Connect the transaxle range switch connector.
51. Connect the solenoid connector.
52. Install the battery carrier.
53. Install the air cleaner assembly.
54. Install the battery and battery cover.
55. Test drive the vehicle. Check for proper operation in all gear ranges.

1993–95 MX3 with FA4A and FA4A-EL Transaxles

1. If equipped with 1.8L engine, relieve the fuel system pressure.
2. Disconnect the negative battery cable. Raise and safely support the vehicle. Drain the transaxle fluid.
3. Remove the wheel and tire assemblies, and splash shields.
4. Disconnect the air flow meter connector, and remove the resonance duct and air cleaner assembly.
5. If equipped with 1.8L engine, remove the strut bar from between the strut towers.
6. Disconnect the positive battery cable and disconnect the wiring harness from the battery tray. Remove the battery and battery tray.
7. Disconnect the speed sensor connector.

8. Remove the clip from the cable housing and the spring clip from the transaxle lever, then remove the shift cable.
9. Label and disconnect the electrical connectors for the inhibitor switch and solenoid valve. Remove the bolt from the harness bracket and position the harness aside.
10. Remove the throttle cable at the throttle body and routing brackets. Disconnect and plug the oil cooler hoses.
11. If equipped with 1.8L engine, disconnect the oxygen sensor connectors and remove the fuel filter and the transverse member. If equipped with 1.6L engine, remove the intake manifold support bracket.
12. Remove the starter, and the front exhaust pipe.
13. Remove the pinch bolt and separate the lower ball joint from the steering knuckle. Remove the cotter pin and nuts and separate the tie rod ends from the knuckles.
14. Disconnect the stabilizer bar from the lower control arms. Remove the brake hose and ABS sensor cable clips.
15. Support the engine using engine support tool 49 G017 5A0 or equivalent. Remove the nuts and bolt and remove the engine mount member.
16. Remove the halfshafts and intermediate shaft. Remove the left side transaxle mount.
17. On 1.8L engine, remove the lower converter housing cover. Hold the flexplate with a small prybar and remove the torque converter nuts.
18. On 1.6L engine, insert a small prybar through the converter housing service hole and hold the flexplate. Remove the cover from the oil pan side service hole and remove the torque converter nuts.
19. Remove the front transaxle mount.
20. Loosen the engine support tool to lean the engine toward the transaxle. Support the transaxle with a jack and remove the transaxle mounting bolts. Carefully lower the transaxle from the vehicle, being careful not to drop the torque converter.

To install:
21. Make sure the torque converter is completely installed in the transaxle. Lay a straightedge across the bellhousing and measure the distance to the torque converter (not the stud); it should be 0.535 in. (13.6mm). If the distance is less than specification, push the converter into

the pump while rotating it, to properly engage the pump drive.
22. Raise the transaxle into position, making sure the converter studs align with the flexplate holes, and install the transaxle-to-engine bolts. Tighten to 59 ft. lbs. (80 Nm).
23. Install the front transaxle mount and tighten the bolt(s) to 38 ft. lbs. (52 Nm).
24. Hold the flexplate with a small prybar and tighten the converter nuts, gradually and evenly to 25–36 ft. lbs. (34–49 Nm). On 1.6L engine, install the service hole cover. On 1.8L engine, install the access plate.
25. Install the left side transaxle mount. Tighten the mount-to-body bolts to 45 ft. lbs. (61 Nm) and the mount-to-transaxle nuts to 69 ft. lbs. (93 Nm).
26. Install the halfshafts.
27. Install the engine mount member. Tighten the mount member-to-mount nuts to 38 ft. lbs. (52 Nm) and the mount member-to-body nuts/bolts to 66 ft. lbs. (89 Nm). Remove the engine support tool.
28. Attach the clip the brake hose and ABS sensor cable. Attach the stabilizer links to the lower control arms and tighten the nuts to 45 ft. lbs. (61 Nm).
29. Connect the tie rod ends to the steering knuckles and tighten the nuts to 42 ft. lbs. (57 Nm). Install new cotter pins. Connect the lower arm ball joints to the steering knuckles and install the pinch bolts and nuts. Tighten to 43 ft. lbs. (59 Nm).
30. Install the front exhaust pipe, using new gaskets. Tighten the pipe-to-converter nuts to 66 ft. lbs. (89 Nm) and the pipe-to-manifold nuts to 34 ft. lbs. (46 Nm). Connect the oxygen sensor connectors on 1.8L engine.
31. On 1.6L engine, install the intake manifold support bracket. On 1.8L engine, install the transverse member and tighten the bolts to 90 ft. lbs. (123 Nm).
32. Install the starter. On 1.8L engine, install the fuel filter and connect the fuel lines.
33. Connect the oil cooler hoses and the throttle cable. Position the harness bracket and secure with the bolt.
34. Connect the solenoid valve and inhibitor switch connectors. Connect the selector cable to the transaxle. Install the cable housing clip and a new manual lever spring clip.
35. Connect the speedometer cable or speed sensor connector, as necessary. Install the battery tray and secure the wiring harness with the bolt. Install the battery.

36. On 1.8L engine, install the strut bar and tighten the nuts to 20 ft. lbs. (26 Nm).

37. Install the air cleaner assembly and resonance duct. Connect the air flow meter connector.

38. Install the splash shields and the wheel and tire assemblies. Lower the vehicle.

39. Connect the battery cables. Fill the transaxle with the proper type and quantity of fluid. Start the engine and bring to normal operating temperature. Check for leaks and proper transaxle operation.

1993–97 626 and MX6 with GFA4-EL Transaxle

1. Disconnect the battery cables and remove the battery and battery tray.
2. Raise and safely support the vehicle. Drain the transaxle fluid.
3. Remove the air ducts and the air cleaner assembly.
4. Remove the housing clip and remove the selector cable from the transaxle.
5. Label and disconnect the electrical connectors for the inhibitor switch, solenoid valve, oxygen sensor, vehicle speed sensor and vehicle pulse generator. Disconnect the ground wires and remove the necessary wiring harness brackets.
6. Remove the fuel filter mounting nuts and position the filter aside, leaving the fuel line attached. Remove the engine mount stay.
7. If equipped with 2.5L engine, remove the starter. Disconnect and plug the oil cooler hoses.
8. Remove the wheel and tire assemblies and the splash shields. Remove the transverse member.
9. If equipped with 2.5L engine, remove the front exhaust pipe.
10. Remove the pinch bolts from the steering knuckles. Pry the lower control arms down to separate the ball joints from the knuckles.
11. Remove the cotter pins and nuts and separate the tie rod ends from the knuckles. Remove the nuts and disconnect the stabilizer links from the lower control arms.
12. Remove the brake hose clips from the struts and remove the ABS speed sensor harness mounting nuts.
13. Separate the right side halfshaft from the intermediate shaft using a hammer and brass drift. Pry the left side shaft from the transaxle using a prybar inserted between the transaxle and the inner CV-joint. Be careful not to damage the oil seal.
14. Suspend the halfshafts with rope, in a level position. Insert plug tools 49 G030 455 or equivalent into the differential side gears to keep them from becoming mispositioned.

15. Remove the intermediate shaft.
16. If equipped with 2.0L engine, remove the intake manifold support bracket and remove the starter.
17. Remove the engine mount rubber from the right side of the engine mount member. Remove the mount-to-transaxle bolts from the rear transaxle mount.
18. Support the engine using engine support tool 49 G017 5A0 or equivalent. Remove the engine mount member nuts and bolts and remove the engine mount member.
19. If equipped with 2.5L engine, remove the torque converter access plate. On 2.0L engine, remove the seal rubber from the transaxle case near the starter mounting hole.
20. Hold the flexplate with a small prybar and remove the torque converter nuts.
21. Remove the left side transaxle mount.
22. Loosen the engine support tool and lean the engine toward the transaxle. Support the transaxle with a jack and remove the transaxle mounting bolts. Carefully lower the transaxle from the vehicle, being careful not to drop the torque converter.
23. Remove the front transaxle mount.

To install:

24. Make sure the torque converter is completely installed in the transaxle. Lay a straightedge across the bellhousing and measure the distance to the torque converter (not the stud); it should be 0.602 in. (15.3mm) on 2.0L engine or 0.551 in. (14mm) on 2.5L engine. If the distance is less than specification, push the converter into the pump while rotating it, to properly engage the pump drive.
25. Install the front transaxle mount.
26. Raise the transaxle into position, making sure the torque converter studs align with the flexplate holes, and install the transaxle mounting bolts.
27. On 2.5L engine, tighten the transaxle mounting bolts to 73 ft. lbs. (99 Nm). On 2.0L engine, tighten bolts **A** to 73 ft. lbs. (99 Nm), bolts **B** to 38 ft. lbs. (51 Nm) and bolt **C** to 18 ft. lbs. (25 Nm).
28. Hold the flexplate with the small prybar and install the torque converter nuts. Tighten the bolts, gradually and evenly, to 45 ft. lbs. (60 Nm). On 2.5L engine, install the tor-

que converter access plate. On 2.0L engine, install the seal rubber.

29. Install the left side transaxle mount and loosely tighten the bolts and nuts.
30. Use the engine support tool to make sure the transaxle bolt holes and the rear transaxle mount align. Install the bolts and tighten to 68 ft. lbs. (93 Nm).
31. Install the engine mount member, making sure the mount rubbers are installed properly.
32. Install the mount member-to-body bolts/nuts and tighten to 68 ft. lbs. (93 Nm). Loosely tighten the mount member-to-mount nuts.
33. Tighten the left transaxle mount mount-to-transaxle nuts and bolt to 68 ft. lbs. (93 Nm). Tighten the mount through-bolt to 86 ft. lbs. (116 Nm).
34. Remove the engine support tool. Tighten the mount member-to-mount nuts to 77 ft. lbs. (104 Nm).
35. Install the mount rubber on the right side of the mount member. Tighten the bolts to 68 ft. lbs. (93 Nm).
36. If equipped with 2.0L engine, install the starter and the intake manifold support bracket.
37. Remove the plug tool from the differential side gear. Install the intermediate shaft into the transaxle. Install the support bearing to the engine and tighten the bolts, in sequence, to 45 ft. lbs. (61 Nm).
38. Remove the plug from the other differential side gear. Install a new circlip on the halfshaft with end gap facing upward. Install the halfshaft into the transaxle, being careful not to damage the seal. Make sure the circlip seats in the differential side gear, by pulling out on the shaft; it must not pull out.
39. Install a new circlip on the end of the other halfshaft and connect it to the intermediate shaft.
40. Connect the lower ball joints to the steering knuckles. Install the pinch bolts and tighten to 43 ft. lbs. (58 Nm).
41. Connect the stabilizer links to the lower control arms and tighten the nuts to 39 ft. lbs. (53 Nm). Connect the tie rod ends to the steering knuckles and tighten the nuts to 32 ft. lbs. (44 Nm). Install new cotter pins.
42. If equipped with 2.5L engine, install the front exhaust pipe, using new gaskets and install the starter.
43. Install the transverse member and tighten the bolts to 96 ft. lbs. (131 Nm). Install the splash shields and the wheel and tire assemblies.

44. Connect the oil cooler hoses. Install the engine mount stay.

45. Install the harness brackets and the ground wires. Connect the electrical connectors.

46. Position the fuel filter and install the mounting bolts. Install the selector cable to the cable bracket and install the clip. Connect the cable to the transaxle lever.

47. Install the air cleaner assembly and ducts. Install the battery and battery tray.

48. Connect the battery cables. Fill the transaxle with the proper type and quantity of fluid. Start the engine and bring to normal operating temperature. Check for leaks and proper transaxle operation.

1995–97 626 and MX6 with LA4A-EL Transaxle

1. Disconnect the battery cables and remove the battery and battery tray.

2. Remove the resonance chamber, fresh air duct and the air cleaner assembly.

3. Label and disconnect the Manual Lever Position (MLP) sensor electrical connector. Remove the two MLP sensor bolts and the sensor.

4. Remove the ground wire bracket and the ground wire. Label and disconnect the solenoid body connector.

5. Disconnect the transaxle electrical connector.

6. Disconnect the Vehicle Speed Sensor (VSS) connector.

7. Remove the oil filler tube.

8. Remove the two ignition coil nuts and position the ignition coil out of the way. Remove the three speed control servo nuts and position the speed control servo out of the way.

9. Remove the three ignition coil mounting strap bolts.

10. Disconnect the wire harness clips from the ignition coil mounting straps. Remove the ignition coil mounting straps.

11. Remove the 2 fuel filter bracket nuts from the left transaxle mount. Position the filter and bracket, aside, without disconnecting the fuel lines.

12. Remove the two emission harness protector nuts.

13. Support the engine. Remove the number one and number four engine mounts.

14. Remove the shift cable from the cable bracket. Remove the two cable bracket bolts and remove the bracket.

15. Raise and safely support the vehicle. Remove the front wheel and tire assemblies and the splash shields.

16. Disconnect the Transmission Speed Sensor (TSS) connector.

17. Remove the 6 transverse member bolts and the transverse member. Remove the 6 transaxle cradle nuts and 2 bolts and remove the transaxle cradle.

18. Remove the engine mounting member (transaxle cradle).

19. Remove the pinch bolts from the steering knuckles. Pry the lower control arms down to separate the ball joints from the knuckles.

20. Remove the cotter pins and nuts and separate the tie rod ends from the knuckles. Remove the nuts and disconnect the stabilizer links from the lower control arms.

21. Remove the brake hose clips from the struts, and (if equipped) remove the ABS speed sensor harness mounting nuts.

22. Separate the right side halfshaft from the intermediate shaft using a hammer and brass drift. Pry the left side shaft from the transaxle using a prybar inserted between the transaxle and the inner CV-joint. Be careful not to damage the oil seal.

23. Suspend the halfshafts with rope, in a level position. Insert plug tools 49 G030 455 or equivalent into the differential side gears to keep them from becoming mispositioned.

24. Remove the intermediate/joint shaft.

25. Remove the two starter motor bolts.

26. Remove the intake manifold support and the starter.

27. Remove the seal rubber located next to the starter opening. Use a small prybar to hold the flexplate and reach through the opening to remove the torque converter nuts.

28. Safely support the transaxle from below. Remove the four front transaxle mount bracket bolts, and remove the bracket.

29. Remove the number two engine mount.

30. Disconnect and plug the oil cooler lines.

31. Remove the top three transaxle-to-engine mounting bolts.

32. Remove the two remaining transaxle-to-engine bolts.

33. Remove the three engine-to-transaxle bolts.

34. Use a small prybar to separate the transaxle from the engine. Slightly tilt the transaxle and engine to ease removal.

35. Remove the transaxle from the engine and lower the transaxle from the vehicle.

To install:

36. Raise the transaxle into position. Align the torque converter studs with the flexplate.

37. Install the transaxle-to-engine and engine-to-transaxle bolts. Tighten bolts **B** to 28–38 ft. lbs. (38–51 Nm). Tighten the transaxle-to-engine bolts **C** to 14–18 ft. lbs. (19–25 Nm).

38. Remove the transaxle support.

39. Install transaxle-to-engine bolts **A** and tighten to 66–86 ft. lbs. (89–117 Nm).

40. Connect the oil cooler hoses.

41. Install the number two engine mount. Install the front transaxle support bracket.

42. Install the torque converter-to-flexplate nuts and tighten to 45 ft. lbs. (60 Nm).

43. Install the torque converter access plug.

44. Install the starter. Do not install the two upper starter motor bolts at this time

45. Install the intake manifold support bracket. Tighten the bolts to 38 ft. lbs. (52 Nm).

46. Install the two upper starter motor bolts and tighten to 23–34 ft. lbs. (31–46 Nm).

47. Remove the plug tool from the differential side gear. Install the intermediate/joint shaft into the transaxle. Install the support bearing to the engine and tighten the bolts, in sequence, to 45 ft. lbs. (61 Nm).

48. Remove the plug from the other differential side gear. Install a new circlip on the halfshaft with end gap facing upward. Install the halfshaft into the transaxle, being careful not to damage the seal. Make sure the circlip seats in the differential side gear, by pulling out on the shaft; it must not pull out.

49. Install a new circlip on the end of the other halfshaft and connect it to the intermediate shaft.

50. Connect the lower ball joints to the steering knuckles. Install the pinch bolts and tighten to 43 ft. lbs. (58 Nm).

51. Connect the stabilizer links to the lower control arms and tighten the nuts to 39 ft. lbs. (53 Nm). Connect the tie rod ends to the steering knuckles and tighten the nuts to 32 ft. lbs. (44 Nm). Install new cotter pins.

52. Install the transaxle cradle. Tighten the cradle-to-body bolts and nuts to 68 ft. lbs. (93 Nm). Tighten the cradle-to-front mount nuts to 77 ft. lbs. (104 Nm) and the cradle-to-rear mount nuts to 44 ft. lbs. (60 Nm).

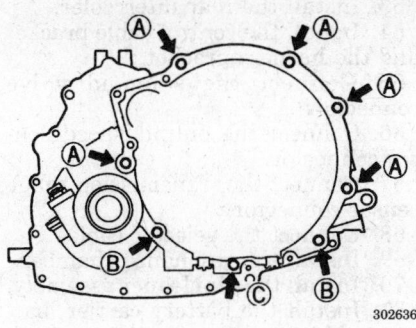

Engine-to-transaxle mounting bolts — 1995–97 626 and MX6 with LA4A-EL transmission

302636

53. Install the transverse member and tighten the bolts to 96 ft. lbs. (131 Nm).

54. Connect the TSS and VSS connectors.

55. Install the splash shields and the wheel and tire assemblies. Lower the vehicle.

56. Install the shift cable bracket.

57. Install the number one and number four engine mounts. Remove the engine support.

58. Install the two emission harness protector nuts.

59. Install the two fuel filter bracket nuts and tighten to 71–97 inch lbs. (8–11 Nm).

60. Position the ignition coil mounting straps, and install the wire harness clips and the three strap bolts.

61. Position the speed control servo, and install and tighten the three nuts.

62. Position the ignition coil and install the two nuts. Tighten the nuts to 71–88 inch lbs (8–10 Nm).

63. Install the oil filler tube.

64. Connect the transaxle electrical connector.

65. Install the MLP sensor and the two MLP sensor bolts. Tighten the bolts to 7 ft. lbs. (10 Nm). Connect the MLP sensor electrical connector.

66. Install the air cleaner assembly, fresh air duct and resonance chamber. Install the battery tray and battery. Connect the battery cables.

67. Install the ground wire bracket and the ground wire. Connect the solenoid body connector.

68. Fill the transaxle with the proper type and quantity of fluid. Run the engine and check for leaks. Road test and check for proper transaxle operation.

1995–97 Millenia with GFA4 Transaxle

1. Remove the battery, battery cover and the carrier.

2. Raise and safely support the vehicle.

3. Drain the transaxle fluid.

4. Remove the air cleaner assembly.

5. Disconnect the selector cable at the transaxle.

6. Disconnect the transaxle range sensor connector and the input/turbine speed sensor connector.

7. Disconnect the solenoid valve connector.

8. Remove the harness bracket and the selector cable bracket.

9. Remove the No. 1 engine mount bracket stay.

10. Remove the starter.

11. Disconnect the speedometer sensor connector.

12. Remove the intake manifold cover.

13. Disconnect the O2 sensor connector.

14. Remove the wheels.

15. Remove the splash shields and the undercover.

16. Remove the front exhaust pipe.

17. Disconnect the upper lateral link (left side).

18. Remove the lower ball joint.

19. Remove the right side halfshaft and the joint shaft.

20. Support the engine from above and remove the No. 4 engine mount.

21. Remove the No. 1 and No. 2 engine mount nuts.

22. Remove the engine mounting member.

23. Disconnect the oil hose.

24. Remove the undercover and the transaxle to engine undercover.

25. Hold the driveplate and remove the torque converter nuts.

26. Remove the No. 2 engine mount and the bracket.

27. Remove the left side upper link and separate the left side halfshaft from the transaxle.

28. Remove the No. 1 engine mount bolts.

29. Loosen the engine support and lean the engine towards the transaxle.

30. Support the transaxle on a jack. Remove the transaxle mounting bolts and remove the transaxle.

To install:

31. Support the transaxle on a jack and place into position at the engine. Install the mounting bolts, torque to 50–73 ft. lbs. (68–99 Nm).

32. Install the torque converter bolts, torque to 28–44 ft. lbs. (38–60 Nm).

33. Loosely tighten the No. 4 engine mount nuts.

34. Use the engine support to make sure that the transaxle bolt holes and the No. 1 engine mount are aligned.

— **WARNING** —

Align the transaxle bolt holes and the engine mount exactly. Any misalignment could result in bolt and bolt holes becoming damaged or stripped during installation.

35. Torque the No. 1 engine mount bolts to 55–77.3 ft. lbs. (75–104 Nm).

36. Install the left side halfshaft.

37. Install the No. 2 engine mount bracket and engine mount.

38. Install the undercover and the engine to trans. undercover.

39. Connect the oil hose.

40. Install the engine mounting member, making sure that the No. 2 engine mount stud bolt passes through the No. 2 engine mount bracket installation hole.

41. Torque the mounting member outer bolts to 50–68 ft. lbs. (67–93 Nm), torque the engine mount to member bolt to 55–77.3 ft. lbs. (75–104 Nm).

42. Install the No. 1 engine mount nut.

43. Install the No. 4 engine mount.

44. Install the joint shaft and the right side halfshaft.

45. Install the lower ball joint.

46. Install the left side upper lateral link.

47. Install the front exhaust pipe.

48. Install the undercover and splash shields.

49. Install the wheels.

50. Connect the O2 sensor connector.

51. Install the intake manifold cover.

52. Connect the speedometer sensor connector.

53. Install the starter.

54. Install the No. 1 engine mount bracket stay.

55. Install the selector cable bracket and the harness bracket.

56. Connect the solenoid valve connector.

57. Connect the input/turbine speed sensor connector.

58. Connect the transmission range sensor connector.

59. Connect the selector cable.

60. Install the air cleaner assembly.

61. Install the battery carrier, battery and battery cover.

62. Fill the transaxle with the proper fluid.

1995–97 Millenia with a LJA4 Transaxle

1. Remove the battery, battery cover and the carrier.

2. Raise and safely support the vehicle.

3. Drain the transaxle fluid.

4. Remove the air cleaner assembly.

5. Remove the resonance chamber.

6. Disconnect the selector cable at the transaxle.

7. Disconnect the transaxle range sensor connector and the output speed sensor connector.

8. Disconnect the solenoid valve connector.

9. Remove the harness bracket and the control cable bracket.

10. Remove the rear intercooler.

11. Remove the bracket and remove the starter.

12. Disconnect the speed sensor connector.

13. Remove the front intercooler.

14. Remove the electric coolant fan assembly.

15. Remove the wheels.

16. Remove the locknut.

17. Remove the splash shields and the undercover.

18. Remove the fresh air duct.

19. Remove the front exhaust pipe.

20. Disconnect the lower ball joint.

21. Disconnect the upper lateral link (left side).

22. Remove the lower arm.

23. Remove the right side halfshaft and the joint shaft.

24. Remove the stabilizer control link.

25. Remove the left side halfshaft.

26. Remove the drive belts and timing belt.

27. Remove the power steering pump and the A/C compressor.

28. Remove the shift control rod and the ATF filler tube.

29. Support the engine from above, loosen the No. 3 engine mount bolt and remove the No. 4 engine mount.

30. Remove the No. 1 engine mount nut an the engine mount damper. Remove the No. 2 engine mount nut.

31. Remove the undercover.

32. Hold the driveplate and remove the torque converter nuts.

33. Remove the No. 1 engine mount bolts.

34. Remove the engine mount member.

35. Loosen the engine support and lean the engine towards the transaxle.

36. Support the transaxle on a jack. Remove the transaxle mounting bolts and remove the transaxle.

To install:

37. Support the transaxle on a jack and place into position at the engine. Install the mounting bolts, torque to 50–73 ft. lbs. (68–99 Nm).

38. Install the torque converter bolts, torque to 28–38 ft. lbs. (38–51 Nm).

39. Install the No. 4 engine mount bracket.

40. Install the engine mounting member making sure that the No. 2 engine mount stud bolt passes through the No. 2 engine mount bracket installation hole.

41. Torque the mounting member outer bolts to 50–68 ft. lbs. (67–93 Nm), torque the engine mount to member bolt to 55–77.3 ft. lbs. (75–104 Nm).

42. Tighten the No. 3 engine mount bolt, torque to 55–77 ft. lbs. (75–104 Nm).

43. Use the engine support to make sure that the transaxle bolt holes and the No. 1 engine mount are aligned.

—— WARNING ——
Align the transaxle bolt holes and the engine mount exactly. Any misalignment could result in the bolt and bolt holes becoming damaged or stripped during installation.

44. Torque the No. 1 engine mount bolts to 55–77.3 ft. lbs. (75–104 Nm).

45. Install the No. 4 engine mount onto the vehicle, torque to 32–44 ft. lbs. (44–60 Nm).

46. Use a jack to make sure that the No. 4 mounting bolt holes and the bracket are aligned, then torque the bolts to 50–68 ft. lbs. (67–93 Nm). Remove the jack.

47. Shift the selector lever to the PARK position and move the manual shaft to the PARK position. Install the shift control rod and torque the mounting nut to 12–16 ft. lbs. (16–22 Nm).

48. Install the ATF filler tube.

49. Install the A/C compressor and the power steering pump.

50. Install the timing belt and the drive belts.

51. Install the left side halfshaft.

52. Install the stabilizer control link, joint shaft and the right side halfshaft.

53. Install the lower arm, upper lateral link and the ball joint.

54. Install the front exhaust pipe.

55. Install the fresh air duct.

56. Install the undercover and the splash shields.

57. Install the locknut.

58. Install the wheels.

59. Install the electric coolant fan assembly.

60. Install the front intercooler.

61. Connect the speed sensor connector.

62. Install the starter and bracket.

63. Install the rear intercooler.

64. Install the control cable bracket and the harness bracket.

65. Connect the solenoid valve connector.

66. Connect the output speed sensor connector.

67. Connect the transmission range sensor connector.

68. Connect the selector cable.

69. Install the resonance chamber.

70. Install the air cleaner assembly.

71. Install the battery carrier, battery and battery cover.

72. Fill the transaxle with the proper fluid.

DRIVE AXLE

Driveshaft

REMOVAL AND INSTALLATION

1993–95 929 and 1993–97 Miata

1. Raise and safely support the vehicle.

2. On the Miata, remove the exhaust pipe and muffler assembly.

3. Mark the position of the driveshaft flange on the rear axle flange for assembly reference.

4. Remove the driveshaft-to-flange bolts and nuts, and remove the driveshaft. Slide the driveshaft yoke from the transmission, being careful not to damage the oil seal. Install a suitable plug in the transmission to prevent fluid leakage.

To install:

5. Remove the plug from the transmission rear housing. Slide the driveshaft yoke into the transmission, being careful not to bottom the yoke against the oil seal.

6. Install the driveshaft on the axle flange, aligning the marks that were made during removal. Install the bolts and nuts. Tighten the nuts and bolts to 37–43 ft. lbs. (50–58 Nm).

7. On the Miata, install the exhaust pipe/muffler assembly with a new gasket. Tighten the pipe nuts to 28–38 ft. lbs. (38–51 Nm).

8. Lower the vehicle.

1993–95 RX7

1. Raise and safely support the vehicle.

2. Remove the right and left undercovers, the center and rear tunnel reinforcements and the catalytic converter.

3. Disconnect the secondary air injection pipe and remove the catalytic converter assembly.

4. Remove the driveshaft cover.

5. Matchmark the driveshaft and companion flange. Remove the mounting nuts and bolts and remove the driveshaft. Cap the transmission to prevent leakage of fluid. If there is excessive play in the U-joints, the driveshaft must be replaced with a new one.

6. Installation is the reverse of removal. Torque the driveshaft mounting nuts and bolts to 37–43 ft. lbs. (50–58 Nm). Torque the tunnel reinforcement bracket mounting bolts to 14–19 ft. lbs. (18–26 Nm). Torque the driveshaft and undercover mounting bolts to 70–95 in. lbs. (7.9–10.7 Nm).

U-Joints

REMOVAL AND INSTALLATION

1993–95 929 and 1993–97 Miata

1. Raise and safely support the vehicle.

2. Remove the driveshaft assembly.

3. Matchmark both the yoke and the driveshaft so that they can be re-turned to their original balancing position during assembly.

4. Remove the bearing snaprings from the yoke.

5. Use a hammer and a brass drift to drive in one of the bearing cups. Remove the cup which is protruding from the other side of the yoke.

6. Remove the other bearing cups by pressing them from the spider.

7. Withdraw the spider from the yoke.

8. Examine the spider journals for rusting or wear. Check the bearing for smoothness or pitting.

NOTE: The spider and bearing are replaced as a complete assembly only.

9. Check the seals and rollers for wear or damage.

To install:

10. Pack the bearing cups with grease.

11. Fit the rollers into the cups and install the dust seals.

12. Place the spider in the yoke and then fit one of the bearing cups into its bore in the yoke.

13. Press the bearing cup home, while guiding the spider into it, so that a snapring can be installed.

14. Press-fit the other bearings into the yoke.

15. Select a snapring to obtain minimum end-play of the spider. Use snaprings of the same thickness on both sides to center the spider.

NOTE: When assembled, the U-joint should have a slight drag but should not bind. If it does bind, use different thickness snaprings. Selective fit snaprings are available in sizes ranging from 1.2–1.4mm

16. Install the spider/yoke assembly and bearings into the driveshaft in the same manner as the spider was assembled to the yoke.

17. Test the operation of the U-joint assembly. The spider should move freely with no binding.

18. Install the driveshaft assembly observing any matchmarks made during removal.

19. Lower the vehicle.

Halfshaft

REMOVAL AND INSTALLATION

1993–94 323, 929, Miata and 1993–95 MX3 and 1993–97 Protege, 626, MX6

1. Raise and safely support the vehicle. Remove the wheel and tire assemblies.

2. Remove the splash shield, and drain the transaxle.

3. Raise the staked portion of the hub locknut with a hammer and chisel. Lock the hub by applying the brakes and remove the nut.

4. Disconnect the stabilizer bar and control link from the lower control arm.

5. Remove the cotter pin and nut from the tie rod end ball stud. Use a suitable tool to separate the tie rod end from the knuckle.

6. Remove the transverse member.

7. Remove the lower ball joint pinch bolt and nut. Use a prybar to pry down the lower control arm and separate the ball joint from the knuckle.

8. If removing the left side shaft with automatic transaxle:

   a. Suspend the engine using engine support tool 49 G017 5A0 or equivalent.

   b. Remove the bolts and nuts and remove the engine mount member.

9. Position a prybar between the inner CV-joint and transaxle case. Carefully pry the halfshaft from the transaxle, being careful not damage the oil seal. If equipped with a right side intermediate shaft, insert the

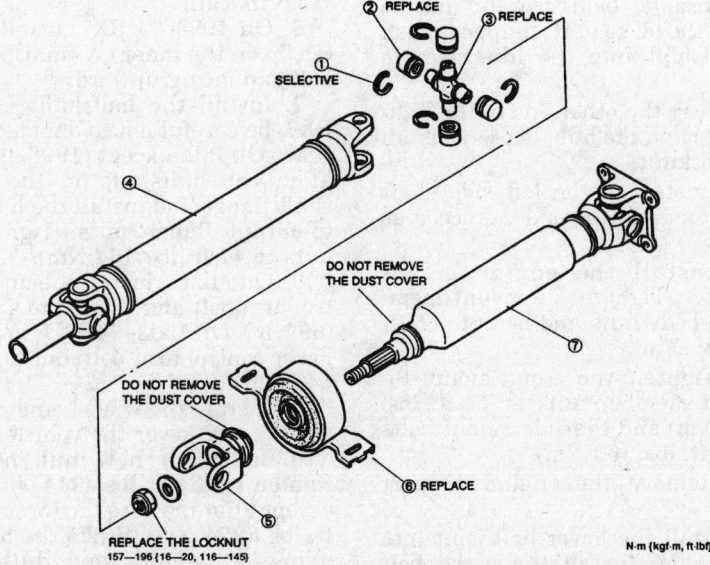

REPLACE THE LOCKNUT
157—196 {16—20, 116—145}

N·m {kgf·m, ft·lbf}

1. Snap ring
2. Bearing cup
3. Spider
4. Front propeller shaft
   Inspect for damage, wear and rough rotation
5. Center yoke
6. Center bearing support assembly
   Inspect for damage and rough rotation
7. Rear propeller shaft

232350

U-joint (exploded view) — 1993–95 929

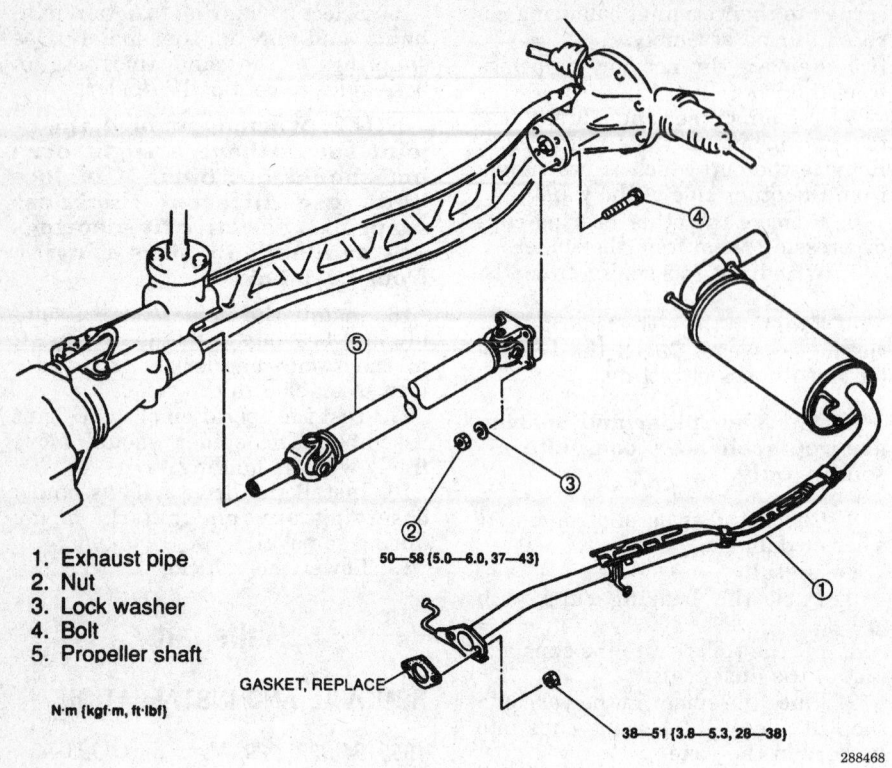

1. Exhaust pipe
2. Nut
3. Lock washer
4. Bolt
5. Propeller shaft

N·m {kgf·m, ft·lbf}

50—58 (5.0—6.0, 37—43)

GASKET, REPLACE

38—51 (3.8—5.3, 28—38)

288468

Driveshaft assembly — 1993-97 Miata

prybar between the halfshaft and intermediate shaft and tap on the bar to uncouple them.

10. Pull outward on the hub/knuckle assembly, push the outer CV-joint stub shaft through the hub, and remove the halfshaft. If the halfshaft is stuck in the hub, install the old hub nut to protect the stub shaft threads. Tap on the nut, with only a soft mallet, and remove the halfshaft.

NOTE: Install plug tool 49 G030 455 or equivalent, into the transaxle after removing the halfshaft, to keep the differential side gear in position. If the gear becomes mispositioned, the differential may have to be removed to realign the gear.

11. If necessary, remove the intermediate shaft by removing the support bearing bolts and pulling the shaft from the transaxle.
To install:
12. If removed, install a new circlip on the end of the intermediate shaft, with the end gap facing upward.
13. Install the intermediate shaft in the transaxle, being careful not to damage the oil seals. Install the support bearing bolts and tighten, in sequence, to 45 ft. lbs. (61 Nm).

14. Install a new circlip on the end of the halfshaft, with the end gap facing upward. Insert the halfshaft into the transaxle, being careful not to damage the oil seal. If equipped, push the halfshaft into the intermediate shaft.

15. Insert the other end of the halfshaft through the hub. Loosely install a new locknut.

16. If installing the left side shaft with automatic transaxle, proceed as follows:

   a. Install the engine mount member. Tighten the mount member-to-body nuts and bolts to 66 ft. lbs. (89 Nm).

   b. Tighten the front mount-to-mount member nuts to 77 ft. lbs. (104 Nm) and the side mount bolts to 44 ft. lbs. (60 Nm).

   c. Remove the engine support tool.

17. Install the lower ball joint into the knuckle. Install the pinch bolt and nut and tighten to 26—41 ft. lbs. (35—56 Nm).

18. Install the transverse member, and tighten the bolts to 69—96.9 ft. lbs. (94—132 Nm).

19. Connect the tie rod end to the steering knuckle and tighten the nut to 24—32 ft. lbs. (32—44 Nm). Install a new cotter pin. Tighten the nut, if

necessary, to align the ball stud hole with the nut castellation.

20. Connect the stabilizer bar to the lower control arm.

21. Install the splash shield and the wheel and tire assemblies. Lower the vehicle.

22. Lock the hub with the brakes. Tighten the new hub nut to 173.6—235 ft. lbs. (236—318 Nm). After torquing, stake the locknut using a hammer and dull bladed chisel.

23. Fill the transaxle with the proper type and quantity of fluid.

RX7

1. Raise and safely support the vehicle. Remove the wheel and tire assembly.

2. Unstake and remove the wheel hub nut.

3. On all except 1993—94 RX7, mark the position of the halfshaft on the output flange and remove the nuts.

4. On Miata, remove the upper control arm bolt. On 1993—94 RX7, remove the lower control arm bolt.

5. Remove the halfshaft.

NOTE: If the halfshaft is stuck in the hub, install the used hub nut until it is flush with the end of the shaft. Tap on the nut, using only a soft mallet, to remove the halfshaft

To install:
6. On 1993—94 RX7, install a new circlip on the inner CV-joint, with the end gap facing upward.

7. Install the halfshaft. On 1993 RX7, be careful not to damage the oil seal. On all except 1993—94 RX7, align the halfshaft on the output shaft flange and install the halfshaft-to-output flange nuts. Tighten the nuts to 47 ft. lbs. (64 Nm).

8. On Miata, install the upper control arm bolt and tighten to 49 ft. lbs. (69 Nm). On 1993—94 RX7, install the lower control arm bolt and tighten to 54 ft. lbs. (73 Nm).

9. Install the wheel and tire assembly and lower the vehicle.

10. Install a new hub nut and tighten to 231 ft. lbs. (314 Nm) on all except Miata, where the torque is 217 ft. lbs. (294 Nm). Stake the hub nut, using a hammer and dull-bladed chisel.

1993-95 929

1. Raise and safely support the vehicle. Remove the wheel and tire assemblies.

2. Remove the exhaust pipe.

3. Remove the rear lower lateral link.

4. Matchmark and remove the driveshaft.

5. Remove the upper and lower lateral links.

6. Remove the stabilizer control link.

7. Raise the staked portion of the hub locknut with a hammer and chisel. Lock the hub by applying the brakes and remove the nut.

8. Matchmark the halfshaft and output shaft for proper alignment before removing the halfshaft.

9. Support the subframe and differential with a jack. Remove the subframe mounting nuts and washers and lower the subframe and differential approximately 4 inches (100mm).

10. Pull outward on the hub/knuckle assembly, push the outer CV-joint stub shaft through the hub, and remove the halfshaft. If the halfshaft is stuck in the hub, install the old hub nut to protect the stub shaft threads. Tap on the nut, with only a soft mallet, and remove the halfshaft.

To install:

11. Position the halfshaft with the matchmarks aligned, and install the mounting bolts. Torque the bolts to 40–47 ft. lbs. (54–63 Nm).

12. Insert the other end of the halfshaft through the hub. Loosely install a new locknut.

13. Raise the subframe and differential and install the washers and mounting nuts.

14. Connect the stabilizer control link, and tighten the nut to 32–39 ft. lbs. (44–53 Nm).

15. Connect the upper and lower lateral links. Tighten the bolts to 69–86 ft. lbs. (94–117 Nm).

16. Install the driveshaft on the axle flange, aligning the marks that were made during removal. Tighten the bolts to 37–43 ft. lbs. (50–58 Nm).

17. Connect the rear lower lateral link. Tighten the mounting nut to 37–47 ft. lbs. (50–64 Nm), and install a new cotter pin.

18. Install the exhaust pipe with new gaskets and locknuts. Tighten the locknuts to 28–38 ft. lbs. (38–51 Nm).

19. Install the wheel and tire assemblies. Lower the vehicle.

20. Lock the hub with the brakes. Tighten the new hub nut to 235 ft. lbs. (318 Nm). After torquing, stake the locknut using a hammer and dull bladed chisel.

1995–97 Millenia

Halfshaft

1. Raise and safely support the vehicle.

2. Drain the transaxle fluid.

3. Remove the stabilizer control link upper nut.

4. Remove the halfshaft locknut.

5. Remove the lower arm ball joint bolt.

6. Disconnect the upper lateral link ball joint. Suspend the axle assembly to prevent damage to the link bushing.

7. Disconnect the tie-rod end.

8. LEFT SIDE: separate the halfshaft from the transaxle by prying with a bar inserted between the outer ring and the transaxle. RIGHT SIDE: separate the halfshaft from the joint shaft by hammering on a brass or copper bar inserted between them.

9. Remove the clip from the halfshaft.

To install:

10. Install a NEW clip on the halfshaft so that the opening is facing upwards.

11. Apply grease to the ends of the halfshaft. Push the halfshaft into the transaxle/joint shaft. After installation, pull the front hub outward to confirm that the halfshaft is securely held by the clip.

12. Connect the tie-rod end and torque the nut to 41–59 ft. lbs. (50–80 Nm).

13. Connect the upper lateral link and torque the nut to 41–59 ft. lbs. (50–80 Nm).

14. Install the lower arm ball joint bolt. Torque to 58–86 ft. lbs. (79–116 Nm).

15. Install a NEW halfshaft locknut. Torque to 174–235 ft. lbs. (236–318 Nm).

16. Install the stabilizer control link upper nut and torque to 32–44 ft. lbs. (44–60 Nm).

17. Fill the transaxle with the proper fluid.

18. Lower the vehicle.

Joint shaft

1. Raise and safely support the vehicle.

2. Drain the transaxle fluid.

3. Disconnect the tie-rod end.

4. Remove the lower arm ball joint bolt.

5. Disconnect the upper lateral link ball joint. Suspend the axle assembly to prevent damage to the link bushing.

6. Separate the right side halfshaft from the joint shaft by hammering on a brass or copper bar inserted between them.

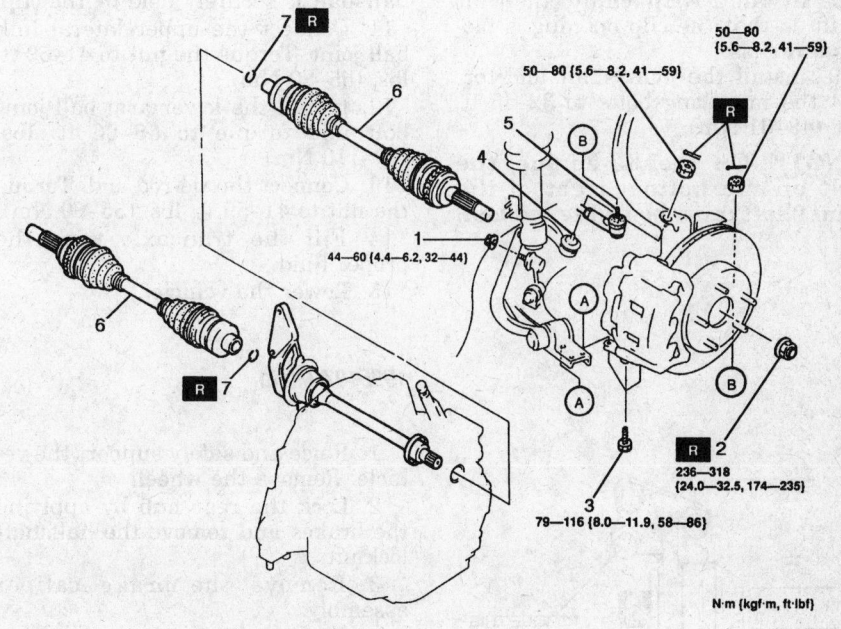

1. Stabilizer control link upper nut
2. Locknut
3. Lower arm ball joint bolt
4. Upper lateral link ball joint
5. Tie rod end ball joint
6. Drive shaft
7. Clip

N·m {kgf·m, ft·lbf}

305887

Halfshaft assembly and related components, exploded view — 1995 Millenia

---

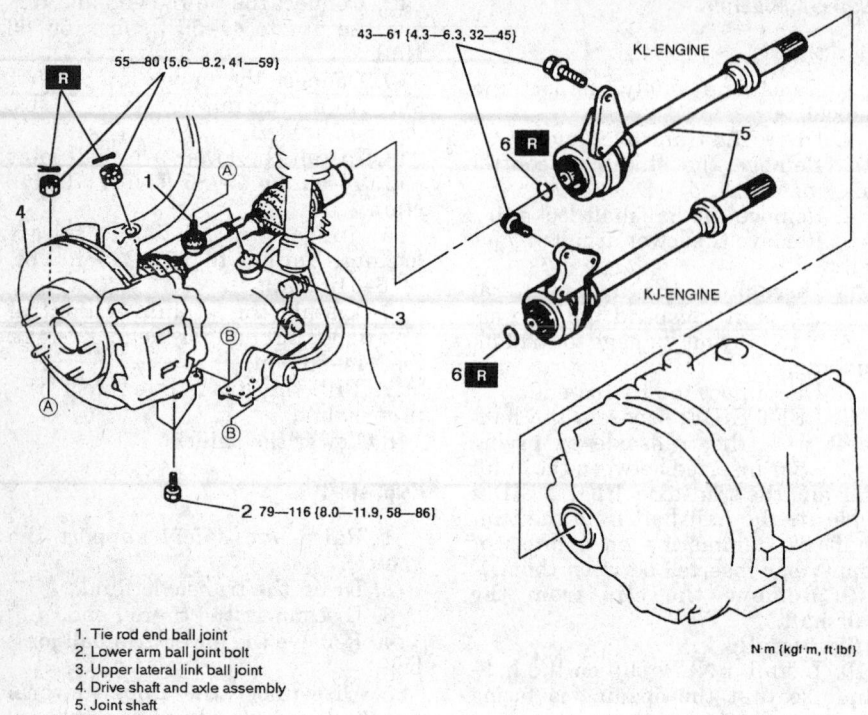

1. Tie rod end ball joint
2. Lower arm ball joint bolt
3. Upper lateral link ball joint
4. Drive shaft and axle assembly
5. Joint shaft
6. Clip

305889

Joint shaft assembly and related components, exploded view — 1995 Millenia

7. Remove the joint shaft. Remove the clip from the joint shaft.

To install:

8. Install a NEW clip to the joint shaft so that the clip opening is facing upwards.

9. Install the joint shaft and torque the mounting bolts to 32–45 ft. lbs. (43–61 Nm).

NOTE: On the KJ-engine, the bolt on the bottom right of the joint shaft must be tightened last.

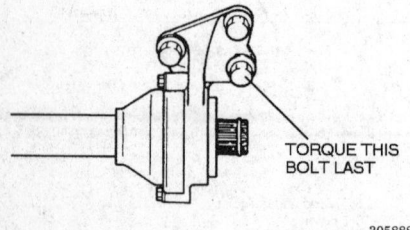

TORQUE THIS BOLT LAST

305888

Joint shaft mount — 1995–97 Millenia

10. Push the halfshaft onto the joint shaft. After installation, pull the front hub outward to confirm that the halfshaft is securely held by the clip.

11. Connect the upper lateral link ball joint. Torque the nut to 41–59 ft. lbs. (55–80 Nm).

12. Install the lower arm ball joint bolt and torque to 58–86 ft. lbs. (79–116 Nm).

13. Connect the tie-rod end. Torque the nut to 41–59 ft. lbs. (55–80 Nm).

14. Fill the transaxle with the proper fluid.

15. Lower the vehicle.

1995–97 Miata

1. Raise and safely support the vehicle. Remove the wheel.

2. Lock the rear hub by applying the brakes and remove the halfshaft locknut.

3. Remove the brake caliper assembly.

4. Remove the lower arm to knuckle bolt.

5. Matchmark the halfshaft and output shaft for proper alignment before removing the halfshaft.

6. Pull outward on the hub/knuckle assembly, push the

outer CV-joint stub shaft through the hub, and remove the halfshaft. If the halfshaft is stuck in the hub, install the old hub nut to protect the stub shaft threads. Tap on the nut, with only a soft mallet, and remove the halfshaft. Remove the halfshaft from the differential by using a prybar.

To install:

7. Install a NEW clip onto the halfshaft. With the ends of the clip facing upward install the halfshaft into the differential with the matchmarks aligned. Verify that the halfshaft is being held in place by the clip.

8. Insert the other end of the halfshaft through the hub. Join the hub/knuckle assembly to the lower arm and install the bolt and nut. Torque the bolt/nut to 40–47 ft. lbs. (54–63 Nm) on 1995 models and 47–54 ft. lbs. (63–74 Nm) on 1996–97 models.

9. Install the brake caliper assembly.

10. Lock the rear hub, install a NEW locknut and torque to 235 ft. lbs. 318 Nm) on 1995 models and 160–216 ft. lbs. (216–294 Nm) on 1996–97 models. Stake the nut after proper torque.

11. Install the wheel. Lower the vehicle.

CV-Joint Boot

REPLACEMENT

1993–94 323, Protege, MX3 and 1993–95 929 and 1993–97 MX6, 626 and Miata

Three types of CV-joints are used. The inboard CV-joints are either double offset type or the tripot type; disassembly procedures differ accordingly. All outboard CV-joints are the Birfield type. The Birfield type CV-joint cannot be disassembled; if an outboard CV-joint boot needs replacement, the inboard CV-joint must be removed.

Double Offset CV-Joint

1. Remove the halfshaft from the vehicle and clamp it in a vise equipped with jaw caps, to prevent damage to the machined surfaces. Do not allow the vise to contact the boot or its clamps.

2. Remove the large boot clamp from the inboard CV-joint, using side

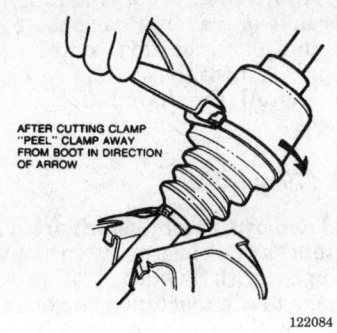

CV-joint boot removal, double offset — 1993-94 323, 626, 929, MX3, MX6, Miata and Protege

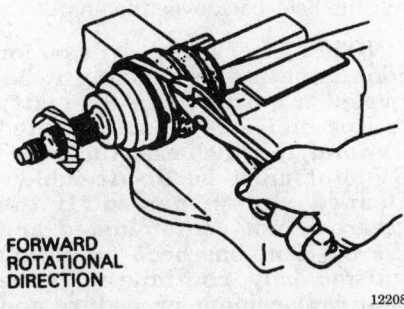

FORWARD ROTATIONAL DIRECTION

CV-joint boot installation, double offset — 1993-94 323, 626, 929, MX3, MX6, Miata and Protege

cutters. After removing the clamp, roll the boot back over the shaft.

NOTE: Check the grease for contamination by rubbing it between 2 fingers. Any gritty feeling indicates a contaminated CV-joint, in which case the entire CV-joint must be disassembled, cleaned and inspected. If the grease is not contaminated and the CV-joint has been operating satisfactorily, continue with the boot replacement procedure and add the required lubricant.

3. Paint alignment marks on the outer race and shaft for assembly reference. Remove the wire ring bearing retainer and remove the outer race.

4. Paint alignment marks on the inner race and shaft for assembly reference. Remove the inner race snapring from the end of the halfshaft and remove the inner race, cage and ball bearings from the shaft as an assembly.

NOTE: Use care to prevent damage to the bearing surfaces and cage.

5. If only the boot is being replaced, go to the next step. If it is

necessary to disassemble the CV-joint further, proceed as follows:

a. Pry the ball bearing out of the bearing cage using a small prybar with blunted edges. Mark the inner race and the bearing cage for proper assembly.

b. Rotate the inner race to align the bearing lands with the windows in the bearing cage. Remove the inner race through the larger end of the cage.

6. Remove the small clamp and remove the inner boot from the halfshaft. If the boot is to be reused, wrap the shaft splines with tape before removing.

7. If the outer CV-joint boot is to be replaced, remove the clamps and slide the boot off the shaft from the inboard side.

To install:

8. If the outboard boot was removed, slide the boot onto the shaft from the inboard side. Wrap tape on the splines before installing to protect the boot.

9. Install the inboard boot and remove the tape from the shaft.

10. Lubricate the inner race, bearing cage and ball bearings with high temperature CV-joint grease.

11. Position the inner race in the bearing cage and align the matchmarks.

NOTE: Install the race with the chamfered splines facing the large end of the cage.

12. Install the ball bearings in the bearing cage. The balls can be pressed into the cage windows with the heel of the hand.

13. Install the inner race, cage and balls on the halfshaft as an assembly. Make sure the chamfer on the bearing cage faces the snapring and the paint marks made during removal line up. Install the inner race snapring.

14. Lubricate the outer race with high temperature CV-joint grease. Install the outer race and add more grease to the outer race. Install the wire ring bearing retainer.

15. Position the CV-joint boot(s). Make sure the boot is fully seated in the grooves in the shaft and outer race.

16. Insert a small prybar with rounded edges between the boot and the outer bearing race to allow trapped air to escape from the boot. Install new boot clamps.

17. Wrap the clamps around the boots in a clockwise direction, pull tight with pliers and bend the locking tabs to secure in position.

18. Work the CV-joint through its full range of travel at various angles. The joint should flex, extend and compress smoothly.

19. Install the halfshaft into the vehicle.

Tripot CV-Joint

1. Remove the halfshaft from the vehicle and clamp it in a vise equipped with jaw caps, to prevent damage to the machined surfaces. Do not allow the vise to contact the boot or its clamps.

2. Remove the large boot clamp from the inboard CV-joint, using side cutters. After removing the clamp, roll the boot back over the shaft.

NOTE: Check the grease for contamination by rubbing it between 2 fingers. Any gritty feeling indicates a contaminated CV-joint, in which case the entire CV-joint must be disassembled, cleaned and inspected. If the grease is not contaminated and the CV-joint has been operating satisfactorily, continue with the boot replacement procedure and add the required lubricant.

3. Paint alignment marks on the outer race and shaft for assembly reference. Remove the wire ring bearing retainer and remove the outer race.

4. Paint alignment marks on the tripot bearing and shaft for assembly reference. Remove the tripot bearing snapring and, using a brass drift and hammer, remove the tripot bearing from the shaft.

5. Remove the small clamp and remove the inner boot from the halfshaft. If the boot is to be reused, wrap the shaft splines with tape before removing.

6. If the outer CV-joint boot is to be replaced, remove the clamps and slide the boot off the shaft from the inboard side.

To install:

7. If the outboard boot was removed, slide the boot onto the shaft from the inboard side. Wrap tape on the splines before installing to protect the boot.

8. Install the inboard boot and remove the tape from the shaft.

9. Install the tripot assembly on the halfshaft. Tap the assembly onto the shaft using a hammer and brass drift. Install the tripot assembly retaining ring.

10. Fill the CV-joint outer race with high temperature CV-joint grease. Install the outer race over the tripot joint and install the wire ring bearing retainer.

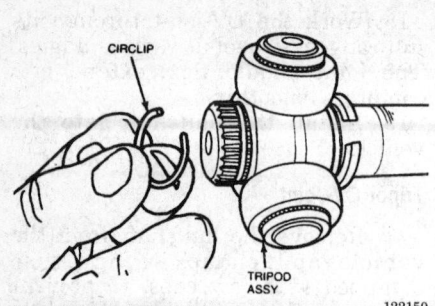

CV-joint, tripod — 1993–94 323, 626, 929, MX3, MX6, Miata and Protege

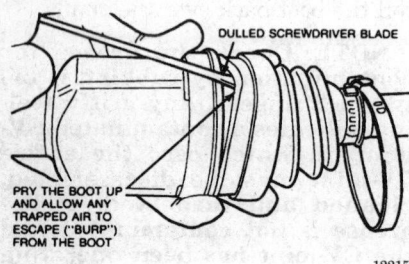

CV-joint boot installation, tripod — 1993–94 323, 626, 929, MX3, MX6, Miata and Protege

11. Position the CV-joint boot(s). Make sure the boot is fully seated in the grooves in the shaft and outer race.

12. Insert a small prybar with rounded edges between the boot and the outer bearing race to allow trapped air to escape from the boot. Install new boot clamps.

13. Wrap the clamps around the boots in a clockwise direction, pull tight with pliers and bend the locking tabs to secure in position.

14. Work the CV-joint through its full range of travel at various angles. The joint should flex, extend and compress smoothly.

15. Install the halfshaft into the vehicle.

1995 MX3 and 1995–97 Protege

Two types of CV-joints are used. The inboard CV-joints are the tripot type. All outboard CV-joints are Birfield type. The Birfield CV-joint cannot be disassembled; if an outboard CV-joint boot needs replacement, the inboard CV-joint must be removed.

1. Remove the halfshaft from the vehicle and clamp it in a vise equipped with jaw caps, to prevent damage to the machined surfaces. Do not allow the vise to contact the boot or its clamps.

2. Remove the large boot clamp from the inboard CV-joint, using side cutters. After removing the clamp, roll the boot back over the shaft.

NOTE: Check the grease for contamination by rubbing it between 2 fingers. Any gritty feeling indicates a contaminated CV-joint, in which case the entire CV-joint must be disassembled, cleaned and inspected. If the grease is not contaminated and the CV-joint has been operating satisfactorily, continue with the boot replacement procedure and add the required lubricant.

3. Paint alignment marks on the outer race and shaft for assembly reference. Remove the wire ring bearing retainer and remove the outer race.

4. Paint alignment marks on the tripot bearing and shaft for assembly reference. Remove the tripot bearing snapring and, using a brass drift and hammer, remove the tripot bearing from the shaft.

5. Remove the small clamp and remove the inner boot from the halfshaft. If the boot is to be reused, wrap the shaft splines with tape before removing.

6. If the outer CV-joint boot is to be replaced, remove the clamps and slide the boot off the shaft from the inboard side.

To install:

7. If the outboard boot was removed, slide the boot onto the shaft from the inboard side. Wrap tape on the splines before installing to protect the boot.

8. Install the inboard boot and remove the tape from the shaft.

9. Install the tripot assembly on the halfshaft. Tap the assembly onto the shaft using a hammer and brass drift. Install the tripot assembly retaining ring.

10. Fill the CV-joint outer race with high temperature CV-joint grease. Install the outer race over the tripot joint and install the wire ring bearing retainer.

11. Position the CV-joint boot(s). Make sure the boot is fully seated in the grooves in the shaft and outer race.

12. Insert a small prybar with rounded edges between the boot and the outer bearing race to allow trapped air to escape from the boot. Install new boot clamps.

13. Wrap the clamps around the boots in a clockwise direction, pull tight with pliers and bend the locking tabs to secure in position.

14. Work the CV-joint through its full range of travel at various angles. The joint should flex, extend and compress smoothly.

15. Install the halfshaft into the vehicle.

1995 929 and Miata

1. Remove the halfshaft from the vehicle and clamp it in a vise equipped with jaw caps to prevent damage to the machined surfaces. Do not allow the vise to contact the boot or its clamps.

2. Remove the large boot clamp from the inboard CV-joint, using side cutters. After removing the clamp, roll the boot back over the shaft.

NOTE: Check the grease for contamination by rubbing it between two fingers. Any gritty feeling indicates a contaminated CV-joint, in which case the entire CV-joint must be disassembled, cleaned and inspected. If the grease is not contaminated and the CV-joint has been operating satisfactorily, continue with the boot replacement procedure and add the required lubricant.

3. Paint alignment marks on the outer race and shaft for assembly reference. Remove the wire ring bearing retainer and remove the outer race.

4. Paint alignment marks on the inner race and shaft for assembly reference. Remove the inner race snapring from the end of the halfshaft and remove the inner race, cage and ball bearings from the shaft as an assembly.

NOTE: Use care to prevent damage to the bearing surfaces and cage.

5. If only the boot is being replaced, go to the next step. If it is necessary to disassemble the CV-joint further, proceed as follows:

a. Pry the ball bearing out of the bearing cage using a small prybar with blunted edges. Mark the inner race and the bearing cage for proper assembly.

b. Rotate the inner race to align the bearing lands with the windows in the bearing cage. Remove the inner race through the larger end of the cage.

6. Remove the small clamp and remove the inner boot from the halfshaft. If the boot is to be reused, wrap the shaft splines with tape before removing.

7. If the outer CV-joint boot is to be replaced, remove the clamps and

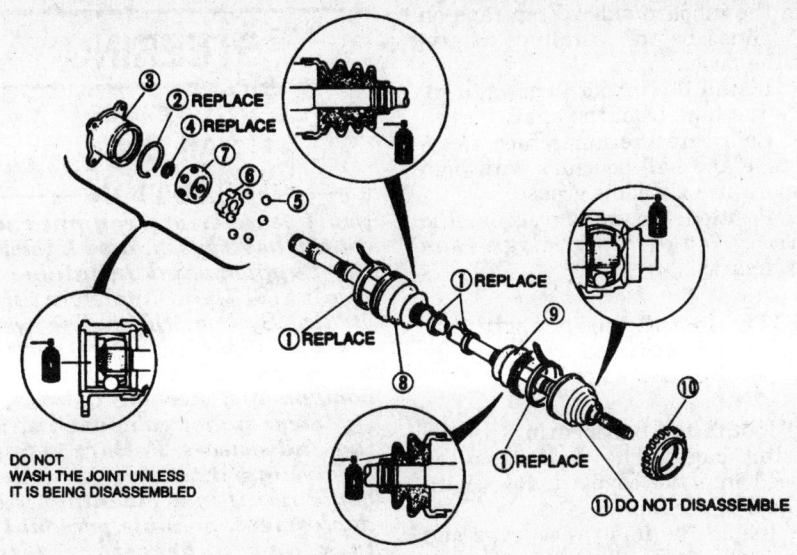

DO NOT
WASH THE JOINT UNLESS
IT IS BEING DISASSEMBLED

1. Boot bands
2. Clip
3. Outer ring
4. Snap ring
5. Balls
6. Inner ring
7. Cage
8. Boot (differential side)
9. Boot (wheel side)
10. ABS sensor rotor
11. Shaft and ball joint assembly

288359

Halfshaft with double offset CV-joint — 1995 929 and Miata

slide the boot off the shaft from the inboard side.

To install:

8. If the outboard boot was removed, slide the boot onto the shaft from the inboard side. Wrap tape on the splines before installing to protect the boot.

9. Install the inboard boot and remove the tape from the shaft.

10. Lubricate the inner race, bearing cage and ball bearings with high temperature CV-joint grease.

11. Position the inner race in the bearing cage and align the matchmarks.

NOTE: Install the race with the chamfered splines facing the large end of the cage.

12. Install the ball bearings in the bearing cage. The balls can be pressed into the cage windows with the heel of the hand.

13. Install the inner race, cage and balls on the halfshaft as an assembly. Make sure the chamfer on the bearing cage faces the snapring and the paint marks made during removal line up. Install the inner race snapring.

14. Lubricate the outer race with high temperature CV-joint grease. Install the outer race and add more

grease to the outer race. Install the wire ring bearing retainer.

15. Position the CV-joint boot(s). Make sure the boot is fully seated in the grooves in the shaft and outer race.

16. Insert a small prybar with rounded edges between the boot and the outer bearing race to allow trapped air to escape from the boot. Install new boot clamps.

17. Wrap the clamps around the boots in a clockwise direction, pull tight with pliers and bend the locking tabs to secure in position.

18. Work the CV-joint through its full range of travel at various angles. The joint should flex, extend and compress smoothly.

19. Install the halfshaft into the vehicle.

1995–97 Millenia

1. Raise and safely support the vehicle.

2. Remove the halfshaft from the vehicle and clamp it in a vise equipped with jaw caps, to prevent damage to the machined surfaces. Do not allow the vise to contact the boot or its clamps.

3. Remove the large boot clamp from the inboard CV-joint, using side

cutters. After removing the clamp, roll the boot back over the shaft.

NOTE: Check the grease for contamination by rubbing it between 2 fingers. Any gritty feeling indicates a contaminated CV-joint, in which case the entire CV-joint must be disassembled, cleaned and inspected. If the grease is not contaminated and the CV-joint has been operating satisfactorily, continue with the boot replacement procedure and add the required lubricant.

4. Paint alignment marks on the outer race and shaft for assembly reference. Remove the wire ring bearing retainer and remove the outer race.

5. Paint alignment marks on the tripod bearing and shaft for assembly reference. Remove the tripod bearing snapring and, using a brass drift and hammer, remove the tripod bearing from the shaft.

6. Remove the small clamp and remove the inner boot from the halfshaft. If the boot is to be reused, wrap the shaft splines with tape before removing.

7. If the outer CV-joint boot is to be replaced, remove the clamps and slide the boot off the shaft from the inboard side.

To install:

8. If the outboard boot was removed, slide the boot onto the shaft from the inboard side. Wrap tape on the splines before installing to protect the boot.

9. Install the inboard boot and remove the tape from the shaft.

10. Install the tripod assembly on the halfshaft. Tap the assembly onto the shaft using a hammer and brass drift. Install the tripod assembly retaining ring.

11. Fill the CV-joint outer race with high temperature CV-joint grease. Install the outer race over the tripod joint and install the wire ring bearing retainer.

12. Position the CV-joint boot(s). Make sure the boot is fully seated in the grooves in the shaft and outer race.

13. Insert a small prybar with rounded edges between the boot and the outer bearing race to allow trapped air to escape from the boot. Install new boot clamps.

14. Wrap the clamps around the boots in a clockwise direction, pull tight with pliers and bend the locking tabs to secure in position.

15. Work the CV-joint through its full range of travel at various angles. The joint should flex, extend and compress smoothly.

16. Install the halfshaft into the vehicle.

1996–97 Miata

1. Remove the halfshaft from the vehicle and clamp it in a vise equipped with jaw caps to prevent damage to the machined surfaces. Do not allow the vise to contact the boot or its clamps.

2. Remove the large boot clamp from the inboard CV-joint, using side cutters. After removing the clamp, roll the boot back over the shaft.

NOTE: Check the grease for contamination by rubbing it between two fingers. Any gritty feeling indicates a contaminated CV-joint, in which case the entire CV-joint must be disassembled, cleaned and inspected. If the grease is not contaminated and the CV-joint has been operating satisfactorily, continue with the boot replacement procedure and add the required lubricant.

3. Paint alignment marks on the outer race and shaft for assembly reference. Remove the wire ring bearing retainer and remove the outer race.

4. Paint alignment marks on the inner race and shaft for assembly reference. Remove the inner race snapring from the end of the halfshaft and remove the inner race, cage and ball bearings from the shaft as an assembly.

NOTE: Use care to prevent damage to the bearing surfaces and cage.

5. If only the boot is being replaced, go to the next step. If it is necessary to disassemble the CV-joint further, proceed as follows:
   a. Pry the ball bearing out of the bearing cage using a small prybar with blunted edges. Mark the inner race and the bearing cage for proper assembly.
   b. Rotate the inner race to align the bearing lands with the windows in the bearing cage. Remove the inner race through the larger end of the cage.

6. Remove the small clamp and remove the inner boot from the halfshaft. If the boot is to be reused, wrap the shaft splines with tape before removing.

7. If the outer CV-joint boot is to be replaced, remove the clamps and slide the boot off the shaft from the inboard side.

To install:

8. If the outboard boot was removed, slide the boot onto the shaft from the inboard side. Wrap tape on the splines before installing to protect the boot.

9. Install the inboard boot and remove the tape from the shaft.

10. Lubricate the inner race, bearing cage and ball bearings with high temperature CV-joint grease.

11. Position the inner race in the bearing cage and align the matchmarks.

NOTE: Install the race with the chamfered splines facing the large end of the cage.

12. Install the ball bearings in the bearing cage. The balls can be pressed into the cage windows with the heel of the hand.

13. Install the inner race, cage and balls on the halfshaft as an assembly. Make sure the chamfer on the bearing cage faces the snapring and the paint marks made during removal line up. Install the inner race snapring.

14. Lubricate the outer race with high temperature CV-joint grease. Install the outer race and add more grease to the outer race. Install the wire ring bearing retainer.

15. Position the CV-joint boot(s). Make sure the boot is fully seated in the grooves in the shaft and outer race.

16. Insert a small prybar with rounded edges between the boot and the outer bearing race to allow trapped air to escape from the boot. Install new boot clamps.

17. Wrap the clamps around the boots in a clockwise direction, pull tight with pliers and bend the locking tabs to secure in position.

18. Work the CV-joint through its full range of travel at various angles. The joint should flex, extend and compress smoothly.

19. Install the halfshaft into the vehicle.

1993–95 RX7

1. Raise and safely support the vehicle.

2. Remove the halfshaft assembly.

3. Remove the boot bands. Slide the boot out of the way and matchmark the outer ring and the shaft.

4. Mark the shaft and tripot joint, then remove the snapring. Drive the joint from the shaft.

5. Installation is the reverse of removal. Once the boot is installed, fill with the grease supplied with the boot kit.

STEERING

Air Bag

——— CAUTION ———

Some vehicles are equipped with an air bag system, also known as the Supplemental Inflatable Restraint (SIR) or Supplemental Restraint System (SRS). The system must be disabled before performing service on or around system components, steering column, instrument panel components, wiring and sensors. Failure to follow safety and disabling procedures could result in accidental air bag deployment, possible personal injury and unnecessary system repairs.

PRECAUTIONS

Several precautions must be observed when handling the inflator module to avoid accidental deployment and possible personal injury.

• Never carry the inflator module by the wires or connector on the underside of the module.

• When carrying a live inflator module, hold securely with both hands, and ensure that the bag and trim cover are pointed away.

• Place the inflator module on a bench or other surface with the bag and trim cover facing up.

• With the inflator module on the bench, never place anything on or close to the module which may be thrown in the event of an accidental deployment.

• An air bag is an explosive device. Handle with extreme caution.

• Always disconnect the battery and the air bag connector before removing the steering wheel or beginning work on the air bag system.

• Air bag components must not be repaired or opened. Always use new parts, including the wiring harness.

• Always place a removed air bag unit with the horn pad facing up. Put it in a safe place where it will not be disturbed.

• The air bag unit must not be exposed to grease, fluids, or cleaning agents.

• The air bag unit must not be exposed to temperatures above 194°F (90°C) at any time. Even the heat of a soldering iron can damage or ignite the charge.

• Storage and transport of air bags is subject to rules governing explosive devices and should be done only in the original package.

• Failure to follow proper safety precautions may result in personal injury through accidental firing of the air bag, or through failure of the air bag in an accident.

DISARMING

1993–94 323, Miata, 1993–95 MX3, Protege, RX7 and 1993–97 MX6 and 626

1. Disconnect the negative battery cable.
2. Locate and unplug the blue and orange connector. Protect this connector from accidental electrical contact, including static electricity.
3. The air bag is now disarmed and the battery can be connected to perform electrical testing on other systems.
4. When ready to reconnect the air bag; disconnect the battery, connect the air bag connector, install the cover or panel and make sure no one is in the vehicle when connecting the battery.

1995–97 Protege

1. Turn the ignition switch to **LOCK**.
2. Disconnect the negative battery cable.
3. Remove the interior right kick panel.
4. Disconnect the SAS unit connector.

1995–97 Miata

Drivers Side Air Bag

─────── CAUTION ───────
The air bag system must be disarmed before removing the steering wheel. Failure to do so may cause accidental deployment, property damage or personal injury.

1. Deactivate the audio anti-theft system.
2. Disconnect the negative battery cable.
3. Remove the lower panel under the steering wheel, then disconnect the orange and blue clock spring connectors.
4. The air bag is now disarmed and the battery can be connected to perform electrical testing on other systems.
5. If removing the unit, remove the mounting nuts, disconnect the support rope, disconnect the electrical connectors and remove the air bag.
6. When ready to reconnect the air bag; disconnect the battery, connect the air bag connectors, connect the support rope and mount the air bag. Torque the nuts in sequence to 35–52 in. lbs. (4.0–5.8 Nm).
7. Connect the negative battery cable. Turn the ignition switch ON. Verify that the air bag system warning light illuminates for four to eight seconds and then goes off.

Passenger Side Air Bag

1. Deactivate the audio anti-theft system.
2. Disconnect the negative battery cable.
3. Remove the glove compartment and the undercover.
4. Disconnect the orange and blue passenger side air bag module connectors.
5. The air bag is now disarmed and the battery can be connected to perform electrical testing on other systems.
6. If removing the unit, remove the mounting nuts, disconnect the electrical connectors and remove the air bag module.
7. When ready to reconnect the air bag; disconnect the battery, connect the air bag connectors and mount the air bag module. Torque the nuts in sequence to 13.1–20.2 ft. lbs. (17.7–27.4 Nm).
8. Connect the negative battery cable. Turn the ignition switch ON. Verify that the air bag system warning light illuminates for four to eight seconds and then goes off.

Millenia

1. Deactivate the audio antitheft system.
2. Turn the ignition switch to LOCK.
3. Disconnect the negative battery cable and wait for more than one minute to allow the backup power supply to deplete its stored power.
4. Remove the driver side undercover and lower dash panel. Disconnect the orange and blue clock spring connectors for the drivers side air bag.
5. Remove the glove compartment and disconnect the orange and blue passenger side air bag module connectors.

1993–95 RX7

1. Deactivate the audio antitheft system, if equipped.
2. Disconnect the negative battery cable.
3. Disconnect the electrical connectors at the diagnostic module and connect special tool 49 H066 004.
4. Arming is the reverse of disarming.

306498

SAS unit and harness connector — 1995–97 Protege

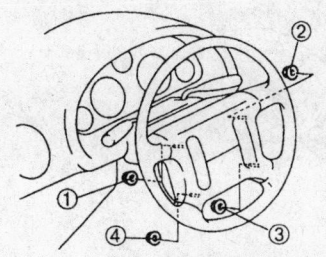

Tighten the nuts in the order shown in the figure.

Tightening torque:
4.0—5.8 N·m { 40—60 kgf·cm , 35—52 in·lbf }

287601

Driver side air bag mounting nut torque sequence — 1996 Miata

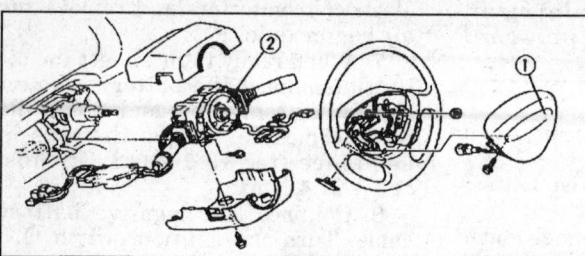

1. Driver-side air bag module
2. Clock spring
3. Passenger-side air bag module
4. SAS unit

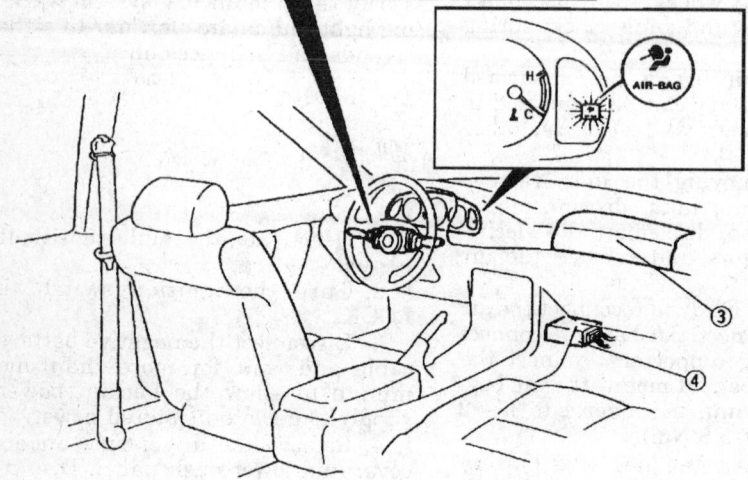

Exploded view of steering wheel components — Millenia

311448

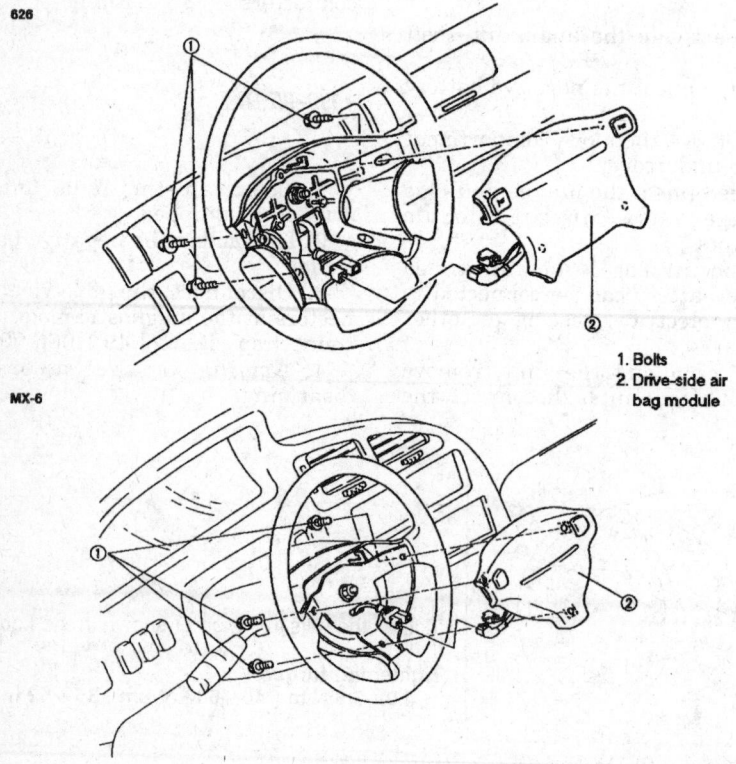

626

MX-6

1. Bolts
2. Drive-side air bag module

Air bag module mounting points — 626 and MX6

302361

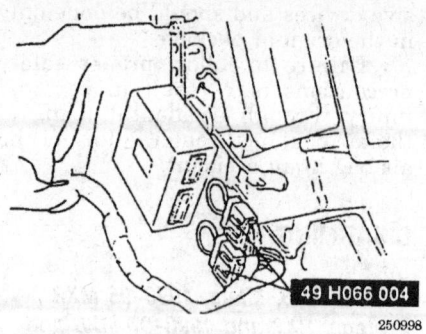

49 H066 004

250998

Air bag diagnostic connector — 1995–97 RX7

Steering Wheel

REMOVAL AND INSTALLATION

— CAUTION —
If equipped with an air bag, the vehicle battery and the system's own back-up battery must be disconnected before removing the steering wheel. Failure to do so may result in deployment of the air bag and possible personal injury. Always carry an air bag assembly with the bag and cover facing away from your body. Store the assembly with the bag and cover facing up; never place the assembly face down on the floor or workbench.

1993–94 323, 626, 929, MX3, MX6, Miata, RX7 and 1994 Protege

Without Air Bag

1. Disconnect the negative battery cable. Remove the horn pad button assembly. If equipped with a 4 spoke steering wheel, pull the center cap toward the wheel top.
2. Make matchmarks on the steering wheel and steering shaft. Never strike the steering shaft with a hammer, as damage to the column may result.
3. Remove the wheel using a suitable puller.
4. Installation is the reverse of removal. Torque the steering wheel nut to 36 ft. lbs. (49 Nm).

With Air Bag

1. Disarm the air bag.
2. At the back of the steering wheel hub, remove the nuts that hold the air bag assembly and remove the air bag. Place it in a safe place, pad side up.
3. Matchmark the wheel to the shaft and remove the nut. Use a puller to remove the wheel.

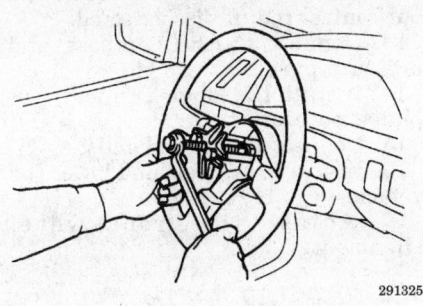

291325

Common method of removing the steering wheel

4. When installing, the clockspring must be reset.

  a. Make sure the front wheels are straight-ahead.

  b. Turn the clockspring all the way to the right.

  c. Turn the clockspring back about 2 ³/₄turns and align the marks.

  d. Connect the wiring and install the steering wheel.

5. Torque the steering wheel nut to 36 ft. lbs. (49 Nm). Install the air bag unit.

1995 MX3, 929 and RX7 and 1995–97 626, MX6, Miata, Millenia and Protege

1. Set the steering wheel so that the front wheels are straight-ahead.

2. Disconnect the negative battery cable.

3. Disarm the air bag.

4. At the back of the steering wheel hub, remove the nuts that hold the air bag assembly and remove the air bag. Place it in a safe place, pad side up.

5. Matchmark the wheel to the shaft and remove the nut. Use a puller to remove the wheel.

To install:

6. Double check that the front wheels are in the straight-ahead position and mount the steering wheel. Make sure the matchmarks are aligned.

7. Torque the steering wheel nut to 29–36 ft. lbs. (40–49 Nm). Install the air bag unit.

8. Connect the negative battery cable.

Tie Rod Ends

REMOVAL AND INSTALLATION

1993–95 RX7, 1995 MX3 and 1995–97 626, MX6, Miata and Protege

1. Raise and support the vehicle safely. Remove the wheel and tire assembly.

2. Remove the cotter pin and loosen the nut on the tie rod end ball stud. With the nut protecting the ball stud, press the stud from the knuckle using a press tool (49 T028 3AO or equivalent). Remove the nut and the tie rod end from the knuckle.

3. Paint a reference mark across the tie rod, jam nut and shaft.

4. Loosen the jam nut and unscrew the tie rod end from the shaft.

To install:

5. Thread the tie rod onto the shaft and align the marks made during removal. If installing a new tie rod end, try to assemble it in the same position as the old one.

6. Install the tie rod end into the knuckle. Install the nut and tighten to 32–41 ft. lbs. (43–56 Nm) on the MX3, and 24–32 ft. lbs. (31–41 Nm) on the RX7, MX6/626 Miata and Protege.

7. Install a new cotter pin. If the cotter pin cannot be installed because the ball stud hole and the nut castellation do not align, tighten the nut further until the cotter pin can be installed. Never loosen the nut to install the cotter pin.

8. Tighten the jam nut to 26–36 ft. lbs. (35–50 Nm).

9. Install the wheel and tire assembly and lower the vehicle. Check the front wheel alignment.

1995 929

1. Raise and support the vehicle safely. Remove the wheel and tire assembly.

2. Remove the cotter pin and loosen the nut on the tie rod end ball stud. With the nut protecting the ball stud, press the stud from the knuckle using a press tools (49 T028 303 and 304 or equivalent). Remove the nut and the tie rod end from the knuckle.

3. Paint a reference mark across the tie rod, jam nut and shaft.

4. Loosen the jam nut and unscrew the tie rod end from the shaft.

To install:

5. Thread the tie rod onto the shaft and align the marks made during removal. If installing a new tie rod end, try to assemble it in the same position as the old one.

6. Install the tie rod end into the knuckle. Install the nut and tighten to 37–47 ft. lbs. (50–63 Nm).

7. Install a new cotter pin. If the cotter pin cannot be installed because the ball stud hole and the nut castellation do not align, tighten the nut further until the cotter pin can be installed. Never loosen the nut to install the cotter pin.

8. Tighten the jam nut to 51–72 ft. lbs. (69–98 Nm).

9. Install the wheel and tire assembly and lower the vehicle. Check the front wheel alignment.

Manual Rack and Pinion

REMOVAL AND INSTALLATION

1993–94 323, MX3 and Protege

Without 4WD

1. Disconnect the negative battery cable. Raise and safely support the vehicle and remove the wheel and tire assemblies.

2. Remove the engine undercover.

3. Remove the cotter pins and nuts and disconnect the tie rod ends.

4. Remove the steering column universal joint bolt. Matchmark the joint to the pinion shaft.

5. If equipped with power steering, disconnect the hydraulic lines and drain the fluid into a container.

6. If equipped with manual transaxle, disconnect the extension bar and shift control rod from the transaxle.

7. Remove the bracket nuts or bolts and remove the steering rack from the vehicle.

To install:

8. When fitting the rack into place, make sure the pinion shaft and universal joint matchmarks are correctly aligned. Fit the shaft into the joint and start the steering rack bracket nuts or bolts.

9. Tighten the bracket nuts or bolts to 38 ft. lbs. (52 Nm). Install the universal joint bolt and torque it to 20 ft. lbs. (26 Nm).

10. Connect the tie rod ends and torque the nuts to 42 ft. lbs. (57 Nm), then tighten as required to install a new cotter pin.

11. If equipped, connect the power steering hydraulic lines and install the undercover.

12. Install the wheel and tire assemblies and lower the vehicle. If equipped with power steering, fill the system with the proper fluid and bleed the air from the system.

With 4WD

1. Remove the battery and battery tray. Raise and safely support the vehicle and remove the front wheels and front crossmember.

2. Disconnect the tie rod ends and remove the bolt from the steering column universal joint. Matchmark the universal joint to the pinion shaft.

3. Disconnect the hydraulic lines and drain the fluid into a container.

4. To remove the steering rack, the front–to–rear engine mount member must be removed:

a. Use the proper lifting equipment and support the engine from above.

b. Disconnect the front and rear engine mounts from the mount member and remove the mount member.

c. Remove the rear mount from the engine.

5. Remove the exhaust pipe and catalytic converter.

6. Matchmark the flanges and remove the driveshaft.

7. Lower the engine gradually until the lower left rack mounting bolt can be reached. Do not lower too far or the halfshaft joints will be damaged.

8. Remove the mount bolts and move the rack to the left to remove the steering rack.

To install:

9. When fitting the rack into place, make sure the pinion shaft and universal joint matchmarks are correctly aligned. Fit the shaft into the joint and start the steering rack bracket bolts. Torque the nuts and bolts to 38 ft. lbs. (52 Nm).

10. Use new gaskets and install the exhaust pipe. Torque the flange nuts to 34 ft. lbs. (46 Nm).

11. Install the driveshaft and torque the flange nuts to 22 ft. lbs. (30 Nm).

12. Install the engine mount and mount member. Torque the mount member nuts to 66 ft. lbs. (89 Nm), the mount–to–engine nuts to 38 ft. lbs. (52 Nm).

13. Connect the fluid lines. Install the universal joint bolt and torque it to 20 ft. lbs. (27 Nm).

14. Connect the tie rod ends and torque the nuts to 42 ft. lbs. (57 Nm) and tighten as required to install a new cotter pin.

15. Install the front crossmember and torque the bolts to 86 ft. lbs. (117 Nm).

16. Fill the pump with fluid and bleed the air from the system.

1993–97 Miata

1. Disconnect the negative battery cable. Raise and safely support the vehicle and remove the wheel and tire assemblies.

2. Remove the engine undercover.

3. Remove the cotter pins and nuts, and disconnect the tie rod ends.

4. Remove the steering column universal joint bolt. Matchmark the joint to the pinion shaft.

5. Remove the bracket bolts, and remove the steering rack from the vehicle.

To install:

6. When fitting the rack into place, make sure the pinion shaft and universal joint matchmarks are correctly aligned. Fit the shaft into the joint and start the steering rack bracket nuts or bolts.

7. Tighten the bracket bolts to 34–43 ft. lbs. (47–58 Nm) on 1995–97 models and 38 ft. lbs. (52 Nm) on 1993–94 models. Install the universal joint bolt, and torque it to 14–19 ft. lbs. (18–26 Nm) on 1995–97 models and 20 ft. lbs. (26 Nm) on 1993–94 models.

8. Connect the tie rod ends, and torque the nuts to 22–32 ft. lbs. (30–44 Nm) on 1995–97 models and 42 ft. lbs. (57 Nm) on 1993–94 models. Install a new cotter pin.

9. Install the undercover.

10. Install the wheel and tire assemblies and lower the vehicle.

11. Adjust the front alignment.

Power Rack and Pinion

REMOVAL AND INSTALLATION

1995–97 Protege

1. Raise and safely support the vehicle. Remove the front wheels.

2. Disconnect the tie-rod ends.

3. Remove the splash shield, set plate and boot.

4. Disconnect the return hose and pressure pipe.

5. (M/T) Disconnect the extension bar/control rod.

6. Safely support the engine and remove the rear engine mount bolt.

7. Remove the rack mounting bracket nuts.

8. Remove the rack through the right side of the vehicle.

To install:

9. Install the rack into the vehicle.

10. Install the mounting bracket nuts. Torque the nuts in sequence to 28–38 ft. lbs. (38–51 Nm).

11. Install the rear engine mount bolt.

12. Connect the extension bar/control rod, if disconnected.

13. Connect the return hose and pressure pipe.

14. Install the splash shield, set plate and boot.

15. Connect the tie-rod ends.

16. Install the wheels and lower the vehicle.

17. Adjust the front wheel alignment.

1993–95 929 and 1995 MX3

1. Disconnect the negative battery cable. Raise and safely support the vehicle.

2. Remove the front wheels.

3. Remove the cotter pins and nuts, and separate the tie rod ends from the knuckles.

4. Disconnect the power steering pressure and return lines.

5. Mark the position of the steering shaft in the column universal joint. Remove the set plate and dust cover. Remove the universal joint pinch bolt.

6. On manual transaxles (MTX) only, remove the control rod.

7. Remove the mounting bolts and remove the steering gear.

To install:

8. Position the steering gear, making sure the marks made on the universal joint and steering gear shaft align, and install the mounting nuts and bolts. Tighten to 28–38 ft. lbs. (38–51 Nm). The bolts and nuts must be torqued in sequence.

9. On MTX equipped vehicles, install the control rod. Tighten the mounting bolt to 12–16 ft. lbs. (16–22 Nm), and the mounting nut to 24–33 ft. lbs. (32–46 Nm).

10. Install the universal joint pinch bolt, and tighten to 14–20 ft. lbs. (18–26 Nm).

11. Connect the power steering lines.

12. Connect the tie rod ends to the knuckles and tighten the nuts to 37–47 ft. lbs. (50–64 Nm). Install new cotter pins.

13. Install the wheel and tire assemblies and lower the vehicle. Fill the power steering system with the proper type and quantity of fluid and bleed the air from the system.

14. Check and adjust the front wheel alignment.

1993–97 626 and MX6

1. Disconnect the negative battery cable. Raise and safely support the vehicle.

2. Remove the front wheels.

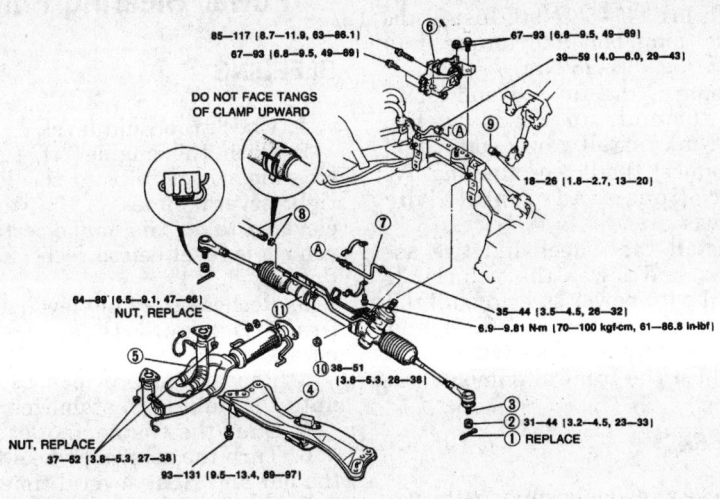

DO NOT FACE TANGS OF CLAMP UPWARD

85—117 [8.7—11.9, 63—86.1]
67—93 [6.8—9.5, 49—69]
67—93 [6.8—9.5, 49—69]
39—59 [4.0—6.0, 29—43]
18—26 [1.8—2.7, 13—20]
64—89 [6.5—9.1, 47—66] NUT, REPLACE
35—44 [3.5—4.5, 26—32]
6.9—9.81 N·m [70—100 kgf·cm, 61—86.8 in·lbf]
35—51 [3.6—5.3, 26—38]
NUT, REPLACE 37—52 [3.8—5.3, 27—38]
93—131 [9.5—13.4, 69—97]
31—44 [3.2—4.5, 23—33]
REPLACE

N·m [kgf·m, ft·lbf]

1. Cotter pin
2. Nut
3. Tie rod end ball joint
4. Transverse member (KL engine)
5. Front exhaust pipe and catalytic converter (KL engine)
6. Engine mount
7. Pressure pipe
8. Return hose and clamp
9. Bolt (Intermediate shaft)
10. Mounting bracket nut and bolt
11. Steering gear and linkage

302363

Rack and pinion assembly and related components — 626 and MX6

3. Remove the cotter pins and nuts, and separate the tie rod ends from the knuckles.

4. If equipped with 2.5L engine, remove the transverse member and the front exhaust pipe.

5. Support the engine with a suitable engine support tool (49 G017 502 or equivalent), and remove the rear engine mount.

6. Disconnect the power steering pressure and return lines.

7. Mark the position of the steering shaft in the column universal joint. Remove the universal joint pinch bolt.

8. Remove the mounting bolts and remove the steering rack.

To install:

9. Position the steering gear, making sure the marks made on the universal joint and steering gear shaft align, and install the mounting nuts and bolts. Tighten to 28–38 ft. lbs. (38–51 Nm). The bolts and nuts must be torqued in sequence.

10. Install the universal joint pinch bolt and tighten to 13–20 ft. lbs. (18–26 Nm).

11. Connect the power steering lines. Tighten the fluid fittings to the proper torque. Hand tighten the clamp bolts until snug, the torque them to 61–89 inch lbs. (7–10 Nm)

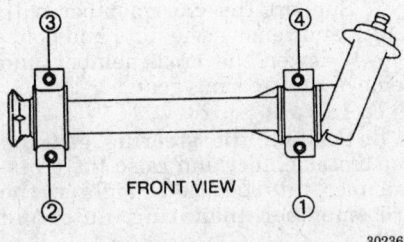

FRONT VIEW

302365

Rack and pinion mounting bolt locations — 1993–97 626 and MX6

12. Install the rear engine mount. Tighten the mounting bolts and nut to 50–68 ft. lbs. (67–93 Nm). Remove the engine support tool.

13. If equipped with 2.5L engine, install the front exhaust pipe, using new gaskets. Tighten the pipe-to-catalytic converter nuts to 47–65 ft. lbs. (64–89 Nm) and the pipe-to-exhaust manifold nuts to 27–38 ft. lbs. (37–52 Nm). Connect the oxygen sensors.

14. If equipped with 2.5L engine, install the transverse member and tighten the bolts to 69–96 ft. lbs. (94–131 Nm).

15. Connect the tie rod ends to the knuckles.

16. Install the wheel and tire assemblies and lower the vehicle. Fill the power steering system with the proper fluid and bleed the air from the system.

17. Check the front wheel alignment.

1995–97 Millenia

1. Raise and safely support the vehicle.

2. Separate the tie-rod end from the steering knuckle.

3. Remove the transverse member.

4. Disconnect the return hose and pressure pipe.

5. Remove the intermediate shaft bolt.

6. Support the engine from the top and remove the engine mount member.

7. Remove the bolts for engine mount No. 1.

8. Remove the upper lateral link bolt and nut.

9. Remove the lower arm bolt (crossmember side).

10. Disconnect the stabilizer control link.

11. Support the crossmember using a jack and remove the nuts and bolts.

12. Lower the crossmember slowly and remove the stabilizer bar from the crossmember.

13. Remove the mounting bracket bolts and remove the steering rack.

To install:

14. Replace the steering rack and loosely install the mounting bracket bolts. Looking straight ahead at the rack mounting bolts torque in the following order: lower left, upper right, upper left and lower right. Torque the bolts to 32–44 ft. lbs. (44–60 Nm).

15. Install the stabilizer and torque the mounting bolts to 32–44 ft. lbs. (44–60 Nm).

16. Install the crossmember.

17. Connect the stabilizer control link, torque the nuts to 32–44 ft. lbs. (44–60 Nm).

18. Install the lower arm bolt, torque to 58–86 ft. lbs. (79–116 Nm).

19. Install the upper lateral link nut and bolt, torque to 58–86 ft. lbs. (79–116 Nm).

20. Install the No. 1 engine mount bolts and torque to 32–44 ft. lbs. (44–60 Nm).

21. Install the engine mount member, torque the bolts to 50–68 ft. lbs. (67–93 Nm) and the nuts to 55–77.3 ft. lbs. (75–104 Nm).

22. Install the intermediate shaft bolt, torque to 14–19 ft. lbs. (18–26 Nm).

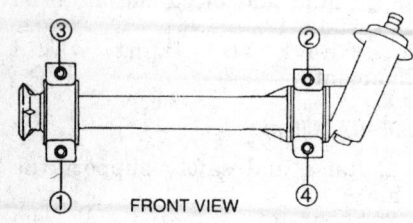

FRONT VIEW

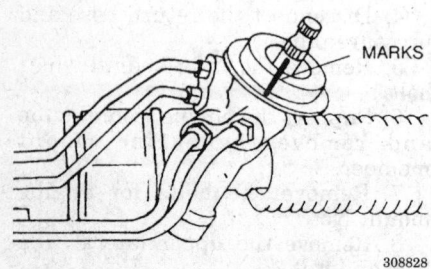

MARKS

308828

Mounting bracket and intermediate shaft alignment marks — 1995-97 Millenia

23. Connect the pressure pipe and return hose.

24. Install the transverse member.

25. Connect the tie-rod end, torque the nut to 41-59 ft. lbs. (55-80 Nm).

26. Lower the vehicle. Check the power steering fluid and fill to proper level.

1995-97 Miata

1. Disconnect the negative battery cable. Raise and safely support the vehicle, and remove the wheel and tire assemblies.

2. Remove the engine undercover.

3. Remove the cotter pins and nuts, and disconnect the tie rod ends.

4. Remove the steering column universal joint bolt. Matchmark the joint to the pinion shaft.

5. Disconnect the hydraulic lines, and drain the fluid into a container.

6. Remove the bracket bolts, and remove the steering rack from the vehicle.

To install:

7. When fitting the rack into place, make sure the pinion shaft and universal joint matchmarks are correctly aligned. Fit the shaft into the joint and start the steering rack bracket nuts or bolts.

8. Tighten the bracket bolts to 34-43 ft. lbs. (47-58 Nm). Install the universal joint bolt and torque it to 14-19 ft. lbs. (18-26 Nm).

9. Connect the tie rod ends, and torque the nuts to 22-33 ft. lbs. (30-44 Nm). Install a new cotter pin.

10. Connect the power steering hydraulic lines and install the undercover.

11. Install the wheel and tire assemblies, and lower the vehicle. If equipped with power steering, fill the system with the proper fluid and bleed the air from the system.

12. Adjust the front alignment.

1993-95 RX7

1. Raise and safely support the vehicle. Remove the wheel and tire assemblies. Remove the undercover.

2. Remove the cotter pins and nuts from the tie rod ends. Separate the tie rod ends from the knuckles using a suitable tool.

3. Remove the stabilizer bar brackets.

4. Disconnect the power steering pressure and return lines.

5. Mark the position of the steering gear shaft in the lower column joint and remove the bolt.

6. Remove the steering gear bracket mounting bolts.

7. Support the crossmember with a jack and remove the nuts and bolts. Slowly lower the crossmember and remove the steering gear.

To install:

8. Position the steering gear on the crossmember and raise the crossmember into position. Tighten the crossmember mounting nuts and bolts.

9. Install the steering gear mounting bracket bolts and tighten, in sequence, to 28-38 ft. lbs. (38-51 Nm).

10. Connect the column joint to the steering shaft, aligning the marks made during removal. Install the bolt and tighten to 14-19 ft. lbs. (18-26 Nm).

11. Connect the power steering lines. When connecting the pressure hose, align the pin on the hose with the hole in the steering gear.

12. Install the stabilizer bar brackets and tighten the bolts to 14-19 ft. lbs. (18-26 Nm).

13. Connect the tie rod ends to the knuckles.

14. Install the wheel and tire assemblies and lower the vehicle. Fill the power steering system with the proper type and quantity of fluid and bleed the air from the system.

Power Steering Pump

BLEEDING

1. Check the fluid level.

2. With the engine off, turn the steering wheel fully to the left and right several times. If the vehicle is elevated to do this, make certain the vehicle is level before rechecking the fluid level.

3. Recheck the fluid level, and add fluid (Dexron® II or M-III) if necessary.

4. Repeat the previous two steps until the fluid level stabilizes.

5. Start the engine and let it idle.

6. Turn the steering wheel fully to the left and right several times.

7. Verify that the fluid level has not dropped, and that the fluid is smooth, not foamy.

8. If necessary, repeat the previous two steps until the fluid level stabilizes.

REMOVAL AND INSTALLATION

1993-94 323, MX3 and Protege

1. Disconnect the negative battery cable. Disconnect and plug the hoses at the pump. If equipped, disconnect the pressure switch connector.

2. Remove the pump drive belt.

3. Support the pump, remove the mounting bolts and lift out the pump.

4. Installation is the reverse of removal. Tighten the pulley nut to 43 ft. lbs. (58 Nm). Adjust the belt tension and fill and bleed the system.

1995-97 Protege

1. Disconnect the negative battery cable.

2. Remove the bottom locknut from the pump bracket.

3. Loosen the adjusting nut and remove the drive belt.

4. Disconnect the pressure pipe and the return hose. Plug the lines to prevent fluid spills.

5. Disconnect the pump switch connector.

6. Remove the pump bracket.

7. Remove the remaining pump mount bolts and remove the pump.

To install:

8. Replace the pump and install the mounting nuts and bolts. Torque the bracket nuts and bolts to 28-38 ft. lbs. (38-51 Nm). Torque the pump to bracket mount bolt to 32-44 ft. lbs. (44-60 Nm).

9. Connect the pump switch connector.

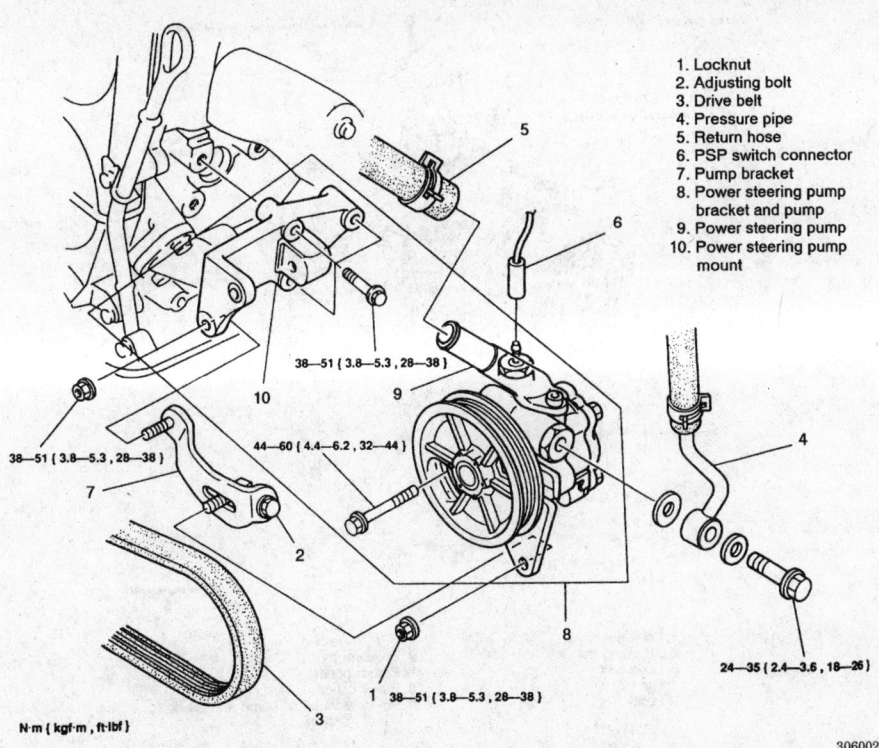

1. Locknut
2. Adjusting bolt
3. Drive belt
4. Pressure pipe
5. Return hose
6. PSP switch connector
7. Pump bracket
8. Power steering pump bracket and pump
9. Power steering pump
10. Power steering pump mount

38—51 { 3.8—5.3 , 28—38 }

38—51 { 3.8—5.3 , 28—38 }

44—60 { 4.4—6.2 , 32—44 }

24—35 { 2.4—3.6 , 18—26 }

1 38—51 { 3.8—5.3 , 28—38 }

N·m { kgf·m , ft·lbf }

306002

Power steering pump removal and installation — 1995–97 Protege

10. Connect the return hose and pressure pipe.

11. Replace the drive belt and properly adjust the tension

12. Install the bottom bracket locknut, torque to 28–38 ft. lbs. (38–51 Nm).

13. Connect the negative battery cable.

1995 MX3

1.6L (B6E) Engine

1. Disconnect the negative battery cable.

2. Remove the bolt located at the pump pulley.

3. Remove the two nuts from the adjusting bracket.

4. Remove the adjusting bolt from the pump bracket.

5. Remove the pump drive belt.

6. Disconnect and plug the hoses at the pump. Remove the O-rings

7. It is necessary to remove the pump pulley before removing the pump. Hold the pulley by inserting a small prybar through one of the holes in the pulley to hold it. Remove the pulley nut and pulley.

8. Disconnect the power steering pressure switch connector.

9. Support the pump, remove the mounting bolts and lift out the pump.

To install:

10. Position the pump, and install the mounting bolts. Tighten the bolts to 28–38 ft. lbs. (38–51 Nm).

11. Unplug and connect the return hose to the pump. Unplug the pressure hose, and install the hose using a new O-rings. Tighten the bolt to 12–17 ft. lbs. (16–23 Nm).

12. Connect the power steering pressure switch connector.

13. Install the pump bracket, and torque the mounting bolts to 24–33 ft. lbs. (31–46 Nm).

14. Position the pulley on the pump shaft, and fasten the washer and nut onto the end of the shaft. Hold the pulley by inserting a small prybar through one of the holes in the pulley to hold it. Tighten the pulley nut to 37–43 ft. lbs. (49–58 Nm).

15. Install the drive belt and the adjusting bolt.

16. Install the nuts onto the adjusting bracket, and the bolt at the pulley.

17. Adjust the belt tension, and fill and bleed the system. Check for leaks.

18. Connect the negative battery cable.

1.8L (FS) Engine

1. Disconnect the negative battery cable.

2. Disconnect the power steering pressure switch connector.

3. Disconnect and plug the hoses at the pump.

4. Remove the pump drive belt.

5. It is necessary to remove the pump pulley before removing the pump. Hold the pulley by inserting a small prybar through one of the holes in the pulley to hold it. Remove the pulley nut and pulley.

6. Support the pump, remove the mounting bolts and lift out the pump along with the bracket as an assembly.

To install:

7. Position the pump, and install the mounting bolts. Tighten the upper bolt to 14–18 ft. lbs. (19–25 Nm). Tighten the lower mounting bolt to 24–33 ft. lbs. (32–46 Nm).

8. Position the pulley on the pump shaft, and fasten the washer and nut onto the end of the shaft. Hold the pulley by inserting a small prybar through one of the holes in the pulley to hold it. Tighten the pulley nut to 37–43 ft. lbs. (49–58 Nm).

9. Install the drive belt.

10. Unplug and connect the return hose to the pump. Unplug the pressure hose, and install the hose using a new washer. Tighten the bolt to 18–26 ft. lbs. (24–35 Nm).

11. Connect the power steering pressure switch connector.

12. Adjust the belt tension, and fill and bleed the system. Check for leaks.

13. Connect the negative battery cable.

1993–97 626 and MX6

2.5L (KL) Engine

1. Disconnect the negative battery cable.

2. Remove the transverse member.

3. Remove the locknut that secures the idler pulley, and remove the idler pulley and bearing.

4. Remove the adjusting bolt from the pump bracket.

5. Disconnect and plug the hoses at the pump.

6. Remove the pump drive belt.

7. It is necessary to remove the pump pulley before removing the pump. Hold the pulley by inserting a small prybar through 1 of the holes in the pulley to hold it. Remove the pulley nut and pulley.

8. Remove the hose and pump brackets. Disconnect the power steering pressure switch connector.

9. Support the pump, remove the mounting bolts and lift out the pump.

To install:

10. Position the pump, and install the mounting bolts. Tighten the bolts to 24–33 ft. lbs. (31–46 Nm).

11. Unplug and connect the return hose to the pump. Unplug the pressure hose, and install the hose using a new washer. Tighten the bolt to 22–32 ft. lbs. (30–44 Nm).

12. Connect the power steering pressure switch connector.

13. Install the pump bracket, and torque the mounting bolts to 24–33 ft. lbs. (31–46 Nm).

14. Install the hose bracket onto the pump bracket and tighten the pressure hose mounting bolt through the bracket at 61–87 inch lbs. (7–10 Nm).

15. Position the pulley on the pump shaft, and fasten the washer and nut onto the end of the shaft. Hold the pulley by inserting a small prybar through 1 of the holes in the pulley to hold it. Tighten the pulley nut to 37–43 ft. lbs. (49–58 Nm).

16. Install the drive belt and the adjusting bolt.

17. Install the idler pulley and bearing. Tighten the locknut to 28–38 ft. lbs. (38–51 Nm).

18. Install the transverse member and tighten the mounting bolts to 69–96 ft. lbs. (94–131 Nm).

19. Adjust the belt tension, and fill and bleed the system. Check for leaks.

20. Connect the negative battery cable.

2.0L (FS) Engine

1. Disconnect the negative battery cable.

2. Remove the transverse member.

3. Remove the lockbolts and the adjusting bolt from the pump bracket.

4. Disconnect and plug the hoses at the pump.

5. Remove the pump drive belt.

6. It is necessary to remove the pump pulley before removing the pump. Hold the pulley by inserting a small prybar through 1 of the holes in the pulley to hold it. Remove the pulley nut and pulley.

7. Disconnect the power steering pressure switch connector.

8. Remove the pulley cover.

9. Support the pump, remove the mounting bolts and lift out the pump along with the bracket as an assembly.

To install:

10. Position the pump, and install the mounting bolts. Tighten the upper bolt to 32–44 ft. lbs. (44–60 Nm).

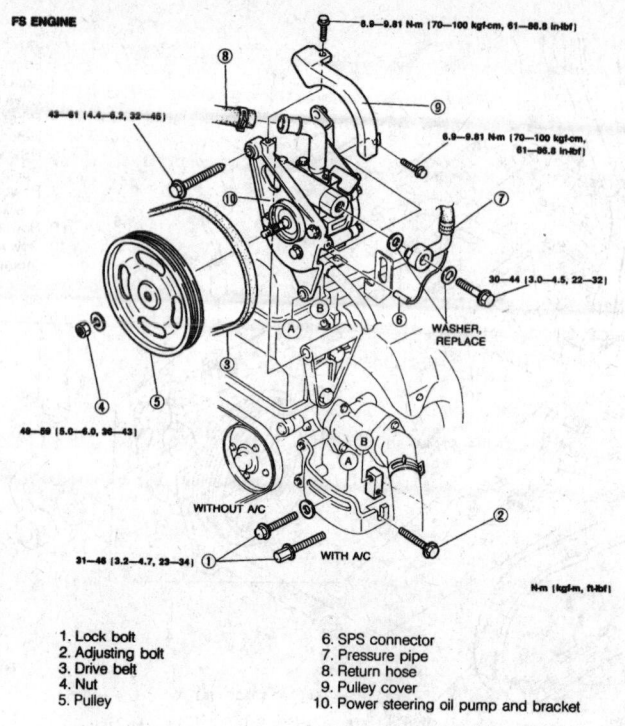

Steering pump, with FS engine — 1995–97 626 and MX6

1. Lock bolt
2. Adjusting bolt
3. Drive belt
4. Nut
5. Pulley
6. SPS connector
7. Pressure pipe
8. Return hose
9. Pulley cover
10. Power steering oil pump and bracket

11. Install the pulley cover to the bracket, and tighten the bolts to 61–87 inch lbs. (7–10 Nm).

12. Unplug and connect the return hose to the pump. Unplug the pressure hose, and install the hose using a new washer. Tighten the bolt to 22–32 ft. lbs. (30–44 Nm).

13. Connect the power steering pressure switch connector.

14. Position the pulley on the pump shaft, and fasten the washer and nut onto the end of the shaft. Hold the pulley by inserting a small prybar through 1 of the holes in the pulley to hold it. Tighten the pulley nut to 37–43 ft. lbs. (49–58 Nm).

15. Install the drive belt.

16. Install the adjusting bolt. Install the lockbolts and tighten to 24–33 ft. lbs. (31–46 Nm).

17. Install the transverse member and tighten the mounting bolts to 69–96 ft. lbs. (94–131 Nm).

18. Adjust the belt tension, and fill and bleed the system. Check for leaks.

19. Connect the negative battery cable.

1993–95 929

1. Disconnect the negative battery cable.

2. Remove the locknut and adjusting bolt.

3. Remove the pump drive belt.

4. It is necessary to remove the pump pulley before removing the pump. Hold the pulley with tool 49 W023 585A or equivalent, or if possible, insert a small prybar through 1 of the holes in the pulley to hold it. Remove the pulley nut and pulley.

5. Disconnect and plug the hoses at the pump. If equipped, disconnect the pressure switch connector.

6. Support the pump, remove the mounting bolts and lift out the pump.

To install:

7. Position the power steering pump, and install the mounting bolts. Tighten the bolts to 12–16 ft. lbs. (16–22 Nm).

8. Unplug and connect the hoses. If equipped, connect the pressure switch connector.

9. Install the pulley. Tighten the pulley nut to 29–43 ft. lbs. (40–58 Nm).

10. Install the drive belt.

11. Install the adjusting bolt and locknut. Tighten the locknut to 28–38 ft. lbs. (38–51 Nm).

12. Adjust the belt tension. Fill and bleed the system.

13. Connect the negative battery cable.

1995–97 Millenia

1. Disconnect the negative battery cable.

2. Remove the idler pulley locknut and remove the pulley bearing.

3. Loosen the adjusting bolt and remove the drive belt.

4. Keeping the pump pulley from moving, remove the pump pulley locknut. Remove the pump pulley.

5. Remove the hose bracket mount nut.

6. Remove the pump bracket.

7. Disconnect the power steering pressure switch connector.

8. Disconnect the pressure pipe and return hose.

9. Remove the power steering pump.

To install:

10. Place the pump to its mounts and install the mounting bolts. Torque to 24–33 ft. lbs. (32–46 Nm).

11. Connect the pressure pipe and return hose. Torque the pressure pipe mount bolt to 22–28 ft. lbs. (30–39 Nm).

12. Connect the switch connector.

13. Install the pump bracket. Torque the mounting bolts to 24–33 ft. lbs. (32–46 Nm).

14. Install the hose bracket and torque the mounting nut to 70–95.4 in. lbs. (7.9–10.7 Nm).

15. Install the pump pulley, torque the locknut to 37–43 ft. lbs. (50–58 Nm).

16. Install the idler pulley and bearing. Torque the nut to 24–33 ft. lbs. (32–46 Nm).

17. Install the drive belt and tighten the adjusting bolt. Make sure that there is proper tension on the belt.

18. Connect the negative battery cable.

Miata

1. Disconnect the negative battery cable.

2. Remove the bolt located at the pump pulley.

3. Remove the nut from the adjusting bracket.

4. Remove the adjusting bolt from the pump bracket.

5. Remove the pump drive belt.

6. Disconnect the power steering pressure switch connector.

7. Remove the pressure hose bracket bolt. Disconnect and plug the hoses at the pump. Remove the O-rings

8. Support the pump, remove the mounting bolts and lift out the pump.

To install:

9. Position the pump, and install the mounting bolts. Tighten the bolts to 28–39 ft. lbs. (37–53 Nm).

10. Unplug and connect the return hose to the pump. Unplug the pressure hose, and install the hose using NEW O-rings. Tighten the fitting to 24–34 ft. lbs. (32–47 Nm). Tighten the bracket bolt to 14–18 ft. lbs. (19–25 Nm).

11. Connect the power steering pressure switch connector.

12. Install the drive belt, and the adjusting bolt.

13. Install the nuts onto the adjusting bracket, and the bolt at the pulley.

14. Adjust the belt tension, and fill and bleed the system. Check for leaks.

15. Connect the negative battery cable.

1993–95 RX7

1. Disconnect the negative battery cable. Disconnect and plug the hoses at the pump.

2. Remove the pump drive belt.

3. It is necessary to remove the pump pulley before removing the pump. Hold the pulley from moving, remove the pulley nut and pulley.

4. Support the pump, remove the mounting bolts and lift out the pump.

5. Installation is the reverse of removal. Tighten the pulley nut to 43 ft. lbs. (58 Nm). Adjust the belt tension and fill and bleed the system.

BRAKES

Anti-Lock Brake System Service

PRECAUTIONS

• Certain components within the Anti-Lock Brake System (ABS) are not intended to be serviced or repaired individually. Only those components with removal and installation procedures should be serviced.

• Do not use rubber hoses or other parts not specifically specified for and ABS system. When using repair kits, replace all parts included in the kit. Partial or incorrect repair may lead to functional problems and require the replacement of components.

• Lubricate rubber parts with clean, fresh brake fluid to ease assembly. Do not use lubricated shop air to clean parts; damage to rubber components may result.

• Use only specified brake fluid from an unopened container.

• If any hydraulic component or line is removed or replaced, it may be necessary to bleed the entire system.

• A clean repair area is essential. Always clean the reservoir and cap thoroughly before removing the cap. The slightest amount of dirt in the fluid may plug an orifice and impair the system function. Perform repairs after components have been thoroughly cleaned; use only denatured alcohol to clean components. Do not allow ABS components to come into contact with any substance containing mineral oil; this includes used shop rags.

• The Anti-Lock control unit is a microprocessor similar to other computer units in the vehicle. Ensure that the ignition switch is **OFF** before removing or installing controller harnesses. Avoid static electricity discharge at or near the controller.

• If any arc welding is to be done on the vehicle, the control unit should be unplugged before welding operations begin.

Master Cylinder

REMOVAL AND INSTALLATION

1993–94 323, 1993–95 MX3, 929 and 1995–97 Protege, Miata, Millenia, 626 and MX6

1. Disconnect the negative battery cable.

2. Disconnect the fluid level sensor connector.

3. Disconnect the brake lines from the master cylinder. Wrap the lines with a rag and plug the lines to prevent drainage.

4. Remove the 2 nuts securing the master cylinder to the brake booster.

5. Remove the proportioning bypass valve and bracket.

6. Remove the master cylinder, and on ABS equipped vehicles remove the O-ring.

To install:

7. Replace the O-ring, if equipped.

8. Position the master cylinder and the proportioning bypass valve onto the mounting studs and install the mounting nuts. Tighten mounting nuts to specifications as follows;

  a. 1993–94 323, 626, 929, MX3, MX6, Miata and Protege; 7–12 ft. lbs. (10–16 Nm).

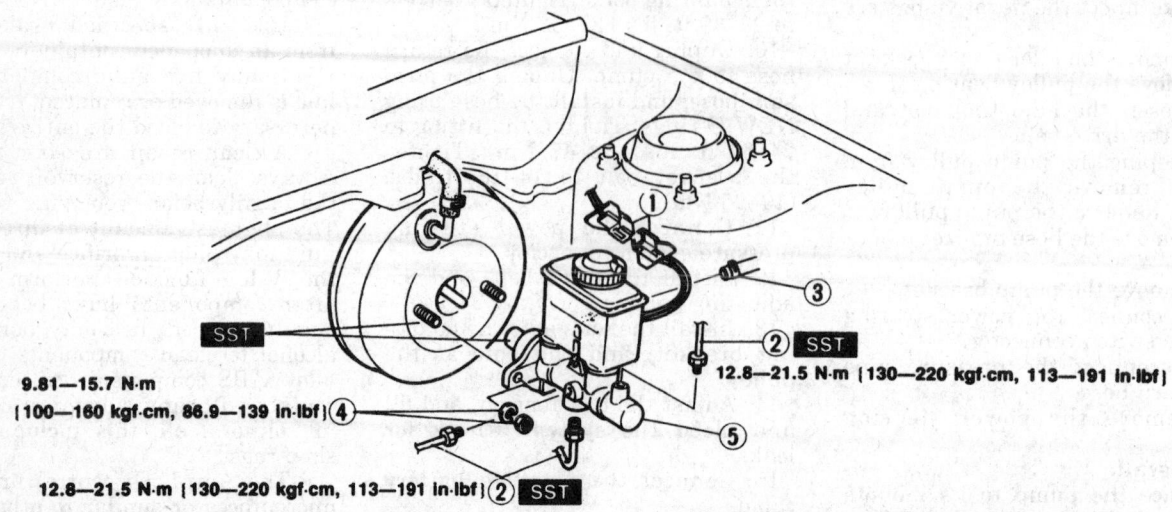

1. Connector
2. Brake pipe
3. Hose (MTX)
4. Nut and washer
5. Master cylinder

9.81—15.7 N·m
(100—160 kgf·cm, 86.9—139 in·lbf)

12.8—21.5 N·m (130—220 kgf·cm, 113—191 in·lbf)

12.8—21.5 N·m (130—220 kgf·cm, 113—191 in·lbf)

SST

328209

Master cylinder removal and installation — 1993-94 323, 626, 929, MX3, MX6, Miata, Protege and 1995 MX3

b. 1995 MX3 ; 9–12 ft. lbs. (12–16 Nm).

c. 1995 929 and 1995–97 Protege; 7–11 ft. lbs. (10–16 Nm).

d. 1995–97 Millenia, 626 and MX6; 87–138 in. lbs. (9–15 Nm)

e. 1995–97 Miata; 7–11 ft. lbs. (10–16 Nm).

9. Fill the reservoir to the proper level with clean DOT 3 brake fluid. Bleed the master cylinder.

10. Replace the brake line washers. Install the brake lines to the master cylinder and tighten to specifications as follows;

a. 1993–94 323, 626, 929, MX3, MX6, Miata and Protege; 9–16 ft. lbs. (13–22 Nm).

b. 1995 MX3 1.8L engine; 15–21 ft. lbs. (20–29 Nm).

c. 1995 MX3 1.6L engine; 10–16 ft. lbs. (13–21 Nm).

d. 1995 929 and 1995–97 Protege; 9–15 ft. lbs. (13–21 Nm).

e. 1995–97 626 and MX6; 113–190 in. lbs. (9–15 Nm).

f. 1995–97 Miata; 15–21 ft. lbs. (20–29 Nm).

11. Apply the brake pedal and check for firmness. If the pedal is spongy, air is present in the system.

If air remains in the system, the entire system must be bled.

12. Connect the negative battery cable and the fluid level sensor connector. Check the brakes for proper operation and leaks.

1993–95 RX7

1. Disconnect the clutch master cylinder hose at the brake master cylinder.

2. Disconnect the brake fluid level sensor connector.

3. Disconnect the brake pipes. Remove the pipe joint and bracket.

4. Remove the mounting nuts and master cylinder bracket. Remove the master cylinder. Remove and discard the O-ring.

5. Installation is the reverse of removal. Properly bench bleed NEW master cylinder and check the piston to pushrod clearance before installation. Torque the master cylinder mounting nuts to 87–138 in. lbs. (9.9–15.6 Nm). Torque the pipe joint mounting bolt to 70–104 in. lbs. (7.9–11.7 Nm).

Brake Caliper

REMOVAL AND INSTALLATION

1993–94 323, 626, 929, MX3, MX6, Miata, Protege and 1995 MX3

Front

1. Raise the vehicle and support safely.

2. Remove the appropriate tire and wheel assembly.

3. Disconnect the flexible brake hose from the caliper.

4. If equipped with one caliper mounting bolt, remove the lower caliper bolt and pivot the caliper upward. Slide the top of the caliper off of the top pin and remove it from the vehicle. If equipped with two caliper mounting bolts, remove both bolts and remove the caliper.

Rear

1. Disconnect the battery negative cable.

2. Raise the vehicle and support safely.

3. Remove the appropriate tire and wheel assemblies. Loosen the parking brake cable adjustment from inside the vehicle.

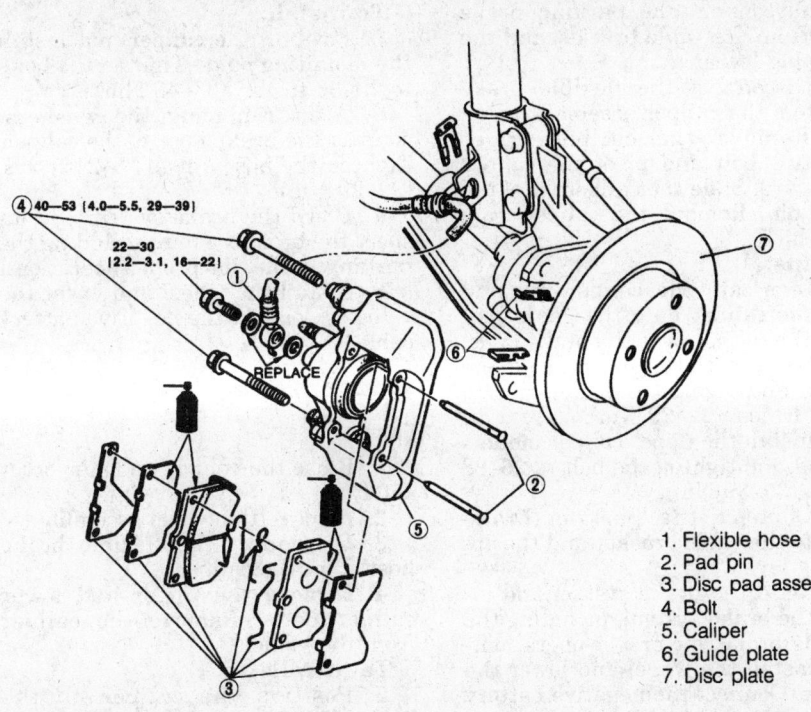

4) 40—53 (4.0—5.5, 29—39)

22—30
(2.2—3.1, 16—22)

REPLACE

1. Flexible hose
2. Pad pin
3. Disc pad assembly
4. Bolt
5. Caliper
6. Guide plate
7. Disc plate

N·m (kgf·m, ft·lbf)
328185

Brake caliper, front — 1993–94 323, 626, 929, MX3, MX6, Miata, Protege and 1995 MX3

4. Disconnect the parking brake cable from the cable bracket and the operating lever.

5. Disconnect the flexible brake line from the caliper assembly.

6. Remove the caliper upper mounting bolt and pivot the caliper downward. Slide the caliper off of the guide pin. Remove the caliper from the vehicle.

To install:

7. Lubricate the caliper pin and slide the caliper onto the guide pin. Pivot the caliper over the brake pads.

8. Connect the brake hose to the caliper and tighten the hose nut to 16–20 ft. lbs. (22–26 Nm).

9. Install the upper caliper mounting bolt and tighten the bolt to 33–49 ft. lbs. (45–67 Nm).

10. Bleed the brake system and inspect the brake system for proper operation.

11. Install the wheel and lower the vehicle. Connect the negative battery cable.

1995–97 Protege

Front

1. Raise the vehicle and support safely.

2. Remove the appropriate tire and wheel assembly.

3. Disconnect the flexible brake hose from the caliper.

4. Remove the disc pad retaining pins. Remove the brake pads.

5. Remove the upper and lower caliper bolts. Remove the caliper from the vehicle.

To install:

6. Position the caliper on the brake disc. Install the caliper mounting bolts and tighten the bolts to 29–36 ft. lbs. (40–49 Nm).

7. Install the disc pad and retaining pins.

8. Replace the washers for the brake line. Connect the brake hose to the caliper and tighten the hose nut to 16–21 ft. lbs. (22–29 Nm).

9. Bleed the brake system and inspect the brake system for proper operation.

10. Install the wheel and lower the vehicle.

Rear

1. Raise the vehicle and support safely. Remove the wheels.

2. Disconnect the parking brake cable from the cable bracket and the operating lever.

3. Disconnect the flexible brake line from the caliper assembly.

4. Turn the manual adjustment gear counterclockwise with an Allen

wrench to pull the caliper piston inward (turn until it stops).

5. Remove the caliper mounting bolts. Remove the caliper from the vehicle.

6. Installation is reverse of removal. Torque the caliper mount bolts to 34–44 ft. lbs. (46–60 Nm). Torque the brake hose line bolt to 16–22 ft. lbs. (22–30 Nm). Properly bleed the brake system.

1995 MX3

Front

1. Raise the vehicle and support safely.

2. Remove the appropriate tire and wheel assembly.

3. Disconnect the flexible brake hose from the caliper.

4. Remove the disc pad retaining pins. Remove the brake pads.

5. Remove the upper and lower caliper bolts. Remove the caliper from the vehicle.

To install:

6. Position the caliper on the brake disc. Install the caliper mounting bolts and tighten the bolts to 29–36 ft. lbs. (40–49 Nm).

7. Install the disc pad and retaining pins.

8. Replace the washers for the brake line. Connect the brake hose to the caliper and tighten the hose nut to 16–21 ft. lbs. (22–29 Nm).

9. Bleed the brake system and inspect the brake system for proper operation.

10. Install the wheel and lower the vehicle. Connect the negative battery cable.

Rear

1. Raise the vehicle and support safely.

2. Remove the wheels. Loosen the parking brake cable adjustment from inside the vehicle.

3. Disconnect the parking brake cable from the cable bracket and the operating lever.

4. Disconnect the flexible brake line from the caliper assembly.

5. Remove the caliper upper mounting bolt and pivot the caliper downward. Slide the caliper off of the guide pin. Remove the caliper from the vehicle.

To install:

6. Lubricate the caliper pin and slide the caliper onto the guide pin. Pivot the caliper over the brake pads.

7. Connect the brake hose to the caliper and tighten the hose nut to 16–21 ft. lbs. (22–29 Nm).

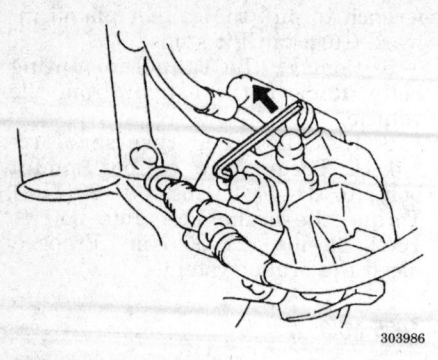

Manual adjustment gear — 1995–97 Protege

8. Install the upper caliper mounting bolt and tighten the bolt to 12–17 ft. lbs. (16–24 Nm).

9. Connect the parking brake cable to the cable bracket and the operating lever.

10. Bleed the brake system and inspect the brake system, including the parking brake, for proper operation.

11. Install the wheel and lower the vehicle. Connect the negative battery cable.

1995–97 626 and MX6

Front

1. Raise the vehicle and support safely.

2. Remove the appropriate tire and wheel assembly.

3. Disconnect the flexible brake hose from the caliper.

4. Remove the lower caliper bolt and pivot the caliper upward. Slide the top of the caliper off of the top pin and remove it from the vehicle.

To install:

5. Lubricate the caliper pin and slide the caliper onto the guide pin. Pivot the caliper over the brake pads.

6. Connect the brake hose to the caliper and tighten the hose nut to 16–21 ft. lbs. (22–29 Nm).

7. Install the caliper mounting bolt and tighten the bolt to 22–29 ft. lbs. (30–40 Nm).

8. Bleed the brake system and inspect the brake system for proper operation.

9. Install the wheel and lower the vehicle. Connect the negative battery cable.

Rear

1. Disconnect the battery negative cable.

2. Raise the vehicle and support safely.

3. Remove the appropriate tire and wheel assemblies. Loosen the parking brake cable adjustment from inside the vehicle.

4. Disconnect the parking brake cable from the cable bracket and the operating lever.

5. Disconnect the flexible brake line from the caliper assembly.

6. Remove the caliper upper mounting bolt and pivot the caliper downward. Slide the caliper off of the guide pin. Remove the caliper from the vehicle.

To install:

7. Lubricate the caliper pin and slide the caliper onto the guide pin. Pivot the caliper over the brake pads.

8. Connect the brake hose to the caliper and tighten the hose nut to 16–21 ft. lbs. (22–29 Nm).

9. Install the upper caliper mounting bolt and tighten the bolt to 26–28 ft. lbs. (35–39 Nm).

10. Connect the parking brake cable to the cable bracket and the operating lever.

11. Bleed the brake system and inspect the brake system, including the parking brake, for proper operation.

12. Install the wheel and lower the vehicle. Connect the negative battery cable.

1995 929

Front

1. Raise the vehicle and support safely.

2. Remove the appropriate tire and wheel assembly.

3. Disconnect the flexible brake hose from the caliper.

4. Remove the upper (marked "L") and lower (marked "G") caliper bolts. Remove the caliper from the vehicle.

To install:

5. Position the caliper on the brake disc. Install the caliper mounting bolts and tighten the bolts to 47–62 ft. lbs. (63–84 Nm).

6. Replace the washers for the brake line. Connect the brake hose to the caliper and tighten the hose nut to 15–21 ft. lbs. (20–29 Nm).

7. Bleed the brake system and inspect the brake system for proper operation.

8. Install the wheel and lower the vehicle. Connect the negative battery cable.

Rear

1. Raise the vehicle and support safely.

2. Remove the wheels.

3. Disconnect the flexible brake line from the caliper assembly.

4. Remove the caliper mounting bolts and remove the caliper from the vehicle.

To install:

5. Position the caliper and install the mounting bolts. Tighten the bolts to 28–36 ft. lbs. (38–49 Nm).

6. After replacing the washers, connect the brake hose to the caliper. Tighten the hose nut to 15–21 ft. lbs. (20–29 Nm).

7. Bleed the brake system and inspect the brake system, including the parking brake, for proper operation.

8. Install the wheel and lower the vehicle. Connect the negative battery cable.

1995–97 Miata

Front

1. Raise the vehicle and support it safely.

2. Remove the wheel assembly.

3. Disconnect the flexible brake hose from the caliper.

4. Remove the upper and lower caliper bolts. Remove the caliper from the vehicle.

To install:

5. Position the caliper on the brake disc. Install the caliper mounting bolts and tighten the bolts to 36–51 ft. lbs. (49–69 Nm).

6. Replace the washers for the brake line. Connect the brake hose to the caliper and tighten the hose nut to 16–22 ft. lbs. (22–29 Nm).

7. Bleed the brake system and inspect the brake system for proper operation.

8. Install the wheel and lower the vehicle. Connect the negative battery cable.

Rear

1. Raise the vehicle and support it safely.

2. Remove the wheels. Loosen the parking brake cable adjustment from inside the vehicle.

3. Disconnect the parking brake cable from the cable bracket and the operating lever.

4. Disconnect the flexible brake line from the caliper assembly.

5. Remove the cover for the manual adjustment gear. Insert an Allen wrench and turn counterclockwise to retract the caliper piston.

6. Remove the caliper mounting bolts and remove the caliper from the vehicle.

To install:

7. Position the caliper and install the mounting bolts. Tighten the upper bolt to 33–36 ft. lbs. (45–49 Nm), and the lower bolt to 25–29 ft. lbs. (34–39 Nm).

8. Turn the manual adjustment gear clockwise to return the caliper until the brake pads just touch the

disc, then turn counterclockwise 1/3 of a turn. Replace the cover.

9. After replacing the washers, connect the brake hose to the caliper. Tighten the hose nut to 16–21 ft. lbs. (22–29 Nm).

10. Connect the parking brake cable to the cable bracket and the operating lever.

11. Bleed the brake system and inspect the brake system, including the parking brake, for proper operation.

12. Install the wheel and lower the vehicle. Connect the negative battery cable.

1995–97 Millenia

Front

1. Raise and safely support the vehicle.
2. Remove the wheels.
3. Disconnect the brake hose and brake pipe. Cap the ends to prevent contamination.
4. Remove the caliper mounting bolts and remove the caliper. Remove the brake rotor.
5. Installation is the reverse of removal. Torque the caliper mounting bolts to 47–62 ft. lbs. (63–84 Nm). Add fluid and properly bleed the brake system.

Rear

1. Raise and safely support the vehicle.
2. Remove the wheels.
3. Disconnect the brake hose and cap the end to prevent contamination.
4. Remove the caliper bracket mounting bolts and remove the caliper. Remove the rotor.
5. Installation is the reverse of removal. Torque the caliper mounting bolts to 12–17 ft. lbs. (16–23 Nm).

1993–95 RX7

Front

1. Raise and safely support the vehicle.
2. Remove the wheels.
3. Disconnect the brake hose and brake pipe. Cap the ends to prevent contamination.
4. Remove the caliper mounting bolts and remove the caliper.
5. Remove the retaining screw and remove the rotor.
6. Installation is the reverse of removal. Torque the rotor retaining screw to 87–130 ft. lbs. (9.9–14.7 Nm). Torque the caliper mounting bolts to 58–72 ft. lbs. (79–98 Nm). Add fluid and properly bleed the brake system

Rear

1. Raise and safely support the vehicle.
2. Remove the wheels.
3. Remove the clip and disconnect the rear parking brake cable.
4. Disconnect the brake hose and cap the end to prevent contamination.
5. Remove the caliper bracket mounting bolts and remove the caliper. Remove the rotor.
6. Installation is the reverse of removal. Torque the caliper mounting bolts to 34–49 ft. lbs. (46–67 Nm).

Disc Brake Pads

REMOVAL AND INSTALLATION

1993–94 323, 626, 929, MX3, MX6, Miata and Protege

------ **CAUTION** ------
Brake pads and shoes may contain asbestos, which has been determined to be a cancer causing agent. Never clean the brake surfaces with compressed air! Avoid inhaling any dust from brake surfaces! When cleaning brakes, use commercially available brake cleaning fluids.

Front

1. Disconnect the negative battery cable.
2. Remove some of the brake fluid from the master cylinder reservoir. The reservoir should be no more than 1/2 full. When the pistons are depressed into the calipers, excess fluid will flow up into the reservoir.
3. Raise the vehicle and support safely.
4. Remove the appropriate tire and wheel assemblies.
5. If equipped with 1 lower caliper mounting bolt, remove the caliper

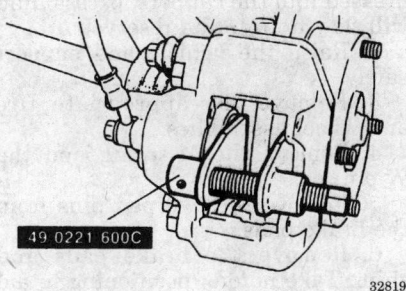

`49 0221 600C`

328196

Compressing the brake caliper

lower mounting bolt and pivot the caliper up and support it. If equipped with 2 caliper mounting bolts, remove both and support the caliper with mechanics wire. Do not kink the brake line or allow the caliper to hang by the brake line.

6. Remove the brake pads, shims and if so equipped, pins. Take note of positioning to aid installation.

7. If the caliper is single piston design, push the caliper back into the bore with a C-clamp or other suitable tool. If the caliper is a 4 piston type caliper, use Mazda tool 49-0221-600C or equivalent and the old inner brake pad, push the caliper piston(s) into the caliper bore.

To install:
8. Install the brake pads and shims to the caliper support. Install the caliper over the brake pads.

NOTE: Be careful that the piston boot does not become caught when lowering the caliper onto the support. Do not twist the brake hose during caliper installation.

9. Install the caliper mounting bolt(s).
10. Install the tire and wheel assemblies. Connect the negative battery cable.
11. Lower the vehicle and test the brakes for proper operation.

Rear

1. Disconnect the battery negative cable.
2. Remove some of the brake fluid from the master cylinder reservoir. The reservoir should be no more than 1/2 full. When the pistons are depressed into the calipers, excess fluid will flow up into the reservoir.
3. Raise the vehicle and support safely.
4. Remove the appropriate tire and wheel assemblies. Loosen the parking brake cable adjustment from inside the vehicle.
5. Disconnect the parking brake cable from the cable bracket and the operating lever.
6. Remove the upper caliper mounting bolt and pivot the caliper downward off of the pads. Do not allow the caliper to hang by the brake line.
7. Remove the brake pads and spring clips from the caliper support. Take note of positioning of each to aid in installation.
To install:
8. Install the brake pads, shims and spring clips to the caliper sup-

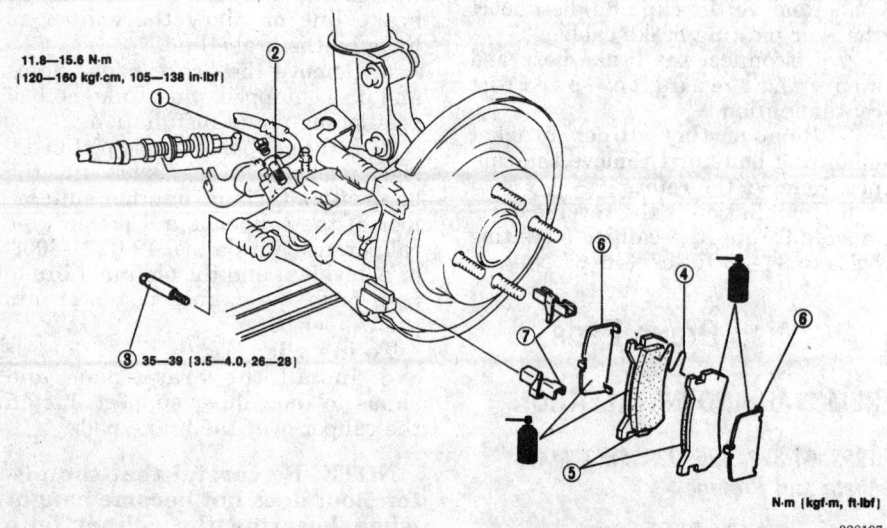

11.8—15.6 N·m
(120—160 kgf·cm, 105—138 in·lbf)

①

②

③ 35—39 (3.5—4.0, 26—28)

④

⑤

⑥

⑦

N·m (kgf·m, ft·lbf)

328197

Brake pads, rear — 1993–94 323, 626, 929, MX3, MX6, Miata, Protege and 1995 MX3

port. Pivot the caliper over the brake pads.

NOTE: Be careful that the piston boot does not become caught when pivoting the caliper onto the support. Do not twist the brake hose during caliper installation.

9. Lubricate and install the top caliper mounting bolt. Tighten the bolt to 12–17 ft. lbs. (16–23 Nm). Attach the parking brake cable to the operating lever and tighten the locknut to 12–17 ft. lbs. (16–23 Nm).

10. Start the engine and forcefully depress the brake pedal 5–6 times. Apply the parking brake and make sure the adjustment is within specifications. Adjust the parking brake cable, as required.

11. Check the disc brake drag by applying the brakes several times and then rotating the wheels to check for excessive dragging.

12. Install the tire and wheel assemblies. Lower the vehicle.

13. Test the brakes for proper operation.

1995–97 Protege

Front

CAUTION

Brake pads and shoes may contain asbestos, which has been determined to be a cancer causing agent. Never clean the brake surfaces with compressed air. Avoid inhaling any dust from brake surfaces. When cleaning brakes, use a commercially available brake cleaning fluids.

1. Remove some of the brake fluid from the master cylinder reservoir. The reservoir should be no more than 1/2 full. When the pistons are depressed into the calipers, excess fluid will flow up into the reservoir.

2. Raise the vehicle and support safely.

3. Remove the appropriate tire and wheel assemblies.

4. Remove the M-spring and the W-pin.

5. Remove the two pad pins from the brake pads.

6. Remove the brake pads and shim. Take note of positioning to aid installation.

7. Push the caliper piston into the caliper bore.

To install:

8. Install the brake pads and shims to the caliper.

9. Insert the pad pins, W-pin and M-spring.

10. Install the tire and wheel assemblies.

11. Lower the vehicle and test the brakes for proper operation.

Rear

1. Remove some of the brake fluid from the master cylinder reservoir. The reservoir should be no more than 1/2 full. When the pistons are depressed into the calipers, excess fluid will flow up into the reservoir.

2. Raise the vehicle and support safely.

3. Remove the wheels. Loosen the parking brake cable adjustment from inside the vehicle.

4. Disconnect the parking brake cable from the cable bracket and the operating lever. Remove the screw plug. Turn the manual adjustment gear counterclockwise with an Allen wrench to pull the brake caliper piston inward (turn until it stops).

5. Remove the upper caliper mounting bolt and pivot the caliper upward off of the pads.

6. Remove the brake pads and spring clips from the caliper support. Take note of positioning of each to aid in installation.

To install:

7. Install the brake pads, shims and spring clips to the caliper support. Pivot the caliper over the brake pads.

NOTE: Be careful that the piston boot does not become caught when pivoting the caliper onto the support. Do not twist the brake hose during caliper installation.

8. Lubricate and install the bottom caliper mounting bolt. Tighten the bolt to 33–44 ft. lbs. (46–60 Nm). Attach the parking brake cable to the operating lever and tighten the screw plug to 9–11 ft. lbs. (12–15 Nm).

9. Start the engine and forcefully depress the brake pedal 5–6 times. Apply the parking brake and make sure the adjustment is within specifications. Adjust the parking brake cable, as required.

10. Check the disc brake drag by applying the brakes several times and then rotating the wheels to check for excessive dragging.

11. Install the tire and wheel assemblies. Lower the vehicle.

12. Test the brakes for proper operation.

1995 MX3

Front

CAUTION

Brake pads and shoes may contain asbestos, which has been determined to be a cancer causing agent. Never clean the brake surfaces with compressed air. Avoid inhaling any dust from brake surfaces. When cleaning brakes, use commercially available brake cleaning fluids.

---

1. Disconnect the negative battery cable.
2. Remove some of the brake fluid from the master cylinder reservoir. The reservoir should be no more than ¹/₂ full. When the pistons are depressed into the calipers, excess fluid will flow up into the reservoir.
3. Raise the vehicle and support safely.
4. Remove the appropriate tire and wheel assemblies.
5. Remove the M-spring and the W-pin.
6. Remove the two pad pins from the brake pads.
7. Remove the brake pads and shim. Take note of positioning to aid installation.
8. Using Mazda tool 49-0221-600C or equivalent and the old inner brake pad, push the caliper piston into the caliper bore.
To install:
9. Install the brake pads and shims to the caliper.
10. Insert the pad pins, W-pin and M-spring.
11. Install the tire and wheel assemblies. Connect the negative battery cable.
12. Lower the vehicle and test the brakes for proper operation.

Rear

1. Disconnect the battery negative cable.
2. Remove some of the brake fluid from the master cylinder reservoir. The reservoir should be no more than ¹/₂ full. When the pistons are depressed into the calipers, excess fluid will flow up into the reservoir.
3. Raise the vehicle and support safely.
4. Remove the wheels. Loosen the parking brake cable adjustment from inside the vehicle.
5. Disconnect the parking brake cable from the cable bracket and the operating lever. Remove the screw plug.
6. Remove the upper caliper mounting bolt and pivot the caliper

downward off of the pads. Do not allow the caliper to hang by the brake line.
7. Remove the brake pads and spring clips from the caliper support. Take note of positioning of each to aid in installation.
To install:
8. Install the brake pads, shims and spring clips to the caliper support. Pivot the caliper over the brake pads.

NOTE: Be careful that the piston boot does not become caught when pivoting the caliper onto the support. Do not twist the brake hose during caliper installation.

9. Lubricate and install the top caliper mounting bolt. Tighten the bolt to 12–17 ft. lbs. (16–24 Nm). Attach the parking brake cable to the operating lever and tighten the screw plug to 105–138 inch lbs. (12–16 Nm).
10. Start the engine and forcefully depress the brake pedal 5–6 times. Apply the parking brake and make sure the adjustment is within specifications. Adjust the parking brake cable, as required.
11. Check the disc brake drag by applying the brakes several times and then rotating the wheels to check for excessive dragging.
12. Install the tire and wheel assemblies. Lower the vehicle.
13. Test the brakes for proper operation.

1995–97 626 and MX6

CAUTION

Brake pads and shoes may contain asbestos, which has been determined to be a cancer causing agent. Never clean the brake surfaces with compressed air. Avoid inhaling any dust from brake surfaces. When cleaning brakes, use commercially available brake cleaning fluids.

Front

1. Remove some of the brake fluid from the master cylinder reservoir. The reservoir should be no more than ¹/₂full. When the pistons are depressed into the calipers, excess fluid will flow up into the reservoir.
2. Raise the vehicle and support safely.
3. Remove the appropriate wheel assemblies.
4. Remove the caliper lower mounting bolt and pivot the caliper up and support it. Do not kink the brake line.

5. Remove the brake pads, shims and pin. Take note of positioning to aid installation.
6. Using Mazda tool 49-0221-600C or equivalent and the old inner brake pad, push the caliper piston into the caliper bore.
To install:
7. Install the brake pads and shims to the caliper support. Install the caliper over the brake pads.

NOTE: Be careful that the piston boot does not become caught when lowering the caliper onto the support. Do not twist the brake hose during caliper installation.

8. Install the caliper mounting bolt and torque to 22–29 ft. lbs. (30–40 Nm).
9. Install the tire and wheel assemblies.
10. Lower the vehicle and test the brakes for proper operation.

Rear

1. Remove some of the brake fluid from the master cylinder reservoir. The reservoir should be no more than ¹/₂ full. When the pistons are depressed into the calipers, excess fluid will flow up into the reservoir.
2. Raise the vehicle and support safely.
3. Remove the appropriate tire and wheel assemblies. Loosen the parking brake cable adjustment from inside the vehicle.
4. Disconnect the parking brake cable from the cable bracket and the operating lever. Remove the screw plug.
5. Remove the upper caliper mounting bolt and pivot the caliper downward off of the pads. Do not allow the caliper to hang by the brake line.
6. Remove the brake pads and spring clips from the caliper support. Take note of positioning of each to aid in installation.
To install:
7. Install the brake pads, shims and spring clips to the caliper support. Pivot the caliper over the brake pads.

NOTE: Be careful that the piston boot does not become caught when pivoting the caliper onto the support. Do not twist the brake hose during caliper installation.

8. Lubricate and install the top caliper mounting bolt. Tighten the bolt to 26–28 ft. lbs. (35–39 Nm). Attach the parking brake cable to the operating lever and tighten the screw

plug to 105–138 inch lbs. (11.8–15.6 Nm).

9. Start the engine and forcefully depress the brake pedal 5–6 times. Apply the parking brake and make sure the adjustment is within specifications. Adjust the parking brake cable, as required.

10. Check the disc brake drag by applying the brakes several times and then rotating the wheels to check for excessive dragging.

11. Install the tire and wheel assemblies. Lower the vehicle.

12. Test the brakes for proper operation.

1995 929

CAUTION

Brake pads and shoes may contain asbestos, which has been determined to be a cancer causing agent. Never clean the brake surfaces with compressed air. Avoid inhaling any dust from brake surfaces. When cleaning brakes, use commercially available brake cleaning fluids.

Front

1. Disconnect the negative battery cable.

2. Remove some of the brake fluid from the master cylinder reservoir. The reservoir should be no more than 1/2 full. When the pistons are depressed into the calipers, excess fluid will flow up into the reservoir.

3. Raise the vehicle and support safely.

4. Remove the appropriate tire and wheel assemblies.

5. Remove the lower lockbolt, and pivot the brake caliper upwards. Secure the caliper with a piece of rope.

6. Remove the brake pads and shim. Take note of positioning to aid installation.

7. Using Mazda tool 49-0221-600C or equivalent and the old inner brake pad, push the caliper piston into the caliper bore.

To install:

8. Install the brake pads and shims onto the caliper.

9. Remove the piece of rope, and pivot the caliper into position. Install the lockbolt and tighten to 47–62 ft. lbs. (63–84 Nm).

10. Install the tire and wheel assemblies. Connect the negative battery cable.

11. Lower the vehicle and test the brakes for proper operation.

Rear

1. Disconnect the battery negative cable.

2. Remove some of the brake fluid from the master cylinder reservoir. The reservoir should be no more than 1/2 full. When the pistons are depressed into the calipers, excess fluid will flow up into the reservoir.

3. Raise the vehicle and support safely.

4. Remove the wheels.

5. Remove the upper caliper mounting bolt and pivot the caliper downward off of the pads. Do not allow the caliper to hang by the brake line.

6. Remove the brake pads and spring clips from the caliper support. Take note of positioning of each to aid in installation.

7. Using Mazda tool 49-0221-600C or equivalent and the old inner brake pad, push the caliper piston into the caliper bore.

To install:

8. Install the brake pads, shims and spring clips to the caliper support. Pivot the caliper over the brake pads.

NOTE: Be careful that the piston boot does not become caught when pivoting the caliper onto the support. Do not twist the brake hose during caliper installation.

9. Lubricate and install the upper caliper mounting bolt. Tighten the bolt to 28–36 ft. lbs. (38–49 Nm).

10. Check the disc brake drag by applying the brakes several times and then rotating the wheels to check for excessive dragging.

11. Install the tire and wheel assemblies. Lower the vehicle.

12. Test the brakes for proper operation.

1995–97 Miata

CAUTION

Brake pads and shoes may contain asbestos, which has been determined to be a cancer causing agent. Never clean the brake surfaces with compressed air. Avoid inhaling any dust from brake surfaces. When cleaning brakes, use commercially available brake cleaning fluids.

Front

1. Remove some of the brake fluid from the master cylinder reservoir. The reservoir should be no more than 1/2 full. When the pistons are depressed into the calipers, excess fluid will flow up into the reservoir.

2. Raise the vehicle and support safely.

3. Remove the appropriate tire and wheel assemblies.

4. Remove the lower lockbolt, and pivot the brake caliper upwards. Secure the caliper with a piece of rope.

5. Remove the brake pads and shim. Take note of positioning to aid installation.

6. Using Mazda tool 49-0221-600C or equivalent and the old inner brake pad, push the caliper piston into the caliper bore.

To install:

7. Install the brake pads and shims onto the caliper.

8. Remove the piece of rope, and pivot the caliper into position. Install the lockbolt and tighten to 58–65 ft. lbs. (78–88 Nm).

9. Install the tire and wheel assemblies. Connect the negative battery cable.

10. Lower the vehicle and test the brakes for proper operation.

Rear

1. Disconnect the battery negative cable.

2. Remove some of the brake fluid from the master cylinder reservoir. The reservoir should be no more than 1/2 full. When the pistons are depressed into the calipers, excess fluid will flow up into the reservoir.

3. Raise the vehicle and support safely.

4. Remove the wheels.

5. Remove the plug for the manual adjustment gear. Using an Allen wrench turn the gear counterclockwise to retract the caliper piston.

6. Remove the lower caliper mounting bolt and pivot the caliper upward off of the pads. Do not allow the caliper to hang by the brake line.

7. Remove the brake pads and spring clips from the caliper support. Take note of positioning of each to aid in installation.

To install:

8. Install the brake pads, shims and spring clips to the caliper support. Pivot the caliper over the brake pads.

NOTE: Be careful that the piston boot does not become caught when pivoting the caliper onto the support. Do not twist the brake hose during caliper installation.

9. Lubricate and install the lower caliper mounting bolt. Tighten the bolt to 25–29 ft. lbs. (34–39 Nm).

10. Turn the manual adjusting gear clockwise until the piston contacts the brake disc, then turn it clockwise 1/3 turn. Replace the plug.

11. Check the disc brake drag by applying the brakes several times and then rotating the wheels to check for excessive dragging.

12. Install the tire and wheel assemblies. Lower the vehicle.

13. Test the brakes for proper operation.

1995–97 Millenia

Front

1. Raise and safely support the vehicle.
2. Remove the wheels.
3. Remove the bottom caliper lock pin and swing the caliper upwards.
4. Remove the V-springs and remove the pads and shims.
5. Press the caliper pistons back into their cylinders. Installation is the reverse of removal. Torque the lock pin to 47–62 ft. lbs. (63–84 Nm). Bleed the brake system if necessary.

Rear

1. Raise and safely support the vehicle.
2. Remove the wheels.
3. Remove the lock pin and rotate the caliper upwards.
4. Remove the V-springs and remove the pads. Remove the shims from the pads.
5. Press the caliper piston back into the cylinder. Installation is the reverse of removal. Torque the lock pin to 37–50 ft. lbs. (50–68 Nm). Bleed the brake system if necessary.

1993–95 RX7

Front

1. Raise and safely support the vehicle.
2. Remove the wheels.
3. Remove the M-clip, the pad pins, and the M-spring. Remove the disc pads. Remove the shims from the pads.
4. Installation is the reverse of removal. Align the outer and inner shim arrows with the disc plate rotation. Bleed the brake system if necessary.

Rear

1. Raise and safely support the vehicle.
2. Remove the wheels.
3. Remove the lock pin and rotate the caliper upwards.
4. Remove the V-spring and remove the pads. Remove the shims from the pads.

5. Installation is the reverse of removal. Before placing the caliper over the pads, rotate the piston clockwise and align the piston grooves to be perpendicular to the lock pin. Bleed the brake system if necessary.

INSPECTION

The front brake pads have built in wear indicators that contact the brake disc when the brake pad thickness becomes too thin and emit a squealing sound to warn the driver.

Inspect the thickness of the brake linings by looking through the brake caliper body check port. The thickness limit of the lining is 0.08 in. (2.0mm) except on 1993–95 RX7, 1995 929, 1995–97 626, MX6, Millenia and Miata. On 1993–95 RX7, 1995 929, 1995–97 626, MX6, Millenia and Miata, the limit is 0.04 in. (1.0mm).

When the limit is exceeded, replace the pads on both sides of the brake disc and also the brake pads on the wheel on the opposite side of the vehicle. Do not replace just one pad on a side without replacing the other pad on the same wheel as well as the brake pads on the other front wheel.

If there is a significant difference in the thickness of the pads on the left and right sides, check the sliding condition of the piston, lock pin sleeve and guide pin sleeve.

Brake Rotor

REMOVAL AND INSTALLATION

1993–94 323, 626, 929, MX3, MX6, Miata, Protege and 1995 MX3

Front

1. Raise the vehicle and support safely. Remove appropriate wheel assembly.
2. Remove the caliper and brake pads. Support the caliper out of the way using wire.
3. The rotor on most models is held to the hub by two small threaded screws. Remove the screws, if equipped, and pull off the rotor.
4. Installation is the reverse of the removal process.

Rear

1. Raise the vehicle and support safely. Remove appropriate wheel assembly.
2. Disconnect the parking brake cable from the rear caliper assembly.

3. Remove the caliper and brake pads. Support the caliper out of the way using wire.
4. Remove the brake rotor (disc) from the rear hub assembly.
5. Installation is the reverse of the removal procedure.

1995 MX3, 929 and 1995–97 626, MX6 and Miata

Front

1. Raise the vehicle and support safely. Remove appropriate wheel assembly.
2. Remove the caliper and brake pads. Support the caliper out of the way using wire.
3. Pull off the rotor.
4. Installation is the reverse of the removal process.

Rear

1. Raise the vehicle and support safely. Remove appropriate wheel assembly.
2. Disconnect the parking brake cable from the rear caliper assembly.
3. Remove the caliper and brake pads. Support the caliper out of the way using wire.
4. Remove the brake rotor (disc) from the rear hub assembly.
5. Installation is the reverse of the removal procedure.

1995–97 Protege

Front

1. Raise the vehicle and support safely.
2. Remove the appropriate tire and wheel assembly.
3. Disconnect the flexible brake hose from the caliper.
4. Remove the disc pad retaining pins. Remove the brake pads.
5. Remove the upper and lower caliper bolts. Remove the caliper from the vehicle.
6. Remove the brake rotor.
   To install:
7. Install the brake rotor.
8. Position the caliper on the brake disc. Install the caliper mounting bolts and tighten the bolts to 29–36 ft. lbs. (40–49 Nm).
9. Install the disc pad and retaining pins.
10. Replace the washers for the brake line. Connect the brake hose to the caliper and tighten the hose nut to 16–21 ft. lbs. (22–29 Nm).
11. Bleed the brake system and inspect the brake system for proper operation.
12. Install the wheel and lower the vehicle.

Rear

1. Raise the vehicle and support safely. Remove the wheels.
2. Disconnect the parking brake cable from the cable bracket and the operating lever.
3. Disconnect the flexible brake line from the caliper assembly.
4. Turn the manual adjustment gear counterclockwise with an Allen wrench to pull the caliper piston inward (turn until it stops).
5. Remove the caliper mounting bolts. Remove the caliper from the vehicle.
6. Remove the small retaining screw holding the disc. Remove the disc.
7. Installation is reverse of removal. Tighten the disc retaining screw to 14 ft. lbs. (19 Nm). Tighten the caliper mount bolts to 34–44 ft. lbs. (46–60 Nm) and the brake hose bolt to 16–22 ft. lbs. (22–30 Nm).
8. Bleed the brake system.

1995–97 Millenia

Front

1. Raise and safely support the vehicle.
2. Remove the wheels.
3. Disconnect the brake hose and brake pipe. Cap the ends to prevent contamination.
4. Remove the caliper mounting bolts and remove the caliper. Remove the brake rotor.
5. Installation is the reverse of removal. Torque the caliper mounting bolts to 47–62 ft. lbs. (63–84 Nm). Add fluid and properly bleed the brake system

Rear

1. Raise and safely support the vehicle.
2. Remove the wheels.
3. Disconnect the brake hose and cap the end to prevent contamination.
4. Remove the caliper bracket mounting bolts and remove the caliper. Remove the rotor.
5. Installation is the reverse of removal. Torque the caliper mounting bolts to 12–17 ft. lbs. (16–23 Nm).

1993–95 RX7

Front

1. Raise and safely support the vehicle.
2. Remove the wheels.
3. Disconnect the brake hose and brake pipe. Cap the ends to prevent contamination.

4. Remove the caliper mounting bolts and remove the caliper.
5. Remove the retaining screw and remove the rotor.
6. Installation is the reverse of removal. Torque the rotor retaining screw to 87–130 ft. lbs. (9.9–14.7 Nm). Torque the caliper mounting bolts to 58–72 ft. lbs. (79–98 Nm). Add fluid and properly bleed the brake system

Rear

1. Raise and safely support the vehicle.
2. Remove the wheels.
3. Remove the clip and disconnect the rear parking brake cable.
4. Disconnect the brake hose and cap the end to prevent contamination.
5. Remove the caliper bracket mounting bolts and remove the caliper. Remove the rotor.
6. Installation is the reverse of removal. Torque the caliper mounting bolts to 34–49 ft. lbs. (46–67 Nm).

INSPECTION

Using a micrometer, measure the disc thickness in at least eight positions, approximately 45 degrees apart and 0.39 in. (10mm) in from the outer edge of the disc. The minimum thickness is 0.87 in. (22mm), with a maximum thickness variation of 0.0006 in. (0.015mm).

If the disc is below limits for thickness, remove it and install a new one. If the thickness variation exceeds the specifications, replace the disc or turn rotor with on the car type brake lathe.

Brake Drums

REMOVAL AND INSTALLATION

1993–94 323, 626, 929, MX3, MX6, Miata, Protege and 1995 MX3

————— CAUTION —————
Brake pads and shoes may contain asbestos, which has been determined to be a cancer causing agent. Never clean the brake surfaces with compressed air. Avoid inhaling any dust from brake surfaces. When cleaning brakes, use commercially available brake cleaning fluids.

1. Disconnect the negative battery cable.
2. Loosen the rear wheel lug nuts. Raise and safely support the vehicle.

3. Remove the rear wheel and remove the center hub cap. Uncrimp the locknut and remove it.
4. Pull the brake drum outward to remove. If the brake drum is difficult to remove, push the operating lever stopper at the backing plate upward to release the operating lever and to increase the shoe clearance.
5. Installation is the reverse of the removal procedure. Install a new locknut and tighten it to 72–130 ft. lbs. (98–177Nm). Crimp the locknut.

1995–97 Protege, 626 and MX6

————— CAUTION —————
Brake pads and shoes may contain asbestos, which has been determined to be a cancer causing agent. Never clean the brake surfaces with compressed air. Avoid inhaling any dust from brake surfaces. When cleaning brakes, use commercially available brake cleaning fluids.

1. Loosen the rear wheel lug nuts. Raise and safely support the vehicle.
2. Remove the rear wheel and remove the center hub cap. Remove the locknut.
3. Remove the screw securing the rear brake drum, and pull the brake drum outward to remove. If the brake drum is difficult to remove, push the operating lever stopper at the backing plate upward to release the operating lever and to increase the shoe clearance.
4. Installation is the reverse of the removal procedure. Install a new locknut and tighten it to 131–173 ft. lbs. (177–235 Nm).

INSPECTION

Inspect the brake drum for any scratches, uneven or abnormal wear. Minor items may be corrected by sanding lightly. The maximum drum inner diameter is 9.06 inch (230.1 mm) on 1993–94 323, 626, 929, MX3, MX6, Miata, Protege and 1995 MX3. It is 7.933 inches (201.5mm) on other models.

Wheel Cylinder

REMOVAL AND INSTALLATION

1993–94 323, 929, Miata, 1993–95 MX3 and 1993–97 Protege, 626 and MX6

1. Raise and support the vehicle safely. Remove the tire and wheel assembly.

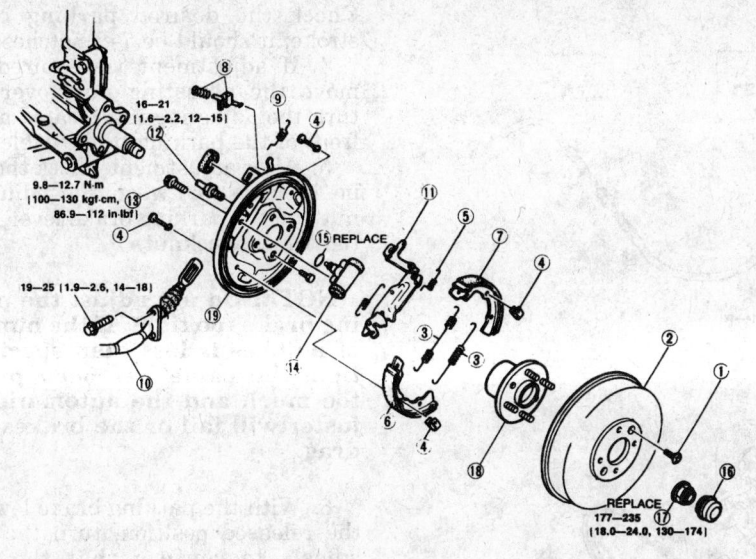

1. Screw
2. Brake drum
3. Return spring
4. Hold pin and spring
5. Anti-rattle spring
6. Brake shoe (leading side)
7. Brake shoe (trailing side)
8. Stopper spring and clip
9. Return spring
10. Parking brake cable
11. Operating lever assembly
12. Brake pipe
13. Bolt
14. Wheel cylinder
15. Wheel cylinder gasket
16. Hub cap
17. Locknut
18. Wheel hub
19. Backing plate

328187

Brake drum assembly — 1993–94 323, 929, Miata, 1993–95 MX3 and 1993–97 626, MX6 and Protege

2. Remove the brake drum and brake shoes.

3. Disconnect and plug the brake line.

4. Remove the stud nuts and bolt attaching the wheel cylinder to the backing plate, and remove the wheel cylinder.

5. Installation is the reverse of removal. Torque the mounting bolts to 86.9–112 in. lbs. (9.81–12.7 Nm). Be sure to bleed the system after installation.

Brake Shoes

INSPECTION

Inspect the brake shoes for peeling, cracking or extremely uneven wear of the brake shoe lining. Measure the brake shoe lining thickness. The minimum thickness is 0.04 inch (1.0mm).

REMOVAL AND INSTALLATION

1993–94 323, 929 Miata and 1993–95 MX3, Protege and 1993–97 626 and MX6

1. Raise and safely support the vehicle.

2. Remove the tire and wheel assembly and remove the brake drum.

3. Disconnect the parking brake cable from the backside of the brake backing plate.

4. From the brake shoe side of the backing plate remove the upper return spring.

5. From the lower (leading side) brake shoe, remove the hold pin and the spring.

6. Remove the lower (leading side) brake shoe and lower return spring and the anti-rattle spring.

7. Remove the upper (trailing side) brake shoe hold pin and spring and remove the upper brake shoe.

To install:

8. Install the upper (trailing side) brake shoe to the operating lever and then to the wheel cylinder and backing plate. Install the brake shoe hold spring and hold pin.

9. Install the anti-rattle spring.

10. Install the lower return spring to both brake shoes.

11. Install the leading side brake shoe to the operating lever and then to the wheel cylinder and anchor plate.

12. Install the hold spring and hold pin to the leading side brake shoe.

13. Install the upper return spring.

14. Install the brake drum.

15. Replace the tire and wheel assembly and safely lower the vehicle.

Parking Brake Cable

ADJUSTMENT

1993–94 323, 929 and Miata and 1993–95 MX3 and 1993–97 626, MX6 and Protege

1. Make sure the parking brake cable is free and is not frozen or sticking. Make sure the brake shoes are properly adjusted.

2. With the engine running, forcefully depress the brake pedal 5–6 times.

3. Apply the parking brake while counting the number of notches. Check the desired parking brake stroke; it should be 5–7 notches.

4. If adjustment is required, remove the adjusting nut clip and turn the adjusting nut which is located at the front of the parking brake cable.

5. After adjustment, check there is no looseness between the adjusting nut and the parking brake lever, then tighten the locknut.

NOTE: Do not adjust the parking brake too tight. If the number of notches is less than specification, the cable has been pulled too much and the automatic adjuster will fail or the brakes will drag.

6. After adjusting the lever stroke, raise the rear of the vehicle and safely support. With the parking brake lever in the released position, turn the rear wheels to confirm that the rear brakes are not dragging.

7. Check that the parking brake holds the vehicle on an incline.

1995 929

1. Make sure the parking brake cable is free and is not frozen or sticking. Make sure the brake shoes are properly adjusted.

2. Apply the parking brake while counting the number of notches. Check the desired parking brake stroke; it should be 5–7 notches.

3. If adjustment is required, turn the adjusting nut located at the base of the parking brake lever.

4. After adjustment, check there is no looseness between the adjusting nut and the parking brake lever, then tighten the locknut.

NOTE: Do not adjust the parking brake too tight. If the number of notches is less than specification, the cable has been pulled too much and the automatic adjuster will fail or the brakes will drag.

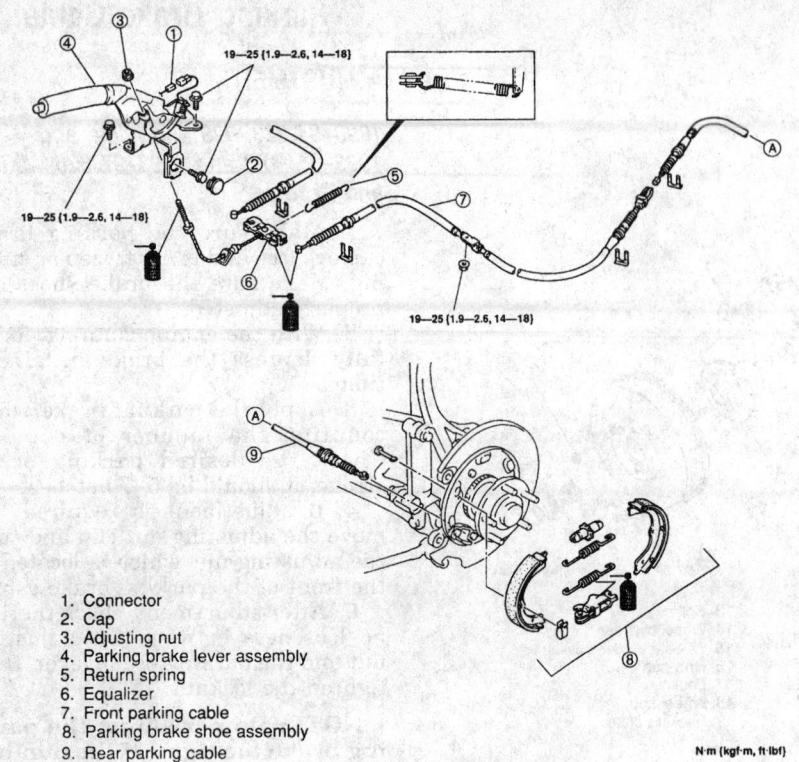

19—25 {1.9—2.6, 14—18}
19—25 {1.9—2.6, 14—18}
19—25 {1.9—2.6, 14—18}

1. Connector
2. Cap
3. Adjusting nut
4. Parking brake lever assembly
5. Return spring
6. Equalizer
7. Front parking cable
8. Parking brake shoe assembly
9. Rear parking cable

N·m {kgf·m, ft·lbf}
335666

Exploded view of the parking brake components

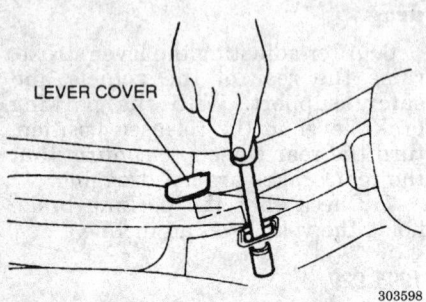

LEVER COVER

303598

Parking brake adjusting cover — 1995 MX3 and 1995–97 626, MX6 and Protege

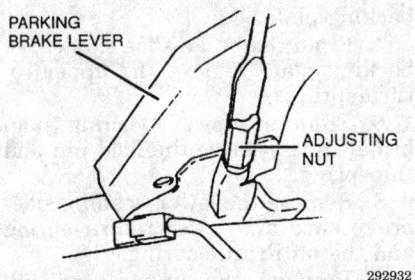

PARKING BRAKE LEVER

ADJUSTING NUT

292932

Parking brake adjusting nut — 1995–97 Millenia

5. With the parking brake lever in the released position, turn the rear wheels to confirm that the rear brakes are not dragging.

6. Check that the parking brake holds the vehicle on an incline.

1995–97 Millenia

1. Make sure the parking brake cable is free and is not frozen or sticking. Make sure the brake shoes are properly adjusted.
2. Raise and safely support the rear wheels.
3. Apply the parking brake while counting the number of notches.

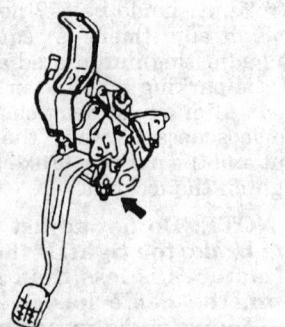

232322

Parking brake adjusting nut — 1995 929

Check the desired parking brake stroke; it should be 7–9 notches.

4. If adjustment is required, remove the adjusting nut cover and turn the adjusting nut located at the front of the parking brake cable.

5. After adjustment, check there is no looseness between the adjusting nut and the parking brake lever, then tighten the locknut.

NOTE: Do not adjust the parking brake too tight. If the number of notches is less than specification, the cable has been pulled too much and the automatic adjuster will fail or the brakes will drag.

6. With the parking brake lever in the released position, turn the rear wheels to confirm that the rear brakes are not dragging.

7. Lower the vehicle.

8. Check that the parking brake holds the vehicle on an incline.

1993–95 RX7

1. Make sure the parking brake cable is free and is not frozen or sticking. Make sure the brake shoes are properly adjusted.

2. With the engine running, forcefully depress the brake pedal 5–6 times.

3. Apply the parking brake while counting the number of notches. Check the desired parking brake stroke; it should be 7–10 notches.

4. If adjustment is required, remove the console panel and remove the adjusting nut clip and turn the adjusting nut which is located at the front of the parking brake cable.

5. After adjustment, check there is no looseness between the adjusting nut and the parking brake lever, then tighten the locknut.

NOTE: Do not adjust the parking brake too tight. If the number of notches is less than specification, the cable has been pulled too much and the automatic adjuster will fail or the brakes will drag.

6. After adjusting the lever stroke, raise the rear of the vehicle and safely support. With the parking brake lever in the released position, turn the rear wheels by hand to confirm that the rear brakes are not dragging.

7. Check that the parking brake holds the vehicle on an incline.

REMOVAL AND INSTALLATION

1993–94 323, 1993–95 MX3 and 1993–97 626, MX6 and Protege

1. Disconnect the negative battery cable.
2. Remove the rear center console as follows:
   a. Remove the screw plugs in the side covers. Remove the retainer screws and the side covers from the vehicle.
   b. Remove the mounting bolts and the floor console from the vehicle.
3. Loosen the cable adjusting nut at the parking brake handle. Disconnect the brake cable from the handle and remove the retaining spring.
4. Raise the vehicle and support safely. Remove the parking brake cable mounting bolts. Disconnect the cable end from the parking brake assembly.
5. Unfasten any remaining frame retainers and remove the cables from the vehicle.

To install:

6. Install the cable to the rear actuator. Secure in place with the parking brake cable mounting bolts.
7. Reattach the parking brake cables to the parking brake handle inside the vehicle. Tighten the ad-

justing nut until the proper tension is placed on the cable.
8. Secure all cable retainers tightening to 14–18 ft. lbs. (19–25 Nm). Apply and release the parking brake a number of times once all adjustments have been made. With the rear wheels raised, make sure the parking brake is not causing drag on the rear wheels.
9. Install the rear center console assembly.
10. Road test the vehicle and check for proper brake operation. Check that the parking brake holds the vehicle on an incline.

1993–95 929

1. Disconnect the negative battery cable.
2. Remove the caliper and brake rotor.
3. Remove the parking brake springs and shoes.
4. Remove the hub dust cover.
5. Loosen the cable adjusting nut at the parking brake lever. Disconnect the brake cable from the handle and remove the clips and retaining spring.
6. Raise the vehicle and support safely. Remove the parking brake cable mounting bolts. Disconnect the

cable end from the parking brake assembly.
7. Unfasten any remaining frame retainers and remove the cables from the vehicle.

To install:

8. Install the cable to the rear actuator. Secure in place with the parking brake cable mounting bolts.
9. Install the retaining spring and clips.
10. Reattach the parking brake cables to the parking brake lever inside the vehicle. Tighten the adjusting nut until the proper tension is placed on the cable.
11. Secure the cable retainer bolts tightening to 14–18 ft. lbs. (19–25 Nm).
12. Install the hub dust cover.
13. Install the parking brake shoes and springs.
14. Install the brake rotor and caliper.
15. Remove the service plug from the rotor. Insert a screwdriver into the hole, and turn the adjuster forward until the disc plate locks. Set the brake shoe clearance to 0.005–0.007 inch (0.12–0.20 mm) by turning the adjuster three to five notches in the opposite direction. Verify that the brakes do not drag by rotating the rotor, and replace the service plug.
16. Apply and release the parking brake a number of times once all adjustments have been made. With the rear wheels raised, make sure the parking brake is not causing drag on the rear wheels.
17. Road test the vehicle and check for proper brake operation. Check that the parking brake holds the vehicle on an incline.

1995–97 Millenia

Front Cable

1. Remove the console around the parking brake lever.
2. Remove the nut securing the front parking brake cable to the lever.
3. Raise and support the vehicle safely.
4. Disconnect the rear parking brake cables from the equalizer.
5. Installation is reverse of removal. Torque the front parking cable nut to 14–18 ft. lbs. (19–25 Nm).

Rear Cable

1. Disconnect the front cable from the rear cables.
2. Remove the clips and mounting bolts and remove the cables.
3. The rear parking brake shoe assembly must be removed for the rear

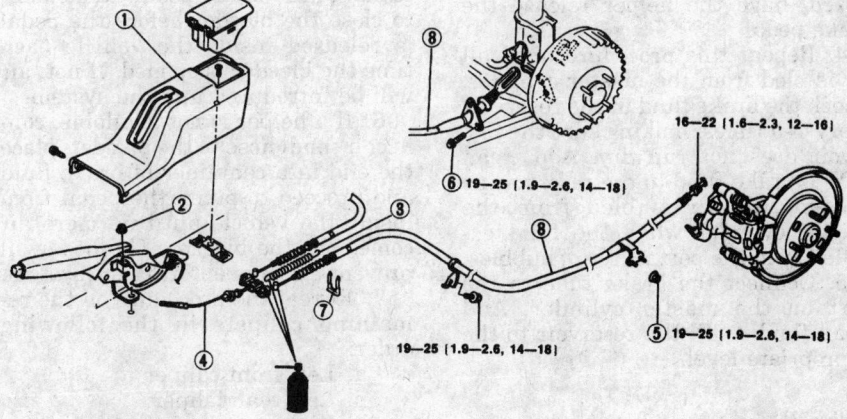

16—22 [1.6—2.3, 12—16]
19—25 [1.9—2.6, 14—18]
19—25 [1.9—2.6, 14—18]
19—25 [1.9—2.6, 14—18]

N·m [kgf·m, ft·lbf]

1. Rear console
2. Adjusting nut
3. Return spring
4. Front parking cable
5. Nut (Disc brake type)
6. Bolt (Drum brake type)
7. Clip
8. Rear parking cable

Parking brake cable — 1993–94 323, 626, MX3, MX6, Protege and 1995 MX3

328746

cable to be disconnected and removed.

4. Installation is the reverse of removal. Torque the mounting screws to 14–18 ft. lbs. (19–25 Nm).

1993–97 Miata

1. Disconnect the negative battery cable.

2. Loosen the cable adjusting nut at the parking brake handle. Disconnect the brake cable from the handle and remove the clips and retaining spring.

3. Raise the vehicle and support safely. Remove the parking brake cable mounting bolts. Disconnect the cable end from the parking brake assembly.

4. Unfasten any remaining frame retainers and remove the cables from the vehicle.

To install:

5. Install the cable to the rear actuator. Secure in place with the parking brake cable mounting bolts.

6. Install the retaining spring and clips.

7. Reattach the parking brake cables to the parking brake handle inside the vehicle. Tighten the adjusting nut until the proper tension is placed on the cable.

8. Secure the cable retainer bolts and tighten to 12–16 ft. lbs. (16–22 Nm).

9. Apply and release the parking brake a number of times once all adjustments have been made. With the rear wheels raised, make sure the parking brake is not causing drag on the rear wheels.

10. Road test the vehicle and check for proper brake operation. Check that the parking brake holds the vehicle on an incline.

1993–95 RX7

1. Remove the adjusting nut at the lever.

2. Disconnect the front cable from the rear cables.

3. Remove the clips and mounting bolts and remove the rear cables.

4. Installation is the reverse of removal. Torque the mounting screws to 14–18 ft. lbs. (19–25 Nm).

Brake System

BLEEDING

NOTE: If using a pressure bleeder, follow the instructions furnished with the unit and choose the correct adaptor for the application. Do not substitute an adapter that "almost fits" as it will not work and could be dangerous.

Master Cylinder

Due to the location of the fluid reservoir, bench bleeding of the master cylinder is not recommended. The master cylinder is to be bled while mounted on the brake booster. If the fluid reservoir runs dry, bleeding of the entire system will be necessary. Two people will be required to bleed the brake system.

1. Fill the brake fluid reservoir with clean brake fluid. Disconnect the brake tube from the master cylinder.

2. Have a helper slowly depress the brake pedal. Once depressed, hold it in that position. Brake fluid will be expelled from the master cylinder.

— CAUTION —
When bleeding the brakes, keep your face away from the area. Spraying fluid may cause facial and/or visual damage. Do not allow brake fluid to spill on the car's finish; it will remove the paint.

3. While the pedal is held down, use a finger to close the outlet port of the master cylinder. While the port is closed, have the helper release the brake pedal.

4. Repeat this procedure until all air is bled from the master cylinder. Check the brake fluid in the reservoir every 4–5 times, making sure the reservoir does not run dry. Add clean DOT 3 brake fluid to the reservoir as needed. All air is bled from the master cylinder when the fluid expelled from the port is free of bubbles.

5. Connect the brake tube to the port on the master cylinder. Add clean fluid to fill the reservoir to the appropriate level.

Calipers

1. Fill the master cylinder with fresh brake fluid. Check the level often during this procedure. Raise and safely support the vehicle.

2. Starting with the wheel furthest from the master cylinder, remove the protective cap from the bleeder and place where it will not be lost. Clean the bleeder screw.

3. Start the engine and run at idle.

— CAUTION —
When bleeding the brakes, keep face away from the brake area. Spewing fluid may cause physical and/or visual damage. Do not allow brake fluid to spill on the car's finish; it will remove the paint.

4. If the system is empty, the most efficient way to get fluid down to the wheel is to loosen the bleeder about 1/2–3/4 turn, place a finger firmly over the bleeder and have a helper pump the brakes slowly until fluid comes out the bleeder. Once fluid is at the bleeder, close it before the pedal is released inside the vehicle.

NOTE: If the pedal is pumped rapidly, the fluid will churn and create small air bubbles, which are almost impossible to remove from the system. These air bubbles will accumulate and a spongy pedal will result.

5. Once fluid has been pumped to the caliper, open the bleed screw again, have the helper press the brake pedal to the floor, lock the bleeder and have the helper slowly release the pedal. Wait 15 seconds and repeat the procedure (including the 15 second wait) until no more air comes out of the bleeder upon application of the brake pedal. Remember to close the bleeder before the pedal is released inside the vehicle each time the bleeder is opened. If not, air will be introduced into the system.

6. If a helper is not available, connect a small hose to the bleeder, place the end in a container of brake fluid and proceed to pump the pedal from inside the vehicle until no more air comes out the bleeder. The hose will prevent air from entering the system.

7. Repeat the procedure on the remaining calipers in the following order:

   a. Left front caliper
   b. Left rear caliper
   c. Right front caliper

8. Hydraulic brake systems must be totally flushed if the fluid becomes contaminated with water, dirt or other corrosive chemicals. To flush, bleed the entire system until all fluid has been replaced with the correct type of new fluid.

9. Install the bleeder cap on the bleeder to keep dirt out. Always road test the vehicle after brake work of any kind is done.

Wheel Speed Sensor

REMOVAL AND INSTALLATION

1993–97 626, MX6 and 1995–97 Protege

1. Raise and safely support the vehicle.
2. Remove the wheel.
3. Disconnect the connector. Remove the mounting bolts/nuts and remove the sensor.
4. Installation is the reverse of removal. Torque the bolts at the sensor to 12–16 ft. lbs. (16–22 Nm).

1995–97 Millenia

Front

1. Disconnect the negative battery cable. Wait at least 90 seconds before performing any work.
2. Working from under the hood disconnect the ABS speed sensor electrical connector.
3. Raise and safely support the vehicle and remove the tire and wheel.
4. Remove the two nuts and the ABS sensor wire bracket from the body.
5. Remove the two nuts and remove the speed sensor wire clamp from the strut bracket.

6. Remove the grommet and draw out the speed sensor wire from the engine compartment.
7. Remove the bolt attaching the speed sensor wire clamp to the knuckle.
8. Remove the mounting bolt and remove the ABS speed sensor.

To install:

9. Install the ABS speed sensor with its mounting bolt and torque to 14–18 ft. lbs. (19–25 Nm).
10. Check the clearance between the wheel-speed sensor and the sensor rotor. Clearance is 0.012–0.043 in. (0.3–1.1 mm). If clearance is not within specifications, replace the

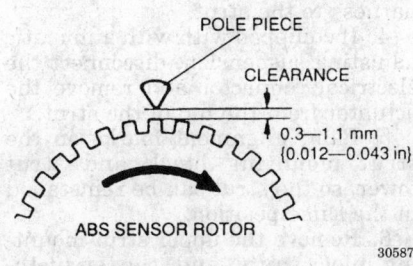

Measuring speed sensor clearance — 1995–97 Millenia and RX7

ABS sensor or the sensor rotor as necessary.

11. Install the speed sensor wire clamp to the steering knuckle with the bolt. Bolt torque is 14–18 ft. lbs. (19–25 Nm).
12. Feed the ABS speed sensor wire into the engine compartment. Reinstall the grommet.
13. Install the wire clamp to the strut bracket with the two nuts and install the wire clamp to body bracket with the two nuts. Torque all the nuts to 70–95.4 in. lbs. (7.9–10.7 Nm).
14. Reinstall the tire and wheel and safely lower the vehicle.
15. Reconnect the speed sensor electrical connector.
16. Reconnect the negative battery cable.

Rear

1. Disconnect the negative battery cable. Wait at least 90 seconds before performing any work.
2. Disconnect the speed sensor electrical connector.
3. Remove the bolts attaching the sensor wire clamps to the body and the strut assembly.
4. Remove the speed sensor mounting bolt and remove the sensor.
5. Installation is the reverse of removal. Mounting bolt and wire clamp mounting bolts are tightened to 14–18 ft. lbs. (19–25 Nm).

RX7

Front

1. Disconnect the negative battery cable. Wait at least 90 seconds before performing any work.
2. Working from under the hood disconnect the ABS speed sensor electrical connector.
3. Raise and safely support the vehicle and remove the tire and wheel.
4. Remove the speed sensor wire clamp from the strut bracket.
5. Remove the grommet and draw out the speed sensor wire from the engine compartment.
6. Remove the bolt attaching the speed sensor wire clamp to the knuckle.
7. Remove the mounting bolt and remove the ABS speed sensor.

To install:

8. Install the ABS speed sensor with its mounting bolt and torque to 12–16 ft. lbs. (16–22 Nm).
9. Check the clearance between the wheel-speed sensor and the sensor rotor. Clearance is 0.012–0.043 in. (0.3–1.1 mm). If clearance is not within specifications, replace the

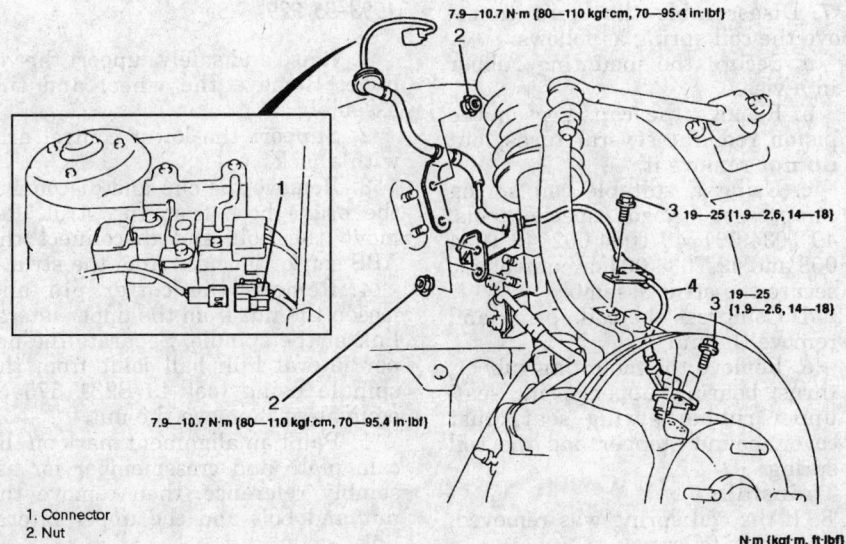

1. Connector
2. Nut
3. Bolt
4. ABS wheel-speed sensor

Front ABS speed sensor and components — 1995–97 Millenia

ABS sensor or the sensor rotor as necessary.

10. Install the speed sensor wire clamp to the steering knuckle.

11. Feed the ABS speed sensor wire into the engine compartment. Reinstall the grommet.

12. Reinstall the tire and wheel and safely lower the vehicle.

13. Reconnect the speed sensor electrical connector.

14. Reconnect the negative battery cable.

Rear

1. Disconnect the negative battery cable. Wait at least 90 seconds before performing any work.

2. Disconnect the speed sensor electrical connector.

3. Remove the bolts attaching the sensor wire clamps to the body and the strut assembly.

4. Remove the speed sensor mounting bolt and remove the sensor.

5. Installation is the reverse of removal. Mounting bolt and wire clamp mounting bolts are tightened to 14–18 ft. lbs. (19–25 Nm).

6. Check the clearance between the wheel-speed sensor and the sensor rotor. Clearance is 0.012–0.043 in. (0.3–1.1 mm). If clearance is not within specifications, replace the ABS sensor or the sensor rotor as necessary.

1993–97 Miata

Front

1. Raise and safely support the vehicle.

2. Remove the wheel.

3. Disconnect the connector. Remove the mounting bolts and the bands. Remove the sensor.

4. Installation is the reverse of removal. Torque the bolts at the sensor to 12–17 ft. lbs. (16–23 Nm).

Rear

1. For the left side, remove the filler pipe protector. Right side, remove the spare tire.

2. Raise and safely support the vehicle.

3. Remove the wheel.

4. Disconnect the connector. Remove the mudguard. Remove the mounting bolts. Remove the sensor.

5. Installation is the reverse of removal. Torque the bolts at the sensor to 12–17 ft. lbs. (16–23 Nm).

FRONT SUSPENSION

Strut

REMOVAL AND INSTALLATION

1993–94 323, MX3, Protege and 1993–97 626 and MX6

1. Raise and safely support the vehicle. Remove the wheel and tire assembly.

2. Support the lower control arm with a jack.

3. Remove the bolt or clip attaching the brake hose and/or ABS sensor harness to the strut.

4. If equipped with with automatic adjusting suspension, disconnect the electrical connector and remove the actuator from the top of the strut.

5. Paint alignment marks on the strut mounting block and strut tower, so the strut can be reinstalled in the same position.

6. Remove the upper strut mounting block nuts and the strut-to-knuckle bolts and remove the strut assembly.

---- CAUTION ----
Serious injury may result from careless action while removing the coil spring.

7. Disassemble the strut to remove the coil spring as follows:

   a. Secure the mounting rubber in a vise.

   b. Remove the cap. Loosen the piston rod nut several turns, but **do not** remove it.

   c. Using a suitable coil spring compressor (Mazda special tools: 49 T034 001, 49 T034 002, 49 T034 003 and 49 T034 004 or equivalent) secure the strut assembly.

   d. Compress the coil spring and remove the nut.

   e. Remove the mounting rubber, thrust bearing, upper spring seat, upper rubber spring seat, dust cover, bound stopper and the coil spring.

To install:

8. If the coil spring was removed, assemble as follows:

   a. Temporarily assemble the upper rubber spring seat, upper spring seat and coil spring to the strut. Mark their locations on the strut for proper reassembly.

   b. Align the marks of the upper spring seat and coil spring. Protect

them with a shop rag, and remove them from the strut.

   c. Using the same coil spring compressor tool that disassembled the strut, compress the spring.

   d. Install the bound stopper and dust cover to the strut.

   e. Install the strut into the coil spring, fitting the end of the coil into the step of the lower seat.

   f. Install the thrust bearing and the mounting rubber. Tighten the piston rod nut several turns.

   g. Remove the compressor tool, and verify that the coil spring is seated on the step of the lower seat.

   h. Secure the mounting rubber in a vise, and tighten the nut to 66–86 ft. lbs. (90–116 Nm). Install the cap.

9. Install the strut into the strut tower, aligning the paint marks made during removal. Install the mounting nuts and tighten to 33–46 ft. lbs. (47–62 Nm).

10. Install the strut-to-knuckle bolts and tighten to 69–86 ft. lbs. (94–117 Nm).

11. If equipped with automatic adjusting suspension, install the actuator and connect the electrical connector.

12. Install the clip or bolt attaching the brake hose and/or ABS sensor harness.

13. Install the wheel and tire assembly and lower the vehicle. Check the front end alignment.

1993–95 929

1. Raise and safely support the vehicle. Remove the wheel and tire assembly.

2. Support the lower control arm with a jack.

3. Remove the clip and disconnect the brake hose from the strut. Remove the bolt and disconnect the ABS sensor harness from the strut.

4. Remove the cotter pin and loosen the nut from the upper lateral link at the spindle. Separate the upper lateral link ball joint from the spindle using tool 49 S231 575 or equivalent. Remove the nut.

5. Paint an alignment mark on the cam plate and crossmember for assembly reference, then remove the nut and bolt and the upper lateral link.

6. Remove the lower strut bolt from the lower control arm.

7. Paint alignment marks on the strut mounting block and strut tower for assembly reference. Remove the nuts and remove the strut and coil assembly.

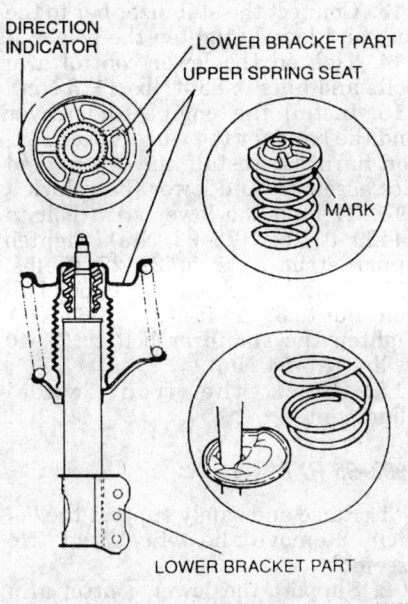

DIRECTION INDICATOR

LOWER BRACKET PART

UPPER SPRING SEAT

MARK

LOWER BRACKET PART

302592

Seating the coil spring — 1993–97 626 and MX6

8. To remove the coil spring from the strut assembly:

a. Safely secure the assembly in a vise. Loosen the piston rod upper nut several turns, but do not remove it.

b. Assembly special tools 49 0223 640B and 49 G030 641. Compress the coil spring using the tools.

c. Remove the upper nut, and remove the coil spring.

To install:

9. Install the coil spring on the strut assembly using the same special tools used to remove it. Tighten the upper nut in a vice with a final torque of 32–44 ft. lbs. (44–60 Nm).

10. Install the strut, making sure the brake hose bracket faces forward.

11. Install the strut into the tower, being sure to align the marks made during removal and loosely install the nuts.

12. Loosely install the lower strut mounting bolt.

13. Install the upper lateral link with the bolt and nut. Align the marks made during removal and tighten the nut to 86 ft. lbs. (117 Nm).

14. Connect the ball joint to the spindle and tighten the nut to 80 ft.

lbs. (108 Nm). Install a new cotter pin.

15. Install the bolt and clip connecting the ABS harness and brake hose to the strut.

16. Install the wheel and tire assembly and lower the vehicle.

17. When the vehicle is on the ground, tighten the strut block-to-tower nuts to 46 ft. lbs. (63 Nm) and the lower strut bolt to 87 ft. lbs. (118 Nm).

18. Check the front wheel alignment.

1995 MX3

1. Raise and safely support the vehicle. Remove the wheel and tire assembly.

2. Support the lower control arm with a jack.

3. Remove the bolt or clip attaching the brake hose and/or ABS sensor harness to the strut.

4. Paint alignment marks on the strut mounting block and strut tower, so the strut can be reinstalled in the same position.

5. Remove the upper strut mounting blocknuts and the strut-to-knuckle bolts and remove the strut assembly.

——— **CAUTION** ———

Serious injury may result from careless action while removing the coil spring.

6. Disassemble the strut to remove the coil spring as follows:

a. Secure the mounting rubber in a vise.

b. Loosen the piston rod nut several turns, but **do not** remove it.

c. Using a suitable coil spring compressor (Mazda special tools: 49 T034 001, 49 T034 002, 49 T034 003 and 49 T034 004 or equivalent) secure the strut assembly.

d. Compress the coil spring and remove the nut.

e. Remove the mounting rubber, thrust bearing, upper spring seat, upper rubber spring seat, dust cover, bound stopper and the coil spring.

To install:

7. If the coil spring was removed, assemble as follows:

a. Temporarily assemble the upper rubber spring seat, upper spring seat and coil spring to the strut. Mark their locations on the strut for proper reassembly.

b. Align the marks of the upper spring seat and coil spring. Protect them with a shop rag, and remove them from the strut.

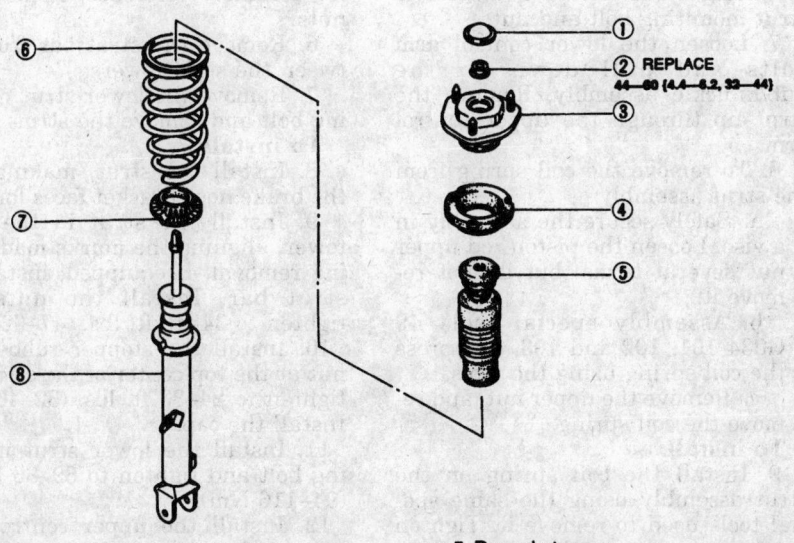

1. Cap
2. Nut
3. Mount
4. Upper spring seat
5. Bound stopper
6. Coil spring
7. Lower spring seat
8. Shock absorber

①
② REPLACE 44—60 {4.4—6.2, 32—44}
③
④
⑤

N·m {kgf·m,ft-lbf}

233007

Strut (exploded view) — 1993–95 929

c. Using the same coil spring compressor tool that disassembled the strut, compress the spring.

d. Install the bound stopper and dust cover to the strut.

e. Install the strut into the coil spring, fitting the end of the coil into the step of the lower seat.

f. Install the thrust bearing and the mounting rubber. Tighten the piston rod nut several turns.

g. Remove the compressor tool, and verify that the coil spring is seated on the step of the lower seat.

h. Secure the mounting rubber in a vise, and tighten the nut to 58–81 ft. lbs. (79–109 Nm).

8. Install the strut into the strut tower, aligning the paint marks made during removal. Install the mounting nuts and tighten to 33–46 ft. lbs. (47–62 Nm).

9. Install the strut-to-knuckle bolts and tighten to 76–93 ft. lbs. (103–126 Nm).

10. Install the clip or bolt attaching the brake hose and/or ABS sensor harness.

11. Install the wheel and tire assembly and lower the vehicle. Check the front end alignment.

Millenia

1. Raise and safely support the vehicle. Remove the wheel and tire assembly.

2. Support the lower control arm with a jack.

3. Remove the nut and disconnect the brake hose from the strut. Remove the nuts and disconnect the ABS sensor cable.

4. Matchmark the strut mounts and remove the mounting nuts, then remove the strut assembly.

5. Remove the cap and the damper fork.

——— CAUTION ———
Removing the piston rod nut is dangerous. The shock absorber and spring could fly off under tremendous pressure and cause serious injury or death. Secure the shock absorber in a compression tool before removing the piston rod nut.

6. Install the assembly into a strut compression tool and remove the top mount nut and mount.

7. Remove the upper spring seat, bound stopper, coil spring and lower spring seat.

To install:

8. Assemble the strut assembly and install the damper fork and cap to the strut.

9. Install the strut, making sure the brake hose bracket faces forward.

10. Install the strut in the strut tower, aligning the marks made during removal. Install the mounting nuts and torque to 34–46 ft. lbs. (47–62 Nm).

11. Install the lower strut damper fork mounting bolt and tighten to 73–101 ft. lbs. (99–137 Nm).

12. Connect the ABS sensor cable mount, tighten the nuts to 70–86 in. lbs. (7.9–9.8 Nm).

13. Attach the brake line mount to the strut and install the mounting nut, torque to 14–18 ft. lbs. (19–25 Nm).

14. Install the wheel and tire assembly and lower the vehicle. Check the front wheel alignment.

1993–97 Miata

1. Raise and safely support the vehicle. Remove the wheel and tire assembly.

2. Remove the engine undercover. Remove the band for the wheel speed sensor harness.

3. Remove the bolt and disconnect the stabilizer bar from the link.

4. Paint alignment marks on the upper strut mounting block and the strut tower, so the strut can be reinstalled in the same position.

5. Remove the cotter pin and the mounting nut for the upper ball joint.

6. Remove the nuts from the strut mounting block. Remove the lower strut mounting bolt and nut.

7. Loosen the lower control arm bolts and pull down on the hub/knuckle assembly. Remove the strut up through the upper control arm.

8. To remove the coil spring from the strut assembly:

a. Safely secure the assembly in a vise. Loosen the piston rod upper nut several turns, but do not remove it.

b. Assembly special tools 49 G034 101, 102 and 103. Compress the coil spring using the tools.

c. Remove the upper nut, and remove the coil spring.

To install:

9. Install the coil spring on the strut assembly using the same special tools used to remove it. Tighten the upper nut in a vice with a final torque of 24–33 ft. lbs. (32–46 Nm).

10. Position the strut and loosely install the lower bolt and nut.

11. Install the strut in the strut tower, making sure the paint marks made during removal are aligned, and loosely install the upper mounting blocknuts.

12. Install the upper ball joint to the knuckle. Use a new cotter pin and loosely install the nut.

13. Connect the stabilizer bar to the link and loosely tighten the bolt.

14. Tighten the lower control arm bolts and nuts to 83 ft. lbs. (113 Nm).

15. Install the engine undercover and the band for the wheel speed sensor harness. Install the wheel and tire assembly and lower the vehicle.

16. Tighten the lower strut bolt to 54–69 ft. lbs. (73–93 Nm). Tighten upper strut nuts to 22–27 ft. lbs. (30–36 Nm). Tighten the upper ball joint nut to 31–44 ft. lbs. (42–60 Nm). Tighten the stabilizer bolt to 27–40 ft. lbs. (37–54 Nm).

17. Check the front wheel alignment.

1993–95 RX7

1. Raise and safely support the vehicle. Remove the wheel and tire assembly.

2. Support the lower control arm with a jack.

3. Remove the clip and disconnect the brake hose from the strut. Remove the bolts and disconnect the ABS sensor and harness.

4. Remove the upper control arm bolts and nuts.

5. Remove the cap, nut and stopper at the top center of the strut. Pain an alignment mark on the upper strut mounting block and strut tower for assembly reference. Remove the nuts.

6. Remove the strut bar from between the strut towers.

7. Remove the lower strut mounting bolt and remove the strut.

To install:

8. Install the strut, making sure the brake hose bracket faces forward.

9. Install the strut in the strut tower, aligning the marks made during removal. If equipped, install the strut bar. Install the nuts and tighten to 34–46 ft. lbs. (47–62 Nm).

10. Install the stopper rubber and nut at the top center of the strut and tighten to 24–33 ft. lbs. (32–46 Nm). Install the cap.

11. Install the lower strut mounting bolt and tighten to 69–86 ft. lbs. (94–116 Nm).

12. Install the upper control arm bolts and nuts and tighten to 44–54 ft. lbs. (59–73 Nm).

13. Connect the ABS sensor and harness. Tighten the bolts to 12–16 ft. lbs. (16–22 Nm). Install the brake hose clip.

14. Install the wheel and tire assembly and lower the vehicle. Check the front wheel alignment.

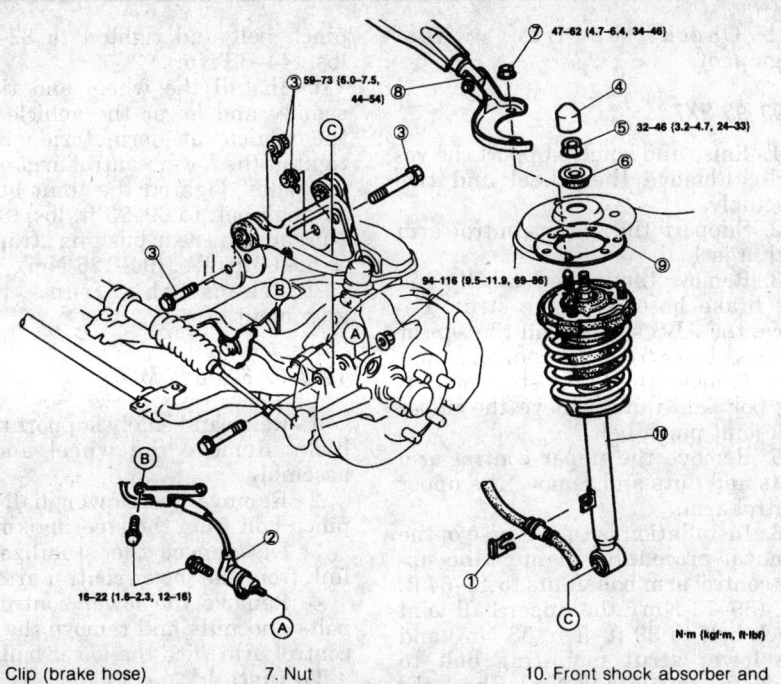

1. Clip (brake hose)
2. ABS wheel-speed sensor
3. Bolt, nut
4. Cap
5. Nut
6. Stopper rubber
7. Nut
8. Front strut bar (R1 vehicle)
9. Insulator
10. Front shock absorber and spring

250008

Strut removal and installation — 1993–95 RX7

Lower Ball Joints

REMOVAL AND INSTALLATION

1993–94 323, MX3 and Protege

1. Raise and safely support the vehicle. Remove the wheel and tire assembly.

2. Remove the ball joint stud pinch bolt and nut from the steering knuckle. Pry the lower control arm down from the knuckle, and separate the ball joint from the knuckle.

3. Remove the bolt and nut and remove the ball joint from the lower control arm.

4. Installation is the reverse of the removal procedure. Tighten the ball joint-to-lower control arm bolt and nut to 86 ft. lbs. (117 Nm). Tighten the ball joint pinch bolt and nut to 43 ft. lbs. (59 Nm). Check the front wheel alignment.

1993–94 Miata

1. Raise and safely support the vehicle. Remove the wheel and tire assembly.

2. Remove the cotter pin and loosen the nut on the lower ball joint. With the nut protecting the ball joint stud, press the stud from the knuckle using press tool 49 0727 575 or

equivalent. Remove the nut from the ball stud.

3. Remove the ball joint-to-lower control arm bolts and nut and remove the lower ball joint.

4. Installation is the reverse of the removal procedure. Tighten the ball joint-to-lower control arm bolts to 69 ft. lbs. (93 Nm). Tighten the ball joint stud nut to 57 ft. lbs. (77 Nm) and install a new cotter pin.

Upper Control Arms

REMOVAL AND INSTALLATION

1993–95 929

NOTE: The curved arm is referred to as the upper lateral link. The straight arm is referred to as the upper leading link.

1. Raise and safely support the vehicle. Remove the wheel and tire assembly.

2. Remove the cotter pin and loosen the lateral link ball joint nut. With the nut protecting the ball joint stud, press the ball joint stud from the knuckle using press tool 49 S231 575 or equivalent. Remove the nut.

3. Paint an alignment mark on the cam plate and crossmember for assembly reference. Remove the nut,

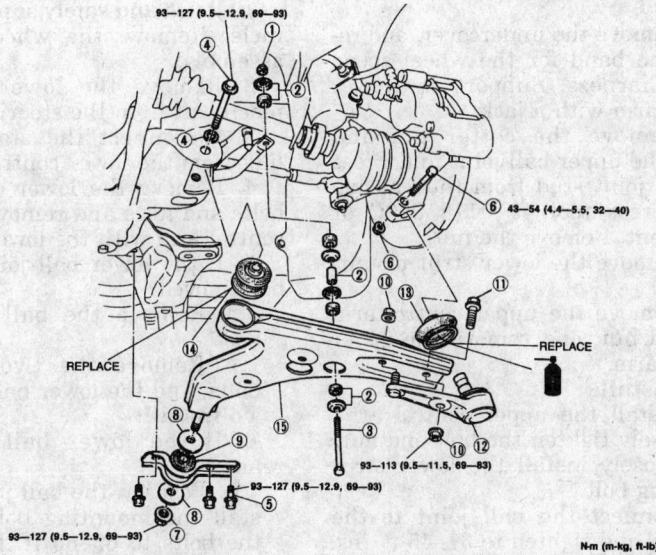

1. Stabilizer nut
2. Retainer, bushing and spacer
3. Stabilizer bolt
4. Bolt and washer
5. Bolt
6. Bolt and nut
7. Nut
8. Washer
9. Lower arm bushing (rear)
   Inspect for wear and damage
10. Nut
11. Bolt
12. Lower arm ball joint
13. Ball joint dust boot
14. Lower arm bushing
15. Front lower arm

128365

Lower control arm, 4WD — 1993–94 323, MX3 and Protege

cam plate, adjusting cam bolt and the upper lateral link.

4. Remove the cotter pin and loosen the leading link ball joint nut. With the nut protecting the ball joint stud, press the ball joint stud from the knuckle using press tool 49 0118 850C or equivalent. Remove the nut.

5. Remove the nuts and remove the upper leading link.

To install:

6. Install the upper leading link and tighten the nuts to 32–40 ft. lbs. (44–54 Nm).

7. Connect the leading link ball joint to the knuckle and tighten the nut to 32–42 ft. lbs. (43–57 Nm). Install a new cotter pin.

8. Install the upper lateral link with the adjusting cam bolt, cam plate and nut. Align the mark made during removal and tighten the nut to 69–86 ft. lbs. (94–117 Nm).

9. Connect the lateral link ball joint to the knuckle, and install the nut. Tighten the nut to 58–80 ft. lbs. (79–108 Nm) and install a new cotter pin.

10. Install the wheel and tire assembly and lower the vehicle. Check the front wheel alignment.

1993–97 Miata

1. Raise and safely support the vehicle. Remove the wheel and tire assembly.

2. Remove the undercover, and remove the band for the wheel speed sensor harness. Support the lower control arm with a jack.

3. Remove the cotter pin and loosen the upper ball joint nut. Press the ball joint stud from the knuckle using press tool 49 0118 850C or equivalent. Remove the nut.

4. Remove the lower strut mounting bolt.

5. Remove the upper control arm bolt and nut, and remove the upper control arm.

To install:

6. Install the upper control arm, and loosely tighten the bolt and nut.

7. Loosely install the lower strut mounting bolt.

8. Connect the ball joint to the knuckle, and tighten to 31–45 ft. lbs. (42–61 Nm). Install a new cotter pin.

9. Install the band for the wheel speed sensor, and install the undercover.

10. Install the wheel and tire assembly and lower the vehicle. When the vehicle is on the ground, tighten the upper control arm bolt and nut to 87–101 ft. lbs. (118–137 Nm), and the lower strut mounting bolt to 54–69 ft. lbs. (73–93 Nm).

11. Check the front wheel alignment.

1993–95 RX7

1. Raise and safely support the vehicle. Remove the wheel and tire assembly.

2. Support the lower control arm with a jack.

3. Remove the brake hose clip and the brake hose from the strut. Remove the ABS sensor and the sensor harness bolts from the control arm.

4. Remove the lower strut mounting bolt and nut. Remove the upper ball joint pinch bolt.

5. Remove the upper control arm bolts and nuts and remove the upper control arm.

6. Installation is the reverse of the removal procedure. Tighten the upper control arm bolts/nuts to 44–54 ft. lbs. (59–73 Nm), the upper ball joint pinch bolt to 39 ft. lbs. (53 Nm) and the lower strut mounting bolt to 69–86 ft. lbs. (94–116 Nm). Check the front wheel alignment.

Lower Control Arms

REMOVAL AND INSTALLATION

1995 MX3 and 1995–97 Protege

1. Raise and safely support the vehicle. Remove the wheel and tire assembly.

2. Remove the lower ball joint pinch bolt from the steering knuckle.

3. Disconnect the stabilizer bar link from the lower control arm.

4. Remove the lower control arm bolts and nuts and remove the lower control arm with the lower ball joint.

5. If the lower ball joint needs to be removed:

  a. Remove the ball joint dust boot.

  b. Remove the two mounting bolts, and the lower ball joint.

To install:

6. If the lower ball joint was removed:

  a. Position the ball joint and install the mounting bolts. Tighten the bolts to 69–86 ft. lbs. (94–126 Nm).

  b. Install a new ball joint dust boot.

7. Install the lower control arm and loosely tighten the mounting nuts and bolts.

8. Connect the stabilizer link to the lower control arm and tighten the nut to 32–44 ft. lbs. (44–60 Nm).

9. Connect the lower ball joint to the steering knuckle. Install the pinch bolt and tighten to 32–43 ft. lbs. (44–58 Nm).

10. Install the wheel and tire assembly and lower the vehicle. With the vehicle at normal ride height, tighten the lower control arm mounting bolts. Tighten the front bushing through-bolt to 69–93 ft. lbs. (94–126 Nm) and the rear bushing strap bolts to 69–93 ft. lbs. (94–126 Nm).

11. Check the front wheel alignment.

1993–97 626 and MX6

1. Raise and safely support the vehicle. Remove the wheel and tire assembly.

2. Remove the lower ball joint pinch bolt from the steering knuckle.

3. Disconnect the stabilizer bar link from the lower control arm.

4. Remove the lower control arm bolts and nuts and remove the lower control arm with the lower ball joint.

To install:

5. Install the lower control arm and loosely tighten the mounting nuts and bolts.

6. Connect the stabilizer link to the lower control arm and tighten the nut to 27–39 ft. lbs. (37–53 Nm).

7. Connect the lower ball joint to the steering knuckle. Install the pinch bolt and tighten to 26–41 ft. lbs. (35–56 Nm).

8. Install the wheel and tire assembly and lower the vehicle. With the vehicle at normal ride height, tighten the lower control arm mounting bolts. Tighten the front bushing through-bolt to 58–78 ft. lbs. (79–106 Nm) and the rear bushing strap bolts to 69–96 ft. lbs. (94–131 Nm).

9. Check the front wheel alignment.

1993–95 929

1. Raise and safely support the vehicle. Remove the wheel and tire assembly.

2. Remove the splash shield.

3. Remove the bolts and disconnect the tension arm from the lower control arm. Remove the tension arm.

4. Remove the stabilizer link bolt.

5. Remove the cotter pin and loosen the nut on the lower ball joint. With the nut protecting the ball joint stud, press the stud from the knuckle using press tool 49 S231 575 or equivalent. Remove the nut.

6. Remove the lower strut mounting bolt.

7. Remove the lower control arm mounting bolt, and remove the control arm.

To install:

8. Install the lower control arm and loosely tighten the bolt.

9. Loosely install the lower strut mounting bolt.

10. Install the ball joint stud in the knuckle. Install the nut and tighten to 87–116 ft. lbs. (118–157 Nm). Install a new cotter pin.

11. Tighten the stabilizer link bolt to 31–40 ft. lbs. (44–54 Nm).

12. Connect the tension arm to the lower control arm. Tighten the bolts to 69–86 ft. lbs. (94–117 Nm). Tighten the tension arm bolt at the chassis to 87–108 ft. lbs. (118–147 Nm).

13. Install the splash shield.

14. Install the wheel and tire assembly and lower the vehicle. With the vehicle at normal ride height, tighten the lower control arm bolt to 86 ft. lbs. (117 Nm), and tighten the lower strut mounting bolt to 87 ft. lbs. (118 Nm).

15. Check the front wheel alignment.

1995–97 Millenia

1. Raise and safely support the vehicle.

2. Remove the transverse member.

3. Disconnect the power steering return hose and pressure pipe.

4. Remove the intermediate steering shaft bolt.

5. Support the engine from the top and remove the engine mount member.

6. Remove the bolts for engine mount No. 1.

7. Remove the lower strut mounting bolt.

8. Disconnect the tie-rod end from the steering knuckle.

9. Remove the upper lateral link ball joint.

10. Remove the lower ball joint.

11. Remove the stabilizer control link nut.

12. Remove the gusset.

13. Support the crossmember using a jack and remove the crossmember mounting nuts. Lower the crossmember to gain clearance and remove the lower arm assembly.

To install:

14. Replace the lower arm assembly to the vehicle and install the crossmember mounting bolts.

15. Install the gusset, torque the gusset mount bolts to 58–86 ft. lbs. (79–116 Nm).

16. Connect the stabilizer control link, torque the nut to 32–44 ft. lbs. (44–60 Nm).

17. Install the lower ball joint, torque the mounting bolts to 58–86 ft. lbs. (79–116 Nm). Torque the retaining nut at the knuckle to 86–115 ft. lbs. (116–156 Nm).

18. Install the upper lateral link nut and bolt, torque to 58–86 ft. lbs. (79–116 Nm).

19. Connect the tie-rod end, torque the nut to 41–59 ft. lbs. (55–80 Nm).

20. Install the strut lower mounting bolt and torque to 73–101 ft. lbs. (98–137 Nm).

21. Install the No. 1 engine mount bolts and torque to 32–44 ft. lbs. (44–60 Nm).

22. Install the engine mount member, torque the bolts to 50–68 ft. lbs. (67–93 Nm) and the nuts to 55–77.3 ft. lbs. (75–104 Nm). Remove the engine support tool.

23. Install the intermediate steering shaft bolt, torque to 14–19 ft. lbs. (18–26 Nm).

24. Connect the power steering pressure pipe and return hose.

25. Install the transverse member.

26. Lower the vehicle. Check the power steering fluid and fill to proper level, bleed if necessary. Check the front end alignment.

1993–94 323, MX3 and Protege

1. Raise and safely support the vehicle. Remove the wheel and tire assembly.

2. Disconnect the stabilizer bar link from the lower control arm.

3. Remove the lower ball joint pinch bolt from the steering knuckle.

4. Remove the lower control arm bolts and nuts and remove the lower control arm.

To install:

5. Install the lower control arm and loosely tighten the mounting nuts and bolts.

6. Connect the lower ball joint to the steering knuckle. Install the pinch bolt and tighten to 40 ft. lbs. (54 Nm).

7. On 323, Protege, assemble the stabilizer bar link. Tighten the top nut until ¾in. of threads are exposed.

8. On MX-3, connect the stabilizer link to the lower control arm and tighten the nut to 39 ft. lbs. (53 Nm).

9. Install the wheel and tire assembly and lower the vehicle. With the vehicle at normal ride height, tighten the lower control arm mounting bolts.

10. On 323, Protege and MX-3, tighten the front bushing through-bolt and the rear bushing strap bolts to 93 ft. lbs. (127 Nm).

1993–97 Miata

1. Raise and safely support the vehicle.

2. Remove the wheel and tire assembly.

3. Remove the undercover.

4. Remove the cotter pin and nut from the tie-rod end. Separate the tie-rod end from the steering knuckle.

5. Remove the stabilizer bar link bolt, and the lower strut mounting bolt.

6. Remove the cotter pin and loosen the nut on the lower ball joint. With the nut protecting the ball joint stud, separate the stud from the knuckle using special tool 49 T028 3A0 or equivalent. Remove the nut.

7. Paint alignment marks on the adjusting cams and the chassis for assembly reference. Remove the bolts, nuts and adjusting cams. Remove the lower control arm.

To install:

8. Install the lower control arm, and loosely tighten the bolts and nuts.

9. Connect the lower ball joint to the knuckle and tighten the nut to 42–57 ft. lbs. (57–77 Nm). Install a new cotter pin.

10. Loosely install the lower strut mounting bolt and stabilizer link bolt.

11. Connect the tie-rod end to the steering knuckle. Tighten the nut to 22–32 ft. lbs. (30–44 Nm), and replace the cotter pin.

12. Install the wheel and tire assembly, and the undercover. Lower the vehicle.

13. With the vehicle at normal ride height, tighten the lower control arm bolts and nuts to 69–83 ft. lbs. (94–113 Nm), being sure to align the marks on the cam plates and chassis made during removal.

14. Tighten the lower strut mounting bolt to 69 ft. lbs. (93 Nm), and the stabilizer link bolt to 40 ft. lbs. (54 Nm).

15. Check the front wheel alignment.

Sway Bar

REMOVAL AND INSTALLATION

1993–323 and Protege

1. Raise and support the vehicle safely. Remove the front wheels.

2. Remove the 2 stabilizer nuts and remove the upper bushing.

3. Remove the stabilizer bolt, bushing and retainer.

4. Remove the bushing, retainer and spacer.

5. Remove the stabilizer center bushings and bracket and remove the

stabilizer bar from the vehicle. Examine the insulators (bushings) carefully for any sign of wear and replace them if necessary.

NOTE: Check the bushings inside the brackets for wear or deformation. A worn bushing can cause a distinct noise as the bar twists during cornering operation.

To install:

6. Install stabilizer bar bushings in the correct position.

7. Temporarily install stabilizer bar brackets. Install the stabilizer bar center bracket bolts and tighten to 27–40 ft. lbs. (38–54 Nm).

8. Install the bushing retainers and spacers. Install the bolt retainer and bushing.

9. Install the upper link nuts. Tighten the link nuts so that there is 0.71–0.87 inches (18.1–22.1mm) of thread exposed.

10. Torque the top link nut to 12–17 ft. lbs. (16–23 Nm).

11. Install front wheels and lower the vehicle.

12. Check front wheel alignment.

1995–97 Protege

1. Disconnect the negative battery cable.

2. Raise and safely support the vehicle.

3. Remove the wheels.

4. Disconnect the tie-rod ends.

5. Remove the stabilizer control link.

6. Remove the transverse member.

7. Remove the three way catalytic converter.

8. Safely support the engine and remove the engine mounting member.

9. (M/T) Remove the change control rod and extension bar.

10. Remove the engine mount bracket.

11. Support the crossmember by using a jack and remove the mounting nuts and bolts.

12. Remove the lower arm, front crossmember and the steering gear assembly.

13. Remove the engine mount.

14. Remove the stabilizer bracket and bushing.

15. Remove the stabilizer bar.

To install:

16. Replace the stabilizer bar to the vehicle and install the bushings and brackets. Torque the bracket mount bolts to 32–44 ft. lbs. (44–60 Nm).

17. Install the engine mount, torque nuts to 50–68 ft. lbs. (67–93 Nm).

18. Replace the steering gear assembly, front crossmember and the lower arm assemblies.

19. Remove the crossmember support.

20. Install the engine mount bracket, torque the bolt to 48–65 ft. lbs. (64–89 Nm).

21. (M/T) Install the extension bar and change control rod.

22. Install the engine mounting member.

23. Install the three way catalytic converter.

24. Install the transverse member.

25. Install the stabilizer control links.

26. Install the tie-rod ends.

27. Connect the negative battery cable.

1993–95 MX3

1. Raise and safely support the vehicle. Remove the front tire/wheels.

2. Remove the engine undercover.

3. Install an engine support device, such as Mazda tool number 49 G017 5A0 or equivalent, to the vehicle.

4. Remove the tie rod end from the steering knuckle assembly.

5. On 1.8L engine only, remove the transverse member from under the engine.

6. Remove the engine mount member.

7. Disconnect the oxygen sensor connectors and remove the front exhaust pipe from the manifold.

8. Remove the stabilizer nuts and insulator pad.

9. Disconnect the steering lines and plug. Remove the steering gear and linkage assembly.

10. Remove the lower arm and front crossmember assembly bolts, and remove the assembly from the vehicle.

11. Remove the remaining stabilizer bar bolts, and remove the stabilizer bar.

To install:

12. Install the stabilizer bar to the vehicle and loosely install the mounting bolts. If the mounting bushings were removed, make sure that they are replaced to the original positions and that the bushings are aligned with the marks on the bar.

13. Install the lower arm and front crossmember assembly to the vehicle. Tighten the mounting bolts to 69–93 ft. lbs. (93–127 Nm).

14. Install the steering gear and linkage. Tighten the mounting nuts to 27–38 ft. lbs. (37–52 Nm), and connect the lines to the steering gear and linkage.

15. Install the insulator plate and install the stabilizer nuts. Tighten the stabilizer nuts to 32–45 ft. lbs. (43–61 Nm).

16. Replace the exhaust pipe gaskets and tighten the nuts to 33 ft. lbs. (44 Nm).

17. Install the engine mount member. Tighten the mounting bolts to 48–65 ft. lbs. (64–89 Nm) and the nuts to 27–38 ft. lbs. (37–52 Nm).

18. On 1.8L engine only, install the transverse member and tighten the bolts to 69–93 ft.lbs. (93–127 Nm).

19. Connect the tie rod end to the steering knuckle. Replace the cotter pin.

20. Remove the engine support device, and install the under engine cover.

21. Install the wheels and lower the vehicle. Tighten the stabilizer bar bolts to 32–43 ft. lbs. (43–59 Nm).

22. Bleed the air from the steering system.

1993–97 626 and MX6

1. Raise and safely support the vehicle. Remove the front tire/wheels.

2. Remove the engine undercover.

3. Install an engine support device, such as Mazda tool number 49 G017 5A0 or equivalent, to the vehicle.

4. Remove the transverse member from under the engine.

5. Remove the engine mount member.

6. Disconnect the oxygen sensor connectors and remove the front exhaust pipe from the manifold.

7. Remove the number one engine mount.

8. Disconnect the tie rod end ball joint from the steering knuckle assembly.

9. Disconnect and plug the pressure and return steering lines. Remove the steering gear and linkage assembly.

10. Support the crossmember, and remove the lower arm and front crossmember assembly mounting bolts.

11. Remove the stabilizer bracket nuts and insulator pad.

12. Lower the crossmember assembly from the vehicle.

13. Remove the remaining stabilizer bar bolts, and remove the stabilizer bar from the crossmember and stabilizer control link.

To install:

14. Install the stabilizer bar to the crossmember, and install the mounting bolts. If the mounting bushings were removed, make sure that they are replaced to the original positions

and that the bushings are aligned with the marks on the bar. Tighten the stabilizer bar bolts to 27–39 ft. lbs. (37–53 Nm).

15. Loosely tighten the stabilizer bracket bolts.

16. Install the lower arm and front crossmember assembly to the vehicle. Tighten the mounting bolts to 69–96 ft. lbs. (93–131 Nm).

17. Install the steering gear and linkage. Tighten the mounting nuts to 28–38 ft. lbs. (38–51 Nm). Unplug and connect the lines to the steering gear and linkage.

18. Connect the tie rod end to the steering knuckle. Replace the cotter pin.

19. Install the number one engine mount.

20. Replace the exhaust pipe gaskets and nuts, and tighten the nuts to 27–38 ft. lbs. (37–52 Nm). Connect the oxygen sensor connectors.

21. Install the engine mount member.

22. Install the transverse member, and tighten the bolts to 69–96 ft. lbs. (94–131 Nm).

23. Remove the engine support device and install the under engine cover.

24. Install the wheels and lower the vehicle.

25. Tighten the stabilizer bracket bolts to 27–39 ft. lbs. (37–53 Nm).

26. Check the front wheel alignment.

27. Bleed the power steering system.

1993–95 929

1. Raise and support the vehicle safely. Remove the front wheels.

2. Remove the 2 stabilizer nuts and remove the upper bushing.

3. Remove the stabilizer bolt, bushing and retainer.

4. Remove the bushing, retainer and spacer.

5. Remove the stabilizer center bushings and bracket and remove the stabilizer bar from the vehicle. Examine the insulators (bushings) carefully for any sign of wear and replace them if necessary.

NOTE: Check the bushings inside the brackets for wear or deformation. A worn bushing can cause a distinct noise as the bar twists during cornering operation.

To install:

6. Install stabilizer bar bushings in the correct position.

7. Temporarily install stabilizer bar brackets. Install the stabilizer

bar center bracket bolts and tighten to 32–40 ft. lbs. (44–54 Nm).

8. Install the bushing retainers and spacers. Install the bolt retainer and bushing.

9. Install the upper link nuts. Tighten the link nuts so that there is 0.71–0.87 inches (18.1–22.1mm) of thread exposed.

10. Torque the stabilizer link bolt to 32–39 ft. lbs. (44–53 Nm).

11. Install front wheels and lower the vehicle.

12. Check front wheel alignment.

1995–97 Millenia

1. Disconnect the negative battery cable and wait at least 90 seconds before performing any work.

2. Raise and safely support the vehicle.

3. Remove the splash shields.

4. Remove the transverse member.

5. Disconnect the power steering return hose and pressure pipe.

6. Remove the intermediate steering shaft bolt.

7. Support the engine from the top and remove the engine mount member.

8. Remove the bolts for engine mount No. 1.

9. Remove the lower strut mounting bolt.

10. Disconnect the stabilizer control link.

11. Support the crossmember using a jack and remove the crossmember mounting nuts. Lower the crossmember to gain clearance and remove the stabilizer bracket and bushing. Remove the stabilizer bar.

To install:

12. Replace the stabilizer bar assembly to the vehicle and install bushings and brackets, torque the nut to 32–44 ft. lbs. (44–60 Nm), then install the crossmember mounting bolts.

13. Connect the stabilizer control link, torque the nut to 32–44 ft. lbs. (44–60 Nm).

14. Install the strut lower mounting bolt and torque to 73–101 ft. lbs. (98–137 Nm).

15. Install the No. 1 engine mount bolts and torque to 32–44 ft. lbs. (44–60 Nm).

16. Install the engine mount member, torque the bolts to 50–68 ft. lbs. (67–93 Nm) and the nuts to 55–77.3 ft. lbs. (75–104 Nm). Remove the engine support tool.

17. Install the intermediate steering shaft bolt, torque to 13–19 ft. lbs. (18–26 Nm).

18. Connect the power steering pressure pipe and return hose.

19. Install the transverse member.

20. Install the splash shields.

21. Lower the vehicle. Reconnect the negative battery cable. Check the power steering fluid and fill to proper level, bleed if necessary. Check the front end alignment.

1993–97 Miata

1. Raise and support the vehicle safely.

2. Remove the splash shield.

3. Remove the two stabilizer nuts, and remove the upper bushing.

4. Remove the stabilizer bolt, bushing and retainer.

5. Remove the stabilizer bar from the vehicle. Examine the insulators (bushings) carefully for any sign of wear and replace them if necessary.

NOTE: Check the bushings inside the brackets for wear or deformation. A worn bushing can cause a distinct noise as the bar twists during cornering operation.

To install:

6. Install stabilizer bar bushings in the correct position.

7. Temporarily install stabilizer bar brackets.

8. Install the bolt retainer and bushing.

9. Install the splash shield.

10. Lower the vehicle.

11. Tighten the stabilizer bar brackets bolts to 14–19 ft. lbs. (18–26 Nm), and the control link mounting bolts to 27–39 ft. lbs. (37–53 Nm).

12. Check front wheel alignment.

1993–95 RX7

1. Raise and support the vehicle safely. Remove the front wheels and the undercover.

2. Remove the stabilizer control link nut and the stabilizer plate mount bolts. Remove the bushings.

3. Remove the stabilizer bar. If necessary, remove the stabilizer control link.

4. Installation is the reverse of removal. Check front wheel alignment.

Front Wheel Bearings

INSPECTION

1993 323, 1993–94 Protege, 1993–95 RX7, MX3 and 1993–97 626 and MX6

1. Raise and support the vehicle safely. Remove the tire and wheel assembly.

1. Stabilizer bracket
2. Stabilizer bushing
3. Stabilizer
4. Stabilizer control link

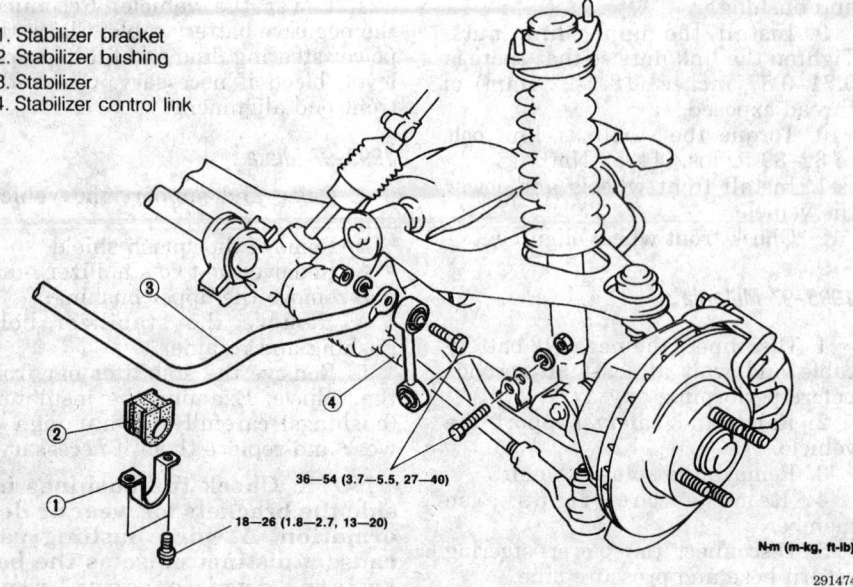

36—54 (3.7—5.5, 27—40)

18—26 (1.8—2.7, 13—20)

N·m (m-kg, ft-lb)

291474

Sway bar (exploded view) — 1993–97 Miata

2. Remove and properly support the caliper assembly.

3. Position a dial indicator gauge against the dust cap. Push and pull the disc brake rotor or brake drum in and out in the axial direction and measure the end-play of the wheel bearing.

4. End-play should not exceed 0.002 in. (0.05 mm) on all other vehicles.

5. If end-play is excessive, check the hub nut torque or replace the bearing.

1995–96 Millenia

1. Raise and support the vehicle safely. Remove the wheel assembly.

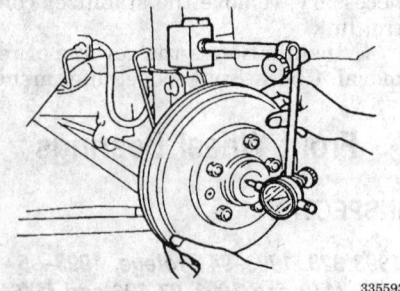

335593

Checking wheel bearing play — 1993 323, 1993–94 Protege, 1993–95 RX7, MX3 and 1993–97 626 and MX6

2. Remove and properly support the caliper assembly.

3. Remove the disc plate.

4. Position a dial indicator gauge against the wheel hub. Push and pull on the hub in the axial direction and measure the end-play of the wheel bearing.

5. Endplay should not exceed 0.002 in. (0.05mm).

6. If end-play is excessive, check the hub nut torque or replace the bearing.

7. Install the disc, caliper, and wheel.

1993–95 MX3, 1995 RX7, 1993–97 626 and MX6

1. Raise and support the vehicle safely. Remove the tire and wheel assembly.

2. Remove and properly support the caliper assembly.

3. Position a dial indicator gauge against the dust cap. Push and pull the disc brake rotor or brake drum in and out in the axial direction and measure the end-play of the wheel bearing.

4. Endplay should not exceed 0.002 in. (0.05 mm) on all other vehicles.

5. If end-play is excessive, check the hub nut torque or replace the bearing.

1993–95 929 and 1993–97 Miata

The wheel bearings on these vehicles are not adjustable. To check if the bearing requires service, remove the wheel and tire assembly, brake caliper and disc brake rotor. Install a dial indicator with the indicator foot resting on the wheel hub. Try to move the hub in and out. If there is more than 0.002 in. (0.05mm) bearing play, check the wheel hub nut torque or replace the hub and bearing assembly.

REMOVAL AND INSTALLATION

1993–94 323, MX3 and Protege

1. Raise and safely support the vehicle. Remove the front wheel and tire assemblies.

2. Uncrimp the tab on the center locknut and remove the locknut. Discard the old locknut.

3. Remove the caliper assembly from the knuckle. Do not disconnect the brake lines. Support the caliper with a piece of wire. Do not allow the caliper to hang by the hose at any time. Remove the brake disc.

4. Remove the tie rod end cotter pin and remove the tie rod end nut. Using Mazda tool 49 0118 850C or equivalent, press the tie rod end out of the knuckle assembly.

5. Remove the stabilizer upper nuts and remove the stabilizer link bolt.

6. Remove the outer lower arm to ball joint mounting bolt and nut. Separate the lower arm from the knuckle assembly.

7. Remove the ABS speed sensor if so equipped.

8. Using a plastic mallet, tap the driveshaft free of the knuckle assembly. Remove the knuckle assembly.

9. Clamp the knuckle in a vise with protected jaws.

10. Remove the inner oil seal from the knuckle.

11. Use Mazda hub puller tools 49 G033 102, 49 G033 104 and 49 G033 105 or equivalent, and remove the front wheel hub from the knuckle assembly.

12. Remove the bearing inner race from the front wheel hub.

13. Remove the retaining ring from within the knuckle and using the hub puller tools, press the front wheel bearing from the knuckle.

14. Remove the brake dust shield.

15. Clean and inspect all parts but do not wash or clean the wheel bearing. The bearing must be replaced.

16. Using Mazda press tools 49 G033 107 and 49 H026 103 or equivalent, install a new dust shield cover assembly to the knuckle.

17. Using the press tools, press a new wheel bearing into the knuckle assembly.

18. Install the wheel bearing retaining ring, and install a new oil seal using installation tool 49 V001 795.

19. Install the front wheel hub by using the Mazda press tools or equivalent.

20. Install the bearing/hub and knuckle assembly in place. Loosely tighten the knuckle to shock absorber bolt.

21. Install the lower arm ball joint to the knuckle and tighten the nut to 27–40 ft. lbs. (36–54 Nm).

22. Install the driveshaft to the knuckle assembly.

23. Install the stabilizer control link and tighten the nuts so that there is 0.71–0.87 inches (18.1–22.1mm) of thread exposed. Torque the top link nut to 12–17 ft. lbs. (16–23 Nm).

24. If equipped with ABS, install the wheel speed sensor and tighten the bolts to 12–17 ft. lbs. (16–23 Nm).

25. Connect the tie rod ends to the knuckle and tighten the nuts to 31–42 ft. lbs. (42–57 Nm). Replace the cotter pins.

26. Install a new wheel hub locknut and tighten the locknut to 174–235 ft. lbs. (235–319 Nm).

27. Check the end play of the wheel bearing by installing a dial indicator against the wheel hub and tire to move the brake disc back and forth. There should be no more than 0.0079 inch (0.2mm) of freeplay present.

28. Stake the locknut into place by bending it into the groove.

29. Install the brake caliper(s) and tighten the bolts to 58–72 ft. lbs. (78–98 Nm).

30. Install the front wheels and lower the vehicle.

31. With the vehicle lowered check all of the bolts and re-torque as is necessary.

32. Inspect the front end alignment and adjust as is necessary.

1995 MX3 and 1995–97 Protege

1. Raise and safely support the vehicle. Remove the front wheels.

2. Uncrimp the tab on the center locknut and remove the locknut. Discard the old locknut.

3. Remove the caliper assembly from the knuckle. Do not disconnect the brake lines. Support the caliper with a piece of wire. Do not allow the caliper to hang by the hose at any time. Remove the brake disc.

4. Remove the ABS speed sensor if so equipped.

5. Remove the tie rod end cotter pin and remove the tie rod end nut. Using Mazda tool 49 T028 3AO or equivalent, press the tie rod end out of the knuckle assembly.

6. Remove the outer lower arm to ball joint mounting bolt and nut. Separate the lower arm from the knuckle assembly.

7. Using a plastic mallet, tap the knuckle assembly free of the driveshaft. Remove the knuckle assembly.

8. Clamp the knuckle in a vise with protected jaws.

9. Remove the inner oil seal from the knuckle.

10. Use Mazda hub-puller tools 49 F026 103, 49 G030 727 and 49 G033 102 or equivalent, and remove the front wheel hub from the knuckle assembly.

11. Remove the bearing inner race from the front wheel hub.

NOTE: If the bearing inner race still remains on the hub assembly, grind a section of the bearing inner race until about 0.02 inch (0.50 mm) remains. Remove with a chisel.

12. Remove the retaining ring from within the knuckle and using the wheel bearing removal tools (49 F027 005 and 49 F027 003 or equivalent), press the front wheel bearing from the knuckle.

13. Clean and inspect all parts but do not wash or clean the wheel bearing. .

14. Using the press tools, press a new wheel bearing into the knuckle assembly.

15. Install the wheel bearing retaining ring.

16. Install the front wheel hub by using the Mazda press tools 49 G033 102 and 49 F027 009 or equivalents.

17. Install a new oil seal using installation tool 49 V001 795 and a hammer. Tap the oil seal in evenly until the special tool contacts the steering knuckle. Coat the lip of the oil seal with grease.

18. Install the bearing/hub and knuckle assembly in place. Loosely

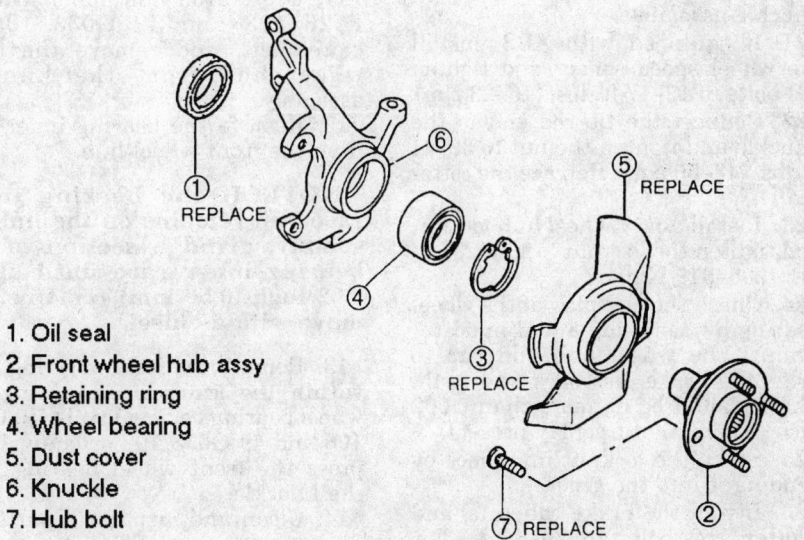

1. Oil seal
2. Front wheel hub assy
3. Retaining ring
4. Wheel bearing
5. Dust cover
6. Knuckle
7. Hub bolt

306047

Steering knuckle and hub disassembly (exploded view) — 1995 MX3 and 1995–97 Protege

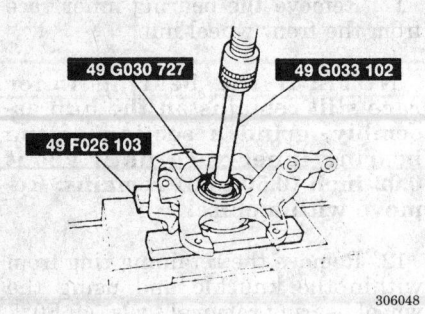

49 G030 727   49 G033 102

49 F026 103

306048

**Removing the hub from the steering knuckle —
1995 MX3 and 1995–97 Protege**

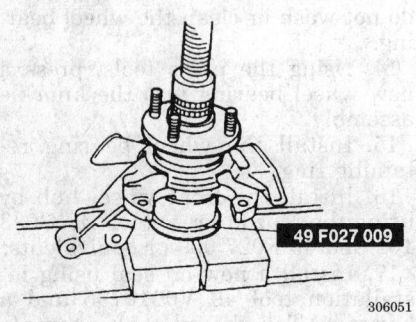

49 F027 009

306051

**Assembling the hub to the knuckle — 1995 MX3
and 1995–97 Protege**

tighten the knuckle to shock absorber
bolt.

19. Install the lower arm ball joint
to the knuckle and tighten the nut to
33–43 ft. lbs. (44–58 Nm).

20. Install the driveshaft to the
knuckle assembly.

21. If equipped with ABS, install
the wheel speed sensor and tighten
the bolts to 12–17 ft. lbs. (16–23 Nm).

22. Connect the tie rod end to the
knuckle and tighten the nut to 32–41
ft. lbs. (43–56 Nm). Replace the cotter
pin.

23. Install a new wheel hub locknut
and tighten the locknut to 174–235 ft.
lbs. (235–319 Nm).

24. Check the end play of the wheel
bearing by installing a dial indicator
against the wheel hub and tire to
move the brake disc back and forth.
There should be no more than 0.002
inch (0.05mm) of freeplay present.

25. Stake the locknut into place by
bending it into the groove.

26. Install the brake caliper(s) and
tighten the bolts to 29–36 ft. lbs.
(40–49 Nm).

27. Install the front wheels and
lower the vehicle.

28. With the vehicle lowered check
all of the bolts and re-torque as
necessary.

29. Inspect the front end alignment
and adjust as is necessary.

1993–97 626 and MX6

1. Raise and safely support the vehicle. Remove the front wheel and
tire assemblies.

2. Uncrimp the tab on the center
locknut and remove the locknut. Discard the old locknut.

3. Remove the caliper assembly
from the knuckle. Do not disconnect
the brake lines. Support the caliper
with a piece of wire. Do not allow the
caliper to hang by the hose at any
time. Remove the brake disc.

4. Remove the ABS speed sensor if
so equipped.

5. Remove the tie rod end cotter
pin and remove the tie rod end nut.
Using Mazda tool 49 T028 3AO or
equivalent, press the tie rod end out
of the knuckle assembly.

6. Remove the stabilizer upper
nuts and remove the stabilizer link
bolt.

7. Remove the outer lower arm to
ball joint mounting bolt and nut. Separate the lower arm from the knuckle
assembly.

8. Using a plastic mallet, tap the
knuckle assembly free of the
driveshaft. Remove the knuckle
assembly.

9. Clamp the knuckle in a vise
with protected jaws.

10. Remove the inner oil seal from
the knuckle.

11. Use Mazda hub puller tools 49
F026 103 and 49 G033 105 or
equivalent, and remove the front
wheel hub from the knuckle
assembly.

12. Remove the bearing inner race
from the front wheel hub.

**NOTE: If the bearing inner
race still remains on the hub assembly, grind a section of the
bearing inner race until about
0.02 inch (0.50 mm) remains. Remove with a chisel.**

13. Remove the retaining ring from
within the knuckle and using the
wheel bearing removal tools (49 G033
106 and 49 G033 102 or equivalent),
press the front wheel bearing from
the knuckle.

14. Clean and inspect all parts but
do not wash or clean the wheel
bearing.

15. Using the press tools, press a
new wheel bearing into the knuckle
assembly.

16. Install the wheel bearing retaining ring.

17. Install the front wheel hub by
using the Mazda press tools 49 G033
105 and 49 F027 009 or equivalent.

18. Install a new oil seal using installation tool 49 V001 795 and a
hammer. Tap the oil seal in evenly
until the special tool contacts the
steering knuckle. Coat the lip of the
oil seal with grease.

19. Install the bearing/hub and
knuckle assembly in place. Loosely
tighten the knuckle to shock absorber
bolt.

20. Install the lower arm ball joint
to the knuckle and tighten the nut to
26–41 ft. lbs. (35–56 Nm).

21. Install the driveshaft to the
knuckle assembly.

22. Install the stabilizer control
link and tighten the nuts so that
there is 0.71–0.87 inches
(18.1–22.1mm) of thread exposed.
Torque the top link nut to 12–17 ft.
lbs. (16–23 Nm).

23. If equipped with ABS, install
the wheel speed sensor and tighten
the bolts to 12–17 ft. lbs. (16–23 Nm).

24. Connect the tie rod ends to the
knuckle and tighten the nuts to
24–32 ft. lbs. (32–44 Nm). Replace the
cotter pin.

25. Install a new wheel hub locknut
and tighten the locknut to 174–235 ft.
lbs. (235–319 Nm).

26. Check the end play of the wheel
bearing by installing a dial indicator
against the wheel hub and tire to
move the brake disc back and forth.
There should be no more than 0.0079
inch (0.2mm) of freeplay present.

27. Stake the locknut into place by
bending it into the groove.

28. Install the brake caliper(s) and
tighten the bolts to 58–72 ft. lbs.
(78–98 Nm).

29. Install the front wheels and
lower the vehicle.

30. With the vehicle lowered check
all of the bolts and re-torque as is
necessary.

31. Inspect the front end alignment
and adjust as is necessary.

1993–95 929

1. Remove the dust cap. Unstake
and loosen the locknut.

2. Raise and safely support the vehicle. Remove the wheel and tire
assembly.

3. Remove the brake caliper and
suspend it from the coil spring.

4. Remove the disc brake rotor.

5. Remove and discard the hub
nut. Remove the wheel hub and bearing assembly. The wheel hub and
bearing cannot be disassembled.

To install:

6. Install the hub over the spindle. Loosely install a new hub nut.

7. Install the brake rotor and the disc brake caliper. Install the wheel and tire assembly, and lower the vehicle.

8. When the vehicle is on the ground, tighten the hub nut. Tighten to 131–173 ft. lbs. (177–235 Nm).

9. Install the dust cap.

1995–97 Millenia

1. Raise and safely support the vehicle.

2. Lock the hub by having a helper apply the brakes firmly. Remove the locknut.

3. Remove the brake caliper assembly and disc plate.

4. Disconnect the ABS wheel speed sensor cable mounting bolts.

5. Disconnect the tie-rod end, upper leading link and upper lateral link.

6. Remove the lower arm ball joint.

7. Remove the front wheel hub and steering knuckle as an assembly.

8. Press out the wheel hub from the knuckle. Grind out a section of the bearing inner race until approx. 0.020 in. (0.5 mm) remains. Then remove it with a chisel.

9. Remove the oil seal from the knuckle.

10. Remove the snapring and press out the wheel bearing using a press.

11. Matchmark the knuckle and dust cover and remove the dust cover.

To install:

12. Install the dust cover to the knuckle.

13. Press the wheel bearing onto the knuckle. Install the snapring.

14. Press the hub assembly onto the knuckle.

15. Install a new oil seal.

16. Replace the knuckle/hub assembly to the vehicle and install the lower arm ball joint. Torque the ball joint mounting bolts to 58–86 ft. lbs. (79–116 Nm). Torque the ball joint nut to 86–115 ft. lbs. (116–156 Nm).

17. Connect the upper lateral link ball joint and torque the nut to 41–59 ft. lbs. (55–80 Nm).

18. Connect the upper leading link ball joint and torque the nut to 28–38 ft. lbs. (38–51 Nm).

19. Connect the tie-rod end and torque the nut to 41–59 ft. lbs. (55–80 Nm).

20. Install the ABS wheel speed sensor and torque the bolts to 14–18 ft. lbs. (19–25 Nm).

21. Install the disc plate and caliper assembly.

22. Install a NEW locknut and torque to 174–235 ft. lbs. (236–318 Nm).

1993–97 Miata

1. Remove the dust cap. Unstake and loosen the locknut.

2. Raise and safely support the vehicle. Remove the wheel and tire assembly.

3. Remove the brake caliper and suspend it from the coil spring.

4. Remove the disc brake rotor.

5. Remove and discard the locknut. Remove the wheel hub and bearing assembly. The wheel hub and bearing cannot be disassembled.

To install:

6. Install the hub over the spindle. Loosely install a new locknut.

7. Install the brake rotor and the disc brake caliper. Install the wheel and tire assembly, and lower the vehicle.

8. When the vehicle is on the ground, tighten the hub nut. Tighten to 159 ft. lbs. (216 Nm).

9. Install the dust cap.

1993–95 RX7

1. Raise and safely support the vehicle. Remove the front wheel and tire assemblies.

2. Uncrimp the tab on the center locknut and remove the locknut. Discard the old locknut.

3. Remove the caliper assembly from the knuckle. Do not disconnect the brake lines. Support the caliper with a piece of wire. Do not allow the caliper to hang by the hose at any time. Remove the brake disc.

4. Remove the tie rod end cotter pin and remove the tie rod end nut. Press the tie rod end out of the knuckle assembly.

5. Remove the stabilizer upper nuts and remove the stabilizer link bolt.

6. Remove the outer lower arm to ball joint mounting bolt and nut. Separate the lower arm from the knuckle assembly.

7. Remove the ABS speed sensor if so equipped.

8. Using a plastic mallet, tap the driveshaft free of the knuckle assembly. Remove the knuckle assembly.

9. Clamp the knuckle in a vise with protected jaws.

10. Remove the inner oil seal from the knuckle.

11. Remove the front wheel hub from the knuckle assembly.

12. Remove the bearing inner race from the front wheel hub.

13. Remove the retaining ring from within the knuckle and using the hub puller tools, press the front wheel bearing from the knuckle.

14. Remove the brake dust shield.

To install:

15. Clean and inspect all parts but do not wash or clean the wheel bearing. The bearing must be replaced.

16. Install a new dust shield cover assembly to the knuckle.

17. Press a new wheel bearing into the knuckle assembly.

18. Install the wheel bearing retaining ring, and install a new oil seal.

19. Install the front wheel hub.

20. Install the bearing/hub and knuckle assembly in place. Loosely tighten the knuckle to shock absorber bolt.

21. Install the lower arm ball joint to the knuckle and tighten the nut to 27–40 ft. lbs. (36–54 Nm).

22. Install the driveshaft to the knuckle assembly.

23. Install the stabilizer control link and tighten the nuts so that there is 0.71–0.87 inches (18.1–22.1mm) of thread exposed. Torque the top link nut to 12–17 ft. lbs. (16–23 Nm).

24. If equipped with ABS, install the wheel speed sensor and tighten the bolts to 12–17 ft. lbs. (16–23 Nm).

25. Connect the tie rod ends to the knuckle and tighten the nuts to 31–42 ft. lbs. (42–57 Nm). Replace the cotter pins.

26. Install a new wheel hub locknut and tighten the locknut to 174–235 ft. lbs. (235–319 Nm).

27. Check the end play of the wheel bearing by installing a dial indicator against the wheel hub and tire to move the brake disc back and forth. There should be no more than 0.0079 inch (0.2mm) of freeplay.

28. Stake the locknut into place by bending it into the groove.

29. Install the brake caliper(s) and tighten the bolts to 58–72 ft. lbs. (78–98 Nm).

30. Install the front wheels and lower the vehicle.

31. With the vehicle lowered check all of the bolts and re-torque as is necessary.

32. Inspect the front end alignment and adjust as is necessary.

REAR SUSPENSION

Strut

REMOVAL AND INSTALLATION

1993–94 323, MX3 and Protege

1. As required, remove the side trim panels from the inside of the trunk or the rear seat and trim.
2. If equipped with Adjustable Shock Absorber (ASA) system, disconnect the wiring and remove the cap. Loosen and remove the top mounting nuts from the strut mounting block assembly.
3. Raise and safely support the vehicle and remove the rear wheels. The suspension will drop when the weight lifts off the wheels.
4. Unclip the brake line or wiring retainers as required and unbolt the bottom strut mount. Remove the strut.
5. Installation is the reverse of removal. Torque the following:
   a. Lower strut mount bolts — 86 ft. lbs (117 Nm).
   b. 323 trailing arm bolt — 50 ft. lbs. (68 Nm).
   c. 323 upper strut mount nuts — 22 ft. lbs. (29 Nm).
   d. MX-3 upper strut mount nuts — 46 ft. lbs. (63 Nm).

1995–97 Protege

1. Remove the trunk side trim.
2. Raise and safely support the vehicle and remove the rear wheels. The suspension will drop when the weight lifts off the wheels.
3. Unclip the brake line and wiring retainers (if equipped with ABS), and unbolt the bottom strut mount.
4. Remove the upper strut mount nuts, and remove the strut from the vehicle.
5. Disassemble the strut to remove the coil spring as follows:
   a. Secure the mounting rubber in a vise.
   b. Remove the cap. Loosen the piston rod nut several turns, but **do not** remove it.
   c. Using a coil spring compressor, (Mazda special tool 49 T034 1A0 or equivalent) secure the strut assembly.
   d. Compress the coil spring and remove the nut.
   e. Remove the mounting rubber, upper spring seat, upper rubber spring seat, dust cover, bound stopper and the coil spring.

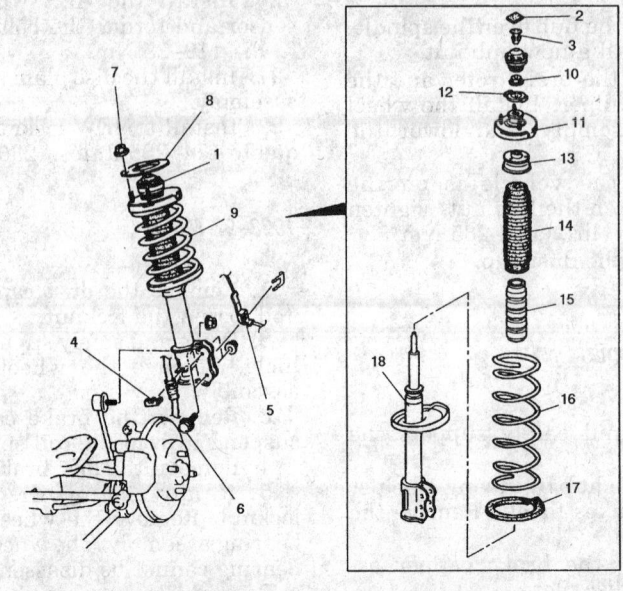

1. Cap
2. Sponge
3. Fastener
4. Clip
5. ABS wheel-speed sensor harness
6. Shock absorber bolt
7. Nut
8. Sheet
9. Rear shock absorber and spring
10. Nut (mounting rubber)
11. Mounting rubber
12. Washer
13. Upper spring seat
14. Dust cover
15. Bound stopper
16. Coil spring
17. Lower spring seat
18. Rear shock absorber

306073

Rear strut and related components (exploded view) — 1995–97 Protege

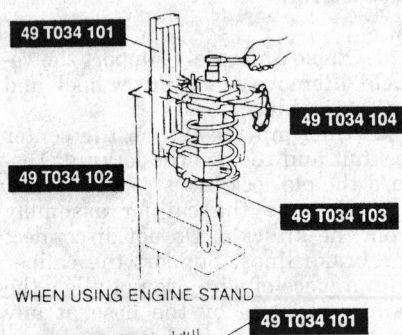

WHEN USING ENGINE STAND

49 T034 101
49 T034 104
49 T034 102
49 T034 103

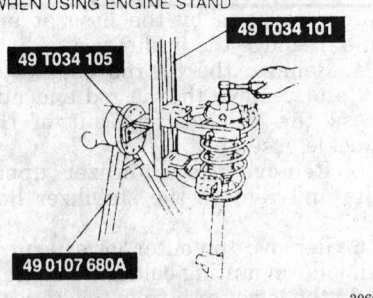

49 T034 101
49 T034 105
49 0107 680A

306076

Compressing the strut — 1995–97 Protege

To install:

6. If the coil spring was removed, assemble as follows:
   a. Temporarily assemble the upper rubber spring seat, upper spring seat and coil spring to the strut. Mark their locations on the strut for proper reassembly.
   b. Align the marks of the upper spring seat and coil spring. Protect them with a shop rag, and remove them from the strut.
   c. Using the same coil spring compressor tool that disassembled the strut, compress the spring.
   d. Install the bound stopper and dust cover to the strut.
   e. Install the strut into the coil spring, fitting the end of the coil into the step of the lower seat.
   f. Install the mounting rubber. Tighten the piston rod nut several turns.
   g. Remove the compressor tool, and verify that the coil spring is seated on the step of the lower seat.
   h. Secure the mounting rubber in a vise, and tighten the nut to 41–49 ft. lbs. (55–67 Nm). Install the cap.

7. Position the strut in the strut tower and install the upper mounting nuts. Tighten the nuts to 34–46 ft. lbs. (47–62 Nm).

8. Install the lower strut mount bolt, and tighten to 69–93.3 ft. lbs. (94–126 Nm).

9. Replace the ABS harness (if equipped), and the brake line clip.

10. Install the wheel assemblies, and lower the vehicle.

11. Install the trunk side trim.

1995 MX3

1. Remove the rear speaker cover and seat belt.

2. Raise and safely support the vehicle and remove the rear wheels. The suspension will drop when the weight lifts off the wheels.

3. Unclip the brake line and wiring retainers (if equipped with ABS), and unbolt the bottom strut mount.

4. Remove the upper strut mount nuts, and remove the strut from the vehicle.

5. Disassemble the strut to remove the coil spring as follows:

   a. Secure the mounting rubber in a vise.

   b. Remove the cap. Loosen the piston rod nut several turns, but **do not** remove it.

   c. Using a suitable coil spring compressor (Mazda special tools: 49 T034 001, 49 T034 002, 49 T034 003 and 49 T034 004 or equivalent) secure the strut assembly.

   d. Compress the coil spring and remove the nut.

   e. Remove the mounting rubber, upper spring seat, upper rubber spring seat, dust cover, bound stopper and the coil spring.

To install:

6. If the coil spring was removed, assemble as follows:

   a. Temporarily assemble the upper rubber spring seat, upper spring seat and coil spring to the strut. Mark their locations on the strut for proper reassembly.

   b. Align the marks of the upper spring seat and coil spring. Protect them with a shop rag, and remove them from the strut.

   c. Using the same coil spring compressor tool that disassembled the strut, compress the spring.

   d. Install the bound stopper and dust cover to the strut.

   e. Install the strut into the coil spring, fitting the end of the coil into the step of the lower seat.

   f. Install the mounting rubber. Tighten the piston rod nut several turns.

   g. Remove the compressor tool, and verify that the coil spring is seated on the step of the lower seat.

   h. Secure the mounting rubber in a vise, and tighten the nut to

41–49 ft. lbs. (55–67 Nm). Install the cap.

7. Position the strut in the strut tower and install the upper mounting nuts. Tighten the nuts to 34–46 ft. lbs. (47–62 Nm).

8. Install the lower strut mount bolt, and tighten to 76–93 ft. lbs. (103–126 Nm).

9. Replace the ABS harness (if equipped), and the brake line clip.

10. Install the wheel assemblies, and lower the vehicle.

11. Install the rear seat belt and speaker cover.

1993–97 626 and MX6

1. Remove the side trim panels from the inside of the trunk, or the rear seat and trim.

2. Raise and safely support the vehicle and remove the rear wheels. The suspension will drop when the weight lifts off the wheels.

3. Unclip the brake line and wiring retainers (if equipped with ABS), and unbolt the bottom strut mount.

4. Remove the upper strut mount nuts, and remove the strut from the vehicle.

5. Disassemble the strut to remove the coil spring as follows:

   a. Secure the mounting rubber in a vise.

   b. Remove the cap. Loosen the piston rod nut several turns, but **do not** remove it.

   c. Using a suitable coil spring compressor (Mazda special tools: 49 T034 001, 49 T034 002, 49 T034 003 and 49 T034 004 or equivalent) secure the strut assembly.

   d. Compress the coil spring and remove the nut.

   e. Remove the mounting rubber, upper spring seat, upper rubber spring seat, dust cover, bound stopper and the coil spring.

To install:

6. If the coil spring was removed, assemble as follows:

   a. Temporarily assemble the upper rubber spring seat, upper spring seat and coil spring to the strut. Mark their locations on the strut for proper reassembly.

   b. Align the marks of the upper spring seat and coil spring. Protect them with a shop rag, and remove them from the strut.

   c. Using the same coil spring compressor tool that disassembled the strut, compress the spring.

   d. Install the bound stopper and dust cover to the strut.

   e. Install the strut into the coil spring, fitting the end of the coil into the step of the lower seat.

   f. Install the mounting rubber. Tighten the piston rod nut several turns.

   g. Remove the compressor tool, and verify that the coil spring is seated on the step of the lower seat.

   h. Secure the mounting rubber in a vise, and tighten the nut to 66–86 ft. lbs. (90–116 Nm). Install the cap.

7. Position the strut in the strut tower and install the upper mounting nuts. Tighten the nuts to 33–46 ft. lbs. (47–62 Nm).

8. Install the lower strut mount bolt, and tighten to 69–96 ft. lbs. (94–116 Nm).

9. Replace the ABS harness (if equipped), and the brake line clip.

10. Install the wheel assemblies, and lower the vehicle.

11. Replace any removed trim from the trunk and rear seat area.

1993–95 929

1. Remove the side trim panels from the inside of the trunk.

2. Remove the rear seat belt cover.

3. Loosen and remove the top mounting nuts from the strut mounting block assembly.

4. Raise and safely support the vehicle and remove the rear wheels. The suspension will drop when the weight lifts off the wheels.

5. Unbolt the bottom strut mount. Remove the strut.

6. To remove the coil spring from the strut assembly:

   a. Safely secure the assembly in a vice. Loosen the piston rod upper nut several turns, but do not remove it.

   b. Assembly special tools 49 0223 640B and 40 G030 641. Compress the coil spring using the tools.

   c. Remove the upper nut, and remove the coil spring.

To install:

7. Install the coil spring on the strut assembly using the same special tools used to remove it. Tighten the upper nut in a vice with a final torque of 26–36 ft. lbs. (35–49 Nm).

8. Install the strut assembly. Torque the upper mount nuts to 34–46 ft. lbs. (47–62 Nm) and the lower mount bolt to 55–68 ft. lbs. (74–93 Nm).

9. Install the wheels, and lower the vehicle.

10. Install the rear seat belt cover and the trunk trim panels.

11. Adjust the rear wheel alignment.

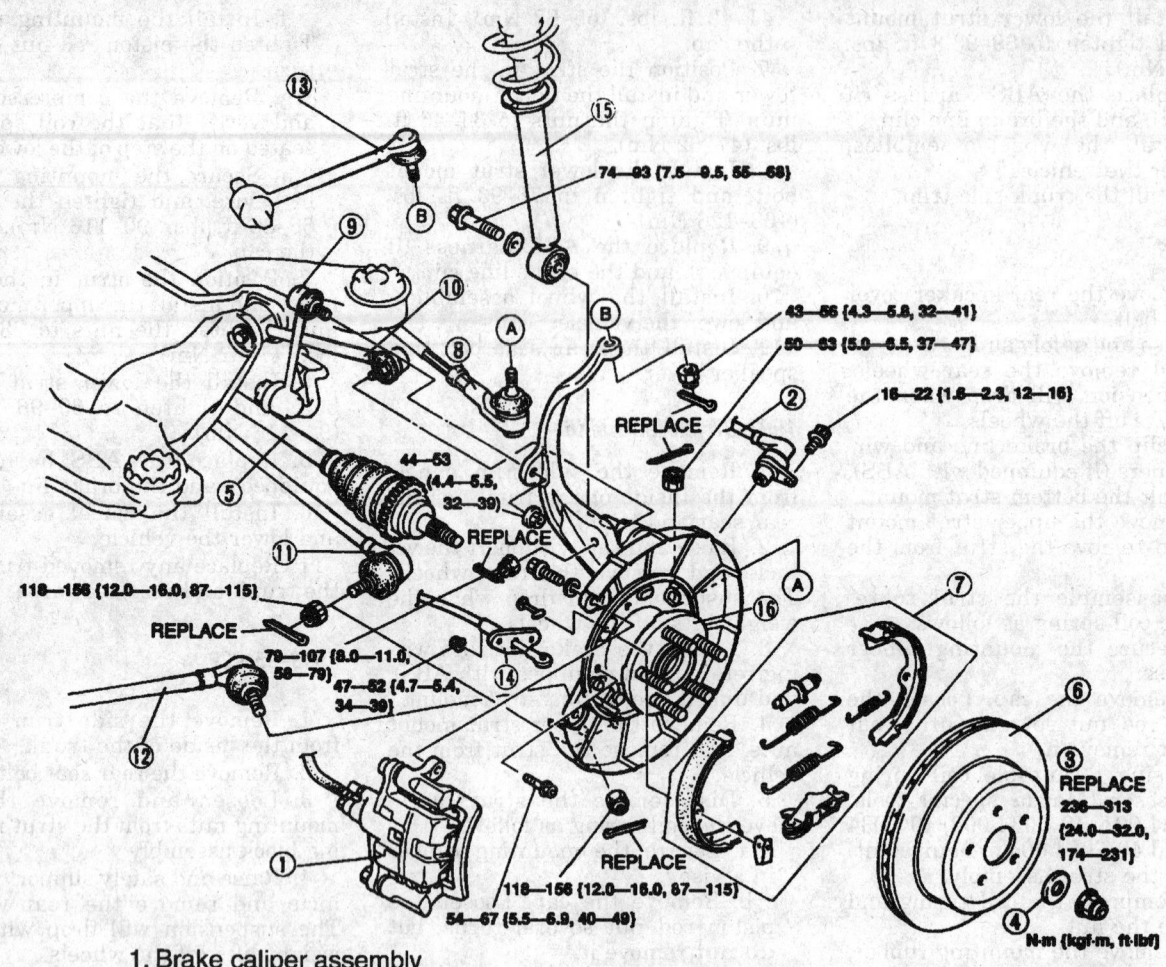

1. Brake caliper assembly
2. ABS wheel speed sensor
3. Wheel Hub nut
4. Washer
5. Drive shaft
6. Disc plate
7. Parking brake shoe assembly
8. Rear lower lateral link
9. Stabilizer link
10. Upper lateral link
11. Lower lateral link
12. Lower trailing link
13. Upper trailing link
14. Parking brake cable
15. Rear shock absorber and spring
16. Rear hub support assembly

Exploded view of rear suspension components — 1993–95 929

1995–97 Millenia

1. Raise and safely support the vehicle. Remove the wheel and tire assembly.

2. Remove the mounting screws and the first aid box. Remove the speaker cover. Pull the speaker cover and rear package front trim upward to disengage the clips and remove the rear package front trim.

3. Remove the bolt and disconnect the ABS sensor cable.

4. Disconnect the lower lateral link ball joint.

5. Matchmark the strut mounts and remove the mounting nuts, then remove the strut assembly.

6. Remove the cap.

— CAUTION —
Removing the piston rod nut is dangerous. The shock absorber and spring could fly off under tremendous pressure and cause serious injury or death. Secure the shock absorber in a compression tool before removing the piston rod nut.

7. Install the assembly into a strut compression tool and remove the top mount nut and mount.

8. Remove the upper spring seat, bound stopper, coil spring and lower spring seat.

To install:

9. Assemble the strut assembly and install the damper fork and cap to the strut.

10. Install the strut assembly.

11. Install the strut in the strut tower, aligning the marks made during removal. Install the upper mounting nuts and torque to 34–46 ft. lbs. (47–62 Nm). Install the lower mounting nut and bolt and torque to 76–101 ft. lbs. (102–137 Nm).

12. Connect the lower lateral link ball joint. Torque the nut to 86–115 ft.lbs. (116–156 Nm).

13. Connect the ABS sensor cable mount, tighten the bolt to 14–18 ft. lbs. (19–25 Nm).

14. Install the rear package front trim.

15. Install the wheel and tire assembly and lower the vehicle. Check the front wheel alignment.

1993–97 Miata

1. Raise and safely support the vehicle and remove the rear wheel(s).

2. On the left side, remove the fuel filler pipe protector panel.

3. Remove the bolt from the lower stabilizer bar connecting link.

4. Loosen the upper arm and adjusting cam nuts.

5. Remove the upper strut mount nuts and lower mount bolt.

6. Lower the upper and lower arms to increase clearance. Lift the spring and strut out as an assembly.

7. To remove the coil spring from the strut assembly:

   a. Safely secure the assembly in a vice. Loosen the piston rod upper nut several turns, but do not remove it.

   b. Assembly special tools 49 G034 101, 102 and 103. Compress the coil spring using the tools.

   c. Remove the upper nut, and remove the coil spring.

To install:

8. Install the coil spring on the strut assembly using the same special tools used to remove it. Tighten the upper nut in a vice with a final torque of 24–33 ft. lbs. (32–46 Nm).

9. Install the strut assembly. Torque the upper mount nuts to 22–27 ft. lbs. (30–36 Nm) and the lower mount bolt to 54–69 ft. lbs. (73–93 Nm).

10. Tighten the upper arm and adjusting cam nuts.

11. Connect the control link to the stabilizer bar, and loosely install the mounting bolt.

12. If removed, install the fuel filler pipe protector panel.

13. Install the wheel(s), and lower the vehicle.

14. Tighten the stabilizer link bolt to 27–40 ft. lbs. (37–54 Nm).

15. Adjust the rear wheel alignment.

1993–95 RX7

1. Raise and safely support the vehicle. Remove the rear wheel and tire assembly.

2. Remove the nut and disconnect the stabilizer bar.

3. Remove the strut upper mount nuts.

4. Remove the rear upper strut bar.

5. Remove the nut and stopper rubber at the top of the strut. Remove the strut and insulator.

To install:

6. Install the insulator on the strut so the notches face the studs.

7. Install the strut so that the identification paint mark at the bottom of the strut faces rearward.

8. Install the stopper rubber and nut. Tighten to 24–33 ft. lbs. (32–46 Nm).

9. Install the upper strut bar, then install the strut upper mount nuts. Tighten to 34–46 ft. lbs. (47–62 Nm).

10. Slide the stabilizer bar link through the control arm and the strut. Install the nut and tighten the 69–81 ft. lbs. (94–110 Nm).

Upper Control Arms

REMOVAL AND INSTALLATION

1993–95 929

1. Raise and safely support the vehicle. Remove the wheel and tire assembly.

2. Remove the cotter pin and loosen the front lower lateral link ball joint nut. With the nut protecting the ball joint stud, press out the stud with press tool 49 S231 575 or equivalent. Remove the nut.

3. Remove the mounting bolt, and remove the lower lateral link.

4. Remove the cotter pin and loosen the upper lateral link ball joint nut. With the nut protecting the ball joint stud, press out the stud with press tool 49 S231 575 or equivalent. Remove the nut.

5. Paint an alignment mark on the cam plate and crossmember for assembly reference, before removing the upper lateral link bolt. Remove the mounting bolt, and the upper lateral link.

6. Remove the cotter pin and loosen the rear lower lateral link ball joint nut. With the nut protecting the ball joint stud, press out the stud with press tool 49 0118 850C or equivalent. Remove the nut.

7. Remove the nuts, and remove the rear lower lateral link.

To install:

8. Install the rear lower lateral link and tighten the nuts, in sequence, to 28–38 ft. lbs. (38–52 Nm).

9. Install the rear lower lateral link inner ball joint, and tighten to 80–87 ft. lbs. (108–118 Nm).

10. Connect the rear lower lateral link outer ball joint to the knuckle and tighten the nut to 37–47 ft. lbs. (50–64 Nm). Install a new cotter pin.

11. Install the upper lateral link, and loosely tighten the bolt. Connect the ball joint to the knuckle, and

tighten the nut to 58–72 ft. lbs. (79–98 Nm). Install a new cotter pin.

12. Install the front lower lateral link and loosely install the bolt. Connect the ball joint to the knuckle and tighten the nut to 58–80 ft. lbs. (79–108 Nm). Install a new cotter pin.

13. Install the wheel and tire assembly and lower the vehicle. With the vehicle at normal ride height, tighten the upper and front lower lateral link bolts. Align the paint marks on the cam plate and crossmember, and tighten the upper lateral link bolt to 86 ft. lbs. (117 Nm). Tighten the lower lateral link bolt to the same specification.

1995–97 Millenia

Lower Lateral Link

1. Raise and safely support the vehicle.

2. Remove the tire and wheel assembly.

3. Remove the cotter pin and nut and disconnect the the lower lateral link ball joint stud from the rear knuckle assembly, using the proper separator tool.

4. Remove the nut and bolt and remove the lower lateral link from the rear crossmember.

5. If the lower lateral link ball joint is to be replaced, the lower lateral link must be replaced.

To install:

6. Install the lower lateral link to the rear crossmember with the nut and the bolt. Do not tighten the bolt and nut at this time.

7. Install a new dust boot and reinstall the lower lateral link ball joint to the rear knuckle. Torque the ball joint nut to 86–115 ft. lbs. (116–156 Nm) and install a new cotter pin.

8. Install the tire and wheel assembly and safely lower the vehicle.

9. Torque the lower lateral link to crossmember bolt and nut to 58–86 ft. lbs. (79–116 Nm).

10. Check the rear wheel alignment.

Upper Lateral Link

1. Raise and safely support the vehicle.

2. Remove the tire and wheel assembly.

3. Remove the cotter pin and nut connecting the upper lateral link to the knuckle. Remove the ball joint dust boot.

4. Remove the adjusting bolt, cam plate, and nut and remove the upper lateral link from the crossmember mount.

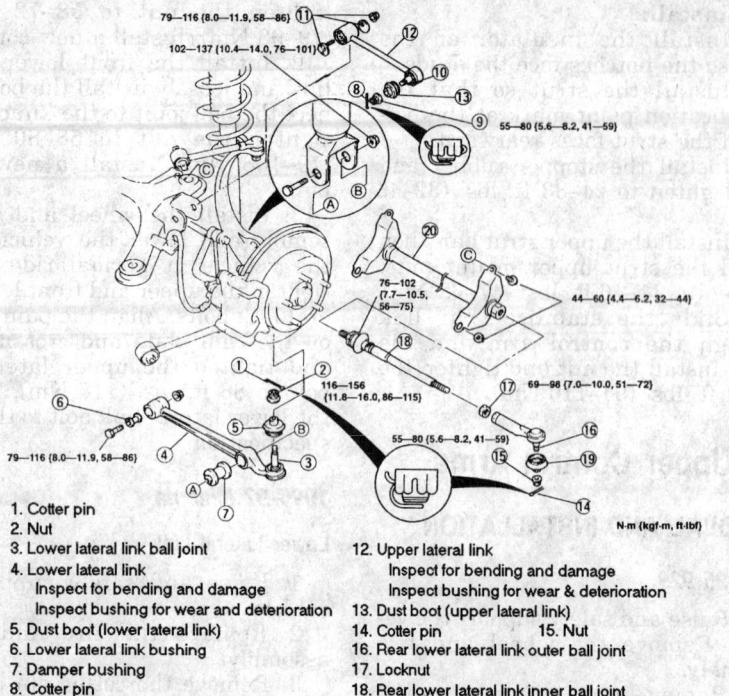

79—116 (8.0—11.9, 58—86)
102—137 (10.4—14.0, 76—101)
55—80 (5.6—8.2, 41—59)
76—102 (7.7—10.5, 56—75)
44—60 (4.4—6.2, 32—44)
116—156 (11.8—16.0, 86—115)
69—98 (7.0—10.0, 51—72)
79—116 (8.0—11.9, 58—86)
55—80 (5.6—8.2, 41—59)

N·m (kgf·m, ft-lbf)

1. Cotter pin
2. Nut
3. Lower lateral link ball joint
4. Lower lateral link
   Inspect for bending and damage
   Inspect bushing for wear and deterioration
5. Dust boot (lower lateral link)
6. Lower lateral link bushing
7. Damper bushing
8. Cotter pin
9. Nut
10. Upper lateral link ball joint
11. Nut, cam plate, and adjusting cam bolt
12. Upper lateral link
    Inspect for bending and damage
    Inspect bushing for wear & deterioration
13. Dust boot (upper lateral link)
14. Cotter pin   15. Nut
16. Rear lower lateral link outer ball joint
17. Locknut
18. Rear lower lateral link inner ball joint
    Inspect for bending and damage
19. Dust boot (rear lower lateral link)
20. Rear lower lateral link

310176

Rear suspension components — 1995–97 Millenia

To install:

5. Install the upper lateral link to the crossmember mount with the adjusting bolt, cam plate, and locknut. Do not tighten at this time.

6. Reinstall a new dust boot to the upper lateral link and reattach the link to the knuckle with the nut. Torque the nut to 41–59 ft. lbs. (55–80 Nm). Install a new cotter pin.

7. Reinstall the tire and wheel assembly and safely lower the vehicle.

8. Check the wheel alignment.

9. Torque the upper lateral link adjusting bolt and nut to 58–86 ft. lbs. (79–116 Nm).

1993–97 Miata

1. Raise and safely support the vehicle, and remove the rear wheel(s).

2. Remove the nuts, washers and eccentric bolts, and the outer through-bolt. Remove the upper control arm. The bushings can be replaced separately.

To install:

3. Position the upper control arm, and loosely install the mounting bolts.

4. Install the rear wheel(s), and lower the vehicle.

5. Torque the upper arm through-bolts to 34–49 ft. lbs. (47–67 Nm).

1993–95 RX7

1. Raise and safely support the vehicle and remove the rear wheels.

2. Disconnect the stabilizer bar link and the lower shock absorber mounting bolt from the control arm.

3. Remove the nuts, through-bolts and remove the control arm.

4. Installation is the reverse of removal. Torque the control arm through-bolts to 44–54 ft. lbs. (59–73 Nm). Torque the strut mount bolt to 69–81 ft. lbs. (94–110 Nm).

Lower Control Arms

REMOVAL AND INSTALLATION

1993–94 323, MX3 and Protege

1. Raise and safely support the vehicle and remove the wheels.

2. Before disconnecting the stabilizer bar link, on all except MX-3, measure the length of the threads protruding above the locknut. Remove the nuts and through-bolt to disconnect the stabilizer bar.

3. Remove the nuts and bolts as required and remove the control arms.

4. Installation is the reverse of removal. Make sure to properly adjust the stabilizer bar before tightening the locknuts and check rear wheel alignment.

5. Torque the following:
   a. Inner control arm through-bolt to 70 ft. lbs. (95 Nm).
   b. Outer control arm through-bolt to 86 ft. lbs. (117 Nm).
   c. Trailing arm-to-body bolt to 69 ft. lbs. (93 Nm).
   d. Trailing arm-to-knuckle bolt to 86 ft. lbs. (117 Nm).

1993–94 626 and MX6

1. Raise and safely support the vehicle. Remove the wheel and tire assemblies.

2. Remove the access cap from the underside of the rear crossmember.

3. Remove the bolts and nuts and remove the lateral links and trailing link. If removing the rear lateral link, paint and alignment mark on the cam plate and crossmember for assembly reference.

4. Installation is the reverse of the removal procedure. Tighten the inner and outer lateral link bolts/nuts to 86 ft. lbs. (116 Nm). Tighten the trailing link-to-body bolt to 68 ft. lbs. (93 Nm) and the trailing link-to-knuckle bolt to 86 ft. lbs. (116 Nm).

5. Do not final tighten the bolts until the vehicle is on the ground and at normal ride height.

1993–97 Miata

1. Raise and safely support the vehicle, and remove the rear wheels.

2. Disconnect the stabilizer bar link, and the lower strut mounting bolt from the lower arm.

3. Remove the nuts, eccentric bolts, and the outer through-bolt. Remove the lower control arm. The bushings can be replaced separately.

WARNING
If removing the upper control arm, either re-install the lower control arm or remove the brake caliper, halfshaft and hub carrier. Do not allow the carrier to be supported only by the halfshaft.

To install:

4. Position the lower control arm, and loosely install the mounting bolts.

5. Loosely install the lower strut mounting bolt.

6. Connect the stabilizer bar link, and loosely tighten the nut and bolt.

7. Install the rear wheels, and lower the vehicle.

8. After the vehicle is lowered, torque the lower arm eccentric bolts to 54–70 ft. lbs. (73–95 Nm). Torque the

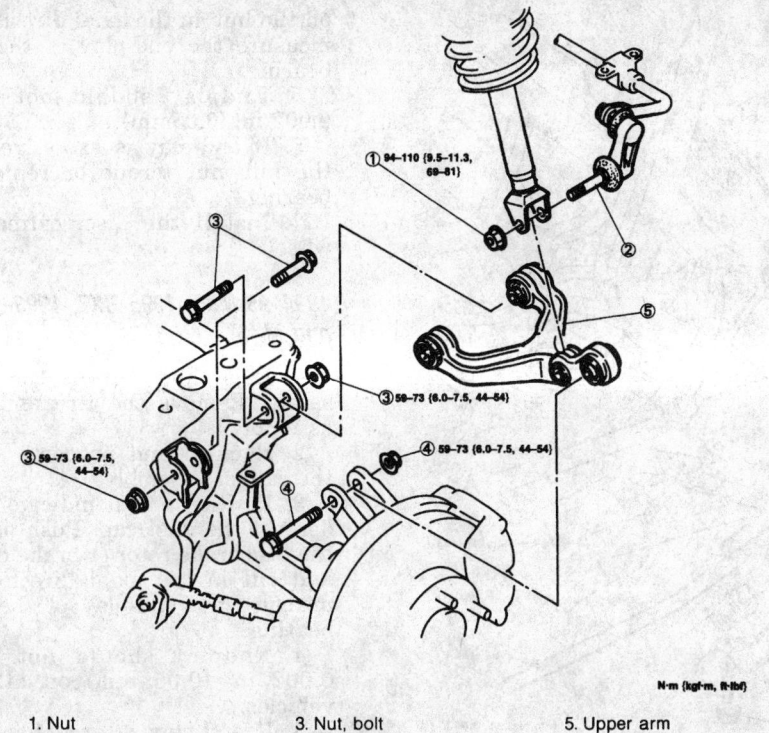

1. Nut
2. Stabilizer control link
3. Nut, bolt
4. Nut, bolt
5. Upper arm

N·m (kgf·m, ft-lbf)

250136

Upper control arm removal and installation — 1993–95 RX7

lower arm outer through-bolt to 47–54 ft. lbs. (63–74 Nm). Tighten the strut bolt to 54–68 ft. lbs. (73–93 Nm), and the link bolt to 27–39 ft. lbs. (37–53 Nm).

1993–95 RX7

1. Raise and safely support the vehicle. Remove the wheel.
2. Remove the nuts and bolt for the trailing link and remove the trailing link.
3. Remove the adjusting cam bolt assembly.
4. Remove the nut and bolt for the I-arm assembly and remove the I-arm.
5. Installation is the reverse of removal. Torque the I-arm mounting bolt and the trailing link bolt to 44–54 ft. lbs. (59–73 Nm). Torque the adjusting cam bolt to 69–86 ft. lbs. (94–116 Nm).

Sway Bar

REMOVAL AND INSTALLATION

1993–94 323, 1993–95 MX3 and 1993–97 Protege

1. Raise and safely support the rear of the vehicle.

2. Remove the rear wheels from the vehicle.
3. Remove the stabilizer bar to link mounting nut and protectors.
4. Remove the stabilizer bracket, and remove the stabilizer bar. Remove the stabilizer bushings and inspect the bushing for deterioration or wear. Replace if necessary.
To install:
5. Apply rubber grease to the inside of the bushing.
6. Position the stabilizer bar, and loosely tighten the mounting bolts.
7. Install the rear wheels, and lower the vehicle.
8. Tighten the stabilizer bar to link nuts and the bracket bolts to 32–44 ft. lbs. (44–60 Nm).

1993–97 626 and MX6

1. Raise and safely support the rear of the vehicle.
2. Remove the rear wheels from the vehicle.
3. Remove the stabilizer bar to link mounting nut and protectors.
4. Remove the stabilizer bracket, and remove the stabilizer bar. Remove the stabilizer bushings and inspect the bushing for deterioration or wear. Replace if necessary.

To install:
5. Position the stabilizer bar, and loosely tighten the mounting bolts.
6. Install the rear wheels, and lower the vehicle.
7. Tighten the stabilizer bar to link nuts and the bracket bolts to 27–39 ft. lbs. (37–53 Nm).

1993–95 929

1. Raise and support the vehicle safely. Remove the front wheels.
2. Remove the stabilizer control link.
3. Remove the 2 stabilizer nuts, and remove the upper bushing and bracket from each side of the vehicle.
4. Remove the stabilizer bar from the vehicle. Examine the insulators (bushings) carefully for any sign of wear and replace them if necessary.

NOTE: Check the bushings inside the brackets for wear or deformation. A worn bushing can cause a distinct noise as the bar twists during cornering operation.

To install:
5. Install the stabilizer bar bushings in the correct position.
6. Loosely install the stabilizer bar brackets.
7. Position the stabilizer control link. Install the upper link nuts. Tighten the link nuts so that there is 0.71–0.87 inches (18.1–22.1mm) of thread exposed.
8. Torque the stabilizer link bolt to 32–39 ft. lbs. (44–53 Nm).
9. Install front wheels and lower the vehicle.
10. Tighten the stabilizer bar bracket bolts to 24–33 ft. lbs. (32–46 Nm). Check front wheel alignment.

1995–97 Millenia

1. Raise and safely support the vehicle.
2. Disconnect the stabilizer control link.
3. Remove the stabilizer brackets and bushings.
4. Remove the rear stabilizer.
5. Installation is the reverse of removal. Torque the stabilizer bracket mounting bolts and the control link mount nut to 32–44 ft. lbs. (44–60 Nm).

1993–97 Miata

1. Raise and safely support the rear of the vehicle.
2. Remove the rear wheels from the vehicle.
3. Remove the stabilizer bracket and the stabilizer bushing, inspect

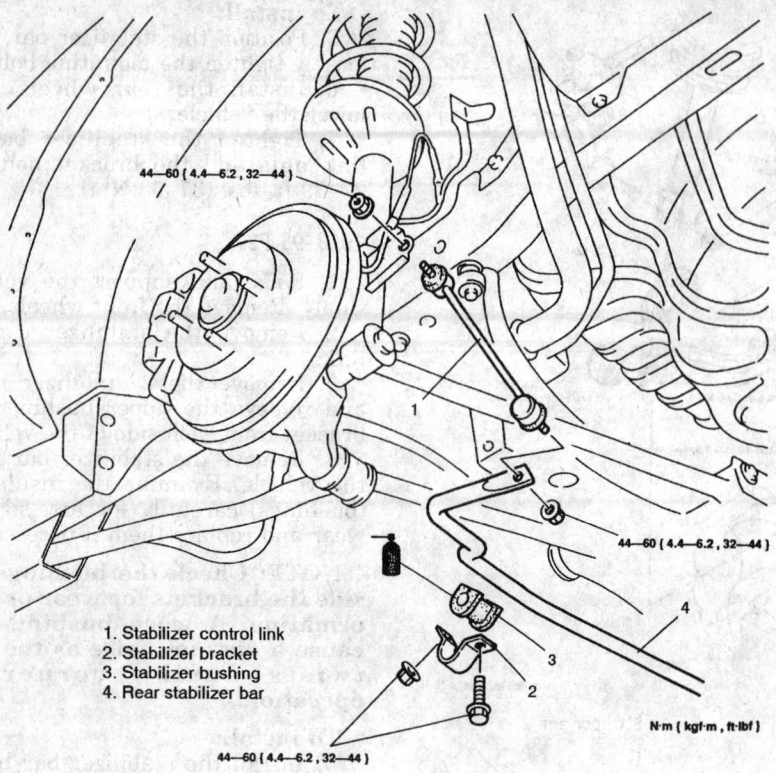

1. Stabilizer control link
2. Stabilizer bracket
3. Stabilizer bushing
4. Rear stabilizer bar

44—60 (4.4—6.2 , 32—44)

44—60 (4.4—6.2 , 32—44)

44—60 (4.4—6.2 , 32—44)

N·m (kgf·m , ft·lbf)

306088

Rear sway bar and related components, exploded view — 1995-97 Protege

the bushing for deterioration or wear and replace if necessary.

4. Remove the stabilizer to control link bolt, and remove the stabilizer bar.

To install:

5. Position the stabilizer bar. Align the bushing on the stabilizer bar with the painted marking. Loosely install the mounting bolts.

6. Install the rear wheels, and lower the vehicle.

7. Tighten the control link to stabilizer bar bolts to 27-40 ft. lbs. (36-54 Nm), and the stabilizer bracket bolts to 14-21 ft. lbs. (20-28 Nm).

1993-95 RX7

1. Raise and support the vehicle safely. Remove the rear wheels and the undercover.

2. Remove the stabilizer control link nut and the stabilizer plate mount bolts. Remove the bushings.

3. Remove the stabilizer bar. If necessary, remove the stabilizer control link.

4. Installation is the reverse of removal. Torque the stabilizer plate mount bolts to 14-19 ft. lbs. (18-26 Nm). Torque the stabilizer bar to control link mount nut to 27-39 ft. lbs.

(37-53 Nm). Check front wheel alignment.

Wheel Bearings

ADJUSTMENT

1993-94 323, MX3 and Protege

1. Raise and support the vehicle safely. Remove the tire and wheel assembly.

2. Remove and properly support the caliper assembly.

3. Position a dial indicator gauge against the dust cap. Push and pull the disc brake rotor or brake drum in and out in the axial direction and measure the end-play of the wheel bearing.

4. Endplay should not exceed 0.002 in. (0.05mm).

5. If end-play is excessive, check the hub nut torque or replace the bearing.

1995-97 Millenia and Protege

1. Raise and support the vehicle safely. Remove the wheel assembly.

2. Remove and properly support the caliper assembly.

3. Remove the disc plate.

4. Position a dial indicator gauge against the wheel hub. Push and pull

on the hub in the axial direction and measure the end-play of the wheel bearing.

5. Endplay should not exceed 0.002 in. (0.05mm).

6. If end-play is excessive, check the hub nut torque or replace the bearing.

7. Install the disc, caliper, and wheel.

1994-95 MX3, 1995 RX7, 1993-97 626 and MX6

1. Raise and support the vehicle safely. Remove the tire and wheel assembly.

2. Remove and properly support the caliper assembly.

3. Position a dial indicator gauge against the dust cap. Push and pull the disc brake rotor or brake drum in and out in the axial direction and measure the end-play of the wheel bearing.

4. Endplay should not exceed 0.002 in. (0.05 mm) on all other vehicles.

5. If end-play is excessive, check the hub nut torque or replace the bearing.

1993-95 929 and 1993-97 Miata

The wheel bearings on these vehicles are not adjustable. To check if the bearing requires service, remove the wheel and tire assembly, brake caliper and disc brake rotor. Install a dial indicator with the indicator foot resting on the wheel hub. Try to move the hub in and out. If there is more than 0.002 in. (0.05mm) bearing play, check the wheel hub nut torque or replace the hub and bearing assembly.

REMOVAL AND INSTALLATION

1993-94 323, 1993-95 MX3 and 1993-97 Protege, 626 and MX6

1. Raise and safely support the vehicle and remove the rear wheels.

2. Remove the hubcap. Hold the brake to remove the center axle nut.

3. If equipped with drum brakes, remove the drum.

4. If equipped with disc brakes, without disconnecting the hydraulic hose, remove the disc brake caliper and hang it from the body. Do not let it hang by the hose. Slide the disc off the spindle.

5. Slide the hub and bearing assembly off the spindle. The hub and bearing cannot be separated and must be replaced as one piece.

To install:

6. Install the hub and drum or rotor. If equipped, install the brake caliper.

7. Install a new spindle nut and tighten to 130–173 ft. lbs. (177–235 Nm). Stake the nut into place. Replace the hubcap.

8. Install the wheel and tire assembly and lower the vehicle.

1993–95 929

1. Loosen the hub nut and raise and safely support the vehicle. Remove the wheel and tire assembly and the nut.

2. Remove the brake caliper and suspend it from the coil spring; do not let the caliper hang by the brake hose.

3. Remove the ABS speed sensor and the disc brake rotor. Remove the parking brake shoe assembly.

4. Disconnect the rear lower lateral link and the stabilizer bar link from the knuckle.

5. Disconnect the upper and lower lateral links and trailing links from the knuckle.

6. Remove the brake backing plate mounting bolts, remove the parking brake cable mounting bolts and remove the cable.

7. Remove the lower strut bolt and remove the hub/knuckle assembly. If the halfshaft sticks to the hub, install the old hub nut until it is flush with the end of the shaft. Tap out the shaft, using a soft mallet only.

8. Properly support the hub/knuckle assembly and press the hub from the knuckle. If the bearing race sticks to the knuckle, grind the race in 1 spot until only 0.20 in. (0.5mm) thickness remains. Cut the race with a chisel and remove it.

9. Remove the brake backing plate from the knuckle. Remove the bearing snapring.

10. Properly support the knuckle and press out the bearing.

To install:

11. Inspect the knuckle and hub for cracks or other damage. Replace parts as necessary.

12. Properly support the knuckle and press in a new bearing. Install the snapring and the brake backing plate.

13. Properly support the knuckle and press on the hub.

14. Position the hub/knuckle assembly and install the lower strut bolt. Tighten to 69 ft. lbs. (93 Nm).

15. Connect the parking brake cable and install the backing plate bolts.

16. Connect the upper trailing link and tighten the nut to 42 ft. lbs. (57 Nm). Connect the lower trailing link and tighten the nut to 116 ft. lbs. (157 Nm). Install new cotter pins.

17. Connect the upper lateral link and tighten the nut to 80 ft. lbs. (108 Nm). Connect the lower lateral link and tighten the nut to 116 ft. lbs. (157 Nm). Install new cotter pins.

18. Connect the stabilizer link and tighten the nut to 40 ft. lbs. (54 Nm). Connect the rear lower lateral link and tighten the nut to 47 ft. lbs. (64 Nm). Install a new cotter pin.

19. Install the parking brake shoe assembly and the brake rotor. Connect the ABS speed sensor and tighten the bolt to 17 ft. lbs. (23 Nm).

20. Install the brake caliper, and tighten the bolts to 50 ft. lbs. (68 Nm).

21. Loosely install a new hub nut and washer. Install the wheel and tire assembly and lower the vehicle.

22. Tighten the hub nut to 174–231 ft. lbs. (236–314 Nm). After torquing, stake the nut with a hammer and dull-bladed chisel.

23. Check the wheel alignment.

1995–97 Millenia

1. Raise and safely support the vehicle.

2. Remove the wheel.

3. Remove the brake caliper assembly.

4. Remove the locknut dust cap and the locknut.

5. Remove the brake disc plate and the hub assembly.

6. Remove the parking brake shoe assembly and disconnect the parking brake cable.

7. Remove the backing plate and the ABS wheel speed sensor from the spindle.

8. Disconnect the rear lower lateral link ball joint, upper and lower trailing link ball joints, and the upper and lower lateral link ball joints.

9. Remove the spindle.

To install:

10. Replace the spindle to the vehicle.

11. Connect the lower lateral link ball joint, torque the nut to 85–115 ft. lbs. (116–156 Nm).

12. Connect the upper lateral link ball joint, torque the nut to 41–59 ft. lbs. (55–80 Nm).

13. Connect the upper trailing link ball joint, torque the nut to 28–38 ft. lbs. (38–51 Nm).

14. Connect the lower trailing link ball joint, torque the nut to 55–77 ft. lbs. (75–104 Nm).

15. Connect the rear lower lateral link ball joint, torque the nut to 41–59 ft. lbs. (55–80 Nm).

16. Install the ABS wheel speed sensor, torque the mounting bolt 14–18 ft. lbs. (19–25 Nm).

17. Install the backing plate, torque the mounting nuts to 21–29 ft. lbs. (28–40 Nm).

18. Connect the parking brake cable and install the parking brake shoe assembly.

19. Install the wheel hub assembly.

20. Replace the disc plate and install the locknut, torque to 131–173 ft. lbs. (177–235 Nm).

21. Install the hub nut dust cap and install the brake caliper.

22. Install the wheel.

1993–97 Miata

1. Loosen the axle nut, then raise and safely support the vehicle.

2. Remove the wheel and tire assembly and the axle nut.

3. Remove the brake caliper assembly and hang it from the body. Do not let it hang by the hose.

4. Remove the disc and the ABS speed sensor. Remove the sensor bracket.

5. Remove the lower and upper control arm bolts and remove the knuckle and hub assembly from the vehicle.

6. Be careful not to distort the back plate. If damaged, a new one must be pressed onto the knuckle.

7. To remove the hub, remove the seal and properly support the assembly to press the hub out from the back. If the inner bearing race stays on the hub, use a chisel to move it far enough to grab it with a bearing puller.

8. Remove the snapring and press the bearing out of the knuckle.

To install:

9. Carefully inspect all parts for wear or damage, replace as necessary. Always install a new seal and bearing.

10. Apply grease to the wheel bearing. When pressing the new bearing into the knuckle, make sure to press only on the outer race.

11. Apply grease to the hub. Install the snapring, support the inner race and press the hub into the bearing. Failure to properly support the bearing races for pressing will ruin the bearing.

12. Lubricate and install a new seal, and fit the knuckle into place on the suspension. Torque the upper control arm bolt to 49 ft. lbs. (67 Nm) and the lower bolt to 55 ft. lbs. (75 Nm).

13. Install the sensor bracket and the ABS speed sensor.

14. Install the brake disc. Install the brake caliper, and torque the bolts to 51 ft. lbs. (69 Nm).

15. Install the washer and a new axle nut. With all 4 wheels on the ground, torque the nut to 160–217 ft. lbs. (216–294 Nm). Stake the nut into place.

16. Check and adjust the rear wheel alignment.

1993–95 RX7

1. Loosen the hub nut and raise and safely support the vehicle. Remove the wheel and tire assembly.

2. Disconnect the ABS sensor. Remove the caliper and suspend it from the coil spring; do not let it hang by the brake hose.

3. Remove the hub nut and washer. Remove the disc brake rotor.

4. Remove the upper and lower control arm bolts and the toe control link bolt. Remove the hub/knuckle assembly.

NOTE: If the halfshaft is stuck in the hub, install the old hub nut until it is flush with the end of the shaft. Tap out the shaft, using a soft mallet only.

5. Properly support the hub/knuckle assembly and press the hub from the knuckle. If the bearing race sticks to the knuckle, grind the race in 1 spot until only 0.02 in. (0.5mm) thickness remains. Cut the race with a chisel and remove it.

6. Remove the snapring from the knuckle. Properly support the knuckle and press out the bearing.

7. Inspect the hub and knuckle for cracks or other damage and replace as necessary.

8. Remove the axle shaft from the differential.

To install:

9. Install the axleshaft into the differential.

10. Properly support the knuckle and press in the new bearing. Install a new snapring.

11. Properly support the knuckle and press on the hub.

12. Install the hub knuckle assembly. Tighten the upper and lower control arm bolts to 44–54 ft. lbs. (59–73 Nm) and the toe control link bolt to 48–57 ft. lbs. (64–78 Nm).

13. Install the brake rotor and loosely install a new hub nut and washer.

14. Install the brake caliper and tighten the bolts to 34–49 ft. lbs. (46–67 Nm). Connect the ABS sensor and tighten the bolt to 12–16 ft. lbs. (16–22 Nm).

15. Install the wheel and tire assembly and lower the vehicle.

16. Tighten the wheel hub nut to 174–231 ft. lbs. (236–313 Nm). After torquing, stake the hub nut using a hammer and dull-bladed chisel.

17. Check the wheel alignment.

FIRING ORDERS

NOTE: To avoid confusion, always replace spark plug wires one at a time.

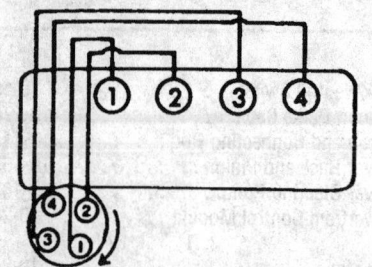

178430

1.5L (G4DJ) Engine
Engine Firing Order: 1–3–4–2
Distributor Rotation: Clockwise

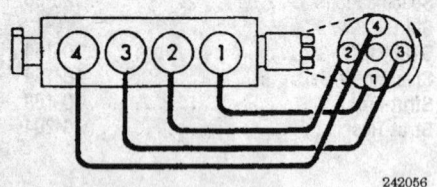

242056

1.5L (4G15) and 1.8L (4G93)
Engine Firing Order: 1-3-4-2
Distributor Rotation: Counterclockwise

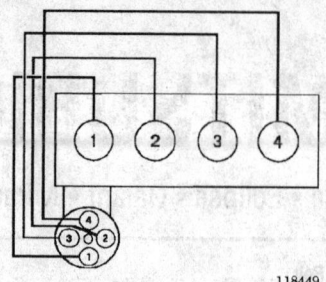

118449

1.8L (4G37) and 2.0L (4G63) SOHC Engines
Engine Firing Order: 1-3-4-2
Distributor Rotation: Clockwise

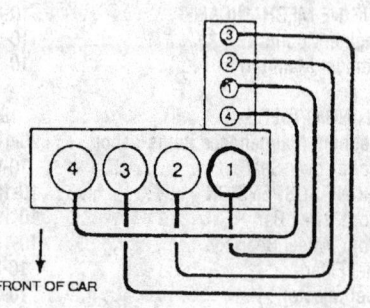

167847

2.0L (4G63) DOHC Engine
Engine Firing Order: 1-3-4-2
Distributorless Ignition System

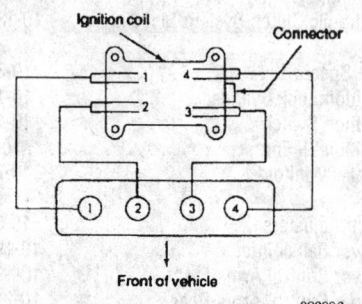

323996

2.0L (420A) Engine
Engine Firing Order: 1-3-4-2
Distributorless Ignition System

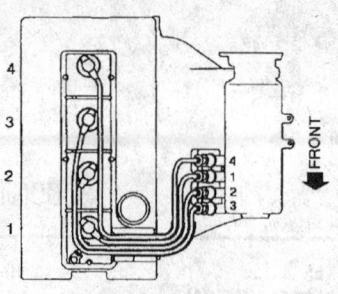

242167

2.4L (4G64) DOHC Engines
Engine Firing Order: 1-3-4-2
Distributorless Ignition System

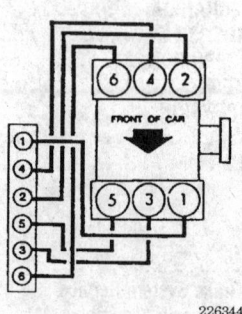

226344

3.0L (6G72) Engine DOHC
Engine Firing Order:
1-2-3-4-5-6
Distributorless Ignition System

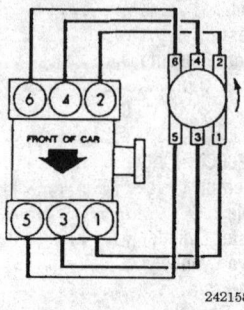

242158

3.0L (6G72) Engine SOHC
Engine Firing Order:
1-2-3-4-5-6
Distributor Rotation:
Counterclockwise

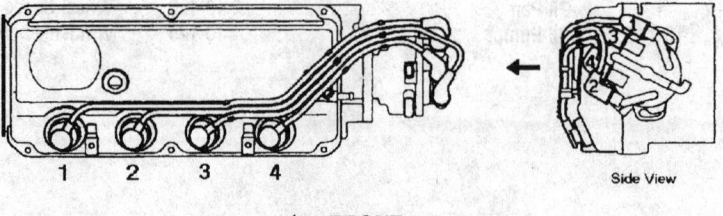

167853

2.4L (4G64) SOHC and 1993 Galant 2.0L 16 Valve (4G63) Engines
Engine Firing Order: 1-3-4-2
Distributor Rotation: Counterclockwise

ENGINE ELECTRICAL

NOTE: Disconnecting the negative battery cable on some vehicles may interfere with the functions of the on board computer systems and may require the computer to undergo a relearning process, once the negative battery cable is reconnected.

Distributor

REMOVAL AND INSTALLATION

Before removing the distributor, position No. 1 cylinder at TDC on the compression stroke and align the timing marks.

1. Disconnect the negative battery cable. Remove the ignition wire cover, if equipped.
2. If necessary for access, tag and disconnect the spark plug wires from the distributor cap.
3. Detach the distributor harness connector.

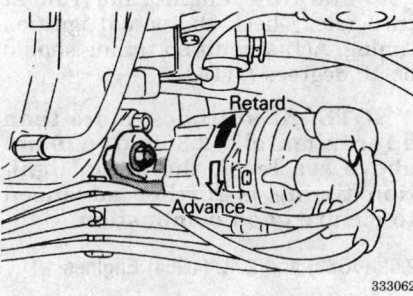

Adjusting distributor — Mirage shown

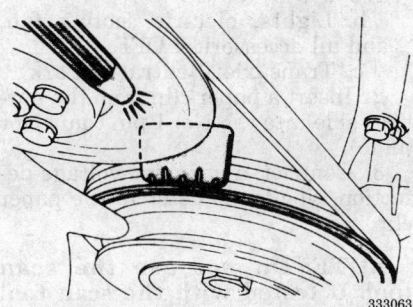

Checking ignition timing — Mirage shown

4. Unscrew the distributor cap hold-down screws or release the clips and lift off the distributor cap with all ignition wires still connected. Remove the coil wire, if necessary.
5. Matchmark the rotor to the distributor housing and the distributor housing to the engine.

NOTE: Do not crank the engine during this procedure. If the engine is cranked, the matchmark must be disregarded.

6. Remove the hold-down nut.
7. Carefully remove the distributor from the engine.

NOTE: Some engines may be sensitive to the routing of the distributor sensor wires. If routed near the high-voltage coil wire or the spark plug wires, the electromagnetic field surrounding the high voltage wires could generate an occasional disruption of the ignition system operation.

To install:

Timing Not Disturbed

1. Install a new distributor housing O-ring and lubricate with clean oil.
2. Install the distributor in the engine so the rotor is aligned with the matchmark on the housing and the housing is aligned with the matchmark on the engine. Make sure the distributor is fully seated and the distributor shaft is fully engaged.
3. Install the hold-down nut.
4. Connect the distributor harness connectors.
5. Make sure the sealing O-ring is in place, install the distributor cap and tighten the screws or secure the clips.
6. Connect the negative battery cable.
7. Adjust the ignition timing and tighten the hold-down nut to 8 ft. lbs. (11 Nm).

Timing Disturbed

1. Install a new distributor housing O-ring and lubricate with clean oil.
2. Position the engine so the No. 1 piston is at TDC of its compression stroke and the mark on the vibration damper is aligned with **0** on the timing indicator.
3. Align the distributor housing and gear mating marks. Install the distributor in engine so the slot or groove of the distributor's installation flange aligns with the distributor

installation stud in the engine block. Make sure the distributor is fully seated. Inspect alignment of the distributor rotor making sure the rotor is aligned with the position of the No. 1 ignition wire in the distributor cap.

NOTE: Make sure the rotor is pointing to where the No. 1 runner originates inside the cap, if equipped, and not where the No. 1 ignition wire plugs into the cap.

4. Install the hold-down nut.
5. Connect the distributor harness connectors.
6. Make sure the sealing O-ring is in place, install the distributor cap and tighten the screws or secure the clips.
7. Connect the negative battery cable.
8. Adjust the ignition timing and tighten the hold-down nut to 8 ft. lbs. (11 Nm).

Ignition Timing

ADJUSTMENT

Mirage

1. Attach the timing light according to the manufacturer's instructions.
2. Locate the timing tab line on the front of the engine and the notch on the crankshaft pulley. Mark them so they are easily recognizable with the timing light. Connect a tachometer to the engine.
3. **START** the engine and allow it to reach operating temperature.
4. Check that the engine idle is at specifications.
5. Turn the engine **OFF**.
6. Locate the ignition timing adjustment connector (brown) and connect a jumper wire to from the connector to a good ground connection.

NOTE: Grounding the ignition timing adjustment connector will set the engine to basic ignition timing.

7. Point the timing light at the crankshaft pulley marks and inspect the ignition timing.
8. If the timing is not within specifications, loosen the distributor mounting nut and rotate the distributor slowly, in either direction, to align the timing marks.
9. Tighten the distributor mounting nut when the ignition timing is correct. Stop the engine and remove the timing light.
10. Adjust engine idle if needed.

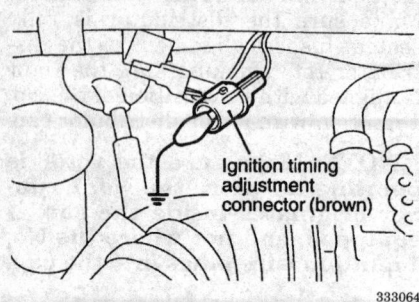

Timing connector identification — Mirage

333064

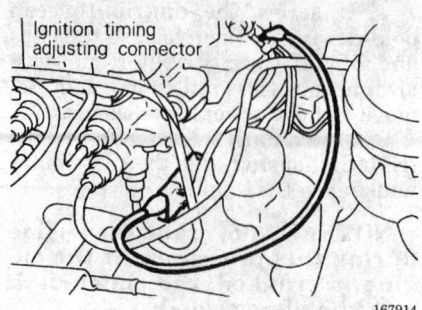

Ignition timing adjusting connector location — 1993 Galant 2.0L (4G63) engine shown

167914

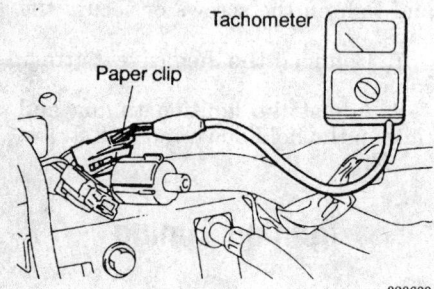

Connecting a tachometer — Eclipse 2.0L (4G63) and 2.4L (4G64) engines

323633

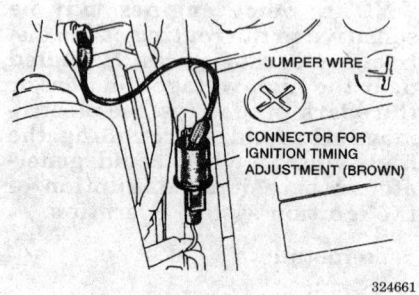

Ignition timing connector — Diamante SOHC and 3000GT engines

324661

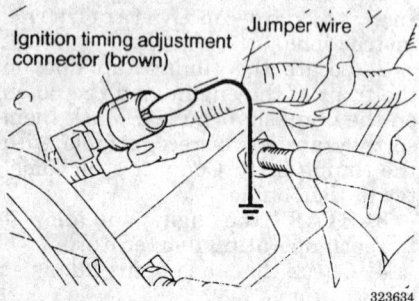

Grounding the ignition timing adjustment connector — Eclipse 2.0L (4G63) and 2.4L (4G64) Engines

323634

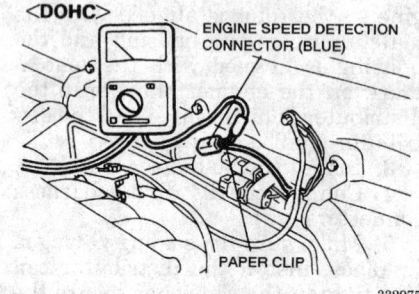

Tachometer connection point — Diamante

332975

Eclipse

1.8L (4G37) Engine

1. Apply the parking brake and block the wheels. Run the engine until the coolant reaches normal operating temperature.
2. Make certain all lights, cooling fan and accessories are OFF.
3. Position the steering wheel in straight ahead position and the gear selector lever in P or N.
4. Connect a timing light to the engine.

5. Insert a paper clip into the CRC filter connector (3-pole connector), located in the engine compartment of the vehicle.
6. Connect a tachometer to the inserted clip.

NOTE: During installation of the paper clip, do not separate the connector.

7. Check the curb idle speed. It should be 600–800 rpm.
8. Turn the engine OFF. Connect a jumper wire to the terminal for ig-

nition-timing adjustment (located in the engine compartment), and ground it.
9. Start and run the engine at curb idle speed.
10. Check the basic ignition timing and adjust, if necessary. Basic ignition timing should be 5 degrees BTDC.

NOTE: Ignition timing can vary depending on equipment and options. Always check and use underhood decal for specification when available.

11. If the timing is not within specifications, loosen the distributor hold-down bolt and turn the distributor to bring the timing within specifications. Turning the distributor to the right retards timing while to the left will advance timing.
12. Tighten the hold-down bolt after adjustment. Recheck the timing and adjust if necessary.
13. Stop the engine and remove the ground for the ignition timing connector.

NOTE: Actual ignition timing may vary, depending on the control mode of the engine control unit. In such case, recheck the basic ignition timing. If there is no deviation, the ignition timing is functioning normally.

14. Start the engine and run at curb idle. Check the actual ignition timing. Actual ignition timing should be 10 degrees BTDC.

NOTE: At altitudes, more than approximately 2,300 ft. (701m) above sea level, the actual ignition timing is further advanced to ensure good combustion.

2.0L (4G63) and 2.4L (4G64) Engines

1. Before inspection and adjustment, set the vehicle to the following conditions:
   a. Engine coolant temperature: 176–203° F (80–95° C).
   b. Lights, electric cooling fan, and all accessories: OFF.
   c. Transaxle: Neutral or Park.
2. Insert a paper clip from the harness side into the No. 1 pin connector (blue).
3. Connect a primary-voltage-detection type tachometer to the paper clip.

NOTE: Do not use the scan tool. If tested with the scan tool connected to the data link connector, the ignition timing will not be the basic timing.

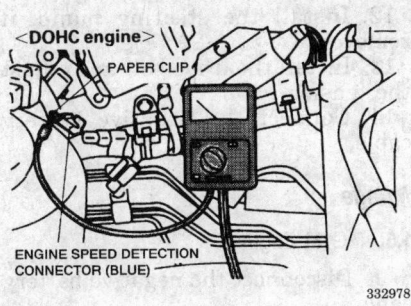

ENGINE SPEED DETECTION
CONNECTOR (BLUE)

332978

Tachometer connection point — 3000GT

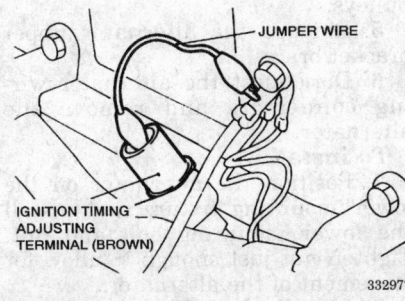

JUMPER WIRE

IGNITION TIMING
ADJUSTING
TERMINAL (BROWN)

332977

Ignition timing connector — Diamante DOHC engines

4. Set up a timing light. Start the engine and run at idle. Idle specification is: 750 ±100 rpm.

NOTE: For rpm, ½ of the actual engine rpm is indicated, so that the actual engine rpm is two times the indicated value shown by the tachometer.

5. Turn the ignition switch to **OFF**.

6. Remove the waterproof connector from the ignition timing adjustment connector (brown). Connect the jumper wire with the clip to the ignition timing adjustment terminal, and ground this to the body.

NOTE: Grounding this terminal sets the engine to the basic ignition timing.

7. Start the engine and run it at idle. Check that the basic ignition timing is within specifications. Specification is: 5° ±3°.

8. If the basic ignition timing is not within specifications, inspect the MFI components.

9. Stop the engine, remove the jumper wire from the ignition timing adjustment connector (brown), and reconnect the connector.

10. Start the engine and check that the actual ignition timing is within

specification. Specification is: 8° BTDC.

NOTE: Ignition timing is variable within ±7°, even under normal operating conditions. Timing is automatically advanced by about 5° from 8° BTDC at higher altitudes.

2.0L (420A) Engine

It is not necessary to check ignition timing using a timing light, because the crankshaft position is detected directly and ignition timing is controlled electronically.

Galant

1. Set the parking brake, start and run the engine until normal operating temperature is obtained. Keep all lights and accessories off and the front wheels straight-ahead. Place the transaxle in **P** for automatic transaxle or neutral for manual transaxle.

NOTE: On Canadian vehicles the lights will remain on when the vehicle is running; this will not be a problem.

2. Locate the wire connector on the ignition coil connector. Insert a paper clip behind the TACH terminal connector to act as a tachometer adapter. Connect a tachometer to the paper clip. If not at specification, set the idle speed at the correct level.

3. Turn the engine **OFF** and remove the waterproof cover from the ignition timing adjusting connector. This connector is a brown connector located near the center of the firewall. Connect a jumper wire from this terminal to a good ground.

4. Connect a conventional power timing light to the No. 1 cylinder spark plug wire. Start the engine and run at idle. Check the ignition timing by aiming the timing light at the timing scale located near the crankshaft pulley. Ignition timing should be 5° BTDC.

5. If ignition timing is not within specification, loosen the distributor hold-down nut just enough so the housing can be rotated.

6. Turn the housing in the proper direction until the specified timing is reached. Tighten the hold-down nut and recheck the timing. Turn the engine **OFF**.

7. Remove the jumper wire from the ignition timing adjusting terminal and install the waterproof cover.

8. Start the engine and check the actual timing without the terminal grounded. This reading should be approximately 5 degrees more than the

basic timing. Actual timing may increase according to altitude. Also, actual timing may fluctuate because of slight variations accomplished by the ECU. As long as the basic timing is correct, the engine is timed correctly.

9. Turn the engine **OFF**. Disconnect the timing equipment and tachometer.

3000GT and Diamante

3.0L (6G72) SOHC Engine

1. Set the parking brake, start and run the engine until normal operating temperature is obtained. Keep all lights and accessories OFF and the front wheels straight-ahead. Place the transaxle in **P** for automatic transaxle or Neutral for manual transaxle.

2. If not at specification, set the idle speed to the correct level.

3. Turn the engine **OFF**. Remove the water-proof cover from the ignition timing adjusting connector. This connector is located slightly left of center on the firewall.

4. Connect a conventional powered timing light to the No. 1 cylinder spark plug wire. Start the engine and run at idle.

5. Aim the timing light at the timing scale located near the crankshaft pulley.

6. Loosen the distributor hold-down nut (SOHC) or the crankshaft position sensor nut (DOHC) just enough so the housing can be rotated.

7. Turn the housing in the proper direction until the specified timing is reached. Tighten the hold-down nut and recheck the timing. Turn the engine **OFF**.

8. Remove the jumper wire from the ignition timing adjusting terminal and install the water-proof cover.

9. Start the engine and check the actual timing (the timing without the terminal grounded). This reading should be approximately 5 degrees more than the basic timing. Actual timing may increase according to altitude. Also, actual timing may fluctuate because of slight variation accomplished by the ECU. As long as the basic timing is correct, the engine is timed correctly.

10. Turn the engine **OFF**. Disconnect the timing light and the tachometer.

3.0L (6G72) DOHC Engine

1. Set the parking brake, start and run the engine until normal operating temperature is obtained. Keep all lights and accessories OFF and the front wheels straight-ahead. Place the transaxle in **P** for automatic tran-

saxle or Neutral for manual transaxle.

2. Connect a tachometer to the single pin connector terminal under the hood. Check the idle speed and if not at specification, set the idle speed to the correct level.

NOTE: The tachometer reading will be $1/3$ of the actual engine speed. Multiply the reading by 3 to figure the actual engine speed.

3. Turn the engine **OFF**. Remove the water-proof cover from the ignition timing inspecting connector. This connector is located:
- **3000GT**: on the firewall just behind the battery.
- **Diamante**: slightly left of center on the firewall.

4. Connect a conventional powered timing light to the No. 1 cylinder spark plug wire. Start the engine and run at idle.

5. Aim the timing light at the timing scale located near the crankshaft pulley. If timing marks do not align, the ignition timing is not within specification.

NOTE: The ignition timing is not adjustable; if timing is not within specification, check the Engine Control System operation and perform the required service.

6. Remove the jumper wire from the ignition timing inspection and terminal and install the water-proof cover.

7. Start the engine and check the actual timing (the timing without the terminal grounded). This reading should be approximately 5 degrees more than the basic timing. Actual timing may increase according to altitude. Also, actual timing may fluctuate because of slight variation accomplished by the ECU. As long as the basic timing is correct, the engine is timed correctly.

8. Turn the engine **OFF**. Disconnect the timing apparatus and tachometer.

Alternator

PRECAUTIONS

Several precautions must be observed with alternator equipped vehicles to avoid damage to the unit.
- If the battery is removed for any reason, make sure it is reconnected with the correct polarity. Reversing the battery connections may result in damage to the 1-way rectifiers.

- When utilizing a booster battery as a starting aid, always connect the positive to positive terminals and the negative terminal from the booster battery to a good engine ground on the vehicle being started.
- Never use a fast charger as a booster to start vehicles.
- Disconnect the battery cables when charging the battery with a fast charger.
- Never attempt to polarize the alternator.
- Do not use test lights of more than 12 volts when checking diode continuity.
- Do not short across or ground any of the alternator terminals.
- The polarity of the battery, alternator and regulator must be matched and considered before making any electrical connections within the system.
- Never separate the alternator on an open circuit. Make sure all connections within the circuit are clean and tight.
- Disconnect the battery ground terminal when performing any service on electrical components.
- Disconnect the battery if arc welding is to be done on the vehicle.

REMOVAL AND INSTALLATION

Precis

1. Disconnect the negative battery cable.
2. If equipped with power steering, remove the oil pump and position the pump above the bracket without disconnecting the fluid lines.
3. Loosen the alternator mounting bolts to relieve the belt tension and remove the drive belt from the alternator pulley.
4. Raise the front of the vehicle and support safely.
5. Remove the left side mud guard.
6. Disconnect the B+ terminal wire from the alternator.
7. Support the alternator by hand, finish removing the mounting bolts and lift the unit up and out of the engine compartment.
To install:
8. Position the alternator onto its mounting and install the mounting bolts.
9. Connect the B+ wire to the alternator terminal. Make sure the wire nut is tight and the protective cap is firmly placed over the wire connection.
10. Install the left side mud guard.

11. Lower the vehicle.
12. Install the steering pump, if removed.
13. Install the drive belt and adjust the tension.
14. Connect the negative battery cable.

Mirage

1.5L (4G15) Engine

1. Disconnect the negative battery cable.
2. Remove the left side cover panel under the vehicle.
3. Remove the drive belts.
4. Remove the water pump pulleys.
5. Remove the alternator upper bracket/brace.
6. Disconnect the alternator wiring connectors and remove the alternator.
To install:
7. Position the alternator on the lower mounting fixture and install the lower mounting bolt and nut. Tighten nut just enough to allow for movement of the alternator.
8. Install the alternator upper bracket/brace and connect the alternator electrical harness.
9. Install the water pump pulleys.
10. Install the drive belts and adjust to the proper tension.
11. Install the left side cover panel under the vehicle as required.
12. Connect the negative battery cable and check for proper operation.

1.8L (4G93) Engine

1. Disconnect negative battery cable. Remove the accessory drive belts.
2. Disconnect the electrical harness from the alternator.
3. Remove the alternator mounting nut, bolt and upper brace assembly from the vehicle.
To install:
4. Install the alternator and secure using mounting nuts. Make sure the upper brace assembly is in place.
5. Install and adjust drive belts to the proper tension. Secure all mounting hardware.
6. Reconnect the negative battery cable and check system operation.

Eclipse

1.8L (4G37) and 2.0L (4G63) Engines

1. Disconnect the negative battery cable. Remove the left side undercover from the vehicle.
2. If equipped with air conditioning, remove the condenser electric fan motor and shroud assembly.

3. Remove alternator, water pump and air conditioner compressor drive belts.

4. Remove both water pump pulleys and the alternator top brace.

5. Disconnect the alternator wiring and remove the alternator from the vehicle.

To install:

6. Install the alternator in position and connect the electrical harness.

7. Install the alternator top brace to the engine and tighten the mounting bolt to 20 ft. lbs. (27 Nm).

8. Install the alternator mounting bolt loosely.

9. Install the water pump pulleys and tighten retainer bolts to 7 ft. lbs. (10 Nm).

10. Install the drive belts and adjust until the proper tension is achieved. Secure the lower alternator through-bolt nut to 18 ft. lbs. (25 Nm) and the upper alternator lock bolt to 11 ft. lbs. (15 Nm).

11. Install the condenser electric fan motor and shroud assembly. Connect all wiring.

12. Install the left side undercover from the vehicle, if removed.

13. Connect the negative battery cable. Start the engine and check the alternator for proper operation.

1995–97 2.0L and 2.4L (4G64) Engines

1. Disconnect the negative battery cable.

2. Remove the left side undercover from the vehicle.

3. Remove the alternator and power steering pump drive belts.

4. Remove the harness connector from the rear of the alternator.

5. Remove the oil pressure switch connector.

6. Remove the alternator mounting bolts and remove the alternator from the underside of the vehicle.

To install:

7. Install the alternator from the underside of the vehicle with the mounting bolts.

8. Reconnect the oil pressure switch.

9. Connect the harness connector to the rear of the alternator.

10. Install and adjust the drive belts.

11. Install the left side undercover.

12. Connect the negative battery cable.

Galant

1. Disconnect the negative battery cable.

2. Remove the left side undercover from the vehicle.

3. On turbocharged models, remove the air intake hose.

4. Remove alternator, water pump and air conditioner compressor drive belts.

5. Remove both water pump pulleys and the alternator top brace.

6. Disconnect the alternator wiring and remove the alternator from the vehicle.

To install:

7. Install the alternator in position and connect the electrical harness.

8. Install the alternator top brace to the engine and tighten the mounting bolt to 18 ft. lbs. (24 Nm).

9. Install the alternator mounting bolt loosely.

10. Install the water pump pulleys and tighten retainer bolts to 7 ft. lbs. (10 Nm).

11. Install the drive belts and adjust until the proper tension is achieved. Secure the lower alternator through-bolt to 15–18 ft. lbs. (20–25 Nm) and the upper alternator lock bolt to 10 ft. lbs. (14 Nm).

12. On turbocharged models, install the air intake hose.

13. Install the left side undercover.

14. Connect the negative battery cable. Start the engine and check the alternator for proper operation.

Diamante

3.0L (6G72) DOHC Engine

1. Disconnect the negative battery cable.

2. Remove the headlamp washer reservoir tank.

3. Remove the condenser fan and upper radiator insulator.

4. Loosen the tensioner pulley and remove the alternator drive belt.

5. Remove the alternator upper and lower mounting bolts.

6. Remove the alternator support bracket mounting bolts.

7. Remove the alternator support bracket from the vehicle.

8. Disconnect the alternator wiring harness.

9. Remove the alternator from the vehicle.

To install:

10. Install the alternator to the vehicle and connect the wiring harness.

11. Install the alternator support bracket to the vehicle and tighten the bracket mounting bolts to specifications.

12. Position the alternator on the mounting bracket. Install and tighten the mounting bolt and nut to 17 ft. lbs. (24 Nm).

13. Reinstall the drive belt and adjust the tensioner until the proper belt tension is achieved.

14. Install the upper radiator insulator and condenser fan.

15. Install the headlamp washer reservoir tank.

16. Connect the negative battery cable and check the charging system for proper operation.

3.0L (6G72) SOHC Engine

1. Disconnect the negative battery cable.

2. Disconnect and remove the air intake hose.

3. Loosen the tensioner pulley and remove the alternator drive belt.

4. On California models, remove the rear bank converter assembly.

5. Remove the engine roll stopper stay bracket assembly.

6. Disconnect the EGR temperature sensor wire and remove the EGR pipe assembly.

7. Remove the intake plenum stay bracket assembly.

8. Disconnect the alternator wiring harness connectors.

9. Remove the alternator upper and lower mounting bolts.

10. From beneath the vehicle, remove the alternator from the transaxle side of the vehicle.

To install:

11. Position the alternator on the lower mounting fixture. Install and tighten the mounting bolt and nut to 14–18 ft. lbs. (20–25 Nm).

12. Connect the alternator wiring harness.

13. Install the intake plenum stay bracket and tighten the mounting bolt to 13 ft. lbs. (18 Nm).

14. Install the EGR pipe and tighten the fitting connections to 43 ft. lbs. (60 Nm).

15. Connect the EGR temperature sensor wire.

16. Connect the engine roll stopper stay and tighten the mounting bolt to 35 ft. lbs. (45 Nm) and the nut to 36–43 ft. lbs. (50–60 Nm).

17. Install the rear converter assembly, if removed.

18. Reinstall the drive belt and adjust the tensioner until the proper belt tension is achieved.

19. Connect the air intake hose.

20. Connect the negative battery cable and check the charging system for proper operation.

3000GT

1. Disconnect the negative battery cable.

CAUTION

Wait at least 90 seconds after the negative (-) battery cable is disconnected to prevent possible deployment of the air bag.

2. On turbocharged models, remove the air intake pipe and hoses.

3. Remove the radiator surge tank.

4. If equipped with air conditioning, remove the clamp nut that secures the air conditioning hose.

5. Raise the air conditioning suction hose and suspend it from the engine hood using a cord.

6. Loosen the tensioner pulley and remove the alternator drive belt.

7. Disconnect the oxygen sensor connector.

8. Disconnect the alternator wiring harness.

9. Remove the alternator bracket to engine block mounting bolts and remove the bracket and alternator as an assembly.

10. Separate the alternator from the mounting bracket on a workbench.

To install:

11. Install the alternator onto the bracket and install bracket assembly to the engine.

12. Connect the oxygen sensor connector.

13. Connect wiring harness to the rear of the alternator.

14. Install the drive belt and adjust to proper tension using the tensioner pulley.

15. Install air conditioning suction hose to its original position and secure using clamp nut.

16. Install the radiator surge tank.

17. Install the air intake delivery hoses and air pipe.

18. Reconnect the negative battery cable and check the charging system for proper operation.

Drive Belt

REMOVAL AND INSTALLATION

Precis and Mirage

Alternator and Water Pump Belt

1. Loosen the alternator support nut.

2. Loosen the adjuster lock bolt.

3. Rotate the adjuster bolt counterclockwise to release the tension on the belt.

4. Remove the belt.

To install:

5. Install the belt on the pulleys.

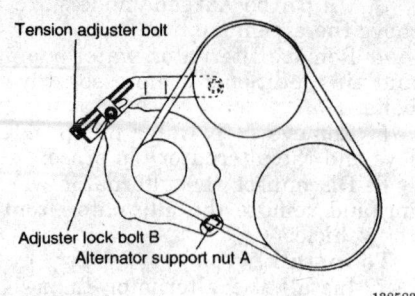

Alternator and water pump belt routing — Precis

138583

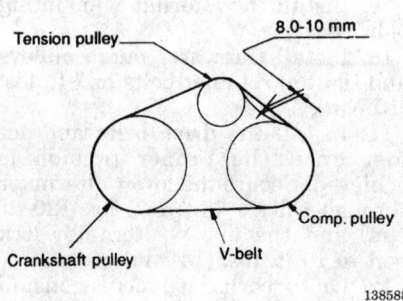

A/C Compressor drive belt routing — Precis

138588

6. Rotate the adjuster bolt clockwise until the proper tension is reached.

7. Tighten the adjuster lock bolt and the alternator support nut.

A/C Compressor Belt

1. Loosen the tension pulley and remove the belt.

2. Reverse the procedure to install the belt.

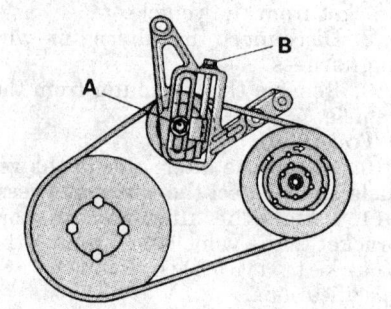

Example of air conditioner belt tensioner adjustment procedure — Mirage

329911

Eclipse with 2.0L (4G63) DOHC and 2.4L (4G64) Engines

Alternator and Water Pump Belt

1. Disconnect the negative battery cable.

2. Loosen the lock bolt.

3. Loosen the pivot bolt.

4. Turn the adjusting bolt to loosen the tension on the belt.

5. Remove the belt.

To install:

6. Install the belt on the pulleys.

7. Turn the adjusting bolt until a tension of 110.2–154.3 lbs.(490–686 N) for a new belt or 88.2 lbs. (392 N) for a used belt is indicated on the tension gauge. Another method is to apply a force of 22 lbs. (98 N) to the belt midway between the water pump and the alternator and measure the belt deflection. Belt deflection should be 0.30–0.35 in. (7.5–9.0 mm) for a new belt or 0.39 in. (10.0 mm) for a used belt.

8. Torque the pivot bolt to 15–18 ft. lbs. (20–25 Nm) for 2.0L engines or to 17 ft. lbs. (23 Nm) for 2.4L engines.

9. Torque the lock bolt to 8.7–11 ft. lbs. (12–15 Nm) for 2.0L engines or to 17 ft. lbs. (23 Nm) for 2.4L engines.

10. Now torque the adjusting bolt to 7.2 ft. lbs. (9.8 Nm).

11. Connect the negative battery cable.

Power Steering Belt

1. Disconnect the negative battery cable.

2. Remove the alternator belt.

3. Loosen the four power steering pump mounting bolts A, B, C and D.

4. Pivot the power steering pump towards the water pump and remove the belt.

To install:

5. Install the belt on the proper pulleys.

6. Pry the power steering pump away from the water pump pulley until a tension of 110.2–154.3 lbs.(490–686 N) for a new belt or 77.2–99.2 lbs. (343–441 N) for a used belt is indicated on the tension gauge. Another method is to apply a force of 22 lbs. (98 N) to the belt midway between the water pump and the power steering pulley and measure the belt deflection. Belt deflection should be 0.18–0.22 in. (4.5–5.5 mm) for a new belt or 0.24–0.28 in. (6.0–7.0 mm) for a used belt.

7. Torque mounting bolt A to 21 ft. lbs. (28 Nm).

8. Torque mounting bolts B and D to 21 ft. lbs. (28 Nm).

9. Torque mounting bolt C to 16 ft. lbs. (22 Nm).

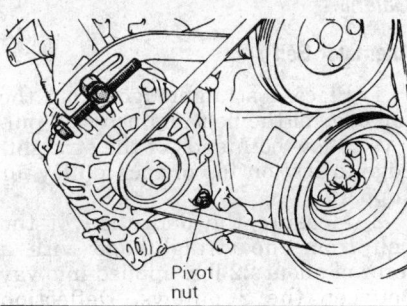

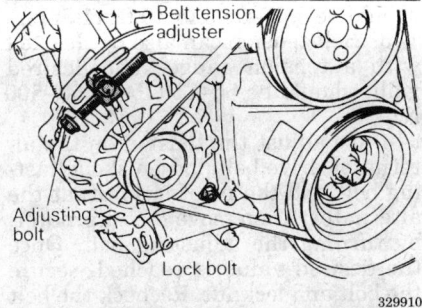

Example of alternator adjustment and pivot points — Mirage

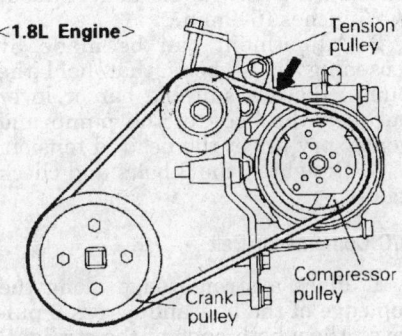

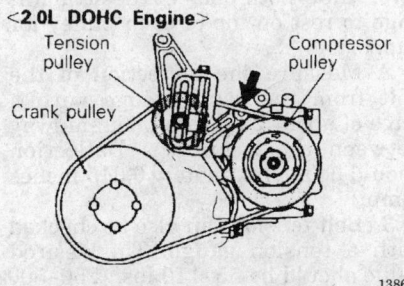

Air conditioner compressor belt adjustment — 1993–94 Eclipse 4G37 and 4G63 DOHC engines

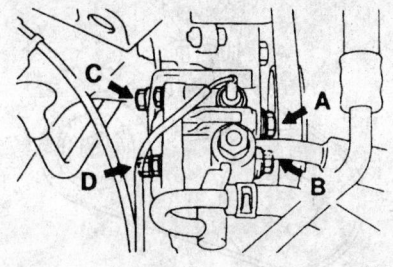

Power steering pump mounting bolts — Eclipse and Diamante shown, Galant similar

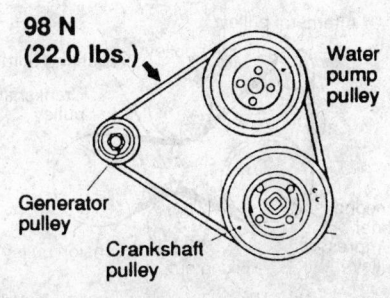

Measuring the belt deflection on the alternator — 1995–97 Eclipse 2.0L (4G63) DOHC and 2.4L (4G64) Engines

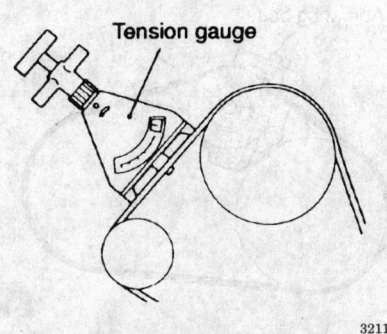

Using a belt tension gauge

10. Install and adjust the alternator belt.
11. Connect the negative battery cable.

A/C Compressor Belt

1. Disconnect the negative battery cable.
2. Remove the alternator belt.
3. Loosen the lock bolt (A) on the tension pulley.
4. Turn the adjusting bolt (B) to loosen the belt.
5. Remove the belt.

To install:
6. Install the belt on the proper pulleys.
7. Turn the adjusting bolt (B) until a tension of 86.0–99.2 lbs.(382–411 N) for a new belt or 57.3–75.0 lbs. (255–333 N) for a used belt is indicated on the tension gauge. Another method is to apply a force of 22 lbs. (98 N) to the belt midway between the A/C compressor and the crankshaft pulley or between the A/C compressor and the tension pulley and measure the belt deflection. Belt deflection should be 0.22–0.24 in. (5.5–6.0 mm) for a new belt or 0.26–0.30 in. (6.5–7.5 mm) for a used belt.

WARNING

Measure the belt tension only after one full rotation of the crankshaft in the forward direction (right).
---

8. Torque the lock bolt A to 17–20 ft. lbs. (23–26 Nm).
9. Install and adjust the alternator belt.
10. Connect the negative battery cable.

1995–97 Eclipse 2.0L (420A) Engine

Alternator Belt

1. Disconnect the negative battery cable.
2. Remove the power steering pump and A/C compressor belt.
3. Loosen the locknut.
4. Loosen the pivot bolt.
5. Turn the adjusting bolt to loosen the tension on the belt.
6. Remove the belt.

To install:
7. Install the belt on the proper pulleys.
8. Turn the adjusting bolt until a tension of 110–160 lbs.(490–712 N) for a new belt or 90–110 lbs. (400–490 N) for a used belt is indicated on the tension gauge. Another method is to apply a force of 22 lbs. (98 N) to the belt midway between the water pump and the alternator and measure the belt deflection. Belt deflection should be 0.30–0.41 in. (7.5–10.5 mm) for a new belt or 0.35–0.47 in. (9.0–12.0 mm) for a used belt.
9. Torque the pivot bolt to 40 ft. lbs. (54 Nm).
10. Torque the locknut to 45 ft. lbs. (61 Nm).
11. Install and adjust the power steering pump and A/C compressor belt.
12. Connect the negative battery cable.

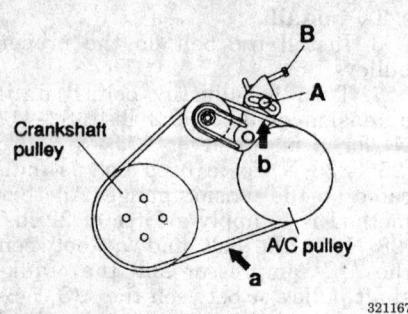

A/C compressor belt adjustment — 1995–97 Eclipse with 2.0L (4G63) DOHC and 2.4L (4G64) Engines

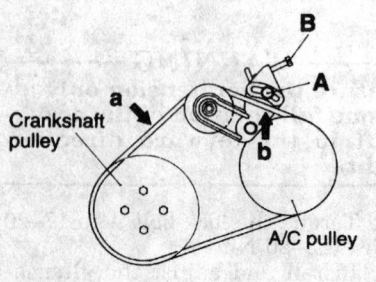

A/C compressor drive belt adjustment and belt deflection measurement points — 1994–97 Galant 2.4L (4G64) engines

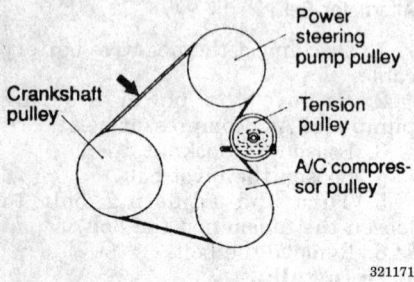

Measuring the belt deflection: Power steering pump and A/C belt — 1995–97 Eclipse 2.0L (420A) and 2.4L (4G64) Engines

Power Steering Pump and A/C Compressor Belt

1. Disconnect the negative battery cable.
2. Loosen the nut in the center of the tension pulley.
3. Turn the adjusting bolt to loosen the belt.
4. Remove the belt.
To install:
5. Install the belt on the proper pulleys.
6. Turn the adjusting bolt until a tension of 136.7–158.7 lbs. (608–706

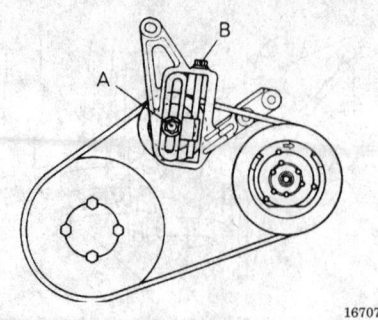

A/C compressor drive belt adjustment points — 1993 Galant 2.0L Engine

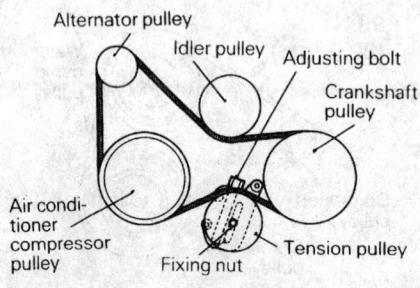

Alternator and A/C compressor drive belt — 3000GT and Diamante 3.0L (6G72) DOHC engine

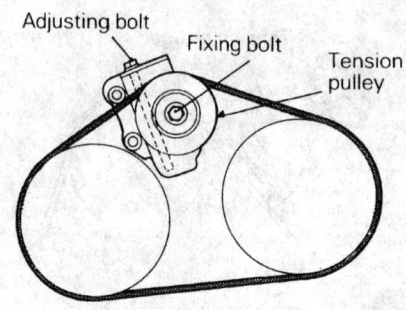

Air conditioning compressor drive belt — Diamante 3.0L (6G72) SOHC engine

N) for a new belt or 92.62–114.6 lbs. (412–510 N) for a used belt is indicated on the tension gauge. Another method is to apply a force of 22 lbs. (98 N) to the belt midway between the crankshaft pulley and the power steering pulley and measure the belt deflection. Belt deflection should be 0.32–0.35 in. (8.0–9.0 mm) for a new belt or 0.39–0.43 in. (10.0–11.0 mm) for a used belt.

NOTE: Make sure the power steering pump is mounted at the most forward position in the mounting bracket.

7. Tighten the tension pulley nut.

Galant

Alternator Belt

1. Place a straightedge along the top edge of the belt and across 2 pulleys. Allow both ends of the straightedge to rest on top of each pulley for support.
2. Measure the deflection of the belt from the straightedge with a force of about 22 lbs. applied midway between the 2 pulleys. Deflection should be 0.35–0.45 in. (9.0–11.5mm).
3. Belt tension can also be checked with a tension gauge. The desired value should be 55–110 lbs. (250–500 N).
4. To adjust the tension on the alternator drive belt, loosen the adjusting bolt and the pivot locknut, at the alternator. Then move the alternator, by turning the adjusting bolt. Once the desired value is reached, secure the bolt and locknut. Recheck the belt tension.

Power Steering Pump Belt

1. Press the belt in about the center between the power steering pump pulley and the pulley it shares, usually the water pump pulley. With reasonable pressure applied (about 22 lbs.), the belt should deflect about $1/4$–$3/8$ inches (6–9mm).
2. Adjustment can be made by loosening the 3 bolts that hold the pump. Place a suitable bar or lever between the body of the pump and gently pry to get the desired tension.
3. Retighten the 3 bolts and check again.

A/C Compressor Belt

1. Place a straightedge along the top edge of the belt and across 2 pulleys. Allow both ends of the straightedge to rest on top of each pulley for support.
2. Measure the deflection of the belt from the straightedge with a force of about 22 lbs. applied midway between the 2 pulleys. Deflection should be approximately 0.315 inches (8mm)
3. Belt tension can also be checked with a tension gauge. The desired value should be 55–110 lbs. (250–500 N).
4. To adjust the tension on the power steering drive belt, loosen the locknut on the tension pulley. Then move the adjusting bolt to adjust belt tension. Once the desired value is reached, secure the locknut and recheck tension.

Items		Check value	Adjustment value	
			New belt	Used belt
For generator and P/S pump	Tension N (lbs.)	350–600 (77–132)	700–900 (155–198)	450–600 (99–132)
	Deflection mm (in.) (Reference value)	6.0–9.0 (.24–.35)	4.0–5.0 (.16–.20)	6.0–8.0 (.24–.32)
For A/C compressor	Tension N (lbs.)	250–500 (55–110)	500–600 (110–132)	320–400 (70–88)
	Deflection mm (in.) (Reference value)	7.5–9.5 (.28–.37)	6.5–7.0 (.26–.28)	7.5–8.5 (.28–.34)

323201

Drive belt tension adjustments — Diamante 3.0L (6G72) SOHC Engine

<Vehicle without air conditioning>

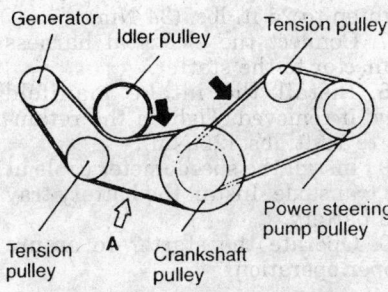

<Vehicle with air conditioning>

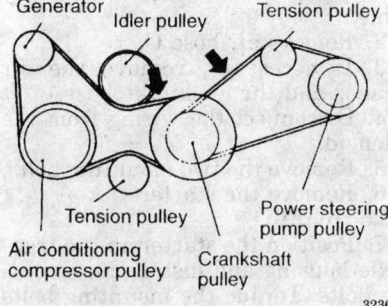

323657

Drive belts — 3000GT 3.0L (6G72) DOHC engine

3000GT and Diamante 3.0L (6G72) DOHC Engine

Alternator and A/C compressor belt

1. Disconnect the negative battery cable.

── CAUTION ──
Wait at least 90 seconds after the negative (-) battery cable is disconnected to prevent possible deployment of the air bag.

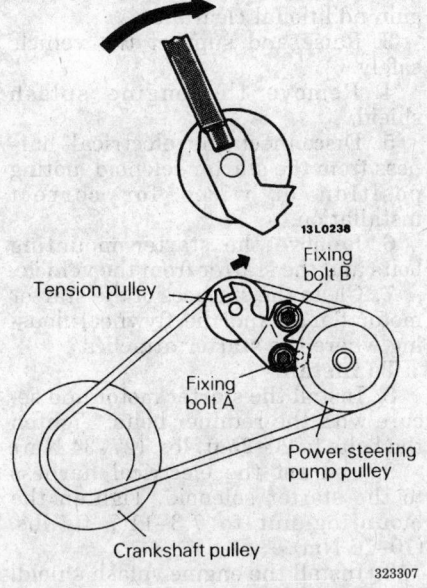

13L0238

Power steering pump drive belt — 3000GT and Diamante 3.0L (6G72) DOHC engine

2. Raise and safely support the vehicle and remove the front undercover.
3. Loosen the tension pulley fixing nut and relieve the tension on the belt by turning the adjusting bolt.
4. Remove the belt.
   To install:
5. Install the belt on the crankshaft and alternator pulleys.
6. Using the adjusting bolt on the tensioner, tighten the belt to the desired tension.
7. Tighten the fixing nut to hold the adjustment.

8. Install the undercover and lower the vehicle to the floor.
9. Connect the negative battery cable.

Power Steering Belt

1. Disconnect the negative battery cable.

── CAUTION ──
Wait at least 90 seconds after the negative (-) battery cable is disconnected to prevent possible deployment of the air bag.

2. Raise and safely support the vehicle and remove the undercover.
3. Remove the alternator and A/C compressor belt.
4. Lower the vehicle and remove the cruise control pump link assembly.
5. Place the power steering hose under the oil reservoir.
6. Loosen the tension pulley fixing bolts and remove the power steering pump drive belt.
   To install:
7. Install the power steering pump drive belt.
8. Insert an extension bar or equivalent into the opening at the end of the tension pulley bracket and pivot the pulley to apply tension to the belt.
9. Tighten the fixing bolts.
10. Raise the vehicle and install the alternator and A/C compressor belt.
11. Install the undercover and lower the vehicle.
12. Connect the negative battery cable.

Diamante 3.0L (6G72) SOHC Engine

1. Disconnect the negative battery cable.
2. Loosen the lockbolt on the face of the A/C tensioner pulley.
3. Turn the adjusting bolt of the A/C tensioner pulley to loosen the tension of the A/C belt.
4. Remove the A/C compressor belt.
5. Loosen the locknut on the face of the power steering/alternator tensioner pulley.
6. Turn the adjusting bolt of the tensioner pulley to loosen the tension of the belt.
7. Remove the power steering/alternator belt.
   To install:
8. Install the power steering/alternator belt first and then the A/C compressor drive belt.
9. Adjust the belts to the proper tension by turning the adjusting bolts and tighten pulley fixing nut/bolt.

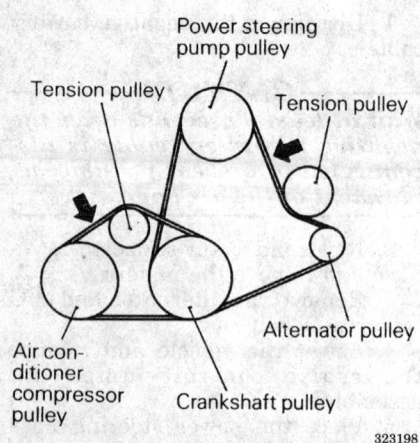

Drive belt routing — Diamante 3.0L (6G72) SOHC Engine

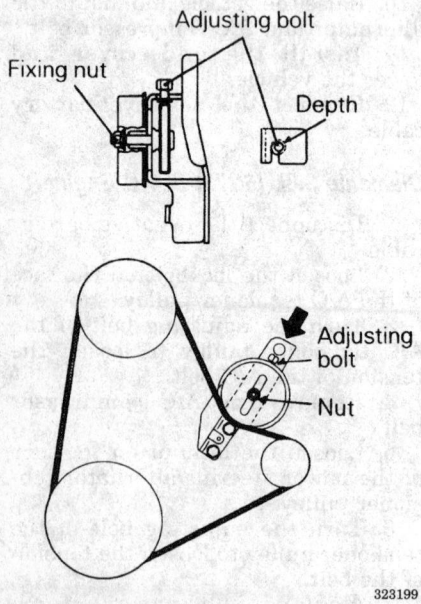

Alternator and power steering pump drive belt — Diamante 3.0L (6G72) SOHC Engine

10. Tighten the mounting nut of the power steering/alternator tensioner pulley to 36 ft. lbs. (50 Nm).

NOTE: The manufacturer does not provide a torque specification for the bolt that secures A/C tensioner pulley.

11. Connect the negative battery cable.

Starter

REMOVAL AND INSTALLATION

Precis

1. Disconnect the negative battery cable.
2. Remove the battery and tray to gain additional clearance.
3. Raise and support the vehicle safely.
4. Remove the engine splash shield.
5. Disconnect the electrical harness from the starter solenoid, noting position of wires for correct installation.
6. Remove the starter mounting bolts and the starter from the vehicle.
7. Clean the surfaces of the starter motor flange and the flywheel housing where the starter attaches.
To install:
8. Install the starter motor and secure with the retainer bolts. Tighten the bolts to 20–25 ft. lbs. (27–34 Nm).
9. Connect the electrical harness to the starter solenoid. Tighten the mounting nut to 7.3–11.7 ft. lbs. (10–16 Nm).
10. Install the engine splash shield.
11. Install the battery and tray, if removed.
12. Reconnect the negative battery cable.

Mirage

1. Disconnect the speedometer cable connector at the transaxle end.
2. Disconnect the starter motor electrical connections.
3. Remove the starter motor mounting bolts and remove the starter.
To install:
4. Install the starter motor and secure with the retainer bolts. Tighten starter mounting bolts to 22 ft. lbs. (31 Nm.)
5. Connect the electrical harness connector to the starter.
6. Connect the negative battery cable and check the starter for proper operation.

Eclipse

1.8L (4G37) and 2.0L (4G63) DOHC Engines

1. Remove the battery and battery tray from the engine compartment.
2. Disconnect the speedometer cable connector at the transaxle end.
3. If equipped with 1.8L engine, remove the bracket on the lower side if the intake manifold.
4. Disconnect the starter motor electrical connections.
5. Remove the starter motor mounting bolts and remove the starter.
To install:
6. Clean both surfaces of starter motor flange and rear plate. Install the starter motor onto the engine and secure with the retainer bolts. Tighten to 25 ft. lbs. (34 Nm).
7. Connect the electrical harness connector to the starter.
8. Install the intake manifold stay, if removed. Tighten the retainers to 18 ft. lbs. (25 Nm).
9. Install the speedometer cable at the transaxle. Install the battery tray and battery.
10. Operate the starter to assure proper operation.

1995–97 2.0L (4G63) and 2.4L (4G64) Engines

1. Disconnect the negative battery cable.
2. Remove air hose C.
3. If necessary, remove the air cleaner and the air intake hose.
4. Disconnect the wiring from the solenoid.
5. Remove the two mounting bolts.
6. Remove the starter.
To install:
7. Position the starter on the transaxle housing and install the mounting bolts. Torque the mounting bolts to 22 ft. lbs. (30 Nm).
8. Connect the wiring to the solenoid.
9. If equipped, install the air intake hose and the air cleaner.
10. Install air hose C.
11. Connect the negative battery cable.

1995 2.0L (420A) Engine

1. Disconnect the negative battery cable.
2. If equipped with manual transaxle, disconnect the aspirator valve.
3. Disconnect the wiring from the starter solenoid.
4. Remove the starter mounting bolts and remove the starter.
To install:
5. Install the starter assembly.

6. Connect the wiring to the solenoid.

7. Connect the aspirator valve if disconnected.

8. Connect the negative battery cable and check the operation of the starter.

Galant

1. Disconnect the negative battery cable.

2. Remove the air cleaner and air intake hoses.

3. If equipped with Active-ECS suspension, remove the air compressor as follows:

   a. Disconnect the two electrical connectors, from the compressor.

   b. Disconnect the air line at the compressor.

   c. Remove the three mounting bolts, securing the compressor to the chassis.

4. Disconnect the starter motor electrical connections.

5. Remove the starter motor mounting bolts and remove the starter.

To install:

6. Clean both surfaces of starter motor flange and rear plate. Install the starter motor onto the engine and secure with the retainer bolts. Tighten bolts to 22 ft. lbs. (30 Nm).

7. Connect the electrical harness connectors to the starter.

8. If equipped with Active-ECS, mount and connect the compressor. Tighten the mounting bolts to 8 ft. lbs. (11 Nm) and the air fitting to 7 ft. lbs. (10 Nm).

9. Install the air cleaner and air intake hoses.

10. Connect the negative battery cable and operate the starter to assure proper operation.

11. If equipped with Active-ECS, check the compressor connection for leaks; using a soapy water solution, wet the fitting and check for signs of leakage.

3000GT and Diamante

1. Disconnect the negative battery cable.

---- **CAUTION** ----
Wait at least 90 seconds after the negative (-) battery cable is disconnected to prevent possible deployment of the air bag.
----

2. Raise the vehicle and support safely.

3. Remove the engine undercover.

4. Disconnect the wiring from the starter.

5. Remove the mounting bolts and remove the starter from the vehicle.

To install:

6. Position the starter and install the mounting bolts.

7. Tighten starter mounting bolts to 20–25 ft. lbs. (27–34 Nm.)

8. Connect the wiring for the starter.

9. Connect the negative battery cable and check the starter for proper operation.

CHASSIS ELECTRICAL

Blower Motor

REMOVAL AND INSTALLATION

Precis

1. Disconnect the negative battery cable.

2. Remove the glove box housing cover.

3. Remove the lower crash pad, crash pad center fascia and lower crash pad center skin.

4. Disconnect the blower motor and blower resistor connectors.

5. Disconnect the blower motor cooling tube.

6. Remove the screws that attach the blower motor to the lower case and lower the blower motor far enough so that the FRESH/RECIRC vacuum connector can be disconnected. Remove the blower motor from the car.

To install:

7. Inspect the blower wheel for damage and replace it as necessary. Replace the blower motor mounting seal.

8. Raise the blower motor and seal up onto the lower case half and connect the FRESH/RECIRC vacuum connector. Mount the blower onto the lower case half and install the three retaining screws.

9. Connect the blower motor cooling tube.

10. Connect the blower motor and resistor wire connectors.

11. Install the lower crash pad center skin, crash pad center fascia and lower crash pad.

12. Install the glove box housing cover.

13. Connect the negative battery cable.

14. Check the blower for proper operation at all speeds.

Mirage

1. Disconnect the negative battery cable.

2. Remove the right side dashboard undercover panel.

3. Remove the glove box panel and frame.

4. Disconnect the blower motor electrical connection.

5. Disconnect and remove the resistor.

6. Disconnect the blower motor ventilation tube.

7. Remove the blower motor mounting bolts, remove the blower motor.

To install:

8. Replace the blower motor and install the mounting bolts.

9. Connect the blower motor electrical connection.

10. Connect the blower motor ventilation tube.

11. Install the resistor and the glove box assembly.

12. Install the right side dashboard undercover panel.

13. Connect the negative battery cable.

1993–94 Eclipse

Heater Motor

1. Disconnect battery negative cable.

2. Remove the right side duct, if equipped.

3. Remove the molded hose from the blower assembly.

4. Remove the blower motor assembly.

5. Remove the packing seal.

6. Remove the fan retaining nut and fan in order to renew the motor.

To install:

7. Check that the blower motor shaft is not bent and that the packing (sealing material) is in good condition. Clean all parts of dust, etc.

8. Assemble the motor and fan. Install the blower motor then connect the motor terminals to battery voltage. Check that the blower motor operates smoothly. Then, reverse the polarity and check that the blower motor operates smoothly in the reverse direction.

9. Install the molded hose and duct, if removed.

10. Connect the negative battery cable and check the entire climate control system for proper operation.

A/C Motor

1. Disconnect the negative battery cable.
2. Remove the shower duct on the right hand side of the dashboard.
3. Remove the blower motor ventilation hose.
4. Disconnect the blower motor electrical connection.
5. Remove the blower motor mounting bolts.
6. Remove the blower motor.
7. Remove the fan installation nut, remove the fan.

To install:
8. Replace the fan on the blower motor and install the fan installation nut.
9. If the blower motor packing is cracked, replace it.
10. Mount the blower motor and install the blower motor mounting bolts.
11. Connect the blower motor ventilation hose.
12. Connect the blower motor electrical connection.
13. Install the shower duct assembly.
14. Connect the negative battery cable.

1995–97 Eclipse

1. Disconnect the negative battery cable.
2. Remove the right instrument panel lower cover.
3. Disconnect the harness from the blower motor.
4. On vehicles with A/C and non-turbo engines, disconnect and remove the automatic compressor ECM.
5. Disconnect the blower motor connector.
6. Remove the blower motor mounting screws and remove the blower motor.
7. Remove the blower fan from the motor.

To install:
8. Install the blower motor and fan assembly to the housing.
9. Connect the harness connector to the blower motor.
10. On vehicles with A/C and non-turbo engines, reinstall and reconnect the automatic compressor ECM.
11. Install the instrument panel lower cover.
12. Connect the negative battery cable.

1993 Galant

1. Disconnect the negative battery cable.
2. Remove the glove box stopper.

3. Swing the glove box door open all the way and remove the bottom retaining screws. Remove the glovebox.
4. Remove the dash undercover. Note that some of the screws and retainers are concealed behind small covers which must be removed.
5. Remove the heater duct for the passenger's feet.
6. Carefully disconnect the 10-pin connector running to the back of the glove box frame. Disconnect the single wire (glove box switch) running to the back of the glove box frame.
7. Remove the four bolts holding the glove box frame and remove the frame.
8. Disconnect the small air hose running from the fan motor to the fan housing.
9. Disconnect the electrical connector for the fan motor.
10. Remove the three small bolts holding the motor to the housing and remove the motor and fan.

To install:
11. Check the inside of the case carefully; any debris can snag the fan and cause noise or poor airflow.
12. Inspect the gasket (packing) under the motor and replace it if cracked or damaged. Reinstall the fan and motor to the case and install the retaining bolts.
13. Connect the air hose and electrical connector.
14. Install the glove box frame and connect both the 10-pin and single pin connectors properly.
15. Install the heater duct
16. Install the undercover, taking care to insure it is in place and all the fasteners are secure.
17. Install the glove box and its stopper.
18. Connect the negative battery cable.

1994–97 Galant

1. Disconnect the negative battery cable.
2. Remove the dash undercover three mounting screws and remove the cover.
3. If equipped with A/C, unplug and remove the compressor module.
4. Disconnect the electrical connector from the fan motor.
5. Remove the three small bolts holding the motor to the housing and remove the motor and fan.

To install:
6. Check the inside of the case carefully; any debris can snag the fan and cause noise or poor airflow.

7. Install the blower motor, in the blower case and secure with the three mounting bolts.
8. Connect the blower motor electrical connector.
9. Install the compressor module, if removed.
10. Install the undercover, taking care to insure it is in place and all the fasteners are secure.
11. Connect the negative battery cable.

Diamante

1. Disarm the air bag as follows:
   a. Position the front wheels in the straight-ahead position and place the key in the **LOCK** position. Remove the key from the ignition lock cylinder.
   b. Disconnect the negative battery cable and insulate the cable end with high-quality electrical tape or similar non-conductive wrapping.
   c. Wait at least 1 minute before working on the vehicle. The air bag system is designed to retain enough voltage to deploy the air bag for a short period of time even after the battery has been disconnected.
2. Remove the passenger side lower instrument panel and shower duct.
3. Remove the glove box striker, glove box, glove box outer casing and the screw below the assembly.
4. Remove the evaporator case mounting bolt and nut.
5. Remove the inside/outside air changeover damper motor assembly.
6. Remove the ECU, mounting bracket and MFI control relay.
7. Remove the instrument panel passengers side lower bracket.
8. Remove the molded hose from the blower assembly.
9. Remove the blower motor assembly.
10. Remove the fan retaining nut and fan in order to replace the motor.

To install:
11. Check that the blower motor shaft is not bent and that the packing is in good condition. Clean all parts of dust, etc.
12. Assemble the motor and fan. Install the blower motor then connect the connector.
13. Install the molded hose. Install the duct or undercover.
14. Install the evaporator case mounting bolt and nut.
15. Install the instrument panel passengers side lower bracket.
16. Install the ECU, mounting bracket and MFI control relay.

17. Install the inside/outside air changeover damper motor assembly.

18. Install the screw below the glove box assembly, and the entire glove box unit.

19. Install the lower instrument panel and shower duct.

20. Connect the negative battery cable and check the entire climate control system for proper operation.

3000GT

1. Disconnect battery negative cable.

2. Remove the instrument panel undercover.

3. Remove the glove box and the glove box outer case assembly.

4. Remove the molded hose from the blower assembly.

5. Remove the blower motor assembly.

6. Remove the packing seal.

7. Remove the fan retaining nut and fan in order to replace the motor.

To install:

8. Check that the blower motor shaft is not bent and that the packing (sealing material) is in good condition. Clean all parts of dust, etc.

9. Assemble the motor and fan. Install the blower motor then connect the connector.

10. Install the molded hose. Install the duct or undercover.

11. Install the glove box outer case and the glove box.

12. Connect the negative battery cable and check the entire climate control system for proper operation.

Windshield Wiper Motor

REMOVAL AND INSTALLATION

Precis

Front

1. Disconnect the negative battery cable.

2. Remove the wiper arm and blade assemblies. Take care not to scratch the paint on the cowl.

3. Remove the air inlet and cowl front center trim panels. Remove the 3 pivot shaft mounting nuts and push the pivot shafts into the area under the cowl.

4. Remove the motor mounting bolts. Pull the motor into the best possible position for access and remove the linkage off the motor crank arm. Remove the motor and then the linkage as required.

To install:

5. Connect the wiper motor to the link securely then install the motor attaching bolts.

6. Install the cowl top cover.

7. Install the cowl top sealing caps.

8. Install the wiper arm and blade assemblies to the pivot shafts. Torque the pivot shaft nuts to 2.9 to 4.3 ft. lbs. (4 to 6 Nm).

9. Position the wiper arms so the blades are about 30mm above the lower windshield molding.

10. Connect the negative battery cable.

Rear

1. Remove the wiper blade and arm by lifting the wiper blade locknut cover and removing the locknut. Then, pull the arm from the shaft.

2. Remove the lift gate trim panel and disconnect the wiring harness connector.

3. Remove the inside and outside motor mounting nuts and remove the motor.

To install:

4. Install the motor assembly to the tailgate.

5. Connect the electrical connector to the wiper motor.

6. Install the tail gate trim.

7. Install the wiper arm.

Mirage

1. Disconnect the negative battery cable.

2. Remove the windshield wiper arms by unscrewing the cap nuts and lifting the arms from the linkage posts.

3. Remove the front deck garnish panel.

4. Remove both windshield holders.

5. Remove the clips that hold the deck cover. If they are the pin type, they may be removed using the following procedure:

   a. Remove the clip by pressing down on the center pin with a suitable blunt pointed tool. Press down a little more than $^1/_{16}$ in. (2mm). This releases the clip. Pull the clip outward to remove it.

   b. Do not push the pin inward more than necessary because it may damage the grommet or if pushed too far, the pin may fall in. Once the clips are removed, use a plastic trim stick to pry the deck cover loose.

6. Remove the air intake screen.

7. Loosen the wiper motor assembly mounting bolts and remove the

windshield wiper motor. Disconnect the linkage from the motor assembly. If necessary, remove the linkage from the vehicle.

NOTE: The installation angle of the crank arm and motor has been factory set. Do not remove unless necessary. If arm must be removed, remove only after marking mounting positions.

To install:

8. Install the windshield wiper motor and connect the linkage. Connect the electrical harness to the motor.

9. When installing the trim and garnish pieces and reusing pin type clips, use the following procedure:

   a. With the pin pulled out, insert the trim clip into the hole in the trim.

   b. Push the pin inward until the pin's head is flush with the grommet.

   c. Check that the trim is secure.

10. Install the wiper arms and tighten nuts.

11. Connect the negative battery cable and check the wiper system for proper operation.

Eclipse and 3000GT

Front

1. Disconnect the negative battery cable.

CAUTION
Wait at least 90 seconds after the negative battery cable is disconnected to prevent possible deployment of the air bag.

NOTE: Disconnect the linkage at the motor.

2. Remove the windshield wiper arms by unscrewing the cap nuts and lifting the arms from the linkage post.

3. Remove the front deck garnish panel.

4. Remove the air inlet trim pieces.

5. Remove the access hole cover.

6. Remove the wiper motor by loosening the mounting bolts, removing the motor assembly, then disconnecting the linkage.

NOTE: The installation angle of the crank arm and motor has been factory set. Do not remove unless necessary. If removal is required, remove only after marking the mounting positions.

To install:

7. Install the windshield wiper motor and connect the linkage.

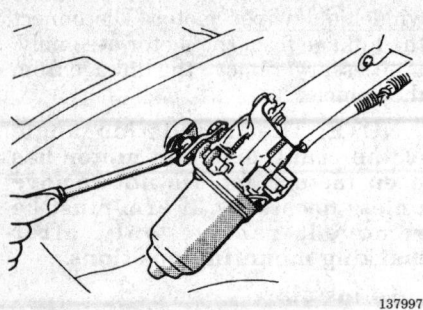

Front wiper motor linkage connection — 1993-94 Eclipse shown

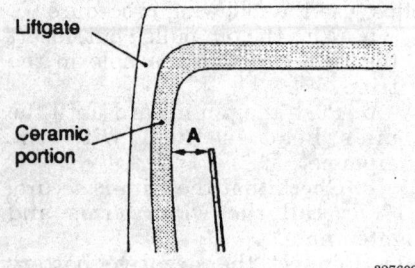

Rear wiper blade parked position — 1995-97 Eclipse shown

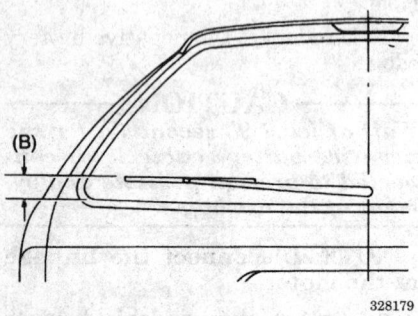

Rear wiper arm installation position — Diamante Wagon

8. Reinstall all trim pieces.

NOTE: Note that the driver's side wiper arm should be marked D or Dr and the passenger's marked A or As at the base of the arm, near the pivot.

9. Reinstall the wiper blades. Adjust the wiper blade assembly so that the clearances between the wiper blade edges and the ceramic line are within specification. **Specification is: A - 2.07 ±0.20 in. B - .77 ±.20 in.**
10. Connect the negative battery cable and check the wiper system for proper operation.

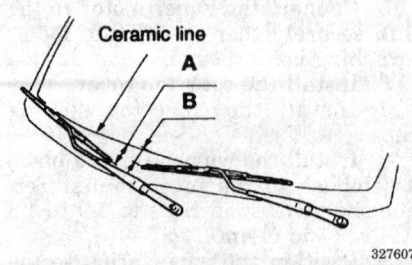

Front wiper blade parked position — 1995-97 Eclipse

Rear

1. Disconnect the negative battery cable.
2. Remove the rear wiper arm by removing the cover, unscrewing the nut and lifting the arm from the linkage post.
3. Remove the large interior trim panel. Use a plastic trim stick to unhook the trim clips of the liftgate trim.
4. For the Eclipse, if equipped with rear air spoiler, remove the wiper grommet.
5. For the 3000GT, remove the rear spoiler, center brace and center brake light. Lift the small cover, remove the retaining nut, then remove the wiper arm and spacer.
6. If necessary, unfasten the mounting bolts, then remove the rear wiper assembly. Do not loosen the grommet for the wiper post.

To install:
7. Install the motor and retaining bolts, and/or grommet. Mount the grommet so the arrow on the grommet is pointing upward.
8. Install the wiper arm. Adjust the wiper arm so that the clearance between the wiper blade edge and the ceramic portion is within specification. **Specification is: A - 4.61 ±.20 in.**
9. Install the rear spoiler and related parts.
10. Connect the negative battery cable and check the rear wiper for proper operation.
11. If operation is satisfactory, fit the tabs on the upper part of the liftgate trim into the liftgate clips and secure the liftgate trim.

Diamante and Galant

Front

1. Disconnect the negative battery cable.
2. Matchmark the wiper arms to the shaft and mark the arms to the

proper side for reinstallation purposes.
3. Remove the windshield wiper arms by unscrewing the cap nuts and lifting the arms from the linkage posts.
4. Remove the front deck garnish assembly.
5. Remove the air inlet cover.
6. Disconnect the electrical harness plug from the wiper motor.
7. Remove the access hole cover.
8. Remove the wiper motor mounting bolts.
9. Detach the motor crank arm from the wiper linkage and remove the motor.

NOTE: The installation angle of the crank arm and motor has been factory set. Do not remove them unless necessary. If they must be removed, remove them only after marking their mounting positions.

To install:
10. Install the windshield wiper motor and connect the linkage.
11. Attach the electrical harness plug.
12. Install the access hole cover.
13. Install the air inlet cover.
14. Install the front deck garnish assembly.
15. Reinstall the wiper arm and tighten the mounting nuts to 14 ft. lbs. (19 Nm). Install the arms so the blades are parallel to the garnish molding when parked.
16. Connect the negative battery cable and check the wiper system for proper operation.

Wagon Rear

1. Disconnect the negative battery cable.
2. Remove the liftgate lower trim.
3. Lift the small cover, remove the retaining nut and remove the wiper arm.
4. Remove the mounting bolts and remove the wiper motor.

To install:
5. Install the motor and install the retaining bolts.
6. Install the wiper arm so that the arm is 3.35 inches (85 mm) between the measurement points, when parked. Secure the wiper arm with the retaining nut.

NOTE: Before proceeding, connect the battery and check the operation of the motor. If satisfactory, disconnect the cable and complete the installation.

7. Install the interior trim piece.

8. Connect the negative battery cable and recheck the system for proper operation.

Headlight Switch

REMOVAL AND INSTALLATION

The headlights, turn signals, dimmer switch, horn switch, windshield wiper/washer, intermittent wiper switch and on some models, the cruise control function are all built into 1 multi-function Combination Switch that is mounted on the steering column.

Combination Switch

REMOVAL AND INSTALLATION

——— CAUTION ———
The air bag system (SRS or SIR) must be disarmed before removing the steering wheel. Failure to do so may cause accidental deployment, property damage or personal injury.

1. Disconnect the negative battery cable.
2. If equipped, disable the air bag system.
3. Remove the steering wheel as follows:
  a. If not equipped with an air bag, remove the horn pad attaching screw on the under side of the steering wheel and remove the horn pad by pushing the pad upward.
  b. If equipped, remove the air bag module mounting nut from behind the steering wheel.
  c. To disconnect the connector of the clockspring from the air bag module, press the air bag's lock toward the module to spread the lock open. While holding lock in this position, use a small tipped prying tool to gently pry the connector from the module.
  d. If equipped, store the air bag module in a clean, dry place with the pad cover facing up.
  e. Remove the steering wheel retaining nut and use a steering wheel puller to remove the wheel. Do not use a hammer or the collapsible mechanism in the column could be damaged.
4. For Diamante, remove the hood lock release handle.
5. Remove the knee protector panel under the steering column,

then the upper and lower column covers.
6. For Diamante and 3000GT, remove the lap cooler and foot blower duct work as necessary. Gently disconnect the combination switch connectors.
7. For Mirage, disconnect all connectors, remove the wiring clip and remove the column switch assembly.
8. For the Galant, if equipped, remove the four screws retaining the cruise control slip ring to the switch.
9. For Galant, remove the two retaining screws from the combination switch and remove the switch from the column.

To install:
10. Install the switch assembly and secure all harness connectors with clips if needed. Make sure the wires are not pinched or out of place.
11. Install the column covers and knee protector and all connectors.
12. If removed, install the foot blower duct work and lap cooler.
13. Confirm that the front wheels are in a straight-ahead position. Center the clockspring by aligning the **NEUTRAL** mark on the clockspring with the mating mark on the casing. Then install the steering wheel and torque the retaining nut to 29 ft. lbs. (40 Nm).
14. Install the air bag module or horn pad, as applicable.
15. Connect the negative battery cable, turn the key to the **ON** position, the SRS warning light should illuminate for seven seconds and go out. If the warning light is not functioning properly, refer to SRS system diagnosis.
16. Check all functions of the combination switch for proper operation.

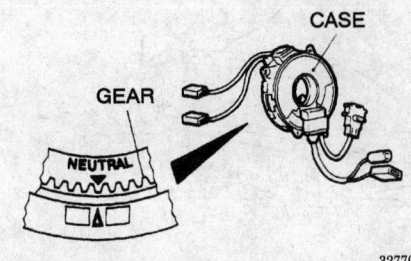

Clock spring mating mark alignment — 1994–97 Galant shown

Ignition Lock Cylinder

REMOVAL AND INSTALLATION

——— CAUTION ———
The air bag system (SRS or SIR) must be disarmed before removing the steering wheel. Failure to do so may cause accidental deployment, property damage or personal injury.

1. If equipped, properly disarm the air bag system.
2. Disconnect the negative battery cable. Remove the hood lock release lever from the lower panel.
3. Remove the lower instrument panel knee protector.
4. On all except 3000GT and Precis, remove the ductwork with lower knee panel and remove the steering wheel assembly.

NOTE: Use proper steering wheel puller equipment when removing the steering wheel. The use of a hammer for removal could damage the collapsible mechanism within the column.

5. Remove the lower and upper steering column covers.
6. For Diamante, remove the lap cooler and foot shower duct work.
7. If necessary, remove the clip that holds the wiring harness against the steering column.
8. Detach all necessary connectors.
9. For Galant and Diamante, remove the retaining screws, then remove the entire column switch/clockspring assembly from the left side of the steering column.
10. For Diamante, remove the mounting screws from the ignition switch and pull the switch from the interlock cylinder.
11. Insert the key into the steering lock cylinder and turn to the **ACC** position.
12. With a small, pointed tool, push the lock pin of the steering lock cylinder inward and pull the lock cylinder out.
13. Remove the key reminder switch, if equipped.
To install:
14. For 3000GT and Precis, using the key, install the cylinder lock into the housing until the retaining tab locks it in place.
15. For Mirage and Eclipse, install the lock cylinder in the lock cylinder bracket. Make sure the cylinder operates properly before breaking off the heads of the special bolts if the bracket was replaced.

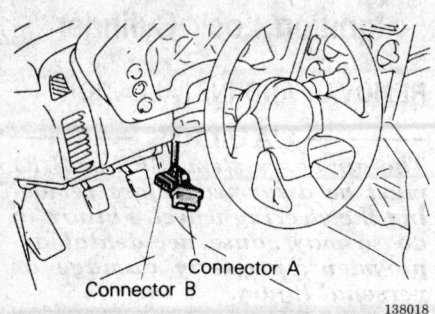

Combination switch connectors — 1993–94 Eclipse shown

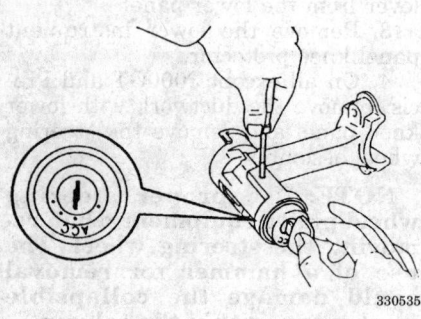

Lock cylinder — 3000GT and Precis

16. For Galant and Diamante, install the lock cylinder into the interlock housing. Be sure the lock pin snaps into place. Install the ignition switch into the interlock housing. Align the keyway of the ignition switch with the lock cylinder and secure with the mounting screws.

17. For Galant and Diamante, install the column switch/clockspring assembly to the and connect the harness.

18. Install the key reminder switch if equipped.

19. Connect harness connections and install the wiring clip.

20. If removed, install the lap cooler and foot shower duct work.

21. Install the steering column upper and lower covers.

22. If necessary, install the knee protector and the steering wheel.

23. Install the hood lock release lever.

24. Connect the negative battery cable, enable the air bag system, and check the ignition switch and lock for proper operation.

Ignition Switch

REMOVAL AND INSTALLATION

Precis

—— CAUTION ——
The air bag system (SRS or SIR) must be disarmed before removing the air bag module or horn pad. Failure to do so may cause accidental deployment, property damage or personal injury.

1. Disconnect the negative battery cable.

2. Mark the steering wheel and the shaft so it can be installed in the original position. Remove the steering wheel.

3. Remove the lower crash pad.

4. Remove the upper and lower steering column covers.

5. Disconnect the ignition switch electrical connector.

6. Remove the combination switch mounting screws and remove the combination switch.

7. Using a hacksaw, cut a slot in the head of the screw and back it out using a flat blade screwdriver.

8. Remove the ignition switch assembly.

To install:

9. Align the steering lock with the column boss and insert the key to confirm the steering lock operation before tightening the switch assembly.

NOTE: Special screws should be used to secure the ignition switch assembly.

10. Install the combination switch.

11. Connect the electrical connectors.

12. Install the steering column covers.

13. Install the lower crash pad.

14. Install the steering wheel in the original position.

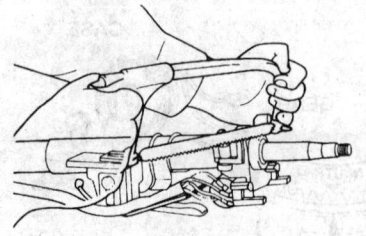

Removing the ignition switch — Precis shown

15. Connect the negative battery cable.

Eclipse, Mirage and Galant

1. Disconnect the negative battery cable. If equipped, disarm the air bag system.

2. Remove the hood lock release lever from the lower panel.

3. Remove the lower instrument panel knee protector.

4. Remove ductwork with lower knee panel and remove the steering wheel assembly.

NOTE: Use proper steering wheel puller equipment when removing the steering wheel. The use of a hammer for removal could damage the collapsible mechanism within the column.

5. Remove the lower steering column cover.

6. Remove the upper steering column cover.

7. Remove the clip that holds the wiring harness against the steering column.

8. For Eclipse and Galant, insert the key into the steering lock cylinder and turn to the ACC position.

9. For Eclipse and Galant, with a small pointed tool, push the lock pin of the steering lock cylinder inward and pull the lock cylinder out.

10. For Eclipse and Galant, remove the key reminder switch, if equipped.

11. Unplug the ignition switch harness connector. Remove the ignition switch mounting screws and pull the switch from the steering lock cylinder.

NOTE: Vehicles equipped with automatic transaxle, have safety-lock systems and will have a key interlock cable installed in a slide lever on the side of the key cylinder. Carefully unhook the interlock cable from the lock cylinder while withdrawing cylinder from lock housing.

To install:

NOTE: To remove the steering lock, use a hacksaw or equivalent to cut the special bolts through the bracket. The bracket and bolts must be replaced with new ones.

12. With the ignition key removed, install the slide lever and the interlock cable to the steering lock cylinder. Apply grease to the interlock cable and install cylinder into the lock housing. Check for normal operation of the interlock system.

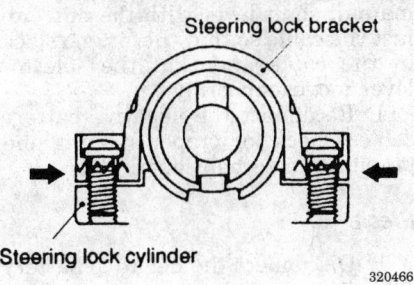

Steering lock bracket

Steering lock cylinder

320466

**Use a hacksaw to cut the special bolts —
1995–97 Eclipse shown**

13. Install the ignition switch into the rear of the lock cylinder housing. Be sure to align the keyway of the ignition switch with interlock cylinder.

14. Connect harness connections and install the wiring clip.

15. Install the steering column upper and lower covers.

16. Install the knee protector.

17. Connect the negative battery cable, enable the air bag system (if equipped) and check the ignition switch and lock for proper operation.

Diamante

---------- CAUTION ----------

Work MUST NOT be started until at least 90 seconds after the ignition switch is turned to the LOCK position and the negative battery cable is disconnected from the battery. This will allow time for the air bag system backup power supply to deplete its stored energy preventing accidental air bag deployment which could result in unnecessary air bag system repairs and/or personal injury.

1. Disconnect the negative battery cable.

2. Disarm the air bag system.

3. Remove the air bag module as follows:

   a. Remove the air bag module mounting nuts from behind the steering wheel.

   b. To disconnect the connector of the clockspring from the air bag module, press the air bag's lock towards the module to spread the lock open. While holding lock in this position, use a small tipped prying tool to gently pry the connector from the module.

   c. Store the air bag module in a clean, dry place with the pad cover facing up.

4. Remove the steering wheel retaining nut and use a steering wheel

puller to remove the wheel. Do not use a hammer or the collapsible mechanism in the column could be damaged.

5. Remove the hood lock release handle.

6. Remove the switches from the knee protector below the steering column and remove the exposed retaining screws. Then remove the knee protector.

7. Remove the steering column upper and lower covers. Use care removing covers to prevent breakage of alignment tabs.

8. Remove lap cooler and foot shower duct work. Disconnect the ignition switch harness connectors.

9. Remove mounting screws from the ignition switch and pull switch from interlock cylinder.

To install:

10. Install ignition switch into interlock housing. Align keyway of ignition switch with lock cylinder and secure with mounting screws.

11. Connect the ignition switch electrical harness plug.

12. Install lap cooler and foot shower duct work.

13. Install the upper and lower steering column covers.

14. Install the knee protector and switches.

15. Install the hood release handle.

16. Center the clockspring by aligning the **NEUTRAL** mark on the clockspring with the mating mark on the casing. Install the steering wheel and torque the retaining nut to 29 ft. lbs. (40 Nm) for 1993 vehicles or to 33 ft. lbs. (45 Nm) for 1994–97 vehicles.

17. Attach air bag module wiring connector to clockspring connection. Install air bag module and tighten mounting nuts to 40–43 inch lbs. (4.5–5 Nm).

18. Connect the negative battery cable and check all functions of column-mounted switches and the ignition switch for proper operation.

3000GT

1. Disconnect the negative battery cable.

---------- CAUTION ----------

Wait at least 90 seconds after the negative (-) battery cable is disconnected to prevent possible deployment of the air bag.

2. Remove the steering column upper and lower covers. Use care removing covers to prevent breakage of alignment tabs.

3. Remove the knee protector.

4. Disconnect the wiring connector from the combination switch.

5. Remove the ignition switch.

To install:

6. Install the ignition switch into the interlock housing.

7. Connect the harness connections.

8. Install the knee protector.

9. Install the upper and lower steering column covers.

10. Connect the negative battery cable and check all functions of column-mounted switches and the ignition switch for proper operation.

Park/Neutral Switch

ADJUSTMENT

Mirage, 1993 Galant and 1993–94 Eclipse

1. Disconnect the negative battery cable and locate the neutral safety switch on the top of the transaxle.

NOTE: Apply parking brake and chock wheels before placing transaxle into the N position

2. At the transmission, loosen the shift cable adjustment nut. Inside the vehicle place the gearshift selector lever in **N**.

3. Place the manual shift control lever in **N**.

4. Loosen neutral safety switch mounting screws and rotate switch body so the manual control lever 0.20 in. (5mm) hole and the switch body 0.20 in. (5mm) holes are aligned.

5. Tighten switch body mounting bolts to 84–96 inch lbs. (10–12 Nm).

6. At the shift cable adjusting nut, gently pull cable to remove any slack. Tighten locknut to 8 ft. lbs. (12 Nm).

7. Verify that the switch lever moves to positions corresponding to each position of the selector lever. Connect the negative battery terminal.

8. Make sure the engine only starts in the **P** and **N** positions. Also make sure the reverse lights operate only in **R** selection.

REMOVAL AND INSTALLATION

Mirage, 1993 Galant, and 1993–94 Eclipse

1. Disconnect the negative battery cable.

2. Disconnect the selector cable from the lever.

3. Remove the two retaining screws and lift off the switch.

To install

4. Mount and position new switch. Do not tighten the bolts until the switch is adjusted.

5. Connect selector cable and adjust switch.

6. After installation and adjustment make sure the engine only starts in the **P** and **N** selections. Also check that the reverse lights operate only in the **R** selection.

1995–97 Eclipse and 1994–97 Galant

1. Disconnect the negative battery cable.

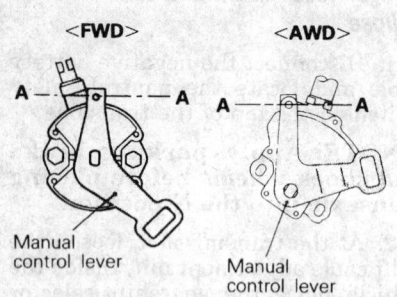

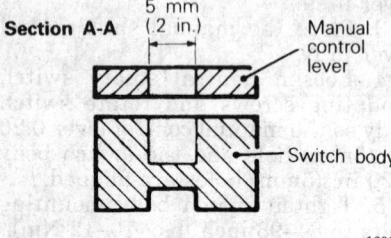

Park neutral switch adjustment points — 1993 Galant with F4A22 and W4A32 automatic transaxles

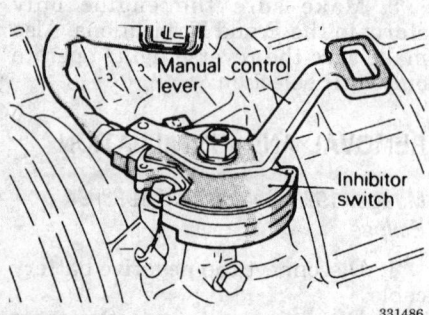

Park/Neutral switch location — 1993–94 Eclipse shown

2. Remove the nut attaching the shift control cable from the transaxle manual shaft lever. Position the control cable out of the way.

3. Place the manual shaft lever in the Neutral position, remove the nut and the manual shaft lever.

4. Disconnect the park/neutral switch electrical connector.

5. Remove the park/neutral switch mounting bolts and remove the switch from the transaxle manual shaft.

To install:

6. Install the park/neutral switch to the transaxle manual shaft and install the switch mounting bolts. Do not tighten the mounting bolts until the switch is adjusted.

7. Install the manual shaft lever to the park/neutral switch with the nut. Make sure that the shaft lever is in the Neutral position.

8. Adjust the switch in the following manner: turn the switch body until the hole in the body of the switch aligns with the hole in the manual shaft lever. Insert a drill bit or equivalent into the holes. Tighten the switch mounting bolts to 8 ft. lbs. (11 Nm).

9. Reconnect the electrical connector.

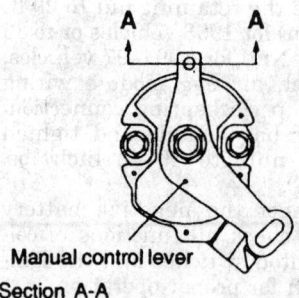

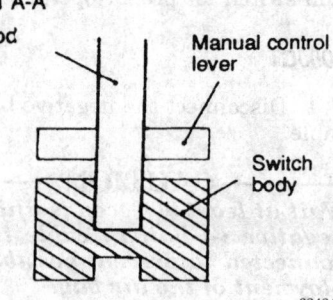

Adjusting the park/neutral switch — 1994–97 Galant

10. Install the control cable to the manual shaft lever with the nut. Adjust the cable so that there is no slack in the cable and that the selector lever moves smoothly.

11. Reconnect the negative battery cable. Check for proper starting and proper reverse light operation.

Precis

1. Disconnect the negative battery cable.

2. Apply the parking brake and place the selector lever in the neutral position.

3. Remove the manual control lever.

4. Disconnect the connector from the park/neutral switch.

5. Remove the 2 mounting screws from the switch and remove the switch.

To install:

6. Position the switch on the transaxle and install the 2 mounting screws finger tight.

7. Install the manual control lever making sure the transaxle is still in neutral.

8. Turn the switch body so the flat on the control lever aligns with the flange on the switch body.

9. Connect the electrical connector and the negative battery cable.

3000GT and Diamante

1. Disconnect the negative battery cable.

— **CAUTION** —
Wait at least 90 seconds after the negative (-) battery cable is disconnected to prevent possible deployment of the air bag.

2. Disconnect the selector cable from the lever.

3. Remove the two retaining screws and lift off the switch.

To install:

4. Install the lever, tighten the bolts only hand tight.

5. Rotate switch body so the manual control lever 0.20 inch (5mm) hole and the switch body 0.20 inch (5mm) holes are aligned.

6. Tighten the mounting bolts to 7–8 ft. lbs. (10–12 Nm).

7. Connect the selector cable to the lever.

8. Connect the negative battery cable.

9. After installation and adjustment make sure the engine only starts in the **P** and **N** selections. Also check that the reverse lights operate only in the **R** selection.

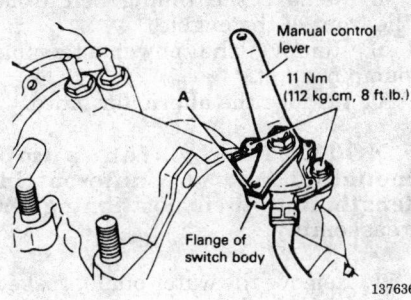

Park/neutral switch mounting and adjustment — 1993–94 Precis

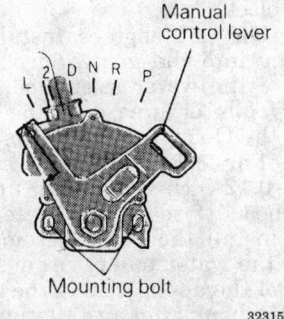

Park/Neutral switch (inhibitor switch) — 3000GT and Diamante

Powertrain Control Module

REMOVAL AND INSTALLATION

Precis

1. Disconnect the negative battery cable. Remove the lower left trim panel.
2. Disconnect the electrical connectors at the control module.
3. Remove the control module from the vehicle.
To install:
4. Install the control module into the mounting.
5. Attach the control module connectors.
6. Install the lower trim panel.
7. Connect the negative battery cable.

Mirage

NOTE: The electronic control unit is located above the passenger side kickpanel.

1. Disconnect the negative battery cable.
2. Remove the glove box, right side kickpanel and lower panel assemblies.

3. Unplug the connectors and remove fasteners. Remove the control unit.
4. Installation is the reverse of the removal procedure.

Eclipse and 1994–97 Galant

1. Disconnect negative battery cable.

---WARNING---

A grounded wrist strap should be used to prevent static discharge to the ECM. Static discharge can easily destroy the electronic components inside the ECM.

2. Remove both center console side panels.
3. Unplug the wiring connector and remove the mounting hardware. Slide the control unit out the side.
4. Installation is the reverse of the removal procedure.

Diamante and 1993 Galant

The engine control module is located behind the glove box assembly.
1. If equipped, disarm the air bag system.
2. Remove the passenger side lower instrument panel and shower duct.
3. Remove the glove box striker, glove box, glove box outer casing and the screw below the assembly.
4. Unplug wiring connector and remove mounting hardware. Slide out the control unit.
To install:
5. Install the control unit with the mounting hardware.
6. Connect the wire connector.
7. Install the glove box striker, the glove box, the glove box casing and the screw below the assembly.
8. Install the passenger side lower instrument panel and the shower duct.
9. Reconnect the negative battery cable.

3000GT

1. Disconnect the negative battery cable and wait at least 1 minute for the airbag to disarm.
2. Remove the center console left side cover to access the electronic control unit.
3. Disconnect the electrical connectors to the ECU and remove the mounting bolts.
4. Remove the ECU.
To install:
5. Position the ECU and install the mounting bolts.
6. Connect the electrical connectors to the ECU.

7. Install the side cover to the console.
8. Connect the negative battery cable.

ENGINE COOLING

Radiator

REMOVAL AND INSTALLATION

---CAUTION---

Allow the cooling system to completely cool before attempting any repair or draining the system. Injury from scalding could result if radiator cap or hose connections are removed while system is hot.

1. Disconnect the negative battery cable.
2. Drain the cooling system when safe.
3. Disconnect the overflow tube. Some vehicles may also require removal of the overflow tank.
4. Disconnect upper and lower radiator hoses.
5. Disconnect electrical connectors for cooling fan and air conditioning condenser fan, if equipped.
6. For Mirage, Eclipse, Diamante and 3000GT, remove the fan assembly.
7. Disconnect thermo sensor wires.
8. Disconnect and plug automatic transaxle cooler lines, if equipped with automatic transaxle.
9. Remove the upper radiator mounts and lift out the radiator assembly or radiator/fan assembly.
10. Service the lower mounts, as required.
To install:
11. Install the radiator and fan assembly, if removed as an assembly.
12. Connect the automatic transaxle cooler lines, if disconnected.
13. Connect the thermo wires.
14. Install the fan if removed separately.
15. Install the radiator hoses.
16. Install the air cleaner support bracket, if removed.
17. Install the overflow tube and reservoir, if removed.
18. Fill the system with coolant.
19. Connect the negative battery cable, run the vehicle until the thermostat opens, fill the radiator completely and check the automatic transaxle fluid level, if equipped.

20. Once the vehicle has cooled, recheck the coolant level.

Water Pump

REMOVAL AND INSTALLATION

Precis

1. Disconnect the negative battery cable. Drain the cooling system.

2. Loosen the four bolts attaching the water pump pulley to the pulley flange. Loosen the alternator mounting bolts and remove the alternator belt.

3. Remove the water pump pulley.

4. Remove the upper and lower timing covers.

5. Rotate the crankshaft to align the camshaft sprocket and crankshaft sprocket timing marks.

6. Remove the timing belt and tensioner. Mark the timing belt's direction of rotation if it is to be reinstalled. Do not disturb the engine after the timing marks have been aligned.

7. Remove the water pump mounting bolts, noting the three different lengths and locations. Remove the pump and gasket, disconnecting the water pipe from the coolant outlet pipe.

To install:

8. Clean the gasket surfaces and coat a new gasket with sealant. Then, position the gasket on the front of the block with all bolt holes aligned. Install a new O-ring in the coolant outlet pipe port.

9. Install the pump and connect the coolant outlet pipe. Install the bolts with the shortest at the bottom; two just slightly longer at the 1 and 4 o'clock positions on the right side of the pump; next-to-longest bolt at the 8 o'clock position, just under the outlet; and the longest bolt at the 11 o'clock position and also attaching the alternator brace.

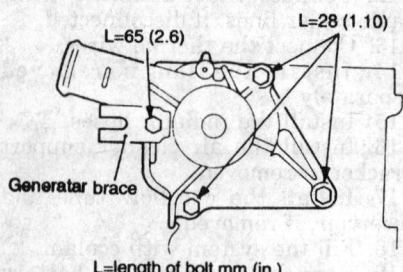

L=65 (2.6)
L=28 (1.10)
Generatar brace
L=length of bolt mm (in.)
257128

Water pump bolt locations — 1993–94 Precis 1.5L (G4DJ) engines

10. Torque the bolts with a head mark 4, to 9–11 ft. lbs. (12–15 Nm); those with a head mark 7, to 14–20 ft. lbs. (20–26 Nm).

11. Install the timing belt and tensioner and upper and lower timing covers.

12. Install the water pump pulley.

13. Install and adjust the engine accessory drive belts.

14. Tighten the water pump pulley bolts to 6–7 ft. lbs. (8–10 Nm) after the drive belts have been installed and tensioned. Recheck belt tension after the pulley bolts are tightened.

15. Refill the cooling system.

16. Connect the negative battery cable.

Mirage

1. Disconnect the negative battery cable.

2. Rotate the engine and position the No. 1 piston to TDC of its compression stroke.

3. Drain the cooling system.

4. Remove the engine undercover.

5. Disconnect the clamp bolt from the power steering hose.

6. Remove the engine drive belts.

7. Support the engine with the appropriate equipment and remove the engine mount bracket.

8. Remove the timing belt from the front of the engine.

9. Remove the power steering pump bracket.

10. Remove the alternator brace.

NOTE: The water pump mounting bolts are different in length, note their positioning for reassembly.

11. Remove the water pump, gasket and O-ring where the water inlet pipe(s) joins the pump.

To install:

12. Thoroughly clean and dry both gasket surfaces of the water pump and block.

13. For 1.5L engines, install a new O-ring into the groove on the front end of the water inlet pipe. Do not apply oils or grease to the O-ring. Wet the O-ring with water only.

14. For 1.8L engines, apply a 0.09–0.12 inch (2.5–3.0mm) continuous bead of sealant to water pump and install the pump assembly. Install the water pump within 15 minutes of the application of the sealant. Wait 1 hour after installation of the water pump to refill the cooling system or starting the engine.

15. Install the gasket and pump assembly and tighten the bolts to specifications. Use care when aligning the

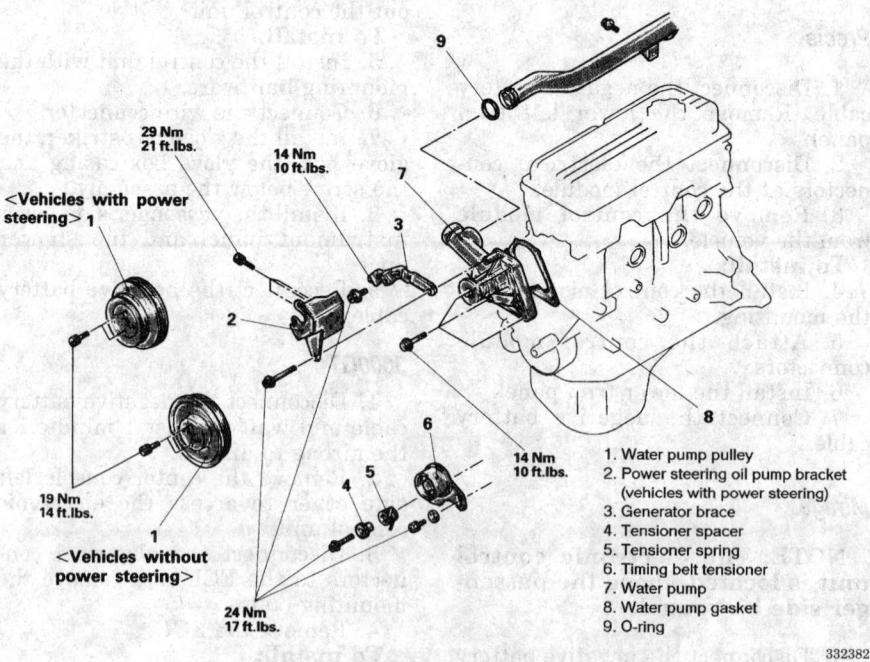

29 Nm / 21 ft.lbs.
14 Nm / 10 ft.lbs.
<Vehicles with power steering> 1
19 Nm / 14 ft.lbs.
<Vehicles without power steering>
24 Nm / 17 ft.lbs.
14 Nm / 10 ft.lbs.

1. Water pump pulley
2. Power steering oil pump bracket (vehicles with power steering)
3. Generator brace
4. Tensioner spacer
5. Tensioner spring
6. Timing belt tensioner
7. Water pump
8. Water pump gasket
9. O-ring

332382

Water pump and related components — Mirage with 1.5L (4G15) engine shown, Precis similar

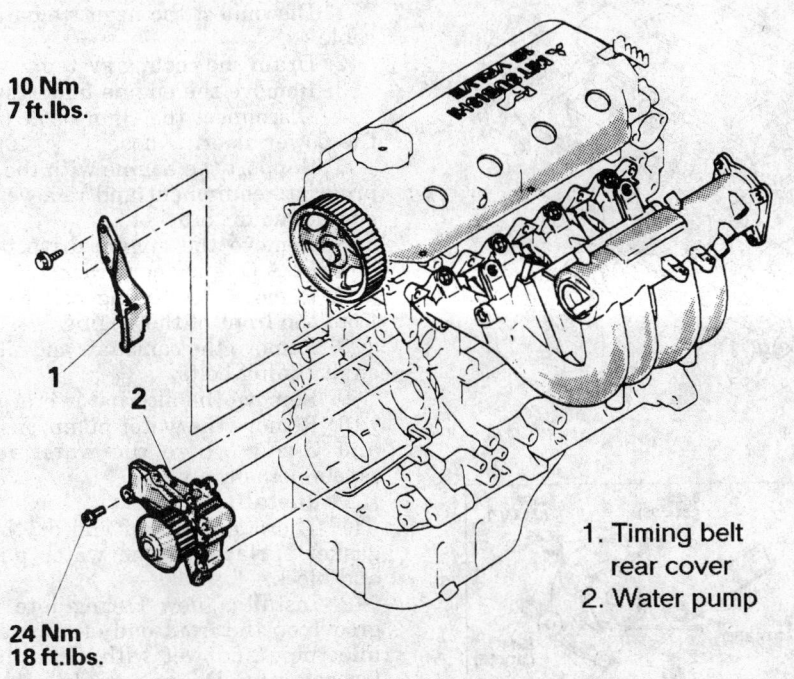

**10 Nm
7 ft.lbs.**

1

2

**24 Nm
18 ft.lbs.**

1. Timing belt
   rear cover
2. Water pump

332344

Water pump and related components — Mirage with 1.8L (4G93) engines

water pump with the water inlet pipe.

16. Install the remaining components in the reverse order of removal.

17. Fill the system with coolant.

18. Connect the negative battery cable, run the vehicle until the thermostat opens and fill the radiator completely.

19. Once the vehicle has cooled, recheck the coolant level.

Eclipse

1993–94 Vehicles

1. Disconnect the negative battery cable.

2. Drain the cooling system. Remove the accessory drive belt.

3. Remove the engine undercover.

4. Disconnect the clamp bolt from the power steering hose. Remove the tensioner pulley bracket.

5. Support the engine with the appropriate equipment and remove the engine mount bracket.

6. Remove both the outer and the inner timing belts from the front of the engine. If reusing old belt, mark the direction of rotation on the outer belt surface. This will assure belt rotation in the original direction and extend belt life.

7. Remove the alternator brace from the front of the water pump.

8. Remove the water pump and gasket from the engine.

9. Remove the O-ring where the water inlet pipe(s) joins the pump. Clean all mating surfaces and inspect for cracks or other damage. Replace components that are damaged or cracked.

To install:

10. Thoroughly clean and dry gasket surfaces of the water pump and block.

11. Install a new O-ring into the groove on the front end of the water inlet pipe. Do not apply oils or grease to the O-ring. Wet with clean antifreeze only.

12. Install the gasket and pump assembly and tighten the bolts. Note the marks on the bolt heads. Those marked **4** should be torqued to 9–11 ft. lbs. (12–15 Nm). Those bolts marked **7** should be torqued from 15–19 ft. lbs. (20–27 Nm).

13. Connect the hoses to the pump.

14. Reinstall the timing belt and related parts.

15. Install the engine undercover.

16. Fill the system with coolant.

17. Connect the negative battery cable, run the vehicle until the thermostat opens and fill the radiator completely.

18. Once the vehicle has cooled, recheck the coolant level.

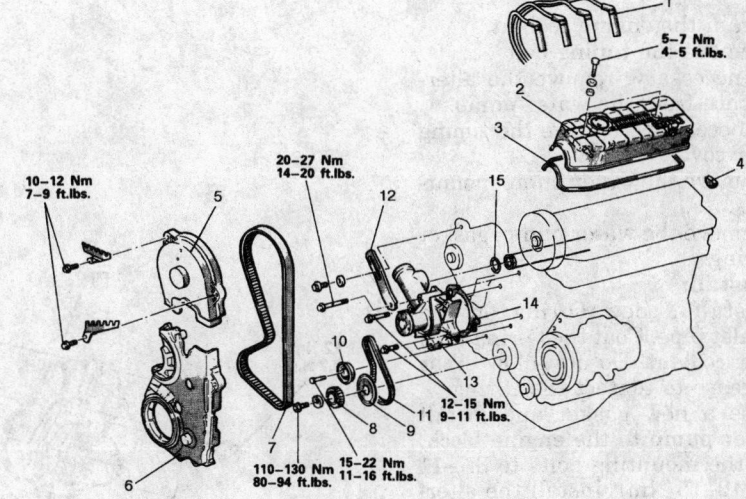

1. Spark plug cable
2. Rocker cover
3. Rocker cover gasket
4. Semi-circular packing
5. Timing belt front upper cover
6. Timing belt front lower cover
7. Timing belt
8. Crankshaft sprocket
9. Flange
10. Timing belt B tensioner
11. Timing belt B
12. Generator brace
13. Water pump
14. Water pump gasket
15. O-ring

46834

Water pump and related components — 1.8L (4G37) engine

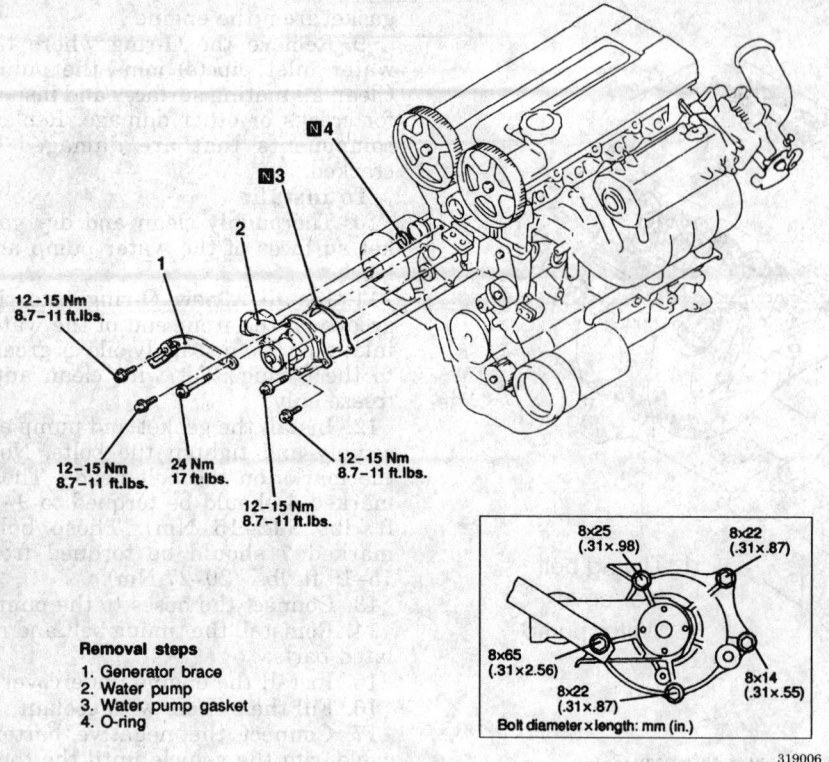

12–15 Nm
8.7–11 ft.lbs.

12–15 Nm
8.7–11 ft.lbs.

24 Nm
17 ft.lbs.

12–15 Nm
8.7–11 ft.lbs.

12–15 Nm
8.7–11 ft.lbs.

Removal steps
1. Generator brace
2. Water pump
3. Water pump gasket
4. O-ring

8×25
(.31×.98)

8×22
(.31×.87)

8×65
(.31×2.56)

8×22
(.31×.87)

8×14
(.31×.55)

Bolt diameter × length: mm (in.)

319006

Water pump mounting — 1995–97 2.0L engines shown

1995–97 Vehicles

1. Disconnect the negative battery cable.

2. Drain the engine coolant.

3. Remove the timing belt.

4. If necessary, remove the alternator brace from the water pump.

5. If necessary, remove the timing belt rear cover.

6. Remove the water pump mounting bolts.

7. Remove the water pump, gasket and O-ring.

To install:

8. Install a new O-ring on the water inlet pipe. Coat the O-ring with water or coolant. Do not allow oil or other grease to contact the O-ring.

9. Use a new gasket and install the water pump to the engine block. Torque the mounting bolts to 8.7–11 ft. lbs. (12–15 Nm). Install the alternator brace on the water pump. Torque the brace pivot bolt to 17 ft. lbs. (24 Nm).

10. If removed, install the timing belt rear cover.

11. Install the timing belt.

12. Install the remaining components.

13. Refill the engine with coolant.

14. Connect the negative battery cable, start the engine and check for leaks.

Galant

1. Disconnect the negative battery cable.

2. Drain the cooling system.

3. Remove the engine undercover.

4. Disconnect the clamp bolt from the power steering hose.

5. Support the engine with the appropriate equipment and remove the engine mount bracket.

6. Remove the engine drive belts and the A/C tensioner bracket.

7. Remove the timing belt covers from the front of the engine.

8. Remove the camshaft and silent shaft timing belts.

9. Remove the alternator brace.

10. Remove the water pump, gasket and O-ring where the water inlet pipe(s) joins the pump.

To install:

11. Thoroughly clean and dry both gasket surfaces of the water pump and block.

12. Install a new O-ring into the groove on the front end of the water inlet pipe and wet with clean antifreeze only. Do not apply oils or grease to the O-ring.

13. Using a new gasket, install the water pump assembly. Tighten bolts with the head mark **4** to 10 ft. lbs. (14 Nm) and bolts with the head mark **7** to 18 ft. lbs. (24 Nm).

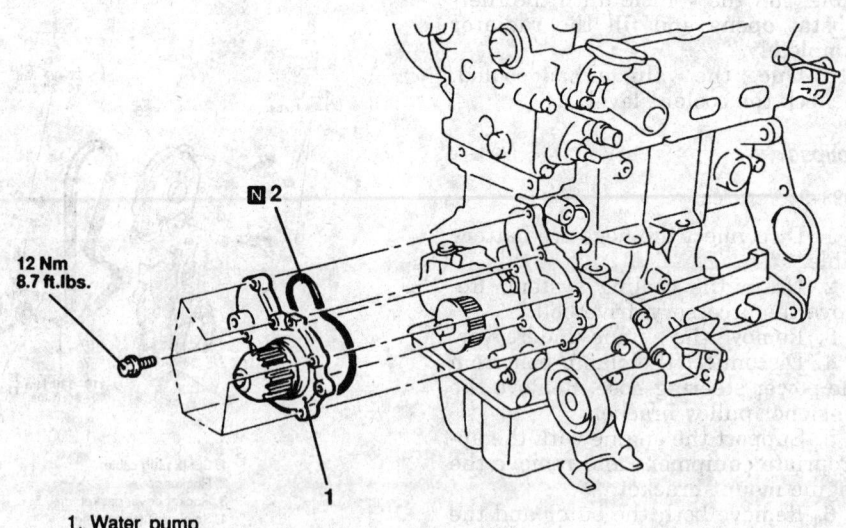

12 Nm
8.7 ft.lbs.

1. Water pump
2. O-ring

319003

Water pump mounting — 1995–97 Eclipse 2.0L (420A) engines

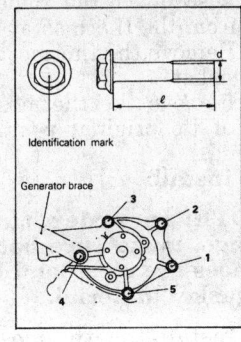

165238

Water pump bolt identification — Galant

No.	Identification mark	Bolt diameter (d) x length (ℓ) mm (in.)	Torque Nm (ft.lbs.)
1	4	8 x 14 (.31 x .55)	
2	4	8 x 22 (.31 x .87)	12–15 (9–10)
3	4	8 x 30 (.31 x 1.18)	
4	7	8 x 65 (.31 x 2.56)	20–27 (15–19)
5	4	8 x 28 (.31 x 1.10)	12–15 (9–10)

14. Reinstall the timing belt and related parts.

15. Install the engine drive belts and adjust.

16. Install the engine undercover.

17. Fill the system with coolant.

18. Connect the negative battery cable, run the vehicle until the thermostat opens and fill the radiator completely.

19. Once the vehicle has cooled, recheck the coolant level.

Diamante and 3000GT

1. Disconnect the negative battery cable. Drain the cooling system.

2. Remove the engine undercover.

3. Disconnect the clamp bolt from the power steering hose.

4. Support the engine with the appropriate equipment and remove the engine mount bracket.

5. Remove the timing belt from the front of the engine.

6. Disconnect the coolant hoses from the pump, if equipped.

7. Remove the alternator brace.

NOTE: The water pump bolts are different in size. Be sure to pay special attention to the bolts during the removal procedure.

8. Remove the water pump, gasket and O-ring where the water inlet pipe joins the pump.

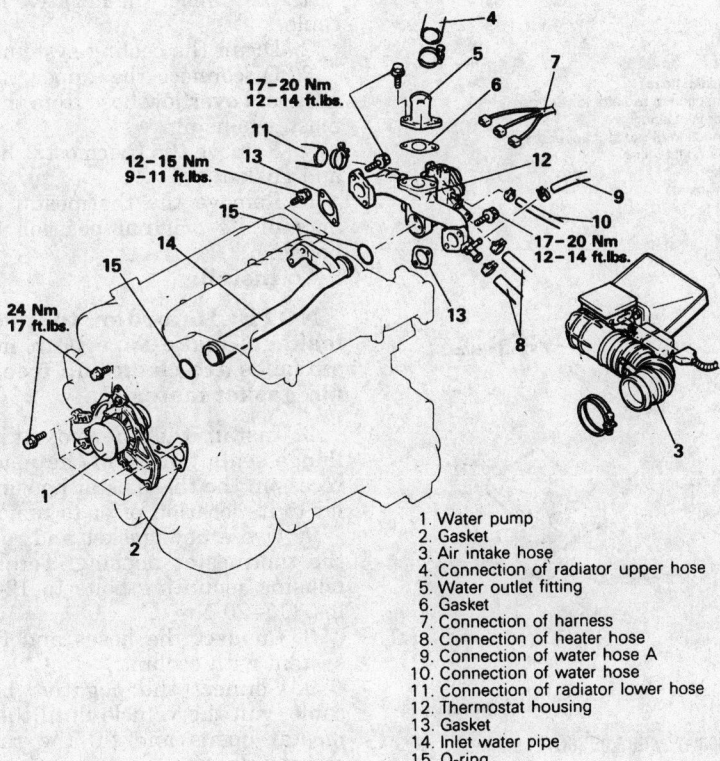

17–20 Nm
12–14 ft.lbs.

12–15 Nm
9–11 ft.lbs.

24 Nm
17 ft.lbs.

17–20 Nm
12–14 ft.lbs.

1. Water pump
2. Gasket
3. Air intake hose
4. Connection of radiator upper hose
5. Water outlet fitting
6. Gasket
7. Connection of harness
8. Connection of heater hose
9. Connection of water hose A
10. Connection of water hose
11. Connection of radiator lower hose
12. Thermostat housing
13. Gasket
14. Inlet water pipe
15. O-ring

328128

Water pump and related components — DOHC Diamante shown, 3000GT similar

To install:

9. Thoroughly clean and dry both gasket surfaces of the water pump and block.

10. Install a new O-ring into the groove on the front end of the water inlet pipe. Do not apply oils or grease to the O-ring. Wet with water only.

NOTE: Use care when aligning the water pump with the inlet water pipe.

11. Using a new gasket, install the water pump assembly to the engine block. Torque the mounting bolts to 17 ft. lbs. (24 Nm).

12. Connect the hoses to the pump.

13. Reinstall the timing belt and related parts.

14. Install the engine drive belts and adjust.

15. Fill the system with coolant.

16. Connect the negative battery cable, run the vehicle until the thermostat opens and fill the radiator completely.

17. Once the vehicle has cooled, recheck the coolant level.

Thermostat

REMOVAL AND INSTALLATION

Precis

1. Disconnect the negative battery cable. Remove the air cleaner.

2. Drain the cooling system to a point below the level of the tubes in the top tank of the radiator.

3. Disconnect the hose at the thermostat water pipe.

4. Remove the water pipe support bracket nut.

5. Unbolt and remove the thermostat housing and pipe.

6. Lift out the thermostat. Discard the gasket.

To install:

7. Clean the mating surfaces of the housing and manifold thoroughly.

8. Install the thermostat with the spring facing downward and position a new gasket over the thermostat. The jiggle valve in the thermostat should be on the manifold side.

9. Install the housing and pipe assembly. Torque the housing bolts to 12–14 ft. lbs. (17–20 Nm).

10. Refill the cooling system. Connect the negative battery cable.

Mirage

1. Disconnect the negative battery cable. Drain the cooling system.

2. Remove the air cleaner assembly.

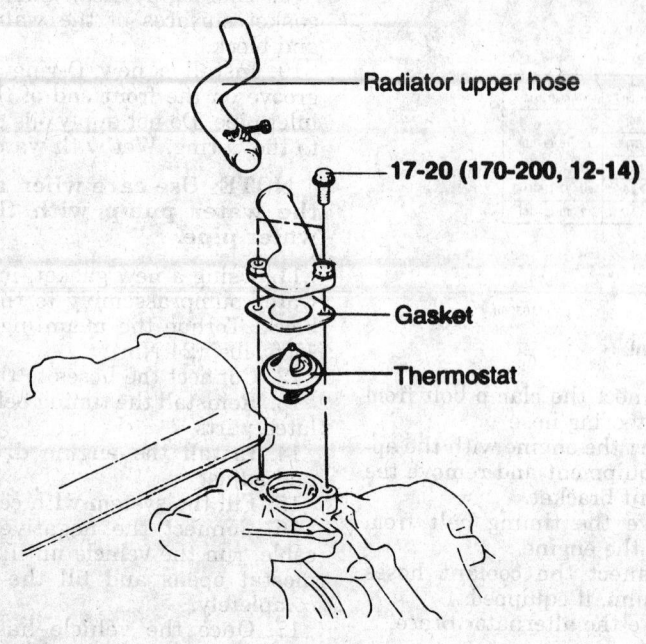

Thermostat mounting — Precis shown, Mirage similar

334090

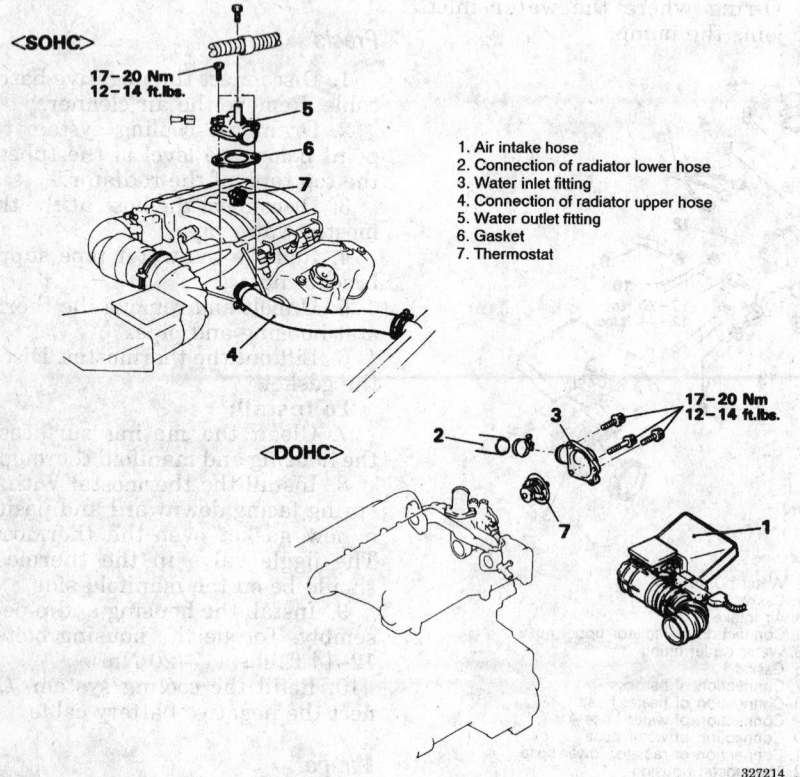

1. Air intake hose
2. Connection of radiator lower hose
3. Water inlet fitting
4. Connection of radiator upper hose
5. Water outlet fitting
6. Gasket
7. Thermostat

Thermostat and related components — Diamante and 3000GT with 3.0L (6G72) engine

327214

3. Disconnect the upper radiator hose from the thermostat housing.

4. Remove the thermostat housing and gasket.

5. Remove the thermostat taking note of its original position in the housing.

To install:

NOTE: In order to prevent leakage, make sure both mating surfaces are clean and free of all old gasket material.

6. Install the thermostat so its flange seats tightly in the machined recess in the thermostat housing. Refer to its location prior to removal.

7. Use a new gasket and reinstall the thermostat housing. Torque the housing mounting bolts to 16 ft. lbs. (22 Nm).

8. Connect the hoses and fill the system with coolant.

9. Install the air cleaner assembly.

10. Connect the negative battery cable, run the vehicle until the thermostat opens and fill the radiator completely.

11. Bleed the cooling system.

12. Once the vehicle has cooled, recheck the coolant level.

Eclipse

1993–94 Vehicles

1. Disconnect the negative battery cable.

2. Drain the cooling system.

3. Disconnect the upper radiator hose and overflow hose from the thermostat housing.

4. Remove the thermostat housing and gasket.

5. Remove the thermostat taking note of its original position in the housing.

To install:

NOTE: In order to prevent leakage, make sure both mating surfaces are clean and free of all old gasket material.

6. Install the thermostat so its flange seats tightly in the machined recess in the thermostat housing. Refer to its location prior to removal.

7. Use a new gasket and reinstall the thermostat housing. Torque the housing mounting bolts to 12–14 ft. lbs. (17–20 Nm).

8. Connect the hoses and fill the system with coolant.

9. Connect the negative battery cable, run the vehicle until the thermostat opens and fill the radiator completely.

10. Once the vehicle has cooled, recheck the coolant level.

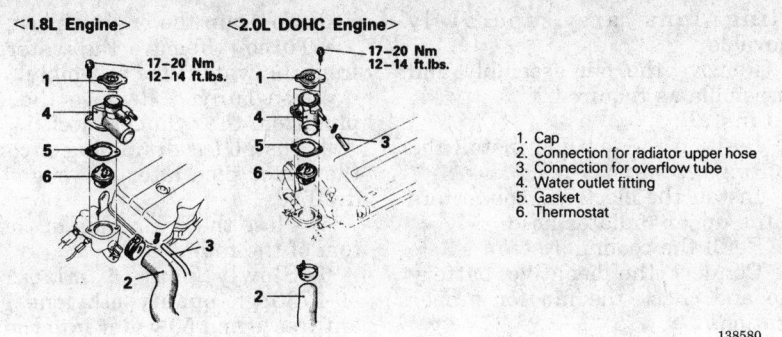

1. Cap
2. Connection for radiator upper hose
3. Connection for overflow tube
4. Water outlet fitting
5. Gasket
6. Thermostat

138580

Thermostat and related parts exploded view — 1993–94 Eclipse with 1.8L (4G37) and 2.0L (4G63) engines

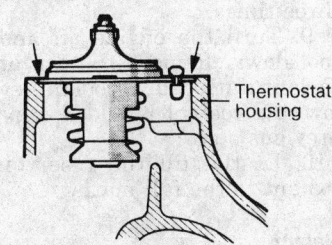

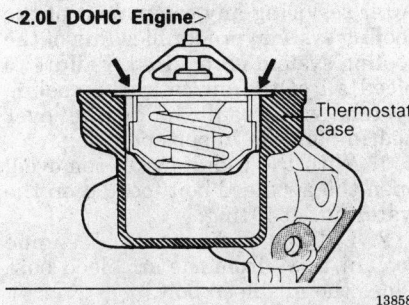

138581

Thermostat positioning in housing — Eclipse 1.8L (4G37), and 2.0L (4G63) engines

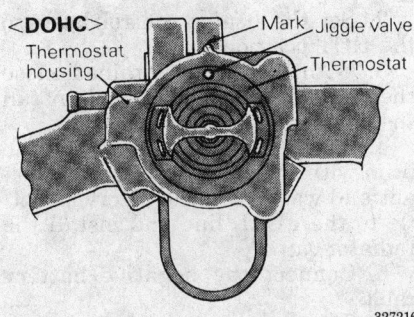

327216

Thermostat mounting — Diamante and 3000GT with 3.0L (6G72) engine

1995–97 Vehicles

1. Disconnect the negative battery cable.
2. Drain the cooling system.
3. Matchmark the clamp to the hose so it can be installed in the original position and disconnect the upper radiator hose from the water outlet.
4. Remove the water outlet or inlet, as applicable.
5. Remove the thermostat and gasket.

To install:

6. Use a new rubber gasket and install the thermostat with the jiggle valve facing up. Torque the water outlet bolts to 16 ft. lbs. (22 Nm) for 2.0L non-turbo engines or to 9.4 ft. lbs. (13 Nm) for turbo and 2.4L engines.
7. Connect the upper radiator hose to the water outlet. Install the clamp in the original position.
8. Refill the cooling system and connect the negative battery cable.
9. Start the engine and check for leaks.

1993 Galant 2.0L (4G63) Engine

1. Disconnect the negative battery cable.
2. Drain the cooling system.
3. Disconnect the lower radiator hose, from the water inlet or outlet, at the thermostat housing.
4. Remove the water inlet, or outlet, as applicable, from the thermostat housing.
5. Remove the thermostat taking note of its original position in the housing.

To install:

6. Install the thermostat so its flange seats tightly in the machined recess in the thermostat housing. Refer to its location prior to removal.
7. Install water inlet or outlet, to the thermostat housing. Torque the housing mounting bolts to 7–10 ft. lbs. (10–15 Nm).
8. Connect the lower hose and fill the system with coolant.
9. Connect the negative battery cable, run the vehicle until the thermostat opens and fill the radiator completely.
10. Once the vehicle has cooled, recheck the coolant level.

Diamante and 3000GT

1. Disconnect the negative battery cable.
2. Drain the cooling system.
3. Remove necessary air intake plumbing.
4. Disconnect the upper radiator hose and overflow hose from the thermostat housing.
5. Remove the thermostat housing and gasket.
6. Remove the thermostat taking note of its original position in the housing or intake manifold.

To install:

7. Install the thermostat so its flange seats tightly in the machined groove in the intake manifold or thermostat case. Refer to its location prior to removal. Align the jiggle valve with the alignment mark on the thermostat housing.
8. Use a new gasket and reinstall the thermostat housing. Torque the housing mounting bolts to 12–14 ft. lbs. (17–20 Nm).
9. Fill the system with coolant.
10. Install the removed air intake plumbing.
11. Connect the negative battery cable, run the vehicle until the thermostat opens and fill the radiator completely.
12. Once the vehicle has cooled, recheck the coolant level.

Electric Cooling Fan

REMOVAL AND INSTALLATION

Except Diamante and 3000GT

1. Disconnect the negative battery cable.
2. If necessary to remove hoses for radiator removal, drain the cooling system.
3. If necessary, remove the overflow hose and disconnect the upper radiator hose.

NOTE: It is recommended that each clamp be matchmarked to the hose. Observe the marks and reinstall the clamps exactly when reinstalling the radiator.

4. If equipped with an automatic transaxle, remove and plug the fluid cooler hoses.
5. Disconnect the electrical connector from the coolant fan motor.
6. Remove the mounting bolts, fan and shroud assembly from the vehicle.
7. Remove the fan blade retainer nut from the shaft on the fan motor and separate the fan from the motor.
8. Remove the motor to shroud attaching screws and the motor from the shroud.
To install:
9. Install the motor to the shroud and secure with the mounting bolts.
10. Install the remaining components in the reverse order of removal.
11. Fill the cooling system. Connect the negative battery cable and check the cooling fan for proper operation.

Diamante and 3000GT

1. Disconnect the negative battery cable.

———— CAUTION ————
Wait at least 90 seconds after the negative (-) battery cable is disconnected to prevent possible deployment of the air bag.

2. Drain the cooling system only when the radiator and the engine are at safe temperatures.
3. Unplug the cooling fan and radiator sensor connector(s). Most of these connectors employ a waterproof connector. When disconnecting, make sure all parts of the connector remain intact.
4. Disconnect the upper radiator hose from the radiator and remove overflow tank.
5. Remove the fan mounting screws. The radiator and condenser

cooling fans are separately removable.
6. Remove the fan assembly and disassemble as required.
To install:
7. Position the fan and install the mounting screws.
8. Install the electrical connectors and the upper radiator hose.
9. Refill the cooling system.
10. Connect the negative battery cable and check the fan for proper operation.

Cooling System Bleeding

PROCEDURE

———— CAUTION ————
Allow the cooling system to completely cool before attempting any repair or draining the system. Injury from scalding could result if radiator cap or hose connections are removed while system is hot.

Mirage

1. Disconnect the negative battery cable.
2. Set the heater control lever to the HOT position.
3. With the coolant drained, close the radiator drain plug and install the engine drain plugs.
4. Slowly fill the radiator to the brim with the proper mixture of coolant and water. Fill the reservoir bottle to the FULL line and install the radiator cap.
5. Connect the negative battery cable.
6. Start the engine and allow it to reach normal operating temperature. Watch the coolant temperature gauge for signs of overheating. Race the engine two or three times with no-load.
7. Shut the engine off and let it cool down completely.
8. After the engine has cooled down, remove the radiator cap and fill the radiator to the brim. Fill the reservoir to the FULL line on the bottle.
9. Check the radiator drain plug and the engine drain plugs for leakage.
10. Install the radiator cap.

Eclipse

1. Remove the cooling system cap and open the drain plug at the bottom of the radiator to drain the coolant.
2. Remove the reservoir to drain the coolant from the tank.

3. To drain the engine block;
• Turbo - Remove the water hose from the water pipe assembly
• Non-Turbo - Remove the drain plug from the cylinder block
4. Install the drain plug or connect the water pipe after the engine has drained.
5. Close the drain plug at the bottom of the radiator.
6. Slowly pour a mixture of 50-17368gh quality ethylene glycol antifreeze and 50 water into the cooling system and install the cap when full.
7. Fill the reservoir tank to the full mark.
8. Start the engine and let it run until it reaches normal operating temperature and the thermostat opens. Race the engine at 3,000 rpm three times.
9. Turn the engine off and let it cool down. Remove the cap and add coolant if needed. If coolant level was low and coolant is added, repeat the previous step.
10. Lastly, fill the reservoir with coolant to the full mark.

Galant

After servicing any component of the cooling system, proper bleeding of the cooling system is required. Failure to bleed all of the air from the cooling system will result in engine overheating and improper operation.

1. With the radiator cap removed, open the air bleed bolt located on the water outlet fitting.
2. Fill the radiator until engine coolant flows from the air bleed bolt. Close the air bleed bolt.
3. Slowly pour the coolant into the radiator until full. Also fill the coolant reservoir to the FULL line.

NOTE: Recommended antifreeze is DIA-QUEEN LONG-LIFE COOLANT (Part No. 0103044) or high quality ethylene-glycol coolant. Capacity is: 7.4 qts.

4. Install the radiator cap securely.
5. Start the engine and warm the engine until the thermostat opens. (Carefully feel the radiator hose with your hand to check that warm coolant is flowing). After the thermostat has opened, race the engine at 3,000 rpm three times.
6. After the engine is stopped, wait until it has cooled down. Remove the radiator cap and check the level. If the level is low, add coolant and repeat the procedure. If the level has not dropped, replace the radiator cap

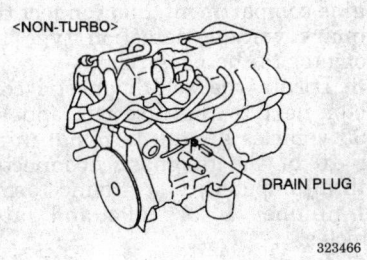

Drain plug location (non-turbo) — Eclipse

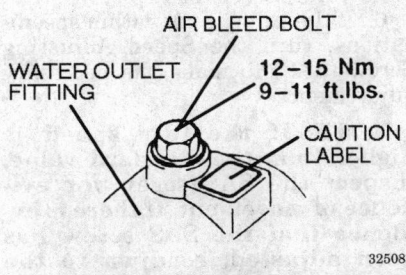

Cooling system air bleed — Galant

and make sure the coolant reservoir is at the FULL mark.

Diamante

1. Disconnect the negative battery cable.
2. With the coolant drained, close the radiator drain plug and install the engine drain plugs.
3. Slowly fill the radiator to the brim with the proper mixture of coolant and water. Fill the reservoir bottle to the FULL line and install the radiator cap.
4. Connect the negative battery cable.
5. Start the engine and allow it to reach normal operating temperature. Watch the coolant temperature gauge for signs of overheating. Race the engine two or three times with no-load.
6. Shut the engine off and let it cool down completely.
7. After the engine has cooled down, remove the radiator cap and fill the radiator to the brim. Fill the reservoir to the FULL line on the bottle.
8. Check the radiator drain plug and the engine drain plugs for leakage.
9. Install the radiator cap.

3000GT

After working on the cooling system, even to replace the thermostat, the system must be bled. Air trapped in the system will prevent proper filling and leave the radiator coolant level low, causing a risk of overheating.

To bleed the system, start with the system cool, the radiator cap off and the radiator filled to about an inch below the filler neck.

1. Start the engine and run it at slightly above normal idle speed. This will insure adequate circulation. If air bubbles appear and the coolant level drops, fill the system with an antifreeze/water mixture to bring the level back to the proper level.
2. Run the engine this way until the thermostat opens. When this happens, coolant will move abruptly across the top of the radiator and the temperature of the radiator will suddenly rise.
3. At this point, air is often expelled and the level may drop quite a bit. Keep refilling the system until the level is near the top of the radiator and remains constant.
4. If the vehicle has an overflow tank, fill the radiator right up to the filler neck. Replace the radiator filler cap.

FUEL SYSTEM

Fuel System Service Precautions

Safety is the most important factor when performing not only fuel system maintenance but any type of maintenance. Failure to conduct maintenance and repairs in a safe manner may result in serious personal injury or death. Maintenance and testing of the vehicle's fuel system components can be accomplished safely and effectively by adhering to the following rules and guidelines.

• To avoid the possibility of fire and personal injury, always disconnect the negative battery cable unless the repair or test procedure requires that battery voltage be applied.

• Always relieve the fuel system pressure prior to disconnecting any fuel system component (injector, fuel rail, pressure regulator, etc.), fitting or fuel line connection. Exercise extreme caution whenever relieving fuel system pressure to avoid exposing skin, face and eyes to fuel spray. Please be advised that fuel under pressure may penetrate the skin or any part of the body that it contacts.

• Always place a shop towel or cloth around the fitting or connection prior to loosening to absorb any excess fuel due to spillage. Ensure that all fuel spillage (should it occur) is quickly removed from engine surfaces. Ensure that all fuel soaked cloths or towels are deposited into a suitable waste container.

• Always keep a dry chemical (Class B) fire extinguisher near the work area.

• Do not allow fuel spray or fuel vapors to come into contact with a spark or open flame.

• Always use a backup wrench when loosening and tightening fuel line connection fittings. This will prevent unnecessary stress and torsion to fuel line piping. Always follow the proper torque specifications.

• Always replace worn fuel fitting O-rings with new. Do not substitute fuel hose or equivalent, where fuel pipe is installed.

Fuel System Pressure

RELIEVING

———— **CAUTION** ————

Fuel injection systems remain under pressure after the engine has been turned OFF. Properly relieve fuel pressure before disconnecting any fuel lines. Failure to do so may result in fire or personal injury.

1. Turn the ignition to the **OFF** position.
2. Loosen the fuel filler cap to release fuel tank pressure.
3. For the Mirage and 1995–97 Eclipse, remove the rear seat cushion, then remove the service cover and disconnect the fuel pump harness connector.
4. For the Precis, FWD Galant, 1993–94 Eclipse, Diamante and 3000GT, detach the fuel pump harness connector located in the area of the fuel tank. It may be necessary to raise the vehicle to access the connector.
5. For the AWD Galant, remove the carpet from the trunk, locate the fuel tank wiring at the pump access cover, then detach the wiring.
6. Start the vehicle and allow it to run until it stalls from lack of fuel. Turn the key to the **OFF** position.

7. Disconnect the negative battery cable, then reconnect the fuel pump connector. Install the access cover, cushion or carpet as necessary.

8. Wrap shop towels around the fitting that is being disconnected to absorb residual fuel in the lines.

9. Place shop towels into proper safety container.

Idle Speed

ADJUSTMENT

Precis

1. Warm the engine to operating temperature, leave lights, electric cooling fan and accessories **OFF**. The transaxle should be in **N**. The steering wheel in a neutral position for vehicles with power steering.

2. Loosen the accelerator cable.

3. Turn the ignition switch **ON** but do not start the engine. Leave the key in this position for at least 15 seconds. Check to see that the ISC servo is fully retracted to the curb idle position.

NOTE: When the ignition switch is turned to the ON position, the ISC plunger extends to the fast idle position opening. After 15 seconds, it retracts to the fully closed (curb idle) position.

4. Turn the ignition switch **OFF**.

5. Disconnect the ISC motor connector and secure the ISC motor at the fully retracted position.

6. In order to prevent the throttle valve from sticking, open it 2 or 3 times, then allow it to click shut and loosen the fixed SAS sufficiently.

7. Start the engine and allow it to run at idle speed. Ensure that the engine is running at idle speed of 700 ±100 rpm. Adjust as necessary by turning the ISC adjust screw.

8. Tighten the fixed SAS until the engine speed starts to increase. Then,

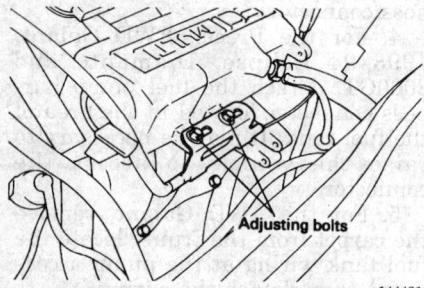

Accelerator cable adjusting bolts — Precis

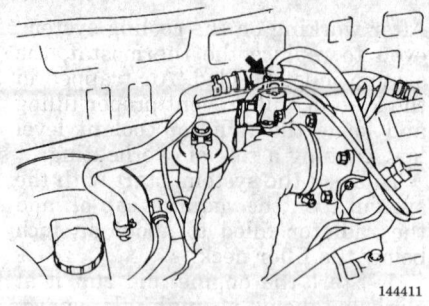

ISC Motor connector — Precis

loosen the screw until the engine speed ceases to drop (touch point) and loosen an additional ½ turn.

9. Turn the ignition switch to the **OFF** position.

10. Adjust the accelerator cable to 0 to 0.04 in. (0 to 1mm) for a manual transaxle and 0.08 to 0.12 in. (2 to 3mm) for an automatic transaxle.

11. Connect the ISC motor connector.

12. Start the engine and check to be sure that the idle speed is correct.

13. Turn the ignition switch **OFF** and disconnect the negative battery cable for 15 seconds and re-connect. This will erase the data stored in memory during ISC adjustment.

Mirage

NOTE: The idle speed is controlled electronically and adjustment is usually unnecessary. However, the idle speed may be checked using the following procedures.

1. Warm the engine to operating temperature, leave lights, electric cooling fan and accessories **OFF**. The transaxle should be in **N** or **P** for automatic transaxle. The steering wheel in a neutral position for vehicles with power steering.

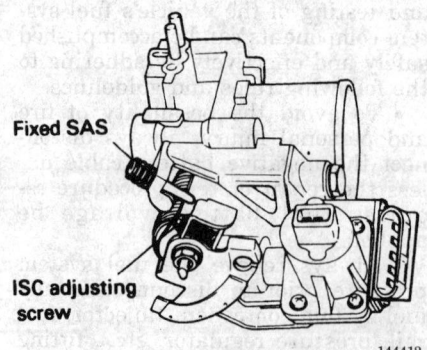

Fixed SAS and ISC adjusting screws — Precis

2. Insert the paper clip into the single terminal rpm connector in the engine compartment, and connect the primary voltage detection type tachometer to the paper clip.

3. Locate the self-diagnostic connector next to the fuse box and for 1993 vehicles ground terminal number 10 of the diagnostic connector with a jumper wire or ground terminal number 1 for 1994 and later vehicles.

4. Remove the waterproof female connector from the ignition timing adjustment connector. Ground the ignition timing adjustment terminal.

5. Start the engine and run at idle. Check the basic idle speed, the desired value is 700–800 rpm.

6. If the value is not within specifications, turn the Speed Adjusting Screw (SAS) to make the necessary adjustment.

NOTE: If the idle speed is higher than the standard value, inspect the SAS screw for evidence of movement. If there is evidence that the SAS screw has been adjusted, readjust to the proper setting. If the screw does not look as though it has been adjusted, it is possible that there is leakage as a result of deterioration of the Fast Idle Air Valve (FIAV) and, if so the throttle body should be replaced.

7. Turn the ignition **OFF**. Disconnect and remove the jumper wires from the diagnosis control terminal and the ignition timing adjustment terminal.

8. Start the engine and let run at idle speed for about 10 minutes, check to be sure the idling condition is normal.

1993–94 Eclipse

1.8L (4G37) Engine

1. Warm the engine to operating temperature, leave lights, electric cooling fan and accessories **OFF**. The transaxle should be in **N** or **P** for automatic transaxle. The steering wheel in a neutral position for vehicles with power steering.

2. Check the ignition timing and adjust, if necessary. Be sure to ground the ignition timing adjustment terminal.

3. Connect a tachometer to the CRC filter connector. Use a paper clip for a tach adapter.

4. Run the engine for more than 10 seconds at 2000–3000 rpm. Allow the engine to idle for 2 minutes. Check the idle rpm. Curb idle should be 750 rpm.

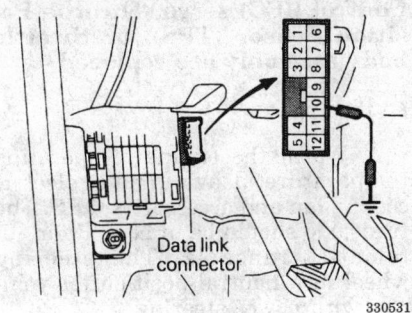

Diagnostic connector and terminal identification — 1993 Mirage

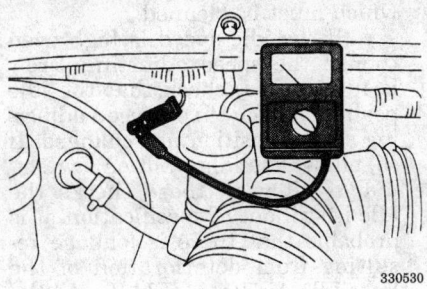

Late style tachometer pickup location — Mirage engines

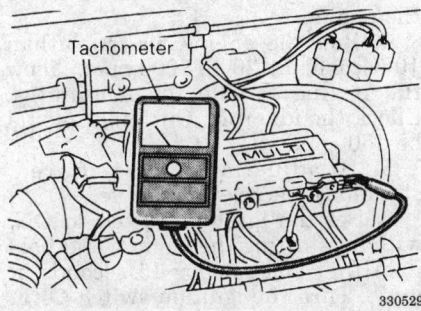

Early style tachometer pickup location — Mirage

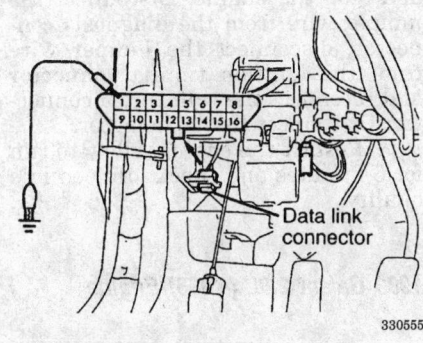

Diagnostic connector and terminal identification — Mirage

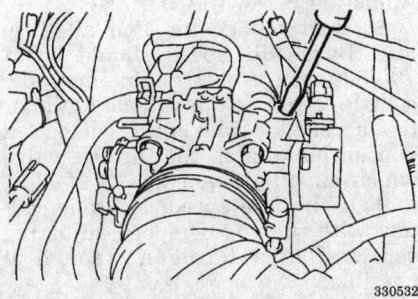

Engine speed adjusting screw location — Mirage

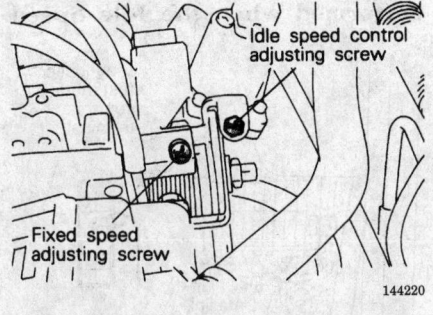

Idle speed adjusting screw location — Eclipse 1.8L (4G37) engine

5. If adjustment is required, slacken the accelerator cable.

6. Connect a digital voltmeter between terminal **19** throttle position sensor output voltage) of the engine control unit and terminal **24** (ground).

7. Set the ignition switch to **ON**, without starting the engine, and hold it in that position for 15 seconds or more. Turn the ignition switch **OFF**.

8. Disconnect the connectors of the idle speed control servo and lock the idle speed control plunger at the initial position. Back out the fixed Speed Adjusting Screw (SAS).

9. Start the engine and allow to idle. Basic idle speed should be at specification. A new engine may idle a little lower. If the vehicle stalls or has a very low idle speed, suspect a deposit buildup on the throttle valve which must be cleaned.

10. If the idle speed is wrong, adjust with the idle speed control adjusting screw. Use a hexagon wrench if possible. Turn in the fixed SAS until the engine speed rises. Then back out the fixed SAS until the Touch Point where the engine speed does not fall any longer, is found. Back out

the fixed SAS an additional ½ turn from the touch point.

11. Stop the engine. Turn the ignition switch to **ON** but do not start engine. Check that the output voltage from the throttle position sensor is 0.48–0.52 volts. If it is out of specification, adjust by loosening the throttle position sensor mounting screws and rotating the throttle position sensor. Turning the throttle position sensor clockwise increases the output voltage. After adjustment, tighten screws firmly.

12. Turn the ignition switch **OFF**.

13. Adjust the free-play of the accelerator cable, reconnect the connectors of the idle speed control servo and remove the voltmeter.

14. Start the engine and check the curb idle. It should be 700 rpm.

15. Turn the ignition switch to **OFF**, disconnect the negative battery cable for more than 10 seconds and reconnect. This clears any trouble codes introduced during testing.

16. Restart the engine, allow to run for 5 minutes and check for good idle quality.

2.0L (4G63) Engine

1. Warm the engine to operating temperature, leave lights, electric cooling fan and accessories **OFF**. The transaxle should be in **N**. The steering wheel in a neutral position for vehicles with power steering.

2. Check the ignition timing and adjust, if necessary. Be sure to ground ignition timing adjustment connector.

3. Connect a tachometer to the single pin connector terminal under the hood.

4. Run the engine for more than 10 seconds at 2000–3000 rpm. Allow the engine to idle for 2 minutes. Check the idle rpm. Curb idle should be 750 rpm.

5. If adjustment is required, disconnect the waterproof female connector used for ignition timing adjustment. Connect this terminal to ground using a jumper wire.

6. Locate the self-diagnosis terminal under the dashboard and connect terminal No. **10** to ground with a jumper wire.

7. Start the engine and allow to idle. Check that the basic idle speed is at specification. If the idle speed deviates from this speed, check the following:

a. A new engine will idle more slowly. Break-in should take approximately 300 miles.

b. If the vehicle stalls or has a very low idle speed, suspect a de-

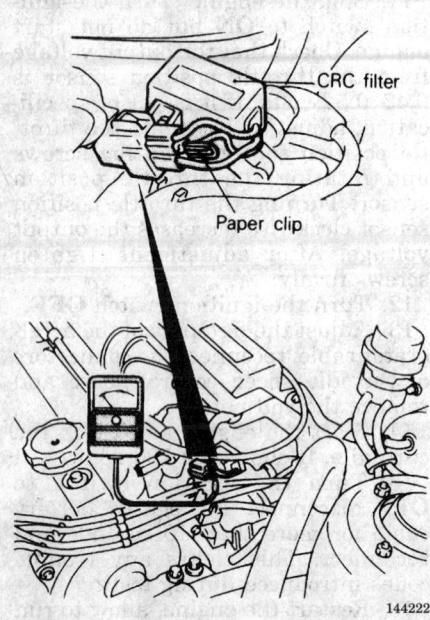

Connecting a tachometer to the engine — Eclipse 1.8L (4G37) engine

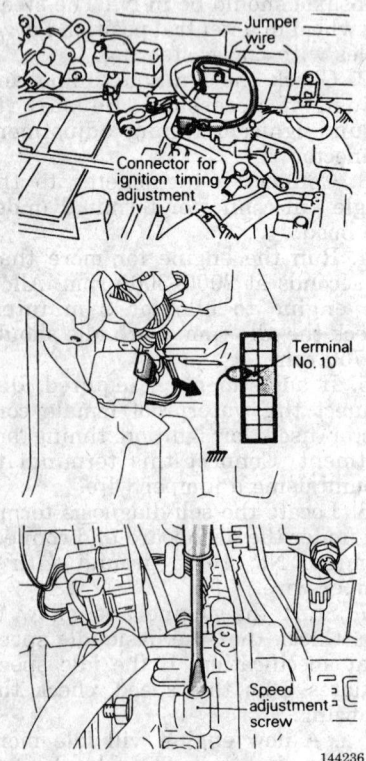

Service points for idle speed adjustment — 1993–94 Eclipse 2.0L (4G63) engine

posit buildup on the throttle valve which must be cleaned.

c. If the idle speed is high even though the speed adjusting screw is fully closed, check that the idle position switch (fixed speed adjusting screw) position has changed. If so, adjust the idle position switch.

d. If after all these checks the idle is still out of specification, it is probable that there is leakage resulting from deterioration of the Fast-Idle Air Valve (FIAV) and the throttle body will need to be replaced.

8. Turn the ignition switch **OFF** and stop the engine. Disconnect the jumper wire from the diagnosis connector, disconnect the jumper wire from the ignition timing connector and reconnect the waterproof connector. Disconnect the tachometer.

9. Restart the engine, allow to run for 5 minutes and check for good idle quality.

1993 Galant 2.0L (4G63) Engine

NOTE: The idle speed is controlled electronically and adjustment is usually unnecessary. However, this procedure must be performed when the Idle Speed

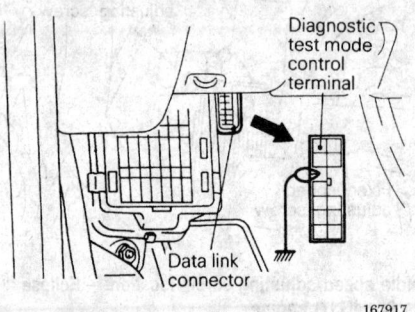

Diagnostic connector and terminal identification (16 valve engines) — 1993 Galant 2.0L (4G63) engines

Engine speed adjusting screw identification (16 valve engines) — 1993 Galant 2.0L (4G63) engine

Control (ISC) servo, Throttle Position Sensor (TPS), or throttle body assembly are replaced.

8 Valve Engine

1. Warm the engine to operating temperature, leave lights, electric cooling fan and accessories **OFF**. The transaxle should be in **N** or **P** for automatic transaxle. The steering wheel in a neutral position for vehicles with power steering.

2. Check the ignition timing and adjust, if necessary. Be sure to ground the ignition timing adjustment terminal.

3. Connect a tachometer to the to the engine.

4. Run the engine for more than 10 seconds at 2000–3000 rpm. Allow the engine to idle for 2 minutes. Check the idle rpm. Curb idle should be 750 rpm.

5. If adjustment is required, slacken the accelerator cable.

6. Set the ignition switch to **ON**, without starting the engine, and hold it in that position for 15 seconds or more. Turn the ignition switch **OFF**.

7. Disconnect the connectors of the idle speed control servo and lock the idle speed control motor at the initial position. Back out the fixed Speed Adjusting Screw (SAS).

8. Start the engine and allow to idle. Basic idle speed should be at 700-800 rpm. A new engine may idle a little lower. If the vehicle stalls or has a very low idle speed, suspect a deposit buildup on the throttle valve which must be cleaned.

9. If the idle speed is wrong, adjust with the engine speed adjusting screw. Use a hexagon wrench if possible.

10. Once the engine speed adjustment is correct, adjust the fixed SAS as follows:

a. Turn in the fixed SAS until the engine speed rises.

b. Back out the fixed SAS until the "Touch Point", where the engine speed does not fall any longer.

c. Back out the fixed SAS an additional $\frac{1}{2}$ turn from the touch point.

11. Turn the ignition switch **OFF**.

12. Adjust the free-play of the accelerator cable, reconnect the connectors of the idle speed control servo and remove the voltmeter.

13. Start the engine and check the curb idle. It should be 700–800 rpm.

14. Turn the ignition switch to **OFF**, disconnect the negative battery cable for more than 10 seconds and reconnect. This clears any trouble codes introduced during testing.

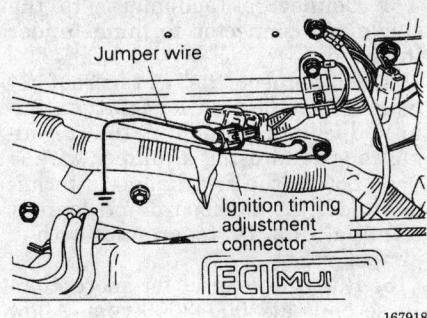

Ignition timing adjusting connector identification — 1993 Galant 2.0L (4G63) engine

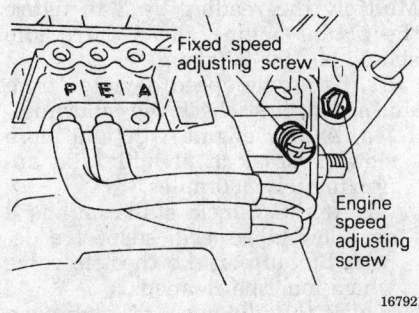

Engine speed adjusting screw identification (8 valve engines) — 1993 Galant 2.0L (4G63) engine

15. Restart the engine, allow to run for 5 minutes and check for good idle quality.

16 Valve Engines

1. Warm the engine to operating temperature, leave lights, electric cooling fan and accessories **OFF**. The transaxle should be in **N**. The steering wheel in a neutral position for vehicles with power steering.
2. Check the ignition timing and adjust, if necessary. Be sure to ground ignition timing adjustment connector.
3. Connect a tachometer to the single pin connector terminal under the hood.
4. Run the engine for more than 10 seconds at 2000–3000 rpm. Allow the engine to idle for 2 minutes.
5. Disconnect the waterproof female connector used for ignition timing adjustment. Connect this terminal to ground using a jumper wire.
6. Locate the self-diagnosis terminal under the dashboard and connect terminal No. **10** to ground with a jumper wire.
7. Start the engine and allow to idle. Check that the basic idle speed is between 700–800 rpm. If the idle

speed deviates from this speed, check the following:

   a. A new engine will idle more slowly. Break-in should take approximately 300 miles.
   b. If the vehicle stalls or has a very low idle speed, suspect a deposit buildup on the throttle valve which must be cleaned.
   c. If the idle speed is high even though the speed adjusting screw is fully closed, check that the idle position switch (fixed speed adjusting screw) position has changed. If so adjust the fixed SAS,
   d. If the fixed SAS screw requires adjustment, turn in the fixed SAS until the engine speed rises. Then back out the fixed SAS until the "Touch Point", where the engine speed does not fall any longer, is found. Back out the fixed SAS an additional ½ turn from the touch point.
   e. If all prior checks are okay, adjust the engine speed to specification, using the engine speed adjusting screw.
   f. If after all these checks the idle is still out of specification, it is probable that there is leakage resulting from deterioration of the Fast-Idle Air Valve (FIAV), and the throttle body will need to be replaced.
8. Turn the ignition switch **OFF** and stop the engine. Disconnect the jumper wire from the diagnosis connector, disconnect the jumper wire from the ignition timing connector and reconnect the waterproof connector. Disconnect the tachometer.
9. Restart the engine, allow to run for 5 minutes and check for good idle quality.

1994–97 Galant 2.4L (4G64) Engine

NOTE: The idle speed is controlled electronically and adjustment is usually unnecessary. However, this procedure must be

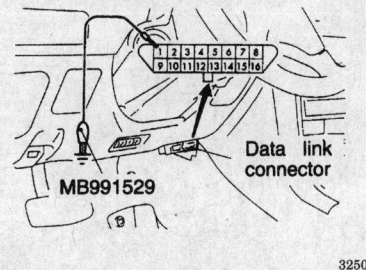

Diagnostic connector and terminal identification — 1994–97 Galant 2.4L (4G64) engine

performed when the Idle Speed Control (ISC) motor, Throttle Position Sensor (TPS), or throttle body assembly are replaced.

1. Warm the engine to operating temperature, leave lights, electric cooling fan and accessories **OFF**. The transaxle should be in **N**. The steering wheel in a neutral position for vehicles with power steering.
2. Check the ignition timing and adjust, if necessary. Be sure to ground ignition timing adjustment connector.
3. Connect a tachometer to the single pin connector terminal under the hood.
4. Run the engine for more than 10 seconds at 2000–3000 rpm. Allow the engine to idle for 2 minutes.
5. Disconnect the waterproof female connector used for ignition timing adjustment. Connect this terminal to ground using a jumper wire.
6. Locate the self-diagnosis terminal under the dashboard and connect terminal No. **1** to ground with a jumper wire.
7. Start the engine and allow to idle. Check that the basic idle speed is between 700–800 rpm on SOHC engines, and 750–850 on DOHC engines. If the idle speed deviates from this speed, check the following:
   a. A new engine will idle more slowly. Break-in should take approximately 300 miles.
   b. If the vehicle stalls or has a very low idle speed, suspect a deposit buildup on the throttle valve which must be cleaned.
   c. If the idle speed is high even though the speed adjusting screw is fully closed, check that the idle position switch (fixed speed adjusting screw) position has changed. If so adjust the fixed SAS.

NOTE: The fixed SAS adjustment is preset at the factory and should not be tampered with during normal adjustments.

8. If the fixed SAS screw requires adjustment, adjust as follows:
   a. Loosen the SAS locknut.
   b. Turn the SAS adjusting screw counterclockwise, until the throttle valve is fully closed.
   c. Turn the SAS adjusting screw clockwise until the throttle valve just begins to open. From this point, turn the SAS adjusting screw an additional 1–¼ turns clockwise.
   d. Hold the SAS adjusting screw and tighten the locknut.
9. If all prior checks are okay, adjust the engine speed to specification,

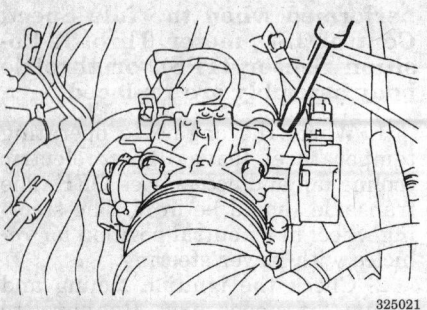

Engine speed adjusting screw location — 1994–97 Galant 2.4L (4G64) engine

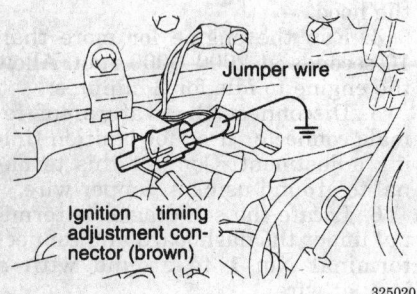

Ignition timing adjusting connector identification — 1994–97 Galant 2.4L (4G64) engine

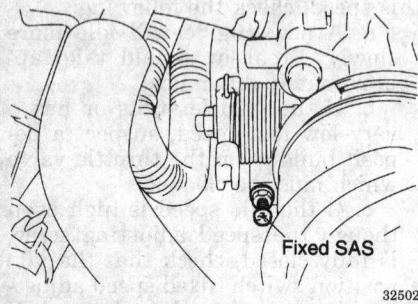

Fixed SAS adjusting screw location — 1994–97 Galant 2.4L (4G64) engine

using the engine speed adjusting screw.

10. If after all these checks the idle is still out of specification, it is probable that there is leakage resulting from deterioration of the Fast-Idle Air Valve (FIAV), and the throttle body will need to be replaced.

11. Turn the ignition switch **OFF** and stop the engine. Disconnect the jumper wire from the diagnosis connector, disconnect the jumper wire from the ignition timing connector and reconnect the waterproof connector. Disconnect the tachometer.

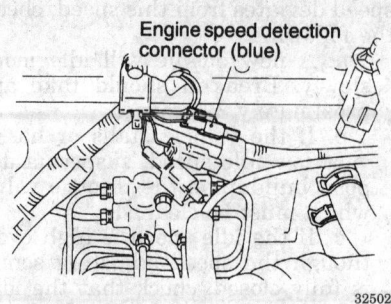

Tachometer connection point — 1994–97 Galant 2.4L (4G64) engine

12. Restart the engine, allow to run for 5 minutes and check for good idle quality.

Diamante and 3000GT

1993 Vehicles

1. Warm the engine to operating temperature, leave lights, electric cooling fan and accessories **OFF**. The transaxle should be in **N**. The steering wheel in a neutral position for vehicles with power steering.

2. Check the ignition timing and adjust, if necessary. Be sure to ground ignition timing adjustment connector.

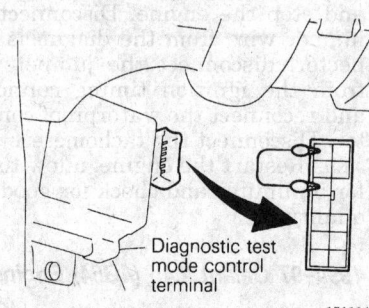

Diagnostic connector location — Diamante 3.0L (6G72) engine

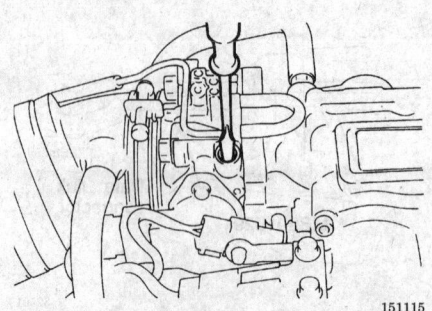

Idle speed adjustment point — Diamante and 3000GT with 3.0L (6G72) engine

3. Connect a tachometer to the single pin connector terminal under the hood.

4. Disconnect the waterproof female connector used for ignition timing adjustment. Connect this terminal to ground using a jumper wire.

5. Locate the self-diagnosis terminal under the dashboard and connect terminal No. **10** to ground with a jumper wire.

6. Run the engine for more than 10 seconds at 2000–3000 rpm. Allow the engine to idle for 2 minutes. Check that the basic idle speed is at specification. The tachometer reading will be $\frac{1}{3}$ of the actual engine speed. Multiply the reading by 3 to figure the actual engine speed Curb idle should be 700–800 rpm.

7. If the idle speed deviates from this speed, first check the following:
   a. A new engine will idle more slowly. Break-in should take approximately 300 miles.
   b. If the vehicle stalls or has a very low idle speed, suspect a deposit buildup on the throttle valve which must be cleaned.
   c. If the idle speed is high even though the speed adjusting screw is fully closed, check that the idle position switch (fixed speed adjusting screw) position has changed. If so, adjust the idle position switch.
   d. If after all these checks the idle is still out of specification, adjust the idle speed using the engine speed adjusting screw.

8. Turn the ignition switch **OFF** and stop the engine. Disconnect the jumper wire from the diagnosis connector, disconnect the jumper wire from the ignition timing connector and reconnect the waterproof connector. Disconnect the tachometer.

9. Restart the engine, allow to run for 5 minutes and check for good idle quality.

1994–97 Vehicles

1. Warm the engine to operating temperature, leave lights, electric cooling fan and accessories **OFF**. The transaxle should be in **N**. The steering wheel in a neutral position for vehicles with power steering.

2. Check the ignition timing and adjust, if necessary. Be sure to ground ignition timing adjustment connector.

3. Connect a tachometer to the single pin connector terminal under the hood.

4. Disconnect the waterproof female connector used for ignition timing adjustment. Connect this terminal to ground using a jumper wire.

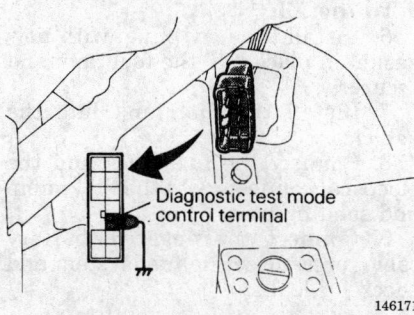

Tachometer connection point — Diamante 3.0L (6G72) DOHC engine

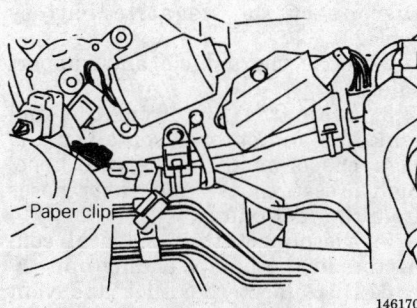

Diagnostic terminal identification — 1993 3000GT 3.0L (6G72) engines

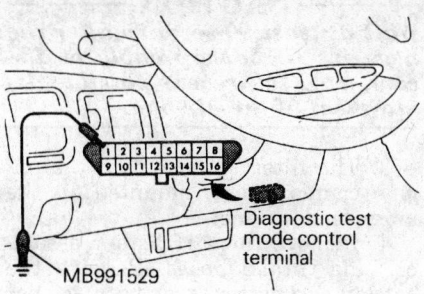

Diagnostic terminal identification — 1994–97 3000GT 3.0L (6G72) engine

Tachometer connector location — 3000GT 3.0L (6G72) engines

5. Locate the self-diagnosis terminal under the dashboard and connect terminal No. **1** to ground with a jumper wire.

6. Run the engine for more than 10 seconds at 2000–3000 rpm. Allow the engine to idle for 2 minutes. Check the idle rpm. Curb idle should be 700–800 rpm.

7. If the idle speed deviates from this speed, first check the following:

a. A new engine will idle more slowly. Break-in should take approximately 300 miles.

b. If the vehicle stalls or has a very low idle speed, suspect a de-

posit buildup on the throttle valve which must be cleaned.

c. If the idle speed is high even though the speed adjusting screw is fully closed, check that the idle position switch (fixed speed adjusting screw) position has changed. If so, adjust the idle position switch.

d. If after all these checks the idle is still out of specification, adjust the idle speed using the engine speed adjusting screw.

8. Turn the ignition switch **OFF** and stop the engine. Disconnect the jumper wire from the diagnosis connector, disconnect the jumper wire from the ignition timing connector and reconnect the waterproof connector. Disconnect the tachometer.

9. Restart the engine, allow to run for 5 minutes and check for good idle quality.

Mixture

ADJUSTMENT

Air/Fuel mixture is controlled by the ECU and is not adjustable.

Fuel Filter

REMOVAL AND INSTALLATION

——— **CAUTION** ———
Fuel injection systems remain under pressure, after the engine has been turned OFF. Properly relieve fuel pressure before disconnecting any fuel lines. Failure to do so may result in fire or personal injury.

——— **CAUTION** ———
Do not use conventional fuel filters, hoses or clamps when servicing fuel injection systems. They are not compatible with the injection system and could fail, caus-

ing personal injury or damage to the vehicle. Use only hoses and clamps specifically designed for fuel injection.

Precis, Mirage, Galant and 1993–95 Eclipse

A replaceable fuel filter is located in the engine compartment.

1. Properly relieve the fuel system pressure.

2. Disconnect the negative battery cable.

3. If necessary, remove the air intake hose and the battery.

NOTE: Wrap shop towels around the fitting that is being disconnected to absorb residual fuel in the lines.

4. Hold the fuel filter nut securely with a backup or spanner wrench. Cover the hoses with shop towels and remove the eye bolt. Discard the gaskets.

5. Separate the flare nut connection at the filter. Discard the gaskets.

6. Remove the mounting bolts and the fuel filter from the vehicle.

To install:

7. If equipped with flare fitting, tighten the fitting by hand before installing the filter to the vehicle.

8. Install the filter to its bracket only finger-tight. Movement of the filter will ease attachment of the fuel lines.

9. Install new gaskets and connect the high pressure hose and eye bolt, then the main pipe. While holding the fuel filter nut, tighten the eye bolts to 22 ft. lbs. (30 Nm). Tighten the flare nut to 27 ft. lbs. (37 Nm).

10. Tighten the filter mounting bolts fully.

11. Install the air intake hose and the battery.

12. Connect the negative battery cable, install the fuel filler cap, turn the key to the **ON** position to pressurize the fuel system and check for leaks.

13. Release the fuel pressure and repair leaks as required.

1996–97 Eclipse

1. Properly relieve the fuel system pressure.

2. Disconnect the negative battery cable. Wait at least 90 seconds before performing any work.

3. On turbo models and models equipped with the 2.4L engine, remove the battery and the air intake hose.

4. Raise and safely support the vehicle.

--- **WARNING** ---

As there will be some pressure remaining in the fuel pipe line, cover it with a shop towel to prevent fuel from spraying out.

5. While holding the fuel filter with a suitable wrench, remove the eye bolt attaching the high pressure fuel line to the fuel filter. Discard the two washers.

6. Hold the fuel filter with a suitable wrench and remove the main fuel pipe from the fuel filter. On 2.0L non-turbo models, remove the fuel pipe from the connector.

7. On the 2.0L non-turbo engine models, remove the clamp and the hose from the fuel pressure regulator.

8. Remove the fuel filter mounting bracket bolts and remove the fuel filter.

9. Remove the bracket screw and remove the fuel filter from the mounting bracket.

10. On 2.0L non-turbo models, remove the following from the filter:
   a. The eye bolt and washer.
   b. The fuel connector and washer with the fuel pressure regulator.

To install:

11. On 2.0L non-turbo models, install the fuel connector with the fuel pressure regulator to the filter with two new washers and torque the eye bolt to 22 ft. lbs. (36 Nm).

12. Install the fuel filter to the mounting bracket with the screw.

13. Reinstall the fuel filter to the vehicle with the bracket mounting bolts.

14. Reconnect the main fuel pipe to the fuel filter connector or the filter itself. Torque the flare nut to 27 ft. lbs. (36 Nm).

15. On the 2.0L non-turbo engine models reconnect the hose and clamp to the fuel pressure regulator.

16. Reconnect the high pressure fuel hose, using two new washers, to the fuel filter with the eye bolt. Torque the eye bolt to 22 ft. lbs. (29 Nm).

17. Safely lower the vehicle.

18. On 2.0L turbo and 2.4L engine models, reinstall the battery and the air intake hose.

19. Reconnect the negative battery cable, start the engine and check for fuel leaks.

Diamante and 3000GT

1. Properly relieve the fuel pressure.

2. Disconnect the negative battery cable.

--- **CAUTION** ---

Wait at least 90 seconds after the negative (-) battery cable is disconnected to prevent possible deployment of the air bag.

3. The filter is located in the engine compartment, mounted on the inner fender panel.

4. Remove the air cleaner assembly and intake hoses. Remove the battery and battery tray with washer tank.

5. Separate the flare nut connection at the line.

6. Hold the fuel filter nut securely with a backup or spanner wrench. Cover the hoses with shop towels and remove the eye bolts. Discard the gaskets.

7. Remove the mounting bolts and remove the fuel filter from the vehicle.

To install:

8. Install a new gaskets or O-rings whenever fuel connections have been disassembled.

9. Install the filter to its bracket only finger-tight. Movement of the filter will ease attachment of the fuel lines.

10. Install new gaskets and connect the high pressure hose and eye bolt, then the main pipe and eye bolt. While holding the fuel filter nut, tighten the eye bolts to 22 ft. lbs. (30 Nm). Tighten the flare nut to 25 ft. lbs. (35 Nm).

11. Tighten the mounting bolts fully.

12. Install the air cleaner assembly, battery and battery tray with washer tank, if removed.

13. Connect the negative battery cable, install the fuel filler cap, turn the key to the **ON** position to pressurize the fuel system and check for leaks. Release the fuel pressure and repair leaks as required.

Fuel Pump

REMOVAL AND INSTALLATION

--- **CAUTION** ---

Fuel injection systems remain under pressure after the engine has been turned OFF. Properly relieve fuel pressure before disconnecting any fuel lines. Failure to do so may result in fire or personal injury.

NOTE: Cover all fuel hose connections with a shop towel, prior to disconnecting, to prevent

splash of fuel that could be caused by residual pressure remaining in the fuel line.

Precis

1. Properly relieve fuel system pressure.

2. Disconnect the negative battery cable.

3. Raise the vehicle and support it safely.

4. Remove the fuel tank from the vehicle. Disconnect the hoses at the pump.

5. Unbolt and remove the pump from the tank.

To install:

6. Install the new pump with new gasket in place into the fuel tank and secure.

7. Install the fuel tank into the vehicle.

8. Connect the fuel lines and the electrical connectors to the fuel pump and sending unit.

9. Connect the negative battery cable, pressurize the fuel system and check for leaks.

Mirage

1. Properly relieve the fuel system pressure using proper procedures. Disconnect the negative battery cable.

2. Raise the vehicle and support safely.

3. Drain the fuel from the fuel tank into an approved container.

4. Disconnect the return hose, high pressure hose and vapor hoses from the fuel pump.

5. Disconnect the electrical connectors at the pump/sending unit.

6. Disconnect the filler and vent hoses.

7. Place a transmission jack under the center of the fuel tank and apply a slight upward pressure. Remove the fuel tank strap retaining nut.

8. Lower the tank slightly and disconnect any remaining electrical or hose connectors at the fuel tank.

9. Remove the fuel tank from the vehicle.

10. Remove the five nuts securing the access plate to the fuel tank and remove the pump assembly.

To install:

11. Install fuel pump into fuel tank, with new packing gasket, and tighten mounting nuts.

12. Install the fuel tank onto the transmission jack. Raise the tank in position under the vehicle. Leave enough clearance to attach the electrical and hose connections to the top of the fuel pump.

13. Attach all connections to the top of the tank.

14. Raise the tank completely and position the retainer straps around the fuel tank. Install new fuel tank self-locking nuts and tighten to 22 ft. lbs. (31 Nm).

15. Connect the return hose and high pressure hoses.

16. Install the vapor hose and the filler hose. Install the filler hose retainer screws to the fender, if removed.

17. Lower the vehicle and pour the drained fuel into the gas tank.

18. Connect the negative battery cable. Check the fuel pump for proper pressure and inspect the entire system for leaks.

1993–94 Eclipse and 1993 Galant

Front Wheel Drive (FWD)

1. Relieve the fuel system pressure using proper procedure. Disconnect negative battery cable.

2. Raise and safely support the vehicle.

3. Drain the fuel from the fuel tank into an approved gasoline container.

Remove the electrical connectors at the fuel pump. Make sure there is enough slack in the electrical harness of the fuel gauge unit to allow for the fuel tank to be lowered slightly. If not, label and disconnect the electrical harness at the fuel gauge unit.

4. Disconnect the high pressure fuel line connector at the pump.

5. Loosen self-locking nuts on tank support straps to the end of the stud bolts.

6. Remove the right side lateral rod attaching bolt and disconnect the arm from the right body coupling. Lower the lateral rod and suspend from the axle beam using wire.

7. Remove the holding bolt and gasket from the base of the tank.

8. Remove the fuel pump assembly.

To install:

9. Align the 3 projections on packing with the holes on the fuel pump and the nipples on the pump facing the same direction as before removal.

10. Install the holding bolt through the bottom of the tank. Make sure the gasket on the bolt is replaced and is not pinched during installation. Torque to 10 ft. lbs. (14 Nm).

11. Install the right side lateral rod and attaching bolt into the right body coupling. Tighten loosely only, at this time.

12. Tighten self-locking nuts on tank support straps until tank is

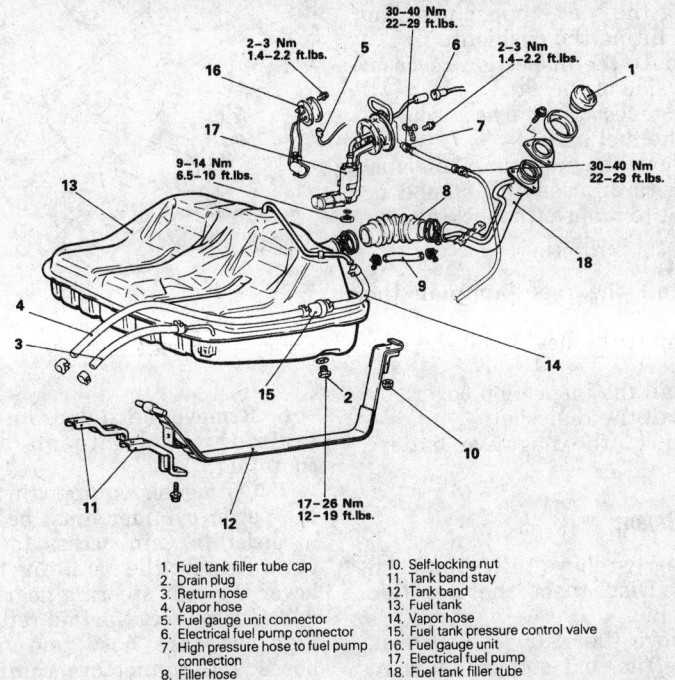

1. Fuel tank filler tube cap
2. Drain plug
3. Return hose
4. Vapor hose
5. Fuel gauge unit connector
6. Electrical fuel pump connector
7. High pressure hose to fuel pump connection
8. Filler hose
9. Vapor hose
10. Self-locking nut
11. Tank band stay
12. Tank band
13. Fuel tank
14. Vapor hose
15. Fuel tank pressure control valve
16. Fuel gauge unit
17. Electrical fuel pump
18. Fuel tank filler tube

165509

Exploded view of the fuel tank components — 1993 Galant FWD with 2.0L (4G64) SOHC engine

seated fully. Torque nuts to 22 ft. lbs. (31 Nm).

13. Install the high pressure fuel hose connector and tighten to 29 ft. lbs. (40 Nm).

14. Install the electrical connectors onto the fuel pump and gauge unit assemblies.

15. Lower the vehicle so the suspension supports the weight of the vehicle. Tighten the lateral rod attaching bolt to 58–72 ft. lbs. (80–100 Nm).

16. Refill the fuel tank with fuel drained during this procedure.

17. Connect the negative battery cable and check the entire system for proper operation and leaks.

All Wheel Drive (AWD)

1. Relieve the fuel system pressure using proper procedure. Disconnect negative battery cable.

2. The fuel pump is located in the fuel tank. Remove the hole cover located in the rear floor pan.

3. Partially drain the fuel tank into an approved gasoline container.

4. Remove the electrical connector from the fuel pump.

5. Remove the overfill limiter (two-way valve), as required.

6. Remove the high pressure fuel hose connector.

7. Remove the fuel pump and gauge assembly from the tank. Note positioning of pump prior to removal from tank.

To install:

8. Align the 3 projections on the packing with the holes on the fuel pump and the nipples on the pump facing the same direction as before removal. Install the retainers and tighten to 2 ft. lbs. (3 Nm).

9. Install the high pressure hose connection and tighten to 29 ft. lbs. (40 Nm).

10. Install the overfill limiter (two-way valve) and the electrical connector to the fuel pump.

11. Fill the fuel tank with the gasoline removed during this procedure.

12. Reconnect the negative battery cable and check the entire system for leaks.

13. Install MOPAR Rope Caulk Sealer part 4026044 or equivalent, to the rear floor pan and install the cover into place.

1995–97 Eclipse

1. Relieve the fuel system pressure.

2. Disconnect the negative battery cable.

3. Remove the rear seat cushion by pulling the seat stopper near the floor and lifting the cushion up.

4. Remove the inspection cover on the right side of the car.

5. Disconnect the harness connector and the fuel lines.

6. Remove the fuel pump assemble from the tank. Use MB991480 or equivalent to remove the locking ring on the AWD model.

To install:

7. Install the fuel pump in the tank.

8. Connect the hoses and the harness connector.

9. Install the inspection cover.

10. Install the rear seat.

11. Connect the negative battery cable.

1994–97 Galant

1. Properly relieve the fuel system pressure. Disconnect the negative battery cable.

2. Remove the rear seat cushion, by pulling the seat stopper outward and lifting the lower cushion upward.

3. Remove the access cover.

4. Disconnect the fuel pump wiring.

5. Disconnect the return hose and the high pressure fuel hose.

6. Remove the pump mounting nuts and remove the pump assembly.

To install:

7. Install the fuel pump assembly to the tank and tighten the retaining nuts to 22 inch lbs. (2.5 Nm).

NOTE: Tilt the float to the left of the vehicle, when installing the pump assembly.

8. Connect the high pressure hose, return hose and the fuel tank wiring.

9. Connect the negative battery cable.

10. Check the fuel pump for proper pressure and inspect the entire system for leaks.

11. Apply sealant to the access cover and install the cover.

12. Install the rear seat cushion.

Diamante

1. Properly relieve the fuel system pressure.

2. Disconnect the negative battery cable.

3. Raise the vehicle and support it safely.

4. Remove the left rear wheel well liner, if equipped.

5. Disconnect the center exhaust system from the main muffler. Disconnect the rear exhaust hangers, lower the system and secure aside.

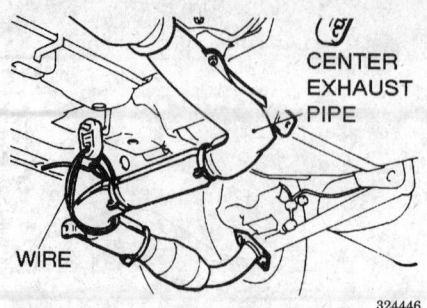

CENTER EXHAUST PIPE

WIRE

324446

Proper method of supporting rear exhaust system — Diamante 3.0L (6G72) engine

6. Remove the tank drain plug and drain the fuel into an approved container.

7. On models equipped with 4WS, the power cylinder must be lowered, in order to gain access to the fuel tank. Remove the mounting bolts and lower the rear steering gear.

8. Disconnect the fuel return hose, high pressure hose and all other hoses and connectors connected to the pump/sending unit.

— **CAUTION** —
Cover all fuel hose connections with a shop towel, prior to disconnecting, to prevent splash of fuel that could be caused by residual pressure remaining in the fuel line.

9. Disconnect the filler and vent hoses. Place a support under the tank and remove the retaining nuts.

10. Lower the tank from the vehicle.

11. Remove the fuel pump retaining nuts and remove the assembly from the tank.

To install:

12. Install the pump assembly to the tank and tighten the retaining nuts to 2 ft. lbs. (3 Nm).

13. Install the fuel tank and connect the filler and vent hoses.

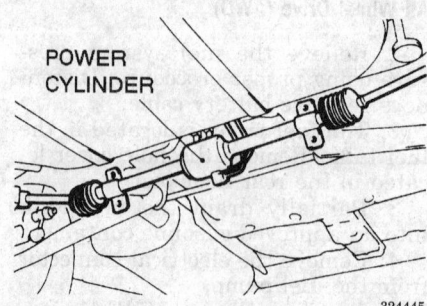

POWER CYLINDER

324445

Power cylinder identification — Diamante 3.0L (6G72) engine

Tighten the tank retaining nuts and bolts to 19 ft. lbs. (26 Nm).

14. Connect the return hose, high pressure hose and all other hoses and connectors connected to the pump/sending unit.

15. If equipped with 4WS, install the power cylinder unit to the crossmember and tighten the mounting bolts to 31 ft. lbs. (43 Nm).

16. Connect the exhaust pipe and secure in place with rear hangers.

17. Install the left rear wheel well liner, if removed.

18. Lower the vehicle and return fuel to the gas tank.

19. Connect the negative battery cable and check the entire system for proper operation and leaks.

3000GT

1. Relieve fuel system pressure. Remove the fuel filler cap.

2. Disconnect the negative battery cable.

— **CAUTION** —
Wait at least 90 seconds after the negative (-) battery cable is disconnected to prevent possible deployment of the air bag.

3. The fuel pump is located in the fuel tank. Drain the fuel from the fuel tank.

4. Remove the fuel gauge cover located in the rear floor pan.

5. Remove the fuel pump and gauge electrical connector. Remove the overfill limiter (two-way valve).

6. Disconnect both sides of the high pressure fuel hose. When disconnecting the fuel pump side of the hose, hold the pump side nut with a wrench while turning the nut on the hose side. This will prevent any damage that will occur to the fittings and the hoses if two wrenches are not used.

7. Remove the fuel pump and gauge assembly from the tank.

To install:

8. Align the three projections on the packing with the holes on the fuel pump and the nipples on the pump facing the same direction as before removal.

9. Temporarily tighten the flare nut on the high pressure hose by hand. Making sure the hose does not twist, tighten body side nut to 22 ft. lbs. (30 Nm) and the fuel pump side nut to 25 ft. lbs. (35 Nm).

10. Install the overfill limiter (2-way valve) with the long shouldered side of the valve facing the canister.

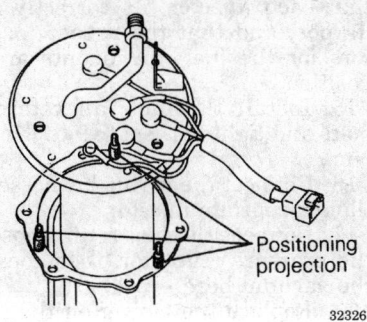

Align the positioning projections — 3000GT 3.0L (6G72) engine

11. Connect the electrical connector to the pump assembly.

12. Reconnect the negative battery cable and check the entire system for leaks.

13. Install sealer to the rear floor pan and install the cover into place.

Fuel Injector

REMOVAL AND INSTALLATION

—————— **CAUTION** ——————
Fuel injection systems remain under pressure, after the engine has been turned OFF. Properly relieve fuel pressure before disconnecting any fuel lines. Failure to do so may result in fire or personal injury.

Precis

1. Relieve the fuel system pressure.
2. Disconnect the negative battery cable.
3. Disconnect and remove the air intake hoses as required.
4. Wrap the connection with a shop towel and disconnect the high pressure fuel line at the fuel rail.
5. Disconnect the fuel return hose.
6. Disconnect the accelerator cable connection from the throttle body and position aside.
7. Disconnect the vacuum connection from the fuel pressure regulator.
8. Disconnect the electrical harness connector from each fuel injector.
9. Remove the injector rail retaining bolts. Make sure the rubber mounting insulators do not get lost.
10. Lift the rail assembly up and away from engine.
11. Remove the injectors from the rail by pulling gently. Discard the lower insulator. Check the resistance through the injector. The specification is 13–16 ohms at 70°F (20°C).

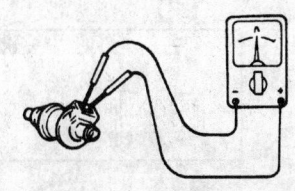

Check the resistance of the injector — 1993–94 Precis 1.5L (G4DJ) engine

To install:

12. Install a new grommet and O-ring to the injector. Coat the O-ring with light weight oil or gasoline.
13. While turning the injector to the left and right, install the injector to the fuel rail.
14. Install the fuel rail and injectors to the manifold. Make sure the rubber bushings are in place before tightening the mounting bolts.
15. Tighten the retaining bolts to 9 ft. lbs. (12 Nm).
16. Connect the electrical connectors to the injectors.
17. Replace the O-ring on the fuel pressure regulator, lightly lubricate and install on the delivery pipe. Connect the vacuum hose to the fuel pressure regulator.
18. Connect the fuel return hose.
19. Replace the O-ring on high pressure fuel line, lightly lubricate it and connect to delivery pipe.
20. Reconnect the accelerator cable to the throttle body and adjust to specifications.
21. Connect the negative battery cable and check the entire system for proper operation and leaks.

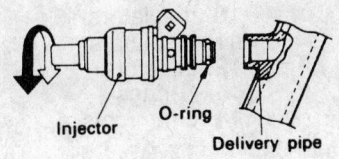

Install the injector to the fuel rail — 1993–94 Precis 1.5L (G4DJ) engine

Mirage 1.5L (4G15), 1.8L (4G93), Eclipse 1.8L (4G37), 2.0L (4G63) and 1993 Galant 2.0L (4G63) Engines

1. Relieve the fuel system pressure following proper procedures.
2. Disconnect the PCV hose from the valve cover. Also disconnect the breather hose at the opposite end of the valve cover.
3. Remove the bolts holding the high pressure fuel line to the fuel rail and disconnect the line. Be prepared to contain fuel spillage; plug the line to keep out dirt and debris.
4. Remove the vacuum hose from the fuel pressure regulator.
5. Disconnect the fuel return hose from the pressure regulator. Remove the fuel pressure regulator mounting bolts and remove from the fuel rail.
6. Label and disconnect the electrical connector from each injector.
7. Remove the bolt(s) holding the fuel rail to the manifold. Carefully lift the rail up and remove it with the injectors attached. Take great care not to drop an injector. Place the rail and injectors in a safe location on the workbench; protect the tips of the injectors from dirt and/or impact.
8. Remove and discard the injector insulators from the intake manifold. The insulators are not reusable.
9. Remove the injectors from the fuel rail by pulling gently in a straight outward motion. Make certain the grommet and O-ring come off with the injector.

To install:

10. Install a new insulator in each injector port in the manifold.
11. Remove the old grommet and O-ring from each injector. Install a new grommet and O-ring; coat the O-ring lightly with clean, thin oil.
12. If the fuel pressure regulator was removed, replace the O-ring with a new one and coat it lightly with clean, thin oil. Insert the regulator straight into the rail, then check that it can be rotated freely. If it does not rotate smoothly, remove it and inspect the O-ring for deformation or jamming. When properly installed, align the mounting holes and tighten the retaining bolts to 7 ft. lbs. (9 Nm). This procedure must be followed even if the fuel rail was not removed.
13. Install the injector into the fuel rail, constantly turning the injector left and right during installation. When fully installed, the injector should still turn freely in the rail. If it does not, remove the injector and inspect the O-ring for deformation or damage.
14. Install the delivery pipe and injectors to the engine. Make certain

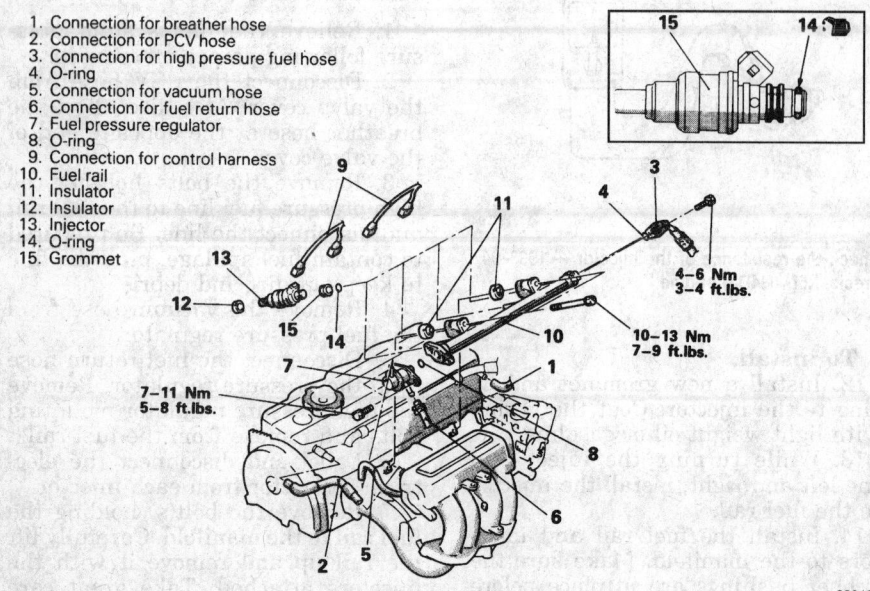

1. Connection for breather hose
2. Connection for PCV hose
3. Connection for high pressure fuel hose
4. O-ring
5. Connection for vacuum hose
6. Connection for fuel return hose
7. Fuel pressure regulator
8. O-ring
9. Connection for control harness
10. Fuel rail
11. Insulator
12. Insulator
13. Injector
14. O-ring
15. Grommet

4–6 Nm
3–4 ft.lbs.

10–13 Nm
7–9 ft.lbs.

7–11 Nm
5–8 ft.lbs.

330405

Fuel injector and related parts — Mirage 1.5L (4G15), 1.8L (4G93), Eclipse 1.8L (4G37), 2.0L (4G63) and 1993 Galant 2.0L (4G63) engines

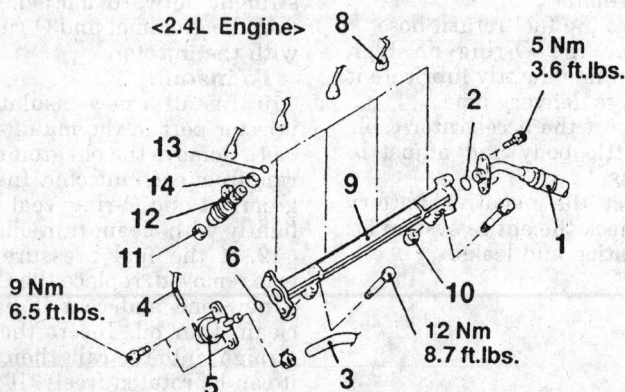

<2.4L Engine>

5 Nm
3.6 ft.lbs.

9 Nm
6.5 ft.lbs.

12 Nm
8.7 ft.lbs.

1. High-pressure fuel hose connection
2. O-ring
3. Fuel return hose connection
4. Vacuum hose connection
5. Fuel pressure regulator
6. O-ring
7. PCV hose

8. Injector connectors
9. Fuel rail
10. Insulators
11. Insulators
12. Injectors
13. O-rings
14. Grommets

319749

Fuel injectors — 1995–97 2.0L (4G63) and 2.4L (4G64) engines

that each injector fits correctly into its port and that the rubber insulators for the fuel rail mounts are in position.

15. Install the fuel rail retaining bolts and tighten them to 9 ft. lbs. (12 Nm).

16. Connect the wiring harnesses to the appropriate injector.

17. Connect the fuel return hose to the pressure regulator, then connect the vacuum hose.

18. Replace the O-ring on the high pressure fuel line, coat the O-ring lightly with clean, thin oil and install the line to the fuel rail. Tighten the mounting bolts to specifications.

19. Connect the PCV hose and the breather hose if they were disconnected.

20. Connect the negative battery cable. Pressurize the fuel system and inspect all connections for leaks.

1995–97 Eclipse

1. Relieve the fuel system pressure.

2. Disconnect the negative battery cable.

3. Label (if needed) and disconnect the cables from the spark plugs.

4. If necessary, remove the battery and air intake hose.

5. Disconnect the high pressure line from the fuel rail.

6. Disconnect the fuel return hose from the fuel rail.

7. Disconnect the vacuum hose from the fuel pressure regulator.

8. On 2.0L turbo engines, remove the PCV hose.

9. Disconnect the harness connectors from the fuel injectors.

10. Remove the fuel rail mounting bolts and pull the rail with the injectors out of the intake manifold.

11. Remove the fuel injector(s) from the rail.

To install:

12. Install new grommets and O-rings on the fuel injector(s). Apply a small amount of engine oil to the grommets and O-ring and install the injector in the fuel rail by turning the injector to the left and right.

—————— **WARNING** ——————
Do not let the engine oil get into the fuel rail.

13. Install the fuel rail on the engine. Turn the injectors to the left and right as needed to get the injectors into the intake manifold.

14. Connect the harness connectors to the fuel injectors.

15. On 2.0L turbo engines, reconnect the PCV hose.

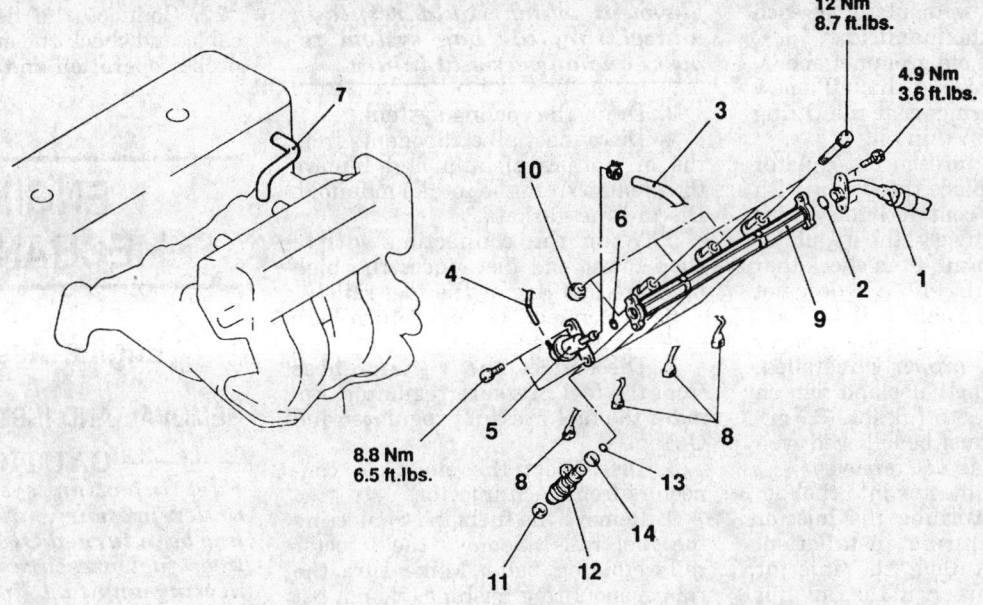

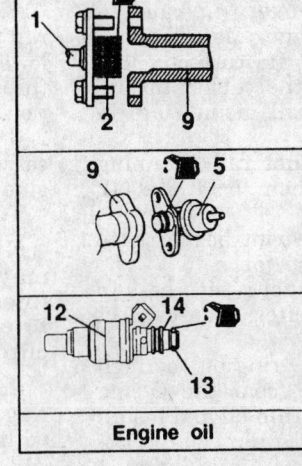

Engine oil

1. Fuel high pressure hose connection
2. O-ring
3. Fuel return hose connection
4. Vacuum hose connection
5. Fuel pressure regulator
6. O-ring
7. PCV hose
8. Injector connectors
9. Fuel rail
10. Insulators
11. Insulators
12. Injectors
13. O-rings
14. Grommets

319733

Fuel injectors and related components — 1995–97 Eclipse 2.0L (4G63) and 2.4L (4G64) engines

16. Connect the vacuum hose to the regulator.

17. Use a new O-ring and connect the high pressure hose to the fuel rail.

18. Connect the fuel return hose to the fuel rail.

19. If necessary, install the battery and the air intake hose.

20. Connect the plug wires to the plugs.

21. Reconnect the negative battery cable.

22. Pressurize the fuel system and check for leaks.

1994–97 Galant

1. Relieve the fuel system pressure following proper procedures.

2. Label and disconnect the spark plug wires. Position the wires aside.

3. Disconnect the PCV hose from the valve cover.

4. Remove the bolts holding the high pressure fuel line to the fuel rail and disconnect the line. Be prepared to contain fuel spillage; plug the line to keep out dirt and debris.

5. Remove the vacuum hose from the fuel pressure regulator.

6. Disconnect the fuel return hose from the pressure regulator. Remove the fuel pressure regulator mounting bolts and remove from the fuel rail.

7. Label and disconnect the electrical connector from each injector.

8. Remove the bolt(s) holding the fuel rail to the manifold. Carefully lift the rail up and remove it with the injectors attached. Take great care not to drop an injector. Place the rail and injectors in a safe location on the workbench; protect the tips of the injectors from dirt and/or impact.

9. Remove and discard the injector insulators from the intake manifold. The insulators are not reusable.

10. Remove the injectors from the fuel rail by pulling gently in a straight outward motion. Make certain the grommet and O-ring come off with the injector.

To install:

11. Install a new insulator in each injector port in the manifold.

12. Remove the old grommet and O-ring from each injector. Install a new grommet and O-ring; coat the O-ring lightly with clean, thin oil.

13. If the fuel pressure regulator was removed, replace the O-ring with a new one and coat it lightly with clean, thin oil. Insert the regulator straight into the rail, then check that it can be rotated freely. If it does not rotate smoothly, remove it and inspect the O-ring for deformation or jamming. When properly installed, align the mounting holes and tighten the retaining bolts to 7 ft. lbs. (9 Nm). This procedure must be followed even if the fuel rail was not removed.

14. Install the injector into the fuel rail, constantly turning the injector left and right during installation. When fully installed, the injector should still turn freely in the rail. If it does not, remove the injector and inspect the O-ring for deformation or damage.

15. Install the delivery pipe and injectors to the engine. Make certain that each injector fits correctly into its port and that the rubber insulators for the fuel rail mounts are in position.

16. Install the fuel rail retaining bolts and tighten them to 9 ft. lbs. (12 Nm).

17. Connect the wiring harnesses to the appropriate injector.

18. Connect the fuel return hose to the pressure regulator, then connect the vacuum hose.

19. Replace the O-ring on the high pressure fuel line, coat the O-ring lightly with clean, thin oil and install the line to the fuel rail. Tighten the mounting bolts to 4 ft. lbs. (6 Nm).

20. Connect the PCV hose and spark plug wires.

21. Connect the negative battery cable. Pressurize the fuel system and inspect all connections for leaks.

Diamante and 3000GT

1. Relieve the fuel system pressure.

2. Disconnect the negative battery cable.

— **CAUTION** —
Work MUST NOT be started until at least 90 seconds after the ignition switch is turned to the LOCK position and the negative battery cable is disconnected from the battery. This will allow time for the air bag system backup power supply to deplete its stored energy preventing accidental air bag deployment which could result in unnecessary air bag system repairs and/or personal injury.

3. Drain the cooling system.

4. Disconnect all components from the air intake plenum and remove the plenum from the intake manifold. Discard the gaskets.

5. Wrap the connection with a shop towel and disconnect the high pressure fuel line at the fuel rail.

6. Disconnect the fuel return hose and remove the O-ring.

7. Disconnect the vacuum hose from the fuel pressure regulator. Remove the fuel pressure regulator and O-ring.

8. Disconnect the electrical connectors from each injector.

9. Remove the fuel pipe connecting the fuel rails. Remove the injector rail retaining bolts. Make sure the rubber mounting bushings do not get lost.

10. Lift the rail assemblies up and away from the engine.

11. Remove the injectors from the rail by pulling gently. Discard the lower insulator. Check the resistance through the injector. The specification for 3.0L turbocharged engine is 2–3 ohms at 68°F (20°C). The specification for non-turbocharged 3.0L engine is 13–16 ohms at 68°F (20°C).

To install:

NOTE: Some of the vehicles may have a clip that secures the injector to the fuel rail. Be sure to remove or install the injector clip where necessary.

12. Install a new grommet and O-ring to the injector. Coat the O-ring with light oil.

13. Install the injector to the fuel rail.

14. Replace the seats in the intake manifold. Install the fuel rails and injectors to the manifold. Make sure the rubber bushings are in place before tightening the mounting bolts.

15. Tighten the retaining bolts to 7–9 ft. lbs. (10–13 Nm). Install the fuel pipe with new gasket.

16. Connect the electrical connectors to the injectors.

17. Replace the O-ring, lightly lubricate it and connect the fuel pressure regulator.

18. Connect the fuel return hose.

19. Replace the O-ring, lightly lubricate it and connect the high pressure fuel line.

20. Using new gaskets, install the intake plenum and all related items. Torque the plenum mounting bolts to 13 ft. lbs. (18 Nm).

21. Fill the cooling system.

22. Connect the negative battery cable and check the entire system for proper operation and leaks.

ENGINE MECHANICAL

Engine Assembly

REMOVAL AND INSTALLATION

— **CAUTION** —
Fuel injection systems remain under pressure, after the engine has been turned OFF. Properly relieve fuel pressure before disconnecting any fuel lines. Failure to do so may result in fire or personal injury.

Precis

The manufacturer recommends that the engine and transaxle be removed as a unit on all models.

1. Properly relieve the fuel system pressure.

2. Matchmark the hood and hinges and remove the hood assembly. Remove the air cleaner assembly and all adjoining air intake duct work.

3. Remove the undercover if equipped.

4. Disconnect the purge control vacuum hose from the purge valve. Remove the purge control valve mounting bracket. Remove the windshield washer reservoir, radiator tank and carbon canister.

5. Drain the cooling system.. Disconnect the upper and lower radiator hoses and then remove the radiator and fan assembly. Be sure to disconnect the fan wiring harness prior to removal.

6. Detach the connectors for the back-up lights and engine harness, located near the battery tray. If equipped with a 5 speed transaxle, disconnect the select control valve connector. Disconnect the alternator harness connectors and the oil pressure sending unit.

7. Disconnect and cap the automatic transaxle oil cooler hoses.

8. Label and disconnect all low tension wires and the one high tension wire going to the coil from the distributor. Disconnect the engine ground.

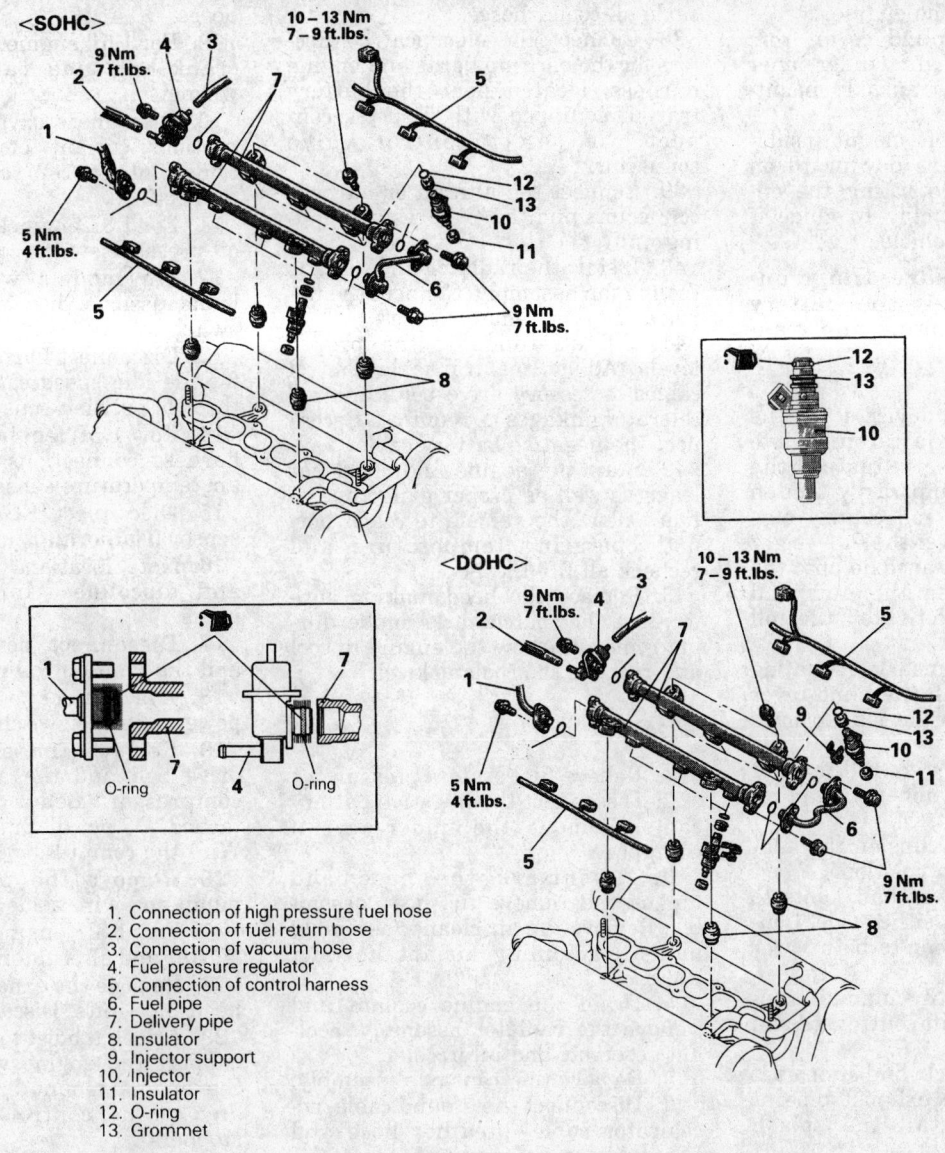

<SOHC>

9 Nm
7 ft.lbs.

10 – 13 Nm
7 – 9 ft.lbs.

5 Nm
4 ft.lbs.

9 Nm
7 ft.lbs.

<DOHC>

O-ring

O-ring

1. Connection of high pressure fuel hose
2. Connection of fuel return hose
3. Connection of vacuum hose
4. Fuel pressure regulator
5. Connection of control harness
6. Fuel pipe
7. Delivery pipe
8. Insulator
9. Injector support
10. Injector
11. Insulator
12. O-ring
13. Grommet

323222

Fuel rail assembly — Diamante and 3000GT 3.0L (6G72) engine

9. Disconnect the brake booster vacuum hose at the intake manifold.

10. Wrap shop towels around the fuel fittings, then disconnect the fuel supply, return and vapor hoses at the side of the engine.

11. Disconnect the heater hoses from the side of the engine. Disconnect the accelerator cable at the engine side.

12. Disconnect the clutch control cable for manual transaxle or transaxle shifter control cable for automatic transaxle from the transaxle.

13. Unscrew and disconnect the speedometer cable at the transaxle.

14. Remove the air conditioner drive belt and the air conditioning compressor. Leave the hoses attached. Do not discharge the system. Wire the compressor aside.

15. Raise and safely support the vehicle. Remove the splash shield. Remove the drain plug and drain the transaxle fluid. Disconnect the exhaust pipe at the manifold and suspend the pipe securely with wire.

16. If equipped with a manual transaxle, remove the shift control rod and extension rod.

17. Disconnect the stabilizer bar at both lower control arms. Remove the bolts that attach the lower ball joints to the control arms on either side.

18. Remove the front halfshafts. Then, seal off the openings in the transaxle to prevent damage caused by the introduction of foreign substances into the transaxle.

19. Attach an engine lift, via chains or cables, to both the engine lifting hooks. Put just a little tension on the cables. Then, remove the nut and bolt from the front roll stopper; unbolt the brace from the top of the engine damper.

20. Separate the rear roll stopper from the No. 2 crossmember. Remove the attaching nut from the left mount insulator bolt, but do not remove the bolt.

21. Raise the engine just enough that the lifting device is supporting

its weight. Check that everything is disconnected from the engine.

22. Remove the blind cover from the inside of the right fender inner shield. Remove the transaxle mounting bracket bolts.

23. Remove the left mount insulator bolt. Then, press downward on the transaxle while lifting the engine/transaxle assembly to guide it up and out of the vehicle.

NOTE: Make sure the transaxle does not hit the battery bracket during engine and transaxle removal.

To install:

24. Using a lifting device, lower the engine and transaxle carefully into position and loosely install the mounting bolts. Temporarily tighten the front and rear roll control rods mounting bolts. Lower the full weight of the engine and transaxle onto the mounts and tighten the nuts and bolts. Loosen and retighten the roll control rods.

25. Install the transaxle mounting bracket bolts. Install the blind cover to the inside of the right fender inner shield.

26. Assemble the rear roll stopper to the No. 2 crossmember. Install retaining nut and bolt.

27. Install new circlips on the half-shafts and install in position.

28. Attach the lower ball joints to the control arms on either side. Connect the stabilizer bar to both lower control arms.

29. If equipped with a manual transaxle, install the shift control rod and extension rod.

30. Raise the vehicle and support it safely. Connect the exhaust pipe at the manifold. Install the splash shield.

31. Connect the speedometer cable at the transaxle. Connect the air conditioning compressor to the mounting bracket.

32. Install the clutch control cable, for manual transaxle, or shifter control cable, for automatic transaxle, to the transaxle.

33. Lower the vehicle. Connect the heater hoses to the engine. Connect the accelerator cable at the engine side.

34. Connect the fuel supply, return and vapor hoses with new O-rings installed.

35. Connect the brake booster vacuum hose at the intake manifold.

36. Connect all low tension wires and the high tension wire going to the coil from the distributor. Connect the engine ground and the fuel injectors.

37. Connect the automatic transaxle oil cooler hoses.

38. Connect the electrical connectors for the back-up lights and engine harness, located near the battery tray. If equipped with 5 speed, connect the select control valve connector.

39. Connect the alternator harness connectors and the oil pressure sending unit.

40. Install the radiator and electric cooling fan assembly. Connect the upper and lower radiator hoses.

41. Fill all fluids to the proper levels. Adjust the transaxle control cables, accessory drive belts and accelerator linkages as required. Reconnect the negative battery cable.

42. Start the engine and check for leaks as well as proper gauge operation. Allow the vehicle to reach normal operating temperature and recheck all fluid levels.

43. Replace the hood making sure to align the matchmarks made during removal. Allow the engine to cool and recheck the coolant level.

Mirage

1. Relieve fuel system pressure.

2. Disconnect the negative battery cable. Remove the undercover if equipped.

3. Matchmark the hood and hinges and remove the hood assembly. Remove the air cleaner assembly and all adjoining air intake duct work.

4. Drain the engine coolant and remove the radiator assembly, coolant reservoir and intercooler.

5. Remove the transaxle assembly.

6. Disconnect the ground cable, accelerator cable, breather hose and heater hose connections from the engine.

7. Note locations and remove vacuum hoses from engine. Be sure to disconnect brake booster vacuum supply.

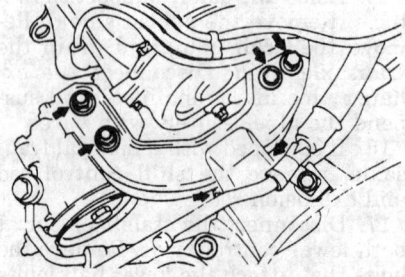

Engine mounting bracket and insulator bolts — Precis

8. Disconnect fuel feed and return hoses.

9. For 1.5L engines, disconnect the crankshaft and camshaft sensor wiring.

10. Disconnect oxygen sensor connections, coolant temperature gauge and coolant temperature sensor connections.

11. For 1.8L engines, disconnect the oil pressure switch connection.

12. On models with automatic transmissions, disconnect the thermo switch.

13. Disconnect harness connections for the idle speed control motor and throttle position sensor.

14. For 1.5L engines, disconnect harness connections for the intake air temperature sensor.

15. Disconnect EGR temperature sensor (California).

16. Note locations for reassembly and disconnect injector harness plugs.

17. Disconnect power transistor and the ignition coil connections.

18. Disconnect alternator and power steering switch wiring.

19. Remove the air conditioner drive belt and the air conditioning compressor. Leave the hoses attached. Do not discharge the system. Wire the compressor aside.

20. Remove the power steering pump and wire aside.

21. For 1.8L engines, remove the starter and alternator harness clamp.

22. Remove the exhaust manifold to head pipe nuts. Discard the gasket.

23. Attach a hoist to the engine and support the engine weight. Remove the engine mount bracket. Remove any torque control brackets (roll stoppers).

24. Remove the engine assembly from the vehicle.

To install:

25. Install the engine and secure in position. The front lower mount through-bolt nut should not be tightened until the full weight of the engine is on the mount. Tighten through-bolt to 72 ft. lbs. (100 Nm) and bracket mounting bolts to 42 ft. lbs. (58 Nm). Tighten bracket mounting nut to 38 ft. lbs. (53 Nm).

26. Using a new gasket, position exhaust pipe onto the manifold and tighten the flange nuts to 36 ft. lbs. (50 Nm).

27. Install power steering pump, alternator and air conditioner compressor. Install and adjust drive belts, tighten all mounting bolts.

28. Connect alternator and power steering wiring.

29. For 1.8L engines, secure the alternator and starter harness clamp.

30. Connect the ignition coil and power transistor connections.

31. Connect fuel injector harness connections.

32. On California models, connect EGR temperature sensor plug

33. For 1.5L engines, connect harness wiring for the intake air temperature sensor.

34. Connect wiring for idle speed control motor and throttle position sensor.

35. On automatic transmission models, connect the thermo switch.

36. For 1.8L engines, connect the oil pressure switch wiring.

37. Connect oxygen sensor, coolant temperature gauge and coolant temperature sensor.

38. For 1.5L engines, connect the crankshaft and camshaft sensor wiring.

39. Using new O-rings, connect fuel feed hose and tighten bolts to 44 inch lbs. (5 Nm).

40. Using a new hose clamp, connect the fuel return hose.

41. Connect noted vacuum hoses and connect brake booster vacuum supply.

42. Connect the breather hose, heater hoses, accelerator cable and ground cables. Inspect accelerator cable for proper adjustment.

43. Install the transaxle assembly.

44. Install radiator assembly and refill the cooling system, engine oil and transmission oil.

45. Install air cleaner and hood assembly.

46. Connect negative battery cable and run engine.

47. Inspect all connections and check all fluid levels.

1993–94 Eclipse

1.8L (4G37) and 2.0L (4G63) Engines

The following procedure can be used on all models. Slight differences may occur, due to options and other equipment variations, but the basic procedure should cover all models.

1. Properly relieve the fuel system pressure as follows.

2. Disconnect the negative battery cable.

3. Remove the engine undercover, if equipped.

4. Matchmark the hood and hinges and remove the hood assembly. Remove the air cleaner assembly and all adjoining air intake duct work.

5. Drain the engine coolant and remove the radiator assembly, coolant reservoir and intercooler.

6. Remove the transaxle and transfer case if equipped with AWD.

7. Tag and disconnect the accelerator cable, heater hoses, brake vacuum hose, connection for vacuum hoses, high pressure fuel line, fuel return line, oxygen sensor connection, coolant temperature gauge connection, coolant temperature sensor connector, connection for thermo switch sensor, if equipped with automatic transaxle, the connection for the idle speed control, the motor position sensor connector, the throttle position sensor connector, the EGR temperature sensor connection (California vehicles), the fuel injector connectors, the power transistor connector, the ignition coil connector, the condenser and noise filter connector, the distributor and control harness, the connections for the alternator and oil pressure switch wires.

8. Remove the air conditioner drive belt and the air conditioning compressor. Leave the hoses attached. Do not discharge the system. Wire the compressor aside.

9. Remove the power steering pump and wire aside.

10. Remove the exhaust manifold to head pipe nuts. Discard the gasket.

11. Attach a hoist to the engine and take up the engine weight. Remove the engine mount bracket. Remove any torque control brackets (roll stoppers). Note that some engine mount pieces have arrows on them for proper assembly. Double check that all cables, hoses, harness connectors, etc., are disconnected from the engine. Lift the engine slowly from the engine compartment.

To install:

12. Install the engine and secure in position. The front lower mount through-bolt nut should not be tightened until the full weight of the engine is on the mount. Tightening the engine mount bolts as followings:

• Upper mount to engine nuts and bolts — 36–47 ft. lbs. (50–65 Nm)

• Upper mount through-bolt nut — 43–58 ft. lbs. (60–80 Nm)

• Lower mount through-bolt nut — 33–47 ft. lbs. (45–65 Nm)

13. Install the exhaust pipe, power steering pump and air conditioning compressor.

14. Checking the tags installed during removal, reconnect all electrical and vacuum connections.

15. Install the transaxle to the vehicle and tighten the upper mounting bolts to 65 ft. lbs. (90 Nm). Install the starter assembly and tighten both mounting bolts to 54–65 ft. lbs. (75–90 Nm).

16. Install the radiator assembly and intercooler.

17. Install the air cleaner assembly. Install all control brackets, if not already done.

18. Fill the engine with the proper amount of engine oil. Connect the negative battery cable.

19. Refill the cooling system. Start the engine, allow it to reach normal operating temperature. Check for leaks.

20. Check the ignition timing and adjust, if necessary.

21. Install the hood making sure to align the matchmarks made during disassembly.

22. Road test the vehicle and check all functions for proper operation.

1995–97 Eclipse

2.0L (4G63) Engine

1. Relieve the fuel system pressure.

2. Disconnect the negative battery cable.

3. Matchmark the hood to the hinges and remove the hood.

4. Remove the intake air duct.

5. Drain the engine coolant.

6. Remove the radiator.

7. Remove the engine undercover.

8. Attach an engine lifting fixture to the engine and remove the transaxle assembly.

9. Disconnect the power steering pressure switch, oil pressure switch, oil pressure gauge sender and the alternator wiring connectors.

10. Remove the alternator.

11. Remove the power steering pump from the bracket and position the pump out of the way. It is not necessary to disconnect the fluid lines.

12. Remove the A/C compressor from the bracket and position it out of the way. Do not disconnect the hoses.

13. Disconnect the accelerator cable from the throttle body and mounting bracket.

14. Disconnect the following connectors;

• Idle air control (IAC) motor
• Knock sensor
• Heated oxygen sensor
• Engine coolant temperature gauge sender
• Engine coolant temperature sensor
• Ignition module (power transistor)
• Throttle position sensor
• Condenser
• Manifold differential pressure sensor
• Injectors

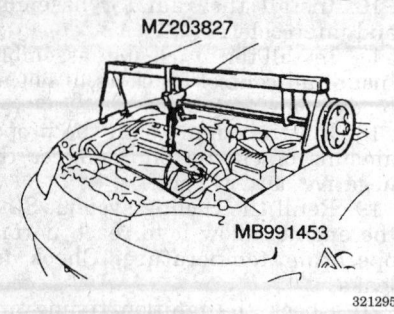

MZ203827

MB991453

321295

Common method of supporting the engine, using an support fixture

- Ignition coil
- Camshaft position sensor
- Crankshaft position sensor
- A/C compressor connector
- Engine control wiring harness

15. Disconnect the brake booster vacuum hose.

16. Disconnect the fuel lines from the fuel supply rail.

17. Disconnect the A and B water hose connections.

18. Label and disconnect the vacuum hoses.

19. Disconnect the front exhaust pipe from the turbocharger.

20. Support the engine under the oil pan with a floor jack and a piece of wood.

21. Remove the engine support fixture and replace it with a hoist. lift up the engine to take the weight off of the engine mount bracket.

22. Remove the engine mount bracket.

23. After checking that all cables, hoses and harness connectors are disconnected from the engine, raise the engine up slowly out of the engine compartment.

To install:

24. Slowly lower the engine assembly into the vehicle. Make sure that any wires, cables or hoses do not get damaged.

25. Position the floor jack under the oil pan with a piece of wood in between. Use the floor jack to adjust the height of the engine while installing the engine mount bracket.

26. Remove the chain hoist and install the engine support fixture.

27. Connect the front exhaust pipe to the turbocharger.

28. Connect the vacuum hoses.

29. Connect the A and B water hoses.

30. Install a new O-ring on the high pressure fuel line. Apply a small amount of clean engine oil to the O-ring and connect the fuel lines to the fuel supply rail.

31. Connect the brake booster vacuum hose.

32. Connect the following connectors;
- Idle air control (IAC) motor
- Knock sensor
- Heated oxygen sensor
- Engine coolant temperature gauge sender
- Engine coolant temperature sensor
- Ignition module (power transistor)
- Throttle position sensor
- Condenser
- Manifold differential pressure sensor
- Injectors
- Ignition coil
- Camshaft position sensor
- Crankshaft position sensor
- A/C compressor connector
- Engine control wiring harness

33. Install the A/C compressor and the power steering pump in their brackets.

34. Install the alternator.

35. Connect the oil pressure gauge sender, oil pressure switch and the power steering pressure switch connectors.

36. Install the transaxle assembly and remove the engine support fixture.

37. Install the engine undercover.

38. Install the radiator and connect the hoses.

39. Install the intake air duct.

40. Refill the engine with coolant.

41. Align the matchmarks and install the hood.

42. Connect the negative battery cable.

2.0L (420A) and 2.4L (4G64) Engines

1. Relieve the fuel system pressure.

2. Disconnect the negative battery cable.

3. Matchmark the hood to the hinges and remove the hood.

4. Remove the intake air duct.

5. Drain the engine coolant.

6. Matchmark the hose clamps to the hoses and remove the radiator.

7. Remove the engine undercover.

8. Attach an engine lifting fixture to the engine and remove the transaxle assembly.

9. Disconnect the following connectors;
- A/C compressor
- Power steering pressure switch
- Heated oxygen sensor
- Engine coolant temperature gauge sender
- Engine coolant temperature sensor
- MAP sensor
- Intake air temperature sensor

10. Remove the power steering pump from the bracket and position the pump out of the way. It is not necessary to disconnect the fluid lines.

11. Remove the A/C compressor from the bracket and position it out of the way. Do not disconnect the hoses.

12. Disconnect the accelerator cable from the throttle body and mounting bracket.

13. Disconnect the following connectors:
- Idle air control (IAC) motor
- Knock sensor
- Ignition module (power transistor)
- EGR solenoid
- Oil pressure switch
- Throttle position sensor
- Condenser
- Manifold differential pressure sensor
- Injectors
- Ignition coil
- Camshaft position sensor
- Crankshaft position sensor
- Engine control wiring harness

14. Disconnect the heater hoses from the engine.

15. Disconnect the fuel lines from the fuel supply rail.

16. Disconnect the purge air hose and the brake booster vacuum hose.

17. Disconnect the front exhaust pipe from the manifold.

18. Place a floor jack against the oil pan with a piece of wood in between to protect the oil pan.

19. Raise the engine with the jack and remove the engine support fixture.

20. Install a chain hoist to the top of the engine.

21. Remove the engine mount bracket.

22. After checking that all cables, hoses and harness connectors are disconnected from the engine, raise the engine up slowly out of the engine compartment.

To install:

23. Slowly lower the engine assembly into the vehicle. Make sure that any wires, cables or hoses do not get damaged.

24. Position the floor jack under the oil pan with a piece of wood in between. Use the floor jack to adjust the height of the engine while installing the engine mount bracket.

25. Remove the chain hoist and install the engine support fixture.

26. Connect the front exhaust pipe to the manifold.

27. Connect the brake booster vacuum hose.

28. Install a new O-ring on the high pressure fuel line. Apply a small amount of clean engine oil to the O-ring and connect the fuel lines to the fuel supply rail.

29. Connect the following connectors;
- Idle air control (IAC) motor
- Knock sensor
- Ignition module (power transistor)
- EGR solenoid
- Oil pressure switch
- Throttle position sensor
- Condenser
- Manifold differential pressure sensor
- Injectors
- Ignition coil
- Camshaft position sensor
- Crankshaft position sensor
- Engine control wiring harness

30. Connect and adjust the accelerator cable.

31. Install the A/C compressor and the power steering pump in their brackets.

32. Connect the intake air temperature sensor, MAP sensor, engine coolant temperature sensor, engine coolant temperature gauge sender, heated oxygen sensor, power steering pressure switch and the A/C compressor harness connectors.

33. Install the radiator and connect the hoses. Position the hose clamps in the original position.

34. Install the transaxle and remove the engine support fixture.

35. Install the engine undercovers.

36. Install the intake air duct.

37. Connect the negative battery cable.

38. Align the matchmarks and install the hood.

39. Refill the engine with the proper amount of coolant.

Galant

1993 2.0L (4G63) Engine

The transaxle must be removed from the car before removing the engine; they will not come out as a unit.

1. Disconnect the negative battery cable.

2. Drain the engine coolant.

3. Drain the engine oil and the transmission oil.

4. Safely relieve the pressure within the fuel injection system.

5. Matchmark the hood to the hinges and remove the hood.

6. Remove the transaxle assembly.

7. Remove the radiator, disconnecting the hoses at the engine.

8. Disconnect the accelerator cable and remove the bracket.

9. Disconnect the heater hoses.

10. Disconnect the brake booster vacuum hose at the engine.

11. Label and disconnect the vacuum hoses running to the firewall.

12. Disconnect the high pressure fuel line and discard the O-ring; it is not reusable.

13. Remove the fuel return hose.

14. Disconnect the electrical connectors to the engine components. All wires and connectors should be labeled at the time of removal. The amount of time saved during reassembly makes the extra effort well worthwhile. Disconnect the:
- oxygen sensor
- coolant temperature gauge
- coolant temperature sensor
- idle speed control
- EGR temperature sensor (Calif. only)
- each injector
- power transistor
- condenser
- throttle position sensor (TPS)
- motor position sensor (MPS)
- distributor connector
- control harness
- alternator
- oil pressure switch
- coolant temperature switch
- crankshaft angle sensor
- body ground
- power steering oil pressure switch

15. Loosen the power steering drive belt and remove it. Remove the bolts holding the pump to its bracket and hang the pump out of the way. Do not disconnect the hoses and do not allow the pump to hang by the hoses.

16. Loosen the adjuster and remove the air conditioning drive belt. Remove the compressor from its mount and hang it from stiff wire out of the way. Note that the hoses are still attached; do not loosen them or discharge the system.

17. Remove the self-locking nuts and bolt at the exhaust system joint just below the manifold. Separate the

exhaust pipes; discard the gasket and the two nuts.

18. Remove the clamp holding the power steering and air conditioning hoses to the top of the left engine mount. Move the hoses out of the way.

19. Elevate the car and support it safely. Install the engine hoist equipment and make certain the attaching points on the engine are secure. Draw tension on the hoist just enough to support the engine's weight but no more. Do not disturb the placement of the car on the stands.

20. Remove the through-bolt from the rear (firewall side) roll stopper. Remove the through-bolt from the front engine roll stopper.

21. Remove the nuts and bolts holding the upper (left side) engine mount to the engine. Remove the through-bolt and remove the mount assembly. Also remove the support bracket below the mount.

22. Double check for any remaining cables, wires or hoses running to the engine. Elevate the hoist and remove the engine from the car. Immediately place it on an engine stand or support it with wooden blocks. Do not allow it to rest on the oil pan or lie on its side. Never leave an engine hanging from a hoist.

To install:

After repairs, make certain the engine is fully reassembled before installation. All components removed with the engine out of the car should be in place before reinstallation.

23. Install the engine into the car and lower it until the bolt holes for the mounts and roll stoppers align with the brackets. Install the through-bolts and new self-locking nuts, tightening them just snug; they will be final tightened later.

24. Install the upper (left side) mount to the engine, tightening the nuts and bolts snug. Install the through-bolt and tighten it snug.

25. Slowly release tension on the hoist, allowing the weight of the engine to bear fully on the mounts. Once the hoist is slack, remove the lifting apparatus from the engine.

26. Connect the exhaust system to the manifold, using a new gasket and new locking nuts. Tighten the nuts and the small bolt to 25 ft. lbs. (34 Nm)

27. Final tighten the engine mount nuts and bolts. Correct torque values are:
- Nuts and bolts holding left mount to engine: 41 ft. lbs. (56 Nm)
- Left mount through-bolt: 50 ft. lbs. (68 Nm)

• Small nuts on head of through-bolt: 25 ft. lbs. (34 Nm)

• Left mount support nuts and bolts 15 ft. lbs. (20 Nm)

• Rear roll stopper through-bolt: 33 ft. lbs. (45 Nm)

• Front roll stopper through-bolt: 41 ft. lbs. (56 Nm)

28. Install the air conditioning compressor, tightening the mounting bolts to 18 ft. lbs. (25 Nm). Install the belt and adjust it.

29. Install the power steering pump, tightening the bolts to 36 ft. lbs. (49 Nm). Install and adjust the belt.

30. Connect the wiring and harness connectors to the engine. Make certain each terminal is clean and the connector is firmly seated to its mate. Do not route wires near hot surfaces or moving parts. Connect the:

• oxygen sensor
• coolant temperature gauge
• coolant temperature sensor
• idle speed control
• EGR temperature sensor (Calif. only)
• each injector
• power transistor
• condenser
• throttle position sensor (TPS)
• motor position sensor (MPS)
• distributor connector
• control harness
• alternator
• oil pressure switch
• coolant temperature switch
• crankshaft angle sensor
• body ground
• power steering oil pressure switch

31. Install the fuel return hose.

32. Using a new O-ring, connect the high pressure fuel line and tighten the bolts to 48 inch lbs. (6 Nm).

33. Connect the vacuum lines running to the firewall. Install the brake booster vacuum hose to the engine.

34. Connect the heater hoses.

35. Install the accelerator cable bracket, tightening the bolts to 48 inch lbs. (6 Nm), and connect the accelerator cable.

36. Install the radiator and connect the hoses.

37. Install the transaxle.

38. Check the engine oil drain plug and secure it if necessary. Install the proper amount of engine oil.

39. Check the transaxle drain plug, tightening it if needed, and install the proper amount of transmission oil.

40. Check the radiator and engine drain cocks, closing them if necessary and refill the coolant system.

41. Double check all installation items, paying particular attention to loose hoses or hanging wires, untightened nuts, poor routing of hoses and wires (too tight or rubbing) and tools left in the engine area.

42. Connect the negative battery cable. Start the engine and check for leaks.

43. Attend to all leaks immediately, remembering that fluids and metal surfaces may be hot. Adjust the drive belts to the correct tension. Adjust all cables (transmission, throttle, shift selector) and check the fluid levels. Check the operation of all gauges and dashboard lights.

44. With the help of an assistant, install the hood and align it for proper body fit and latching.

45. In a safe location at low speed, road test the car for correct operation of steering brakes, transaxle, clutch and speedometer.

2.4L (4G64) Engine

The transaxle must be removed from the car before removing the engine; they will not come out as a unit.

1. Disconnect the negative battery cable.

2. Drain the engine coolant.

3. Drain the engine oil and the transmission oil.

— CAUTION —

Used motor oil may cause skin cancer if repeatedly left in contact with the skin for prolonged periods.

4. Safely relieve the pressure within the fuel injection system.

5. Matchmark the hood to the hinges and remove the hood.

6. Remove the transaxle assembly.

7. Remove the radiator, disconnecting the hoses at the engine.

8. Disconnect the accelerator cable and remove the bracket.

9. Disconnect the air intake and breather hoses.

10. Disconnect the heater hoses.

11. Disconnect the brake booster vacuum hose at the engine.

12. Label and disconnect the vacuum hoses at the throttle body.

13. Disconnect the high pressure fuel line and discard the O-ring; it is not reusable.

14. Remove the fuel return hose.

15. Disconnect the electrical connectors to the engine components. All wires and connectors should be labeled at the time of removal. The amount of time saved during reas-

sembly makes the extra effort well worthwhile. Disconnect the:

• power steering pressure switch
• alternator
• oil pressure switch
• A/C compressor
• each injector
• power transistor
• ignition coil
• throttle position sensor (TPS)
• idle air control motor (IAC)
• coolant temperature switch
• coolant temperature sensor
• EGR temperature sensor
• control wiring harness
• oxygen sensor
• crankshaft angle sensor
• camshaft position sensor
• refrigerant temperature switch
• condenser connection

16. Loosen the power steering drive belt and remove it. Remove the bolts holding the pump to its bracket and hang the pump out of the way. Do not disconnect the hoses and do not allow the pump to hang by the hoses.

17. Loosen the adjuster and remove the air conditioning drive belt. Remove the compressor from its mount and hang it from stiff wire out of the way. Note that the hoses are still attached; do not loosen them or discharge the system.

18. Remove the self-locking nuts and bolt at the exhaust system joint just below the manifold. Separate the exhaust pipes; discard the gasket and the two nuts.

19. Elevate the car and support it safely. Install the engine hoist equipment and make certain the attaching points on the engine are secure. Draw tension on the hoist just enough to support the engine's weight but no more. Do not disturb the placement of the car on the stands.

20. Remove the through-bolt from the rear (firewall side) roll stopper. Remove the through-bolt from the front engine roll stopper.

21. Remove the nuts and bolts holding the upper (left side) engine mount to the engine. Remove the through-bolt and remove the mount assembly. Also remove the support bracket below the mount.

22. Double check for any remaining cables, wires or hoses running to the engine. Elevate the hoist and remove the engine from the car. Immediately place it on an engine stand or support it with wooden blocks. Do not allow it to rest on the oil pan or lie on its side. Never leave an engine hanging from a hoist.

To install:

After repairs, make certain the engine is fully reassembled before installation. All components removed with the engine out of the car should be in place before reinstallation.

23. Install the engine into the car and lower it until the bolt holes for the mounts and roll stoppers align with the brackets. Install the through-bolts and new self-locking nuts, tightening them just snug; they will be final tightened later.

24. Install the upper (left side) mount to the engine, tightening the nuts and bolts snug. Install the through-bolt and tighten it snug.

25. Slowly release tension on the hoist, allowing the weight of the engine to bear fully on the mounts. Once the hoist is slack, remove the lifting apparatus from the engine.

26. Connect the exhaust system to the manifold, using a new gasket and new locking nuts. Tighten the nuts and the small bolt to 33 ft. lbs. (44 Nm)

27. Final tighten the engine mount nuts and bolts. Correct torque values are:
- Upper mount to engine nuts 42 ft. lbs. (57 Nm)
- Upper mount to engine bolt 9 ft. lbs. (12 Nm)
- Upper mount through-bolt 72–87 ft. lbs. (98–118 Nm)
- Rear roll stopper through-bolt 32 ft. lbs. (44 Nm)
- Front roll stopper through-bolt 41 ft. lbs. (57 Nm)

28. Install the air conditioning compressor, tightening the mounting bolts to 18 ft. lbs. (25 Nm). Install the belt and adjust it.

29. Install the power steering pump, tightening the front bolts to 21 ft. lbs. (28 Nm) and the rear bolt to 16 ft. lbs. (22 Nm). Install and adjust the belt.

30. Connect the wiring and harness connectors to the engine. Make certain each terminal is clean and the connector is firmly seated to its mate. Do not route wires near hot surfaces or moving parts.

31. Install the fuel return hose and secure with the retaining clamp.

32. Using a new O-ring, connect the high pressure fuel line and tighten the bolts to 48 inch lbs. (6 Nm).

33. Connect the vacuum lines running to the throttle body.

34. Connect the heater hoses.

35. Install the accelerator cable bracket, tightening the bolts to 48 inch lbs. (6 Nm), and connect the accelerator cable.

36. Install the radiator and connect the hoses.

37. Install the transaxle.

38. Check the engine oil drain plug and secure it if necessary. Install the proper amount of engine oil.

39. Check the transaxle drain plug, tightening it if needed, and install the proper amount of transmission oil.

40. Check the radiator and engine drain cocks, closing them if necessary and refill the coolant system.

41. Double check all installation items, paying particular attention to loose hoses or hanging wires, untightened nuts, poor routing of hoses and wires (too tight or rubbing) and tools left in the engine area.

42. Connect the negative battery cable. Start the engine and check for leaks.

43. Attend to all leaks immediately, remembering that fluids and metal surfaces may be hot. Adjust the drive belts to the correct tension. Adjust all cables (transmission, throttle, shift selector) and check the fluid levels. Check the operation of all gauges and dashboard lights.

44. With the help of an assistant, install the hood and align it for proper body fit and latching.

45. In a safe location at low speed, road test the car for correct operation of steering brakes, transaxle, clutch and speedometer.

Diamante

1. Matchmark the hood and hinges and remove the hood assembly.

2. Relieve fuel system pressure.

3. Disconnect the negative then the positive battery cable. Remove the battery from the vehicle.

4. Remove the air cleaner assembly and all adjoining air intake duct work.

5. Drain the engine coolant and remove the radiator assembly and coolant reservoir (and bracket).

6. Remove the engine undercover, if equipped.

7. Disconnect and remove the front exhaust pipe, main muffler and catalytic converter for 1993–94 Federal models and 1993 California models.

8. Disconnect and remove the front exhaust pipe, catalytic converter (front bank side) and catalytic converter (rear bank side) for 1995–96 Federal models and 1994–96 California models.

9. Remove the transaxle assembly.

10. Disconnect the accelerator cable from the throttle body.

11. Disconnect the brake booster vacuum hose from the booster.

12. Note the locations and disconnect the vacuum hoses.

13. Disconnect the high pressure fuel line and the fuel return line.

14. Note the locations and remove the vacuum hoses from the solenoid valves.

15. Remove the vacuum hoses from the purge canister.

16. Disconnect the heater hose connections from the engine.

17. If equipped, unplug the harness for the EGR temperature sensor connection.

18. Remove the engine drive belts.

19. Remove the power steering pump oil pressure switch connection from the pump.

20. Remove the power steering pump and wire aside. Do not disconnect the fluid hoses.

21. Remove the air conditioning compressor. Wire the compressor aside. Do not discharge or disconnect the A/C lines.

22. Disconnect the wiring to the alternator.

23. Note the locations and disconnect the harness plugs for the atmospheric pressure sensor, idle speed control motor, throttle position sensor connector, fuel injectors and knock sensor.

24. Note the locations and disconnect the harness plugs for the air conditioning engine coolant temperature switch, coolant temperature sensor and coolant temperature gauge.

25. Note the locations and disconnect the harness plugs for the ignition coil, condenser and ignition power transistor.

26. Note the locations and disconnect the harness plugs for the variable induction control motor and the manifold absolute pressure sensor.

27. Note the locations and disconnect the harness plugs for the crankshaft and camshaft position sensors.

28. Remove the radiator overflow tank and remove the mounting bracket.

29. Remove the ground cable connections.

30. Attach a hoist to the engine and take up the engine weight. Remove the engine mount bracket. Remove any torque control brackets (roll stoppers).

NOTE: Note that some engine mount components have arrows on them for proper assembly.

31. Double check that all cables, hoses, harness connectors, etc., are disconnected from the engine. Lift the engine slowly and remove from the engine compartment.

To install:

32. Install the engine and secure all control brackets. Be sure to properly position the brackets.

33. Install the transaxle assembly.

34. Attach the engine ground cable connections.

35. Connect the harness plugs for the crankshaft and camshaft position sensors.

36. Connect the harness plugs for the variable induction control motor and the manifold absolute pressure sensor.

37. Connect the harness plugs for the ignition coil, condenser and ignition power transistor.

38. Connect the harness plugs for the air conditioning engine coolant temperature switch, coolant temperature sensor and coolant temperature gauge.

39. Connect the harness plugs for the atmospheric pressure sensor, idle speed control motor, throttle position sensor , fuel injectors and knock sensor.

40. Connect the wiring to the alternator.

41. Install the air conditioning compressor assembly.

42. Install the power steering pump assembly.

43. Connect the power steering pump oil pressure switch harness plug to the pump.

44. Install and adjust the engine drive belts.

45. If unplugged, connect the harness for the EGR temperature sensor.

46. Using new hose clamps, attach the heater hose connections to the engine.

47. Attach the vacuum hoses to the purge canister.

48. Connect the vacuum hoses to the solenoid valves.

49. Using new clamps or O-rings, connect the high pressure fuel line and the fuel return line.

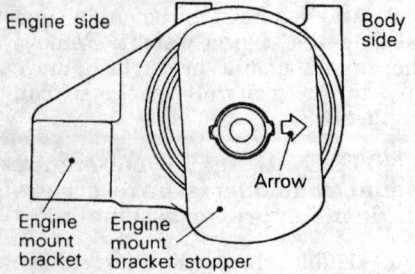

Alignment of the engine mount stopper bracket — Diamante shown

50. Using the noted locations, connect the vacuum hoses.

51. Connect the brake booster vacuum hose to the booster.

52. Connect and adjust the accelerator cable to the throttle body.

53. Install the air cleaner assembly and all adjoining air intake duct work.

54. Install the radiator and coolant reservoir assembly.

55. Install the transaxle assembly.

56. Using new gaskets, connect the front exhaust pipe, main muffler and catalytic converter for 1993–94 Federal models and 1993 California models.

57. Using new gaskets, connect the front exhaust pipe, catalytic converter (front bank side) and catalytic converter (rear bank side) for 1995–96 Federal models and 1994–96 California models.

58. Tighten the nuts that secure the front exhaust pipe to the manifold to 36 ft. lbs. (50 Nm).

59. Tighten the bolts that secure the front exhaust pipe to the catalytic converter to 36 ft. lbs. (50 Nm).

60. Install the battery to the vehicle. Connect the positive and then the negative battery cables.

61. Install the engine undercover, if equipped.

62. Fill the engine with the proper amount of engine oil and coolant.

63. Start the engine, allow it to reach normal operating temperature. Check for leaks.

64. Check the ignition timing and adjust if necessary.

65. Align the matchmarks and install the hood.

66. Road test the vehicle and check all fluid levels and functions for proper operation.

3000GT

1. Relieve fuel system pressure.
2. Disconnect the negative battery cable.

CAUTION

Wait at least 90 seconds after the negative (-) battery cable is disconnected to prevent possible deployment of the air bag.

3. Matchmark the hood and hinges and remove the hood assembly. Remove the air cleaner assembly and all adjoining air intake duct work.

4. Disconnect and remove the cruise control linkage and actuator assemblies.

5. Drain the engine coolant and remove the radiator assembly, coolant reservoir and intercooler.

6. Disconnect the heated oxygen sensor connection at the front exhaust pipe.

7. Unbolt and remove the front exhaust pipe assembly, discard gaskets.

8. Remove the transaxle assembly.

9. Disconnect the accelerator cable, breather hose and heater hose connections from the engine

10. Note locations and remove the vacuum hoses from the engine. Be sure to disconnect the brake booster vacuum supply.

11. Disconnect the fuel feed and return hoses.

12. Remove the solenoid valve assembly and disconnect ground cable.

13. Disconnect the purge hose and EGR temperature sensor, if equipped.

14. Remove the air conditioning and power steering drive belts.

15. Unbolt and remove the air conditioning compressor and the power steering pump assemblies.

NOTE: When removing the power steering pump and a/c compressor, it is not necessary to disconnect the hoses. Position the units aside and use rope or wire to secure.

16. Disconnect the harness connections for the idle speed control, motor position sensor and throttle position sensor.

17. Disconnect the EGR temperature sensor (California).

18. For the turbocharger, disconnect the following:

  a. Connection for the booster vacuum hose.

  b. Connections for the oil cooler lines and discard the sealing rings.

  c. Connection for the oxygen sensor.

19. Disconnect the wiring at the oil pressure switch and oil pressure gauge unit.

20. Disconnect the fuel injection wiring harness plug.

21. Disconnect the wiring from the knock sensor and the crankshaft angle sensor.

22. Disconnect the coolant temperature switch, coolant temperature sensor and the coolant temperature gauge unit connections.

23. Disconnect the wiring to the ignition coil, condenser and the power transistor.

24. Disconnect the variable induction motor connection.

25. Open the cover of relay box and disconnect the alternator wiring.

26. Attach a hoist to the engine and support the engine weight. Remove the engine mount bracket.

27. Remove the front and rear roll stopper bracket mounting bolts.

28. Remove the engine assembly from the vehicle.

To install:

29. Install the engine and secure into position. Secure the engine mount bracket to block and tighten bolts to 72–87 ft. lbs. (100–120 Nm). Install through-bolt and tighten bolt to 51 ft. lbs. (70 Nm).

30. Install the front and rear roll stopper through-bolt and tighten to 36–43 ft. lbs. (50–60 Nm).

31. Open the cover of relay box and connect alternator wiring.

32. Connect the variable induction motor connection.

33. Connect the fuel feed and return hoses. Using a new sealing ring, tighten pressure hose connection to 4 ft. lbs. (5 Nm).

34. Connect the wiring to the ignition coil, condenser and power transistor.

35. Connect the coolant temperature switch, coolant temperature sensor and the coolant temperature gauge unit connections.

36. Connect the wiring from the knock sensor and the crankshaft angle sensor.

37. Connect the fuel injection wiring harness plug.

38. Connect the wiring at the oil pressure switch and oil pressure gauge unit.

39. Connect the following for the turbocharger:

    a. Connection for booster vacuum hose.

    b. Connections for oil cooler lines using new sealing rings. Tighten fittings to 29–33 ft. lbs. (40–49 Nm).

    c. Connection for oxygen sensor.

40. Connect EGR temperature sensor (California).

41. Connect harness connections for the idle speed control, motor position sensor and throttle position sensor.

42. Install the air conditioning compressor and the power steering pump assemblies.

43. Install the engine drive belts.

44. Connect the purge hose and the EGR temperature sensor, if equipped.

45. Install the solenoid valve assembly and connect ground cable to engine block.

46. Reconnect the vacuum hoses to the engine. Be sure to connect the brake booster vacuum supply.

47. Connect the accelerator cable, breather hose and heater hose connections to the engine.

48. Install the transaxle assembly.

49. Install the front exhaust pipe assembly, using new gaskets. Tighten manifold mounting bolts to 36 ft. lbs. (50 Nm).

50. Connect the heated oxygen sensor connection at the front exhaust pipe.

51. Replace the radiator assembly, coolant reservoir and intercooler. Refill the cooling system.

52. Install and connect the cruise control linkage and the actuator assemblies.

53. Install the hood assembly, air cleaner assembly and all adjoining air intake duct work.

54. Connect the negative battery cable and run engine.

55. Inspect all connections and check all fluid levels.

Engine Mounts

REMOVAL AND INSTALLATION

Precis

1. Disconnect the negative battery cable.

2. Attach an engine hoist to the engine hooks and raise the engine just enough to take the pressure off of the engine mounts.

3. The mounts can now be removed and serviced as needed.

4. If working under the vehicle, an adjustable stand may be used to raise the engine and take the pressure off of the engine mounts.

5. Connect the negative battery cable.

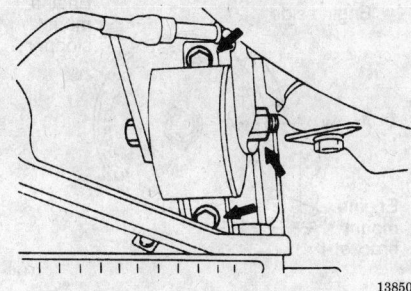

138509

Front roll stopper and bracket — Precis

Mirage, 1993 Galant and 1993–94 Eclipse

Upper Mounts

1. Disconnect the negative battery cable. Remove the air cleaner and all necessary air duct work.

2. Raise and safely support the engine so it is not resting on the engine mount. One suggested way is a block of wood between a floor jack and the oil pan. Use care not to bend or damage any components.

3. Remove the retainer bolt from the clamp securing the power steering pressure hose and the air conditioning low pressure hose.

4. Remove the engine mount bracket and body connection through-bolt. Take note of the position of the arrow on the oval shaped mounting stopper plate. This is important.

5. Remove the engine mounting bracket and stopper plate.

To install:

6. Install the engine mounting bracket and stopper plate. Note the arrows on the stopper plates and make sure they are installed properly. On most engines the arrows will face the towards the center of the engine. Torque upper mount to engine nuts and bolts to 36–47 ft. lbs. (50–65 Nm) and the upper mount through-bolt nut to 43–58 ft. lbs. (60–80 Nm).

Lower Mounts

1. Disconnect the negative battery cable.

2. Raise and safely support the engine so it is not resting on the engine mount. One suggested way is a block of wood between a floor jack and the oil pan. Use care not to bend or damage any components.

3. Remove the stopper through-bolt.

4. Remove the stopper mount frame bolts and pry mount out.

To install

5. Position the lower front roll stopper so the part of the bracket with the hole in it is facing the front of the vehicle. Install the frame mounting bolts and tighten.

6. Install the front lower mount through-bolt nut, but do not fully tighten until the full weight of the engine is on the mount. Torque lower mount through-bolt nut 36–47 ft. lbs. (50–65 Nm).

7. Connect the negative battery cable.

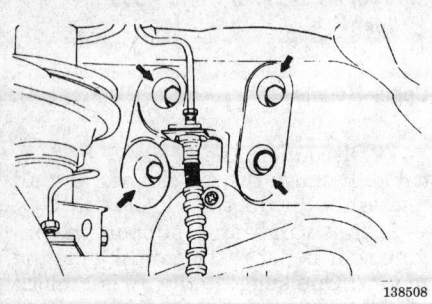

Transaxle bracket mounting bolts — Precis

138508

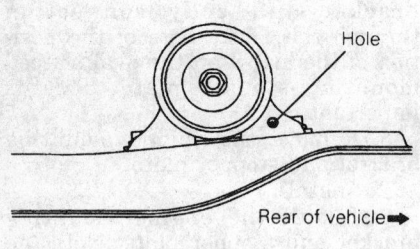

Engine roll stopper service points of installation — Mirage 1.5L (4G15) and 1.8L (4G93) engine

330042

1995–97 Eclipse

Engine Mount

1. Disconnect the negative battery cable.
2. Place a block of wood on a floor jack. Jack up the engine under the oil pan to take the weight off of the engine mount.
3. Remove the engine mount through-bolt.
4. Remove the bolt and two nuts securing the mount to the engine and remove the mount.

To install:
5. Install the engine mount so the arrow points toward the engine.
6. Line up the holes and install the through-bolt.
7. Remove the jack from under the oil pan.

Transaxle Mount

1. Remove the air cleaner assembly.
2. Remove the battery and the battery bracket.
3. Use a jack with a block of wood to support the weight of the engine and remove the engine roll stopper.
4. Remove the engine mount.
5. Remove the charcoal canister and bracket.
6. Remove the four nuts and the transaxle mount bracket.
7. Remove the through-bolt and the transaxle mount.

To install:
8. Install the transaxle mount.
9. Install the charcoal canister and bracket.
10. Install the engine mount and the engine roll stopper.
11. Install the battery tray and bracket.
12. Install the battery.
13. Install the air cleaner assembly.

Engine Roll Stopper

1. Disconnect the negative battery cable.
2. Safely raise and support the vehicle.
3. Remove the engine undercovers.
4. Carefully raise the engine with an adjustable stand to take the weight off of the roll stopper.
5. Remove the roll stopper through-bolt.
6. Remove the roll stopper mounting bolts.
7. Remove the roll stopper.

To install:
8. Install the roll stopper and the mounting bolts. Torque the through-

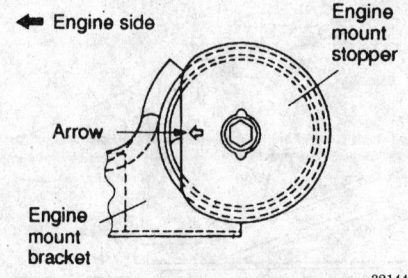

Correct engine mount insulator installation — 1995–97 Eclipse 2.0L and 2.4L engines

321446

bolt to specification after the vehicle is on the floor.

NOTE: Measure the distance from the center member to the through-bolt on the front roll stopper. If the dimension is greater than 1.69 ±0.12 in. (43 ±3 mm), replace the front roll stopper bracket assembly.

9. Remove the stand or jack from under the engine.
10. Install the engine undercovers.
11. Lower the vehicle to the floor.
12. Connect the negative battery cable.

1994–97 Galant

Upper Mounts

1. Disconnect the negative battery cable. Remove the air cleaner and all necessary air duct work.
2. Raise and safely support the engine so it is not resting on the engine mount. One suggested way is a block of wood between a floor jack and the oil pan. Use care not to bend or damage any components.
3. Remove the engine mount bracket and body connection through-bolt. Take note of the position of the arrow on the oval shaped mounting stopper plate. This is important.
4. Remove the engine mounting bracket and stopper plate.

To install:
5. Install the engine mounting bracket and stopper plate. Note the arrows on the stopper plates and make sure they are installed properly. On most engines the arrows will face the towards the center of the engine. Torque upper mount to engine nuts and bolt to 41 ft. lbs. (57 Nm) and the upper mount through-bolt nut to 71–85 ft. lbs. (98–118 Nm).

Lower Mounts

1. Disconnect the negative battery cable.
2. Raise and safely support the engine so it is not resting on the engine mount. One suggested way is a block of wood between a floor jack and the oil pan. Use care not to bend or damage any components.
3. Remove the stopper through-bolt.
4. Remove the stopper mount frame bolts and pry mount out.

To install
5. Position the lower front roll stopper so the part of the bracket with the hole in it is facing the front of the vehicle. Install the frame mounting bolts and tighten.

<2.0L Engine (Non-turbo)>

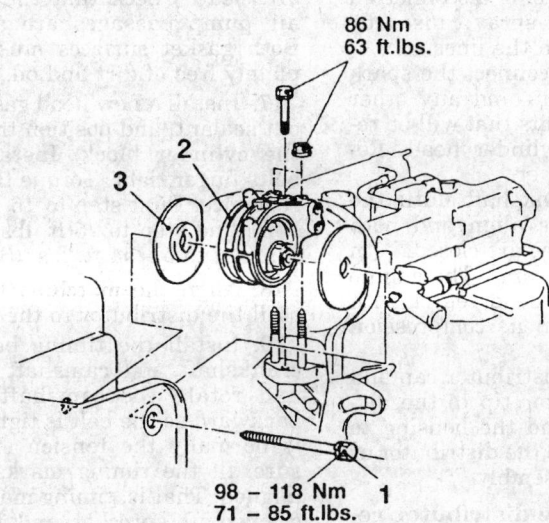

86 Nm
63 ft.lbs.

98 – 118 Nm
71 – 85 ft.lbs.

1

2

3

<2.0L Engine (Turbo) and 2.4L Engine>

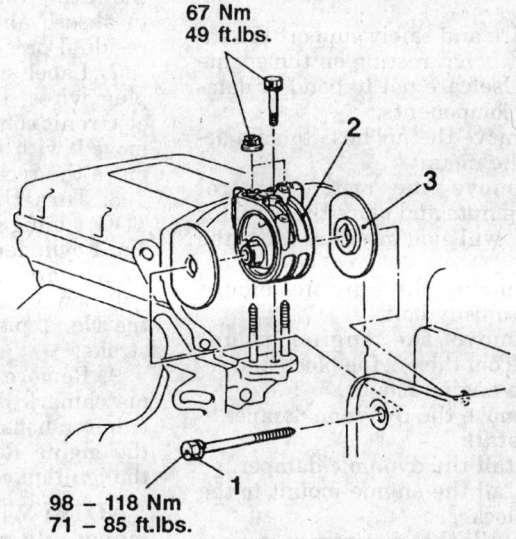

67 Nm
49 ft.lbs.

98 – 118 Nm
71 – 85 ft.lbs.

1

2

3

321443

Removal steps

1. Engine mount insulator mounting bolt
2. Engine mount bracket
3. Engine mount stopper

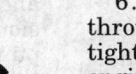

Transaxle mount stopper

Transaxle side ➡

Arrow

Transaxle mount bracket

321460

Correct transaxle mount insulator installation — 1995 Eclipse 2.0L (4G63 and 420A) engines

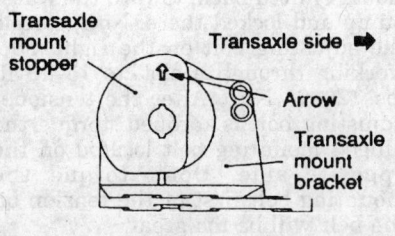

Front roll stopper bracket

A

Center member

321462

Measuring the front roll stopper — 1995 Eclipse 2.0L (4G63 and 420A) engines

Engine mounts — 1995–97 2.0L and 2.4L engines

6. Install the lower mount through-bolt nut, but do not fully tighten until the full weight of the engine is on the mount. Torque lower mount through-bolt nut 32 ft. lbs. (44 Nm).

Diamante

1. Disconnect the negative battery cable. Remove the air cleaner and all necessary duct work.
2. Raise and safely support the engine so it is not resting on the engine mount. One suggested way is a block of wood between a floor jack and the oil pan. Use care not to bend or damage any components.
3. Remove the retainer bolt from the clamp securing the power steering pressure hose and the air conditioning low pressure hose.
4. Remove the coolant reservoir and bracket if they are preventing access to the engine mount bracket. It is imperative to take note of the position of the arrow on the oval shaped mounting stopper plate prior to disassembly. Remove the engine mount bracket and body connection through-bolt.
5. Remove the engine mounting bracket and stopper plate.

6. Lower mounts (roll stoppers) are removed by removing the through-bolt, then the frame bolts.

To install:

7. Place the mount into position. Note the arrows on the stopper plates, and make sure they are installed properly; on most engines the arrows will face towards the center of the engine. When installing the lower front roll stopper, install the stopper bracket so the part of the bracket with the hole in it is facing the front of the vehicle.
8. The front lower mount through-bolt nut should not be tightened until the full weight of the engine is on the mount. Torque specifications are as follows:
• Upper mount to engine nuts and bolts — 72–87 ft. lbs. (100–120 Nm)
• Upper mount through-bolt nut — 51 ft. lbs. (70 Nm)
• Lower mount through-bolt nut — 36–43 ft. lbs. (47–60 Nm)
• Front roll stopper to crossmember bolt — 43–51 ft. lbs. (60–70 Nm)
• Rear roll stopper to crossmember bolt — 33 ft. lbs. (45 Nm)
9. Install the retainer bolt from the clamp securing the power steering pressure hose and the air conditioning low pressure hose.

3000GT

1. Disconnect the negative battery cable.

2. Raise and safely support the engine so it is not resting on the engine mount. Use care not to bend or damage any components.

3. Remove the air hose to gain access to the mount.

4. Remove the cruise control mounting nuts and place the actuator where it will not interfere with the work.

5. Remove the engine mount bracket to body bolt.

6. Remove the engine mount bracket from the engine block and remove the engine mount.

7. Remove the dynamic damper.

To install:

8. Install the dynamic damper.

9. Install the engine mount to the engine block.

10. Install the mounting stopper and the bracket to body bolt and remove the floor jack from under the engine.

NOTE: It may be necessary to move the engine up or down to align the bolt holes.

11. Install the cruise control actuator assembly.

12. Reconnect the air hose and connect the negative battery cable.

Cylinder Head

REMOVAL AND INSTALLATION

Precis

——— WARNING ———
Do not remove the cylinder head unless the engine is cold, a hot cylinder head will warp.

1. Release the fuel system pressure on fuel injected engines and disconnect the negative battery cable.

2. Drain the cooling system and then disconnect the upper radiator hose.

3. Remove the PCV hose that runs between the air cleaner and the rocker cover.

4. Remove the air cleaner. Label and disconnect any vacuum lines running to the cylinder head, manifold or carburetor from other parts of the engine compartment. Disconnect the intake pipe on turbocharged engines.

5. Disconnect the heater hoses going to the head.

6. Disconnect the fuel supply line, return line and vent hoses. Prior to disconnecting the fuel supply or return lines, wrap shop towels around the fitting that is being disconnected to absorb any fuel spray caused by residual pressure in the lines.

7. Label and disconnect the spark plug wires, injectors and any other electronic components that will be removed with the cylinder head. Remove the rocker cover.

8. Turn the crankshaft until the TDC timing marks align and both No. 1 cylinder valves are closed, both rockers are off the cams. The engine will now be at top dead center with the No. 1 piston on its compression stroke.

9. Remove the distributor cap and matchmark the rotor tip to the distributor housing and the housing to the engine. Remove the distributor or the ignition coil assembly.

NOTE: With the distributor removed, do not crank the engine. If rotated with the distributor removed, the engine will have to be positioned with the No. 1 cylinder at top dead center of its compression stroke prior to installing the distributor.

10. Remove the carburetor, if equipped.

11. Remove the intake and the exhaust manifolds.

12. Remove the timing belt cover. Note the location of the camshaft sprocket timing mark. Loosen both timing belt tensioner mounting bolts and move tensioner toward the water pump as far as it will go. Retighten the adjusting bolt to hold the tensioner in this position. Remove the timing belt.

13. Loosen the head bolts in the proper sequence. When all have been loosened, remove them. Then, pull the head off the engine block, rocking it slightly to break it loose.

14. Inspect the head with a straightedge and a flat feeler gauge of 0.002 in. (0.05mm) thickness. The tolerance for warping of a used head is 0.002 in. (0.05mm). The block deck must be flat within the same tolerance. The height of the head should be 3.5 in. (89mm) with a maximum machining limit of 0.012 in. (0.3mm).

To install:

15. Clean the combustion chambers of carbon with a scraper that is not excessively sharp, being careful not to damage the aluminum surface. Sharp edges in the combustion chambers can cause detonation. Clean the gasket mating surfaces with a scraper and solvent.

16. The oil and water passages should be cleaned thoroughly. Also, blow compressed air through all the small oil passages to ensure that they are clear. Check that the EGR and air pump passages are also clear. Both gasket surfaces must be completely free of dirt and oil.

17. Install a new head gasket, without sealant, and position the head on the cylinder block. Install all the bolts finger tight. Torque the bolts in sequence. First step to 15 ft. lbs. (20 Nm), 2nd step to 25 ft. lbs. (35 Nm). Finally, to 51–54 ft. lbs. (69–74 Nm).

18. Align the matchmarks and install the distributor to the engine.

19. Install the timing belt on the crankshaft and camshaft sprockets and rotate the camshaft sprocket backward so the belt is tight on what is normally the tension side. Make sure all the timing marks are now aligned. That is, timing marks on the crankshaft sprocket and front case must align; and the marks on the camshaft sprocket and the tab on the cylinder head must be simultaneously aligned with the side of the belt away from the tensioner under tension. Now, loosen the timing belt tensioner adjusting bolt and allow spring tension to tension the belt. Make sure all timing marks are still aligned. If not, the belt is out of time and must be shifted with the tensioner shifted back, toward the water pump and locked there. Now, torque the adjusting bolt on the right side, working through a slot, to 15–18 ft. lbs. (20–25 Nm). After the tensioner adjusting bolt is torqued, torque the hinged mounting bolt located on the opposite side. Don't torque the mounting bolt first or the tension on the belt will be too great.

20. Turn the crankshaft one full turn in the normal direction of rotation. Loosen first the tensioner pivot bolt and then the adjusting bolt. Now torque them exactly as before. This extra step is necessary to ensure the timing belt is properly seated before final tension is adjusted.

21. Install the cylinder head cover and tighten the bolts to 13–16 inch lbs. (1.5–2.0 Nm).

22. Install the timing belt cover.

23. Install the intake manifold using a new gasket and tighten the bolts and nut to 12–14 ft. lbs. (16–19 Nm).

24. Install the exhaust manifold using a new gasket and tighten the nuts to 12–14 ft. lbs. (16–19 Nm).

25. Install the carburetor or throttle body assembly.

26. Install the distributor. Connect all hoses, lines, and the air cleaner.

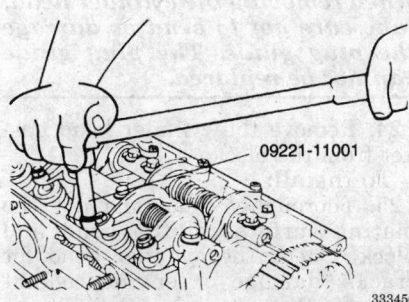

Cylinder head bolt removal sequence — Precis

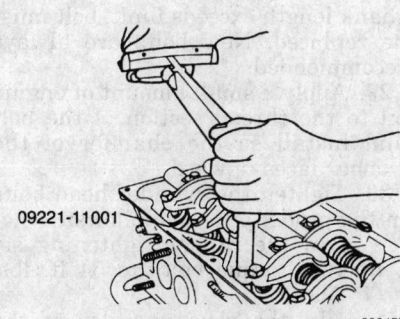

Cylinder head bolt installation torque sequence — Precis

Reconnect the intake pipe on turbocharged engines.

27. Refill and bleed the cooling system. Operate the engine and check for leaks. After the engine has reached normal operating temperature, turn it off and remove the air cleaner and rocker cover. Retighten the cylinder head bolts to 58–61 ft. lbs. (78–83 Nm), in the proper sequence.

28. Reinstall the rocker cover and the air cleaner.

Mirage

1.5L (4G15) Engine

1. Relieve the fuel system pressure. Disconnect the negative battery cable.

2. Position the No. 1 piston to TDC of its compression stroke.

3. Drain the cooling system.

4. Remove the air intake hose and the air cleaner assembly.

5. Disconnect the ground cable connection and the accelerator cable. There will be 2 cables, if equipped with cruise control.

6. Disconnect the PCV and the breather hose connection.

7. Note the locations and disconnect the vacuum hoses from the intake and throttle body.

8. Disconnect the vacuum line for the brake booster.

9. Remove the upper radiator hose, throttle body hoses, bypass hose and heater hose connections.

10. Place a shop towel around the high pressure fuel line to absorb any residual fuel remaining in the system. Disconnect the high pressure fuel line.

11. Disconnect and plug the fuel return line.

12. Remove the spark plug cables.

13. Disconnect the electrical harness plugs from the crankshaft and camshaft position sensors.

14. Disconnect the electrical harness plugs from the oxygen sensor, engine coolant temperature gauge unit and the water temperature sensor.

15. Disconnect the electrical harness plugs from the idle speed control motor, throttle position sensor and air intake temperature sensor.

16. Disconnect electrical harness plugs from the ignition distributor, fuel injectors, EGR temperature sensor, power transistor and ground cable.

17. Disconnect the engine control wiring harness.

18. Remove the clamp that holds the power steering pressure hose to the engine mounting bracket.

19. Place a jack and wood block under the oil pan and carefully lift just enough to take the weight off the engine mounting bracket, remove the bracket.

20. Remove the valve cover and gasket.

21. Remove the timing belt front upper cover.

22. If not already done, rotate the crankshaft clockwise and align the timing marks.

23. While securing the sprocket, remove the sprocket bolt and remove the sprocket with the timing belt attached. Be sure to keep the sprocket and belt wired or tied together as one unit.

24. Remove the timing belt rear upper cover.

——— **CAUTION** ———
The crankshaft must always be rotated clockwise. Do not rotate engine by turning the camshaft.

25. Remove the exhaust pipe self-locking nuts and separate the exhaust pipe from the exhaust manifold. Discard the gasket.

26. Loosen the cylinder head mounting bolts in sequence using three steps. Lift off the cylinder head assembly and remove the head gasket.

To install:

27. Thoroughly clean and dry the mating surfaces of the head and block. Check the cylinder head for cracks, damage or engine coolant leakage. Remove scale, sealing compound and carbon. Clean oil passages thoroughly. Check the head for flatness. End to end, the head should be within 0.002 inch with 0.008 inch the maximum allowed out of true. The total thickness allowed to be removed from the head and block is 0.008 in. maximum.

28. Place a new head gasket on the cylinder block with the identification marks facing upward. Make sure the gasket has the proper identification mark for the engine. Do not use sealer on the gasket.

NOTE: The cylinder head torque specifications are for a cold cylinder head.

29. Carefully install the cylinder head on the block. Using 3 even steps, torque the head bolts in sequence to 53 ft. lbs. (73 Nm).

30. Install a new exhaust pipe gasket and connect the exhaust pipe to the manifold.

31. Install the upper rear timing cover.

Intake side

Front of engine ⇨

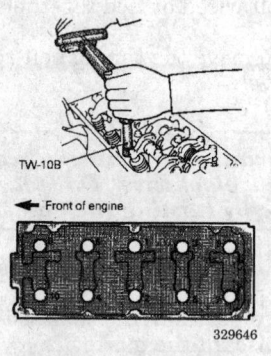

Exhaust side

329710

Cylinder head bolt loosening sequence — Mirage with 1.5L (4G15) engine

TW-10B

← Front of engine

329646

Cylinder head bolt torque sequence — Mirage with 1.5L (4G15) engine

32. Align the timing marks and install the cam sprocket. Torque the retaining bolt to 51 ft. lbs. (70 Nm). Check the belt tension and adjust, if necessary. Install the outer timing cover.

33. Install a new valve cover gasket. Install the valve cover and torque the retaining bolts to 16 inch lbs. (1.8 Nm).

34. Install the engine mount bracket and remove the support jack.

35. Install the clamp that holds the power steering pressure hose to the engine mounting bracket.

36. Connect or install all previously disconnected hoses, cables and electrical connections. Adjust the throttle cable(s).

37. Replace the O-rings and connect the fuel lines.

38. Install the air cleaner assembly. Connect the breather hose.

39. Change the engine oil and oil filter.

40. Fill the system with coolant.

41. Connect the negative battery cable, run the vehicle until the thermostat opens, fill the radiator completely.

42. Check and adjust the idle speed and ignition timing.

43. Once the vehicle has cooled, recheck the coolant level.

1.8L (4G93) Engine

1. Relieve fuel system pressure. Disconnect the negative battery cable.

2. Position the No. 1 piston to TDC of its compression stroke.

3. Remove the air cleaner assembly.

4. Drain the cooling system.

5. Disconnect the brake booster vacuum hose and PVC valve connection.

6. Note the locations and disconnect the vacuum hoses from the intake and throttle body.

7. Remove the upper radiator hose, overflow tube and the water hose from the thermostat to the throttle body.

8. Wrap the connection with a shop towel and disconnect the high pressure fuel line at the fuel rail. Discard the O-ring.

9. Disconnect the fuel return hose from the fuel pressure regulator.

10. Disconnect the accelerator cable connection from the throttle body.

11. Disconnect the electrical harnesses at the oil pressure switch, oxygen sensor, water temperature sensor connectors and distributor.

12. Disconnect the electrical harnesses at the idle air control motor and the EGR temperature sensor.

13. Disconnect the wiring from condenser, idle speed control, throttle position sensor and knock sensor.

14. Note harness plug connections for reassembly and disconnect fuel injectors.

15. Disconnect the spark plug cables from each spark plug.

16. Unbolt the control harness assembly and position aside.

17. Remove the thermostat housing, thermostat and the thermostat case with O-ring from the engine.

18. Remove the rocker cover.

19. Remove the timing belt upper cover.

20. If not already done, rotate the crankshaft in the clockwise direction to align the camshaft timing marks. Matchmark the camshaft sprocket and the timing belt. Tie the camshaft sprocket and the timing belt together so the sprocket will not move with respect to the timing belt.

21. While holding the camshaft sprocket in position using the appropriate wrench, remove the camshaft sprocket and with the belt attached. Wire the sprocket and belt aside making sure constant tension is maintained on the belt. Do not allow

the belt to slacken or engine timing may be altered.

NOTE: When removing the camshaft sprocket, do not allow the crankshaft to rotate. Confirm proper engine timing during installation.

22. Loosen the cylinder head bolts in two or three steps in the proper sequence.

23. Remove the cylinder head from the engine.

--- CAUTION ---
When removing the cylinder head, take care not to bend or damage the plug guide. The plug guide can not be replaced.

24. Remove the cylinder head gasket from the block.

To install:

25. Thoroughly clean and dry the mating surfaces of the head and block. Check the cylinder head for cracks, damage or engine coolant leakage. Remove scale, sealing compound and carbon. Clean oil passages thoroughly. Check the head for flatness. end to end, the head should be within 0.002 inch with 0.008 inch the maximum allowed out of true. The total thickness allowed to be removed from the head and block is 0.008 in. maximum.

26. Place a new head gasket on the cylinder block with the identification marks facing upward. Make sure the gasket has the proper identification mark for the engine. Do not use sealer on the gasket.

27. Carefully install the cylinder head on the block.

28. Inspect the cylinder head bolt prior to installation. The length below the head of the bolts should not exceed 3.795 inches (96.4mm). If bolt shank length exceeds limit, bolt must be replaced. New bolts are always recommended.

29. Apply a small amount of engine oil to the thread section of the bolt and install so the chamfer of the washer faces upward.

30. Tighten the cylinder head bolts in the proper order as follows:

 a. In the proper tightening sequence, torque bolts to 54 ft. lbs. (75 Nm).

 b. In the reverse order of the tightening sequence, fully loosen bolts.

 c. In the proper tightening sequence, torque bolts to 14 ft. lbs. (20 Nm).

 d. In the proper tightening sequence, tighten bolts an additional ¼ turn (90 degrees).

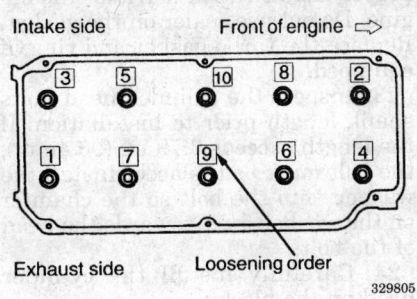

Cylinder head bolt loosening sequence — Mirage and Eclipse with 1.8L engine

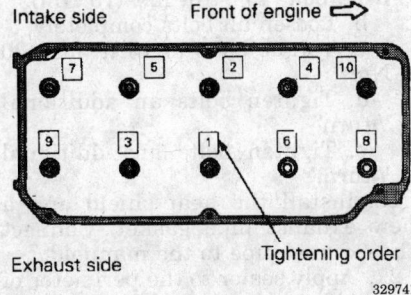

Cylinder head bolt torque sequence — Mirage and Eclipse with 1.8L engine

e. In the proper tightening sequence, tighten bolts an additional 1/4 turn (90 degrees).

31. Install the camshaft sprocket and tighten bolt to 65 ft. lbs. (90 Nm), while holding the sprocket in place using the appropriate wrench. Confirm proper timing mark alignment.

32. Install the upper timing belt cover and rocker cover. Torque the rocker cover bolts to 29 inch lbs. (3.3 Nm).

33. Loosen the water pipe mounting bolt for ease of thermostat housing installation.

34. Apply a thin bead of sealant MD970389 or equivalent, to the water tube connection on the thermostat case.

35. Apply a small amount of water to the O-ring of the water inlet pipe and press the thermostat case assembly onto the water inlet pipe. Install the thermostat case assembly mounting bolt tightening to 16 ft. lbs. (22 Nm).

36. Tighten the water pipe mounting bolt.

37. Install the thermostat into the housing so the jiggle valve is located at the top. Tighten the housing bolts to 10 ft. lbs. (14 Nm).

38. Connect the upper radiator hose to the thermostat housing.

39. Connect or install all previously disconnected hoses, cables and electrical connections. Adjust the throttle cable(s).

40. Replace the O-ring for the high pressure hose and install a new clamp on the return hose and reconnect the fuel lines.

41. Install the air intake hose. Connect the breather hose and air cleaner case cover.

42. Reconnect the brake booster and the PCV vacuum hoses.

43. Change the engine oil and oil filter.

44. Fill the system with coolant.

45. Connect the negative battery cable, run the vehicle until the thermostat opens, fill the radiator completely.

46. Check and adjust the idle speed and ignition timing.

47. Check all systems for leaks. Allow the engine to cool and recheck the coolant level.

Eclipse

1.8L (4G37) Engine

1. Relieve the fuel system pressure using proper procedure.

2. Drain the cooling system.

3. Remove the air intake hose and the breather hose.

4. Disconnect the accelerator cable. There will be 2 cables, if equipped with cruise-control.

5. Place a shop towel around the high pressure fuel line to absorb any residual fuel remaining in the system. Disconnect the high pressure fuel line.

6. Remove the upper radiator hose, the water breather hose, the water bypass hose and the heater hose.

7. Disconnect the PCV hose.

8. Label and remove the spark plug cables. Make sure not to pull on the cable, when removing the spark plug wire.

9. Disconnect and plug the fuel return line.

10. Disconnect the vacuum line for the brake booster.

11. Disconnect the electrical connections for the oxygen sensor, engine coolant temperature gauge unit and the water temperature sensor.

12. Disconnect the electrical connections for the idle speed control motor, Throttle Position Sensor (TPS), distributor, motor position sensor connector, fuel injectors, EGR temperature sensor (California vehicles), power transistor, condenser and engine ground cable.

13. Disconnect the engine control wiring harness.

14. Remove the clamp that holds the power steering pressure hose to the engine mounting bracket.

15. Place a jack and wood block under the oil pan and carefully lift just enough to take the weight off the engine mounting bracket. Then remove the bracket.

16. Remove the valve cover, gasket and half-round seal. Remove the timing belt front upper cover.

17. If possible, rotate the crankshaft clockwise until the timing marks on the cam sprocket and belt align. This should position the engine so No. 1 piston is at top dead center of it's compression stroke.

18. Remove required parts for access and remove timing belt. Remove the timing belt rear upper cover.

19. Remove the exhaust pipe self-locking nuts and separate the exhaust pipe from the exhaust manifold. Discard the gasket.

20. Loosen the cylinder head mounting bolts in 3 steps, starting from the outside and working inward. Lift off the cylinder head assembly and remove the head gasket.

To install:

21. Thoroughly clean and dry the mating surfaces of the head and block. Check the cylinder head for cracks, damage or engine coolant leakage. Remove scale, sealing compound and carbon. Clean oil passages thoroughly. Check the head for flatness. End to end, the head should be within 0.002 in. normally with 0.008 in. the maximum allowed out of true. The total thickness allowed to be removed from the head and block is 0.008 in. maximum.

22. Place a new head gasket on the cylinder block with the identification marks facing upward. Make sure the gasket has the proper identification mark for the engine. Do not use sealer on the gasket.

23. Carefully install the cylinder head on the block. Using 3 even steps, torque the head bolts in sequence to 51–54 ft. lbs. (70–75 Nm).

24. Install a new exhaust pipe gasket and connect the exhaust pipe to the manifold.

25. Install the upper rear timing cover. Align the timing marks and install the timing belt. Install the outer timing cover.

26. Apply sealer to the perimeter of the half-round seal. Install a new valve cover gasket. Install the valve cover.

27. Install the engine mount bracket. Once secure, remove the jack.

28. Install the clamp that holds the power steering pressure hose to the engine mounting bracket.

29. Connect or install all previously disconnected hoses, cables and electrical connections. Adjust the throttle cable(s).

30. Replace the O-rings and connect the fuel lines.

31. Install the air intake hose. Connect the breather hose.

32. Change the engine oil and oil filter.

33. Fill the system with coolant.

34. Connect the negative battery cable, run the vehicle until the thermostat opens, fill the radiator completely.

35. Check and adjust the idle speed and ignition timing.

36. Once the vehicle has cooled, recheck the coolant level.

1993–94 2.0L (4G63) Engine

1. Following proper procedure, relieve the fuel system pressure.
2. Drain the cooling system.
3. Disconnect the accelerator cable. There will be 2 cables if equipped with cruise-control.
4. Remove the air cleaner with the air intake hose.

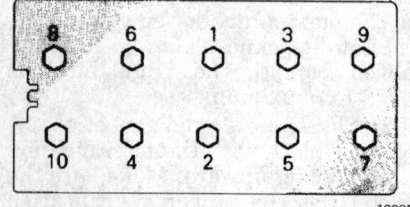

Cylinder head bolt torque sequence — 1993–94 Eclipse 2.0L (4G63) engine

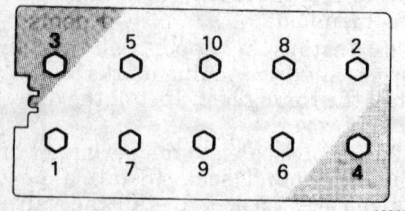

Cylinder head bolt removal sequence — 1993–94 Eclipse 2.0L (4G63) engine

5. Disconnect the oxygen sensor, engine coolant temperature sensor, the engine coolant temperature gauge unit and the engine coolant temperature switch on vehicles with air conditioning.

6. Disconnect the ISC motor, throttle position sensor, crankshaft angle sensor, fuel injectors, ignition coil, power transistor, noise filter, knock sensor on turbocharged engines, EGR temperature sensor (California vehicles), ground cable and engine control wiring harness.

7. Remove the upper radiator hose and the overflow tube.

8. Remove the spark plug cable center cover and remove the spark plug cable.

9. Disconnect and plug the high pressure fuel line.

10. Disconnect the small vacuum hoses.

11. Remove the heater hose and water bypass hose.

12. Remove the PCV hose.

13. Remove the vacuum hoses, water line and eyebolt connection for the oil line for the turbocharger.

14. Disconnect and plug the fuel return hose.

15. Disconnect the brake booster vacuum hose.

16. Remove the timing belt.

17. Remove the valve cover and the half-round seal.

18. Remove the sheet metal heat protector and remove the bolts that attach the turbocharger to the exhaust manifold.

19. Remove the exhaust pipe self-locking nuts and separate the exhaust pipe from the exhaust manifold. Discard the gasket.

20. Loosen the cylinder head mounting bolts in 3 steps, starting from the outside and working inward. Lift off the cylinder head assembly and remove the head gasket.

To install:

21. Thoroughly clean and dry the mating surfaces of the head and block. Check the cylinder head for cracks, damage or engine coolant leakage. Remove scale, sealing compound and carbon. Clean oil passages thoroughly. Check the head for flatness. End to end, the head should be within 0.002 in. normally with 0.008 in. the maximum allowed out of true. The total thickness allowed to be removed from the head and block is 0.008 in. maximum.

22. Place a new head gasket on the cylinder block with the identification marks at the front top (upward) position. Make sure the gasket has the

proper identification mark for the engine. Do not use sealer on the gasket. Replace the turbo gasket and ring, if equipped.

23. Inspect the cylinder head bolts shank length prior to installation. If the length exceeds 3.79 in. (96.4mm), the bolt must be replaced. Install the washer onto the bolt so the chamfer on the washer faces towards the head of the bolt.

24. Carefully install the cylinder head on the block.

25. Tighten the cylinder head bolts as follows:

    a. Following the proper tightening sequence, tighten the cylinder head bolts to 54 ft. lbs. (75 Nm).

    b. Loosen all bolts completely.

    c. Torque bolts to 15 ft. lbs. (20 Nm).

    d. Tighten bolts an additional ¼ turn.

    e. Tighten bolts an additional ¼ turn.

26. Install the heat shield and a new exhaust pipe gasket. Connect the exhaust pipe to the manifold.

27. Apply sealer to the perimeter of the half-round seal and to the lower edges of the half-round portions of the belt-side of the new gasket. Install the valve cover.

28. Install the timing belt and all related items.

29. Connect or install all previously disconnected hoses, cables and electrical connections. Adjust the throttle cable(s).

30. Install the spark plug cable center cover.

31. Replace the O-rings and connect the fuel lines.

32. Install the air cleaner and intake hose. Connect the breather hose.

33. Change the engine oil and oil filter.

34. Fill the system with coolant.

35. Connect the negative battery cable, run the vehicle until the thermostat opens, fill the radiator completely.

36. Check and adjust the idle speed and ignition timing.

37. Once the vehicle has cooled, recheck the coolant level.

1995–97 2.0L (4G63) Engine

1. Relieve the fuel system pressure.
2. Disconnect the negative battery cable.
3. Drain the engine coolant.
4. Drain the engine oil.
5. Disconnect the accelerator cable and remove the mounting bracket.
6. Disconnect the intake air duct (hose) from the throttle body.

7. Disconnect the following connectors;
- Idle air control (IAC) motor
- Knock sensor
- Heated oxygen sensor
- Engine coolant temperature gauge sender
- Engine coolant temperature sensor
- Ignition module (power transistor)
- Throttle position sensor
- Condenser
- Manifold differential pressure sensor
- Injectors
- Ignition coil
- Camshaft position sensor
- Crankshaft position sensor
- A/C compressor
- Engine control wiring harness

8. Remove the engine center cover.

9. Disconnect the spark plug wires.

10. Disconnect the brake booster vacuum hose.

11. Disconnect the fuel lines from the fuel supply rail.

12. Disconnect the by-pass hose and the water hose connections.

13. Disconnect the vacuum hoses, breather hose and the PCV hose.

14. Remove the timing belt.

15. Remove the power steering pump.

16. Remove the cylinder head cover and the semi-circular packing.

17. Remove the heat protector.

18. Mark the position of the hose clamps on the hoses and disconnect the water hoses and the radiator hoses.

19. Remove the thermostat housing and the O-ring.

20. Remove the intake manifold stay.

21. Remove the turbocharger assembly from the exhaust manifold.

22. Gradually loosen the cylinder head bolts in two or three steps using the specified sequence and remove the bolts.

23. Remove the cylinder head and the gasket.

To install:

24. Thoroughly clean the deck surface of the engine block and the sealing surface of the cylinder head. Check the cylinder head for warpage.

25. Measure the length of the cylinder head bolts from below the head to the end, if the bolt measures more than 3.913 in. (99.4 mm), replace the bolt.

26. Install a new gasket on the engine block with the identification mark facing upwards.

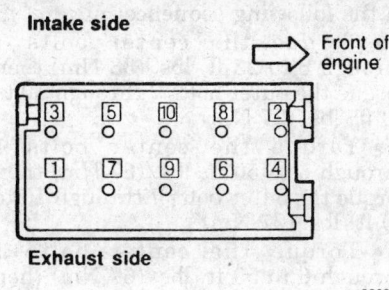

Intake side

Front of engine

Exhaust side

320957

Cylinder head bolt removal sequence — 1995–97 Eclipse 2.0L (4G63) engine

27. Carefully place the cylinder head on the engine. Apply clean engine oil to the bolts and install the bolts finger tight.

28. Torque the bolts using the following procedure;

a. Torque the bolts in sequence to 58 ft. lbs. (78 Nm).

b. Loosen the bolts completely in the reverse order.

c. Torque the bolts in sequence to 15 ft. lbs. (20 Nm).

d. Make a paint mark on the head of the bolt and the cylinder head at the same spot. Tighten the bolt 90° or ¼ turn from the mark.

e. Tighten the bolt an additional 90° or ¼ turn so that the mark on the head of the bolt is opposite the original mark on the cylinder head.

29. Use a new gasket and install the turbocharger to the exhaust manifold.

30. Install the intake manifold stay.

31. Install the thermostat housing and connect the hoses. Align the matchmarks and install the clamps.

32. Install the heat protector.

33. Apply sealant to the semi-circular packing and install it on the cylinder head.

34. Apply sealant at the front of the cylinder head where the camshaft oil seal retainer and the cylinder head

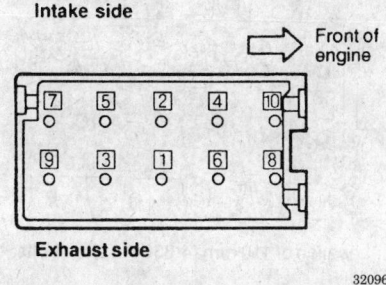

Intake side

Front of engine

Exhaust side

320960

Cylinder head bolt torque sequence — 1995–97 Eclipse 2.0L (4G63) engine

come together and install the cylinder head cover using a new gasket.

35. Install the power steering pump.

36. Install the timing belt.

37. Connect the PCV, breather and vacuum hoses.

38. Connect the water hose and the by-pass hose.

39. Use a new O-ring and connect the lines to the fuel supply rail. Apply a small amount of engine oil to the new O-ring.

40. Connect the brake booster vacuum hose.

41. Connect the spark plug wires and install the center cover.

42. Connect the following connectors;
- Idle air control (IAC) motor
- Knock sensor
- Heated oxygen sensor
- Engine coolant temperature gauge sender
- Engine coolant temperature sensor
- Ignition module (power transistor)
- Throttle position sensor
- Condenser
- Manifold differential pressure sensor
- Injectors
- Ignition coil
- Camshaft position sensor
- Crankshaft position sensor
- A/C compressor
- Engine control wiring harness

43. Connect the intake air hose to the throttle body.

44. Connect and adjust the accelerator cable.

45. Refill the engine with coolant.

46. Replace the oil filter and refill the engine with the proper amount of oil.

47. Connect the negative battery cable, start the engine and check for fuel, coolant and oil leaks.

1995–97 2.0L (420A) Engine

1. Relieve the fuel system pressure.

2. Disconnect the negative battery cable.

3. Drain the engine coolant.

4. Drain the engine oil.

5. Remove the air cleaner and air intake duct.

6. Disconnect the following connectors;
- A/C compressor
- Power steering pressure switch
- Heated oxygen sensor
- Engine coolant temperature gauge sender
- Engine coolant temperature sensor
- MAP sensor

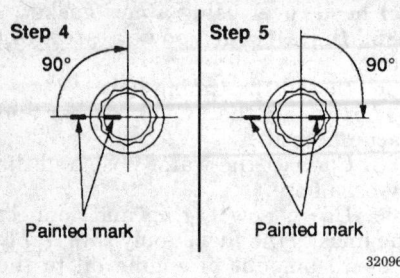

Mark the head bolt and cylinder head —
1995–97 Eclipse 2.0L (4G63) engine

- Intake air temperature sensor
- Throttle position sensor
- Idle air control (IAC) motor
- Injector harness
- Ignition coil
- Camshaft position sensor
- EGR solenoid valve

7. Disconnect the accelerator cable from the throttle body.

8. Disconnect the heater hoses from the rear of the engine.

9. Disconnect the fuel lines from the fuel supply rail.

10. Disconnect the purge air hose and the brake booster vacuum hose connections.

11. Disconnect the overflow tube connection.

12. Mark the position of the clamp on the hose and disconnect the upper radiator hose and the water hose connections.

13. Remove the timing belt.

14. Remove the intake manifold stay.

15. Remove the intake and exhaust camshafts.

16. Disconnect the exhaust pipe connection from the exhaust manifold.

17. Remove the 10 cylinder head mounting bolts and the cylinder head.

To install:

18. Thoroughly clean the cylinder head and engine block sealing surfaces. Check the deck and the cylinder head for warpage.

19. Clean the cylinder head bolts and inspect them for stretching. If the bolt appears to by stretched, replace it.

20. Place a new head gasket on the engine block and carefully place the cylinder head on the engine.

21. Coat the threads of the bolts with clean engine oil and install the bolts finger tight in the engine block. The short bolts go in the corners.

22. Torque the cylinder head bolts in the following sequence;
- Torque the center bolts 1 through 6 to 25 ft. lbs. (33 Nm) then torque the outer bolts 7 through 10 to 20 ft. lbs. (27 Nm)
- Torque the center bolts 1 through 6 to 50 ft. lbs. (67 Nm) then torque the outer bolts 7 through 10 to 20 ft. lbs. (27 Nm)
- Torque the center bolts 1 through 6 to 50 ft. lbs. (67 Nm) then torque the outer bolts 7 through 10 to 20 ft. lbs. (27 Nm)
- Turn all fasteners 1 through 10 ¼ turn (90°) more in sequence. Do not use a torque wrench for this step.

23. Use a new gasket and connect the front exhaust pipe to the exhaust manifold.

24. Install the camshafts.

25. Install the timing belts.

26. Install the intake manifold stay.

27. Connect the upper radiator hose. Install the clamp in the original position.

28. Connect the water hose to the water pipe.

29. Connect the overflow tube.

30. Connect the brake booster vacuum hose and purge air hose connection.

31. Use a new O-ring and connect the fuel lines to the fuel supply rail.

32. Connect the heater hose.

33. Connect the following connectors;
- A/C compressor
- Power steering pressure switch
- Heated oxygen sensor
- Engine coolant temperature gauge sender
- Engine coolant temperature sensor
- MAP sensor
- Intake air temperature sensor
- Throttle position sensor
- Idle air control (IAC) motor

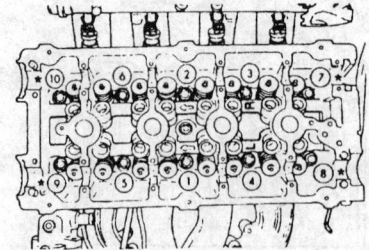

*Location of 110 mm (4.330 in.) short bolts.

Cylinder head bolt torque sequence — 1995–97
Eclipse 2.0L (420A) engine

- Injector harness
- Ignition coil
- Camshaft position sensor
- EGR solenoid valve

34. Install and adjust the accelerator cable.

35. Install the air intake duct and the air cleaner assembly.

36. Refill the engine with oil and coolant. Replace the oil filter.

37. Turn the ignition to the **ON** position and check for fuel leaks. Then start the engine and check for coolant leaks and proper operation.

1996–97 2.4L (4G64) Engine

1. Relieve the fuel system pressure.

2. Disconnect the negative battery cable.

3. Remove the air cleaner with all air intake hoses.

4. Drain the cooling system.

5. Drain the engine oil.

6. Disconnect the accelerator cable. Remove cable mounting brackets and position the cable aside.

7. Remove the breather hose.

8. At the throttle body, disconnect the three small vacuum hoses, the coolant hoses, and the brake booster vacuum hose.

9. Disconnect and plug the high pressure fuel line.

10. Disconnect and plug the fuel return hose.

11. Disconnect the oxygen sensor, engine coolant temperature sensor, the engine coolant temperature gauge unit and the engine coolant temperature switch on vehicles with air conditioning.

12. Disconnect the ISC motor, throttle position sensor, distributor, fuel injectors, noise filter, EGR temperature sensor, ground cable and engine control wiring harness.

13. Remove the spark plug cables.

14. At the thermostat case assembly, remove the coolant hoses and unbolt the thermostat case from the engine.

15. Remove the timing belt:
   a. Remove the upper timing belt cover.
   b. Align all timing marks.
   c. Secure the timing belt to the camshaft sprocket.
   d. Remove the camshaft sprocket.

16. Remove the valve cover and the half-round seal.

17. Disconnect the intake manifold stay bracket from the intake manifold.

18. Remove the exhaust pipe self-locking nuts and separate the exhaust pipe from the exhaust manifold. Discard the gasket.

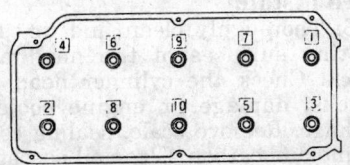

Intake side

Front of engine ⇨

Exhaust side

320988

Cylinder head bolt removal sequence —
1996–97 Eclipse 2.4L (4G64) engine

19. Loosen the cylinder head mounting bolts in 3 steps, starting from the outside and working inward. Lift off the cylinder head assembly and remove the head gasket.

To install:

20. Thoroughly clean and dry the mating surfaces of the head and block. Check the cylinder head for cracks, damage or engine coolant leakage. Remove scale, sealing compound and carbon. Clean oil passages thoroughly. Check the head for flatness.

21. Place a new head gasket on the cylinder block with the identification marks at the front top (upward) position. Make sure the gasket has the proper identification mark for the engine. Do not use sealer on the gasket.

22. Inspect the cylinder head bolts shank length prior to installation. If the length exceeds 3.91 in. (99.4mm), the bolt must be replaced. Install the washer onto the bolt so the chamfer on the washer faces towards the head of the bolt.

23. Carefully install the cylinder head on the block and tighten the cylinder head bolts as follows:

   a. Following the proper tightening sequence, tighten the cylinder head bolts to 58 ft. lbs. (78 Nm).

   b. Loosen all bolts completely.

Intake side

Front of engine ⇨

Exhaust side

320989

Cylinder head bolt installation sequence —
1996–97 Eclipse 2.4L (4G64) engine

c. Torque bolts to 15 ft. lbs. (20 Nm).

   d. Tighten bolts an additional ¼turn.

   e. Tighten bolts an additional ¼turn.

24. Install the new exhaust pipe gasket and connect the exhaust pipe to the manifold. Tighten the self-locking bolts to 33 ft. lbs. (44 Nm).

25. Install the thermostat case and tighten the mounting bolts to 18 ft. lbs. (24 Nm).

26. Connect the coolant hoses to the thermostat case.

27. Apply sealer to the perimeter of the half-round seal and to the lower edges of the half-round portions of the belt-side of the new gasket. Install the valve cover.

28. Connect the intake manifold stay and tighten the mounting bolts to 22 ft. lbs. (30 Nm).

29. Install the timing belt and all related items.

30. Connect or install all previously disconnected hoses, cables and electrical connections.

31. Install the spark plug cable center cover.

32. Replace the O-rings and connect the fuel lines.

33. Install the air cleaner and intake hose. Connect the breather hose.

34. Replace the engine oil and oil filter.

35. Fill the system with coolant.

36. Connect the negative battery cable, run the vehicle until the thermostat opens, fill the radiator completely.

37. Check and adjust the idle speed and ignition timing.

38. Check for leaks and road test for proper operation.

Galant

2.0L (4G63) Engine

1. Following proper procedure, relieve the fuel system pressure.

2. Remove the air cleaner with all air intake hoses.

3. Drain the cooling system and remove the radiator assembly.

4. Disconnect the accelerator cable.

5. Disconnect the small vacuum hose at the pressure regulator.

6. Disconnect the oxygen sensor, engine coolant temperature sensor, the engine coolant temperature gauge unit and the engine coolant temperature switch on vehicles with air conditioning.

7. Disconnect the ISC motor, throttle position sensor, crankshaft angle sensor, fuel injectors, ignition

coil, power transistor, noise filter, knock sensor on turbocharged engines, EGR temperature sensor, ground cable and engine control wiring harness.

8. Remove the spark plug cable center cover and remove the spark plug cable.

9. Disconnect and plug the high pressure fuel line.

10. Remove the heater hose and water bypass hose.

11. Remove the PCV hose.

12. Disconnect and plug the fuel return hose.

13. Disconnect the brake booster vacuum hose.

14. Remove the timing belt.

15. Remove the valve cover and the half-round seal.

16. Remove the exhaust pipe self-locking nuts and separate the exhaust pipe from the exhaust manifold. Discard the gasket.

17. Loosen the cylinder head mounting bolts in 3 steps, starting from the outside and working inward. Lift off the cylinder head assembly and remove the head gasket.

To install:

18. Thoroughly clean and dry the mating surfaces of the head and block. Check the cylinder head for cracks, damage or engine coolant leakage. Remove scale, sealing compound and carbon. Clean oil passages thoroughly. Check the head for flatness. End to end, the head should be within 0.002 in. normally, with 0.008 in. the maximum allowed out-of-true. The total thickness allowed to be removed from the head and block is 0.008 in. maximum.

19. Place a new head gasket on the cylinder block with the identification marks at the front top (upward) position. Make sure the gasket has the proper identification mark for the engine. Do not use sealer on the gasket. Replace the turbo gasket and ring, if equipped.

20. Inspect the cylinder head bolts shank length prior to installation. If the length exceeds 3.91 in. (99.4mm), the bolt must be replaced. Install the washer onto the bolt so the chamfer on the washer faces towards the head of the bolt.

21. Carefully install the cylinder head on the block.

22. On 1993 models, tighten the cylinder head bolts as follows:

   a. Following the proper tightening sequence, tighten the cylinder head bolts to 58 ft. lbs. (80 Nm).

   b. Loosen all bolts completely.

   c. Torque bolts to 14 ft. lbs. (20 Nm).

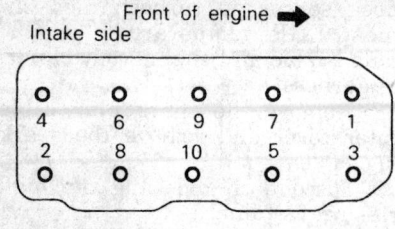

Front of engine ➡
Intake side

```
 4   6   9   7   1
 2   8  10   5   3
```

Exhaust side

164578

Cylinder head bolt removal sequence — 1993 Galant 2.0L (4G63) engine

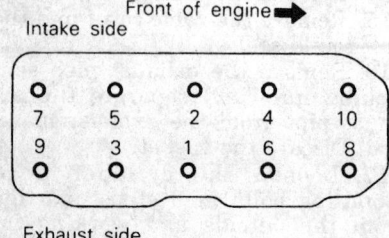

Front of engine ➡
Intake side

```
 7   5   2   4  10
 9   3   1   6   8
```

Exhaust side

164580

Cylinder head bolt installation sequence — 1993 Galant 2.0L (4G63) engine

d. Tighten bolts an additional ¼ turn.

e. Tighten bolts an additional ¼ turn.

23. Install the new exhaust pipe gasket and connect the exhaust pipe to the manifold. Tighten the self-locking bolts to 29 ft. lbs. (40 Nm).

24. Apply sealer to the perimeter of the half-round seal and to the lower edges of the half-round portions of the belt-side of the new gasket. Install the valve cover.

25. Install the timing belt and all related items.

26. Connect or install all previously disconnected hoses, cables and electrical connections. Adjust the throttle cable(s).

27. Install the spark plug cable center cover.

28. Replace the O-rings and connect the fuel lines.

29. Install the air cleaner and intake hose. Connect the breather hose.

30. Change the engine oil and oil filter.

31. Install the radiator and fill the system with coolant.

32. Connect the negative battery cable, run the vehicle until the thermostat opens, fill the radiator completely.

33. Check and adjust the idle speed and ignition timing.

34. Once the vehicle has cooled, recheck the coolant level.

2.4L (4G64) Engine

1. Following proper procedure, relieve the fuel system pressure.

2. Remove the air cleaner with all air intake hoses.

3. Drain the cooling system.

4. Disconnect the accelerator cable. Remove cable mounting brackets and position the cable aside.

5. Remove the breather hose.

6. At the throttle body, disconnect the three small vacuum hoses, the coolant hoses, and the brake booster vacuum hose.

——— CAUTION ———
Do not allow fuel spray or fuel vapors to come in contact with a spark or open flame. Keep a dry chemical fire extinguisher nearby. Never store fuel in an open container due to risk of fire or explosion.

7. Disconnect and plug the high pressure fuel line.

8. Disconnect and plug the fuel return hose.

9. Disconnect the oxygen sensor, engine coolant temperature sensor, the engine coolant temperature gauge unit and the engine coolant temperature switch on vehicles with air conditioning.

10. Disconnect the ISC motor, throttle position sensor, distributor, fuel injectors, noise filter, EGR temperature sensor, ground cable and engine control wiring harness.

11. Remove the spark plug cables.

12. At the thermostat case assembly, remove the coolant hoses and unbolt the thermostat case from the engine.

13. Remove the timing belt:

a. Remove the upper timing belt cover.

b. Align all timing marks.

c. Secure the timing belt to the camshaft sprocket.

d. Remove the camshaft sprocket.

14. Remove the valve cover and the half-round seal.

15. Disconnect the intake manifold stay bracket from the intake manifold.

16. Remove the exhaust pipe self-locking nuts and separate the exhaust pipe from the exhaust manifold. Discard the gasket.

17. Loosen the cylinder head mounting bolts in 3 steps, starting from the outside and working inward. Lift off the cylinder head assembly and remove the head gasket.

To install:

18. Thoroughly clean and dry the mating surfaces of the head and block. Check the cylinder head for cracks, damage or engine coolant leakage. Remove scale, sealing compound and carbon. Clean oil passages thoroughly. Check the head for flatness.

19. Place a new head gasket on the cylinder block with the identification marks at the front top (upward) position. Make sure the gasket has the proper identification mark for the engine. Do not use sealer on the gasket. Replace the turbo gasket and ring, if equipped.

20. Inspect the cylinder head bolts shank length prior to installation. If the length exceeds 3.91 in. (99.4mm), the bolt must be replaced. Install the washer onto the bolt so the chamfer on the washer faces towards the head of the bolt.

21. Carefully install the cylinder head on the block and tighten the cylinder head bolts as follows:

a. Following the proper tightening sequence, tighten the cylinder head bolts to 58 ft. lbs. (78 Nm).

b. Loosen all bolts completely.

c. Torque bolts to 15 ft. lbs. (20 Nm).

d. Tighten bolts an additional ¼ turn.

e. Tighten bolts an additional ¼ turn.

22. Install the new exhaust pipe gasket and connect the exhaust pipe to the manifold. Tighten the self-locking bolts to 33 ft. lbs. (44 Nm).

23. Install the thermostat case and tighten the mounting bolts to 18 ft. lbs. (24 Nm).

24. Connect the coolant hoses to the thermostat case.

25. Apply sealer to the perimeter of the half-round seal and to the lower edges of the half-round portions of the belt-side of the new gasket. Install the valve cover.

26. Connect the intake manifold stay and tighten the mounting bolts to 22 ft. lbs. (30 Nm).

27. Install the timing belt and all related items.

28. Connect or install all previously disconnected hoses, cables and electrical connections.

29. Install the spark plug cable center cover.

30. Replace the O-rings and connect the fuel lines.

31. Install the air cleaner and intake hose. Connect the breather hose.

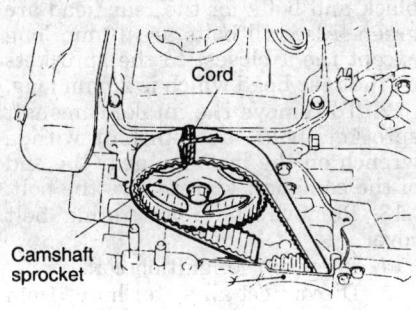

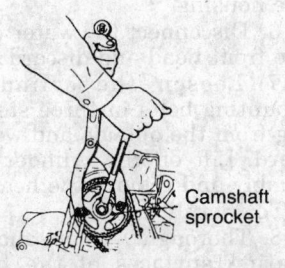

327730

Secure the timing belt to the camshaft sprocket and remove the sprocket — Galant 2.4L (4G64) engine

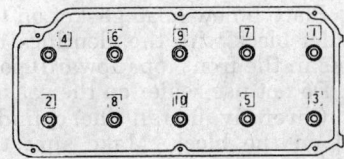

327724

Cylinder head bolt removal sequence — Galant 2.4L (4G64) engine

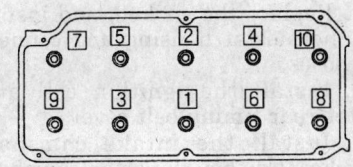

327725

Cylinder head bolt installation sequence — Galant 2.4L (4G64) engine

32. Change the engine oil and oil filter.

33. Fill the system with coolant.

34. Connect the negative battery cable, run the vehicle until the thermostat opens, fill the radiator completely.

35. Check and adjust the idle speed and ignition timing.

36. Once the vehicle has cooled, recheck the coolant level.

Diamante

3.0L (6G72) DOHC Engine

1. Relieve fuel system pressure. Disconnect the negative battery cable.

2. Drain the cooling system.

3. Remove the air intake hoses.

4. Remove air intake plenum and intake manifold.

5. Remove the exhaust manifold.

6. Remove the timing belt.

— WARNING —
Once the timing belt is removed, DO NOT rotate the engine or camshafts. Internal engine damage will result.

7. Remove the breather hose.

8. Remove the spark plug cable center cover and remove the spark plug cables.

9. Remove the rocker covers.

10. Holding the flats of the camshaft, remove the intake camshaft sprocket holding bolts and remove the sprockets.

11. Remove the rear timing belt cover.

12. Remove the ignition coil assembly.

13. Disconnect all water hoses from the thermostat housing and remove the housing.

14. Disconnect the water inlet from the front head and discard O-ring.

15. Loosen the cylinder head mounting bolts in the reverse of the torque sequence and loosen the bolts in three steps. Lift off the cylinder head assembly and remove the head gasket.

To install:

16. Thoroughly clean and dry the mating surfaces of the head and block. Check the cylinder head for cracks, damage or engine coolant leakage. Remove scale, sealing compound and carbon. Clean oil passages thoroughly. Check the head for flatness. End to end, the head should be within 0.002 inches. The total thickness allowed to be removed from the head and block is 0.008 inches maximum.

17. Place a new head gasket on the cylinder block with the identification marks in the front top (upward) position. Do not use sealer on the gasket.

18. Carefully install the cylinder head on the block. Make sure the head bolt washers are installed with the chamfered edge upward. Using three even steps, torque the head bolts in sequence, to 76–83 ft. lbs. (105–115 Nm) for cold engine.

19. Install new O-ring and connect the water inlet to the head. Tighten the mounting bolt to 9–11 ft. lbs. (12–15 Nm).

20. Replace the gaskets and install the thermostat housing. Tighten the mounting bolts to 12–14 ft. lbs. (17–20 Nm).

21. Using new hose clamps, connect the hoses to the thermostat housing.

22. Install the ignition coil and torque the mounting bolts to 7 ft. lbs. (10 Nm).

23. Install the rear timing belt cover and torque the mounting bolts to 17 ft. lbs. (24 Nm).

24. Install the intake camshaft sprocket. Use the hex flange on camshaft to secure and torque the retaining bolt to 65 ft. lbs. (90 Nm).

25. Apply sealer to the lower edges of the valve cover. Tighten the bolts in the proper sequence to 44–51 inch lbs. (5–6 Nm).

26. Connect the spark plug cables and install the center cover. Tighten the bolts that secure the center cover to 27 inch lbs. (3 Nm)

27. Install the breather hose.

28. Install the timing belt assembly.

29. Install the exhaust manifold assembly.

30. Using all new gaskets, install the intake manifold and air intake plenum.

31. Install the air intake hoses.

32. Change the engine oil and oil filter.

33. Fill the system with coolant.

34. Connect the negative battery cable, run the vehicle until the thermostat opens, fill the radiator completely.

35. Adjust the accelerator cable. Check and adjust the idle speed and ignition timing.

36. Once the vehicle has cooled, recheck the coolant level.

3.0L (6G72) SOHC Engine

1. Relieve the fuel system pressure. Disconnect the negative battery cable.

2. Drain the cooling system.

3. Remove the air intake hose.

4. Remove the exhaust manifold.

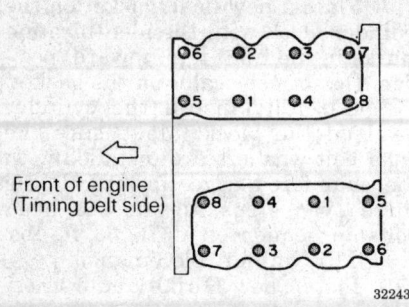

Cylinder head bolt torque sequence — Diamante and 3000GT

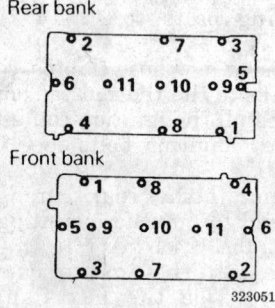

Rocker cover bolt torque sequence — Diamante 3.0L (6G72) DOHC engine

5. Remove the air intake plenum and intake manifold.

6. Remove the timing belt.

7. Remove the camshaft sprocket and the rear timing belt cover.

8. Remove the power steering pump bracket. If removing the rear head, remove the alternator brace.

9. Disconnect the water inlet pipe.

10. Remove the purge pipe assembly.

11. Remove the valve cover.

12. Using the reverse sequence of the tightening torque, loosen the cylinder head mounting bolts in three steps. Lift off the cylinder head assembly and remove the head gasket.

To install:

13. Thoroughly clean and dry the mating surfaces of the head and block. Check the cylinder head for cracks, damage or engine coolant leakage. Remove scale, sealing compound and carbon. Clean oil passages thoroughly. Check the head for flatness. End to end, the head should be within 0.002 inches of true. The total thickness allowed to be removed from the head and block is 0.008 in. maximum.

14. Place a new head gasket on the cylinder block making sure the identification mark on the cylinder head

gasket is in the front top (upward) location. Do not use sealer on the gasket. Make sure the gasket has the proper identification mark for the engine.

15. Carefully install the cylinder head on the block. Make sure the head bolt washers are installed with the chamfered edge upward. Using three even steps, torque the head bolts in sequence, to 76–83 ft. lbs. (105–115 Nm). This torque specification is for a cold engine.

16. Apply sealer to the lower edges of the half-round portions and install the valve cover. Tighten valve cover bolts to 7 ft. lbs. (9 Nm).

17. Install the purge pipe assembly.

18. Connect the water inlet pipe.

19. Install the power steering pump bracket and alternator brace.

20. Install the rear timing belt cover and cam sprocket. Torque the retaining bolt to 65 ft. lbs. (90 Nm).

21. Install the timing belt and all related items.

22. Using all new gaskets, install the intake manifold, air intake plenum and exhaust manifold, following the proper torque sequences.

23. Install the air intake hose.

24. Change the engine oil and oil filter.

25. Fill the system with coolant.

26. Connect the negative battery cable, run the vehicle until the thermostat opens, fill the radiator completely.

27. Check and adjust the idle speed and ignition timing.

28. Once the vehicle has cooled, recheck the coolant level.

3000GT Engine

1. Relieve fuel system pressure. Disconnect the negative battery cable.

―――――― CAUTION ――――――
Wait at least 90 seconds after the negative (-) battery cable is disconnected to prevent possible deployment of the air bag.

2. Drain the cooling system.

3. Remove the air intake hoses.

4. Remove air intake plenum and intake manifold.

5. Remove the turbocharger, if equipped.

6. Remove the exhaust manifold.

7. Remove the timing belt.

8. Remove the triple pipe assembly across the top of the engine.

9. Remove the breather hose.

10. Remove the spark plug cable center cover and remove the spark plug cables.

11. When removing the valve cover, note that bolts for the front head are black and bolts for the rear head are green. Also, all bolts are 10mm long except the 1 closest to the sprockets on the rear head which is 20mm long.

12. To remove the intake camshaft sprocket, hold the camshaft with a wrench on the hexagon near the end of the camshaft and remove the bolt.

13. Remove the rear timing belt cover.

14. Remove the ignition coil.

15. Disconnect all water hoses from the thermostat housing and remove the housing.

16. Disconnect the water inlet from the front head and discard O-ring.

17. Loosen the cylinder head mounting bolts in three steps, starting from the outside and working inward. Lift off the cylinder head assembly and remove the head gasket.

To install:

18. Thoroughly clean and dry the mating surfaces of the head and block. Check the cylinder head for cracks, damage or engine coolant leakage. Remove scale, sealing compound and carbon. Clean oil passages thoroughly. Check the head for flatness. End to end, the head should be within 0.002 in. normally with 0.008 in. the maximum allowed out of true. The total thickness allowed to be removed from the head and block is 0.008 in. maximum.

19. Place a new head gasket on the cylinder block with the identification marks in the front top (upward) position. Do not use sealer on the gasket.

20. Carefully install the cylinder head on the block. Make sure the head bolt washers are installed with the chamfered edge upward. Using three even steps, torque the head bolts in sequence, to 76–83 ft. lbs. (105–115 Nm) for non-turbocharged cold engine or 87–94 ft. lbs. (120–130 Nm) for turbocharged cold engine.

21. On turbocharged models, loosen all cylinder head bolts and retighten in sequence to 87–94 ft. lbs. (120–130 Nm).

22. Install new O-ring and connect the water inlet to the front head.

23. Replace the gaskets and install the thermostat housing and connect the hoses.

24. Install the ignition coil and center rear timing belt cover.

25. Install the intake camshaft sprocket. Use hex flange on camshaft to secure and torque the retaining bolt to 65 ft. lbs. (90 Nm).

26. Apply sealer to the lower edges of the half-round portions of the belt-side of the new gasket and install the

valve cover. Make sure green bolts are installed on the rear head and black bolts are installed on the front head. Also, make sure the longest bolt is installed in its proper location closest to the sprockets on the rear head. Tighten the bolts in the proper sequence to 26 inch lbs. (3 Nm). Then retighten bolts No. 1–6 to 35 inch lbs. (4 Nm).

27. Connect the spark plug cables and install the center cover.

28. Install the breather hose.

29. Install the triple pipe assembly across the top of the engine and torque the retaining bolts to 7 ft. lbs. (10 Nm).

30. Install the timing belt and all related items.

31. Using all new gaskets, install the intake manifold, air intake plenum, turbocharger and exhaust manifold, following the proper torque sequences.

32. Install the air intake hoses.

33. Change the engine oil and oil filter.

34. Fill the system with coolant.

35. Connect the negative battery cable, run the vehicle until the thermostat opens, fill the radiator completely.

36. Adjust the accelerator cable. Check and adjust the idle speed and ignition timing.

37. Once the vehicle has cooled, recheck the coolant level.

Lash Adjusters

BLEEDING

Precis, Diamante, Galant and Eclipse

NOTE: The hydraulic lash adjuster is a precision component that relies on a clean operating environment. When handling the lash adjuster, make certain that no dirt or foreign particles are allowed to get inside the unit. DO NOT try to take the unit apart as it is not rebuildable. The adjuster is filled with diesel fuel. Hold the adjuster in the upright position so that the diesel fuel does not spill out. When cleaning the unit, only use clean diesel fuel.

To bleed the hydraulic lash adjusters when disassembling the rocker arms or as part of an engine overhaul, perform the following:

1. Mount the adjuster into special bleeding tool 09246–32100 or equivalent and immerse the tool and adjuster in a container of clean diesel fuel.

2. With air bleed wire 09246–32200 or equivalent inserted in the adjuster, lightly press down on the steel ball and compress the plunger four or five times.

3. Remove the air bleed wire from the adjuster and push down firmly on the plunger. If the plunger moves even slightly, repeat Steps 1 and 2 until the plunger stops moving. If the plunger continues to move, replace it.

REMOVAL AND INSTALLATION

Eclipse 1.8L (4G37) Engine

1. Disconnect the negative battery cable.

2. Remove the valve cover. Install lash adjuster retainer tools MD998443 or equivalent, to the rocker arm.

3. Remove the distributor extension, if necessary.

4. Remove camshaft timing belt.

5. Working in a crisscross pattern from the center outward, loosen the camshaft bearing caps in gradual steps.

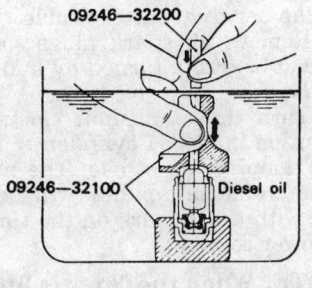

Bleed the lash adjuster in clean diesel fuel

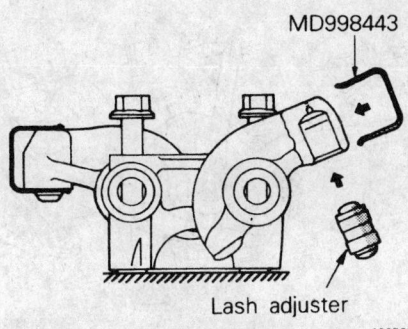

Valve lash adjuster identification — Eclipse 1.8L (4G37) engine

6. Remove the rocker arms, shafts and bearing caps as an assembly.

NOTE: It is essential that all parts being reused be kept in the same order and orientation for reinstallation. Be sure to mark and separate parts, so parts will not be mixed during reassembly. Otherwise, valvetrain noise and excessive wear could result.

7. Disassemble rocker shaft assembly. Starting at rear bearing cap, slide each piece off shafts.

NOTE: Inspect the roller surfaces of the rockers. Replace if there are any signs of damage or if the roller does not turn smoothly. Check the inside bore of the rockers and lifter for wear. It is also recommended that all rocker arms and lash adjusters be replaced together.

To install:

8. Apply a drop of sealant to the rear edges of the end caps.

9. Install the assembly into the front bearing cap making sure the notches in the rocker shafts are facing up. Insert the installation bolt but do not tighten at this point.

10. Install the remaining cap bolts. Tighten all bolts evenly and gradually to 15 ft. lbs. (20 Nm). Remove the lash adjuster retainers.

11. Install the timing belt as required.

12. Install the distributor extension, if removed.

13. Install the valve cover, with a new gasket and semi-circular packing in place.

14. Connect the negative battery cable.

15. Run engine and check ignition timing.

Eclipse 2.0L (4G63 and 420A) and All 1993 Galant Engines

1. Relieve the fuel system pressure following proper procedure. Disconnect negative battery cable.

2. Disconnect the accelerator cable, PCV hoses, breather hoses, spark plug cables and the remove the valve cover.

NOTE: Always rotate the crankshaft in a clockwise direction. Make a mark on the back of the timing belt indicating the direction of rotation so it may be reassembled in the same direction if it is to be reused.

3. Rotate the crankshaft clockwise and align the timing marks so No. 1 piston will be at TDC of the compres-

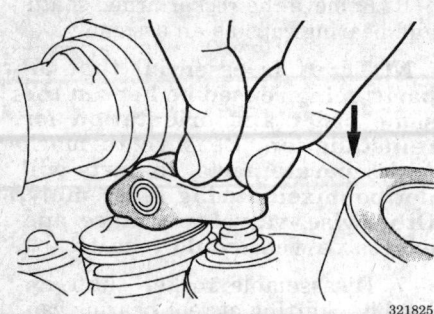

Valve lash adjuster location — 1993 Galant and Eclipse with 2.0L (4G63 and 420A) engines

sion stroke. At this time the timing marks on the camshaft sprocket and the upper surface of the cylinder head should coincide, and the dowel pin of the camshaft sprocket should be at the upper side.

4. Remove the timing belt upper and lower covers.

5. Remove the timing belt.

6. Remove the crank angle sensor.

NOTE: It is essential that all parts be kept in the same order and orientation for reinstallation. In order to prevent confusion during installation, be sure to mark and separate all parts.

7. Remove the camshafts, rocker arms and lash adjusters.

8. Visually inspect the rocker arm roller and replace if dent, damage or seizure is evident. Check the roller for smooth rotation. Replace if excess play or binding is present. Also, inspect valve contact surface for possible damage or seizure. It is recommended that all rocker arms and lash adjusters be replaced together.

To install:

9. Install the lash adjusters and rocker arms into the cylinder head. Lubricate lightly with clean oil prior to installation.

10. Apply engine oil to the lobes and journals of each camshaft. Install the camshafts into the cylinder head taking care not to confuse the intake and the exhaust camshaft; the intake camshaft has a slit on its rear end for driving the crank angle sensor. Align shafts so dowel pins on camshaft sprocket end are located on the top.

11. Install and tighten the camshaft bearing caps in the proper sequence torquing to specifications in three even progressions.

12. Replace the camshaft oil seals and install the sprockets.

13. Locate the dowel pin on the sprocket end of the intake camshaft at the top position, if not already done.

14. Align the punch mark on the crank angle sensor housing with the notch on the sensor plate. Install the crank angle sensor into the cylinder head.

15. Install the timing belt, covers and related components.

16. Install the valve cover using new gasket. Reconnect all related components.

17. Reconnect the negative battery cable.

Valve Lash

ADJUSTMENT

Except Mirage

Valve clearance is not adjustable on these vehicles. If the valve makes noise, look for a leaky lash adjuster or excessive camshaft or rocker arm wear.

Mirage

NOTE: Incorrect valve clearances will cause unsteady engine operation, excessive noise and reduced engine output. Check the valve clearances and adjust as required while the engine is hot.

1. Warm the engine to operating temperature, turn **OFF** and disconnect the negative battery cable.

2. Remove all spark plugs so engine can be easily turned by hand.

3. Remove the valve cover.

4. Turn the crankshaft clockwise to position the No. 1 cylinder at TDC of its compression stroke. The notch on the crankshaft pulley will be aligned with the **T** mark on the timing belt lower cover.

NOTE: When the No. 1 cylinder at TDC of its compression stroke, the No. 1 cylinder intake and exhaust valves will have valve lash.

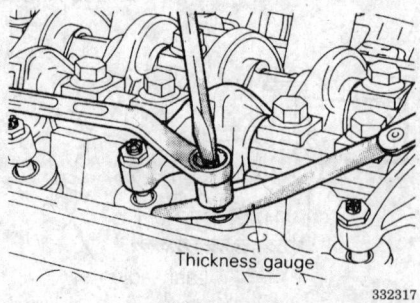

Thickness gauge

332317

Proper method of adjusting valve clearance — Mirage

5. Check the valve lash at cylinder No. 1 intake, cylinder No. 1 exhaust, cylinder No. 2 intake and cylinder No. 3 exhaust valves.

6. Rotate the crankshaft clockwise 1 complete turn and align the **T** mark. This will position the No. 4 cylinder at TDC of its compression stroke.

NOTE: When the No. 4 cylinder at TDC of its compression stroke, the No. 4 cylinder intake and exhaust valves will have valve lash.

7. Check the valve lash at cylinder No. 2 exhaust, cylinder No. 3 intake and cylinder No. 4 intake and exhaust valves.

8. If the valve clearances are out of specification, loosen the rocker arm locknut and adjust the clearance using a feeler gauge while turning the adjusting screw. Be sure to hold the screw to prevent it from turning when tightening the locknut.

9. After adjusting the valves, install the valve cover with new gasket and spark plugs, and connect the negative battery cable.

Rocker Arms

REMOVAL AND INSTALLATION

Precis

1. Disconnect the negative battery cable.

2. Remove the PCV hose running from the rocker cover and the air cleaner. Remove the air cleaner.

3. Remove the upper timing belt cover. Remove the rocker cover.

4. Loosen the rocker shaft mounting bolts but do not remove them. Remove each rocker shaft, rocker arms and springs as an assembly. Disassemble the whole assembly by progressively removing each bolt and then the associated springs and rockers, keeping all parts in the exact order of disassembly. The left and right springs have different tension ratings and free length. Observe the location of the rocker arms as they are removed. Exhaust and intake, right and left are different. Do not mix them up.

5. Check the rocker arm face contacting the cam lobe and the adjusting screw that contacts the valve stem for excess wear. Inspect the fit of the rockers on the shaft. Replace adjusting screws, rockers, and/or shafts that show excessive wear. Pay special attention to the contact pad ends of the rocker arms and the ball surface of the adjusting studs. Check

the diameter of the shaft at the rocker mounting points and subtract that number from the measured inside diameter of the corresponding rocker arm. Clearance should be 0.0005–0.0017 in. (0.013–0.043mm). The service limit is 0.004 in. (0.1mm). Check the rocker shaft bend. Total rocker shaft bend should be 0.002 in. (0.05mm). Check the spring free length. Maximum free length should be 2.1 in. (53.3mm) for the exhaust side springs; 2.6 in. (66mm) for intake side springs.

To install:

6. Assemble all the parts, noting the differences between intake and exhaust parts. The intake rocker shaft is much longer; the intake rocker shaft springs are over 3 in. long, while those for the exhaust side are less than 2 in. long; rockers are labeled 1–3 and 2–4 for the cylinder with which they are associated. Torque the rocker shaft mounting bolts to 15–19 ft. lbs. (20–26 Nm).

7. Adjust the valve clearances. This step may be omitted only if all parts are being reused.

8. Install the rocker cover with a new gasket, torquing the bolts to 12–18 inch lbs. (1.5–2.0 Nm).

9. Install the air cleaner and PCV valve. Connect the battery cable.

10. Run the engine at idle speed until it is hot. Then, unless valves did not require adjustment, remove the valve cover again and adjust the valve clearances with the engine hot.

11. Replace the rocker cover and timing belt cover, air cleaner and PCV valve.

Mirage

1. Disconnect the negative battery cable.

2. Rotate the engine and position the No. 1 piston to TDC of its compression stroke.

3. For 1.8L engines, label and disconnect the spark plug cables.

4. For 1.8L engines, disconnect the air flow sensor connector and remove the air cleaner case cover.

5. Disconnect the accelerator cable, breather hose and PCV hose connections.

6. Remove the rocker cover and discard the gasket.

7. Loosen both rocker arm shaft assemblies gradually and evenly and remove the rocket shafts from the vehicle. Do not disassembly rocker arms and rocker arm shaft assemblies.

8. If disassembly is required, keep all parts in the exact order of removal. Inspect the roller surfaces of the rockers. Replace if there are any signs of damage or if the roller does not turn smoothly. Check the inside bore of the rockers and the adjuster tip for wear.

To install:

9. Lubricate the rocker shaft with clean engine oil and install the rockers and springs in their proper places.

10. Install the rocker arm and shaft assemblies. Tighten the rocker arm shaft retainer bolts to 23 ft. lbs. (32 Nm).

11. Check valve adjustment and install valve cover with a new gasket. Tighten the valve cover bolts to 16 inch lbs. (1.8 Nm) for the 1.5L engine or to 29 inch lbs. (3.3 Nm) for the 1.8L engine.

12. If detached, connect the spark plug cables.

13. Connect the accelerator cable, breather hose and PCV hose.

14. For the 1.8L engines, connect the air flow sensor connector and install the air cleaner case cover.

15. Connect the negative battery cable. Run the engine at idle until normal operating temperature is reached. Check idle speed and ignition timing and adjust as required.

Eclipse

1.8L (4G37) Engine

1. Disconnect the negative battery cable.

2. Remove the valve cover. Install lash adjuster retainer tools MD998443 or equivalent, to the rocker arm.

3. Remove the distributor extension, if necessary.

4. Remove camshaft timing belt.

5. Working in a crisscross pattern from the center outward, loosen the camshaft bearing caps in gradual steps.

6. Remove the rocker arms, shafts and bearing caps as an assembly.

NOTE: It is essential that all parts be kept in the same order and orientation for reinstallation. Be sure to mark and separate parts to keep them from getting mixed. This will aid assembly.

7. Disassemble rocker shaft assembly. Starting at rear bearing cap, slide each piece off shafts.

NOTE: Inspect the roller surfaces of the rockers. Replace if there are any signs of damage or if the roller does not turn smoothly. Check the inside bore of the rockers and lifter for wear.

COMPONENTS [FBC]

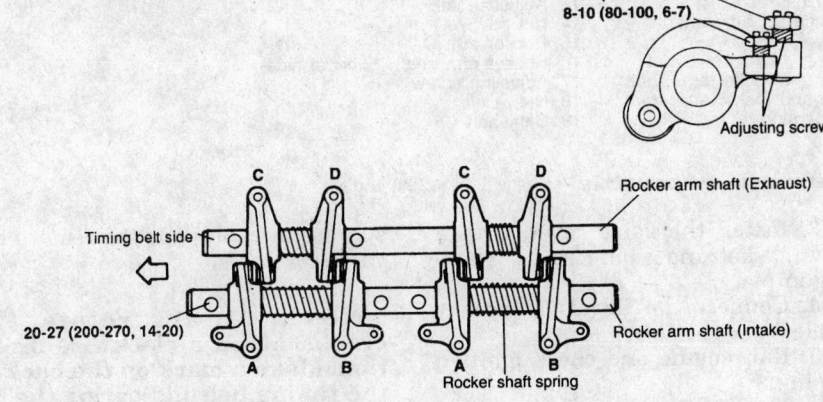

333746

Rocker arm and shaft assembly with jet valves — Precis

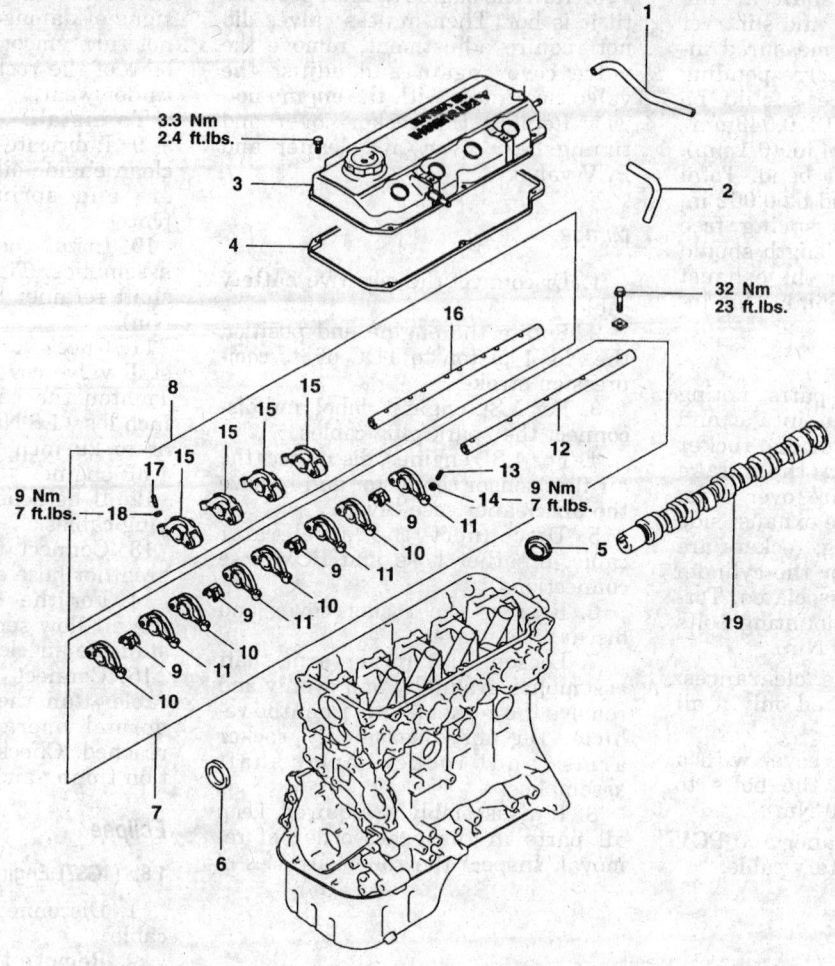

1
2

3.3 Nm
2.4 ft.lbs.

3

4

32 Nm
23 ft.lbs.

16

8
15   15

15   15
17   15
9 Nm
7 ft.lbs.   18

13
14

9 Nm
7 ft.lbs.

12

11
9   11
10
9   11
10
9   11
10
9   11
10
9   11
10

5

19

7

6

1. Breather hose
2. P.C.V. hose
3. Rocker cover
4. Rocker cover gasket
   Valve clearance pre-adjustment
5. Oil seal
6. Oil seal
7. Rocker arms and rocker arm shaft
8. Rocker arms and rocker arm shaft
9. Rocker shaft spring

10. Rocker arm A
11. Rocker arm B
12. Rocker arm shaft (Intake side)
13. Adjusting screw
14. Nut
15. Rocker arm C
16. Rocker arm shaft (Exhaust side)
17. Adjusting screw
18. Nut
19. Camshaft

331925

Camshaft, rocker arm and shaft assemblies — Mirage 1.8L (4G93) engine

To install:

8. Apply a drop of sealant to the rear edges of the end caps.

9. Install the assembly into the front bearing cap making sure the notches in the rocker shafts are facing up. Insert the installation bolt but do not tighten at this point.

10. Install the remaining cap bolts. Tighten all bolts evenly and gradually to 15 ft. lbs. (20 Nm). Remove the lash adjuster retainers.

11. Install the timing belt as required.

12. Install the distributor extension, if removed.

13. Install the valve cover, with a new gasket and semi-circular packing in place.

14. Connect the negative battery cable.

15. Run engine and check ignition timing.

2.0L (4G63) Engine

1. Relieve the fuel system pressure following proper procedure. Disconnect negative battery cable.

2. Disconnect the accelerator cable, PCV hoses, breather hoses,

spark plug cables and the remove the valve cover.

NOTE: Always rotate the crankshaft in a clockwise direction. Make a mark on the back of the timing belt indicating the direction of rotation so it may be reassembled in the same direction if it is to be reused.

3. Rotate the crankshaft clockwise and align the timing marks so No. 1 piston will be at TDC of the compression stroke. At this time the timing marks on the camshaft sprocket and the upper surface of the cylinder head should coincide, and the dowel

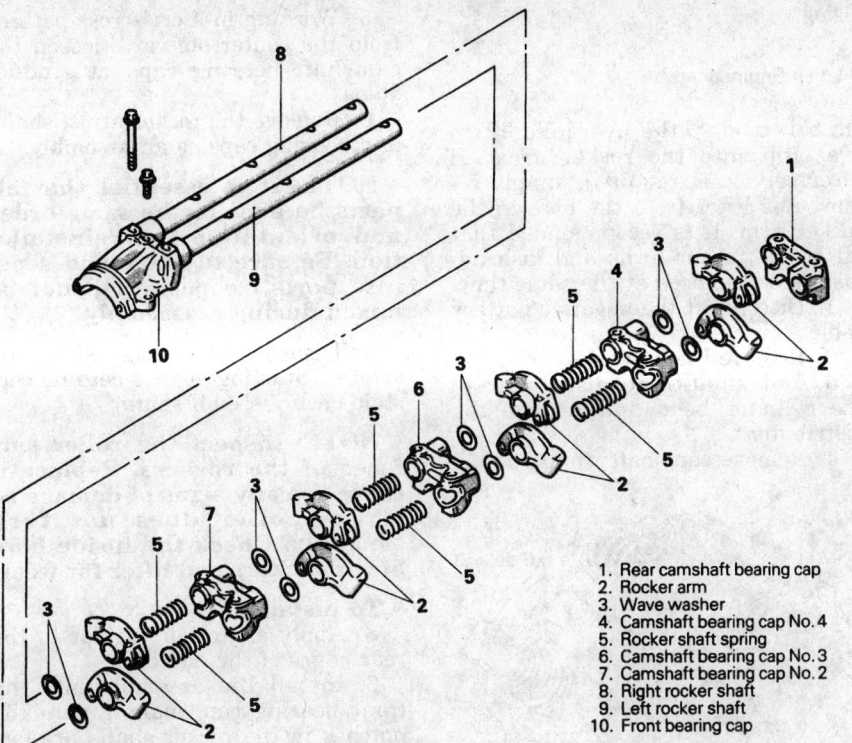

1. Rear camshaft bearing cap
2. Rocker arm
3. Wave washer
4. Camshaft bearing cap No. 4
5. Rocker shaft spring
6. Camshaft bearing cap No. 3
7. Camshaft bearing cap No. 2
8. Right rocker shaft
9. Left rocker shaft
10. Front bearing cap

176628

Rocker arm shaft exploded view — Eclipse 1.8L (4G37) engine

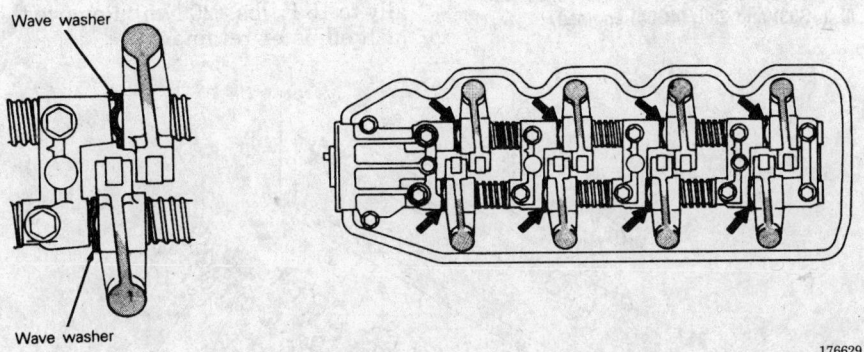

176629

Wave washer locations — Eclipse 1.8L (4G37) engine

pin of the camshaft sprocket should be at the upper side.

4. Remove the timing belt upper and lower covers.

5. Remove the timing belt.

6. Remove the crank angle sensor.

NOTE: It is essential that all parts be kept in the same order and orientation for reinstallation. Mark and separate all parts. Do not mix parts. Valve train components that are to be reused must be installed in the same locations from which they were removed.

7. Remove the camshafts, rocker arms and lash adjusters.

8. Visually inspect the rocker arm roller and replace if dent, damage or seizure is evident. Check the roller for smooth rotation. Replace if excess play or binding is present. Also, inspect valve contact surface for possible damage or seizure. It is recommended that all rocker arms and lash adjusters be replaced together.

To install:

9. Install the lash adjusters and rocker arms into the cylinder head. Lubricate lightly with clean oil prior to installation.

10. Apply engine oil to the lobes and journals of each camshaft. Install the camshafts into the cylinder head taking care not to confuse the intake and the exhaust camshaft; the intake camshaft has a slit on its rear end for driving the crank angle sensor. Align shafts so dowel pins on camshaft sprocket end are located on the top.

11. Install and tighten the camshaft bearing caps in the proper sequence torquing to specifications in three even steps.

12. Replace the camshaft oil seals and install the sprockets.

13. Locate the dowel pin on the sprocket end of the intake camshaft at the top position, if not already done.

14. Align the punch mark on the crank angle sensor housing with the notch on the sensor plate. Install the crank angle sensor into the cylinder head.

15. Install the timing belt, covers and related components.

16. Install the valve cover using new gasket. Reconnect all related components.

17. Reconnect the negative battery cable.

2.0L (420A) Engine

1. Relieve the fuel system pressure following proper procedure. Disconnect negative battery cable.

2. Disconnect the accelerator cable, PCV hoses, breather hoses, spark plug cables and the remove the cylinder head cover.

NOTE: Always rotate the crankshaft in a clockwise direction. Make a mark on the back of the timing belt indicating the direction of rotation so it may be reassembled in the same direction if it is to be reused.

3. Rotate the crankshaft clockwise and align the timing marks so No. 1 piston will be at TDC of the compression stroke. At this time the timing marks on the camshaft sprocket and the upper surface of the cylinder head should coincide, and the dowel pin of the camshaft sprocket should be at the upper side.

4. Remove the timing belt upper and lower covers.

5. Remove the timing belt.

6. Remove the crank angle sensor.

NOTE: It is essential that all parts be kept in the same order and orientation for reinstallation. Mark and separate all parts. Do not mix parts. Valve train components that are to be reused must be installed in the same locations from which they were removed.

7. Remove the camshafts, rocker arms (cam followers) and lash adjusters.

8. Visually inspect the rocker arm roller and replace if dent, damage or seizure is evident. Check the roller for smooth rotation. Replace if excess play or binding is present. Also, inspect valve contact surface for possible damage or seizure. It is recommended that all rocker arms and lash adjusters be replaced together.

To install:

9. Install the lash adjusters and rocker arms into the cylinder head. Lubricate lightly with clean oil prior to installation.

10. Apply engine oil to the lobes and journals of each camshaft. Install the camshafts into the cylinder head taking care not to confuse the intake and the exhaust camshaft; the intake camshaft has a slit on its rear end for driving the crank angle sensor. Align shafts so dowel pins on camshaft sprocket end are located on the top.

------ **WARNING** ------
Piston should not be at TDC when installing the camshaft.

11. Install and tighten the four center camshaft bearing caps in the proper sequence torquing to specifications in three even steps. Torque the bolts to 9 ft. lbs. (12 Nm).

12. Apply Loctite 51817® or equivalent to the front and rear bearing caps. Install the bearing caps and torque the bolts to 20 ft. lbs. 28 Nm.

13. Replace the camshaft oil seals and install the sprockets.

14. Locate the dowel pin on the sprocket end of the intake camshaft at the top position, if not already done.

15. Align the punch mark on the crank angle sensor housing with the notch on the sensor plate. Install the crank angle sensor into the cylinder head.

16. Install the timing belt, covers and related components.

17. Install the valve cover using new gasket. Reconnect all related components.

18. Reconnect the negative battery cable.

Galant

8 Valve Engines

On this engine, the hydraulic lifters are built into the rocker arms. If lifter service is required, simply remove the lifter from the bore in the rocker arm. It is recommended that all of the rocker arms and lash adjusters be replaced at the same time.

1. Disconnect the negative battery cable.
2. Remove the valve cover.
3. Matchmark the distributor to the cylinder head and remove the distributor.
4. Remove camshaft timing belt.

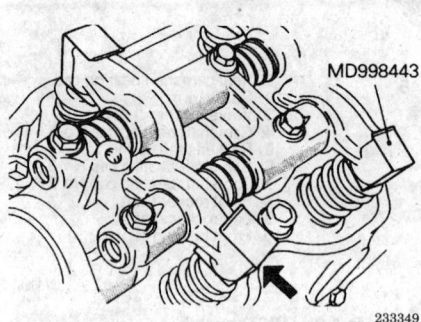

Auto lash adjuster holder — 1993–95 Galant 2.0L (4G63) and 2.4L (4G64) engines

5. Working in a crisscross pattern from the center outward, loosen the camshaft bearing caps in gradual steps.
6. Remove the rocker arms, shafts and bearing caps as an assembly.

NOTE: It is essential that all parts be kept in the same order and orientation for reinstallation. Be sure to mark and separate parts, so parts will not be mixed during reassembly.

7. Disassemble rocker shaft assembly. Starting at rear bearing cap, slide each piece off shafts.

NOTE: Inspect the roller surfaces of the rockers. Replace if there are any signs of damage or if the roller does not turn smoothly. Check the inside bore of the rockers and lifter for wear.

To install:

8. Apply a drop of sealant to the rear edges of the end caps.

9. Install the assembly into the front bearing cap, making sure the notches in the rocker shafts are facing up. Insert the installation bolt but do not tighten at this point.

10. Install the remaining cap bolts. Tighten all bolts evenly and gradually to 15 ft. lbs. (20 Nm). Remove the lash adjuster retainers.

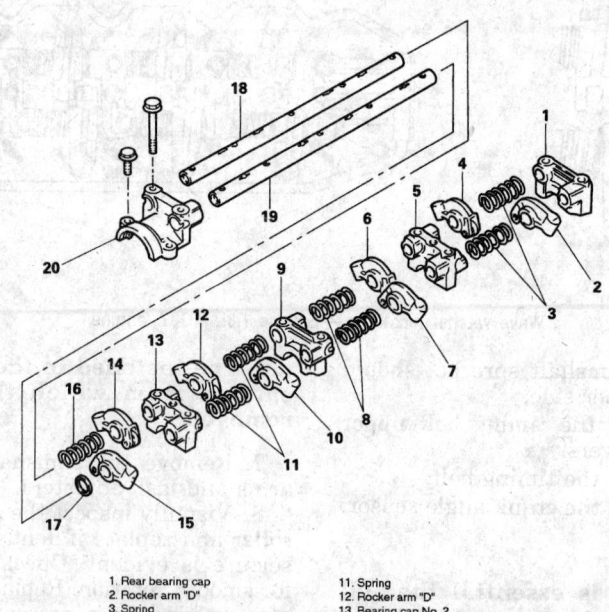

1. Rear bearing cap	11. Spring
2. Rocker arm "D"	12. Rocker arm "D"
3. Spring	13. Bearing cap No. 2
4. Rocker arm "D"	14. Rocker arm "C"
5. Bearing arm No. 4	15. Rocker arm "C"
6. Rocker arm "C"	16. Spring
7. Rocker arm "C"	17. Wave washer
8. Spring	18. Right rocker arm shaft
9. Bearing cap No. 3	19. Left rocker arm shaft
10. Rocker arm "D"	20. Front bearing cap

Exploded view of the rocker shaft assembly (8 valve engine) — 1993–95 Galant 2.0L (4G63) and 2.4L (4G64) engines

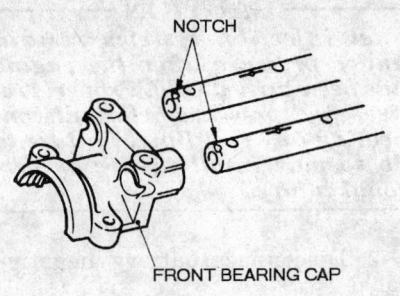

Aligning the front bearing cap to the rocker arm shaft (8 valve engine) — 1993–95 Galant 2.0L (4G63) and 2.4L (4G64) engines

11. Install the timing belt as required.

12. Align the matchmarks and install the distributor.

13. Remove the lash adjuster retaining tools.

14. Install the valve cover, with a new gasket and semi-circular packing in place.

15. Connect the negative battery cable.

16. Run the engine and check ignition timing.

16 Valve Engine

1. Disconnect the negative battery cable.

2. Remove the valve cover and discard the gasket.

3. Install lash adjuster retainer tools MD998443 or equivalent, to the rocker arm.

4. Remove the rocker shaft hold-down bolts gradually and evenly and remove the rocker shaft/arm assemblies.

5. If disassembly is required, keep all parts in the exact order of removal. Inspect the roller surfaces of the rockers. Replace if there are any signs of damage or if the roller does not turn smoothly. Check the inside bore of the rockers and the adjuster tip for wear.

To install:

6. Lubricate the rocker shaft with clean engine oil and install the rockers and springs in their proper places.

7. Install the rocker shaft assemblies on the engine. Tighten the bolts gradually and evenly to 21–25 ft. lbs. (29–35 Nm).

NOTE: When installing the rocker arm shaft, make certain the notch is properly located.

8. Remove the lash adjuster retaining tools.

9. Install the valve cover with a new gasket.

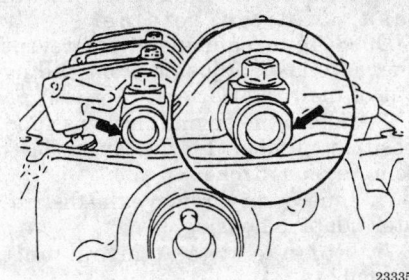

Rocker arm shaft installed position (16 valve engine) — 1993–95 Galant 2.0L (4G63) and 2.4L (4G64) engines

10. Connect the negative battery cable.

1996–97 Eclipse and Galant 2.4L (4G64) Engines

1. Disconnect the negative battery cable. Wait at least 90 seconds before performing any work.

2. Remove the battery.

3. Disconnect the accelerator cable, remove the cable clamp mounting screws and position the accelerator cable out of the way.

4. Remove the air intake hose.

5. Disconnect the breather hose and the PCV hose.

6. Disconnect the spark plug cables from the spark plugs.

7. Remove the rocker cover and gasket.

8. Install lash adjuster retainer tools MD998443 or equivalent, to the rocker arm.

9. Remove the rocker shaft hold-down bolts gradually and evenly and remove the rocker shaft/arm assemblies.

10. Disassemble the rockers and the rocker shaft springs from the rocker shafts. Remove the lash adjuster holding tool. Inspect the roller surfaces of the rockers. Replace if there are any signs of damage or if the roller does not turn smoothly. Check the inside bore of the rockers and the adjuster tip for wear. Check the rocker shafts for scoring. Inspect the lash adjusters for wear and smooth operation.

To install:

11. Immerse the lash adjusters in clean diesel fuel, and using a small wire, move the plunger up and down four or five times. while pushing down lightly on the check ball in order to bleed the air from the adjuster.

12. Install the lash adjusters to the rocker arms and attach the special holding tool. Be careful not to spill the diesel fuel from the adjuster.

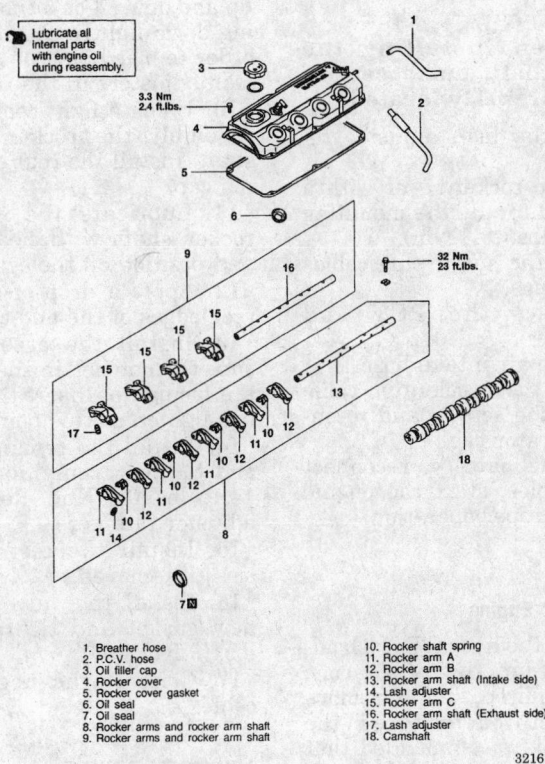

1. Breather hose
2. P.C.V. hose
3. Oil filler cap
4. Rocker cover
5. Rocker cover gasket
6. Oil seal
7. Oil seal
8. Rocker arms and rocker arm shaft
9. Rocker arms and rocker arm shaft
10. Rocker shaft spring
11. Rocker arm A
12. Rocker arm B
13. Rocker arm shaft (Intake side)
14. Lash adjuster
15. Rocker arm C
16. Rocker arm shaft (Exhaust side)
17. Lash adjuster
18. Camshaft

Rocker arm shafts and components — 1996–97 Eclipse and Galant 2.4L (4G64) engines

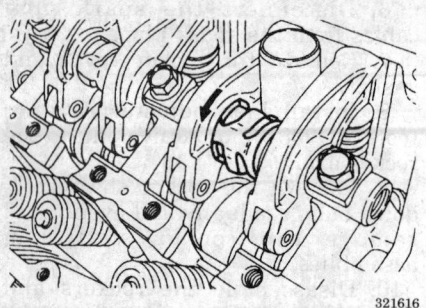

Installing the rocker shaft springs — 1996–97 Eclipse and Galant 2.4L (4G64) engines

13. Lubricate the rocker shaft with clean engine oil and install the rocker arms.

14. Temporarily tighten the rocker shaft assembly with the mounting bolts so that all rocker arms on the inlet valve side do not push on the valves.

15. Fit the rocker shaft springs from above and position them so that they are at right angles to the plug side. Install the rocker springs before installing the exhaust side rocker shaft and rocker arm assembly.

16. Install the exhaust side rocker shaft assembly in the engine. Tighten all the rocker shaft mounting bolts gradually and evenly to 23 ft. lbs. (32 Nm).

NOTE: When installing the rocker arm shaft, make certain the notch is properly located.

17. Remove the lash adjuster retaining tools.

18. Install the rocker cover with a new gasket and torque the mounting bolts to 2.4 ft. lbs. (3.3 Nm).

19. Reinstall the spark plug cables to the spark plugs.

20. Reconnect the PCV and breather hoses.

21. Install the air intake hose.

22. Reattach the accelerator cable brackets with the screws and reconnect the accelerator cable.

23. Install the battery, reconnect the battery cables, start the engine and check for proper operation.

Diamante

3.0L (6G72) SOHC Engine

On this engine, the hydraulic lash adjusters are built into the rocker arms. If service is required, simply remove the lash adjuster from the bore in the rocker arm. It is recommended that all of the rocker arms and lash adjusters are replaced at the same time.

1. Disconnect the negative battery cable.

2. Remove the valve cover. Install lash adjuster retainer tools MD998443 or equivalent, to prevent the auto-lash adjuster from falling out of the rocker arm.

3. Rotate the engine clockwise and position number 1 cylinder at TDC compression stroke.

4. If necessary, remove the distributor adapter housing.

5. Remove the timing belt assembly.

6. Loosen rocker arm and shaft assembly evenly in several steps. Remove the rocker arm and shaft assembly as a complete unit.

7. Remove the rear camshaft bearing cap and slide rocker arms, springs and washers from shaft. Note location and positioning of all rocker shaft components.

8. Visually inspect the rocker arm roller and replace if damage or seizure is evident. Check the roller for smooth rotation. Replace if excess play or binding is present. Also, inspect valve contact surface for possible damage or seizure. It is recommended that all rocker arms and lash adjusters be replaced together.

To install:

9. Immerse the lash adjusters in clean diesel fuel. Using a small wire, move the plunger of the lash adjuster up and down 4 or 5 times while pushing down lightly on the check ball in order to bleed out the air. Install the lash adjusters in the rocker arms.

10. Using a light coat of engine oil, assembly the rocker arms to the shaft. Install the rear camshaft bearing cap.

11. Lubricate the camshaft and rocker shaft with heavy engine oil and position on the cylinder head.

12. Apply a drop of sealant to the rear edges of the end caps.

13. Install the assembly making sure the notches in the rocker shafts are facing up. Insert the bolts but do not tighten at this point.

14. Install the remaining cap bolts and tighten evenly and gradually to 14 ft. lbs. (20 Nm). Remove the lash adjuster retainers.

15. Install the distributor extension, if removed.

16. Install the valve cover with a new gasket and tighten to 84 inch lbs. (9 Nm).

17. Connect the negative battery cable.

3.0L (6G72) DOHC Engine

1. Relieve the fuel system pressure.

CAUTION

Fuel injection systems remain under pressure after the engine has been turned OFF. Properly relieve fuel pressure before disconnecting any fuel lines. Failure to do so may result in fire or personal injury.

2. Disconnect battery negative cable.

3. Remove the timing belt cover and timing belt.

4. Remove the center cover, breather and PCV hoses, and spark plug cables.

5. Remove the rocker cover, semi-circular packing, throttle body stay, both camshaft sprockets, and oil seals.

6. Remove the crank angle sensor and adapter from the rear of the camshaft.

7. Remove the intake and exhaust camshafts.

8. Remove rocker arms and lash adjusters from the head. It is recommended that all lash adjusters and rockers be replaced as a complete set.

To install:

9. Immerse the lash adjusters in clean diesel fuel. Using a small wire, move the plunger of the lash adjuster up and down four or five times while pushing down lightly on the check ball in order to bleed out the air. Lubricate and install the lash adjusters in the cylinder head.

10. Lubricate the camshafts with heavy engine oil and position the camshafts on the cylinder head.

NOTE: Do not confuse the intake camshaft with the exhaust camshaft. On 1992 models, the intake camshaft has a B stamped on the hexagon and the exhaust camshaft has a D. On 1993–96 models, the intake camshaft has a J stamped on the hexagon and the exhaust camshaft has a K.

11. Make sure the dowel pin on both camshaft sprocket ends are positioned properly.

12. Install the bearing caps. Tighten the caps in sequence and in 2 or 3 steps. Caps 2, 3 and 4 have a front mark. Install with the mark aligned with the front mark on the cylinder head. Intake caps have an **I** stamped on the cap and exhaust caps have **E**. Also, make sure the rocker arm is correctly mounted on the lash adjuster and the valve stem end. Torque the front and rear retaining cap bolts to 14 ft. lbs. (20 Nm) and

13. Bearing cap
14. Rocker arm
15. Spring
16. Rocker arm
17. Spring
18. Bearing cap no. 3
19. Rocker arm
20. Spring
21. Rocker arm
22. Spring
23. Bearing cap no. 2
24. Rocker arm
25. Spring
26. Rocker arm
27. Spring
28. Rocker arm shaft
29. Rocker arm shaft
30. Bearing cap no. 1

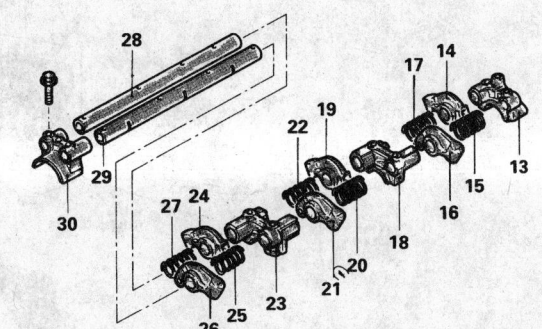

326969

Rocker arm assembly — Diamante 3.0L (6G72) SOHC engine

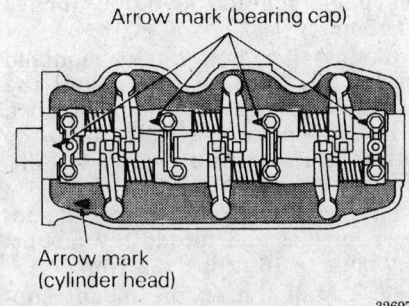

326970

Rocker arm shaft installation — Diamante 3.0L (6G72) SOHC engine

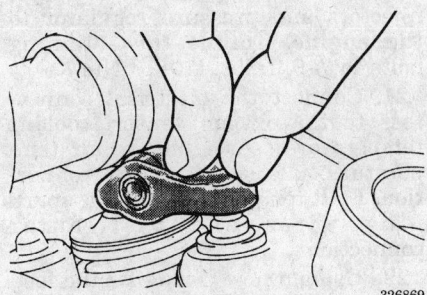

326869

Engine rocker arm — Diamante 3.0L (6G72) DOHC engine

tighten the center 3 retaining cap bolts to 8 ft. lbs. (11 Nm).

NOTE: If installing the camshaft to a cylinder head that is positioned on a workbench, the valves will protrude.

13. Apply a coating of engine oil to the oil seals and install.
14. Install the timing belt, valve cover and all related parts.
15. Connect the negative battery cable and check for leaks.

Intake Manifold

REMOVAL AND INSTALLATION

——————— **CAUTION** ———————
Fuel injection systems remain under pressure after the engine has been turned OFF. Properly relieve fuel pressure before disconnecting any fuel lines. Failure to do so may result in fire or personal injury.

Precis

1. Disconnect the negative battery cable.
2. Disconnect the air intake hose from the throttle body. Position the

hose off to the side and out of the way.
3. Disconnect the accelerator cable from the throttle lever. To do this loosen the two cable adjusting bracket bolts so that there is enough play in the cable to disengage the cable end from the throttle lever. Move the cable out of the way.
4. Disconnect the water hose from the water outlet fitting.
5. Remove the throttle body and gasket.
6. Disconnect the PCV hose from the rocker cover and disconnect the brake vacuum hoses.
7. Disconnect all vacuum hoses from their respective connections on the intake manifold. Make sure that you label them to avoid confusion during installation. On turbocharged engines, disconnect the air intake pipe from the turbocharger.
8. Relieve the fuel system pressure, then disconnect the high pressure fuel hose connection from the fuel delivery pipe. Cover the connection joint with a rag to absorb the excess fuel.
9. Unbolt and remove the air intake surge tank (and gasket) from the intake manifold.
10. Disconnect the fuel injector harness connectors.
11. Unbolt and remove the fuel delivery pipe (with pressure regulator) from the intake manifold. Take care not to drop the injectors when removing the delivery pipe.
12. Disconnect the wiring harness that runs between the water temperature gauge and the water temperature sensor assembly.
13. Remove the water outlet fitting and gasket from the intake manifold. Remove the thermostat.
14. Disconnect the spark plug wires from the distributor cap. Make sure that you label them first.
15. Remove the distributor and the ignition coil.
16. Remove the intake manifold stay.
17. Remove the intake manifold retaining nuts. Rock the intake manifold back and forth to separate it from the cylinder head. Remove the intake manifold gasket.
18. Clean all the intake manifold and cylinder head gasket surfaces making sure that all existing gasket material is removed. Do the same for the surge tank and all other gasket mating surfaces. Inspect the intake manifold and surge tank for cracks and check the coolant passages for restrictions. Replace all gaskets.

To install:

19. Install the intake manifold gasket. Position the intake manifold to the cylinder head studs and lower it onto the gasket. Install the intake manifold retaining nuts and torque them to 11–14 ft. lbs. (15–19 Nm), starting from the center and working outwards.

20. Install the intake manifold stay and torque the retaining bolts to 13–18 ft. lbs. (18–24 Nm).

21. Install the ignition coil, distributor, and spark plug wires.

22. Install the thermostat into the intake manifold.

23. Install the water outlet fitting with a new gasket. Torque the retaining bolts to 12–14 ft. lbs. (16–19 Nm).

24. Connect the wiring harness that runs between the temperature sensor and temperature gauge.

25. Install the fuel delivery pipe onto the intake manifold. Torque the retaining bolts to 7 to 9 ft. lbs.

26. Connect the fuel injector harness connectors.

27. Place the surge tank and gasket onto the intake manifold. Install the retaining bolts and torque them to 11–14 ft. lbs. (15–19 Nm).

28. Connect the high pressure hose to the fuel delivery pipe.

29. Connect the intake manifold, brake and PCV vacuum hoses.

30. Install the throttle body with a new gasket. Torque the throttle body bolts to 7–9 ft. lbs. (9–12 Nm).

31. Connect the water hose to the outlet fitting.

32. Connect the accelerator cable to the throttle lever and tighten the cable bracket adjusting bolts.

33. Connect the air intake hose to the throttle body.

34. Start the engine and check/adjust the ignition timing and the idle speed. Check for fuel and coolant leaks.

Mirage

1.5L (4G15) Engine

1. Relieve the fuel system pressure.

2. Disconnect battery negative cable and drain the cooling system.

3. Disconnect the upper radiator hose, heater hose and water bypass hose.

4. Remove the thermostat housing from intake manifold.

5. Disconnect the accelerator cable, breather hose and air intake hose.

6. Remove all vacuum hoses and pipes as necessary, including the brake booster vacuum line.

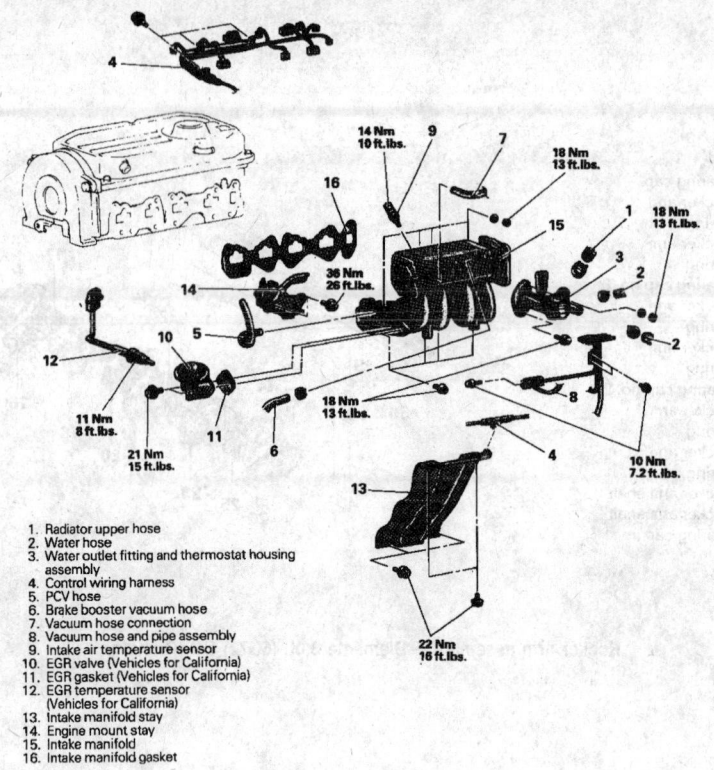

1. Radiator upper hose
2. Water hose
3. Water outlet fitting and thermostat housing assembly
4. Control wiring harness
5. PCV hose
6. Brake booster vacuum hose
7. Vacuum hose connection
8. Vacuum hose and pipe assembly
9. Intake air temperature sensor
10. EGR valve (Vehicles for California)
11. EGR gasket (Vehicles for California)
12. EGR temperature sensor (Vehicles for California)
13. Intake manifold stay
14. Engine mount stay
15. Intake manifold
16. Intake manifold gasket

330777

Intake manifold and related components — Mirage 1.5L (4G15) engine shown

7. Remove the throttle body assembly.

8. Disconnect the high pressure fuel line and the fuel return hose.

9. Tag and disconnect the electrical connectors from the oxygen sensor, coolant temperature sensor, intake air temperature, idle speed control connection, EGR temperature sensor, spark plug wires and distributor connectors.

10. Remove the fuel rail, fuel injectors, pressure regulator and insulators.

11. Remove the EGR valve from the intake manifold.

12. Remove the intake manifold support bracket and remove the engine mount support bracket.

13. Remove the intake manifold mounting bolts and remove the intake manifold assembly.

To install:

14. Clean all gasket material from the cylinder head intake mounting surface and intake manifold assembly. Check both surfaces for cracks or other damage. Check the intake manifold water passages and jet air passages for clogging. Clean if necessary.

15. Using a straight edge, measure the distortion of the intake manifold-to-cylinder head. Total distortion or warpage should be 0.006 inches (0.15mm or less).

16. Install a new intake manifold gasket to the head and install the manifold. Torque the manifold in a criss-cross pattern, starting from the inside and working outwards to 13 ft. lbs. (18 Nm).

17. Install the intake manifold support bracket and torque the mounting bolts to 16 ft. lbs. (22 Nm).

18. Install the engine mount support bracket and torque the mounting bolts to 26 ft. lbs. (36 Nm).

19. Using a new gasket, install the EGR valve and torque the mounting bolts to 15 ft. lbs. (21 Nm).

20. Using new insulators and O-rings, install the fuel delivery pipe, injectors and pressure regulator to the engine. Torque the retaining bolts to 7–9 ft. lbs. (10–13 Nm).

21. Connect the electrical connectors to the oxygen sensor, coolant temperature sensor, intake air temperature, idle speed control connection, EGR temperature sensor, spark plug wires and distributor connections.

22. Using a new O-ring for the feed pipe and a new clamp for the return pipe, install the fuel hoses.

23. Install the throttle body assembly.

24. Install the vacuum hoses and pipes as necessary, including the brake booster vacuum line.

25. Install and adjust the accelerator cable. Install the breather and air intake hose.

26. Using a new gasket, install the thermostat housing to the intake manifold and tighten the mounting bolts to 13 ft. lbs. (18 Nm).

27. Connect the upper radiator hose, heater hose and water bypass hose. Be sure to use new hose clamps.

28. Fill the system with coolant.

29. Connect the negative battery cable, run the vehicle until the thermostat opens, fill the radiator completely.

30. Check and adjust the idle speed and ignition timing.

31. Once the vehicle has cooled, recheck the coolant level.

1.8L (4G93) Engine

1. Relieve the fuel system pressure.

2. Disconnect battery negative cable and drain the cooling system.

3. Disconnect the accelerator cable and the air intake hose.

4. Tag and disconnect the electrical connectors from the oxygen sensor, coolant temperature sensor, idle speed control connection, EGR temperature sensor, oil pressure switch, spark plug wires and distributor connectors.

5. Disconnect the wiring from the throttle position sensor, fuel injectors and disconnect the ground cables.

6. Remove all vacuum hoses and pipes as necessary, including the brake booster and PCV vacuum lines.

7. Disconnect the upper radiator hose, heater hose and water bypass hose.

8. Disconnect the high pressure fuel line and the fuel return hose.

9. Remove the fuel rail, fuel injectors, pressure regulator and insulators.

10. Remove the intake manifold support bracket.

11. If the thermostat housing is preventing removal of the intake manifold, remove it.

12. Remove the intake manifold mounting bolts/nuts and remove the intake manifold assembly.

To install:

13. Clean all gasket material from the cylinder head intake mounting surface and intake manifold assembly. Check both surfaces for cracks or other damage. Check the intake manifold water passages and jet air passages for clogging. Clean if necessary.

14. Using a straight edge, measure the distortion of the intake manifold-to-cylinder head. Total distortion or warpage should be 0.006 inches (0.15mm or less).

15. Install a new intake manifold gasket to the head and install the manifold. Torque the manifold in a criss-cross pattern, starting from the inside and working outwards to 14 ft. lbs. (20 Nm).

16. If removed, install the thermostat housing.

17. Install the intake manifold brace bracket.

18. Install the fuel delivery pipe, injectors and pressure regulator to the engine. Torque the retaining bolts to 108 inch lbs. (12 Nm).

19. Using a new O-ring for the feed pipe and a new clamp for the return pipe, install the fuel hoses.

20. Connect the upper radiator hose, heater hose and water bypass hoses.

21. Install the vacuum hoses and pipes as necessary. Be sure to connect the brake booster and PCV vacuum lines.

22. Connect the wiring to the throttle position sensor, fuel injectors and connect the ground cables.

23. Connect the electrical wiring to the oxygen sensor, coolant temperature sensor, idle speed control connection, EGR temperature sensor, oil pressure switch, spark plug wires and distributor connectors.

24. Connect and adjust the accelerator cable and install the air intake hose.

25. Fill the system with coolant.

26. Connect the negative battery cable, run the vehicle until the thermostat opens, fill the radiator completely.

27. Check and adjust the idle speed and ignition timing.

28. Once the vehicle has cooled, recheck the coolant level.

Eclipse

1.8L (4G37) Engine

1. Relieve the fuel system pressure using proper procedure.

2. Disconnect negative battery cable.

3. Drain the cooling system.

4. Disconnect the accelerator cable, breather hose and air intake hose.

5. Disconnect the upper radiator hose, heater hose and water bypass hose.

6. Remove all vacuum hoses and pipes as necessary, including the brake booster vacuum line.

7. Disconnect the high pressure fuel line, fuel return hose and remove throttle control cable brackets.

8. Tag and disconnect the electrical connectors from the oxygen sensor, coolant temperature sensor, thermo switch, idle speed control connection, EGR temperature sensor, spark plug wires, etc. that may interfere with the manifold removal procedure.

9. Remove the fuel rail, fuel injectors, pressure regulator and insulators from the engine.

NOTE: If the distributor passes through the manifold, it will have to be removed. Matchmark the distributor shaft to the housing and the housing to the head or nearest accessory prior to removal.

10. Remove the intake manifold bracket.

11. Disconnect the water hose connections at the throttle body, water inlet, and heater assembly.

12. If the thermostat housing is preventing removal of the intake manifold, remove it.

13. Disconnect the vacuum connection at the power brake booster and the PCV valve if still connected.

14. Remove the intake manifold mounting bolts and remove the intake manifold assembly. Disassemble manifold from the intake plenum on a work bench as required.

To install:

15. Assemble the intake manifold assembly using all new gaskets. Torque air intake plenum bolts to 11–14 ft. lbs. (15–19 Nm).

16. Clean all gasket material from the cylinder head intake mounting surface and intake manifold assembly. Check both surfaces for cracks or other damage. Check the intake manifold water passages and jet air passages for clogging. Clean if necessary.

17. Install a new intake manifold gasket to the head and install the manifold. Torque the manifold in a crisscross pattern, starting from the inside and working outwards to 11–14 ft. lbs. (15–19 Nm).

18. Install the fuel delivery pipe, injectors and pressure regulator, lubricating all seals lightly with oil. Torque the retaining bolts to 7–9 ft. lbs. (10–13 Nm).

19. Install the thermostat housing, intake manifold brace bracket, distributor (if removed) and throttle body stay bracket.

20. Connect or install all hoses, cables and electrical connectors that

were removed or disconnected during the removal procedure.

21. Fill the system with coolant.

22. Connect the negative battery cable, run the vehicle until the thermostat opens, fill the radiator completely. Check for leaks.

23. Adjust the accelerator cable. Check and adjust the idle speed as required. Check ignition timing, if distributor was removed.

24. Once the vehicle has cooled, recheck the coolant level.

1993–94 2.0L (4G63) Engine

1. Relieve the fuel system pressure using proper procedure. Disconnect negative battery cable.

2. Drain the cooling system.

3. Disconnect the accelerator cable, breather hose and air intake hose.

4. Disconnect the upper radiator hose, heater hose and water bypass hose.

5. Remove all vacuum hoses and pipes as necessary, including the brake booster vacuum line.

6. Disconnect the high pressure fuel line, fuel return hose and remove throttle control cable brackets.

7. Tag and disconnect the electrical connectors from the oxygen sensor, coolant temperature sensor, thermo switch, idle speed control connection, EGR temperature sensor, spark plug wires, etc. that may interfere with the manifold removal procedure.

8. Remove the fuel rail, fuel injectors, pressure regulator and insulators from the engine.

9. Remove the ignition coil. Remove the intake manifold bracket.

10. Disconnect the water hose connections at the throttle body, water inlet, and heater assembly.

11. If the thermostat housing is preventing removal of the intake manifold, remove it.

12. Disconnect the vacuum connection at the power brake booster and the PCV valve if still connected.

13. Remove the intake manifold mounting bolts and remove the intake manifold assembly. Disassemble manifold from the intake plenum on a work bench as required.

To install:

14. Assemble the intake manifold assembly using all new gaskets. Torque air intake plenum bolts to 11–14 ft. lbs. (15–19 Nm).

15. Clean all gasket material from the cylinder head intake mounting surface and intake manifold assembly. Check both surfaces for cracks or other damage. Check the intake manifold water passages and jet air

passages for clogging. Clean if necessary.

16. Install a new intake manifold gasket to the head and install the manifold. Torque the manifold in a crisscross pattern, starting from the inside and working outwards to 11–14 ft. lbs. (15–19 Nm).

17. Install the fuel delivery pipe, injectors and pressure regulator from the engine. Torque the retaining bolts to 7–9 ft. lbs. (10–13 Nm).

18. Install the thermostat housing, intake manifold brace bracket, and throttle body stay bracket.

19. Connect or install all hoses, cables and electrical connectors that were removed or disconnected during the removal procedure.

20. Fill the system with coolant.

21. Connect the negative battery cable, run the vehicle until the thermostat opens, fill the radiator completely. Check for leaks.

22. Adjust the accelerator cable. Check and adjust the idle speed as required.

23. Once the vehicle has cooled, recheck the coolant level.

1995–97 2.0L (4G63) Engine

1. Relieve the fuel system pressure.

2. Remove the battery.

3. Drain the engine coolant.

4. Disconnect the accelerator cable.

5. Remove the air intake hose.

6. Disconnect the ignition coil and the module wiring connectors.

7. Disconnect the Manifold Differential Pressure sensor.

8. Disconnect the condenser.

9. Disconnect the TPS and the IAC motor connectors.

10. Disconnect the knock sensor and the ECT sensor connectors.

11. Disconnect the camshaft position sensor and the crankshaft position sensor connectors.

12. Disconnect the A/C compressor connector.

13. Remove the engine control wiring harness retaining bracket and position the harness out of the way.

14. Label and disconnect the vacuum hoses.

15. Disconnect the spark plug wire from the ignition coil.

16. Disconnect the fuel lines from the fuel rail.

17. Disconnect the heater hoses.

18. Remove the fuel rail assembly and insulators.

19. Remove the ignition coil and module.

20. Remove the EGR valve assembly.

21. Remove the intake manifold stay and the engine hanger.

22. Remove the intake manifold and gasket.

To install:

23. Use a new gasket and install the intake manifold.

24. Install the intake manifold stay and the engine hanger.

25. Use a new gasket and install the EGR assembly.

26. Install the ignition coil and module.

27. Install the fuel rail and insulators.

28. Connect the heater hoses and fuel lines.

29. Connect the spark plug wires to the coil towers.

30. Connect the vacuum hoses.

31. Install the engine harness in the proper position.

32. Connect the harness to the proper sensors.

33. Connect the IAC motor.

34. Connect the ignition condenser.

35. Connect the ignition coil and the module connectors.

36. Install and adjust the accelerator cable.

37. Install the battery.

38. Refill the engine with coolant.

39. Start the engine and check for leaks.

1995–97 Eclipse 2.0L (420A) Engine

1. Disconnect the negative battery cable.

2. Drain the engine coolant.

3. Remove the vacuum reservoir if equipped with cruise control.

4. Remove the air intake hose and breather hose.

5. Remove the accelerator cable from the bracket.

6. Disconnect the engine harness retaining clips.

7. Disconnect the MAP sensor connector.

8. Disconnect the charge temperature sensor connector.

9. Disconnect the vacuum hose connection.

10. Disconnect the TPS and the AIS motor connectors.

11. Position the engine control wiring harness out of the way.

12. Disconnect the alternator wiring harness connector.

13. Remove the PCV hose assembly.

14. Label and disconnect the vacuum hoses.

15. Disconnect the EGR pipe connection.

16. Disconnect the fuel lines from the fuel rail.

17. Remove the intake manifold stay and the engine hanger.

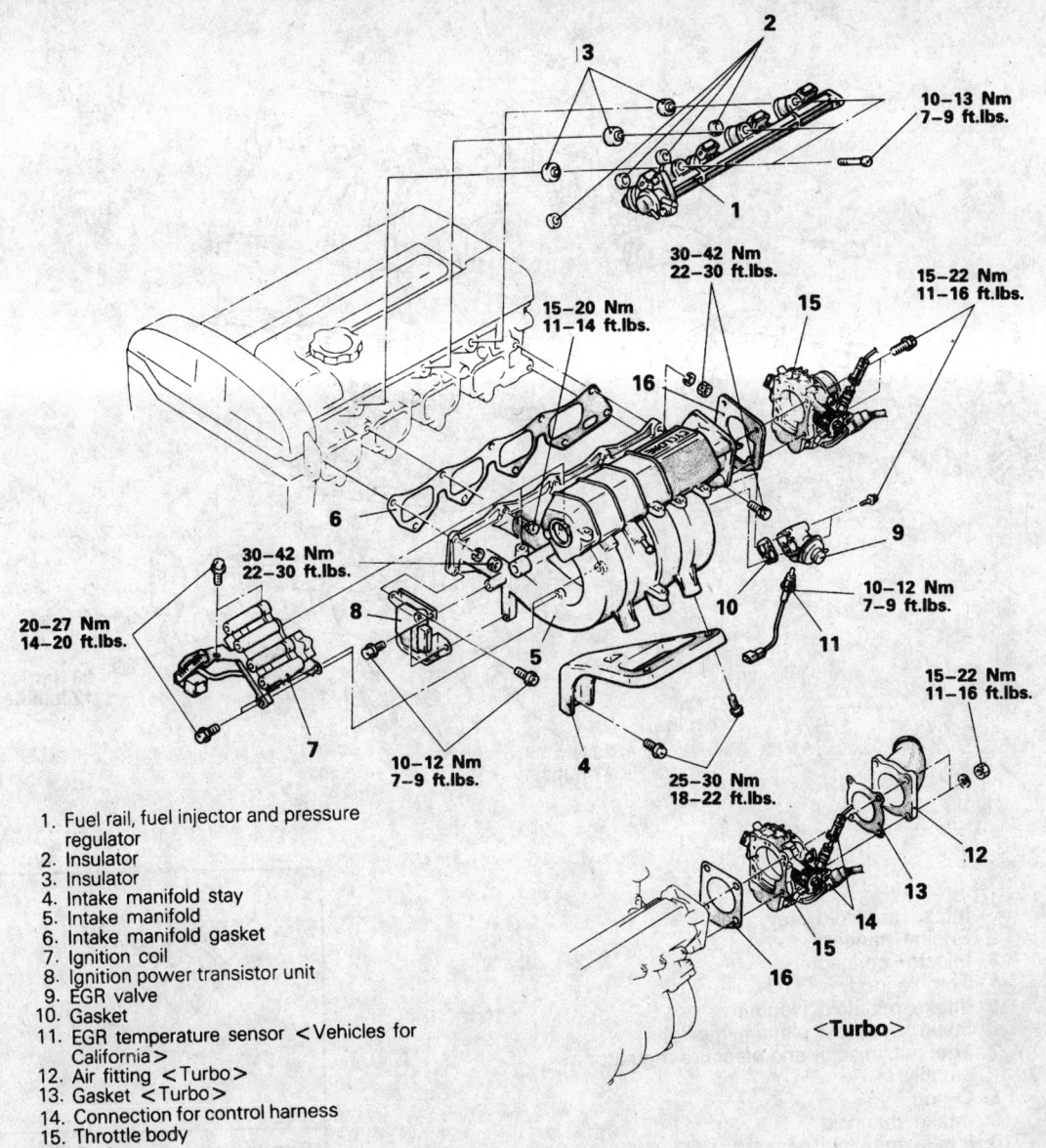

**10–13 Nm
7–9 ft.lbs.**

**30–42 Nm
22–30 ft.lbs.**

**15–22 Nm
11–16 ft.lbs.**

**15–20 Nm
11–14 ft.lbs.**

**30–42 Nm
22–30 ft.lbs.**

**10–12 Nm
7–9 ft.lbs.**

**20–27 Nm
14–20 ft.lbs.**

**15–22 Nm
11–16 ft.lbs.**

**10–12 Nm
7–9 ft.lbs.**

**25–30 Nm
18–22 ft.lbs.**

1. Fuel rail, fuel injector and pressure regulator
2. Insulator
3. Insulator
4. Intake manifold stay
5. Intake manifold
6. Intake manifold gasket
7. Ignition coil
8. Ignition power transistor unit
9. EGR valve
10. Gasket
11. EGR temperature sensor <Vehicles for California>
12. Air fitting <Turbo>
13. Gasket <Turbo>
14. Connection for control harness
15. Throttle body
16. Gasket

139321

<Turbo>

Intake manifold exploded view — 1993–94 Eclipse 2.0L (4G63) engine

18. Remove the throttle body.
19. Remove the intake manifold plenum and gasket.
20. Disconnect the injector connectors.
21. Remove the fuel rail with the injectors.
22. Remove the intake manifold.

To install:

23. Use a new gasket and install the intake manifold.
24. Install the fuel rail assembly and connect the injectors.
25. Use a new gasket and install the intake plenum.
26. Use a new gasket and install the throttle body.

27. Install the intake manifold stay and the engine hanger.
28. Use a new O-ring and connect the fuel lines to the fuel rail.
29. Connect the EGR pipe.
30. Connect the vacuum hoses and the PCV hose assembly.
31. Connect the alternator wiring harness connector.
32. Reposition the engine control wiring harness and install the brackets and clips.
33. Connect the AIS motor and the TPS sensor connectors.
34. Connect the vacuum hose to the throttle body.

35. Connect the MAP and the charge temperature sensor connectors.
36. Install the accelerator cable in the bracket and connect it to the throttle body.
37. Connect the breather hose and the air intake hose.
38. Install the vacuum reservoir, if equipped.
39. Connect the negative battery cable.
40. Refill the engine with coolant.
41. Adjust the accelerator cable.

1996–97 2.4L (4G64) Engine

1. Relieve the fuel system pressure.

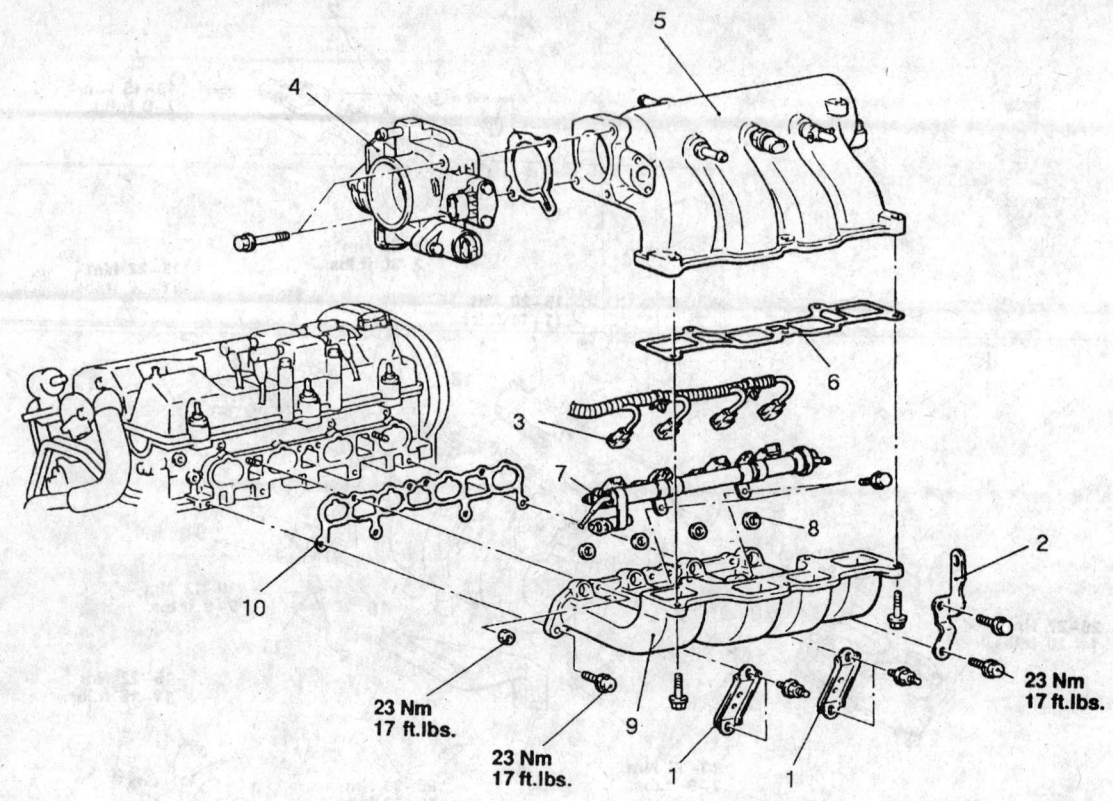

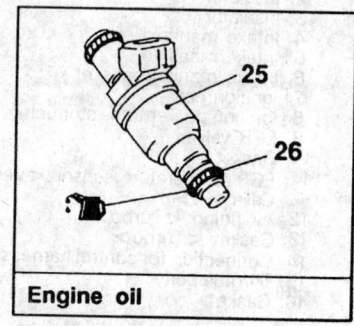

Engine oil

320468

1. Intake manifold stay
2. Engine hanger
3. Injector connector
4. Throttle body
5. Intake manifold plenum
6. Intake manifold plenum gasket
7. Fuel rail, injector and pressure regulator assembly
8. O-ring
9. Intake manifold
10. Intake manifold gasket

Intake manifold and related components — 1995–97 Eclipse 2.0L (420A) engine

2. Remove the battery.

3. Drain the engine coolant.

4. Disconnect the accelerator cable.

5. Remove the air intake hose.

6. Disconnect the ignition coil and the module wiring connectors.

7. Disconnect the Manifold Differential Pressure sensor.

8. Disconnect the condenser.

9. Disconnect the TPS and the IAC motor connectors.

10. Disconnect the heated oxygen sensor connector.

11. Disconnect the crankshaft position sensor connectors.

12. Disconnect the A/C compressor connector.

13. Remove the engine control wiring harness retaining bracket and position the harness out of the way.

14. Label and disconnect the vacuum hoses.

15. Disconnect the spark plug wire from the ignition coil.

16. Disconnect the fuel lines from the fuel rail.

17. Disconnect the heater hoses.

18. Disconnect the high-pressure fuel hose connection and remove the fuel rail assembly and insulators.

19. Remove the manifold differential pressure sensor.

20. Remove the ignition coil and module.

21. Remove the EGR valve assembly.

22. Remove the intake manifold stay and the engine hanger.

23. Remove the intake manifold and gasket.

24. Remove the throttle body assembly and gasket from the intake manifold.

To install:

25. Install the throttle body assembly with a new gasket to the intake manifold and torque the mounting bolts to 11–16 ft. lbs. (15–22 Nm).

26. Use a new gasket and install the intake manifold. Torque the intake manifold bolts to 15 ft. lbs. (20 Nm).

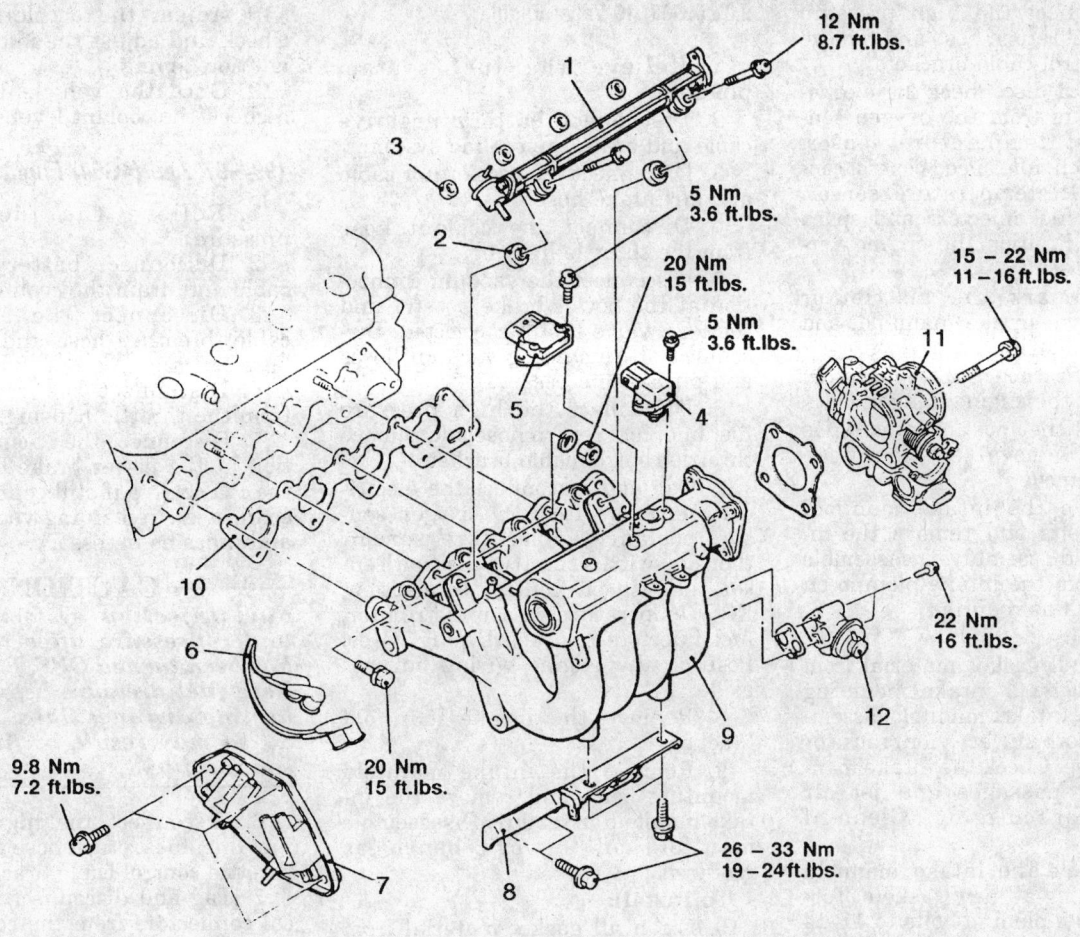

12 Nm
8.7 ft.lbs.

5 Nm
3.6 ft.lbs.

20 Nm
15 ft.lbs.

5 Nm
3.6 ft.lbs.

15 – 22 Nm
11 – 16 ft.lbs.

22 Nm
16 ft.lbs.

9.8 Nm
7.2 ft.lbs.

20 Nm
15 ft.lbs.

26 – 33 Nm
19 – 24 ft.lbs.

1. Fuel rail, fuel injector and pressure regulator assembly
2. Insulator
3. Insulator
4. Manifold differential pressure sensor
5. Ignition power transistor
6. Spark plug cable connection
7. Ignition coil
8. Intake manifold stay
9. Intake manifold
10. Intake manifold gasket
11. Throttle body
12. EGR valve assembly

320493

Intake manifold and related components — 1996–97 Eclipse 2.4L (4G64) engine

27. Install the intake manifold stay and the engine hanger. Torque the mounting bolts to 19–24 ft. lbs. (26–33 Nm).

28. Use a new gasket and install the EGR assembly. EGR bolt torque is 16 ft. lbs. (22 Nm).

29. Install the ignition coil and module.

30. Install the fuel rail and insulators and reconnect the high-pressure fuel hose.

31. Connect the heater hoses and fuel lines.

32. Connect the spark plug wires to the coil towers.

33. Connect the vacuum hoses.

34. Install the engine harness in the proper position.

35. Connect the harness to the proper sensors.

36. Connect the IAC motor.

37. Connect the ignition condenser.

38. Connect the ignition coil and the module connectors.

39. Install and adjust the accelerator cable.

40. Install the battery.

41. Refill the engine with coolant.

42. Start the engine, check for leaks, and check for proper operation.

1993 Galant

2.0L (4G63) 8 Valve Engines

1. Relieve the fuel system pressure.

2. Disconnect the battery negative cable.

3. Drain the cooling system.

4. Disconnect the accelerator cable and air intake hose.

5. Disconnect the upper radiator hose, heater hose and water bypass hose.

6. Disconnect the vacuum connection at the power brake booster and the PCV valve if still connected. Disconnect all remaining vacuum hoses and pipes as necessary.

7. Disconnect the high pressure fuel line, fuel return hose and remove throttle control cable brackets.

8. Tag and disconnect the electrical connectors from the oxygen sensor, coolant temperature sensor, thermo switch, idle speed control connection, EGR temperature sensor, distributor, fuel injectors and spark plug wires. Position the engine wiring harness aside.

9. Matchmark the distributor housing to the intake manifold, and remove the distributor.

10. Remove the intake manifold bracket and the engine hanger.

11. If the thermostat housing is preventing removal of the intake manifold, remove it.

12. Remove the intake manifold mounting bolts and remove the intake manifold assembly. Disassemble manifold from the intake plenum on a work bench as required.

To install:

13. Clean all gasket material from the cylinder head intake mounting surface and intake manifold assembly. Check both surfaces for cracks or other damage. Check the intake manifold water passages and jet air passages for clogging. Clean if necessary.

14. Assemble the intake manifold assembly using all new gaskets. Torque air intake plenum bolts to 11–14 ft. lbs. (15–19 Nm).

15. Install a new intake manifold gasket to the head and install the manifold. Torque the manifold in a criss-cross pattern, starting from the inside and working outwards to 11–14 ft. lbs. (15–19 Nm).

16. Install the fuel delivery pipe, injectors and pressure regulator to the engine. Torque the retaining bolts to 4 ft. lbs. (6 Nm).

17. Install the thermostat housing, intake manifold brace bracket, and engine hanger bracket.

18. Connect or install all hoses, cables and electrical connectors that were removed or disconnected during the removal procedure.

19. Align the distributor matchmarks and install the distributor.

20. Fill the system with coolant.

21. Connect the negative battery cable, run the vehicle until the thermostat opens, fill the radiator completely.

22. Adjust the accelerator cable. Check and adjust the idle speed and ignition timing.

23. Once the vehicle has cooled, recheck the coolant level.

2.0L (4G63) 16 Valve Engine

1. Relieve the fuel system pressure.

2. Disconnect battery negative cable and drain the cooling system.

3. Disconnect the accelerator cable and air intake hose.

4. Disconnect the coolant hose from the throttle housing.

5. Disconnect the vacuum connection at the power brake booster and the PCV valve if still connected. Disconnect all remaining vacuum hoses and pipes as necessary.

6. Disconnect the high pressure fuel line, fuel return hose and remove throttle control cable brackets.

7. Tag and disconnect the electrical connectors from the oxygen sensor, coolant temperature sensor, thermo switch, throttle position sensor, idle speed control connection, EGR temperature sensor, distributor, fuel injectors and spark plug wires. Position the engine wiring harness aside.

8. Remove the intake manifold bracket.

9. Remove the intake manifold mounting bolts and remove the intake manifold assembly. Disassemble manifold on a work bench as required.

To install:

10. Clean all gasket material from the cylinder head intake mounting surface and intake manifold assembly. Check both surfaces for cracks or other damage. Check the intake manifold water passages and jet air passages for clogging. Clean if necessary.

11. Assemble the intake manifold assembly using all new gaskets.

12. Install a new intake manifold gasket to the head and install the manifold. Torque the manifold in a criss-cross pattern, starting from the inside and working outwards to 11–14 ft. lbs. (15–20 Nm).

13. Install the fuel delivery pipe, injectors and pressure regulator to the engine. Torque the retaining bolts to 4 ft. lbs. (6 Nm).

14. Install the intake manifold brace bracket and tighten bolts to 13–18 ft. lbs. (18–25 Nm).

15. Connect or install all hoses, cables and electrical connectors that were removed or disconnected during the removal procedure.

16. Fill the system with coolant.

17. Connect the negative battery cable, run the vehicle until the thermostat opens, fill the radiator completely.

18. Adjust the accelerator cable. Check and adjust the idle speed and ignition timing.

19. Once the vehicle has cooled, recheck the coolant level.

1994–97 2.4L (4G64) Engine

1. Relieve the fuel system pressure.

2. Disconnect battery negative cable and drain the cooling system.

3. Disconnect the accelerator cable, breather hose and air intake hose.

4. Disconnect the coolant hose from the throttle housing.

5. Disconnect the vacuum connection at the power brake booster and the PCV valve if still connected. Disconnect all remaining vacuum hoses and pipes as necessary.

— CAUTION —
Fuel injection systems remain under pressure after the engine has been turned OFF. Properly relieve fuel pressure before disconnecting any fuel lines. Failure to do so may result in fire or personal injury.

---

6. Disconnect the high pressure fuel line, fuel return hose and remove throttle control cable brackets.

7. Tag and disconnect the electrical connectors from the coolant temperature sensor, coolant temperature gauge, IAC valve, ignition coil, EGR temperature sensor, knock sensor, oxygen sensor, throttle position sensor, distributor, A/C temperature sensor, fuel injectors and ignition power transistor. Position the engine wiring harness aside.

8. Label and disconnect the spark plug wires, from the spark plugs.

9. Remove the intake manifold stay bracket.

10. Remove the intake manifold mounting bolts and remove the intake manifold assembly. Disassemble manifold on a work bench as required.

To install:

11. Clean all gasket material from the cylinder head intake mounting surface and intake manifold assembly. Check both surfaces for cracks or other damage. Check the intake manifold water passages and jet air passages for clogging. Clean if necessary.

12. Assemble the intake manifold assembly using all new gaskets.

13. Install a new intake manifold gasket to the head and install the manifold. Torque the manifold in a criss-cross pattern, starting from the

inside and working outwards to 15 ft. lbs. (20 Nm) for bolts, and 26 ft. lbs. (35 Nm) for nuts on the 1994 and 15 ft. lbs. (20Nm) on the 1995–96.

14. Install the fuel delivery pipe, injectors and pressure regulator to the engine. Torque the retaining bolts to 4 ft. lbs. (6 Nm).

15. Install the intake manifold brace bracket and tighten bolts to 21 ft. lbs. (29 Nm).

16. Connect or install all hoses, cables and electrical connectors that were removed or disconnected during the removal procedure.

17. Fill the system with coolant.

18. Connect the negative battery cable, run the vehicle until the thermostat opens, fill the radiator completely.

19. Adjust the accelerator cable. Check and adjust the idle speed and ignition timing.

20. Once the vehicle has cooled, recheck the coolant level.

Diamante

1. Relieve the fuel system pressure.

2. Disconnect battery negative cable and drain the cooling system.

3. Remove the air intake hose(s).

4. Disconnect the accelerator control cables from the throttle body.

5. Tag and disconnect the vacuum hoses including the brake booster hose.

6. Tag and disconnect the wire harness connectors.

7. Disconnect the high pressure and return fuel hoses.

8. Disconnect EGR pipe and remove the EGR valve and EGR temperature sensor from the intake plenum assembly.

9. If equipped, remove the manifold pressure sensor.

10. Remove the plenum retaining bracket.

11. Remove the plenum retaining nuts and bolts and remove the air intake plenum from the intake manifold. Discard the gasket.

12. Remove the upper timing belt covers.

13. Remove the water pump stay bracket.

NOTE: It is not necessary to remove the fuel injectors from the intake unless the manifold assembly is being replaced.

14. Remove the fuel rail with the injectors attached.

15. Disconnect the coolant hoses from the intake manifold. Be sure to note the connections.

16. Remove the intake manifold mounting nuts and remove the intake manifold.

17. Clean the gasket mounting surfaces.

To install:

18. Check all items for cracks, clogging and warpage. Maximum warpage is 0.0059 inches (0.15mm). Replace any questionable parts.

19. Thoroughly clean and dry the mating surfaces of the heads, intake manifold and air intake plenum.

20. Install new intake manifold gaskets to the cylinder heads with the adhesive side facing up.

21. Place the manifold on the cylinder heads.

22. Lubricate the studs lightly with oil and install the nuts.

23. For vehicles produced up to and including November, 1993, tighten the mounting nuts as follows:
   a. Front bank nuts to 27–43 inch lbs. (3–5 Nm).
   b. Rear bank nuts to 9–11 ft. lbs. (12–15 Nm).
   c. Front bank nuts to 9–11 ft. lbs. (12–15 Nm).

24. For vehicles produced after November, 1993, tighten the mounting nuts as follows:
   a. Front bank nuts to 48–72 inch lbs. (5–8 Nm).
   b. Rear bank nuts to 14–17 ft. lbs. (20–23 Nm).
   c. Front bank nuts to 14–17 ft. lbs. (20–23 Nm).

25. Using new clamps, connect the coolant hoses to the intake manifold.

26. Using new O-rings, install the fuel rail assembly, if removed. Tighten the mounting bolts to 7–9 ft. lbs. (10–13 Nm).

27. Install a new intake air plenum gasket and install the plenum. Tighten the retaining nuts and bolts evenly and gradually to 13 ft. lbs. (18 Nm).

28. Install the retaining bracket and tighten the retaining bolts to 13 ft. lbs. (18 Nm).

29. If removed, install the manifold pressure sensor.

30. Using a new gasket, install the EGR valve and tighten the bolts to 16 ft. lbs. (22 Nm).

31. Install the EGR temperature sensor and tighten the fitting to 7–9 ft. lbs. (10–12 Nm).

32. Connect the EGR pipe and tighten the fittings to 43 ft. lbs. (60 Nm).

33. Replace the O-ring and connect the high pressure fuel hose. Tighten the retaining bolts to 48 inch lbs. (5 Nm).

34. Using a new hose clamp, connect the fuel return hose.

35. Install the water pump stay bracket.

36. Install the upper timing belt covers.

37. Connect the harness connector and vacuum hoses.

38. Connect and adjust the accelerator cables.

39. Install the air intake hose(s).

40. Fill the system with coolant.

41. Connect the negative battery cable, run the vehicle until the thermostat opens, fill the radiator completely.

42. Check and adjust the idle speed and ignition timing.

43. Once the vehicle has cooled, recheck the coolant level.

3000GT

1. Relieve the fuel system pressure.

2. Disconnect battery negative cable and drain the cooling system.

3. Remove the air intake hose(s).

4. Disconnect the accelerator control cables from the throttle body.

5. Matchmark and disconnect the vacuum hoses including the brake booster hose.

6. Disconnect the clutch booster vacuum hose connection, if equipped.

7. Disconnect all harness connectors.

8. Disconnect EGR components on California vehicles.

9. Remove the plenum retaining bracket.

10. Remove the plenum retaining nuts and bolts and remove the air intake plenum. Discard the gasket.

11. Disconnect the high pressure and return fuel hoses.

12. Matchmark and disconnect the vacuum hoses.

13. Disconnect the wire harness connectors.

14. Remove the fuel rail with the injectors attached.

15. Remove the timing belt upper cover.

16. Remove the intake manifold mounting nuts; turbocharged engines have cone disc springs under some of the nuts which should be removed. Remove the intake manifold and discard the gaskets.

To install:

17. Check all items for cracks, clogging and warpage. Maximum warpage is 0.008 in. (0.2mm). Replace any questionable parts.

18. Thoroughly clean and dry the mating surfaces of the heads, intake manifold and air intake plenum.

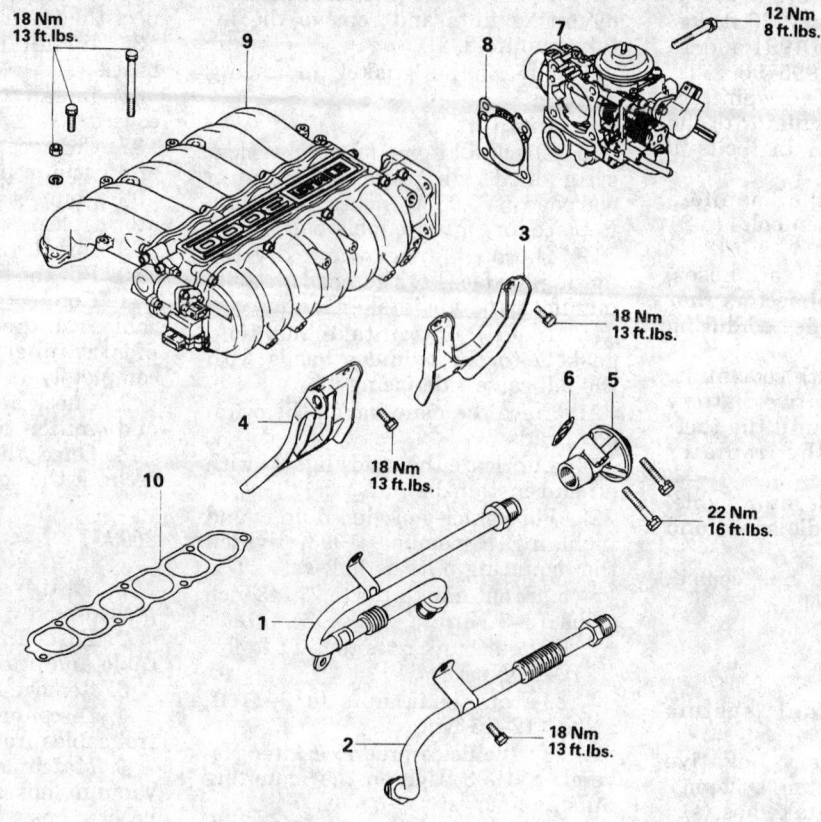

1. EGR pipe – Up to 1993 <California> model
2. EGR pipe – From 1994 <California> model
3. Intake manifold plenum stay, rear
4. Intake manifold plenum stay, front
5. EGR valve
6. EGR valve gasket } <For California>
7. Throttle body
8. Throttle body gasket
9. Intake manifold plenum
10. Intake manifold plenum gasket

325372

Exploded view of air intake plenum assembly — Diamante shown, 3000GT similar

19. Install new intake manifold gaskets to the heads with the adhesive side facing up.

20. Place the manifold on the heads and install the cone disc springs and/or lock washers.

21. For Turbo engines up to November 1993: Lubricate the studs lightly with oil, then install the nuts following this procedure:

  a. Tighten the nuts on the front bank to 2.2–3.6 ft. lbs. (3–5 Nm).

  b. Tighten the nuts on the rear bank to 9–11 ft. lbs. (12–15 Nm).

  c. Tighten the nuts on the front bank to 9–11 ft. lbs. (12–15 Nm).

  d. Repeat Steps B and C.

22. For Turbo engines after November 1993: Lubricate the studs lightly with oil, then install the nuts following this procedure:

  a. Tighten the nuts on the front bank to 4–6 ft. lbs. (3–5 Nm).

  b. Tighten the nuts on the rear bank to 14–17 ft. lb. (20–23 Nm).

  c. Tighten the nuts on the front bank to 14–17 ft. lbs. (20–23 Nm).

  d. Repeat Steps B and C.

23. On non-turbocharged engines only, tighten the nuts to a final torque of 14 ft. lbs. (18 Nm).

24. Install the timing belt upper cover.

25. Install the fuel rail assembly.

26. Connect the harness connector and vacuum hoses.

27. Replace the O-ring and connect the fuel hoses.

28. Install a new intake air plenum gasket and install the plenum. Tighten the retaining nuts and bolts evenly and gradually to 13 ft. lbs. (18 Nm).

29. Install the retaining bracket.

30. Connect EGR components on California vehicles.

31. Connect the harness connectors and vacuum hoses.

32. Connect and adjust the accelerator cables.

33. Install the air intake hose(s).

34. Fill the system with coolant.

1. Connection for high-pressure fuel hose
2. O-ring
3. Connection for fuel return hose
4. Connection for vacuum hoses
5. Wiring harness connector
6. Oxygen sensor <For California from 1994 models>
7. Fuel rail (with injectors)
8. Insulators
9. Timing belt upper cover
10. Water pump stay mounting bolt
11. Intake manifold mounting nut
12. Intake manifold mounting nut
13. Cone disc spring
14. Intake manifold
15. Intake manifold gasket

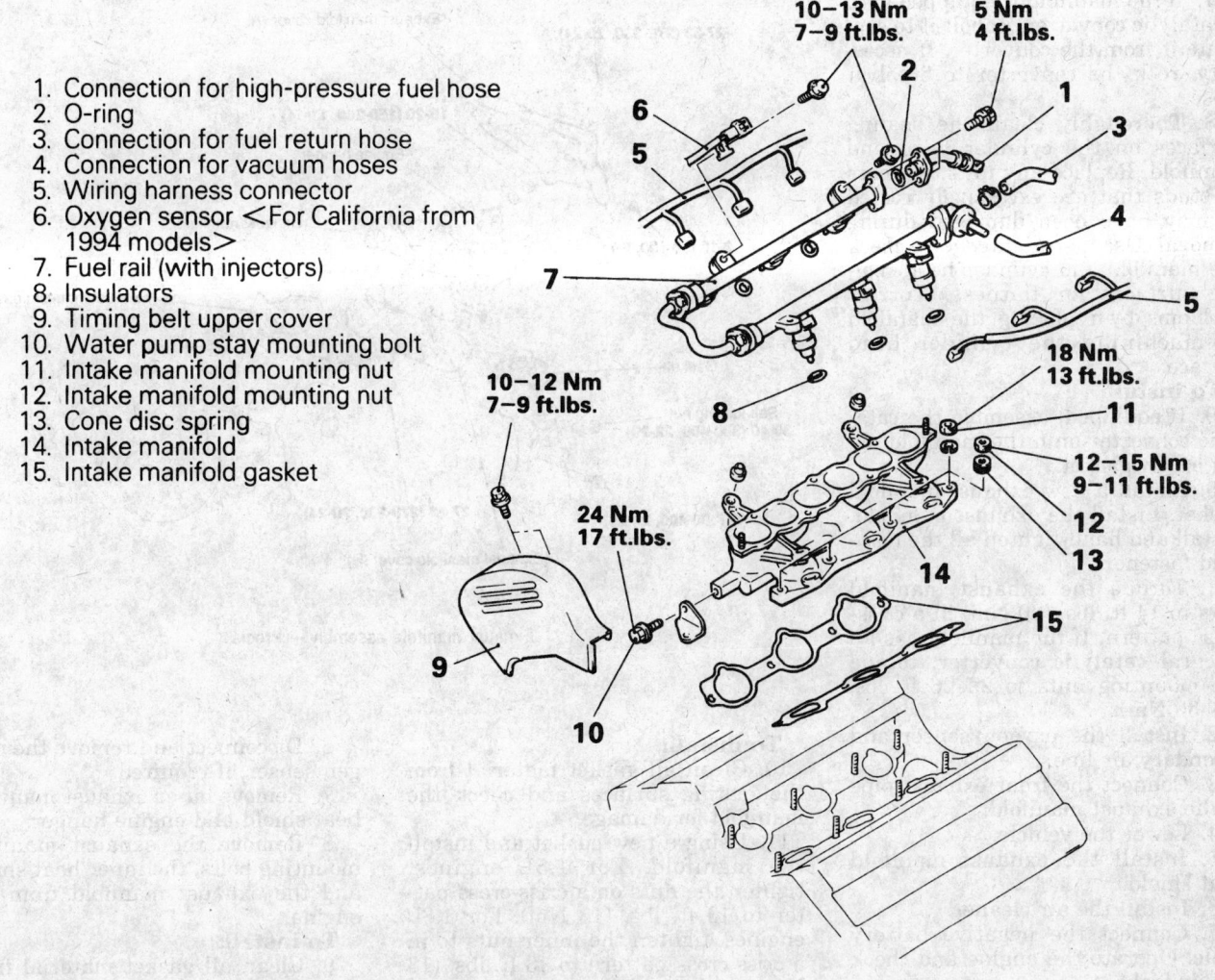

Intake manifold and related components — Diamante shown, 3000GT similar

325373

---

35. Connect the negative battery cable, run the vehicle until the thermostat opens, fill the radiator completely.
36. Check and adjust the idle speed and ignition timing.
37. Once the vehicle has cooled, recheck the coolant level.

Exhaust Manifold

REMOVAL AND INSTALLATION

Precis

NOTE: A hot engine should be allowed to cool before the manifold is removed.

1. Disconnect the negative battery cable. Remove the air cleaner.

2. Remove the exhaust manifold heat shield. Soak all manifold nuts and studs with a liquid penetrant.

NOTE: Only 6-point sockets should be used on the exhaust manifold bolts to prevent the bolt heads from being rounded off.

3. Raise and safely support the vehicle.
4. Disconnect the front exhaust pipe from the exhaust manifold.
5. Disconnect and remove the oxygen sensor. If there is a secondary air line connected to the exhaust manifold, disconnect it.
6. Loosen the manifold nuts in a crisscross pattern. Support the mani-

fold and remove all attaching nuts and washers. Slide the manifold from the cylinder head.

7. If the manifold is equipped with a catalytic converter, unbolt it to separate it from the converter. If necessary, rock the converter to break it loose.

8. Thoroughly clean the sealing surfaces on the cylinder head and manifold. Replace any nuts, washers or studs that are excessively rusted or may have been damaged during removal. Use a straightedge to check the manifold and cylinder head sealing surfaces for flatness. Correct problems by replacing the manifold or machining the cylinder head surface.

To install:

9. If equipped, assemble the catalytic converter onto the manifold using a new gasket.

10. Install a new exhaust manifold gasket. Install the exhaust manifold. Install and hand-tighten all the manifold fasteners.

11. Torque the exhaust manifold nuts to 14 ft. lbs. (20 Nm) in a criss-cross pattern. If the manifold has an integral catalytic converter, torque the mounting nuts to 22–26 ft. lbs. (30–35 Nm).

12. Install the oxygen sensor and secondary air line.

13. Connect the front exhaust pipe to the exhaust manifold.

14. Lower the vehicle.

15. Install the exhaust manifold heat shield.

16. Install the air cleaner.

17. Connect the negative battery cable. Operate the engine and check for air leaks.

Mirage

1. Disconnect battery negative cable.

2. Raise the vehicle and support safely.

3. Remove the exhaust pipe to exhaust manifold nuts and separate exhaust pipe. Discard gasket.

4. Lower vehicle.

5. Remove electric cooling fan assembly, if necessary.

6. If the oxygen sensor is located in the manifold, remove the sensor.

7. Disconnect necessary EGR components.

8. Remove outer exhaust manifold heat shield and engine hanger. Disconnect the electrical connector and remove the oxygen sensor.

9. Remove the exhaust manifold mounting bolts, the inner heat shield and the exhaust manifold.

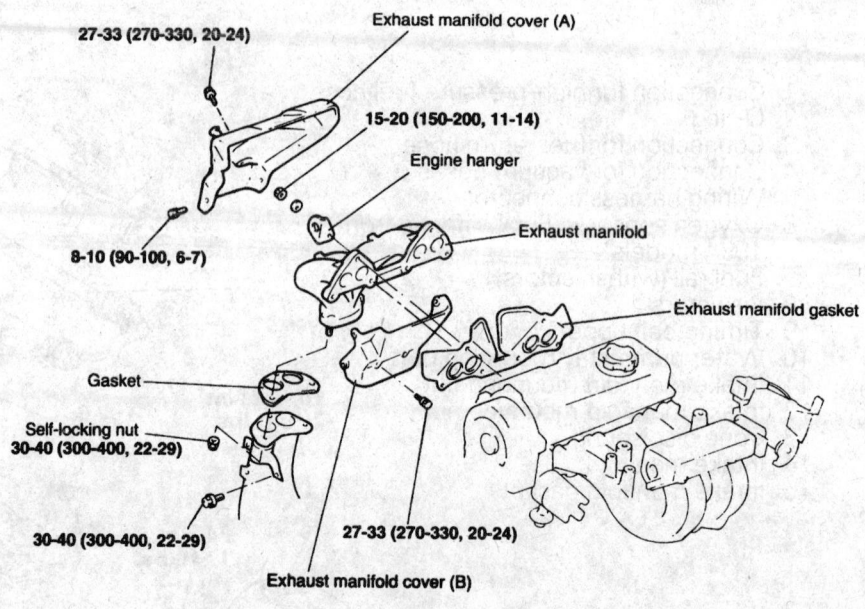

27-33 (270-330, 20-24)

Exhaust manifold cover (A)

15-20 (150-200, 11-14)

Engine hanger

Exhaust manifold

8-10 (90-100, 6-7)

Exhaust manifold gasket

Gasket

Self-locking nut
30-40 (300-400, 22-29)

30-40 (300-400, 22-29)

27-33 (270-330, 20-24)

Exhaust manifold cover (B)

260526

Exhaust manifold assembly — Precis

To install:

10. Clean all gasket material from the mating surfaces and check the manifold for damage.

11. Using a new gasket and install the manifold. For 1.5L engines, tighten the nuts on a criss-cross patter to 13 ft. lbs. (18 Nm). For 1.8L engines, tighten the inner nuts to in a criss-cross pattern to 13 ft. lbs. (18 Nm) and tighten the two outer (larger) nuts to 22 ft. lbs. (30 Nm).

12. Install the heat shields.

13. Connect EGR components.

14. If removed, install the oxygen sensor.

15. Install the electric cooling fan assembly as required.

16. Install a new flange gasket and connect the exhaust pipe.

17. Connect the negative battery cable and check for exhaust leaks.

Eclipse

1.8L (4G37) Engine

1. Disconnect battery negative cable.

2. Raise the vehicle and support safely.

3. Remove the exhaust pipe to exhaust manifold nuts and separate exhaust pipe. Discard gasket.

4. Lower vehicle.

5. Remove the oil dipstick tube.

6. Disconnect and remove the oxygen sensor, if required.

7. Remove outer exhaust manifold heat shield and engine hanger.

8. Remove the exhaust manifold mounting bolts, the inner heat shield and the exhaust manifold from the engine.

To install:

9. Clean all gasket material from the mating surfaces and check the manifold for damage or cracking.

10. Install a new gasket and install the manifold. Tighten the nuts to in a crisscross pattern to 11–14 ft. lbs. (15–20 Nm).

11. Install the heat shields.

12. Install the dipstick tube and alternator as required.

13. Install a new flange gasket and connect the exhaust pipe.

14. Connect the negative battery cable and check for exhaust leaks.

1993–94 2.0L (4G63) Engine

1. Disconnect battery negative cable.

2. Raise the vehicle and support safely.

3. Remove the exhaust pipe to exhaust manifold nuts and separate exhaust pipe. Discard gasket.

4. Lower vehicle.

5. Remove electric cooling fan assembly. Remove the oil dipstick tube.

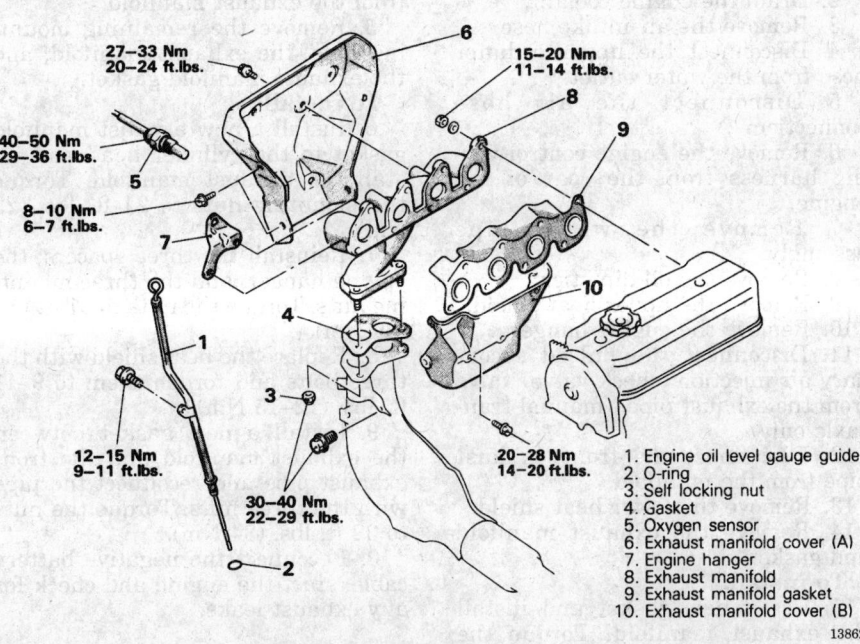

27–33 Nm
20–24 ft.lbs.

15–20 Nm
11–14 ft.lbs.

40–50 Nm
29–36 ft.lbs.

8–10 Nm
6–7 ft.lbs.

12–15 Nm
9–11 ft.lbs.

30–40 Nm
22–29 ft.lbs.

20–28 Nm
14–20 ft.lbs.

1. Engine oil level gauge guide
2. O-ring
3. Self locking nut
4. Gasket
5. Oxygen sensor
6. Exhaust manifold cover (A)
7. Engine hanger
8. Exhaust manifold
9. Exhaust manifold gasket
10. Exhaust manifold cover (B)

138625

Exhaust manifold and related parts — Eclipse 1.8L (4G37) engine

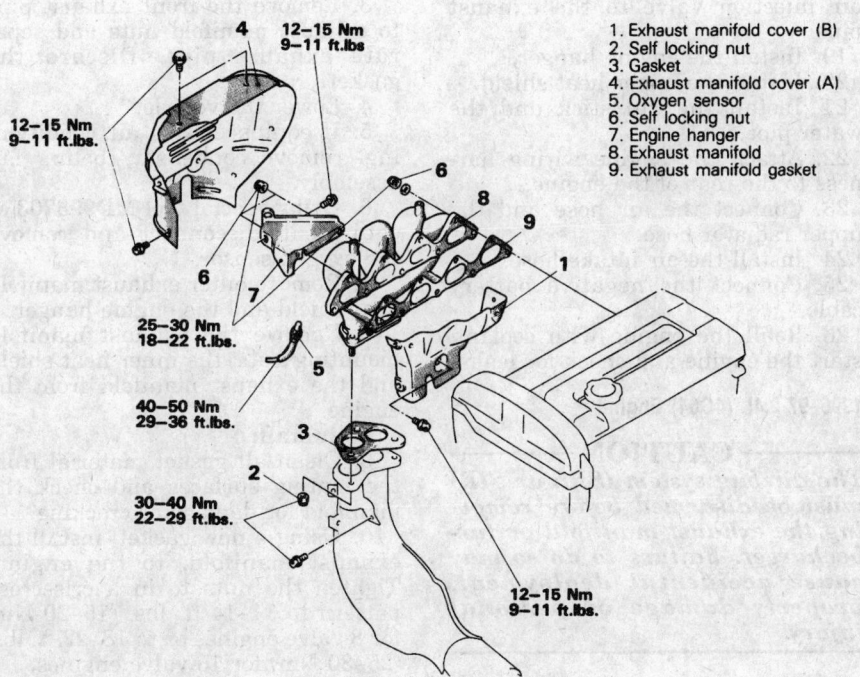

12–15 Nm
9–11 ft.lbs.

12–15 Nm
9–11 ft.lbs.

25–30 Nm
18–22 ft.lbs.

40–50 Nm
29–36 ft.lbs.

30–40 Nm
22–29 ft.lbs.

1. Exhaust manifold cover (B)
2. Self locking nut
3. Gasket
4. Exhaust manifold cover (A)
5. Oxygen sensor
6. Self locking nut
7. Engine hanger
8. Exhaust manifold
9. Exhaust manifold gasket

12–15 Nm
9–11 ft.lbs.

139303

Exhaust manifold and related parts — 1993 Galant and 1993–94 Eclipse 2.0L (4G63) engine

6. Disconnect necessary EGR components. Disconnect and remove the oxygen sensor, as required.

7. Remove outer exhaust manifold heat shield and engine hanger.

8. Remove the exhaust manifold mounting bolts, the inner heat shield and the exhaust manifold from the engine.

To install:

9. Clean all gasket material from the mating surfaces and check the manifold for damage or cracking.

10. Install a new gasket and install the manifold. Tighten the nuts to in a crisscross pattern to 18–22 ft. lbs. (25–30 Nm).

11. Install the heat shields.

12. Connect EGR components.

13. Install the electric cooling fan assembly, dipstick tube and alternator, as required.

14. Install a new flange gasket and connect the exhaust pipe.

15. Connect the negative battery cable and check for exhaust leaks.

1995–97 2.0L (4G63) Engine

------- **CAUTION** -------

The air bag system (SRS or SIR) must be disarmed before removing the exhaust manifold or turbocharger. Failure to do so may cause accidental deployment, property damage or personal injury.

1. Disconnect the negative battery cable.

2. Drain the engine coolant.

3. Remove the condenser fan motor assembly if equipped with air conditioning.

4. Remove the heated oxygen sensor.

5. Remove the dipstick and tube assembly.

6. Remove the air cleaner and air intake hose assembly.

7. Disconnect the air intake hose from the turbocharger.

8. Disconnect the engine coolant hoses from the turbocharger.

9. Disconnect the oil supply pipe connection. Do not let dirt of foreign particles enter the oil pipe.

10. Remove the heat shields.

11. Remove the engine hanger.

12. Disconnect the front exhaust pipe from the turbocharger.

13. Remove the oil return pipe and gaskets.

14. Remove the flange bolts, (washers if equipped), and nuts that attach the turbo to the exhaust manifold. Take note of the positions of the coned disc springs and the washers.

15. Remove the turbocharger, gasket and ring.

16. Remove the exhaust manifold and gasket.

To install:

17. Use a new gasket and install the exhaust manifold.

18. Use a new gasket and install the turbo to the exhaust manifold. Make sure the coned disc spring and the washers are installed in their original positions. Torque the bolts, (if equipped with washers, install them), and nuts to specification, loosen them and torque them again.

19. Use a new gasket and connect the exhaust pipe to the turbo.

20. Using new gaskets, install the oil return pipe.

21. Install the engine hanger.

22. Install the heat shields.

23. Connect the oil supply pipe.

24. Connect the engine coolant hoses to the turbo.

25. Connect the air hose.

26. Install the air cleaner and duct assembly.

27. Position a new O-ring on the dipstick tube and install the tube and dipstick.

28. Install the heated oxygen sensor.

29. Install the condenser fan assembly if removed.

30. Change the engine oil and filter.

31. Connect the negative battery cable and refill the engine with coolant.

32. Start the engine and let it idle. Do not race the engine until the oil reaches the turbo. Check for leaks.

1995–97 2.0L (420A) Engine

-------------- CAUTION --------------

The air bag system (SRS or SIR) must be disarmed before removing the exhaust manifold. Failure to do so may cause accidental deployment, property damage or personal injury.

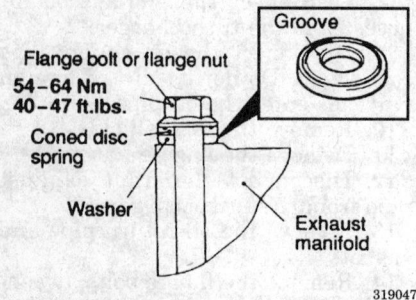

Flange bolt or flange nut
54–64 Nm
40–47 ft.lbs.

Coned disc spring

Washer

Exhaust manifold

Groove

319047

Install the groove of the coned disc spring toward the flange bolt or nut — 1995–97 Eclipse 2.0L (4G63) engine

1. Disconnect the negative battery cable.

2. Drain the engine coolant.

3. Remove the air intake hose.

4. Disconnect the upper radiator hose from the water outlet.

5. Disconnect the air hose connection.

6. Remove the engine control wiring harness from the rear of the engine.

7. Remove the water pipe assembly.

8. Remove the oil dip stick.

9. Remove the upper heat shield.

10. Remove the engine hanger.

11. Disconnect the pulsed secondary air injection (check valve) valve from the exhaust pipe. (manual transaxle only)

12. Disconnect the front exhaust pipe from the manifold.

13. Remove the lower heat shield.

14. Remove the exhaust manifold and gasket.

To install:

15. Use a new gasket and install the exhaust manifold. Torque the nuts and bolts to 17 ft. lbs. (23 Nm).

16. Install the lower heat shield.

17. Use a new gasket and connect the front exhaust pipe to the manifold.

18. On vehicles with manual transaxles, connect the pulsed secondary air injection valve to the exhaust pipe.

19. Install the engine hanger.

20. Install the upper heat shield.

21. Install the dip stick and the water pipe.

22. Attach the engine wiring harness to the rear of the engine.

23. Connect the air hose and the upper radiator hose.

24. Install the air intake hose.

25. Connect the negative battery cable.

26. Refill the engine with coolant, start the engine and check for leaks.

1996–97 2.4L (4G64) Engine

-------------- CAUTION --------------

The air bag system (SRS or SIR) must be disarmed before removing the exhaust manifold or turbocharger. Failure to do so may cause accidental deployment, property damage or personal injury.

1. Disconnect the negative battery cable.

2. Remove the nuts and the gasket and disconnect the front exhaust pipe from the exhaust manifold.

3. Remove the three bolts and the heat shield.

4. Remove the three nuts, the engine hanger, and the three spacers from the exhaust manifold.

5. Remove the remaining mounting nuts, the exhaust manifold, and the exhaust manifold gasket.

To install:

6. Install a new exhaust manifold gasket to the cylinder head and install the exhaust manifold. Torque the mounting nuts to 21 ft. lbs. (29 Nm).

7. Reinstall the three spacers, the engine hanger, and the three mounting nuts. Torque the nuts to 21 ft. lbs. (29 Nm).

8. Replace the heat shield with the three bolts and torque them to 9–11 ft. lbs. (12–15 Nm).

9. Install a new gasket between the exhaust manifold and the front exhaust pipe and reconnect the pipe with the three nuts. Torque the nuts to 32 ft. lbs. (34 Nm).

10. Reconnect the negative battery cable, start the engine and check for any exhaust leaks.

Galant

1993 2.0L (4G63) Engine

1. Disconnect the negative battery cable.

2. Raise the vehicle and support safely.

3. Remove the front exhaust pipe to exhaust manifold nuts and separate exhaust pipe. Discard the gasket.

4. Lower the vehicle.

5. If equipped with air conditioning, remove condenser cooling fan assembly.

6. Using special tool MD998703 or equivalent, disconnect and remove the oxygen sensor.

7. Remove outer exhaust manifold heat shield and the engine hanger.

8. Remove the exhaust manifold mounting nuts, the inner heat shield and the exhaust manifold from the engine.

To install:

9. Clean all gasket material from the mating surfaces and check the manifold for damage or cracking.

10. Using a new gasket, install the exhaust manifold, to the engine. Tighten the nuts to in a crisscross pattern to 11–14 ft. lbs. (15–20 Nm) for 8 valve engines or to 18–22 ft. lbs. (25–30 Nm) for 16 valve engines.

11. Install the outer heat shield and tighten the mounting bolts to 10 ft. lbs. (14 Nm).

12. Install the electric cooling fan assembly, if removed.

13. Using a new flange gasket, connect the exhaust pipe and tighten the

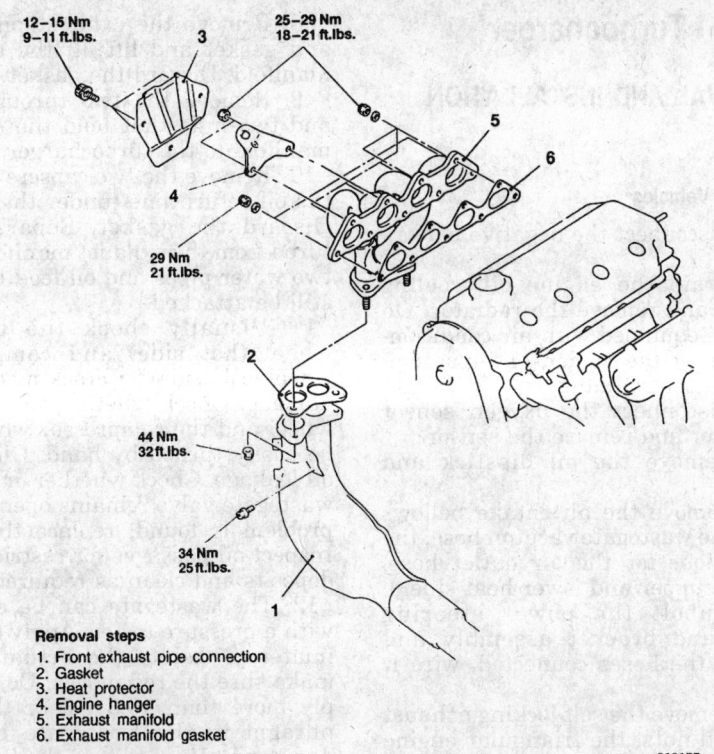

12–15 Nm
9–11 ft.lbs.
3
25–29 Nm
18–21 ft.lbs.
5
6
4
29 Nm
21 ft.lbs.
2
44 Nm
32 ft.lbs.
34 Nm
25 ft.lbs.
1

319177

Removal steps
1. Front exhaust pipe connection
2. Gasket
3. Heat protector
4. Engine hanger
5. Exhaust manifold
6. Exhaust manifold gasket

Exhaust manifold removal and installation — Eclipse 2.4L (4G64) engine shown, Galant similar

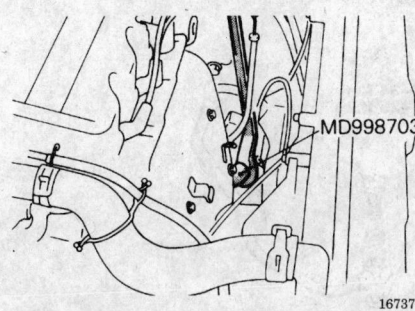

MD998703

167370

Removal of the oxygen sensor — 1993 Galant 2.0L (4G63) engine

mounting nuts to 29–36 ft. lbs. (40–50 Nm).

14. Connect the negative battery cable and check for exhaust leaks.

1994–97 2.4L (4G64) Engine

1. Disconnect battery negative cable.
2. Raise the vehicle and support safely.
3. Remove the exhaust pipe to exhaust manifold nuts and separate exhaust pipe. Discard the gasket.
4. Remove the outer exhaust manifold heat shield and engine hanger.

5. Remove the exhaust manifold mounting nuts and the exhaust manifold from the engine.

To install:
6. Clean all gasket material from the mating surfaces and check the manifold for damage or cracking.
7. Install a new gasket and install the manifold. Tighten the nuts to in a crisscross pattern to 18–21 ft. lbs. (25–29 Nm).
8. Install the heat shields and tighten the mounting bolts 10 ft. lbs. (14 Nm).
9. Install a new flange gasket and connect the exhaust pipe. Tighten the mounting nuts to 32 ft. lbs. (44 Nm).
10. Connect the negative battery cable and check for exhaust leaks.

Diamante

CAUTION

Do not attempt the work on the exhaust system until it has completely cooled.

1. Disconnect battery negative cable.
2. Raise the vehicle and support safely.
3. Remove the exhaust pipe to exhaust manifold nuts and remove the front exhaust pipe.
4. Lower the vehicle.

5. If removing the front manifold, remove condenser electric cooling fan assembly.
6. For the 6G72 DOHC engine, if removing the front manifold, remove the alternator and mounting bracket from the vehicle.
7. For the 6G72 DOHC engine, separate the A/C compressor from the mounting bracket. Leaving the hoses connected, position the compressor aside.
8. If removing the front manifold, remove the oil dipstick and tube from the engine.
9. For the 6G72 DOHC engine, if removing the front manifold, remove the heat protector.
10. If removing the rear manifold, disconnect the EGR tube.
11. For the 6G72 SOHC engine, if removing the rear manifold, remove the intake plenum stay and the roll stopper bracket.
12. Disconnect the electrical connector and remove the oxygen sensor.
13. Remove the exhaust manifold mounting bolts the manifold.

To install:
14. Clean all gasket material from the mating surfaces and check the manifold for damage.
15. Install a new gasket and install the manifold. Tighten the nuts in a criss-cross pattern to 21 ft. lbs. (30 Nm) for the J-6G72 engine or to 14 ft. lbs. (19 Nm) for the H-6G72 engine.
16. Install the heat shields.
17. Connect the EGR tube and intake plenum stay and roll stopper bracket, if removed.
18. Install the oxygen sensor.
19. Install the electric cooling fan assembly, A/C compressor, dipstick tube and alternator, as required.
20. Install a new flange gasket and connect the exhaust pipe or converter assembly.
21. Install the drive belt(s) and adjust for proper tension.
22. Connect the negative battery cable and check for exhaust leaks.

3000GT

Non-Turbocharged Engines

1. Disconnect battery negative cable.
2. Raise the vehicle and support safely.
3. Remove the exhaust pipe to exhaust manifold nuts and separate exhaust pipe. Discard gasket.
4. Lower vehicle.
5. Remove electric cooling fan assembly, if necessary. If removing the front manifold, remove the dipstick tube. If removing the front manifold

from 3.0L DOHC engine, remove the alternator.

6. Disconnect necessary EGR components.

7. Disconnect the electrical connector and remove the oxygen sensor.

8. Remove the exhaust manifold mounting bolts, the inner heat shield and the exhaust manifold.

To install:

9. Clean all gasket material from the mating surfaces and check the manifold for damage.

10. Install a new gasket and install the manifold. Tighten the nuts in a criss-cross pattern to 22 ft. lbs. (30 Nm).

11. Install the heat shields.

12. Connect EGR components.

13. Install the oxygen sensor.

14. Install the electric cooling fan assembly, dipstick tube and alternator, as required.

15. Install a new flange gasket and connect the exhaust pipe.

16. Connect the negative battery cable and check for exhaust leaks.

Turbocharged Engines

1. Disconnect the negative battery cable.

2. Drain the engine coolant.

3. Remove the turbocharger assembly.

4. Remove the heat shield.

5. Remove the mounting nuts and remove the exhaust manifold. Note that the cone disc springs are installed at all lower mounting points.

To install:

6. Clean all gasket material from the mating surfaces and check the manifold for damage.

7. Install new gaskets and install the manifold. Make sure all cone disc springs are in their original locations with the grooved side facing the nut. Tighten the manifold nuts using the following procedure:

   a. Tighten all but the outer two nuts to 22 ft. lbs. (30 Nm).

   b. Tighten the outer two nuts to 34–38 ft. lbs. (47–53 Nm).

   c. Loosen the outer two nuts, then torque them to 22 ft. lbs. (30 Nm).

8. Install the heat shield.

9. Install the turbocharger assembly.

10. Fill the cooling system.

11. Connect the negative battery cable and check for exhaust leaks.

Turbocharger

REMOVAL AND INSTALLATION

Eclipse

1993–94 Vehicles

1. Disconnect the negative battery cable.

2. Drain the engine oil, cooling system and remove the radiator. On vehicles equipped with air conditioning, remove the condenser fan assembly with the radiator.

3. Disconnect the oxygen sensor connector and remove the sensor.

4. Remove the oil dipstick and tube.

5. Remove the air intake bellows hose, the wastegate vacuum hose, the connections for the air outlet hose, and the upper and lower heat shield.

6. Unbolt the power steering pump and bracket assembly and leaving the hoses connected, wire it aside.

7. Remove the self-locking exhaust manifold nuts, the triangular engine hanger bracket, the eyebolt and gaskets that connect the oil feed line to the turbo center section and the water cooling lines. The water line under the turbo has a threaded connection.

8. Remove the exhaust pipe nuts and gasket and lift off the exhaust manifold. Discard the gasket.

9. Remove the two through-bolts and two nuts that hold the exhaust manifold to the turbocharger.

10. Remove the two capscrews from the oil return line (under the turbo). Discard the gasket. Separate the turbo from the exhaust manifold. The two water pipes and oil feed line can still be attached.

11. Visually check the turbine wheel (hot side) and compressor wheel (cold side) for cracking or other damage. Check whether the turbine wheel and the compressor wheel can be easily turned by hand. Check for oil leakage. Check whether or not the wastegate valve remains open. If any problem is found, replace the part. Inspect oil passages for restriction or deposits and clean as required.

12. The wastegate can be checked with a pressure tester. Apply approximately 9 psi to the actuator and make sure the rod moves. Do not apply more than 10.3 psi or the diaphragm in the wastegate may be damaged. Vacuum applied to the wastegate actuator should be maintained, replace if leaks vacuum. Do not attempt to adjust the wastegate valve.

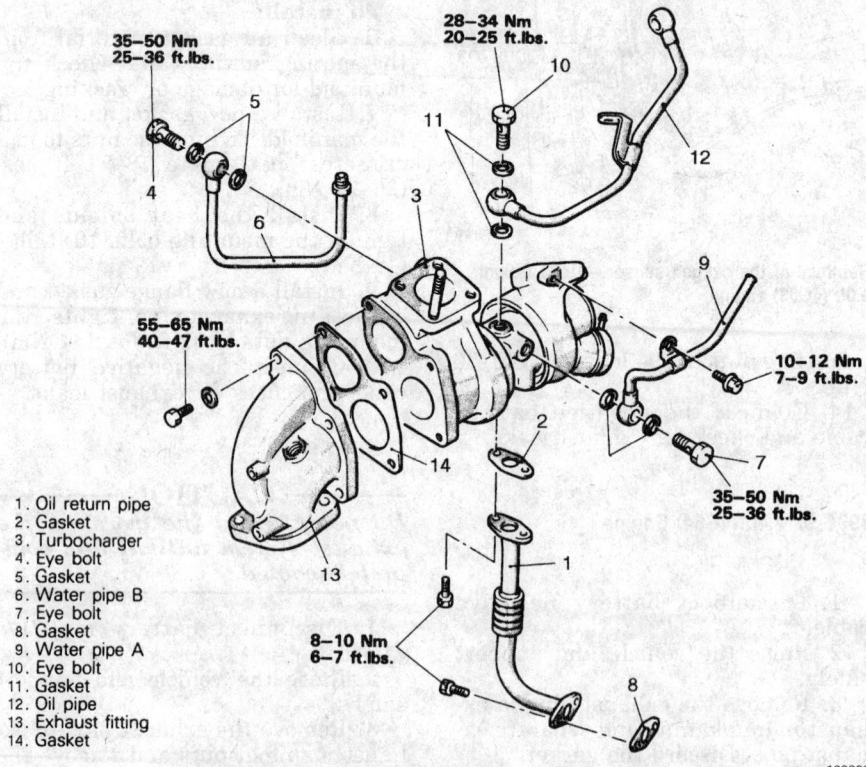

1. Oil return pipe
2. Gasket
3. Turbocharger
4. Eye bolt
5. Gasket
6. Water pipe B
7. Eye bolt
8. Gasket
9. Water pipe A
10. Eye bolt
11. Gasket
12. Oil pipe
13. Exhaust fitting
14. Gasket

Turbocharger assembly and related components — 1993–94 Eclipse shown

139299

To install:

13. Prime the oil return line with clean engine oil. Replace all locking nuts. Before installing the threaded connection for the water inlet pipe, apply light oil to the inner surface of the pipe flange. Assemble the turbocharger and exhaust manifold.

14. Install the exhaust manifold using a new gasket.

15. Connect the water cooling lines, oil feed line and engine hanger.

16. If removed, install the power steering pump and bracket.

17. Install the heat shields, air outlet hose, wastegate hose and air intake bellows.

18. Install the oil dipstick tube and dipstick. Install the oxygen sensor.

19. Install the radiator assembly.

20. Fill the engine with oil, fill the cooling system and reconnect the negative battery cable.

21. Run engine until thermostat opens. Check engine for leaks and top off coolant.

1995–97 Vehicles

—— CAUTION ——

The air bag system (SRS or SIR) must be disarmed before removing the turbocharger. Failure to do so may cause accidental deployment, property damage or personal injury.

1. Disconnect the negative battery cable.
2. Drain the engine coolant.
3. Remove the condenser fan motor assembly if equipped with air conditioning.
4. Remove the heated oxygen sensor.
5. Remove the dipstick and tube assembly.
6. Remove the air cleaner and air intake hose assembly.
7. Disconnect the air intake hose from the turbocharger.
8. Disconnect the engine coolant hoses from the turbocharger.
9. Disconnect the oil supply pipe connection. Do not let dirt of foreign particles enter the oil pipe.
10. Remove the heat shields.
11. Remove the engine hanger.
12. Disconnect the front exhaust pipe from the turbocharger.
13. Remove the oil return pipe and gaskets.
14. Remove the flange bolts and nut that attach the turbo to the exhaust manifold. Take note of the positions of the coned disc springs and the washers.
15. Remove the turbocharger, gasket and ring.

To install:

16. Use a new gasket and install the turbo to the exhaust manifold. Make sure the coned disc spring and the washers are installed in their original positions. Torque the bolts and nut to 20–23 ft. lbs. (27–31 Nm). Further tighten the bolts and nuts 60°–70°.

17. Use a new gasket and connect the exhaust pipe to the turbo. Torque the mounting bolts to 40–47 ft. lbs. (54–64 Nm).

18. Using new gaskets, install the oil return pipe.

19. Install the engine hanger.

20. Install the heat shields.

21. Connect the oil supply pipe. Torque the flare nut fittings to 14 ft. lbs. (19 Nm).

22. Connect the engine coolant hoses to the turbo.

23. Connect the air hose.

24. Install the air cleaner and duct assembly.

25. Position a new O-ring on the dipstick tube and install the tube and dipstick.

26. Install the heated oxygen sensor.

27. Install the condenser fan assembly if removed.

28. Change the engine oil and filter.

29. Connect the negative battery cable and refill the engine with coolant.

30. Start the engine and let it idle. Do not race the engine until the oil reaches the turbo. Check for leaks.

3000GT

—— CAUTION ——

Work must be started after 90 seconds from the time the ignition switch is turned to the LOCK position and the negative (-) battery cable is disconnected.

Right Side (Front) Turbocharger

1. Disconnect the negative battery cable.

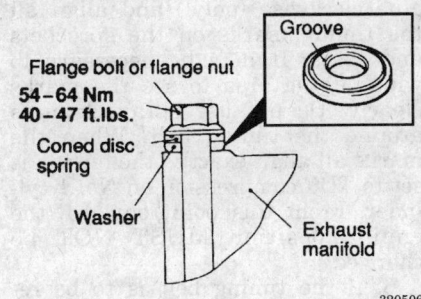

Install the groove of the coned disc spring toward the flange bolt or nut — 1995–97 Eclipse

2. Remove the radiator.
3. Remove the right side transaxle bracket.
4. Remove the front exhaust pipe.
5. Carefully matchmark, diagram or photograph all air intake hoses and pipes along the front of the engine. It is imperative that all of these pieces are installed in the exact same positions when assembling. Remove the hoses and pipes and keep covered in a clean area.
6. Remove the alternator.
7. Remove the oil dipstick tube.
8. Remove the turbocharger heat protector.
9. Remove the water feed pipes.
10. Remove the oxygen sensor.
11. Remove the oil return line.
12. Remove the exhaust extension fitting and bracket.
13. Remove all air conditioning components preventing removal of the turbocharger.
14. Remove the oil feed tube.
15. Remove the turbocharger to exhaust manifold bolts and remove the turbocharger assembly.

To install:

16. Visually check the turbine wheel (hot side) and compressor wheel (cold side) for cracking or other damage. Check whether the turbine wheel and the compressor wheel can be easily turned by hand. Check for oil leakage. Check whether or not the wastegate valve remains open. If any problem is found, replace the part.

17. Clean all mating surfaces. Pour clean engine oil through the oil pipe feed hole in the turbocharger.

18. Install a new gasket and ring a install the turbocharger to the manifold. Torque the bolts to 40–47 ft. lbs. (55–65 Nm).

19. Replace the eye-bolt rings and install the oil feed pipe.

20. Install the removed air conditioning components.

21. Install the exhaust extension fitting and bracket with a new gasket. Torque the nuts to 40–47 ft. lbs. (55–65 Nm).

22. Install the oil return line with new gaskets.

23. Install the oxygen sensor.

24. Replace the eye-bolt rings and install the water feed pipes.

25. Install the turbocharger heat protector.

26. Install the dipstick tube.

27. Install the alternator.

28. Install all air intake hoses and pipes along the front of the engine. Make sure all are in their proper positions.

29. Install a new gasket and connect the front exhaust pipe.

30. Install the right side transaxle bracket.

31. Install the radiator.

32. Fill the system with coolant.

33. Connect the negative battery cable and check for exhaust leaks.

Left Side (Rear) Turbocharger

1. Remove the battery.

2. Drain the coolant.

3. Remove the front exhaust pipe.

4. Disconnect the accelerator cable from the throttle body.

5. Remove the intake air hose, the air pipe across the top of the engine and its heat shield.

6. Remove the clutch booster vacuum hose and disconnect the accelerator cable from the pedal.

7. Remove the air intake hoses coming from the air cleaner box.

8. Remove the oxygen sensor and the turbocharger heat protector.

9. Remove the EGR pipe, if equipped.

10. Remove the oil feed pipe.

11. Remove the EGR valve, if equipped.

12. Remove the water feed pipes.

13. Remove the exhaust extension fitting and bracket.

14. Remove the inner heat protector.

15. Remove the oil return tube.

16. Remove the turbocharger to exhaust manifold nuts and remove the turbocharger assembly.

To install:

17. Visually check the turbine wheel (hot side) and compressor wheel (cold side) for cracking or other damage. Check whether the turbine wheel and the compressor wheel can be easily turned by hand. Check for oil leakage. Check whether or not the wastegate valve remains open. If any problem is found, replace the part.

18. Clean all mating surfaces. Pour clean engine oil through the oil pipe feed hole in the turbocharger.

19. Install a new gasket and ring a install the turbocharger to the manifold. Torque the nuts to 40–47 ft. lbs. (55–65 Nm).

20. Install the oil return line with new gaskets.

21. Install the inner heat protector.

22. Install the exhaust extension fitting and bracket with a new gasket. Torque the nuts to 40–47 ft. lbs. (55–65 Nm).

23. Replace the eye-bolt rings and install the water feed pipes.

24. Install the EGR valve, if equipped.

25. Replace the eye-bolt rings and install the oil feed pipe.

26. Install the EGR pipe if equipped.

27. Install the turbocharger heat protector and oxygen sensor.

28. Install the air intake hoses coming from the air cleaner box. Make sure the triangular aligning marks are engaged.

29. Connect the accelerator cable to from the pedal and install the clutch booster vacuum hose.

30. Install the heat shield, the air pipe across the top of the engine and the air intake hose.

31. Connect the accelerator cable to the throttle body.

32. Install a new gasket and connect the front exhaust pipe.

33. Fill the system with coolant.

34. Install the battery.

35. Connect the negative battery cable and check for exhaust leaks.

Front Cover Seal

REMOVAL AND INSTALLATION

Precis

NOTE: These engines have two timing belts. One connects the crankshaft and camshaft, the second one drives the right silent shaft.

1. With the vehicle in neutral or park, make certain the parking brake is set and the wheels are blocked. Disconnect the negative battery cable.

2. Remove the water pump drive belt and water pump pulley. If the vehicle is equipped with air conditioning, loosen the compressor belt tensioner and remove the compressor belt.

3. Remove the bolts holding the crankshaft pulley(s) and remove the pulley(s).

4. Remove the upper and lower timing belt covers and their gaskets.

5. Use a socket wrench on the projecting crankshaft bolt to turn the engine clockwise (only!) and align all the timing marks on the sprockets and cases. It may be necessary to wipe off the area to see the marks clearly. Do not use spray cleaners around the timing belt. When the marks all align exactly, the engine is set to TDC/compression on No. 1 cylinder. From this point onward, the engine position MUST NOT be changed.

6. If the timing belt is to be reused, make a chalk or crayon arrow on the belt showing the direction of rotation so that it may be reinstalled correctly.

7. Loosen the bolts holding the tensioner and pivot the tensioner towards the water pump. Temporarily tighten the bolts to hold the tensioner in its slack position.

8. Carefully slide the belt off the sprockets. Place the belt in a clean, dry, protected location away from the work area.

9. Remove the crankshaft sprocket retaining bolt. Remove the sprocket and flange.

10. Remove the tensioner for the silent shaft belt. (Tensioner B).

11. Remove the silent shaft belt (timing belt B). Remove the crankshaft pulley for the silent shaft belt.

———— WARNING ————
Do not spray or immerse the sprockets or tensioners in cleaning solvent. The sprocket may absorb the solvent and transfer it to the belt. The tensioners are internally lubricated and the solvent will dilute or dissolve the lubricant.

12. Carefully pry out the oil seal with a screwdriver being careful not to damage the crankshaft.

To install:

13. Coat the lip of the new seal with oil and install the seal using the proper seal driver.

14. To reassemble, install the sprocket for the silent shaft belt onto the crankshaft. Make certain it is installed correctly.

15. If the sprocket for the right silent shaft was removed, coat the spacer with a light coating of clean engine oil and install the spacer to the shaft. Be sure to install it in the correct direction. Install the silent shaft sprocket and tighten the bolt finger tight.

16. Double check the timing marks on the silent shaft sprocket and crankshaft sprocket. Carefully align the marks if necessary.

17. Install the silent shaft belt, observing the direction of rotation mark made earlier. Handle the belt carefully and do not use metal tools to guide or force the belt into place. When installing the belt, make sure the tension side has no slack in it.

18. Install the tensioner (B) with the center of the pulley located to the left side of the mounting bolt and with the pulley flange to the front of the engine.

19. Lift the tensioner with your hand so that the belt becomes taut. Hold the tensioner in this position and tighten its bolt. Use care that only the bolt and not the tensioner shaft is turned during tightening.

The bolt should be tightened to 13 ft. lbs.

20. Check that the timing marks are still aligned. Push down on the center of the tension side of the belt with your finger. Correct belt deflection is 5–7mm or approximately ¼ inch. If the deflection is not correct, the tensioner must be released fully and the belt re-tensioned.

21. Install the flange and the crankshaft sprocket onto the crankshaft. Make certain the flange is installed in the correct direction. If it is put on incorrectly, the belt will wear and break.

22. Install the special washer and the sprocket retaining bolt to the crankshaft. Tighten the bolt to 88 ft. lbs.

23. Install the spacer, tensioner and tensioner spring if they were removed. Install the lower end of the spring to its position on the tensioner, then place the upper end in position at the water pump. Move the tensioner towards the water pump and temporarily tighten it in this position.

24. Double check the alignment of the timing marks for the cam, crank and oil pump sprockets.

25. Install the timing belt onto the crankshaft sprocket, the oil pump sprocket and the cam sprocket in that order. Keep the belt taut between sprockets. If reusing an old belt, make certain that the direction of rotation arrow is properly oriented.

26. Loosen the tensioner mounting nut and bolt. The spring will move the tensioner against the belt and tension it.

27. Check the belt as it passes over the camshaft sprocket. The belt may tend to lift in the area to the left of the sprocket. (Roughly 7 o'clock to 12 o'clock when viewed from the pulley end.) Make certain the belt is well seated and not rubbing on any flanges or nearby surfaces.

28. At the tensioner, tighten the bolt in the slotted hole first, then tighten the nut on the pivot. If this order is not followed, the belt will become too tight and break.

29. Once again, check all the timing marks for alignment. Nothing should have changed; check anyway.

30. Remove the screwdriver blocking the left side silent shaft. Using a socket on the crankshaft bolt, turn the crankshaft smoothly one full turn (360°). Do NOT turn the engine backwards.

31. Loosen the tensioner nut and bolt. The spring will allow the tensioner to tighten a little bit more be-cause of the slack picked up during engine rotation. Tighten the bolt, then the nut to 36 ft. lbs.

32. Check the deflection of the belt. At the middle of the right (tension) side of the belt, deflect the belt outward with your finger, toward the timing case. The distance between the belt and the line of the cover seal should be about 14mm (⁹⁄₁₆ in.)

33. Install the lower and upper timing belt covers. Make certain the gaskets are properly seated in the covers and that they don't come loose during installation. Tighten the bolts to 8 ft. lbs.

34. Install the crankshaft pulley(s) and tighten the retaining bolts to 18 ft. lbs.

35. Install the water pump pulley, tightening the bolts to 6 ft. lbs. and install the belt. Adjust the belt to the correct tension.

36. Connect the negative battery cable. Start the engine and let it idle, listening for any unusual noises from the area of the timing belt. Possible causes of noise are the belt rubbing against the covers or a sprocket flange, the belt being too loose and slapping, or a tensioner binding. Do not accelerate the engine if abnormal noises are heard from the timing belt train — severe damage can result.

Mirage

1. Disconnect the negative battery cable. Drain the engine oil.
2. Remove the timing belt.
3. Remove the crankshaft pulley retainer bolts and remove the pulley.
4. Remove the vibration damper retainer bolt and washer and remove damper. If difficult to remove, the appropriate puller may be used.
5. Remove the crankshaft sprocket. If sprocket is difficult to remove, the appropriate puller may be used.
6. Pry out the oil seal from front of engine.

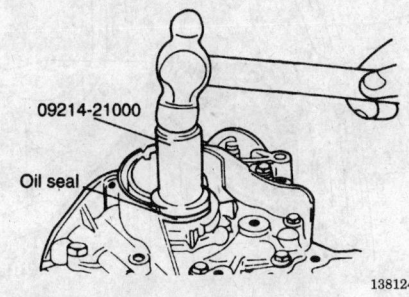

09214-21000

Oil seal

138124

Crankshaft oil seal installation — 1993–94 Precis 1.5L (G4DJ) engine

To install:
7. Using proper size driver, install new front seal.

---------- **WARNING** ----------

Small nicks and burrs on crankshaft surface will damage the oil seal. Use care when installing the oil seal not to damage crankshaft surface.

8. Lubricate the lips of the new seal with clean engine oil.
9. Install the crankshaft sprocket and vibration damper. Torque the retaining bolt to specifications.
10. Install the timing belt, timing covers, valve cover and remaining components.
11. Install the engine undercover and connect the negative battery cable.
12. Fill engine oil, start the engine and check for leaks.

Eclipse

2.0L (4G63) Engine

1. Disconnect the negative battery cable.
2. Remove the valve covers.
3. Remove the timing belts as required.

NOTE: When reusing the timing belts, be sure to mark the direction of rotation on the belt. This will extend belt life, ensuring the same direction of rotation.

4. Remove the crankshaft pulley retainer bolts and remove the pulley.
5. Remove the crankshaft sprocket retainer bolt and washer from the sprocket, if used, and remove sprocket. If sprocket is difficult to remove, the appropriate puller may be used. If no bolts are used on the sprocket, use the appropriate puller to remove.
6. Pry the seal from the bore and replace using the proper installation tools.
To install:
7. Using proper size driver, install new seal
8. Install the crankshaft sprocket and torque the retaining bolt to 80–94 ft. lbs. (110–130 Nm).
9. Install the timing belt(s), timing covers, valve cover(s) and remaining components.
10. Install the engine undercover. Connect the negative battery cable, start the engine and check for leaks.

1995–97 2.0L and 2.4L Engines

1. Disconnect the negative battery cable.

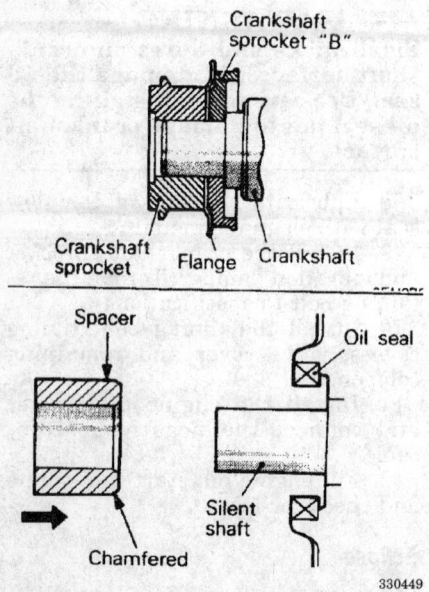

Front crankshaft oil seal — Eclipse 2.0L (4G63) engine

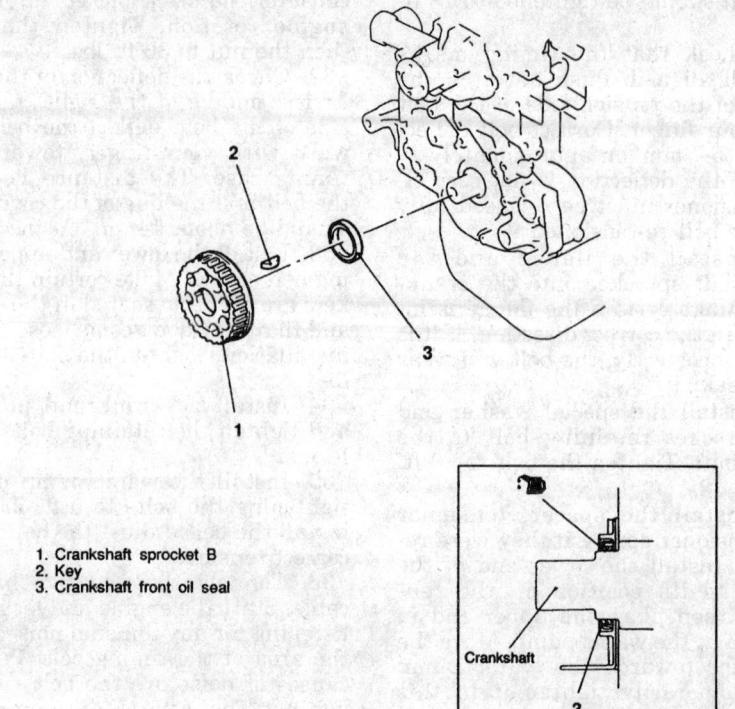

1. Crankshaft sprocket B
2. Key
3. Crankshaft front oil seal

Front crankshaft oil seal — 1995–97 Eclipse 2.0L (4G63) engine shown

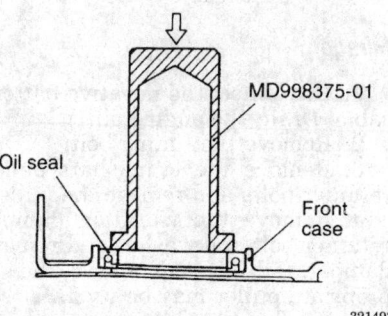

Front crankshaft oil seal installation — 1995–97 Eclipse 2.0L (4G63) engine shown

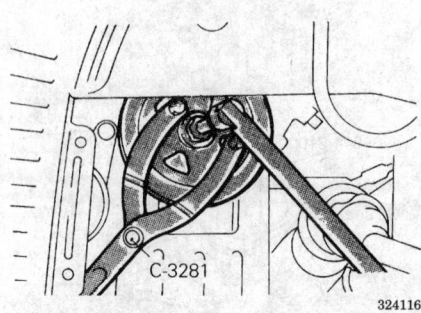

Crankshaft pulley pin spanner tool is used to hold pulley while bolt is removed — Diamante sand 3000GT

2. Remove the timing belt.

3. Remove the crankshaft sprocket using MB995027 or equivalent proper puller.

4. Carefully pry the oil seal out of the front case assembly using MB995020 or equivalent. Be careful not to damage the oil seal bore or the crankshaft sealing surface.

To install:

5. Apply clean engine oil to the oil seal lip. Using MB995022 or equivalent seal driver, install the oil seal.

6. Install the crankshaft sprocket using MB995035 and MB995026 or equivalent, if necessary. If equipped, torque the crankshaft bolt to 87 ft. lbs. (118 Nm).

7. Install the timing belt.

8. Connect the negative battery cable.

Diamante

1. Disconnect the negative battery cable.

2. Position the engine to TDC of No. 1 cylinder compression stroke.

3. Remove the drive belts.

4. Using a pin spanner, if available, hold the crankshaft pulley and remove the retaining bolt. Remove the pulley.

5. Remove the timing belt covers and the timing belt .

6. If necessary, remove the crankshaft position sensor.

7. Remove the crankshaft sprocket, and if necessary, the sensing blade, spacer and woodruff key®.

NOTE: If sprocket is difficult to remove, an appropriate puller may be used.

8. Pry the seal from the bore, using a suitable tool.

———— **WARNING** ————
Use care not to nick or scratch the crankshaft, when removing the seal.

To install:

9. Using driver tool MD998717 or equivalent, install the new crankshaft seal. Lubricate the lips of the seal with clean engine oil.

10. Install the Woodruff key®, spacer, sensing blade (if necessary) and the crankshaft sprocket.

11. If removed, install the crankshaft position sensor and torque the retaining bolts to 7 ft. lbs. (9 Nm).

12. Install the timing belt and the timing belt cover(s).

13. Install the crankshaft pulley and retaining bolt. Torque the retaining bolt to 130–137 ft. lbs. (180–190

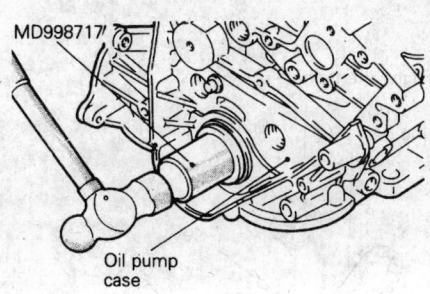

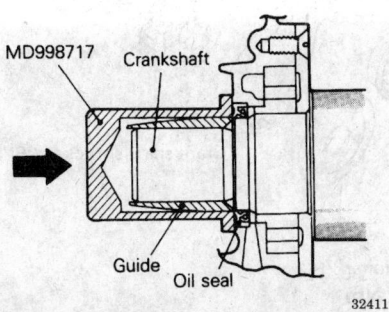

Crankshaft seal installation — Diamante and 3000GT

324117

Nm) for the 6G72 DOHC engine or to 108–116 ft. lbs. (150–160 Nm) for the 6G72 SOHC engine.

14. Install and adjust the drive belts.

15. Connect the negative battery cable and check for leaks.

3000GT

1. Disconnect the negative battery cable.

2. Remove the undercover.

3. Remove the cruise control pump and link assembly.

4. Remove the alternator assembly.

5. Raise and support the engine to take the weight off of the engine mount.

6. Remove the air hose and the air pipe.

7. Remove the power steering tensioner and drive belt.

8. Using a pin spanner, if available, hold the crankshaft pulley from turning and remove the crankshaft pulley bolt.

9. Remove the brake fluid level sensor and the upper timing belt cover.

10. Remove the engine mount bracket and the idler pulley for the alternator and A/C compressor drive belt.

11. Remove the engine support bracket and the lower timing belt cover.

12. Remove the timing belt and the auto tensioner.

13. Remove the crankshaft sprocket.

14. Pry out the crankshaft seal using a suitable tool.

To install:

15. Using a driver tool, install the new crankshaft seal.

16. Install the crankshaft sprocket, timing belt and the belt auto tensioner.

17. Install the lower timing belt cover and the engine support bracket.

18. Install the idler pulley for the alternator and A/C compressor and the engine mount bracket.

19. Install the upper timing belt cover and the brake fluid sensor.

20. Using the spanner tool, if available, install the crankshaft pulley.

21. Install the power steering drive belt and tensioner.

22. Install the air hose and the air pipe.

23. Install the alternator and the cruise control link and pump assembly.

24. Install the undercover and the negative battery cable.

Timing Belt, Sprockets, Tensioner and Front Cover

REMOVAL AND INSTALLATION

1.5L (G4DJ) Engine

Precis

1. Disconnect the negative battery cable. Remove the engine accessory drive belts.

2. Remove the water pump pulley.

3. Remove the timing belt cover.

4. Rotate the crankshaft clockwise and align the timing marks so the No. 1 piston will be at TDC of the compression stroke. Loosen the tensioning bolt in the slotted portion of the tensioner, and the pivot bolt on the timing belt tensioner. Move the tensioner as far as it will go toward the water pump. Tighten the adjusting bolt. If the belt is to be reused, mark its direction of rotation with an arrow.

5. Remove the timing belt.

6. Remove the camshaft sprocket as required.

7. Remove the crankshaft sprocket bolts and remove the crankshaft sprocket and flange, noting the direction of installation for each. Remove the timing belt tensioner.

8. Inspect the belt thoroughly and replace if necessary.

9. Inspect the tensioner for grease leaking from the grease seal and any roughness in rotation. Replace a tensioner if either condition is present.

10. The sprockets should be inspected and replaced if there is any sign of damaged teeth or cracking. Do not immerse the sprockets or tensioner in solvent.

To install:

11. Install the flange and crankshaft sprocket. The flange must go on first with the chamfered area outward. The sprocket is installed with the boss forward and the studs for the fan belt pulley outward. Install and torque the crankshaft sprocket bolt to 51–72 ft. lbs. (69–98 Nm). Install the camshaft sprocket and bolt, torquing it to 47 to 54 ft. lbs. (64 to 74 Nm).

12. Align the camshaft sprocket timing marks. Check that the crankshaft timing marks are still in alignment (the locating pin on the front of the crankshaft sprocket is aligned with a mark on the front case).

13. Mount the spring and spacer onto the tensioner. Then, install the bolts and tighten the adjusting bolt slightly with the tensioner moved as far as possible away from the water pump. Install the free end of the spring into the locating tang on the front case. Position the belt over the crankshaft sprocket and then over the camshaft sprocket. Slip the back of the belt over the tensioner wheel. Turn the camshaft sprocket in the opposite of its normal direction of rotation until the straight side of the belt is tight and the timing marks align.

14. Loosen the tensioner mounting bolts so the tensioner works under spring pressure without the interference of any friction.

15. Torque the tensioner adjusting bolt to 15–18 ft. lbs. (20–26 Nm). Then, torque the tensioner pivot bolt to 15–18 ft. lbs. (20–26 Nm). The bolts must be tightened in order or the tension won't be correct.

16. Turn the crankshaft one turn clockwise until the timing marks again align to seat the belt. Loosen both tensioner attaching bolts and let the tensioner position itself under spring tension as before. Torque the bolts to 15–18 ft. lbs. (20–26 Nm) in order. Check belt tension by putting your finger on the water pump side of the tensioner wheel and pulling the belt toward . Under moderate pressure, the belt should move toward the pump until the teeth are about $1/4$–$1/2$

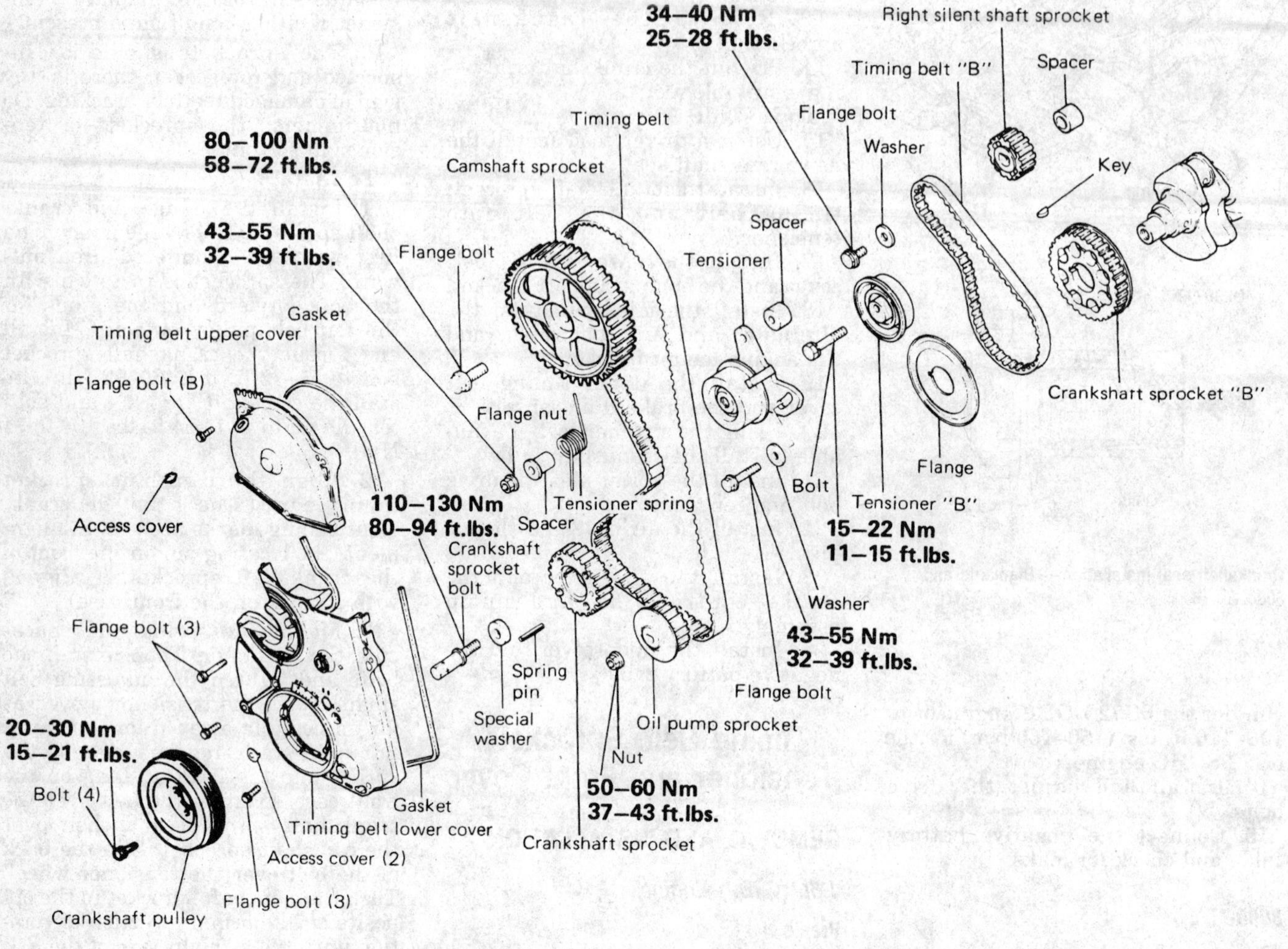

34–40 Nm
25–28 ft.lbs.

Right silent shaft sprocket

Timing belt "B"

Spacer

Timing belt

Camshaft sprocket

Flange bolt

Washer

Key

80–100 Nm
58–72 ft.lbs.

Spacer

Tensioner

43–55 Nm
32–39 ft.lbs.

Flange bolt

Crankshaft sprocket "B"

Timing belt upper cover

Gasket

Flange bolt (B)

Flange nut

Flange

Access cover

Tensioner spring

Bolt

Tensioner "B"

110–130 Nm
80–94 ft.lbs.

Spacer

15–22 Nm
11–15 ft.lbs.

Crankshaft sprocket bolt

Flange bolt (3)

Washer

Spring pin

43–55 Nm
32–39 ft.lbs.

Flange bolt

20–30 Nm
15–21 ft.lbs.

Special washer

Oil pump sprocket

Bolt (4)

Nut

50–60 Nm
37–43 ft.lbs.

Gasket

Timing belt lower cover

Crankshaft sprocket

Access cover (2)

Flange bolt (3)

Crankshaft pulley

151887

Timing belts and related components — Precis 1.5L (G4DJ) engine

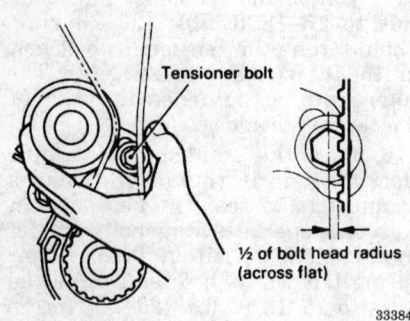

Tensioner bolt

½ of bolt head radius (across flat)

333849

Checking the tension of the timing belt — Precis 1.5L (G4DJ) engine

of the way across the head of the tensioner adjusting bolt.

17. Install the timing belt covers.

18. Install the crankshaft pulley, making sure the pin on the crankshaft sprocket fits through the hole in the rear surface of the pulley. Install the bolts and torque to 7.5–8.5 ft. lbs. (9–12 Nm).

19. Install the water pump pulley.

20. Install and tension the accessory drive belts.

21. Connect the negative battery cable.

1.5L (4G15) and 1.8L (4G93) Engines

Mirage

1. Disconnect the negative battery cable. Remove the engine undercover.

2. Rotate crankshaft clockwise and position engine at TDC compression stroke.

3. Raise and safely support the weight of the engine using the appropriate equipment. Remove the A/C clamp, front engine mount bracket and accessory drive belts.

4. Remove the crankshaft pulley.

5. Remove timing belt upper and lower covers.

6. Make a mark on the back of the timing belt indicating the direction of

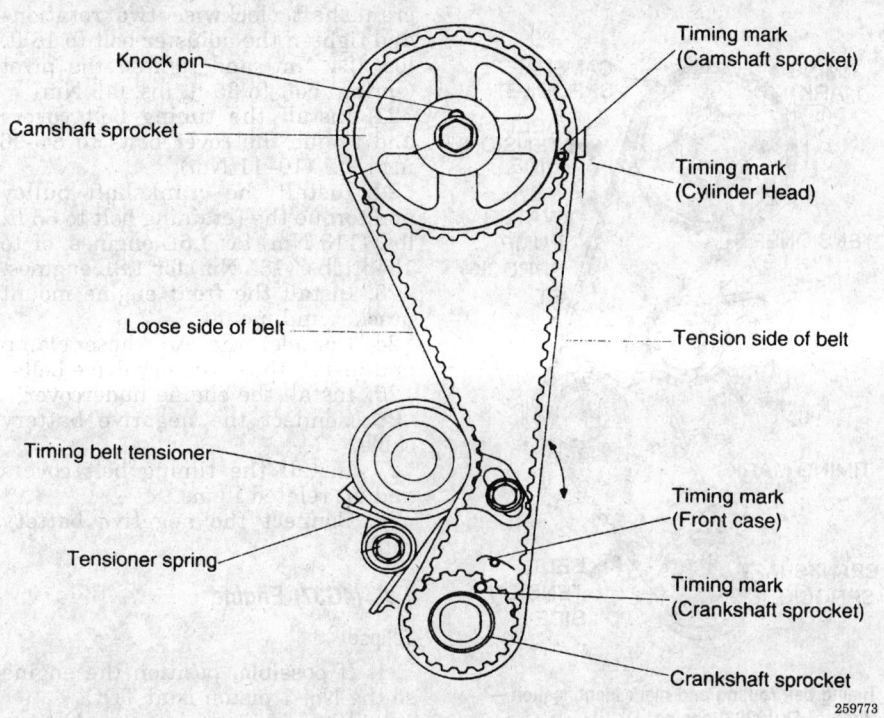

Timing belt components — Precis 1.5L (G4DJ) engine

Knock pin

Camshaft sprocket

Timing mark (Camshaft sprocket)

Timing mark (Cylinder Head)

Loose side of belt

Tension side of belt

Timing belt tensioner

Tensioner spring

Timing mark (Front case)

Timing mark (Crankshaft sprocket)

Crankshaft sprocket

259773

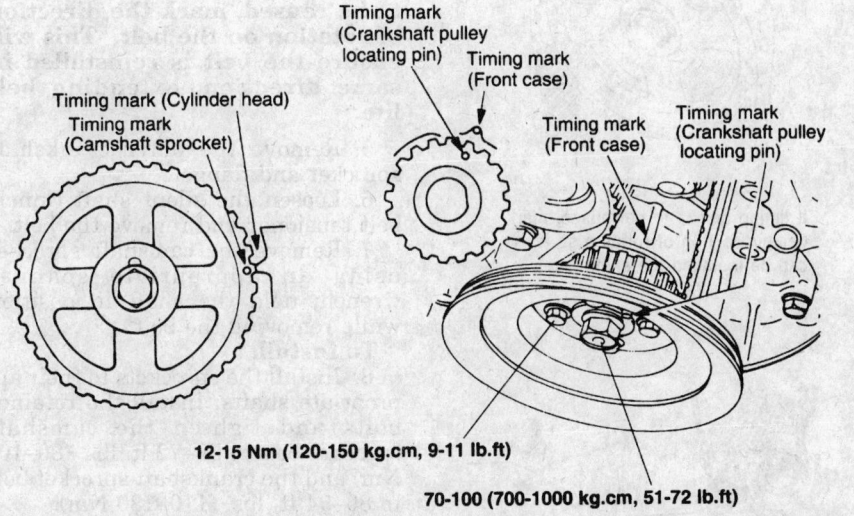

Timing mark (Cylinder head)
Timing mark (Camshaft sprocket)

Timing mark (Crankshaft pulley locating pin)

Timing mark (Front case)

Timing mark (Front case)

Timing mark (Crankshaft pulley locating pin)

12-15 Nm (120-150 kg.cm, 9-11 lb.ft)

70-100 (700-1000 kg.cm, 51-72 lb.ft)

259072

Crankshaft pulley and sprocket installation — Precis 1.5L (G4DJ) engine

rotation so it may be reassembled in the same direction if it is to be re-used. Loosen the timing belt tensioner and move the tensioner to provide slack to the timing belt. Tighten the tensioner in this position.

7. Remove the timing belt.

NOTE: Coolant and engine oil will damage the rubber in the timing belt, drastically reducing its life. Do not allow engine oil or coolant to contact the timing belt, the sprockets or tensioner assembly.

8. Remove the tensioner spacer, tensioner spring and tensioner assembly.

9. Remove the crankshaft sprocket retaining bolt and washer from the sprocket, if used, and remove the sprocket. If it is difficult to remove, a suitable puller may be used.

10. Remove the camshaft sprocket by using the appropriate spanner tool or unfastening the retaining bolt.

NOTE: It is recommended that the timing belt is replaced every 60,000 miles (96,000km).

11. Inspect the timing belt for cracks or wear. Check the tensioner pulley for smooth rotation.

To install:

12. Install the camshaft and crankshaft sprockets.

13. Position the tensioner, tensioner spring and tensioner spacer on engine block.

14. Align the timing marks on the camshaft sprocket and crankshaft sprocket. This will position No. 1 piston on TDC on the compression stroke.

15. Position the timing belt on the crankshaft sprocket and keeping the tension side of the belt tight, set it on the camshaft sprocket and then the tensioner.

16. Apply slight counterclockwise force to the camshaft sprocket to give tension to the belt and make sure all timing marks are aligned.

17. Loosen the pivot side tensioner bolt and the slot side bolt. Allow the spring to remove the slack.

18. Tighten the slot side tensioner bolt and then the pivot side bolt. If the pivot side bolt is tightened first, the tensioner could turn with bolt, causing over tension.

19. For 1.5L engines, turn the crankshaft clockwise. Loosen the pivot side tensioner bolt and then the slot side bolt to allow the spring to take up any remaining slack. Tighten the slot bolt and then the pivot side bolt to 17 ft. lbs. (24 Nm).

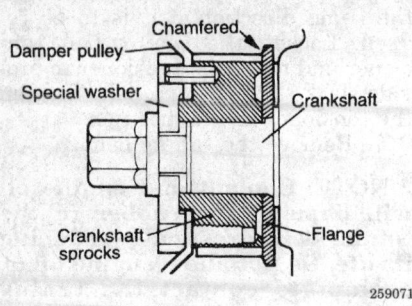

Crankshaft sprocket installation — Precis 1.5L (G4DJ) engine

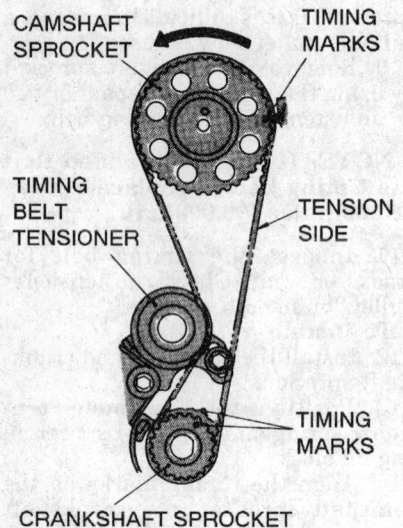

Timing belt routing and mark identification — Mirage with 1.5L (4G15) engine shown

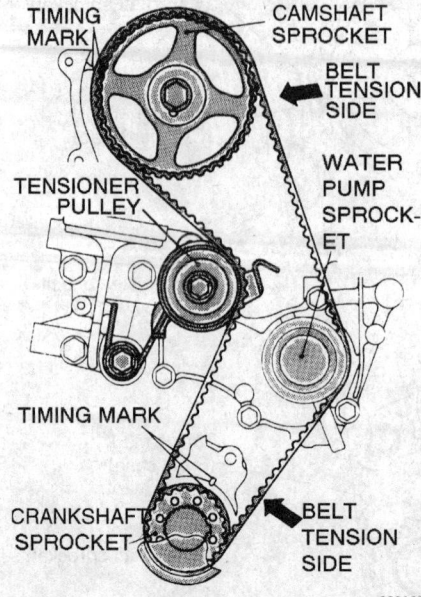

Timing belt routing and mark identification — Mirage 1.8L (4G93) engine

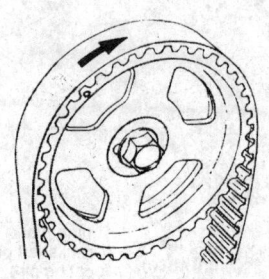

If timing belt is to be reused, mark belt's direction of rotation so it can be installed the same way

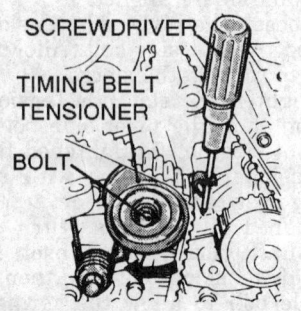

Proper method of releasing belt tension — Mirage 1.8L (4G93) engine

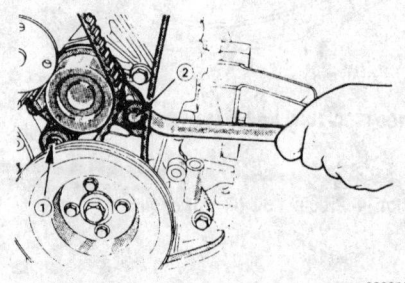

Loosen pivot bolt #1 then slot side bolt #2. Tighten #2 first, then #1 — Mirage with 1.5L (4G15) engine

20. For 1.8L engines, turn the crankshaft clockwise two rotations and tighten the adjuster bolt to 18 ft. lbs. (24 Nm) and tighten the pivot (spring) bolt to 35 ft. lbs. (45 Nm).

21. Install the timing belt covers and torque the cover bolts to 84–96 inch lbs. (10–11 Nm).

22. Install the crankshaft pulley and torque the retaining bolt to 83 ft. lbs. (113 Nm) for 1.5L engines, or to 134 ft. lbs. (185 Nm) for 1.8L engines.

23. Install the front engine mount bracket and mount.

24. Connect the A/C hose clamp and install the accessory drive belts.

25. Install the engine undercover.

26. Connect the negative battery cable.

27. Install the timing belt covers and all related items.

28. Connect the negative battery cable.

1.8L (4G37) Engine

Eclipse

1. If possible, position the engine so the No. 1 piston is at TDC.

2. Disconnect the negative battery cable.

3. Remove the timing belt covers.

4. Loosen the timing (outer) belt tensioner, move tensioner toward the water pump and lightly tighten bolt to hold position. Remove the outer timing belt.

NOTE: If timing belts are going to be reused, mark the direction of rotation on the belt. This will ensure the belt is reinstalled in same direction, extending belt life.

5. Remove the outer crankshaft sprocket and flange.

6. Loosen the silent shaft (inner) belt tensioner and remove the belt.

7. Remove the camshaft sprocket using an appropriate spanner wrench; hold the shaft in position while removing the bolt.

To install:

8. Install the sprockets to their appropriate shafts. Install the retainer bolts and tighten the camshaft sprocket bolt to 58–72 ft. lbs. (80–100 Nm) and the crankshaft sprocket bolt to 80–94 ft. lbs. (110–130 Nm).

9. Turn both tensioner pulleys and check for any signs of bearing wear.

10. Align the timing marks of the silent shaft sprockets and the crankshaft sprocket with the timing marks on the front case. Wrap the timing belt around the sprockets so there is no slack in the upper span of the belt and the timing marks are still aligned.

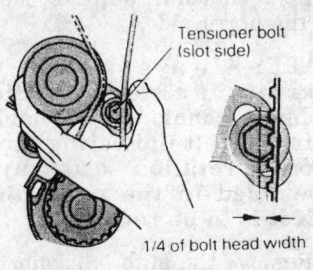

Belt is correctly tensioned when thumb pressure deflects belt as shown — Mirage with 1.5L (4G15) engine

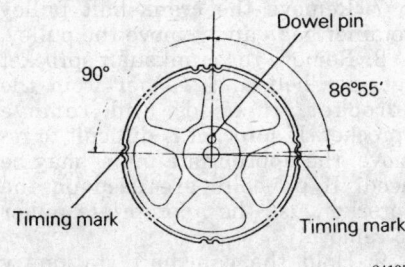

Same sprocket is used for exhaust and intake, but two timing marks are provided — 1993–95 Mirage

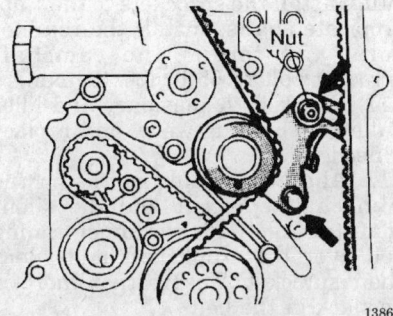

Timing belt tensioner identification and adjustment — Eclipse 1.8 (4G37) engine

11. Install the tensioner pulley and move the pulley by hand so the long side of the belt deflects about ¼ inch.

12. Hold the pulley tightly so the pulley cannot rotate when the bolt is tightened. Tighten the bolt to 15 ft. lbs. (20 Nm) and recheck the deflection amount.

13. Install the timing belt tensioner fully toward the water pump and tighten the bolts. Place the upper end of the spring against the water pump body.

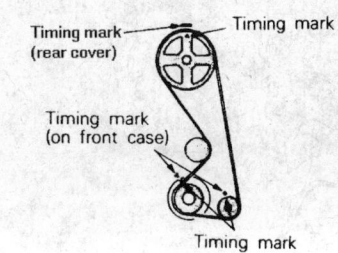

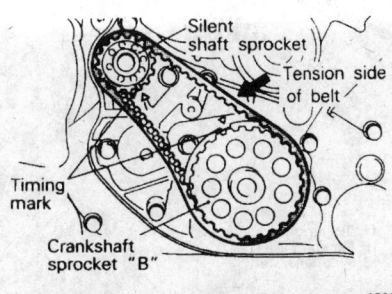

Timing mark locations and alignment — Eclipse 1.8 (4G37) engine

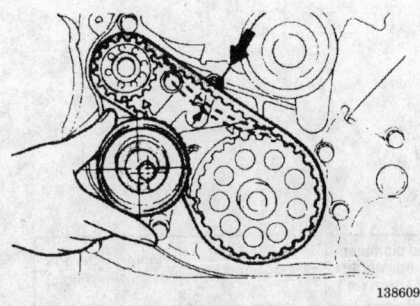

Silent shaft belt adjustment procedure — Eclipse 1.8 (4G37) engine

14. Align the timing marks of the camshaft, crankshaft and oil pump sprockets with their corresponding marks on the front case or rear cover.

NOTE: There is a possibility of aligning all timing marks and have the oil pump sprocket and silent shaft out of time, causing an engine vibration during operation. If the following step is not followed exactly, there is a 50 percent chance that the silent shaft alignment will be 180 degrees off.

15. Before installing the timing belt, ensure that the left side (rear)

silent shaft (oil pump sprocket) is in the correct position as follows:

a. Remove the plug from the rear side of the block and insert a tool with shaft diameter of 0.31 in. (8mm) into the hole.

b. With the timing marks still aligned, the shaft of the tool must be able to go in at least 2 ½ in. If the tool can only go in about 1 in., the shaft is not in the correct orientation and will cause a vibration during engine operation. Remove the tool from the hole and turn the oil pump sprocket 1 complete revolution. Realign the timing marks and insert the tool. The shaft of the tool must go in at least 2 ⅓ in.

c. Recheck and realign the timing mark.

d. Leave the tool in place to hold the silent shaft while continuing.

16. Install the belt to the crankshaft sprocket, oil pump sprocket, then camshaft sprocket, in that order. While doing so, make sure there is no slack between the sprocket except where the tensioner is installed.

17. Recheck the timing marks alignment. If all are aligned, loosen the tensioner mounting bolt and allow the tensioner to apply tension to the belt.

18. Remove the tool that is holding the silent shaft and rotate the crankshaft a distance equal to 2 teeth on the camshaft sprocket. This will allow the tensioner to automatically apply the proper tension on the belt. Do not manually overtighten the belt or it will howl.

19. Tighten the lower mounting bolt first, then the upper spacer bolt.

20. To verify correct belt tension, check that the deflection at the longest span of the belt is about ½ inch.

21. Install the timing belt covers and all related items.

22. Connect the negative battery cable.

23. Run engine until thermostat opens. Check and adjust ignition timing.

2.0L (4G63) Engine

1993–94 Eclipse

1. Disconnect the negative battery cable.

2. Remove the timing belt upper and lower covers.

3. Rotate the crankshaft clockwise and align the timing marks so No. 1 piston will be at TDC of the compression stroke. At this time the timing marks on the camshaft sprocket and the upper surface of the cylinder head should coincide, and the dowel

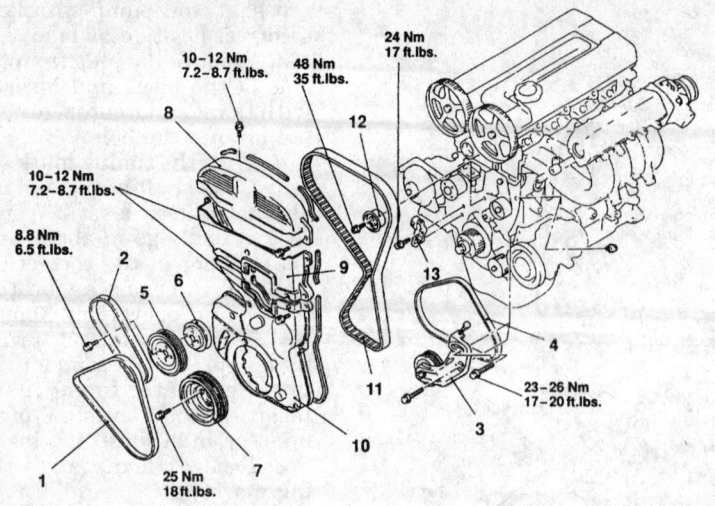

1. Drive belt (Generator)
2. Drive belt (Power steering)
3. Tensioner pulley bracket
4. Drive belt (A/C)
5. Water pump pulley
6. Water pump pulley (Power steering)
7. Crankshaft pulley
8. Timing belt front upper cover
9. Timing belt front center cover
10. Timing belt front lower cover

11. Timing belt
12. Tension pulley
13. Auto tensioner

321727

Timing belt and related components — 1995-97 Eclipse 2.0L (4G63) engine

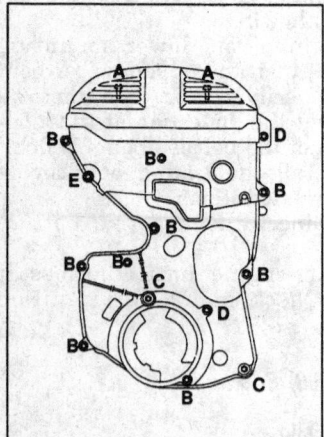

	Thread diameter × thread length mm (in.)	Bolt classification	Tightening torque Nm (ft.lbs.)
A	6 × 16 (.24 × .63)	Flange bolt	10–12 (7.2–8.7)
B	6 × 18 (.24 × .71)	Flange bolt	10–12 (7.2–8.7)
C	6 × 25 (.24 × .98)	Washer assembled bolt	8.8 (6.5)
D	6 × 25 (.24 × .98)	Flange bolt	10–12 (7.2–8.7)
E	6 × 45 (.24 × 1.77)	Flange bolt	10–12 (7.2–8.7)

321737

Timing belt cover bolt locations — 1995-97 Eclipse 2.0L (4G63) engine

pin of the camshaft sprocket should be at the upper side.

NOTE: Always rotate the crankshaft in a clockwise direction. Make a mark on the back of the timing belt indicating the direction of rotation so it may be reassembled in the same direction if it is to be reused.

4. Remove the auto tensioner and remove the outermost timing belt.

5. Remove the timing belt tensioner pulley, tensioner arm, idler pulley, oil pump sprocket, special washer, flange and spacer.

6. Remove the silent shaft (inner) belt tensioner and remove the belt.

7. Remove the crankshaft pulley retainer bolts and remove the pulley.

8. Remove the crankshaft sprocket retainer bolt and washer from the sprocket, if used, and remove sprocket. If sprocket is difficult to remove, the appropriate puller may be used. If no bolts are used on the sprocket, use the appropriate puller to remove.

9. Hold the camshaft stationary using the hexagon cast between journals No. 2 and 3 and remove the retainer bolt. Remove the sprocket from the camshaft.

To install:

10. Install the sprockets to their appropriate shafts. Install the retainer bolts and torque the camshaft sprocket bolt to 65 ft. lbs. (90 Nm)

11. Check both tensioner and idler pulley for bearing wear, and replace if needed.

12. Align the timing marks on the crankshaft sprocket and the silent shaft sprocket. Fit the inner timing belt over the crankshaft and silent shaft sprocket. Ensure that there is no slack in the belt.

13. While holding the inner timing belt tensioner with your fingers, adjust the timing belt tension by applying a force towards the center of the belt, until the tension side of the belt is taut. Tighten the tensioner bolt.

NOTE: When tightening the bolt of the tensioner, ensure that the tensioner pulley shaft does not rotate with the bolt. Allowing it to rotate with the bolt can cause excessive tension on the belt.

14. Check belt for proper tension by depressing the belt on its long side with your finger and noting the belt deflection. The desired reading is 0.20–0.28 in. (5–7mm). If tension is not correct, readjust and check belt deflection.

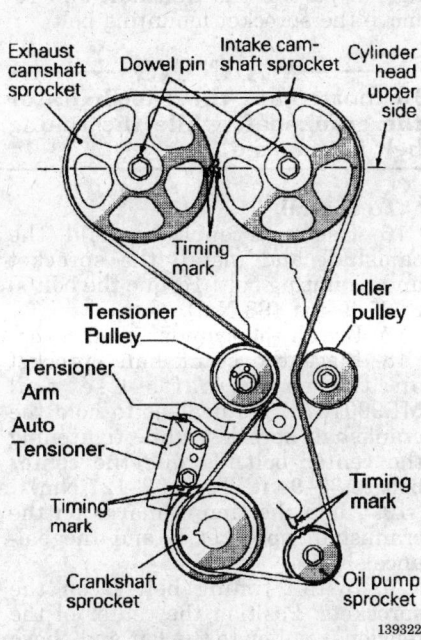

Timing belt routing and mark identification — All 2.0L (4G63) engines

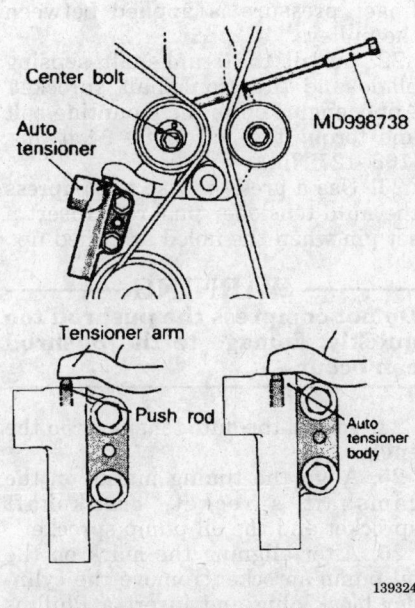

Auto tensioner adjustment — All 2.0L (4G63) engines

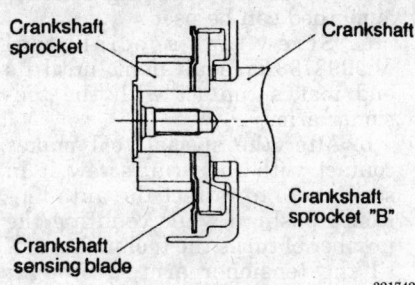

Correct crankshaft sensing blade installation — 1995–97 Eclipse 2.0L (4G63) engine

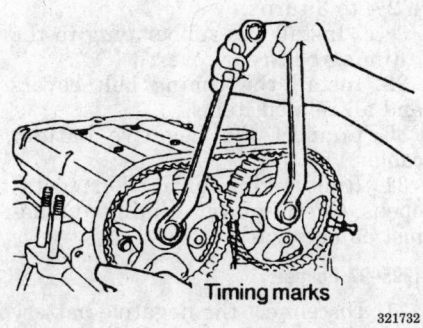

Use two wrenches to position the camshaft sprockets — 1995–97 Eclipse 2.0L (4G63) engine

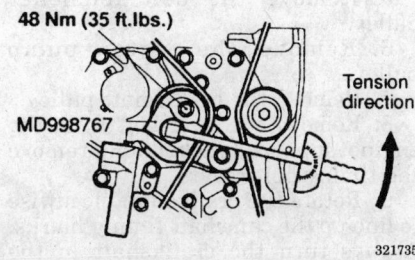

Use a torque wrench and the adapter to apply tension to the timing belt — 1995–97 Eclipse 2.0L (4G63) engine

15. Install the flange, crankshaft and washer to the crankshaft. The flange on the crankshaft sprocket must be installed towards the inner timing belt sprocket. Tighten bolt to 80–94 ft. lbs. (110–130 Nm).

16. To install the oil pump sprocket, insert a Phillips screwdriver with a shaft 0.31 in. (8mm) in diameter into the plug hole in the left side of the cylinder block to hold the left silent shaft. Tighten the nut to 36–43 ft. lbs. (50–60 Nm).

17. Using a wrench, hold the camshaft at its' hexagon between

journal No. 2 and 3 and tighten camshaft sprocket mounting bolt, if removed, to 58–72 ft. lbs. (80–100 Nm). If no hexagon is present between journal No. 2 and 3, hold the sprocket stationary with a spanner wrench while tightening the sprocket retainer bolt.

18. Carefully push the auto tensioner rod in until the set hole in the rod aligned up with the hole in the cylinder. Place a wire into the hole to retain the rod.

19. Install the tensioner pulley onto the tensioner arm. Locate the pinhole in the tensioner pulley shaft to the left of the center bolt. Then, tighten the center bolt finger-tight.

20. When installing the timing belt, turn the 2 camshaft sprockets so their dowel pins are located on top. Align the timing marks facing each other and with the top surface of the cylinder head. When you let go of the exhaust camshaft sprocket, it will rotate 1 tooth in the counterclockwise direction. This should be taken into account when installing the timing belts on the sprocket.

NOTE: Both camshaft sprockets are used for the intake and exhaust camshafts and are provided with 2 timing marks. When the sprocket is mounted on the exhaust camshaft, use the timing mark on the right with the dowel pin hole on top. For the intake camshaft sprocket, use the 1 on the left with the dowel pin hole on top.

21. Align the crankshaft sprocket and oil pump sprocket timing marks.

22. After alignment of the oil pump sprocket timing marks, remove the plug on the cylinder block and insert a Phillips screwdriver with a shaft diameter of 0.31 in. (8mm) through the hole. If the shaft can be inserted 2.4 in. deep, the silent shaft is in the correct position. If the shaft of the tool can only be inserted 0.8–1.0 in. (20–25mm) deep, turn the oil pump sprocket 1 turn and realign the marks. Reinsert the tool making sure it is inserted 2.4 in. deep. Keep the tool inserted in hole for the remainder of this procedure.

NOTE: The above step assures that the oil pump socket is in correct orientation to the silent shafts. This step must not be skipped or a vibration may develop during engine operation.

23. Install the timing belt as follows:

a. Install the timing belt around the intake camshaft sprocket and

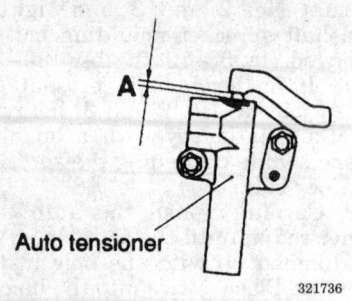

A

Auto tensioner

321736

Measure the pushrod protrusion from
the auto tensioner — 1995-97 Eclipse
2.0L (4G63) engine

retain it with 2 spring clips or
binder clips.

b. Install the timing belt around
the exhaust sprocket, aligning the
timing marks with the cylinder
head top surface using 2 wrenches.
Retain the belt with 2 spring clips.

c. Install the timing belt around
the idler pulley, oil pump sprocket,
crankshaft sprocket and the tensioner pulley. Remove the 2 spring
clips.

d. Lift upward on the tensioner
pulley in a clockwise direction and
tighten the center bolt. Make sure
all timing marks are aligned.

e. Rotate the crankshaft ¼ turn
counterclockwise. Then turn clockwise until the timing marks are aligned again.

24. To adjust the timing (outer)
belt, turn the crankshaft ¼ turn counterclockwise, then turn clockwise to
move No. 1 cylinder to TDC.

25. Loosen the center bolt. Using
tool MD998738 or equivalent and a
torque wrench, apply a torque of
1.88–2.03 ft. lbs. (2.6–2.8 Nm).
Tighten the center bolt.

26. Screw the special tool into the
engine left support bracket until its
end makes contact with the tensioner
arm. At this point, screw the special
tool in some more and remove the set
wire attached to the auto tensioner, if
the wire was not previously removed.
Then remove the special tool.

27. Rotate the crankshaft 2 complete turns clockwise and let it sit for
approximately 15 minutes. Then,
measure the auto tensioner protrusion (the distance between the tensioner arm and auto tensioner body)
to ensure that it is within 0.15–0.18
in. (3.8–4.5mm). If out of specification, repeat Step 1–4 until the specified value is obtained.

28. If the timing belt tension adjustment is being performed with the
engine mounted in the vehicle, and
clearance between the tensioner arm
and the auto tensioner body cannot

be measured, the following alternative method can be used:

a. Screw in special tool
MD998738 or equivalent, until its
end makes contact with the tensioner arm.

b. After the special tool makes
contact with the arm, screw it in
some more to retract the auto tensioner pushrod while counting the
number of turns the tool makes until the tensioner arm is brought
into contact with the auto tensioner body. Make sure the number
of turns the special tool makes conforms with the standard value of
2½ to 3 turns.

c. Install the rubber plug to the
timing belt rear cover.

29. Install the timing belt covers
and all related items.

30. Connect the negative battery
cable.

31. Run engine until thermostat
opens. Check ignition timing and adjust as necessary.

1995–97 Eclipse

1. Disconnect the negative battery
cable.

2. Remove the engine undercover.

3. Remove the engine mount
bracket.

4. Remove the drive belts.

5. Remove the belt tensioner
pulley.

6. Remove the water pump
pulleys.

7. Remove the crankshaft pulley.

8. Remove the stud bolt from the
engine support bracket and remove
the timing belt covers.

9. Rotate the crankshaft clockwise
to line up the camshaft timing marks.
Always turn the crankshaft in the
forward direction only.

10. Loosen the tension pulley
center bolt.

**NOTE: If the timing belt is to
be reused, mark the direction of
rotation on the flat side of the
belt with an arrow.**

11. Move the tension pulley towards the water pump and remove
the timing belt.

12. Remove the crankshaft sprocket
center bolt using special tool
MB990767 to hold the crankshaft
sprocket while removing the center
bolt. Then use MB998778 or
equivalent puller to remove the
sprocket.

13. Mark the direction of rotation
on the timing belt B with a arrow.

14. Loosen the center bolt on the
tensioner and remove the belt.

15. To remove the camshaft
sprocket, remove the cylinder head

cover. Use a wrench to hold the hexagonal part of the camshaft and remove the sprocket mounting bolt.

---— **WARNING** ———
**Do not rotate the camshafts or
the crankshaft while the timing
belt is removed.**

To install:

16. Use a wrench to hold the
camshaft and install the sprocket
and mounting bolt. Torque the bolt(s)
to 65 ft. lbs. (88 Nm).

17. Install the cylinder head cover.

18. Place the crankshaft sprocket
on the crankshaft. Use tool
MB990767 or equivalent to hold the
crankshaft sprocket while tightening
the center bolt. Torque the center
bolt to 80–94 ft. lbs. (108–127 Nm).

19. Align the timing marks on the
crankshaft sprocket B and the balance shaft.

20. Install timing belt B on the
sprockets. Position the center of the
tensioner pulley to the left and above
the center of the mounting bolt.

21. Push the pulley clockwise toward the crankshaft to apply tension
to the belt and tighten the mounting
bolt to 14 ft. lbs. (19 Nm). Do not let
the pulley turn when tightening the
bolt because it will cause excessive
tension on the belt. The belt should
deflect 0.20–0.28 in. (5–7 mm) when
finger pressure is applied between
the pulleys.

22. Install the crankshaft sensing
blade and the crankshaft sprocket.
Apply engine oil to the mounting bolt
and torque the bolt to 80–94 ft. lbs.
(108–127 Nm).

23. Use a press or vise to compress
the auto tensioner pushrod. Insert a
set pin when the holed are lined up.

---— **WARNING** ———
**Do not compress the pushrod too
quickly, damage to the pushrod
can occur.**

24. Install the auto tensioner on the
engine.

25. Align the timing marks on the
camshaft sprocket, crankshaft
sprocket and the oil pump sprocket.

26. After aligning the mark on the
oil pump sprocket, remove the cylinder block plug and insert a Phillips
screwdriver in the hole to check the
position of the counter balance shaft.
The screwdriver should go in at least
2.36 in. or more, if not, rotate the oil
pump sprocket once and realign the
timing mark so the screwdriver goes
in. Do not remove the screwdriver
until the timing belt is installed.

27. Install the timing belt on the intake camshaft and secure it with a clip.

28. Install the timing belt on the exhaust camshaft. Align the timing marks with the cylinder head top surface using two wrenches. Secure the belt with a clip.

29. Install the belt around the idler pulley, oil pump sprocket, crankshaft sprocket and the tensioner pulley.

30. Turn the tension pulley so the pinholes are at the bottom. Press the pulley lightly against the timing belt.

31. Screw the special tool into the left engine support bracket until it contacts the tensioner arm, then screw the tool in a little more and remove the pushrod pin from the auto tensioner. Remove the special tool and torque the center bolt to 35 ft. lbs. (48 Nm).

32. Turn the crankshaft ¼ turn counterclockwise and then clockwise until the timing marks are aligned.

33. Loosen the center bolt. Install special tool MD998767 on the tension pulley. Turn the tension pulley counterclockwise with a torque of 2.6 ft. lbs. (3.5 Nm) and tighten the center bolt to 35 ft. lbs. (48 Nm). Do not let the tension pulley turn with the bolt.

34. Turn the crankshaft two revolutions to the right and align the timing marks. After 15 minutes, measure the protrusion of the pushrod on the auto tensioner. The standard measurement is 0.150–0.177 in (3.8–4.5 mm). If the protrusion is out of specification, loosen the tension pulley, apply the proper torque to the belt and retighten the center bolt.

35. Install the crankshaft pulley. Torque the mounting bolts to 18 ft. lbs. (25 Nm).

36. Install the water pump. Torque the mounting bolts to 6.5 ft. lbs. (8.8 Nm).

37. Install and adjust the drive belts.

38. Install the engine mount bracket.

39. Install the engine undercovers.

40. Connect the negative battery cable.

Galant

1. If possible, position the engine so the No. 1 piston is at TDC.

2. Disconnect the negative battery cable.

3. Remove the splash shield under the engine.

4. Remove the bolt holding the two hoses to the top of the left engine mount.

5. Place a floor jack and a broad piece of lumber under the oil pan and remove the front engine mount.

6. Remove the timing belt covers.

NOTE: If timing belts are going to be reused, mark the direction of rotation on the belt. This will ensure the belt is reinstalled in same direction, extending belt life.

7. To loosen the timing (outer) belt tensioner, install special tool MD998738 or equivalent, to the slot and screw inward to move tensioner toward the water pump. Once the tension has been relieved, remove the outer timing belt.

8. If tensioner replacement is required, align the pin hole in the tensioner rod to the hole in the tensioner cylinder. Insert a 0.055 inch (1.4 mm) wire in the hole and remove the special tool from the slot. With the cylinder tension relieved, remove the auto tensioner cylinder assembly two mounting bolts.

9. Remove the outer crankshaft sprocket and flange.

10. Loosen the silent shaft (inner) belt tensioner and remove the belt. If pulley replacement is required, remove the center adjusting bolt.

To install:

Inspect the timing belts in detail for any flaw or wear. Check the sprockets and tensioner for wear. The sprocket teeth should be well defined, not rounded and the valleys between the teeth should be clean. Turn both tensioner pulleys and check for any signs of bearing wear. If sprockets or pulleys show any sign of wear, they must be replaced.

--- **WARNING** ---

Do not spray or immerse the sprockets or tensioners in cleaning solvent. The sprocket may absorb the solvent and transfer it to the belt. The tensioners are internally lubricated and the solvent will dilute or dissolve the lubricant.

11. Align the timing marks of the silent shaft sprockets and the crankshaft sprocket with the timing marks on the front case. Wrap the timing belt around the sprockets so there is no slack in the upper span of the belt and the timing marks are still aligned.

12. Install the tensioner pulley and move the pulley by hand so the long side of the belt deflects about ¼ inch.

13. Hold the pulley tightly so the pulley cannot rotate when the bolt is tightened. Tighten the bolt to 14 ft. lbs. (19 Nm) and recheck the deflection amount.

14. Align the timing marks of the camshaft, crankshaft and oil pump sprockets with their corresponding marks on the front case or rear cover.

NOTE: It is possible to align all timing marks and still have the oil pump sprocket and silent shaft out of time; this will cause an engine vibration during operation. If the following step is not followed exactly, there is a 50 percent chance that the silent shaft alignment will be 180 degrees off.

15. Before installing the timing belt, ensure that the left side (rear) silent shaft (oil pump sprocket) is in the correct position as follows:

a. Remove the plug from the rear side of the block and insert a tool with shaft diameter of 0.31 inches. (8mm) into the hole.

b. With the timing marks still aligned, the shaft of the tool must be able to go in at least 2 ½ inches. If the tool can only go in about 1 in., the shaft is not in the correct orientation and will cause a vibration during engine operation. Remove the tool from the hole and turn the oil pump sprocket 1 complete revolution. Realign the timing marks and insert the tool. The shaft of the tool must go in at least 2 ½ inches.

c. Recheck and realign the timing mark.

d. Leave the tool in place to hold the silent shaft while continuing.

16. If the camshaft belt tensioner was removed, use a vise to carefully push the auto tensioner rod in until the set hole in the rod aligned up with the hole in the cylinder. Place a wire into the hole to retain the rod. Mount the tensioner to the engine block and tighten the mounting bolt to 17 ft. lbs. (23 Nm).

17. Install the belt to the crankshaft sprocket, oil pump sprocket, then camshaft sprocket, in that order. While doing so, make sure there is no slack between the sprocket except where the tensioner is installed.

18. To adjust the timing (outer) belt perform the following steps:

a. Turn the crankshaft ¼ turn counterclockwise, then turn it clockwise to move No. 1 cylinder to TDC.

b. Loosen the center bolt. Using tool MD998752 or equivalent and a torque wrench, apply a torque of 2.6 ft. lbs. (3.6 Nm). Tighten the center bolt.

c. Screw the special tool into the engine left support bracket until its end makes contact with the tensioner arm. At this point, screw the special tool in some more and remove the set wire attached to the auto tensioner, if the wire was not previously removed. Then remove the special tool.

d. Rotate the crankshaft two complete turns clockwise and let it sit for approximately 15 minutes. Then, measure the auto tensioner protrusion (the distance between the tensioner arm and auto tensioner body) to ensure that it is within 0.15–0.18 inch (3.8–4.5mm). If out of specification, repeat these steps until the specified value is obtained.

NOTE: Do not manually overtighten the belt or it will howl during operation.

19. Install the upper and lower timing belt covers. Tighten the bolts to 8 ft. lbs.

20. Install the engine mount bracket to the engine. Tighten the mounting nuts and bolts to 42 ft. lbs.

21. Install the splash shield.

22. Position the high pressure hose and secure its bracket.

23. Connect the negative battery cable. Start the engine and let it idle.

24. Run engine until thermostat opens. Check and adjust ignition timing.

2.4L (4G64) Engine

Eclipse

1. If possible, position the engine so the No. 1 piston is at TDC.

2. Disconnect the negative battery cable.

3. Remove the splash shield under the engine.

4. Safely support the weight of the engine and remove the engine mount and bracket assembly.

5. Remove the drive belts and the timing belt covers.

NOTE: If timing belts are going to be reused, mark the direction of rotation on the belt. This will ensure the belt is reinstalled in same direction, extending belt life.

6. To loosen the timing (outer) belt tensioner, install special tool MD998738 or equivalent, to the slot and screw inward to move tensioner toward the water pump. Once the tension has been relieved, remove the outer timing belt.

7. If tensioner replacement is required, align the pin hole in the tensioner rod to the hole in the tensioner cylinder. Insert a 0.055 inch (1.4 mm) wire in the hole and remove the spe-

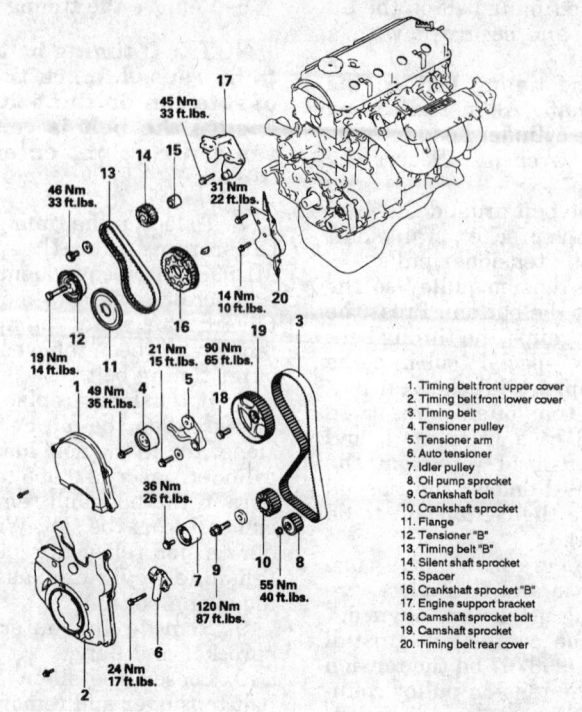

1. Timing belt front upper cover
2. Timing belt front lower cover
3. Timing belt
4. Tensioner pulley
5. Tensioner arm
6. Auto tensioner
7. Idler pulley
8. Oil pump sprocket
9. Crankshaft bolt
10. Crankshaft sprocket
11. Flange
12. Tensioner "B"
13. Timing belt "B"
14. Silent shaft sprocket
15. Spacer
16. Crankshaft sprocket "B"
17. Engine support bracket
18. Camshaft sprocket bolt
19. Camshaft sprocket
20. Timing belt rear cover

321775

Exploded view of the timing belt components — 1996–97 Eclipse 2.4L (4G64) engine

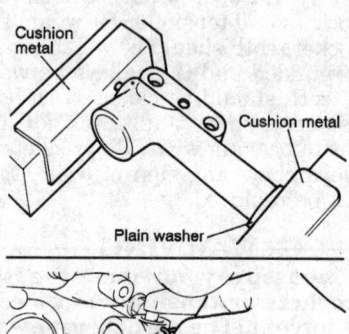

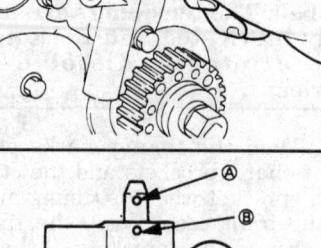

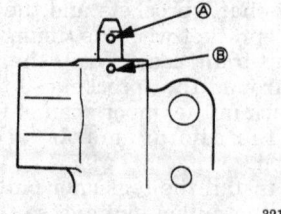

321778

Service points of tensioner installation — Eclipse and Galant 2.4L (4G64) and Galant 2.0L (4G63) engines

cial tool from the slot. With the cylinder tension relieved, remove the auto tensioner cylinder assembly two mounting bolts.

8. Remove the outer crankshaft sprocket and flange.

9. Loosen the silent shaft (inner) belt tensioner and remove the belt. If pulley replacement is required, remove the center adjusting bolt.

10. Remove the crankshaft pulley retainer bolts and remove the pulley.

11. Remove the crankshaft sprocket retainer bolt and washer from the sprocket, if used, and remove sprocket. If sprocket is difficult to remove, the appropriate puller may be used. If no bolts are used on the sprocket, use the appropriate puller to remove.

12. Hold the camshaft stationary using the hexagon cast between journals No. 2 and 3 and remove the retainer bolt. Remove the sprocket from the camshaft.

To install:

13. Install the sprockets to their appropriate shafts. Install the retainer bolts and torque the camshaft sprocket bolt to 65 ft. lbs. (90 Nm)

Inspect the timing belts in detail for any flaw or wear. Check the sprockets and tensioner for wear. The sprocket teeth should be well defined, not rounded and the valleys between

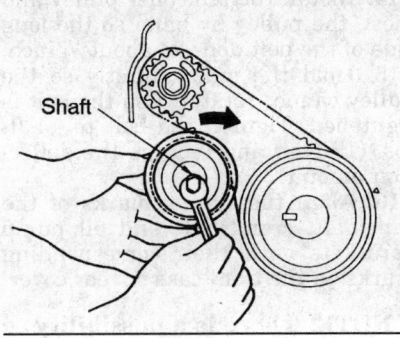

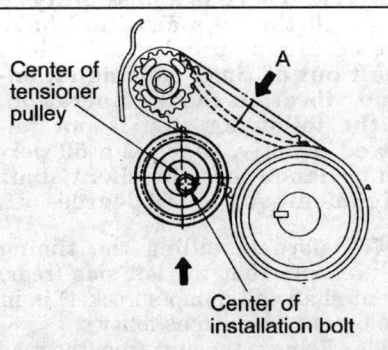

Adjusting the silent shaft belt tension — Eclipse and Galant 2.4L (4G64) and Galant 2.0L (4G63) engines

321771

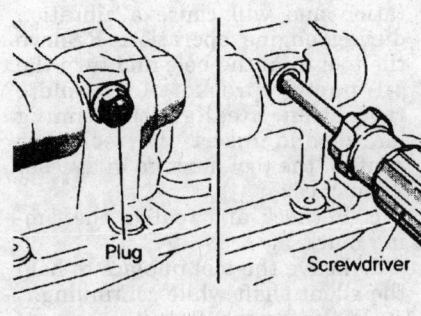

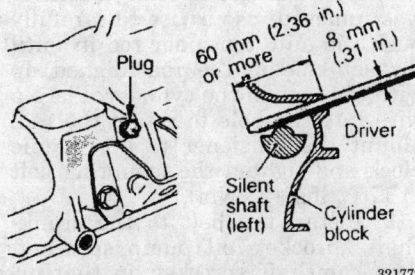

Silent shaft alignment procedure — Eclipse and Galant 2.4L (4G64) and Galant 2.0L (4G63) engines

321772

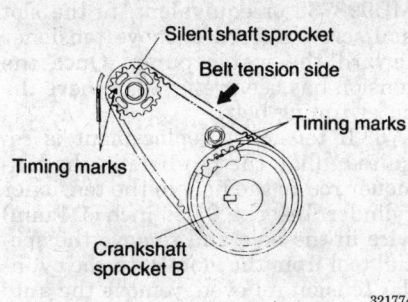

Silent shaft timing mark identification — Eclipse and Galant 2.4L (4G64) engines

321774

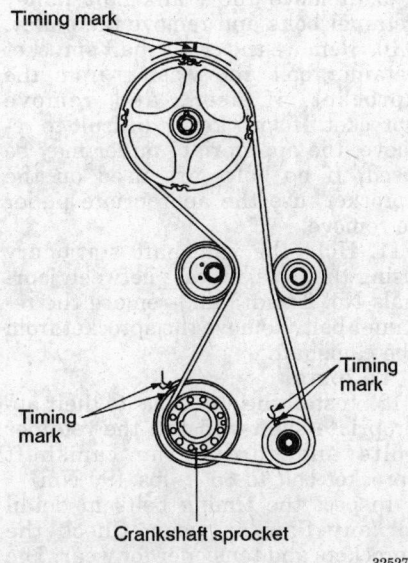

Camshaft belt timing mark locations — 1994–97 Galant 2.4L (4G64) engines

325277

the teeth should be clean. Turn both tensioner pulleys and check for any signs of bearing wear. If sprockets or pulleys show any sign of wear, they must be replaced.

WARNING

Do not spray or immerse the sprockets or tensioners in cleaning solvent. The sprocket may absorb the solvent and transfer it to the belt. The tensioners are internally lubricated and the solvent will dilute or dissolve the lubricant.

14. Align the timing marks of the silent shaft sprockets and the crankshaft sprocket with the timing marks on the front case. Wrap the timing belt around the sprockets so there is no slack in the upper span of the belt and the timing marks are still aligned.

15. Install the tensioner pulley and move the pulley by hand so the long side of the belt deflects about ¼ inch.

16. Hold the pulley tightly so the pulley cannot rotate when the bolt is tightened. Tighten the bolt to 14 ft. lbs. (19 Nm) and recheck the deflection amount.

17. Align the timing marks of the camshaft, crankshaft and oil pump sprockets with their corresponding marks on the front case or rear cover.

NOTE: There is a possibility to align all timing marks and have the oil pump sprocket and silent shaft out of time, causing an engine vibration during operation. If the following step is not followed exactly, there is a 50 percent chance that the silent shaft alignment will be 180 degrees off.

18. Before installing the timing belt, ensure that the left side (rear) silent shaft (oil pump sprocket) is in the correct position as follows:

  a. Remove the plug from the rear side of the block and insert a tool with shaft diameter of 0.31 inches. (8mm) into the hole.

  b. With the timing marks still aligned, the shaft of the tool must be able to go in at least 2 ½ inches. If the tool can only go in about 1 in., the shaft is not in the correct orientation and will cause a vibration during engine operation. Remove the tool from the hole and turn the oil pump sprocket 1 complete revolution. Realign the timing marks and insert the tool. The shaft of the tool must go in at least 2 ¼ inches.

  c. Recheck and realign the timing mark.

  d. Leave the tool in place to hold the silent shaft while continuing.

19. If the camshaft belt tensioner was removed, use a vise to carefully push the auto tensioner rod in until the set hole in the rod aligned up with the hole in the cylinder. Place a wire into the hole to retain the rod. Mount the tensioner to the engine block and tighten the mounting bolt to 17 ft. lbs. (23 Nm).

20. Install the belt to the crankshaft sprocket, oil pump sprocket, then camshaft sprocket, in that order. While doing so, make sure there

is no slack between the sprocket except where the tensioner is installed.

21. To adjust the timing (outer) belt perform the following steps:

a. Turn the crankshaft ¼ turn counterclockwise, then turn it clockwise to move No. 1 cylinder to TDC.

b. Loosen the center bolt. Using tool MD998752 or equivalent and a torque wrench, apply a torque of 2.6 ft. lbs. (3.6 Nm). Tighten the center bolt.

c. Screw the special tool into the engine left support bracket until its end makes contact with the tensioner arm. At this point, screw the special tool in some more and remove the set wire attached to the auto tensioner, if the wire was not previously removed. Then remove the special tool.

d. Rotate the crankshaft two complete turns clockwise and let it sit for approximately 15 minutes. Then, measure the auto tensioner protrusion (the distance between the tensioner arm and auto tensioner body) to ensure that it is within 0.15–0.18 inch (3.8–4.5mm). If out of specification, repeat Step 1–4 until the specified value is obtained.

NOTE: Do not manually overtighten the belt or it will howl.

22. Install the upper and lower timing belt covers. Tighten the bolts to 8 ft. lbs. (11 Nm)

23. Install the drive belts and properly adjust.

24. Reinstall the engine mount and bracket assembly and safely lower the engine.

25. Install the splash shield.

26. Connect the negative battery cable. Start the engine and let it idle.

27. Run engine until thermostat opens. Check and adjust ignition timing.

Eclipse

1. If possible, position the engine so the No. 1 piston is at TDC.

2. Disconnect the negative battery cable.

3. Remove the splash shield under the engine.

4. Remove the drive belts and the timing belt covers.

NOTE: If timing belts are going to be reused, mark the direction of rotation on the belt. This will ensure the belt is reinstalled in same direction, extending belt life.

5. To loosen the timing (outer) belt tensioner, install special tool MD998738 or equivalent, to the slot and screw inward to move tensioner toward the water pump. Once the tension has been relieved, remove the outer timing belt.

6. If tensioner replacement is required, align the pin hole in the tensioner rod to the hole in the tensioner cylinder. Insert a 0.055 inch (1.4 mm) wire in the hole and remove the special tool from the slot. With the cylinder tension relieved, remove the auto tensioner cylinder assembly two mounting bolts.

7. Remove the outer crankshaft sprocket and flange.

8. Loosen the silent shaft (inner) belt tensioner and remove the belt. If pulley replacement is required, remove the center adjusting bolt.

9. Remove the crankshaft pulley retainer bolts and remove the pulley.

10. Remove the crankshaft sprocket retainer bolt and washer from the sprocket, if used, and remove sprocket. If sprocket is difficult to remove, the appropriate puller may be used. If no bolts are used on the sprocket, use the appropriate puller to remove.

11. Hold the camshaft stationary using the hexagon cast between journals No. 2 and 3 and remove the retainer bolt. Remove the sprocket from the camshaft.

To install:

12. Install the sprockets to their appropriate shafts. Install the retainer bolts and torque the camshaft sprocket bolt to 65 ft. lbs. (90 Nm)

Inspect the timing belts in detail for any flaw or wear. Check the sprockets and tensioner for wear. The sprocket teeth should be well defined, not rounded and the valleys between the teeth should be clean. Turn both tensioner pulleys and check for any signs of bearing wear. If sprockets or pulleys show any sign of wear, they must be replaced.

------ **WARNING** ------
Do not spray or immerse the sprockets or tensioners in cleaning solvent. The sprocket may absorb the solvent and transfer it to the belt. The tensioners are internally lubricated and the solvent will dilute or dissolve the lubricant.

13. Align the timing marks of the silent shaft sprockets and the crankshaft sprocket with the timing marks on the front case. Wrap the timing belt around the sprockets so there is no slack in the upper span of the belt and the timing marks are still aligned.

14. Install the tensioner pulley and move the pulley by hand so the long side of the belt deflects about ¼ inch.

15. Hold the pulley tightly so the pulley cannot rotate when the bolt is tightened. Tighten the bolt to 14 ft. lbs. (19 Nm) and recheck the deflection amount.

16. Align the timing marks of the camshaft, crankshaft and oil pump sprockets with their corresponding marks on the front case or rear cover.

NOTE: There is a possibility to align all timing marks and have the oil pump sprocket and silent shaft out of time, causing an engine vibration during operation. If the following step is not followed exactly, there is a 50 percent chance that the silent shaft alignment will be 180 degrees off.

17. Before installing the timing belt, ensure that the left side (rear) silent shaft (oil pump sprocket) is in the correct position as follows:

a. Remove the plug from the rear side of the block and insert a tool with shaft diameter of 0.31 inches. (8mm) into the hole.

b. With the timing marks still aligned, the shaft of the tool must be able to go in at least 2 ½ inches. If the tool can only go in about 1 in., the shaft is not in the correct orientation and will cause a vibration during engine operation. Remove the tool from the hole and turn the oil pump sprocket 1 complete revolution. Realign the timing marks and insert the tool. The shaft of the tool must go in at least 2 ¼inches.

c. Recheck and realign the timing mark.

d. Leave the tool in place to hold the silent shaft while continuing.

18. If the camshaft belt tensioner was removed, use a vise to carefully push the auto tensioner rod in until the set hole in the rod aligned up with the hole in the cylinder. Place a wire into the hole to retain the rod. Mount the tensioner to the engine block and tighten the mounting bolt to 17 ft. lbs. (23 Nm).

19. Install the belt to the crankshaft sprocket, oil pump sprocket, then camshaft sprocket, in that order. While doing so, make sure there is no slack between the sprocket except where the tensioner is installed.

20. To adjust the timing (outer) belt perform the following steps:

a. Turn the crankshaft ¼ turn counterclockwise, then turn it clockwise to move No. 1 cylinder to TDC.

b. Loosen the center bolt. Using tool MD998752 or equivalent and a torque wrench, apply a torque of 2.6 ft. lbs. (3.6 Nm). Tighten the center bolt.

c. Screw the special tool into the engine left support bracket until its end makes contact with the tensioner arm. At this point, screw the special tool in some more and remove the set wire attached to the auto tensioner, if the wire was not previously removed. Then remove the special tool.

d. Rotate the crankshaft two complete turns clockwise and let it sit for approximately 15 minutes. Then, measure the auto tensioner protrusion (the distance between the tensioner arm and auto tensioner body) to ensure that it is within 0.15–0.18 inch (3.8–4.5mm). If out of specification, repeat Step 1–4 until the specified value is obtained.

NOTE: Do not manually overtighten the belt or it will howl.

21. Install the upper and lower timing belt covers. Tighten the bolts to 8 ft. lbs. (11 Nm)
22. Install the drive belts and properly adjust.
23. Install the splash shield.
24. Connect the negative battery cable. Start the engine and let it idle.
25. Run engine until thermostat opens. Check and adjust ignition timing.

2.0L (420A) Engine

Eclipse

1. Disconnect the negative battery cable.
2. Remove the drive belts.
3. Remove the power steering pump from the bracket and position it out of the way. Do not disconnect the hoses.
4. Remove the power steering pump bracket from the engine.

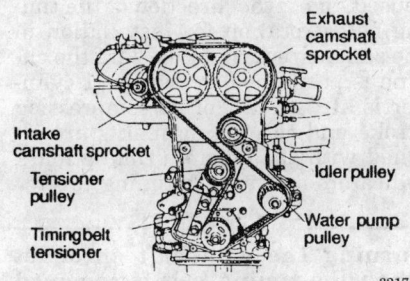

Engine timing system parts identification — 1995–97 Eclipse 2.0L (420A) engine

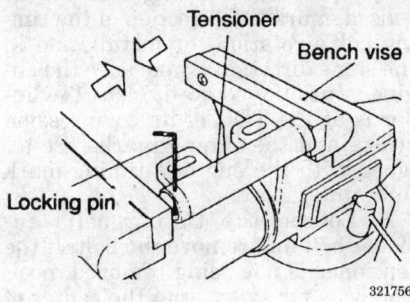

Compress the tensioner in a vise — 1995–97 Eclipse 2.0L (420A) engine

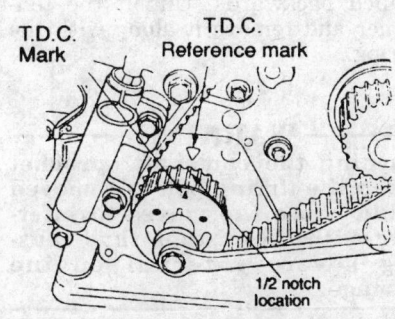

Engine timing marks — 1995–97 Eclipse 2.0L (420A) engine

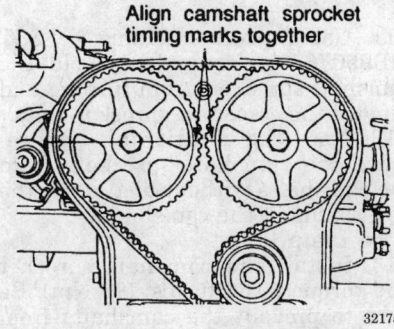

Align camshaft sprocket timing marks together

5. Use a floor jack with a piece of wood on it and jack up the engine to take the weight off of the engine mount.
6. Remove the engine mount and bracket.
7. Remove the crankshaft pulley.
8. Remove the front timing belt cover.

NOTE: If the timing belt is going to be reused, mark the direction of rotation on the belt with an arrow. Install the belt in the same direction.

9. Rotate the crankshaft sprocket clockwise until the timing marks are aligned. Loosen the timing belt tensioner and remove the timing belt.
10. Use tool MB995027 or equivalent to hold the camshaft sprocket while removing the mounting bolt in the center.
11. Use tool MB990767 or equivalent to hold the camshaft sprocket while removing the mounting bolt in the center.

——— WARNING ———
Do not rotate the crankshaft or the camshafts while the belt is removed.

To install:
12. Use the special tool to hold the camshaft sprocket and install the center bolt. Torque the bolt to 75 ft. lbs. (101 Nm).
13. Use tool MB995038 and MB995026 or equivalent, to install the crankshaft sprocket.
14. Using a vise, slowly compress the plunger into the body of the tensioner and install a pin through the body of the tensioner to retain the plunger.
15. Make sure the timing marks are still aligned, if not, align the camshaft sprocket timing marks facing each other. Align the crankshaft sprocket timing mark with the mark on the oil pump housing, then turn the crankshaft sprocket backward ½ notch.
16. Install the timing belt starting at the crankshaft, go around the water pump sprocket, idler pulley, camshaft sprockets and then around the tensioner pulley.
17. Turn the crankshaft sprocket ½ notch to TDC to take up the slack in the belt.
18. Install the tensioner on the engine but do not tighten the bolts.
19. Place a torque wrench on the tensioner pulley and apply 21 ft. lbs. (28 Nm) of torque in the direction of the water pump. Push the tensioner up against the tensioner pulley and torque the mounting bolts to 23 ft. lbs. (31 Nm).
20. Pull the pin out of the tensioner. Belt tension is correct when the pin can be removed and installed.
21. Rotate the crankshaft two revolutions and check the timing marks for alignment. Repeat the previous steps if necessary.
22. Install the timing belt cover.
23. Install the crankshaft pulley.
24. Install the engine mount and bracket. Remove the jack from under the engine.
25. Install the power steering pump bracket and pump.
26. Install the drive belts.

27. Connect the negative battery cable.

3.0L (6G72) SOHC Engine

Diamante

1. Position the engine so the No. 1 cylinder is at TDC of its compression stroke.
2. Disconnect the negative battery cable.

— CAUTION —

Wait at least 90 seconds after the negative (-) battery cable is disconnected to prevent possible deployment of the air bag.

---

3. Remove the engine undercover.
4. Remove the cruise control pump and the link assembly.
5. Raise and suspend the engine so that force is not applied to the engine mount.
6. Remove the engine mount and support bracket from the engine.
7. Remove the timing covers from the engine.

NOTE: It is recommended to change the timing belt at 60,000 mile intervals.

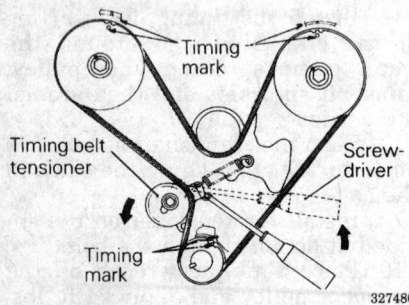

Timing mark

Timing belt tensioner

Screw-driver

Timing mark

327480

Timing marks and belt removal — Diamante 3.0L (6G72) SOHC engine

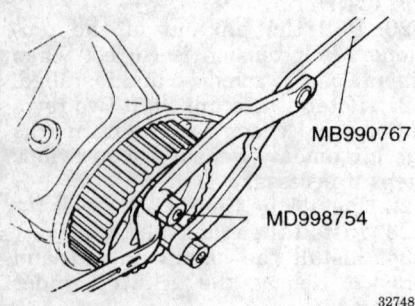

MB990767

MD998754

327481

Removing the camshaft sprocket — Diamante 3.0L (6G72) SOHC engine

8. If the same timing belt will be reused, mark the direction of the timing belt's rotation for installation in the same direction. Make sure the engine is positioned so the No. 1 cylinder is at the TDC of its compression stroke and the timing marks are aligned with the engine's timing mark indicators.
9. Loosen the timing belt tensioner bolt and remove the belt. If the tensioner is not being removed, position it as far away from the center of the engine as possible and tighten the bolt.
10. If the tensioner is being removed, mark the outside of the spring to ensure that it is not installed backwards. Unbolt the tensioner and remove it along with the spring.

— WARNING —

Turning the camshaft sprocket when the timing belt is removed could cause the valves to interfere with the pistons thus causing severe internal engine damage.

---

11. Using sprocket holding tool SST MB990767 or equivalent, hold the camshaft sprockets from turning and remove the camshaft sprocket bolts.
12. Note the positioning and remove the camshaft sprockets from the camshafts without disturbing the positioning of the camshafts.
    To install:
13. Install the camshaft sprocket and torque to 66 ft. lbs. (90 Nm). Be sure to prevent the camshafts from turning.
14. Install the tensioner, if removed, and hook the upper end of the spring to the water pump pin and the lower end to the tensioner in exactly the same position as originally installed.
15. Position both camshafts so the marks align with those on the rear timing covers. Rotate the crankshaft so the timing mark aligns with the mark on the front cover.
16. Install the timing belt on the crankshaft sprocket and while keeping the belt tight on the tension side, install the belt on the front (left) camshaft sprocket.
17. Install the belt on the water pump pulley, then the rear (right) camshaft sprocket and the tensioner.
18. Loosen the bolt that secures the adjustment of the tensioner and lightly press the tensioner against the timing belt.

19. Check that the timing marks are in alignment.
20. Rotate the crankshaft 2 full turns in the clockwise direction only. Align the timing marks.
21. Torque the bolt that secures the tensioner to 19 ft. lbs. (26 Nm).
22. Install the lower and the upper timing belt covers.
23. Install the engine support bracket and torque the bolts to specifications.
24. Install the engine mount and remove the support jack.
25. Install the cruise control pump and the link assembly.
26. Install the engine undercover.
27. Connect the negative battery cable.
28. Perform all necessary engine adjustments and road test the vehicle.

3.0L (6G72) DOHC Engine

Diamante

1. Position the engine so the No. 1 cylinder is at TDC of its compression stroke.
2. Disconnect the negative battery cable.

— CAUTION —

Wait at least 90 seconds after the negative (-) battery cable is disconnected to prevent possible deployment of the air bag.

---

3. Remove the engine undercovers and the front wheel and inner wheel covering.
4. Raise and suspend the engine so that force is not applied to the engine mount.
5. Remove the engine mount and support bracket from the engine.
6. Remove the timing covers from the engine.

NOTE: It is recommended to change the timing belt at 60,000 mile intervals.

7. If the same timing belt will be reused, mark the direction of the timing belt's rotation for installation in the same direction. Make sure the engine is positioned so the No. 1 cylinder is at the TDC of its compression stroke and the timing marks are aligned with the engine's timing mark indicators on the rear timing covers.

— WARNING —

Turning the camshaft sprocket when the timing belt is removed could cause the valves to contact with the pistons, resulting in severe engine damage.

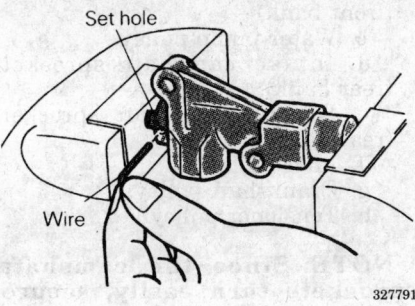

Timing belt tensioner — Diamante and 3000GT

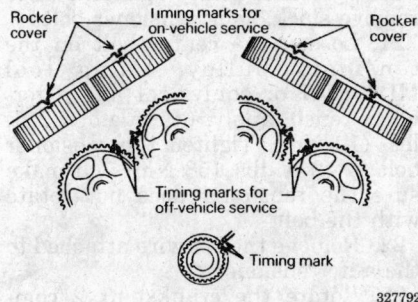

Timing marks — Diamante and 3000GT

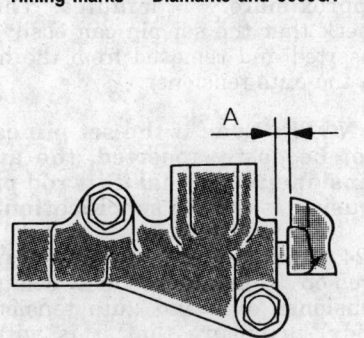

Auto tensioner protrusion — Diamante and 3000GT

8. Remove the bolts that secure the auto tensioner to the engine block and remove the tensioner.

NOTE: The auto-tensioner assembly must be reset to correctly adjust belt tension.

9. Loosen the center bolt of tensioner pulley to provide timing belt slack. Remove the timing belt assembly.
10. Position the auto-tensioner into a vise with soft jaws. The plug at the rear of tensioner protrudes, be sure to use a washer as a spacer to protect the plug from contacting vise jaws.

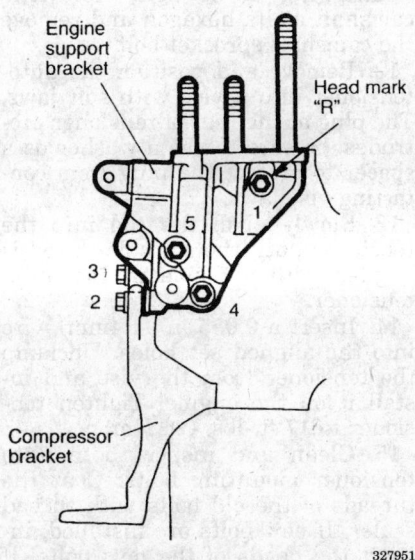

Torque sequence for the engine support bracket — Diamante 3.0L (6G72) DOHC engine

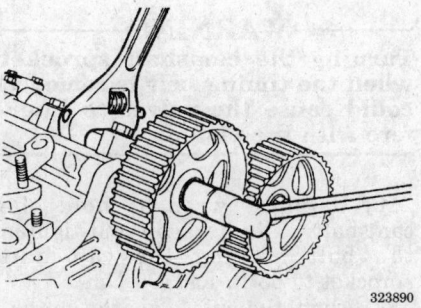

Removing the camshaft sprocket — 3000GT 3.0L (6G72) engine

11. Slowly push the rod into the tensioner until the set hole in rod is aligned with set hole in the auto-tensioner.
12. Insert a 0.055 inch (1.4mm) wire into the aligned set holes. Unclamp the tensioner from the vise and install it on the engine. Tighten tensioner mounting bolts to 17 ft. lbs. (24 Nm).

————— WARNING —————
DO NOT rotate or turn the camshafts when removing the sprockets or severe engine damage will result from internal component interference.

---

13. If removing the camshaft sprockets perform the following:
   a. Remove the valve covers.
   b. Holding the flats of the camshafts, loosen the sprocket retaining bolt.
   c. Note the positioning of the sprocket and remove the sprocket.
To install:
14. If removed, install the camshaft sprockets and tighten the mounting bolts to 66 ft. lbs. (90 Nm). Be sure and secure and prevent the camshafts from turning when tightening the mounting bolts.
15. Align the mark on the crankshaft sprocket with the mark on the front case. Then move the crankshaft sprocket 1 tooth counterclockwise.
16. Align the timing marks of the camshafts with the marks on the rear covers.
17. Using large paper clips to secure the timing belt to sprockets, install the timing belt in the following order. Be sure camshafts to cylinder heads and crankshaft to front cover timing marks are aligned. Install the timing belt around the pulleys in the following order:
   a. Exhaust camshaft sprocket (front bank).
   b. Intake camshaft sprocket (front bank).
   c. Water pump pulley.
   d. Intake camshaft sprocket (rear bank).
   e. Exhaust camshaft sprocket (rear bank).
   f. Tensioner pulley.
   g. Crankshaft pulley.
   h. Idler pulley.

NOTE: Since the camshaft sprockets turn easily, secure them with box wrenches to install timing belt.

18. Align all timing mark on the crankshaft and raise tensioner pulley against belt to remove slack, snug tensioner bolt.
19. Check the alignment of all the timing marks and remove the clips that secure the timing belt to the camshaft sprockets.
20. Rotate the engine ¼ turn counterclockwise then rotate the engine clockwise to align the timing marks. Check that all the timing marks are in alignment.
21. Loosen the center bolt on the tensioner pulley. Using tool MD998752 or equivalent and a torque wrench, apply a torque of 84 inch lbs. (10 Nm) to the tool on the tensioner. Tighten the tensioner bolt to 35 ft. lbs. (49 Nm) and make sure the tensioner does not rotate with the bolt.

22. Rotate the crankshaft two complete turns clockwise and let it sit for approximately five minutes. Then check that the set pin can easily be inserted and removed from the hole in the auto tensioner.

23. Remove the set wire attached to the auto tensioner.

24. Measure the auto tensioner protrusion (the distance between the tensioner arm and auto tensioner body) to ensure that it is within 0.15–0.18 inches (3.8–4.5mm). If out of specification, repeat adjustment procedure until the specified value is obtained.

25. Check again that the timing marks on all sprockets are in proper alignment.

26. Install the timing belt covers.

27. Install the engine support bracket and tighten the bolts to specifications in the proper sequence.

28. Install the engine mount and tighten mounting bolt.

29. Install the engine covers.

30. Connect the negative battery cable and perform all necessary adjustments.

3000GT

1. Position the engine so the No. 1 cylinder is at TDC of its compression stroke.

2. Disconnect the negative battery cable.

--------- CAUTION ---------
Wait at least 90 seconds after the negative (-) battery cable is disconnected to prevent possible deployment of the air bag.

3. Remove the engine undercover.

4. Remove the front undercover panel.

5. Remove the cruise control pump and the link assembly.

6. Remove the alternator.

7. Raise and suspend the engine so that force is not applied to the engine mount.

8. Remove the timing covers from the engine.

9. If the same timing belt will be reused, mark the direction of the timing belt's rotation for installation in the same direction. Make sure the engine is positioned so the No. 1 cylinder is at the TDC of its compression stroke and the timing marks are aligned with the engine's timing mark indicators on the valve covers or head.

10. Loosen the center bolt of tensioner pulley and unbolt auto-tensioner assembly. The auto-tensioner

assembly must be reset to correctly adjust belt tension. Remove the timing belt.

11. Using a wrench, hold the camshaft at its hexagon and remove the camshaft sprocket bolt.

12. Remove and position the auto-tensioner into a vise with soft jaws. The plug at the rear of tensioner protrudes, be sure to use a washer as a spacer to protect the plug from contacting vise jaws.

13. Slowly push the rod into the tensioner until the set hole in rod is aligned with set hole in the auto-tensioner.

14. Insert a 0.055 in. (1.4mm) wire into the aligned set holes. Unclamp the tensioner from the vise and install it on the engine. Tighten tensioner to 17 ft. lbs. (24 Nm).

15. Clean and inspect both auto tensioner mounting bolts. Coat the threads of the old bolts with thread sealer. If new bolts are installed, inspect the heads of the new bolts. If there is white paint on the bolt head, no sealer is required. If there is no paint on the head of the bolt, apply a coat of thread sealer to the bolt. Install both bolts and torque to 17 ft. lbs. (24 Nm).

To install:

--------- WARNING ---------
Turning the camshaft sprocket when the timing belt is removed could cause the valves to interfere with the pistons.

16. Using a wrench, hold the camshaft at its hexagon and tighten the bolt holding the camshaft sprocket to 65 ft. lbs. (90 Nm).

17. Align the mark on the crankshaft sprocket with the mark on the front case. Then move the sprocket 3 teeth clockwise to lower the piston so the valves do not touch the piston if the camshafts are being moved.

18. Turn each camshaft sprocket 1 at a time to align the timing marks with the mark on the valve cover or head. If the intake and exhaust valves of the same cylinder are opened simultaneously, they could interfere with each other. Therefore, if any resistance is felt, turn the other camshaft to move the valve.

19. Using large paper clips to secure the timing belt to sprockets, install the timing belt in the following order. Be sure camshafts to cylinder heads and crankshaft to front cover timing marks are aligned.

   a. Exhaust camshaft sprocket (front bank).

   b. Intake camshaft sprocket (front bank).

   c. Water pump pulley.

   d. Intake camshaft sprocket (rear bank).

   e. Exhaust camshaft sprocket (rear bank).

   f. Idler pulley.

   g. Crankshaft pulley.

   h. Tensioner pulley.

NOTE: Since the camshaft sprockets turn easily, secure them with box wrenches to install timing belt.

20. Align all timing marks and raise tensioner pulley against belt to remove slack, snug tensioner bolt.

21. Loosen the center bolt on the tensioner pulley. Using tool MD998767 or equivalent and a torque wrench, apply a torque of 7.2 ft. lbs. (10 Nm). Tighten the tensioner bolt to 42 ft. lbs. (58 Nm) and make sure the tensioner does not rotate with the bolt.

22. Remove the set wire attached to the auto tensioner.

23. Rotate the crankshaft 2 complete turns clockwise and let it sit for approximately 5 minutes. Then, check that the set pin can easily be inserted and removed from the hole in the auto tensioner.

NOTE: Even if the set pin cannot be easily inserted, the auto tensioner is normal if its rod protrusion is within specification.

24. Measure the auto tensioner protrusion (the distance between the tensioner arm and auto tensioner body) to ensure that it is within 0.15–0.18 in. (3.8–4.5mm). If out of specification, repeat adjustment procedure until the specified value is obtained.

25. Check again that the timing marks on all sprockets are in proper alignment.

26. Lower the engine so that the weight is again applied to the engine mount.

27. Make any necessary engine adjustments.

28. Install the alternator.

29. Install the cruise control pump and the link assembly.

30. Install the timing belt covers and all related items.

31. Install the front undercover panel.

32. Install the engine undercover.

33. Connect the negative battery cable.

34. Start the engine and check the timing.

Camshaft

REMOVAL AND INSTALLATION

Precis

1. Disconnect the negative battery cable.
2. Disconnect the breather hose and the secondary air hose.
3. Remove the air cleaner.
4. Remove the rocker cover.
5. Remove the timing belt cover.
6. Remove the distributor.
7. Loosen the two bolts and move the timing belt tensioner toward the water pump as far as it will go, then retighten the timing belt tensioner adjusting bolt. Disengage the timing belt from the camshaft sprocket and unbolt and remove the sprocket. The timing belt may be left engaged with the crankshaft sprocket and tensioner.
8. Remove the rocker shaft assembly. Remove the small, square cover that sits directly behind the camshaft on the transaxle side of the head. Remove the camshaft thrust case tightening bolt that sits on the top of the head right near that cover.
9. Carefully, slide the camshaft out of the head through the hole in the transaxle side of the head, being careful that the cam lobes do not strike the bearing bores in the head.

To install:
10. Lubricate all journal and thrust surfaces with clean engine oil.
11. Carefully insert the camshaft into the engine. Make sure the camshaft goes in with the threaded hole in the top of the thrust case straight upward.
12. Align the threaded hole with the bolt hole in the cylinder head surface.
13. Install the thrust case bolt and tighten firmly.

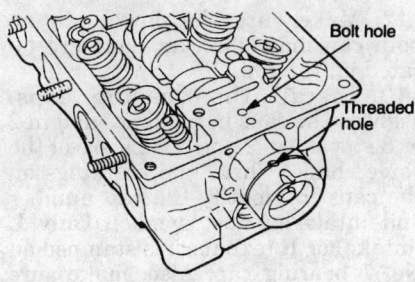

333407

Align the threaded hole with the hole in the cylinder head — Precis

14. Install the rear cover with a new gasket and install and tighten the bolts to 5.8–7.2 ft. lbs. (8–10 Nm).
15. Coat the external surface of the front oil seal with engine oil.
16. Using special installer tool MD 998306-01 or equivalent, drive a new front camshaft oil seal into the clearance between the cam and head at the forward end, making sure the seal seats fully.
17. Install the camshaft sprocket and torque the bolt to 47–54 ft. lbs. (64–74 Nm).
18. Reconnect the timing belt, check the timing and adjust the belt tension.
19. Reinstall the rocker shaft assembly. Adjust the valves and install the rocker and timing belt covers.

Mirage

1.5L (4G15) Engine

1. Disconnect the negative battery cable.
2. Rotate the engine and position the No. 1 piston to TDC of its compression stroke.
3. Disconnect the accelerator cable, breather hose and PCV hose connections.
4. Matchmark the positioning of the distributor housing and the positioning of the distributor rotor to the engine block and remove the distributor.
5. Remove the valve cover and discard the gasket.
6. Loosen both rocker arm assemblies gradually and evenly and remove the rocket shafts from the vehicle.
7. Remove the timing belt covers.

NOTE: DO NOT allow the camshaft or the crankshaft to rotate after the timing belt is removed.

8. Remove the timing belt assembly.
9. Holding the camshaft sprocket from turning, loosen and remove the bolt that secures the sprocket.
10. Remove the camshaft sprocket from the camshaft. Note the positioning of the dowel pin at the end of the camshaft.
11. Remove the camshaft oil seal from the front of the cylinder head.
12. Remove the camshaft from the head.
13. Carefully check all parts for damage and wear.

To install:
14. Lubricate the camshaft with heavy engine oil and slide it into the head. Be sure to position the dowel pin at the 12 o'clock position.

15. Check the camshaft end-play between the thrust case and camshaft. The camshaft end-play should be 0.002–0.008 in. (0.05–0.20mm). If the end-play is not within specification, replace the camshaft thrust bearing.
16. Install a new camshaft oil seal. Be sure to lubricate the lips of the seal with clean engine oil.
17. Install the camshaft sprocket and install the mounting bolt. Tighten the bolt to 51 ft. lbs. (70 Nm) while holding the camshaft from turning.
18. Install the timing belt assembly.
19. Install the timing belt covers.
20. Install the rocker shaft assemblies. Torque the bolts gradually and evenly to 23 ft. lbs. (32 Nm).
21. Check valve adjustment and install the valve cover with a new gasket. Tighten the valve cover bolt to 16 inch lbs. (1.8 Nm).
22. Align the distributor marks and install the distributor.
23. Connect the accelerator cable, breather hose and PCV hose.
24. Connect the negative battery cable and check the ignition timing.

1.8L (4G93) Engine

1. Disconnect the negative battery cable.
2. Rotate the engine and position the No. 1 piston to TDC of its compression stroke.
3. Label and disconnect the spark plug cables.
4. Matchmark the positioning of the distributor housing and the positioning of the distributor rotor to the engine block and remove the distributor.
5. Disconnect the air flow sensor connector and remove the air cleaner case cover.
6. Disconnect the accelerator cable, breather hose and PCV hose connections.
7. Remove the rocker cover and discard the gasket.
8. Loosen both rocker arm shaft assemblies gradually and evenly and remove the rocket shafts from the vehicle. Do not disassembly rocker arms and rocker arm shaft assemblies.
9. Remove the timing belt covers.

NOTE: DO NOT allow the camshaft or the crankshaft to rotate after the timing belt is removed.

10. Remove the timing belt assembly.

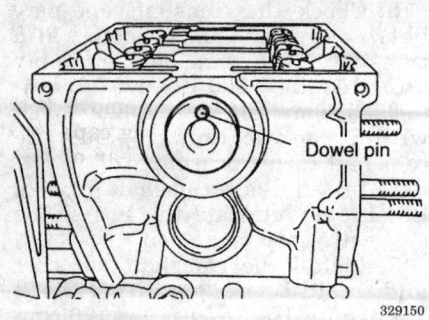

Positioning of the camshaft dowel pin — Mirage 1.5L (4G15) engine

329150

11. Holding the camshaft sprocket from turning, loosen and remove the bolt that secures the sprocket.

12. Remove the camshaft sprocket from the camshaft. Note the positioning of the dowel pin at the end of the camshaft.

13. Remove the camshaft oil seal from the front of the cylinder head.

14. Remove the camshaft from the head.

15. Carefully check all parts for damage and wear.

To install:

16. Lubricate the camshaft journals and camshaft with clean engine oil and install the camshaft in the cylinder head. Be sure to position the dowel pin at the end of the camshaft as noted during the removal procedure.

17. Check the camshaft end-play between the thrust case and camshaft. The camshaft end-play should be 0.002–0.008 in. (0.05–0.20mm). If the end-play is not within specification, replace the camshaft thrust bearing.

18. Install a new camshaft oil seal. Be sure to lubricate the lips of the seal with clean engine oil.

19. Install camshaft sprocket and torque the retainer bolt to 65 ft. lbs. (90 Nm). Be sure to secure the sprocket while tightening the bolt.

20. Install the timing belt assembly.

21. Install the timing belt covers.

22. Install the rocker arm and shaft assemblies. Tighten the rocker arm shaft retainer bolts to 23 ft. lbs. (32 Nm).

23. Check valve adjustment and install valve cover with a new gasket. Tighten the valve cover bolts to 29 inch lbs. (3.3 Nm).

24. Align the distributor marks and install the distributor.

25. Connect the spark plug cables.

26. Connect the accelerator cable, breather hose and PCV hose.

27. Connect the air flow sensor connector and install the air cleaner case cover.

28. Connect the negative battery cable. Run the engine at idle until normal operating temperature is reached. Check idle speed and ignition timing and adjust as required.

Eclipse

1.8L (4G37) Engine

1. Disconnect the negative battery cable. Remove the air intake hose and the PCV hose.

2. Remove the valve cover.

3. Remove distributor extension housing if necessary.

4. Remove the outer timing belt.

5. Install auto lash adjuster retainer tools MD998443 or equivalent, on the rocker arm.

6. Remove the camshaft bearing caps but do not remove the bolts from the carrier.

7. Remove the rocker arms, rocker shafts and bearing caps from the engine as an assembly.

8. Remove the camshaft from the cylinder head.

9. Inspect the bearing journals on the camshaft for excess wear or damage. Measure the cam lobe height and compare to the desired readings. Inspect the bearing surfaces in the cylinder head. Replace any components that is damaged or shows signs of excess wear.

To install:

10. Lubricate the camshaft journals and camshaft with clean engine oil and install the camshaft in the cylinder head.

11. Align the camshaft bearing caps with the arrow mark depending on cylinder numbers and install in numerical order.

12. Apply sealer at the ends of the bearing caps and install the assembly.

13. Torque the bearing cap bolts in the following sequence: No. 3, No. 2, No. 1 and No. 4 to 85 inch lbs. (10 Nm).

14. Repeat the sequence increasing the torque to 15 ft. lbs. (20 Nm).

15. Install the distributor extension if it was removed.

16. Install the timing belt, valve cover and all related parts.

17. Connect the negative battery cable and check for leaks.

1993–94 2.0L (4G63) Engine

1. Relieve the fuel system pressure following proper procedure. Disconnect negative battery cable.

2. Disconnect the accelerator cable, PCV hoses, breather hoses, spark plug cables and remove the valve cover.

NOTE: Always rotate the crankshaft in a clockwise direction. Make a mark on the back of the timing belt indicating the direction of rotation so it may be reassembled in the same direction if it is to be reused.

3. Rotate the crankshaft clockwise and align the timing marks so No. 1 piston will be at TDC of the compression stroke. At this time the timing marks on the camshaft sprocket and the upper surface of the cylinder head should coincide, and the dowel pin of the camshaft sprocket should be at the upper side.

4. Remove the timing belt upper and lower covers.

5. Remove the timing belt.

6. Matchmark the crank angle sensor to the cylinder head and remove the crank angle sensor.

7. Remove the rocker cover, semi-circular packing, throttle body stay, both camshaft sprockets, and oil seal.

8. Loosen the bearing cap bolts in 2–3 steps. Label and remove all camshaft bearing caps.

NOTE: If the bearing caps are difficult to remove, use a plastic hammer to gently tap the rear part of the camshaft.

9. Remove the intake and exhaust camshafts.

10. Check the camshaft journals for wear or damage. Check the cam lobes and rocker rollers for damage. Also, check the cylinder head oil holes for clogging.

To install

11. Lubricate the camshafts with heavy engine oil and position the camshafts on the cylinder head.

NOTE: Do not confuse the intake camshaft with the exhaust camshaft. The intake camshaft has a split on its rear end for driving the crank angle sensor.

12. Make sure the dowel pin on both camshaft sprocket ends are located on the top.

13. Install the bearing caps. Tighten the caps in sequence and in 2 or 3 steps. No. 2 and 5 caps are of the same shape. Check the markings on the caps to identify the cap number and intake/exhaust symbol. Only **L** (intake) or **R** (exhaust) is stamped on No. 1 bearing cap. Also, make sure the rocker arm is correctly mounted on the lash adjuster and the valve stem end. Torque the retaining bolts to 15 ft. lbs. (20 Nm).

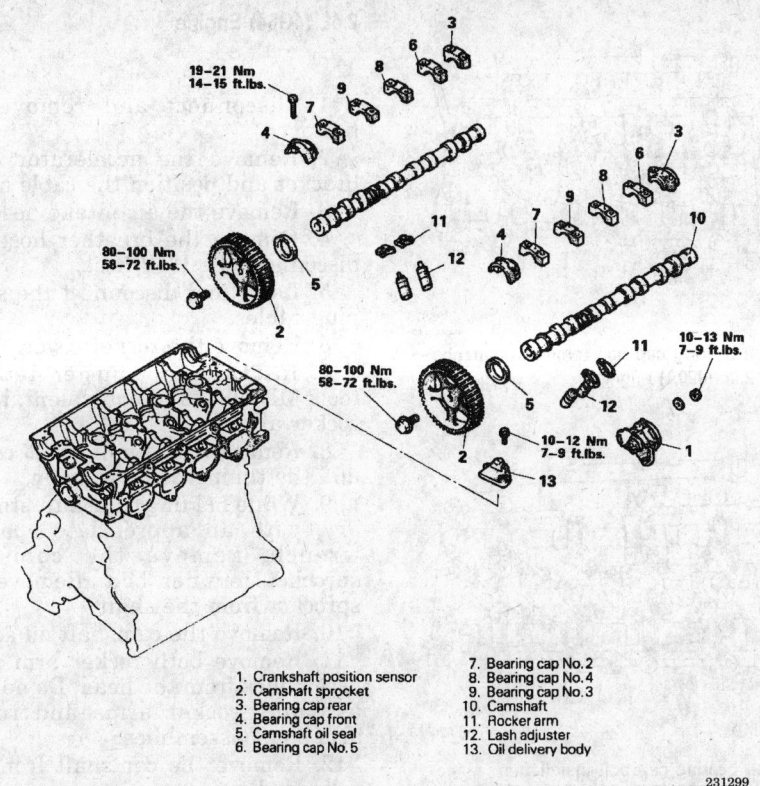

19–21 Nm
14–15 ft.lbs.

80–100 Nm
58–72 ft.lbs.

80–100 Nm
58–72 ft.lbs.

10–12 Nm
7–9 ft.lbs.

10–13 Nm
7–9 ft.lbs.

1. Crankshaft position sensor
2. Camshaft sprocket
3. Bearing cap rear
4. Bearing cap front
5. Camshaft oil seal
6. Bearing cap No.5
7. Bearing cap No.2
8. Bearing cap No.4
9. Bearing cap No.3
10. Camshaft
11. Rocker arm
12. Lash adjuster
13. Oil delivery body

231299

Camshaft, rocker and lifter assemblies — 1993–94 Eclipse 2.0L (4G63) engine shown

Intake side   Exhaust side

Slits

231300

Camshaft identification — Eclipse and Galant

14. Apply a coating of engine oil to the oil seal. Using proper size driver, press-fit the seal into the cylinder head.

15. Install the camshaft sprockets. While holding the camshaft at its hexagon, between number 2 and 3 journals tighten sprocket bolts to 58–72 ft. lbs. (80–100 Nm).

16. Align the punch mark on the crank angle sensor housing with the notch in the plate. With the dowel pin on the sprocket side of the intake camshaft at top, install the crank angle sensor on the cylinder head.

NOTE: Do not position the crank angle sensor with the punch mark positioned opposite the notch; this position will result in incorrect fuel injection and ignition timing.

17. Install the timing belt, covers and related components.

18. Install the valve cover using new gasket. Reconnect all related components.

19. Reconnect the negative battery cable.

1995–97 2.0L (4G63) Engine

1. Disconnect the negative battery cable.

2. Disconnect the accelerator cable from the throttle body and remove the cable bracket from the intake plenum.

3. Remove the engine center cover.

4. Disconnect the spark plug cables from the spark plugs. Label them if necessary.

5. Disconnect the breather hose and the PCV hose from the rocker cover.

6. Remove the rocker cover.

7. Position the No. 1 cylinder at TDC on compression.

8. Remove the timing belt.

9. Use a wrench on the hex shaped part of the camshaft to hold the cam and remove the camshaft sprockets.

10. Loosen the bearing cap bolts in two or three steps and remove the bearing caps. If the bearing caps are hard to remove, tap the rear of the camshaft with a plastic hammer.

11. Remove the camshaft(s) and the oil seals.

To install:

12. Apply engine oil or assembly lube to the camshafts and install them on the cylinder head.

――――――― WARNING ―――――――
If new camshaft(s) are being installed, remove the rocker arms and install the camshaft(s) and the bearing caps. Make sure the camshaft(s) can be turned by hand. After checking, remove the camshafts and install the rocker arms.
―――――――――――――――――――――

NOTE: Bearing caps and rocker arms must be installed in the same location that they were remove from.

13. Install the bearing caps and torque the bolts evenly in two or three steps to specifications.

14. Apply engine oil to the lip of the seal. Using MB998713, install the front oil seal.

15. Install the camshaft sprockets.

16. Install the timing belt.

17. Apply sealant to the semi-circular packing and install it in the cylinder head.

18. Apply sealant to the lower part of the front and rear bearing caps where they meet the cylinder head. Use a new gasket and install the rocker cover.

19. Connect the PCV hose and the breather hose.

20. Connect the spark plug wires.

21. Install the center cover.

22. Install the adjust the accelerator cable.

23. Connect the negative battery cable.

2.0L (420A) Engine

1. Disconnect the negative battery cable.

2. Remove the ignition coil pack.

3. Disconnect the PCV hose and the breather hose from the cylinder head cover.

4. Remove the semi-circular packing from the rear of the head.

5. Remove the camshaft position sensor.

6. Remove the timing belt.

7. Use tool MB990767 and MB998719 or equivalent to hold the

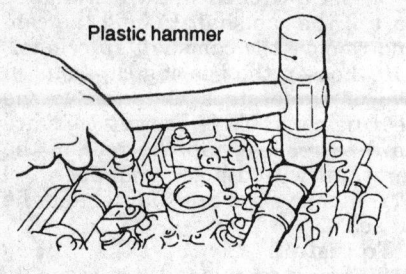

Tap the camshaft with a plastic hammer to loosen the bearing caps — 1995–97 Eclipse 2.0L (4G63) engine

320860

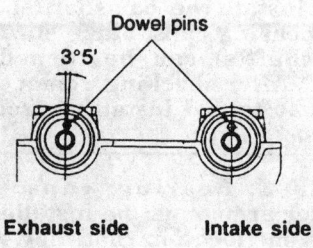

Position the camshafts with the dowels facing up — 1995–97 Eclipse 2.0L (4G63) engine

320862

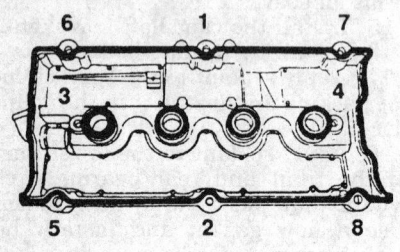

Cylinder head cover bolt installation sequence — Eclipse 2.0L (420A) engine

320879

camshaft sprockets and remove the sprocket mounting bolt and the sprocket.

8. Remove the bracket and the rear timing belt cover.

9. Remove the outside camshaft bearing cap.

10. Gradually loosen the camshaft bearing caps in sequence, one camshaft at a time and remove the bearing caps.

NOTE: Keep the bearing caps in order. They must be installed in the location that they were removed from.

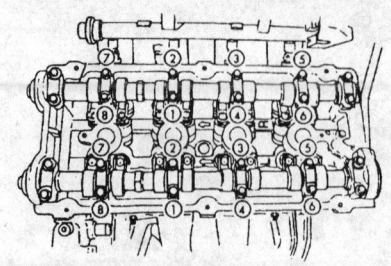

Camshaft bearing cap bolt removal sequence — Eclipse 2.0L (420A) engine

320876

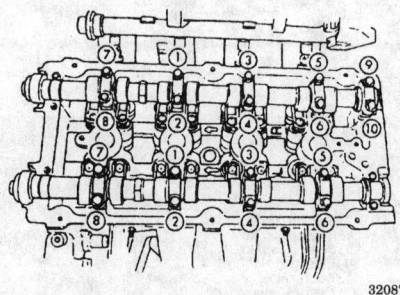

Camshaft bearing cap bolt installation sequence — 1995–97 Eclipse 2.0L (420A) engine

320878

11. Mark the camshafts for later identification and remove the camshafts. The camshafts are not interchangeable.

To install:

12. Apply engine oil or assembly lube to the camshaft and install the camshafts.

13. Install the bearing caps. Torque the bolts evenly and in sequence.

14. Apply Loctite 518® to the outside camshaft bearing caps and install them.

15. Install the camshaft oil seal.

16. Install the rear timing belt cover and the bracket.

17. Use the special tools and install the camshaft sprockets.

18. Install the timing belt.

19. Apply Loctite 5699® or equivalent to the semi-circular packing and install it in the rear of the cylinder head.

20. Install the camshaft position sensor.

21. Install the cylinder head cover. Torque the bolts evenly in three steps in the proper sequence.

22. Install the air, breather and PCV hoses.

23. Install the coil pack.

24. Connect the negative battery cable.

2.4L (4G64) Engine

1. Disconnect and remove the battery.

2. Remove the accelerator cable bracket and position the cable aside.

3. Remove the air intake hose.

4. Remove the breather hose and disconnect the PCV hose.

5. Label and disconnect the spark plug cables.

6. Remove the rocker cover.

7. Install lash adjuster retainer tools MD998443 or equivalent, to the rocker arm.

8. Remove the timing belt covers and the timing belt assembly.

9. While holding camshaft stationary, with an appropriate spanner wrench, remove the camshaft sprocket retainer bolt. Remove the sprocket from the shaft.

10. Remove the camshaft oil seal.

11. Remove both rocker arm shaft assemblies from the head. Do not disassembly rocker arms and rocker arm shaft assemblies.

12. Remove the camshaft from the cylinder head.

13. Inspect the bearing journals on the camshaft, cylinder head, and bearing caps.

To install:

14. Lubricate the camshaft journals and camshaft with clean engine oil and install the camshaft in the cylinder head.

15. Install the rocker arm and shaft assemblies. Tighten the rocker arm shaft retainer bolts to 21–25 ft. lbs. (29–35 Nm).

16. Apply a coating of engine oil to the oil seal. Using the proper size driver, press-fit the seal into the cylinder head.

17. Install camshaft sprocket and retainer bolt torquing to 65 ft. lbs. (90 Nm).

18. Install the timing belt and belt covers.

19. Remove the lash adjuster retaining tools.

20. Install the rocker cover using new gasket material on mating surfaces.

21. Connect the spark plug cables.

22. Reinstall the air intake hose.

23. Install the breather hose and connect the PCV hose.

24. Install the battery.

25. Run the engine at idle until normal operating temperature is reached. Check idle speed and ignition timing; adjust as required.

1993 Galant 2.0L (4G63) Engine

8 Valve Engines

1. Disconnect the negative battery cable.
2. Disconnect the breather and the PCV hoses.
3. Remove the accelerator cable bracket and position the cable aside.
4. Install lash adjuster retainer tools MD998443 or equivalent, to the rocker arm.
5. Remove the valve cover and semi-circular packing.
6. Matchmark the distributor housing to the cylinder head, and remove the distributor.
7. Remove the timing belt covers and the timing belt.
8. Remove the camshaft sprocket.
9. Remove the carrier bolts and remove the rocker arms, rocker shafts and bearing caps from the engine as an assembly.
10. Remove the camshaft from the cylinder head.
11. Inspect the bearing journals on the camshaft for excess wear or damage. Measure the cam lobe height and compare to the desired readings. Inspect the bearing surfaces in the cylinder head. Replace any components that are damaged or show signs of excess wear.

To install:
12. Lubricate the camshaft journals and camshaft with clean engine oil and install the camshaft in the cylinder head.
13. Align the camshaft bearing caps with the arrow marks (depending on cylinder numbers) and install in numerical order.
14. Install the rocker shaft assembly to the cylinder head. Torque the bearing cap bolts from the center outward, in three steps, until a final torque of 15 ft. lbs. (20 Nm) is reached.
15. Apply a coating of engine oil to the oil seal. Using the proper size driver, press-fit the seal into the cylinder head.
16. Install the camshaft sprocket and torque retaining bolt to 65 ft. lbs. (90 Nm).
17. Install the timing belt.
18. Align the matchmarks and install the distributor.
19. Remove the lash adjuster retaining tools.
20. Install the valve cover and all related parts.
21. Connect the negative battery cable and run engine to check for leaks.
22. Check and adjust ignition timing, if necessary.

16 Valve Engines

1. Disconnect the negative battery cable.
2. Remove the accelerator cable bracket and position the cable aside.
3. Remove the breather hose and disconnect the PCV hose.
4. Label and disconnect the spark plug cables.
5. Matchmark the distributor housing, to the cylinder head, and remove the distributor.
6. Remove the rocker cover.
7. Install lash adjuster retainer tools MD998443 or equivalent, to the rocker arm.
8. Remove the timing belt covers and the timing belt assembly.
9. Remove the camshaft sprocket retainer bolt while holding shaft stationary with an appropriate wrench. Remove the sprocket from the shaft.
10. Remove the camshaft oil seal.
11. Remove both rocker arm shaft assemblies from the head. Do not disassembly rocker arms and rocker arm shaft assemblies.
12. Remove the camshaft from the cylinder head.
13. Inspect the bearing journals on the camshaft, cylinder head, and bearing caps.

To install:
14. Lubricate the camshaft journals and camshaft with clean engine oil and install the camshaft in the cylinder head.
15. Install the rocker arm and shaft assemblies. Tighten the rocker arm shaft retainer bolts to 21–25 ft. lbs. (29–35 Nm).
16. Apply a coating of engine oil to the oil seal. Using the proper size driver, press-fit the seal into the cylinder head.
17. Install camshaft sprocket and retainer bolt torquing to 65 ft. lbs. (90 Nm).
18. Install the timing belt and belt covers.
19. Align the matchmarks and install the distributor.
20. Remove the lash adjuster retaining tools.
21. Install the rocker cover using new gasket material on mating surfaces.
22. Connect the spark plug cables.
23. Install the breather hose and connect the PCV hose.
24. Connect the negative battery cable. Run the engine at idle until normal operating temperature is reached. Check idle speed and ignition timing; adjust as required.

1994–97 Galant 2.4L (4G64) Engine

1. Relieve the fuel system pressure following proper procedure and disconnect negative battery cable and
2. Disconnect the accelerator cable, PCV hoses, breather hoses, spark plug cables and the remove the valve cover.

NOTE: Always rotate the crankshaft in a clockwise direction. Make a mark on the back of the timing belt indicating the direction of rotation so it may be reassembled in the same direction if it is to be reused.

3. Rotate the crankshaft clockwise and align the timing marks so No. 1 piston will be at TDC of the compression stroke. At this time the timing marks on the camshaft sprocket and the upper surface of the cylinder head should coincide, and the dowel pin of the camshaft sprocket should be at the upper side.
4. Remove the timing belt upper and lower covers.
5. Remove the camshaft timing belt.
6. Use a wrench between No. 2 and No. 3 journals to hold the camshaft; remove the camshaft sprockets.
7. Loosen the bearing cap bolts in 2–3 steps. Label and remove all camshaft bearing caps.

NOTE: If the bearing caps are difficult to remove, use a plastic hammer to gently tap the rear part of the camshaft.

8. Remove the intake and exhaust camshafts.
9. Remove the rocker arms and lash adjusters.

NOTE: It is essential that all parts be kept in the same order and orientation for reinstallation. In order to prevent confusion during installation, be sure to mark and separate all parts.

To install:
10. Install the lash adjusters and rocker arms into the cylinder head. Lubricate lightly with clean oil prior to installation.
11. Lubricate the camshafts with heavy engine oil and position the camshafts on the cylinder head.
12. Check the camshaft journals and lobes for wear or damage. Also, check the cylinder head oil holes for clogging. Visually inspect the rocker arm roller and replace if dent, damage or seizure is evident. Check the roller for smooth rotation. Replace if excess play or binding is present.

Also, inspect valve contact surface for possible damage or seizure. It is recommended that all rocker arms and lash adjusters be replaced together.

NOTE: Do not confuse the intake camshaft with the exhaust camshaft. The intake camshaft has a split on the rear face for driving the crank angle sensor.

13. Make sure the dowel pin on both camshaft sprocket ends are located on the top.

14. Install the bearing caps. Tighten the caps in sequence and in 2 or 3 steps. No. 2 and 5 caps are of the same shape. Check the markings on the caps to identify the cap number and intake/exhaust symbol. Only **L** (intake) or **R** (exhaust) is stamped on No. 1 bearing cap. Also, make sure the rocker arm is correctly mounted on the lash adjuster and the valve stem end. Torque the retaining bolts to 15 ft. lbs. (20 Nm).

15. Apply a coating of engine oil to the oil seal. Using the proper size driver, press-fit the seal into the cylinder head.

16. Install the camshaft sprockets. While holding the camshaft at its hexagon, between number 2 and 3 journals tighten sprocket bolts to 58–72 ft. lbs. (80–100 Nm).

17. Install the timing belt, covers and related components.

18. Install the valve cover, using new gasket, and reconnect all related components.

19. Reconnect the negative battery cable.

Diamante

3.0L (6G72) DOHC Engine

1. Relieve the fuel system pressure.

2. Disconnect negative battery cable.

3. Remove the intake manifold plenum.

4. Remove the timing belt cover and the timing belt.

------ **WARNING** ------

DO NOT rotate the crankshaft or camshafts after the timing belt has been removed. If rotated, severe internal engine damage will result from the pistons hitting the valves.

5. Remove the center cover, breather, PCV hoses, and the spark plug cables.

6. Remove the rocker cover and the semi-circular packing.

7. Matchmark the positioning of the crankshaft position sensor at the

rear of the camshaft and remove the sensor.

8. If equipped with a camshaft sensor, remove the sensor from the front of the engine.

9. Being sure to hold the flats of the camshaft, loosen the camshaft sprocket bolts.

10. Noting the positioning and location of the sprockets, remove the sprockets from the camshafts.

NOTE: Be sure to note the positioning of the knock pin at the end of the camshafts for reinstallation purposes.

NOTE: Be sure to keep the valve train components labeled and in proper order for reassembly.

11. Loosen the bearing cap bolts in 2–3 steps. Label and remove all camshaft bearing caps.

NOTE: If the bearing caps are difficult to remove, use a plastic hammer to gently tap the components.

12. Mark the components and remove the intake and the exhaust camshafts.

13. Remove the rocker arms and the lash adjusters. Be sure to note the location of the valve train components for reinstallation purposes.

14. Check the camshaft journals for wear or damage. Check the cam lobes for damage. Also, check the cylinder head oil holes for clogging.

To install:

NOTE: Lubricate the valve train components with clean engine oil.

15. Bleed and install the lash adjusters to the to the original bores in the cylinder head.

16. Install the rocker arms to the cylinder head.

17. Lubricate the camshafts with clean engine oil and position the camshafts on the cylinder head.

------ **WARNING** ------

Be sure to properly position the knock pins of the camshaft to prevent valve to piston interference.

NOTE: Do not confuse the intake camshaft with the exhaust camshaft. The intake camshaft on the Diamante has a B or J stamped on the hexagon depending on the application. The exhaust camshaft on the Diamante has a D or K stamped on the hexagon depending on application.

NOTE: Install the bearing caps according to the identification mark and cap number. Bearing caps No. 2, 3 and are marked as such. The caps also are marked I for intake or E for exhaust.

18. Install the bearing caps. Tighten the caps in sequence and in 2 or 3 steps. Caps 2, 3 and 4 have a front mark. Install with the mark aligned with the front mark on the cylinder head. Torque the retaining bolts for caps No. 2, 3 and 4 to 8 ft. lbs. (11 Nm) and torque the retaining bolts for the front and rear caps to 14 ft. lbs. (20 Nm).

19. Apply a coating of engine oil to the oil seals and install the oil seals to the front and rear of the camshafts.

20. Holding the flats of the camshaft, install and tighten the sprocket bolts to 65 ft. lbs. (90 Nm).

21. If removed, install the camshaft position sensor and tighten the mounting bolts to 78 inch lbs. (9 Nm).

22. Aligning the matchmark, install the crankshaft position sensor at the rear of the camshaft and tighten the mounting nut to 7 ft. lbs. (12 Nm).

23. Align the marks on the camshaft and crankshaft sprockets. Install the timing belt assembly.

24. Install the rocker cover and the semi-circular packing.

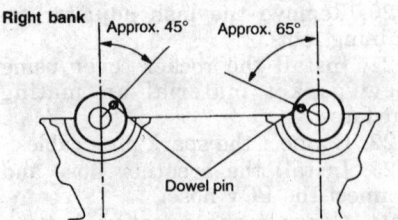 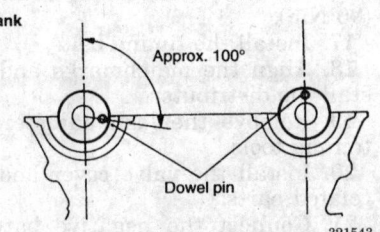

Proper positioning of the camshaft knock pins — Diamante 3.0L (6G72) DOHC engine

25. Install the intake manifold plenum.

26. Install the spark plug cables, center cover, breather and PCV hoses.

27. Connect the negative battery cable and check for leaks.

3.0L (6G72) SOHC Engine

1. Disconnect the negative battery cable.

2. Rotate and position the engine to TDC of compression stroke.

3. If removing the right side (front) camshaft, matchmark the distributor rotor and distributor housing to the engine block and remove the distributor.

4. Remove the intake manifold plenum stay bracket.

5. Remove the distributor housing adapter and discard the O-ring.

6. Remove the valve covers and the timing belt.

7. Using camshaft sprocket holding tool MB990767 and MD998719 or equivalent, hold the sprocket and loosen the bolt.

8. Remove the bolt and note the positioning of the of the knock pin at the end of the camshaft and remove the sprocket.

9. Install auto lash adjuster retainer tools MD998443 or equivalent, on the rocker arms.

NOTE: Be sure to note the position of the rocker arms, rocker shafts and bearing caps for reinstallation purposes.

10. Remove the camshaft bearing caps but do not remove the bolts from the caps.

11. Remove the rocker arms, rocker shafts and bearing caps, as an assembly.

12. Remove the camshaft from the cylinder head.

13. Inspect the bearing journals on the camshaft, cylinder head, and bearing caps.

To install:

NOTE: The right bank camshaft is identified by a 4mm slit at the rear end of the camshaft.

14. Lubricate the camshaft journals and camshaft with clean engine oil and install the camshaft in the cylinder head. Be sure to properly position the knock pin of the camshaft as noted during removal.

15. Apply sealer at the ends of the bearing caps and install the rocker arms, rocker shafts and bearing caps

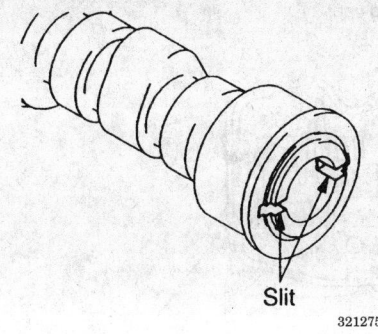

Right bank camshaft identification — Diamante 3.0L (6G72) SOHC engine

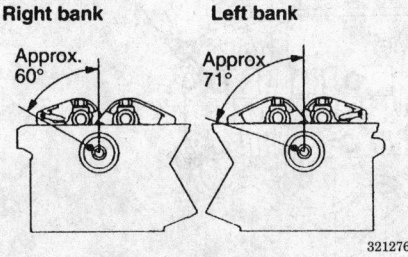

Proper positioning of the camshafts — Diamante 3.0L (6G72) SOHC engine

as an assembly. Properly position the arrows on the bearing caps.

16. Torque the bearing cap bolts in the following sequence: No. 3, No. 2, No. 1 and No. 4 to 85 inch lbs. (10 Nm).

17. Repeat the sequence increasing the torque to 14 ft. lbs. (20 Nm).

18. Remove the auto lash adjuster retainer tools from the rocker arms.

19. Install the camshaft sprocket and bolt.

20. Using camshaft sprocket holding tool MB990767 and MD998719 or equivalent, hold the sprocket and tighten the bolt to 65 ft. lbs. (90 Nm).

21. Install the timing belt and valve covers.

22. Using a new O-ring, install the distributor extension housing.

23. Install the intake manifold plenum stay bracket.

24. Install the distributor assembly. Be sure to align the rotor and distributor housing matchmarks.

25. Connect the negative battery cable and check for leaks.

3000GT (6G72) Engine

1. Relieve the fuel system pressure.

> **── CAUTION ──**
> *Fuel injection systems remain under pressure after the engine has been turned OFF. Properly relieve fuel pressure before disconnecting any fuel lines. Failure to do so may result in fire or personal injury.*

2. Disconnect battery negative cable.

> **── CAUTION ──**
> *Wait at least 90 seconds after the negative (-) battery cable is disconnected to prevent possible deployment of the air bag.*

3. Remove the timing belt cover and timing belt.

4. Remove the center cover, breather and PCV hoses, and spark plug cables.

5. Remove the rocker cover, semi-circular packing, throttle body stay, both camshaft sprockets, and oil seals.

6. Remove the crank angle sensor and adapter.

7. Remove the intake and exhaust camshafts.

To install:

8. Lubricate the camshafts with heavy engine oil and position the camshafts on the cylinder head.

NOTE: Do not confuse the intake camshaft with the exhaust camshaft. The intake camshaft has a J stamped on the hexagon and the exhaust camshaft has a K or N.

9. Make sure the dowel pins on both camshaft sprocket ends are positioned properly.

10. Install the bearing caps. Tighten the caps in sequence and in 2 or 3 steps. Caps 2, 3 and 4 have a front mark. Install with the mark aligned with the front mark on the cylinder head. Intake caps have **I** stamped on the cap and exhaust caps have **E**. Also, make sure the rocker arm is correctly mounted on the lash adjuster and the valve stem end. Torque the front and rear retaining cap bolts to 15 ft. lbs. (20 Nm) and tighten the center 3 retaining cap bolts to 8 ft. lbs. (11 Nm).

NOTE: If installing the camshaft to a cylinder head that is positioned on a workbench, the valves will protrude.

11. Apply a coating of engine oil to the oil seals and install.

12. Install the timing belt, valve cover and all related parts.

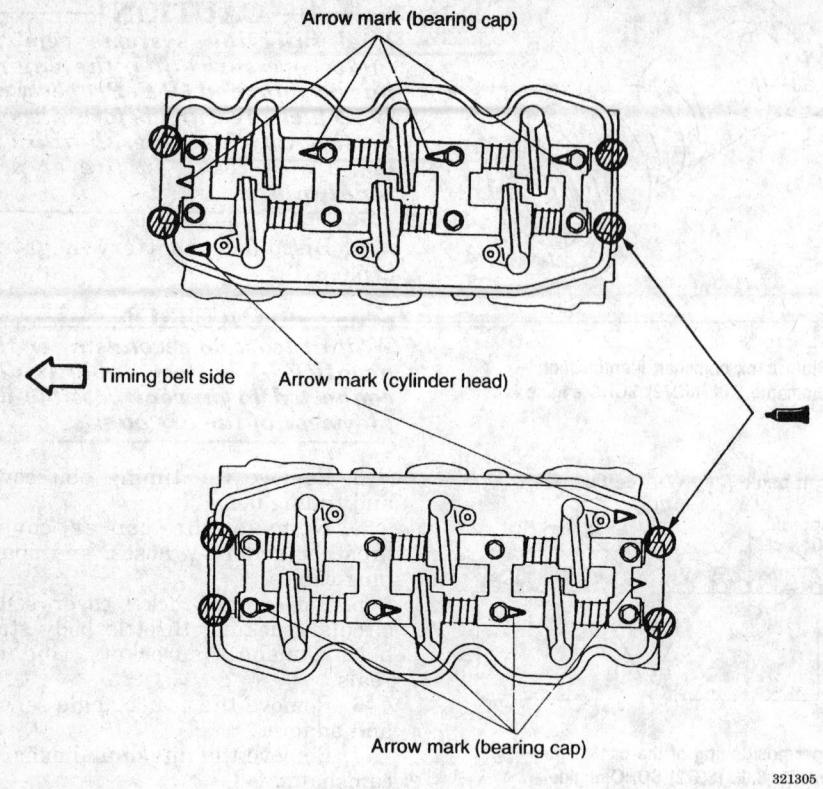

Alignment of the rockershafts and application of sealant — Diamante 3.0L (6G72) SOHC engine

front case. The specifications for oil clearances are as follows:
- **Right shaft**
- Front — 0.0008–0.0024 in. (0.02–0.06mm)
- Rear — 0.0020–0.0036 in. (0.05–0.09mm)
- **Left shaft**
- Front — 0.0008–0.0021 in. (0.02–0.05mm)
- Rear — 0.0020–0.0036 in. (0.05–0.09mm)

To install:

NOTE: Whenever the oil pump is disassembled or the cover is removed, the gear cavity must be filled with petroleum jelly. This seals the pump and acts as a prime so the oil pump will draw oil as soon as the engine begins to turn. Do not use grease

9. Lubricate the bearing surface of the shaft and the bearing journals with clean engine oil. Carefully install the silent shafts to the block.
10. Clean the gasket material from the mating surface of the cylinder block and the engine front cover. Install new gasket in place.
11. Using proper size driver, install the crankshaft oil seal into the front engine cover.
12. Using the proper size socket wrench, press in the silent shaft oil seal into the front case.
13. Place pilot tool MD998285 or equivalent, onto the nose of the crankshaft. Apply clean engine oil to the outer circumference of the pilot tool.
14. Install the front case onto the engine block and install the retainer bolts in their original positions. Tighten retainers evenly to 12 ft. lbs. (17 Nm).
15. Install the oil pump relief plunger and spring into the bore in the front case and tighten to 33 ft. lbs. (45 Nm). Make sure a new gasket is in place.
16. Install the oil pimp drive gear and driven gear to the front case, lining up the timing marks. Lubricate the gears with clean engine oil.
17. Insert the Phillips screwdriver into the hole on the side of the engine block, to hold the silent shaft. Install and tighten the flange bolt to 27 ft. lbs. (37 Nm).
18. Install a new oil pump cover gasket in the groove of the front case. When installing the gasket, make sure the round side of the gasket is towards the oil pump cover.
19. Install the oil pump seal into the oil pump cover, making sure the lip is facing the correct direction. The

13. Connect the negative battery cable and check for leaks.

Balance Shaft

REMOVAL AND INSTALLATION

Eclipse 1.8L (4G37) Engine

NOTE: A special oil seal guide MD998285 or equivalent, is needed to complete this operation.

1. Disconnect the negative battery cable.
2. Remove the oil filter, oil pressure switch, oil gauge sending unit, oil filter mounting bracket and gasket.
3. Raise and safely support the vehicle. Drain engine oil. Remove engine oil pan, oil screen and gasket.
4. Lower the vehicle. Remove the timing belts.
5. Remove the front engine cover. Different length bolts are used. Take note of their locations. If the cover sticks to the block, look for a special slot provided and pry with a flat bladed tool. Discard the shaft seal and gasket.

6. Remove the oil pump driven gear flange bolt. When loosening this bolt, first insert a tool approximately 3/8 in. diameter into the plug hole on the left side of the cylinder block to hold the silent shaft. Remove the oil pump gears and remove the front case assembly. Remove the threaded plug, the oil pressure relief spring and plunger.
7. Remove the silent shaft oil seals, the crankshaft oil seal and front case gasket.
8. Remove the silent shafts and inspect as follows:
   a. Check the oil holes in the shaft for clogging.
   b. Check journals of the shaft for seizure, damage and contact with bearing. If there is anything wrong with the journal, replace the silent shaft bearing, silent shaft or front case.
   c. Check the silent shaft oil clearance. If the clearance is beyond the specifications, replace the silent shaft bearing, silent shaft or

lip of the seal should be installed against the oil it is to stop.

NOTE: The timing of the oil pump sprocket and connected silent shaft can be incorrect, even with the timing mark aligned. Incorrect orientation of the silent shaft will result in engine vibration during operation. Follow the alignment procedure in the timing belt section of this chapter.

20. Install the timing belts and all related items. Make sure the timing and the orientation of the silent shafts is correct, using alignment tool in the hole in the left side of the engine block.

21. Install the oil pan, oil filter mounting bracket, oil switches and new oil filter to the engine. Fill the crankcase to the proper level with clean engine oil.

22. Connect the negative battery cable and start the engine. Check for proper timing and inspect for leaks.

Eclipse 2.0L (4G63) and All 1993 Galant Engines

NOTE: A special oil seal guide MD998285 and a plug cap socket tool MD998162 or exact equivalents are needed to complete this operation.

1. Disconnect the negative battery cable.
2. Remove the oil filter, oil pressure switch, oil gauge sending unit, oil filter mounting bracket and gasket. Remove the oil cooler bolt and oil cooler from the oil filter bracket.
3. Raise and safely support the vehicle.
4. Drain the engine oil. Remove engine oil pan, oil screen and gasket. Remove the relief plug, gasket, relief spring and relief plunger.
5. Lower the vehicle. Using the proper equipment, support the weight of the engine.

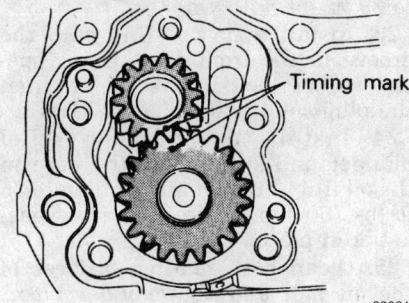
Aligning oil pump timing marks

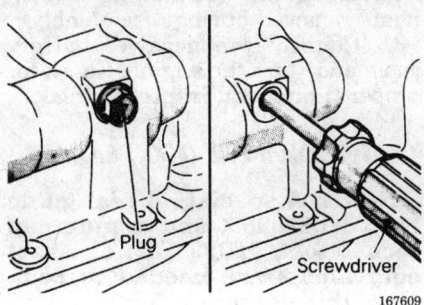
Alignment of oil pump driveshaft

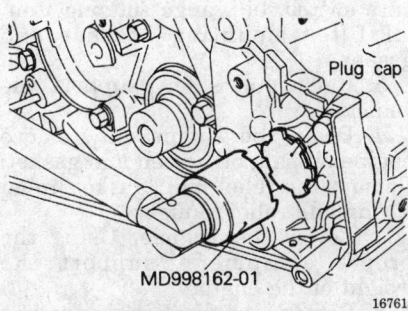
Removal and installation of the front case plug — 1993 Galant 2.0L (4G63) engine

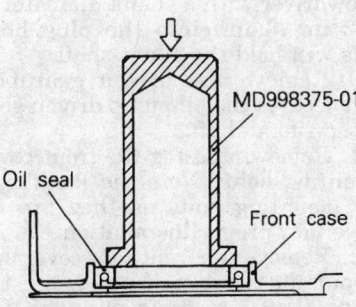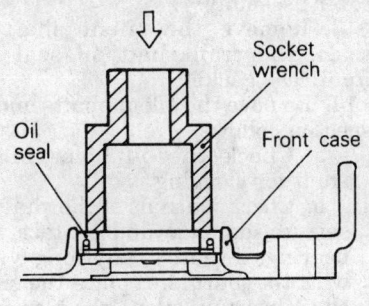
Installation of the front case and seal to the engine — 1993 Galant 2.0L (4G63) engine

6. Remove the front engine mount bracket and accessory drive belt.
7. Using special tool MD998162, remove the plug cap in the engine front cover.
8. Remove the plug on the side of the engine block. Insert a Phillips screwdriver with a shank diameter of 0.32 in. (8mm) into the plug hole. This will hold the silent shaft.
9. Remove the driven gear bolt that secures the oil pump driven gear to the silent shaft.
10. Remove and tag the front cover mounting bolts. Note the lengths of the mounting bolts as they are removed for proper installation.
11. Remove the front case cover and oil pump assembly. If necessary, the silent shaft can come out with the cover assembly.
12. Remove the silent shaft oil seals, the crankshaft oil seal and front case gasket.
13. Remove the silent shafts and inspect as follows:
   a. Check the oil holes in the shaft for clogging.
   b. Check journals of the shaft for seizure, damage and contact with bearing. If there is anything wrong with the journal, replace the silent shaft bearing, silent shaft or front case.
   c. Check the silent shaft oil clearance. If the clearance is beyond the specifications, replace the silent shaft bearing, silent shaft or front case. The specifications for oil clearances are as follows:
   • **Right shaft**
   • Front — 0.0012–0.0024 in. (0.03–0.06mm)
   • Rear — 0.0008–0.0021 in. (0.02–0.05mm)
   • **Left shaft**
   • Front — 0.0020–0.0036 in. (0.05–0.09mm)
   • Rear — 0.0017–0.0033 in. (0.04–0.08mm)

To install:
14. Lubricate the bearing surface of the shaft and the bearing journals with clean engine oil. Carefully install the silent shafts to the block.
15. Clean the gasket material from the mating surface of the cylinder block and the engine front cover. Install new gasket in place.
16. Install the oil pump drive gear and driven gear to the front case, lining up the timing marks. Lubricate the gears with clean engine oil. Install the oil pump cover, with new gasket in place and tighten the mounting bolts to 13 ft. lbs. (18 Nm).

17. Using proper size driver, install the crankshaft oil seal into the front engine case.

18. Using the proper size socket wrench, press in the silent shaft oil seal into the front case.

19. Place pilot tool MD998285 or equivalent, onto the nose of the crankshaft. Apply clean engine oil to the outer circumference of the pilot tool.

20. Install the front case onto the engine block and temporarily tighten the flange bolts (other than those for tightening the filter bracket). Mount the oil filter bracket with new gasket in place. Install the 4 bolts with washers and tighten to 16 ft. lbs. (22 Nm).

21. Insert the Phillips screwdriver into the hole on the side of the engine block. Secure the oil pump driven gear onto the left silent shaft by tightening the driven gear flange bolt to 29 ft. lbs. (40 Nm).

22. Install a new O-ring onto the groove in the front case. Using special socket tool, install and tighten the plug cap to 20 ft. lbs. (27 Nm).

23. Install the oil pump relief plunger and spring into the bore in the oil filter bracket and tighten to 36 ft. lbs. (50 Nm). Make sure a new gasket is in place.

24. Clean both mating surfaces of the oil pan and the cylinder block. Apply sealant in the groove in the oil pan flange, keeping towards the inside of the bolt holes. The width of the sealant bead applied is to be about 0.16 in. (4mm) wide.

NOTE: After applying sealant to the oil pan, do not exceed 15 minutes before installing the oil pan.

25. Install the oil pan to the engine and secure with the retainers. Tighten bolts to 6 ft. lbs. (8 Nm).

26. Install the oil pressure gauge unit and the oil pressure switch. Connect the electrical harness connector.

27. Install the oil cooler secure with oil cooler bolt tightened to 33 ft. lbs. (45 Nm).

28. Install new oil filter and fill engine with clean engine oil.

29. Install the timing belts and all related items.

NOTE: The timing of the oil pump sprocket and connected silent shaft can be incorrect, even with the timing mark aligned. Make certain that all special timing belt installation procedures are followed to ensure proper orientation of the silent shafts.

30. Install any remaining components removed during disassembly.

31. Connect the negative battery cable and start the engine. Check for proper timing and inspect for leaks.

1994–97 Galant 2.4L (4G64) Engine

NOTE: A special oil seal guide tool, MD998285, and a plug cap socket tool, MD998162, or exact equivalents are needed to complete this operation.

1. Disconnect the negative battery cable.

2. Remove the oil filter, oil pressure switch, oil gauge sending unit, oil filter mounting bracket and gasket.

3. Raise and safely support the vehicle.

4. Drain the engine oil. Remove engine oil pan, oil screen and gasket. Remove the relief plug, gasket, relief spring and relief plunger.

5. Lower the vehicle. Using the proper equipment, support the weight of the engine.

6. Remove the front engine mount bracket and accessory drive belt.

7. Remove the timing belts and sprockets.

8. Using special tool MD998162, remove the plug cap in the engine front cover.

9. Remove the plug on the side of the engine block. Insert a Phillips screwdriver with a shank diameter of 0.32 in. (8mm) into the plug hole. This will hold the silent shaft.

10. Remove the driven gear bolt that secures the oil pump driven gear to the silent shaft.

11. Remove and tag the front cover mounting bolts. Note the lengths of the mounting bolts as they are removed for proper installation.

12. Remove the front case cover and oil pump assembly. If necessary, the silent shaft can come out with the cover assembly.

13. Remove the silent shaft oil seals, the crankshaft oil seal and front case gasket.

14. Remove the silent shafts and inspect as follows:
   a. Check the oil holes in the shaft for clogging.
   b. Check journals of the shaft for seizure, damage and contact with bearing. If there is anything wrong with the journal, replace the silent shaft bearing, silent shaft or front case.
   c. Check the silent shaft oil clearance. If the clearance is beyond the specifications, replace the silent shaft bearing, silent shaft or

front case. The specifications for oil clearances are as follows:
   - **Right shaft**
   - Front — 0.0012–0.0024 in. (0.03–0.06mm)
   - Rear — 0.0008–0.0021 in. (0.02–0.05mm)
   - **Left shaft**
   - Front — 0.0020–0.0036 in. (0.05–0.09mm)
   - Rear — 0.0017–0.0033 in. (0.04–0.08mm)

To install:

15. Lubricate the bearing surface of the shaft and the bearing journals with clean engine oil. Carefully install the silent shafts to the block.

16. Clean the gasket material from the mating surface of the cylinder block and the engine front cover. Install new gasket in place.

17. Install the oil pump drive gear and driven gear to the front case, lining up the timing marks. Lubricate the gears with clean engine oil. Install the oil pump cover, with new gasket in place and tighten the mounting bolts to 13 ft. lbs. (18 Nm).

18. Using proper size driver, install the crankshaft oil seal into the front engine case.

19. Using the proper size socket wrench, press in the silent shaft oil seal into the front case.

20. Place pilot tool MD998285 or equivalent, onto the nose of the crankshaft. Apply clean engine oil to the outer circumference of the pilot tool.

21. Install the front case onto the engine block and temporarily tighten the flange bolts (other than those for tightening the filter bracket). Mount the oil filter bracket with new gasket in place. Install the 4 bolts with washers and tighten to 16 ft. lbs. (22 Nm).

22. Insert the Phillips screwdriver into the hole on the side of the engine block. Secure the oil pump driven gear onto the left silent shaft by tightening the driven gear flange bolt to 29 ft. lbs. (40 Nm).

23. Install a new O-ring onto the groove in the front case. Using special socket tool, install and tighten the plug cap to 20 ft. lbs. (27 Nm).

24. Install the oil pump relief plunger and spring into the bore in the oil filter bracket and tighten to 36 ft. lbs. (50 Nm). Make sure a new gasket is in place.

25. Clean both mating surfaces of the oil pan and the cylinder block. Apply sealant in the groove in the oil pan flange, keeping towards the inside of the bolt holes. The width of

the sealant bead applied is to be about 0.16 in. (4mm) wide.

NOTE: After applying sealant to the oil pan, do not exceed 15 minutes before installing the oil pan.

26. Install the oil pan to the engine and secure with the retainers. Tighten bolts to 6 ft. lbs. (8 Nm).
27. Install the oil pressure gauge unit and the oil pressure switch. Connect the electrical harness connector.
28. Install new oil filter and fill engine with clean engine oil.
29. Install the timing belts and all related items.

NOTE: The timing of the oil pump sprocket and connected silent shaft can be incorrect, even with the timing mark aligned. Make certain that all special timing belt installation procedures are followed to ensure proper orientation of the silent shafts.

30. Install any remaining components removed during disassembly.
31. Connect the negative battery cable and start the engine. Check for proper timing and inspect for leaks.

Piston and Connecting Rod

POSITIONING

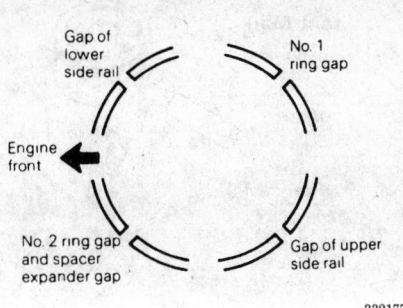

Piston ring end gap positions — Precis

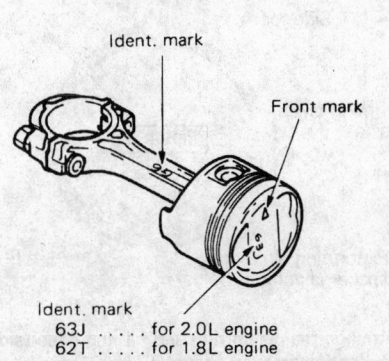

Piston identification marks — Precis

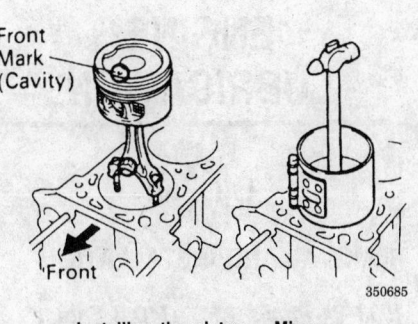

Installing the piston — Mirage

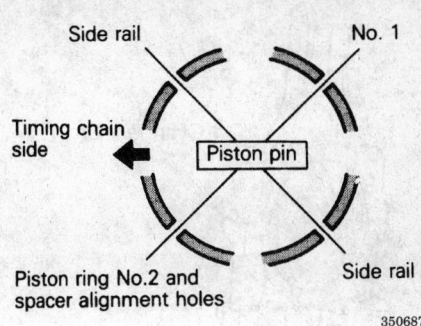

Arrange the piston ring and oil ring gaps (side rail and spacer) as shown — Mirage

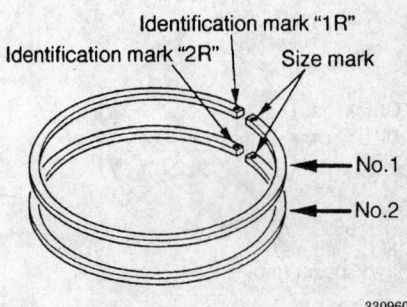

Piston ring identification — Eclipse

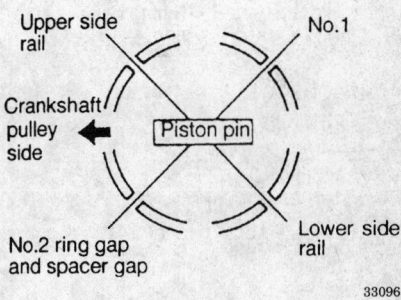

Piston ring end gap — Eclipse

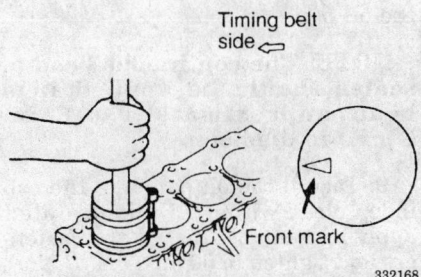

Position of the piston in the cylinder block — Precis and Eclipse

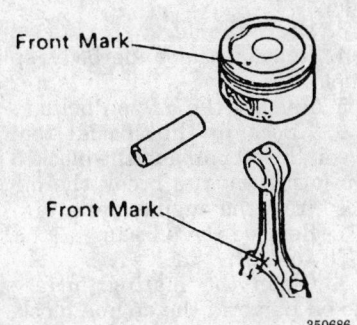

Aligning the piston and the connecting rod — Mirage

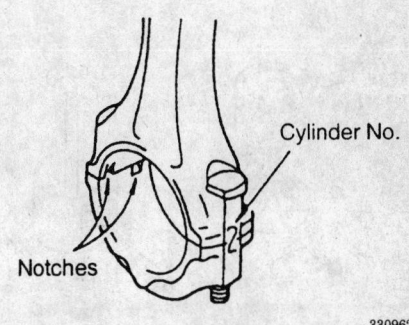

Correct connecting rod installation — Precis and Eclipse

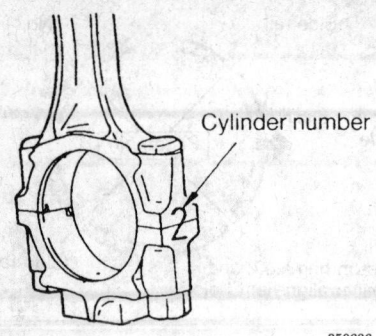

Connecting rod matchmarks — Galant

350636

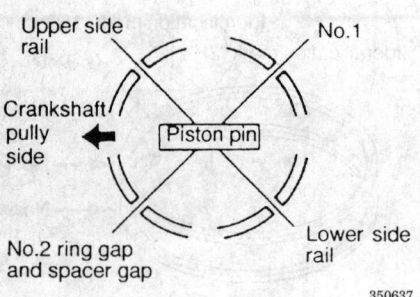

Piston ring positioning — Galant

350637

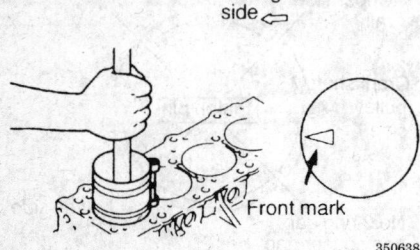

Position of the piston in the cylinder block — Galant

350638

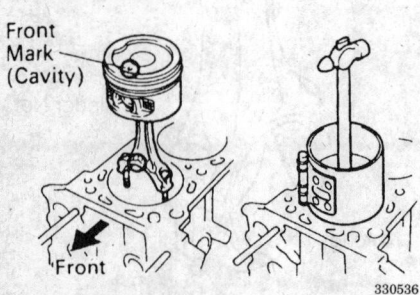

Installing the piston — 3000GT and Diamante

330536

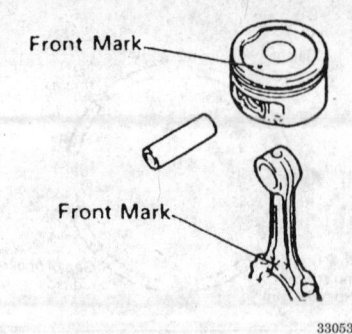

Aligning the piston and the connecting rod — 3000GT and Diamante

330537

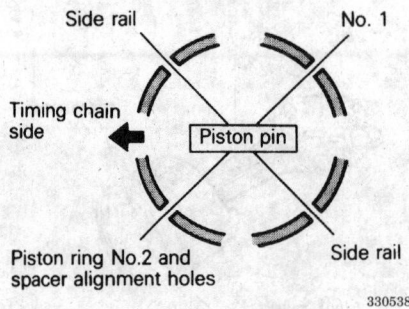

Arrange the piston ring and oil ring gaps (side rail and spacer) as shown — 3000GT and Diamante

330538

ENGINE LUBRICATION

Oil Pan

REMOVAL AND INSTALLATION

1993–94 Precis 1.5L (G4DJ) Engine

1. Disconnect the negative battery cable.
2. Raise the vehicle and support it safely.
3. Drain the oil.
4. Remove the underbody splash shield.
5. Remove the oil pan bolts.
6. Knock in the special tool between the oil pan and the block. Then tap it sideways to break the oil pan loose from the engine block.
7. Remove the oil pan.

To install:

8. Clean the mating surfaces of the oil pan and the engine block.
9. Apply a ⅛ in. (3mm) bead of RTV sealer along the groove in the oil pan.

10. Install the oil pan. Hand tighten the retaining bolts.
11. Starting on one side of the pan, gradually tighten the retaining bolts to 4–6 ft. lbs. (6–8 Nm), in a crisscross pattern switching from one side to the other.
12. Lower the engine and tighten the mount retaining nuts.
13. Install the oil pan drain plug.
14. Install the splash shield and lower the vehicle.
15. Refill the crankcase with oil. Start the engine and check for leaks.

Mirage 1.5L (4G15) Engine

1. Disconnect the negative battery cable.
2. Raise the vehicle and support safely.
3. Remove the oil pan drain plug and drain the engine oil.
4. Remove the bell housing lower cover.
5. Remove the oil pan retainer bolts. Tap the oil pan with a rubber mallet to break seal.

NOTE: Do not use a chisel, screwdriver or similar tool when removing the oil pan. Damage to engine components may occur. If available, oil pan remover tool MD998727 or equivalent may be used break the seal.

6. Inspect the oil pan for damage and cracks. Replace if faulty. While the pan is removed, inspect the oil screen for clogging, damage and cracks. Replace if faulty.

To install:

7. Using a wire brush or other tool, scrape clean all gasket surfaces of the cylinder block and the oil pan so that all loose material is removed. Clean sealing surfaces of all dirt and oil.
8. Apply sealant around the gasket surfaces of the oil pan in such a manner that all bolt holes are circled and there is a continuous bead of sealer around the entire perimeter of the oil pan.

NOTE: The continuous bead of sealer should be applied in a bead approximately 0.16 in. (4mm) in diameter.

9. Install the oil pan onto the cylinder block within 15 minutes after applying sealant. Install the fasteners and tighten to 60 inch lbs. (7 Nm).
10. Install the bellhousing cover.
11. Install the oil drain plug with a new seal and tighten to 29 ft. lbs. (40 Nm).

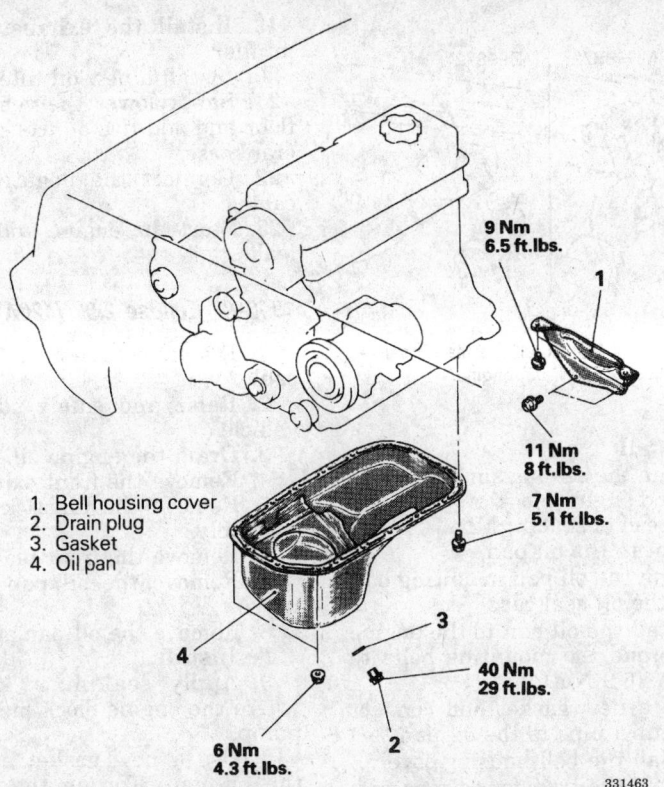

9 Nm
6.5 ft.lbs.

11 Nm
8 ft.lbs.

7 Nm
5.1 ft.lbs.

40 Nm
29 ft.lbs.

6 Nm
4.3 ft.lbs.

1. Bell housing cover
2. Drain plug
3. Gasket
4. Oil pan

331463

Oil pan and related components — Mirage 1.5L (4G15) engine

12. Lower the vehicle and fill the crankcase to the proper level with clean engine oil.

13. Connect the negative battery cable. Start the engine and check for leaks.

Mirage 1.8L (4G93) Engine

1. Disconnect the negative battery cable.

2. Raise the vehicle and support safely.

3. Remove the oil pan drain plug and drain the engine oil.

4. Disconnect and lower the exhaust pipe from the engine manifold.

5. Remove the bell housing lower cover.

6. Remove the oil pan retainer bolts. Tap the oil pan with a rubber mallet to break the seal.

NOTE: Do not use a chisel, screwdriver or similar tool when removing the oil pan. Damage to engine components may occur. If available, oil pan remover tool MD998727 or equivalent may be used break the seal.

7. Inspect the oil pan for damage and cracks. Replace if faulty. While the pan is removed, inspect the oil screen for clogging, damage and cracks. Replace if faulty.

To install:

8. Using a wire brush or other tool, scrape clean all gasket surfaces of the cylinder block and the oil pan so that all loose material is removed. Clean sealing surfaces of all dirt and oil.

9. Apply sealant around the gasket surfaces of the oil pan in such a manner that all bolt holes are circled and there is a continuous bead of sealer around the entire outside edge of the oil pan.

NOTE: The continuous bead of sealer should be applied in a bead approximately 0.16 in. (4mm) in diameter.

10. Install the oil pan onto the cylinder block within 15 minutes after applying sealant. Install the fasteners and tighten to 60 inch lbs. (5 Nm).

11. Install the bellhousing cover.

12. Connect the exhaust pipe from the engine manifold with new gasket in place. Tighten the exhaust pipe to manifold flange nuts to 33 ft. lbs. (45 Nm). Install and tighten the support bolt to 18 ft. lbs. (25 Nm).

13. Install the oil drain plug and tighten to 29 ft. lbs. (40 Nm).

14. Lower the vehicle and fill the crankcase to the proper level with clean engine oil.

15. Connect the negative battery cable. Start the engine and check for leaks.

Eclipse 1.8L (4G37) and 2.0L (4G63) Engines

1. Disconnect the negative battery cable.

2. Raise the vehicle and support safely.

3. Remove the oil pan drain plug and drain the engine oil.

4. Disconnect and lower the exhaust pipe from the engine manifold.

5. Using the appropriate equipment, support the weight of the engine.

6. On FWD models, remove the retainer bolts and the center crossmember and AWD models, remove the left member.

7. Remove the oil pan bolts. Tap in thin prybar between the engine block and the oil pan.

NOTE: Do not use a chisel, screwdriver or similar tool when removing the oil pan. Damage to engine components may occur.

8. Inspect the oil pan for damage and cracks. Replace if faulty. While the pan is removed, inspect the oil screen for clogging, damage and cracks. Replace if faulty.

To install:

9. Using a wire brush or other tool, scrape clean all gasket surfaces of the cylinder block and the oil pan so that all loose material is removed. Clean sealing surfaces of all dirt and oil.

10. Apply sealant around the gasket surfaces of the oil pan in such a manner that all bolt holes are circled and there is a continuous bead of sealer around the entire perimeter of the oil pan.

NOTE: The continuous bead of sealer should be applied in a bead approximately 0.16 in. (4mm) in diameter.

11. Install the oil pan onto the cylinder block within 15 minutes after applying sealant. Install the fasteners and tighten to 4–6 ft. lbs. (6–8 Nm).

12. On FWD models, install the crossmember and tighten the crossmember mounting bolts to 72 ft. lbs. (100 Nm).

13. On AWD models, install the left member and tighten the forward retainer bolts to 72 ft. lbs. (100 Nm). Tighten the rearward left member bolts to 58 ft. lbs. (80 Nm).

14. Connect the exhaust pipe from the engine manifold with new gasket

in place. Tighten the exhaust pipe to manifold flange nuts to 29 ft. lbs. (40 Nm). Install and tighten the support bolt to 29 ft. lbs. (40 Nm).

15. Install the oil drain plug and tighten to 33 ft. lbs., If not already done.

16. Lower the vehicle and fill the crankcase to the proper level with clean engine oil.

17. Connect the negative battery cable. Start the engine and check for leaks.

1995–97 Eclipse 2.0L (4G63) Engine

1. Disconnect the negative battery cable.

2. Safely raise and support the vehicle.

3. Remove the front exhaust pipe.

4. Remove the exhaust pipe and muffler assembly.

5. Drain the engine oil.

6. Remove the dipstick and tube.

7. Remove the transfer case assembly (AWD).

8. Remove the bell housing cover.

9. Disconnect the oil return pipe from the oil pan.

10. Remove the oil pan mounting bolts. Tap the oil pan seal breaker MB998727 or equivalent between the oil pan and the engine block to break the seal and remove the oil pan.

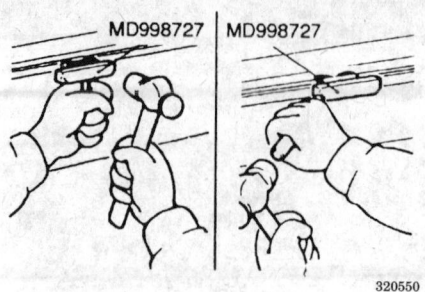

Using the special tool to remove the oil pan — 1995–97 Eclipse 2.0L (4G63) engine

To install:

11. Clean the sealing surface on the oil pan and engine block. Apply a continuous bead of sealant MD970389 or equivalent to the oil pan.

12. Clean the oil pan mounting bolt holes in the oil seal case.

13. Install the oil pan to the engine block. Torque the mounting bolts to 5.1 ft. lbs. (6.9 Nm).

14. Use a new gasket and connect the oil return pipe to the oil pan.

15. Install the bell housing cover.

16. Install the transfer case assembly if equipped.

17. Install the dipstick and tube assembly.

18. Install the front exhaust pipe.

19. Install the exhaust pipe and muffler.

20. Install a new oil filter.

21. Safely lower the vehicle to the floor and add five quarts of oil to the crankcase.

22. Connect the negative battery cable.

23. Start the engine and check for leaks.

1995–97 Eclipse 2.0L (420A) Engine

1. Disconnect the negative battery cable.

2. Raise and safely support the vehicle.

3. Drain the engine oil.

4. Remove the front exhaust pipe.

5. Remove the dipstick and tube assembly.

6. Remove the front plate.

7. Remove the oil pan mounting bolts.

8. Remove the oil pan and gasket.

To install:

9. Apply sealant at the point where the engine block meets the oil pump.

10. Use a new gasket and install the oil pan. Torque the mounting bolts to 8.9 ft. lbs. (12 Nm).

11. Install the front plate.

12. Install the front exhaust pipe.

13. Install the dipstick and tube assembly.

14. Safely lower the vehicle to the floor.

15. Refill the crankcase with oil to the proper level.

16. Connect the negative battery cable.

17. Start the engine and check for leaks.

1996–97 Eclipse 2.4L (4G64) Engine

1. Remove the negative battery cable. Wait at least 90 seconds before performing any work.

2. Drain the engine oil.

3. Remove the engine dipstick and tube assembly.

4. Remove the front exhaust pipe.

5. Remove the bell housing inspection cover.

6. Remove the bolts attaching the oil pan to the cylinder block.

7. Using the special tool remove the oil pan assembly.

To install:

8. Make sure that the oil pan and cylinder block mating surfaces are free of any old sealing material. Apply the specified sealant or the equivalent.

9. Install the oil pan to the cylinder block and torque the bolts to 5 ft. lbs. (7 Nm).

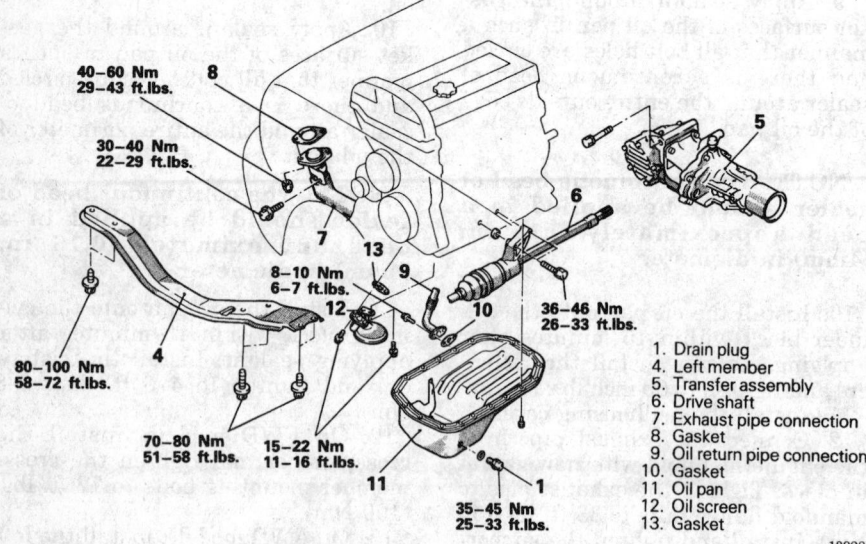

40–60 Nm
29–43 ft.lbs.

30–40 Nm
22–29 ft.lbs.

8–10 Nm
6–7 ft.lbs.

80–100 Nm
58–72 ft.lbs.

70–80 Nm
51–58 ft.lbs.

15–22 Nm
11–16 ft.lbs.

36–46 Nm
26–33 ft.lbs.

35–45 Nm
25–33 ft.lbs.

1. Drain plug
2. —
3. —
4. Left member
5. Transfer assembly
6. Drive shaft
7. Exhaust pipe connection
8. Gasket
9. Oil return pipe connection
10. Gasket
11. Oil pan
12. Oil screen
13. Gasket

AWD oil pan removal — 1993–94 Eclipse 2.0L (4G63) engine

10. Reinstall the bell housing inspection cover. Torque the bolts to 7 ft. lbs. (9 Nm).

11. Install the front exhaust pipe.

12. Reinstall the engine dipstick and tube assembly using a new O-ring.

13. Refill the engine with oil. Reconnect the negative battery cable. Start the engine and check for leaks.

1993 Galant 2.0L (4G63) Engine

FWD

1. Disconnect the negative battery cable.

2. Raise the vehicle and support safely.

3. Remove the oil pan drain plug and drain the engine oil.

4. Disconnect and lower the exhaust pipe from the engine manifold.

5. Using the appropriate equipment, support the weight of the engine.

6. Remove the retainer bolts and the center crossmember.

7. Remove the oil pan bolts. Using special tool MD998727, tap in between the engine block and the oil pan.

NOTE: Do not use a chisel, screwdriver or similar tool when removing the oil pan. Damage to engine components may occur.

8. Inspect the oil pan for damage and cracks. Replace if faulty. While the pan is removed, inspect the oil screen for clogging, damage and cracks. Replace if faulty.

To install:

9. Using a wire brush or other tool, scrape clean all gasket surfaces of the cylinder block and the oil pan so that all loose material is removed. Clean sealing surfaces of all dirt and oil.

10. Apply sealant around the gasket surfaces of the oil pan in such a manner that all bolt holes are circled and there is a continuous bead of sealer around the entire perimeter of the oil pan.

NOTE: The continuous bead of sealer should be applied in a bead approximately 0.16 in. (4mm) in diameter.

11. Install the oil pan onto the cylinder block within 15 minutes after applying sealant. Install the fasteners and tighten to 4–6 ft. lbs. (6–8 Nm).

12. Install the crossmember and tighten the mounting bolts to 72 ft. lbs. (100 Nm).

13. Connect the exhaust pipe from the engine manifold with new gasket in place. Tighten the exhaust pipe to manifold flange nuts to 29 ft. lbs. (40 Nm).

14. Install the oil drain plug and tighten to 33 ft. lbs.

15. Lower the vehicle and fill the crankcase to the proper level with clean engine oil.

16. Connect the negative battery cable. Start the engine and check for leaks.

AWD

1. Disconnect the negative battery cable.

2. Raise the vehicle and support safely.

3. Remove the oil pan drain plug and drain the engine oil.

4. Disconnect and lower the exhaust pipe from the engine manifold.

5. Remove the transfer assembly and right driveshaft.

6. Using the appropriate equipment, support the weight of the engine and remove the center crossmember.

7. Disconnect the return pipe for the turbocharger from the side of the oil pan.

8. Remove the oil pan bolts. Using special tool MD998727, tap in between the engine block and the oil pan.

NOTE: Do not use a chisel, screwdriver or similar tool when removing the oil pan. Damage to engine components may occur.

9. Inspect the oil pan for damage and cracks. Replace if faulty. While the pan is removed, inspect the oil screen for clogging, damage and cracks. Replace if faulty.

To install:

10. Using a wire brush or other tool, scrape clean all gasket surfaces of the cylinder block and the oil pan so that all loose material is removed. Clean sealing surfaces of all dirt and oil.

11. Apply sealant around the gasket surfaces of the oil pan in such a manner that all bolt holes are circled and there is a continuous bead of sealer around the entire perimeter of the oil pan.

NOTE: The continuous bead of sealer should be applied in a bead approximately 0.16 in. (4mm) in diameter.

12. Install the oil pan onto the cylinder block within 15 minutes after applying sealant. Install the fasteners and tighten to 4–6 ft. lbs. (6–8 Nm).

13. Install the oil return pipe using a new gasket, if removed. Tighten retainers to 5–7 ft. lbs. (7–10 Nm).

14. Install the left member and tighten the forward retainer bolts to 72 ft. lbs. (100 Nm). Tighten the rearward left member bolts to 58 ft. lbs. (80 Nm).

15. Install the transfer assembly and right driveshaft.

16. Connect the exhaust pipe from the engine manifold with new gasket in place. Tighten the exhaust pipe to manifold flange nuts to 29 ft. lbs. (40 Nm).

17. Install the oil drain plug and tighten to 33 ft. lbs.

18. Lower the vehicle and fill the crankcase to the proper level with clean engine oil.

19. Connect the negative battery cable. Start the engine and check for leaks.

1994–97 Galant 2.4L (4G64) Engine

1. Disconnect the negative battery cable.

2. Raise the vehicle and support safely.

3. Remove the oil pan drain plug and drain the engine oil.

4. Remove the oil dipstick and tube assembly.

5. Disconnect the oxygen sensor connector. Remove the nuts and disconnect the front exhaust pipe from the exhaust manifold. Disconnect the front exhaust pipe from the catalytic converter and remove the front exhaust pipe from the vehicle.

6. Remove the four bolts securing the bellhousing cover, to the engine and transmission.

7. Remove the oil pan retainer bolts. Using special tool MD998727 or equivalent, tap in between the engine block and the oil pan.

NOTE: Do not use a chisel, screwdriver or similar tool when removing the oil pan. Damage to engine components may occur.

8. Inspect the oil pan for damage and cracks. Replace if faulty. While the pan is removed, inspect the oil screen for clogging, damage and cracks. Replace if faulty.

To install:

9. Using a wire brush or other tool, scrape clean all gasket surfaces of the cylinder block and the oil pan so that all loose material is removed. Clean sealing surfaces of all dirt and oil.

10. Apply sealant around the gasket surfaces of the oil pan in such a manner that all bolt holes are circled and there s a continuous bead of

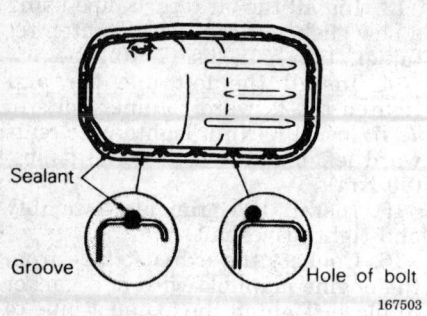

Sealant

Groove

Hole of bolt

167503

Applying sealant to oil pan — 1993 Galant 2.0L (4G63) engine

sealer around the entire perimeter of the oil pan.

NOTE: The continuous bead of sealer should be applied in a bead approximately 0.16 in. (4mm) in diameter.

11. Install the oil pan onto the cylinder block within 15 minutes after applying sealant. Install the fasteners and tighten to 6 ft. lbs. (8 Nm).

12. Install the oil drain plug and tighten to 29 ft. lbs. (39 Nm), If not already done.

13. Install the bellhousing cover, and tighten the mounting bolts to 7 ft. lbs. (9 Nm).

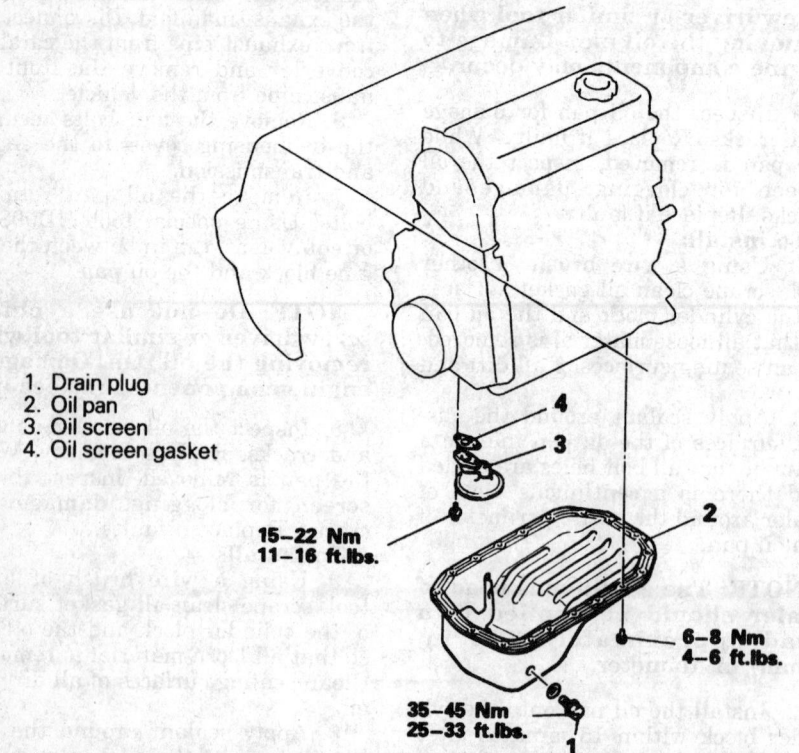

1. Drain plug
2. Oil pan
3. Oil screen
4. Oil screen gasket

15–22 Nm
11–16 ft.lbs.

6–8 Nm
4–6 ft.lbs.

35–45 Nm
25–33 ft.lbs.

167510

Oil pan and related components — 1993 Galant 2.0L (4G63) engine

14. Reconnect the front exhaust pipe to the catalytic converter with a new gasket and torque the bolts to 36 ft. lbs. (49 Nm). Reinstall the front exhaust pipe with a new gasket to the exhaust manifold and torque the nuts to 32 ft. lbs. (44 Nm). Reconnect the oxygen sensor connector.

15. Lower the vehicle and fill the crankcase to the proper level with clean engine oil.

16. Connect the negative battery cable. Start the engine and check for leaks.

Diamante 3.0L (6G72) Engines

1. Disconnect the negative battery cable.

2. Raise the vehicle and support safely.

3. Remove the oil pan drain plug and drain the engine oil.

4. Remove the left side crossmember. If equipped with 4WS, it will also be necessary to remove the right side crossmember.

5. Remove the starter motor.

6. Disconnect the roll stopper stay bracket, from the rear transaxle stay bracket. Remove the both transaxle stay brackets.

7. Remove the bellhousing lower cover.

8. Remove the oil pan mounting bolts. Using special tool MD998727 or equivalent, separate and remove the engine oil pan.

To install:

9. Thoroughly clean and dry the oil pan, cylinder block bolts and bolt holes.

10. Apply a 0.16 inch (4mm) continuous bead of sealer around the surface of the oil pan.

NOTE: Assemble the oil pan to the cylinder block within 15 minutes after applying the sealant.

11. Install the oil pan mounting bolts. Following proper sequence, torque mounting bolts to 48 inch lbs. (6 Nm).

12. Install lower bellhousing cover and the starter motor.

13. Install the transaxle stay brackets and connect the roll stopper bracket.

14. Install the crossmember(s) and tighten the mounting bolts to 43–51 ft. lbs. (60–70 Nm).

15. Fill the engine with the proper amount of oil.

16. Connect the negative battery cable and check for leaks.

3000GT 3.0L (6G72) Engines

1. Disconnect the negative battery cable.

2. Raise the vehicle and support safely.

3. Remove the oil pan drain plug and drain the engine oil.

4. On vehicles equipped with AWD, remove the transfer assembly.

5. Disconnect and lower the exhaust pipe and on turbocharged engines, disconnect the return pipe for the turbocharger from the side of the oil pan.

6. Remove the oil pan mounting bolts.

7. Using the special tool, separate and remove the engine oil pan.

To install:

8. Thoroughly clean and dry the oil pan, cylinder block bolts and bolt holes.

9. Apply a thin bead of sealer around the surface of the oil pan.

10. Assemble the oil pan to the cylinder block within 15 minutes after applying the sealant.

11. Install the oil pan mounting bolts and torque to 4–6 ft. lbs. (6–8 Nm).

12. Fill the engine with the proper amount of oil.

13. Connect the negative battery cable and check for leaks.

14. Safely lower the vehicle to the floor.

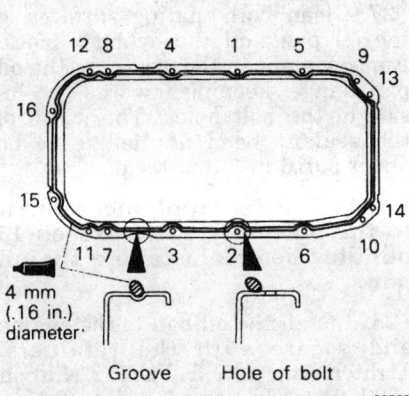

Oil pan bolt tightening sequence and application of sealant to the pan — Diamante 3.0L (J-6G72 and H-6G72) engines

Oil Pump

REMOVAL AND INSTALLATION

1993–94 Precis 1.5L (G4DJ) Engine

NOTE: Whenever the oil pump is disassembled or the cover removed, the gear cavity must be filled with petroleum jelly for priming purposes. Do not use grease.

1. Disconnect the negative battery cable. Remove the timing belt.
2. Remove the oil pan.
3. Remove the oil screen.
4. Unbolt and remove the front case assembly.
5. Remove the oil pump cover.
6. Remove the inner and outer gears from the front case.

NOTE: The outer gear has no identifying marks to indicate direction of rotation. Clean the gear and mark it with an indelible marker.

7. Remove the plug, relief valve spring and relief valve from the case.
To install:
8. Check the front case for damage or cracks. Replace the front seal. Replace the oil screen O-ring. Clean all parts thoroughly with a safe solvent.

9. Check the pump gears for wear or damage. Clean the gears thoroughly and place them in their original position in the case to check the clearances. There is a crescent-shaped piece between the 2 gears.
10. Check that the relief valve can slide freely in the case.
11. Check the relief valve spring for damage. The relief valve free length should be 1.8 in. (47mm). Load length should be 13.4 lbs. at 1.6 in. (40mm).
12. Thoroughly coat both oil pump gears with clean engine oil and install them in the correct direction of rotation.
13. Install the pump cover and torque the bolts to 6–8 ft. lbs. (8–12 Nm).
14. Coat the relief valve and spring with clean engine oil, install them and tighten the plug to 30–36 ft. lbs. (39–49 Nm).
15. Position a new front case gasket, coated with sealer, on the engine and install the front case. Torque the bolts to 8–11 ft. lbs. (12–15 Nm). Note that the bolts have different shank lengths. Use the following guide to determine which bolts go where. Bolts marked:
   - A: 0.8 in. (20mm)
   - B: 1.2 in. (30mm)
   - C: 2.4 in. (60mm)
16. Coat the lips of a new seal with clean engine oil and slide it along the crankshaft until it touches the front case. Drive it into place with a seal driver.
17. Install the sprocket, timing belt and pulley.
18. Install the oil screen.
19. Thoroughly clean both the oil pan and engine mating surfaces. Apply a 1/8 in. (3mm) wide bead of RTV sealer in the groove of the oil pan mating surface.
20. Tighten the oil pan bolts to 60–72 inch lbs. Connect the negative battery cable.

Mirage 1.5L (4G15) and 1.8L (4G93) Engine

NOTE: Whenever the oil pump is disassembled or the cover removed, it is suggested the gear cavity is filled with petroleum jelly for priming purposes. Do not use grease.

1. Disconnect the negative battery cable.
2. Remove the front engine mount bracket and accessory drive belts.
3. Remove timing belt upper and lower covers.
4. Remove the timing belt and crankshaft sprocket.

5. Remove the oil pan and remove the oil screen.
6. Remove and tag the front cover mounting bolts. Note the lengths of the mounting bolts as they are removed for proper installation.
7. Remove the front case assembly and oil pump assembly.
8. Remove the oil pump cover.
9. Remove the inner and outer gears from the front case.

NOTE: The outer gear has no identifying marks to indicate direction of rotation. Clean the gear and mark it with an indelible marker.

10. Check the front case for damage or cracks. Replace the front seal. Replace the oil screen O-ring. Clean all parts thoroughly with a safe solvent. Check the pump gears for wear or damage, and ensure that the relief valve can slide freely in the case.
To install
11. Remove all gasket material from the mating surfaces and clean all parts.
12. Thoroughly coat both oil pump gears with clean engine oil and install them in the correct direction of rotation.
13. Install the pump cover and torque the bolts to 84 inch lbs. (10 Nm).
14. Coat the relief valve and spring with clean engine oil, install them and tighten the plug to 33 ft. lbs. (45 Nm).
15. Install a new front crankshaft seal and coat the lips of the seal with clean engine oil.
16. Install the front case and oil pump assembly to the engine block using a new gasket. Use the noted locations of the mounting bolts for proper positioning and tighten the bolts to 10 ft. lbs. (14 Nm).
17. Install the oil screen with new gasket. Torque the screen bolts to 14 ft. lbs. (19 Nm).
18. Install the oil pan.
19. Install the sprocket, timing belt and pulley.
20. Connect the battery cable refill the engine oil, run the engine and check for leaks.

Eclipse 2.0L (4G63), 2.4L (4G64) and 1995–97 Galant 2.4L (4G64) Engines

NOTE: Whenever the oil pump is disassembled or the cover removed, the gear cavity must be filled with petroleum jelly. This seals the pump and acts like a primer so the oil pump draws oil as soon as the engine turns. Do not use grease.

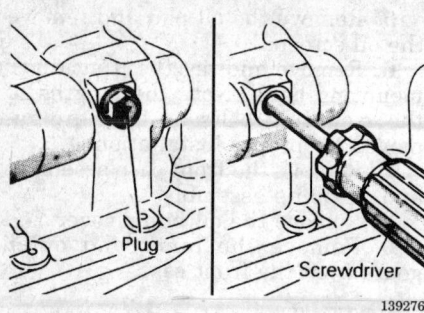

Holding silent shaft for oil pump gear removal — 1993–94 Eclipse 2.0L (4G63 and 4G63) engine

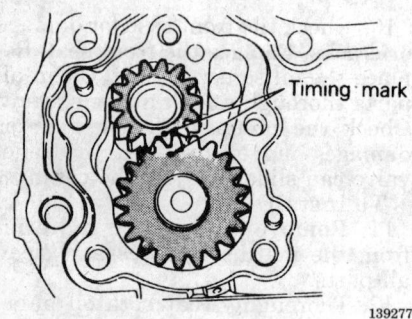

Timing mark

Aligning oil pump timing marks — 1993–94 Eclipse 2.0L (4G63 and 4G63) engine

1. Disconnect the negative battery cable. Rotate the engine so No. 1 cylinder is on Top Dead Center (TDC) of its compression stroke. The timing marks should be aligned at this point.

2. Raise and safely support the vehicle.

3. Drain the engine oil. Lower the vehicle.

4. Using the proper equipment, support the weight of the engine. Remove the front engine mount bracket and accessory drive belts.

5. Remove timing belt upper and lower covers.

6. Remove the timing belt and crankshaft sprocket.

7. Disconnect the electrical connector from the oil pressure sending unit and remove the oil pressure sensor. Remove the oil filter and the oil filter bracket.

8. Remove the oil pan, oil screen and gasket.

9. Using special tool MD998162, remove the plug cap in the engine front cover.

10. Remove the plug on the side of the engine block. Insert a Phillips screwdriver with a shank diameter of 0.32 in. (8mm) into the plug hole. This will hold the silent shaft.

11. Remove the driven gear bolt that secures the oil pump driven gear to the silent shaft.

12. Remove and tag the front cover mounting bolts. Note the lengths of the mounting bolts as they are removed for proper installation.

13. Remove the front case cover and oil pump assembly. If necessary, the silent shaft can come out with the cover assembly.

14. Remove the oil pump cover, located on the back of the engine front cover. Remove the oil pump drive and driven gears.

15. After disassembling the oil pump, clean all components and remove gasket material from mating surfaces.

16. Assemble the oil pump gears into the front case and rotate it to ensure smooth rotation and no looseness. Make sure there is no ridge wear on the contact surface between the front case and the gear surface of the oil pump front cover.

To install

17. Align the timing mark on the oil pump drive gear with that on the driven gear and install them into the engine front case. Apply engine oil to the gears.

18. Install the oil pump cover and tighten the retainer bolts to 13 ft. lbs. (18 Nm) on Eclipse models and 17 ft. lbs. (24 Nm) on Galant models.

19. Using the appropriate driver, install a new crankshaft seal into the front case.

20. Position new front case gasket in place. Set seal guide tool MD998285 on the front end of the crankshaft to protect the seal from damage. Apply a thin coat of oil to the outer circumference of the seal pilot tool.

21. Install the front case assembly through a new front case gasket and temporarily tighten the flange bolts.

22. Mount the oil filter on the bracket with new oil filter bracket gasket in place. Install the bolts with washers and tighten to 25 ft. lbs. (34 Nm) on 1993–94 models and 14 ft. lbs. (19 Nm) on 1995–97 models

23. Insert a Phillips screwdriver into a hole in the left side of the engine block to lock the silent shaft in place.

24. Secure the oil pump drive gear onto the left silent shaft by installing and tightening the driven gear bolt to 29 ft. lbs. (40 Nm) on 1993–94 models and 27 ft. lbs. (37 Nm) on 1995–97 models.

25. Install a new O-ring to the groove in the front case and install the plug cap. Using the special tool

MD998162, tighten the cap to 20 ft. lbs. (27 Nm) on 1993–94 models and 17 ft. lbs. (24 Nm) on 1995–97 models.

26. Install the oil screen in position with new gasket in place.

27. Clean both mating surfaces of the oil pan and the cylinder block. Apply sealant in the groove in the oil pan flange, keeping towards the inside of the bolt holes. The width of the sealant bead applied is to be about 0.016 in. (4mm) wide.

NOTE: After applying sealant to the oil pan, do not exceed 15 minutes before installing the oil pan.

28. Install the oil pan to the engine and secure with the retainers. Tighten bolts to 9 ft. lbs. (12 Nm) on 1993–94 models and 5 ft. lbs. (7 Nm) on 1995–97 models.

29. Install the oil pressure gauge unit and the oil pressure switch. Connect the electrical harness connector.

30. Install the oil cooler. Secure with oil cooler bolt tightened to 33 ft. lbs. (45 Nm) on 1993–94 models and 31 ft. lbs. (43 Nm) on 1995–97 models.

31. Refill the crankcase. Install new oil filter.

32. Connect the negative battery cable and start the engine. Verify oil pressure. Inspect for leaks.

To install

1995–97 Eclipse 2.0L (420A) Engine

1. Disconnect the negative battery cable.

2. Raise the safely support the vehicle.

3. Drain the engine oil.

4. Remove the rear plate.

5. Remove the oil filter and adapter.

6. Remove the oil pan.

7. Remove the oil pick-up tube.

8. Remove the timing belt.

9. Using tool MB995027 or equivalent, remove the crankshaft sprocket.

—— **WARNING** ——
Do not nick the crankshaft sealing surface or the seal bore.

10. Using tool MB995020 or equivalent, remove the crankshaft oil seal.

11. Remove the oil pump mounting bolts.

12. Remove the oil pump.

To install:

13. Apply a bead of the specified sealant to the sealing surface of the oil pump and install a new O-ring

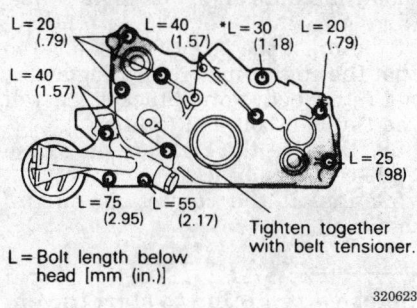

L = Bolt length below head [mm (in.)]

Front case bolt identification — 1995–97 Eclipse 2.0L (4G63) and 2.4L (4G64) engines

into the counter bore on the oil pump discharge passage.

14. Carefully install the oil pump on the crankshaft until seated to the engine block. Torque the bolts to 17 ft. lbs. (23 Nm).

15. Install a new crankshaft oil seal in the oil pump.

16. Install the crankshaft sprocket using the proper installation tools.

17. Install the timing belt and related components.

18. Install the oil pick-up tube.

19. Apply Loctite®18718 or equivalent at the point where the oil pump meets the engine block.

20. Install the oil pan using a new gasket. Torque the mounting bolts to 9 ft. lbs. (12 Nm).

21. Use a new O-ring and install the oil filter adapter to the engine. Made sure the roll pin aligns with the hole. Torque the assembly to 40 ft. lbs. (55 Nm).

22. Install a new oil filter.

23. Install the rear plate.

24. Safely lower the vehicle to the floor.

25. Refill the engine with the proper amount of oil.

26. Start the engine and check for leaks.

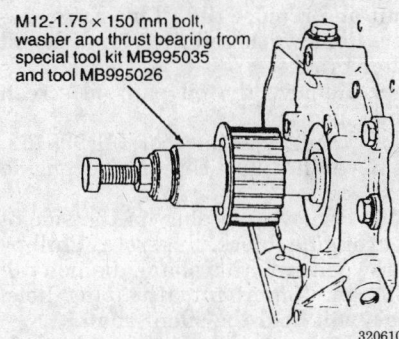

M12-1.75 × 150 mm bolt, washer and thrust bearing from special tool kit MB995035 and tool MB995026

Crankshaft sprocket installation — 1995–97 Eclipse 2.0L (420A) engine

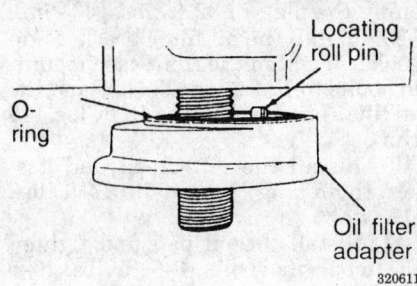

Oil filter adapter installation — 1995–97 Eclipse 2.0L (420A) engine

1993 Galant 2.0L (4G63) Engines

1. Disconnect the negative battery cable.

2. Remove the splash shield under the engine.

3. Remove the bolt holding the two hoses to the top of the left engine mount.

4. Place a floor jack and a broad piece of lumber under the oil pan.

— WARNING —
Elevate the jack just enough to support the engine without raising it.

5. Remove the nuts holding the through-bolt for the left engine bracket and remove the bolt. It may be necessary to adjust the jack tension slightly to allow the bolt to come free. Use the jack to keep the engine in its normal position while the mount is disconnected.

6. Remove the nuts and bolts holding the engine bracket to the engine and remove the bracket.

7. Remove the engine drive belts.

8. Remove the power steering and water pump pulleys.

9. Remove the air conditioner pulley from the crankshaft and remove the crankshaft pulley.

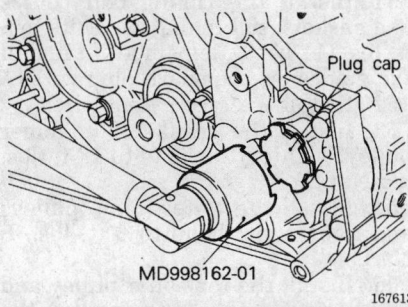

Removal and installation of the front case plug — 1993 Galant 2.0L (4G63 and 4G63) engine

10. Remove the timing belt covers and its gaskets.

NOTE: Before removing the timing belt, mark it with an arrow to show the direction of rotation. If the belt is to be reused, it must be reinstalled so that it rotates in the same direction as before.

11. Rotate the crankshaft clockwise until the timing marks are aligned and remove both the camshaft and silent shaft timing belts.

— WARNING —
Once the engine is set to this position, the crankshaft and camshaft(s) must not be moved out of place. Severe damage can result.

12. Remove the oil filter. Drain the engine oil.

13. Remove the oil pressure switch.

14. Remove the oil pan and the oil pick up screen.

15. Remove the oil filter bracket and its gasket.

16. Remove the oil relief plunger plug and gasket from the oil filter bracket assembly. Carefully remove the relief spring and valve from the bore.

17. Using special tool MD 998162 or equivalent, remove the plug cap.

18. Remove the crankshaft sprocket and the flange.

19. Remove the plug from the left side of the cylinder block and insert a screwdriver with a 8mm (0.31 in.) shank in the hole. The screwdriver shaft must go in at least 58mm (2.3 in.). This blocks the silent shaft and keeps it from turning.

20. Remove the bolt holding the oil pump driven gear to the silent shaft and remove the gear.

21. Remove the front case mounting bolts; remove the front case assembly and its gasket.

— WARNING —
The front case bolts are of several different lengths and must be replaced exactly as removed. Labeling or diagramming their placement during removal is very important.

22. Remove the screwdriver from the left silent shaft. Remove the silent shafts from the engine.

23. Remove the oil pump cover from the front case.

24. Remove the oil pump gears.

25. Visually check the contact surfaces of the oil pump cover and the front case for wear. If excessive wear

is present, replace the components. Inspect the oil passages for clogging and clean them as needed. The silent shafts should be inspected for any of wear or seizure. Closely inspect the oil seals on the front case, replacing any with worn or damaged lips.

To install:

26. Apply engine clean engine oil to the oil pump gears and install them in the front case. With a straight edge and feeler gauge, check the clearance between the side of the gears and the case. Maximum allowable clearance for either gear is 0.25mm (0.0098 in.). With the feeler gauge, check the gear-to-case clearance. Maximum allowable clearance is 0.25mm (0.0098 in.).

27. Insert the relief plunger into the oil filter bracket and check for smooth operation.

28. Align the mating marks on the pump gears.

29. Install the oil pump cover to the front case and tighten the bolts to 12 ft. lbs. (16 Nm).

30. Carefully install the left and right silent shafts into the engine.

───── WARNING ─────
Use care when installing the silent shafts, the bearings can be damaged by careless installation.

31. Using the proper size driver, install a new crankshaft seal into the front case.

NOTE: The front oil seal should be replaced every time the case is removed.

32. Install a seal guide tool MD 998285–01 or equivalent, to the crankshaft. Note that the tapered end should face away from the block. Coat the surface of the guide with clean engine oil.

33. Install a new front case gasket, carefully aligning all the bolt holes.

34. Install the front case to the block and lightly tighten the eight bolts. When the case is properly seated, the seal guide may be removed.

35. Insert a screwdriver through the left side access hole and block the left silent shaft.

36. Install the oil pump driven gear onto the left silent shaft and tighten the bolt to 26 ft. lbs. (35 Nm).

37. Install a new O-ring and, using the special tool, install the plug cap. Tighten the cap plug to 18 ft. lbs. (24 Nm).

38. Coat the relief plunger and spring in clean engine oil and install them into the oil filter bracket. In-

stall the relief plug and gasket, tightening the plug to 32 ft. lbs. (43 Nm).

39. Install the oil filter bracket and gasket. Tighten the front case mounting bolts to 17 ft. lbs. (23 Nm) and the oil filter bracket bolts to 13 ft. lbs. (18 Nm).

40. Install the oil screen and gasket, tightening the bolts to 14 ft. lbs. (19 Nm).

41. Install the oil pan and tighten the mounting bolts 4–6 ft. lbs. (6–8 Nm).

42. Coat the threads of the oil pressure switch with sealant such as 3M ART 8660® or equivalent, and using the special socket, install the switch. Tighten the switch to 8 ft. lbs. (11 Nm).

43. Install a new oil filter.

44. Align the silent shaft timing marks. Install the silent shaft timing belt and properly adjust.

45. Install the flange on the crankshaft sprocket and the outer crankshaft sprocket and/or the oil pump sprocket.

46. Make certain that all the timing marks for the camshaft sprocket, the crankshaft sprocket and the oil pump sprocket are correctly aligned.

47. Once the oil pump sprocket is aligned, remove the plug in the left side of the block. Insert a Phillips screwdriver with a shaft diameter of 8mm (0.31 in.) into the plug hole. It should enter 58mm (2.3 in.) or more. Do not remove the screwdriver until the timing belt is completely installed. If the screwdriver shaft only enters about 25mm (1 inch), turn the oil pump sprocket one rotation and align the timing mark again; insert the screwdriver to block the silent shaft from turning.

48. Observe the directional arrow made during removal and install the camshaft timing belt.

49. Remove the screwdriver from the silent shaft.

50. Tighten the crankshaft sprocket bolt to 88 ft. lbs. (120 Nm).

51. Install the timing belt covers and gaskets, tightening the bolts to 8 ft. lbs. (11 Nm). Remember to install the guides for the spark plug wires.

52. Install the spark plug wires.

53. Install the crankshaft damper pulley. Tighten the bolts to 18 ft. lbs. (24 Nm).

54. Install the water pump pulleys and tighten the bolts to 6 ft. lbs. (8 Nm).

55. Install the tensioner pulley and the engine drive belts; adjust the belts to the correct tension.

56. Install the engine mount bracket to the engine. Tighten the

mounting nuts and bolts to 42 ft. lbs. (57 Nm).

57. Adjust the jack (if necessary) so that the engine mount bushing aligns with the bodywork bracket. Install the through-bolt and tighten.

58. Position the high pressure hose and secure its bracket.

59. Install the correct amount of motor oil.

───── WARNING ─────
Before attempting to start the engine, make certain that the screwdriver has been removed from the silent shaft access hole.

60. Connect the negative battery cable and start the engine. Verify correct oil pressure and inspect for any signs of leakage.

1995–97 Galant 2.4L (4G64) Engines

NOTE: Whenever the oil pump is disassembled or the cover removed, the gear cavity must be filled with petroleum jelly. This seals the pump and acts lime a prime so the oil pump draws oil as soon as the engine starts to turn. Do not use grease.

1. Disconnect the negative battery cable. Rotate the engine so No. 1 cylinder is on Top Dead Center (TDC) of its compression stroke. The timing marks should be aligned at this point.

2. Raise and safely support the vehicle.

3. Drain the engine oil. Lower the vehicle.

4. Using the proper equipment, support the weight of the engine. Remove the front engine mount bracket and accessory drive belts.

5. Remove timing belt upper and lower covers.

6. Remove the timing belt and crankshaft sprocket.

7. Disconnect the electrical connector from the oil pressure sending unit and remove the oil pressure sensor. Remove the oil filter and the oil filter bracket.

8. Remove the oil pan, oil screen and gasket.

9. Using special tool MD998162, remove the plug cap in the engine front cover.

10. Remove the plug on the side of the engine block. Insert a Phillips screwdriver with a shank diameter of 0.32 in. (8mm) into the plug hole. This will hold the silent shaft.

11. Remove the driven gear bolt that secures the oil pump driven gear to the silent shaft.

12. Remove and tag the front cover mounting bolts.

NOTE: The mounting bolts are different lengths, make certain to identify their original location as they are removed, for proper installation.

13. Remove the front case cover and oil pump assembly. If necessary, the silent shaft can come out with the cover assembly.

14. Remove the oil pump cover, located on the back of the engine front cover. Remove the oil pump drive and driven gears.

15. After disassembling the oil pump, clean all components and remove gasket material from mating surfaces.

16. Assemble the oil pump gears into the front case and rotate it to ensure smooth rotation and no looseness. Make sure there is no ridge wear on the contact surface between the front case and the gear surface of the oil pump front cover.

To install

17. Align the timing mark on the oil pump drive gear with that on the driven gear and install them into the engine front case. Apply engine oil to the gears.

18. Install the oil pump cover and tighten the retainer bolts to 17 ft. lbs. (24 Nm).

19. Using the appropriate driver, install a new crankshaft seal into the front case.

20. Position a new front case gasket in place. Set seal guide tool MD998285 on the front end of the crankshaft to protect the seal from damage. Apply a thin coat of oil to the outer circumference of the seal pilot tool.

21. Install the front case assembly through a new front case gasket and temporarily tighten the flange bolts.

22. Mount the oil filter on the bracket with new oil filter bracket gasket in place. Install the 3 bolts with washers and tighten to 16 ft. lbs. (22 Nm).

23. Insert a Phillips screwdriver into a hole in the left side of the engine block to lock the silent shaft in place.

24. Secure the oil pump drive gear onto the left silent shaft by installing and tightening the driven gear bolt to 29 ft. lbs. (40 Nm).

25. Install new O-ring to the groove in the front case and install the plug cap. Using the special tool MD998162, tighten the cap to 20 ft. lbs. (27 Nm).

26. Install the oil screen in position with new gasket in place.

27. Clean both mating surfaces of the oil pan and the cylinder block. Apply sealant in the groove in the oil pan flange, keeping towards the inside of the bolt holes. The width of the sealant bead applied is to be about 0.016 in. (4mm) wide.

NOTE: After applying sealant to the oil pan, do not exceed 15 minutes before installing the oil pan.

28. Install the oil pan to the engine and secure with the retainers. Tighten bolts to 6 ft. lbs. (8 Nm).

29. Install the oil pressure gauge unit and the oil pressure switch. Connect the electrical harness connector.

30. Refill crankcase with oil. Install new oil filter.

31. Install the timing belts and timing belt covers. Assemble the remaining components to the front of the engine.

32. Connect the negative battery cable and start the engine. Verify oil pressure. Inspect for leaks.

Diamante and 3000GT 3.0L (6G72) Engines

NOTE: Whenever the oil pump is disassembled or the cover removed, the gear cavity must be filled with petroleum jelly to seal the pump and act as a prime. This allows the pump to draw oil as soon as the engine starts. Do not use grease.

1. Disconnect the negative battery cable.

---- **CAUTION** ----
Wait at least 90 seconds after the negative (-) battery cable is disconnected to prevent possible deployment of the air bag.

2. Remove the front engine mount bracket and accessory drive belts.

3. Remove timing belt upper and lower covers.

4. Remove the timing belt and crankshaft sprocket.

5. Remove the oil pan.

6. Remove the oil screen and gasket.

7. Remove and tag the front cover mounting bolts. Note the lengths of the mounting bolts as they are removed for proper installation.

8. Remove the front case cover and oil pump assembly. Disassemble as required.

To install:

9. Thoroughly clean all gasket material from all mounting surfaces.

10. Apply engine oil to the entire surface of the gears or rotors.

11. Assemble the front case cover and oil pump assembly to the engine block using a new gasket.

12. Install the oil screen with new gasket.

13. Install the oil pan and timing belts.

14. Install the timing belt covers.

15. Install the drive belts and the front engine mount bracket.

16. Connect the negative battery cable, refill the crankcase and check for adequate oil pressure.

TRANSAXLE

Manual Transaxle Assembly

REMOVAL AND INSTALLATION

1993–94 Precis with KM200 and KM201 Transaxles

NOTE: If the vehicle is going to be rolled while the halfshafts are out of the vehicle, obtain 2 outer CV-joints or proper equivalent tools and install to the hubs. If the vehicle is rolled without the proper torque applied to the front wheel bearings, the bearings will no longer be usable.

1. Disconnect the negative battery cable. Remove the air cleaner assembly, battery and battery tray as required.

2. On 5-speed transaxle, disconnect the electrical connector for the selector control valve.

NOTE: The actuator-to-shaft coupling pin collar is not reusable; replace it.

3. Disconnect and remove the speedometer cable.

4. If equipped with a cable operated clutch, disconnect the clutch cable from the transaxle assembly.

5. If equipped with a hydraulically operated clutch, remove the clevis pin connecting the slave cylinder to the release fork shaft and remove the slave cylinder mounting bolts. Remove the bolts attaching the hydraulic line support bracket to the transaxle. Remove and support the slave cylinder assembly out of the way with a length of mechanics wire.

6. Disconnect the back-up lamp electrical connector. Remove the starter motor electrical harness.

7. Remove the transaxle mounting bolts accessible from the top side of the transaxle.

8. Unbolt and remove the starter motor.

9. Raise the vehicle and support it safely. Then, remove the splash shield from under the engine. Drain the transaxle fluid.

10. Disconnect the extension rod and the shift rod at the transaxle end and lower them.

11. Disconnect the stabilizer bar at the lower control arm.

12. Remove the halfshafts from the transaxle assembly.

13. Support the transaxle from below with a floor jack. Make sure the support is widely enough spread that the transaxle pan will not be damaged. Then, remove the attaching bolts and the bell housing cover.

14. Remove the lower bolts attaching the transaxle to the engine.

15. Remove the transaxle insulator mount bolt. Remove the cover from inside the right fender shield and remove the transaxle support bracket.

16. Remove the transaxle mount bracket.

17. Pull the assembly away from the engine and lower it from the vehicle.

To install:

18. Install the transaxle to the engine and install the mounting bolts. Tighten the mounting bolts as follows:

- M10—7T engine-to-transaxle bolts — 35 ft. lbs. (48 Nm)
- M8—10T engine-to-transaxle bolts — 25 ft. lbs. (34 Nm)

19. When installing the halfshafts, use new circlips on the axle ends. Take care to get the inboard joint parts straight, not bent relative to the axle. Care must be taken to ensure that the oil seal lip of the transaxle is not damaged by the serrated part of the driveshaft.

20. Install the undercover.

21. Install the mounting brackets and torque the mounting bracket bolt to 40 ft. lbs. (54 Nm).

22. Install the starter making sure to fasten the ground wire with the upper fastener and the harness fastener with the lower fastener.

23. Connect the back-up light switch connector and speedometer cable.

24. Install the clutch and shifter actuation components. If the hydraulic

system was opened, it should be bled after installation.

25. Install the air cleaner and battery.

26. Make sure the vehicle is level when refilling the transaxle. Use Hypoid gear oil or equivalent, GL-4 or higher.

27. Connect the negative battery cable and check the transaxle for proper operation. Make sure the reverse lights come on when in R.

Mirage with F5M21 and F5M22 Transaxles

NOTE: If the vehicle is going to be rolled while the halfshafts are out of the vehicle, obtain 2 outer CV-joints or proper equivalent tools and install to the hubs. If the vehicle is rolled without the proper torque applied to the front wheel bearings, the bearings will no longer be usable.

NOTE: The suspension components should not be tightened until the vehicles weight is resting on the ground.

1. Disconnect the negative battery cable.

2. Remove the front wheels and the inner wheel panels.

3. Remove the air cleaner assembly and vacuum hoses.

4. Note the locations and disconnect the shifter cables.

5. Disconnect the backup lamp switch connector, speedometer cable connection and remove the starter motor.

6. Remove the upper transaxle-to-engine mounting bolts.

7. Raise the vehicle and support safely.

8. Remove the undercover and splash pan.

9. Drain the transaxle oil.

10. Support the engine and remove the crossmember.

11. Remove the upper transaxle mounting bolt and bracket.

12. Disconnect the stabilizer bar, tie rod ends and the lower ball joint connections.

13. Remove the clutch release cylinder and clutch oil line bracket. Do not disconnect the fluid lines and secure the slave cylinder with wire. Disconnect the clutch cable, if equipped with cable controlled clutch system.

14. Remove the halfshafts by inserting a prybar between the transaxle case and the driveshaft and

prying the shaft from the transaxle. Do not pull on the driveshaft. Doing so damages the inboard joint. Do not insert the prybar so far the oil seal in the case is damaged.

NOTE: It is not necessary to disconnect the halfshafts from the steering knuckle. Remove the shaft with the hub and knuckle as an assembly. Tie the shafts aside. Note the circle clip on the end of the inboard shafts should not be reused.

15. Remove the bellhousing lower cover.

16. Remove the transaxle to engine bolts and lower the transaxle from the vehicle.

To install:

NOTE: When installing the transaxle, be sure to align the splines of the transaxle with the clutch disc.

17. Install the transaxle to the engine and install the mounting bolts. Torque the bolts to specifications.

18. Install the bellhousing cover.

NOTE: When installing the halfshafts, use new circlips on the axle ends. Care must be taken to ensure that the oil seal lip of the transaxle is not damaged by the serrated part of the driveshaft.

19. Install and fully seat the halfshafts into the transaxle.

20. Install the slave cylinder.

21. Connect the ball joints, tie rod ends and the stabilizer bar connections.

22. Install the upper transaxle mounting bracket and bolt.

23. Install the crossmember.

24. Install the undercover.

25. Install the upper transaxle-to-engine mounting bolts.

26. Install the starter motor.

27. Connect the backup light switch connector and speedometer cable.

28. Connect and adjust the shifter cables.

29. Install the air cleaner assembly.

30. Install the front wheels.

31. Make sure the vehicle is level when refilling the transaxle. Use Hypoid gear oil or equivalent, GL-4 or higher.

32. Connect the negative battery cable and check the transaxle for proper operation. Make sure the reverse lights operate when in reverse.

1993 Galant and 1993–94 Eclipse with F5M31, F5M22, F5M33, W5M31 and W5M33 Transaxles

NOTE: If the vehicle is going to be rolled on its wheels while the halfshafts are out of the vehicle, obtain two outer CV-joints or proper equivalent tools and install to the hubs. If the vehicle is rolled without the proper torque applied to the front wheel bearings, the bearings will no longer be usable.

1. Remove the battery and the air intake hoses.
2. If equipped with Active-ECS, unplug the compressor wiring.
3. Remove the auto-cruise actuator and underhood bracket, located on the passenger side inner fender wall.
4. Drain the transaxle and transfer case fluid, if equipped, into a suitable waste container.
5. Remove the retainer bolt and pull the speedometer cable from the transaxle assembly.
6. Remove the cotter pin securing the select and shift cables and remove the cable ends from the transaxle.
7. Remove the connection for the clutch release cylinder and without disconnecting the hydraulic line, secure aside.
8. Disconnect the backup light switch harness and position aside.
9. Disconnect the starter electrical connections, if necessary, remove the starter motor and position aside.
10. Remove the transaxle mount bracket. Remove the upper transaxle mounting bolts.
11. Raise the vehicle and support safely on jackstands. Remove the undercover and the front wheels.
12. Remove the cotter pin and disconnect the tie rod end from the steering knuckle.
13. Remove the self-locking nut from the halfshafts. Disconnect the lower arm ball joint from the steering knuckle.
14. Remove the halfshafts from the transaxle.
15. On AWD vehicle, disconnect the front exhaust pipe.
16. On AWD vehicle, remove the transfer case by removing the attaching bolts, moving the transfer case to the left and lowering the front side. Remove it from the rear driveshaft. Be careful of the oil seal. Do not allow the driveshaft to hang; once the front is removed from the transfer, tie it up. Cover the transfer case openings to keep out dirt.
17. Remove the cover from the transaxle bellhousing. On AWD, also remove the crossmember and the triangular gusset.
18. Remove the transaxle lower coupling bolt. It is just above the halfshaft opening on FWD or transfer case opening on AWD.
19. Support the weight of the engine from above (chain hoist). Support the transaxle using a transmission jack and remove the remaining lower mounting bolts.
20. On turbocharged vehicle, be careful not to damage the lower radiator hose with the transaxle housing during removal. Wrap tape on both the lower hose and the transaxle housing to prevent damage. Move the transaxle assembly to the right and carefully lower it from the vehicle.

To install:
21. Install the transaxle to the engine and install the mounting bolts. Install the transaxle lower coupling bolt.
22. Install the underpan, crossmember and the triangular gusset.
23. Install the transfer case on AWD vehicles and connect the exhaust pipe.
24. Install the halfshafts, using new circlips on the axle ends. Try to keep the inboard joint straight in relation to the axle. Be careful not to damage the oil seal lip of the transaxle with the serrated part of the halfshaft.
25. Connect the tie rod and ball joint to the steering knuckle.
26. Install the transaxle mount bracket.
27. Install wheels and lower vehicle. Retorque axle shaft nuts to 145–188 ft. lbs.
28. Install the starter motor.
29. Connect the backup light switch and the speedometer cable.
30. Install the clutch release cylinder.
31. Connect the select and shift cables and install new cotter pins.
32. Install the air intake hose.
33. Install the auto-cruise actuator and bracket.
34. Install the battery.
35. If equipped with Active-ECS, connect the air compressor.
36. Make sure the vehicle is level when refilling the transaxle. Use Hypoid gear oil or equivalent, GL-4 or higher.
37. Check the transaxle and transfer case for proper operation. Make sure the reverse lights come on when in reverse.

1995–97 Eclipse with F5M31, F5M33, F5MC1 and W5M33 Transaxles

1. Remove the battery and the air intake hoses.
2. Remove the battery tray and support.
3. If equipped with cruise control, remove the auto-cruise actuator and bracket.
4. Drain the transaxle and transfer case fluid, if equipped, into a suitable container.
5. Remove the charcoal canister and bracket.
6. Disconnect the shift and select cables from the transaxle.
7. Disconnect the back-up light switch and the vehicle speed sensor connectors.
8. Remove the starter assembly.
9. Attach an engine support fixture to the engine and remove the transaxle mounting bolts.
10. Remove the rear roll stopper bracket mounting bolts.
11. Remove the transaxle mounting bracket mounting nuts.
12. Raise the vehicle and remove the engine undercovers.
13. If equipped with all wheel drive, remove the transfer case assembly.

——— **WARNING** ———
Do not remove or install the axle shaft nut when the vehicle is on the floor or damage to the bearings will occur.

14. Remove the axle shafts.
15. Remove the slave cylinder from the bell housing but do not disconnect the fluid line. Position it out of the way.
16. Remove the bell housing cover and the right hand center member stay (support).
17. Remove the center member.
18. Place a transmission jack under the transaxle and remove the transaxle mounting bolt.
19. Remove the transaxle mounting and lower the transaxle.

To install:
20. Raise the transaxle into position and install the transaxle mounting. Torque the through-bolt to 50 ft. lbs. (69 Nm).
21. Install the transaxle assembly mounting bolt. Torque the bolt to 22–25 ft. lbs. (30–34 Nm).
22. Install the center member assembly and the right hand stay.
23. Install the bell housing cover and the slave cylinder.
24. Install the axle shafts. Make sure to install the washer in the proper direction.

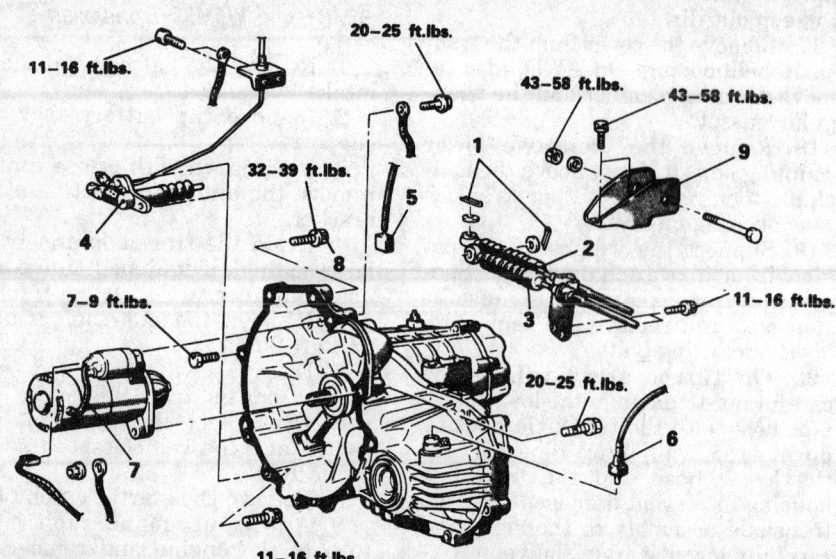

11–16 ft.lbs.
20–25 ft.lbs.
43–58 ft.lbs.
43–58 ft.lbs.
32–39 ft.lbs.
7–9 ft.lbs.
11–16 ft.lbs.
20–25 ft.lbs.
11–16 ft.lbs.

1. Cotter pin
2. Connection for select cable
3. Connection for shift cable
4. Connection for clutch release cylinder
5. Backup light switch connector
6. Connection for speedometer cable
7. Starter
8. Transaxle assembly upper part coupling bolt
9. Transaxle mount bracket

138285

Transaxle assembly and related components — 1993–94 Eclipse with F5M22, F5M33 and W5M33 manual transaxles

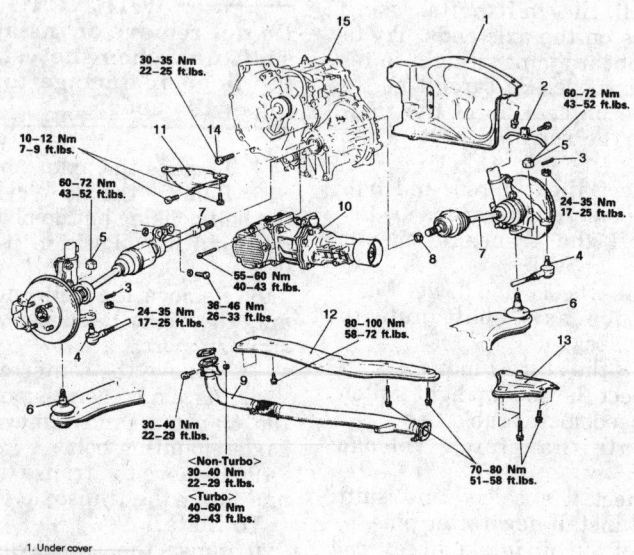

30–35 Nm
22–25 ft.lbs.
10–12 Nm
7–9 ft.lbs.
60–72 Nm
43–52 ft.lbs.
60–72 Nm
43–52 ft.lbs.
24–35 Nm
17–25 ft.lbs.
55–60 Nm
40–43 ft.lbs.
36–46 Nm
26–33 ft.lbs.
24–35 Nm
17–25 ft.lbs.
80–100 Nm
58–72 ft.lbs.
30–40 Nm
22–29 ft.lbs.
70–80 Nm
51–58 ft.lbs.

<Non-Turbo>
30–40 Nm
22–29 ft.lbs.
<Turbo>
40–60 Nm
29–43 ft.lbs.

1. Under cover
2. Speed sensor <Vehicles with Anti-lock Braking System>
3. Cotter pin
4. Connection for tie rod end
5. Self-locking nut
6. Connection for lower arm ball joint
7. Connection for drive shaft
8. Circlip
9. Front exhaust pipe
10. Transfer assembly
11. Bell housing cover
12. Right member
13. Gusset
14. Transaxle assembly lower part coupling bolt
15. Transaxle assembly

162788

Transaxle lower connection points — 1993 Galant AWD models with F5M22, F5M31, W5M31 and W5M33 manual transaxles

25. Install the engine undercovers and lower the vehicle.
26. Install the transfer case assembly if removed.
27. Install the transaxle mounting bracket mounting nuts.
28. Install the rear roll stopper bracket mounting bolts.
29. Install the transaxle assembly mounting bolts. Torque the mounting bolts to 35 ft. lbs. (48 Nm).
30. Remove the engine support fixture.
31. Install the starter assembly.
32. Connect the vehicle speed sensor and the back-up light connectors.
33. Install the cruise control actuator if removed.
34. Install the battery tray support and the tray.
35. Install the charcoal canister bracket and the canister.
36. Install the air duct and the air cleaner assembly.
37. Refill the transaxle and the transfer case if equipped with oil.

1994–97 Galant with an F5M31 Transaxle

1. Disconnect the negative battery cable and wait at least 90 seconds before performing any work.
2. Remove the air cleaner and intake hoses.
3. Drain the transaxle into a suitable waste container.
4. Remove the cotter pins and clips securing the select and shift cables and remove the cable ends from the transaxle.
5. If equipped with Active-ECS, disconnect the air compressor.
6. Disconnect the backup light switch harness and position aside.
7. Disconnect the speedometer electrical connector, from the transaxle assembly.
8. Remove the starter motor and position aside.
9. Using special tool MZ203827 or equivalent, support the engine assembly.
10. Remove the rear roll stopper mounting bracket.
11. Remove the transaxle mount bracket.
12. Remove the upper transaxle mounting bolts.
13. Raise and safely support the vehicle.
14. Remove the front wheel assemblies.
15. Remove the right hand undercover.
16. Remove the cotter pin and disconnect the tie rod end, from the steering knuckle.

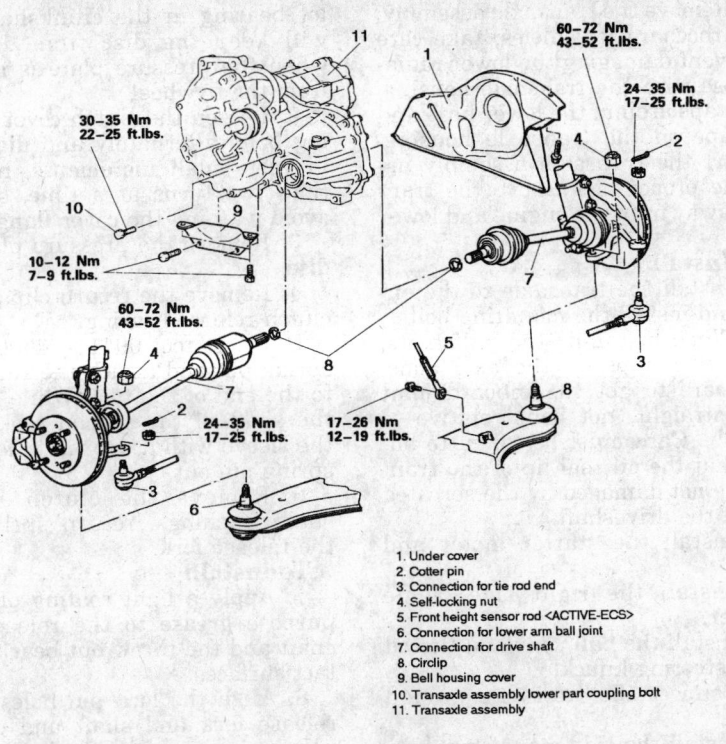

1. Under cover
2. Cotter pin
3. Connection for tie rod end
4. Self-locking nut
5. Front height sensor rod <ACTIVE-ECS>
6. Connection for lower arm ball joint
7. Connection for drive shaft
8. Circlip
9. Bell housing cover
10. Transaxle assembly lower part coupling bolt
11. Transaxle assembly

162786

Transaxle lower connection points — 1993 Galant FWD models with F5M22, F5M31, W5M31 and W5M33 manual transaxles

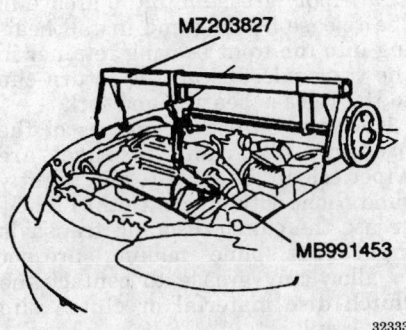

MZ203827

MB991453

323334

Proper method of supporting the engine assembly for transaxle removal

17. Disconnect the stabilizer bar link, from the damper fork.
18. Disconnect the damper fork, from the lateral lower control arm.
19. Disconnect the later lower arm, and the compression arm, lower ball joints, from the steering knuckle.
20. Pry the halfshafts from the transaxle, and secure aside.
21. Remove the connection for the clutch release cylinder and without disconnecting the hydraulic line, secure aside.

22. Remove the cover from the transaxle bellhousing.
23. Remove the engine front roll stopper through-bolt.
24. Remove the crossmember and the triangular right hand stay.
25. Support the transaxle, using a transmission jack, and remove the transaxle lower coupling bolt.

NOTE: The coupling bolt threads from the engine side, into the transaxle, and is located just above the halfshaft opening.

26. Slide the transaxle rearward and carefully lower it from vehicle.
To install:
27. Install the transaxle to the engine and install the mounting bolts and tighten to 35 ft. lbs. (48 Nm). Install the transaxle lower coupling bolt and tighten to 22–25 ft. lbs. (30–34 Nm).
28. Install the cover to the transaxle bellhousing and tighten the mounting bolts to 7 ft. lbs. (9 Nm).
29. Install the crossmember and tighten the front mounting bolts to 65 ft. lbs. (88 Nm) and the rear bolt to 54 ft. lbs. (73 Nm). Install the front en-

gine roll stopper through-bolt and lightly tighten. Once the full weight of the engine is on the mounts, tighten the bolt to 42 ft. lbs. (57 Nm).
30. Install the triangular stay bracket and tighten the mounting bolts to 65 ft. lbs. (88 Nm).
31. Connect the clutch release cylinder.
32. Install the halfshafts, using new circlips on the axle ends.

— WARNING —
When installing the axleshaft, keep the inboard joint straight in relation to the axle, so not to damage the oil seal lip of the transaxle, with the serrated part of the halfshaft.

33. Connect the tie rod and ball joints to the steering knuckle. Tighten the ball joint self-locking nuts to 48 ft. lbs. (65 Nm). Tighten the tie rod end nut to 21 ft. lbs. (28 Nm) and secure with a new cotter pin.
34. Connect the damper fork to the lower control arm and tighten the through-bolt to 65 ft. lbs. (88 Nm).
35. Connect the stabilizer link to the damper fork, and tighten the self-locking nut to 29 ft. lbs. (39 Nm).
36. Install the underpan.
37. Install wheels and lower vehicle.
38. Install the transaxle mount bracket, to the transaxle, and tighten the mounting nuts to 32 ft. lbs. (43 Nm).
39. Install the rear roll stopper mounting bracket.
40. Remove the engine support. Tighten the transaxle mount through-bolt to 51 ft. lbs. (69 Nm) and tighten the front engine roll stopper through-bolt.
41. Install the upper transaxle mounting bolts and tighten to 35 ft. lbs. (48 Nm).
42. Install the starter motor.
43. Connect the backup light switch and the speedometer connector.
44. Connect the select and shift cables and install new cotter pins.
45. Install the air cleaner and the air intake hose.
46. Connect the negative battery cable.
47. Make sure the vehicle is level, and refill the transaxle.
48. Check the transaxle for proper operation. Make sure the reverse lights come on when in reverse.

3000GT with F5M33, W5MG1 and W6MG1 Transaxles

NOTE: If the vehicle is going to be rolled on its wheels while the halfshafts are out of the vehicle, obtain two outer CV-joints or proper equivalent tools and install to the hubs. If the vehicle is rolled without the proper torque applied to the front wheel bearings, the bearings will no longer be usable.

--- **CAUTION** ---

Wait at least 90 seconds after the negative (-) battery cable is disconnected to prevent possible deployment of the air bag.

1. Remove the battery and battery tray. Raise the vehicle and support safely. Drain the transaxle oil and the oil from the transfer case.
2. If equipped with AWD, disconnect the exhaust pipe. Remove the mounting bolts and lower the transfer case from the vehicle.
3. Remove the left side splash shield and engine undercover.
4. Remove the air cleaner assembly and all adjoining duct work.
5. Disconnect the shifter control cables and speedometer connector.
6. Remove the clutch release cylinder.
7. Disconnect the reverse light switch.
8. Support the weight of the transaxle and remove the transaxle mount through-bolt. Remove the access plug, remove the bolts for the bracket and remove the brackets.
9. Disconnect the transaxle ground cable.
10. Disconnect the tie rod end and ball joint from the steering knuckle.
11. Remove the right frame member.
12. Remove the starter motor.
13. Remove the halfshafts by inserting a prybar between the transaxle case and the driveshaft and prying the shaft from the transaxle. Do not pull on the driveshaft. Doing so damages the inboard joint. Use the prybar. Do not insert the prybar so far the oil seal in the case is damaged. On AWD, remove the right side shaft as just described. The left side shaft can be removed by tapping with a plastic hammer. Remove the shaft with the hub and knuckle as an assembly. Don't tap on the center bearing or it will be damaged. Tie the shafts aside. Note the circle clip on the end of the inboard shafts. These should not be reused.
14. Remove the transaxle brackets.

15. Remove the transaxle assembly. On turbocharged vehicles, take care to prevent damaging the lower radiator hose with the transaxle housing. Wind tape around the lower hose and put tape on the transaxle housing. Support the transaxle assembly using the proper jack, move the transaxle away from the engine and lower it.

To install:

16. Install the transaxle to the engine and install the mounting bolts.
17. When installing the halfshafts, use new circlips on the axle ends. Take care to get the inboard joint parts straight, not bent relative to the axle. Care must be taken to ensure that the oil seal lip of the transaxle is not damaged by the serrated part of the driveshaft.
18. Install the starter motor and cover.
19. Install the right side frame member.
20. Install the ball joint and tie rod to the steering knuckle.
21. Connect the transaxle ground cable.
22. Install the side mount brackets and install the access plug.
23. Connect the reverse light switch.
24. Install the clutch release cylinder.
25. Connect the shifter control cables and speedometer connector.
26. Install the transfer case and related items on AWD vehicles.
27. Install the air cleaner assembly and all adjoining duct work.
28. Install the left side splash shield.
29. Install the battery tray and battery.
30. Make sure the vehicle is level when refilling the transaxle. Use Hypoid gear oil or equivalent, GL-4 or higher.
31. Connect the negative battery cable and check the transaxle and transfer case for proper operation. Make sure the reverse lamps come ON when in reverse.

Clutch Assembly

REMOVAL AND INSTALLATION

1993–94 Precis with KM200 and KM201 Transaxles

1. Remove the transaxle. Insert the forward end of an old transaxle input shaft or a clutch disc guide tool through the splined center of the clutch disc, pressure plate and the pi-

lot bearing in the crankshaft. This will keep the disc from dropping when the pressure plate is removed from the flywheel.
2. Loosen the clutch cover mounting bolts alternately and diagonally in very small increments, no more than two turns at a time, so as to avoid warping the cover flange.
3. Remove the pressure plate and disc.
4. Remove the return clip and the clutch release bearing.
5. Insert tool 09414–24000 in the spring pin and attach the round nut to the end of the tool. While holding the shaft of the special tool, rotate the sleeve with a wrench to force the spring pin out.
6. Remove the clutch release shaft, packings, return spring and the release fork.

To install:

7. Apply a light coating of multipurpose grease to the release fork shaft and the throw out bearing contact surfaces.
8. Align the lock pin holes of the release fork and shaft and drive 2 new spring pins into the holes. Make sure the spring pin slot is at right angles to the centerline of the control shaft.
9. Apply grease into the groove in the release bearing and install bearing into the front bearing retainer in the transaxle. Install the return clip to the release bearing and fork.
10. Make sure the surfaces of the pressure plate and flywheel are wiped clean of grease and lightly sand them with crocus cloth. Lightly grease the clutch disc and transaxle input shaft splines making sure not to allow any grease to contact the clutch disc material or clutch slip may result.
11. Locate the clutch disc on the flywheel with the stamped mark facing outward. Use a clutch disc guide or old input shaft to center the disc on the flywheel and then install the pressure plate over it. Install the bolts and tighten them evenly. Tighten them in increments of 2 turns or less to avoid warping the pressure plate. Torque to 11 to 15 ft. lbs. (15 to 21 Nm).
12. Remove the clutch disc centering tool. Install the transaxle. Adjust the clutch free-play.

Eclipse and 1993 Galant

1. Disconnect the negative battery cable. Raise and safely support the vehicle.
2. Remove the transaxle assembly from the vehicle.

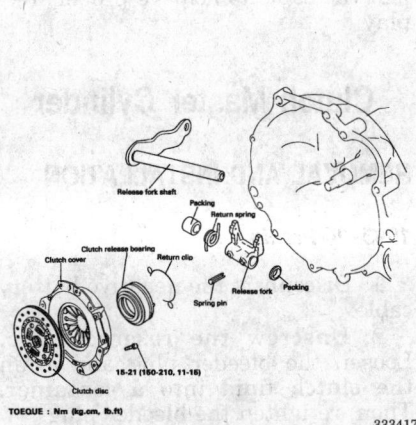

Clutch and pressure plate assembly — 1993–94 Precis with KM200 and KM201 manual transaxles

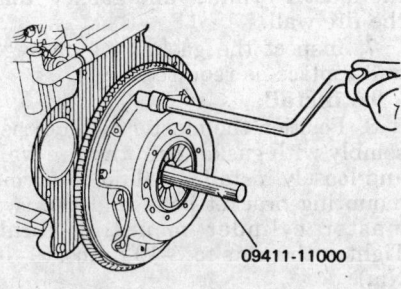

09411-11000

Removing or installing the pressure plate — 1993–94 Precis with KM200 and KM201 manual transaxles

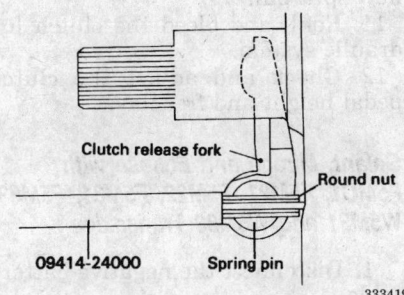

Clutch release fork — 1993–94 Precis with KM200 and KM201 manual transaxles

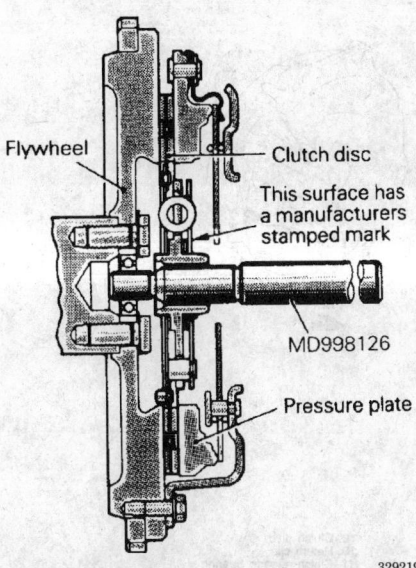

Clutch plate and disk alignment procedure — Mirage

3. Remove the pressure plate attaching bolts, pressure plate and clutch disc. If the pressure plate is to be reused, loosen the bolts in a diagonal pattern, 1 or 2 turns at a time. This will prevent warping the clutch cover assembly.

4. Remove the return clip and the pressure plate release bearing. Do not use solvent to clean the bearing.

5. Inspect the clutch release fork and fulcrum for damage or wear. If necessary, remove the release fork and unthread the fulcrum from the transaxle.

6. Carefully inspect the condition of the clutch components and replace any worn or damaged parts.

To install:

7. Inspect the flywheel for heat damage or cracks. Resurface or replace the flywheel as required. Install the flywheel using new bolts.

8. Install the fulcrum and tighten to 25 ft. lbs. (35 Nm). Install the release fork. Apply a coating of multi-purpose grease to the point of contact with the fulcrum and the point of contact with the release bearing. Apply a coating of multi-purpose grease to the end of the release cylinder's

pushrod and the pushrod hole in the release fork.

NOTE: When installing the clutch, apply grease to each part, but be careful not to apply excessive grease. Excessive grease will cause clutch slippage and shudder.

9. Apply multi-purpose grease to the clutch release bearing. Pack the bearing inner surface and the groove with grease. Do not apply grease to the resin portion of the bearing. Place the bearing in position and install return clip.

10. Apply a coating of grease to the clutch disc splines and then use a brush to rub it in the grooves. Using a universal clutch disc alignment tool, position the clutch disc on the flywheel. Install the retainer bolts and tighten a little at a time, in a diagonal sequence. Tighten them to a final torque of 14 ft. lbs. (19 Nm) on 1994–97 Galant models and 16 ft. lbs. (22 Nm) on all other models. Remove the aligning tool.

11. Install the transaxle assembly and check fluid level.

12. Check for proper clutch operation.

3000GT

1. Disconnect the negative battery cable.

— CAUTION —

Wait at least 90 seconds after the negative (-) battery cable is disconnected to prevent possible deployment of the air bag.

2. Raise and safely support the vehicle.

3. Remove the transaxle assembly from the vehicle.

4. Remove the pressure plate attaching bolts. If the pressure plate is to be reused, loosen the bolts in succession, one or two turns at a time to prevent warping the cover flange.

5. Remove the pressure plate release bearing assembly and the clutch disc. Do not use solvent to clean the bearing.

6. Inspect the condition of the clutch components and replace any worn parts.

To install:

7. Inspect the flywheel for heat damage or cracks. Resurface or replace the flywheel as required, using new bolts.

8. Using the proper alignment tool, install the clutch disc to the flywheel. Install the pressure plate assembly and tighten the pressure plate bolts evenly to 11–15 ft. lbs.

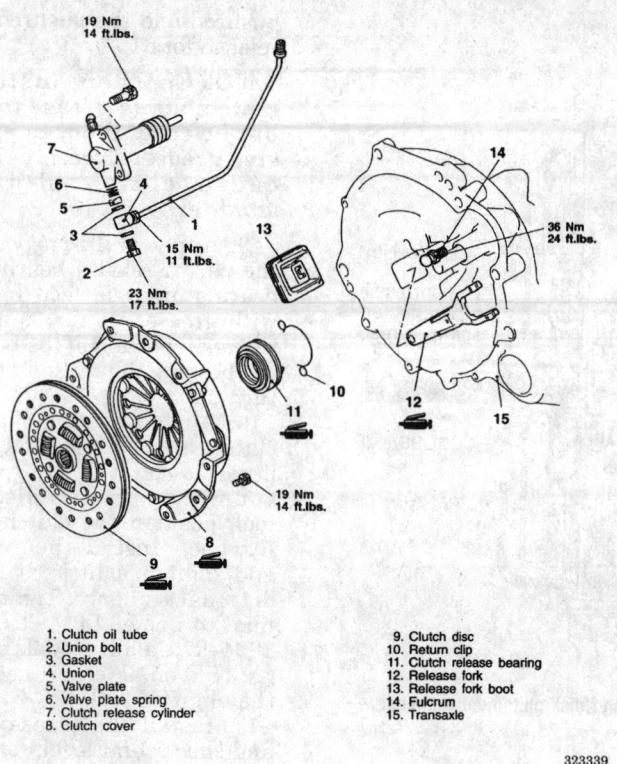

1. Clutch oil tube
2. Union bolt
3. Gasket
4. Union
5. Valve plate
6. Valve plate spring
7. Clutch release cylinder
8. Clutch cover

9. Clutch disc
10. Return clip
11. Clutch release bearing
12. Release fork
13. Release fork boot
14. Fulcrum
15. Transaxle

323339

Exploded view of non-turbo clutch assembly — 1995–97 Eclipse with F5M31, F5M33, F5Mc1 and W5M33 manual transaxles

(15–21 Nm). Remove the alignment tool.

9. Apply a very light coat of high temperature grease to the clutch fork at the ball pivot and where the fork contacts the bearing. Also a little bit of grease can be applied to end of the release cylinder's pushrod and to the pushrod hole on the fork. Apply a light coat of grease on the transaxle input shaft splines.

10. Install a new clutch release bearing. Pack its inner surface with high temperature grease.

11. Install the transaxle assembly.

12. Lower the vehicle and connect the negative battery cable.

13. Check the clutch for proper operation.

Clutch Cable

ADJUSTMENT

Mirage

1. Measure the clutch pedal height (measurement A). The specification is 6.38–6.50 inches (162–165mm).

NOTE: The clutch pedal height is not adjustable. If not within specifications, part replacement is required.

2. Depress clutch pedal several times and check the pedal freeplay (measurement B).

3. If measurement is not 0.67–0.87 inches (17–22mm), adjustment is required.

4. To adjust turn the outer cable adjusting nut, located at the firewall, until freeplay is within range.

5. Depress clutch pedal several times and recheck measurement.

REMOVAL AND INSTALLATION

Mirage

1. Rotate the adjusting wheel counterclockwise to loosen the cable.

2. Remove the cable retaining clamps.

3. Remove the cotter pin from the clutch actuating arm at the transaxle and disconnect the cable.

4. Disconnect the cable at the pedal and remove the cable from the vehicle.

NOTE: In order to prevent cable binding or abrasion, be sure to take note of cable routing, so that it can be reinstalled in the same position.

To install

5. Route cable and make connection at clutch pedal.

6. Make connection at transaxle and secure cable with retaining clamp. Install a new cotter pin.

7. Lubricate all pivot points. Adjust the cable to achieve proper freeplay.

Clutch Master Cylinder

REMOVAL AND INSTALLATION

1993–94 Precis

1. Disconnect the negative battery cable.

2. Unscrew the reservoir cap. Loosen the bleeder plug and drain the clutch fluid into a container. Then, retighten the bleeder plug.

3. From inside the vehicle; remove the cotter pin, clevis pin and washer from the clutch pedal to release the master cylinder pushrod.

4. Unbolt the clutch reservoir bracket, if equipped.

5. Loosen the clutch reservoir hose clamps to permit movement of the clutch hose. Then, disconnect the clutch hose from the master cylinder.

6. Remove the two nuts and pull the master cylinder and gasket from the fire wall.

7. Inspect the gasket for damage and replace as required.

To install:

8. Position the master cylinder assembly with gasket onto the fire wall and loosely install the fluid reservoir mounting bracket bolts and the two master cylinder mounting nuts. Tighten the nuts to 7–10 ft. lbs. (9–14 Nm).

9. Connect the clutch tube to the master cylinder and tighten the tube clamps. Tighten the clutch reservoir bracket bolts.

10. Connect the pushrod to the clutch pedal and insert the clevis pin. Lubricate the clevis pin with multi-purpose grease and secure it with a new split pin.

11. Refill and bleed the clutch hydraulic system.

12. Check and adjust the clutch pedal height and freeplay.

Galant, Mirage and Eclipse with F5MC1, F5M21, F5M22, F5M31, F5M33, W5M31 and W5M33 Transaxles

1. Disconnect the negative battery cable.

2. Remove necessary underhood components in order to gain access to the clutch master cylinder.

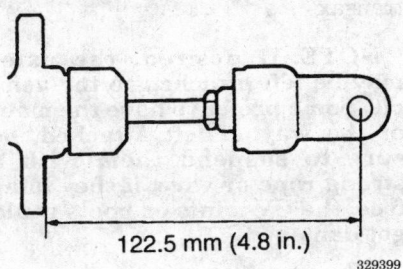

122.5 mm (4.8 in.)

329399

Clutch cylinder pushrod adjustment — Mirage with F5M21 and F5M22 manual transaxles

—————— WARNING ——————
Clutch hydraulic system uses brake fluid. Use care; brake fluid is harmful to painted surfaces.

3. Loosen the clutch fluid line at the cylinder and allow the fluid to drain.
4. Remove the clevis pin retainer at the clutch pedal and remove the washer and clevis pin.
5. Remove the 2 nuts and pull the cylinder from the firewall. A seal should be between the mounting flange and firewall. This seal should be replaced.

To install

NOTE: The length of the clutch master cylinder pushrod should be 4.8 inches (122.5mm) from the flange to the center of the clevis pin hole.

6. Mount master cylinder on firewall studs, using new seal, and torque nuts to 7–11 ft. lbs. (10–15 Nm) on Galant and Eclipse models and 9 ft. lbs. (13 Nm) on Mirage models.
7. Lubricate all pivot points with grease and install the clevis pin.
8. Connect hydraulic line and bleed the system at the slave cylinder using fresh DOT 3 brake fluid.
9. Check the adjustment of the clutch pedal for proper freeplay.
10. Connect the negative battery cable.

3000GT

1. Disconnect the negative battery cable.
2. Remove the power brake booster and any other underhood components necessary in order to gain access to the clutch master cylinder.
3. Loosen the line at the cylinder and allow the fluid to drain. Use care, brake fluid damages paint.
4. Remove the clevis pin retainer at the clutch pedal and remove the

washer and clevis pin. AWD vehicles have a clutch pedal booster which directly activates the master cylinder.
5. Remove the two nuts and pull the cylinder from the firewall. A seal should be between the mounting flange and firewall. This seal should be replaced.

To install:
6. Using a new seal, mount the clutch master cylinder to the firewall.
7. Install the clevis pin on FWD vehicles.
8. Lubricate all pivot points with grease.
9. Install the power brake booster.
10. Bleed the system at the slave cylinder using DOT 3 brake fluid and check the adjustment of the clutch pedal.

Clutch Slave Cylinder

REMOVAL AND INSTALLATION

1993–94 Precis

1. Disconnect the negative battery cable.
2. Drain the fluid from the clutch master cylinder reservoir.
3. Disconnect the clutch hose from the slave cylinder.
4. Unbolt the cylinder from the clutch housing and remove it.
5. Inspect the cylinder for leakage or torn boots and repair or replace as required.

To install:
6. Apply a thin coating of grease to the contact points of the release fork and the cylinder and install the slave cylinder to the clutch housing. Tighten the cylinder retainer bolts to 11–16 ft. lbs. (15–22 Nm).
7. Connect the fluid line to the cylinder and tighten the fitting to 7–10 ft. lbs. (13–17 Nm).
8. Refill and bleed the clutch hydraulic system.
9. Reconnect the negative battery cable.

Eclipse, Mirage, Galant and 3000GT

1. Disconnect the negative battery cable. Remove necessary underhood components in order to gain access to the clutch release cylinder.
2. Remove the hydraulic line and allow the system to drain.
3. Remove the bolts and pull the cylinder from the transaxle housing.

To install
4. Lubricate all pivot points with grease.

5. Mount slave cylinder to transaxle and tighten bolts to 11–16 ft. lbs.
6. Connect hydraulic line and fill the system with clean brake fluid meeting DOT 3 specifications.
7. Bleed the system and adjust the clutch pedal height and the clevis pin play.

Hydraulic Clutch System Bleeding

PROCEDURE

Galant, Eclipse, 3000GT and Mirage

—————— CAUTION ——————
The clutch hydraulic system uses brake fluid. Use care; brake fluid is harmful to painted surfaces.

1. Fill the reservoir with clean brake fluid meeting DOT 3 specifications.
2. Press the clutch pedal to the floor then open the bleeder screw on the slave cylinder.
3. Tighten the bleed screw and release the clutch pedal.
4. Repeat the procedure until the fluid is free of air bubbles.

NOTE: It is suggested that a hose be attached to the bleeder with the other end immersed in a container at least half full of brake fluid during the bleeding operation. Do not allow the reservoir to run out of fluid during bleeding.

Automatic Transaxle Assembly

REMOVAL AND INSTALLATION

1993–94 Precis with an KM176 Transaxle

1. Shift the transaxle into the **N** position.
2. Disconnect the negative battery cable.
3. Drain the fluid from the transaxle.
4. Remove the entire air cleaner assembly including the air intake and breather hoses. Disconnect the air flow sensor connector and the purge control solenoid valve connector and hoses. Label all the connections to avoid confusion during assembly.
5. Disconnect the cooler supply and return lines from the transaxle.

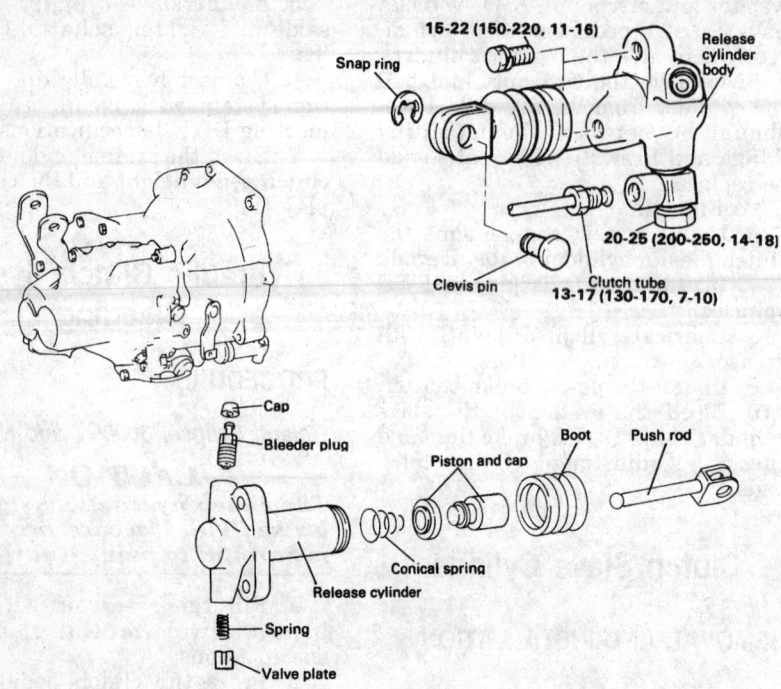

15-22 (150-220, 11-16)

Snap ring

Release cylinder body

20-25 (200-250, 14-18)

Clevis pin

Clutch tube
13-17 (130-170, 7-10)

Cap

Bleeder plug

Piston and cap

Boot

Push rod

Conical spring

Release cylinder

Spring

Valve plate

TORQUE : Nm (kg.cm, lb.ft)

258063

Clutch slave cylinder — 1993–94 Precis with KM200 and KM201 manual transaxles

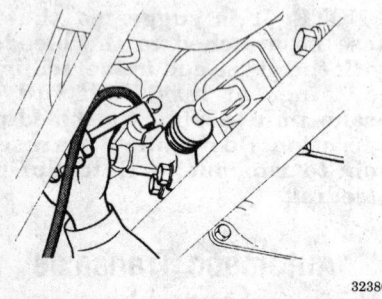

323802

Bleeding the clutch hydraulic system — Eclipse with F5M22, F5M31, F5M33, F5MC1 and W5M33 manual transmission

Plug the line openings to prevent the entry of dirt and foreign matter.

6. Disconnect the control cable from the shift lever.

7. Disconnect the speedometer cable from the transaxle. Route the cable off to the side and out of the way.

8. Unplug the connectors to the solenoid valves, inhibitor switch, pulse generators, kickdown servo switch and oil temperature sensor.

9. Working from the top of the transaxle, remove the upper transaxle mounting bolts.

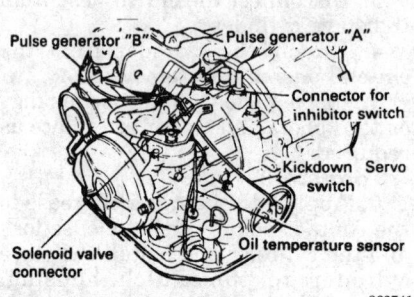

Pulse generator "B"   Pulse generator "A"

Connector for inhibitor switch

Kickdown   Servo
switch

Solenoid valve connector

Oil temperature sensor

263741

Transaxle electrical connectors — 1993–94 Precis with an KM176 automatic transaxle

10. Remove the transaxle mounting bracket.

11. Support the transaxle with a jack and remove the center member mounting bolts.

NOTE: To avoid placing excessive pressure on the oil pan, support the transaxle in a wide area.

12. Raise and safely support the vehicle.

13. Remove both front wheels and both axle-end cotter pins, nuts and washers.

14. Remove the ball joint castle nuts and separate the ball joints from the knuckles.

15. Remove the halfshafts from the transaxle.

NOTE: If desired, the axles may be left attached to the vehicle. Some prefer to have them out of the way. If left attached, be sure to suspend them with a strong rope or wire; if they hang free, the CV-joints or boots could get damaged.

16. Unbolt and remove the bell housing cover.

17. Remove the three special bolts attaching the torque converter to the drive plate. To remove these bolts, engage the crankshaft pulley nut with a socket and turn the crankshaft until each bolt comes into view. Remove each bolt with the proper size box end wrench by working through the bell housing cover access hole.

18. After the drive plate bolts are removed, reinstall the center member temporarily.

19. Remove the lower transaxle mounting bolts and lower the transaxle from the engine.

To install:

20. Place the transaxle on the transaxle jack securely and install onto the engine, using the dowels as guides. Make sure all heater tube clamps, wires, etc. are out of the way or it will not be possible to install the transaxle flush with the block.

21. Install the lower transaxle mounting bolts. Torque the 2.4 in. (60mm) bolt to 22–26 ft. lbs. (30–35 Nm). Torque the 2.6 in. (65mm) bolt to 31–40 ft. lbs. (43–55 Nm).

22. Remove the center member.

23. Install the torque converter-to-drive plate bolts and torque them to 34–39 ft. lbs. (46–53 Nm).

24. Install the bell housing cover and torque the bolts to 7–9 ft. lbs. (10–12 Nm).

25. Install the halfshafts.

26. Connect the ball joints to the steering knuckles. Tighten the castle nuts to 43–53 ft. lbs. (60–72 Nm), and then only tighten them enough to install new cotter pins.

27. Install the front wheels.

28. Lower the vehicle.

29. Install the center member.

30. Install the transaxle mounting bracket and torque the bolts to 43–58 ft. lbs. (60–80 Nm).

31. Install the upper transaxle mounting bolts. Torque the upper two bolts to 31–40 ft. lbs. (43–55 Nm). Torque the two bolts at the starter motor housing to 20–25 ft. lbs. (27–34 Nm).

32. Plug in all the transaxle electrical connectors.

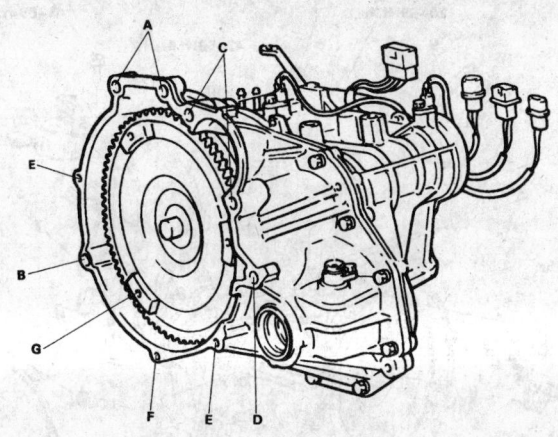

	Nm	Kg.cm	lb.ft	O.D. x Length mm (in.)	Bolt identification
A	43—55	430—550	31—40	10 x 40 (1.6)	A X B
B	43—55	430—550	31—40	10 x 65 (2.6)	
C	27—34	270—340	20—25	10 x 55 (2.2)	
D	30—35	300—350	22—26	8 x 60 (2.4)	
E	10—12	100—120	7—9	6 x 12 (0.5)	
F	30—35	300—350	22—26	8 x 12 (0.5)	
G	46—53	460—530	34—39		

263739

Transaxle assembly and bolt locations — 1993–94 Precis with an KM176 automatic transaxle

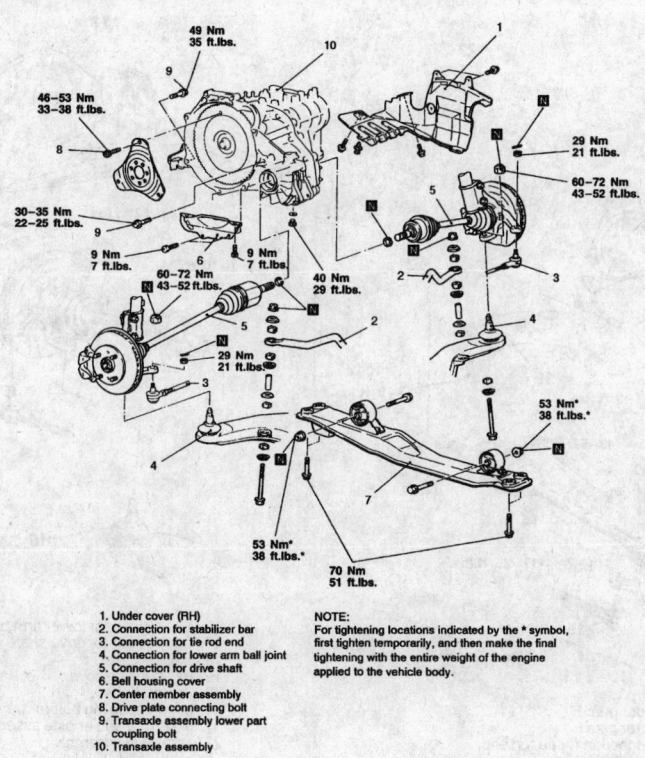

1. Under cover (RH)
2. Connection for stabilizer bar
3. Connection for tie rod end
4. Connection for lower arm ball joint
5. Connection for drive shaft
6. Bell housing cover
7. Center member assembly
8. Drive plate connecting bolt
9. Transaxle assembly lower part coupling bolt
10. Transaxle assembly

NOTE:
For tightening locations indicated by the * symbol, first tighten temporarily, and then make the final tightening with the entire weight of the engine applied to the vehicle body.

332232

Automatic transaxle components-bottom view — Mirage with F3A21 and F4A22 automatic transaxles

33. Connect the speedometer cable to the transaxle.

34. Connect the control cable to the shift lever. Adjust the control cable.

35. Connect the cooler supply and return lines to the transaxle.

36. Install the air cleaner assembly.

37. Reconnect the negative battery cable.

38. Fill the transaxle to the proper level.

Mirage with F3A21 and F4A22 Transaxles

NOTE: If the vehicle is going to be rolled on its wheels while the halfshafts are out of the vehicle, obtain two outer CV-joints or proper equivalent tools and install to the hubs. If the vehicle is rolled without the proper torque applied to the front wheel bearings, the bearings will no longer be usable.

1. Disconnect the negative battery cable.

2. Remove the battery and battery tray.

3. Remove the air hose and air cleaner assembly.

4. Raise the vehicle and support safely.

5. Remove the under guard pan.

6. Drain the transaxle oil.

7. Disconnect the control cable and cooler lines.

8. Disconnect the shift control solenoid valve connector.

9. Disconnect the inhibitor switch, kickdown servo switch, the pulse generator and oil temperature sensor, if equipped.

10. Disconnect the speedometer cable and remove the starter.

11. Remove the transaxle mounting bolts and bracket.

12. Disconnect the stabilizer bar from the lower control arm.

13. Disconnect the steering tie rod end and the ball joint from the steering arm.

14. Remove the halfshafts at the inboard side from the transaxle. Tie the joint assembly aside.

NOTE: It is not necessary to disconnect the halfshafts from the wheel hubs.

15. Support the engine and remove the center member.

16. Remove the bellhousing cover and remove the driveplate bolts.

17. Remove the transaxle assembly lower connecting bolt, located just over the halfshaft opening.

18. Properly support the transaxle assembly and lower it moving it to the right for clearance.

To install:

19. After the torque converter has been mounted on the transaxle, install the transaxle assembly on the engine. Install the mounting bolts and torque to specifications.

20. Tighten the driveplate bolts to 33–38 ft. lbs. (46–53 Nm). Install the bellhousing cover.

21. Install the center member and torque the bolts to specifications.

22. Replace the circlips and install the halfshafts to the transaxle.

23. Install the tie rods, ball joints and stabilizer links to the steering arm.

24. Install the transaxle mounting bracket and bolts.

25. Install the starter.

26. Connect the speedometer cable.

27. Connect the inhibitor switch, kickdown servo switch, the pulse generator and oil temperature sensor, if disconnected.

28. Connect the shift control solenoid valve connector.

29. Connect the control cables and oil cooler lines.

30. Install the tension rod, if removed.

31. Install the air cleaner assembly.

32. Install the battery tray and battery. Connect the positive then the negative terminal.

33. Refill with Dexron® II, Mopar ATF Plus type 7176 or equivalent, automatic transaxle fluid.

34. Start the engine and allow to idle for two minutes. Apply parking brake and move selector through each gear position, ending in **N**. Recheck fluid level and add if necessary. Fluid level should be between the marks in the **HOT** range.

1993 Galant and 1993–94 Eclipse with F4A22, F4A33, W4A32 and W4A33 Transaxles

1. Remove the battery and battery tray.

2. On vehicles equipped with autocruise, remove the control actuator and bracket.

3. If equipped with an active ECS, disconnect the air compressor.

4. Drain the transaxle fluid.

5. Remove the air cleaner assembly, intercooler and air hose, as required.

6. Mark the shift cable. Remove the adjusting nut and disconnect the shift cable.

7. Disconnect and tag the electrical connectors for the solenoid, neutral safety switch (inhibitor switch), the pulse generator kickdown servo switch and oil temperature sensor.

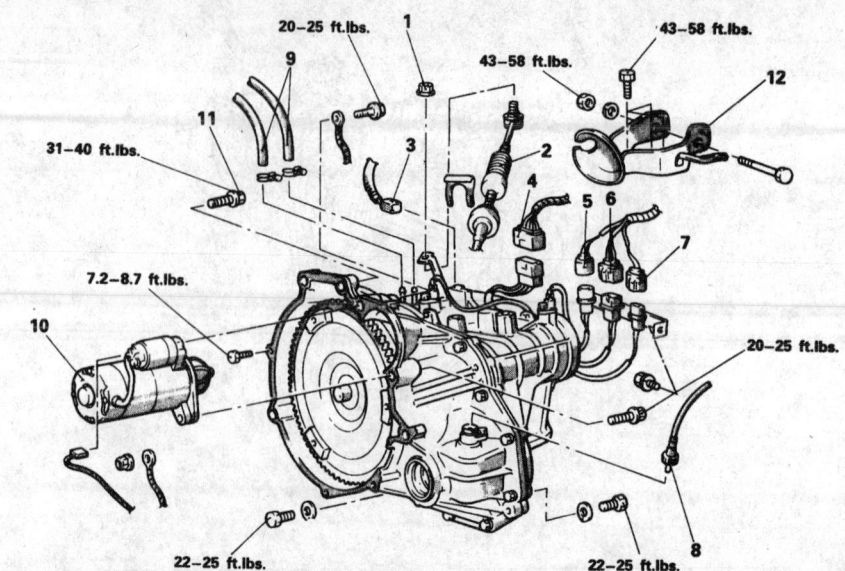

1. Adjusting nut
2. Connection for transaxle control cable
3. Connection for solenoid connector
4. Connection for park/neutral position switch connector
5. Connection for pulse generator connector
6. Connection for kickdown servo switch connector
7. Connection for oil temperature sensor connector
8. Connection for speedometer cable
9. Connection for oil cooler hose
10. Starter motor
11. Upper coupling bolt for transaxle assembly and engine assembly
12. Transaxle mount bracket

138144

Automatic transaxle removal and installation view A — 1993–94 Eclipse with F4A22, F4A33 and W4A33 with automatic transaxles

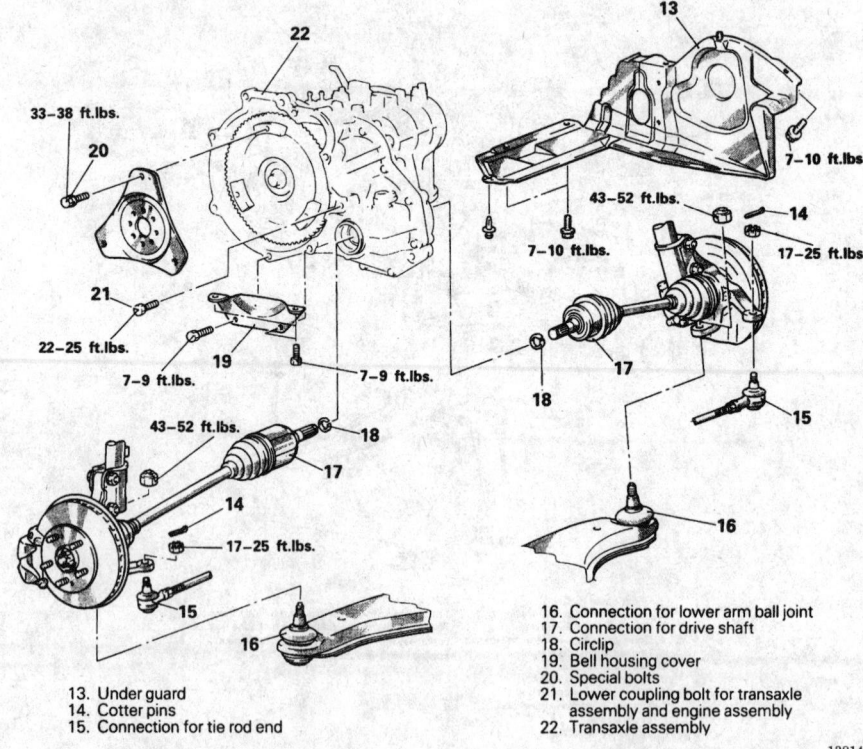

13. Under guard
14. Cotter pins
15. Connection for tie rod end
16. Connection for lower arm ball joint
17. Connection for drive shaft
18. Circlip
19. Bell housing cover
20. Special bolts
21. Lower coupling bolt for transaxle assembly and engine assembly
22. Transaxle assembly

138148

Automatic transaxle removal and installation view B — 1993–94 Eclipse with F4A22, F4A33 and W4A33 with automatic transaxles

8. Disconnect the speedometer cable and oil cooler lines.

9. Disconnect the wires to the starter motor and remove the starter.

10. Remove the upper transaxle to engine bolts.

11. Support the transaxle and remove the transaxle mounting bracket.

12. Raise the vehicle and support safely. Remove the sheet metal under guard.

13. Remove the tie rod ends and the ball joints from the steering knuckle.

14. Remove the halfshafts by inserting a prybar between the transaxle case and the driveshaft and prying the shaft from the transaxle. Do not pull on the driveshaft. Doing so damages the inboard joint. Use the prybar. Do not insert the prybar so far the oil seal in the case is damaged. Tie the halfshafts aside.

15. On AWD vehicles, disconnect the exhaust pipe and remove the transfer case.

16. Remove the lower bellhousing cover and remove the special bolts holding the flexplate to the torque converter. To remove, turn the engine crankshaft with a box wrench and bring the bolts into a position appropriate for removal, one at a time. After removing the bolts, push the torque converter toward the transaxle so it doesn't stay on the engine allowing oil to pour out the converter hub or cause damage to the converter.

17. Remove the lower transaxle to engine bolts and remove the transaxle assembly.

To install:

18. After the torque converter has been mounted on the transaxle, install the transaxle assembly on the engine. Tighten the driveplate bolts to 34–38 ft. lbs. (46–53 Nm). Install the bellhousing cover.

19. On AWD, install the transfer case and frame pieces. Connect the exhaust pipe using a new gasket.

20. Replace the circlips and install the halfshafts to the transaxle.

21. Install the tie rods and ball joint to the steering arm.

22. Install the transaxle mounting bracket.

23. Install the under guard.

24. Install the starter.

25. Connect the speedometer cable and oil cooler lines.

26. Connect the solenoid, neutral safety switch (inhibitor switch), the pulse generator kickdown servo switch and oil temperature sensor.

27. Install the shift control cable.

28. Install the air hose, intercooler and air cleaner assembly.

29. If equipped with an active ECS, connect the air compressor.

30. If equipped with auto-cruise, install the control actuator and bracket.

31. Refill with Dexron II, Mopar ATF Plus type 7176, Mitsubishi Plus ATF or equivalent, automatic transaxle fluid. If vehicle is AWD check and fill the transfer case.

32. Start the engine and allow to idle for 2 minutes. Apply parking brake and move selector through each gear position, ending in **N**. Recheck fluid level and add if necessary. Fluid level should be between the marks in the **HOT** range.

1995–97 Eclipse with F4A23, F4A33, F4AC1 and W4A33 Transaxles

1. Remove the battery and the air intake hoses.

2. Remove the battery tray and support.

3. If equipped with cruise control, remove the auto-cruise actuator and bracket.

4. Drain the transaxle and transfer case fluid, if equipped, into a suitable container.

5. Remove the charcoal canister and bracket.

6. Disconnect the shift and select cables from the transaxle.

7. Disconnect the back-up light switch and the vehicle speed sensor connectors.

8. Remove the dipstick and tube assembly.

9. Remove the starter assembly.

10. Disconnect the park/neutral switch, oil temperature sensor, kick down servo switch, solenoid valve, pulse generator and speedometer connections.

11. Attach an engine support fixture to the engine and remove the transaxle mounting bolts.

12. Remove the rear roll stopper bracket mounting bolts.

13. Remove the transaxle mounting bracket mounting nuts.

14. Raise the vehicle and remove the engine undercovers.

15. Remove the front exhaust pipe.

16. If equipped with all wheel drive, remove the transfer case assembly.

WARNING ---
Do not remove or install the axle shaft nut when the vehicle is on the floor or damage to the bearings will occur.

---

17. Remove the axle shafts.

18. Remove the slave cylinder from the bell housing but do not disconnect the fluid line. Position it out of the way.

19. Remove the bell housing cover and the right hand center member stay (support).

20. Remove the center member.

21. Remove the drive plate connecting bolts.

22. Place a transmission jack under the transaxle and remove the transaxle mounting bolt.

23. Remove the transaxle mounting and lower the transaxle.

To install:

24. Raise the transaxle into position and install the transaxle mounting bracket. Torque the through-bolt to 51 ft. lbs. (69 Nm).

25. Install the transaxle assembly mounting bolt. Torque the bolt to 22–25 ft. lbs. (29–34 Nm).

26. Install the drive plate connecting bolts. Torque the bolts to 33–38 ft. lbs. (45–52 Nm).

27. Install the center member assembly and the right hand stay.

28. Install the bell housing cover and the slave cylinder.

29. Install the axle shafts. Make sure to install the washer in the proper direction.

30. Install the front exhaust pipe.

31. Install the engine undercovers and lower the vehicle.

32. Install the transfer case assembly if removed.

33. Install the transaxle mounting bracket mounting nuts.

34. Install the rear roll stopper bracket mounting bolts.

35. Install the transaxle assembly mounting bolts. Torque the bolts to 35 ft. lbs. (48 Nm).

36. Remove the engine support fixture.

37. Connect the park/neutral switch, oil temperature sensor, kick down servo switch, solenoid valve, pulse generator and speedometer connections.

38. Install the starter assembly.

39. Install the dipstick and tube assembly.

40. Connect the vehicle speed sensor and the back-up light connectors.

41. Install the cruise control actuator if removed.

42. Install the battery tray support and the tray.

43. Install the charcoal canister bracket and the canister.

44. Install the air duct and the air cleaner assembly.

45. Refill the transaxle and the transfer case if equipped with the proper fluid.

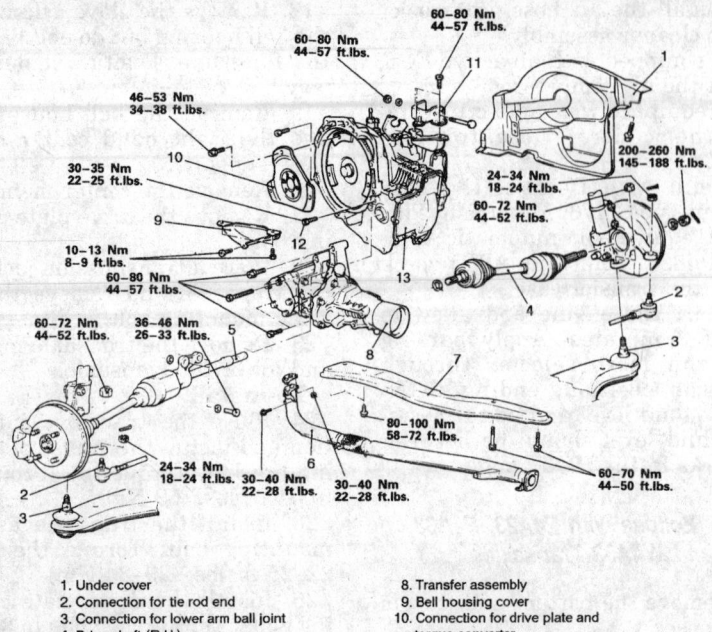

1. Under cover
2. Connection to tie rod end
3. Connection for lower arm ball joint
4. Drive shaft (R.H.)
5. Connection for drive shaft (L.H.)
6. Front exhaust pipe
7. Right member
8. Transfer assembly
9. Bell housing cover
10. Connection for drive plate and torque converter
11. Transaxle mounting bracket
12. Lower coupling bolt for transaxle assembly and engine assembly
13. Transaxle assembly

162838

Transaxle lower connection points (AWD Models) — 1993 Galant with F4A22 and W4A32 automatic transaxle

1994–97 Galant with F4A23 and F4A33 Transaxles

1. Disconnect the negative battery cable.
2. Remove the air cleaner and intake hoses.
3. Drain the transaxle into a suitable waste container.
4. Remove the nut securing the shifter lever to the transaxle. Remove the cable retaining clip and remove the cable from the transaxle.
5. Remove the shifter cable mounting bracket.
6. Disconnect and tag the electrical connectors for the speedometer, solenoid, neutral safety switch (inhibitor switch), the pulse generator, kickdown servo switch, and the oil temperature sensor.
7. Disconnect and tag the oil cooler lines, at the transaxle.
8. Remove the bolt securing the fluid dipstick tube, to the transaxle. Remove the dipstick and tube from the transaxle.
9. Remove the starter motor and position it aside.
10. Using special tool MZ203827 or equivalent, support the engine assembly.
11. Remove the rear roll stopper mounting bracket.

12. Remove the transaxle mount bracket.
13. Remove the upper transaxle mounting bolts.
14. Raise and safely support the vehicle.
15. Remove the front wheel assemblies.
16. Remove the right hand undercover.
17. Remove the cotter pin and disconnect the tie rod end, from the steering knuckle.
18. Disconnect the stabilizer bar link, from the damper fork.
19. Disconnect the damper fork, from the lateral lower control arm.
20. Disconnect the later lower arm, and the compression arm, lower ball joints, from the steering knuckle.
21. Pry the halfshafts from the transaxle, and secure aside.
22. Remove the cover from the transaxle bellhousing.
23. Remove the engine front roll stopper through-bolt.
24. Remove the crossmember and the triangular right hand stay.
25. Remove the bolts holding the flexplate to the torque converter with a box wrench. Rotate the engine to bring the bolts into a position appropriate for removal, one at a time. After removing the bolts, push the torque converter toward the transaxle.

This will prevent the converter from remaining intact with the engine, possibly damaging the converter.
26. Support the transaxle, using a transmission jack, and remove the transaxle lower coupling bolt.

NOTE: The coupling bolt threads from the engine side, into the transaxle, and is located just above the halfshaft opening.

27. Slide the transaxle rearward and carefully lower it from the vehicle.
To install:
28. After the torque converter has been mounted on the transaxle, install the transaxle assembly to the engine. Install the mounting bolts and tighten to 35 ft. lbs. (48 Nm). Install the transaxle lower coupling bolt and tighten to 21–25 ft. lbs. (29–34 Nm).
29. Connect the torque converter to the flexplate and tighten the bolts to 33–38 ft. lbs. (45–52 Nm).
30. Install the cover to the transaxle bellhousing and tighten the mounting bolts to 7 ft. lbs. (9 Nm).
31. Install the crossmember and tighten the front mounting bolts to 65 ft. lbs. (88 Nm) and the rear bolt to 54 ft. lbs. (73 Nm). Install the front engine roll stopper through-bolt and lightly tighten. Once the full weight of the engine is on the mounts, tighten the bolt to 42 ft. lbs. (57 Nm).
32. Install the triangular stay bracket and tighten the mounting bolts to 65 ft. lbs. (88 Nm).
33. Install the halfshafts, using new circlips on the axle ends.

—— **WARNING** ——

When installing the axleshaft, keep the inboard joint straight in relation to the axle, so as not to damage the oil seal lip of the transaxle, with the serrated part of the halfshaft.

34. Connect the tie rod and ball joints to the steering knuckle. Tighten the ball joint self-locking nuts to 48 ft. lbs. (65 Nm). Tighten the tie rod end nut to 21 ft. lbs. (28 Nm) and secure with a new cotter pin.
35. Connect the damper fork to the lower control arm and tighten the through-bolt to 65 ft. lbs. (88 Nm).
36. Connect the stabilizer link to the damper fork, and tighten the self-locking nut to 29 ft. lbs. (39 Nm).
37. Install the underpan.
38. Install wheels and lower vehicle.
39. Install the transaxle mount bracket, to the transaxle, and tighten

the mounting nuts to 32 ft. lbs. (43 Nm).

40. Install the rear roll stopper mounting bracket.

41. Remove the engine support. Tighten the transaxle mount through-bolt to 51 ft. lbs. (69 Nm) and tighten the front engine roll stopper through-bolt.

42. Install the upper transaxle mounting bolts and tighten to 35 ft. lbs. (48 Nm).

43. Install the starter motor.

44. Install the dipstick tube and the dipstick.

45. Install the shifter cable mounting bracket.

46. Connect the shifter lever and tighten the retaining nut to 14 ft. lbs. (19 Nm).

47. Connect the oil cooler lines and secure with clamps.

48. Connect the electrical connectors for the speedometer, solenoid, neutral safety switch (inhibitor switch), the pulse generator, kickdown servo switch and oil temperature sensor.

49. Install the air cleaner and the air intake hose.

50. Connect the negative battery cable.

51. Make sure the vehicle is level, and refill the transaxle. Start the engine and allow to idle for 2 minutes. Apply parking brake and move selector through each gear position, ending in **N**. Recheck fluid level and add if necessary. Fluid level should be between the marks in the **HOT** range.

52. Check the transaxle for proper operation. Make sure the reverse lights come on when in reverse and the engine starts only in **P** or **N**.

Diamante with a F4A33 Transaxle

NOTE: If the vehicle is going to be rolled while the halfshafts are out of the vehicle, obtain 2 outer CV-joints or proper equivalent tools and install to the hubs. If the vehicle is rolled without the proper torque applied to the front wheel bearings, the bearings will no longer be usable.

1. Properly disarm the SRS system (air bag).

2. Raise and safely support the vehicle.

3. Remove the front wheels.

4. Remove the engine side and undercovers.

5. Drain the transaxle assembly.

6. If equipped, remove the front catalytic converter and exhaust pipe.

7. Remove the exhaust pipe, main muffler and catalytic converter.

8. Disconnect the tie rod end and ball joint from the steering knuckle.

NOTE: It will be necessary to unbolt the support bearing for the left side halfshaft.

9. Remove the halfshafts by inserting a prybar between the transaxle case and the driveshaft and prying the shaft from the transaxle. Do not pull on the driveshaft. Doing so damages the inboard joint. Do not insert the prybar so far that the oil seal in the case is damaged. Tie the halfshafts aside.

10. Remove the air cleaner assembly and adjoining duct work.

11. Detach the engine harness connection.

12. If the vehicle is equipped with active electronically controlled suspension (ACTIVE-ECS), remove the compressor assembly from the transaxle and suspend with wire. DO NOT disconnect the air hose from the compressor.

13. If equipped, remove the roll stopper stay bracket.

14. Disconnect the speedometer cable from the transaxle.

15. Remove the clip that secures the shifter and disconnect the shifter control cable from the transaxle.

16. Disconnect and plug the oil cooler hoses from the transaxle.

17. Disconnect the park/neutral switch electrical harness.

18. Disconnect the kickdown servo switch, pulse generator, oil temperature sensor electrical harness.

19. Disconnect the shift control solenoid valve harness.

20. Support the transaxle and remove the connection for the transaxle mounting bracket.

21. Remove the three upper transaxle-to-engine mounting bolts.

22. For vehicles with 4WS remove the following

   a. Remove the heat shield for the 4WS oil pump.

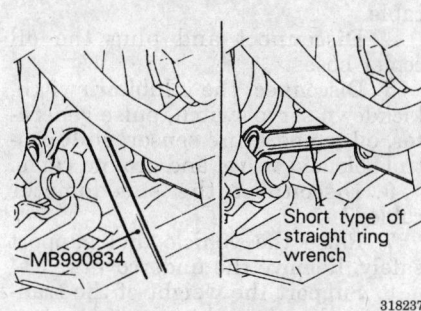

Location of 4-wheel steering oil pump mounting bolts — Diamante with a F4A33 automatic transaxle

b. Without removing the oil hoses, remove the bolts that secure the oil pump and remove the pump.

   c. Secure the oil pump with a piece of wire.

23. For vehicles equipped with ACTIVE-ECS, disconnect the height sensor rod from the lower control arm.

24. Remove the bolt that secures the oxygen sensor harness to the right side crossmember.

25. Remove the starter assembly.

26. Remove the mounting brackets for access to the bell housing cover.

27. Remove the bell housing/oil pan covers assembly.

28. Remove the four bolts holding the flexplate to the torque converter. It will be necessary to rotate the engine by the front crankshaft bolt for access to all the torque converter bolts.

29. After removing the bolts, push the torque converter toward the transaxle so it does not stay on the engine side of the vehicle.

30. Remove the lower transaxle to engine bolts and remove the transaxle assembly.

To install:

NOTE: Be sure the torque converter is fully seated into the front of the transaxle before installing the transaxle.

31. Install the transaxle assembly to the engine block and install the mounting bolts. Tighten the mounting bolts to specifications.

32. Install the four bolts that secure the torque convert to the driveplate. Tighten the driveplate bolts to 34–38 ft. lbs. (46–53 Nm).

33. Install the bell housing/oil pan covers.

34. Install the transaxle stay brackets that were removed for access to the bell housing cover.

35. Install the starter assembly and connect the wiring.

36. Install the bolt that secures the oxygen sensor harness to the right side crossmember and tighten the bolt to 7–9 ft. lbs. (10–12 Nm).

37. For vehicles equipped with ACTIVE-ECS, connect the height sensor rod from the lower control arm. Check the height sensor rod for a length (A) of 10.59–10.63 inches (269–270mm).

NOTE: Be sure to keep the height sensor rod (B) equal on both sides of the adjuster and the ball joint at the tip of the rod should centered on the fulcrum.

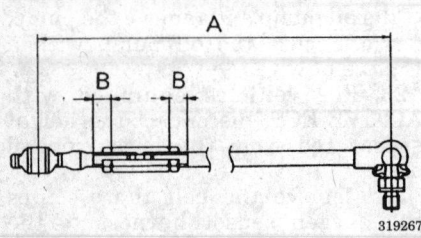

Height sensor rod adjustment — Diamante with a F4A33 automatic transaxle

38. If removed, install the 4WS oil pump and tighten the mounting bolts to 17 ft. lbs. 24 Nm).

39. If removed, install the 4WS oil pump heat shield and tighten the mounting bolts to 17 ft. lbs. 24 Nm).

40. Install the three upper transaxle-to-engine mounting bolts. Tighten the mounting bolts to 54 ft. lbs. (75 Nm).

NOTE: One of the upper bolts has a grounding strap to secure under the bolt, DO NOT forget this strap.

41. Install and connect the transaxle mounting bracket. Tighten the mounting nut and bolts to 51 ft. lbs. (70 Nm).

42. Connect the shift control solenoid valve harness.

43. Connect the kickdown servo switch, pulse generator and oil temperature sensor electrical harness.

44. Connect the park/neutral switch electrical harness.

45. Using new hose clamps, install the oil cooler hoses to the transaxle.

46. Install shifter control cable to the transaxle and secure the cable with clip.

47. Connect the speedometer cable to the transaxle.

48. If removed, install the roll stopper stay bracket and tighten the one through nut and bolt to 36–43 ft. lbs. (50–60 Nm). Tighten the two mounting bolts to 16 ft. lbs. (22 Nm).

49. If removed, install the ACTIVE-ECS compressor assembly. Tighten the mounting bolts to 48 inch lbs. (5 Nm) and connect the electrical harness.

50. Attach the engine harness connection.

51. Install the air cleaner assembly and adjoining duct work.

52. Using new circlips, install the halfshafts and seat halfshafts into the transaxle. Install the bolt that secure the left side support bearing and

tighten the bolts to 33 ft. lbs. (45 Nm).

53. Connect the ball joint and tie rod end to the steering knuckle. Using new nuts, tighten the ball joint castle nut to 43–52 ft. lbs. (60–72 Nm) and tighten the tie rod castle nut to 22 ft. lbs. (30 Nm). Install new cotter pins to both connections.

54. Using new gaskets, install the exhaust system.

55. If removed, install front catalytic converter and exhaust pipe. Be sure to use new gaskets.

56. Install the engine undercovers.

57. Connect the negative battery cable.

58. Refill with Dexron®II.

59. Start the engine and allow to idle for 2 minutes. Apply parking brake and move selector through each gear position, ending in **N**. Recheck fluid level and add if necessary. Fluid level should be between the marks in the **HOT** range.

60. Road test the vehicle.

3000GT with a F4A33 Transaxle

NOTE: If the vehicle is going to be rolled on its wheels while the halfshafts are out of the vehicle, obtain two outer CV-joints or proper equivalent tools and install to the hubs. If the vehicle is rolled without the proper torque applied to the front wheel bearings, the bearings will no longer be usable.

───── CAUTION ─────
Wait at least 90 seconds after the negative (-) battery cable is disconnected to prevent possible deployment of the air bag.

1. Disarm the air bag, if equipped. Remove the battery, battery tray and washer tank.

2. Remove the air cleaner assembly and adjoining duct work.

3. Disconnect the shifter control cable.

4. Disconnect and plug the oil cooler hoses.

5. Disconnect the inhibitor switch, kickdown servo switch, pulse generator, oil temperature sensor, shift control solenoid valve, and ground cable.

6. Disconnect the speedometer cable.

7. Raise the vehicle and support safely. Remove the undercovers.

8. Support the weight of the transaxle and remove the mount bracket. Remove the upper bellhousing bolts.

9. Disconnect the tie rod end and ball joint from the steering knuckle.

10. Remove the right frame member.

11. Remove the starter.

12. Remove the halfshafts by inserting a prybar between the transaxle case and the driveshaft and prying the shaft from the transaxle. Do not pull on the driveshaft. Doing so damages the inboard joint. Use the prybar. Do not insert the prybar so far the oil seal in the case is damaged. Tie the halfshafts aside.

13. Remove the remaining mounting brackets.

14. Remove the bellhousing cover plate.

15. Remove the special bolts holding the flexplate to the torque converter.

16. After removing the bolts, push the torque converter toward the transaxle so it doesn't stay on the engine side and allow oil to pour out the converter hub.

17. Remove the lower transaxle to engine bolts and remove the transaxle assembly.

To install:

18. After the torque converter has been mounted on the transaxle, install the transaxle assembly on the engine. Tighten the driveplate bolts to 34–38 ft. lbs. (46–53 Nm). Install the bellhousing cover.

19. Install the mounting brackets.

20. Replace the circlips and install the halfshafts to the transaxle.

21. Install the starter and frame member.

22. Install the tie rods and ball joint to the steering arm.

23. Install the upper bellhousing bolts.

24. Install the transaxle mounting bracket.

25. Install the undercovers.

26. Connect the speedometer cable.

27. Connect the inhibitor switch, kickdown servo switch, pulse generator, oil temperature sensor, shift control solenoid valve, and ground cable.

28. Connect the oil cooler hoses.

29. Connect the shifter control cable.

30. Install the air cleaner assembly and adjoining duct work.

31. Install the washer tank, battery tray and battery.

32. Refill with DEXRON II, Mopar ATF Plus type 7176 or equivalent, automatic transaxle fluid.

33. Start the engine and allow it to idle for two minutes. Apply parking brake and move selector through each gear position, ending in **N**. Recheck fluid level and add if necessary. Fluid level should be between the marks in the **HOT** range.

Throttle Valve Cable

ADJUSTMENT

Mirage and 1993–94 Eclipse

1. Place selector lever and manual control lever in **N** position.
2. Loosen adjusting nut. While lightly pulling on control cable tighten mounting nut to 7–10 ft. lbs. (10–14 Nm).
3. When adjustment is complete, be sure selector lever is still in the **N** position. Verify all functions correspond to the position indicated on the selector lever.

Transfer Case Assembly

REMOVAL AND INSTALLATION

Eclipse

1. Raise and safely support the vehicle.
2. Remove the engine undercovers.
3. Remove the front exhaust pipe.
4. Drain the transfer case fluid.
5. Remove the transfer case mounting bolts.
6. Support the driveshaft with wire or string and remove the transfer case from the transaxle.

To install:

7. Slide the driveshaft into the transfer case and install the transfer case to the transaxle. Torque the bolts to 40–44 ft. lbs. (54–59 Nm).
8. Install the front exhaust pipe.
9. Refill the transfer case with the proper gear oil.
10. Install the engine undercover.
11. Safely lower the vehicle to the floor.

3000GT

1. Disconnect the negative battery cable.

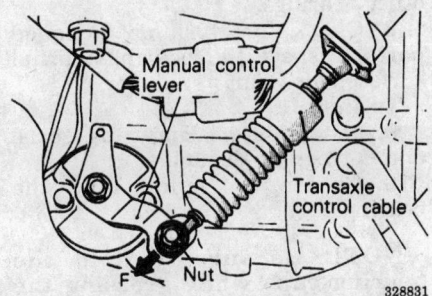

Adjusting nut location — Mirage and 1993–94 Eclipse with F3A21, F4A22, F4A23, F4A33, W4A32 and W4A33 automatic transaxles

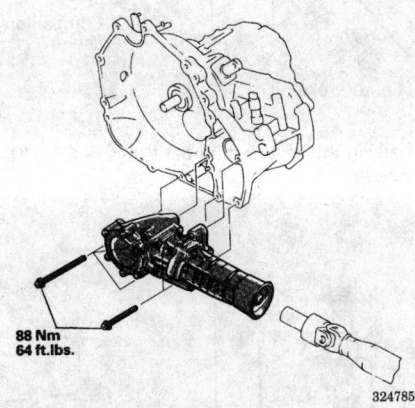

Transfer case assembly — 3000GT with F5M33, W5MG1 and W6MG1 transfer case

---

— **CAUTION** —

Wait at least 90 seconds after the negative (-) battery cable is disconnected to prevent possible deployment of the air bag.

---

2. Raise the vehicle and support safely. Drain the transfer assembly oil.
3. Remove necessary front bumper components.
4. Disconnect the front exhaust pipe.
5. Unbolt the transfer case assembly and remove by sliding it off the rear driveshaft. Be careful not to damage the oil seal in the transfer case output housing. Do not let the rear driveshaft hang; suspend it from a frame piece. Cover the opening in the transaxle and transfer case to keep oil from dripping and to keep dirt out.

To install:

6. Lubricate the driveshaft sleeve yoke and oil seal lip on the transfer extension housing. Install the transfer case assembly to the transaxle. Use care when installing the rear driveshaft to the transfer case output shaft.

7. Tighten the transfer case to transaxle bolts to:
- 1993: 61–65 ft. lbs. (85–90 Nm)
- 1994–96: 18–22 ft. lbs. (25–29 Nm)
8. Install the exhaust pipe using a new gasket. Install removed bumper components.
9. Refill the transfer case and check the oil levels in the transaxle and transfer case.
10. Safely lower the vehicle and connect the negative battery cable.

DRIVE AXLE

Driveshaft

REMOVAL AND INSTALLATION

Eclipse, 3000GT and 1993 Galant

1. Disconnect the negative battery cable. Raise the vehicle and support safely.
2. The rear driveshaft is a 3-piece unit, with a front, center and rear propeller shaft. Remove the nuts and insulators from the center support bearing. Work carefully. There will be a number of spacers which will differ from vehicle to vehicle. Check the number of spacers and write down their locations for reference during reassembly.
3. Matchmark the rear differential companion flange and the rear driveshaft flange yoke. Remove the companion shaft bolts and remove the driveshaft, keeping it as straight as possible so as to ensure that the boot is not damaged or pinched. Use care to keep from damaging the oil seal in the output housing of the transfer case.

NOTE: Damage to the boot can be avoided and work will be easier if a piece of cloth or similar material is inserted in the boot.

4. Do not lower the rear of the vehicle or oil will flow from the transfer case. Cover the opening to keep dirt out.

To install:

5. Install the driveshaft to the vehicle and align the matchmarks at the rear yoke.
6. Install the bolts at the rear differential flange and torque to 36–43 ft. lbs. (50–60 Nm) on 3000GT models and 22–25 ft. lbs. (30–35 Nm) on all other models.

7. Install the center support bearing with all spacers in place. Torque the retaining nuts to 22–25 ft. lbs. (30–35 Nm).

8. Check the fluid levels in the transfer case and rear differential case.

U-Joints

REMOVAL AND INSTALLATION

Eclipse

1. Raise and properly support vehicle. Remove driveshaft and secure in vise.

2. Make mating marks on the yoke and the universal joint that is to be disassembled. Remove the snaprings from the yoke with snapring pliers.

3. Force out the bearing journals from the yoke using a large C-clamp. Install a collar on the fixed side of the C-clamp. Press the journal bearing into the collar by applying pressure with the C-clamp, on the opposite side.

4. Pull the journal bearing from the yoke.

NOTE: If the journal bearing is hard to remove, strike the yoke with a plastic hammer.

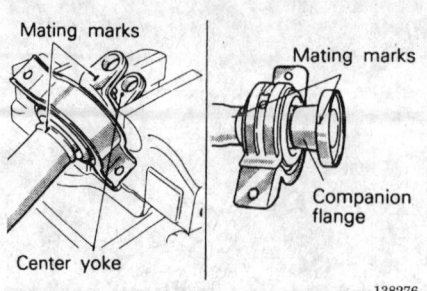

Marking the driveshaft to the flange — 1993–94 Eclipse

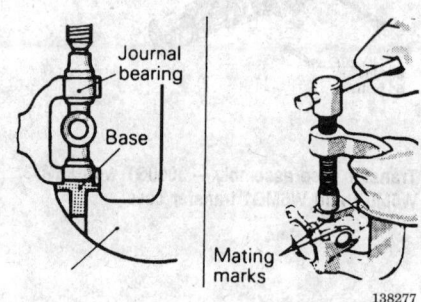

Pressing out universal joint — 1993–94 Eclipse

5. Press the journal shaft using C-clamp or similar tool, to remove the remaining bearings.

6. Once all bearings are removed, remove the journal.

To install:

7. Apply multi-purpose grease to the shafts, grease sumps, dust seal lips and needle roller bearings of the replacement U-joint. Do not apply excessive grease. Otherwise, it may be difficult to install the bearing caps and errors in selection of snaprings may result.

8. Press fit the journal bearings to the yoke using a C-clamp as follows:

   a. Install a solid base onto the bottom of the C-clamp.

   b. Insert both bearings into the yoke. Hold and press fit them by tightening the C-clamp.

   c. Install snaprings of the same thickness onto both sides of each yoke.

   d. Press the bearing and journal into 1 side by using a brass bar with diameter of 0.59 in. (15mm).

9. Measure the clearance between the snapring and the groove wall of the yoke with a feeler gauge. If clearance exceeds 0.0008–0.0024 in. (0.02-0.06mm) on 1993–94 models and 0.0004–0.0012 in. (0.01–0.03mm) on 1995–97 models, the snaprings should be replaced.

3000GT

1. Remove the driveshaft from the vehicle.

2. Matchmark the yoke and the driveshaft.

3. Using a brass bar and hammer, slightly tap in the bearing outer race.

4. Remove the four snaprings from the bearings.

5. Push out the bearing from the flange.

6. Clamp the bearing outer race in a vise and tap off the flange with a hammer.

7. Repeat Steps 3, 4, and 5 for the other bearings.

8. Check for worn or damaged parts. Inspect the bearing journal surfaces for wear.

To install:

9. Install the bearing cups, seals, and O-rings on the spider.

10. Grease the spider and the bearings.

NOTE: Be sure to hold the bearing caps while greasing the U-joints. The grease will force the bearing caps off the spider when they are not secured in the driveshaft yoke.

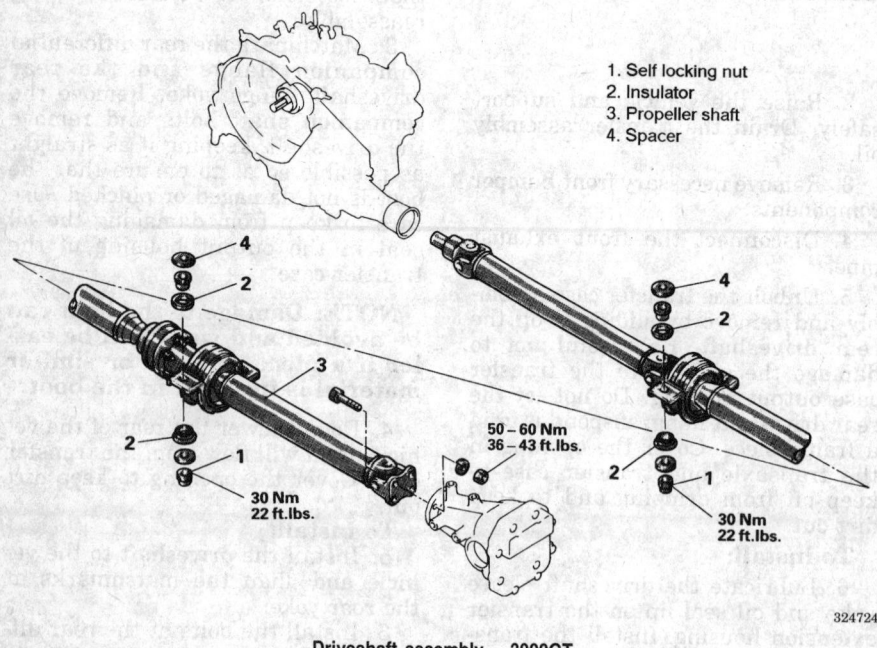

1. Self locking nut
2. Insulator
3. Propeller shaft
4. Spacer

50 – 60 Nm
36 – 43 ft.lbs.

30 Nm
22 ft.lbs.

30 Nm
22 ft.lbs.

Driveshaft assembly — 3000GT

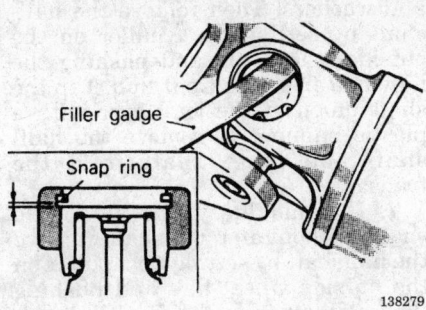

Checking universal joint installation — 1993–94 Eclipse

Snap ring thickness mm (in.)	Identification color
1.28 (.0503)	–
1.31 (.0516)	Yellow
1.34 (.0528)	Blue
1.37 (.0539)	Purple
1.40 (.0551)	Brown

Snapring identification chart — 1995–97 Eclipse

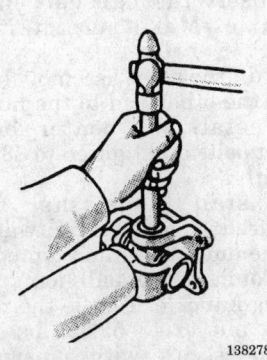

Installing new universal joint — 1993–94 Eclipse

11. Remove the bearing cups from the spider.

12. Position the spider in the yoke.

13. Start the bearings in the yoke and then press them into place, using a vise. Install the snaprings to hold the bearing cups in place.

14. Make sure the bearings and snaprings are fully seated by lightly tapping on the yoke with a hammer.

15. If the axial play of the spider is greater than 0.024 inch (0.06 mm), select snaprings which will provide the correct play. Be sure that the snaprings are the same size on both

sides or driveshaft noise and vibration will result.

16. Install the driveshaft in the vehicle.

Halfshaft

REMOVAL AND INSTALLATION

1993–94 Precis

1. Remove the hub center cap and loosen the driveshaft (axle) nut.

2. Loosen the wheel lug nuts.

3. Raise and support the front of the vehicle safely.

4. Remove the front wheels.

5. Remove the engine splash shield.

6. Remove the lower ball joint and strut bar from the lower control arm.

NOTE: Place the lower arm ball joint on the lower arm to prevent damage to the ball joint dust boot.

7. Drain the transaxle fluid into a suitable waste container.

8. Insert a prybar between the transaxle case (on the raised rib) and the driveshaft inner joint case. Move the bar to the right to withdraw the left driveshaft; left, to remove the right driveshaft.

NOTE: Do not insert the pry bar too deeply (7mm) or you will damage the oil seal.

9. Plug the transaxle case with a clean rag to prevent dirt from entering the case.

10. Use a puller/driver mounted on the wheel studs to push the driveshaft from the front hub. Take care to prevent the spacer shims from falling out of place.

To install:

11. To install, insert the driveshaft into the hub first, then install the transaxle end.

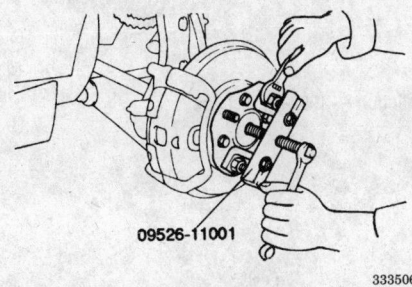

Removing the halfshaft from the hub — 1993–94 Precis

12. Install the lower ball joint mounting bolts and the strut bar to the lower control arm.

13. Install the wheel and tire assembly.

14. Install the splash shield if equipped.

15. Lower the vehicle to the floor.

16. Install the hub nut washer. Torque the axle shaft hub nut to 148–192 ft. lbs.; the lower arm-to-ball joint nuts to 69–87 ft. lbs.

NOTE: Always use a new inner joint retaining ring every time you remove the driveshaft.

17. Check and add transaxle fluid if necessary.

1993–94 Eclipse

1. Raise the vehicle and support safely.

2. Remove the bolts that attach the rear halfshaft to the companion flange.

3. Use a suitable tool to pry the inner shaft out of the differential case. Don't insert the bar too far or the seal could be damaged.

4. Remove the rear driveshaft from the vehicle.

To install:

5. Replace the circlip and install the rear driveshaft to the differential case. Make sure it snaps in place.

6. Install the companion flange bolts and tighten to 40–47 ft. lbs. (55–65 Nm).

7. Check the fluid level in the rear differential.

Diamante and Mirage

NOTE: If the vehicle is going to be rolled while the halfshafts are out of the vehicle, obtain 2 outer CV-joints or proper equivalent tools and install to the hubs. If the vehicle is rolled without the proper torque applied to the front wheel bearings, the bearings will no longer be usable.

1. Raise the vehicle and support it safely.

2. Remove the cotter pin, halfshaft nut and washer.

3. If equipped with ABS, remove the front wheel speed sensor.

4. If equipped with Active Electronic Control Suspension, disconnect the front height sensor from the lower control arm.

5. Disconnect the lower ball joint and the tie rod end from the steering knuckle.

6. If removing the left side axle with an inner shaft, remove the center support bearing bracket bolts

and washers. Then remove the half-shaft by setting up a puller on the outside wheel hub and pushing the halfshaft from the front hub. Tap the shaft union at the joint case with a plastic hammer to remove the halfshaft and inner shaft from the transaxle.

7. If removing right side axle shafts without an inner shaft, remove the halfshaft by setting up a puller on the outside wheel hub and pushing the halfshaft from the front hub. After pressing the outer shaft, insert a prybar between the transaxle case and the halfshaft and pry the shaft from the transaxle.

NOTE: Do not pull on the shaft; doing so damages the inboard joint. Do not insert the prybar too far or the oil seal in the case may be damaged.

To install:

8. Inspect the halfshaft boot for damage or deterioration. Check the ball joints and splines for wear.

9. Replace the circlips on the ends of the halfshafts.

10. Insert the halfshaft into the transaxle. Make sure it is fully seated.

11. Pull the strut assembly out and install the other end to the hub.

12. Install the center bearing bracket bolts and tighten to 33 ft. lbs. (45 Nm).

13. Install the washer so the chamfered edge faces outward. Install the nut and tighten temporarily.

14. Connect the ball joint to the steering knuckle. Torque the new retaining nut to 43–52 ft. lbs. (60–72 Nm) and secure with a new cotter pin.

15. Connect the tie rod end to the steering knuckle. Torque the retaining nut to 21 ft. lbs. (29 Nm) and secure with a new cotter pin.

16. If equipped with ABS, install the front wheel speed sensor.

17. If equipped with Active Electronic Control Suspension, connect the front height sensor to the lower control arm.

18. Install the wheel and lower the vehicle to the floor.

19. Tighten the axle nut with the brakes applied, to a maximum torque of 145–188 ft. lbs. (200–260 Nm) and secure with a new cotter pin.

20. Check the transaxle fluid and fill if necessary.

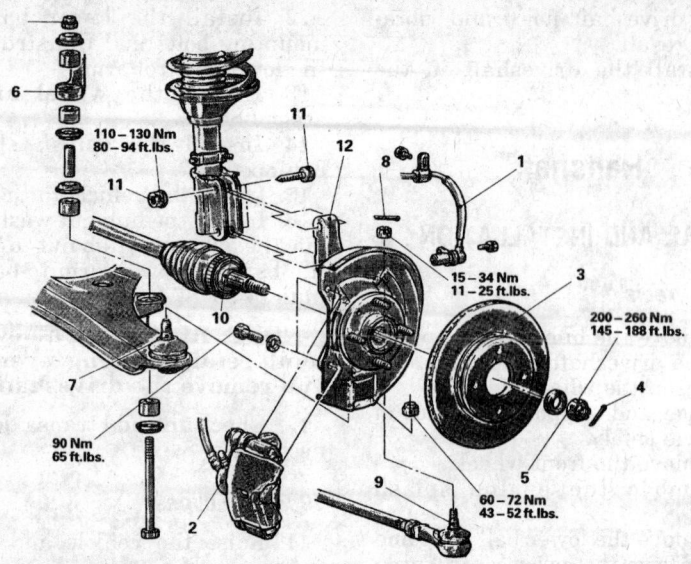

1. Front speed sensor <Vehicle with ABS>
2. Caliper assembly
3. Brake disc
4. Cotter pin
5. Drive shaft nut
6. Connection for stabilizer bar
7. Connection for lower arm ball joint
8. Cotter pin
9. Connection for tie rod end
10. Drive shaft
11. Front strut mounting bolt and nut
12. Hub and knuckle

330135

Front axle shaft and related parts — Mirage

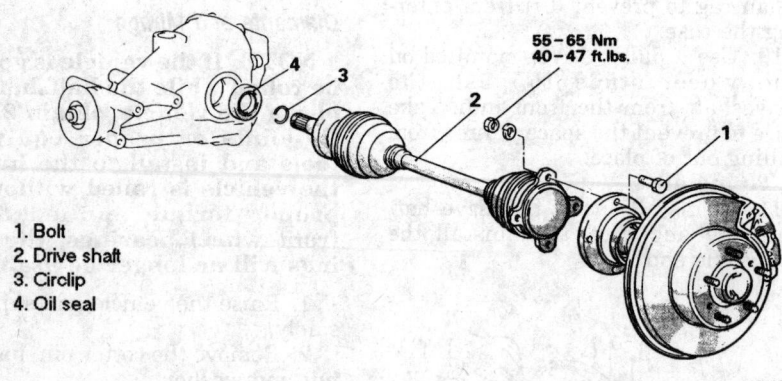

1. Bolt
2. Drive shaft
3. Circlip
4. Oil seal

138236

Rear axle shaft removal and installation — 1993–94 Eclipse

3000GT

NOTE: If the vehicle is going to be rolled on its wheels while the halfshafts are out of the vehicle, obtain two outer CV-joints or proper equivalent tools and install to the hubs. If the vehicle is rolled without the proper torque applied to the front wheel bearings, the bearings will no longer be usable.

Front Halfshaft

1. Disconnect the negative battery cable.

2. With the vehicle on the floor and the brakes applied, remove the cotter pin, halfshaft nut and the washer.

3. Raise the vehicle and support safely. Remove the lower ball joint and the tie rod end from the steering knuckle.

4. On vehicles with an inner shaft, remove the center support bearing bracket bolts and washers.

5. On vehicles with an inner shaft, remove the halfshaft by setting up a puller on the outside wheel hub and pushing the halfshaft from the front hub. Then tap the shaft union at the joint case with a plastic hammer to remove the halfshaft shaft and inner shaft from the transaxle.

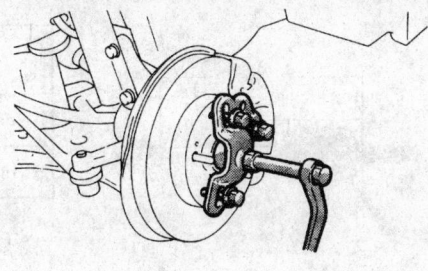

324727

Front halfshaft removal — 3000GT

6. On vehicles without an inner shaft, remove the halfshaft by setting up a puller on the outside wheel hub and pushing the halfshaft from the front hub. After pressing the outer shaft, insert a prybar between the transaxle case and the halfshaft and pry the shaft from the transaxle. Do not pull on the shaft. Doing so damages the inboard joint. Do not insert the prybar too far or the oil seal in the case may be damaged.

To install:

7. Inspect the halfshaft boot for damage or deterioration. Check the ball joints and splines for wear.

8. Replace the circlips on the ends of the halfshafts.

9. Insert the halfshaft into the transaxle. Make sure it is fully seated.

10. Pull the strut assembly out and install the other end to the hub.

11. Install the center bearing bracket bolts and tighten to 33 ft. lbs. (45 Nm).

12. Install the washer so the chamfered edge faces outward. Install the nut and tighten temporarily.

13. Install the tie rod end and ball joint.

14. Install the wheel and lower the vehicle to the floor. Tighten the axle nut with the brakes applied. Tighten the nut to a maximum torque of 188 ft. lbs. (260 Nm). Install the cotter pin and bend to secure.

Rear Halfshaft

NOTE: On vehicles with Limited Slip Differential, the right and left halfshafts are not the same. If both halfshafts are to be removed, be sure to mark one of the halfshafts (left or right) for proper installation.

1. Disconnect the negative battery cable. Raise the vehicle and support it safely.

2. Matchmark the halfshaft and the companion flange.

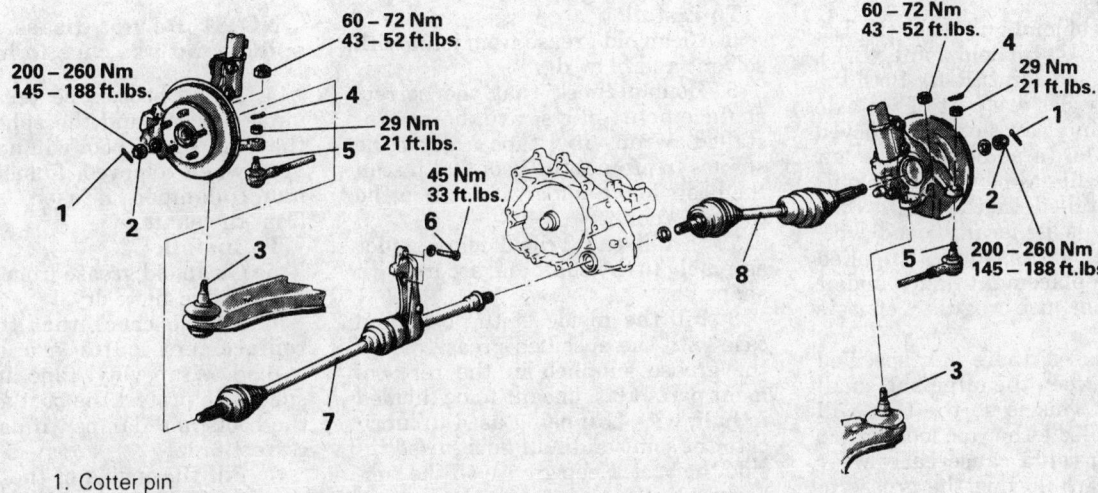

1. Cotter pin
2. Drive shaft nut
3. Lower arm ball joint connection
4. Cotter pin
5. Tie rod end connection
6. Center bearing bracket installation bolt
7. Drive shaft and inner shaft assembly (L.H.)
8. Drive shaft (R.H.)
9. Circlip

Caution
In the case of AWD-vehicles with A.B.S., take care not to damage the rotor for A.B.S. installed to the B.J. outer race.

324725

Front halfshafts and related components — 3000GT

3. Remove the bolts that attach the rear halfshaft to the companion flange.

4. Use a prybar to pry the inner shaft out of the differential case. Don't insert the prybar too far or the seal could be damaged.

5. Remove the rear halfshaft from the vehicle.

To install:

6. Install a new circlip on the halfshaft and install it into the differential. Make sure it is fully seated.

7. Align the matchmarks and attach the halfshaft to the companion flange.

8. Safely lower the vehicle and connect the negative battery cable.

CV-Joint Boot

REPLACEMENT

These vehicles use several different types of joints. Engine size, transaxle type, whether the joint is an inboard or outboard joint, even which side of the vehicle is being serviced could make a difference in joint type. Be sure to properly identify the joint before attempting joint or boot replacement. Look for identification numbers at the large end of the boots and/or on the end of the metal retainer bands.

The types of joints used are the Tripod Joint (T.J.), Birfield Joint, (B.J.), and the Rzeppa Joint (R.J.). Both the outer joint styles are part of the axle assembly, and can not be removed from the axle. In addition, some left side shafts will have a round dynamic damper installed on the shaft. Special grease is generally used with these joints and is often supplied with the replacement joint and/or boot. Do not use regular chassis grease.

In most cases, there is a specified distance between the large and small boot bands. This is so the boot will not be installed either too loose or too tight, which could cause early wear and cracking, allowing the grease to get out and water and dirt in leading to early joint failure.

Mirage

1. Raise and properly support the vehicle. Remove the halfshaft.

2. Remove the T.J. boot bands from the boot. Side cutter pliers can be used to cut off the metal retaining bands. Remove the T.J. case from the halfshaft.

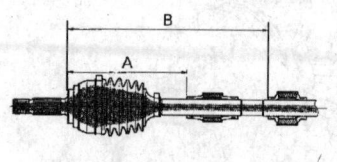

Items		1.5L Engine	1.8L Engine	
			L.H.	R.H.
A	mm (in.)	351 ± 3 (13.82 ± .12)	365 ± 3 (14.37 ± .12)	200.5 ± 3*² (7.89 ± .12)
B	mm (in.)	481 ± 3*¹ (18.94 ± .12)	–	–

NOTE
*¹: A/T
*²: M/T

329602

Dynamic damper installation position — Mirage

3. Remove the snapring next to the Tripod joint spider assembly from the halfshaft with snapring pliers and remove the spider assembly from the shaft. Do not disassemble the spider and use care in handling.

NOTE: Both of the halfshaft boots are going to be removed from the T.J. case side of the halfshaft.

4. If the boot is be reused, wrap vinyl tape around the spline part of the shaft so the boot will not be damaged when removed. Remove the dynamic damper, if used, and boots from the shaft.

To install:

5. Clean old grease from joint with solvent and blow dry.

6. Double check that the correct replacement parts are being installed. Wrap vinyl tape around the splines to protect the boot and install the boots and damper, if used, in the correct order.

7. Install the Tripod joint spider assembly to the shaft and secure with snapring.

8. Fill the inside of the boot and case with the specified grease. Often the grease supplied in the replacement parts kit is meant to be divided in half, with half being used to lubricate the joint and half being used inside the boot. Keep grease off the rubber part of the dynamic damper (if used).

9. Install the T.J. case to the spider assembly.

10. Secure the boot bands with the halfshaft in a horizontal position. Make sure the boot span on the halfshaft is 3.25 ±0.12 in. (80 ±3mm) in length.

11. Check dynamic damper for proper positioning and adjust if necessary.

12. Install the halfshaft.

Eclipse and 1993 Galant

Front — FWD Vehicles

Both of the halfshaft boots are going to be removed from the T.J. case side of the halfshaft.

1. Disconnect the negative battery cable. Remove the halfshaft from the vehicle.

2. Remove the T.J. boot bands from the boot. Side cutter pliers can be used to cut off the metal retaining bands. Remove the T.J. case from the halfshaft.

3. Remove the snapring next to the Tripod joint spider assembly from the halfshaft with snapring pliers. Remove the spider assembly from the shaft.

NOTE: Do not disassemble the spider and use care in handling.

4. If the boot is be reused, wrap vinyl tape around the spline part of the shaft so the boot will not be damaged when removed. Remove the dynamic damper, if used, and boots from the shaft.

To install:

5. Clean old grease from joint with solvent and blow dry.

6. Double check that the correct replacement parts are being installed. Wrap vinyl tape around the splines to protect the boot and install the boots and damper, if used, in the correct order.

7. Fill the inside of the boot with the specified grease. Often the grease supplied in the replacement parts kit is meant to be divided in half, with half being used to lubricate the joint and half being used inside the boot. Keep grease off the rubber part of the dynamic damper (if used).

8. Secure the boot bands with the halfshaft in a horizontal position. Make sure the boot span on the halfshaft is 3.15 ±0.12 in. (80 ±3mm) in length.

9. Install halfshaft into vehicle.

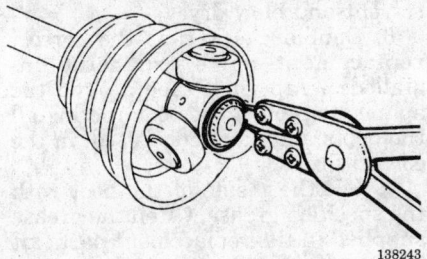

Front inner joint disassembly — Eclipse
138243

1. Circlip
2. Boot band (B)
3. Boot band (C)
4. Circlip
5. D.O.J. outer race
6. Snap ring
7. D.O.J. inner race, cage and ball assembly
8. Balls
9. D.O.J. inner race
10. D.O.J. cage
11. D.O.J. boot
12. Boot band (A)
13. Boot band (C)
14. B.J. boot
15. Dust cover
16. Drive shaft and B.J.

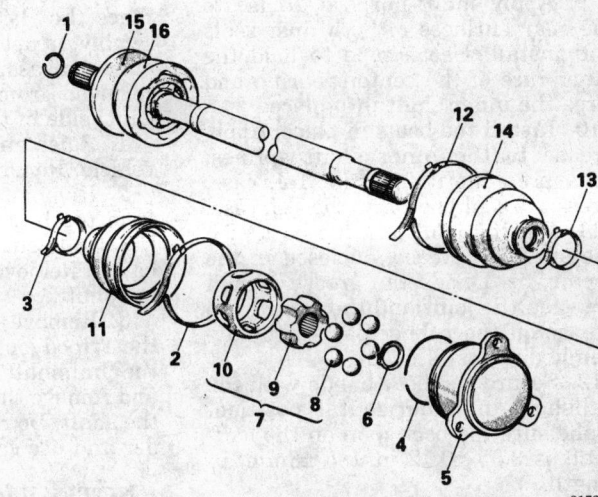

Rear driveshaft exploded view — 1995–97 Eclipse
317298

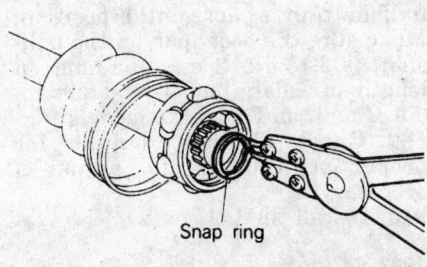

Snap ring
138246

Rear axle disassembly — Eclipse

1. T.J. boot band
2. Boot band (small)
3. T.J. case
4. Snap ring
5. Spider assembly
6. T.J. boot
7. B.J. boot band
8. Boot band (small)
9. B.J. boot
10. B.J. assembly
11. Dust cover
12. Circlip

Front axle shaft exploded view — 1995–97 Eclipse
317295

10. Check transaxle fluid level and fill if necessary.

Front — AWD Vehicles

1. Raise and properly support vehicle. Remove the halfshaft from the vehicle.

2. Remove the T.J. large and small boot bands. Remove the T.J. case from inner shaft assembly.

3. Remove the snapring next to the Tripod joint spider assembly from the halfshaft with snapring pliers. Remove the spider assembly from the shaft.

NOTE: Do not disassemble the spider and use care in handling.

4. If the boot is be reused, wrap vinyl tape around the spline part of the shaft so the boot will not be damaged when removed.

5. Remove the inner and the outer dust seals from the center support bearing assembly. Remove the center bearing from the shaft.

6. Remove the inner shaft assembly, together with the seal plate, from the T.J. case. Using puller tool, remove the inner shaft from the center bearing bracket.

To install:

7. Clean old grease from joint(s) with solvent and blow dry.

8. Apply multi-purpose grease to the center bearing and inside the center bearing bracket. Using proper size driver, press fit the center bearing into the center bearing bracket.

9. Apply multi-purpose grease to the rear surfaces of both dust seals and install. Use a pipe to hold the inner race of the center bearing and force the inner shaft into place.

10. Install the boots in place. Apply grease to the inner shaft splines, then press fit it into the T.J. case. Press the seal plate into the T.J. case.

11. Fill the joint and the boot with the specified grease, enclosed in the repair kit. Divide the grease in half between the joint and the boot. Keep grease off the rubber part of the dynamic damper (if used).

12. Secure the boot bands with the halfshaft in a horizontal position. Make sure the boot span on the halfshaft is 3.35 ±0.12 in. (85 ±3mm) in length.

13. Install halfshaft into vehicle.

14. Check transaxle fluid level and fill if necessary.

Rear — AWD Vehicles

Both boots are removed from the DOJ (outer) side of the halfshaft.

1. Raise and properly support vehicle. Remove the halfshaft from the vehicle.

2. Remove the outer joint large and small boot bands, and slide boot back out of way.

3. Remove outer case circlip and slide case from inner shaft assembly.

4. Remove the inner snapring, located on the end of the axle shaft. Slide ball and cage assembly from the halfshaft.

NOTE: Do not disassemble the cage and ball assembly.

5. Remove inner boot clamps and slide boot from axle.

To install:

6. Clean old grease from joint(s) with solvent and blow dry.

7. Pack joint(s) with specified grease, enclosed in repair kit. Divide the grease in half between the joint and the boot. Slide boots and clamps onto axle shaft.

8. Reinstall the ball and cage assembly onto the axleshaft. Secure with snapring.

9. Slide outer cage assemble onto shaft and secure with circlip.

10. Tighten the boot bands with the halfshaft in a horizontal position. Make sure the boot span on the halfshaft is 3.00 ±0.23 in. (76 ±3mm) in length.

11. Install halfshaft into vehicle.

12. Check differential fluid level and fill if necessary.

Diamante and 1994–97 Galant

Since the B.J. joint (outer joint) assembly, is not serviceable and should not be disassembled, both joints are removed from the T.J. joint (inner joint) side of the axle shaft.

1. Raise and properly support vehicle. Remove the halfshaft.

2. Remove the T.J. boot bands from the boot. Side cutter pliers can be used to cut off the metal retaining bands. Remove the T.J. case from the halfshaft.

3. Remove the snapring next to the Tripod joint spider assembly from the halfshaft with snapring pliers and remove the spider assembly from the shaft. Do not disassemble the spider and use care in handling.

NOTE: Both of the halfshaft boots are going to be removed from the T.J. case side of the halfshaft.

4. If the boot is be reused, wrap vinyl tape around the spline part of the shaft so the boot will not be damaged when removed. Remove the dynamic damper, if used, and boots from the shaft.

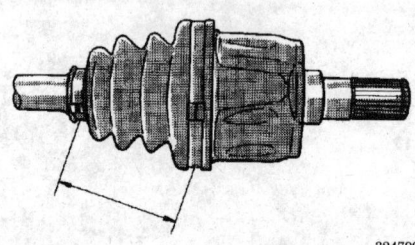

Checking the installed length of the CV boot — 1994–97 Galant

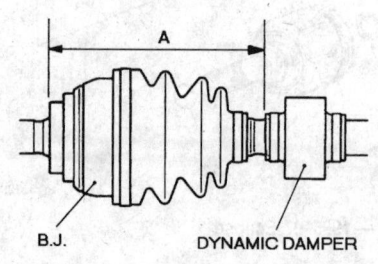

Dynamic damper installation — 1994–97 Galant

To install:

5. Clean old grease from joint with solvent and blow dry.

6. Double check that the correct replacement parts are being installed. Wrap vinyl tape around the splines to protect the boot and install the boots and damper, if used, in the correct order.

7. Fill the inside of the boot with the specified grease. Often the grease supplied in the replacement parts kit is meant to be divided in half, with half being used to lubricate the joint and half being used inside the boot. Keep grease off the rubber part of the dynamic damper (if used).

8. Secure the boot bands with the halfshaft in a horizontal position. Make sure the boot span on the halfshaft is 3.15 ±0.12 in. (80 ±3mm) in length on Galant models and 2.95 in. (75 ±3mm) on Diamante models.

9. Check dynamic damper for proper positioning and adjust if necessary.

10. Install the halfshaft.

1993–94 Precis

NOTE: Do not disassemble a Birfield joint. Service with a new joint or clean and repack using a new boot kit. The driveshaft joints use special grease, do not add any grease other than that supplied with the kit.

The Double Offset Joint (D.O.J.) is bigger than other joints and in these applications, is normally used as an inboard joint.

1. Remove the halfshaft from the vehicle.

2. Side cutter pliers can be used to cut the metal retaining bands. Remove the boot from the joint outer race.

3. Locate and remove the large circlip at the base of the joint. Remove the outer race (the body of the joint).

4. Remove the small snapring and take off the inner race, cage and balls as an assembly. Clean the inner race, cage and balls without disassembling.

5. If the boot is to be reused, wipe the grease from the splines and wrap the splines in vinyl tape before sliding the boot from the shaft.

6. Remove the inner (D.O.J.) boot from the shaft. If the outer (B.J.) boot is to be replaced, remove the boot retainer rings and slide the boot down and off of the shaft at this time.

To install:

7. Be sure to tape the shaft splines before installing the boots. Fill the inside of the boot with the specified grease. Often the grease supplied in

boot. Do not use regular chassis grease.

NOTE: If the Birfield joint boot is being replaced, the Tripod joint will have to be disassembled and the new boot can then be installed from the Tripod joint side of the shaft. The Birfield joint cannot be disassembled.

1. Disconnect the negative battery cable.

CAUTION
Wait at least 90 seconds after the negative (-) battery cable is disconnected to prevent possible deployment of the air bag.

2. Raise and safely support the vehicle and remove the halfshaft.
3. Remove the bands securing the boot to the CV-joint and remove the boot. The damaged boot may be cut off if needed.
4. Remove the Tripod joint case from the spider assembly. Do not disassemble the spider and use care in handling.
5. Remove the snapring and the spider assembly from the shaft.
6. If the boot is be reused, wrap vinyl tape around the spline part of the shaft so the boot will not be damaged when removed. Remove the dynamic damper, if used, and boots from the shaft.
7. Thoroughly clean the old grease out of the CV-joint and blow dry it with compressed air.
To install:
8. Double check that the correct replacement parts are being installed. Wrap vinyl tape around the splines to protect the boot and install the boots and damper, if used, in the correct order.
9. Fill the inside of the boot with the specified grease. Often the grease supplied in the replacement parts kit is meant to be divided in half, with half being used to lubricate the joint and half being used inside the boot. Keep grease off the rubber part of the dynamic damper (if used).
10. Secure the boot bands with the halfshaft in a horizontal position.
11. Install the halfshaft.
12. Install the wheel and tire assembly if removed.
13. Safely lower the vehicle and connect the negative battery cable.

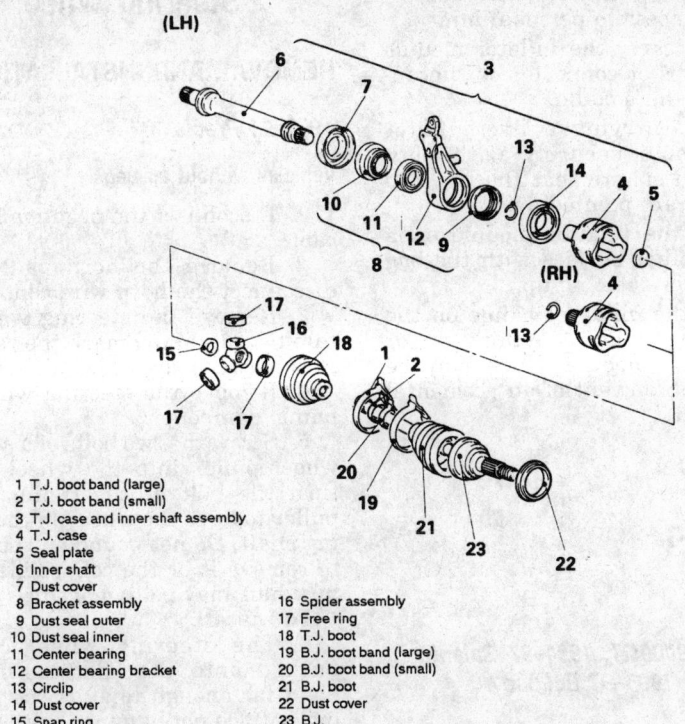

1 T.J. boot band (large)
2 T.J. boot band (small)
3 T.J. case and inner shaft assembly
4 T.J. case
5 Seal plate
6 Inner shaft
7 Dust cover
8 Bracket assembly
9 Dust seal outer
10 Dust seal inner
11 Center bearing
12 Center bearing bracket
13 Circlip
14 Dust cover
15 Snap ring
16 Spider assembly
17 Free ring
18 T.J. boot
19 B.J. boot band (large)
20 B.J. boot band (small)
21 B.J. boot
22 Dust cover
23 B.J.

Exploded view of front axle shaft assembly — Diamante

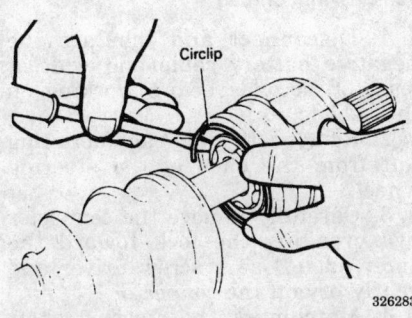

Remove the large circlip from the outer race — 1993–94 Precis

Tape the splines to protect the boot — 1993–94 Precis

the replacement parts kit is meant to be divided in half, with half being used to lubricate the joint and half being used inside the boot.

8. Install the cage onto the halfshaft so the small diameter side of the cage is installed first. With a brass drift pin, tap lightly and evenly around the inner race to install the race until it comes into contact with the rib of the shaft. Apply the specified grease to the inner race and cage and fit them together. Insert the balls into the cage.

9. Install the outer race (the body of the joint) after filling with the specified grease. The outer race should be filled with this grease.

10. Tighten the boot bands securely. Make sure the distance between the boot bands is correct.

11. Install the halfshaft to the vehicle.

3000GT

The two types of joints used are the Birfield Joint, (B.J.), the Tripod Joint (T.J.). In addition, some left side shafts will have a center bearing bracket installed on the shaft. Special grease is generally used with these joints and is often supplied with the replacement joint and/or

Remove the spider assembly — 3000GT

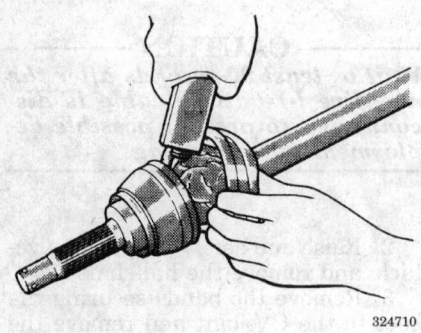

Apply the grease to the inside of the boot — 3000GT

STEERING

Air Bag

CAUTION

Some vehicles are equipped with an air bag system, also known as the Supplemental Inflatable Restraint (SIR) or Supplemental Restraint System (SRS). The system must be disabled before performing service on or around system components, steering column, instrument panel components, wiring and sensors. Failure to follow safety and disabling procedures could result in accidental air bag deployment, possible personal injury and unnecessary system repairs.

PRECAUTIONS

Several precautions must be observed when handling the inflator module to avoid accidental deployment and possible personal injury.

• Never carry the inflator module by the wires or connector on the underside of the module.

• When carrying a live inflator module, hold securely with both hands, and ensure that the bag and trim cover are pointed away.

• Place the inflator module on a bench or other surface with the bag and trim cover facing up.

• With the inflator module on the bench, never place anything on or close to the module which may be thrown in the event of an accidental deployment.

DISARMING

Diamante, 3000GT, 1994–97 Galant, Mirage and 1995–97 Eclipse

1. Position the front wheels in the straight-ahead position and place the key in the **LOCK** position. Remove the key from the ignition lock cylinder.

2. Disconnect the negative battery cable and insulate the cable end with high-quality electrical tape or similar non-conductive wrapping.

3. Wait at least one minute before working on the vehicle. The air bag system is designed to retain enough voltage to deploy the air bag for a short period of time after the battery has been disconnected.

To arm:

4. Connect the negative battery cable, turn the ignition switch to the **ON** position and check the SRS warning light for proper operation.

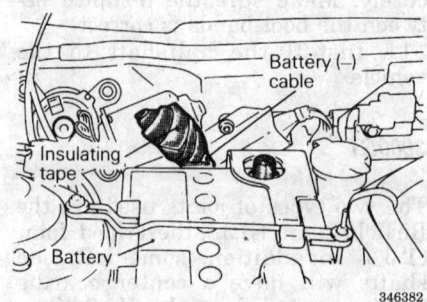

Insulate the negative battery cable to prevent accidental deployment of the air bag

Steering Wheel

REMOVAL AND INSTALLATION

1993–94 Precis

Vehicles without air-bag

1. Disconnect the negative battery cable.

2. Remove the horn pad. Then, disconnect the horn wire connector.

3. Remove the steering wheel retaining nut. Matchmark the relationship between the wheel and shaft.

4. Remove the steering wheel dynamic dampener.

5. Screw the two bolts of a steering wheel puller into the wheel. Then, turn the bolt at the center of the puller to force the wheel off the steering shaft. Do not pound on the wheel to remove it or the collapsible steering shaft may be damaged.

To install:

6. The steering wheel can be pushed onto the shaft splines by hand far enough to start the retaining nut. Do not hammer on the steering wheel. Install the retaining nut and torque it to 26–32 ft. lbs. (34–44 Nm).

Vehicles with air-bag

1. Disconnect and insulate the negative battery cable and wait at least 30 seconds before working on the vehicle.

2. Remove the air-bag mounting nut from the back of the steering wheel.

3. Carefully remove the connector by spreading the lock toward the outer edge. Use a screw driver and gently pry off the connector.

4. Matchmark the steering shaft and the steering wheel so it can be installed in the original position.

5. Using a steering wheel puller, remove the steering wheel.

To install:

6. Install the steering wheel. Torque the nut to 26–32 ft. lbs. (34–44 Nm).

7. Reconnect and install the air-bag module.

8. Connect the negative battery cable.

9. Turn the ignition switch on and check for proper operation of the SRS warning light.

1993 Mirage

1. Disconnect the negative battery cable.

2. Remove the horn pad from the steering wheel by, pulling the lower

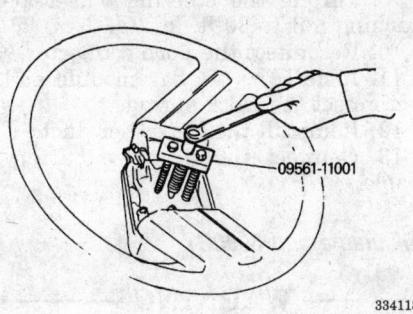

Typical method of removing the steering wheel

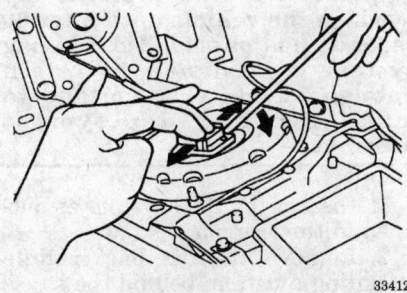

Remove the connector from the air-bag module — 1993–94 Precis

end of the pad upward. Disconnect horn button connector.

3. Remove steering wheel retaining nut.

4. Matchmark the steering wheel to the shaft.

5. Use a steering wheel puller to remove the steering wheel.

— **WARNING** —

Do not hammer on steering wheel to remove it. The collapsible column mechanism may be damaged.

To install:

6. Line up the matchmarks and install the steering wheel to the shaft.

7. Torque the steering wheel attaching nut to 29 ft. lbs. (40 Nm).

8. Reconnect the horn connector and install the horn pad.

9. Connect the negative battery cable.

1994–97 Mirage

— **CAUTION** —

If equipped with an air bag, be sure to disarm it before starting repairs on the vehicle. Failure to do so could result in severe personal injury and damage to vehicle.

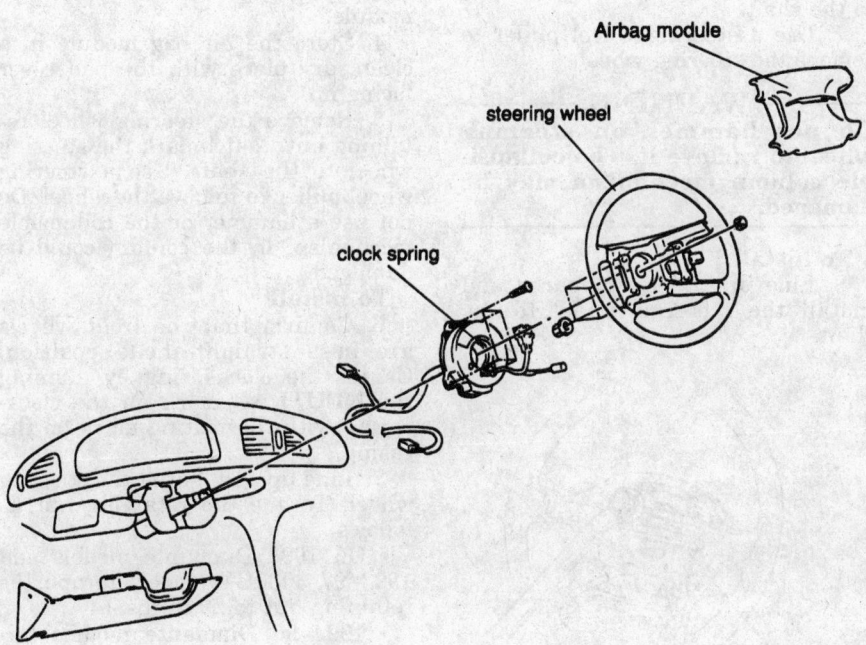

Steering wheel assembly with air-bag — 1993–94 Precis

1. Disarm the air bag as outlined in this section.

2. Remove the air bag module mounting nut from behind the steering wheel.

3. To disconnect the connector of the clockspring from the air bag module, press the air bag's lock toward the module to spread the lock open. While holding lock in this position, use a small tipped prying tool to gently pry the connector from the module.

4. Remove the air bag module and store in a clean, dry place with the pad cover facing up.

5. Matchmark the steering wheel to the shaft.

6. Remove the steering wheel retaining nut and use a steering wheel puller to remove the wheel. Do not use a hammer or the collapsible mechanism in the column could be damaged.

To install:

7. Confirm that the front wheels are in a straight-ahead position. Center the clockspring by aligning the **NEUTRAL** mark on the clockspring with the mating mark on the casing. Then install the steering wheel and torque the new retaining nut to 29 ft. lbs. (40 Nm).

8. Install the air bag module.

9. Connect the negative battery cable, turn the key to the **ON** position, the SRS warning light should illuminate for seven seconds and go out. If the warning light is not functioning properly, refer to SRS system diagnosis.

1993–94 Eclipse

1. Disconnect the negative battery cable.

2. Remove the horn pad from the steering wheel. Remove the retainers in the horn pad. Push pad upward to remove. Disconnect horn button connector.

3. Remove steering wheel retaining nut.

4. Matchmark the steering wheel to the shaft.

5. Use a steering wheel puller to remove the steering wheel.

— **WARNING** —

Do not hammer on steering wheel to remove it. The collapsible column mechanism may be damaged.

To install:

6. Line up the matchmarks and install the steering wheel to the shaft.

7. Torque the steering wheel attaching nut to 33 ft. lbs. (45 Nm).

8. Reconnect the horn connector and install the horn pad.

9. Connect the negative battery cable.

1993 Galant

1. Disconnect the negative battery cable.

2. Remove the horn pad from the steering wheel as follows:

   a. On models equipped with a Type 1 steering wheel, the horn pad is removed by pushing the pad upward, to release the pad from the retaining clips. Disconnect horn button connector and remove the pad.

   b. On models equipped with a Type 2 steering wheel, remove the screw from the bottom of the pad and push the pad upward, to release the pad from the retaining clips. Disconnect horn button connector and remove the pad.

3. Remove steering wheel retaining nut.

4. Matchmark the steering wheel to the shaft.

5. Use a steering wheel puller to remove the steering wheel.

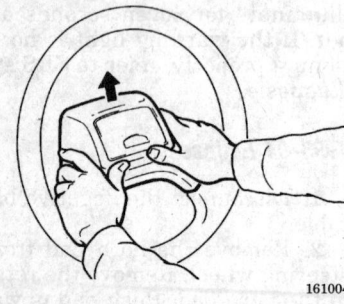

161004

Horn pad removal for Type 1 steering wheel — 1993 Galant

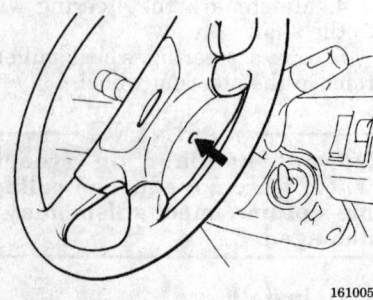

161005

Horn pad removal for Type 2 steering wheel — 1993 Galant

— **WARNING** —
Do not hammer on steering wheel to remove it. The collapsible column mechanism may be damaged.

To install:

6. Line up the matchmarks and install the steering wheel to the shaft.

7. Torque the steering wheel attaching nut to 25–33 ft. lbs. (35–45 Nm).

8. Reconnect the horn connector and install the horn pad.

9. Connect the negative battery cable.

1995–97 Eclipse

— **CAUTION** —
The air bag system (SRS or SIR) must be disarmed before removing the steering wheel. Failure to do so may cause accidental deployment, property damage or personal injury.

1. Disconnect the negative battery cable.

2. Remove the lap cooler duct.

3. Remove the air bag module and disconnect the clock spring from the steering wheel.

4. Disconnect the horn connector.

5. Remove steering wheel retaining nut.

6. Matchmark the steering wheel to the shaft.

7. Use a steering wheel puller to remove the steering wheel.

— **WARNING** —
Do not hammer on steering wheel to remove it. The collapsible column mechanism may be damaged.

To install:

8. Line up the matchmarks and install the steering wheel to the shaft.

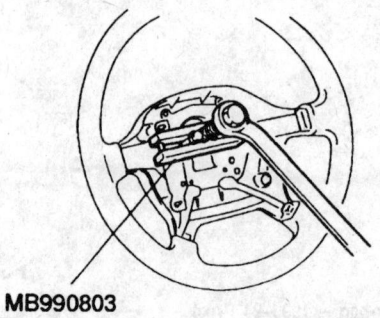

MB990803

327627

Steering wheel removal — 1995–97 Eclipse

---

9. Torque the steering wheel attaching nut to 30 ft. lbs. (41 Nm).

10. Reconnect the horn connector.

11. Install the air bag module and reconnect the clock spring.

12. Reinstall the lap cooler duct.

13. Connect the negative battery cable.

Diamante and 3000GT

— **WARNING** —
Be sure to disarm the SRS (air bag) system, before starting repairs on the vehicle. Failure to do so could result in personal injury or death. DO NOT perform any work on the vehicle until after 90 seconds has passed. The air bag system is designed to retain enough short term voltage to make air bag deployment possible.

1. Disarm the SRS system as outlined in this section.

2. Remove the air bag module mounting nut from behind the steering wheel. Matchmark the steering wheel.

3. Disconnect the connector of the clockspring from the air bag module, press the air bag's lock towards the module to spread the lock open. While holding lock in this position, use a small tipped prying tool to gently pry the connector from the module.

4. Store the air bag module in a clean, dry place with the pad cover facing up.

5. Remove the steering wheel retaining nut. Matchmark the steering wheel to the shaft. Use a steering wheel puller to remove the wheel. Do not use a hammer or the collapsible mechanism in the column could be damaged.

To install:

6. Confirm that the front wheels are in a straight-ahead position. Center the clockspring by aligning the **NEUTRAL** mark on the clockspring with the mating mark on the casing.

7. Line up and install the steering wheel. Torque the retaining nut as follows:

• On 1993 Diamante models and 1993–97 3000GT models torque the retaining nut to 29 ft. lbs. (40 Nm)

• 1994–96 Diamante models torque the retaining nut to 33 ft. lbs. (45 Nm)

8. Install the air bag module and torque the retaining nuts to 48 inch lbs. (5 Nm).

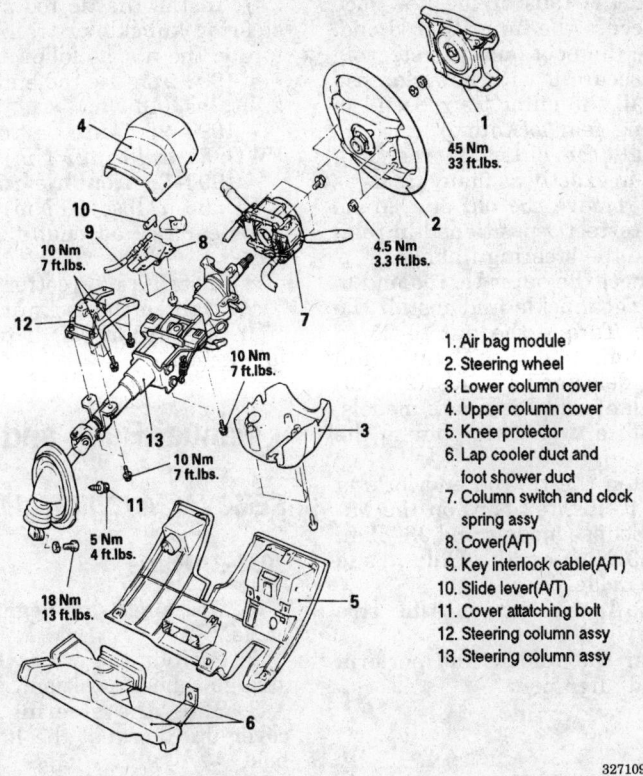

1. Air bag module
2. Steering wheel
3. Lower column cover
4. Upper column cover
5. Knee protector
6. Lap cooler duct and foot shower duct
7. Column switch and clock spring assy
8. Cover(A/T)
9. Key interlock cable(A/T)
10. Slide lever(A/T)
11. Cover attatching bolt
12. Steering column assy
13. Steering column assy

327109

Exploded view of steering wheel and shaft — Diamante

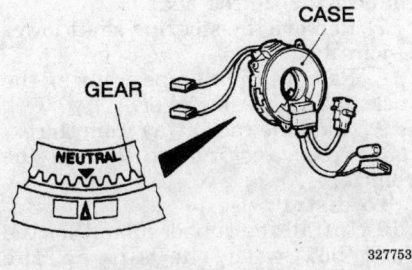

327753

Clock spring mating mark alignment — 1994–97 Galant

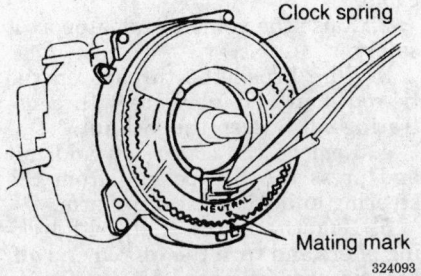

324093

Clockspring mating marks — 3000GT

9. Connect the negative battery cable and check the SRS warning light operation.

Tie Rod Ends

REMOVAL AND INSTALLATION

1993–94 Precis

1. Raise and safely support the vehicle.
2. Remove the front wheels.
3. Loosen the locknut on the tie rod end to be removed.

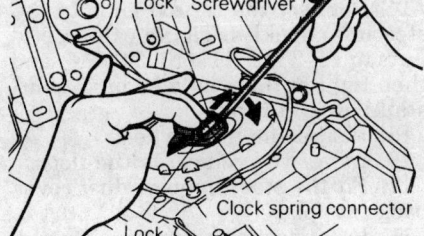

327752

Airbag module connection — 1994–97 Galant

4. Remove the cotter pin and the castellated nut.
5. Use a ball joint separator to separate the tie rod end from the knuckle.
6. Remove the tie rod end counting the number of turns it takes to remove it from the tie rod.

To install:

7. Install the new tie rod end on the shaft the same amount of turns that as it took to remove the old one.
8. Reconnect the tie rod end to the steering knuckle. Install the castle nut and tighten it to 11–25 ft. lbs. (15–34 Nm). Then, tighten the castle nut only enough to install a new cotter pin.
9. Tighten the tie rod end locknut making sure the tie rod end is perpendicular to the steering knuckle mounting surface.
10. Install the front wheels.
11. Lower the vehicle to the floor.
12. Check and adjust the front wheel alignment.

Mirage and Eclipse

Outer

1. Raise the front of the vehicle and support it on jackstands. Remove the wheel.
2. Remove the cotter pin and the tie rod ball joint stud nut. Note the position of the steering linkage.
3. Wire brush the threads on the tie rod shaft and lubricate with penetrating oil.
4. Using a suitable ball joint separator tool, remove the tie rod ball joint from the steering knuckle.
5. Loosen the locknut and remove the outer tie rod end from the tie rod. Count the number of complete turns it takes to completely remove it.

To install:

6. Install the new tie rod end, turning it in exactly as many turns as it was to remove the old one. Make sure it is correctly positioned in relationship to the steering linkage.
7. Connect the outer tie rod end to the steering knuckle and install the castle nut. Torque the nut to 25 ft. lbs. (34 Nm) on Mirage and Eclipse models and 21 ft. lbs. (29 Nm) on the Galant rear outer and Diamante models. Torque the Galant rear outer tie rod end castle nut to 37–47 ft. lbs. (50–65 Nm).
8. Install a new cotter pin to the castle nut.
9. Tighten the tie rod end locking nut to 30 ft. lbs. (42 Nm) on Mirage and Eclipse models and 36–39 ft. lbs. (49–53 Nm).
10. Install the wheel and tire assembly.

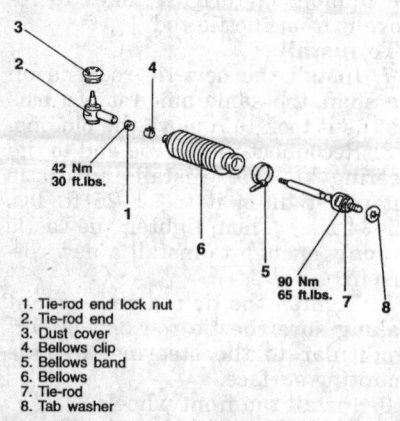

1. Tie-rod end lock nut
2. Tie-rod end
3. Dust cover
4. Bellows clip
5. Bellows band
6. Bellows
7. Tie-rod
8. Tab washer

331469

Tie rod assembly — Mirage and Eclipse

11. Lower the vehicle and perform a front end alignment.

Inner

1. Raise the front of the vehicle and support it on jackstands. Remove the wheel.
2. Remove the cotter pin and the outer tie rod ball joint stud nut. Note the position of the steering linkage.
3. Wire brush the threads on the tie rod shaft and lubricate with penetrating oil.
4. Using a suitable ball joint separator tool, remove the tie rod ball joint from the steering knuckle.
5. Loosen the locknut and remove the tie rod end from the tie rod. Count the number of complete turns it takes to completely remove it.
6. Remove the tie rod-to-steering gear locknut.
7. Remove the clamps that secure the flexible boot to the steering gear.
8. Slide the boot from the inner tie rod and remove the boot.
9. Bend the lock plate tabs from the inner tie rod end nut.
10. Loosen the inner tie rod end nut from the steering gear and remove the inner tie rod end.
To install:
11. Using a new lock plate, install the tie rod end and torque the tie rod to 65 ft. lbs. (90 Nm).

12. Bend the tabs of the new lock plate to secure the inner tie rod end.
13. Slide the boot onto the steering gear and secure it with new clamps.
14. Install the outer tie rod end to the steering gear locknut.
15. Install the outer tie rod end, turning it in exactly as many turns as it was to remove the old one. Make sure it is correctly positioned in relationship to the steering linkage.
16. Connect the outer tie rod end to the steering knuckle and install the castle nut. Torque the nut to 25 ft. lbs. (34 Nm.) on the Mirage and Eclipse models and 21 ft. lbs. (29 Nm) on the Galant and Diamante models.
17. Install a new cotter pin to the castle nut.
18. Tighten the tie rod end locking nut to 30 ft. lbs. (42 Nm) on the Mirage and Eclipse models and 36–39 ft. lbs. (49–53 Nm) on the Galant and Diamante models.
19. Install the wheel and tire assembly.
20. Lower the vehicle and perform a front end alignment.

3000GT

1. Disconnect the battery negative cable.
2. Raise the vehicle and support it safely.
3. Wire brush the threads on the tie rod shaft and lubricate with penetrating oil. Loosen the locknut.
4. Remove the cotter pin and nut and press the tie rod end from the steering knuckle or trailing arm.
5. Hold the tie rod shaft with locking pliers and turn the tie rod end off, counting the number of turns for installation purposes.
To install:
6. Thread on the new tie rod end the same number of turns required to remove the old one.

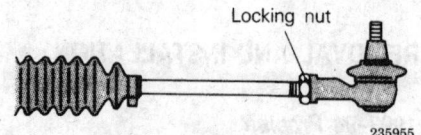

Common outer tie rod assembly

7. Install the tie rod stud into the steering knuckle or trailing arm and torque the nut as follows:
• 1993 front tie rod end with FWD: 29 ft. lbs. (40 Nm)
• 1994–96 front tie rod end with FWD: 21 ft. lbs. (29 Nm)
• 1993–96 front tie rod end with AWD: 36 ft. lbs. (50 Nm)
• Rear tie rod end: 42 ft. lbs. (58 Nm)
8. Install a new cotter pin.
9. Tighten the locknut
10. Perform a four wheel alignment.

Manual Rack and Pinion

REMOVAL AND INSTALLATION

1993–94 Precis

1. Disconnect the negative battery cable.
2. Position the front wheels in the straight-ahead position.
3. Slide the steering joint dust cover up to reveal the lower U-joint bolt.
4. Raise the vehicle and support it safely.
5. Remove the front wheels.
6. Use a ball joint separator to disconnect the tie rod ends.
7. Remove the steering shaft lower U-joint bolt.
8. Remove the clamps securing the rack to the crossmember.
9. Slide the rack away from the exhaust pipe and remove it from the vehicle.
To install:
10. Install the rubber mount for the gear box with the slit on the downside.
11. Fit the rack into the vehicle and install the mounting bolts. Tighten the bolts to 44–59 ft. lbs. (60–80 Nm).
12. Center the rack ends within their strokes.
13. Install the steering shaft coupling bolt and tighten it to 11–15 ft. lbs. (15–20 Nm).
14. Connect the tie rod ends to the steering knuckles. Tighten the castle nuts to 11–25 ft. lbs. (15–34 Nm), and then tighten them only enough to install new cotter pins.
15. Install the front wheels.
16. Lower the vehicle to the floor.
17. Fit the steering joint dust cover back into place.
18. Reconnect the negative battery cable.
19. Check and adjust the front wheel alignment and the steering wheel spoke angle.

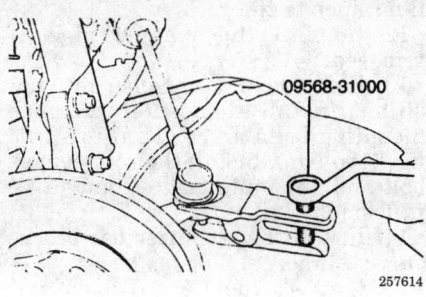

09568-31000

257614

Proper tie-rod end removal method

Mirage

NOTE: If equipped with air bag, prior to removal of the steering rack, center the front wheels and remove the ignition key. Failure to do so may damage the SRS clockspring and render SRS system inoperative, risking serious driver injury. Be sure to properly disarm the air bag system.

1. Disconnect the battery negative cable. Raise the vehicle and support safely and remove the wheels.
2. Disconnect the oxygen sensor and remove the front exhaust pipe.

NOTE: It may be easier on some vehicles, to disconnect all hangers and lower the complete exhaust system.

3. Properly support the engine. Remove both roll stopper mounting bolts and the four center member installation bolts.
4. Remove the center member.

NOTE: Matchmark the pinion input shaft of the rack to the lower steering column joint for installation purposes.

5. Remove the pinch bolt holding the lower steering column joint to the rack and pinion input shaft.
6. Remove the cotter pins and disconnect the tie rod ends from the steering knuckle.
7. Remove the rack and pinion steering assembly and its rubber mounts from the right side of the vehicle.
To install:
8. Align the matchmarks of the input shaft and install the rack to the vehicle.
9. Secure the rack using the retainer clamps and bolts. Torque the bolts to 51 ft. lbs. (70 Nm).
10. Torque the steering column pinch bolt to 13 ft. lbs. (18 Nm).
11. Install the center member.

12. Install the front exhaust pipe.
13. Connect the tie rod ends to the steering knuckles and torque the castle nuts to 25 ft. lbs. (34 Nm). Install new cotter pins.
14. Install the wheels and connect the negative battery cable.
15. Perform a front end alignment.

1993–94 Eclipse

1. Position the wheels in a straight ahead position. Disconnect the negative battery cable. Raise the vehicle and support safely.
2. Remove the bolt holding lower steering column joint to the rack and pinion input shaft.
3. Remove the cotter pins and, using the proper separating tools, disconnect the tie rod ends from the knuckle.
4. Locate the triangular brace near the stabilizer bar brackets on the crossmember and remove both the brace and the stabilizer bar bracket.
5. Place a jack under the center member. Remove the through-bolt from the round roll stopper. Remove the rear bolts from the center crossmember.
6. Disconnect the front exhaust pipe and tie out of the way. Lower the center member slightly.
7. Remove the rack and pinion steering assembly and its rubber mounts. Move the rack to the right to remove from the crossmember. While tilting downward, remove the rack assembly from the left side of the vehicle. Use caution to avoid damaging the boots.
To install:
8. Install the rack and mounting bolts, torquing bolts to 43–58 ft. lbs. (60–80 Nm). When installing the rubber rack mounts, align the projection of the mounting rubber with the indentation in the crossmember.
9. Raise the center member using the jack and install the center support rear bolts. Tighten to 72 ft. lbs. (100 Nm).
10. Install the roll stopper bolt and new nut. Tighten nut to 47 ft. lbs. (65 Nm). Remove the jack supporting the center member.
11. Install the joint assembly and gear box connecting bolt and tighten to 14 ft. lbs. (20 Nm).
12. Reposition the exhaust pipe and connect to the manifold.
13. Install the stabilizer bar brackets and brace.
14. Connect the tie rod ends to the steering knuckles. Install the retaining nuts.

15. Perform a front end alignment.

Power Rack and Pinion

REMOVAL AND INSTALLATION

1993–94 Precis

1. Disconnect the negative battery cable.
2. Position the front wheels in the straight-ahead position.
3. Drain the fluid from the power steering reservoir.
4. Slide the steering joint dust cover up to reveal the lower U-joint bolt.
5. Raise the vehicle and support it safely.
6. Remove the front wheels.
7. Use a ball joint separator to disconnect the tie rod ends.
8. Remove the steering shaft lower U-joint bolt.
9. Disconnect the fluid lines from the rack valve body. Plug the lines to prevent fluid loss and contamination.
10. If equipped, unbolt and remove the stabilizer bar.
11. If necessary, remove the rear roll stopper from the center member and move the rear roll stopper forward.
12. Remove the rack unit mounting clamp bolts. Move the rack to the side to clear the exhaust pipe and take it out the right side of the vehicle.
To install:
13. Install the rack with the projection on the rubber mount aligned with the hole in the clamp. Apply adhesive to the cylinder side of the mount so that the slit part will not open. Tighten the rack mounting bolts to 44–59 ft. lbs. (60–80 Nm).
14. Install the steering shaft lower U-joint bolt.
15. Connect the fluid pressure and return lines.
16. Connect the tie rod ends to the steering knuckles. Tighten the castle nuts to 11–25 ft. lbs. (15–34 Nm), and then tighten them only enough to install new cotter pins.
17. Install the stabilizer bar if it was removed. Reconnect the rear roll stopper if it was disconnected.
18. Install the front wheels.
19. Lower the vehicle to the floor.
20. Fit the steering joint dust cover back into place.
21. Refill and bleed the power steering system.
22. Check and adjust the front wheel alignment and the steering wheel spoke angle.

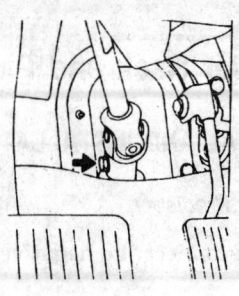

257634

Steering shaft coupling bolt — 1993-94 Precis

Mirage

NOTE: If equipped with air bag, prior to removal of the steering gear box, center the front wheels and remove the ignition key. Failure to do so may damage the SRS clockspring and render SRS system inoperative, risking serious driver injury.

1. Drain power steering system:
   a. Disconnect the return hose at the reservoir and place into a suitable container.
   b. Disable the ignition system. While cranking the engine, turn the wheels several times, until system has been drained.
2. Disconnect the battery negative cable. Raise the vehicle and support safely.
3. Disconnect the oxygen sensor and remove the front exhaust pipe.

NOTE: It may be easier on some vehicles, to disconnect all hangers and lower the complete exhaust system.

4. Properly support the engine. Remove both roll stopper mounting bolts and the four center member installation bolts. Remove the center member.
5. Remove the center member.

NOTE: Matchmark the pinion input shaft of the rack to the lower steering column joint for installation purposes.

6. Remove the pinch bolt holding the lower steering column joint to the rack and pinion input shaft.
7. Remove the cotter pins and disconnect the tie rod ends from the steering knuckle.
8. Disconnect the power steering fluid pressure pipe and return hose from the rack fittings.
9. Remove the rack and pinion steering assembly and its rubber mounts from the right side of the vehicle.

To install:
10. Align the matchmarks of the input shaft and install the rack to the vehicle.
11. Secure the rack using the retainer clamps and bolts. Torque the bolts to 51 ft. lbs. (70 Nm).
12. Torque the steering column pinch bolt to 13 ft. lbs. (18 Nm).
13. Using new O-rings, connect the power steering fluid lines to the rack fittings.
14. Install the center member.
15. Install the front exhaust pipe.
16. Connect the tie rod ends to the steering knuckles and torque the castle nuts to 25 ft. lbs. (34 Nm). Install new cotter pins.
17. Install the wheels and connect the negative battery cable.
18. Refill the reservoir and bleed the system.
19. Perform a front end alignment.

1993 Galant and 1993-94 Eclipse

1. Disconnect the negative battery cable. Drain the power steering fluid. Raise the vehicle and support safely.
2. Remove the bolt holding lower steering column joint to the rack and pinion input shaft.
3. Remove the transfer case, if equipped.
4. Remove the cotter pins and using the proper tools, separate the tie rod ends from the steering knuckle.
5. Locate the triangular brace near the stabilizer bar brackets on the crossmember and remove both the brace and the stabilizer bar bracket.
6. Support the center crossmember. Remove the through-bolt from the round roll stopper and remove the rear bolts from the center crossmember.
7. Disconnect the front exhaust pipe, if equipped with FWD.
8. Disconnect the power steering fluid pressure pipe and return hose from the rack fittings. Plug the fittings to prevent excess fluid leakage.
9. Lower the crossmember slightly. Remove the rack and pinion steering assembly and its rubber mounts. Move the rack to the right to remove from the crossmember. Tilt the assembly downward and remove from the left side of the vehicle. Use caution to avoid damaging the boots.

To install:
10. Install the rack and install the mounting bolts. Torque the mounting bolts to 43-58 ft. lbs. (60-80 Nm). When installing the rubber rack mounts, align the projection of the mounting rubber with the indentation in the crossmember.
11. Connect the power steering fluid lines to the rack.
12. Connect the exhaust pipe, if removed.
13. Raise the crossmember into position. Install the center member mounting bolts and tighten to 72 ft. lbs. (100 Nm). Install the roll stopper bolt and new nut. Tighten nut to 47 ft. lbs. (65 Nm).
14. Install the stabilizer bar brackets and brace.
15. Connect the tie rod ends and tighten nuts to 25 ft. lbs. (34 Nm).
16. Install the transfer case, if removed. Check and fill fluid.
17. Refill the reservoir with power steering fluid and bleed the system.
18. Perform a front end alignment.

1995-97 Eclipse

—————— CAUTION ——————
The air bag system (SRS or SIR) must be disarmed before removing the rack and pinion. Failure to do so may cause accidental deployment, property damage or personal injury.

Non-Turbo

1. Center the front wheel and remove the ignition key from the switch.
2. Disconnect the negative battery cable.
3. Drain the power steering fluid.
4. Raise and safely support the vehicle.
5. Remove the stabilizer bar.
6. Remove the windshield washer reservoir.
7. Remove the pinch bolt from the joint assembly.
8. Disconnect the fluid lines from the steering rack.
9. Using the proper tools, disconnect the tie rod ends from the steering knuckles.
10. Remove the left and right stays (supports).
11. Support the engine and remove the center member.
12. Remove the clamp and the mounting bolts.
13. Disconnect the left lower compression arm from the body side of the vehicle and support it with wire or string.
14. Disconnect the steering rack from the joint assembly and remove the rack from the left side of the vehicle.

To install:
15. Position the steering rack in the vehicle and install the clamp and the mounting bolts. Make sure the rack

is centered before connecting it to the joint assembly.

16. Install the left lower compression arm to the body.

17. Install the center member.

18. Install the left and right stays and remove the engine support fixture or jack.

19. Connect the tie rods to the steering knuckles.

20. Connect the fluid lines to the steering rack. Torque to specifications.

21. Install the pinch bolt in the joint assembly.

22. Install the stabilizer bar and the windshield washer reservoir.

23. Safely lower the vehicle.

24. Connect the negative battery cable.

25. Refill and bleed the power steering system.

26. Check wheel alignment.

Turbo

1. Center the front wheel and remove the ignition key from the switch.

2. Disconnect the negative battery cable.

3. Drain the power steering fluid.

4. Raise and safely support the vehicle.

5. Remove the stabilizer bar.

6. Disconnect the fluid level sensor and remove the brake fluid reservoir and position it out of the way. Do not disconnect the brake hose.

7. Disconnect the electrical connector from the A/C compressor.

8. Remove the A/C compressor from the bracket and position it out of the way. Do not disconnect the hoses.

9. Remove the pinch bolt from the joint assembly.

10. Disconnect the fluid lines from the steering rack.

11. Using the proper tools, disconnect the tie rod ends from the steering knuckles.

12. Remove the left and right stays (supports).

13. Support the engine and remove the center member assembly.

14. Remove the clamp and the mounting bolts.

15. Disconnect the left lower compression arm from the body side of the vehicle and support it with wire or string.

16. Disconnect the steering rack from the joint assembly and remove the rack from the left side of the vehicle.

To install:

17. Position the steering rack in the vehicle and install the clamp and the mounting bolts. Make sure the rack

is centered before connecting it to the joint assembly.

18. Install the left lower compression arm to the body.

19. Install the center member assembly.

20. Install the left and right stays.

21. Connect the tie rod ends to the steering knuckles.

22. Connect the fluid lines to the steering rack.

23. Install the pinch bolt in the joint assembly.

24. Install the stabilizer bar.

25. Safely lower the vehicle.

26. Install the A/C compressor and connect the harness connector.

27. Install the brake fluid reservoir and connect the fluid level sensor.

28. Connect the negative battery cable.

29. Refill and bleed the power steering system.

30. Check wheel alignment.

1994–97 Galant

— **WARNING** —

Prior to removal of the steering gear box, center the front wheels and remove the ignition key. Failure to do so may damage the SRS clock spring and render SRS system inoperative, risking serious driver injury.

1. Drain the power steering fluid as follows:

a. Disconnect the power steering return (low side) hose.

b. Connect a suitable container to the hose.

c. Properly disable the ignition system, by disconnecting the ignition coil wire and connecting it to a suitable ground.

d. While cranking the engine, turn the wheels, several times, from side to side, until the fluid is removed.

2. Disarm the SRS system as outlined in this section.

3. Raise and properly support the vehicle.

4. Remove both front wheel assemblies.

5. Remove the bolt holding lower steering column joint to the rack and pinion input shaft.

6. Remove the stabilizer bar.

7. Remove the cotter pins and using joint separator MB991113, disconnect the tie rod ends, from the steering knuckle.

8. On vehicles equipped with Electronic Control Power steering (EPS), disconnect the wiring harness, from the solenoid connector.

A13X0236

```
57 Nm*
42 ft.lbs.*

23–26 Nm
17–19 ft.lbs.

78–88 Nm
58–65 ft.lbs.        69–78 Nm
                     51–58 ft.lbs.
```

```
12
15 Nm
11 ft.lbs.    18 Nm
              13 ft.lbs.

28 Nm
21 ft.lbs.

69 Nm        28 Nm
51 ft.lbs.   21 ft.lbs.
```

322147

1. Brake fluid reservoir assembly
2. A/C compressor
3. Joint assembly and gear box connecting bolt
4. Power steering pipe connection
5. Cotter pin
6. Tie-rod end and knuckle connection
7. Stay (L.H.)
8. Stay (R.H.)
9. Centermember assembly
10. Clamp
11. Gear box assembly
12. Return tube

NOTE
The fasteners marked * should be temporarily tightened before they are finally tightened once the total weight of the engine has been placed on the vehicle body.

Power steering rack assembly and related components — 1995–97 Eclipse

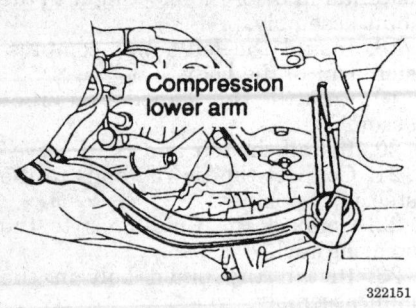

Compression lower arm

322151

Disconnect the lower compression arm from the body — 1995–97 Eclipse

9. Locate the two triangular braces near the crossmember and remove both.

10. Support the center crossmember. Remove the through-bolt from the front round roll stopper and remove the three bolts securing the center crossmember.

11. Remove the center crossmember.

12. Properly support the engine and remove the rear roll stopper through-bolt. Lower the engine slightly.

---- WARNING ----
In order to prevent damage to the engine, when supporting and jacking the engine, place a block of wood between the engine and the oil pan.

13. Disconnect the power steering fluid pressure pipe and return hose from the rack fittings. Plug the fittings to prevent excessive fluid leakage.

14. Remove the clamp bolts and the two bolts securing the rack assembly to the chassis.

15. Remove the rack and pinion steering assembly and its rubber mounts.

NOTE: When removing the rack and pinion assembly, tilt the assembly to the vehicle side of the compression lower arm, and remove from the left side of the vehicle. Use caution to avoid damaging the boots.

To install:

16. Align the rack assembly so the splines are inserted into the steering column shaft.

17. Install the rack and with the mounting bolts. Torque the mounting bolts to 51 ft. lbs. (69 Nm).

18. Install the pinch bolt and torque the bolt to 13 ft. lbs. (18 Nm).

19. Connect the power steering fluid lines to the rack and tighten to

high side fitting to 11 ft. lbs. (15 Nm). Secure the low side hose with the clamp.

20. Raise the engine into position. Install the rear roll stopper through-bolt and tighten to 32 ft. lbs. (43 Nm).

21. Raise the crossmember into position. Install the center member mounting bolts and tighten the front bolts to 58–65 ft. lbs. (78–88 Nm) and the rear bolt to 51–58 ft. lbs. (69–78 Nm).

22. Install the front roll stopper bolt and tighten the nut to 32 ft. lbs. (43 Nm).

23. Install the two triangular braces and tighten the mounting bolts to 50–56 ft. lbs. (69–78 Nm).

24. Install the stabilizer bar.

25. Connect the tie rod ends and tighten nuts to 20 ft. lbs. (27 Nm).

26. On vehicles equipped with EPS, connect the wiring harness to the solenoid connector.

27. Install the wheel assemblies and lower the vehicle.

28. Refill the reservoir with power steering fluid and bleed the system.

29. Perform a front end alignment.

3000GT and Diamante

NOTE: Prior to removal of the steering gear box, center the front wheels and remove the ignition key. Failure to do so may damage the SRS clock spring and render SRS system inoperative, risking serious driver injury.

1. Disconnect the negative battery cable. Disarm the air bag.

---- CAUTION ----
Work must be started after 90 seconds from the time the ignition switch is turned to the LOCK position and the negative (-) battery cable is disconnected.

2. Disconnect the front exhaust pipe.

3. If equipped with AWD, remove the transfer case assembly.

4. Remove the bolt holding the lower steering column joint to the rack and pinion input shaft.

5. Remove the cotter pins and disconnect the tie rod ends.

6. Remove the left and right frame members.

7. Remove the stabilizer bar bracket.

8. If equipped with four-wheel steering, disconnect the lines going to the rear pump.

9. Remove the rack and pinion steering assembly and its rubber mounts. Move the rack to the right to

remove it from the crossmember. Use caution to avoid damaging the boots.

To install:

10. Install the rack and install the mounting bolts, tightening bolts to 51 ft. lbs. (70 Nm). When installing the rubber rack mounts, align the projection of the mounting rubber with the indentation in the crossmember. Install the pinch bolt.

11. Connect the lines going to the four-wheel steering rear pump and to the rack itself.

12. Install the frame members and torque the bolts to 43–51 ft. lbs. (60–70 Nm).

13. Connect the tie rods and Install new cotter pins.

14. Install the transfer case and front exhaust pipe.

15. Refill the reservoir and bleed the system.

16. Perform a front end alignment.

Power Steering Pump

BLEEDING

3000GT and Diamante

Front

1. Raise the vehicle and support it safely.

2. Manually turn the pump pulley a few times.

3. Turn the steering wheel all the way to the left and to the right 5 or 6 times.

NOTE: If bleeding is attempted with the engine running, the air will be absorbed in the fluid. Bleed only while cranking.

4. Disconnect the ignition high tension cable, and, while operating the starter motor intermittently, turn the steering wheel all the way to the left and right 5-6 times for 15–20 seconds. During bleeding, make sure the fluid in the reservoir never falls below the lower position of the filter.

5. Connect the ignition high tension cable, start the engine and allow to idle.

6. Turn the steering wheel left and right until there are no air bubbles in the reservoir. Confirm that the fluid is not milky and the level is up to the specified position on the gauge. Confirm that there is very little change in the fluid level when the steering wheel is turned. If the fluid level changes more than 0.2 in., the air has not been completely bled. Repeat the process.

Rear

1. Bleed the front steering system.

2. Start the engine and let it idle.

3. Loosen the bleeder screw on the left side of the control valve and install a plastic tube on the bleeder.

4. Turn the steering wheel all the way to the left, then immediately turn it half way back. Confirm that air has discharged with the fluid.

5. Repeat Step 4 two or three times as required, to remove all of the air from the rear system. Stop the engine.

6. Loosen the power cylinder (rear steering gear) bleeder screw about 1/8 turn and install the same special tool with the rotation prevention metal fixtures to prevent the bleeder from opening more.

7. Start the engine and run to 50 mph to circulate the fluid.

8. Maintain a speed of 20 mph and turn the steering wheel back and forth. Air should be discharged through the tube of the special tool and into the oil reservoir.

9. Repeat until all air is removed from the power cylinder.

1993–94 Precis

1. Ensure that the reservoir is full of DEXRON II automatic transmission fluid.

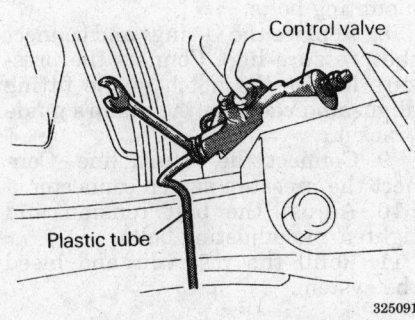

Rear control valve — 3000GT

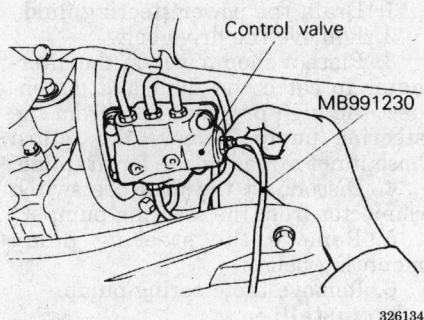

Control valve bleeder screw location — Diamante

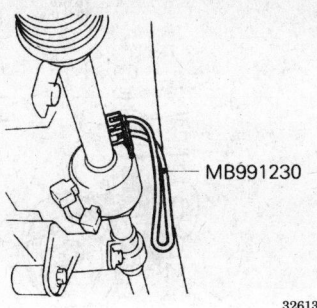

Power valve bleeder screw location — Diamante

2. Raise and safely support the front wheels of the vehicle.

3. Turn the steering wheel from lock to lock 5 or 6 times.

4. Disconnect the coil wire and connect to a solid ground. Operate the starter motor intermittently for 15 to 20 seconds and turn the steering wheel from lock to lock 5 or 6 times.

NOTE: Ensure that the reservoir is full during air bleeding to prevent the fluid level from falling below the lower position of the filter.

5. Connect the coil wire and start the engine.

6. Turn the steering wheel from lock to lock until no more air bubbles are visible in the reservoir.

7. Confirm that the oil is not milky and that the fluid level is correct.

8. Confirm that there is little change in the fluid level when the steering wheel is turned to the left and right.

NOTE: An abrupt rise in the fluid level after stopping the engine is a sign of incomplete bleeding. If this occurs, repeat the bleeding procedure.

Mirage

1. Raise and support the front of the vehicle safely.

2. Check and add fluid to the reservoir, if necessary.

3. Disconnect the ignition coil wire so the engine will not start.

4. Turn the steering wheel (all the way), right and left, just touching the stops while cranking the key. Continue turning the steering wheel until the fluid level in the reservoir no longer decreases. Do not crank the engine for longer than 15–20 seconds.

5. Check the fluid level and add fluid as required.

6. Attach the coil wire.

7. Start the engine and continue turning the steering wheel from right to left, lightly touching the stops until the fluid level in the reservoir no longer decreases.

8. Stop the engine and check the fluid level, add fluid as required.

9. Repeat the steps until all of the air is bled from the system.

10. If the air cannot be bled from the system, turn and hold the steering wheel at each stop for at least 5 seconds but never more than 15 seconds.

Eclipse and 1994–97 Galant

1. Raise the vehicle and support safely.

2. Manually turn the pump pulley a few times.

3. Turn the steering wheel all the way to the left and to the right 5 or 6 times.

4. Disconnect the ignition high tension cable and, while operating the starter motor intermittently, turn the steering wheel all the way to the left and right 5–6 times for 15–20 seconds. During bleeding, make sure the fluid in the reservoir never falls below the lower position of the filter. If bleeding is attempted with the engine running, the air will be absorbed in the fluid. Bleed only while cranking.

5. Connect ignition high tension cable, start engine and allow to idle.

6. Turn the steering wheel left and right until there are no air bubbles in the reservoir. Confirm that the fluid is not milky and the level is up to the specified position on the gauge. Confirm that there is very little change in the fluid level when the steering wheel is turned. If the fluid level changes more than 0.2 in., the air has not been completely bled. Repeat the process.

REMOVAL AND INSTALLATION

1993–94 Precis

1. Disconnect the negative battery cable.

2. Drain the fluid from the power steering pump reservoir.

3. Loosen the pump mounting bolts and remove the drive belt.

4. Disconnect the suction hose from the pump and plug it.

5. Disconnect the pressure hose from the pump and the return hose from the reservoir. Plug both hoses to prevent fluid loss and contamination

6. Remove the pump mounting bracket bolts and lift out the pump. If

necessary, unbolt and remove the pump reservoir.

To install:

7. Install the pump reservoir, if removed. Position the pump onto the mounting bracket and torque the mounting bolts to 15–20 ft. lbs. (20–27 Nm).

8. Connect the suction hose to the pump. Install the hose so that the painted portion is toward the pump body.

9. Connect the pressure hose to the pump and the return hose to the fluid reservoir. Use new crush washers and tighten the pressure hose banjo bolt to 41–44 ft. lbs. (55–60 Nm). When installing the return line, make sure you push it at least 25–30mm onto the return tube. The hoses should not be twisted or allowed to come in contact with another component in the engine compartment.

10. Install the drive belt and adjust the tension. Tighten the tension adjusting bolt to 18–24 ft. lbs. (25–33 Nm).

11. Connect the negative battery cable.

12. Fill the system with DEXRON II ATF and bleed the system. Check for leaks and the hydraulic fittings.

Mirage

1. Drain power steering system:
   a. Disconnect the return hose at the reservoir and place into a suitable container.
   b. Disable the ignition system. While cranking the engine, turn the wheels several times, until system has been drained.

2. Disconnect the battery negative cable.

3. Remove the pressure switch connector from the side of the pump.

4. If the alternator is located under the oil pump, cover it with a shop towel to protect it from oil.

5. Disconnect the pressure line.

6. Remove the steering pump drive belt and the water pump pulley drive belt.

7. Remove the water pump pulley.

8. Remove the bolts that secure the oil pump the remove the pump from its bracket.

To install:

9. Install the power steering pump and torque the mounting bolts to specifications.

10. Install the water pump pulley and torque the mounting bolts 14 ft. lbs. (19 Nm).

11. Install and adjust the drive belts.

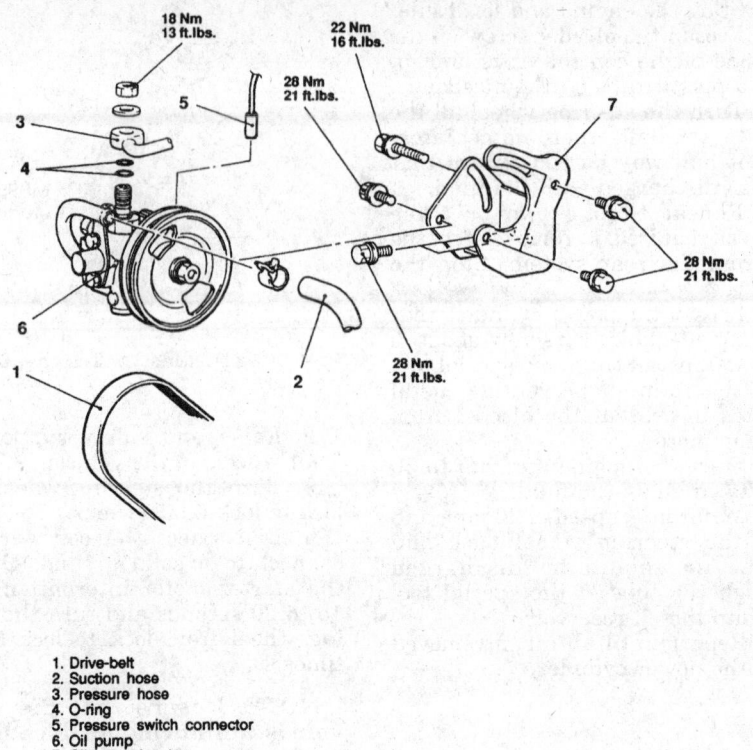

1. Drive-belt
2. Suction hose
3. Pressure hose
4. O-ring
5. Pressure switch connector
6. Oil pump
7. Oil pump bracket

322222

Power steering pump and related parts — 1996–97 Eclipse

12. Replace the O-rings and connect the pressure line. Connect the pressure line so the notch in the fitting aligns and contacts the pump's guide bracket. Tighten the nut that secures the pressure line to 13 ft. lbs. (18 Nm).

13. Connect the return line. Connect the pressure switch connector.

14. Refill the reservoir and bleed the system.

1993–94 Eclipse

1. Disconnect the battery negative cable.

2. Remove the pressure switch connector from the side of the pump.

3. If the alternator is located under the oil pump, cover it with a shop towel to protect it from oil.

4. Disconnect the return fluid line. Remove the reservoir cap and allow the return line to drain the fluid from the reservoir. If the fluid is contaminated, disconnect the ignition high tension cable and crank the engine several times to drain the fluid from the gearbox.

5. Disconnect the pressure line.

6. Remove the pump drive belt and unbolt the pump from its bracket.

To install:

7. Install the pump, wrap the belt around the pulley and tighten the mounting bolts.

8. Replace the O-rings and connect the pressure line. Connect the pressure line so the notch in the fitting aligns and contacts the pump's guide bracket.

9. Connect the return line. Connect the pressure switch connector.

10. Adjust the belt tension and tighten the adjusting bolts.

11. Refill the reservoir and bleed the system.

1995–97 Eclipse

Non-Turbo

1. Drain the power steering fluid.

2. Remove the drive belt.

3. Place a shop towel on the alternator to catch the fluid and disconnect the high pressure hose from the steering pump. Remove the return hose if not already done for draining.

4. Disconnect the pressure switch connector from the steering pump.

5. Remove the steering pump mounting bolts.

6. Remove the steering pump.

To install:

7. Install the power steering pump in the bracket as forward as possible. Torque the bolts to specifications.

8. Connect the pressure switch connector.

9. Use new gaskets and connect the pressure line to the steering pump.

10. Connect the return hose to the steering pump.

11. Install and adjust the drive belt.

12. Connect the negative battery cable.

13. Refill and bleed the power steering system.

Turbo

1. Disconnect the negative battery cable.

2. Remove the drive belt.

3. Place a shop towel on the alternator to catch the fluid and remove the suction (return) hose if not already done for draining.

4. Disconnect the pressure hose from the pump.

5. Disconnect the pressure switch connector.

6. Remove the steering pump mounting bolts.

7. Remove the steering pump.

To install:

8. Install the steering pump in the bracket.

9. Connect the pressure switch connector.

10. Connect the pressure hose and the suction hose to the steering pump. The notch in the pressure hose should contact the suction connector.

11. Install and adjust the drive belt.

12. Connect the negative battery cable.

13. Refill and bleed the power steering system.

1993 Galant

Front

1. Disconnect the battery negative cable.

2. Remove the pressure switch connector from the side of the pump.

3. If the alternator is located under the oil pump, cover it with a shop towel to protect it from oil.

4. Disconnect the return fluid line. Remove the reservoir cap and allow the return line to drain the fluid from the reservoir. If the fluid is contaminated, disconnect the ignition high tension cable and crank the engine several times to drain the fluid from the gearbox.

5. Disconnect the pressure line.

6. Remove the pump drive belt.

7. Remove the pump mounting bolts and remove the pump from the engine.

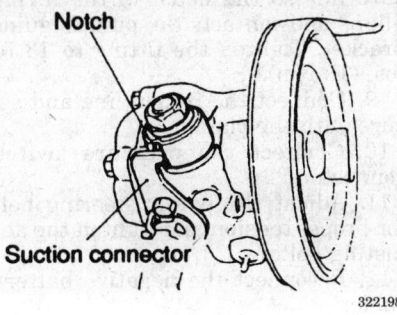

The notch in the pressure hose should contact the suction connector — 1995 Eclipse

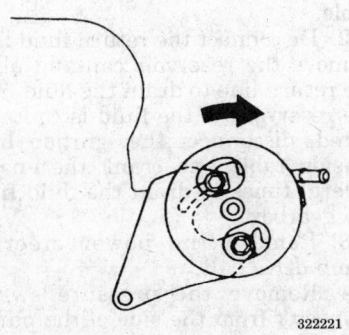

Install the pump towards the front of the vehicle — 1996–97 Eclipse

To install:

8. Install the pump, wrap the belt around the pulley and loosely tighten the mounting bolts.

9. Replace the O-rings and connect the pressure line. Connect the pressure line so the notch in the fitting aligns and contacts the pump's guide bracket.

10. Connect the return line. Connect the pressure switch connector.

11. Adjust the belt tension and tighten the adjusting bolts to 25–33 ft. lbs. (35–45 Nm).

12. Refill the reservoir and bleed the system.

Rear

1. Raise the vehicle and support safely.

2. Drain the differential gear oil.

3. Matchmark and remove the rear driveshaft.

4. Remove the rear halfshafts.

5. Remove the center exhaust pipe and muffler assembly, as required.

6. Disconnect the pressure and suction hoses from the fittings on the pump.

7. The large mounting bolts that hold the differential carrier support plate to the underbody may use self-locking nuts. Before removing them, support the rear axle assembly in the middle with a transaxle jack. Remove

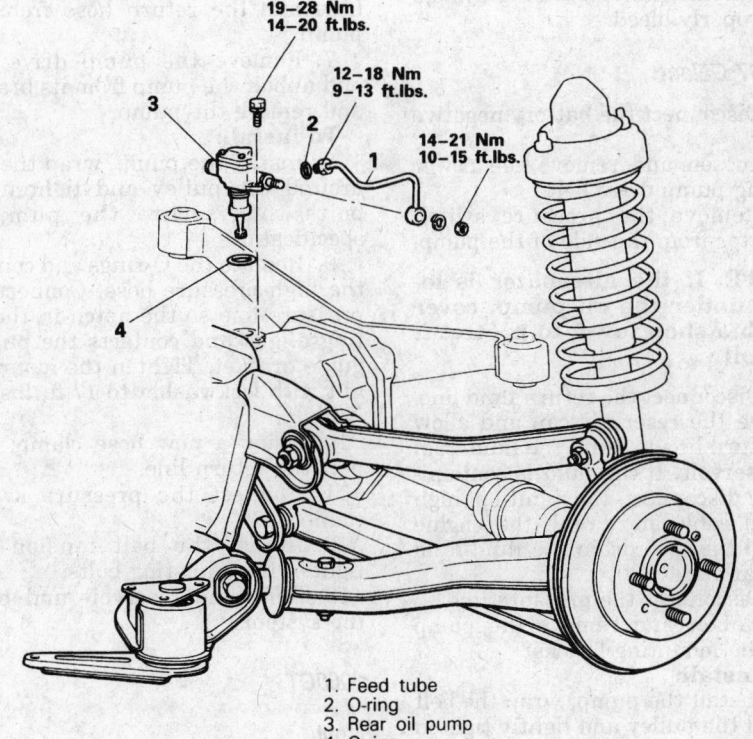

1. Feed tube
2. O-ring
3. Rear oil pump
4. O-ring

Rear steering pump and related components — 1993 Galant

the nuts, then remove the support plate(s) and the square dynamic damper from the rear of the carrier.

8. Lower the differential carrier and remove from the vehicle.

9. Remove the pump retaining bolt and remove the pump from the rear differential assembly.

To install:

10. Install the pump and tighten the mounting bolt to 14–20 ft. lbs. (19–28 Nm).

11. Raise the rear differential carrier into position and install support member bolts. Replace all self-locking nuts. Tighten all mounting nuts and bolts as follows:

• Upper support plate to carrier bolts — 72–87 ft. lbs. (100–120 Nm)

• Support member/dynamic damper to carrier bolts — 58–72 ft. lbs. (80–100 Nm)

• Differential support member mounting bolt nuts — 80–94 ft. lbs. (110–130 Nm)

12. Connect the pressure and suction lines to the pump.

13. Install new circlips on both rear driveshafts and install.

14. Install the propeller shaft and tighten mounting hardware to 22–25 ft. lbs. (30–35 Nm).

15. Install the center exhaust pipe and muffler.

16. Lower the vehicle. With the vehicle level, fill the rear differential.

17. Fill the power steering system and properly bleed.

1994–97 Galant

1. Disconnect the battery negative cable.

2. Loosen and remove the power steering pump drive belt.

3. Remove the pressure switch connector from the side of the pump.

NOTE: If the alternator is located under the oil pump, cover it with a shop towel to protect it from oil.

4. Disconnect the return fluid line. Remove the reservoir cap and allow the return line to drain the fluid from the reservoir. If the fluid is contaminated, disconnect the ignition high tension cable and crank the engine several times to drain the fluid from the gearbox.

5. Disconnect the pressure line.

6. Unbolt and remove the pump from the mounting bracket.

To install:

7. Install the pump, wrap the belt around the pulley and lightly tighten the mounting bolts.

8. Replace the O-rings and connect the pressure line. Connect the pres-

sure line so the notch in the fitting aligns and contacts the pump's guide bracket. Tighten the fitting to 13 ft. lbs. (18 Nm).

9. Connect the return line and secure with the clamp.

10. Connect the pressure switch connector.

11. Adjust the power steering belt for proper tension and tighten the adjusting bolts.

12. Reconnect the negative battery cable.

13. Refill the reservoir and bleed the system.

Diamante

1. Disconnect the battery negative cable.

2. Disconnect the return fluid line. Remove the reservoir cap and allow the return line to drain the fluid from the reservoir. If the fluid is contaminated, disconnect the ignition high tension cable and crank the engine several times to drain the fluid from the gearbox.

3. Remove the power steering pump drive belt.

4. Remove the pressure switch connector from the side of the pump.

5. If the alternator is located under the oil pump, cover it with a shop towel to protect it from oil.

6. Disconnect the high pressure hose and the return hose from the pump.

7. Remove the pump drive belt and unbolt the pump from its bracket and remove the pump.

To install:

8. Install the pump, wrap the belt around the pulley and tighten the bolts that secure the pump to specifications.

9. Replace the O-rings and connect the high pressure hose. Connect the pressure line so the notch in the fitting aligns and contacts the pump's guide bracket. Tighten the mounting nut with lockwasher to 17 ft. lbs. (24 Nm).

10. Using a new hose clamp, connect the return line.

11. Connect the pressure switch connector.

12. Adjust the belt tension and tighten the adjusting bolts.

13. Refill the reservoir and bleed the system.

3000GT

Front

1. Disconnect the negative battery cable.

Wait at least 90 seconds after the negative (-) battery cable is disconnected to prevent possible deployment of the air bag.

2. Remove the pressure switch connector from the side of the pump.

3. If the alternator is located under the oil pump, cover it with a shop towel to protect it from oil.

4. Disconnect the return fluid line. Remove the reservoir cap and allow the return line to drain the fluid from the reservoir. If the fluid is contaminated, disconnect the ignition high tension cable and crank the engine several times to drain the fluid from the gearbox.

5. Disconnect the pressure line.

6. Remove the pump drive belt and unbolt the pump from its bracket.

To install:

7. Install the pump, wrap the belt around the pulley and tighten the bolts.

8. Replace the O-rings and connect the pressure line. Connect the pressure line so the notch in the fitting aligns and contacts the pump's guide bracket.

9. Connect the return line.

10. Connect the pressure switch connector.

11. Adjust the belt tension and tighten the adjusting bolts.

12. Refill the reservoir and bleed the system.

Rear

1. Disconnect the negative battery cable. Raise the vehicle and support safely.

2. Drain the power steering fluid.

3. Remove the main muffler assembly.

4. Remove the rear shock absorber lower mounting bolts.

5. Remove the 2 small crossmember brackets.

6. Using the proper equipment, support the weight of the rear differential. Remove the large self-locking crossmember mounting nuts on the differential side.

7. Disconnect the pressure and suction hoses from the fittings on the pump.

8. Remove the pump retaining bolt and remove the pump from the rear differential assembly. Do not attempt to disassemble the pump; it is not serviceable.

To install:

9. Replace the O-ring and install the pump assembly to the differential. Make sure the housing is fully

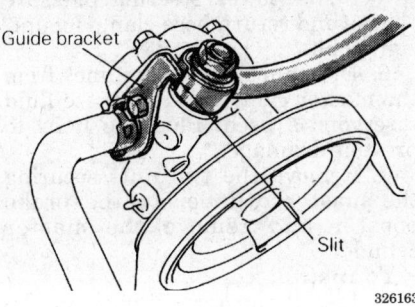

Guide bracket

Slit

326163

Pressure hose installation — Diamante

seated and the gear is fully engaged. Install the retaining bolt.

10. Replace the O-ring and connect the fluid lines to the pump.

11. Install the large self-locking crossmember mounting nuts on the differential side. Torque to 80–94 ft. lbs. (110–130 Nm). Remove the support equipment.

12. Install the 2 small crossmember brackets.

13. Install the shock mounting bolts.

14. Install the muffler assembly.

15. Refill the reservoir and bleed the system.

─────── **CAUTION** ───────

Extreme caution should be taken when testing the rear steering pump. Ensure that the vehicle is supported safely and that all components are torqued to specification prior be testing.

16. To check and see if the system is functioning:

a. Raise the vehicle safely so all 4 wheels turn freely.

b. Run the vehicle at 50 mph.

c. Turn the steering wheel quickly to the left and right and make sure the rear wheels steer in the same direction as the front wheels.

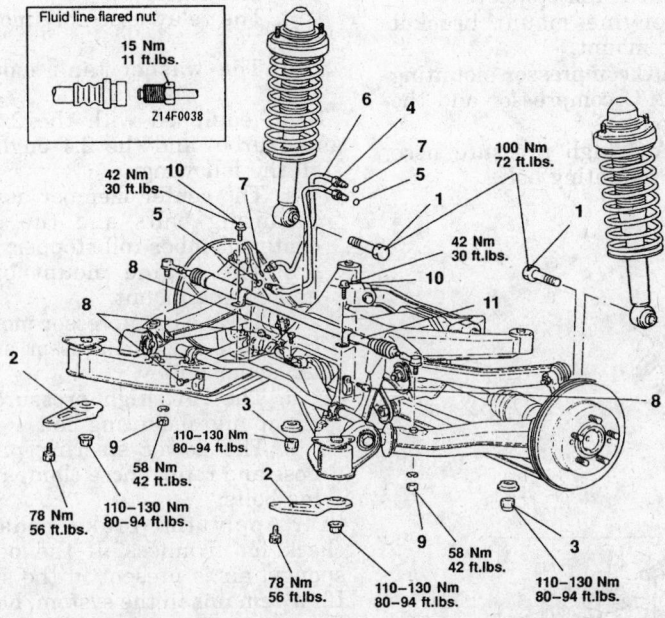

Fluid line flared nut
15 Nm
11 ft.lbs.
Z14F0038

42 Nm
30 ft.lbs.

100 Nm
72 ft.lbs.

42 Nm
30 ft.lbs.

110–130 Nm
80–94 ft.lbs.

58 Nm
42 ft.lbs.

78 Nm
56 ft.lbs.

110–130 Nm
80–94 ft.lbs.

78 Nm
56 ft.lbs.

110–130 Nm
80–94 ft.lbs.

58 Nm
42 ft.lbs.

110–130 Nm
80–94 ft.lbs.

1. Rear shock absorber lower mounting bolt
2. Crossmember bracket
3. Crossmember mounting nut (on differential side)
4. Pressure tube (RL)
5. O-ring
6. Pressure tube (RR)
7. O-ring
8. Oil line clamp bolt
9. Tie rod end nut
10. Power cylinder installation bolt
11. Power cylinder

324060

Rear power steering pump — 3000GT

BRAKES

Anti-Lock Brake System Service

PRECAUTIONS

• Certain components within the Anti-Lock Brake System (ABS) are not intended to be serviced or repaired individually. Only those components with removal and installation procedures should be serviced.

• Do not use rubber hoses or other parts not specifically specified for and ABS system. When using repair kits, replace all parts included in the kit. Partial or incorrect repair may lead to functional problems and require the replacement of components.

• Lubricate rubber parts with clean, fresh brake fluid to ease assembly. Do not use lubricated shop air to clean parts; damage to rubber components may result.

• Use only specified brake fluid from an unopened container.

• If any hydraulic component or line is removed or replaced, it may be necessary to bleed the entire system.

• A clean repair area is essential. Always clean the reservoir and cap thoroughly before removing the cap. The slightest amount of dirt in the fluid may plug an orifice and impair the system function. Perform repairs after components have been thoroughly cleaned; use only denatured alcohol to clean components. Do not allow ABS components to come into contact with any substance containing mineral oil; this includes used shop rags.

• The Anti-Lock control unit is a microprocessor similar to other computer units in the vehicle. Ensure that the ignition switch is **OFF** before removing or installing controller harnesses. Avoid static electricity discharge at or near the controller.

• If any arc welding is to be done on the vehicle, the control unit should be unplugged before welding operations begin.

DEPRESSURIZING

Master Cylinder

REMOVAL AND INSTALLATION

Mirage, Galant, Diamante, 3000GT and 1993–94 Precis

1. Disconnect the negative battery cable.
2. Disconnect the fluid level sensor.
3. Disconnect the brake lines from the master cylinder.
4. Remove the two nuts securing the master cylinder to the brake booster.
5. Slide the proportioning valve assembly off the mounting studs and remove the master cylinder.

To install:

6. Install master cylinder and proportioning valve to the mounting studs. Install the mounting nuts and tighten to 7 ft. lbs. (10 Nm).
7. Fill the reservoir to the proper level with clean DOT 3 brake fluid. Bleed the master cylinder.
8. Connect the brake lines to the master cylinder.
9. Apply the brake pedal and check for firmness. If the pedal is spongy, air is present in the system

and bleeding the entire system is required.

10. Check the brake system for leakage and proper operation.

Eclipse

1. Disconnect the negative battery cable.
2. Disconnect the fluid level sensor connector, if equipped.
3. If equipped with M/T, remove the clutch fluid reservoir bracket.
4. If equipped with the 2.0L engine (non-turbo), remove the following:
   a. The battery.
   b. The relay assembly mounting bolts.
   c. The washer tank mounting bolts.
5. If equipped with the 2.0L engine (turbo) and the 2.4 engine, remove the following:
   a. The center member assembly mounting bolts and the engine center member roll stopper.
   b. The engine mount bracket and engine mount.
   c. The A/C compressor mounting bolts, the A/C compressor and the tensioner pulley.
   d. The A/C high pressure hose clamp and mounting bolt.

e. The power steering pressure hose and return hose clamp mounting bolts.

6. Disconnect the brake lines from the master cylinder. A separate fluid reservoir is used. Plug the lines to prevent drainage.
7. Remove the two nuts securing the master cylinder to the brake booster and remove the master cylinder.

To install:

8. Install master cylinder to the mounting studs and install the mounting nuts. Tighten mounting nuts to 7 ft. lbs. (10 Nm).
9. Connect reservoir hoses to master cylinder and secure with clamps.
10. Fill the reservoir to the proper level with clean DOT 3 brake fluid. Bleed the master cylinder.
11. Connect the brake lines to the master cylinder.
12. If equipped with the 2.0L engine (non-turbo), install the following:
   a. The battery.
   b. The relay assembly mounting bolts.
   c. The washer tank mounting bolts.
13. If equipped with the 2.0L engine (turbo) and the 2.4 engine, install the following:
   a. The center member assembly mounting bolts and the engine center member roll stopper.
   b. The engine mount bracket and engine mount.
   c. The A/C compressor mounting bolts, the A/C compressor and the tensioner pulley.
   d. The A/C high pressure hose clamp and mounting bolt.
   e. The power steering pressure hose and return hose clamp mounting bolts.
14. Apply the brake pedal and check for firmness. If the pedal is spongy, air is present in the system. If air remains in the system, bleeding the entire system is required.
15. Check the brakes for proper operation and leaks.

Brake Caliper

REMOVAL AND INSTALLATION

Mirage, Eclipse and 1993 Galant

Front

1. Raise the vehicle and support safely.

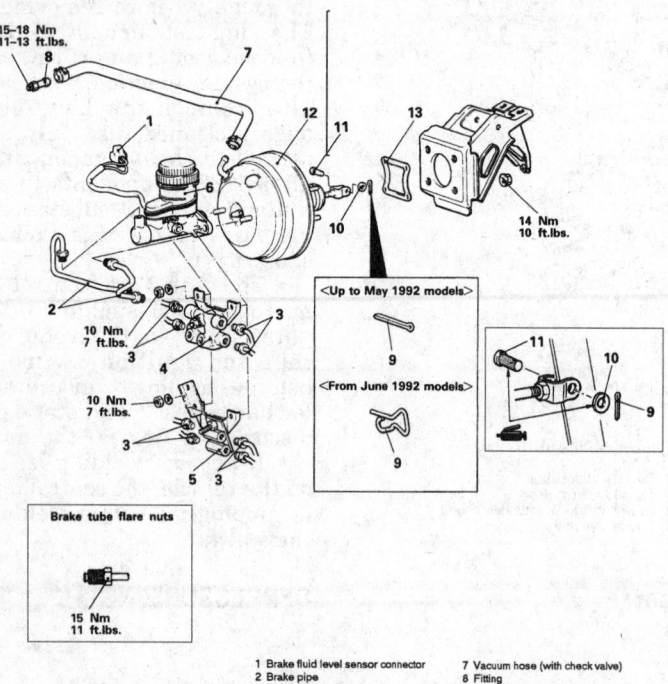

1 Brake fluid level sensor connector
2 Brake pipe
3 Brake pipe connection
4 Proportioning valve
5 Connector assembly
  <Vehicles with ABS>
6 Master cylinder
7 Vacuum hose (with check valve)
8 Fitting
9 Cotter pin*1 or snap pin*2
10 Washer
11 Clevis pin
12 Brake booster
13 Sealer

325675

Brake master cylinder and related components — Diamante

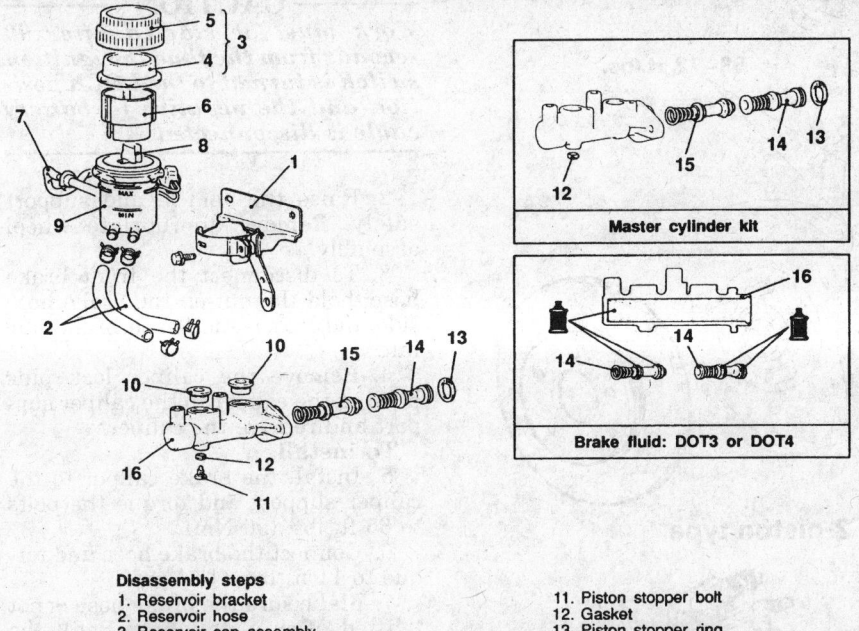

Disassembly steps
1. Reservoir bracket
2. Reservoir hose
3. Reservoir cap assembly
4. Diaphragm
5. Reservoir cap
6. Filter
7. Brake fluid level sensor
8. Float
9. Reservoir
10. Reservoir seal

11. Piston stopper bolt
12. Gasket
13. Piston stopper ring
14. Primary piston assembly
15. Secondary piston assembly
16. Master cylinder body

Master cylinder kit

Brake fluid: DOT3 or DOT4

Caution
Do not disassemble the primary and secondary piston assembly.

324468

Master cylinder and related components — 1994-97 Galant

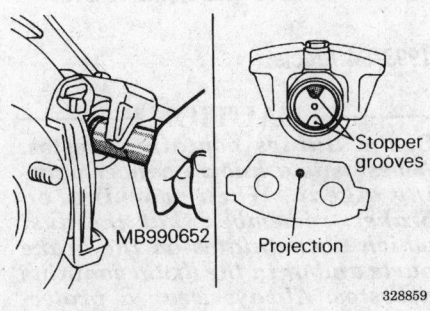

Retracting brake caliper piston and aligning pad to piston — Mirage

2. Remove the appropriate tire and wheel assembly.

NOTE: Do not let air into the master cylinder by allowing the reservoir to empty or complete system bleeding will be required.

3. To disconnect the front brake hose, hold the nut on the brake hose side and loosen the flared brake line nut. With the hose disconnected at the line, remove the brake hose from the caliper.
4. Remove the caliper guide and lock pins. Remove the caliper assembly from the caliper support.

To install
5. Position the brake caliper onto the caliper support. Install and tighten the guide and lock pins.
6. Reconnect the brake hose.

NOTE: Use caution not to twist brake hose during installation.

7. Bleed the brake system.
8. Apply brake pedal and inspect the system for proper operation and no leakage.
9. Install tire and wheel assembly.

Rear

1. Disconnect the battery negative cable.
2. Raise the vehicle and support safely.
3. Remove the appropriate tire and wheel assemblies.
4. Loosen the parking brake cable adjustment from inside the vehicle.
5. Remove the retaining clips, and disconnect the parking brake cable from the rear brake caliper assembly.

NOTE: Do not let air into the master cylinder by allowing the reservoir to empty or complete system bleeding will be required.

6. On FWD models, to disconnect the brake hose from the caliper, remove the banjo bolt from the brake

caliper. On AWD models, hold the nut on the brake hose side and loosen the flared brake line nut. With the hose disconnected at the line, remove the brake hose from the caliper.
7. Remove the caliper lock pin. Pivot the caliper upward, and slide the caliper assembly from the caliper support.

To install:
8. On FWD models, install the rear brake hose onto the caliper with new washers in place and tighten the brake hose retainer. On AWD models, connect the brake hose to the caliper and tighten the fitting to 9–12 ft. lbs. (13–17 Nm). Then connect the hose at the bracket to the steel line and tighten the fitting to 9–12 ft. lbs. (13–17 Nm).

NOTE: Do not twist the brake hose during installation.

9. Install the caliper over the brake pads, making sure stopper grove lines up with pad projection.
10. Lubricate and install the lock pin and tighten to 23 ft. lbs. (32 Nm).
11. Bleed the brake system.
12. Inspect the brake system for leaks and ensure proper operation.
13. Install tire and wheel assemblies.
14. Properly adjust parking brake cable.

Diamante and Galant

Unlike most rear disc brake designs, this system does not incorporate the parking brake system into the rear brake caliper. Therefore, the rear brake system is serviced the same as the front system.
1. Raise the vehicle and support safely.
2. Remove the appropriate tire and wheel assembly.

NOTE: Do not allow the master cylinder reservoir to empty. An empty reservoir will allow air to enter the entire brake system and complete system bleeding will be required.

3. To disconnect the brake hose on models with a banjo-bolt connecting the brake hose to the caliper assembly, simply remove the bolt at the hose connection. To disconnect the brake hose on all other systems, hold the nut on the brake hose side and loosen the flared brake line nut. Once the hose has been disconnected from the line, remove the brake hose from the caliper.
4. Remove the caliper guide and lock pins and lift the caliper assembly from the caliper support.

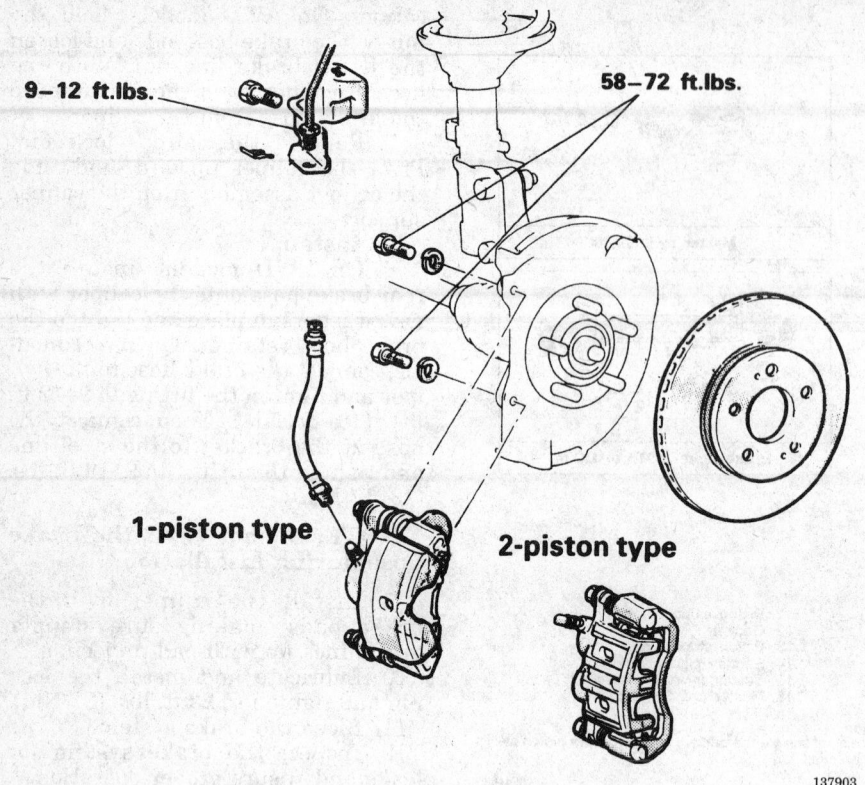

9–12 ft.lbs.

58–72 ft.lbs.

1-piston type

2-piston type

137903

Front disc brake and caliper removal and installation — 1993–94 Eclipse shown, others similar

To install

5. Position the caliper onto the caliper support. Install the guide pin and lock pin. Tighten to specification.

6. Reconnect the brake hose or install the banjo bolt with new washers.

NOTE: Use caution not to twist the brake hose during installation.

7. Bleed the brake system.

8. Apply brake pedal and inspect the system for leaks. Ensure proper operation and no leakage.

9. Install tire and wheel assembly. Torque lug nuts to 87–101 ft. lbs. (120–140 Nm).

1993–94 Precis

CAUTION

Brake linings contain asbestos. Asbestos is a known cancer-causing agent. When working on brakes, remember that the dust which accumulates on the brake parts and/or in the drum contains asbestos. Always wear a protective face covering, such as a painter's mask, when working on the brakes. NEVER blow the dust from the brakes or drum.

1. Raise and support the front end on jackstands.

2. Remove the front wheels.

3. Loosen the brake line at the caliper and disconnect it.

NOTE: Have some kind of capping device handy to plug the brake line once it is disconnected.

4. Remove the brake pads.

5. Remove the pin and sleeve boots.

6. Remove the lower caliper bolt and raise the caliper up and out to remove it.

To install:

7. Position the caliper onto its mounting and install the lower mounting bolt. Torque the bolt to 16 to 24 ft. lbs. (22 to 32 Nm).

8. Install the pin boots, sleeve boots and brake pads.

9. Connect the brake line to the caliper with two new metal gaskets. Torque the brake line union bolt to 18–22 ft. lbs. (25–30 Nm).

10. Bleed the system.

11. Mount the front wheels and lower the vehicle.

3000GT

1. Disconnect the negative battery cable.

CAUTION

Work must be started after 90 seconds from the time the ignition switch is turned to the LOCK position and the negative (-) battery cable is disconnected.

2. Raise the vehicle and support safely. Remove appropriate wheel assembly.

3. To disconnect the front brake hose, hold the nut on the brake hose side and loosen the flared brake line nut.

4. Remove the caliper lock pins holding the caliper to the caliper support and remove the caliper.

To install:

5. Install the brake caliper to the caliper support and torque the bolts to 65 ft. lbs. (90 Nm).

6. Connect the brake hose and torque to 11 ft. lbs. (15 Nm).

7. Make sure the brake hose is not twisted after installation. Refill the brake fluid as required and bleed the brakes.

Disc Brake Pads

REMOVAL AND INSTALLATION

1993–94 Precis

CAUTION

Brake linings contain asbestos. Asbestos is a known cancer-causing agent. When working on brakes, remember that the dust which accumulates on the brake parts and/or in the drum contains asbestos. Always wear a protective face covering, such as a painter's mask, when working on the brakes. NEVER blow the dust from the brakes or drum! There are solvents made for the purpose of cleaning brake parts.

1. Raise the vehicle and support it safely.

2. Remove the front wheels.

3. Remove the lower caliper mounting bolt and rotate the caliper upward. Secure the caliper with a wire or heavy string.

4. Remove the pads from the caliper.

To install:

5. Install the pad clips.

6. Install the pads onto the pad clips.

7. Compress the caliper piston using a C-clamp.

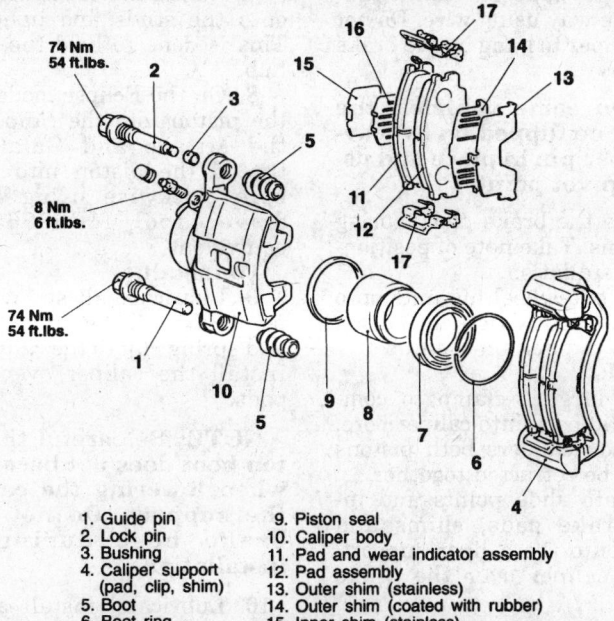

74 Nm
54 ft.lbs.

8 Nm
6 ft.lbs.

74 Nm
54 ft.lbs.

1. Guide pin
2. Lock pin
3. Bushing
4. Caliper support
   (pad, clip, shim)
5. Boot
6. Boot ring
7. Piston boot
8. Piston
9. Piston seal
10. Caliper body
11. Pad and wear indicator assembly
12. Pad assembly
13. Outer shim (stainless)
14. Outer shim (coated with rubber)
15. Inner shim (stainless)
16. Inner shim (coated with rubber)
17. Clip

324462

Exploded view of the front brake pads and related components — 1996–97 Galant, others similar

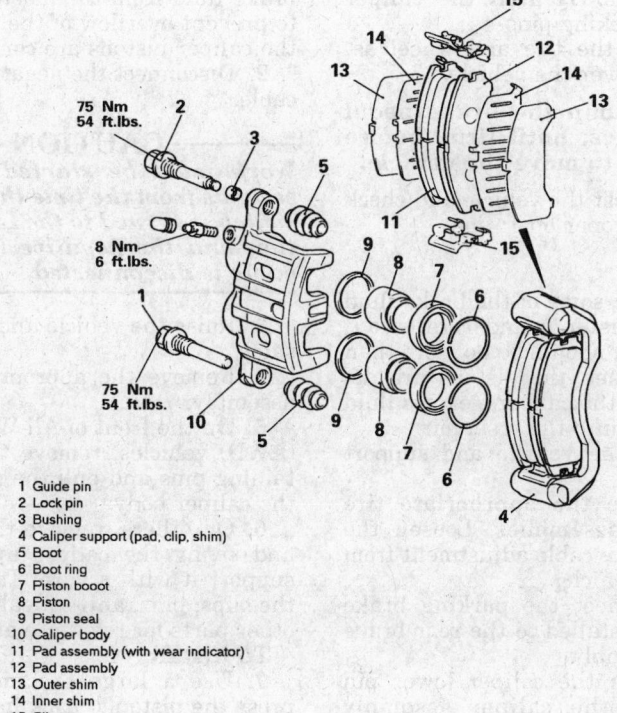

75 Nm
54 ft.lbs.

8 Nm
6 ft.lbs.

75 Nm
54 ft.lbs.

1 Guide pin
2 Lock pin
3 Bushing
4 Caliper support (pad, clip, shim)
5 Boot
6 Boot ring
7 Piston booot
8 Piston
9 Piston seal
10 Caliper body
11 Pad assembly (with wear indicator)
12 Pad assembly
13 Outer shim
14 Inner shim
15 Clip

320927

Front dual piston caliper exploded view — Diamante

8. Rotate the caliper downward and install the mounting bolt.

NOTE: Replace all brake pads at the same time. Never replace the pads on 1 wheel only.

9. Install the wheels and lower the vehicle.
10. Depress the brake pedal a few times. The first couple of strokes on the pedal will feel overly long. However, the pads will set themselves and the stroke will return to normal.

Diamante and 1994–97 Galant

——— **CAUTION** ———
Brake pads and shoes contain asbestos, which has been determined to be a cancer causing agent. Never clean the brake surfaces with compressed air! Avoid inhaling any dust from brake surfaces! When cleaning brakes, use commercially available brake cleaning fluids.

Unlike most rear disc brake designs, this system does not incorporate the parking brake system, into the rear brake caliper, therefore, the rear brake system is serviced the same as the front system.

1. Remove some of the brake fluid from the master cylinder reservoir. The reservoir should be no more than ½ full. When the pistons are depressed into the calipers, excess fluid will flow up into the reservoir.
2. Raise the vehicle and support safely.
3. Remove the appropriate tire and wheel assemblies.
4. Remove the caliper guide and lock pins and lift the caliper assembly from the caliper support. Tie the caliper out of the way using wire. Do not allow the caliper to hang by the brake line.

NOTE: On some vehicles, the caliper can be flipped up by leaving the upper pin in place and using it as a pivot point.

5. Remove the brake pads, spring clip and shims. Take note of positioning to aid installation.
6. Install the wheel lug nuts onto the studs and lightly tighten. This is done to hold the disc on the hub.
To install:
7. Use a large C-clamp to compress the piston(s) back into caliper bore.
8. Lubricate slide points and install the brake pads, shims and spring clip onto the caliper support.

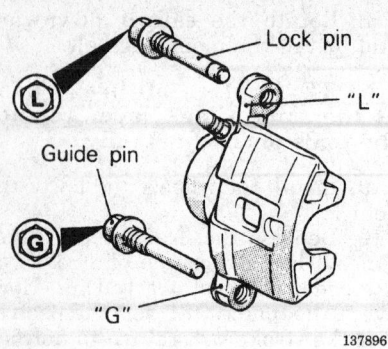

Caliper locking pin and guide pin description — 1993-94 Eclipse

Install the caliper over the brake pads.

NOTE: Be careful that the piston boot does not become caught when lowering the caliper onto the support. Do not twist the brake hose during caliper installation.

9. Lubricate and install the caliper guide and lock pins in their original positions. Tighten guide and locking pins to 54 ft. lbs. (75 Nm) on the front, and 20 ft. lbs. (27 Nm) on the rear.
10. Install the tire and wheel assemblies. Lower the vehicle.

— **WARNING** —
Pump brake pedal several times, until firm, before attempting to move vehicle.

11. Road test the vehicle and check brakes for proper operation.

Mirage, Eclipse and 1993 Galant

— **CAUTION** —
Brake pads and shoes contain asbestos, which has been determined to be a cancer causing agent. Never clean the brake surfaces with compressed air! Avoid inhaling any dust from brake surfaces! When cleaning brakes, use commercially available brake cleaning fluids.

Front

1. Remove some of the brake fluid from the master cylinder reservoir. The reservoir should be no more than half full. When the pistons are pressed into the calipers, excess fluid will flow up into the reservoir.
2. Raise the vehicle and support safely.
3. Remove the appropriate tire and wheel assemblies.
4. Remove the caliper guide and lock pins and lift the caliper assembly

from the caliper support. Tie the caliper out of the way using wire. Do not allow the caliper to hang by the brake line.

NOTE: On some vehicles, the caliper can be flipped up by leaving the upper pin in place and using it as a pivot point.

5. Remove the brake pads, spring clip and shims. Take note of positioning to aid installation.
6. Install two wheel lug nuts onto the studs and lightly tighten. This is done to hold the disc on the hub.
To install:
7. Use a large C-clamp to compress piston(s) back into caliper bore. On two piston calipers both pistons will have to be retracted together.
8. Lubricate slide points and install the brake pads, shims and spring clip onto the caliper support. Install the caliper over the brake pads.

NOTE: Be careful that the piston boot does not become caught when lowering the caliper onto the support. Do not twist the brake hose during caliper installation.

9. Lubricate and install the caliper guide and lock pins in their original positions. Tighten the caliper guide and locking pins.
10. Install the tire and wheel assemblies. Lower the vehicle.

NOTE: Pump the brake pedal several times, until firm, before attempting to move the vehicle.

11. Road test the vehicle and check brakes for proper operation.

Rear

1. Remove some of the brake fluid from the master cylinder reservoir. The reservoir should be no more than half full. When the pistons are depressed into the calipers, excess fluid will flow up into the reservoir.
2. Raise the vehicle and support safely.
3. Remove the appropriate tire and wheel assemblies. Loosen the parking brake cable adjustment from inside the vehicle.
4. Disconnect the parking brake cable end installed to the rear brake caliper assembly.
5. Remove the caliper lower pin and swing the caliper assembly upwards. Tie the caliper out of the way using wire.
6. Remove the outer shim, brake pads and spring clips from the caliper support. Take note of positioning of each to aid in installation.

7. Install two of the wheel lug nuts onto the studs and lightly tighten. This is done to hold the disc on the hub.
8. On the Eclipse model, compress the piston into the caliper bore. On the Mirage and Galant models, thread the piston into the caliper bore clockwise using disc brake driver tool MB990652 or its equivalent.
To install:
9. Lubricate all sliding and pivot points. Install the brake pads, shims and spring clip to the caliper support. Install the caliper over the brake pads.

NOTE: Be careful that the piston boot does not become caught when lowering the caliper onto the support. Do not twist the brake hose during caliper installation.

10. Lubricate, install and tighten the lower pin.
11. Install the tire and wheel assemblies. Lower the vehicle.
12. Test the brakes for proper operation.

3000GT

1. Remove approximately ⅓ of the brake fluid from the master cylinder to prevent overflow of the fluid when the caliper pistons are compressed.
2. Disconnect the negative battery cable.

— **CAUTION** —
Work must be started after 90 seconds from the time the ignition switch is turned to the LOCK position and the negative (-) battery cable is disconnected.

3. Raise the vehicle and support it safely.
4. Remove the appropriate wheel assembly.
5. On the front of All Wheel Drive (AWD) vehicles, remove the pad retaining pins and pull the pads out of the caliper body.
6. On others, remove the lock pin and swing the caliper upward and support it with a wire. Take note of the clips, pins, anti-squeal shims and other parts for reference at assembly.
To install:
7. Use a large C-clamp to compress the piston(s) back into the caliper bore.
8. Install the pads and all other small parts.
9. On AWD front disc brakes, apply a small amount of multi-purpose grease to the inner pad shims.

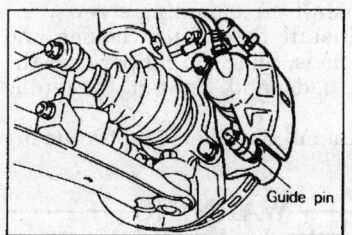

Guide pin

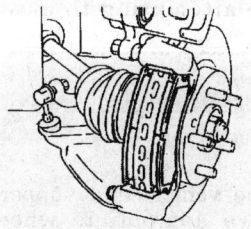

① Pad & wear indicator assembly
② Pad assembly
③ Clip
④ Inner shim
⑤ Outer shim

1-piston type   **2-piston type**

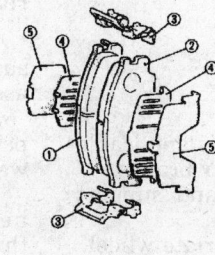

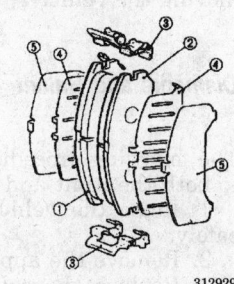

312929

Front brake pad installation — 1995–97 Eclipse

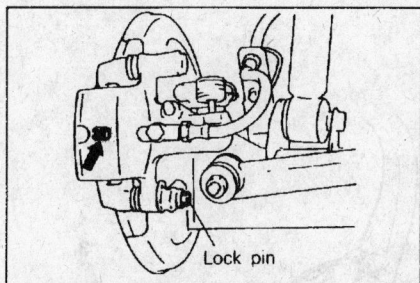

Lock pin

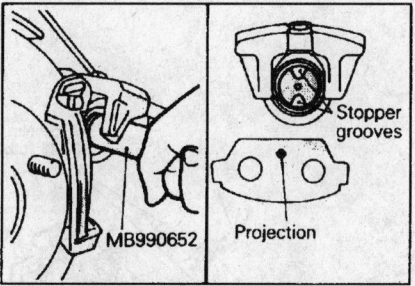

MB990652   Stopper grooves

Projection

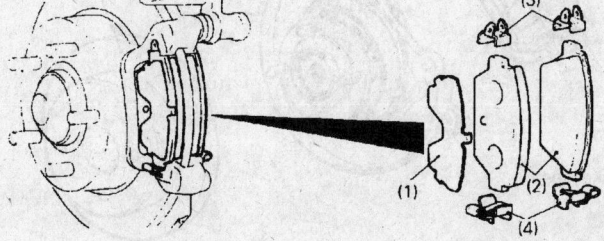

(1) Outer shim
(2) Pad assembly
(3) Pad clips C
(4) Pad clips B

137944

Rear brake pad installation — 1993–94 Eclipse shown, others similar

10. Install the caliper to the caliper support. Make sure the brake hose is not twisted after installation if both guide pins were removed.

11. Install the tire and wheel assembly and torque to 87–101 ft. lbs. (120–140 Nm).

12. Connect the negative battery cable.

13. Refill the master cylinder if needed and pump the brake pedal until firm before putting transaxle in gear or moving vehicle.

Brake Rotor

REMOVAL AND INSTALLATION

1993–94 Precis

——————— **CAUTION** ———————
Brake linings contain asbestos. Asbestos is a known cancer-causing agent. When working on brakes, remember that the dust which accumulates on the brake parts and/or in the drum contains asbestos. Always wear a protective face covering, such as a painter's mask, when working on the brakes. NEVER blow the dust from the brakes or drum! There are solvents made for the purpose of cleaning brake parts.

Front

1. Remove the center hub cap and halfshaft nut. Then raise the vehicle and support it safely. Allow the wheels to hang freely. Then remove the front wheel.

2. Remove the brake caliper without disconnecting the hydraulic line and suspend it out of the way with a piece of wire.

3. Disconnect the stabilizer bar and strut bar from the lower control arm.

4. Disconnect the ball joint stud from the knuckle.

5. Remove the halfshaft from the transaxle and press the halfshaft out of the hub using tool 09526-11001 or equivalent.

6. Unbolt and remove the hub and knuckle from the bottom of the strut and remove the hub and knuckle assembly from the vehicle.

7. Several special tools are required to press the hub and disc from the steering knuckle and to remount them. Use 09517-21600 or equivalent. Do not attempt to hammer the parts apart, or the bearing will be damaged. Install the arm of the special tool then the body onto the knuckle and tighten the nut man-

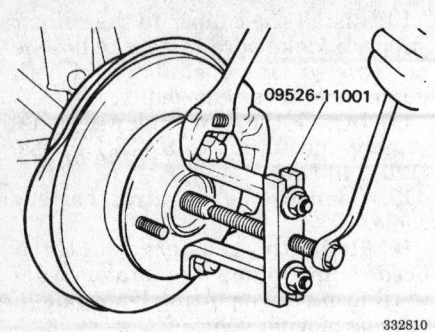

Remove the halfshaft from the hub — 1993–94 Precis

ually. Using special tool 09517–21500, separate the hub from the knuckle. Pull the bearings out, noting their positions and direction of installation (smaller diameter inward).

8. Matchmark the relationship between the brake disc and hub. Then place the knuckle in a vise and separate the rotor from the hub by removing the attaching bolts.

To install:

9. Install the hub to the rotor.

10. Install the hub and rotor to the knuckle.

11. Install the hub and knuckle to the bottom of the strut.

12. Install the halfshaft to the transaxle.

13. Connect the stabilizer bar and strut bar to the lower control arm, if disconnected.

14. Install the lower ball joint stud to the knuckle.

15. Install the brake caliper.

16. Install the halfshaft nut and center hub cap and halfshaft nut.

17. Install the front wheel and lower the vehicle.

18. Torque the hub nut to 145 to 188 ft. lbs. (200 to 260 Nm).

Rear

1. Raise and safely support the vehicle.

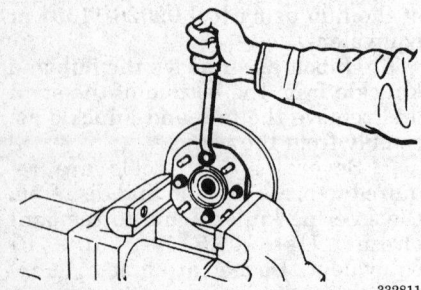

Remove the rotor from the hub — 1993–94 Precis

2. Remove the wheel and tire assembly.

3. Remove the brake caliper and support it with and wire to prevent damage to the brake hose.

4. Remove the retaining screw and pull the rotor from the hub.

To install:

5. Position the rotor on the hub and install the retaining screw.

6. Install the brake caliper on the rotor assembly.

7. Install the wheel and tire assembly.

8. Lower the vehicle to the floor and apply the brakes a few times to insure good brake operation before moving the vehicle.

Diamante and Galant

The following procedure is applicable to both the front and rear brakes.

1. Raise the vehicle and support safely.

2. Remove the appropriate wheel.

3. Remove the caliper and brake pads. Support the caliper out of the way using a wire.

4. On some models the rotor is held to the hub by two small threaded screws. Remove the screws and pull off the rotor.

To install

5. Position the rotor on the hub and install the mounting screws.

6. Install the caliper holder and brake pads. Slide the caliper over the brake pads and tighten the guide pins.

7. Install the wheel and torque lug nuts.

—— WARNING ——
Pump the brake pedal several times before attempting to move vehicle.

Mirage and 1993–94 Eclipse

Front

1. Raise the vehicle and support safely. Remove appropriate wheel assembly.

2. Remove the caliper and brake pads. Support the caliper out of the way using wire.

3. The rotor on most models is held to the hub by two small threaded screws. Remove screws, if equipped, and pull off the rotor.

To install

4. Position the rotor on the hub and install mounting screws.

5. Install caliper holder and brake pads. Slide caliper over brake pads and tighten guide pins.

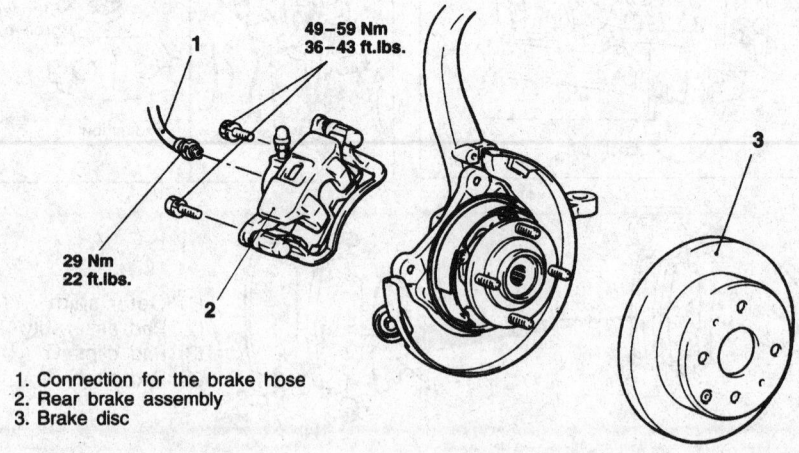

1. Connection for the brake hose
2. Rear brake assembly
3. Brake disc

Rear brake system component identification — 1994–95 Galant

6. Install wheel and torque lug nuts. Check brake pedal before attempting to move vehicle.

Rear

1. Raise the vehicle and support safely. Remove appropriate wheel assembly.

2. Disconnect the parking brake connection at the rear caliper assembly.

3. Remove the caliper and brake pads. Support the caliper out of the way using wire.

4. Remove the brake rotor from the rear hub assembly.

To install

5. Position the rotor on the hub. Install a couple of lug nuts and lightly tighten to hold rotor on hub.

6. Install the caliper holder and place brake pads in holder. Slide caliper over brake pads and install guide pins. Once caliper is secured, lug nuts can be removed.

7. Reconnect parking brake cable and install wheel(s).

1995–97 Eclipse

1. Raise the vehicle and support safely. Remove appropriate wheel assembly.

2. Remove the caliper, brake pads and caliper support. Support the caliper out of the way using wire.

3. Matchmark the rotor to the hub and remove the rotor. It may be necessary to loosen the parking brake adjusting nut to remove the rear rotor.

To install:

4. Align the matchmark and install the rotor on the hub.

5. Install the caliper support, brake pads and caliper.

6. Install the wheel and tire assembly.

7. Lower the vehicle to the floor.

1993–95 3000GT

1. Raise and safely support the vehicle.

2. Remove the appropriate wheel and tire assembly.

3. Remove the caliper with the caliper support attached and using a wire, support the caliper from the coil spring.

4. Match mark the rotor and the hub and remove the brake rotor.

To install:

5. Align the match mark and install the rotor on the hub.

6. Using a large C-clamp, compress the caliper pistons enough to allow the rotor to fit between the pads and install the caliper assembly. Torque the bolts to 65 ft. lbs. (90 Nm).

7. Install the wheel and tire assembly and torque the lug nuts to 87–101 ft. lbs. (120–140 Nm).

> ——— CAUTION ———
> *Apply the brake pedal several times to insure good brake operation before moving the vehicle.*

Brake Drums

REMOVAL AND INSTALLATION

1993–94 Precis

> ——— CAUTION ———
> *Brake linings contain asbestos. Asbestos is a known cancer-causing agent. When working on brakes, remember that the dust which accumulates on the brake parts and/or in the drum contains asbestos. Always wear a protective face covering, such as a painter's mask, when working on the brakes. NEVER blow the dust from the brakes or drum.*

1. Raise the vehicle and support safely.

2. Remove the wheel and tire assembly.

3. Remove the dust cap, cotter pin, nut lock, wheel bearing nut and washer from the spindle. Remove the outer wheel bearing. Remove the drum with the inner wheel bearing from the spindle. If the drum is difficult to remove, remove the plug from the rear of the backing plate and push the self-adjuster lever away from the star wheel. Rotate the star wheel to retract the shoes.

4. Remove the brake drum and the grease seal.

To install:

5. Lubricate and install the inner wheel bearing. Install a new grease seal. Install the drum to the spindle. Lubricate and install the outer wheel bearing, washer and nut. Adjust the bearing preload as required. When the bearing preload is properly set, install the nut lock and a new cotter pin. Install the grease cap.

6. Install the wheel and tire assembly. Adjust the rear brakes as required.

7. Apply the brakes until a firm pedal is obtained, prior to moving the vehicle.

Mirage

1. Raise the vehicle and support safely.

2. Remove the wheel and tire assembly.

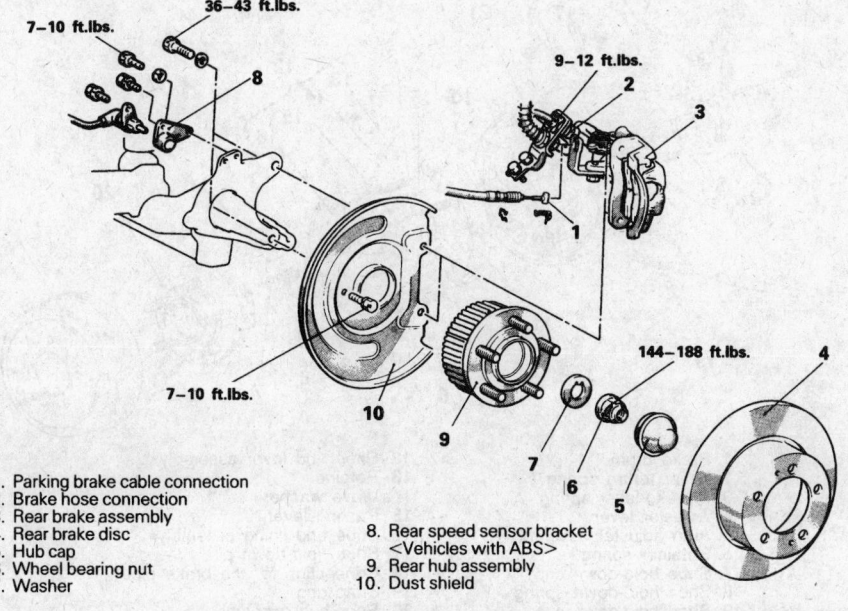

36–43 ft.lbs.
7–10 ft.lbs.
9–12 ft.lbs.
7–10 ft.lbs.
144–188 ft.lbs.

1. Parking brake cable connection
2. Brake hose connection
3. Rear brake assembly
4. Rear brake disc
5. Hub cap
6. Wheel bearing nut
7. Washer
8. Rear speed sensor bracket <Vehicles with ABS>
9. Rear hub assembly
10. Dust shield

137904

Rear disc brake exploded view — 1993–94 FWD Eclipse

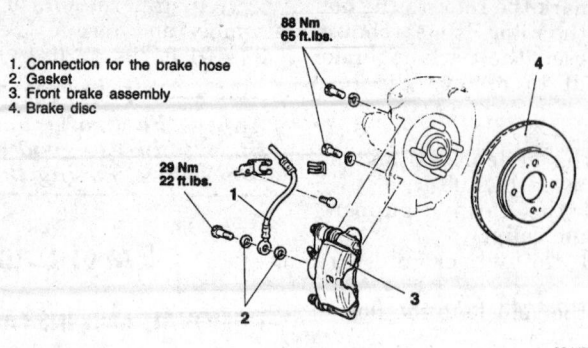

1. Connection for the brake hose
2. Gasket
3. Front brake assembly
4. Brake disc

88 Nm
65 ft.lbs.

29 Nm
22 ft.lbs.

324454

Front brake system component identification — 1996–97 Galant

3. Remove the dust cap.

4. Remove the self-locking nut.

5. Remove the outer wheel bearing.

6. Remove the drum with the inner wheel bearing from the spindle. Remove the grease seal.

To install:

7. To determine if the self-locking nut is reusable:

   a. Screw in the self-locking nut until about 1/8 in. of the spindle is showing.

   b. Measure the torque required to turn the self-locking nut counterclockwise.

   c. The lowest allowable torque is 48 inch lbs. (5.5 Nm). If the measured torque is less than the specification, replace the nut.

8. Lubricate and install the inner wheel bearing. Install a new grease seal.

9. Install the drum to the spindle.

10. Lubricate and install the outer wheel bearing.

11. Torque the self-locking nut to 108–145 ft. lbs. (150–200 Nm).

12. Install the grease cap.

Galant and 1995–97 Eclipse

1. Raise and safely support the vehicle.

2. Remove the rear wheel.

3. Loosen the parking brake adjusting nut.

4. Pull the drum from the rear hub assembly. Tap the drum with a soft mallet if necessary.

To install:

5. Install the drum on the rear hub assembly.

6. Install the wheel and adjust the parking brake.

7. Lower the vehicle to the floor.

Brake Shoes

REMOVAL AND INSTALLATION

Mirage

—————— **WARNING** ——————

Brake shoes contain asbestos, which has been determined to be a cancer causing agent. Never clean the brake surfaces with compressed air! Avoid inhaling any dust from brake surfaces! When cleaning brakes, use commercially available brake cleaning fluids.

1. Raise vehicle and support safely. Remove the tire and wheel assembly.

NOTE: Note the location of all springs and clips for proper reassembly.

2. Remove the brake drum. Remove the front shoe to rear shoe spring.

3. Remove the shoe to lever spring and remove the adjuster assembly.

4. Remove the shoe hold-down clips and the brake shoes.

5. Disconnect the parking brake cable from the rear shoes by spreading the horseshoe clip apart.

To install

6. Thoroughly clean and dry the backing plate. Ensure the backing plate bosses are smooth, so shoes won't bind.

7. Lubricate backing plate bosses, anchor pin, and parking brake actuating mechanism with a lithium-based grease.

8. Remove, clean and dry all remaining parts. Apply anti-seize to the star wheel threads and transfer all parts to the new shoes.

9. Connect the parking brake arm to the appropriate brake shoe. Attach shoes to backing plate and install all remaining hardware in the reverse order it was removed.

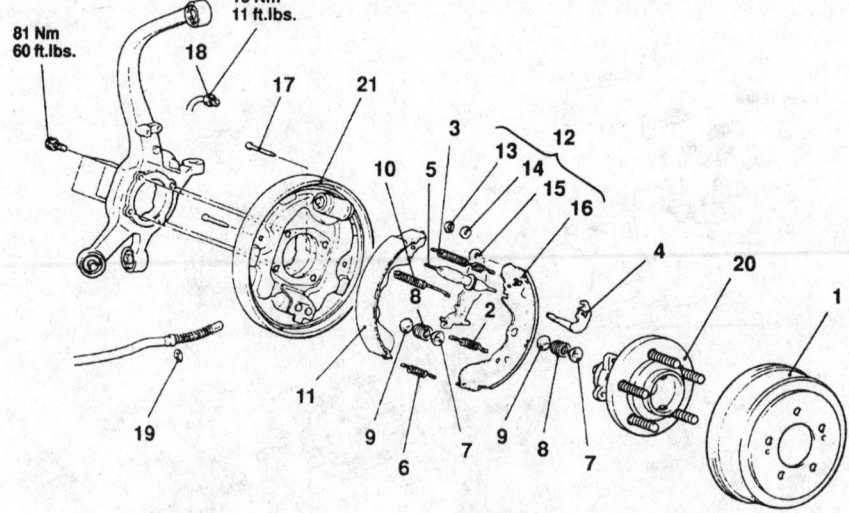

81 Nm
60 ft.lbs.

15 Nm
11 ft.lbs.

1. Brake drum
2. Lever return spring
3. Shoe-to-lever spring
4. Adjuster lever
5. Auto adjuster assembly
6. Retainer spring
7. Shoe hold-down cup
8. Shoe hold-down spring
9. Shoe hold-down cup
10. Shoe-to-shoe spring
11. Shoe and lining assembly
12. Shoe and lever assembly
13. Retainer
14. Wave washer
15. Parking lever
16. Shoe and lining assembly
17. Shoe hold-down pin
18. Connection for the brake pipe
19. Snap ring
20. Rear hub assembly
21. Backing plate

312908

Rear drum brake components — Galant and 1995–97 Eclipse

10. Preadjust the shoes so the drum slides on with a light drag and install brake drum.

11. Install a new wheel bearing self-locking nut and torque to 130 ft. lbs. (180 Nm).

12. Install wheel bearing dust cap and adjust rear brake shoes.

1993–94 Precis

1. Raise and support the vehicle.
2. Remove the wheel and tire assembly.
3. Remove the brake drum.
4. Remove the adjuster spring and the lever.
5. Spread the shoes apart and remove the adjuster strut.
6. Remove the shoe to shoe springs.
7. Remove the hold down springs and the brake shoes.
8. Remove the parking brake lever or the cable from the secondary shoe.
To install:
9. Clean the backing plate with brake cleaning solvent.
10. Apply grease to the friction areas on the backing plate.
11. Transfer the parking brake lever to the new secondary brake shoe.
12. Position the brake shoe on the backing plate and install the hold down pin and spring.
13. Position the primary shoe on the backing plate and install the pin and hold down spring.
14. Install the shoe to shoe springs.
15. Install the adjuster lever and spring.
16. Install the drum and adjust the brake shoes.
17. Install the wheel and tire assembly and lower the vehicle to the floor.

Galant and 1995–97 Eclipse

1. Raise and safely support the vehicle.
2. Remove the rear wheels and drums.
3. Remove the lever return spring.
4. Remove the shoe-to-lever spring.
5. Remove the adjuster lever.
6. Remove the auto-adjuster assembly.
7. Remove the retainer spring.
8. Remove the brake shoe hold-down springs and spring cups.
9. Remove the shoe-to-shoe spring.
10. Remove the brake shoes.
11. Disconnect the parking brake cable from the lever on the rear shoe.
To install:
12. Remove the parking brake lever from the used shoe and install it on

the new brake shoe. Make sure the wave washer is installed in the proper direction.

13. Clean the backing plate and lightly apply brake grease to the six shoe support pads.

14. Clean the adjuster assembly and apply brake grease to the threads, do not use more grease than necessary.

15. Connect the parking brake cable to the lever on the rear shoe. Position the rear shoe on the backing plate and install the hold-down spring and pin.

16. Position the front shoe on the backing plate and install the hold-down spring and pin.

17. Position the adjuster assembly between the two shoes.

18. Install the shoe-to-shoe spring.
19. Install the retainer spring.
20. Install the adjuster lever.
21. Install the shoe-to-lever spring.
22. Install the lever return spring.
23. Adjust the brake shoes and install the drum.
24. Install the wheel and tire assembly.
25. Lower the vehicle to the floor.

Wheel Cylinder

REMOVAL AND INSTALLATION

NOTE: It is important not let the master cylinder reservoir run dry, at any time during this procedure, or the entire system will have to be bleed.

1993–94 Precis

1. Raise and support the vehicle.
2. Remove the wheel and tire assembly.
3. Remove the brake drum.
4. Remove the brake shoes.
5. Place a container or some old rags under the brake backing plate to catch the brake fluid that will run out of the wheel cylinder.

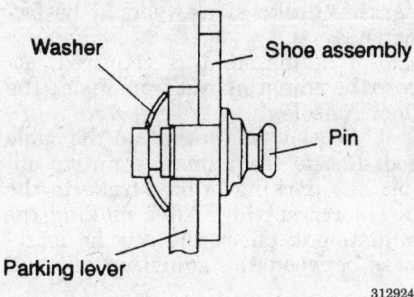

Correct wave washer position — Galant and 1995–97 Eclipse

312924

6. Disconnect the brake line(s) and remove the cylinder mounting bolts. Remove the cylinder from the backing plate.
To install:
7. Install the wheel cylinder to the brake backing plate. Torque the wheel cylinder mounting bolts to 4–9 ft. lbs. (5.5–12 Nm).
8. Connect the brake fluid line to the wheel cylinder.
9. Install the brake shoes and the drum.
10. Install the wheel and tire assembly.
11. Adjust the brakes.
12. Refill and bleed the brake system.

Mirage

1. Raise the vehicle and support safely.
2. Remove the wheel and the brake drum.
3. Remove the shoe-to-lever spring and the upper shoe-to-shoe spring. Spread the upper portion of the brake shoes slightly.
4. Remove and plug the brake line from the wheel cylinder.
5. Remove the wheel cylinder retaining bolts and remove the cylinder from the backing plate.
To install:
6. Apply a very thin coating of silicone sealer to the cylinder mounting surface, install the cylinder to the backing plate and install the retaining bolts.
7. Connect the brake line to the wheel cylinder.
8. Install brake springs and the brake drum. If the wheel cylinder was removed to correct a fluid leakage complaint, carefully clean and inspect all components. Brake shoe linings contaminated by brake fluid must not be reused. Always install new brake shoes as an axle set, both rear wheels at once. New brake return springs and hardware is always recommended. Make sure the adjuster is free and operates correctly. This is also a good time to check the parking brake cable assemblies and operation. Frozen or sticking brake cables must be replaced. Install all brake components that were removed, using new replacement parts as required. Inspect the inside of the brake drum. Most brake drum wear can be removed by proper machining. If the drum inside diameter is beyond limits, usually marked on the drum, a replacement drum must be installed.
9. Torque the new self-locking nut to 130 ft. lbs. (180 Nm).

10. Install the grease cap.

11. Install the tire and wheel assembly.

12. Fill the system with clean brake fluid and bleed the brake system.

Galant and 1995–97 Eclipse

1. Safely raise and support the vehicle.

2. Remove the rear wheel and the drum.

3. Remove the spring on the adjuster lever.

4. Remove the shoe-to-shoe spring.

5. Remove the auto adjuster assembly.

NOTE: Special flare nut wrenches should be used on all line fittings to prevent damage to the flats on the nut.

6. Place a drain pan under the wheel to catch the brake fluid and disconnect the brake line from the rear of the wheel cylinder.

7. Remove the wheel cylinder mounting bolts and the wheel cylinder.

To install:

8. Install the wheel cylinder on the backing plate. Torque the mounting bolts to 7 ft. lbs. (10 Nm).

9. Connect the brake line the wheel cylinder. Torque the line fitting to 11 ft. lbs. (15 Nm).

10. Add brake fluid to the reservoir and leave the wheel cylinder bleeder screw loose. Brake fluid will start the flow into the wheel cylinder and may save time when bleeding the brake system later.

11. Install the auto adjuster assembly.

12. Install the shoe-to-shoe spring.

13. Install the spring on the adjuster lever.

14. Install the drum and the wheel assembly.

15. Lower the vehicle and bleed the brake system.

Parking Brake Cable

ADJUSTMENT

Mirage

NOTE: If the vehicle is equipped with rear drum brakes, make certain that the brake shoes are properly adjusted before attempting to adjust the parking brake.

1. Make sure the parking brake cable is free and is not frozen or sticking.

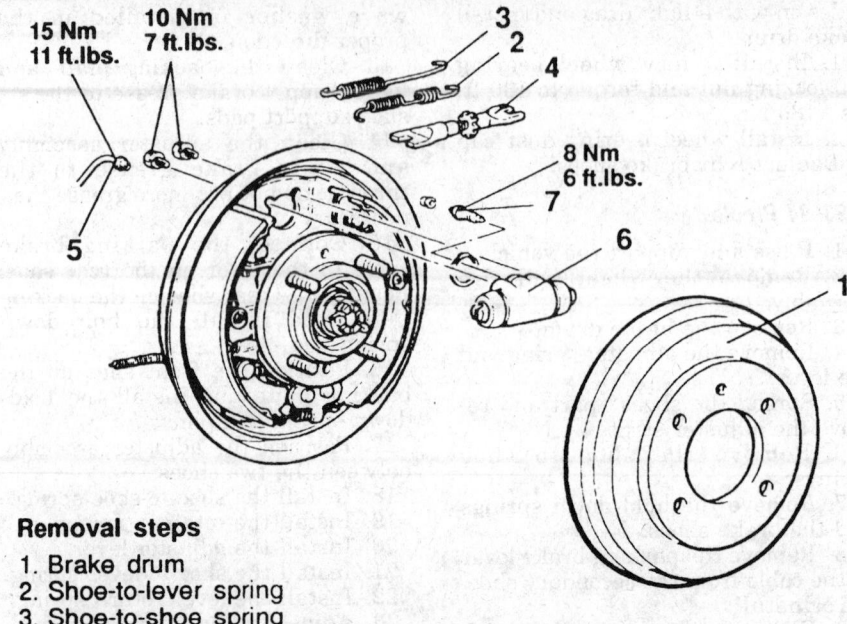

Removal steps

1. Brake drum
2. Shoe-to-lever spring
3. Shoe-to-shoe spring
4. Auto adjuster assembly
5. Connection for the brake pipe
6. Wheel cylinder
7. Bleeder screw

313219

Wheel cylinder removal — Galant and 1995–97 Eclipse

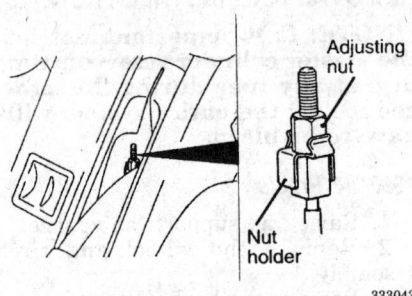

333043

Parking brake adjusting nut — Mirage

2. Apply the parking brake with 45 lbs. (200 N) of force while counting the number of notches. The desired parking brake stroke should be 5–7 notches.

3. If adjustment is required, access the adjusting nut from inside the floor console.

4. Loosen the locknut on the cable rod. Rotate the adjusting nut to adjust the parking brake stroke to the 5–7 notch setting. After making the adjustment, check there is no looseness between the adjusting nut and

the parking brake lever, then tighten the locknut.

NOTE: Do not adjust the parking brake too tight. If the number of notches is less than specification, the cable has been pulled too much and the automatic adjuster will fail or the brakes will drag.

5. After adjusting the lever stroke, raise the rear of the vehicle and safely support. With the parking brake lever in the released position, turn the rear wheels to confirm that the rear brakes are not dragging.

6. Check that the parking brake holds the vehicle on an incline.

Eclipse

1. Pull the parking brake lever with a force of approx. 45 lbs. (196 N) and count the number of notches. **Standard value is: Vehicles with drum-in-disc — 3–5 notches. Vehicles with drum brakes — 5–7 notches.**

━━━ **CAUTION** ━━━
The 45 lbs. (196 N) force of the parking brake lever must be strictly observed.

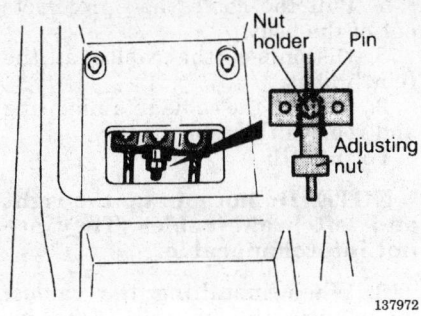

Parking brake cable adjusting nut location — 1993–94 Eclipse

2. If the parking brake lever is not the standard value, adjust in the following manner:

a. Remove the inner compartment mat of the floor console

b. Loosen the adjusting nut at the end of the cable rod, freeing the parking brake.

c. Vehicles with drum brakes - with the engine idling, forcefully depress the brake pedal five or six times and confirm that the pedal stroke stops changing. If the pedal stroke stops changing, the automatic-adjustment mechanism is functioning normally, and the clearance between the shoe and the drum is correct.

d. Vehicles with drum-in-disc brakes, remove the adjustment hole plug, and then with a suitable tool, turn the adjuster to expand the shoes against the brake drum so that the rotor will not turn. Return the adjuster five notches in the opposite direction.

e. Turn the parking brake adjusting nut and adjust the parking brake lever stroke to within the standard value.

— CAUTION —

If the number of brake lever notches engaged is less than the standard value, the cable has been pulled excessively. Be sure to adjust it within the standard value.

f. After making the adjustment, check to be sure that there is no play between the adjusting nut and the pin. Also check that the adjusting nut is securely held at the nut holder.

g. After adjusting the parking brake lever stroke, safely raise and support the rear of the vehicle and with the parking brake lever in the released position, turn the rear wheels to confirm that there is no brake drag.

1993 Galant

1. Pull the parking brake lever up with a force of about 45 lbs. (61 Nm). The total number of clicks heard should be 5 to 7 clicks. If the number of clicks was not within that range, system requires adjustment.

NOTE: The parking brake shoes must be adjusted before attempting to adjust the cable mechanism

2. To adjust the parking brake shoes perform the following steps.

a. remove the floor console, release the lever and back off the cable adjuster locknut at the base of the lever.

b. Raise the vehicle, support safely and remove the wheel. Remove the hole plug in the brake rotor.

c. Remove the brake caliper and hang out of the way with wire.

d. Use a suitable prybar to pry up on the self-adjuster wheel until the rotor will not turn.

e. Return the adjuster 5 notches in the opposite direction. Make sure the rotor turns freely with a slight drag.

f. Install the caliper and check operation.

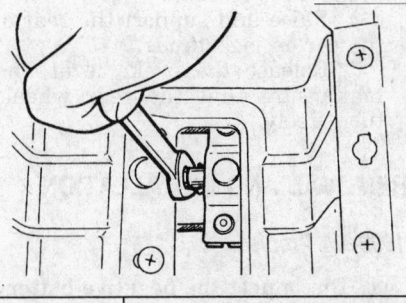

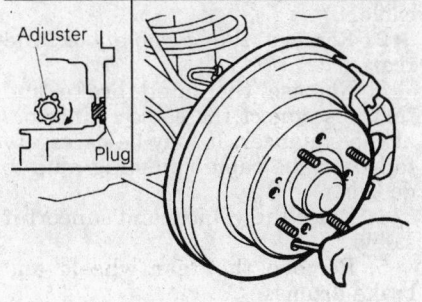

Parking brake system adjustment points — 1993 Galant

3. Once the parking brake shoes have been properly adjusted, adjust the cable mechanism, by performing the following steps:

a. Turn the adjusting nut to give the proper number of clicks when the lever is raised full travel.

b. Raise and support the rear of the car on jackstands.

c. Release the brake lever and make sure that the rear wheels turn freely. If not, back off on the adjusting nut until they do.

1996–97 Galant

1. Pull the parking brake lever with a force of approx. 45 lbs. (196 N) and count the number of notches. **Standard value is: 5–7 notches.**

— CAUTION —

The 45 lbs. (196 N) force of the parking brake lever must be strictly observed.

2. If the parking brake lever is not the standard value, adjust in the following manner:

a. Remove the inner compartment mat of the floor console

b. Loosen the adjusting nut at the end of the cable rod, freeing the parking brake.

c. With the engine idling, forcefully depress the brake pedal five or six times and confirm that the pedal stroke stops changing. If the pedal stroke stops changing, the automatic-adjustment mechanism is functioning normally, and the clearance between the shoe and the drum is correct.

d. After adjusting the parking brake lever stroke, safely raise and support the rear of the vehicle and with the parking brake lever in the released position, turn the rear wheels to confirm that there is no brake drag.

3000GT

With Drum Brakes

NOTE: Make certain that the brake shoes are properly adjusted before attempting to adjust the parking brake.

1. Pull the parking brake lever up with a force of about 45 lbs. If that value cannot be determined, just pull it up as far as possible. The total number of clicks heard should be 5–7.

2. If the number of clicks was not within that range, release the lever and back off the cable adjuster locknut at the base of the lever and tighten the adjusting nut until there is no more slack in the cable.

3. Operate the lever and brake pedal several times, until no more clicks are heard from the automatic adjuster.

4. Turn the adjusting nut to give the proper number of clicks when the lever is raised full travel.

5. Raise and support the rear of the car on jackstands.

6. Release the brake lever and make sure that the rear wheels turn freely. If not, back off on the adjusting nut until they do.

With Disc Brakes

1. Pull the parking brake lever up with a force of about 45 lbs. (61 N). The total number of clicks heard should be 3–5. If the number of clicks was not within that range, system requires adjustment.

NOTE: The parking brake shoes must be adjusted before attempting to adjust the cable mechanism

2. To adjust the parking brake shoes perform the following steps.

a. remove the floor console, release the lever and back off the cable adjuster locknut at the base of the lever.

b. Raise the vehicle, support safely and remove the wheel. Remove the hole plug in the brake rotor.

c. Remove the brake caliper and hang out of the way with wire.

d. Use a suitable prybar to pry up on the self-adjuster wheel until the rotor will not turn.

e. Return the adjuster 5 notches in the opposite direction. Make sure the rotor turns freely with a slight drag.

f. Install the caliper and check operation.

3. Once the parking brake shoes have been properly adjusted, adjust the cable mechanism, by performing the following steps:

a. Turn the adjusting nut to give the proper number of clicks when the lever is raised full travel.

b. Raise and support the rear of the car on jackstands.

c. Release the brake lever and make sure that the rear wheels turn freely. If not, back off on the adjusting nut until they do.

Diamante

1. Pull the parking brake lever up with a force of about 45 lbs. (200 N). The total number of clicks heard should be 3–5. If the number of clicks

was not within that range, system requires adjustment.

NOTE: The parking brake shoes must be adjusted before attempting to adjust the cable mechanism

2. To adjust the parking brake shoes perform the following steps.

a. Remove the floor console, release the lever and back off the cable adjuster locknut at the base of the lever.

b. Raise the vehicle, support safely and remove the wheel. Remove the hole plug in the brake rotor.

c. Remove the brake caliper and hang out of the way with wire.

d. Use a suitable prybar to pry up on the self-adjuster wheel until the rotor will not turn.

e. Return the adjuster 5 notches in the opposite direction. Make sure the rotor turns freely with a slight drag.

f. Install the caliper and check operation.

3. Once the parking brake shoes have been properly adjusted, adjust the cable mechanism, by performing the following steps:

a. Pull the parking brake lever up with a force of 45 lbs. (200 N). The total number of clicks heard should be 3–5.

b. Turn the adjusting nut to give the proper number of clicks when the lever is raised.

c. Raise and support the rear of the car on jackstands.

d. Release the brake lever and make sure that the rear wheels turn freely.

REMOVAL AND INSTALLATION

1993–94 Precis

1. Disconnect the negative battery cable.

2. Remove the console box and rear seat.

3. Release the hand brake and then disconnect the cable connectors at the equalizer. It may be necessary to loosen the cable adjusting nuts to do this.

4. Raise the vehicle and support it safely.

5. Remove the rear wheels and brake drums.

6. Disconnect all cable clamps from the body. Remove the mounting bolts for the large mounting clamp located just forward of where the cables pass through the body grommets.

7. Pull the cables and grommets out of the body.

8. Disconnect the cables at the rear brakes.

9. Remove the cable retaining ring and remove the cable.

To install:

NOTE: Do not mix up the right and left brake cables. They are not interchangeable.

10. When installing the cables, make sure the grommets are installed in the body completely and that the concave side faces to the rear.

11. Install the cable, and the cable retaining ring.

12. Connect the cables to the rear brakes.

13. Install the drums and the wheels.

14. Install the rear seat and console box.

15. Connect the negative battery cable.

16. Adjust the hand brake mechanism.

17. Adjust the switch for the indicator light so the light comes on when the lever is pulled one notch.

Mirage

With Drum Rear Brakes

NOTE: If equipped with an air bag (SRS system), be sure to disarm system before starting repairs on the vehicle.

1. Disconnect the negative battery cable.

2. Remove the screws from the center section and remove the rear part of the console.

NOTE: If equipped with SRS, when removing the floor console, don't allow any impact or shock to the SRS diagnostic unit.

3. Remove the rear seat cushion.

4. Remove the center cable clamp and grommet.

5. Raise the vehicle and support safely.

6. At the rear wheel, remove the brake drum and shoes. Disconnect the cable end from the parking brake strut lever. Compress the retaining strips to remove the cable from the backing plate.

7. Unfasten any other frame retainers and remove the cables.

To install

8. The parking brake cables may be color coded to indicate side. Check the parking brake cables for an identification mark.

9. Install the cable to the rear actuator. Secure in place with the parking brake cable clip and retainer spring.

10. Install the brake shoes and drum.

11. Position the cable in the under the vehicle and install retainers loose.

12. Reattach the parking brake cables to the actuator inside the vehicle.

13. Adjust the rear brake shoes.

14. Tighten the adjusting nut until the proper tension is placed on the cable. Adjust the parking brake stroke using appropriate method.

15. Secure all cable retainers. Apply and release the parking brake a number of times once all adjustments have been made.

16. With the rear wheels raised, make sure the parking brake is not causing excess drag on the rear wheels.

17. Install the floor console and rear seat assembly.

18. Connect the negative battery cable.

19. Check that the parking brake holds the vehicle on an incline.

With Rear Disc Brakes

NOTE: If equipped with an air bag (SRS system), be sure to disarm system before starting repairs on the vehicle.

1. Disconnect the negative battery cable.

2. Remove the screws from the center section and remove the rear part of the console.

NOTE: If equipped with SRS, when removing the floor console, don't allow any impact or shock to the SRS diagnostic unit.

3. Remove the rear seat cushion.

4. Loosen the cable adjusting nut and disconnect the rear brake cables from the actuator. Remove the center cable clamp and grommet.

5. Raise the vehicle and support safely. Remove the parking brake cable clip and retainer spring. Disconnect the cable end from the parking brake assembly.

6. Unfasten any remaining frame retainers and remove the cables from the vehicle.

To install:

7. The parking brake cables may be color coded to indicate side. Check the parking brake cables for an identification mark.

8. Install the cable to the rear actuator. Secure in place with the park-

ing brake cable clip and retainer spring.

9. Position the cable in the under the vehicle and install retainers loose.

10. Reattach the parking brake cables to the actuator inside the vehicle.

11. Tighten the adjusting nut until the proper tension is placed on the cable. Adjust the parking brake stroke using appropriate method.

12. Secure all cable retainers. Apply and release the parking brake a number of times once all adjustments have been made. With the rear wheels raised, make sure the parking brake is not causing excess drag on the rear wheels.

13. Install the floor console and rear seat assembly.

14. Connect the negative battery cable and check console electrical components for proper operation.

15. Road test the vehicle and check for proper brake operation. Check that the parking brake holds the vehicle on an incline.

Eclipse

——— CAUTION ———
The air bag system (SRS or SIR) must be disarmed before removing the center floor console. Fail-

ure to do so may cause accidental deployment, property damage or personal injury.

1. Disconnect the negative battery cable.

2. Remove the floor console from the vehicle as follows:

a. Remove the screw plugs in the side covers. Remove the retainer screws and the side covers from the vehicle.

b. Remove the front mounting screw cover from the floor console. Remove the manual transaxle shift lever knob.

c. Remove the cup holder and the carpet inserts from the floor console assembly.

d. Label and disconnect the electrical wire harness connections for the floor console.

e. Remove the mounting bolts and the floor console from the vehicle.

3. Loosen the cable adjusting nut and disconnect the rear brake cables from the actuator. Remove the center cable clamp and grommet.

4. Raise the vehicle and support safely. Remove the parking brake cable clip and retainer spring. Disconnect the cable end from the parking brake assembly.

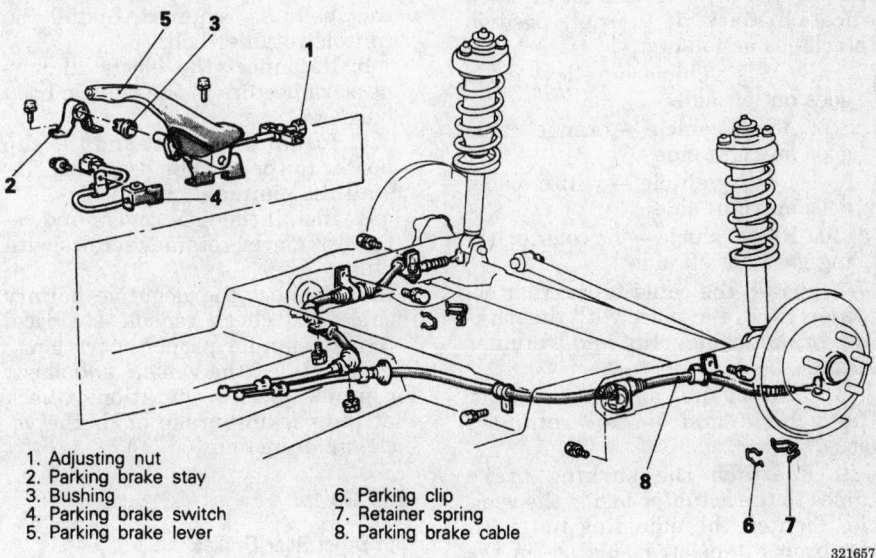

1. Adjusting nut
2. Parking brake stay
3. Bushing
4. Parking brake switch
5. Parking brake lever
6. Parking clip
7. Retainer spring
8. Parking brake cable

321657

Parking brake system (front wheel drive) — 1993–94 Eclipse

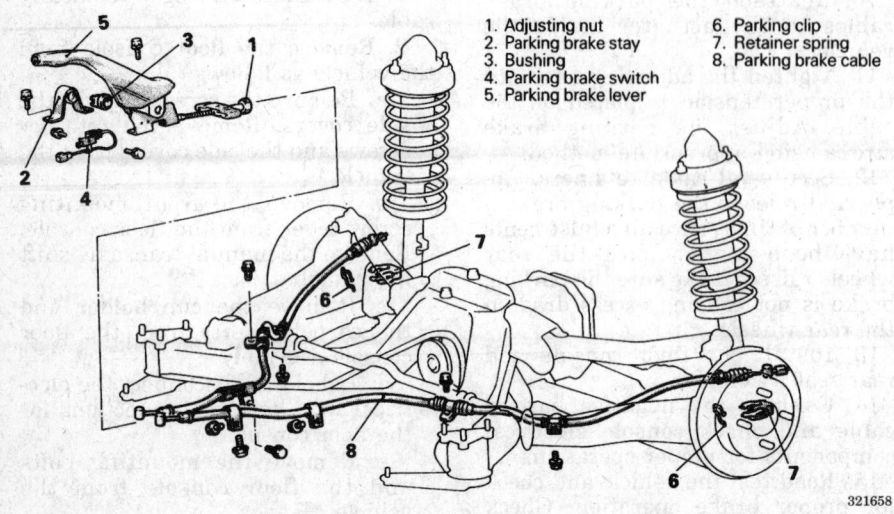

1. Adjusting nut
2. Parking brake stay
3. Bushing
4. Parking brake switch
5. Parking brake lever
6. Parking clip
7. Retainer spring
8. Parking brake cable

321658

Parking brake system — 1993–94 Eclipse with AWD

5. Unfasten any remaining frame retainers and remove the cables from the vehicle.

To install:

6. The parking brake cables may be color coded to indicate side. Check the parking brake cables for an identification mark. If present, position the cables as follows:

   a. AWD vehicle — yellow cable goes on left side

   b. AWD vehicle — orange cable goes on right side

   c. FWD vehicle — white cable goes on right side

   d. FWD vehicle — no color marking goes on left side

7. Install the cable to the rear actuator. Secure in place with the parking brake cable clip and retainer spring.

8. Position the cable in the under the vehicle and install retainers loose.

9. Reattach the parking brake cables to the actuator inside the vehicle. Tighten the adjusting nut until the proper tension is placed on the cable. Adjust the parking brake stroke using appropriate method.

10. Secure all cable retainers. Apply and release the parking brake a number of times once all adjustments have been made. With the rear

wheels raised, make sure the parking brake is not causing excess drag on the rear wheels.

11. Install the floor console assembly as follows:

   a. Install the floor console in position in the vehicle. Position the seat belts as required. Install the console retainer bolts.

   b. Reconnect the electrical harness connectors to the vehicle body harness.

   c. Install the carpet and the cup holder to the console assembly. Install the manual shift knob.

   d. Install the side covers and retainers. Cover retainer screws with plugs.

   e. Connect the negative battery cable and check console electrical components for proper operation.

12. Road test the vehicle and check for proper brake operation. Check that the parking brake holds the vehicle on an incline.

1993 Galant

With Drum Rear Brakes

1. Disconnect the negative battery cable.

2. Remove the center console as follows:

   a. Remove both side cover panels.

   b. Remove the sifter knob on manual transmission models. Remove the spacer trim piece on automatic transmission models.

   c. Remove the switch panel/box and remove the two screws beneath the panel/box.

   d. Remove the radio trim panel.

   e. Remove the radio and tape player.

   f. Remove the console inner panel and remove the two screws for beneath the panel.

   g. Remove the remaining screws from the sides of the console.

   h. Remove the console assembly from the vehicle.

3. While pressing downward on the front of the rear seat cushion, release the locking levers and remove the seat cushion.

4. Loosen the cable adjustment at the cable equalizer.

5. Remove the center cable clamp and grommet.

6. Raise the vehicle and support safely.

7. At the rear wheel, remove the brake drum and shoes.

8. Disconnect the cable end from the parking brake strut lever.

9. Remove the snapring securing the cable to the backing plate.

10. Unfasten any other frame retainers and remove the cables.

To install

NOTE: The parking brake cables may be color coded to indicate side. Check the parking brake cables for an identification mark.

11. Install the cable to the rear actuator. Secure in place with the parking brake cable clip and snapring.

12. Install the brake shoes and drum.

13. Position the cable in the under the vehicle and install retainers loose.

14. Reattach the parking brake cables to the actuator inside the vehicle. Tighten the adjusting nut until the proper tension is placed on the cable. Adjust the parking brake stroke using appropriate method.

15. Secure all cable retainers. Apply and release the parking brake a number of times once all adjustments have been made.

16. Adjust the rear brakes and parking brake cables.

17. Check the rear wheels to confirm that the rear brakes are not dragging.

18. Install the center console and rear seat cushion.

19. Connect the negative battery cable and check console electrical components for proper operation.

20. Check that the parking brake holds the vehicle on an incline.

With Rear Disc Brakes

1. Disconnect the negative battery cable.

2. Remove the center console as follows:

  a. Remove both side cover panels.

  b. Remove the sifter knob on manual transmission models. Remove the spacer trim piece on automatic transmission models.

  c. Remove the switch panel/box and remove the two screws beneath the panel/box.

  d. Remove the radio trim panel.

  e. Remove the radio and tape player.

  f. Remove the console inner panel and remove the two screws for beneath the panel.

  g. Remove the remaining screws from the sides of the console.

  h. Remove the console assembly from the vehicle.

3. While pressing downward on the front of the rear seat cushion, release the locking levers and remove the seat cushion.

4. Loosen the cable adjusting nut and disconnect the rear brake cables from the actuator.

5. Remove the center cable clamp and grommet.

6. Raise the vehicle and support it safely.

7. At the rear caliper assembly, remove the parking brake cable clip and retainer spring. Disconnect the cable end from the caliper.

8. Unfasten any remaining frame retainers and remove the cables from the vehicle.

  To install:

  NOTE: The parking brake cables may be color coded to indicate side. Check the parking brake cables for an identification mark.

9. Connect the cable to the actuator at the brake caliper. Secure in place with the parking brake cable clip and retainer spring.

10. Position the cable in the under the vehicle and install retainers loose.

11. Reattach the parking brake cables to the actuator inside the vehicle. Tighten the adjusting nut until the proper tension is placed on the cable. Adjust the parking brake stroke using appropriate method.

12. Secure all cable retainers. Apply and release the parking brake a number of times once all adjustments have been made. With the rear wheels raised, make sure the parking brake is not causing excess drag on the rear wheels.

13. Install the floor console assembly and rear seat cushion.

14. Connect the negative battery cable and check console electrical components for proper operation.

15. Road test the vehicle and check for proper brake operation. Check that the parking brake holds the vehicle on an incline.

1994–97 Galant

With Rear Drum Brakes

1. Disconnect the negative battery cable.

—————— **WARNING** ——————
The SRS control unit is mounted beneath the center console. Use care when working with the center console assembly not to impact or shock the control unit.

2. Remove the center floor console assembly as follows:

  a. Remove the shifter knob on models equipped with a manual transmission.

  b. Remove the shifter trim panel.

  c. Remove the center instrument panel.

  d. Remove the panel box from the console assembly.

  e. Remove the two screws from the center of the console.

  f. Remove the four side panel screws and remove the console from the vehicle.

3. Loosen the cable adjuster nut and then remove the parking brake cable, by pulling it from the passenger compartment.

4. Raise the vehicle and support safely.

5. At the rear wheel, remove the brake drum and shoes.

6. Disconnect the cable end from the parking brake strut lever. Compress the retaining strips to remove the cable from the backing plate.

7. Unfasten any other frame retainers and remove the cables.

  To install

8. Install the cable to the rear actuator. Secure in place with the parking brake cable clip and retainer spring.

9. Install the brake shoes and drum.

10. Position the cable in the under the vehicle and install retainers loose.

11. Reattach the parking brake cables to the actuator inside the vehicle. Tighten the adjusting nut until the proper tension is placed on the cable. Adjust the parking brake stroke using appropriate method.

12. Secure all cable retainers. Apply and release the parking brake a number of times once all adjustments have been made.

13. Assemble the interior components which were removed.

14. Adjust the rear brakes and parking brake cables.

15. Connect the negative battery cable and check the rear wheels to confirm that the rear brakes are not dragging.

16. Check that the parking brake holds the vehicle on an incline.

With Rear Disc Brakes

Unlike conventional rear disc brake systems, the parking brake operation is **not** incorporated into the brake caliper. This system, uses a separate set of brake shoes, located behind the brake rotor.

1. Disconnect the negative battery cable.

—————— **WARNING** ——————
The SRS control unit is mounted beneath the center console. Use care when working with the center console assembly not to impact or shock the control unit.

2. Remove the center floor console assembly as follows:

  a. Remove the shifter knob on models equipped with a manual transmission.

  b. Remove the shifter trim panel.

  c. Remove the center instrument panel.

  d. Remove the panel box from the console assembly.

  e. Remove the two screws from the center of the console.

  f. Remove the four side panel screws and remove the console from the vehicle.

3. Loosen the cable adjuster nut and then remove the parking brake cable, by pulling it from the passenger compartment.

4. Raise the vehicle and support safely.

5. At the rear wheel, remove the brake caliper and rotor.

6. Remove the parking brake shoes as follows:

  a. Remove the upper shoe to anchor springs.

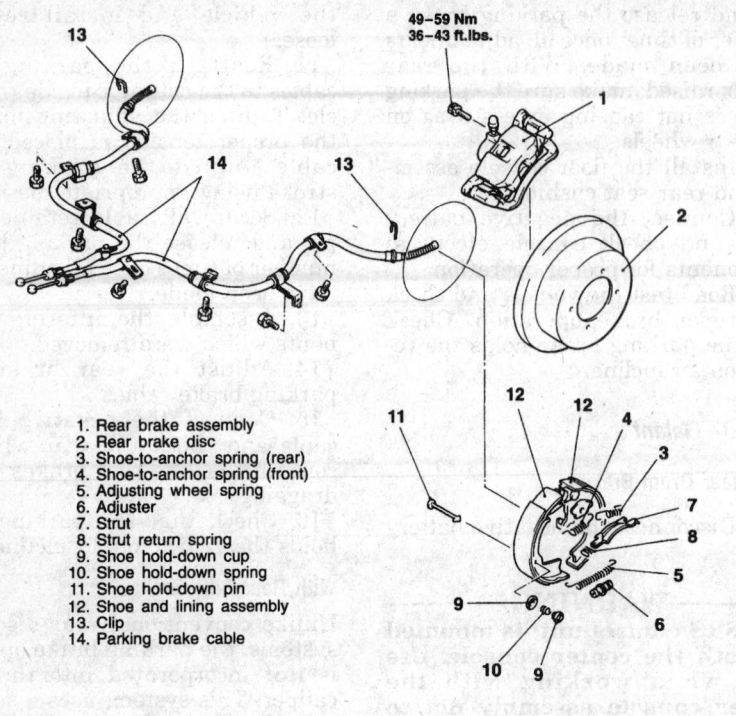

49–59 Nm
36–43 ft.lbs.

1. Rear brake assembly
2. Rear brake disc
3. Shoe-to-anchor spring (rear)
4. Shoe-to-anchor spring (front)
5. Adjusting wheel spring
6. Adjuster
7. Strut
8. Strut return spring
9. Shoe hold-down cup
10. Shoe hold-down spring
11. Shoe hold-down pin
12. Shoe and lining assembly
13. Clip
14. Parking brake cable

233118

Exploded view of parking brake components with disc brake rear system) — 1994–95 Galant

b. Remove the lower shoe to shoe spring.

c. Remove the brake shoe hold-down springs.

d. Disconnect the parking brake cable from the actuating lever.

7. Disconnect the cable end from the parking brake strut lever. Compress the retaining strips to remove the cable from the backing plate.

8. Unfasten any other frame retainers and remove the cables.

To install

9. Install the cable to the rear actuator. Secure in place with the parking brake cable clip and retainer spring.

10. Install the parking brake shoes.

11. Install the brake rotor and caliper assembly.

12. Position the cable in the under the vehicle and install retainers loose.

13. Reattach the parking brake cables to the actuator inside the vehicle. Tighten the adjusting nut until the proper tension is placed on the cable. Adjust the parking brake stroke using appropriate method.

14. Secure all cable retainers. Apply and release the parking brake a number of times once all adjustments have been made.

15. Assemble the interior components which were removed.

16. Adjust the parking brake shoes and parking brake cables.

17. Connect the negative battery cable and check the rear wheels to confirm that the rear brakes are not dragging.

18. Check that the parking brake holds the vehicle on an incline.

3000GT and Diamante

Unlike conventional rear disc brake systems, the parking brake operation is **not** incorporated into the brake caliper. This system, uses a separate set of brake shoes, located behind the brake rotor.

1. Disconnect the negative battery cable.

——— **CAUTION** ———
Work must be started after 90 seconds from the time the ignition switch is turned to the LOCK position and the negative (-) battery cable is disconnected.

NOTE: If equipped with an air bag, be sure to disarm it before starting repairs on the vehicle.

2. On Diamante models, remove the center floor console assembly as follows:

a. Remove the ashtray or console switch panel from the console assembly.

b. Remove the two screws from the center of the console.

c. Remove the four side panel screws and remove the console from the vehicle.

3. On 3000GT models, remove the front and rear floor consoles, by prying out the coin holder, box tray and remote mirror switch, if equipped, or the cover. Remove the small cover around the seat belt from the console side. Remove the screws from the center section and remove the rear part of the console.

NOTE: If equipped with SRS, when removing the floor console, don't allow any impact or shock to the SRS diagnostic unit.

4. Loosen the cable adjuster nut and then remove the parking brake cable, by pulling it from the passenger compartment.

5. Raise the vehicle and support it safely.

6. At the rear wheel, remove the brake caliper and rotor.

7. Remove the parking brake shoes, following the same procedures as conventional drum brake shoes.

8. Disconnect the cable end from the parking brake strut lever. Compress the retaining strips to remove the cable from the backing plate.

9. Unfasten any other frame retainers and remove the cables.

To install

10. Install the cable to the rear actuator. Secure in place with the parking brake cable clip and retainer spring.

11. Install the parking brake shoes.

12. Install the brake rotor and caliper assembly.

13. Position the cable in the under the vehicle and install retainers loose.

14. Reattach the parking brake cables to the actuator inside the vehicle. Tighten the adjusting nut until the proper tension is placed on the cable. Adjust the parking brake stroke using appropriate method.

15. Secure all cable retainers. Apply and release the parking brake a number of times once all adjustments have been made.

16. Assemble the interior components which were removed.

17. Adjust the parking brake shoes and parking brake cables.

18. Connect the negative battery cable and check the rear wheels to

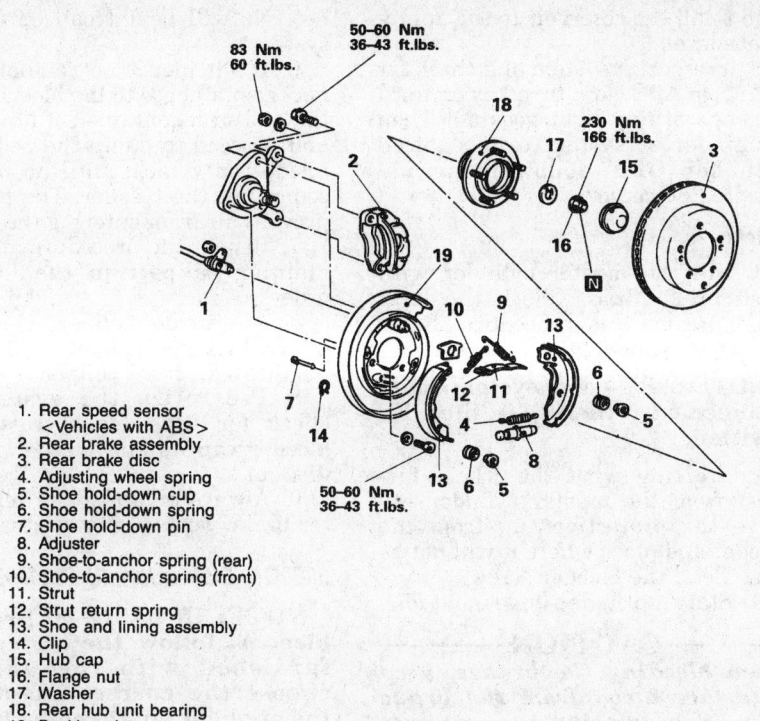

83 Nm
60 ft.lbs.

50–60 Nm
36–43 ft.lbs.

230 Nm
166 ft.lbs.

50–60 Nm
36–43 ft.lbs.

1. Rear speed sensor
   <Vehicles with ABS>
2. Rear brake assembly
3. Rear brake disc
4. Adjusting wheel spring
5. Shoe hold-down cup
6. Shoe hold-down spring
7. Shoe hold-down pin
8. Adjuster
9. Shoe-to-anchor spring (rear)
10. Shoe-to-anchor spring (front)
11. Strut
12. Strut return spring
13. Shoe and lining assembly
14. Clip
15. Hub cap
16. Flange nut
17. Washer
18. Rear hub unit bearing
19. Backing plate

325964

Parking brake shoes and related components — Diamante

confirm that the rear brakes are not dragging.

19. Check that the parking brake holds the vehicle on an incline.

Brake System Bleeding

PROCEDURE

1993–94 Precis

The brakes should be bled whenever a brake line, caliper, wheel cylinder, or master cylinder has been removed or when the brake pedal is low or soft.

NOTE: If using a pressure bleeder, follow the instructions furnished with the unit and choose the correct adapter for the application. Do not substitute an adapter that "almost fits" as it will not work and could be dangerous.

1. Fill the master cylinder with fresh brake fluid. Check the level often during the procedure.
2. Starting with the right rear wheel, remove the protective cap from the bleeder, if equipped, and place where it will not be lost. Clean the bleed screw.

—— CAUTION ——
When bleeding the brakes, keep face away from the brake area. Spewing fluid may cause facial and/or visual injury. Do not allow brake fluid to spill on the vehicle's finish; it will remove the paint.

3. If the system is empty, the most efficient way to get fluid down to the wheel is to loosen the bleeder about ½ to ¾ turn, place a finger firmly over the bleeder and have a helper pump the brakes slowly until fluid comes out the bleeder. Once fluid is at the bleeder, close it before the pedal is released inside the vehicle.

NOTE: If the pedal is pumped rapidly, the fluid will churn and create small air bubbles, which are almost impossible to remove from the system. These air bubbles will eventually congregate and a spongy pedal will result.

4. Once fluid has been pumped to the caliper or wheel cylinder, open the bleed screw again, have an assistant press the brake pedal to the floor, lock the bleeder and have an assistant slowly release the pedal. Wait 15 seconds and repeat the procedure, including the 15 second wait, until no more air comes out of the

bleeder upon application of the brake pedal. Remember to close the bleeder before the pedal is released inside the vehicle each time the bleeder is opened. If not, air will be induced into the system.

5. If a helper is not available, connect a small hose to the bleeder, place the end in a container of brake fluid and proceed to pump the pedal from inside the vehicle until no more air comes out the bleeder. The hose will prevent air from entering the system.
6. Repeat the procedure on remaining wheel cylinders in order:
   a. Left front caliper
   b. Left rear wheel cylinder or caliper
   c. Right front caliper
7. Hydraulic brake systems must be totally flushed, if the fluid becomes contaminated with water, dirt or other corrosive chemicals. To flush, bleed the entire system until all fluid has been replaced with the correct type of new fluid.
8. Install the bleeder cap(s) on the bleeder to keep dirt out. Always road test the vehicle.

Mirage, Eclipse and 1993 Galant

Bleeding the brake system is required anytime the normally closed system has been opened to the atmosphere. When bleeding the system, keep the brake fluid level in the master cylinder reservoir above half full. If the reservoir is empty, air will be pushed through the system. Hydraulic brake systems must be totally flushed if the fluid becomes contaminated with water, dirt or other corrosive chemicals. To flush, bleed the entire system until all fluid has been replaced with the correct type of new fluid.

NOTE: If using a pressure bleeder, follow the instructions furnished with the unit and choose the correct adapter for the application. Do not substitute an adapter that "almost fits" as it will not work and could be dangerous.

Master Cylinder

Due to the location of the fluid reservoir, bench bleeding of the master cylinder is not recommended. The master cylinder is to be bled while mounted on the brake booster. If the fluid reservoir runs dry, bleeding of the entire system will be necessary. Two people will be required to bleed the brake system.

1. Fill the brake fluid reservoir with clean brake fluid. Disconnect

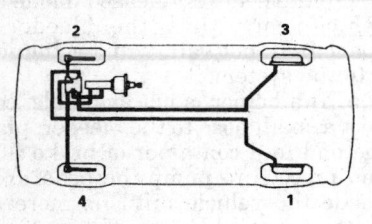

Brake bleeding sequence — 1993–94 Precis

the brake tube from the master cylinder.

2. Have a helper slowly depress the brake pedal. Once depressed, hold it in that position. Brake fluid will be expelled from the master cylinder.

— CAUTION —

When bleeding the brake system, keep your face away from the area. Spewing brake fluid may cause facial and/or visual damage. Do not allow brake fluid to spill onto the car's finish; it will remove the paint.

3. While the pedal is held down, use a finger to close the outlet port of the master cylinder. While the port is closed, have the helper release the brake pedal.

4. Repeat this procedure until all air is bled from the master cylinder. Check the brake fluid in the reservoir every 4–5 times, making sure the reservoir does not run dry. Add clean DOT 3 brake fluid to the reservoir as needed. All air is bled from the master cylinder when the fluid expelled from the port is free of bubbles.

5. Connect the brake tube to the port on the master cylinder and torque to 10 ft. lbs.(13.5 Nm). Add clean

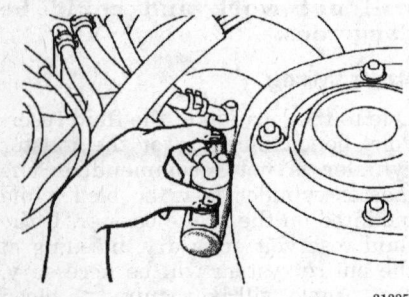

Master cylinder bleeding technique — Mirage, Galant and Eclipse

fluid to fill the reservoir to the appropriate level.

6. Pressurize system and check for leaks. On ABS cars, turn key on until system warning light goes off. Then remainder of system can be bleed with key **OFF**, following normal bleeding procedure.

Calipers

1. Fill the master cylinder with fresh brake fluid. Check the level often during this procedure. Raise and safely support the vehicle.

NOTE: ABS cars, system must be bleed with the key in the OFF position.

2. Starting with the wheel farthest from the master cylinder, remove the protective cap from the bleeder and place where it will not be lost. Clean the bleeder screw.

3. Start the engine and run at idle.

— CAUTION —

When bleeding the brakes, keep your face away from the brake area. Spewing fluid may cause physical and/or visual damage. Do not allow brake fluid to spill onto the car's finish; it will remove the paint.

4. If the system is empty, the most efficient way to get fluid down to the wheel is to loosen the bleeder about ½–¾ turn, place a finger firmly over the bleeder and have a helper pump the brakes slowly until fluid comes out the bleeder. Once fluid is at the bleeder, close it before the pedal is released inside the vehicle.

NOTE: If the pedal is pumped rapidly, the fluid will churn and create small air bubbles, which are almost impossible to remove from the system. These air bubbles will accumulate and a spongy pedal will result. Also note, it is important not to exceed normal pedal travel during bleeding procedure. This will prevent possible master cylinder piston(s) damage, due to build-up on the bore walls.

5. Once fluid has been pumped to the caliper, open the bleed screw again, have a helper press the brake pedal, lock the bleeder and have the helper slowly release the pedal. Wait 15 seconds and repeat the procedure (including the 15 second wait) until no more air comes out of the bleeder upon application of the brake pedal. Remember to close the bleeder before the pedal is released inside the vehicle each time the bleeder is opened. If

not, air will be introduced into the system.

6. If a helper is not available, connect a small hose to the bleeder, place the end in a container of brake fluid and proceed to pump the pedal from inside the vehicle until no more air comes out the bleeder. The hose will prevent air from entering the system.

7. Repeat the procedure on the remaining calipers in the following order:

    a. Left front caliper

    b. Left rear caliper

    c. Right front caliper

8. Pressurize the system and check for fluid leaks. Install the bleeder cap on the bleeder to keep dirt out.

9. Always road test the vehicle after brake work of any kind is done.

3000GT, Diamante and 1994–97 Galant

NOTE: If using a pressure bleeder, follow the instructions furnished with the unit and choose the correct adapter for the application. Do not substitute an adapter that almost fits as it will not work and could be dangerous.

Master Cylinder

If the master cylinder is off the vehicle it can be bench bled.

1. Connect two short pieces of brake line to the outlet fittings, bend them until the free end is below the fluid level in the master cylinder reservoir.

2. Fill the reservoir with fresh brake fluid. Pump the piston slowly until no more air bubbles appear in the reservoirs.

3. Disconnect the two short lines, refill the master cylinder and securely install the cylinder caps.

4. If the master cylinder is on the vehicle, it can still be bled, using a flare nut wrench.

5. Open the brake lines slightly with the flare nut wrench while pressure is applied to the brake pedal by a helper inside the vehicle.

6. Be sure to tighten the line before the brake pedal is released.

7. Repeat the process with both lines until no air bubbles come out.

8. Refill master cylinder and always bleed the complete brake system.

Calipers

1. Fill the master cylinder with fresh brake fluid. Check the level often during the procedure.

2. Starting with the wheel farthest from the master cylinder, re-

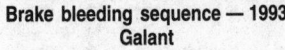

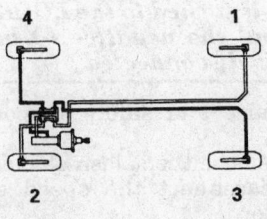

Brake bleeding sequence — 1993 Galant

Brake bleeding sequence — Mirage, Eclipse and Diamante

move the protective cap from the bleeder and place it where it will not be lost. Clean the bleeder screw.

— CAUTION —

When bleeding the brakes, keep face away from the brake area. Spewing fluid may cause facial and/or visual damage. Do not allow brake fluid to spill on the car's finish. It will remove the paint.

3. If the system is empty, the most efficient way to get fluid down to the wheel is to use a pressure bleeder tool. Open the bleeder until brake fluid flows without signs of air bubbles, close bleeder.

NOTE: If the pedal is pumped rapidly, the fluid will churn and create small air bubbles, which are almost impossible to remove from the system. These air bubbles will accumulate and a spongy pedal will result.

4. If the manual procedure is to be used to pump brake fluid to the caliper or wheel cylinder, open the bleed screw and have an assistant press the brake pedal to the floor.

5. Close the bleeder bleeder screw before releasing the brake pedal, have the helper slowly release the pedal. Wait 15 seconds and repeat the procedure until no more air comes out of the bleeder upon application of the brake pedal. Remember to close the bleeder before the pedal is released inside the vehicle each time the bleeder is opened. If not, air will be introduced into the system.

6. Repeat the procedure on remaining wheel cylinders in the following order:
- Right rear caliper.
- Left front caliper.
- Left rear caliper.
- Right front caliper.

7. Hydraulic brake systems must be totally flushed if the fluid becomes contaminated with water, dirt or other corrosive chemicals. To flush, bleed the entire system until all fluid has been replaced with the correct type of new fluid.

8. Install the bleeder cap on the bleeder to keep dirt out and refill master cylinder. Always road test the vehicle after brake work of any kind is done.

Wheel Speed Sensor

REMOVAL AND INSTALLATION

Mirage

Front

1. Disconnect the negative battery cable.

— CAUTION —

Wait at least 90 seconds after the negative (-) battery cable is disconnected to prevent possible deployment of the air bag.

2. Raise and safely support the vehicle.
3. Remove the splash shield.

4. Disconnect the speed sensor connector.
5. Remove the clips holding the sensor harness.
6. Remove the speed sensor from the bracket.
To install:
7. Install the speed sensor to the bracket and secure with mounting bolt.
8. Install the clips holding the sensor harness.
9. Connect the speed sensor connector.
10. Connect the negative battery cable.

Rear

1. Disconnect the negative battery cable.

— CAUTION —

Wait at least 90 seconds after the negative (-) battery cable is disconnected to prevent possible deployment of the air bag.

2. Raise and safely support the vehicle.
3. Disconnect the speed sensor connector.
4. Remove the clips holding the sensor harness.
5. Remove the mounting bolt and the speed sensor.
To install:
6. Install the speed sensor and loosely tighten the mounting bolt.
7. Install the clips holding the sensor harness.
8. Install the sensor wire harness and attach the connector.
9. Using a non-magnetic feeler gauge, adjust the sensor so the clearance between the rotor and the sensor is 0.012–0.035 inches (0.3–0.9mm).
10. Tighten the sensor bracket and recheck the clearance.
11. Connect the negative battery cable.

Eclipse and 1994–97 Galant

Front

1. Disconnnect the negative battery cable. Wait at least 90 seconds before performing any work.
2. Raise and safely support the vehicle. Remove the necessary tire and wheel assembly.
3. Remove the fender splash shield.
4. Disconnect the ABS speed sensor connector.
5. Remove the sensor harness clamp bolts and clamps.
6. Remove the ABS speed sensor mounting bolt and the sensor.

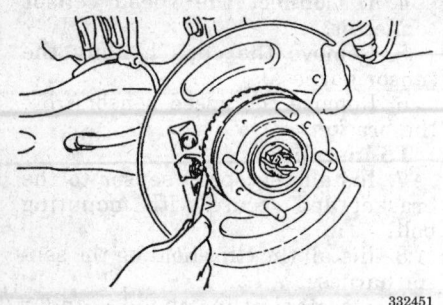

Adjusting the rear speed sensor air gap — Mirage

To install:

7. Install the ABS speed sensor with its mounting bolt.

NOTE: The clearance between the wheel speed sensor and the rotor's toothed surface is not adjustable, but measure the distance between the sensor installation surface and the rotor's toothed surface. Standard value is: 1.11–1.12 in. If not within specifications, replace the speed sensor or the toothed rotor.

8. Reinstall the sensor harness with its clamps and bolts.

9. Reconnect the speed sensor connector.

10. Install the fender splash shield.

11. Reinstall the tire and wheel, safely lower the vehicle, and reconnect the negative battery cable.

Rear

1. Disconnnect the negative battery cable. Wait at least 90 seconds before performing any work.

2. Raise and safely support the vehicle. Remove the necessary tire and wheel assembly.

3. Disconnect the ABS speed sensor connector.

4. Remove the sensor harness clamp bolts and clamps.

5. Remove the ABS speed sensor mounting bolt and the sensor.

To install:

6. Install the ABS speed sensor with its mounting bolt.

NOTE: The clearance between the wheel speed sensor and the rotor's toothed surface is not adjustable, but measure the distance between the sensor installation surface and the rotor's toothed surface. Standard value is: 1.11–1.12 in. If not within specifications, replace the speed sensor or the toothed rotor.

7. Reinstall the sensor harness with its clamps and bolts.

AWD

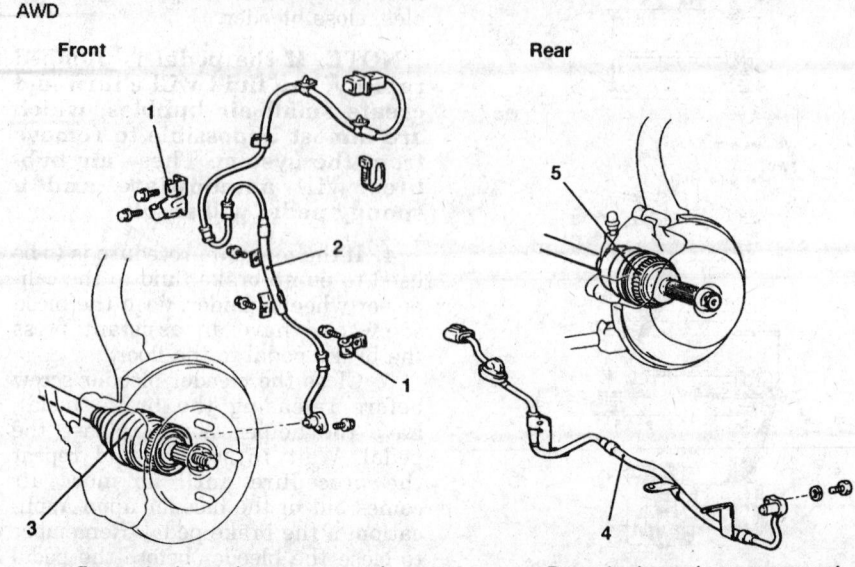

Front wheel speed sensor removal steps
1. Clip
2. Front wheel speed sensor

Rear wheel speed sensor removal
4. Rear wheel speed sensor

NOTE
The toothed rotor is integrated with the drive shaft and is not disassembled.

Typical wheel speed sensors

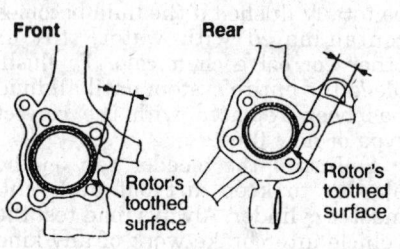

ABS speed sensor clearance — Eclipse

8. Reconnect the speed sensor connector.

9. Reinstall the tire and wheel, safely lower the vehicle, and reconnect the negative battery cable.

3000GT and Diamante

Front

1. Disconnect the negative battery cable.

— CAUTION —
Work must be started after 90 seconds from the time the ignition switch is turned to the LOCK position and the negative (-) battery cable is disconnected.

2. Raise and safely support the vehicle.

3. Remove the splash shield.

4. Disconnect the speed sensor connector.

5. Remove the clips holding the sensor harness.

6. Remove the speed sensor.

To install:

7. Install the speed sensor and torque to 9 ft. lbs. (12 Nm).

8. Install the clips holding the sensor harness.

9. Connect the speed sensor connector.

10. Connect the negative battery cable.

Rear

1. Disconnect the negative battery cable.

— CAUTION —
Work must be started after 90 seconds from the time the ignition switch is turned to the LOCK position and the negative (-) battery cable is disconnected.

2. Raise and safely support the vehicle.

3. Disconnect the speed sensor connector.

4. Remove the clips holding the sensor harness.

5. On AWD models, remove the cable band.

6. Remove the mounting bolt and the speed sensor with the O-ring.

To install:

7. Install the speed sensor with the O-ring and torque to 9 ft. lbs. (12 Nm).

8. Install the clips holding the sensor harness.

9. Install the cable band.

10. Install the sensor wire harness and connect the connector.

11. Connect the negative battery cable.

FRONT SUSPENSION

Strut

REMOVAL AND INSTALLATION

1993–94 Precis

1. Remove the front wheels.

2. Raise and safely support the vehicle.

3. Detach the brake hose from the clip on the strut.

4. Unbolt the strut from the knuckle.

5. Remove the four strut-to-fender apron nuts.

6. Pull the strut away from the steering knuckle and wheelhouse and out from the car.

To install:

NOTE: **Before installing the strut, make sure the surface where the strut attaches to the knuckle is clean. This ensures a good connection.**

7. Install the strut to the inner fender. Tighten the mounting nuts to 11–14 ft. lbs. (15–20 Nm).

8. Install the lower strut to the steering knuckle. Tighten the mounting nuts to 65–76 ft. lbs. (90–105 Nm).

9. Attach the brake line the bracket on the strut body.

10. Install the wheel and tire assembly.

11. Install the wheel and tire assembly, and lower the vehicle to the floor.

12. Check and adjust the front wheel alignment as needed.

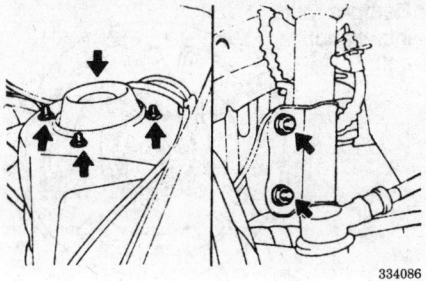

Strut mounting points — 1993–94 Precis

Mirage and 1993–94 Eclipse

1. Disconnect the negative battery cable.

2. Raise and safely support vehicle.

3. Remove the brake hose and tube bracket retainer bolt and bracket from the front strut. Do not pry the brake hose and tube clamp away when removing.

4. If equipped with ABS, disconnect the front speed sensor mounting clamp from the strut.

5. Support the lower arm using floor jack or equivalent. Remove the lower strut to knuckle bolts. Once the mounting bolts have been removed, jack up the lower arm. Use a piece of wire to attach the brake hose, tube and driveshaft to the knuckle and to help keep the weight off. These components are not to be pulled.

6. Before removing the top bolts, make matchmarks on the body and the strut insulator for proper reassembly. If this plate is installed improperly, the wheel alignment will be wrong. Remove the strut upper mounting bolts. Remove the strut assembly from the vehicle.

To install:

7. Install the strut to the vehicle and install the top mounting bolts. Make sure the insulator is installed so the matchmarks made during disassembly are in alignment. Tighten the mounting bolts to 36 ft. lbs. (50 Nm) on Eclipse models and 29 ft. lbs. (40 Nm) on Mirage models.

8. Position the strut on the knuckle and install the mounting bolts. While holding the head of the lower mounting bolt, tighten the nuts to 80–101 ft. lbs. (110–140 Nm) on Eclipse models and 80–94 ft. lbs. (110–130 Nm) on Mirage models.

9. Install the brake hose bracket and the ABS clamp, if equipped.

10. Install the wheel and tire assembly. Perform a front end alignment.

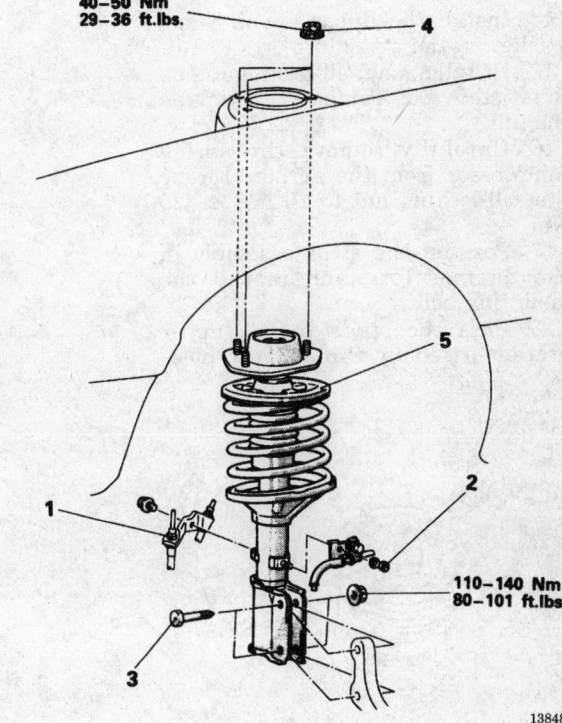

1. Brake hose and tube bracket
2. Front speed sensor cramp
3. Strut lower mounting bolt
4. Strut upper mounting nut
5. Strut assembly

Front strut assembly and related parts — 1993–94 Eclipse shown, others similar

1995–97 Eclipse

1. Raise and safely support the vehicle.

2. Remove the front wheel.

3. Remove the three upper strut mounting nuts. Do not remove the larger nut in the center of the strut at this time.

4. Disconnect the stabilizer link from the damper fork.

5. Remove the damper fork mounting bolt.

6. Remove the strut assembly from the vehicle.

7. Use a coil spring compressor and compress the coil spring. An air tool should not be used to tighten the spring compressor.

8. While holding the piston rod, remove the self-locking nut.

9. Remove the upper bracket assembly and spring pad.

10. Remove the collar, upper bushing, cup assembly, bump rubber and dust cover.

11. Remove the coil spring from the strut.

To install:

12. Align the end of the coil spring with the stepped part of the spring seat and install the compressed coil spring on the strut.

13. Install the dust cover, bump rubber, cup assembly, upper bushing, collar, upper spring pad and bracket assembly on the strut.

14. Install the upper bushing and washer on the piston rod.

15. Install a new self-locking nut on the piston rod. Temporarily tighten the nut.

16. Carefully remove the spring compressor from the spring. Torque the self-locking nut to 16 ft. lbs. (25 Nm).

17. Position the strut assembly in the damper fork and install the mounting bolt.

18. Pass the studs in the upper bracket assembly through the holes

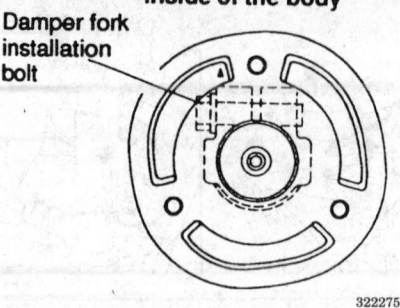

Inside of the body

Damper fork installation bolt

322275

Upper bracket assembly alignment — 1995–97 Eclipse

in the inner fender and install the three mounting nuts.

19. Connect the stabilizer link to the damper fork.

20. Install the wheel assembly.

21. Safely lower the vehicle to the floor.

22. Check and adjust the front wheel alignment if necessary.

Diamante and 1993 Galant

1. Disconnect the negative battery cable.

2. Raise and safely support the vehicle.

3. Remove the brake hose and the tube bracket.

NOTE: Do not pry the brake hose and tube clamp away when removing it.

4. If equipped with ABS, disconnect the front speed sensor mounting clamp from the strut.

5. Support the lower arm and remove the strut to knuckle bolts. Use a piece of wire to suspend the knuckle to keep the weight off the brake hose.

6. If equipped with Active-ECS, disconnect the air tubes and remove the O-rings from the actuator on top of the strut. Once the air line is removed, disconnect the ECS connector at the top of the strut assembly.

NOTE: Before removing the top bolts, make matchmarks on the body and the strut insulator for proper reassembly. If this plate is installed improperly, the wheel alignment will be wrong.

7. Remove the strut upper nuts and remove the strut assembly from the vehicle.

To install:

8. Install the strut to the vehicle and tighten the upper mounting nuts to 33 ft. lbs. (45 Nm).

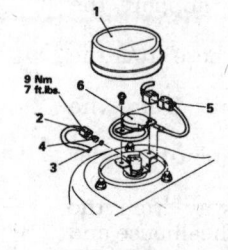

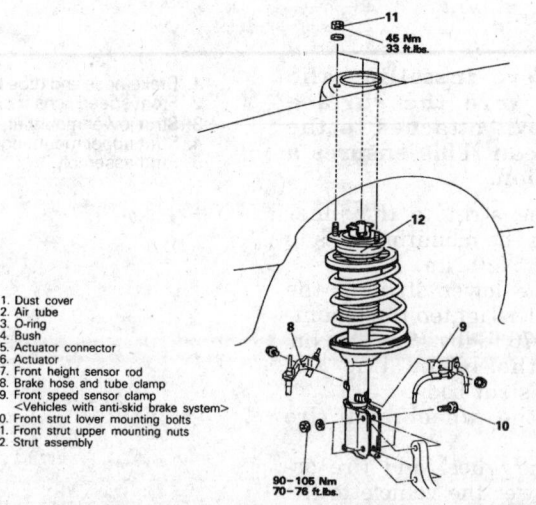

1. Dust cover
2. Air tube
3. O-ring
4. Bush
5. Actuator connector
6. Actuator
7. Front height sensor rod
8. Brake hose and tube clamp
9. Front speed sensor clamp
   <Vehicles with anti-skid brake system>
10. Front strut lower mounting bolts
11. Front strut upper mounting nuts
12. Strut assembly

324339

Front strut assembly with Active-ECS — Diamante and 1993 Galant

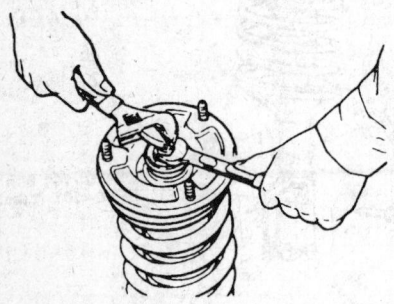

322273

Removing the self-locking nut — 1995–97 Eclipse

9. Align the strut to the knuckle and connect with the mounting bolts. Torque the mounting bolts to 70–76 ft. lbs. (90–105 Nm).

10. If equipped with Active-ECS, lubricate the air tube and new O-ring. Connect the air line to the actuator and tighten the fitting to 7 ft. lbs. (9 Nm).

11. If equipped with Active-ECS, connect the actuator wiring.

12. Install the brake hose bracket and the ABS clamp, if equipped.

13. Install the wheel and tire assembly.

14. Perform a front end alignment.

1994–97 Galant

1. Disconnect the negative battery cable.

2. Raise and safely support vehicle.

3. Remove the appropriate wheel assembly.

4. Disconnect the sway bar link from the damper fork.

5. Remove the damper fork lower through-bolt and upper pinch bolt. Remove the damper fork assembly.

6. Remove the strut upper nuts and remove the strut assembly from the vehicle.

7. Compress the coil spring with a special compression tool.

——————— CAUTION ———————
Make sure that the coil spring compression tool is of an approved design. Great care should be utilized in the removal and installation of the coil spring. If the coil spring is not handled safely, personal injury and/ or property damage could result.

8. Remove the self-locking nut and washer. Remove the upper bushing, upper bracket assembly, the upper spring pad, and the collar.

9. Remove the other upper bushing, cup assembly, bump rubber, dust cover, and the coil spring. Carefully remove the coil spring compression tool.

To install:

10. Install the compressed coil spring to the strut assembly. Make sure to align the edge of coil spring to the stepped part of the strut spring seat. Install the dust cover, bump rubber, cup assembly, upper bushing, collar, and upper spring pad.

11. Install the upper bracket assembly and position it so that the three bolts are in the correct position.

12. Install the upper bushing, washer, and locknut. Torque the locknut to 18 ft. lbs.

13. Install the strut to the vehicle and tighten the upper mounting nuts to 32 ft. lbs. (44 Nm).

14. Align the strut to the damper fork and install the damper fork. Tighten the lower through-bolt/nut to 65 ft. lbs. (88 Nm) and the upper pinch bolt to 76 ft. lbs. (103 Nm).

15. Connect the sway bar link to the damper fork and tighten the link nut to 29 ft. lbs. (39 Nm).

16. Install the wheel and tire assembly.

17. Perform a front end alignment.

3000GT

1. Disconnect the negative battery cable.

——————— CAUTION ———————
Work must be started after 90 seconds from the time the ignition switch is turned to the LOCK position and the negative (-) battery cable is disconnected.

2. Raise and safely support vehicle.

3. Remove the brake hose and tube bracket. Do not pry the brake hose and tube clamp away when removing it.

4. If equipped with ABS, disconnect the front speed sensor mounting clamp from the strut.

5. Support the lower arm and remove the strut to knuckle bolts. Use a piece of wire to suspend the knuckle to keep the weight off the brake hose.

6. Disconnect the ECS connector.

7. Before removing the top bolts, make matchmarks on the body and the strut insulator for proper reassembly. If this plate is installed improperly, the wheel alignment will be wrong. Remove the strut upper bolts and remove the strut assembly from the vehicle.

To install:

8. Install the strut to the vehicle and install the top bolts.

9. Install the strut to the knuckle and install the bolts.

10. Connect the ECS connector.

11. Install the brake hose bracket and the ABS clamp.

12. Install the daytime running lamp delay and control unit to the mounting bracket located on top of the left strut tower.

13. Install the auto-cruise control actuator.

14. Install the wheel and tire assembly.

15. Connect the negative battery cable.

16. Perform a front end alignment.

Coil Spring

REMOVAL AND INSTALLATION

1993–94 Precis

1. Remove the strut from the car and clamp in a protected jaw vise.

2. Gently pry the dust cover from the strut insulator with a flat-blade screwdriver.

NOTE: Matchmark the upper end of the coil spring and bearing plate to avoid confusion during reassembly.

3. Using a spring compressor, fully compress the spring.

4. Using the special tools shown, remove the self-locking gland nut.

5. Remove the rubber insulator, support, spring seat, rubber bumper and the spring.

To install:

6. Install the compressed spring, rubber bumper, dust cover, support and insulator onto the strut.

7. Tighten the nut to 29–36 ft. lbs. (40–50 Nm). Install the dust cover.

8. Slowly release the tension on the spring and remove the spring compressor.

9. Install the assembled strut into the vehicle.

1994–97 Mirage

1. Disconnect the negative battery cable.

2. Raise and safely support vehicle.

3. Remove the brake hose and tube bracket retainer bolt and bracket from the front strut. Do not pry the brake hose and tube clamp away when removing.

4. If equipped with ABS, disconnect the front speed sensor mounting clamp from the strut.

5. Support the lower arm using a floor jack. Remove the lower strut to knuckle bolts. Use a piece of wire to support the brake hose, tube and driveshaft. These components are not to be pulled or stretched.

6. Before removing the top bolts, make matchmarks on the body and the strut insulator for proper reassembly. If this plate is installed improperly, the wheel alignment will be wrong. Remove the strut upper mounting bolts. Remove the strut assembly from the vehicle.

NOTE: Matchmark the coil spring to the strut for reassembly purposes.

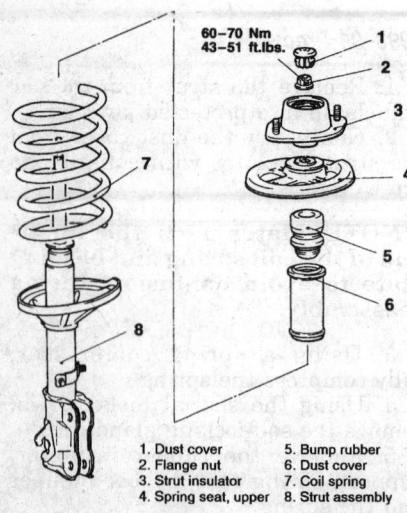

60–70 Nm
43–51 ft.lbs.

1. Dust cover
2. Flange nut
3. Strut insulator
4. Spring seat, upper
5. Bump rubber
6. Dust cover
7. Coil spring
8. Strut assembly

330180

Exploded view of the strut assembly — 1994–97 Mirage

7. Compress the coil spring of the strut using the proper spring compressor.

8. Hold the spring upper seat and remove the self-locking nut at the top of the strut.

9. Slowly release the coil spring tool and remove the coil spring from the strut.

To install:

10. Position and place the spring in the spring compressor and compress the spring.

11. Place the spring onto the strut and install the dust cover, bumper, upper spring seat and the upper insulator.

12. Temporarily tighten a new self-locking nut on the piston shaft.

13. Line up the holes in the upper and lower spring seats and carefully release the spring from the compressor.

14. Torque the self-locking nut to 43–51 ft. lbs. (60–70 Nm).

15. Install the strut to the vehicle and install the top mounting bolts. Make sure the insulator is installed so the matchmarks made during disassembly are in alignment. Tighten the mounting bolts to 29 ft. lbs. (40 Nm).

16. Position the strut on the knuckle and install the mounting bolts. While holding the head of the

lower mounting bolt, tighten the nuts to 80–94 ft. lbs. (110–130 Nm).

17. Install the brake hose bracket and the ABS clamp, if equipped.

18. Install the wheel and tire assembly. Perform a front end alignment.

3000GT and Diamante

1. Disconnect the negative battery cable.

---— **CAUTION** ---—
DO NOT start work until 90 seconds has passed from the time the ignition switch is turned to the LOCK position and the negative (-) battery cable is disconnected.

2. Raise and safely support vehicle.

3. Remove the brake hose and tube bracket. Do not pry the brake hose and tube clamp away when removing it.

4. If equipped with ABS, disconnect the front speed sensor mounting clamp from the strut.

5. Support the lower arm and remove the strut to knuckle bolts. Use a piece of wire to suspend the knuckle to keep the weight off the brake hose.

6. Disconnect the ECS connector.

7. Before removing the top bolts, make matchmarks on the body and the strut insulator for proper reassembly. If this plate is installed improperly, the wheel alignment will be wrong. Remove the strut upper bolts and remove the strut assembly from the vehicle.

8. Place the strut assembly in a suitable vise or holding fixture.

9. Remove the dust shield and loosen the self-locking nut.

---— **CAUTION** ---—
The self-locking nut should be loosened only, not removed until the spring is properly compressed with a suitable coil spring compressor.

10. Install a spring compressor and remove the self-locking nut and upper seat assembly.

11. Remove the spring pad and remove the coil spring.

12. Carefully remove the spring from the spring compressor.

To install:

13. Compress the new spring and place it on the strut.

14. Install the spring pad and the upper seat assembly.

15. Install a new self-locking nut and carefully release the spring from the spring compressor.

16. Install the strut to the vehicle and install the top bolts.

17. Install the strut to the knuckle and install the bolts.

18. Connect the ECS connector.

19. Install the brake hose bracket and the ABS clamp.

20. Install the daytime running lamp delay and control unit to the mounting bracket located on top of the left strut tower.

21. Install the auto-cruise control actuator.

22. Install the wheel and tire assembly.

23. Perform a front end alignment.

Lower Ball Joints

REMOVAL AND INSTALLATION

1993–94 Precis

1. Raise the vehicle and support it safely.

2. Remove the wheel and tire assembly.

3. Unbolt the ball joint from the control arm.

4. Remove the stud retaining nut.

5. Use a ball joint removing tool and separate the ball joint from the steering knuckle.

To install:

6. Replace the ball joint and tighten the ball joint to control arm nut to 69–87 ft. lbs. (99–118 Nm). Torque the ball joint stud nut 43–52 ft. lbs. (59–71 Nm).

7. Install the wheel and tire assembly.

8. Lower the vehicle to the floor.

Mirage, Eclipse and Diamante

The lower ball joint is an integral part of the lower control arm assembly, and can not be serviced separately. A worn or damaged ball joint, requires replacement of lower control arm assembly.

Upper Control Arms

REMOVAL AND INSTALLATION

1995–97 Eclipse

The upper ball joint is an integral part of the upper control arm. If the upper ball joint is to be serviced, the upper control arm will have to be replaced.

1. Raise and safely support the vehicle.

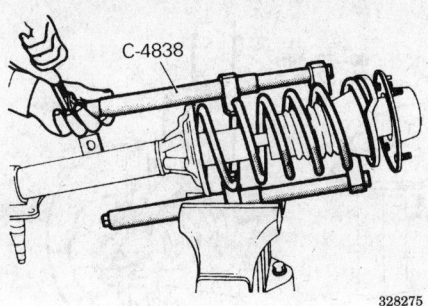

C-4838

328275

Typical method of compressing the coil spring

2. Remove the front wheel(s).
3. Disconnect the upper arm ball joint from the steering knuckle.
4. Remove the upper arm shaft mounting nuts from the body.
5. Remove the upper arm.
6. Remove the through-bolts that attach the upper arm to the shafts.

To install:

7. Assembly the upper arm to the shafts at the proper angle. Torque the through-bolts and nuts to 41 ft. lbs. (57 Nm). The proper angle is 85 °±1°. After the arm and the shafts are connected at the right angle, measure

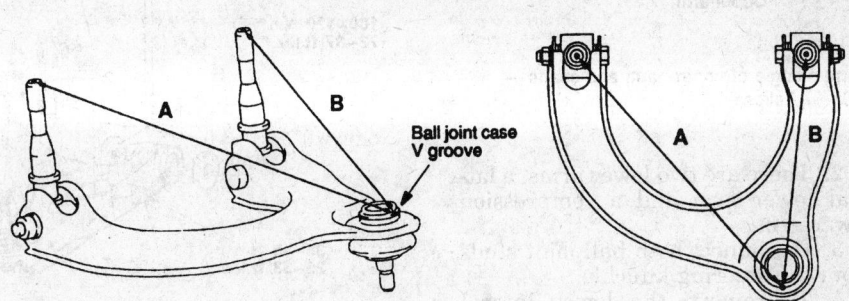

A : 299.9 mm (11.8 in.)
B : 234.0 mm (9.2 in.)

Ball joint case
V groove

322379

Measuring dimensions A and B — 1995–97 Eclipse

dimensions A and B to insure correct assembly.
• A O-ring 11.8 in. (299.9mm)
• B O-ring 9.2 in. (234.0 mm)
8. Install the control arm assembly to the body with new self-locking nuts. Torque the self-locking nuts to 62 ft. lbs. (86 Nm).
9. Connect the upper arm ball joint to the steering knuckle with a new self-locking nut. Torque the locking nut to 20 ft. lbs. (28 Nm).
10. Install the front wheel(s).
11. Perform front wheel alignment and adjust if necessary.
12. Safely lower the vehicle to the floor.

Lower Control Arms

REMOVAL AND INSTALLATION

Galant

The lower lateral arm ball joint and the compression arm ball joint are integral components of the lateral arm and the compression arm. If the ball joints are to be serviced, the arms must be replaced.
1. Raise and support the vehicle safely.

86 Nm
62 ft.lbs.

2

Dust cover

Lips

57 Nm
41 ft.lbs.

4   4

57 Nm
41 ft.lbs.

3

5

28 Nm
20 ft.lbs.

1

1. Upper arm ball joint and knuckle connection
2. Self-locking nut for upper arm installation
3. Upper arm assembly
4. Upper arm shaft assembly
5. Dust cover

322377

Upper control arm and related components — 1995–97 Eclipse

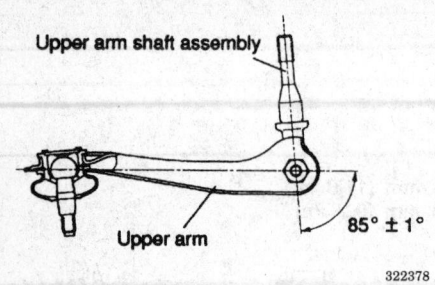

Upper arm shaft assembly

Upper arm

85° ± 1°

322378

**Correct angle of control arm and shafts —
1995–97 Eclipse**

2. There are two lower arms, a lateral lower arm and a compression lower arm.

3. Disconnect both ball joint studs from the steering knuckle.

4. To remove the lower lateral arm, remove the crossmember brackets.

5. Remove the inner lateral arm mounting bolts and nut.

6. Remove the arm from the vehicle.

7. Remove the two bolts holding the compression arm.

8. Remove the compression arm.

To install:

9. Assemble the control arms and bushings.

10. Install the lateral control arm to the vehicle and install the inner mounting bolts. Install a new nut and snug temporarily.

11. Install the compression arm to the vehicle with the two bolts.

12. Connect the ball joint studs to the knuckle. Install new nuts and torque to 43–51 ft. lbs. (59–71 Nm).

13. Lower the vehicle to the floor for the final torquing.

14. Once the full weight of the vehicle is on the suspension, torque the lateral arm rear bolt to 71–85 ft. lbs. (98–118 Nm) and the front bolt to the damper fork to 64 ft. lbs. (88 Nm).

15. Torque the bolts for the compression arm to 60 ft. lbs. (83 Nm).

16. Reinstall the crossmember brackets with their mounting bolts. Torque the mounting bolts to 51–58 ft. lbs. (69–78 Nm).

17. Inspect all suspension bolts, making sure they all have been fully tightened.

18. Perform an alignment on the vehicle.

1993–94 Precis

1. Raise the vehicle and support it safely.

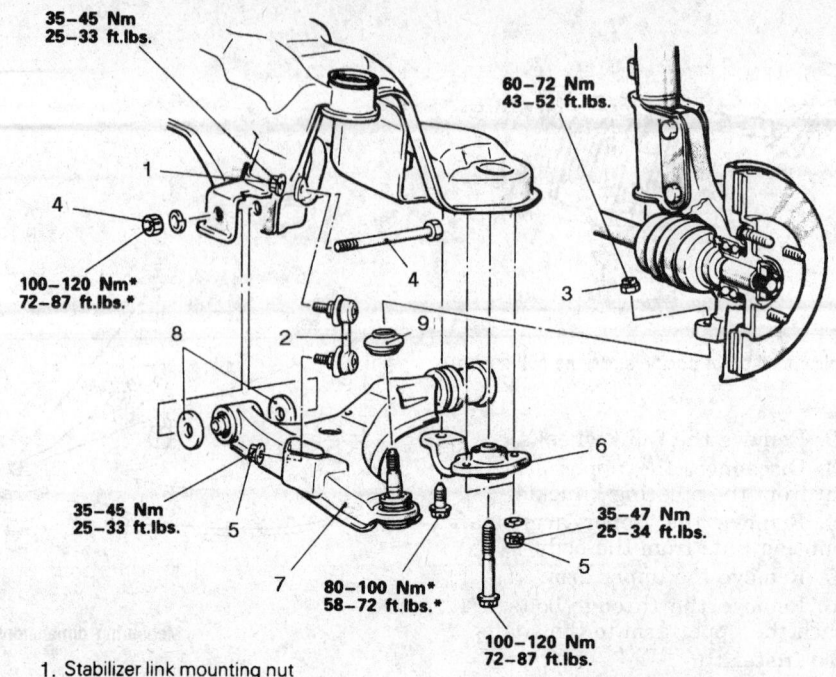

35–45 Nm
25–33 ft.lbs.

60–72 Nm
43–52 ft.lbs.

100–120 Nm*
72–87 ft.lbs.*

1
4
4
3
8
2
9
6
5
7

35–45 Nm
25–33 ft.lbs.

35–47 Nm
25–34 ft.lbs.

80–100 Nm*
58–72 ft.lbs.*

100–120 Nm
72–87 ft.lbs.

1. Stabilizer link mounting nut
2. Stabilizer link
3. Self-locking nut
4. Lower arm mounting nut and bolt
5. Self-locking nut

6. Clamp
7. Lower arm
8. Stopper
9. Ball joint dust cover

160939

Lower control arm with pillow-ball type stabilizer exploded view — 1993 Galant

2. Remove the front wheel and tire assembly.

3. Remove the undercover.

4. Remove the ball joint stud nut and the ball joint stud from the lower control arm.

5. Disconnect the stabilizer bar from the lower arm. Remove the nut from under the control arm and take off the washer and spacer.

6. Remove the lower control arm brackets from the body, and remove the lower control arm from the vehicle.

To install:

7. Install the lower control arm to the body. Tighten the mounting bracket bolts to 43–58 ft. lbs. (60–80 Nm), and the lower arm mounting bolts to 116–137 ft. lbs. (160–190 Nm).

8. Install the stabilizer bar to the control arm. The nut on the stabilizer bar bolt must be torqued until the link shows 0.9–1.0 inch (24–26mm) of threads below the bottom of the nut.

9. Install the ball joint to the steering knuckle. Tighten the nut to 43–52 ft. lbs. (60–72 Nm).

10. Install the undercover.

11. Install the wheel and tire assembly.

12. Lower the vehicle.

Mirage and 1993–94 Eclipse

NOTE: The suspension components should not be tightened until the vehicle's weight is resting on its wheels.

1. Raise the vehicle and support safely.

2. Remove the wheel and tire assembly.

3. Remove sway bar links or mounting nuts and bolts from lower control arm. Remove the joint cups and bushings.

4. Disconnect the ball joint stud from the steering knuckle.

5. Remove the inner lower arm mounting bolt and nut.

6. Remove the rear mount bolts from the retaining clamp. Remove the rear retainer clamp if equipped.

7. Remove the arm from the vehicle.

To install:

8. Install the control arm to the vehicle and install the inner mounting bolt. Install new nut and torque to 78 ft. lbs. (108 Nm).

9. Install the rear mount clamp and bolts. Torque the clamp mounting bolts to 65 ft. lbs. (90 Nm) on Mirage models and 34 ft. lbs. (47 Nm) on Eclipse models.

GALANT/MIRAGE/PRECIS **MITSUBISHI** **10**

10. Connect the ball joint stud to the knuckle. Install a new nut and torque to 43–52 ft. lbs. (60–72 Nm).

11. Install the sway bar and links.

12. Lower the vehicle to the floor for the final torquing of the inner frame mount bolt.

13. Inspect all suspension bolts, making sure they all have been fully tightened.

14. Install the wheel and tire assembly.

1995–97 Eclipse

The compression lower arm ball joint and the lateral lower arm ball joint are integral parts of the arms. If the ball joints are to be replaced, the arms must be replaced.

Compression Lower Arm

1. Raise and safely support the vehicle.

2. Remove the wheel assembly.

3. Use the proper tool and disconnect the compression lower arm ball joint from the steering knuckle.

4. Remove the two mounting bolts and remove the compression lower arm.

To install:

5. Install the compression lower arm. Torque the two mounting bolts to 60 ft. lbs. (81 Nm).

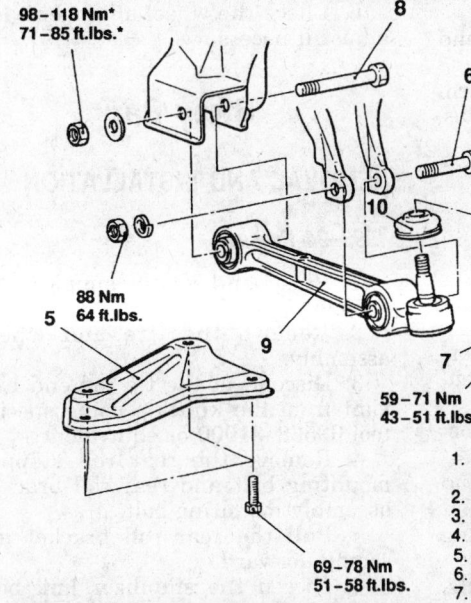

1. Compression lower arm ball joint and knuckle connection
2. Compression lower arm mounting bolts
3. Compression lower arm assembly
4. Dust cover
5. Stay
6. Shock absorber lower mounting bolt
7. Lateral lower arm ball joint and knuckle connection
8. Lateral lower arm mounting bolt
9. Lateral lower arm assembly
10. Dust cover

Caution
*: Indicates parts which should be temporarily tightened, and then fully tightened with the vehicle on the ground in the unladen condition.

333198

Compression lower arm and lateral lower arm components — 1995–97 Eclipse and 1994–97 Galant

6. Using a new self-locking nut, connect the ball joint to the knuckle assembly. Torque the nut to 43–51 ft. lbs. (59–71 Nm).

7. Install the wheel assembly and safely lower the vehicle to the floor.

8. Check and adjust the front wheel alignment if necessary.

Lateral Lower Arm

1. Raise and safely support the vehicle.

2. Remove the wheel assembly.

3. Remove the stay (bracket).

4. Remove the shock absorber lower mounting bolt.

5. Use the proper tool and disconnect the lateral lower arm from the knuckle assembly.

6. Remove the lateral lower arm mounting bolt and the lateral arm.

To install:

7. Install the lateral lower arm and temporarily install the mounting bolt. Do not torque the bolt until the vehicle is on the floor at normal riding height.

8. Use a new self-locking nut and connect the ball joint to the knuckle assembly. Torque the self-locking nut to 43–51 ft. lbs. (59–71 Nm).

9. Install the shock absorber lower mounting bolt, secure with the nut and torque to 64 ft. lbs. (88 Nm).

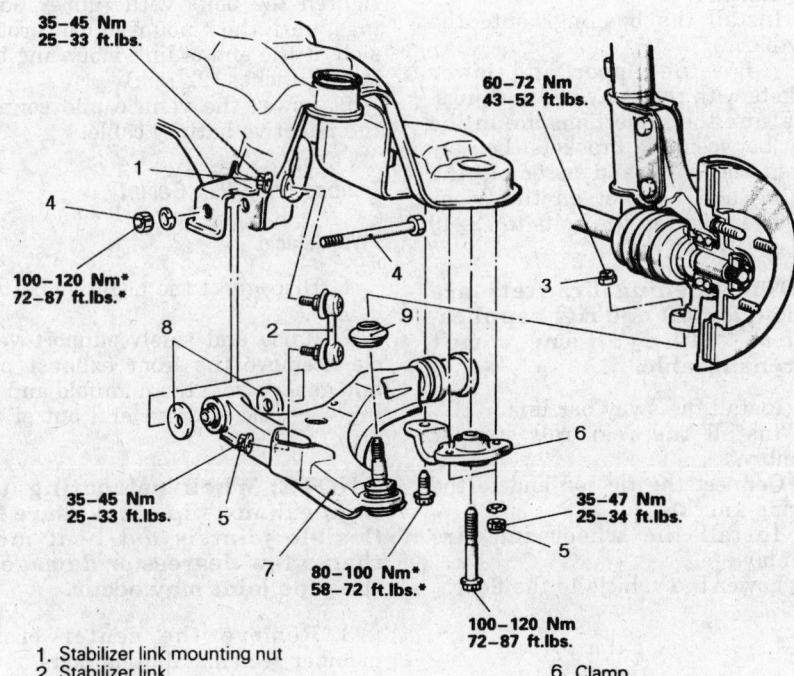

1. Stabilizer link mounting nut
2. Stabilizer link
3. Self-locking nut
4. Lower arm mounting nut and bolt
5. Self-locking nut
6. Clamp
7. Lower arm
8. Stopper
9. Ball joint dust cover

138495

Lower control arm with pillow-ball type stabilizer exploded view — 1993–94 Eclipse shown, Galant, Diamante and 3000GT similar

<section>10-195</section>

10. Install the stay with the bolts and torque them to 51–58 ft. lbs. (69–78 Nm).

11. Install the wheel assembly and lower the vehicle to the floor.

12. Torque the lateral lower arm through-bolt and nut to 71–85 ft. lbs. (98–118 Nm).

13. Check and adjust the front wheel alignment if necessary.

3000GT and Diamante

1. Disconnect the negative battery cable.

2. Raise the vehicle and support safely allowing wheels and suspension to hang freely.

3. Remove the sway bar links from the lower control arm.

4. Disconnect the ball joint stud from the steering knuckle.

5. Remove the inner mounting frame through-bolt and nut.

6. Remove the rear mount bolts. Remove the clamp if equipped.

7. Remove the rear rod bushing if servicing.

To install:

8. Assemble the control arm and bushing.

9. Install the control arm to the vehicle and install the through-bolt. Replace the nut and snug temporarily.

10. Install the rear mount clamp, bolts and replacement nuts. Torque the bolts to 72–87 ft. lbs. (100–120 Nm). Torque the nuts to 29 ft. lbs. (40 Nm).

11. Connect the ball joint stud to the knuckle. Install a new nut and torque to 43–52 ft. lbs. (60–72 Nm).

12. Install the sway bar and links.

13. Lower the vehicle to the floor for the final torquing of the frame mount through-bolt.

14. Once the full weight of the vehicle is on the floor, torque the frame mount through-bolt nuts to 75–90 ft. lbs. (102–122 Nm).

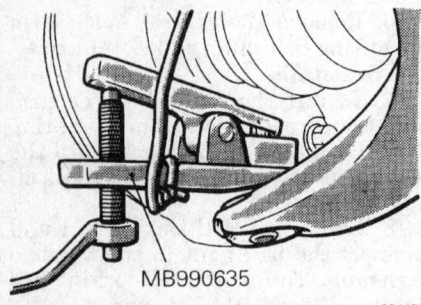

MB990635

324451

Separate the ball joint from the steering knuckle — 3000GT and Diamante

15. Connect the negative battery cable.

16. Check the wheel alignment and adjust if necessary.

Sway Bar

REMOVAL AND INSTALLATION

1993–94 Precis

1. Raise and safely support the vehicle.

2. Remove the tire and wheel assembly.

3. Disconnect the tie rod end ball joint from the knuckle using special tool 09568–31000 or equivalent.

4. Remove the rear roll stopper mounting bolt and rear roll bracket assembly mounting bolt.

5. Pull the rear roll bracket assembly forward.

6. Loosen the stabilizer link bolt and nut, then separate the stabilizer bar from the lower arm.

7. Loosen the stabilizer bar mounting bolts through the steering gear box access opening provided on the vehicle body.

8. Remove the stabilizer through the access opening.

9. Detach the upper and lower bracket, then remove the bushing.

To install:

10. Install the bushings onto the sway bar.

11. Align the upper and lower brackets with the sway bar bushings. Make sure the projections are in the space between the brackets. Loosely tighten the bolts and assemble the bushing and bracket on the other side. Then tighten the bolts to 12–19 ft. lbs. (19–26 Nm).

NOTE: Bushing brackets are marked for left and right applications. They are not interchangeable.

12. Install the sway bar link.

13. Install the rear roll stopper assembly.

14. Connect the tie rod end to the steering knuckle.

15. Install the wheel and tire assembly.

16. Lower the vehicle to the floor.

Mirage

1. Disconnect the negative battery cable.

2. Raise and safely support vehicle.

3. Disassemble the links, remove the locknut, joint cup, bushing, and

collar. Remove the stabilizer link bolts.

4. It will be necessary to remove the center crossmember in order to remove the sway bar. The following steps are required to remove the crossmember.

  a. Remove the front exhaust pipe.

  b. Properly support the engine, remove the engine roll stopper bolts. Remove the four center member mounting bolts and remove the center member assembly.

  c. Remove both steering rack mounts.

  d. Disconnect the lower control arm from the crossmember.

  e. Support the crossmember, remove the mounting bolts and lower the crossmember for access.

5. Remove the stabilizer bar mounts and remove the bar from the vehicle.

To install

NOTE: Note that the bar brackets are marked left and right.

6. Position the stabilizer bar in the vehicle, install the crossmember in the reverse order it was removed.

7. Install the sway bar mount brackets, and tighten the mounting bolts to 16 ft. lbs. (22 Nm).

8. Connect the stabilizer links and tighten the bolts with rubber bushings, until the amount of bolt protrusion at the end of link mounting bolt is 0.87 inches (22 mm).

9. Lower the vehicle and connect the negative battery cable.

Eclipse and 1993 Galant

FWD Vehicle

1. Disconnect the negative battery cable.

2. Raise and safely support vehicle. Remove the front exhaust pipe and gasket from the manifold and using wire, tie it down and out of the way.

NOTE: When relocating the front exhaust pipe, make sure the flexible joint is not bent more than a few degrees or damage to the pipe joint may occur.

3. Remove the center crossmember rear installation bolts.

4. Remove the stabilizer link bolts. On the pillow-ball type, hold ball stud with a hex wrench and remove the self-locking nut with a box wrench.

5. Remove the stabilizer bar bolts and mounts.

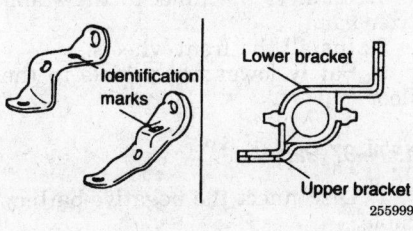

Correct bushing installation — 1993–94 Precis

255999

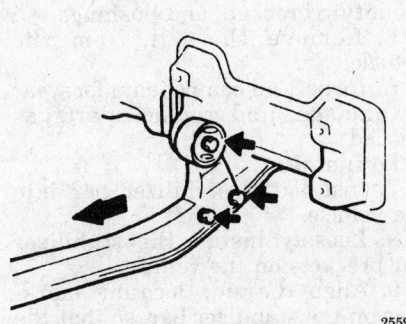

255997

Rear roll stopper mounting — 1993–94 Precis

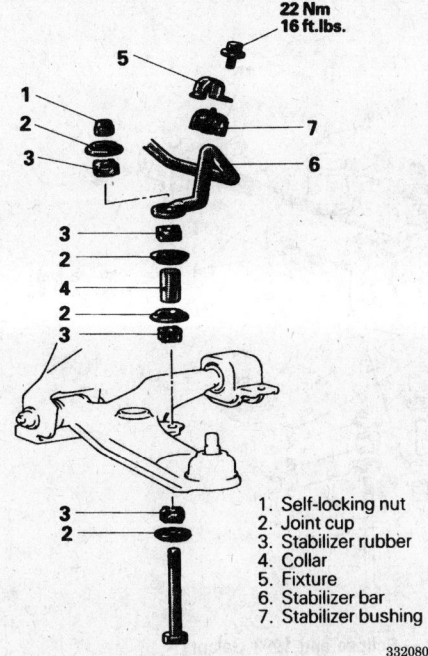

5 — 22 Nm 16 ft.lbs.

1. Self-locking nut
2. Joint cup
3. Stabilizer rubber
4. Collar
5. Fixture
6. Stabilizer bar
7. Stabilizer bushing

332080

Front sway bar and related components — Mirage

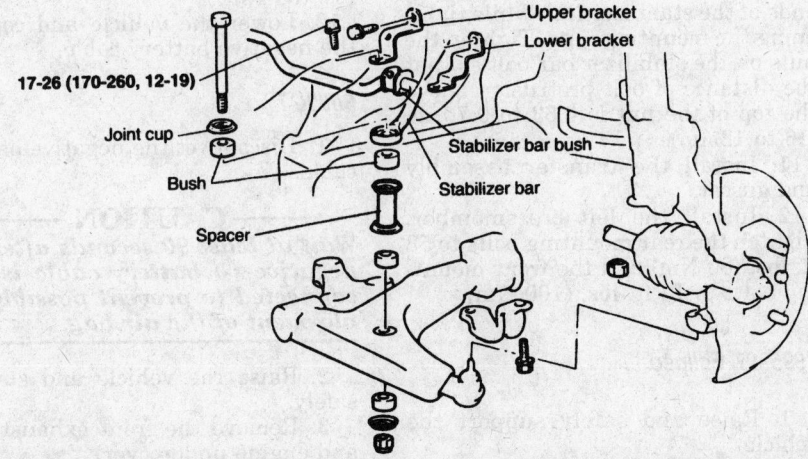

Upper bracket
Lower bracket

17-26 (170-260, 12-19)

Joint cup

Bush

Spacer

Stabilizer bar bush
Stabilizer bar

TORQUE : Nm (Kg.cm, lb.ft)

255996

Sway bar mounting — 1993–94 Precis

6. Remove the bar from the vehicle.

a. Pull both ends of the stabilizer bar toward the rear of the vehicle.

b. Move the right stabilizer bar end until the end clears the lower arm.

c. Remove the stabilizer bar out the right side of the vehicle.

7. Inspect all bushings for wear and deterioration and replace as required. Check the stabilizer bar for damage, and replace as required.

To install:

8. Install the stabilizer bar into the vehicle.

9. Install the stabilizer bar brackets on the vehicle, following any side locating markings on the brackets. Temporarily tighten the stabilizer bar bracket. Align the bushing end with the marked part of the stabilizer bar and then fully tighten the stabilizer bar bracket.

10. If equipped with the pillow-ball type mounting, install the stabilizer bar links and link mounting nuts. Using a wrench, secure the ball studs at both ends of the stabilizer link while tightening the mounting nuts. Tighten the nuts on the stabilizer bar bolt so that the distance of bolt protrusion above the top of the nut is 0.63 to 0.70 in. (16 to 18mm).

11. Install the front exhaust pipe with new gasket in place. Tighten new self-locking nuts to 29 ft. lbs. (40 Nm).

12. Connect the negative battery cable.

AWD Vehicle

1. Disconnect the negative battery cable.

2. Remove the front exhaust pipe.

3. Remove the center gusset and transfer assembly.

4. Using a wrench to secure the ball studs at both ends of the stabilizer link, remove the stabilizer link mounting nuts. Remove the stabilizer link.

5. Remove the stabilizer bar bracket installation bolt and the stabilizer bar bracket and bushing.

6. Disconnect the stabilizer bar coupling at the right lower control arm. Pull out the left side stabilizer edge, pulling it out between the driveshaft and the lower arm. Pull out the right side bar below the lower arm.

To install:

7. Install the bar into the vehicle in the same manner as removal.

8. Temporarily tighten the stabilizer bar bracket. Align the bushing end with the marked part of the sta-

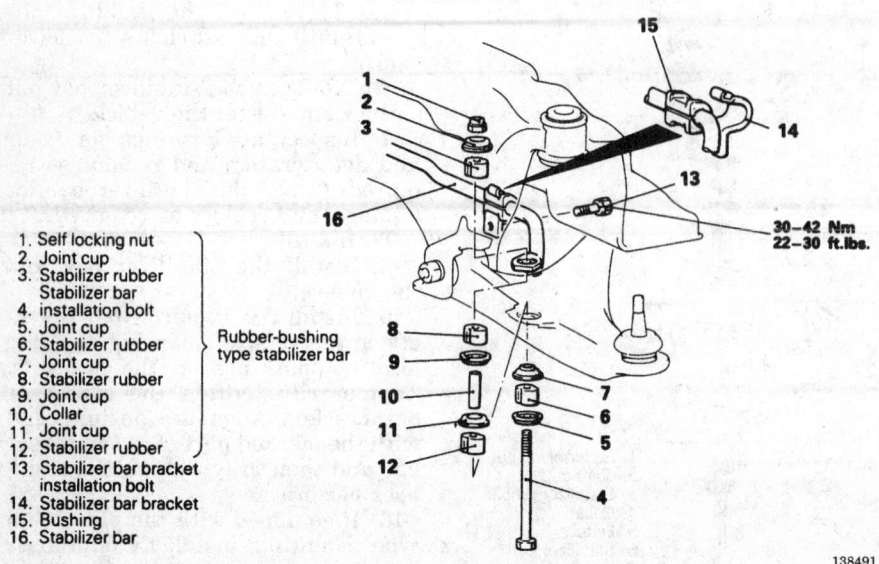

1. Self locking nut
2. Joint cup
3. Stabilizer rubber
   Stabilizer bar
4. installation bolt
5. Joint cup
6. Stabilizer rubber
7. Joint cup
8. Stabilizer rubber
9. Joint cup
10. Collar
11. Joint cup
12. Stabilizer rubber
13. Stabilizer bar bracket
    installation bolt
14. Stabilizer bar bracket
15. Bushing
16. Stabilizer bar

Rubber-bushing type stabilizer bar

30–42 Nm
22–30 ft.lbs.

138491

Rubber bushing type stabilizer bar — Eclipse and 1993 Galant

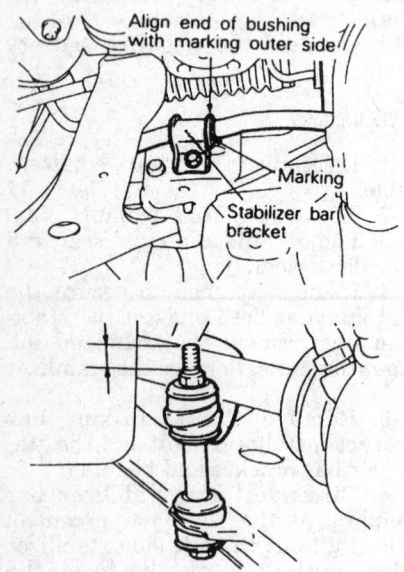

Align end of bushing with marking outer side

Marking

Stabilizer bar bracket

138493

Areas to check on installation — Eclipse and 1993 Galant

bilizer bar and then fully tighten the stabilizer bar bracket.

9. Install and tighten the stabilizer bar bracket bolt.

10. Install the stabilizer bar links and link mounting nuts. Using a wrench, secure the ball studs at both ends of the stabilizer link while tightening the mounting nuts. Tighten the nuts on the stabilizer bar bolt so that the distance of bolt protrusion above the top of the nut is 0.63 to 0.70 in. (16 to 18mm).

11. Install the transfer assembly and gusset.

12. Install the left crossmember. Tighten the rear mounting bolts to 58 ft. lbs. (80 Nm) and the front mounting bolts to 72 ft. lbs. (100 Nm).

1995–97 Eclipse

1. Raise and safely support the vehicle.
2. Remove the front wheels.
3. Disconnect the stabilizer link from the bar.
4. Remove the stabilizer bar bracket mounting bolts.
5. Remove the stabilizer bar.
   To install:
6. Position the stabilizer bar so the distance from the marking on the bar and the edge of the bracket is

0.39 in. (10 mm). Torque the mounting bolts to specifications.

7. Connect the links to the stabilizer bar.
8. Install the front wheels.
9. Safely lower the vehicle to the floor.

1994–97 Galant

1. Disconnect the negative battery cable.
2. Raise and safely support the vehicle.
3. Disconnect the stabilizer links by removing the self-locking nuts.
4. Remove the stabilizer bar mounting brackets and bushings.
5. Remove the bar from the vehicle.
6. Inspect all components for wear or damage, and replace parts as needed.
   To install:
7. Install the stabilizer bar into the vehicle.
8. Loosely install the stabilizer bar brackets on the vehicle.
9. Align the side locating markings on the stabilizer bar, so that the marking on the bar, extends approximately 0.40 inches (10 mm) from the inner edge of the mounting bracket, on both sides.
10. With the stabilizer bar properly aligned, tighten the mounting bracket bolts to 28 ft. lbs. (39 Nm).
11. Connect the stabilizer links to the damper fork and the stabilizer bar. Tighten the locking nuts to 28 ft. lbs. (38 Nm).
12. Lower the vehicle and connect the negative battery cable.

3000GT

1. Disconnect the negative battery cable.

CAUTION

Wait at least 90 seconds after the negative (-) battery cable is disconnected to prevent possible deployment of the air bag.

2. Raise the vehicle and support safely.
3. Remove the front exhaust pipe and engine undercover.
4. Remove the left and right frame members.
5. If necessary, disconnect the harness plugs for the oxygen sensor. Remove the bolt that secures the harness to the frame.
6. On AWD vehicles with automatic transaxle, remove the transfer case bracket and transfer case.
7. Remove the sway bar link.

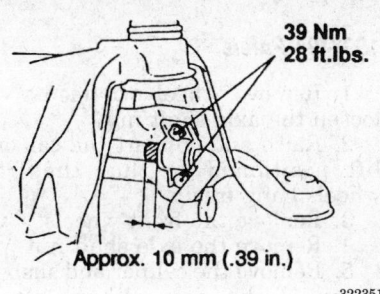

39 Nm
28 ft.lbs.

Approx. 10 mm (.39 in.)

322351

Correct position of the stabilizer bar in the bracket — 1995–97 Eclipse

8. Remove the sway bar brackets and remove the sway bar from the vehicle.

To install:

9. Note that the bar brackets are marked left and right. Lubricate all rubber parts and install the bushings, the sway bar and brackets.

10. Install the sway bar link and torque to 29 ft. lbs. (40 Nm).

11. Install the transfer case and bracket.

12. Install the frame members and torque to 29 ft. lbs. (40 Nm) on 3000GT models and 43–51 ft. lbs. (60–70 Nm) on Diamante models.

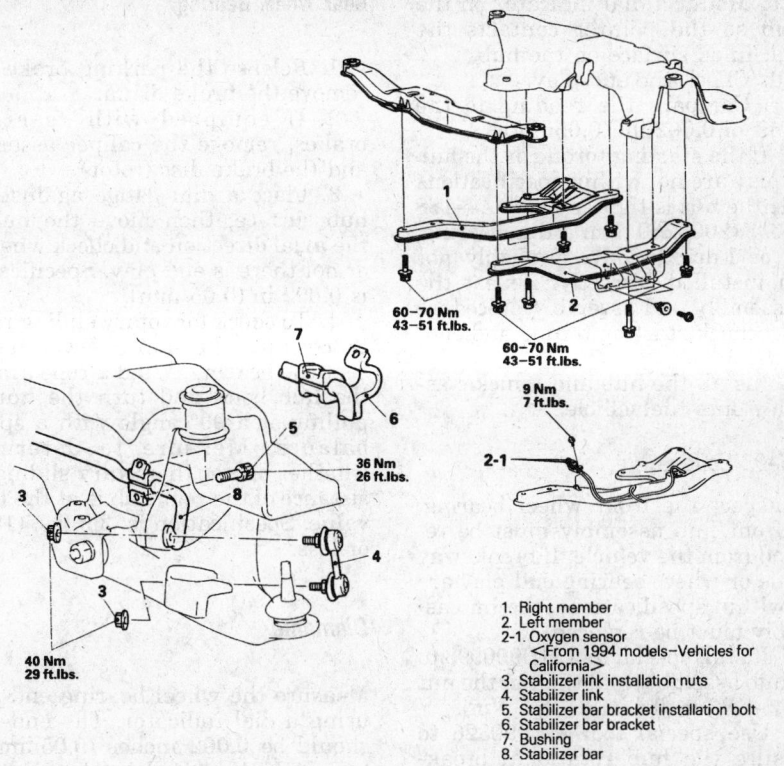

60–70 Nm
43–51 ft.lbs.

60–70 Nm
43–51 ft.lbs.

9 Nm
7 ft.lbs.

36 Nm
26 ft.lbs.

40 Nm
29 ft.lbs.

1. Right member
2. Left member
2-1. Oxygen sensor
    <From 1994 models–Vehicles for California>
3. Stabilizer link installation nuts
4. Stabilizer link
5. Stabilizer bar bracket installation bolt
6. Stabilizer bar bracket
7. Bushing
8. Stabilizer bar

324367

Front sway bar and related components — Diamante

13. Install the engine undercover and exhaust pipe.

14. If disconnected, attach the harness plugs for the oxygen sensor.

15. Connect the negative battery cable.

Front Wheel Bearings

ADJUSTMENT

Mirage

1. Remove the hub, knuckle and bearing assembly from the vehicle.

2. Using pressing tool MB990998 or equivalent, mount the front hub assembly into the knuckle. Tighten the nut of the pressing tool to 144–188 ft. lbs. (200–260 Nm). Rotate the hub to seat the bearing.

3. Mount the knuckle assembly in a vise. Check the hub assembly turning torque and end-play as follows:

   a. Using a torque wrench and socket MB990998 or equivalent, turn the hub in the knuckle assembly. Note the reading on the torque wrench and compare to the desired reading of 16 inch lbs. (1.8 Nm) or less. This is known as the breakaway torque.

   b. Check for roughness when turning the bearing.

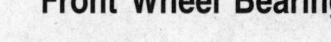

39 Nm
28 ft.lbs.

39 Nm
28 ft.lbs.

39 Nm
28 ft.lbs.

39 Nm
28 ft.lbs.

1. Stabilizer link mounting nut
2. Stabilizer link
3. Stabilizer bar bracket
4. Bushing
5. Stabilizer bar

325729

Stabilizer bar and related components — 1994–97 Galant and 1995–97 Eclipse

c. Mount a dial indicator on the hub so the pointer contacts the machined surface on the hub.

d. Check the end-play.

e. Compare the reading to the limit of 0.002 in. (0.05mm).

4. If the starting torque or the hub end-play are not within specifications while the nut is tightened to 144–188 ft. lbs. (200–260 Nm), the bearing, hub or knuckle have probably not been installed correctly. Repeat the disassembly and assembly procedure and recheck starting torque and end-play.

5. Install the hub and knuckle assembly onto the vehicle.

1995–97 Eclipse

To inspect the front wheel bearing, the front hub assembly must be removed from the vehicle. If breakaway torque or wheel bearing end play are not within specifications, the hub assembly must be replaced.

1. Install special tool MB990998 to the hub assembly and tighten the nut to 145–188 ft. lbs. (196–255 Nm).

2. Use special tool MB990326 to measure the hub rotational breakaway torque. The hub rotation breakaway torque is: 9 in. lbs. (1.0 Nm) or less. There should be no engagement or feeling of roughness.

3. To check wheel bearing end play: Install special tool MB990998 to the hub assembly and torque the nut to 145–188 ft. lbs. (196–255 Nm).

4. Measure the play in the hub axial direction. Specification limit is: 0.002 in. (0.05 mm).

Galant

To inspect the front wheel bearing, the front hub assembly must be removed from the vehicle. If breakaway torque or wheel bearing end play are not within specifications, the hub assembly must be replaced.

Front Wheel Bearing

1. Install special tool MB990998 to the hub assembly and tighten the nut to 145–188 ft. lbs. (196–255 Nm).

2. Use special tool MB990326 to measure the hub rotational breakaway torque. The hub rotation breakaway torque is: 16 in. lbs. (1.8 Nm) or less. There should be no engagement or feeling of roughness.

3. To check wheel bearing end play: Install special tool MB990998 to the hub assembly and torque the nut to 145–188 ft. lbs. (196–255 Nm).

4. Measure the play in the hub axial direction. Specification limit is: 0.002 in. (0.05 mm).

Rear Wheel Bearing

1. Release the parking brake and remove the brake drum.

2. If equipped with rear disc brakes, remove the caliper assembly and the brake disc (rotor).

3. Place a dial gauge against the hub surface; then move the hub in the axial direction and check whether or not there is end play. Specification is 0.002 in (0.05 mm).

4. To check for rotary sliding resistance: turn the hub a few times to seat the bearing. Wind a rope around the hub bolts and turn the hub by pulling at a 90° angle with a spring balance. Measure to determine whether or not the rotary sliding resistance of the rear hub is at the limit value. Specification is: 3.9 lbs. (18 N) or less.

Diamante

Measure the wheel bearing end-play using a dial indicator. The end-play should be 0.002 inches (0.05mm) or less with the wheel bearing locknut torqued to specifications. The wheel bearings are sealed units and are not adjustable. If defective, replacement is the only option.

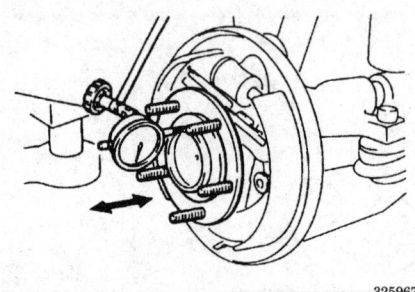

325967

Inspecting rear wheel bearing end play — Galant

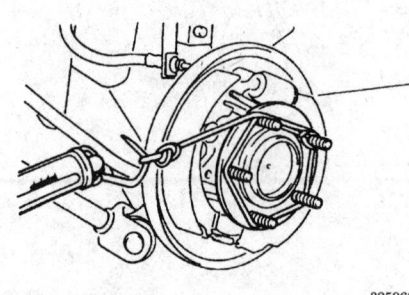

325968

Inspecting rear wheel bearing sliding resistance — Galant

REMOVAL AND INSTALLATION

1993–94 Precis

1. Remove wheel ornaments and loosen the axle shaft nut.

2. Raise and support the car on a lift positioned so that the front wheels hang freely.

3. Remove the front wheels.

4. Remove the axle shaft nut.

5. Remove the caliper and suspend it out of the way without disconnecting the brake hose.

6. Disconnect the lower ball joint from the knuckle.

7. Disconnect the tie rod end from the knuckle using the proper tool.

8. Using a two-jawed puller, press the axle shaft from the hub.

9. Unbolt the strut from the knuckle. Remove the hub and knuckle assembly from the car.

10. Install first the arm, then the body of special tool 09517-21600 on the knuckle and tighten the nut.

11. Using special tool 09517-21500, separate the hub from the knuckle.

NOTE: Prying or hammering will damage the bearing. Use these special tools, or their equivalent to separate the hub and knuckle.

To install:

12. Mount the knuckle in a vise. Position the hub and knuckle together. Install tool 09517-21500 and tighten the tool to 145–188 ft. lbs. (200–260 Nm). Rotate the hub to seat the bearing.

13. With the knuckle still in the vise, measure the hub starting torque with an inch lbs. torque wrench and tool 09517-215000. Starting torque should be 11.5 inch lbs. If the starting torque is 0, measure the hub bearing axial play with a dial indicator. If axial play exceeds 0.11mm while the nut is tightened to specification, the procedure has not been done correctly. Disassemble the knuckle and hub and start again.

14. Remove the special tool.

15. Place the outer bearing in the hub and drive the seal into place.

16. Position the knuckle and hub assembly onto the strut and install the attaching bolts. Torque the bolts to 54–65 ft. lbs. (74–88 Nm) on 1986–89 models and 65–76 ft. lbs. (90–105 Nm) on 1990–94 models.

17. Insert the axle shaft through the hub.

18. Connect the tie rod end to the steering knuckle. Install the tie rod end nut and torque it to 11–25 ft. lbs. (15–34 Nm).

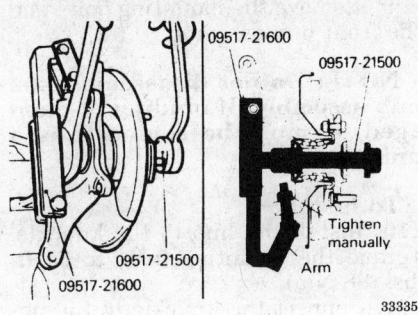

Removing the hub from the knuckle — 1993–94 Precis

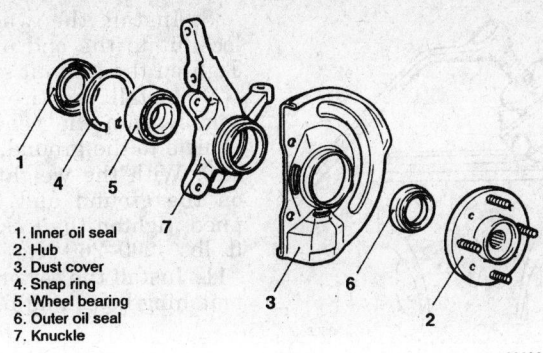

1. Inner oil seal
2. Hub
3. Dust cover
4. Snap ring
5. Wheel bearing
6. Outer oil seal
7. Knuckle

Wheel bearing assembly exploded view — Mirage, Diamante, 1993 Galant and 1993–94 Eclipse

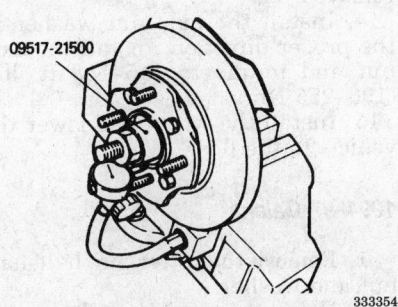

Measuring the axial play — 1993–94 Precis

19. Connect the lower ball joint to the steering knuckle. Install the attaching bolt and snug it. The bolt will be torqued properly when the wheels are on the ground.

20. Mount the brake caliper assembly and brake hose.

21. Mount the front wheels and lower the vehicle to the ground. Now torque the lower ball joint bolt.

22. Install the axleshaft nut and torque the nut to 145–188 ft. lbs. (200–260 Nm). Secure the nut with a new cotter pin.

Mirage, Diamante, 1993 Galant and 1993–94 Eclipse

Removal

1. Disconnect the negative battery cable.

2. Remove the cotter pin from the driveshaft nut. With the brakes applied, loosen the halfshaft nut.

3. Raise the vehicle and support safely. Remove the halfshaft nut.

4. If equipped with ABS, remove the front wheel speed sensor.

5. If equipped with Active-ECS, disconnect the height sensor from the lower control arm.

6. Remove the caliper assembly and brake pads. Suspend the caliper with a wire.

7. Using tool MB991113 or equivalent, disconnect the ball joint and tie rod end from the steering knuckle.

NOTE: It is important to use proper methods of joint separation. Use of unapproved techniques can result in damage to joint and possible failure.

8. Remove the halfshaft by setting up a puller on the outside wheel hub and pushing the halfshaft from the front hub. After pressing the outer shaft, insert a prybar between the transaxle case and the halfshaft and pry the shaft from the transaxle.

9. Unbolt the lower end of the strut and remove the hub and steering knuckle assembly from the vehicle.

10. Install the hub/knuckle assembly in a vise. Using puller MB991056 or equivalent, remove the hub from the knuckle.

Disassembly

NOTE: Do not use a hammer to accomplish this or the bearing will be damaged.

1. Remove the oil seal from the axle side of the knuckle using a small prying tool.

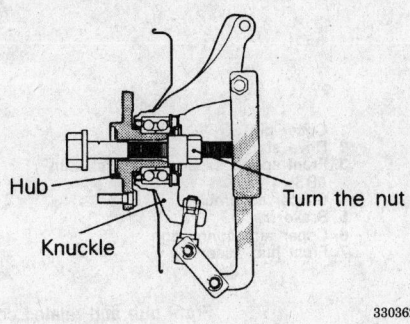

Use of press tool for hub removal — Mirage, 1993 Galant and 1993–94 Eclipse

2. Remove the wheel bearing inner race from the front hub using a puller.

NOTE: Be careful that the front hub does not fall when the inner race is removed.

3. Remove the snapring from the axle side of the knuckle. Remove the bearing from the knuckle using a puller.

4. Once the bearing is removed, the bearing outer race can be removed by tapping out with a brass drift pin and a hammer.

Assembly

1. Fill the wheel bearing with multipurpose grease. Apply a thin coating of multipurpose grease to the knuckle and bearing contact surfaces.

2. Press the wheel bearing into the knuckle using appropriate pressing tool. Once the bearing is installed, install the inner race using the proper driving tool.

3. Drive the oil seal into the knuckle by using the proper size driver. Drive seal into knuckle until it is flush with the knuckle end surface.

4. Using pressing tool MB990998 or equivalent, mount the front hub assembly into the knuckle. Tighten the nut of the pressing tool to 144–188 ft. lbs. (200–260 Nm). Rotate the hub to seat the bearing.

5. Mount the knuckle assembly in a vise. Check the hub assembly turning torque and end-play as follows:

a. Using a torque wrench and socket MB990998 or equivalent, turn the hub in the knuckle assembly. Note the reading on the torque wrench and compare to the desired reading of 16 inch lbs. (1.8 Nm) or less. This is known as the breakaway torque.

b. Check for roughness when turning the bearing.

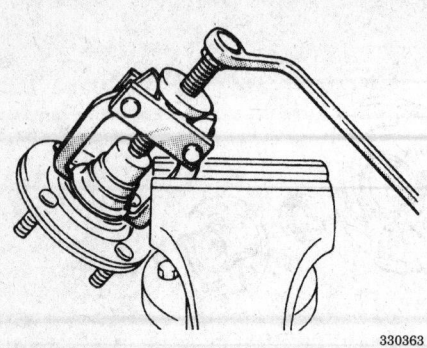

Removing inner race from hub — Mirage, 1993 Galant and 1993–94 Eclipse

c. Mount a dial indicator on the hub so the pointer contacts the machined surface on the hub.

d. Check the end-play.

e. Compare the reading to the limit of 0.002 in. (0.05mm).

6. If the starting torque or the hub end-play are not within specifications while the nut is tightened to 144–188 ft. lbs. (200–260 Nm), the bearing, hub or knuckle have probably not been installed correctly. Repeat the disassembly and assembly procedure and recheck starting torque and end-play.

Installation

1. Install the hub and knuckle assembly onto the vehicle. Install the lower ball joint stud into the steering knuckle and install new nut. Tighten to 52 ft. lbs. (72 Nm).

2. Install the halfshaft into the transaxle extension housing and guide the outer end through the hub/knuckle assembly.

3. Install the two front strut lower mounting bolts and tighten to 80–94 ft. lbs. (110–130 Nm) on all models except Diamante and 65–76 ft. lbs. (90–105 Nm) on Diamante models.

4. Install the connection for the tie rod end and tighten nut to 25 ft. lbs. (34 Nm) on all models except Diamante and 21 ft. lbs. (29 Nm) on Diamante models. Install new cotter pin and bend to locknut in position.

5. Install the brake disc and caliper assembly.

6. If equipped with Active-ECS, connect the height sensor and tighten the mounting bolt to 15 ft. lbs. (20 Nm).

7. Install the front speed sensor, if removed.

NOTE: When installing front speed sensor, make sure harness is routed in the original position and that it is not twisted.

8. Install the washer and new locknut to the end of the halfshaft. Tighten the locknut snugly.

9. Install the tire and wheel assembly onto the vehicle. Lower the vehicle to the ground.

10. With the weight of the vehicle on the ground and the brakes applied, tighten the locknut to 144–188 ft. lbs. (200–260 Nm).

11. Install the cotter pin in the first matching holes and bend it securely.

1995–97 Eclipse

1. Raise and safely support the vehicle.

2. Remove the front wheel.

3. Use tool MB990767 or equivalent to hold the hub assembly while removing the axle nut.

4. On vehicles with ABS, remove the wheel speed sensor.

5. Remove the caliper and suspend it out of the way with wire or string.

6. Remove the brake rotor.

7. Disconnect the steering knuckle from the upper arm.

8. Pull the knuckle away from the vehicle to access the hub mounting bolts on the inboard side of the hub. Be careful not to damage the ball joint boot or the ABS rotor if equipped.

9. Remove the mounting bolts and the front hub assembly.

NOTE: Do not disassemble the hub assembly. If binding or damaged, it must be replaced as a unit.

To install:

10. Install the hub to the knuckle. Torque the mounting bolts to 65 ft. lbs. (88 Nm).

11. Connect the knuckle to the upper arm.

12. Install the brake rotor and the caliper.

13. Install the wheel speed sensor if removed.

14. Install the axle nut washer in the proper direction. Install the axle nut and torque to 145–188 ft. lbs. (196–255 Nm).

15. Install the wheel and lower the vehicle to the floor.

1994–97 Galant

1. Remove the cotter pin, halfshaft nut and washer.

2. Raise the vehicle and support safely.

3. Remove the appropriate wheel assembly.

4. If equipped with ABS, remove the vehicle speed sensor.

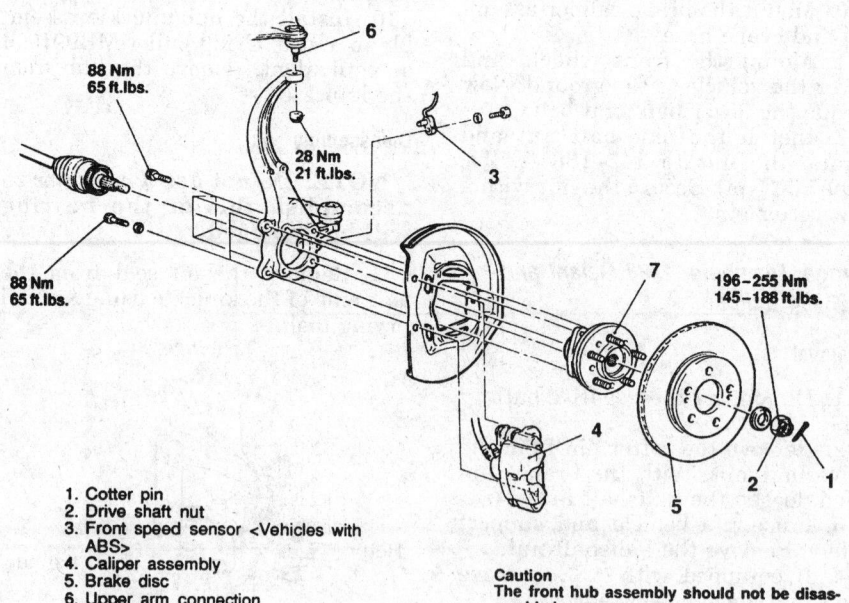

1. Cotter pin
2. Drive shaft nut
3. Front speed sensor <Vehicles with ABS>
4. Caliper assembly
5. Brake disc
6. Upper arm connection
7. Front hub assembly

Caution
The front hub assembly should not be disassembled.

Front hub and related components — 1995–97 Eclipse

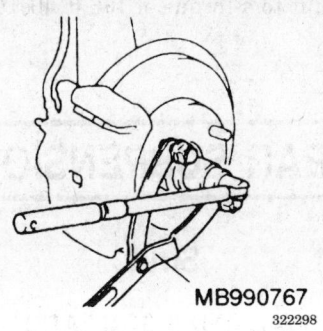

Hold the hub while removing the axle nut — 1995–97 Eclipse

MB990767
322298

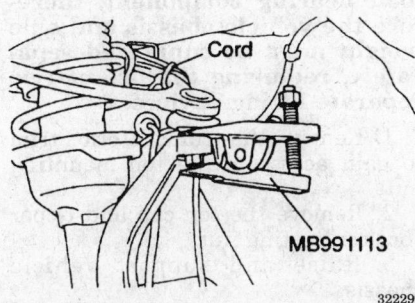

Cord

MB991113

322299

Disconnect the knuckle from the upper arm — 1995–97 Eclipse

5. Remove the caliper and brake pads. Support the caliper out of the way using wire.

6. Remove the brake rotor from the hub assembly.

7. Disconnect the upper ball joint from the steering knuckle and pull the knuckle outward.

———— WARNING ————
Use of improper methods of joint separation can result in damage to joint, leading to possible failure.

8. From the back of the knuckle, remove the four bolts securing the hub to the knuckle.

9. Remove the hub and bearing assembly from the knuckle.

NOTE: The hub assembly is not serviceable and should not be disassembled.

To install

10. Install the hub to the steering knuckle and tighten the mounting bolts to 65 ft. lbs. (88 Nm).

11. Connect the upper ball joint to the steering knuckle and tighten the self-locking nut to 21 ft. lbs. (28 Nm).

12. Position the rotor on the hub. Install a couple of lug nuts and lightly tighten to hold rotor on hub.

13. Install the caliper holder and place brake pads in holder. Slide caliper over brake pads and install guide pins. Once caliper is secured, lug nuts can be removed.

14. If equipped with ABS, install the vehicle speed sensor.

15. Install the wheel assembly and lower the vehicle.

16. Install the wheel and lower the vehicle to the floor. Tighten the axle nut with the brakes applied. Tighten the nut to torque of 145–188 ft. lbs. (200–260 Nm).

17. Install a new cotter pin and bend to secure.

———— WARNING ————
Pump the brake pedal until hard, before attempting to move the vehicle.

3000GT

1. Disconnect the negative battery cable.

———— CAUTION ————
Work must be started after 90 seconds from the time the ignition switch is turned to the LOCK position and the negative (-) battery cable is disconnected.

2. Remove the cotter pin, halfshaft nut and washer.

3. Raise the vehicle and support safely. If equipped with ABS, remove the front wheel speed sensor. Remove the ball joint and tie rod end from the steering knuckle.

4. Remove the caliper and brake pads and suspend with a wire.

5. On vehicles with an inner shaft, remove the center support bearing bracket bolts and washers. Remove the halfshaft by setting up a puller on the outside wheel hub and pushing the halfshaft from the front hub. Then tap the joint case with a plastic hammer to remove the halfshaft shaft and inner shaft from the transaxle.

6. On vehicles without an inner shaft, remove the halfshaft by setting up a puller on the outside wheel hub and pushing the halfshaft from the front hub. After pressing the outer shaft, insert a prybar between the transaxle case and the halfshaft and pry the shaft from the transaxle.

7. On vehicles with AWD, the front hub/bearing assembly can be serviced at this point as a unit. If the knuckle is being removed, proceed.

8. Unbolt the lower end of the strut and remove the hub and steering knuckle assembly.

9. Set up a puller with the knuckle/hub in a vise and pull the hub from the knuckle. Do not use a

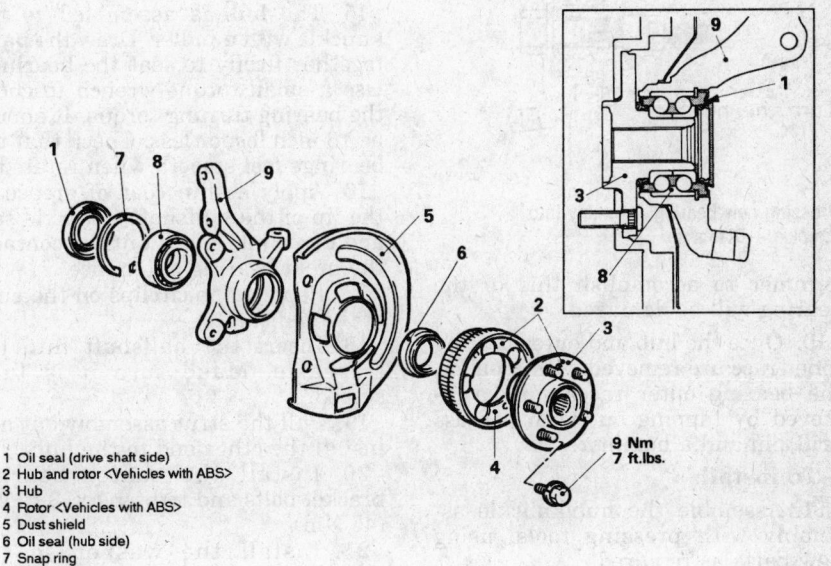

1 Oil seal (drive shaft side)
2 Hub and rotor <Vehicles with ABS>
3 Hub
4 Rotor <Vehicles with ABS>
5 Dust shield
6 Oil seal (hub side)
7 Snap ring
8 Wheel bearing
9 Knuckle

9 Nm
7 ft.lbs.

324385

Exploded view of the wheel bearing assembly — Diamante

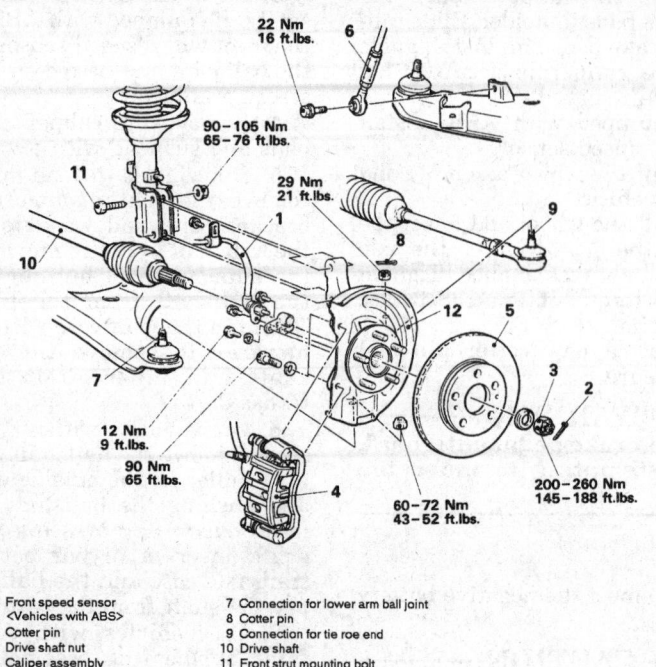

22 Nm 16 ft.lbs.
90–105 Nm 65–76 ft.lbs.
29 Nm 21 ft.lbs.
12 Nm 9 ft.lbs.
90 Nm 65 ft.lbs.
60–72 Nm 43–52 ft.lbs.
200–260 Nm 145–188 ft.lbs.

1 Front speed sensor <Vehicles with ABS>
2 Cotter pin
3 Drive shaft nut
4 Caliper assembly
5 Brake disc
6 Front height sensor <Vehicles with ACTIVE-ECS>
7 Connection for lower arm ball joint
8 Cotter pin
9 Connection for tie roe end
10 Drive shaft
11 Front strut mounting bolt
12 Hub and knuckle

324386

Steering knuckle and related components — Diamante

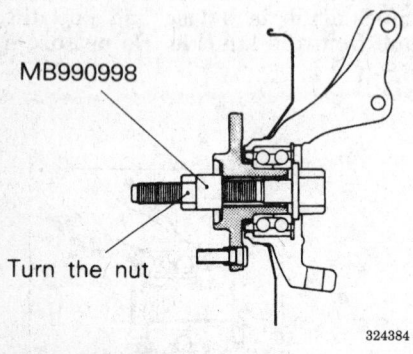

MB990998

Turn the nut

324384

Pressing new bearing assembly into knuckle — Diamante

hammer to accomplish this or the bearing will be damaged.

10. Once the hub and outer bearing inner race are removed with a puller, the bearing outer races can be removed by tapping out with a brass drift pin and a hammer.

To install:

11. Assemble the hub/knuckle assembly with pressing tools, using new parts as required.

12. Install the knuckle assembly to the vehicle and install the strut bolts.

13. Apply a thin coat of grease to the outside of the outer races and install into the hub with a bearing driver.

14. Apply multi-purpose grease to the bearings, inside surface of the hub and the lip of the grease seal. Place the outside bearing into the knuckle and install the seal with a driver.

15. The hub is assembled to the knuckle with a puller. Draw the parts together firmly to seat the bearings. Use a small torque wrench to check the bearing turning torque. It should be 16 inch lbs. or less. Check that the bearings feel smooth when rotated.

16. Apply a thin coat of grease to the lip of the halfshaft side axle seal and drive into place until it contacts the inner bearing outer race.

17. Replace the circlips on the ends of the halfshafts.

18. Insert the halfshaft into the transaxle. Make sure it is fully seated.

19. Pull the strut assembly out and install the other end to the hub.

20. Install the center bearing bracket bolts and tighten to 33 ft. lbs. (45 Nm).

21. Install the washer so the chamfered edge faces outward. Install the nut and tighten temporarily.

22. Install the tie rod end and ball joint.

23. Install the wheel and lower the vehicle to the floor. Tighten the axle nut with the brakes applied. Tighten

the nut to a torque of 166 ft. lbs. (230 Nm).

REAR SUSPENSION

Strut

REMOVAL AND INSTALLATION

Mirage

NOTE: The strut assembly is a load bearing component, therefore the vehicle chassis and axle weight must be supported separately, requiring the use of two separate lifting devices.

1. Remove the trunk interior trim to gain access to the top mounting nuts.

2. Remove the top cap and upper shock mounting nuts.

3. Raise and support vehicle chassis.

4. Raise and support the trailing arm assembly slightly.

NOTE: Matchmark the upper spring plate to the vehicle chassis for reassembly.

5. Remove the shock lower mounting bolt and remove the assembly from the vehicle.

6. Compress the coil spring using the proper spring compressor.

—— **CAUTION** ——
Do not use air tools to tighten the spring compressor.

7. Hold the piston rod with a wrench and remove the self-locking nut.

8. Remove the washer, upper bushing A, bracket, spring pad, upper bushing B, collar, cup, dust cover and bump rubber.

NOTE: Align the stepped part of the spring pad with the end of the spring.

9. Remove the coil spring.

To install

10. Install the coil spring on the strut.

11. Instal the bump rubber, dust cover, cup, collar, upper bushing A, spring pad, bracket, upper bushing B and the washer.

12. Temporarily install a new self-locking nut, carefully release the spring from the compressor and tighten the self-locking nut to specifications.

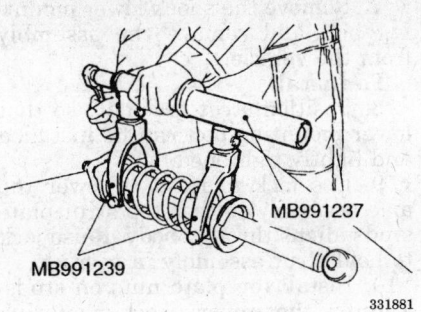

Compress the coil spring using the proper tool — Mirage

13. Position strut assembly so that lower mounting bolt can be installed and lightly tightened.

14. Use jack to raise or lower the axle assembly so that top strut plate studs aligns through body. Raise jack to hold strut assembly in position. Be sure to properly position the upper spring plate.

15. Install top plate nuts on studs. Tighten the upper shock mounting nuts to 20 ft. lbs. (28 Nm) and the lower mounting bolt to 65 ft. lbs. (90 Nm).

16. Lower the vehicle. Install top cap and interior trim.

1993–94 Eclipse

NOTE: The strut assembly is a load bearing component, therefore the vehicle chassis and axle weight must be supported separately, requiring the use of two separate lifting devices.

1. Disconnect the negative battery cable.
2. Raise and support vehicle chassis.
3. Raise and support torsion axle and arm assembly slightly. Make sure the jack does not contact the lateral rod.

NOTE: Always use a wooden block between the jack receptacle and the axle beam. Place the jack at the center of the axle beam.

4. Remove the trunk interior trim to gain access to the top mounting nuts.
5. Remove the top cap and upper shock mounting nuts.
6. Remove the brake tube bracket bolt.
7. Remove the shock lower mounting bolt and remove the assembly from the vehicle.

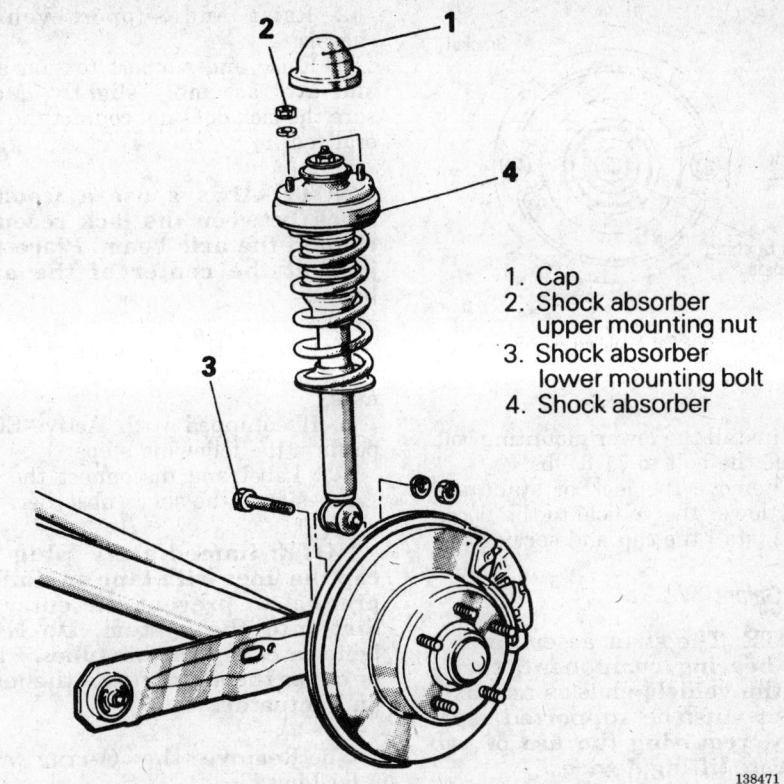

1. Cap
2. Shock absorber upper mounting nut
3. Shock absorber lower mounting bolt
4. Shock absorber

Rear strut assembly and related components — 1993–94 Eclipse

To install

8. Position strut assembly so that lower mounting bolt can be installed and lightly tightened.
9. Use jack to raise or lower the axle assembly so that top strut plate studs aligns through body. Raise jack to hold strut assembly in position.
10. Install top plate nuts on studs. Tighten the upper shock mounting nuts to 29 ft. lbs. (40 Nm) and the lower mounting bolt to 72 ft. lbs. (100 Nm).
11. Connect brake bracket and lower vehicle.
12. Install top cap and interior trim.

1995–97 Eclipse

1. Remove the service lid in the luggage compartment.
2. Remove the cap and flange nuts securing the upper mounting bracket to the body of the vehicle. Do not remove the larger nut in the center of the shock absorber.
3. Raise and safely support the vehicle.
4. Remove the bolt attaching the lower end of the shock to the knuckle and remove the shock absorber from the vehicle.
5. Use a coil spring compressor and compress the coil spring. An air

tool should not be used to tighten the spring compressor.

6. While holding the piston rod, remove the self-locking nut.
7. Remove the upper bracket assembly and spring pad.
8. Remove the collar, upper bushing, cup assembly, bump rubber and dust cover.
9. Remove the coil spring from the strut.

To install:

10. Align the end of the coil spring with the stepped part of the spring seat and install the compressed coil spring on the strut.
11. Install the dust cover, bump rubber, cup assembly, upper bushing, collar, upper spring pad and bracket assembly on the strut.
12. Install the upper bushing and washer on the piston rod.
13. Install a new self-locking nut on the piston rod. Temporarily tighten the nut.
14. Carefully remove the spring compressor from the spring. Torque the self-locking nut to 16 ft. lbs. (25 Nm).
15. Install the upper bracket of the shock to the vehicle. Torque the mounting nuts to 32 ft. lbs. (44 Nm).
16. Raise the suspension up with a jack or adjustable stand to align the shock absorber lower mounting holes.

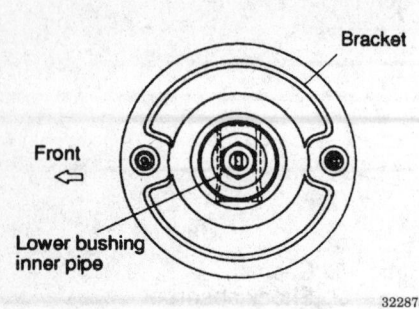

Correct upper bracket installed position — 1995–97 Eclipse

Front ⟸
Bracket
Lower bushing inner pipe

322875

17. Install the lower mounting bolt. Torque the bolt to 71 ft. lbs.
18. Remove the jack or stand and safely lower the vehicle to the floor.
19. Install the cap and service lid.

1993 Galant

NOTE: The strut assembly is a load bearing component, therefore the vehicle chassis and axle weight must be supported separately, requiring the use of two separate lifting devices.

FWD Vehicles

1. Disconnect the negative battery cable.

2. Raise and support vehicle chassis.
3. Raise and support torsion axle and arm assembly slightly. Make sure the jack does not contact the lateral rod.

NOTE: Always use a wooden block between the jack receptacle and the axle beam. Place the jack at the center of the axle beam.

4. Remove the trunk interior trim to gain access to the top mounting nuts.
5. If equipped with Active-ECS, perform the following steps:
  a. Label and disconnect the air tubes from the shock absorber.

NOTE: Immediately plug or cap the lines with tape or similar product to prevent the entry of dirt into the system. Do NOT bend or crimp the air tubes. Plug or cover the air ports on the joint and actuator.

  b. Remove the O-ring and bushing.
  c. Remove the actuator assembly.
6. Remove the top cap and upper shock mounting nuts.

7. Remove the shock lower mounting bolt and remove the assembly from the vehicle.
  To install
8. Position strut assembly so that lower mounting bolt can be installed and lightly tightened.
9. Use jack to raise or lower the axle assembly so that top strut plate studs aligns through body. Raise jack to hold strut assembly in position.
10. Install top plate nuts on studs. Tighten the upper shock mounting nuts to 29 ft. lbs. (40 Nm).
11. With the car on the ground, tighten the lower mounting bolt to 72 ft. lbs. (100 Nm).
12. If equipped with Active-ECS, assemble the actuator components in the following order:
  a. Install the adapter to the mounting bracket.
  b. Using new O-rings, connect the air line to the actuator. Tighten the locknuts to 6 ft. lbs. (9 Nm).
  c. Swab the air connectors with a solution of soapy water. Start the engine, cycle the suspension controls and observe the joints for any sign of air leaks.
13. Install top cap and interior trim.

AWD Vehicles

1. Disconnect the negative battery cable.
2. Raise and support vehicle chassis.
3. Raise and support trailing arm assembly slightly.
4. Remove the trunk interior trim to gain access to the top mounting nuts.
5. Remove the top cap and upper shock mounting nuts.
6. Remove the brake tube bracket bolt.
7. Remove the shock lower mounting bolt and remove the assembly from the vehicle.
  To install
8. Position strut assembly so that lower mounting bolt can be installed and lightly tightened.
9. Use jack to raise or lower the trailing arm, so that top strut plate studs aligns through body. Raise jack to hold strut assembly in position.
10. Install top plate nuts on studs. Tighten the upper shock mounting nuts to 29 ft. lbs. (40 Nm).
11. With the car on the ground, tighten the lower mounting bolt to 65–80 ft. lbs. (90–110 Nm).
12. Connect brake bracket and tighten the mounting bolt to 12 ft. lbs. (17 Nm).
13. Install top cap and interior trim.

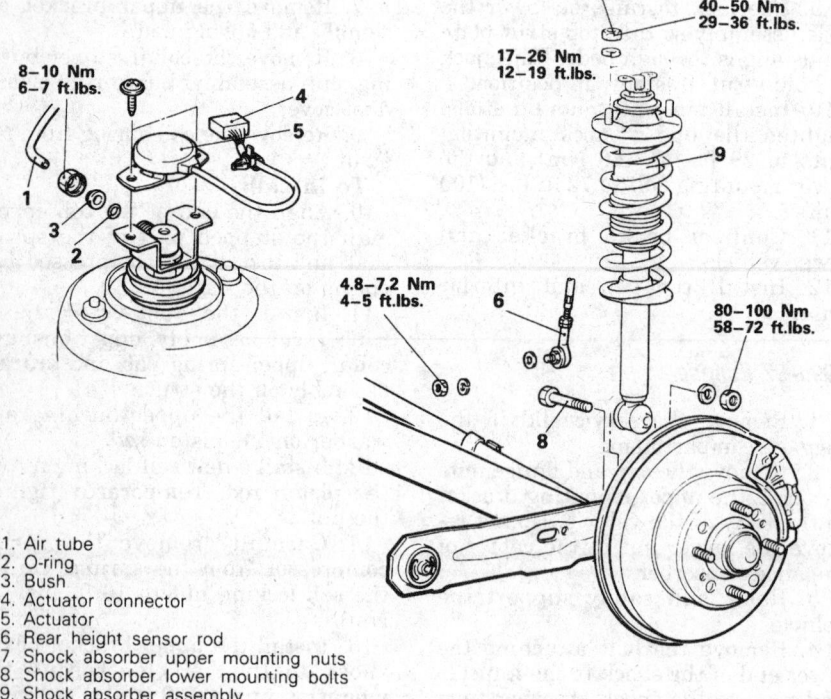

8–10 Nm
6–7 ft.lbs.

7   40–50 Nm
     29–36 ft.lbs.

17–26 Nm
12–19 ft.lbs.

9

4.8–7.2 Nm
4–5 ft.lbs.

6

80–100 Nm
58–72 ft.lbs.

1. Air tube
2. O-ring
3. Bush
4. Actuator connector
5. Actuator
6. Rear height sensor rod
7. Shock absorber upper mounting nuts
8. Shock absorber lower mounting bolts
9. Shock absorber assembly

162179

Active-ECS strut assembly component identification — 1993 Galant

1994–97 Galant

NOTE: The strut assembly is a load bearing component, therefore the vehicle chassis and axle weight must be supported separately, requiring the use of two separate lifting devices.

1. Raise and support the vehicle chassis.
2. Raise and support the lower control arm assembly slightly.
3. In order to gain access to the top mounting nuts, remove the rear seat as follows:
   a. While pulling the rear seat stopper outward, lift the lower cushion upward. Remove the lower cushion.
   b. Remove the seat back mounting bolts.
   c. Lift the seat back upward and remove the seat.
4. Remove the strut upper mounting nuts.
5. Remove the shock lower mounting bolt and remove the assembly from the vehicle.
6. Use a coil spring compressor and compress the coil spring. An air tool should not be used to tighten the spring compressor.
7. Remove the strut cap.
8. While holding the piston rod, remove the self-locking nut.
9. Remove the upper bracket assembly and spring pad.
10. Remove the collar, upper bushing, cup assembly, bump rubber and dust cover.
11. Remove the coil spring from the strut.
   To install
12. Align the end of the coil spring with the stepped part of the spring seat and install the compressed coil spring on the strut.
13. Install the dust cover, bump rubber, cup assembly, upper bushing, collar, upper spring pad and bracket assembly on the strut.
14. Install the upper bushing and washer on the piston rod.
15. Install a new self-locking nut on the piston rod. Temporarily tighten the nut.
16. Carefully remove the spring compressor from the spring. Torque the self-locking nut to 16 ft. lbs. (25 Nm).
17. Install the strut cap.
18. Position the strut assembly so that the lower mounting bolt can be installed and lightly tightened.
19. Use a jack to raise or lower the lower control arm, so that the top strut plate studs align through the

body. Raise the jack to hold the strut assembly in position.
20. Install the top plate nuts on the studs and tighten the mounting nuts to 32 ft. lbs. (44 Nm).
21. With the car on the ground, tighten the lower mounting bolt to 71 ft. lbs. (98 Nm).
22. Install the rear seat back and cushion.

Diamante

NOTE: For rear strut replacement the vehicle chassis and axle weight must be supported separately, requiring the use of two separate lifting devices.

1. Disconnect the negative battery cable.
2. Raise and properly support vehicle. Remove both rear wheels.
3. Working on one side at a time, lightly jack under the trailing arm for support.
4. Matchmark the positioning of the upper spring plate to the vehicle for reinstallation purposes.
5. If equipped with electronic control suspension (ECS) perform the following:
   a. Loosen the nut that secures the air line to the to the top of the strut and discard the O-ring.
   b. Remove the bolts that secure the actuator to the top of the strut and remove the component. Disconnect the wiring harness.
6. Remove the strut lower mounting bolt and remove the two nuts that secure the strut upper plate to the vehicle.
7. Lower the support jack and remove the strut from the vehicle.
   To install
8. Position the upper spring plate and install the strut. Use the support jack to assist with installation.
9. Tighten the upper strut mounting nuts to 33 ft. lbs. (45 Nm).
10. Tighten the lower strut mounting bolt to 65 ft. lbs. (90 Nm).
11. If equipped with electronic control suspension (ECS) perform the following:
   a. Using a new O-ring, tighten the nut that secures the air line to the to the top of the strut to 84 inch lbs. (9 Nm).
   b. Install the actuator to the top of the strut and secure with mounting bolts. Connect the wiring harness.
12. Remove the support jack, install wheels and lower vehicle.
13. Connect the negative battery cable.

3000GT

1. Disconnect the negative battery cable and wait one minute for the air bag to disarm before working on the vehicle.
2. Raise and safely support the vehicle.
3. Remove the rear side trim in the luggage compartment and remove the ECS connector and cap.
4. Support the suspension and remove the upper strut mounting bolts.
5. Remove the wheel and tire assembly and the lower strut mounting bolt.
6. Remove the strut from the vehicle.
   To install:
7. Position the strut in the trailing arm and torque the lower mounting bolt to 72 ft. lbs. (100 Nm).
8. Guide the upper mounting studs through the body and torque the upper mounting nuts to 33 ft. lbs. (45 Nm).
9. Install the cap and the ECS connector.
10. Install the wheel and tire assembly and connect the negative battery cable.
11. Lower the vehicle to the floor.

Shock Absorber

REMOVAL AND INSTALLATION

1993–94 Precis

1. Raise the vehicle and support it safely.
2. Remove the wheel.
3. Jack up the control arm slightly, and remove the upper mounting bolt and nut.
4. Remove the lower mounting bolt and nut.
5. Remove the shock absorber.
6. Check the shock for:
   a. Excessive oil leakage, some minor weeping is permissible;
   b. Bent center rod, damaged outer case, or other defects.
   c. Pump the shock absorber several times, if it offers even resistance on full strokes it may be considered serviceable.
   To install:
7. Install the upper shock mounting nut and bolt. Hand-tighten the nut.
8. Install the bottom eye of the shock in the mounting bracket. Tighten the nut to 47–58 ft. lbs. (64–78 Nm).
9. Tighten the upper fasteners to 47–58 ft. lbs. (64–78 Nm).

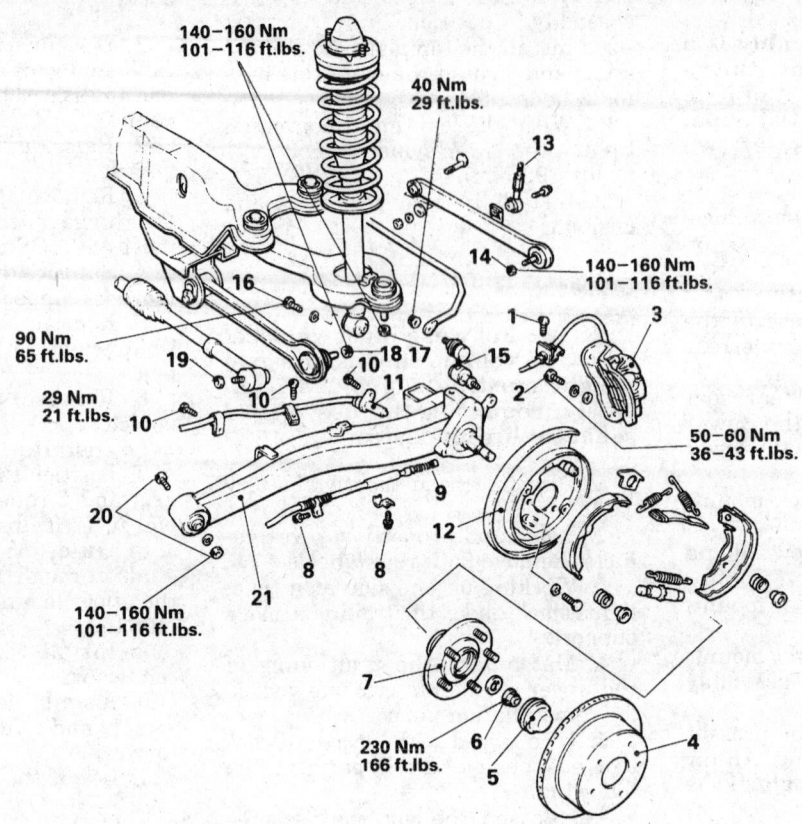

140–160 Nm
101–116 ft.lbs.

40 Nm
29 ft.lbs.

13

14

140–160 Nm
101–116 ft.lbs.

90 Nm
65 ft.lbs.

16

19

29 Nm
21 ft.lbs.

10

10   18 17

11

10

1

2

15

3

50–60 Nm
36–43 ft.lbs.

20

8   8

9

12

21

140–160 Nm
101–116 ft.lbs.

7

6

5

230 Nm
166 ft.lbs.

4

1 Brake line clamp bolt
2 Brake caliper mounting bolts
3 Brake caliper
4 Brake disk
5 Hub cap
6 Flange nut
7 Hub assembly
8 Bolt
9 Parking brake cable end
10 ABS speed sensor clamp bolts
   <Vehicles equipped with ABS>
11 ABS speed sensor
   <Vehicles equipped with ABS>
12 Backing plate

13 ECS height sensor rod
   <Vehicles equipped with ABS>
14 Self lock nut
15 Stabilizer link
16 Shock absorber mounting bolt
17 Self lock nut
18 Self lock nut
   <Vehicles not equipped with 4WS>
19 Power cylinder tie-rod connection
   nut <Vehicles equipped with 4WS>
20 Trailing arm mounting bolt and nut
21 Trailing arm

326621

Exploded view of rear suspension — Diamante

10. Remove the jack from under the control arm.

11. Install the wheel and tire assemblies.

12. Lower the vehicle to the floor.

Coil Spring

REMOVAL AND INSTALLATION

1993–94 Precis

1. Raise the vehicle and support it safely.

2. Remove the rear wheels.

3. Support the rear suspension arm with a floor jack. Then, remove the lower shock absorber attaching bolt, nut and lock washer.

4. Slowly, lower the jack just to the point where the spring can be removed and remove the spring.

To install:

5. If the spring is being replaced, transfer the spring seat to the new spring.

6. When installing the coil spring, make sure the smaller diameter is upward. Make sure the spring identification and load markings match up.

7. Install and torque the lower shock mounting nut/bolt to 47–58 ft. lbs. (64–78 Nm).

8. Install the rear wheels.

9. Lower the vehicle.

3000GT and Diamante

1. Disconnect the negative battery cable.

────── **CAUTION** ──────
Wait at least 90 seconds after the negative (-) battery cable is disconnected to prevent possible deployment of the air bag.
───────────────────────

2. Raise and safely support the vehicle.

3. Remove the wheel and tire assembly.

4. Remove the rear strut assembly from the vehicle by loosening the bracket fasteners, not the center piston rod nut.

—————— **CAUTION** ——————

Do not remove the piston rod nut until the coil spring is properly compressed and secure.

5. Using a suitable coil spring compressor, compress the coil spring and remove the piston rod nut and the upper bracket assembly.

6. Remove the coil spring.

7. If a new spring is being installed, slowly and evenly release the spring from the spring compressor and compress the new spring in the compressor.

To install:

8. Position the compressed coil spring on the strut and install the upper bracket assembly and new locknut.

9. Slowly and evenly release the coil spring.

10. Install the strut assembly in the vehicle.

11. Install the wheel and tire assembly and safely lower the vehicle to the floor.

12. Connect the negative battery cable.

Upper Control Arms

REMOVAL AND INSTALLATION

1993 Galant and 1993–94 Eclipse

1. Disconnect the negative battery cable. Raise and safely support vehicle. Remove the tire and wheel assembly.

2. Support the rear lower control arm. Remove the brake line clamp bolt.

3. Remove the nut and separate the upper ball joint, using tool MB990635 or equivalent, from the rear trailing arm/steering knuckle.

NOTE: It is important to use proper method of joint separation when reusing joint. Damage can result from unapproved methods, resulting in possible joint failure.

4. Matchmark the eccentric on the upper installation bolt and remove from the control arm.

5. Remove the upper arm from the vehicle.

To install:

6. Install the arm to the vehicle and install the upper arm installation bolt. Align the matchmarks and tighten the nut snugly only.

7. Install the upper arm ball joint to the rear spindle assembly and install new nut. Tighten to 52 ft. lbs. (72 Nm) torque.

8. Install the tire and wheel assembly.

9. Lower the vehicle until the suspension supports its weight. Tighten the upper arm installation bolt to 116 ft. lbs. (160 Nm).

10. Check the rear wheel alignment.

1995–97 Eclipse and 1994–97 Galant

1. Raise and safely support the vehicle.

2. Remove the wheel and tire assembly.

3. Remove the through-bolt securing the upper arm to the knuckle.

4. Remove the four bolts securing the upper arm to the vehicle.

5. Remove the upper arm assembly.

6. Remove the through-bolts and nuts and remove the upper arm to body brackets.

To install:

7. Install the upper arm to body brackets to the upper arm. Torque the bolts and nuts to 41 ft. lbs. (57 Nm).

8. Install the upper arm to the vehicle and torque the four bolts to 28 ft. lbs. (39 Nm).

9. Install the through-bolt securing the upper arm to the knuckle. Do not tighten the nut until the vehicle is on the floor at normal riding height.

10. Install the wheel and tire assembly.

11. Safely lower the vehicle to the floor and torque the nut for the through-bolt to 71 ft. lbs. (98 Nm).

12. Check and adjust wheel alignment if necessary.

Diamante

1. Raise and properly support vehicle. Remove appropriate wheel assembly.

2. Disconnect the sway bar and remove the strut assembly.

3. Support trailing arm assembly and remove the self-locking nut connecting the control arm to the trailing arm.

4. Using a joint separation tool, disconnect the upper ball joint from the trailing arm.

5. Disconnect the control arm from the subframe and remove the assembly.

To install

6. Install the control arm to the subframe and lightly tighten the mounting bolt.

7. Connect the control arm to the trailing arm and tighten the self-locking nut to 54–61 ft. lbs. (75–89 Nm).

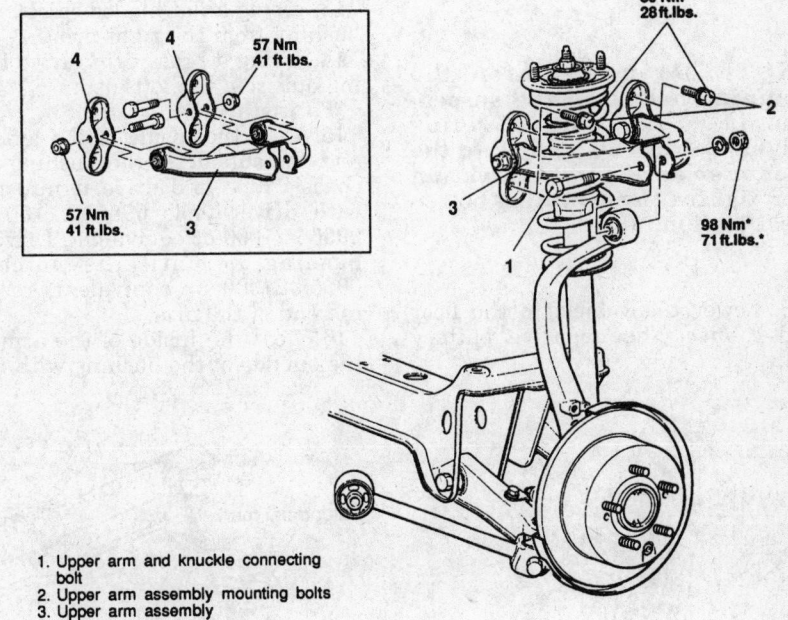

1. Upper arm and knuckle connecting bolt
2. Upper arm assembly mounting bolts
3. Upper arm assembly
4. Upper arm bracket

Caution
* : Indicates parts which should be temporarily tightened, and then fully tightened with the vehicles on the ground in the unladen condition.

323000

Upper control arm and related components — 1995–97 Eclipse

8. Install the strut assembly and connect the sway bar.

9. Install wheel and lower vehicle.

10. With full weight of vehicle on the ground, torque the control arm to subframe bolt to 54–61 ft. lbs. (75–89 Nm).

11. Check rear wheel alignment and adjust if necessary.

3000GT

1. Disconnect the negative battery cable and wait one minute for the air bag to disarm before working on the vehicle.

2. Raise and safely support the vehicle.

3. Remove the wheel and tire assembly.

4. Remove the self-locking nut on the upper control arm and using the special tool, separate the ball stud from the knuckle.

5. Remove the upper arm mounting bolt and nut.

6. Remove the upper control arm from the vehicle.

To install:

7. Position the upper arm on the vehicle and torque the mounting bolt to the knuckle to 54–64 ft. lbs. (75–89 Nm) and the bolt to the body to 101–116 ft. lbs. (140–160 Nm).

NOTE: Do not tighten the mounting bolt until the suspension is at the normal riding height. Lower the vehicle to the floor or on an alignment rack and then tighten the mounting bolt to specification.

8. Lower the vehicle to the floor and connect the negative battery cable.

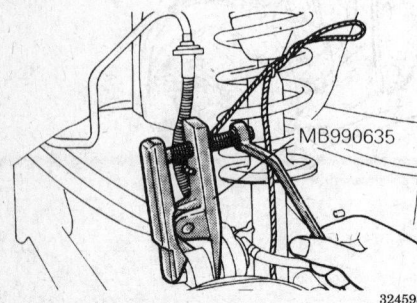

Removing the ball stud from the knuckle — 3000GT

Lower Control Arms

REMOVAL AND INSTALLATION

1993–94 Precis

1. Raise the vehicle and support it safely.

2. Remove the muffler assembly.

3. Raise the suspension arm assembly slightly and keep in this position.

4. Disconnect the parking brake cable from the arm.

5. Remove the shock absorber.

6. Disconnect the brake hoses from their clips on the suspension members.

7. Lower the suspension slightly and carefully remove the coil spring.

8. Remove the rear suspension from the vehicle as an assembly.

9. Before removing any fixtures from the suspension arm matchmark all parts for assembly reference; this is extremely important! If equipped with stabilizer bars, make a mark on the bar in line with the punch mark on the bracket.

10. Remove the dust cover clamp.

11. Remove the nuts securing the control arms and pull them apart. Leave the dust cover attached to the right arm.

12. Remove the rubber stopper from the right arm.

13. Using a flat bladed chisel, drive bushing from the right arm.

14. Using a brass drift, drive bushing out from the left arm.

To install:

15. Coat the inside of the left arm and the outside of the bushing with chassis lube and drive it into place with driver tools 09555-21100 and 09555-21000 or equivalent. Drive the bushing in until the notch on 09555-21000 or equivalent, reaches the end of the arm.

16. Coat the inside of the arm and the outside of the bushing with chas-

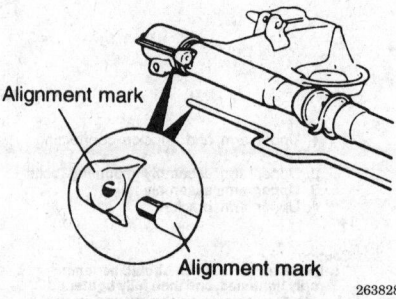

Mark the sway bar for proper installation — 1993–94 Precis

sis lube and drive it into the arm until it is fully seated.

17. Install the dust cover to the center position of the right arm, about 400mm.

18. Apply chassis lube to the surface of the right arm and install the rubber stopper.

19. Align all matchmarks, including the stabilizer bar and slowly push the suspension halves together.

20. Install all remaining bushing, washers and attaching parts.

NOTE: The toothed sides of the washers face the bushings.

21. Install the end nuts and torque them loosely at this time.

22. Jack the assembly into position and torque the suspension-to-body bolts to 50 ft. lbs. (68 Nm).

23. Install the coil springs and loosely install the shock absorbers.

24. Install the rear brake assembly and related components.

25. Attach the parking brake cable and brake hoses to their clips on the suspension arm.

26. Install the tire and wheel assemblies.

27. Lower the vehicle and tighten the suspension arm end nuts to specifications. Tighten the shock mounting bolts to 47–58 ft. lbs. (63–79 Nm).

28. Adjust the rear brake shoe clearance.

29. Install the wheels and tires.

30. Lower the vehicle to the floor.

1993 Galant and 1993–94 Eclipse

1. Disconnect the negative battery cable.

2. Raise the vehicle and support safely.

3. Remove sway bar links or mounting nuts and bolts from lower control arm. Remove the joint cups and bushings, if equipped.

4. Disconnect the ball joint stud from the steering knuckle, using tool MB990635 or equivalent.

NOTE: It is important to use proper method when separating joints. Damage to joint could occur, resulting in possible failure.

5. Remove the inner lower arm mounting bolts and nut.

6. Remove the rear mount bolts. Remove the rear retainer clamp if equipped.

7. Remove the arm from the vehicle.

8. Remove the rear rod bushing, if service is required.

To install:

9. Assemble the control arm and bushing. Install the control arm to

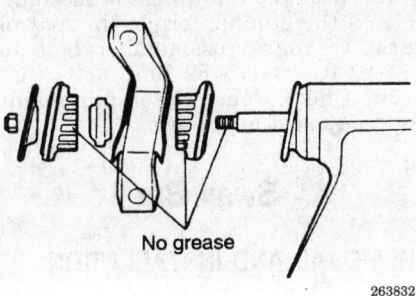

No grease

263832

Install the mounting fixtures in the original positions — 1993–94 Precis

the vehicle and install the inner mounting bolts. Install new nut and snug temporarily.

10. Install the rear mount clamp, bolts and replacement nuts. Torque the clamp mounting nuts to 34 ft. lbs. (47 Nm). Temporarily tighten the clamp mounting bolt. Once the weight of the vehicle is on the suspension, the bolt will be tightened to 72 ft. lbs. (100 Nm).

11. Connect the ball joint stud to the knuckle. Install a new nut and torque to 43–52 ft. lbs. (60–72 Nm).

12. Install the sway bar and links.

13. Lower the vehicle to the floor for the final torquing of the inner frame mount bolt.

14. Once the full weight of the vehicle is on the suspension, torque the inner lower arm mounting bolt nuts to 87 ft. lbs. (120 Nm). Tighten the inner clamp mounting bolt to 72 ft. lbs. (100 Nm).

15. Inspect all suspension bolts, making sure they all have been fully tightened.

16. Connect the negative battery cable.

1995–97 Eclipse

1. Raise and safely support the vehicle.

2. Remove the wheel and tire assembly.

3. If equipped with a lower arm cover, remove the cover.

4. Disconnect the stabilizer link from the lower arm.

5. If equipped with ABS, remove the wheel speed sensor clamp bolts.

6. Remove the through-bolt securing the lower arm to the knuckle.

7. Remove the through-bolt securing the lower arm to the suspension crossmember.

8. Remove the lower control arm.

To install:

9. Install the lower arm to the suspension crossmember and the knuckle but do not tighten the

through-bolts until the vehicle is on the floor at normal riding height.

10. Install the ABS wheel speed sensor clamp bolts to the lower arm, if removed.

11. Connect the stabilizer link to the lower arm.

12. Install the lower arm cover if removed.

13. Install the wheel and tire assembly.

14. Safely lower the vehicle to the floor and torque the through-bolts to 71 ft. lbs. (98 Nm).

1994–97 Galant

Lower Control Arm

1. Raise and support the vehicle safely.

2. Remove the appropriate wheel assembly.

3. If equipped with ABS, disconnect the speed sensor harness brackets, from the lower control arm.

4. Disconnect the stabilizer bar link from the lower control arm.

5. Remove the through-bolt, connecting the knuckle assembly to the lower control arm.

6. Remove the mounting bolt connecting the lower control arm to the suspension crossmember.

7. Remove the lower control arm from the vehicle.

To install:

NOTE: The control arm mounting bolts must not be fully tightened until the full weight of the vehicle is on the ground.

8. Install the control arm to the suspension crossmember and temporarily tighten the mounting bolt.

9. Connect the knuckle to the lower control arm and lightly tighten the through-bolt.

10. Connect the stabilizer bar link to the control arm and torque the nut to 28 ft. lbs. (39 Nm).

11. Install the wheels and lower the vehicle to the floor.

12. Once the full weight of the vehicle is on the suspension, torque the lower arm mounting bolt nuts to 71 ft. lbs. (98 Nm).

13. Check rear wheel alignment and adjust if necessary.

Toe Control Lower Arm

The lower ball joint is integral with the lower toe control arm. They are removed and replaced as an assembly.

1. Raise and support the vehicle safely.

2. Remove the appropriate wheel assembly.

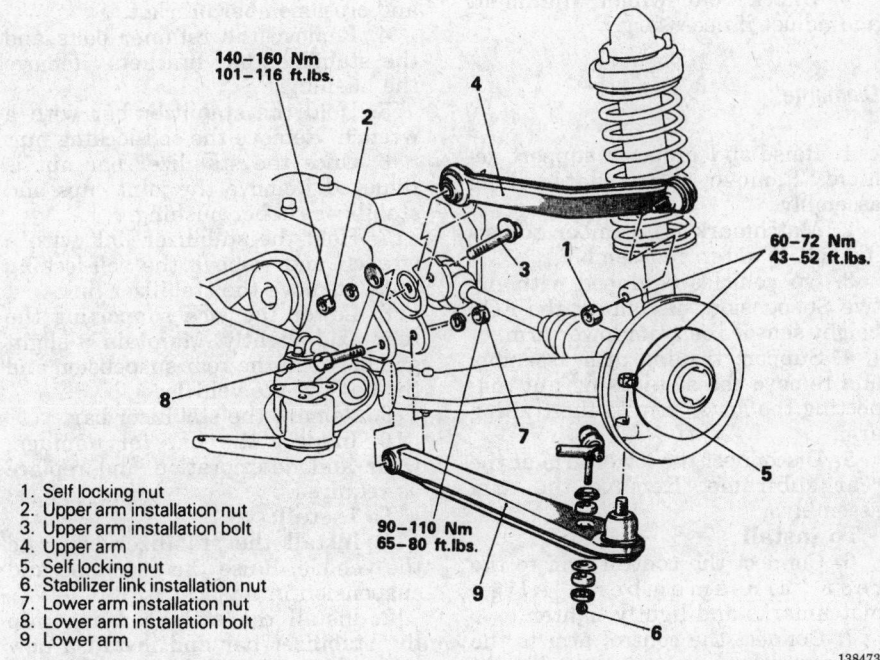

140–160 Nm
101–116 ft.lbs.

60–72 Nm
43–52 ft.lbs.

90–110 Nm
65–80 ft.lbs.

1. Self locking nut
2. Upper arm installation nut
3. Upper arm installation bolt
4. Upper arm
5. Self locking nut
6. Stabilizer link installation nut
7. Lower arm installation nut
8. Lower arm installation bolt
9. Lower arm

138473

Rear suspension exploded view — AWD Galant and Eclipse shown, 3000GT similar

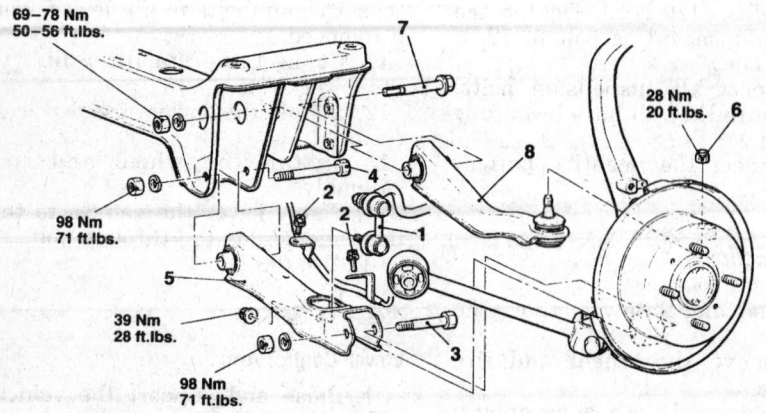

**69–78 Nm
50–56 ft.lbs.**

**98 Nm
71 ft.lbs.**

**39 Nm
28 ft.lbs.**

**98 Nm
71 ft.lbs.**

**28 Nm
20 ft.lbs.**

1. Stabilizer link
2. ABS speed sensor clamp bolts <Vehicles with ABS>
3. Lower arm assembly and knuckle connection
4. Lower arm assembly mounting bolt
5. Lower arm assembly

6. Connection for toe control arm ball joint and knuckle
7. Toe control arm assembly mounting bolt
8. Toe control arm assembly

326004

Lower control arms and related components — 1994–97 Galant

3. Matchmark the control arm adjusting bolt to aid in reassembly.

4. Using joint separator MB991113, disconnect the ball joint stud from the steering knuckle.

5. Remove the mounting bolts connecting the lower control arm to the suspension crossmember.

To install:

6. Connect the control arm to the suspension crossmember. Align the matchmarks on the adjustment bolt and lightly tighten the bolt.

7. Connect the ball joint stud to the knuckle and torque the nut to 20 ft. lbs. (28 Nm).

8. Install the wheels and lower the vehicle to the floor.

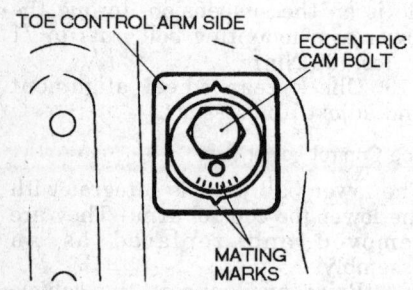

TOE CONTROL ARM SIDE

ECCENTRIC CAM BOLT

MATING MARKS

326005

Removal and installation of the toe control arm — 1994–97 Galant

9. With the full weight of the vehicle is on the ground, torque the control arm through-bolt to 50–56 ft. lbs. (69–78 Nm).

10. Check rear wheel alignment and adjust if necessary.

Diamante

1. Raise and properly support vehicle. Remove appropriate wheel assembly.

2. Matchmark the camber adjusting bolt to aid in reassembly.

3. On vehicles equipped with Active Suspension, disconnect the ECS height sensor rod from lower arm.

4. Support trailing arm assembly and remove the self-locking nut connecting the lower arm to the trailing arm.

5. Disconnect the lower arm at the rear subframe. Remove the arm assembly.

To install

6. Connect the control arm to the rear crossmember. Align matchmarks and lightly tighten.

7. Connect the control arm to the trailing arm and torque the self-locking nut to 54–61 ft. lbs. (75–89 Nm).

8. Connect the ECS height sensor rod to the lower arm, if equipped.

9. Install wheel and lower vehicle.

10. With the full weight of the vehicle on the ground, torque the control arm to rear crossmember bolt to 54–61 ft. lbs. (75–89 Nm).

11. Check rear wheel alignment and adjust if necessary.

Sway Bar

REMOVAL AND INSTALLATION

1993–94 Precis

1. Raise and safely support the vehicle.

2. Remove the rear wheels.

3. Remove the rear axle assembly.

4. Mark the sway bar for proper installation.

5. Pull apart the suspension arms slightly and remove the sway bar from the axle assembly.

To install:

6. Install the sway bar on the axle assembly.

7. Install the axle assembly into the vehicle.

8. Install the rear wheels, and lower the vehicle to the floor.

1993 Galant and 1993–94 Eclipse

1. Raise and support the vehicle safely.

2. Place a jack under the rear axle and suspension assembly.

3. Remove the self-locking nuts and crossmember bracket.

4. Remove the retainer bolts and the stabilizer bar brackets. Remove the bushing.

5. Hold the stabilizer bar with a wrench. Remove the self-locking nut.

6. Once the stabilizer bar nut is removed, remove the joint cups and stabilizer rubber bushing.

7. Hold the stabilizer link with a wrench and remove the self-locking nuts. Remove the stabilizer link.

8. Lower the jack supporting the rear axle slightly. Maintain a slight gap between the rear suspension and the body of the vehicle.

9. Remove the stabilizer bar.

10. Inspect the bar for damage, wear and deterioration and replace as required.

To install:

11. Install the stabilizer bar into the vehicle. Raise the rear axle and suspension into place.

12. Install the stabilizer link into the stabilizer bar and install a new self-locking nut. Tighten the nut to 33 ft. lbs. (45 Nm).

13. Install the joint cups and stabilizer rubber to the link. Install a new self-locking nut onto the link. While

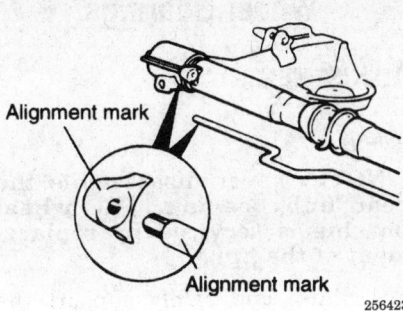

Mark the sway bar for installation — 1993–94 Precis

holding the stabilizer link ball studs with a wrench, tighten the self-locking nut so the protrusion of the stabilizer link is within 0.354–0.433 in. (9–11mm).

14. Install the center stabilizer bar bushings, brackets and bolts. Tighten the bolts to 10 ft. lbs. (14 Nm).

15. Install the parking brake cable and rear speed sensor installation bolt.

16. Install the crossmember bracket and tighten the bolt to 61 ft. lbs. (85 Nm). Tighten the crossmember bracket mounting nut to 94 ft. lbs. (130 Nm).

17. Install the rubber insulators and new self-locking nuts onto the

crossmember brackets. Tighten the nuts to 80–94 ft. lbs. (110–130 Nm).

18. Lower the vehicle.

1995–97 Eclipse

1. Raise and safely support the vehicle.

2. Remove the rear wheels.

3. If equipped with lower control arm covers, remove them.

4. Remove the stabilizer link mounting nuts and remove the link.

5. Remove the bolts securing the stabilizer bar brackets.

6. Remove the stabilizer bar from the vehicle.

To install:

7. Install the stabilizer bar so the identification mark is on the left and install the bushing and brackets. Torque the mounting bolts to 7–10 ft. lbs. (9–14 Nm).

8. Install the stabilizer links and torque the nuts to 28 ft. lbs. (39 Nm).

9. Install the lower control arm cover if removed.

10. Install the wheels and lower the vehicle to the floor.

1994–97 Galant and Diamante

1. Disconnect the negative battery cable.

2. Raise and safely support the vehicle.

3. Disconnect the stabilizer links by removing the self-locking nuts.

4. Remove the stabilizer bar mounting brackets and bushings.

5. Remove the bar from the vehicle.

6. Inspect all components for wear or damage, and replace parts as needed.

To install:

7. Install the stabilizer bar into the vehicle.

8. Loosely install the stabilizer bar brackets on the vehicle.

9. Align the side locating markings on the stabilizer bar, so that the marking on the bar, extends approximately 0.39 inches (10 mm) from the outer edge of the mounting bracket, on both sides.

10. With the stabilizer bar properly aligned, tighten the mounting bracket bolts to 28 ft. lbs. (39 Nm).

11. Connect the stabilizer links to the damper fork and the stabilizer bar. Tighten the locking nuts to 28 ft. lbs. (38 Nm).

12. Lower the vehicle and connect the negative battery cable.

3000GT

1. Disconnect the negative battery cable and wait one minute for the air bag to disarm before working on the vehicle.

2. Remove both wheel and tire assemblies.

3. Raise and safely support the vehicle.

4. Remove the self-locking nuts securing the sway bar link to the sway bar and the control arm.

5. Disconnect the power cylinder tie rod ends from the trailing arms.

6. Remove the lower shock absorber mounting bolts.

7. Remove the parking brake cable mounting bolt.

8. Remove the 4WS (Four Wheel Steering) pipe bracket bolts and the power cylinder brackets.

9. Support the rear suspension assembly and remove the crossmember brackets and mounting nuts.

10. Remove the sway bar brackets and remove the sway bar.

To install:

11. Position the sway bar on the crossmember and install the brackets.

12. Install the crossmember brackets and nuts and torque to 29 ft. lbs. (40 Nm).

13. Install the power cylinder brackets and the 4WS pipe brackets and bolts.

14. Install the parking brake cable mounting bolt.

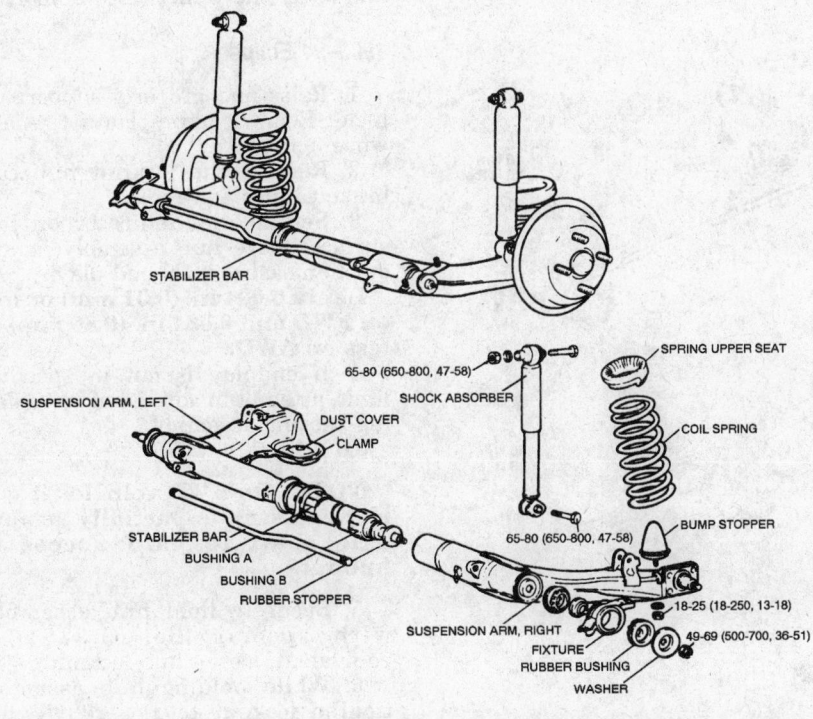

Rear axle assembly — 1993–94 Precis

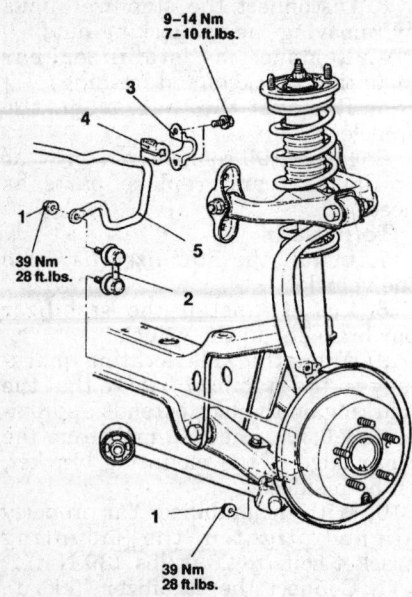

1. Stabilizer link mounting nuts
2. Stabilizer link
3. Stabilizer bar brackets
4. Bushing
5. Stabilizer bar

322944

Stabilizer bar and related components — 1995–97 Eclipse

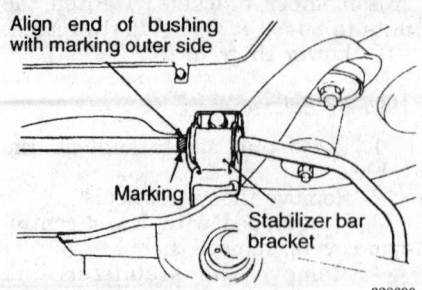

328690

Aligning the marks with the sway bar brackets — FWD 3000GT

15. Install the lower shock absorber mounting bolts.

16. Using new self-locking nuts, install the sway bar links. Make sure that the link protrudes .197–.276 in. (5–7 mm) on the lower arm side.

17. Install the wheels and lower the vehicle to the floor.

18. Connect the negative battery cable.

1. Stabilizer link mounting nuts
2. Bolts
3. Stabilizer bar brackets
4. Bush
5. Stabilizer bar

326865

Rear sway bar assembly and related components — Diamante

Wheel Bearings

ADJUSTMENT

Mirage

NOTE: Never disassemble the rear hub bearing. The wheel bearing is serviced by replacement of the hub.

1. Raise and safely support the vehicle.

2. Remove the rear wheel.

3. Remove the caliper and brake disc or brake drum.

4. Remove the dust cap and torque the flange nut to 130 ft. lbs. (180 Nm).

5. Using a dial indicator, measure wheel bearing end-play. The maximum limit for end-play is 0.0020 inches (0.05mm).

6. Using a spring scale and a rope wrapped around the bolts, measure the rotary sliding resistance of the bearing/hub. The maximum limit for resistance is 4 lbs. (19 N).

7. If any of the readings exceed the specifications, replacement of the hub is required.

8. Install the dust cap.

9. Install the brake disc and caliper, or brake drum.

10. Install the rear wheel assembly and lower the vehicle to the floor.

1993–94 Eclipse

1. Raise and properly support vehicle. Remove appropriate tire and wheel assembly.

2. Remove the locknut hub cap, brake caliper and rotor.

3. Set up dial indicator on hub surface; move hub assembly in and out to measure total end play. **Limit: 0.004 in. (0.01 mm) or less on FWD and 0.031 in. (0.80 mm) or less on AWD.**

4. If endplay is not in specified limit, attempt to adjust before replacing bearing assembly. **To adjust**

NOTE: On AWD vehicles it will be necessary to partially remove axle shaft to gain access to hubnut.

5. Securely hold hub assembly, with special tool C-3281 or equivalent, loosen hub locknut.

6. While holding hub assembly tighten locknut to 144–188 ft. lbs. (200–260 Nm).

7. Recheck endplay, if still not within specifications, replace bearing assembly.

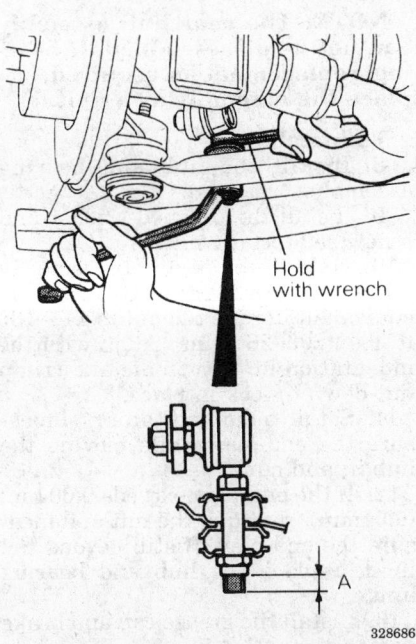

Sway bar link installation — AWD 3000GT

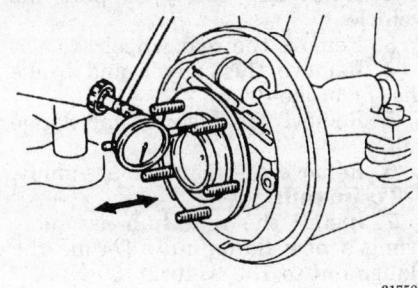

Inspecting wheel bearing end play — 1995–97 Eclipse

1995–97 Eclipse

The wheel bearing play is not adjustable. If the wheel bearing play is not within specifications, the hub assembly must be replaced.

FWD Vehicles

1. Release the parking brake and remove the brake drum.
2. If equipped with rear disc brakes, remove the caliper assembly and the brake disc (rotor).
3. Place a dial gauge against the hub surface; then move the hub in the axial direction and check whether or not there is end play. Specification is 0.002 in (0.05 mm).
4. To check for rotary sliding resistance: turn the hub a few times to seat the bearing. Wind a rope around the hub bolts and turn the hub by pulling at a 90° angle with a spring balance. Measure to determine whether or not the rotary sliding resistance of the rear hub is at the limit value. Specification is: 3.9 lbs. (18 N) or less.

AWD Vehicles

1. Release the parking brake and remove the brake drum.
2. If equipped with rear disc brakes, remove the caliper assembly and the brake disc (rotor).

3. Place a dial gauge against the hub surface; then move the hub in the axial direction and check whether or not there is end play. Specification is 0.002 in (0.05 mm).

1993 Galant

NOTE: Vehicles equipped with rear disc brakes use a sealed hub and bearing assembly, which requires no adjustment. Drum brake models are adjusted using the following procedure.

1. Raise the vehicle and support it safely.
2. Remove the wheel and tire assemblies.
3. If equipped with rear disc brakes, remove the caliper assembly.
4. Remove the grease cap and the hub nut.
5. Tighten the wheel bearing nut to 20 ft. lbs. (27 Nm) while rotating the drum/hub.
6. Back off the adjusting nut to remove the preload and then tighten it to 7 ft. lbs. (10 Nm).
7. Install the nut lock and a new cotter pin.
8. If brake caliper was removed, reinstall.
9. Install the wheel and lower the vehicle.

Diamante

Measure the wheel bearing end-play using a dial indicator. The end-play should be 0.002 inches (0.05mm) or less with the wheel bearing locknut torqued to specifications. The wheel bearings are sealed units and are not adjustable. If defective, replacement is the only option.

REMOVAL AND INSTALLATION

1993–94 Precis

1. Safely raise and support the rear of the vehicle.

2. Remove the tire and wheel assembly.
3. Remove the grease cap, cotter pin, serrated nut cap, axle shaft nut and washer from the spindle.
4. Pull outward on the brake drum slightly to remove the outer wheel bearing.
5. Slide the drum down the spindle and remove assembly from the vehicle.
6. Pry the inner grease seal from the rear hub of the drum and discard.
   To install:
7. Coat the new races with wheel bearing grease and drive them into the hub, making sure they are fully and squarely seated.
8. Pack the hub cavity with wheel bearing grease.
9. Install the inner bearing and drive a new grease seal into place. Make sure to pack the bearings completely with grease prior to installing into the drum.
10. Install the brake drum onto the spindle. Install the outer bearing, washer and shaft nut onto spindle.
11. Install and tighten bearing locknut as follows:
   a. Prior to installation, inspect the rear bearing nut by threading the nut onto the spindle until the distance between the shoulder of the spindle and the inner flat on the nut is 0.07–0.11 inch (2.0–3.0mm).
   b. Install the brake drum and outer bearing onto the spindle. Install and torque the nut to 108–145 ft. lbs. (150–200 Nm).
   c. Check for correct bearing endplay by placing a dial indicator on the hub surface and moving the drum outward. Note the movement of the gauge and compare to the desired reading of 0.0043 in. or less (0.11mm or less). If end-play exceeds the desired reading, retighten the rear hub bearing nut and recheck end-play. If reading is still excessive, replace the hub assembly.
12. If end-play is correct, check the starting torque by attaching a spring balance to the hub lug bolts and pulling at a 90 degree angle while noting the required force to turn the hub. If the torque required is above the desired reading of 4.9 lbs. or less (22 N or less), loosen the nut and again tighten to the desired torque. Recheck the starting torque. If torque is still above the desired reading, replace the rear bearing.
13. After final tightening the wheel bearing nut, align with the spindle's

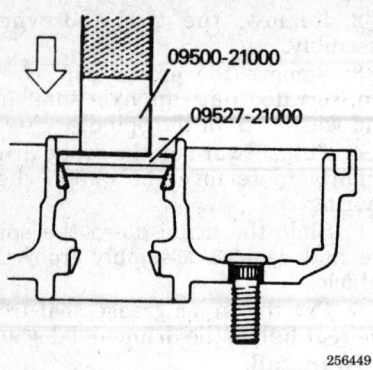

Inner race installation — 1993–94 Precis

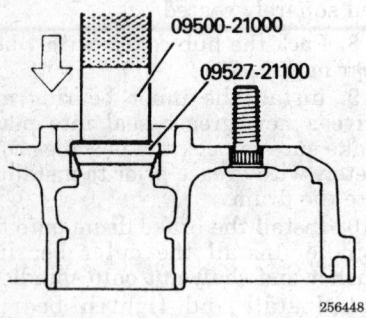

Outer race installation — 1993–94 Precis

indentation and crimp the edge of the nut to wedge in position.

14. Install the tire and wheel assembly, and lower the vehicle.

15. Prior to moving the vehicle, pump the brakes until a firm pedal is obtained.

Mirage

NOTE: Never disassemble the rear hub bearing. The wheel bearing is serviced by replacement of the hub.

1. If equipped with ABS, remove the wheel speed sensor.

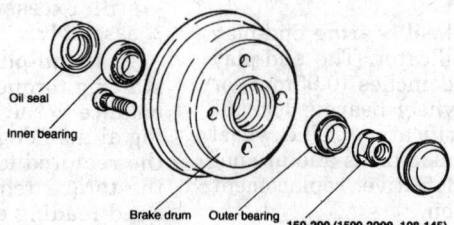

Drum and bearing assembly — 1993–94 Precis

2. Raise and safely support the vehicle.

3. Remove the rear wheel.

4. Remove the caliper and brake disc or brake drum.

5. Remove the dust cap and flange nut.

6. Remove the rear hub assembly.

To install:

7. Install the rear hub assembly using a new flange nut. Torque the flange nut to 130 ft. lbs. (180 Nm).

8. Install the dust cap.

9. Install the wheel speed sensor if removed. The air gap should be 0.012–0.035 in. (0.3–0.9 mm).

10. Install the brake disc and caliper, or brake drum.

11. Install the rear wheel assembly and lower the vehicle to the floor.

1993–94 Eclipse

FWD Vehicles

1. Raise the vehicle and support safely.

2. Remove the tire and wheel assembly.

3. Remove the bolt(s) holding the speed sensor bracket to the knuckle and remove the assembly from the vehicle.

NOTE: The speed sensor has a pole piece projecting from it. This exposed tip must be protected from impact or scratches. Do not allow the pole piece to contact the toothed wheel during removal or installation.

4. Remove the caliper from the brake disc and suspend with a wire.

5. Remove the brake disc.

6. Remove the grease cap, self-locking nut and tongued washer.

7. Remove the rear hub and bearing assembly.

NOTE: The rear hub assembly can not be disassembled. If bearing replacement is required, replace the assembly as a unit.

To install:

8. Install the hub and bearing assembly.

9. Install the tongued washer and a new self-locking nut.

10. Hold hub assembly securely with special tool C-3281 or equivalent, torque the nut to 144–188 ft. lbs. (200–260 Nm). Align with the indentation in the spindle and crimp down to lock in place.

11. Set up a dial indicator and measure the end-play while moving the hub in and out.

12. If the end-play exceeds 0.004 in. (0.01mm), retorque the nut and measure the end-play. If still beyond the limit, replace the hub and bearing unit.

13. Install the grease cap and brake parts.

14. Temporarily install the speed sensor to the knuckle; tighten the bolts only finger-tight.

15. Route the speed sensor cable correctly and loosely install the clips and retainers. All clips must be in their original position and the sensor cable must not be twisted. Improper installation may cause cable damage or system failure.

NOTE: The wiring in the harness is easily damaged by twisting and flexing. Use the white stripe on the outer insulation as a guide to keep the sensor harness properly placed.

16. Use a brass or other non-magnetic feeler gauge to check the air gap between the tip of the pole piece and the toothed wheel. Correct gap is 0.012–0.035 in. (0.3–0.9mm). Tighten the 2 sensor bracket bolts to 10 ft. lbs. (14 Nm) with the sensor located so the gap is the same at several points on the toothed wheel. If the gap is incorrect, it is likely that the toothed wheel is worn or improperly installed.

17. Install the tire and wheel assembly. Be sure to pump brake pedal until firm before moving vehicle

AWD Vehicles

1. Disconnect the negative battery cable.

2. Raise and support the vehicle safely.

3. Remove the tire and wheel assembly from the vehicle.

4. If equipped with ABS, remove the rear wheel speed sensor.

NOTE: Be cautious to ensure that the tip of the pole piece on the rear speed sensor does not come in contact with other parts during removal. Sensor damage could occur.

5. Remove the rear caliper and support assembly out of the way. Remove the brake disc.

6. Remove the driveshaft and companion flange installation bolts, nuts and washers. Move the end of shaft slightly to access the self-locking nut.

7. Using axle holding tool C-3281 or equivalent, secure the rear axle shaft in position, then remove the self-locking nut.

8. Using a slide hammer with hub adapter, remove the rear axle shaft from the trailing arm.

9. If equipped with ABS, remove the rear rotor from the axle assembly using collar and press. The rotor is a press fit.

10. Remove the outer bearing and dust cover concurrently from the axle shaft using a press.

11. Using puller, remove the oil seal and inner bearing from the trailing arm.

12. Inspect the companion flange and axle shaft for wear or damage. Inspect the dust cover for deformation or damage. Inspect the bearings for burning or declaration. Replace components as required.

To install:

13. Using the proper driver, press fit the inner bearing onto the trailing arm. Press fit the oil seal onto the trailing arm with the depression in the oil seal facing upward, and until it contacts the shoulder on the inner arm.

NOTE: When tapping the oil seal in, use a plastic hammer to lightly tap the top and circumfer-

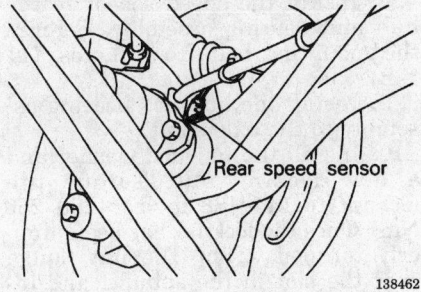

Rear speed sensor

138462

ABS rear wheel sensor location — 1993-94 Eclipse

ence of the seal installation tool, press fitting gradually and evenly.

14. Press fit the dust covers onto the axle until it contacts the axle shaft shoulder. Install the innermost cover so the depression is facing upward.

15. Apply multi-purpose grease around the entire circumference of the inner side of the outer bearing seal lip. Press fit the outer bearing to the axle shaft so that the bearing seal lip surface is facing towards the axle shaft flange.

16. Press fit the rear rotor to the axle shaft with the rear rotor groove surface towards the axle shaft flange.

17. Install the rear axle shaft to the trailing arm temporarily. Install the companion flange to the rear axle shaft, then install a new self-locking nut.

18. While holding the rear axle shaft in position using holding tool, tighten a new self-locking nut to 159 ft. lbs. (220 Nm).

19. Install the driveshaft nuts, washers and bolts. Tighten to 47 ft. lbs. (65 Nm).

20. Install the rear brake disc, caliper assembly and parking brake.

21. Install the tire and wheel assembly and lower the vehicle. Check the parking brake stroke and adjust as required.

22. Before moving the vehicle, pump the brakes until a firm pedal is achieved.

1995–97 Eclipse

The rear wheel bearing is not serviceable. If the wheel bearing must be replaced for any reason, the hub assembly must be replaced.

NOTE: The radio may contain a coded theft protection circuit. Always obtain the code number from the customer before disconnecting the battery.

1. Disconnect the negative battery cable.

2. Raise and safely support the vehicle.

3. Remove the wheel and tire assembly.

4. Remove the rear wheel speed sensor if equipped with ABS.

5. Remove the caliper assembly and rotor, or drum. Suspend the caliper out of the way with wire.

6. On vehicles with rear disc brakes, remove the parking brake shoes.

7. On vehicles equipped with AWD, remove the axle shaft locking

nut, and using a suitable tool, separate the hub from the axle shaft.

8. Remove the hub mounting bolts from behind the backing plate and remove the hub.

NOTE: The rotor for the ABS must be removed and installed using a press.

9. Remove the through-bolt, lock washer and nut, and disconnect the trailing arm.

10. Remove the bolt, washer, and locknut, and disconnect the lower control arm from the knuckle.

11. Remove the locknut and the toe control arm ball joint from the knuckle.

12. Remove the lower strut mounting bolt and disconnect the strut from the knuckle.

13. Remove the through-bolt, washer and locknut and disconnect the upper control arm from the knuckle. Remove the knuckle assembly.

To install:

14. Install the knuckle assembly to the upper control arm with the through-bolt, washer and locknut. Torque is 71 ft. lbs. (98 Nm).

15. Connect the lower strut mount to the knuckle and torque the bolt to 71 ft. lbs. (98 Nm).

16. Install the toe control arm ball joint to the knuckle and torque the mounting locknut to 20 ft. lbs. (28 Nm).

17. Reconnect the lower control arm to the knuckle with the through-bolt, washer, and locknut. Torque the bolt and nut to 71 ft. lbs.

18. Install the lower trailing arm to the knuckle with the bolt, washer, and locknut. Torque the nut and bolt to 85–99 ft. lbs. (118–137 Nm).

19. Press the rotor (ABS) to the hub.

20. On vehicles with AWD, engage the splines of the axle shaft with the hub assembly and torque the axle shaft locking nut to 145–188 ft. lbs. (196–255 Nm).

21. Install the hub and torque the mounting bolts to 54–65 ft. lbs. (74–88 Nm).

22. Install the parking brake shoes if equipped.

23. Install the rotor and caliper or drum.

24. Install the speed sensor if equipped.

25. Install the wheel and tire assembly.

26. Lower the vehicle to the floor.

27. Connect the negative battery cable.

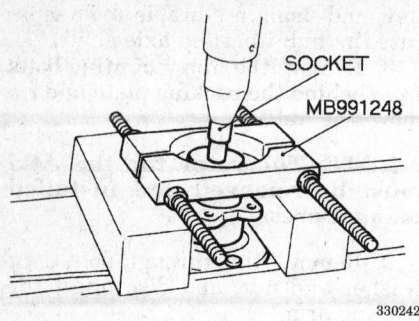

Removal of the speed sensor rotor from the hub — Galant

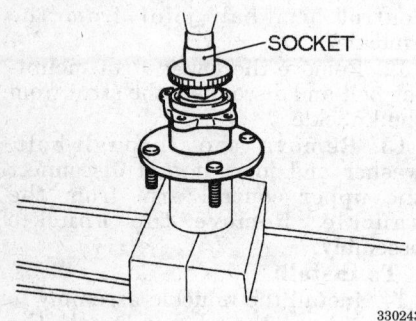

Installation of the speed sensor rotor to the hub — Galant

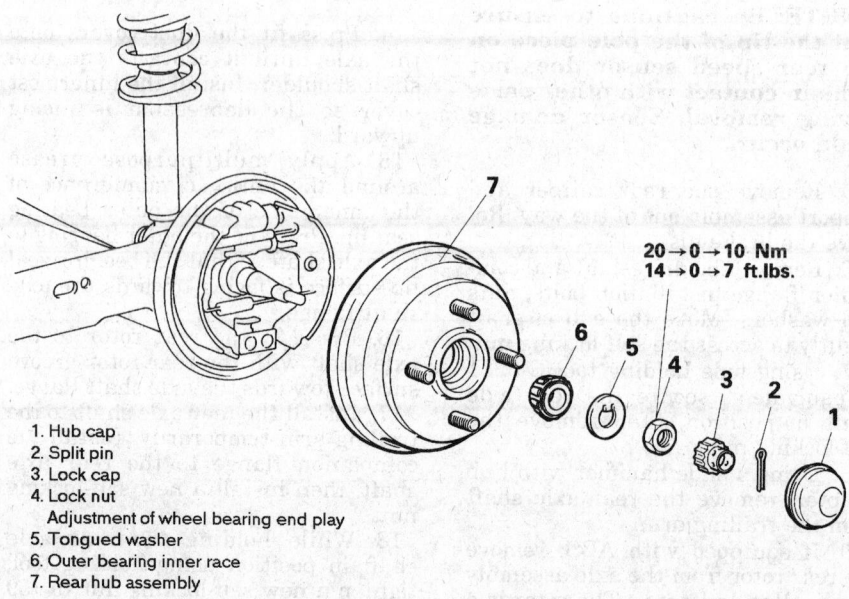

1. Hub cap
2. Split pin
3. Lock cap
4. Lock nut
   Adjustment of wheel bearing end play
5. Tongued washer
6. Outer bearing inner race
7. Rear hub assembly

$20 \rightarrow 0 \rightarrow 10$ Nm
$14 \rightarrow 0 \rightarrow 7$ ft.lbs.

Exploded view of rear bearing assembly (Drum brakes) — 1993 Galant

1993 Galant

Drum Brake Vehicles

1. Raise the vehicle and support it safely.
2. Remove the wheel and tire assemblies.
3. Remove the grease cap and the hub nut.
4. Remove the brake drum. The outer bearing will fall out while the drum is coming off. Do not drop it. Remove the hub and rotor assembly.
5. Pry out and discard the oil seal.
6. Remove the inner bearing.

NOTE: Check the bearing races. If any scoring, heat checking or damage is noted, they should be replaced. When bearing or races need replacement, replace them as a set.

7. If the bearings and races are to be replaced, drive out the race with a brass drift.
To install:
8. Before installing new races, coat them with wheel bearing grease. Drive into place with proper size driver. Make sure they are fully seated.
9. Thoroughly pack the bearings and lubricate the hubs with wheel bearing grease. Install the inner bearing and coat the lip and rim of the grease seal with grease. Drive the seal into place with a seal driver.
10. Install the drum assembly on the axle.
11. Lubricate and install the outer wheel bearing, washer and nut. To properly adjust the wheel bearing preload:
   a. Tighten the wheel bearing nut to 20 ft. lbs. (27 Nm) while rotating the drum.
   b. Back off the adjusting nut to remove the preload, then tighten it to 7 ft. lbs. (10 Nm).
   c. Install the nut lock and a new cotter pin.
12. Install the wheel and lower the vehicle.

Disc Brake Vehicles

1. Raise the vehicle and support safely.
2. Remove the tire and wheel assembly.
3. Remove the bolt(s) holding the speed sensor bracket to the knuckle and remove the assembly from the vehicle.

——— WARNING ———
The speed sensor has a pole piece projecting from it. This exposed tip must be protected from impact or scratches. Do not allow the pole piece to contact the toothed wheel during removal or installation.

4. Remove the caliper from the brake disc and suspend with a wire. Remove the brake rotor.
5. Remove the grease cap, locking nut and tongued washer.
6. Remove the rear hub and bearing assembly.

NOTE: The rear hub assembly can not be disassembled. If bearing replacement is required, replace the assembly as a unit.

7. If replacing the hub assembly, remove the two bolts securing the speed sensor ring to the hub.
To install:
8. Install the speed sensor to the hub and bearing assembly. Tighten the mounting bolts to 8 ft. lbs. (11 Nm).
9. Install the hub and bearing assembly to the axle shaft.
10. Install the tongued washer and a new locking nut. Torque the locknut to 144–188 ft. lbs. (200–260 Nm). Once the locknut has been properly torqued, crimp the nut flange over the slot in the spindle, and install the grease cap.
11. Install the brake caliper and rotor.

12. Install the speed sensor and tighten the mounting bolt to 8 ft. lbs. (11 Nm).

13. Install the tire and wheel assembly.

— WARNING —
Be sure to pump brake pedal until firm before moving vehicle.

1994–95 Galant

Drum Brake Vehicles

1. Raise the vehicle and support safely.

2. Remove the appropriate wheel assembly.

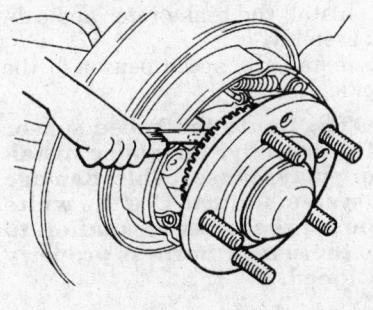

150490

Measurement point for checking speed sensor clearance — Diamante and 3000GT

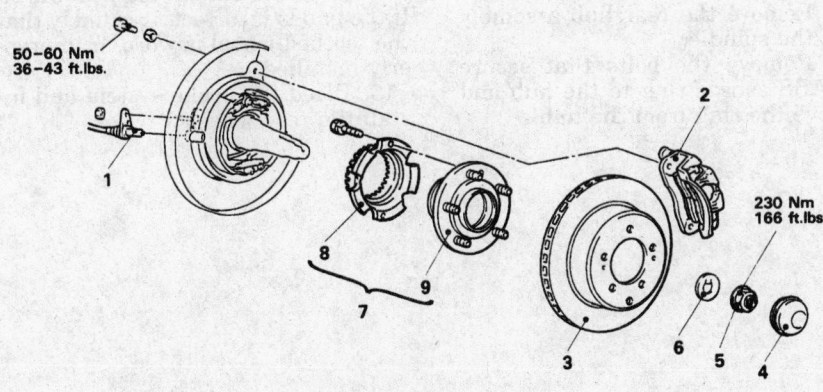

150465

1. Rear speed sensor <Vehicles with ABS>
2. Caliper assembly
3. Brake disc
4. Hub cap
5. Flange nut
6. Tongued washer
7. Rear hub assembly
8. Rear rotor <Vehicles with ABS>
9. Rear hub unit bearing

Exploded view of rear bearing and hub assembly — 1993–94 Diamante

<Vehicles with drum brake>
74–88 Nm
54–65 ft.lbs

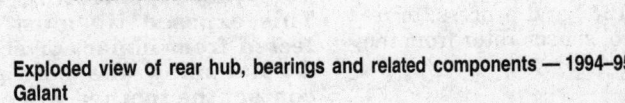

<Vehicles with disc brake>
74–88 Nm
54–65 ft.lbs.

49–59 Nm
36–43 ft.lbs

1. Rear speed sensor <Vehicles with A.B.S.>
2. Caliper assembly
3. Brake drum
4. Brake disc
5. Clip mounting bolt
6. Shoe and lining assembly
7. Rear hub assembly
8. Rotor<Vehicles with A.B.S.>

233484

Exploded view of rear hub, bearings and related components — 1994–95 Galant

3. If equipped with ABS, remove the vehicle speed sensor.

4. Remove the brake drum from the hub assembly.

5. From the back of the knuckle, remove the four bolts securing the hub to the knuckle.

6. Remove the hub and bearing assembly from the knuckle.

NOTE: The hub assembly is not serviceable and should not be disassembled.

7. If replacing the hub, use special socket MB991248 and a press, to remove the wheel sensor rotor from the hub.

To install

8. Press the wheel sensor rotor onto the hub.

9. Install the hub to the knuckle and tighten the mounting bolts to 54–65 ft. lbs. (74–88 Nm).

10. Install the brake drum on the hub.

11. If equipped with ABS, install the vehicle speed sensor.

12. Install the wheel assembly and lower the vehicle.

Disc Brake Vehicles

1. Remove the cotter pin, halfshaft nut and washer.

2. Raise the vehicle and support safely.

3. Remove the appropriate wheel assembly.

4. If equipped with ABS, remove the vehicle speed sensor.

5. Remove the caliper and brake pads. Support the caliper out of the way using wire.

6. Remove the brake rotor from the hub assembly.

7. Remove the parking brake shoes as follows:

   a. Remove the upper shoe to anchor springs.

   b. Remove the lower shoe to shoe spring.

   c. Remove the brake shoe hold-down springs.

   d. Disconnect the parking brake cable from the actuating lever.

8. From the back of the knuckle, remove the four bolts securing the hub to the knuckle.

9. Remove the hub and bearing assembly from the knuckle.

NOTE: The hub assembly is not serviceable and should not be disassembled.

10. If replacing the hub, use special socket MB991248 and a press, to remove the wheel sensor rotor from the hub.

To install

11. Press the wheel sensor rotor onto the hub.

12. Install the hub to the knuckle and tighten the mounting bolts to 54–65 ft. lbs. (74–88 Nm).

13. Install the parking brake shoes.

14. Position the rotor on the hub. Install a couple of lug nuts and lightly tighten to hold rotor on hub.

15. Install the caliper holder and place brake pads in holder. Slide caliper over brake pads and install guide pins. Once caliper is secured, lug nuts can be removed.

16. If equipped with ABS, install the vehicle speed sensor.

17. Install the wheel assembly and lower the vehicle.

3000GT and Diamante

NOTE: The hub assembly is not repairable, if defective replacement is the only option. If the hub is removed for any reason it must be replaced.

1. Raise and support vehicle safely.

2. Remove the both of the rear wheels.

3. Remove the caliper and the brake disc. Support the caliper with wire to prevent stress to the brake hose.

4. If equipped with ABS, remove the bolt holding the speed sensor to the trailing arm and remove the sensor.

NOTE: The speed sensor has a pole piece projecting from it. This exposed tip must be protected from impact or scratches. Do not allow the pole piece to contact the toothed wheel during removal or installation.

5. Remove the grease cap, self-locking nut and tongued washer.

NOTE: Do not use an air gun to remove the hub locknut.

6. Remove the rear hub assembly from the spindle.

7. Remove the bolts that secure the ABS sensor ring to the hub and remove the ring from the hub.

To install

8. Secure the sensor ring to the hub assembly and tighten the mounting bolts.

9. Install the hub assembly, tongued washer and a new self-locking nut. Torque the nut to 166 lbs. (230 Nm), align with the indentation in the spindle, and crimp.

10. Using a rope around the hub bolts and a spring balance, measure the resistance necessary to rotate the hub. If the resistance exceeds 7 lbs. (31 N), loosen and retighten the locknut. If the resistance still exceeds the specification, the hub must be replaced.

11. Using a dial indicator, measure the hub endplay. The endplay should be 0.002 inches (0.05mm) or less.

12. Install the brake rotor and caliper assembly.

13. Install the speed sensor to the knuckle.

NOTE: Route the speed sensor cable correctly. Improper installation may cause cable damage and system failure. Use the white stripe on the outer insulation to keep the sensor harness properly positioned.

14. Use a brass or other non-magnetic feeler gauge to check the air gap between the tip of the pole piece and the toothed wheel. Correct gap is 0.008–0.028 in. (0.2–0.7mm). Tighten the sensor bracket nut with the sensor located so the gap is the same at several points on the toothed wheel. If the gap is incorrect, it is likely that the toothed wheel is worn or improperly installed.

15. Bleed the brake system and install the rear wheels.

FIRING ORDERS

NOTE: To avoid confusion, always replace spark plug wires one at a time.

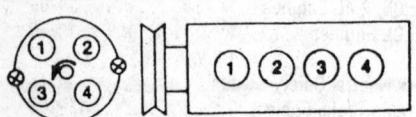

202698

1993–94 1.6L (GA16DE) and 2.0L (SR20DE) Sentra Engines
Engine Firing Order: 1–3–4–2
Distributor Rotation: Counterclockwise

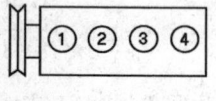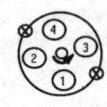

285516

1.6L (GA16DE) 200SX Engine
Engine Firing Order: 1–3–4–2
Distributor Rotation: Counterclockwise

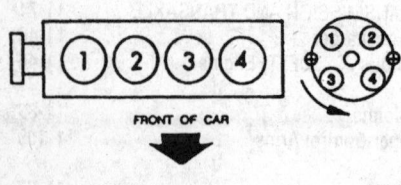

285499

2.4L (KA24DE) Altima Engine
Engine Firing Order: 1–3–4–2
Distributor Rotation: Counterclockwise

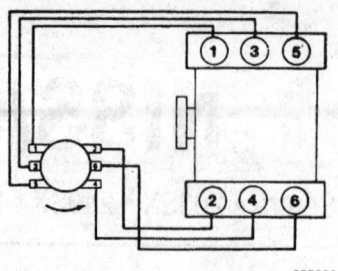

255980

1993–94 3.0L (VE30DE) and (VG30E) Maxima Engines
Engine Firing Order: 1–2–3–4–5–6
Distributor Rotation : Counterclockwise

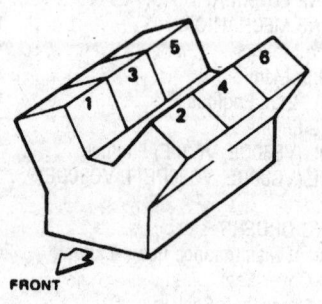

285508

3.0L Maxima and 300ZX Engines
Engine Firing Order: 1–2–3–4–5–6
Distributorless Ignition System

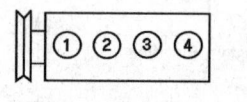

285524

1995–97 2.0L (SR20DE) 200SX Engine
Engine Firing Order: 1–3–4–2
Distributor Rotation: Counterclockwise

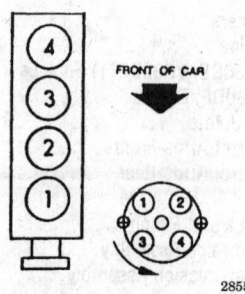

285500

2.4L (KA24DE) 240SX Engine
Engine Firing Order:1–3–4–2
Distributor Rotation: counterclockwise

ENGINE ELECTRICAL

NOTE: Disconnecting the negative battery cable on some vehicles may interfere with the functions of the on board computer systems and may require the computer to undergo a relearning process, once the negative battery cable is reconnected.

Distributor

REMOVAL AND INSTALLATION

1993–94 Sentra with 2.0L (SR20DE) Engine

1. Disconnect the negative battery cable.

2. Release the retaining clips and lift the distributor cap straight up. It will be easier to install the distributor if the wiring is not disconnected from the cap. If the wires must be removed from the cap, label the wires according to cylinder number to aid in installation and avoid confusion.

3. Disconnect the distributor wiring harness.

4. If equipped, disconnect and label the vacuum lines.

5. Note the position of the rotor in relation to the base. Scribe a mark on the base of the distributor and on the engine block to facilitate reinstallation. Align the marks with the direction the rotor is pointing.

6. Remove the bolt(s) which hold the distributor to the engine.

7. Lift the distributor assembly from the engine.

NOTE: Once the distributor is removed, do not to disturb the position of the rotor, camshaft or crankshaft.

To install:

Engine Not Disturbed

1. Insert the distributor shaft and assembly into the engine.

2. Align the distributor and engine block matchmarks with the rotor. Make sure the housing is pointed in the same direction as it was pointed originally. This will be done automatically if the marks on the engine and the distributor are lined up with the rotor.

3. Install the distributor hold-down bolt(s) and if equipped, the hold-down clamp. Leave the screw

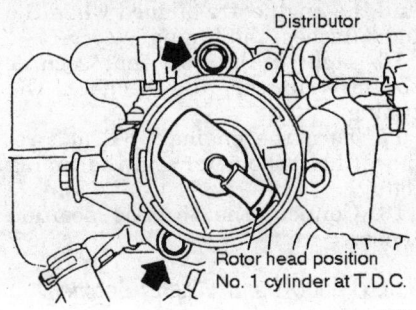

Distributor mounting and alignment points — 1993–94 Sentra with 2.0L (SR20DE) engine

loose enough so the distributor can be moved with moderate hand pressure.

4. If equipped, connect the vacuum lines.

5. Connect the primary wire to the coil.

6. Install the distributor cap on the distributor housing. Secure the distributor cap with the spring clips.

7. Install the spark plug wires if removed. Make sure the wires are pressed all the way into the top of the distributor cap and firmly onto the spark plug.

8. Connect the negative battery cable.

9. Start the engine and set the ignition timing.

10. After setting the ignition timing, tighten the distributor hold-down bolt(s) completely.

Engine Disturbed

NOTE: If the crankshaft has been turned or the engine disturbed in any manner (i.e., disassembled and rebuilt) while the distributor was removed or if the marks were not drawn, it will be necessary to initially time the engine. Follow the procedure given below.

1. It is necessary to place the No. 1 cylinder in the firing position to correctly install the distributor. To locate this position, the ignition timing marks on the crankshaft front pulley are used.

2. Remove the No. 1 cylinder spark plug. Turn the crankshaft until the piston in the No. 1 cylinder is moving up on the compression stroke. This can be determined by placing a thumb over the spark plug hole and feeling the air being forced out of the cylinder. Stop turning the crankshaft when the timing marks are aligned.

3. Oil the distributor housing lightly where the distributor mounts to the block.

4. Install the distributor so the rotor, which is mounted on the shaft, points toward the No. 1 spark plug terminal tower position when the cap is installed.

NOTE: The distributor rotor will be facing the 5 o'clock position with the engine at number one cylinder TDC.

5. When the distributor shaft has reached the bottom of the hole, move the rotor back and forth slightly until the driving lug on the end of the distributor shaft enters the slots cut in the end of the oil pump shaft and the distributor assembly slides down into place.

6. When the distributor is correctly installed, the reluctor teeth should be aligned with the pickup coil. This can be accomplished by rotating the distributor body after it has been installed in the engine. Once again, line up the marks made before the distributor was removed.

7. Install the distributor hold-down bolt(s) and if equipped, the hold-down clamp. Leave the screw loose enough so the distributor can be moved with moderate hand pressure.

8. Install the spark plug into the No. 1 spark plug hole.

9. If equipped, connect the vacuum lines.

10. Connect the primary wire to the coil.

11. Install the distributor cap on the distributor housing. Secure the distributor cap with the spring clips.

12. Install the spark plug wires if removed. Make sure the wires are pressed all the way into the top of the distributor cap and firmly onto the spark plug.

13. Connect the negative battery cable.

14. Start the engine and set the ignition timing.

15. After setting the ignition timing, tighten the distributor hold-down bolt(s) completely.

Altima and 240SX with 2.4L (KA24DE), Sentra and 200SX with 1.6L (GA16DE), 2.0L (SR20DE) and 1993–94 Maxima with 3.0L (VG30E) Engines

1. Disconnect the negative battery cable.

2. Set the engine to TDC. (top dead center) with the number 1 piston on compression stroke

3. Remove and label the distributor spark plug wires from the distributor cap.

4. Remove the distributor cap and scribe a mark on the engine block to show the rotor and distributor position prior to removal.

5. Disconnect and label the wiring connections to the distributor.

6. Remove the bolt(s) holding distributor to engine.

7. Pull the distributor upward to remove from cylinder block.

NOTE: Do not disturb the camshaft or crankshaft position after the distributor is removed from the engine. If any of these components are moved, TDC on cylinder 1 will have to be found again before reinstalling the distributor.

To install:

Engine Not Disturbed

1. Install a new distributor housing O-ring.

2. Install the distributor in the engine so the rotor is aligned with the matchmark on the housing and the housing is aligned with the matchmark on the engine. Make sure the distributor is fully seated and the distributor gear is fully engaged.

3. Install and snug the hold-down bolt.

4. Connect the distributor pickup lead wires.

5. Install the distributor cap and tighten the screws. Install the splash shield.

6. Install the spark plug wires.

7. Connect the negative battery cable.

8. After the ignition timing has been adjusted, tighten the hold-down bolt(s) to 80–104 inch lbs. (9–11 Nm) for GA16DE and VG30E engines or to 108–144 inch lbs. (13–16 Nm) for SR20DE and the Altima (KA24DE) engines. On the 240SX (KA24DE) engine, tighten to 34–39 inch lbs. (4–5 Nm).

Engine Disturbed

1. Install a new distributor housing O-ring.

2. Position the engine so the No. 1 piston is at TDC of its compression stroke and the mark on the vibration damper is aligned with **0** on the timing indicator.

3. Install the distributor in the engine so the rotor is aligned with the position of the No. 1 ignition wire on the distributor cap. Make sure the distributor is fully seated and that the distributor shaft is fully engaged.

4. Install and snug the hold-down bolt.

5. Connect the distributor pickup lead wires.

6. Install the distributor cap and tighten the screws. Install the splash shield, if equipped.

7. Install the spark plug wires.

8. Connect the negative battery cable.

9. After the ignition timing has been adjusted, tighten the hold-down bolt(s) to 80–104 inch lbs. (9–11 Nm) for GA16DE and VG30E engines or to 108–144 inch lbs. (13–16 Nm) for SR20DE and the Altima (KA24DE) engines. On the 240SX (KA24DE) engine, tighten to 34–39 inch lbs. (4–5 Nm).

Ignition Timing

ADJUSTMENT

1.6L (GA16DE), 2.0L (SR20DE) and 2.4L (KA24DE) Engines

Visually check the air cleaner, intake hoses, ducts, EGR valve operation and electrical connections prior to the adjustment of the ignition timing. Correct or repair any problem as required. Be sure to inspect the throttle valve and the throttle position sensor for proper operation.

1. Locate the timing marks on the crankshaft pulley and the front of the engine.

2. Clean the timing marks.

3. Using chalk or white paint, color the mark on the crankshaft pulley and the mark on the scale which will indicate the correct timing when aligned with the notch on the crankshaft pulley.

4. Attach a tachometer to the engine.

5. Attach a timing light to the engine, to No.1 cylinder's ignition wire.

6. Check to make sure all of the wires clear the fan; start the engine and allow it to reach normal operating temperatures.

7. Block the front wheels and set the parking brake. Shift the trans-

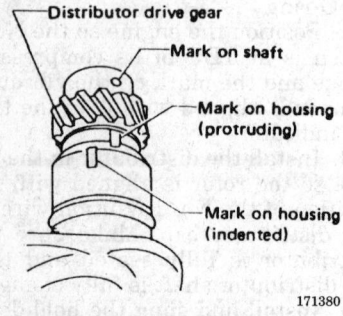

Distributor shaft and housing alignment marks — 1993–94 Maxima with 3.0L (VG30E) engine

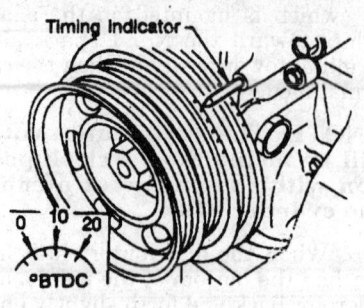

340844

Timing indicator — 1.6L (GA16DE) engine

mission into **NEUTRAL** for automatic and manual transaxles; do not stand in front of the vehicle when making adjustments.

8. Perform the following procedures:

a. Race the engine at 2000 rpm for about two minutes under a no-load condition; make sure all of the accessories are turned off.

b. Perform on board engine diagnostics and repair any fault code.

c. Race the engine 2–3 times under no-load then run the engine it for one minute at idle.

d. Stop the engine and disconnect the throttle position sensor.

e. Race the engine at 2000 rpm for about two minutes under a no-load condition; make sure all of the accessories are turned **OFF**.

f. Run the engine at idle speed.

1.6L (GA16DE) — 6–10 degrees BTDC

2.0L (SR20DE) — 13–17 degrees BTDC

2.4L (KA24DE) — 18–22 degrees BTDC

9. Aim the timing light at the timing marks. If the marks on the pulley and the engine are aligned when the light flashes, the timing is correct. Turn the engine **OFF** and remove the tachometer and the timing light. If the marks are not in alignment, proceed with the following steps.

10. Turn the engine **OFF**.

11. Loosen the bolts that secure the distributor just enough so it can be turned.

12. Start the engine. Keep the wires of the timing light clear of the cooling fan.

13. With the timing light aimed at the pulley and the marks on the engine, turn the distributor for the proper adjustment.

14. Race the engine 2–3 times under no-load then run the engine it for one minute at idle.

15. Aim the timing light at the timing marks. If the marks on the pulley

and the engine are aligned when the light flashes, the timing is correct.

16. Tighten the bolt that secures the distributor and recheck the timing.

17. Turn the engine **OFF** and remove the tachometer and the timing light.

18. Connect the throttle position sensor.

3.0L (VE30DE and VG30E) Engines

Before the ignition timing can be checked, there must be no trouble codes stored in the ECM. Also, the physical condition of the ignition components must be inspected and any suspect parts replaced.

1. Run the engine until the water temperature indicator points to the middle of the gauge.

2. Run the engine for 1–2 minutes with no load at 2000 rpm; all electrical accessories in the off position.

3. Race the engine 2 or 3 times and allow to idle for a minute.

4. Connect a timing light to cylinder 1 and check timing. The timing must be 13–17 degrees BTDC.

5. To adjust timing, stop the engine and loosen the distributor hold-down bolt just enough to allow the distributor to be turned by hand.

6. Start the engine and race it 2–3 times with no load and then allow the engine to run at idle speed.

7. Adjust the ignition timing by rotating the crank angle sensor located at the front of the cylinder head either clockwise or counterclockwise. Adjust the ignition timing to 15 degrees BTDC at idle speed.

8. Tighten the crank angle mounting bolts and recheck the timing.

3.0L (VQ30DE) Engine

NOTE: The ignition timing is not adjustable. If not within specifications, further diagnostic inspection is required. The following procedure is for viewing the ignition timing setting.

Visually check the air cleaner, intake hoses, ducts, EGR valve operation and electrical connections prior to the adjustment of the ignition timing. Correct or repair any problem as required. Be sure inspect the throttle valve and throttle position sensor for proper operation.

1. Locate the timing marks on the crankshaft pulley and the front of the engine.

2. Clean the timing marks.

NOTE: The ignition timing specification is 15 degrees ±2 BTDC.

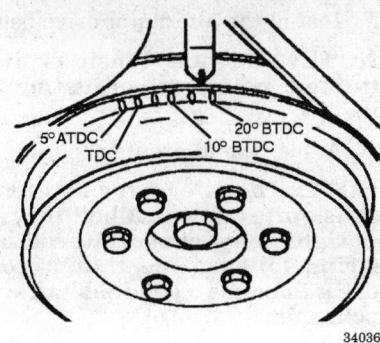

View of the timing marks — 2.4L (KA24DE) engine

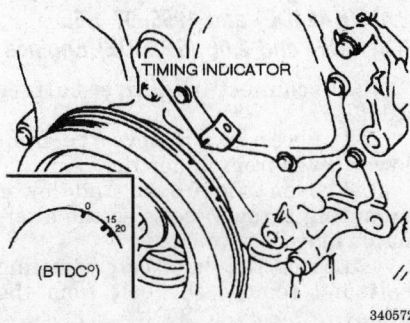

View of the timing marks — 3.0L (VQ30DE) engine

3. Using chalk or white paint, color the mark on the crankshaft pulley and the mark on the scale which will indicate the correct timing when aligned with the notch on the crankshaft pulley.

4. Attach a tachometer to the engine.

5. Attach a timing light to the engine to number one cylinder ignition wire.

6. Turn all electrical equipment and accessories off.

7. Check to make sure all of the wires clear the fan, then, start the engine and allow it to reach normal operating temperatures.

8. Block the front wheels and set the parking brake. Shift the transmission into **NEUTRAL** for manual transmission and automatic transmissions. Do not stand in front of the vehicle when making adjustments.

9. Perform the following procedures:

a. Race the engine at 2000 rpm for about two minutes under a no-load condition; make sure all of the accessories are turned off.

b. Perform on board engine diagnostics and repair any fault code.

c. Race the engine at 2000 rpm for about two minutes under a no-load condition.

d. Turn the engine **OFF** and disconnect the throttle position sensor.

e. Start and race the engine 2–3 times under no-load then run the engine at idle speed.

NOTE: The ignition timing specification is 15 degrees ±2 BTDC.

10. Aim the timing light at the timing marks. If the marks on the pulley and the engine are aligned when the light flashes, the timing is correct. Turn the engine **OFF** and remove the tachometer and the timing light. If the marks are not in alignment, proceed with the following steps.

11. Turn the engine **OFF**.

12. Check the camshaft position sensor (PHASE), crankshaft position sensor (REF) and crankshaft position sensor (POS). Replace if necessary.

13. If the ignition timing is still not correct, substitute a known good ECM.

NOTE: The ECM may be the cause of the problem but this is rarely the case.

14. Turn the engine **OFF** and remove the tachometer and the timing light.

3.0L (VG30DETT and VG30DE) Engines

1993–94 Models

Visually check the air cleaner, intake hoses, ducts, EGR valve operation and electrical connections prior to the adjustment of the ignition timing. Correct or repair any problem as required.

1. Locate the timing marks on the crankshaft pulley and the front of the engine.

2. Clean the timing marks.

NOTE: The ignition timing specification is 15 degrees ±2 BTDC.

3. Using chalk or white paint, color the mark on the crankshaft pulley and the mark on the scale which will indicate the correct timing when aligned with the notch on the crankshaft pulley.

4. Attach a tachometer to the engine.

5. Attach a timing light to the engine to number one cylinder ignition wire.

6. Turn all electrical equipment and accessories off.

7. Check to make sure all of the wires clear the fan, then, start the engine and allow it to reach normal operating temperatures.

8. Block the front wheels and set the parking brake. Shift the transmission into **NEUTRAL** for manual transmission and automatic transmissions. Do not stand in front of the vehicle when making adjustments.

9. Perform the following procedures:

a. Race the engine at 2000 rpm for about two minutes under a no-load condition; make sure all of the accessories are turned off.

b. Perform on board engine diagnostics and repair any fault code.

c. Run the engine at idle speed.

d. Race the engine 2–3 times under no-load then run the engine it for one minute at idle.

NOTE: The ignition timing specification is 15 degrees ±2 BTDC.

10. Aim the timing light at the timing marks. If the marks on the pulley and the engine are aligned when the light flashes, the timing is correct. Turn the engine **OFF** and remove the tachometer and the timing light. If the marks are not in alignment, proceed with the following steps.

11. Turn the engine **OFF**.

12. Loosen the camshaft position sensor securing bolts just enough so it can be turned.

13. Start the engine. Keep the wires of the timing light clear of the cooling fan.

14. With the timing light aimed at the pulley and the marks on the engine, turn the crankshaft position sensor for the proper adjustment.

15. Race the engine 2–3 times under no-load then run the engine it for one minute at idle.

16. Aim the timing light at the timing marks. If the marks on the pulley and the engine are aligned when the light flashes, the timing is correct.

17. Tighten the bolts securing the sensor and recheck the timing.

18. Turn the engine **OFF** and remove the tachometer and the timing light.

1995–97 Vehicles

NOTE: The engine should be in good mechanical condition and all electrical connectors and vacuum hoses connected before making this adjustment.

1. Start the engine and let it warm up to normal operating temperature.

2. Open the hood and run the engine under no load at about 2,000 rpm for about two minutes.

3. Perform Diagnostic Test Mode II and repair any causes of trouble codes as needed.

4. Run the engine under no load at 2,000 rpm for about two minutes. Rev the engine two or three times and let it idle for one minute.

5. Turn off the engine and disconnect the throttle position sensor connector. Remove the No. 1 ignition coil. Connect the coil to the spark plug using a spare piece of high-tension wire so you have a place to connect your timing light. Start the engine.

6. Run the engine under no load at 2,000 rpm for about two minutes. Rev the engine two or three times and let it idle.

7. Check the ignition timing and adjust if needed. Correct ignition timing is 10° ±2° BTDC on non-turbocharged vehicles; 15° ±2° on turbocharged vehicles. Adjustment is made by loosening the screws and turning the camshaft position sensor. Tighten the mounting screws and confirm ignition timing has not changed.

8. Turn the engine **OFF** and connect the throttle position sensor connector.

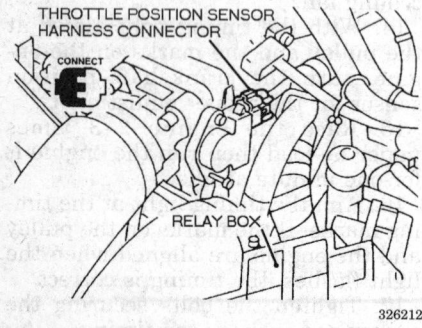

Throttle position sensor connector — 3.0L (VG30DE and VG30DETT) engines

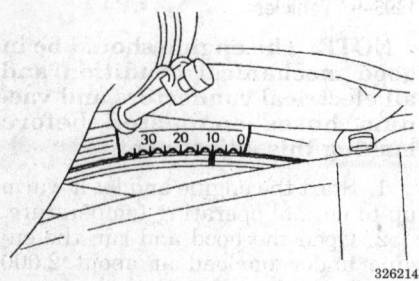

Location of timing marks — 3.0L (VG30DE and VG30DETT) engines

Alternator

PRECAUTIONS

Several precautions must be observed with alternator equipped vehicles to avoid damage to the unit.

• If the battery is removed for any reason, make sure it is reconnected with the correct polarity. Reversing the battery connections may result in damage to the 1-way rectifiers.

• When utilizing a booster battery as a starting aid, always connect the positive to positive terminals and the negative terminal from the booster battery to a good engine ground on the vehicle being started.

• Never use a fast charger as a booster to start vehicles.

• Disconnect the battery cables when charging the battery with a fast charger.

• Never attempt to polarize the alternator.

• Do not use test lights of more than 12 volts when checking diode continuity.

• Do not short across or ground any of the alternator terminals.

• The polarity of the battery, alternator and regulator must be matched and considered before making any electrical connections within the system.

• Never separate the alternator on an open circuit. Make sure all connections within the circuit are clean and tight.

• Disconnect the battery ground terminal when performing any service on electrical components.

• Disconnect the battery if arc welding is to be done on the vehicle.

REMOVAL AND INSTALLATION

1993–94 1.6L (GA16DE), 2.0L (SR20DE) and 3.0L (VG30E and VE30DE) Engines

1. Disconnect the negative battery terminal.
2. Disconnect the lead wires and connector from the alternator.
3. Loosen the drive belt adjusting bolt locknut and remove the belt.
4. Unscrew the alternator attaching bolts and remove the alternator from the vehicle.
To install:
5. Mount the alternator to the engine and partially tighten the attaching bolts.
6. Reconnect the lead wires and connector to the alternator.

7. Install the alternator drive belt.

NOTE: The belt tensions are adjusted using an adjustment bolt.

8. Adjust the alternator belt.

NOTE: The alternator belt tension is quite critical. A belt that is too tight may cause alternator bearing failure; one that is too loose will cause a gradual battery discharge.

9. Connect the battery cable. Start the engine and check for proper operation.

2.4L (KA24DE) and 1995–97 1.6L (GA16DE) and 2.0L (SR20DE) Engines

1. Disconnect negative battery cable.
2. If necessary, remove the front lower cover from under the vehicle.
3. Loosen the upper and lower mounting through bolts so the alternator may be pivoted.
4. Loosen the drive belt adjusting bolt and remove the belt from the pulley.
5. Remove the upper and lower mounting bolts from the alternator.

NOTE: Be sure to note the positioning of the electrical connections for reinstallation.

6. Position the alternator so the lead wires and harness connector can be removed from the alternator.
7. Remove the alternator from the vehicle.
To install:
8. Connect the electrical wiring to the rear of the alternator.
9. Position the alternator on the mounting brackets.
10. Install the attaching bolts, but do not fully tighten the bolts at this time.
11. Install the drive belt and adjust to the proper tension.
12. Tighten the lower mounting through bolt to 27–37 ft. lbs. (37–50 Nm).
13. Tighten the upper adjusting lock bolt to 12–15 ft. lbs. (16–21 Nm).
14. Reconnect negative battery cable.
15. Check for proper operation.

3.0L (VQ30DE) Engine

1. Disconnect the negative battery cable.
2. Remove the engine RH undercover and remove the RH side inspection cover.
3. Loosen the nut that secures the alternator belt idler pulley and turn

the adjusting bolt to loosen the alternator belt.

4. Remove the alternator belt from the alternator pulley.

5. Remove the 4 air conditioner compressor mounting bolts.

6. Remove the cooling fan and the fan shroud.

7. Slide the A/C compressor forward and disconnect the alternator connectors. Be sure to note electrical connections for installation purposes.

8. Remove the alternator upper and lower mounting bolts.

9. Carefully remove the alternator from the vehicle.

To install:

10. Install the alternator to the vehicle and install the mounting bolts.

11. Tighten the upper mounting bolt to 12–15 ft. lbs. (16–21 Nm).

12. Tighten the lower mounting bolt to 33–38 ft. lbs. (44–52 Nm).

13. Connect the harness electrical connections and secure the wiring harness.

14. Install the cooling fan and the cooling fan shroud.

15. Slide the A/C compressor rearward and install the 4 mounting bolts. Tighten the mounting bolts to 33–44 ft. lbs. (45–60 Nm).

16. Install the alternator drive belt and tighten the idler pulley adjusting screw to 35–61 inch lbs. (4–7 Nm).

17. Tighten the idler pulley mounting nut to 19–24 ft. lbs. (25–32 Nm).

18. Install the engine RH inspection cover and install the RH undercover.

19. Connect the negative battery cable.

1993–94 3.0L (VG30DE and VG30DETT) Engines

NOTE: The alternator is removed from the bottom of the vehicle.

1. Disconnect the negative battery cable.

2. Raise and safely support the vehicle.

3. Disconnect the 2 lead wires and harness connector from the alternator.

4. Loosen the drive belt adjusting bolt and remove the belt.

5. Remove the lower radiator hose bracket and pull the hose upward to allow clearance to remove the alternator.

NOTE: Do not disconnect or remove the lower hose.

6. Remove the front stabilizer bar mounting bolts.

NOTE: It is not necessary to remove the sway bar, only to disconnect the chassis mounts.

7. Remove the alternator mounting bolts and lower the alternator. While pulling the stabilizer bar down, remove the alternator from the vehicle.

To install:

8. Using the same method as removal, install the alternator in the vehicle and connect to the mounting bracket.

9. Connect the wiring to the alternator.

10. Install the remaining components in the opposite order from which they were removed.

11. Start the engine and check alternator operation.

1995–97 3.0L (VG30DE and VG30DETT) Engines

1. Disconnect negative battery cable.

2. Remove the front lower cover from under the vehicle.

3. Remove the stabilizer bar bracket mounting bolts and allow the stabilizer bar to hang. It is not necessary to disconnect the stabilizer links and completely remove stabilizer bar from the vehicle.

4. Remove the lower radiator hose support bracket and push the lower hose upward to make room for alternator removal.

5. Loosen the upper and lower mounting through bolts so the alternator may be pivoted.

6. Loosen the drive belt adjusting bolt and remove the belt from the pulley.

7. Remove the upper and lower mounting bolts from the alternator.

NOTE: Be sure to note the positioning of the electrical connections for reinstallation.

8. Position the alternator so the lead wires and harness connector can be removed from the alternator.

9. Remove the alternator from the vehicle.

To install:

10. Connect the electrical wiring to the rear of the alternator.

11. Position the alternator on the mounting brackets.

12. Install the attaching bolts, but do not fully tighten the bolts at this time.

13. Install the remaining components in the reverse order of removal.

14. Tighten the lower adjusting lock bolt to 12–15 ft. lbs. (16–21 Nm).

15. Tighten the stabilizer bar mounting bolts to 29–36 ft. lbs. (39–49 Nm). Be sure to align the mounting brackets with the paint marks on the stabilizer bar.

16. Reconnect negative battery cable.

17. Check for proper operation.

Drive Belt

REMOVAL AND INSTALLATION

1.6L (GA16DE) Engine

———— **WARNING** ————

If a removed belt is to be reused, be certain to mark the direction of rotation on the belt, this will extend the belt's life.

Alternator

1. Disconnect the negative battery cable.

2. Loosen the pivot and mounting bolts of the alternator.

3. Loosen the locking bolt on the alternator adjusting bolt. The alternator adjusting bolt will loosen and tighten the alternator drive belt tension.

4. When there is enough slack in the belt, remove the belt from the alternator pulley.

To install:

5. Verify that the new belt and the old belt have the same length and width. These measurements must be the same or problems will occur when the new belt is adjusted.

6. Correctly route the belt around the pulleys.

7. After new belt is installed correctly, adjust the tension of the new belt.

8. Tighten the mounting, pivot, and lock bolts.

9. Reconnect the negative battery cable.

Air Conditioning Compressor

1. Disconnect the negative battery cable.

2. Loosen the lock bolt for the idler pulley.

3. Loosen the idler pulley adjusting bolt. When the idler pulley adjustment bolt is loosened, the drive belt tension will slowly be released.

4. When there is enough slack in the belt, remove the belt from the pulleys.

To install:

5. Verify that the new belt and the old belt have the same length and

width. These measurements must be the same or problems will occur when the new belt is adjusted.

6. Correctly route the belt around the pulleys.

7. After new belt is installed correctly, adjust the tension of the new belt.

8. Tighten the mounting and pivot bolts.

9. Reconnect the negative battery cable.

Power Steering

1. Disconnect the negative battery cable.

2. Loosen the power steering oil pump mounting and pivot bolts.

3. Loosen the drive belt adjustment locking bolt. The drive belt adjustment bolt is located on the power steering oil pump. The drive belt adjustment bolt will move the power steering oil pump and increase or decrease belt tension.

4. Turn the power steering oil pump adjusting bolt until there is enough slack in the drive belt to remove it.

5. Remove the drive belt from around the pulleys.

To install:

6. Verify that the new belt and the old belt have the same length and width. These measurements must be the same or problems will occur when the new belt is adjusted.

7. Correctly route the belt around the pulleys.

8. After the drive belt is installed correctly, adjust the drive belt tension.

9. Tighten the mounting, pivot, and lock bolts.

10. Reconnect the negative battery cable.

2.0L (SR20DE) Engine

Alternator and Air Conditioning Compressor

1. Disconnect the negative battery cable.

2. Loosen the pivot and mounting bolts of the alternator.

3. Loosen the locking bolt on the alternator adjusting bolt. The alternator adjusting bolt will loosen and tighten the alternator/air conditioning compressor drive belt tension.

4. When there is enough slack in the belt, remove the belt from the alternator and air conditioning compressor pulleys.

To install:

5. Verify that the new belt and the old belt have the same length and

width. These measurements must be the same or problems will occur when the new belt is adjusted.

6. Correctly route the belt around the pulleys.

7. After the belt is installed correctly, adjust the tension by applying 22 lbs. (98 N) of pressure to the belt, then measuring the deflection. Compare the deflection with the following specifications:

• Alternator with air conditioner compressor — 0.256–0.295 in. (6.5–7.5mm)

• Alternator without air conditioner compressor — 0.28–0.31 in. (7–8mm)

8. Tighten the mounting, pivot, and lock bolts.

9. Reconnect the negative battery cable.

Power Steering

1. Disconnect the negative battery cable.

2. Loosen the power steering oil pump mounting and pivot bolts.

3. Loosen the drive belt adjustment locking bolt. The drive belt adjustment bolt is located on the power steering oil pump. The drive belt adjustment bolt will move the power steering oil pump and increase or decrease belt tension.

4. Turn the power steering oil pump adjusting bolt until there is enough slack in the drive belt to remove it.

5. Remove the drive belt from around the pulleys.

To install:

6. Verify that the new belt and the old belt have the same length and width. These measurements must be the same or problems will occur when the new belt is adjusted.

7. Correctly route the belt around the pulleys.

8. After the drive belt is installed correctly, adjust the tension by applying 22 lbs. (98 N) of pressure to the belt, then measuring the deflection. Compare the deflection with the following specifications:

• Power steering pump — 0.138–0.177 in. (3.5–4.5mm)

9. Tighten the mounting, pivot, and lock bolts.

10. Reconnect the negative battery cable.

Altima with 2.4L (KA24DE) Engine

Alternator and Power Steering

1. Disconnect the negative battery cable.

2. Loosen the pivot and mounting bolts of the alternator.

3. Loosen the locking bolt on the alternator adjusting bolt. The alternator adjusting bolt will loosen and tighten the alternator drive belt tension.

4. When there is enough slack in the belt, remove the belt from the alternator pulley.

To install:

5. Verify that the new belt and the old belt have the same length and width. These measurements must be the same or problems will occur when the new belt is adjusted.

6. Correctly route the belt around the pulleys.

7. After the new belt is installed correctly, adjust the tension by applying 22 lbs. (98 N) of pressure to the belt, then measuring the deflection. Compare the deflection with the following specifications:

• Alternator and power steering pump — 0.20–0.24 in. (5–6mm)

8. Tighten the alternator adjustment locking bolt to 12–14 ft. lbs. (16–19 Nm) and recheck the drive belt deflection.

9. Tighten the mounting and pivot bolts.

10. Connect the negative battery cable.

Air Conditioning Compressor

1. Disconnect the negative battery cable.

2. Loosen the locknut for the idler pulley.

3. Loosen the idler pulley adjusting bolt. When the idler pulley adjustment bolt is loosened, the drive belt tension will slowly be released.

4. When there is enough slack in the belt, remove the belt from the pulleys.

To install:

5. Verify that the new belt and the old belt have the same length and width. These measurements must be the same or problems will occur when the new belt is adjusted.

6. Correctly route the belt around the pulleys.

7. After the new belt is installed correctly, adjust the tension by applying 22 lbs. (98 N) of pressure to the belt, then measuring the deflection. Compare the deflection with the following specifications:

• Air conditioner compressor — 0.24–0.28 in. (6–7mm)

8. Tighten the idler pulley locknut on the face of the idler pulley to 24–28 ft. lbs. (32–38 Nm).

9. Tighten the mounting and pivot bolts.

10. Connect the negative battery cable.

3.0L (VE30DE and VG30E) Engines

Alternator — with drive belt adjusting bolt

1. Disconnect the negative battery cable.
2. Loosen the pivot and mounting bolts of the alternator.
3. Loosen the locking bolt on the alternator adjusting bolt. The alternator adjusting bolt will loosen and tighten the alternator drive belt tension.
4. When there is enough slack in the belt, remove the belt from the alternator pulley.

To install:

5. Verify that the new belt and the old belt have the same length and width. These measurements must be the same or problems will occur when the new belt is adjusted.
6. Correctly route the belt around the pulleys.
7. After new belt is installed correctly, adjust the tension of the new belt.
8. Tighten the mounting, pivot, and lock bolts.
9. Reconnect the negative battery cable.

Alternator — without drive belt adjusting bolt

1. Disconnect the negative battery cable.
2. Loosen the pivot and mounting bolts of the alternator.
3. Using a proper tool, pry the component inward to relieve the tension on the drive belt.

NOTE: Always be careful where using the pry bar not to damage the alternator or surrounding components.

4. When there is enough slack in the belt, remove the belt from the alternator pulley.

To install:

5. Verify that the new belt and the old belt have the same length and width. These measurements must be the same or problems will occur when the new belt is adjusted.
6. Correctly route the belt around the pulleys.
7. After new belt is installed correctly, adjust the tension by applying 22 lbs. (98 N) of pressure to the belt, then measuring the deflection. Compare the deflection with the following specifications:

- 3.0L (VG30E) engine — 0.24–0.31 in. (6–8mm)
- 3.0L (VE30DE) engine — 0.256–0.295 in. (6.5–7.5mm)

8. Tighten the mounting and pivot bolts.
9. Reconnect the negative battery cable.

Air Conditioning Compressor

1. Disconnect the negative battery cable.
2. Loosen the lock bolt for the idler pulley.
3. Loosen the idler pulley adjusting bolt. When the idler pulley adjustment bolt is loosened, the drive belt tension will slowly be released.
4. When there is enough slack in the belt, remove the belt from the pulleys.

To install:

5. Verify that the new belt and the old belt have the same length and width. These measurements must be the same or problems will occur when the new belt is adjusted.
6. Correctly route the belt around the pulleys.
7. After new belt is installed correctly, adjust the tension by applying 22 lbs. (98 N) of pressure to the belt, then measuring the deflection. Compare the deflection with the following specifications:

- 3.0L (VG30E) engine — 0.16–0.24 in. (4–6mm)
- 3.0L (VE30DE) engine — 0.20–0.24 in. (5–6mm)

8. Tighten the mounting and pivot bolts.
9. Reconnect the negative battery cable.

Power Steering — with adjustable idler pulley

1. Disconnect the negative battery cable.
2. Loosen the lock bolt for the idler pulley.
3. Loosen the idler pulley adjusting bolt. When the idler pulley adjustment bolt is loosened, the drive belt tension will slowly be released.
4. When there is enough slack in the belt, remove the belt from the pulleys.

To install:

5. Verify that the new belt and the old belt have the same length and width. These measurements must be the same or problems will occur when the new belt is adjusted.
6. Correctly route the belt around the pulleys.
7. After new belt is installed correctly, adjust the tension by applying 22 lbs. (98 N) of pressure to the belt, then measuring the deflection. Compare the deflection with the following specifications:

- 3.0L (VG30E) engine — 0.31–0.39 in. (8–10mm)
- 3.0L (VE30DE) engine — 0.256–0.295 in. (6.5–7.5mm)

8. Tighten the mounting, pivot, and lock bolts.
9. Reconnect the negative battery cable.

Power Steering — with non-adjustable idler pulley

1. Disconnect the negative battery cable.
2. Loosen the power steering oil pump mounting and pivot bolts.
3. Loosen the drive belt adjustment locking bolt. The drive belt adjustment bolt is located on the power steering oil pump. The drive belt adjustment bolt will move the power steering oil pump and increase or decrease belt tension.
4. Turn the power steering oil pump adjusting bolt until there is enough slack in the drive belt to remove it.
5. Remove the drive belt from around the pulleys.

To install:

6. Verify that the new belt and the old belt have the same length and width. These measurements must be the same or problems will occur when the new belt is adjusted.
7. Correctly route the belt around the pulleys.
8. After the drive belt is installed correctly, adjust the drive belt tension.
9. Tighten the mounting, pivot, and lock bolts.
10. Reconnect the negative battery cable.

3.0L (VQ30DE) Engine

Air Conditioning Compressor and Alternator

1. Disconnect the negative battery cable.
2. Loosen the lock bolt for the idler pulley.
3. Loosen the idler pulley adjusting bolt. When the idler pulley adjustment bolt is loosened, the drive belt tension will slowly be released.
4. When there is enough slack in the belt, remove the belt from the pulleys.

To install:

5. Verify that the new belt and the old belt have the same length and width. These measurements must be the same or problems will occur when the new belt is adjusted.

6. Correctly route the belt around the pulleys.

7. After new belt is installed correctly, adjust the tension by applying 22 lbs. (98 N) of pressure to the belt, then measuring the deflection. Compare the deflection with the following specifications:

• Alternator with air conditioner compressor — 0.150–0.161 in. (3.8–4.1mm)

• Alternator without air conditioner compressor — 0.228–0.244 in. (5.8–6.2mm)

8. Tighten the mounting nut to 19–24 ft. lbs. (25–32 Nm).

9. Connect the negative battery cable.

Power Steering

1. Disconnect the negative battery cable.

2. Loosen the power steering oil pump mounting and pivot bolts.

3. Loosen the drive belt adjustment locking bolt. The drive belt adjustment bolt is located on the power steering oil pump. The drive belt adjustment bolt will move the power steering oil pump and increase or decrease belt tension.

4. Turn the power steering oil pump adjusting bolt until there is enough slack in the drive belt to remove it.

5. Remove the drive belt from around the pulleys.

To install:

6. Verify that the new belt and the old belt have the same length and width. These measurements must be the same or problems will occur when the new belt is adjusted.

7. Correctly route the belt around the pulleys.

8. After the drive belt is installed correctly, adjust the tension by applying 22 lbs. (98 N) of pressure to the belt, then measuring the deflection. Compare the deflection with the following specifications:

• Alternator with air conditioner compressor — 0.256–0.276 in. (6.5–7mm)

9. Tighten the mounting nut to 12–15 ft. lbs. (16–15 Nm). Also tighten the pivot bolts.

10. Connect the negative battery cable.

240SX with 2.4L (KA24DE) Engine

Alternator

1. Disconnect the negative battery cable.

2. Loosen the pivot and mounting bolts of the alternator.

3. Loosen the locking bolt on the alternator adjusting bolt. The alter-

nator adjusting bolt will loosen and tighten the alternator drive belt tension.

4. When there is enough slack in the belt, remove the belt from the alternator pulley.

To install:

5. Verify that the new belt and the old belt have the same length and width. These measurements must be the same or problems will occur when the new belt is adjusted.

6. Correctly route the belt around the pulleys.

7. After the new belt is installed correctly, adjust the tension by applying 22 lbs. (98 N) of pressure to the belt, then measuring the deflection. Compare the deflection with the following specifications:

• Alternator — 0.240–0.280 in. (6–7mm)

8. Tighten the mounting, pivot, and lock bolts.

9. Connect the negative battery cable.

Air Conditioning Compressor

1. Disconnect the negative battery cable.

2. Loosen the locknut for the idler pulley.

3. Loosen the idler pulley adjusting bolt. When the idler pulley adjustment bolt is loosened, the drive belt tension will slowly be released.

4. When there is enough slack in the belt, remove the belt from the pulleys.

To install:

5. Verify that the new belt and the old belt have the same length and width. These measurements must be the same or problems will occur when the new belt is adjusted.

6. Correctly route the belt around the pulleys.

7. After the new belt is installed correctly, adjust the tension by applying 22 lbs. (98 N) of pressure to the belt, then measuring the deflection. Compare the deflection with the following specifications:

• Air conditioner compressor — 0.256–0.295 in. (6.5–7.5mm)

8. Tighten the idler pulley locknut.

9. Connect the negative battery cable.

Power Steering

1. Disconnect the negative battery cable.

2. Loosen the power steering oil pump mounting and pivot bolts.

3. Loosen the drive belt adjustment locking bolt. The drive belt adjustment bolt is located on the power steering oil pump. The drive belt ad-

justment bolt will move the power steering oil pump and increase or decrease belt tension.

4. Turn the power steering oil pump adjusting bolt until there is enough slack in the drive belt to remove it.

5. Remove the drive belt from around the pulleys.

To install:

6. Verify that the new belt and the old belt have the same length and width. These measurements must be the same or problems will occur when the new belt is adjusted.

7. Correctly route the belt around the pulleys.

8. After the drive belt is installed correctly, adjust the tension by applying 22 lbs. (98 N) of pressure to the belt, then measuring the deflection. Compare the deflection with the following specifications:

• Power steering pump without SUPER HICAS — 0.256–0.295 in. (6.5–7.5mm)

• Power steering pump with SUPER HICAS — 0.217–0.256 in. (5.5–6.5mm)

9. Tighten the mounting, pivot, and lock bolts.

10. Connect the negative battery cable.

1993–94 3.0L (VG30DE and VG30DETT) Engines

Alternator — with drive belt adjusting bolt

1. Disconnect the negative battery cable.

2. Loosen the pivot and mounting bolts of the alternator.

3. Loosen the locking bolt on the alternator adjusting bolt. The alternator adjusting bolt will loosen and tighten the alternator drive belt tension.

4. When there is enough slack in the belt, remove the belt from the alternator pulley.

To install:

5. Verify that the new belt and the old belt have the same length and width. These measurements must be the same or problems will occur when the new belt is adjusted.

6. Correctly route the belt around the pulleys.

7. After the new belt is installed correctly, adjust the tension by applying 22 lbs. (98 N) of pressure to the belt, then measuring the deflection. Compare the deflection with the following specifications:

• Alternator — 0.256–0.295 in. (6.5–7.5mm)

8. Tighten the mounting, pivot, and lock bolts.

9. Connect the negative battery cable.

Alternator — without drive belt adjusting bolt

1. Disconnect the negative battery cable.
2. Loosen the pivot and mounting bolts of the alternator.
3. Using a proper tool, pry the component inward to relieve the tension on the drive belt.

NOTE: Always be careful where using the pry bar not to damage the alternator or surrounding components.

4. When there is enough slack in the belt, remove the belt from the alternator pulley.
   To install:
5. Verify that the new belt and the old belt have the same length and width. These measurements must be the same or problems will occur when the new belt is adjusted.
6. Correctly route the belt around the pulleys.
7. After the new belt is installed correctly, adjust the tension of the new belt.
8. Tighten the mounting and pivot bolts.
9. Connect the negative battery cable.

Air Conditioning Compressor

1. Disconnect the negative battery cable.
2. Loosen the lock bolt for the idler pulley.
3. Loosen the idler pulley adjusting bolt. When the idler pulley adjustment bolt is loosened, the drive belt tension will slowly be released.
4. When there is enough slack in the belt, remove the belt from the pulleys.
   To install:
5. Verify that the new belt and the old belt have the same length and width. These measurements must be the same or problems will occur when the new belt is adjusted.
6. Correctly route the belt around the pulleys.
7. After the new belt is installed correctly, adjust the tension by applying 22 lbs. (98 N) of pressure to the belt, then measuring the deflection. Compare the deflection with the following specifications:
   • Power steering pump without SUPER HICAS — 0.28–0.31 in. (7–8mm)
8. Tighten the idler pulley locknut.

9. Tighten the mounting and pivot bolts.
10. Connect the negative battery cable.

Power Steering — with adjustment bolt

1. Disconnect the negative battery cable.
2. Loosen the power steering oil pump mounting and pivot bolts.
3. Loosen the drive belt adjustment locking bolt. The drive belt adjustment bolt is located on the power steering oil pump. The drive belt adjustment bolt will move the power steering oil pump and increase or decrease belt tension.
4. Turn the power steering oil pump adjusting bolt until there is enough slack in the drive belt to remove it.
5. Remove the drive belt from around the pulleys.
   To install:
6. Verify that the new belt and the old belt have the same length and width. These measurements must be the same or problems will occur when the new belt is adjusted.
7. Correctly route the belt around the pulleys.
8. After the drive belt is installed correctly, adjust the tension by applying 22 lbs. (98 N) of pressure to the belt, then measuring the deflection. Compare the deflection with the following specifications:
   • Power steering pump without turbocharger — 0.413–0.453 in. (10.5–11.5mm)
   • Power steering pump with turbocharger — 0.35–0.39 in. (9–10mm)
9. Tighten the mounting, pivot, and lock bolts.
10. Connect the negative battery cable.

Power Steering — with adjustable idler pulley

1. Disconnect the negative battery cable.
2. Loosen the lock bolt for the idler pulley.
3. Loosen the idler pulley adjusting bolt. When the idler pulley adjustment bolt is loosened, the drive belt tension will slowly be released.
4. When there is enough slack in the belt, remove the belt from the pulleys.
   To install:
5. Verify that the new belt and the old belt have the same length and width. These measurements must be the same or problems will occur when the new belt is adjusted.
6. Correctly route the belt around the pulleys.

7. After the new belt is installed correctly, adjust the tension by applying 22 lbs. (98 N) of pressure to the belt, then measuring the deflection. Compare the deflection with the following specifications:
   • Power steering pump without turbocharger — 0.413–0.453 in. (10.5–11.5mm)
   • Power steering pump with turbocharger — 0.35–0.39 in. (9–10mm)
8. Tighten the mounting, pivot, and lock bolts.
9. Connect the negative battery cable.

1995–97 3.0L (VG30DETT and VG30DE) Engines

Belt deflection is measured by applying pressure on the drive belt at the mid points between the component pulleys. Deflection is the distance the belt moves when pressure is applied at the midpoint. Always measure belt deflection when the drive belts are cold.

Alternator

1. Disconnect the negative battery cable.
2. Loosen the pivot mounting bolt of the alternator.
3. Loosen the locknut on the alternator adjusting bolt.
4. Turn the alternator adjusting bolt to loosen the drive belt.
5. Remove the alternator drive belt from the pulleys.
   To install:
6. Verify that the new belt and the old belt have the same length and width. These measurements must be the same or problems will occur when the new belt is adjusted.
7. Correctly route the belt around the pulleys.
8. After new belt is installed correctly, adjust the tension of the new belt.
9. Tighten the alternator adjustment locknut to 12–15 ft. lbs. (16–21 Nm) and check the drive belt deflection.
10. Tighten the alternator pivot and mounting bolts.
11. Apply a pushing force of 22 lb. (98 N) to the alternator belt. The belt deflection should be 0.28–0.31 in. (7–8mm) for used belts or 0.256–0.295 in. (6.5–7.5mm) for new belts.
12. After all the drive belts have been installed, run the engine and recheck the drive belt deflection.

Power Steering

The belt tension is adjusted with an adjusting bolt. The adjusting bolt is located on the power steering oil pump.

1. Loosen the power steering oil pump mounting(pivot) bolts.

2. Loosen the drive belt adjustment locking nut.

3. Turn the power steering oil pump adjusting bolt until the drive belt tension has loosened.

4. Remove the power steering drive belt from the pulley and remove the belt from the vehicle.

To install:

5. Verify that the new belt and the old belt have the same length and width. These measurements must be the same or problems will occur when the new belt is adjusted.

6. Install the power steering belt around the crankshaft and power steering pump pulley.

7. Turn the power steering oil pump adjusting bolt until the desired drive belt tension is achieved.

8. Tighten the adjusting bolt locknut to 12–15 ft. lbs. (16–21 Nm).

9. Tighten the power steering oil pump mounting and pivot bolts.

10. Apply a pushing force of 22 lb. (98 N) to the power steering belt. The belt deflection should be 0.472–0.531 in. (12–13.5mm) for used belts or 0.413–0.453 in. (10.5–11.5mm) for new belts.

11. Reconnect the negative battery cable.

12. After all the drive belts have been installed, run the engine and recheck the drive belt deflection.

Air Conditioning Compressor

The idler pulley is the smallest of all the pulleys. There is a bolt, at the top of the slotted bracket holding the idler pulley, which is used to either raise or lower the pulley.

1. Disconnect the negative battery cable.

2. Loosen the locknut in the face of the idler pulley.

3. Loosen the idler pulley by turning the adjusting bolt on the idler pulley.

4. When there is enough slack in the belt, remove the belt from the pulleys.

To install:

5. Verify that the new belt and the old belt have the same length and width. These measurements must be the same or problems will occur when the new belt is adjusted.

6. Correctly route the belt around the pulleys.

7. Turn the air conditioner belt adjusting bolt until the desired drive belt tension is achieved.

8. After adjusting the belt tension, tighten the locknut on the face of the idler pulley to 23 ft. lbs. (31 Nm).

9. Apply a pushing force of 22 lb. (98 N) to the air conditioning compressor belt. The belt deflection should be 0.31–0.35 in. (8–9mm) for used belts or 0.28–0.31 in. (7–8mm) for new belts.

10. After all the drive belts have been installed, run the engine and recheck the drive belt deflection.

Starter

REMOVAL AND INSTALLATION

Sentra, 200SX and 300ZX

1. Disconnect the negative battery cable from the battery.

2. If equipped, remove the starter heat shield.

3. Disconnect the starter wiring at the starter, taking note of the positions for correct reinstallation.

4. Remove the bolts attaching the starter to the engine and remove the starter from the vehicle.

To install:

5. Install the starter motor to the engine and install and tighten the mounting bolts to 23–31 ft. lbs. (31–42 Nm) on GA16DE engines or 30–38 ft. lbs. (41–52 Nm) on SR20DE engines.

6. Reconnect the starter wiring.

7. If removed, install the starter heat shield.

8. Connect the negative battery cable.

9. Check the starter motor for correct operation.

Altima, Maxima and 1993–94 240SX

1. Disconnect the negative battery cable from the battery.

2. Remove the air duct assembly.

3. Disconnect the wiring harness connector at the starter.

4. Remove the starter-to-engine bolts and remove the starter from the vehicle.

NOTE: The starter mounting bolts are two different lengths. Note the positioning during removal.

To install:

5. Install the starter to the bell housing and install the starter bolts to the proper locations.

6. Tighten the long starter mounting bolt to 58–72 ft. lbs. (78–98 Nm), then the short mounting bolt to 23–30 ft. lbs. (31–41 Nm).

7. Reconnect the starter wiring harness connection.

8. Install the air duct assembly.

9. Connect the negative battery cable.

10. Start the engine and verify proper starter operation.

1995–97 240SX

Manual Transmission

1. Disconnect the negative battery cable from the battery.

2. If necessary, remove the air duct and its connector brackets.

3. Disconnect the wiring at the starter, taking note of the positions for correct installation.

4. Remove the starter-to-engine bolts and remove the starter from the vehicle.

To install:

5. Install the starter to the bell housing.

6. Tighten the starter mounting bolts to 22–29 ft. lbs. (29–39 Nm). Be careful not to overtighten the mounting bolts and crack the nose of the starter case.

7. Connect the starter wiring that was removed from the starter.

8. If removed, install the air duct and its connector brackets.

9. Connect the negative battery cable.

10. Start the engine and verify proper starter operation.

Automatic Transmission

1. Disconnect the negative battery cable from the battery.

2. Support the automatic transmission with a jack.

3. Remove the four rear mounting bracket bolts.

4. Slightly lower the transmission for access.

5. Remove the dipstick from the automatic transmission and remove the dipstick tube from the transmission.

6. Remove the connector bracket from the front mount bracket.

7. Disconnect the electrical harness connector from the starter.

8. Remove the bolts that secure the starter and remove the starter assembly.

To install:

9. Install the starter assembly and tighten the mounting bolts to 22–29 ft. lbs. (29–39 Nm).

10. Connect the electrical harness connector to the starter.

11. Install the dipstick tube to the automatic transmission and install the dipstick.

12. Raise the transmission and connect the bolts that secure the front mount.

13. Install the four rear mounting bracket bolts.

14. Remove the jack, connect the negative battery cable, run the engine and check the fluid level.

CHASSIS ELECTRICAL

Blower Motor

REMOVAL AND INSTALLATION

Altima, Sentra, 200SX and 240SX

NOTE: The blower motor is located behind the glove box.

1. Disconnect the negative battery cable.

2. Remove the glove box assembly.

3. Disconnect the electrical harness from the blower motor.

4. Remove the retaining bolts from the bottom of the blower unit, then lower the blower motor from the case. It may be necessary to remove a hose for access. Have a rag handy.

To install:

5. With the blower motor assembly removed, check the case for any debris or signs of fan contact. Inspect the fan for wear spots, cracked blades or hub problems.

6. Install the blower motor assembly and secure with retaining bolts.

7. Connect the electrical harness to the blower motor.

8. Install the glove box assembly.

9. Connect the negative battery cable.

Maxima and 300ZX

NOTE: It may be necessary to remove the glove box assembly to gain clearance for the blower motor removal and installation. The blower motor is located behind the glove box, facing the floor.

1. Disconnect the negative battery cable.

2. Remove all panels and ducting necessary to gain access to the blower motor.

3. Disconnect the blower motor harness wiring connectors.

4. Remove the blower motor retaining screws and lower the motor/wheel from the intake housing.

On some models, release the clips that attach the blower casing to the intake housing to remove the blower motor.

To install:

5. Transfer the old blower wheel to the shaft of the new motor.

6. Raise the blower/wheel assembly up and onto the intake housing. Use a new gasket, if required.

7. Install the blower motor retaining screws or lock the clips.

8. Connect the blower motor wiring.

9. Install all ducting and panels.

10. Connect the negative battery cable.

11. Check the blower for proper operation at all speeds.

Windshield Wiper Motor

REMOVAL AND INSTALLATION

The wiper motor is on the firewall under the hood. The operating linkage is located under the cowl panel.

Front Wiper Motor

1. Disconnect the negative battery cable.

2. Disconnect the wiper motor electrical connection.

3. Make a chalk mark on the windshield on each side to mark the wiper blade position.

4. Lift the wiper arms. Remove the securing nuts and detach the arms.

5. Remove the cowl cover or air intake grille.

6. Remove the bolts securing the wiper motor to the bulkhead.

7. Turn the wiper motor so that the wiper motor link comes through the oblong hole in the cowl top panel. Disconnect the ball joint which connects the motor link and wiper link by pulling the wiper motor straight out.

8. Remove the nuts holding the wiper pivots to the body.

9. Remove the wiper linkage and pivots as an assembly through the oblong hole in the cowl top panel.

To install:

10. Install the wiper linkage and pivots through the oblong hole in the cowl top panel.

11. Apply some white lithium grease to the wiper link ball joint. Press the wiper motor link ball joint into the fitting on the wiper arm linkage.

12. Attach the wiper pivots to the body. Tighten the pivot mounting nuts to 3–9 ft. lbs. (4–12 Nm).

13. Install the wiper motor mounting bolts. Tighten the wiper motor mounting bolts to 3–9 ft. lbs. (4–12 Nm).

14. Install the cowl cover panel or air intake grill.

15. To reduce wiper arm looseness, prior to connecting the wiper arm, make sure the motor spline shaft and pivot area is completely free of debris and corrosion. Wire brush as necessary.

16. When installing the wiper arms, make sure the blades align with the chalk marks made during disassembly before attaching the wiper arm to the pivot. Install the wiper arms and tighten the mounting nuts to 9–17 ft. lbs. (13–23 Nm).

17. The distance from the bottom of the windshield glass to the middle of the wiper blades in the rest position should be as follows:

   a. 240SX — 0.689–1.280 in. (17.5–32.5 mm)

   b. Altima — 1.06–1.61 in. (27–41 mm)

   c. Maxima — 0.20–0.79 in. (5–20 mm)

18. Connect the wiper motor electrical connection and negative battery cable.

19. Verify that the wipers operate correctly.

20. Clean chalk marks from the windshield.

Rear Wiper Motor

1. Disconnect the negative battery cable.

2. Lift up the cover and remove the nut attaching the rear wiper arm to the pivot. Remove the rear wiper arm.

3. Open the rear hatch and remove the door trim panel.

4. Disconnect the wiper motor electrical connection.

5. Remove the wiper motor mounting bolts.

6. Remove the rear wiper motor.

To install:

7. Install the wiper motor and its mounting bolts. Tighten the mounting bolts to 3–4 ft. lbs. (4–5 Nm).

8. Connect the wiper motor electrical connection.

9. Install the rear door trim panel.

10. Install the wiper arm. The distance between the edge of the wiper blade in the rest position and the edge of the black print on the rear glass should be:

   a. Altima and 240SX — 0.98–1.38 in. (25–35 mm)

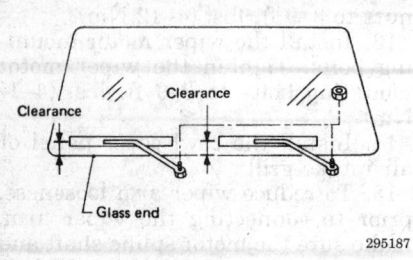

Adjusting the front windshield wiper blade clearance — Altima and 240SX

b. Maxima — 1.18–1.97 in. (30–50 mm)

11. Tighten the wiper blade mounting nut to 9–13 ft. lbs. (13–18 Nm).

12. Verify that the wipers operate correctly.

Combination Switch

REMOVAL AND INSTALLATION

Except 300ZX

———— CAUTION ————
The air bag system (SRS or SIR) must be disarmed before removing the combination switch. Failure to do so may cause accidental deployment, property damage, or personal injury.

1. Disarm the air bag, if equipped and disconnect the battery ground cable.

2. Remove the steering wheel. On Maxima with Sonar suspension, remove the steering angle sensor from the steering column.

3. Remove the steering column covers.

NOTE: At this point, the individual switch assemblies can be removed without removing the combination switch base assembly. To service an individual switch/stalk assembly, disconnect the electrical lead and remove the two stalk-to-base mounting screws. If the switch base must be removed, proceed with the remainder of the removal procedure.

4. Disconnect the electrical plugs from the switch.

5. Remove the retaining screws, push down on the base of the switch with moderate pressure and twist the switch and pull it from the steering wheel shaft.

To install:

6. Install the remaining components in the reverse order of removal.

7. Check the switch functions for proper operation. Many vehicles have turn signal switches that have a tab which must fit into a hole in the steering shaft. This fit is necessary in order for the system to return the switch to the neutral position after the turn has been made. Be sure to align the tab and the hole when installing the combination switch.

300ZX

———— CAUTION ————
The air bag system (SRS or SIR) must be disarmed before removing the steering wheel. Failure to do so may cause accidental deployment, property damage or personal injury.

1. Disarm air bag, if equipped, refer to the necessary service procedure. Disconnect the battery ground cable.

2. Remove the steering wheel.

3. Remove the steering column covers.

NOTE: At this point, the individual switch assemblies can be removed without removing the combination switch base assembly. To service an individual switch/stalk assembly, disconnect the electrical lead and remove the two stalk-to-base mounting screws. If the switch base must be removed, proceed with the remainder of the removal procedure.

4. Disconnect the electrical plugs from the switch.

5. Remove the retaining screws, push down on the base of the switch with moderate pressure and twist the switch from the steering wheel shaft.

To install:

6. Reinstall the switch base and retaining screws.

7. Reconnect the wiring to the combination switch.

8. Reinstall the steering wheel covers.

NOTE: Many vehicles have turn signal switches that have a tab which must fit into a hole in the steering shaft. This fit is necessary in order for the system to return the switch to the neutral position after the turn has been made. Be sure to align the tab and the hole when installing.

9. Reinstall the steering wheel.

10. Rearm the air bag system.

11. Reconnect the negative battery cable.

Ignition Switch and Cylinder

REMOVAL AND INSTALLATION

The steering lock/ignition switch/warning buzzer switch assembly is attached to the steering column by special bolts whose heads shear off upon installation. The bolts must be drilled out to remove the assembly or removed with an appropriate tool. The bolts may also be removed with a hammer and chisel by notching the bolts and tapping them counterclockwise with the hammer and chisel.

NOTE: The ignition switch or warning switch can be replaced without removing the steering lock assembly. The ignition switch is on the back of the assembly and the warning switch on the side.

1. Disconnect the negative battery cable and insulate the terminal end.

———— CAUTION ————
The air bag system must be disarmed before removing the steering wheel. Failure to do so may cause accidental deployment, property damage or personal injury.

2. Disarm the air bag (SRS) system.

3. Remove the steering wheel and the steering column upper and lower covers.

4. Remove the spiral cable assembly.

5. Remove the combination switch from the steering column.

6. Remove steering column support nuts and lower the steering column.

7. Disconnect the ignition switch wiring.

8. Remove the bolts that secure the steering lock and remove steering lock assembly.

To install:

9. Install the steering lock assembly and secure with new shear type bolts.

10. Connect the ignition switch wiring harness.

11. Raise the steering column and secure with mounting nuts. Tighten the mounting nuts to 11–14 ft. lbs. (15–19 Nm).

12. Install the spiral cable assembly.

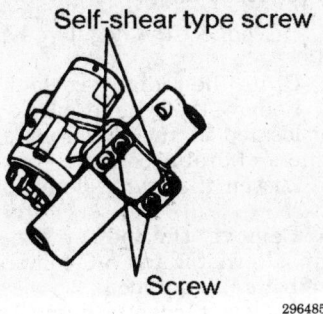

Ignition switch

296485

13. Install the combination switch, steering column covers, and the steering wheel.

14. Connect the negative battery cable and enable the air bag system.

Park/Neutral Safety Switch

REMOVAL AND INSTALLATION

1. Raise and support the vehicle safely.

2. Disconnect the manual control linkage from the manual shaft.

3. Disconnect the harness connectors from the park/neutral safety switch.

4. Remove the bolts that secure the switch to the side of the transfer case and remove the switch.

To install:

5. Install the switch and secure with mounting bolts.

6. Connect the harness connectors to the park/neutral switch.

7. Set the manual shaft on the side of the transaxle to the **N** position.

8. Loosen the park/neutral switch mounting screws enough to allow for movement of the switch.

9. Insert a 0.16 in. (4 mm) diameter pin and move the switch until the pin falls through the locating hole in the inhibitor switch and manual shaft. Tighten the switch screws equally.

10. Remove the pin and connect the manual control linkage to the shaft.

11. Check for continuity at the park/neutral switch in the **N**, **P** and **R** ranges.

12. Make sure while holding the brakes on, that the engine will start only in **P** or **N** and that the backup lights only illuminate in reverse.

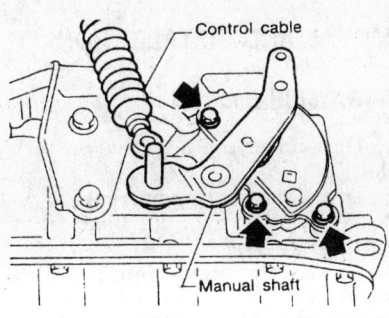

Control cable and adjustment pin

291510

Powertrain Control Module

REMOVAL AND INSTALLATION

NOTE: It is recommended that a grounding strap be worn when handling the ECM. A grounding strap will prevent a static electric discharge from damaging the component.

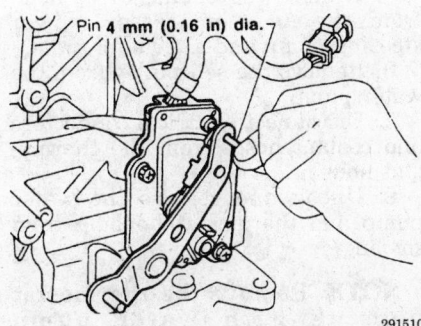

Except 300ZX

The ECM is located under the radio at the front of the console on most of the models. The console does not have to be removed to gain access to the control module. On the other models, the ECM is located between the front center console and is accessible through the passengers side.

1. Disconnect negative battery cable.

2. Remove the console side covers.

3. Remove electronic control module (ECM) bracket bolts.

4. Swing the ECM out from under the console.

5. Disconnect wiring harness by removing the connector hold down bolt.

6. Unplug the connector.

7. Remove the ECM.

To install:

— WARNING —

Before installing a new ECM, perform input/output signal inspection at the ECM connector terminals to prevent damage to the new ECM.

---

8. Plug the wiring harness into the ECM.

9. On the Altima, Maxima and 240SX models, install the ECM into position, then tighten the securing bolt until the gap between the orange indicator disappears.

10. On the Sentra and 200SX models, tighten the hold down bolt until the red projector is in line with the connector face. Install the ECM into its proper location.

11. If necessary, reattach the ECM bracket bolts.

12. If removed, install the console side covers.

13. Reconnect the negative battery cable.

300ZX

The electronic control module is located behind the glove compartment. The glove compartment must be removed to gain access to the ECM.

1. Disconnect negative battery cable.

2. Remove the glove box.

3. Remove electronic control module (ECM) bracket bolts.

4. Swing the ECM through the glove box opening.

5. Disconnect wiring harness by removing the connector hold down bolt.

6. Unplug the connector.

7. Remove the ECM.

To install:

8. Plug the wiring harness into the ECM.

9. Tighten the hold down bolt until the red projection is in line with the connector face.

10. Reinstall the ECM in to the dashboard.

11. Reinstall the ECM bracket bolts.

12. Install the glove box assembly

13. Reconnect the negative battery cable.

ENGINE COOLING

Radiator

REMOVAL AND INSTALLATION

1. Disconnect the negative battery cable.
2. Remove the cap and drain the radiator via the drain plug in the bottom tank. Be sure to only do this when the engine is cold.
3. If necessary to gain access, unbolt and set aside the power steering pump.

NOTE: Do not disconnect the power steering pressure hoses or drain the system.

4. Disconnect the upper and lower radiator hoses, along with the coolant overflow reservoir hose.
5. On the 300ZX models, raise and safely support the vehicle, then remove the engine undercover.
6. If necessary, disconnect the water temperature switch electrical wiring.
7. Disconnect the fan motor wires and remove the fan/shroud assembly.
8. If equipped with an automatic transaxle, disconnect and cap the A/T oil cooling lines at the radiator.
9. Unbolt the radiator support brackets and remove the brackets.
10. Remove the radiator assembly from the vehicle.
To install:

NOTE: Be sure the rubber mounting bushings are in position before the radiator is installed.

11. Install the radiator to the vehicle.
12. Install radiator support brackets and tighten mounting nuts evenly.
13. Install the fan motor and blade to the shroud assembly. Tighten the fan shroud mounting bolts to 29–44 inch lbs. (3–5 Nm).
14. Install the remaining components in the opposite order from which they were removed.
15. Operate the engine until warm and then check for leaks. Check the coolant level and the fan operation.
16. If the A/T oil cooling lines were disconnected, check transaxle fluid level and add fluid as necessary.

Water Pump

REMOVAL AND INSTALLATION

1.6L (GA16DE) Engine

1. Disconnect the negative battery cable.
2. Drain the cooling system.
3. Remove the cylinder head front mounting bracket.
4. Loosen the water pump pulley bolts.
5. Remove the engine drive belts from the A/C compressor, power steering pump and alternator.
6. Remove the belt pulley from the water pump.
7. Disconnect electrical connectors and coolant hoses from the thermostat housing.
8. Unbolt and remove the water pump and thermostat housing from the engine.

NOTE: Remove the thermostat housing with water pump assembly.

9. Remove the bolts that secure the thermostat housing to the water pump.
10. Remove all traces of gasket material from sealing surfaces.
To install:
11. Apply a continuous bead of liquid sealer to the sealing surface of the thermostat housing. The sealant should be 0.079–0.118 in. (2–3mm) diameter.
12. Install the thermostat housing to the water pump and tighten mounting bolts to 56–73 inch lbs. (7–8 Nm).
13. Apply a continuous bead of liquid sealer to the sealing surface of the water pump. The sealant should be 0.079–0.118 in. (2–3mm) diameter.
14. Install the water pump on the engine and tighten mounting bolts to 56–73 inch lbs. (7–8 Nm).
15. Install the pulley to the water pump and tighten the mounting bolts to 56–73 inch lbs. (7–8 Nm).
16. Connect electrical connectors and coolant hoses to the thermostat housing.
17. Install and adjust the alternator, power steering and A/C compressor drive belts.
18. Refill the cooling system and connect the negative battery cable.
19. Start the engine, bleed the cooling system, warm the engine to full operating temperature, and check for leaks.
20. If necessary, refill the cooling system when the engine has cooled.

2.0L (SR20DE) Engine

1. Disconnect the negative battery cable.
2. Drain the radiator coolant.
3. Remove the cylinder block drain plug located at the left front of the engine and drain coolant.
4. Loosen the water pump pulley bolts.
5. Remove the power steering pump, alternator and A/C compressor drive belts (if equipped).
6. Remove the water pump pulley.
7. Note positioning of power steering pump adjusting bracket and remove the power steering pump adjusting bracket from the water pump. If necessary, remove the power steering pump for access to bracket.

NOTE: When removing the power steering pump, it is not necessary to disconnect the pressure hoses or drain the system. Position or tie the pump aside.

8. Support the engine and remove the front engine mount.
9. Remove the mounting bolts from the water pump and remove the water pump.
10. Remove all traces of liquid gasket material from sealing surfaces.
To install:
11. Apply a continuous bead of liquid sealer to the mating surface of the water pump. Sealer should be 0.079–0.118 in. (2–3mm) wide.
12. Install the water pump assembly and tighten mounting bolts to 12–15 ft. lbs. (16–21 Nm).

NOTE: Be sure to properly position the adjusting bracket that was noted during removal.

13. Install the front engine mount.
14. Install the power steering pump adjusting bracket and install the power steering pump if removed.
15. Install the water pump pulley and tighten mounting bolts to 55–73 inch lbs. (6–8 Nm).
16. Install and adjust the power steering pump, alternator and A/C compressor drive belts (if equipped).
17. Install the cylinder block drain plug located at the left front of the engine and tighten drain plug to 70–104 inch lbs. (8–12 Nm).
18. Refill the cooling system and connect the negative battery cable.
19. Start the engine, bleed the cooling system, and check for leaks.

2.4L (KA24DE) Engine

Altima

1. Disconnect the negative battery cable.

2. Drain the cooling system and water pipe, using the drain plugs.

3. Remove the upper radiator hose to provide working room and remove the drive belt(s) from the pulleys.

4. Remove the alternator and the A/C compressor.

NOTE: Do not disconnect the A/C compressor lines. Unbolt the compressor and lay it off to the side.

5. Remove the water pump pulley.

6. Remove the mounting bolts and remove the water pump from the engine.

NOTE: The mounting bolts are different sizes and must be reinstalled in the correct location, therefore it is a good idea to arrange the bolts so that they can be easily identified during installation.

To install:

7. Make sure all gasket surfaces are clean and properly apply a continuous bead of silicone sealer to the pump.

8. Install the pump to the engine and torque the 6mm bolts to 57–66 inch lbs. (6–8 Nm) and the 8mm bolts to 12–14 ft. lbs. (16–19 Nm).

9. Install the water pump pulley and torque the bolts to 57–66 inch lbs. (6–8 Nm).

10. Install the remaining components in the opposite order from which they were removed.

11. Fill and bleed the cooling system.

12. Start the engine and check for leaks.

240SX

1. Disconnect the negative battery cable.

2. Drain the cooling system and the engine block, using the radiator petcock and cylinder block drain plug.

3. Remove the upper radiator hose to provide working room and remove the drive belt(s) from the pulleys.

4. Remove the retaining screws and lift the fan shroud from the engine.

5. While holding the pulley, remove the nuts retaining the cooling fan and pulley to the water pump.

6. Remove the mounting bolts and remove the water pump from the engine.

To install:

7. Make sure all gasket surfaces are clean and properly apply liquid sealer to the pump.

8. Install the pump to the engine and torque the bolts to 12–14 ft. lbs. (16–19 Nm).

9. Install the remaining components in the reverse order of removal.

10. Tighten the fan clutch, fan, and pulley mounting nuts to 66 inch lbs. (8 Nm).

11. Start the engine and check for leaks.

3.0L (VE30DE) Engine

1. Disconnect the negative battery cable.

2. Drain the cooling system to a level below the water pump.

3. Remove the drive belts from the front of the engine.

4. Remove the water pump pulley-to-water pump bolts.

5. Remove the water pump-to-engine bolts.

6. Remove the water pump from the engine. If the water pump has excessive end play or rough operation, replace it.

To install:

7. Clean the gasket mounting surfaces and grooves.

8. Using liquid sealant, apply a continuous bead of it to the gasket mounting surfaces of the water pump.

9. Install the water pump and torque water pump-to-engine bolts to 12–15 ft. lbs. (16–21 Nm).

10. Install the remaining components in the opposite order from which they were removed.

11. Tighten the water pump pulley-to-water pump bolts to 5–6 ft. lbs. (6–8 Nm).

12. Start the engine, run to normal operating temperature and check for the correct coolant level and for leaks.

3.0L (VG30E) Engine

1. Disconnect the negative battery cable.

2. Drain the cooling system through the cock at the bottom of the radiator and the drain plug, on the left side of the block, behind the alternator.

3. Remove the upper radiator hose and thermostat housing.

4. Loosen and remove the drive belt(s) from the water pump pulley.

5. If necessary, remove the drive belt tensioner.

6. Remove the water pump pulley.

7. Remove the timing belt cover.

NOTE: Be careful to keep coolant off the timing belt.

8. Remove the water pump-to-engine bolts (noting different lengths)

and remove the pump. If the water pump has excessive end-play or rough operation, replace it.

To install:

9. Clean the gasket mounting surfaces. If originally equipped with a gasket, use a new gasket and coat the gasket with sealant. If not originally equipped with a gasket, use liquid silicone gasket applied in an even bead around the mounting surfaces.

10. Reinstall the pump and torque the water pump-to-engine bolts evenly to 12–15 ft. lbs. (16–21 Nm).

11. The remaining components are installed in the reverse order from which they were removed.

12. Start the engine, run to normal operating temperature and check for the correct coolant level and for leaks.

3.0L (VQ30DE) Engine

1. Disconnect the negative battery cable.

2. Drain the coolant from the plugs on the radiator and both sides of the engine block.

3. Position a jack under the oil pan for support. Be sure to place a block of wood on the jack for protection to the engine parts.

4. Remove the right side engine mount and engine mounting bracket.

5. Remove the drive belts and the idler pulley bracket.

6. Remove the chain tensioner cover and the water pump cover.

7. Push the timing chain tensioner sleeve and apply a stopper pin so it does not return.

8. Remove the timing chain tensioner assembly.

9. Remove the three bolts that secure the water pump.

10. Rotate the crankshaft 20 degrees counterclockwise to provide timing chain slack.

11. Put M8 bolts to two M8 threaded holes of the water pump.

12. Tighten each bolt by turning alternately ½ turn until they reach the timing chain rear case. Be sure to turn each bolt ½ turn at a time to prevent damage.

13. Lift up the water pump and remove it.

14. When removing the water pump, do not allow the water pump gear to hit the timing chain.

15. Remove and discard the O-rings from the water pump.

16. Clean all traces of liquid gasket from the water pump and covers.

To install:

17. Using new O-rings, install the water pump to the engine block.

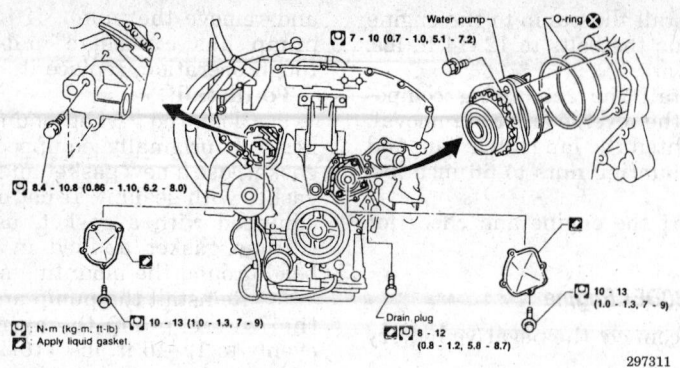

N·m (kg-m, ft-lb)
Apply liquid gasket

Water pump and timing cover assembly — 3.0L (VQ30DE) engine

18. Tighten the three water pump mounting bolts evenly to 62–86 inch lbs. (7–10 Nm).

19. Rotate the crankshaft pulley to its original position by turning it 20 degrees clockwise.

20. Install the timing chain tensioner and tighten the mounting bolts to 75–96 inch lbs. (9–10 Nm).

21. Remove the stopper pin from the timing chain tensioner.

22. Apply a continuous 0.091–0.130 inch (2.3–3.3mm) bead of liquid sealant to the mating surfaces of the timing chain tensioner and water pump covers.

23. Install the timing chain tensioner and water pump covers to the engine block. Tighten the cover mounting bolts to 84–108 inch lbs. (10–13 Nm).

24. Install the drive belts and the idler pulley bracket.

25. Install the right side engine mounting bracket and the engine mount.

26. Remove the jack from under the engine and install the drain plugs to the cylinder block.

27. Connect the negative battery cable and refill the cooling system.

28. Start the engine, bleed the cooling system, and check for leaks.

3.0L (VG30DETT and VG30DE) Engines

1. Disconnect negative battery cable.

2. Drain coolant from radiator and engine block.

3. Remove the undercover and the radiator.

4. Remove cooling fan assembly, water inlet, outlet and drive belts.

5. Remove crankshaft pulley and timing belt cover.

6. Remove water pump.

To install:

7. Remove all traces of gasket material.

8. Apply a continuous bead of liquid gasket to water pump mating surface.

9. Install water pump to engine block.

10. Torque water pump bolts to 12–15 ft. lbs. (16–21 Nm).

11. Reinstall timing belt cover and tighten the cover bolts to 26–43 inch lbs. (3–5 Nm).

12. Replace crankshaft pulley and tighten the mounting bolt to 159–174 ft. lbs. (216–235 Nm).

13. Reinstall the drive belts, water inlet, water outlet, and cooling fan assembly.

14. Torque cooling fan nuts to 51–86 inch lbs. (6–10 Nm).

15. Reinstall the radiator and undercover.

16. Refill and bleed the cooling system.

17. Reconnect the negative battery cable

18. Start engine and check for proper operation.

Thermostat

REMOVAL AND INSTALLATION

1.6L (GA16DE) and 2.0L (SR20DE) Engines

NOTE: The thermostat is housed in the water outlet casting on the cylinder head on most models. On the 1995–97 GA16DE engines, the thermostat is housed in the water outlet casting attached to the water pump.

1. Disconnect the negative battery cable.

2. When cool, open the drain cock on the radiator and drain the coolant into a suitable drain pan.

3. Remove the upper radiator hose from the water outlet side and remove the bolts securing the water outlet to the cylinder head and any other components in the way.

4. Remove the water outlet and thermostat.

To install:

5. Clean off the old gasket from the mating surfaces of the thermostat housing and engine block.

6. When installing the thermostat, be sure to install a new gasket and be sure the air bleed hole in the thermostat is facing upward and that the spring is toward the inside of the engine.

7. Install the water outlet and tighten the bolts.

• GA16DE — 13–16 ft. lbs. (18–22 Nm)

• SR20DE — 5–6 ft. lbs. (6–8 Nm)

8. If removed, install the exhaust air induction tube clamp bolts.

9. Install the upper radiator hose, close drain cock and refill cooling system.

10. Connect the negative battery cable.

11. Start the engine and check for leaks.

2.4L (KA24DE) Engine

Altima

1. Disconnect the negative battery cable and drain the cooling system.

2. Remove the lower hose from the thermostat housing at the rear of the engine.

3. Remove the bolts that secure the thermostat housing and remove the housing from the engine.

4. Remove the thermostat from the engine.

5. Clean the gasket mating surfaces.

To install:

6. Clean the gasket surfaces on both the cylinder block and the thermostat housing.

7. Apply liquid gasket to the thermostat housing. The gasket material should be 0.079–0.118 in. (2.0–3.0mm) wide.

8. Install the thermostat to the housing. Be sure to position the thermostat with the jiggle valve to the top of the housing.

9. Attach the housing to the engine.

10. Tighten the housing mounting bolts as follows:

a. 6mm — 57–66 inch lbs. (6–8 Nm)

b. 8mm — 12–14 ft. lbs. (16–19 Nm)

11. Using a new hose clamp, connect the lower hose to the thermostat housing.

12. Connect the negative battery cable and refill the cooling system.

13. Start the engine, check for leaks, and bleed the cooling system.

240SX

1. Disconnect the negative battery cable.

2. Drain the engine coolant into a clean container so that the level is below the thermostat housing.

3. Disconnect the upper radiator hose at the water outlet.

4. Loosen the three securing bolts and remove the water outlet, gasket, and the thermostat from the thermostat housing.

To install:

NOTE: When installing the thermostat, be sure to position the jiggle valve upward.

5. Install the thermostat to the engine with the thermostat spring toward the inside of the engine.

6. Using liquid sealer, install the thermostat housing and tighten the mounting bolts to 61–66 inch lbs. (7–8 Nm).

7. Using a new hose clamp, connect the upper radiator hose to the thermostat housing.

8. Connect the negative battery cable.

9. Refill and bleed the cooling system. Start the engine and check for leaks. Allow the engine reach normal operating temperature then check coolant for the correct level.

3.0L (VE30DE) Engine

The thermostat is located at the rear of the engine, directly above the transaxle.

1. Disconnect the negative battery cable.

2. Drain the cooling system to a level below the thermostat housing.

3. Disconnect the radiator hose from the water inlet housing.

4. Remove the water inlet housing-to-thermostat housing bolts. Separate the water inlet housing from the thermostat housing. Remove the thermostat from the thermostat housing.

To install:

5. Clean the gasket mounting surfaces.

6. Using liquid sealant, apply a bead to the water inlet housing.

7. Install the thermostat into the thermostat housing with the thermostat spring facing the thermostat housing and the thermostat pintle in the upward direction.

8. Install the water inlet housing and torque the water inlet housing-to-thermostat housing bolts to 12–15 ft. lbs. (16–21 Nm).

9. Reconnect the radiator hose to the water inlet housing.

10. Refill the cooling system.

11. Connect the negative battery cable.

12. Start the engine and allow it to reach normal operating temperatures; then check for leaks. After cooling, recheck the coolant level.

3.0L (VG30E) Engine

The thermostat is located at the front of the engine, directly above the water pump.

1. Disconnect the negative battery cable.

2. Drain the cooling system to a level below the thermostat housing.

3. Disconnect the water hose from the water outlet housing.

4. Remove the water outlet housing-to-thermostat housing bolts. Separate the water outlet housing from the thermostat housing. Remove the thermostat from the thermostat housing.

To install:

5. Clean the gasket mounting surfaces.

6. Using liquid sealant, apply a bead to the water outlet housing.

7. Install the thermostat into the thermostat housing with the thermostat spring facing the hose and the thermostat pintle in the upward direction.

8. Install the water outlet housing and torque the water outlet housing-to-thermostat housing bolts to 12–15 ft. lbs. (16–21 Nm).

9. Reconnect the hose to the front housing.

10. Refill the cooling system.

11. Connect the negative battery cable.

12. Start the engine and allow it to reach normal operating temperatures; then check for leaks. After cooling, recheck the coolant level.

3.0L (VQ30DE) Engine

1. Disconnect the negative battery cable.

2. Drain the cooling system from the radiator and both sides of the engine block.

3. Disconnect the radiator hose from the water inlet housing.

4. Remove the water inlet housing mounting nuts. Separate the water inlet housing from the engine.

5. Remove the thermostat from the thermostat housing.

To install:

6. Clean the gasket mounting surfaces.

7. Install the thermostat into the thermostat housing with the thermo-

stat spring facing toward the engine. Be sure to position the jiggle valve facing upward.

8. Using a new gasket, install the water inlet housing and torque the water inlet housing mounting nuts to 74–99 inch lbs. (9–11 Nm).

9. Connect the radiator hose to the water inlet housing. Be sure to use a new clamp.

10. Refill the cooling system.

11. Connect the negative battery cable.

12. Start the engine, bleed the cooling system, and allow it to reach normal operating temperatures. After cooling, check for leaks and recheck the coolant level.

1995–97 2.0L (SR20DE) Engine

NOTE: The engine thermostat is housed in the water inlet housing where the lower hose attaches to the engine block.

1. Disconnect the negative battery cable.

2. When cool, open the drain cock on the radiator and drain the coolant into a suitable drain pan.

3. Remove the lower radiator hose from the thermostat housing.

4. Remove the bolts that secure the water inlet to the thermostat housing and remove the thermostat.

To install:

5. Clean off the old gasket from the mating surfaces of the thermostat housing.

6. When installing the thermostat, be sure to install a new gasket or apply liquid gasket and be sure the air bleed hole in the thermostat is facing upward and that the spring is toward the inside of the engine.

7. Install the thermostat and the water inlet housing.

8. Install the mounting bolts and tighten to 56–74 inch lbs. (6–8 Nm).

9. Install the lower radiator hose, close drain cock and refill cooling system.

10. Connect the negative battery cable.

11. Start the engine, check for leaks, and bleed the cooling system.

3.0L (VG30DE and VG30DETT) Engines

1. Disconnect the negative battery cable.

2. Drain the coolant from both sides of the engine block and drain the radiator.

3. Remove the engine undercover.

4. Remove the lower radiator hose.

5. Remove the radiator shroud.

6. Remove the engine drive belts.

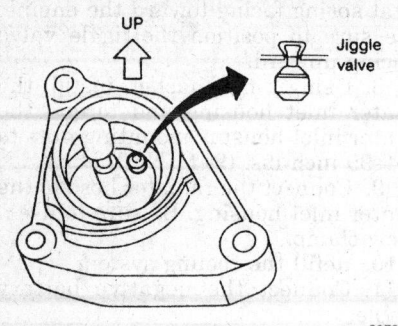

297214

Positioning of the jiggle valve — 3.0L (VQ30DE) engine

7. Remove the cooling fan, the fan coupling, and the water pump pulley.

8. Remove the nuts and bolts that secure the water inlet and remove the inlet assembly.

9. Remove the thermostat assembly from the water inlet.

10. Clean the gasket mounting surfaces.

To install:

11. Using a new gasket, install the thermostat into the water inlet housing.

NOTE: Be sure to position the thermostat with the jiggle valve facing upward.

12. Install the water housing and secure with mounting bolts/nuts. Tighten the nuts and bolts to 12–14 ft. lbs. (16–19 Nm).

13. Install the water pump pulley, cooling fan, and the fan coupling .

14. Install and adjust the engine drive belts.

15. Install the radiator shroud.

16. Using new clamps, install the lower radiator hose.

17. Install the engine undercover.

18. Refill the cooling system and connect the negative battery cable.

19. Start the engine, bleed the cooling system, and check for leaks.

Electric Cooling Fan

REMOVAL AND INSTALLATION

Sentra and 200SX

1. Disconnect the negative battery cable.

2. Drain the cooling system.

3. If necessary to gain access, unbolt and set aside the power steering pump assembly.

NOTE: Do not disconnect the power steering pressure hoses or drain the system.

4. Disconnect the upper radiator hose.

5. Unbolt the fan shroud assembly from the radiator and/or support.

NOTE: On some vehicles, it may be necessary to remove the radiator first before removing the fan and shroud assembly.

6. Disconnect the fan motor wires and remove the fan and shroud assembly.

7. If necessary to remove radiator first, follow the proper procedures for radiator removal first.

To install:

8. Install the fan motor and blade to the shroud assembly and tighten the bolts.

9. Install the radiator in the vehicle and torque the mounting bolts evenly.

10. The remaining components are installed in the reverse order from which they were removed.

11. Connect the fan motor electrical connectors.

12. Connect the upper radiator hose and all other cooling system hoses, that may have been removed, to the radiator.

13. If removed, install the power steering pump.

14. Refill the radiator with proper coolant / water mixture and follow the appropriate system bleeding procedures.

15. Connect the negative battery cable.

16. Operate the engine until warm and then check for leaks, coolant level, and fan operation.

Altima and 240SX with 2.4L (KA24DE) and Maxima with 3.0L (VG30E) Engines

1. Disconnect the negative battery cable.

2. On the some models it may be necessary to remove the radiator cap and drain the radiator via the drain plug in the bottom tank. Remove the upper radiator hose.

3. Disconnect the cooling fan electrical connectors.

4. Remove the bolts securing the fan/shroud assembly from the radiator.

5. Remove the fan and shroud assembly from the vehicle.

To install:

6. Install the cooling fan and shroud assembly. Tighten the mounting bolts to 29–37 inch lbs. (3–5 Nm).

7. Using new clamps, install the upper hose if removed.

8. Refill the cooling system with 50/50 antifreeze/water mix and bleed the system.

9. Connect the cooling fan electrical connectors and connect the negative battery cable.

10. Run engine and check the cooling system for leaks and verify the operation of the cooling fans.

300ZX

1. Disconnect negative battery cable.

2. Remove drive belts.

3. Remove fan bolts from fan coupling.

4. Remove fan assembly.

5. Remove nuts that attach fan coupling to fan pulley.

6. Remove fan coupling.

To install:

NOTE: Access to the cooling fan may not be possible from above with the fan shroud installed. Removing the lower fan shroud from underneath the vehicle could allow access to fan clutch nuts. If the fan nuts are not accessible with the lower fan shroud removed, removal of radiator upper fan shroud will be necessary.

To install:

7. Install the fan blade assembly to the fan clutch and tighten the mounting bolts to 52–86 inch lbs. (6–10 Nm).

8. Install the fan clutch and fan blade assembly to the water pump. Tighten the mounting bolts to 52–86 inch lbs. (6–10 Nm).

9. Install the drive belts and adjust.

10. Reinstall the fan shrouds, if removed.

11. Connect the negative battery cable.

Cooling System Bleeding

PROCEDURE

1.6L (GA16DE) And 2.0L (SR20DE) Engines

1. Turn the heater temperature control to the HOT position and make sure the A/C is OFF.

2. Remove the air relief plug and air bleeder cap.

3. Fill the radiator with the proper mix of coolant and water until the coolant starts to spill out of the relief plug.

4. Install the air relief plug and air bleeder cap.

5. Fill the coolant reservoir up to the H or MAX line.

6. Install a steel wire between the negative pressure valve and its seat on the radiator cap. This allows the air and coolant to be directed into the coolant reservoir regardless of system pressure.

7. Install the radiator cap.

8. Run the engine until it reaches normal operating temperature.

9. Raise the engine speed to 2500 RPM for ten seconds, then let the engine return to idle speed.

10. Repeat step 9 two or three times.

NOTE: Watch the coolant temperature gauge to avoid overheating the engine.

11. Stop engine and allow it to cool.

NOTE: A fan can be used to expedite cool down time.

12. Remove the radiator cap and check coolant level.

13. Top off the coolant level in the radiator and in the coolant reservoir.

14. Reinstall the radiator cap and repeat steps 9 through 13 two more times.

15. Stop engine and allow it to cool.

16. Remove the radiator cap and remove the steel wire from negative pressure valve.

17. Reinstall the cap and warm engine to normal operating temperature.

18. Raise the engine speed to 4000 RPM and listen for coolant flow. Move the temperature control lever to several positions between cool and hot while running engine at 4000 RPM. The sound may be most noticeable at the heater water cock.

19. If water sounds are heard, use the following procedures:

   a. On 1.6L (GA16DE) engines, bleed the air from the cooling system by repeating steps 6 through 10

   b. On 2.0L (SR20DE) engines, follow steps C through M

   c. Allow the engine to cool down and remove the air bleeder cap from the heater inlet hose.

   d. Attach a suitable clear hose at the bleeder pipe and put the opposite end in the coolant overflow reservoir.

   e. Install a steel wire between the negative pressure valve and its seat on the radiator cap.

   f. Install the radiator cap and start the engine.

   g. Look for bubbles in the coolant reservoir tank.

   h. Set the heater temperature control lever to its MAX. cool position in order to bypass coolant through the clear hose.

   i. Run engine at 2300 RPM until all bubbles disappear from the clear hose.

NOTE: DO NOT run engine at higher than 2300 RPM, because engine damage may result from the reduced coolant flow.

   j. When the bubbles are gone, move heater control lever to the HOT position and check for the sound of coolant flow. If sound is heard, repeat steps H through J.

   k. Stop the engine and allow it to cool.

   l. Remove the cap and remove the steel wire. If necessary, top off the coolant and reinstall the radiator cap.

   m. Remove the clear hose and reinstall the air bleeder cap.

20. Check for leaks.

2.4L (KA24DE) and 3.0L (VE30DE and VG30E) Engines

1. Move the heater temperature control lever to the hot position.

2. Open the air relief plug.

3. Fill the radiator with coolant until coolant starts to spill from the air relief plug.

4. Close the relief plug and start the engine.

5. Run the engine until warm.

6. Race the engine 2 or 3 times under no load.

7. Stop the engine and allow it to cool.

8. Check the coolant level in the radiator and add if necessary.

9. Fill the reservoir to the MAX line.

10. Reinstall the radiator cap, start the engine, and check for leaks.

3.0L (VQ30DE, VG30DE and VG30DETT) Engines

1. Disconnect the negative battery cable.

2. With the coolant drained, close the radiator drain plug and install the engine drain plugs.

3. Slowly fill the radiator with the proper mixture of coolant and water. Fill the reservoir and install the radiator cap.

4. Connect the negative battery cable.

5. Start the engine and allow it to reach normal operating temperature, watch the coolant temperature gauge for signs of overheating. Race the engine two or three times with no-load.

6. Shut the engine **OFF** and let it cool down completely.

7. After the engine has cooled down remove the radiator cap and fill the radiator to the filler opening. Fill the reservoir to the **H** level.

8. Check the radiator drain plug and the engine drain plugs for leakage.

9. Install the radiator cap.

FUEL SYSTEM

Fuel System Service Precautions

Safety is the most important factor when performing not only fuel system maintenance but any type of maintenance. Failure to conduct maintenance and repairs in a safe manner may result in serious personal injury or death. Maintenance and testing of the vehicle's fuel system components can be accomplished safely and effectively by adhering to the following rules and guidelines.

• To avoid the possibility of fire and personal injury, always disconnect the negative battery cable unless the repair or test procedure requires that battery voltage be applied.

• Always relieve the fuel system pressure prior to disconnecting any fuel system component (injector, fuel rail, pressure regulator, etc.), fitting or fuel line connection. Exercise extreme caution whenever relieving fuel system pressure to avoid exposing skin, face and eyes to fuel spray. Please be advised that fuel under pressure may penetrate the skin or any part of the body that it contacts.

• Always place a shop towel or cloth around the fitting or connection prior to loosening to absorb any excess fuel due to spillage. Ensure that all fuel spillage (should it occur) is quickly removed from engine surfaces. Ensure that all fuel soaked cloths or towels are deposited into a suitable waste container.

• Always keep a dry chemical (Class B) fire extinguisher near the work area.

• Do not allow fuel spray or fuel vapors to come into contact with a spark or open flame.

• Always use a backup wrench when loosening and tightening fuel line connection fittings. This will prevent unnecessary stress and torsion to fuel line piping. Always follow the proper torque specifications.

• Always replace worn fuel fitting O-rings with new. Do not substitute fuel hose or equivalent, where fuel pipe is installed.

Fuel System Pressure

RELIEVING

Except 300ZX

The fuel pump fuse is located in the dash fuse box or in the engine compartment fuse box. Check the lid of the fuse box for exact location.
1. Remove the fuel pump fuse.
2. Start the engine.
3. Start the engine and run until the engine stalls.
4. After the engine stalls, try to restart the engine; if the engine will not start, the fuel pressure has been released.
5. Turn the ignition switch OFF. Reinstall the fuel pump fuse into the fuse block.

NOTE: Do not crank the engine or turn the ignition switch ON after the fuel pump fuse has been reinstalled, or the fuel pressure will be re-established.

300ZX

The fuel pump relay is located in the drivers side kick panel near the fuse box.
1. Remove the fuel pump relay.
2. Start the engine.
3. The engine should run and then stall when the fuel in the lines is exhausted. When the engine stops, crank the starter a few times for about 5 seconds to make sure all pressure in the fuel lines is released.
4. Turn the ignition OFF and re-install the fuel pump relay.

NOTE: Do not crank the engine or turn the ignition switch to the ON position after reinstalling the relay, or fuel pressure will be reestablished.

5. On some models, the Check Engine Light will stay on after test has been completed. This is caused by the computer sensing a open fuel pump circuit. The memory code in the control unit must be erased. To erase the code disconnect the battery cable for 30 seconds then reconnect the cable.

Idle Speed

ADJUSTMENT

NOTE: The engine should be in good mechanical condition and all electrical connectors and vacuum hoses connected before making this adjustment.

1. Start the engine and let it warm up to normal operating temperature.
2. Check the ignition timing and adjust if needed.
3. Run the engine under no load at 2,000 rpm for about two minutes. Rev the engine two or three times and let it idle for one minute.
4. Check the idle speed and compare to the specifications:
5. If adjustment is required, disconnect the throttle position sensor connector for Non-turbo or disconnect the idle air control valve for turbocharged vehicles.
6. Run the engine under no load at 2,000 rpm for about two minutes. Rev the engine two or three times and let it idle.
7. Adjust the idle speed using the idle speed adjusting screw to the specifications earlier.
8. Connect the throttle position sensor for Non-turbo or connect the idle air control valve for turbocharged vehicles.
9. Run the engine under no load at 2,000 rpm for about two minutes. Rev the engine two or three times and let it idle.
10. Check the idle speed for the proper specification. If not, check the IACV-AAC (idle air control valve) valve or related wiring.
11. If the idle speed is still not correct, substitute a known good ECM.

NOTE: The ECM may be the cause of the problem, however this is rarely the case.

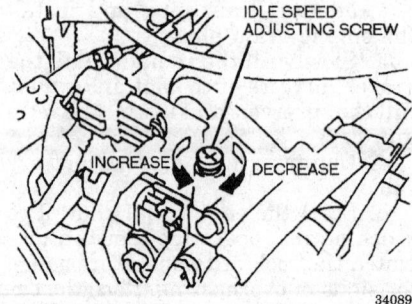

340890

Idle speed adjusting screw — 2.0L (SR20DE) engines

Mixture

ADJUSTMENT

All Engines

The air/fuel mixture is automatically adjusted by the engine control system. If the mixture is too rich or too lean, use the appropriate diagnostic procedure to locate the problem.

Fuel Filter

REMOVAL AND INSTALLATION

All Models

——— CAUTION ———
Make sure to relieve the fuel system pressure on fuel injected engines before replacing the fuel filter.

1. Properly relieve fuel system pressure.
2. Disconnect the negative battery cable.
3. Loosen the fuel hose clamps and disconnect the hoses from the fuel filter.
4. Remove the bolt securing the filter to the bracket or just remove the filter from the bracket clips.
5. Remove the filter.
To install:
6. Install the new filter and secure the filter in the bracket.
7. If necessary, replace the fuel line hoses and hose clamps. Reconnect the fuel hoses and tighten the clamps.
8. Reconnect the negative battery cable.
9. Install the fuel pump fuse.
10. Start the engine and check for leaks.

Fuel Pump

REMOVAL AND INSTALLATION

1993–94 Sentra

——— CAUTION ———
Never smoke when working around gasoline! Avoid all sources of sparks or ignition. Gasoline vapors are EXTREMELY volatile!

The fuel pump is located in the fuel tank on all vehicles. In-tank fuel pumps are accessible either by lifting up the rear seat or through an opening (access or inspection cover) in the

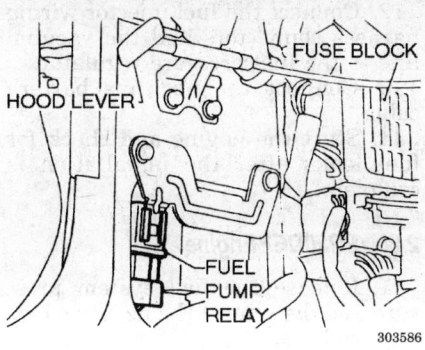

Fuel pump relay location — 300ZX

303586

trunk compartment. If the vehicle has no fuel tank inspection cover, the fuel tank assembly must be lowered or removed to gain access to the in-tank fuel pump.

1. Relieve the fuel system pressure.

2. Disconnect the negative battery cable.

3. Disconnect the electrical connector and the fuel gauge connector. Remove the inspection cover.

4. Disconnect the inlet and outlet fuel lines from the fuel pump assembly. Remove the fuel tank if necessary.

5. Remove the locking ring or unbolt and remove the fuel pump from the top of the fuel tank.

NOTE: When removing or installing the locking ring, be sure to use a brass drift pin to unseat or seat the ring.

6. Discard the O-ring seal or gasket.

To install:

7. Install the pump with a new gasket or O-ring seal. Tighten the pump retaining bolts or seat locking ring and connect the fuel hoses. Be sure to use new clamps and verify that all hoses are properly seated on the fuel tank. If removed, install the fuel tank.

8. Reconnect the fuel gauge connector.

9. Install the fuel pump access cover.

10. Reconnect the pump wiring harness connector.

11. Reconnect the negative battery cable.

12. Start the engine and check for leaks.

NOTE: On some models, the Check Engine Light will stay on after the repair is complete. The memory code in the ECU must be

erased. To erase the code, disconnect the battery cables for 1 minute then reconnect.

1995–97 Sentra and 200SX

The fuel pump is located in the fuel tank on all vehicles. In-tank fuel pumps are accessible by lifting up the rear seat to gain access to the inspection cover.

1. Relieve the fuel system pressure.

2. Disconnect the negative battery cable.

3. Remove the rear seat from the vehicle.

4. Remove the inspection cover that is located under the rear seat.

5. Disconnect the inlet and outlet fuel lines from the fuel pump assembly.

6. Disconnect the fuel pump and gauge wiring connections.

7. Remove the six mounting bolts that secure the fuel pump assembly to the top of the fuel tank.

8. Raise up the fuel pump assembly and disconnect the fuel tubes and connector. Remove the fuel gauge assembly.

9. Remove the fuel pump with the fuel chamber.

10. Pull up the front of the fuel pump chamber and slide the chamber forward.

11. Remove the fuel pump from the chamber.

12. Discard the O-ring seal or gasket.

To install:

13. Install the fuel pump to the fuel pump chamber and slide chamber rearward.

14. Install the fuel pump with the fuel pump chamber.

15. Using a new O-ring, install the fuel gauge assembly and connect the fuel tubes and connector. Use new hoses and clamps.

16. Install the six mounting bolts to the top of the fuel gauge unit. Tighten the bolts to 28–37 inch lbs. (3–4 Nm).

17. Connect the fuel pump and gauge wiring connections.

18. Using new hoses and clamps, connect the inlet and outlet fuel lines to the fuel pump assembly.

19. Install the inspection cover and install the rear seat.

20. Connect the negative battery cable, start the engine, and check for leaks.

NOTE: On some models, the Check Engine Light will stay on after the repair is complete. This is caused by an open fuel pump

circuit when relieving the fuel pump pressure The memory code in the ECU must be erased. To erase the code, disconnect the battery cables for 1 minute then reconnect.

Altima

1. Relieve the pressure from the fuel system.

2. Disconnect the negative battery cable.

3. Remove the rear seat and remove the access cover.

4. Disconnect the fuel pump electrical connector.

5. Disconnect the fuel lines from the fuel pump assembly.

6. Remove the locking ring using special tool J38879 or equivalent.

7. Remove the fuel gauge assembly; disconnect the fuel tube and connector from the fuel gauge.

NOTE: When the fuel sending unit needs to be removed, pull the tab upwards. The tab is located on the sending unit, opposite the end of the float. After the tab is pulled, the sending unit will lift straight out of the tank bracket.

8. Remove the fuel pump by pinching the two locking tabs together. Lift the fuel pump assembly straight upward and out of fuel tank.

9. Remove the O-ring and discard. Place a clean rag in the hole to keep out dirt.

To install:

10. Remove the rag and install a new O-ring.

11. Install the fuel pump.

12. Connect the electrical connection and fuel tube to the fuel gauge sending unit.

13. Install the fuel sending unit into the tank.

NOTE: Verify that the mark on the fuel tank and the components are aligned when installing the pump and fuel gauge sending unit.

14. Install the locking ring and tighten ring to 22–26 ft. lbs. (30–35 Nm).

15. Connect the fuel lines and fuel pump electrical connector. Always install new clamps on the fuel lines.

16. Install the fuel pump access cover.

17. Install the rear seat.

18. Connect the negative battery cable.

19. Start the engine and check for leaks.

Maxima, 300ZX and 240SX

1. Relieve the fuel system pressure

2. Disconnect the negative battery cable.

3. Remove the rear seat or open the access panel in the trunk.

4. Disconnect the fuel gauge electrical connector and pump electrical connector.

5. Disconnect the fuel outlet and the return hoses. If necessary, remove the fuel tank.

6. On some 240SX models you need to remove the fuel pump assembly-to-fuel tank bolts and lift the fuel pump assembly from the fuel tank.

7. On other models you need to remove the locking ring with tool SST J38879-A or equivalent and raise the fuel pump from the tank. Disconnect the feed tube while raising the pump.

8. Discard the O-ring. Plug the fuel tank opening with a clean rag to prevent dirt from entering the system.

NOTE: When removing or installing the fuel pump assembly, be careful not to damage or deform it and always install a new O-ring.

To install:

9. Remove the rag; using a new O-ring, install fuel pump assembly into the fuel tank.

10. Install the fuel pump assembly-to-fuel tank bolts and torque the bolts to 17–22 inch lbs. (2.0–2.5 Nm).

11. Install the locking ring assembly and tighten.

12. If removed, install the fuel tank assembly.

13. Connect the fuel lines and the electrical connectors. Always use new clamps when reconnecting fuel line hoses.

NOTE: When installing the upper plate, be sure to align the mark with the center marks on the fuel tank.

14. Install the fuel pump access cover.

15. Connect the negative battery cable.

16. Start the engine and check for fuel leaks.

NOTE: On some models, the Check Engine Light will stay ON after installation is completed. The memory code in the control unit must be erased. This code is stored for an open fuel pump circuit, this is caused when the fuel pressure is released. To erase the code, disconnect the battery cable for 10 seconds then reconnect after installation of fuel pump.

Fuel Injector

REMOVAL AND INSTALLATION

1.6L (GA16DE) Engine

1. Relieve the fuel system pressure and disconnect the negative battery cable.

2. Disconnect the fuel injector wiring harness connectors and the vacuum line from the fuel pressure regulator.

3. Disconnect the fuel hoses from the fuel rail assembly.

4. Remove the bolts that secure the entire fuel rail to the intake.

5. Remove the injectors with the fuel rail assembly.

6. Remove the individual injector by removing the bolts that secure the cap to the injector.

7. Remove the injector from the fuel rail by pushing the injector from the bottom and discard the O-rings.

——— WARNING ———
DO NOT remove the injector by pulling on the electrical connector.

To install:

8. Install new O-rings onto the injectors and coat the O-rings with clean fuel.

9. Install the injector into the fuel rail assembly and install the cap that secures the injector. Tighten the cap bolts to 26–33 inch lbs. (3–4 Nm).

10. Install the insulators, injectors, and the fuel rail assembly to the intake manifold. Torque the retaining bolts in sequence and in two (2) steps. Torque the bolts to 15–20 ft. lbs. (20–27 Nm).

11. Install the fuel hoses with new clamps to the fuel rail.

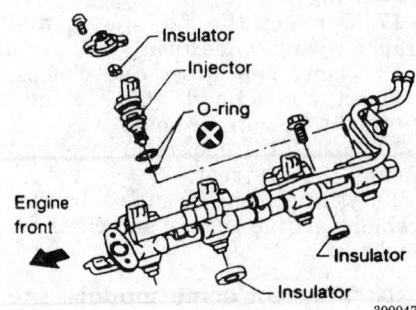

Fuel injector assembly — 1.6L (GA16DE) engine

12. Connect the fuel injector wiring harness connectors and the vacuum line to the fuel pressure regulator.

13. Connect the negative battery cable.

14. Start the engine and check for fuel leaks after the installation is complete.

2.0L (SR20DE) Engine

1. Release the fuel system pressure and disconnect the negative battery cable.

2. Disconnect all fuel and vacuum hoses and electrical harnesses from the pressure regulator and fuel rail.

3. Remove the injector/fuel rail assembly.

4. Push out the injector from the bottom of the fuel rail assembly.

——— WARNING ———
DO NOT remove the injector by pulling on the electrical connector.

To install:

5. Always replace the O-rings and insulators with new. Lubricate the O-rings with fuel.

6. Install the injectors to the fuel rail and tighten the cap that secures the injector to 26–33 inch lbs. (3–4 Nm). Install the rail onto the intake manifold.

7. Torque the fuel rail bolts as follows:
   a. Tighten all bolts in sequence to 8 ft. lbs. (11 Nm).
   b. Tighten all bolts in sequence to 15–20 ft. lbs. (21–26 Nm).

8. Connect the fuel hoses with new clamps and the connect vacuum lines. Connect the injectors to the harness.

9. Connect the battery cable, start the engine, and check for leaks before road testing.

2.4L (KA24DE) Engine

1. Relieve the fuel system pressure.

——— CAUTION ———
Fuel injection systems remain under pressure after the engine has been turned OFF. Properly relieve the fuel pressure before disconnecting any fuel lines. Failure to do so may result in fire or personal injury.

2. Disconnect the negative battery cable.

3. Note the locations of the vacuum hoses and disconnect the hoses from the intake manifold.

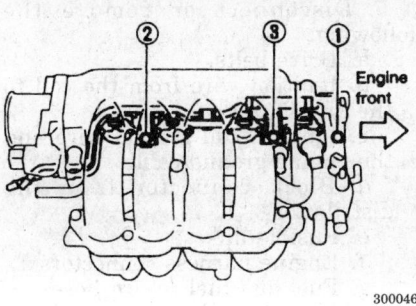

Fuel rail torque sequence — 1.6L (GA16DE) engine

300046

4. Note the connector locations and remove the injector electrical harness connections.

5. Disconnect the fuel feed and return lines from the fuel rail.

6. Remove the bolts that secure the fuel rail to the intake and remove the rail with injectors.

7. Remove the screws that secure the cap to the individual injectors and remove the cap.

8. Push the injector tail piece to remove the injector from the tube assembly. Do not pull on the connector.

To install:

NOTE: When replacing the fuel injectors always use new O-rings.

9. Clean the exterior of the injector tailpiece.

10. Push the injectors into the fuel tube and secure with cap.

11. Install the fuel tube and injectors assembly into the engine.

12. Using new clamps, reconnect the fuel lines to the fuel rail.

13. Reconnect the injectors electrical harness and the vacuum hoses.

14. Reconnect the negative battery cable.

15. Start the engine and check for leaks.

3.0L (VE30DE) Engine

1. Relieve the fuel system pressure.

2. Remove the intake manifold collector.

3. If necessary, disconnect the electrical connectors from the ignition coils.

4. If necessary, disconnect the electrical connector from the crank angle sensor and the power transistor.

5. Remove the fuel injector assembly by performing the following procedures:

   a. Disconnect the electrical connectors from the fuel injectors.

   b. Disconnect the fuel injector assembly from the fuel lines.

   c. Remove the fuel rail-to-cylinder head bolts.

   d. Remove the fuel rail assembly from the engine.

6. To remove the fuel injector from the fuel rail, remove the fuel injector-to-fuel rail bolts and remove the fuel injector from the fuel rail; discard the O-rings.

To install:

7. To install the fuel injector to the fuel rail, perform the following procedures:

   a. Install new O-rings onto the fuel injector.

   b. Wet the new O-rings with fuel and press the injector into the fuel rail.

   c. Install the bolts and tighten the fuel injector retainer.

8. Clean the gasket mounting surfaces.

9. Install the fuel injector assembly by performing the following procedures:

   a. Install the fuel rail assembly to the engine.

   b. Install the fuel rail-to-cylinder head bolts and torque the bolts to 12–14 ft. lbs. (16–20 Nm).

   c. Connect the fuel injector assembly to the fuel lines.

   d. Connect the electrical connectors to the fuel injectors.

10. The remaining components are installed in the reverse order from which they were removed.

11. Refill the cooling system and connect the negative battery cable.

12. Start the engine, bleed cooling system and check for leaks.

3.0L (VG30E) Engine

1. Relieve the fuel system pressure.

2. Remove the upper and lower intake manifold collector.

3. Remove the fuel injector assembly by performing the following procedures:

   a. Disconnect the electrical connectors from the fuel injectors.

   b. Disconnect the fuel injector assembly from the fuel lines.

   c. Remove the fuel rail-to-cylinder head bolts.

   d. Remove the fuel rail assembly from the engine.

4. To remove the fuel injector from the fuel rail, remove the fuel injector-to-fuel rail bolts and the fuel injector from the fuel rail; discard the O-ring.

To install:

5. To install the fuel injector to the fuel rail, perform the following procedures:

   a. Install a new O-ring onto the fuel injector.

   b. Wet the new O-ring with fuel and press the injector into the fuel rail.

   c. Install the bolts and tighten the fuel injector retainer.

6. Clean the gasket mounting surfaces.

7. Install the fuel injector assembly by performing the following procedures:

   a. Install the fuel rail assembly to the engine.

   b. Install the fuel rail-to-cylinder head bolts and torque the bolts to 1.8–2.4 ft. lbs. (2–3 Nm).

   c. Connect the fuel injector assembly to the fuel lines.

   d. Connect the electrical connectors to the fuel injectors.

8. Use a new gasket and install the lower intake manifold collector.

9. balance of installation is the reverse of the removal procedure.

10. Connect the negative battery cable.

11. Start the engine and check for leaks.

3.0L (VQ30DE) Engine

1. Release the fuel system pressure.

─────── **CAUTION** ───────

The fuel injection system remains under pressure after the engine has been turned OFF. Properly relieve the fuel pressure before disconnecting any fuel lines. Failure to do so may result in fire or personal injury.

2. Disconnect the negative battery cable and drain the cooling system.

3. Remove the intake manifold collector.

4. Remove the fuel injector assembly by performing the following procedures:

   a. Disconnect the electrical connectors from the fuel injectors.

   b. Disconnect the fuel lines from the fuel injector assembly.

   c. Remove the fuel rail-to-cylinder head bolts.

   d. Remove the fuel rail assembly from the engine.

5. To remove the fuel injector from the fuel rail, remove the fuel injector-to-fuel rail bolts and remove the fuel injector from the fuel rail; discard the O-rings.

To install:

6. To install the fuel injector to the fuel rail, perform the following procedures:

a. Install new O-rings onto the fuel injector.

b. Wet the new O-rings with fuel and press the injector into the fuel rail.

c. Install the bolts and tighten the fuel injector retainer to 27–33 inch lbs. (3–4 Nm).

7. Clean the gasket mounting surfaces.

8. Install the fuel injector assembly by performing the following procedures:

a. Install the fuel rail assembly to the engine.

b. Install the fuel rail-to-cylinder head bolts and torque the bolts to 15–20 ft. lbs. (21–26 Nm) in two progressive steps.

c. Connect the fuel lines to the fuel injector assembly.

d. Connect the electrical connectors to the fuel injectors.

9. Using a new gasket, install the intake manifold collector.

10. Install the remaining components in the opposite order from which they were removed.

11. Refill the cooling system and connect the negative battery cable.

12. Start the engine, bleed cooling system, and check for leaks.

3.0L (VG30DE and VG30DETT) Engines

CAUTION

Fuel injection systems remain under pressure after the engine has been turned OFF. Properly relieve fuel pressure before disconnecting any fuel lines. Failure to do so may result in fire or personal injury.

1. Disconnect the negative battery cable, drain the cooling system, and release the fuel system pressure.

2. Remove the intake manifold collector.

3. Disconnect the electrical connectors from the ignition coils.

4. Disconnect the electrical connector from the crank angle sensor and the power transistor.

5. Disconnect the electrical connectors from the fuel injectors.

6. Disconnect the fuel injector assembly from the fuel lines.

7. Remove the fuel rail-to-cylinder head bolts.

8. Remove the fuel rail assembly from the engine.

9. Remove the bolts that secure fuel injector to the fuel rail.

10. Gently pull the fuel injector from the fuel rail and disregard the O-ring.

To install:

NOTE: When the fuel injector is remove the injector sealing O-rings must be replaced.

11. Clean the gasket mounting surfaces.

12. Using new O-rings, install the fuel injector to the fuel rail and torque the mounting bolts to 22–28 inch lbs. (2–3 Nm).

13. Install and tighten the fuel rail-to-cylinder head bolts. Be sure to install the fuel rail insulators.

14. Using new clamps, connect the fuel lines to the fuel rail assembly.

15. Connect the electrical connectors to the fuel injectors.

16. Connect the electrical connectors to the ignition coils.

17. Using a new gasket, install the intake manifold collector.

18. The remaining components are installed in the reverse order from which they were removed.

19. Connect the negative battery cable.

20. Refill the cooling system.

21. Start the engine, bleed the cooling system, and check for leaks.

ENGINE MECHANICAL

Engine Assembly

REMOVAL AND INSTALLATION

1993–94 Sentra

NOTE: On the 1.6L (GA16DE) engines, the engine cannot be removed separately from the transaxle. Remove the engine and the transaxle as a unit. If equipped with 4WD, remove the engine, transaxle and transfer case together.

1. Mark the hood hinge relationship and remove the hood.

2. Release the fuel system pressure, disconnect the negative battery cable and raise and support the vehicle safely.

3. Drain the cooling system and the engine oil.

4. Remove the air cleaner and disconnect the throttle cable.

5. Disconnect or remove the following:

a. Drive belts.

b. Ignition wire from the coil to the distributor

c. Ignition coil ground wire and the engine ground cable

d. Block connector from the distributor

e. Fusible links

f. Engine harness connectors

g. Fuel and fuel return hoses

h. Upper and lower radiator hoses

i. Heater inlet and outlet hoses

j. Engine vacuum hoses

k. Carbon canister hoses and the air pump air cleaner hose

l. Any interfering engine accessories: power steering pump, air conditioning compressor or alternator

m. Driveshaft from transfer unit for 4WD vehicles.

NOTE: Make sure to match-mark flanges of driveshafts

6. Remove the air pump air cleaner.

7. Remove the carbon canister.

8. Remove the auxiliary fan, washer tank, grille and radiator (with fan assembly).

9. Remove the clutch cylinder from the clutch housing for manual transaxles.

10. Remove both buffer rods without altering the length of the rods. Disconnect the speedometer cable.

11. Remove the spring pins from the transaxle gear selector rods.

12. Install engine slingers to the block and connect a suitable lifting device to the slingers. Do not tension the lifting device at this point.

13. Disconnect the exhaust pipe at both the manifold connection and the clamp holding the pipe to the engine.

14. Remove the lower ball joints.

15. Drain the transaxle gear oil.

16. Disconnect the right and left side halfshafts from their side flanges and remove the bolt holding the radius link support.

NOTE: When drawing out the halfshafts, it is necessary to loosen the strut head bolts.

17. Lower the shifter and selector rods and remove the bolts from the motor mount brackets. Remove the nuts holding the front and rear motor mounts to the frame.

18. Disconnect the clutch and accelerator wires and remove the speedometer cable with its pinion from the transaxle.

19. Lift the engine/transaxle assembly up and away from the vehicle.

To install:

20. Lower the engine/transaxle assembly into the vehicle. When lowering the engine onto the frame, make sure to keep it as level as possible.

21. Check the clearance between the frame and clutch housing and make sure the engine mount bolts are seated in the groove of the mounting bracket.

22. After installing the motor mounts, adjust and install the buffer rods. The front should be 3.50–3.58 in. (89–91mm), and the rear 3.90–3.98 in. (99–101mm).

23. Tighten the engine mount bolts first, then apply a load to the mounting insulators before tightening the buffer rod and sub-mounting bolts.

24. Install the remaining components in the opposite order from which they were removed.

25. Fill the transaxle and cooling system to the proper levels.

26. Install the hood and connect the negative battery cable.

27. Make all the necessary engine adjustments. Charge the air conditioning system. Road test the vehicle for proper operation.

1995–97 Sentra and 200SX

1.6L (GA16DE) and 2.0L (SR20DE) Engines

NOTE: The engine and transaxle are removed as one unit from the underside of the vehicle.

────── CAUTION ──────
Fuel injection systems remain under pressure after the engine has been turned OFF. Properly relieve fuel pressure before disconnecting any fuel lines. Failure to do so may result in fire or personal injury.

1. Relieve the fuel system pressure.

2. Disconnect the negative and positive battery cables.

3. Remove the battery and battery tray from the vehicle.

4. Raise and safely support the vehicle.

5. Remove both front wheels.

6. Remove the engine undercovers and remove the engine side covers.

7. Drain the coolant from the radiator and the engine block.

8. Drain the engine oil.

9. Remove the air cleaner assembly and remove air duct.

10. Note the locations and remove the vacuum hoses.

11. Disconnect the heater hoses from the engine.

12. If equipped, disconnect the A/T cooler hoses from the transaxle.

13. Disconnect the fuel hoses from the engine.

14. Note the locations and disconnect the harness and wiring connections.

15. Disconnect the throttle cable and the cruise control cable.

16. If equipped with automatic transmission, disconnect the control cable.

17. Remove the cooling fans, radiator, and the recovery tank.

18. Remove the front driveshafts from the vehicle.

19. Remove the front exhaust pipe.

20. On the 1.6L engine, disconnect the control rod and support rod from the transaxle.

21. Remove the starter motor and intake manifold support brackets.

22. Remove the engine drive belts.

23. Remove the alternator and adjusting brackets.

24. Remove the power steering pump and A/C compressor. It is not necessary to disconnect the lines.

25. On the 1.6L engine, remove the cylinder head front mounting bracket.

26. Position a transmission jack under the transaxle and support the engine with engine slinger.

27. Remove the center crossmember.

28. On some models it may be necessary to remove the front stabilizer bar.

29. Remove the engine mounting bolts from both sides of the engine.

30. Slowly lower the jacking devices and remove the engine and transaxle from the vehicle.

To install:

31. Install the engine and transaxle assembly.

32. Install the mounting bolts to both sides of the engine.

33. For vehicles with manual transaxles, adjust the height of the mounting bracket (buffer rod). The distance between the two through bolts should be 2.126–2.205 in. (54–56mm).

34. Install the center crossmember and remove the engine support jacks. Remove the engine slinger.

35. Install the remaining components in the reverse order of removal.

36. Tighten the control rod bolt to 10–13 ft. lbs. (14–18 Nm). Tighten the support rod bolt to 26–35 ft. lbs. (35–47 Nm).

37. Install the battery tray and install the battery.

38. Connect the positive and then the negative battery cables.

39. Start the engine and check for leaks. Make all the necessary adjustments.

Altima

────── CAUTION ──────
Release the fuel pressure in the system before disconnecting the fuel lines. The fuel system will remain under pressure after the ignition has been turned OFF. Failure to do so may result in fire or personal injury.

NOTE: The engine and transaxle must be removed as a single unit. The engine and transaxle are removed from under the vehicle.

1. Mark the location of the hinges on the hood and remove the hood from the vehicle.

2. Release fuel system pressure.

3. Disconnect the battery cables and remove the battery and battery tray.

4. Drain the coolant from the plug on the water pipe and drain the radiator.

5. If equipped with automatic transaxle, disconnect the cooler lines from the radiator.

6. Remove the upper and lower hoses from the radiator and then remove the radiator assembly.

7. Disconnect the heater hoses from the engine.

8. Disconnect the throttle cable and cruise control cable (if equipped).

9. Remove the air cleaner, air box, and the intake hose.

10. Disconnect the fuel feed and return hoses.

11. Disconnect and label all the necessary vacuum hoses and electrical connectors.

12. Disconnect the wiring from starter motor.

13. If equipped, disconnect the slave cylinder from the transaxle. It is not necessary to disconnect the hydraulic hose.

14. Remove the engine drive belts. Be sure to mark belts for reinstallation.

15. Remove the alternator, A/C compressor, and the power steering pump from the engine.

16. Remove the right and left driveshafts from the transaxle.

17. Disconnect the exhaust pipe from the exhaust manifold.

18. Disconnect the crankshaft position sensor from the engine block.

19. Support the engine with slinger and support the transaxle with proper jack.

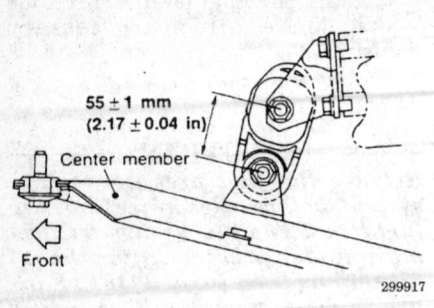

55 ± 1 mm
(2.17 ± 0.04 in)

Center member

Front

299917

**Height adjustment of the buffer rod — 1995–97
Sentra and 200SX with 1.6L (GA16DE) engine**

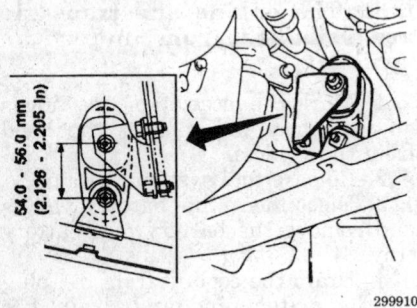

54.0 – 56.0 mm
(2.126 – 2.205 in)

299910

**Adjusting the height of the engine mount for
M/T vehicles — 1995–97 2.0L (SR20DE) engine**

20. Disconnect the left and right engine mounting through bolts.
21. Remove the bolts that secure the crossmember to the vehicle and remove the crossmember.
22. Remove the front and rear engine mounts.
23. Lower the transaxle and engine assembly from the vehicle.

NOTE: The engine and transaxle assembly should be removed through the bottom of the vehicle. Do not attempt to remove the assembly from above.

To install:
24. Raise the transaxle and engine assembly to the vehicle.
25. Install the front and rear engine mounts. Tighten the mounting bolts to specifications.
26. Install the crossmember and tighten the mounting bolts to 57–72 ft. lbs. (77–98 Nm).
27. Connect the left and right engine mounting through bolts. Tighten the bolts to specifications.
28. Remove the engine and transaxle support jacks.
29. The balance of installation is the reverse of the removal procedure.
30. Refill the cooling system and the engine oil. Check all the fluid levels.

31. Start the engine, bleed the cooling system, and check for leaks. Make all the necessary adjustments.
32. Install the hood.

Maxima

3.0L (VQ30DE and VE30DE) and 1993–94 (VG30E) Engines

It is recommended the engine and transaxle be removed as a single unit. If need be, the units may be separated after removal.

NOTE: The engine and transaxle assembly must be removed from the underside of the vehicle.

1. Matchmark the hood hinge relationship and remove the hood.
2. Release the fuel system pressure. Disconnect the negative battery cable and raise and safely support the vehicle.
3. Drain the coolant from the cylinder block and the radiator. Drain the crankcase and the automatic transaxle, if equipped.
4. Remove the air cleaner, the air intake tube, the air flow meter and disconnect the throttle linkage.
5. Disconnect and/or remove the following:
   • Drive belts
   • Engine ground cable
   • Electrical connector from the crank angle sensor
   • Engine electrical harness connectors

----- **CAUTION** -----

The fuel injection system remains under pressure after the engine has been turned OFF. Properly relieve the fuel pressure before disconnecting any fuel lines. Failure to do so may result in fire or personal injury.

   • Fuel feed and fuel return hoses
   • Upper and lower radiator hoses
   • Heater inlet and outlet hoses
   • Electrical connector from the crank angle sensor
   • Engine vacuum hoses
   • Carbon canister hoses
   • Any interfering engine accessory: power steering pump, air conditioning compressor, and the alternator
6. Remove the carbon canister.
7. Remove the auxiliary fan, washer tank, and the radiator (with the fan assembly).
8. If equipped with a manual transaxle, remove the clutch release cylinder from the clutch housing.
9. On the VG30E engine, remove the buffer rods without altering the

length of the rods. Disconnect the speedometer cable.
10. On the VG30E engine, remove the spring pins from the transaxle gear selector rods.
11. On some models with a manual transaxle, disconnect the shift control rod and disconnect the shift support rod. On others with an automatic transaxle, disconnect the control cable from the transaxle.
12. Install engine slingers to the block and connect a suitable lifting device to the slingers. Do not tension the lifting device at this point.
13. Disconnect the exhaust pipe at both the manifold connections and remove the front exhaust pipe from the vehicle.
14. If equipped with a manual transaxle, drain the transaxle gear oil.
15. Support the engine and transaxle assembly with proper jack.
16. Disconnect the right and left side halfshafts from their side flanges and remove the bolt holding the radius link support.
17. Lower the shifter and selector rods and remove the bolts from the motor mount brackets. Remove the nuts holding the front and rear motor mounts to the frame.
18. On some models it will be necessary to remove the center crossmember assembly from the vehicle.
19. Lower the engine/transaxle assembly down and onto an engine stand.

To install:
20. Raise the engine/transaxle assembly into the vehicle. When raising the engine onto the mounts, make sure to keep it as level as possible.
21. After installing the motor mounts, adjust and install the buffer rods; the front should be 3.50–5.58 in. (89–91mm) and the rear should be 3.90–3.98 in. (99–101mm).
22. Check the clearance between the frame and clutch housing and make sure the engine mount bolts are seated in the groove of the mounting bracket.
23. Remove the transaxle and engine jack assembly.
24. The remaining components are installed in the reverse order from which they were removed.
25. Fill the transaxle, the engine, and the cooling system to the proper levels with the appropriate fluids.
26. Install the hood and connect the negative battery cable.
27. Make all the necessary engine adjustments. Charge the air conditioning system, if discharged. Road test the vehicle for proper operation.

240SX

NOTE: The engine assembly is removed from the top of the vehicle.

1. Make sure it is on a flat and level surface and that wheels are tightly chocked.
2. Allow the exhaust system to cool completely before starting work to prevent burns and possible fire as fuel lines are disconnected.

CAUTION

The fuel injection system remains under pressure after the engine has been turned OFF. Properly relieve the fuel pressure before disconnecting any fuel lines. Failure to do so may result in fire or personal injury.

3. Release fuel pressure from the fuel system before attempting to disconnect any fuel lines.
4. Mark the location of the hinges on the hood. Unbolt and remove the hood.
5. Disconnect the battery cables and remove the battery. Be sure to disconnect the negative cable first.
6. Remove the engine undercover.
7. Remove the transmission from the vehicle.
8. Drain the coolant from the radiator and the engine block.
9. Drain the engine oil from the vehicle.
10. Remove the radiator and radiator shroud after disconnecting the automatic transmission to radiator cooling tubes.
11. Remove the air cleaner.
12. Remove the engine drive belts.
13. Remove the fan and pulley.
14. Disconnect the electrical harness connectors at the water temperature sensor, oil pressure sending unit, and the starter motor. Disconnect the primary ignition wires.
15. Disconnect the fuel hoses.

CAUTION

On all fuel injected models, the fuel pressure must be released before the fuel lines can be disconnected.

16. Disconnect the electrical connections at the alternator. Disconnect the heater hoses and throttle connections.
17. Disconnect the engine ground cable, thermal transmitter wire, wire to the fuel cut-off solenoid and the vacuum cut solenoid wire.
18. Remove the front exhaust pipe from the vehicle.

CAUTION

On models with air conditioning, it is necessary to remove the compressor from the vehicle. DO NOT ATTEMPT TO UNFASTEN ANY OF THE AIR CONDITIONER HOSES BEFORE PROPERLY EVACUATING THE SYSTEM.

19. Remove the A/C compressor from the vehicle.
20. Remove the power steering pump from the engine.
21. Disconnect the power brake booster hose from the engine.
22. Attach a hoist to the lifting hooks on the engine (at either end of the cylinder head). Support the engine.
23. Remove the engine mounting nuts from both lower sides of the engine mounts.

NOTE: When lifting the engine out, guide it carefully to avoid hitting parts such as the master cylinder.

24. Remove the engine from the vehicle.

To install:
25. With the engine assembly safely secured to the hoist, lower the assembly into the vehicle.
26. Tighten the engine mounting nuts to 51–58 ft. lbs. (69–78 Nm). It may be necessary to lower or raise the engine hoist to correctly position the engine assembly to line up with the mount holes. Remove the engine hoist.
27. Install the power steering pump to the engine. Install the power brake booster.
28. The balance of installation is the reverse of the removal procedure.
29. Refill the engine, transmission, and coolant levels with the proper type and amount of fluid.
30. Check all fluids to assure they are at the correct level. Start the engine and run at idle until normal operating temperature is reached. Check for fluid leaks and repair as required.
31. Road test the vehicle after you are sure there are no leaks. Check vehicle for proper operation. Recheck all fluid levels once road test is completed.

300ZX

1993–94 Vehicles

1. Mark the hood hinge relationship and remove the hood.
2. Release the fuel system pressure, disconnect the negative battery cable and raise and safely support the vehicle.

3. Remove the undercover.
4. Drain the coolant from both sides of the block and from the radiator.
5. Drain the oil pan.
6. Disconnect and label all engine vacuum hoses, fuel piping, harnesses and connectors.
7. Disconnect and remove the front exhaust tube sections.
8. Mark the relationship of the flanges and disconnect the driveshaft.
9. Remove the radiator.
10. Remove the drive belts.
11. Remove the cooling fan and coupling.
12. Remove the power steering pump, alternator, starter and clutch operating cylinder (if so equipped).
13. Discharge the air conditioning system and remove the compressor from the engine.
14. Disconnect the air conditioning tube clamps.
15. Disconnect the steering column lower joint from the steering rack.
16. Remove the tension rod retaining bolts on both sides.
17. Loosen the transverse link bolts on both sides.
18. Support the rear suspension member using the proper equipment.
19. Install engine slingers to the block and connect a suitable lifting device to the slingers. Tension the lifting device slightly.
20. Remove the rear suspension member retaining bolts and center nut.
21. Remove the engine mount bracket bolts from both sides and slowly lower the transmission jack. Lift the engine and transmission from the vehicle.

To install:
22. Lower the engine and transmission into the vehicle and slowly raise the transmission jack. Install the engine mount bracket bolts. Torque the bolts to 30–38 ft. lbs. (40–42 Nm).
23. Install the rear suspension bolts and center nut. Torque the bolts to 38–48 ft. lbs. (51–65 Nm) and the center nut to 26–33 ft. lbs. (35–45 Nm).
24. Remove the jack and disconnect the engine hoist.
25. Torque the transverse link bolts to 80–94 ft. lbs. (108–127 Nm).
26. Install the tension rod retaining bolts and torque them to 80–94 ft. lbs. (108–127 Nm).
27. Connect the steering column lower joint to the steering rack. Torque the lower joint bolt to 17–22 ft. lbs. (24–29 Nm).

28. Install the remaining components in the reverse order of removal.
29. Install the driveshaft. Make sure the flanges are aligned properly. On non-turbo models, torque the flange bolts to 29–33 ft. lbs. (39–45 Nm) or 40–47 ft. lbs. (54–64 Nm) on turbocharged models.
30. Fill the transmission and cooling system to the proper levels.
31. Install the hood and connect the negative battery cable.
32. Make all the necessary engine adjustments. Charge the air conditioning system.

1995–97 Models — Manual transmission

1. Release the fuel system pressure and disconnect the negative battery cable.

—— CAUTION ——
Fuel injection systems remain under pressure after the engine has been turned OFF. Properly relieve fuel pressure before disconnecting any fuel lines. Failure to do so may result in fire or personal injury.

2. Matchmark the hood hinge relationship and remove the hood.
3. Drain the coolant from the cylinder block and the radiator. Drain the crankcase and the transaxle.
4. Remove the air cleaner, the air intake tube, the air flow meter and disconnect the throttle linkage.
5. Note the locations and disconnect the vacuum hoses from the engine.
6. Disconnect the fuel feed and return hoses.
7. Note the wiring locations and disconnect the harness connections.
8. Raise and safely support the vehicle.
9. Remove the engine undercover(s).
10. Disconnect the front exhaust pipes from the engine.
11. Disconnect and remove the driveshaft from the rear of the transmission.
12. Remove the upper and lower radiator hoses.
13. Remove the radiator shroud and remove the radiator from the vehicle.
14. Remove the engine drive belts.
15. Remove the cooling fan and the cooling fan coupling.
16. Disconnect and remove the power steering pump, A/C compressor, and the alternator.
17. Disconnect the wiring to the starter motor and remove the starter assembly.

18. Remove the clutch slave cylinder from the transmission and position aside. It is not necessary to disconnect the fluid line.
19. Disconnect the A/C tube clamps from the A/C line.
20. Disconnect the steering column lower joint.
21. Remove the tension rod to lower control arm mounting nuts from both sides and disconnect the rod at the lower control arm.
22. Loosen but do not remove the lower control arm to crossmember bolts.
23. Position a suitable transmission jack under the crossmember and support the engine with slinger and hoist.
24. Remove the crossmember mounting bolts.

NOTE: The crossmember will remain in the vehicle during engine removal.

25. Remove the engine mounting bolts from both sides of the engine and then slowly lower the jack.
26. Remove the engine and transmission from the vehicle.

NOTE: When lifting the engine from vehicle, be careful not to strike adjacent parts. Pay special attention to the accelerator wire casing, brake lines, and the master cylinder.

To install:
27. Install the engine and transmission assembly.
28. Slowly raise the floor jack and install the engine mounting bolts.
29. Install the crossmember mounting bolts.
30. Remove the engine slinger and hoist.

NOTE: All suspension bolts must be tightened with the full weight of the vehicle on the ground.

31. Tighten the lower control arm to crossmember bolts.
32. Install the remaining components in the opposite order from which they were removed.
33. Tighten the traverse link and tension rod mounting bolts and nuts to 80–94 ft. lbs. (108–127 Nm).
34. Fill the engine oil, transmission, and the cooling system with the proper type and amount of fluid.
35. Install the hood and connect the negative battery cable.
36. Perform all necessary adjustments.
37. Start the engine and warm to full operating temperature, bleed cooling system, and check for leaks.

1995–97 Models — Automatic transmission

1. Release the fuel system pressure and disconnect the negative battery cable.

—— CAUTION ——
Fuel injection systems remain under pressure after the engine has been turned OFF. Properly relieve fuel pressure before disconnecting any fuel lines. Failure to do so may result in fire or personal injury.

2. Matchmark the hood hinge relationship and remove the hood.
3. Drain the coolant from the cylinder block and the radiator. Drain the crankcase and the automatic transaxle.
4. Remove the air cleaner, the air intake tube, the air flow meter and disconnect the throttle linkage.
5. Note the locations and disconnect the vacuum hoses from the engine.
6. Disconnect the fuel feed and return hoses.
7. Note the wiring locations and disconnect the harness connections.
8. Raise and safely support vehicle.
9. Remove the engine undercover(s).
10. Disconnect the front exhaust pipes from the engine.
11. Disconnect and remove the driveshaft from the rear of the transmission.
12. Remove the upper and lower radiator hoses. Disconnect the heater core hoses.
13. Remove the radiator shroud and remove the radiator from the vehicle.
14. Remove the engine drive belts.
15. Disconnect and remove the power steering pump, A/C compressor, and the alternator.
16. Disconnect the wiring to the starter motor and remove the starter assembly.
17. Remove the transmission from the vehicle.
18. Install an engine slinger and hoist.
19. Support the weight of the engine with the hoist and disconnect the engine mounting bolts.
20. Lift the engine from the mounts and remove the engine assembly.

NOTE: When lifting the engine from vehicle, be careful not to strike adjacent parts. Pay special attention to the accelerator wire casing, brake lines, and the master cylinder.

To install:

21. Install the engine assembly to the engine compartment.

22. Install the engine mounting bolts.

23. Install the transmission assembly.

24. Install the starter motor and connect the wiring.

25. Install the remaining components in the reverse order of removal.

26. Refill the cooling system, transmission, and the crankcase with the proper type and amount of fluid.

27. Install the hood and connect the negative battery cable.

28. Perform all necessary adjustments.

29. Start the engine and warm it to full operating temperature, bleed cooling system, and check for leaks.

Engine Mounts

REMOVAL AND INSTALLATION

1. Disconnect negative battery cable.

2. Raise and support the vehicle.

━━━━━━━ **WARNING** ━━━━━━━
Make certain vehicle is properly supported when raising engine.

3. Raise the engine slightly to take the tension off the motor engine mounts.

NOTE: When raising the engine to replace the engine mounts, DO NOT place the jack directly under the oil pan or the crankshaft torsional damper. Use a block of wood between engine and jack to prevent damage at jacking point.

4. Remove the through bolt nut(s) and through bolt(s).

5. Remove the engine mount-to-engine block bolts.

6. Raise the engine enough to remove the engine mount(s) and remove the engine mount(s).

NOTE: Only raise the engine enough to remove the mounts. Raising the engine too far could result in damage to some engine components.

To install:

7. Install the engine mount(s) and engine mount-to-engine block bolts. Tighten the bolts to correct specification.

8. Lower the engine enough to install the engine mount through bolt(s) and nut(s). Tighten the through bolts to correct specification.

9. Lower the vehicle and connect the negative battery cable.

Cylinder Head Cover

REMOVAL AND INSTALLATION

1.6L (GA16DE) Engine

1. Disconnect the negative battery cable.

2. Remove the rocker arm cover mounting bolt in the reverse order of the tightening sequence and remove the cover.

To install:

3. Using liquid sealant or equivalent, install the rocker cover and torque the bolts in sequence to 18–34 inch lbs. (2–4 Nm).

4. Install the remaining components. Refill and check all fluid levels.

5. Connect negative battery cable.

6. Start the engine and check for leaks.

2.0L (SR20DE) Engine

1. Disconnect the negative battery cable.

2. Remove the rocker cover and oil separator from the engine.

To install:

3. Clean the rocker cover and mating surfaces and apply a continuous bead of liquid gasket to the mating surface.

4. Install the rocker cover and oil separator. Tighten the rocker cover bolts as follows:

   a. Tighten nuts 1, 10, 11 and 8, in that order to 3 ft. lbs. (4 Nm).

   b. Tighten nuts 1–13 as indicated in the figure to 6–7 ft. lbs. (8–10 Nm).

5. Connect the negative battery cable.

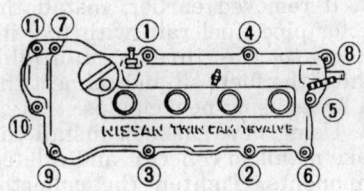

Tighten in numerical order.

302342

Rocker arm cover torque sequence, loosen in the reverse order — 1.6L (GA16DE) engine

2.4L (KA24DE) Engine

1. Disconnect the negative battery cable.

2. Remove any vacuum hoses, fuel hoses, wires, harness, and connectors that are necessary for removal of the cylinder head cover.

NOTE: If any fuel lines must be disconnected in order to remove the cylinder head cover, depressurize the fuel system first.

3. Disconnect the spark plug wires from the spark plugs.

4. Remove the cylinder head cover by loosening the bolts in the reverse order of installation.

To install:

5. Install the cylinder head rubber plugs and the cylinder head cover. Tighten nuts No. 1–5–6–4, in that order, to 35 inch lbs. (4 Nm). Then tighten nuts No. 1–11 in sequence to 70–95 inch lbs. (8–11 Nm).

6. Reconnect the spark plug wires.

7. Install any vacuum hoses, fuel lines, wires and the harness connectors removed earlier.

8. Connect the negative battery cable.

3.0L (VG30E) Engine

1. Disconnect the negative battery cable.

2. Remove the intake collector assembly (upper intake manifold).

3. Remove the cylinder head cover attaching bolts, then remove the cover from the cylinder head.

To install:

4. Install the cylinder head cover onto the engine using a new gasket.

5. Install the intake manifold and collector assembly.

6. Connect the negative battery cable.

3.0L (VE30DE and VQ30DE) Engines

1. Disconnect the negative battery cable.

2. Remove the left side rocker cover ornament.

NOTE: Before disconnecting any hoses or connectors, note the locations for reassembly.

3. Remove the upper intake manifold assembly.

4. Disconnect and remove all six ignition coils from the spark plugs.

5. Remove the spark plugs.

6. Remove the cylinder head covers from the cylinder head.

To install:

7. Apply a 0.12 inch (3mm) continuous bead of liquid gasket to the rocker covers and install the covers.

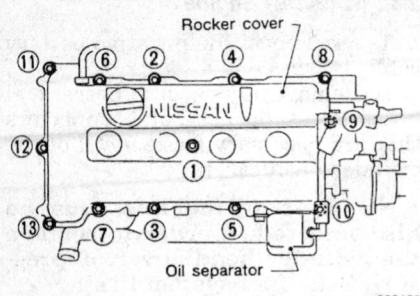

Rocker arm cover torque sequence, loosen in the reverse order — 2.0L (SR20DE) engine

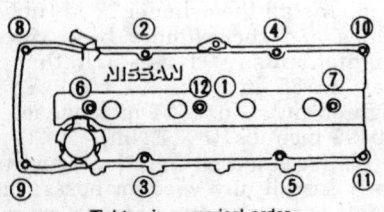

Rocker cover tightening sequence — Altima

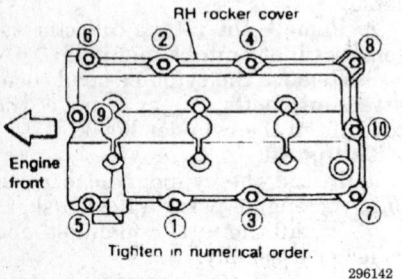

Right rocker cover torque sequence — 3.0L (VQ30DE) engine

Tighten the mounting bolts in sequence as follows:

   a. Bolts No. 1–10 — Tighten to 9–26 inch lbs. (1–3 Nm).

   b. Bolts No. 1–10 — Tighten to 52–69 inch lbs. (6–8 Nm)

8. Install the intake manifold collector gasket with the arrow facing forward.

9. Install the intake manifold collector assembly and related components.

10. Using new gaskets, install the cylinder head covers.

11. Install the spark plugs.

Left rocker cover torque sequence — 3.0L (VQ30DE) engine

12. Install the ignition coils and tighten the mounting bolts to 27–33 inch lbs. (2.9–3.8 Nm).

13. Install the rocker cover ornament on the left side.

14. Connect the negative battery cable.

3.0L (VG30DE and VG30DETT) Engines

1. Disconnect the negative battery cable.

2. Unbolt and remove the intake manifold collector from the intake manifold.

NOTE: After the intake manifold collector has been removed, cover the openings on the intake manifold to prevent foreign objects from entering the combustion chamber.

3. Remove the ignition coils and spark plugs from the engine.

4. The fuel rail may need to be removed to gain access to the cylinder head cover mounting bolts.

NOTE: If the fuel rail is to be removed, first depressurize the fuel system.

5. Remove the valve covers.
To install:
6. Install the cylinder head covers. Use liquid sealant on the exhaust side of the valve cover.

7. If removed earlier, install the injector pipe (fuel rail) with fuel injectors to the intake manifold. Tighten the fuel rail and connect the fuel hoses using new clamps.

8. Using new gaskets, install the intake manifold collector and related components. Tighten the collector mounting bolts to 12–15 ft. lbs. (16–21 Nm).

9. Install the spark plugs and ignition coils.

10. Connect the negative battery cable.

Cylinder Head

REMOVAL AND INSTALLATION

1.6L (GA16DE) Engine

— **CAUTION** —
Fuel injection systems remain under pressure after the engine has been turned OFF. Properly relieve fuel pressure before disconnecting any fuel lines. Failure to do so may result in fire or personal injury.

1. Disconnect the negative battery cable, drain the cooling system, and relieve the fuel system pressure.

2. Remove all engine drive belts.

3. Remove the cylinder head cover and any related components.

4. Remove the distributor, plug wires and spark plugs.

5. Remove the spark plugs.

6. Remove the intake manifold and all related components.

7. Remove the idler pulley, camshaft sprockets and timing chains.

8. Remove the camshafts.

9. Loosen the cylinder head bolts in 2–3 steps in the reverse order of the tightening sequence to prevent warpage or cracking of the cylinder head assembly.

10. Carefully remove the cylinder head from the block, pulling the head up evenly from both ends. If the head seems stuck, do not pry it off. Tap lightly around the lower perimeter of the head with a rubber mallet to help break the seal. The cylinder head and the intake and exhaust manifolds are removed together. Remove the cylinder head gasket.

To install:
11. Thoroughly clean both the cylinder block and head mating surfaces. Avoid scratching either surface.

12. Coat the threads and the seating surface of the head bolts with clean engine oil. Install the cylinder head assembly (always replace the head gasket). Install head bolts (with washers) in their proper locations and tighten in sequence as follows:

   a. Tighten bolts No. 1 through 10 in sequence to 22 ft. lbs. (29 Nm).

   b. Tighten bolts No. 1 through 10 in sequence to 43 ft. lbs. (59 Nm).

   c. Loosen bolts completely.

   d. Tighten bolts No. 1 through 10 in sequence to 22 ft. lbs. (29 Nm).

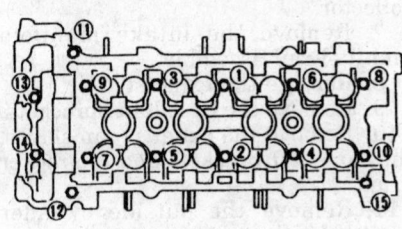

Tighten in numerical order.

299660

Torque sequence of cylinder head bolts — 1.6L (GA16DE) engine

e. Tighten bolts No. 1 through 10 to 50–55 degrees clockwise in sequence or if angle wrench is not available, tighten bolts to 40–47 ft. lbs. (54–64 Nm) in sequence.

f. Finally, tighten bolts No. 11 through 15 to 56–73 inch lbs. (6.3–8.3 Nm).

13. Install the camshafts.

14. Install the idler pulley, camshaft sprockets and timing chains.

15. Install the remaining components in the reverse order of removal. Refill and check all fluid levels.

16. Connect negative battery cable.

17. Start the engine, check for leaks, and adjust the ignition timing to specification.

18. Road test the vehicle for proper operation.

2.0L (SR20DE) Engine

1. Release the fuel pressure. Disconnect the negative battery cable.

2. Drain the cooling system. Remove the radiator assembly.

3. Remove the cylinder head cover and oil separator.

4. On some models it may be necessary to disconnect the front exhaust pipe from exhaust manifold.

5. Remove the intake manifold.

6. Remove the distributor assembly.

7. Remove the timing chain, tensioner, chain guide and camshaft sprockets.

8. Remove the camshafts.

9. Remove the water hose from the cylinder block and water hose from the heater.

10. Remove the starter motor. Remove the water pipe bolt.

11. Remove the knock sensor harness connector and remove the EGR tube.

12. Remove the cylinder outside bolts. Remove the cylinder head bolts in 2 or 3 steps. Remove the cylinder head completely with manifolds attached.

To install:

13. Check all components for wear. Replace as necessary. Clean all mating surfaces and replace the cylinder head gasket.

NOTE: If the length of any cylinder head bolt exceeds 6.228 in. (158.2mm), replace the bolt.

14. Install cylinder head. Tighten cylinder head in the following sequence:

a. Tighten all bolts in sequence to 29 ft. lbs. (39 Nm).

b. Tighten all bolts in sequence to 58 ft. lbs. (78 Nm).

c. Loosen all bolts in sequence completely.

d. Tighten all bolts in sequence to 25–33 ft. lbs. (34–44 Nm).

e. Tighten all bolts to 90–100 degrees clockwise in sequence.

f. Tighten all bolts additional 90–100 degrees clockwise in sequence. Do not turn any bolt 180–200 degrees clockwise all at once.

15. Install the starter motor and connect the wiring.

16. Install the remaining components in the opposite order from which they were removed.

17. Install and connect the intake air hose.

18. Connect the negative battery cable.

19. Refill and check all fluid levels. Road test the vehicle for proper operation.

2.4L (KA24DE) Engine

Altima

1. Disconnect the negative battery cable.

2. Drain coolant from the engine and radiator.

3. Relieve the fuel system pressure.

⇦ ENGINE FRONT

Tighten in numerical order

299640

Cylinder head bolt tightening sequence — 2.0L (SR20DE) engines

─── **CAUTION** ───

Fuel injection systems remain under pressure after the engine has been turned OFF. Properly relieve fuel pressure before disconnecting any fuel lines. Failure to do so may result in fire or personal injury.

4. Remove the intake manifold collector, exhaust manifold and all related components.

5. Remove the distributor assembly.

6. Using a block of wood, set a jack under the aluminum oil pan and remove the front engine mount.

7. Remove the cylinder head cover.

8. Remove the timing chain and camshaft sprockets.

9. Remove the camshafts.

NOTE: The valve train components must be reassembled in their original positions.

10. Loosen the cylinder head bolts in reverse order of tightening.

NOTE: A warped or cracked cylinder head could result from loosening in incorrect order. The cylinder head bolts should be loosened in two or three steps.

11. Remove the cylinder head and the intake manifold. Remove the cylinder head gasket. The lower timing chain will not be disengaged from crankshaft sprocket.

To install:

12. Clean the gasket surfaces.

13. Install new cylinder head gasket.

14. Install the cylinder head and temporarily tighten the cylinder head bolts. This is necessary to avoid damaging the cylinder head gasket. Be sure to install washers between the bolts and cylinder head.

15. Install the idler shaft assembly.

16. Install the upper timing chain and cover.

17. Tighten the cylinder head bolts in the reverse order of removal as follows:

a. Tighten all the bolts to 22 ft. lbs. (29 Nm).

b. Tighten all the bolts to 59 ft. lbs. (79 Nm).

c. Loosen all the bolts completely.

d. Tighten all the bolts to 18–25 ft. lbs. (25–34 Nm).

e. Turn all the bolts 86–91 degrees clockwise.

18. Install the camshafts.

19. Install the timing chains, chain tensioner, and camshaft sprockets.

20. The balance of installation is the reverse of the removal procedure.

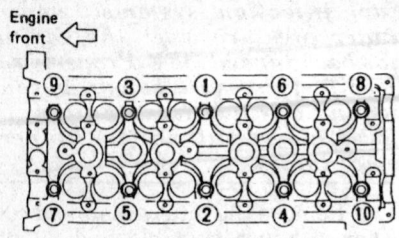

Engine front ⟸

Tighten in numerical order.

289271

Cylinder head bolt tightening sequence — 2.4L (KA24DE) engines

21. Refill the engine coolant.
22. Connect the negative battery cable.
23. Start the engine and make the necessary adjustments. Check for proper operation and leaks.

240SX

1. Disconnect negative battery cable.
2. Relieve the fuel system pressure.

────── **CAUTION** ──────
Fuel injection systems remain under pressure after the engine has been turned OFF. Properly relieve fuel pressure before disconnecting any fuel lines. Failure to do so may result in fire or personal injury.
──────────────────

3. Disconnect and tag all vacuum hoses, fuel lines, and electrical connections.
4. Remove all spark plugs.
5. Remove the fuel injector tube assembly with injectors.
6. Remove all accessory drive belts.
7. Remove the distributor assembly and related components.
8. Remove the cylinder head cover.
9. Remove the front engine mount.
10. Remove the timing chain and camshaft sprockets.
11. Remove the camshafts.
12. Loosen the cylinder head bolts the reverse order of tightening.

NOTE: Head warpage or cracking could result from removing the head bolts in the incorrect order. Cylinder head bolts should be loosened in two or three steps.

13. Remove the cylinder head with the intake manifold and the exhaust manifold assembly.
14. Remove the cylinder head gasket.

To install:
15. Check all components for wear and replace any worn components as necessary. Clean all mating surfaces and replace the cylinder head gasket.
16. Install a new cylinder head gasket.
17. Install the cylinder head and temporarily tighten the cylinder head bolts when installing the front cover. This is necessary to avoid damaging the cylinder head gasket. Be sure to install washers between the bolts and cylinder head.
18. Install the idler shaft assembly.
19. Install the upper timing chain and camshaft sprocket cover.
20. Tighten cylinder head bolts, in sequence as follows:
   a. Tighten all bolts to 22 ft. lbs. (29 Nm).
   b. Tighten all bolts to 59 ft. lbs. (79 Nm).
   c. Loosen all bolts completely.
   d. Tighten all bolts to 18–25 ft. lbs. 25–34 Nm).
   e. Turn all bolts 86–91 degrees (90° preferred) clockwise, or if an angle wrench is not available, tighten bolts to 55–62 ft. lbs. (75–84 Nm).
21. The remaining components are installed in the reverse order from which they were removed.
22. Install the air duct and engine undercover.
23. Refill the coolant and engine oil.
24. Connect the negative battery cable.

3.0L (VE30DE)

NOTE: To remove or install the cylinder head, you'll need a cylinder head bolt wrench tool No. ST10120000 (J24239–01) or equivalent.

1. Release the fuel pressure. Disconnect the negative battery cable.
2. Rotate the crankshaft to position the No. 1 piston on TDC of it's compression stroke.
3. Drain the cooling system. Disconnect all the electrical connectors, vacuum hoses and water hoses connected to the intake manifold collector.
4. The crank angle sensor is located in the rear of the left cylinder head. Mark its position, disconnect the electrical connector and remove the sensor.
5. Remove the timing chain assembly.

NOTE: Do not rotate either the crankshaft or camshaft from this point on or the valves could be bent by hitting the pistons.

6. Remove the intake manifold collector.
7. Remove the intake manifold and fuel rail assembly.
8. Remove the exhaust.
9. Remove the camshaft sprockets.
10. Remove the exhaust camshafts, the camshaft brackets and the rocker arms.
11. Remove the outside cylinder head bolts.
12. Loosen the cylinder head-to-engine bolts, in sequence, using 2–3 steps.
13. Lift the cylinder heads from the engine and discard the gaskets.
14. Unbolt camshaft cap bolts and remove the camshafts.

To install:
15. Install the camshafts, position the dowels to 12 o'clock and secure with bearing caps. Tighten the bearing caps to 7–9 ft. lbs. (9–12 Nm).
16. Clean the gasket mounting surfaces. Inspect the cylinder heads for warpage, wear, cracks and/or damage.
17. Using liquid gasket sealant, apply a continuous bead to the mating surface of the cylinder block.
18. Install the cylinder heads and torque the head bolts in the reverse of tightening sequence by performing the following procedures:
   a. Torque all bolts, in sequence, to 29 ft. lbs. (39 Nm).
   b. Torque all bolts, in sequence, to 90 ft. lbs. (123 Nm).
   c. Loosen all bolts.
   d. Torque all bolts, in sequence, to 25–33 ft. lbs. (34–44 Nm).
   e. Torque all bolts, in sequence, to 87–94 ft. lbs. (118–127 Nm).
19. The balance of installation is the reverse of the removal procedure.
20. Connect all the electrical connectors, vacuum hoses and water hoses to the intake manifold collector. Refill the cooling system.
21. Connect the negative battery cable. Start the engine check the engine timing. After the engine reaches the normal operating temperature check for the correct coolant level.
22. Road test the vehicle for proper operation.

3.0L (VG30E) Engine

NOTE: To remove or install the cylinder head, you will need a cylinder head bolt wrench No. ST10120000 (J24239–01) or equivalent. The collector assembly and intake manifold have special bolt sequence for removal and installation.

1. Release the fuel pressure. Disconnect the negative battery cable.

2. Remove the timing belt assembly.

3. Remove the intake manifold and fuel rail assembly.

4. Remove the cylinder head covers.

5. Remove the exhaust manifold components.

6. Remove the camshaft pulleys and the rear timing cover securing bolts.

7. Loosen the cylinder head bolts, in the reverse order of the tightening sequence, using 2–3 steps.

8. Remove the cylinder head with the exhaust manifold attached, if not already removed.

9. Check the positions of the timing marks and camshaft sprockets to make sure they have not shifted.

To install:

10. Clean the gasket mounting surfaces. Inspect the cylinder head(s) for warpage, wear, cracks and/or damage.

11. Make the following checks by performing the following procedures:

   a. If the engine was disturbed, rotate the crankshaft to position the No. 1 piston on the TDC of its compression stroke.

   b. Align the mark on the crankshaft sprocket with the mark on the oil pump body.

   c. Make sure the knock pin on the camshaft is set at the top.

12. Using a new gasket, install the cylinder head. Apply clean engine oil to the threads and seats of the bolts and install the bolts with washers in the correct position. Note that bolts 4, 5, 12, and 13 are 5.00 in. (127mm) long. The other bolts are 4.17 in. (106mm) long.

13. Torque the bolts according to the pattern for the cylinder head on each side, in the following stages:

   a. Torque all bolts, in order, to 22 ft. lbs. (29 Nm).

   b. Torque all bolts, in order, to 43 ft. lbs. (59 Nm).

   c. Loosen all bolts completely.

   d. Torque all bolts, in order, to 22 ft. lbs. (29 Nm).

   e. Torque all bolts, in order, to 40–47 ft. lbs. (54–64 Nm). If you have a special wrench available that torques bolts to a certain angle, torque them 60–65° tighter rather than going to 40–47 ft. lbs.

14. Install the remaining components in the reverse order of removal.

15. Refill the cooling system. Connect the negative battery cable. Start the engine check the engine timing. After the engine reaches the normal operating temperature check for the correct coolant level.

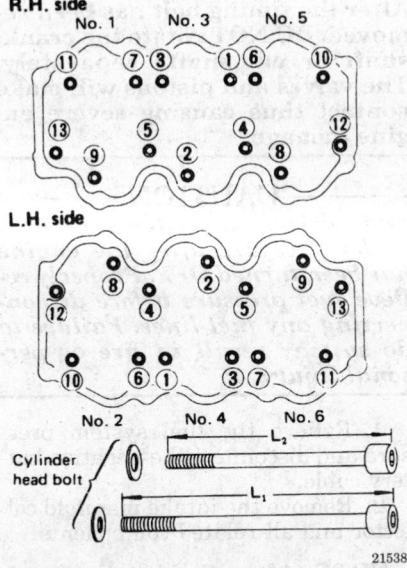

Cylinder head bolt torque sequence — 3.0L (VG30E) engine

16. Road test the vehicle for proper operation.

3.0L (VQ30DE) Engine

1. Relieve the fuel system pressure.

— **CAUTION** —

Fuel injection systems remain under pressure after the engine has been turned OFF. Properly relieve fuel pressure before disconnecting any fuel lines. Failure to do so may result in fire or personal injury.

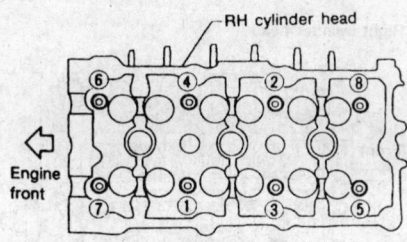

Right cylinder head torque sequence — 3.0L (VQ30DE) engine

2. Disconnect the negative battery cable.

3. Drain the engine oil and the cooling system. Be sure to drain the engine block and the radiator.

NOTE: Before disconnecting any hoses or connectors, note the locations for reassembly.

4. Remove the intake manifold collector.

5. Remove the fuel tube.

6. Remove the intake manifold.

7. Remove the cylinder head covers.

8. Remove the drive belts and idler pulley.

9. Remove the steel (lower) and aluminum (upper) oil pans.

10. Remove the water pump cover.

11. Remove the timing chain case cover.

12. Remove the timing chains, camshaft sprockets and related components.

13. Remove the crankshaft sprocket.

14. Loosen the bolts that secure the rear timing chain case. The bolts must be loosened in the reverse order of installation sequence.

15. Using seal cutter tool, remove the rear timing case cover.

NOTE: Remove the O-rings from the front of the engine block.

16. Remove the camshafts.

17. Remove the cylinder head bolts in the reverse order of the tightening sequence. The bolts should be loosened in 2–3 steps.

NOTE: A warped or cracked cylinder head could result from removing the bolts in incorrect order.

18. Remove the cylinder heads from the vehicle.

19. Remove and discard the head gaskets.

20. Remove all traces of liquid gasket from the timing chain case and from the water pump covers.

21. Remove all traces of liquid gasket from the engine block.

22. Inspect the timing chain for excessive wear or damage and replace as necessary.

To install:

23. Turn the crankshaft until the No. 1 piston is set 240 degrees before TDC on compression stroke.

24. Using new head gaskets, install the cylinder heads.

NOTE: If possible, replacement of the head bolts is suggested.

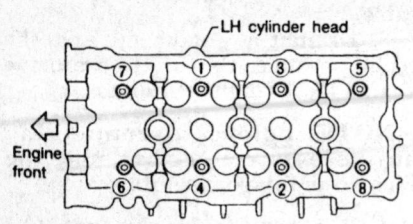

Tighten in numerical order.

296157

Left cylinder head torque sequence — 3.0L (VQ30DE) engine

25. If replacement of the head bolts is not possible, perform the following bolt measurement:

    a. Measure the diameter of the head bolt 0.43 in. (11mm) from the bottom of the bolt.

    b. Measure the diameter of the head bolt 1.89 in. (48mm) from the bottom of the bolt.

    c. Whenever the size difference between the two measurements exceeds 0.0043 in. (0.11mm) the head bolts must be replaced.

26. Install the cylinder head bolts and tighten in sequence as follows:

    a. Tighten all bolts in sequence to 72 ft. lbs. (98 Nm).

    b. Completely loosen all bolts.

    c. Tighten all bolts in sequence to 25–33 ft. lbs. (24–44 Nm).

    d. Turn all bolts in sequence 90–95 degrees clockwise.

    e. Turn all bolts in sequence 90–95 degrees clockwise.

27. Install the camshafts and related components.

28. Install new O-rings to the front of the engine block.

29. Apply sealant to the hatched portion of the of the rear timing chain case.

30. Align the rear timing chain case with the dowel pins and install onto the cylinder heads and engine block.

31. Tighten the rear timing chain case mounting bolts in sequence to 105–121 inch lbs. (11.8–13.7 Nm).

32. The remaining components are installed in the reverse order from which they were removed.

33. Connect the negative battery cable.

34. Start the engine and run at 3000 RPM under no load to purge the air from the high pressure chamber. The engine may produce a rattling noise. This indicates that air still remains in the chamber and is not a matter of concern.

35. Verify that there are no leaks.

3.0L (VG30DE and VG30DETT) Engines

— WARNING —
After the timing belt has been removed, DO NOT rotate the crankshaft or camshafts separately. The valves and pistons will make contact thus causing severe engine damage.

— CAUTION —
Fuel injection systems remain under pressure after the engine has been turned OFF. Properly relieve fuel pressure before disconnecting any fuel lines. Failure to do so may result in fire or personal injury.

1. Relieve the fuel system pressure and disconnect the negative battery cable.

2. Remove the intake manifold collector and all related components.

NOTE: After the intake manifold collector has been removed, cover the openings on the intake manifold to prevent foreign objects from entering the combustion chamber.

3. Remove the fuel rail and injectors.

4. Remove the cylinder head covers.

5. Remove the accessory drive belts.

6. Remove the radiator and cooling fans.

7. Remove the timing belt, automatic tensioner and idler pulley assemblies.

8. Remove the intake manifold assembly.

9. Disconnect the exhaust pipe from the exhaust manifold.

10. Loosen the cylinder head bolts (in reverse order of installation sequence) in 2–3 stages. Lift the cylinder head off the engine block with the

Right cylinder head

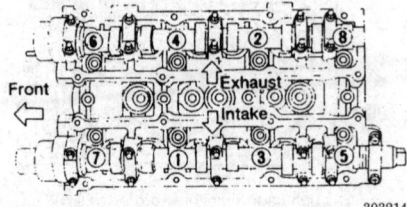

303214

Right cylinder head torque sequence — 3.0L (VG30DE and VG30DETT) engines

exhaust manifolds attached. It may be necessary to tap the head lightly with a rubber mallet to loosen it.

To install:

11. Thoroughly clean the engine block and cylinder head surfaces.

12. Make sure the No. 1 cylinder is set at TDC on its compression stroke as follows:

    a. Align the crankshaft timing mark with the mark on the oil pump housing.

    b. Align camshaft sprocket timing mark with the mark on the rear timing belt cover.

13. Install the cylinder head with a new gasket. Apply clean engine oil to the threads and seats of the bolts and install the bolts with washers in the correct position. Be sure to position the head bolt washers with the flat side toward the cylinder head.

NOTE: There is one special 6mm bolt for each cylinder head. Follow the proper torque specifications for these bolts.

14. Torque the bolts in the proper sequence as follows:

    a. Torque all bolts, in sequence, to 29 ft. lbs. (39 Nm).

    b. Torque all bolts, in sequence, to 90 ft. lbs. (123 Nm).

    c. Loosen all bolts completely.

    d. Torque all bolts, in sequence, to 25–33 ft. lbs. (34–44 Nm).

    e. Torque all bolts, in sequence, to 90 ft. lbs. (123 Nm). If using an angle torque wrench, torque them 70 degrees tighter rather than going to 90 ft. lbs. (123 Nm).

    f. Torque the 6mm **X** bolts to 7–9 ft. lbs. (10–12 Nm). There is one of these bolts per head.

15. Install the remaining components in the opposite order from which they were removed.

16. Fill the cooling system to the proper level and connect the negative battery cable.

17. Run the engine and make all the necessary engine adjustments.

18. Road test the vehicle for proper operation.

Lash Adjusters

BLEEDING

2.0L (SR20DE) Engine

200SX

NOTE: The hydraulic lifters must be bled to remove the air from them. Air can not be bled from this type of lifter by running the engine.

Left cylinder head

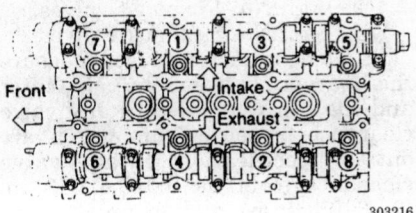

303216

Left cylinder head torque sequence — 3.0L (VG30DE and VG30DETT) engines

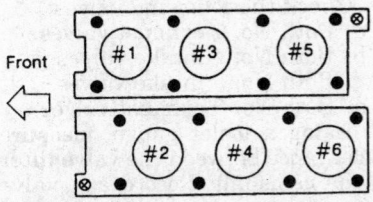

⊗ : M6 bolt

303217

Location of cylinder head 6mm bolt — 3.0L (VG30DE and VG30DETT) engines

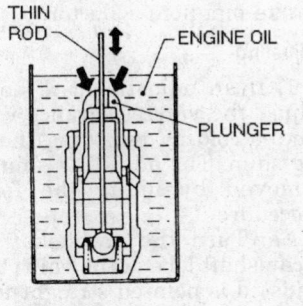

298695

Bleeding air from hydraulic lifter — 200SX with 2.0L (SR20DE) engines

1. Bleed the lifters as follows:
   a. Submerse the lifter into a container of clean engine oil.
   b. While pushing the plunger, insert a thin rod into the check ball and lightly push the check ball.
   c. Air is completely bled when the plunger no longer moves.

3.0L (VG30DE and VG30DETT) Engines

When removed, the hydraulic lash adjusters should be immersed in a container of clean engine oil in the upright position. No other specific procedures are required

3.0L (VE30DE) Engine

1. Remove the rocker arms and lash adjusters from the cylinder heads.
2. Push down on the lash adjuster. If the adjuster can be depressed 0.04 in. (1.0 mm) or more, the adjuster must be bled. Place the adjuster in a bath of engine oil and insert a wire through the hole in the top of the plunger and depress the check ball. Work the plunger until no air remains in the lash adjuster.
3. Install the rocker arms and the camshaft.
4. Install the rocker covers, start the engine and check operation.

Valve Lash

ADJUSTMENT

1.6L (GA16DE) and 2.4L (KA24DE) Engines

Checking

1. Run the engine until it reaches normal operating temperature and shut if off.
2. Remove the cylinder head cover and all the spark plugs.
3. Set the No. 1 cylinder at TDC on its compression stroke. Align the pointer with the TDC mark on the crankshaft pulley. Check that the valve lifters on the No. 1 cylinder are loose and valve lifters on the No. 4 cylinder are tight. If not, turn the crankshaft one revolution (360 degrees) and align the pointer with the TDC mark on the crankshaft pulley.
4. Check the following valves:
   a. Both No. 1 intake valves.
   b. Both No. 1 exhaust valves.

Push

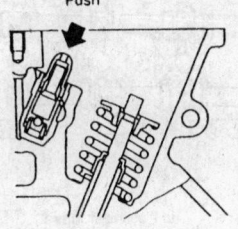

167392

Push on the lash adjuster to check for air. If the adjuster can be pushed down more than 0.04 in. (1.0 mm), it must be bled — 3.0L (VE30DE) engine

   c. Both No. 2 intake valves.
   d. Both No. 3 exhaust valves.
5. Using a feeler gauge, measure the clearance between the valve lifter and the camshaft. Record any valve clearance measurements which are out of specification.
6. Turn the crankshaft one revolution (360 degrees) and align the mark on the crankshaft pulley with the pointer. Check the following valves:
   a. Both No. 2 exhaust valves.
   b. Both No. 3 intake valves.
   c. Both No. 4 intake valves.
   d. Both No. 4 exhaust valves.
7. Using a feeler gauge, measure the clearance between the valve lifter and the camshaft. Record any valve clearance measurements which are out of specification.
8. If all the valve clearances are within specification, install the cylinder head cover and the spark plugs.

Adjusting

1. If an adjustment is necessary, adjust the valve clearance while engine is cold by removing the adjusting shim. The adjusting shim can be removed by using the following procedures:
   a. Turn the crankshaft so the camshaft lobe of the valve to be adjusted is pointed straight up.
   b. Turn the lifter so the notch is pointed towards the center of the cylinder head; this will facilitate the shim removal process.
   c. Using a depressor tool, push down on the lifter and insert a keeper tool on the edge of the lifter to keep the lifter in the depressed position.
   d. Remove the depressor tool and remove the shim with a magnet.
2. Determine the replacement adjusting shim size by using the following procedures and formula:
   a. Using a micrometer determine thickness of the removed shim.
   b. Calculate the thickness of a new adjusting shim so valve clearance is within the specified values.
   c. R = thickness of the removed shim.
   d. N = thickness of the new shim.
   e. M = measured valve clearance.
   1.6L engine — Intake shim determination formula: $N = R + (M - 0.0146$ in. or 0.37mm$)$
   1.6L engine — Exhaust shim determination formula: $N = R + (M - 0.0157$ in. or 0.40mm$)$

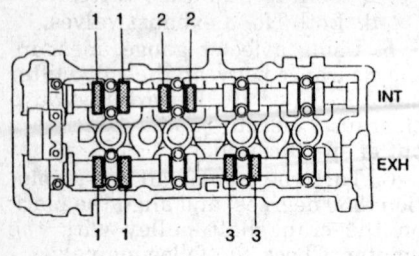

Valve lash checking sequence when the engine is at TDC of cylinder No. 1 — 1.6L and 2.4L engines

302436

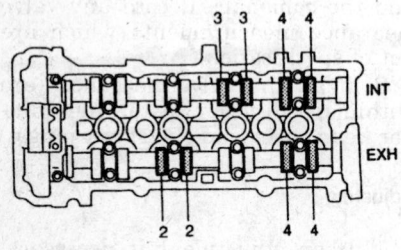

Valve lash checking sequence when the engine is at TDC of cylinder No. 4 — 1.6L and 2.4L engines

302437

2.4L engine — Intake shim determination formula: N = R + (M - 0.0138 in. or 0.35mm)

2.4L engine — Exhaust shim determination formula: N = R + (M - 0.0146 in. or 0.37mm)

3. Shims are available different sizes from 0.0772–0.1055 in. (1.96–2.68mm) in increments of 0.0008 in. (0.02mm). The thickness is stamped on the shim; this side is always installed facing down. Select new shims with thickness as close as possible to calculated valve and install it in the lifter.

4. Install the new shim onto the lifter.

5. Depress the lifter and remove the keeper tool. Remove the depressor tool and recheck the valve clearance. Repeat this procedure for any other valves requiring adjustment.

6. Install the cylinder head cover and spark plugs when all valve adjustments are finished.

3.0L (VQ30DE) Engine

NOTE: Check and adjust the valve clearances while the engine is cold and not running.

Checking

1. Remove the intake manifold collector.

2. Remove the left and right rocker covers.

3. Remove the spark plugs.

4. Set the No. 1 cylinder at TDC on its compression stroke. Align the pointer with the TDC mark on the crankshaft pulley. Check that the valve lifters on the No. 1 cylinder are loose and valve lifters on the No. 4 cylinder are tight. If not, turn the crankshaft one revolution (360 degrees) and align the pointer with the TDC mark on the crankshaft pulley.

5. Check the following valves:
   a. Both No. 1 intake valves.
   b. Both No. 2 exhaust valves.
   c. Both No. 3 exhaust valves.
   d. Both No. 6 intake valves.

6. Using a feeler gauge, measure the clearance between the valve lifter and the camshaft. Record any valve clearance measurements which are out of specification. Intake valve clearance (cold) is 0.010–0.013 in. (0.26–0.34mm) and exhaust valve clearance (cold) is 0.011–0.015 in. (0.29–0.37mm).

7. Turn the crankshaft 240 degrees and set the No. 3 cylinder to TDC of its compression stroke.

RH cylinder head

Engine front

LH cylinder head

289216

Valve lash checking sequence at TDC of cylinder No. 1 — 3.0L (VQ30DE) engine

8. Check the following valves:
   a. Both No. 2 intake valves.
   b. Both No. 3 intake valves.
   c. Both No. 4 exhaust valves.
   d. Both No. 5 exhaust valves.

9. Using a feeler gauge, measure the clearance between the valve lifter and the camshaft. Record any valve clearance measurements which are out of specification. Intake valve clearance (cold) is 0.010–0.013 in. (0.26–0.34mm) and exhaust valve clearance (cold) is 0.011–0.015 in. (0.29–0.37mm).

10. Turn the crankshaft 240 degrees and set the No. 5 cylinder to TDC of its compression stroke.

11. Check the following valves:
   a. Both No. 1 exhaust valves.
   b. Both No. 4 intake valves.
   c. Both No. 5 intake valves.
   d. Both No. 6 exhaust valves.

12. Using a feeler gauge, measure the clearance between the valve lifter and the camshaft. Record any valve clearance measurements which are out of specification. Intake valve clearance (cold) is 0.010–0.013 in. (0.26–0.34mm) and exhaust valve clearance (cold) is 0.011–0.015 in. (0.29–0.37mm).

13. If all the valve clearances are within specification, install the cylinder head cover, spark plugs, and the intake manifold collector.

Adjusting

1. If an adjustment is necessary, adjust the valve clearance while engine is cold by removing the adjusting shim. The adjusting shim can be removed by using the following procedures:
   a. Turn the crankshaft so the camshaft lobe of the valve to be adjusted is pointed straight up.
   b. Turn the lifter so the notch is pointed towards the center of the cylinder head; this will facilitate the shim removal process.
   c. Using a depressor tool No. KV10115110 or equivalent, push down on the lifter and insert a keeper tool on the edge of the lifter to keep the lifter in the depressed position.
   d. Remove the depressor tool and remove the shim with a magnet.

NOTE: Compressed air can be blown into the hole of the lifter to separate the adjusting shim from the lifter.

2. Determine the replacement adjusting shim size by using the following procedures and formula:
   a. Using a micrometer determine thickness of the removed shim.

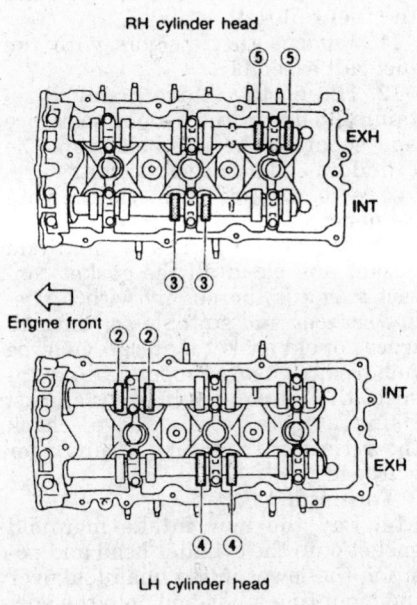

Valve lash checking sequence at TDC of cylinder No. 3 — 3.0L (VQ30DE) engine

289217

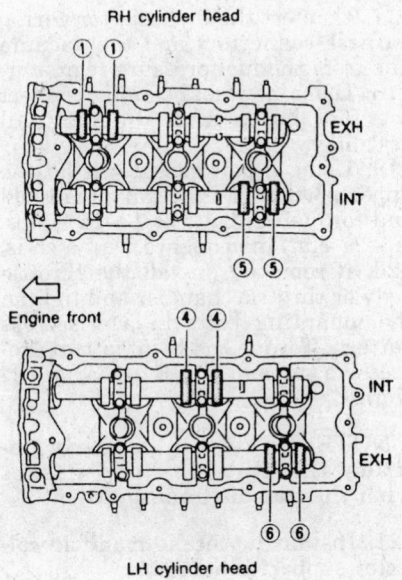

Valve lash checking sequence at TDC of cylinder No. 5 — 3.0L (VQ30DE) engine

289218

b. Calculate the thickness of a new adjusting shim so valve clearance is within the specified values.

c. R = thickness of the removed shim.

d. N = thickness of the new shim.

e. M = measured valve clearance.

Intake shim determination formula: N = R + (M - 0.0118 in. or 0.30mm)

Exhaust shim determination formula: N = R + (M - 0.0130 in. or 0.33mm)

3. Shims are available in 64 sizes from 0.0913–0.1161 in. (2.32–2.95mm) in steps of 0.004 in. (0.01mm). The thickness is stamped on the shim; this side is always installed facing down. Select new shims with thickness as close as possible to calculated valve and install it in the lifter.

4. Install the new shim onto the lifter.

5. Depress the lifter and remove the keeper tool. Remove the depressor tool and recheck the valve clearance. Repeat this procedure for any other valves requiring adjustment.

6. When all valve adjustments are finished, install the cylinder head cover, spark plugs, and the intake manifold collector.

3.0L (VG30DE and VG30DETT) Engines

The engine is equipped with Hydraulic Lash Adjusters. The valve lash is not adjustable. If the valves make noise, check the cylinder head for mechanical damage or excessive wear of the camshaft or lash adjusters.

Rocker Arms

REMOVAL AND INSTALLATION

2.0L (SR20DE) Engine

1. Release the fuel pressure following the proper procedure.

2. Disconnect the negative battery cable.

3. Remove the rocker arm cover, gasket, and the oil separator.

4. Remove the intake manifold supports, oil filter bracket, and the power steering pump.

5. Set the No. 1 cylinder at TDC on the compression stroke.

6. Remove the timing chain tensioner from the side of the head.

7. Matchmark the position of the rotor and housing and remove the distributor.

8. Remove the timing chain guide. Remove the camshaft sprockets while holding the camshaft stationary with a large wrench. Secure the timing chain with wire so the timing is not lost. The front cover will have to be removed if the chain timing is lost.

NOTE: When removing the camshafts, loosen the journal caps in the opposite sequence of tightening. Camshaft damage may result if this step is not followed.

9. Remove the camshafts, brackets, oil tubes, and the baffle plate. Label all components for proper installation.

NOTE: It is essential that all parts be kept in the same order and orientation for reinstallation. Be sure to mark and separate parts to keep them from getting mixed. This will aid assembly.

10. Remove the rocker arms, shims, rocker arm guides, and the hydraulic lash adjusters. Label all components for proper installation.

NOTE: The valve lifters must be stored in the vertical position or submersed in clean oil to prevent air from entering the lifters.

11. Inspect the surfaces of the rockers and replace if there are any signs of damage.

To install:

12. Lubricate the rocker arms, shims, rocker arm guides, and the hydraulic lash adjusters. Install them in their original locations.

13. Install the camshafts, brackets, oil tubes, and the baffle plate in the proper location.

14. Torque the bolts in sequence as follows:

a. Tighten right camshaft bolts No. 9 and No. 10 to 17 inch lbs. (2 Nm).

b. Tighten right camshaft bolts No. 1–No. 8 to 17 inch lbs. (2 Nm).

c. Tighten left camshaft bolts No. 11 and No. 12 to 17 inch lbs. (2 Nm).

d. Tighten left camshaft bolts No. 1–No. 10 to 17 inch lbs. (2 Nm).

e. Tighten all camshaft bolts in numerical sequence to 52 inch lbs. (6 Nm).

f. Tighten all camshaft bolts in numerical sequence to 87–104 inch lbs. (10–11 Nm).

g. Tighten the rear two bolts of the LH camshaft to 13–19 ft. lbs. (18–25 Nm).

15. Install the camshaft sprockets while holding the camshaft stationary with a large wrench.

16. The remaining components are installed in the reverse order from which they were removed.

17. Connect the negative battery cable.

18. Check and adjust the ignition and valve timing. If there is air in the lifters, bleed the air by running the engine at 1000 rpm for 10 minutes.

Intake Manifold

REMOVAL AND INSTALLATION

1.6L (GA16DE) Engine

1993–95 Models

————— CAUTION —————
Fuel injection systems remain under pressure after the engine has been turned OFF. Properly relieve fuel pressure before disconnecting any fuel lines. Failure to do so may result in fire or personal injury.

1. Relieve the fuel system pressure, disconnect the negative battery cable, and drain the cooling system.

2. Remove the air cleaner assembly.

3. Disconnect and tag the throttle linkage, electrical connections, fuel and vacuum lines from the throttle body or throttle chamber.

4. The throttle body/throttle chamber can be removed from the manifold at this point or can be removed as an assembly with the intake manifold.

5. Remove the bolts holding the upper portion of the intake to the lower portion. Remove the bolts in reverse order of the tightening sequence.

6. Remove the upper portion of the intake.

7. Loosen the intake manifold retaining bolts in the proper sequence and separate the manifold from the cylinder head. Remove the bolts in reverse order of the tightening sequence

8. Remove the intake manifold gasket and clean all the gasket contact surfaces thoroughly with a gasket scraper and suitable solvent. All traces of old gasket material must be removed to ensure proper sealing. Inspect the intake manifold for cracks. Using a metal straightedge, check the surface of the intake manifold for warpage.

To install:

9. Lay the new intake manifold gasket onto the cylinder head and position the lower intake manifold over the mounting studs and onto the gasket. Install the mounting nuts and torque them to 12–15 ft. lbs. (16–21 Nm) in sequence.

10. Using a new gasket, install the upper portion of the intake manifold and torque the bolts to 12–15 ft. lbs. (16–21 Nm) in sequence.

11. If removed, install the throttle body or throttle chamber and tighten the mounting bolts in a crisscross pattern. Torque the bolts in two progressive steps to 15 ft. lbs. (21 Nm).

NOTE: Be sure the gasket for the throttle body is positioned properly.

12. Install the remaining components in the opposite order from which they were removed.

13. Fill the cooling system to the proper level and connect the negative battery cable.

14. Start the engine, bleed the cooling system, and check for leaks.

15. Road test the vehicle for proper operation.

1996–97 Vehicles

————— CAUTION —————
Fuel injection systems remain under pressure after the engine has been turned OFF. Properly relieve fuel pressure before disconnecting any fuel lines. Failure to do so may result in fire or personal injury.

1. Relieve the fuel system pressure, disconnect the negative battery cable, and drain the cooling system.

2. Remove the air cleaner assembly.

3. Disconnect and tag the throttle linkage, electrical connections and vacuum lines from the throttle body.

4. Remove the intake manifold collector support brackets.

5. The throttle body can be removed from the manifold at this point or can be removed as an assembly with the intake manifold.

6. Remove the bolts holding the upper portion of the intake to the lower portion. Remove the bolts in reverse order of the tightening sequence.

7. Remove the upper portion of the intake.

8. Disconnect the fuel injector wiring harness connectors and the vacuum line from the fuel pressure regulator.

9. Disconnect the fuel hoses from the fuel rail assembly.

10. Remove the bolts that secure the fuel rail to the intake.

11. Remove the injectors with the fuel rail assembly.

12. Loosen the intake manifold retaining bolts in the proper sequence and separate the manifold from the cylinder head. Remove the bolts in reverse order of the tightening sequence

13. Remove the intake manifold gasket and clean all the gasket contact surfaces thoroughly with a gasket scraper and suitable solvent. All traces of old gasket material must be removed to ensure proper sealing. Inspect the intake manifold for cracks. Using a metal straightedge, check the surface of the intake manifold for warpage.

To install:

14. Lay the new intake manifold gasket onto the cylinder head and position the lower intake manifold over the mounting studs and onto the gasket. Install the mounting nuts and bolts; torque them to 13–15 ft. lbs. (18–21 Nm) in sequence.

15. Install the injectors with the fuel rail assembly. Be sure to install the fuel rail insulators.

16. Install the bolts that secure the fuel rail to the intake. Torque the bolts in two steps to 13–15 ft. lbs. (18–21 Nm).

17. Connect the fuel injector wiring harness connectors and the vacuum line from the fuel pressure regulator.

18. Using new hose clamps, connect the fuel hoses from the fuel rail assembly.

19. Using a new gasket, install the upper portion of the intake manifold and torque the bolts to 13–15 ft. lbs. (18–21 Nm) in sequence.

20. If removed, install the throttle body or throttle chamber and tighten the mounting bolts in a crisscross pattern. Torque the bolts in two progressive steps to 13–16 ft. lbs. (18–22 Nm).

NOTE: Be sure to properly position the throttle body gasket with the cut out facing down.

21. Install the intake manifold collector support brackets.

22. Connect the throttle linkage, electrical connections and vacuum lines.

23. Install the air cleaner.

24. Fill the cooling system to the proper level and connect the negative battery cable.

25. Start the engine, bleed the cooling system, and check for leaks.

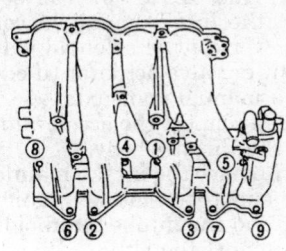

Tighten in numerical order.

300855

**Lower intake manifold tightening sequence —
1.6L (GA16DE) engines**

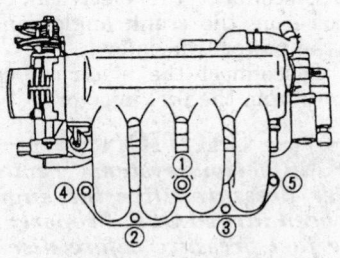

Tighten in numerical order.

300854

**Upper intake manifold tightening sequence —
1.6L (GA16DE) engines**

26. Road test the vehicle for proper operation.

2.0L (SR20DE) Engine

1993–94 Sentra

1. Relieve the fuel system pressure, disconnect the negative battery cable and drain the cooling system.
2. Remove the air cleaner assembly.
3. Disconnect the throttle linkage, electrical connections, fuel and vacuum lines from the throttle body.
4. Remove the EGR tube from the manifold.

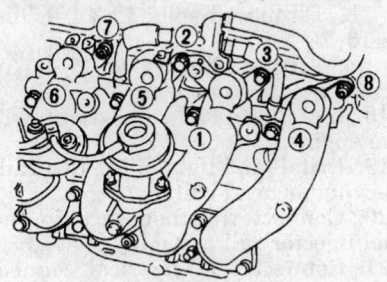

Tighten in numerical order

137152

**Intake manifold assembly bolt torque
sequence — 1993–95 Sentra models with
2.0L (SR20DE) engines**

5. Remove the drive belts and water pump pulley.
6. Remove the alternator and power steering pump.
7. Remove the intake manifold supports, oil filter bracket and power steering bracket.
8. Loosen the intake manifold retaining bolts in the proper sequence and separate the manifold from the cylinder head.
9. If necessary, remove the fuel rail and separate the intake collector assembly from the intake manifold.
10. Remove the intake manifold gasket and clean all the gasket contact surfaces thoroughly with a gasket scraper and suitable solvent. All traces of old gasket material must be removed to ensure proper sealing. Inspect the intake manifold for cracks. Using a metal straightedge, check the surface of the intake manifold for warpage.

To install:
11. Assemble the collector assembly to the manifold. Torque nuts and bolts in sequence to 13–15 ft. lbs. (18–21 Nm).
12. Lay the new intake manifold gasket onto the cylinder head and position the intake manifold over the mounting studs and onto the gasket. Install the mounting nuts and torque them to 13–15 ft. lbs. (18–21 Nm), in the proper sequence.
13. The balance of installation is the reverse of the removal procedure.
14. Fill the cooling system to the proper level and connect the negative battery cable.
15. Road test the vehicle for proper operation.

1995–97 200SX

----- CAUTION -----
Fuel injection systems remain under pressure after the engine has been turned OFF. Properly relieve fuel pressure before disconnecting any fuel lines. Failure to do so may result in fire or personal injury.

1. Relieve the fuel system pressure, disconnect the negative battery cable, and drain the cooling system.
2. Remove the air cleaner assembly.
3. Disconnect the manifold support brackets.
4. Disconnect the throttle linkage, electrical connections, and vacuum lines from the throttle body. Be sure to note the locations of all connections.
5. Remove the EGR tube from the manifold.

6. Unbolt and remove the fuel rail assembly.
7. Remove the drive belts and water pump pulley.
8. Remove the alternator and power steering pump.
9. Remove the oil filter bracket and power steering bracket.
10. Loosen the intake manifold collector retaining bolts in the reverse of installation sequence and separate the collector from the manifold.
11. Loosen the intake manifold assembly retaining bolts in the reverse order of the tightening sequence and separate the manifold from the cylinder head.
12. Remove all gasket material and clean all the gasket contact surfaces thoroughly with a gasket scraper and a suitable solvent. All traces of old gasket material must be removed to ensure proper sealing. Inspect the intake manifold for cracks. Using a metal straightedge, check the surface of the intake manifold for warpage.

To install:
13. Using new gaskets, install the intake manifold assembly to the cylinder head and torque the mounting bolts in sequence to 13–15 ft. lbs. (18–21 Nm).
14. Using new gaskets, install the intake manifold collector to the intake manifold assembly torque the mounting bolts and nuts in sequence to 13–15 ft. lbs. (18–21 Nm).
15. Install the remaining components in the reverse order of removal.
16. Fill the cooling system to the proper level and connect the negative battery cable.
17. Bleed the cooling system and road test the vehicle for proper operation.

2.4L (KA24DE) Engine

1. Relieve the fuel system pressure, disconnect the negative battery cable and drain the cooling system.

----- CAUTION -----
Fuel injection systems remain under pressure after the engine has been turned OFF. Properly relieve the pressure before disconnecting the fuel lines. Failure to do so may result in fire or personal injury.

2. Remove the air duct between the air flow meter and the throttle body.
3. Disconnect the throttle cable and the cruise control cable, if equipped.
4. Disconnect the fuel supply and return lines from the fuel injector as-

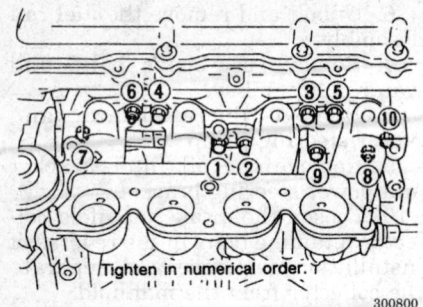

Lower intake manifold assembly torque sequence — 1995–97 200SX with 2.0L (SR20DE) engines

300800

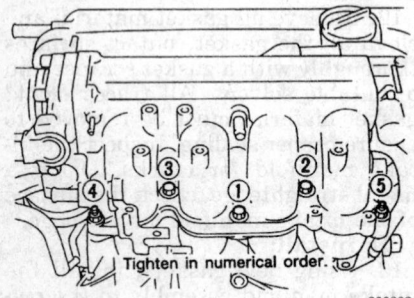

Upper intake manifold collector torque sequence — 1995–97 200SX with 2.0L (SR20DE) engines

300802

sembly. Plug the lines to prevent leakage.

5. Disconnect and tag the electrical connectors and the vacuum hoses to the throttle body and intake manifold/collector assembly.

6. Remove the spark plug wires from the spark plugs.

7. Remove the throttle body assembly from the intake manifold.

8. Disconnect the EGR valve tube from the exhaust manifold.

9. Remove the intake manifold mounting brackets.

10. Remove the intake manifold collector to the intake manifold bolts/nuts in the reverse sequence of the tightening procedure and separate the intake manifold from the intake manifold collector.

11. Remove the bolts that secure the intake manifold to the cylinder head and remove the manifold. Be sure to loosen the bolts in the reverse sequence of the tightening procedure.

12. Using a putty knife or equivalent, clean the gasket mounting surfaces. Check the intake manifold/collector for cracks and warpage.

To install:

13. Using new gaskets, install the intake manifold to the cylinder head and tighten the mounting bolts in sequence to 12–14 ft. lbs. (16–19 Nm).

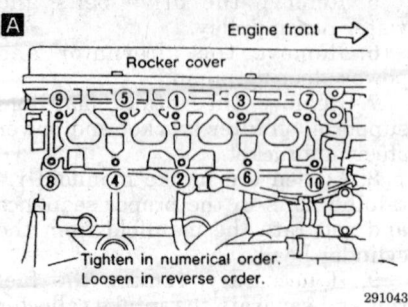

Intake manifold torque sequence — 2.4L (KA24DE) engines

291043

14. Using new gaskets, install the intake manifold collector to the intake manifold and tighten the mounting bolts/nuts in sequence to 12–14 ft. lbs. (16–19 Nm).

15. Install intake manifold mounting brackets.

16. Connect the EGR valve tube to the exhaust manifold.

17. Using a new gasket, install the throttle body and tighten the mounting bolts in a crisscross pattern to 13–16 ft. lbs. (18–22 Nm). Be sure to tighten the bolts in two progressive steps.

18. The balance of installation is the reverse of the removal procedure.

19. Fill the cooling system to the proper level and connect the negative battery cable.

20. Make all the necessary engine adjustments. Road test the vehicle for proper operation.

3.0L (VE30DE) Engine

1. Disconnect the negative battery cable, drain the cooling system, and release the fuel system pressure.

2. Disconnect the electrical connectors from the throttle position sensor, exhaust gas temperature sensor, coolant temperature sensor, etc.

3. Label and disconnect the hoses from the throttle body, the EGR

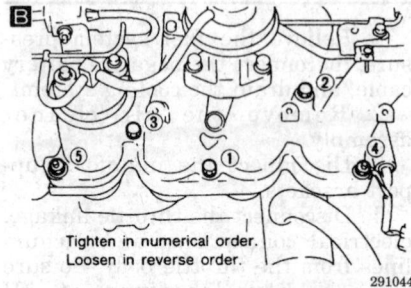

Intake manifold collector torque sequence — 2.4L (KA24DE) engines

291044

valve, the EGR control solenoid valve, the intake manifold collector, the power control solenoid valve and the power valve actuator (if equipped with a manual transaxle).

4. Disconnect the accelerator cable from the throttle body.

5. Remove the intake manifold collector support-to-intake manifold collector and the intake manifold collector support-to-cylinder head bolts and remove the support.

6. Remove the intake manifold collector-to-intake manifold bolts and remove the intake manifold collector.

7. Disconnect the electrical connectors from the ignition coils.

8. Disconnect the electrical connector from the crank angle sensor and the power transistor.

9. Disconnect the electrical connectors from the fuel injectors.

— CAUTION —

Fuel injection systems remain under pressure after the engine has been turned OFF. Properly relieve fuel pressure before disconnecting any fuel lines. Failure to do so may result in fire or personal injury.

10. Disconnect the fuel injector assembly from the fuel lines.

11. Remove the fuel rail-to-cylinder head bolts.

12. Remove the fuel rail assembly from the engine.

13. Remove the intake manifold-to-engine bolts, in sequence, by reversing the tightening sequence.

14. Lift the intake manifold from the engine and discard the gasket.

To install:

15. Clean the gasket mounting surfaces.

16. Install the intake manifold and torque the intake manifold-to-engine bolts and nuts in sequence.

17. Torque the intake manifold bolts and nuts in 2 steps as follows:

   a. Torque all bolts and nuts to 26–43 inch lbs. (3–5 Nm).

   b. Torque all bolts 12–14 ft. lbs. (16–20 Nm).

   c. Torque all nuts to 17–20 ft. lbs. (24–27 Nm).

18. Install the fuel rail assembly to the engine.

19. Install and tighten the fuel rail-to-cylinder head bolts.

20. Connect the fuel lines to the fuel injector rail assembly.

21. Connect the electrical connectors to the fuel injectors.

22. Connect the electrical connectors to the ignition coils.

23. Using a new gasket, install the intake manifold collector and torque

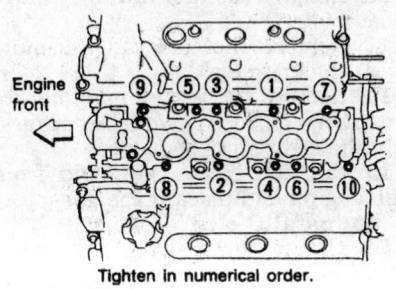

Intake manifold torque sequence — 3.0L (VE30DE) engine

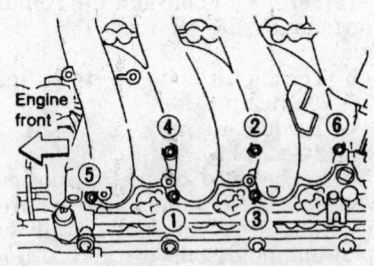

Intake collector torque sequence — 3.0L (VE30DE) engine

the intake manifold collector-to-intake manifold bolts, in sequence, to 13–16 ft. lbs. (18–22 Nm).

24. Install the intake manifold collector support and torque the bolts to 12–15 ft. lbs. (16–21 Nm).

25. The remaining components are installed in the reverse order from which they were removed.

26. Connect the negative battery cable.

27. Refill the cooling system.

28. Start the engine and check for leaks.

3.0L (VG30E) Engine

1. Disconnect the negative battery cable and relieve the fuel pressure.

2. Disconnect the air intake duct from the dual duct housing.

3. Disconnect the electrical connectors from the throttle body, the step motor AAC valve and the exhaust gas temperature sensor.

4. Disconnect and label the hoses from the rocker arm cover, the throttle body, the step motor AAC valve, the EGR control valve and the air cut valve.

5. If the spark plug wires are in the way, disconnect them and move them aside. Disconnect the accelerator cable from the throttle body.

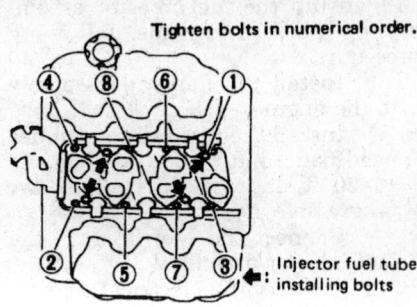

Intake manifold bolt tightening sequence — 3.0L (VG30E) engine

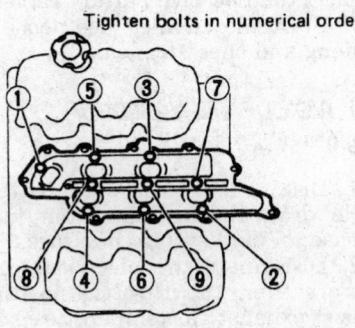

Lower intake collector torque sequence — 3.0L (VG30E) engine

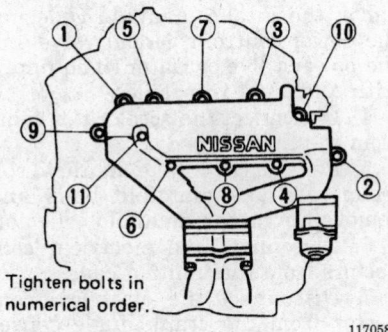

Upper intake collector torque sequence — 3.0L (VG30E) engine

6. Remove the upper intake manifold collector-to-intake manifold bolts, in sequence, and lift the assembly from the intake manifold. Discard the gasket.

7. Remove the lower intake manifold collector-to-intake manifold bolts, in sequence, and lift the assembly from the intake manifold. Discard the gasket.

8. If the fuel injector assembly is in the way, perform the following procedures:

   a. Disconnect the electrical connectors from the fuel injectors.

b. Disconnect the fuel injector assembly from the fuel lines.

   c. Remove the fuel rail-to-cylinder head bolts.

   d. Remove the fuel rail assembly from the engine.

9. Remove the intake manifold-to-engine bolts, in sequence, lift the intake manifold from the engine and discard the gasket.

To install:

10. Clean the gasket mounting surfaces.

11. Use a new gasket and install the intake manifold by performing the following procedures:

   a. Torque the intake manifold-to-engine nuts/bolts, in sequence, to 2.2–3.6 ft. lbs. (3–5 Nm).

   b. Retighten the intake manifold-to-engine nuts/bolts, in sequence, to 12–14 ft. lbs. (16–20 Nm).

   c. Finally, torque the intake manifold-to-engine nuts/bolts, in sequence, to 17–20 ft. lbs. (24–27 Nm).

12. If the fuel injector assembly was removed, perform the following procedures:

   a. Install the fuel rail assembly to the engine.

   b. Install the fuel rail-to-cylinder head bolts and torque the bolts to 1.8–2.4 ft. lbs. (2.5–3.2 Nm).

   c. Connect the fuel injector assembly to the fuel lines.

   d. Connect the electrical connectors to the fuel injectors.

13. Install the remaining components in the opposite order from which they were removed.

14. Connect the negative battery cable.

15. Start the engine and check for oil leaks.

3.0L (VQ30DE) Engine

1. Disconnect the negative battery cable and drain the cooling system.

2. Release the fuel system pressure.

—————— CAUTION ——————
The fuel injection system remains under pressure after the engine has been turned OFF. Properly relieve the fuel pressure before disconnecting any fuel lines. Failure to do so may result in fire or personal injury.

3. Remove the throttle body coolant hoses.

4. Label and disconnect the electrical connectors from the throttle position sensor.

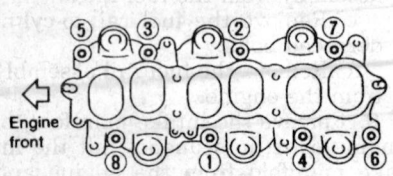

Tighten in numerical order.

288334

Intake manifold tightening sequence — 3.0L (VQ30DE) engine

5. Label and disconnect the hoses from the throttle body, the EGR valve, intake manifold collector, IAC valve, and the fuel pressure regulator.

6. Disconnect the canister purge hose and blow-by hose.

7. Disconnect the EGR guide tube.

8. Disconnect the accelerator cable from the throttle body.

9. Remove the intake manifold collector support brackets.

10. Disconnect the right side electrical connectors from the ignition coils.

11. If necessary, disconnect the electrical connector from the crank angle sensor and the power transistor.

12. Remove the intake manifold collector-to-intake manifold bolts/nuts and remove the intake manifold collector.

13. Remove the fuel injector assembly by performing the following procedures:

   a. Disconnect the electrical connectors from the fuel injectors.

   b. Disconnect the fuel lines from the fuel injector assembly.

   c. Remove the fuel rail-to-cylinder head bolts.

   d. Remove the fuel rail assembly from the engine.

14. Remove the intake manifold bolts/nuts in the reverse sequence of the torque procedure.

15. Remove the intake manifold from the engine and discard the gaskets.

16. Clean all gasket mounting surfaces.

To install:

17. Using new gaskets, install the intake manifold to the engine.

18. Tighten the bolts/nuts in sequence as follows:

   a. Tighten nuts and bolts to 44–86 inch lbs. (5–10 Nm).

   b. Tighten nuts and bolts to 20–23 ft. lbs. (26–31 Nm).

19. Install the fuel injector assembly by performing the following procedures:

   a. Install the fuel rail assembly to the engine.

   b. Install the fuel rail-to-cylinder head bolts and torque the bolts to 15–20 ft. lbs. (21–26 Nm) in two progressive steps.

   c. Connect the fuel lines to the fuel injector assembly.

   d. Connect the electrical connectors to the fuel injectors.

20. Install the remaining components in the reverse order of removal.

21. Refill the cooling system and connect the negative battery cable.

22. Start the engine, bleed cooling system, and check for leaks.

3.0L (VG30DE and VG30DETT) Engines

1. Disconnect the negative battery cable, drain the cooling system, and release the fuel system pressure.

2. Disconnect the electrical connectors from the throttle position sensor, exhaust gas temperature sensor, coolant temperature sensor, etc.

3. Label and disconnect the hoses from the throttle body, the EGR valve, the EGR control solenoid valve, the intake manifold collector, the power control solenoid valve and the power valve actuator (if equipped with a manual transaxle).

4. Disconnect the accelerator cable from the throttle body.

5. Remove the intake manifold collector-to-intake manifold bolts and remove the intake manifold collector.

6. Disconnect the electrical connectors from the ignition coils.

7. Disconnect the electrical connector from the crank angle sensor and the power transistor.

8. Disconnect the electrical connectors from the fuel injectors.

9. Disconnect the fuel injector assembly from the fuel lines.

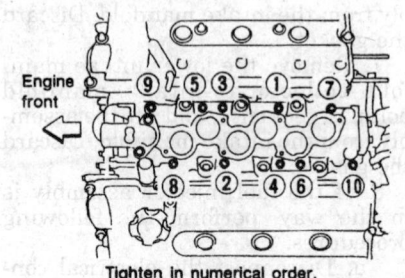

Engine front

Tighten in numerical order.

303668

Intake manifold tightening sequence — 3.0L (VG30DE and VG30DETT) engines

10. Remove the fuel rail-to-cylinder head bolts.

11. Remove the fuel rail assembly from the engine.

12. Remove the intake manifold-to-engine bolts, in sequence, by reversing the tightening sequence.

13. Lift the intake manifold from the engine and discard the gasket.

To install:

14. Clean the gasket mounting surfaces.

15. Install the intake manifold and torque the intake manifold-to-engine bolts and nuts in sequence.

16. Torque the intake manifold bolts and nuts in two steps as follows:

   a. Torque all bolt-and-nut combinations to 26–43 inch lbs. (3–5 Nm).

   b. Torque all bolts 12–14 ft. lbs. (16–20 Nm).

   c. Torque all nuts to 17–20 ft. lbs. (24–27 Nm).

17. The balance of installation is the reverse of the removal procedure.

18. Using a new gasket, install the intake manifold collector and torque the intake manifold collector-to-intake manifold bolts to 12–15 ft. lbs. (16–21 Nm).

19. Reconnect the negative battery cable.

20. Refill the cooling system.

21. Start the engine and check for leaks.

Exhaust Manifold

REMOVAL AND INSTALLATION

1.6L (GA16DE) and 2.0L (SR20DE) Engine

1993–94 Sentra

NOTE: If any fuel system components must be removed, be sure to relieve the fuel system pressure first.

1. Disconnect the negative battery cable. Raise and support the vehicle safely.

2. Remove the undercover and dust covers, if equipped.

3. Remove the air cleaner or collector assembly, if necessary for access.

4. Remove the heat shield(s), if equipped.

5. Disconnect the exhaust pipe from the exhaust manifold.

6. Remove or disconnect the temperature sensors, oxygen sensors, air induction pipes.

7. Remove manifold support brackets.

8. Disconnect the EAI and EGR tubes from their fittings if so equipped.

9. Loosen and remove the exhaust manifold attaching nuts and remove the manifold(s) from the block. Discard the exhaust manifold gaskets and replace with new.

10. Clean the gasket surfaces and check the manifold for cracks and warpage.

To install:

11. Install the exhaust manifold with a new gasket. Torque the manifold fasteners from the center outward in several stages.

12. Connect the EAI and EGR tubes to the connections on the manifold as necessary.

13. The remaining components are installed in the reverse order from which they were removed.

14. Install the undercovers and dust covers.

15. Connect the negative battery cable.

1995–97 Sentra and 200SX

1. Disconnect the negative battery cable. Raise and support the vehicle safely.

2. Remove the engine undercovers.

3. Remove the air cleaner or collector assembly.

4. Remove the heat shields from the manifold and front exhaust pipe.

5. Disconnect the front exhaust pipe from the exhaust manifold.

6. Remove or disconnect the temperature sensors, oxygen sensors, air induction pipes from the manifold.

7. Remove manifold support brackets.

8. Loosen and remove the exhaust manifold attaching nuts and remove the manifold from the block. Discard the exhaust manifold gaskets.

9. Clean the gasket surfaces and check the manifold for cracks and warpage.

To install:

10. Install the exhaust manifold with a new gasket. Torque the manifold fasteners as follows from the center outward in several stages.

   a. GA16DE engines — Tighten mounting nuts with washers to 12–15 ft. lbs. (16–21 Nm).

   b. SR20DE engines — Tighten mounting nuts with washers to 27–35 ft. lbs. (37–48 Nm).

11. Install or connect the temperature sensors, oxygen sensors and air induction pipes.

12. Install the manifold support brackets.

13. Connect the exhaust pipe to the manifold using a new gasket. Tighten

the manifold nuts to 21–25 ft. lbs. (28–33 Nm) for GA16DE engine models or 32–37 ft. lbs. (43–50 Nm) for SR20DE engine models.

14. Install the heat shields.

15. Install the air cleaner or collector assembly.

16. Install the engine undercovers.

17. Connect the negative battery cable.

18. Start the engine and check for exhaust leaks.

2.4L (KA24DE) Engine

1. Disconnect the negative battery cable.

2. Raise and safely support the vehicle.

NOTE: Before loosening the exhaust pipe retaining nuts or bolts, soak the retaining nuts or bolts with penetrating oil.

3. Disconnect the exhaust pipe from the exhaust manifold.

NOTE: On California models equipped with A/T and on all M/T equipped models, disconnect the exhaust pipe at the exhaust manifold collector.

4. Disconnect the oxygen sensor electrical connector.

5. Remove the exhaust manifold cover.

6. Remove the EGR tube from the exhaust manifold.

7. Remove the exhaust manifold-to-engine bolts and nuts and discard the gaskets. Remove the retaining bolts and nuts in reverse of the tightening sequence.

8. Remove the exhaust manifold from the vehicle.

To install:

9. Clean all gasket mounting surfaces and install new gaskets.

10. Install the exhaust manifold to the engine and torque the new nuts to 27–35 ft. lbs. (37–48 Nm). Tighten the bolts and nuts evenly in sequence

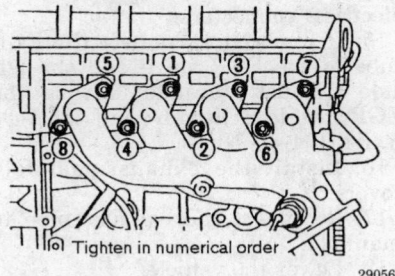

Tighten in numerical order

290569

Exhaust manifold bolt tightening sequence — 2.4L (KA24DE) engine (except California models)

until snug; now torque the bolts and nuts in sequence to specification.

11. Install the EGR tube to the exhaust manifold, tighten the EGR tube nuts to 29–36 ft. lbs. (39–49 Nm).

12. Install the exhaust manifold cover.

13. Torque the exhaust manifold cover bolts to 12–14 ft. lbs. (16–19 Nm) for 1993–94 vehicles. Torque the exhaust manifold cover bolts to 46–57 inch lbs. (5–7 Nm) for 1995–96 vehicles.

14. Connect the oxygen sensor electrical connector.

15. Using a new gasket, connect the exhaust pipe to the exhaust manifold and tighten the mounting nuts/bolts to 33–44 ft. lbs. (45–60 Nm).

16. Connect the negative battery cable.

17. Start the engine and check for exhaust leaks.

3.0L (VE30DE) Engine

1. Disconnect the negative battery cable.

2. Raise and safely support the vehicle.

NOTE: If necessary, soak the exhaust pipe retaining bolts with penetrating oil to loosen them.

3. Disconnect the exhaust manifolds from the exhaust pipes.

4. Remove the exhaust manifold-to-engine bolts, in the reverse sequence of tightening. Discard the gaskets.

To install:

5. Clean all gasket mounting surfaces. Install new gaskets.

6. Install the exhaust manifold to the engine and perform the following procedure:

   a. Torque the exhaust manifold-to-engine bolts, in sequence, to 13–16 ft. lbs. (16–22 Nm).

   b. Torque the exhaust manifold-to-engine bolts, in sequence, to 17–20 ft. lbs. (24–27 Nm).

7. Install the exhaust manifolds to the exhaust pipes.

8. Connect the negative battery cable. Start the engine and check for exhaust leaks.

3.0L (VG30E) Engine

1. Disconnect the negative battery cable.

2. Raise and safely support the vehicle.

NOTE: If necessary, soak the exhaust pipe retaining bolts with penetrating oil to loosen them.

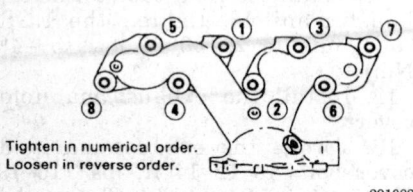

Tighten in numerical order.
Loosen in reverse order.

291082

Exhaust manifold bolt tightening sequence — 2.4L (KA24DE) engine (California models)

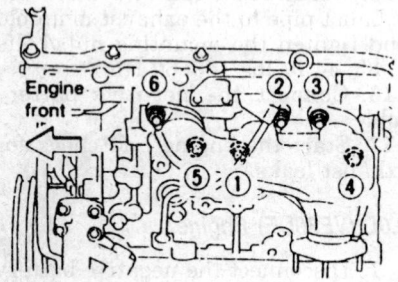

Engine front

Tighten in numerical order.

156966

Left exhaust manifold tightening sequence — 3.0L (VE30DE) engine

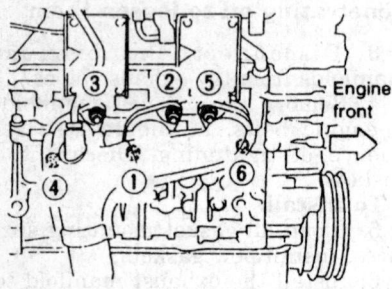

Engine front

Tighten in numerical order.

156967

Right exhaust manifold tightening sequence — 3.0L (VE30DE) engine

3. Disconnect the exhaust manifolds from the exhaust pipes.

4. Remove both exhaust manifolds-to-engine bolts, in sequence. Discard the gaskets.

To install:

5. Clean all gasket mounting surfaces. Install new gaskets.

6. Install the exhaust manifolds to the engine and torque the exhaust manifold-to-engine bolts, alternately in 2 stages, in the exact reverse order of removal to 13–16 ft. lbs. (16–22 Nm).

7. Install the exhaust manifolds to the exhaust pipes.

8. Connect the negative battery cable. Start the engine and check for exhaust leaks.

3.0L (VG30DE and VG30DETT) Engines

1. Disconnect the negative battery cable.

2. Raise and safely support the vehicle.

3. Disconnect the exhaust manifolds from the exhaust pipe.

4. Disconnect the oxygen sensor(s) electrical connections.

5. If equipped, remove the turbocharger assembly and three-way warm up catalyst from the exhaust manifold.

6. Remove the exhaust manifold cover(s).

7. If removing the left side exhaust manifold, remove the EGR tube.

NOTE: Before loosening the exhaust manifold retaining nuts, soak the retaining nuts with penetrating oil.

8. Remove the retaining nuts from the exhaust manifold. Loosen all the retaining nuts evenly

9. Remove the manifold from the vehicle.

To install:

10. Clean all gasket mounting surfaces and install new gaskets.

11. Install the exhaust manifold to the engine and torque the exhaust manifold-to-engine retaining nuts to 17–20 ft. lbs. (24–27 Nm) on non turbo engines, and torque retaining nuts to 20–23 ft. lbs. (27–31 Nm) on turbo engines.

12. If equipped, install the turbocharger and three-way catalyst assembly to the exhaust manifold.

13. Torque the turbocharger bolts to 32–40 ft. lbs. (43–54 Nm). Torque the three-way catalyst bolts and nuts to 18–22 ft. lbs. (25–29 Nm).

14. Reconnect the oxygen sensor(s) electrical connectors.

15. If removed, reconnect the EGR tube to the EGR valve and the left side exhaust manifold. Torque the EGR tube to EGR valve nut to 25–33 ft. lbs. (34–44 Nm).

16. Install the exhaust manifold covers.

17. Reconnect the exhaust pipe to manifold.

18. Lower the vehicle.

19. Connect the negative battery cable.

20. Start the engine and check for exhaust leaks.

3.0L (VQ30DE) Engine

1. Disconnect the negative battery cable.

2. Raise and safely support the vehicle.

NOTE: If necessary, soak the exhaust pipe retaining nuts with penetrating oil to loosen them.

3. Disconnect the exhaust manifolds from the exhaust pipes.

4. Remove the protective covers from the manifolds.

5. Remove the exhaust manifold-to-engine mounting nuts. Remove the manifolds from the engine an discard the gaskets.

To install:

6. Clean all gasket mounting surfaces. Install new gaskets.

7. Install the exhaust manifold to the engine and tighten the mounting nuts in two progressive steps to 22–24 ft. lbs. (30–32 Nm).

8. Install the protective shields and tighten the mounting bolts in two progressive steps to 46–57 inch lbs. (5–7 Nm).

9. Install the exhaust manifolds to the exhaust pipes and tighten the mounting nuts to 32–37 ft. lbs. (43–50 Nm).

10. Connect the negative battery cable, start the engine, and check for exhaust leaks.

Turbocharger

REMOVAL AND INSTALLATION

3.0L (VG30DETT) Engine

Right Side

1. Remove the right side of the cowl panel and remove the battery.

2. Remove the air inlet hoses and pipes from the turbocharger.

3. Disconnect the lower pipe from the turbocharger.

4. Remove the automatic speed control bracket with wiper motor and solenoid valves.

5. Disconnect the oxygen sensor harness connector.

6. Disconnect the water tubes and oil inlet from the turbocharger. Cap the tubes to prevent contaminants from entering the system

7. Remove the front exhaust tube and the three-way catalyst assembly.

8. Remove the oil pressure switch.

9. Remove the oil filter.

10. Remove the turbocharger oil return tube. Cap the tube to prevent contaminants from entering the oil.

11. Disconnect the oil hose from the oil filter bracket.

12. Remove the rod pin from the wastegate valve actuator.

13. Remove the oil filter bracket.

14. Unbend the locking plates for turbocharger fastening nuts.

15. Remove the turbocharger fastening nuts and remove the turbocharger from the vehicle.

To install:

16. Using new gaskets, install the turbocharger unit. Tighten the turbocharger attaching nuts to 20–23 ft. lbs. (27–31 Nm).

17. Bend the locking tabs around the nuts that secure the turbocharger.

18. Reconnect the oil filter bracket. Tighten the bracket attaching bolts to 12–15 ft. lbs. (16–21 Nm).

19. Install the rod pin of wastegate valve actuator.

20. Using new gaskets, reconnect the oil hose to the turbocharger.

21. Using new gaskets, reconnect the oil return and water return tubes. Tighten the tube attaching bolts and nuts to 12–15 ft. lbs. (16–21 Nm).

22. Install a new oil filter.

23. Install the oil pressure switch.

24. Using a new gasket, install the front exhaust tube and three-way catalyst assembly. Tighten the catalyst-to-turbocharger bolts to 18–22 ft. lbs. (25–29 Nm).

25. Using new gaskets, install the water and oil inlet tubes. Tighten the nuts to 11–14 ft. lbs. (15–20 Nm).

26. The balance of installation is the reverse of the removal procedure.

27. Check the oil and coolant levels; add oil or coolant if necessary.

28. Start the engine and check the turbocharger for proper operation.

Left Side

1. Remove the master cylinder and power brake booster.

2. Disconnect the oxygen sensor electrical connector.

3. Remove the air inlet hose and pipe.

4. Disconnect the lower pipe from the turbocharger.

5. Disconnect the water tubes and oil inlet tube from the turbocharger. Cap the tube to prevent contaminants from entering the system

6. Remove the front exhaust pipe and three way catalyst assembly.

7. Disconnect the lower steering joint.

8. Disconnect the turbocharger oil return tube. Cap the tube to prevent contaminants from entering the oil.

9. Disconnect the EGR tube and the turbocharger wastegate valve actuator bracket.

10. Remove the exhaust manifold heat shield.

11. Remove the left side exhaust manifold attaching nuts.

12. Remove the left side exhaust manifold and turbocharger as an assembly.

To install:

13. Using new gaskets, install the left side exhaust manifold and turbocharger assembly. Tighten the exhaust manifold bolts to 17–20 ft. lbs. (24–27 Nm).

14. Install the exhaust manifold heat shield.

15. Reconnect the EGR tube and the turbocharger wastegate valve actuator bracket.

16. Using new gaskets, reconnect the oil return tube. Tighten the tube attaching bolts to 12–15 ft. lbs. (16–21 Nm).

17. Reconnect the lower steering joint.

18. Using new gaskets, install the front exhaust pipe and three way catalyst assembly. Tighten the catalyst to turbocharger bolts to 18–22 ft. lbs. (25–29 Nm).

19. Using new gaskets, reconnect the water tubes and oil inlet tube to the turbocharger. Tighten the nuts to 11–14 ft. lbs. (15–20 Nm).

20. Install the remaining components in the reverse order of removal.

21. Start the engine and check the turbocharger for proper operation.

Front Cover Seal

REMOVAL AND INSTALLATION

1993–94 1.6L (GA16DE) Engine

1. Disconnect the negative battery cable.

2. Remove the drive belts.

3. Remove the crankshaft pulley.

4. Remove the timing chain cover.

5. Carefully pry the oil seal from the front cover with a small flat-bladed tool.

NOTE: When removing the oil seal, be careful not the gouge or scratch the seal bore or crankshaft surfaces.

6. Wipe the seal bore with a clean rag.

To install:

7. Lubricate the lip of the new seal with clean engine oil.

8. Install the seal into the front cover with a suitable seal installer.

9. Install the timing chain cover.

10. Install the crankshaft pulley.

11. Install the drive belts

12. Connect the negative battery cable.

13. Start the engine and check for oil leaks.

2.0L (SR20DE), 2.4L (KA24DE), 3.0L (VE30DE and VQ30DE) and 1995–97 1.6L (GA16DE) Engines

1. Disconnect the negative battery cable.

2. Raise and safely support the vehicle.

3. Remove the right front tire.

4. Remove the engine undercover and right side inner wheel cover.

5. Remove the engine drive belts.

6. For 1.6L (GA16DE) and 2.0L (SR20DE) engines, use a suitable puller to remove the crankshaft pulley.

7. For 2.4L (KA24DE) engines, remove the bolt that secures the crankshaft pulley and remove the pulley.

8. Using a suitable tool, pry the oil seal from the front cover.

NOTE: When removing the oil seal, be careful not the gouge or scratch the seal bore or crankshaft surfaces.

9. Wipe the seal bore with a clean rag.

To install:

10. Lubricate the lip of the new seal with clean engine oil.

11. Install the seal into the front cover with a suitable seal installer.

NOTE: Be sure to install the seal with the oil seal lip facing the engine block.

12. Install the crankshaft pulley.

   a. For the 2.0L (SR20DE), 2.4L (KA24DE) and 1995–97 1.6L (GA16DE) engines, tighten the mounting bolt to 105–112 ft. lbs. (142–152 Nm).

   b. For 3.0L (VE30DE and VQ30DE) engines, tighten the mounting bolt to 14–22 ft. lbs. (20–29 Nm). Turn the crankshaft bolt an additional 60–66 degrees clockwise.

13. Install and adjust the engine drive belts.

14. Install the inner wheel cover and engine undercover.

15. Install the right front wheel and lower the vehicle.

16. Connect the negative battery cable.

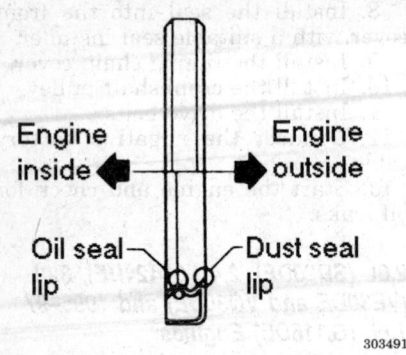

Correct seal position

Front Crankshaft Seal

REMOVAL AND INSTALLATION

3.0L (VG30E) Engine

NOTE: The front oil seal is a part of the oil pump body.

1. Disconnect the negative battery cable.
2. Remove the timing belt and the crankshaft sprocket.
3. Remove the oil pump assembly.
4. Remove the oil seal from the oil pump body using a pry tool. Be careful not to damage the oil pump body during seal removal.
To install:
5. Apply clean engine oil to the new oil seal. Install the seal using proper size driver.
6. Install the oil pump assembly to the engine.
7. Install the remaining components in reverse order of removal.
8. Connect the negative battery cable.

3.0L (VG30DE and VG30DETT) Engines

1. Disconnect the negative battery cable.
2. Position the engine to TDC of compression stroke.
3. Remove the engine undercover.
4. Remove all accessory drive belts and remove the alternator from the vehicle.
5. Remove the crankshaft damper-to-crankshaft bolt and remove the damper.
6. Remove the timing belt covers and remove the timing belt.
7. Using a proper prying tool, remove the seal from the front oil pump housing.

WARNING
Use care when prying the seal front the front cover so as not to damage the cover or crankshaft.

To install:
8. Using a proper seal driver, install the front crankshaft oil seal to the oil pump.
9. Install the timing belt and the timing belt covers.
10. Install the crankshaft damper and tighten the mounting bolt to 159–174 ft. lbs. (216–235 Nm).
11. Install the alternator assembly and install the engine drive belts.
12. Install the engine undercover.
13. Connect the negative battery cable and check all fluid levels.
14. Start engine, check ignition timing, and check for oil leaks.

Timing Chain, Sprockets and Front Cover

REMOVAL AND INSTALLATION

1.6L (GA16DE) Engine

1. Disconnect the negative battery cable. Relieve the fuel pressure.
2. Drain the coolant from the radiator and cylinder block.
3. Remove the upper radiator hose from the engine.
4. Remove the engine drive belts.
5. Remove the power steering pulley and remove the pump with bracket.
6. Remove the air duct from the intake manifold collector.
7. Disconnect the vacuum hoses, wiring, and harness connectors.
8. Remove the right front wheel and remove the inner wheel covers.
9. Remove the engine undercovers.
10. Remove the front exhaust pipe.
11. Remove the cylinder head front mounting bracket.

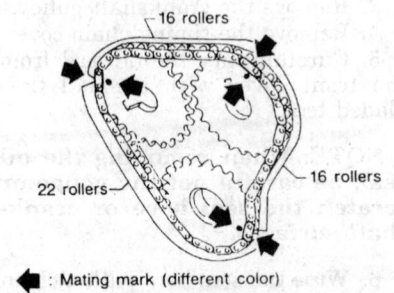

: Mating mark (different color)

299666

Alignment of the camshaft sprockets with the timing chain — 1.6L (GA16DE) engine

12. Remove the cylinder head cover from the engine.
13. Remove the distributor.
14. Remove the water pump pulley from water pump.
15. Remove the complete thermostat housing assembly from the engine.
16. Remove the lower timing chain tensioner.
17. Remove the upper timing chain tensioner and slack side timing chain guide.
18. Loosen the idler sprocket bolt.
19. Remove the camshaft sprocket bolts and remove the sprockets from the camshafts. Be sure to mark the sprockets for proper reinstallation.
20. Remove the camshaft mounting caps by loosening the bolts in two or three steps. Remove the camshafts from the engine.
21. Remove the idler sprocket bolt.
22. Remove the cylinder head with the manifolds.
23. Remove the idler sprocket shaft from the rear side.
24. Remove the upper timing chain.
25. Support the engine assembly and remove the center crossmember.
26. Remove the oil pan and strainer assembly.
27. Remove the crankshaft pulley.
28. Remove the engine front mount.
29. Remove the engine front mount bracket.
30. Remove the bolts that secure the front timing cover and remove the cover from the engine.
31. Remove the idler sprocket and remove the lower timing chain.
32. Remove the oil pump drive spacer and remove the crankshaft sprocket.
33. Remove the timing chain guide.
To install:
34. Confirm that No. 1 piston is set at TDC on compression stroke.
35. Install the crankshaft sprocket with the marks of the sprocket facing the front of the engine.
36. Install the oil pump drive spacer and install the chain guide.
37. Install the lower timing chain. Set the chain by aligning its mating mark with the one on the crankshaft sprocket. Make sure the sprocket's mating mark faces the front of the engine.

NOTE: The number of links between the alignment marks are the same for the left and the right side.

38. Install the crankshaft sprocket and the lower timing chain. Set the timing chain by aligning its mating mark with the one on the crankshaft sprocket. Make sure sprocket's mat-

ing mark faces engine front. The number of links between the alignment marks are the same for the left and right side.

39. Using liquid gasket, install the front cover assembly.

40. Install engine front mounting bracket and install the engine mount.

41. Install the oil strainer, oil pan assembly, and the crankshaft pulley.

42. Install the center crossmember.

43. Set the idler sprocket by aligning the mating mark on the larger sprocket with the silver mating mark on the lower timing chain.

44. Install the upper timing chain and set it by aligning the mating mark on the smaller sprocket with the silver mating marks on the upper timing chain. Make sure sprocket marks face engine front.

45. Install the idler sprocket shaft from the rear side.

46. Install the cylinder head assembly.

47. Install the idler sprocket bolt. Be sure to lubricate the bolt with clean engine oil.

48. Install the exhaust and intake camshafts. The camshafts and marked **I** for intake and **E** for exhaust.

49. Position the intake camshaft knock pin at the 9 o'clock position and the exhaust camshaft knock pin at the 12 o'clock position.

50. Install the camshaft bearing caps and distributor bracket. Apply liquid sealant to the distributor bracket.

51. Tighten the mounting bolts in sequence as follows:

   a. Tighten bolts 11–15, then bolts 1–10 to 18 inch lbs. (2.0 Nm)

   b. Tighten bolts 1–15 to 52 inch lbs. (5.9 Nm)

   c. Tighten bolts 1–14 to 81–104 inch lbs. (11 Nm)

   d. Tighten bolt 15 to 55–73 inch lbs. (11 Nm)

52. Install the camshaft sprockets with timing chain. Set the camshaft sprockets by aligning the mating marks of the timing chain with the marks on the camshaft sprockets.

53. Install the camshaft sprocket bolts and torque to 86 ft. lbs. (117 Nm). Be sure to lubricate the bolts with clean engine oil.

54. Install the upper timing chain tensioner. Before installation of the tensioner, install a suitable pin to hold the tensioner in the relaxed position. After installing the chain tensioner, remove the pin.

55. Install the lower timing chain tensioner. Be sure the notch of the gasket is positioned down.

56. Install the remaining components in the opposite order from which they were removed.

57. Connect the negative battery cable. Refill all fluid levels. Road test the vehicle for proper operation.

2.0L (SR20DE) Engine

> **CAUTION**
> *Fuel injection systems remain under pressure after the engine has been turned OFF. Properly relieve the fuel pressure before disconnecting any fuel lines. Failure to do so may result in fire or personal injury.*

1. Relieve the fuel system pressure and remove the negative battery cable.

2. Drain the coolant from the radiator and engine block. Remove the radiator.

3. Remove the right front wheel and engine side cover.

4. Remove the spark plugs.

5. Rotate the engine and position the No. 1 cylinder to TDC.

6. Remove the air duct to the intake manifold.

7. Remove the drive belts and the water pump pulley.

8. Remove the alternator and the power steering pump from the engine.

9. Label and remove the vacuum hoses, fuel hoses, and the wire harness connectors.

10. Remove the cylinder head cover.

11. Remove the intake manifold supports.

12. Unbolt and remove the oil filter bracket and the power steering pump bracket.

13. Remove the timing chain tensioner.

14. Remove the distributor.

15. Remove the timing chain guide.

16. Holding the flats of the camshaft sprockets, remove the bolts that secure the sprockets.

17. Remove the timing chain sprockets from the camshafts.

18. Remove the oil tubes, baffle plate, camshaft brackets and remove the camshafts from the cylinder head.

19. Remove the starter motor.

20. Remove the coolant hoses from the engine block.

21. Remove the knock sensor harness connector.

22. Remove the EGR tube.

23. Remove the cylinder head assembly.

24. Raise and support the vehicle safely.

25. Remove the oil pan, oil strainer, and the baffle plate.

26. Remove the crankshaft pulley using a suitable puller.

27. Remove the engine front mount.

28. Remove the front cover and oil pump drive spacer.

29. Remove the timing chain guides and timing chain. Check the timing chain for excessive wear at the roller links. Replace the chain if necessary.

To install:

30. Clean all gasket mating surfaces.

31. Install the crankshaft sprocket. Position the crankshaft so that No. 1 piston is set at TDC (keyway at 12 o'clock, mating mark at 4 o'clock) fit timing chain to crankshaft sprocket so the mating mark is in line with mating mark on crankshaft sprocket. The mating marks on timing chain for the camshaft sprockets should be silver. The mating mark on the timing chain for the crankshaft sprocket should be gold.

32. Install the timing chain to the crankshaft sprocket and install the timing chain guides. Tighten the timing chain guides to 9–14 ft. lbs. (13–19 Nm). Drape the timing chain over the left chain guide.

33. Install the oil pump drive spacer to the crankshaft.

34. Apply a continuous bead of liquid sealant to the front timing cover

---

Intake   Exhaust

10° – 11°   Knock pin

Knock pin

302335

Positioning of camshaft knock pins — 1.6L (GA16DE) engine

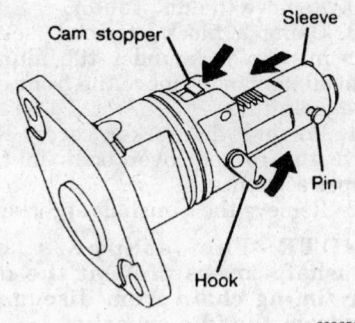

Cam stopper   Sleeve

Pin

Hook

299650

Timing chain tensioner — 2.0L (SR20DE) engines

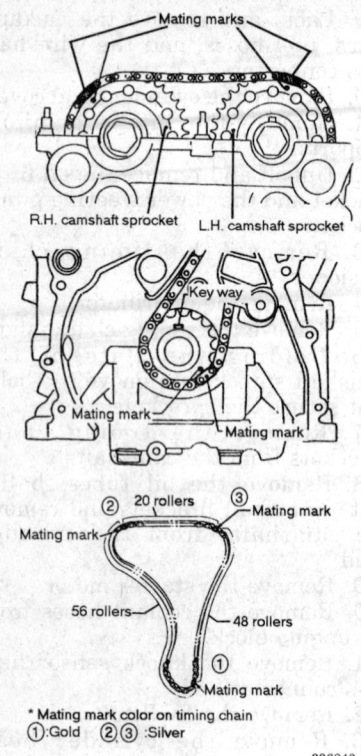

Timing chain sprocket alignment marks — 2.0L (SR20DE) engines

and install the cover. Torque the front cover mounting bolts to 57–66 inch lbs. (6.4–7.5 Nm).

35. Install right front engine mount.

36. Install the crankshaft pulley and tighten the mounting bolt to 105–112 ft. lbs. (142–152 Nm). Be sure the No. 1 piston is at TDC.

37. Install the oil strainer, baffle plate, and the oil pan assembly.

38. Install the cylinder head assembly. Be sure to apply a bead of sealant to the joint of the block and front timing cover.

39. Install the EGR tube.

40. Install the knock sensor harness connector.

41. Using new hose clamps, install the coolant hoses to the engine block.

42. Install the starter motor.

43. Install the camshafts, camshaft bearing caps, oil tubes, and the baffle plate.

NOTE: When installing the camshafts, be sure to position the LH and RH camshaft keys at 12 o'clock. Also make sure the camshaft brackets are facing in the correct direction.

44. Install the camshaft sprockets by lining up the mating marks on the timing chain with the mating marks

on the camshaft sprockets. Tighten the camshaft sprocket bolts to 101–116 ft. lbs. (137–157 Nm).

45. Install the timing chain guide and distributor.

46. Install the chain tensioner. Press the cam stopper down and the press-in sleeve until the hook can be engaged on the pin. When tensioner is bolted in position the hook will release automatically.

NOTE: Ensure the arrow on the outside of the tensioner faces the front of the engine.

47. Install the remaining components in the reverse order of removal.

48. Connect the negative battery cable, Refill fluid levels, start the engine, and bleed the cooling system. Check for leaks and road test the vehicle for proper operation.

2.4L (KA24DE) Engine

Altima

1. Disconnect the negative battery cable.

2. Drain the coolant from the engine and radiator.

3. Drain the engine oil.

4. Remove the engine undercover.

---------------- CAUTION ----------------

Fuel injection systems remain under pressure after the engine has been turned OFF. Properly relieve fuel pressure before disconnecting any fuel lines. Failure to do so may result in fire or personal injury.

----

5. Remove the vacuum hoses, fuel hoses, wires, harness, and connectors.

6. Remove the alternator and bracket, the upper radiator hose, the air duct, and the front exhaust tube.

7. Remove the intake manifold collector supports, intake manifold collector, and the exhaust manifold.

8. Set the No. 1 piston at TDC on its compression stroke.

9. Remove the distributor.

10. Using a block of wood, set a transmission jack under the aluminum oil pan and remove the front engine mounting.

11. Remove the rocker cover. Remove the rocker cover bolts in the proper sequence.

12. Remove the camshaft sprockets.

NOTE: The stoppers on camshaft covers prevent the upper timing chain from disengaging from the idle sprocket.

13. Remove the cam bearing caps in sequence and remove the camshafts.

The camshaft brackets must be loosened in reverse order of tightening to prevent damage to the camshaft.

NOTE: These parts must be reassembled in their original positions.

14. Loosen the cylinder head bolts in the reverse order of installation.

NOTE: A warped or cracked cylinder head could result from loosening in incorrect order. The cylinder head bolts should be loosened in two or three steps.

15. Remove the cam sprocket cover.

16. Remove the upper chain tensioner and upper chain guides.

17. Remove the upper timing chain.

18. Remove the idler sprocket bolt.

19. Remove the cylinder head and the intake manifold. Remove the cylinder head gasket. The lower timing chain will not be disengaged from crankshaft sprocket.

NOTE: The cast portion of the front cover is located on the lower side of the crankshaft sprocket, so the lower timing chain need not be disengaged from idler sprocket.

20. Remove the steel oil pan bolts in the reverse sequence of the tightening procedure.

21. Install a seal cutter between the steel oil pan and the aluminum oil pan.

22. Tapping the cutter with a hammer, slide it around the entire edge of the oil pan. Take care not to damage the aluminum oil pan.

23. Remove the steel oil pan.

24. Remove the baffle plate, oil strainer, and the front plate.

25. Support the transaxle with a jack and the engine with a engine hoist. Remove the front suspension member.

26. Remove the A/C compressor gussets.

27. Remove the rear cover plate.

28. Remove the aluminum oil pan retaining bolts in sequence.

29. Insert a seal cutter between the oil pan and the cylinder block.

30. Tapping the cutter with a hammer, slide it around the entire edge of the oil pan. Take care not to damage the aluminum oil pan.

31. Lower the oil pan from the cylinder block and remove it from the engine.

32. Remove the crankshaft pulley.

33. Remove the front timing chain cover.

34. Remove the oil pump drive spacer.

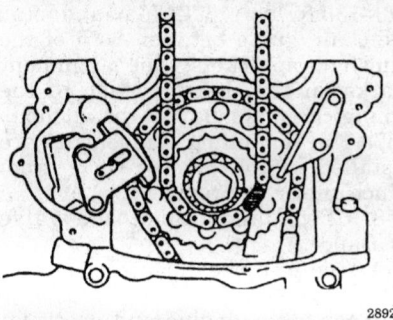

289270

Lower portion of the upper timing chain with mating marks — Altima with 2.4L (KA24DE) engines

35. Remove the lower timing chain tensioner, tensioner arm, and lower timing chain guide.

36. Remove the lower timing chain and idler sprocket.

To install:

37. Install the crankshaft sprocket and oil pump drive spacer.

38. Install the idler sprocket and lower timing chain.

39. Set the lower timing chain on the sprockets, aligning the mating marks. The mating marks on the timing chain assembly will be silver.

40. Install the chain tension arm and chain guide.

41. Install the lower timing chain tensioner.

42. Apply a continuous bead of liquid gasket to the front cover and install the front cover. Install a new oil seal.

43. Install the crankshaft pulley and torque bolt to 105–112 ft. lbs. (142–152 Nm).

44. Carefully scrape the old gasket material away from the pan and cylinder block mounting surfaces and then apply a continuous bead (3.5–4.5 mm) of liquid gasket around the oil pan and the cylinder block.

45. Install the aluminum oil pan and tighten the mounting bolts in sequence, to 13 ft. lbs. (17.5 Nm).

46. Install the baffle plate, oil strainer, and the front tube.

47. Install the steel oil pan and torque in sequence to 61 inch lbs. (7 Nm).

48. Reinstall the rear cover plate.

49. Install the A/C compressor gussets.

50. Reinstall the front suspension member.

51. Install the front engine mounting.

52. Remove the engine hoist and transaxle support.

53. Install new cylinder head gasket.

54. Reinstall the cylinder head and temporarily tighten the cylinder head

bolts when installing the front cover. This is necessary to avoid damaging the cylinder head gasket. Be sure to install washers between the bolts and cylinder head.

55. Install the upper timing chain, chain tensioner and chain guide.

56. Set the upper timing chain on idler sprockets, aligning the mating marks.

57. Install cam sprocket cover. Apply a continuous bead of liquid gasket to front cover. Be careful not to damage the cylinder head gasket. Be careful that the upper timing chain does not slip or jump when installing cam sprocket cover.

58. Tighten cylinder head bolts.

59. Install the camshafts and camshaft bearing caps.

60. Install the camshaft sprockets, then tighten the sprocket bolts to 123–130 ft. lbs. (167–176 Nm). Install the chain guide between both camshaft sprockets. The alignment marks on the upper portion of the timing chain should now be aligned.

61. The balance of installation is the reverse of the removal procedure.

62. Connect the negative battery cable.

63. Start the engine and make the necessary adjustments. Check for proper operation and leaks.

240SX

1. Disconnect the negative battery cable.

2. Relieve the fuel system pressure.

3. Drain the coolant from engine and radiator.

4. Drain the engine oil.

5. Remove the engine undercover.

6. Remove the air duct assembly.

7. Remove the fan shroud and the cooling fan.

8. Remove the exhaust manifold cover.

9. Remove the front exhaust tube and if equipped, the A.I.V. pipe.

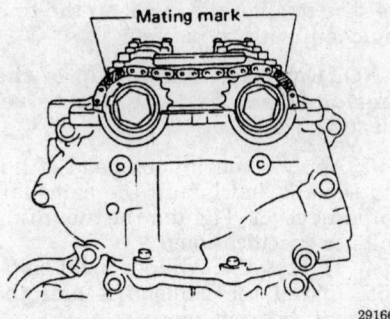

291663

Upper portion of the upper timing chain and mating marks — 240SX and Altima

- CAUTION -

Fuel injection systems remain under pressure after the engine has been turned OFF. Properly relieve the fuel pressure before disconnecting any fuel lines. Failure to do so may result in fire or personal injury.

10. Disconnect and tag all of the vacuum hoses, fuel lines, and electrical connections.

11. Remove all of the spark plugs.

12. Remove the distributor cap with the spark plug wires attached.

13. Remove the injector tube assembly with the injectors.

14. Remove the rocker cover bolts, in reverse order of installation.

15. Set the No. 1 piston at TDC (top dead center) on its compression stroke.

16. Remove the distributor. Be sure to note the positioning of the distributor rotor before removing the distributor.

17. Remove the camshaft sprockets. Be sure to hold the flats of the camshafts when removing the sprocket bolts.

NOTE: The stoppers on the inside of the camshaft covers prevent the upper timing chain from disengaging from the idle sprocket.

18. Remove the camshaft bearing caps, in reverse order of installation, then remove both of the camshafts.

NOTE: The camshaft bearing caps and camshafts should be kept in their original position for reassembly.

19. Loosen the cylinder head bolts in the reverse order of installation.

NOTE: Head warpage or cracking could result from removing the head bolts in the incorrect order. The cylinder head bolts should be loosened in two or three steps.

20. Remove the camshaft sprocket cover.

21. Remove the upper chain tensioner and upper chain guides.

NOTE: Compress the piston of the tensioner and insert a suitable pin into the pin hole.

22. Remove the upper timing chain.

23. Remove the idler sprocket bolt.

24. Remove the cylinder head with the intake manifold and exhaust manifold assembly.

25. Remove the cylinder head gasket.

26. Remove the oil pan bolts.

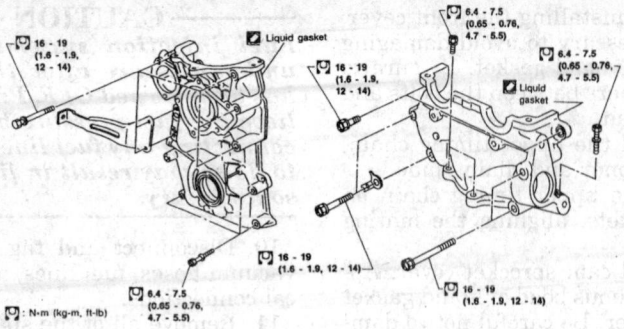

Timing chain front cover — 240SX with 2.4L (KA24DE) engine

27. Install a seal cutter between the oil pan and the engine block.
28. Tapping the cutter with a hammer, slide it around the entire edge of the oil pan. Take care not to damage the oil pan.
29. Remove the oil pan assembly.
30. Remove the baffle plate, oil strainer, and the front tube.
31. Remove the engine drive belts.
32. Remove the A/C compressor idler pulley.
33. Remove the crankshaft pulley.
34. Remove the front timing chain cover.
35. Remove the oil pump drive spacer.
36. Remove the lower timing chain tensioner, tensioner arm, and the lower timing chain guide.

NOTE: Compress the piston of the tensioner and insert a suitable pin into the pin hole.

37. Remove the lower timing chain and crankshaft sprocket.
   To install:
38. Check all components for wear. Replace as necessary. Clean all mating surfaces and replace the cylinder head gasket.
39. Install the crankshaft sprocket and the oil pump drive spacer.
40. Install the idler sprocket and the lower timing chain.

NOTE: Make sure that the mating marks of the crankshaft sprocket are facing front of engine.

41. Set the lower timing chain on the sprockets, aligning the mating marks. The mating marks on the timing chain assembly will be silver.
42. Install the chain tension arm and the chain guide.
43. Install the lower timing chain tensioner.

NOTE: After installation of the tensioner, remove the pin to release the piston.

44. Apply a continuous bead of liquid gasket to the front cover and install the front cover. Install a new oil seal.
45. Install the crankshaft pulley and torque the bolt to 105–112 ft. lbs. (142–152 Nm).
46. Carefully scrape the old gasket material away from the pan and cylinder block mounting surfaces and then apply a continuous bead (3.5–4.5 mm) of liquid gasket around the oil pan and the cylinder block.
47. Install the baffle plate, oil strainer, and the front tube.
48. Install the oil pan and tighten the mounting bolts, in sequence, to 57–66 inch lbs. (6.4–7.5 Nm).
49. Install the A/C compressor idler pulley.
50. Install a new cylinder head gasket and install the idler shaft.
51. Install the cylinder head and temporarily tighten the cylinder head bolts when installing the front cover. This is necessary to avoid damaging the cylinder head gasket. Be sure to install washers between the bolts and cylinder head.

—— CAUTION ——
Do not fully tighten any of the cylinder head bolts at this time. tighten the bolts finger-tight only.

52. Install the upper timing chain, chain tensioner, and the chain guide. Be sure to align the mark on the timing chain with the idler.

NOTE: After installation of the tensioner, remove the pin to release the piston.

53. Apply a continuous bead of liquid sealant and install the camshaft sprocket cover. Tighten the mounting bolts to specifications.
54. Tighten the cylinder head bolts.
55. Install the camshafts and the camshaft bearing caps.
56. Install the camshaft sprockets and torque the sprocket bolts to

123–130 ft. lbs. (167–176 Nm). Install the chain guide between both of the camshaft sprockets. The alignment marks on the upper portion of the timing chain should now be aligned.
57. The remaining components are installed in the reverse order from which they were removed.
58. Check and adjust the valve clearance.
59. Connect the negative battery cable.
60. Start the engine and check for leaks. Bleed the cooling system.
61. Install the engine undercover.
62. Check the ignition timing; adjust the timing as necessary.

3.0L (VE30DE) Engine

1. Release the fuel system pressure. Disconnect the negative battery cable.
2. Rotate the crankshaft to position the No. 1 cylinder on TDC of its compression stroke.
3. Raise and safely support the vehicle. Remove the undercovers from the vehicle.
4. Remove the front right wheel and engine side cover.
5. Drain the cooling system by removing the engine drain plugs and opening the radiator drain cock.
6. Remove the radiator.
7. Remove the air duct from the intake manifold.
8. Remove the blow-by pipe.
9. Remove the vacuum hoses, the fuel lines, electrical connectors.
10. From the right rear of the engine, remove the EGR valve-to-exhaust manifold tube.
11. Remove the intake manifold collector supports.
12. Remove the fuel pressure regulator vacuum hose.
13. Remove the throttle chamber water hoses.
14. Remove the canister purge hoses.
15. Remove the blow-by hoses.
16. Remove the intake manifold collector.
17. Remove the ignition coils and the spark plugs.
18. Remove the IAC unit and heater pipe. Remove the fuel injector rail assembly.
19. From the front of the intake manifold, remove the right VTC solenoid valve.
20. Remove the intake manifold.
21. Remove the drive belts, air conditioning compressor and alternator.
22. Remove the air conditioning compressor bracket and the alternator bracket.

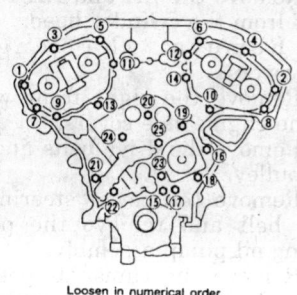

Loosen in numerical order.

295972

Rear timing chain case loosening sequence — 3.0L (VE30DE and VQ30DE) engines

23. Remove the idler pulley bracket and the dipstick tube.

24. Remove the left exhaust manifold. Remove the power steering pump and its bracket.

25. Remove the right exhaust manifold. Remove the rocker arm covers.

26. Working between the intake and exhaust camshafts, of each cylinder head, remove the upper timing chain guides.

27. Rotate the crankshaft to position the No. 1 cylinder on TDC of it's compression stroke.

28. Remove the crank angle sensor. Remove the timing chain tensioners from each upper timing chain.

29. Using a backup wrench to secure the intake camshaft, remove the intake camshaft sprocket-to-camshaft bolts, the VTC assemblies and the intake camshaft sprockets.

30. Remove the both cylinder heads. Remove the water pipe.

31. Remove the water pump pulley and the water pump.

32. Remove the oil pan, the crankshaft pulley, the oil strainer and the oil filter bracket.

33. Remove the front cover, the alternator adjusting bar and the upper timing chains.

34. Remove the lower timing chain guides, the idler sprockets and the lower timing chain.

35. Inspect the timing chains for cracks and/or excessive wear at the roller links; if necessary, replace the timing chain(s).

To install:

36. Install the crankshaft sprocket on the crankshaft.

37. Make sure the No. 1 cylinder is at the TDC of it's compression stroke.

38. Install the right idler sprocket and timing chain guides.

39. Position the lower timing chain on the right idler sprocket by aligning the mating mark on the right idler sprocket with the silver mating mark on the lower timing chain.

40. Install the left idler sprocket with the lower timing chain; align the mating marks on the lower timing chain with the mating marks on the left idler sprocket and crankshaft sprocket. Install the chain guide and the chain tensioner.

41. Position the upper timing chains on the idler sprockets by aligning the mating marks on the idler sprockets with the gold mating marks on the upper timing chains.

42. Install the oil pump drive spacer, front cover and the alternator adjusting bar.

43. Install the oil filter bracket, the oil strainer, the crankshaft pulley and the oil pan.

44. Make sure the No. 1 cylinder is on the TDC of its compression stroke.

45. Install the water pump.

46. Install the water pump pulley. Install the water pipe and tighten bolts to 12–15 ft. lbs. (16–21 Nm).

47. Rotate the crankshaft counterclockwise, until the No. 1 piston is set at approximately 120° before TDC on the compression stroke, to prevent interference of the valves and pistons.

48. Install the cylinder heads.

49. Install the camshafts, the camshaft brackets and the rocker arms. Position the right side exhaust camshaft at about 10 o'clock position and the left side exhaust camshaft at about 12 o'clock position.

50. Install the right side VTC assembly and the right camshaft sprocket. Align the mating marks on the right upper timing chain with the mating marks on the right VTC assembly and the right camshaft sprocket.

NOTE: There are 2 types of chain tensioners; be careful not to install the left chain tensioner onto the right cylinder head or right tensioner in left head.

51. Install the right timing chain tensioner. Before installing the chain tensioner, press-in the sleeve until the hook can be engaged on the pin; make sure the hook used to retain the chain tensioner is released for proper timing chain tension.

——— **WARNING** ———

The timing chain tensioner hooks must be released after installation for proper timing chain tension and operation.

52. Rotate the crankshaft clockwise to set the No. 1 piston at the TDC of its compression stroke.

53. Install the left side VTC assembly and the left camshaft sprocket. Align the mating marks on the left upper timing chain with the mating marks on the left VTC assembly and the left camshaft sprocket.

54. Install the left timing chain tensioner. Before installing the chain tensioner, press-in the sleeve until the hook can be engaged on the pin; make sure the hook used to retain the chain tensioner is released for proper timing chain tension.

——— **WARNING** ———

The timing chain tensioner hooks must be released after installation for proper timing chain tension and operation.

55. Install the remaining components in the opposite order from which they were removed.

56. Install the upper front covers. Install the upper chain guides on both cylinder heads. Install the rocker arm covers.

57. Install the engine undercovers and connect the negative battery cable.

58. Refill the cooling system and engine oil to proper specified levels.

59. Run engine and recheck fluid levels.

3.0L (VQ30DE) Engine

1. Disconnect the negative battery cable.

2. Drain the engine oil and the cooling system. Be sure to drain the engine block and the radiator.

3. Relieve the fuel system pressure.

4. Remove the left side rocker cover ornament.

NOTE: Before disconnecting any hoses or connectors, note the locations for reassembly.

5. Remove the air duct to intake manifold hose, collector hose, blow-by hose, and vacuum hoses.

——— **CAUTION** ———

Fuel injection systems remain under pressure after the engine has been turned OFF. Properly relieve fuel pressure before disconnecting any fuel lines. Failure to do so may result in fire or personal injury.

6. Remove the fuel hoses and disconnect the harness connections.

7. Disconnect the canister purge hoses.

8. Remove the water hoses from the cylinder head and intake manifold.

9. Disconnect and remove all six ignition coils from the spark plugs.

10. Remove the spark plugs.

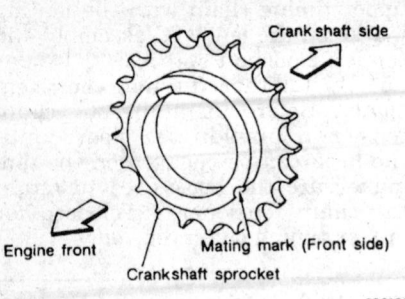

Crankshaft sprocket with mating marks — 3.0L (VQ30DE) engine

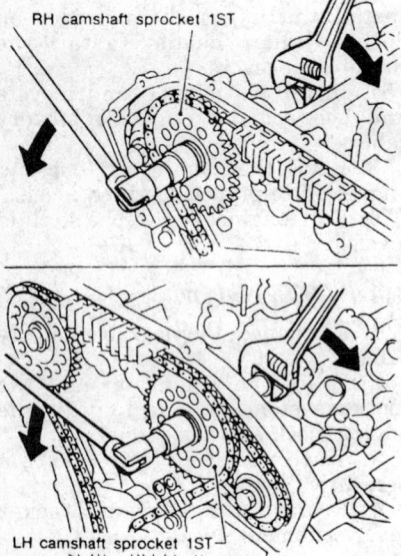

Removing the intake camshaft sprocket bolts — 3.0L (VQ30DE) engine

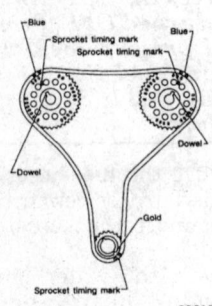

Lower timing chain alignment marks — 3.0L (VQ30DE) engine

11. Remove the bolts that secure the EGR tube and remove the tube.

12. Remove the intake manifold collector supports and remove the collector.

13. Remove the bolts that secure the fuel tube and remove the fuel tube from the vehicle.

14. Remove the bolts that secure the intake manifold to the engine block and remove the manifold. Loosen the bolts in the reverse sequence of the tightening procedure.

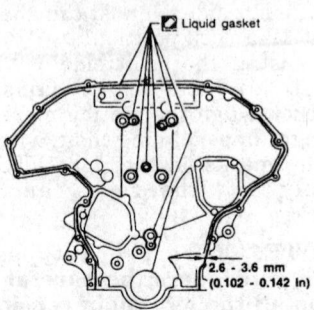

Rear timing chain case torque sequence — 3.0L (VE30DE and VQ30DE) engines

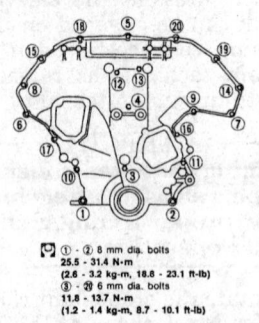

Application of liquid gasket to the front timing case — 3.0L (VE30DE and VQ30DE) engines

Front timing case bolt torque sequence — 3.0L (VE30DE and VQ30DE) engines

15. Remove the LH and RH rocker covers from the cylinder head.

16. Remove the engine undercovers.

17. Remove the right front wheel and the engine side covers.

18. Remove the drive belts and the idler pulley.

19. Remove the power steering oil pump belt and remove the power steering oil pump assembly.

20. Remove the camshaft position sensor (PHASE) and crankshaft position sensors (REF)/(POS).

21. Set the No. 1 piston to TDC of compression stroke by rotating the crankshaft.

22. Remove the ring gear cover access plate.

23. Loosen the crankshaft pulley bolt while securing the ring gear so the crankshaft cannot rotate.

NOTE: Use care not to damage the ring gear teeth.

24. Using a suitable puller, remove the crankshaft pulley.

25. Remove the A/C compressor and bracket.

26. Remove the front exhaust pipe and its support.

27. Hang the engine at the right and left side engine slingers with a suitable hoist.

28. Support the transaxle with jack.

29. Remove the right side engine mounting, mounting bracket, and nuts.

30. Remove the center crossmember assembly.

31. Remove the steel (lower) oil pan bolts in the reverse sequence of the torque sequence.

32. Insert a seal cutter between the steel and aluminum oil pan.

33. Tapping the cutter with a hammer, slide it around the entire edge of the oil pan. Be careful not to damage the aluminum mating surface of the upper oil pan.

34. Remove the steel oil pan and the oil strainer.

35. Remove the aluminum (upper) oil pan bolts in the reverse sequence of the torque sequence.

36. Remove the transaxle bolts that secure the oil pan.

37. Insert a seal cutter between the aluminum oil pan and the engine block.

38. Tapping the cutter with a hammer, slide it around the entire edge of the oil pan. Be careful not to damage the mating surfaces of the oil pan or engine block.

39. Remove the oil pan from the vehicle.

40. Remove the water pump cover and remove the bolts that secure the front timing chain case.

41. Using the seal cutter, remove the timing chain case cover.

42. Remove the internal timing chain guide and the upper chain guide.

43. Remove the timing chain tensioner and slack side chain guide.

44. Remove the left and right intake camshaft sprockets first. Be sure to hold the flats of the camshafts while removing the sprocket bolts.

45. Remove the lower timing chain assembly. Be sure to note the aligning marks of the chain before removal.

46. Insert a suitable stopper pin for the left and right camshaft tensioners.

47. Remove the left and right exhaust camshaft sprocket bolts. Be sure to hold the flats of the camshafts while removing the sprocket bolts.

48. Remove the upper timing chain assembly. Be sure to note the aligning marks of the chain before removal.

49. Remove the lower timing chain guide.

50. Remove the crankshaft sprocket.

51. Remove all traces of liquid gasket from the front timing chain case and from the water pump.

52. Inspect the timing chain for excessive wear or damage and replace as necessary.

To install:

53. Install the crankshaft sprocket with the mating mark facing out.

54. Position the crankshaft to TDC of compression stroke and align the dowels of the camshaft sprockets to the 12 o'clock position in respect to the cylinder head.

55. Install the lower timing chain guide. The front mark on the guide should face upwards.

56. On a work bench, align the marks on the intake and exhaust camshaft sprockets with the marks of the chain.

57. Put the exhaust camshaft sprockets onto the dowel pin and tighten the mounting bolts to 88–95 ft. lbs. (119–128 Nm). Be sure to secure the camshafts while tightening the bolts.

58. Install the timing chains and sprockets to the intake camshafts. Be sure to align the timing chain and sprocket mating marks.

59. Remove the left and right camshaft tensioner stopper pins.

60. Align the mating mark on the crankshaft with the matchmark (gold link) on the lower timing chain.

61. Attach the lower timing chain to the water pump sprocket.

62. Working counterclockwise, install the lower timing chain camshaft sprockets. Be sure to align the sprocket marks with the blue links of the timing chain during installation.

63. Install the intake sprocket bolts and tighten to 88–95 ft. lbs. (119–128 Nm). Be sure to secure the camshafts while tightening the bolts.

64. Install the internal timing chain guide, upper timing chain guide, lower timing chain tensioner and slack side timing chain guide.

65. Tighten the tensioner mounting bolt to 75–96 inch lbs. (8.4–10.8 Nm) and tighten the guide bolts to 108–168 inch lbs. (13–19 Nm).

66. Apply a 0.102–0.142 inch (2.6–3.6mm) continuous bead of liquid gasket to all necessary areas as shown on the front timing cover.

67. Install the timing cover evenly and gently. Be sure to align the dowel pin holes.

68. Tighten the mounting bolts in sequence as follows:
  a. Bolts No. 1 and 2 — Tighten to 19–23 ft. lbs. (26–31 Nm)
  b. Bolts No. 3–20 — Tighten to 105–121 inch lbs. (11.8–13.7 Nm)

NOTE: Leave the bolts unattended for 30 minutes or more after tightening. This will allow the liquid gasket to cure sufficiently.

69. Apply a 0.091–0.130 inch (2.3–3.3mm) continuous bead of liquid gasket to the water pump cover and install the cover. Tighten the bolts to 84–108 inch lbs. (10–13 Nm).

70. The balance of installation is the reverse of the removal procedure.

71. Connect the negative battery cable.

72. Start the engine and run at 3000 RPM under no load to purge the air from the high pressure chamber. The engine may produce a rattling noise. This indicates that air still remains in the chamber and is not a matter of concern.

Timing Belt, Sprockets, Tensioner and Front Cover

ADJUSTMENT

3.0L (VG30DE and VG30DETT) Engines

If Auto-Tensioner Was Removed

1. Prepare the automatic tensioner for installation:
  a. Remove the bolt holding the tensioner position.
  b. Use a vise to adjust the gap between the tensioner arm and pusher body to 0.160 in. (4mm).
  c. Install the bolt again to hold the arm in position. Do not try to use the bolt to adjust the gap or the threads will be damaged.

2. Install the automatic tensioner, push it towards the belt to just take up the slack, then tighten bolts finger tight.

3. Before adjusting timing belt tension, the slack must be properly distributed:
  a. Turn the crankshaft 10 degrees clockwise and tighten the tensioner bolts and nut to 12–15 ft. lbs. (16–21 Nm). Do not push the automatic tensioner hard or the belt will be adjusted too tight.
  b. Turn the crankshaft 120 degrees counterclockwise.
  c. Loosen the tensioner bolts and nut ½ turn and move the tensioner body away from the timing belt as far as it will move.
  d. Turn the crankshaft clockwise to TDC again.
  e. Push the tensioner against the belt with a force of 13 lbs. (59 N) using a spring scale or similar tool and tighten the bolts again to 12–15 ft. lbs. (16–21 Nm). The pressure specification is important and a special spring scale tool, J-38387, is available to measure the tensioner force.

With Auto-Tensioner Installed

1. To check timing belt tension:
  a. Turn the crankshaft 120 degrees clockwise, then turn counterclockwise and return to TDC.
  b. Prepare a steel plate that is about ³/₈ in. (10mm) wide and longer than the width of the belt.
  c. Set the plate on the timing belt between two camshaft sprockets and push against the plate with a force of 11 lbs. (49 N). Note the belt deflection.
  d. Repeat the procedure between the other camshaft sprockets and between the exhaust sprockets and

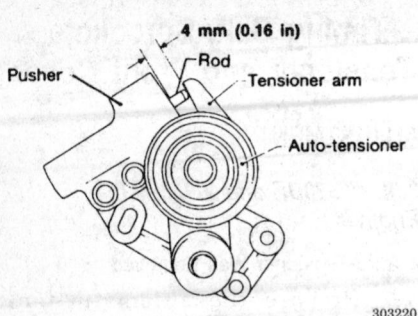

Measuring clearance of the automatic tensioner — 3.0L (VG30DE and VG30DETT) engines

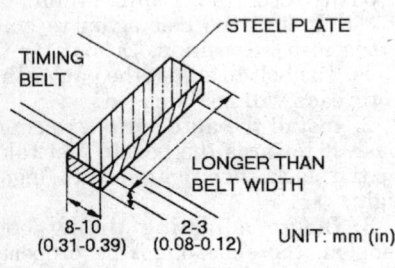

Fabrication of the steel plate — 3.0L (VG30DE and VG30DETT) engines

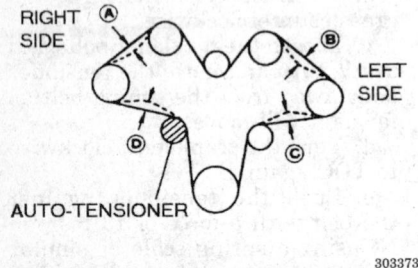

Belt deflection measurement points — 3.0L (VG30DE and VG30DETT) engines

idler/tensioner pulleys. There will be a total of four measurements.

e. Add the deflection measurements and divide by 4. The average deflection must be 0.240–0.280 in. (6–7mm). If belt tension is not correct, start the entire adjustment procedure again.

2. Confirm the automatic tensioner mounting nuts are torqued to 12–15 ft. lbs. (16–21 Nm) and remove the automatic tensioner stopper bolt.

3. After 5 minutes, measure the clearance between the tensioner arm and the pusher. It should be 0.138–0.205 in. (3.5–5.2mm).

4. Make sure all the sprocket timing marks are correctly aligned. Install the timing belt covers and torque the bolts to 24–38 inch lbs. (3–5 Nm).

REMOVAL AND INSTALLATION

3.0L (VG30E) Engine

1. Raise and safely support the vehicle.

2. Remove the engine undercovers and drain engine coolant from the radiator; be careful not to allow coolant to contact drive belts.

3. Remove the front right side wheel and tire assembly. Remove the engine side cover.

4. Remove all the drive belts from the engine.

5. Rotate the crankshaft to position the No. 1 cylinder at the TDC of it's compression stroke.

6. Remove the upper radiator hose and the water inlet hose. Remove the water pump pulley.

7. Remove the idler bracket of the compressor drive belt and crankshaft pulley.

8. Remove the upper and lower timing belt covers.

NOTE: Make sure the punch marks on the camshaft sprockets align with the punch marks on the rear timing belt cover and the punch mark on the crankshaft sprocket aligns with the punch mark on the oil pump.

9. Loosen the timing belt idler pulley bolt. Using a hexagon wrench, rotate the idler pulley to release its tension and remove the timing belt.

NOTE: After removing timing belt, do not rotate crankshaft or camshaft because valves will hit piston heads.

10. Using an adjustable spanner wrench (to hold the camshaft sprockets) and a socket wrench, remove the camshaft sprocket bolt and washer.

11. Pull the camshaft sprockets from the camshafts. Be careful not to lose the key.

NOTE: The right and left camshaft pulleys are different parts. Install them in their correct positions. The right pulley has an R3 identification mark and the left pulley has an L3.

12. Crankshaft sprocket can pulled off crankshaft. If sprocket is difficult to remove, use an appropriate puller to remove.

To install:

13. Install the camshaft sprockets. Hold the sprocket and torque mounting bolt to 65 ft. lbs. (88 Nm).

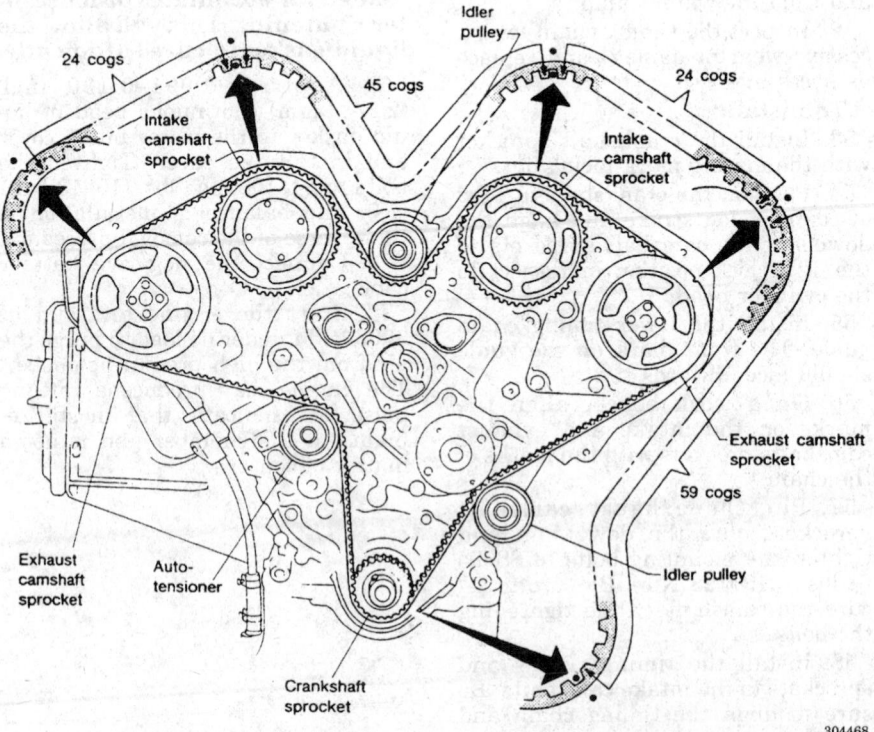

Timing belt alignment on sprockets — 3.0L (VG30DE and VG30DETT) engines

14. Align Woodruff key and tap crankshaft sprocket onto crankshaft, using a plastic hammer or mallet.

NOTE: Be careful not to bend the new belt installing it.

15. Confirm that No. 1 cylinder is at TDC on its compression stroke. Install tensioner and tensioner spring. If stud is removed apply locking sealant to threads before installing.
16. Swing tensioner fully clockwise with hexagon wrench and temporarily tighten locknut.
17. Install the timing belt assembly and be sure the marks on all 3 pulleys are aligned.

NOTE: There are 133 total timing belt teeth. If timing belt is installed correctly there will be 40 teeth between left and right camshaft sprocket timing marks. There will be 43 teeth between left camshaft sprocket and crankshaft sprocket timing marks.

18. Loosen tensioner locknut, keeping tensioner steady with a hexagon wrench.
19. Swing tensioner 70–80° clockwise with hexagon wrench and temporarily tighten locknut.
20. Install all the spark plugs. Turn crankshaft clockwise 2–3 times, then slowly set No. 1 cylinder at TDC on its compression stroke.
21. Push middle of timing belt between right camshaft sprocket and tensioner pulley with a force of 22 lbs. (98 N).
22. Loosen tensioner locknut, keeping tensioner steady with a hexagon wrench.
23. Using a feeler gauge or equivalent, which is 0.0138 in. (0.35mm) thick and 0.500 in. (13mm) wide, set gauge at the bottom of tensioner pulley and timing belt. Turn crankshaft clockwise and position the gauge completely between tensioner pulley and timing belt. The timing belt will move about 2.5 teeth.
24. Tighten tensioner locknut, keeping tensioner steady with a hexagon wrench.
25. Turn crankshaft clockwise or counterclockwise and remove the gauge.
26. Rotate the engine 3 times, then set No. 1 at TDC on its compression stroke. Check alignment of the timing marks.
27. Check timing belt deflection. Timing belt deflection is 13.0–15mm at 22 lbs. of pressure. If it is out of specified range, readjust the timing belt.

28. The remaining components are installed in the reverse order from which they were removed.

3.0L (VG30DE and VG30DETT) Engines

1. Remove the spark plugs and position the engine so that No. 1 piston is at TDC of the compression stroke.
2. Disconnect the negative battery cable.
3. Remove the engine undercover.
4. Drain the cooling system.
5. Remove the radiator.
6. Remove the drive belts.
7. Remove the cooling fan and cooling fan clutch assembly from the front of the water pump.
8. Remove the water pump pulley.
9. Remove the starter and lock the flywheel ring gear using a suitable locking device.
10. Remove the crankshaft pulley bolt.

NOTE: The engine is locked to prevent the crankshaft gear from turning during removal and installation. Remove the lock during engine timing portion of this procedure.

11. Remove the crankshaft pulley using a suitable puller.
12. Remove the water inlet and outlet housings.
13. Remove the timing belt covers and gaskets.

───── **WARNING** ─────
After the timing belt has been removed, DO NOT rotate the camshafts or the crankshaft. Severe internal engine damage will result from piston and valve contact.

14. Install a suitable 6mm stopper bolt in the tensioner arm of the auto tensioner so the length of the pusher does not change.
15. Remove the automatic tensioner and the timing belt.
16. Check the automatic tensioner for oil leaks in the pusher rod and diaphragm. If oil is evident, replace the automatic tensioner assembly.
17. Inspect the timing gear teeth for wear and if the sprockets must be removed, perform the following:
   a. On the exhaust camshaft, use a suitable spanner wrench to hold the sprocket in place and remove the camshaft pulley bolts. Remove the washer, and pull the sprocket from the camshaft.
   b. On the intake camshaft, use a suitable spanner wrench to hold the sprocket in place, and carefully

remove the front cover bolts, O-ring, and spring. Then remove the center retaining bolt and the pull the sprocket, from the camshaft.
   c. Using a suitable puller, remove the crankshaft gear and timing belt plates from the crankshaft. Be careful not to gouge or scratch the surface of the crankshaft when removing the gear.
To install:
18. If the sprockets were removed, install the crankshaft gear with new Woodruff key and torque the pulley mounting bolt to 159–174 ft. lbs. (216–235 Nm). Install the camshaft sprockets and torque to specifications.
19. Verify that the No. 1 piston is at TDC of the compression stroke.
20. Align the timing marks on the camshaft and crankshaft sprockets with the timing marks on the rear timing belt cover and the oil pump housing.
21. With a feeler gauge, check the clearance between the tensioner arm and the pusher of the automatic tensioner. The clearance should be 0.16 in. (4mm) with a slight drag on the feeler gauge. If the clearance is not as specified, mount the tensioner in a vise and adjust the clearance. When the clearance is set, insert the stopper bolt into the tensioner arm to retain the adjustment.

NOTE: When adjusting the clearance, do not push the tensioner arm with the stopper bolt fitted, because damage to the threaded portion of the bolt will result.

22. Mount the automatic tensioner and tighten nuts and bolts by hand.
23. Install the timing belt. Ensure the timing sprockets are free of oil and water. Do not bend or twist the timing belt. Align the white lines on the belt with the timing marks on the camshaft and crankshaft sprockets. Point the arrow on the belt towards the front.
24. Push the automatic tensioner slightly towards the timing belt to prevent the belt from slipping. At the same time, turn the crankshaft 10 degrees clockwise and torque the tensioner fasteners to 12–15 ft. lbs. (16–21 Nm).

NOTE: Do not push the tensioner too hard because it will create excessive tension on the belt.

25. Turn the crankshaft 120 degrees counterclockwise.
26. Back off on the automatic tensioner fasteners ½ turn.

27. Turn the crankshaft clockwise and set the No. 1 piston at TDC of the compression stroke.

28. Using push-pull gauge No. EG1486000 (J–38387) or equivalent, apply approximately 15.2–18.3 lbs. (67.7–81.4 N) of force to the tensioner.

29. Tighten the tensioner mounting bolts to 12–15 ft. lbs. (17–21 Nm).

30. Turn the crankshaft 120 degrees clockwise.

31. Turn the crankshaft 120 degrees counterclockwise and set the No. 1 piston at TDC of the compression stroke.

NOTE: If the timing belt deflection exceeds the specifications, change the applied pushing force.

32. Fabricate a 0.35 in. (9mm) wide x 0.10 in. (2mm) steel plate. The length of the plate should be slightly longer than the width of the belt.

33. Set the steel plate at positions mid-way between the camshaft sprockets on each head, between the left exhaust camshaft sprocket and the idler pulley, and between the right exhaust camshaft sprocket and the tensioner.

34. Using the push-pull gauge or equivalent, apply approximately 11 lbs. (49 N) of force to the tensioner.

35. Check and record the belt deflection at each position with the steel plate in place. The timing belt deflection at each position should be 0.24–0.28 in. (6–7mm). Another means of determining the belt deflection is to add all deflection readings and divide them by 4. This average deflection should be 0.24–0.28 in. (6–7mm).

36. If the belt deflection is not as specified, repeat the timing belt adjusting procedure until the belt deflection is correct.

37. Once the belt is properly tensioned, torque the automatic tensioner fasteners to 12–15 ft. lbs. (16–21 Nm).

38. Remove the stopper bolt from the tensioner and wait 5 minutes. After 5 minutes, check the clearance between the tensioner arm and the pusher of the automatic tensioner. The clearance should remain at 0.138–0.205 in. (3.5–5.2mm).

39. Make sure the belt is installed and aligned properly on each pulley and the timing sprocket. There must be no slippage or misalignment.

40. Install the timing belt covers with new gaskets. Torque the covers bolts to 2–4 ft. lbs. (3–5 Nm).

41. Install the remaining components in the reverse order of removal.

42. Connect the negative battery cable.

43. Start the engine and check the ignition timing.

Camshaft

REMOVAL AND INSTALLATION

1.6L (GA16DE) Engine

─────── **CAUTION** ───────
Fuel injection systems remain under pressure after the engine has been turned OFF. Properly relieve fuel pressure before disconnecting any fuel lines. Failure to do so may result in fire or personal injury.

1. Disconnect the negative battery cable, drain the cooling system, and relieve the fuel system pressure.

2. Remove all engine drive belts. Disconnect the exhaust pipe from the exhaust manifold.

3. Remove the power steering pulley and pump with the mounting bracket.

4. Remove the cylinder head cover.

5. Remove the distributor assembly.

6. Remove the timing chain tensioners and camshaft sprocket.

NOTE: Before the camshafts are removed from the cylinder head, note the positioning of the pins at the end of the camshafts for reassembly purposes.

7. Remove the camshaft bearing caps in sequence and remove the camshafts from the cylinder head. Remove the idler sprocket bolt. These parts should be reassembled in their original position.

8. Remove the shims from the tops of the lifters. Be sure to note the position of each shim.

9. Remove the valve lifters from the bores in the cylinder head. Note

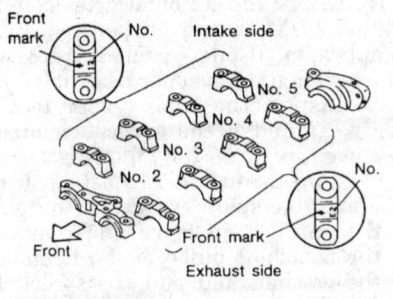

Installation of camshaft bearing caps — 1.6L (GA16DE) engine

302337

the positioning of the lifters for reassembly.

10. Measure the diameter of the lifters. The diameter should be 1.1795–1.1801 in. (29.960–29.975mm).

11. Measure the diameter of the lifter bores. The diameter should be 1.1811–1.1819 in. (30.000–30.021mm).

12. Clearance between the lifter and bore should be 0.0010–0.0024 in. (0.025–0.061mm).

To install:

13. Install the lifters and shims to the cylinder head in the proper locations as noted during removal.

NOTE: The exhaust and intake camshafts are marked with identification stamps. (E for exhaust and I for intake).

14. Install the camshafts to the cylinder head and position the intake camshaft knock pin at the 9 o'clock position and the exhaust camshaft at the 12 o'clock position.

15. Install the camshaft bearing caps and tighten the mounting bolts as follows:

    a. Tighten bolts 11 through 15, then bolts 1 through 10 to 18 inch lbs. (2 Nm)

    b. Tighten bolts 1 through 15 to 53 inch lbs. (6 Nm)

    c. Tighten bolts 1 through 14 to 87–105 inch lbs. (10–12 Nm)

    d. Tighten bolt 15 to 56–73 inch lbs. (7–8 Nm)

NOTE: If any part of the valve train has been has been replaced, the valve adjustment must be checked. DO NOT adjust the valves or rotate the camshafts at this point. Internal engine damage will result.

16. Install the camshaft sprockets with timing chains.

17. Install distributor assembly.

18. Check and adjust the valve clearance.

19. Install the cylinder head cover.

20. Install the remaining components. Refill and check all fluid levels.

21. Connect negative battery cable.

22. Start the engine, check for leaks, and adjust the ignition timing to specification.

23. Road test the vehicle for proper operation.

2.0L (SR20DE) Engine

Sentra

1. Disconnect the negative battery cable. Remove the rocker cover and oil separator.

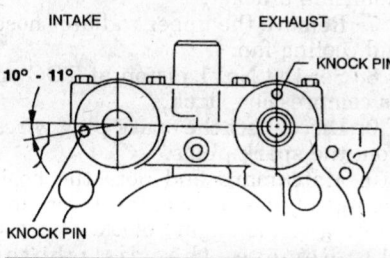

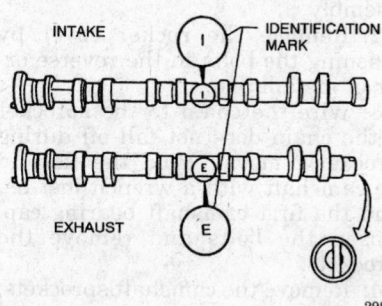

298717

Positioning and identification of camshafts — 1.6L (GA16DE) Engine

2. Rotate the crankshaft until the No.1 piston is at TDC on the compression stroke. Then rotate the crankshaft until the mating marks on the camshaft sprockets line up with the mating marks on the timing chain.

3. Remove the timing chain tensioner.

4. Remove the distributor.

5. Remove the timing chain guide.

6. Remove the camshaft sprockets. Use a wrench to hold the camshaft while loosening the sprocket bolt.

7. Loosen the camshaft bracket bolts in the opposite order of the tightening sequence.

8. Remove the camshaft.

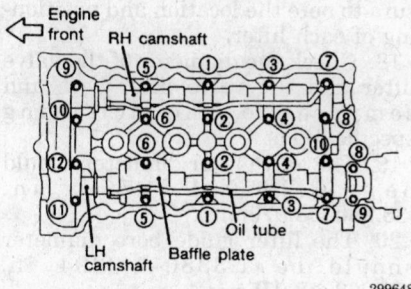

299648

Camshaft bearing cap torque sequence — 2.0L (SR20DE) engines

To install:

9. Clean the left hand camshaft end bracket and coat the mating surface with liquid gasket. Install the camshafts, camshaft brackets, oil tubes and baffle plate. Ensure the left camshaft key is at 12 o'clock and the right camshaft key is at 10 o'clock.

10. The procedure for tightening camshaft bolts must be followed exactly to prevent camshaft damage. Tighten bolts as follows:

    a. Tighten right camshaft bolts 9 and 10 (in that order) to 1.5 ft. lbs. (2 Nm), then tighten bolts 1–8 (in that order) to the same specification.

    b. Tighten left camshaft bolts 11 and 12 (in that order) to 1.5 ft. lbs. (2 Nm), then tighten bolts 1–10 (in that order) to the same specification.

    c. Tighten all bolts in sequence to 4.5 ft. lbs. (6 Nm).

    d. Tighten all bolts in sequence to 6.5–8.5 ft. lbs. (9–12 Nm) for type A, B and C bolts and 13–19 ft. lbs. (18–25 Nm) for type D bolts.

11. Line up the mating marks on the timing chain and camshaft sprockets and install the sprockets. Tighten sprocket bolts to 101–116 ft. lbs. (137–157 Nm).

12. The remaining components are installed in the reverse order from which they were removed.

13. Connect the negative battery cable. Refill all fluid levels. Road test the vehicle for proper operation.

200SX

― **CAUTION** ―
Fuel injection systems remain under pressure after the engine has been turned OFF. Properly relieve fuel pressure before disconnecting any fuel lines. Failure to do so may result in fire or personal injury.

1. Release the fuel system pressure and disconnect the negative battery cable.

2. Raise and safely support the vehicle. Remove the engine undercovers.

3. Remove the right front wheel and the engine side cover.

4. Drain the cooling system and remove the radiator assembly.

5. Remove the air duct from the intake manifold.

6. Remove the drive belts and water pump pulley.

7. Remove the alternator and power steering pump from the engine.

8. Note the locations for reassembly and remove the vacuum hoses, fuel hoses, wires, and the electrical connections.

9. Remove the distributor cap and ignition wires from the engine.

10. Remove all spark plugs.

11. Remove the rocker cover nuts in sequence.

12. Remove the rocker cover and oil separator.

13. Remove the intake manifold supports, oil filter bracket, and the power steering bracket.

14. Rotate the engine and set the No. 1 piston at TDC on the compression stroke. Rotate crankshaft until mating marks on camshaft sprockets are in the correct position.

15. Remove the timing chain tensioner.

16. Matchmark the position of the distributor rotor and housing to the engine block for reinstallation purposes. Unbolt and remove the distributor assembly.

17. Remove the timing chain guide.

NOTE: Wire the camshaft sprockets to the timing chain to maintain proper timing chain position.

18. Holding the flats of the camshaft sprockets, remove the mounting bolts. Remove the sprockets from the camshafts.

NOTE: Note the positioning of the pins on the end of the camshafts for installation purposes.

19. Remove the oil tubes, baffle plate, camshaft brackets and the camshafts. It is important that all parts are kept in order for correct installation.

NOTE: It is essential that the valve train components are kept in specific order for reassembly.

20. Remove the rocker arms, shims, rocker arm guides and the hydraulic lash adjusters.

NOTE: When the lifters are removed, keep them straight up or soak them in clean engine oil to prevent air from entering them.

21. Measure the diameter of the lifters. The diameter should be 0.6685–0.6690 in. (16.980–16.993mm).

22. Measure the diameter of the lifter bores. The diameter should be 0.6693–0.6701 in. (17.000–17.020mm).

23. Standard clearance between the lash adjuster and the guide hole should be 0.0003–0.0016 in. (0.007–0.040mm).

To install:

NOTE: The hydraulic lifters must be bled to remove the air from them. Air can not be bled from this type of lifter by running the engine.

24. Bleed the lifters as follows:
    a. Submerse the lifter into a container of clean engine oil.
    b. While pushing the plunger, insert a thin rod into the check ball and lightly push the check ball.
    c. Air is completely bled when the plunger no longer moves.
25. Check all components for wear, replace as necessary and clean all mating surfaces.

NOTE: Apply clean engine oil to all components prior to installation. Always replace the rocker arm guide with a new one.

26. At this point it is necessary to perform valve adjustment. Adjust the valves as follows:

NOTE: It will be necessary to determine the proper shim size when replacing the valve, cylinder head, shim, rocker arm guide, or the valve seat.

27. Insert tool KV10115700 (J38957) with dial gauge into the lifter bore.
28. Before measuring, make sure the following parts are installed in the cylinder head.
    • Valve
    • Valve spring
    • Collet
    • Retainer
    • Rocker arm guide (except shim)
29. On the shim side, measure the difference between contact surfaces of the rocker arm guide and the valve stem end.

NOTE: When measuring, lightly pull dial indicator rod toward you to eliminate play in tool.

30. Using this reading, select the proper shim size.
31. Shims are available in thickness' from 0.1102–0.1260 in. (2.800–3.200mm) in steps of 0.0010 in. (0.025mm).
32. Measure all the valves and select the proper shim sizes.
33. Remove the tool from cylinder head.
34. Install the valve lifters to the bores in the cylinder head.
35. Install the rocker arm guides, shims, and the rocker arms to the cylinder head.
36. Clean the left hand camshaft end bracket and coat the mating surface with liquid gasket. Install the camshafts, camshaft brackets, oil tubes, and the baffle plate.

--- WARNING ---
Ensure that the left camshaft key is at the 12 o'clock position and the right camshaft key is also at the 12 o'clock position.

37. The procedure for tightening camshaft bolts must be followed exactly to prevent camshaft damage. Tighten bolts as follows:
    a. Tighten the right camshaft bolts No. 9 and 10 (in that order) to 18 inch lbs. (2 Nm), then tighten bolts 1–8 (in that order) to the same specification.
    b. Tighten the left camshaft bolts No. 11 and 12 (in that order) to 18 inch lbs. (2 Nm), then tighten bolts 1–10 (in that order) to the same specification.
    c. Tighten all bolts in sequence to 51 inch lbs. (6 Nm).
    d. Tighten all bolts in sequence to 78–102 inch lbs. (9–12 Nm) and then tighten the two bolts that secure the distributor housing cap to 13–19 ft. lbs. (18–25 Nm).
38. Line up the mating marks on the timing chain and camshaft sprockets and install the timing chain and sprockets. Tighten sprocket bolts to 101–116 ft. lbs. (137–157 Nm). Be sure to hold the flats of the camshaft when tightening the sprocket bolts.
39. The balance of installation is the reverse of the removal procedure.
40. Run the engine and reset the ignition timing.
41. Road test the vehicle for proper operation.

2.4L (KA24DE) Engine

Altima

1. Relieve the fuel system pressure.

--- CAUTION ---
Fuel injection systems remain under pressure after the engine has been turned OFF. Properly relieve fuel pressure before disconnecting any fuel lines. Failure to do so may result in fire or personal injury.

2. Disconnect the negative battery cable.
3. Drain coolant from the engine and radiator.
4. Remove the air intake ducts and the air cleaner assembly.
5. Remove the vacuum hoses, fuel hoses, wires, harness, and connectors that are necessary for removal of the rocker cover.
6. Remove the alternator and mounting bracket.
7. Remove the upper radiator hose and cooling fan.
8. Set the No. 1 piston at TDC on its compression stroke.
9. Disconnect the spark plug wires from the spark plugs.
10. Matchmark and note the positioning of the distributor rotor and housing to the engine block.
11. Remove the distributor assembly.
12. Remove the rocker cover by loosening the bolts in the reverse order of installation.
13. Wire the chain to the sprocket so the chain does not fall off during sprocket removal. Hold the flats of the camshaft with a wrench just behind the first camshaft bearing cap. Loosen the bolts and remove the sprockets.
14. Remove the camshaft sprockets.

NOTE: The stoppers on camshaft covers prevent the upper timing chain from disengaging from the idle sprocket. Also, after removal of the camshaft sprockets, note the positioning of the pins at the end of the camshafts for reinstallation purposes.

15. Remove the camshaft bearing caps in reverse order of installation, then remove the camshafts. The camshaft brackets must be loosened in the correct sequence to prevent damage to the camshaft.

NOTE: All the valve train components must be reassembled in their original positions.

16. Remove the valve lifter adjusting shims from the tops of the of the lifters. Be sure to note the location and positioning of each shim.
17. Remove the valve lifters from the bores in the cylinder heads. Be sure to note the location and positioning of each lifter.
18. Check the diameter of the valve lifter and the valve lifter bore and compare to the following specifications.
19. The valve lifter diameter should be 1.3370–1.3376 in. (33.960–33.975mm).
20. The lifter guide bore diameter should be 1.3386–1.3394 in. (33.960–33.975mm).
21. The valve lifter to lifter guide bore clearance should be 0.0010–0.0024 in. (0.025–0.061mm).

To install:

NOTE: When installing the valve components, apply a coat of clean engine oil to the component.

22. Install the lifters into the lifter bores from which they were removed.

23. Install the valve shims to the lifters from which they came.

24. Install the camshafts in the same position as noted during removal and camshaft bearing caps; tighten cap bolts in the proper sequence as follows:

    a. Tighten all bolts to 17 inch lbs. (2 Nm).

    b. Tighten all bolts to 81–104 inch lbs. (9–12 Nm).

NOTE: When installing the timing chain and sprockets, align the marks on the sprockets with the colored links of the chain.

25. Install the camshaft sprockets with the timing chain and torque the sprocket bolts to 123–130 ft. lbs. (167–177 Nm). Install the chain guide between both camshaft sprockets. The alignment marks on the upper portion of the timing chain should now be aligned with the marks on the sprockets.

26. Install the remaining components in the reverse order of removal.

27. Refill engine coolant and check the engine oil level.

28. Connect the negative battery cable.

29. Start the engine and make the necessary adjustments. Check for proper operation and leaks.

240SX

1. Disconnect negative battery cable.

2. Release fuel system pressure.

———— **CAUTION** ————
Fuel injection systems remain under pressure after the engine has been turned OFF. Properly relieve fuel pressure before disconnecting any fuel lines. Failure to do so may result in fire or personal injury.

3. Remove the air intake ducts and the air cleaner assembly.

4. Remove and tag vacuum hoses from valve cover. The fuel rail may need to be removed to access valve cover nuts.

5. Remove the cooling fan and the radiator fan shroud.

6. Remove spark plug wires.

7. Remove the valve cover and turn the crankshaft to align the timing marks at TDC on No. 1 cylinder.

8. Matchmark the position of the distributor rotor and the distributor. Remove the distributor assembly from the vehicle.

9. Remove the upper timing chain cover.

10. Remove the upper timing chain guide and tensioner.

11. Wire the chain to the sprocket so the chain does not fall off during sprocket removal. Hold the flats of the camshaft with a wrench just behind the first camshaft bearing cap. Loosen the bolts and remove the sprockets.

NOTE: The stoppers on camshaft covers prevent the upper timing chain from disengaging from the idle sprocket.

NOTE: After removal of the camshaft sprockets, note the positioning of the pins at the end of the camshafts for reinstallation purposes.

12. Remove the cam bearing caps in reverse order of the tightening sequence and remove the camshafts. The camshaft brackets must be loosened in the correct sequence to prevent damage to the camshaft.

NOTE: These parts should be reassembled in their original position. Bolts should be loosened in 2–3 steps (loosen all bolts in the reverse of the tightening order).

13. Remove the valve lifter adjusting shims from the tops of the of the lifters.

14. Remove the valve lifters from the bores in the cylinder heads.

15. Check the diameter of the valve lifter and the valve lifter bore and compare to the following specifications.

16. The valve lifter diameter should be 1.3370–1.3376 in. (33.960–33.975mm).

17. The lifter guide bore diameter should be 1.3386–1.3394 in. (33.960–33.975mm).

18. The valve lifter to lifter guide bore clearance should be 0.0010–0.0024 in. (0.025–0.061mm).

To install:

NOTE: Apply a clean coat of new engine oil to the valve train components.

19. Install the lifters into the lifter bores from which they were removed.

20. Install the valve shims to the lifters from which they came.

21. Install the camshafts in the same position as noted during removal and install the camshaft bearing caps.

22. Tighten bearing cap bolts in the proper sequence as follows:

    a. Tighten all bolts to 17 inch lbs. (2 Nm).

    b. Tighten all bolts to 81–104 inch lbs. (9–12 Nm).

NOTE: When installing the timing chain and sprockets, align the marks on the sprockets with the colored links of the chain.

23. Install the camshaft sprockets and torque the bolts to 123–130 ft. lbs. (167–177 Nm).

24. Install the timing chain, then install the remaining components in the opposite order from which they were removed.

25. Connect negative battery cable.

26. Start the engine and make the necessary adjustments. Check for proper operation and leaks.

3.0L (VG30E) Engine

1. Disconnect the negative battery cable.

2. Drain the coolant from the cooling system and the engine. Coolant from the engine can be drained by removing the drain plug on the cylinder block.

3. Remove the timing belt assembly.

4. Remove the collector assembly and intake manifold.

5. Remove the cylinder head from the engine.

6. With cylinder head mounted on a suitable workbench, remove the rocker shafts with rocker arms. Bolts should be loosened in 2–3 steps.

7. Remove hydraulic valve lifters and lifter guide.

8. Hold hydraulic valve lifters with wire so they will not drop from lifter guide.

9. Remove the camshaft front oil seal and slide camshaft out the front of the cylinder head assembly.

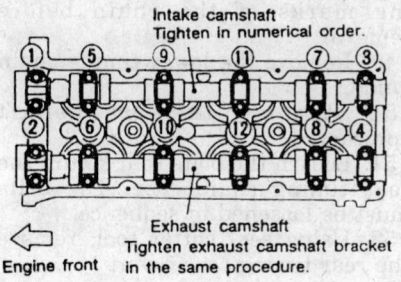

Intake camshaft
Tighten in numerical order.

Exhaust camshaft
Tighten exhaust camshaft bracket in the same procedure.

Engine front

289272

Camshaft bearing cap tightening sequence — 2.4L (KA24DE) engines

To install:

10. Install camshaft, locate plate, cylinder head rear cover and front oil seal. Set camshaft knock pin at 12 o'clock position. Install cylinder head with new gasket to engine.

11. Install valve lifter guide assembly. Assemble valve lifters in their original position. After installing them in the correct location remove the wire holding them in lifter guide.

12. Install rocker shafts in correct position with rocker arms. Tighten bolts in 2–3 stages to 13–16 ft. lbs. Before tightening, be sure to set camshaft lobe at the position where lobe is not lifted or the valve closed. You can set each cylinder one at a time or follow the procedure below (timing belt must be installed in the correct position):

   a. Set No. 1 piston at TDC on its compression stroke and tighten rocker shaft bolts for No. 2, No. 4 and No. 6 cylinders.

   b. Set No. 4 piston at TDC on its compression stroke and tighten rocker shaft bolts for No. 1, No. 3 and No. 5 cylinders.

   c. Torque specification for the rocker shaft retaining bolts is 13–16 ft. lbs.

13. Install the intake manifold and collector assembly.

14. Install the timing belt cover and camshaft sprocket. The left and right camshaft sprockets are different parts. Install the correct sprocket in the correct position.

15. Install the timing belt assembly, fill coolant and set engine timing to specifications.

3.0L (VE30DE and VQ30DE) Engines

1. Relieve the fuel system pressure.

─────── **CAUTION** ───────
Fuel injection systems remain under pressure after the engine has been turned OFF. Properly relieve fuel pressure before disconnecting any fuel lines. Failure to do so may result in fire or personal injury.
───────────────────────

2. Disconnect the negative battery cable.

3. Drain the engine oil and the cooling system. Be sure to drain the engine block and the radiator.

4. Remove the left side rocker cover ornament.

NOTE: Before disconnecting any hoses or connectors, note the locations for reassembly.

5. Remove the air duct to intake manifold hose, collector hose, blow-by hose, and vacuum hoses.

6. Remove the fuel hoses and disconnect the harness connections.

7. Disconnect the canister purge hoses.

8. Remove the water hoses from the cylinder head and intake manifold.

9. Disconnect and remove all six ignition coils from the spark plugs.

10. Remove the spark plugs.

11. Remove the bolts that secure the EGR tube and remove the tube.

12. Remove the intake manifold collector supports and remove the collector.

13. Remove the bolts that secure the fuel tube and remove the fuel tube from the vehicle.

14. Remove the bolts that secure the intake manifold to the engine block and remove the manifold. Loosen the bolts in the reverse sequence of the tightening procedure.

15. Remove the LH and RH rocker covers from the cylinder head.

16. Remove the engine undercovers.

17. Remove the right front wheel and engine side covers.

18. Remove the drive belts and idler pulley.

19. Remove the power steering oil pump belt and remove the power steering oil pump assembly.

20. Remove the camshaft position sensor (PHASE) and crankshaft position sensors (REF)/(POS).

21. Set the No. 1 piston to TDC of compression stroke by rotating the crankshaft.

22. Remove the ring gear cover access plate.

23. Loosen the crankshaft pulley bolt while securing the ring gear so the crankshaft cannot rotate.

NOTE: Use care not to damage the ring gear teeth.

24. Using a suitable puller, remove the crankshaft pulley.

25. Remove the A/C compressor and bracket.

26. Remove the front exhaust pipe and its support.

27. Hang the engine at the right and left side engine slingers with a suitable hoist.

28. Support the transaxle with jack.

29. Remove the right side engine mounting, mounting bracket and nuts.

30. Remove the center crossmember assembly.

31. Remove the steel (lower) oil pan bolts in the reverse sequence of the torque sequence.

32. Insert a seal cutter between the steel and aluminum oil pan.

33. Tapping the cutter with a hammer, slide it around the entire edge of the oil pan. Be careful not to damage the aluminum mating surface of the upper oil pan.

34. Remove the steel oil pan and the oil strainer.

35. Remove the aluminum (upper) oil pan bolts in the reverse sequence of the torque sequence.

36. Remove the transaxle bolts that secure the oil pan.

37. Insert a seal cutter between the aluminum oil pan and the engine block.

38. Tapping the cutter with a hammer, slide it around the entire edge of the oil pan. Be careful not to damage the mating surfaces of the oil pan or engine block.

39. Remove the oil pan from the vehicle.

40. Remove the water pump cover and remove the bolts that secure the front timing chain case cover.

41. Using the seal cutter, remove the timing chain case cover.

42. Remove the internal timing chain guide and the upper chain guide.

43. Remove the timing chain tensioner and slack side chain guide.

44. Remove the left and right intake camshaft sprockets first. Be sure to hold the flats of the camshafts while removing the sprocket bolts.

45. Remove the lower timing chain assembly. Be sure to note the aligning marks of the chain before removal.

46. Insert a suitable stopper pin for the left and right camshaft tensioners.

47. Remove the left and right exhaust camshaft sprocket bolts. Be sure to hold the flats of the camshafts while removing the sprocket bolts.

48. Remove the upper timing chain assembly. Be sure to note the aligning marks of the chain before removal.

49. Remove the lower timing chain guide.

50. Remove the crankshaft sprocket.

51. Loosen the bolts that secure the rear timing chain case. The bolts must be loosened in sequence.

52. Using seal cutter tool, remove the rear timing case cover.

NOTE: Remove the O-rings from the front of the engine block.

53. Loosen the camshaft bearing caps in several steps. The bearing caps MUST be loosened in sequence.

NOTE: Keep all bearing caps and camshafts in proper order for reinstallation.

54. Remove the LH and RH camshaft tensioners from the cylinder head.

55. Remove the camshafts from the cylinder heads.

NOTE: The valve lifters have a replaceable shim on the top of the lifter. Note the proper locations of each shim to lifter and remove the shims from the lifters.

56. Using a magnet, remove the valve adjusting shim from the lifter.

57. Remove the lifter assembly from the bore. Be sure to note the locations from where each lifter came.

58. Check the diameter of the valve lifter and the valve lifter guide bore.

59. The diameter of the lifter should be 1.3764–1.3770 in. (34.960–34.975mm) and the diameter of the bore should be 1.3780–1.3788 in. (35.000–35.021mm).

60. Remove all traces of liquid gasket from the timing chain case and from the water pump covers.

61. Remove all traces of liquid gasket from the engine block.

62. Inspect the camshafts for excessive wear or damage and replace as necessary.

To install:

63. Lubricate the valve lifters with clean engine oil and install the lifters into the bore from which they were removed.

64. Lubricate the valve lifter shims with clean engine oil and install the shims into the lifter from which they were removed.

65. Turn the crankshaft clockwise until the No. 1 piston is set 240 degrees before TDC on compression stroke.

66. Install the camshaft tensioners on both sides of the cylinder heads. Tighten the tensioner mounting bolts to 75–96 inch lbs. (8.4–10.8 Nm).

NOTE: The camshafts can be identified by the paint marks on the camshaft. The left cylinder head camshafts have a YELLOW paint mark and the right cylinder head camshafts have a WHITE paint mark.

67. Install the exhaust and intake camshafts and install the bearing caps. Before installing the No. 1 bear-

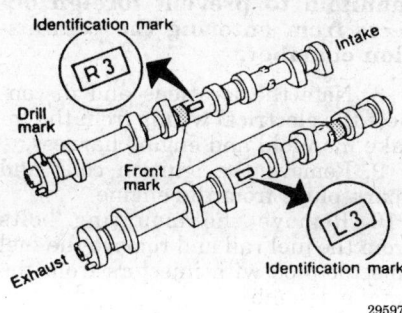

Camshaft identification — 3.0L (VE30DE and VQ30DE) engines

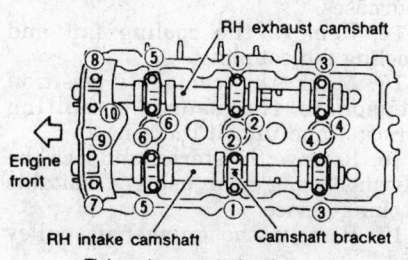

Right intake and exhaust camshaft bearing cap torque sequence — 3.0L (VE30DE and VQ30DE) engines

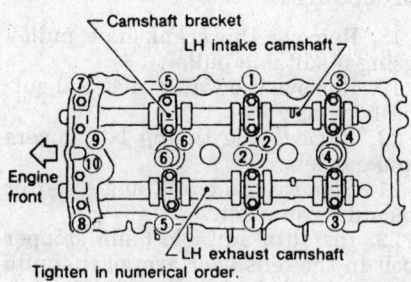

Left intake and exhaust camshaft bearing cap torque sequence — 3.0L (VE30DE and VQ30DE) engines

ing cap, apply liquid gasket to the corners of the cap.

NOTE: When installing the camshafts, position the camshaft keys at the 12 o'clock position in respect to the cylinder head angle.

68. Tighten the camshaft bearing caps as follows:
   a. Tighten bolts No. 7–10 to 17 inch lbs. (2 Nm).
   b. Tighten bolts No. 1–6 to 17 inch lbs. (2 Nm).
   c. Tighten bolts No. 1–10 to 52 inch lbs. (6 Nm).

   d. Tighten bolts No. 1–10 to 81–104 inch lbs. (9–11 Nm).

69. Install new O-rings to the front of the engine block.

70. Apply sealant to the hatched portion of the of the rear timing chain case.

71. Align the rear timing chain case with the dowel pins and install onto the cylinder heads and engine block.

72. Tighten the rear timing chain case mounting bolts in sequence to 105–121 inch lbs. (11.8–13.7 Nm).

73. Install the crankshaft sprocket with the mating mark facing out.

74. Rotate the crankshaft clockwise and position the crankshaft to TDC of compression stroke and align the dowels of the camshaft sprockets to the 12 o'clock position in respect to the cylinder head.

75. Install the lower chain guide on the dowel pin with the front mark on the guide facing upward.

76. On a work bench, align the marks on the intake and exhaust camshaft sprockets with the marks of the chain.

77. Put the exhaust camshaft sprockets onto the dowel pin and tighten the mounting bolts to 88–95 ft. lbs. (119–128 Nm). Be sure to secure the camshafts while tightening the bolts.

78. Install the timing chains, sprockets and related components.

79. Install the transaxle bolts that secure the oil pan.

80. Install the oil pan strainer and tighten the mounting bolts to 12–14 ft. lbs. (16–19 Nm).

81. Apply a 0.177–0.217 inch (4.5–5.5mm) continuous bead of liquid gasket to the lower oil pan mating surface and install the oil pan. Tighten the mounting bolts in sequence to 57–66 inch lbs. (6.4–7.5 Nm).

82. Tighten the oil pan drain plug to 22–29 ft. lbs. (29–39 Nm).

83. Install the center crossmember assembly.

84. Install the right side engine mounting bracket and mount assembly.

85. Remove the engine slinger assembly.

86. Install the front exhaust pipe and its support.

87. Install the A/C compressor and bracket.

88. Install the crankshaft pulley to the crankshaft and install the mounting bolt.

89. Tighten the mounting bolt to 14–22 ft. lbs. (20–29 Nm). Tighten the crankshaft bolt an additional 60–66 degrees clockwise. This is

about the angle from one hexagon bolt head corner to another.

90. Install the remaining components in the opposite order from which they were removed.

91. Refill the engine oil and coolant with the proper type and amount of fluid.

92. Connect the negative battery cable.

93. Start the engine and run at 3000 RPM under no load to purge the air from the high pressure chamber. The engine may produce a rattling noise. This indicates that air still remains in the chamber and is not a matter of concern.

3.0L (VG30DE and VG30DETT) Engines

———— WARNING ————
After the timing belt has been removed, DO NOT rotate the crankshaft or camshafts. The valves and pistons will make contact thus causing severe engine damage.

1. Position the engine so that No. 1 piston is at TDC of the compression stroke.

2. Relieve the fuel system pressure and disconnect the negative battery cable.

3. Drain the cooling system. Be sure to drain the engine block.

4. Disconnect the accelerator cable from the throttle body linkage.

5. Disconnect the cruise control cable from the throttle body linkage.

6. Note the locations and disconnect the vacuum lines from the intake manifold collector.

7. Unbolt and remove the intake manifold collector from the intake manifold.

NOTE: After the intake manifold collector has been removed, cover the openings on the intake

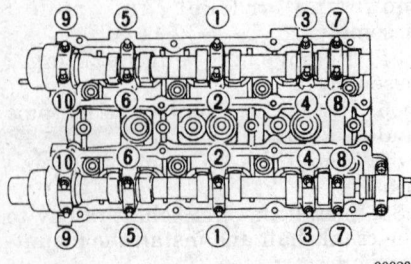

303231

Camshaft bearing cap bolt tightening sequence — 3.0L (VG30DE and VG30DETT) engines

manifold to prevent foreign objects from entering the combustion chamber.

8. Note the locations and disconnect the electrical wiring from the intake manifold and engine harness.

9. Remove the ignition coils and spark plugs from the engine.

10. Remove the mounting bolts from the fuel rail and remove the fuel injector pipe with injectors from the intake assembly.

11. Remove the valve covers.

12. Remove the radiator.

13. Remove the drive belts. Mark the drive belts for reinstallation purposes.

14. Remove the cooling fan and cooling fan coupling.

15. Mark the camshaft position sensor to the sensor mounting bracket and unbolt the sensor.

16. Remove the starter and lock the flywheel ring gear using a suitable locking device.

17. Remove the crankshaft pulley bolt.

NOTE: The engine is locked to prevent the crankshaft gear from turning during removal and installation. Remove the lock during engine timing portion of this procedure.

18. Remove the crankshaft pulley using a suitable puller.

19. Remove the water inlet and outlet housings.

20. Remove the timing belt covers and gaskets.

21. Remove the crank angle sensor mounting bracket.

22. Install a suitable 6mm stopper bolt in the tensioner arm of the auto tensioner so the length of the pusher does not change.

NOTE: Mark the timing belt direction of rotation for reinstallation purposes.

23. Remove the automatic tensioner and the timing belt.

24. Check the automatic tensioner for oil leaks in the pusher rod and diaphragm. If oil is evident, replace the automatic tensioner assembly.

25. Inspect the timing gear teeth for wear and if the sprockets show signs of wear they must be replaced.

26. Remove the idler pulley mounting nut, remove the idler pulley and remove the idler pulley stud.

27. Unbolt and remove the intake manifold assembly.

28. Disconnect the exhaust pipe from the exhaust manifold.

29. Loosen the cylinder head bolts (in reverse order of installation se-

quence) in 2–3 stages. Lift the cylinder head off the engine block with the exhaust manifolds attached. It may be necessary to tap the head lightly with a rubber mallet to loosen it.

30. Position the cylinder head on blocks of wood to protect the surface of the head and to protect the any valves that may extend below the surface of the cylinder head.

31. Remove the exhaust manifold from the cylinder head.

32. While holding the flats of the camshaft in position with a wrench, remove the mounting bolts from the camshaft sprockets.

NOTE: Remove the front plate, O-ring and spring from the right (intake) camshaft to gain access to the sprocket bolt. The left camshaft sprocket is held in place by plate and 4 bolts.

33. Note the positioning of the sprockets to the camshaft and remove the sprocket from the camshafts.

34. To remove the camshaft you will need to remove the rear timing belt cover which will be accessible after the camshaft sprockets are removed.

35. Mount a dial indicator and set the stylus of the indicator on the end of the camshaft. Zero the indicator and measure the camshaft end-play by moving the camshaft back and forth. End-play should be within 0.0012–0.0031 in. (0.03–0.08mm).

36. Remove the camshaft bearing caps. Loosen the bolts in the proper sequence (reverse the installation torque sequence) gradually in 2–3 stages.

37. Gently pry the camshaft oil seals from the cylinder head.

38. Remove the timing control solenoid valves.

39. Remove the camshafts from the cylinder head.

NOTE: Be sure to note positioning of the exhaust and the intake camshafts before removing them.

To install:

40. Install the camshafts so the knock pins are aligned properly. The exhaust side camshaft (left side) has a spline that accepts the crank angle sensor.

41. Install the timing control solenoid valves. Torque the bracket bolts to 12–18 ft. lbs. (16–25 Nm). Apply liquid gasket to the valve seating surface before installation.

42. Install the camshaft bearing caps. Torque the bracket bolts in sequence to 7–9 ft. lbs. (9–12 Nm).

Tighten the bolts gradually in 2–3 stages. When installing the front camshaft brackets, apply liquid gasket to the bracket seating surface.

43. Coat the lips of the new camshaft seals with clean engine oil and install the seals into the cylinder head.

44. Install the rear timing belt covers. Torque the cover bolts to 5–6 ft. lbs. (6–8 Nm).

45. Install the camshaft sprockets.

46. Install the cylinder head with a new gasket.

47. Connect the exhaust tube to the exhaust manifold.

48. Install the remaining components in the reverse order of removal.

49. Fill the cooling system to the proper level and connect the negative battery cable.

50. Run the engine and make all the necessary engine adjustments.

51. Road test the vehicle for proper operation.

Piston and Connecting Rod

POSITIONING

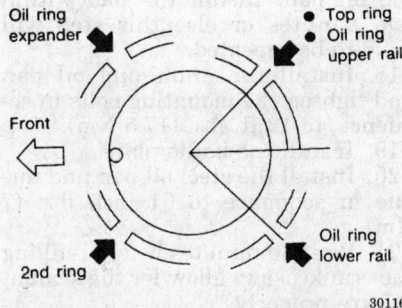

Piston ring positioning — 1.6L (GA16DE), 2.0L (SR20DE) and 3.0L (VQ30DE) engines

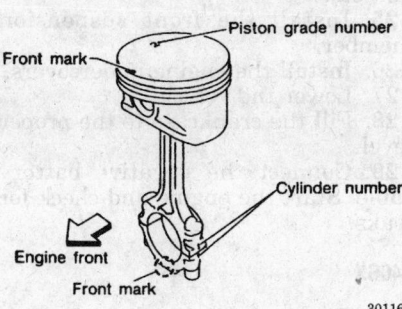

Piston and connecting rod front marks — all engines except 3.0L (VG30E and VE30DE) engines

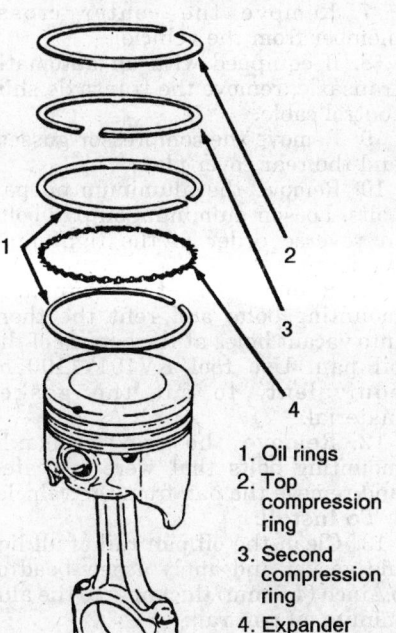

1. Oil rings
2. Top compression ring
3. Second compression ring
4. Expander

Piston, connecting rod and piston rings — 3.0L (VG30DE and VG30DETT) engines

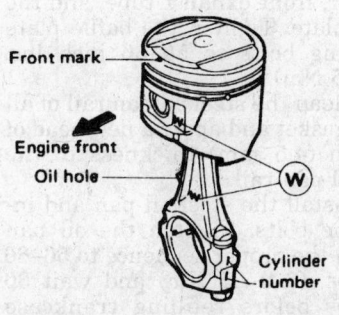

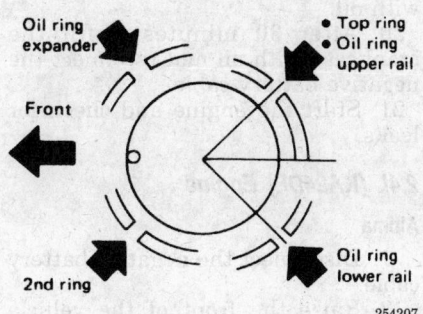

Piston and ring alignment — 3.0L (VG30E and VE30DE) engines

ENGINE LUBRICATION

Oil Pan

REMOVAL AND INSTALLATION

1.6L (GA16DE) Engine

1. Disconnect the negative battery cable.

2. Raise and safely support the vehicle.

3. Remove the undercovers.

4. Remove the oil pan plug and drain the oil into a container.

5. Remove the front exhaust tube.

6. Remove center crossmember assembly.

7. Remove the support brackets from the sides of the oil pan.

8. Remove the oil pan mounting bolts. Using the oil pan seal cutter tool KV10111100 or equivalent, separate the oil pan from the engine.

WARNING

Do not drive the seal cutter into the oil pump or rear oil seal retainer portion, for the aluminum mating surfaces will be damaged. Do not use a prybar to remove the oil pan; the flange will be deformed.

9. Clean all the sealing surfaces.
To install:

10. Apply sealant to the rear oil seal retainer.

11. Apply a 0.128–0.177 in. (3.5–4.5 mm) continuous bead of liquid gasket to the oil pan mating surface.

12. Install the oil pan and torque the oil pan mounting bolts/nuts in sequence to 56–73 inch lbs. (6.3–8.3 Nm).

13. Install the oil pan support brackets.

14. Install the center crossmember.

15. Install the front exhaust tube.

16. Install the undercovers.

17. Using a new gasket, install the oil pan plug and tighten the plug to 21–28 ft. lbs. (7–8 Nm).

18. After 30 minutes of gasket curing time, refill the oil pan with the specified quantity of clean oil. Operate the engine and check for leaks.

19. Connect the negative battery cable.

2.0L (SR20DE) Engine

1. Disconnect the negative battery cable.

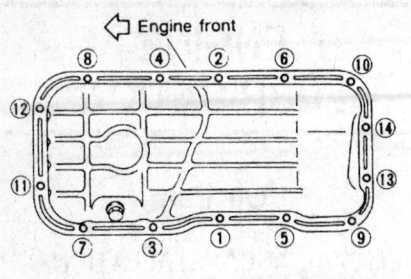

Oil pan torque sequence — 1.6L (GA16DE) engine

301008

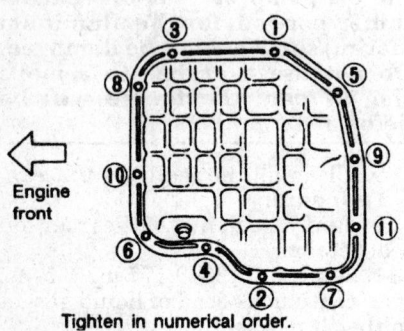

Tighten in numerical order.

300998

Aluminum oil pan bolt torque sequence — 2.0L (SR20DE) engine

Tighten in numerical order.

300996

Steel oil pan bolt torque sequence — 2.0L (SR20DE) engine

2. Raise and support the vehicle safely.

3. Remove the engine undercover and drain the oil.

4. Remove the lower steel oil pan bolts in the reverse of installation sequence. Remove the steel oil pan. Insert tool KV10111100 or equivalent, between steel oil pan and aluminum oil pan. Tap the tool around the perimeter of the pan to cut the gasket material.

5. Remove the oil baffle bolts and oil baffle. Remove the front exhaust tube.

6. Set a suitable jack under the transaxle and raise the engine.

7. Remove the center cross-member from the vehicle.

8. If equipped with an automatic transaxle, remove the transaxle shift control cable.

9. Remove the compressor gussets and the rear cover plate.

10. Remove the aluminum oil pan bolts. Loosen aluminum oil pan bolts in reverse order of the tightening sequence.

11. Remove the two transaxle mounting bolts and refit the them into vacant holes at the bottom of the oil pan. Use tool KV10111100 or equivalent, to cut the gasket material.

12. Remove the two transaxle mounting bolts that were relocated and remove the pan from the vehicle.

To install:

13. Clean the oil pan rail of all liquid gasket and apply a new bead of 5/32 inch (4.5 mm) thickness to the aluminum oil pan rail.

14. Install the aluminum oil pan and torque bolts No. 1–16 to 12–14 ft. lbs. (16–19 Nm) and bolts No. 17–18 to 5–6 ft. lbs. (6–8 Nm) in the opposite order of removal.

15. Install the two transaxle mounting bolts, rear cover plate, and the compressor gussets.

16. Install the automatic transmission shift control cable (if equipped).

17. Install the center crossmember member, front exhaust tube, and the baffle plate. Tighten the baffle plate mounting bolts to 56–66 inch lbs. (6.4–7.5 Nm).

18. Clean the steel oil pan rail of all liquid gasket and apply a new bead of 5/32 inch (4.5 mm) thickness to the steel oil pan rail.

19. Install the steel oil pan and install the bolts. Tighten the oil pan bolts in the proper sequence to 56–66 inch lbs. (6.4–7.5 Nm) and wait 30 minutes before refilling crankcase with oil.

20. After 30 minutes, refill the crankcase with oil and reconnect the negative battery cable.

21. Start the engine and check for leaks.

2.4L (KA24DE) Engine

Altima

1. Disconnect the negative battery cable.

2. Raise the front of the vehicle and support it safely.

3. Drain the oil from the oil pan.

4. Remove the engine undercover.

5. Remove the bolts securing the steel oil pan to the aluminum oil pan

in reverse order of the tightening sequence.

6. Install a seal cutter between the steel oil pan and the aluminum oil pan

7. Tapping the cutter with a hammer, slide it around the entire edge of the oil pan. Take care not to damage the aluminum oil pan.

8. Remove the steel oil pan.

9. Remove the baffle plate and oil strainer.

10. Remove the front suspension member.

11. Remove the A/C compressor gussets.

12. Remove the rear cover plate.

13. Remove the aluminum oil pan retaining bolts in reverse order of the tightening sequence.

14. Insert a seal cutter between the oil pan and the cylinder block.

15. Tapping the cutter with a hammer, slide it around the entire edge of the oil pan. Take care not to damage the aluminum oil pan.

16. Lower the oil pan from the cylinder block and remove it from the engine.

To install:

17. Carefully scrape the old gasket material away from the pan and cylinder block mounting surfaces and then apply a continuous bead (3.5–4.5mm) of liquid gasket around the oil pan. Install the pan within five minutes or else this step will have to be repeated.

18. Install the aluminum oil pan and tighten the mounting bolts in sequence, to 13 ft. lbs. (17.5 Nm).

19. Install the baffle plate

20. Install the steel oil pan and torque in sequence to 61 inch lbs. (7 Nm).

21. Wait 30 minutes before refilling the crankcase to allow for the sealant to cure properly.

22. Install the rear cover plate.

23. Install the front suspension member.

24. Install the A/C compressor gussets.

25. Install the front suspension member.

26. Install the engine undercovers.

27. Lower the vehicle.

28. Fill the crankcase to the proper level.

29. Connect the negative battery cable. Start the engine and check for leaks.

240SX

1. Disconnect the negative battery cable.

2. Raise the front of the vehicle and support it safely.

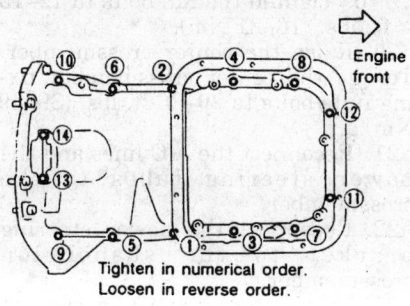

Tighten in numerical order.
Loosen in reverse order.

291657

**Oil pan loosening and tightening sequence —
2.4L (KA24DE) engine**

3. Position a hoist on the engine and support the engine.

4. Drain the oil pan.

5. Disconnect the tension rod bolts at the transverse link.

6. Separate the front stabilizer bar from the side member.

7. Remove the left and right engine mounting bolts.

8. Disconnect the lower steering joint.

9. Disconnect the power steering tube bracket at the left tension rod.

10. Remove the bolts and lower the front suspension member while supporting it with a jack. It is only necessary to lower the suspension member 2.36 in. (60mm).

11. Remove the oil pan retaining bolts.

12. Insert a seal cutter between the oil pan and the cylinder block.

13. Tapping the cutter with a hammer, slide it around the entire edge of the oil pan. Do not drive the seal cutter into the oil pump or rear seal retainer portion or the aluminum mating surface will be deformed.

14. Lower the oil pan from the cylinder block and remove it from the front side of the engine.

To install:

15. To install, carefully scrape the old gasket material away from the pan and cylinder block mounting surfaces and then apply a continuous bead 0.138–0.177 in. (3.5–4.5mm) of liquid gasket around the oil pan. Install the pan within 5 minutes or this step will have to be repeated.

16. Install the oil pan and tighten the mounting bolts. Start tightening the bolts from the center and work towards the ends.

17. Torque the oil pan bolts to 57–61 inch lbs. (6.4–7.5 Nm) for 1993–94 vehicles or 12–14 ft. lbs. (16–19 Nm) for 1995–96 vehicles. Wait 30 minutes before refilling the crankcase to allow for the sealant to cure properly.

18. The remaining components are installed in the reverse order from which they were removed.

19. Fill the crankcase to the proper level.

20. Connect the negative battery cable. Start the engine and check for leaks.

21. Recheck the oil level.

3.0L (VE30DE and VG30E) Engines

1. Disconnect the negative battery cable.

2. Raise and safely support the vehicle.

3. Connect a lifting device to the engine and apply upward pressure to take the engine weight off the engine mounts.

4. Remove the oil pan plug and drain the oil into proper safety container.

5. Remove the undercovers from under the engine.

6. Remove the exhaust front pipe mounting nuts and bolts.

7. Lower the front exhaust pipe assembly.

8. Remove the engine mount insulator-to-crossmember nuts and bolts.

9. Remove the center cross-member-to-chassis bolts and the crossmember.

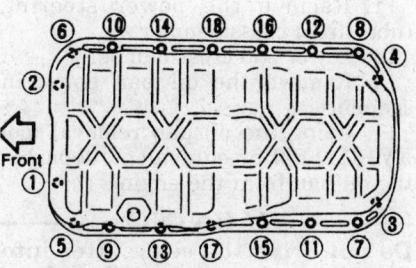

Loosen in numerical order.

159485

Oil pan bolt removal sequence — 3.0L (VE30DE and VG30E) engines

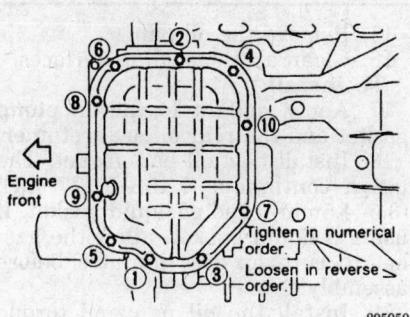

Tighten in numerical order.
Loosen in reverse order.

295950

Steel oil pan tightening sequence — 3.0L (VQ30DE) engines

10. Remove the oil pan-to-engine bolts, in sequence. Using the oil pan removal tool KV10111100 or equivalent, separate the oil pan from the engine.

—————— WARNING ——————
Do not drive the seal cutter into the oil pump or rear oil seal retainer portion, for the aluminum mating surfaces will be damaged. Do not use a prybar, for the oil pan flange will be deformed.

To install:

11. Clean all the sealing surfaces. Apply sealant to the ends of the front cover/oil pump gasket and to the ends of the rear oil seal retainer gasket.

12. Apply a continuous bead of liquid sealer to the mating surface of the oil pan.

NOTE: Be sure the liquid sealer is 0.138–0.177 in. (3.5–4.5mm) wide.

13. Tighten the oil pan mounting bolts to 5.1–5.8 ft. lbs. (7–8 Nm) by reversing the removal sequence.

14. Install the exhaust pipe connection.

15. Install the center crossmember. Torque the center crossmember-to-chassis nuts/bolts and the engine mount bolts to 57–72 ft. lbs. (77–98 Nm).

16. Install the undercovers to the engine. After 30 minutes of gasket curing time, refill the oil pan with the specified quantity of clean oil. Operate the engine and check for leaks.

3.0L (VQ30DE) Engine

1. Disconnect the negative battery cable.

2. Drain the engine oil and remove the engine undercovers.

3. Remove the steel (lower) oil pan bolts in the reverse sequence of the torque sequence.

4. Insert a seal cutter between the steel and aluminum oil pan.

5. Tapping the cutter with a hammer, slide it around the entire edge of the oil pan. Be careful not to damage the aluminum mating surface of the upper oil pan.

6. Remove the steel oil pan and the oil strainer.

7. Remove the front exhaust pipe and its support.

8. Hang the engine at the right and left side engine slingers with a suitable hoist.

9. Position a suitable jack under the transaxle.

10. Remove the crankshaft position sensors (REFERENCE and POSITION) from the oil pan.

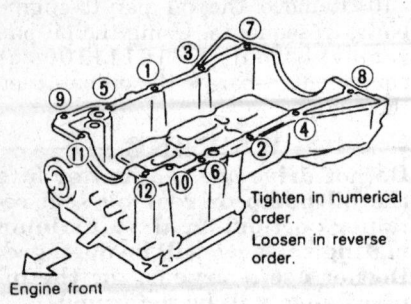

Tighten in numerical order.
Loosen in reverse order.

Engine front

295953

Aluminum oil pan torque sequence — 3.0L (VQ30DE) engines

11. Remove the front and rear engine mounting nuts and bolts.
12. Remove the center crossmember assembly.
13. Remove the engine drive belts.
14. Remove the air conditioner compressor and the compressor mounting bracket.
15. Remove the rear cover plate and the lower transaxle bolts.
16. Remove the aluminum (upper) oil pan bolts in the reverse sequence of the torque sequence.
17. Insert a seal cutter between the aluminum oil pan and the engine block.
18. Tapping the cutter with a hammer, slide it around the entire edge of the oil pan. Be careful not to damage the mating surfaces of the oil pan or engine block.
19. Remove the oil pan assembly.
20. Remove the bolts that secure the baffle plate and remove the baffle plate.
21. Remove the O-rings from the cylinder block and oil pump body.
   To install:
22. Install the baffle plate to the oil pan and tighten the mounting bolts to 22–27 inch lbs. (2.5–3.1 Nm).
23. Apply sealant to the front and rear seal of the oil pan.
24. Install new O-rings to the cylinder block and the oil pump body.
25. Apply a 0.177–0.217 inch (4.5–5.5mm) continuous bead of liquid gasket to the upper oil pan mating surface and install the oil pan. Tighten the mounting bolts in sequence to 12–14 ft. lbs. (16–19 Nm).
26. Install the oil pan strainer and tighten the mounting bolts to 12–14 ft. lbs. (16–19 Nm).
27. The balance of installation is the reverse of the removal procedure.
28. Apply a 0.177–0.217 inch (4.5–5.5mm) continuous bead of liquid gasket to the lower oil pan mating surface and install the oil pan. Tighten the mounting bolts in se-

quence to 57–66 inch lbs. (6.4–7.5 Nm).
   NOTE: Wait at least 30 minutes before refilling the engine oil.
29. Tighten the oil pan drain plug to 22–29 ft. lbs. (29–39 Nm).
30. Install the engine undercovers.
31. Refill the engine oil with the proper type and amount of fluid.
32. Start the engine and check for leaks.

3.0L (VG30DE and VG30DETT) Engine

1. Raise and safely support the vehicle.
2. Remove the oil pan plug and drain the oil into a container.
3. Remove the engine undercover.
4. Remove the engine oil filter and oil filter housing assembly.
5. Disconnect the A/C tube clamps.
6. Remove the tension rod securing bolts on both sides and loosen the traverse link bolts on both sides.
7. Disconnect the steering column lower joint.
8. Disconnect the engine mounts and raise engine using a suitable lifting device.
9. Remove the power steering gear bracket from the suspension crossmember.
10. Support the crossmember and remove the crossmember fixing bolts.
11. Remove the power steering tubes from crossmember.
12. Lower the crossmember.
13. Remove the oil pan bolts, in sequence.
14. Using the oil pan removal tool KV10111100 or equivalent, separate the oil pan from the engine.

—————— WARNING ——————
Do not drive the seal cutter into the oil pump or rear oil seal retainer portion, for the aluminum mating surfaces will be damaged. Do not use a prybar or screwdriver, for the oil pan flange will be deformed.
———————————————————

15. Remove the oil pan.
16. Clean all the sealing surfaces.
   To install:
17. Apply sealant to the oil pump gasket and the rear oil seal retainer.
18. Install new oil pan gasket. The use a continuous 0.138–0.177 inch (3.5–4.5mm) bead of liquid gasket. If using a liquid gasket, allow the gasket to cure for five minutes before assembly.
19. Install the oil pan and torque the bolts in sequence as follows:
   a. Tighten the M6 bolts to 55–73 inch lbs. (6.3–8.3 Nm).

   b. Tighten the M8 bolts to 12–15 ft. lbs. (16–21 Nm).
20. Raise the center crossmember. Torque the center crossmember fixing nuts/bolts to 29–36 ft. lbs. (39–49 Nm).
21. Reconnect the A/C lines and the power steering tubes to the crossmember.
22. Reconnect the power steering bracket to the suspension crossmember.
23. Lower the engine and connect the motor mounts.
24. Using a new washer, install the oil pan drain plug and tighten the plug to 22–29 ft. lbs. (29–39 Nm).
25. Install the tension rod bolts on both sides and tighten the mounting through bolt to 80–100 ft. lbs. (108–135 Nm).
26. Connect the tension rod to the lower control arm and tighten mounting nuts as follows:
   a. Convertible models torque to 69–83 ft. lbs. (93–113 Nm).
   b. Non-convertible models torque to 80–94 ft. lbs. (108–127 Nm).
27. Connect the stabilizer links and tighten the mounting nuts to 41–47 ft. lbs. (56–64 Nm).
28. Reconnect the steering column lower joint and tighten the lower pinch bolt to 17–22 ft. lbs. (24–29 Nm).
29. Using a new gasket, install the oil filter and oil filter housing assembly. Tighten the oil filter housing bolts to 12–15 ft. lbs. (16–21 Nm).
30. Install the engine undercover.
31. Lower the vehicle.
32. Refill the oil pan with the specified quantity of clean oil.

   NOTE: If a liquid gasket was used, wait at least 30 minutes for material to cure before refilling engine with oil.

33. Start the engine and check for leaks.

Oil Pump

REMOVAL AND INSTALLATION

1.6L (GA16DE) Engine

—————— CAUTION ——————
Fuel injection systems remain under pressure after the engine has been turned OFF. Properly relieve fuel pressure before disconnecting any fuel lines. Failure to do so may result in fire or personal injury.
———————————————————

1. Disconnect the negative battery cable.

2. Raise and safely support the vehicle.

3. Remove the engine undercovers and drain the oil.

4. Drain the cooling system.

5. Remove the engine center member and the front exhaust pipe.

6. Remove the oil pan and the oil strainer.

7. Lower the vehicle and remove the drive belts.

8. Remove the cylinder head and the upper timing chain.

9. Support the engine assembly with a jack and remove the front engine mount.

10. Remove the crankshaft pulley.

11. Remove the front cover mounting bolts, the front cover, and the oil pump assembly from the inside of the front cover.

To install:

12. Coat the oil pump gears with oil. Install the oil pump to the front cover. Torque the oil pump mounting bolts to 56–73 inch lbs. (7–8 Nm). Torque the oil pump mounting screws to 33–44 inch lbs. (4–5 Nm).

13. Clean the mating surfaces of liquid gasket and apply a fresh bead of 1/8 in. (3mm) thickness.

14. Using a new oil seal and O-ring, install the front cover assembly.

15. Install the oil strainer and the oil pan.

16. The balance of installation is the reverse of the removal procedure.

17. Prime the oil pump prior to initial engine start-up. The oil pump can be primed by disabling the ignition system and cranking the engine for a few seconds.

18. Start the engine, check for leaks, and verify that the oil pressure is within specifications.

19. Bleed the cooling system, check and/or adjust the ignition timing, and road test the vehicle.

2.0L (SR20DE) Engine

1. Disconnect the negative battery cable.

2. Drain the engine oil and coolant.

3. Remove the cylinder head assembly.

4. Raise and support the vehicle safely.

5. Remove the oil pan, oil strainer, and the baffle plate.

6. Remove the crankshaft pulley using a suitable puller.

7. Support the engine and remove the front engine mount.

8. Loosen the front cover bolts in 2 or 3 steps and remove the front cover.

9. Remove the oil pump mounting hardware, from the front cover.

To install:

10. Install the outer gear into the cavity. Make sure the gears mesh properly and prime the oil pump with clean engine oil.

11. Install the oil pump cover. Torque the retaining screws to 33–44 inch lbs. (4–5 Nm) and the bolts to 56–66 inch lbs. (7 Nm).

12. Clean all mating surfaces of liquid gasket material.

13. Apply a continuous bead of liquid gasket to the mating surface of the timing cover. Install the oil pump drive spacer and front cover. Tighten front cover bolts (in steps) to 5–6 ft. lbs. (6–8 Nm). Wipe excess liquid gasket material.

14. Install front engine mount.

15. Install crankshaft pulley and tighten bolt to 105–112 ft. lbs. (142–152 Nm).

16. Install the oil strainer, baffle, and the oil pan.

17. Install the cylinder head assembly.

18. Lower the vehicle and connect the negative battery cable. Refill the fluid levels, start the engine, and check for leaks. Remember to bleed the cooling system. Verify correct oil pressure and road test the vehicle for proper operation.

2.4L (KA24DE) Engine

1. Disconnect the negative battery cable.

2. Drain the engine oil and coolant.

3. Remove the drive belts.

4. Remove the alternator and mounting bracket.

5. Remove the water pump assembly.

6. Remove the breather separator assembly from the timing cover.

7. Remove the oil pan, oil strainer, and the baffle plate.

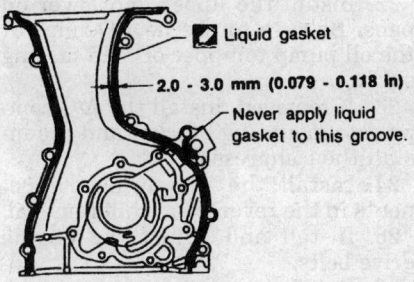

301028

Applying sealant to the front cover — 2.0L (SR20DE) engine

8. Remove the crankshaft pulley using a suitable puller.

9. Support the engine and remove the front engine mount.

10. Remove the valve cover assembly.

11. Remove the camshaft sprocket cover.

12. Remove the front timing cover assembly (oil pump assembly is mounted in the front cover).

13. Remove the oil pump cover and remove the gears from the front cover housing.

To install:

NOTE: When installing the timing cover, be sure to replace the O-ring seal for the oil hole in the front cover.

14. Install the outer gear into the cavity in the front cover housing. Make sure the gears mesh properly and prime the oil pump with clean engine oil.

15. Install the oil pump cover. Torque the retaining screws to 33–44 inch lbs. (3.7–5.0 Nm) and torque the bolts to 12–15 inch lbs. (16–21 Nm).

16. Clean the mating surfaces of liquid gasket and apply a continuous bead of 0.079–0.118 in. (3mm) thickness of liquid sealer to the mating surfaces of the timing cover.

17. Install the timing cover assembly.

18. Torque the M6 mounting bolts to 56–66 inch lbs. (6.4–7.5 Nm) and torque the M8 mounting bolts to 12–14 ft. lbs. (15–19 Nm).

19. Clean the mating surfaces of liquid gasket and apply a continuous bead of 0.079–0.118 in. (3mm) thickness of liquid sealer to the mating surfaces of the camshaft sprocket cover.

20. Install the camshaft sprocket cover.

21. Torque the M6 mounting bolts to 56–66 inch lbs. (6.4–7.5 Nm) and torque the M8 mounting bolts to 12–14 ft. lbs. (15–19 Nm).

22. Install the front engine mount.

23. Install crankshaft pulley and tighten bolt to 105–112 ft. lbs. (142–152 Nm).

24. The remaining components are installed in the reverse order from which they were removed.

25. Refill the engine with oil and coolant. Start the engine and check for leaks.

26. Remember to bleed the cooling system. Verify correct oil pressure and road test the vehicle for proper operation.

1993–94 3.0L (VE30DE, VG30DE, VG30E and VG30DETT) Engines

The oil pump is an integral part of the front cover.

1. Remove all accessory drive belts and the alternator.
2. Remove the cylinder heads.
3. Unbolt the engine from its mounts and raise the engine up from the unibody.
4. Remove the crankshaft damper-to-crankshaft bolt and the damper.
5. Drain the engine oil and remove the oil pan.
6. Remove the oil strainer-to-engine bolt, oil strainer-to-oil pump bolts and the strainer.
7. Remove the front cover-to-engine bolts and the front cover.
8. Remove the oil pump assembly-to-engine bolts, along with the oil strainer, and remove the assembly from the engine.
9. Clean the gasket mating surfaces.

To install:

10. Pack the oil pump full of petroleum jelly to prevent the pump from cavitating when the engine is started.
11. Apply a bead of liquid sealant to the front cover mating surfaces.
12. Install the front cover and torque the front cover-to-engine bolts to 4.6–6.1 ft. lbs. (6.3–8.3 Nm) except for bolt above oil filter housing and 12–15 ft. lbs. (16–21 Nm) for bolt above oil filter housing.
13. Using a new gasket, install the strainer and torque the strainer-to-front cover bolts to 12–15 ft. lbs. (16–21 Nm) and the strainer-to-engine bolt to 4.6–6.1 ft. lbs. (6.3–8.3 Nm).
14. Install the oil pan.
15. Install the cylinder heads.
16. Install the alternator and all drive belts. Reconnect the negative battery cable.
17. Start engine, check ignition timing and check for oil leaks.

3.0L (VQ30DE) Engine

NOTE: The oil pump bolts to the front of the engine block and is driven by the crankshaft. Removal of the timing cover and chains are necessary for oil pump service.

1. Disconnect the negative battery cable and drain the engine oil.
2. Rotate the engine and position it to TDC compression stroke of cylinder No. 1.
3. Remove the drive belts.

4. Remove the camshaft position sensor (PHASE) and the crankshaft position sensor (REF/POS).
5. Remove the right front wheel and inner fender cover.
6. Remove the engine undercovers.
7. Remove the bolt that secures the crankshaft pulley and remove the pulley.
8. Remove the front exhaust pipe and its support.
9. Support the engine at the left and right side slingers with a suitable hoist.
10. Remove the engine right side mounting insulator and bracket nuts and bolts.
11. Remove the center crossmember assembly.
12. If equipped, remove the A/C compressor and the mounting bracket.
13. Remove the lower and upper oil pans.
14. Remove the oil strainer from the oil pump.
15. Remove the water pump cover and remove the front cover assembly.
16. Remove the lower timing chain assembly.
17. Remove the bolts that secure the oil pump to the engine block and remove the oil pump.

To install:

NOTE: When installing the oil pump, be sure to apply engine oil to the gears.

18. Install the oil pump to the engine block. Tighten the mounting bolts to 57 inch lbs. (6.5 Nm) and tighten the mounting screws to 33–44 inch lbs. (4–5 Nm).
19. Install the lower timing chain assembly.
20. Install the front timing cover and water pump covers.
21. Using a new gasket, install the oil strainer and tighten the mounting bolts to 12–14 ft. lbs. (16–19 Nm).
22. Install the upper and lower oil pans. Be sure to use new O-rings at the oil pump to upper oil pan mating surface.
23. If removed, install the A/C compressor mounting bracket and the install the compressor.
24. Install the remaining components in the reverse order of removal.
25. Install and adjust the engine drive belts.
26. Refill the engine oil with the proper type and amount.
27. Start the engine, check the oil pressure, and check for oil leaks.

1995–97 3.0L (VG30DE and VG30DETT) Engine

1. Disconnect the negative battery cable.
2. Position the engine to TDC of compression stroke.
3. Remove the engine undercover.
4. Remove all accessory drive belts and remove the alternator from the vehicle.
5. Remove the crankshaft damper bolt and remove the damper.
6. Remove the timing belt covers and remove the timing belt.
7. Remove the crankshaft sprocket and the timing belt plates.
8. Drain the engine oil and remove the oil pan.
9. Remove the oil strainer (pickup) support bracket bolts and remove the bolts that secure to strainer to the oil pump.
10. Remove the strainer from the oil pump and discard the O-ring.
11. Remove the oil pump bolts and remove the pump assembly from the engine.
12. Clean the gasket mating surfaces.

To install:

NOTE: Lubricate the oil pump prior to installation and prime the oil pump before initial engine start-up.

13. Using a new gaskets, install the oil pump and torque the oil pump mounting bolts to 55–73 inch lbs. (6.3–8.3 Nm) except for the two bolts closest to the oil pan. Torque the two longer bolts closest to the oil pan to 16–22 ft. lbs. (16–21 Nm).
14. Using a new O-ring, install the strainer and torque the strainer-to-front cover (oil pump) bolts to 12–15 ft. lbs. (16–21 Nm) and the strainer-to-engine bolt to 55–73 inch lbs. (6.3–8.3 Nm).
15. Using liquid gasket, install the oil pan and tighten the mounting bolts to 22–29 ft. lbs. (29–39 Nm).
16. Install the crankshaft sprocket and the timing belt plates.
17. Install the timing belt and the timing belt covers.
18. Install the crankshaft damper and tighten the mounting bolt to 159–174 ft. lbs. (216–235 Nm).
19. Install the remaining components in the opposite order from which they were removed.
20. After verifying there is oil pressure, enable the ignition system.
21. Start the engine, check the ignition timing, and check for oil leaks. Verify proper oil pressure.

TRANSMISSION AND TRANSAXLE

Manual Transmission Assembly

REMOVAL AND INSTALLATION

240SX

1. Disconnect the negative battery cable.

2. Raise and support the vehicle safely.

3. Place the shift lever in the **N** position and disconnect the shifter lever from the transmission.

4. Remove the shifter lever with control housing from the transmission.

5. Remove the crankshaft position sensor from the upper side of the transmission housing.

6. Remove the clutch operating (slave cylinder) cylinder from the clutch housing.

7. Disconnect the electrical harness connections from the transmission and disconnect the rear heated oxygen sensor connection.

8. Matchmark then unbolt the driveshaft at the rear and remove. If equipped with a center bearing, unbolt it from the crossmember. Plug the end of the transmission extension to prevent leakage.

9. Support the engine with a large wood block and a jack under the oil pan. Do not place the jack under the oil pan drain plug.

10. Unbolt the transmission from the crossmember. Support the transmission with a jack and remove the crossmember.

11. Lower the rear of the engine to allow clearance.

12. Disconnect the wiring to the starter motor and remove the starter from the transaxle.

13. Remove the exhaust tube mounting bracket from the transmission.

14. Unbolt the transmission assembly from the engine. Lower and move it to the rear.

NOTE: Tagging the different length transmission bolts upon removal is necessary to ensure that they are installed in their original position.

To install:

15. Clean the engine and transmission mating surfaces. Lightly lubri-cate the clutch disc splines and main drive gear splines. Also lubricate the control lever sliding surfaces with grease.

16. Properly support the transmission and install the transmission to the rear of the engine.

17. Use the following torque specifications to bolt the transmission to the engine:

    a. Torque bolts No. 1 and 2 to 29–36 ft. lbs. (39–49 Nm).

    b. Torque bolts No. 3, 4 and 5 to 22–29 ft. lbs. (29–39 Nm).

18. Install the exhaust tube mounting bracket to the transmission.

19. Install the starter assembly.

20. Raise the rear of the engine to its original position.

21. Bolt the crossmember in place and remove the jack.

22. Install the remaining components in the reverse order of removal.

23. Connect the negative battery cable and check the transmission fluid level.

24. Lower the vehicle and road test the vehicle for proper operation.

300ZX

1. Disconnect the negative battery cable.

2. Raise and support the vehicle safely.

3. Remove the exhaust pipe section from the manifold and remove the support bracket from the transmission.

4. Matchmark the driveshaft and unbolt the shaft from the rear housing. Unbolt the driveshaft at the center bearing and remove the shaft from the rear of the transmission. Install a sealing plug in the end of the transmission extension housing to prevent leakage.

5. Disconnect the speedometer drive cable from the transmission.

6. Disconnect the shifter lever.

7. Without disconnecting the hydraulic line, remove the clutch oper-ating cylinder from the clutch housing and position it aside.

8. Support the engine with a large wood block and a jack under the oil pan. Do not place the jack under the oil pan drain plug.

9. Unbolt the transmission from the crossmember. Support the transmission with a jack and remove the crossmember.

10. Lower the rear of the engine to allow clearance.

11. Unplug the backup light, neutral and overdrive switch connectors.

12. Unbolt the transmission. Lower and remove it to the rear.

NOTE: The transmission bolts are different lengths. Tagging the transmission-to-engine bolts upon removal will facilitate proper positioning during installation.

To install:

13. Lubricate the output shaft of the transmission with high temperature grease.

14. Raise the transmission onto the engine and install the mounting bolts. Tighten bolts to correct torque specification.

NOTE: The bolts that secure the transmission to the engine are of different length and require different torque. Be sure to properly position all bolts.

15. Connect the backup light, and the neutral and overdrive switch connectors.

16. Install the crossmember.

17. Install the clutch slave cylinder and tighten the mounting bolts to 22–30 ft. lbs. (30–40 Nm).

18. Install the shifter lever, shift knob and console boot finisher or control rod.

19. Connect the speedometer drive cable.

20. On 1993–94 models, install the driveshaft. Torque the flange bolts to 29–33 ft. lbs. (34–44 Nm). Torque the

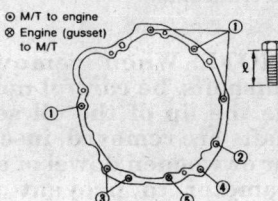

Bolt No.	Tightening torque N·m (kg-m, ft-lb)
①	39 - 49 (4.0 - 5.0, 29 - 36)
②	39 - 49 (4.0 - 5.0, 29 - 36)
③*	29 - 39 (3.0 - 4.0, 22 - 29)
④*	29 - 39 (3.0 - 4.0, 22 - 29)
⑤	29 - 39 (3.0 - 4.0, 22 - 29)
Gusset to engine	29 - 39 (3.0 - 4.0, 22 - 29)

*: With nut.

306521

Transmission bolt tightening specifications — 240SX

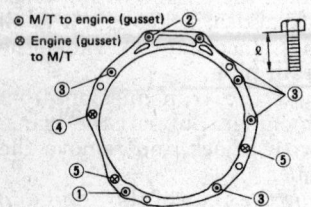

Bolt No.	Tightening torque N·m (kg-m, ft-lb)	ℓ mm (in)
①	39 - 49 (4.0 - 5.0, 29 - 36)	100 (3.94)
②	39 - 49 (4.0 - 5.0, 29 - 36)	65 (2.56)
③	39 - 49 (4.0 - 5.0, 29 - 36)	60 (2.36)
④	29 - 39 (3.0 - 4.0, 22 - 29)	55 (2.17)
⑤	29 - 39 (3.0 - 4.0, 22 - 29)	25 (0.98)

132612

Manual transmission mounting bolt identification and tightening torque — 300ZX without turbocharger

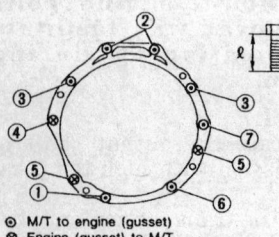

● Tighten all transmission bolts.

Bolt No.	Tightening torque N·m (kg-m, ft-lb)	ℓ mm (in)
①	39 - 49 (4.0 - 5.0, 29 - 36)	100 (3.94)
②	39 - 49 (4.0 - 5.0, 29 - 36)	55 (2.17)
③	39 - 49 (4.0 - 5.0, 29 - 36)	60 (2.36)
④	29 - 39 (3.0 - 4.0, 22 - 29)	55 (2.17)
⑤	29 - 39 (3.0 - 4.0, 22 - 29)	25 (0.98)
⑥	29 - 39 (3.0 - 4.0, 22 - 29)	60 (2.36)
⑦	39 - 49 (4.0 - 5.0, 29 - 36)	65 (2.56)

303698

Manual transmission mounting bolt identification and tightening torque — 300ZX with turbocharger

center bearing bracket nuts to 19–29 ft. lbs. (25–39 Nm).

21. On 1995–97 models, install the driveshaft. Torque the flange bolts to 41–48 ft. lbs. (55–65 Nm) for non-turbo models or 47–54 ft. lbs. (64–74 Nm) for turbo models. Torque the center bearing bracket mounting bolts to 43–58 ft. lbs. (59–78 Nm).

22. Connect the exhaust tube section to the manifolds and attach the support bracket to the transmission.

23. Remove the jack from underneath the engine.

24. Lower the vehicle and connect the negative battery cable. Road test the vehicle for proper operation.

Transaxle Assembly

REMOVAL AND INSTALLATION

Sentra and 200SX

1. Disconnect the negative and positive battery cables.
2. Remove the battery and battery bracket.
3. Remove the air duct, air cleaner box and the air flow meter.
4. Raise the front of the vehicle and support it safely.

5. Remove the clutch control cable from the operating lever.
6. Disconnect the speedometer cable from the transaxle.
7. Disconnect the wires from the reverse (back-up), neutral and ground harness connectors.
8. Disconnect the vehicle speed sensor and the crankshaft position sensor from the transaxle.
9. Remove the starter motor from the transaxle.
10. Remove the through bolts that secure shift control rod and the support rod from the transaxle. Disconnect the rods from the transaxle.
11. Drain the transaxle oil.
12. Remove the exhaust front tube.
13. Withdraw the halfshafts from the transaxle.

NOTE: When removing the halfshafts, be careful not to damage the lip of the oil seal. After shafts are removed, insert a steel bar or wooden dowel of a suitable diameter to prevent the side gears from rotating and falling into the differential case.

14. Support the engine by placing a jack under the oil pan, with a wooden block placed between the jack and pan for protection.

15. Support the transaxle with a jack.
16. Remove the rear and left side engine mounts.

NOTE: Most of the transaxle mounting bolts are different lengths. Tagging the bolts upon removal will facilitate proper tightening during installation.

17. Remove the bolts attaching the transaxle to the engine.
18. Slide the transaxle away from the engine and carefully lower the transaxle down and away from the engine.

To install:
19. Before installing, clean the mating surfaces on the engine rear plate and clutch housing.
20. Apply a light coat of a lithium-based grease to the spline parts of the clutch disc and the transaxle input shaft.
21. Raise the transaxle into place and bolt it to the engine. Install the engine mounts. Torque the transaxle mounting bolts to specifications.
22. Install the halfshafts.
23. The balance of installation is the reverse of the removal procedure.
24. Connect the battery cables.
25. Remove the filler plug and fill the transaxle to the proper level with fluid that meets API GL–4 specifications. Fill to the level of the plug hole. Apply a thread sealant to the threads of the filler plug and install the plug in the transaxle case.

Altima

1. Remove the battery and the battery bracket.
2. Remove the air cleaner housing with mass air flow sensor.
3. Remove the air duct.
4. Raise and safely support the vehicle so there is clearance to remove the transaxle from underneath. Securely support the engine with a jack. Place a wooden block between the jack and the oil pan. This will protect the oil pan from being damaged.
5. Remove the clutch release cylinder; do not disconnect the hydraulic line from the cylinder.
6. Disconnect the back-up lamp switch harness connectors.
7. Disconnect the speedometer cable, inhibitor switch, and the ground connections.
8. Remove the starter motor assembly.
9. Remove the crankshaft position sensor from the transaxle.
10. Disconnect the shift linkage.
11. Drain the fluid from the transaxle.

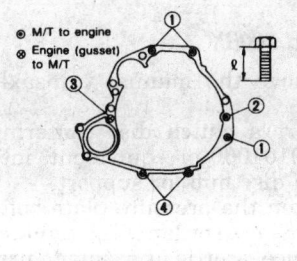

⊛ M/T to engine
⊗ Engine (gusset) to M/T

● GA engine models

Bolt No.	Tightening torque N·m (kg-m, ft-lb)	"ℓ" mm (in)
1	30 - 40 (3.1 - 4.1, 22 - 30)	70 (2.76)
2	30 - 40 (3.1 - 4.1, 22 - 30)	85 (3.35)
3	30 - 40 (3.1 - 4.1, 22 - 30)	30 (1.18)
4	16 - 21 (1.6 - 2.1, 12 - 15)	25 (0.98)
Front gusset to engine	30 - 40 (3.1 - 4.1, 22 - 30)	20 (0.79)
Rear gusset to engine	16 - 21 (1.6 - 2.1, 12 - 15)	16 (0.63)

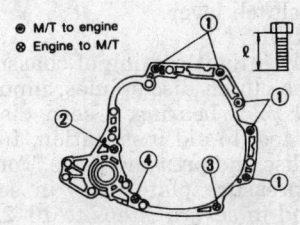

⊛ M/T to engine
⊗ Engine to M/T

● SR engine models

Bolt No.	Tightening torque N·m (kg-m, ft-lb)	"ℓ" mm (in)
1	70 - 79 (7.1 - 8.1, 51 - 59)	55 (2.17)
2	70 - 79 (7.1 - 8.1, 51 - 59)	65 (2.56)
3	30 - 40 (3.1 - 4.1, 22 - 30)	35 (1.38)
4	30 - 40 (3.1 - 4.1, 22 - 30)	45 (1.77)

302616

Transaxle bolt torque specifications — 1994–97 Sentra and 200SX

12. Remove both halfshafts from the transaxle assembly. Securely support the transaxle with another jack.

13. Remove the rear and left side mounts.

14. Raise the jack and remove the lower housing bolts. Lower the jack, but continue to support the transaxle and engine.

15. Remove the transaxle mounting bolts. Remove the transaxle from the vehicle by sliding the transaxle input shaft out of the clutch, lowering the rear of the transaxle and then lowering the transaxle from of the vehicle.

To install:

16. Install the transaxle in the correct position. Torque the transaxle bolts to specification.

17. Securely support the transaxle. Install the lower housing bolts and the halfshafts.

18. The remaining components are installed in the reverse order from which they were removed.

19. Install the battery and the battery bracket.

20. Road test the vehicle and verify proper shift and clutch operation.

1993–94 Maxima

1. Remove the battery and battery bracket.

2. Remove the air cleaner and the air flow meter.

3. Raise and safely support the vehicle so there is clearance to remove the transaxle from underneath. Securely support the engine via the oil pan using a cushioning wooden block and a floor jack.

4. Drain the fluid from the transaxle.

5. Remove both driveshafts from the transaxle assembly. Securely support the transaxle with another jack.

6. Remove the clutch operating cylinder; do not disconnect the hydraulic line from the cylinder.

7. Disconnect the speedometer pinion and the position switch connectors..

8. Disconnect the back-up lamp switch, the neutral position switch and the ground harness connectors.

9. Remove the starter motor assembly.

10. Disconnect the shift linkage.

11. Remove the rear and the LH engine mounts.

12. Remove the transaxle bolts. Remove the transaxle from the vehicle by sliding the transaxle input shaft out of the clutch, lowering the rear of the transaxle and then lowering transaxle from of the vehicle.

To install:

13. Install the transaxle in the correct position. Torque the transaxle bolts to specification in the correct sequence.

14. Install the rear engine mount and torque the through bolt to 65–72 ft. lbs. 88–98 Nm). Install the LH engine mount and torque the through bolt to 47–58 ft. lbs. (64–78 Nm).

15. Install the remaining components in the opposite order from which they were removed.

16. Install the battery and the battery bracket.

17. Road test the vehicle for proper shift operation.

1995–97 Maxima

1. Disconnect the negative and the positive battery cables.

2. Remove the battery and battery bracket.

3. Remove the air cleaner assembly with the mass air flow sensor.

4. Raise and safely support the vehicle so there is clearance to remove the transaxle from underneath. Securely support the engine via the oil pan using a cushioning wooden block and a floor jack.

5. Remove the clutch operating cylinder; do not disconnect the hydraulic line from the cylinder.

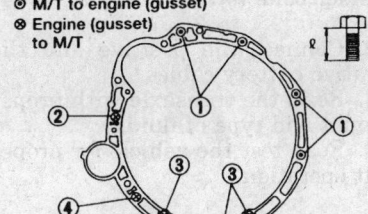

⊛ M/T to engine (gusset)
⊗ Engine (gusset) to M/T

Bolt No.	Tightening torque N·m (kg-m, ft-lb)
1	39 - 49 (4.0 - 5.0, 29 - 36)
2	39 - 49 (4.0 - 5.0, 29 - 36)
3	30 - 40 (3.1 - 4.1, 22 - 30)
4	30 - 40 (3.1 - 4.1, 22 - 30)

294978

Transaxle bolt tightening specifications — Altima

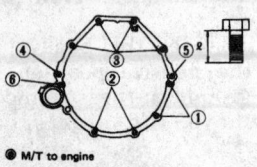

Bolt No.	Tightening torque N·m (kg-m, ft-lb)	ℓ mm (in)
1	16 - 21 (1.6 - 2.1, 12 - 15)	25 (0.98)
2	30 - 40 (3.1 - 4.1, 22 - 30)	28 (1.10)
3	39 - 49 (4.0 - 5.0, 29 - 36)	57 (2.24)
4	39 - 49 (4.0 - 5.0, 29 - 36)	57 (2.24)
5	39 - 49 (4.0 - 5.0, 29 - 36)	64 (2.52)
6	30 - 40 (3.1 - 4.1, 22 - 30)	25 (0.98)
Front gusset to engine	30 - 40 (3.1 - 4.1, 22 - 30)	25 (0.98)
Rear gusset to engine	30 - 40 (3.1 - 4.1, 22 - 30)	25 (0.98)

● M/T to engine
⊗ Engine (gusset) to M/T

205706

Transaxle bolt tightening specifications — 1993–94 Maxima

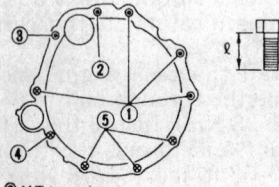

Bolt No.	Tightening torque N·m (kg-m, ft-lb)	"ℓ" mm (in)
①	70 - 79 (7.1 - 8.1, 51 - 59)	52 (2.05)
②	70 - 79 (7.1 - 8.1, 51 - 59)	65 (2.56)
③	70 - 79 (7.1 - 8.1, 51 - 59)	124 (4.88)
④	35.1 - 47.1 (3.58 - 4.80, 25.89 - 34.74)	40 (1.57)
⑤	35.1 - 47.1 (3.58 - 4.80, 25.89 - 34.74)	40 (1.57)

● M/T to engine
⊗ Engine to M/T

③ with starter
④ with support rod bracket

289089

Transaxle bolt tightening specifications — 1995–97 Maxima

6. Remove the clutch hose clamp.

7. Disconnect the speedometer pinion and the neutral position switch connectors and the ground harness connectors.

8. Remove the starter motor assembly from the transaxle.

9. Disconnect the back-up lamp switch and the neutral position switch.

10. Remove the crankshaft position sensor (POS) from the transaxle front side.

11. Remove the shifter control rod and the support rod bracket from the transaxle.

12. Drain the fluid from the transaxle.

13. Remove both driveshafts from the transaxle assembly. Securely support the transaxle with another jack.

14. Support the engine of the transaxle by placing a jack under the oil pan. Be sure to use a block of wood between the oil pan and jack.

15. Remove the bolts that secure the center crossmember.

16. Remove the LH engine mounts.

NOTE: The transaxle bolts are of different lengths, be sure to note the location of the bolts for reassembly.

17. Remove the transaxle bolts. Remove the transaxle from the vehicle by sliding the transaxle input shaft out of the clutch, lowering the rear of the transaxle and then lowering the transaxle from of the vehicle.

To install:

18. Install the transaxle assembly to the bell housing while aligning the output shaft of the transaxle with the clutch disc. Torque the transaxle bolts to specifications.

19. Install the LH engine mount and torque the through bolt to 32–41 ft. lbs. (43–55 Nm), then install the center crossmember assembly and tighten the mounting bolts to 57–72 ft. lbs. (77–98 Nm).

20. Install the remaining components in the reverse order of removal.

21. Tighten the clutch operating cylinder bolts to 22–30 ft. lbs. (30–40 Nm).

22. Connect the positive and the negative battery cables.

23. Refill the transaxle with proper amount and type of fluid.

24. Road test the vehicle for proper shift operation.

Clutch Assembly

REMOVAL AND INSTALLATION

Sentra and 200SX

1. Remove the manual transaxle assembly.

2. Insert a clutch disc centering tool KV30101000 or equivalent, into the clutch disc hub for support.

3. Loosen the pressure plate bolts evenly in reverse order of the tightening sequence, a little at a time to prevent distortion.

4. Remove the clutch assembly.

5. Remove the throw-out bearing from the clutch lever.

To install:

6. Apply a light coating of chassis lube to the clutch disc splines, input shaft and pilot bearing. Use a disc centering tool to aid installation. Install the disc and pressure plate. Torque the pressure plate bolts in sequence and in several steps to 16–22 ft. lbs. (20–26 Nm).

7. Install the new throw-out bearing in the clutch release lever. Remove the clutch disc centering tool.

8. Install the transaxle into the vehicle. If the mating surfaces will not come together, do not force the units together. Remove the transaxle and recheck that the disc is centered.

NOTE: DO NOT draw the transaxle to the engine with the bolts. This may damage the clutch and/or transaxle. Also, be careful not to move the throw-out bearing when installing the transaxle.

9. After the transaxle is installed, connect the clutch cable and check operation before complete reassembly.

10. Adjust the clutch pedal as necessary.

Altima

1. Disconnect the negative battery cable.

2. Raise and safely support the vehicle.

3. Remove the slave cylinder from the transaxle. It is not necessary to disconnect the fluid hose from the slave cylinder.

4. Remove transaxle assembly from the vehicle.

5. Insert clutch disc alignment tool into the clutch disc.

6. Remove bolts holding pressure plate to flywheel.

7. Remove pressure plate and clutch disc from the flywheel.

8. Inspect the flywheel for cracks, scoring, or signs of overheating (blue

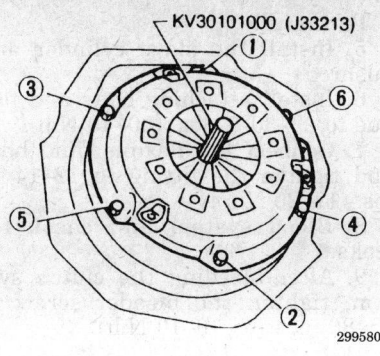

Pressure plate torque sequence — Sentra and 200SX

299580

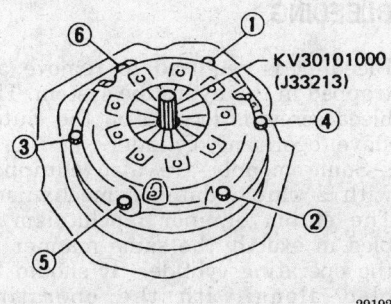

Pressure plate bolt torque sequence — Altima

291928

marks). If necessary, resurface the flywheel.

9. Install a dial indicator gauge to the surface of the flywheel. Zero the gauge and rotate the flywheel one complete revolution.

10. The flywheel run-out should be less than 0.0059 inches (0.15mm).

11. Inspect the throw-out bearing; the bearing should spin freely and be free from noise, pits, cracks or wear. Replace the bearing as necessary.

12. Inspect the pilot bearing in the end of the crankshaft. This bearing also be free from noise or wear. The bearing should be replaced as necessary.

NOTE: Factory replacement pilot bearings are pre-lubricated. Some aftermarket bearings may require lubrication before being installed.

13. Check the rear of the engine or front of the transaxle for fluid leaks. Replace leaking seals before installing the clutch assembly.

To install:

14. Lightly lubricate the throw-out bearing sleeve and the clutch lever pivot point.

15. Lubricate the input shaft and clutch disc splines with lubriplate®.

Slide the clutch disc back and forth over the input shaft several times. Wipe off excess grease and lubricant from the disc or splines. Do not get grease or lube on the flywheel, pressure plate, or the clutch disc.

16. Install clutch disc against the pressure plate.

NOTE: The clutch disc will have a marking for which side should be positioned against the flywheel. Be sure to properly install the disc during assembly.

17. Align the clutch disc with an alignment tool.

18. Hold pressure plate and clutch disc against flywheel and install bolts.

19. Tighten the bolts in sequence to 16–22 ft. lbs. (22–29 Nm) and remove the alignment tool.

20. Install the transaxle assembly and the slave cylinder.

21. Connect the negative battery cable.

22. Verify proper clutch operation and adjust the clutch pedal as necessary.

Maxima, 240SX and 300ZX

1. Remove the transaxle/transmission from the vehicle.

2. Insert a clutch aligning bar or similar tool all the way into the clutch disc hub. This must be done so as to support the weight of the clutch disc during removal. Mark the clutch assembly-to-flywheel relationship with paint or a center punch so the clutch assembly can be assembled in the same position from which it is removed.

3. Loosen the bolts in reverse order of tightening sequence, a turn at a time. Remove the bolts.

4. Remove the pressure plate and clutch disc.

5. Remove the release mechanism from the transaxle housing.

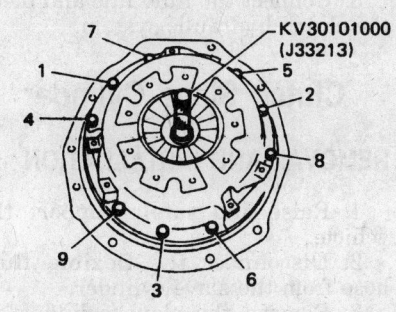

Pressure plate bolt tightening sequence — Maxima

296014

6. Inspect the pressure plate for wear, scoring, etc., and resurface or replace, as necessary.

7. Measure the thickness of the clutch plate lining to the rivet heads; if the it is worn to a minimum of 0.012 in. (0.3mm), replace the clutch plate.

8. Inspect the release bearing and replace as necessary.

9. Using a dial indicator, mount it to the engine and inspect the flywheel runout; if the runout exceeds 0.0059 in. (0.15mm), replace it.

To install:

10. Apply a small amount of grease to the transaxle input shaft splines. Install the disc on the splines and slide back and forth a few times. Remove the disc and remove excess grease on hub. Be sure no grease contacts the disc or pressure plate.

11. Apply lithium based molybdenum disulfide grease to the bearing sleeve inside groove, the contact point of the withdrawal lever and bearing sleeve, the contact surface of the lever ball pin and lever.

12. Install the release mechanism and release bearing.

13. Install the pressure plate and clutch disc, aligning it with a splined dummy shaft tool KV301010000 or equivalent.

14. Torque the pressure plate bolts in sequence to 16–22 ft. lbs. (22–29 Nm) on 240SX models or 25–33 ft. lbs. (34–44 Nm) on Maxima and 300ZX models.

15. Remove the dummy shaft.

16. Install the transaxle/transmission in the vehicle.

Clutch Cable

ADJUSTMENT

Sentra and 200SX

NOTE: When the clutch cable is replaced, depress the clutch pedal 50 times as a break in procedure before adjusting the free-play.

1. Loosen the adjusting knob at the transaxle as much that is needed to obtain a proper adjustment.

2. Adjust the pedal freeplay by turning the adjusting knob at the transaxle.

3. Measure the distance from the pedal resting position to the end of the free travel.

4. The free-play travel should be 0.42–0.59 inches (10.5–15mm).

5. Road test vehicle for proper operation.

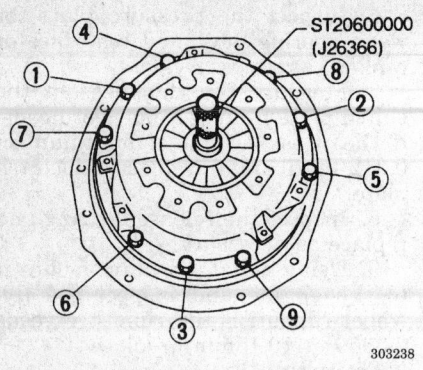

303238

Pressure plate bolt torque sequence — 300ZX

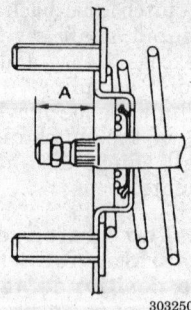

303250

Measuring the output rod length — 300ZX (VG30DETT) engine

Clutch Master Cylinder

REMOVAL AND INSTALLATION

Altima, Maxima and 240SX

1. Disconnect the clutch pedal arm from the pushrod.
2. Disconnect the clutch hydraulic line from the master cylinder.

—————— WARNING ——————
Take precautions to keep brake fluid from coming in contact with any painted surfaces.

3. Remove the nuts attaching the master cylinder and remove the master cylinder and pushrod toward the engine compartment side.
4. If equipped, remove the dust boot from the pushrod. The boot will not fit through the cowl without tearing.
To install:
5. Install the master cylinder assembly into the fire wall.
6. Install the attaching nuts and torque to 70–95 in. lbs. (8–11 Nm).
7. Connect the hydraulic line.
8. Connect clutch pedal.
9. Bleed system and check for leaks.
10. Adjust the clutch pedal free-play as follows:
   a. Loosen the locknut to the clutch master cylinder pushrod.
   b. Move the clutch pedal in a downward location and measure only the free-play.
   c. The free-play should be 0.04–0.12 inches (1–3mm).
   d. Rotate the clutch master rod to obtain the proper free-play.
   e. Tighten the locknut to 69–104 inch lbs. (8–12 Nm) and recheck the free-play.

300ZX

The non-turbocharged model uses a conventional release mechanism. The turbocharged engine model uses a clutch booster assembly similar a vacuum operated power brake booster.

—————— WARNING ——————
The clutch hydraulic system uses brake fluid which is extremely harmful to painted surfaces.

1. Disconnect the negative battery cable.
2. On VG30DE engines, disconnect the clutch pedal arm from the pushrod.
3. Disconnect the clutch hydraulic line from the master cylinder. Plug the end of line to prevent leakage.
4. Remove the nuts attaching the master cylinder to the firewall (VG30DE engines) or to the clutch booster (VG30DETT engines), then remove the master cylinder.
To install:
5. On the VG30DETT model, measure the length of the output rod from the face of the booster to the end of the output rod. The length of the rod should be 0.5256–0.5354 inches (13.35–13.60mm).
6. Install the master cylinder and tighten mounting nuts to 6–8 ft. lbs. (8–11 Nm).
7. For VG30DE engines, connect the pedal arm to the pushrod, then secure it with a snap-pin.
8. Connect the fluid line and bleed the clutch hydraulic system.

Clutch Slave Cylinder

REMOVAL AND INSTALLATION

1. Raise and safely support the vehicle.
2. Disconnect the flexible fluid hose from the slave cylinder.
3. Remove the slave cylinder attaching bolts.
4. Remove slave cylinder and pushrod.

To install:
5. Install the slave cylinder and pushrod.
6. Install attaching bolts and torque to 22–30 ft. lbs. (30–40 Nm).
7. Connect the flexible fluid hose and tighten the fitting to 12–14 ft. lbs. (17–20 Nm).
8. Bleed system and check for leaks.
9. After bleeding the clutch system, tighten the bleeder screw to 52–86 inch lbs. (6–10 Nm).

Hydraulic Clutch System

BLEEDING

Bleeding is required to remove air trapped in the hydraulic system. The bleed screw is located on the clutch slave (operating) cylinder.
Some models are also equipped with a clutch damper mechanism. The clutch damper mechanism is bled in exactly the same manner as the operating cylinder. It should be bled along with the operating cylinder.
1. Remove the bleed screw dust cap.
2. Attach a transparent vinyl tube to the bleed screw, immersing the free end in a clean container of clean brake fluid.
3. Fill the master cylinder with the proper fluid.
4. Open the bleed screw about ¾ turn.
5. Depress the clutch pedal quickly. Hold it down. Have an assistant tighten the bleed screw. Allow the pedal to return slowly.
6. Repeat the above procedure until no more air bubbles are seen in the fluid container.
7. Remove the bleed tube.
8. Replace the dust cap and refill the master cylinder.
9. Bleed the clutch damper, if equipped.

Automatic Transmission Assembly

REMOVAL AND INSTALLATION

240SX

1. Disconnect the negative battery terminal.
2. Raise and safely support the vehicle.
3. Drain the fluid from the transmission.

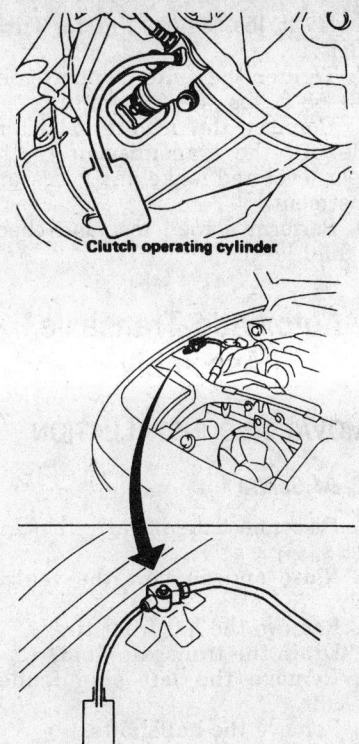

Clutch operating cylinder

291981

Clutch system bleeding points — Altima, Maxima, 240SX and 300ZX

4. Disconnect the cooler lines from the transmission and plug the openings.

NOTE: It may be necessary disconnect to the left side cooler line after the transmission has been slightly lowered.

5. Disconnect the electrical harness connections from the transmission.
6. Disconnect the shift linkage from the transmission.
7. Matchmark the driveshaft to the rear flange and disconnect the driveshaft. Unbolt the driveshaft center support and remove the shaft from the vehicle.

NOTE: Insert a plug into the rear of the transmission to prevent oil leakage.

8. Remove the heat insulator from the catalytic converter.
9. Remove the exhaust tube bracket and separate rear exhaust tube from the catalytic converter.
10. Disconnect the wiring connections and remove the starter motor assembly.
11. Remove the transmission support brackets and remove the torque converter cover plate.
12. Mark the relationship between the torque converter and the drive

plate. Remove the bolts that secure the torque converter to the drive plate. Gain access to the bolts by turning the front crankshaft bolt.
13. Support the transmission with a jack under the transmission oil pan. Also, place a jack, with a piece of wood, under the oil pan to support the engine.
14. Remove the rear crossmember.
15. Unbolt the transmission from the engine and remove it. Be careful not to damage the steering gear and the tubes.

NOTE: The transmission bolts are different lengths. Tag each bolt according to location to ensure proper installation.

To install:

16. Using a dial indicator, measure the drive plate runout. The maximum allowable runout should be less than 0.020 inches (0.5mm).
17. If the torque converter was removed from the transmission for any reason, after it is installed, the distance from the face of the converter bolt mounts to the edge of the converter housing must be checked prior to installing the transmission. This is done to ensure proper installation of the torque converter. The dimension should be 1.024 inches (26mm) or more.
18. Raise the transmission and bolt the transmission to the engine. Torque the transmission mounting bolts as follows:
   a. Tighten bolts No. 1 and No. 2 to 29–36 ft. lbs. (39–49 Nm)
   b. Tighten bolt No. 3 to 22–29 ft. lbs. (29–39 Nm)
   c. Tighten the gusset mounting bolts to 22–29 ft. lbs. (29–39 Nm).
19. Torque the torque converter mounting bolts to 33–43 ft. lbs. (44–59 Nm).

NOTE: After the converter is installed, rotate the crankshaft several times to make sure the transmission rotates freely and does not bind.

20. Install the remaining components in the reverse order of removal.
21. Connect the negative battery cable, fill the transmission to the proper level, and make any necessary adjustments.
22. Perform a road test and check the fluid level.

300ZX

1. Disconnect the negative battery cable.
2. Remove the accelerator linkage.
3. Detach the shift linkage.

4. Disconnect the neutral safety switch and downshift solenoid wiring.
5. Remove the crankshaft position sensor.
6. Raise and safely support the vehicle.
7. Remove the drain plug and drain the torque converter. If there is no converter drain plug, remove the transmission drain plug located in the pan.
8. Remove the front exhaust pipe.
9. Remove the vacuum tube and speedometer cable.
10. Disconnect the fluid cooler and dipstick tubes. Plug the tube ends to prevent leakage.
11. Matchmark the driveshaft to the companion flange at the center support bearing and unbolt the center support bearing from the vehicle.

NOTE: The rear section of the driveshaft will remain in the vehicle. Be sure to support the shaft with wire. DO NOT allow the driveshaft to hang by the rear joint.

12. Remove the driveshaft from the rear of the transmission and install a plug to the rear of the transmission to prevent fluid leakage.
13. Disconnect the wiring from the starter motor and remove the starter. Be sure to note the location of the starter wiring for reassembly.
14. Remove the bracket (gusset) that secures the engine to the transmission.
15. Support the transmission with a jack under the oil pan. Support the engine also.
16. Remove the rear crossmember.
17. Mark the relationship between the torque converter and the drive plate.
18. Remove the bolts holding the torque converter to the drive plate through the access hole. Rotate the crankshaft to gain access to the additional bolts.
19. Remove the transmission-to-engine mounting bolts and lower the transmission assembly.

NOTE: The transmission bolts are different lengths. Tag each bolt according to location to ensure proper installation.

20. Check the drive plate runout with a dial indicator. Maximum allowable runout is 0.0059 inch (0.15mm).

To install:

If the torque converter was removed from the transmission for any reason, after it is installed, the dis-

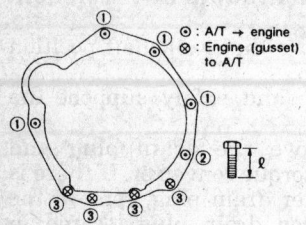

Bolt No.	Tightening torque N·m (kg-m, ft-lb)	Bolt length "ℓ" mm (in)
1	39 - 49 (4.0 - 5.0, 29 - 36)	40 (1.57)
2	39 - 49 (4.0 - 5.0, 29 - 36)	50 (1.97)
3	29 - 39 (3.0 - 4.0, 22 - 29)	25 (0.98)
Gusset to engine (4 bolts)	29 - 39 (3.0 - 4.0, 22 - 29)	20 (0.79)

306627

Transmission bolt torque sequence — 240SX

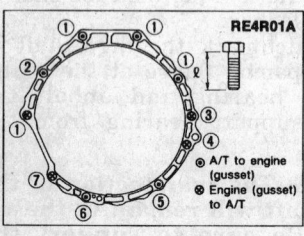

RE4R01A

Bolt No.	Tightening torque N·m (kg-m, ft-lb)	Bolt length "ℓ" mm (in)
1	39 - 49 (4.0 - 5.0, 29 - 36)	45 (1.77)
2	39 - 49 (4.0 - 5.0, 29 - 36)	50 (1.97)
3	39 - 49 (4.0 - 5.0, 29 - 36)	60 (2.36)
4	29 - 39 (3.0 - 4.0, 22 - 29)	25 (0.98)
5	29 - 39 (3.0 - 4.0, 22 - 29)	60 (2.36)
6	39 - 49 (4.0 - 5.0, 29 - 36)	20 (0.79)
7	39 - 49 (4.0 - 5.0, 29 - 36)	25 (0.98)
Gusset to engine	29 - 39 (3.0 - 4.0, 22 - 29)	20 (0.79)

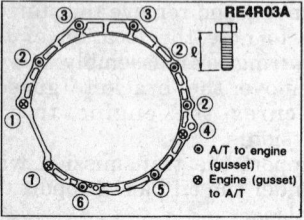

RE4R03A

Bolt No.	Tightening torque N·m (kg-m, ft-lb)	Bolt length "ℓ" mm (in)
1	39 - 49 (4.0 - 5.0, 29 - 36)	65 (2.56)
2	39 - 49 (4.0 - 5.0, 29 - 36)	60 (2.36)
3	39 - 49 (4.0 - 5.0, 29 - 36)	55 (2.17)
4	29 - 39 (3.0 - 4.0, 22 - 29)	25 (0.98)
5	29 - 39 (3.0 - 4.0, 22 - 29)	60 (2.36)
6	39 - 49 (4.0 - 5.0, 29 - 36)	20 (0.79)
7	39 - 49 (4.0 - 5.0, 29 - 36)	25 (0.98)
Gusset to engine	29 - 39 (3.0 - 4.0, 22 - 29)	20 (0.79)

214547

Transmission mounting bolt identification and torque specification — 1993–95 300ZX

tance from the face of the converter bolt mounting lug to the edge of the converter housing must be checked prior to installing the transmission. This is done to ensure proper installation of the torque converter.

21. The converter measured distance should be:

a. 0.984 in. (25mm) for RE4R03A transmissions (turbo models)

b. 1.024 inches (26mm) for RE4R01A transmissions (non-turbo models)

22. Raise the transmission and tighten the transmission to the engine mounting bolts to the proper tor-

que. Be sure to position the bolts in the proper locations.

23. Tighten the drive plate to the converter mounting bolts to 33–43 ft. lbs. (44–59 Nm).

NOTE: After the torque converter is installed, rotate the crankshaft several times to make sure the transmission rotates freely and does not bind.

24. Install the rear crossmember.

25. The balance of installation is the reverse of the removal procedure.

26. Tighten all driveshaft flange mounting bolts to 29–33 ft. lbs. (34–44 Nm) for Non-turbo models, or

to 40–47 ft. lbs. (54–64 Nm) for Turbo models.

27. Tighten the center support bolts to 43–58 ft. lbs. (59–78 Nm).

28. Connect the negative battery cable, fill the transmission to the proper level and make any necessary adjustment.

29. Perform a road test and check the fluid level.

Automatic Transaxle Assembly

REMOVAL AND INSTALLATION

1993–94 Sentra

1. Disconnect the negative battery cable.

2. Raise and support the vehicle safely.

3. Remove the left front tire.

4. Drain the transaxle fluid.

5. Remove the left side fender protector.

6. Remove the halfshafts.

NOTE: Be careful not to damage the oil seals when removing the halfshafts. After removing the halfshafts, install a suitable bar so the side gears will not rotate and fall into the differential case.

7. On the Sentra Wagon, disconnect and remove the forward exhaust pipe.

8. Disconnect the speedometer cable.

9. Disconnect the throttle wire (cable) connection.

10. Remove the control cable rear end from the unit and remove the oil level gauge tube.

11. Place a suitable jack under the transaxle and engine. Do not place the jack under the oil pan drain plug. Support the engine with wooden blocks placed between the engine and the center member.

12. Disconnect the oil cooler and charging tubes. Plug the tube ends to prevent leakage.

13. Remove the engine motor mount securing bolts, as required.

14. Remove the starter motor and disconnect all electrical wires from the transaxle.

15. Loosen and remove all but 3 of the bolts holding the transaxle to the engine. Leave the 3 bolts in to support the weight of the transaxle while removing the converter bolts.

16. Remove the dust covers.

17. Remove the bolts holding the torque converter to the drive plate.

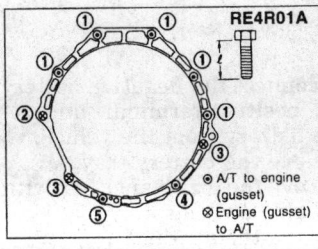

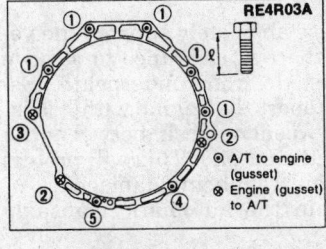

RE4R01A

Bolt No.	Tightening torque N·m (kg-m, ft-lb)	Bolt length "ℓ" mm (in)
①	39 - 49 (4.0 - 5.0, 29 - 36)	47.5 (1.870)
②	39 - 49 (4.0 - 5.0, 29 - 36)	58 (2.28)
③	29 - 39 (3.0 - 4.0, 22 - 29)	25 (0.98)
④	29 - 39 (3.0 - 4.0, 22 - 29)	60 (2.36)
⑤	29 - 39 (3.0 - 4.0, 22 - 29)	65 (2.56)
Gusset to engine	29 - 39 (3.0 - 4.0, 22 - 29)	20 (0.79)

RE4R03A

Bolt No.	Tightening torque N·m (kg-m, ft-lb)	Bolt length "ℓ" mm (in)
①	39 - 49 (4.0 - 5.0, 29 - 36)	65 (2.56)
②	29 - 39 (3.0 - 4.0, 22 - 29)	25 (0.98)
③	39 - 49 (4.0 - 5.0, 29 - 36)	58 (2.28)
④	29 - 39 (3.0 - 4.0, 22 - 29)	62 (2.44)
⑤	29 - 39 (3.0 - 4.0, 22 - 29)	100 (3.94)
Gusset to engine	29 - 39 (3.0 - 4.0, 22 - 29)	20 (0.79)

302962

Transmission mounting bolt identification and torque specification — 1996–97 300ZX

Rotate the crankshaft to gain access to each bolt. Before separating the torque converter, place chalk marks on 2 parts for alignment purposes during installation.

NOTE: The transaxle bolts are different lengths. Tag each bolt according to location to ensure proper installation.

18. Remove the 3 remaining bolts. Move the jack gradually until the transaxle can be lowered and removed from the vehicle through the left side wheel housing.
19. Check the drive plate runout with a dial indicator. Runout must be no more than 0.020 in. (0.5mm).
   To install:
20. If the torque converter was removed from the transaxle for any reason, after it is installed, the distance from the face of the converter to the edge of the converter housing must be checked prior to installing the transaxle. This is done to ensure proper installation of the torque converter. Check the distance and make sure it is as follows:
   a. RL3F01A transaxles should be 0.831 in. (21mm) or more.
   b. RL4F03A and RL4F03V transaxles should be 0.748 in. (19mm) or more.

21. Raise the transaxle onto the engine and install the torque converter bolts. Torque the bolts to specification. Install 3 bolts to support the transaxle while tightening the converter bolts.

NOTE: After the converter is installed, rotate the crankshaft several times to make sure the transaxle rotates freely and does not bind.

22. Install the dust covers.
23. Install the transaxle mounting bolts and tighten as shown (see illustration).
24. Install the remaining components in the opposite order from which they were removed.
25. Road test the vehicle and check for leaks.

1995–97 Sentra and 200SX

1. Disconnect the negative and positive battery cables.
2. Remove the battery and bracket from the vehicle.
3. Remove the air duct between the throttle body and the air cleaner.
4. Disconnect the solenoid harness connector, inhibitor switch harness connector and the revolution speed sensor harness connector.
5. Disconnect the torque converter clutch solenoid harness connector

and the vehicle speed sensor harness connector.
6. Remove the crankshaft position sensor from the transaxle.
7. For RL4F03A transaxles, disconnect the throttle wire at the engine side.
8. Drain the fluid from the transaxle.
9. Disconnect the control cable from the transaxle.
10. Disconnect the oil cooler lines from the transaxle.
11. Remove the halfshafts from the transaxle.
12. Remove the intake manifold support brackets.
13. Remove the starter motor from the transaxle.
14. Remove the upper bolts that secure the transaxle to the engine.
15. Using a block of wood, support the transaxle with a jack.
16. Remove the center crossmember.
17. On vehicles with GA16DE engines, remove the front and rear gussets.
18. Remove the rear plate cover.
19. Remove the torque converter mounting bolts. It will be necessary to rotate the engine by hand to gain access to all bolts.
20. Remove the rear transaxle to engine bracket.
21. Remove the rear transaxle mount.
22. Remove the lower transaxle to engine mounting bolts.
23. Slide the transaxle away from the engine and lower the transaxle assembly.
   To install:
When connecting the torque converter to the transaxle, be sure to measure the distance between the mounting lug of the converter and the front edge of the transaxle.
24. The measured distance between the converter and the front of the transaxle should be:
   a. 0.831 inches (21.1mm) or more for GA16DE engine vehicles
   b. 0.626 inches (15.9mm) or more for SR20DE engine vehicles
25. Raise the transaxle and install to engine drive plate.
26. Install the transaxle mounting bolts in the proper location as noted during removal. Torque the bolts to specifications.
27. The remaining components are installed in the reverse order from which they were removed.
28. Tighten the torque converter mounting bolts to 33–43 ft. lbs. (44–59 Nm).

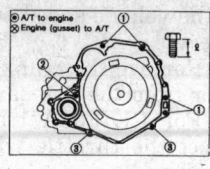

RL3F01A

Bolt No.	Tightening torque N·m (kg-m, ft-lb)	Bolt length "ℓ" mm (in)
①	30 - 40 (3.1 - 4.1, 22 - 30)	50 (1.97)
②	30 - 40 (3.1 - 4.1, 22 - 30)	30 (1.18)
③	16 - 21 (1.6 - 2.1, 12 - 15)	20 (0.79)
Front gusset to engine	30 - 40 (3.1 - 4.1, 22 - 30)	20 (0.79)
Rear gusset to engine	16 - 21 (1.6 - 2.1, 12 - 15)	16 (0.63)

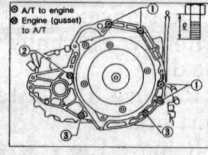

RL4F03A

Bolt No.	Tightening torque N·m (kg-m, ft-lb)	Bolt length "ℓ" mm (in)
①	30 - 40 (3.1 - 4.1, 22 - 30)	50 (1.97)
②	30 - 40 (3.1 - 4.1, 22 - 30)	30 (1.18)
③	16 - 21 (1.6 - 2.1, 12 - 15)	25 (0.98)
Front gusset to engine	30 - 40 (3.1 - 4.1, 22 - 30)	20 (0.79)
Rear gusset to engine	16 - 21 (1.6 - 2.1, 12 - 15)	16 (0.63)

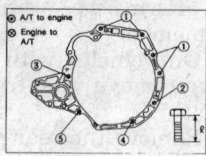

RL4F03V

Bolt No.	Tightening torque N·m (kg-m, ft-lb)	Bolt length "ℓ" mm (in)
①	70 - 79 (7.1 - 8.1, 51 - 59)	55 (2.17)
②	70 - 79 (7.1 - 8.1, 51 - 59)	50 (1.97)
③	70 - 79 (7.1 - 8.1, 51 - 59)	65 (2.56)
④	16 - 21 (1.6 - 2.1, 12 - 15)	35 (1.38)
⑤	16 - 21 (1.6 - 2.1, 12 - 15)	45 (1.77)

204456

Transaxle mounting bolts specifications — 1993-94 Sentra

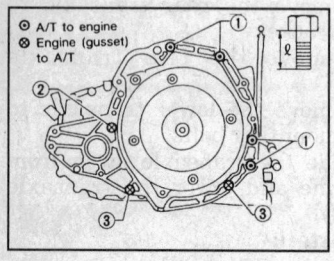

RL4F03A

Bolt No.	Tightening torque N·m (kg-m, ft-lb)	Bolt length "ℓ" mm (in)
①	30 - 40 (3.1 - 4.1, 22 - 30)	50 (1.97)
②	30 - 40 (3.1 - 4.1, 22 - 30)	30 (1.18)
③	16 - 21 (1.6 - 2.1, 12 - 15)	25 (0.98)
Front gusset to engine	30 - 40 (3.1 - 4.1, 22 - 30)	20 (0.79)
Rear gusset to engine	16 - 21 (1.6 - 2.1, 12 - 15)	16 (0.63)

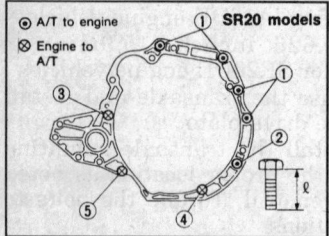

RE4F03V

Bolt No.	Tightening torque N·m (kg-m, ft-lb)	Bolt length "ℓ" mm (in)
①	70 - 79 (7.1 - 8.1, 51 - 59)	55 (2.17)
②	70 - 79 (7.1 - 8.1, 51 - 59)	50 (1.97)
③	70 - 79 (7.1 - 8.1, 51 - 59)	65 (2.56)
④	16 - 21 (1.6 - 2.1, 12 - 15)	35 (1.38)
⑤	16 - 21 (1.6 - 2.1, 12 - 15)	45 (1.77)

302653

Transaxle bolt torque specifications — 1995-97 Sentra and 200SX

29. Fill the transaxle to the proper level, start the engine, and recheck the fluid level.

30. Road test the vehicle and verify proper operation.

Altima

1. Disconnect the negative battery terminal, positive terminal, and remove the battery from the vehicle.

2. Remove the battery tray.

3. Remove the air cleaner box with the mass air flow sensor.

4. Remove the air duct.

5. Remove the vehicles front wheels.

6. Raise and safely support the vehicle so there is clearance to remove the transaxle from underneath. Securely support the engine with a jack. Place a wooden block between the jack and the oil pan. This will protect the oil pan from being damaged.

7. Drain the automatic transaxle fluid.

8. Disconnect the wiring harness connectors from the side of the transaxle.

9. Disconnect the wiring harness connectors from the vehicle speed sensor and the revolution sensor.

10. Remove the crankshaft position sensor from the transaxle.

11. Remove the left hand mounting bracket from the transaxle and body.

12. Disconnect the control cable from the side of the transaxle.

13. Remove both halfshafts from the transaxle assembly. Securely support the transaxle with another jack.

14. Disconnect the oil cooler lines.

15. Remove the starter motor from the transaxle.

16. Remove the center crossmember from the vehicle.

17. Remove the rear cover plate for access to the torque converter bolts.

18. Remove the torque converter to drive plate bolts.

NOTE: When removing the torque converter, turn the crankshaft for access to the bolts. Place alignment marks on the converter and drive plate, so the converter can be installed in its original position.

19. Remove the transaxle mounting bolts, pull the transaxle away from the engine and lower it from the vehicle.

NOTE: The transaxle mounting bolts are different lengths. Tagging the bolts upon removal will facilitate proper tightening during installation.

To install:

NOTE: When installing the torque converter to the transaxle, measure the depth of the converter to ensure proper installation.

20. Using a straight edge across the mounting flange, measure the depth of the converter. The measurement is to the bolt mounting flange of the converter.

21. The depth measurement of the converter should be 0.75 inches (19mm) or more.

22. Install the transaxle assembly into the vehicle and tighten the mounting bolts to specifications.

NOTE: The transaxle mounting bolts are different lengths and require special torque specifications. Use care when installing and tightening these bolts.

23. Torque the bolts holding the converter to the drive plate to 33–43 ft. lbs. (44–59 Nm).

24. Install the remaining components in the opposite order from which they were removed.

25. Connect the positive and then the negative battery terminals.

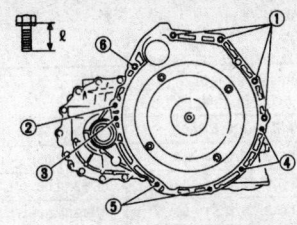

Bolt No.	Tightening torque N·m (kg-m, ft-lb)	ℓ mm (in)
1	39 - 49 (4.0 - 5.0, 29 - 36)	60 (2.36)
2	39 - 49 (4.0 - 5.0, 29 - 36)	60 (2.36)
3	30 - 40 (3.1 - 4.1, 22 - 30)	25 (0.98)
4	30 - 40 (3.1 - 4.1, 22 - 30)	25 (0.98)
5*	30 - 40 (3.1 - 4.1, 22 - 30)	–
6	43 - 58 (4.4 - 5.9, 32 - 43)	115 (4.53)
Front gusset or Rear gusset to engine	30 - 40 (3.1 - 4.1, 22 - 30)	25 (0.98)

*: Nuts and washers.

125287

Transaxle-to-engine bolt torque — 1993–94 Maxima (VE30DE) engines

26. Refill the transaxle with the proper type and amount of fluid.

27. Road test the vehicle for proper operation and recheck the fluid level.

1993–94 Maxima

1. Disconnect the negative battery cable. If necessary, remove the battery and the battery tray.

2. Raise and safely support vehicle.

3. Remove the oil cooler lines from the transaxle and drain the fluid from the transaxle and the radiator.

Plug the transaxle fittings and the oil cooler lines.

4. Disconnect the shift linkage from the transaxle.

5. If equipped, disconnect the electrical harness connector from the transaxle.

6. Disconnect and remove the drive axles from the transaxle.

7. Remove the starter assembly.

8. Matchmark the torque converter-to-driveplate position.

9. Rotate the crankshaft and remove the torque converter-to-flywheel bolts.

10. Using transmission jack, support the transaxle and remove the bolts that secure the transaxle to the frame.

11. Remove the transaxle assembly-to-engine bolts and pull the transaxle assembly rearward from the engine; be careful the torque converter does not fall off the transaxle's shaft.

12. To inspect the drive plate for runout, perform the following procedures:

a. Mount a dial indicator to the rear of the engine and zero the indicator.

b. Rotate the crankshaft by hand and measure the runout at various positions on the drive plate.

c. The runout should be less than 0.020 in. (0.5mm); if not, replace the drive plate and ring gear.

13. Using a straight-edge and a ruler, measure the torque converter-to-transaxle recessed depth, it should be for VG30E engine—0.71 in. (18mm) or more. For VE30DE engine, it should be 0.55 in. (14mm) or more

To install:

14. Install the transaxle assembly to the engine. Torque the transaxle assembly-to-engine bolts, as follows for vehicles with VG30E engine:

Bolt 1: 22–30 ft. lbs. (30–40 Nm)
Bolt 2: 29–36 ft. lbs. (39–49 Nm)
Bolt 3: 22–30 ft. lbs. (30–40 Nm)

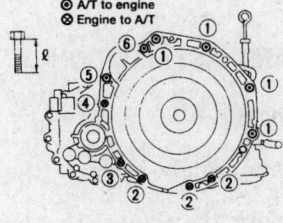

Bolt No.	Tightening torque N·m (kg-m, ft-lb)	ℓ mm (in)
1	39 - 49 (4.0 - 5.0, 29 - 36)	45 (1.77)
2	30 - 36 (3.1 - 3.7, 22 - 27)	30 (1.18)
3	30 - 36 (3.1 - 3.7, 22 - 27)	40 (1.57)
4	74 - 83 (7.5 - 8.5, 54 - 61)	45 (1.77)
5	30 - 36 (3.1 - 3.7, 22 - 27)	80 (3.15)
6	30 - 36 (3.1 - 3.7, 22 - 27)	65 (2.56)

295024

Transaxle bolt torque specifications — Altima

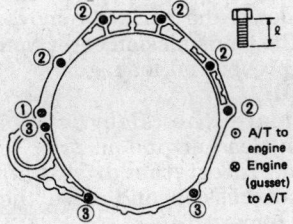

Bolt No.	Tightening torque N·m (kg-m, ft-lb)	ℓ mm (in)
1	30 - 40 (3.1 - 4.1, 22 - 30)	60 (2.36)
2	39 - 49 (4.0 - 5.0, 29 - 36)	45 (1.77)
3	30 - 40 (3.1 - 4.1, 22 - 30)	25 (0.98)

125246

Transaxle-to-engine bolt torque — 1993–94 Maxima (VG30E) engine

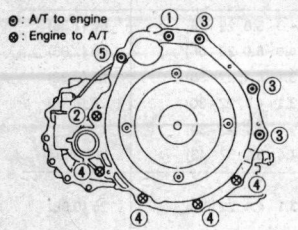

Bolt No.	Tightening torque N·m (kg-m, ft-lb)	mm (in)
1	70 - 79 (7.1 - 8.1, 51 - 59)	65 (2.56)
2	70 - 79 (7.1 - 8.1, 51 - 59)	52 (2.05)
3	70 - 79 (7.1 - 8.1, 51 - 59)	52 (2.05)
4	70 - 79 (7.1 - 8.1, 51 - 59)	40 (1.57)
5	70 - 79 (7.1 - 8.1, 51 - 59)	124 (4.88)

289091

Transaxle bolt torque specifications — 1995–97 Maxima

15. For vehicles with a VE30DE engine, the tightening torque should be:
Bolt 1: 29–36 ft. lbs. (39–49 Nm)
Bolt 2: 29–36 ft. lbs. (39–49 Nm)
Bolt 3: 22–30 ft. lbs. (30–40 Nm)
Bolt 4: 22–30 ft. lbs. (30–40 Nm)
Bolt 5: 22–30 ft. lbs. (30–40 Nm)
Bolt 6: 32–43 ft. lbs. (43–58 Nm)

16. Align the torque converter-to-driveplate matchmarks, apply locking sealant to the torque converter-to-driveplate bolts and torque the torque converter-to-driveplate bolts to 33–43 ft. lbs. (44–59 Nm).

17. The balance of installation is the reverse of the removal procedure.

18. With the parking brake applied and the engine idling, move the selector lever through **N** to **D**, to **2**, to **1** and to **R**.

NOTE: When moving the selector lever, a slight notch or bump should be felt by the hand gripping the selector lever each time the transaxle is shifted.

19. Test drive the vehicle and check for proper operation.

1995–97 Maxima

1. Disconnect the negative and positive battery cables.
2. Remove the battery and the battery tray.
3. Remove the air cleaner and the resonator.
4. Raise and safely support the vehicle so there is clearance to remove the transaxle from underneath. Securely support the engine via the oil pan using a cushioning wooden block and a jack.
5. Disconnect the terminal cord assembly wiring harness and the inhibitor switch.
6. Disconnect the wiring harness for the revolution sensor and the vehicle speed sensor.
7. Remove the crankshaft position sensor (POS) from the transaxle.

8. Remove the LH mounting bracket from the transaxle and body.
9. Disconnect the control cable at the transaxle side.
10. Drain the fluid from the transaxle.
11. Remove both driveshafts from the transaxle assembly.
12. Disconnect and plug the oil cooler lines.
13. Remove the starter motor from the transaxle.
14. Support the transaxle with proper safety jack.
15. Remove the center crossmember assembly.
16. Remove the rear cover or access plate.
17. Remove the bolts that secure the flywheel to the torque converter. Rotate the flywheel to gain access to the three converter mounting bolts.

NOTE: When removing the torque converter, turn the crankshaft for access to the bolts. Place alignment marks on the converter and drive plate, so the converter can be installed in its original position.

18. Remove the bolts that secure the transaxle to the engine block.

NOTE: The transaxle bolts are of different lengths, be sure to note the location of the bolts for reassembly.

19. Lower the transaxle while supporting it with a jack. Be sure the torque converter remains with the transaxle.
20. Using a dial indicator measure the drive plate run out while rotating the crankshaft.
21. The runout must be less than 0.0059 inches (0.15mm).
To install:

NOTE: When installing the torque converter to the transaxle, measure the depth of the converter to ensure proper installation.

22. Using a straight edge across the mounting flange, measure the depth of the converter. The measurement is to the bolt mounting flange of the converter.
23. The depth measurement of the converter should be 0.55 inches (14mm) or more.
24. Install the transaxle assembly to the engine block while aligning the torque converter. Torque the transaxle bolts to specifications.
25. Install the converter mounting bolts and tighten the bolts to 33–43 ft. lbs. (44–59 Nm).
26. Install the rear access cover plate to the transaxle.
27. Install the center crossmember assembly and tighten the mounting bolts to 57–72 ft. lbs. (77–98 Nm).
28. Install the LH engine mount and torque the through bolt to 32–41 ft. lbs. (43–55 Nm)
29. Install the remaining components in the reverse order of removal.
30. Lower the vehicle.
31. Refill the transaxle with the proper type and amount of fluid.

DRIVELINE

Driveshaft

REMOVAL AND INSTALLATION

240SX and 300ZX

1. Raise and safely support the vehicle.
2. Matchmark the flanges on the driveshaft and differential so the driveshaft can be reinstalled in its original position; this will help maintain drive line balance.
3. Unbolt the rear flange and the center bearing bracket.
4. Withdraw the driveshaft from the transmission and pull the driveshaft down and back to remove.
5. Plug the transmission extension housing to prevent oil leakage.
To install:
6. Lubricate the sleeve yoke splines with clean engine oil prior to installation. Insert the driveshaft into the transmission and align the flange matchmarks.
7. For 240SX models, perform the following:
  a. Install the flange and the center bearing bolts.

b. Tighten the center bearing mounting bolts to the following specifications:
- 1993–94 vehicles — 19–29 ft. lbs. (25–39 Nm).
- 1995–97 vehicles — 32–41 ft. lbs. (43–55 Nm).

8. For 240SX models, perform the following:

a. Tighten the driveshaft rear flange-to-differential mounting bolts to the following specifications:
- 1993–94 vehicles — 29–33 ft. lbs. (39–44 Nm).
- 1995–97 vehicles — 41–48 ft. lbs. (55–65 Nm).

9. For 300ZX models, perform the following:

a. Position the driveshaft and center support bearing. Start but do not tighten the center support bearing bolts.

b. Align the driveshaft rear flange with the matchmarks and install the mounting bolts. Tighten the rear flange mounting bolts as follows:
- Turbocharged models — 47–54 ft. lbs. (64–74 Nm)
- Non-turbocharged models — 41–48 ft. lbs. (55–65 Nm)

10. For 300ZX models, tighten the center support bearing mounting bolts to 43–58 ft. lbs. (59–78 Nm)

11. Lower the vehicle and check the transmission fluid level. Add the appropriate type of fluid as necessary.

U-Joints

REMOVAL AND INSTALLATION

240SX and 300ZX

The driveshaft must be removed from the vehicle to inspect the U-joints correctly. Mount the driveshaft in a bench-mounted vise and mount a dial indicator so that it reads forward/rearward motion (axial end-play) of the U-joint. Manually push and pull the U-joint frontward and rearward while reading the dial indicator. The axial end-play should be zero. Next, mount the dial indicator so that it will register any side-to-side movement of the U-joint. There should also be no side-to-side movement. If any front/rear or side-to-side movement is exhibited by the U-joint, the entire driveshaft must be replaced as an assembly. The U-joints are not individually serviceable.

Halfshaft

REMOVAL AND INSTALLATION

1993–94 Sentra

NOTE: On some models, installation of the halfshafts will require a special tool for the spline alignment of the halfshaft end and the transaxle case. Do not perform this procedure without access to this tool. The KentMoore tool Number is J–34296 and J–34297

1. Raise the front of the vehicle and support it on jackstands, then remove the wheel and the tire assembly.
2. Remove the caliper assembly.
3. Remove the cotter pin from the drive axle.
4. Using a bar to hold the wheel from turning, loosen the hub nut.
5. Using the removal tool HT72520000 or equivalent, remove the tie rod ball joint from the steering knuckle.
6. Remove the control arm-to-steering knuckle, ball joint mounting nuts and separate the ball joint from the control arm.
7. Drain the lubricant from the transaxle.
8. Pull the hub/steering knuckle assembly away from the vehicle, to disconnect the driveshaft from the transaxle.

NOTE: When removing the driveshaft from the transaxle, do not pull on the driveshaft. The driveshaft will separate at the sliding joint (damaging the boot). Use a small pry bar to remove it from the transaxle. Be sure to replace the oil seal in the transaxle. After removing the driveshaft from the transaxle, be sure to install a holding tool to hold the side gear in place while the axle is removed.

9. Support the engine properly and remove support bracket if so equipped.
10. Use a wheel puller tool to press the driveshaft from the hub/steering knuckle assembly.
To install:
11. Use a new circlip on the driveshaft and install a new oil seal to the transaxle.

12. Install the driveshaft assembly and connect the lower control arm-to-ball joint.

NOTE: When installing the driveshaft into the transaxle, use oil seal protector tool KV38106700 or equivalent to protect the oil seal from damage; after installation, remove the tool.

13. Torque the control arm-to-ball joint to 40–47 ft. lbs. (54–64 Nm), the lower ball joint stud nut to 25–36 ft. lbs. (34–49 Nm).
14. Install the tie rod to the steering knuckle and torque the tie rod stud nut to 22–36 ft. lbs. (30–49 Nm).
15. Install then hub nut and tighten to 87–145 ft. lbs. (114–197 Nm).
16. Install new cotter pin in drive axle and mount the caliper assembly.
17. Install the wheel and tire assembly.
18. Road test for proper operation.

1995–97 Sentra and 200SX

NOTE: The halfshafts will require a special tool for the spline alignment of the halfshaft end into the transaxle case. Do not perform this procedure without access to this tool. The KentMoore tool Number is J–34296 and J–34297

1. Raise the front of the vehicle and support it on jackstands, then remove the wheel and the tire assembly.
2. Using a bar to hold the wheel from turning, loosen and remove the hub nut.
3. Remove the clip and separate the brake hose from the strut.
4. Remove the caliper assembly and support it with a wire. Do not allow the caliper to hang from the brake hose.
5. Remove the bolts that secure the strut to the steering knuckle.

NOTE: Cover the halfshaft boots with shop towels to protect them during removal of the shaft.

6. Separate the halfshaft from the knuckle by lightly tapping it with a hammer. If it is hard to remove, use a puller.
7. Remove the halfshaft from the transaxle as follows:

a. Models without support bearing — Pry the halfshaft from the transaxle.

b. Models with support bearing — Remove the support bearing

bolts and pull the halfshaft from transaxle.

NOTE: When removing the halfshaft from the transaxle, do not pull on the halfshaft. The halfshaft will separate at the sliding joint (damaging the boot). Use a small pry bar to remove it from the transaxle. Be sure to replace the oil seal in the transaxle.

8. Remove the halfshaft from the vehicle.

To install:

9. Use a new circlip on the halfshaft and install a new oil seal to the transaxle.

NOTE: When installing the halfshaft into the transaxle, use oil seal protector tool KV38106700 (J34296) for the left side or tool KV38106800 (J34297) for the right side to protect the oil seal from damage; after installation of the shaft remove the tool.

10. Install the halfshaft assembly into the transaxle.

NOTE: After installation of the halfshaft, try to pull the flange out by hand. If it pulls out, the circular clip is not locked into the transaxle.

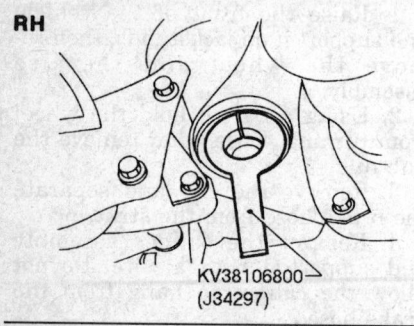

RH

KV38106800
(J34297)

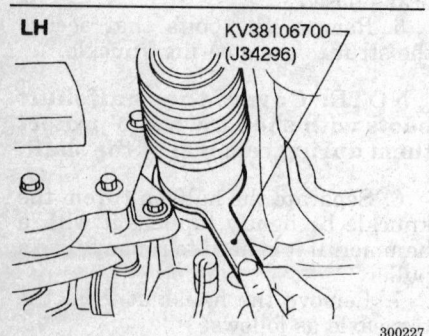

LH

KV38106700
(J34296)

300227

Halfshaft installation tools — 1995–97 Sentra and 200SX

11. If removed, install the support bearing bracket bolts and tighten the mounting bolts to 19–26 ft. lbs. (25–35 Nm).

12. Lubricate the splines of the halfshaft and insert the shaft through the steering knuckle.

13. Align the steering knuckle with the lower strut mount and install the mounting bolts. Tighten the mounting bolts to 68–82 ft. lbs. (92–111 Nm).

14. Install the disc brake caliper and connect the brake hose to the strut with the clip.

15. Install the washer and hub nut to the halfshaft. Tighten the hub nut to 145–202 ft. lbs. (197–274 Nm).

16. Install the adjusting cap and a new cotter pin in drive axle.

17. Install the wheel and tire assembly and lower the vehicle.

18. Road test the vehicle for proper operation.

Altima

1. Raise and safely support the vehicle with the front wheels hanging freely.

2. Remove the front wheels from the vehicle.

NOTE: The brake caliper does not need to be disconnected from the knuckle.

3. Pull out the cotter pin from the castellated nut on the wheel hub and then remove the wheel bearing locknut.

NOTE: Cover the CV-joint boots with a shop towel or waste cloth so not to damage them when removing the halfshaft.

4. Remove the cotter pin and castle nut from the lower ball joint.

5. Strike the knuckle with a hammer and pull down on the transverse link to separate the lower ball joint from the knuckle.

6. Disconnect the tie rod end from the steering knuckle.

7. Separate the halfshaft from the steering knuckle by tapping it with a block of wood and a mallet.

8. Using a prybar, reach through the engine crossmember and carefully pry the right inner CV-joint from the transaxle.

9. If equipped with manual transaxle, carefully pry the left inner CV-joint from the transaxle.

10. If equipped with automatic transaxle, insert a long tool into the opening for the right halfshaft and strike the tool to with a hammer.

11. Remove the left halfshaft from the transaxle.

To install:

NOTE: Whenever the halfshafts are removed, the axle seals should be replaced.

12. When installing the shafts into the transaxle, use a new oil seal and then install an alignment tool, KV38106700 for the left side or KV38106800 for the right side, along the inner circumference of the oil seal.

13. Insert the halfshaft into the transaxle, align the serrations and then remove the alignment tool.

14. Push the halfshaft, then press-fit the circular clip on the shaft into the clip groove on the side gear.

NOTE: After insertion, attempt to pull the flange out of the side joint to make sure the circular clip is properly seated in the side gear and will not come out.

15. Insert the halfshaft into the steering knuckle.

16. Connect the lower ball joint and tie rod end in the correct position. Torque the lower ball joint-to-control arm nuts to 52–64 ft. lbs. (71–86 Nm) and the tie rod end-to-steering knuckle nut to 22–29 ft. lbs. (29–39 Nm). Install new cotter pins to the castle nuts.

17. Install the axle nut and tighten the locknut to 174–231 ft. lbs. (235–314 Nm).

18. Install a new cotter pin on the wheel hub and install the wheel.

19. Install the front wheels to the vehicle.

20. Road test the vehicle for proper operation.

21. Check the transaxle fluid level and top off as necessary.

1993–94 Maxima

1. Raise and safely support the vehicle with the front wheels hanging freely.

2. Remove the wheel(s). Remove the brake caliper assembly. The brake hose does not need to be disconnected from the caliper. Be careful not to depress the brake pedal or twist the brake hose.

3. Support the caliper with a wire. Do not allow the caliper to hang from the brake line.

4. Pull out the cotter pin from the castellated nut on the wheel hub and then remove the wheel bearing locknut.

NOTE: Cover the CV-joint boots with a shop towel or waste cloth so not to damage them when removing the halfshaft.

5. Disconnect the tie rod from the steering knuckle. Remove the three lower ball joint-to-lower control arm nuts/bolts and then pull the control arm down.

NOTE: Always use a new nut(s) when reinstalling the tie rod end or lower ball joint.

6. Separate the halfshaft from the steering knuckle by tapping it with a block of wood and a mallet.

NOTE: It may be necessary to loosen (do not remove) the strut-to-chassis nuts to gain clearance for steering knuckle separation from the halfshaft. If these bolts are loosened, the alignment will have to be checked when the job is finished.

7. Using a prybar, reach through the engine crossmember and carefully pry the inner CV-joint from the transaxle or the support bearing flange.

8. Using a block of wood on an hydraulic floor jack, support the engine under the oil pan.

9. Withdraw the shaft from the steering knuckle and remove it from the vehicle. If necessary on the right side, remove the support bearing bracket from the engine and then withdraw the right halfshaft.

To install:

10. When installing the shafts into the transaxle, use a new oil seal and then install an alignment tool, KV38106700 for the left side or KV38106800 for the right side, along the inner circumference of the oil seal.

11. Insert the halfshaft into the transaxle, align the serrations and then remove the alignment tool.

12. Push the halfshaft, then press-fit the circular clip on the shaft into the clip groove on the side gear.

NOTE: After insertion, attempt to pull the flange out of the side joint to make sure the circular clip is properly seated in the side gear and will not come out.

13. Install the support bearing bracket bolts and tighten the bolts to 26 ft. lbs. (35 Nm). Insert the halfshaft in the steering knuckle. If loosened, tighten the strut mounting bolts.

14. Connect the lower ball joint and tie rod end in the correct position. Torque the lower ball joint-to-control arm nuts to 56–80 ft. lbs. (76–109 Nm) and the tie rod end-to-steering knuckle nut to 22–29 ft. lbs. (29–39 Nm).

15. Install the caliper assembly and the wheel bearing locknut. Tighten the locknut to 174–231 ft. lbs. (235–314 Nm).

16. Install a new cotter pin on the wheel hub and install the wheel.

17. If the brake hose was disconnected, bleed the brake system.

18. Road test the vehicle for proper operation.

19. Check the transaxle fluid level and top off as necessary.

1995–97 Maxima

1. Raise and support the front of the vehicle safely and remove the wheels.

2. Remove the ABS wheel sensor and move it out of the way.

3. Remove the brake hose from the strut.

4. Remove wheel bearing locknut.

5. Matchmark and remove the bolts attaching the steering knuckle to the strut.

NOTE: Cover axle boots with waste cloth or equivalent so as not to damage them when removing halfshaft.

6. Separate the halfshaft from the knuckle by slightly tapping it.

7. Loosen the bolts attaching the support bearing to the support bearing bracket.

8. If equipped with a manual transaxle, pry the halfshaft from the transaxle with a flat bladed tool.

9. If equipped with a automatic transaxle perform the following:

   a. Remove the right halfshaft from the vehicle.

   b. Insert a flat bladed tool into the transaxle where the right halfshaft was, place the end of the tool on the halfshaft, then drive the left shaft from the pinion side gear.

10. Remove the support bearing bolts and remove the halfshaft from the vehicle.

11. Remove and discard the circlip on the end of the halfshaft. Remove the seal from the transaxle.

To install:

12. Install a new seal into the transaxle and install a halfshaft alignment tool KV38106700 into the transaxle seal.

13. Install a new circlip to the halfshaft, then insert the halfshaft into the transaxle.

14. With the serrations aligned remove the alignment tool.

15. Push the halfshaft fully into the transaxle to seat the circlip. Try to pull the halfshaft from the transaxle by hand to verify that the circlip is properly seated.

16. Install the support bearing bolts and torque the bolts to 9–14 ft. lbs. (13–19 Nm).

17. Insert the halfshaft into the steering knuckle and install the hub locknut, do not tighten the hub nut at this time.

18. Connect the steering knuckle to the strut.

19. Install the strut mounting bolts and align the matchmarks. Torque the bolts to 103–117 ft. lbs. (140–159 Nm).

20. Install the brake hose to the strut.

21. Install the ABS wheel sensor and torque the attaching bolt to 13–17 ft. lbs. (18–24 Nm).

22. Install the front wheels, lower the vehicle and torque hub locknut to 174–231 ft. lbs. (235–314 Nm).

23. Check and/or adjust the wheel alignment as necessary.

240SX

Rear

1. Raise and safely support the vehicle.

2. Remove the rear wheel(s).

3. Loosen the wheel bearing locknut and lightly tap on the axle shaft to loosen it from the steering knuckle. Remove the nut from the axle shaft.

4. Remove the brake caliper assembly.

NOTE: Support the brake caliper, do not allow the caliper to hang freely from the brake hose.

5. Remove the ABS sensor from the steering knuckle.

6. Disconnect and separate the lower ball joint.

7. Pull the hub assembly outward.

8. Remove the shaft from hub assembly.

9. Unbolt and remove axle shaft from the differential.

To install:

10. Reconnect the shaft to the differential and tighten the mounting bolts to 25–33 ft. lbs. (34–44 Nm).

11. Insert the shaft into the hub assembly.

12. Reconnect the lower control arm and ball joint. Tighten the mounting nut to 52–64 ft. lbs. (71–86 Nm) and install a new cotter pin.

13. Replace the wheel bearing nut and torque it to 152–202 ft. lbs. (206–274 Nm).

14. Install the ABS sensor to the steering knuckle and tighten the mounting bolt to 13–17 ft. lbs. (18–24 Nm).

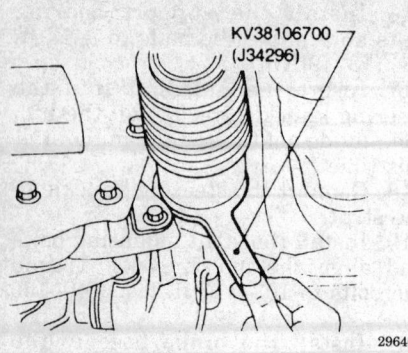

KV38106700
(J34296)

296409

Left halfshaft alignment tool — 1995–97 Maxima

15. Reinstall the brake caliper assembly. If the brake line was disconnected, bleed the brake system.

16. Install the wheel assembly, lower the vehicle, and perform a road test.

1995–97 300ZX

Rear

NOTE: When removing the driveshaft assembly, disconnect the ABS wheel sensor to prevent damage to the wheel sensor. When removing the driveshaft, cover the axle boots with a shop towel for protection.

1. Raise and safely support the vehicle.
2. Remove the rear wheel(s).
3. Unbolt the halfshaft from the flange at the rear housing.
4. Remove the cotter pin, adjusting cap, insulator, wheel bearing locknut and washer.
5. Remove the axle shaft by lightly tapping it with a copper hammer.

WARNING

To avoid damaging the threads of the axle shaft, install the axle nut to protect the threads.

---

6. Remove the shaft from the hub assembly.
To install:
7. Insert the shaft into the hub assembly.
8. Install the washer and axle nut but do not tighten at this time.
9. Reconnect the shaft to the differential.
10. Tighten the mounting bolts and nuts with washers to:
   a. 51–58 ft. lbs. (69–78 Nm) for turbo models
   b. 47–58 ft. lbs. (64–78 Nm) for non-turbo models
11. Replace the wheel bearing nut and torque it to 152–203 ft. lbs. (206–275 Nm).

12. Install the insulator, adjusting cap and install a new cotter.
13. Install the wheel assembly and lower the vehicle.

CV-Joint Boot

REPLACEMENT

Altima and 1993–94 Sentra

NOTE: When installing the new boots, tape the splines of the axle shaft to protect the boots.

Transaxle Side Tripod Joint (TS79C and TS70C)

1. Remove the halfshaft from the vehicle.
2. Place the halfshaft securely in a soft-jawed vise.
3. Remove the boot bands.
4. Matchmark the slide joint housing to the halfshaft and remove the slide joint housing.
5. Remove the snapring and matchmark the spider assembly to the halfshaft.
6. Press the spider assembly from the halfshaft.
7. Remove the CV-joint boot.
To install:
8. Install the boot and a new small clamp on the halfshaft.
9. Place the halfshaft in a soft-jawed vise.
10. Align the matchmarks and using a suitable driver, press the spider assembly onto the halfshaft. Install the spider assembly with the serrated chamfered edge facing the shaft.
11. Install a new snapring. The round surface should face the spider assembly.
12. Pack the slide joint housing with 7.94–8.29 oz. (225–235g) of grease.
13. Align the matchmarks and install the slide joint housing.
14. Slide the CV-joint boot over the slide joint housing and install the large boot clamp. Secure the boot with both clamps.
15. Install the halfshaft into the vehicle.

Transaxle Side Double Offset Joint (DS76, DS80, DS86 and DS90)

1. Remove the halfshaft assembly from the vehicle.
2. Remove the boot bands.
3. Before separating the joint assembly from the halfshaft, matchmark the slide joint housing and inner race.

4. Pry off the outer snapring (**A**) and pull out the slide joint housing.

NOTE: Cover driveshaft serration with tape so not to damage the boot.

5. Place matchmarks on the inner race and the halfshaft.
6. Pry off the middle snapring (**C**).
7. Remove the ball cage, the inner race, and the balls as a unit.
8. Pry off the rear snapring (**B**) and remove the CV-joint boot.
To install:
9. Install the boot and a new rear snapring (**B**).
10. Install the ball cage, the inner race, and the balls as a unit. Be sure that the matchmarks on the inner race and halfshaft are aligned; install a new middle snapring (**C**).
11. Pack the CV-joint housing and boot with the following specified amounts:
   a. 1993–94 Sentra — 4.94–5.64 oz. (140–160g) of grease.
   b. 1993–97 Altima — 5.11–5.82 oz. (145–164g) of grease.
12. Install the slide joint housing and align the matchmarks made during disassembly.
13. Install a new outer snapring (**A**).
14. Install the boot and secure with the boot bands.
15. Install the halfshaft in the vehicle.

Wheel Side — Rzeppa joint (ZF80, ZF90 and ZF100)

NOTE: The joint on the wheel side cannot be disassembled.

1. Remove the halfshaft from the vehicle.
2. Matchmark the halfshaft and the joint assembly.
3. Separate the joint assembly from the halfshaft by placing the halfshaft securely in a vise and place the locknut on the end of the joint. Attach a slide hammer to the locknut and pull the joint from the halfshaft.
4. Remove the boot bands.
To install:
5. Install the boot with new boot bands.
6. Pack CV-joint and boot with the following specified amounts:
   a. 1993–94 Sentra/NX Coupe — 5.47–6.17 oz. (155–175g) of grease.
7. Align the matchmarks and lightly tap the joint assembly onto the shaft.
8. Install the boot and secure it with both boot band clamps.
9. Install the halfshaft in the vehicle.

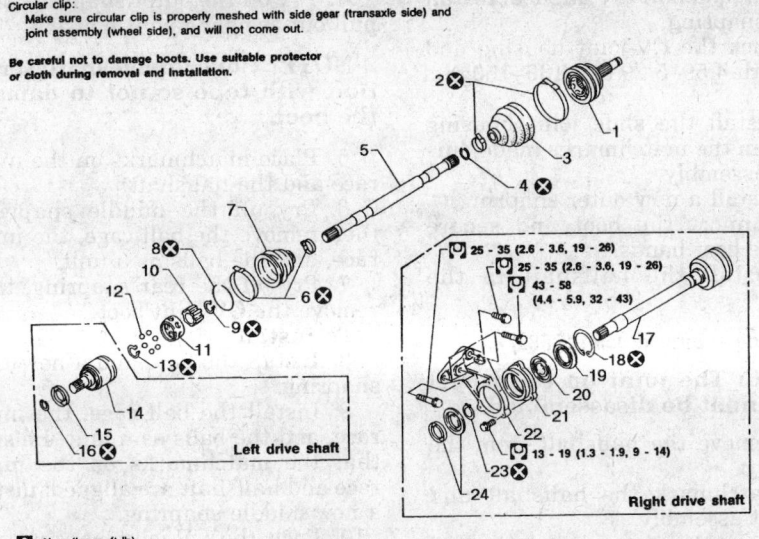

Circular clip:
Make sure circular clip is properly meshed with side gear (transaxle side) and joint assembly (wheel side), and will not come out.

Be careful not to damage boots. Use suitable protector or cloth during removal and installation.

25 - 35 (2.6 - 3.6, 19 - 26)
25 - 35 (2.6 - 3.6, 19 - 26)
43 - 58 (4.4 - 5.9, 32 - 43)
13 - 19 (1.3 - 1.9, 9 - 14)

Left drive shaft

Right drive shaft

: N·m (kg-m, ft-lb)

1 Joint assembly	9 Snap ring B	17 Slide joint housing with extension shaft
2 Boot band	10 Inner race	18 Snap ring E
3 Boot	11 Cage	19 Dust shield
4 Circular clip B	12 Ball	20 Support bearing
5 Drive shaft	13 Snap ring C	21 Support bearing retainer
6 Boot band	14 Slide joint housing	22 Bracket
7 Boot	15 Dust shield	23 Snap ring D
8 Snap ring A	16 Circular clip A	24 Dust shield

293123

Exploded view of halfshafts with Birfield and Double Offset CV-joints — 1993–94 Sentra

Wheel Side — Birfield Joint (BF76, BF80, BF86 and BF90)

NOTE: The joint on the wheel side cannot be disassembled.

1. Remove the halfshaft from the vehicle.
2. Matchmark the halfshaft and the joint assembly.
3. Separate the joint assembly from the halfshaft by placing the halfshaft securely in a vise and place the locknut on the end of the joint. Attach a slide hammer to the locknut and pull the joint from the halfshaft.
4. Remove the boot bands.
To install:
5. Install the boot with new boot bands.
6. Pack CV-joint and boot with the following specified amounts:
    a. 1993–94 Sentra — 3.70–4.41 oz. (105–125g) of grease.
    b. 1993–97 Altima — 3.53–4.23 oz. (100–120g) of grease.
7. Align the matchmarks and lightly tap the joint assembly onto the shaft.
8. Install the boot and secure it with both boot band clamps.
9. Install the halfshaft in the vehicle.

1995–97 Sentra and 200SX (GA16DE) Engine

Transaxle Side — Tripod Joint (TS79C)

1. Remove the halfshaft from the vehicle.
2. Place the halfshaft securely in a soft-jawed vise.
3. Remove the boot bands.
4. Matchmark the slide joint housing to the halfshaft and remove the slide joint housing.
5. Remove the snapring and matchmark the spider assembly to the halfshaft.
6. Press the spider assembly from the halfshaft.
7. Remove the CV-joint boot.
To install:
8. Install the boot and a new small clamp on the halfshaft.

NOTE: Cover the driveshaft serration with tape so as not to damage the new boot during installation.

9. Place the halfshaft in a soft-jawed vise.
10. Align the matchmarks and using a suitable driver, press the spider assembly onto the halfshaft. Install the spider assembly with the serrated chamfered edge facing the shaft.
11. Install a new snapring.

12. Pack the slide joint housing with 5.47–5.82 oz. (155–165g) of grease.
13. Align the matchmarks and install the slide joint housing.
14. Slide the CV-joint boot over the slide joint housing and install the large boot clamp. Secure the boot with both clamps.
15. Set the axle boot so that it does not swell or deform when stretched to a length of 4.00–4.07 inches.
16. Install the halfshaft into the vehicle.

Wheel Side — Rzeppa Joint (ZF90)

NOTE: The joint on the wheel side cannot be disassembled, if defective it must be replaced.

1. Remove the halfshaft from the vehicle.
2. Matchmark the halfshaft and the joint assembly.
3. Separate the joint assembly from the halfshaft by placing the halfshaft securely in a vise and place the locknut on the end of the joint. Attach a slide hammer to the locknut and pull the joint from the halfshaft.
4. Remove the boot bands.
To install:
5. Install the boot with new boot bands to the axle shaft.

NOTE: Cover the driveshaft serration with tape so as not to damage the new boot during installation.

6. Pack CV-joint and boot with 4.06–4.41 oz. (115–125g) of grease.
7. Align the matchmarks and lightly tap the joint assembly onto the shaft until the joint is locked on the circular clip.

NOTE: Be sure to position the axle nut on the end of the CV-joint to protect the threads when tapping the joint on to the axle.

8. Install the boot and secure it with the boot band clamps.
9. Set the axle boot so that it does not swell or deform when stretched to a length of 3.78–3.86 inches.
10. Install the halfshaft in the vehicle.

1995–97 200SX (SR20DE) Engine

Transaxle Side — Tripod Joint (TS83)

1. Remove the halfshaft from the vehicle.
2. Place the halfshaft securely in a soft-jawed vise.
3. Remove the boot bands.
4. Matchmark the slide joint housing to the halfshaft and remove the slide joint housing.

5. Remove the snapring and matchmark the spider assembly to the halfshaft.

6. Press the spider assembly from the halfshaft.

7. Remove the CV-joint boot.

To install:

8. Install the boot and a new small clamp on the halfshaft.

NOTE: Cover the driveshaft serration with tape so as not to damage the new boot during installation.

9. Place the halfshaft in a soft-jawed vise.

10. Align the matchmarks and using a suitable driver, press the spider assembly onto the halfshaft. Install the spider assembly with the serrated chamfered edge facing the shaft.

11. Install a new snapring.

12. Pack the slide joint housing with 4.59–5.29 oz. (130–150g) of grease.

13. Align the matchmarks and install the slide joint housing.

14. Slide the CV-joint boot over the slide joint housing and install the large boot clamp. Secure the boot with both clamps.

15. Set the axle boot so that it does not swell or deform when stretched to a length of 3.90 inches.

16. Install the halfshaft into the vehicle.

Transaxle Side — Double Offset Joint (DS83)

1. Remove the halfshaft assembly from the vehicle.

2. Remove the boot bands.

3. Before separating the joint assembly from the halfshaft, matchmark the slide joint housing and inner race.

4. Pry off the outer snapring and pull out the slide joint housing.

5. Place matchmarks on the inner race and the halfshaft.

6. Pry off the snapring that retains the ball cage.

7. Remove the ball cage, the inner race, and the balls as a unit.

8. Remove the boot and bands.

To install:

9. Install the boot assembly with new bands.

NOTE: Cover the halfshaft serration with tape so as not to damage the new boot during installation.

10. Install the ball cage, the inner race, and the balls as a unit. Be sure that the matchmarks on the inner

race and halfshaft are aligned; install a new snapring.

11. Pack the CV-joint housing and boot with 4.59–5.29 oz. (130–150g) of grease.

12. Install the slide joint housing and align the matchmarks made during disassembly.

13. Install a new outer snapring.

14. Connect the boot and secure with the boot bands.

15. Install the halfshaft in the vehicle.

Wheel Side — Birfield Joint (BF83)

NOTE: The joint on the wheel side cannot be disassembled.

1. Remove the halfshaft from the vehicle.

2. Matchmark the halfshaft and the joint assembly.

3. Separate the joint assembly from the halfshaft by placing the halfshaft securely in a vise and place the locknut on the end of the joint. Attach a slide hammer to the locknut and pull the joint from the halfshaft.

4. Remove the boot bands.

To install:

5. Install the axle boot with new boot bands.

NOTE: Cover the halfshaft serration with tape so as not to damage the new boot during installation.

6. Pack CV-joint and boot with 3.70–4.41 oz. (105–125g) of grease.

7. Align the matchmarks and lightly tap the joint assembly onto the shaft until the joint is locked on the circular clip.

NOTE: Be sure to position the axle nut on the end of the CV-joint to protect the threads when tapping the joint on to the axle.

8. Install the boot and secure it with both boot band clamps.

9. Set the axle boot so that it does not swell or deform when stretched to a length of 3.74 inches.

10. Install the halfshaft in the vehicle.

Maxima

Transaxle Side — Double Offset Joint (DS90)

1. Remove the halfshaft assembly from the vehicle.

2. Remove the boot bands and pull the boot back to expose the CV-joint components.

3. Before separating the joint assembly from the halfshaft, matchmark the slide joint housing and inner race.

4. Pry off the outer snapring, then pull out the slide joint housing.

NOTE: Cover driveshaft serration with tape so not to damage the boot.

5. Place matchmarks on the inner race and the halfshaft.

6. Pry off the middle snapring, then remove the ball cage, the inner race, and the balls as a unit.

7. Pry off the rear snapring, then remove the CV-joint boot.

To install:

8. Install the boot and a new rear snapring.

9. Install the ball cage, the inner race, and the balls as a unit. Be sure that the matchmarks on the inner race and halfshaft are aligned; install a new middle snapring.

10. Pack the CV-joint housing and boot with 6.0 oz. (170g) of grease.

11. Install the slide joint housing and align the matchmarks made during disassembly.

12. Install a new outer snapring.

13. Install the boot and secure with the boot bands.

14. Install the halfshaft in the vehicle.

Wheel Side — Rzeppa Joint (ZF100)

NOTE: The joint on the wheel side cannot be disassembled.

1. Remove the halfshaft from the vehicle.

2. Matchmark the halfshaft and the joint assembly.

3. Separate the joint assembly from the halfshaft by placing the halfshaft securely in a vise and place the locknut on the end of the joint. Attach a slide hammer to the locknut and pull the joint from the halfshaft.

4. Remove the boot bands.

To install:

5. Install the boot with new boot bands.

6. Pack CV-joint and boot with 7.58 oz. (215g) of grease.

7. Align the matchmarks and lightly tap the joint assembly onto the shaft.

8. Install the boot and secure it with both boot band clamps.

9. Install the halfshaft in the vehicle.

240SX

Differential side

1. Remove the halfshaft assembly.

2. Remove the plug seal from the joint assembly by tapping lightly around the flange of the slide joint housing.

3. Remove the spring and spring cap.

4. Remove the CV-joint boot bands and slide the boot towards the center of the halfshaft.

5. Put matchmarks on the slide joint housing and the halfshaft before separating the joint assembly.

6. Put matchmarks on the spider assembly and the halfshaft.

7. Remove the snapring and pull the spider assembly from the halfshaft.

NOTE: Do not disassemble the spider assembly.

8. Remove the slide joint housing from the halfshaft.

9. Remove the boot.

To install:

10. Install the boot onto the halfshaft with the boot bands. Do not crimp the boot bands at this time.

11. Align the matchmarks during disassembly and install the slide joint housing onto the halfshaft.

12. Align the matchmarks made during disassembly and reinstall the spider assembly. Install the spider assembly into place with the chamfered edge facing the shaft. Install a new snapring.

13. Install the spring cap and spring.

14. Apply sealant to the mating surface of the plug seal.

15. Install the plug seal using a suitable press.

   a. When installing the plug seal into place, hold the seal horizontal so that the spring inside it does not tilt or fall down.

   b. Move the shaft in axial (right to left) direction to ensure that the spring is installed correctly. If the shaft drags, binds, or the spring is not installed properly, remove the plug seal and install a new one. Discard the plug seal after it is removed.

16. Pack the slide joint housing and spider assembly with the following amounts of grease:

   a. 6.52–6.88 oz. (185–195 g) of grease for 1993–94 vehicles.

   b. 5.47–5.82 oz. (155–165 g) of grease for 1995 vehicles.

   c. 3.60–3.77 oz. (102–107 g) of grease for 1997 vehicles.

17. Slide the boot over the slide joint housing. The small part of the boot should be firmly seated in the halfshaft groove.

18. Set the boot so the that it does not swell or be deformed when the measured length of the boot is as follows:

   a. 4.35–4.43 inches (110.5–112.5mm) for 1993–94 vehicles.

   b. 3.74–3.82 inches (95–97mm) for 1995–97 vehicles.

19. Place the boot bands in their proper location and crimp the boot bands at this time.

20. Install the halfshaft in the vehicle.

Wheel side

1. Remove the halfshaft from the vehicle.

2. Remove the CV-joint boot bands and pull the boot away from the joint housing.

3. Put matchmarks on the housing and the halfshaft.

4. Remove the snapring from the joint housing.

5. Remove the housing from the halfshaft.

6. Put matchmarks on the spider assembly and the halfshaft.

7. Remove the snapring from the spider assembly and remove the spider assembly from the halfshaft.

8. Remove the boot.

To install:

9. Install the new boot and new small boot clamp onto the driveshaft. Do not crimp the boot band at this time.

10. Align the matchmarks made during disassembly and install the spider assembly. Install a new snapring.

11. Pack the slide joint housing and spider assembly with the following amounts of grease:

   a. 5.11–5.47 oz. (145–155 g) of grease for 1993–94 vehicles.

   b. 4.76–5.11 oz. (135–145 g) of grease for 1995 vehicles.

   c. 4.06–4.41 oz. (115–125 g) of grease for 1997 vehicles.

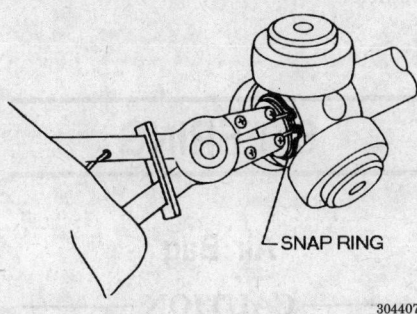

Removing the spider assembly snapring — 240SX

12. Align the matchmarks made during disassembly and install the housing onto the driveshaft.

13. Install a new housing snapring.

14. Slide the boot onto the housing and verify that the small end of the boot is firmly seated in the halfshaft groove.

15. Set the boot so the that it does not swell or be deformed when the measured length of the boot is as follows:

   a. 4.35–4.43 in. (110–112mm) for 1993–94 vehicles.

   b. 3.74–3.82 in. (95–97mm) for 1995–97 vehicles.

16. Install the large boot band and make sure the small boot band is in the correct position. Crimp the CV-joint boot bands at this time.

17. Install the halfshaft in the vehicle.

300ZX

NOTE: After it is removed from the vehicle, the actual halfshaft is referred to as a driveshaft during disassembly and reassembly.

Differential side

1. Remove the halfshaft assembly.

2. Remove the plug seal from the joint assembly by tapping lightly around the flange of the slide joint housing.

3. Remove the CV-joint boot bands and slide the boot towards the center of the driveshaft.

4. Put matchmarks on the slide joint housing and the driveshaft before separating the joint assembly.

5. Remove the large snapring from the outside of the slide joint housing.

6. Remove the slide joint housing from the driveshaft.

7. Put matchmarks on the ball and cage assembly and the driveshaft.

8. Remove the snapring and pull the ball and cage assembly from the driveshaft.

NOTE: Do not disassemble the ball and cage assembly.

9. Remove the boot.

To install:

NOTE: Cover the driveshaft serration with tape to prevent damage to the new CV-joint boot.

10. Install the boot onto the driveshaft with the boot bands. Do not crimp the boot bands at this time.

11. Align the matchmarks made during disassembly and reinstall the ball and cage assembly. Install a new snapring to secure the ball and cage assembly.

12. Fill the slide joint housing with the following specified amount of grease:

    a. Non-Turbocharged vehicles — 5.82–6.17 oz. (165–175 g)

    b. 1994–95 Turbocharged vehicles — 6.35–7.05 oz. (180–200 g)

    c. 1996 Turbocharged vehicles — 5.47–6.17 oz. (155–175 g)

13. Align the matchmarks during disassembly and install the slide joint housing onto the driveshaft.

14. Apply sealant to the mating surface of the plug seal.

15. Install the plug seal using a suitable press.

    a. When installing the plug seal into place, hold the seal horizontally so that the spring inside it does not tilt or fall down.

    b. Move the shaft in axial (right to left) direction to ensure that the joint is installed correctly. If the shaft drags or binds, it is not installed properly.

16. Slide the boot over the slide joint housing. The small part of the boot should be firmly seated in the driveshaft groove.

17. Set the boot so the that it does not swell or become deformed when the measured length of the boot is as follows:

    a. Non-Turbocharged vehicles — 3.66–3.74 inches (93–95mm)

    b. 1994–95 Turbocharged vehicles — 4.04–4.11 inches (102.5–104.5mm)

    c. 1996 Turbocharged vehicles — 3.76–3.84 inches (95.5–97.5mm)

18. Place the boot bands in their proper location and crimp the boot bands at this time.

19. Install the halfshaft in the vehicle.

Wheel side

After it is removed from the vehicle, the actual halfshaft is referred to as a driveshaft during disassembly and reassembly.

NOTE: The joint on the wheel side can not be disassembled. If defective it can only be replace.

1. Remove the driveshaft from the vehicle.

2. Position the driveshaft in a vise.

3. Remove the CV-joint boot bands and pull the boot away from the joint housing.

4. Put matchmarks on the housing and the driveshaft.

5. Install the axle shaft nut and position a slide hammer tool onto the nut. Remove the CV-joint using the hammer tool.

6. Remove the circular clip from the driveshaft.

NOTE: The circular clips must be replace whenever the CV-joint is removed.

7. Remove the boot and discard the circular clips.

To install:

NOTE: Cover the driveshaft serration with tape to prevent damage to the new CV-joint boot.

8. Install the new boot and new clamps onto the driveshaft. Do not crimp the boot band at this time.

9. Install new circular clips.

10. Pack the CV-joint housing assembly with the following amount of grease:

    a. 1994 Non-Turbocharged vehicles — 5.29–5.64 oz. (150–160 g)

    b. 1995 Non-Turbocharged vehicles — 3.99–4.34 oz. (113–123 g)

    c. 1994–95 Turbocharged vehicles — 6.00–6.70 oz. (170–190 g)

    d. 1996 Turbocharged vehicles — 4.41–5.11 oz. (125–145 g)

11. Align the matchmarks made during disassembly and install the housing onto the driveshaft.

12. With the axle shaft nut on the CV-joint, tap on the end of the shaft to seat the joint.

13. Set the boot so the that it does not swell or be deformed when the measured length of the boot is as follows:

    a. Non-Turbocharged vehicles — 3.78–3.86 inches (96–98mm)

    b. Turbocharged vehicles — 3.98–4.06 inches (102.5–104.5mm)

14. Install the large boot band and make sure the small boot band is in the correct position. Crimp the CV-joint boot bands.

15. Install the halfshaft in the vehicle.

STEERING

Air Bag

——— CAUTION ———
Some vehicles are equipped with an air bag system, also known as the Supplemental Inflatable Restraint (SIR) or Supplemental Restraint System (SRS). The system must be disabled before performing service on or around system components, steering column, instrument panel components, wiring and sensors. Failure to follow safety and disabling procedures could result in accidental air bag deployment, possible personal injury and unnecessary system repairs.

PRECAUTIONS

Several precautions must be observed when handling the inflator module to avoid accidental deployment and possible personal injury.

• Never carry the inflator module by the wires or connector on the underside of the module.

• When carrying a live inflator module, hold securely with both hands, and ensure that the bag and trim cover are pointed away.

• Place the inflator module on a bench or other surface with the bag and trim cover facing up.

• With the inflator module on the bench, never place anything on or close to the module which may be thrown in the event of an accidental deployment.

DISARMING

All Models

NOTE: All SRS electrical wiring harnesses and connectors are covered with YELLOW outer insulation. Do not use electrical test equipment on any circuit related to the SRS (air bag) sensors. When installing SRS components, always install with the arrow marks facing the front of the vehicle.

To disarm the SRS system turn the ignition switch to **OFF** position. Then disconnect the both battery cables starting with the negative cable first and wait at least 10 minutes after the cables are disconnected. Be sure to insulate the battery terminal ends.

Arming

To arm the SRS system turn the ignition switch to **OFF** position. Connect the both battery cables starting with the positive cable first.

NOTE: The SRS or air bag system is equipped with a self-diagnostic operation. After turning the ignition key to the ON or START position, the AIR BAG warning lamp will illuminate for 7 seconds. After 7 seconds, the AIR BAG lamp will extinguish if no malfunction is detected. If the AIR BAG lamp does not extinguish after 7 seconds, check the SRS self diagnostic system for a malfunction.

Steering Wheel

REMOVAL AND INSTALLATION

Sentra, 200SX and 300ZX

Without Air Bag

1. Position the wheels in the straight ahead direction. The steering wheel should be right side up and level.
2. Disconnect the battery ground cable.
3. Some vehicles have countersunk screws on the back of the steering wheel, remove the screws and pull off the horn pad.

NOTE: Some vehicles have a horn wire running from the pad to the steering wheel; disconnect it.

4. Remove the rest of the horn switching mechanism, noting the relative location of the parts. Remove the mechanism only if it hinders subsequent wheel removal procedures.
5. Matchmark the top of the steering column shaft and the steering wheel flange.
6. Remove the attaching nut. Using the steering wheel remover tool ST27180001 or equivalent, pull the steering wheel from the steering column.

NOTE: Do not strike the shaft with a hammer, which may cause the column to collapse.

To install:

7. Prior to installing the steering wheel, apply multi-purpose grease to the entire surface of the turn signal cancel pins and the horn contact ring. Install the steering wheel in the reverse order of removal and align the punch marks. DO NOT drive or hammer the wheel.
8. Tighten the steering wheel nut to 29–40 ft. lbs. (39–54 Nm).

9. Install the horn button, pad, or the ring.

With Air Bag (SRS)

——————— **CAUTION** ———————
To avoid rendering the SRS (Supplemental Restraint System) inoperative, which could lead to personal injury or death in the event of a severe frontal collision, extreme caution must be taken when servicing the electrical related systems. All SRS electrical wiring harnesses and connectors are covered with YELLOW outer insulation. Do not use electrical test equipment on any circuit related to the SRS (air bag).

NOTE: On vehicles equipped with an air bag, turn the ignition switch to OFF position and properly disable the air bag system.

1. Disconnect the negative battery cable and wait for 10 minutes. The 10 minute wait is to allow the SRS system to de-energize. Always install SRS components with the arrow facing the front of the vehicle.
2. Remove the lower lid from the steering wheel and disconnect the air bag module connector.
3. Remove the steering column side lid. Using a T50H Torx® bit, remove the left and right special screws. The air bag module can be removed by disconnecting the spiral cable connector.
4. Set the steering wheel straight ahead and disconnect the horn connector and remove the nuts.

NOTE: Do not strike the shaft with a hammer, which may cause the column to collapse.

5. Using a steering wheel puller, remove the steering wheel.

To install:

6. Prior to installing the steering wheel, apply multi-purpose grease to the entire surface of the turn signal cancel pins and the horn contact ring. Install the steering wheel and torque the nut to 22–29 ft. lbs. (29–39 Nm).
7. Connect the spiral cable connector. Install the air bag module and tighten the special Torx® screws to 24 inch. lbs. (4.0 Nm) on 300ZX models and 11–18 ft. lbs. (15–25 Nm) on all other models.

NOTE: The Torx® bolts must be replaced with new bolts whenever removed.

8. Enable the supplemental restraint system.

9. Install the cover lids.
10. Connect the negative battery cable and check operation.

NOTE: When the ignition switch is first turned on, the AIR BAG lamp should illuminate for approximately seven seconds. After seven seconds, the lamp should extinguish. If the lamp does not extinguish or illuminate, check the air bag self diagnostic system for a malfunction.

240SX

1. Disconnect the negative battery cable.
2. Disarm the air bag system.

——————— **CAUTION** ———————
The air bag system must be disarmed before removing the steering wheel. Failure to do so may cause accidental deployment, property damage or personal injury.

3. Remove the access cover on the side of the steering wheel.
4. Remove and discard the special bolt from the side of the steering column.
5. Remove the steering wheel pad/air bag module assembly from the steering wheel.

——————— **CAUTION** ———————
Always carry an air bag assembly with the bag and trim cover away from your body. Store the assembly facing upward; never place the assembly face down on any surface.

6. Matchmark the steering wheel to the steering column shaft.
7. Remove the steering wheel hold down nut.
8. Remove the steering wheel with a suitable puller tool.

To install:

9. Install the steering wheel and align the matchmark located on the steering column shaft.
10. Install the steering wheel hold down nut.
11. Torque the nut to 22–29 ft. lbs. (29–39 Nm.).
12. Install the steering wheel pad/air bag module.
13. Install a new steering wheel pad/air bag module assembly securing bolt and tighten the bolt to 11–18 ft. lbs. (15–25 Nm).
14. Install the access cover to the side of the steering column.
15. Connect negative battery cable and re-arm the air bag system.

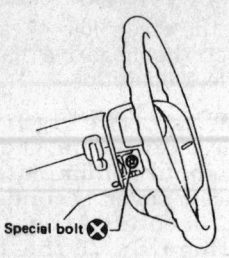

Special bolt ⊗

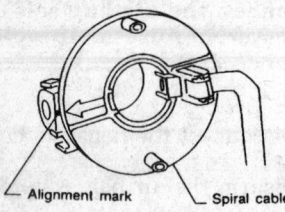

Alignment mark — Spiral cable

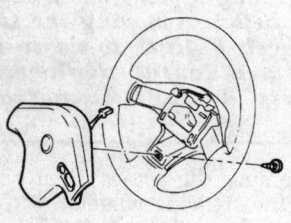

304172

Service points of air bag
removal — 300ZX

Altima

1. Disconnect the negative battery cable.
2. Disable the air bag system.

----- **CAUTION** -----
Be sure to follow proper procedures for disabling the air bag. Failure to do so may result in bodily injury and damage to the vehicle.

NOTE: The air bag module mounting bolts are coated with a special bonding agent. Replace with new bolts whenever they are removed.

3. Remove the air bag left and right retaining bolts and remove the air bag assembly.
4. Remove the steering wheel hold-down nut.
5. Verify that the wheels are in the straight ahead position. Make a matchmark on the steering shaft and a matching one on the steering wheel.
6. Remove the steering wheel with the proper pulling tool.

To install:
7. Install the steering wheel, be sure to align the matchmarks.

8. Install the steering wheel hold-down nut.
9. Torque the hold-down nut to 25–29 ft. lbs. (33–39 Nm.).
10. With the front wheels raised off the floor, verify that the steering wheel turns freely and an equal number of turns to the left and the right from the center.
11. Install the air bag and tighten the mounting bolts to 11–18 ft. lbs. (15–25 Nm).
12. Enable the air bag system.
13. Connect the negative battery cable.

Maxima

Without Air Bag

1. Position the wheels in the straight-ahead direction. The steering wheel should be right side up and level.
2. Disconnect the negative battery cable.
3. Pull out the horn pad. If the vehicle has a horn wire running from the pad to the steering wheel, disconnect it.
4. Remove the rest of the horn switching mechanism, noting the relative location of the parts. Remove the mechanism only if it hinders subsequent wheel removal procedures.
5. Matchmark the top of the steering column shaft and the steering wheel flange.
6. Remove the steering wheel-to-steering column nut. Using the steering wheel puller tool ST27180001, or equivalent, pull the steering wheel from the steering column.

NOTE: Do not strike the shaft with a hammer, which may cause the column to collapse.

To install:
7. Apply multi-purpose grease to the entire surface of the turn signal cancel pin (both portions) and the horn contact slip ring.
8. Install the steering wheel by aligning the matchmarks. Do not drive or hammer the steering wheel into place or the steering column will be damaged.
9. Tighten the steering wheel-to-steering column nut to 22–29 ft. lbs. (29–39 Nm).
10. Install the horn pad and connect the negative battery cable.

With Air Bag

1. Position the wheels in the straight-ahead direction. The steering wheel should be right side up and level.
2. Disconnect the negative battery cable.

----- **CAUTION** -----
Wait 10 minutes after the battery cable has been disconnected, before attempting to work on the air bag unit. The air bag unit is still armed and can inflate, during the 10 minute period, and possibly causing bodily injury.

3. Remove the lower lid from the steering wheel and disconnect the air bag unit electrical connector.
4. Remove both side lids from the steering wheel.
5. Using a T50H Torx® bit, remove the air bag-to-steering wheel bolts from both sides of the steering wheel; discard the bolts.
6. Lift the air bag unit upward and place it in a safe, clean, dry place with the pad side facing upward; be sure the temperature in the area will not exceed 212°F (100°C).
7. Disconnect the horn connector and remove the nuts.
8. Matchmark the top of the steering column shaft and the steering wheel flange.
9. Remove the steering wheel-to-steering column nut. Using the steering wheel puller tool ST27180001, or equivalent, pull the steering wheel from the steering column.

NOTE: Do not strike the shaft with a hammer, which may cause the column to collapse.

To install:
10. Apply multi-purpose grease to the entire surface of the turn signal cancel pin (both portions) and the horn contact slip ring.
11. Install the steering wheel by aligning the matchmarks. Do not drive or hammer the steering wheel into place or you may cause the (collapsible) steering column to collapse, in which case you'll have to buy a whole new steering column unit.
12. Torque the steering wheel-to-steering column nut to 25–29 ft. lbs. (33–39 Nm).
13. Connect the horn electrical connector and install the nuts.
14. Align the spiral cable correctly with the alignment mark and install the air bag unit into the steering wheel.
15. Using new air bag-to-steering wheel Torx® bolts, install them into both sides of the steering wheel and torque to 11–18 ft. lbs. (15–25 Nm).
16. Install both side lids to the steering wheel.
17. Connect the air bag unit electrical connector and install the lower lid to the steering wheel.
18. Connect the negative battery cable and make sure the air bag **red**

indicator light turns ON. The AIR BAG light should extinguish after about seven seconds. If the light does not extinguish, check the air bag self diagnostic system for a malfunction.

Tie Rod Ends

REMOVAL AND INSTALLATION

1993–94 Sentra and Maxima

1. Raise the front of the vehicle and support it on jackstands. Remove the wheel.
2. Remove the cotter pin and the tie rod ball joint stud nut. Note the position of the steering linkage.
3. Loosen the tie rod-to-steering gear locknut.
4. Using a suitable ball joint separator tool remove the tie rod ball joint from the steering knuckle.
5. Loosen the locknut and remove the tie rod end from the tie rod, counting the number of complete turns it takes to completely free it.
To install:
6. Install the new tie rod end, turning it in exactly as far as you screwed out the old one. Make sure it is correctly positioned in relationship to the steering linkage.
7. Fit the ball joint and nut, tighten them and install a new cotter pin. Torque the ball joint stud nut to specifications. Check front end alignment.
8. The outer tie rod end-to-steering knuckle torque specification is 22–29 ft. lbs. (29–39 Nm.).

NOTE: Always replace the cotter pins and if necessary replace the retaining nut.

9. Check and adjust the front end alignment as necessary.

1995–97 Sentra and 200SX

1. Raise the front of the vehicle and support it on jackstands. Remove the wheel.
2. Remove the cotter pin and the tie rod ball joint stud nut. Note the position of the steering linkage.
3. Using a suitable ball joint separator tool, remove the tie rod ball joint from the steering knuckle.
4. Loosen the locknut and remove the tie rod end from the tie rod. Count the number of complete turns it takes to completely remove it.
To install:
5. Install the new tie rod end, turning it in exactly as many turns as it was to remove the old one. Make sure it is correctly positioned in relationship to the steering linkage.
6. Connect the outer tie rod end-to-steering knuckle and install the castle nut. Torque the nut to 22–29 ft. lbs. (29–39 Nm.).
7. Install a new cotter pin to the castle nut.
8. Tighten the tie rod end locking nut to 27–34 ft. lbs. (37–46 Nm.).
9. Install the wheel and tire assembly.
10. Lower the vehicle and perform a front end alignment.

240SX, 300ZX and Altima

1. Raise and safely support the vehicle. Remove the wheel(s).
2. Remove the cotter pin and the tie rod ball joint stud nut. Note the position of the steering linkage.
3. Loosen the tie rod-to-steering gear locknut.
4. Using a suitable ball joint separator tool, remove the tie rod ball joint from the steering knuckle.
5. Loosen the locknut and remove the tie rod end from the inner tie rod, counting the number of complete turns it takes to completely free it.
To install:
6. Thread the new tie rod end onto the inner tie rod, turning it in the

exact number of turns as removal. Make sure it is correctly positioned in relationship to the steering linkage.
7. Connect the tie rod end to the knuckle assembly and torque the retaining nut to 22–29 ft. lbs. (29–39 Nm). Install a new cotter pin.
8. Install the wheel(s) and lower the vehicle.
9. Check front end alignment.
10. Once adjustment is complete, tighten tie rod locknut to 58–72 ft. lbs. (78–98 Nm).

1995–97 Maxima

1. Raise the front of the vehicle and support it on jackstands. Remove the wheel.
2. Remove the cotter pin and the tie rod ball joint stud nut. Note the position of the steering linkage.
3. Using a suitable ball joint separator tool, remove the tie rod ball joint from the steering knuckle.
4. Loosen the locknut and remove the tie rod end from the tie rod. Count the number of complete turns it takes to completely remove it.
To install:
5. Install the new tie rod end, turning it in exactly as many turns as it was to remove the old one. Make sure it is correctly positioned in relationship to the steering linkage.
6. Connect the outer tie rod end-to-steering knuckle and install the castle nut. Torque the nut to 46–54 ft. lbs. (63–73 Nm.).
7. Install a new cotter pin to the castle nut.
8. Tighten the tie rod end locking nut to 58–72 ft. lbs. (78–98 Nm).
9. Install the wheel and tire assembly.
10. Lower the vehicle and perform a front end alignment.

Sentra and 200SX

1. Raise and support the car on jackstands.
2. Remove the front wheels.
3. Remove the both tie rod ends from the steering knuckles.
4. Matchmark the steering column shaft to the lower joint and remove the pinch bolt from the joint.
5. Loosen and remove the steering gear mounting bolts.
6. Remove the mounting clamps from the steering gear and remove the steering gear by sliding it off the steering shaft.
7. Remove the steering gear from the vehicle.
To install:
8. Install the steering gear assembly to the vehicle. Be sure to align the

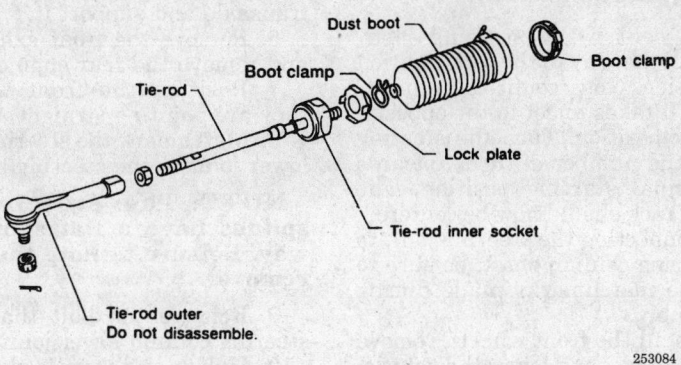

253084

This style of tie rod is common to all the vehicles covered in this section

matchmarks of the rack with the marks on the steering shaft.

9. Install the steering gear mounting clamps and tighten the bolts to 58 ft. lbs. (78 Nm).

10. Install the lower joint-to-steering column pinch bolt and tighten to 22 ft. lbs. (29 Nm).

11. Connect the tie rod end to the steering knuckle and install castle nut. Torque the nut to 29 ft. lbs. (39 Nm) and install a new cotter pin.

NOTE: If installing a new rack and pinion assembly, transfer the lower steering joint to the new rack and pinion prior to installation. When installing the lower steering joint to the steering gear, make sure that the wheels are aligned with the vehicle (straight ahead position). Turn the steering gear all the way to the lock position on one side. Now, count the number of turns it takes to get to the opposite side lock position. Turn the steering gear ½ the number of turns towards the original starting position. The steering rack should now be centered. When connecting the steering joint to the steering column shaft, be sure to align the matchmarks made during disassembly.

12. Install the front wheels, remove the jackstands, and lower the vehicle.

13. Check the vehicle's alignment.

Manual Rack and Pinion

REMOVAL AND INSTALLATION

Sentra and 200SX

1. Raise and support the car on jackstands.

2. Remove the front wheels.

3. Remove the both tie rod ends from the steering knuckles.

4. Matchmark the steering column shaft to the lower joint and remove the pinch bolt from the joint.

5. Loosen and remove the steering gear mounting bolts.

6. Remove the mounting clamps from the steering gear and remove the steering gear by sliding it off the steering shaft.

7. Remove the steering gear from the vehicle.

To install:

8. Install the steering gear assembly to the vehicle. Be sure to align the matchmarks of the rack with the marks on the steering shaft.

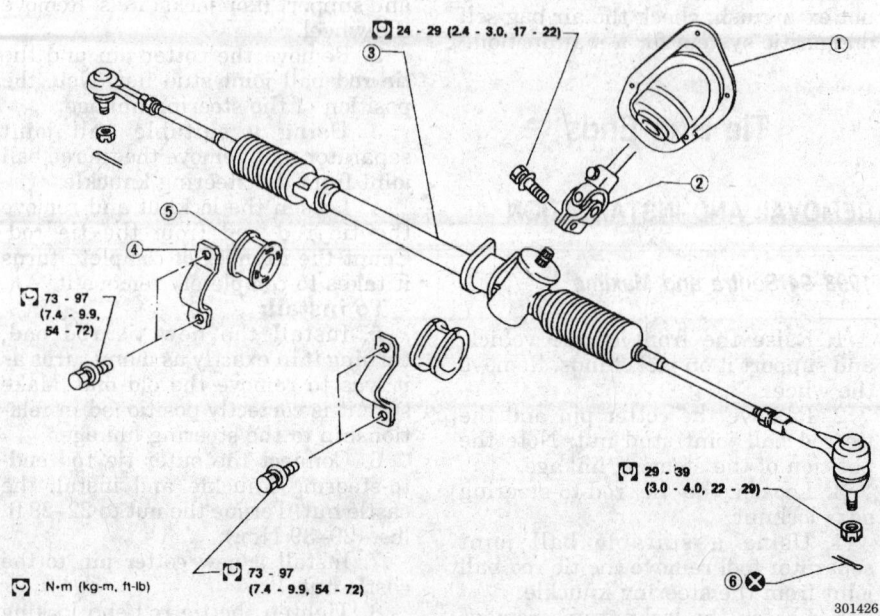

9. Install the steering gear mounting clamps and tighten the bolts to 58 ft. lbs. (78 Nm).

10. Install the lower joint-to-steering column pinch bolt and tighten to 22 ft. lbs. (29 Nm).

11. Connect the tie rod end to the steering knuckle and install castle nut. Torque the nut to 29 ft. lbs. (39 Nm) and install a new cotter pin.

NOTE: If installing a new rack and pinion assembly, transfer the lower steering joint to the new rack and pinion prior to installation. When installing the lower steering joint to the steering gear, make sure that the wheels are aligned with the vehicle (straight ahead position).

12. To center the steering gear, turn it all the way to the lock position on one side. Now, count the number of turns it takes to get to the opposite side lock position. Turn the steering gear ½ the number of turns towards the original starting position. The steering rack should now be centered. When connecting the steering joint to the steering column shaft, be sure to align the matchmarks made during disassembly.

13. Install the front wheels, remove the jackstands, and lower the vehicle.

14. Check the vehicle's alignment.

Power Rack and Pinion

REMOVAL AND INSTALLATION

Sentra and 200SX

1. Raise and support the vehicle safely.

2. Disconnect the low pressure hose clamp and remove the low pressure hose at the steering gear. Be sure to use a pan to catch the fluid.

3. Disconnect the flare nut and the high pressure tube at the steering gear, then drain the fluid from the gear.

4. Remove the tie rod ends from the steering knuckle.

5. Place a floor jack under the transaxle and support it.

6. Remove the front exhaust pipe and remove the rear engine mount.

7. Position the front wheels so they are pointing straight ahead.

8. Matchmark the steering column lower joint to the steering gear.

NOTE: The steering gear splines have a flat spot or key way. Be sure to note this during removal.

9. Remove the bolt that secures steering column lower joint.

10. Unbolt and remove the steering gear unit and the linkage.

Caption for figure: Manual rack and pinion assembly — Sentra and 200SX

Figure callout text:
24 - 29 (2.4 - 3.0, 17 - 22)
73 - 97 (7.4 - 9.9, 54 - 72)
73 - 97 (7.4 - 9.9, 54 - 72)
N·m (kg-m, ft-lb)
29 - 39 (3.0 - 4.0, 22 - 29)
301426

To install:

11. Install the power steering gear assembly to the vehicle. Align the steering column to the steering gear.

NOTE: Be sure to align the flat spot or keyway during installation.

12. Install the steering gear mounts and tighten the mounting bolts in sequence to 54–72 ft. lbs. (73–97 Nm).

13. Install the pinch bolt for the steering column-to-gear connection and tighten the bolt to 17–22 ft. lbs. (24–29 Nm).

14. Connect the tie rod ends to the steering knuckle and tighten the mounting nut to 22–29 ft. lbs. (29–39 Nm).

15. Tighten the tie rod mounting nut further so the groves in the nut align with first cotter pin hole. Install a new cotter pin.

16. Connect the power steering low pressure hose to the steering gear and tighten the fitting to 20–29 ft. lbs. (27–39 Nm).

17. Connect the power steering high pressure hose to the steering gear and tighten the fitting to 11–18 ft. lbs. (15–25 Nm).

18. Install the rear engine mount and remove the floor jack.

19. Using new gaskets, install the front exhaust pipe assembly.

20. Fill the power steering system and start the engine.

21. Bleed the power steering system and check the wheel alignment.

Altima and 240SX

——————— **CAUTION** ———————

The air bag system must be disarmed before removing the rack and pinion. Failure to do so may cause accidental deployment, property damage or personal injury.

1. Disconnect the negative battery cable and disarm the air bag.

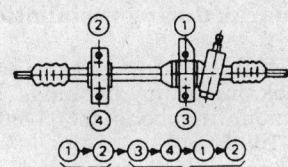

Temporary tightening   Secure tightening

301431

Mounting bolt torque sequence — 1995–97 Sentra and 200SX

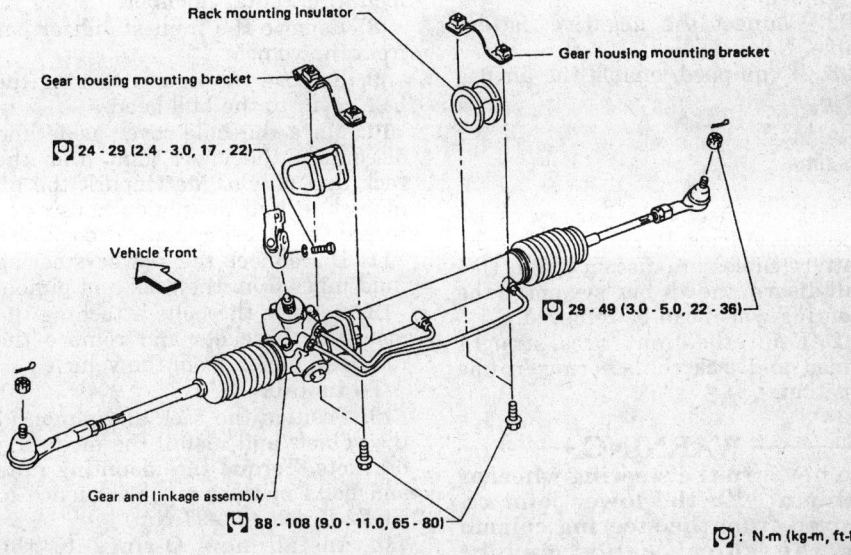

⊙ 24 - 29 (2.4 - 3.0, 17 - 22)

Vehicle front

⊙ 29 - 49 (3.0 - 5.0, 22 - 36)

Rack mounting insulator

Gear housing mounting bracket

Gear housing mounting bracket

Gear and linkage assembly

⊙ 88 - 108 (9.0 - 11.0, 65 - 80)

⊙ : N·m (kg-m, ft-lb)

294444

Power steering gear assembly — 240SX

2. Raise and safely support the vehicle as necessary.

3. Remove the bolt securing the lower steering column shaft to the power steering gear assembly. Be sure to matchmark the shaft from the steering gear to the steering column joint for correct installation.

4. Disconnect the hoses from the power steering gear and plug the hoses to prevent leakage.

5. Remove the cotter pins and castle nuts from the tie rod ends.

6. Using a ball joint separating tool, remove tie rod ends from the steering knuckle.

7. Remove the front exhaust pipe mounting nuts and bolts. Remove the front exhaust pipe from the vehicle.

8. If necessary, disconnect the control cable or linkage from the transmission and position it out of the way.

9. Remove the power steering gear mounting bolts or nuts.

10. Remove the steering gear from the vehicle. Use care when separating the steering column joint.

11. Inspect the steering gear mount bushings and replace as necessary.

To install:

12. Align the steering column-to-steering gear matchmark and install the steering gear to the vehicle. Be

sure to properly install the mounting bushings and hand tighten the mounting nuts or bolts.

NOTE: When installing the lower steering joint to the steering gear, make sure that the wheels are aligned straight and the steering joint slot is aligned.

13. Tighten the steering gear mounts as follows:
   a. 1993–97 240SX — Tighten to 65–80 ft. lbs. (88–108 Nm)
   b. 1993–96 Altima — Tighten to 54–72 ft. lbs. (73–97 Nm)

14. Install the pinch bolt securing the lower steering column shaft to the power steering gear assembly. Tighten the pinch bolt to 17–22 ft. lbs. (24–29 Nm)

15. Install the tie rod end to steering knuckle and tighten the castle nut to 22–29 ft. lbs. (29–39 Nm)

16. Tighten the castle nut further to align the slot in the castle nut with the cotter pin hole and install a new cotter pin.

17. If removed, connect the control cable or linkage to the transmission and position it out of the way.

18. Using new gaskets, install the front exhaust pipe assembly.

19. Connect the power steering hoses to the steering gear and fill the power steering reservoir.

20. Start the engine and refill the power steering reservoir.
21. Bleed the power steering system and perform a front end alignment.
22. Connect the negative battery cable.
23. If equipped, enable the air bag system.

Maxima

1. Disconnect both battery cables and wait at least 10 minutes after the battery cables are disconnected. This will disarm the air bag system so the steering wheel can be removed.
2. Point the front tires straight ahead and lock the steering in this position.

WARNING
Do not turn the steering wheel or column with the lower joint removed from the steering column or the spiral cable may be damaged.

3. Remove the steering wheel.

CAUTION
The air bag system must be disarmed before removing the steering wheel. Failure to do so may cause accidental deployment, property damage or personal injury.

NOTE: The steering wheel must be removed before disconnecting the steering column lower joint to avoid damaging the SRS spiral cable.

4. Raise and support the vehicle safely and remove the front wheels.
5. Disconnect the tie rod ends from the steering knuckles.
6. Remove the carbon canister from the vehicle.

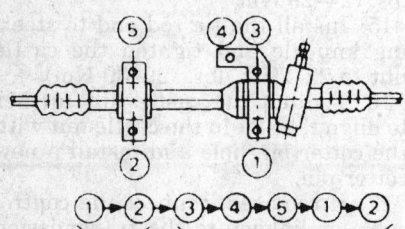

Rack and pinion bolt tightening sequence — Maxima

288833

Temporary tightening   Secure tightening

7. Support the engine then remove the bolts attaching the engine mounts to the engine mounting center member. Remove the engine mounting center member.
8. Remove the front stabilizer bar from the vehicle.
9. Remove the nuts attaching the hole cover to the bulkhead.
10. Move the hole cover aside and disconnect the lower joint from the rack and pinion. Matchmark the pinion shaft and the pinion housing to record the steering neutral position.
11. Disconnect the power steering fluid pipes from the rack and pinion.
12. Remove the bolts attaching the mounting brackets and remove the rack and pinion from the vehicle.
To install:
13. Position the rack and pinion in the vehicle and install the mounting brackets. Torque the mounting nuts and bolts in the proper sequence to 54–72 ft. lbs. (73–97 Nm).
14. Install new O-rings to the power steering fluid pipes and connect them to the rack and pinion. Torque the low pressure line 20–29 ft. lbs. (27–39 Nm). Torque the high pressure line to 11–18 ft. lbs. (15–25 Nm).
15. Align the lower steering joint to the pinion shaft and install the joint onto the pinion shaft. Install the bolt and torque the bolt to 17–22 ft. lbs. (24–29 Nm).
16. Properly position the hole cover and install the attaching nuts, torque the nuts to 2.9–3.6 ft. lbs. (4–5 Nm).
17. Install the front stabilizer.
18. Install the engine mounting center member and torque the attaching bolts to 57–72 ft. lbs. (77–98 Nm). Attach the engine mounts to the center member and torque the bolts to 57–72 ft. lbs. (77–98 Nm). Remove the support from the engine.
19. Install the remaining components in the reverse order of removal.
20. Tighten the tie rod end nuts to 46–61 ft. lbs. (63–82 Nm), then install a new cotter pin.
21. Fill the power steering reservoir with fluid and bleed the air from the power steering system.
22. Check the vehicle front end alignment and adjust as necessary.

300ZX

WARNING
On air bag equipped vehicles, the rotation of the spiral cable is limited. If the steering gear must be removed, set the front wheels in the straight-ahead direction. Do not rotate the steering column while the steering gear is removed.

1. Raise and support the vehicle safely.
2. Disconnect the electrical connector from the solenoid valve.
3. Position an oil catch pan under the power steering gear, remove the hydraulic lines from the gear and drain the oil. Plug the lines to prevent leakage and discard the copper washers.
4. Loosen the steering column lower joint shaft bolt.
5. Before disconnecting the lower ball joint set the steering gear assembly in neutral by making the wheels straight. Loosen the bolt and disconnect the lower joint. Matchmark the pinion shaft to the pinion housing to record the neutral gear position.
6. Remove the tie rod end-to-knuckle arm cotter pins and castle nuts.
7. Separate the tie rods from the knuckle using a suitable puller.
8. Remove the steering gear housing mounting bolts.
9. Position a jack under the engine and raise it just enough to support the engine. Loosen the engine mounting bolts and raise the engine about ½ in. (12mm).

NOTE: To prevent damage to engine components, place a block of wood between the engine and the jack, when raising the engine.

10. Remove the steering gear and linkage from the vehicle.
To install:
11. Install the power steering gear and linkage into the vehicle.
12. Install the lower steering column joint to the steering gear pinion and torque bolt to 17–22 ft. lbs. (22–29 Nm).

NOTE: Be sure to align the matchmarks during installation.

13. Install steering gear clamps with rack mounting bushings and torque mounting bolts to 65–80 ft. lbs. (88–108 Nm).
14. Connect the tie rod ends to the steering knuckles and tighten nuts to 22–29 ft. lbs. (29–39 Nm). Install new cotter pins.
15. Using new copper washers, connect the power steering lines and torque the high pressure hydraulic line fitting to 22–26 ft. lbs. (36–40 Nm)

and lower pressure fitting to 27–30 ft. lbs. (36–40 Nm).

NOTE: The copper washers in the lower pressure hydraulic line fitting are larger than the washers in the high pressure line. Make sure the washers are installed in the proper fittings. Observe the torque specification given for the hydraulic line fittings. Over-tightening will cause damage to the fitting threads and washers.

16. Connect the electrical connector to the solenoid valve.

17. Lower the engine and tighten the mounting bolts.

18. Refill the power steering pump, start the engine and bleed the system.

19. Perform front end alignment.

Power Steering Pump

BLEEDING

All Models

1. Raise and support the front of the vehicle safely.

2. Check and add fluid to the reservoir, if necessary.

3. Turn the steering wheel quickly (all the way), right and left, just touching the stops. Continue turning the steering wheel until the fluid level in the reservoir no longer decreases.

4. Check the fluid level, add fluid as required.

5. Start the engine and continue turning the steering wheel from right to left, lightly touching the stops until the fluid level in the reservoir no longer decreases.

6. Stop the engine and check the fluid level, add fluid as required.

NOTE: When bleeding the system, make sure the temperature of the fluid reaches 140–176°F (60–80°C).

7. Repeat the steps until all of the air is bled from the system.

8. If the air cannot be bled from the system, turn and hold the steering wheel at each stop for at least 5 seconds but never more than 15 seconds.

REMOVAL AND INSTALLATION

Sentra and 200SX

1. Disconnect the negative battery cable.

2. Remove the air cleaner duct and air cleaner.

3. Loosen the idler pulley locknut and turn the adjusting nut counterclockwise, in order to remove the power steering belt.

4. Remove the drive belt from the air conditioning compressor.

5. Loosen the power steering hoses at the pump and remove the bolts holding the power steering pump to the bracket.

6. Disconnect and plug the power steering hoses and remove the pump from the vehicle.

To install:

7. Install the power steering pump and connect hoses. Hand-tighten the power steering hoses at this time.

8. Tighten the power steering pump mounting bolts to specifications.

9. Install the A/C and power steering drive belts. Adjust the drive belt tension.

10. Finish tightening the power steering hoses.

11. Connect the negative battery cable and fill the power steering reservoir with Dexron II® automatic transmission fluid.

12. Bleed the power steering system by turning the steering wheel from side-to-side with the engine running. Verify that there are no power steering system fluid leaks.

Maxima, Altima, 300ZX and 240SX

1. Disconnect the negative battery cable.

2. If necessary, remove the air cleaner duct and the air cleaner.

3. Remove the drive belt from the air conditioning compressor, if equipped.

4. Loosen power steering pump belt adjustment as follows:

   a. Loosen the pivot and mounting bolts.

   b. Loosen the idler pulley locknut and turn the adjusting nut counterclockwise to remove the power steering belt.

5. Loosen the power steering hoses at the pump and remove the bolts holding the power steering pump to the bracket.

6. Disconnect and plug the power steering hoses and remove the pump from the vehicle.

To install:

7. Using new O-rings, connect the power steering hoses to the steering pump.

8. Install the power steering pump and secure it with its mounting bolts.

Tighten the front and rear mounting bolts to specifications.

9. Install the remaining components in the reverse order from which they were removed.

10. Connect the negative battery cable.

11. Fill the power steering system, start the engine and turn the steering wheel from side-to-side to bleed air from system.

BRAKES

Anti-Lock Brake System Service

PRECAUTIONS

• Certain components within the Anti-Lock Brake System (ABS) are not intended to be serviced or repaired individually. Only those components with removal and installation procedures should be serviced.

• Do not use rubber hoses or other parts not specifically specified for and ABS system. When using repair kits, replace all parts included in the kit. Partial or incorrect repair may lead to functional problems and require the replacement of components.

• Lubricate rubber parts with clean, fresh brake fluid to ease assembly. Do not use lubricated shop air to clean parts; damage to rubber components may result.

• Use only specified brake fluid from an unopened container.

• If any hydraulic component or line is removed or replaced, it may be necessary to bleed the entire system.

• A clean repair area is essential. Always clean the reservoir and cap thoroughly before removing the cap. The slightest amount of dirt in the fluid may plug an orifice and impair the system function. Perform repairs after components have been thoroughly cleaned; use only denatured alcohol to clean components. Do not allow ABS components to come into contact with any substance containing mineral oil; this includes used shop rags.

• The Anti-Lock control unit is a microprocessor similar to other computer units in the vehicle. Ensure that the ignition switch is **OFF** before removing or installing controller harnesses. Avoid static electricity discharge at or near the controller.

• If any arc welding is to be done on the vehicle, the control unit should be unplugged before welding operations begin.

DEPRESSURIZING

All Nissan vehicles equipped with an Anti-lock Brake System (ABS) utilize a low-pressure system. Special depressurization procedures are therefore NOT necessary prior to servicing the ABS brakes on vehicles so-equipped.

Master Cylinder

REMOVAL AND INSTALLATION

Sentra and 200SX

NOTE: Never reuse any brake fluid. Any brake fluid that is removed from the system should be properly discarded.

1. Clean the outside of the master cylinder thoroughly, particularly around the cap and fluid lines.
2. If equipped with a fluid level sensor, disconnect the wiring harness from the master cylinder.
3. Disconnect the brake fluid lines, then plug the openings to prevent dirt from entering the system.
4. Remove the mounting bolts at the firewall or the brake booster (if equipped) and remove the master cylinder from the vehicle.
To install:
5. Bench bleed the master cylinder.
6. Install the master cylinder to the vehicle.
7. Torque the master cylinder mounting nuts to 9 ft. lbs. (12 Nm).
8. Connect all brake lines and fluid level sensor wiring if so equipped.
9. Tighten the brake line fittings to 12–14 ft. lbs. (17–20 Nm).
10. Refill the reservoir with brake fluid and bleed the system.
NOTE: Use DOT 3 brake fluid in the brake systems.
11. Check for leaks and verify proper brake system operation.

Altima, Maxima and 1993–94 240SX and 300ZX

1. Disconnect the negative battery terminal and the electrical connector from the reservoir.
2. Clean the outside of the master cylinder thoroughly, particularly around the cap and fluid lines.

3. Disconnect the brake fluid tubes, then plug the openings to prevent dirt from entering the system.
4. Remove the master cylinder mounting nuts and remove the master cylinder from the vehicle.
To install:

NOTE: At this point it will be necessary to adjust the output rod length.

5. Apply -19.69 in. Hg of vacuum to the brake booster using a vacuum pump.
6. Measure the output rod length from the end of the rod to the front face of the booster.
7. The adjustment of the rod should be 0.4045–0.4144 in. (10.275–10.525mm).
8. Bench bleed the master cylinder assembly prior to installation.
9. Install the master cylinder and torque the master cylinder mounting nuts to the following specifications:
    a. 1993–97 Maxima, 1997 Altima and 1993–94 300ZX — 9–11 ft. lbs. (12–15 Nm).
    b. All other vehicles — 5.8–8.0 ft. lbs. (8–11 Nm).
10. Connect the brake lines.
11. Refill the reservoir with brake fluid and bleed the system. Adjust the brake system if necessary.
12. Connect the negative battery cable and the electrical to the bottom of the reservoir.

NOTE: The adjustable pushrod is used to adjust the brake pedal free height.

13. Check for fluid leaks and verify proper brake system operation.

1995–97 240SX and 300ZX

1. Disconnect the negative battery terminal.
2. Clean the outside of the master cylinder thoroughly, particularly around the cap and fluid lines.

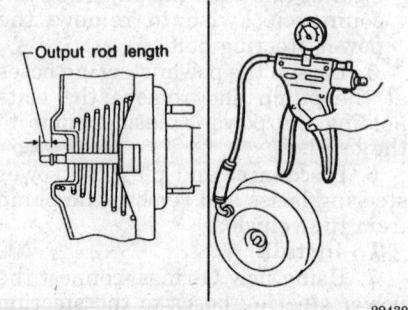

Measuring the output rod length — Altima, Maxima and 1993–94 300ZX

3. Disconnect the brake fluid tubes, then plug the openings to prevent dirt from entering the system.
4. Remove the master cylinder mounting nuts and remove the master cylinder from the vehicle.
To install:
5. Bench bleed the master cylinder assembly prior to installation.
6. Install the master cylinder and torque the master cylinder mounting nuts to 9–11 ft. lbs. (12–15 Nm).
7. Connect the brake lines and tighten the flare fittings to 11–13 ft. lbs. (15–18 Nm).
8. Refill the reservoir with brake fluid and bleed the system. Adjust the brakes, if necessary.
9. Connect the negative battery cable and bleed the entire brake system.

NOTE: Ordinary brake fluid will boil and cause brake failure under the high temperatures developed in disc brake systems; use DOT 3 brake fluid in the brake systems.

10. Check for fluid leaks and verify proper brake system operation.

Master Cylinder Bench Bleeding

1. Place the master cylinder in a vise.
2. Connect 2 lines to the fluid outlet orifices and place them into the reservoir.
3. Fill the reservoir with brake fluid.
4. Using a wooden dowel, depress the pushrod slowly, allowing the pistons to return. Do this several times until the air bubbles in the brake fluid are gone.
5. Remove the bleeding tubes from the master cylinder, plug the outlets, and install the caps.

Brake Caliper

REMOVAL AND INSTALLATION

300ZX

——— **CAUTION** ———
Brake pads may contain asbestos, which has been determined to be a cancer causing agent. Never clean the brake surfaces with compressed air! Avoid inhaling any dust from any brake surface! When cleaning brake surfaces, use a commercially available brake cleaning fluid.

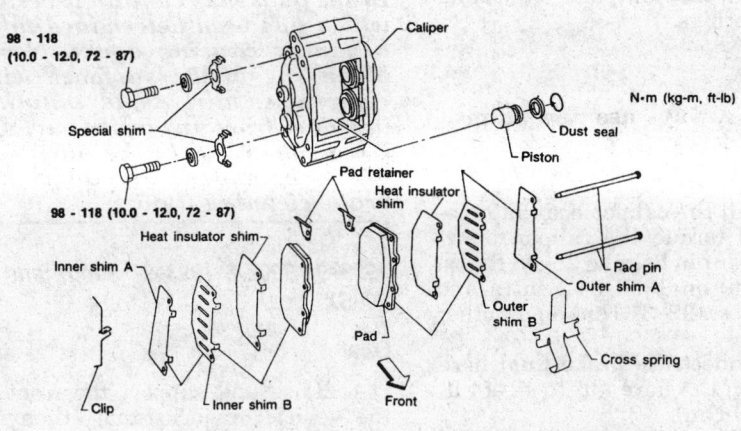

98 - 118
(10.0 - 12.0, 72 - 87)

Caliper

Special shim

N•m (kg-m, ft-lb)

Dust seal

Piston

98 - 118 (10.0 - 12.0, 72 - 87)

Pad retainer

Heat insulator shim

Heat insulator shim

Inner shim A

Pad pin

Outer shim A

Outer shim B

Cross spring

Pad

Inner shim B

Front

Clip

303199

Front brake caliper assembly — 1995–96 Sentra and 200SX

Front

1. Raise and support the vehicle on jackstands, then remove the front wheels.
2. Remove the disc brake pads.
3. Using a suitable tool, retract the disc brake pistons by prying them back into the disc caliper.

NOTE: When prying disc caliper piston, observe the other pistons and prevent them from coming out of their bores.

4. Disconnect and plug the brake tube at the brake hose connection.
5. Remove the brake caliper mounting bolts, then remove the caliper assembly from the vehicle.

───── **WARNING** ─────
Never remove the caliper joining bolts that hold the two caliper halves together.

To install:
6. Install the brake caliper mounting bolts. Tighten the mounting bolts to 72–87 ft. lbs. (98–118 Nm).
7. Reconnect the brake hose.
8. Install the disc brake pads, shims, and the pad springs. Always use new shims.

9. Bleed the brake system. Top off the master cylinder fluid level as necessary
10. Install the wheel assembly and lower the vehicle.

Rear

1. Raise and support the vehicle safely.
2. Remove the rear wheels.
3. Release the parking brake and remove the cable and bracket assembly.
4. Remove the disc brake pads.
5. Using a suitable tool, retract the disc brake pistons by prying them back into the disc caliper.

NOTE: When prying disc caliper piston, observe the other pistons and prevent them from coming out of their bores.

6. Disconnect the brake hose from the caliper and plug it.
7. Remove the caliper mounting bolts and remove the caliper assembly.
To install:
8. Clean the piston end of the caliper assembly and the area around the pin holes. Be careful not to get oil on the piston boot.
9. Position the caliper assembly in the mounting support and tighten

the mounting bolts to 28–38 ft. lbs. (38–52 Nm).
10. Install the disc brake pads, shims, and the pad springs. Always use new shims.
11. Reconnect the rear brake hose to the caliper.
12. Bleed the brake system and top of the master cylinder fluid level as necessary.
13. Mount the wheels and lower the vehicle.

───── **WARNING** ─────
Never remove the caliper joining bolts that hold the two caliper halves together.

Altima, Maxima, 240SX, 1995–97 Sentra and 200SX

───── **CAUTION** ─────
Brake pads may contain asbestos, which has been determined to be a cancer causing agent. Never clean the brake surfaces with compressed air! Avoid inhaling any dust from any brake surface! When cleaning brake surfaces, use a commercially available brake cleaning fluid.

Front Caliper

1. Raise and safely support vehicle.
2. Remove the front wheels.
3. Remove both guide pin bolts and brake fluid hose connector; be sure to plug the openings to prevent dirt from entering the system.
4. Remove the caliper assembly from the vehicle.
To install:
5. Check the level of fluid in the master cylinder. If the fluid is near the maximum level, use a clean syringe to remove fluid until the level is down well below the lip of the reservoir.
6. Use a large C-clamp to press the caliper piston back into the caliper, to allow room for the installation of the thicker new pads.
7. Install the new pads, new shims and pad retainer(s).
8. Install the brake caliper and torque the caliper-to-torque member pin bolts to 40–47 ft. lbs. (54–84 Nm) on the 1993–94 Sentra and 23 ft. lbs. (31 Nm) on all other models.
9. Using new copper washers, install the brake line to the brake caliper and torque the connecting bolt to 12–14 ft. lbs. (17–20 Nm).
10. Install the wheels and tighten the lug nuts to the proper specification.

11. Bleed the brake system and top off the master cylinder as necessary.

NOTE: Make sure you pump the brakes and get a hard pedal before moving the vehicle!

Rear Caliper

NOTE: All rear brake pads must always be replaced as a complete set.

1. Raise and safely support the vehicle. Remove the rear wheels.
2. Remove the parking brake cable stay fixing bolt and the lock spring.
3. Remove the brake fluid hose from the caliper.
4. Remove the lower caliper-to-torque member pin bolt and raise the caliper.
5. Remove the pad retainers, the pads and the shims.
6. Remove the upper caliper-to-torque member pin bolt and remove the caliper.
To install:
7. Clean the piston end of the caliper body and the area around the pin holes. Be careful not to get oil on the rotor.
8. Using a pair of needle nose pliers, carefully turn the piston clockwise back into the caliper body; re-

move some brake fluid from the master cylinder, if necessary. Take care not to damage the piston boot.
9. Coat the pad contact area on the mounting support with a silicone based grease.
10. Install the new pads, shims and the pad springs.

NOTE: Always use new shims.

11. Install the caliper body into position and torque the caliper-to-torque member pin bolts to 28–38 ft. lbs. (38–52 Nm) on 1993–94 Sentra and 16–23 ft. lbs. (22–31 Nm) on all other models.
12. Reconnect the brake fluid hose and tighten the flare nut to 12–14 ft. lbs. (17–20 Nm).
13. Install the lock spring and the parking brake stay fixing bolt.
14. Bleed the brake system and top off the master cylinder as necessary.
15. Replace the wheel, lower the vehicle.

NOTE: Make sure you pump the brakes and get a hard pedal before moving the vehicle!

Disc Brake Pads

REMOVAL AND INSTALLATION

—— CAUTION ——
Brake pads may contain asbestos, which has been determined to be a cancer causing agent. Never clean the brake surfaces with compressed air! Avoid inhaling any dust from any brake surface! When cleaning brake surfaces, use a commercially available brake cleaning fluid.

1993–94 300ZX, 1993–97 Sentra and 200SX

Front

1. Raise and support the front of the vehicle on jackstands, then remove the wheels.
2. Remove the bottom guide pin from the caliper and swing the caliper cylinder body upward as hang by a wire.
3. Remove the brake pad retainers and the pads.
To install:
4. Compress the piston of the disc brake caliper.
5. Install the brake pads and caliper assembly. Torque the guide pin to 23–30 ft. lbs. (31–41 Nm).
6. Install the wheels.
7. Apply the brakes a few times to seat the pads. Check the master cylinder and add fluid if necessary. Bleed the brakes, if necessary.

Rear

NOTE: Do not press the piston into the bore as performed on the front disc brakes. Due to the parking brake mechanism, the caliper piston must be turned into the bore using a special tool.

1. Raise and support the vehicle safely.
2. Remove the rear wheels.
3. Release the parking brake and remove the cable bracket bolt.
4. Remove the pin bolts and lift off the caliper body.
5. Pull out the pad springs and then remove the pads and shims.
To install:
6. Clean the piston end of the caliper body and the area around the pin holes. Be careful not to get oil on the rotor.
7. Using the proper tool, carefully turn the piston clockwise back into the caliper body. Take care not to damage the piston boot.

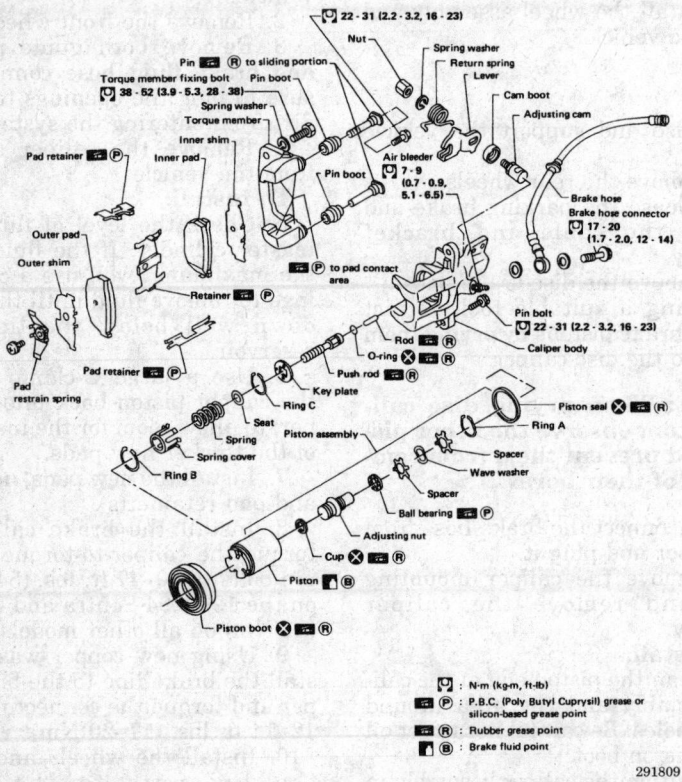

291809

Rear disc brake caliper assembly — 200SX shown, others similar

8. Coat the pad contact area on the mounting support with a silicone based grease.

9. Install the pads, shims and the pad springs. Always use new shims.

10. Position the caliper body in the mounting support and tighten the pin bolts to 28–38 ft. lbs. (38–52 Nm).

11. Mount the wheels, lower the vehicle and bleed the system if necessary.

1995–97 300ZX

WARNING

When the disc brake pads are removed, do not depress the brake pedal because the piston will pop out.

Front

1. Raise and support the front of the vehicle, then remove the wheels.

2. Remove the clip from the pad pins and then remove the pins.

3. Remove the cross spring from the disc caliper.

4. Pull out the outer disc brake pad and insert it temporarily between the lower piston and disc rotor.

5. Push the upper piston back using a suitable tool and insert the new brake pad so it contacts the upper piston only.

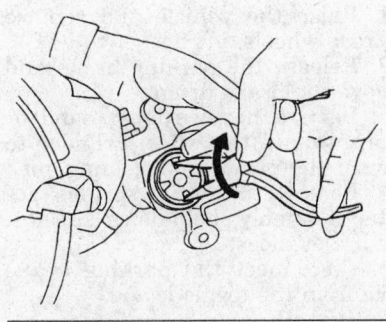

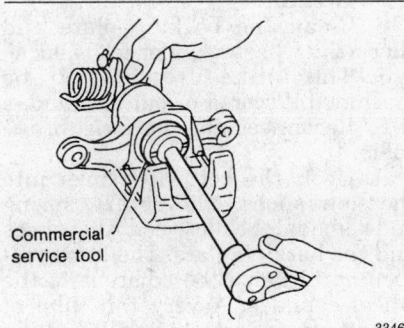

Commercial service tool

334658

Using service tools to turn piston back into caliper — Altima, Maxima, 240SX and 1995–97 Sentra and 200SX

6. Pull the old brake pad from the lower piston.

7. Push the lower piston back using a suitable tool.

8. Pull out the new disc brake pad and reinstall it in the proper position.

NOTE: Be sure to install the disc brake pad shims.

9. Repeat steps 3 to 7 for the inner disc brake pad.

10. Install the cross spring, pad retaining pins and retaining pin securing clip.

11. Install the wheels.

12. Be sure to pump the brake pedal several times prior to moving the vehicle.

Rear

NOTE: The parking brake assembly is located inside the disc rotor and has no affect on the rear disc caliper.

1. Raise and support the vehicle safely.

2. Remove the rear wheels.

3. Remove the clip from the pad pins and then remove the pins.

4. Remove the cross spring from the disc caliper.

5. Pull the outer disc brake pad from the disc caliper.

6. Push back the outer piston using a suitable tool and install a new disc brake pad with shims.

7. Pull the inner disc brake pad from the disc caliper.

8. Push back the inner piston using a suitable tool and install a new disc brake pad with shims.

9. Install the cross spring, pad pins and pad pin clips.

10. Install the rear wheels.

11. Be sure to pump the brake pedal several times prior to moving the vehicle.

Altima, Maxima and 200SX

Front

1. Raise and support the front of the vehicle, then remove the wheels.

2. Remove the bottom guide pin from the caliper and swing the caliper cylinder body upward; support the caliper with a wire.

3. Remove the brake pad retainers and the pads.

To install:

4. Compress the piston of the disc brake caliper.

5. Install the brake pads and caliper assembly. Torque the guide pin to 16–23 ft. lbs. (22–31 Nm).

6. Install the wheels.

7. Apply the brakes a few times to seat the pads. Check the master cylinder and add fluid if necessary. Bleed the brakes, if necessary.

Rear

NOTE: Do not press the piston into the bore as performed on the front disc brakes. Due to the parking brake mechanism, the caliper piston must be turned into the bore using a special tool.

1. Raise and support the vehicle safely.

2. Remove the rear wheels.

3. Release the parking brake and remove the cable bracket bolt.

4. Remove the pin bolts and lift off the caliper body.

5. Pull out the pad springs and then remove the pads and shims.

To install:

6. Clean the piston end of the caliper body and the area around the pin holes. Be careful not to get oil on the rotor.

7. Using the proper tool, carefully turn the piston clockwise back into the caliper body. Take care not to damage the piston boot.

8. Coat the pad contact area on the mounting support with a silicone based grease.

9. Install the pads, shims, and the pad springs. Always use new shims.

10. Position the caliper body in the mounting support and tighten the pin bolts to 16–23 ft. lbs. (22–31 Nm) on Altima and Maxima and 28–38 ft. lbs. (38–52 Nm) on 200SX.

11. Install the wheels and lower the vehicle.

12. Apply the brakes a few times to seat the pads. Check the master cylinder and add fluid if necessary. Bleed the brakes, if necessary.

Brake Rotor

REMOVAL AND INSTALLATION

Sentra and 200SX

CAUTION

Brake pads may contain asbestos, which has been determined to be a cancer causing agent. Never clean the brake surfaces with compressed air! Avoid inhaling any dust from any brake surface! When cleaning brake surfaces, use a commercially available brake cleaning fluid.

1. Raise and safely support vehicle. Remove the wheels from vehicle.

2. Remove the disc brake caliper assembly and support with wire.

NOTE: Do not disconnect the brake hose (if possible).

3. Remove the disc from the hub without removing the halfshaft nut and bearings.
To install:
4. Install the disc rotor to the hub assembly.
5. Install the disc brake caliper assembly and brake pads.
6. Install the wheel and tire assembly.
7. Bleed brake system if necessary.
8. Pump the brake pedal several times prior to moving the vehicle.

Altima, Maxima and 240SX

1. Raise and support the vehicle safely.
2. Remove the wheel assembly.
3. Remove the disc brake caliper assembly. Using a wire, support the caliper assembly; do not disconnect the brake line from the caliper.
4. Remove the caliper mounting bolts and remove the mount.
5. Remove the brake disc (rotor) from the wheel hub.

NOTE: On some rotors there are two threaded holes in the rotor. Two 8mm·1.25 bolts can be placed in these holes. When the bolts are turned evenly, the rotor is pressed away from the hub.

6. Inspect the disc for wear, cracks, runout and thickness; if necessary, replace the disc.
To install:
7. Install disc onto the wheel hub.
8. Install the disc caliper mount and torque the torque mounting bolts to the following specifications:
   a. Front wheel — torque the mounting bolts to: 53–72 ft. lbs.
   b. Rear wheel — torque the mounting bolts to: 28–38 ft. lbs. (38–52 Nm) for all vehicles except 1995–97 Maxima.
   c. Rear wheel — torque the mounting bolts to: 16–23 ft. lbs. (22–31 Nm) 1995–97 Maxima.
9. Install the caliper assembly to the caliper mounting bracket. Tighten the brake caliper mounting pin bolts to 23 ft. lbs. (31 Nm).
10. Check the level of fluid in the master cylinder. Top off fluid as necessary.
11. Install the wheels and torque the wheel lug nuts to the proper specification.

12. If the brake system was opened, bleed the brake system.

NOTE: Make sure to pump the brakes and get a hard pedal before driving the vehicle!

300ZX

NOTE: When removing the disc brake rotor it is not necessary to remove the axle shaft nut.

1. Raise and support the vehicle safely.
2. Remove the wheel and tire assembly.
3. Remove the disc brake pad pins and remove the disc brake pads.
4. Unbolt and remove the disc brake caliper, but do not remove the brake hose. Support the caliper with a wire; do not allow the caliper to hang from the hose.

── WARNING ──
DO NOT remove the 4 bolts that hold the halves of the disc brake caliper together. Only remove the 2 bolts that secure the caliper to the steering knuckle.

5. Matchmark the disc brake rotor to the wheel studs.
6. Remove the brake rotor by pulling it from the hub.
7. If the disc brake rotor is hard to remove, install 2 M8 · 1.25 bolts into the holes on the face of the disc brake rotor and gradually tighten the bolts. This will press the disc rotor from the hub.
To install:
8. Align the matchmarks and install the disc brake rotor.
9. Install the brake caliper and tighten front disc brake caliper mounting bolts to 72–87 ft. lbs. (98–118 Nm). Tighten the rear disc brake caliper mounting bolts to 28–38 ft. lbs. (38–52 Nm.)
10. Install the disc brake pads and install the brake pad mounting pins.
11. Install the wheel and tire assembly.

Brake Drums

REMOVAL AND INSTALLATION

Altima, Maxima, Sentra and 200SX

── CAUTION ──
Brake shoes may contain asbestos, which has been determined to be a cancer causing agent. Never clean the brake surfaces with compressed air! Avoid inhaling any dust from any brake surface!

When cleaning brake surfaces, use a commercially available brake cleaning fluid.

1. Raise the rear of the vehicle and support it on jackstands.
2. Remove the wheels.
3. Release the parking brake.
4. Remove the brake drum from the brake shoes. Using two 8mm x 1.25 bolts, tighten the bolts evenly to remove the drum.
5. If necessary, back off the brakes to remove the drum.
To install:
6. Install the drum assembly to the vehicle.
7. Install the wheels.
8. Adjust the rear brakes.

Brake Shoes

REMOVAL AND INSTALLATION

Altima, Maxima, Sentra and 200SX

── CAUTION ──
Brake shoes may contain asbestos, which has been determined to be a cancer causing agent. Never clean the brake surfaces with compressed air! Avoid inhaling any dust from any brake surface! When cleaning brake surfaces, use a commercially available brake cleaning fluid.

1. Raise the vehicle and remove the rear wheels.
2. Release the parking brake and remove the brake drum.
3. Place a heavy rubber band or clamp around the wheel cylinder to prevent the piston from coming out.
4. Remove the return springs, adjuster assembly, hold-down springs, and brake shoes.
5. Disconnect the parking brake cable from the toggle lever.
To install:
6. Clean the backing plate and check the wheel cylinder for leaks.
7. The brake drums must be machined if scored or out of round.
8. Reconnect the parking brake cable.
9. Hook the return springs into the new shoes. The return spring ends should be between the shoes and the backing plate. The longer return spring must be adjacent to the wheel cylinder. A very thin film of lithium grease may be applied to the pivot points at the ends of the brake shoes. Grease the shoe locating buttons on the backing plate, also. Be careful not to get grease on the linings or drums.

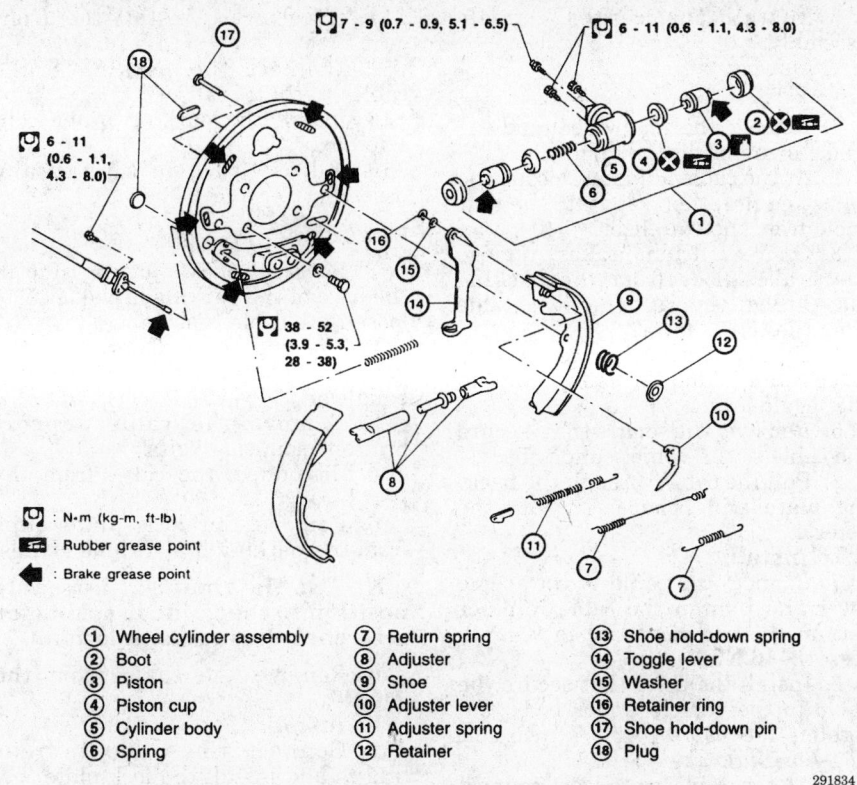

7 - 9 (0.7 - 0.9, 5.1 - 6.5)   6 - 11 (0.6 - 1.1, 4.3 - 8.0)

6 - 11 (0.6 - 1.1, 4.3 - 8.0)

38 - 52 (3.9 - 5.3, 28 - 38)

: N·m (kg-m, ft-lb)
: Rubber grease point
: Brake grease point

① Wheel cylinder assembly
② Boot
③ Piston
④ Piston cup
⑤ Cylinder body
⑥ Spring
⑦ Return spring
⑧ Adjuster
⑨ Shoe
⑩ Adjuster lever
⑪ Adjuster spring
⑫ Retainer
⑬ Shoe hold-down spring
⑭ Toggle lever
⑮ Washer
⑯ Retainer ring
⑰ Shoe hold-down pin
⑱ Plug

291834

Brake shoe and wheel cylinder assembly — Altima and Maxima

10. Install the adjuster assembly (rotate nut until adjuster rod is at its shortest point) between brake shoes. Place one shoe in the adjuster and piston slots and pry the other shoe into position. Install hold-down springs and reconnect the return springs.

11. Install the drums (adjust wheel bearings if necessary) and wheels. Adjust the brakes and bleed the hydraulic system, if necessary.

12. Make sure the shoes do not drag when the parking brake is applied and then is released.

Wheel Cylinder

REMOVAL AND INSTALLATION

Altima, Maxima, Sentra and 200SX

————— **CAUTION** —————
Brake shoes may contain asbestos, which has been determined to be a cancer causing agent. Never clean the brake surfaces with compressed air! Avoid inhaling any dust from any brake surface! When cleaning brake surfaces, use a commercially available brake cleaning fluid.

1. Raise and safely support vehicle.

2. Remove the rear wheels and the brake drums.

3. Disconnect the flare nut and the brake tube from the wheel cylinder, then plug the line to prevent dirt from entering the system.

4. Remove the brake shoes from the backing plate.

5. Remove the wheel cylinder-to-backing plate bolts and the wheel cylinders.

NOTE: If the wheel cylinder is difficult to remove, bump it with a soft hammer to release it from the backing plate.

To install:

6. Install the wheel cylinder assembly to the backing plate. Torque the wheel cylinder-to-backing plate bolts to 60–72 inch lbs. (6–9 Nm).

7. Connect the brake line(s) and install the rear brake shoes.

8. Install the brake drum.

9. Bleed the brake system.

10. Adjust the rear brakes as necessary.

Parking Brake Cable

ADJUSTMENT

Sentra and 200SX

NOTE: Make sure the rear brakes are properly adjusted prior to making any adjustments to the parking brake.

1. Make sure that the rear brakes are in good condition.

2. Carefully remove the dust boot from around the parking brake lever.

3. The adjustment is made by determining the amount of force needed to pull up on the lever. A force of 44 lbs. (20 kg.) should be needed to raise the lever 7–8 notches for drum brake vehicles or 8–9 notches for disc brake vehicles.

4. To adjust the pull, raise and support the vehicle safely on jackstands. There is an adjusting nut on the hand brake clevis rod. Loosen the locknut and turn the adjusting nut to establish the correct pull.

5. Tighten the adjuster locknut.

6. With the parking brake released, the rear wheels should turn freely. When the parking brake lever is pulled up 6–8 notches, the rear wheels should not turn.

7. Bend the parking brake warning lamp switch plate so that the warning lamp illuminates when the parking brake lever is raised 1 notch. The warning lamp should extinguish after the parking brake lever is completely released.

Altima and Maxima

NOTE: Make sure the rear brakes are properly adjusted prior to adjusting the parking brake.

1. Pull the parking brake lever with 44 lbs. of force and note the number of notches.

2. The parking brake lever should raise:
   a. 6–7 notches on the 300ZX models
   b. 7–8 notches for Altima
   c. 8–11 notches for Maxima and Stanza

3. Locate the parking brake adjuster at the base of the hand lever and rotate the adjuster nut on the threaded rod to obtain the proper parking brake lever adjustment.

4. Bend the parking brake warning lamp switch plate so that brake warning light comes on when parking brake lever is pulled up 1 notch. The brake light should turn off when the lever is fully released.

240SX

1. Raise parking brake lever 4 to 5 notches.

2. Insert a offset box end wrench or ratchet and socket assembly into the opening in the control lever and loosen the self-lock adjusting nut. Remove the wrench and push the lever completely down.

3. Depress the brake pedal about five times. This will automatically set the rear caliper in the proper position.

NOTE: Use 44 lb. (196 N) of force to raise the lever for adjustment purposes.

4. Adjust parking brake cable until the proper stroke is achieved. The proper stroke is 6 to 8 notches for 1993–94 vehicles or 7 to 9 notches for 1995–97 vehicles. When the parking brake lever is pulled the specific number of notches, the rear wheels should not turn. When the parking brake lever is released, the rear wheels should turn freely.

5. Bend the parking brake warning lamp switch plate so that brake warning light comes on when parking brake lever is pulled up 1 notch. The brake light should be extinguished when the lever is pushed completely down.

6. After the proper stroke is achieved, tighten the adjusting nut.

REMOVAL AND INSTALLATION

1993–94 Altima, Sentra, 200SX, Maxima and 1993–97 300ZX

Front Cable

1. Remove the center console assembly.

2. Loosen and remove the parking brake cable adjusting nut at the base of the parking brake lever.

3. From under the vehicle, disconnect the parking brake cables at the equalizer.

4. Unbolt the parking brake lever from the center console.

5. Unbolt and remove the front parking brake cable from the vehicle.

To install:

6. Install the front parking brake cable into the parking brake lever and start the adjusting nut on the front cable.

7. Install the lever and cable assembly into the vehicle. Tighten the mounting bolts to 9–12 ft. lbs. (13–16 Nm).

8. From under the vehicle, connect the parking brake cables at the equalizer.

9. Install the center console assembly.

Rear Cable

1. Remove the rear wheel and the brake drum or disc assembly.

2. At the cable adjuster, loosen the adjusting nut, then separate the rear cable from the equalizer.

3. On rear drum brakes, remove the brake shoes from the backing plate, then separate the rear cable from the toggle lever.

4. On rear disc brakes, remove the cable retainer and the cable end from the toggle lever.

5. Remove the bolts that secure the cable to the vehicle under body.

6. Pull the cable through the backing plate and remove it from the vehicle.

To install:

7. Connect the cable to the toggle lever and tighten the cable retainer nut on the backing plate to 9–12 ft. lbs. (13–16 Nm).

8. Install the bolts that secure the cable to the vehicle under body and tighten the bolt to 45–57 inch lbs. (5.1–6.5 Nm).

9. At the cable equalizer, connect the parking brake cable and tighten the bolt or mounting nut to 9–12 ft. lbs. (13–16 Nm).

10. Install the disc brake rotor and wheel assembly.

11. Adjust the parking brake cable.

240SX

Front Cable

1. Remove the parking brake console box.

2. Fully loosen the parking brake cable adjustment at the base of the parking brake lever and remove the adjusting nut.

3. Disconnect the parking brake cables from the equalizer.

4. If necessary, remove the front seat.

5. Disconnect the warning lamp switch plate connector.

6. Using a chisel and hammer, break the clinched portion of the control lever at the front cable.

7. Remove the bolts that secure the front cable to the floor and remove the front cable.

To install:

8. Thread the front cable into the parking brake lever and start the adjusting nut by hand.

9. Reinstall the front cable through the floor and secure with mounting bolts.

10. Connect the warning lamp switch plate connector.

11. If removed, install the front seat.

12. Connect the parking brake cables to the equalizer.

13. Install the parking brake console box.

14. Adjust the parking brake assembly.

Rear Cable

1. Back off the adjusting nut in the base of the parking brake lever to loosen the cable tension.

2. Working from underneath the vehicle, disconnect the cable at the equalizer.

3. Remove the cable support bracket mounting bolts.

4. Disconnect the cable from the rear brakes.

5. Remove the lock plate clips from the parking brake cable guides.

NOTE: Matchmark the cable position to the guide brackets for reference during installation.

6. Remove the cable from the vehicle.

To install:

7. Connect the cable to rear brakes and install the lock plate.

8. Connect the cable to the equalizer.

9. Connect the cable support brackets.

10. Adjust the cable.

Brake System Bleeding

ALL MODELS

The purpose of bleeding the brakes is to expel air trapped in the hydraulic system. The system must be bled whenever the pedal feels spongy, indicating that compressible air has entered the system. It must also be bled whenever the system has been opened or repaired. You will need a helper for this job.

WARNING
Be careful! Brake fluid is extremely harmful to painted surfaces. Brake fluid picks up moisture from the air. Don't leave the master cylinder or the fluid container uncovered any longer than necessary. Never reuse brake fluid which has been bled from the brake system.

1. The sequence for bleeding is as follows:
• Models not equipped with ABS: left rear wheel cylinder, right rear wheel cylinder, left front caliper, right front caliper.

• Model equipped with ABS: left rear caliper or wheel cylinder, right rear caliper or wheel cylinder, left front caliper, right front caliper, ABS actuator.

NOTE: On models with ABS, be sure to turn the ignition OFF and disconnect the actuator connector.

2. Clean all the bleeder screws. You may want to give each one a shot of a penetrating lubricant to loosen it up; seizure is a common problem with bleeder screws, which then break off, usually requiring replacement of the part to which they are attached.
3. Fill the master cylinder with DOT 3 brake fluid.

NOTE: Check the level of the fluid often when bleeding, and refill the reservoirs as necessary. Don't let them run dry, or you will have to repeat the process.

4. Attach a length of clear vinyl tubing to the bleeder screw on the wheel cylinder (or master cylinder). Insert the other end of the tube into a clear, clean jar half filled with brake fluid.
5. Have you helper slowly depress the brake pedal. As this is done, open the bleeder screw ⅓–½ of a turn, and allow the fluid to run through the tube. Then close the bleeder screw before the pedal reaches the end of its travel. Have you assistant slowly release the pedal. Repeat this process until no air bubbles appear in the expelled fluid.

NOTE: If the brake pedal is depressed too fast, small air bubbles will form in the brake fluid.

6. Repeat the procedure on the other remaining bleeder screws, checking the level of fluid in the cylinder reservoirs often.
7. When all the air has been bleed from the system, perform the following steps:
   a. If disconnected, reconnect the actuator.
   b. Pressurize the system and check for leaks.
   c. Check and fill the fluid reservoir.

Wheel Speed Sensor

REMOVAL AND INSTALLATION

Sentra and 200SX

——— **WARNING** ———
Be careful not to damage the sensor edge or the teeth on the rotor.

1. Raise the safely support the vehicle.
2. Disconnect the harness connector to the wheel speed sensor.
3. Remove the wheel speed sensor mounting bolt and carefully remove the sensor.
4. Installation is the reverse of removal.
5. On 1993–94 Sentra models, torque both speed sensor mounting bolts to 8–11 ft. lbs. (11–15 Nm).
6. On all other models, torque the bolt on the front speed sensor to 13–17 ft. lbs. (18–24 Nm) and the bolt on the rear speed sensor to 18–25 ft. lbs. (25–33 Nm).

NOTE: Always install the sensor by hand, never force it into place.

Altima and Maxima

——— **WARNING** ———
Be careful not to damage the sensor edge or the teeth on the rotor.

1. Raise the safely support the vehicle.
2. Disconnect the harness connector to the wheel speed sensor.
3. Remove the wheel speed sensor mounting bolt and carefully remove the sensor.
4. Installation is the reverse of removal. Torque the bolt on the front speed sensor to 13–17 ft. lbs. (18–24

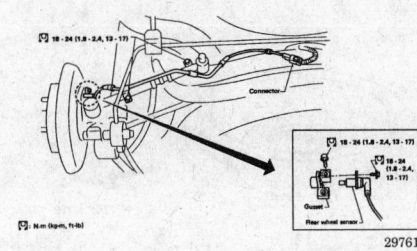

Rear wheel speed sensor mounting locations — 1993–94 Maxima shown, other models are similar

Nm) and the bolt on the rear speed sensor to 18–25 ft. lbs. (25–33 Nm).

240SX and 300ZX

——— **WARNING** ———
Be careful not to damage the sensor edge or the teeth on the rotor.

1. Raise the safely support the vehicle.
2. Disconnect the harness connector to the speed sensor.
3. Remove the wheel speed sensor mounting bolt and carefully remove the sensor.
4. Installation is the reverse of removal.
5. Torque the bolt on the front speed sensor to 8–12 ft. lbs. (11–16 Nm) and the bolt on the rear speed sensor to 13–20 ft. lbs. (18–26 Nm).

FRONT SUSPENSION

Strut

REMOVAL AND INSTALLATION

Sentra and 200SX

1. Raise and support the vehicle on jackstands.
2. Remove the wheel.
3. Detach the brake tube from the strut. If equipped with ABS, disconnect the ABS wiring from the strut.
4. Support the transverse link with a jackstand.
5. Detach the steering knuckle from the strut.

NOTE: Note the positioning of the strut alignment mark for reassembly purposes.

6. Support the strut and remove the three upper attaching nuts. Remove the strut from the vehicle.

——— **CAUTION** ———
Never loosen the center spring retaining nut until the coil spring is compressed, or serious injury or vehicle damage may occur.

7. Place the strut assembly in a vise with the special holding tool ST35652000 or in a spring compressor.
8. Loosen the piston rod locknut.

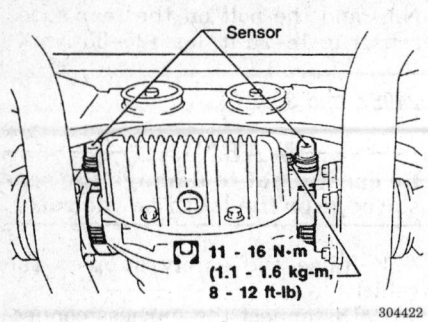

Rear wheel speed sensor mounting — 1996–97 300ZX

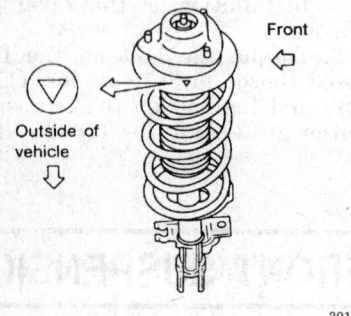

Positioning of strut alignment mark — 1995–97 Sentra and 200SX

9. Compress the spring with the spring compressor then remove the piston rod locknut.

NOTE: Before removing the strut from the coil spring, note the positioning of the strut in relationship to the coil spring for reassembly.

10. Remove the strut mounting insulator bracket, strut mounting bearing, upper spring seat, and the upper spring rubber seat.
11. Remove the strut, leaving the coil spring compressed.
12. Remove the piston boot and rebound bumper from the strut.
To install:
13. Install the rebound bumper and the boot to the strut piston.
14. Install the strut into the coil spring, make sure the strut and spring are properly positioned.
15. Install the upper spring rubber seat, upper spring seat, strut mounting bearing, and the strut mounting insulator bracket. Make sure that the cutout on the upper spring seat is facing the outside of the vehicle.
16. Install the piston rod locknut then remove the spring compressor.

17. Torque the piston rod locknut to 43–54 ft. lbs. (59–74 Nm).

NOTE: When installing the strut, be sure to position the alignment mark toward the outside of the vehicle.

18. Position the strut to the vehicle and install the 3 upper attaching nuts. Tighten the upper mounting nuts to 18–22 ft. lbs. (25–29 Nm).
19. Connect the steering knuckle to the strut and tighten mounting nuts of the mounting bolts to 68–82 ft. lbs. (92–111 Nm).
20. Connect the brake tube to the strut and connect the ABS wiring to the strut, if it was removed.
21. Bleed the brake system and install the wheel.
22. Perform a front end alignment.

Altima

1. Raise and support the vehicle on jackstands.
2. Remove the wheel.
3. Detach the brake tube from the strut. If equipped with ABS, disconnect the ABS wiring from the strut.
4. Support the transverse link with a jackstand.
5. Detach the steering knuckle from the strut.
6. Support the strut and remove the three upper attaching nuts. Remove the strut from the vehicle.

—— **WARNING** ——
Never loosen the center spring retaining nut until the coil spring is compressed, or serious injury or vehicle damage may occur.

To install:

NOTE: When installing the strut, be sure to position the alignment mark toward the outside of the vehicle.

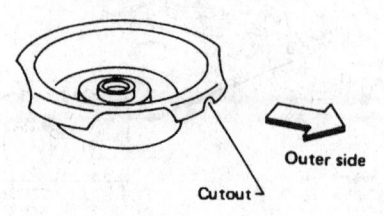

Positioning of strut alignment mark — Altima

7. Position the strut to the vehicle and install the three upper attaching nuts. Tighten the upper mounting nuts to 29–40 ft. lbs. (39–54 Nm).
8. Connect the steering knuckle to the strut and tighten mounting nuts to 87–108 ft. lbs. (118–147 Nm).
9. Connect the brake tube to the strut and connect the ABS wiring to the strut, if it was removed.
10. Bleed the brake system and install the wheel.
11. Lower the vehicle and perform a front end alignment.

1993–94 Maxima

1. Disconnect the negative battery cable. If equipped with adjustable struts, disconnect the sub-harness connector at the strut tower.
2. Raise and safely support the vehicle.
3. Remove the wheel. Mark the position of the strut-to-steering knuckle location.
4. Detach the brake tube from the strut.
5. Support the control arm.
6. Remove the strut-to-steering knuckle bolts.

NOTE: On models equipped with adjustable struts, remove the shock absorber actuator-to-plate bolts and separate the actuator from the plate; the plate is located on top of the strut.

7. Support the strut and remove the 3 upper strut-to-chassis nuts. Remove the strut from the vehicle.

—— **WARNING** ——
Never loosen the strut center nut until the spring is compressed or serious injury or vehicle damage may occur.

To install:
8. Install the strut assembly onto the vehicle and torque the strut-to-body nuts to 29–40 ft. lbs. (39–54 Nm). Tighten the strut-to-knuckle bolts to 116–123 ft. lbs. (157–167 Nm).
9. If equipped with a shock absorber actuator, install the actuator and torque the actuator-to-plate bolts to 2.2–2.8 ft. lbs. (2.9–3.8 Nm).
10. If the brake hose was disconnected from the brake caliper, bleed brakes and install the wheel. Reattach the brake hose to the strut.
11. Connect the negative battery cable and the adjustable strut electrical connectors, if equipped.
12. Check and/or adjust the wheel alignment.

1995–97 Maxima

1. Raise and safely support the vehicle.
2. Remove the wheel. Matchmark the position of the strut-to-steering knuckle location.
3. Disconnect the brake hose from the strut.
4. Remove the ABS wheel sensor and move it out of the way.
5. Matchmark and remove the bolts attaching the steering knuckle to the strut.
6. Open the hood and remove the strut attaching nuts while holding the strut.

—————— CAUTION ——————

Do not remove the center locknut from the strut assembly until the strut is safely compressed.

7. Remove the strut from the vehicle.
8. Place the strut assembly in a vise with the special holding tool ST35652000 or in a spring compressor.
9. Loosen the piston rod locknut.

—————— CAUTION ——————

Do not remove the piston rod locknut, the spring is under tension and can cause serious personal injury.

10. Compress the spring with the spring compressor then remove the piston rod locknut.

NOTE: Before removing the strut from the coil spring, note the positioning of the strut in relationship to the coil spring for reassembly.

11. Remove the strut mounting insulator bracket, strut mounting bearing, upper spring seat, and the upper spring rubber seat.
12. Remove the strut, leaving the coil spring compressed.
13. Remove the piston boot and rebound bumper from the strut.

To install:

14. Install the rebound bumper and the boot to the strut piston.
15. Install the strut into the coil spring, make sure the strut and spring are properly positioned.
16. Install the upper spring rubber seat, upper spring seat, strut mounting bearing, and the strut mounting insulator bracket. Make sure that the cutout on the upper spring seat is facing the outside of the vehicle.
17. Install the piston rod locknut then remove the spring compressor.

18. Torque the piston rod locknut to 43–58 ft. lbs. (59–76 Nm).
19. Install the strut into the strut tower and install new attaching nuts. Torque the nuts to 29–40 ft. lbs. (39–54 Nm).
20. Install the bolts attaching the steering knuckle to the strut and align the matchmarks. Torque the bolts to 103–117 ft. lbs. (140–159 Nm).
21. Install the ABS wheel sensor and torque the attaching bolt to 13–17 ft. lbs. (18–24 Nm).
22. Install the brake hose to the strut.
23. Install the front wheels and lower the vehicle.
24. Check and/or adjust the wheel alignment as necessary.

240SX

1. Raise and support the vehicle on jackstands.
2. Remove the wheel.
3. Detach the brake tube from the strut. If equipped with ABS, disconnect the ABS wiring from the strut.
4. Support the transverse link with a jackstand.
5. Detach the steering knuckle from the strut by removing the two through bolts.
6. Support the strut and remove the three upper attaching nuts. Remove the strut from the vehicle.

—————— WARNING ——————

Never loosen the center spring retaining nut until the coil spring is compressed, or serious injury or vehicle damage may occur.

7. Compress the strut coil spring with a spring compressor.

—————— CAUTION ——————

If coil spring is not properly compressed serious injury could result.

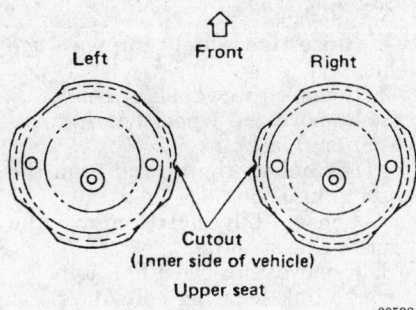

Positioning of strut alignment mark — 240SX

8. Remove the strut assembly center locknut.

NOTE: Before removing the strut from the coil spring, note the positioning of the strut in relationship to the coil spring for reassembly.

9. Separate the strut from the coil spring. Keep the coil spring compressed.

To install:

10. Install the strut into the coil spring.
11. Install and torque the center locknut to 43–58 ft. lbs. (59–78 Nm).

NOTE: When installing the strut, be sure to position the alignment mark toward the outside of the vehicle.

12. Position the strut to the vehicle and install the three new upper attaching nuts. Tighten the upper mounting nuts to 29–40 ft. lbs. (39–54 Nm).
13. Connect the steering knuckle to the strut and tighten mounting through bolts to 84–98 ft. lbs. (114–133 Nm) for 1993–94 vehicles or 90–112 ft. lbs. (123–152 Nm) for 1995–96 vehicles.
14. Connect the brake tube to the strut and if it was removed, connect the ABS wiring to the strut.
15. Bleed the brake system and install the wheel.
16. Perform a front end alignment.

300ZX

1. Raise and support the vehicle safely.
2. Remove the front wheels.
3. Lightly jack under the lower arm to unload the suspension.
4. Remove the nut connecting the lower portion of the strut to the third link.
5. From under the hood and remove the two nuts holding the top of the strut to the chassis.

—————— CAUTION ——————

Do not remove the center nut from the strut piston. The center nut retains the coil spring, which is contained under considerable pressure. Improper servicing can lead to vehicle damage and serious injury.

6. If equipped with adjustable or Sonar suspension shocks, disconnect the electrical lead from the actuating unit.
7. Lower the jack slowly and cautiously until the strut assembly can be removed.

To install:

NOTE: On vehicles with Sonar suspension, before installing the actuator ensure the output shaft of the actuating unit is aligned with the shock absorber control rod. If this is not done, the actuator will be damaged.

8. Position the strut in the vehicle and using 2 new self-locking nuts, torque the nuts holding the top of the strut to 25–33 ft. lbs. (34–44 Nm).

9. If equipped with adjustable or Sonar suspension shocks, connect the electrical lead to the actuating unit.

10. Connect the strut assembly to the knuckle arm and torque the mounting nut to 72–87 ft. lbs. (98–118 Nm).

11. Install the front wheel and lower the vehicle.

12. Check front wheel alignment and adjust if necessary.

Shock Absorber

REMOVAL AND INSTALLATION

Altima and 240SX

1. Raise and safely support the vehicle.
2. Remove the strut assembly.

———— CAUTION ————
Coil springs are under extreme loads when compressed. Be sure to properly align the spring with the compressing tool to prevent personal injury from the spring releasing unexpectedly.

NOTE: Mark the coil spring position to the strut assembly for reinstallation purposes.

3. Compress the coil spring with the proper compressor tool.

4. Remove the center retaining nut holding strut mounting insulator.

5. Slowly decompress the coil spring.

6. Remove the strut mounting insulator.

7. Remove coil spring.

To install:

8. Install the coil spring onto the strut assembly. Be sure to align the matchmarks made during the removal procedure.

9. Install the strut mounting insulator.

10. Compress the coil spring assembly.

NOTE: It will be necessary to use a new locknut for the center retaining nut of the coil spring.

11. Install the center retaining nut and tighten the nut to 43–56 ft. lbs. (59–76 Nm). Make sure the spring is seated properly on the strut and in the mounting insulator.

12. Slowly remove the spring compressor tool.

13. Install the strut assembly on the vehicle.

14. Perform a vehicle alignment and road test.

300ZX

1. Remove the strut from the vehicle.

2. Use the strut holding fixture and mount the strut in a vise.

3. **Loosen** but do not remove the piston rod locknut.

4. Use a coil spring compressor and compress the coil spring until the upper spring seat can be turned by hand.

5. Remove the piston rod locknut.

6. Remove the upper spring seat and the coil spring.

To install:

7. Install the compressed coil spring on the strut.

8. Install the upper spring seat on the coil spring.

9. Install a new self locking nut on the piston rod. Torque the nut to 13–17 ft. lbs. (18–24 Nm).

10. Carefully remove the spring compressor from the spring.

11. Install the strut assembly in the vehicle.

Lower Ball Joints

REMOVAL AND INSTALLATION

Sentra, Altima, 200SX and 1995–97 Maxima

The lower ball joint is not replaceable, if the ball joint is defective the lower control arm or transverse link must be replaced.

1993–94 Maxima

1. Raise and safely support the vehicle.

2. Remove the wheel assembly.

3. Remove the wheel bearing cotter pin and locknut.

4. Disconnect the tie rod from the steering knuckle.

5. Loosen but not remove the strut-to-chassis nuts.

6. Remove the lower ball joint-to-traverse link securing bolts/nuts.

7. Remove the drive axle.

8. Remove the ball joint cotter pin and castle nut.

9. Using the Ball Joint Remover tool HT72520000 or equivalent, separate the ball joint from the steering knuckle.

10. Remove the ball joint.

To install:

11. Install the ball joint in the spindle and tighten the ball joint-to-steering knuckle nut to 52–64 ft. lbs. (71–86 Nm). Install a new cotter pin.

12. Install the drive axle

13. Install the ball joint-to-traverse link securing bolts. Tighten the ball joint-to-traverse link bolts/nuts to 56–80 ft. lbs. (76–109 Nm).

14. Tighten the strut-to-chassis nuts to 116–137 ft. lbs. (157–186 Nm).

15. Reconnect the tie rod to the steering knuckle.

16. Install the wheel bearing locknut and tighten to 174–231 ft. lbs. (235–314 Nm). Install a new cotter pin.

17. Install the wheel assembly.

18. Lower the vehicle.

19. Check the vehicle for proper alignment.

300ZX

NOTE: The lower ball joint is integral with the lower control arm (transverse link). They are removed and replaced as an assembly.

1. Raise and support the vehicle safely.

2. Remove the front wheels.

3. Disconnect the ball joint stud from the steering knuckle.

4. Disconnect the tension rod from the lower control arm.

5. Remove the mounting through bolt connecting the lower control arm to the suspension crossmember.

NOTE: It may be necessary to turn the steering wheel to move the steering rack for removal of the through bolt.

To install:

NOTE: When installing the control arm, temporarily tighten the nuts and/or bolts securing the control arm to the suspension crossmember. Tighten them fully only after the vehicle's weight is resting on the wheels.

6. Install the lower control arm and install the through bolt with securing nut. At this time only snug the mounting nut.

7. Connect the ball joint stud to the knuckle and torque the nut to 65–80 ft. lbs. (88–108 Nm). Install a new cotter pin.

8. Connect the tension rod to the control arm and torque the nuts as follows:

a. Convertible models torque to 69–83 ft. lbs. (93–113 Nm).

b. Non-convertible models torque to 80–94 ft. lbs. (108–127 Nm).

9. Install the wheels and lower the vehicle to the floor.

10. Once the full weight of the vehicle is on the suspension, torque the inner lower arm mounting bolt nut to 80–94 ft. lbs. (108–127 Nm).

Upper Control Arms

REMOVAL AND INSTALLATION

300ZX

1. Raise and properly support vehicle.

2. Remove appropriate wheel assembly.

3. Support the lower control arm using a separate jack.

4. Remove the two mounting bolts connecting the upper control arm and remove the assembly.

To install

5. Install the control arm (link), with the **A** identification mark toward the axle side and facing upward. Lightly tighten the mounting bolts.

6. Install wheel and lower vehicle.

7. With full weight of vehicle on the ground, torque the mounting bolts to 65–80 ft. lbs. (88–108 Nm).

Lower Control Arms

REMOVAL AND INSTALLATION

Sentra and 1995–97 200SX

1. Raise the vehicle and support it safely.

2. Remove the front wheels.

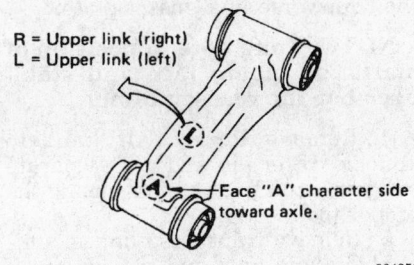

R = Upper link (right)
L = Upper link (left)

Face "A" character side toward axle.

304351

Upper control arm identification marks — 300ZX

3. Remove the disc brake caliper from the steering knuckle.

———— **WARNING** ————
DO NOT allow the disc brake caliper to hang from the brake hose. Support the disc caliper with safety wire.

4. Remove the cotter pin and loosen the wheel bearing locknut.

5. Remove cotter pin and the castle nut from the tie rod ball joint. Separate the tie rod with a suitable puller.

6. Remove the two bolts that secure the lower portion of the strut to the steering knuckle.

7. Using a plastic or rubber mallet, tap on the loosened wheel bearing locknut to loosen the halfshaft in the knuckle. Remove the locknut and remove the halfshaft from the steering knuckle. Be sure to cover the CV-joints with a shop rag.

NOTE: Support the halfshaft assembly with wire. Do not allow the halfshaft to hang by the inner joint.

8. Remove the nut that secures the stabilizer link to the lower control arm and disconnect the link from the control arm. Note the positioning of the washers and spacers for reassembly.

9. Remove the cotter pin and castle nut from the lower ball joint.

10. Separate the lower ball joint from the knuckle and remove the knuckle from the vehicle.

11. Remove the mounting nuts/bolts that secure the lower control arm to the frame and remove the control arm from the vehicle.

To install:

NOTE: Final tightening of all suspension components should take place with the weight of the vehicle on the wheels.

12. Install the lower control arm assembly and tighten mounting bolts/nuts as follow:

a. Tighten the through bolt and nut to 76–87 ft. lbs. (103–118 Nm).

b. Tighten the two saddle bracket mounting bolts to 58–72 ft. lbs. (78–98 Nm).

13. Connect the steering knuckle to the lower ball joint and tighten the lower ball joint castle nut to 43–54 ft. lbs. (59–74 Nm). Install a new cotter pin.

14. Connect the stabilizer link to the lower control arm and tighten the

link mounting nut to 12–16 ft. lbs. (16–22 Nm).

NOTE: When installing cotter pins to castle nuts, always tighten the castle nut further to align the cotter pin hole. DO NOT loosen the castle nut to align the cotter pin hole.

NOTE: When installing the stabilizer link be sure to properly align link.

15. Insert the halfshaft through the wheel bearing and install the wheel bearing locknut. Do not torque the locknut at this time.

16. Install the two bolts that secure the steering knuckle to the strut and tighten the bolts as follows:

a. 1993–94 vehicles — Tighten to 84–98 ft. lbs. (114–133 Nm).

b. 1995–96 vehicles — Tighten to 68–82 ft. lbs. (92–111 Nm).

17. Connect the tie rod end and tighten the tie rod castle nut to 22–29 ft. lbs. (29–39 Nm). Install a new cotter pin.

18. Install the disc brake caliper to the steering knuckle.

19. Tighten the halfshaft mounting nut (hub nut) and torque the nut to 145–202 ft. lbs. (196–274 Nm). It may be necessary to have an assistant hold the brake pedal while tightening the locknut. Install the adjusting cap and a new cotter pin.

20. Install the front wheels, lower the vehicle, and perform a front end alignment.

Maxima

1. Raise and safely support the vehicle.

2. Remove the front wheels.

3. Remove the ABS wheel sensor and move it out of the way.

4. Remove the wheel bearing locknut.

5. Disconnect the tie rod from the steering knuckle.

6. Matchmark then remove the bolts attaching the strut to the steering knuckle.

7. Separate the halfshaft from the steering knuckle by lightly tapping the end of the shaft.

8. Separate the steering knuckle and the lower ball joint.

9. Disconnect the stabilizer bar from the lower control arm.

10. Remove the bolts attaching the link bushing pin to the chassis. If necessary, remove the nut attaching the link to the control arm and remove the link.

11. Remove the bolts attaching the compression rod bushing clamp and

remove the lower control arm/traverse link.

To install:

12. Install the lower control arm and the compression rod bushing clamp into the vehicle.

13. Install the link bushing pin, if removed from the control arm.

14. Tighten all bolts and nuts until they are snug enough to support the weight of the vehicle but not fully tight, the bolts should be torqued to specification with the vehicle on the floor.

15. Install the steering knuckle to the lower control arm and connect the ball joint. Torque the ball joint nut to 46–56 ft. lbs. (62–76 Nm).

NOTE: Always use a new nut when installing the ball joint to the control arm.

16. Connect the steering knuckle to the strut and to the halfshaft.

17. Install the strut mounting bolts and align the matchmarks. Torque the bolts to 103–117 ft. lbs. (140–159 Nm).

18. Install the tie rod ball joint and torque the nut to 46–54 ft. lbs. (63–73 Nm).

19. Install the wheel bearing locknut

20. Install the ABS wheel sensor and torque the attaching bolt to 13–17 ft. lbs. (18–24 Nm).

21. Install the front wheels, lower the vehicle and torque hub locknut to 174–231 ft. lbs. (235–314 Nm).

22. Torque the bolts attaching the compression rod bushing clamp and the link bushing pin, in the proper sequence to 87–108 ft. lbs. (118–147 Nm).

23. If the link bushing pin was removed from the control arm torque the attaching nut to 87–108 ft. lbs. (118–147 Nm).

24. Torque the sway bar attaching nut to 30–35 ft. lbs. (41–47 Nm).

25. Check the vehicle alignment.

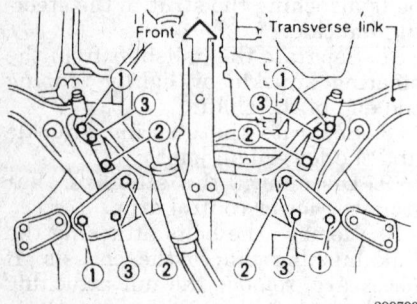

Bolt tightening sequence for the lower control arms — 1995–97 Maxima

299793

300ZX

NOTE: The lower ball joint is integral with the lower control arm (transverse link). They are removed and replaced as an assembly.

1. Raise and support the vehicle safely.

2. Remove the front wheels.

3. Disconnect the ball joint stud from the steering knuckle.

4. Disconnect the tension rod from the lower control arm.

5. Remove the mounting through bolt connecting the lower control arm to the suspension crossmember.

NOTE: It may be necessary to turn the steering wheel to move the steering rack for removal of the through bolt.

To install:

NOTE: When installing the control arm, temporarily tighten the nuts and/or bolts securing the control arm to the suspension crossmember. Tighten them fully only after the vehicle's weight is resting on the wheels.

6. Install the lower control arm and install the through bolt with securing nut. At this time only snug the mounting nut.

7. Connect the ball joint stud to the knuckle and torque the nut to 65–80 ft. lbs. (88–108 Nm). Install a new cotter pin.

8. Connect the tension rod to the control arm and torque the nuts as follows:

   a. Convertible models torque to 69–83 ft. lbs. (93–113 Nm).

   b. Non-convertible models torque to 80–94 ft. lbs. (108–127 Nm).

9. Install the wheels and lower the vehicle to the floor.

10. Once the full weight of the vehicle is on the suspension, torque the inner lower arm mounting bolt nut to 80–94 ft. lbs. (108–127 Nm).

240SX

1. Raise and support the vehicle safely. Allow the lower control arm to hang free.

2. Remove the front wheels.

3. Remove the cotter pin and castle nut from the tie rod end. Disconnect the tie rod end from the steering knuckle.

4. Remove the cotter pin and castle nut from the ball joint stud.

5. Using the proper tools, separate the knuckle from the ball joint.

6. Remove the stabilizer bar link from the lower arm.

7. Remove the nuts and bolts that connect the lower control arm to the tension rod.

8. Remove the bolt that connects the lower control arm to the crossmember and remove the lower control arm from the vehicle.

To install:

NOTE: Torquing of the suspension components should be performed with the full weight of the vehicle resting on the suspension.

9. Connect control arm to the crossmember and install the through bolt, lockwasher, and nut. Tighten the nut to 65–80 ft. lbs. (88–108 Nm) for 1993–94 vehicles or 80–94 ft. lbs. (108–127 Nm) for 1995–96 vehicles.

10. Install the nuts that connect the lower control arm to the tension rod. Secure the bolts and tighten the mounting nuts to 69–80 ft. lbs. (93–108 Nm).

11. Install the stabilizer bar transverse link to the lower arm and tighten the link to 14–22 ft. lbs. (20–29 Nm).

12. Connect the knuckle to the lower ball joint. Tighten the castle nut to 71–88 ft. lbs. (96–120 Nm) and install a new cotter pin.

13. Connect the tie rod end to the knuckle and tighten the castle nut to 22–36 ft. lbs. (29–49 Nm). Install a new cotter pin.

14. Install the front wheels and lower the vehicle.

15. Tighten all nuts and bolts to specification and perform a wheel alignment.

Altima

Transverse Link

1. Raise and safely support the vehicle.

2. Remove the front wheels.

3. Unbolt and disconnect the stabilizer bar. The bar is removed by unfastening the nut that hold the bar to the transverse link gusset plate.

NOTE: Take note of position of marks on clamp face and stabilizer bar for reassembling.

4. Remove lower ball joint to knuckle cotter pin and nut. Separate ball joint stud from knuckle using the proper tool.

5. Remove transverse link mounting bolts and nuts and remove link.

To install:

6. Install transverse link with mounting bolts and nuts. Tighten

nuts and bolts to 87–108 ft. lbs. (118–147 Nm).

NOTE: The final tightening of suspension components must be done with wheels on the ground and vehicle at curb weight.

7. Install the lower ball joint to the knuckle, tighten the nut to 52–64 ft. lbs. (71–86 Nm) and install a new cotter pin.

8. Reattach the stabilizer bar link to the transverse link. Tighten the stabilizer link nuts to 30–35 ft. lbs. (41–47 Nm).

9. Install wheels and safely lower vehicle to ground.

10. Check the front end alignment.

Sway Bar

REMOVAL AND INSTALLATION

Sentra and 200SX

1. Raise and support the vehicle safely.

2. Disconnect the front exhaust pipe at the manifold and position it out of the way.

3. Remove the stabilizer bar-to-transverse link (lower control arm) mounting nuts. Note location of the washers and bushings for reinstallation.

4. Matchmark the stabilizer bar to the mounting clamps. Take note to the paint marks on the stabilizer bar for reassembly.

5. Supporting the stabilizer bar, remove the stabilizer bar mounting clamp bolts. Remove the bar from the vehicle while pushing down on the exhaust pipe.

To install:

NOTE: When installing suspension components, do not fully tighten the mounting bolts unless the vehicle is resting on the ground with full vehicle weight upon the wheels.

6. Install the stabilizer bar and mounting brackets. Tighten the mounting bracket bolts to 23–31 ft. lbs. (31–42 Nm).

NOTE: When installing the stabilizer bar be sure to align the paint marks with the mounting bushings and brackets and properly align the sway bar links.

7. Connect the stabilizer link to the lower control arm and tighten the mounting nuts to 12–16 ft. lbs. (16–22 Nm).

8. Using a new gasket, connect the front exhaust pipe at the manifold.

9. Final tightening of all bolts should take place with the weight of the vehicle on the wheels.

Altima

1. Raise and support the vehicle safely.

2. Remove the stabilizer bar-to-transverse link (lower control arm) mounting nuts. Note location of the washers and bushings for reinstallation.

3. Matchmark the stabilizer bar to the mounting clamps and note the positioning of the mounts to the paint marks on the stabilizer bar.

4. Supporting the stabilizer bar, remove the stabilizer bar mounting clamp bolts. Remove the bar from the vehicle.

To install:

5. Install and support the stabilizer bar.

NOTE: The slits on the stabilizer bushings should face the front of the vehicle.

6. Install the mounting clamps with bushings and tighten the mounting nuts or bolts as follows:

    a. 1993–95 Altima — Tighten bolts to 12–16 ft. lbs. (16–22 Nm).

    b. 1996 Altima — Tighten bolts to 30–36 ft. lbs. (40–49 Nm).

7. Connect the stabilizer link to the lower control arm and tighten the link nut to 30–35 ft. lbs. (41–47 Nm).

Maxima

1. Raise and safely support the vehicle.

2. Remove the ball joint socket nuts connecting the stabilizer bar to the lower control arm.

3. Remove the 4 stabilizer bar bracket bolts and then pull the bar from the vehicle.

To install:

4. Make sure the stabilizer ball joint socket is positioned properly.

5. Install the stabilizer bar and mounting brackets. Never fully tighten the mounting bolts unless the vehicle is resting on the ground with normal weight upon the wheels. Be sure the stabilizer bar ball joint socket is properly positioned.

NOTE: When installing the stabilizer bar, make sure the paint mark and the bushings are aligned. Make sure the clamp is facing in the right direction.

6. Lower the vehicle.

7. Bounce the vehicle to stabilize the suspension. Torque the stabilizer bar bracket bolts to 36–43 ft. lbs. (49–59 Nm)

8. Torque the nuts connecting the stabilizer bar to the control arm to 30–35 ft. lbs. (40–47 Nm)

240SX

1. Safely raise and support the vehicle.

2. Remove the nuts holding the stabilizer bar to the sway bar links.

3. Remove the stabilizer brackets bolts/nuts and note the positioning of the brackets for reinstallation.

4. Remove the stabilizer bar.

To install:

NOTE: Never tighten any bolts or nuts to their final torque unless the vehicle is resting on the wheels (Unladen condition).

5. Install the stabilizer bar.

6. Install the stabilizer bar to the sway bar links and torque the new nuts to 34–38 ft. lbs. (46–52 Nm) for 1993–94 vehicles or 30–35 ft. lbs. (41–47 Nm) for 1995–96 vehicles . Be sure to install new nuts and properly position the links to the sway bar.

7. Install the stabilizer brackets and torque the bolts/nuts to 29–36 ft. lbs. (39–49 Nm).

8. Lower the vehicle and tighten stabilizer bolts and nuts to specification.

300ZX

1. Raise and support the vehicle safely.

2. Note the location and the direction in which the sway bar clamps face. There is a paint mark that needs to be aligned when reinstalling the sway bar.

3. Disconnect the sway bar connecting rod from the sway bar and the third link.

4. Remove the 4 sway bar bracket bolts and remove the sway bar.

To install:

NOTE: All bolts should not be torqued until the full weight of the vehicle is resting on the ground.

5. Position the sway bar in the vehicle and tighten the sway bar bracket bolts to 29–36 ft. lbs. (39–49 Nm).

NOTE: Be sure to align the sway bar clamps with the paint marks and be sure the clamps face in the proper direction.

6. Connect the sway bar to the connecting rod and to the third link. Tighten mounting nuts to 41–47 ft. lbs. (56–64 Nm).

7. Lower the vehicle and properly torque all hardware.

Front Wheel Bearings

ADJUSTMENT

Sentra, 200SX, Altima and Maxima

NOTE: Whenever the hub or bearing assemblies are removed, the wheel bearing must be replaced. Never reuse the old bearing assembly.

The wheel bearings are sealed and are not adjustable. If defective, replacement is the only option.

240SX

1. Raise and safely support the front wheels.
2. Remove the front wheels.
3. Remove the hub cap and wheel bearing nut cotter pin.
4. Torque wheel bearing locknut to 151–210 ft. lbs. (206–284 Nm).
5. Check that wheel bearings operate smoothly.
6. Check axial end-play. Axial end play must be 0.0020 inches (0.05 mm) or less. If axial end play is not within specification or wheel bearing does not turn smoothly, replace the wheel bearing assembly.
7. Replace the wheel bearing locknut cotter pin and reinstall the hub cap.
8. Install the wheels and safely lower the vehicle.

300ZX

The front wheel bearing is not adjustable. Replace the bearing assembly if a growling noise is emitted during operation or if the bearing drags or turns roughly.

REMOVAL AND INSTALLATION

Sentra and 200SX

NOTE: Whenever the hub or bearing assembly is removed, the wheel bearing assembly must be replaced. Never reuse the old bearing assembly.

1. Raise the vehicle and support it safely. Remove the front wheel.
2. Remove the wheel bearing/axleshaft locknut while depressing the brake pedal.
3. Remove the brake caliper and support it with a piece of wire. It is not necessary to disconnect the brake line from the caliper.

4. Remove the ABS sensor from the steering knuckle.

NOTE: Do not depress the brake pedal or twist the brake line.

5. Remove the tie rod end using a tie rod removing tool J25730A or equivalent.
6. Separate the halfshaft from the knuckle by slightly tapping with a soft hammer. Position the axle shaft nut on the threads of the shaft to protect them when lightly tapping.
7. Loosen the lower ball joint nut and separate using a ball joint separator J25730A or equivalent.
8. Remove the two strut-to-knuckle retaining bolts and separate.
9. Remove the steering knuckle from the vehicle.
10. Place the assembly in a vise. Drive the hub with the inner race from the knuckle with a suitable tool. Remove the inner and outer grease seals.
11. Remove the bearing inner race and outer grease seal from the hub.
12. Remove the snapring and press the bearing outer race to remove the bearing from the steering knuckle.
   To install:
13. Press a new wheel bearing into the knuckle assembly not exceeding 3.3 tons (3,000 kg) pressure.
14. Install the snapring and pack the grease seal lips with chassis grease.
15. Install the inner and outer grease seals.
16. Press the wheel hub into the knuckle not exceeding 3.3 tons (3,000 kg) pressure.
17. Check bearing operation and by applying 3.9–5.5 tons of pressure to the hub assembly. Spin the hub several times in both directions.
18. Make sure the bearings rotate freely. If the bearings do not rotate freely, replace the bearings.
19. Install the knuckle and wheel hub assembly.
20. Install the lower ball joint and torque the nut to 43–54 ft. lbs. (59–74 Nm). Install a new cotter pin.
21. Install the strut bolts and torque them to 84–98 ft. lbs. (114–133 Nm) for 1991–94 vehicles or 68–82 ft. lbs. (92–118 Nm) for 1995–96 vehicles.
22. Connect the tie end and tighten the nut to 22–29 ft. lbs. (29–39 Nm). Install a new cotter pin
23. Install the disc brake caliper.
24. Install and torque the wheel bearing locknut to 145–203 ft. lbs. (196–275 Nm). Install a new cotter pin.

25. Install the front wheels and lower the vehicle.
26. Check the vehicle's alignment.
27. Road test the vehicle and verify proper operation.

Maxima and Altima

NOTE: Whenever the hub or bearing assembly is removed, the wheel bearing assembly must be replaced. Never reuse the old bearing assembly.

1. Remove the knuckle assembly from the vehicle.
2. Using a shop press and a suitable tool, press the hub with the inner race from the steering knuckle.
3. Using a shop press and a suitable tool, press the bearing inner race from the hub and remove the outer grease seal.
4. Using a prybar, pry the inner grease seal from the steering knuckle.
5. Using snapring pliers to remove the inner and outer snaprings from the steering knuckle.
6. Using a shop press and a suitable tool, press the sealed bearing assembly from the steering knuckle.
7. Inspect the hub, steering knuckle and snaprings for cracks and/or wear; if necessary, replace the damaged part(s).
   To install:
8. Install the inner snapring in the steering knuckle groove.
9. Using a shop press and a suitable tool, press the new wheel bearing assembly into the steering knuckle, until it seats, using a maximum pressure of 3 tons.
10. Install the outer snapring.
11. Pack the new grease seal lips with multi-purpose grease.
12. Using a shop press and a suitable tool, press the new outer grease seal into the steering knuckle.
13. Using a shop press and a suitable tool, press the new inner grease seal into the steering knuckle.
14. Using a shop press and a suitable tool, press the hub into the steering knuckle, until it seats, using a maximum pressure of 5.5 tons; be careful not to damage the grease seal.
15. To check the bearing operation, perform the following procedures:
   a. Increase the press pressure to 3.5–5.0 tons.
   b. Spin the steering knuckle, several turns, in both directions.
   c. Make sure the wheel bearings operate smoothly.
16. If the wheel bearings do not operate smoothly, replace the wheel bearing assembly.
17. Install the knuckle assembly.

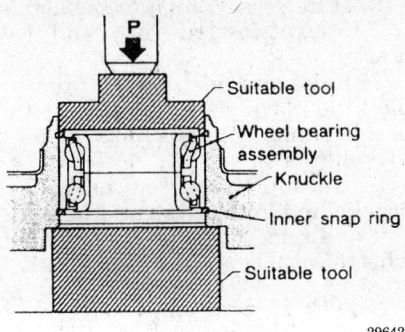

Typical method of installing the wheel bearing

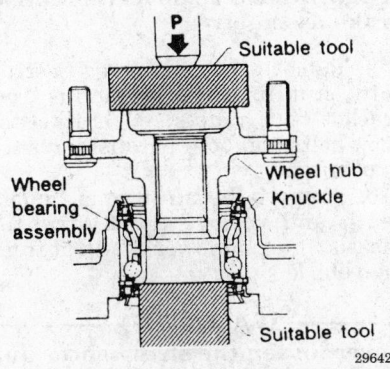

Typical hub installation

18. Install the halfshaft into the hub and tighten the locknut to 174–231 ft. lbs. (235–314 Nm).
19. Install the wheel assembly and lower the vehicle.
20. Road test the vehicle and verify proper operation.

240SX and 300ZX

NOTE: If defective, the wheel bearing assembly can only be serviced by replacement of the hub.

1. Raise and support the front of the vehicle safely and remove the wheels.
2. Remove the brake caliper assembly. Make sure not to twist the brake hose.

NOTE: When removing the brake caliper it is not necessary to disconnect the brake hose. Be sure to support the brake caliper after removal.

3. If applicable, remove the ABS brake sensor from the steering knuckle.
4. Remove the brake rotor.
5. Remove the wheel bearing dust cap.
6. Remove the locking nut.

7. Remove the wheel bearing locknut and washer.

NOTE: The hub bearing is a sealed type and will be removed with the hub assembly.

8. Pull hub assembly toward you to remove the hub assembly from the spindle.
9. Remove the tie rod joint and lower ball joint.
10. Disconnect the knuckle from the strut.
   To install:
11. Connect the knuckle to the strut.
12. Install the tie rod joint and lower ball joint.
13. Lubricate the spindle and slide hub assembly onto spindle.
14. Install the washer and wheel bearing locknut.
15. On 1993–94 models, torque the wheel bearing locknut to 147 to 216 Nm (108 — 159 ft. lbs.).
16. On 1995–97 models, torque the wheel bearing locknut to 152–210 ft. lbs. (206–284 Nm.).
17. Turn the hub assembly several times in both directions to seat the hub.
18. Measure the wheel bearing axial end-play.
   a. Axial end-play — 0.0020 inches (0.05mm) or less
19. Clinch the locking nut with a hammer and chisel.
20. Install a new dust cap.
21. Install the ABS sensor and tighten the mounting bolt to 96–144 in. lbs. (11–16 Nm).
22. Install the brake rotor and caliper assembly.
23. Install the front wheels.
24. If the brake line was disconnected, bleed the brake system.

REAR SUSPENSION

Strut

REMOVAL AND INSTALLATION

1993–94 Sentra

1. Raise the vehicle and support it safely.
2. Remove the rear wheel and brake caliper.
3. Remove the rear seat and trim panel.
4. Remove the upper strut nuts.

---- **WARNING** ----
Never remove the center strut nut until the strut is removed from the vehicle and the spring is compressed.

5. Remove the lower strut bolts and remove the strut from the vehicle.
   To install:
6. Perform final tightening on all suspension bolts when the vehicle's weight is resting on the suspension.
7. Install the strut and torque the upper mounting nuts to 18–22 ft. lbs. (25–29 Nm).
8. Torque the lower strut bolts to 72–98 ft. lbs. (98–133 Nm).
9. Install the rear seat and trim panel.
10. Lower the vehicle and have the rear end aligned.

1995–97 Sentra and 200SX

1. Raise the vehicle and support it safely.
2. Remove the rear wheel.
3. Remove the trim panel from the trunk to gain access to the upper mounting nuts of the strut.
4. Remove the protective cap from the upper portion of the strut.
5. Position a floor jack under the rear axle for support.

NOTE: Note and mark the positioning of the upper strut plate to the vehicle body.

---- **CAUTION** ----
Never remove the center strut nut until the strut is removed from the vehicle and the spring is safely compressed.

6. Remove the lower strut mounting through bolt.
7. Remove the 2 upper strut mounting nuts and remove the strut from the vehicle.
8. Place the strut assembly in a vise with the special holding tool ST35652000 or in a spring compressor.
9. Loosen the piston rod locknut.
10. Compress the spring with the spring compressor then remove the piston rod locknut.

NOTE: Before removing the strut from the coil spring, note the positioning of the strut in relationship to the coil spring for reassembly.

11. Remove the strut mounting insulator bracket, strut mounting bearing, upper spring seat, and the upper spring rubber seat.

12. Remove the strut, leaving the coil spring compressed.

13. Remove the piston boot and rebound bumper from the strut.

To install:

14. Install the rebound bumper and the boot to the strut piston.

15. Install the strut into the coil spring, make sure the strut and spring are properly positioned.

16. Install the upper spring rubber seat, upper spring seat, strut mounting bearing, and the strut mounting insulator bracket. Make sure that the cutout on the upper spring seat is facing the outside of the vehicle.

17. Install the piston rod locknut and torque the piston rod locknut to 13–17 ft. lbs. (18–24 Nm).

18. Remove the spring compressor from the coil spring.

19. Install the strut to the vehicle and torque the 2 upper mounting nuts to 12–14 ft. lbs. (16–19 Nm).

20. Install the upper mount protective cap.

21. Install the through bolt to the lower mount of the strut. Torque the lower strut bolt to 72–87 ft. lbs. (98–118 Nm).

22. Install the trunk trim panel.

23. Install the rear wheel.

24. Lower the vehicle and perform an alignment.

Altima

1. Raise and safely support the vehicle.

2. Remove the rear wheels from the vehicle.

3. Support the rear axle with a jack.

4. Remove the strut lower mounting through bolts.

NOTE: Be sure to note the position the strut upper plate to the vehicle for reinstallation purposes.

5. Remove the two nuts from the top of the strut and remove the strut as an assembly.

— CAUTION —

Do not remove the center locknut from the strut assembly until the strut is safely compressed.

6. Compress the strut coil spring with a spring compressor.

7. Remove the strut assembly center locknut.

NOTE: Before removing the strut from the coil spring, note the positioning of the strut in relationship to the coil spring for reassembly.

8. Remove the strut leaving the coil spring compressed.

NOTE: Mark the coil spring position to the strut assembly for reinstallation purposes.

9. To remove the spring from the strut assembly, perform the following steps:

   a. Compress the coil spring with the proper compressor tool.

   b. Remove the center retaining nut holding strut mounting insulator.

   c. Slowly decompress the coil spring.

   d. Remove the strut mounting insulator.

   e. Remove coil spring.

To install:

10. Install the coil spring onto the strut assembly. Be sure to align the matchmarks made during the removal procedure.

11. Install the strut mounting insulator.

12. Compress the coil spring assembly.

NOTE: It will be necessary to use a new locknut for the center retaining nut of the coil spring.

13. Install the center retaining nut and tighten the nut to 43–58 ft. lbs. (59–78 Nm). Make sure the spring is seated properly on the strut and in the mounting insulator.

14. Slowly remove the spring compressor tool.

15. Install the strut assembly into the vehicle.

16. Torque the upper strut mounting nuts to 31–40 ft. lbs. (42–54 Nm).

17. Torque the lower strut through bolt to 87–108 ft. lbs. (118–147 Nm).

NOTE: Be sure to hold the through bolt and tighten the nuts.

18. Install the wheels, lower the vehicle and perform a front end alignment.

1993–94 Maxima

1. Disconnect the negative battery cable.

2. Remove the rear seat and the parcel shelf.

3. If equipped with an adjustable shock absorber, disconnect the electrical connector from the sonar unit.

4. Remove the strut mounting cap from the strut.

5. If equipped, remove the shock absorber actuator bracket bolts and the shock absorber actuator.

6. If necessary, remove the actuator bracket-to-strut nut and the bracket.

7. Raise and safely support the vehicle; do not raise the vehicle at the parallel links or radius links. Remove the wheel(s).

8. Uncouple the rear brake hydraulic line from the strut. Then, unbolt and remove the brake assembly, wheel bearings and backing plate.

NOTE: If equipped with disc brakes, suspend the brake caliper so the hydraulic line will not be stressed. If equipped with drum brakes, remove the entire brake assembly

9. Remove the radius rod-to-strut bolt, stabilizer bar-to-radius rod bracket bolt, radius rod bracket-to-strut bolts and both parallel links-to-strut nut/bolt.

10. Support the strut from underneath and remove the 3 strut-to-chassis nuts and lower the strut from the vehicle.

— WARNING —

Never loosen the strut center nut until the spring is compressed or serious injury or vehicle damage may occur.

To install:

11. Install in the strut onto the vehicle and support it with the 3 mounting nuts.

12. Install both parallel links-to-strut nut and bolt, radius rod bracket-to-strut bolts, stabilizer bar-to-radius rod bracket bolt and the radius rod-to-strut bolt.

13. If removed, install the brake assembly.

14. Tighten all bolts sufficiently to safely support the vehicle; then, lower the vehicle to the ground. The final tightening of all the suspension components must be carried at with the vehicle in an unladened condition.

15. Final torque the nuts and bolts, as follows:

- Upper strut-to-chassis nuts: 23–31 ft. lbs. (31–41 Nm)
- Radius rod bracket-to-strut bolts: 43–58 ft. lbs. (59–78 Nm)
- Radius rod-to-bracket bolt: 65–80 ft. lbs. (88–108 Nm)
- Parallel links-to-strut nut/bolt: 65–87 ft. lbs. (88–118 Nm)
- Stabilizer bar-to-radius rod bracket bolt: 43–58 ft. lbs. (59–78 Nm)

16. Connect the brake line to the strut housing.

17. If equipped, connect the electrical harness connector and the shock absorber actuator.
18. Install the strut mounting cap.
19. If the brake line was disconnected, bleed the brake system.
20. Install the rear seat and the parcel shelf.
21. Install the wheels and connect the negative battery cable.
22. Check and/or adjust the wheel alignment.

1995–97 Maxima

1. Raise and safely support the vehicle.
2. Remove the rear wheels.
3. Support the rear torsion beam assembly with a jack.
4. Open the trunk and remove the two nuts attaching the strut to the vehicle.

— CAUTION —
Do not remove the center locknut from the strut assembly until the strut is safely compressed.

5. Remove the bolt attaching the strut to the rear torsion beam assembly and remove the strut.
6. Place the strut assembly in a vise with the special holding tool HT71780000 or in a spring compressor.
7. Loosen the piston rod locknut.

— CAUTION —
Do not remove the piston rod locknut, the spring is under tension and can cause serious personal injury.

8. Compress the spring with the spring compressor then remove the piston rod locknut.

NOTE: Before removing the strut from the coil spring, note the positioning of the strut in relationship to the coil spring for reassembly.

9. Remove the bushing, strut mounting bracket, and the upper spring seat rubber.
10. Remove the strut, leaving the coil spring compressed.
11. Remove the bushing, bound bumper cover, and the bound bumper.
To install:
12. Install the bound bumper, bound bumper cover, and the bushing.
13. Install the strut into the coil spring, make sure the strut and spring are properly positioned.

14. Install the upper spring seat rubber, strut mounting bracket, and the bushing. Make sure that the mounting bracket is properly positioned.
15. Install the piston rod locknut then remove the spring compressor.
16. Torque the piston rod locknut to 13–17 ft. lbs. (18–24 Nm).
17. Install the strut into the vehicle and install new attaching nuts. Torque the nuts to 12–14 ft. lbs. (16–19 Nm).
18. Position the strut on the rear torsion beam and install the bolt. Torque the bolt attaching the strut to the torsion beam assembly to 72–87 ft. lbs. (98–118 Nm).
19. Remove the support from the rear torsion beam.
20. Install the rear wheels and lower the vehicle.
21. Check the vehicle's alignment and adjust as necessary.

240SX

1. Raise and safely support the vehicle.
2. Remove the rear wheels from the vehicle.
3. Support the rear control arm with a jack.
4. Remove the strut lower mounting bolt.
5. Mark the upper spring plate to the vehicle for reinstallation purposes.
6. Remove the two strut upper mounting nuts.

— WARNING —
Do not remove the center piston locknut until the spring has been compressed.

7. Remove the strut from the vehicle.
8. Mark the positioning of the coil spring to the strut assembly.
9. Using a coil spring compressor, compress the spring.

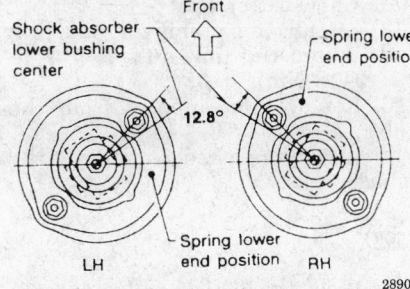

Positioning of the strut mounting brackets — 1995–97 Maxima

— CAUTION —
If coil spring is not properly compressed serious injury could result.

10. Remove the strut assembly center locknut.

NOTE: Before removing the strut from the coil spring, note the positioning of the strut in relationship to the coil spring for reassembly.

11. Remove the upper strut plate.
12. Remove the strut leaving the coil spring compressed.
13. Note the positioning of the spring and slowly release the spring compressor. Remove the coil spring assembly.
To install:
14. Properly position and compress the coil spring.
15. Install the strut to the coil spring.
16. Install the upper strut plate and tighten the center locknut to 13–17 ft. lbs. (18–24 Nm). Verify that the coil spring is seated properly on the strut.
17. Remove the coil spring compressor.
18. Install the strut assembly to the vehicle.
19. Using new locking nuts, install the top mounting nuts and torque to 12–14 ft. lbs. (16–19 Nm).

NOTE: Make sure that all final torques are done with the full weight of the vehicle on the ground.

20. Install the lower mounting nut and torque to 65–80 ft. lbs. (88–108 Nm) for 1993–94 vehicles.
21. Install the lower mounting bolt and torque to 72–87 ft. lbs. (98–118 Nm) for 1995–97 vehicles.
22. Install the rear wheels.
23. Check the vehicle alignment and adjust it as necessary.

300ZX

1. Block the front wheels.
2. Raise and support the vehicle safely. Remove the rear wheel(s).

NOTE: The vehicle should be far enough off the ground so the rear spring does not support any weight.

3. Working inside the luggage compartment, turn and remove the caps above the strut mounts.
4. Disconnect the shock actuator sub-harness connector.
5. Remove the actuator fixing bolts.

6. Remove the strut mounting nuts.

WARNING

Do not remove the center spring retaining nut until the coil spring is compressed or serious personal injury may result.

7. Remove the mounting bolt for the strut at the lower arm (transverse link) and then lift out the strut.

8. Place the strut in a spring compressor and compress the spring.

9. Remove the center spring retaining nut. Decompress the spring and remove the spring seat, bushing, plate, and the spring.

10. Remove the strut from the spring compressor.

To install:

11. Install the strut into the spring compressor. Install the coil spring, plate, bushing, and upper spring seat.

12. Compress the coil spring and install the center spring retaining nut. Decompress the spring and tighten the nut to 13–17 ft. lbs. (18–24 Nm).

13. Remove the strut assembly from the coil spring compressor.

14. Install the strut upper end first and secure with the nuts snugged down, but not fully tightened. Attach the lower end of the strut to the transverse link and tighten the upper nuts to 12–14 ft. lbs. (16–19 Nm). Tighten the lower mounting bolt to 57–72 ft. lbs. (77–98 Nm).

15. Install the shock absorber actuator fixing bolts, torque to 26–34 inch lbs. (3–4 Nm)

16. Reconnect the sub-harness connector.

17. Install the wheel assembly and lower the vehicle.

18. Install the strut mount caps in the trunk.

19. Inspect the vehicle alignment and adjust as necessary.

Lower Control Arms

REMOVAL AND INSTALLATION

Altima

1. Raise and safely support the vehicle.

2. Each link or rod can be removed individually as required by removing the nuts and bolts.

3. To remove the rear parallel link, remove the bolts that secure the link and remove the link.

To install:

4. The bushings cannot be removed from the links. If the bushings are faulty, the link must be replaced.

5. Install the link and secure with mounting bolts but do not tighten it yet.

NOTE: Final tightening must be done with the vehicle at normal ride height, tires on the ground and the chassis loaded.

6. When all link bolts are installed, lower the vehicle so it is resting on all four wheels. Roll it forward and back several times to settle the bushings.

7. Torque the inner parallel link nuts to 72–87 ft. lbs. (98–118 Nm) and tighten the outer bolts/nuts to 62–72 ft. lbs. (84–98 Nm).

240SX

1. Raise and safely support the vehicle.

2. Remove the wheel assembly.

3. Remove the stabilizer bar links.

4. Support the hub with jack.

5. Remove the lower ball joint nut and cotter pin. Separate the lower ball joint from the knuckle assembly.

6. Remove the control arm support bolts.

7. Remove the control arm.

To install:

8. Install the control arm,

NOTE: Install new nuts to the lower control arm mounting bolts and the ball joint. The final tightening of the mounting bolts should be performed with the vehicle on the ground.

9. Install the control arm support bolts. Torque bolts to 57–72 ft. lbs. (77–98 Nm).

10. Raise the control arm and reconnect the lower ball joint to the knuckle assembly.

11. Tighten the lower ball joint nut to 52–64 ft. lbs. (71–86 Nm) and install a new cotter pin.

12. Install the stabilizer bar link and tighten the link nuts to 7–9 ft. lbs. (9–12 Nm).

13. Remove the support from the hub.

14. Install the wheels and lower the vehicle.

300ZX

1. Raise the vehicle and support it safely.

2. Disconnect the sway bar links from the sway bar assembly.

3. Disconnect the ball joint stud from the knuckle assembly.

NOTE: It is important to use proper method when separating joints. Damage to joint could occur resulting in possible joint failure.

4. Remove the 2 lower arm mounting bolts and nut.

5. Remove the arm from the vehicle.

To install:

6. Install the control arm to the vehicle. Temporarily snug the mounting bolts.

7. Connect the ball joint stud to the knuckle and torque the nut to 58–69 ft. lbs. (78–93 Nm). Install a new cotter pin.

8. Install the sway bar and links. Tighten the link mounting nuts to 6.5–8.7 ft. lbs. (9–12 Nm).

9. Once the weight of the vehicle is on the suspension, tighten the lower control arm nuts to 57–72 ft. lbs. (77–98 Nm).

Sway Bar

REMOVAL AND INSTALLATION

Sentra

1. Raise and support the vehicle safely.

2. Remove the self-locking nuts at the sway bar link. Once the stabilizer bar nut is removed, remove the joint cups and stabilizer rubber bushings.

3. Remove the retainer bolts and the stabilizer bar brackets. Remove the bushings.

4. Remove the stabilizer bar.

5. Inspect the bar for damage, wear, and deterioration. Replace as required.

To install:

NOTE: Do not tighten the mounting hardware until the full weight of the vehicle is on the ground.

6. Install the stabilizer bar into the vehicle.

7. Install the center stabilizer bar bushings and brackets. Torque the bolts to 23–31 ft. lbs. (31–42 Nm)

8. Install the stabilizer bar to the stabilizer connecting rod and torque the nut to 34–38 ft. lbs. (46–52 Nm). Make sure that the ball joint socket of the connecting rod is properly placed during installation.

9. Lower the vehicle and perform the final tightening of the sway bar fasteners. Road test the vehicle.

Altima

1. Raise and support the vehicle safely.

2. Remove the stabilizer bar-to-strut mounting link. Note location of the washers for reinstallation.

3. Matchmark the stabilizer bar to the mounting clamps and note the positioning of the mounts to the paint marks on the stabilizer bar.

4. Supporting the stabilizer bar, remove the stabilizer bar mounting clamp nuts. Remove the bar from the vehicle.

To install:

NOTE: All final torque should be performed with the vehicle resting on the ground.

5. Install and support the stabilizer bar.

6. Install the mounting clamps with bushings and tighten the mounting nuts to 30–35 ft. lbs. (41–47 Nm).

NOTE: Be sure to align the mounting clamps with the paint marks and properly position the stabilizer links.

7. Connect the stabilizer link to the strut assembly and tighten the nuts to 30–35 ft. lbs. (41–47 Nm).

Maxima

1. Raise and safely support the vehicle.

2. Remove the mounting nuts connecting the stabilizer bar to the radius rod brackets.

3. Remove the 4 bolts holding the stabilizer bar bracket to the chassis and then pull the bar from the vehicle.

To install:

4. Install the stabilizer bar and mounting brackets. Never fully tighten the mounting bolts unless the

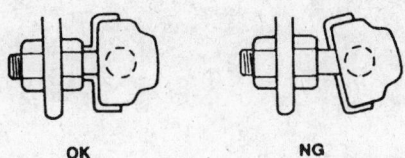

OK   NG
294910

Proper positioning of the stabilizer links — Altima, Sentra and 1995–97 Maxima and 200SX

vehicle is resting on the ground with normal weight upon the wheels.

NOTE: When installing the stabilizer bar, make sure the paint mark is aligned with the bushing and the clamp is facing in the correct direction.

5. Torque the bolts connecting the stabilizer bar bracket to the chassis to 23–31 ft. lbs. (31–42 Nm).

6. Torque the nuts connecting the stabilizer bar to the radius rod brackets to 43–58 ft. lbs. (59–78 Nm).

7. On all models, make sure the stabilizer bar connecting rods are installed on their correct sides.

8. Lower the vehicle.

240SX and 300ZX

1. Raise and support the vehicle safely.

2. Remove the self-locking nuts from the stabilizer link. Disconnect the stabilizer link from the from the stabilizer bar.

NOTE: Be sure to note the positioning of the stabilizer link bushings and washers for reinstallation.

3. Remove the retainer bolts and the stabilizer bar brackets. Remove the bushings and note the positioning of the mounting brackets.

4. Remove the stabilizer bar.

5. Inspect the bar for damage, wear and deterioration. Replace the sway bar as required.

To install:

NOTE: Do not tighten mounting hardware until the full weight of the vehicle is on the ground.

6. Install the stabilizer bar into the vehicle.

7. Install the two center stabilizer bar bushings, brackets and bolts. Tighten the bolts to 32–41 ft. lbs. (43–55 Nm).

8. Assemble stabilizer links and tighten the new self-locking nuts to 78–104 inch lbs. (9–12 Nm).

9. Lower the vehicle.

Wheel Bearings

ADJUSTMENT

Sentra, 200SX, Altima and Maxima

If the wheel hub bearing assembly is removed, it must be replaced.

NOTE: The wheel hub bearing assembly is not repairable; it must be replaced when defective.

1. Torque the wheel bearing locknut to 138–188 ft. lbs. (187–255 Nm).

2. Verify that the wheel bearings operate smoothly.

3. Install a new cotter pin into the spindle to hold the wheel bearing locknut.

4. Install a dial indicator to the rear wheel hub bearing assembly and check the axial end-play; it should be less than 0.0020 in. (0.05 mm).

5. Install the grease cap.

6. If the axial end-play exceeds specifications, the wheel bearing must be replaced.

REMOVAL AND INSTALLATION

Sentra, 200SX, Altima and Maxima

If the wheel hub bearing assembly is removed, it must be replaced.

NOTE: If the vehicle is equipped with ABS, the sensor must be removed to protect the sensor and its wiring.

1. Raise and safely support the vehicle. Remove the rear wheel(s).

2. If equipped with disc brakes, perform the following procedures:
   a. Remove the brake caliper and hang it by a piece of wire.
   b. Remove the brake caliper support.
   c. Remove the disc brake pads.
   d. Remove the brake disc.

3. If equipped with drum brakes, perform the following procedures:
   a. Remove the brake drum.
   b. If necessary, remove the brake shoe assembly.

4. Remove the grease cap.

5. Remove the cotter pin, wheel bearing locknut, washer, and the wheel hub bearing assembly. A slide hammer may be needed to remove the hub bearing assembly.

NOTE: The wheel hub bearing assembly is not repairable; it must be replaced when defective.

To install:

NOTE: If the vehicle is equipped with ABS, the sensor ring must be removed and installed on the new hub.

6. Install the wheel hub bearing assembly, the washer and the wheel bearing locknut. Torque the wheel bearing locknut to 138–188 ft. lbs. (187–255 Nm).

7. Verify that the wheel bearings operate smoothly.

8. Install a new cotter pin into the spindle to hold the wheel bearing locknut.

9. Install a dial micrometer to the rear wheel hub bearing assembly and check the axial end-play; it should be less than 0.0020 in. (0.05 mm).

10. Install the grease cap.

11. If removed, install the ABS sensor and its wiring.

12. Install the brake assembly and the wheels.

240SX and 300ZX

1. Raise and support the rear of the vehicle and remove the rear wheels. Remove the cotter pin, adjusting cap and insulator.

2. Apply the parking brake firmly to hold the rear halfshaft while removing the wheel bearing locknut. Remove the wheel bearing locknut.

3. Unbolt the caliper and move it aside. Do not disconnect the hose from the caliper. Do not allow the caliper to hang by the hose; support the caliper with a length of wire or rest it on a suspension member. Remove the brake disc.

4. Separate the halfshaft from the axle housing by lightly tapping it. Cover the driveshaft boots with a shop towel to prevent damage.

5. Remove the 4 bolts on the rear of the axle housing and that hold the wheel bearing, flange, and hub to the axle housing. Remove the wheel bearing, flange, and hub assembly.

6. Press the wheel bearing from the axle hub.

7. Mount the hub in a vise and remove the inner race, using bearing puller tool ST30031000, or equivalent. Discard the inner race and grease seals.

8. Clean all parts in a suitable solvent. Check the wheel hub and axle housing for cracks, preferably using the dye penetrant method. Check the wheel bearing seating surface for roughness, seizure or other damage that may interfere with proper bearing function. Check the rubber bushing for wear.

To install:

9. Place the hub on a block of wood and seat the inner race using a suitable drift. Be careful not to damage the grease seals during installation of the inner race.

10. Press the bearing into the hub using a suitable drift.

11. Mount the bearing/hub assembly to the axle housing and torque the mounting bolts to 58–72 ft. lbs. (78–98 Nm).

12. Lubricate the halfshaft splines. Properly align the splines and insert the halfshaft into the wheel hub.

13. Install the wheel bearing locknut and torque the nut to 152–203 ft. lbs. (206–275 Nm). Install the insulator and fit adjusting cap. Install a new cotter pin.

14. Check the axial end-play as follows before mounting the rear wheels. Mount a dial indicator so the stylus of the dial rests on the face of the hub and check the wheel bearing axial end-play by attempting to rock the wheel hub in and out. The end-play should be 0.0020 in. (0.05mm) or less.

15. Install the rotor and caliper assembly.

16. Mount the rear wheels and lower the vehicle.

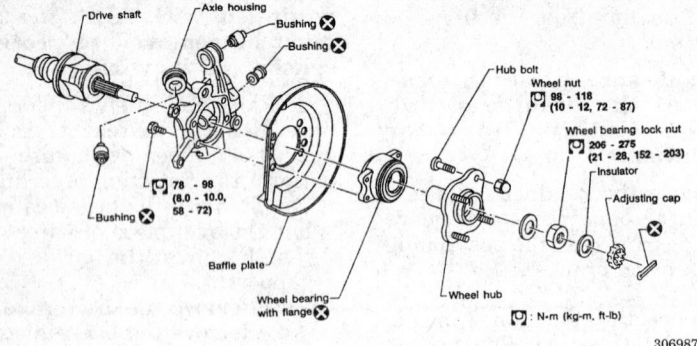

Exploded view of rear wheel bearing assembly — 240SX

FIRING ORDERS

NOTE: To avoid confusion, always replace spark plug wires one at a time.

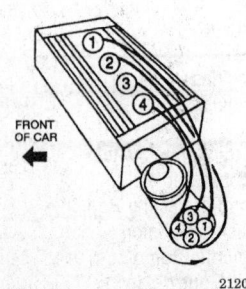

212017

900 Series 2.0L (B202L) and all 2.1L & 2.3L Engines
Firing Order: 1–3–4–2
Distributor Rotation: Counterclockwise

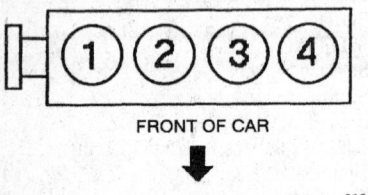

FRONT OF CAR

212010

900 Series 2.0L (B204L) and all 9000 Series 2.3L Engines
Engine Firing Order: 1–3–4–2
Direct Ignition: uses 1 coil at each cylinder

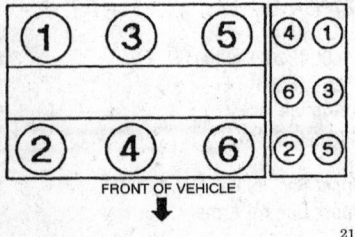

FRONT OF VEHICLE

211986

900 Series 2.5L (B258I) and 9000 Series 3.0L (B308I) Engines
Engine Firing Order: 1–2–3–4–5–6
Distributorless Ignition System

ENGINE ELECTRICAL

NOTE: Disconnecting the negative battery cable on some vehicles may interfere with the functions of the on board computer systems and may require the computer to undergo a relearning process, once the negative battery cable is reconnected.

Distributor

REMOVAL AND INSTALLATION

900 Series

2.0L (B202L) and 2.1L (B212I) Engines

1. Disconnect the negative battery cable.
2. Remove the distributor cap by releasing the retaining clamp.
3. Disconnect the Hall effect sensor and vacuum hose.
4. Mark the position of the distributor housing in relation to the engine, for installation reference.
5. Unscrew the distributor holddown clamp. Remove the distributor.

NOTE: Do not rotate the engine while the distributor is removed from the engine.

To install:

6. Install the distributor. Rotate the distributor shaft until the shaft tab seats into the offset slot in the camshaft. Align the distributor housing with the mark made during removal.
7. Tighten the distributor holddown clamp.
8. Reconnect the Hall effect sensor and vacuum hose.
9. Install the distributor cap.
10. Reconnect the negative battery cable. Check the ignition timing.

2.3L (B234I) Engine

1. Disconnect the negative battery cable.
2. Remove the distributor cap and disconnect the Hall effect sensor connector.
3. Remove the distributor holddown bolt. Remove the distributor from the engine.

NOTE: Do not rotate the engine while the distributor is removed from the engine.

To install:

4. Install the distributor onto the engine. Turn the distributor shaft so that the tab in the distributor shaft inserts into the offset slot in the camshaft.
5. Tighten the distributor holddown bolt.
6. Reconnect the Hall effect sensor connector and install the distributor cap.
7. Reconnect the negative battery cable.

Alternator

PRECAUTIONS

Several precautions must be observed with alternator equipped vehicles to avoid damage to the unit.

• If the battery is removed for any reason, make sure it is reconnected with the correct polarity. Reversing the battery connections may result in damage to the 1-way rectifiers.
• When utilizing a booster battery as a starting aid, always connect the positive to positive terminals and the negative terminal from the booster battery to a good engine ground on the vehicle being started.
• Never use a fast charger as a booster to start vehicles.
• Disconnect the battery cables when charging the battery with a fast charger.
• Never attempt to polarize the alternator.
• Do not use test lights of more than 12 volts when checking diode continuity.
• Do not short across or ground any of the alternator terminals.
• The polarity of the battery, alternator and regulator must be matched and considered before making any electrical connections within the system.
• Never separate the alternator on an open circuit. Make sure all connections within the circuit are clean and tight.
• Disconnect the battery ground terminal when performing any service on electrical components.
• Disconnect the battery if arc welding is to be done on the vehicle.

REMOVAL AND INSTALLATION

900 Series

2.0L (B202L) and 2.1L (B212I) Engines

1. Disconnect the negative battery cable.
2. Remove the retaining bolts. Loosen the drive belt, then remove it. Disconnect the alternator leads.

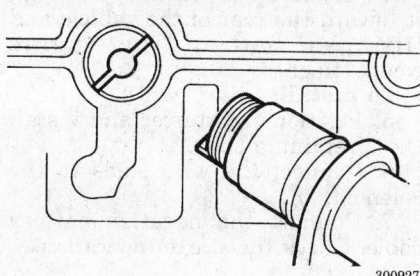

Aligning the distributor shaft tab to the offset slot in the camshaft — 2.3L (B234I) engine

3. Remove the alternator mounting and the alternator.

4. Remove the 2 screws holding the voltage regulator / brush assembly. Remove the assembly.

To install:

5. Compress the brushes to slide the voltage regulator into place and tighten the 2 retaining screws.

6. Connect the wire leads to the rear of the alternator.

7. Install the alternator and mounting.

8. Install and tighten the drive belt.

9. Connect the negative battery cable. Start the vehicle and check the alternator output.

2.0L (B204L) and 2.3L (B234I) Engines

1. Disconnect the negative battery cable.

2. Remove the air cleaner and air induction hose assemblies.

3. Remove the belt tensioner.

4. Raise the vehicle and rest it on safety stands. Remove the right front wheel.

5. Remove the belt cover.

6. Disconnect the wire leads from the rear of the alternator.

7. Remove the 2 alternator retaining bolts.

8. Remove the 3 exhaust pipe to manifold bolts.

9. Remove the catalytic converter bracket.

10. Position the catalytic converter to the left for clearance to remove the alternator.

11. To remove the voltage regulator, remove the rear cover on the alternator and remove the 2 retaining screws.

To install:

12. Position the alternator and connect the wire leads. Tighten the retaining bolts.

13. Position the catalytic converter and secure the converter bracket.

14. Tighten the 3 exhaust pipe to manifold bolts.

15. Install the belt cover.

16. Install the right front wheel. Lower the vehicle from the stands.

17. Install the belt tensioner.

18. Install the air cleaner and air induction hose assemblies.

19. Connect the negative battery cable.

2.5L (B258I) Engine

1. Disconnect the negative battery cable.

2. Remove the air cleaner and air induction hose assemblies.

3. Relieve the pressure on the tensioner and remove the drive belt.

4. With all four wheels on the ground, loosen the right side axle nut.

5. Properly raise and support the vehicle on safety stands. Remove the right front wheel.

6. Remove the cover for the drive belt.

7. Remove the drive axle nut and remove the outboard drive axle.

8. Disconnect the wire leads from the rear of the alternator.

9. Remove the cooling duct to the alternator.

10. Remove the 2 retaining bolts to the alternator. Remove the alternator.

11. If removal of the voltage regulator is required, remove the rear cover from the alternator and remove the 2 retaining screws from the regulator.

To install:

12. Install the regulator by compressing the brushes and installing the regulator. Tighten the 2 retaining screws.

13. Position the alternator in position, connect the wire leads from the rear of the alternator and tighten the retaining bolts.

14. Connect the cooling duct to the alternator.

15. Install the drive axle and nut but do not tighten the nut yet.

16. Install the cover for the drive belt.

17. Install the right front wheel. Remove the safety stands and lower the vehicle.

18. With all four wheels on the ground, torque the axle nut to 214 ft. lbs. (290 Nm). Install the center hubcap.

19. Install the drive belt.

20. Install the air cleaner and air induction hose assemblies.

21. Connect the negative battery cable.

9000 Series

2.3L (B234L, B234R, B234E, B234I) Engines

1. Disconnect the negative battery cable.

2. Raise the vehicle and support it on safety stands. Remove the right front wheel and inner wheel well.

3. Remove the serpentine belt from the alternator and power steering pump and fit a tensioner lock tool 83–94–488 or equivalent in place on the tensioner.

4. Remove the power steering pump bracket.

5. Remove the 2 securing bolts and drop the alternator down and towards the rear.

6. Disconnect the electrical leads from the alternator and remove it by lifting it towards the front of the vehicle.

To install:

7. Connect the leads to the alternator and place it in position in the engine compartment.

8. Install the bracket for the power steering pump.

9. Install the top bolt in the alternator, leaving it loose. Align the alternator, secure it to the bracket and tighten the bolts.

10. Install the power steering pump and serpentine belt. Remove the lock bar from the tensioner.

11. Install the wheel well and wheel. Lower the vehicle.

12. Reconnect the negative battery cable.

3.0L (B308I) Engine

1. Disconnect the negative battery cable.

2. Disconnect the rear oxygen sensor wiring.

3. Remove the drive belt.

4. Remove the drive belt tensioner.

5. Raise vehicle and support with safety stands. Remove right front wheel and wheel well liner.

6. Disconnect the front oxygen sensor wiring.

7. Remove the front exhaust pipe section.

8. Remove the alternator retaining bolts and lift alternator so leads can be disconnected.

9. Lift alternator out of vehicle.

To install:

10. Connect lead to back of alternator.

11. Place alternator in bracket and secure retaining bolts.

12. Fit front exhaust pipe section.

13. Connect front oxygen sensor wiring.

14. Install right front wheel well liner and fit wheel.

15. Install drive belt tensioner.

16. Install drive belt.

17. Connect rear oxygen sensor wiring.

18. Connect negative battery cable.

Drive Belt

REMOVAL AND INSTALLATION

2.0L (B202L and B204L), 2.1L (B212I) and 2.3L (B234I, B234L, B234R and B234E) Engines

1. Remove the air filter assembly.

2. Release the pressure on the belt tensioner using a 3/8 inch ratchet extension and insert a 6mm bit into the access hole to hold in position to remove the belt.

3. Remove the drive belt.

To install:

4. Fit drive belt over pulley assemblies.

5. Release tensioner using extension, remove the 6mm bit from the access hole and slowly remove ratchet extension.

6. Fit air filter assembly.

2.5L (B258I) and 3.0L (B308I) Engines

1. To release tensioner on the drive belt use a long pattern 15mm box wrench on the belt tensioner.

2. Turn the wrench toward the rear of the vehicle and hold there while remove belt from top pulleys.

3. When belt is off the pulley slow release the belt tensioner.

To install:

4. Fish belt around proper pulleys until you need to move tensioner.

5. Using a long pattern 15mm box wrench turn the tensioner toward rear of the vehicle allowing belt to slacken enough to install over all pulleys.

6. When the belt is properly routed slowly release pressure on the belt tensioner.

Starter

REMOVAL AND INSTALLATION

2.0L (B202L) and 2.1L (B212I) Engines

1. Disconnect the negative battery cable.

2. Remove the 2 retaining bolts from the starter.

3. Disconnect the wire leads to the solenoid.

4. Remove the starter by moving it toward the rear of the vehicle and lifting it out of the engine compartment.

To install:

5. Position the starter and install the 2 retaining bolts.

6. Connect the wire leads to the solenoid.

7. Connect the negative battery cable. Check the starter operation.

2.0L (B204L) and 2.3L (B234I, B234L, B234E, B234R) Engines

900 Series

1. Disconnect the negative battery cable.

2. Remove the upper retaining bolt from the starter.

3. Raise the vehicle and support safely.

4. Disconnect the wires at the solenoid.

5. Remove the lower bolt securing the starter.

6. Remove the starter. It may be necessary to move the exhaust.

To install:

7. Position the starter and install the lower retaining nut. Install the exhaust pipe if necessary.

8. Connect the wiring to the solenoid.

9. Lower the vehicle and install the upper retaining bolt.

10. Connect the negative battery cable. Check the starter operation.

9000 Series

1. Disconnect the negative battery cable.

—————— **WARNING** ——————
Do not disconnect the battery cable while the engine is running. Serious damage to the alternator could result.
——————————————————————

2. Disconnect the harness connections at the solenoid.

3. Loosen but do not remove the upper mounting bolt of the support brace by the starter.

4. Remove the starter retaining bolts.

5. Lower the starter away from the block and remove from the vehicle.

To install:

6. Lower the starter into the vehicle and up into position on the block.

7. Install and tighten the starter retaining bolts.

8. Tighten the upper support brace bolt.

9. Connect the harness connections at the solenoid.

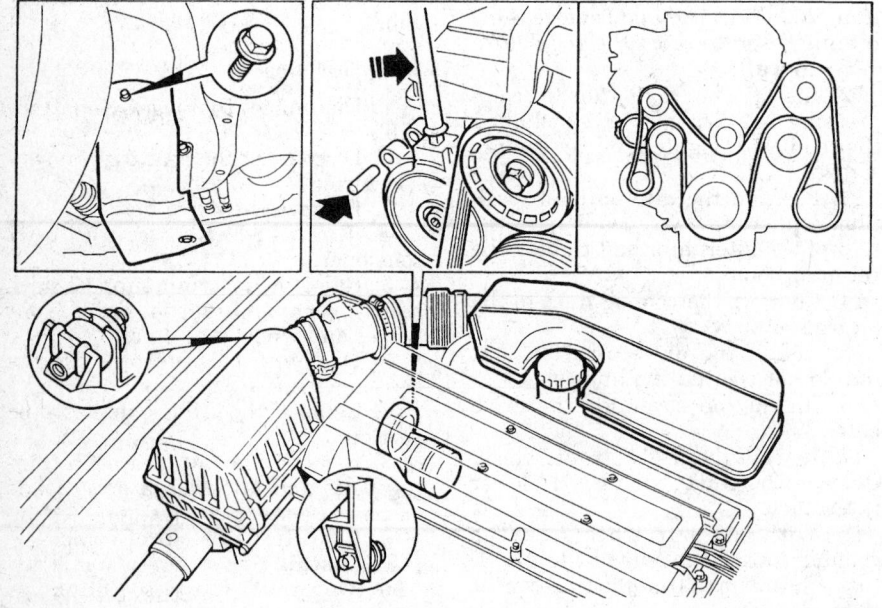

311538

Drive belt removal and installation — 2.0L (B202L and B204L), 2.1L (B212I) and 2.3L (B234I, B234L, B234R and B234E) engines

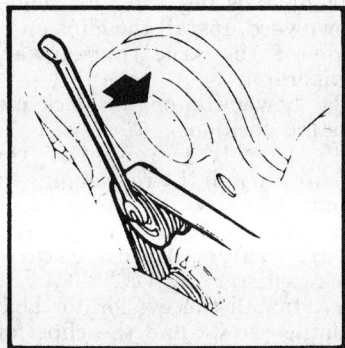

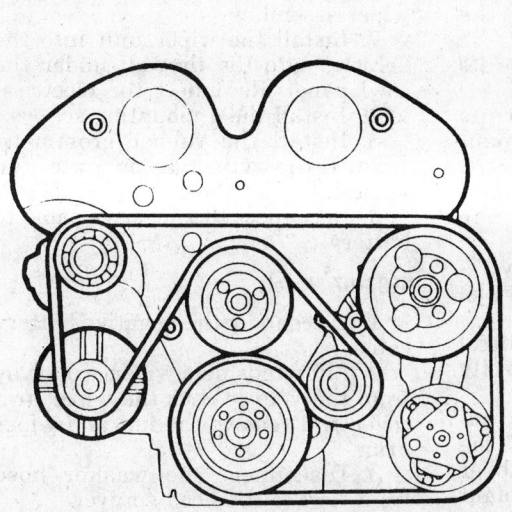

Drive belt routing and tension relief point — 2.5L and 3.0L engines

305191

10. Connect the negative battery cable. Check starter operation.

2.5L (B258I) and 3.0L (B308I) Engines

1. Disconnect the negative battery cable.
2. Raise the vehicle and support the vehicle with safety stands.
3. Cut the plastic cable ties holding the wiring at the starter.
4. Remove the wire leads from the solenoid.
5. Remove the 2 starter retaining bolts.

NOTE: The starter retaining bolts are Torx® head bolts.

6. Remove the starter from under the vehicle.
   To install:
7. Position the starter and tighten the 2 retaining bolts.
8. Connect the wire leads to the solenoid.
9. Secure the wire loom with new cable ties.
10. Lower the vehicle to the ground.
11. Connect the negative battery cable. Check starter operation.

CHASSIS ELECTRICAL

Blower Motor

REMOVAL AND INSTALLATION

900 Series

1993 Models

1. Disconnect the negative battery cable.
2. Remove the dash panel and the upper section of the instrument panel. The screws retaining the switch panel are of different lengths. The screws are marked with grooves. Note the placement of the screws for reassembly.
3. Disconnect the electrical leads to the fan motor.
4. Remove the retaining screws for the right defroster valve housing.
5. Remove the fan retaining screws. Remove the fan from its housing.
   To install:
6. Position the fan in the housing and tighten the retaining screws.

7. Secure the right defroster valve housing.
8. Connect the electrical leads to the fan motor.
9. Install the switch panel and the upper section of the instrument panel. Be sure to use the correct length screws during assembly.
10. Connect the negative battery cable.

1994–97 Models

1. Disconnect the negative battery cable.
2. Remove the windshield wiper arms.
3. Remove the cover over the bulkhead.
4. Remove the wiper unit.
5. Remove the fresh-air filter.
6. Unplug the wiring connector to the fan motor.
7. Remove the filter housing.
8. Remove the ventilation fan cover.
9. Remove the blower motor retaining screws and remove the motor.
   To install:
10. Position the motor and tighten the retaining screws.
11. Install the ventilation fan cover.
12. Install the filter housing fresh-air filter.
13. Install the cover over the bulkhead.
14. Install the windshield wiper arms.
15. Connect the negative battery cable.

9000 Series

1. Disconnect the negative battery cable.
2. Remove the hood assembly.
3. Disconnect the wiper arms. Remove the covers on the evaporator and wiper motor.
4. On vehicles with automatic climate control, unplug the connector for the fan control unit.
5. Remove the false bulkhead panel.
6. Drain the radiator.
7. Remove the plastic drainage tube molding below the windshield molding.
8. Remove the bolts securing the Electronic Control Unit and position it aside.
9. Remove the clip and unplug the connectors. Remove the complete wiper assembly.
10. Remove the rubber lead through panel for the coolant hoses. Disconnect the quick release couplings for the coolant hoses at the heat exchanger.

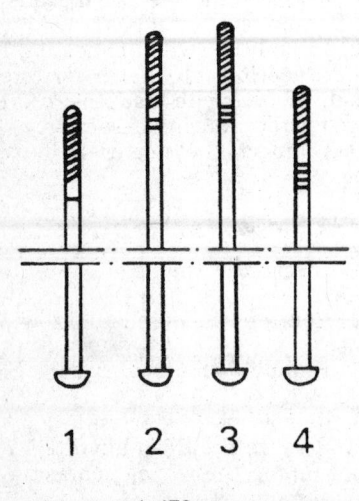

1 176 mm
2 205 mm
3 210 mm
4 189 mm

213265

Dashboard assembly screws are marked with grooves to designate length — 1993 900 Series

11. Move the cruise control vacuum pump assembly out of the way.

12. Remove the evaporator body retaining screws and the clips for the refrigerant hoses.

13. Remove the lock washer and disconnect the cable for the temperature valve.

14. Remove the engine torque mount. Remove the rear engine mount nuts.

15. Connect the engine to an engine lift and raise the engine enough to gain clearance to the evaporator housing.

16. Carefully lift the evaporator and remove the clips on either side of the fan. Remove the complete fan assembly by twisting the fan diagonally upwards.

17. Remove the screw in the center of the casing. Release the clips and the grille at the discharge duct.

18. Separate the fan housing and remove the securing screw for the fan motor.

19. Lift the cover upward and remove the motor complete with the impeller.

To install:

20. Install the motor complete with the impeller, then install the cover.

21. Install the securing screw for the fan motor, then install the fan housing.

22. Attach the clips and the grille at the discharge duct. Install the screw in the center casing.

23. Install the complete fan assembly by twisting the fan diagonally downward. Install the clips on either side of the fan, then lower the evaporator.

24. Lower the engine back into its resting position.

25. Install the engine torque mount. Install the rear engine mount nuts.

26. Attach the cable for the temperature valve, then install the lockwasher.

27. Install the evaporator body retaining screws and the clips for the refrigerant hoses.

28. Reposition the cruise control vacuum pump assembly into its place.

29. Connect the quick release couplings for the coolant hoses at the heat exchanger. Install the rubber lead through the panel for the coolant hoses.

30. Install the complete wiper assembly. Install the clip and attach the connectors.

31. Install the bolts securing the Electronic Control Unit.

32. Install the plastic drainage tube molding below the windshield molding.

33. Fill and bleed the cooling system.

34. Install the false bulkhead panel.

35. On vehicles with automatic climate control, attach the connector for the fan control unit.

36. Connect the wiper arms. Install the covers on the evaporator and wiper motor.

37. Install the hood assembly.

38. Connect the negative battery cable.

Windshield Wiper Motor

REMOVAL AND INSTALLATION

900 Series

1993 Models

1. Disconnect the negative battery cable.

2. Lift windshield wiper arm away from the vehicle. Lift the wiper arm cover and remove the nut and wiper arm. Do this to both wiper arms.

3. Remove the rubber grommets located at the cowl panel.

4. The wiper unit is located on the firewall under the cowl panel. Remove the 4 mounting screws. Discon-nect the electrical lead. Remove the wiper unit from the vehicle.

5. Separate the wiper motor from the wiper assembly.

To install:

6. Install the wiper motor onto the wiper assembly.

7. Install the wiper unit into the vehicle, onto the firewall under the cowl panel. Reconnect the electrical lead. Install the 4 mounting screws.

8. Install the rubber grommets. Install the wiper arms onto the vehicle.

9. Reconnect the negative cable to battery.

1994–97 Models

1. Disconnect the negative battery cable.

2. Lift windshield wiper arm away from the vehicle. Lift the wiper arm cover and remove the nut and wiper arm.

3. Disconnect the washer hose from cowl to the hood sprayer.

4. The wiper unit is located on the firewall under the cowl panel. Remove the four mounting screws, disconnect the wiring and remove the wiper unit from the vehicle.

5. Separate the wiper motor from the wiper assembly.

To install:

6. Install the wiper motor onto the wiper assembly.

7. Install the wiper unit into the vehicle, onto the firewall under the cowl panel. Reconnect the wiring and install the mounting screws.

8. Install washer hose to hood sprayer.

9. Reconnect the negative cable to battery.

10. Turn the ignition switch **ON** and test the motor. With the ignition still **ON**, turn the wiper switch **OFF** and allow the motor to stop in the park position.

11. Install the wiper arms onto the vehicle.

9000 Series

1. Disconnect the negative battery cable.

2. Remove the windshield wipers and the protective rubber covers.

3. Remove the 2 upper screws that hold the wiper motor mounting plate.

4. Remove the hood seal and panel cover and disconnect the washer hose.

5. Carefully unscrew the ABS electronic unit, if equipped, and the LH electronic unit, then remove the bracket.

6. Disconnect the 24-pin connector and remove the bracket.

7. Disconnect the wiper motor connector.

8. Remove the 2 lower screws holding the wiper motor mounting plate.

9. Carefully push the wiper spindles, slightly move the unit to the right and remove it, left-hand wiper spindle first.

10. Carefully separate the wiper linkage from the wiper motor using a suitable prying tool.

11. Remove the wiper motor from the wiper motor mounting bracket.

To install:

12. Install the wiper motor onto the wiper motor mounting bracket and reconnect the wiper linkage to the motor.

13. Carefully install the wiper unit into the car, inserting the right-hand wiper spindle first, then install the upper mounting bolts.

14. Tighten the lower bracket mounting bolts.

15. Reconnect the wiper motor connector, making sure that the colors of the leads match each other.

16. Reconnect the 24-pin connector.

17. Fit the bracket, LH electronic unit, and the ABS electronic unit, if equipped.

18. Reconnect the washer hose. Install the panel cover and hood seal.

19. Install the 2 upper screws to the upper motor mounting plate.

20. Reconnect the negative cable to the battery.

21. Turn the ignition **ON** and operate the wiper motor. With the ignition still **ON**, turn the wiper switch **OFF** to be sure the motor is in the park position before installing the protective rubber covers and wiper arms.

22. Install the protective rubber wiper covers and install the wiper arms. Torque the wiper arm nuts to 10–16 ft. lbs. (14–22 Nm).

Headlight Switch

REMOVAL AND INSTALLATION

All Models

1. Disconnect the negative battery cable.

2. Carefully pull the switch from its mounting on the instrument panel assembly.

3. Disconnect the electrical connectors from the switch.

4. Remove the switch from the vehicle.

To install:

5. Reconnect the electrical connectors to the headlight switch.

6. Install the switch in its mounting on the instrument panel assembly.

7. Reconnect the negative cable to battery.

Turn Signal Switch

REMOVAL AND INSTALLATION

900 Series

1993 Models

The turn signal switch on the 1993 vehicles of the 900 series is incorporated in the combination switch. For removal and installation instructions, please refer to the combination switch procedures.

1994–97 Models

———— **CAUTION** ————
On vehicles equipped with an air bag, the negative battery cable must be disconnected for 20 minutes before working on the system. Failure to do so may result in deployment of the air bag and possible personal injury.

1. Disconnect the negative battery cable.

2. On vehicles equipped with adjustable steering wheels, make sure steering wheel is fully extended.

3. To remove the air bag:

 a. Remove the 2 screws securing the air bag.

 b. Remove the air bag and disconnect the air bag connectors.

 c. Store the air bag facing up.

 d. Remove the horn contact.

4. Remove four screws from the underside of the steering wheel.

5. Unplug horn connector.

6. Make sure wheels are heading straight ahead and the steering wheel is straight. Loosen steering wheel nut but do not remove.

7. Shake steering wheel back and forth to loosen from shaft. When loose, remove steering wheel nut and steering wheel.

8. Remove cowl mounting screws and pull cowl from mounting bracket.

9. Remove turn signal or wiper switch by pressing in the two clips on each mount. Pull switch out and unplug harness connector.

To install:

10. Plug in electrical harness connector and press switch into mount, make sure mounting clips are secure.

11. Install steering column cowl and tighten all four mounting screws.

12. Place steering wheel on shaft, making sure steering wheel is straight. Tighten the nut to 20 ft. lbs. (28 Nm).

13. To install the air bag:

 a. Install the contact assembly. Make sure it is centered in its travel.

 b. Install the air bag and connect the wiring.

 c. Install the 2 screws securing the air bag and tighten to 60 inch lbs. (7 Nm). These screws must be tightened evenly for proper air bag deployment.

14. With no one in the vehicle, connect the negative battery cable.

9000 Series

1. Disconnect the negative battery cable.

2. Pull out the telescoping steering wheel as far as it will go by moving the lever under the steering column. Do not remove the steering wheel.

3. Remove the steering column bearing bracket covers (4 screws).

4. On the turn signal switch, unplug the upper connector.

5. Remove the two switch retaining bolts.

6. Withdraw the switch and then unplug the connector.

To install:

7. Plug in the connector and install the switch onto the steering column bearing bracket. Tighten the retaining bolts.

8. Plug in the upper turn signal switch connector.

9. Install the steering column bearing bracket covers.

10. Reconnect the negative cable to battery. Check the turn signal switch for correct operation.

Combination Switch

REMOVAL AND INSTALLATION

900 Series

1993 Models

1. Disconnect the negative battery cable.

———— **CAUTION** ————
Before performing any repairs, if equipped with an air bag, disconnect the negative battery cable and wait 20 minutes before performing any procedures. Failure to do so may result in deployment of the air bag and possible personal injury.

2. Disconnect the horn connections in the steering wheel. Discon-

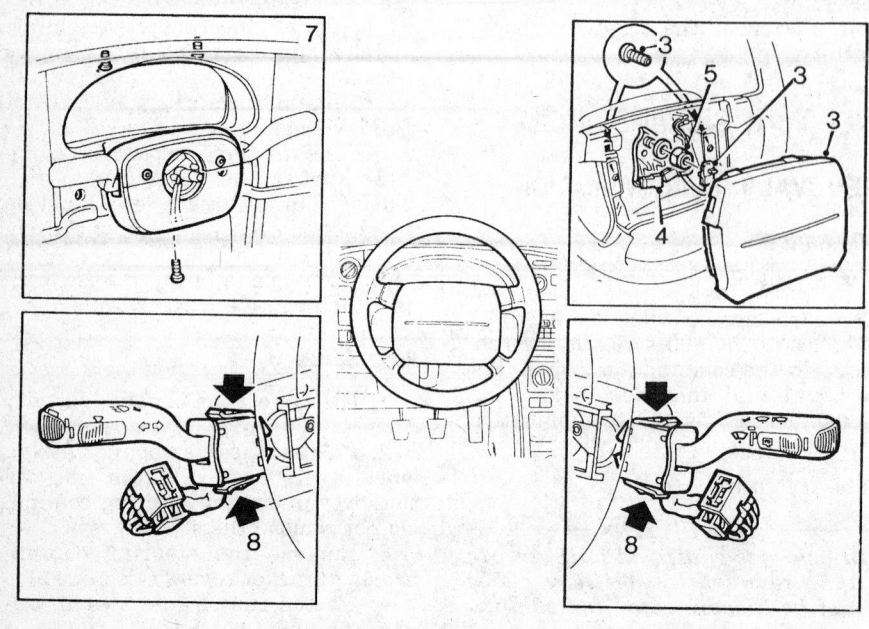

Remove air bag, steering wheel and column cover to access the turn signal and wiper switches — 1994–97 900 Series

289114

nect and remove the air bag, if equipped.

3. Remove the steering wheel using a proper steering wheel removal tool.

4. Remove the cover under the bearing support.

5. Remove the combination switch retaining bolts and electrical connections.

6. Remove the switch from the vehicle.

To install:

7. Install the switch onto the steering column bearing support and reconnect the electrical connections. Tighten the retaining bolts.

8. Install the cover under the bearing support.

9. Install the steering wheel and reconnect the horn connections. Install and reconnect the air bag unit.

10. Reconnect the negative battery cable. Test combination switch for operation.

Ignition Lock Cylinder

REMOVAL AND INSTALLATION

All Models

1. Disconnect negative battery cable.

2. Remove the protective cover around the lock cylinder.

3. Remove the three bolts securing the center console on 900 Series, and remove the bolt securing the ignition lock assembly on 9000 Series.

4. Turn the key to the straight ahead position and remove the lock cylinder. Use a narrow screwdriver to release the catch, insert it at an angle to the right in the hole. Withdraw the lock cylinder straight up.

To install:

5. Install the lock cylinder with the key inserted and turned to the first position.

6. Push the lock cylinder down until it is locked into position.

7. Fit the three bolts securing the center console.

8. Install the protective cover around the lock cylinder.

9. Connect the negative battery cable.

Ignition Switch

REMOVAL AND INSTALLATION

900 Series

2.0L (B202L) and 2.1L (B212I) Engines With Manual Transaxles

1. Disconnect the negative battery cable.

2. Move the left front seat back rearward as far as it will go. Remove the mounting bolts.

3. Disconnect the electrical connector underneath the seat.

4. Move the seat as far as it will go and release the front catch.

5. Tip back the seat. Lift the seat upwards and forwards out of the vehicle.

6. Apply the parking brake. Remove the bellows from between the center and rear consoles.

7. Place the shift lever in **R** and remove the ignition key.

8. Remove the console top cover retaining screws. Lift the cover so that the interior lighting switch can be disconnected. Remove the top cover from the shift lever.

9. Remove the rear console ashtray and remove the console mounting screws.

10. Lift the console and remove the lamp for the ignition switch lighting. Disconnect the electric window switch connector, if equipped. Remove the rear console from the vehicle.

11. Remove the mounting screws at the bottom edge of the center console. Disconnect the electric sun roof switch connector, if equipped.

12. Remove the ashtray. Remove the mounting screw behind the ashtray.

13. Lift out the center console, remove the ashtray light and remove the center console from the vehicle.

14. Pull back the carpet and remove the ventilation air duct.

15. Disconnect and label the ignition switch connectors. Disconnect the reversing light switch connector.

16. Disconnect the spring from the guide bracket and remove the rear console support bracket. Use special spanner socket tool 87–90–370 or equivalent, to remove the mounting bolts for the shift lever housing.

17. Lift and turn the shift lever housing on its side. Remove the cover plate bolts and cover plate underneath the shift lever housing.

18. Remove the ignition switch retaining screws and remove the ignition switch.

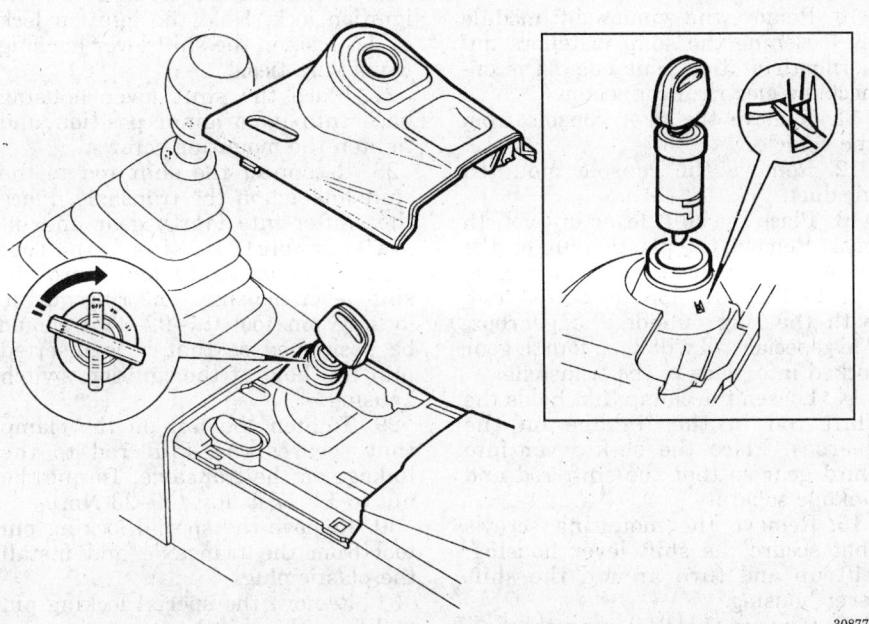

To remove the lock cylinder, align the switch and pry as indicated — 900 Series

308776

To install:

19. Before installing the new ignition switch, insert a screwdriver into the drive slot of the ignition switch and turn it so that the setting mark on the drive slot lines up with the arrow on the ignition switch.

20. Turn the ignition key to the **LOCK** position. Install the ignition switch into the shift lever housing. Be sure that the locating pad of the switch fits into the corresponding slot in the shift lever housing.

21. Turn the ignition switch back and forth several times to make sure that there is no binding or jamming in the shift lever locking mechanism.

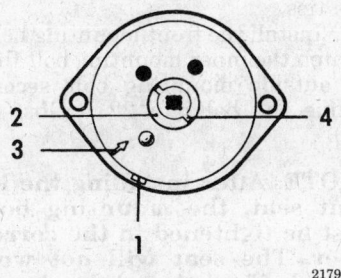

217983

Ignition switch: locating pad (1), setting mark (2), arrow (3) and drive slot (4) — 2.0L (B202L) and 2.1L (B212l) engines with manual transaxles

22. Install the cover plate back underneath the shift lever housing.

23. Position the shift lever housing back into its original place and tighten the shift lever housing mounting bolts using special spanner socket tool 87–90–370 or equivalent.

24. Install the rear console support bracket and reconnect the spring to the guide bracket.

25. Reconnect the ignition switch connectors and reversing light switch connector.

26. Install the ventilation air duct and place the carpet back to its original position.

27. Install the center console and ashtray light. Reconnect the electric sun roof switch connector, if equipped.

28. Install the center console mounting screws. Install the center console ashtray.

29. Install the rear console into the vehicle. Install the lamp for the ignition switch lighting and reconnect the electric window switch connector, if equipped.

30. Tighten the center console mounting screws and install the rear console ashtray.

31. Reconnect the interior lighting switch connector.

32. Install the rear console shift lever top cover. Install the console

bellows between the center and rear consoles.

33. Install the left front seat into the vehicle. Reconnect the electrical connector under the seat.

34. Tighten the mounting bolts. Move the seat forward and backward to engage the front catch mechanism.

35. Reconnect the negative battery cable.

2.0L (B202L) and 2.1L (B212l) Engines With Automatic Transaxles

1. Disconnect the negative battery cable.

2. Move the left front seat back rearward as far as it will go. Remove the mounting bolts.

3. Disconnect the electrical connectors underneath the seat.

4. Move the seat as far as it will go and release the front catch.

5. Tip back the seat. Lift the seat upwards and forwards out of the vehicle.

6. Remove the bellows from between the center and rear consoles.

7. Place the shift lever into the **P** position and remove the ignition key.

8. Loosen the locknut under the shift lever knob and remove the knob and locknut.

9. Remove the shift lever indicator cover.

10. Remove the light for the shift lever indicator.

11. Remove the mounting screws in the front of the rear console. Lift up the cover and disconnect the interior lighting switch connector.

12. Remove the ashtray in the rear of the console. Remove the mounting screws in the rear of the console and remove the console from the vehicle.

13. Move the carpet out of the way. Remove the set screw securing the shift cable to the shift lever.

14. Disconnect and label the ignition switch connectors. Disconnect the park/neutral switch connector.

15. Remove the shift lever housing mounting bolts using special spanner socket tool 87–90–370 or equivalent.

16. Lift up the shift lever housing and turn it to one side. Remove the housing cover underneath the shift lever housing and remove the ignition switch.

To install:

17. Install the ignition switch into the shift lever housing. Install the housing cover underneath the housing.

18. Place the shift lever housing back to its original position and tighten the mounting bolts using the special spanner socket tool 87–90–370 or equivalent.

19. Install the set screw securing the shift cable to the shift lever. The shift cable should be fully withdrawn in the **P** position. Move the shift cable in 2 notches to the **N** position. Place the shift lever into the **N** position and tighten the set screw.

20. Reconnect the ignition switch connectors and the park/neutral switch connector.

21. Place the carpet back down to its original position.

22. Install the rear console into the vehicle. Tighten the console mounting screws in the front and rear and install the rear console ashtray.

23. Reconnect the interior lighting switch connector. Install the shift lever indicator light and shift lever indicator cover.

24. Install the shift lever knob and tighten the locknut.

25. Install the bellows back between the rear and center consoles.

26. Install the left front seat into the vehicle. Reconnect the electrical connector under the seat.

27. Tighten the mounting bolts. Move the seat forward and backward to engage the front catch mechanism.

28. Reconnect the negative battery cable.

2.3L (B234I) and 2.5L (B258I) Engines With Manual Transaxles

1. Disconnect the negative battery cable.

2. Lower the left front seat as far as it will go. On five-door vehicles only, disconnect the seat belt from its anchoring point on the seat by pressing in the catch using a curved screwdriver.

3. Push the seat up forward and remove the rear mounting bolts. Push the seat back rearwards and remove the front mounting bolts.

4. Cut off the wire tie on the wiring harness and disconnect the seat connector. Remove the left front seat out of the vehicle.

5. Apply the parking brake. Remove the ignition switch cover by loosening the front part of the left-hand side and then loosening the rear edge of the cover. Unhook the front edge.

6. Disconnect the ignition switch lighting connector. Remove the floor console mounting screws.

7. Remove the rear ashtray cover plate. Remove the rear console ashtray housing.

8. Detach the console by pulling it straight back and slightly lifting it up.

9. Using a small screwdriver, remove the interior lighting switch by pushing it straight up from underneath. Disconnect the switch connector.

10. Remove the window lift module by loosening the snap fasteners underneath at the front edge. Disconnect the electrical connector.

11. Remove the floor console from the vehicle.

12. Remove the console mounted air duct.

13. Place the shift lever into fourth gear. Remove the plastic plug on the transaxle and insert special locking pin tool 87–92–335 or equivalent, with the ring outside the gearbox. This special tool will keep fourth gear locked into place in the transaxle.

14. Loosen the clamp that holds the shift rod in the linkage on the gearbox. Place the shift lever into third gear so that the shift rod and linkage separate.

15. Remove the mounting screws that secure the shift lever housing. Lift up and turn around the shift lever housing.

16. Remove the 2 clamps that secure the ignition lock cables in the shift lever housing. Unplug the terminal on the ignition lock.

17. Pull back and lift the shift lever housing and shift rod out of the vehicle.

18. Remove the locking plate holder. Remove the spring that holds the locking plate.

19. Using a screwdriver, push up the rider which is controlled from the ignition lock. Lift up the locking plate along with its plastic fastener.

20. Remove the ignition switch and ignition lock.

To install:

21. Install the ignition lock and ignition switch back into the shift lever housing.

22. Install the locking plate. Be sure that the 2 locking plate bushings are in place.

23. Clip the ignition lock's rider into the locking plate. Install the locking plate holder. Be sure to lightly coat the screws which secure the locking plate holder with thread-locking fluid.

24. Install the locking plate spring. Adjust the locking plate position with the screw on the locking plate holder. Be sure that the locking plate is level with the heel in the shift lever housing.

NOTE: During adjustment, the stop may not touch the locking plate.

25. Install the shift lever housing and shift rod into the vehicle. Guide the shift rod through the hole in the firewall.

26. Reconnect the terminal on the ignition lock. Hold the ignition lock cables back in the shift lever housing using wire ties.

27. Place the shift lever housing back into its original position and tighten the mounting screws.

28. Reconnect the shift rod to the shift linkage on the transaxle. Place the shifter into fourth gear and install special locking pin tool 87–92–335 or equivalent, into the shift lever housing. The ring on the locking pin tool (87–92–335) should be positioned so that it is inserted into the hole of the ignition switch housing.

29. Tighten the nut on the clamp that secures the shift rod to the linkage on the transaxle. Torque the nut to 14–17 ft. lbs. (19–23 Nm).

30. Remove the special locking pin tool from the transaxle and install the plastic plug.

31. Remove the special locking pin tool from the shift lever housing and try changing the gears several times.

32. Install the console mounted air duct.

33. Position the floor console back into place in the vehicle. Install the window lift module and interior lighting switch into the floor console. Reconnect the electrical connectors to the interior light switch and window lift module.

34. Install the rear console ashtray casing and rear ashtray cover plate.

35. Reconnect the ignition switch lighting connector. Be sure that the floor console is secured in position and tighten the floor console mounting screws.

36. Install the ignition switch cover.

37. Place the left front seat into position in the vehicle. Be sure it is positioned onto the inner bracket's locating lugs.

38. Reconnect the seat connector and secure the wiring harness with wire ties.

39. Install the front mounting bolts. Tighten the inner mounting bolt first, the outside mounting bolt second. Torque the bolts to 22 ft. lbs. (30 Nm).

NOTE: After installing the left front seat, the mounting bolts must be tightened in the correct order. The seat will not work properly if this is not done.

40. Move the seat forward and install the rear mounting bolts. Tighten the inner mounting bolt first, the

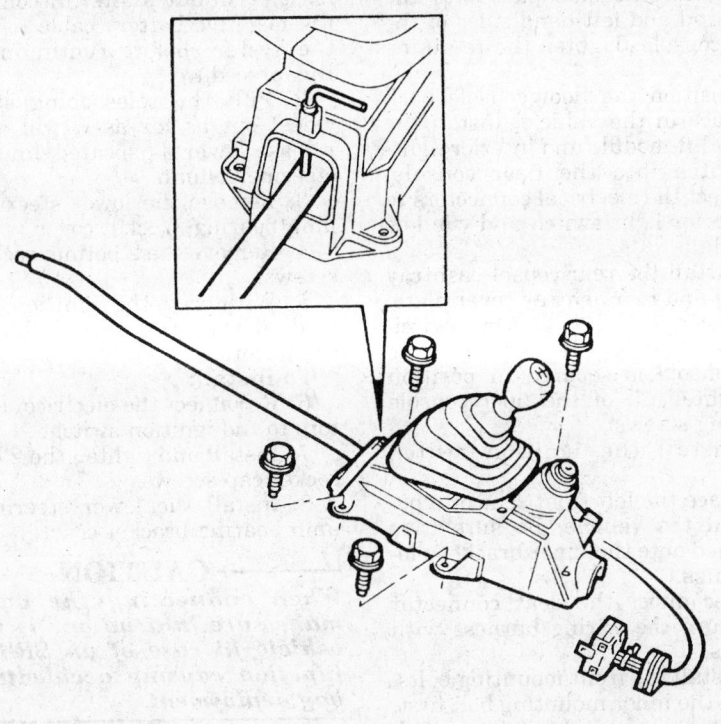

Install the special locking pin tool in the shift lever housing as shown — 900 Series 2.3L (B234I) and 2.5L (B258I) engines with manual transaxles

300714

outside mounting bolt second. Torque the bolts to 22 ft. lbs. (30 Nm).

41. Check to make sure that the seat works properly. The seat should lock into place on both sides simultaneously.

42. On five-door vehicles only, push the seat belt firmly down into its anchoring point. Be sure that the catch engages properly.

43. Reconnect the negative battery cable.

2.3L (B234I) and 2.5L (B258I) Engines With Automatic Transaxle

1. Disconnect the negative battery cable.

2. Lower the left front seat as far as it will go. On five-door vehicles only, disconnect the seat belt from its anchoring point on the seat by pressing in the catch using a curved screwdriver.

3. Push the seat up forward and remove the rear mounting bolts. Push the seat back rearwards and remove the front mounting bolts.

4. Cut off the wire tie on the wiring harness and disconnect the seat connector. Remove the left front seat out of the vehicle.

5. Apply the parking brake. Remove the ignition switch cover by loosening the front part of the left-hand side and then loosening the rear edge of the cover. Unhook the front edge.

6. Disconnect the ignition switch lighting connector. Remove the floor console mounting screws.

7. Remove the rear ashtray cover plate. Remove the rear console ashtray housing.

8. Detach the console by pulling it straight back and slightly lifting it up.

9. Using a small screwdriver, remove the interior lighting switch by pushing it straight up from underneath. Disconnect the switch connector.

10. Remove the window lift module by loosening the snap fasteners underneath at the front edge. Disconnect the electrical connector.

11. Remove the floor console from the vehicle.

12. Remove the left-hand and right-hand side panel retaining screws and remove the side panels out of the center console.

13. If equipped with Automatic Climate Control (ACC), push out the ACC module from behind the center console and unplug the connector.

14. If the vehicle does not have Automatic Climate Control (ACC), push out the heating control panel from behind the center console.

15. Separate the air distribution shaft from the air distribution control.

16. Disconnect the connectors to the heated rear window, fan speed, air recirculation and air distribution control lighting.

17. Disconnect the heating control cable from the air distribution control module.

18. Remove the heating control panel and disconnect all the center console connectors.

19. Remove the expanding rivet that secures the center console to the dashboard. To withdraw the expanding rivet, push in the center pin approximately 1/8 in. (3mm) and pull out the rivet.

20. Remove the center console out of the vehicle.

21. Remove the console mounted air ducts. Place the shift lever in the **N** position.

22. Release the gear shift indicator cover and its rubber seal from the shift lever housing. Turn it to one side so that the securing point for the shift lever cable becomes more accessible.

23. Loosen the locking screw at the securing point of the shift lever cable. Release the front cable securing point.

24. Cut the holding straps and cable holder that secure the wiring harness to the right-hand side of the shift lever housing by the ignition lock.

25. Remove the four mounting bolts that secure the shift lever housing to the floor of the vehicle.

26. Disconnect the electrical connectors on the right-hand side of the shift lever housing.

27. Remove the locking ring which locks the shift lever cable at the shift lever.

28. Disconnect the ignition switch connector and bushing which holds the cable contact to the shift lever housing.

29. Remove the shift lever housing from the vehicle.

30. Place the shift lever in the **P** position and remove the key. Remove the ignition switch.

To install:

31. Install the ignition switch into the shift lever housing. Install the shift lever housing into the vehicle.

32. Reconnect the ignition switch connector to the ignition switch and position the rubber bushing onto the shift lever housing.

33. Place the shift lever into the **N** position. Reconnect the shift lever

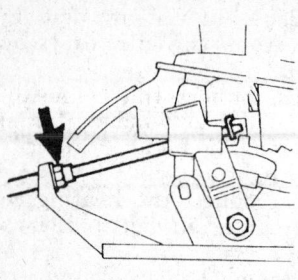

300731

The shift lever cable adjusting nut location — 900 Series 2.3L (B234I) and 2.5L (B258I) engines with automatic transaxles

cable at the shift lever and install the locking ring.

34. Reconnect all the electrical connectors on the right-hand side of the shift lever housing.

35. Be sure that the shift lever housing is positioned correctly on the floor of the vehicle and tighten the shift lever housing mounting bolts.

36. Install 2 new clips around the wire harness and install the cable holder to the housing next to the ignition lock.

37. With the transaxle and the shift lever in the **N** position, screw in the shift lever cable front securing point. Tighten the adjustment locking nut at the shift lever cable front securing point. Be sure that the shift lever has the same amount of play in the **D** and **R** positions as it does in the **N** position.

38. Install the shift indicator plate and rubber seal.

39. Install the console mounted air ducts.

40. Position the center console back into place on the dashboard. Install the expanding rivets that secure the center console to the dashboard by first pushing in the rivet and then pressing in the center pin.

41. Reconnect all the center console connectors. The orange connector is for heating in the right-hand seat and the yellow connector is for heating in the left-hand seat.

42. Reconnect the ACC module connectors and press the module into place, if necessary.

43. If necessary, reconnect the heating control cable to the valve control module. This cable is secured with a locking washer.

44. Reconnect the heated rear window, air recirculation, fan speed and valve control lighting connectors.

45. Reconnect the air distribution shaft to the air distribution control. Install the heating control panel.

46. Install the side panels to the right-hand and left-hand sides of the center console. Tighten the retaining screws.

47. Position the floor console back into place in the vehicle. Install the window lift module and interior lighting switch into the floor console. Reconnect the electrical connectors to the interior light switch and window lift module.

48. Install the rear console ashtray housing and rear ashtray cover plate.

49. Reconnect the ignition switch lighting connector. Be sure that the floor console is secured in position and tighten all of the floor console mounting screws.

50. Install the ignition switch cover.

51. Place the left front seat into position in the vehicle. Be sure it is positioned onto the inner bracket's locating lugs.

52. Reconnect the seat connector and secure the wiring harness with wire ties.

53. Install the front mounting bolts. Tighten the inner mounting bolt first, the outside mounting bolt second. Torque the bolts to 22 ft. lbs. (30 Nm).

NOTE: After installing the left front seat, the mounting bolts must be tightened in the correct order. The seat will not work properly if this is not done.

54. Move the seat forward and install the rear mounting bolts. Tighten the inner mounting bolt first, the outside mounting bolt second. Torque the bolts to 22 ft. lbs. (30 Nm).

55. Check to make sure that the seat works properly. The seat should lock into place on both sides simultaneously.

56. On five-door vehicles only, push the seat belt firmly down into its anchoring point. Be sure that the catch engages properly.

57. Reconnect the negative battery cable.

9000 Series

--- **CAUTION** ---
On vehicles equipped with a Supplemental Restraint System (SRS), the system must be properly disarmed and the air bag must be removed and properly stored. Unintended air bag deployment can cause serious or fatal injury.

1. With the ignition switch **OFF**, disconnect the negative battery cable. If equipped with an air bag, wait at

least 20 minutes after disconnecting the negative battery cable to disarm the system before continuing with this procedure.

2. Pull the telescoping steering wheel out as far as it will go. The release lever is located under the steering column.

3. Remove the lower steering column bearing bracket cover.

4. Remove the 2 bottom socket cap screws.

5. Withdraw the ignition switch and disconnect the electrical connector.

To install:

6. Reconnect the electrical connectors to the ignition switch.

7. Install and tighten the 2 bottom socket cap screws.

8. Install the lower steering column bearing bracket cover.

--- **CAUTION** ---
When connecting the battery, make sure that no one is in the vehicle in case of an SRS malfunction causing accidental air bag deployment.

9. Reconnect the negative battery cable.

10. Check for proper switch operation.

Park/Neutral Safety Switch

REMOVAL AND INSTALLATION

1. Put selector in the park position.

2. Remove battery.

3. Take out dip stick.

4. Remove servo pump pipe's securing console to release the transmission range switch.

5. Remove the lever.

6. Remove the transmission range switch screw.

7. Remove the transmission range switch nut.

8. Release the contacts and lift up the sensor from the shaft.

To install:

9. Reconnect the contacts and position switch on shaft.

10. Install the transmission range switch nut.

11. Screw in the lever and move the select into neutral position.

12. Tighten the switch so that the markings on the transmission range switch corresponds with the selector position.

13. Install the servo pump pipe's securing console.

14. Insert dipstick.

15. Refit battery.

Powertrain Control Module

REMOVAL AND INSTALLATION

--- **WARNING** ---

All control modules are sensitive to static electricity and must be handled with extreme care. When removing or installing a control module, it is important to properly ground oneself. Never allow any foreign objects or fingers to come into contact with the control module connecting pins.

900 Series

1993 Models

1. Disconnect the negative battery cable. Make sure that the ignition switch is **OFF**.
2. Remove the door sill scuff plates from the right-hand side of the vehicle. Move away the bottom front door weather-strip.
3. Remove the right-hand side kick panel and pull back the carpet.
4. Unplug the module connector by releasing the clip and pulling it diagonally up and out.
5. Remove the engine control module mounting screws from the wheel arch. Remove the engine control module from the vehicle.

To install:

6. Install the engine control module into the vehicle and tighten the mounting screws.
7. Plug the module connector back into the engine control module, making sure that the clip snaps the connector secure into place.
8. Reposition the carpet back into place and install the kick panel.
9. Reposition the bottom front door weather-strip back into place. Install the door sill scuff plates on the right-hand side of the vehicle.
10. Reconnect the negative battery cable.

1994–97 Models

1. Disconnect the negative battery cable. Make sure the ignition switch is **OFF**.
2. Open the glove box lid. Remove the retaining screw covers and remove the retaining screws.
3. Remove the retaining bolt and expanding rivet in the front edge of glove box.
4. Remove the catch from the firewall bracket.
5. Slightly pull out the glove box and unplug the connector for the glove box light. Remove the glove box from the vehicle.

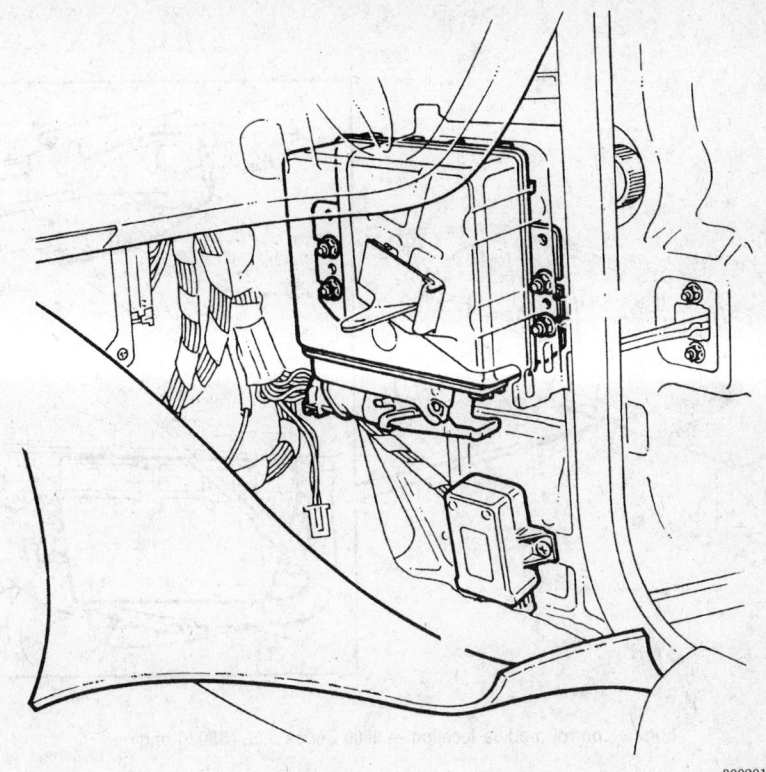

300201

Engine control module location — 1994–97 900 Series

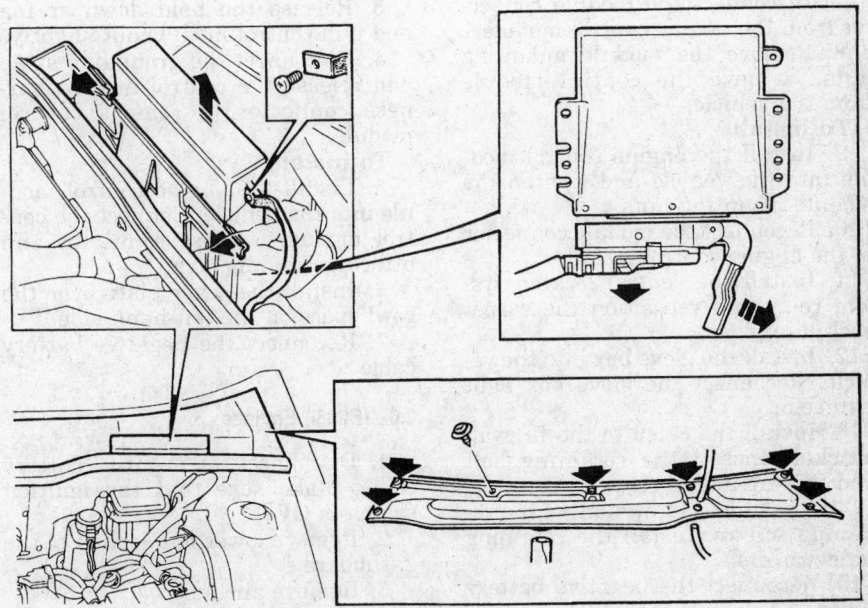

303593

Engine control module location — 9000 Series 2.3L engines

9. Place lower dash in position and tighten mounting screws.
10. Connect negative battery cable, and test drive vehicle.

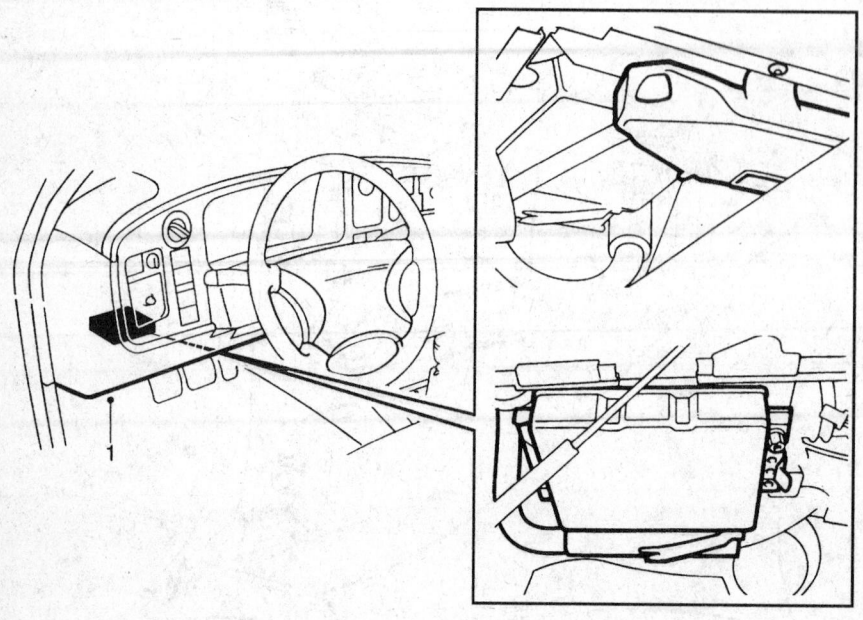

Engine control module location — 9000 Series 3.0L (B308I) engine

303591

6. Move the carpet out of the way and remove the central locking system relay.
7. Disconnect the module connector from the engine control module.
8. Remove the module mounting nuts. Remove the control module from the vehicle.
To install:
9. Install the engine control module into the vehicle and tighten the module mounting nuts.
10. Reconnect the module connector to the engine control module.
11. Install the central locking system relay and reposition the carpet back into place.
12. Install the glove box into the vehicle. Reconnect the glove box light connector.
13. Install the catch to the firewall bracket. Install the retaining bolt and the expanding rivet.
14. Install and tighten all of the retaining screws. Install the retaining screw covers.
15. Reconnect the negative battery cable.

9000 Series

All 2.3L Engines

1. Disconnect the negative battery cable. Make sure that the ignition switch is **OFF**.

2. Access the engine control module by removing the cover plate over the cowl space on the left-hand side.
3. Release the hold down spring and pull control module out slightly.
4. Disconnect the grounding strap and release the control module harness connector to remove control module.
To install:
5. Install the engine control module into the vehicle. Connect the control module harness connector and attach grounding strap.
6. Install the cover plate over the cowl space on the left-hand side.
7. Reconnect the negative battery cable.

3.0L (B308I) Engines

1. Disconnect the negative battery cable. Make sure that the ignition switch is **OFF**.
2. Remove lower left section of the dashboard.
3. Remove air duct.
4. Unscrew nuts to the control module and remove.
5. Disconnect harness connector.
To install:
6. Connect harness connector.
7. Fit control module to mounts and tighten nuts.
8. Install air duct.

ENGINE COOLING

Radiator

REMOVAL AND INSTALLATION

900 Series

2.0L (B202L) and 2.1L (B212I) Engines

1. Disconnect the negative battery cable.
2. Drain the radiator. Disconnect and plug the transmission cooling lines, as required.
3. Disconnect the hoses from the radiator. Disconnect the electrical leads to the radiator fan and the auxiliary fan, if equipped.
4. Disconnect the electrical lead to the thermal switch and solenoid valve. Remove the ignition coil and solenoid valve from the bracket. Remove the oil cooler.
5. Remove the 2 bolts from the upper radiator support. Lift the radiator out of the vehicle by pulling the top of the radiator slightly backwards.
To install:
6. Guide the radiator studs into the holes in the radiator support and install the retaining bolts.
7. Install the oil cooler. Install the ignition coil and solenoid valve on the bracket and connect the electrical lead to the thermal switch and solenoid valve.
8. Connect the electrical leads to the radiator fan and the auxiliary fan, if equipped.
9. Properly fill and bleed the cooling system.
10. Connect the negative battery cable.
11. Start the engine, allow it to reach normal operating temperature and check for proper cooling system operation.

2.0L (B204L) Engine

1. Disconnect and remove battery and battery shelf.
2. Remove grille and horn.
3. Removal headlights and direction indicators, separate electrical connectors.
4. Remove spoiler to gain access to outside temperature sensor's electri-

cal connection and washer pump hose, disconnect both units.

5. Remove bumper securing screws and pull bumper forward.

6. Disconnect hoses and remove charge air cooler.

7. Release engine oil cooler and hang up out of the way of radiator.

8. Remove A/C condenser from radiator.

9. Drain coolant from radiator.

10. Remove bypass hose together with intake hose.

11. Disconnect upper radiator hose.

12. Disconnect the electrical connector to the radiator fan.

13. Disconnect lines and clips for radiator breather hose.

14. Remove the radiator fan cowling attaching screws and remove the fan and cowling as an assembly.

15. Remove the hold down screws on the radiator for the oil cooler lines.

16. Remove lower radiator hose.

17. Remove radiator clips take hold it to the radiator cross member.

18. Remove radiator.

To install:

19. Install radiator.

20. Connect clips to radiator cross member and radiator.

21. Connect lower radiator hose, and install hold down screw for oil cooler lines to radiator.

22. Connect lines and clips to radiator breather hose.

23. Install the fan and cowling as an assembly.

24. Plug in electrical connector for fan.

25. Install upper radiator hose, bypass hose and intake hose.

26. Install A/C condenser and oil cooler to radiator support.

27. Install charge air cooler and connect lines.

28. Put bumper into place and secure.

29. Connect outside air temperature sensor's electrical connection and washer pump hose. Install spoiler.

30. Connect and install directional indicators and headlights.

31. Install grille and horn.

32. Install battery shelf and connect battery cables.

2.3L (B234I) and 2.5L (B258I) Engines

1. Disconnect and remove battery and battery shelf.

2. Remove grille and horn.

3. Release engine oil cooler and hang up out of the way of radiator.

4. Remove A/C condenser from radiator.

5. Drain coolant from radiator.

6. Remove bypass hose together with intake hose.

7. Disconnect upper radiator hose.

8. Disconnect the electrical connector to the radiator fan.

9. Disconnect lines and clips for radiator breather hose.

10. Remove the radiator fan cowling attaching screws and remove the fan and cowling as an assembly.

11. Remove the hold down screws on the radiator for the oil cooler lines.

12. Remove lower radiator hose.

13. Remove radiator clips take hold it to the radiator cross member.

14. Remove radiator.

To install:

15. Install radiator.

16. Connect clips to radiator cross member and radiator.

17. Connect lower radiator hose, and install hold down screw for oil cooler lines to radiator.

18. Connect lines and clips to radiator breather hose.

19. Install the fan and cowling as an assembly.

20. Plug in electrical connector for fan.

21. Install upper radiator hose, bypass hose and intake hose.

22. Install A/C condenser and oil cooler to radiator support.

23. Install grille and horn.

24. Install battery shelf and connect battery cables.

9000 Series

1. Loosen the expansion tank pressure cap.

2. Raise and safely support the vehicle.

3. Remove the 2 lower retaining bolts from the A/C condenser.

4. Lower the vehicle.

5. Disconnect the negative battery cable.

6. Drain the radiator.

7. Disconnect the hoses from the radiator.

8. Disconnect the electrical leads to the radiator fan and the auxiliary fan, if equipped.

9. Disconnect the electrical lead to the boost pressure valve and bracket. Remove the ignition coil and solenoid valve from the bracket.

10. Remove the oil cooler, if equipped.

11. Remove the 2 bolts from the upper radiator support. Lift the radiator out of the vehicle by pulling from its rubber mountings.

To install:

12. Fit radiator into rubber supports and install the 2 mounting bolts.

13. Install oil cooler, if equipped.

14. Install ignition coil and solenoid valve on bracket. Connect electrical

harness connector to the boost pressure valve.

15. Install electrical harness connector to radiator cooling fans.

16. Install radiator hoses.

17. Fill vehicle with coolant.

18. Install negative battery cable.

19. Raise and safely support the vehicle.

20. Install A/C condenser support bolts.

21. Lower vehicle and test drive, top off any lost coolant.

Water Pump

REMOVAL AND INSTALLATION

2.0L (B202L) and 2.1L (B212I) Engines

1. Disconnect the negative battery cable.

2. Remove the pressure cap from the expansion tank.

3. Drain the coolant into a suitable container.

4. Remove the water pump drive belt.

5. Remove the pulley.

6. Remove the bolts and lift out the water pump.

To install:

7. Thoroughly clean all mating surfaces between the engine and water pump.

8. Apply a suitable gasket sealant and install the new pump with a new gasket.

9. Install the bolts and tighten them to 15 ft. lbs. (20 Nm).

10. Properly fill the cooling system.

11. Connect the negative battery cable.

12. Start the engine, allow it to reach normal operating temperature and check for proper cooling system operation.

2.0L (B204L) and All 2.3L Engines

1. Disconnect the negative battery cable.

2. Loosen the expansion tank pressure cap.

3. Raise and safely support the vehicle.

4. Remove the center air deflector and properly drain the coolant.

5. Remove the right front wheel assembly and remove the front section of the inner fender panel.

6. Lower the vehicle.

7. Remove the expansion tank, then remove the coolant hoses and electrical connector. Pull hard on the drive belt and lock the tensioner using tool 83-94-488 or equivalent. Re-

move the belt from the water pump and the air conditioning compressor.

8. Protect the engine oil cooler and place a protective cover over the upper radiator crossmember.

9. Unplug the electrical connector to the air conditioning compressor and detach the compressor with lines attached and position it to the side. Remove the air conditioning compressor bracket.

10. Disconnect the coolant hoses from the water pump.

11. Remove the oxygen sensor wire from the clips.

12. Remove the coolant pipe from the turbocharger.

13. Raise and safely support the vehicle.

14. Remove the coolant pipe from the water pump and lower the vehicle.

15. Remove the 3 retaining screws securing the water pump to the timing cover.

16. Carefully pry the water pump loose. Start at the sleeve in the cylinder block and remove the water pump.

—————— WARNING ——————
Use care not to damage the oxygen sensor.

17. Remove the water pump pulley and take the pump out of its housing.
To install:
18. Clean the sealing surfaces and install the water pump in the pump housing with a new gasket.

19. Install the water pump pulley and torque the retaining bolts to 72 inch lbs. (8 Nm).

20. Lubricate the new O-rings with acid free petroleum jelly and fit the sleeve together with the water pump. Tighten the water pump bolts to 15 ft. lbs. (20 Nm).

21. Raise the vehicle and attach the coolant pipe to the water pump.

22. Lower the vehicle and attach the coolant pipe to the turbocharger.

23. Install the oxygen sensor wire in the clips.

24. Connect the hoses to the water pump.

25. Install the air conditioning compressor bracket. Install the air conditioning compressor and connect the electrical connector.

26. Remove the protective covers from the engine oil cooler and the upper radiator crossmember.

27. Install the belt, pull on it firmly and remove the locking pin from the tensioner. Check for proper belt alignment and tension.

28. Install the coolant hoses and electrical connector to the expansion tank, then install the expansion tank.

29. Check that the radiator drain plug is tightened and install the center air deflector.

30. Install the front section of the inner fender panel and install the wheel.

31. Properly fill the cooling system.

32. Connect the negative battery cable. Start the engine and check for proper cooling system operation.

2.5L (B258I) Engine

1. Disconnect the negative battery cable.

2. Safely raise and support the vehicle.

3. Remove the lower center air deflector. Drain the engine coolant into a catch pan.

4. Lower the vehicle and remove the top engine covers, air filter assembly and mass air flow sensor.

5. Loosen the power steering pump and water pump pulley bolts. Remove the drive belt and tensioner. Remove the power steering pump, water pump pulley and timing cover.

6. Check the timing belt for damage and replace if necessary.

7. Remove the water pump.
To install:
8. Clean water pump mounting surface. Coat the water pump O-ring and sealing surface with acid-free petroleum jelly.

9. Install water pump with new O-ring and torque water pump bolts to 18 ft. lbs. (25 Nm)

10. Install the timing cover.

11. Install the water pump pulley. Torque water pump pulley bolts to 6 ft. lbs. (8 Nm).

12. Install power steering pump, drive belt tensioner and drive belt.

13. Install air filter assembly, mass air flow sensor. Install the upper engine covers and safely raise and support the vehicle.

14. Make sure the radiator drain plug is tightened and install the lower center air deflector.

15. Fill the radiator with clean coolant and check the cooling system for leaks

16. Bleed the cooling system.

17. Test drive the vehicle and check for any leaks.

3.0L (B308I) Engines

1. Disconnect the negative battery cable.

2. Safely raise and support the vehicle.

3. Remove the lower center air deflector. Drain the engine coolant into a catch pan.

4. Lower the vehicle and remove the top engine covers.

5. Lift up the power steering reservoir.

6. Disconnect the connection on the torque arm engine mount and remove the torque arm.

7. Remove the power steering line clamp from the torque arm engine mount and remove the engine mount.

8. Remove the upper coolant hose.

9. Disconnect the hose from the coolant expansion tank and remove the upper alternator air intake.

10. Loosen the power steering pump and water pump pulley bolts. Remove the drive belt and tensioner. Remove the power steering pump, water pump pulley and timing cover.

11. Check the timing belt for damage and replace if necessary.

12. Remove the water pump.
To install:
13. Clean water pump mounting surface. Coat the water pump O-ring and sealing surface with acid-free petroleum jelly.

14. Install water pump with new O-ring and torque water pump bolts to 18 ft. lbs. (25 Nm)

15. Install the timing cover.

16. Install the water pump pulley. Torque water pump pulley bolts to 6 ft. lbs. (8 Nm).

17. Install power steering pump, drive belt tensioner and drive belt.

18. Install the upper alternator air intake.

19. Install the torque arm engine mount and attach the power steering line clamp.

20. Install the torque arm and bolt the connection to the torque arm engine mount.

21. Install the upper coolant hose.

22. Connect the upper hose to the coolant expansion tank and set the power steering reservoir back into position.

23. Install the upper engine covers and safely raise and support the vehicle.

24. Make sure the radiator drain plug is tightened and install the lower center air deflector.

25. Fill the radiator with clean coolant and check the cooling system for leaks

26. Bleed the cooling system.

27. Test drive the vehicle and check for any leaks.

Thermostat

REMOVAL AND INSTALLATION

900 Series

2.0L, 2.1L (B212I) and 2.3L (B234I) Engines

1. Open the drain plug and drain coolant. Remove overflow tank cap to help speed process.
2. Remove the throttle body cover.
3. Remove the hoses to the charge air cooler and the by-pass hose.
4. Remove the coolant pipe coming from the coolant pump.
5. Remove the pipe retaining bolt that holds the coolant pipe in the front left side of the engine.
6. Take off the preheater hose and the hose into the thermostat housing.
7. Take off the thermostat lid to remove the thermostat.
   To install:
8. Install thermostat and fit lid over housing.
9. Fit hose into thermostat housing and install preheater hose.
10. Install coolant pipe into left front side of engine with mounting bolt.
11. Reconnect heater by-pass hose and check all hose mounting points.
12. Tighten coolant drain and fill system with coolant.
13. Install hoses to the charge air cooler and start engine. Check for leaks.

2.5L (B258I) Engine

1. Open the drain plug and drain coolant. Remove overflow tank cap to help speed process.
2. Remove the upper intake manifold.
3. Disconnect and swing inlet pipe to one side and cover engine's intake ports.
4. Take off middle intake pipes and move slightly.
5. Disconnect upper radiator hose and remove from engine.
6. Pull out dipstick.
7. Remove coolant pipe.
8. Take off the thermostat lid to remove the thermostat.
   To install:
9. Install thermostat and fit lid over housing, tighten lid to 15 ft. lbs. (20 Nm).
10. Fit coolant pipe to engine.
11. Install the dipstick.
12. Reconnect upper radiator hose to engine.
13. Install intake manifold and tighten bolts to 15 ft. lbs. (20 Nm). Install inlet pipe.

14. Fill with coolant and run engine to check for leaks. As vehicle warms up, top off coolant.

9000 Series

2.3L Engines

1. Disconnect the negative battery cable.
2. Loosen the expansion tank pressure cap.
3. Raise and safely support the vehicle.
4. Remove the center air deflector, properly drain the coolant and lower the vehicle.
5. On vehicles without a turbocharger, remove the retaining bolts from the fuel pressure regulator. Unplug the connection between the fuel pressure regulator and the fuel injection rail. Move the pressure regulator to the side.
6. On vehicles equipped with a turbocharger, remove the stay from the thermostat housing lid and the hose that goes to the throttle body preheater.
7. Remove the bolts from the thermostat housing.
8. Remove the housing and the thermostat.
   To install:
9. Thoroughly clean all mating surfaces between the engine and thermostat housing.
10. Properly install the thermostat (spring facing toward the engine) and install the housing with a new gasket.
11. Install the bolts and tighten them to 15 ft. lbs. (20 Nm).
12. On vehicles with a turbocharger, connect the hose to the throttle body preheater and install the stay to the thermostat housing.
13. On vehicles without a turbocharger, position the fuel pressure regulator for installation. Plug in the connection between fuel pressure regulator and the fuel injection rail and install the retaining bolts for the fuel pressure regulator.
14. Tighten the coolant drain plug and install the center air deflector.
15. Properly fill and bleed the cooling system.
16. Connect the negative battery cable.
17. Start the engine, allow it to reach normal operating temperature and check for proper cooling system operation.

3.0L (B308I) Engine

1. Open the drain plug and drain coolant. Remove overflow tank cap to help speed process.

2. Remove the upper intake manifold.
3. Disconnect and swing inlet pipe to one side and cover engine's intake ports.
4. Take off middle intake pipes and move slightly.
5. Disconnect upper radiator hose and remove from engine.
6. Pull out dipstick.
7. Remove coolant pipe.
8. Take off the thermostat lid to remove the thermostat.
   To install:
9. Install thermostat and fit lid over housing, tighten lid to 15 ft. lbs. (20 Nm).
10. Fit coolant pipe to engine.
11. Install dipstick.
12. Reconnect upper radiator hose to engine.
13. Install intake manifold and tighten bolts to 15 ft. lbs. (20 Nm). Install inlet pipe.
14. Fill with coolant and run engine to check for leaks. As vehicle warms up, top off coolant.

Electric Cooling Fan

REMOVAL AND INSTALLATION

900 Series

2.0L (B202L) and 2.1L (B212I) Engines

1. Disconnect the negative battery cable.
2. Disconnect the electrical connector to the radiator fan.
3. Remove the radiator fan cowling attaching screws and remove the fan and cowling as an assembly.
4. Remove the nuts securing the radiator fan to the fan cowling, detach the resistor (2 speed fan), remove the switches from the fan cowling and remove the radiator fan.
   To install:
5. Fit the fan, press the switches onto the fan cowling, fit the resistor (2 speed fan) and tighten the retaining nuts securing the fan to the cowling.
6. Install the fan and cowling as an assembly.
7. Connect the negative battery cable.
8. Start the engine, allow it to reach normal operating temperature and check for proper fan operation.

2.0L (B204L), 2.3L (B234I) and 2.5L (B258I) Engines

1. Disconnect the battery cables and remove battery.
2. Disconnect the electrical connector to the radiator fan.

3. Disconnect lines and clips for radiator breather hose.

4. Remove the radiator fan cowling attaching screws and remove the fan and cowling as an assembly.

5. Remove the nuts securing the radiator fan to the fan cowling, detach the resistor (2 speed fan), remove the switches from the fan cowling and remove the radiator fan.

To install:

6. Fit the fan, press the switches onto the fan cowling, fit the resistor (2 speed fan) and tighten the retaining nuts securing the fan to the cowling.

7. Connect lines and clips to radiator breather hose.

8. Install the fan and cowling as an assembly.

9. Install battery and connect the battery cables.

10. Start the engine, allow it to reach normal operating temperature and check for proper fan operation.

9000 Series

1. Disconnect the negative battery cable.

2. Disconnect the electrical connector to the radiator fan and, if there is an ignition coil, remove the wire from the coil to the distributor.

3. Remove the upper retaining screws from the fan cowling, carefully pull the cowling away enough to remove the clip on the radiator fan cable assembly.

4. Raise and safely support the vehicle.

5. Remove the center air deflector and remove the lower retaining screw from the fan cowling.

6. Lower the vehicle and lift the fan cowling complete with the fan.

7. Remove the nuts securing the radiator fan to the fan cowling, detach the resistor (2 speed fan), remove the switches from the fan cowling and remove the radiator fan.

To install:

8. Fit the fan, press the switches onto the fan cowling, fit the resistor (2 speed fan) and tighten the retaining nuts securing the fan to the cowling.

9. Install the fan and cowling as an assembly.

10. Raise and safely support the vehicle.

11. Tighten the lower fan cowling retaining screw, fit the center air deflector and lower the vehicle.

12. Press down the radiator fan cable assembly clip and install the other retaining screws.

13. Connect the negative battery cable.

14. Start the engine, allow it to reach normal operating temperature and check for proper fan operation.

Cooling System Bleeding

PROCEDURE

900 SERIES

2.0L, 2.1L (B212I) and 2.3L Engines

The bleeder nipple, located in the thermostat housing, should be opened when adding coolant to the system. The nipple should not be opened when the engine is running.

2.5L (B258I) Engine

1. Fill the coolant to the MAX level, install the pressure cap.
2. Start the engine and warm it up at varying rpm until the cooling fan turns **ON**.
3. Refill the coolant to the MAX level and run the engine at varying rpm until the cooling fan turns **ON** three more times.
4. Turn the engine **OFF** and if needed, fill the coolant to the MAX level.
5. Test drive the vehicle and check for any leaks.

9000 SERIES

2.3L Engines

9000 Series 2.3L engines do not require a special bleeding procedure. Fill the engine with coolant, start the engine and allow it to reach operating temperature (thermostat open) and check the coolant level. Add coolant as necessary.

3.0L (B308I) Engine

1. Fill the coolant to the MAX level, install the pressure cap.
2. Start the engine and warm it up at varying rpm until the cooling fan turns **ON**.
3. Refill the coolant to the MAX level and run the engine at varying rpm until the cooling fan turns **ON** three more times.
4. Turn the engine **OFF** and if needed, fill the coolant to the MAX level.
5. Test drive the vehicle and check for any leaks.

FUEL SYSTEM

Fuel System Service Precautions

Safety is the most important factor when performing not only fuel system maintenance but any type of maintenance. Failure to conduct maintenance and repairs in a safe manner may result in serious personal injury or death. Maintenance and testing of the vehicle's fuel system components can be accomplished safely and effectively by adhering to the following rules and guidelines.

• To avoid the possibility of fire and personal injury, always disconnect the negative battery cable unless the repair or test procedure requires that battery voltage be applied.

• Always relieve the fuel system pressure prior to disconnecting any fuel system component (injector, fuel rail, pressure regulator, etc.), fitting or fuel line connection. Exercise extreme caution whenever relieving fuel system pressure to avoid exposing skin, face and eyes to fuel spray. Please be advised that fuel under pressure may penetrate the skin or any part of the body that it contacts.

• Always place a shop towel or cloth around the fitting or connection prior to loosening to absorb any excess fuel due to spillage. Ensure that all fuel spillage (should it occur) is quickly removed from engine surfaces. Ensure that all fuel soaked cloths or towels are deposited into a suitable waste container.

• Always keep a dry chemical (Class B) fire extinguisher near the work area.

• Do not allow fuel spray or fuel vapors to come into contact with a spark or open flame.

• Always use a backup wrench when loosening and tightening fuel line connection fittings. This will prevent unnecessary stress and torsion to fuel line piping. Always follow the proper torque specifications.

• Always replace worn fuel fitting O-rings with new. Do not substitute fuel hose or equivalent, where fuel pipe is installed.

Fuel System Pressure

RELIEVING

All Engines

─────── CAUTION ───────
Fuel injection systems remain under pressure, even after the engine has been turned OFF. The fuel system pressure must be relieved before disconnecting any fuel lines. Failure to do so may result in fire and/or personal injury.

1. Disconnect the negative battery cable.

2. Place clean shop rags under and around the fuel line banjo coupling to absorb any released fuel or fuel spray.

3. With the ignition switch in the **OFF** position, carefully relieve the fuel system pressure by firmly cracking open the banjo coupling at the inlet to the fuel injection manifold.

4. Be sure to always replace the banjo coupling washers whenever you loosen or remove the banjo couplings.

Idle Speed

ADJUSTMENT

1993 Models

2.0L (B202L) and 2.1L (B212I) Engines

The engine idle speed is controlled by the Engine Control Module (ECM), which actuates the Idle Air Control valve (IAC valve). The ECM automatically adjusts the idle rpm taking into account for engine wear and accessory loads. If needed, the base idle

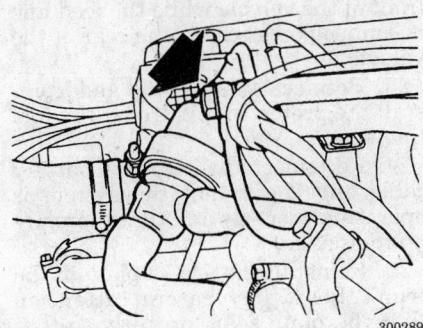

Location of fitting on fuel rail to relieve fuel pressure — 900 Series

can be checked and adjusted as follows:

1. Clean the throttle body of sludge deposited from normal engine operation.

2. Allow the engine to warm up to normal operating temperature. The transmission oil must also be at normal operating temperature.

3. Shut the engine **OFF**.

4. Loosen the dashpot and lower it to clear the throttle lever.

5. Unplug the electrical harness connector at the Throttle Position sensor (TP sensor).

6. Connect a jumper wire across pins 1 and 2 at the rear of the harness connector. The jumper temporarily disengages the Engine Control Module (ECM) from the idle circuit.

7. Loosen the 2 screws that secure the TP sensor and rotate the sensor clear of the throttle plate. Make sure that the throttle cable has some slack in it.

8. Start the engine making sure that all accessories are turned **OFF** and that the engine cooling fan is not running during the test.

9. Connect a tachometer.

10. If the throttle body has an air bleed screw, the locknut must be loosened and the screw turned in until it is fully seated. Tighten the locknut.

11. Loosen the locknut on the throttle plate adjusting screw. Using Saab nut driver 83–94–322 or equivalent, turn the adjusting screw until the throttle plate is fully closed.

12. Locate the green/red wire of the single-pin test socket found between the fresh-air vent and the right-hand wheel arch in the engine compartment.

13. Place a jumper wire from the test socket (green/red wire) to ground.

14. Adjust the base idle to 775 ± 25 rpm. using the throttle plate adjust-

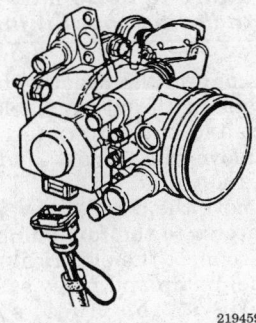

Throttle position sensor connector location — 1993 900 Series

ing screw on the side of the throttle body and tighten the locknut while holding the adjusting screw.

15. Recheck the base idle rpm before removing the tachometer.

16. Remove the jumper wire and the idle speed should return to the specified rpm of 850 ± 75 rpm.

17. If no change in rpm was detected, check the operation of the IAC valve.

18. Turn the engine **OFF**.

19. Adjust the TP switch so that it is closed in the idling position and tighten the 2 screws. The TP sensor should make a clicking sound when the throttle lever is moved from the idle position.

20. Adjust the throttle dashpot so that it closes within a period of 2–3 seconds and tighten the locknut.

21. Remove the jumper wire from the TP sensor harness connector and reconnect it to the TP sensor.

2.3L (B234I and B234L) Engines

The engine idle speed is controlled by the Idle Air Control valve (IAC valve). The Engine Control Module automatically adjusts the idle rpm taking into account for engine wear and accessory loads. If needed, the base idle can be checked and adjusted as follows:

1. Allow the engine to warm up to normal operating temperature.

2. Connect a tachometer and check the idle rpm with all accessories turned **OFF**.

3. Check that the engine idle is 875 ± 75 rpm as controlled by the IAC valve.

4. If the idle is not within parameters, the base idle will need to be reset.

5. Locate the green/red wire of the single pole test socket on the left-hand side of the engine compartment.

6. With the engine running, place a jumper wire from the single pin socket (green/red wire) to ground.

7. Adjust the base idle to 725–775 rpm. using the idle adjusting screw on the side of the throttle body being sure to first loosen the locknut.

8. Turn the adjusting screw to achieve the specified rpm.

9. Tighten the locknut while holding the adjusting screws position.

10. Remove the jumper wire and check the idle speed. It should return to the specified rpm of 875 ± 75 rpm.

11. If no change was noticed in the idle speed, check the operation of the IAC valve.

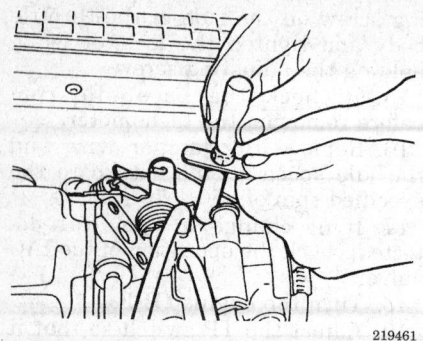

Throttle plate adjustment point — 1993 900 Series

219461

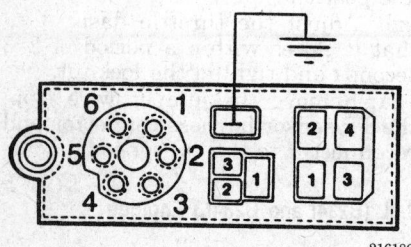

216186

Test socket grounding point for base idle adjustment — 1993 9000 Series 2.3L (B234I and B234L) engines

1994-97 Models

All Engines

Engine idle speed is not adjustable. The engine management system controls the engine idle speed through an Idle Air Control (IAC) valve. The idle speed control system adapts continuously to various changes, such as engine load or wear. The control module is programmed to maintain a constant idle speed when the engine is at normal operating temperature.

Fuel Filter

REMOVAL AND INSTALLATION

— CAUTION —
Fuel injection systems remain under pressure, even after the engine has been tuned OFF. The fuel system pressure must be relieved before disconnecting any fuel lines. Failure to do so may result in fire and/or personal injury.

1. Raise and safely support the vehicle.

2. Locate the fuel filter which is mounted under the vehicle, forward of the fuel tank.
3. Make sure that the ignition switch is in the OFF position.
4. Thoroughly clean the area around the banjo fittings before continuing.
5. Relieve the fuel pressure by placing a shop towel around one of the banjo fittings on the fuel filter and crack open the fitting until the pressure is relieved.
6. Once relieved, continue to remove the banjo fittings on both sides of the fuel filter, containing the fuel with the shop towel. Properly dispose of the towel once the job is complete.
7. Loosen the band-type clamp so the old filter can be removed.
To install:
8. Position the new fuel filter, making sure that the arrow is pointing in the correct direction of fuel flow.
9. Properly position the new filter and tighten the clamp.
10. Reinstall the banjo fitting to both ends of the filter. Use new sealing washers and tighten the fittings.
11. Wipe the area clean of any remaining fuel and start the engine. Check for leakage and immediately correct if any leaks are found.
12. Lower the vehicle.

Fuel Pump

REMOVAL AND INSTALLATION

900 Series

— CAUTION —
Fuel injection systems remain under pressure, even after the engine has been tuned OFF. The fuel system pressure must be relieved before disconnecting any fuel lines. Failure to do so may result in fire and/or personal injury.

1. Disconnect the negative battery cable. Tape or tie off the cable so it cannot contact the battery terminals.
2. Remove the carpet and floor panel in the trunk.
3. Remove the circular cover plate to gain access to the fuel pump.
4. Disconnect the electrical wiring at the fuel pump and move aside.
5. Relieve the fuel system pressure.
6. Remove the screw and protective plate over the fuel line fittings.
7. Remove the fuel lines at the fuel pump using angled needle-nose pliers

or similar tool. Use a shop towel to absorb any fuel released.
8. Move the fuel lines and wiring harness to the side.
9. Place Saab tool 83-94-397 or equivalent over the fuel pump. Place the chain supplied with the tool through the two load securing brackets on the trunk floor and secure each end with a screwdriver or similar tool.
10. Loosen the large threaded ring that holds the pump in the tank using a wrench on the shaft of the tool.
11. Remove the chains and tool. Finish removal of the ring by hand.
12. Remove the rubber seal and remove the pump by tilting the top part of it forward and to the left. Allow the fuel to run out of the pump before removing it from the trunk.
13. Wipe up any fuel spilled.
To install:
14. Wipe clean the sealing area of the opening in the fuel tank.
15. Carefully place the pump into the tank to prevent damage to the fuel level float arm.
16. Once in the tank, line up the marks on the tank and pump.
17. Place a new seal with a thin coat of petroleum jelly into the large threaded ring and place on top of the fuel pump with the mark on the ring lined up with the mark on the fuel tank.
18. Press down hard and rotate the ring 90°.
19. Reinstall special tool 83-94-397 or equivalent, with the chain situated for tightening.
20. Tighten the ring until it will no longer turn.
21. Check that the pump did not rotate more than 30° from the alignment marks.
22. Connect the fuel lines using new O-rings. The return line is connected to the nipple closest to the front of the vehicle while the feed line is connected closest to the rear of the vehicle.
23. Connect the wiring and reinstall the protective plate over the fuel line fittings.
24. Reconnect the negative battery cable and check that the system is operating properly and that there are no fuel leaks.
25. Reinstall the access plate in the trunk floor. Pay careful attention that the plate seals properly so that no exhaust fumes can enter the cabin.
26. Reinstall the floor panel and carpet.

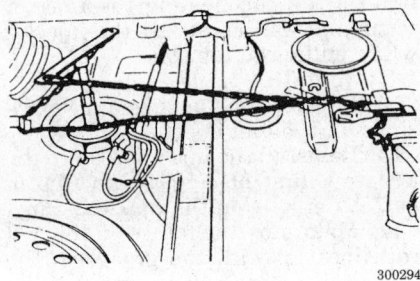

Positioning of tool and chains for tightening
the retaining ring — 900 Series

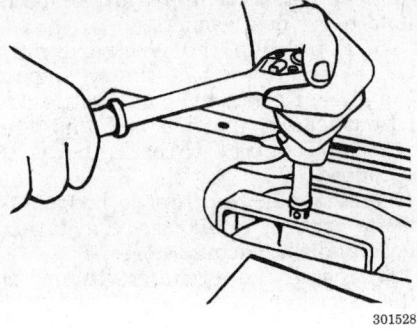

Removing threaded ring using Saab tool —
9000 Series

9000 Series

——— CAUTION ———

Fuel injection systems remain under pressure, even after the engine has been tuned OFF. *The fuel system pressure must be relieved before disconnecting any fuel lines. Failure to do so may result in fire and/or personal injury.*

1. Disconnect the negative battery cable. Tape or tie off the cable so that it cannot be reconnected during this procedure.
2. Relieve the fuel system pressure.
3. Gain access to the trunk floor. Remove the carpet in the trunk.
4. Remove the 2 screws securing the floor panel and remove the panel.
5. Remove the oblong cover plate to gain access to the fuel pump.
6. Release the clip and remove the electrical harness connector. Move the cover aside.
7. Place a shop towel around the fuel lines at the fuel pump and disconnect the lines. Place the fuel lines off to the side.
8. Soak up any fuel released with the shop towel.
9. Using Saab tool 83–94–462 or similar device, remove the large threaded ring that secures the pump in the tank.
10. Lift the fuel pump out of the tank by tilting the top to the right, being careful not to damage the arm for the fuel level sender.
11. Soak up any spilled fuel.
To install:
12. Clean the groove on the tank and place a new O-ring in the groove.
13. Place the fuel pump into the tank.
14. Line up the mark on the pump with the mark on the tank. This must be done for the fuel sender to operate properly.
15. Place the large threaded ring over the pump and tighten using tool

83–94–462 or equivalent, to 40 ft. lb. (55 Nm).
16. Check that the alignment of the marks are still in line. A tolerance of 5° is permissible.
17. Replace the O-rings on the inside of each fuel line and then reconnect to the fuel pump.
18. Reinstall the electrical connector and the clip.
19. Connect the negative battery cable. Check that the pump is operating properly and there are no leaks.
20. Reinstall the access plate in the trunk floor. Pay careful attention that the plate seals properly so that no exhaust fumes can enter the cabin.
21. Reinstall the floor panel, carpet and any accessories removed.

Fuel Injector

REMOVAL AND INSTALLATION

900 Series

Except 2.5L (B258I) Engines

——— CAUTION ———

Fuel injection systems remain under pressure, even after the engine has been tuned OFF. *The fuel system pressure must be relieved before disconnecting any fuel lines. Failure to do so may result in fire and/or personal injury.*

1. Locate the fuel injectors and fuel rail which is mounted to the top of the intake manifold.
2. Clean the area around the injectors at the intake manifold and at the fuel rail and fuel lines. If available, blow the remaining dirt away with compressed air.
3. Make sure that the ignition switch is in the OFF position.
4. Disconnect the battery.
5. Remove the crankcase ventilation hose from the valve cover.

6. Unplug the electrical connectors at each injector and move the wiring loom out of the way.
7. Relieve the fuel system.
8. Once relieved, continue to remove the banjo fittings, containing the fuel with the shop towel, which should be properly disposed of once the job is complete.
9. Remove the two bolts securing the fuel rail to the intake manifold.
10. Carefully lift off the fuel rail with the injectors.
11. To remove individual injectors, remove the clip holding the injector to the fuel rail and twist the injector while pulling to remove.
To install:
12. Replace the O-rings on both ends of the fuel injector(s).
13. Reinstall the injector(s) onto the fuel rail, first placing a light coating of petroleum jelly on the O-rings that fit into the fuel rail.
14. Place a light coating of petroleum jelly onto the lower O-rings and carefully refit the fuel rail/injector assembly back into the intake manifold.
15. Make sure that the injectors are in the correct position and gently push down on the fuel rail to fully seat the injector assembly.
16. Make sure that the flange on the fuel rail lines up with the groove in each injector and reinstall the clips.
17. Reconnect the fuel line(s) using new washers and tighten the fittings, making sure to hold the fuel rail from twisting with a wrench.
18. Reconnect the electrical connectors to each fuel injector.
19. Install the crankcase ventilation hose to the valve cover.
20. Once the job is complete, wipe the area clean of any remaining fuel.
21. Reconnect the battery and start the engine.
22. Check for leakage and immediately correct if any leaks are found.

2.5L (B258I) Engine

——— CAUTION ———

Fuel injection systems remain under pressure, even after the engine has been tuned OFF. *The fuel system pressure must be relieved before disconnecting any fuel lines. Failure to do so may result in fire and/or personal injury.*

1. Disconnect the negative cable from the battery.
2. Remove the engine cover, if equipped.
3. Disconnect the cruise control cable and the throttle cable and set aside.

4. Disconnect the control rod and bracket and move aside.

5. Disconnect the air intake tubing to the throttle body and move aside.

6. Drain the coolant to a level below the intake manifold.

7. Disconnect the throttle body preheater hoses, crankcase ventilation hose and the vacuum hoses from the intake manifold, noting their locations for reassembly.

8. Disconnect the wiring to the idle air control valve, throttle position sensor, ignition coil and the traction control throttle body connector if equipped. Move the harnesses aside.

9. Remove the bolts securing the upper intake manifold and remove. Cover the openings in the lower manifold to keep dirt out.

10. Clean the area around the injectors located in the middle of the center intake manifold and at the fuel rail and fuel lines. If available, blow the remaining dirt away with compressed air.

11. Unplug the electrical connectors at each injector and move the wiring loom out of the way.

12. Relieve the fuel system.

13. Once relieved, continue to remove the banjo fittings, containing the fuel with the shop towel which should be properly disposed of once the job is complete.

14. Remove the bolts securing the fuel rail to the center intake manifold.

15. Carefully lift off the fuel rail with the injectors.

16. The injectors may be serviced as necessary.

To install:

17. Replace the O-rings on both ends of the fuel injector(s).

18. Reinstall the injector(s) removed onto the fuel rail, first placing a light coating of petroleum jelly on the O-rings that fit into the fuel rail.

19. Place a light coating of petroleum jelly onto the lower O-rings and carefully refit the fuel rail/injector assembly back into the intake manifold.

20. Make sure the injectors are in the correct position and gently push down on the fuel rail to fully seat the injector assembly.

21. Make sure the flange on the fuel rail lines up with the groove in each injector.

22. Reconnect the fuel line(s) using new washers and tighten the fittings, making sure to hold the fuel rail from twisting.

23. Reconnect the electrical connectors to each fuel injector.

24. Reinstall the upper intake manifold using new gaskets.

25. Reposition the electrical harnesses and reconnect the wiring to the idle air control valve, throttle position sensor, ignition coil and the traction control throttle body if equipped.

26. Reinstall the throttle body preheater hoses, crankcase ventilation hose and the vacuum hoses.

27. Install the air intake tubing to the throttle body.

28. Reinstall the control rod and bracket.

29. Reinstall the cruise control cable if equipped and the throttle cable.

30. Reinstall the engine cover.

31. Refill the engine coolant.

32. Wipe the area clean of any remaining fuel.

33. Reconnect the battery negative cable and start the engine.

34. Check for leakage and immediately correct if any leaks are found.

9000 Series

2.3L Engines

—— CAUTION ——

Fuel injection systems remain under pressure, even after the engine has been tuned OFF. The fuel system pressure must be relieved before disconnecting any fuel lines. Failure to do so may result in fire and/or personal injury.

1. Locate the fuel injectors and fuel rail, mounted to the bottom of the intake manifold.

2. Clean the area around the injectors at the intake manifold and at the fuel rail. If available, blow the remaining dirt away with compressed air.

3. Make sure the ignition switch is in the **OFF** position.

4. Disconnect the battery.

5. Remove the false bulkhead panel to gain access to the fuel rail area.

6. Unplug the electrical connectors at each injector and move the wiring loom out of the way.

7. Relieve the fuel system pressure.

8. Once relieved, continue to remove the banjo fittings from both ends of the fuel rail while containing the fuel with the shop towel, which should be properly disposed of once the job is complete.

9. Remove the two bolts securing the fuel rail to the intake manifold.

10. Carefully lift off the fuel rail with the injectors.

11. To remove individual injectors, remove the clip holding the injector to the fuel rail and twist the injector while pulling to remove.

To install:

12. Replace the upper and lower O-rings on the fuel injector(s).

13. Reinstall the injector(s) onto the fuel rail, first placing a thin film of petroleum jelly on the upper O-ring.

14. Make sure the flange on the fuel rail lines up with the groove in the injector(s) before reinstalling the clip.

15. Place a small amount of petroleum jelly onto the lower O-rings and carefully refit the fuel rail/injector assembly back into the intake manifold.

16. Make sure the injectors are in the correct position and gently push down on the fuel rail to fully seat the injector assembly.

17. Reconnect the fuel lines using new washers and tighten the fittings, making sure to hold the fuel rail from twisting with a wrench.

18. Reconnect the wiring to each fuel injector.

19. Reinstall the false bulkhead panel.

20. Once the job is complete, wipe the area clean of any remaining fuel.

21. Reconnect the battery and start the engine.

22. Check for leakage and immediately correct if any leaks are found.

3.0L (B308I) Engine

—— CAUTION ——

Fuel injection systems remain under pressure, even after the engine has been tuned OFF. The fuel system pressure must be relieved before disconnecting any fuel lines. Failure to do so may result in fire and/or personal injury.

1. Disconnect the negative cable from the battery.

2. Remove the engine cover, if equipped.

3. Disconnect the cruise control cable and the throttle cable and set aside.

4. Disconnect the control rod and bracket and move aside.

5. Disconnect the air intake tubing to the throttle body and move aside.

6. Drain the coolant to a level below the intake manifold.

7. Disconnect the throttle body preheater hoses, crankcase ventilation hose and the vacuum hoses from the intake manifold, noting their locations for reassembly.

8. Disconnect the wiring to the idle air control valve, throttle position sensor, ignition coil and the trac-

tion control throttle body connector if equipped. Move the harnesses aside.

9. Remove the bolts securing the upper intake manifold and remove. Cover the openings in the lower manifold to keep dirt out.

10. Clean the area around the injectors located in the middle of the center intake manifold and at the fuel rail and fuel lines. If available, blow the remaining dirt away with compressed air.

11. Unplug the electrical connectors at each injector and move the wiring loom out of the way.

12. Relieve the fuel system pressure.

13. Once relieved, continue to remove the banjo fittings, containing the fuel with the shop towel which should be properly disposed of once the job is complete.

14. Remove the bolts securing the fuel rail to the center intake manifold.

15. Carefully lift off the fuel rail with the injectors.

16. The injectors may be serviced as necessary.

To install:

17. Replace the O-rings on both ends of the fuel injector(s).

18. Reinstall the injector(s) removed onto the fuel rail, first placing a light coating of petroleum jelly on the O-rings that fit into the fuel rail.

19. Place a light coating of petroleum jelly onto the lower O-rings and carefully refit the fuel rail/injector assembly back into the intake manifold.

20. Make sure the injectors are in the correct position and gently push down on the fuel rail to fully seat the injector assembly.

21. Make sure the flange on the fuel rail lines up with the groove in each injector.

22. Reconnect the fuel line(s) using new washers and tighten the fittings, making sure to hold the fuel rail from twisting.

23. Reconnect the electrical connectors to each fuel injector.

24. Reinstall the upper intake manifold using new gaskets.

25. Reposition the electrical harnesses and reconnect the wiring to the idle air control valve, throttle position sensor, ignition coil and the traction control throttle body if equipped.

26. Reinstall the throttle body preheater hoses, crankcase ventilation hose and the vacuum hoses.

27. Install the air intake tubing to the throttle body.

28. Reinstall the control rod and bracket.

29. Reinstall the cruise control cable if equipped and the throttle cable.

30. Reinstall the engine cover.

31. Refill the engine coolant.

32. Wipe the area clean of any remaining fuel.

33. Reconnect the battery negative cable and start the engine.

34. Check for leakage and immediately correct if any leaks are found.

ENGINE MECHANICAL

Engine Assembly

REMOVAL AND INSTALLATION

900 Series

2.0L (B202L) and 2.1L (B212I) Engines

NOTE: The engine and transaxle are removed together as an assembly.

CAUTION

Fuel injection systems remain under pressure, even after the engine has been turned OFF. The fuel system pressure must be relieved before disconnecting any fuel lines. Failure to do so may result in fire and/or personal injury.

1. Remove the hood, after scribing lines around the mounting bolt positions for installation reference.
2. Properly relieve the fuel system pressure.
3. Install tool 83-93-209, or equivalent, under the upper control arm on the right side.
4. Disconnect and remove the battery.
5. Disconnect the exhaust pipe flange bolts.
6. Drain the engine coolant.
7. Loosen the wheel nuts on the right front wheel.
8. Raise and safely support the vehicle on jackstands.
9. Put the transmission selector into **R**.
10. Under the vehicle, remove the taper pin from the gearshift rod joint.
11. Disconnect the speedometer cable.
12. Remove the bolt securing the exhaust pipe to the clamp bracket on the transaxle.

13. Loosen the clamps around the rubber boots on the CV-joints and slide the boots clear, this operation can also be done from above.
14. On the right side of the vehicle, remove the front wheel.
15. Disconnect the hydraulic hose from the slave cylinder.
16. Separate the steering knuckle from the lower control arm.
17. Separate the universal joint and position the knuckle in front of the driver. Support the end piece against the outer end of the control arm.
18. Disconnect the positive battery cable from the clips holding it to the body. Disconnect the ground cable from the transaxle.
19. Disconnect the starter motor wiring.
20. Disconnect the pressure and return pipe from the steering servo pump and have a plug handy to prevent oil escaping from the pipe. Take care not to drip oil onto the engine mounting and control arm rubbers.
21. From the left side of the vehicle, disconnect the cooling system hoses at the following connections:
   a. the heater control valve.
   b. the expansion tank.
   c. the bottom of the radiator.
   d. the thermostat housing.
22. Disconnect the left fuel injection system cable harness as follows:
   a. At the air mass meter sensor
   b. At the throttle switch.
   c. At the A.I.C. actuator.
   d. At the injectors.
   e. At the NTC resistor (thermostatic switch).
   f. The ground points on the front lifting lug.

NOTE: Use the proper tool to release the tension in the springs on the terminal blocks.

23. Label as necessary and disconnect the following:
   a. Lead at the alternator.
   b. Green/white cable to the positive terminal on the regulator.
   c. Ground (black) cable.
   d. Black cable from the oil pressure switch.
   e. Cable for the A.I.C. actuator.
   f. Yellow/white cable from the temperature transmitter.
   g. Gray cable from the knock detector.
   h. Cable harnesses from the clip on the fuel injection manifold, from the rear of the engine and from the coolant hose between the engine and expansion tank.
24. Withdraw the loose cables and guide the harness unit out of the en-

gine compartment. Place it on top of the power distribution unit.

25. Remove the adjusting bolt in the alternator bracket, remove the drive belts and lift off the alternator.

26. Disconnect the brake servo hose from the intake manifold. Disconnect the throttle cable and sheath.

27. Remove the air conditioner compressor and bracket from the block. Place them on the filter housing for the heater system. Secure the alternator so it will not drop or become damaged.

28. Disconnect the fuel lines at their connections at the front of the fuel injection manifold and on the fuel pressure regulator.

29. Remove the coil.

30. If equipped, disconnect the turbocharger pressure line from the turbocharger compressor and the intercooler/throttle housing.

31. Remove the auxiliary fan.

32. Remove the air mass meter together with the suction pipe for the turbocharger unit. Disconnect the hoses at the solenoid valve and the crankcase ventilation at the suction pipe.

33. Disconnect the cables from the Hall transmitter and coil in the distributor. Free the Hall transmitter cable from the clips on the clutch cover.

34. Disconnect the solenoid valve hoses from the connections on the turbocharger unit and charging pressure regulator.

35. Remove the engine mounting bolts.

36. Attach suitable lifting equipment to the engine lifting hooks. Raise the engine until the left, inner CV-joint can be separated.

37. Raise the engine to enable the hoses on the oil cooler to be disconnected.

NOTE: When lifting the engine out of the vehicle, keep it close to the bulkhead to prevent the radiator and solenoid valve from being damaged by the front engine mounting.

38. Disconnect the hose to the power steering pump and drain the oil in the system.

To install:

39. Before installation, check that the inner CV-joint boots are packed with the correct grease.

40. Suspend the engine from the lifting gear. Adjust the lifting gear so the front engine mounting is slightly lower than the rear mounting.

41. Lower the engine into the engine compartment until the hoses to

the oil cooler and servo pump can be connected.

42. Guide the engine into position, attending to the following items in order:
   a. Front engine mounting.
   b. Left inner CV-joint.
   c. Right inner CV-joint.

43. Lower the engine until it rests on the rear engine mountings and install the mounting bolts. Unhook the lifting gear and unbolt the lifting lug from the water pump.

44. Connect the clutch master cylinder and all turbocharger unit connections.

45. Install the distributor, coil and all wiring for the electronic ignition.

46. Install the air mass meter, auxiliary fan, fuel lines, air conditioner, throttle cable and brake booster.

47. Install the alternator, drive belts, all electrical connections, cooling system hoses, exhaust system, battery and cables.

48. Assemble the steering knuckle and halfshafts.

49. Fill the cooling system. Fill the engine with the proper type and quantity of engine oil. Check the transmission fluid level.

50. Start the engine and allow it to reach operating temperature. Check the timing and recheck all fluid levels.

51. Test drive the vehicle.

2.0L (B204L), All 2.3L and 2.5L (B258I) Engines

NOTE: The engine and transaxle are removed together as an assembly.

1. With all four wheels on the ground, loosen the axle hub nuts.

2. Drain the engine coolant and remove the battery.

3. Remove air cleaner assembly and attached hoses. Take off the resonator.

4. Disconnect throttle cable and move it to one side.

5. Disconnect cruise control wiring harness and cables from throttle body. Loosen retaining nuts on cruise control unit and remove.

6. Relieve the fuel system pressure.

CAUTION
Fuel injection systems remain under pressure even after the engine is stopped. The fuel system pressure must be relieved before disconnecting any fuel lines. Failure to do so may result in fire and/or personal injury.

---

7. Disconnect the fuel lines at their connections at the front of the fuel injection manifold and on the fuel pressure regulator.

8. If equipped, disconnect the turbocharger pressure line from the turbocharger compressor and the intercooler/throttle housing.

9. Disconnect the vacuum hose from the secondary injection and disconnect the brake booster vacuum hose from the intake manifold.

10. If equipped, remove the boost pressure control unit.

11. Using a 3/8 inch ratchet extension relieve belt tensioner of pressure. Remove drive belt.

12. Disconnect the pressure and return pipe from the steering servo pump and have a plug handy to prevent oil escaping from the pipe.

13. Disconnect the cooling system hoses at the following connections:
   a. the heater control valve.
   b. the expansion tank.
   c. the bottom of the radiator.
   d. the thermostat housing.

14. Without disconnecting any hoses, remove the air conditioner compressor and it mounting bracket. Place them on the filter housing for the heater system. Secure the alternator so it will not drop or become damaged.

15. If not already done, raise and safely support the vehicle.

16. Disconnect the positive battery cable from the clips holding it to the body. Disconnect the ground cable from the transaxle.

17. Remove ignition cable and disconnect electrical connections from ignition coil.

18. Unplug the wiring connectors from the transaxle.

19. Disconnect the oxygen sensor wiring and the catalytic converter temperature sensor connector.

20. Under the vehicle, remove the taper pin from the gearshift rod joint. If equipped, disconnect the clutch cable and clutch pipe.

21. Inside the vehicle, pull back carpet under glove box to gain access to the central locking system and disconnect the control module. Feed the wires to the engine compartment through the grommet.

22. Remove hub nuts and remove both wheels. Remove front splash shield.

23. Remove both ball joint nuts and disconnect the ball joints from the struts. Make sure the halfshafts can slide out of the hubs. If necessary, use a wheel puller to push them out now.

24. Remove front pipe from exhaust manifold and take off catalytic converter.

25. Disconnect oil cooler lines and all lines connected to transaxle housing.

26. Position lifting table under vehicle so it is directly under front engine mount and gearbox. Take off subframe bolts and front engine mounts.

27. Lower lifting table slightly and separate the halfshafts from the hubs.

28. Lower lifting table fully and remove subframe.

29. Place engine on engine stand.

To install:

30. Place engine/transaxle on lifting table.

31. Install subframe on rear engine mount and tighten to 35 ft. lbs. Fit halfshafts into hubs as engine is raised into engine compartment.

32. Connect engine mount and subframe bolts and tighten in sequence:
- Front bolts–85 ft. lbs.
- Middle subframe bolts–141 ft. lbs.
- Rear subframe bolts — to 90 ft. lbs.
- Rear engine mount–54 ft. lbs.

33. Connect all oil cooler and hydraulic lines.

34. Reconnect all charge air cooler lines between the turbocharger and the cooler.

35. Install the front exhaust pipe to the exhaust manifold. Secure the catalytic converter to the rear exhaust section.

36. Fit engine shield to under side of vehicle.

37. Install hub nuts and wheels. Tighten wheel nuts to 89 ft. lbs. but do not tighten hub nuts yet.

38. Feed the wiring through the grommet and reconnect the central locking system control module.

39. As equipped, hook the clutch cable in place or bleed the clutch hydraulic cylinder.

40. Reconnect the taper pin to the gearshift selector.

41. Connect the oxygen and catalytic converter temperature sensor.

42. Connect all transaxle wiring.

43. Install A/C compressor, servo pump and the drive belt.

44. Reconnect power brake booster vacuum hose and pressure sensor connections.

45. Install cruise control including cables and electrical harness connections.

46. Install the battery and connect the cables.

47. With all four wheels on the ground, torque the hub nuts to 215 ft. lbs. (290 Nm).

48. Fill coolant and oil to specified level and check.

49. Start engine and test drive vehicle.

9000 Series

All Engines

NOTE: The engine and transaxle assembly are removed together.

1. Raise the vehicle and support it safely.

2. Drain the cooling system.

3. Disconnect the battery cables and remove the battery and tray.

4. Remove the connecting bolt for the expansion tank, disconnect the tank from the suction and remove the overflow hoses from the radiator.

5. Loosen the drive belt for the compressor by loosening the locknut, and loosening the adjusting nut under the locknut.

6. Disconnect the upper connection on the oil cooler, loosen the pipe clip on the radiator and slide the pipe down behind the radiator.

7. Unplug the connector to the electromagnetic clutch on the compressor and loosen the compressor mounting complete with the belt tensioner.

8. Place a protective cloth over the radiator member and rest the compressor on the radiator member. Secure the compressor to the radiator member.

9. If turbocharged, remove the turbocharger pressure pipe, situated between the turbocharger unit and the intercooler.

10. Disconnect the Lambda probe connector leads and disconnect them from the clips.

11. From the engine compartment, unbolt the flange joint between the exhaust pipe and the exhaust manifold. If turbocharged, disconnect the exhaust pipe coming from the turbocharger. Push the exhaust pipe to one side and unhook the rubber hangers from the exhaust system. Disconnect the bottom coolant hose from the water pump.

12. From under the vehicle, remove the bottom retaining bolt for the radiator fan.

13. Disconnect the speedometer drive from the gearbox.

14. If equipped with manual transmission, select the 4th gear and separate the rubber joint in the gear selector linkage.

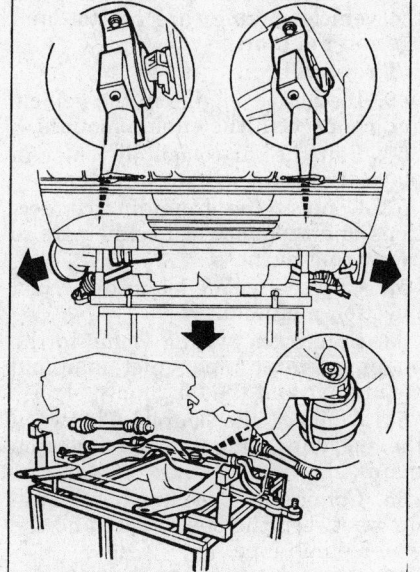

Lowering the entire powertrain assembly — 900 Series 2.0L (B204L) and 2.3L engines shown

291702

15. If equipped with an automatic transmission, disconnect the gear selector cable from the selector lever. Do not separate at the ball joint.

16. Remove the clips on the rubber covers over the inboard universal joints and slide the covers off the drive axles.

17. Disconnect the electrical leads from the alternator and the starter motor. Unplug the connector for the oil pressure switch.

18. Remove the clips and remove the top radiator hose.

19. Disconnect the top radiator hose at the cylinder head.

20. Disconnect the high tension lead from the ignition coil at the distributor cap.

21. Remove the solenoid valve from the bracket on the radiator and unplug the electrical connections.

22. Remove the bolts from the top of the radiator fan. Disconnect the wiring loom and lift out the fan.

23. Pull the connector off the air mass meter. Disconnect the air mass meter from the air intake duct socket connector and the air cleaner. Leave the rubber socket connector attached to the turbocharger unit.

24. Remove the air intake duct by pulling it out of the aperture in the wing and twisting the ends inwards.

25. Remove the air cleaner top section first, then the remaining section.

26. Disconnect the relief valve hose from the turbocharger pressure pipe and remove the pipe, if equipped.

27. Disconnect the Hall Effect transducer, the earth lead from the gear box and the electrical connector for the back-up lights.

28. Disconnect the end of the throttle cable and disconnect the throttle linkage.

29. Install a clamp to the hydraulic line to the slave cylinder and pinch the line tightly. With proper wrenches, open the line to the clutch slave cylinder.

30. Remove the front wheels.

31. From both sides of the vehicle, slacken the lower bolts retaining the steering swivel member to the strut assembly. Remove the 2 upper bolts.

32. Pivot the steering swivel member outwards to pull the inboard universal joint out of the halfshaft. Position dust covers over the exposed halfshaft cups.

33. Remove the engine mount bolt.

34. Remove the steering reservoir for the servo and position it within the engine compartment. Drain the fluid from the container.

35. Disconnect the large bore hose and the delivery hose from the steering servo pump and plug the open ends.

36. Disconnect the fuel return line from the pressure regulator.

37. Remove the nut from the rear engine mounting and back off the front mount bolts a few turns.

38. Attach the lifting sling (Saab 83–92–409) or equivalent to the rear lifting lug.

39. Lift the engine sufficiently to provide access for the removal of the components located between the engine and the firewall.

40. Disconnect and tag the vacuum hoses from the inlet manifold.

41. Remove the coolant hoses running between the heater core and the water pump pipe.

42. Separate the coupling between the fuel pipe and the fuel injection manifold. Do not allow the fuel to spill or collect.

43. Cut the clips securing the wiring looms to the oil pipe, water pipe. Disconnect and tag the vacuum hoses from the inlet manifold.

44. Unclip the wiring loom to the fuel injection manifold.

45. Disconnect the grounding connections and the electrical connectors from the wiring harness.

46. Unbolt the air cooled oil cooler and place it on top of the engine. The 2 lower bolts need only be loosened.

47. Disconnect the hood lifts and install hood extenders for extra clearance.

48. Carefully lift the engine from the vehicle, taking care not to damage the radiator.

To install:

49. Install the engine in the vehicle and connect all the engine mounts.

50. Bolt the air cooled oil cooler in place. Tighten the 2 lower bolts.

51. Connect the grounding connections and the electrical connectors to the wiring harness.

52. Clip the wiring loom to the fuel injection manifold.

53. Secure the wiring looms to the oil pipe, water pipe, inlet manifold steady bar and the oil supply pipe.

54. Connect the coupling between the fuel pipe and the fuel injection manifold.

55. Connect the coolant hoses running between the heater core and the water pump pipe.

56. Connect the vacuum hoses to the correct connections on the inlet manifold.

57. Before removing the lifting sling, connect the components located between the engine and the firewall.

58. Attach the nut from the rear engine mounting and tighten the front mount bolts.

59. Connect the fuel return line to the pressure regulator.

60. Connect the large bore hose and the delivery hose to the steering servo pump.

61. Connect the steering reservoir for the servo. Fill the fluid container.

62. Connect the engine stay bolt.

63. Pivot the steering swivel member outwards to install the inboard universal joint into the halfshaft. Position dust covers over the exposed halfshaft cups and install clamps.

64. If equipped with manual transmission, select the 4th gear and connect the rubber joint to the gear selector linkage.

65. If equipped with an automatic transmission, connect the gear selector cable to the selector lever.

66. Connect the lower bolts retaining the steering swivel member to the strut assembly. Tighten the 2 upper bolts.

67. Install the front wheels.

68. With line wrenches, tighten the line to the clutch slave cylinder.

69. Connect the end of the throttle cable and the throttle linkage.

70. Connect the Hall Effect transducer, the negative battery cable to the gear box and the electrical connector for the back-up lights.

71. Connect the relief valve hose to the turbocharger pressure pipe, if equipped.

72. Install the air cleaner.

73. Install the air intake duct.

74. Push the connector on the air mass meter. Connect the air mass meter to the air intake duct socket connector and the air cleaner.

75. Connect the high tension lead to the ignition coil at the distributor cap.

76. Connect the top radiator hose at the cylinder head.

77. Connect the electrical leads to the alternator and the starter motor. Plug in the connector for the oil pressure switch.

78. Connect the speedometer drive to the gearbox.

79. From the engine compartment, connect the flange joint between the exhaust pipe and the exhaust manifold. If turbocharged, connect the exhaust pipe coming from the turbocharger. Hook the rubber hangers to the exhaust system.

80. Connect the bottom coolant hose to the water pump.

81. Connect the Lambda probe connector leads.

82. Plug the connector into the electromagnetic clutch on the compressor and tighten the compressor mounting complete with the belt tensioner.

83. Connect the upper connection on the oil cooler, tighten the pipe clip on the radiator.

84. Tighten the drive belt for the compressor .

85. Install the expansion tank, connect the tank to the suction and overflow hoses from the radiator.

86. Install the battery and tray. Connect the battery cables.

87. Remove the lift supports and lower the vehicle.

88. Fill the radiator with coolant, the engine with oil and the transmission with fluid. Start the engine and allow it to reach operating temperature. Check the ignition timing and all fluid levels. Test drive the vehicle.

Engine Mounts

REMOVAL AND INSTALLATION

2.0L (B202L) and 2.1L (B212I) Engines

Front Engine Mount

1. Disconnect the negative battery cable.

2. Using an engine lifting device, raise the engine just enough to re-

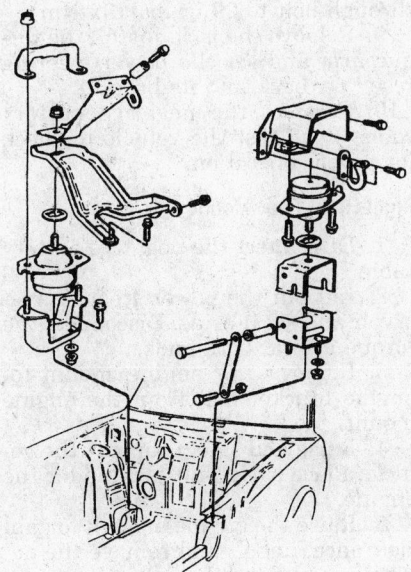

213810

Exploded view of the mount assemblies — 2.0L (B202L) and 2.1L (B212I) engines

lease pressure from the engine mounts.

3. Remove the mounting bolts from underneath the radiator support.

4. Remove the mount-to-engine bolts and the mount.

To install:

5. Position the engine mount and secure it to the engine.

6. Lower the engine until the mount contacts the radiator support.

7. Install the mounting bolts from underneath.

8. Install the mounting nut and washer and remove the engine lifting device.

9. Connect the negative battery cable.

Left Side Mount

1. Disconnect the negative battery cable.

2. Using an engine lifting device, raise the engine just enough to release pressure from the engine mounts.

3. Remove the mounting nut and washer.

4. Raise the engine enough to gain access and remove the two retaining bolts and the mount.

To install:

5. Install the new mount and secure it with the two retaining bolts.

6. Lower the engine slowly and position the center mounting bolt so it slides into the hole in the mounting bracket and tighten the center mounting nut.

7. Lower the vehicle and remove the engine lifting device.

8. Connect the negative battery cable.

Right Side Mount

1. Disconnect the negative battery cable.

2. Using an engine lifting device, raise the engine just enough to release pressure from the engine mounts.

3. Remove the mounting nut and washer.

4. Raise the engine enough to gain access and remove the two mount-to-bracket retaining nuts and the mount.

To install:

5. Install the new mount and secure it with the two retaining nuts.

6. Lower the engine slowly and position the center mounting bolt so it slides into the hole in the mounting bracket and tighten the center mounting nut.

7. Lower the vehicle and remove the engine lifting device.

8. Connect the negative battery cable.

2.0L (B204L), 2.3L and 2.5L Engines

Front Engine Mounts

1. Disconnect the negative battery cable.

2. Raise and safely support the vehicle. Remove the wheel, fender liner and spoiler half from the side of the vehicle where the mount is being replaced.

3. Place a jack under the engine or transmission.

4. Remove the bolts securing the engine mount to the body.

5. Lower the engine slightly and unbolt the mount from the engine mount bracket.

To install:

6. Bolt the mount to the engine mount bracket and torque to 29 ft. lbs. (39 Nm).

7. Lift up the engine and install the mount-to-body mounting bolts. Torque the bolts to 54 ft. lbs. (73 Nm).

8. Install the spoiler half, fender liner and wheel on the side of the vehicle where the mount is being replaced. Lower the vehicle and torque the wheel bolts to 89 ft. lbs. (120 Nm).

9. Connect the negative battery cable. Road test the vehicle to check for proper operation.

Rear Engine Mount

1. Disconnect the negative battery cable.

2. Install engine lifting beam 83–94–850 (or equivalent) and raise the engine just enough to take the weight of the engine off the mounts.

3. Disconnect the oxygen sensor connector(s) and if necessary, remove the cable clamp and clip.

4. Safely raise and support the vehicle. Remove the front wheels and front spoiler.

5. Disconnect the front and intermediate exhaust pipes. Unbolt the front exhaust pipe from the exhaust manifold(s) or turbocharger. Remove the catalytic converter from it's bracket and remove the front exhaust pipe.

6. Disconnect the ball joints from the struts on both sides. Remove the rear mount-to-subframe nuts.

7. Position a lifting table (or suitable equivalent) under the engine. Raise the table, unbolt the subframe and lower the subframe from the vehicle.

8. Remove the rear engine mount-to-engine bolts and remove the mount.

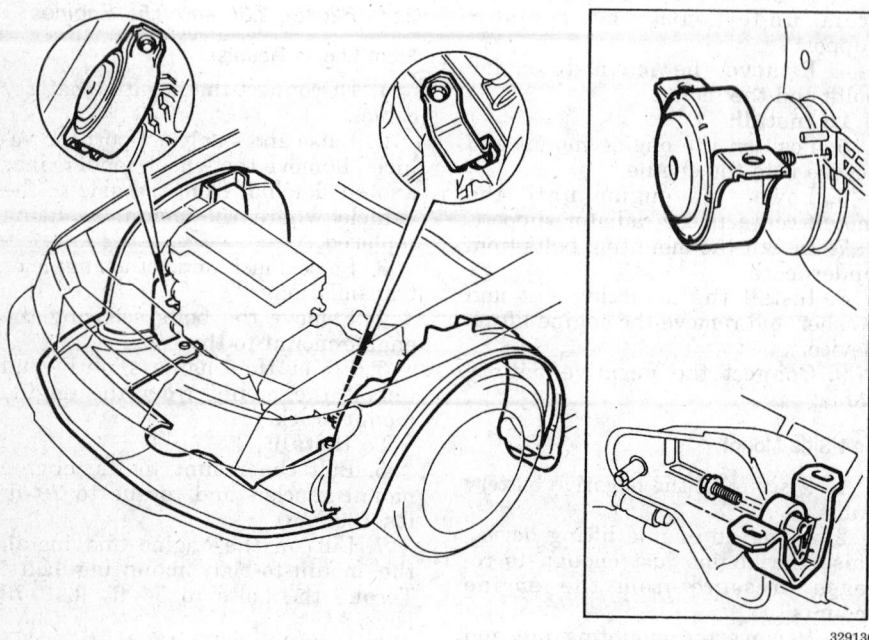

Front engine mount locations — 900 Series 2.5L engine shown

329136

To install:

9. Install the mount to the engine and torque the bolts to 29 ft. lbs. (39 Nm).

10. Lift the subframe into place and install the subframe mounting bolts. Torque the bolts as follows:
- Front bolts to 85 ft. lbs. (115 Nm)
- Middle bolts to 140 ft. lbs. (190 Nm)
- Rear bolts to 81 ft. lbs. (110 Nm) plus 75° additional rotation

11. Install the engine mount-to-subframe nuts and torque to 29 ft. lbs. (39 Nm).

12. Connect the ball joints to the strut assemblies and torque the nuts to 55 ft. lbs. (75 Nm).

13. Attach the catalytic converter to it's bracket but do not tighten the bolt. Connect the front exhaust pipe to the exhaust manifold(s) or turbocharger.

NOTE: Before assembling turbocharged vehicles, apply Molykote® 1000 to the threads on the turbocharger studs.

14. Tighten the bolt on the converter bracket and reattach the front and intermediate exhaust pipes.

15. Install the front spoiler and front wheels. Torque the wheel bolts to 89 ft. lbs. (120 Nm).

16. Lower the vehicle, connect the oxygen sensor connector(s) and if necessary, install the cable clamp and clip.

17. Remove the engine lifting beam.

18. Connect the negative battery cable. Road test the vehicle to check for proper operation.

3.0L (B308I) Engine

Right-Hand Front Mount

1. Disconnect the negative battery cable.

2. Lift out the power steering reservoir and set it aside. Disconnect the torque arm at both ends.

3. Remove the engine mount-to-engine bracket bolt from the engine mount.

4. Raise and safely support the vehicle. Place a high-lift jack under the engine.

5. Raise the engine enough to gain clearance, unbolt and remove the engine mount from the subframe.

To install:

6. Install the front mount and torque the mounting bolts to 29 ft. lbs. (39 Nm).

7. Lower the engine and install the engine mount-to-engine bracket bolt. Torque the bolt to 29 ft. lbs. (39 Nm).

8. Remove the high-lift jack and lower the vehicle.

9. Install the torque arm and set the power steering reservoir back into its bracket.

10. Connect the negative battery cable. Road test the vehicle to check for proper operation.

Left-Hand Front Mount

1. Disconnect the negative battery cable.

2. Lift out the power steering reservoir and set it aside. Disconnect the torque arm at both ends.

3. Unbolt the mount-to-subframe through bolt and keep the washers for reassembly.

4. Place a jack under the transmission and raise the transmission enough to remove the mount.

5. Unbolt the mount cap from the transmission and remove the mount.

To install:

6. Loosely install the mount-to-subframe through bolt in the mount.

7. Set the mount into position on the transmission and install the mount cap. Torque the mount cap bolts to 18 ft. lbs. (24 Nm).

NOTE: Make sure the positioning mark on the mount is correct when installing the mount onto the transmission.

8. Lower the transmission to set the mount onto the subframe. Install the washers saved at removal and torque the mount-to-subframe through bolt to 29 ft. lbs. (39 Nm).

9. Remove the jack, install the torque arm and set the power steering reservoir back into its bracket.

10. Connect the negative battery cable. Road test the vehicle to check for proper operation.

Right-Hand Rear Mount

1. Disconnect the negative battery cable.

2. Lift out the power steering reservoir and set it aside. Disconnect the torque arm at both ends.

3. Remove the engine mount-to-engine bracket bolt from the engine mount.

4. Raise and safely support the vehicle. Place a high-lift jack under the engine.

5. Raise the engine enough to gain clearance, unbolt and remove the engine mount from the subframe.

To install:

6. Align the rear mount and bolt it to the subframe. Torque the mounting bolts to 29 ft. lbs. (39 Nm).

7. Lower the engine and install the engine mount-to-engine bracket

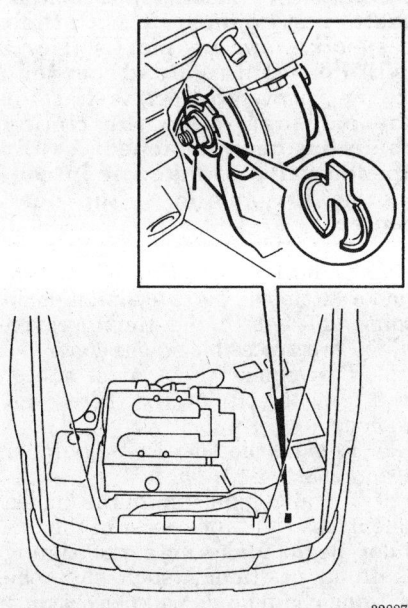

Location of the left-hand front mount — 3.0L engine

328879

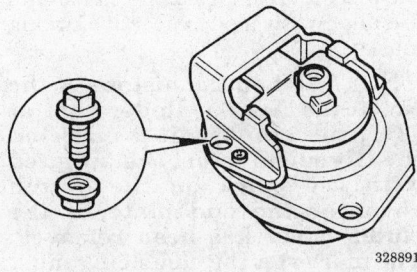

328891

Right-hand rear mount — 3.0L engine

bolt. Torque the bolt to 29 ft. lbs. (39 Nm).

8. Remove the high-lift jack and lower the vehicle.

9. Install the torque arm and set the power steering reservoir back into its bracket.

10. Connect the negative battery cable. Road test the vehicle to check for proper operation.

Cylinder Head

REMOVAL AND INSTALLATION

All 2.0L and 2.1L (B212I) Engines

——————— CAUTION ———————
Fuel injected systems remain under pressure, even after the engine has been turned OFF. The fuel system pressure must be relieved before disconnecting any fuel lines. Failure to do so may result in personal injury.

1. Remove the hood after scribing reference marks next to the mounting bolts to aid in reinstallation.

2. Remove the battery.

3. Drain the coolant from the radiator and cylinder block. Drain the oil from the crankcase.

4. Remove the exhaust manifold and turbocharger unit, if equipped.

5. Remove the tensioning pulley and drive belt for the air conditioner compressor.

6. Slacken the securing bolts for the steering pump bracket, remove the drive belt and push the pump aside.

7. Detach the wiring harness clips on the cylinder head.

8. Remove the 2 bolts in the timing cover, which are screwed into the cylinder head from underneath.

9. Remove the bolts in the right-hand engine mounting which are screwed into the cylinder head, together with the spacer sleeves.

10. Disconnect the hose between the thermostat housing and the radiator at the thermostat housing.

11. Remove the fuel pressure regulator and disconnect the ground leads for the fuel injection system.

12. Remove the air conditioning compressor with lines still attached and carefully put it on the air intake for the heating system.

13. Remove the auxiliary air valve (not applicable to vehicles with a catalytic converter). Remove the bracket for the air conditioning compressor from the cylinder head.

14. Remove the intake manifold complete with injectors.

15. Disconnect the lead from the temperature sensor.

16. Remove the lid on the valve cover and the ignition cables together with the distributor cap.

17. Remove the valve cover. Disconnect the crankcase ventilation hose and remove the semicircular rubber plug halves from the cylinder head.

18. Remove the timing chain, tensioner and camshaft sprockets.

NOTE: Do not rotate the camshaft or crankshaft after the sprockets have been removed, as rotation of one of the shafts can result in damage to the valves.

19. Remove the cylinder head bolts and siphon off the oil from the cylinder head.

20. Install a guide pin (Saab special tool 83–92–128 or equivalent) in one of the bolt holes and lift off the cylinder head, making sure the pivoting guide for the timing chain is not damaged.

To install:

21. Align the **0** mark on the flywheel with the timing mark on the housing. Align the marks on the camshafts with their respective timing marks.

22. Install the cylinder head gasket, making sure it is held in position by the guide sleeves in the cylinder head flange.

23. Position the timing chain and pivoting guide for cylinder head installation.

24. With the guide pin still installed, carefully position the cylinder head. Use the guide pin as a pivot for the head, which must be turned slightly to enable it to pass the pivoting guide. Thereafter, alignment will be determined by the guide sleeves.

NOTE: Remember to install the 2 M8 sized bolts in the underside of the cylinder head.

25. Install the cylinder head bolts and tighten them in sequence and in 3 stages. Stage 1, torque to 44 ft. lbs. (60 Nm) evenly. Stage 2, torque to 59 ft. lbs. (80 Nm) evenly. Stage 3, another 90 degrees (¼ turn).

26. Unless the head is fitted with a new, special gasket which does not require retorquing, the cylinder head bolts must be retorqued. Run the engine for 30 minutes allow to cool and then retighten the head.

27. In the same sequence as before, slacken and retighten each bolt to 59 ft. lbs. (80 Nm).

28. Install the camshaft sprockets, timing chain and tensioner.

29. Install the semi-circular rubber plug halves in the cylinder head and install the valve cover torque the retaining bolts to 11 ft. lbs. (14 Nm). Install the crankcase ventilation hose.

30. Install the spark plug wires and distributor cap in their original position and install the lid on the valve cover.

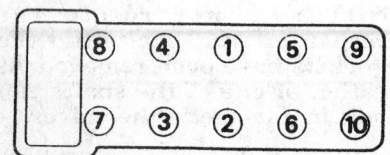

214105

Cylinder head bolt torque sequence — all 2.0L, 2.1L and 2.3L engines

31. Connect the electrical lead to the temperature sensor.

32. Install the intake manifold complete with injectors.

33. Install the bracket for the air conditioning compressor and install the compressor.

34. Install the auxiliary air valve, if applicable.

35. Install the fuel pressure regulator and connect the ground leads for the fuel injection system.

36. Connect the upper radiator hose to the thermostat housing.

37. Install the bolts in the right-hand engine mounting. Screw them into the cylinder head, together with the spacer sleeves.

38. Connect the wiring harness clips on the cylinder head.

39. Install the power steering belt.

40. Install the tensioning pulley and drive belt for the air conditioner compressor.

41. Install the exhaust manifold and turbocharger unit, if equipped.

42. Fill the crankcase with new oil and properly fill the cooling system.

43. Install the battery.

44. Using the reference marks made earlier, align and install the hood.

45. Start the engine and check for proper operation. Check for leaks.

All 2.3L Engines

1. Disconnect the negative battery cable.

2. Raise and safely support the vehicle.

3. Remove the right front wheel assembly and the inner fender panel.

4. Drain the coolant. Remove the radiator expansion tank. Disconnect the power steering reservoir and set aside. Leave the hoses attached.

5. Loosen the compressor drive belt and remove the belt.

6. Disconnect the electrical leads from the air compressor.

7. Unbolt the compressor from its mounting bracket. Disconnect the top pipe connected to the air cooled oil cooler and push the pipe to one side. Carefully rest the compressor on the radiator crossmember. Unbolt the compressor mounting bracket and remove it.

8. Unbolt the front exhaust pipe flange and unhook the rubber hangers.

9. Remove the stay bar for the turbocharger unit and the oil return pipe (if vehicle is equipped with a turbocharger).

10. Disconnect the hose from the intercooler at the turbocharger unit. Disconnect the oil supply pipe from the turbocharger (if applicable).

11. Disconnect the hose between the air mass meter and the turbocharger unit (if applicable). Disconnect the coolant hose from the thermostat housing and the hose from the cylinder head.

12. Disconnect the oil supply hose or pipe so as not to obstruct the removal of the exhaust manifold. If necessary, remove the clip holding the pipe to the cylinder head and slave cylinder.

13. Unbolt and lift off the exhaust manifold complete with the turbocharger unit (if applicable), pushing the oil supply pipe aside at the same time.

14. Disconnect the lead to the temperature transducer.

15. Remove the engine stay bracket from its attachment point on the wing.

16. Remove the bolt securing the engine stay bracket to the cylinder head. Remove the intake manifold from the cylinder head.

17. Disconnect the breather hose for the crankcase ventilation from the camshaft cover.

18. Disconnect the vacuum hose and the Hall Effect transducer lead from the distributor and remove the distributor cap complete with the spark plug wires (if applicable).

19. Unscrew and remove the spark plug inspection plate and the clips for the spark plug wires.

20. Remove the camshaft cover.

21. Align the crankshaft with the 0 timing mark and check that the camshaft timing marks also coincide.

22. Using a wrench installed over the flats on the exhaust camshaft, hold the camshaft and remove the center bolt securing the camshaft sprocket. Repeat this step for the intake camshaft.

23. Remove the timing chain tensioner. Remove the 2 cylinder head bolts adjacent to the timing cover, which are accessible from below.

24. Disconnect the starter motor lead from the clip on the thermostat housing.

25. Remove the Torx® type cylinder head bolts.

26. Install a guide pin in the drilled hole in the right top corner of the cylinder head. Make sure the timing chain is positioned such that the pivoting chain guide will not obstruct the cylinder head and carefully lift the cylinder head from the engine block.

To install:

27. Before installation, clean both the cylinder head and the engine block surfaces. Install a new gasket. Be sure the crankshaft is aligned in the **0** position and that the camshafts are align with their respective timing marks.

NOTE: When the pistons of the No. 1 and No. 4 cylinders are at TDC, the crankshaft 0 mark on the flywheel must be aligned with the mark on the clutch cover or the end plate, if the clutch cover has been removed. The marks on the camshafts must be aligned with those on the cam bearing caps. This indicates the exhaust valves for No. 1 and No. 4 cylinders are closed.

28. Install a guide pin in the drilled hole in the top of the right corner of the cylinder head and lower the cylinder head carefully into position on the engine block. Locate the cylinder head on the guide sleeves.

29. Install the cylinder head bolts and tighten as follows:

    a. Tighten, in sequence, to 44 ft. lbs. (60 Nm).

    b. Tighten, in sequence, to 59 ft. lbs. (80 Nm).

    c. Tighten, in sequence, an additional 90 degrees (1/4 turn).

    d. At the completion of cylinder head installation, run the engine until it reaches normal operating temperature, then shut the engine

OFF and allow to cool for 30 minutes.

e. Slacken, then retorque the cylinder head bolts, in sequence, to 59 ft. lbs. (80 Nm).

f. Tighten each bolt, in sequence, an additional 90 degrees (¼ turn).

30. To install the timing chain, run the chain up through the opening in the cylinder head. Install the chain and sprocket on the exhaust cam first. Make sure the chain is taut between the crankshaft and camshaft sprockets. Snug the bolt but do not tighten.

31. Install the chain and sprocket to the intake cam. Keep the chain taut between the cam sprockets while it is being installed. Snug the bolt but do not tighten. Make sure the chain is seated in the guide tensioner grooves.

32. Tension the chain tensioner by fully depressing the piston and then rotating it to the locked position.

33. Install the chain tensioner with the piston under tension. Make sure the copper gasket is in good condition and that the sealing surface is clean and free from burrs. Tighten the tensioner to 47 ft. lbs. (63 Nm).

34. Trigger the chain tensioner by pressing the pivoting chain guide against it, thereafter, press the pivoting guide against the chain to give the chain its basic tension. Check that the chain tensioner maintains tension on the chain when the pressure on the chain guide is released and that the basic setting stop for the tensioner holds the chain guide tight against the chain. A limited amount of play will be present until the hydraulic pressure takes over once the engine is running.

35. Check the setting by rotating the crankshaft 2 complete turns in its normal direction of rotation around to the timing mark. The basic setting of the cams should remain unaltered.

36. Lock the exhaust cam by using a wrench installed over the flats on the camshaft and torque the sprocket bolt to 48 ft. lbs. (65 Nm). Repeat this step on the intake cam.

NOTE: The accuracy of the timing chain adjustment will depend on the condition of the chain.

37. Install both halves of the split seal and the camshaft cover. Install the bolt at the distributor end and the middle bolt at the other end first. Tighten the bolts in sequence to 16 ft. lbs. (22 Nm).

38. Check that the timing marks for the distributor rotor are aligned,

Install the distributor cap and connect the lead for the Hall Effect transducer. Connect all vacuum hoses.

39. Connect the spark plug wires. Secure the leads in the clips. Install the inspection plate and tighten the retaining screws.

40. Install the clip securing the starter motor lead to the thermostat housing.

41. Install a new gasket on the inlet manifold and install the manifold in place. Install the top securing bolts first and then install the lower bolts, using an extension bar.

42. Install the bolt for the engine stay bracket to the cylinder head and position the stay bracket in place. Install a new gasket onto the exhaust manifold and position the exhaust manifold to the cylinder head.

43. Install the oil supply pipe. Install the clip and the slave cylinder bolt. Install the oil return line and the steady bar for the turbocharger unit (if applicable).

44. Connect the hose between the turbocharger unit and the intercooler (if applicable). Connect the cooler hose to the thermostat housing and the hose to the cylinder head.

45. Install the air mass meter socket connector into the turbocharger unit and tighten the clip. Connect the hose between the intercooler and the turbocharger unit (if applicable).

46. Install and tighten the nuts securing the front section of the exhaust pipe to the turbocharger compressor (if applicable). Bolt the air conditioning compressor mounting bracket onto the cylinder head and engine block.

47. Install the air conditioning compressor. Leave the coolant hose in the bracket when installing the compressor.

48. Connect the electrical leads and make sure the lead is clear of the compressor pulley. Install the steering servo reservoir. Install the coolant expansion tank and tighten the hose clip.

49. Connect the top pipe to the air cooled oil cooler and secure the cooler to the radiator. Install the overflow line between the expansion tank and the radiator.

50. Install the compressor belt, adjust the tension and tighten the belt tensioner bolt. Install the inner right wheel arch, and install the wheel.

51. Lower the vehicle and tighten the wheel. Connect the negative battery cable and properly fill the cooling system.

52. Start the engine.
53. Test the engine operation and check for leaks.

2.5L (B258I) Engine

———— WARNING ————
To avoid damage to the valves, DO NOT rotate the camshafts. The crankshaft may only be turned between 0° and 60° BTDC when the camshafts are locked in position with the appropriate locking tool.

1. Disconnect the negative battery cable and drain the coolant from the radiator.

2. Safely raise and support vehicle and disconnect the rearward flange of the front exhaust pipe from the exhaust manifold.

3. Lower vehicle, remove engine covers, air filter assembly and mass air flow sensor.

4. Disconnect and label all hoses and electrical harness connections at the intake manifold. Remove the upper intake plenum and plug the center intake runner ports with paper.

5. Disconnect and plug the fuel lines.

6. Remove the center intake runners complete with fuel rail.

7. Remove lower section of the intake manifold and cover intake port openings.

8. Disconnect electrical connectors and upper radiator hose at the coolant bridge. Remove the coolant bridge.

9. Remove the spark plug wires.

10. Remove the oxygen sensor from it's holder and remove crankcase breather. Cover holes with paper.

11. Remove the ignition coil. Remove the bracket for the ignition coil, camshaft position sensor and oxygen sensor connectors and lay the bracket aside.

12. Disconnect the upper radiator hose from the engine. Remove the lifting eye.

13. Remove the right-front wheel and forward cover in the wheel housing.

14. Loosen the power steering pump, water pump pulley bolts and 6 outer crankshaft pulley bolts. Remove the drive belt and tensioner. Remove the power steering pump pulley, water pump pulley and timing cover.

NOTE: When removing the crankshaft pulley, remove the 6 outer bolts only, DO NOT remove the center bolt.

15. Remove the crankshaft pulley.

16. Set the engine at TDC No. 1 cylinder. The timing marks on both camshafts and crankshaft should be aligned. Install locking tools (83–94–926 or equivalent) between the camshaft sprockets to lock the camshafts in position.

17. Remove the timing belt, tensioner and camshaft sprockets.

—————— WARNING ——————
The wrench used to hold the camshaft must not have jaws that are too long. A wrench that is too long could damage the casting and lock the tappets.

18. Remove the bearing caps on the exhaust camshaft. Note the position markings on the bearing caps and loosen the bearing cap bolts in stages of ½ to 1 turn at a time. Remove the front camshaft seal and remove the camshaft.

NOTE: The bearing caps located where valve tappets are compressed, should be removed last. Also, it is not necessary to remove the intake camshaft for cylinder head removal.

19. Loosen the cylinder head bolts ¼ turn at a time in sequence, then ½ turn at a time. Remove cylinder head bolts and lift cylinder head off the engine.

—————— WARNING ——————
Be careful when setting down the cylinder head, since the intake camshaft is still installed, the valve stems could be accidentally bent.

20. With the cylinder head removed, the spark plugs, intake camshaft, tappets and exhaust manifold can be removed.

NOTE: When removing the intake camshaft, use the same procedures used for the exhaust camshaft.

To install

21. Make sure the crankshaft is at 60° BTDC.

22. Make sure all gasket contact surfaces are clean. Note the positioning of the cylinder head gasket and make sure the words OBEN/TOP marked on the gasket are facing up.

23. Install the exhaust manifold and tappets, if removed from the cylinder head.

24. Install the cylinder head onto the engine using NEW cylinder head bolts.

25. Step torque the cylinder head bolts in sequence as follows:
Step 1: 18.5 ft. lbs. (25 Nm)
Step 2: Tighten bolts 90°
Step 3: Tighten bolts 90°
Step 4: Tighten bolts 90°

26. Thoroughly lubricate the camshafts, install the camshafts and new front camshaft seals. Make sure the locating pins are properly positioned. Install the bearing caps in their proper location and position. Tighten the bearing cap bolts in sequence ½ to 1 turn at a time, to a torque of 6 ft. lbs. (8 Nm).

27. Install the timing cover and 2 retaining screws into cylinder head. Tighten retaining screws to 6 ft. lbs. (8 Nm).

28. Check to make sure that the camshaft locating pins are in the proper position.

NOTE: On the rear or left bank cylinder head, the locating pins of both camshafts should be point towards the inboard bolts of the camshaft bearing caps. On the front or right bank cylinder head, the locating pin for the intake camshaft should be pointing downward in line with the inboard bolt of the bearing caps. The locating pin for the exhaust camshaft should be pointing upward in line with the edge of the camshaft sensor.

29. Install the camshaft sprockets, the timing belt tensioner and timing belt.

30. Install crankshaft locking tool (83–94–926 or equivalent). Install tool 83–93–985 with a cut piece from an old timing belt, to measure belt tension.

31. Snug the center bolts of the adjusting rollers. Turn the lower adjusting roller counterclockwise, until a belt tension of 275–300 Nm is reached. Tighten adjusting roller center bolts to 30 ft. lbs. (40 Nm).

NOTE: This adjustment of the timing belt is just preparation and should not be used as a final check.

32. Adjust the tensioner pulley until the marks are aligned. Tighten the tensioner pulley to 15 ft. lbs. (20 Nm).

33. Remove camshaft locking tool on camshaft sprockets 1 & 2. Adjust the upper adjusting roller until sprocket No. 2 moves 1–2mm clockwise. Tighten the upper adjusting roller to 30 ft. lbs. (40 Nm) and remove the upper locking tool.

34. Rotate the engine two complete revolutions to just before 0° TDC and install the locking tool on the crank-shaft. Carefully turn the crankshaft until the arm of the locking tool is against the water pump flange and tighten the locking tool. Set tool (83–94–926 or equivalent) into position on the front of the camshaft sprockets. Make sure that the timing marks on the camshaft sprockets are aligned with the marks on the tool and that the edge of the timing belt is flush with the edge of the camshaft sprockets.

NOTE: Also check that the alignment marks on the tensioner pulley are still aligned.

35. Install the crankcase ventilation housing and set the oxygen sensor connector back in it's holder. Make sure the valve cover O-rings are still in position and clean. Lubricate the O-rings with soapy water, apply (81–52–381 or equivalent) sealer at the corners of the large end bearing caps and install the valve cover.

36. Install the spark plug wires.

37. Install the crankshaft pulley and torque bolts to 15 ft. lbs. (20 Nm).

38. Install the timing cover. Install power steering pump pulley, water pump pulley, drive belt tensioner and drive belt.

39. Install the forward cover in the wheel housing and the right-front wheel. Torque wheel bolts to 89 ft. lbs. (120 Nm).

40. Install coolant bridge and torque the bolts to 22 ft. lbs. (30 Nm). Install the radiator hose and electrical wiring harness connectors.

41. Install a new gasket, the lower intake manifold and torque the bolts to 15 ft. lbs. (20 Nm).

42. Install the center intake runners complete with fuel rail and torque the bolts to 15 ft. lbs. (20 Nm).

43. Connect the fuel lines.

44. Connect all hoses and electrical harness connections at the intake manifold, which should have been labeled at disassembly.

NOTE: Make sure the pressure regulator's vacuum hose is connected to front nipple on the throttle housing, in front of the butterfly.
Install the upper intake plenum and tighten the bolts at the mating surface. Starting in the middle and working alternately outward to the ends. Torque upper intake plenum bolts to 15 ft. lbs. (20 Nm).

45. Install the mass air flow sensor, air filter assembly and engine covers.

46. Safely raise and support vehicle and connect the rearward flange of the front exhaust pipe to the exhaust manifold.

47. Connect the negative battery cable.

48. Check all other oil and fluids for condition and proper level. Replace fluids and filters or top off as necessary.

49. Fill the radiator with clean coolant and check the cooling system for leaks

50. Bleed the cooling system.

51. Test drive the vehicle and check for any leaks.

3.0L (B308I) Engine

— **WARNING** —
To avoid damage to the valves, DO NOT rotate the camshafts. The crankshaft may only be turned between 0° and 60° BTDC when the camshafts are locked in position with the appropriate locking tool.

1. Disconnect the negative battery cable.

2. Safely raise and support the vehicle. Disconnect the front exhaust pipe from the exhaust manifolds. Disconnect the bracket for the check valves.

3. Remove the lower center air deflector. Drain the engine coolant into a catch pan.

4. Lower the vehicle and remove the top engine covers.

5. Disconnect the cruise control cable, throttle cable. Unclip the throttle control rod from the bracket. Remove the bracket mounting bolt and carefully set the bracket with cables attached to the side.

6. Loosen the clamps on the air intake pipes at the intake manifold and the mass air flow meter. Disconnect the pipes and raise them slightly. Disconnect the vacuum hoses and

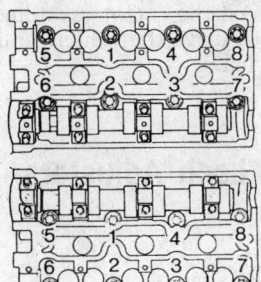

320710

Cylinder head bolt torque sequence — 2.5L and 3.0L engines

electrical connection. Lift the pipes with resonator attached carefully out.

7. Disconnect and label all electrical and vacuum connections on the intake manifold. Remove the intake plenum bolts.

8. Disconnect the IAC connector and fuel pressure regulator hose and remove the wiring harness from under the throttle body.

9. Disconnect the throttle position indicator, ignition coil and TCS connectors. Lift off the intake plenum and plug the intake runners with paper.

10. Disconnect the fuel injector and camshaft position connectors.

11. Disconnect and plug the fuel line connections.

12. Remove the center intake manifold, with fuel rails and set aside.

13. Mark the position of the lower intake manifold. Remove the lower intake manifold and plug the engine intake ports with paper.

14. Unbolt the coolant bridge and carefully bend it aside.

15. Disconnect the spark plug wires and the ignition coil. Bend the ignition coil aside and disconnect the ignition coil bracket.

16. Remove the lifting eye and the heat shield over the exhaust manifold.

17. Unbolt the resonator bracket and secondary air injection pipe from the exhaust manifold.

18. Lift up the power steering reservoir.

19. Disconnect the connection on the torque arm engine mount and remove the torque arm.

20. Remove the power steering line clamp from the torque arm engine mount and remove the engine mount.

21. Disconnect the hose from the coolant expansion tank and remove the upper alternator air intake.

22. Loosen the power steering pump, water pump pulley and 6 outer crankshaft pulley bolts. Remove the drive belt and tensioner. Remove the power steering pump, water pump pulley and timing cover.

NOTE: When removing the crankshaft pulley, remove the 6 outer bolts only, DO NOT remove the center bolt.

23. Remove the crankshaft pulley.

24. Remove the timing belt, tensioner and camshaft sprockets.

25. Remove the valve cover. Make sure the O-rings stay in position and do not fall into the engine.

26. Remove the bearing caps on the exhaust camshaft. Note the position

markings on the bearing caps and loosen the bearing cap bolts in stages of ½ to 1 turn at a time.

NOTE: The bearing caps located where valve tappets are compressed, should be removed last. Also, it is not necessary to remove the intake camshaft for cylinder head removal.

27. Loosen the cylinder head bolts ¼ turn at a time in sequence, then ½ turn at a time. Remove cylinder head bolts and lift cylinder head off the engine.

— **WARNING** —
Be careful when setting down the cylinder head, since the intake camshaft is still installed, the valve stems could be accidentally bent.

28. Once the cylinder head is removed from the engine assembly, the spark plugs, camshafts, tappets and exhaust manifold can be removed.

To install:

29. Make sure the crankshaft is still positioned at 60° BTDC.

30. Make sure all gasket contact surfaces are clean. Note the positioning of the cylinder head gasket and make sure the words OBEN/TOP marked on the gasket are facing up.

31. Install the exhaust manifold and tappets, if removed from the cylinder head.

32. Install the cylinder head onto the engine using NEW cylinder head bolts.

Step torque the cylinder head bolts in sequence as follows:
Step 1: 18.5 ft. lbs. (25 Nm)
Step 2: Tighten bolts 90°
Step 3: Tighten bolts 90°
Step 4: Tighten bolts 90°

NOTE: On the front or right bank cylinder head, the camshaft bearing caps are marked R1–R8 and the camshaft bearing seats in the head are numbered 1–8. On the rear or left bank cylinder head, the camshaft bearing caps are marked L1–L8, and the camshaft bearing seats in the head are numbered 1–8.

33. Thoroughly lubricate the camshafts and install the camshafts with new front camshaft gaskets. Make sure the locating pins are properly positioned. Install the bearing caps in their proper location and position. Tighten the bearing cap bolts in sequence ½ to 1 turn at a time, to a torque of 6 ft. lbs. (8 Nm).

34. Install the timing cover and 2 retaining bolts.

35. Check to make sure that the camshaft locating pins are in the proper position. Check the locating pins, if they are hollow, replace them with solid pins.

NOTE: On the front or right bank cylinder head, the locating pin for the intake camshaft should be pointing downwards, in line with the inboard bolt of the bearing caps. The locating pin for the exhaust camshaft should be pointing upwards, in line with the edge of the camshaft sensor. On the rear or left bank cylinder head, the locating pins of both camshafts should point toward the inboard bolts of the camshaft bearing caps.

36. Install the camshaft sprockets, tensioner and camshaft sprockets.

37. Install crankshaft locking tool KM-800-10 (or equivalent). Install tool 83-93-985 with a cut piece from an old timing belt, to measure belt tension.

38. Snug the center bolts of the adjusting rollers. Turn the lower adjusting roller counterclockwise, until a belt tension of 275–300 Nm is reached. Tighten adjusting roller center bolts to 30 ft. lbs. (40 Nm).

NOTE: This adjustment of the timing belt is just preparation and should not be used as a final check.

39. Adjust the tensioner pulley until the marks are aligned. Tighten the tensioner pulley to 15 ft. lbs. (20 Nm).

40. Remove camshaft locking tool KM-800-1 (or equivalent) on camshaft sprockets 1 & 2. Adjust the upper adjusting roller until sprocket No. 2 moves 1–2mm clockwise. Tighten the upper adjusting roller to 30 ft. lbs. (40 Nm) and remove the upper locking tool.

41. Rotate the engine two complete revolutions to just before 0° TDC and install the locking tool on the crankshaft. Carefully turn the crankshaft until the arm of the locking tool is against the water pump flange and tighten the locking tool. Set tool KM-800-20 (or equivalent) into position. Make sure that the timing marks on the camshaft sprockets are aligned with the marks on the tool and that the edge of the timing belt is flush with the edge of the camshaft sprockets.

NOTE: Also check that the alignment marks on the tensioner pulley are still aligned.

42. Install the crankcase ventilation housing and set the oxygen sensor connector back in it's holder. Make sure the valve cover O-rings are still in position and clean. Lubricate the O-rings with soapy water, apply 81-52-381 (or equivalent) sealer at the corners of the large end bearing caps and install the valve cover.

43. Install the timing cover.

44. Install the crankshaft pulley and torque bolts to 15 ft. lbs. (20 Nm).

45. Install the water pump pulley, power steering pump, drive belt tensioner and drive belt.

46. Install the upper alternator air intake.

47. Install the torque arm engine mount and attach the power steering line clamp.

48. Install the torque arm and bolt the connection to the torque arm engine mount.

49. Connect the upper hose to the coolant expansion tank and set the power steering reservoir back into position.

50. Install the lifting eye and secondary air injection pipe to the exhaust manifold. Install the exhaust manifold heat shield.

51. Install the resonator bracket.

52. Install the ignition coil bracket and ignition coil.

53. Install the spark plugs (if removed) and torque to 18.5 ft. lbs. (25 Nm). Install the spark plug wires.

54. Carefully bend the coolant bridge back into position. Apply 74-96-284 (or equivalent) to the mounting bolts. Install the mounting bolts and torque to 22 ft. lbs. (30 Nm).

55. Apply an acid-free petroleum jelly to the lower intake manifold and set the lower intake manifold onto the engine, noting the position marks made at disassembly. Apply thread locker 74-96-268 (or equivalent) to the lower intake manifold bolts. Install the lower intake manifold bolts and torque to 15 ft. lbs. (20 Nm).

56. Install the center intake manifold with fuel rail. Apply thread locker 74-96-268 (or equivalent) to the lower intake manifold bolts. Install the center intake manifold bolts and torque to 15 ft. lbs. (20 Nm).

57. Connect the fuel lines, crankshaft position sensor and fuel injector connectors.

58. Set the intake plenum into position and connect the TCS throttle body.

59. Tighten the intake plenum bolts to 15 ft. lbs. (20 Nm).

60. Connect the throttle position indicator, ignition coil and TCS connectors.

61. Route the wiring harness under the throttle body and connect the fuel pressure regulator hose and IAC connector.

62. Connect all labeled electrical and vacuum connections on the intake manifold.

63. Set the air intake pipes with resonator attached, carefully into position. Connect the vacuum hoses and electrical connection. Lower the air intake pipes and connect them to the mass air flow meter and intake manifold. Install and tighten the clamps on the air intake pipes at the intake manifold and the mass air flow meter.

64. Carefully set the throttle control bracket with cables attached into position and install the bracket mounting bolt. Clip the throttle control rod to the bracket. Connect the throttle cable and cruise control cable. Adjust the kick-down cable and the throttle cable.

65. Install the upper engine covers and safely raise and support the vehicle.

66. Make sure the radiator drain plug is tightened and install the lower center air deflector.

67. Bolt on the bracket for the check valves and connect the front exhaust pipe to the exhaust manifolds.

68. Connect the front exhaust pipe to the exhaust manifolds and lower the vehicle.

69. Connect the negative battery cable.

70. Check all other oil and fluids for condition and proper level. Replace fluids and filters or top off as necessary.

71. Fill the radiator with clean coolant and check the cooling system for leaks

72. Bleed the cooling system.

73. Test drive the vehicle and check for any leaks.

Lash Adjusters

BLEEDING

The hydraulic cam followers, which are used in all 1993–97 Saab engines, do not require bleeding.

Valve Lash

ADJUSTMENT

The hydraulic cam followers, which are used in all 1993–97 Saab engines, do not require adjusting. The cam followers keep the valve clearance within 0.7382–0.8189 in. (18.75–20.8mm). However, if the cam followers are making excessive noise or are diagnosed to be defective, perform the following procedure:

1. Disconnect the negative battery cable.

2. If a cam follower is noisy, it can be found by removing the valve cover and using a screw driver, gently pushing down on each cam follower until the defective follower(s) is found by exhibiting a spongy feeling.

3. Replace the defective cam follower(s); first removing the camshaft(s).

4. Reinstall the camshaft(s) and the valve cover.

5. Reconnect the negative battery cable.

Intake Manifold

REMOVAL AND INSTALLATION

2.0L (B202L) and 2.1L (B212I) Engines

— CAUTION —

Fuel injected systems remain under pressure, even after the engine has been turned OFF. The fuel system pressure must be relieved before disconnecting any fuel lines. Failure to do so may result in personal injury.

1. Disconnect the negative battery cable.

2. Label and disconnect the necessary hoses and wiring.

3. Remove the turbocharger pressure pipe, the lubricating oil pressure pipe and the return oil pipe.

4. Remove the intake manifold retaining bolts. Remove the intake manifold along with the injection manifold, injectors and the AIC regulator.

To install:

5. Thoroughly clean all gasket mating surfaces.

6. Be sure the proper gasket is used. A coolant leak could occur if the wrong one is used.

7. Install intake manifold and tighten bolts to 16 ft. lbs. (22 Nm).

8. Install the turbocharger pressure pipe, the lubricating oil pressure pipe and the return oil pipe.

9. Connect all hoses and wiring to their proper locations.

10. Connect the negative battery cable.

11. Start the engine and check for leaks.

2.0L (B204L) and All 2.3L Engines

— CAUTION —

Fuel injected systems remain under pressure, even after the engine has been turned OFF. The fuel system pressure must be relieved before disconnecting any fuel lines. Failure to do so may result in fire or personal injury.

1. Disconnect the negative battery cable. Drain the engine coolant.

2. If vehicle is equipped with a turbocharger, disconnect and remove the rubber elbow running between the throttle housing and the turbocharger.

3. Unplug the throttle position sensor. Disconnect the hoses at the throttle housing, remove the three nuts and lift out the housing.

4. Unbolt the oil filler pipe bracket at the manifold and carefully position it out of the way.

5. Tag and disconnect all hoses and lines attached to the manifold.

6. Remove the AIC valve. Disconnect the fuel line from the pressure regulator.

7. Loosen the banjo fitting connecting the fuel line to the fuel rail. Cut the plastic tie and move the fuel line and pulsator out of the way. Be careful not to lose the seals.

8. Unplug each fuel injector electrical lead.

9. Disconnect the temperature sensor and the ground wires at the manifold.

10. Loosen the two screws on the cable clip underneath the manifold and move the harness assembly out of the way.

11. Disconnect the EGR pipe and all connectors.

12. Remove the eight mounting bolts and lift off the intake manifold.

To install:

13. Scrape off any excess gasket material, install a new gasket and install the manifold. Tighten the bolts in a crisscross pattern to 16 ft. lbs. (22 Nm).

14. Reposition the wire bundle and reconnect the EGR pipe.

15. Connect the ground wires and the temperature sensor.

16. Reconnect all injector leads.

17. Connect the fuel line to the pressure regulator. Connect the fuel line/pulsator to the fuel rail and secure it with a plastic tie.

18. Connect the oil filler pipe bracket to the manifold and install the AIC valve.

19. Install the throttle housing and all its attachments.

20. If vehicle is equipped with a turbocharger connect the rubber elbow between the turbocharger and the intake manifold.

21. Fill the cooling system.

22. Connect the negative battery cable.

23. Start the engine and check for leaks.

2.5L (B258I) and 3.0L (B308I) Engines

1. Remove the engine covers.

2. Disconnect the cruise control cable, detach the throttle cable and unclip the control rod and move it to aside.

3. Remove the hose clamps and intake tube from the mass air flow sensor assembly to the throttle plate housing. Disconnect the vacuum hoses and electrical connectors and remove complete assembly.

4. Disconnect all hoses and wiring from the upper intake manifold.

5. Remove the upper intake manifold bolts and remove the upper manifold.

6. Unplug the idle air control valve and the fuel pressure regulator hose. Remove harness from under the throttle plate housing.

7. Unplug the injector electrical harness connector and camshaft position sensor connector.

8. Disconnect and plug the fuel line connectors.

9. Remove the bolts from the lower intake manifold and lift off engine.

To install:

10. Install the lower intake manifold with new gaskets and tighten retaining bolts to 15 ft. lbs. (20 Nm).

11. Connect the fuel lines, injector wiring and the camshaft position sensor.

12. Install the upper intake manifold with new gaskets and connect the throttle plate housing. Tighten the manifold bolts to 15 ft. lbs. (20 Nm).

13. Connect all wiring and hoses.

14. Install the idle air control valve and the hose to the fuel pressure regulator.

15. Install the mass air flow sensor assembly and hoses.

16. Connect the throttle control rod and associated cruise control cable.

17. Fit engine covers.

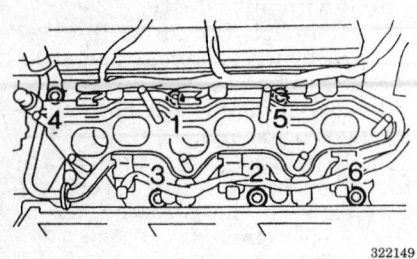

322149

Intake manifold bolt torque sequence — 2.5L (B258I) and 3.0L (B308I) engines

Exhaust Manifold

REMOVAL AND INSTALLATION

2.0L (B202L) Engine

1. Remove the battery and the heat shield.
2. Remove the heat shield from the turbocharger unit.
3. Remove the suction and pressure connections from the compressor and loosen the preheating hose.
4. Remove the exhaust elbow between the exhaust manifold and the compressor.
5. Disconnect the oil supply line and the oil return line at the turbocharger unit.
6. Remove the nuts, distance pieces and washers from the exhaust manifold.
7. Slide the exhaust manifold clear of the studs and remove it complete with the turbocharger unit.
8. Separate the exhaust manifold from the turbocharger unit.
   To install:
9. Position the turbocharger unit on the exhaust manifold and install the retaining bolts.
10. Slide the exhaust manifold, complete with the turbocharger unit and a new gasket onto the studs and secure it.
11. Install the washers, distance pieces and nuts and tighten the nuts to 19 ft. lbs. (25 Nm).
12. Connect the oil return line with a new gasket. Fill the lubricating inflow of the turbocharger unit with engine oil, then connect the oil supply line with a new gasket.
13. Connect the exhaust elbow and the exhaust manifold.
14. Install the suction and pressure connections to the compressor and tighten the preheating hose.
15. Crank the engine for about 30 seconds with terminal 15 on the ignition coil disconnected. This will prime the lubricating system of the turbocharger before the engine is started.
16. Install the battery and the heat shield.
17. Start the engine and check for leaks and proper engine operation.

2.1L (B212I) Engine

1. Disconnect the negative battery cable.
2. Remove the transmission oil filler tube, if equipped.
3. Move aside the preheater hose and remove the preheater cover.
4. Label and disconnect all necessary hoses and wiring.
5. Unbolt the exhaust pipe at the connecting flange.
6. The manifold is a 2 piece unit; remove the exhaust manifold nuts and distance pieces and remove the inner unit first. Remove the upper stud from the outboard flange of the outer exhaust manifold section. Use a second nut to lock the securing nut when unscrewing the stud.
7. Remove the remaining exhaust manifold nuts and distance pieces and then free the exhaust manifold from the studs by raising the leading end to clear the engine mountings.
   To install:
8. Install the outer exhaust manifold first, then the inner, with new gaskets. Apply an anti-seize compound to the manifold studs and tighten the nuts to 13 ft. lbs. (18 Nm).
9. Connect the exhaust pipe at the connecting flange.
10. Connect all wires and hoses previously disconnected. Check that none of the wires or hoses make contact with the exhaust manifold.
11. Connect the negative battery cable.
12. Start the engine and check for leaks.

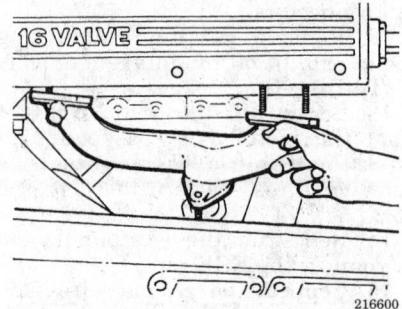

216600

When removing the exhaust manifold, raise the front end to clear the engine mounting — 2.1L (B212I) engine

2.0L (B204L) Engine

1. Drain coolant from cooling system.
2. Raise and safely support the vehicle and remove the hose to the charge air cooler.
3. Remove the front exhaust pipe from the engine and be careful not to damage the oxygen sensor electrical wiring harness.
4. Break the seal and remove locking ring on the wastegate.
5. Remove the oil return pipe from the turbocharger.
6. Disconnect the oil pipe from the oil filter housing.
7. Remove the hose to the boost control valve.
8. Disconnect the intake pipe and the bypass hose, take off the nuts to the wastegate diaphragm and remove the intake pipe together as one unit.
9. Remove the turbocharger coolant pipe with the oil pipe.
10. Remove the turbocharger.
11. Using a ³⁄₈ inch ratchet extension, relieve the pressure on the drive belt tensioner. Remove the drive belt.
12. Remove the power steering pump bracket and pump without disconnecting the hoses.
13. Remove the exhaust manifold studs, take off exhaust manifold and seal.
    To install:
14. Clean mounting surface and install exhaust manifold seal and the manifold. Torque the nuts to 18 ft. lbs. (25 Nm).
15. Fit power steering pump bracket with pump.
16. Install drive belt and release the pressure on the drive belt tensioner.
17. Install turbocharger tighten retaining nuts to 16 ft. lbs. (22 Nm).
18. Install turbocharger coolant and oil pipes.
19. Install the intake pipe for the turbocharger and mount the wastegate with the boost pressure control valve and pipes.
20. Fit the intake hose with the bypass hose.
21. Raise and safely supported the vehicle.
22. Install the locking ring for the wastegate.
23. Install the front pipe to the turbocharger.
24. Install the hose connecting the charge air cooler with the turbocharger.
25. Fill the coolant system with coolant and check oil level and quality.
26. Reseal the boost pressure control rod.

27. Test drive and check the boost pressure.

2.3L Engines

900 Series

1. Unplug the oxygen sensor connector.
2. Safely raise and support the vehicle.
3. Disconnect the front pipe from exhaust manifold and lower the vehicle.
4. Using a ⅜ in. ratchet extension, relieve the drive belt tensioner and remove the belt.
5. Remove the power steering pump pulley and disconnect the pump bracket. Move pump to one side without disconnecting the hoses.
6. Remove the middle part of the exhaust manifold first, then remove the outer tubes.
To install:
7. Fit the outer tubes first, then mount middle part of exhaust manifold. Tighten the nuts to 18 ft. lbs. (25 Nm).
8. Install the power steering pump bracket and install pump and pulley.
9. Fit drive belt onto the pulleys and release the belt tensioner.
10. Raise the vehicle to connect front pipe to the exhaust manifold.
11. Connect the oxygen sensor.

9000 Series

1. Disconnect the negative battery cable.
2. Raise and safely support the vehicle. Remove the center air deflector and properly drain the cooling system.
3. Disconnect the top pipe coupling on the air cooled oil cooler and disconnect the clips securing the pipe to the radiator.
4. Remove the solenoid valve from its mounting on the radiator and disconnect the electrical leads.
5. Disconnect the electrical leads at the radiator fan. Unbolt and remove the fan.
6. Remove the oxygen sensor cable clamps.
7. Disconnect the exhaust pipe from the turbocharger and bend it down slightly.
8. Unplug the electrical connectors for the air mass meter. Disconnect the toggle fasteners securing the air mass meter to the air cleaner cover and pull the rubber socket connector off of the turbocharger unit.
9. Disconnect the turbocharger pressure pipe from the turbocharger compressor.

10. Remove the oil pipe to the turbocharger unit and remove the clip securing the oil pipe to the cylinder head. Disconnect the oil pipe banjo coupling from the block and remove the clip on the intake manifold.
11. Remove the coolant pipes from the turbocharger unit.
12. Remove the steady bar bracket between the engine block and the turbocharger compressor. Remove the securing bolts and loosen the oil return lines. Cap the opening to prevent washers or nuts from the exhaust manifold dropping inside during the removal.
13. Remove the turbocharger unit from the exhaust manifold.
14. Release the tension on the air conditioning compressor belt by pulling hard upwards on the belt and installing 83–94–488 belt tensioner locking clamp tool or equivalent, on the tensioner and remove the belt.
15. Remove the air conditioning compressor mounting bolts. Insert a sheet of metal to protect the oil cooler and lift the air conditioning compressor towards the expansion tank.
16. Remove the nuts securing the exhaust manifold to the cylinder head.
17. Lift the exhaust manifold from the cylinder head complete with the gasket.
To install:
18. Clean the mating surfaces and install a new gasket over the studs for the exhaust manifold and install the manifold. Tighten the nuts to 19 ft. lbs. (25 Nm).
19. Install the air conditioning compressor and belt, remove the locking tool and check for proper belt tension.
20. Position the turbocharger unit to the exhaust manifold. Using new locknuts, tighten the retaining nuts to 16 ft. lbs. (22 Nm).
21. Fill the turbocharger interchamber passage with engine oil.

─── **WARNING** ───
It is very important that there is oil in the turbocharger when the engine is started to avoid damage to the unit.

22. Install the clip holding the turbocharger oil supply pipe to the intake manifold. Connect and tighten the banjo coupling to the engine block. Make sure the copper washers are in good condition. Secure the pipe to the turbocharger unit.
23. Install the return oil pipe and the steady bar bracket between the turbocharger unit and the crankcase.

Connect the rubber hangers for the front exhaust hanger.
24. Lubricate the 3 studs on the turbocharger with Molykote 1000® Connect the exhaust pipe to the turbocharger compressor with new locknuts and tighten the nuts to 19 ft. lbs. (25 Nm).
25. Install the coolant pipes to the turbocharger unit.
26. Install the turbocharger pressure pipe to the turbocharger compressor and assemble the air mass meter and rubber socket connector between the air cleaner body and the inlet side of the turbocharger compressor.
27. Assemble the fan and solenoid valve, securing the electrical leads into their clips. Connect the return hose to the solenoid valve. Install the air conditioning compressor.
28. Reconnect the oil pipe to the oil cooler and secure the pipe clip to the radiator.
29. Tighten the radiator drain plug and install the center air deflector.
30. Properly fill the cooling system and check the oil level and quality.
31. Connect the negative battery cable.
32. Start the engine and check for leaks and proper turbocharger system operation.

2.5L (B258I) and 3.0L (B308I) Engines

1. Disconnect oxygen sensor cable.
2. Loosen top nuts on rear exhaust manifold.
3. Raise and safely support vehicle.
4. Disconnect catalytic converter and remove nuts on front pipe and remove pipe from exhaust manifold.
5. Remove lower nuts on rear exhaust manifold and remove.
6. Lower vehicle.
7. Remove dipstick holder.
8. Remove nuts on front exhaust manifold and remove.
To install:
9. Install front exhaust manifold and tighten nuts to 15 ft. lbs. (20 Nm).
10. Install the dipstick holder.
11. Raise and safely support vehicle.
12. Install rear exhaust manifold and tighten lower nuts to 15 ft. lbs. (20 Nm).
13. Install front pipe and tighten nuts to 30 ft. lbs. (40 Nm).
14. Install catalytic converter onto the front pipe.
15. Lower car and tighten rear exhaust manifold top nuts to 15 ft. lbs. (20 Nm).
16. Connect oxygen sensor cable.

Turbocharger

REMOVAL AND INSTALLATION

2.0L (B202L) Engine

1. Disconnect the battery cables from the battery, negative cable first.
2. Remove the battery and the heat shield.
3. Remove the heat shield from the turbocharger unit.
4. Remove the suction and pressure connections from the compressor and loosen the preheating hose.
5. Remove the exhaust elbow between the exhaust manifold and the compressor.
6. Disconnect the oil supply line and the oil return line at the turbocharger unit.
7. Remove the retaining bolts from the turbocharger flange on the exhaust manifold and remove the turbocharger.

To install:

8. Position the turbocharger unit and install the retaining bolts.
9. Connect the oil return line with a new gasket. Fill the lubricating inflow of the turbocharger unit with engine oil, then connect the oil supply line with a new gasket.
10. Connect the exhaust elbow and the exhaust manifold.
11. Install the suction and pressure connections to the compressor and tighten the preheating hose.
12. Install the battery and the heat shield.
13. Crank the engine for about 30 seconds with terminal 15 on the ignition coil disconnected. This will prime the lubricating system of the turbocharger before the engine is started.
14. Start the engine and check for leaks and proper engine operation.

2.0L (B204L) Engine

1. Drain coolant from cooling system.
2. Raise and safely support the vehicle. Remove hose to the charge air cooler.
3. Remove the front exhaust pipe from the engine and be careful not to damage the oxygen sensor electrical wiring harness.
4. Break seal and remove locking ring on the wastegate.
5. Remove the oil return pipe from the turbocharger.
6. Disconnect the oil pipe from the oil filter housing.
7. Remove the hose to the boost control valve.

8. Disconnect the intake pipe and the bypass hose, take off the nuts to the wastegate diaphragm and remove the intake pipe together as one unit.
9. Remove the turbocharger coolant pipe with the oil pipe.
10. Remove the turbocharger.

To install:

11. Install the turbocharger with a new gasket and tighten the retaining nuts to 16 ft. lbs. (22 Nm).
12. Install turbocharger coolant and oil pipes.
13. Install the intake pipe for the turbocharger and mount the wastegate with the boost pressure control valve and pipes.
14. Fit the intake hose with the bypass hose.
15. Raise and safely support the vehicle.
16. Install the locking ring for the wastegate.
17. Install the front pipe to the turbocharger.
18. Install the hose connecting the charge air cooler with the turbocharger.
19. Fill the cooling system and check oil level.
20. Reseal the boost pressure control rod.
21. Test drive and check the boost pressure.

All 2.3L Engines

1. Disconnect the negative battery cable.
2. Raise and safely support the vehicle. Remove the center air deflector and properly drain the cooling system.
3. Disconnect the top pipe coupling on the air cooled oil cooler and disconnect the clips securing the pipe to the radiator.
4. Remove the solenoid valve from its mounting on the radiator and disconnect the electrical leads.
5. Disconnect the electrical leads at the radiator fan. Unbolt and remove the fan.
6. Remove the oxygen sensor cable clamps.
7. Disconnect the exhaust pipe from the turbocharger and bend it down slightly.
8. Unplug the electrical connectors for the air mass meter. Disconnect the toggle fasteners securing the air mass meter to the air cleaner cover and pull the rubber socket connector off the turbocharger unit.
9. Disconnect the turbocharger pressure pipe from the turbocharger compressor.
10. Remove the oil pipe to the turbocharger unit and remove the clip

securing the oil pipe to the cylinder head. Disconnect the oil pipe banjo coupling from the block and remove the clip on the intake manifold.
11. Remove the coolant pipes from the turbocharger unit.
12. Remove the steady bar bracket between the engine block and the turbocharger compressor. Remove the securing bolts and loosen the oil return lines. Cap the opening to prevent washers or nuts from the exhaust manifold dropping inside during the removal.
13. Remove the turbocharger unit from the exhaust manifold.

To install:

14. Position the turbocharger unit to the exhaust manifold. Using new locknuts, tighten the retaining nuts to 16 ft. lbs. (22 Nm).
15. Fill the turbocharger interchamber passage with engine oil.

————— **WARNING** —————
It is very important that there is oil in the turbocharger when the engine is started to avoid damage to the unit.

16. Install the clip holding the turbocharger oil supply pipe to the intake manifold. Connect and tighten the banjo coupling to the engine block. Make sure the copper washers are in good condition. Secure the pipe to the turbocharger unit.
17. Install the return oil pipe and the steady bar bracket between the turbocharger unit and the crankcase. Connect the rubber hangers for the front exhaust hanger.
18. Lubricate the 3 studs on the turbocharger with Molykote 1000® and connect the exhaust pipe to the turbocharger compressor with new locknuts. Tighten the nuts to 19 ft. lbs. (25 Nm).
19. Install the coolant pipes to the turbocharger unit.
20. Install the turbocharger pressure pipe to the turbocharger compressor and assemble the air mass meter and rubber socket connector between the air cleaner body and the inlet side of the turbocharger compressor.
21. Assemble the fan and solenoid valve, securing the electrical leads into their clips. Connect the return hose to the solenoid valve. Install the air conditioning compressor.
22. Reconnect the oil pipe to the oil cooler and secure the pipe clip to the radiator.
23. Tighten the radiator drain plug and install the center air deflector.
24. Properly fill the cooling system and check the oil level and quality.

25. Connect the negative battery cable.
26. Start the engine and check for leaks and proper turbocharger system operation.

Front Crankshaft Seal

REMOVAL AND INSTALLATION

All 2.0L, 2.1L and 2.3L Engines

1. Disconnect the negative battery cable.
2. Raise and safely support the vehicle.

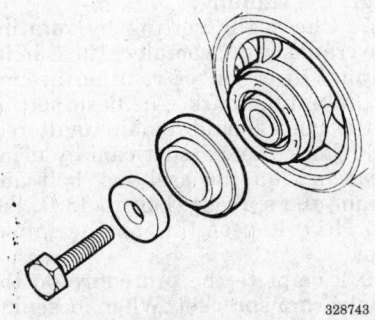

328743

Installing front crankshaft seal using installation tool 83–94–876 — all 2.0L, 2.1L and 2.3L engines

3. Remove the right front wheel and remove the inner front fender panel.
4. Loosen and remove the drive belts.
5. Remove the retaining bolt for the crankshaft pulley.
6. Remove the crankshaft pulley.
7. Using a small prybar, carefully remove the oil seal without marring the crankshaft stub end.

To install:
8. Install a new, oiled seal, using the seal installer tool 83–94–876, or the equivalent.
9. Install the pulley and tighten the retaining bolt to 140 ft. lbs. (190 Nm).
10. Install the drive belts.
11. Install the forward section of the inner fender panel.
12. Replace the front wheel and lower the vehicle.
13. Start the engine and check for leaks.

2.5L (B258I) and 3.0L (B308I) Engines

1. Disconnect the negative battery cable.
2. Remove the air cleaner complete with hoses.
3. Using a 15mm wrench turn the serpentine drive belt tensioner towards the front of the vehicle, remove

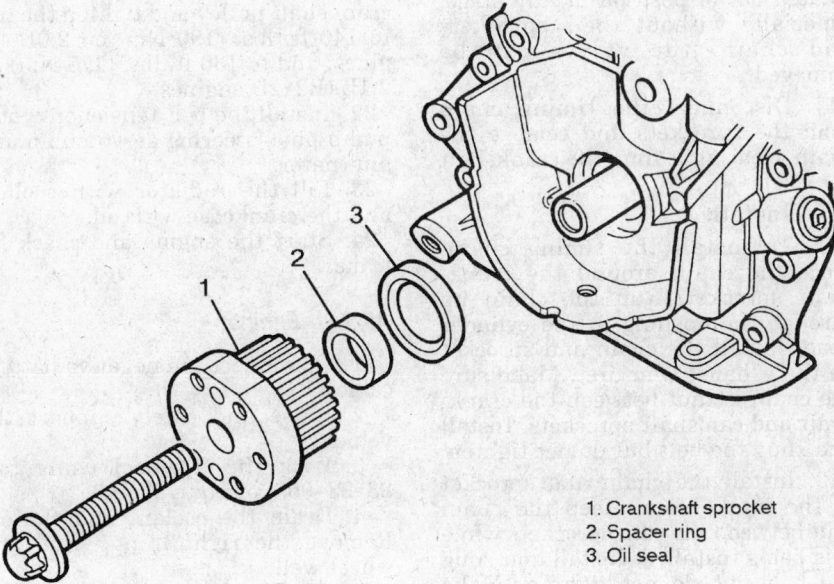

1. Crankshaft sprocket
2. Spacer ring
3. Oil seal

295860

Remove the crankshaft sprocket for access to the front seal assembly — 2.5L and 3.0L engines

the belt from the water pump pulley and slowly release the belt tensioner.
4. Remove the belt tensioner and the power steering pump pulley.
5. Remove the water pump pulley. To do this, carefully pry the engine towards the left of the vehicle, using the engine bracket as a fulcrum.
6. Remove the timing cover.
7. Remove the right front wheel and the cover in the wheel well.
8. Remove the 6 bolts holding the crankshaft pulley and remove the pulley. Do not remove the center bolt.
9. Remove the tensioning roller and the two adjusting rollers and remove the timing belt.
10. Remove the crankshaft sprocket and spacer ring.
11. Using a small prybar, carefully remove the oil seal without marring the crankshaft stub end.
To install:
12. Lubricate and install a new oil seal using an appropriate seal installer.
13. Install the timing belt and belt tensioner.
14. Install the crankshaft pulley and tighten the retaining bolts to 15 ft. lbs. (20 Nm).
15. Install the timing belt cover tighten the bolts to 6 ft. lbs. (8 Nm).
16. Install the water pump pulley by prying the engine to the left, using the engine bracket as a fulcrum.
17. Install the power steering pump pulley and tighten the retaining bolts to 6 ft. lbs. (8 Nm).
18. Using a 15mm wrench turn the serpentine drive belt tensioner towards the front of the vehicle, position the belt and slowly release the belt tensioner.
19. Install the air cleaner complete with hoses.
20. Connect the negative battery cable.
21. Start the engine and check the ignition timing.

Timing Chain, Sprockets and Front Cover

REMOVAL AND INSTALLATION

All 2.0L and 2.1L (B212I) Engines

1. Disconnect the negative battery cable.
2. Drain the engine oil and the coolant.
3. Remove the bracket for the steering servo pump, complete with the pump and alternator.
4. Remove the belt tensioner and the water pump pipe.

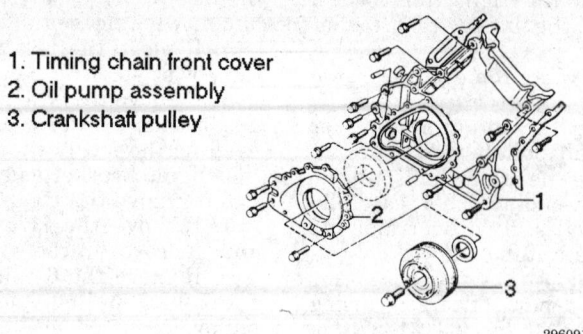

1. Timing chain front cover
2. Oil pump assembly
3. Crankshaft pulley

296091

Timing chain front cover assembly exploded view — all 2.0L engines

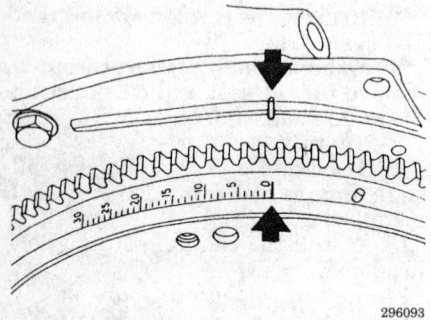

296093

Crankshaft timing mark — all 2.0L and 2.1L (B212l) engines

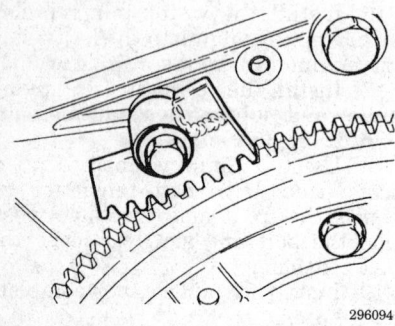

296094

Secure the engine in position using flywheel locking tool 83-92-987 — all 2.0L and 2.1L (B212l) engines

5. Secure the flywheel with a flywheel locking segment, tool 83–92–987, and loosen the crankshaft pulley bolt.

6. Using a suitable puller remove the crankshaft pulley.

7. Remove the oil pipes and the water pump pulley.

8. Remove the bolts and timing chain cover.

9. The cam chain and crankshaft timing sprocket should now both be visible. From above, release the timing chain tensioner by pressing the pivoting guide firmly against it. Remove the chain tensioner.

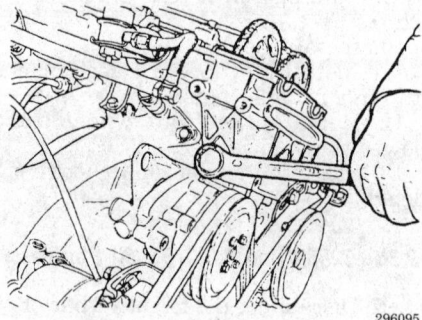

296095

Timing chain tensioner location — all 2.0L and 2.1L (B212l) engines

10. Using a wrench to hold the cast hex bolt on the camshafts, remove the center bolts securing the camshaft sprockets. Throughout this procedure, keep the camshafts in their basic correct setting. If they are rotated out of position at any stage, especially without their sprockets and chain, the valves can be damaged.

11. Disconnect the timing chain from the sprockets and remove the chain, clearing it from the crankshaft sprockets.

To install:

12. To install the timing chain, place the chain around the crankshaft sprocket. Run the chain up through the opening in the cylinder head. Install the chain and sprocket on the exhaust cam first. Make sure the chain is taut between the crankshaft and camshaft sprockets. Install and snug the bolt but do not tighten.

13. Install the chain and sprocket to the intake cam. Keep the chain taut between the cam sprockets while it is being installed. Install and snug the bolt but do not tighten. Make sure the chain is seated in the guide tensioner grooves.

14. Tension the chain tensioner by fully depressing the piston and then rotating it to the locked position.

15. Install the chain tensioner with the piston under tension. Make sure the copper gasket is in good condition and that the sealing surface is clean and free from burrs. Tighten the tensioner to 47 ft. lbs. (63 Nm).

16. Trigger the chain tensioner by pressing the pivoting chain guide against it, thereafter, press the pivoting guide against the chain to give the chain its basic tension. Check that the chain tensioner maintains tension on the chain when the pressure on the chain guide is released and that the basic setting stop for the tensioner holds the chain guide tight against the chain. A limited amount of play will be present until the hydraulic pressure takes over once the engine is running.

17. Check the setting by rotating the crankshaft 2 complete turns in its normal direction of rotation around to the timing mark. The basic setting of the cams should remain unaltered.

18. Lock the exhaust cam by using a wrench on the cast hex bolt and torque the sprocket bolt to 48 ft. lbs. (65 Nm). Repeat this on the intake cam.

19. Complete the procedure on the intake cam sprocket. When loosening or torquing the sprocket center bolts, hold the cam still using a wrench installed over the flats on the camshaft.

20. Install the timing cover and tighten the bolts to 15 ft. lbs. (20 Nm).

21. Install the oil pump and pipes and water pump pulley. Install the crankshaft pulley and tighten the nut to 140 ft. lbs. (190 Nm) on 2.0L engines and to 130 ft. lbs. (175 Nm) on 2.1L (B212l) engines.

22. Install the belt tensioner, water pump pipe, steering servo pump and alternator.

23. Fill the radiator with coolant and the crankcase with oil.

24. Start the engine and check for leaks.

All 2.3L Engines

1. Disconnect the negative battery cable.

2. Raise and safely support the vehicle.

3. Lock the flywheel using tool 83–93–993, or equivalent.

4. Drain the coolant and the oil. Remove the right front wheel and wheel well.

5. Remove the serpentine belt and belt tensioner. Remove the tie bar between the wheel arch and the subframe.

6. Remove the steering servo pump, pump bracket and the alterna-

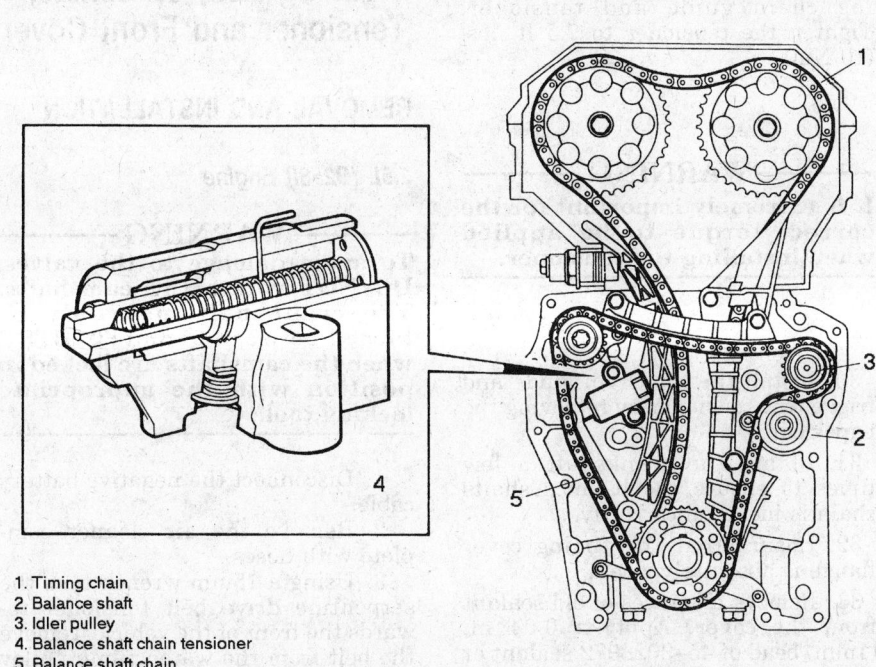

1. Timing chain
2. Balance shaft
3. Idler pulley
4. Balance shaft chain tensioner
5. Balance shaft chain

296077

Timing and balance shaft chain assemblies — all 2.3L engines

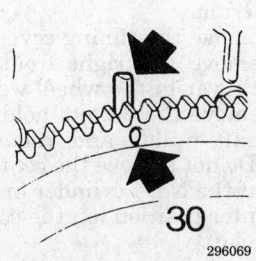

296069

Flywheel timing mark — all 2.3L engines

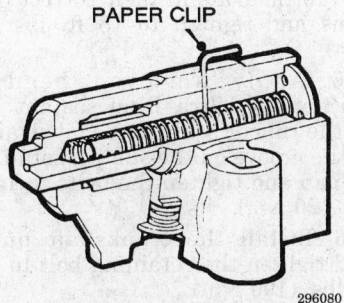

PAPER CLIP

296080

Preparing the balance shafts tensioner for installation — all 2.3L engine

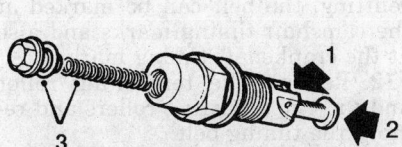

1. Push down on the catch
2. Push in on tensioner arm
3. Tensioner plug, push rod and spring

296078

Timing chain tensioner components — all 2.3L engines

tor. Remove the top engine mounting bracket. Remove the torque arm.

7. Remove the top engine mounting bracket. Note the location of the bolts as the bolts are of different lengths.

8. Remove the top belt tensioner bracket, Without disconnecting any hoses, remove the air conditioner compressor and compressor bracket. Use a suitable rigid cover to prevent the oil cooler from being damaged.

9. Disconnect the coolant hoses and remove the water pump. Remove the crankshaft pulley and position the crankshaft sensor out of the way.

10. Move the coolant pipe aside and remove the oil pan. Remove the timing cover securing bolts. Note the locations of all bolts as they are of different lengths. Remove the bolts securing the timing cover to the cylinder head.

11. Carefully tap the cover off of the guide pin and remove the timing cover.

12. Remove the lid on the valve cover and remove the ignition wires. Remove the valve cover.

NOTE: To remove or refit the timing chain, the camshafts and crankshaft must be lined up with their respective timing marks (No. 1 cylinder at top dead center).

13. Remove the top chain guide and chain tensioner for the balance shaft chain.

14. Remove the idler sprocket and balance shaft chain.

— **WARNING** —
Throughout this procedure, keep the camshafts in their basic correct setting. If they are rotated out of position at any stage, especially without their sprockets and chain, the valves can be damaged.

15. The camshaft chain and crankshaft timing sprocket should both be visible. From above, release the timing chain tensioner by pressing the pivoting guide firmly against it. Remove the chain tensioner.

16. Using a wrench installed over the flats on the exhaust camshaft, hold the camshaft and remove the center bolt securing the camshaft sprocket. Repeat this step for the intake camshaft.

17. Disconnect the timing chain from the sprockets and remove the chain, clearing it from the crankshaft sprockets.

To install:

18. To install the timing chain, place the chain around the crankshaft sprocket. Run the chain up through the opening in the cylinder head. Install the chain and sprocket on the exhaust cam first. Make sure the chain is taut between the crankshaft and camshaft sprockets. Snug the bolt but do not tighten.

19. Install the chain and sprocket to the intake cam. Keep the chain taut between the cam sprockets while it is being installed. Snug the bolt but do not tighten. Make sure the chain is seated in the guide tensioner grooves.

20. Tension the chain tensioner by fully depressing the piston and then rotating it to the locked position.

21. Install the chain tensioner with the piston under tension. Make sure the copper gasket is in good condition and that the sealing surface is clean and free from burrs.

22. Trigger the chain tensioner by pressing the pivoting chain guide against it, thereafter, press the pivoting guide against the chain to give the chain its basic tension. Check that the chain tensioner maintains tension on the chain when the pressure on the chain guide is released and that the basic setting stop for the tensioner holds the chain guide tight against the chain. A limited amount of play will be present until the hydraulic pressure takes over once the engine is running.

23. Check the setting by rotating the crankshaft 2 complete turns in its normal direction of rotation around to the timing mark. The basic setting of the cams should remain unaltered.

24. Lock the exhaust cam by using a wrench installed over the flats on the camshaft and torque the sprocket bolt to 48 ft. lbs. (65 Nm). Repeat this step on the intake cam.

NOTE: The accuracy of the timing chain adjustment will depend on the condition of the chain.

25. Install the balance shaft chain and sprocket on the crankshaft.

26. Install the oil pump drive dog.

27. Install the balance shaft chain and idler wheel sprocket, ensuring that the aligning marks on the bearing housing and sprocket are in line. When installing, leave some slack in the chain in line with the tensioner, and keep the chain reasonably taut by means of the top chain guide.

NOTE: There is an alternate way of installing the balance shaft chain. Install the top chain guide first and then adjust the run of the chain around the sprockets. Adjusting the chain is easier this way, although it will be more awkward to install the idler wheel sprocket.

28. Cock the balance shaft chain tensioner and insert a paper clip through the hole in the cylinder to prevent the tensioner being triggered. Before installing the tensioner, make sure that the plunger is turned to the position in which the spring acts fully on it.

29. Install the balance shafts pivoting chain guide and tensioner. Tighten the tensioner to 7.5 ft. lbs. (10 Nm).

——— WARNING ———
It is extremely important for the correct torque to be applied when installing the tensioner.

30. Install the top chain guide and trigger the tensioner by removing the paper clip.

31. Rotate the crankshaft a few times to ensure the balance shafts chain is installed correctly.

32. Ensure that the timing cover flange is absolutely clean.

33. Remove all traces of old sealant from the cover. Apply a 0.04 in. (1mm) bead of 45–3028972 sealant or equivalent to the flanges of the cover. Use sealant sparingly as excess sealant can get into the oil ways and do serious damage to the engine.

34. Install the timing cover taking care not to damage the head gasket. Install the bolts in their correct positions and tighten to 15 ft. lbs. (20 Nm).

35. Apply an even bead of Permatex® Ultra Blue sealant and fit the rubber seal for the oil strainer in the groove on the sump. Install the oil pan and tighten the bolts to 15 ft. lbs. (20 Nm).

36. Install the crankshaft pulley and tighten the retaining bolt to 140 ft. lbs. (190 Nm).

37. Install the coolant pipe and secure the crankshaft sensor.

38. Install the engine mountings, water pump and cooling hoses, air conditioner compressor, steering servo pump, pump bracket and the alternator. Install the top engine mounting bracket and torque arm.

39. Install the serpentine belt and belt tensioner. Install the tie bar between the wheel arch and the subframe.

40. Install the wheel well and wheel.

41. Fill the radiator with coolant and the engine with oil.

42. Connect the negative battery cable.

43. Start the engine and allow it to reach normal operating temperature and check for leaks.

Timing Belt, Sprockets, Tensioner and Front Cover

REMOVAL AND INSTALLATION

2.5L (B258I) Engine

——— WARNING ———
To avoid damage to the valves, DO NOT rotate the camshafts. The crankshaft may only be turned between 0° and 60° BTDC when the camshafts are locked in position with the appropriate locking tool.

1. Disconnect the negative battery cable.

2. Remove the air cleaner complete with hoses.

3. Using a 15mm wrench turn the serpentine drive belt tensioner towards the front of the vehicle, remove the belt from the water pump pulley and slowly release the belt tensioner.

4. Remove the belt tensioner and the power steering pump pulley.

5. Remove the water pump pulley. To do this you will have to carefully press the engine towards the left in the vehicle, using the engine bracket as a fulcrum.

6. Remove the timing cover.

7. Remove the right front wheel and the cover in the wheel well.

8. Remove the 6 bolts holding the crankshaft pulley and remove the pulley. Do not remove the center bolt.

9. Put the No. 1 cylinder in the top dead center position (on the compression stroke).

10. The timing marks on the crankshaft and camshafts should be in alignment with their respective marks on the engine. Insert the camshaft locking tool 83–94–926 and install tool 83–94–868, the locking tool for the crankshaft.

11. If reusing the belt, mark the direction of its rotation. To help with refitting, the belt can be marked at the camshaft timing marks and also at the crankshaft timing mark.

12. Remove the tensioning roller and the two adjusting rollers and remove the timing belt.

13. Release tension from and remove the timing belt. Loosen the timing belt adjuster bolts.

14. Rotate the crankshaft back to 60° BTDC, to prevent damage to the valves.

15. Remove the bracket with the upper timing belt adjuster and tensioner rollers.

16. Remove the valve covers. Make sure the O-rings stay in position and do not fall into the engine.

17. Note the positioning of the camshaft sprockets. Use an open ended wrench on the hex flats of the camshaft to hold the camshaft in position when removing the timing sprockets. Remove the camshaft sprocket Torx® bolts and sprockets (the sprockets are marked 1, 2, 3 & 4).

—————— **WARNING** ——————
The wrench used to hold the camshaft must not have jaws that are too long. A wrench that is too long could damage the casting and lock the tappets.

18. Remove the crankshaft sprocket and spacer ring.
To install:
19. Install the crankshaft sprocket and center bolt. Torque the crankshaft bolt to 184 ft. lbs. (250 Nm) + 45° additional torque.
20. Install the camshaft sprockets. Use an open ended wrench on the hex flats of the camshaft to hold the camshaft in position when removing the timing sprockets. Check the proper position of the sprockets by locating pins and timing marks. Torque the camshaft Torx® bolts to 37 ft. lbs. (50 Nm) + 60° additional torque.

—————— **WARNING** ——————
The wrench used to hold the camshaft must not have jaws that are too long. A wrench that is too long could damage the casting and lock the tappets.

21. Install the bracket with the upper timing belt adjuster and tensioner pulleys.
22. Install and locking tools 83–94–926 (or equivalent) between the camshaft sprockets to lock the camshafts of both heads in position.
23. Rotate the crankshaft forward to just before 0° TDC and install crankshaft locking tool 83–94–926 (or equivalent) on the crankshaft. Carefully rotate the engine until the arm of the tool is against the water pump flange. Make sure the crankshaft is at 0° TDC and all timing marks are aligned. Remove the locking tool.
24. Install the timing belt, noting the direction of rotation marked at disassembly.
   a. Adjust the tensioner lightly by hand counterclockwise to keep the timing belt from falling off. Make sure the crankshaft is a 0° TDC and all timing marks are aligned.

   b. Install the crankshaft locking tool (83–94–926 or equivalent).
   c. Install tool (83–93–985) with a cut piece from an old timing belt, to measure belt tension.
   d. Snug the center bolts of the adjusting rollers. Turn the lower adjusting roller counterclockwise, until a belt tension of 202–221 ft. lbs. (275–300 Nm) is registered on the torque wrench.
   e. Tighten adjusting roller center bolts to 30 ft. lbs. (40 Nm).

NOTE: This adjustment of the timing belt is only preparatory and should, therefore, not be used as a final check.

25. Adjust the tensioner pulley until the marks are aligned. Tighten the tensioner pulley to 15 ft. lbs. (20 Nm). Remove the camshaft locking tool on camshaft sprockets 1 & 2. Adjust the upper adjusting roller until sprocket No. 2 moves 1–2mm clockwise. Tighten the upper adjusting roller to 30 ft. lbs. (40 Nm) and remove the upper locking tool.
26. Rotate the engine two complete revolutions to just before 0° TDC and install the locking tool on the crankshaft. Carefully turn the crankshaft until the arm of the locking tool is against the water pump flange and tighten the locking tool. Set tool 83–94–926 (or equivalent) into position on the front of the camshaft sprockets. Make sure that the timing marks on the camshaft sprockets are aligned with the marks on the tool and that the edge of the timing belt is flush with the edge of the camshaft sprockets.

NOTE: Also check that the alignment marks on the tensioner pulley are still aligned.

27. Make sure the valve cover O-rings are still in position and clean. Lubricate the O-rings with soapy water, apply 81–52–381 (or equivalent) sealer at the corners of the large end bearing caps and install the valve covers.
28. Install the spark plug wires.
29. Install the crankshaft pulley and tighten the retaining bolts to 15 ft. lbs. (20 Nm).
30. Install the timing belt cover tighten the bolts to 6 ft. lbs. (8 Nm).
31. Install the water pump pulley. To do this you will have to carefully press the engine towards the left in the vehicle, using the engine bracket as a fulcrum.
32. Install the power steering pump pulley and tighten the retaining bolts to 6 ft. lbs. (8 Nm).

33. Using a 15mm wrench turn the serpentine drive belt tensioner towards the front of the vehicle, position the belt and slowly release the belt tensioner.
34. Install the air cleaner complete with hoses.
35. Connect the negative battery cable.
36. Start the engine and check the ignition timing.

3.0L (B308I) Engine

—————— **WARNING** ——————
To avoid damage to the valves, DO NOT rotate the camshafts. The crankshaft may only be turned between 0° and 60° BTDC when the camshafts are locked in position with the appropriate locking tool.

1. Disconnect the negative battery cable.
2. Lift up the power steering reservoir.
3. Disconnect the connection on the torque arm engine mount and remove the torque arm.
4. Remove the power steering line clamp from the torque arm engine mount and remove the engine mount.
5. Disconnect the hose from the coolant expansion tank and remove the upper alternator air intake.
6. Remove the right front wheel and the cover in the wheel well.
7. Loosen the power steering pump, water pump pulley and 6 outer crankshaft pulley bolts.
8. Using a 15mm wrench turn the serpentine drive belt tensioner towards the front of the vehicle, remove the belt from the water pump pulley and slowly release the belt tensioner. Remove the belt tensioner.
9. Remove the power steering pump and water pump pulley.
10. Remove the timing cover.

NOTE: When removing the crankshaft pulley, remove the 6 outer bolts only, DO NOT remove the center bolt.

11. Remove the crankshaft pulley.
12. Rotate the crankshaft to TDC of No. 1 cylinder.
13. The timing marks on the crankshaft and camshafts should be in alignment with their respective marks on the engine. Insert tool KM-800–1 (or equivalent) between the camshaft sprockets 1 & 2 and KM-800–2 between camshaft sprockets 3 & 4 and install tool 83–94–868 or an equivalent flywheel locking tool.

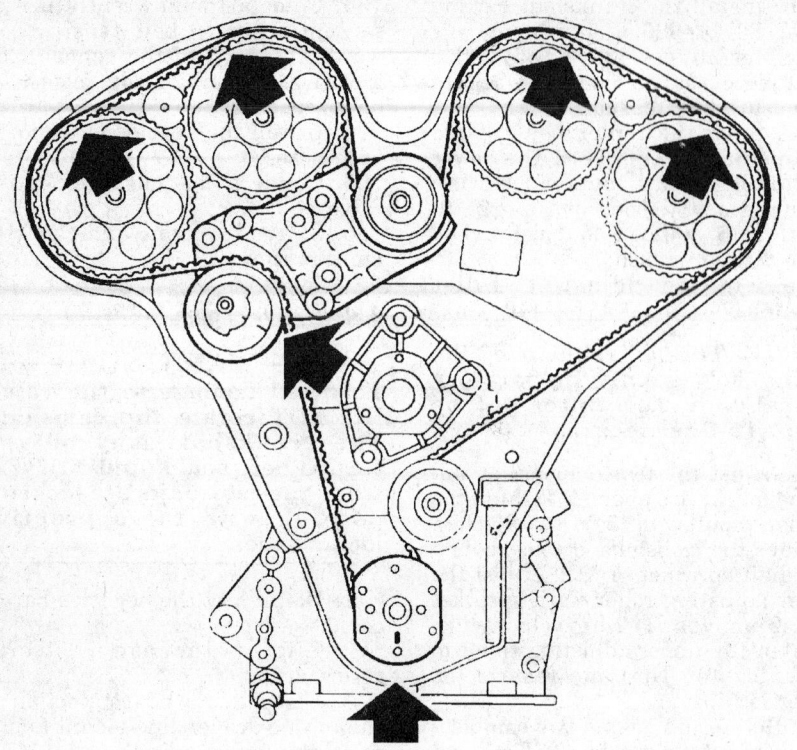

329871

Timing mark alignment with the engine at TDC — 2.5L and 3.0L engines

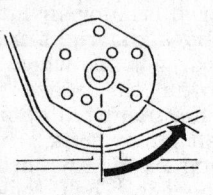

329872

The crankshaft must not be rotated beyond the points shown — 2.5L and 3.0L engines

14. Mark the direction of rotation of the timing belt for reassembly.

15. Release tension from and remove the timing belt. Loosen the timing belt adjuster bolts.

16. Rotate the crankshaft back to 60° BTDC, to prevent damage to the valves.

17. Remove the bracket with the upper timing belt adjuster and tensioner rollers.

18. Remove the valve covers. Make sure the O-rings stay in position and do not fall into the engine.

19. Note the positioning of the camshaft sprockets. Use an open ended wrench on the hex flats of the

camshaft to hold the camshaft in position when removing the timing sprockets. Remove the camshaft sprocket Torx® bolts and sprockets (the sprockets are marked 1, 2, 3 & 4).

WARNING ---
The wrench used to hold the camshaft must not have jaws that are too long. A wrench that is too long could damage the casting and lock the tappets.

---

20. Remove the crankshaft center bolt, sprocket and spacer ring.

To install:

21. Install the crankshaft sprocket and center bolt. Torque the crankshaft bolt to 184 ft. lbs. (250 Nm) plus an additional 1/8 (45°) turn.

22. Remove the flywheel locking tool and install the flywheel inspection cover.

23. Install the camshaft sprockets. Use an open ended wrench on the hex flats of the camshaft to hold the camshaft in position when removing the timing sprockets. Check the proper position of the sprockets by locating pins and timing marks. Torque the camshaft Torx® bolts to 37 ft. lbs. (50 Nm) plus an additional 1/6 (60°) turn.

WARNING ---
The wrench used to hold the camshaft must not have jaws that are too long. A wrench that is too long could damage the casting and lock the tappets.

---

24. Install the bracket with the upper timing belt adjuster and tensioner pulleys.

25. Install and locking tools KM-800–1 (or equivalent) between the camshaft sprockets 1 & 2 and KM-800–2 (or equivalent) between the camshaft sprockets 3 & 4 to lock

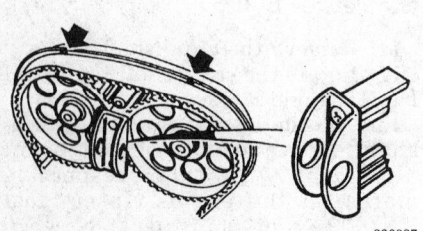

330087

Install tool as shown to lock the camshafts in position — 2.5L and 3.0L engines

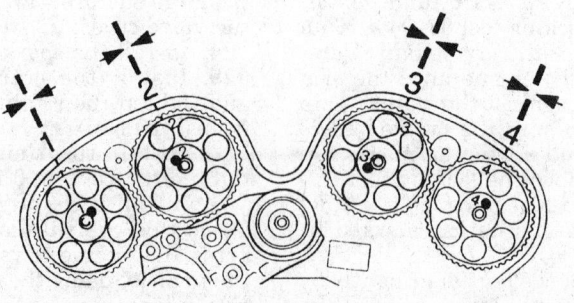

329874

Camshaft sprocket identification and positioning — 2.5L and 3.0L engines

the camshafts of both heads in position.

26. Rotate the crankshaft forward to just before 0° TDC and install crankshaft locking tool KM-800–10 (or equivalent) on the crankshaft. Carefully rotate the engine until the arm of the tool is against the water pump flange. Make sure the crankshaft is at 0° TDC and all timing marks are aligned. Remove the locking tool.

27. If reusing the belt, fit the timing belt according to its marked direction of rotation and timing marks. Adjust the tensioning roller loosely by hand to prevent the belt from slipping out of the cogs. Always adjust counterclockwise.

28. Measure the belt tension with tool 83–93–985.

29. Snug the center bolts of the adjusting rollers. Turn the lower adjusting rollers counterclockwise, until a belt tension of 275–300 Nm is reached. Tighten adjusting roller center bolts to 30 ft. lbs. (40 Nm).

NOTE: This is a preliminary adjustment of the belt tension and must not be used as a check when the belt is finally adjusted.

30. Continue to carry out the adjustment by means of the tensioning roller, mark against mark. Remove the locking tool for camshaft sprockets 1 and 2. Carry out the final adjustment with the upper center adjusting roller until camshaft sprocket No. 2 moves 0.04–0.08 in. (1–2mm) forward.

31. Remove the locking tool for camshaft sprockets 3 and 4 and also remove the crankshaft locking tool.

32. Tighten the tensioning roller to 15 ft. lbs. (20 Nm). Tighten the upper adjusting roller to 30 ft. lbs. (40 Nm) and tighten the lower adjusting roller to 15 ft. lbs. (20 Nm).

33. Rotate the engine two complete revolutions to just before 0° TDC and install the locking tool on the crankshaft. Carefully turn the crankshaft until the arm of the locking tool is against the water pump flange and tighten the locking tool. Set tool KM-800–20 (or equivalent) into position. Make sure that the timing marks on the camshaft sprockets are aligned with the marks on the tool and that the edge of the timing belt is flush with the edge of the camshaft sprockets.

NOTE: Also check that the alignment marks on the tensioner pulley are still aligned.

34. Install the crankshaft pulley and tighten the retaining bolts to 15 ft. lbs. (20 Nm).

35. Install the timing belt cover tighten the bolts to 6 ft. lbs. (8 Nm).

36. Install the water pump pulley. Torque water pump pulley bolts to 6 ft. lbs. (8 Nm).

37. Install power steering pump.

38. Install the power steering pump pulley and tighten the retaining bolts to 6 ft. lbs. (8 Nm).

39. Using a 15mm wrench turn the serpentine drive belt tensioner towards the front of the vehicle, position the belt and slowly release the belt tensioner.

40. Install the upper alternator air intake.

41. Install the torque arm engine mount and attach the power steering line clamp.

42. Install the torque arm and bolt the connection to the torque arm engine mount.

43. Connect the upper hose to the coolant expansion tank and set the power steering reservoir back into position.

44. Connect the negative battery cable.

45. Start the engine and check the ignition timing.

Camshaft

REMOVAL AND INSTALLATION

All 2.0L, 2.1L (B212l) and 2.3L Engines

1. Disconnect the negative battery cable.

2. Remove the engine from the vehicle.

3. Remove the lid on the valve cover. Disconnect the spark plug wires and vacuum hose from the distributor and remove the distributor cap (if applicable).

4. Remove the valve cover and position the crankshaft for TDC. The **0** mark on the flywheel should align with the timing mark on the bell housing end plate. The camshafts should be lined up with their respective timing marks.

5. Remove the distributor (if applicable) and remove the oil pipes.

6. Remove the center bolts securing the camshaft sprockets. Use a proper holding tool to hold the camshafts from rotating. Always keep the camshafts in their correct basic setting. If the setting of the crankshaft or camshafts is altered at this stage the valves can be damaged.

7. Remove the timing chain tensioner. Remove the camshaft sprockets.

8. Mark the bearing cap positions and relation to the front of the engine. The caps must be installed in their original location.

9. Loosen the camshaft bearing cap bolts one turn at a time to avoid uneven valve spring pressure on the camshafts. When all bolts are loose, remove the bearing caps and lift out the camshafts.

To install:

10. Place the camshafts in their proper positions and install the bearing caps in their original location and position.

11. When installing the bolts, tighten them one turn at a time to draw the camshaft down evenly against the valve springs. Torque the bearing cap bolts to 11 ft. lbs. (15 Nm).

NOTE: The black bolts have an oiling passage and must be installed on the spark plug side.

12. Install the camshaft sprockets, fitting the sprocket for the exhaust cam first. Make sure the chain between the crankshaft sprocket and the camshaft sprocket is kept tight. Next install the intake cam sprocket. Keep the chain tight between the sprockets. Hand tighten the center bolts securing the camshaft sprockets. .

13. Install the chain tensioner with the piston under tension. Make sure the copper gasket is in good condition and that the sealing surface is clean and free from burrs. Tighten the tensioner to 47 ft. lbs. (63 Nm).

14. Trigger the chain tensioner by pressing the pivoting chain guide against it, thereafter, press the pivoting guide against the chain to give the chain its basic tension. Check that the chain tensioner maintains tension on the chain when the pressure on the chain guide is released and that the basic setting stop for the tensioner holds the chain guide tight against the chain. A limited amount of play will be present until the hydraulic pressure takes over once the engine is running.

15. Depress the pivoting guide to check that the tensioner is working. Rotate the crankshaft 2 complete turns clockwise, viewed from the transmission end. Make sure the crankshaft and camshaft timing marks still align properly.

16. Hold the camshafts in their proper position and tighten the cam sprocket bolts to 49 ft. lbs.

17. Install the oil pipes and the distributor (if applicable).

18. Install the semi-circular rubber plug halves in the cylinder head and install the valve cover torque the retaining bolts to 11 ft. lbs. (14 Nm). Install the crankcase ventilation hose.

19. Install the spark plug wires and distributor cap in their original position (if applicable) and install the lid on the valve cover.

20. Install the engine assembly and connect the negative battery cable.

21. Start the engine and check for proper operation and leaks.

2.5L (B258I) Engine

------ WARNING ------
To avoid damage to the valves, DO NOT rotate the camshafts. The crankshaft may only be turned between 0° and 60° BTDC when the camshafts are locked in position with the appropriate locking tool.

1. Disconnect the battery and drain the coolant.

2. Remove engine covers, air filter assembly and mass air flow sensor.

3. Disconnect and label all hoses and electrical harness connections at the intake manifold. Remove the upper intake manifold plenum.

4. Remove spark plug wires.

5. Remove the right-front wheel and forward cover in the wheel housing.

6. Loosen the power steering pump, water pump pulley bolts and 6 outer crankshaft pulley bolts. Remove the drive belt and tensioner. Remove the power steering pump pulley, water pump pulley and timing cover.

NOTE: When removing the crankshaft pulley, remove the 6 outer bolts only, DO NOT remove the center bolt.

7. Remove the crankshaft pulley.

8. Remove the valve covers. Make sure the O-rings stay in position and do not fall into the engine.

9. Remove the timing belt, tensioner and camshaft sprockets.

10. Rotate the crankshaft back to 60° BTDC, to prevent damage to the valves.

11. Remove the bearing caps on the camshafts. Note the position markings on the bearing caps and loosen the bearing cap bolts in stages of ½ to 1 turn at a time. Remove the front camshaft seal and remove the camshaft.

NOTE: The bearing caps located where valve tappets are compressed, should be removed last.

To install:

12. Make sure the crankshaft is at 60° BTDC.

13. Thoroughly lubricate the camshafts, install the camshafts and new front camshaft seals. Make sure the locating pins are properly positioned. Install the bearing caps in their proper location and position. Tighten the bearing cap bolts in sequence ½ to 1 turn at a time, to a torque of 6 ft. lbs. (8 Nm).

14. Install the timing cover and retaining screws into cylinder head. Tighten retaining screws to 6 ft. lbs. (8 Nm).

15. Check to make sure that the camshaft locating pins are in the proper position.

NOTE: The locating pins of both camshafts 1 & 2 should be point towards the inboard bolts of the camshaft bearing caps. The locating pin for the intake camshaft 3 should be pointing downwards, in line with the inboard bolt of the bearing caps. The locating pin for the exhaust camshaft 4 should be pointing upwards, in line with the edge of the camshaft sensor.

16. Install the camshaft sprockets. Use an open ended wrench on the hex flats of the camshaft to hold the camshaft in position when removing the timing sprockets. Check the proper position of the sprockets by locating pins and timing marks. Torque the camshaft Torx® bolts to 37 ft. lbs. (50 Nm) + 60°.

------ WARNING ------
The wrench used to hold the camshaft must not have jaws that are too long. A wrench that is too long could damage the casting and lock the tappets.

17. Install the tensioner and timing belt.

18. Make sure the valve cover O-rings are still in position and clean. Lubricate the O-rings with soapy water, apply 81-52-381 (or equivalent) sealer at the corners of the large end bearing caps and install the valve covers.

19. Install the spark plug wires.

20. Install crankshaft pulley and tighten retaining bolts to 15 ft. lbs. (20 Nm).

21. Install the outer timing cover. Install power steering pump pulley, water pump pulley, drive belt tensioner and drive belt.

22. Install the forward cover in the wheel housing and the right front wheel. Torque the wheel bolts to 89 ft. lbs. (120 Nm).

23. Install the upper intake plenum and tighten the bolts at the mating surface. Starting in the middle and working alternately outward to the ends. Torque upper intake plenum bolts to 15 ft. lbs. (20 Nm).

24. Connect all hoses and electrical harness connections at the intake manifold, which should have been labeled at disassembly.

NOTE: Make sure the pressure regulator's vacuum hose is connected to front nipple on the throttle housing, in front of the butterfly.

25. Install the mass air flow sensor, air filter assembly and engine covers.

26. Connect the battery cable.

27. Fill coolant system and check all other fluids as necessary.

28. Start engine and check for leaks.

3.0L (B308I) Engine

------ WARNING ------
To avoid damage to the valves, DO NOT rotate the camshafts. The crankshaft may only be turned between 0° TDC and 60° BTDC when the camshafts are locked in position with the appropriate locking tool.

1. Disconnect the negative battery cable.

2. Safely raise and support the vehicle. Disconnect the front exhaust pipe from the exhaust manifolds. Disconnect the bracket for the check valves.

3. Remove the lower center air deflector. Drain the engine coolant into a catch pan.

4. Lower the vehicle and remove the top engine covers.

5. Disconnect the cruise control cable, throttle cable. Unclip the throttle control rod from the bracket. Remove the bracket mounting bolt and carefully set the bracket with cables attached to the side.

6. Loosen the clamps on the air intake pipes at the intake manifold and the mass air flow meter. Disconnect the pipes and raise them slightly. Disconnect the vacuum hoses and electrical connection. Lift the pipes with resonator attached carefully out.

7. Disconnect and label all electrical and vacuum connections on the intake manifold. Remove the intake plenum bolts.

8. Disconnect the IAC connector and fuel pressure regulator hose and remove the wiring harness from under the throttle body.

9. Disconnect the throttle position indicator, ignition coil and TCS connectors. Lift off the intake plenum and plug the intake runners with paper.

10. Disconnect the fuel injector and camshaft position connectors.

11. Disconnect and plug the fuel line connections.

12. Remove the center intake manifold, with fuel rails and set aside.

13. Disconnect the spark plug wires and the ignition coil. Bend the ignition coil aside and disconnect the ignition coil bracket.

14. Remove the lifting eye and the heat shield over the exhaust manifold.

15. Unbolt the resonator bracket and secondary air injection pipe from the exhaust manifold.

16. Lift up the power steering reservoir.

17. Disconnect the connection on the torque arm engine mount and remove the torque arm.

18. Remove the power steering line clamp from the torque arm engine mount and remove the engine mount.

19. Disconnect the hose from the coolant expansion tank and remove the upper alternator air intake.

20. Loosen the power steering pump, water pump pulley and 6 outer crankshaft pulley bolts. Remove the drive belt and tensioner. Remove the power steering pump, water pump pulley and timing cover.

NOTE: When removing the crankshaft pulley, remove the 6 outer bolts only, DO NOT remove the center bolt.

21. Remove the crankshaft pulley.

22. Remove the timing belt, tensioner and camshaft sprockets.

23. Rotate the crankshaft back to 60° BTDC, to prevent damage to the valves.

24. Remove the valve cover. Make sure the O-rings stay in position and do not fall into the engine.

25. Remove the bearing caps on the camshafts. Note the position markings on the bearing caps and loosen the bearing cap bolts in stages of ½ to 1 turn at a time.

NOTE: The bearing caps located where valve tappets are compressed, should be removed last.

26. Carefully lift the camshafts off of the cylinder head.

To install:

27. Make sure the crankshaft is still positioned at 60° BTDC.

28. Make sure all gasket contact surfaces are clean.

NOTE: The camshaft bearing caps are marked L1–L8 for the Front or Left bank and R1–R8 for the Rear or Right bank and the camshaft bearing seats in each head are numbered 1–8.

29. Thoroughly lubricate the camshafts and install the camshafts with new front camshaft gaskets. Make sure the locating pins are properly positioned. Install the bearing caps in their proper location and position. Tighten the bearing cap bolts in sequence ½ to 1 turn at a time, to a torque of 6 ft. lbs. (8 Nm).

30. Install the timing covers and bolt to the head.

31. Check to make sure that the camshaft locating pins are in the proper position. Check the locating pins, if they are hollow, replace them with solid pins.

NOTE: The locating pins of both camshafts 1 & 2 should be point towards the inboard bolts of the camshaft bearing caps. The locating pin for the intake camshaft No, 3 should be pointing downwards, in line with the inboard bolt of the bearing caps. The locating pin for the exhaust No. 4 camshaft should be pointing upwards, in line with the edge of the camshaft sensor.

32. Install the camshaft sprockets, tensioner and timing belt.

33. Install the crankcase ventilation housing and set the oxygen sensor connector back in it's holder. Make sure the valve cover O-rings are still in position and clean. Lubricate the O-rings with soapy water, apply (81–52–381 or equivalent) sealer at the corners of the large end bearing caps and install the valve cover.

34. Install the timing cover.

35. Install the crankshaft pulley and torque bolts to 15 ft. lbs. (20 Nm).

36. Install the water pump pulley, power steering pump, drive belt tensioner and drive belt.

37. Install the upper alternator air intake.

38. Install the torque arm engine mount and attach the power steering line clamp.

39. Install the torque arm and bolt the connection to the torque arm engine mount.

40. Connect the upper hose to the coolant expansion tank and set the power steering reservoir back into position.

41. Install the lifting eye and secondary air injection pipe to the exhaust manifold. Install the exhaust manifold heat shield.

42. Install the resonator bracket.

43. Install the ignition coil bracket and ignition coil.

44. Install the spark plugs (if removed) and torque to 19 ft. lbs. (25 Nm). Install the spark plug wires.

45. Install the center intake manifold with fuel rail. Apply thread locker (74–96–268 or equivalent) to and install the center intake manifold bolts and torque to 15 ft. lbs. (20 Nm).

46. Connect the fuel lines, crankshaft position sensor and fuel injector connectors.

47. Set the intake plenum into position and connect the TCS throttle body.

48. Tighten the intake plenum bolts to 15 ft. lbs. (20 Nm).

49. Connect the throttle position indicator, ignition coil and TCS connectors.

50. Route the wiring harness under the throttle body and connect the fuel pressure regulator hose and IAC connector.

51. Connect all labeled electrical and vacuum connections on the intake manifold.

52. Set the air intake pipes with resonator attached, carefully into position. Connect the vacuum hoses and electrical connection. Lower the air intake pipes and connect them to the mass air flow meter and intake manifold. Install and tighten the clamps on the air intake pipes at the intake manifold and the mass air flow meter.

53. Carefully set the throttle control bracket with cables attached into position and install the bracket mounting bolt. Clip the throttle control rod to the bracket. Connect the throttle cable and cruise control cable. Adjust the kick-down cable and the throttle cable.

54. Install the upper engine covers and safely raise and support the vehicle.

55. Make sure the radiator drain plug is tightened and install the lower center air deflector.

56. Bolt on the bracket for the check valves and connect the front exhaust pipe to the exhaust manifolds. Lower the vehicle.

57. Connect the negative battery cable.

58. Check all other oil and fluids for condition and proper level. Replace fluids and filters or top off as necessary.

59. Fill the radiator with clean coolant and check the cooling system for leaks

60. Bleed the cooling system.

61. Test drive the vehicle and check for any leaks.

Balance Shaft

REMOVAL AND INSTALLATION

All 2.3L Engines

1. Disconnect the negative battery cable.

2. Remove the engine assembly and place it in a suitable engine stand.

3. Rotate the crankshaft to bring the No. 1 piston to TDC on the compression stroke.

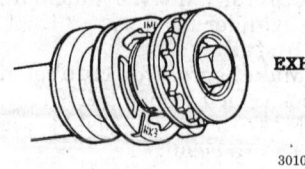

301023

Balance shaft identification — all 2.3L engines

4. Secure the flywheel with a flywheel locking segment, tool 83–92–987, and remove the crankshaft pulley bolt.

5. Using a suitable puller, remove the crankshaft pulley.

6. Remove the timing cover noting the different bolt lengths.

7. Remove the top guide from the balance shaft chain.

8. Remove the chain tensioner and pivoting guide. Remove the idler wheel sprocket and the balance shaft chain.

9. Remove the oil pump drive dog and the sprocket from the crankshaft.

10. Remove the balance shafts taking care not to damage the inner bearing shells.
   To install:

NOTE: The shaft with the smaller thrust ring, marked INL, is the one for the intake side of the engine. The shaft with the larger thrust ring, marked EXH, is the one for the exhaust side of the engine.

11. Lubricate the balance shaft journals and bearing housings. Insert the balance shafts into their respective tunnels, taking care not to damage the inner bearing shells. Tighten the bearing bolts to 9 ft. lbs. (12 Nm).

12. Install the chain and sprocket on the crankshaft.

13. Install the oil pump drive dog.

14. Install the balance shaft chain and idler wheel sprocket, ensuring that the aligning marks on the bearing housing and sprocket are in line. When installing, leave some slack in the chain in line with the tensioner, and keep the chain reasonably taut by means of the top chain guide.

NOTE: There is an alternate way of installing the balance shaft chain. Install the top chain guide first and then adjust the run of the chain around the sprockets. Adjusting the chain is easier this way, although it will be more awkward to install the idler wheel sprocket.

15. Cock the chain tensioner and insert a paper clip through the hole in the cylinder to prevent the tensioner being triggered. Before installing the tensioner, make sure that the plunger is turned to the position in which the spring acts fully on it.

16. Install the pivoting chain guide and the tensioner. Tighten the tensioner to 7.5 ft. lbs. (10 Nm).

—— WARNING ——
It is extremely important for the correct torque to be applied when installing the tensioner.

17. Install the top chain guide and trigger the tensioner by removing the paper clip.

18. Rotate the crankshaft a few times to ensure the chain is installed correctly.

19. Ensure that the timing cover flange is absolutely clean.

20. Install the timing cover, be sure all the bolts are in their proper location. Tighten all timing cover bolts to 16 ft. lbs. (22 Nm).

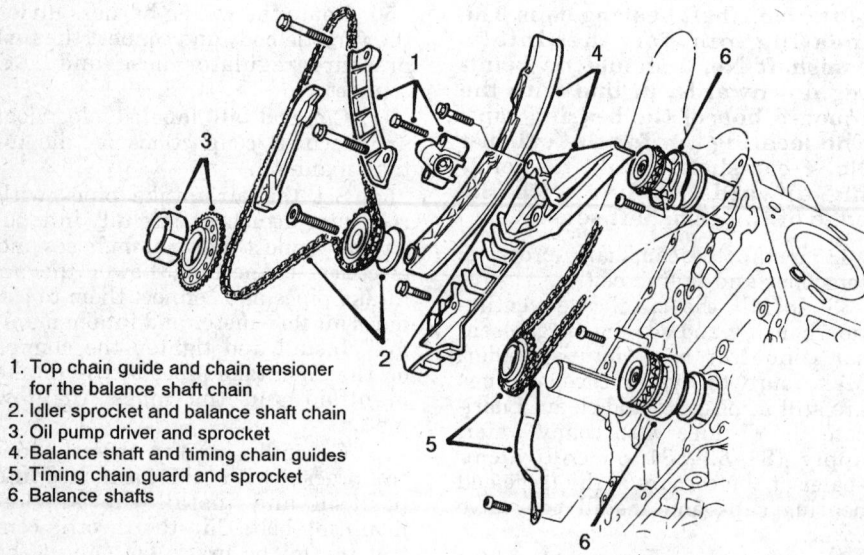

1. Top chain guide and chain tensioner for the balance shaft chain
2. Idler sprocket and balance shaft chain
3. Oil pump driver and sprocket
4. Balance shaft and timing chain guides
5. Timing chain guard and sprocket
6. Balance shafts

301022

Balance shafts and timing chains exploded view — all 2.3L engines

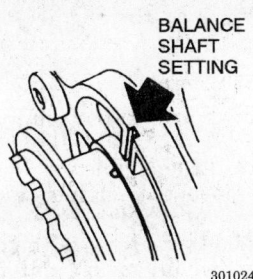

BALANCE
SHAFT
SETTING

301024

**Balance shaft aligning
marks — all 2.3L engines**

21. Install the crankshaft pulley and tighten the retaining bolt to 130 ft. lbs. (175 Nm).

22. Install the engine assembly.

23. Connect the negative battery cable.

24. Start the engine and check the timing.

25. Check for leaks and test drive the vehicle.

Piston and Connecting Rod

POSITIONING

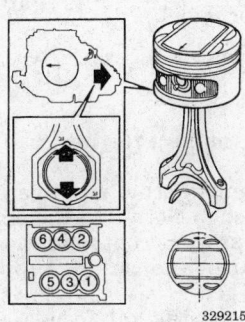

329215

Piston and connecting rod positioning — 2.5L and 3.0L engines

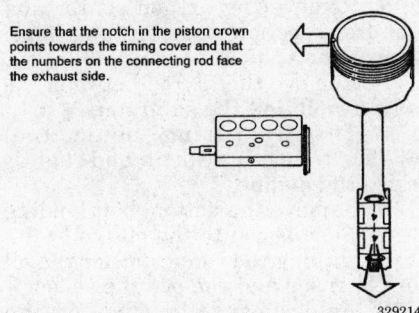

Ensure that the notch in the piston crown points towards the timing cover and that the numbers on the connecting rod face the exhaust side.

329214

Piston and connecting rod positioning — 2.0L, 2.1L and 2.3L engines

ENGINE LUBRICATION

Oil Pan

REMOVAL AND INSTALLATION

900 Series

On 900 Series vehicles the top of the transaxle is used as the oil pan (sump). To remove the oil pan, remove the engine and transaxle as an assembly. Then, separate the engine and transaxle once removed from the vehicle.

2.0L (B202L) and 2.1L (B212l) Engines With Manual Transmissions

1. After the engine and transmission assembly has been removed from the vehicle, clean their external surfaces.

2. Remove the clutch cover and dipstick tube. If turbocharged, remove the oil return line from the turbocharger unit.

3. Remove the starter and the steady bar from the transmission.

4. Withdraw the transmission input shaft using a slide hammer and joint 87 90 529.

5. Remove the slave cylinder.

6. Remove all of the bolts from the mating flanges between the engine and transmission. Remove the bracket for the oil filler pipe at the throttle control on the intake manifold.

7. Carefully lift the engine away from the transmission, removing the release bearing sleeve at the same time.

WARNING

If the engine and transmission fail to separate, do not attempt to pry them apart without first checking that all bolts have been removed.

---

8. Remove the oil pump suction line complete with the O-ring.

To install:

9. Remove all traces of oil and dirt from the flanges.

10. Install the oil pump suction line complete with the O-ring.

11. Make sure that the 2 guide sleeves are fitted in the transmission.

12. Fit a new gasket onto the transmission flange and apply a sealing compound to the slots.

13. Apply a thread sealing compound to the 6 bolt holes shown in the illustration.

NOTE: The release bearing and sleeve must be held in position against the clutch when the engine is placed onto the transmission.

14. Carefully place the engine onto the transmission and position the release bearing sleeve at the same time.

15. Install all of the bolts for the mating flanges between the engine and transmission and tighten them to 20 ft. lbs. (25 Nm).

16. Install the input shaft and the slave cylinder.

17. Install the starter and install the steady bar to the transmission.

18. If turbocharged, install the oil return line to the turbocharger unit. Install the clutch cover and dipstick tube.

19. Install the engine and transmission assembly.

2.0L (B202L) and 2.1L (B212l) Engines With Automatic Transmissions

1. After the engine and transmission assembly has been removed from

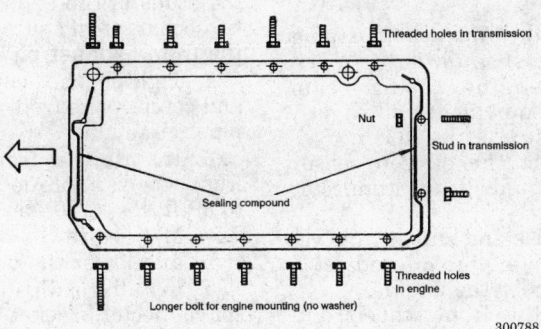

Threaded holes in transmission

Nut

Stud in transmission

Sealing compound

Longer bolt for engine mounting (no washer)

Threaded holes in engine

300788

Engine-to-transmission bolt locations — 2.0L (B202L) and 2.1L (B212l) engines

the vehicle, clean their external surfaces.

2. Remove the flywheel cover.

3. Remove all of the bolts from the mating flanges between the engine and transmission. Remove the hoses for the transmission fluid cooler.

4. Remove the bolts securing the torque converter to the flywheel.

5. Rotate the flywheel to bring the angle irons horizontal. Carefully lift the engine off of the transmission.

--- WARNING ---
If the engine and transmission fail to separate, do not attempt to pry them apart without first checking that all bolts have been removed.

6. Fit the torque converter support bracket 87–90–255 or equivalent.
To install:
7. Remove all traces of oil and dirt from the flanges.

8. Make sure that the 2 guide sleeves are fitted in the transmission.

9. Fit a new gasket onto the transmission flange and apply a sealing compound to the slots.

10. Apply a thread sealing compound to the 6 bolt holes shown in the illustration.

11. Install the torque converter support bracket.

12. Carefully place the engine onto the transmission. Take care not to damage the torque converter spindle.

13. Install the bolts securing the torque converter to the flywheel.

14. Install all of the bolts for the mating flanges between the engine and transmission and tighten them to 20 ft. lbs. (25 Nm).

15. Install the hoses for the transmission fluid cooler.

16. Install the flywheel cover.

17. Install the engine and transmission assembly.

2.0L (B204L) and 2.3L Engines

1. Disconnect the negative battery cable.

2. Remove the dipstick and plug the hole with a shop towel.

3. Install engine lifting beam 83–94–850 or equivalent and slightly raise the engine.

4. Disconnect the oxygen sensor connector and unbolt the connector bracket.

5. Safely raise and support the vehicle, drain the engine oil and set a lifting table under the engine.

6. Remove the front wheels and front spoiler. Disconnect the front exhaust pipe at the exhaust manifold and intermediate pipe.

7. If equipped with turbocharger, unbolt the front exhaust pipe from the turbocharger. Disconnect the front pipe catalytic converter bracket and remove the front pipe.

8. Disconnect the lower ball joints from the steering knuckles. Unbolt the rear engine mount.

9. Raise the lifting table into position. Unbolt and remove the subframe.

10. Disconnect the oil level sensor and pull the sensor harness from it's clips.

11. Remove the flywheel inspection cover and oil pan.

12. Do not remove the guide sleeve from the block. Wipe off excess oil.

To install:
13. Thoroughly clean the flanges on the sump and block using a suitable solvent. Apply an even bead of Loctite® 518 sealant, or equivalent, along the oil pan flange.

14. Install the oil pan and loosely install the bolts.

15. Tighten the bolts, starting in the center on both sides, then crisscross out to the ends until snug,

16. Torque the bolts in the same fashion to 16 ft. lbs. (22 Nm).

17. Install the flywheel inspection cover.

18. Set the oil level sensor harness back into it's clips and connect the oil level sensor.

19. Raise the subframe and install the subframe bolts. Torque the subframe bolts as follows:
    a. Front bolts to 85 ft. lbs. (115 Nm).
    b. Center bolts to 140 ft. lbs. (190 Nm).
    c. Rear bolts to 81 ft, lbs. (110 Nm) +75° additional torque.

20. Connect the lower ball joints to the steering knuckles and torque the nuts to 55 ft. lbs. (75 Nm).

21. Loosely connect the front exhaust pipe catalytic converter bracket. Apply Molykote 1000 ® to the studs of the exhaust manifold or turbocharger (if equipped) and install the front exhaust pipe.

22. Tighten the converter bracket and connect the front pipe to the mid pipe.

23. Install the front spoiler and front wheels. Torque the wheel bolts to 89 ft. lbs. (120 Nm).

24. Make sure the oil drain plug is tight and lower the vehicle.

25. Install the dipstick, oxygen sensor connector bracket and connect the oxygen sensor.

26. Fill the engine with oil and remove the engine lifting beam.

27. Connect the negative battery cable. Warm up the engine and check for leaks.

2.5L (B258I) Engine

1. Raise and safely support the vehicle.

2. Drain the engine oil.

3. Remove the engine cradle sub frame.

4. Unbolt and remove the oil pan.

5. Remove all gasket material and clean sealing surfaces.

To install:
6. Apply a sealing compound to the areas where the engine block mate to the oil pump and rear main bearing cap.

7. Install a new oil pan gasket. Install the oil pan to the engine block.

8. Using a bolt locking compound install oil pan retaining bolts and tighten to 11 ft. lbs. (15 Nm).

9. Install oil pan drain plug and tighten to 41 ft. lbs. (55 Nm).

10. Install the engine cradle sub frame and tighten bolts as follows:
    a. Front bolts to 85 ft. lbs. (115 Nm).
    b. Center bolts to 140 ft. lbs. (190 Nm).
    c. Rear bolts to 81 ft. lbs. (110 Nm) + 75° additional torque.

11. Lower the vehicle. Fill the engine with oil. Start the engine and check for leaks.

9000 Series

2.3L Engines

1. Disconnect the negative battery cable.

2. Remove the oil dipstick and place a shop cloth into the end of the tube to prevent contamination.

3. Raise and safely support the vehicle.

4. Drain the oil into a suitable container.

5. Remove the right front wheel and inner wheel well.

6. Unbolt the front and rear engine mounts.

7. Remove the oxygen sensor and the front section of the exhaust pipe. Lower the vehicle.

8. Remove the tie rod between the wheel well and the subframe.

9. Install lifting beam tool 83–93–977 or equivalent, and slightly raise the engine.

10. Remove the bottom bolt holding the transmission to the oil pan.

11. Unplug the connector for the oil level sensor and remove the sensor.

12. Fold down the edge of the splash plate and remove the two rubber plugs in the back of the transmission case.

13. Remove the two bolts securing the oil pan to the block under the plugs.

14. Tap the guide sleeve into the block.

15. Remove the remaining oil pan bolts and remove the oil pan from the back first.

16. Remove the guide sleeve from the cylinder block.

To install:

17. Thoroughly clean the flanges on the sump and block using a suitable solvent. Apply an even bead of Permatex® Ultra Blue sealant, or equivalent, along the oil pan flange.

18. Install the rubber seal for the oil strainer in the groove on the oil pan.

19. Install the oil pan, front edge first and then the back.

NOTE: The longer bolt with the washer should be installed in the middle on the right hand side.

20. Install the bolts loosely, then, starting in the middle, tighten the bolts to 15 ft. lbs. (20 Nm).

21. Install the two rubber plugs in the back of the transmission and return the edge of the splash plate to its original position.

22. Install the bolt securing the oil pan to the transmission case at the bottom and install the oil level sensor.

23. Align the engine over the mounts and lower it into position. Install the tie rod between the wheel well and the subframe. Install the dipstick.

24. Install the bolts in the front and rear engine mounts. Install the oxygen sensor and exhaust pipe.

25. Install the inner wheel well and wheel. Torque wheel bolts to 96 ft. lbs. (130 Nm).

26. Lower the vehicle, fill with oil and run the engine to normal operating temperature to check for leaks.

3.0L (B308I) Engine

1. Raise and safely support the vehicle.

2. Drain the engine oil.

3. Unbolt front exhaust pipe from the exhaust manifolds, pull down slightly to allow for room to remove the oil pan.

4. Remove the flywheel inspection cover and remove the oil pan.

5. Remove all gasket material and clean sealing surfaces.

To install:

6. Apply a sealing compound to the areas where the engine block mate to the oil pump and rear main bearing cap.

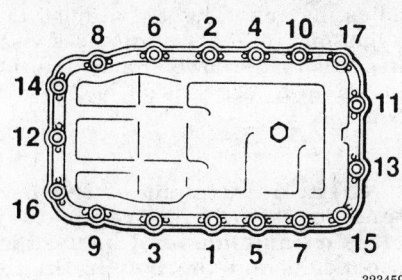

Oil pan bolt torque sequence — 2.5L and 3.0L engines

323459

7. Install a new oil pan gasket. Install the oil pan to the engine block.

8. Using a bolt locking compound install oil pan retaining bolts and tighten to 11 ft. lbs. (15 Nm).

9. Install oil pan drain plug and tighten to 41 ft. lbs. (55 Nm).

10. Install the flywheel inspection cover and tighten bolts to 6 ft. lbs. (8 Nm).

11. Lower the vehicle. Fill the engine with oil. Start the engine and check for leaks.

Oil Pump

REMOVAL AND INSTALLATION

900 Series

All 2.0L, 2.1L (B212I) and 2.3L Engines

1. Disconnect the negative battery cable.

2. Drain the engine oil and the coolant.

3. Remove the bracket for the steering servo pump, complete with the pump and alternator.

4. Remove the belt tensioner and the water pump pipe.

5. Secure the flywheel with a flywheel locking segment, tool 83–92–987, and loosen the crankshaft pulley bolt.

6. Using a suitable puller remove the crankshaft pulley.

7. Remove the oil pipes and the water pump pulley.

8. Remove the oil pump.

To install:

9. Install the oil pump and tighten the retaining bolts to 6 ft. lbs. (8 Nm). Install the oil pipes and water pump pulley.

10. Install the crankshaft pulley and tighten the nut to 140 ft. lbs. (190 Nm).

11. Install the belt tensioner, water pump pipe, steering servo pump and alternator.

12. Fill the radiator with coolant and the crankcase with oil.

13. Start the engine and check for leaks.

2.5L (B258I) Engine

1. Disconnect the negative battery cable.

2. Remove the air cleaner complete with hoses.

3. Using a 15mm wrench turn the serpentine drive belt tensioner towards the front of the vehicle, remove the belt from the water pump pulley and slowly release the belt tensioner.

4. Remove the belt tensioner and the power steering pump pulley.

5. Remove the water pump pulley. To do this you will have to carefully press the engine towards the left in the vehicle, using the engine bracket as a fulcrum.

6. Remove the timing cover.

7. Remove the right front wheel and the cover in the wheel well.

8. Remove the 6 bolts holding the crankshaft pulley and remove the pulley. Do not remove the center bolt.

9. Put the No. 1 cylinder in the top dead center position (on the compression stroke).

10. The timing marks on the crankshaft and camshafts should be in alignment with their respective marks on the engine. Insert the camshaft locking tool 83–94–926 and install, 83–94–868 the locking tool for the crankshaft.

11. If reusing the belt, mark the direction of its rotation. To help with refitting, the belt can be marked at the camshaft timing marks and also at the crankshaft timing mark.

12. Remove the tensioning roller and the two adjusting rollers and remove the timing belt.

13. Rotate the crankshaft back to 60° BTDC, to prevent damage to the valves.

14. Remove the bracket with the upper timing belt adjuster and tensioner rollers.

15. Remove inner timing cover bolts.

16. Install an engine lifting beam.

17. Raise and safely support the vehicle. Remove the front spoiler and disconnect the front exhaust pipe.

18. Remove the flywheel inspection cover and install a flywheel stop.

19. Remove the crankshaft sprocket.

20. Remove the engine subframe.

21. Remove the alternator.

22. Remove the AC compressor and hang to aside.

23. Drain the engine oil.

24. Remove the oil pan and strainer.

25. Disconnect the oil pump pressure sensor wiring.

26. Remove the AC compressor bracket.

27. Remove the oil pump housing. Extract the oil pump impellers.

To install:

28. Install the oil pump impellers and fit the oil pump cover.

29. Install a new oil pump housing gasket with the housing.

30. Torque bolts to 53 inch lbs. (6 Nm).

31. Fit the AC compressor bracket.

32. Connect the oil pump pressure sensor wiring.

33. Install the oil strainer and fit the oil pan.

34. Torque oil pan bolts to 11 ft. lbs. (15 Nm).

35. Install the A/C compressor and alternator.

36. Install the subframe and bolts. Torque the subframe bolts as follows:

 a. Front bolts to 85 ft. lbs. (115 Nm).

 b. Center bolts to 140 ft. lbs. (190 Nm).

 c. Rear bolts to 81 ft, lbs. (110 Nm) + 75° additional torque.

37. Install crankshaft sprocket and torque center crankshaft bolt to 184 ft. lbs. (250 Nm) + 45° additional torque.

38. Remove flywheel stop.

39. Install the front spoiler and connect the front exhaust pipe to the exhaust manifolds.

40. Lower the vehicle and remove the engine lifting beam.

41. Install the inner timing cover bolts.

42. Install the bracket with the upper timing belt adjuster and tensioner pulleys.

43. Install and locking tools 83–94–926 (or equivalent) between the camshaft sprockets to lock the camshafts of both heads in position.

44. Rotate the crankshaft forward to just before 0° TDC and install crankshaft locking tool 83–94–926 (or equivalent) on the crankshaft. Carefully rotate the engine until the arm of the tool is against the water pump flange. Make sure the crankshaft is at 0° TDC and all timing marks are aligned. Remove the locking tool.

45. If reusing the belt, install the timing belt according to its marked direction of rotation and timing marks. Adjust the tensioning roller loosely by hand to prevent the belt from slipping out of the cogs. Always adjust counterclockwise.

46. Measure the belt tension with tool 83–93–985.

47. Tighten the center bolts of the adjusting roller lightly. Adjust the adjusting rollers counterclockwise. Begin with the lower roller and adjust it to a belt tension of 275–300 Nm.

NOTE: Adjustment of the belt tension is only a preparatory measure and must not be used as a check when the belt is finally adjusted.

48. Continue to carry out the adjustment by means of the tensioning roller, mark against mark. Remove the locking tool for camshaft sprockets 1 and 2. Carry out the final adjustment with the upper center adjusting roller until camshaft sprocket No. 2 moves 0.04–0.08 in. (1–2mm) forward.

49. Remove the locking tool for camshaft sprockets 3 and 4 and also remove the crankshaft locking tool.

50. Tighten the tensioning roller to 15 ft. lbs. (20 Nm). Tighten the upper adjusting roller to 30 ft. lbs. (40 Nm) and tighten the lower adjusting roller to 15 ft. lbs. (20 Nm).

51. Turn the engine over two revolutions to the zero mark and refit the locking tool on the crankshaft. Check that the markings on the camshaft sprockets are in alignment with the markings on the timing cover. Check the positioning by installing the two camshaft locking tools, which should fit, and also by fitting tool 83–94–926 on camshaft sprockets 1 and 2 and 3 and 4. Also check the tensioning roller to ensure that the marks are still in alignment.

52. Install the crankshaft pulley and tighten the retaining bolts to 15 ft. lbs. (20 Nm).

53. Install the timing belt cover tighten the bolts to 6 ft. lbs. (8 Nm).

54. Install the water pump pulley and tighten the retaining bolts to 6 ft. lbs. (8 Nm).

55. Install the power steering pump pulley and tighten the retaining bolts to 6 ft. lbs. (8 Nm).

56. Install the forward cover in the wheel housing and the right-front wheel. Torque wheel bolts to 89 ft. lbs. (120 Nm).

57. Using a 15mm wrench turn the serpentine drive belt tensioner towards the front of the vehicle, position the belt and slowly release the belt tensioner.

58. Install the air cleaner complete with hoses.

59. Connect the negative battery cable.

60. Change the engine oil filter and fill the engine with oil.

61. Check all fluid levels and top off or change as necessary.

62. Start the engine and check the ignition timing.

9000 Series

2.3L Engines

1. Remove air filter assembly.

2. Using a ⅜ inch ratchet extension release the pressure on the drive belt. Place a short drill bit in tensioner housing to hold in the non-tension position.

3. Raise the vehicle and support on safety stands remove left front wheel.

4. Remove crankshaft access cover. Remove crankshaft pulley.

5. Remove circlip to release oil pump. Lift off oil pump cover and extract oil pump gears.

To install:

6. Insert oil pump gears making sure that the marking on the oil pump ring gear faces outward.

7. Install the oil pump cover and fit a new circlip with the chamfer facing outward.

8. Install crankshaft pulley and tighten bolt to 130 ft. lbs. (175 Nm).

9. Install crankshaft cover and fit left front wheel.

10. Lower vehicle and release tensioner to fit drive belt.

11. Install air filter assembly.

12. Test drive car and check drive belt is seated properly.

3.0L (B308I) Engine

1. Disconnect the negative battery cable.

2. Safely raise and support the vehicle.

3. Remove the lower center air deflector. Drain the engine coolant into a catch pan.

4. Lower the vehicle and remove the top engine covers.

5. Lift up the power steering reservoir.

6. Disconnect the connection on the torque arm engine mount and remove the torque arm.

7. Remove the power steering line clamp from the torque arm engine mount and remove the engine mount.

8. Disconnect the hose from the coolant expansion tank and remove the upper alternator air intake.

9. Loosen the power steering pump, water pump pulley and 6 outer crankshaft pulley bolts. Remove the drive belt and tensioner. Remove the

power steering pump, water pump pulley and timing cover.

NOTE: When removing the crankshaft pulley, remove the 6 outer bolts only, DO NOT remove the center bolt.

10. Remove the crankshaft pulley.
11. Remove the timing belt. Loosen the timing belt adjuster bolts.
12. Rotate the crankshaft back to 60° BTDC, to prevent damage to the valves.
13. Remove the bracket with the upper timing belt adjuster and tensioner pulleys.
14. Remove the water pump.

NOTE: Keep a drain pan under the engine to catch any coolant draining out.

15. Install camshaft locking tools (KM-800-1 and KM-800-2 or equivalents) between the camshaft sprockets, then remove the camshaft sprockets and inner timing covers.
16. Safely raise and support the vehicle. Disconnect the front exhaust pipe from the exhaust manifolds.
17. Remove the flywheel inspection cover and install holding tool (83-95-063 or equivalent) onto the flywheel.
18. Remove the crankshaft center bolt and crankshaft sprocket.
19. Remove the alternator.
20. Remove the A/C compressor and hang it aside.
21. Drain the engine oil. Remove the oil pan and strainer.
22. Disconnect the oil pressure switch.
23. Lower the vehicle and place a stand and wood block under the front edge of the engine block.
24. Remove the A/C compressor bracket.
25. Remove the oil pump housing.
26. Remove the oil pump housing cover and two oil pump impellers.
27. Inspect the oil pressure relief valve and oil pressure control valve.
To install:
28. Install the oil relief and control valves. Torque the valves as follows:
   a. With a copper gasket, tighten to 21 ft. lbs. (28 Nm).
   b. With an aluminum gasket, tighten to 15 ft. lbs. (20 Nm).
29. Install the oil pump impellers into the housing. Apply a thread locking compound to the cover bolts and

tighten the cover bolts to 53 inch lbs. (6 Nm).

NOTE: Make sure the alignment marks on the impellers are properly aligned.

30. Position a new gasket and install the oil pump housing. Tighten the housing bolts to 53 inch lbs. (6 Nm).
31. Install the A/C compressor bracket.
32. Raise the vehicle and remove the stand and wood block from under the engine block.
33. Connect the oil pressure switch.
34. Apply a sealing compound to the areas where the engine block mate to the oil pump and rear main bearing cap.
35. Install the oil strainer. Apply a thread locking compound to the bolts and tighten the bolts to 6 ft. lbs. (8 Nm).
36. Install a new oil pan gasket. Install the oil pan to the engine block.
37. Using a bolt locking compound install oil pan retaining bolts and tighten to 11 ft. lbs. (15 Nm).
38. Install oil pan drain plug and tighten to 41 ft. lbs. (55 Nm).
39. Install the A/C compressor.
40. Install the alternator.
41. Install the crankshaft sprocket and torque the crankshaft center bolt to 184 ft. lbs. (250 Nm) plus 45° additional bolt rotation.
42. Remove the crankshaft holding tool and install the flywheel inspection cover.
43. Connect the front exhaust pipe to the exhaust manifolds.
44. Make sure the radiator drain plug is installed properly. Install the lower center air deflector and lower the vehicle.
45. Install the inner timing covers and camshaft sprockets.
46. Install the bracket with the upper timing belt adjuster and tensioner pulleys.
47. Install the timing belt.
48. Clean water pump mounting surface. Coat the water pump O-ring and sealing surface with acid-free petroleum jelly.
49. Install water pump with new O-ring and torque water pump bolts to 18 ft. lbs. (25 Nm)
50. Install the crankshaft pulley and torque the bolts to 15 ft. lbs. (20 Nm).
51. Install the timing cover.
52. Install the water pump pulley. Torque water pump pulley bolts to 6 ft. lbs. (8 Nm).

53. Install power steering pump.
54. Install the drive belt tensioner and torque bolts to 30 ft. lbs. (40 Nm).
55. Install the upper alternator air intake.
56. Install the torque arm engine mount and attach the power steering line clamp.
57. Install the torque arm and bolt the connection to the torque arm engine mount.
58. Connect the upper hose to the coolant expansion tank and set the power steering reservoir back into position.
59. Change the oil filter and fill the engine with oil.
60. Make sure the radiator drain plug is tightened and install the lower center air deflector.
61. Fill the radiator with clean coolant and check the cooling system for leaks
62. Check all other fluids and top off or change as necessary.
63. Bleed the cooling system.
64. Test drive the vehicle and check for any leaks.

TRANSAXLE

Manual Transaxle Assembly

REMOVAL AND INSTALLATION

900 Series

AX4, FA44, FA46, FM51, FM54, FM55 and MX5 Transaxles

NOTE: The engine and transaxle are removed together as an assembly.

1. With all four wheels on the ground, loosen the axle hub nuts.
2. Drain the engine coolant and remove the battery.
3. Remove air cleaner assembly and attached hoses. Take off the resonator.
4. Disconnect throttle cable and move it to one side.
5. Disconnect cruise control wiring harness and cables from throttle body. Loosen retaining nuts on cruise control unit and remove.
6. Properly relieve the fuel system pressure.

CAUTION

Fuel injection systems remain under pressure even after the engine is stopped. The fuel system pressure must be relieved before disconnecting any fuel lines. Failure to do so may result in fire and/or personal injury.

7. Disconnect the fuel lines at their connections at the front of the fuel injection manifold and on the fuel pressure regulator.

8. If equipped, disconnect the turbocharger pressure line from the turbocharger compressor and the intercooler/throttle housing.

9. Disconnect the vacuum hose from the secondary injection and disconnect the brake booster vacuum hose from the intake manifold.

10. If equipped, remove the boost pressure control unit.

11. Using a ⅜ inch ratchet extension relieve belt tensioner of pressure. Remove drive belt.

12. Disconnect the pressure and return pipe from the steering servo pump and have a plug handy to prevent oil escaping from the pipe.

13. Disconnect the cooling system hoses at the following connections:
   a. the heater control valve.
   b. the expansion tank.
   c. the bottom of the radiator.
   d. the thermostat housing.

14. Without disconnecting any hoses, remove the air conditioner compressor and it mounting bracket. Place them on the filter housing for the heater system. Secure the alternator so it will not drop or become damaged.

15. If not already done, raise and safely support the vehicle.

16. Disconnect the positive battery cable from the clips holding it to the body. Disconnect the ground cable from the transaxle.

17. Remove ignition cable and disconnect electrical connections from ignition coil.

18. Unplug the wiring connectors from the transaxle.

19. Disconnect the oxygen sensor wiring and the catalytic converter temperature sensor connector.

20. Under the vehicle, remove the taper pin from the gearshift rod joint. If equipped, disconnect the clutch cable and clutch pipe.

21. Inside the vehicle, pull back carpet under glove box to gain access to the central locking system and disconnect the control module. Feed the wires to the engine compartment through the grommet.

22. Remove hub nuts and remove both wheels. Remove front splash shield.

23. Remove both ball joint nuts and disconnect the ball joints from the struts. Make sure the halfshafts can slide out of the hubs. If necessary, use a wheel puller to push them out now.

24. Separate front pipe from exhaust manifold and remove the catalytic converter.

25. Disconnect oil cooler lines and all lines connected to transaxle housing.

26. Position lifting table under vehicle so it is directly under front engine mount and gearbox. Take off subframe bolts and front engine mounts.

27. Lower lifting table slightly and separate the halfshafts from the hubs.

28. Lower lifting table fully and remove subframe.

29. Place engine on engine stand.

To install:

30. Place engine/transaxle on lifting table.

31. Install subframe on rear engine mount and tighten to 35 ft. lbs. Fit halfshafts into hubs as engine is raised into engine compartment.

32. Connect engine mount and subframe bolts and tighten in sequence:
   - Front bolts–85 ft. lbs.
   - Middle subframe bolts–141 ft. lbs.
   - Rear subframe bolts– to 90 ft. lbs.
   - Rear engine mount–54 ft. lbs.

33. Connect all oil cooler and hydraulic lines.

34. Reconnect all charge air cooler lines between the turbocharger and the cooler.

35. Install the front exhaust pipe to the exhaust manifold. Secure the catalytic converter to the rear exhaust section.

36. Fit engine shield to under side of vehicle.

37. Install hub nuts and wheels. Tighten wheel nuts to 89 ft. lbs. but do not tighten hub nuts yet.

38. Feed the wiring through the grommet and reconnect the central locking system control module.

39. As equipped, hook the clutch cable in place or bleed the clutch hydraulic cylinder.

40. Reconnect the taper pin to the gearshift selector.

41. Connect the oxygen and catalytic converter temperature sensor.

42. Connect all transaxle wiring.

43. Install A/C compressor, servo pump and the drive belt.

44. Reconnect power brake booster vacuum hose and pressure sensor connections.

45. Install cruise control including cables and electrical harness connections.

46. Install the battery and connect the cables.

47. With all four wheels on the ground, torque the hub nuts to 215 ft. lbs. (290 Nm).

48. Fill coolant and oil to specified level and check.

49. Start engine and test drive vehicle.

9000 Series

GM65103 and GM75701 Transaxles

1. Disconnect the battery cables from the battery, negative cable first.

2. Remove the battery, washer fluid container and connectors, terminal blocks and battery tray. Release the stay for the hydraulic unit on ABS equipped vehicles.

3. Remove the 8 bolts for the bulkhead cover. Remove the rubber strip, lift the cover and disconnect the washer hoses from the nozzle and remove the cover.

4. Separate the speedometer cable connector by first removing the washer hose and then the speedometer cable through the rubber grommet.

5. Disconnect the electrical connector from the air mass meter. Disconnect the connector for the intake air temperature sensor and disconnect the hose on the delivery pipe from the by-pass valve.

6. Remove the delivery pipe between the throttle housing and the intercooler. Also, remove the nuts retaining the starter motor. Remove the starter and place it on the steering gear.

7. Separate the gear selector rod universal joint and selector rod. Disconnect the slave cylinder pressure hose.

8. Remove the upper bolts for the stay at the wheel housing. Release the left-hand engine mounting. Attach a sling to the engine lifting beam. Raise and support the vehicle safely.

9. Remove the left wheel and wheel housing liner. Disconnect the reverse light connector from the transaxle.

10. Separate the suspension arm from the ball joint. Remove the sway bar. Remove the lower bolt for the stay at the wheel housing and the 3 bottom bolts holding the engine to the transaxle.

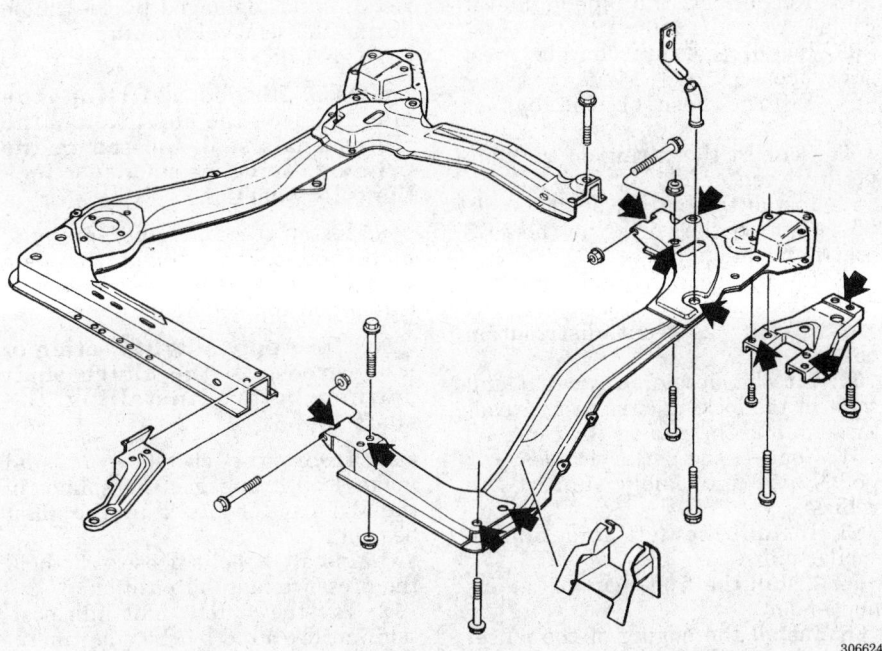

306624

Subframe connection points — 9000 Series GM65103 and GM75701 Transaxles

11. Remove the center and left skirts under the spoiler. Separate the subframe at the front and rear. Lower the subframe. Remove the universal joint and lower the vehicle.

12. Sling the transaxle from the workshop hoist and remove the top nut and bolt. Remove the transaxle and separate the halfshafts. Carefully lower the transaxle to the floor.

To install:

13. Prior to installation, ensure that the halfshaft is in position and the aluminum tube is pressed into the seal.

14. Slide the transaxle into position, guiding the driver and input into place. Install the top bolt and nut into the engine/transaxle joint face. Release the transaxle from the hoist. Raise and support the vehicle safely.

15. Fit the 3 bottom bolts into the joint face and tighten to 40–74 ft. lbs. (54–100 Nm). Install the universal joint. Raise the subframe into position and secure to 31–42 ft. lbs. (43–57 Nm).

16. Install the sway bar and tighten to 30–40 ft. lbs. (40–52 Nm).

17. Secure the suspension arm to the ball joint and torque to 14–20 ft. lbs. (20–27 Nm). Start the bolts to the bracket for the wheel housing stay. Do not tighten the wheel housing

stay bolt at this time. Lower the vehicle.

18. Remove the lifting beam. Install the starter and the top mounting of the wheel housing. Now tighten all wheel housing bolts.

19. Tighten the left engine mount 37–67 ft. lbs. (49–91 Nm). Raise and support the vehicle safely.

20. Connect the negative battery cable and reverse light switch. Install the wheel housing liners, left wheel, under car skirts, selector rod universal joint, slave cylinder pressure pipe (remove the clamping tongs), speedometer cable, washer hose and the remainder of the components removed.

21. Bleed the slave cylinder, check the fluid level and road test the vehicle.

FM51001, FM54001, FM57001, FM57101 and MX5 Transaxles

1. Place vehicle in 4th gear.

2. Disconnect and remove battery.

3. Remove locking clips which hold the accelerator cable in the leadthough and turn cable to one side.

4. Disconnect and remove front distribution box.

5. Remove the positive junction (without removing the cables).

6. Remove the positive cable's two clamps on battery shelf.

7. Remove the connect from the control module ABS.

8. Remove battery shelf.

9. Remove the bypass valve from the turbocharger pressure pipe.

10. Disconnect the temperature sensor from the turbocharger pressure pipe.

11. Remove the hose clips from the turbocharger pressure pipe.

12. Pull apart the connector on the cable for the speedometer sensor.

13. Remove the reverse light switch connector.

14. Disconnect the clutch slave cylinder hydraulic line.

15. Remove all the engine-to-transaxle bolts accessible from the top side, except the top bolt.

16. Place lifting beam on the edges of the engine compartment. Insert the hook into the engine's lifting eye and tighten the lifting beam wing nut slightly to support the engine and transaxle from above.

17. Safely raise and support the vehicle.

18. Remove left front wheel and the edging of the wheel housing.

19. Remove the front part of the inner fender and both left and middle spoiler units.

20. Remove the ground cable from the gear box.

21. Remove the two nuts on the gear selector universal joint and disconnect the shift linkage.

22. Remove the three bolts connecting the ball joint to the lower control arm.

23. Remove the anti-roll bar from the lower control arm.

24. Disconnect the front engine mounting nut from the bolt.

25. Remove the screw from the wheel housing stay and put the two washers in a safe place.

26. The transaxle side of the subframe must be folded down out of the way:

   a. Remove the bolt in the front link of the subframe assembly and remove the two bolts that hold the front link.

   b. Remove the bolt in the rear link of the subframe assembly.

   c. Remove the two bolts that hold the rear link, one of which also holds the steering rack.

   d. Remove the two bolts in the front corners of the subframe, then the four bolts in the rear corners.

   e. Carefully fold down the subframe. Put the plate between the subframe and the chassis in a safe place.

27. Remove the clamp around the CV-joint boot and disconnect the CV-

joint. With the ball joint disconnected, the strut will move with the halfshaft but be careful not to damage the boot.

28. Remove the protective plate for the transaxle.

29. Lower the vehicle.

30. Support the transaxle and remove the remaining engine-to-transaxle bolts. Carefully lower the transaxle out of the vehicle.

To install:

31. Fit two locating pins into the engine block.

32. Lift transaxle into position and onto the locating pins. Push the transaxle against the engine, making sure to align the clutch with the crankshaft, then start some of the bolts. Use the bolts to draw the transaxle up against the engine block.

33. Remove the locating pins and safely raise and support the vehicle.

34. Install the protection plate for the transaxle and install the remaining lower bolts. Tighten the bolts to 50 ft. lbs. (70 Nm).

35. Hang the subframe and secure each link with one bolt. Make sure the plate in the rear corner is positioned correctly.

36. Install the bolt in the front engine mount, making sure the washer is in the correct position.

37. Install all the remaining subframe bolts but do not tighten any of them yet.

38. When all bolts are installed, tighten all the bolts in the subframe to 41 ft. lbs. (55 Nm).

39. Install the bolt to the stay in the wheel housing, tighten to 37 ft. lbs. (50 Nm).

40. Install the bolts which hold the anti-roll bar bearing.

41. Install the nut which holds the anti-roll bar mount in the lower control arm.

42. Connect the halfshaft.

43. Install the three bolts of the lower ball joint to the control arm tighten to 25 ft. lbs. (34 Nm).

44. Install the CV-joint boot clamp.

45. Make sure the transaxle is in fourth gear, fit the two screws in the selector universal joint.

46. Fit the battery grounding cable to the gearbox.

47. Lower the vehicle and remove the lifting beam.

48. Install and tighten all the remaining engine-to transaxle bolts to 50 ft. lbs. (70 Nm).

49. Check the position of the washer in the engine mount and tighten nut to 50 ft. lbs. (70 Nm).

50. Connect the slave cylinder hydraulic line.

51. Reconnect the reverse light connector.

52. Reconnect the speedometer cable.

53. Install the turbocharger pressure pipe.

54. Fit the hose on the bypass valve.

55. Connect the wiring on the temperature sensor.

56. Install the battery shelf.

57. Connect the wiring on the ABS control module.

58. Fit the positive cable and the two clamps on the battery shelf.

59. Install the front distribution box.

60. Fit the accelerator cable and connect the locking clips in the lead-through and the clip on the cable.

61. Connect the battery cables.

62. Raise and safely support the vehicle.

63. Install the left and middle spoiler units.

64. Install the front part of the inner fender.

65. Install the edging of the wheel housing.

66. Install the left front wheel.

67. Bleed the clutch.

68. Check for any leakage. Check the oil level in the gearbox.

69. Test drive vehicle.

Clutch Assembly

REMOVAL AND INSTALLATION

900 Series

1993 Models

1. Scribe the hood position and remove the hood.

2. Disconnect the negative battery cable.

3. Remove the clutch housing cover.

4. Install the spacer 83–90–023, or equivalent, between the clutch cover and the diaphragm spring. Keep the clutch pedal depressed when the ring is being installed.

5. Unhook the spring clip and remove the cover located in front of the clutch shaft. Remove the clutch shaft plastic oil slinger propeller.

6. Remove the clutch shaft by means of an M8 bolt installed in the shaft end and tool 83–93–175, or equivalent. Withdraw the shaft as far as possible.

7. Remove the clutch slave cylinder retaining bolts.

8. Remove the clutch retaining bolts and remove the clutch pressure plate, clutch disc and the slave cylin-

der complete with the clutch release bearing. Be sure the slave cylinder sleeve is not damaged by the clutch during the removal procedure.

To install:

NOTE: Before refitting the clutch, check the condition of the clutch shaft seal, located in the primary drive case, also check the pilot bearing.

9. Install the clutch disc, pressure plate and slave cylinder complete with the release bearing into the clutch housing as one unit.

NOTE: Apply a light coating of moly grease to the clutch shaft splines before installing the shaft.

10. Insert the clutch shaft and make sure it engages the splines of the disc and centers in the pilot bearing.

11. Install 2 bolts loosely to hold the pressure plate in position.

12. Tap the clutch shaft into position until secured by the snapring in the primary gear.

13. Bolt the slave cylinder into place on the primary gear case and tighten to 5–10 ft. lbs. (6–14 Nm). Be sure to coat the bolts with thread sealing compound.

14. Position the plastic propeller on the end of the clutch shaft.

15. Install and tighten the remaining pressure plate bolts 17–21 ft. lbs.(23–28 Nm) to the flywheel.

16. Have an assistant depress the clutch pedal only far enough to remove spacer tool 83–90–023.

NOTE: Do not depress the clutch pedal completely, this could damage the lip seal causing hydraulic fluid loss.

17. With the pedal still depressed, move the old style sliding lock ring up to the slave cylinder. If the vehicle has the new version slave cylinder without the lock ring, press the plastic sleeve up against the release bearing.

18. Replace the clutch cover.

19. Install the preheater hose.

20. Install the hood.

21. Check for proper pedal operation.

1994–97 Models

1. Place car on lift and engage in 4th gear.

2. Disconnect battery and remove from vehicle.

3. Remove ground strap from transaxle housing.

4. Disconnect the positive cable routing straps.

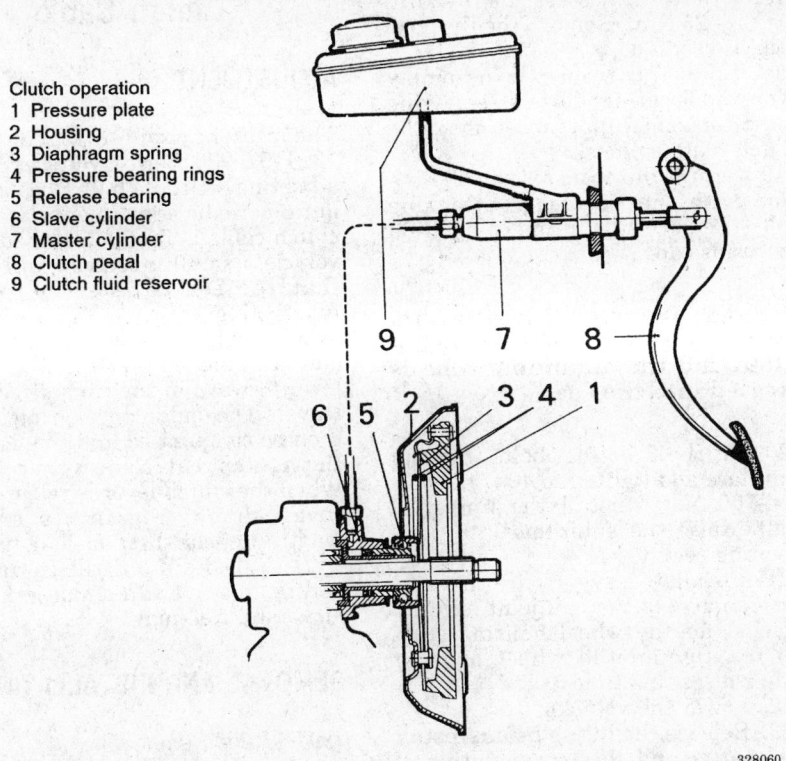

Clutch operation
1 Pressure plate
2 Housing
3 Diaphragm spring
4 Pressure bearing rings
5 Release bearing
6 Slave cylinder
7 Master cylinder
8 Clutch pedal
9 Clutch fluid reservoir

328060

Hydraulic clutch assembly component identification — 1993 900 Series

5. Disconnect the electrical harness connector for the rear light switch.

6. Disconnect the clutch cable from the clutch lever and release the cable's rubber damper from the fastener on the transaxle.

7. Disconnect the oxygen sensor connectors and remove the securing straps .

8. Install the engine holder. Insert the lifting bar in the eyelet's and connect to the engine holder.

9. Install transaxle locking pin to ensure the car does not move out of gear.

10. Disconnect the shift rod and linkage from the transaxle.

11. Safely raise vehicle.

12. Remove front wheels.

13. Remove front exhaust pipes.

14. Remove front spoilers.

15. Disconnect the ball joints on both sides.

16. Using the proper jack remove the carrying frame assembly to gain access to transaxle housing.

17. Drain the transaxle.

18. Remove the left and right front halfshafts and suspend shafts with securing straps.

19. Remove intermediate shaft bearing bracket and pull out shaft.

20. Remove flywheel cover.

21. Remove transaxle mounting bracket.

22. Remove three transaxle bracket bolts on the transaxle

23. Remove the engine/transaxle surface mounting bolts.

24. Lower vehicle.

25. Remove the two outer bolts holding gear box to engine mounting surface.

26. Install lifting eye to transaxle and connect hoist to transaxle.

27. Remove last bolt and pull out transaxle assembly.

28. Place manual transaxle assembly on proper stand.

29. Lock flywheel in position and remove pressure plate retaining nuts.

30. Remove pressure plate and clutch disc.

To install:

31. Using a clutch disc alignment tool install clutch disc with pressure plate.

32. Tighten pressure plate retaining nuts to 16 ft. lbs. (22 Nm).

33. Using a hoist lift transaxle into position.

34. To help with installation use a guide pin to line up transaxle.

35. When transaxle is in position fit the three upper bolts into engine housing.

36. Tighten transaxle/engine surface mounting bolts to 65 ft. lbs. (90 Nm).

37. Safely raise vehicle.

38. Fit transaxle/engine surface mounting bolts from under car and tighten to 65 ft. lbs. (90 Nm).

39. Install transaxle mounting bracket to the transaxle housing and the frame assembly.

40. Install flywheel cover.

41. Install intermediate driveshaft with bracket bearing assembly.

42. Install both halfshafts.

43. Using a proper jack install subframe assembly. Tighten ball joints to 55 ft. lbs. (75 Nm). Fit sub frame bolts and tighten front bolts to 85 ft. lbs. (11 Nm), tighten middle bolts to 141 ft. lbs. (190 Nm), and tighten rear bolts to 81 ft. lbs. (110 Nm) + 75 degrees additional torque.

44. Install front exhaust pipes.

45. Install front wheels.

46. Lower vehicle.

47. Install shift rod and engage 4th gear, install shift rod mounting bolt to secure linkage.

48. Remove the engine holder and lifting rods.

49. Remove the transaxle locking pin.

50. Fill transaxle with fluid.

51. Connect the oxygen sensor electrical harness connectors.

52. Fit the positive battery cable.

53. Connect clutch cable, and connect rear light switch harness connector.

54. Install battery and connect ground cable.

55. Test drive vehicle.

9000 Series

1. Disconnect the battery cables from the battery, negative cable first.

2. Remove the battery, washer fluid container and connectors, terminal blocks, battery tray and release the stay for the hydraulic unit on ABS equipped vehicles.

3. Remove the 8 bolts for the bulkhead cover. If equipped, remove the rubber strip, lift the cover and disconnect the washer hoses from the nozzle and remove the cover.

4. If equipped, separate the speedometer cable connector by first removing the washer hose and then the speedometer cable through the rubber grommet.

5. Disconnect the electrical connector from the air mass meter. Disconnect the connector for the intake air temperature sensor and disconnect the hose on the delivery pipe from the bypass valve.

6. If equipped, remove the delivery pipe between the throttle housing and the intercooler. Also, remove the nuts retaining the starter motor. Remove the starter and place it on the steering gear.

7. Separate the selector rod universal joint and selector rod. Pinch the slave cylinder pressure hose with clamping tongs and disconnect the pressure line.

8. Remove the upper bolts for the stay at the wheel housing. Remove the left-hand engine mounting. Attach a sling to the engine lifting beam. Raise and support the vehicle safely.

9. Remove the left wheel and wheel housing liner. Disconnect the reverse light connector from the transaxle.

10. Separate the suspension arm from the ball joint. Remove the sway bar. Remove the lower bolt for the stay at the wheel housing and the 3 bottom bolts from the joint between the engine and the transaxle.

11. Disconnect the speedometer cable at the transaxle.

12. Remove the center and left skirts under the spoiler. Separate the subframe at the front and rear. Lower the subframe. Remove the universal joint and lower the vehicle.

13. Sling the transaxle from a suitable hoist and remove the top nut and bolt from the clutch housing face. Remove the transaxle by lowering to the floor.

14. Install a flywheel locking tool, if available, and remove the clutch assembly from the flywheel.

15. Inspect the flywheel for wear, scoring, cracking or other damage. Replace or machine, as necessary.

To install:

16. Use a centering arbor type tool or an appropriate input shaft to center the clutch disc to the flywheel.

17. Install the clutch pressure plate and tighten the bolts, alternately and evenly, in several steps, to 10–20 ft. lbs. (13–25 Nm). Remove the flywheel lock, if used.

18. Slide the transaxle assembly over the locating dowels, engaging the transaxle input shaft into the clutch plate splines.

19. Secure the transaxle to the engine with the necessary attaching bolts. Remove the lifting sling from the transaxle.

NOTE: Prior to installation, ensure that the halfshaft is in position and the aluminum tube is pressed into the seal.

20. Fit the 3 bottom bolts into the joint face and tighten to 40–74 ft. lbs. (54–100 Nm). Install the universal joint. Raise the subframe into position and secure.

21. Install the sway bar, the suspension arm to the ball joint and the bracket for the wheel housing stay. Do not tighten the wheel housing stay bolt at this time.

22. Lower the vehicle.

23. Remove the lifting beam. Install the starter and the top mounting of the wheel housing. Now tighten all wheel housing bolts to 32–43 ft. lbs. (43–57 Nm).

24. Tighten the left engine mount to 37–67 ft. lbs. (41–91 Nm). Raise and support the vehicle safely.

25. Connect the negative battery cable.

26. Connect the reverse light switch.

27. Install the wheel housing liners, left wheel, under car skirts and selector rod universal joint.

28. Connect the slave cylinder pressure pipe (remove the clamping tongs), speedometer cable, washer hose and the remainder of the components removed.

29. Bleed the slave cylinder, check the transaxle fluid level and road test the vehicle.

Clutch Cable

ADJUSTMENT

The clutch cable utilized in the 1994–97 900 Series vehicles is self-adjusting. Although it does not require periodic adjustments, when the clutch cable is first reinstalled in the vehicle a small procedure will insure that the self-adjuster is properly functioning.

1. Check the functioning of the clutch cable by moving the clutch lever forward in the car's direction of travel. The balancing spring should then be compressed and the length of the clutch cable's cover reduced. When the clutch lever is released, the cover should regain its original length. Repeat three or four times.

2. Grip hold of the balancing spring and give it a small jerk to remove any free-play.

REMOVAL AND INSTALLATION

1994–97 900 Series

1. Disconnect and remove battery.

2. Remove clutch cable from gearbox connection.

3. Disconnect clutch cable from rubber dampener.

4. Remove the power distribution box.

5. Remove all routing fasteners on clutch cable.

6. From inside the vehicle, remove the data link connector from under the dash.

7. Remove lower dash panel.

8. Remove air duct for floor heating.

9. Remove knee shield.

10. Remove fuse box and swing out of way.

11. Remove the Integrated Central Electronics (ICE) unit and swing out.

12. Remove the air duct to the air distributor on the instrument panel.

13. Push the pedal spring to one side and remove the cable's plastic eye.

14. Hold the self adjusting mechanism and pull cable through the firewall access.

To install:

15. Install the self adjusting unit through the firewall.

16. Hook the plastic eye of the cable on the pedal and replace the pedal return spring.

17. Press the rubber dampener in the gearbox holder.

18. Connect the cable to the stop on the clutch lever.

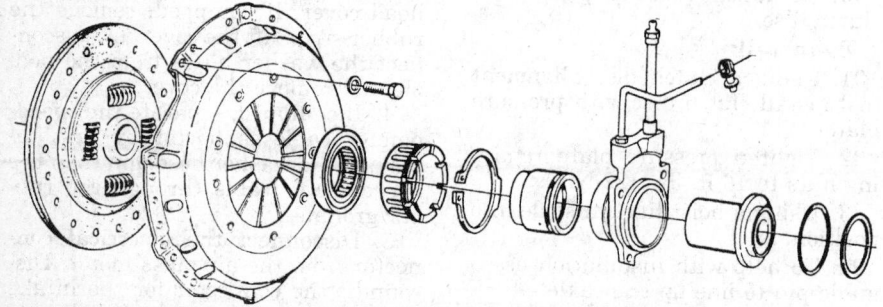

306668

Exploded view of the clutch assembly — 9000 Series

19. Install all fasteners to clutch cable.
20. Check the function of the clutch lever and fit clutch lever return spring.
21. Install the power distribution box.
22. Install the air duct to the air distributor on the instrument panel.
23. Fit the control unit.
24. Install the fuse box.
25. Install the knee shield.
26. Fit the floor directed air ducts.
27. Install the data link connector.
28. Install battery and test drive vehicle.

Clutch Master Cylinder

REMOVAL AND INSTALLATION

900 Series

1. Disconnect the negative battery cable.
2. Remove the left finish panel under the instrument panel.
3. Place a suitable drip tray under the hose connections. Pinch the fluid supply hose with a small clamp and disconnect the hose and pressure line from the master cylinder.
4. Remove the clevis pin holding the pushrod to the clutch pedal.
5. Remove the clutch cylinder bolts inside the dash panel.
6. Remove the clutch cylinder from inside the engine compartment.
7. Remove the hose from the fluid container and hang it aside so the fluid does not come out.
   To install:
8. Position the replacement cylinder assembly. Tighten bolts to 6–7 ft. lbs. (8–10 Nm).
9. Connect the pushrod and insert the clevis pin.
10. Connect the fluid hose and pressure line and tighten. Tighten the clamp securing the line.
11. Install the finish panel.
12. Connect the negative battery cable.
13. Bleed the system.

9000 Series

1. Remove the underdash trim panel.
2. Remove the clevis pin from the pushrod.

   NOTE: Before disconnecting the fluid hoses from the cylinder, place a drain pan and protective cover over the carpet to protect against leaking fluid.

3. From under the hood, disconnect the supply hose from the cylinder and close off the end.
4. Disconnect the pressure hose from the master cylinder.
5. Remove the 2 retaining bolts and remove the master cylinder.
   To install:
6. Position the master cylinder and tighten the 2 retaining bolts 16 ft. lbs. (22 Nm).
7. Connect the pressure line and tighten to 10–13 ft. lbs. (14–18 Nm).
8. Connect the supply hose.
9. Connect the linkage to the pedal.
10. Install the lower trim panel.
11. Refill the fluid reservoir and pump the pedal.
12. Bleed the hydraulic system.

Clutch Slave Cylinder

REMOVAL AND INSTALLATION

900 Series

1. Disconnect the negative battery cable.
2. Remove the clutch assembly.
3. Remove the clutch release bearing together with the clutch slave cylinder.
4. Installation is the reverse of the removal procedure.

9000 Series

1. Disconnect the negative battery cable.
2. Remove the transaxle from the vehicle.
3. Remove the clutch release bearing. Disconnect the pressure pipe. Remove the bleed nipple.
4. Remove the retaining bolts that hold the slave cylinder in place.
5. Remove the clutch slave cylinder.
6. Installation is the reverse of the removal procedure.

Hydraulic Clutch System Bleeding

1. Connect a hose to the slave cylinder bleeder valve. Place the other end of the hose in a suitable jar partially filled with brake fluid.
2. Fill the master cylinder with brake fluid.
3. Open the bleeder valve on the slave cylinder ½ turn.
4. Place a cooling system pressure tester gauge over the opening of the master cylinder.

5. Pump the tester until all air is expelled from the hydraulic clutch system.
6. Close the slave cylinder bleeder valve.
7. Check that all air was removed from the system and the clutch is functioning properly. Adjust the fluid level, as required.

Automatic Transaxle Assembly

REMOVAL AND INSTALLATION

900 Series

1. Disconnect battery and remove from vehicle.
2. Remove ground strap from transaxle housing.
3. Remove dipstick and sleeve, plug access hole in transaxle housing.
4. Remove vent hose from transaxle housing.
5. Disconnect the positive cable routing straps.
6. Remove the transaxle selector lever from housing.
7. Disconnect electrical harness connector on transaxle housing.
8. Disconnect the oxygen sensor connectors and remove the securing straps .
9. Install the engine lifting beam. Insert the lifting bar in the eyelet's and connect to the engine holder.
10. Safely raise vehicle.
11. Remove front wheels.
12. Remove front exhaust pipes.
13. Remove front spoilers.
14. Disconnect the ball joints on both sides.
15. Using the proper jack remove the subframe assembly to gain access to transaxle housing.
16. Remove transaxle cooler lines from the transaxle.
17. Remove the left and right front driveshafts and suspend shafts with securing straps.
18. Remove intermediate shaft bearing bracket and pull out shaft.
19. Remove actuator cover.
20. Release the torque converter's securing point on the actuator.
21. Remove transaxle mounting bracket.
22. Remove transaxle mounting bolts that secure the transaxle to the engine.
23. Remove the engine/transaxle surface mounting bolts.
24. Lower vehicle.
25. Remove the two outer bolts holding gear box to engine mounting surface.

26. Install lifting eye to transaxle and connect hoist to transaxle.

27. Remove last bolt and pull out transaxle assembly.

To install:

28. Using a hoist lift transaxle into position.

29. To help with installation use a guide pin to line up transaxle.

30. When transaxle is in position fit the three upper bolts into engine housing.

31. Tighten upper transaxle-to-engine mounting bolts to 55 ft. lbs. (75 Nm).

32. Install transaxle/engine lifting beam to hold unit in place.

33. Safely raise vehicle.

34. Fit transaxle/engine surface mounting bolts from under car and tighten to 55 ft. lbs. (75 Nm).

35. Install transaxle mounting bracket to the transaxle housing and the frame assembly.

36. Install torque converter to the actuator with retaining bolts, use locking fluid and tighten to 37 ft. lbs. (55 Nm).

37. Install actuator cover.

38. Install the intermediate half-shaft with the bracket bearing assembly.

39. Install the outer halfshafts.

40. Install transaxle cooler lines to transaxle assembly.

41. Using a proper jack install sub frame assembly. Fit sub frame bolts and tighten front bolts to 85 ft. lbs. (11 Nm), middle bolts to 141 ft. lbs. (190 Nm), and the rear bolts to 81 ft. lbs. (110 Nm) + 75° bolt rotation. Tighten ball joints to 55 ft. lbs. (75 Nm).

42. Install front exhaust pipes.

43. Install front wheels.

44. Lower vehicle.

45. Fill transaxle with fluid.

46. Install selector lever cable and selector lever to the transaxle housing.

47. Connect the oxygen sensor electrical harness connectors.

48. Install dipstick and sleeve to transaxle housing.

49. Fit the positive battery cable.

50. Install battery and connect ground cable.

51. Test drive vehicle.

9000 Series

NOTE: The engine and transaxle assembly are removed together.

1. Raise the vehicle and support it safely.

2. Drain the cooling system.

3. Disconnect the battery cables and remove the battery and tray.

4. Remove the connecting bolt for the expansion tank, disconnect the tank from the suction and remove the overflow hoses from the radiator.

5. Loosen the drive belt for the compressor by loosening the locknut, and loosening the adjusting nut under the locknut.

6. Disconnect the upper connection on the oil cooler, loosen the pipe clip on the radiator and slide the pipe down behind the radiator.

7. Unplug the connector to the electromagnetic clutch on the compressor and loosen the compressor mounting complete with the belt tensioner.

8. Place a protective cloth over the radiator member and rest the compressor on the radiator member. Secure the compressor to the radiator member.

9. If turbocharged, remove the turbocharger pressure pipe, situated between the turbocharger unit and the intercooler.

10. Disconnect the Lambda probe connector leads and disconnect them from the clips.

11. From the engine compartment, unbolt the flange joint between the exhaust pipe and the exhaust manifold. If turbocharged, disconnect the exhaust pipe coming from the turbocharger. Push the exhaust pipe to one side and unhook the rubber hangers from the exhaust system. Disconnect the bottom coolant hose from the water pump.

12. From under the vehicle, remove the bottom retaining bolt for the radiator fan.

13. Disconnect the speedometer drive from the gearbox.

14. Disconnect the gear selector cable from the selector lever. Do not separate at the ball joint.

15. Remove the clips on the rubber gaiters over the inboard universal joints and slide the gaiters off the drive axles.

16. Disconnect the electrical leads from the alternator and the starter motor. Unplug the connector for the oil pressure switch.

17. Remove the clips and remove the top radiator hose.

18. Disconnect the top radiator hose at the cylinder head.

19. Disconnect the high tension lead from the ignition coil at the distributor cap.

20. Remove the solenoid valve from the bracket on the radiator and unplug the electrical connections.

21. Remove the bolts from the top of the radiator fan. Disconnect the wiring loom and lift out the fan.

22. Pull the connector off the air mass meter. Disconnect the air mass meter from the air intake duct socket connector and the air cleaner. Leave the rubber socket connector attached to the turbocharger unit.

23. Remove the air intake duct by pulling it out of the aperture in the wing and twisting the ends inwards.

24. Remove the air cleaner top section first, then the remaining section.

25. Disconnect the relief valve hose from the turbocharger pressure pipe and remove the pipe, if equipped.

26. Disconnect the Hall effect transducer, the earth lead from the gear box and the electrical connector for the back-up lights.

27. Disconnect the end of the throttle cable and disconnect the throttle linkage.

28. Install a clamp to the hydraulic line to the slave cylinder and pinch the line tightly. With proper wrenches, open the line to the clutch slave cylinder.

29. Remove the front wheels.

30. From both sides of the vehicle, slacken the lower bolts retaining the steering swivel member to the strut assembly. Remove the 2 upper bolts.

31. Pivot the steering swivel member outwards to pull the inboard universal joint out of the halfshaft. Position dust covers over the exposed halfshaft cups.

32. Remove the engine mount bolt.

33. Remove the steering reservoir for the servo and position it within the engine compartment. Drain the fluid from the container.

34. Disconnect the large bore hose and the delivery hose from the steering servo pump and plug the open ends.

35. Disconnect the fuel return lie from the pressure regulator.

36. Remove the nut from the rear engine mounting and back off the front mount bolts a few turns.

37. Attach the lifting sling (Saab 83–92–409) or equivalent to the rear lifting lug.

38. Lift the engine sufficiently to provide access for the removal of the components located between the engine and the firewall.

39. Disconnect and tag the vacuum hoses from the inlet manifold.

40. Remove the coolant hoses running between the heater core and the water pump pipe.

41. Separate the coupling between the fuel pipe and the fuel injection manifold. Do not allow the fuel to spill or collect.

42. Cut the clips securing the wiring looms to the oil pipe, water pipe.

Disconnect and tag the vacuum hoses from the inlet manifold.

43. Unclip the wiring loom to the fuel injection manifold.

44. Disconnect the grounding connections and the electrical connectors from the wiring harness.

45. Unbolt the air cooled oil cooler and place it on top of the engine. The 2 lower bolts need only be loosened.

46. Disconnect the hood lifts and install hood extenders for extra clearance.

47. Carefully lift the engine from the vehicle, taking care not to damage the radiator.

To install:

48. Install the engine in the vehicle and connect all the engine mounts.

49. Bolt the air cooled oil cooler in place. Tighten the 2 lower bolts.

50. Connect the grounding connections and the electrical connectors to the wiring harness.

51. Clip the wiring loom to the fuel injection manifold.

52. Secure the wiring looms to the oil pipe, water pipe, inlet manifold steady bar and the oil supply pipe.

53. Connect the coupling between the fuel pipe and the fuel injection manifold.

54. Connect the coolant hoses running between the heater core and the water pump pipe.

55. Connect the vacuum hoses to the correct connections on the inlet manifold.

56. Before removing the lifting sling, connect the components located between the engine and the firewall.

57. Attach the nut from the rear engine mounting and tighten the front mount bolts.

58. Connect the fuel return line to the pressure regulator.

59. Connect the large bore hose and the delivery hose to the steering servo pump.

60. Connect the steering reservoir for the servo. Fill the fluid container.

61. Connect the engine stay bolt.

62. Pivot the steering swivel member outwards to install the inboard universal joint into the halfshaft. Position dust covers over the exposed halfshaft cups and install clamps.

63. Connect the gear selector cable to the selector lever.

64. Connect the lower bolts retaining the steering swivel member to the strut assembly. Tighten the 2 upper bolts.

65. Install the front wheels.

66. With line wrenches, tighten the line to the clutch slave cylinder.

67. Connect the end of the throttle cable and the throttle linkage.

68. Connect the Hall Effect transducer, the negative battery cable to the gear box and the electrical connector for the back-up lights.

69. Connect the relief valve hose to the turbocharger pressure pipe, if equipped.

70. Install the air cleaner.

71. Install the air intake duct.

72. Push the connector on the air mass meter. Connect the air mass meter to the air intake duct socket connector and the air cleaner.

73. Connect the high tension lead to the ignition coil at the distributor cap.

74. Connect the top radiator hose at the cylinder head.

75. Connect the electrical leads to the alternator and the starter motor. Plug in the connector for the oil pressure switch.

76. Connect the speedometer drive to the gearbox.

77. From the engine compartment, connect the flange joint between the exhaust pipe and the exhaust manifold. If turbocharged, connect the exhaust pipe coming from the turbocharger. Hook the rubber hangers to the exhaust system.

78. Connect the bottom coolant hose to the water pump.

79. Connect the Lambda probe connector leads.

80. Plug the connector into the electromagnetic clutch on the compressor and tighten the compressor mounting complete with the belt tensioner.

81. Connect the upper connection on the oil cooler, tighten the pipe clip on the radiator.

82. Tighten the drive belt for the compressor .

83. Install the expansion tank, connect the tank to the suction and overflow hoses from the radiator.

84. Install the battery and tray. Connect the battery cables.

85. Remove the lift supports and lower the vehicle.

86. Fill the radiator with coolant, the engine with oil and the transmission with fluid. Start the engine and allow it to reach operating temperature. Check the ignition timing and all fluid levels. Test drive the vehicle.

Throttle Valve Cable

ADJUSTMENT

900 Series

1. Remove the screw for the pressure tap on the transaxle and connect a pressure gauge. Block the drive wheels and apply the parking brake.

2. Start the engine and check that the idle speed is 850 rpm in **P**.

3. Disconnect the throttle cable from the spindle lever and ensure that it is not binding. If so, clean the throttle cable thoroughly and reconnect.

4. With the gear selector in **D**, check that the cable is released and adjust the throttle cable to obtain the lowest possible pressure.

5. Readjust the cable so the pressure increases to 1.4 psi.

6. With the gear selector in **P**, check that the pressure is now 59–69 psi. Pressure should not be allowed to exceed 69 psi.

7. Tighten the cable locknuts.

9000 Series

1. With the engine idling, check the clearance between the cable stop and the end of the throttle cable. The clearance should be 0.08–0.10 in. (2.0–2.5mm).

2. If clearance is not within specification, loosen the locknuts and adjust the cable.

3. Tighten the locknuts and recheck the clearance.

DRIVE AXLE

Halfshaft

REMOVAL AND INSTALLATION

900 Series

1993 Models

1. Disconnect the negative battery cable.

2. Loosen the hub center nut.

3. Support the upper control arm with special tool 83–93–209 or equivalent.

4. Raise the vehicle and support it safely. Remove the wheel.

5. If equipped with ABS, unbolt the ABS wheel sensor and move it out of the way.

6. Remove the brake housing and position it in the wheel housing to avoid damage to the brake hose. Remove the brake disc and parking brake assembly along with the cable, if equipped.

7. Remove the large clamp from the CV-joint boot on the inner universal joint. To separate the inner universal joint, install a cover over the rubber bellows to stop the needle bearings from falling out and to keep

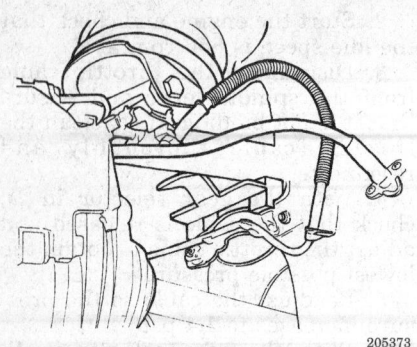

205373

Supporting the upper control arm using special tool 83-93-209 — 1993 900 Series

dirt from entering. Install the protective cap on the inner driver.

8. Remove the securing bolts for the upper and lower ball joints on the upper and lower control arms. Separate the tie rod end from the knuckle.

9. Remove the halfshaft and steering knuckle assembly through the wheel housing. Separate the halfshaft from the hub.

To install:

10. Install the halfshaft through the wheel housing. Before connecting the halfshaft to the inner driveshaft, pack the CV-joint boot with fresh, clean CV-joint grease and install a new boot clip.

11. Install the outer halfshaft end into the hub on the steering knuckle.

12. Install the upper and lower control arm ball joints and torque the securing bolts at each control arm to 30-40 ft. lbs. (40-54 Nm). Install the tie rod to the steering knuckle and torque the tie rod nut to 37-44 ft. lbs. (50-60 Nm). Check the upper and lower ball joint nuts to make sure that they are torqued to 26-37 ft. lbs. (35-50 Nm).

13. Install a new hub center nut. The center nut should be finger tight only.

14. If equipped with ABS, Install the wheel sensor, gently pressing the

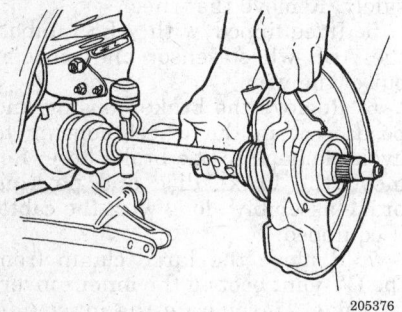

205376

Removing the halfshaft and steering knuckle assembly from the wheel housing — 1993 900 Series

sensor body against the sensor wheel. Tighten the retaining screw.

15. Install the brake rotor and brake caliper assembly. Torque the brake caliper mounting bolts to 44-74 ft. lbs. (60-100 Nm).

16. Install the wheel and tighten the lug nuts. Torque the lug nuts to 80-90 ft. lbs. (108-122 Nm). Carefully lower the vehicle to the ground.

17. Torque the hub center nut to 214-229 ft. lbs. (290-310 Nm). Remove spacer tool no. 83-93-209 or equivalent, out from under the upper control arm. Using a hammer and punch, stake the hub center nut at the groove in the end of the shaft.

18. Reconnect the negative battery cable. Check front wheel alignment and test drive the vehicle.

1994-97 Models

1. Disconnect the negative battery cable.

2. With the vehicle sitting on all four wheels, remove the hubcap and loosen the hub center nut. Do not remove the center nut at this time.

3. Carefully raise and support the vehicle safely. Remove the wheel.

4. Remove the hub center nut.

5. Remove the ball joint nut. Separate the ball joint from the knuckle using special tool 89-96-696 or equivalent.

6. Remove the sway bar nut. Remove the washer and rubber bushing.

7. Push down on the lower control arm.

8. Using a rubber mallet, tap the halfshaft out of the hub.

9. Move the strut to one side and withdraw the halfshaft. Be extremely careful not to stretch or break the ABS sensor cables or brake hoses. Place a drain pan under the transaxle to catch any fluid spillage.

10. Remove the inner halfshaft joint from the transaxle on the left-hand side. Remove the halfshaft joint from the intermediate shaft on the right-hand side. Use special tool 89-96-654 or equivalent to pry the joints out.

To install:

11. Place the halfshaft into the intermediate shaft on the right-hand side. Reposition the halfshaft into the transaxle on the left-hand side. If necessary, use a rubber mallet to secure the halfshaft into place.

12. Insert the halfshaft into the center of the hub. Install a new hub center nut but do not tighten yet.

13. Insert the ball joint into the bottom of the steering knuckle. Torque the ball joint nut to 55 ft. lbs. (75 Nm).

14. Reconnect the sway bar to the lower control arm. Install the rubber bushing, a new washer and new retaining nut. Torque the nut to 84 inch lbs. (10 Nm).

15. Install the wheels and lug nuts. Torque the lug nuts to 80-90 ft. lbs. (108-122 Nm).

16. Lower the vehicle to the ground. With all four wheels on the ground, torque the hub center nut to 214 ft. lbs. (290 Nm). Install the center hubcap.

17. Check the transaxle fluid level and top off if necessary.

18. Reconnect the negative battery cable.

9000 Series

1. Disconnect the negative battery cable.

2. With all four wheels on the ground, loosen the center axle nut.

3. Raise and support the vehicle safely.

4. Remove the front wheel.

5. Remove the inner fender panel for working access.

6. Unbolt the MacPherson strut from the steering swivel member and detach the flexible brake hose from the clip on the strut. If equipped with ABS, remove the ABS sensor and lead.

7. Loosen the clip on the rubber boot on the inboard CV-joint.

8. Separate the 2 halves of the joint. Install protective covers over the rubber boot and the drive axle.

9. Remove the hub center nut and withdraw the halfshaft from the steering swivel member.

To install:

10. Install the halfshaft and a new hub center nut. Do not tighten the nut yet.

11. Join the two halves of the joint and install a new rubber boot along with a new clamp.

12. Install the MacPherson strut and tighten the strut-to-steering swivel bolts to 58-77 ft. lbs. (78-105 Nm).

13. If equipped with ABS, clean the end of the sensor of any debris. Clean the sensor wheel of any debris. Install the sensor unit gently against the sensor wheel and tighten the retaining screw.

14. Install the brake hose to the clip on the strut and install the inner fender panel. Install the wheel and lug nuts.

15. Lower the vehicle and connect the negative battery cable.

16. With all four wheels on the ground, torque the hub center nut to 207-221 ft. lbs. (280-300 Nm). Check

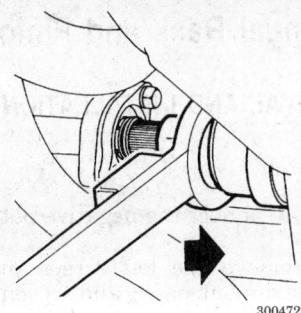

Removing the halfshaft joint from the intermediate shaft using special tool 89–96–654 — 1994–97 900 Series

300472

the alignment and test drive the vehicle.

CV-Joint Boot

REPLACEMENT

All Models

NOTE: Be sure to clean all grease, dust and dirt particles from the rubber boot and halfshaft assembly prior to beginning this procedure. After disassembling the CV-joint, place plastic covers over both ends of the joint to keep out dust and dirt.

1. Disconnect the negative battery cable.
2. On 1993 900 vehicles only, place upper control arm support tool no. 83–93–209, or equivalent, between the underside of the upper control arm and the body.
3. Raise and support the vehicle safely. Remove the front wheel.

NOTE: Remove CV-joint boot clips of the non-screw type using a pair of pliers. Unless you are replacing the CV-joint boot with a new one, be careful not to damage the rubber.

4. Remove the CV-joint boot clamps and slide the boot back along the shaft.
5. Remove the halfshaft and steering knuckle assembly. Separate the halfshaft from the steering knuckle assembly.
6. Spread open the circlip using a pair of snapring pliers, and pull the shaft out of the CV-joint. Slide the boot off of the intermediate shaft. Clean the CV-joint and shaft of all the old grease along with any dirt.

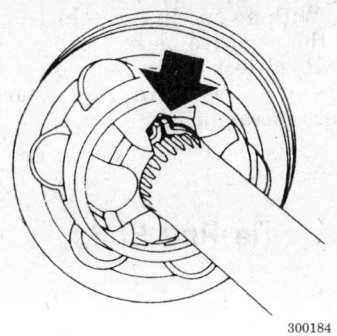

Intermediate shaft snapring located in the CV-joint

300184

To install:

— WARNING —
When installing the CV-joint boot onto the halfshaft, be careful not to cut the boot when sliding it over the splines of the shaft. This could result in loss of CV-joint grease due to damage of the boot.

7. Install the new boot on the intermediate shaft. Install the intermediate shaft into the CV-joint and make sure that the circlip has snapped back into its groove securing the CV-joint assembly. Install the halfshaft and steering knuckle assembly back into the vehicle.
8. Fully pack the CV-joint and boot with clean, fresh grease only and be sure that the inner and outer CV-boots are correctly positioned over the joints.
9. Complete installation of the halfshaft and steering knuckle assembly. Tighten the new CV-joint boot clamps. The non-screw type clamps require the use of special boot clamp pliers.
10. Install the front wheel, lower the vehicle and remove the spacer tool.
11. Reconnect the negative battery cable.

STEERING

Air Bag

— CAUTION —
Some vehicles are equipped with an air bag system, also known as the Supplemental Inflatable Re-

straint (SIR) or Supplemental Restraint System (SRS). The system must be disabled before performing service on or around system components, steering column, instrument panel components, wiring and sensors. Failure to follow safety and disabling procedures could result in accidental air bag deployment, possible personal injury and unnecessary system repairs.

PRECAUTIONS

Several precautions must be observed when handling the inflator module to avoid accidental deployment and possible personal injury.
- Never carry the inflator module by the wires or connector on the underside of the module.
- When carrying a live inflator module, hold securely with both hands, and ensure that the bag and trim cover are pointed away.
- Place the inflator module on a bench or other surface with the bag and trim cover facing up.
- With the inflator module on the bench, never place anything on or close to the module which may be thrown in the event of an accidental deployment.

DISARMING

All Models

— CAUTION —
The air bag system must be disarmed before performing service around air bag system components or system wiring. Failure to do so may cause accidental deployment of the air bag, resulting in unnecessary air bag system repairs and/or personal injury.

Always disconnect the negative battery cable and wait 20 minutes prior to performing service around air bag system components or system wiring. Do not use any diagnostic instruments that are battery powered, such as buzzers, ohmmeters or diode testers, to diagnose faults in the steering wheel or electronic control unit. Using such devices may trigger the air bag.

Steering Wheel

REMOVAL AND INSTALLATION

All Models

─────── CAUTION ───────
On vehicles equipped with an air bag, the negative battery cable must be disconnected for 20 minutes before working on the system. Failure to do so may result in deployment of the air bag and possible personal injury.

1. Disconnect the negative battery cable.
2. Air bag equipped steering wheels:
   a. Remove the 2 screws securing the air bag.
   b. Remove the air bag and disconnect the air bag connectors.
   c. Store the air bag facing up.
   d. Remove the horn contact.
3. Without air bag, remove the steering wheel emblem and remove the horn contact.
4. Remove the steering wheel nut and washer.
5. Remove the steering wheel using a puller if necessary.
To install:
6. Install the steering wheel on the shaft and tighten the retaining nut 18–20 ft. lbs. (25–28 Nm).
7. Without air bag, install the horn contact and the steering wheel emblem.
8. Air bag equipped steering wheels:
   a. Install the horn contact.
   b. Install the air bag and connect the air bag connectors.
   c. Install the 2 screws securing the air bag and tighten to 60 inch lbs. (7 Nm). These screws must be tightened evenly for proper air bag deployment.

9. With no one in the vehicle, connect the negative battery cable.
10. Road test the vehicle on a flat highway to be sure the steering wheel is straight.

Tie Rod Ends

REMOVAL AND INSTALLATION

1. Raise and support the vehicle safely.
2. Remove the tire and wheel assembly. Remove the nut from the tie rod end.
3. Disconnect the tie rod end from the steering knuckle using tool 89–96–696 or equivalent.
4. On 900 Series vehicles, back off the nut that locks the end assembly to the adjusting rod.
5. On 9000 Series vehicles, measure the length of the thread from the tie rod end locknut to the inner end of the threads on the tie rod. Loosen the locknut.
6. Unscrew the assembly from the adjusting rod.
To install:
7. On 900 Series vehicles, screw the tie rod end on to the adjusting rod to the preset positions.
8. For 9000 Series vehicles, position the locknut where it was before removal. Screw the tie rod end onto the rod up to the locknut, then tighten the locknut against the tie rod end.
9. Connect the end joint to the steering knuckle and tighten to 44–59 ft. lbs. (60–80 Nm).
10. Install the wheels and lower the vehicle. Check and adjust the toe-in as required. Tighten the rod end locknut to 16 ft. lbs. (22 Nm).

Manual Rack and Pinion

REMOVAL AND INSTALLATION

900 Series

1. Disconnect the negative battery cable.
2. Remove the left screen under the instrument panel and loosen the rubber bellows at the body lead through for the steering gear intermediate shaft, if required.
3. Raise and support the vehicle safely.
4. Remove the bolt holding the joint to the steering gear pinion or intermediate shaft.
5. Loosen the steering column tube from the body and separate the steering column joint from the pinion. Position the steering column so the wiring harness is not damaged.
6. Remove both tire and wheel assemblies. Remove the tie rod ends at the steering arms with the proper removal tool. Remove the 2 steering gear clamps.
7. Move the rack to the right as far as possible. Lift the steering gear to the right so the tie rod can be bent down in the opening of the engine compartment floor.
8. Pull the rack to the left and lift the steering gear down through the opening in the engine compartment floor.
To install:
9. Carefully lift the rack assembly in through the right side wheel arch. Install and tighten the mounting bolts.
10. Secure the pinion to intermediate shaft.
11. Screw the tie-rod ends back into the track rods and measure the distance between the tie-rod end and groove in the rack to ensure that both tie-rod ends are back in the same position as before removal.

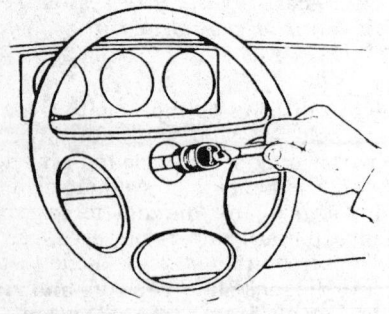

296481

Access to wheel nut with emblem removed

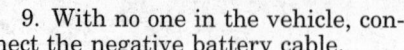

A = 100mm (3.94 in.) maximum - manual steering
A = 125mm (4.92 in.) maximum - power-assisted steering

296613

Tie rod end maxium installed length — 1993 900 Series vehicles

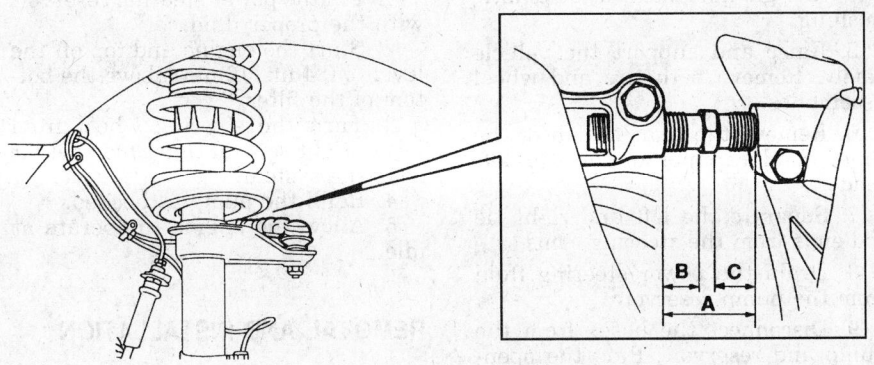

A: Measurement must not exceed 2.03 in. (52mm)
B,C: Measurement between A + B and A + C is 3mm

296614

Tie rod end maxium installed length — 1994–97 900 Series vehicles

To install:

10. Maneuver the rack to the left and lift the steering gear into the opening in the engine compartment floor. Be sure the tie rod sleeves clear the openings on each end.

11. Rotate the rack assembly to connect the pinion shaft to the column lock pin. Tighten the clamp bolt to 26–30 ft. lbs. (35–42 Nm).

12. Center the rack and connect the retaining bolts. Tighten to 44–60 ft. lbs. (60–80 Nm).

13. Replace the hydraulic line O-rings and tighten the lines to 15–25 ft. lbs. (20–34 Nm).

14. Install the return pipe straps.

15. Install the tie rod ends with the same amount of turns as was required to remove them. Tighten the locking nuts to 37–44 ft. lbs. (50–60 Nm) after the toe adjustment is checked.

16. Install the front wheels.

17. Bleed the steering system by filling the reservoir and turning the steering wheel with the engine off, stop to stop, 3 or 4 times.

18. Lower the vehicle. Connect the negative battery cable.

19. Start the engine and turn the wheel 2 more times to be sure the system is working properly and there are no leaks.

1994–97 Models

1. Disconnect the hydraulic fluid return line and plug. Connect a length of hose to the fluid line and a suitable container to recover the fluid. Turn the steering wheel, stop to stop, until all the fluid is out of the system.

2. Disconnect the negative battery cable.

3. Remove the lower left dash finish panel. Remove the fuse box.

4. Remove the steering column to pinion shaft locknut.

5. Straighten the steering wheel and mark it with chalk to center.

6. Sperate the column shaft and pinion shaft.

7. Disconnect the tracking rods from the center of the rack assembly.

8. Raise and safely support the vehicle.

9. Remove the front wheels.

10. Disconnect the tie rods from the strut/knuckle using tool 89–96–696 or equivalent.

11. Disconnect the rack assembly retaining clamps.

12. Disconnect the pressure and return lines using special wrench 87–91–287 or equivalent, from the rack assembly.

12. Tighten the nut securing each tie-rod end to the steering swivel members. Tighten the clamp bolt to 26–30 ft. lbs. (35–42 Nm); the body to steering gear bolts to 44–60 ft. lbs. (60–80 Nm); the tie rod end nuts to 37–44 ft. lbs. (50–60 Nm).

13. Install both tire and wheel assemblies.

14. Lower the vehicle.

15. Refit the left screen and rubber bellows under the instrument panel.

16. Connect the negative battery cable.

Power Rack and Pinion

REMOVAL AND INSTALLATION

900 Series

1993 Models

1. On vehicles with a remote power steering fluid reservoir, disconnect the 2 hoses, drain the reservoir and plug the hoses.

2. On pumps with built in reservoir, disconnect the return hose and place it in a suitable container to recover the fluid. Start the engine and turn the steering wheel stop to stop until the fluid is pumped out. Do not allow the pump to run without fluid.

3. Disconnect the negative battery cable.

4. Remove the left finish panel under the instrument panel and loosen the rubber bellows at the bulkhead lead through for the steering gear intermediate shaft. Disconnect and plug the power steering fluid lines.

5. Raise and support the vehicle safely. Remove the bolt holding the joint to the steering gear pinion or intermediate shaft.

6. Loosen the steering column tube from the body and separate the steering column joint from the pinion. Position the steering column so the wiring harness is not damaged.

7. Remove both tire and wheel assemblies. Loosen the locknuts and unscrew the tie rod ends at the steering arms. Count the number of turns it takes to remove them. Remove the return pipe retaining straps from the rack assembly.

8. Move the rack to the right as far as possible. Lift the steering gear to the right so the tie rod can be bent down in the opening of the engine compartment floor.

9. Pull the rack to the left and lift the steering gear down through the opening in the engine compartment floor.

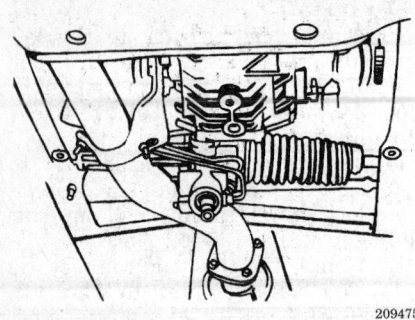

Rotating the rack is required during removal and installation — 1993 900 Series

13. Maneuver the rack assembly out of the vehicle through the left wheel housing.

To install:

NOTE: If a replacement rack assembly is being used, transfer all lines and bellows to the replacement unit as required.

14. Maneuver the rack assembly in through the left wheel housing.
15. Connect the pressure and return lines using specialty wrench 87–91–287 from the rack assembly.
16. Connect the rack assembly retaining clamps.
17. Connect the tie rods to the strut/knuckle. Tighten to 44 ft. lbs. (60 Nm)
18. Connect the tracking rods to the center of the rack assembly. Tighten to 63–74 ft. lbs. (85–100 Nm).
19. Connect the column shaft and pinion shaft. Be sure the steering wheel was aligned with the chalk center mark.
20. Install the lower left dash finish panel. Install the fuse box.
21. Connect the negative battery cable.
22. Connect the hydraulic fluid return line. Fill the reservoir and turn the steering wheel, stop to stop, until all the air is out of the system.
23. Lower the vehicle.
24. Check the toe-in and steering wheel alignment.

9000 Series

1. Disconnect the negative battery cable.
2. Remove the padding from under the instrument panel and the trim on the left side of the center tunnel, as required. Fold back the carpet where the steering column passes through the bulkhead. Remove the rubber boot from the intermediate shaft.

3. Remove the pinch bolt in the lower clamp, loosen the bolt in the upper clamp and remove the intermediate shaft.
4. Remove the cover panel from the bulkhead. Take care not to damage the gasket, seal and plastic bushing.
5. Raise and support the vehicle safely. Remove both tire and wheel assemblies.
6. Remove the rear section of the inner fender panel under the left fender.
7. Separate the left and right tie rod ends from the steering arms.
8. Drain the power steering fluid from the pump reservoir.
9. Disconnect the hoses from the pump and reservoir. Plug the openings to prevent fluid from leaking out and dirt from entering.
10. Remove the retaining bolts from the rack and pinion assembly.
11. Remove the vertical brace between the engine subframe and the body.
12. Lift out the rack and pinion unit through the left fender inner panel opening. Do not damage the rubber boots or brake hose.

To install:

13. Maneuver the rack and pinion unit through the left fender inner panel opening. Do not damage the rubber boots or brake hose. Tighten the rack and pinion gear securing bolts to 46–56 ft. lbs. (60–80 Nm).
14. Install the vertical brace between the engine subframe and the body.
15. Connect the hoses from the pump and reservoir.
16. Connect the left and right tie rod ends to the steering arms.
17. Assemble the rear section of the inner fender panel under the left fender.
18. Install the wheel assemblies and lower the vehicle.
19. Install the cover panel to the bulkhead, do not damage the gasket, seal and plastic bushing.
20. Install the pinch bolt in the lower clamp. Tighten the steering column pinch bolt to 27–32 ft. lbs. (35–42 Nm).
21. Connect the negative battery cable.
22. Fill the reservoir and bleed the system by allowing the engine to run at idle and turning the steering wheel, stop to stop, two or more times.

Power Steering Pump

BLEEDING

1. Fill the power steering reservoir with the proper fluid.
2. Start the engine and top off the level to 0.4 in. (10mm) above the bottom of the filter.
3. Turn the steering wheel from left to right several times to expel air from the system.
4. Refill the pump as needed.
5. Allow the engine to operate at idle.

REMOVAL AND INSTALLATION

900 Series

1993 Models

1. Disconnect the negative battery cable.
2. Drain the fluid from the power steering pump by disconnecting the return hose and connecting a drain hose. Start the engine and turn the steering wheel stop to stop 2 times. Do not allow the pump to run dry. Dispose of the used fluid.
3. Drain the coolant from the drain cock on the engine block and disconnect the hose from between the expansion tank and the water pump.
4. Disconnect the power steering pump hoses. Grip the hexagonal nipple on the pump when removing the delivery line.
5. Unbolt the pump unit from the bracket and the engine mounting.
6. Remove the power steering belt.
7. Remove the pump complete with its mounting bracket.

To install:

8. Install the pump complete with its mounting bracket.
9. Connect the power steering pump hoses. Grip the hexagonal nipple on the pump when tightening the delivery line.
10. Install and adjust the tension of the drive belt.
11. Connect the hose between the expansion tank and the water pump.
12. Connect the negative battery cable.
13. Fill the cooling system.
14. Fill the power steering reservoir with fresh fluid. Start the vehicle and turn the steering wheel stop to stop to work out any air in the system. Recheck the fluid, do not overfill.

1994–97 Models

1. Drain the hydraulic fluid from the pump, as follows:
   a. Disconnect the return hose and connect a drain hose.
   b. Start the vehicle.
   c. Turn the steering wheel stop-to-stop to pump out the fluid.

NOTE: Do not allow the pump to run once it is empty.

2. Remove the air cleaner.
3. Remove the belt from the pump pulley.
4. Disconnect the hoses from the pump.
5. Remove both retaining bolts. Aligning the hole in the pulley is required to access the front retaining bolt. Remove the pump.

To install:
6. Install the pump, tighten the retaining bolts to 15 ft. lbs. (20 Nm).
7. Install the drive belt.
8. Connect the pressure and return hoses.
9. Install the air cleaner.
10. Fill the reservoir with hydraulic fluid. Start the vehicle and turr. the steering wheel stop-to-stop to purge air from the system. Recheck the fluid level.

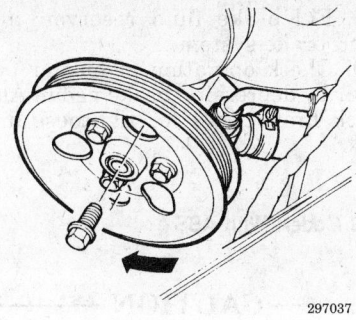

297037

Access to retaining bolt through hole in pulley — 1994–97 900 Series with 4-cylinder engines

9000 Series

1. Disconnect the negative battery cable.
2. Remove the fluid from the pump reservoir.
3. Raise and support the vehicle safely. Remove the right front wheel and the right inner fender panel.
4. Remove the drive belt. Remove the bracket for the engine oil filler pipe. Remove the engine stay bracket. Disconnect the hoses from the pump and plug the openings.
5. Remove the pump retaining bolts and remove the pump. Note that 1 bolt is located behind the

pump pulley and is accessible only through the hole in the pulley.
To install:
6. Install the pump and tighten the retaining bolts 15 ft. lbs. (20 Nm).
7. Connect the lines to the pump and secure the bracket bolts.
8. Install the drive belt.
9. Install the inner fenderwell and wheel. Lower the vehicle to the ground.
10. Fill the reservoir with fluid. Start the vehicle and turn the steering wheel, stop to stop, a few times to purge the air from the system.

BRAKES

Anti-Lock Brake System Service

PRECAUTIONS

• Certain components within the Anti-Lock Brake System (ABS) are not intended to be serviced or repaired individually. Only those components with removal and installation procedures should be serviced.
• Do not use rubber hoses or other parts not specifically specified for and ABS system. When using repair kits, replace all parts included in the kit. Partial or incorrect repair may lead to functional problems and require the replacement of components.
• Lubricate rubber parts with clean, fresh brake fluid to ease assembly. Do not use lubricated shop air to clean parts; damage to rubber components may result.
• Use only specified brake fluid from an unopened container.
• If any hydraulic component or line is removed or replaced, it may be necessary to bleed the entire system.
• A clean repair area is essential. Always clean the reservoir and cap thoroughly before removing the cap. The slightest amount of dirt in the fluid may plug an orifice and impair the system function. Perform repairs after components have been thoroughly cleaned; use only denatured alcohol to clean components. Do not allow ABS components to come into contact with any substance containing mineral oil; this includes used shop rags.
• The Anti-Lock control unit is a microprocessor similar to other computer units in the vehicle. Ensure

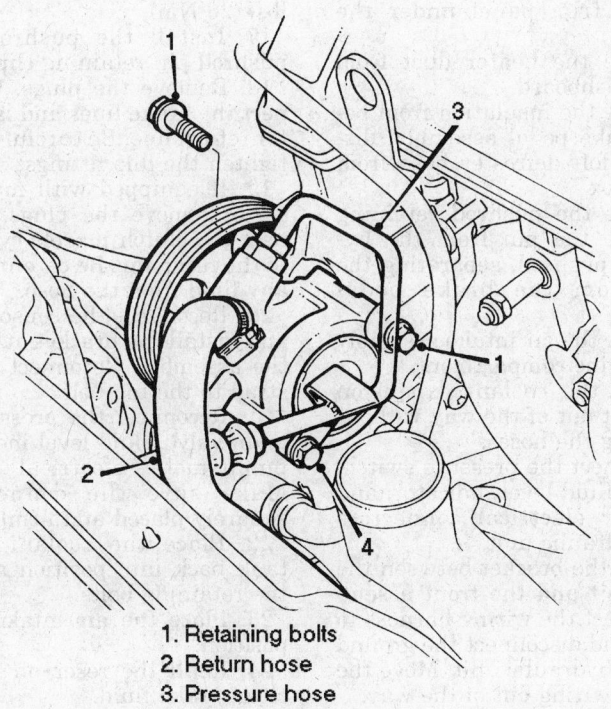

1. Retaining bolts
2. Return hose
3. Pressure hose
4. Hose retaining bolt

297034

Power steering pump mounting points — 1994–97 900 Series with 4-cylinder engines

that the ignition switch is **OFF** before removing or installing controller harnesses. Avoid static electricity discharge at or near the controller.

• If any arc welding is to be done on the vehicle, the control unit should be unplugged before welding operations begin.

DEPRESSURIZING

All Models

— CAUTION —

The hydraulic accumulator contains brake fluid and nitrogen gas at extremely high pressures. Certain portions of the hydraulic system also contain brake fluid at high pressure. It is mandatory that the system be depressurized before disconnecting any hoses, lines or fittings, or personal injury may result.

1. Turn the ignition switch **OFF** during the procedure.
2. Step firmly on the brake pedal at least 20–30 times. The system is not completely discharged until a distinct change is felt in the brake pedal. The effort needed to press the pedal will clearly increase when the system pressure is released.
3. Once the pedal feel changes, pump the pedal a few more times.
4. Perform tests and/or repairs as necessary with the ignition **OFF** at all times. If the ignition is turned **ON**, the pump will run, repressurizing the system.

Master Cylinder

REMOVAL AND INSTALLATION

900 Series

1993 Models Without ABS

1. Remove the brake fluid from the brake fluid reservoir.
2. Remove coolant expansion tank to gain easier access to the master cylinder.
3. Disconnect and plug brake line connection at the master cylinder.
4. Remove the two mounting nuts and lift master cylinder out of engine compartment.
   To install:
5. Remove brake fluid reservoir and install on new master cylinder.
6. Install master cylinder to the brake booster and tighten retaining nuts.
7. Connect brake lines to the master cylinder ports.

8. Fill brake fluid reservoir and bleed brake system.
9. Check operation of master cylinder by depressing brake pedal. Also check brake lines for tightness and leaks.

1993 Models With ABS

— CAUTION —

This vehicle is equipped with an ABS system. Certain portions of the hydraulic system contain brake fluid at high pressure. It is mandatory that the system be depressurized before disconnecting any hoses, lines or fittings, or personal injury may result.

1. With the ignition switch **OFF**, depressurize the hydraulic system by pushing down on the brake pedal approximately 20 times until positive resistance is felt in the brake pedal.
2. Disconnect the negative battery cable.
3. Thoroughly clean the unit, connections and surrounding surfaces to prevent dirt from entering the hydraulic system.
4. Remove the center console and the padded trim panel under the dash.
5. Remove the heater duct from under the dashboard.
6. Remove the insulation from behind the brake pedal assembly; disconnect the left defroster hose from the heater box.
7. Remove the pushrod retaining clip. Remove the pin from the hydraulic unit pushrod, separating the pushrod from the brake pedal assembly.
8. Remove the air intake assembly from the engine compartment.
9. Unbolt the coolant expansion tank; move it out of the way without disconnecting the hoses.
10. Disconnect the pressure switch, main valve, fluid level indicator, and pump motor electrical connectors from the hydraulic unit.
11. Unbolt the bracket between the hydraulic unit and the front assembly. Disconnect the wiring harness to the sensor and disconnect the ground strap at the hydraulic unit. Move the bracket and wiring out of the way.
12. Use a syringe or similar tool to remove as much fluid as possible from the hydraulic unit reservoir. It will not be possible to remove all the fluid.

— WARNING —

Be careful not to spill any brake fluid on the vehicle since brake fluid damages body paint.

13. If equipped with manual transaxle, disconnect the hose running to the clutch system from the reservoir. Immediately plug the hose end; do not allow air to enter the clutch system.
14. Label or diagram the placement of the brake lines and large-diameter return line at the hydraulic unit. Disconnect the lines together to avoid undue strain on any one line. Immediately plug the lines and the ports on the valve block to prevent dirt from entering the system.
15. Remove the hydraulic unit mounting nuts located on the firewall behind the brake pedal assembly and remove the hydraulic unit from the vehicle.
    To install:
    NOTE: During installation, it is possible for the brake switch and/or the cruise control switch to be pressed in inadvertently. If this occurs, the switch may be reset with a pair of pliers.
16. Install the hydraulic unit into the vehicle and align the pushrod to the brake pedal assembly. Torque the hydraulic unit retaining nuts to 19 ft. lbs. (26 Nm).
17. Install the pushrod pin and pushrod pin retaining clip.
18. Remove the plugs, then reconnect the brake lines and large-diameter return line. Be careful not to overtighten the line fittings.
19. If equipped with manual transaxle, remove the plugs and reconnect the clutch master cylinder hose to the reservoir. Be careful not to lose any fluid from the hose.
20. Reconnect the sensor connector and install the bracket at the front of the assembly. Reconnect the ground strap to the top bolt.
21. Reconnect the pressure switch, main valve, fluid level indicator, and pump motor electrical connectors. Make sure the connectors are squarely placed and firmly fitted.
22. Place the coolant expansion tank back into position and tighten the retaining bolts.
23. Place the air intake back into position.
24. Refill the reservoir with fresh, clean brake fluid.
25. Bleed the brake system.
26. Reconnect the negative battery cable.
27. Turn the ignition **ON** and check that the pump is working. The pump

should shut off within 60–90 seconds when correct pressure is achieved. If the pump does not shut off within 120 seconds, shut the ignition **OFF** immediately and allow the pump at least 10 minutes to cool.

28. Fully inspect the brake system for any leaking components or line fittings. Be sure that the BRAKE and ABS warning lights on the instrument cluster go out.

29. Reconnect the left side defroster hose. Install the underdash insulation back behind the brake pedal assembly.

30. Install the heater duct and the padded trim panel under the dashboard.

31. Install the center console.

32. Test drive the vehicle to confirm correct operation of the service brakes, the ABS and the clutch, if equipped.

1994–97 Models With ABS

—————— **CAUTION** ——————
This vehicle is equipped with an Anti-lock braking system. The hydraulic system contains brake fluid at extremely high pressure that must be depressurized before disconnecting any hoses, lines or fittings, or personal injury may result.

1. To depressurize the system, with the ignition **OFF**, depress the brake pedal approximately 20 times until positive resistance in the brake pedal is felt.

2. Disconnect the negative battery cable. Disconnect the electrical connection to the brake warning switch.

3. Remove as much of the brake fluid as possible from the brake fluid reservoir using a fluid siphoning tool. It will not be possible to remove all of the brake fluid.

—————— **WARNING** ——————
Painted surfaces can be damaged by brake fluid. If fluid is spilled, immediately clean the affected surface with soap and water.

4. Place a rag or paper underneath the master cylinder to absorb any spilled brake fluid. Before removing the master cylinder, be sure to thoroughly clean the connections and surrounding surfaces to prevent system contamination.

5. Disconnect and plug the brake lines from the master cylinder. Use a line wrench to disconnect the brake line fittings.

6. Remove the master cylinder retaining nuts and remove the master cylinder out of the vehicle.

To install:

7. Remove the fluid reservoir from the master cylinder and install it onto the new master cylinder. Be sure to use new rubber seals for the reservoir.

8. Position the master cylinder up against the power brake booster and tighten the retaining nuts.

9. Connect the brake lines to the master cylinder and tighten the brake line fittings.

10. Refill the fluid reservoir with clean, fresh brake fluid. Use only DOT 4 brake fluid.

11. Without running the engine, bleed the brake system. Top off the brake fluid to the proper level, if necessary.

12. Reconnect the electrical connection to the brake warning switch. Reconnect the negative battery cable.

13. Check the brake line connections for leaks. Check that the brake system operates properly.

9000 Series

Mark II ABS System

—————— **CAUTION** ——————
The hydraulic accumulator contains brake fluid and nitrogen gas at extremely high pressures. Certain portions of the hydraulic system also contain brake fluid at high pressure. It is mandatory that the system be depressurized before disconnecting any hoses, lines or fittings, or personal injury may result.

NOTE: This hydraulic control unit is part of the Mark II ABS system. This procedure refers to all 1993 and later Saab 9000 Turbocharged vehicles with manual transaxle and traction control.

1. To depressurize the system, with the ignition **OFF**, depress the brake pedal approximately 20 times until positive resistance is felt in the brake pedal.

2. Remove the lower dash panel located under the steering wheel.

3. Remove the hydraulic control unit pushrod retaining clip and pin.

4. Disconnect the battery cables from the battery, negative cable first. Remove the battery from the vehicle.

5. Remove the clips for the positive leads at the bottom of the battery tray.

6. Remove the positive cables and the terminal block on the front of the battery tray.

7. Remove bracket and any electrical connectors at the rear of the battery tray. Remove the battery tray.

8. Separate the fuel filter from the battery tray and carefully move the fuel filter out of the way.

9. Remove as much brake fluid out of the reservoir as possible using a fluid siphoning tool. It will not be possible to remove the brake fluid entirely.

—————— **WARNING** ——————
Be careful not to allow brake fluid to spill, as brake fluid will damage body paint. If spillage occurs, be sure to clean any affected area immediately with soap and water.

10. Disconnect the pump motor, fluid-level indicator, main valve, pressure switch and valve block connectors from the hydraulic control unit.

11. Remove the clip and disconnect the ground connection.

12. If equipped with a manual transaxle, disconnect the clutch master cylinder hose and plug the end so that moisture does not contaminate the system.

13. Carefully raise and support the vehicle safely. Remove the left front wheel and inner fender liner.

14. Remove the steady bar between the wheel well and the subframe.

15. Using a degreasing agent, clean the brake lines to the valve block, then dry with compressed air. Disconnect the brake lines from the valve block using a line wrench. Be sure to plug the ends of the brake lines and the ports in the valve block.

16. Install the left front wheel and tighten the lug nuts. Lower the vehicle to the ground.

17. Remove the accumulator using an 8mm hex bit.

18. Remove the hydraulic control unit mounting nuts located on the firewall behind the brake pedal assembly. Remove the hydraulic control unit from the vehicle.

To install:

19. Install the hydraulic control unit into the vehicle. Torque the mounting bolts to 16–22 ft. lbs. (22–30 Nm). Reconnect the hydraulic control unit pushrod to the brake pedal assembly and install the retaining pin and clip. If the stoplight switch and/or cruise control switch are accidentally pushed in, reset them using a pair of pliers.

20. Install the accumulator. Torque to 25–34 ft. lbs. (34–46 Nm).

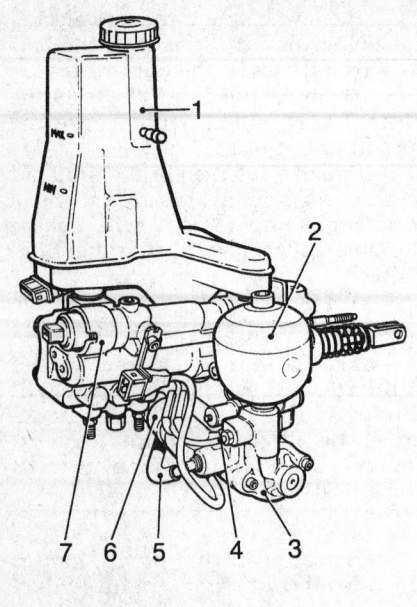

1 Brake fluid reservoir
2 Accumulator
3 Hydraulic pump
4 Pump delivery pipe
5 Pump inlet hose
6 Electric motor
7 Main valve

301619

Hydraulic control unit component identification — 9000 Series with Mark II ABS System

21. Carefully raise and support the vehicle safely. Remove the left front wheel.

22. Reconnect the brake lines and tighten the fittings to the valve block.

23. Install the steady bar between the wheel well and the subframe.

24. Install the inner fender liner and the left front wheel. Torque the wheel lug nuts to 80–90 ft. lbs. (108–122 Nm). Lower the vehicle to the ground.

25. Be sure there is brake fluid in the clutch master cylinder hose and reconnect the hose to the brake fluid reservoir.

26. Install the clip and reconnect the ground connection.

27. Reconnect the pump motor, fluid-level indicator, main valve, pressure switch and valve block connectors to the hydraulic control unit.

28. Install the battery tray. Install the clips for the positive lead at the bottom of the battery tray.

29. Install the positive cables and terminal block to the front of the battery tray.

30. Install the bracket and any connectors to the rear of the battery tray.

31. Carefully reposition the fuel filter and secure it to the battery tray.

32. Install the battery and reconnect the positive battery cable, then the negative battery cable.

33. Refill the reservoir with fresh, clean DOT 4 brake fluid.

34. Turn the ignition **ON** to make sure the pump is working. Bleed the brake system and top off the brake fluid if necessary.

35. Check for leaks in the brake system. Check to see if the brake-warning and the ABS warning lights go out after the ignition has been turned **ON**.

36. Install the lower dash panel.

37. Test drive the vehicle to verify proper brake system operation and clutch system operation if equipped with a manual transaxle.

Mark IV ABS System

NOTE: This procedure applies only to the Mark IV ABS system fitted to most 1993 and later Saab 9000 vehicles. Vehicles equipped with a turbocharger and a manual transaxle and Traction Control are equipped with the Mark II ABS system used in previous vehicles.

——————— CAUTION ———————
Anti-Lock Brake Systems contain brake fluid at extremely high pressures. It is mandatory that the system be depressurized before disconnecting any hoses, lines or fittings, or personal injury may result.

1. To depressurize the system, with the engine **OFF**, depress the brake pedal approximately 20 times until positive resistance in the brake pedal is felt. Remove the key from the ignition to prevent accidentally running the pump and pressurizing the system.

2. Disconnect the battery cables, then remove the battery from the vehicle.

3. Disconnect the main fuse box by pulling it upwards. Disconnect the ABS wiring loom from the battery tray.

4. Remove the mounting bolts and remove the battery tray.

5. Remove as much brake fluid from the reservoir as possible using a siphoning tool. It may not be possible to remove the brake fluid entirely. Be careful not to spill any brake fluid on the body paint.

6. Disconnect the brake lines from the master cylinder using a line wrench. Plug the ends of the brake lines and the openings in the master cylinder to prevent dirt from entering the system.

7. Remove the securing bolt under the master cylinder.

8. Remove the mounting nuts securing the master cylinder to the power brake booster, then remove the master cylinder and brake fluid reservoir.

9. Disconnect the hoses from the master cylinder.

To install:

10. Fit the reservoir-to-master cylinder hoses firmly onto the new master cylinder.

11. Install the master cylinder and reservoir onto the power brake booster. Tighten the mounting nuts and the securing bolt under the master cylinder, then secure the reservoir back into its original position.

12. Reconnect the brake lines to the master cylinder and tighten the brake line fittings.

13. Refill the brake fluid reservoir with fresh, clean brake fluid. Use only the manufacturer's recommended brake fluid type.

14. Install the battery tray. Tighten the mounting bolts and screws. Reattach the ABS wiring loom to the battery tray.

15. Install the battery and connect the cables.

16. Install the main fuse box back in its bracket in the original position.

17. Without running the engine, bleed the brake system. Top off the brake fluid to the proper level. Check the brake system for leaks.

18. Test drive the vehicle and be sure that there is proper brake system operation.

Brake Caliper

REMOVAL AND INSTALLATION

Front

1993 900 Series

1. Carefully raise and support the vehicle safely. Remove the front wheel.

——————— CAUTION ———————
If equipped with an ABS system, you must first depressurize the system before disconnecting any hoses, lines or fittings, or personal injury may result. With the ignition OFF, push down on the brake pedal approximately 20 times until you start to feel positive resistance in the brake pedal.

2. Be sure to clean the caliper and brake hose fittings before removal, so dirt does not get into the system.

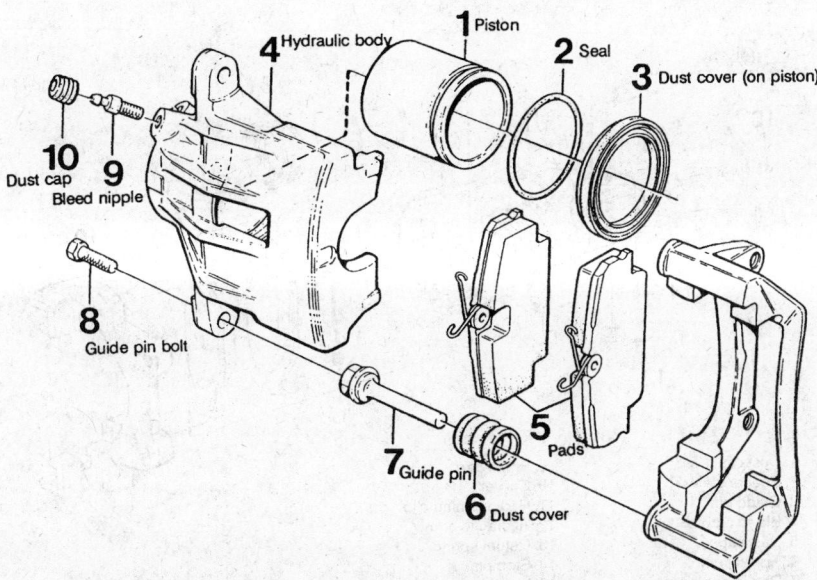

Exploded view of front brake components — 1993 900 Series

184577

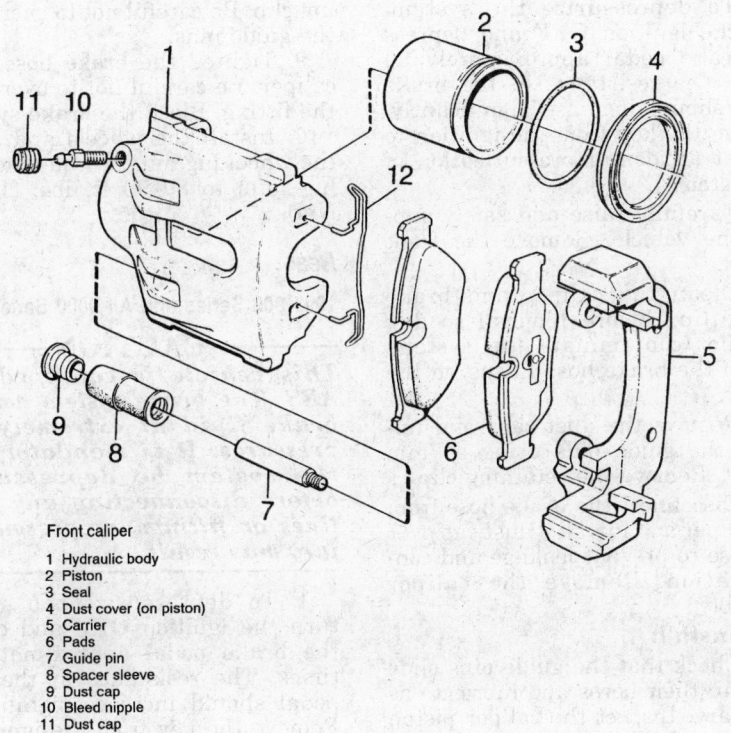

Front caliper

1 Hydraulic body
2 Piston
3 Seal
4 Dust cover (on piston)
5 Carrier
6 Pads
7 Guide pin
8 Spacer sleeve
9 Dust cap
10 Bleed nipple
11 Dust cap
12 Retaining clip

301351

Exploded view of front brake components — 9000 Series

3. Carefully and slowly retract the brake pads, using a pair of slip-joint pliers. Be careful not to damage the disc brake pad material or the caliper piston.

4. Use a brake hose clamp to pinch off the brake hose.

5. Disconnect the brake hose from the brake caliper and install a dust cap over the hose to prevent leakage or contamination.

6. Cover the hole in the caliper and remove the bracket for the brake hose from the caliper.

7. Remove both brake caliper guide pin bolts. Be careful not to damage the dust covers for the guide pin bolts and the caliper piston. Remove the caliper assembly from the steering knuckle.

To install:

NOTE: Before starting the installation procedure, use a wire brush to thoroughly clean the surfaces of the caliper that come in contact with the brake pads.

8. Install the brake caliper onto the steering knuckle. Torque the brake caliper-to-steering knuckle bolts to 52–82 ft. lbs. (71–111 Nm). Torque the guide pin bolts to 22–26 ft. lbs. (30–35 Nm). Be sure to lubricate the guide pins lightly with special grease before installing them into the caliper assembly.

9. Install the brake line fitting and brake line bracket onto the brake caliper. Be careful not to overtighten.

10. Remove the brake hose clamp tool.

11. Bleed the brake system then top off the brake fluid to proper level, if necessary.

12. Install the wheel back onto the car and tighten the wheel lug nuts. Torque the wheel lug nuts to 80–90 ft. lbs. (108–122 Nm). Lower the vehicle to the ground and pump the brake pedal several times to seat the brake pads.

1994–97 900 Series

─── **CAUTION** ───
This vehicle is equipped with ABS. The brake system contains brake fluid at extremely high pressures. It is mandatory that the system be depressurized before disconnecting any hoses, lines or fittings, or personal injury may result.

1. To depressurize the system, turn the ignition **OFF** and depress the brake pedal approximately 20 times. The resistance in the brake pedal should increase significantly.

Remove the key from the ignition to prevent accidental pressurization of the system.

2. Carefully raise and support the vehicle safely. Remove the appropriate wheel.

3. Using a pair of water-pump pliers, press the piston back into the caliper.

4. Using a brake clamp to push down on the brake pedal will minimize fluid loss.

5. Disconnect the brake hose from the brake caliper. Plug the brake line so that dirt cannot enter the hydraulic system. Pay attention to how the brake hose was positioned on the brake caliper.

6. Remove the brake caliper mounting bolts and remove the brake caliper.

To install:

7. Fit the brake caliper onto the strut. Lightly apply thread locking compound to the mounting bolts and tighten the caliper mounting bolts to 59 ft. lbs. (80 Nm).

8. Install the brake hose onto the caliper. Be careful to install the brake hose in its original position.

9. Remove the brake clamp. Bleed the brake system and fill the fluid to the proper level.

10. Install the wheel and lug nuts. Lower the vehicle to the ground.

11. Check that the brake pedal does not drop and that the brake system operates properly.

9000 Series

—— **CAUTION** ——

This vehicle is equipped with ABS. The brake system contains brake fluid at extremely high pressures. It is mandatory that the system be depressurized before disconnecting any hoses, lines or fittings, or personal injury may result.

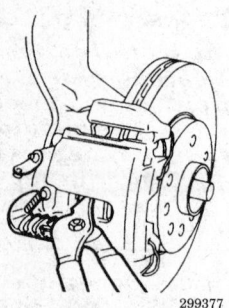

299377

Using water pump pliers to press in the caliper piston — all 900 Series

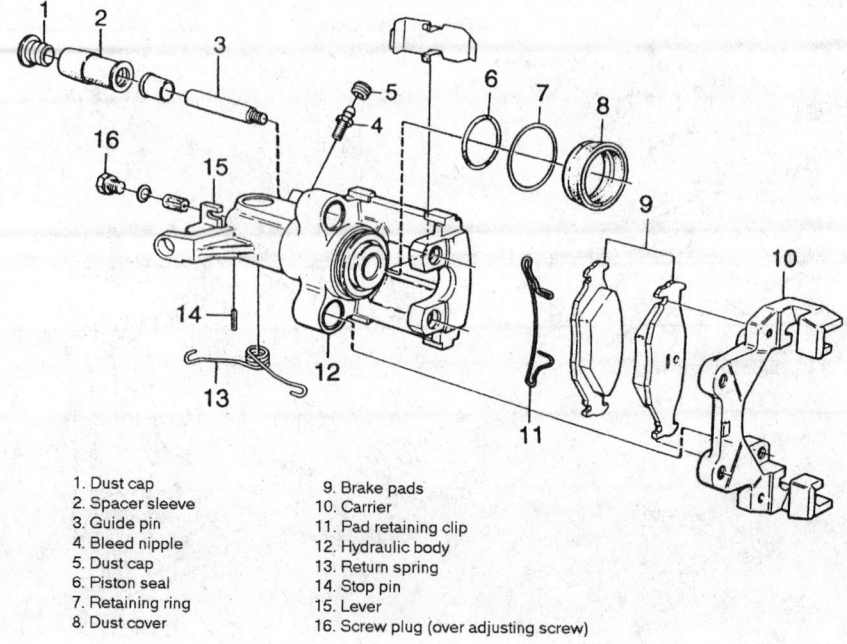

1. Dust cap	9. Brake pads
2. Spacer sleeve	10. Carrier
3. Guide pin	11. Pad retaining clip
4. Bleed nipple	12. Hydraulic body
5. Dust cap	13. Return spring
6. Piston seal	14. Stop pin
7. Retaining ring	15. Lever
8. Dust cover	16. Screw plug (over adjusting screw)

302105

Rear brake caliper component identification — 9000 and 1993 900 Series

1. To depressurize the system, turn the ignition **OFF** and depress the brake pedal approximately 20 times. The resistance in the brake pedal should increase significantly. Remove the key from the ignition to prevent accidental pressurization of the system.

2. Carefully raise and safely support the vehicle. Remove the front wheels.

3. Clean the caliper and brake hose fitting before removal so dirt does not contaminate the system. Loosen the brake hose fitting on the caliper.

4. Remove the dust caps and unscrew the guide pins using a 7mm hex bit. Remove the retaining clip.

5. Disconnect the brake hose from the caliper and install a dust cap over the hose to prevent leakage and contamination. Remove the caliper assembly.

To install:

6. Check that the guide pins slide freely in their bores and lubricate as necessary. Inspect the caliper piston dust cover for damage and replace as necessary.

7. Reconnect the brake hose to the caliper but leave the fitting loose.

8. Tighten the guide pins and install the dust caps. Install the retaining clip. Be careful not to overtighten the guide pins.

9. Tighten the brake hose at the caliper. Be careful not to overtighten the fitting. Bleed the brake system.

10. Install the wheels and tighten the wheel lug nuts. Torque the wheel lug nuts to 80–90 ft. lbs. (108–122 Nm).

Rear

1993 900 Series and All 9000 Series

—— **CAUTION** ——

This vehicle is equipped with ABS. The brake system contains brake fluid at extremely high pressures. It is mandatory that the system be depressurized before disconnecting any hoses, lines or fittings, or personal injury may result.

1. To depressurize the system, turn the ignition **OFF** and depress the brake pedal approximately 20 times. The resistance in the brake pedal should increase significantly. Remove the key from the ignition to prevent accidental pressurization of the system.

2. Raise and safely support the vehicle. Remove the rear wheels.

3. Disconnect the parking brake cable from the parking brake lever

located on the back of the caliper assembly.

4. Remove the disc brake pad retaining clip.

5. Thoroughly clean the brake caliper assembly so that no dirt contaminates the hydraulic system.

6. Remove the screw plug from the adjusting screw and unscrew the adjusting screw using a 4mm hex key tool.

7. Clamp the brake hose closed using hose clamp pliers.

8. Loosen the brake line fitting slightly. Remove the dust caps and the caliper guide pins using a 7mm hex key tool.

9. Remove the caliper assembly and remove the brake pads. Fully disconnect the brake line from the brake caliper. Be sure to plug the brake line opening and the brake line connection on the caliper so brake fluid does not escape or contamination enter the hydraulic system.

To install:

10. Before installing the brake caliper, make sure that the guide pins are clean and slide freely in the caliper unit. Use a wire brush to clean them if needed, but do not lubricate them. Using a wire brush, clean the brake pad contact surface of the caliper.

11. Reconnect the brake line to the brake caliper making sure that the line is not twisted. Do not tighten the brake line fitting.

12. Install the brake pads. Be sure that the pad spring is against the caliper piston.

13. Install the brake caliper into position. Install and tighten the caliper guide pins. Install the guide pin dust caps.

14. Tighten the brake line fitting on the brake caliper. Remove the hose clamp pliers.

15. Install the brake pad retaining clip into place.

16. Reconnect the parking brake cable to the parking brake lever on the back of the caliper.

17. Adjust the parking brake to proper specifications. Bleed the brake system. Top off the brake fluid, if needed.

18. Install the wheel and lower the vehicle to the ground. Torque the wheel lug nuts to 80–90 ft. lbs. (108–122 Nm).

19. Pump the brake pedal several times to be sure that the brake pads are seated up against the brake rotors.

1994–97 900 Series

---- **CAUTION** ----

This vehicle is equipped with ABS. The brake system contains brake fluid at extremely high pressures. It is mandatory that the system be depressurized before disconnecting any hoses, lines or fittings, or personal injury may result.

---

1. To depressurized the system, turn the ignition **OFF** and remove the key to prevent accidental activation of the pump. Depress the brake pedal approximately 20 times until positive resistance is felt in the brake pedal.

2. Carefully raise and support the vehicle safely. Remove the appropriate wheel.

3. Using a pair of water-pump pliers, carefully push the caliper pistons back into the caliper.

4. Using a brake pedal clamp, push down the brake pedal to limit fluid loss.

5. Disconnect the brake line from the caliper using a line wrench. Be sure to plug the brake line and the opening in the caliper so dirt does not enter the hydraulic system.

6. Remove the locking spring and caliper pins.

7. Remove the brake pads.

8. Remove the brake caliper mounting bolts and remove the brake caliper.

To install:

9. Install the brake caliper. Apply a light coating of thread locking compound and tighten the brake caliper mounting bolts to 59 ft. lbs. (80 Nm).

10. Reconnect the brake lines to the caliper and tighten the fittings.

11. Install the brake pads, caliper pins and locking spring.

12. Remove the brake pedal clamp.

13. Bleed the brake system.

14. Install the wheels and tighten the lug nuts. Lower the vehicle to the ground.

15. Check the brake line fittings for any leaks. Depress the brake pedal several times to seat the pads against the rotor. Check the brakes for proper operation.

Disc Brake Pads

REMOVAL AND INSTALLATION

Front

1993 900 Series

1. Carefully raise and safely support the vehicle. Remove the front wheels.

2. Using a pair of slip-joint pliers, slowly compress the brake piston.

3. Remove the lower guide pin bolt.

4. Swing the brake caliper upwards. Remove the brake pads.

To install:

5. Before installing the brake pads, thoroughly clean the surfaces of the caliper that the brake pads contact, using a wire brush. Clean the guide pins and lightly lubricate them using special grease. Also inspect the guide pin and brake piston dust covers for damage and replace if necessary.

6. Install the brake pads and reposition the brake caliper.

7. Install the lower guide pin bolt. Torque the lower guide pin bolts to 22–26 ft. lbs. (30–35 Nm).

8. Install the wheels and lower the vehicle to the ground. Torque the wheel lug nuts to 80–90 ft. lbs. (108–122 Nm).

9. Before test driving the vehicle be sure to pump the brake pedal several times to snug the brake pads to the rotor.

1994–97 900 Series

1. Disconnect the negative battery cable.

2. Raise and support the vehicle and support it safely. Remove the wheel.

3. Using a pair of water pump pliers, carefully press back the brake piston.

4. Remove the retaining clip from the brake caliper.

5. Remove the caliper guide pin dust caps and the caliper guide pin bolts.

6. Lift off the brake caliper and remove the pads. Hang the brake caliper securely off the strut using a piece of strong wire.

7. Check that the guide pins slide freely and the dust covers are in good condition.

8. Clean the surfaces between the pads and the carrier.

To install:

9. Before installing the new brake pads, be sure to lightly coat the backing plate of the brake pads with a disc brake anti-squeal lubricant. In-

stall the new brake pads in the caliper and place the brake caliper back into its original position.

10. Install the caliper guide pins. Torque the guide pin bolts to 78 ft. lbs. (105 Nm).

11. Install the guide pin dust caps and the caliper retaining clip.

12. Install the wheel and tighten the wheel lug nuts. Lower the vehicle to the ground.

13. Pump the brake pedal to move the pads to their operating positions.

14. Connect the negative battery cable.

9000 Series

1. Carefully raise and safely support the vehicle. Remove the front wheels.

2. Remove the dust caps on the caliper guide pins and remove the caliper guide pins using a 7mm hex bit.

3. Remove the brake pad retaining clip.

4. Lift off the caliper assembly and remove the brake pads.

5. Support the caliper assembly by hanging it off the strut with a piece of strong wire.

To install:

6. Before installing the brake pads, be sure to clean the contact surface between the brake pads and the caliper.

7. Compress the brake pad and piston into the caliper unit using a pair of water pump pliers. Use the old brake pad for this procedure so that you do not damage the new brake pad.

8. Install the new brake pads onto the caliper assembly along with the anti-rattle spring.

9. Install the caliper assembly, lubricate, install and tighten the guide pins. Install the dust caps and brake pad retaining clip.

10. Install the wheels and wheel lug nuts. Torque the lug nuts to 80–90 ft. lbs. (108–122 Nm). Carefully lower the vehicle to the ground.

11. Pump the brakes several times to seat the brake pads to the brake rotors.

Rear

1993 900 Series and All 9000 Series

1. Carefully raise and safely support the vehicle. Remove the rear wheels.

2. Disconnect the parking brake cable from the parking brake lever on the caliper.

3. Remove the screw plug to access the parking brake adjusting screw.

4. Move the brake caliper piston back against its stop by turning in the adjusting screw using a 4mm hex key tool.

5. Remove the dust caps covering the caliper guide pins. Using a 7mm hex key tool, unscrew the caliper guide pins.

6. Remove the brake pad retaining clip.

7. Move the brake caliper out of the way and remove the brake pads. Inspect the caliper piston dust cover for wear or damage and replace, if necessary.

To install:

8. Before installing the brake pads, thoroughly clean the sliding surface between the caliper and brake pad, then lightly coat the backing plate of the new pads with a disc brake pad anti-squeal lubricant. Install the new brake pads, fitting the brake pad with the clip against the piston.

9. Install the caliper into position and install the caliper guide pins. Torque the guide pins to 18–22 ft. lbs. (25–30 Nm).

10. Install the guide pin dust caps. Install the brake pad retaining clips.

11. Rotate the adjusting screw all the way in and then back it out $1/4$–$1/2$ turn. Check that the brake rotor freely rotates. Install the adjusting screw plug.

12. Reconnect the parking brake cable to the parking brake lever on the caliper and adjust the parking brake.

13. Install the wheels and lug nuts. Carefully lower the vehicle to the ground. Torque the wheel lug nuts to 80–90 ft. lbs. (105–125 Nm).

1994–97 900 Series

1. Carefully raise and support the vehicle safely. Remove the rear wheels.

2. If the fluid reservoir is full, carefully remove some brake fluid with a suction bulb. Discard the old fluid.

3. Using suitable pliers, gently press the caliper piston into its caliper bore.

4. Remove the brake pad retaining pins and lock spring.

5. Remove the brake pads.

To install:

6. Before installing the new brake pads, be sure to first clean the contact surfaces between the brake caliper and the brake pads.

7. Install the brake pads into the brake caliper.

8. Install the lock spring and the brake pad retaining pins.

9. Install the wheels and tighten the lug nuts. Torque the lug nuts to 80–90 ft. lbs. (108–122 Nm).

10. Carefully lower the vehicle to the ground.

11. Depress the brake pedal several times to seat the brake pads evenly against the brake rotor. Check the fluid level and add new fluid as needed.

CAUTION

Do not attempt to move the vehicle until a firm pedal is obtained.

Brake Rotor

REMOVAL AND INSTALLATION

Front

1993 900 Series

1. Raise and support the vehicle safely. Remove the appropriate wheel.

2. Spread the brake pads from the brake disc by using a pair of slip-joint pliers.

3. Remove the brake caliper-to-steering knuckle bolts and hang the caliper out of the way using a piece of wire. Do not allow the caliper to hang by the brake hose.

4. Remove the locating stud and the servicing screw. Remove the rotor.

To install:

5. Before installing the rotor, check that the hub contact surface to the rotor is free of corrosion, burrs, or any dirt.

6. Install the front rotor, locating stud, and tighten the service screw.

7. Check the rotor run-out. If not within 0.003 in. (0.08mm) specification, have the rotor resurfaced.

8. Install the caliper. Torque the front caliper-to-frame bolts to 52–81 ft. lbs. (70–110 Nm).

9. Install the wheel and lower the vehicle to the ground. Torque the lug nuts to 80–90 ft. lbs. (108–122 Nm).

10. Pump the brakes several times until the brake pads are positioned against the rotor.

1994–97 900 Series

1. Raise and support the vehicle safely. Remove the appropriate wheel.

2. Using a pair of water pump pliers, carefully press in the caliper piston.

3. Remove the two caliper mounting bolts, remove the caliper and hang it out of the way using a piece of

strong wire. Do not allow the caliper to hang by the brake hose.

4. Remove the brake rotor locking bolt. Remove the rotor.

To install:

5. Check the rotor run-out. Maximum run-out allowed is 0.003 in. (0.08mm).

6. Before installing the brake rotor, be sure that the hub contact surface is free of any dirt, corrosion or burrs.

7. Install the brake rotor and install the locking bolt. Lightly coat the locking bolt with a thread locking compound. Torque the locking bolt to 3 ft. lbs. (4 Nm).

8. Install the brake caliper. Lightly apply a thread locking compound to the caliper bolts. Torque the bolts to 78 ft. lbs. (105 Nm).

9. Install the wheel and lug nuts. Lower the vehicle to the ground.

10. Depress the brake pedal several times to evenly seat the brake pads against the rotors.

9000 Series

1. Carefully raise and safely support the vehicle. Remove the front wheels.

2. Loosen the rotor retaining screw and the locating stud.

3. Slowly and carefully push in the brake pads using a pair of water pump pliers.

4. Remove the brake caliper retaining bolts.

5. Move the brake caliper out of the way and hang it from the strut using a piece of strong wire.

6. Remove the brake rotor retaining screw and locating stud. Remove the brake rotor.

To install:

7. Before installing the rotor, clean any dirt, burrs, or corrosion from the contact surface of the hub. Install the rotor, retaining screw and locating stud. After installing the rotor, check the rotor for run-out. Maximum allowable run-out is 0.003 in. (0.08mm).

8. Install the brake caliper and retaining bolts. Torque the retaining bolts to 52–81 ft. lbs. (70–110 Nm).

9. Install the front wheels and wheel lug nuts. Torque the lug nuts to 80–90 ft. lbs. (108–122 Nm). Carefully lower the vehicle to the ground.

10. Pump the brake pedal several times until the brake pads have seated against the rotor.

Rear

1993 900 Series

1. Carefully raise and safely support the vehicle. Remove the rear wheels.

2. Disconnect the parking brake cable from the parking brake lever on the brake caliper.

3. Remove the screw plug from the adjusting screw.

4. Turn the adjusting screw back slightly using a 4mm hex key tool so the brake pads can be pressed in.

5. Remove the brake caliper-to-backing plate retaining bolts. Move the brake caliper out of the way and support it by hanging it with a piece of wire.

6. Remove the locating stud and pull off the brake rotor.

To install:

7. Before installing the brake rotor, clean the contact surface of the wheel hub from any burrs or corrosion. Install the brake rotor and locating stud.

8. Install the rear brake caliper assembly. Torque the retaining bolts to 30–40 ft. lbs. (40–45 Nm).

9. Reconnect the parking brake cable to the parking brake cable lever on the back of the caliper. Adjust the parking brake cable.

10. Install the rear wheels and wheel lug nuts. Carefully lower the vehicle to the ground. Torque the lug nuts to 80–90 ft. lbs. (105–125 Nm).

11. Before driving the vehicle, be sure to pump the brakes several times to seat the pads against the rotor.

1994–97 900 Series

1. Carefully raise and support the vehicle safely.

2. If the brake fluid reservoir is full, remove some brake fluid with a suction bulb and discard the old fluid.

3. Using suitable pliers, gently press the caliper piston back into the caliper bore.

4. Remove the brake caliper mounting bolts. Disconnect the brake line from the retaining clip. Be careful not to damage the brake line when disconnecting it from the clip. Lift the brake caliper up off of the brake rotor and hang it out of the way with a strong piece of wire; do not let the caliper hang by the brake hose.

5. Loosen the parking brake adjustment.

6. Remove the brake rotor retaining bolt, then remove the brake rotor.

To install:

7. Before installing the new brake rotor, be sure that it is clean of all oil

and dirt. Use a non-petroleum cleaning solvent that leaves no residue.

8. Install the brake rotor and tighten the rotor retaining bolt.

9. Adjust the parking brake shoes. Rotate the adjusting nut until the rotor is locked, then unscrew the adjusting nut until the rotor rotates freely. This can be done by inserting a suitable tool through a hole in the brake rotor.

10. Install the brake caliper back to its original position and tighten the mounting bolts. Be careful when repositioning the brake line to its retaining clip.

11. Install the wheels and tighten the lug nuts. Carefully lower the vehicle to the ground.

--- CAUTION ---

Do not attempt to move the vehicle until a firm pedal is obtained.

12. Depress the brake pedal several times to seat the brake pads against the rotor.

9000 Series

1. Carefully raise and support the vehicle safely. Remove the rear wheels.

2. Loosen the locating stud and retaining screw on the rotor.

3. If necessary, retract the brake pads and caliper piston into the caliper bore using a pair of water pump pliers.

4. Remove the brake rotor backing plate retaining bolts.

5. Remove the caliper mounting bolts.

6. Remove the caliper assembly out of the way and support it off the strut using a piece of strong wire.

7. Unscrew the locating stud, rotor retaining screw, and remove the brake rotor.

To install:

8. Before installing the rotor, be sure to clean the contact surface of the hub of any dirt, corrosion or burrs. Install the new rotor along with the locating stud and retaining screw.

9. Install the brake caliper and mounting bolts. Torque the caliper mounting bolts to 52–67 ft. lbs. (70–90 Nm).

10. Install the rotor backing plate retaining bolts.

11. Adjust the parking brake, if necessary.

12. Install the wheels and lug nuts. Torque the lug nuts to 80–90 ft. lbs. (108–122 Nm). Lower the vehicle and pump the brake pedal to seat the brake pads against the rotor.

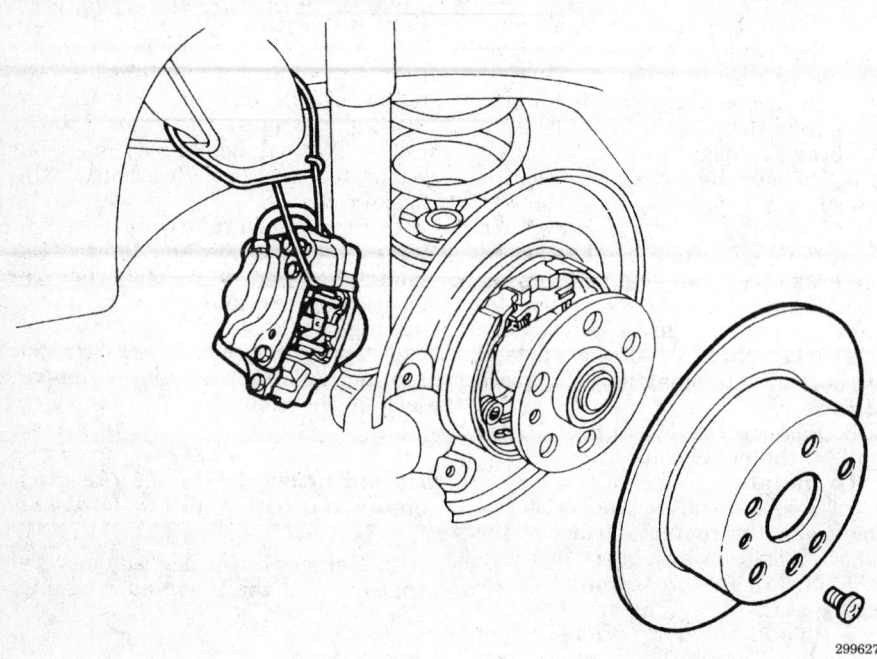

Support the brake caliper when removing the brake rotor — 1994–97 900 Series

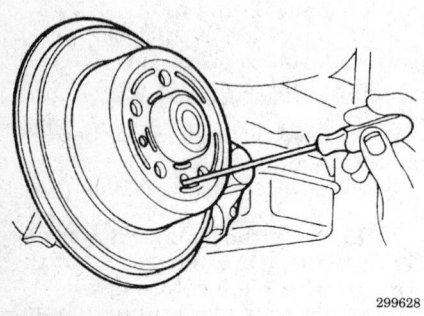

Making the parking brake shoe adjustment — 1994–97 900 Series

Parking Brake Cable

ADJUSTMENT

900 Series

1993 Models

1. Raise the rear seat of the vehicle. On the 900 convertible, remove the rear seat cushion by gripping the front edge of the seat and lifting it straight up.
2. Pry apart the adjusting device.
3. Insert an 0.08 in. (2.0mm) feeler gauge between the parking brake lever and the stop on the back of the rear brake caliper.
4. Rotate the locknut against the adjusting sleeve until the feeler gauge drops out.
5. To settle the adjusting device, apply and release the parking brake several times.
6. Be sure that the clearance between the parking brake lever and stop is 0.02–0.08 in. (0.5–2.0mm).
7. Lower the rear seat cushion back into place. On the 900 convertible, install the rear seat cushion, pushing on the front edge of the seat to secure it in position.

1994–97 Models

1. Raise and support the vehicle safely. Remove the rear wheels.
2. Using a screwdriver, rotate the brake shoe adjuster until it locks up the brake rotor.
3. Loosen back the adjuster until the brake rotor drags slightly but can be turned. Make this adjustment to the other rear wheel.
4. Install the rear wheels. Torque the lug nuts to 80–90 ft. lbs. (108–122 Nm).
5. Pull the hand parking brake lever up to the 6th notch. This should lock up the rear wheels.
6. If the parking brake cable equalization also needs adjustment, release the parking brake lever and set it to the second notch.
7. Turn the adjusting nut on the parking brake cable equalizer until the parking brake starts to engage.
8. Lower the vehicle slightly. Release the parking brake. The rear wheels should be fully locked up when the parking brake lever is pulled up 6 notches. Continue the adjustment procedure if necessary.
9. Lower the vehicle safely to the ground.

9000 Series

1. Remove the console brush seal off of the hand brake lever.
2. Remove the locking plate off of the adjusting nuts under the hand brake lever.
3. Raise and safely support the vehicle.
4. Remove the screw plug from the adjusting screw at the rear caliper. Rotate the adjusting screw inward until it bottoms and then back it out by ¼–½ turn. Be sure that the brake rotor turns freely. Install the screw plug.
5. To adjust the parking brake cable: insert a 0.040 in. (1.0mm) feeler gauge between the parking brake lever and stop on the back of the caliper.
6. Rotate the adjusting nut under the hand brake lever until the feeler gauge falls out. Clearance is 0.040±0.020 in. (1.0 ± 0.5mm).
7. Carefully lower the vehicle to the ground.
8. Install the plastic locking plate over the adjusting nuts. Install the console brush seal over the hand brake lever and back into position on the console.

REMOVAL AND INSTALLATION

900 Series

1993 Models

1. Remove the driver's seat and disconnect the electrical connector under the driver's seat.
2. Remove the rubber boot between the shift lever console and the center console.
3. Remove the switch cluster from the shift lever console and disconnect the connector by pressing in the tab.
4. On manual transaxle, pull up the shift lever rubber boot. On automatic transaxle, remove the shift lever knob and shift cover.
5. Remove the 4 shift lever console retaining screws.
6. Disconnect the electrical connector for the interior lighting and re-

move the ignition switch illumination bulb.

7. On manual transaxle, Place the shift lever into reverse gear and remove the ignition key.

8. Pull up the hand brake lever. Lift up the console. On manual transaxle, insert the ignition key and place the shift lever into third gear.

9. Remove the shift lever console.

10. On both sides, remove the rear sill scuff plates.

11. The rear seat can be removed by first removing the hinge pins. On the convertible, remove the molding retaining screws under the front of the rear seat, remove the molding and the rear seat.

12. Push up the rear of the carpet.

13. Push back the floor insulation and remove the retaining screws for the metal cover over the parking brake cables.

14. Remove the clamp that holds down the cables.

15. On cars equipped with an ABS system, carefully cut through the cable tie holding the parking brake cable and ABS sensor cable together. Be very careful not to damage the ABS sensor cable.

16. Release the hand brake lever. Remove the cable retaining nut at the hand brake lever.

17. Remove the locking plate holding the cables to the cable bracket and then withdraw the cables.

18. Carefully raise and safely support the vehicle. Remove the rear wheels.

19. Disconnect the cable from the rear brake caliper parking brake lever.

20. Remove the cable guide from the rear suspension spring link.

21. Withdraw the brake cable from the vehicle.

To install:

NOTE: Be aware that the left-side brake cable is 20mm shorter than the right-side cable.

22. Loosen the locknut from the adjusting sleeve on the new cable until the parking brake covering can be pressed to the bottom of the adjusting sleeve. Tighten the locknut 3 turns against the adjusting sleeve.

23. Push the new cable through the hole in the floor under the rear seat and hold it down in the front retaining bracket.

24. Install and hand tighten the retaining nut onto the brake cable at the hand brake lever as far as possible.

25. Install the locking plate onto the cables. Install the clamp holding down the cables.

26. Install the metal cover over the cables. Fit a new cable tie around the ABS sensor cable and the parking brake cable.

27. Fit the rubber cable grommet into the hole in the floor.

28. Connect the brake cable to the parking brake lever on the rear caliper.

29. Attach the cable guide to the rear suspension spring link. Install the rear wheels and safely lower the vehicle to the ground. Torque the wheel lug nuts to 80–90 ft. lbs. (109–122 Nm).

30. To adjust the parking brake cable, insert a 0.080 in. (2.0mm) feeler gauge between the stop and the parking brake lever on the rear brake caliper and adjust the nut on the cable until the feeler gauge falls out.

31. To settle the adjusting device located under the rear seat, apply and release the parking brake several times.

32. Be sure that the clearance between the parking brake lever and the stop is 0.020–0.080 in. (0.5–2.0mm).

33. Install the shift lever console and install the switch cluster. Reconnect all electrical connectors and install the ignition switch illumination bulb.

34. On manual transaxle, reposition the shift lever boot to the console. On automatic transaxle, install the shift lever cover and knob.

35. Install the rubber boot between the shift lever console and the center console.

36. Install the driver seat back into place and reconnect the electrical connector under the seat.

37. Reposition the rear carpet back into place and install the rear seat. The rear seat may require pressing on its front edge to secure it into position.

38. Install the rear sill scuff plates and all moldings back into place.

1994–97 Models

1. Carefully raise and support the vehicle safely.

2. Disconnect the return springs from the levers on the right and left-hand sides.

3. Push the rear guide sleeve out of the brackets on the right and left-hand sides.

4. Push the front guide sleeve out of the bracket on the right side.

5. Disconnect the cable from the levers.

6. Remove the rubber bushings from the equalizer and disconnect the cables from the equalizer.

7. Remove the retaining units from the exhaust pipe heat shield above the rear axle and the 2 front retaining nuts. Move the heat shield panel to the side and remove the cable guide sleeve from the bracket.

8. Remove the retaining nuts of the center heat shield on the left side and slightly bend the heat shield down.

9. Remove the parking brake cable from the bracket.

10. Pull the parking brake cable out on the left side in front of the fuel tank.

To install:

11. Route the new cable into the bracket in front of the fuel tank and connect the cable to the hook.

12. Fit the clip onto the hook.

13. Place the other part of the parking brake cable to the side of the fuel tank and onto the rear axle. Push the cable guide into the bracket along the side of the fuel tank.

14. Install the parking brake cable equalizer onto the cable and thread on the rubber bushing.

15. Install the rubber bushings.

16. Pull the cables down to the rear wheels and install the front cable guide onto the right-hand side.

17. Connect the cables to the parking brake levers.

18. Install the rear guide sleeves. Install the lever return springs.

19. Reposition the heat shields and tighten the retaining nuts.

20. Set the parking brake at the second notch.

21. Tighten the adjusting nut on the cable equalizer until the brake starts to engage.

22. Release the parking brake and tighten the adjusting nut again.

23. The wheels must lock at the sixth notch when setting the parking brake.

24. Carefully lower the vehicle to the ground.

9000 Series

CAUTION
Due to the parking brake cable routing, it is necessary to lower the fuel tank slightly in order to correctly place the parking brake cable. The procedure does not require disconnecting the fuel system, however all applicable safety precautions must be observed. Keep a dry chemical (Class B) fire extinguisher near the work area.

1. Disconnect any electrical connectors under the front passenger seat.

2. Push the passenger seat up as far as it will go to access and remove the seat rail mounting bolts. Push the passenger seat back as far as it will go and remove the front seat rail mounting bolts.

3. Disconnect the seat-belt anchorage point on the seat and remove the seat from the vehicle.

4. Slide the brush seal off the parking brake lever inside the car.

5. Remove the plastic locking plate from the adjusting nuts.

6. Remove the shift lever console by removing the mounting screws at the rear, front, and either sides of the console. Remove the console bezel as well as the gear selector cover.

7. Move back the interior carpet after removing the door sill scuff plates.

8. Remove the mounting screws and cover over the parking brake cables.

9. Unscrew the adjusting nuts from the ends of the cables at the parking brake lever.

10. Unhook the parking brake cable from the slot at the parking brake lever on the brake caliper. Remove the rubber boot and remove the cable.

11. Remove the cable guide from the suspension link. Withdraw the parking brake cable from the vehicle.

To install:

12. Slightly lower the fuel tank, route the cable through the grommet and secure the fuel tank back to its original position.

13. Screw the cable guide onto the suspension link.

14. Pull the parking brake cable up to the parking brake lever inside the car. Install the adjusting nuts onto the ends of the cables at the brake lever.

15. At the brake caliper, pack the rubber boot with grease and slide it over the end of the cable.

16. Remove the adjusting screw's plug at the caliper. Rotate the adjusting screw in as far as it will go and then back it out ¼–½ turn. Be sure the brake rotor rotates freely. Install the screw plug.

17. Insert the cable end into the slot on the parking brake lever at the caliper. Apply the parking brake several times to stretch the cable.

18. To adjust the brake cable, insert a 0.040 in. (1.0mm) feeler gauge between the parking brake lever and the stop on the caliper.

19. Rotate the adjusting nut at the parking brake lever until the feeler gauge slips out. Clearance is 0.040–0.060 in. (0.5–1.5mm).

20. Install the cover over the cables. Place the interior carpet back into its correct position and install the door sill scuff plate.

21. Install the shift lever console, bezel, and gear selector cover.

22. Install the plastic locking plate back over the adjusting nuts.

23. Place the brush seal over the parking brake lever and back into position.

24. Fit the front passenger seat into the correct position. Install and tighten the mounting bolts.

25. Reconnect the electrical connectors under the seat and the seat-belt anchorage point on the seat itself.

Brake System Bleeding

900 SERIES

1993 Models

NOTE: This procedure applies to both ABS and Non-ABS equipped 1993 900 models.

——— **CAUTION** ———
Turn the ignition switch OFF and remove the key to prevent accidentally running the ABS high pressure pump.

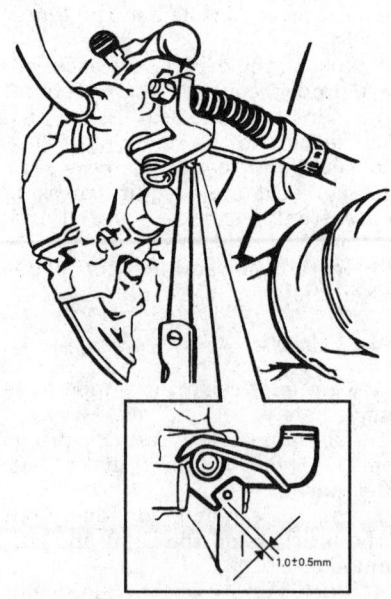

Inspect the parking brake adjustment using a feeler gauge — 9000 Series

1.0±0.5mm

301973

1. Remove the cap off of the brake fluid reservoir and top off with clean DOT 4 brake fluid. Check the fluid level frequently during the bleeding procedure.

2. Raise and support the vehicle safely.

3. Connect a brake bleeder hose to the brake bleeder screw behind the caliper at the right front wheel. Place the free end of the hose into a container partially filled with clean DOT 4 brake fluid. You must keep the end of the hose below the surface of the fluid in the bottle at all times.

NOTE: The brake bleeder screw need only be opened ⅓–½ turn.

4. With the aid of an assistant and with the ignition **OFF**, depress the brake pedal several times until the pedal feel changes to a noticeably stiffer resistance. Slowly pump the pedal a few more times; with the pedal depressed, open the bleeder screw. With the pedal depressed, tighten the bleeder screw.

5. Release the pedal, pump again slowly, hold pedal down and repeat the procedure until no air bubbles are seen in the fluid.

6. Perform this bleeding procedure next on the left front wheel.

7. When bleeding the rear wheel circuits, turn the ignition switch **ON** and depress the brake pedal. You must not run the pump motor for more than 2 minutes at a time. After the motor has been operating, you must allow it to cool for 10 minutes before starting it again.

8. Connect the brake bleeding hose to the right rear wheel. Perform the bleeding procedure to the right rear wheel caliper and then to the left rear wheel caliper.

9. Top off the brake fluid to the MAX level on the side of the reservoir.

1994–97 Models

——— **CAUTION** ———
The hydraulic accumulator on vehicles equipped with ABS contains brake fluid and nitrogen gas at extremely high pressures. Certain portions of the hydraulic system also contain brake fluid at high pressure. It is mandatory that the system be depressurized before disconnecting any hoses, lines or fittings, or personal injury may result. When the ignition is turned ON, the pump will run, repressurizing the system.

1. For vehicles equipped with ABS, perform the following:

   a. Turn the ignition switch **OFF** and remove the key.

   b. Step firmly on the brake pedal at least 20–30 times. The system is not completely discharged until a distinct change is felt in the brake pedal. The effort needed to press the pedal will clearly increase when the system pressure is released.

   c. Once the pedal feel changes, pump the pedal a few more times.

2. Remove the cap off of the brake fluid reservoir and top off with clean DOT 4 brake fluid. Check the fluid level frequently during the bleeding procedure.

3. Raise and support the vehicle safely.

4. Connect a brake bleeder hose to the brake bleeder screw behind the caliper at the left front wheel. Place the free end of the hose into a container partially filled with clean DOT 4 brake fluid. The end of the hose must be kept below the surface of the fluid in the bottle at all times.

NOTE: The brake bleeder screw need only be opened 1/3–1/2 turn.

5. With the ignition **OFF**, depress the brake pedal several times until pedal feel changes to a noticeably stiffer resistance. Slowly pump the pedal a few more times; with the pedal depressed, open the bleeder screw. Tighten the bleeder screw before the fluid pressure is gone.

6. Release the pedal, pump again slowly, hold the pedal down and repeat the procedure until no air bubbles are seen in the fluid.

7. Perform the brake bleeding procedure in the following sequence: left front wheel, right front wheel, left rear wheel, right rear wheel.

8. Lower the vehicle to the ground.

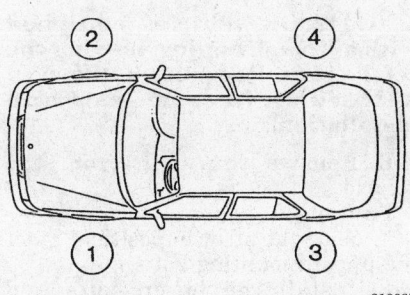

310087

Brake bleeding sequence — 1994–97 900 Series

9. Top off the brake fluid reservoir to the proper level using clean DOT 4 brake fluid only.

9000 SERIES

───────── **CAUTION** ─────────

On vehicles equipped with ABS, the hydraulic accumulator contains brake fluid and nitrogen gas at extremely high pressures. Certain portions of the hydraulic system also contain brake fluid at high pressure. It is mandatory that the system be depressurized before disconnecting any hoses, lines or fittings, or personal injury may result. When the ignition is turned ON, the pump will run, repressurizing the system.

1. Turn the ignition switch **OFF** and remove the key.

2. To depressurize the anti-lock brake system, step firmly on the brake pedal at least 20–30 times. The effort needed to press the pedal will clearly increase when the system pressure is released. Continue to pump the pedal a few more times.

3. Remove the cap off of the brake fluid reservoir and top off to the proper level with clean DOT 4 brake fluid. Check the fluid level frequently during the bleeding procedure.

4. Raise and support the vehicle safely.

5. Connect a brake bleeder hose to the brake bleeder screw behind the caliper at the right front wheel. Place the free end of the hose into a container partially filled with clean DOT 4 brake fluid. The end of the hose must be kept below the surface of the fluid in the container at all times.

NOTE: The brake bleeder screw need only be opened 1/3–1/2 turn.

6. With the ignition **OFF**, depress the brake pedal several times until the pedal feel changes to a noticeably stiffer resistance. Slowly pump the pedal a few more times; with the pedal depressed, open the bleeder screw. Tighten the bleeder screw before the fluid pressure is gone.

7. Release the pedal, pump again slowly, hold the pedal down and repeat the procedure until no air bubbles are seen in the fluid.

8. Perform the brake bleeding procedure in the following sequence: right front wheel, left rear wheel, left front wheel, right rear wheel.

9. Lower the vehicle to the ground.

10. Top off the brake fluid reservoir to the proper level using clean DOT 4 brake fluid only.

11. If applicable, turn the ignition switch **ON** and check that the ABS warning light, on the dashboard, goes out.

Wheel Speed Sensor

REMOVAL AND INSTALLATION

900 Series

Front Sensors

1. Raise vehicle and safety support with jack stands.

2. Remove wheel from hub assembly.

3. Clean area surrounding speed sensor using bristle brush.

4. Disconnect speed sensor electrical harness connector.

5. Remove speed sensor retaining bolt and remove speed sensor.

To install:

6. Install speed sensor into its holder and tighten retaining bolt.

7. Connect speed sensor electrical harness connect and attach cable to its retainer.

8. Install wheel and tighten to 89 ft. lbs.

9. Lower and test drive vehicle.

Rear Sensors

1. Raise and safely support the vehicle. Remove rear wheel.

2. Using slip joint pliers, press brake pistons back into calipers.

3. Remove brake caliper mounting bolts and safety hang caliper out of the way. Do not let the caliper hang by the brake hose.

4. Screw brake shoe adjuster back to allow rotor to come off easily.

5. Detach the hand brake cable and remove the disc brake retaining screw.

6. Remove rotor.

7. Disconnect the speed sensor electrical harness connection.

8. Remove the four hub retaining nuts.

9. Remove the wheel hub, backing plate and spacer.

10. Unplug speed sensor from inner hub mount.

To install:

11. Install speed sensor into inner hub mount.

12. Fit spacer backing plate and wheel hub on mount and tighten retaining nuts to 40 ft. lbs. (54 Nm).

13. Install the brake rotor and tighten the locking bolt using a locking compound to 96 inch lbs. (11 Nm).

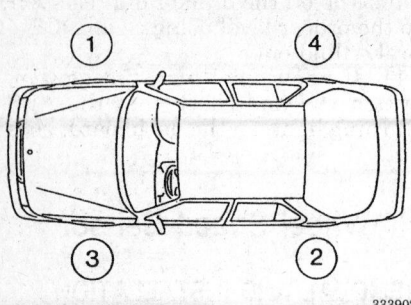

Brake bleeding sequence — 1993–97 9000 Series

14. Connect hand brake cable.
15. Install brake caliper and tighten retaining bolts to 59 ft. lbs. (80 Nm).
16. Install brake adjuster. Check brake pedal travel is proper and wheel turns freely with foot off brake pedal.
17. Install rear wheel and tighten mounting nuts to 89 ft. lbs. (120 Nm).

9000 Series

Front Sensors

1. Raise and safely support the vehicle.
2. Remove the wheel.
3. Clean area surrounding speed sensor using bristle brush.
4. Disconnect speed sensor electrical connector.
5. Remove speed sensor retaining bolt and remove speed sensor.
To install:
6. Install speed sensor into its holder and tighten retaining bolt.
7. Connect speed sensor wiring and attach cable to its retainer.
8. Install wheel and tighten to 89 ft. lbs.
9. Lower car and test drive vehicle.

Rear Sensors

1. From inside of vehicle move rear seat forward and remove cover on the appropriate side floor.
2. Unplug the sensor lead from the bracket by snipping through the tie on the top bracket.
3. Raise and safely support the vehicle. Remove rear wheel.
4. Remove the clip and pull the sensor lead through the rubber grommet in the floor.
5. Disconnect the securing bolt and remove sensor from wheel bearing assembly.
To install:
6. Install speed sensor into wheel bearing assembly.

7. Refit the clip for the sensor lead and run the lead through the rubber grommet.
8. Install rear wheel and torque to 89 ft. lbs. (120 Nm).
9. Plug in connect beneath rear seat.
10. Install cover.
11. Reposition rear seat.

FRONT SUSPENSION

Strut

REMOVAL AND INSTALLATION

900 Series

1994–97 Models

1. With all four wheels on the ground, loosen the front hub nut.
2. Raise the vehicle and support it safely.
3. Remove the front wheel.
4. Remove the hub nut and remove the wheel speed sensor.
5. Retract the caliper piston, unbolt and remove the caliper. Be sure the caliper is properly supported and not hanging from the brake hose.
6. Remove the rotor and backing plate.
7. Using tool 89–96–696 or equivalent, remove the tie rod end from the knuckle.
8. Disconnect the sway bar from the lower control arm.
9. Using tool 89–96–696 or equivalent, press the ball joint out of the steering knuckle.
10. Remove the three upper strut mounting bolts. Lift out the strut housing.
11. Place the strut housing in a vise.
12. Compress the spring with tool 89–18–809 or equivalent.
13. Remove the self-locking nut from the bearing plate and discard the nut.
14. Remove the coil spring, bellows and rubber snubber.
15. Remove the housing nut with spanner wrench 89–96–670 or equivalent.
16. Remove the strut cartridge from the housing.
To install:
17. Install the new strut cartridge into the housing and tighten the spanner nut to 159 ft. lbs. (215 Nm).

18. Position the coil spring in place. Be sure the end of the spring is up against the spring stop.
19. Position the upper spring seat with the notches properly aligned. Place the upper bearing assembly on the upper seat and secure it using a new self locking nut.
20. Remove the spring compressor and be sure the coil spring stays properly seated.
21. Position the strut assembly on the vehicle and tighten the upper mounting bolts 13 ft. lbs. (18 Nm).
22. Position the ball joint and install a new self locking nut and tighten it to 55 ft. lbs. (75 Nm).
23. Connect the tie rod end.
24. Tighten the sway bar at the lower control arm link to 8 ft. lbs. (10 Nm).
25. Install the backing plate.
26. Reassemble the brakes.
27. Install the wheel speed sensor.
28. Start the center hub nut.
29. Install the wheels and tighten the lugs to 77–95 ft. lbs. (105–130 Nm).
30. Lower the vehicle and with all four wheels on the ground, tighten the hub nut to 215 ft. lbs. (290 Nm).
31. Pump the brake pedal to position the brake caliper piston.

— CAUTION —
Do not attempt to move the vehicle until a firm pedal is obtained.

32. Check front wheel alignment.

9000 Series

1993 Models

1. Disconnect the negative battery cable.
2. Raise and support the vehicle safely. Remove the front wheel.
3. Remove the front brake hose from the retaining clip on the strut assembly.
4. Unbolt the strut from the steering knuckle.
5. Remove the 3 retaining bolts from the top of the strut.

NOTE: On vehicles equipped with air conditioning, moving the refrigerant line out of the way will ease in removal and installation.

6. Remove the strut from the vehicle.
To install:
7. Slide the strut in position. Start the upper mounting bolts.
8. Install the lower bolts and tighten to 58–77 ft. lbs. (78–105 Nm).
9. Tighten upper mounting bolts to 30–40 ft. lbs. (40–54 Nm).

10. Connect the brake hose to the strut housing.

11. Install the wheel and lower the vehicle.

12. Check the wheel alignment.

1994–97 Models

1. With all four wheels on the ground, loosen the front hub nut.

2. Raise the vehicle and support it safely.

3. Remove the front wheel.

4. Remove the hub nut and remove the wheel speed sensor.

5. Retract the caliper piston, unbolt and remove the caliper. Be sure the caliper is properly supported and not hanging from the brake hose.

6. Remove the rotor and backing plate.

7. Using tool 89–96–696 or equivalent, press the tie rod end from the knuckle.

8. Disconnect the sway bar from the lower control arm.

9. Unbolt strut from lower steering swivel and pull strut from swivel assembly.

10. Remove the three upper strut mounting bolts and lower the strut out of the vehicle.

11. To remove the spring and shock absorber cartridge:

    a. Place the strut in a vise.

    b. Compress the spring with tool 89–18–809 or equivalent.

    c. Remove the self-locking nut from the bearing plate and discard the nut.

    d. Remove the coil spring, bellows and rubber snubber.

    e. Remove the upper flange nut with spanner wrench 89–96–670 or equivalent.

    f. Remove the strut cartridge from the housing.

To install:

12. To assemble the strut:

    a. Install the new strut cartridge into the housing and tighten the spanner nut to 159 ft. lbs. (215 Nm).

    b. Position the compressed coil spring in place. Be sure the end of the spring is up against the spring stop.

    c. Position the upper spring seat with the notches properly aligned. Place the upper bearing assembly on the upper seat and secure it using a new self locking nut.

    d. Remove the spring compressor and be sure the coil spring stays properly seated.

13. Position the strut assembly on the vehicle and tighten the upper mounting bolts 13 ft. lbs. (18 Nm).

14. Install mounting bolts to connect strut to steering swivel assembly.

15. Connect the tie rod end.

16. Tighten the sway bar at the lower control arm link to 96 inch lbs. (10 Nm).

17. Install the backing plate.

18. Reassemble the brakes.

19. Install the wheel sensor.

20. Start the center hub nut.

21. Install the wheel and tighten the lugs to 77–95 ft. lbs. (105–130 Nm).

22. Lower the vehicle. With all four wheels on the ground, tighten the hub nut to 215 ft. lbs. (290 Nm).

23. Pump the brake pedal to position the brake caliper piston.

─────── **CAUTION** ───────

Do not attempt to move the vehicle until a firm pedal is obtained.

24. Check front wheel alignment.

Shock Absorber

REMOVAL AND INSTALLATION

900 Series

1993 Models

1. Open the hood. Locate and remove the upper shock absorber nuts.

2. Raise the vehicle and support it safely.

3. Remove the tire and wheel assembly.

4. Remove the bottom shock absorber retaining bolt. Remove the shock from the vehicle.

To install:

5. Extend the shock and install. Install the bottom bolt to hold the shock in place.

NOTE: The manufacturer recommends using only replacement shocks with a built in rubber buffer to prevent front suspension damage.

6. Install the upper bushing and position through the wheel well hole.

7. Tighten the lower shock bolt to 65 ft. lbs. (95 Nm).

8. Install the tire and wheel assemblies.

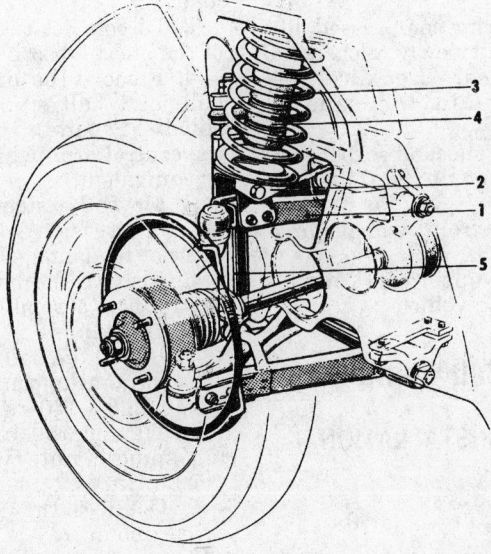

Front suspension
1. Upper control arm
2. Lower spring support
3. Coil spring
4. Rubber buffer
5. Shock absorber

204366

Front suspension component identification — 1993 900 Series

9. Lower the vehicle and tighten the upper shock nut until the shock bushing is compressed even with shock washers.

10. Road test vehicle.

Coil Spring

REMOVAL AND INSTALLATION

900 Series

1993 Models

1. Working under the hood, remove the upper shock absorber retaining nuts.

2. Raise and support the vehicle safely. Remove the tire and wheel assembly.

3. Install a spring compressor tool, engaging the upper coil at the second free turn from the top and the lower coil around the spring caps. Alignment shanks are located on the last turn of the spring with the color coded cup right beside the end of the coil.

4. Compress the spring at the top end, approximately 1 ½ in. (38mm). If the upper spring attachment of the steel cone is left behind in the wheel housing, remove it.

5. Remove the spring and the steel cone from the vehicle.

To install:

6. With the spring compressed, lift it into position and slowly release the spring compressor. Be sure the spring ends are in the saddles properly.

7. Position the shock absorber and tighten the lower bolt to 66–74 ft. lbs. (90–100 Nm)

8. Install the front wheels and lower the vehicle.

9. Connect the upper shock nuts.

10. Road test the vehicle.

Upper Ball Joints

REMOVAL AND INSTALLATION

900 Series

1993 Models

1. Raise and support the vehicle safely. Remove the tire and wheel assembly.

2. Place a floor jack under the lower control arm and raise it slightly to relieve pressure from the ball joints.

3. Disconnect the lower mounting of the shock absorber.

4. Remove the nut that holds the ball joint ball stud to the steering spindle. Separate the ball joint from the control arm using tool 89–95–409 or equivalent.

5. Move the steering spindle and position it aside so the brake hose will not be damaged.

6. Remove the ball joint from the control arm assembly.

To install:

7. Install the ball joint and tighten the ball joint-to-control arm bolts to 30–40 ft. lbs. (40–54 Nm).

8. Tighten the ball joint-to-steering spindle nuts to 26–37 ft. lbs. (35–50 Nm).

9. Connect the lower shock absorber mount.

10. Mount the wheels and lower the vehicle.

11. Check front wheel alignment.

Lower Ball Joints

REMOVAL AND INSTALLATION

900 Series

1993 Models

1. Raise and support the vehicle safely. Remove the tire and wheel assembly.

2. Place a floor jack under the lower control arm and raise it slightly to relieve pressure from the ball joints.

3. Disconnect the lower mounting of the shock absorber.

4. Remove the nut that holds the ball joint ball stud to the steering spindle. Separate the ball joint from the control arm using tool 89–95–409 or equivalent.

5. Move the steering spindle and position it aside so the brake hose will not be damaged.

6. Remove the ball joint from the control arm assembly.

To install:

7. Install the ball joint and tighten the ball joint-to-control arm bolts to 30–40 ft. lbs. (40–54 Nm).

8. Tighten the ball joint-to-steering spindle nuts to 26–37 ft. lbs. (35–50 Nm).

9. Connect the lower shock absorber mount.

10. Mount the wheels and lower the vehicle.

11. Check front wheel alignment.

1994–97 Models

NOTE: The ball joint cannot be removed from the control arm. To replace the ball joint, the control arm must be replaced.

1. Raise and support the vehicle safely. Remove the wheel.

2. Remove the sway bar link bolt.

3. Loosen the ball joint nut. Use tool 89-96-696 or equivalent to press the joint out of the steering knuckle.

4. Remove the retaining nut at the support arm.

5. Remove the retaining bolt at the subframe.

To install:

6. Install the arm and install the bolt at the subframe.

7. Install the bolt at the support arm and tighten to 68 ft. lbs. (92 Nm).

8. Tighten the retaining bolt at the subframe to 85 ft. lbs. (115 Nm).

9. Connect the sway bar link and tighten to 7 ft. lbs. (10 Nm).

10. Connect the ball joint and tighten to 55 ft. lbs. (75 Nm)

11. Install the wheels and lower the vehicle to the ground.

12. Check front wheel alignment.

9000 Series

1. Raise vehicle and support on safety stands.

2. Remove wheel.

3. Disconnect the three mounting bolts securing the ball joint to the lower control arm.

4. Remove the lock washer from atop the steering swivel member.

5. Remove the bolt securing the ball joint to the steering swivel member.

6. Remove the ball joint.

To install:

7. Place the ball joint in the steering member socket. Install the bolt and tighten to 20 ft. lbs. (28 Nm).

8. Secure the locking ring atop the ball joint.

9. Fit the ball joint to the lower control arm and tighten the three mounting bolts to 22 ft. lbs. (30Nm).

10. Fit the wheel and tighten to 89 ft. lbs. (120 Nm).

11. Lower the vehicle and test drive.

Upper Control Arms

REMOVAL AND INSTALLATION

900 Series

1993 Models

1. Disconnect the negative battery cable.

2. Raise the vehicle and support it safely.

3. Remove the tire and wheel assembly.

4. Remove the shock absorber.

5. Compress the coil spring, using a suitable spring compression tool.

Lower Control Arms

REMOVAL AND INSTALLATION

900 Series

1993 Models

1. Raise the vehicle and support it safely. Remove the tire and wheel assembly.
2. Disconnect the lower end of the shock absorber.
3. Remove the 2 bolts that attach the ball joint to the control arm.
4. Remove the 6 lower control arm attaching bolts from under the engine compartment floor.
5. Remove the control arm and its attaching brackets from the vehicle.
6. Remove the control arm bearing nuts and remove the bearings from the control arm.
To install:
7. Install the rubber bushings with the aid of tool 78–41–349, or equivalent. Soap the bushings prior to installation. When both nuts are tightened, the angle between the control arm and bearing should be 16–20°. Tighten the bolts to 70–77 ft. lbs. (100–120 Nm).
8. Install the control arm and secure the ball joint. Raise the control arms slightly and fit the shock absorber.
9. Install the wheel and lower the vehicle. Check the wheel alignment and test drive the vehicle.

1994–97 Models

1. Raise and support the vehicle safely. Remove the wheel.
2. Remove the sway bar link bolt.
3. Loosen the ball joint nut. Use tool 89–96–696 or equivalent to press the joint out of the steering knuckle.
4. Remove the retaining nut at the support arm.
5. Remove the retaining bolt at the subframe.
To install:
6. Install the arm and install the bolt at the subframe.
7. Install the bolt at the support arm and tighten to 68 ft. lbs. (92 Nm).
8. Tighten the retaining bolt at the subframe to 85 ft. lbs. (115 Nm).
9. Connect the sway bar link and tighten to 7 ft. lbs. (10 Nm).
10. Connect the ball joint and tighten to 55 ft. lbs. (75 Nm)
11. Install the wheels and lower the vehicle to the ground.
12. Check front wheel alignment.

1. Sway bar nut
2. Ball joint nut
3. Ball joint press tool
4. Support arm connection
5. Subframe connection
6. Lower control arm
7. Support arm

334351

Lower control arm connection points — 1994–97 900 Series

6. Remove the 2 bolts attaching the upper ball joint and lower spring seat to the upper control arm.
7. From under the hood, move the alternator and master cylinder, as required, to gain access to the left side rear upper control arm bolt. On the right side, the power steering pump may have to be moved to access the right rear upper control arm bolt.

— **CAUTION** —

To properly handle the removal and installation of the compressed coil spring, utilize the correct coil spring compressor tool. Exercise care when handling a compressed coil spring, the spring is powerful and can easily cause physical injury.

8. Remove the compressed coil spring from the vehicle.
9. Remove the bolts from both upper control arm bearing brackets.
10. Remove the control arm and bearings from the vehicle. Save the spacers under the bearings and record the number of spacers used under each bearing.

11. Remove both of the bearing nuts. Now the bearings and bushings can be removed from the control arm.
To install:
12. Install the rubber bushings with tool 78–41–331 or equivalent. Spray the bushings with soapy water prior to installation. Do not use oil or grease to aid installation.
13. Install the bearing to the control arm. Once the 2 nuts are tight, the angle between the control arm and the bearing should be 50–54°.
14. Install the control arm and tighten the bolts to 54–66 ft. lbs. (75–90 Nm). Install the upper ball joint to the control arm and tighten the bolts to 30–40 ft. lbs. (40–54 Nm).
15. Install the coil spring. Check that the spring bushing is seated in the upper spring pocket.
16. Raise the outer end of the lower control arm slightly and install the shock absorber. Install the wheel and lower the vehicle.
17. Reposition the alternator, master cylinder or power steering pump, if disturbed.
18. Check the wheel alignment and test drive the vehicle.

9000 Series

1. Raise the vehicle and support it safely. Remove the tire and wheel assembly.

2. Remove the bolts securing the suspension arm to the ball joint.

3. Remove the nut from the bolt securing the suspension arm to the sway bar link. Remove the upper securing bolt for the link.

4. Press down on the suspension arm and withdraw the sway bar link.

5. Remove the nuts at the front of the suspension arm from the bolts securing the arm to the frame.

6. Remove the rear bolts securing the reinforcement member to the frame.

7. Remove the bolts securing the control arm rear pivot to the frame. Remove the control arm.

To install:

8. Install the control arm. Tighten the sway bar link and ball joint connection but leave the nuts for the bushings in the suspension arm rear pivot loose.

9. After the arm is installed and the remaining bolts in place, tighten the rear pivot bolts. Tighten the bolts to 33–40 ft. lbs. (45–54 Nm).

10. Check the wheel alignment and adjust as required, after the vehicle has been allowed to settle by bouncing or driving.

Sway Bar

REMOVAL AND INSTALLATION

900 Series

1. Raise the hood. Disconnect the negative battery cable.

2. Place the engine support I-beam tool in position above the engine compartment. Connect the engine to the I-beam tool.

3. Raise and safely support the vehicle to access the undercarriage.

4. Disconnect the front exhaust pipe at the engine and before the muffler. Secure the pipe so that it is not hanging free.

5. Place support stands under the subframe and loosen the 2 front retaining bolts.

6. Remove the 4 retaining bolts and 2 retaining nuts at the rear of the subframe.

7. Raise the vehicle enough to allow the rear section of the subframe to lower.

8. Disconnect the sway bar links.

9. Disconnect the sway bar U-clamps and remove the sway bar.

To install:

10. Position the sway bar and connect the sway bar U-clamps. Tighten the bolts to 7 ft. lbs. (10 Nm).

11. Connect the sway bar links and tighten to 19 ft. lbs. (26 Nm).

12. Lower the vehicle enough to allow the rear section of the subframe to align with the bolt holes in the underbody.

13. Install and tighten the subframe bolts to 85 ft. lbs. (115 Nm).

14. Connect the exhaust.

15. Remove the support stands and lower the vehicle.

16. Disconnect the engine from the I-beam. Remove the I-beam from the engine compartment.

17. Connect the negative battery cable.

9000 Series

1. Raise vehicle and support on safety stands.

2. Remove front wheels.

3. Disconnect the sway bar link nut, washer and bushing on both sides.

4. Remove the upper securing nut on both sides.

5. Disconnect the sway bar U-clamp and bushing on both sides.

6. Lift out sway bar.

To install:

7. Place sway bar in mounts and loosely fit U-clamps with bushings.

8. Install sway bar links tighten top nut to 22 ft. lb. (30 Nm).

9. Tighten U-clamp on both sides to 18 ft. lbs. (24 Nm).

10. Install lower link nut on both sides and tighten to 18 ft. lbs. (24 Nm).

11. Install wheels.

12. Lower vehicle and test drive.

Front Wheel Bearings

ADJUSTMENT

The front wheel bearings found in all 1993–97 Saab vehicles are sealed units requiring no adjustment.

REMOVAL AND INSTALLATION

900 Series

1993 Models

1. With all four wheels on the ground, loosen the center hub nut.

2. Place the spring spacer tool 83–93–209 or equivalent under the upper control arm.

3. Raise and safely support the vehicle.

4. Remove the appropriate wheels.

5. Remove the ABS speed sensor, if equipped.

6. Align the disc access grooves to remove the brake pads and caliper assembly. Do not let the caliper hang by the brake hose.

7. Remove the brake disc.

8. Disconnect the outer tie rod end using tool 89–95–409 or equivalent.

9. Remove the ball joint retaining bolts at the control arms or the ball joint-to-steering knuckle nuts.

10. Remove the steering knuckle assembly. If required lightly tap the halfshaft with a soft head hammer to free it from the hub.

11. Place the knuckle assembly in a press and fit the correct press tools. Press the hub from the bearing.

NOTE: Pressing or removing the hub will destroy the bearing: always replace with a new bearing.

12. Remove the 2 retaining clips and press the bearing from the knuckle.

To install:

13. Lightly coat the bearing housing in the knuckle with molybdenum grease.

14. Install the retaining clip in the inboard groove. Press the bearing in the knuckle until it stops against the retaining clip. Make sure the press tool contacts only the outer bearing race or the bearing will be damaged.

15. Install the retaining clip in the outboard groove.

16. Press the hub into the new bearing. Make sure the inner bearing race is properly supported or set the hub on the press table and press the inner bearing race onto the hub.

17. Position the knuckle assembly and guide the outer CV-joint shaft through the hub bearing. Start the center hub nut.

18. Reconnect the upper and lower ball joints. Reconnect the tie rod end.

19. Reassemble the brakes. Connect the ABS speed sensor.

20. Install the wheels and lower the vehicle.

21. With all four wheels on the ground, tighten the center hub nut to 220 ft. lbs. (300 Nm) and stake the locking collar.

22. Remove the spacer tool from under the upper control arm.

23. Check front wheel alignment.

1994–97 Models

1. With all four wheels on the ground, loosen the front hub nut.

2. Raise and safely support the vehicle.

3. Remove the front wheel.

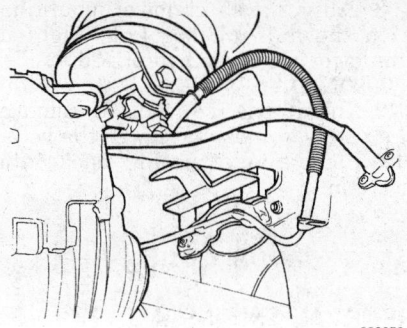

Supporting the upper control arm using special tool 83–93–209 — 1993 900 Series

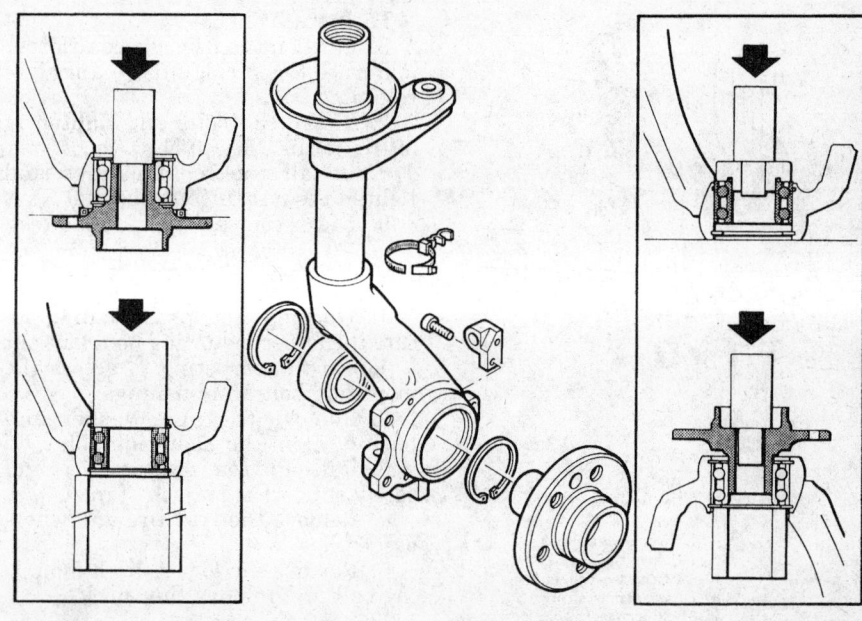

Bearing races must be properly supported during hub and wheel bearing installation — 1994–97 900 Series

4. Remove the hub nut and remove the wheel speed sensor.

5. Retract the brake caliper piston, unbolt and remove the caliper. Be sure the caliper is properly supported and not hanging from the brake hose.

6. Remove the rotor and backing plate.

7. Using tool 89–96–696 or equivalent, press the tie rod end from the knuckle.

8. Disconnect the sway bar from the lower control arm.

9. Using tool 89–96–696 or equivalent, press the ball joint out of the steering knuckle.

10. Remove the three upper strut mounting bolts. Lift out the strut housing.

11. Using the proper arbor press adapters, press the hub from the wheel bearing.

12. Remove the two circlips on each side of the wheel bearing.

13. Press out the wheel bearing using the proper size press adapters.

To install:

14. Install one circlip and press in wheel bearing until it contacts the circlip. Make sure the press tool contacts only the outer bearing race.

15. Install other circlip.

16. Clean the bearing surface of the hub on a wire wheel. If the hub is pitted or damaged, it should be replaced.

17. Support the inner bearing race and press hub into wheel bearing. If the inner race is not properly supported, the bearing will fail quickly.

18. Position the strut assembly on the vehicle and tighten the upper mounting bolts 13 ft. lbs. (18 Nm).

19. Position the ball joint, install a new self locking nut and tighten it to 55 ft. lbs. (75 Nm).

20. Connect the tie rod end.

21. Tighten the sway bar at the lower control arm link to 96 inch lbs. (10 Nm).

22. Install the backing plate.

23. Reassemble the brakes.

24. Install the wheel sensor.

25. Start the center hub nut.

26. Install the wheel assemblies and tighten the lugs to 77–95 ft. lbs. (105–130 Nm).

27. Lower the vehicle. With all four wheels on the ground, tighten the hub nut to 215 ft. lbs. (290 Nm).

28. Pump the brake pedal to position the brake caliper piston.

— CAUTION —
Do not attempt to move the vehicle until a firm pedal is obtained.

29. Check front wheel alignment.

9000 Series

1. With all four wheels on the ground, loosen the hub center nut and the wheel bolts.

2. Raise the vehicle and support it safely.

3. Remove the tire and wheel assembly. Remove the hub center nut and thrust washer.

4. Remove the flexible brake hose from its support clip.

5. Unbolt the caliper and secure it to the suspension arm; do not let the caliper hang by the brake hose.

6. Unscrew the locating stud for the disc and remove it from the hub.

7. Push in on the halfshaft. If the CV-joint shaft does not push in, use puller 87–91–287 or equivalent to break it loose.

NOTE: Do not allow the puller to push the halfshaft in more the 2 in. or damage to the inboard joint may occur.

8. Remove the 4 bolts securing the hub to the knuckle. All vehicles use socket head screws to retain the hub assembly. For easier removal, cut an Allen wrench to fit the head and turn it with an 8mm wrench.

9. Lift the hub and disc backing plate from the knuckle assembly.

10. Remove the two bolts connecting the knuckle to the strut.

To install:

11. Clean the bearing seat and lightly coat with molybdenum grease. Install the hub assembly. Draw the hub in by tightening the four retaining screws.

12. Tighten the hub retaining screws to 41–44 ft. lbs. (55–60 Nm).

13. Lift the knuckle into place and connect the two strut bolts with the nuts facing to the front of the vehicle. Tighten to 58–78 ft. lbs. (78–105 Nm).

14. Assemble the disc and brake assembly. Torque the caliper bolts to 52–80 ft. lbs. (70–110 Nm). Connect the ABS speed sensor, if equipped.

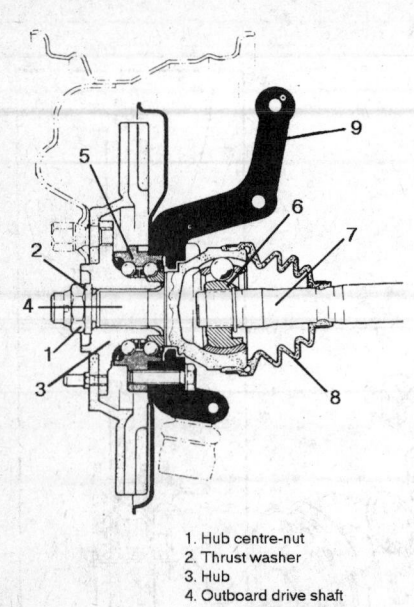

1. Hub centre-nut
2. Thrust washer
3. Hub
4. Outboard drive shaft
5. Bearings and seals
6. Constant-velocity joint
7. Inboard drive shaft
8. Rubber gaiter
9. Spindle

313989

Hub and knuckle component identifcation — 9000 Series

15. Install the wheels and lower the vehicle.

16. With all four wheels on the ground, tighten the center hub nut to 207–221 ft. lbs. (280–300 Nm).

17. Pump the brakes to seat the brakes before road testing.

REAR SUSPENSION

Shock Absorber

REMOVAL AND INSTALLATION

900 Series

1993 Models

1. Raise the vehicle and support it safely.

2. Place a suitable jackstand under the rear axle to prevent it from dropping and stretching the brake lines.

3. Position a jack at the rear of the spring link. Remove the upper and lower shock absorber fasteners.

4. Remove the bolts in the spring link mounting on the rear axle.

5. Lower the spring link so the shock absorber can be removed from the vehicle.

To install:

6. Position the new shock absorber and raise the spring link to align the retaining bolts.

7. Install the bolts and tighten to 59–67 ft. lbs. (80–90 Nm).

8. Install the top bushings and tighten the nut to 15 ft. lbs. (20 Nm).

9. Lower the vehicle and road test.

1994–97 Models

1. Working inside the trunk, locate the upper mounting bolt. Cut out a flap in the carpeting to access the mounting bolt and bushings.

2. Remove the nut, washer and bushing from the mounting point.

3. Raise the rear of the vehicle and safely support it on jack stands.

4. Remove the rear tire and wheel assemblies.

5. Remove the lower shock mounting bolt and remove the shock.

To install:

6. Install the shock into the upper mounting. Be sure to install the lower part of the bushing on the shock.

7. Install the lower shock mounting bolt and tighten to 46 ft. lbs. (62 Nm).

8. Install the rear wheels and tighten the nuts to 77–96 ft. lbs. (105–130 Nm).

9. Lower the vehicle to the ground.

10. Install the upper bushing, washer and nut. Tighten to 15 ft. lbs. (20 Nm).

11. Reposition the carpeting.

9000 Series

1. Disconnect the negative battery cable.

2. Raise and safely support the vehicle at the rear jacking point and remove the rear wheels.

3. Place jackstands under the rear end of the trailing arms where the they mount to the rear axle.

4. Position a jack at the rear jacking point. Raise the rear of the vehicle enough to relieve the load on the shock absorbers and anti-roll bar. Remove the lower shock absorber bolts.

5. From the trunk, pull back the carpet to locate the upper shock bolts. Remove the nut washer and bushing.

6. Remove the shock absorber from the vehicle.

To install:

7. Install the shock with the bushings in the proper orientation and tighten the upper shock bolts to 7–16 ft. lbs. (10–20 Nm).

8. Align the shock lower mounting with the anti-roll bar link. Tighten the lower shock bolt to 59–66 ft. lbs. (80–90Nm).

9. Install the rear wheels, remove the safety stands and lower the vehicle. Place the carpeting back into position.

Coil Spring

REMOVAL AND INSTALLATION

900 Series

1993 Models

1. Apply the hand brake.

2. Safely raise the rear of the vehicle from the rear jacking point and place jack stands under the rear axle.

NOTE: Do not raise the vehicle with the jack under the rear axle.

3. Remove the rear wheels.

4. Place a jack under the lower arm and disconnect the lower shock mounting bolt and the 2 locknuts that secure the front spring link bearing to the body.

5. Slowly lower the jack to remove the spring and spring seats.

To install:

6. Install the spring, making sure the rubber spring cushions are properly positioned.

7. Raise the lower control arm into position and secure the front spring link.

8. Secure the lower shock mount bolt to 70 ft. lbs. (95 Nm).

9. Install the rear wheels and lower the vehicle.

10. Road test the vehicle.

1994–97 Models

1. Raise the vehicle and safely support it on jack stands. Do not place the stands under the rear axle assembly.

2. Remove the rear wheels.

3. Place a floor jack under the lower control arm and raise upward slightly.

NOTE: If the same spring is to be reinstalled, mark the rear of the spring to be sure it is reinstalled in the proper position.

4. Remove the lower mounting bolt from the shock absorber.

5. Slowly lower the floor jack until the lower arm is relaxed.

6. Use a pry bar to bring the lower arm down far enough to remove the spring.

To install:

7. Pry down the lower arm to install the spring.

8. Install the spring with the rubber cushions in place.

9. Raise the control arm with the floor jack, secure the lower shock mount and tighten to 46 ft. lbs. (62 Nm).

10. Install the wheels and torque the nuts to 77–96 ft. lbs.105–130 Nm).

11. Lower the vehicle and road test.

9000 Series

1. Raise and safely support the vehicle on jack stands.

2. Remove the rear wheels.

3. Disconnect the hand brake cable from the retaining bracket connected to the lower control arm.

4. Remove the ABS sensor wiring, if necessary, by remove the clip and releasing the cable.

5. Place a floor jack under the lower arm and disconnect the trailing-end bolt and lower shock absorber bolt.

6. Lower the jack slowly and remove the coil spring.

To install:

7. Position the coil spring in the vehicle.

8. Raise the lower arm until the bolt holes align. Install and tighten the bolts.

9. Install ABS wiring, if necessary, by connecting the clip and cable.

10. Install the wheels and lower the vehicle.

11. Road test the vehicle.

Sway Bar

REMOVAL AND INSTALLATION

900 Series

1994–97 Models

1. Raise and safely support the vehicle.

2. Remove all bolts holding outer anti-roll bar in position.

3. Lower outer anti-roll bar.

4. Remove one wheel.

5. Remove all bolts from inner anti-roll bar.

6. Slide bar out from one side.

To install:

7. Slide inner anti-roll bar into position.

8. Tighten inner anti-roll bar retaining bolts to 45 ft. lbs. (62 Nm).

9. Install wheel and tighten wheel nuts to 89 ft. lbs. (72 Nm).

10. Position outer anti-roll bar.

11. Tighten outer anti-roll bar retaining bolts to 18 ft. lbs. (24 Nm).

9000 Series

1. Raise the vehicle and support it with jack stands placed under the control arms at the rear axle mounts.

2. Remove the rear wheels.

3. Using a floor jack, raise the vehicle at the rear jacking point to relieve the load on the anti-roll bar.

4. Remove the bolts at the anti-roll bar end mountings. Separate the anti-roll bar from the support and remove.

To install:

5. Install the anti-roll bar to the support. Install the bolts in the end mountings and tighten to 58–65 ft. lbs. (80–90 Nm).

6. Install the wheels and lower the vehicle.

Rear Wheel Bearings

ADJUSTMENT

The rear wheel bearings found in all 1993–97 Saab vehicles are sealed units requiring no adjustment.

REMOVAL AND INSTALLATION

900 Series

1993 Models

NOTE: The wheel bearing and seal are an integral part of the wheel hub and can not be services separately. If the wheel bearing and/or seal are damaged, the complete hub assembly must be replaced.

1. Raise and safely support the vehicle. Remove the wheel assembly.

2. Disconnect the parking brake cable from the caliper and rest the cable on the rear axle.

3. If equipped with ABS brakes, remove the wheel sensor from the hub and disconnect the sensor harness from the bracket on the trailing end of the spring link.

4. Remove the screw plug and unscrew the adjusting screw in the caliper.

5. Unbolt the brake caliper and wire it to the torque arm.

NOTE: Do not let the caliper hang from the brake hose.

6. Remove the locating stud from the brake disc and remove the disc.

7. Remove the dust cap from the hub and remove the center nut. Pull

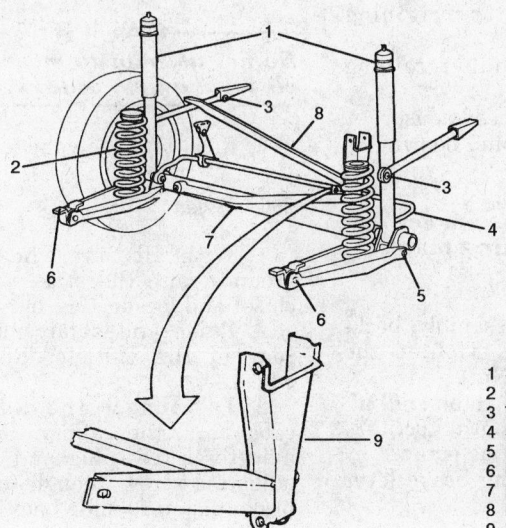

1 Shock absorber
2 Coil spring
3 Torque arm
4 Sway (anti-roll) bar
5 Lower control arm
6 Lower control arm bushing
7 Rear axle tube
8 Panhard rod
9 Frame support

305721

Rear suspension component identification — 9000 Series

the hub and bearing assembly off the stub axle.

To install:

8. Lightly lubricate the stub axle with oil. Hold the hub with both hands while applying pressure with the thumbs to the bearing race. Gently push the hub onto the stub axle, making sure it is centered on the stub axle.

9. Install a new center nut and tighten to 207–221 ft. lbs. (280–300 Nm).

10. Install the hub assembly dust cap.

11. Install the brake disc and locating stud.

12. Install the brake caliper.

13. Reconnect the parking brake cable and adjust, as required.

14. Install the wheel sensor and secure the wiring harness.

15. Install the tire and wheel assembly. Torque the wheel bolts to 78–92 ft. lbs. (105–125 Nm) Lower the vehicle.

1994–97 Models

1. Raise the vehicle and support it safely.

2. Remove the rear wheels.

3. Compress the caliper piston. Unbolt the caliper and secure it out of the way. Do not let the caliper hang by the brake hose.

4. Back off the adjuster on the parking brake shoes.

5. Remove the parking brake return spring and lever.

6. Remove the rotor retaining screws and the rotor.

7. Remove the 4 hub retaining nuts.

8. Disconnect the speed sensor.

9. Remove the hub and bearing as an assembly.

NOTE: There is a spacer behind the brake backing plate.

To install:

10. Install the hub assembly, backing plate and spacer. Tighten the nuts to 37 ft. lbs. (50 Nm).

11. Install the brake rotor and, after applying low strength Loctite®, install the rotor retaining screw.

12. Install the parking brake lever and return spring.

13. Connect the speed sensor.

14. Install the Caliper and retaining bolts with Loctite®, applied.

15. Screw in the brake shoe adjusting screw until the rotor cannot turn.

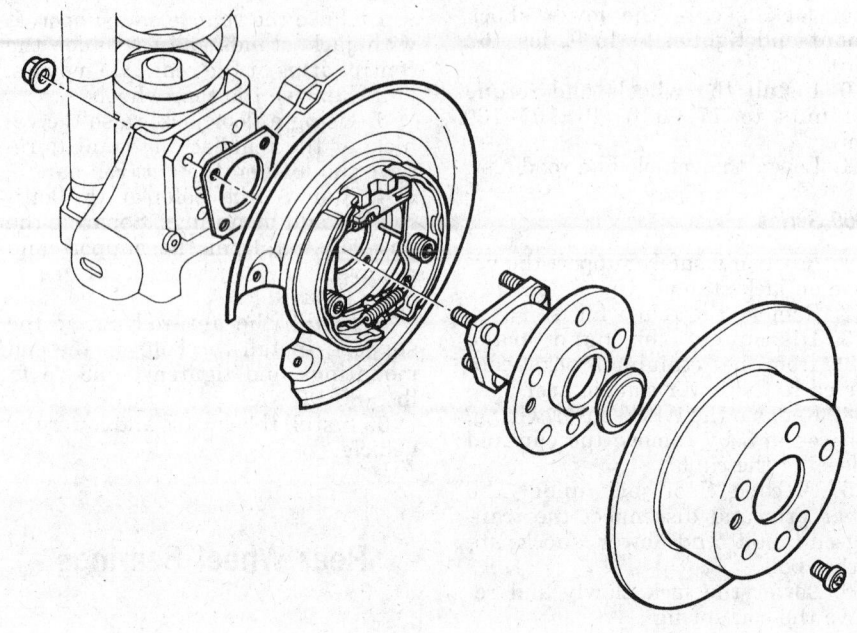

Rear hub and bearing components — 1994–97 900 Series

297008

Back off the screw until the rotor can rotate freely.

16. Install the wheels and lower the vehicle. Tighten the wheel lugs to 75–95 ft. lbs. (105–130 Nm).

17. Pump the brakes to position the caliper piston.

CAUTION
Do not attempt to move the vehicle until a firm pedal is obtained.

---

18. Road test the vehicle.

9000 Series

1. With all four wheels on the ground, raise the stake with a cold chisel and loosen the rear hub nut.

2. Raise and safely support the rear of the vehicle. Remove the wheel.

3. Disconnect the hand brake cable from the caliper. Remove the adjuster screw plug and loosen the adjusting screw enough to allow the brake piston to slide back.

4. Unbolt and remove the brake caliper and backing plate. Remove the ABS sensor, if equipped. Do not allow the caliper to hang by the brake hose.

5. Remove the bolt and pull off the brake disc.

6. Pry off the hub nut dust cap. Remove the hub nut and thrust washer and pull off the hub assembly. Discard the used hub nut.

NOTE: Whenever the hub nut is removed, a new one must always be used because the locking device on the nut becomes ineffective.

To install:

7. Check the spindle for damage and repair or replace as required. Install the hub assembly.

8. Install the thrust washer and nut but do not tighten the nut yet.

9. Install the brake disc. Install the caliper and tighten the bolts to 51–65 ft. lbs. (70–90 Nm).

10. Install the wheel and lower the vehicle.

11. With all four wheels on the ground, tighten the hub nut to 195–208 ft. lbs. (270–290 Nm). Stake the nut with a cold chisel and press the dust cap into place.

12. Pump the brake pedal several times before driving the vehicle.

FIRING ORDERS

NOTE: To avoid confusion, always replace the spark plug wires one at a time.

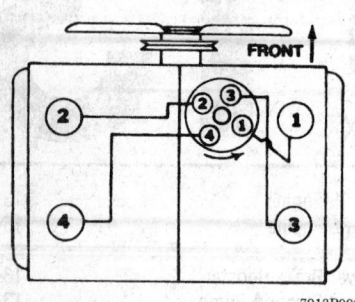

1.8L Engine
Engine Firing Order: 1–3–2–4
Distributor Rotation: Counterclockwise

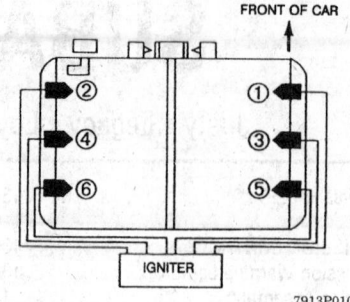

3.3 Engine
Engine Firing Order: 1–6–3–2–5–4
Distributorless Ignition System

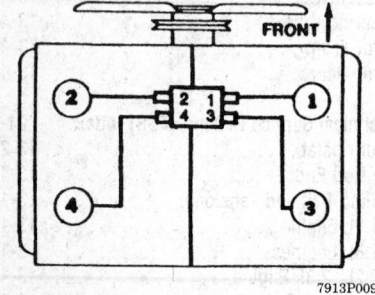

1.2L Engine
Engine Firing Order: 1–3–2
Distributor Rotation: Counterclockwise

2.2L Engine
Engine Firing Order: 1–3–2–4
Distributorless Ignition System

SERIAL NUMBER IDENTIFICATION

Vehicle Identification Plate

The vehicle identification plate is located on the bulkhead in the engine compartment.

Engine Number

The engine serial number is stamped on the front right side of the crankcase, on all engines except the 1.2L engine. On the 1.2L engine, the serial number is stamped at the right rear side of the engine below the cylinder head.

Vehicle Identification Number is located on the left side of the dash

Vehicle Identification Number

The Vehicle Identification Number (VIN) is stamped on a plate located on the top of the dashboard on the drivers side and is visible through the windshield.

Transaxle Number

The transaxle serial number is located on a sticker fixed to the upper surface of the main case (manual transaxle) or to the converter housing (automatic transaxle).

ENGINE MECHANICAL

NOTE: Disconnecting the negative battery cable on some vehicles may interfere with the functions of the on board computer systems and may require the computer to undergo a relearning process, once the negative battery cable is reconnected.

Engine Assembly

REMOVAL AND INSTALLATION

1.2L Engine

1. Disconnect the negative battery terminal from the battery.
2. Raise and support the front of the vehicle on jackstands.
3. Raise and support the hood with the stay so it opens wider than usual.
4. Position a drain pan under the radiator, remove the drain plug and the radiator cap, then drain the cooling system.
5. Remove the bumper and the grille.
6. Disconnect the electrical connectors and the hoses from the radiator and remove the radiator.
7. Disconnect the hood release cable and remove the radiator upper member.
8. Label, then disconnect all hoses, wires and cables.
9. Disconnect the pitching stopper from the bracket.

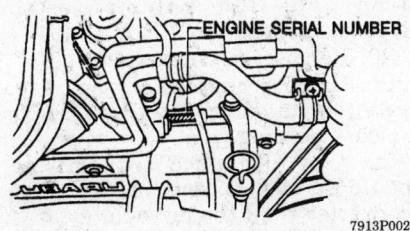

The engine number is stamped on the front right-side of the crankcase — except 1.2L engine

The engine number is stamped on the rear-side of the engine, below the cylinder head — 1.2L engine

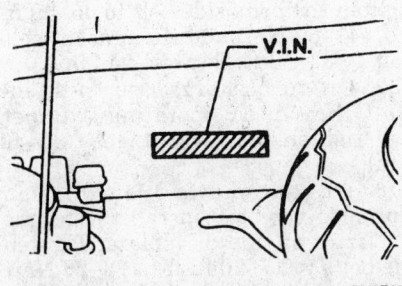

Vehicle Identification Plate is located on a plate attached to the bulkhead panel in the engine compartment

10. Remove the engine splash covers and the exhaust pipes.

11. Disconnect the gearshift rod and stay from the transmission.

12. Remove the transverse link. Using a rod, remove the spring pin and separate the front axle shaft.

13. Remove the engine/transmission mounting brackets.

14. Using an engine hoist and a cable, attach it to the engine and lift it slightly.

15. Remove the center member and crossmember from the vehicle.

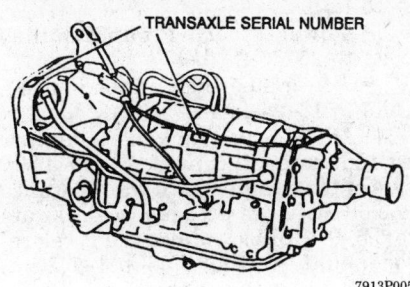

Transaxle identification plate location

16. Lift the engine/transmission assembly carefully and remove it from the vehicle.

17. Remove the engine from the transmission, then secure the engine to a workstand.

To install:

18. Attach the engine to the transmission and attach them to an engine hoist and a cable.

19. Lower the engine/transmission slowly and carefully into the vehicle.

20. With the engine/transmission assembly slightly raised, install the center member and crossmember to the vehicle.

21. Completely lower the engine/transmission and install the mounting brackets.

22. Install the front axle shaft, spring pin and transverse link.

23. Install the gearshift rod and stay into the transmission.

24. Install the exhaust pipes and the engine splash cover.

25. Install the pitching stopper.

26. Install all removed hoses.

27. Install the hood release cable.

28. Install the radiator and connect the hoses and electrical connectors.

29. Install the grille and bumper.

30. Make sure the drain has been placed in the radiator, and fill the radiator with coolant.

31. Lower the vehicle, install the battery and close the hood.

1.8L Engine

1. Open the hood and prop it, securely. Remove the spare tire and the spare tire bracket.

2. If equipped with Turbo or MPFI, perform the following procedures to reduce the fuel pressure:

    a. From under the vehicle, disconnect the fuel pump electrical harness connector.

    b. Crank the engine for at least 5 seconds. If the engine starts, allow it to run until it stalls.

c. Reconnect the fuel pump connector.

3. Remove the negative battery terminal from the battery.

4. Disconnect the air temperature sensor plug from the engine compartment.

5. Label and disconnect the fuel system hoses and the evaporative emissions system hoses.

6. Label and disconnect the vacuum hoses from the cruise control, the Master-Vac®, the air intake shutter and the heater air intake door.

7. Disconnect the electrical wiring connectors from the the alternator, the EGI, the thermoswitch, the electric fan, the A/C condenser and the ignition coil, then disconnect the main engine harness.

8. Label and disconnect the spark plug wires, the engine ground strap and the fusible link assembly.

9. Disconnect the accelerator linkage. Remove the windshield washer reservoir and position it behind the right strut tower.

10. To remove the power steering pump, perform the following procedures:

    a. Loosen the alternator pivot and mounting bolts, then shift the alternator to loosen the drive belt and remove the belt.

    b. Remove the pulley from the power steering pump.

    c. Remove the power steering pump-to-engine bolts and clamp.

    d. Remove the engine oil filler pipe brace.

    e. Remove the power steering pump and secure it to the bulkhead without disturbing the pressure lines.

11. Loosen the air intake duct hose clamps and remove the duct; seal the openings to keep dirt out of the air intake passages. Remove the upper cover.

12. Remove the air intake-to-flow meter line and cover the openings.

13. Remove the horizontal damper and clip.

14. To remove the center section of the exhaust pipe, perform the following procedures:

    a. Disconnect the temperature sensor connector.

    b. If equipped, disconnect the exhaust pipe-to-turbocharger bolts.

    c. Remove the rear cover.

    d. Remove the center exhaust section-to-transmission bolt.

    e. Remove the hanger bolts, then carefully remove the exhaust pipe (clearance is tight) to avoid damage.

f. Slightly loosen the attaching bolts, then remove the torque converter cover.

15. If equipped, disconnect the turbocharger oil supply and drain lines. Remove the turbo-to-exhaust bolts, the turbo assembly, the lower cover and the gasket.

16. Disconnect the electrical connector from the O_2 sensor. Remove the torque converter-to-driveplate bolts.

17. Using a chain hoist, connect it to the crankshaft damper bracket and support the engine. Remove the upper engine-to-transmission bolts; leave the starter in place.

18. Drain the engine coolant, using a hose to lead coolant to a clean container. Disconnect the upper/lower radiator hoses, the oil cooler lines, the ground wire and the radiator.

19. Disconnect the oil cooler lines from the engine and drain the oil into a clean container. Disconnect the heater hoses from the side of the engine.

20. Remove the front engine mount, then the lower engine-to-transmission nuts.

21. Position a floor jack under the transmission, then raise the engine/transmission slightly. Pull the engine forward until the transmission shaft clears the clutch, then carefully raise the engine out of the engine compartment.

To install:

22. To install, use new gaskets and observe the following points:

a. After installing all major mounting nuts and bolts finger-tight, tighten the upper transmission-to-engine bolts just snug, then, remove the engine/transmission support. Tighten the lower transmission-to-engine bolts, then tighten the engine-to-mount nuts.

b. When torquing the turbocharger (if equipped) and the exhaust system bolts, be sure to go back and forth, tighten the bolts evenly.

23. To complete the installation, reverse the removal process. Observe the following torques:

• Transmission-to-engine bolts to 14–17 ft. lbs.
• Torque converter-to-driveplate bolts to 17–20 ft. lbs.
• Turbocharger-to-exhaust system bolts to 31–38 ft. lbs.
• Exhaust system-to-transmission bolt to 18–25 ft. lbs.
• Exhaust system hanger bolts to 7–13 ft. lbs.

• Rear exhaust pipe joint nuts to 7–13 ft. lbs.
• Power steering pump pulley bolts to 25–30 ft. lbs.
• Power steering pump mounting bolts to 18–25 ft. lbs.

24. Adjust the crankshaft damper by tightening the nuts on the body side of the damper until the clearance is 0.08 in. (2mm); torque the locknuts to 6.5–9.4 ft. lbs. Adjust the accelerator pedal so there is 0.4–1.2 in. (10–30mm) between the pin and stop. Adjust the cable for an endplay of 0–0.08 in. (0–2mm) on the actuator side. Replenish all of the fluids. Run the engine to normal operating temperatures and check for leaks in oil cooler and lines.

2.2L Engine

1. Raise and support the vehicle safely.
2. Release the fuel system pressure.
3. Disconnect the battery cables and remove the battery.
4. Drain the coolant.
5. On non-turbo engines, remove the manifold cover.
6. Remove the cooling system, radiator and fan assembly, and reservoir tank.
7. On air conditioning equipped models, discharge the air conditioning system and remove the high pressure hoses.
8. Remove the air intake system. Remove the air intake duct from non-turbo engines and the resonator chamber from turbo engines.
9. Remove the air cleaner upper cover and element.
10. On turbo engines, remove the turbocharger cooling duct and air inlet and outlet ducts.
11. Remove the evaporative canister and bracket.
12. Label and disconnect all electrical connectors, cables and vacuum hoses.
13. Remove the power steering pump.
14. On turbocharged engines, remove the turbocharger unit from the center exhaust pipe.
15. Remove the exhaust system from the engine.
16. On turbocharged engines, remove the clutch damper with bracket.
17. Remove the nut which holds the power side of the starter.
18. Remove the nuts which hold the lower side of the transmission to the engine.

19. Remove the nuts which hold the front cushion rubber to the crossmember.
20. Remove the starter.
21. On turbocharged engines, separate the clutch release fork from the release bearing.
22. On automatic transmission equipped models, separate the torque converter from the driveplate.
23. Remove the pitching stopper and bracket.
24. Disconnect the fuel delivery hose, return hose and evaporation hose.
25. Support the engine with a lifting device and the transmission with a floor jack.
26. Remove the bolt which holds the upper side of the transmission to the engine.
27. Remove the engine.

To install:

28. Install the clutch release fork and bearing onto the transmission.
29. Install the engine to the transmission and tighten the bolts which holds the right upper side of the transmission to 34–40 ft. lbs. (46–54 Nm).
30. Remove the lifting device from the engine and remove the floor jack.
31. Install the pitching stopper and tighten the body side bolt to 35–49 ft. lbs. (47–67 Nm) and the bracket side bolt to 33–40 ft. lbs. (44–54 Nm).
32. On turbocharged engines, install the clutch operating cylinder and tighten to 25–30 ft. lbs. (34–40 Nm).
33. On automatic transmission equipped vehicles, install the torque converter on the driveplate. Tighten the bolts to 17–20 ft. lbs. (23–26 Nm).
34. Install the canister and bracket. Install the power steering pump and tighten the bolts to 22–36 ft. lbs. (29–49 Nm).
35. Install the starter and tighten the bolts to 22–36 ft. lbs. (29–49 Nm).
36. Tighten the nuts which hold the lower side of the transmission to the engine to 34–40 ft. lbs. (46–54 Nm).
37. Tighten the nuts which hold the front cushion rubber to the crossmember to 40–61 ft. lbs. (54–83 Nm).
38. Install the exhaust system.
39. On turbocharged engines, install the air inlet and outlet ducts.
40. Connect all hoses, electrical connectors and cables previously disconnected.
41. On turbocharged engines, install the turbo cooling duct.
42. Install the air intake system.
43. If equipped with air conditioning, install the air conditioner high

pressure hoses. Tighten to 13–23 ft. lbs. (18–31 Nm).

44. Install the cooling system. Tighten bolts to 9–11 ft. lbs. (12–15 Nm).

45. Install the manifold cover.

46. Install the battery and connect the battery cables.

47. Fill the radiator with coolant.

48. Check the level of the transmission fluid and add as necessary.

49. Check the level of the engine oil and add as necessary.

50. Start the engine and allow it to reach operating temperature. Check for leaks. Test drive the vehicle.

3.3L Engine

1. Raise and support the vehicle safely.

2. Release the fuel system pressure.

3. Disconnect the negative battery cable.

4. Remove the under body cover and drain the engine coolant.

5. Remove the radiator and all coolant hoses.

6. Discharge the air conditioning system. Disconnect and plug the air conditioning lines.

7. Remove the air intake system.

8. Disconnect the accelerator cable.

9. Disconnect the cruise control cable.

10. Label and disconnect all wiring harness connectors and cables.

11. Remove the evaporation canister, vacuum hoses and bracket.

12. Remove the exhaust system from the engine.

13. Disconnect the power steering hoses from the gear box.

14. Disconnect the automatic transmission cooler lines.

15. Remove the nuts which hold the lower side of the engine to the transmission and attach the lower side of starter.

16. Remove the nuts which attach the front cushion rubber to the subframe.

17. Separate the torque converter from the driveplate.

18. Remove the pitching stopper and bracket.

19. Disconnect the fuel delivery hose, return hose and evaporation hoses.

20. Support the engine with a lifting device and the transmission with a transmission jack.

21. Remove the bolts which hold the upper side of the engine to the transmission.

22. Remove the engine from the vehicle.

To install:

23. Install the engine to the transmission and tighten the bolts which hold the right upper side of the engine to 34–40 ft. lbs. (46–54 Nm).

24. Remove the lifting device and transmission jack.

25. Install the pitching stopper and tighten to 33–40 ft. lbs. (44–54 Nm).

26. Install the torque converter to driveplate bolts and tighten to 17–20 ft. lbs. (23–26 Nm).

27. Connect all hoses previously disconnected.

28. Install the evaporation canister and bracket.

29. Install the cooling system.

30. Install the nuts which hold the lower side of the engine to the transmission and attach the lower side of the starter. Tighten to 34–40 ft. lbs. (54–83 Nm).

31. Install the nuts which hold the front cushion rubber to the subframe. Tighten to 40–61 ft. lbs.(54–83 Nm).

32. Connect the power steering hoses to the gear box and the automatic transmission cooler lines.

33. Install the exhaust system.

34. Install the engine under cover.

35. Connect all electrical harness connectors.

36. Connect the accelerator cable.

37. Connect the cruise control cable.

38. Connect the high pressure hoses to the air conditioner compressor. Tighten to 13–23 ft. lbs. (18–31 Nm).

39. Install the air intake system

40. Connect the negative battery cable.

41. Fill the radiator with coolant.

42. Check the automatic transmission oil level and add as required.

43. Check the power steering fluid level. Add as necessary and bleed all air from the system.

44. Check the engine oil level.

45. Start the engine and allow it to reach operating temperature. Check for leaks. Test drive the vehicle.

Cylinder Head

REMOVAL AND INSTALLATION

1.2L Engine

1. Disconnect the negative battery cable. Drain the cooling system.

2. Remove the air cleaner assembly. Remove the drive belts. Remove the spark plug wires.

3. Position the engine at TDC with No. 1 cylinder on the compression stroke. Matchmark and remove the distributor assembly.

4. Remove the crankshaft pulley, using pulley removal tool 499205500 or equivalent. Remove the outer front timing belt cover.

5. Loosen the tensioner bolt and position it in the direction that loosens the belt. Tighten the tensioner bolt in that position.

6. Remove the camshaft driveplate. Mark the timing belt, in the direction of rotation, for reinstallation and than remove it from the engine.

7. Remove the tensioner and spring. Remove the camshaft pulley, using pulley removal tool 499205500 or equivalent. Remove the inner belt cover and cover mount.

8. Remove the PCV hose from the rocker arm cover. Remove the rocker arm cover retaining bolts. Remove the rocker arm cover from the engine. Remove the rocker arm assembly.

9. Remove the exhaust manifold retaining bolts. Remove the exhaust manifold from the engine. Discard the gasket.

10. Disconnect all required electrical wiring and vacuum lines. Remove the air suction valve and pipe, if equipped.

11. Disconnect the accelerator linkage. Remove the intake manifold retaining bolts. Remove the intake manifold. Discard the gasket.

12. Be sure the engine is cold before removing the cylinder head bolts. Loosen, than remove the cylinder head bolts. Carefully remove the cylinder head from the engine.

To install:

13. Installation is the reverse of the removal procedure. Be sure to use new gaskets or RTV sealant, as required.

14. Install the cylinder head and tighten the bolts as follows:

   a. Step 1 — Torque all bolts in sequence to 29 ft. lbs. (39 Nm).

   b. Step 2 — Torque all bolts in sequence to 54 ft. lbs. (73 Nm).

   c. Step 3 — Loosen bolts 90 degrees or more in reverse order of tightening sequence.

   d. Step 4 — Torque all bolts in sequence to 54 ft. lbs. (73 Nm).

15. Adjust the valves to specification, as required. Install the camshaft pulley, timing belt, tensioner and driveplate.

16. Install the distributor and all other components previously removed. Fill the cooling system with coolant and check the engine oil.

17. Start the engine, check the ignition timing, check for leaks and test drive the vehicle.

1.8L

1. Disconnect the negative battery cable.

2. Remove the timing belt, belt cover and related components.

3. On turbocharged engines, remove the turbo cooling pipe together with the union screws and gaskets from the cylinder head.

4. Remove the camshaft cases, lash adjusters and related components.

5. On turbocharged engines, remove the EGR pipe.

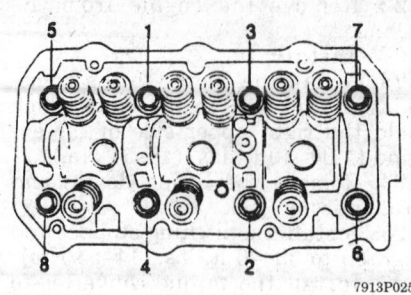

Cylinder head bolt torque sequence — 1.2L engine

6. Remove the accessory drive belts, alternator and air conditioner compressor if not already removed. Remove the bolt attaching the alternator bracket to the cylinder head.

7. On fuel injected engines, relieve the fuel system pressure.

8. Remove the bolts attaching the intake manifold to the cylinder head and remove the intake manifold.

9. Remove the bolt attaching the water bypass pipe bracket to the cylinder head.

10. Remove the spark plugs.

To install:

11. Clean all gasket mating surfaces thoroughly. Inspect the cylinder

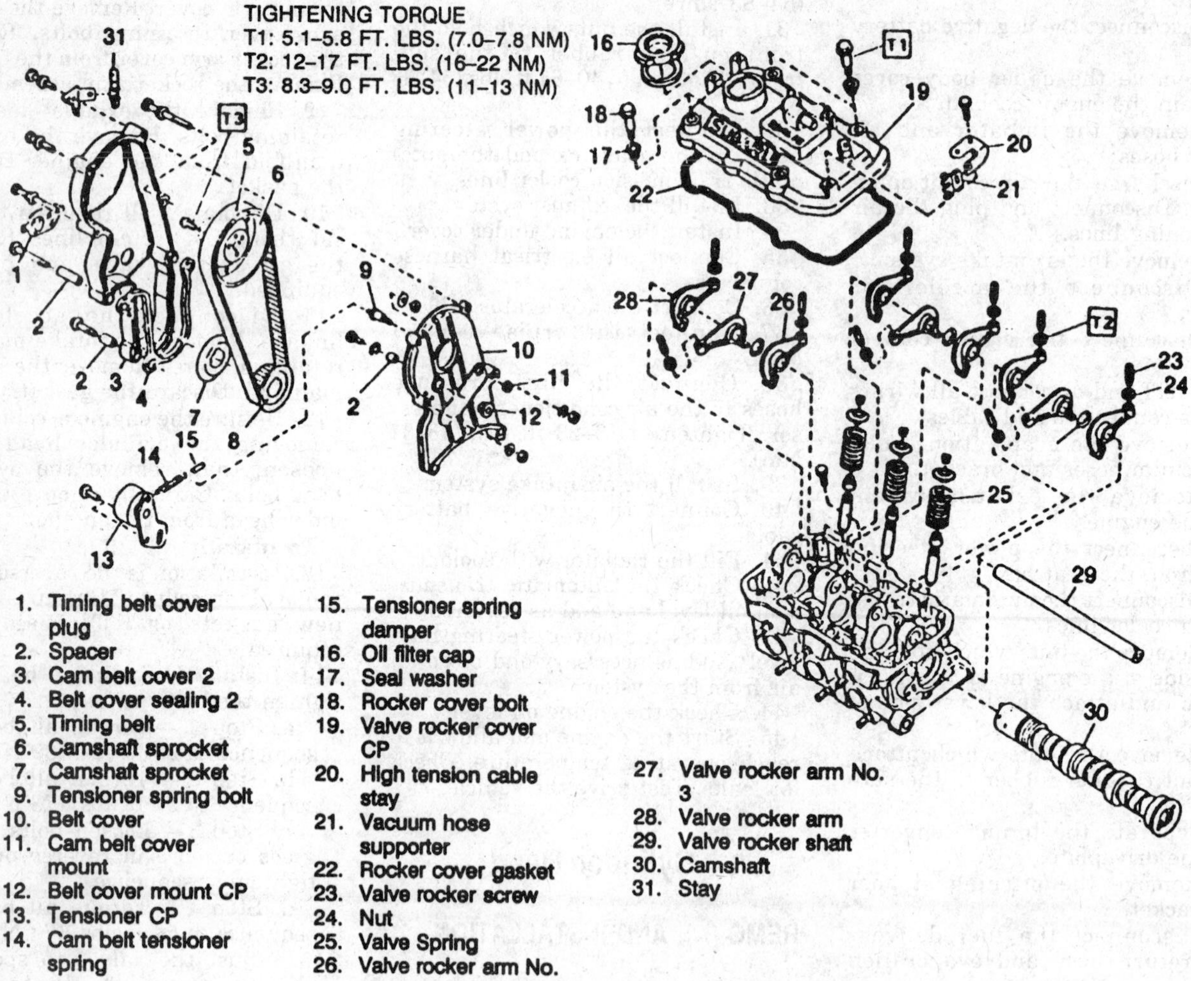

TIGHTENING TORQUE
T1: 5.1–5.8 FT. LBS. (7.0–7.8 NM)
T2: 12–17 FT. LBS. (16–22 NM)
T3: 8.3–9.0 FT. LBS. (11–13 NM)

1. Timing belt cover plug
2. Spacer
3. Cam belt cover 2
4. Belt cover sealing 2
5. Timing belt
6. Camshaft sprocket
7. Camshaft sprocket
9. Tensioner spring bolt
10. Belt cover
11. Cam belt cover mount
12. Belt cover mount CP
13. Tensioner CP
14. Cam belt tensioner spring
15. Tensioner spring damper
16. Oil filter cap
17. Seal washer
18. Rocker cover bolt
19. Valve rocker cover CP
20. High tension cable stay
21. Vacuum hose supporter
22. Rocker cover gasket
23. Valve rocker screw
24. Nut
25. Valve Spring
26. Valve rocker arm No.
27. Valve rocker arm No. 3
28. Valve rocker arm
29. Valve rocker shaft
30. Camshaft
31. Stay

Cylinder head and related components — 1.2L engine

7913P024

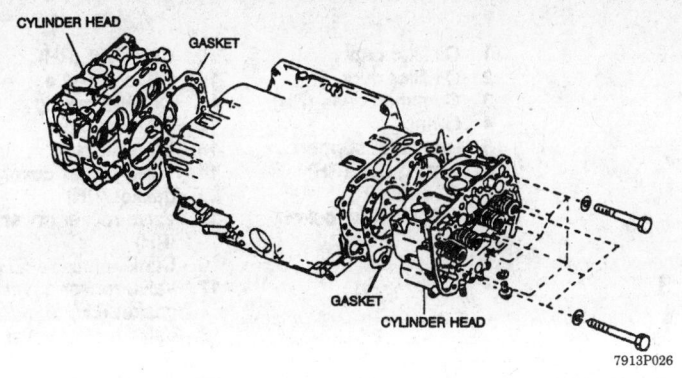

CYLINDER HEAD   GASKET

GASKET
CYLINDER HEAD

7913P026

Cylinder head and related components — 1.8L engine

head for warpage. Warpage should not exceed 0.0020 in. (0.05mm).

12. Install the cylinder heads using new gaskets.

13. Tighten the cylinder head bolts as follows:

    a. Tighten all bolts in sequence to 22 ft. lbs. (29 Nm).

    b. Tighten all bolts in sequence to 43 ft. lbs. (59 Nm).

    c. Tighten all bolts in sequence to 47 ft. lbs. (64 Nm).

14. Install the spark plugs. Install the water bypass pipe bracket.

15. Install the intake manifold and tighten the bolts to 13–16 ft. lbs. (18–22 Nm).

16. Install the alternator and bracket, air conditioner compressor and accessory drive belt.

17. On turbocharged engines, install the EGR pipe. Tighten the bolts to 23–27 ft. lbs. (31–37 Nm).

18. Install the camshaft cases, lash adjusters and related components.

19. On turbocharged engines, install the turbo cooling pipe. 16–18 ft. lbs. (21–24 Nm).

20. Install the timing belt, belt cover and related components.

21. Connect the negative battery cable.

22. Adjust the valve lash, as required. Start the engine and allow it

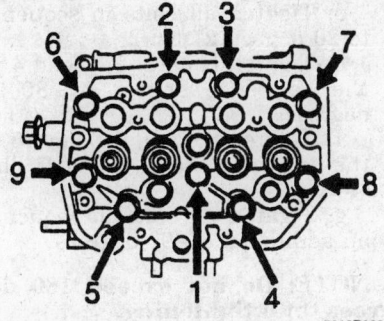

7913P029

Cylinder head bolt torque sequence — 1.8L engine

to reach operating temperature. Adjust the ignition timing.

23. Check for leaks and test drive the vehicle.

NOTE: Depending on the type of cylinder heads used, retightening of the bolts may be necessary after the vehicle has be running.

2.2L Engine

1. Disconnect the negative battery cable.

2. Remove the V-belt, power steering pump, alternator and bracket.

3. Remove the valve rocker cover.

4. Disconnect the PCV hose and spark plug wires.

5. Remove the connector bracket attaching bolt.

6. Remove the crank angle sensor and cam angle sensor.

7. Disconnect the oil pressure switch. Remove the knock sensor.

8. Disconnect the blowby hose.

9. Relieve the fuel system pressure and disconnect the fuel pipes.

10. Remove the intake manifold and gasket. Remove the water pipe.

11. Remove the timing belt, camshaft sprocket and related components.

12. Remove the oil level gauge guide attaching bolt on the left cylinder head.

13. Remove the cylinder head bolts in the proper sequence. Leave bolts 1 and 3 installed loosely to prevent the cylinder head from falling.

14. While tapping the cylinder head with a plastic hammer, separate it from the cylinder block.

15. Remove bolts 1 and 3. Remove the cylinder head and gasket.

   To install:

16. Clean all gasket mating surfaces thoroughly. Inspect the cylinder head for warpage. Warpage should not exceed 0.0020 in. (0.05mm).

17. Install the cylinder heads on the block using new gaskets.

18. Tighten the cylinder head bolts, after lubricating them with oil, to the following specifications:

    a. Tighten all bolts in sequence to 22 ft. lbs. (29 Nm).

    b. Tighten all bolts in sequence to 51 ft. lbs. (69 Nm).

    c. Loosen all bolts by 180 degrees, then loosen an additional 180 degrees.

    d. Tighten bolts 1 and 2 to 25 ft. lbs. (24 Nm) for non-turbo engines or 27 ft. lbs. (37 Nm) for turbo engines.

    e. Tighten bolts 3, 4, 5 and 6 to 11 ft. lbs. (15 Nm) for non-turbo engines or 14 ft. lbs. (20 Nm) for turbo engines.

    f. Tighten all bolts in sequence by 80–90 degrees.

    g. Tighten all bolts in sequence an additional 80–90 degrees.

NOTE: Do not exceed 180 degrees total tightening.

19. Install the oil level gauge guide attaching bolt on the left cylinder head.

20. Install the timing belt, camshaft sprocket and related components.

21. Install the water pipe.

22. Install the intake manifold and tighten bolts to 21–25 ft. lbs. (28–34 Nm). Connect the fuel delivery pipes.

23. Connect the blowby hose. Install the knock sensor.

24. Connect the oil pressure switch connector.

25. Install the crank and cam angle sensors.

26. Install the connector bracket attaching bolt.

27. Connect the spark plug wires. Connect the PCV hose.

28. Install the valve rocker cover and tighten bolts to 4 ft. lbs. (9 Nm).

29. Install the alternator, power steering pump and accessory drive belt.

30. Connect the negative battery cable. Start the engine and allow it to reach operating temperature. Check for leaks and test drive the vehicle.

3.3L Engine

1. Disconnect the negative battery cable.

2. Remove the timing belt, camshaft sprockets and related components.

3. Remove the EGR valve, EGR pipe and BPT.

4. Disconnect the auxiliary air control valve connector.

5. Disconnect the blowby hoses and auxiliary air valve hose.

6. Disconnect the PCV hose.

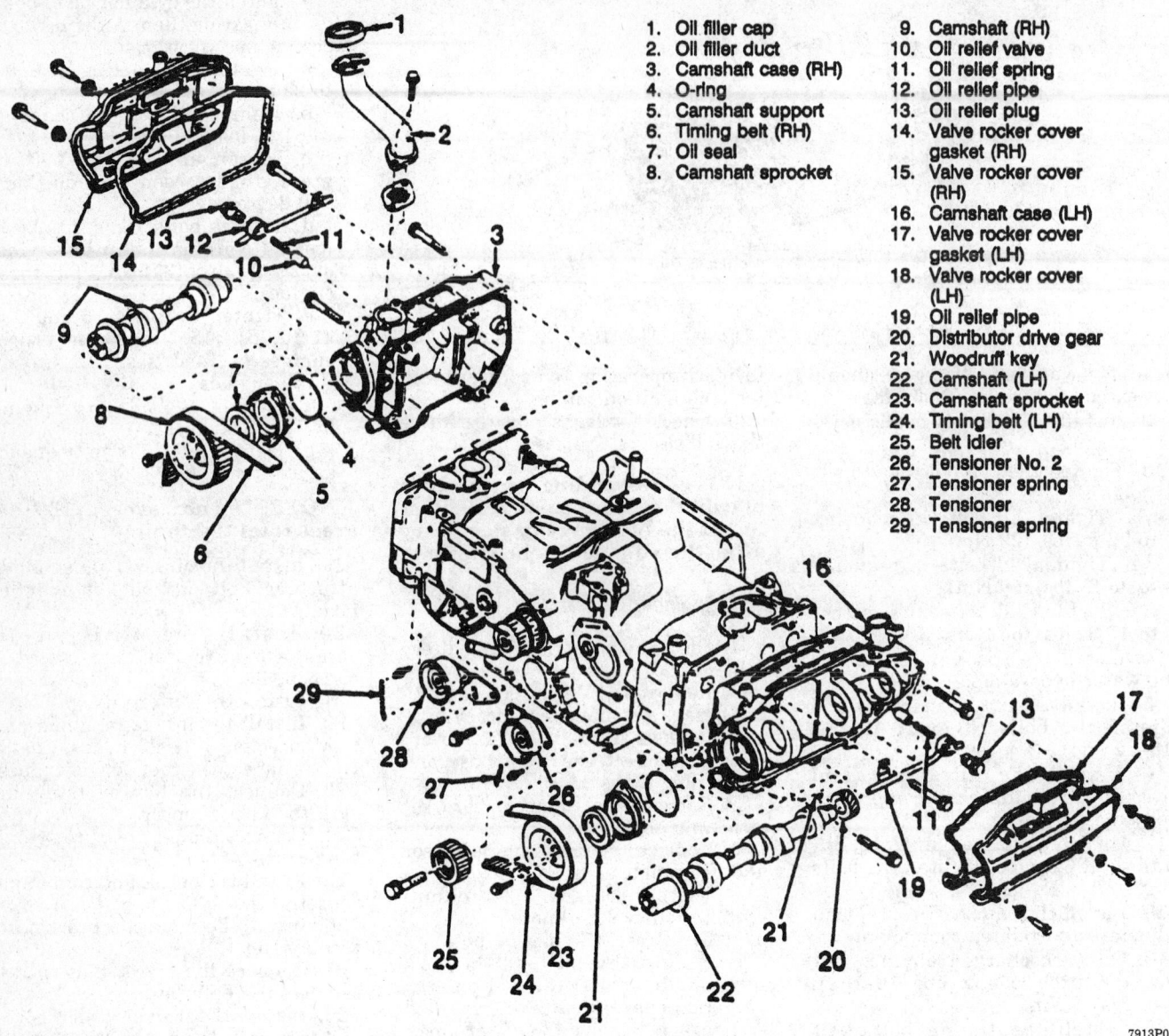

1. Oil filler cap
2. Oil filler duct
3. Camshaft case (RH)
4. O-ring
5. Camshaft support
6. Timing belt (RH)
7. Oil seal
8. Camshaft sprocket
9. Camshaft (RH)
10. Oil relief valve
11. Oil relief spring
12. Oil relief pipe
13. Oil relief plug
14. Valve rocker cover gasket (RH)
15. Valve rocker cover (RH)
16. Camshaft case (LH)
17. Valve rocker cover gasket (LH)
18. Valve rocker cover (LH)
19. Oil relief pipe
20. Distributor drive gear
21. Woodruff key
22. Camshaft (LH)
23. Camshaft sprocket
24. Timing belt (LH)
25. Belt idler
26. Tensioner No. 2
27. Tensioner spring
28. Tensioner
29. Tensioner spring

7913P030

Cylinder head assembly — 1.8L engine

7. Disconnect the water hoses from the throttle body.

8. Relieve the fuel system pressure. Remove the collector and intake manifold assembly with the gaskets.

9. Remove the exhaust manifold and gasket.

10. Remove the cylinder head covers, camshafts and related components.

11. Remove the oil level guide and heater pipe.

12. Remove the cylinder head bolts in the proper sequence. Leave bolts 5 and 8 loosely installed to prevent the cylinder head from falling.

13. While tapping the cylinder head with a plastic hammer, separate it from the cylinder block. Remove bolts 5 and 8 to remove the cylinder head and gasket.

To install:

14. Clean all gasket mating surfaces thoroughly. Inspect the cylinder head for warpage. Warpage should not exceed 0.0020 in. (0.05mm).

15. Install the cylinder head, using new gaskets, on the cylinder block.

16. Tighten the cylinder head bolts, after lubricating them with oil, to the following specifications:

a. Tighten all bolts in sequence to 22 ft. lbs. (29 Nm).

b. Tighten all bolts in sequence to 51 ft. lbs. (69 Nm).

c. Loosen all bolts by 180 degrees, then loosen an additional 180 degrees.

d. Tighten all bolts in sequence to 20 ft. lbs. (27 Nm).

e. Tighten bolts 1, 2, 3 and 4 in the sequence shown by 80–90 degrees.

f. Tighten bolts 5, 6, 7 and 8 in the sequence shown to 33 ft. lbs. (44 Nm).

g. Tighten all bolts in sequence an additional 80–90 degrees.

NOTE: Do not exceed 180 degrees total tightening.

17. Install the heater pipe and oil level gauge.

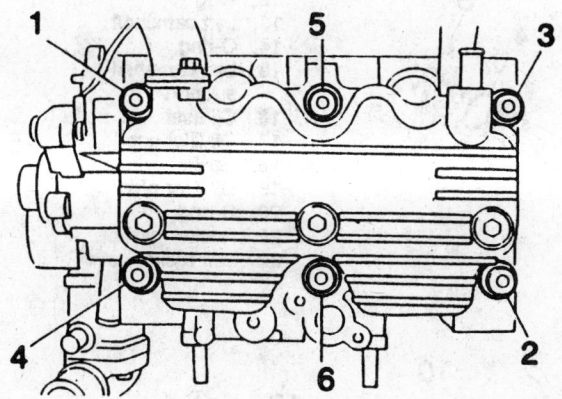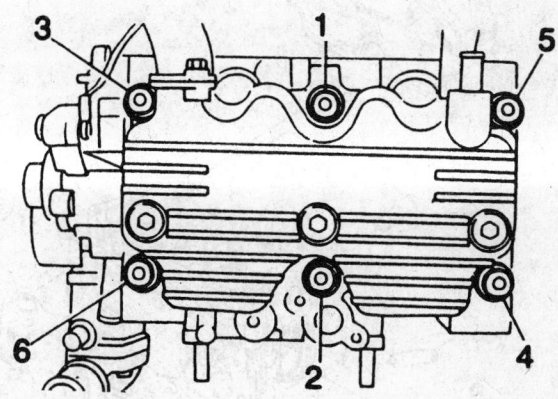

7913P031

Cylinder head torque sequence — 2.2L engine

18. Install the camshafts, cylinder head covers and related components. Tighten the cylinder head cover bolts to 3–4 ft. lbs. (4–5 Nm).

19. Install the exhaust manifold using a new gasket. Tighten the bolts to 25–33 ft. lbs. (29–39 Nm).

20. Install the collector and intake manifold assembly using new gaskets. Tighten intake manifold bolts to 17–20 ft. lbs. (23–26 Nm). Connect the fuel pipes.

21. Connect the water hoses to the throttle body.

22. Connect the PCV hose, auxiliary air control valve hose, blowby hoses and auxiliary air control valve connector.

23. Install the EGR valve, EGR pipe and BPT.

24. Install the camshaft sprockets, timing belt and related components.

25. Connect the negative battery cable. Start the engine and allow it to reach operating temperature.

26. Check for leaks and test drive the vehicle.

Valve Lash Adjusters

REMOVAL AND INSTALLATION

NOTE: The rocker arms ride directly on the camshaft. All other engines use hydraulic valve lash adjusters. No service adjustment is possible. If valve clatter or tapping is noticed, inspect for worn valve train components.

1.8L and 3.3L Engines

1. Disconnect the negative battery cable.

2. Remove the distributor, timing belt, belt cover and related components.

3. Remove the rocker covers. Remove the camshaft cases, camshaft support and camshaft as a unit. When removing the camshaft cases, place a rag under the cylinder head to catch any valve rockers that may be knocked loose.

4. Remove the rocker arms and valve lash adjusters from the cylinder head. Do not lay the adjusters down. Keep all components in the order of removal.

To install:

5. With the lash adjuster in a vertical position, push the adjuster pivot inward with a quick and hard motion. If the pivot is depressed more than 0.020 in. (0.5mm), put the adjuster in a container of oil and pump the plunger until the depression is within specification. If the lash adjuster is still not within specification, replace it.

6. Insert the lash adjusters into the cylinder head. Apply grease to the valve rocker and install the rocker on the cylinder head.

7. Install the camshaft case assembly.

8. Install the rocker covers.

9. Install timing belt, belt cover, distributor and related parts.

10. Connect the negative battery cable. Start the engine and check the ignition timing.

Valve Lash

ADJUSTMENT

1.2L Engine

1. Remove the valve covers. Before starting the valve adjustment procedure, retorque the cylinder head bolts to 51–57 ft. lbs. (70–77 Nm).

2. Set the cylinder that is to be adjusted to the TDC (Top Dead Center) position of the compression stroke.

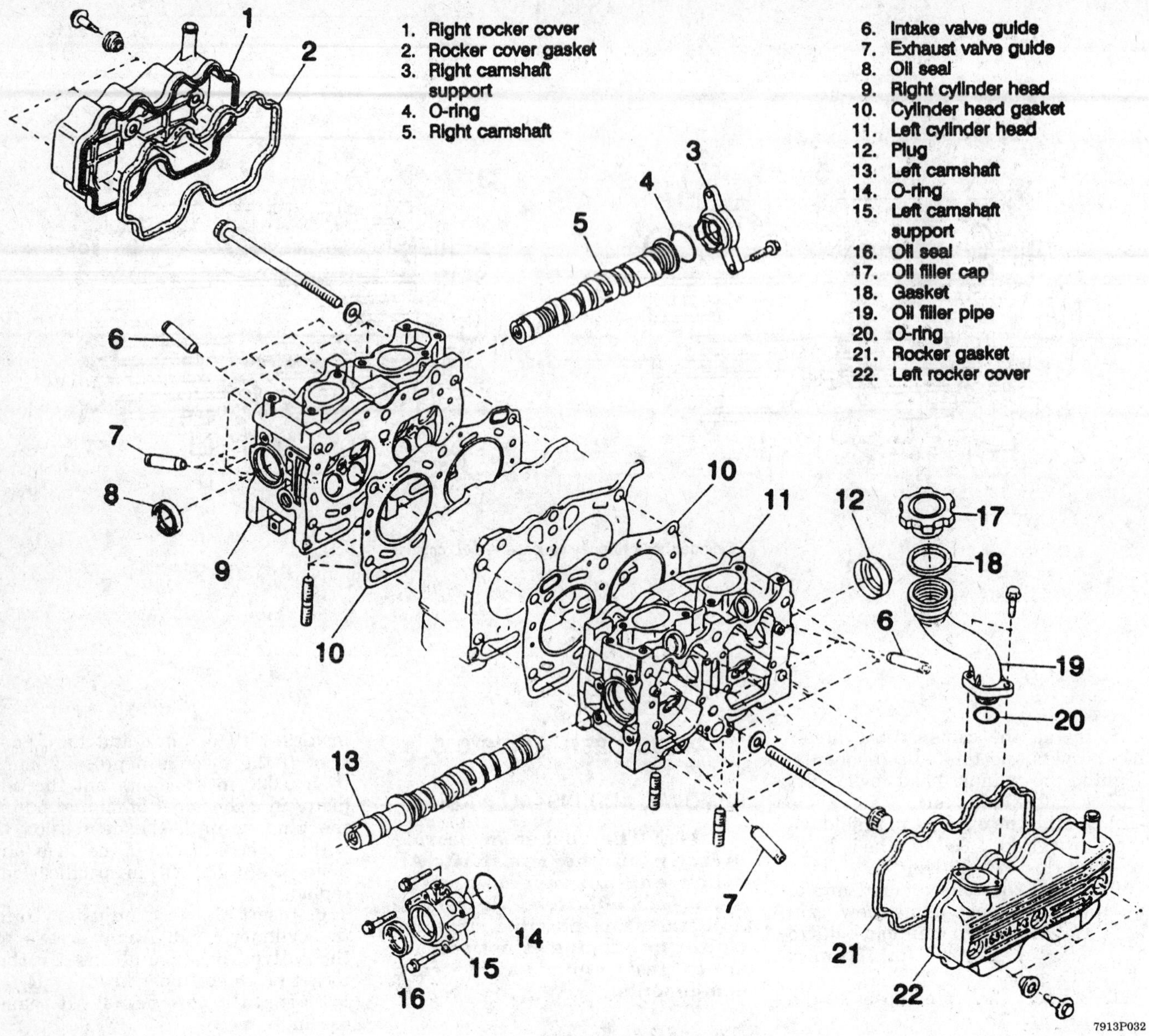

1. Right rocker cover
2. Rocker cover gasket
3. Right camshaft support
4. O-ring
5. Right camshaft
6. Intake valve guide
7. Exhaust valve guide
8. Oil seal
9. Right cylinder head
10. Cylinder head gasket
11. Left cylinder head
12. Plug
13. Left camshaft
14. O-ring
15. Left camshaft support
16. Oil seal
17. Oil filler cap
18. Gasket
19. Oil filler pipe
20. O-ring
21. Rocker gasket
22. Left rocker cover

2.2L engine cylinder head — exploded view

7913P032

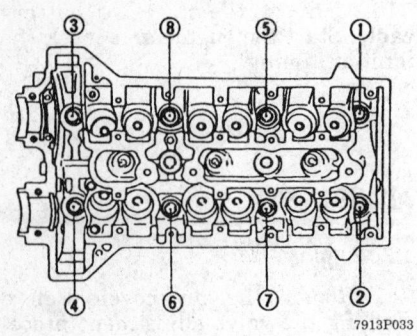

Cylinder head torque sequence — 3.3L engine

7913P033

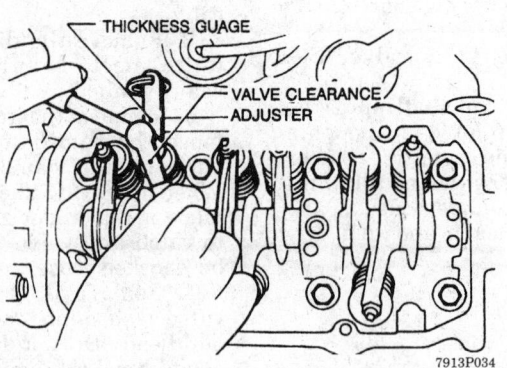

THICKNESS GUAGE

VALVE CLEARANCE ADJUSTER

Justy 1.2L engine valve adjustment procedure

7913P034

3. Using a thickness gauge, measure the clearance between the valve and the valve rocker arm.

4. The valve clearance should be as follows:
   a. Intake valves — 0.0051–0.0067 inch (0.13–0.17mm)
   b. Exhaust valves — 0.0091–0.0106 inch (0.23–0.27mm)

5. If the valves are not within the allowable range, adjust the clearance using Subaru valve adjusting tool 49876700 or an equivalent valve adjuster wrench.

6. After the adjustments are made, tighten the valve adjuster to 12–17 ft. lbs. (17–23Nm).

Rocker Shafts

REMOVAL AND INSTALLATION

1.2L Engine

1. Remove the engine from the vehicle.

2. Using the engine stand tool 499815500 or equivalent, attach to the engine.

3. Remove the suction and the air cleaner hoses from the air cleaner.

4. Loosen the alternator-to-engine bolts, reduce the drive belt tension and remove the drive belt.

5. Using the crank/camshaft pulley wrench tool 499205500 or equivalent, secure the crankshaft pulley and remove the pulley bolt.

6. Remove the timing belt cover. Rotate the crankshaft until the alignment mark on the camshaft sprocket aligns with the pointer on the timing belt housing.

7. Remove the timing belt cover. Loosen the timing belt tensioner bolt ½ turn, move the tensioner to relax the belt tension, remove the timing belt and tighten the bolt.

NOTE: Before removing the timing belt, be sure to mark it for the direction of rotation.

8. Using the crank/camshaft pulley wrench tool 499205500 or equivalent, secure the camshaft sprocket, then remove the sprocket-to-camshaft bolts and the sprocket.

9. Remove the timing belt housing-to-engine bolts and the housing.

10. Remove the throttle body bolts and remove the throttle body.

11. Remove the rocker arm cover-to-cylinder head bolts and the cover from the cylinder head.

12. Using the valve clearance adjuster tool 498767000 or equivalent, loosen the valve adjuster locknuts and back-off the adjusting screw.

13. Remove the valve rocker arm shaft-to-journal bolt, pull out the rocker arm shaft, then remove the spring washers, the valve rocker arms. Keep the rocker arms and the spring washers in order to make the installation easier.

To install:

14. Using a putty knife, clean the gasket mounting surfaces. Inspect the rocker arm for wear; the clearance between the rocker arm and the shaft should be 0.0006–0.0022 in. (0.016–0.057mm).

15. To install, use new gaskets, sealant (where necessary) and reverse the removal procedures. Torque the camshaft sprocket-to-camshaft bolts to 8.3–9.0 ft. lbs., the crankshaft pulley-to-crankshaft bolt to 47–54 ft. lbs. Refill the cooling system and the crankcase. Operate the engine until normal operating temperature is reached and check for leaks.

2.2L Engine

1. Disconnect the PCV hose and remove the rocker cover.

2. Remove the valve rocker assembly by removing bolts 2 through 4 in numerical sequence.

3. Loosen bolt 1 but leave engaged to retain valve rocker assembly.

4. Remove bolts 5 through 8, taking care not to gouge the dowel pin.

5. Remove the valve rocker assembly.

6. Place the valve rocker assembly with the air vent on the rocker arm facing upward into clean engine oil until ready to install. This is done to prevent damaging the hydraulic lash adjuster.

To install:

7. Install the valve rocker assembly on the cylinder head.

8. Temporarily tighten bolts 1 through 4 equally.

NOTE: Do not allow the valve rocker assembly to gouge the dowel pins.

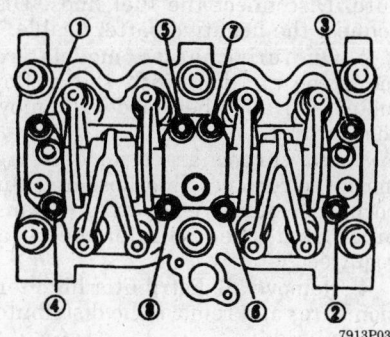

Rocker arm assembly torque sequence — 2.2L engine

9. Tighten bolts 5 through 8 to 9 ft. lbs. (12 Nm).

10. Tighten bolts 1 through 4 to 9 ft. lbs. (12 Nm).

11. Install the rocker cover and connect the PCV hose.

Intake Manifold

REMOVAL AND INSTALLATION

1.2L Engine

1. Disconnect the negative battery cable. Drain the cooling system. Remove the air cleaner assembly. On fuel injected models, relieve the fuel system pressure.

2. Disconnect the fuel line and accelerator cable. Label and disconnect the required vacuum lines. Label and disconnect all required electrical connections. Remove the upper radiator hose.

3. Remove the necessary components in order to gain access to the intake manifold retaining bolts.

4. Remove the intake manifold retaining bolts. Remove the intake manifold and discard the gasket.

To install:

5. Clean all gasket material from the intake manifold and cylinder head.

6. Install the intake manifold using a new gasket. Tighten the intake manifold bolts to 14–22 ft. lbs. (19–30 Nm).

7. Reconnect all vacuum and electrical connections. Install all components previously removed.

8. Reconnect the radiator hose and fill the cooling system with coolant.

9. Reconnect the fuel lines and then the negative battery cable. Start the engine and check for leaks.

1.8L SPFI Engine

1. Disconnect the negative battery cable.

2. Drain the cooling system and remove the radiator hose.

3. Remove all distributor high tension wires and remove the distributor as required.

4. Label and disconnect all applicable vacuum hoses.

5. Remove the alternator as required to gain clearance.

6. Remove the silencers and silencer hoses. Remove the air cleaner assembly.

7. Remove the air suction valves and hoses.

8. Remove the EGR cover and EGR pipe.

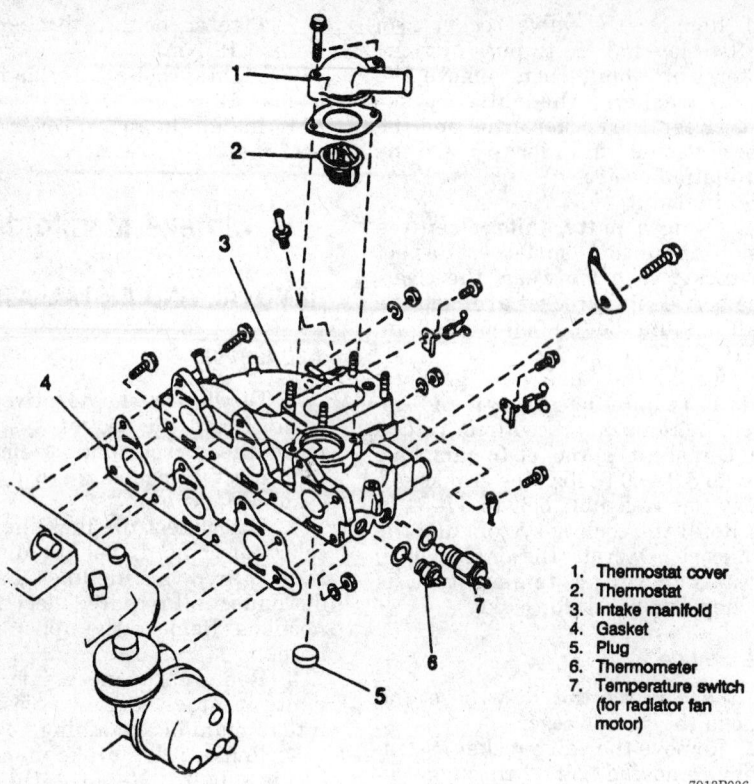

1. Thermostat cover
2. Thermostat
3. Intake manifold
4. Gasket
5. Plug
6. Thermometer
7. Temperature switch (for radiator fan motor)

7913P036

Intake manifold assembly — 1.2L engine

9. Remove the PCV valve and blowby hoses.

10. Remove the air bleed from the thermostat case.

11. Label and disconnect all applicable electrical harnesses.

12. Disconnect the fuel lines and accelerator cable.

13. Remove the intake manifold bolts and carefully lift the intake manifold off the engine.

To install:

14. Clean the gasket mating surfaces thoroughly. Using a straight-edge and a feeler gauge, inspect the intake manifold for flatness. Distortion should not exceed 0.020 in. (0.5mm).

15. Install the intake manifold using new gaskets and tighten the bolts to 13–16 ft. lbs. (18–22 Nm).

16. Inspect all electrical connectors for damage and replace as necessary. Connect all electrical connectors.

17. Install the air bleed on the thermostat.

18. Install the PCV valve and hoses. Install the EGR cover and pipe, and tighten bolts to 23–27 ft. lbs. (31–37 Nm). Install the air suction valve and hoses.

19. Install the silencers and hoses. Install the alternator if removed.

20. Inspect all vacuum lines for damage and replace as necessary. Install all vacuum lines.

21. Install the distributor if removed. Connect all distributor high tension wires.

22. Install the radiator hose and fill the cooling system.

23. Connect all fuel lines and install the air cleaner assembly.

24. Connect the negative battery cable, start the engine and allow it to reach operating temperature. Check for leaks. Test drive the vehicle.

1.8L MPFI Engines

1. Relieve the fuel system pressure. Disconnect the fuel lines. Disconnect the negative battery cable.

2. On turbocharged models, remove the air duct with the airflow meter. On all other models, remove the air duct.

3. On turbocharged models, remove the turbo cooling hose and turbocharger. Remove the front exhaust pipe from the cylinder head as required.

4. Remove all distributor high tension wires and remove the distributor as required.

5. Remove the alternator as required to gain clearance.

6. Label and remove all applicable electrical connectors.

7. Label and remove all applicable vacuum hoses.

8. Remove the EGR cover and pipe.

9. Disconnect the accelerator linkage. Remove the intake manifold assembly.

To install:

10. Clean the gasket mating surfaces thoroughly. Using a straight-edge and a feeler gauge, inspect the intake manifold for flatness. Distortion should not exceed 0.020 in. (0.5mm).

11. Install the intake manifold using new gaskets and tighten the bolts to 13–16 ft. lbs. (18–22 Nm). Connect the fuel lines.

12. Install the EGR pipe and tighten to 23–27 ft. lbs. (31–37 Nm). Install the EGR cover.

13. Check all vacuum lines for deterioration and replace as necessary. Install all previously removed vacuum lines.

14. Check all electrical connectors for deterioration and replace as necessary. Install all previously disconnected electrical connectors.

15. Install the alternator if removed.

16. Inspect all distributor high tension wires and replace as necessary. Install the high tension wires and distributor.

17. On turbocharged models, install the turbo cooling hose and turbocharger. Install the front exhaust pipe from the cylinder head if removed.

18. On turbocharged models, install the air duct with the airflow meter. On all other models, install the air duct.

19. Connect the negative battery cable. Start the engine and allow it to reach operating temperature. Check for leaks and test drive the vehicle.

2.2L AND 3.3L Engines

1. Release the fuel system pressure. Disconnect the negative battery cable and remove the engine cover.

2. Drain the cooling system and remove the water pipes as required.

3. Remove power steering pump, alternator and bracket as necessary to gain clearance.

4. Label and disconnect all electrical connectors leading to the intake manifold.

5. Label and disconnect all vacuum hoses leading to the intake manifold. Disconnect the PCV and blowby hoses.

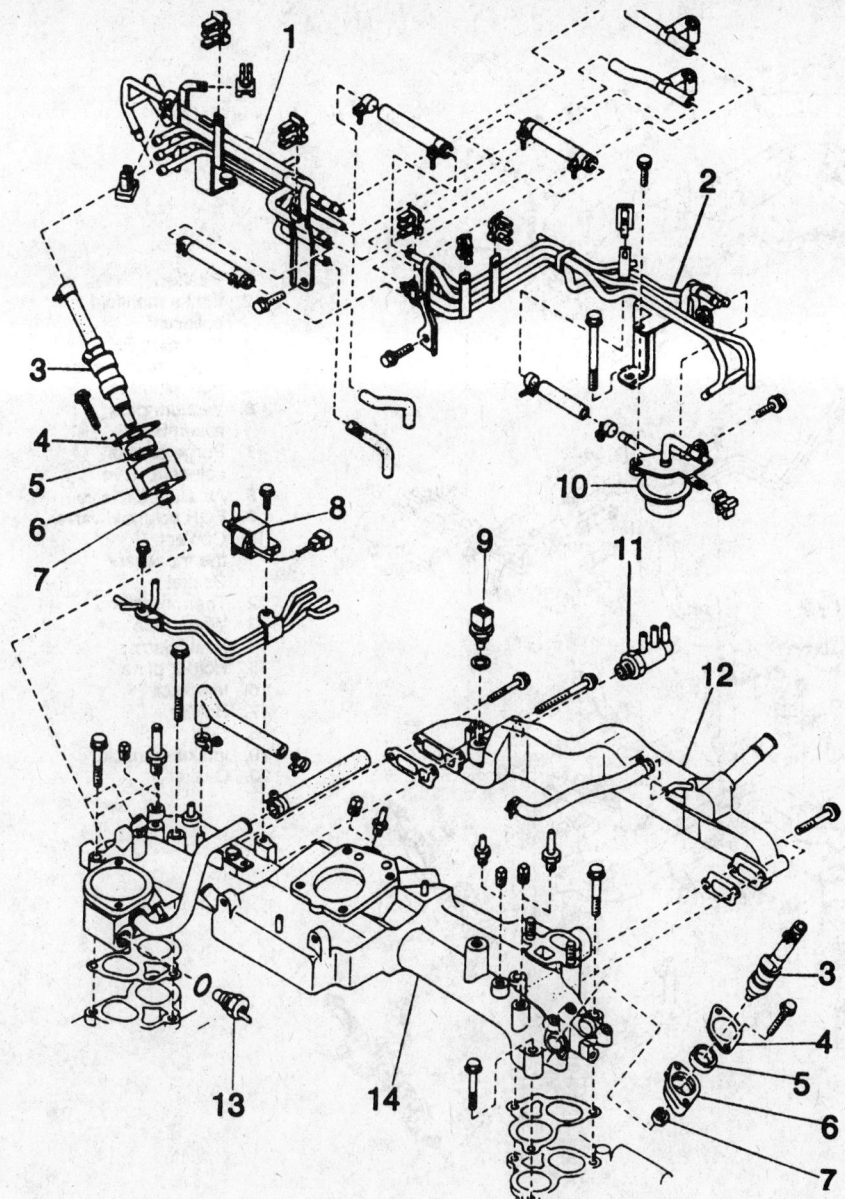

1. Fuel pipe assembly RH
2. Fuel pipe assembly LH
3. Fuel injector
4. Holder plate
5. Insulator
6. Holder
7. Seal
8. EGR solenold valve
9. Coolant thermosensor
10. Pressure regulator
11. Thermo valve
12. Water pipe
13. Thermometer
14. Intake manifold

7913P037

Intake manifold assembly — 1.8L engine with TBI

6. Label and disconnect the ignition high tension wires at the spark plugs and lay them aside.

7. Disconnect the air intake duct.

8. On turbocharged models, disconnect the turbo from the intake manifold and remove as required.

9. Disconnect the fuel supply lines and accelerator linkage.

10. Remove the intake manifold assembly.

To install:

11. Clean the gasket mating surfaces thoroughly. Using a straight edge and a feeler gauge, inspect the intake manifold for flatness. Distor-

tion should not exceed 0.020 in. (0.5mm).

12. Install the intake manifold and tighten the bolts to specification. On the 2.2L engine tighten the short bolts to 21–25 ft. lbs. (28–34 Nm); the long bolts to 4–5 ft. lbs. (6–7 Nm). On the 3.3L engine, tighten all bolts to 17–20 ft. lbs. (23–26 Nm).

13. Install the fuel lines and accelerator linkage.

14. Install the turbocharger assembly.

15. Install the air intake duct.

16. Check the ignition high tension wires for damage and install on the spark plugs.

17. Check all vacuum lines for deterioration and replace as necessary. Install the vacuum lines.

18. Check all electrical connectors for damage and replace as necessary. Install the electrical connectors.

19. Install the PCV valve and blowby hose.

20. Install the power steering pump and alternator if removed.

21. Install the water pipes and fill the cooling system.

22. Install the engine cover.

23. Start the engine and allow it to reach operating temperature. Check for leaks and test drive the vehicle.

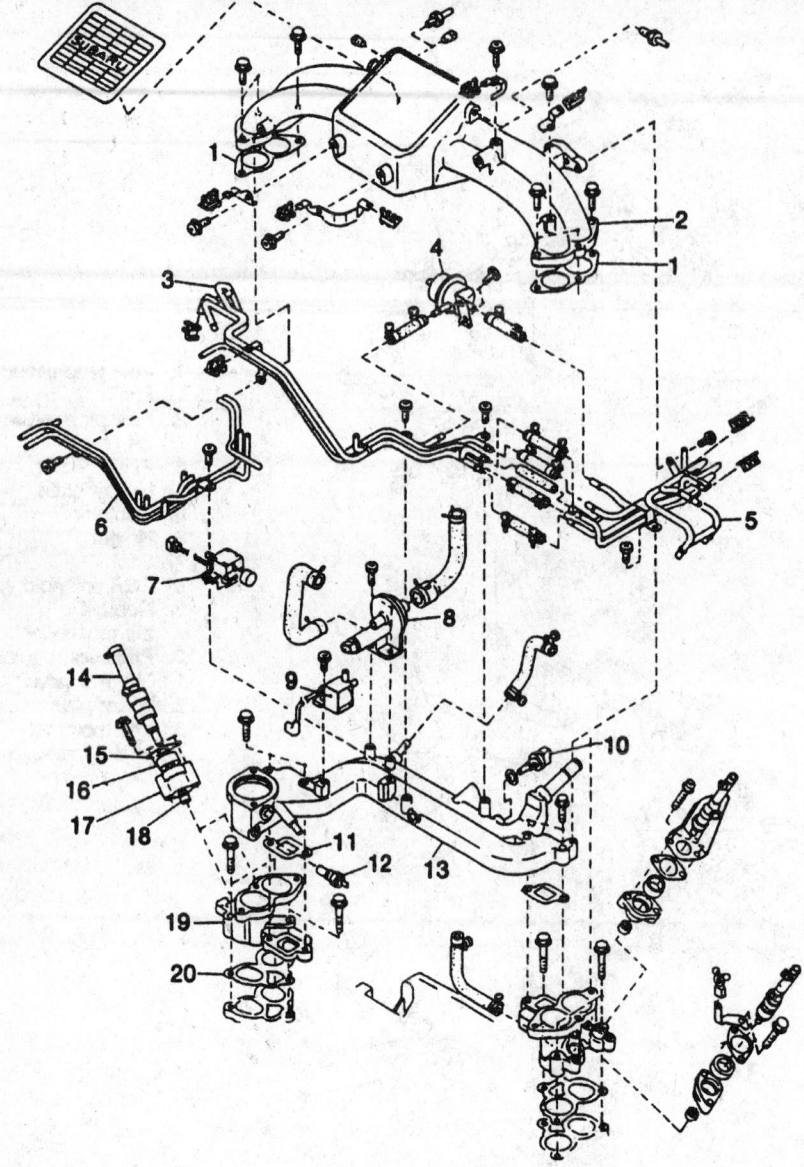

1. Gasket
2. Intake manifold collector
3. Fuel pipe RH
4. Pressure regulator
5. Fuel pipe LH
6. Vacuum pipe assembly
7. Purge control solenoid valve
8. Auxiliary air valve
9. EGR solenoid valve
10. Coolant thermosensor
11. Gasket
12. Thermometer
13. Water pipe
14. Fuel injector
15. Holder plate
16. Insulator
17. Holder
18. Seal
19. Intake manifold
20. Gasket

7913P038

Intake manifold assembly — 1.8L engine with MPFI

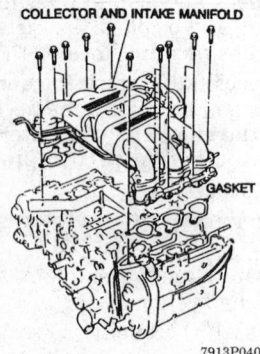

COLLECTOR AND INTAKE MANIFOLD

GASKET

7913P040

Removing the intake manifold — 3.3L engine

Exhaust Manifold

REMOVAL AND INSTALLATION

1.2L Engine

1. Disconnect the negative battery cable. Remove the air cleaner assembly.

2. Raise and support the vehicle safely. Disconnect the exhaust manifold from the exhaust pipe. Lower the vehicle.

3. Disconnect the oxygen sensor electrical connector. Remove the exhaust manifold cover plate assembly.

4. Remove the exhaust manifold retaining bolts. Remove the exhaust manifold from the engine. Discard the gasket.

To install:

5. Clean the mating surfaces of the exhaust manifold and cylinder head thoroughly. Install the exhaust manifold and tighten the bolts to 14–22 ft. lbs. (19–30 Nm).

6. Install the oxygen sensor electrical connector and manifold cover plate. Raise the vehicle and install the exhaust pipe. Lower the vehicle.

7. Install the oxygen sensor electrical connector and air cleaner. Reconnect the negative battery cable.

1.8L, 2.2L Engines

On these engines the front exhaust pipe bolts directly to the under side of the cylinder head. No exhaust manifold is used.

1. Disconnect the negative battery cable.

2. Raise and safely support the vehicle. Remove the engine under cover, if equipped.

3. Disconnect the electrical lead from the oxygen sensor. Remove the front exhaust pipe-to-cylinder head bolts. If equipped with a turbocharger, remove the exhaust pipe to turbocharger assembly.

4. Remove the front exhaust pipe-to-rear exhaust pipe nuts, then separate the pipes.

5. To install, reverse the removal procedures. Replace all gaskets and tighten the retaining bolts to 19–22 ft. lbs. (26–30 Nm).

3.3L Engine

1. Raise and support the vehicle safely.

2. Disconnect the oxygen sensor harness.

3. Remove the front under cover.

4. Remove the exhaust manifold covers.

5. Remove the front exhaust pipes.

6. Disconnect the EGR pipe from the right exhaust manifold.

7. Remove the exhaust manifolds.

To install:

8. Clean all gasket mating surfaces thoroughly.

9. Install the exhaust manifolds using new gaskets. Tighten the exhaust manifold-to-cylinder head nuts to 25–33 ft. lbs. (34–44 Nm) and the exhaust manifold-to-front exhaust pipe nuts to 22–29 ft. lbs. (29–39 Nm).

10. Install the exhaust manifold covers and tighten the bolts to 13–15 ft. lbs. (17–20 Nm).

11. Install the front under cover.

12. Connect the oxygen sensor harness.

13. Lower the vehicle. Start the engine and check for leaks.

Turbocharger

REMOVAL AND INSTALLATION

1. Disconnect the negative battery cable. Drain the cooling system. Remove the air cleaner.

2. Disconnect the airflow meter-to-turbocharger inlet clamp, then remove the air intake duct. Cover the airflow meter and turbocharger openings.

3. Loosen the turbocharger-to-air outlet hose clamp and the throttle body inlet to air inlet hose clamp. Remove the turbocharger-to-throttle body hose. Plug all of the openings.

4. Remove the turbocharger-to-center exhaust pipe nuts and the front exhaust pipe to turbocharger nuts.

5. Disconnect and plug the coolant lines.

6. Remove the oil feed line to turbocharger bolt and disconnect the turbocharger to oil return hose clamp and the return hose.

7. Remove the turbocharger from the exhaust manifold.

NOTE: When removing the turbocharger from the vehicle, disconnect the oil return hose.

To install:

8. Installation is the reverse of the removal procedure. Be sure to fill the turbocharger assembly with clean engine oil prior to installation.

9. Be sure to use new gaskets, as required. Start the engine and check for leaks, correct as necessary.

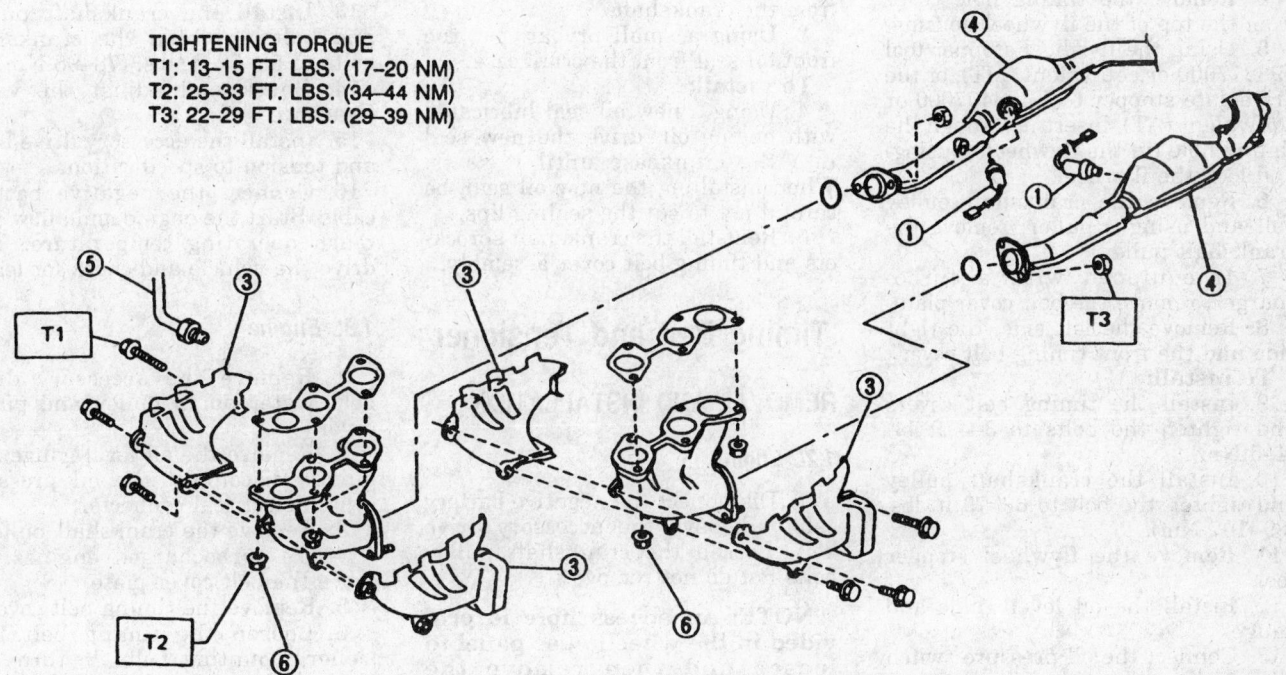

TIGHTENING TORQUE
T1: 13–15 FT. LBS. (17–20 NM)
T2: 25–33 FT. LBS. (34–44 NM)
T3: 22–29 FT. LBS. (29–39 NM)

7913P041

Removing the exhaust manifold — 3.3L engine

Timing Belt Front Cover

REMOVAL AND INSTALLATION

1.2L Engine

1. Loosen the alternator-to-engine bolts, relax the drive belt tension and remove the drive belt from the front of the engine.
2. Using a socket wrench (through the hole in the right fender) and the crank/camshaft pulley wrench tool 499205500 or equivalent (to hold the crankshaft pulley), remove the crankshaft pulley-to-crankshaft bolt and the pulley from the crankshaft.
3. Remove the timing belt cover-to-engine bolts and the cover from the engine.
4. Install the timing belt cover and tighten the bolts securely.
5. Install the crankshaft pulley and tighten the bolt to 58–72 ft. lbs. (78–98 Nm).
6. Install the accessory drive belt and tighten to the proper tension.

1.8L Engines

1. Loosen the water pump pulley nut/bolts and the alternator-to-engine bolts, the remove the drive belt.
2. Disconnect the electrical connector from the oil pressure switch.
3. Remove the oil level gauge guide with the gauge.
4. Remove the timing hole cover from the top of the flywheel housing.
5. Using the flywheel stopper tool 498277000 or equivalent (MT), or the driveplate stopper tool 498407000 or equivalent (AT), insert it through the timing hole (in the flywheel housing) and lock the flywheel.
6. Remove the crankshaft pulley bolt and using a puller, remove the crankshaft pulley.
7. If equipped with a turbocharger, remove the belt cover plate.
8. Remove the left side, the right side and the front timing belt cover.
   To install:
9. Install the timing belt covers and tighten the bolts to 3–4 ft.lbs. (4–5 Nm).
10. Install the crankshaft pulley and tighten the bolt to 66–79 ft. lbs. (89–107 Nm).
11. Remove the flywheel stopper tool.
12. Install the oil level guide and gauge.
13. Connect the oil pressure switch electrical connector.
14. Install the water pump pulley and drive belt. Tighten the drive belt to the proper tension.

2.2L and 3.3L Engines

1. Remove the accessory drive belt.
2. As required, remove the power steering pump, alternator, air conditioner compressor and associated brackets.
3. Remove the crankshaft pulley bolt and remove the crankshaft pulley.
4. Remove the belt covers.
   To install:
5. Install the belt covers and tighten the bolts to 3–4 ft. lbs. (4–5 Nm).
6. Install the crankshaft pulley and tighten the bolt to 69–76 ft. lbs. (93–103 Nm) on the 2.2L engine or 108–123 ft. lbs. (147–167 Nm) on the 3.3L engine.
7. Install the power steering pump, alternator, air conditioner compressor and associated brackets.
8. Install the accessory drive belt and tension to specification.

OIL SEAL REPLACEMENT

1. Remove the timing belt cover assembly.
2. On the 1.8L engine, slide both the No. 1 and No. 2 crankshaft sprockets from the crankshaft. On the 1.2L engine, slide crankshaft sprocket from the crankshaft. When removing the crankshaft sprockets, be sure to remove the Woodruff® key from the crankshaft.
3. Using a small prybar, pry the front oil seal from the crankcase.
   To install:
4. Using a new oil seal lubricated with engine oil, drive the new seal into the crankcase until it seats. When installing the new oil seal, be careful not to cut the sealing lips.
5. Reinstall the crankshaft sprockets and timing belt cover assembly.

Timing Belt and Tensioner

REMOVAL AND INSTALLATION

1.2L Engine

1. Disconnect the negative battery cable. Remove the accessory drive belt. Loosen the crankshaft pulley bolts but do not remove.

NOTE: An access hole is provided in the wheelhouse panel to loosen and then remove the crankshaft pulley bolts.

2. Position the crankshaft with No. 3 cylinder at TDC.

3. Remove the crankshaft bolts and pulley. Remove the outer front timing belt cover.
4. Loosen the tensioner bolt and position it in the direction that loosens the belt. Tighten the tensioner bolt in that position.
5. Remove the camshaft drive pulley plate. Mark the timing belt, if to be used again, in the direction of rotation for reinstallation. Remove the belt from the sprockets.
6. If necessary, remove the tensioner and spring. Remove the camshaft pulley. Remove the inner belt cover and cover mount, only as required.
   To install:
7. When installing the timing belt, rotate and align the matchmark of the camshaft driven pulley 0.120 in. (3mm) diameter hole with the matchmark of the cam belt side cover.
8. Align the matchmark of the camshaft drive pulley and the crankshaft cover. Install the camshaft drive belt.
9. Ensure that each rocker arm can be moved. Loosen the tensioner bolt ½ turn.
10. Tighten the tensioner bolt below the adjusting wheel first. Tighten the other bolt. Check to be sure all sprocket and housing matching marks are in agreement.
11. Install the camshaft drive pulley plate.
12. Install the cam belt cover.
13. Install the crankshaft pulley and bolts. Tighten the crankshaft bolts to 58–72 ft. lbs. (78–98 Nm).
14. Check and adjust the valve clearance.
15. Install the accessory drive belts and tension to specification.
16. Connect the negative battery cable. Start the engine and allow it to reach operating temperature. Test drive the vehicle and check for leaks.

1.8L Engine

1. Remove the accessory drive belt, water pump pulley and pulley cover.
2. Remove the oil level gauge and guide. Disconnect the oil pressure switch electrical connector.
3. Remove the crankshaft pulley.
4. On turbocharged engines, remove the belt cover plate.
5. Remove the timing belt covers.
6. Loosen the timing belt tensioner mounting bolts ½ turn and slacken the timing belt. Tighten the mounting bolts.
7. Mark the rotating direction of the timing belt, then remove the belt.

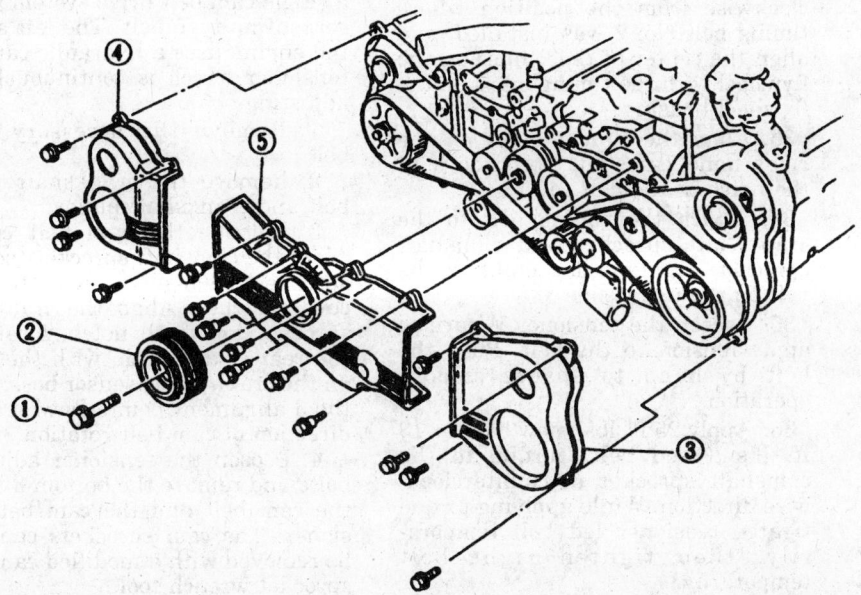

1. Water pump pulley bolt
2. Crankshaft pulley
3. Timming belt side cover
4. Timming belt side cover
5. Front timming belt cover

7913P042

Removing the front engine covers — 3.3L engine

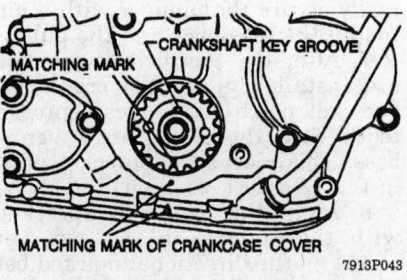

MATCHING MARK

CRANKSHAFT KEY GROOVE

MATCHING MARK OF CRANKCASE COVER

7913P043

Crankshaft gear alignment — 1.2L engine

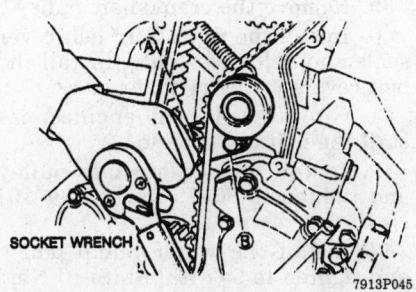

SOCKET WRENCH

7913P045

Timing belt tension adjustment — 1.2L engine

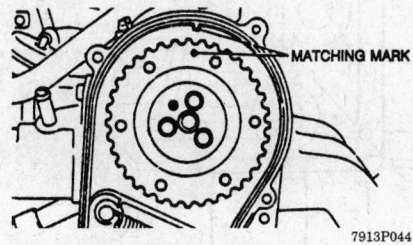

MATCHING MARK

7913P044

Camshaft gear alignment — 1.2L engine

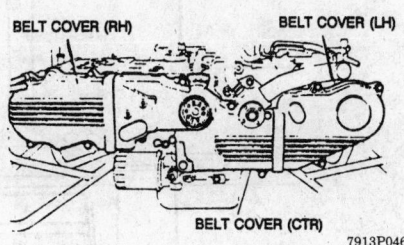

BELT COVER (RH)

BELT COVER (LH)

BELT COVER (CTR)

7913P046

Left, center and right belt covers — typical of 1.8L, 2.2L engines

8. Perform the same procedure for the No. 2 timing belt. Remove the crankshaft sprockets.

9. Remove both tensioners together with the tensioner springs.

10. Remove the belt idler. Remove the camshaft sprockets.

11. Remove the No. 2 belt covers.

To install:

12. Inspect the timing belt for breaks, cracks and wear. Replace as required.

13. Check the belt tensioner and idler for smooth rotation. Replace if noisy or excessive play is noticed.

14. Install the left hand belt cover seal No. 3 to the cylinder block.

15. Install the left hand belt cover seal, left hand belt cover seal No. 4, and belt cover mount to the right rear belt cover, then install the assembly on the cylinder block. Tighten to 3–4 ft. lbs. (4–5 Nm).

16. Install the left hand belt cover seal No. 2 and belt cover mounts to left hand belt cover No. 2, then install to the cylinder head and camshaft case. Tighten to 3–4 ft. lbs. (4–5 Nm).

17. Install the right hand belt cover seal, belt cover seal No. 2 and belt cover mounts to the right hand belt cover No. 2, then install to the cylinder head and camshaft case. Tighten to 3–4 ft. lbs. (4–5 Nm).

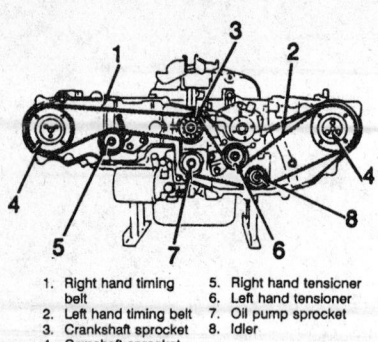

1. Right hand timing belt
2. Left hand timing belt
3. Crankshaft sprocket
4. Camshaft sprocket
5. Right hand tensioner
6. Left hand tensioner
7. Oil pump sprocket
8. Idler

7913P047

Timing belt configuration — 1.8L engines

18. Install the camshaft sprockets to the right and left camshafts. Tighten the bolts gradually in 2–3 steps to 6–7 ft. lbs. (9–10 Nm).

19. Attach the tensioner spring to the tensioner, then install to the right side of the cylinder block. Tighten the bolts temporarily by hand.

20. Attach the tensioner spring to the bolt, tighten the right side bolt and then loosen ½ turn.

21. Push down tensioner until it stops, then temporarily tighten the left bolt.

22. Install the left side tensioner in the same manner.

23. Install the belt idler to the cylinder block using care not to turn the seal. Tighten to 29–35 ft. lbs. (39–47 Nm).

24. Install sprockets to the crankshaft. Install the crankshaft pulley and tighten the bolt temporarily.

25. Align the center of the three lines scribed on the flywheel with the timing mark on the flywheel housing.

26. Align the timing mark on the left hand camshaft sprocket with notch on the belt cover.

27. Attach timing belt No. 2 to the crankshaft sprocket No. 2, oil pump sprocket, belt idler, and camshaft sprocket in that order. Avoid downward slackening of the belt.

28. Loosen tensioner No. 2 lower bolt ½ turn to apply tension. Push timing belt by hand to ensure smooth movement of tensioner.

29. Apply 25 ft. lbs. (new belt) or 18 ft. lbs. (used belt) torque to the camshaft sprocket in counterclockwise direction. While applying torque tighten tensioner No. 2 lower bolt temporarily, then tighten upper bolt temporarily.

30. Tighten the lower bolt, then the upper bolt to 13–15 ft. lbs. (17–20 Nm) in that order.

31. Check that the flywheel timing mark and left hand camshaft sprocket mark are in their proper positions.

32. Turn the crankshaft 1 turn clockwise from the position where timing belt No. 2 was installed, and align the center of the 3 lines on the flywheel with the timing mark on the flywheel housing.

33. Align the timing mark on the right hand camshaft sprocket with the notch in the belt cover.

34. Attach the timing belt to the crankshaft sprocket and camshaft sprocket, avoiding slackening of the belt on the upper side.

35. Loosen the tensioner ½ turn to apply tension to the belt. Push the belt by hand to ensure smooth operation.

36. Apply 25 ft. lbs. (new belt) or 18 ft. lbs. (used belt) torque to the camshaft sprocket in counterclockwise direction. While applying torque tighten tensioner left bolt temporarily, then tighten right bolt temporarily.

37. Tighten the left bolt, then the right bolt to 13–15 ft. lbs. (17–20 Nm) in that order.

38. Check that the flywheel timing mark and left hand camshaft sprocket mark are in their proper positions.

39. Remove the crankshaft pulley.

40. Install the right front belt cover seals and belt cover plug. Install the belt covers to the cylinder block.

41. On turbocharged engines, install the belt cover plate.

42. Install the crankshaft pulley and tighten to 66–79 ft. lbs. (89–107 Nm).

43. Install the water pump pulley and tighten to 6–7 ft. lbs. (9–10 Nm). Install the pulley cover, oil level guide and gauge and oil pressure switch connector.

44. Install and properly tension the accessory drive belt.

2.2L and 3.3L Engines

The 2.2L and 3.3L OHC engines use a single cam belt drive system with a serpentine type belt. The left side of the engine uses a hydraulic cam belt tensioner which is continuously self adjusting.

1. Remove the accessory drive belt.

2. Remove the crankshaft pulley bolt and crankshaft pulley.

3. Remove the cam belt covers. Align the camshaft sprockets so each sprocket notch aligns with the cam cover notches. Align the crankshaft sprocket top tooth notch, located at the rear of the tooth, with the notch on the crank angle sensor boss. Mark the 3 alignment points as well as the direction of cam belt rotation.

4. Loosen the tensioner adjusting bolts and remove the bottom 3 idlers, the cam belt and the cam belt tensioner. The cam sprockets can then be removed with a modified camshaft sprocket wrench tool.

5. If the sprockets are removed, note the reference sensor at the rear of the left cam sprocket.

To install:

6. Install the crankshaft sprocket and all of the idlers except for the lower right. Compress the hydraulic tensioner in a vise slowly and temporarily secure the plunger with a pin. Install the tensioner and the pulley.

7. After the cam belt components are installed, align the crankshaft sprocket notch on the rear sprocket tooth with the crank angle sensor boss. This places the sprocket notch in the 12 o'clock position.

8. Align the camshaft sprockets with the notches in the cam belt cover. As the directional marked belt is installed, align the marks on the belt with the crankshaft sprocket and the left camshaft sprocket. Install the lower right idler.

9. Load the tensioner by pushing it towards the crankshaft with a

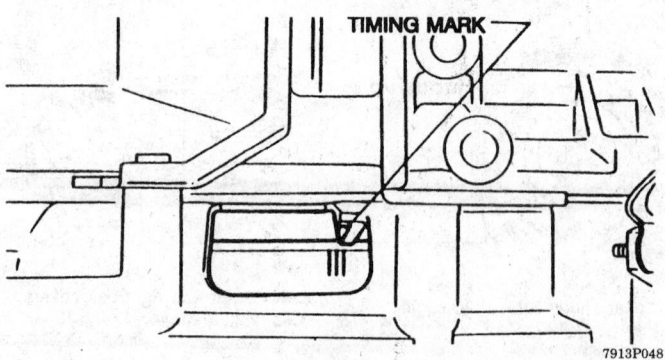

TIMING MARK

7913P048

Flywheel alignment marks for timing belt servicing — 1.8L engine

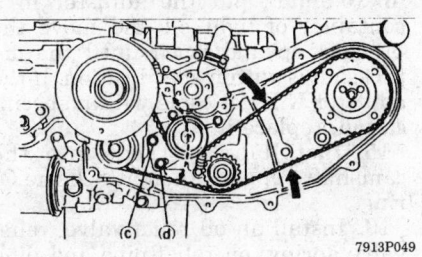

Left timing belt tensioner servicing — 1.8L engine

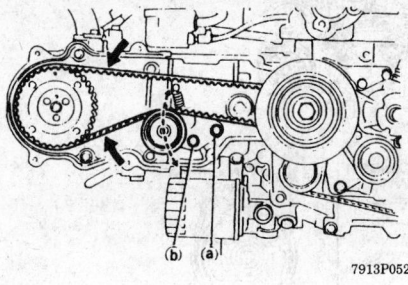

Right timing belt tensioner servicing — 1.8L engine

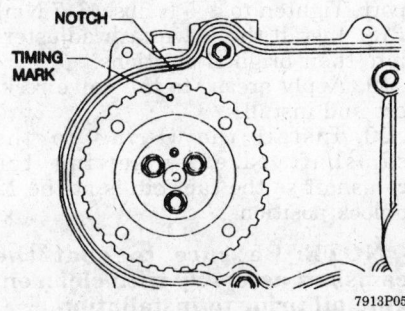

Right camshaft gear alignment — 1.8L engine — after turning engine 1 complete rotation with the left timing belt installed

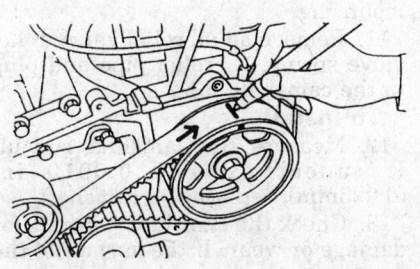

Mark the timing belt for easy reference on installation

Left camshaft gear alignment — 1.8L engine

prybar and tighten the bolts. Remove the tensioner retention pin and the belt tension is automatically set. Rock the crankshaft back and forth 1 time to distribute the belt tension.

10. Verify the correctness of the timing by noting that the notches on the 2 cam pulleys and the notch on the crankshaft pulley all point to the 12 o'clock position when the belt is properly installed.

11. Complete the engine component assembly by installing the cam belt covers, the crankshaft pulley bolt and pulley and the remaining components.

Camshaft

REMOVAL AND INSTALLATION

1.2L Engine

1. Disconnect the negative battery cable.

2. Remove the timing belt. Remove the camshaft driven pulley.

3. Remove the valve rocker cover and slacken the valve rocker adjustment.

4. Remove the bolt from the valve rocker shaft journal and pull the valve rocker shaft out from the cylin-

der head. Remove the spring washer and valve rockers.

5. Remove the camshaft taking care not to damage the bearings.

To install:

6. Clean and check the disassembled components for damage and replace as necessary.

7. Measure the rocker arm-to-shaft clearance. If clearance is greater than 0.0022 in. (0.057mm), replace the components as required.

8. Replace the rocker shaft spring if a permanent set is noted.

9. Lubricate and install the camshaft.

10. Measure the thrust clearance of the camshaft with the breaker case installed. If greater than 0.020 in. (0.5mm), grind the breaker surface of the cylinder head until the thrust clearance fall within specification.

11. Install the valve rocker assembly but do not adjust the valve clearance.

12. Install the camshaft pulley and tighten to 8–9 ft. lbs. (11–12 Nm).

13. Align the marks for the camshaft pulley and the crankshaft sprocket with their respective matchmarks.

14. Install the timing belt assembly.

15. Adjust the valve rocker clearance. Install the valve rocker cover.

16. Start the engine and allow it to reach operating temperature. Check for leaks. Set the ignition timing.

1.8L Engines

1. Disconnect the negative battery cable.

2. Drain the cooling system.

3. Remove the timing belt, belt covers and related components.

4. Remove the distributor.

5. Remove the water pipe assembly and oil filler duct.

6. On turbocharged engines, remove the EGR pipe cover, pipe clamps and EGR pipe.

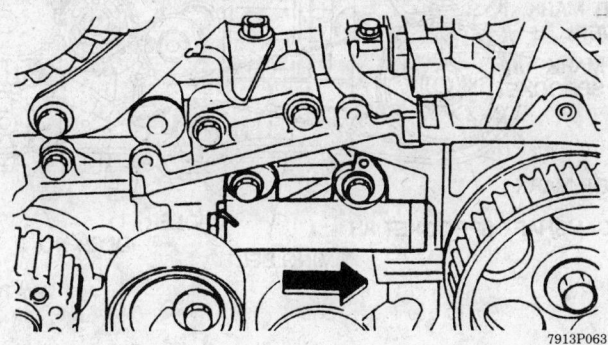

Temporarily move the tension adjuster aside with snugged bolts — 2.2L engine

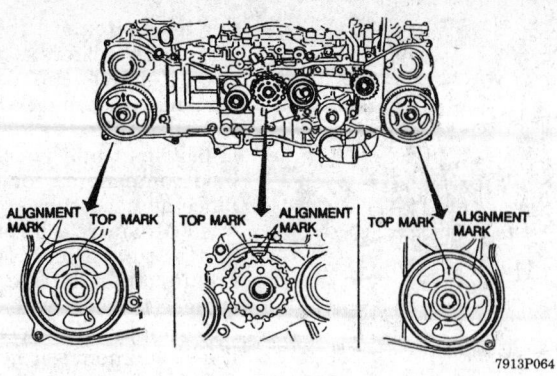

Alignment marks for timing belt installation — 3.3L engine

7913P064

7. Remove the valve rocker covers.

8. Remove the camshaft case, camshaft support and camshaft as a complete unit.

NOTE: When removing the camshaft case, the valve rockers may come off their mounting. To prevent them from being damaged, place a rag under the head.

9. Remove the valve lash adjusters from the cylinder head.

NOTE: Be sure to keep the adjusters and the rockers in the proper order for reinstallation.

10. Remove the camshaft support from the camshaft case. Carefully remove the camshaft from its mounting.

11. Remove an oil relief valve, relief valve spring, oil relief pipe and plug to the camshaft case.

To install:

12. Measure the camshaft runout. If runout exceeds 0.0010 in. (0.025mm), replace the camshaft.

13. Check the camshaft journals for damage or wear. If the surface of the camshaft or valve rocker is damaged or worn, repair by removing the minimum necessary amount, otherwise replace the damaged components.

14. With the valve lash adjuster in a vertical position, push the adjuster pivot quick and hard by hand. If the pivot is depressed more than 0.020 in. (0.5mm), put the adjuster in a container of light oil and move the plunger up and down until the depression is within specification. If the adjuster will not come within specification, replace it.

15. Install an oil seal into the camshaft support, then attach the O-ring.

16. Install an oil relief valve, relief valve spring, oil relief pipe and plug to the camshaft case.

17. Install the Woodruff key on the camshaft and press fit the distributor gear. Insert the camshaft into the case and install the camshaft support. Tighten to 4–5 ft. lbs. (6–7 Nm).

18. Install the valve lash adjusters into their original positions.

19. Apply grease to the valve rockers and install.

20. Install the O-ring to the camshaft case by setting the camshaft so the cam pin is at the 12 o'clock position.

NOTE: Be sure to coat the camshaft assembly with clean engine oil prior to installation.

21. Install the camshaft case to the cylinder head, using sealing compound 1207B or equivalent. Torque the retaining bolts 17–20 ft. lbs. (23–27 Nm).

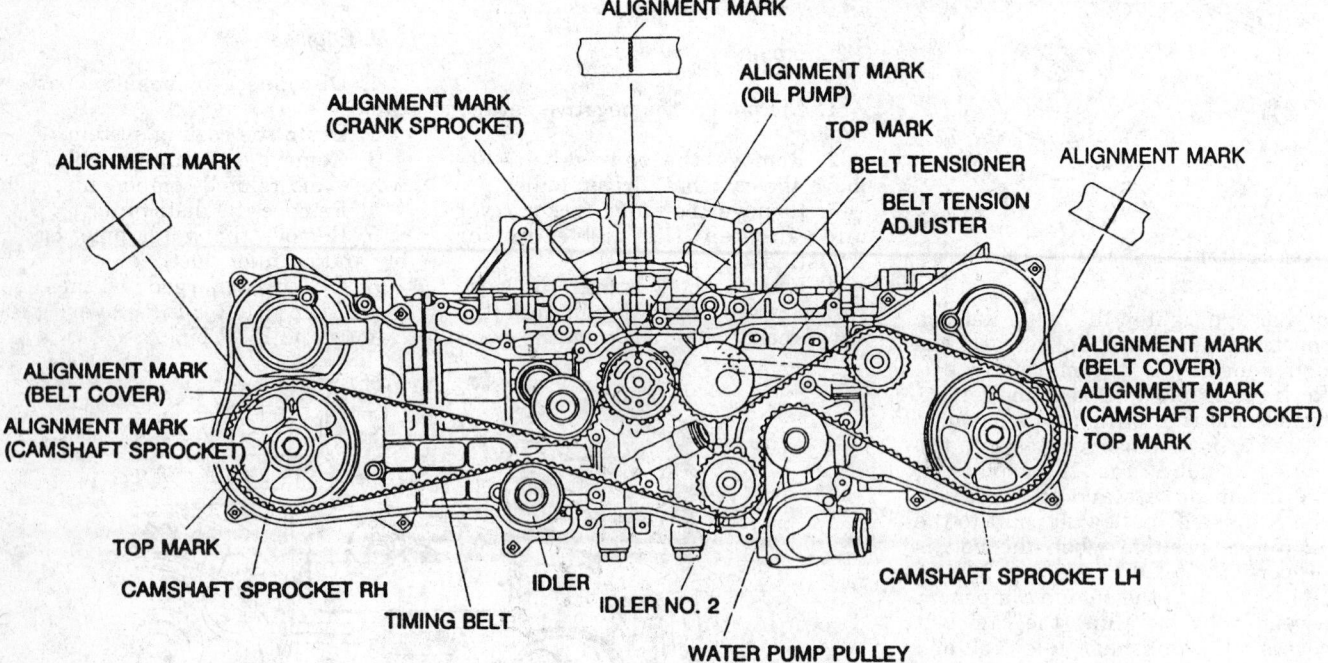

Timing belt routing and position — 3.3L engine

7913P065

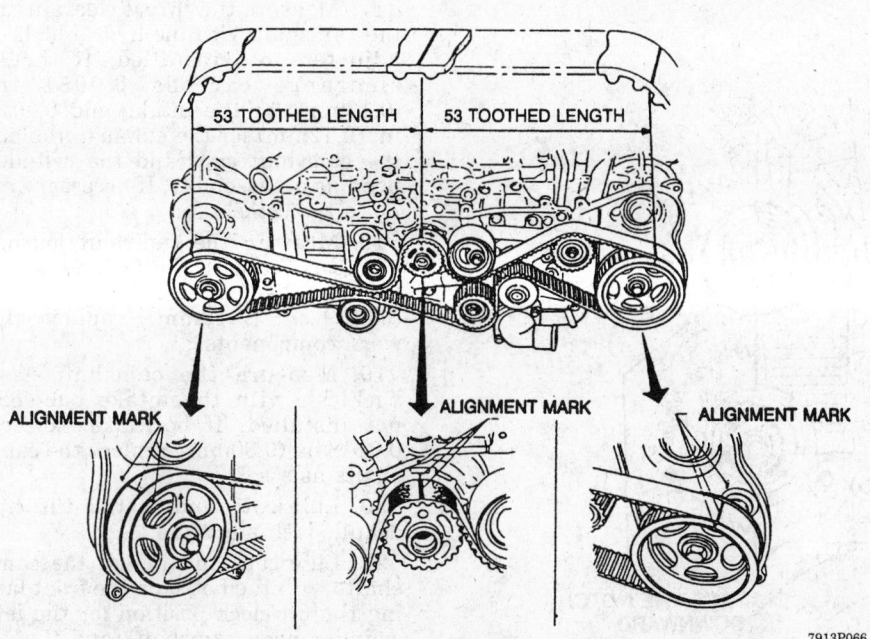

Proper installation of the timing belt — 3.3L engine

7913P066

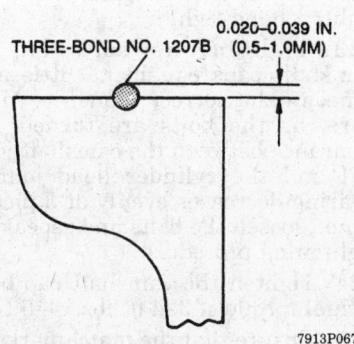

THREE-BOND NO. 1207B

0.020–0.039 IN.
(0.5–1.0MM)

7913P067

Camshaft case to cylinder head installation — 1.8L engine

22. Install the valve rocker cover assemblies and tighten to 3–4 ft. lbs. (4–5 Nm).

23. Install the PCV hoses, and on turbocharged engines install the EGR pipe, clamps and pipe cover. Tighten the EGR pipe to 23–27 ft. lbs. (31–37 Nm).

24. Install the oil filler duct and water pipe.

25. Install the timing belt, belt cover and related parts.

26. Fill the radiator with coolant. Start the vehicle and adjust the ignition timing. Test drive the vehicle and check for leaks.

2.2L Engine

1. Remove the timing belt covers, the timing belt, camshaft sprockets and related components necessary to expose the camshaft.

2. Remove the valve rocker covers. Remove the rocker arm assemblies.

3. To remove the left camshaft, perform the following procedures:
   a. Remove the cam angle sensor.
   b. Remove the oil dipstick tube attaching bolt.
   c. Remove the camshaft support on the left side.
   d. Remove the O-ring.
   e. Remove the camshaft and oil seal (as necessary) from the left side.

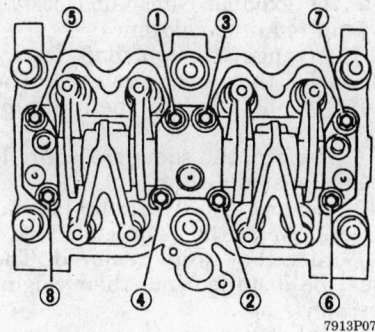

7913P070

Rocker arm assembly bolt removal sequence — 2.2L engine

4. To remove the right camshaft, perform the following procedures:
   a. Remove the camshaft support on the right side.
   b. Remove the O-ring.
   c. Remove the camshaft and seal (rear) from the right side. Remove the oil seal from the camshaft support.

To install:

5. To install the left camshaft, perform the following procedures:
   a. Lubricate the camshaft journals, install the oil seal (rear) and install the camshaft into the cylinder head.
   b. Install the O-ring into the camshaft support and install the support.
   c. Install oil seal into the camshaft support.
   d. Install the bolt into the dipstick tube and install the camshaft sensor.

6. To install the right camshaft, perform the following procedures:
   a. Lubricate the camshaft journals and install the right camshaft.
   b. Install the O-ring into the camshaft support and install the support.
   c. Install a new oil seal in the rear of the cylinder head.

7. Install the valve rocker covers and tighten to 4 ft. lbs. (5 Nm).

8. Install the timing belt covers, the timing belt, camshaft sprockets and related components.

9. Start the engine and check the ignition timing. Allow the engine to reach operating temperature and check for leaks.

3.3L ENGINE

1. Remove the timing belt, camshaft sprockets and related components necessary to gain access to the camshaft.

2. Disconnect the cam angle sensor and bracket.

3. Disconnect the ignition coil connectors and coils.

4. Disconnect the blowby hose and remove the cylinder head cover and gasket.

5. Remove the front camshaft cap.

6. Remove the camshaft oil seal and plug.

NOTE: Since the camshaft thrust clearance is small, the camshaft must be removed by holding it parallel to the cylinder head. If the camshaft is not parallel to the cylinder head, the cylinder head thrust bearing portion may be damaged.

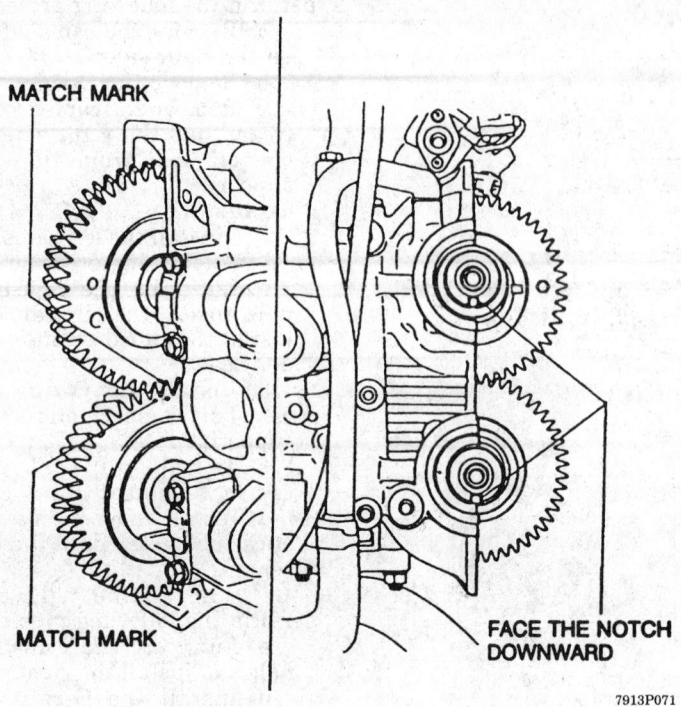

Aligning the camshafts for removal — 3.3L engine

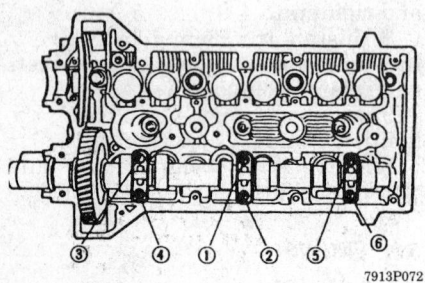

Removing and installing the camshaft retaining caps — 3.3L engine

7. Remove the left cylinder camshafts as follows:

8. Rotate the intake (upper) and exhaust (lower) camshafts to that the notch at the front of the camshafts faces the 6 o'clock position on the left cylinder head and the 12 o'clock position on the right cylinder head.

9. Inspect the rear of the camshaft and check that the matchmarks on the rear gears are aligned.

10. Install a service bolt to the sub-gear mounting bolt hole of the intake camshaft gear to secure the sub-gear and driven gear.

NOTE: When removing the camshafts, ensure that the torsional spring force of the sub-gear has been eliminated.

11. Loosen the intake camshaft bolt caps in sequence. Make sure, as the bolts are turned, the clearance between the camshaft journal and the cylinder head journal bearing increases evenly at 3 places. If not, tighten the bolts and repeat the loosening procedure.

12. Remove the camshaft cap while holding the intake camshaft with 1 hand, then remove the camshaft. Rotate the exhaust camshaft clockwise to gain required clearance.

13. Arrange the camshaft caps in the order they were removed. They must be installed to their original positions.

14. Perform the same procedure for the exhaust camshaft.

15. Remove the hydraulic lash adjusters. Keep the lash adjusters in the order they were removed. They must be installed into their original positions.

To install:

16. Inspect the camshafts for scratches, flaking and wear. Measure the camshaft runout. If it exceeds 0.0008 in. (0.02mm), replace it.

17. Measure the thrust clearance of the camshaft with the hydraulic lash adjusters not installed. If thrust clearance exceeds 0.0051 in. (0.13mm) for the intake and 0.0047 in. (0.12mm) for the exhaust, replace the camshaft caps and the cylinder head as an assembly. If necessary replace the camshaft.

18. Measure the camshaft journal oil clearance with the lash adjusters not installed. If clearance exceeds 0.0039 in. (0.10mm), replace the worn components.

19. Measure the camshaft gear backlash with the intake sub-gear not installed. If backlash exceeds 0.0118 in. (0.30mm), replace the camshafts as a set.

20. Lubricate and install the hydraulic lash adjusters.

21. Lubricate and install the camshafts with the notch on the front facing the 6 o'clock position for the left cylinder head camshafts and the 12 o'clock position for the right cylinder head camshafts. Ensure that the marks for both camshafts are facing the same position.

22. Install the camshaft caps and tighten hand-tight.

23. Tighten the bolts on the camshaft caps equally, a little at a time, in the correct sequence. Make sure, as the bolts are turned, the clearance between the camshaft journal and the cylinder head journal bearing decreases evenly at 3 places. If not, loosen the bolts and repeat the tightening procedure.

24. Tighten the camshaft cap bolts a final torque of 3–4 ft. lbs. (4–5 Nm).

25. Ensure that the matchmarks on the rear side of the camshaft gears are aligned. If not, disassemble the camshaft and perform the installation procedure again.

26. Remove the sub-gear securing bolt from the camshaft.

27. Install the front camshaft cover using fluid packing.

28. Lubricate and install new oil seals.

29. Install the camshaft plug.

30. Install the camshaft cover and connect the blowby hose.

31. Connect the ignition coil connectors and coils.

32. Connect the cam angle sensor and bracket.

33. Install the timing belt, camshaft sprockets and related components.

ENGINE LUBRICATION

Oil Pan

REMOVAL AND INSTALLATION

NOTE: In most cases it is not necessary to remove the engine from the vehicle to remove the oil pan.

1. Disconnect the negative battery cable. Drain the engine oil.
2. Raise and support the vehicle safely. Remove the required components in order to gain access to the oil pan retaining bolts.
3. Loosen the engine mounts and raise the engine, as necessary, to gain access to the oil pan retaining bolts.
4. Remove the oil pan retaining bolts. Remove the oil pan from the engine.
To install:
5. Clean all mating surfaces thoroughly. Using a new gasket, install the oil pan and tighten the retaining bolts to 3–4 ft. lbs (4–5 Nm).
6. Lower the engine and tighten the mounting bolts. Reinstall all components previously removed. Lower the vehicle.
7. Fill the crankcase with oil. Connect the negative battery cable. Start the engine and check for leaks.

Rear Main Bearing Oil Seal

REMOVAL AND INSTALLATION

1. Remove the engine from the vehicle and position it in a suitable holding fixture.
2. Remove the clutch assembly and the flywheel (manual transaxle) or the torque converter and flexplate (automatic transaxle) from the crankshaft.
3. Using a small prybar, pry the rear oil seal from the crankcase. Be careful not to damage the crankshaft or the crankcase housing.
To install:
4. Lubricate a new crankshaft oil seal with engine oil.
5. Using the crankshaft rear oil seal guide tool or equivalent, and the rear oil seal press tool or equivalent, drive the new oil seal into the housing until it seats.
6. Install the clutch assembly and flywheel (manual transaxle) or tor-

que converter and flexplate (automatic transaxle). Install the engine in the vehicle.

Oil Pump

REMOVAL AND INSTALLATION

1.2L Engine

1. Disconnect the negative battery cable. Drain the engine oil. Drain the cooling system.
2. Remove the oil dipstick, dipstick guide and guide sealing.
3. Remove the alternator. Remove the timing belt.
4. Raise and support the vehicle safely. Remove the oil pan. Lower the vehicle.
5. Remove the water pump cover. Remove the water pump impeller. When removing the impeller, lock the balance shaft using the proper tool.
6. Remove the crankcase cover retaining bolts. Remove the crankcase cover along with the oil pump assembly.
To install:
7. Disassemble the oil pump. Check the tip clearance of the rotors. Clearance should be 0.0008–0.0059 in. (0.02–0.15mm). Check the side clearance between the oil pump inner rotor and the pump cover. Clearance should be 0.0020–0.0063 in. (0.05–0.16mm). If not within specification, replace the oil pump.
8. Install the oil pump and crankcase. Install the water pump assembly.
9. Install the oil pan and lower the vehicle. Install the timing belt, alternator and oil dipstick.
10. Fill the engine with oil, start the engine and check for proper oil pressure.

1.8L Engines

1. Disconnect the negative battery cable. Drain the engine oil.

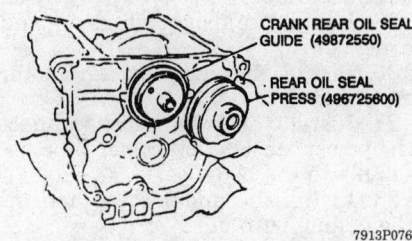

Rear main seal installation — 1.2L engine

2. Remove the timing belts. Before removing the camshaft drive belts, loosen the oil pump pulley mounting nut.
3. Remove the oil pump retaining bolts. Remove the oil pump along with the oil filter.
4. Remove the oil pump outer rotor from the cylinder block. Remove the oil filter from the oil pump.
To install:
5. Disassemble the oil pump. Check the case clearance by measuring the clearance between the outer rotor and the cylinder block rotor housing. The clearance should be 0.0039–0.0071 in. (0.10–0.18mm). Check the side clearance as follows:
   a. Measure the depth of the rotor housing bore in the cylinder block (L).
   b. Measure the total height of the case projection (H1) plus oil pump inner and outer rotors (H2).
   c. Calculate the side clearance using the formula: $C = L - (H1 + H2)$.
6. The clearance should be 0.05–0.16 in. (0.0020–0.0063mm). If not within specification, replace the oil pump rotors.
7. Install the oil pump assembly on the cylinder block. Install the timing belts and camshaft drive belts.
8. Fill the crankcase with oil. Start the engine and check for correct oil pressure.

2.2L Engine

1. Drain the engine oil.
2. Drain coolant from engine.
3. Remove all belt covers, drive belts and other necessary components.
4. Remove the belt tensioner bracket.
5. Remove the water pump.
6. Remove the oil pump.
To install:
7. Disassemble the oil pump and scribe alignment marks on the inner and outer rotors for ease of reassembly.
8. Measure the top clearance of the rotors. Clearance should be 0.0016–0.0055 in. (0.04–0.14mm). Measure the clearance between the outer rotor and the cylinder block housing. Clearance should be 0.0039–0.0069 in. (0.10–0.175mm). Measure the clearance between the oil pump inner rotor and the pump cover. Clearance should be 0.0008–0.0028 in. (0.02–0.07mm). If not within specification, replace the pump body or rotors.
9. Assemble the oil pump and install on the cylinder block using three

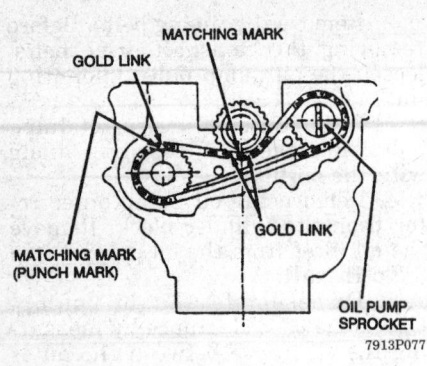

Oil pump sprocket alignment — 1.2L engine

bond 1215 or equivalent. Tighten bolts to specification.

10. Install the water pump, drive belts and tensioner. Fill the crankcase with oil and the radiator with coolant.

11. Start the engine and check for oil pressure and/or leaks.

3.3L Engine

1. Disconnect the negative battery cable.
2. Drain the engine oil.
3. Remove the under cover.
4. Remove the bolts which connect the power steering oil cooler pipe assembly to the body.

NOTE: Do not remove the pipe assembly.

5. Remove the radiator fan motor assemblies.
6. Remove the drive belt cover and drive belts.
7. Remove the air conditioner belt idler assembly.

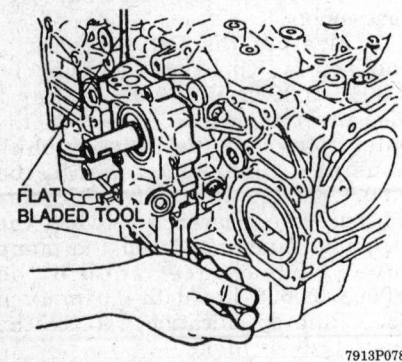

Removing the oil pump — 3.3L engine

8. Remove the power steering pump bracket.

NOTE: Do not remove the power steering pump.

9. Remove the crank angle sensors and crankshaft pulley.
10. Remove the timing belt covers, timing belt and related components.
11. Remove the crankshaft sprocket, belt idlers, belt tensioner and tensioner bracket.
12. Remove the oil pump by prying it from the engine with 2 prybars.

NOTE: Take care not to scratch the gasket mating surfaces.

13. Disassemble the oil pump.
To install:
14. Measure the tip clearance of the rotors. If clearance is greater than 0.0071 in. (0.18mm), replace the rotors.
15. Measure the clearance between the outer rotor and the cylinder block rotor housing. If clearance exceeds 0.0079 in. (0.20mm), replace the rotor.
16. Measure the side clearance between the oil pump inner rotor and the pump cover. If clearance exceeds 0.0047 in. (0.12mm), replace the rotor or pump body.
17. Install a new front oil seal on the pump cover using a driver.
18. Assemble the oil pump.
19. Install the oil pump and tighten the bolts to 4–5 ft. lbs. (5–6 Nm).
20. Install the crankshaft sprocket, belt idlers, belt tensioner and tensioner bracket.
21. Install the timing belt covers, timing belt and related components.
22. Install the crank angle sensors and crankshaft pulley.
23. Install the power steering pump bracket.
24. Install the air conditioner belt idler assembly.
25. Install the drive belt cover and drive belts.
26. Install the radiator fan motor assemblies.
27. Install the bolts which connect the power steering oil cooler pipe assembly to the body.
28. Install the under cover and fill the engine with oil.
29. Connect the negative battery cable.
30. Start the engine and allow it to reach operating temperature. Check

for adequate oil pressure and check for leaks.

ENGINE COOLING

Radiator

REMOVAL AND INSTALLATION

1. Disconnect the negative battery cable. Remove the under cover as required and drain the cooling system. Drain the cooling system.
2. Loosen the hose clamps and remove the inlet (upper) and outlet (lower) hoses from the radiator. Disconnect inlet and outlet oil cooler lines (automatic transmission).
3. Remove the overflow hose and tank as required.
4. Disconnect the thermoswitch, and electric fan motor electrical connectors.
5. Remove the V-belt cover as required to gain clearance.
6. Remove the radiator attaching bolts and lift the radiator from the vehicle with the fan attached.
7. Remove the fan assembly to service the radiator.
To install:
8. Inspect the radiator cushions and replace as necessary.
9. Install the fan assembly if removed.
10. Install the radiator and tighten the attaching bolts to 9–17 ft. lbs. (13–23 Nm).
11. Install the V-belt cover, overflow hose and tank.
12. Inspect the electrical connectors for damage and replace as necessary. Install the electrical connectors.
13. Inspect the radiator hoses for deterioration and replace as necessary. Install the radiator hoses.
14. Install the transmission cooler lines on automatic transmission equipped vehicles.
15. Install the under cover if removed.
16. Fill the radiator with coolant. Start the engine and allow it to reach operating temperature.
17. Stop the engine and allow it to cool. Remove the radiator cap and add coolant as required.

Heater Core

REMOVAL AND INSTALLATION

─────── CAUTION ───────
Properly disarm the air bag on vehicles equipped with the SRS system. Failure to do so can cause serious injury.

Except Justy

NOTE: Depending upon working clearance the air conditioning system may have to be discharged in order to service the blower motor. If this is the case, be sure to observe all the required safety precautions when discharging and recharging the air conditioning system.

1. Disconnect the negative battery cable. Drain the cooling system.
2. Disconnect the heater hoses from the heater core assembly.
3. Remove the instrument panel assembly.
4. Disconnect the electrical harness connector from the blower motor assembly. Disconnect the temperature control cable.
5. Remove the heater unit retaining bolts. Remove the heater unit from the vehicle.
6. Remove the heater core retaining connectors. Remove the heater core from its mounting.
To install:
7. Pressure test the heater core prior to installation. Install the heater core in the heater unit, and the heater unit in the vehicle.
8. Reconnect all electrical connections. Install the heater hoses.
9. Fill the cooling system with coolant. Connect the negative battery cable. Start the engine and bring to operating temperature. Check for leaks.

Justy

1. Disconnect the negative battery cable. Drain the cooling system.
2. Disconnect the heater hoses from the heater core assembly.
3. Pull off the right and left defroster ducts from the defroster nozzles. Pull the ducts from the heater unit.
4. Disconnect the electrical wires from the fan switch and the blower motor.
5. Disconnect the air mix cable from the heater unit. Disconnect the mode cable from the heater unit.

6. Remove the bolts that retain the heater unit to the instrument panel.
7. As required, for working clearance remove the glove box door assembly.
8. Disconnect the inside/outside air control cable from the blower assembly.
9. Remove the instrument panel assembly.
10. Remove the heater unit retaining bolts. Remove the heater unit from the vehicle.
11. Remove the heater core cushion. Loosen the heater core holder and than remove it. Pull the heater core from its mounting and remove it from the heater case.
To install:
12. Pressure test the new heater core prior to installation.
13. Install the heater core in the heater cushion, then install the heater core/unit assembly in the vehicle.
14. Install the instrument panel, control cables and glove box door. Connect the heater mode and control cables to the heater box. Connect all electrical connections previously disconnected.
15. Install the defroster nozzles and the heater hoses. Fill the cooling system with coolant.
16. Connect the negative battery cable, start the engine and check for leaks.

Water Pump

REMOVAL AND INSTALLATION

1.2L Engine

1. Disconnect the negative battery cable.
2. Drain the engine oil and coolant.
3. Remove the oil level gauge, oil level gauge guide and level gauge guide sealing.
4. Disconnect the connector from the alternator and remove the alternator and V-belt.
5. Remove the crankshaft pulley and camshaft belt cover.
6. Remove the camshaft belt tensioner spring and tensioner.

NOTE: Prior to removing the camshaft belt, scribe a mark indicating the drive direction of the belt for installation reference.

7. Remove the camshaft driveplate and drive belt.
8. Remove the camshaft drive pulley and driven pulley.

9. Remove the camshaft belt cover and cover mount.
10. Remove the flywheel housing.
11. Remove the oil pan and pan gasket.

NOTE: The water pump is part of the crankcase cover and must be disassembled for rebuilding.

12. Remove the crankcase cover and disassemble the water pump.
To install:
13. Clean all gasket mating surfaces thoroughly. Use new gaskets during installation.
14. Install the crankcase cover. Install the oil pan and gasket. Install the flywheel housing.
15. Install the camshaft drive belt pulleys, drive belt and drive belt covers.
16. Install the alternator and V-belt. Install the oil level gauge assembly.
17. Fill the engine with oil and the radiator with coolant.
18. Connect the negative battery cable. Start the engine and allow it to reach operating temperature. Check for leaks and test drive the vehicle.

1.8L Engines

1. Drain the coolant.
2. Disconnect the radiator outlet hose and water bypass hose from the water pump.
3. Loosen the pulley nuts. Loosen the alternator assembly and remove the drive belt.
4. Remove the front belt cover.
5. Remove the water pump.
To install:
6. Clean the gasket mating surfaces thoroughly. Always use new gaskets during installation.
7. Install the water pump and pulley.
8. Install the alternator, drive belt and drive belt cover. Adjust the drive belt to the proper tension. Tighten the water pump pulley bolts.
9. Inspect the coolant hoses and replace as necessary. Install the radiator outlet hose and water bypass hose on the water pump.
10. Fill the radiator with coolant. Start the engine and allow it to reach operating temperature. Check for leaks.

2.2L and 3.3L Engines

1. Disconnect the negative battery cable.
2. Drain the coolant.
3. Disconnect the radiator outlet hose.
4. Remove the radiator fan motor assembly

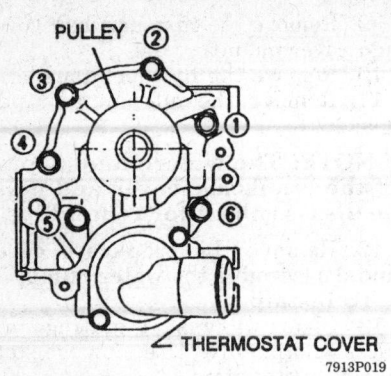

PULLEY ②

THERMOSTAT COVER

7913P019

Removing the water pump — 3.3L engine

5. Remove the accessory drive belts.

6. Remove the timing belt, tensioner and camshaft angle sensor.

7. Remove the left side camshaft pulley and left side rear timing belt cover. Remove the tensioner bracket.

8. Disconnect the radiator hose and heater hose from the water pump.

9. Remove the water pump.

To install:

10. Clean the gasket mating surfaces thoroughly. Always use new gaskets during installation.

11. Install the water pump and tighten the bolts, in sequence, to 7–10 ft. lbs. (10–14 Nm). After tightening the bolts once, retighten to the same specification again.

12. Inspect the radiator hoses for deterioration and replace as necessary. Install the radiator hose and heater hose on the water pump.

13. Install the left side rear timing belt cover, left side camshaft pulley and tensioner bracket.

14. Install the camshaft angle sensor, tensioner and timing belt.

15. Install the accessory drive belts.

16. Install the radiator fan motor assembly.

17. Install the radiator outlet hose. Fill the system with coolant.

18. Connect the negative battery cable. Start the engine and allow it to reach operating temperature. Check for leaks.

Thermostat

REMOVAL AND INSTALLATION

1.2L Engine

1. Disconnect the negative battery cable. Drain the cooling system.

2. Remove the thermostat housing retaining bolts. Remove the thermostat housing and cover assembly.

3. Remove the thermostat from the intake manifold.

4. Installation is the reverse of the removal procedure. Be sure to use a new gasket or RTV sealant, as required.

1.8L Engines

1. Disconnect the negative battery cable. Drain the cooling system.

2. Remove the thermostat housing retaining bolts. Remove the thermostat housing and cover assembly.

3. Remove the thermostat from the intake manifold.

4. Installation is the reverse of the removal procedure. Be sure to use a new gasket or RTV sealant, as required.

2.2L and 3.3L Engines

1. Disconnect the negative battery cable. Drain the cooling system.

2. Remove the thermostat case cover and gasket. Pull out the thermostat.

To install:

3. Clean the mating surface of the thermostat case cover thoroughly.

4. Install the thermostat using a new gasket. The thermostat must be installed with the jiggle pin upward and to the front.

5. Fill the cooling system with coolant. Star the vehicle and allow it to reach operating temperature. Check the coolant level.

ENGINE ELECTRICAL

NOTE: Disconnecting the negative battery cable on some vehicles may interfere with the functions of the on board computer systems and may require the computer to undergo a relearning process, once the negative battery cable is reconnected.

Distributor

NOTE: The 1.8L engines with both the SPFI and MPFI systems use an LED and photodiode pulse pick-up in the distributor for cylinder and crankshaft position determination. The electronic ignition circuit operates in the same basic manner as the standard electronic distributors used on the remaining engines.

REMOVAL

1. Disconnect the negative battery cable. Remove the air cleaner assembly. If equipped, label and disconnect the hose from the distributor.

2. Disconnect the primary wire from the coil. If equipped with a breakerless ignition, disconnect the distributor electrical wiring connector from the vehicle wiring harness.

3. Disconnect the distributor cap retaining clamps or remove the screws and the cap from the distributor. Position the cap and ignition wires aside.

NOTE: If necessary to remove the ignition wires from the cap to provide room to remove the distributor, be sure to label the wires and the cap terminals for easy and accurate installation.

4. Position the engine at TDC with No. 1 cylinder on the compression stroke or using chalk, mark the distributor rotor to distributor housing and the distributor housing to engine relationships.

5. Remove the distributor to engine hold-down bolt.

6. Remove the distributor from the engine, taking care not to damage or lose the O-ring.

NOTE: Do not disturb the engine while the distributor is removed. If the engine cranked or rotated while the distributor is removed, the engine will have to be retimed.

INSTALLATION

Timing Not Disturbed

1. Position the distributor in the block, make sure the O-ring is in place, align the distributor rotor to housing marks and the distributor housing to engine marks.

NOTE: If equipped with an octane selector, install and tighten the hold-down bolt finger-tight.

2. To complete the installation, reverse the removal procedures. Recheck the ignition timing.

Timing Disturbed

1. If equipped, remove the plastic dust cover from the timing port on the flywheel housing.

2. Remove the No. 1 spark plug. Use a wrench on the crankshaft pulley bolt and place the transaxle in the N position and slowly rotate the engine until the TDC 0 degree mark on the flywheel aligns with the pointer.

3. If Step 2 is impractical for any reason, the following method can be used to get the No. 1 piston on TDC. Remove the 2 bolts that hold the right valve cover and remove the cover to expose the valves on No. 1 cylinder. Rotate the engine so the valves in No. 1 cylinder are closed and the TDC **0** degree mark on the flywheel aligns with the pointer.

4. Align the small depression on the distributor drive pinion with the mark on the distributor housing; this will align the rotor with the No. 1 spark plug terminal on the distributor cap.

NOTE: If equipped with an octane selector, set the pointer midway between the A and R. Make sure the O-ring is located in the proper position.

5. Align the distributor housing to engine matchmarks and install the distributor into the engine. Make sure the drive is engaged. Install the hold-down bolt finger-tight. Using a timing light, perform the ignition timing procedures.

6. To complete the installation, remove the timing light and reverse the removal procedures.

Distributorless Ignition System

REMOVAL AND INSTALLATION

Ignition Coil

2.2L AND 3.3L ENGINES

1. Disconnect the battery negative terminal.
2. Remove the intake manifold cover.
3. Disconnect the wires from the ignition coil.
4. Remove the ignition coil.
5. To install, reverse the installation procedure.

Ignition Timing

ADJUSTMENT

NOTE: There are no timing procedures available for the 2.2L or 3.3L engines. The ignition systems are distributorless and are operated via crankshaft and camshaft sensors, using the "'waste type" spark system.

Justy

1. Connect test mode connectors (2 pin type, Green in color), located beneath the left side of the instrument panel.

2. Allow the engine to reach operating temperature. Adjust the idle speed to specification. Connect a timing light, according to manufacturer's instructions.

3. Start the engine and check the ignition timing. If timing is not within specification, loosen the distributor hold-down bolt.

4. Rotate the distributor until the correct timing specification is reached. Tighten the distributor hold-down bolt.

5. Disconnect the test mode connector.

Except Justy

1. Allow the engine to reach operating temperature. Adjust the idle speed to specification. Connect a timing light, according to manufacturer's instructions.

2. Be sure the idle switch is in the engaged position. Connect the test mode connectors, located under the left side of the dash for the MPFI system or on the left side of the engine compartment for the SPFI system. The connectors will be located side-by-side.

NOTE: The check engine warning light will come ON; this does not indicate that there is a problem. The ignition timing must not be adjusted and cannot be checked while the idle switch is inoperative or the test mode connectors disconnected.

3. If timing is not within specification, loosen the distributor hold-down bolt.

4. Rotate the distributor until the correct timing specification is reached. Tighten the distributor hold-down bolt.

Alternator

PRECAUTIONS

Observing these precautions will ensure safe handling of the electrical system components and will avoid damage to the vehicle's electrical system.

• Be absolutely sure of the polarity of a booster battery before making connections. Connect the cables positive to positive and negative to negative. If jump starting, connect the positive cables first and the last connection to a ground on the body of the booster vehicle, so arcing cannot ignite the hydrogen gas that may have accumulated near the battery. Even a momentary connection of a booster battery with polarity reserved may damage the alternator diodes.

• Disconnect both vehicle battery cables before attempting to charge the battery.

• Never ground the alternator output or battery terminal. Be cautious when using metal tools around a battery to avoid creating a short circuit between the terminals.

• Never run an alternator without a load unless the field circuit is disconnected.

• Never attempt to polarize an alternator.

• Never disconnect any electrical components with the ignition switch turned **ON**.

BELT TENSION ADJUSTMENT

1. To adjust the belt tension, first loosen the alternator to bracket adjusting bolt.

2. Lift the alternator to increase the tension on the belt. When it takes moderate thumb pressure to move the longest span of belt ½ in., the tension adjustment is correct.

3. Tighten the adjusting bolt so the alternator will not move in the adjusting bracket.

REMOVAL AND INSTALLATION

1. Disconnect the negative battery cable.
2. Label and disconnect the wiring from the alternator. Remove the necessary components in order to gain access to the alternator retaining bolts.
3. Remove the alternator retaining bolts.
4. Remove the drive belt. Remove the alternator from the vehicle.
5. Installation is the reverse of the removal procedure.

Starter

REMOVAL AND INSTALLATION

1. Remove the spare tire from the engine compartment, as required.
2. Disconnect the negative battery cable. As required, raise and support the vehicle safely.
3. Disconnect the wiring harness from the starter.
4. Remove the starter retaining bolts. Remove the starter from its mounting.
5. Installation is the reverse of the removal procedure.

EMISSION CONTROLS

Emission Warning Lamps

RESETTING

Some vehicles are equipped with an EGR light that illuminates when the vehicle attains 60,000 miles (96,000 km). In order to reset the light for another 60,000 mile increment, the following procedure must be done:

1. Remove the lower instrument panel cover, exposing the fuel panel.
2. Directly behind or along the side of the fuse panel, a blue 2 piece connector will be noted. Disconnect the 2 blue connectors.
3. Near the blue connectors will be a green connector that is not connected to any other wire.
4. Connect the green connector into the matching blue connector, thus resetting the emission light and recycling the system for another 60,000 mile increment.
5. Be sure the indicator light is out and reinstall the lower instrument panel cover.

FUEL SYSTEM

RELIEVING FUEL SYSTEM PRESSURE

Fuel Injected Engines

1. Disconnect the electrical wiring connector from the fuel pump.
2. Start the engine. Once the engine has stopped, crank the engine for 5 seconds or more. If the engine starts, let the engine run until it stops.
3. Turn the ignition switch **OFF**.
4. Reconnect the electrical wiring connector of the fuel pump.

Fuel Tank

REMOVAL AND INSTALLATION

Justy

1. Release the fuel system pressure.

2. Remove the rear seat and package shelf.
3. Disconnect the rollover valve and separator.
4. Remove the access hole lid and disconnect the wiring harness.
5. Disconnect the filler hose and air vent hose.
6. Raise and support the vehicle safely.
7. Drain the fuel through the delivery pipe.
8. Support the fuel tank with a floor jack.
9. Remove the parking brake cable and fuel filter bracket.
10. Disconnect the fuel hoses leading to the tank.
11. Remove the tank attaching bolts and lower the tank.
    To install:
12. Raise the tank and install the attaching bolts. Tighten to 9–17 ft. lbs. (13–23 Nm).
13. Connect the fuel hoses leading to the tank.
14. Install the parking brake cable and fuel filter bracket.
15. Lower the vehicle.
16. Connect the filler hose and air vent hose.
17. Connect the wiring harness and install the access hole lid.
18. Connect the rollover valve and separator.
19. Install the rear seat and package shelf.

Except Justy

1. Release the fuel system pressure.
2. Remove the muffler and rear differential assembly (4WD). On SVX, remove the rear sub-frame.
3. Remove the fuel filler cap and drain the fuel.
4. Remove the fuel filler pipe protector.
5. Remove the fuel filler hose, air vent and delivery hose.
6. Raise and support the vehicle safely.
7. Support the fuel tank with a floor jack.
8. Loosen the attaching bolts and lower the fuel tank.
9. Disconnect the harness connector.
    To install:
10. Connect the harness connector.
11. Raise the fuel tank and install the attaching bolts. Tighten to 9–17 ft. lbs. (13–23 Nm).
12. Lower the vehicle.
13. Install the fuel filler hose, air vent and delivery hose.

14. Install the fuel filler pipe protector.
15. Install the fuel filler cap.
16. Install the muffler and rear differential assembly (4WD).

Fuel Filter

REMOVAL AND INSTALLATION

Justy

1. Carefully relieve the fuel pump pressure. As required, raise and support the vehicle safely.
2. Remove the flange bolts and remove the lower fuel pump bracket assembly.
3. Disconnect and plug the fuel lines at the fuel filter.
4. Remove the fuel filter retaining bolts. Remove the fuel filter assembly from its mounting.
5. Installation is the reverse of the removal procedure.

Except Justy

The fuel filter is located inside the engine compartment on the left fender assembly.

1. Properly relieve the fuel system pressure.
2. Disconnect the fuel lines from the fuel filter.
3. Pull the fuel filter from the bracket and remove it from the vehicle.
4. Installation is the reverse of the removal procedure. Start the engine and check for leaks.

Electric Fuel Pump

PRESSURE TESTING

1. Raise and support the vehicle safely.
2. Using a fuel pressure gauge, connect into the fuel line.
3. Turn the ignition switch to the **ON** position. Observe the fuel pressure, it should be:
   • 61–71 psi for MPFI equipped vehicle.
   • 36–50 psi for SPFI equipped vehicle.
4. If the fuel pump does not meet specification, replace it.
5. After testing, disconnect the pressure gauge and reconnect the fuel line.

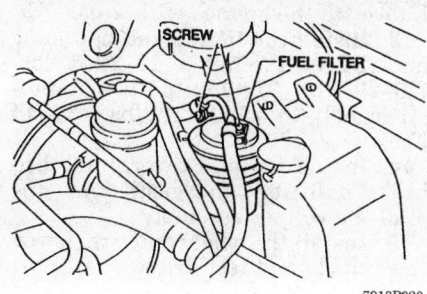

Removing the fuel filter

REMOVAL AND INSTALLATION

Justy

1. Carefully relieve the fuel pump pressure. As required, raise and support the vehicle safely.
2. Remove the flange bolts and remove the lower fuel pump bracket assembly.
3. Disconnect and plug the fuel lines at the fuel pump assembly. Disconnect the fuel pump electrical connector.
4. Remove the fuel pump assembly retaining bolts. Remove the fuel pump assembly from its mounting.
5. Installation is the reverse of the removal procedure.

Except Justy

1.8L SPFI ENGINE

The pump is at the rear of the vehicle, bolted to the vehicle frame.
1. Release the fuel system pressure and disconnect the negative battery cable.
2. Keep the fuel pump harness disconnected after releasing the fuel system pressure.
3. Raise and support the vehicle safely.
4. Clamp the middle portion of the thick hose connecting the pipe and the pump to prevent fuel from flowing out of the tank.
5. Loosen the hose clamp and disconnect the hose.
6. Remove the 3 pump bracket mounting bolts and remove the pump together with the pump damper.
To install:
7. If the pump and damper have been removed from the bracket, reinstall and tighten the bolts securely.
8. Install the hose and tighten the clamp screw to 0.7–1.1 ft. lbs. (1.0–1.5 Nm).

9. Install the pump bracket in position to the vehicle body and secure it with the bolts.

NOTE: Take care to position the rubber cushion properly.

10. Connect the pump harness connector.
11. Connect the negative battery cable and test the fuel pump for proper operation.

EXCEPT 1.8L SPFI ENGINE

The fuel pump is located in the fuel tank and is part of the fuel sender assembly.
1. Relieve the fuel system pressure. Disconnect the negative battery cable.
2. Remove the floor mat or carpet from the luggage compartment.
3. Remove the access lid and disconnect the wiring connector from the sender assembly.
4. Loosen the hose clamps and disconnect the lines from the sender assembly.
5. Remove the retaining bolts and remove the pump unit from the fuel tank.
6. Use a new gasket and install the assembly in the reverse of the removal procedure. Tighten the retaining bolts to 2–4 ft. lbs. (3–6 Nm).

Fuel Injector

REMOVAL AND INSTALLATION

SPFI

1. Disconnect the negative battery cable.

NOTE: This procedure may be performed with the throttle body mounted on the intake manifold or removed. If the throttle body is mounted on the intake manifold during servicing, ensure that

debris does not fall into the intake manifold through the throttle body.

2. Remove the air intake boot.
3. Remove the injector cap and gasket.
4. Disconnect the injector electrical connector.
5. Hold the injector using pliers, then pull the injector from throttle body.
6. Remove the O-ring and discard.
To install:
7. Using a new O-ring, install the injector in the throttle body.
8. Connect the injector electrical connector.
9. Install the injector cap using a new gasket.
10. Install the air intake boot.
11. Connect the negative battery cable.

MPFI

EXCEPT 1.8L ENGINES

1. Relieve the fuel system pressure.
2. Disconnect the negative battery cable.
3. Label and disconnect the fuel injector electrical connectors.
4. Label and remove the hoses attached to the fuel rail.
5. Remove the fuel rail attaching bolts and gently lift the fuel rail from the engine.

NOTE: Some fuel injectors may be removed with the fuel rail while others may remain in the engine.

6. Remove the fuel injectors by pulling with a slight twist. Discard all gaskets and O-rings.
To install:
7. Install the fuel injectors using new gaskets and O-rings. Lubricate the O-rings prior to installation.
8. Install the fuel rail and tighten the attaching bolts securely.

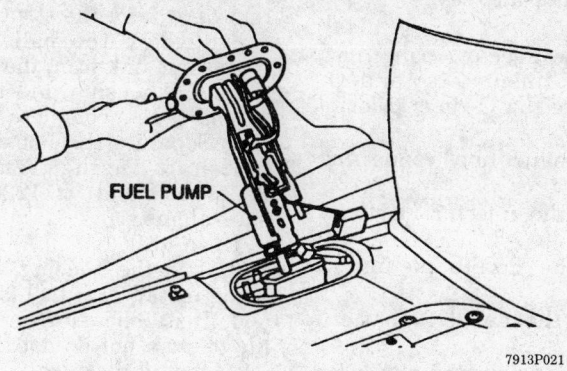

Removing the fuel sending/pump assembly from the fuel tank

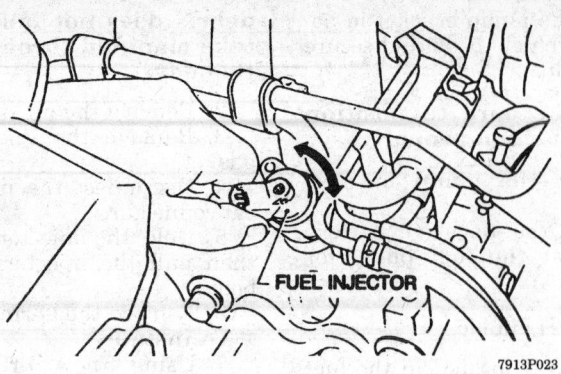

Removing the fuel injector-MPFI vehicles.

9. Install all hoses attached to the fuel rail.

10. Install the fuel injector electrical connectors.

11. Connect the negative battery cable.

1.8L ENGINES

1. Relieve the fuel system pressure.

2. Disconnect the negative battery cable.

NOTE: On some engines, it may be necessary to remove the intake plenum to gain access to the fuel lines connecting the injectors.

3. Remove the fuel lines connecting the injectors.

4. Disconnect the fuel injector electrical connectors.

5. Remove the injectors by pulling with a slight twist. Discard the gaskets and O-rings.

6. Remove the injector holder plate, insulator, holder and seal. Discard the insulator and seal.

To install:

7. Install the injectors using new gaskets, seals (insulators) and O-rings. Lubricate the O-rings prior to installation.

8. Install the fuel lines connecting the injectors.

9. Connect the injector electrical connectors.

10. Install the intake plenum if removed.

11. Connect the negative battery cable.

12. Start the engine and check for leaks.

DRIVE AXLE

Halfshaft

REMOVAL AND INSTALLATION

Justy

1. Raise and support the vehicle safely. Remove the tire and wheel assembly.

2. Remove the disc brake assembly. Remove the dust cover, cotter pin, castle nut, conical spring. Remove the center piece, using the proper tools.

3. Pull the hub and disc assembly from the halfshaft. Remove the disc cover from the housing.

4. Drive out the spring pin connecting the halfshaft to the differential, using the proper tool.

5. Remove the cotter pin and the castle nut from the tie rod end ball joint.

6. Remove the tie rod end ball joint from the knuckle arm, using the proper puller.

7. Remove the bolt that retains the housing to the strut. Carefully push down the housing in order to remove it from the strut.

8. Remove the ball joint of the transverse link from the housing. Remove the housing and the halfshaft assembly as a complete unit.

9. Separate the housing from the halfshaft, using removal tools 922493000 and 921122000 or their equivalents.

To install:

10. Join the housing and halfshaft using installation tool 927210000.

11. Install housing and axle assembly to strut but do not tighten.

12. Install the dust seal on the spindle. Insert the halfshaft into the dif-

ferential and install the spring pin. Lubricate the splines with grease.

13. Install the tie rod end ball joint and tighten the nut to 18–22 ft. lbs. (25–29 Nm). Tighten the housing-to-strut bolt to 25–33 ft. lbs. (34–44 Nm).

14. Install the disc cover, hub, disc brake assembly and castle nut. Install the caliper assembly.

15. Install the wheel and tire, lower the vehicle and test drive.

Except Justy

1. Release the parking brake. Raise and support the vehicle safely. Remove the tire and wheel assembly.

2. Pull out the parking brake cable outer clip from the caliper. Disconnect the parking brake cable end from the caliper lever.

3. Drive out the double offset joint spring pin, using the proper tools.

4. Loosen the 2 retaining bolts and remove the disc brake assembly from the housing. Remove the 2 bolts that connect the housing and the damper strut.

5. Remove the dust cover, cotter pin. Disconnect the tie rod end ball joint from the housing knuckle arm, using the proper puller tool.

6. Remove the halfshaft from the differential spindle along with the housing assembly.

7. Remove the housing from the halfshaft, using tool 926470000 or equivalent.

To install:

8. Install the halfshaft into the hub using installer 922431000, or equivalent. Take care not to damage the inner oil seal lip. Tighten the axle nut temporarily.

9. Install the double offset joint on the spindle and drive a new spring pin into place. Install the tie rod and tighten the nut to 61–83 ft. lbs. (83–113 Nm).

10. Install the axle nut and tighten to 137 ft. lbs. (186 Nm).

11. Install the disc brake assembly and parking brake. Install the wheel and tire.

12. Lower the vehicle and test drive.

Driveshaft

REMOVAL AND INSTALLATION

4WD

1. Raise and support the vehicle safely.

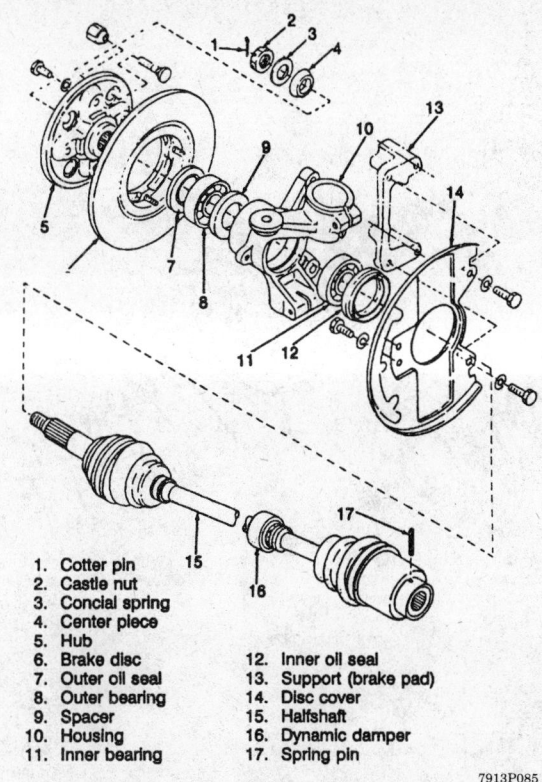

1. Cotter pin
2. Castle nut
3. Concial spring
4. Center piece
5. Hub
6. Brake disc
7. Outer oil seal
8. Outer bearing
9. Spacer
10. Housing
11. Inner bearing
12. Inner oil seal
13. Support (brake pad)
14. Disc cover
15. Halfshaft
16. Dynamic damper
17. Spring pin

7913P085

Front halfshaft assembly and related components — Justy

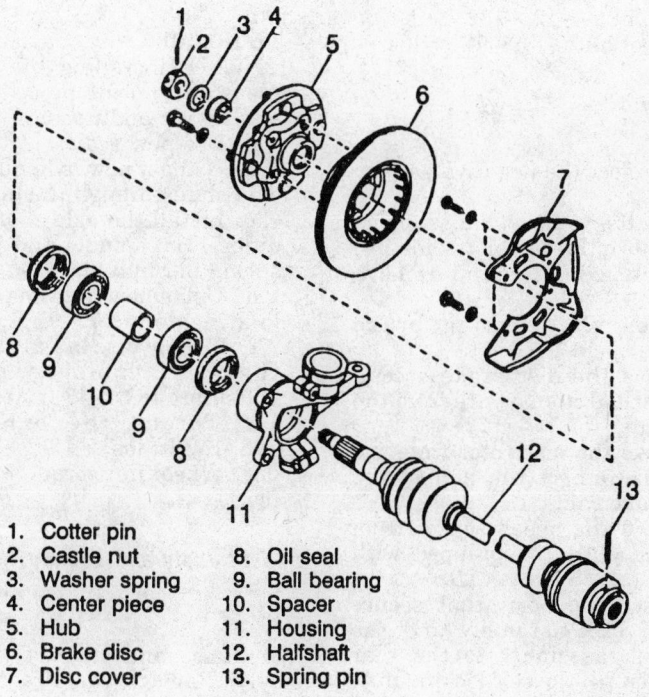

1. Cotter pin
2. Castle nut
3. Washer spring
4. Center piece
5. Hub
6. Brake disc
7. Disc cover
8. Oil seal
9. Ball bearing
10. Spacer
11. Housing
12. Halfshaft
13. Spring pin

7913P086

Front halfshaft assembly and related components — except Justy

2. Remove the driveshaft flange to rear differential flange bolts.

NOTE: If equipped with a center bearing, remove the center bearing to chassis bolts and lower the assembly from the vehicle.

3. Position a drain pan under the rear of the transaxle. Remove the driveshaft from the vehicle.
To install:
4. Install the driveshaft and tighten the flange bolts to 17–24 ft. lbs. (24–32 Nm).
5. If equipped with a center bearing, raise the assembly and install the center bearing bolts. Tighten to attaching bolts to 25–33 ft. lbs. (34–44 Nm).
6. Lower the vehicle, check the transaxle fluid level and test drive.

Rear Axle Shafts

REMOVAL AND INSTALLATION

Justy

2WD

1. Raise and support the vehicle safely. Remove the tire and wheel assembly.
2. Remove the dust cap. Straighten the locking washer edge. Remove the nut, lock washer and washer.
3. Remove the brake drum. Be sure not to drop the outer bearing.
4. Remove the brake line bracket from the spindle housing.
5. Loosen the bolts and remove the brake assembly. Suspend the assembly aside with wire.
6. Using the proper tools, drive out the spring pin connecting the halfshaft assembly to the differential.
7. Remove the strut, lower link and trailing link. Pull the housing along with the halfshaft from its mounting.
8. Separate the housing from the halfshaft, using removal tools 922493000 and 921122000 or their equivalent.
To install:
9. Join the housing and halfshaft. Install the strut, lower link and trailing link. Install the spring pin connecting the halfshaft assembly to the differential.
10. After tightening the rear axle halfshaft-to-axle housing nut, tighten the axle shaft nut 30 degrees further.
11. Install the brake assembly, brake line bracket and brake drum.

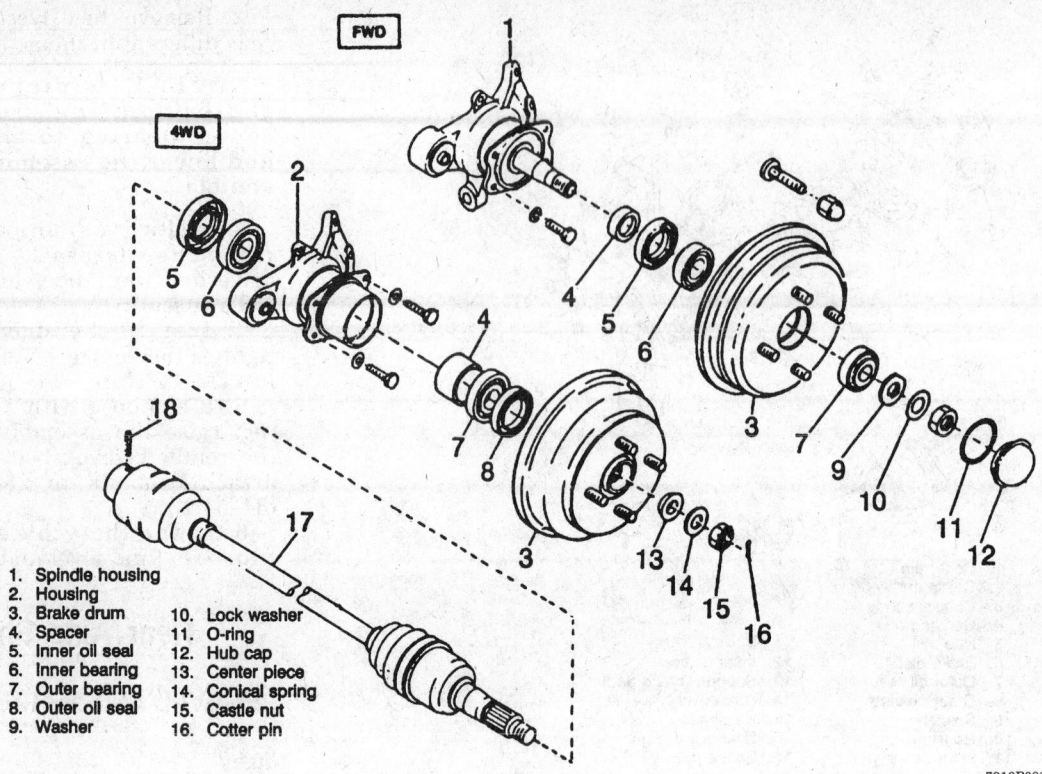

1. Spindle housing
2. Housing
3. Brake drum
4. Spacer
5. Inner oil seal
6. Inner bearing
7. Outer bearing
8. Outer oil seal
9. Washer
10. Lock washer
11. O-ring
12. Hub cap
13. Center piece
14. Conical spring
15. Castle nut
16. Cotter pin

7913P088

Rear axle assembly — Justy

12. Install the wheel and tire, lower the vehicle and test drive.

Legacy and SVX

2WD LEGACY ONLY

1. Disconnect the negative battery cable.
2. Raise the vehicle and support safely.
3. Remove the wheels and unlock the axle nut. Remove the axle nut.
4. Loosen the parking brake adjuster. Remove the disc brake assembly from the backing plate and suspend it with a wire from the strut.
5. Remove the disc brake rotor from the hub and disconnect the end of the parking brake cable.
6. Remove the bolts that retain the lateral link, trailing link and the strut to the rear spindle.
7. Remove the rear spindle, backing plate and hub as a unit.
To install:
8. The installation is the reverse of the removal procedure. Use the following torque values during installation.
  a. Rear spindle to strut assembly — 98–119 ft. lbs.
  b. Rear spindle assembly to trailing link — 72–94 ft. lbs.
  c. Rear spindle to lateral link — 87–116 ft. lbs.

  d. Disc brake assembly to backing plate — 34–43 ft. lbs.
  e. Axle nut — 123–152 ft. lbs.
  f. Wheel nuts — 58–72 ft. lbs.

4WD

1. Disconnect the negative battery cable.
2. Raise the vehicle and support safely. Remove the wheel assemblies.
3. Unlock axle nut and remove from axle.
4. Loosen the parking brake adjuster.
5. Remove the disc brake assembly and suspend it on a wire from the body or strut.
6. Remove the disc rotor from the hub and disconnect the end of the parking brake cable.
7. Remove the speed sensor from the backing plate, if equipped with Anti-lock Brake System (ABS).
8. Remove the bolts that secure the lateral link assembly and the trailing link assembly to the rear housing. Discard the self-locking nuts and replace with new nuts.
9. Remove the spring pin that secures the rear differential spindle to the inner CV-joint.
10. Remove the inner CV-joint and shaft from the differential spindle.

11. Disengage the rear driveshaft from the rear hub and remove the shaft.
To install:
12. When installing the shaft, reverse the removal procedures with the following additions:
  a. Use new seals.
  b. Using a new axle nut, pull the axle shaft through the hub splines.
  c. Install the axle shaft onto the differential spindle and install the spring pin into place.
  d. Using new nuts on the trailing link, tighten to 72–94 ft. lbs.
  e. Torque disc brake assembly to the rear housing assembly bolts/nuts to 34–43 ft. lbs.
  f. Torque the axle nut to 123–152 ft. lbs.
  g. Wheel nut torque to 58–72 ft. lbs.

Except Justy and Legacy

2WD

1. Raise and support the vehicle safely. Remove the tire and wheel assembly.
2. Remove the dust cap. Straighten the lock washer. Remove the nut, lock washer and washer.
3. Remove the brake drum. Be sure not to drop the outer bearing.

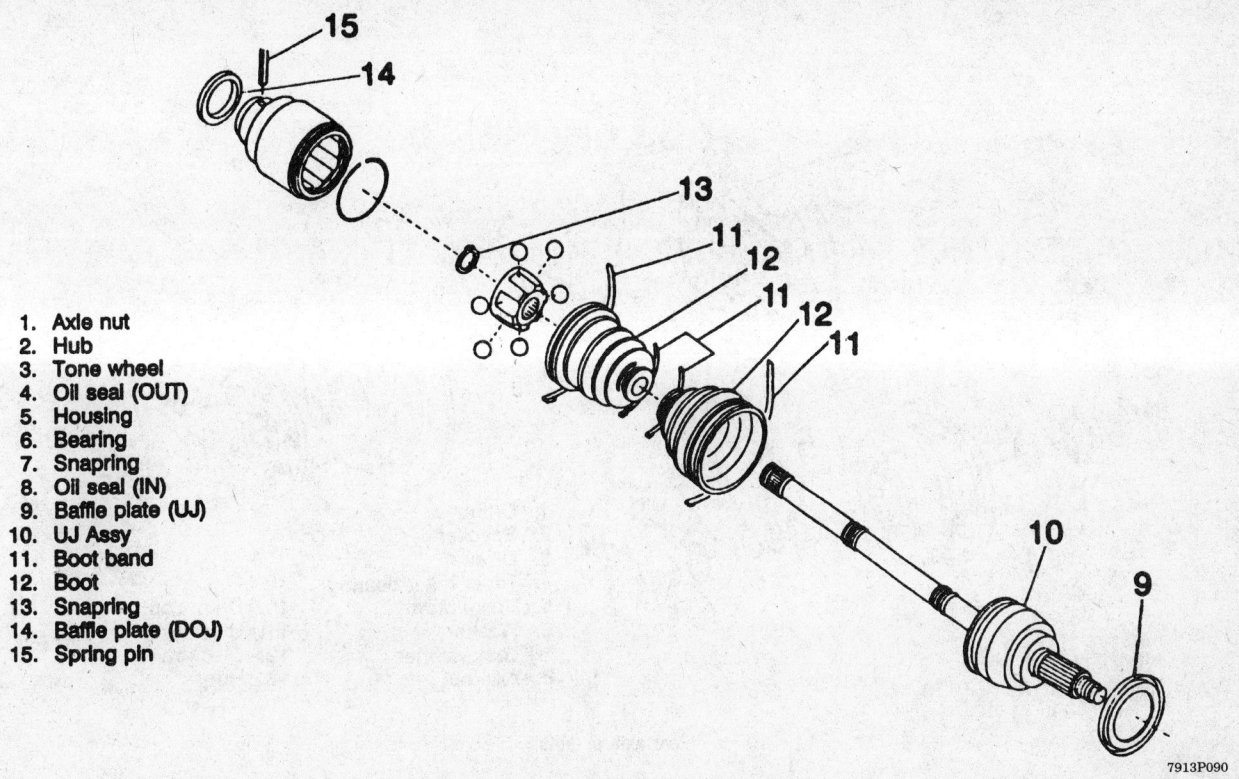

1. Axle nut
2. Hub
3. Tone wheel
4. Oil seal (OUT)
5. Housing
6. Bearing
7. Snapring
8. Oil seal (IN)
9. Baffle plate (UJ)
10. UJ Assy
11. Boot band
12. Boot
13. Snapring
14. Baffle plate (DOJ)
15. Spring pin

7913P090

Front axle and hub assembly — Legacy

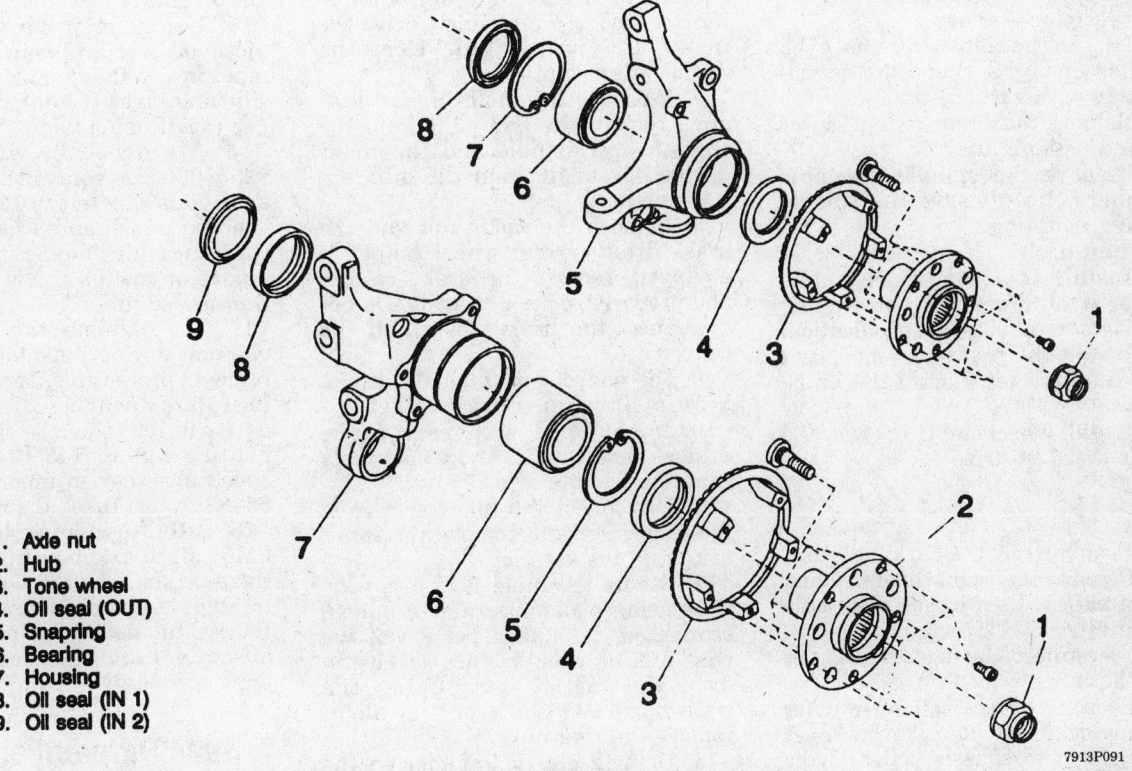

1. Axle nut
2. Hub
3. Tone wheel
4. Oil seal (OUT)
5. Snapring
6. Bearing
7. Housing
8. Oil seal (IN 1)
9. Oil seal (IN 2)

7913P091

Rear hub assembly — Legacy with 4WD

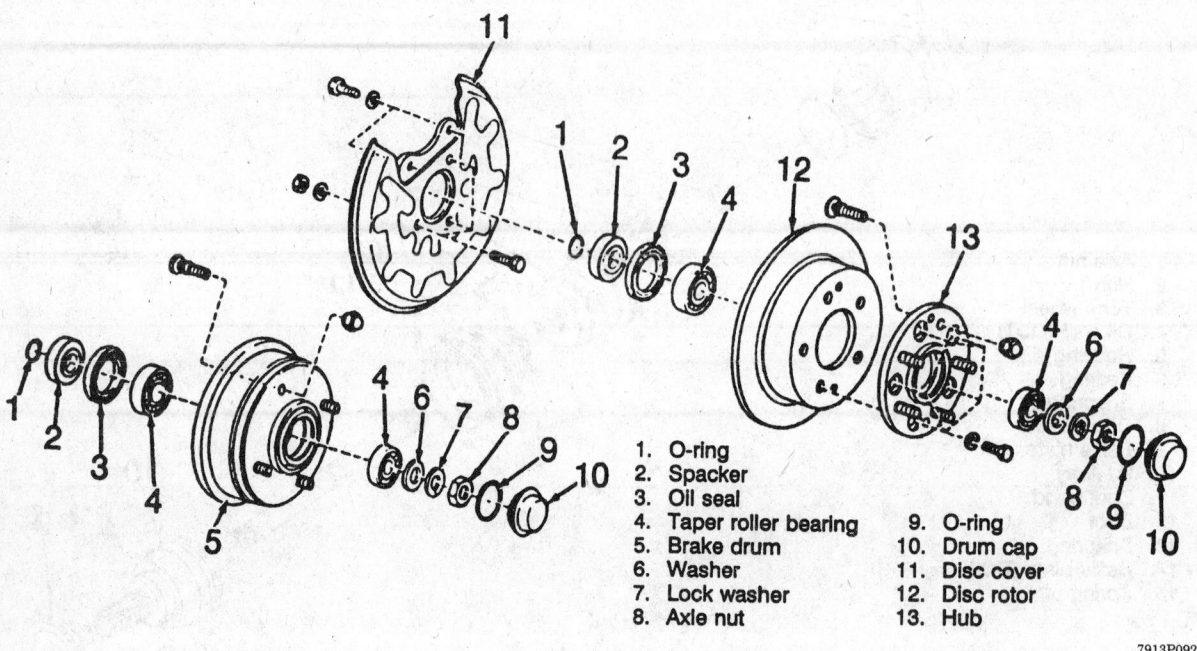

1. O-ring
2. Spacer
3. Oil seal
4. Taper roller bearing
5. Brake drum
6. Washer
7. Lock washer
8. Axle nut
9. O-ring
10. Drum cap
11. Disc cover
12. Disc rotor
13. Hub

7913P092

Rear axle assembly

4. Remove the brake line bracket from the spindle housing.

5. Loosen the bolts and remove the brake assembly. Suspend the assembly aside with wire.

6. Remove the damper strut, lower link and trailing link.

7. Remove the spindle assembly retaining bolts. Remove the spindle from its mounting.

To install:

8. Install the spindle assembly, damper strut, lower link and trailing link. Tighten all bolts to specification.

9. Install the brake assembly and brake line bracket. Install the brake drum.

10. Install wheel and tire, lower the vehicle and test drive.

4WD

1. Firmly apply the parking brake.

2. Remove the rear wheel cap and the cotter pin, then loosen the castle nut.

3. Disconnect the shock absorber from the inner arm.

4. Loosen the crossmember outer bushing lock bolts. Remove the inner trailing arm to chassis bolt and the inner arm.

5. Raise and support the vehicle safely. Remove the rear wheel assemblies.

6. Using a 0.24 in. (6mm) diameter steel rod or a pin punch, drive the inner/outer spring pins from the double offset joints.

7. With the trailing arm fully lowered, remove the ball joint from the trailing arm spindle and the inner double offset joint and the differential spindle.

8. Remove the castle nut and the brake drum or rear wheel caliper If equipped, remove the brake caliper and properly position it aside. Do not disconnect the brake hose from the caliper.

9. Disconnect and plug the brake hose from the inner arm bracket.

10. If equipped with rear brake drums, remove the brake assembly from the trailing arm.

11. Disconnect the inner arm from the outer arm and remove the inner arm from the vehicle.

12. Secure the inner arm in a vise, then using a hammer and a punch, straighten the staked portion of the ring nut or remove the cotter pin from the castled nut. Using the wrench tool 925550000 or equivalent, remove the ring nut.

13. Using a plastic hammer on the outside of the spindle, drive it inward to remove it.

14. Clean, inspect and replace the necessary parts.

To install:

15. Using an arbor press and a piece of 1.38 in. (35mm) diameter pipe, insert the spindle from the inside and press the outer bearing's inner race from outside.

16. Using the wrench tool 925550000 or equivalent, torque the axle shaft ring nut to 127–163 ft. lbs. Using a punch and a hammer, stake the ring nut, facing the ring nut groove or install a new cotter pin in the castled nut.

17. To complete the installation, use new spring pins and reverse the removal procedures. Torque the backing plate to axle housing bolts to 34–43 ft. lbs., the axle spindle to axle housing nut to 145 ft. lbs. and the shock absorber to inner arm bolt to 65–87 ft. lbs. Bleed the brake system.

18. After tightening the rear axle halfshaft to axle housing nut, tighten the axle shaft nut 30 degrees further to align cotter pin holes as required. Be careful not to install the double offset joint and the constant velocity joint oppositely.

Rear Differential Carrier

REMOVAL AND INSTALLATION

1. Drain the differential oil.

2. Raise the rear wheels and support the car on jackstands.

3. Remove the exhaust pipe and muffler.

4. Remove the driveshaft and axle shafts.

5. Support the differential carrier with a jack.

6. Remove the 2 nuts securing the differential to the rear mounting bracket.

7. Remove the 2 bolts securing the differential to the front mounting bracket.

8. Lower the jack and remove the differential.

9. To install reverse Steps 1 through 8. Observe the following torque specifications; rear mounting nuts: 53 ft. lbs. (72 Nm), front mounting bolts: 53 ft. lbs. (72 Nm)

Knuckle and Spindle

REMOVAL AND INSTALLATION

1. Disconnect the ground cable from the battery.

2. Apply the parking brake.

3. Remove the front wheel cap and cotter pin, and loosen the castle nut and wheel nuts.

4. Raise the vehicle, support it with jackstands and remove the front tires and wheels.

5. Release the parking brake.

6. Pull out the parking brake cable outer clip from the caliper.

7. Disconnect the parking brake cable end from the caliper lever.

8. Loosen the 2 nuts and remove the disc brake assembly from the housing.

9. Remove the 2 nuts which connect the housing and damper strut.

10. Remove the cotter pin and castle nut, and disconnect the tie rod end ball joint from the housing knuckle arm by using a puller.

11. Disconnect the strut from the housing by opening the slit of the housing and lowering the housing gradually, being careful not to damage the ball joint boot.

NOTE: Do not expand the slit of the housing more than 4mm. If the housing is hard to remove from the strut, lightly tap the hub and disc with a large rubber mallet.

12. Remove the castle nut, washer spring center piece on the axle shaft, and take out the hub and disc assembly.

13. Remove the disc cover.

14. Attach puller 926470000 to the housing and drive the axle shaft out of the housing toward the engine at the bearing location.

NOTE: If the inner bearing and/or oil seal are left on the axle shaft, remove them with a puller.

15. Disconnect the transverse link from the housing, and detach the housing.

To install:

16. Fit the housing onto the axle shaft and attach spacer installer 925130000 or 922430000, on the outer bearing inner race taking care not to damage the oil seal lip. Then, connect the rod of the installer to the thread of the axle shaft so the housing does not drop off the axle shaft.

17. Install the transverse link ball joint to the housing.

18. Turn the handle while holding the rod end, by means of a spanner, thus pushing in the housing.

19. Connect the damper strut to the housing.

20. Connect the tie rod end ball joint and housing knuckle arm. Torque the castle nut. After tightening the nut to the specified torque, further torque the nut just enough to align the holes of the nut and ball stud. Then insert a cotter pin into the ball stud and bend it around the castle nut.

21. Install the disc cover to the housing.

22. Install the hub and disc assembly onto the axle shaft.

NOTE: Be sure to press the hub and disc assembly onto the axle shaft until the end surface of the hub contacts the ball bearing. If the assembly is hard to press, rotate it to locate the point where it is easily pressed.

23. Install the brake caliper to the housing assembly.

24. Connect the parking brake cable to the brake assembly.

25. Apply the parking brake.

26. Position the center piece, washer spring and castle nut in this order onto the axle shaft and tighten the castle nut to 145 ft. lbs. (196 Nm) on all models except the SVX and Legacy or on the SVX and Legacy tighten the castle nut to 123–152 ft. lbs. (167–206 Nm) then insert a new cotter pin and bend it around the nut.

NOTE: After tightening the nut to the specified torque, retighten it further until a slot of the castle nut is aligned to the hole in the axle shaft.

27. Install the wheel and hub cap.

28. Remove the jack stands, lower the vehicle and reconnect the negative battery cable.

Front Wheel Hub and Bearing

REMOVAL AND INSTALLATION

Justy

1. Raise and safely support the vehicle. Remove the halfshaft from the vehicle. Remove the steering knuckle from the vehicle.

2. Using a finger, move the spacer (inside the steering knuckle) in the radial direction.

3. Using a brass bar, insert it through the inner race of the outer bearing, then tap the bar with a hammer to drive out the bearing (with the oil seal); discard the bearing and the oil seal.

4. Remove the spacer and the inner bearing.

5. Position the brass bar through the outer race of the inner bearing, then using a hammer, drive the out the bearing (with the oil seal); discard the bearing and the oil seal.

To install:

6. Clean and inspect the parts for wear, cracks and/or damage; if necessary, replace the damaged parts.

7. Using new bearings, pack them with grease.

8. Using the stand tool 922441000 or equivalent, install the steering knuckle onto the stand.

9. Using the bearing installer tool 922470000 or equivalent, and the handle tool 498477000 or equivalent, press the inner bearing into the housing until it contacts the housing stopper.

10. Using 1/4 oz. of wheel bearing grease, pack the inside of the housing.

11. Invert the housing on the stand tool 922441000 or equivalent, and install the spacer.

12. Using the bearing installer tool 922470000 or equivalent, and the handle tool 498477000 or equivalent, press the outer bearing into the housing until it contacts the housing stopper.

13. Using a press and the oil seal installer tool 922450000 or equivalent, press the new outer oil seal into the steering knuckle housing, until it comes in contact with the bearing end face.

14. Invert the steering knuckle housing.

15. Using a press and the oil seal installer tool 922460000 or equivalent, press the new inner oil seal into the steering knuckle housing, until it comes in contact with the bearing end face.

16. To complete the installation, reverse the remaining removal procedures.

Except Justy

1. Raise and safely support the vehicle. Remove the halfshaft from the vehicle; be sure to remove the steering knuckle from the vehicle.

2. Using a finger, move the spacer (inside the steering knuckle) in the radial direction.

3. Using a brass bar, insert it through the inner race of the outer bearing, then tap the bar with a hammer to drive out the bearing (with the oil seal); discard the bearing and the oil seal.

4. Remove the spacer and the inner bearing.

5. Position the brass bar through the outer race of the inner bearing, then using a hammer, drive the out the bearing (with the oil seal); discard the bearing and the oil seal.

To install:

6. Clean and inspect the parts for wear, cracks and/or damage; if necessary, replace the damaged parts.

7. Using new bearings, pack them with wheel bearing grease.

8. Using the die tool 926490000 or equivalent, install the steering knuckle onto the die.

9. Using a press and the punch tool 926490000 or equivalent, press the outer bearing into the housing until it contacts the housing stopper.

10. Using ½ oz. of wheel bearing grease, pack the inside of the housing.

11. Invert the housing on the die tool 926490000 or equivalent, and install the spacer.

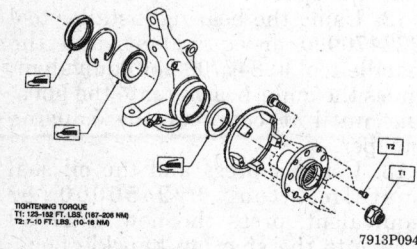

TIGHTENING TORQUE
T1: 123-152 FT. LBS. (167-206 NM)
T2: 7-10 FT. LBS. (10-16 NM)

7913P094

Front knuckle and hub assembly — Legacy and SVX

12. Using a press and the punch tool 926490000 or equivalent, press the inner bearing into the housing until it contacts the housing stopper.

13. Using a press and the punch tool 926490000 or equivalent, position the new outer oil seal in the punch tool so the lip faces the groove, then press it into the steering knuckle housing, until it comes in contact with the bearing end face.

14. Invert the steering knuckle housing onto the punch tool 926490000 or equivalent, so the seal lip faces the groove.

15. Using a press and the die tool 926490000 or equivalent, press the new inner oil seal into the steering knuckle housing, until it comes in contact with the bearing end face.

16. To install the steering knuckle, fit the housing onto the axle shaft and attach a spacer of the installation tool 922430000 or equivalent, on the outer bearing inner race; be careful not to damage the oil seal. Thread the axle shaft onto the installation tool, then turn the handle to draw the axle into the housing, until it is seated.

17. Install the remaining components in reverse order.

MANUAL TRANSAXLE

REMOVAL AND INSTALLATION

Justy

1. Disconnect the negative battery cable. Remove the air cleaner assembly. Raise and support the vehicle safely.

2. Disconnect the electrical wiring connectors from the starter. Remove the starter to transaxle bolts and the starter from the vehicle.

3. From the transaxle, disconnect the speedometer cable, the backup light switch connector and the ground cable. If equipped with 4WD, remove the activation hoses from the actuator.

4. Disconnect the electrical connector between the ignition coil and the distributor.

5. Disconnect the clutch cable and the bracket from the transaxle. In place of the clutch cable bracket, install the lifting hook, or equivalent.

6. Removing the pitching stopper and brackets between the transaxle and chassis.

7. Install engine supporter tool 921540000 or equivalent.

8. Install the vertical hoist to T000100 transaxle lifting hook and raise the transaxle slightly.

9. From under the vehicle, remove the under covers.

10. Disconnect the rear exhaust pipe from the front exhaust pipe and the vehicle.

11. Remove the center crossmember to engine/transaxle assembly bolts.

12. Using a pin punch and a hammer, drive out the axle shaft to driveshaft spring pin. Discard the spring pin and separate the axle shaft.

13. Remove the transaxle mounting bracket.

14. Disconnect the gearshift rod and stay from the transaxle.

15. Properly support the engine assembly. Remove the transaxle to engine bolts.

16. Using the vertical hoist, lift the transaxle from the vehicle.

To install:

17. Install the transaxle assembly in the vehicle and install the transaxle-to-engine bolts. Install the gearshift rods on the transaxle.

18. Join the axle shaft and the differential. Install a new axle shaft spring pin. Install the center crossmember and tighten bolts to 27–49 ft. lbs. (37–67 Nm).

19. Install the rear exhaust pipe, engine under covers, pitching stopper and brackets, clutch cable, electrical connectors, speedometer cable, 4WD activation hoses, starter wires and starter-to-transaxle bolts.

20. Lower the vehicle, connect the negative battery cable, check the transaxle fluid and test drive the vehicle.

Legacy

1. Disconnect the negative battery cable. If equipped with a turbocharger, remove the manifold cover.

2. Remove the air intake duct. If equipped with a turbocharger, discharge the air conditioning and disconnect the air conditioner pressure hose. Remove the resonator chamber, air inlet and outlet ducts.

3. Disconnect all cables and harness connectors attached to the transaxle. Remove the starter.

4. Remove the pitching stopper rod and bracket. On turbocharged models, remove the clutch operating cylinder assembly and free the re-

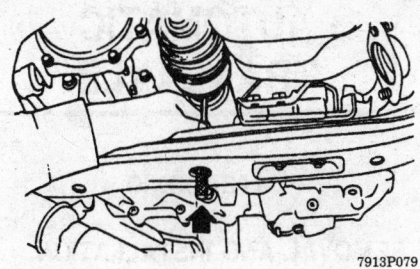

**Removing the spring pin from the axle shaft —
Justy**

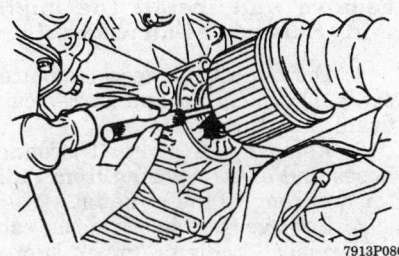

Separating the axle shaft from the driveshaft

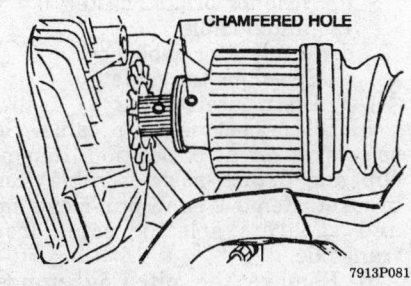

**Aligning the chamfered holes of the axle shaft
with the driveshaft**

lease fork. Remove the transaxle oil level gauge.

5. Remove the connector holder bracket. Remove the turbocharger cooling ducts and disconnect the center exhaust pipe from the turbocharger.

6. Remove the upper transaxle attaching bolts. Remove the driveshaft, gearshift system, front stabilizer and halfshafts.

7. Remove the nuts holding the lower side of the engine to the transaxle. Install a transaxle jack. Remove the rear cushion rubber mount-

ing nuts and rear crossmember. Remove the transaxle.

8. On turbocharged models, remove the release bearing from the clutch cover.

To install:

9. Install the transaxle and temporarily tighten bolts. Install the clutch release assembly on Turbo models.

10. Install the rear cushion and crossmember. Tighten cushion bolts to 20–35 ft. lbs. (27–47 Nm); crossmember front bolts to 87–116 ft. lbs. (118–157 Nm), rear 40–61 ft. lbs. (54–83 Nm). Tighten the transaxle to engine bolts to 34–40 ft. lbs. (46–54 Nm).

11. After tightening all bolts check that the release fork is in the proper position. Install the halfshafts, and temporarily install the transverse link and stabilizer.

12. Install the gear shift system and driveshaft. Lower the vehicle and tighten the transverse link to 43–51 ft. lbs. (59–69 Nm); stabilizer to 14–2 ft. lbs. (20–29 Nm).

13. Install the connector holder bracket, pitching stopper, turbocharger cooling duct, clutch operating cylinder, air conditioner hoses, resonator, air inlet and outlet.

14. Install the starter assembly. Connect all previously disconnected harnesses and connectors.

15. Connect the negative battery cable, check the transaxle fluid level and test drive the vehicle.

Except Legacy and Justy

1. Disconnect the negative battery cable. Remove the air cleaner assembly.

2. Remove the clutch cable and the hill holder cable. Remove the speedometer cable.

3. Remove the oxygen sensor electrical connector and the neutral switch connector.

4. If equipped with 4WD, remove the disconnect the electrical connections at the backup light and differential lock indicator switch assembly. Disconnect the differential lock vacuum hose.

5. Disconnect the starter electrical connections. Remove the starter retaining bolts. Remove the starter from the transaxle case.

6. Remove the air intake boot. Disconnect the pitching stopper rod from its mounting bracket. Remove the right side engine to transaxle mounting bolt.

7. Install engine support bracket 927160000 and engine support tool 927150000 or their equivalents. Re-

move the buffer rod from the engine and body side bracket.

NOTE: Before attaching the special engine support tools, connect the adjuster to the buffer rod assembly on the right side of the engine.

8. Raise and support the vehicle safely.

9. Disconnect the exhaust pipes at the exhaust manifold flange. Remove the exhaust system up to the rear exhaust pipe assembly.

10. If equipped with 4WD, matchmark and remove the driveshaft. Remove the complete gear shift assembly.

11. Loosen the upper bolt and nut from the plate that secures the transverse link to the stabilizer. Remove the lower bolt and separate the link from the stabilizer.

12. Remove the right brake cable bracket from the transverse link. Remove the bolt retaining the link to the crossmember on each side.

13. Lower the transverse link. Using tool 398791700 or equivalent, remove the spring pin and separate the axle shaft from the driveshaft on each side of the assembly by pushing the rear of the tire outward.

14. Remove the engine to transaxle mounting bolts. Position the proper transaxle jack under the transaxle assembly.

15. Remove the rear cushion rubber mounting bolts. Remove the rear crossmember assembly.

16. Turn the engine support tool adjuster counterclockwise in order to slightly raise the engine.

17. Move the transaxle jack toward the rear of the vehicle until the mainshaft is withdrawn from the clutch cover.

18. Carefully remove the transaxle assembly from the vehicle.

To install:

19. Carefully raise the transaxle until the mainshaft is aligned with the clutch side. Install the engine to the transaxle and temporarily tight the mounting bolts.

20. Install the rear crossmember rubber cushion and tighten nuts to 20–35 ft. lbs. (27–47 Nm). Install the rear crossmember and tighten front bolts to 65–87 ft. lbs. (88–118 Nm); rear bolts to 27–49 ft. lbs. (37–67 Nm).

21. Tighten the engine to transaxle nuts to 34–40 ft. lbs. (46–54 Nm). Remove the transaxle jack.

22. Install the halfshaft into the differential and spring pin into place. Install the transverse link and stabilizer temporarily to the front cross-

member. Install the brake cable bracket. Lower the vehicle and tighten transverse link bolt to 43–51 ft. lbs. (59–69 Nm); stabilizer bolts to 14–22 ft. lbs. (20–29 Nm).

23. Install the gearshift system. Install the driveshaft (4WD vehicles). Install the starter, pitching stopper, timing hole plug, air intake boot and speedometer cable. Reconnect all electrical and vacuum connectors.

24. Connect the clutch cable and hill holder. Install the front exhaust pipe.

25. Connect the negative battery cable, check the transaxle fluid level and test drive the vehicle.

CLUTCH

REMOVAL AND INSTALLATION

1. Remove the transaxle from the vehicle.

2. Gradually loosen the pressure plate to flywheel assembly bolts. Loosen the bolts 1 turn at a time, working around the pressure plate.

3. Remove the clutch plate and the disc from the vehicle.

To install:

4. Inspect the parts for wear or damage and replace any parts, as necessary.

5. Installation is the reverse of the removal procedure.

6. Use clutch disc guide tool 499747000 or equivalent, to align the clutch on the non-turbocharged 1.8L engines. Use tool 499747100 or equivalent, to align the clutch on the turbocharged 1.8L engine. Use tool 499745500 or equivalent, to align the clutch on the 1.2L engine.

7. When installing the clutch pressure plate assembly, make sure the marks on the flywheel and the clutch pressure plate assembly are at least 120 degrees apart. This is for purposes of balance. Also, make sure the clutch disc is installed properly, noting the **FRONT** and **REAR** markings.

FREE-PLAY ADJUSTMENT

1. Remove the clutch release fork return spring.

2. Loosen the cable locknut, then adjust the spherical nut so there is

the following play between the spherical nut and the release fork seat.

- 1.8L engines, 2WD except turbocharger — 0.08–0.12 in.
- 2WD/4WD turbocharged, 1.8L engine — 0.12–0.16 in.
- 1.2L engine — 0.08–0.16 in.

3. Tighten the locknut and reconnect the release spring.

Clutch Cable

REMOVAL AND INSTALLATION

The clutch cable is connected to the clutch pedal at 1 end and to the clutch release lever at the other end. The cable conduit is retained by a bolt and clamp on a bracket mounted on the flywheel housing.

1. If necessary, raise and support the vehicle safely.

2. Disconnect both ends of the cable and the conduit, then remove the assembly from under the vehicle.

3. Using engine oil, lubricate the clutch cable. If the cable is defective, replace it.

4. Installation is the reverse the removal procedure.

ADJUSTMENT

The clutch cable can be adjusted at the cable bracket where the cable is attached to the side of the transaxle housing.

1. Remove the circlip and clamp.

2. Slide the cable end in the direction desired and then replace the circlip and clamp into the nearest gutters on the cable end.

NOTE: The cable should not be stretched out straight nor should it have right angle kinks in it. Any straightening should be gradual.

3. Check the clutch for proper operation.

AUTOMATIC TRANSAXLE

Transaxle

REMOVAL AND INSTALLATION

Justy

WITH ECVT

NOTE: When removing and installing ECVT transaxle, always remove and install the engine and transaxle as an assembly.

1. Disconnect the negative battery cable. Drain the coolant by removing drain plug from radiator.

2. Remove the grille. Disconnect hoses and electric wiring from radiator and remove the radiator.

3. Remove front hood release cable and remove radiator upper support member. Disconnect horn and remove the air cleaner assembly.

4. Disconnect the following hoses and cables:
   a. Hoses from the heater unit
   b. Hose for brake booster
   c. Clutch cable
   d. Accelerator cable
   e. Speedometer cable
   f. Distributor wiring

5. Disconnect selector cable. Set selector lever at **N** position. Remove clip and detach selector cable from bracket. Remove snap pin, clevis pin and separate selector cable from transaxle.

6. Remove the pitching stopper from the bracket.

7. Disconnect the starter cable, engine wiring harness connectors, ground lead terminals and brush holder harness connector.

8. Remove the hanger from the rear of transaxle.

9. Remove under covers and remove the exhaust system.

10. Remove the driveshaft from transaxle.

11. Remove transverse link.

12. Remove the spring pin retaining the axle shaft by using a suitable tool and separate front axle shaft from the transaxle.

13. Remove engine and transaxle mounting brackets.

14. Raise the engine and remove center member and crossmember.

15. Lift the engine/transaxle assembly carefully and remove it from the vehicle.

To install:

16. Position the engine/transaxle assembly in the vehicle. Install engine and transaxle mounting brackets.

17. Install center member and crossmember.

18. Install the axle shaft to transaxle with new spring pin.

19. Install gearshift rod and stay to transaxle.

20. Install the exhaust system. Connect driveshaft to transaxle.

21. Install transverse link and under covers to the vehicle.

22. Reconnect the pitching stopper to bracket.

23. Reconnect the following hoses and cables:
   a. Hoses to the heater unit
   b. Hose to brake booster
   c. Clutch cable to transaxle
   d. Accelerator cable
   e. Speedometer cable
   f. Distributor wiring

24. Reconnect the starter cable, engine wiring harness connectors, ground lead terminals and brush holder harness connector. Install the air cleaner assembly.

25. Install radiator upper member and connect hood release cable to lock assembly. Reconnect the horn.

26. Install the radiator and connect hoses and electric wiring. Attach grille to the vehicle.

27. Refill the coolant. Reconnect the battery cable.

28. Check all fluid levels. Road test vehicles for proper operation in all driving ranges.

Loyale

2WD AND 4WD NON-ELECTRONIC 3- AND 4-SPEED TRANSAXLES

1. Disconnect the negative battery cable.

2. Remove clamp from spare tire supporter and remove the spare tire.

NOTE: Use care when removing spare tire assembly from the vehicle.

3. Remove spare tire supporter and battery clamp.

4. Remove speedometer cable and retaining clip. Before disconnecting speedometer cable, remove front exhaust pipe on 4 speed automatic transaxle.

5. Disconnect the following electrical harness connections on the 3 speed automatic transaxle:
   a. Oxygen sensor connector
   b. ATF temperature switch connector

   c. Kickdown solenoid valve connector
   d. 4WD solenoid valve connector on 4WD equipped vehicles

6. Disconnect the following electrical harness connections on the 4 speed automatic transaxle:
   a. Oxygen sensor connector
   b. Transaxle harness connector
   c. Inhibitor switch connector
   d. Revolution sensor connector on 4WD equipped vehicles

7. Disconnect the diaphragm vacuum hose on 3 speed automatic transaxle and 4WD vacuum hose on 4WD equipped vehicles.

8. Remove clip band which secures air breather hose to pitching stopper.

9. Remove the pitching stopper rod. Remove the starter.

10. Remove timing hole inspection plug and remove the 4 bolts which hold torque converter to driveplate.

11. Support the engine assembly with special engine support tool 926610000 or equivalent.

12. Remove engine-to-transaxle mounting nut and bolt on the right side.

13. Remove the exhaust system.

NOTE: Apply a penetrating oil or equivalent to all exhaust retaining nuts in advance to facilitate removal.

14. On turbocharged vehicles, remove accelerator cable cover and upper and lower turbocharger covers. Remove the center exhaust pipe at turbocharger location and at rear exhaust pipe. Remove any exhaust brackets or hangers that attach to the transaxle, as necessary.

15. On non-turbocharged vehicles, disconnect front exhaust pipe from the engine and from the rear exhaust pipe. Remove any exhaust brackets or hangers that attach to the transaxle as necessary.

16. Drain all transaxle fluid from the oil pan.

17. Remove the driveshaft on 4WD vehicles. Plug the opening at the rear of extension housing to prevent oil from flowing out.

18. Disconnect the linkage rod for a 3 speed or cable for a 4 speed. from the select lever.

19. Remove stabilizer from transverse link by loosening (not removing) nut and bolt on the lower side of plate.

20. Remove parking brake cable bracket from transverse link and bolt holding transverse link to crossmember on each side. Lower the transverse link.

21. Remove spring pin and separate axle shaft from transaxle on each side.

NOTE: Use a suitable tool to remove spring pin. Discard old spring pin and always install a new pin.

22. Disconnect the axle shaft from transaxle on each side. Be sure to remove axle shaft from transaxle by pushing the rear of tire outward.

23. Remove engine-to-transaxle mounting nuts.

24. Disconnect oil cooler hoses and oil supply pipe. Be careful not to damage the oil supply pipe O-ring.

25. Place transaxle jack or equivalent under transaxle. Always support transaxle case with a transaxle jack.

NOTE: Do not place jack under oil pan otherwise oil pan may be damaged.

26. Remove rear cushion rubber mounting nuts and rear crossmember. Move torque converter and transaxle as a unit away from the engine. Remove the transaxle.

To install:

27. Install transaxle to engine and temporarily tighten engine-to-transaxle mounting nuts.

28. Install rear crossmember to rear cushion rubber mounts. Align rear cushion guide with rear crossmember guide hole and tighten nuts.

29. Install rear crossmember to chassis. Be careful not to damage threads. Torque rear crossmember bolts to 39–49 ft. lbs.

30. Tighten engine to transaxle nuts on the lower side to 34–40 ft. lbs. Remove transaxle jack from the vehicle.

31. Install axle shaft to transaxle and install spring pin into place.

NOTE: Always use new spring pin. Be sure to align the halfshaft and shaft from the transaxle at chamfered holes and engage shaft splines correctly.

32. Install transverse link temporarily to front crossmember by using bolt and self-locking nut. Do not complete final torque at this point.

33. Install stabilizer temporarily to transverse link. Install parking brake cable bracket to transverse link.

34. Connect the linkage rod for a 3 speed or cable for a 4 speed to the select lever. Make sure the lever operates smoothly all across the operating range.

35. Install propeller shaft on 4WD vehicles. Torque propeller shaft to

rear differential retaining bolts to 13–20 ft. lbs. and center bearing location retaining bolts to 25–33 ft. lbs.

36. Connect oil cooler hoses and oil supply pipe. Lower vehicle to floor.

37. Tighten transverse link to front crossmember mounting bolts and transverse link to stabilizer mounting bolts with the tires placed on the ground when the vehicle is not loaded. Tightening torque for transverse link to front crossmember (self-locking nuts) 43–51 ft. lbs. and transverse link to stabilizer 14–22 ft. lbs.

38. Tighten engine to transaxle nuts on the upper side to 34–40 ft. lbs.

39. Raise vehicle and safely support. Install exhaust system.

NOTE: Before installing exhaust system, connect speedometer cable on 4 speed vehicles.

40. On turbocharged vehicles, install the center exhaust pipe at turbocharger location and at rear exhaust pipe. Install any exhaust brackets or hangers that attach to the transaxle as necessary. Install upper and lower turbocharger covers and accelerator cable cover.

41. On non-turbocharged vehicles, connect front exhaust pipe to the engine and rear exhaust pipe. Install any exhaust brackets or hangers that attach to the transaxle as necessary.

42. Remove the special engine support tool. Install and tighten torque converter to driveplate mounting bolts to 17–20 ft. lbs.

43. Install timing hole inspection plug.

44. Install starter.

45. Install pitching stopper. Be sure to tighten the bolt for the body side first and then the 1 for engine or transaxle side. Tightening torque for chassis side is 27–49 ft. lbs. and for engine or transaxle side is 33–40 ft. lbs.

46. Reconnect the following electrical harness connections on the 3 speed automatic transaxle:
   a. Oxygen sensor connector
   b. ATF temperature switch connector
   c. Kickdown solenoid valve connector
   d. 4WD solenoid valve connector on 4WD equipped vehicles

47. Reconnect the following electrical harness connections on the 4 speed automatic transaxle:
   a. Oxygen sensor connector
   b. Transaxle harness connector
   c. Inhibitor switch connector
   d. Revolution sensor connector on 4WD equipped vehicles

48. Reconnect the diaphragm vacuum hose on 3 speed automatic transaxle and 4WD vacuum hose on 4WD equipped vehicles.

49. Secure air breather hose to pitching stopper with a clip band.

50. Reconnect the speedometer cable. Manually tighten cable nut all the way and then turn it approximately 30 degrees more with a tool.

51. Connect the battery ground cable. Refill and check transaxle oil level.

52. Install spare tire supporter and battery clamp. Install spare tire.

53. Road test vehicle for proper operation across all operating ranges.

Legacy and SVX

4-SPEED ELECTRONIC TRANSAXLE

1. Disconnect the negative battery cable.

2. Remove speedometer cable or electronic wiring connector from speed sensor.

3. Disconnect the following electrical harness connections on the automatic transaxle:
   a. Oxygen sensor connector
   b. Transaxle harness connector
   c. Inhibitor switch connector
   d. Revolution sensor connector on 4WD equipped vehicles
   e. Crankshaft and camshaft angle sensor connector on Legacy vehicles
   f. Knock sensor connectors and transaxle ground terminal on Legacy vehicles

4. Remove clip band which secures air breather hose to pitching stopper.

5. Remove the starter and air intake boot.

6. Remove timing hole inspection plug and remove the 4 bolts which hold torque converter to driveplate.

7. Disconnect pitching stopper rod from bracket.

8. Remove engine to transaxle mounting nut and bolt on the right side.

9. Remove the buffer rod from the vehicle. Support the engine assembly with special engine support tool or equivalent.

10. Remove the exhaust system. Remove exhaust brackets or hangers that attach to the transaxle, as necessary.

11. Matchmark and remove the driveshaft on 4WD vehicles. Plug the opening at the rear of extension housing to prevent oil from flowing out.

12. Disconnect the gear shift cable from the transaxle select lever.

13. Remove stabilizer from transverse link.

14. Remove parking brake cable bracket from transverse link and bolt holding transverse link to crossmember on each side. Lower the transverse link.

15. Remove spring pin and separate halfshaft from transaxle on each side.

NOTE: Use a suitable tool to remove spring pin. Discard old spring pin and always install a new pin.

16. Disconnect the halfshaft from transaxle on each side. Be sure to remove axle shaft from transaxle by pushing the rear of tire outward.

17. Remove engine to transaxle mounting nuts.

18. Disconnect oil cooler hoses.

19. Place transaxle jack or equivalent, under transaxle. Always support transaxle case with a transmission jack.

NOTE: Do not place jack under oil pan otherwise oil pan may be damaged.

20. Remove rear cushion rubber mounting nuts and rear crossmember.

21. Move torque converter and transaxle as a unit away from the engine. Remove the transaxle.

To install:

22. Install transaxle to engine and temporarily tighten engine to transaxle mounting nuts.

23. Install rear crossmember to rear cushion rubber mounts. Align rear cushion guide with rear crossmember guide hole and tighten nuts.

24. Install rear crossmember to chassis; be careful not to damage threads. Torque rear crossmember bolts to 39–49 ft. lbs.

25. Tighten engine to transaxle retaining nuts to 34–40 ft. lbs. Remove transaxle jack from the vehicle.

26. Remove the engine support tool and install buffer rod.

27. Install axle shaft to transaxle and install spring pin into place.

NOTE: Always use new spring pin. Be sure to align the axle shaft and shaft from the transaxle at chamfered holes and install shaft splines correctly.

28. Install transverse link temporarily to front crossmember by using bolt and self locking nut. Do not complete final torque at this point.

29. Install stabilizer temporarily to transverse link. Install parking brake cable bracket to transverse link.

30. Lower vehicle to floor. Tighten transverse link to front crossmember mounting bolts and transverse link to

stabilizer mounting bolts with the tires placed on the ground when the vehicle is not loaded. Tightening torque for transverse link to front crossmember (self locking nuts) 43–51 ft. lbs. and transverse link to stabilizer 14–22 ft. lbs.

31. Raise and safely support the vehicle. Reconnect the gear shift cable to the select lever. Make sure the lever operates smoothly all across the operating range.

32. Install propeller shaft on 4WD vehicles. Torque propeller shaft-to-rear differential retaining bolts to 17–24 ft. lbs. and center bearing location retaining bolts to 25–33 ft. lbs.

33. Connect oil cooler hoses.

34. Tighten engine to transaxle bolts to 34–40 ft. lbs.

35. Install starter.

36. Install pitching stopper. Be sure to tighten the bolt for the body side first and then the 1 for engine or transaxle side. Tightening torque for chassis side is 27–49 ft. lbs. and for engine or transaxle side is 33–40 ft. lbs.

37. Install and tighten torque converter-to-driveplate mounting bolts to 17–20 ft. lbs.

38. Install timing hole inspection plug, air intake boot and air breather hose to pitching stopper.

39. Reconnect the following electrical harness connections on the automatic transaxle:

    a. Oxygen sensor connector

    b. Transaxle harness connector

    c. Inhibitor switch connector

    d. Revolution sensor connector on 4WD equipped vehicles

    e. Crankshaft and camshaft angle sensor connector on Legacy

    f. Knock sensor connectors and transaxle ground terminal on Legacy

40. Reconnect the speedometer cable. Manually tighten cable nut all the way and then turn it approximately 30 degrees more with a tool.

41. Install exhaust system and exhaust brackets or hangers that attach to the transaxle, as necessary.

42. Connect the battery ground cable. Refill and check transaxle oil level.

43. Road test vehicle for proper operation across all operating ranges.

SHIFT LINKAGE ADJUSTMENT

1. Loosen the clamp nuts on the shifting rod at the bottom of the shift lever on the transaxle.

2. Place the selector lever in **N** and hold it forward against the detent.

3. Check that the transaxle shift lever is in the **N** position by pulling it all the way back into **P** and then pushing it forward 2 positions.

4. Tighten the clamp nuts.

KICKDOWN SOLENOID ADJUSTMENT

If used, an audible click should be heard from the solenoid on the right side of the transaxle, when the accelerator pedal is pushed down all the way with the engine OFF and the ignition switch in the **ON** position. The switch is operated by the upper part of the accelerator lever inside the vehicle. The position of the switch can be varied to give quicker or slower kickdown response.

TRANSFER CASE

REMOVAL AND INSTALLATION

1. Disconnect the negative battery cable.

2. Raise and support the vehicle safely.

3. Remove the transaxle assembly from the vehicle.

4. Position the assembly in a suitable holding fixture.

5. Disassemble the transfer case from the transaxle.

6. Installation is the reverse of removal procedure. Tighten the transfer case-to-transaxle nuts to 35–40 ft. lbs.

FRONT SUSPENSION

MacPherson Strut

REMOVAL AND INSTALLATION

Justy

1. Disconnect the negative battery cable. Remove the bolts that retain the strut assembly to the body.

2. Raise and support the vehicle safely. Remove the tire and wheel assembly.

3. Remove the brake hose from the brake hose bracket on the strut assembly. Remove the retaining bolt that retains the brake hose bracket to the strut.

4. Properly support the hub and disc assembly. Remove the retaining bolt from the strut to the housing.

5. Fit the proper tool into the housing slit and pull the strut assembly from the housing.

6. Remove the strut from the vehicle.

To install:

7. Install the strut and tighten the upper attaching nuts to 29–43 ft. lbs. (39–59 Nm); lower attaching bolts to 25–40 ft. lbs. (34–54 Nm).

8. Install the brake hose and bracket assembly. Install the wheel and lower the vehicle.

Except Justy

1. Disconnect the negative battery cable. If equipped with air suspension, remove the cover and the air line assembly.

2. Remove the bolts that retain the strut assembly to the body.

3. Raise and support the vehicle safely. Remove the tire and wheel assembly.

4. Disconnect the brake hose from the caliper body. Pull the brake hose retaining clip and remove the brake hose from the damper strut bracket.

5. Remove the bolt that retains the damper strut to the housing. Remove the bolt that retains the damper strut bracket to the housing.

6. Pull the strut assembly from the housing gradually and carefully, with the housing assembly in the downward position.

7. Remove the strut assembly from the vehicle.

To install:

8. Install the strut assembly on the vehicle. Tighten the strut attaching bolts to 28–37 ft. lbs. (38–50 Nm).

9. Install the brake hose on the caliper and bleed the brake system. If equipped with air suspension, install the air line assembly.

10. Install the wheel and tire. Lower the vehicle, connect the negative battery cable and test drive the vehicle.

Ball Joints

INSPECTION

1. Raise and support the vehicle safely.

2. Using a prybar, position it under the wheel, then pry upward on the wheel several times. If more than 0.012 in. (3mm) of movement is no-

ticed at the ball joint it should be replaced.

3. Inspect the dust seal, if damaged it should be replaced.

REMOVAL AND INSTALLATION

1. Raise and support the vehicle safely. Remove the tire and wheel assembly.

2. Properly support the lower control arm assembly. Remove the cotter pin and castle nut from the ball joint.

3. Disconnect the ball joint from the lower control arm assembly.

4. Remove the bolt retaining the ball joint to the housing. Remove the ball joint from the housing.

To install:

5. Install the ball joint into the housing and tighten the nut to 28–37 ft. lbs. (38–50 Nm).

6. Connect ball joint to the transverse link and tighten the castle nut to 29 ft. lbs. (39 Nm). Install the cotter pin in the castle nut.

7. Install the front wheels and lower the vehicle.

Lower Control Arm

REMOVAL AND INSTALLATION

Justy

1. Raise and support the vehicle safely. Remove the tire and wheel assembly.

2. Properly support the lower control arm. Remove the brake hoses as necessary. Remove the bolt that retains the lower control arm to the crossmember.

3. Remove the ball joint retaining bolt and the stabilizer tension rods. Remove the ball joint, if required.

4. Remove the lower control arm from the vehicle.

To install:

5. Install the ball joint, if removed. Install the lower control arm. Tighten the crossmember-to-control arm bolt to 43–58 ft. lbs. (59–78 Nm) only after the vehicle is on the ground with the chassis loaded.

6. Install the castle nut on the ball joint and tighten to 29 ft. lbs. (39 Nm).

7. Temporarily install the tension rod-to-control arm bolt. Then, install the tension rod-to-bracket bolt and tighten to 40–54 ft. lbs. (54–74 Nm). Now, tighten the tension rod-to-con-

trol arm bolt to 54–69 ft. lbs. (74–93 Nm).

NOTE: It is very important that the tension rod bolts be tightened in the order given in the text. If the tightening sequence is reversed, the tension rod will interfere with the bracket causing unusual noise.

8. Install the wheels. If components were replaced, have the alignment checked.

Loyale

1. Raise and support the vehicle safely. Remove the tire and wheel assembly.

2. As required, remove the parking brake cable from the lower control arm assembly.

3. Remove the bolt that retains the stabilizer assembly to the lower control arm.

4. Remove the front exhaust pipe, as necessary to gain working clearance.

5. Properly support the lower control arm assembly. Remove the ball joint from its mounting.

6. Remove the lower control arm to crossmember retaining bolt. Remove the lower control arm from the vehicle.

To install:

7. Install the lower control arm. Tighten the retaining bolt to 43–51 ft. lbs. (59–69 Nm) only after the vehicle is on the ground with the chassis loaded.

8. Install the ball joint and tighten the castle nut to 18–22 ft. lbs. (25–29 Nm). Install any exhaust system components previously removed.

9. Install the stabilizer assembly and tighten the bolts to 14–22 ft. lbs. (20–29 Nm). Install the parking brake cable bracket.

10. Install the wheels, lower the vehicle and check the alignment.

Legacy

1. Raise and safely support the front of the vehicle.

2. Remove the wheel and tire assembly.

3. Disconnect the stabilizer link and the transverse link.

4. Remove the ball joint-to-housing retaining bolt and remove the ball joint end from the housing.

5. Remove the nuts (not the bolts) that retain the transverse link to the crossmember.

6. Remove the 2 bolts securing the rear of the transverse link to the chassis.

7. Remove the bolts from the transverse link and lower it from the vehicle.

To install:

8. Install the transverse link into position and loosely install the 2 bolts that retain the link to the chassis.

--- CAUTION ---

All of the retaining nuts used are of the self-locking type and must be replaced when they are removed. Failure to replace the bolts may cause the retaining bolts to work loose and cause loss of vehicle control and personal injury

9. Install the bolts used to retain the transverse link to the crossmember and loosely install the nuts.

10. Install the ball joint into the housing. Tighten the ball joint pinch bolt to 29–43 ft. lbs. (39–59 Nm).

11. Connect the stabilizer link to the transverse link and loosely install the bolts.

12. The suspension bolts must be tightened in the following order:

    a. Transverse link-to-stabilizer: 14–22 ft. lbs. (20–29 Nm).

    b. Transverse link-to-crossmember: 61–83 ft. lbs. (83–113 Nm).

    c. Transverse link rear bushing-to-chassis: 145–217 ft. lbs. (196–294 Nm).

13. Move the transverse link back and forth until the clearance between the link and bushing is 1–1.5mm. The torque for the transverse link end bushing is 152–195 ft. lbs. (206–265 Nm).

14. Install the wheel and lower the vehicle.

SVX

1. Raise and safely support the front of the vehicle.

2. Remove the front wheel.

3. Separate the ball joint from the housing.

4. Remove the rear control arm support bolts.

5. Remove the left and right retaining bolts and lower the arm from the vehicle.

To install:

--- CAUTION ---

All of the retaining nuts used are of the self-locking type and must be replaced when they are removed. Failure to replace the bolts may cause the retaining bolts to work loose and cause loss of vehicle control and personal injury

6. Install the lower control arm assembly and the rear support bolts. Keep all of the bolts loose.

7. Install the ball joint to the housing and loosely install the retaining bolt.

NOTE: The suspension bolts must be tightened with the vehicle on the ground and the weight of the vehicle on the suspension.

8. Install the wheel and lower the vehicle to the ground.

9. Tighten the ball joint retaining bolt to 33–43 ft. lbs. (45–59 Nm). Tighten the rear support-to-sub frame bolts to 93–123 ft. lbs. (127–167 Nm). Tighten the lower arm rear retaining bolts to 56–73 ft. lbs. (76–100 Nm).

Stabilizer Bar

REMOVAL AND INSTALLATION

1. Raise and safely support the front of the vehicle.

2. Remove the wheels.

3. Mark the location and direction of the stabilizer mountings. This will ensure proper installation. Remove the right side ABS sensor clamp, if equipped.

4. Remove the stabilizer-to-crossmember bolts. On the SVX, remove the 2 stabilizer link bolts and remove the stabilizer link.

5. Remove the bolts that secure the stabilizer to the front transverse link.

6. Remove the jack-up plate from the crossmember and remove the stabilizer from the vehicle. On the SVX it will have to be removed from the right side.

To install:

7. Check all of the bushings for deformity or tears. Replace any bushing that shows signs of deterioration.

8. Install the bushings and housings on the stabilizer, aligning any marks before removal.

9. Install the stabilizer into position and install the retaining bolts. Tighten bolts to the following torque:
- Except SVX:
- Jack-up plate-to-crossmember — 17–31 ft. lbs. (23–42 Nm).
- Stabilizer link-to-transverse link — 18–25 ft. lbs. (25–34 Nm).
- Stabilizer-to-crossmember — 15–21 ft. lbs. (21–28 Nm).
- SVX:
- Stabilizer link-to-stabilizer lever — 23–31 ft. lbs. (32–42 Nm).
- Stabilizer bar-to-stabilizer lever — 33–43 ft. lbs. (45–59 Nm).
- Stabilizer bar-to-crossmember — 15–21 ft. lbs. (21–28 Nm).

10. Install the right side ABS sensor clamp, if removed.

11. Install the wheels and lower the vehicle.

REAR SUSPENSION

Shock Absorbers

REMOVAL AND INSTALLATION

1. Raise and support the vehicle safely. Remove the tire and wheel assembly.

2. Properly support the rear axle assembly. Loosen the upper shock absorber to chassis nuts.

3. Remove the washer and the bushing, being sure to note their correct assembly sequence for installation.

4. Remove the shock absorber to trailing arm retaining bolt. Remove the shock absorber from its mounting.

To install:

5. Install the shock absorber and tighten the bolts to specification. Be sure to properly install the washers.

6. Do not fully tighten the upper mounting nuts until the lower shock nut has been installed with the washer and the pin shoulder contracting each other.

7. Install the wheels and lower the vehicle.

MacPherson Strut

REMOVAL AND INSTALLATION

Justy

1. Raise and support the vehicle safely. Remove the tire and wheel assembly. Properly support the rear axle assembly.

2. From the upper portion of the strut mount, remove the trim cover.

3. Remove the strut to body retaining nut. Push the lower arm downward, and remove the coil spring.

4. Remove the strut to axle housing bolts. Remove the strut from the vehicle.

To install:

5. When installing coil spring, fit the lower rubber seat and coil spring end face in the coil spring seat mounting recess of the lower control arm.

6. Install the strut-to-housing bolt and tighten to 25–40 ft. lbs.

7. Install the wheels and remove the supports from under the rear axle. Lower the vehicle and test drive.

Except Justy

1. Raise and support the vehicle safely. Remove the tire and wheel assembly.

2. If equipped with air suspension, remove the cover and disconnect the air line.

3. Properly support the rear axle assembly. On Loyale, remove the upper strut retaining bracket mounting bolts. On Legacy and SVX models, remove the upper strut mounting nut.

4. Remove the lower strut retaining bolts.

5. Remove the strut assembly from the vehicle.

To install:

6. On Loyale, install the strut assembly and tighten the lower attaching bolts to 51–87 ft. lbs. (69–118 Nm), tighten the upper retaining bolts to 65–94 ft. lbs. (88–127 Nm). On Legacy, tighten the lower mounting bolts to 137–174 ft. lbs. (186–235 Nm) and the upper nut to 36–51 ft. lbs. (49–69 Nm). On SVX, tighten the lower mounting bolts to 98–127 ft. lbs. (132–172 Nm).

7. If equipped with air suspension, reconnect the air line. Install the wheel and tire assembly and remove the supports under the rear axle.

8. Lower the vehicle and test drive.

Springs

REMOVAL AND INSTALLATION

Justy

1. Raise and support the vehicle safely. Remove the tire and wheel assembly. Properly support the rear axle assembly.

2. Remove the strut bolt trim cover and remove the strut upper bolts.

NOTE: The rear spring is under extreme tension. Serious injury can result if the spring should fly out of the vehicle.

3. Place a floorjack under the control arm to prevent the spring from expanding. Remove the rear spindle to control arm bolt. Slowly lower the control arm until all spring pressure is released. Push the control arm downward, and remove the coil spring.

To install:

4. Place the spring in the holder cups. Ensure that the spring insulators are installed. Using a floor jack, lift up on the lower control arm to compress the spring. Install the control arm bolts and tighten to 54–69 ft. lbs. (74–93 Nm).

5. Install the rear strut and tighten the lower bolts to 72–87 ft. lbs. (98–118 Nm) and the upper nuts to 40–54 ft. lbs. (54–74 Nm).

6. Remove the rear axle supports and lower the vehicle.

Rear Control Arms

REMOVAL AND INSTALLATION

Justy

1. Raise and support the vehicle safely. Remove the tire and wheel assembly.

2. Properly support the rear axle assembly. Remove the coil spring assembly.

3. Remove the control arm to crossmember bolt. Separate the control arm from the crossmember.

4. Remove the control arm to axle housing bolt Separate the control arm from the axle housing.

5. Remove the assembly from the vehicle.

To install:

6. Install the control arm on the axle housing and tighten the bolts to 54–69 ft. lbs. (74–93 Nm).

7. Install the control arm to crossmember bolt and tighten to 43–58 ft. lbs. (59–78 Nm).

8. Install the coil spring. Install the wheels, remove the rear axle supports and lower the vehicle.

Legacy

TRAILING LINK

1. Loosen the rear wheel lugs, raise and safely support the vehicle and remove the wheel assemblies.

2. Remove the rear parking brake clamps and the ABS sensors, as required.

3. Remove the bolts retaining the trailing link to the body.

4. Remove the bolts retaining the trailing link to the rear housing.

5. Remove the trailing link from the vehicle.

6. To install the trailing link, place in position and install the bolts at each end.

7. Torque the bolts to 72–94 ft. lbs.

8. Complete the assembly.

LATERAL LINK

1. Remove the stabilizer from the lateral link.

2. Remove the parking brake cable and the ABS sensor clamp from the trailing link, as required.

3. Loosen the bolts that secure the trailing link to the bracket and remove the bolts that retain the trailing link to the rear housing.

4. If equipped with 4WD, remove the Double Offset Joint (DOJ) pin and axle shaft to provide working space.

5. Remove the front lateral link from the rear crossmember.

6. Temporarily install front lateral link to the rear crossmember and remove the rear lateral link from the crossmember.

7. To install the link, reverse the removal procedure. Torque the bolts to the following specifications:
   a. 4WD — 61–83 ft. lbs.
   b. FWD — 87–116 ft. lbs.

Loyale

1. Raise and support the vehicle safely. Remove the tire and wheel assembly.

2. Properly support the rear axle assembly.

3. Remove the strut to lower control arm bolt and separate the strut from the lower control arm.

4. If equipped with 4WD, use a 0.24 in. (6mm) pin punch and drive the spring pins from the halfshaft-to-axle shaft and the halfshaft to differential assembly. While pushing downward on the inner arm, separate the halfshaft from the axle shaft. Pull the halfshaft from the differential and position it aside.

5. Disconnect and plug the brake hose from the brake line at the lower control arm.

6. Remove the outer arm-to-lower control arm bolts, then separate the lower control arm from the outer arm. Properly support the inner arm.

7. Remove the inner arm-to-crossmember bolt. Remove the lower control arm from the vehicle.

To install:

8. Install the inner arm and tighten the inner arm to crossmember bolt to 51–65 ft. lbs. (69–88 Nm). Install the outer arm and

tighten the attaching bolts to 94–108 ft. lbs. (127–147 Nm).

9. Install the brake hose and line. If equipped with 4WD, reassemble the halfshaft to differential assembly using new spring pins.

10. Install the strut, remove the rear axle supports and lower the vehicle. As required, bleed the brake system.

SVX

TRAILING LINK

1. Loosen the rear wheel lugs, raise and safely support the vehicle and remove the wheel assemblies.

2. Remove the rear parking brake clamps and the ABS sensors, as required.

3. Remove the bolts retaining the trailing link to the body.

4. Remove the bolts retaining the trailing link to the rear housing.

5. Remove the trailing link from the vehicle.

To install:

6. To install the trailing link, place in position and install the bolts at each end.

7. Torque the rear link-to-housing through bolt to 80–101 ft. lbs. (137–177 Nm) and the front link through bolt to 101–130 ft. lbs. (137–177 Nm).

8. Complete the assembly by connecting the remaining components, installing the wheel and lowering the vehicle.

LATERAL LINK

1. Raise and safely support the vehicle.

2. Remove the wheel and tire assemblies. Remove the rear exhaust pipe.

3. Remove the stabilizer bar from the rear suspension.

4. Remove the parking brake cable and ABS sensor harness brackets.

5. Disconnect the parking brake cable clamp.

6. Disconnect the trailing link at the housing.

7. Disconnect the lateral links at the housing assembly.

8. Using a suitable halfshaft removing tool, carefully pry the halfshaft from the rear differential and support it with a wire from the body.

9. Place an alignment mark on the lateral link-to-crossmember bolt. This bolt must be installed in the same position or rear wheel alignment will be incorrect.

10. Remove the later link-to-rear crossmember bolts and remove the link.

To install:

11. Install the lateral link to the crossmember and loosely install the bolts.

NOTE: All of the fasteners must be tightened with the weight of the vehicle on the suspension. Be sure to align the marks made during removal on the rear lateral link bolts.

12. Connect the lateral link at the housing assembly. Loosely install the bolts.

13. Connect The trailing link at the housing and loosely install the bolts.

14. Reposition the ABS harness and the parking brake cable clips.

15. Install the rear exhaust pipe and the rear tire assemblies.

16. Lower the vehicle to the ground and tighten all fasteners to the following torques:

　　a. Front lateral link-to-sub frame through bolt (bolt without cap for the end) — 72–101 ft. lbs. (98–137 Nm).

　　b. Rear lateral link-to-sub frame through bolt (bolt with cap, closest to the differential) — 61–83 ft. lbs. (83–113 Nm).

　　c. Trailing link-to-housing bolt — 80–101 ft. lbs. (108–137 Nm).

　　d. Stabilizer link nut — 12–17 ft. lbs. (16–14 Nm).

Rear Wheel Bearings

ADJUSTMENT

2WD

1. Raise and support the vehicle safely. Remove the rear wheel assembly.

2. Temporarily tighten the axle nut to 36 ft. lbs. on all vehicles except Justy or to 29 ft. lbs. for Justy.

3. Turn the drum or disc back and forth several times to ensure that bearings are properly seated.

4. Turn the nut backwards 1/8–1/4 turn in order to obtain the correct starting point.

5. Using a spring gauge at 90 degrees to the wheel lug, check the rotating force. Specifications should be 1.9–3.2 lbs. for all vehicles except Justy or for Justy are 3.1–4.4 lbs.

6. After the adjustment is completed, bend the lock washer. After installing a new O-ring to the grease cap, install the cap.

REMOVAL AND INSTALLATION

1. Raise and support the vehicle safely. Remove the rear tire and wheel assembly.

2. If equipped with rear disc brakes, remove the caliper and properly support it.

3. Using a small prybar, remove the rear wheel grease cap.

4. Using a hammer and a punch, flatten the lock washer and loosen the axle nut. Remove the lock washer and the thrust plate. When removing the drum or disc, be careful not to drop the inner race from the outer bearing.

NOTE: If the brake drum on the Justy is difficult to remove, use wheel puller tool 9224930000 or equivalent, to remove the brake drum.

5. Using a gear puller, remove the spacer and the inner race of the inner bearing.

6. Using a brass drift and a hammer, drive the outer race of the inner bearing from the drum or disc.

7. Using a brass drift and a hammer, drive the outer race of the outer bearing from the drum or disc.

To install:

8. Clean and inspect the parts for damage, replace defective parts, if necessary.

9. Using bearing installation tool 925220000 or equivalent, for all vehicles except Justy or tool 922111000 or equivalent, for Justy, press the outer race of the inner bearing into the drum or disc until it seats against the shoulder.

10. When pressing the bearing, be sure not to exceed the load to the bearing, so as not to damage it.

11. Apply a small amount of grease to the oil seal lips, then install the oil seal until it is flush with the drum or disc.

12. Using bearing installation tool 921130000 or equivalent, for all vehicles except Justy or tool 922111000 or equivalent, for Justy, press the outer race of the outer bearing into the drum or disc until it seats against the shoulder.

13. Apply approximately 1/8 oz. of wheel bearing grease to the inner and the outer bearings. Fill the disc or drum hub with 1 oz. of wheel bearing grease.

14. Install a new spacer O-ring, the spacer and the inner race of the inner bearing onto the trailing arm spindle.

15. When installing the spacer, be sure to face the stepped surface toward the bearing. Use a new thrust plate and lock washer.

16. To complete the installation, reverse the removal procedure. Adjust the wheel bearing.

STEERING

— CAUTION —
Properly disarm the air bag on vehicles equipped with the SRS system. Failure to do so can cause serious injury.

Steering Wheel

REMOVAL AND INSTALLATION

Except legacy and SVX

1. Disconnect the negative battery cable.

2. Disconnect the horn lead from the wiring harness, located beneath the instrument panel.

NOTE: If equipped with telescopic steering wheel, remove the telescopic lever assembly.

3. Working behind the steering wheel, remove the steering wheel cover to steering wheel screws. It may be necessary to lower the column from the dash by removing the screws.

4. Lift the crash pad assembly from the front of the wheel.

5. Matchmark the steering wheel and the column for installation.

6. Remove the steering wheel retaining nut. Using a steering wheel puller tool, remove the steering wheel from the column.

To install:

7. Install the steering wheel on the column in the same position as removed. Tighten the center nut to 36–43 ft. lbs. (49–59 Nm) on Justy and 22–29 ft. lbs. (29–39 Nm) on Loyale.

NOTE: Do not hammer on the steering wheel or the steering column, as damage to the collapsible column could result.

8. Install the crash pad assembly and wheel cover.

9. If the column was lowered, tighten the steering column-to-dash screws.

10. Install the telescopic lever, if removed. Connect the horn lead and install the horn pad, if removed.

Legacy and SVX

> **CAUTION**
> *Properly disarm the air bag on vehicles equipped with the SRS system. Failure to do so can cause serious injury.*

1. Properly disarm the air bag system, on models equipped. Disconnect the negative battery cable.

> **CAUTION**
> *Wait at least 10 minutes after disarming the air bag to avoid accidental deployment.*

2. Disconnect the horn lead from the wiring harness, located beneath the instrument panel. On models without air bag, remove the horn pad by pulling it off.

NOTE: If equipped with telescopic steering wheel, remove the telescopic lever assembly.

3. On models with an air bag, working behind the steering wheel, remove the steering column covers. Use a No. 30 Torx® bit and remove the air bag module retaining bolts.

4. Disconnect the air bag module connector and remove the module from the steering wheel. Place the module face up on a flat surface.

5. Matchmark the steering wheel and the column for installation.

6. Remove the steering wheel retaining nut. Using a steering wheel puller tool, remove the steering wheel from the column.

To install:

7. Install the steering wheel on the column in the same position as removed. Tighten the center nut to 22–29 ft. lbs. (29–39 Nm).

NOTE: Do not hammer on the steering wheel or the steering column, as damage to the collapsible column could result.

8. Install the crash pad or air bag assembly and the column covers.

9. Install the telescopic lever, if removed. Connect all of the electrical leads. Re-arm the air bag.

Manual Steering Rack

REMOVAL AND INSTALLATION

Justy

1. Disconnect the negative battery cable. Raise and support the vehicle safely. Remove the front tire and wheel assemblies.

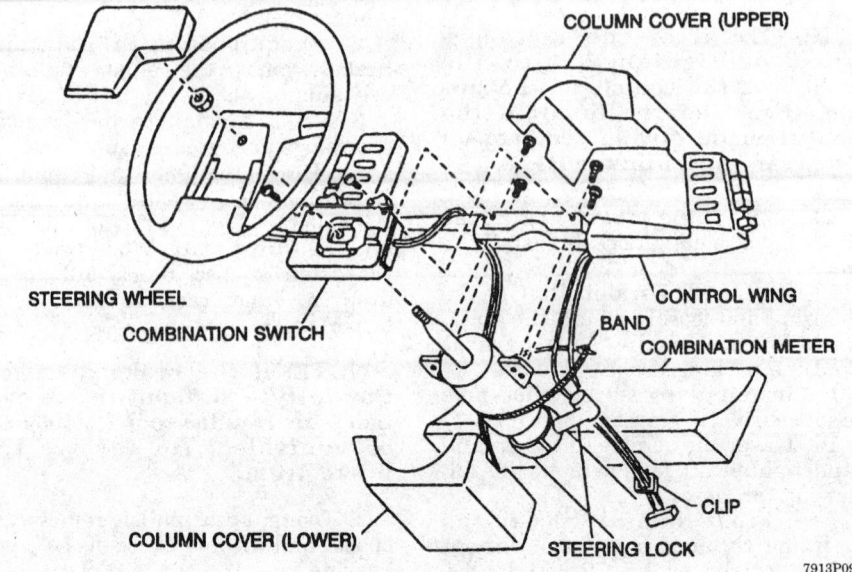

Typical steering wheel and related components

2. Disconnect the universal joint coupling bolts. Remove the dust seal.

3. Using the proper tools, disconnect the tie rod ends from the knuckle arms.

4. Remove the steering rack retaining bolts. Lower the assembly and pull the pinion from the dust seal toward the engine compartment.

5. Remove the steering rack from the vehicle.

To install:

6. Installation is the reverse of the removal procedure. Tighten the rack mounting bolts to 33–43 ft. lbs. (44–59 Nm).

7. Adjust the toe-in and the turning angles to specifications.

8. Tighten the tie rod end to 18–22 ft. lbs. (25–29 Nm).

Except Justy

1. Disconnect the negative battery cable.

2. Raise and support the vehicle safely. Remove the front tire and wheel assemblies.

3. Remove the tie rod end cotter pin and loosen the castle nut. Using a ball joint puller, separate the tie rod ends from the housing knuckle arm.

4. If necessary, disconnect the hand brake cable hanger from the tie rod.

5. Remove the pinch bolt from the torque rod universal joint. Disconnect the pinion with the gearbox from the steering column.

6. If equipped with an hot air pipe, disconnect it.

7. Disconnect the exhaust manifold to engine bolts, pull downward on the exhaust manifold.

8. Remove the boot from the steering rack.

9. Remove the steering rack to crossmember bolts, pull downward on the steering rack to disconnect the pinion flange. Turn the gearbox rearward and remove it toward the left side.

10. When removing the gearbox, be careful not to damage the gearbox boot. Inspect the removed parts for wear or damage and if necessary, replace the parts.

To install:

11. To install, reverse the removal procedures. Torque the steering gearbox to crossmember bolts to 35–52 ft. lbs. (48–66 Nm).

12. Torque the pinch bolt to universal joint to 15–20 ft. lbs.

13. Torque the exhaust manifold to engine bolts to 19–22 ft. lbs. (25–29 Nm).

14. Torque the rubber coupling to steering rack bolts to 7–14 ft. lbs. (10–20 Nm).

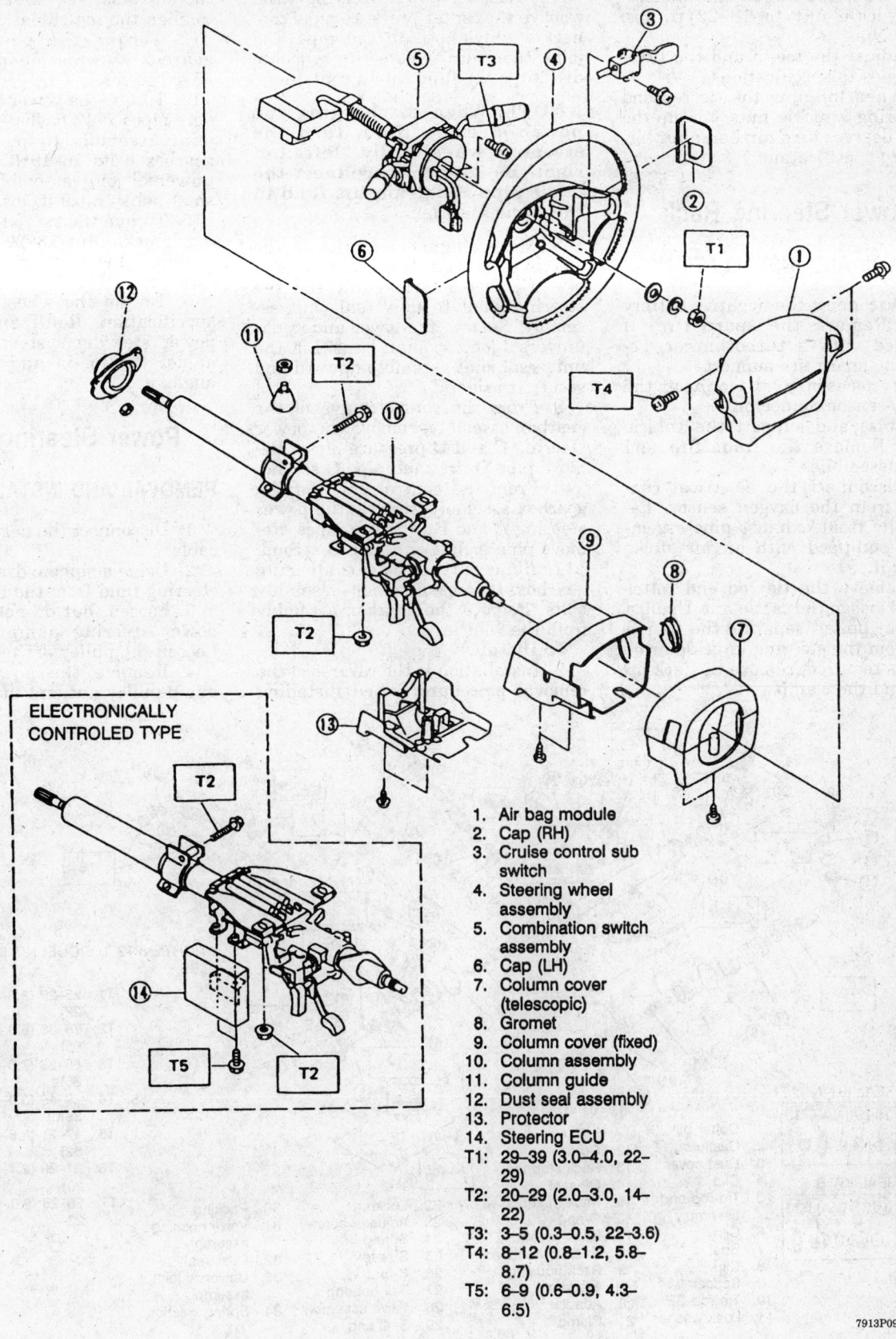

1. Air bag module
2. Cap (RH)
3. Cruise control sub switch
4. Steering wheel assembly
5. Combination switch assembly
6. Cap (LH)
7. Column cover (telescopic)
8. Gromet
9. Column cover (fixed)
10. Column assembly
11. Column guide
12. Dust seal assembly
13. Protector
14. Steering ECU
T1: 29–39 (3.0–4.0, 22–29)
T2: 20–29 (2.0–3.0, 14–22)
T3: 3–5 (0.3–0.5, 22–3.6)
T4: 8–12 (0.8–1.2, 5.8–8.7)
T5: 6–9 (0.6–0.9, 4.3–6.5)

ELECTRONICALLY CONTROLED TYPE

7913P096

Exploded view of the steering column — SVX — Legacy similar

15. Torque the tie rod end to steering knuckle nut to 18–22 ft. lbs. (26–29 Nm).

16. Adjust the toe-in and the turning angles to specifications.

17. When torquing the tie rod end to steering knuckle nuts, torque the nut 60 degrees turn further, after torquing to specification.

Power Steering Rack

REMOVAL AND INSTALLATION

1. Disconnect the negative battery cable. Remove the spare tire. If equipped with a turbocharger, remove the spare tire support.

2. If necessary, disconnect the thermo-sensor connector.

3. Raise and support the vehicle safely. Remove the front tire and wheel assemblies.

4. Disconnect the electrical connector from the oxygen sensor. Remove the front exhaust pipe assembly. If equipped with an air stove, remove it.

5. Remove the tie rod end cotter pin and loosen the castle nut. Using a ball joint puller, separate the tie rod ends from the steering knuckle arm.

6. As required, remove the jack up plate and the clamp.

7. From the power steering rack, remove the center pressure pipe, connect a vinyl hose to the pipe and joint, then turn the steering wheel to discharge the fluid into a container.

NOTE: When discharging the power steering fluid, turn the steering wheel fully, left and right. Be sure to disconnect the other pipe and drain the fluid in the same manner.

8. Make alignment marks on the steering shaft universal joint assembly to power steering unit and the steering shaft to universal joint assembly. Remove the lower and upper universal joint to shaft bolts. Lift the universal joint assembly upward and secure it aside.

9. From the control valve of the gearbox assembly, remove the power steering **C** and **D** pressure pipes. Remove pipe **D** first and pipe **C** second.

10. From the control valve of the gearbox assembly, remove the power steering **A** and **B** pressure pipes. Remove pipe **A** first and pipe **B** second.

11. Remove the power steering gearbox to crossmember assembly bolts. Remove the gearbox assembly from the vehicle.

To install:

12. Installation is the reverse of the removal procedure. When installing the universal joint assembly, be sure to align the matchmarks.

13. Torque the power steering gearbox to crossmember bolts to 35–52 ft. lbs.

14. Torque the power steering pressure pipes 7–12 ft. lbs., the universal joint assembly to power steering gearbox bolts 16–19 ft. lbs. and the universal joint assembly to steering shaft bolts 16–19 ft. lbs.

15. Torque the tie rod end to steering knuckle nut 18–22 ft. lbs. After torquing this nut, turn it 60 degrees further.

16. Torque the wheel lug nuts to specification. Refill and bleed the power steering system. Check and adjust the toe-in and the steering angle.

Power Steering Pump

REMOVAL AND INSTALLATION

1. Disconnect the negative battery cable.

2. Using a siphon, drain the power steering fluid from the reservoir.

3. Loosen, but do not remove the power steering pump pulley nut. Loosen the pulley drive belts.

4. Remove the power steering pump pulley nut and the pulley.

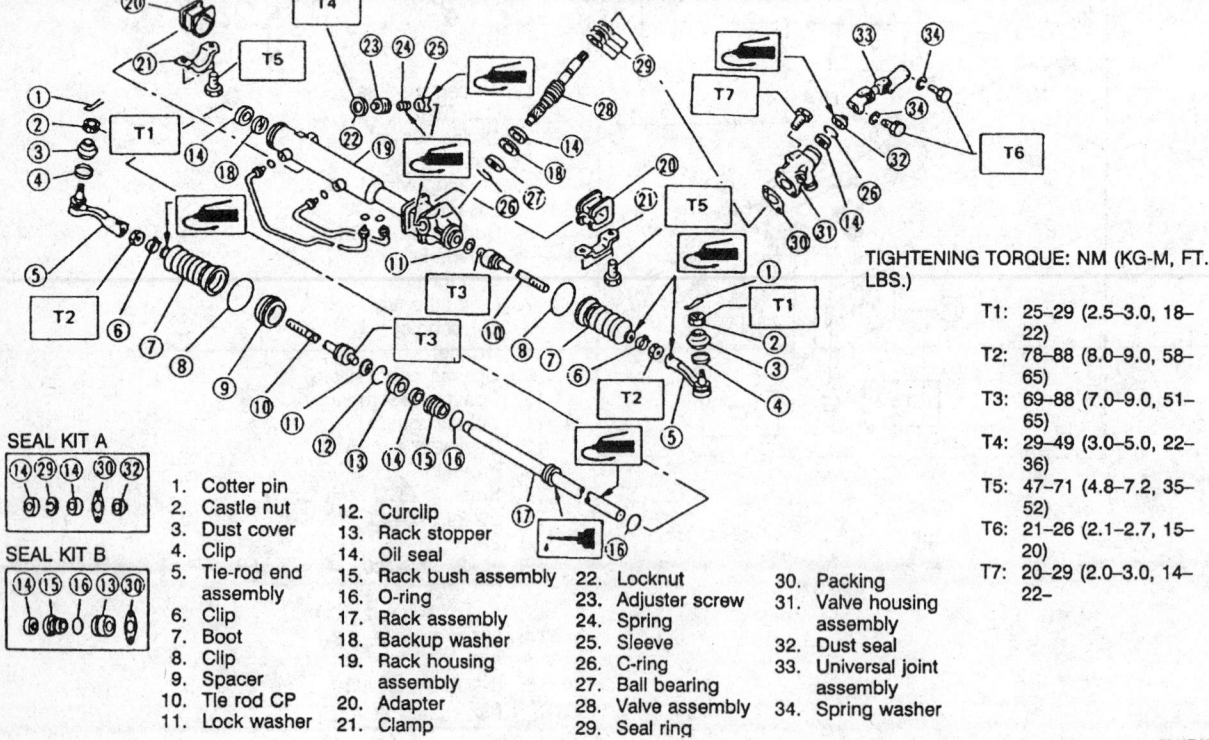

SEAL KIT A

SEAL KIT B

1. Cotter pin
2. Castle nut
3. Dust cover
4. Clip
5. Tie-rod end assembly
6. Clip
7. Boot
8. Clip
9. Spacer
10. Tie rod CP
11. Lock washer
12. Curclip
13. Rack stopper
14. Oil seal
15. Rack bush assembly
16. O-ring
17. Rack assembly
18. Backup washer
19. Rack housing assembly
20. Adapter
21. Clamp
22. Locknut
23. Adjuster screw
24. Spring
25. Sleeve
26. C-ring
27. Ball bearing
28. Valve assembly
29. Seal ring
30. Packing
31. Valve housing assembly
32. Dust seal
33. Universal joint assembly
34. Spring washer

TIGHTENING TORQUE: NM (KG-M, FT. LBS.)

T1: 25–29 (2.5–3.0, 18–22)
T2: 78–88 (8.0–9.0, 58–65)
T3: 69–88 (7.0–9.0, 51–65)
T4: 29–49 (3.0–5.0, 22–36)
T5: 47–71 (4.8–7.2, 35–52)
T6: 21–26 (2.1–2.7, 15–20)
T7: 20–29 (2.0–3.0, 14–22—

7913P097

Exploded view of the steering rack — Legacy, Loyale

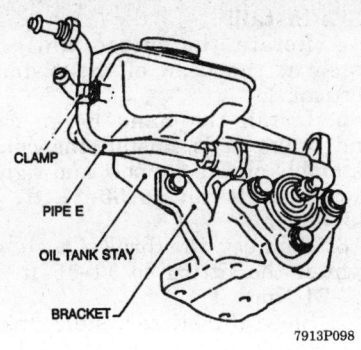

View of the power steering pump with reservoir attached

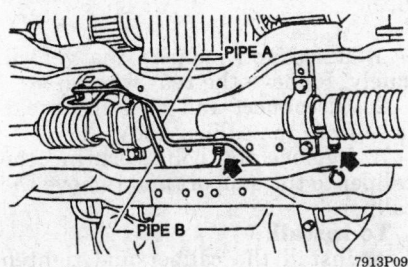

Remove the pressure lines to drain the power steering fluid

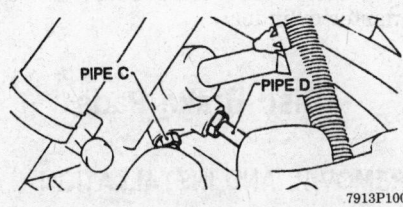

View of the power steering gear pressure lines

5. Disconnect and plug the **A** pressure hose from the **E** pipe. Disconnect the **B** pressure hose from the oil tank.

6. When disconnecting the **A** hose, use wrenches to prevent the **E** pipe from twisting.

7. Remove the **E** hose to reservoir clamp. Loosen the reservoir to bracket bolt, then remove the **A** and **B** bolts on the upper part of the reservoir, this will allow the fluid to run out.

NOTE: To minimize the fluid loss from the reservoir, remove both bolts while the reservoir is

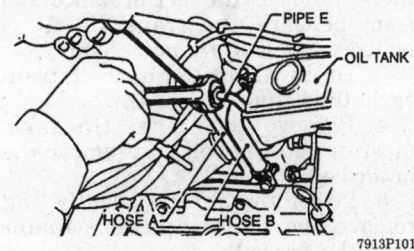

Disconnect the power steering pump hoses from the pressure lines

pressed against the oil pump, then quickly remove the reservoir. It is a good idea to remove the pump and the reservoir as a unit, then separate the reservoir from the pump on a bench.

8. Remove the power steering pump to bracket bolts. Remove the pump from the vehicle.

To install:

9. Installation is the reverse of the removal procedure; be sure to use new O-rings.

10. Torque the power steering pump to bracket bolts to 22–36 ft. lbs. (29–49 Nm).

11. Torque the reservoir stay to bracket bolts to 14–17 ft. lbs. (20–24 Nm).

12. Torque the reservoir to pump bolts to 14–22 ft. lbs. (20–29 Nm).

13. Torque the pulley nut to pump nut to 31–46 ft. lbs. (42–62 Nm).

14. Refill the power steering reservoir. Bleed the power steering system.

DRIVE BELT ADJUSTMENT

1. Using a pair of adjustable jawed pliers, with a piece of rag between the jaws, remove the idler cover cap by turning and pulling.

2. Turn the adjusting bolt until the correct belt tension is obtained. If removing the belt, loosen the adjusting bolt until the drive belt can be removed.

NOTE: The correct belt tension is obtained when the belt can be flexed 6–8 mm by applying finger pressure to the midpoint of the longest span.

3. After a new belt is installed and the correct tension obtained, replace the idler cap cover by pushing in and turning.

SYSTEM BLEEDING

1. Be sure the power steering reservoir is filled with fluid. Raise and support the vehicle safely.

2. With the engine running, turn the steering wheel back and forth, from lock to lock, until the air is removed from the fluid.

3. Lower the vehicle, recheck the reservoir fluid level and correct, as required.

Tie Rod Ends

REMOVAL AND INSTALLATION

1. Raise and support the vehicle safely.

2. Remove the front tire and wheel assemblies.

3. Remove the cotter pin and castle nut from the tie rod end stud.

4. Using a ball joint puller, separate the tie rod end from the steering knuckle.

To install:

5. Install the tie rod and tighten the castle nut to 18–22 ft. lbs. Install a new cotter pin.

6. Install the front tire and lower the vehicle.

BRAKES

Master Cylinder

REMOVAL AND INSTALLATION

1. Disconnect the negative battery cable. Disconnect and plug the brake lines at the master cylinder.

2. It is advised to thoroughly drain the fluid from the master cylinder before performing any removal procedures.

3. If equipped with fluid level indicator, disconnect the electrical harness connector from the master cylinder.

4. Remove the master cylinder to power brake booster retaining nuts. Remove the master cylinder from its mounting.

To install:

5. Bench bleed the master cylinder prior to installation.

6. Install the master cylinder on the power booster and tighten the nuts to 7–13 ft. lbs. (10–18 Nm).

7. Connect the fluid level indicator. Connect the brake lines and

tighten the flarenut to 9–13 ft. lbs. (13–18 Nm).

8. Bleed the brake system as required.

Proportioning Valve

The proportioning valve is attached to a bracket and is located directly under the master cylinder. It's purpose is to provide even braking pressure to all of the wheels.

REMOVAL AND INSTALLATION

1. Disconnect the negative battery cable. Disconnect and plug the brake tubes from the proportioning valve. If equipped with an electrical connector, disconnect it.

2. Remove the proportioning valve-to-bracket bolts. Remove the valve from the vehicle.

3. Installation is the reverse of the removal procedure. Tighten the flarenuts to 9–13 ft. lbs. (13–18 Nm); the proportioning valve attaching nuts to 15–21 ft. lbs. (20–29 Nm).

Power Brake Booster

REMOVAL AND INSTALLATION

Except SVX

1. Disconnect the negative battery cable. Disconnect the vacuum hose from the power brake booster. If equipped, disconnect the connector for the brake fluid level indicator.

2. Remove the master cylinder from the brake booster. Depending upon the vehicle, it may not be necessary to completely remove the master cylinder. It may be possible to remove the retaining bolts and position the assembly aside.

3. Remove the brake pedal pushrod to power booster spring pin and clevis pin, then disconnect the pushrod from the brake pedal.

4. From under the dash, remove the power booster to firewall bolts.

5. Remove the brake booster assembly from the vehicle.

6. Installation is the reverse of the removal procedure. Tighten the attaching nuts to 9–17 ft. lbs. (13–23 Nm). Bleed the brake system, as required.

SVX

1. Disconnect the negative battery cable. Properly discharge the air conditioning system.

2. Raise and safely support the vehicle. Remove the performance rod from beneath the transaxle. It is bolted to the sub frame assembly.

3. Drain about 1 quart of transaxle fluid from the transaxle.

4. Remove the upper transaxle dipstick housing bolt and remove the lower bolt.

5. Lower the vehicle slightly and remove the cruise control actuator from the firewall.

6. Disconnect the positive battery wire from the starter.

7. Disconnect and plug the low pressure air conditioning line.

8. Disconnect the vacuum hose from the brake booster.

9. Remove the master cylinder from the booster. Inside of the vehicle, remove the snap pin and the clevis from the actuator rod.

10. Inside of the vehicle, remove the 4 nuts that secure the booster. Remove the brake booster from the engine compartment.

To install:

11. Install the brake booster in position and install the mounting nuts. Tighten the mounting bolts to 7–13 ft. lbs. (13–18 Nm).

12. Reconnect the actuator rod at the pedal assembly. Install the master cylinder onto the booster.

13. Connect the vacuum hose at the booster. Connect the refrigerant low pressure line.

14. Connect the positive battery cable at the starter. Install the cruise control actuator.

15. Raise and safely support the vehicle. Install the transaxle dipstick housing. Install the performance rod.

16. Lower the vehicle. Properly recharge the air conditioning system.

17. Fill the transaxle to the proper level. Connect the negative battery cable.

Brake Caliper

REMOVAL AND INSTALLATION

Front

1. Raise and support the vehicle safely. Remove the front wheels.

2. Remove the brake hose from the caliper body and plug the hose to prevent the entrance of dirt or moisture.

3. Remove the hand brake cable and brake pads. Remove the caliper assembly by pulling it out of the support. Do not remove the guide pin unless it it damaged.

To install:

4. Rotate the piston until the notch at the head of the piston is vertical.

5. Install the hand brake cable and brake pads. Install the caliper assembly on the support and tighten the support bolt to 36–51 ft. lbs. (49–69 Nm).

6. Connect the brake hose and tighten the fitting to 11–15 ft. lbs. (15–21 Nm).

7. Bleed the brake system. Install the wheels and lower the vehicle. Check the fluid level in the master cylinder.

Rear

1. Raise and support the vehicle safely. Remove the rear wheels.

2. Disconnect and plug the brake hose from the caliper body.

3. Remove the bolts securing the caliper to the support and remove the caliper.

To install:

4. Install the caliper and tighten the attaching bolts to 34–43 ft. lbs. (46–58 Nm).

5. Connect the brake hose and tighten the fitting to 12–14 ft. lbs. (16–20 Nm).

6. Bleed the brake system. Install the rear wheels and lower the vehicle. Check the fluid level in the master cylinder.

Disc Brake Pads

REMOVAL AND INSTALLATION

Front

1. Raise and support the vehicle safely. Remove the wheel assemblies.

2. Release the parking brake and disconnect the cable from the caliper lever.

3. Remove the lock pin bolts from the lower front of the caliper.

4. Rotate the caliper on the support, swinging it upward and aside.

5. Remove the brake disc pads, noting the position of the shim pads and pad clips.

To install:

6. Inspect the brake rotor, calipers and retaining components. Correct as necessary.

7. Remove a small portion of brake fluid from the master cylinder reservoir. With an appropriate tool, turn the caliper piston clockwise into the cylinder bore and align the notches.

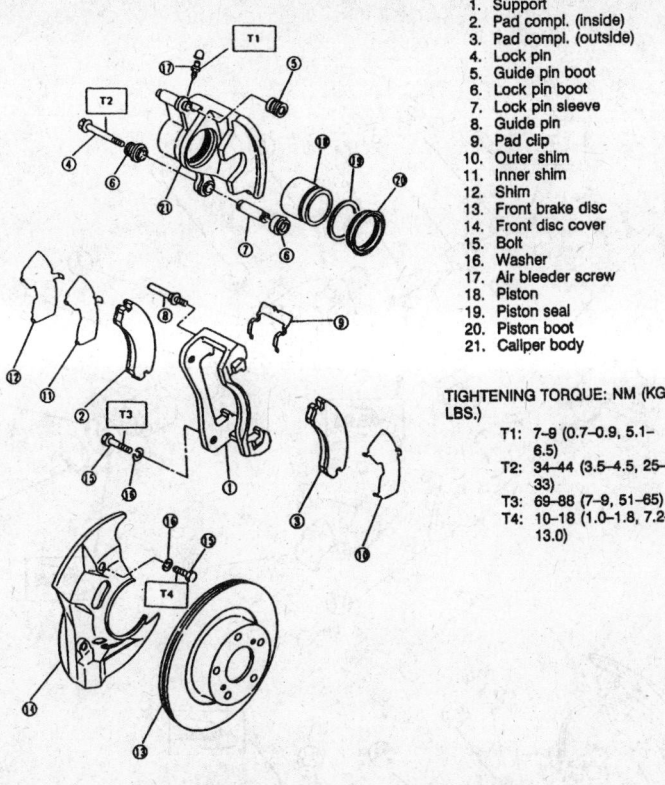

1. Support
2. Pad compl. (inside)
3. Pad compl. (outside)
4. Lock pin
5. Guide pin boot
6. Lock pin boot
7. Lock pin sleeve
8. Guide pin
9. Pad clip
10. Outer shim
11. Inner shim
12. Shim
13. Front brake disc
14. Front disc cover
15. Bolt
16. Washer
17. Air bleeder screw
18. Piston
19. Piston seal
20. Piston boot
21. Caliper body

TIGHTENING TORQUE: NM (KG-M, FT. LBS.)

T1: 7–9 (0.7–0.9, 5.1–6.5)
T2: 34–44 (3.5–4.5, 25–33)
T3: 69–88 (7–9, 51–65)
T4: 10–18 (1.0–1.8, 7.2–13.0)

7913P102

Exploded view of the front disc brake assembly

Be sure the boot is not twisted or pinched.

NOTE: Do not force the piston straight into the caliper bore. The piston is mounted on a threaded spindle which will bend under pressure.

8. Install the new pads into the calipers, being sure all shims and clips are in their original positions.
9. Swing the calipers down into position and install the lock pin bolts. Tighten the lock pin bolt to the following specifications:
    a. Justy — 16–23 ft. lbs. (22–31 Nm).
    b. Loyale — 33–40 ft. lbs. (44–54 Nm).
    c. Legacy — 25–33 ft. lbs. (34–44 Nm).
    d. SVX — 25–33 ft. lbs. (34–44 Nm).
10. Reconnect the parking brake cable and fill the master cylinder reservoir.
11. Install the wheel assembly. Bleed the brakes as required and lower the vehicle. Road test the vehicle.

Rear

1. Raise and safely support the vehicle. Remove the wheel assemblies.

2. Disconnect the brake pad lining wear indicator, if equipped. Remove any anti-rattle springs or clips, if equipped.
3. Pull the caliper away from the center of the vehicle to push piston into caliper bore. Remove the caliper guide pins and remove the caliper from the rotor. Hang the caliper from the body with a support wire.
4. Slide the disc pads from the caliper, noting any shims or shields behind the pad.

NOTE: If equipped with parking brake, use a suitable tool to rotate the piston back into the caliper bore. If not equipped with parking brake, the piston can be pushed straight back into the bore.

5. Push the piston into the caliper bore. To install the pads, position any shims or shields in place and reverse the removal procedure.
6. Tighten the lower caliper bolt to the following torque:
    a. Loyale — 16–23 ft. lbs. (22–31 Nm).
    b. Legacy — 12–17 ft. lbs. (16–24 Nm).
    c. SVX — 12–17 ft. lbs. (16–24 Nm).

Brake Rotor

REMOVAL AND INSTALLATION

1. Raise and support the vehicle safely. Remove the front wheels.
2. Remove the brake caliper assembly and suspend out of the way.
3. Remove the castle nut, conical spring and center piece. Remove the center piece by inserting a prybar into the center slit and lightly tapping.
4. Pull the hub and disc assembly from the half axle shaft by hand.
To install:
5. Install the hub and disc assembly on the axle shaft. Install the conical spring and castle nut and temporarily tighten. Ensure that the conical spring is installed in the correct direction.
6. Install the caliper assembly and tighten the bolts to specification.
7. Tighten the castle nut to 145 ft. lbs. (196 Nm). Install the wheels and lower the vehicle.

Brake Drums

REMOVAL AND INSTALLATION

1. Raise and support the vehicle safely. Remove the wheels.
2. Pry off the center cap using an appropriate tool. Remove the castle nut, conical spring and center piece. Remove the center piece by inserting a prybar into the center slit and lightly tapping.
3. Pull brake drum off by hand.
To install:
4. Install the brake drum on the axle. Install the center piece, conical spring and castle nut. Ensure that the conical spring is installed in the correct direction. The word **OUT** should be facing you. Tighten the castle nut to 108 ft. lbs. (147 Nm).
5. Install the center cap and wheels. Lower the vehicle.

Brake Shoes

REMOVAL AND INSTALLATION

1. Raise and safely support the vehicle. Remove the rear wheels.
2. Remove the brake drums.
3. Remove the adjusting wedge spring and the upper and lower return springs.
4. Remove the hold-down springs.
5. Lift the brake shoes from the backing plate and disconnect the parking brake, if equipped.

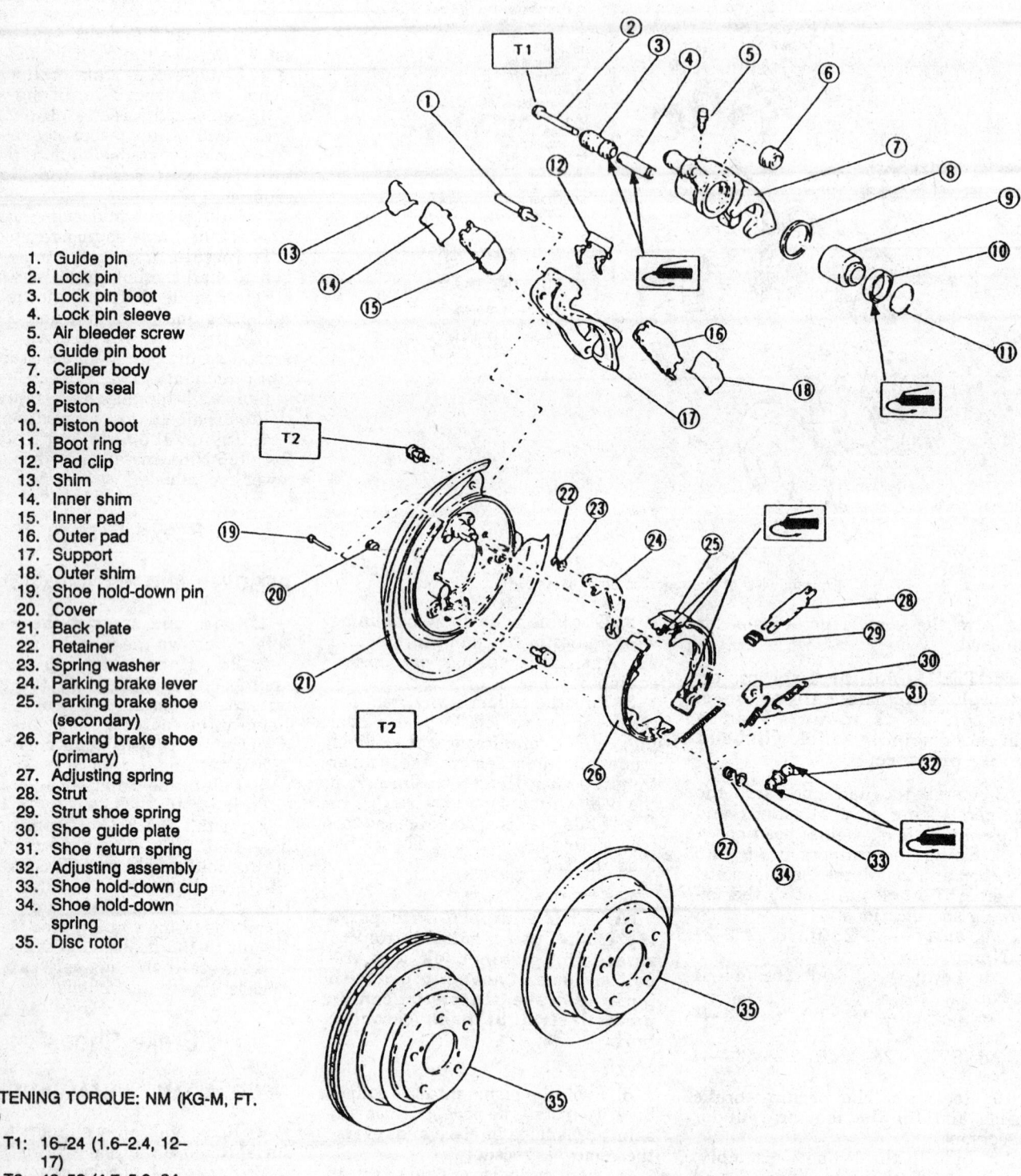

1. Guide pin
2. Lock pin
3. Lock pin boot
4. Lock pin sleeve
5. Air bleeder screw
6. Guide pin boot
7. Caliper body
8. Piston seal
9. Piston
10. Piston boot
11. Boot ring
12. Pad clip
13. Shim
14. Inner shim
15. Inner pad
16. Outer pad
17. Support
18. Outer shim
19. Shoe hold-down pin
20. Cover
21. Back plate
22. Retainer
23. Spring washer
24. Parking brake lever
25. Parking brake shoe (secondary)
26. Parking brake shoe (primary)
27. Adjusting spring
28. Strut
29. Strut shoe spring
30. Shoe guide plate
31. Shoe return spring
32. Adjusting assembly
33. Shoe hold-down cup
34. Shoe hold-down spring
35. Disc rotor

TIGHTENING TORQUE: NM (KG-M, FT. LBS.)

T1: 16–24 (1.6–2.4, 12–17)

T2: 46–58 (4.7–5.9, 34–43)

7913P103

Exploded view of the rear disc brake assembly

6. Disconnect the rear shoe from the push bar.

7. Clamp the push bar in a vise and remove the tension spring and adjusting wedge.

To install:

8. The installation is the reverse of the removal procedure.

9. Center the brake shoes on the backing plate, making sure the adjusting wedge is fully released before installing the drum.

10. Install the drums and the wheel assemblies.

11. Apply the brakes several times to automatically adjust the shoes. If necessary, bleed the brake system to obtain proper brake operation.

12. Road test the vehicle as required.

Wheel Cylinder

REMOVAL AND INSTALLATION

1. Raise and support the vehicle safely. Remove the wheel and tire assembly.

2. Remove the brake drum and the brake shoes from the backing plate.

3. Disconnect and plug the brake line at the back of the wheel cylinder.

4. Remove the wheel cylinder to backing plate bolts. Remove the wheel cylinder from the backing plate.

5. Installation is the reverse of the removal procedure. Tighten the wheel cylinder attaching nuts to 6–7 ft. lbs. (8–10 Nm). Be sure to bleed the brake system, as required.

Brake System Bleeding

There are some general rules for effectively bleeding the brakes. First, start with the brakes connected to the secondary (rear) chamber of the master cylinder. Second, the time interval between pumping the brake pedal should be approximately 3 seconds. Finally, the air bleeder on each brake should be opened only for 1–2 seconds.

STANDARD BRAKES

1. Fit one end of a vinyl tube into the air bleeder and put the other end into a brake fluid container.

2. Starting with the wheel that is farthest from the master cylinder, slowly depress the brake pedal and keep it depressed. Then, open the air bleeder to discharge air together with the fluid. Keep the bleeder open only 1–2 seconds.

3. With the bleeder closed, slowly release the brake pedal and repeat the procedure in 3–4 seconds.

4. When all air has been released from the system, check and fill the master cylinder with fluid.

ANTI-LOCK BRAKES

1. Bleed the brakes at all 4 wheels as described above.

2. Attach the vinyl hose to the bleeders on top of the hydraulic unit. Bleed this port in the same fashion as the wheels. Move the hose to the other bleeder and repeat. These ports bleed the primary circuit.

3. Remove the cone screw from the secondary bleeder port and install a bleeder screw. Install the clear vinyl hose on the bleeder.

4. Open the bleeder and depress the brake pedal. Keep the bleeder open and intermittently apply an electrical signal to the solenoid valve. To apply the signal:

a. Disconnect both battery terminals.

b. Disconnect the 2-pin and 12-pin connectors at the hydraulic assembly.

c. At the 12-pin connector, connect terminals 1 and 3 to battery ground and terminals 5 and 7 to the positive battery terminal. Take care not to short the terminals.

d. When the last connection is made, the electrical signal is transmitted to the solenoids. Do not send the signal for more than 5 seconds. Break the connections at the positive terminal after 2–3 seconds.

5. When the brake pedal moves to the end of its stroke, close the bleeder and allow the pedal to return slowly. If the electrical signal is not transmitted for any reason, the bleeder need not be closed before returning the pedal.

6. Repeat the above steps until the fluid tube contains no air.

7. With the electrical signal disconnected and the brake pedal released, remove the bleeder fitting and re-install the cone screw. Tighten to 6 ft. lbs. (8 Nm).

8. Repeat the procedure for the other secondary bleeder port. Both secondary ports must be bled.

9. Carefully remove the jumper wires and reconnect the connectors to the hydraulic unit. Connect the battery cables with the ignition **OFF**.

Anti-Lock System Brake Service

PRECAUTIONS

• Certain components within the ABS system are not intended to be serviced or repaired individually. Only those components with removal and Installation procedures should be serviced.

• Do not use rubber hoses or other parts not specifically specified for the ABS system. When using repair kits, replace all parts included in the kit. Partial or incorrect repair may lead to functional problems and require the replacement of components.

• Lubricate rubber parts with clean, fresh brake fluid to ease assembly. Do not use lubricated shop air to clean parts; damage to rubber components may result.

• Use only DOT 3 brake fluid from an unopened container.

• If any hydraulic component or line is removed or replaced, it may be necessary to bleed the entire system.

• A clean repair area is essential. Always clean the reservoir and cap thoroughly before removing the cap. The slightest amount of dirt in the fluid may plug an orifice and impair the system function. Perform repairs after components have been thoroughly cleaned; use only denatured alcohol to clean components. Do not allow ABS components to come into contact with any substance containing mineral oil; this includes used shop rags.

• The Anti-Lock control unit is a microprocessor similar to other computer units in the vehicle. Ensure that the ignition switch is **OFF** before removing or installing controller harnesses. Avoid static electricity discharge at or near the controller.

• If any arc welding is to be done on the vehicle, the Anti-Lock Control Unit (ALCU) connectors should be disconnected before welding operations begin.

Hydraulic Unit

REMOVAL AND INSTALLATION

1. Disconnect the negative battery cable. Disconnect the harness connectors at the hydraulic unit.

2. Remove the emission canister from the engine compartment.

3. Disconnect the inlet and outlet lines from the top of the actuator. Label the lines for installation. Immedi-

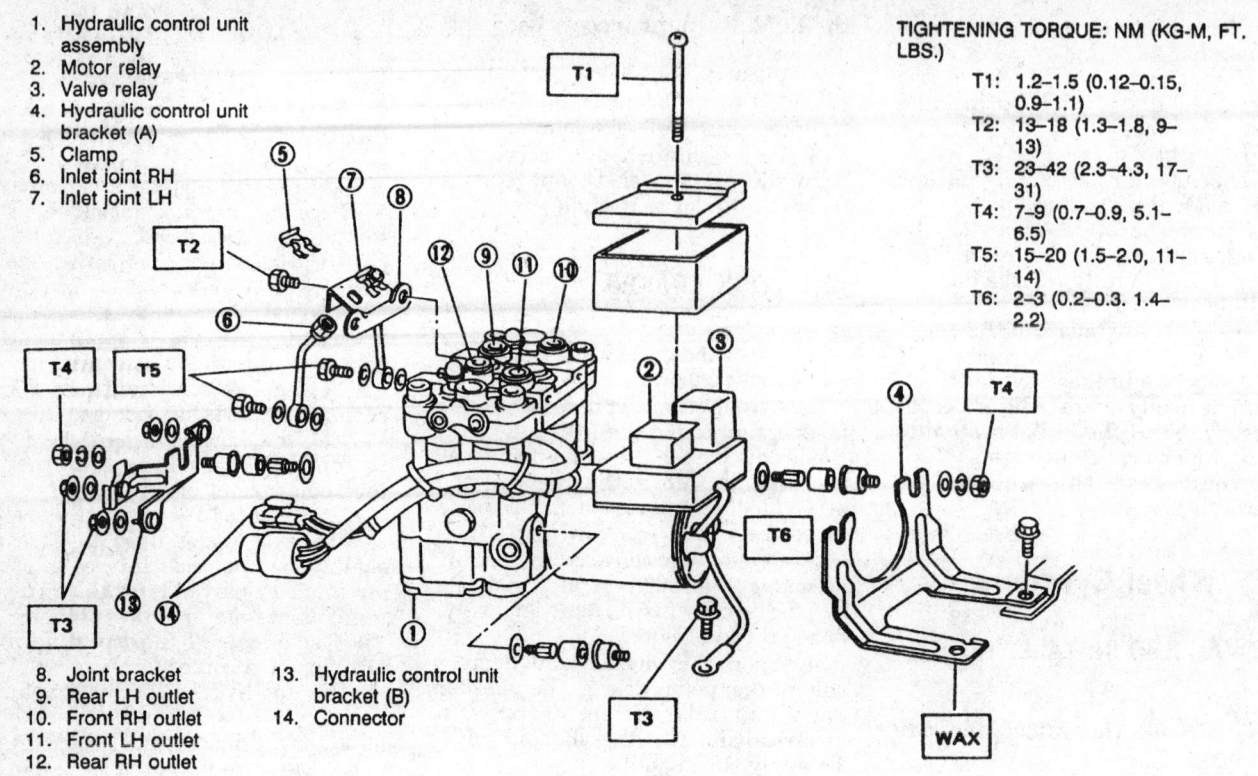

1. Hydraulic control unit assembly
2. Motor relay
3. Valve relay
4. Hydraulic control unit bracket (A)
5. Clamp
6. Inlet joint RH
7. Inlet joint LH

8. Joint bracket
9. Rear LH outlet
10. Front RH outlet
11. Front LH outlet
12. Rear RH outlet

13. Hydraulic control unit bracket (B)
14. Connector

TIGHTENING TORQUE: NM (KG-M, FT. LBS.)

T1: 1.2–1.5 (0.12–0.15, 0.9–1.1)
T2: 13–18 (1.3–1.8, 9–13)
T3: 23–42 (2.3–4.3, 17–31)
T4: 7–9 (0.7–0.9, 5.1–6.5)
T5: 15–20 (1.5–2.0, 11–14)
T6: 2–3 (0.2–0.3. 1.4–2.2)

7913P104

Removing the ABS hydraulic unit

ately plug the lines and ports to prevent the entry of dirt.

4. Remove the screw holding the ABS relay cover and remove the cover. Remove the bolts holding the hydraulic unit bracket to the body. Note that one of these bolts has the pump motor ground attached.

5. Lift the actuator and bracket clear of the vehicle. Keep the unit upright at all times. The brackets and relays may be removed for transfer to a replacement unit.

6. Except for the 2 relays, the hydraulic unit contains no replaceable components. Never attempt to disassemble the unit.

To install:

7. Install the relays and brackets. The nuts on the bushing bolts holding the hydraulic unit to the brackets should be tightened to 6 ft. lbs. (8 Nm).

8. Install the hydraulic unit and brackets and tighten the nuts to 25 ft. lbs. (34 Nm). Make sure the ground is attached.

9. Check that the relays are firmly seated and install the relay cover. Connect the brake lines and tighten to 11 ft. lbs. (15 Nm).

10. Install the canister in the engine compartment. Bleed all 4 wheels, then bleed the hydraulic ac-

tuator primary and secondary circuits.

Wheel Speed Sensor

REMOVAL AND INSTALLATION

Front

1. Disconnect the negative battery cable. Disconnect the speed sensor harness in the engine compartment.

2. Remove the bolts holding the sensor harness brackets. Take careful note of placement and location of the harness retainers.

3. Remove the sensor retaining bolt at the front hub. Remove the front wheel speed sensor by lifting it straight out of the housing. Do not damage the tip of the sensor. Clean or replace as necessary.

To install:

4. Place the sensor into the mount and install the retaining bolts. Tighten to 10 ft. lbs. (14 Nm).

5. Remove the caliper and brake disc. Use a non-ferrous feeler gauge to check the clearance between the top of the sensor and the tone wheel. Check the clearance at several locations on the hub. Clearance should be 0.039–0.059 in. (1.0–1.5mm).

6. If the air gap is too small, the sensor may be raised using ABS sensor shims. Remove the sensor, install the shim and recheck. If the clearance is too large, check the tone wheel, hub and sensor for damage.

7. Once the air gap is correct, reset the retaining bolt to the correct torque.

8. Working from the sensor end, install each harness clip and retainer exactly as removed. Connect the sensor to the ABS harness in the engine compartment.

Rear

1. Remove the rear seat. Disconnect the ABS harness connector.

2. Remove the rear sensor harness retaining bracket from the trailing link. Remove any other retainers and take careful note of the cable routing.

3. Remove the retaining bolt holding the sensor to the hub. Remove the rear wheel sensor by lifting straight out of the housing. Do not damage the tip of the sensor. Clean or replace as necessary.

To install:

4. Place the sensor into the mount and install the retaining bolts. Tighten to 10 ft. lbs. (14 Nm).

5. Remove the caliper and brake disc. Use a non-ferrous feeler gauge

to check the clearance between the top of the sensor and the tone wheel. Check the clearance at several locations on the hub. Clearance should be 0.031–0.051 in. (0.8.0–1.3mm).

6. If the air gap is too small, the sensor may be raised using ABS sensor shims. Remove the sensor, install the shim and recheck. If the clearance is too large, check the tone wheel, hub and sensor for damage.

7. Once the air gap is correct, reset the retaining bolt to the correct torque.

8. Working from the sensor end, install each harness clip and retainer exactly as removed. Connect the sensor to the ABS harness in the engine compartment.

CHASSIS ELECTRICAL

Air Bag

PRECAUTIONS

—— CAUTION ——

To avoid deployment when servicing the SIR system or components in the immediate area, do not use electrical test equipment such as battery or A.C. powered voltmeter, ohmmeter, etc. or any type of tester other than specified on the air bag system. Do not use a non-powered probe tester. To avoid personal injury all precautions must be strictly adhered to.

• Do not disassemble any air bag system components.
• Always carry an inflator module with the trim cover pointed away.
• Always place an inflator module on the workbench with the pad side facing upward, away from loose objects.
• After deployment, the air bag surface may contain sodium hydroxide dust. Always wear gloves and safety glasses when handling the assembly. Wash hands with mild soap and water afterwards.
• When servicing any SRS parts, discard the old bolts and replace with new ones.
• Always use a fine needle test lead for testing, to prevent damaging the connector terminals.
• Never disconnect any electrical connection with the ignition switch

ON unless instructed to do so in a test.
• Before disconnecting the negative battery cable, make a record of the contents memorized by each memory system like the clock, audio, etc., when service or repairs are completed make certain to reset these memory systems.
• Always wear a grounded wrist static strap when servicing any control module or component labeled with a Electrostatic Discharge (ESD) sensitive device symbol.
• Avoid touching module connector pins.
• Leave new components and modules in the shipping package until ready to install them.
• Always touch a vehicle ground after sliding across a vehicle seat or walking across vinyl or carpeted floors to avoid static charge damage.
• All sensors are specifically calibrated to a particular series.The sensors, mounting brackets and wiring harness must never be modified from original design.
• Never strike or jar a sensor, or deployment could happen.
• The inflator module must be deployed before it is scrapped.
• Any visible damage to sensors requires component replacement.

DISARMING

NOTE: Be sure to properly disconnect and connect the air bag connector.

1. Turn the ignition switch to the **OFF** position.
2. Disconnect the negative battery cable.

—— CAUTION ——

Wait at least 10 minutes after disconnecting the battery cable before doing any further work. The SRS system is designed to retain enough voltage to deploy the air bag for a short time, even after the battery has been disconnected. Serious injury may result from unintended air bag deployment, if work is done on the SRS system immediately after the battery cable is disconnected.

3. Remove the lower lid from the steering column and disconnect the connector between the air bag module and spiral cable.

ENABLING THE SYSTEM

1. Reconnect the connector between the air bag module and spiral

cable. Then install the lower steering column lid.
2. Reconnect the negative battery cable.
3. Turn the ignition switch to the **ON** position and observe the SRS warning light. The SRS warning light should illuminate for approximately 7 seconds, turn OFF and remain OFF for at least 45 seconds.
4. If the SRS warning light function as indicated in Step 3, the SRS system is functioning properly.

Blower Motor

REMOVAL AND INSTALLATION

Justy

1. Disconnect the negative battery cable.
2. Remove the coupler that connects the instrument panel harness to the blower motor.
3. Remove the coupler that connects the resistor to the instrument panel harness.
4. Detach the blower assembly. Remove the screws retaining the blower motor to the blower assembly.
5. Remove the motor assembly. Remove the nut retaining the fan to the motor assembly.
To install:
6. Install the motor assembly and tighten the nuts to specification.
7. Install the couplers that connect the resistor to the instrument panel and the instrument panel to the blower motor.
8. Connect the negative battery cable.

Legacy and SVX

NOTE: Depending upon working clearance the air conditioning system may have to be discharged in order to service the blower motor. If this is the case, be sure to observe all the required safety precautions when discharging and recharging the air conditioning system.

1. Disconnect the negative battery cable.
2. Remove the lower instrument panel cover on the passenger side of the vehicle.
3. Remove the glove box assembly, as required for working clearance.
4. Remove the heater duct, if not equipped with air conditioning.
5. If equipped with air conditioning, separate the evaporator from the blower assembly.

6. Disconnect the blower motor harness and the resistor electrical harness connector.

7. Remove the blower motor retaining bolts. Remove the blower motor assembly from its mounting.

To install:

8. Install the blower motor retaining bolts and tighten to 4–7 ft. lbs. Connect the blower motor harness and the resistor electrical harness connector.

9. Install the evaporator to the blower assembly as required. Install the heater duct as required. Install the glove box.

10. Install the lower instrument panel cover and connect the negative battery cable.

Loyale

NOTE: Depending upon working clearance the air conditioning system may have to be discharged in order to service the blower motor. If this is the case, be sure to observe all the required safety precautions when discharging and recharging the air conditioning system.

1. Disconnect the negative battery cable.

2. Remove the lower instrument panel cover on the passenger side of the vehicle. Remove the glove box assembly, as required for working clearance.

3. If equipped with a vacuum actuator, set the control lever to the **CIRC** position and disconnect the vacuum hose from the assembly. Remove the actuator from its mounting.

4. Remove the heater duct, if not equipped with air conditioning.

5. If equipped with air conditioning, separate the evaporator from the blower assembly.

6. Disconnect the blower motor harness and the resistor electrical harness connector.

7. Remove the blower motor retaining bolts. Remove the blower motor assembly from its mounting. As required, separate the fan from the blower motor.

To install:

8. Install the blower motor and tighten the retaining bolts to specification. Connect the blower motor harness and the resistor electrical harness connector.

9. Install the evaporator to the blower assembly as required. Install the heater duct as required. Install the glove box.

10. Install the vacuum actuator and connect the vacuum line. Install the

lower instrument panel cover and connect the negative battery cable.

Windshield Wiper Motor

REMOVAL AND INSTALLATION

Justy

1. Disconnect the negative battery cable.

2. At the wiper motor, disconnect the electrical connector.

3. Remove the wiper motor to cowl bolts.

4. Separate the wiper link from the motor.

5. If necessary, replace the wiper motor.

To install:

6. Install the wiper motor and tighten the cowl bolts. Install the wiper link on the motor. Tighten the bolts to 65 inch lbs. (7.4 Nm).

7. Connect the electrical connector and the negative battery cable. Check for proper operation.

Except Justy

1. Disconnect the negative battery cable.

2. Remove the wiper blades from the wiper arms by pulling the retaining lever up and sliding the blade away from the arm.

3. Slide the covering boot up the wiper arm.

4. Remove the wiper arms to linkage nuts and the arms.

5. Disconnect the electrical wiring connectors from the wiper motor.

6. Remove the cowl to body screws and the cowl from the vehicle.

7. Find or fabricate a ring which has the same diameter as the outer diameter of the plastic joint that retains the linkage to the wiper motor. Force the ring down over the joint to force the 4 plastic retaining jaws inward, then disconnect and remove the linkage.

8. Remove the wiper motor to firewall bolts and the motor.

To install:

9. Install the wiper motor and tighten the attaching bolts. Tighten the bolts to 65 inch lbs. (7.4 Nm).

10. Install the wiper linkage. Install the cowl. Connect the wiper electrical wiring harness.

11. Connect the negative battery cable and install the wiper arms after the ignition switch has been on for a few seconds to put the linkage in the parked position.

Rear Window Wiper Motor

REMOVAL AND INSTALLATION

1. At the rear window, pull the wiper blade outward from the arm and press down on the clip, then remove the blade from the arm.

2. Remove the wiper arm cover.

3. Loosen the wiper arm-to-wiper assembly nut, then remove the nut and the arm from the assembly.

4. Remove the wiper assembly-to-rear gate cap, nut and cushion.

5. From inside of the rear gate, remove the wiper motor assembly trim panel.

6. Disconnect the electrical connector from the wiper motor assembly.

7. Remove the wiper motor assembly-to-rear gate bolts and the motor assembly from the rear gate.

8. If necessary, replace the wiper motor.

9. To install, reverse the above procedures. Tighten the wiper motor bolts to 65 inch lbs. (7.4 Nm). With the rear wiper motor switch in the OFF position, install the wiper arm blade so it is positioned 25mm above the rear glass molding.

Instrument Cluster

REMOVAL AND INSTALLATION

Justy

1. Disconnect the negative battery cable. Remove the steering wheel.

2. Remove the defroster duct assembly.

3. Disconnect the heater control cable from the inside/outside air selector rod at the heater unit.

4. Disconnect the speedometer cable. Disconnect the electrical harness connectors.

5. Remove the covers for the instrument cluster retaining bolts.

6. Remove the instrument cluster retaining bolts. Remove the instrument cluster from its mounting.

To install:

7. Install the instrument cluster and tighten the retaining bolts securely. Install the screw covers.

8. Connect the speedometer cable and electrical harness.

9. Install the heater control cable and the defroster duct assembly.

10. Install the steering wheel and connect the negative battery cable.

FIRING ORDERS

NOTE: To avoid confusion, always replace spark plug wires one at a time.

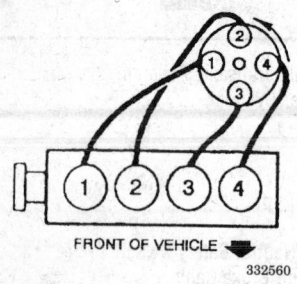

1.3L Engine
Firing order: 1-3-4-2
Distributor rotation:
counterclockwise

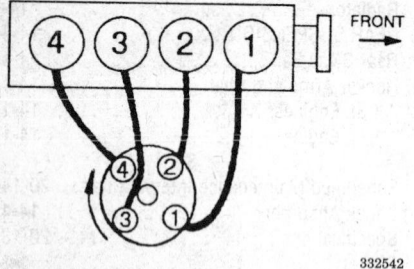

1.6L Engine
Engine Firing Order: 1-3-4-2
Distributor Rotation: Clockwise

ENGINE ELECTRICAL

NOTE: Disconnecting the negative battery cable on some vehicles may interfere with the functions of the on board computer systems and may require the computer to undergo a relearning process, once the negative battery cable is reconnected.

Distributor

REMOVAL AND INSTALLATION

1. Disconnect the negative battery cable.
2. Disconnect the distributor electrical coupler.
3. Remove the distributor cap.

4. Mark the position of the distributor rotor in relation to the distributor body, and the distributor body in relation to the cylinder head.
5. Remove the hold-down bolts and the distributor from the cylinder head. Do not rotate the engine after the distributor has been removed.

To install:

NOTE: Before installing the distributor, check to make sure the O-ring is in good condition. If using a new O-ring, apply a light coat of engine oil to it.

6. Install the distributor into the cylinder head. The tabs of the distributor coupling are offset. Therefore, if the tabs can not be fitted into the slots, turn the distributor shaft 180 degrees and try again.
7. Align the reference marks made during removal.
8. Connect the electrical coupler to the distributor.
9. Install the distributor cap.
10. Connect the negative battery cable.
11. Check and adjust the ignition timing as necessary.

Ignition Timing

ADJUSTMENT

1. Start the engine and allow it to reach normal operating temperature. Prior to any adjustment, be sure all electrical accessories including the A/C are **OFF**.
2. After the engine is reaches normal operating temperature, check and make sure the idle speed is:
   a. a minimum of 800 rpm for automatic transmission equipped models.
   b. a minimum of 750 rpm for manual transmission equipped models.
3. Make sure manual transmission equipped models are in **NEUTRAL** and automatic transmission equipped modes are in **PARK**. Also make sure the parking brake is fully applied.
4. Remove the cap from monitor coupler and connect terminals **D** and **E** with a jumper wire.
5. Connect a timing light according to manufactures instruction, using No. 1 cylinder spark plug wire as an ignition pickup.

NOTE: When terminals D and E are connected, observe if ignition timing is varying. If ignition timing is varying, this indicates ungrounded D terminal which

prevents accurate inspection and adjustment. Make sure to ground the D terminal securely.

6. On 1.6L models, open the air cleaner upper case cover and position the upper case and hose out of the way to observe the timing marks on the crankshaft pulley.
7. With the engine running, direct the timing light to the crankshaft pulley. If the timing mark on the timing tab are aligned with the timing notch on the crankshaft pulley, the ignition is properly timed.
8. Initial ignition timing should be 5 degrees BTDC at 800 rpm for automatic models, 750 rpm for manual models.
9. To adjustment the timing, loosen the distributor flange bolt and turn the distributor housing to advance or retard the timing.
10. After adjusting, tighten the flange bolt and recheck the timing. Torque the flange bolt to 9–13 ft lbs. (12–18 Nm).
11. After checking or adjusting, disconnect the service wire from monitor coupler.
12. With the engine idling and test terminals ungrounded, check that the ignition timing is 12° BTDC. A constant variation of timing is an indication that the computer controlled timing is working correctly.

Alternator

PRECAUTIONS

Several precautions must be observed to avoid damage to the alternator.
- If the battery is removed for any reason, make sure it is reconnected with the correct polarity. Reversing the battery connections may result in damage to the 1-way rectifiers.
- When utilizing a booster battery as a starting aid, always connect the positive to positive terminals and the negative terminal from the booster battery to a good engine ground on the vehicle being started.
- Never use a fast charger as a booster to start vehicles.
- Disconnect the battery cables when charging the battery with a fast charger.
- Never attempt to polarize the alternator.
- Do not use test lights of more than 12 volts when checking diode continuity.
- Do not short across or ground any of the alternator terminals.

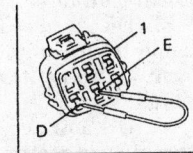

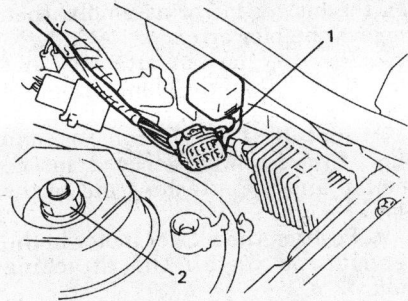

1. Monitor coupler
2. Front strut of left side
D: Ground terminal
E: Test switch terminal

330006

Data link connector terminal location(E and D) — 1.3L engine

• The polarity of the battery, alternator and regulator must be matched and considered before making any electrical connections within the system.
• Never separate the alternator on an open circuit. Make sure all connections within the circuit are clean and tight.
• Disconnect the battery ground terminal when performing any service on electrical components.
• Disconnect the battery if arc welding is to be done on the vehicle.

REMOVAL AND INSTALLATION

1.3L (G13) Engine

1. Disconnect the negative battery cable.
2. Remove the air cleaner case and the air flow meter.
3. Disconnect the terminal wire and electrical coupler from the alternator.
4. Remove the drive belt adjusting bolt and loosen the adjuster arm bolt.
5. Remove the drive belt.
6. Remove the alternator upper cover bolt and the two lower cover bolts, then remove the cover.
7. Remove the alternator lower mounting bolts and nut.

8. Lower the alternator to remove it from the vehicle.
To install:
9. Position the alternator on the lower bracket and loosely install the lower mounting bolts and nut.
10. Install the alternator cover and attaching bolts. Torque the bolts to 3–5 ft. lbs. (4–7 Nm).
11. Install the adjuster bolt finger-tight.
12. Install the drive belt and adjust the tension.
13. Tighten the alternator lower mounting bolts to 14–20 ft. lbs. (18–28 Nm).
14. Install the remaining components.
15. Connect the negative battery cable.
16. Start the vehicle and check the charging system.

1.6L (G16) Engine

1. Disconnect the negative battery cable.
2. Disconnect the wire connector from the alternator by removing the nut.
3. Loosen and remove the alternator drive belt.
4. Remove the alternator cover.
5. Remove the alternator mounting bolt and drive belt adjusting bolt.
6. Remove the alternator from the vehicle.
To install:
7. Install the alternator and loosely install the mounting bolt and drive belt adjusting bolt.
8. Install the drive belt and adjust the tension. Tighten the alternator mounting bolts to 17 ft. lbs. (23 Nm).
9. Install the remaining components.
10. Reconnect the negative battery cable.

Drive Belt

REMOVAL AND INSTALLATION

1.3L (G13) Engine

Power Steering and Air Conditioning Belt

1. Disconnect the negative battery cable.
2. Remove the air cleaner assembly with the Mass Air Flow (MAF) sensor.
3. Safely raise and support the vehicle.
4. Remove the right front wheel and splash shield.
5. Loosen the tensioner pulley nut.
6. Loosen the tension adjusting bolt (if equipped with A/C, turn the

bolt clockwise to loosen the tension; without A/C, turn the bolt counter-clockwise to loosen the tension).
7. Remove the belt from the vehicle.
To install:
8. Install the belt on the power steering pump pulley, then the air conditioning compressor pulley, then the crankshaft.
9. Tighten the adjusting bolt (if equipped with A/C, turn the bolt counter-clockwise; without A/C, turn the bolt clockwise). Adjust the belt until about 22 lbs. (10 kg) of force deflects the belt 0.32–0.39 in. (8–10mm) with the deflection measured between the crankshaft and the power steering pump (or A/C pulley if not equipped with power steering).
10. Tighten the tensioner pulley nut.
11. Install the remaining components.
12. Connect the negative battery cable.

Water Pump and Alternator Belt

1. Disconnect the negative battery cable.
2. Remove the power steering/air conditioning belt, if equipped.
3. Loosen the alternator adjuster bolt (upper bolt).
4. Loosen the lower alternator bolt and move the alternator toward the engine.
5. Remove the drive belt.
To install:
6. Install the drive belt to the alternator, water pump and crankshaft pulleys.
7. Using a suitable pry tool, move the alternator toward the bulkhead and tighten the adjusting bolt.
8. Check that the belt deflection is 0.24–0.31 in. (6–8 mm) for a used belt or 0.20–0.27 in. (5–7 Nm) for a new belt, with 22 lbs. (10kg.) of force applied between the water pump pulley and the crankshaft pulley.
9. Tighten the alternator lower bolt.
10. Install the power steering/air conditioning belt, if equipped.
11. Connect the negative battery cable.

1.6L (G16) Engine

A/C and/or Power Steering Drive Belt

1. Disconnect the negative battery cable.
2. Raise and safely support the front of the vehicle.
3. Remove the engine undercover from the right side.

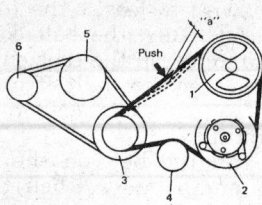

a. Belt deflection
1. P/S pump pulley
2. A/C compressor pulley (if equipped)
3. Crankshaft pulley
4. Tension pulley
5. Water pump pulley
6. Generator

200266

Drive belt routing — 1.3L engine

Vehicle with A/C

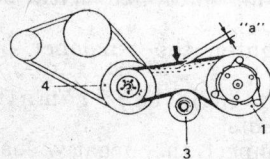

Vehicle with A/C and power steering

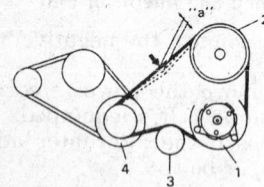

Vehicle with power steering

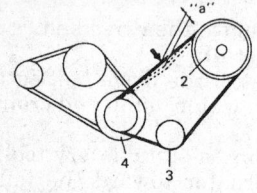

1. A/C compressor pulley
2. Power steering pump pulley
3. Tension pulley
4. Crankshaft pulley

331808

Drive belt routing — 1.6L engine

4. Loosen the tension nut to the idler pulley.
5. Loosen the tension bolt for the idler pulley.
6. Remove the drive belt from the vehicle.
To install:
7. Install the belt on the pulleys, making sure the belt is in the proper grooves and not crossed.
8. Tighten the tension bolt to apply tension to the drive belt. Deflection should be 0.32–0.39 inch (8–10 mm) at 22 lbs. (10 kg) of force.

9. Once the belt is properly adjusted, torque the tension nut to 33 ft. lbs. (45 Nm).
10. Connect the negative battery.

Alternator Drive Belt

1. Disconnect the negative battery cable.
2. Remove the A/C and/or power steering belt from the engine.
3. Loosen the alternator pivot bolt.
4. Loosen the alternator adjusting bolt.
5. Push the alternator towards the vehicle and remove the drive belt.
To install:
6. Install the alternator drive belt on the pulleys.
7. Pull the alternator back and apply tension to the drive belt. Tension should be between 0.20–0.27 inch (5–7 mm) at 22 lbs. (10 kg) of force.
8. Once belt tension is applied, torque the adjusting and pivot bolts to 17 ft. lbs. (23 Nm).
9. Install the A/C and/or power steering belt to the engine.
10. Connect the negative battery cable.

Starter

REMOVAL AND INSTALLATION

1. Disconnect the negative battery cable.
2. Disconnect the switch lead and battery cable from the starter.
3. Remove the two mounting bolts.
4. Remove the starter.
To install:
5. Position the starter on the flywheel and install the mounting bolts. Torque the mounting bolts to 14–20 ft. lbs. (18–28 Nm).
6. Attach the battery cable and the switch lead to the starter.
7. Connect the negative battery cable.

CHASSIS ELECTRICAL

Blower Motor

REMOVAL AND INSTALLATION

1993–94 Models

1. Disconnect the negative battery cable.

2. Disconnect the blower motor electrical couplers and resistor electrical couplers.
3. Disconnect the fresh air control cable from the blower motor unit.
4. Remove the glove box upper panel.
5. Disconnect the air hose from the blower motor.
6. Remove the three bolts mounting the blower motor assembly, then remove the blower motor.
7. Remove the fan attaching nut then the fan from the blower motor, if necessary.
To install:
8. If the fan was removed, install the fan and the attaching nut to the blower motor.
9. Position the blower motor in the vehicle and install the attaching bolts.
10. Attach the air tube to the blower motor.
11. Install the remaining components.
12. Connect the negative battery cable.
13. Test the blower motor operation at all speeds.

1995–97 Models

— WARNING —
When working around the Air Bag system, always disable the Air Bag. Failure to follow the procedures could result in Air Bag deployment or personal injury.

1. Disconnect the negative battery cable.
2. Disable the driver's Air Bag system.
3. Disconnect the passenger's Air Bag connector and remove the Air Bag harness wire from the support member.
4. Remove the glove box assembly.
5. Remove the powertrain control module mounting bolts and remove the module from the vehicle.

— WARNING —
The powertrain control module consists of precision parts. Be careful not to expose it to excessive shock.

6. Release the Air Bag and powertrain control module harness coupler clamp from the PCM bracket.
7. Remove the glove box
8. Disconnect the blower motor electrical couplers.
9. Disconnect the blower motor resistor and electrical couplers.
10. Disconnect the fresh air control cable from the blower motor unit.

11. Remove the three bolts mounting the blower motor assembly, then remove the blower motor unit.

12. Disconnect the air hose from the blower motor.

13. Remove the three blower motor mounting bolts, then remove the blower motor.

14. Remove the fan attaching nut then the fan from the blower motor, if necessary.

To install:

15. If the fan was removed, install the fan and the attaching nut to the blower motor.

16. Place the blower motor in the blower motor assembly and install the mounting bolts.

17. Position the blower motor assembly in the vehicle and install the attaching bolts.

18. Install the remaining components.

19. Enable the Air Bag system using the outlined procedure in the steering portion of this section.

20. Connect the negative battery cable.

21. Test the blower motor operation at all speeds.

Windshield Wiper Motor

REMOVAL AND INSTALLATION

1. Disconnect the negative battery cable.

2. Disconnect the electrical connector from the wiper motor.

3. Remove the three wiper motor mounting bolts and pull the wiper motor away from the bulkhead.

4. Remove the wiper motor linkage retaining nut and disconnect the wiper linkage from the wiper motor.

5. Remove the wiper motor.

To install:

6. Connect the linkage to the wiper motor and install the wiper linkage retaining nut and tighten to 18 ft. lbs. (25 Nm).

7. Install the wiper motor on the cowl and install the wiper motor mounting bolts and tighten to 15 ft. lbs. (20 Nm).

8. Connect the electrical connector to the wiper motor.

9. Connect the negative battery cable.

Combination Switch

REMOVAL AND INSTALLATION

1993–94 Models

1. Disconnect the negative battery cable.

2. Remove the steering wheel.

3. Remove the steering column upper and lower covers.

4. Disconnect the lead wire from the combination switch at the coupler.

5. Loosen the wire bands.

6. Remove the combination switch attaching screws then gently pull the switch off of the steering column.

To install:

7. Feed the wire down the steering column and position the combination switch on the steering column.

8. Install the combination switch attaching screws.

9. Connect the lead wire at the coupler.

10. Install the remaining components.

11. Connect the negative battery cable.

12. Test the operation of the combination switch.

1995–97 Models

———— **CAUTION** ————
The Air Bag system must be disarmed before removing the steering wheel. Failure to do so may cause accidental deployment, property damage or personal injury.

1. Disconnect the negative battery cable.

2. Disable the Air Bag system.

3. Remove the driver side Air Bag and the steering wheel.

4. Remove the steering column covers.

NOTE: If the steering column upper cover cannot be pulled out easily, loosen the steering shaft joint upper side bolt first and then the steering column mounting bolts.

5. Loosen the bands for the contact coil and combination switch wire harness, and unfasten the connectors from the junction/fuse block.

6. Remove the contact coil and combination switch assembly mount-

ing screws then remove the assembly from the steering column.

To install:

7. Check the contact coil and combination switch wire harness for any signs of scorching, melting, or other damage. If any damage is found replace the switch assembly.

8. Install the contact coil and combination switch assembly to the steering column.

NOTE: The new coil and switch assembly is supplied with the contact coil set with a lock pin and seal. Remove this lock pin after installing the coil and switch assembly to the steering column.

9. Check that the front tires are straight ahead and the contact coil is centered. If the contact coil is not centered, perform the following:

a. Check that the ignition is in the **LOCK** position.

b. Turn the contact coil counterclockwise, slowly with a light force until the contact coil will not turn any further.

NOTE: The contact coil can turn about 5 turns maximum. If it is at the center position, the contact coil can turn about two and a half turns clockwise or counterclockwise. Excessive turning will break the coil.

c. From the position where the contact coil stopped, turn it back clockwise about two and a half rotations and align the center mark with the alignment mark.

———— **WARNING** ————
If the front tires are not set straight ahead and the contact coil is not centered, the steering wheel will break the contact coil when turned.

10. Connect the contact coil and combination switch electrical connectors, (but NOT the yellow connector) and tighten the wire harness with bands and clamp.

11. Install the remaining components.

12. Enable the Air Bag system using the steps outlined in the steering portion of this section.

13. Turn the ignition **ON** and verify that the **AIR BAG** warning lamp flashes seven times, then turns off. If the system does not operate as described, diagnosis and repairs to the Air Bag system are necessary.

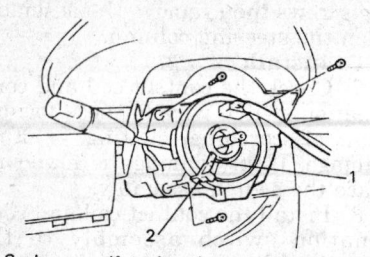

1. Seal — If equipped, remove lock pin after
2. Lock pin — installing contact coil and combination switch assembly.

325525

Combination switch — 1995–97 models

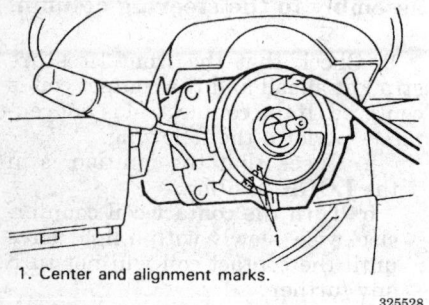

1. Center and alignment marks.

325528

Contact coil aligned — 1995–97 models

Ignition Lock Cylinder and Switch

REMOVAL AND INSTALLATION

——— WARNING ———
**When performing service around
the Air Bag system, disable the
system. Failure to follow the procedures could result in deployment or personal injury.**

1. Disconnect the negative battery
cable.
2. If equipped, disable the Air Bag
system.
3. Remove the steering wheel.
4. Remove the steering column
from the vehicle.
5. Use a sharp center punch, and
turn the mounting bolts out of the
steering column lock mechanism. Be
careful not to damage the aluminum
part of the steering column.
6. Turn the ignition key to the
ACC or **ON** position and remove the
steering lock assembly.
To install:
7. Position the oblong hole in the
steering shaft to the center of the
hole in the column.

8. Turn the ignition key to the
ACC or **ON** position and install the
steering lock assembly to the steering
column.
9. Turn the key to the **LOCK** position and remove the key.
10. Align the ignition switch hub
with the oblong hole in the steering
shaft and rotate the shaft to ensure
that the column shaft is locked.
11. Tighten the two new bolts until
the head of each bolt breaks off.
12. Insert the key into the ignition
and turn the key to the **ACC** or **ON**
position and ensure that the steering
shaft rotates smoothly. Remove the
key and check the operation of the
steering column lock.
13. Install the remaining
components.
14. Connect the negative battery
cable, and enable the Air Bag system.
15. Turn the ignition **ON** and verify
that the **AIR BAG** warning lamp
flashes seven times then turns off. If
the system does not operate as described diagnosis and repairs to the
Air Bag system are necessary.

Park/Neutral Safety Switch

REMOVAL AND INSTALLATION

Swift

1. Disconnect the negative battery
cable.
2. Remove the electrical connector
at the engine wiring harness.
3. Remove the harness from the
retaining clamps.
4. Remove the switch from the
transaxle.
To install:
5. Place the transaxle in the **N**
position.
6. Using a screwdriver, turn the
shift switch assembly joint clockwise
or counterclockwise until a distinct
click noise is heard.
7. Install the switch and tighten
the bolt to 17 ft. lbs. (23 Nm).
8. Install the harness retaining
clips and connect the electrical
connector.
9. Connect the negative battery
cable.
10. To check the park/neutral
switch for proper operation, apply the
parking brake and block the wheels.
Ensure the starter motor operates
only when the transaxle shift lever is
in the **P** or **N** positions and does not
operate when in the **D**, **2**, **L** or **R**
positions.

Esteem

1. Disconnect the negative battery
cable.
2. Remove the shift lever by removing the nut.
3. Pry off the lock washer.
4. Remove the locknut holding the
park/neutral switch to the transaxle
manual valve shaft.
5. Remove the lock washer and
rubber plate from the manual valve
shaft.
6. Remove the two bolts holding
the park/neutral switch to the transaxle and remove the park/neutral
switch.
To install:
7. Install the park/neutral switch
to the transaxle with the two bolts.
Do not torque the bolts at this time.
8. Place a new rubber plate and
new lock washer on the manual valve
shaft.
9. Install the locknut and torque
the nut to 5 ft. lbs. (7 Nm).
10. Adjust the transaxle
park/neutral switch.
11. Torque the two park/neutral
switch bolts to 15 ft. lbs. (20 Nm).
12. Install the shift lever with the
washer and torque the nut to 15 ft.
lbs. (20 Nm).
13. Adjust the shift lever.

Powertrain Control Module

REMOVAL AND INSTALLATION

1993–94 Models

The powertrain control module is located on the driver's side of the vehicle, next to the steering wheel, on the
left.
1. Disconnect the negative battery
cable.
2. Remove the trim panel below
the steering column, to access the
powertrain control module.
3. Remove the PCM mounting
bolts, and lower the PCM.
4. Tag then disconnect the electrical couplers from the PCM.
To install:
5. Connect the electrical couplers
to the PCM.
6. Position the PCM and install
the mounting bolts.
7. Install the trim panel below the
steering wheel.
8. Connect the negative battery
cable.

1995–97 Models

The powertrain control module is located on the passenger side of the vehicle, behind the glove box.

———— **WARNING** ————
When performing service around the Air Bag system, disable the Air Bag. Failure to follow the procedures could result in deployment or personal injury.

1. Disconnect the negative battery cable.
2. Disable the Air Bag system.
3. Remove the glove box.
4. Remove the PCM mounting bolts, and remove the PCM.
5. Release the Air Bag harness coupler clamp from the PCM bracket.
6. Disconnect the electrical couplers from the PCM while releasing the coupler lock.

To install:
7. Connect the electrical couplers to the PCM.
8. Position the PCM and install the mounting bolts.
9. Install the glove box.
10. Enable the Air Bag system.
11. Connect the negative battery cable.
12. Turn the ignition **ON** and verify that the **AIR BAG** warning lamp flashes seven times then turns off. If the system does not operate as described diagnosis and repairs to the Air Bag system are necessary.

ENGINE COOLING

Radiator

REMOVAL AND INSTALLATION

1. Disconnect the negative battery cable.
2. Drain the cooling system by loosening the drain plug to the radiator.
3. Disconnect the electrical connector to the cooling fan.
4. Remove the front grill by removing the clips.
5. Disconnect the cooling fan motor electrical connector.
6. Remove the upper, lower, and reservoir tank hoses from the radiator.
7. If equipped with an automatic transaxle, disconnect and plug the oil cooler lines.

8. Remove the radiator mounting bolts and lift the radiator from the vehicle with the cooling fan attached.
9. Remove the cooling fan and shroud.

To install:
10. Install the cooling fan and shroud on the radiator.
11. Place the radiator in the vehicle and install the mounting bolts. Tighten the bolts to 89 in. lbs. (10 Nm).
12. If equipped with an automatic transaxle, connect the oil cooler lines to the radiator.
13. Install the remaining components.
14. Fill the cooling system and connect the negative battery cable.
15. Start the engine, allow it to reach normal operating temperature and check the cooling system for leaks. Top off the cooling system as necessary.

Water Pump

REMOVAL AND INSTALLATION

1.3L (G13) Engines

1. Disconnect the negative battery cable.
2. Drain the cooling system into a suitable container and tighten the drain plug.
3. Remove the air cleaner assembly and the MAF sensor and outlet hose.
4. Remove the air cleaner bracket.
5. Raise and safely support the vehicle.
6. Remove the right side fender apron clips by pushing the center pin.

NOTE: Do not push the center pin too far in, or it will fall off into the fender.

7. If equipped, remove the power steering and air conditioning belt.
8. Loosen the water pump pulley bolts.
9. Remove the alternator drive belt.
10. Remove the water pump pulley.
11. To remove the crankshaft pulley perform the following:
   a. If equipped with a manual transaxle, insert a suitable flat bladed tool into the hole in the bell housing next to the exhaust pipe. This will lock the crankshaft in place.
   b. If equipped with a automatic transaxle, hold a suitable flat bladed tool in line with the oil pan and insert into the teeth of the

drive plate. This will lock the crankshaft in place.
   c. Loosen the crankshaft pulley bolts.
   d. Remove the crankshaft timing belt pulley bolt with special tool 09919–16020 or a 17 mm socket.
   e. Remove the pulley from the crankshaft.
   f. Install the crankshaft bolt.
   g. Remove the flat bladed tool that was used to lock the crankshaft in place.

NOTE: On 1995–97 models, to remove the crankshaft pulley with the engine assembly mounted on the body, it is necessary to remove the crankshaft timing belt pulley bolt. If the engine assembly is dismounted, the bolt does not need to be removed.

12. On 1995–97 models, remove the resonator and the timing belt outside cover.
13. Loosen the right engine mounting bolt.
14. Remove the timing belt.

———— **CAUTION** ————
After the timing belt is removed never turn the camshafts or the crankshaft. Interference may occur between the pistons and the valves causing component damage.

15. Remove the timing belt inside cover.
16. Remove the water pump belt adjusting arm.
17. Carefully remove the rubber seal between the water and oil pumps, and remove the seal between the water pump and the cylinder head.
18. Remove the water pump bolts and remove the water pump.

To install:
19. Clean the water pump mounting surface of old gasket material.
20. Install a new water pump gasket to the cylinder block.
21. Install the water pump to the cylinder block and torque the bolts to 7–9 ft. lbs. (10–13 Nm).
22. Install the rubber seal between the water pump and the oil pump. Install the seal between the water pump and the cylinder head.
23. Install the water pump belt adjusting arm.
24. Install the timing belt inside cover.
25. On 1995–97 models, with the crankshaft locked in position, remove the crankshaft bolt and install the crankshaft pulley. Torque the crankshaft pulley bolts to 10–13 ft. lbs.

(14–18 Nm). Using special tool 09919–16020 or a 17mm socket, torque the crankshaft timing belt pulley bolt to 76–83 ft. lbs. (105–115 Nm).

26. Install the timing belt.

27. Install the water pump pulley and drive belt. Torque the water pump pulley bolts to 7–8 ft. lbs. (9–12 Nm).

28. Install the remaining components.

29. Fill the cooling system.

30. Connect the negative battery cable.

31. Start the engine and top off the coolant as necessary.

32. Check the cooling system for leaks.

33. Check the ignition timing.

1.6L (G16) Engine

1. Disconnect the negative battery cable.

2. Drain the cooling system into a resealable container and tighten the drain plug.

3. Remove the timing belt.

4. Remove the alternator adjusting shim.

5. Remove the oil dipstick guide and dipstick.

6. Remove the water pump bolts, gasket, the water pump and rubber seal.

To install:

7. Clean the water pump mounting surface of old gasket material.

8. Install a new water pump gasket to the cylinder block.

9. Install the water pump to the cylinder block and torque the bolts to 7–9 ft. lbs. (10–13 Nm).

10. Install the rubber seal between the water pump and the oil pump. Install the seal between the water pump and the cylinder head.

11. Install the timing belt.

12. Install the alternator adjusting arm.

13. Using a new O-ring, install the oil dipstick guide and dipstick.

14. Lower the vehicle.

15. Fill the cooling system with engine coolant.

16. Connect the negative battery cable.

17. Start the engine and top off the coolant as necessary.

18. Check the ignition timing.

Thermostat

REMOVAL AND INSTALLATION

--- **CAUTION** ---

Used antifreeze is considered a hazardous material and should be recycled or disposed of properly. Check with local authorities for the proper procedures in handling used antifreeze.

1. Disconnect the negative battery cable.

2. Drain the cooling system into a suitable container and tighten the drain plug.

3. Disconnect the radiator hose from the thermostat housing.

4. Remove the bolts attaching the thermostat housing to the intake manifold, then remove the thermostat housing.

5. Remove the thermostat.

To install:

6. Clean and inspect the surfaces of the housing and the engine.

7. Install the new thermostat with the spring facing toward the engine. The air bleed valve should be at the top of the engine.

8. Install a new gasket and the thermostat housing. Install the housing bolts and tighten to 10 ft. lbs. (13 Nm).

9. Connect the radiator hose to the thermostat housing.

10. Fill the cooling system.

11. Connect the negative battery cable.

12. Run the engine and check for leaks.

Electric Cooling Fan

REMOVAL AND INSTALLATION

1. Disconnect the negative battery cable.

2. Place a drain pan under the radiator and drain the cooling system.

3. Loosen the hose clamp and unfasten the upper hose from the radiator.

4. Unfasten the electrical connector from the cooling fan motor.

5. Raise and safely support the vehicle.

6. Remove the lower fan shroud-to-radiator frame bolt(s).

7. Lower the vehicle.

8. Remove the upper fan shroud-to-radiator mounting bolts and lift out the fan/shroud assembly.

9. Remove the fan blade-to-motor nut, fan blade and washer.

10. Remove the fan-to-shroud bolts and the fan motor from the shroud.

To install:

11. Install the fan motor in the shroud.

12. Install the fan blade on the motor.

13. Install the cooling fan/shroud assembly to the radiator in the vehicle.

14. Install the upper fan shroud-to-radiator mounting bolts and tighten the bolts to 89 inch lbs. (10 Nm).

15. Raise and safely support the vehicle.

16. Install the lower cooling fan/shroud assembly mounting bolt(s). Torque the mounting bolt(s) to 89 inch lbs. (10 Nm).

17. Install the remaining components.

18. Reconnect the negative battery cable and fill the cooling system.

19. Test the cooling fan for proper operation by running the engine.

20. Turn **OFF** the engine and allow to cool down. Check the coolant level and top off, as necessary.

Cooling System Bleeding

PROCEDURE

1. With the engine **OFF**, fill the radiator to the bottom of the filler neck with a 50/50 mix of antifreeze and water.

2. Start the engine.

3. Allow the engine to reach normal operating temperature at idle.

4. When the thermostat opens (the upper radiator hose gets hot), turn the engine **OFF**. Top off the system with coolant.

5. Install the radiator cap.

6. Fill the coolant recovery reservoir to the full hot mark.

NOTE: When installing the reservoir tank cap, align the arrow marks on the tank and cap.

7. Allow the engine to cool.

8. With the engine cool, check the level of coolant in the recovery bottle and top off as necessary.

FUEL SYSTEM

Fuel System Service Precautions

Safety is the most important factor when performing not only fuel system maintenance but any type of maintenance. Failure to conduct maintenance and repairs in a safe manner may result in serious personal injury or death. Maintenance and testing of the vehicle's fuel system components can be accomplished safely and effectively by adhering to the following rules and guidelines.

• To avoid the possibility of fire and personal injury, always disconnect the negative battery cable unless the repair or test procedure requires that battery voltage be applied.

• Always relieve the fuel system pressure prior to disconnecting any fuel system component (injector, fuel rail, pressure regulator, etc.), fitting or fuel line connection. Exercise extreme caution whenever relieving fuel system pressure to avoid exposing skin, face and eyes to fuel spray. Please be advised that fuel under pressure may penetrate the skin or any part of the body that it contacts.

• Always place a shop towel or cloth around the fitting or connection prior to loosening to absorb any excess fuel due to spillage. Ensure that all fuel spillage (should it occur) is quickly removed from the engine surfaces. Ensure that all fuel soaked cloths or towels are deposited into a suitable waste container.

• Always keep a dry chemical (Class B) fire extinguisher near the work area.

• Do not allow fuel spray or fuel vapors to come into contact with a spark or open flame.

• Always use a backup wrench when loosening and tightening fuel line connection fittings. This will prevent unnecessary stress and torsion to fuel line piping. Always follow the proper torque specifications.

• Always replace worn fuel fitting O-rings with new. Do not substitute fuel hose or equivalent, where fuel pipe is installed.

Idle Speed

ADJUSTMENT

1.3L (G13) Engine

1993–94 Models

1. Place the manual transmission in **NEUTRAL**, or the automatic transmission in **PARK**. Set the parking brake and block the drive wheels.
2. Connect a tachometer to the engine according to manufactures instructions.
3. Warm the engine to normal operating temperature.
4. Use a service wire to ground the diagnosis switch terminal in the monitor coupler. The wire should go from the diagnostic terminal (B) to the ground terminal (D). The **CHECK ENGINE** light should indicate code number 12.
5. Stop the engine, and connect a duty meter between the duty check terminal and ground terminal of the monitor coupler.
6. Check the idle speed. If the speed is not between 700–800 r/min for vehicles with M/T and 850 r/min for vehicles with A/T, adjust by turning the idle speed adjusting screw on the carburetor.

1. Monitor coupler
2. Front strut of left side
B: Diag. switch terminal
D: Ground terminal
F: Duty output terminal

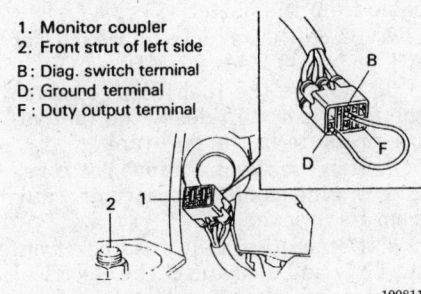

199811

Data link connector terminal locations (B, D and F) — 1993–94 1.3L engine

1. Monitor coupler
2. Duty meter

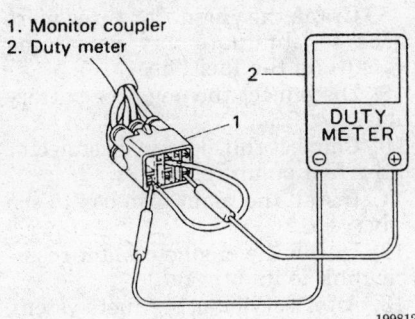

199812

Duty meter terminal hook up — 1993–94 1.3L engine

7. When adjustment is completed, install an adjusting screw cap onto the throttle body.
8. This step is for checking and/or adjusting engine idle speed and IAC duty when the A/C is working. For vehicles without A/C, proceed to step 8. For vehicles with A/C, follow these step:
   a. Turn the A/C switch **ON** and set heater blower switch to the high speed position. Then check that A/C is working.
   b. Check to ensure that idle speed and IAC duty are between 900–1000 rpm with the A/C **ON**.
   c. If the idle speed and/or IAC duty is not within specifications, adjust the idle by turning adjusting screw or A/C solenoid valve.
9. Remove the service wire from the monitor coupler.
10. Install the cap to the monitor coupler.

1995–97 Models

1. Place the manual transmission in **NEUTRAL**, or the automatic transmission in **PARK**. Set the parking brake and block the drive wheels.
2. Connect a tachometer to the engine according to manufactures instructions.
3. Start and warm the engine to normal operating temperature.
4. Check idle speed with the A/C **OFF**. Idle speed should be between 800–900 rpm. If idle speed is out of specification, check idle speed control system and any other system and parts which might affect idle speed.

NOTE: Idle speed is preset at the factory and is not adjustable. If a problem exists, there is a faulty condition within the system.

5. For vehicles with A/C, check idle speed with the A/C **ON**. Idle speed should be between 825–925 rpm. If idle speed is not within specification, check the A/C control signal.
6. When complete, disconnect all testing equipment from the engine.

1.6L (G16) Engine

1. Place the manual transmission in **NEUTRAL**, or the automatic transmission in **PARK**. Set the parking brake and block the drive wheels.
2. Warm the engine to normal operating temperature.
3. Use a service wire to ground the diagnosis switch terminal in the diagnosis connector 1. The **CHECK ENGINE** light should indicate code number 12. The diagnosis connector 1 is located in the relay box.

4. Stop the engine, and connect a duty meter between the duty output terminal and ground terminal of the diagnosis connector 1.

5. Connect a tachometer to the engine according to manufactures instructions.

6. Start and warm the engine to normal operating temperature.

7. Check the idle speed. If the speed is not between 750–850 rpm, adjust by turning the idle speed adjusting screw.

8. When adjustment is completed, install an adjusting screw cap onto the throttle body.

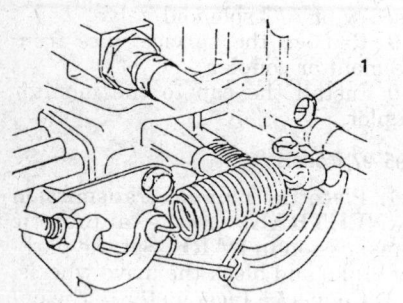

1. Idle speed adjusting screw

199813

Idle speed adjusting screw — 1993–94 1.3L engine

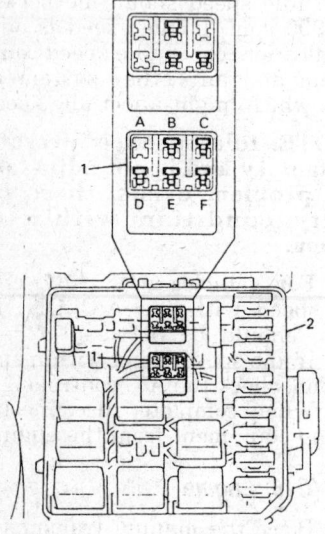

1. Diagnosis connector 1
2. Relay box

A: Blank
B: Diagnosis switch terminal
C: Diagnosis output terminal
D: Ground terminal
E: Test switch terminal
F: Duty output terminal

331859

Grounding the diagnostic switch terminal — 1.6L engine

9. Remove the service wire from the monitor coupler.

10. Install the cap to the monitor coupler.

Mixture

ADJUSTMENT

The air/fuel mixture is not adjustable. The engine control system consists of sensors which detect engine conditions and the Powertrain Control Module (PCM) which controls the fuel mixture based on these sensor signals.

Fuel System Pressure

RELIEVING

— **CAUTION** —
Care should be used when working around the fuel system. DO NOT smoke or expose the fuel system to any open flames. Keep a fire extinguisher handy.

1. Disconnect the negative battery cable from the battery.

2. Place the vehicle in **PARK** for A/T or **NEUTRAL** for M/T.

3. Remove the fuel filler cap from the filler neck to release the fuel vapor pressure in the fuel tank.

4. Remove the main fuse box cover and engine coolant reservoir tank from its bracket.

5. Detach the main fuse box from the body and disconnect the electrical connector from the fuel pump relay.

6. Connect the negative battery cable.

7. Start the vehicle and allow the engine to run until it stalls.

8. Crank the engine for three more times to eliminate any remaining pressure in the fuel lines.

9. Disconnect the negative battery cable.

10. Connect the electrical connector to the fuel pump relay.

11. Install the main fuse box to the body.

12. Install the engine coolant reservoir tank to its bracket.

13. After servicing the fuel system, connect the negative battery cable.

14. Start the engine and check for leaks in the system.

Fuel Filter

REMOVAL AND INSTALLATION

— **CAUTION** —
The fuel system pressure must be relieved before disconnecting any fuel lines. Failure to do so may result in personal injury.

1. Properly relieve the fuel system pressure.

2. Disconnect the negative battery cable.

3. Raise and safely support the vehicle.

4. Place a container under the fuel filter.

5. Disconnect the fuel inlet hose from the fuel filter.

— **CAUTION** —
A small amount of fuel may be released after the fuel hose is disconnected. Cover the hose and pipe with a shop towel.

6. Disconnect the outlet hose from the fuel feed pipe.

7. Remove the two fuel filter mounting bracket bolts and remove the fuel filter from the frame with the outlet hose attached.

8. Remove the fuel filter from the bracket by removing the mounting bolt.

9. Disconnect the outlet hose from the fuel filter.

To install:

10. Install the fuel filter on the bracket and install the mounting bolt.

11. Install the remaining components.

12. Connect the inlet and outlet hoses to the filter.

13. Connect the negative battery cable.

14. With the ignition **ON** and the engine **OFF** check for leaks.

Fuel Pump

REMOVAL AND INSTALLATION

1. Relieve the pressure from the fuel system.

2. Disconnect the negative battery cable.

3. Remove the rear seat cushion by performing the following:

   a. Remove the spare tire.

   b. Remove the seat back by removing the two center mounting nuts and the four mounting screws.

   c. Remove the fitting screws from the rear of the seat cushion.

d. Lift the front of the seat cushion and remove the cushion.

4. Disconnect the fuel level gauge and the fuel pump lead wire couplers and detach the wire tape.

5. Raise and safely support the vehicle on jackstands.

6. Disconnect the fuel filler hose from the fuel tank and disconnect the breather hose from the filler neck.

7. Drain the fuel from the tank by pumping the fuel out through the fuel tank filler.

── **CAUTION** ──

Use a gasoline safe hand operated pump device to drain the fuel tank.

8. Disconnect the fuel hoses from the fuel pipes, located near the fuel filter.

── **CAUTION** ──

A small amount of fuel may be released after the fuel hose is disconnected. Cover the hose and pipe to be disconnected with a shop cloth.

9. Install a support (for example, a transmission jack) under the fuel tank.

10. Remove the fuel tank mounting hardware and remove the tank from the vehicle.

11. Disconnect the fuel lines from the fuel pump and sender assembly.

12. Remove the twelve screws that secure the fuel pump and fuel gauge assembly to the tank and remove the pump and sender assembly.

13. Disconnect the fuel pump electrical connectors.

14. Remove the fuel strainer.

15. Remove the fuel pump.

To install:

16. Install the fuel pump to the fuel gauge assembly.

17. Install the fuel strainer on the fuel pump.

NOTE: Always install a new fuel pump strainer when replacing the fuel pump.

18. Connect the electrical connectors to the fuel pump.

19. Install the remaining components.

20. Connect the negative battery cable.

21. Turn the ignition switch to the **ON** position, but leave the engine **OFF** and check for fuel leaks.

Fuel Injector

REMOVAL AND INSTALLATION

1.3L (G13) Engine

1993–94 Models

1. Properly relieve the fuel system pressure.

2. Disconnect the negative battery cable.

3. Remove the ISC solenoid valve and the EGR modulator bracket from the intake manifold.

4. Disconnect the fuel hoses (feed and return fuel hoses) and the vacuum hose from the delivery pipe.

5. Disconnect the electrical coupler from each injector.

6. Remove the fuel delivery pipe with the fuel injectors.

7. Remove the injectors from the delivery pipe.

To install:

8. Replace the injector O-rings with new O-rings and install the grommet to the injectors.

9. Apply a thin coat of fuel to the injector O-rings and install the injectors to the delivery pipe. Make sure that the injectors rotate smoothly. If the injectors do not rotate smoothly, the O-ring may be improperly installed.

10. Replace any injector insulators if they are scored or damaged. Install the insulators and the fuel delivery pipe spacers to the cylinder head.

11. Install the injectors with the fuel delivery pipe and torque the fuel delivery pipe bolts to 13–20 ft. lbs. (18–28 Nm). Make sure the injectors rotate smoothly.

12. Connect the fuel hoses and the vacuum hose to the delivery pipe.

13. Connect the electrical couplers to the injectors and make sure that the connections are secure.

14. Install the EGR modulator bracket and the ISC solenoid valve to the intake manifold.

15. Connect the negative battery cable.

16. Turn the ignition switch to the **ON** position, leaving the engine **OFF**, and check for fuel leaks.

1995–97 Models

── **CAUTION** ──

The fuel system pressure must be relieved before disconnecting any fuel lines. Failure to do so may result in personal injury.

1. Relieve the fuel pressure from the fuel system.

2. Disconnect the negative battery cable.

3. Remove the air cleaner assembly.

4. Remove the air cleaner mounting stay.

5. Remove the fuel injector cover screws then gently lift the injector cover off of the injector.

── **CAUTION** ──

A small amount of fuel may be released after the fuel hose is disconnected. Cover the hose and pipe that are to be disconnected with a shop towel.

6. Remove the fuel injector from the throttle body.

7. Remove the O-rings from the throttle body.

To install:

8. Apply a thin coat of spindle oil or gasoline to the new upper and lower O-rings. Install the lower O-ring into the throttle body first, then install the upper O-ring.

9. Install the injector by pushing it straight into the fuel injector cavity. Never turn the injector while installing.

NOTE: The positive terminal on the injector should be toward the throttle position sensor.

10. Make sure the fuel injector cover is clean and not damaged then apply a thin coat of spindle oil or gasoline to the O-ring.

11. Install the injector cover, making sure that the electrical terminals fit correctly, then install the mounting screws. Torque the screws to 2–3 ft. lbs. (3–4 Nm).

12. Connect the negative battery cable.

13. Turn the ignition switch to the **ON** position, leaving the engine **OFF**, and check for fuel leaks.

14. Install the air cleaner mounting stay.

15. Install the air cleaner assembly.

1.6L (G16) Engine

1. Release the fuel pressure in the fuel feed line.

2. Disconnect the negative battery cable.

3. Disconnect the four fuel injector connectors.

4. Remove the clamp bolts for the fuel feed pipe and return pipe.

5. Remove the three bolts, then pull the fuel rail up from the intake manifold. Remove the 4 fuel injectors from the intake manifold.

To install:

6. Install new O-rings onto the fuel injectors; be careful not to dam-

age the O-rings. Install the grommets onto the fuel injectors.

NOTE: Inspect the insulators for breakage or scoring. If necessary, replace them.

7. Install the insulators and cushions to the intake manifold.

8. Lubricate the O-rings with fuel and install the injectors onto the fuel rail, then the whole assembly to the intake manifold.

NOTE: Make sure the fuel injectors rotate smoothly. If not, an O-ring may be installed incorrectly; replace it.

9. Install the three fuel rail bolts and torque to 17 ft. lbs. (23 Nm). Make sure the fuel injectors rotate freely.

10. Lubricate the new gaskets with fuel. Install the fuel feed pipe and return pipe.

11. Connect the four fuel injector connectors.

12. Connect the negative battery cable.

13. Start the engine and check for leaks.

ENGINE MECHANICAL

Engine Assembly

REMOVAL AND INSTALLATION

1.3 (G13) Engine

— CAUTION —
Fuel injection systems remain under pressure after the engine has been turned OFF. Properly relieve fuel pressure before disconnecting any fuel lines. Failure to do so may result in fire or personal injury.

1. Properly relieve the fuel system pressure.

2. Disconnect the negative battery cable.

3. Remove the battery and tray.

4. Disconnect the windshield washer hose from the hood. Using a grease pencil or marker, mark the hood hinge to hood outline. With the aid of an assistant, remove the hood.

5. Drain the cooling system.

6. Remove the air cleaner assembly with the MAF sensor outlet hose.

7. Remove the radiator and cooling fan.

8. Disconnect the following electrical wires and release the wire harness from the clamps:
   a. Ignition coil wire from the distributor cap.
   b. Distributor electrical wires.
   c. EGR solenoid vacuum valve.
   d. Radiator fan temperature switch.
   e. Engine coolant temperature gauge sensor.
   f. ECT sensor.
   g. IAC actuator.
   h. Ground wires from the intake manifold.
   i. TP sensor.
   j. Fuel injector.
   k. Oxygen sensor.
   l. Oil pressure gauge sensor.
   m. Alternator.
   n. Starter.
   o. Backup light switch.
   p. Negative battery cable from the transaxle.
   q. Vehicle speed sensor.
   r. Noise filter ground wire.
   s. EGRT sensor, if equipped.
   t. EVAP SP valve.

9. Disconnect the following vacuum hoses:
   a. Brake booster hose from the intake manifold.
   b. Canister purge hose.
   c. Air conditioning SV valve hose.

10. Disconnect the fuel return hose and the fuel feed hose from the throttle body.

11. Disconnect the heater inlet and outlet hoses.

12. Disconnect the following cables:
   a. Accelerator cable from the throttle body.
   b. Clutch cable from the transaxle, if equipped.
   c. Speedometer cable from the transaxle, if equipped.
   d. Shift switch, if equipped with an automatic transaxle.
   e. Vehicle speed sensor, if equipped.

13. Remove the EVAP canister from the vehicle.

14. Safely raise and support the vehicle.

15. Remove the fender apron extensions.

16. Disconnect the exhaust pipe from the exhaust manifold.

17. Disconnect the control shaft and extension rod from the transaxle.

18. Drain the engine and transaxle oil.

19. Remove the left and right halfshafts.

NOTE: For engine and transaxle removal, it is not necessary to remove the halfshafts from the steering knuckles.

20. If equipped with air conditioning, remove the air conditioning compressor from its mounting bracket with the hoses still attached.

NOTE: Suspend the compressor where no damage will occur during engine removal and installation.

21. If equipped with power steering, disconnect the power steering hoses from the power steering pump.

NOTE: Plug the power steering hose, pipe and pump ports to minimize fluid loss.

22. If equipped with a automatic transaxle, remove the rear torque rod bracket from the transaxle.

23. If equipped with a manual transaxle, remove the rear mount from the body.

24. If equipped with a automatic transaxle, remove the rear mounting nut.

25. Lower the vehicle.

26. Install an engine lifting device.

27. Remove the rear mount from the body.

28. Remove the left side engine mounting bracket bolts and bracket.

29. Remove the right side engine mount from its bracket.

30. Before lifting the engine and assembly check to make sure that all the hoses, electric wires and cables are disconnected.

31. Remove the engine with the transaxle from the vehicle.

To install:

32. Lower the engine and transaxle into the engine compartment but do not remove the lifting device.

33. Install the rear mount to the body.

34. Install the left side engine mounting bracket and bolts.

35. Install the right side engine mount to its bracket.

36. If equipped with an automatic transaxle, install the rear mounting nut.

37. Torque the engine mounting nuts and bolts to specification.

38. Remove the lifting device.

39. Safely raise and support the vehicle.

40. If equipped with air conditioning, install the air conditioning compressor to its mounting bracket. Torque the mounting bolts to 13–20 ft. lbs. (18–28 Nm).

41. If equipped with power steering, connect the power steering hose and pipe to the power steering pump.
42. Install the left and right halfshafts.
43. Connect the control shaft and the extension rod to the transaxle. Torque the control shaft nuts and bolts to 11–14 ft. lbs. (15–20 Nm) and torque the extension rod nut to 19–29 ft. lbs. (25–40 Nm).
44. Connect the exhaust pipe to the exhaust manifold. Torque the bolts to 29–36 ft. lbs. (40–50 Nm).
45. Fill the transaxle with gear oil.
46. Install the remaining components.
47. Adjust the clutch pedal freeplay.
48. Adjust the accelerator cable freeplay.
49. Install the air cleaner assembly.
50. Fill the engine with engine oil and the cooling system with coolant.
51. Install the hood and connect the windshield washer hose.
52. Install the battery and tray, and connect the negative battery cable
53. Fill the power steering reservoir and bleed the power steering system.
54. Run the engine and verify that there are no fuel, coolant, transmission or exhaust leaks.

1.6L (G16) Engine

— **CAUTION** —
The fuel system pressure must be relieved before disconnecting any fuel lines. Failure to do so may result in personal injury.

1. Disconnect the battery cables from the battery, negative cable first.
2. Mark the position of the hood on the hinges for installation reference, then remove the hood with the aid of an assistant.
3. Drain the cooling system.
4. Remove the radiator and cooling fan.
5. Remove the air cleaner outlet hose.
6. Remove the air cleaner case by removing the fastening bolts.
7. Disconnect the following cables:
   a. Accelerator cable from the throttle body
   b. If equipped with M/T, disconnect the clutch cable from the transaxle
   c. If equipped with A/T, disconnect the gear select cable from the transaxle.
8. Disconnect the following vacuum hoses:
   a. Brake booster hose from the intake manifold

   b. Canister purge hose from the EVAP canister purge valve
   c. MAP sensor hose from the intake manifold
9. Disconnect the following electrical connectors:
   a. Distributor coil wire
   b. Camshaft position sensor
   c. Engine oil pressure switch
   d. EGR solenoid vacuum valve
   e. EVAP canister purge valve
   f. Engine coolant temperature sensor
   g. Fuel injectors
   h. Power steering pressure switch
   i. Heated oxygen sensor
   j. Back–up light switch (M/T)
   k. Shift switch (A/T)
   l. Forward clutch revolution sensor (A/T)
   m. A/T vehicle speed sensor
   n. Alternator
   o. Starter
   p. Battery negative cable from the transaxle
   q. Vehicle speed sensor
   r. Throttle position sensor
   s. Idle air control valve
   t. Manifold absolute sensor
10. Remove the engine wires from the engine.
11. Disconnect the fuel feed hose from the feed pipe and remove the return hose from the fuel pressure regulator.
12. Disconnect the heater inlet and outlet hoses.
13. Raise and safely support the front of the vehicle.
14. Remove the right and left engine undercovers.
15. Remove the front exhaust pipe from the exhaust manifold and center exhaust pipe.
16. If equipped with a manual transaxle, remove the gear shift control shaft from the transaxle and remove the extension rod.
17. Drain the engine and transaxle oil.
18. Remove the left and right halfshafts.
19. If equipped with A/C, remove the A/C compressor from the compressor bracket with the hoses still attached. Position the A/C out of the way from the engine.
20. If equipped with power steering, drain the power steering pump of fluid.
21. If equipped with power steering, disconnect the power steering hose from the power steering pump.
22. Install a lifting device to the engine.

23. Remove the center member from the vehicle by removing the seven nuts and four bolts.
24. Remove the left engine mount.
25. Remove the right engine mount and bracket.
26. Check to make sure all cooling hoses, vacuum hoses and electrical wires are disconnected from the engine.
27. Lower the engine with the transaxle from the vehicle.
To install:
28. Raise the engine and transaxle into the engine compartment.
29. Install the right engine mount with the bolts and nuts. Torque the bolts and nuts to 40 ft. lbs. (55 Nm).
30. Install the left engine mount and install the bolts and nuts, Torque the bolts and nuts to 40 ft. lbs. (55 Nm).
31. Install the center member using the seven nuts and four bolts. Torque the bolts and nuts as follows:
• Center member to the radiator support to 33 ft. lbs. (45 Nm).
• Center member to crossmember to 33 ft. lbs. (45 Nm).
• Engine mounts to center member nuts to 40 ft. lbs. (55 Nm).
32. Remove the lifting device.
33. Connect the power steering hose to the power steering pump.
34. Install the A/C compressor to the A/C bracket on the engine.
35. Install the left and right halfshafts.
36. If equipped with a manual transaxle, install the gear shift control shaft to the transaxle and install the extension rod.
37. Install the front exhaust pipe to the center exhaust pipe and exhaust manifold.
38. Install the remaining components.
39. Fill the cooling system, engine, transaxle and power steering pump.
40. Adjust all cables and check all connections.
41. Connect the negative battery cable to the battery.
42. Start the vehicle and check for leaks.

Engine Mounts

REMOVAL AND INSTALLATION

1.3L (G13) Engine

Left Engine Mount

1. Disconnect the negative battery cable.
2. Support the engine with a suitable lifting device.

3. Raise and safely support the vehicle.

4. Remove the left engine mount body bracket attaching bolts.

5. Remove the mounting bolt.

6. Remove the nut securing the engine mount to bracket stiffener.

7. Remove the mount from the vehicle.

To install:

8. Position the mount in the vehicle.

9. Install the nut attaching the mount to the mounting bracket, do not tighten the nut at this time.

10. Install the mounting bolt, do not tighten the bolt at this time.

11. Install the engine mounting body bracket bolts, torque the bolts to 14–20 ft. lbs. (18–28 Nm).

12. Torque the mounting bolt and nut to 37–43 ft. lbs. (50–60 Nm).

13. Lower the vehicle and remove the engine support.

14. Connect the negative battery cable.

Right Engine Mount

1. Disconnect the negative battery cable.

2. Support the engine with a suitable lifting device.

3. Raise and safely support the vehicle.

4. Remove the right engine mount bolt.

5. Remove the two bolts attaching the mount to the bracket.

6. Remove the mount from the vehicle.

To install:

7. Position the mount in the vehicle.

8. Install the two bolts attaching the mount to the bracket. Do not tighten the bolts at this time.

9. Install the mounting bolt. Torque to 37–43 ft. lbs. (50–60 Nm).

10. Torque the bolts attaching the mount to the bracket to 37–43 ft. lbs. (50–60 Nm).

11. Lower the vehicle and remove the engine support.

12. Connect the negative battery cable.

Rear Engine Mount With Automatic Transaxle

1. Disconnect the negative battery cable.

2. Support the engine with a suitable lifting device.

3. Raise and safely support the vehicle.

4. Remove the nut attaching the rear engine bracket to the mount.

5. Remove the two nuts securing the mounting body No.1 bracket to the mounting body No.2 bracket.

6. Remove the mount attached to the No.1 bracket from the vehicle.

7. Remove the nut attaching the mount to the No.1 bracket.

To install:

8. Attach the mount to the No.1 bracket, do not tighten the nut at this time.

9. Position the mount in the vehicle.

10. Install the two nuts attaching the mounting body No.1 bracket to the mounting body No.2 bracket. Torque the nuts to 29–36 ft. lbs. (40–50 Nm).

11. Install the nut attaching the rear engine mounting bracket to the mount. Torque the mounting nut to 29–36 ft. lbs. (40–50 Nm).

12. Lower the vehicle and remove the engine support.

13. Connect the negative battery cable.

Rear Mount With Manual Transaxle

1. Disconnect the negative battery cable.

2. Support the engine with a suitable lifting device.

3. Raise and safely support the vehicle.

4. Remove the rear engine mount nut and bolt.

5. Remove the two bolts attaching the mount to the body.

6. Remove the mount from the vehicle.

To install:

7. Position the mount in the vehicle.

8. Install the two bolts attaching the mount to the body and torque the bolts to 37–43 ft. lbs. (50–60 Nm).

9. Install the mounting bolt and torque to 37–43 ft. lbs. (50–60 Nm).

10. Lower the vehicle and remove the engine support.

11. Connect the negative battery cable.

Rear Torque Rod With Automatic Transaxle

1. Disconnect the negative battery cable.

2. Support the engine with a suitable lifting device.

3. Raise and safely support the vehicle.

4. Remove the rear torque rod lower mounting bolt, then the upper mounting bolt.

5. Remove the mount from the vehicle.

To install:

6. Position the mount in the vehicle.

7. Install the upper mounting bolt, do not tighten the bolt at this time.

8. Install the lower mounting bolt.

9. Torque the mounting bolts to 37–43 ft. lbs. (50–60 Nm).

10. Lower the vehicle and remove the engine support.

11. Connect the negative battery cable.

1.6L (G16) Engine

Left Engine Mount

1. Disconnect the negative battery cable.

2. Support the engine with a suitable lifting device.

3. Remove the left engine mounting body bracket attaching bolts.

4. Remove the nut and through-bolt from the left mount.

5. Remove the mount from the vehicle.

To install:

6. Position the mount in the vehicle.

7. Install the through-bolt and nut to the left mount. Do not torque the nut at this time.

8. Install the left engine mount body bracket attaching bolts. Torque the bolts to 40 ft. lbs. (55 Nm).

9. Torque the through-bolt nut to 40 ft. lbs. (55 Nm).

10. Connect the negative battery cable.

Right Engine Mount

1. Disconnect the negative battery cable.

2. Support the engine with a suitable lifting device.

3. Remove the right engine mount bolts and nut from the bracket and stiffener.

4. Remove the right engine mount through-bolt and nut.

5. Remove the engine mount from the vehicle.

To install:

6. Position the engine mount to the engine.

7. Install the nut and two bolts attaching the mount to the bracket. Do not tighten the bolts at this time.

8. Install the through-bolt and nut to the right mount. Torque the nut to 40 ft. lbs. (55 Nm).

9. Torque the two bolts and nut to 40 ft. lbs. (55 Nm).

10. Remove the engine support.

11. Connect the negative battery cable.

Rear Engine Mount

1. Disconnect the negative battery cable.

2. Support the engine with a suitable lifting device.

3. Raise and safely support the vehicle.

4. Remove the two nuts securing the rear engine mount to the center member.

5. Remove the through-bolt holding the engine mount to the bracket.

6. Remove the engine mount from the vehicle.

To install:

7. Attach the engine mount to the bracket on the engine. Install the through-bolt and nut. Do not tighten the nut at this time.

8. Install the two nuts attaching the engine mount to the center member. Torque the nuts to 33 ft. lbs. (45 Nm).

9. Torque the through-bolt nut to 40 ft. lbs. (55 Nm).

10. Lower the vehicle and remove the engine support.

11. Connect the negative battery cable.

Front Engine Mount

1. Disconnect the negative battery cable.

2. Support the engine with a suitable lifting device.

3. Raise and safely support the vehicle.

4. Remove the two nuts securing the engine mount to the center member.

5. Remove the through-bolt holding the front engine mount to the bracket.

6. Remove the engine mount from the vehicle.

To install:

7. Attach the engine mount to the mounting bracket on the engine. Install the through-bolt and nut. Do not tighten the nut at this time.

8. Install the two nuts attaching the engine mount to the center member. Torque the nuts to 33 ft. lbs. (45 Nm).

9. Torque the through-bolt nut to 40 ft. lbs. (55 Nm).

10. Lower the vehicle and remove the engine support.

11. Connect the negative battery cable.

Cylinder Head

REMOVAL AND INSTALLATION

1.3L (G13) Engine

Dual Overhead Cam Engine

1. Relieve the fuel system pressure.

2. Disconnect the negative battery cable.

3. Drain the cooling system into a suitable container.

4. Remove the air cleaner assembly including the MAF sensor outlet hose.

5. Disconnect the following electrical wires:

  a. Throttle position sensor

  b. Fuel injectors

  c. Evaporative canister solenoid purge valve

  d. EGR solenoid vacuum valve (if equipped)

  e. Ground wires from the intake manifold

  f. Idle air control valve

  g. EGR temperature sensor (if equipped)

  h. Ignition coil wire from the distributor

  i. Camshaft position sensor (in distributor)

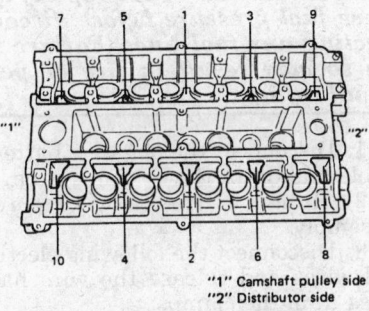

"1" Camshaft pulley side
"2" Distributor side

177079

Cylinder head bolt tightening sequence — 1.3L DOHC engine

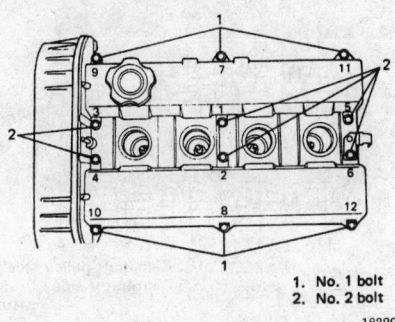

1. No. 1 bolt
2. No. 2 bolt

183297

Cylinder head cover torque sequence — 1.3L DOHC engine

  j. Engine coolant temperature gauge

  k. Radiator fan thermostat switch

  l. Engine coolant temperature sensor

  m. Oxygen sensor

6. Disconnect the brake booster hose.

7. Disconnect the canister purge hose.

8. Disconnect the accelerator cable from the throttle lever and its bracket.

9. Disconnect the radiator inlet hose from the thermostat cap.

10. Disconnect the heater inlet hose from the cylinder head.

11. Disconnect the throttle body coolant inlet hose from the throttle body.

12. Disconnect the fuel feed hose and the return hose from the fuel delivery pipe.

13. Remove the water pump and crankshaft pulleys.

14. Remove the timing belt.

15. Disconnect the exhaust pipe from the exhaust manifold.

16. Remove the exhaust stiffener bolt.

17. Remove the alternator adjusting arm stiffener.

18. Remove the intake manifold bracket.

19. Disconnect the ignition wires from the spark plugs.

20. Disconnect the PCV hose and the breather hose from the cylinder head cover.

21. Remove the cylinder head cover.

22. Remove the cylinder head bolts using a 8 mm hexagon bit. Remove the head bolts starting at the ends and working towards the center.

23. Remove the cylinder head with the intake manifold, exhaust manifold and other mounted components in place.

— **WARNING** —

In this state, one of the valves is out toward the combustion chamber (i.e. it is open). Be sure NOT to place the cylinder head on any flat surface with its mating surface with the engine block facing down. This will cause damage to the open valve.

24. Remove the intake manifold with the throttle body and exhaust manifold from the cylinder head.

25. Remove the distributor and fuel delivery pipe with the fuel injectors from the cylinder head.

26. Remove the camshafts and the valve lash adjusters from the cylinder head.

To install:

27. Thoroughly clean the cylinder block mating surface.

28. Install the cylinder head gasket so the "TOP" mark on the gasket is facing toward the cylinder head and is on the crankshaft pulley side.

29. Install the cylinder head to the cylinder block. Apply engine oil to the cylinder head bolts then install the bolts. Tighten the bolts in two or three steps. The final torque should be 47–50 ft. lbs. (65–70 Nm). Be sure to follow the proper bolt tightening sequence.

30. Reinstall the valve lash adjusters and the camshafts. Be sure to place the camshaft and lash adjusters on the appropriate side, intake or exhaust. Torque the camshaft housings in the proper sequence.

31. Install the cylinder head cover to the cylinder head. Follow the proper sequence for tightening the cylinder head cover bolts. Torque the bolts closest to the intake manifold and the exhaust manifold to 4–5 ft. lbs. (5–7 Nm). Torque the six bolts near the spark plugs to 6–8 ft. lbs. (8–12 Nm).

32. Install the timing belt inside covers and distributor to the cylinder head.

33. Install a new exhaust manifold gasket then install the exhaust manifold on the cylinder head.

34. Install a new gasket to the exhaust pipe and to the exhaust manifold. Torque the three attaching bolts to 29–43 ft. lbs. (40–60 Nm).

35. Install the exhaust stiffener bolt and torque the bolt to 29–43 ft. lbs. (40–60 Nm).

36. Install the exhaust manifold cover.

37. Replace the injector O-rings with new O-rings and install the grommet to the injectors.

38. Install the injectors to the delivery pipe.

39. Install the insulators and the fuel delivery pipe spacers to the cylinder head.

40. Install the injectors with the fuel delivery pipe and torque the fuel delivery pipe bolts to 13–20 ft. lbs. (18–28 Nm). Make sure the injectors rotate smoothly.

41. Install the intake manifold.

42. Install the intake manifold bracket and torque the attaching bolts and nut to 29–43 ft. lbs. (40–60 Nm).

43. Connect the PCV hose and the breather hose to the cylinder head cover.

44. Connect the ignition wires to the proper spark plugs.

45. Install the alternator adjusting arm stiffener bracket.

46. Install the timing belt and covers.

47. Install the water pump and crankshaft pulleys.

48. Connect the fuel feed hose and the fuel return hose to the fuel delivery pipe.

49. Connect the throttle body coolant inlet hose to the throttle body.

50. Connect the heater inlet hose to the cylinder head.

51. Connect the radiator inlet hose to the thermostat cap.

52. Connect the accelerator cable to the throttle lever and its bracket.

53. Connect the canister purge hose.

54. Connect the brake booster hose.

55. Install the remaining components

56. Refill the cooling system.

57. Connect the negative battery cable.

58. Start the engine and adjust the ignition timing.

59. Top off the coolant as necessary.

60. Check to be sure that there are no fuel, coolant, or exhaust leaks.

Single Overhead Cam Engines

CAUTION

Fuel injection systems remain under pressure after the engine has been turned OFF. Properly relieve fuel pressure before disconnecting any fuel lines. Failure to do so may result in fire or personal injury.

---

1. Disconnect the negative battery cable and drain the cooling system.

2. Remove the air cleaner assembly.

3. Disconnect the following electrical wires and release the wire harness from the clamps:

  a. Ignition coil wire from the distributor cap.

  b. Distributor electrical wires.

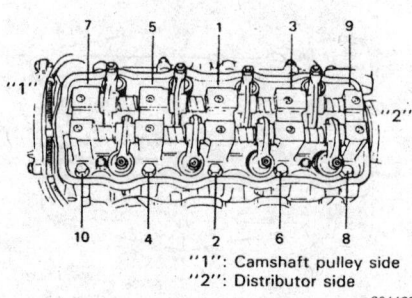

"1": Camshaft pulley side
"2": Distributor side

204468

Cylinder head bolt tightening sequence — 1.3L SOHC engine

  c. EGR solenoid vacuum valve.

  d. Radiator fan thermo switch.

  e. Engine coolant temperature gauge sensor on 1995–97 engines.

  f. Engine coolant temperature gauge sensor.

  g. ECT sensor.

  h. IAC valve.

  i. TP sensor.

  j. Fuel injector.

  k. Ground wires from the intake manifold.

  l. Oxygen sensor.

4. Disconnect the radiator hose from the thermostat housing.

5. Disconnect the heater hose from the intake manifold.

6. Disconnect the throttle body coolant outlet hose from the throttle body.

7. Disconnect the following vacuum hoses:

  a. MAP sensor hose from the intake manifold.

  b. Canister hose from its pipe.

  c. Canister purge hose from the intake manifold.

  d. Brake booster from the intake manifold.

8. Disconnect the fuel return hose and the fuel feed hose from the throttle body.

9. Disconnect the throttle cable from the throttle body.

10. Remove the water pump and crankshaft pulleys. Remove the timing belt.

11. Remove the rubber seal between the cylinder head and the water pump.

12. Disconnect the exhaust pipe from the exhaust manifold.

13. Remove the spark plug wire clamps from the cylinder head cover and disconnect the PCV hose.

14. Remove the cylinder head cover.

15. Loosen all of the valve adjusting screws and allow the valves to close.

16. Remove the cylinder head bolts in the reverse order of the tightening sequence.

17. Remove the cylinder head with the distributor, intake manifold and exhaust manifold.

18. Remove the distributor, intake manifold and exhaust manifold from the cylinder head.

19. Clean the cylinder block mating surface of any old gasket material and clean any engine coolant from the cylinders.

To install:

20. Install the intake and exhaust manifolds.

21. Install the distributor to the cylinder head.

22. Install a new cylinder head gasket with the top mark facing up and toward the crankshaft pulley.

23. Install the cylinder head to the engine block. Coat the cylinder head mounting bolts threads with clean engine oil and install them. Tighten the bolts evenly in 3 equal steps and in sequence to 51–54 ft. lbs. (70–75 Nm).

24. Install the rubber seal between the cylinder head and the water pump.

25. Install the timing belt.

26. Connect the exhaust pipe to the exhaust manifold, torque the attaching bolts to 26–36 ft. lbs. (35–50 Nm).

27. Install the crankshaft and water pump pulleys.

28. Connect the throttle cable to the throttle body.

29. Connect the fuel return hose and the fuel feed hose to the throttle body.

30. Install the remaining components.

31. Refill the cooling system with coolant.

32. Connect the negative battery cable.

33. Adjust the ignition timing.

34. With the engine running, make sure that there are no fuel, coolant or exhaust leaks.

1.6L (G16) Engine

CAUTION

The fuel system pressure must be relieved before disconnecting any fuel lines. Failure to do so may result in personal injury.

1. Disconnect the negative battery cable and drain the cooling system.
2. Remove the air cleaner outlet hose.
3. Remove the intake manifold rear stiffener bolt, alternator adjustment arm reinforcement bolt and right mounting bracket stiffener from the intake manifold.
4. Disconnect the heated oxygen sensor coupler and release its clamps.
5. Remove the exhaust from the manifold.
6. Disconnect the electrical connectors from the following components:
   a. Distributor
   b. Engine coolant temperature sensor and gauge
   c. Engine ground wire from intake manifold
   d. EGR solenoid vacuum valve
   e. Fuel injectors
   f. Throttle position sensor
   g. Idle air control valve
   h. Heated oxygen sensor
   i. Evaporative emissions solenoid purge valve
7. Label and disconnect the vacuum hoses from the following:
   a. EVAP canister purge hose
   b. Brake booster supply hose
8. Disconnect the fuel feed and return hoses from each pipe.
9. Remove the cylinder head cover.
10. Fully loosen all the valve lash adjusting screws.
11. Disconnect the following engine cooling water hose:
   a. Radiator inlet hose
   b. Heater inlet hose
   c. IAC valve outlet
12. Remove the timing belt.
13. Loosen the cylinder head bolts in reverse order of tightening. Once each bolt is loose, remove the bolts from the cylinder head.
14. Check to make sure all components are removed or disconnected before removing the cylinder head.
15. Remove the cylinder head with the intake manifold, exhaust manifold and distributor as an assembly.

To install:
16. Install a new cylinder head gasket and the cylinder head with the distributor case onto the cylinder block. Torque the cylinder head bolts, in sequence, in 3 Steps:
   a. Step 1: 26 ft. lbs. (35 Nm)
   b. Step 2: 41 ft. lbs. (55 Nm)
   c. Step 3: 49 ft. lbs. (68 Nm)
17. Install the timing belt.
18. Connect the engine cooling water hoses.
19. Adjust the valve lash.
20. Install the cylinder head cover.
21. Connect the fuel feed and return hoses to each pipe.
22. Connect the vacuum hoses.
23. Connect the remaining electrical components.
24. Install the remaining components.
25. Fill the engine coolant and check all fluids.
26. Connect the negative battery cable to the battery.
27. Start the engine and check for leaks.

Lash Adjusters

BLEEDING

NOTE: Do not disassemble the Hydraulic Valve Lash (HVL) adjuster for any reason. If the HVL adjuster is removed from the rocker, soak in clean engine oil face up until installation.

The direct-acting hydraulic lash adjuster is located between the camshaft and valve stem. It is not externally adjustable.

If the adjusters have been reinstalled after engine repair, do not start the engine for 30 minutes. This interval allows the adjusters to settle into position. Starting the engine too early may cause piston-to-valve contact and engine damage.

If air is trapped in the adjuster, it will make a tapping noise during operation. To clear the air from the adjuster, run the engine at 2000 rpm for 30 minutes. The air should purge and the tapping sound should cease. If the tapping continues, check for a failed adjuster. To test for a failed adjuster:
1. Stop the engine and remove the valve cover.
2. With the cam lobe away from the suspected adjuster, push the adjuster downward (by hand) with less than 44 lbs. (20 kg) of force. Check for clearance between the cam and the adjuster. If any clearance is present, the adjuster is defective.

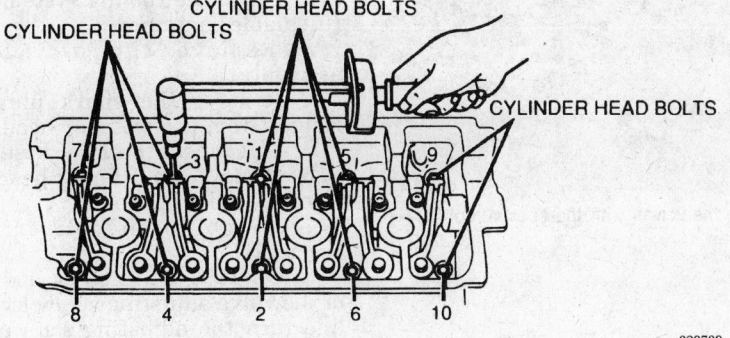

Cylinder head bolt torque sequence — 1.6L engine

Valve Lash

ADJUSTMENT

All Single Overhead Cam Engines Except 1995–97 1.3L

1. Disconnect the negative battery cable.
2. Remove the cylinder head cover.
3. Safely raise and support the vehicle.
4. Remove the right front wheel and fender apron.
5. Turn the crankshaft pulley clockwise until the **V** mark on the pulley is aligned with the 0 calibration on the timing belt cover.
6. Remove the distributor cap and make sure the ignition rotor is aligned with cylinder number one's ignition wire. If the ignition rotor is not aligned, turn the crankshaft one rotation clockwise and realign the timing marks.
7. The valve lash is measured between the rocker arm adjusting screw and the valve stem. Use a thickness gauge to measure the gap.
8. Check the valve lash for the following valves:
   a. Intake valve of cylinder number one (I.D. 1).

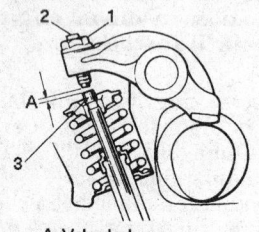

A. Valve lash gap
1. Adjusting screw lock nut
2. Adjusting screw
3. Valve stem

209570

Rocker arm components — 1993–94 1.3L SOHC engine

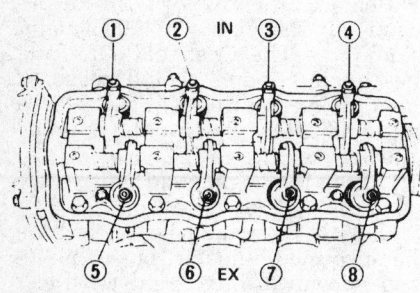

209571

Rocker arm identification — 1993–94 1.3L SOHC engine

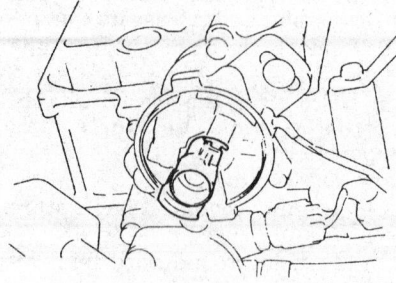

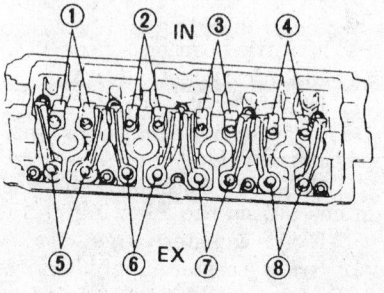

333753

Valve numbered locations — 1.6L engine

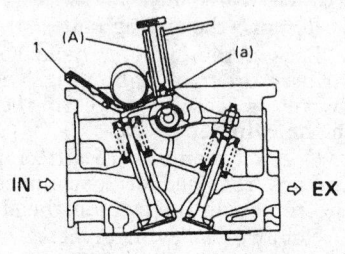

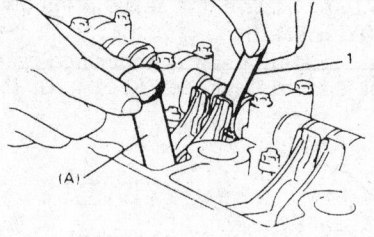

333754

Adjusting the valves with the special tool (A) — 1.6L engine

   b. Intake valve of cylinder number two (I.D. 2).
   c. Exhaust valve of cylinder number one (I.D. 5).
   d. Exhaust valve of cylinder number three (I.D. 7).
9. If the valve lash is out of specification, adjust the specification after loosening the locknut and turning the adjusting screw. Hold the screw stationary while tightening the locknut. Recheck the specification after tightening the locknut.
10. Rotate the crankshaft one rotation clockwise and realign the timing marks.
11. Check the valve lash for the following valves:
   a. Intake valve of cylinder number three (I.D. 3).
   b. Intake valve of cylinder number four (I.D. 4).
   c. Exhaust valve of cylinder number two (I.D. 6).
   d. Exhaust valve of cylinder number four (I.D. 8).
12. Adjust the valves that are out of specification and recheck after tightening the locknut.
13. Install the remaining components.
14. Connect the negative battery cable.

1995–97 1.3L Single Overhead Cam Engines

Hydraulic Valve Lash (HVL) adjusters, located on the valve stem side of the rocker arms, are used to adjust the valve clearance to "0" automatically at all times. Adjustment is not required.

Rocker Arms and Shaft

REMOVAL AND INSTALLATION

1.3L (G13) Engine

1. Disconnect the negative battery cable.
2. Drain the cooling system into a resealable container.
3. Remove the air cleaner assembly.
4. Remove the spark plug wire clamps from the cylinder head.
5. Disconnect the PCV hose.
6. Remove the cylinder head cover from the cylinder head.
7. Remove the distributor assembly.
8. On 1993–94 engines. loosen all of the valve adjusting screw locknuts, and turn the adjusting screws to allow all of the rocker arms to move freely.

9. Remove the rocker arm shaft screws. Be careful not to drop them into the engine.

10. Remove the rocker arm shafts, arms and springs. Keep all of the parts in order so they can be reinstalled in their original locations.

11. On 1993–94 engines, inspect the tip of the adjusting screw and replace if it is badly worn. Inspect the rocker arm's cam riding face and replace if the face is badly worn.

12. Inspect the rocker arms and shafts for wear and/or damage and replace parts as necessary.

13. On 1995–97 engines, remove the hydraulic valve lash adjuster from the rocker arm if necessary.

NOTE: Do not remove the valve lash adjusters unless they need air bleeding or replacement. Be careful not to scratch the valve lash adjuster. Never disassemble the valve lash adjusters. Immerse the removed valve lash adjusters in clean engine oil until reinstallation. If a valve lash adjuster is left in the air, place it with its body facing down. Do not place the valve lash adjusters on their side or the body facing up.

a. If the tip of the valve lash adjuster is badly worn, replace the adjuster. Check the O-ring for breakage or deterioration, replace as necessary.

b. Using a valve lash adjuster air bleeding tool, bleed the air from the adjusters in kerosene.

c. After filling the valve lash adjuster with fresh kerosene, compress the plunger and body with your finger (about 5 kg. or 11 lbs.) for a moment and inspect that its stroke is 0–0.02 inches (0–0.5 mm). If its stroke is more than specified, bleed the air again and recheck the stroke. If the stroke is not within specification, replace the lash adjuster.

To install:

14. For 1995–97 models, install the hydraulic valve lash adjusters, if the valve lash adjusters were removed from the rocker arms.

15. Apply engine oil to the rocker arms and the rocker arm shafts.

16. Install the rocker arms, springs, and the rocker shafts.

NOTE: The two rocker arm shafts are different, they are distinguishable by the stepped ends of the shafts. Looking at the screw holes in the intake shaft, one end of the shaft will have the two sides that are stepped, and

this end is installed toward the camshaft pulley. Looking at the screw holes in the exhaust rocker shaft, one end will have a step on one side only, and this end is installed toward the distributor.

17. For 1993–94 engines, align the screw holes in the rocker arm shafts and install the rocker arm shaft screws. Torque the screws to 7–8 ft. lbs. (9–12 Nm).

18. For 1995–97 engines, Proceed as follows;

a. Install the intake rocker arms, springs, and washers to the intake side rocker shaft. Install the holder, special tool part 09916–56030 to hold the rocker arms, springs, and washers in place on the rocker shaft then install the plate to the holders.

b. Install the rocker arm assembly to the cylinder head and evenly tighten the rocker shaft bolts. Torque the attaching bolts to 16–20 ft. lbs. (22–28 Nm) starting with the center bolt and following the required sequence.

c. Install the exhaust rocker arms, springs, and washers to the exhaust side rocker shaft. Install the holder, special tool part 09916–56030 to hold the rocker

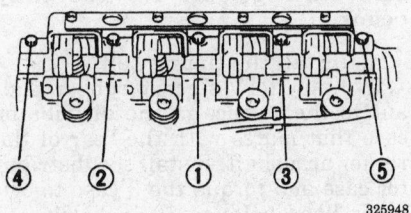

Intake side rocker arm shaft torque sequence — 1995–97 1.3L SOHC engine

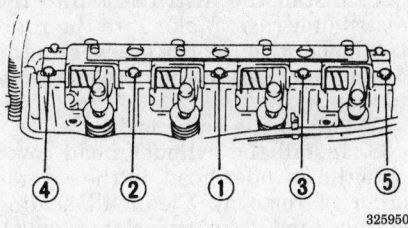

Exhaust side rocker arm shaft torque sequence — 1995–97 1.3L SOHC engine

arms, springs, and washers in place on the rocker shaft then install the plate to the holders.

d. Install the rocker arm assembly to the cylinder head and evenly tighten the rocker arm shaft bolts. Torque the attaching bolts to 16–20 ft. lbs. (22–28 Nm) starting with the center bolt and following the required sequence.

19. On 1993–94 models, adjust the valve lash adjusters.

20. Install the remaining components.

21. Refill the cooling system.

22. Connect the negative battery cable.

23. Adjust the ignition timing.

24. Top off the engine coolant as necessary.

— WARNING —
If air is trapped in a hydraulic valve lash adjuster, the valve may make a tapping sound when the engine is operated after a valve lash adjuster is installed. In such a case, run the engine for about half an hour at 2,000 RPM. The air will be purged and the tapping will stop. If the tapping continues, it is possible that a valve lash adjuster is defective.

1.6L (G16) Engine

1. Disconnect the negative battery cable.

2. Remove the timing belt.

— WARNING —
After the timing belt is removed, never turn the camshaft and crankshaft independently more than ±90°. If turned, interference may occur among the piston and valves causing possible damage to the effected parts.

3. Using camshaft sprocket holding tool (09917–68220) or equivalent, hold the sprocket stationary and remove the camshaft sprocket bolt.

4. Remove the cylinder head cover.

5. Remove the distributor cap and distributor assembly from the engine.

6. Loosen all of the valve adjusting screw locknuts until the rocker arms move freely.

7. Remove the camshaft.

8. Remove the rocker arm shaft plug from the cylinder head.

9. Remove the timing belt inner cover-to-cylinder head bolts and the cover.

10. Remove all intake rocker arms and clips from the rocker shaft. Keep all parts in order so they can be reinstalled in their original locations.

11. Remove the six rocker arm shaft-to-cylinder head bolts. Push the rocker arm shaft through the rear of the cylinder head until the end of the rocker shaft appears. Remove the O-ring from the rear of the rocker arm shaft.

12. Remove the exhaust rocker arms, rocker arm springs and rocker shaft by pulling the rocker arm shaft through the front of the cylinder head. Be sure to keep the parts in order for installation purposes.

13. Clean and inspect all parts for wear and/or damage; replace parts as necessary.

To install:

14. Lubricate the rocker arms and shafts with clean engine oil before installation.

15. Push the rocker shaft into the front of the cylinder head; install the exhaust rocker arms and springs as the rocker arm shaft is being installed into the cylinder head.

16. Push the rocker arm shaft through the rear of the cylinder head. Install a new O-ring onto the rocker shaft.

17. Rotate the rocker arm shaft so the flat machined surface is horizontal and facing downward, parallel with the cylinder head mating surface and slide the shaft back into the cylinder head.

18. Install the 6 rocker arm shaft bolts and torque the rocker arm-to-cylinder head bolts to 89 inch lbs. (10 Nm). Fill the rocker arm shaft bolt holes with clean engine oil.

19. Install the intake rocker arms and clips onto the rocker arm shaft.

NOTE: The camshaft carrier caps are embossed with numbers and arrows to ensure correct assembly. The No. 1 camshaft carrier cap must be installed at the front of the cylinder head with the remaining carrier caps following in numerical order. The directional arrows must always point toward the front of the cylinder head.

20. Install the camshaft.

--- **WARNING** ---
If the camshaft carrier cap bolts are tightened at random, damage to the camshaft may occur.

21. Lubricate the new camshaft seal lip with clean engine oil and install it into the cylinder head until it is flush with the camshaft carrier surface.

22. Install the timing belt inner cover and torque the cover-to-cylin-

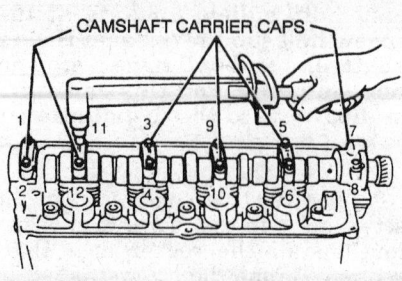

CAMSHAFT CARRIER CAPS

329690

Camshaft carrier cap bolt torque sequence — 1.6L engine

der head bolts to 89 inch lbs. (10 Nm). Install the rocker arm shaft plug into the cylinder head and torque to 24 ft. lbs. (33 Nm).

NOTE: During camshaft timing belt sprocket installation, align the camshaft dowel pin with the slot in the camshaft timing belt gear designated as "E".

23. Install the camshaft sprocket. Using holding tool (09917–68220) or equivalent, to hold the sprocket in place, torque the camshaft sprocket bolt to 44 ft. lbs. (60 Nm).

NOTE: When installing the timing belt, the directional arrows on the timing belt must be matched with the rotation of the crankshaft; if not, excessive wear and timing belt failure may occur.

24. Install the timing belt.

25. Apply RTV silicone rubber sealant to the surface of the distributor case that mates with the rear of the rocker arm shaft. Install the distributor case and torque the 3 case-to-cylinder head bolts to 89 inch lbs. (10 Nm).

NOTE: With the timing marks aligned on the sprockets and the timing belt installed, the number four piston is at TDC of the compression stroke.

26. Install the distributor into the distributor case. Make sure the rotor is aligned with the No. 4 tower on the distributor cap. Install the distributor cap.

27. Adjust the valve lash.

28. Install the cylinder head cover onto the cylinder head, in the reverse order of removal. Clean all sealing surfaces and use a new gasket and O-rings. Tighten the cylinder head cover bolts to 89 inch lbs. (10 Nm).

29. Connect the negative battery cable.

30. Start the engine; allow it to reach normal operating temperature and check for leaks.

31. Check and adjust the ignition timing as necessary.

Intake Manifold

REMOVAL AND INSTALLATION

Dual Overhead Cam Engines

1. Disconnect the negative battery cable.

2. Drain the coolant into a appropriate resealable container.

3. Disconnect the following hoses:
   a. MAF outlet hose.
   b. PCV hose.
   c. Cylinder head cover breather hose.
   d. Air conditioning solenoid vacuum valve hose (if equipped).
   e. Brake booster hose.
   f. Canister purge hose from the canister.
   g. Vacuum hose from the fuel pressure regulator.
   h. Vacuum hose from the EGR valve (if equipped).
   i. Remove the EGR modulator hose (if equipped).

4. Disconnect the following electrical wires:
   a. EGR Temperature sensor (if equipped).
   b. IAC valve.
   c. EVAP solenoid purge valve.
   d. EGR solenoid vacuum valve (if equipped).
   e. Throttle position sensor.
   f. Ground wires from the intake manifold.

5. Disconnect the accelerator cable from the throttle valve lever and its bracket.

6. Disconnect the engine coolant hoses from the fast idle air valve and the throttle body.

7. Remove the alternator adjusting arm stiffener.

8. Remove the intake manifold bracket.

9. Remove the intake manifold mounting nuts and bolts, then remove the intake manifold.

To install:

10. Before installing a new intake manifold gasket clean the mating surfaces.

11. Install a new gasket to the cylinder head.

12. Install the intake manifold to the cylinder head. Be sure to install the clamps to the bottom of the intake manifold and torque the nuts and bolts to 13–20 ft. lbs. (18–28 Nm).

13. Install the intake manifold bracket and torque the nuts and bolts to 29–43 ft. lbs. (40–60 Nm).

14. Install the alternator adjusting arm stiffener and torque the bolts to 13–20 ft. lbs. (18–28 Nm).

15. Install the remaining components.

16. Refill the cooling system.

17. Connect the negative battery cable.

18. Start the engine and check for vacuum and coolant leaks.

Single Overhead Cam Engine

1. Properly relieve the fuel system pressure.

────── **CAUTION** ──────
The fuel system pressure must be relieved before disconnecting any fuel lines. Failure to do so may result in personal injury.

2. Disconnect the negative battery cable.

3. Drain the coolant from the vehicle.

────── **WARNING** ──────
To help avoid the danger of being burned, do not remove the drain plug and the radiator cap while the engine is still hot. Scalding fluid and steam can be blown out under pressure if the plug and cap are taken off too soon.

4. Remove the air cleaner assembly.

5. Disconnect the following electrical wires:
    a. EGR solenoid vacuum valve.
    b. ISC actuator.
    c. Ground wires from the intake manifold.
    d. Fuel injector.
    e. Throttle position sensor.
    f. Early fuel evaporator heater.

6. Disconnect the fuel return and the fuel feed hoses from the fuel pipes.

7. Disconnect the coolant hoses from the throttle body and the intake manifold.

8. Disconnect the following vacuum hoses:
    a. Canister purge hose from the intake manifold.
    b. MAP sensor hose from the intake manifold.
    c. Brake booster hose from the intake manifold.
    d. Disconnect the PCV hose from the PCV valve.

9. Disconnect the accelerator cable from the throttle body.

10. Remove the intake manifold attaching nuts and bolts and remove the intake manifold and throttle body.

To install:

11. Before installing the gasket make sure the mating surfaces of the intake manifold and the cylinder head are clean and undamaged.

12. Install a new intake manifold gasket to the cylinder head.

13. Position the intake manifold and throttle body on the cylinder head and install the mounting nuts and bolts. Torque the nuts and bolts to 13–20 ft. lbs. (18–28 Nm). Make sure that the clamps are properly installed on the lower intake manifold bolts.

14. Install the remaining components.

15. Refill the cooling system.

16. Connect the negative battery cable.

17. Start the engine and check for fuel and cooling system leaks.

Exhaust Manifold

REMOVAL AND INSTALLATION

1.3L (G13) Engine

────── **CAUTION** ──────
To avoid the danger of being burned, do not service the exhaust system while it is hot. Service should be performed only after the system cools down.

1. Disconnect the negative battery cable.

2. Disconnect the oxygen sensor electrical coupler and release the wire from its clamps.

3. Remove the exhaust manifold cover.

4. Remove the exhaust manifold stiffener bolt.

5. On dual overhead cam models, remove the three nuts attaching the exhaust pipe to the exhaust manifold.

6. On single overhead cam models, remove the two bolts attaching the exhaust pipe to the exhaust manifold.

7. Remove the exhaust manifold mounting nuts and bolts.

8. Remove the exhaust manifold and the gasket.

To install:

9. Before installing any components check the exhaust manifold and the engine for deterioration or damage and replace as necessary.

10. Install the manifold gasket and the exhaust manifold to the engine.

Torque the nuts and bolts to 13–20 ft. lbs. (18–28 Nm).

11. Install the exhaust pipe gasket to the exhaust pipe and position the exhaust pipe.

12. On dual overhead cam models, install the three nuts that attach the exhaust pipe to the exhaust manifold and torque the nuts to 29–43 ft. lbs. (40–60 Nm).

13. On single overhead cam models, install the two bolts that attach the exhaust pipe to the exhaust manifold and torque the bolts to 25–36 ft. lbs. (35–50Nm).

14. Install the exhaust manifold stiffener and torque the bolt to 29–43 ft. lbs. (40–60 Nm).

15. Install the remaining components.

16. Connect the negative battery cable.

17. Run the engine and check for exhaust leaks.

1.6L (G16) Engine

1. Disconnect the negative cable.

2. Raise and safely support the vehicle.

3. Disconnect the two exhaust pipe bolts connecting the exhaust pipe to the exhaust manifold. Lower the vehicle after the exhaust manifold is disconnected from the exhaust pipe.

4. Disconnect the oxygen sensor lead wire at the coupler.

5. Disconnect the exhaust manifold heat shield from the manifold by removing the nut and bolt.

6. Remove the exhaust manifold mounting bolts and nuts.

7. Remove the exhaust manifold from the cylinder head.

To install:

8. Clean and inspect the sealing surfaces of the exhaust manifold and the cylinder head.

9. Using new gaskets, install the exhaust manifold to the cylinder head and tighten the mounting bolts and nuts to 17 ft. lbs. (23 Nm).

10. Connect the oxygen sensor lead wire at the coupler.

11. Raise the vehicle. Install the three exhaust pipe bolts connecting the exhaust pipe to the exhaust manifold and torque to 36 ft. lbs. (50 Nm).

12. Connect the manifold heat shield to the exhaust manifold and install the nuts and bolts.

13. Lower the vehicle. Connect negative battery cable.

14. Check for exhaust leaks when finished.

Front Crankshaft Seal

REMOVAL AND INSTALLATION

1.3L (G13) Engine

1. Disconnect the negative battery cable.
2. Remove the timing belt.
3. Remove the crankshaft sprocket bolt using a suitable gear stopper to hold the flywheel. Remove the sprocket bolt, sprocket and key.
4. Use a suitable tool to remove the seal from the oil pump housing.

NOTE: Be careful not to damage the crankshaft or the oil pump sealing surfaces when removing or installing the seal.

5. Clean and inspect the surfaces of the crankshaft and the oil pump assembly.

To install:

6. Lubricate the new seal with clean engine oil.
7. Install the new seal over the crankshaft and into the oil pump, making sure the oil seal lip is not turned up. Use oil seal guide tool 09926–18210 or equivalent.
8. Install the crankshaft sprocket and timing belt.

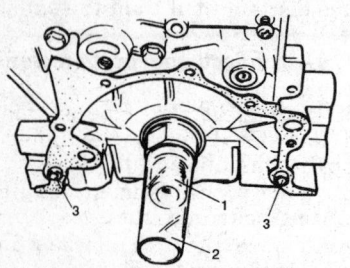

1 Crankshaft
2 Oil seal guide (Vinyl resin) (special tool 09926-18210)
3 Oil pump pin

325892

Oil seal guide tool — 1.3L engine

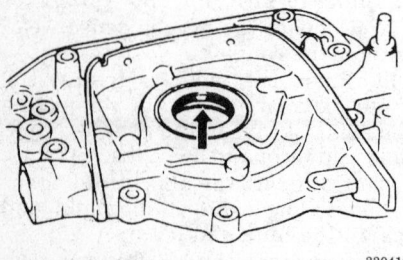

330410

Front crankshaft oil seal location — 1.6L engine

9. Connect the negative battery cable.

1.6L (G16) Engine

NOTE: The front oil seal can be removed from the engine without removing the oil pump.

1. Disconnect the negative battery cable from the battery.
2. Remove the timing belt.
3. Remove the crankshaft timing belt sprocket.

——— **WARNING** ———
When removing the front seal, be extremely careful not to damage the crankshaft.

4. Using a knife, cut off the oil seal lip.
5. Tape the end of a flat bladed tool to avoid damaging the crankshaft. Pry out the oil seal using the taped end of the tool.
6. Inspect the oil seal riding surface on the crankshaft for signs of wear or damage.

To install:

7. Wipe the seal bore with a clean rag.
8. Apply multipurpose grease to the lip of a new oil seal.
9. Drive the oil seal into place using a seal installer tool. Make sure the seal surface is flush with the oil pump case edge. Work from the front of the cover. Be extremely careful not to damage the seal.
10. Install the crankshaft sprocket.
11. Install the timing belt.
12. Connect the negative battery cable to the battery.
13. Start the engine and check for leaks.

Timing Belt, Sprockets and Tensioner

REMOVAL AND INSTALLATION

1.3L (G13) Engine

1. Disconnect the negative battery cable.
2. Remove the air cleaner assembly with the MAF sensor and outlet hose.
3. Remove the air cleaner bracket.
4. Raise and safely support the vehicle.
5. Remove the right side fender apron clips by pushing the center pin.

NOTE: Do not push the center pin too far in, or it will fall off into the fender.

6. If equipped, remove the power steering and air conditioning accessory belt.
7. Loosen the water pump pulley bolts.
8. Remove the alternator drive belt.
9. Remove the water pump pulley.
10. Remove the crankshaft pulley by performing the following:
   a. If equipped with a manual transaxle, insert a suitable flat bladed tool into the hole in the bell housing next to the exhaust pipe. This will lock the crankshaft in place.
   b. If equipped with a automatic transaxle, hold a suitable flat bladed tool in line with the oil pan and insert the flat bladed tool into the teeth of the drive plate.
   c. Loosen the crankshaft pulley bolts. Some vehicles may be equipped with bolts requiring a 5 mm hexagon tool.
   d. Remove the crankshaft timing belt pulley bolt with special tool 09919–16020 or a 17mm socket.
   e. Remove the pulley from the crankshaft.
   f. Install the crankshaft bolt.
   g. Remove the flat bladed tool that was used to lock the crankshaft in place.

NOTE: On single overhead cam models, to remove the crankshaft pulley with the engine assembly mounted on the body, it is necessary to remove the crankshaft timing belt pulley bolt. If the engine assembly is dismounted, the bolt does not need to be removed.

11. Loosen the right engine mounting bolt and push the air cleaner bracket to the right.
12. Remove the upper and lower timing belt outside covers.
13. Set the camshaft timing belt pulley(s) to their timing marks. The crankshaft mark and camshaft marks are straight up.
14. On single overhead cam models. remove the resonator and the timing belt outside cover.
15. Remove the tensioner stud and loosen the tensioner bolt.
16. Remove the tensioner spring and damper, then remove the timing belt.

——— **WARNING** ———
After the timing belt is removed never turn the camshafts or the crankshaft. Interference may occur between the pistons and the valves causing component damage.

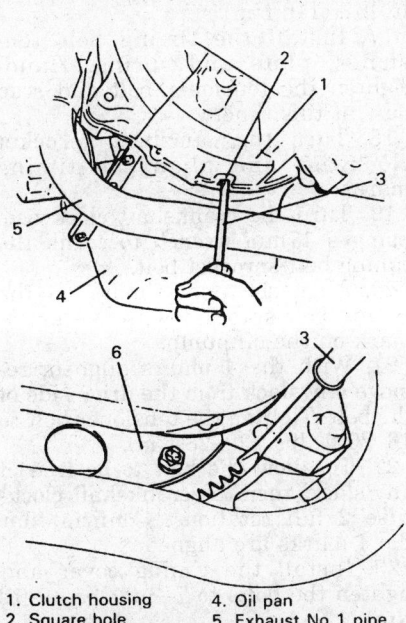

1. Clutch housing
2. Square hole
3. Slotted screwdriver
4. Oil pan
5. Exhaust No.1 pipe
6. Drive plate

177311

Locking the crankshaft — 1.3L engine

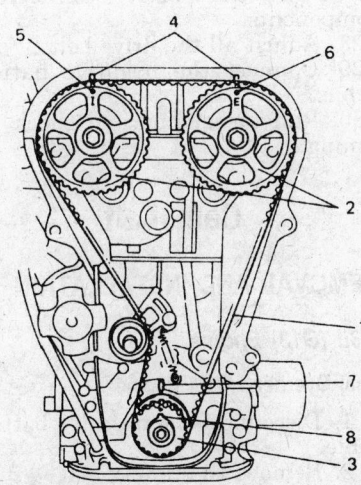

1. Timing belt
2. Camshaft timing belt pulleys
3. Crankshaft timing belt pulleys
4. Marks on cylinder head cover
5. Punch mark by "I" (intake side)
6. Punch mark by "E" (exhaust side)
7. Mark on oil pump case
8. Key on crankshaft

177315

Timing marks — 1.3L DOHC engine

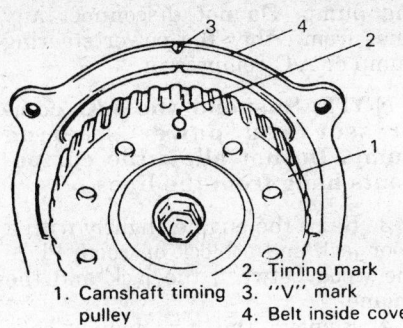

1. Camshaft timing pulley
2. Timing mark
3. "V" mark
4. Belt inside cover

177388

Camshaft sprocket alignment marks — 1993–94 1.3L SOHC engine

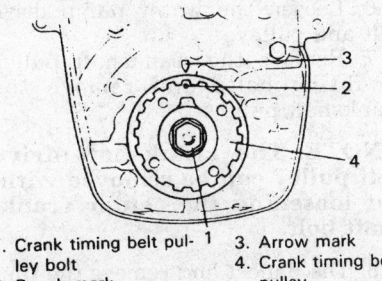

1. Crank timing belt pulley bolt
2. Punch mark
3. Arrow mark
4. Crank timing belt pulley

177389

Crankshaft sprocket alignment marks — 1993–94 1.3L SOHC engine

17. Remove the tensioner and the tensioner plate.
   To install:
18. Install the timing belt tensioner plate and tensioner. Only hand-tighten the tensioner bolt.

NOTE: Make sure that the lug on the tensioner plate is inserted into the hole on the tensioner.

19. Make sure the tensioner plate and the tensioner move uniformly. If they do not move together remove the tensioner and the tensioner plate and reinsert the plate lug into the tensioner hole.
20. Check the camshaft sprockets to verify that they have not moved.

NOTE: On dual overhead cam models, if the sprockets will not stay aligned due to valve spring tensions they can be positioned by using four 8mm bolts and two flanged nuts. Install two bolts into the holes on the head in between the camshafts. Take and install the nuts with the flanges on two other bolts, the flange must face away from the head of the bolt. Position the head of the second pair of bolts on the threaded section of the first bolts, the nut should be facing

up. The nut can be positioned into a groove on the appropriate camshaft sprocket so the sprocket alignment can be adjusted. Turn the flanged nut without having the sprocket resting on the nut so the sprocket is not damaged.

21. Check the crankshaft alignment by verifying that the punch mark on the timing belt pulley(s) is aligned with the arrow on the oil pump case.
22. On dual overhead cam models, install the timing belt on the three pulleys in such a way that there is no slack in the belt. Install the spring and the spring damper and hand-tighten the tensioner stud.

NOTE: If installed, remove the bolts used to secure the camshaft pulleys prior to rotating the engine.

23. On single overhead cam models, remove the cylinder head cover.
24. On 1993–94 single overhead cam models, loosen the all the of the valve adjusting screws on the intake and exhaust rocker arms all of the way, after loosening each locknut.

NOTE: This is to permit the free rotation of the camshaft. When installing the timing belt to the pulleys, the belt should be correctly tensioned by the tensioner spring force. If the camshaft does not rotate freely the belt will not be correctly tensioned.

25. On single overhead cam models, with the timing marks aligned, hold the tensioner plate up by hand and install the timing belt on the pulleys so there is no slack on the drive side of the belt.
26. Turn the crankshaft two rotations clockwise. Confirm that the three sets of timing marks are still properly aligned.
27. If the belt is free of slack and the alignment marks are correct tighten the tensioner stud to 7–8 ft. lbs. (9–12 Nm). Tighten the tensioner bolt to 17–21 ft. lbs. (24–30 Nm).
28. Install the timing belt upper and lower outside covers. Torque the timing cover bolts to 7–8 ft. lbs. (9–12 Nm).
29. Lock the crankshaft in place using a flat bladed tool then remove the crankshaft bolt and install the crankshaft pulley. Torque the crankshaft pulley bolts to 10–13 ft. lbs. (14–18 Nm). Torque the crankshaft timing belt pulley bolt to 76–83 ft. lbs. (105–115 Nm), using special tool

09919–16020 or a 17mm socket. Remove the flat bladed tool that was used to lock the crankshaft.

30. Install the water pump pulley and drive belt.

31. If equipped, install the power steering and air conditioning accessory belt.

32. Reposition the air cleaner bracket then tighten the right side engine mounting bolt.

33. Install the right fender apron.

34. On 1993–94 single overhead cam models, adjust the intake and the exhaust valve lash.

35. Install the cylinder head cover and the air cleaner assembly.

36. Lower the vehicle.

37. Install the air cleaner assembly with the MAF sensor and its outlet hose.

38. Connect the negative battery cable.

39. Start the engine and check the ignition timing.

1.6L (G16) Engine

1. Disconnect the negative battery cable.

2. If equipped with A/C or power steering, remove the drive belt, the A/C compressor, and the power steer-

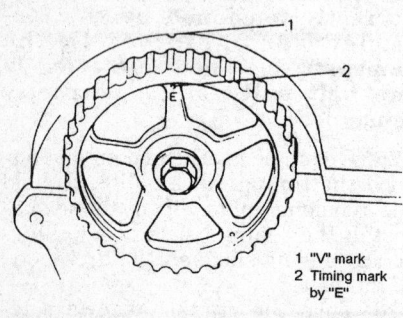

1 "V" mark
2 Timing mark by "E"
327001

Camshaft timing mark alignment — 1995–97 1.3L engine

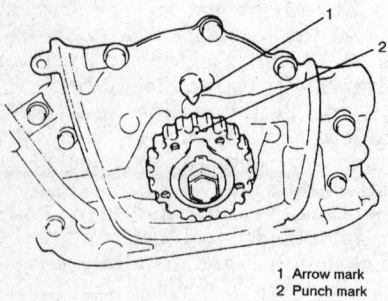

1 Arrow mark
2 Punch mark
327002

Crankshaft timing mark alignment — 1995–97 1.3L engine

ing pump. Do not disconnect any lines from either the power steering pump or A/C compressor.

NOTE: Suspend the A/C compressor and power steering pump. Do not allow the components hang from the lines.

3. Raise the engine slightly with a floor jack and a block of wood. Place the wood between the jack and the engine.

4. Remove the air cleaner case with the air cleaner outlet hose.

5. Remove the engine right mounting bracket and stiffener by removing the nut and bolts.

6. Loosen the water pump drive belt and pulley.

7. Remove the crankshaft pulley mounting bolts and remove the crankshaft pulley.

NOTE: The crankshaft drive belt pulley can be removed without loosening the center crankshaft bolt.

8. Disconnect and remove the timing belt cover bolts and cover. Loosen but do not remove the tensioner bolt.

——— CAUTION ———
After timing belt is removed, never turn the camshaft and crankshaft independently. This engine is an interference engine and if the camshaft or crankshaft is turned beyond a certain point, damage to the valves could occur.

9. Loosen the timing belt tensioner adjusting bolt and pivot nut. Hold pressure on the tensioner to loosen the timing belt and remove the timing belt from the camshaft and crankshaft sprockets.

10. Remove the timing belt tensioner, tensioner plate and tensioner spring.

11. Remove the camshaft timing belt sprocket bolt with tool 09917–68220.

12. Remove the camshaft sprocket mounting bolt and sprocket.

13. Remove the crankshaft sprocket bolt by using a suitable gear stopper to hold the flywheel. Remove the crankshaft timing belt sprocket bolt, sprocket and key.

To install:

14. Install the camshaft sprocket and bolt on the camshaft.

15. Lock the camshaft using the rod and tighten the sprocket bolt to 41–46 ft. lbs. (55–62 Nm). Remove the rod.

16. With the crankshaft locked, install the crankshaft timing belt

sprocket and key. Install the crankshaft sprocket bolt and tighten to 80 ft. lbs. (110 Nm).

17. Install the timing belt tensioner, plate and spring. Hand-tighten the tensioner bolt and stud only at this time.

18. Turn the camshaft sprocket clockwise and align the timing marks.

19. Turn the crankshaft clockwise, using a 17mm wrench to crank the timing belt sprocket bolt.

20. Align the punch mark on the timing belt sprocket with the arrow mark on the oil pump.

21. With the 4 marks aligned, remove any slack from the drive side of the belt. Tighten the tensioner bolt to 16–20 ft. lbs. (22–28 Nm).

22. To allow the belt to be free of any slack, turn the crankshaft clockwise 2 full rotations. Confirm that the 4 marks are aligned.

23. Install the timing cover and tighten the bolts to 7–8 ft. lbs. (9–12 Nm).

24. Install the crankshaft pulley and replace the mounting bolts. Tighten to 10–13 ft. lbs. (14–18 Nm).

25. Install the water pump pulley and water pump drive belt.

26. Instal the engine right mounting bracket and stiffener. Tighten the nuts and bolts to 40 ft. lbs. (55 Nm).

27. Install the remaining components.

28. Adjust all the drive belts.

29. Connect the negative battery cable.

30. Run the engine and check the timing.

Camshaft

REMOVAL AND INSTALLATION

1.3L (G13) Engine

Dual Overhead Cam Engine

1. Disconnect the negative battery cable.

2. Remove the timing belt.

3. Rotate the crankshaft clockwise ninety degrees. This will prevent any interference between the pistons and the valves.

4. Remove the ignition wires from the spark plugs.

5. Disconnect the PCV hose and the breather hose from the cylinder head cover.

6. Remove the bolts from the cylinder head cover and remove the cover.

7. Remove the distributor assembly.

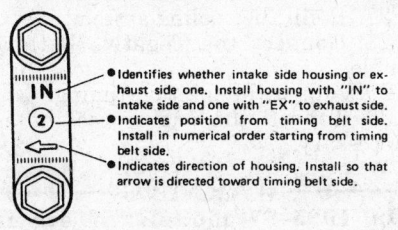

Camshaft housing — 1993–94 1.3L DOHC engine

176681

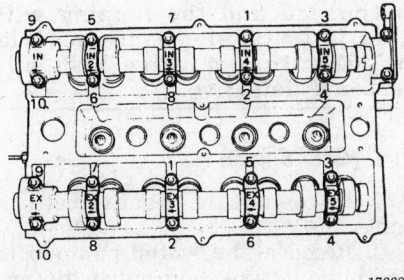

Camshaft housing torque sequence — 1993–94 1.3L DOHC engine

176682

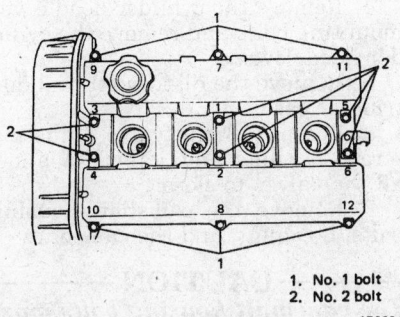

1. No. 1 bolt
2. No. 2 bolt

176684

Cylinder head cover torque sequence — 1993–94 1.3L DOHC engine

8. Remove the camshaft timing belt pulleys. When loosening the attaching bolt, use special tool 09917–68220 for locking the pulley in position is needed.

NOTE: Mark the intake and the exhaust camshaft pulleys so they may be reinstalled on the same camshaft.

9. Remove the camshaft housing bolts in reverse order of the tightening sequence. Remove the housings then remove the camshafts from the cylinder head.

To install:

——— **WARNING** ———

Do not turn the camshafts or start the engine (i.e., the valves should not be operated) for about half an hour after reinstalling the hydraulic valve lash adjusters and the camshafts. It takes time for the valves to settle in place, operating the engine within half an hour after their installation may cause interference to occur between the valves themselves or the valves and the pistons.

10. Fill the oil passage of the cylinder head with engine oil, pour engine oil through the camshaft journal holes. Check to make sure that oil comes out of the oil holes in the sliding part of the valve lash adjuster. Check the intake and the exhaust sides.

11. Apply engine oil around the valve lash adjuster.

12. Apply engine oil to the camshaft journals and set the intake and exhaust camshafts into the cylinder head. Make sure that the intake camshaft is on the intake side of the cylinder head and that the exhaust camshaft is on the exhaust side of the cylinder head.

13. Position the timing belt pulley pin hole in the camshaft at the lower position (6 o'clock position).

NOTE: Embossed marks are on each camshaft housing, these indicate the position and the direction of installation. The embossing identifies IN for intake and EX for exhaust. The number indicates the numbered position starting at the timing belt pulley. The camshaft housings should be installed in numerical order. The arrow indicates the direction of the housing. Install the camshaft housing so the arrow is pointing toward the timing belt.

14. Apply engine oil to the camshaft housing sliding surface of each housing. Apply sealant to the mating surface of the number one camshaft housings, because they will mate with the cylinder head. Make sure that the number one camshaft hous-

ings fit securely since this retains the camshaft in the proper thrust direction.

15. Position the remaining camshaft housings on the camshafts. Apply engine oil to the camshaft housing bolts and start tightening the intake side first. Follow the proper sequence. Tighten the bolts repeating the tightening sequence three or four times before they are tightened to the specified torque. Repeat this tightening procedure for the exhaust side. Torque the camshaft housing bolts to 7–8 ft. lbs. (9–12 Nm).

16. Apply engine oil to the camshaft oil seal lip. Press the camshaft oil seal into the cylinder head till the oil seal surface becomes flush with the housing. This must be done to the intake and the exhaust camshafts.

NOTE: Do not reverse the camshaft sprocket positions for reinstallation. The one that was installed on the intake side should be reinstalled to the same side, and the same with the exhaust side pulley. A new camshaft timing pulley may be used on either side.

17. For the intake camshaft timing belt pulley, fit the pulley pin on the camshaft into the slot marked **I** on the camshaft pulley. Install the timing belt pulley bolt and torque the bolt to 41–46 ft. lbs. (56–64 Nm). Use the special tool for locking the camshaft in place.

18. For the exhaust camshaft timing belt pulley, fit the pulley pin on the camshaft into the slot marked **E** on the camshaft pulley. Install the timing belt pulley bolt and torque the bolt to 41–46 ft. lbs. (56–64 Nm). Use the special tool for locking the camshaft in place.

19. Install the cylinder head cover gasket to the cylinder head cover, replace the gasket if it is damaged or deteriorated.

20. Install the cylinder head cover to the cylinder head. Follow the proper sequence for tightening the cylinder head cover. Torque the bolts closest to the intake manifold and the exhaust manifold to 4–5 ft. lbs. (5–7 Nm). Torque the six bolts near the spark plugs to 6–8 ft. lbs. (8–12 Nm).

21. Install the remaining components.

22. Connect the negative battery cable.

WARNING

If air is trapped in the valve lash adjuster, the valve may make a tapping sound when the engine is operated. In such an case, run the engine for about half an hour at about 2000 RPM, and then the air will be purged and the tapping sound will cease. Should the tapping sound not cease, it is possible that the valve lash adjuster is defective. Replace the valve lash adjuster if necessary.

23. Adjust the ignition timing.

Single Overhead Cam Engines

1. Disconnect the negative battery cable.
2. Drain the cooling system into a resealable container.
3. Remove the air cleaner assembly.
4. Remove the spark plug wire clamps from the cylinder head.
5. Disconnect the PCV hose.
6. Remove the cylinder head cover from the cylinder head.
7. Remove the distributor assembly.
8. On 1993–94 models, loosen all the valve adjusting screw locknuts and screws until the rocker arms move freely.

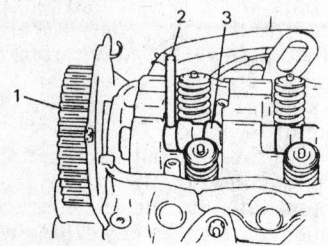

1. Camshaft timing belt pulley   3. Camshaft
2. Proper size rod

325818

Locking the camshaft — 1995–97 1.3L SOHC engine

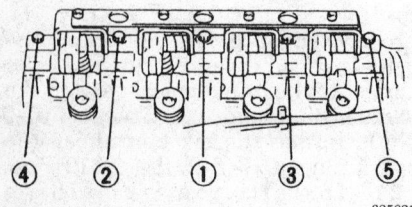

325821

Torque sequence for the rocker arm shaft — 1995–97 1.3L SOHC engine

9. Remove the rocker arm shaft bolts. Be careful not to drop them into the engine.
10. Remove the intake and the exhaust rocker arm shafts.
11. Remove the crankshaft pulley and timing belt.
12. The camshaft must be locked in position to remove the camshaft pulley. Lock the camshaft by inserting a proper sized rod into a hole in the camshaft. The hole for locking the camshaft is located toward the timing belt sprocket.

WARNING

Use care not to bump the rod against the cylinder head when loosening the pulley bolt.

13. With the camshaft locked into position, remove the camshaft pulley bolt and remove the pulley.
14. Remove the distributor case from the cylinder head.
15. Remove the camshaft from the cylinder head.

To install:
16. Apply engine oil to the camshaft lobes and journals and to the camshaft oil seal on the cylinder head.
17. Install the camshaft into the cylinder head from the transmission case side.
18. Install the distributor case to the cylinder head.
19. Lock the camshaft in place.
20. Install the camshaft timing belt pulley by fitting the pulley pin on the camshaft into the slot on the camshaft pulley. Install the attaching bolt and torque the bolt to 41–46 ft. lbs. (56–64 Nm)
21. Install the timing belt.
22. Apply engine oil to the rocker arms and the rocker arm shafts.

NOTE: The two rocker arm shafts are different. They are distinguishable by the stepped ends of the shafts. Looking at the screw holes in the intake shaft, one end of the shaft will have two sides that are stepped; this end goes toward the camshaft pulley. Looking at the screw holes in the exhaust rocker shaft, one end will have a step on one side only; this end goes toward the distributor.

23. Install the intake and the exhaust rocker arms and shafts, Torque the bolts to 16–20 ft. lbs. (22–28 Nm) in the proper sequence.
24. Install the distributor to the camshaft and its case.
25. On 1993–94 models, adjust the intake and exhaust valve lash.

26. Install the cylinder head cover and the air cleaner assembly.
27. Refill the cooling system.
28. Connect the negative battery cable.
29. Adjust the ignition timing.
30. Top off the engine coolant as necessary.

WARNING

On 1995–97 models, if air is trapped in a valve lash adjuster, the valve may make a tapping sound when the engine is operated after a valve lash adjuster is installed. In such a case, run the engine for about half an hour at 2,000 RPM, And then the air will be purged and the tapping will stop. If the tapping continues it is possible that a valve lash adjuster is defective.

1.6L (G16) Engine

1. Disconnect the negative battery cable.
2. Remove the water pump belt and pulley, crankshaft pulley, timing belt cover, and the timing belt.
3. Using tool 09917–68220 or equivalent, remove the camshaft sprocket.
4. Remove the cylinder head cover mounting bolts and remove the cylinder head cover.
5. Remove the distributor and distributor case.
6. Loosen all the valve adjusting screw locknuts and screws to allow all the valves to close.
7. Remove the camshaft housing bolts, housings, and the camshaft.

CAUTION

The camshaft housing bolts must be removed in the reverse order of installation or damage to the camshaft may occur.

To install:
8. Lubricate the lobes and journals of the camshaft with clean engine oil.
9. Install the camshaft on the cylinder head. Install the camshaft housing to the camshaft and cylinder head, starting with the number one housing.

NOTE: Embossed marks are provided on each camshaft housing, indicating position and direction for installation.

10. Apply engine oil to the sliding surface of each housing against the camshaft journal. Apply sealant to the mating surface of the number six

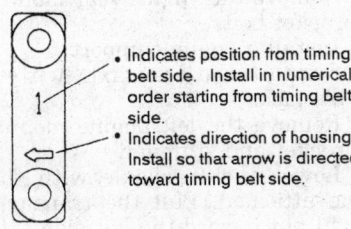

Camshaft housing identification — 1.6L engine

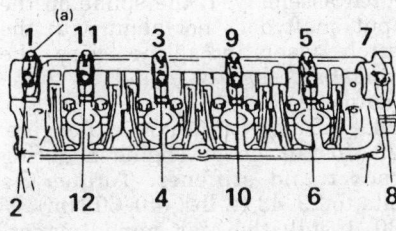

Camshaft housing bolt tightening sequence — 1.6L engine

housing which will mate with the cylinder head.

11. Apply engine oil to the housing bolts, and hand-tighten the bolts into the housing. Follow the tightening sequence in three to four even stages, finishing with a final torque of 7–8 ft. lbs. (9–12 Nm).

————— **CAUTION** —————
The camshaft housing bolts must be tightened in the correct order or damage to the camshaft may occur.

12. Apply engine oil to the camshaft oil seal lip. Install the camshaft oil seal until the surface becomes flush.

13. Reconnect the camshaft sprocket, timing belt, timing belt cover, crankshaft pulley, water pump pulley, and the water pump belt. Make sure the pin on the camshaft fits into the slot at the **E** mark on the camshaft sprocket. Tighten the sprocket bolt to 41–46 ft. lbs. (56–64 Nm).

14. Prior to installation, apply sealant to the area of the distributor housing that covers the rear of the rocker arm shaft on the cylinder head. Install the distributor and distributor housing to the cylinder head.

Make sure the distributor is facing the correct firing position.

15. Adjust the valve lash.

16. Using a new gasket, install the cylinder head cover.

17. Connect the negative battery cable to the battery.

18. Start the engine and check for any water or oil leaks when finished.

19. Check and/or adjust the ignition timing as necessary.

Piston and Connecting Rod

POSITIONING

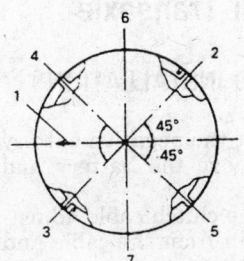

1. Arrow mark
2. 1st ring end gap
3. 2nd ring end gap and oil ring spacer gap
4. Oil ring upper rail gap
5. Oil ring lower rail gap
6. Intake side
7. Exhaust side

Piston ring positioning

ENGINE LUBRICATION

Oil Pan

REMOVAL AND INSTALLATION

1. Disconnect the negative battery cable.

2. Safely raise and support the vehicle.

3. On 1.6L models, remove the engine undercovers.

4. Drain the engine oil into a suitable container.

5. On 1.3L models, remove the lower plate, from the clutch housing if equipped with a manual transaxle, or from the torque converter housing if equipped with a automatic transaxle.

6. On 1.6L models, perform the following steps;
   a. Remove the front exhaust pipe from the vehicle.
   b. Remove the transaxle stiffener plate from the engine and transaxle.

c. Support the transmission and engine.
   d. Remove the vehicle center member by removing the seven nuts and four bolts from the center member.

7. Remove the crankshaft position sensor from the oil pan, if equipped.

8. Remove the oil pan retainer bolts. Remove the oil pan from the cylinder block.

9. Remove the oil pump pickup.

To install:

10. Clean the mating surfaces of the oil pan and the engine block.

11. Install the oil pump strainer. Tighten the strainer bolt first at the bracket. Torque the bolts to 7–8 ft. lbs. (9–12 Nm).

12. Apply silicon sealant to the oil pan mating surface in one continuous bead.

13. Fit the oil pan to the engine block and install the bolts. Start tightening the bolts at the center and move outward. Torque the bolts to 7–8 ft. lbs. (9–12 Nm).

14. Install the crankshaft position sensor to the oil pan.

15. Install the drain plug and drain plug gasket to the oil pan.

16. On 1.3L models, install the lower plate, to the clutch housing if equipped with a manual transaxle, or to the torque converter housing if equipped with an automatic transaxle.

17. On 1.6L models, do the following;
   a. Install the center member to the vehicle. Torque the bolts and nuts at the engine mounts to 40 ft. lbs. (55 Nm) and all other nuts to 33 ft. lbs. (45 Nm).
   b. Install the transaxle stiffener plate and torque the bolts to 37 ft. lbs. (50 Nm).
   c. Install the exhaust pipe and tighten the nuts and bolts to 37 ft. lbs. (50 Nm).
   d. Install the engine undercovers.

18. Lower the vehicle.

19. Refill the engine with oil.

20. Connect the negative battery cable.

21. Start the engine and check for leaks.

Oil Pump

REMOVAL AND INSTALLATION

1. Disconnect the negative battery cable.

2. Safely raise and support the vehicle.

3. Drain the engine oil.

4. Remove the right side fender apron clips by pushing the center pin.

─── **WARNING** ───

Do not push the center pin too far in, or it will fall off into the fender.

5. If equipped, remove the power steering and air conditioning belt.

6. Loosen the water pump pulley bolts.

7. Remove the alternator drive belt.

8. Remove the water pump pulley and alternator bracket.

9. Remove the crankshaft pulley, the timing belt outside covers, timing belt guide and the timing belt.

10. Remove the engine oil level gauge.

11. If equipped, remove the air conditioning compressor bracket bolts.

12. Remove the timing belt and crankshaft pulley.

13. Remove the oil pan and oil pump pickup.

14. Remove the seven bolts securing the oil pump to the engine block and remove the oil pump.

To install:

15. Clean the engine block where the oil pump mounts, then install the oil pump gasket and the two oil pump alignment pins.

16. To prevent damage to the oil seal when installing the oil pump, fit special tool 09926-18210 to the crankshaft and apply a thin coating of engine oil to the special tool.

17. Install the oil pump to the crankshaft and the engine block. Install a long bolt to the lowest bolt hole on the intake manifold side of the engine. Install two long bolts to the two lowest bolt holes on the exhaust manifold side of the engine. Install the four short bolts to the other four bolt holes in the oil pump. Torque all of the bolts to 7-8 ft. lbs. (9-12 Nm). Check that the oil seal lip is not turned up, then remove the special tool.

18. If the of the oil pump gasket bulges where the oil pan attaches cut the excess off with a sharp knife.

19. Install the oil pan and the oil pump pickup.

20. Install the rubber seal between the oil pump and the water pump.

21. Install the crankshaft key, timing belt pulley and the crankshaft pulley pin.

22. With the crankshaft locked, install and tighten the pulley bolt to 80 ft. lbs. (110 Nm).

23. Install the timing belt guide so that the concave side faces the oil

pump then install the crankshaft key and the timing belt pulley.

24. Install the remaining components.

25. Connect the negative battery cable.

26. Start the engine and check the engine oil pressure.

27. Check that no leaks are present.

TRANSAXLE

Manual Transaxle

REMOVAL AND INSTALLATION

1. Disconnect the negative battery cable, then remove the battery and tray.

2. Remove the clutch cable adjusting nut, joint pin from the cable and cable from the bracket.

3. Disconnect and tag all the wiring harness clamps and connectors involved with the transaxle removal.

4. Remove the speedometer cable boot, case clip and cable from the case.

5. On Swift models, remove the radiator outlet pipe from the transmission side cover.

6. Remove the transaxle retaining bolts.

7. Remove the starter.

8. Raise and support the vehicle safely. Drain the transaxle oil.

9. Remove the fender apron extension on the left side.

10. Remove the bolts connecting the exhaust pipe to the exhaust manifold and disconnect the joint.

11. On Esteem models, with the engine supported, remove the vehicle center mounting member by removing the seven nuts and four bolts.

12. Remove the gearshift control shaft nut and bolt, then disconnect the control shaft from the gear shift control shaft.

13. Remove extension rod nut and washers.

14. Remove the clutch housing lower plate.

15. Remove the sway bar.

16. Remove the left and right front wheels.

17. Disconnect the left and right ball joints.

18. Remove the left and right halfshafts.

19. Remove the transaxle stiffener.

20. Remove the transmission to engine bolt and nut.

21. Remove the engine rear mounting bracket bolts.

22. Install a engine support.

23. Support the transaxle with a suitable jack.

24. Remove the left engine mounting bracket and stiffener.

25. Lower the transaxle with the engine attached. Pull the transaxle straight out toward the left side.

26. Lower and remove the transaxle.

To install:

27. Install the transaxle from the left side of the vehicle. Use care when inserting the pilot shaft into the clutch assembly. If the spline on the input shaft does not align with the clutch assembly spline, turn the crankshaft slightly to aid in spline alignment.

28. Raise the transaxle and engine.

29. Install the left engine mounting bracket and stiffener. Torque the bolts to 29-43 ft. lbs. (40-60 Nm).

30. Install the rear engine mounting bracket bolts and torque them to 29-43 ft. lbs. (40-60 Nm).

NOTE: Before installing the bolts into the rear mounting bracket, apply sealant to the bolt threads.

31. Install the transmission-to-engine bolt and nut. Torque the nut and bolt to 29-43 ft. lbs. (40-60 Nm).

32. On Esteem models, install the center member to the vehicle and install the seven nuts and four bolts. Torque the bolts and nuts as follows:

   a. Center member to the radiator support to 33 ft. lbs. (45 Nm).

   b. Center member to cross-member to 33 ft. lbs. (45 Nm).

   c. Engine mounts to center member nuts to 40 ft. lbs. (55 Nm).

33. Install the transaxle stiffener.

34. Lower the transaxle supporting jack.

35. Install the left and right halfshafts.

36. Install the ball joints.

37. Connect the sway bar.

38. Install the clutch housing lower plate.

39. Install the extension rod nut and washers. Torque the rod nut to 18-28 ft. lbs. (25-40 Nm).

40. Install the control shaft to gear shift and install the gear shift control shaft bolt and nut. Torque the gear shift control shaft bolt and nut to 11-14 ft. lbs. (15-20 Nm).

41. Connect the exhaust pipe to the manifold and install the bolts. Torque the bolts to 29-36 ft. lbs. (40-50 Nm).

42. Install the left fender apron extension.

43. Refill the transaxle with the recommended lubricant.

44. Lower the vehicle.

45. Remove the engine support fixture.

46. Install the starter.

47. Install the transaxle retaining bolts. Torque the retaining bolts to 29–43 ft. lbs. (40–60 Nm).

48. Install the remaining components.

49. Install the negative battery cable and the ground strap to the transaxle.

Clutch Assembly

REMOVAL AND INSTALLATION

1. Remove the transaxle.

2. Hold the flywheel stationary with tool 09924–17810 or equivalent.

3. Matchmark the pressure plate and flywheel for installation reference.

4. Loosen the pressure plate attaching bolts, one turn at a time (evenly) until the spring pressure is released.

5. Remove the clutch disc and pressure plate.

To install:

6. Clean the flywheel mating surfaces of all oil, grease and metal deposits. Inspect flywheel for cracks, heat checking or other defects and replace or resurface as necessary.

7. Check the wear on the facings of the clutch disc by measuring the depth of each rivet head depression. Replace clutch disc when rivet heads are 0.02 in. (0.5 mm) below the surface of clutch surface.

8. Check the diaphragm spring and pressure plate for wear or damage. If the spring or plate is excessively worn, replace the pressure plate assembly.

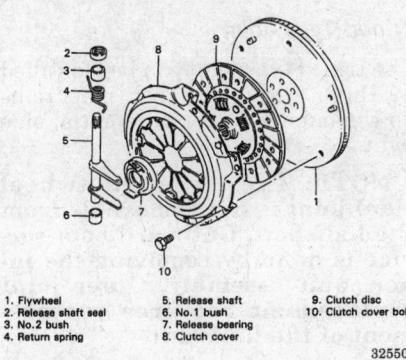

1. Flywheel
2. Release shaft seal
3. No.2 bush
4. Return spring
5. Release shaft
6. No.1 bush
7. Release bearing
8. Clutch cover
9. Clutch disc
10. Clutch cover bolt

325502

Clutch component identification

9. Check the pilot bearing for smooth operation. If the bearing does not spin freely, replace it.

10. Position the clutch disc and pressure plate with the matchmarks aligned and install a clutch alignment tool 09923–36330 or equivalent.

11. Install the pressure plate bolts. Tighten the mounting bolts evenly and in a crisscross pattern to 13–20 ft. lbs. (18–28 Nm). Remove the alignment tool and the flywheel holding tool.

12. Lightly lubricate the transaxle input shaft splines, pilot bearing surface of the input shaft, and the release bearing with grease.

13. Install the transaxle.

14. Adjust the clutch cable.

Clutch Cable

ADJUSTMENT

1. Depress the clutch pedal lightly until tension on the clutch cable can be felt.

2. Measure the clutch pedal free-play; it should be 0.6–0.8 in. (15–20mm).

3. Adjust the clutch pedal free-play by tightening or loosening the clutch cable adjustment nut.

4. Measure the clutch lever free-play; it should be 0.0–0.08 in. (0–2mm). If the clutch release lever free-play exceeds specification, inspect the release shaft return spring for cracks or weakness.

NOTE: Make sure the marks on the clutch release lever and release shaft are aligned. If they are not, remove the lever from the shaft, align the marks and repeat the free-play adjustment procedure.

REMOVAL AND INSTALLATION

1. Disconnect the negative battery cable.

2. Remove the clutch cable joint nut, then disconnect the joint pin from the release arm.

3. Remove the two bolts that attach the clutch cable to the cowl wall.

4. Unhook the cable end from the pedal top. Use a flat bladed tool from the engine compartment.

5. Remove the cable from the vehicle.

To install:

6. Apply grease to the cable end hook and joint pin.

7. Hook the cable end to the pedal with a flat bladed tool or long nose pliers.

8. Connect the inner cable to the joint pin on the release arm.

9. Apply sealant to the mating surface of the cable, where it contacts the cowl wall.

10. Position the cable to the cowl wall and install the two mounting bolts. Torque the bolts to 3–5 ft. lbs. (4–7 Nm).

11. Install the joint nut to the inner cable and adjust the free-play of the pedal to specification.

12. Connect the negative battery cable.

13. Check the clutch for proper operation with the engine running.

Automatic Transaxle Assembly

REMOVAL AND INSTALLATION

1. Disconnect the negative battery cable from the battery and transaxle.

2. Disconnect the speedometer cable.

3. Disconnect the electrical connector for the solenoids, vehicle speed sensor, shift lever switch, forward clutch cylinder revolution sensor and vehicle speed sensor (for A/T).

4. Remove the wire harness from the clamps on the transaxle.

5. Disconnect the select cable from the transaxle.

6. Drain the cooling system.

7. Remove the cooling system pipe from the transaxle.

8. Remove the top transaxle-to-engine bolts.

9. Remove the starter.

10. Remove the exhaust manifold cover. Disconnect the front exhaust pipe from the exhaust manifold.

11. Support the engine.

12. Lift the vehicle and safely support.

13. Drain the transaxle fluid from the transaxle.

14. Remove the engine undercovers, if equipped.

15. Place an oil pan under the transaxle and disconnect the cooler hoses.

16. On Esteem models, with the engine supported, remove the vehicle center mounting member by removing the seven nuts and four bolts.

17. Remove the front exhaust pipe from the vehicle.

18. Remove the transaxle stiffener plate by removing the bolts.

19. Remove the transaxle housing lower plate.

20. Remove the torque converter bolts. To lock the drive plate, engage a flat bladed tool in the flywheel.

21. Disconnect the sway bar from the control arms.

22. Remove the left and right front wheels.

23. Disconnect the left and right ball joints from the steering knuckles.

24. Remove the left and right halfshafts.

25. Remove the engine rear mount and bracket.

26. After removing the rear mount, remove the engine-to-transaxle bolt and nut located behind the rear bracket. Remove all bolts holding the engine to the transaxle.

27. Support the transaxle with a transaxle jack.

28. Remove the bolts from the engine left hand mount.

29. Remove the transaxle with the torque converter from the engine compartment.

NOTE: When removing the transaxle from the engine, move it parallel with the crankshaft and use care so not to apply excessive force to the drive plate and torque converter.

—— CAUTION ——
Be sure to keep the transaxle with the torque converter horizontal or facing up throughout the work. Should it be tilted with converter down, the converter may fall off and cause personal injury.

To install:

30. Install the transaxle to the engine assembly and install the attaching nuts and bolts.

31. Install the left hand mounting bolts. Torque the bolts to 40 ft. lbs. (55 Nm).

32. Install the engine-to-transaxle bolt and nut before installing the rear transaxle mount. Torque the nut to 65 ft. lbs. (90 Nm).

33. Install all bolts for the transaxle. Torque the bolts to 65 ft. lbs. (90 Nm).

34. Install the left halfshafts.

35. Install the ball joints.

36. Connect the sway bar to the control arms.

37. Install the torque converter bolts and torque the bolts to 14 ft. lbs. (19 Nm).

38. Install the transaxle housing lower plate.

39. Install the stiffener plate with the four bolts. Torque the bolts to 40 ft. lbs. (55 Nm).

40. Install the exhaust pipe to the center pipe and torque the bolts to 37 ft. lbs. (50 Nm).

41. On Esteem models, install the center member to the vehicle and install the seven nuts and four bolts. Torque the bolts and nuts as follows:

   a. Center member to the radiator support to 33 ft. lbs. (45 Nm).

   b. Center member to crossmember to 33 ft. lbs. (45 Nm).

   c. Engine mounts to center member nuts to 40 ft. lbs. (55 Nm).

42. Connect the oil hoses for the transaxle.

43. Install the engine undercovers.

44. Lower the vehicle.

45. Remove the engine support.

46. Connect the exhaust pipe to the exhaust manifold and install the nuts.

47. Install the exhaust manifold cover.

48. Install the starter motor.

49. Install the remaining components.

50. Fill the cooling system and the transaxle. Check all fluids.

51. Connect the negative battery cable to the transaxle and battery.

DRIVE AXLE

Halfshaft

REMOVAL AND INSTALLATION

1. Disconnect the negative battery cable.

2. Undo the caulking on the halfshaft nut, then remove the nut and washer.

3. Safely raise and support the vehicle.

4. Drain the oil from the transmission.

5. Use two large prybars, release the snapring fitting on the halfshaft inner joint from the differential.

6. Remove the sway bar attaching nut, washer and bushing from the suspension arm.

7. Remove the ball joint stud bolt and nut.

8. Pull the inboard joint from the differential then disconnect the outer joint from the steering knuckle.

9. Remove the halfshaft from the vehicle.

—— WARNING ——
To prevent breakage of the boots, be careful not to bring the boots in contact with other components when removing the shaft assembly.

10. If the center shaft requires service, drain the transmission oil and remove the support bolts. Remove the center shaft from the differential then from the vehicle.

To install:

11. If the center shaft was removed, install the shaft into the differential then install the support bolts. Torque the support bolts to 29–43 ft. lbs. (40–60 Nm).

12. Clean the grease seal on the steering knuckle and apply a small amount of fresh grease.

13. Install the wheel side joint to the steering knuckle and then the differential side joint to the differential. Seat the differential joint by hand, making sure that the snapring is seated. Install the halfshaft nut to the outer joint loosely, to hold it in position.

—— WARNING ——
Do not hit the joints with a hammer to seat them use your hands only or component damage may occur.

14. Connect the suspension arm to the steering knuckle and install the ball joint stud.

15. Position the sway bar on the suspension arm and install the sway bar bushing, washer and nut. Torque the nut to 17–20 ft. lbs. (23–28 Nm).

16. If equipped with a manual transaxle fill the transaxle with the specified gear oil.

17. Lower the vehicle.

18. Torque the halfshaft nut to 109–145 ft. lbs. (150–200 Nm).

19. Connect the negative battery cable.

20. If equipped with a automatic transaxle, fill the transaxle with specified transmission oil.

CV-Joint Boot

REPLACEMENT

Tripod Type Joint

The tripod type joint can be identified by three cylinder shaped indentions on the outside of the differential side joint.

NOTE: The outboard (wheel side) joint is not removable from the halfshaft. Outboard boot service is done by removing the inner joint assembly. Outer joint replacement requires replacement of the halfshaft.

1. Raise and safely support the vehicle.

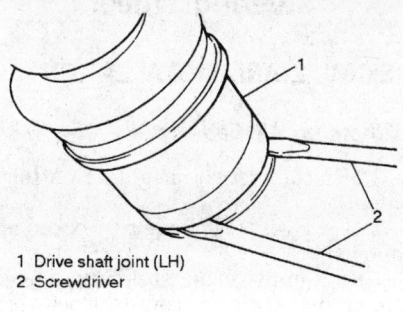

1 Drive shaft joint (LH)
2 Screwdriver

326117

Disconnecting the left inboard joint

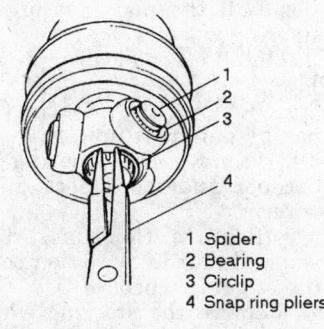

1 Spider
2 Bearing
3 Circlip
4 Snap ring pliers

194315

Circlip removal of the tripod bearing

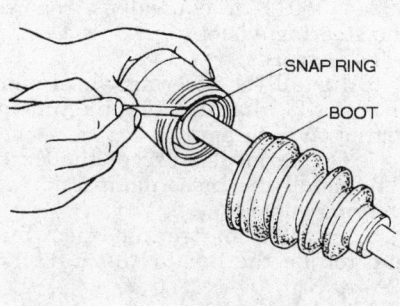

SNAP RING
BOOT

194312

Snapring removal from the outer race

2. Remove the halfshaft from the vehicle and place in a vise equipped with jaw covers.

3. Remove the large boot clamp from the inner joint and discard.

4. Remove the tripod joint housing.

5. Wipe the grease from the spider joint and the end of the shaft.

— **WARNING** —

To prevent the needle bearings of the joint from being damaged, do not wash it if reusing.

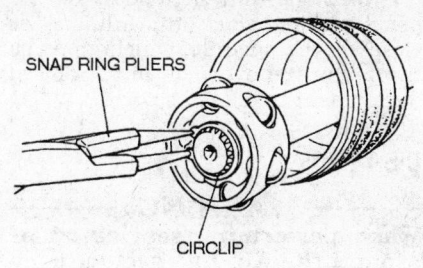

SNAP RING PLIERS

CIRCLIP

194314

Circlip removal for the cage bearing

6. Remove the outer circlip from the shaft.

7. Attach a bearing puller to the tripod bearing and remove the bearing from the shaft.

— **WARNING** —

When removing a tripod bearing, be sure to use a bearing puller with three arms. Removing it otherwise may cause damage to the component. Do not disassemble the tripod joint spider. If any component damage is found, replace it.

8. Remove the small inboard boot clamp and discard.

9. Remove the inboard boot.

10. If removing the outboard CV boot, remove the boot clamps and slide the boot off the inboard end.

To install:

11. Clean the outer race assembly to remove the old grease and any other contaminates.

12. Install the outboard boot over the shaft and slide down. Fill the boot with grease and install the boot over the outer joint. Install the clamps. Use the specified black grease only.

13. Install the small clamp on the shaft and install the boot to the shaft.

14. Install the spider joint with the aid of a piece of pipe with a inner diameter of 0.9 in. (23 mm) or more and a outer diameter of 1.2 in. (32 mm) or less. The chamfered side of the tripod joint should be toward the center of the shaft. Then install the circlip to the ring groove.

15. Install the tripod housing packed with grease. Use the specified yellow grease only.

16. Slide the boot into place. Insert a flat bladed tool into the boot on the outer race side. This will allow air to enter the boot so that the air pressure in the boot becomes the same as atmospheric pressure. Secure the boot with the mounting clamps.

17. Install the halfshaft in the vehicle.

18. Lower the vehicle.

Bearing Cage Type Joint

1. Raise and safely support the vehicle.

2. Remove the halfshaft from the vehicle and place in a vise equipped with jaw covers.

3. Remove the large boot clamp from the inner joint and discard.

4. Slide the boot toward the center of the shaft.

5. Remove the snapring from the outer race then take the shaft out.

6. Wipe the grease off of the bearing cage, then remove the circlip.

7. Remove the bearing cage from the shaft with the aid of a bearing puller.

8. Remove the small inboard boot clamp and discard.

9. Remove the inboard boot.

10. If removing the outboard CV boot, remove the outboard boot clamps and slide the boot off the inboard end.

To install:

11. Clean the outer race assembly to remove the old grease and any other contaminates.

12. Install the outboard boot over the shaft and slide down. Fill the boot with grease and install the boot over the outer joint. Install the clamps. Use the specified black grease only.

13. Install the small clamp on the shaft and push the boot onto the shaft. Fit the small end of the boot to the shaft groove and affix the boot with the boot band.

14. Position the bearing cage on the shaft with the smaller outside diameter end of the cage toward the center of the shaft. Install the bearing cage to the shaft with the aid of a piece of pipe with a inner diameter of 0.9 in. (23 mm) or more and an outer diameter of 1.2 in. (32 mm) or less.

15. Install the circlip in the ring groove. Apply grease to the entire surface of the bearing cage.

16. Install the bearing cage into the outer race packed with grease and fit the snapring into the groove. Position the opening of the snapring so that it will not be lined up with a bearing ball.

17. Apply grease to the inside of the outer race and the boot and slide the boot into place. After fitting the boot in to place, insert a flat bladed tool into the boot on the outer race side. This will allow air to enter the boot so that the air pressure in the boot becomes the same as atmospheric pressure.

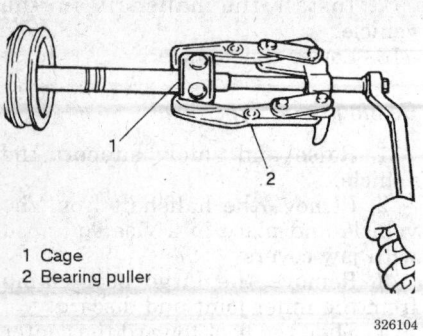

1 Cage
2 Bearing puller

326104

Removal of the bearing cage

18. Fix the boot to the outer race with the boot band.
19. Install the halfshaft in the vehicle.
20. Lower the vehicle.

STEERING

Air Bag

——— CAUTION ———

Some vehicles are equipped with an Air Bag system, also known as the Supplemental Inflatable Restraint (SIR) or Supplemental Restraint System (SRS). The system must be disabled before performing service on or around system components, steering column, instrument panel components, wiring and sensors. Failure to follow safety and disabling procedures could result in accidental Air Bag deployment, possible personal injury and unnecessary system repairs.

PRECAUTIONS

Several precautions must be observed when handling the inflator module to avoid accidental deployment and possible personal injury.

• Never carry the inflator module by the wires or connector on the underside of the module.
• When carrying a live inflator module, hold securely with both hands, and ensure that the bag and trim cover are pointed away.
• Place the inflator module on a bench or other surface with the bag and trim cover facing up.

• With the inflator module on the bench, never place anything on or close to the module which may be thrown in the event of an accidental deployment.

DISARMING

——— WARNING ———

When performing service on or around the Air Bag system components or wiring, disable the Air Bag system. Failure to follow the procedures could result in possible deployment, personal injury or unneeded system repairs.

1. Disconnect the negative battery cable.
2. Turn the steering wheel so the wheels are pointing straight ahead.
3. Turn the ignition switch to the **LOCK** position and remove the key.
4. Remove the **AIR BAG-IG** fuse from the Air Bag fuse box located near the junction/fuse box.
5. Remove the left side steering wheel side cap and disconnect the yellow connector for the driver side Air Bag (inflator) module.
6. Pull out the glove box while pushing in on the stoppers from the left and the right sides. Disconnect the yellow connector for the passenger Air Bag (inflator) module.

ARMING

——— WARNING ———

When performing service on or around the Air Bag system components or wiring, disable the Air Bag system. Failure to follow the procedures could result in possible deployment, personal injury or unneeded system repairs.

1. Connect the negative battery cable.
2. Turn the ignition switch to the **LOCK** position and remove the key.
3. Connect the yellow connector for the passenger side Air Bag (inflator) module and the yellow connector for the driver Air Bag (inflator) module. Be sure to lock each connector with the lock lever.
4. Install the glove box assembly.
5. Install the left side steering wheel side cover.
6. Install the **AIR BAG-IG** fuse to the Air Bag fuse box.
7. Turn the ignition **ON** and verify that the **AIR BAG** warning lamp flashes seven times, then turns off. If the system does not operate as described, diagnosis and repairs to the Air Bag system are necessary.

Steering Wheel

REMOVAL AND INSTALLATION

Without an Air Bag

1. Disconnect the negative battery cable.
2. On Swift GT models, proceed as follows:
   a. Remove the steering wheel pad, by turning it counterclockwise while pressing in. Turn until it stops and remove it.
   b. Disconnect the horn electrical wire from the steering wheel.
   c. Pull the pad seat off of the steering wheel.
3. On all other models, proceed as follows;
   a. Remove the steering wheel pad by pulling it upward.
   b. Remove the steering wheel bumper and disconnect the horn wire.
   c. Remove the mass damper screws and the mass damper from the steering wheel.
4. Remove the steering wheel attaching nut. Make alignment marks on the steering wheel and the shaft for a guide during reinstallation.
5. Use a suitable puller, tool 09944–36011 or equivalent, remove the steering wheel.
   To install:
6. Install the steering wheel onto the shaft, aligning the alignment marks on each part.
7. On all models except the Swift GT, install the mass damper to the steering wheel.
8. Install the steering wheel nut and torque the nut to 18–28 ft. lbs. (25–40 Nm).
9. On Swift GT models, install the steering wheel pad seat to the steering wheel pad assembly.

NOTE: Use special care when installing the pad seat because of the relation of the position between the contact point of the horn switch on the steering wheel pad and boss on the pad seat

10. On Swift GT models, connect the horn wire to the steering wheel pad and then, install the pad with the seat to the steering wheel.

NOTE: The boss on the pad seat should go into the hole in the steering wheel.

11. On all other models, connect the steering wheel bumper horn wire to the steering wheel and install the

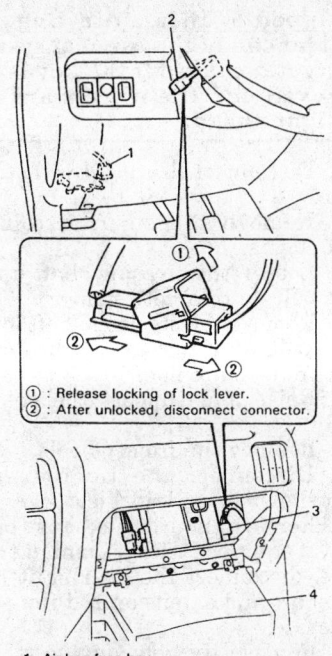

1 Air bag fuse box
2 Yellow connector of driver air bag (inflator) module
3 Yellow connectors of passenger air bag (inflator) module
4 Glove box

329942

Disabling the Air Bag

bumper assembly. Install the steering wheel pad.

12. Connect the negative battery cable.

With an Air Bag

———— WARNING ————
When performing service on or around the Air Bag system components or wiring, disable the Air Bag system. Failure to follow the procedures could result in possible deployment, personal injury, or unneeded system repairs.

1. Disconnect the negative battery cable.

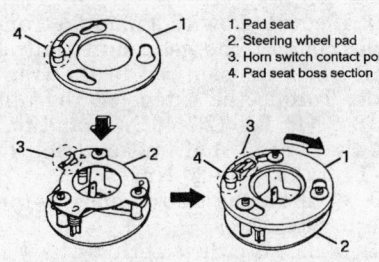

1. Pad seat
2. Steering wheel pad
3. Horn switch contact point
4. Pad seat boss section

196504

Steering wheel pad and seat assembly — Swift

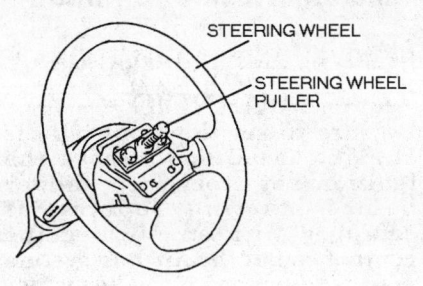

STEERING WHEEL
STEERING WHEEL PULLER

196506

Removal of the steering wheel

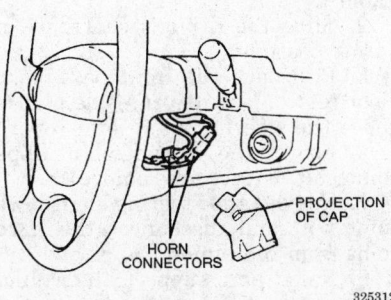

PROJECTION OF CAP
HORN CONNECTORS

325312

Horn wire connectors location and steering wheel cap — Swift

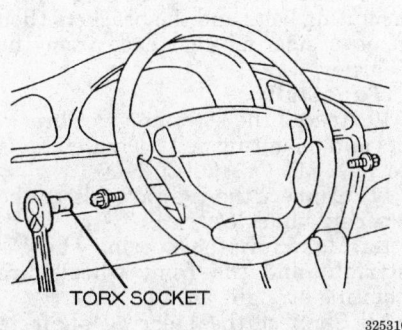

TORX SOCKET

325310

Removal of the Air Bag module mounting bolts — Swift

2. Disable the Air Bag system.

3. Remove the right steering wheel side cap. Disconnect the horn electrical connectors that are attached to the side cap.

4. Remove the two Air Bag module mounting bolts.

5. Remove the Air Bag module from the steering wheel.

———— CAUTION ————
When carrying a live Air Bag, make sure the Air Bag and cover are pointed away from the body. In the event of accidental deploy-

ment, the bag will deploy with the minimal chance of injury. When placing an Air Bag on a bench or other surface, face the Air Bag and trim cover up, away from the surface. This will reduce the motion of the module if it is accidentally deployed.

6. Remove the steering wheel attaching nut. Make alignment marks on the steering wheel and the shaft for a guide during reinstallation.

7. Use a suitable puller, tool 09944–36010 or 36011, or equivalent, remove the steering wheel.

———— WARNING ————
Do not turn the contact coil (on the combination switch) more than the allowable number of turns (about two and a half turns from the center position clockwise or counterclockwise respectively), or the coil will break.

To install:
8. Check that the front tires are in the straight-ahead position and the contact coil is centered.

———— WARNING ————
These two conditions are prerequisite for installation of the steering wheel. If the steering wheel has been installed without these conditions, the contact coil will break when the steering wheel is turned.

9. Install the steering wheel to the steering shaft with the two lugs on the contact coil fitted in the two grooves in the back of the steering wheel and also aligning the marks on the steering wheel and steering shaft.

10. Install the steering wheel nut and torque the nut to 18–28 ft. lbs. (25–40 Nm).

11. Position the contact coil wire through the left inner hole in the steering wheel and position the horn wire through the right inner hole in the steering wheel.

12. After making sure that the horn wire is fitted into the rib on the back of the Air Bag module, install the Air Bag module to the steering wheel. Take care so that no part of the wire harness is caught between the inflator module or the steering wheel.

13. Install the inflator module mounting bolts, torque the left bolt first then the right side bolt, torque the bolts to 13–20 ft. lbs. (18–28 Nm). Make sure that the clearance between the module and steering wheel is uniform all the way around.

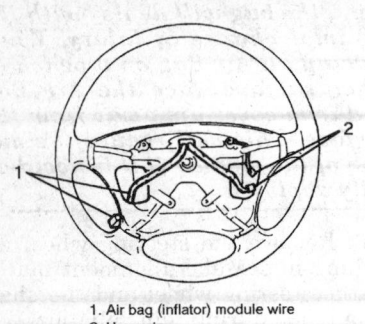

1. Air bag (inflator) module wire
2. Horn wire

325316

Positioning the contact coil lead wires — Swift

14. Fasten the horn connectors securely and fit the connector onto the projection on the right side steering wheel cap.

15. Install the right steering wheel cap with the horn wires attached.

16. Connect the negative battery cable.

17. Enable the Air Bag system.

18. Turn the ignition on and verify that the **AIR BAG** warning lamp flashes seven times then turns off. If the system does not operate as described diagnosis and repairs to the Air Bag system are necessary.

Tie Rod Ends

REMOVAL AND INSTALLATION

1. Raise and support the vehicle safely.

2. Remove the front wheels.

3. Remove the cotter pin and castle nut from the tie rod end ball stud.

4. Using special tool 09913–65210 or equivalent, remove the tie rod end from the knuckle.

5. Loosen the locknut on the threaded end of the tie rod.

6. Unscrew the tie rod end from the tie rod. Count the number of turns necessary for removal.

To install:

7. Screw the tie rod end onto the tie rod the exact number of turns necessary to remove the tie rod end.

8. Install the castle nut and tighten to 22–39 ft. lbs. (30–55 Nm). Install the cotter pin.

9. Tighten the locknut to 26–40 ft. lbs. (35–55 Nm).

10. Install the front wheels.

11. Lower the vehicle.

12. Check and adjust the front wheel alignment.

Manual Rack and Pinion

REMOVAL AND INSTALLATION

——— WARNING ———
Be sure to set the front wheels straight ahead and remove the ignition key from the cylinder before starting repairs. If equipped with an Air Bag the contact coil of the Air Bag system may get damaged if the key is not removed and the wheels are not straight ahead.

1. Disconnect the negative battery cable.

2. Slide the driver's seat back as far as possible.

3. Pull back the front part of the floor mat on the driver's side and remove the steering shaft joint cover.

4. Loosen the steering shaft upper joint bolt, but do not remove.

5. Remove the steering shaft lower joint bolt and disconnect the lower joint from the pinion.

6. Raise and support the vehicle safely.

7. Remove the front wheels.

8. Disconnect the tie rod ends from the steering knuckles.

9. Remove the steering gear mounting bolts and the brackets then remove steering gear case from the vehicle.

To install:

10. Install the steering gear, brackets and mounting bolts. Tighten bolts to 14–21 ft. lbs. (20–30 Nm).

11. Connect the tie rod ends to the steering knuckles.

12. Make sure the steering wheel is straight and the front wheels are pointing straight ahead.

13. Connect the steering shaft to the steering gear. Install the lower steering shaft-to-steering gear clinch bolt and tighten both steering joint bolts (upper and lower) to 14–21 ft. lbs. (20–30 Nm).

14. Install the remaining components.

15. Connect the negative battery cable.

16. Check and adjust the front wheel alignment.

Power Rack and Pinion

REMOVAL AND INSTALLATION

——— WARNING ———
Be sure to set the front wheels straight ahead and remove the ignition key from the cylinder before starting repairs. If equipped with an Air Bag the contact coil of the Air Bag system may get damaged if the key is not removed and the wheels are not straight ahead.

1. Disconnect the negative battery cable.

2. Remove the steering column joint covers.

3. Loosen the steering shaft upper joint bolt, but do not remove.

4. Remove the steering shaft lower joint bolt and disconnect the lower joint from the pinion.

5. Raise and support the vehicle safely.

6. Remove the front wheels.

7. Disconnect the tie rod ends from the steering knuckles.

8. Remove the front exhaust pipe.

9. If equipped with a manual transaxle, disconnect the gear shift control shaft and extension rod from the transaxle.

10. Remove the rear engine mount together with the bracket from the engine and suspension member.

11. Remove the mounting member from the suspension frame by removing the two bolts.

12. Disconnect the high and low pressure lines from the rack and pinion.

NOTE: When the lines are disconnected plug the lines or place a oil pan under the vehicle.

13. Remove the cylinder lines from the rack and pinion.

14. Remove the rack and pinion mounting bolts and brackets.

15. Remove rack and pinion case from the vehicle.

To install:

16. Install the rack and pinion, brackets and mounting bolts. Tighten the bolts to 40 ft. lbs. (55 Nm).

17. Install the cylinder lines to the rack and pinion and torque their fittings to 14–21 ft. lbs. (20–30 Nm).

18. Connect the high and low pressure lines to the rack and pinion. Torque the fittings to 22–28 ft. lbs. (30–40 Nm).

19. If equipped with a manual transaxle, connect the gear shift control shaft and extension rod to the transaxle. Torque the extension rod nut to 18–28 ft. lbs. (25–40 Nm) and torque the control shaft nut and bolt to 11–14 ft. lbs. (15–20 Nm).

20. Connect the tie rod ends to the steering knuckles.

21. Make sure the steering wheel is straight and the front wheels are pointing straight ahead.

22. Connect the steering shaft to the rack and pinion. Install the lower

steering shaft-to-rack and pinion clinch bolt and tighten both steering joint bolts (upper and lower) to 14–21 ft. lbs. (20–30 Nm).

23. Install the front wheels.
24. Connect the negative battery cable.
25. Bleed the power steering system.
26. Lower the vehicle.
27. Check and adjust the front wheel alignment.

Power Steering Pump

BLEEDING

NOTE: Check the power steering system for leaks before starting the bleeding process.

1. Raise and safely support the vehicle. The vehicle only has to be raised up until the front wheels no longer touch the ground.
2. Fill the power steering reservoir with fluid.
3. Start and run the engine. Wait 3–5 seconds and add power steering fluid as necessary.
4. Turn the engine OFF. With the engine stopped, turn the steering wheel all the way to the right and then all the way to the left. Repeat this step three or four times and check the fluid level.
5. Start the engine. With the engine running at idle, turn the steering wheel all the way to the right and then all the way to the left. Repeat turning the wheel three or four times until all of the foam in the reservoir is gone.

NOTE: Make sure to bleed all of the air completely. If any air remains in the fluid, the power steering pump may make a humming noise or the steering wheel may feel heavy.

6. When finished, lower the vehicle.

REMOVAL AND INSTALLATION

1. Disconnect the negative battery cable.
2. Clean the steering hose fittings so they are free of dirt and other contaminates.
3. Raise the vehicle and support safely.
4. Remove the right front wheel and remove the splash shield.
5. Remove the power steering drive belt.

— WARNING —
If equipped with air conditioning, take care not to damage the air conditioning condenser.

6. Place a pan under the vehicle to catch draining power steering oil. Disconnect the power steering high and the low pressure hoses. Plug the hose ports to prevent dust or other foreign matter from entering the pump.

NOTE: If equipped with air conditioning and the high pressure hose cannot be adequately reached, it must be disconnected when the pump is removed.

7. Disconnect the pressure switch lead.
8. If equipped with air conditioning, remove the air conditioning compressor mounting bolts then remove the compressor from the mounting bracket with the hoses attached.

NOTE: Hang the air conditioning compressor with a wire hook to prevent damage to the hoses.

9. Remove the bracket mounting bolts and remove the pump together with its mounting bracket.
10. Remove the pump from its bracket.

To install:
11. Position the pump on its mounting bracket and install the mounting bolts.
12. Install the power steering pump and bracket assembly to the vehicle and install the mounting bolts.
13. Install the air conditioning compressor.
14. Connect the pressure switch lead to the pump.
15. Connect the pressure hoses to the pump assembly.
16. Install the remaining components.
17. Fill the power steering reservoir with fluid and bleed the air from the system.

BRAKES

Anti-Lock Brake System Service

PRECAUTIONS

• Certain components within the Anti-Lock Brake System (ABS) are not intended to be serviced or repaired individually. Only those components with removal and installation procedures should be serviced.
• Do not use rubber hoses or other parts not specifically specified for and ABS system. When using repair kits, replace all parts included in the kit. Partial or incorrect repair may lead to functional problems and require the replacement of components.
• Lubricate rubber parts with clean, fresh brake fluid to ease assembly. Do not use lubricated shop air to clean parts; damage to rubber components may result.
• Use only specified brake fluid from an unopened container.
• If any hydraulic component or line is removed or replaced, it may be necessary to bleed the entire system.
• A clean repair area is essential. Always clean the reservoir and cap thoroughly before removing the cap. The slightest amount of dirt in the fluid may plug an orifice and impair the system function. Perform repairs after components have been thoroughly cleaned; use only denatured alcohol to clean components. Do not allow ABS components to come into contact with any substance containing mineral oil; this includes used shop rags.
• The Anti-Lock control unit is a microprocessor similar to other computer units in the vehicle. Ensure that the ignition switch is OFF before removing or installing control harnesses. Avoid static electricity discharge at or near the controller.
• If any arc welding is to be done on the vehicle, the control unit should be unplugged before welding operations begin.

Master Cylinder

REMOVAL AND INSTALLATION

NOTE: Be careful not to spill brake fluid on the painted surfaces of the vehicle; it will damage the finish.

1. Disconnect the negative battery cable. Clean around the reservoir cap and take some of the fluid out with a syringe.

2. Disconnect and plug the brake tubes from the master cylinder.

3. Remove the mounting nuts and washers.

4. Remove the master cylinder from its mounting studs. On models equipped with ABS, rotate the master cylinder 90° counterclockwise and remove the master cylinder from the brake booster.

5. Remove the master cylinder.

To install:

6. If a new master cylinder is to be installed, bleed before installing.

7. If a new master cylinder is to be installed, adjust the clearance between the brake booster piston rod and the master cylinder primary piston by performing the following procedure:

   a. Disconnect the vacuum hose from the power brake booster.

   b. Place the booster pin gauge 09950–96010 or equivalent, on the master cylinder and push the pin until it contacts the piston.

   c. Turn the booster pin gauge upside down and place it on the booster. Adjust the booster rod length until the rod contacts the pin head.

   d. Adjust the clearance by turning the adjusting screw of the piston rod with tool 09952–16010 or equivalent.

   e. Connect the vacuum hose to the power brake booster.

NOTE: When adjusting the booster piston rod, if negative pressure is applied to the booster with the engine at idle, piston rod clearance should be 0.010–0.020 in. (0.25–0.50 mm).

8. If equipped with ABS, lay the master cylinder on its side with the reservoir toward the passenger side, and insert the master cylinder into

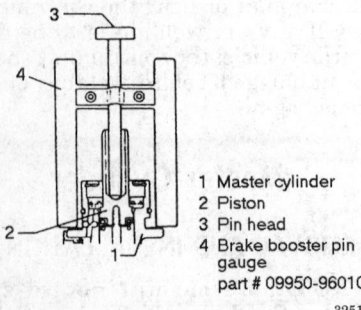

1 Master cylinder
2 Piston
3 Pin head
4 Brake booster pin gauge
part # 09950-96010

325197

Measuring the master cylinder pin depth

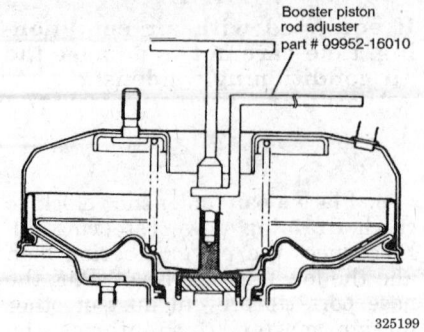

Booster piston rod adjuster part # 09952-16010

325199

Adjusting the booster piston rod length

the brake booster. Then rotate the master cylinder into its proper position on the mounting studs.

9. If not equipped with ABS, mount the assembly on the studs. Tighten the mounting nuts to 8–12 ft. lbs. (10–16 Nm).

10. Connect the brake tubes to the master cylinder. Tighten to 10–13 ft. lbs. (14–18 Nm). Tighten the brake tube flare nuts at the ABS proportioning valve to 10–13 ft. lbs. (14–18 Nm).

11. Fill the master cylinder with clean brake fluid and bleed the brake system.

12. Connect the negative battery cable.

Brake Caliper

REMOVAL AND INSTALLATION

Front Brake Caliper

1. Raise and safely support the front of the vehicle. Set the parking brake and block the rear wheels.

NOTE: Disassemble brakes one wheel at a time. This will prevent parts confusion and also prevent the opposite caliper piston from popping out during installation. Mark the relationship between the wheel and the axle hub before removing the tire and wheel assembly.

2. Remove the brake hose mounting bolt from the brake caliper. Use a pan to catch any spilled fluid and immediately plug the disconnected hose.

3. Remove the 2 caliper mounting bolts, then remove the caliper from the mounting carrier.

To install:

4. Use a caliper piston compressor, C-clamp, or a large pair of pliers to slowly press the caliper piston back into the caliper.

— **WARNING** —
Make sure that the piston is pressed straight into the caliper, so not to damage the piston bore. When compressing the piston any brake fluid in the brake caliper will be forced out.

5. Install the caliper assembly to the mounting carrier. Before installing the retaining bolts, apply a thin, even coating of anti-seize compound to the threads and slide surfaces. Don't use grease or spray lubricants; they will not hold up under the extreme temperatures generated by the brakes. Tighten the mounting bolts to 16–23 ft. lbs. (22–32 Nm).

6. Install the brake hose to the caliper. Always use new washers and tighten the mounting bolt to 14–18 ft. lbs. (20–25 Nm).

7. Bleed the brake system and install the front wheel(s).

8. Lower the vehicle and check the level of the brake fluid in the master cylinder reservoir.

9. Check the brake system for leakage.

Rear Brake Caliper

1. Raise and safely support the rear of the vehicle with the front wheels blocked.

NOTE: Disassemble brakes one wheel at a time. This will prevent parts confusion and also prevent the opposite caliper piston from popping out during installation. Mark the relationship between the wheel and the axle hub before removing the tire and wheel assembly.

2. Remove the brake hose mounting bolt from the brake caliper. Use a pan to catch any spilled fluid and immediately plug the disconnected hose.

3. Release the parking brake lever.

4. Remove the 2 caliper mounting bolts and then remove the caliper from the mounting carrier.

5. Remove the parking brake cable E-ring.

6. Disconnect the parking brake cable from the lever on the caliper.

7. Remove the cover from the brake caliper.

To install:

8. Install the cover to the brake caliper.

9. Connect the parking brake cable to the camshaft lever and the cable bracket. Install the E-ring and the cable clip.

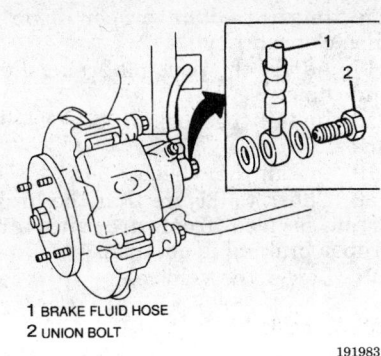

1 BRAKE FLUID HOSE
2 UNION BOLT

191983

Front brake caliper hose attachment location

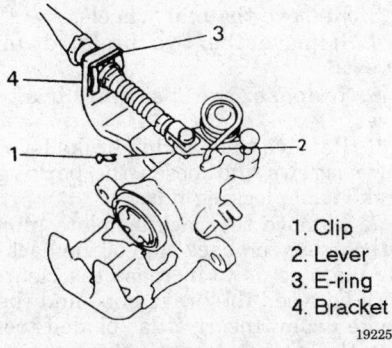

1. Clip
2. Lever
3. E-ring
4. Bracket

192256

Rear brake caliper

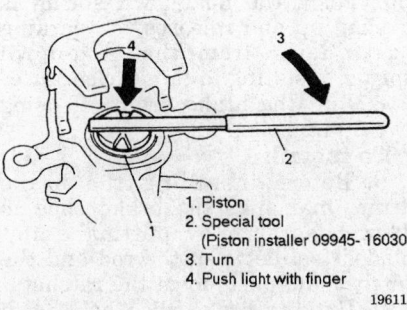

1. Piston
2. Special tool
   (Piston installer 09945- 16030)
3. Turn
4. Push light with finger

196113

Turn the brake caliper piston to obtain additional clearance

10. Install the caliper assembly to the mounting carrier. Before installing the retaining bolts, apply a thin, even coating of anti-seize compound to the threads and slide surfaces. Don't use grease or spray lubricants; they will not hold up under the extreme temperatures generated by the brakes. Tighten the mounting bolts to 16–23 ft. lbs. (22–32 Nm).
11. Install the brake hose to the caliper. Always use new washers and tighten the mounting bolt to 14–18 ft. lbs. (20–25 Nm).
12. Bleed the brake system and install the rear wheel(s).

13. Depress the brake pedal with 66 lbs. (30 kg) of force three to five times to obtain proper disc to pad clearance.
14. Check to ensure that the brake is free from dragging and that proper braking is obtained.
15. Make sure that the proper parking brake lever stroke is as specified.
16. Lower the vehicle and check the level of the brake fluid in the master cylinder reservoir.
17. Check the brake system for leakage.

Disc Brake Pads

REMOVAL AND INSTALLATION

1. Raise and safely support the front of the vehicle.
2. Siphon a sufficient quantity of brake fluid from the master cylinder reservoir. This is necessary as the brake caliper piston must be forced into the cylinder bore to provide sufficient clearance to install the new pads.
3. Remove the wheels.

NOTE: Disassemble brakes one wheel at a time. This will prevent parts confusion and also prevent the opposite caliper piston from popping out during pad installation.

4. Remove the 2 caliper mounting bolts and then remove the caliper from the caliper carrier. Position the caliper out of the way and support it with wire so it doesn't hang by the brake line.
5. Remove the brake pads.
6. **To install:**
6. Clean the caliper thoroughly; remove any dirt or dust. Check the brake rotor for grooves or cracks and machine or replace, as necessary.
7. Install the pads into the caliper carrier.
8. On front calipers, use a caliper compressor, or C-clamp to slowly press the caliper piston back into the caliper. If the piston is frozen, or if the caliper is leaking hydraulic fluid, the caliper must be overhauled or replaced.
9. On rear calipers, use a piston installer 09945–16030 or equivalent, to rotate the caliper piston clockwise into the caliper bore. Apply a small amount of force compressing the piston into the caliper while rotating the piston. Compress the piston enough to enable the caliper to fit over the pads. Lubricate the piston boot with

silicone grease to avoid twisting the piston boot.

---- **WARNING** ----
Make sure that the piston is pressed straight into the caliper, so not to damage the piston bore.

10. Install the caliper assembly to the caliper carrier. Before installing the retaining bolts, apply a thin, even coating of anti-seize compound to the threads and slide surfaces. Don't use grease or spray lubricants; they will not hold up under the extreme temperatures generated by the brakes. Tighten the bolts to 16–23 ft. lbs. (22–32 Nm).
11. Install the wheels.
12. Lower the vehicle to the ground. Check the level of the brake fluid in the master cylinder reservoir; it should be at least to the middle of the reservoir.
13. Depress the brake pedal several times and make sure that the movement feels normal. The first brake pedal application may result in a very LONG pedal due to the pistons being retracted. Always make several brake applications before starting the vehicle.
14. Recheck the fluid level and add if necessary add fluid until it reaches the MAX line.

Brake Rotor

REMOVAL AND INSTALLATION

1. Raise and safely support the front of the vehicle.

NOTE: Disassemble brakes one wheel at a time. This will prevent parts confusion and also prevent the opposite caliper piston from popping out during installation.

2. Remove the two brake caliper carrier bolts and remove the carrier and brake caliper assembly. Do not disconnect the brake line from the caliper. Support the caliper assembly with a piece of wire. Do not allow the caliper assembly to hang from the brake line.

---- **WARNING** ----
Be careful not to damage the brake flexible hose or depress the brake pedal.

3. Remove the two brake rotor screws.
4. Pull the brake rotor off by inserting two 8 mm bolts into the two holes in the hub of the brake rotor.

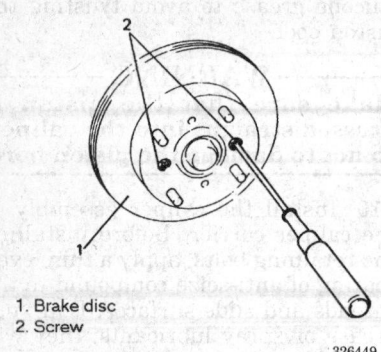

1. Brake disc
2. Screw

326449

Removing the brake rotor screws

Turn the bolts in slowly and evenly to remove the rotor.

5. Remove the two 8 mm bolts from the brake rotor, after removing the brake rotor.

To install:

6. Place the brake rotor on the hub and install the brake rotor screws.

7. Install the brake caliper assembly and the carrier bolts.

8. Install the wheel assembly.

9. Upon completion, depress the brake pedal three to five times to obtain proper disc to pad clearance.

10. Check to ensure that the brakes do not drag and proper braking is obtained.

11. Lower the vehicle.

Brake Drums

REMOVAL AND INSTALLATION

Without Wheel Hubs

1. Apply the parking brake.
2. Raise and safely support the vehicle.
3. Remove the rear wheels.
4. Remove the spindle dust cap.
5. Uncaulk the spindle nut and remove the spindle nut and washer.
6. Release the parking brake lever.

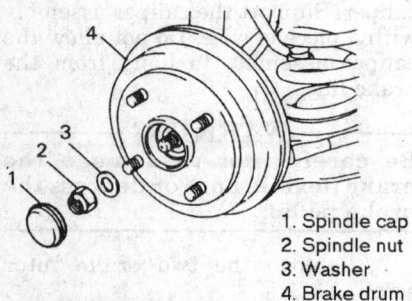

1. Spindle cap
2. Spindle nut
3. Washer
4. Brake drum

326455

Brake drum without wheel hub

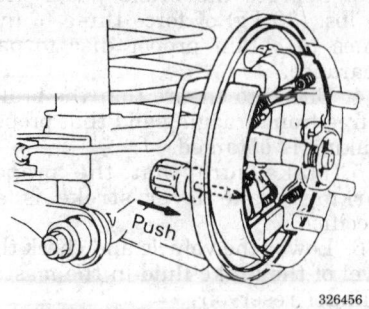

326456

Releasing the parking brake lever from its position

7. Remove the parking brake lever cover screws and loosen the parking brake cable locking nut.

8. Remove the backing plate plug attached to the back side of the backing plate so as to increase the clearance between the brake shoe and the brake drum. Insert a flat bladed tool into the plug hole until the tip contacts the brake shoe, then push the tool in toward the brake drum. With this push, the hold-down spring is pushed up and releases the parking brake lever from the hold-down spring, resulting in larger clearance.

9. Install the brake drum removal tool 09943–17911 or equivalent to the brake drum, then attach a slide hammer to the tool and remove the brake drum.

To install:

10. Before installing the brake drum, maximize the brake shoe to drum clearance by putting a flat bladed tool between the rod and the ratchet and push down the ratchet.

11. Put the brake shoe hold-down spring back to its original position. The parking brake lever should rest against the hold-down spring and not under the spring.

12. Install the brake drum after making sure that the inside of the brake drum and brake shoe are free of dirt and oil.

13. Install the spindle washer and a new spindle nut. Torque the spindle nut to 58–86 ft. lbs. (80–120 Nm).

14. Caulk the spindle nut.

15. Install the spindle dust cap.

NOTE: When installing the spindle cap, hammer lightly several times on the collar of the cap until the collar comes closely into contact with the brake drum. If the fitting part of the cap is deformed or damaged or if it fits loose, replace the cap with a new one.

16. Depress the brake pedal with about 66 lbs. (30 kg) of force three to

five times to obtain proper drum to shoe clearance.

17. Adjust the parking brake shoes and cable.

18. Install the parking brake lever cover.

19. Install the wheels.

20. Check to ensure that the brake drum is free from dragging and proper braking is obtained.

21. Lower the vehicle.

With Wheel Hubs

1. Set the parking brake.
2. Raise and safely support the vehicle.
3. Remove the rear wheels.
4. Remove the two brake drum screws.
5. Release the parking brake lever.
6. Remove the parking brake lever cover screws and loosen the parking brake cable locking nut.

7. Remove the backing plate plug attached to the back side of the backing plate so as to increase the clearance between the brake shoe and the brake drum. Insert a flat bladed tool into the plug hole until the tip contacts the brake shoe and push the tool in toward the brake drum. With this push, the hold-down spring is pushed up and releases the parking brake lever from the hold-down spring, resulting in larger clearance.

8. Pull the brake drum off using two 8 mm bolts.

To install:

9. Before installing the brake drum, maximize the brake shoe to drum clearance by putting a flat bladed tool between the rod and the ratchet and push down the ratchet.

10. Put the brake shoe hold-down spring back to its original position. The parking brake lever should rest against the hold-down spring and not under the spring.

11. Install the brake drum after making sure that the inside of the brake drum and brake shoe are free of dirt and oil.

12. Tighten the brake drum screws.

13. Depress the brake pedal with about 66 lbs. (30 kg) of force three to five times to obtain proper drum to shoe clearance.

14. Adjust the parking brake shoes and cable.

15. Install the parking brake lever cover

16. Install the wheels.

17. Check to ensure that the brake drum is free from dragging and proper braking is obtained.

18. Lower the vehicle.

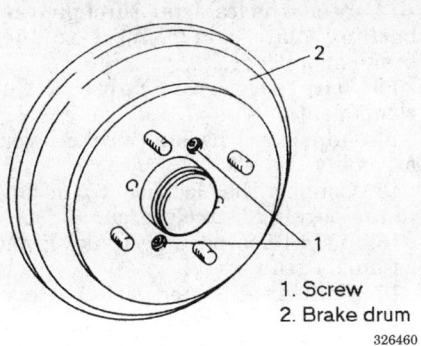

1. Screw
2. Brake drum

326460

Brake drum attaching screws with wheel hub

Brake Shoes

REMOVAL AND INSTALLATION

1. Disconnect the battery negative cable.
2. Apply the parking brake.
3. Raise and support the vehicle safely.
4. Remove the rear wheels.
5. Remove the rear brake drums,
6. Remove the upper and lower springs from the brake shoes.
7. Remove the anti-rattle spring and brake shoe adjustment strut.
8. Remove the primary and secondary brake shoe hold-down springs by turning the hold-down pins and removing the shoes from vehicle.
9. Remove the clip securing the parking brake shoe lever to the secondary shoe.
10. Remove the parking brake cable from the lever.
11. Remove the hold-down pins from the brake backing plate.

To install:

12. Inspect all components for damage from heat and stress. Replace components as necessary. Lubricate all contact points on the backing plate with grease.
13. Install the hold-down pins on the backing plate.
14. Install the parking brake cable to the parking brake shoe lever.
15. Install the clip securing the parking brake shoe lever to the secondary shoe.
16. Install the primary and secondary brake shoes to the vehicle and secure with the hold-down springs. Position the hold-down spring on the hold-down pins and turn the pins downward.
17. Install the brake adjustment strut and anti-rattle spring.
18. Install the upper and lower return springs to the primary and secondary brake shoes.
19. Put the brake shoe hold-down spring back to its original position.

The parking brake lever should rest against the hold-down spring and not under the spring,

20. Install the brake drum
21. Bleed the brake system and top of the master cylinder fluid level as necessary.
22. Install the wheels.
23. Press the brake pedal 3–5 times to adjust the brake shoe clearance.
24. Adjust the parking brake cable.
25. Install the parking brake lever cover,
26. Check to ensure the brake drum is free from dragging and proper braking is obtained.
27. Lower the vehicle.

Wheel Cylinder

REMOVAL AND INSTALLATION

1. Disconnect the battery negative cable.
2. Apply the parking brake.
3. Raise and support the vehicle safely.
4. Remove the rear wheel.
5. Remove brake drums,
6. Remove the brake shoes and hardware.
7. Loosen the brake pipe flare nut enough that fluid does not leak.
8. Remove the wheel cylinder mounting bolts. Disconnect the brake pipe from the wheel cylinder and cap the brake pipe with the bleeder cap to prevent fluid spillage.
9. Remove the wheel cylinder from the backing plate.

To install:

10. Apply water tight sealant to the joint seam of the wheel cylinder and the backing plate, if the wheel cylinder does not have packing on it. Position the wheel cylinder on the backing plate. Take the bleeder plug cap from the brake pipe and connect the pipe to the wheel cylinder just enough to prevent the fluid from leaking.
11. Tighten the mounting bolts to 7–9 ft. lbs. (10–13 Nm).
12. Torque the flare nut of the brake tubing that is connected to the wheel cylinder to 10–13 ft. lbs. (14–18 Nm). Install the bleeder cap that was used to cap the brake pipe.
13. Inspect all components for damage from heat and stress. Replace components as necessary. Lubricate all contact points on the backing plate with grease.
14. Install the brake shoes and hardware.
15. Install the brake drums.

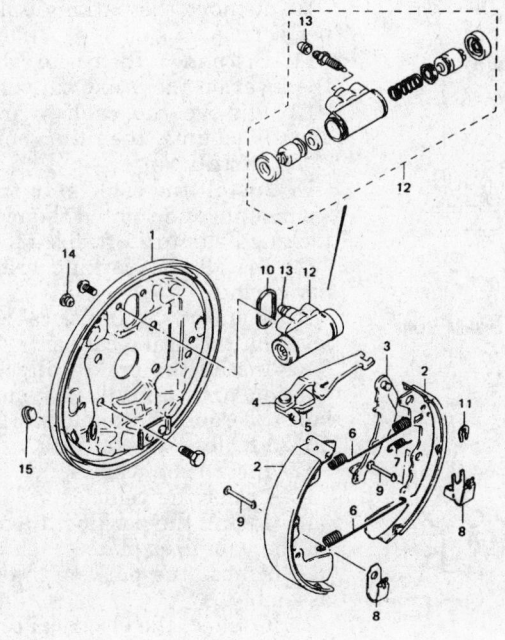

1. Brake back plate
2. Brake shoe
3. Parking brake shoe lever
4. Brake strut
5. Quadrant spring
6. Shoe return spring
7. Antirattle spring
8. Shoe hold down spring
9. Shoe hold down pin
10. Packing
11. Parking lever retainer
12. Wheel cylinder
13. Bleeder plug cap
14. Rubber plug
15. Rubber plug

326477

Brake assembly without a hub

16. Bleed the brake system and top of the master cylinder fluid level as necessary.

17. Install the wheels.

18. Adjust the parking brake shoes and cable.

19. Check to ensure the brake drum is free from dragging and proper braking is obtained.

20. Lower the vehicle.

Parking Brake Cable

ADJUSTMENT

1. Check the following components before adjusting the parking brake cable:

   a. No air is trapped in brake system.

   b. Make sure the rear shoes are within wear limits.

   c. Brake pedal travel is within limits.

   d. Press the brake pedal a few times before adjustment.

   e. Parking brake lever has been pulled up a few times with about 44 ft. lbs. (20 kg) of force.

2. Remove the parking brake cover.

3. After checking the above conditions, adjust the parking brake lever stroke by loosening or tightening the self locking nut. The nut is located at the bottom of the parking brake handle.

4. The parking brake should be fully applied when the lever is raised between 3–6 notches.

5. Replace the parking brake lever cover.

6. Verify that there is no brake drag after the parking brake lever has been released.

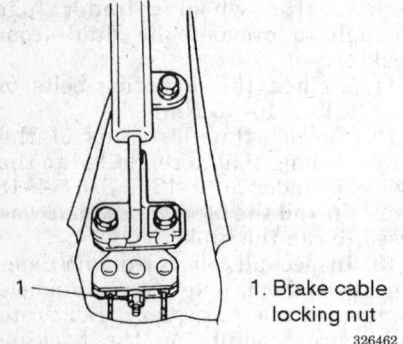

1. Brake cable locking nut

326462

Parking brake lever adjusting locknut

REMOVAL AND INSTALLATION

With Drum Brakes

1. Disconnect the negative battery cable.

2. Remove the parking brake lever assembly trim cover.

3. Disconnect the electrical connector from the parking brake switch.

4. Loosen the adjuster nut on the lever assembly.

5. Remove the parking brake cables from the equalizer plate.

6. Raise and support the vehicle safely.

7. Remove the wheels.

8. Remove the brake drums and shoes.

9. Disconnect the parking brake cable from the brake shoe lever and the backing plate.

10. Remove the cable(s) from the chassis mounts, then the cable from the vehicle.

To install:

11. Install the cable(s) to the chassis mounts and tighten the mount attaching bolts to 11 ft. lbs. (15 Nm).

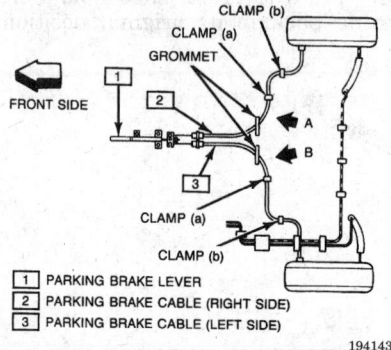

1 PARKING BRAKE LEVER
2 PARKING BRAKE CABLE (RIGHT SIDE)
3 PARKING BRAKE CABLE (LEFT SIDE)

194143

Parking brake cable routing — 1993–94 Swift

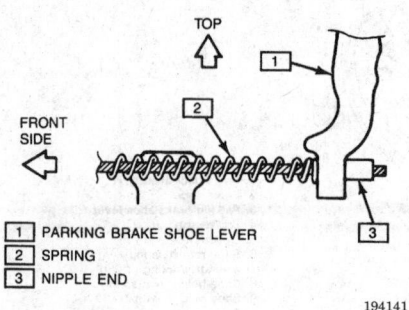

1 PARKING BRAKE SHOE LEVER
2 SPRING
3 NIPPLE END

194141

Parking brake cable to parking lever connection — 1993–97 Swift

12. Feed the cable(s) through the backing plate and connect to the brake shoe lever.

13. Install the remaining components.

14. Adjust the parking brake lever assembly.

15. Connect the electrical connector to the parking brake switch.

16. Install the parking brake lever assembly trim cover.

17. Connect the negative battery cable.

With Disc Brakes

1. Disconnect the negative battery cable.

2. Remove the parking brake lever assembly trim cover.

3. Disconnect the electrical connector from the parking brake switch.

4. Loosen the adjuster nut on the lever assembly.

5. Remove the parking brake cables from the equalizer plate.

6. Raise and support the vehicle safely.

7. Remove the wheels.

8. Remove the rear brake caliper mounting bolts.

9. Remove the brake caliper from the carrier.

10. Remove the parking brake caliper E-ring.

11. Disconnect the brake cable from the lever on the brake caliper.

12. Remove the cable(s) from the chassis mounts, then the vehicle.

To install:

13. Install the cable(s) to the chassis mounts and tighten the mount attaching bolts to 11 ft. lbs. (15 Nm).

14. Install the parking brake caliper E-ring.

15. Connect the brake cable to the lever on the brake caliper.

16. Install the brake caliper to the carrier and install the mounting bolts. Torque the mounting bolts to 16–23 ft. lbs. (22–32 Nm).

17. Install the wheels.

18. Lower the vehicle.

19. Install the parking brake cables on the equalizer plate.

20. Adjust the parking brake lever assembly.

21. Connect the electrical connector to the parking brake switch.

22. Install the parking brake lever assembly trim cover.

23. Connect the negative battery cable.

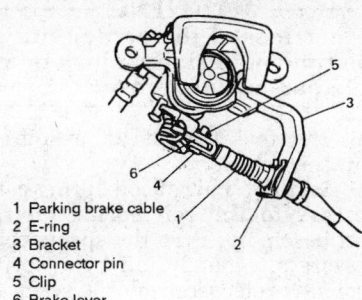

1 Parking brake cable
2 E-ring
3 Bracket
4 Connector pin
5 Clip
6 Brake lever

194162

Rear brake caliper brake cable connection for rear disc brakes — 1993–94 Swift

Brake System Bleeding

PROCEDURE

Master Cylinder

— WARNING —

Do not allow brake fluid to splash or spill onto painted surfaces; the paint will be damaged. If spillage occurs, flush the area immediately with clean water.

1. Fill the master cylinder reservoir to the MAX line with brake fluid and keep it at least half full throughout the bleeding procedure.
2. If the master cylinder is new, it must be bled before installing. Bleed as follows:
   a. Install the master cylinder in a holding fixture, taking precautions to protect the cylinder body from damage.
   b. Fill the fluid reservoir with clean brake fluid meeting DOT 3 recommendations.
   c. Position a container under the outlet fittings. Depress the cylinder piston slowly through it's full travel. Cover both outlets and let the piston return to normal resting position.
   d. Wait 5 seconds, then repeat the operation until the fluid exiting the master cylinder fittings is free of air bubbles.
3. If the master cylinder has been removed or disconnected, it must be bled before any brake unit is bled. To bleed the master cylinder:
   a. Loosen the front brake line from the master cylinder and allow fluid to flow from the front connector port. Make sure to use a suitable container or a shop rag to catch the brake fluid. Dispose of brake fluid properly.
   b. Reconnect the line to the master cylinder and tighten it until it is fluid tight.

c. Have a helper press the brake pedal down one time and hold it down.
   d. Loosen the front brake line connection at the master cylinder. This will allow trapped air to escape, along with some fluid. Do not allow the brake pedal to return upwards before the line is tightened. This will prevent air from reentering the master cylinder.
   e. Release the pedal slowly and repeat the sequence until only fluid runs from the port. No air bubbles should be present in the fluid.
   f. When there is no air left in the master cylinder, tighten the line fitting at the master cylinder to 11 ft. lbs.
   g. After all the air has been bled from the front connection, bleed the master cylinder at the rear connection by repeating the steps.

— WARNING —

Do not reuse brake fluid which has been bled from the brake system.

4. When bleeding is done, top off the master cylinder with brake fluid.
5. Make sure the brakes are firm and there are no leaks.

Brakes

— WARNING —

Do not allow brake fluid to splash or spill onto painted surfaces; the paint will be damaged. If spillage occurs, flush the area immediately with clean water.

1. Fill the master cylinder reservoir to the MAX line with brake fluid and keep it at least half full throughout the bleeding procedure.
2. If the master cylinder has been removed or disconnected, it must be bled before any brake unit is bled.
3. Place the correct size box-end or line wrench over the bleeder valve and attach a tight-fitting transparent hose over the bleeder. Allow the tube to hang submerged in a transparent container of clean brake fluid. The fluid must remain above the end of the hose at all times, otherwise the system will ingest air instead of fluid.
4. Have an assistant pump the brake pedal several times slowly and hold it down.
5. Slowly unscrew the bleeder valve ($\frac{1}{4}$–$\frac{1}{2}$ turn is usually enough). After the initial rush of air and fluid, tighten the bleeder valve. Have the assistant release the pedal.
6. Repeat until no air bubbles are seen in the hose or container. If air is

constantly appearing after repeated bleeding, the system must be examined for the source of the leak or loose fitting.
7. If the entire system must be bled, begin with the right rear, then the left front, left rear and right front brake in that order. After each brake is bled, check and top off the fluid level in the reservoir.

— WARNING —

Do not reuse brake fluid which has been bled from the brake system.

Wheel Speed Sensor

REMOVAL AND INSTALLATION

Swift

Front Sensor

1. Disconnect the negative battery cable.
2. Raise and safely support the vehicle.
3. Remove the front wheel.
4. Disconnect the wheel speed sensor connector.
5. Remove the wheel speed sensor from the steering knuckle by removing the bolt.
   To install:
6. Install the speed sensor to the steering knuckle with the bolt. Torque the bolt to 8 ft. lbs. (12 Nm).
7. Connect the speed sensor connector.
8. Install the wheel.
9. Lower the vehicle.
10. Connect the negative battery cable.
11. Check the speed sensor signal.

Rear Sensor

1. Disconnect the negative battery cable.
2. Raise and safely support the vehicle.
3. Remove the rear wheel.
4. Disconnect the wheel speed sensor connector.
5. Remove the wheel speed sensor from the backing plate by removing the bolt.
   To install:
6. Install the speed sensor to the backing plate with the bolt. Torque the bolt to 8 ft. lbs. (12 Nm).
7. Connect the speed sensor connector.
8. Install the wheel.
9. Lower the vehicle.
10. Connect the negative battery cable.
11. Check the speed sensor signal.

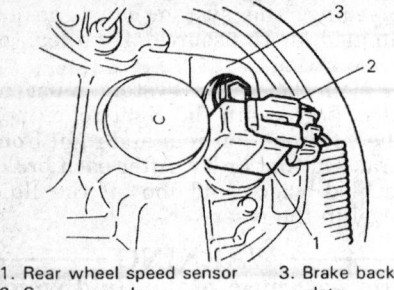

1. Rear wheel speed sensor
2. Sensor coupler
3. Brake back plate

329921

Rear wheel speed sensor — Swift

Esteem

Front Sensor

1. Disconnect the negative battery cable.
2. Raise and safely support the vehicle.
3. Remove the front wheel.
4. Disconnect the speed sensor connector.
5. Push out the grommet of the harness from the inner fender.
6. Remove the bolts holding the speed sensor clamps to the strut.
7. Remove the bolt holding the speed sensor to the steering knuckle and remove the speed sensor from the vehicle.

To install:
8. Install the speed sensor to the vehicle and install the bolt. Torque the bolt to 17 ft. lbs. (23 Nm).
9. Connect the speed sensor wire to the strut with the bolts.
10. Push in the grommet of the harness to the inner fender.
11. Connect the front wheel sensor connector.
12. Install the wheel and lower the vehicle.
13. Connect the negative battery cable.
14. Check the speed sensor signal.

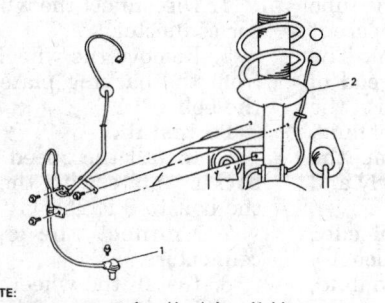

NOTE:
Never remove sensor rotor from drive shaft outside joint. Refer to Section 4 for replacement procedure of joint with sensor.

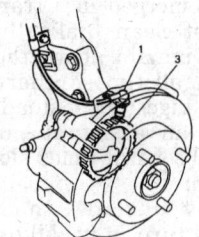

1. Left front wheel speed sensor
2. Left front strut
3. Sensor rotor

329369

Front wheel speed sensor — Esteem

Rear Sensor

1. Disconnect the negative battery cable.
2. Raise the rear of the vehicle.
3. Disconnect the rear wheel sensor connector and detach the connector and wire harness from the suspension frame by removing the bolts.
4. Remove the harness clamp bolt and remove the rear wheel speed sensor from the knuckle.

To install:
5. Install the rear wheel speed sensor to the knuckle and install the bolt. Torque the bolt to 17 ft. lbs. (23 Nm).
6. Route the speed sensor wiring and install the bolts.
7. Connect the speed sensor connector.
8. Lower the vehicle.
9. Connect the negative battery cable.
10. Check the speed sensor signal.

FRONT SUSPENSION

Strut

REMOVAL AND INSTALLATION

1. Raise and support the vehicle safely.
2. Remove the wheels.
3. If equipped with ABS, disconnect the ABS wheel speed sensor.
4. Remove the brake hose clip, then the hose from the strut.
5. Support the lower control arm with a floor jack.
6. Remove the strut bracket bolts.
7. Remove the upper strut support nuts from the engine compartment, hold the strut by hand so it will not fall.

― **WARNING** ―
Do not loosen the center nut at this time or serious injury or vehicle damage may result.

8. Remove the strut assembly from the vehicle.
9. Install a pair of coil spring compressors to the coil spring on the strut assembly. Turn the spring compressors alternately until the spring tension is released from the strut assembly. If the spring can be turned slightly then it has been collapsed enough.

― **CAUTION** ―
This procedure requires the use of a spring compressor; it cannot be performed without one. If you do not access to this special tool, do not attempt to disassemble the strut. The coil spring is retained under considerable pressure. It can exert enough force to cause serious injury. Exercise extreme caution.

10. Keeping the spring collapsed remove the strut center nut and remove the other components from the top of the strut assembly.
11. Remove the spring from the strut.

To install:
12. If installing a new spring, compress the spring with a pair of spring compressors. Make sure that the spring compresses to 9 in. (230 mm) for installation.
13. Position the coil spring on the strut making sure that the end of the spring is mated to the stepped part of the lower seat.
14. Install the bump stop to the strut rod.
15. Install the strut cover, spring seat, upper spring seat, and the bearing spacer. Align the strut bracket to the mark on the upper spring seat.
16. Clean the bearing lower washer and install it to the strut rod.
17. Clean the strut bearing and apply fresh grease to the bearing. Install the bearing to the lower washer.
18. Clean the bearing upper washer and install it.
19. Install these components in the following order: the bearing upper seal, bearing seat, strut support, inner spacer, washer and strut nut. Torque the nut to 29–43 ft. lbs. (40–60 Nm). Apply a water proof coating (paint or lacquer) to the nut and strut rod threads.
20. Loosen the spring compressors alternately, checking that the stepped part of the spring seat and spring are properly positioned.

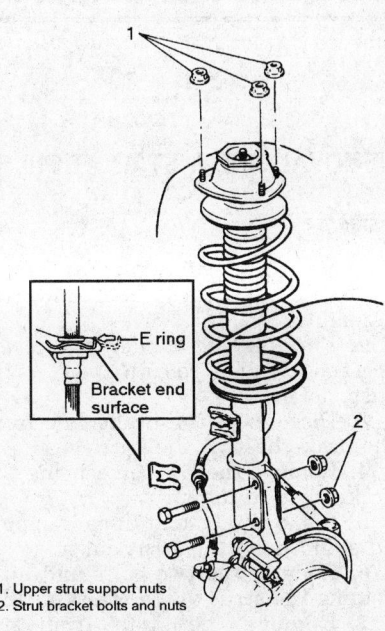

1. Upper strut support nuts
2. Strut bracket bolts and nuts

324747

Strut assembly mounting

21. Install the strut assembly onto the vehicle. Install the upper support nuts loosely. Torque the upper strut support nuts to 16–23 ft. lbs. (22–33 Nm).
22. Install the strut bracket nuts and bolts and torque to 51–65 ft. lbs. (70–90 Nm).
23. If equipped with ABS, connect the ABS wheel speed sensor.
24. Install the brake hose clip.
25. Install the wheels.
26. Lower the vehicle.
27. Check the front end alignment.

Lower Ball Joint and Control Arm

REMOVAL AND INSTALLATION

The lower control arm and ball joint are a complete unit which will not separate.
1. Raise and support the vehicle safely.
2. Remove the front wheels.
3. Remove the sway bar link nut, washer, and cushion.
4. Remove the ball joint stud bolt and nut.
5. Remove the lower control arm front bushing bolt.

6. Remove the two bolts holding the lower control arm bracket to the vehicle.
7. Separate the lower control arm from the steering knuckle and remove the lower control arm from the vehicle.
   To install:
8. Install the ball joint to the knuckle and secure it with the nut and bolt.
9. Install the lower control arm rear bracket and bolts.
10. Install the lower control arm front mounting bolt.
11. Connect the sway bar link to the control arm and install the cushion, washer and nut.
12. Tighten control arm rear mounting bracket bolts to 27 ft. lbs. (37 Nm); front mounting bolt to 65 ft. lbs. (90 Nm); ball joint nut and bolt to 44 ft. lbs. (60 Nm) and the sway bar link nut to 18 ft. lbs. (26 Nm).
13. Install wheels.
14. Lower the vehicle.
15. Check the front wheel alignment.

Sway Bar

REMOVAL AND INSTALLATION

1. Safely raise and support the vehicle.
2. Remove the sway bar link nuts, washers, and cushions.
3. Remove the sway bar mounting brackets by removing the bolts.
4. Remove the sway bar from the vehicle.
5. Remove the sway bar links from the sway bar.
   To install:

NOTE: When installing the sway bar, loosely assemble the components while insuring the sway bar is centered at each side.

6. Install the stabilizer links to the sway bar.
7. Install the sway bar to the vehicle.
8. Install the mounting brackets but do not tighten the bolts.
9. Install the sway bar links cushions, washers and nuts, but do not tighten the nuts.
10. Check the sway bar centering, be sure that the color paint on the sway bar aligns with the mounting bushings.
11. Torque the sway bar mounting bolts to 14–21 ft. lbs. (20–30 Nm). Torque the upper link nuts to 29–43 ft. lbs. (40–60 Nm). Torque the lower link nuts to 16–23 ft. lbs. (22–33 Nm).

1. Nut
2. Stopper
3. Inner spacer
4. Support comp.
5. Bearing seat
6. Bearing upper washer
7. Bearing seal
8. Bearing
9. Bearing lower washer
10. Bearing spacer
11. Coil spring upper seat
12. Coil spring seat
13. Strut cover
14. Bump stopper
15. Coil spring
16. Coil spring lower seat
17. Strut

324749

Exploded view of the strut assembly components

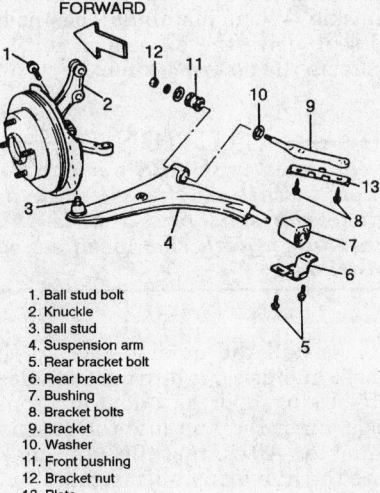

FORWARD

1. Ball stud bolt
2. Knuckle
3. Ball stud
4. Suspension arm
5. Rear bracket bolt
6. Rear bracket
7. Bushing
8. Bracket bolts
9. Bracket
10. Washer
11. Front bushing
12. Bracket nut
13. Plate

324835

Front suspension components — Swift

12. Lower the vehicle.

Front Wheel Bearings

ADJUSTMENT

The front wheel bearings are a cartridge type design and cannot be adjusted.

REMOVAL AND INSTALLATION

NOTE: Always replace bearing races as a complete set.

1. Raise and support the vehicle.
2. Remove the front wheel.
3. Remove the brake caliper, carrier and disc from the steering knuckle.
4. If equipped with ABS, remove the speed sensor.
5. Remove the tie rod from the steering knuckle.
6. Remove the hub from the steering knuckle then remove the steering knuckle.
7. Remove the wheel bearing outside race from the hub using a suitable bearing puller.

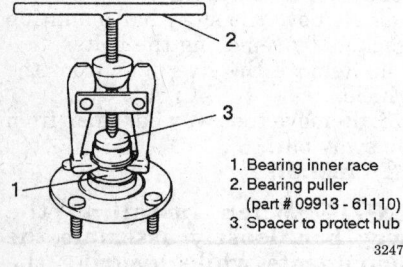

1. Bearing inner race
2. Bearing puller
   (part # 09913 - 61110)
3. Spacer to protect hub

324773

Removal of the bearing race from the hub

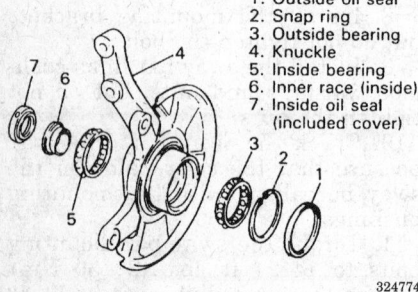

1. Outside oil seal
2. Snap ring
3. Outside bearing
4. Knuckle
5. Inside bearing
6. Inner race (inside)
7. Inside oil seal
   (included cover)

324774

Steering knuckle bearing components — Swift

8. Remove the outside oil seal, snapring, outside bearing, inside oil seal and the inside bearing in that order.

NOTE: Once the bearing outer race is removed, the bearing set (outer race, bearings and inner races) should be replaced.

9. Remove the bearing outer race from the knuckle by pressing the race out of the knuckle with the aid of tool 09913–75520.
To install

NOTE: When installing the oil seals, be careful not to deform or tilt. Damage to the rubber part of the seal may occur.

10. Press the new bearing outer race into the knuckle using a press and the following tools: bearing installer handle 09924–74510, bearing and oil seal installer 09944–68210 and a bearing installer support 0994–78210
11. Apply lithium grease to the bearing races, bearings and oil seal lips.
12. Install the outside bearing to the steering knuckle. Install the snapring to hold the outside bearing in place and then install the outside oil seal.
13. Install the inside bearing to the steering knuckle. Install the inside race and oil seal to the steering knuckle. When installing the inside oil seal, drive the oil seal in until it contacts the steering knuckle.

——— **CAUTION** ———
If equipped with ABS use caution when installing the oil seal, because the seal has a hole that must align with the speed sensor position.

14. Install the outside race to the wheel hub using a bearing installer.
15. Using a press and the proper tools, press the hub into the steering knuckle. After installation, make sure the hub is installed straight and turns freely.
16. Install the steering knuckle in the vehicle.
17. Connect the tie rod end.
18. Install the brake caliper, carrier and disc to the steering knuckle.
19. Install the front wheel.
20. Lower the vehicle.

REAR SUSPENSION

Strut

REMOVAL AND INSTALLATION

Esteem

1. Raise and safely support the vehicle. Remove the wheel and tire assembly.
2. Place a jack under the lower control arm to support the suspension.
3. Disconnect the brake line from the brake hose at the strut.
4. Remove the E ring securing the brake hose to the strut.
5. Remove the strut upper support nuts and push the strut down.
6. Remove the two bolts and nuts holding the strut to the rear knuckle.
7. Remove the strut from the knuckle and then remove the strut from the vehicle.
8. Install a pair of coil spring compressors to the coil spring on the strut assembly. Turn the spring compressors alternately until the spring tension is released from the strut assembly. If the spring can be turned slightly then it has been collapsed enough.

——— **CAUTION** ———
This procedure requires the use of a spring compressor; it cannot be performed without one. If you do not access to this special tool, do not attempt to disassemble the strut. The coil spring is retained under considerable pressure. Id can exert enough force to cause serious injury. Exercise extreme caution.

9. Keeping the spring collapsed remove the strut center nut and remove the other components from the top of the strut assembly.
10. Remove the spring from the strut.
To install:
11. If installing a new spring, compress the spring with a pair of spring compressors, make sure that the spring compresses to 11 in. (290 mm) for installation.
12. Position the coil spring on the strut making sure that the end of the spring is mated to the stepped part of the lower seat.
13. Install the remaining components.

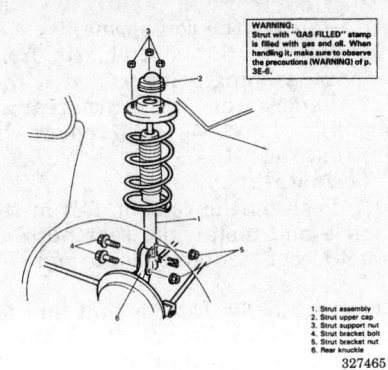

WARNING:
Strut with "GAS FILLED" stamp is filled with gas and oil. When handling it, make sure to observe the precautions (WARNING) of p. 3E-6.

1. Strut assembly
2. Strut upper cap
3. Strut support nut
4. Strut bracket bolt
5. Strut bracket nut
6. Rear knuckle

327465

Strut assembly — Esteem

14. Install the strut support with the center nut. Torque the nut to 40 ft. lbs. (55 Nm).

15. Loosen the spring compressors alternately, checking that the stepped part of the spring seat and spring are properly positioned.

16. Install the strut in the vehicle.

17. Install the two bolts and nuts to hold the strut to the rear knuckle. Torque the nuts 65 ft. lbs. (90 Nm).

18. Fully extend the strut and position the upper part of the strut into the vehicles body. If the upper part of the strut does not reach the vehicle body, raise the jack under the control arm a little.

19. Install the upper support nuts and tighten them to 20–27 ft. lbs. (28–38 Nm).

20. Connect the brake hose to the strut and install the E clip.

21. Connect the brake hose to the brake line.

22. Remove the jack from under the control arm.

23. Install the wheels.

24. Fill the master cylinder with brake fluid and bleed the brake system.

25. Lower the vehicle.

Shock

REMOVAL AND INSTALLATION

Swift

1. Raise and safely support the vehicle. Remove the wheels.

2. Place a jack under the lower control arm to support the suspension.

─── **CAUTION** ───

The coil spring is under extreme pressure. Make sure the lower control arm is firmly supported with a hydraulic jack before con-

tinuing with procedure. If this caution is not observed, serious bodily injury may result.

3. Remove the shock support nuts and push the shock down.

4. Remove the shock lower mounting bolt.

5. Remove the shock from the knuckle. Compress the shock as short as possible for removal. If the shock is hard to remove, open the slit of the knuckle by inserting a wedge.

NOTE: Do not open the knuckle slit wider than necessary. Do not lower the jack more than necessary during the shock removal to prevent the coil spring from coming off, or the brake flexible hose from stretching.

To install:

6. Install the shock in the vehicle. Position the bottom of the alignment projection inside the knuckle opening.

7. Install shock lower mounting bolt and tighten it to 36–50 ft. lbs. (50–70 Nm).

8. Fully extend the shock and position the upper part of the shock into the vehicles body. If the upper part of the shock does not reach the vehicle

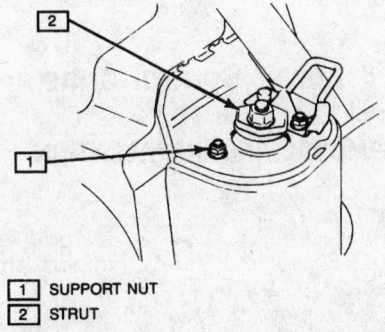

1 SUPPORT NUT
2 STRUT

325026

Rear shock upper mounting nuts — Swift

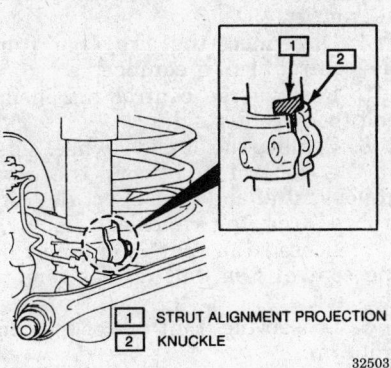

1 STRUT ALIGNMENT PROJECTION
2 KNUCKLE

325031

Shock to steering knuckle alignment — Swift

body raise the jack under the control arm a little.

9. Install the upper support nuts and tighten them to 20–27 ft. lbs. (28–38 Nm).

10. Remove the jack from under the control arm.

11. Install the wheels.

12. Lower the vehicle.

Coil Spring

REMOVAL AND INSTALLATION

Swift

1. Raise and safely support the vehicle.

2. Remove the rear wheels.

NOTE: To facilitate the toe-in adjustment after reinstallation, confirm which one of the lines stamped on the washer is in the closest alignment with the stamped line on the control rod. If not marked, add matchmarks.

3. Remove the control rod inside bolt (body center side).

4. Remove the outside (wheel side) of the control rod from the rear knuckle stud and disconnect the control rod from the knuckle.

5. Remove the nuts, washers, and bushings connecting the rear sway bar to the rear lower control arms.

6. Loosen the rear mount nut on the control arm, but do not remove the bolt.

7. Loosen the front nut of the control arm.

8. If equipped with ABS, disconnect the wheel speed sensor.

─── **CAUTION** ───

The coil spring is under extreme pressure. Make sure control arm is firmly supported with a hydraulic jack before continuing with procedure. If this precaution is not observed, serious bodily injury may result.

9. Loosen the lower mount nut on the knuckle. Place a jack under the control arm to prevent it from lowering and remove the lower mount nut on the knuckle.

10. Raise the jack placed under the control arm enough to allow the removal of the lower mount bolt of the knuckle.

11. Move the brake drum/backing plate toward the outside of the vehicle body so as to separate the lower mount of the knuckle from the control arm. Then lower the jack gradually and remove the coil spring.

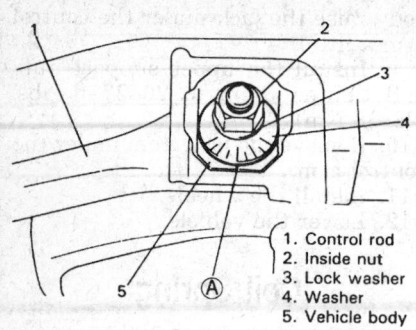

1. Control rod
2. Inside nut
3. Lock washer
4. Washer
5. Vehicle body
A. Alignment lines

324993

Control rod inside mount — Swift

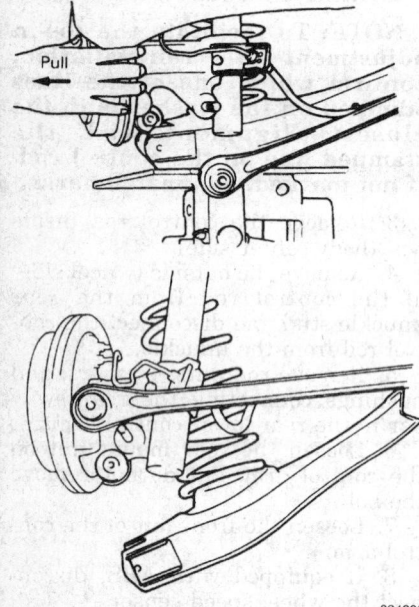

324998

Disconnecting the knuckle and removing the coil spring

To install:

12. Place the jack under the control arm.

13. Install the coil spring on the spring seat of the control arm then raise the control arm. When seating the coil spring, mate the spring end with the stepped part of the control arm.

14. Install the lower knuckle mount bolt. Torque the bolt to 29–33 ft. lbs. (40–45 Nm).

15. Remove the jack from under the suspension arm.

16. Connect the rear sway bar joints to the rear control arms and install the bushings, washers, and nuts. Torque the nuts to 16–20 ft. lbs. (22–28 Nm).

17. Position the control rod and install the inside control rod bolt and the outside control rod nut, but do not tighten them at this time.

18. If equipped with ABS, install the wheel speed sensor.

19. Install the wheels and lower the vehicle.

20. Torque the control rod inside and outside nuts to 51–65 ft. lbs. (70–90 Nm).

NOTE: When tightening the nuts, the vehicle should be off the hoist and in a non-loaded state. Also when tightening the inside nut, align the line stamped on the body with the line on the washer as confirmed before removal or align the matchmarks if marked.

21. Torque the suspension arm front and rear nuts: the front nuts to 36–50 ft. lbs. (50–70 Nm) and the rear nuts to 29–33 ft. lbs. (40–45 Nm). After tightening the suspension arm front nut, make sure that the washer is not tilted.

22. Check the rear wheel alignment.

Lower Control Arms

REMOVAL AND INSTALLATION

Swift

1. Raise and safely support the vehicle.

2. Remove the rear wheels.

3. Scribe alignment marks on the control rod washer and the control rod. This is necessary for proper alignment.

4. Disconnect the sway bar from the control arm, if equipped.

5. Remove the control rod inside bolt (body center side).

6. Remove the outside (wheel side) of the control rod from the rear knuckle stud and disconnect the control rod from the knuckle.

7. Loosen the rear mount nut on the control arm, but do not remove the bolt.

8. Loosen the front nut of the control arm.

9. If equipped with ABS, disconnect the wheel speed sensor.

10. Remove the coil spring.

11. Remove the control arm front mount attaching bolts.

12. Remove the control arm rear attaching nut and bolt, then remove the control arm.

To install:

13. Position the control arm in the vehicle and install the rear nut and bolt. Do not tighten the nut and bolt at this time.

14. Install the front mount and install the bolts. Torque the bolts to 25–39 ft. lbs. (35–55 Nm).

NOTE: Make sure that the bushing is installed properly and that the washer is in the proper position.

15. Install the coil spring.

16. Connect the rear sway bar joints.

17. Position the control rod and install the inside control rod bolt and the outside control rod nut, but do not tighten them at this time.

18. If equipped with ABS, install the wheel speed sensor.

19. Install the wheels and lower the vehicle.

20. Torque the control rod inside and outside nuts to 51–65 ft. lbs. (70–90 Nm).

NOTE: When tightening the nuts, it is most desirable to have the vehicle off the hoist and in a non-loaded state. Also when tightening the inside nut, align the line stamped on the body with the line on the washer as confirmed before removal or align the matchmarks if marked.

21. Torque the suspension arm front and rear nuts: the front nuts to 36–50 ft. lbs. (50–70 Nm) and the rear nuts to 29–33 ft. lbs. (40–45 Nm). After tightening the suspension arm front nut, make sure that the washer is not tilted.

22. Check the rear wheel alignment.

Esteem

1. Disconnect the negative battery cable from the battery.

2. Raise the rear of the vehicle and safely support.

3. Remove the wheel.

NOTE: There are two lower control arms on each side of the suspension. The No. 1 lower control arm is to the front of the vehicle and the No. 2 lower control arm is towards the rear.

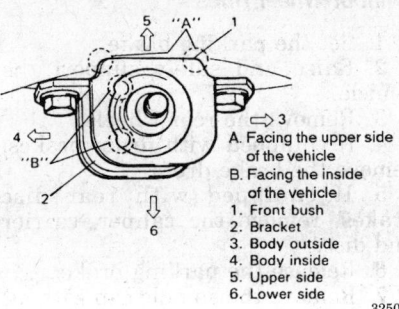

A. Facing the upper side of the vehicle
B. Facing the inside of the vehicle
1. Front bush
2. Bracket
3. Body outside
4. Body inside
5. Upper side
6. Lower side

325013

Proper bushing installation — Swift

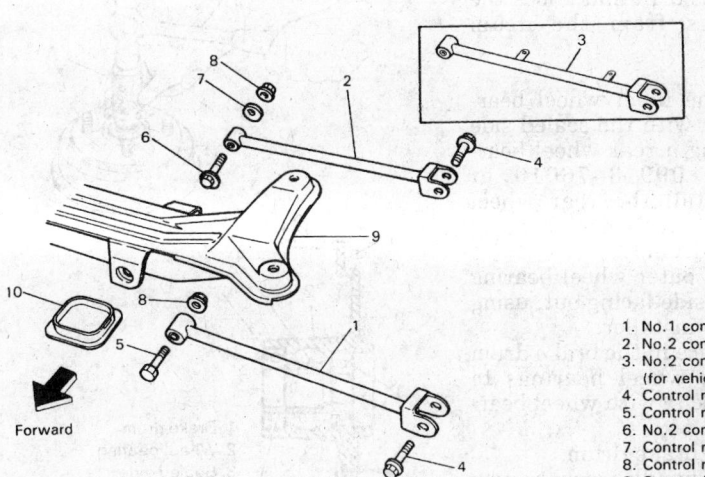

1. No.1 control rod
2. No.2 control rod
3. No.2 control rod (for vehicle equipped with ABS)
4. Control rod outer bolt
5. Control rod inner bolt (inner bolt)
6. No.2 control rod bolt (inner bolt)
7. Control rod washer
8. Control rod inner nut
9. Suspension frame
10. Suspension frame cap

327721

Lower control arm assembly — Esteem

4. If equipped with ABS brakes, disconnect the wheel sensor.

NOTE: For the No. 2 control arm washer: To facilitate the toe-in adjustment after reinstallation, confirm which one of the lines stamped on the washer is in the closest alignment with the stamped line on the suspension frame. If not marked, add matchmarks.

5. Remove the suspension frame cap.
6. Remove the No. 2 control arm nut, washer and bolt from the suspension frame.
7. Remove the No. 1 control arm nut and bolt from the suspension frame.
8. Remove the bolts holding the No. 1 and No. 2 control arms to the knuckle.
9. Remove the No. 1 and No. 2 control arms from the vehicle.

NOTE: Mark the No. 1 and No. 2 control arms to aid in correct installation.

To install:
10. Install the No. 1 and No. 2 control arms to the vehicle.
11. Install the bolts to hold the No. 1 and No. 2 lower control arms to the rear knuckle. The No. 1 lower control arm bolt is installed from the front of the vehicle. The No. 2 lower control arm bolt is installed from the rear of the vehicle. Do not torque the bolts at this time.
12. Connect the No. 1 lower control arm to the suspension frame. Install the bolt and nut to hold the lower control arm to the suspension frame. The No. 1 control arm bolt is installed from the outside of the suspension frame. Do not torque the nut at this time.
13. Connect the No. 2 lower control arm to the suspension frame. The No. 2 lower control arm bolt is installed from inside the suspension frame. In-

stall the No. 2 lower control arm washer and nut. Align the matchmarks on the washer and frame. Do not torque the nut at this time.
14. For vehicles equipped with ABS brakes, connect the ABS speed sensor.
15. Install the wheel and lower the vehicle to the ground.
16. Bounce the vehicle up and down to stabilize the suspension.
17. Tighten the lower control arm bolts and nuts to 65 ft. lbs. (90 Nm). When tightening the No. 2 lower control arm, make sure the washer is aligned with the matchmarks.
18. Install the suspension cap.
19. Check the rear wheel alignment.

Sway Bar

REMOVAL AND INSTALLATION

1. Safely raise and support the vehicle.
2. Remove the nuts, washers and bushing attaching the sway bar to the control arm.
3. Remove the sway bar mounting brackets from the body.
4. Remove the sway bar from the vehicle.

To install:
5. Position the sway bar under the vehicle and install the mounts. Do not tighten the mounting bolts at this time.
6. Connect the sway bar to the control arms and install the bushings, washers and nuts. Do not tighten the nuts at this time.
7. Check the sway bar position by verifying that the paint marks on the sway bar align with the mounting bushings.
8. Torque the mount bolts to 14–20 ft. lbs. (20–28 Nm) and torque the nuts attaching the sway bar to the control arms to 16–20 ft. lbs. (22–28 Nm).
9. Lower the vehicle.

Wheel Bearings

ADJUSTMENT

The rear wheel bearings are a cartridge type design and cannot be adjusted.

REMOVAL AND INSTALLATION

With Wheel Hubs

1. Set the parking brake.

2. Raise and safely support the vehicle.

3. Remove the rear wheels.

4. Remove the rear brake drums.

5. Use a brass drift and knock the wheel bearings from the drum assembly.

To install:

6. Position the inner wheel bearing on the drum with the sealed side facing out. Using a rear wheel bearing installer 09913–76010 or equivalent, install the rear wheel bearing. Install the wheel bearing spacer into the drum.

7. Install the outer wheel bearing with the sealed side facing out, using a wheel bearing installer.

8. Fill the space in the brake drum in between the wheel bearings to about 40% capacity with wheel bearing grease.

9. Install the brake drum.

10. Install the spindle washer and a new spindle nut. Torque the spindle nut to 58–86 ft. lbs. (80–120 Nm).

11. Caulk the spindle nut and the spindle dust cap.

NOTE: When installing the spindle cap, hammer lightly several times on the collar of the cap

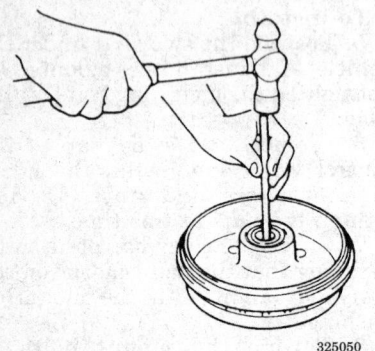

325050

Wheel bearing removal — Swift

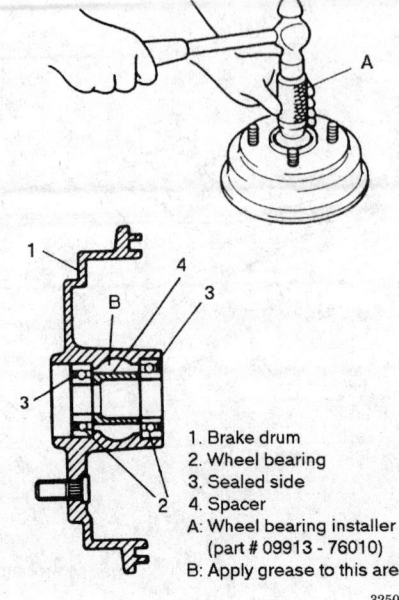

1. Brake drum
2. Wheel bearing
3. Sealed side
4. Spacer
A: Wheel bearing installer (part # 09913 - 76010)
B: Apply grease to this area

325051

Wheel bearing installation — Swift

until the collar comes closely into contact with the brake drum. If the fitting part of the cap is deformed or damaged or if it fits loose, replace the cap with a new one.

12. Depress the brake pedal with about 66 lbs. (30 kg) of force three to five times to obtain proper drum to shoe clearance.

13. Install the wheels.

14. Check to ensure that the brake drum is free from dragging and proper braking is obtained.

15. Lower the vehicle.

Without Wheel Hubs

1. Set the parking brake.

2. Raise and safely support the vehicle.

3. Remove the rear wheels.

4. If equipped with drum brakes, remove the brake drum.

5. If equipped with rear disc brakes, remove the caliper, carrier and disc.

6. Release the parking brake.

7. Remove the spindle cap without deforming it.

8. Uncaulk the spindle nut then remove the spindle nut and washer.

9. Using a brake hub removal tool 09943–17911 or equivalent, and a slide hammer remove the hub from the spindle.

NOTE: The wheel bearing and hub are a solid unit. When the wheel bearing is found defective and it is necessary to replace it, replace the hub assembly.

To install:

10. Install the wheel hub, washer and a new spindle nut. Torque the spindle nut to 108–144 ft. lbs. (150–200 Nm).

11. Caulk the spindle nut and install the spindle cap.

12. If equipped with rear drum brakes, install the brake drums.

13. If equipped with rear disc brakes, install the brake caliper carrier and disc.

14. Depress the brake pedal with about 66 lbs. (30 kg) of force three to five times to obtain proper drum/rotor to shoe/pad clearance.

15. Install the wheel and tighten the lug nuts to 36–58 ft. lbs. (50–80 Nm).

16. Check to ensure that the brakes are free from dragging and that proper braking is obtained.

17. Lower the vehicle.

FIRING ORDERS

NOTE: To avoid confusion, always replace spark plug wires one at a time.

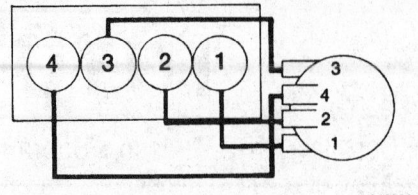

249290

5S-FE, 3S-GTE, 3S-GTE and 5S-FE Engines
Firing Order: 1–3–4–2
Distributor Rotation: Counterclockwise

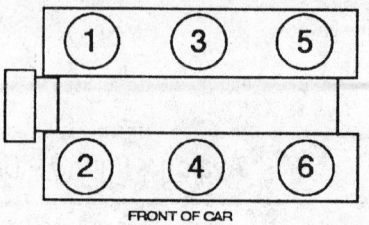

7920ej03

1MZ-FE Engine
Firing Order: 1–2–3–4–5–6
Distributorless Ignition System

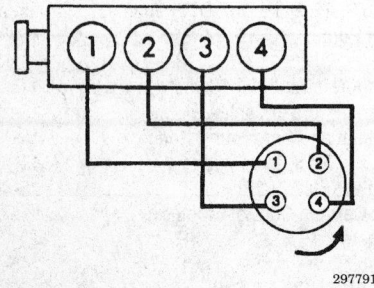

297791

3E-E Engine
Firing Order: 1–3–4–2
Distributor Rotation: Counterclockwise

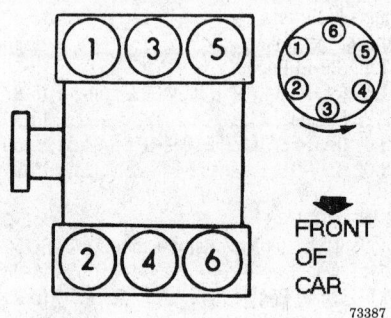

73387

3VZ-FE Engine
Firing Order: 1–2–3–4–5–6
Distributor Rotation: Counterclockwise

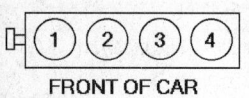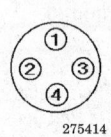

275414

5E-FE Engine
Firing Order: 1–3–4–2
Distributor Rotation: Clockwise

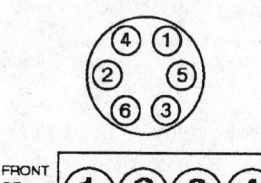

7920ej01

2JZ-GE Engine
Firing Order: 1–5–3–6–2–4
Distributor Rotation: Clockwise

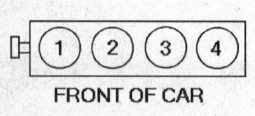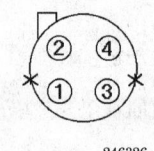

246326

4A-FE and 7A-FE Engines
Firing Order: 1–3–4–2
Distributor Rotation: Counterclockwise

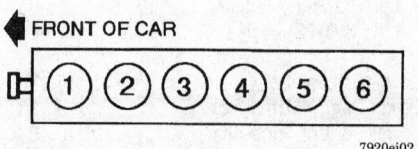

7920ej02

2JZ-GTE Engine
Firing Order: 1–5–3–6–2–4
Direct Ignition: uses one coil at each cylinder

ENGINE ELECTRICAL

NOTE: Disconnecting the negative battery cable on some vehicles may interfere with the functions of the on board computer systems and may require the computer to undergo a relearning process, once the negative battery cable is reconnected.

Distributor

REMOVAL AND INSTALLATION

1. Disconnect the negative battery cable. On vehicles equipped with an air bag, wait at least 90 seconds before proceeding.

2. If equipped, disconnect the electrical connector from the air flow meter. Disconnect the air cleaner hose from the throttle body and remove the air cleaner cover, air flow meter and air duct as one unit.

3. If equipped, remove the air intake to provide clearance.

4. If equipped, remove the intercooler to provide clearance. Disconnect the air temperature sensor, cruise control actuator cable, and the air cleaner hose, if more clearance is necessary.

5. Disconnect the spark plug wires from the distributor cap. Be careful to pull the spark plug wires from the boot, not the wire. Using a suitable tool, lift up the lock claw and disconnect the holder from the distributor cap.

6. Disconnect the distributor connector.

7. Remove the distributor mounting bolt and pull the distributor out.

NOTE: The marks on the distributor drive gear and distributor housing should be aligned. If the marks are not aligned, mark the distributor housing and rotor position.

8. Remove the O-ring from the distributor housing.

To install:

Engine Disturbed

1. On the 2JZ-GE engine, remove the No. 3 timing belt cover.

2. Turn the crankshaft clockwise, and position the slit of the intake camshaft as required to bring No. 1 cylinder to TDC on the compression stroke.

3. Apply a light coat of engine oil to a new O-ring and install it to the distributor housing.

4. Align the cut out portion of the coupling with the groove of the housing. Insert the distributor, aligning the bolt hole of the flange with that of the bolt hole on the cylinder head. Tighten the mounting bolts.

5. If removed, reinstall the intercooler or the air intake and associated parts.

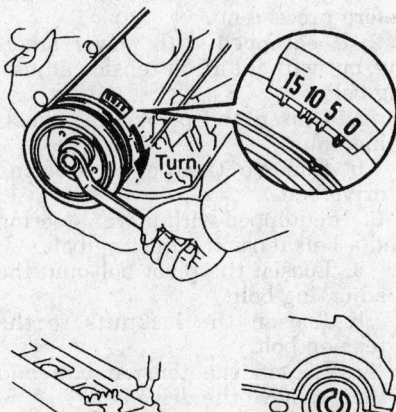

Common method of properly positioning the distributor drive mechanism to TDC

310365

6. Reinstall the spark plug wires to the distributor cap, making sure the firing order is correct.

7. On the 2JZ-GE engine, reinstall the No. 3 timing belt cover.

8. Install the air flow meter, air hose and the air cleaner cover if removed.

9. Connect the distributor and air flow meter connectors.

10. Install the air flow meter, air hose and the air cleaner cover.

11. Connect the negative battery cable.

12. Start the engine and check the ignition timing.

Engine Not Disturbed

1. Apply a light coat of engine oil to a new O-ring and install it to the distributor housing.

2. Align the cut out portion of the coupling with the groove of the housing. Insert the distributor, aligning the bolt hole of the flange with that of the bolt hole on the cylinder head. Tighten the mounting bolts.

3. Make sure the mark on the engine aligns with the mark on the distributor made during removal. Also make sure the rotor is in the same position as removal.

4. If removed, reinstall the intercooler or the air intake and associated parts.

5. Install the distributor cap.

6. Reinstall the spark plug wires to the distributor cap, making sure the firing order is correct.

7. Install the air flow meter, air hose and the air cleaner cover if removed.

8. Connect the distributor and air flow meter connectors.

9. Connect the negative battery cable.

10. Start the engine and check the ignition timing.

Ignition Timing

ADJUSTMENT

1. Start the engine and run it at idle until the engine reaches normal operating temperature. Remove the cap from the check connector (DLC1) which is usually located in the engine compartment.

2. Connect the tachometer and timing light to the engine. Connect the tachometer tester probe to terminal IG (-) of the DLC1.

NOTE: Never allow the tachometer test probe to touch ground as it could result in damage to the igniter and or ignition coil.

NOTE: Not all tachometers are compatible with this system. Make sure to confirm compatibility of your unit before use.

3. Using a jumper connector, connect terminals TE_1 and E_1 of the DLC1. On the 3E-E engines connect terminals T and E_1

4. Check the idle speed.

5. Aim the timing light at the timing indicator and check the ignition timing. Timing should be 10° BTDC at idle.

6. If adjustment is necessary, loosen the distributor hold-down bolt and adjust by turning. Tighten the hold-down bolt and recheck the timing.

7. Remove the jumper connector. Check that the ignition timing advances.

8. Disconnect the tachometer and timing light from the engine.

Alternator

PRECAUTIONS

Several precautions must be observed with alternator equipped vehicles to avoid damage to the unit.

• If the battery is removed for any reason, make sure it is reconnected with the correct polarity. Reversing the battery connections may result in damage to the 1-way rectifiers.

• When utilizing a booster battery as a starting aid, always connect the positive to positive terminals and the negative terminal from the booster battery to a good engine ground on the vehicle being started.

• Never use a fast charger as a booster to start vehicles.

• Disconnect the battery cables when charging the battery with a fast charger.

• Never attempt to polarize the alternator.

• Do not use test lights of more than 12 volts when checking diode continuity.

• Do not short across or ground any of the alternator terminals.

• The polarity of the battery, alternator and regulator must be matched and considered before making any electrical connections within the system.

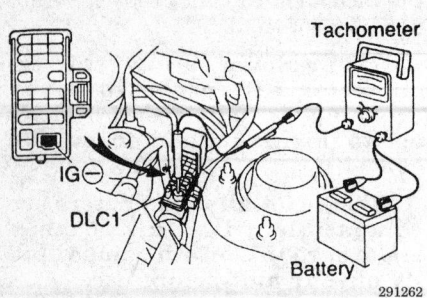

Attach the tachometer test probe to the IG-terminal of the data link connector

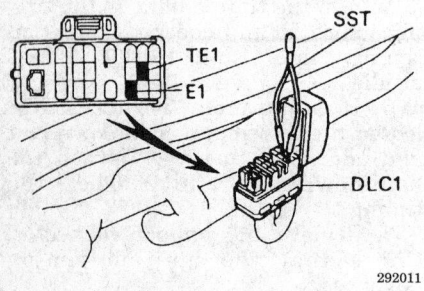

A jumper connector is used to connect terminals TE1 and E1 of the DLC

• Never separate the alternator on an open circuit. Make sure all connections within the circuit are clean and tight.
• Disconnect the battery ground terminal when performing any service on electrical components.
• Disconnect the battery if arc welding is to be done on the vehicle.

Except Supra

NOTE: It may be necessary to remove the gravel shield and work from underneath the car in order to gain access to the alternator retaining bolts.

1. Disconnect the negative battery terminal from the battery.
2. Loosen the pivot bolt and the lock bolt.
3. Loosen the adjusting bolt to release tension on the drive belt.

NOTE: It may be necessary to remove other belts such as power steering or A/C to access the alternator belt.

4. Remove the drive belt from the alternator.
5. Unplug the wiring connector from the back of the alternator.

6. Remove the nut and alternator wire.
7. Remove the wire harness from the 2 clips.
8. Remove the pivot bolt, nut, and the lock bolt from the alternator.
9. Remove the alternator from the mounting bracket.

To install:
10. Place the alternator in its brackets and install the pivot bolt, nut, and the lock bolt. Do not tighten them at this time.
11. Leave the bolts finger tight so that the belt may be adjusted to the correct tension.
12. Make sure that the electrical plugs and connectors are properly seated and secure in their mounts. Adjust belt tension and tighten all necessary hardware. If equipped with A/C, tighten the adjusting bolt to apply tension to the belt.
13. Tighten the pivot bolt to 30–45 ft. lbs. (41–61 Nm) and the adjusting bolt to 9–13 ft. lbs. (12–18 Nm).
14. Connect the negative battery cable.
15. Start the engine and verify proper charging system operation.

Supra

1. Disconnect the negative battery cable. Wait at least 90 seconds before performing any other work.
2. Remove the engine undercover.
3. Remove the No. 2 air tube for the charge air cooler.
4. Remove the lower fan shroud.
5. If equipped with manual transmission, remove the drive belt tensioner damper by removing the two nuts.
6. Turn the drive belt tensioner clockwise and remove the drive belt from the engine.
7. Disconnect the alternator electrical connector.
8. Remove the rubber cap and nut and then disconnect the alternator wire.
9. Disconnect the alternator wire clamp from the wire clip on the alternator.
10. Disconnect the A/T coolant hoses and bracket from the alternator by removing the bolt and pipe clamp.
11. Remove the alternator bolt, nut, pipe bracket, and then remove the alternator from the engine.

To install:
12. Install the alternator and pipe bracket. Tighten the bolt and nut to 27 ft. lbs. (37 Nm).

13. Install the A/T oil cooler pipes to the alternator and install the bolt and pipe clamp.
14. Connect the alternator wire clamp to the wire clip on the alternator.
15. Connect the electrical wire to the alternator and install the nut.
16. Install the alternator electrical connector.
17. Install the drive belt.
18. If equipped with manual transmission, install the drive belt tensioner damper with the two nuts. Tighten the nuts to 14 ft. lbs. (20 Nm).
19. Install the lower fan shroud.
20. Install the No. 2 air tube for the charge air cooler.
21. Install the lower engine cover.
22. Connect the negative battery cable and check the vehicle's charging system for proper operation.

Drive Belt

REMOVAL AND INSTALLATION

1.5L (3E-E and 5E-FE) Engines

NOTE: In order to remove the alternator belt from the engine, the power steering belt must be also be removed.

1. Disconnect the negative battery cable. On vehicles equipped with an air bag, wait at least 90 seconds before proceeding.
2. If equipped with power steering, but without a belt tension adjusting bolt:
   a. Loosen the pivot and adjusting bolts.
   b. Remove the power steering drive belt.
3. If equipped with power steering and a belt tension adjusting bolt:
   a. Loosen the pivot bolt and the adjusting bolt.
   b. Loosen the locknuts to the tension bolt.
   c. Loosen the tension bolt and then remove the drive belt.
4. If equipped with A/C, but without power steering, loosen the idler pulley mounting nut and adjusting bolt. Remove the drive belt.
5. To remove the alternator belt, loosen the pivot nut and adjusting bolt. Remove the drive belt from the engine.
6. Visually check the belt for separation of the ribs, torn or worn ribs, or cracks in the inner ridges of the ribs. If necessary, replace the drive belt.

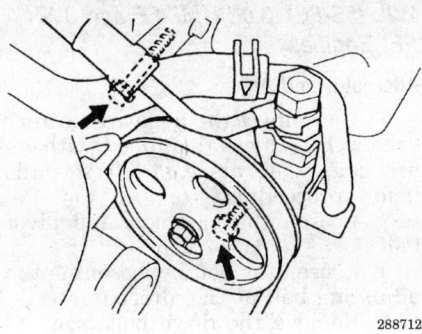

288712

Power steering bolts (without a tension bolt) — 5E-FE engine

To install:

7. To install the alternator drive belt:

a. Position the alternator drive belt over the component pulleys and drive pulley.

b. Adjust the alternator to tighten the alternator belt.

c. Tighten the pivot bolt to 30 ft. lbs. (41 Nm) and the adjusting bolt to 9 ft. lbs. (12 Nm).

8. If equipped with power steering drive and a belt tension adjusting bolt:

a. Install drive belt over component pulleys and drive pulley.

b. Push up on power steering pump and then tighten the tension bolt until the correct tension is applied to the drive belt.

c. Tighten the pivot and adjusting bolts to 32 ft. lbs. (43 Nm).

9. If equipped with power steering, but without a belt tension adjusting bolt:

a. Install the power steering drive belt.

b. Tighten the power steering belt.

c. Tighten the pivot and adjusting bolts to 32 ft. lbs. (43 Nm).

10. Check the drive belt tension using a belt tension gauge.

NOTE: After installing the drive belt(s), check that the belt does not touch the bottom of the pulley groove (conventional type belts) or that it fits properly in the ribbed grooves (V-ribbed type belts).

11. Connect the negative battery cable.

12. If a new belt was installed, run the engine for approximately 5 minutes and then recheck the tension.

13. Drive belt tension:

- New alternator belt–160 ±20 lbs.
- Used alternator belt–100 ±20 lbs.
- New A/C belt–165 ±15 lbs.
- Used A/C belt–110 ±20 lbs.

- New power steering belt–175 ±5 lbs.
- Used power steering belt–115 ±20 lbs.
- New power steering and A/C belt–165 ±15 lbs.
- Used power steering and A/C belt–110 ±20 lbs.

1.6L (4A-FE) and 1.8L (7A-FE) Engines

1. Disconnect the negative battery cable. On vehicles equipped with an air bag, wait at least 90 seconds before proceeding.

2. Loosen the lock, pivot and adjusting bolts of the component. Loosen the idler pulley, if equipped.

3. Remove the drive belt(s) from its driven accessory.

4. Visually check the belt for separation of the ribs, torn or worn ribs or cracks in the inner ridges of the ribs. If necessary, replace the drive belt.

To install:

5. Position the drive belt(s) over the component pulleys and drive pulley.

6. Partially tighten the pivot and adjusting bolts.

7. Check the drive belt(s) tension using a belt tension gauge. Fully tighten the lock, pivot and adjusting bolts.

NOTE: After installing the drive belt(s), check that the belt does not touch the bottom of the pulley groove (conventional type belts) or that it fits properly in the ribbed grooves (V-ribbed type belts).

8. Connect the negative battery cable.

9. If a new belt was installed, run the engine for approximately 5 minutes and then recheck the tension.

10. Celica drive belt tension:

- New alternator belt–160 ±20 lbs.
- Used alternator belt–130 ±20 lbs.
- New A/C belt, 130 ±10 lbs.
- Used A/C belt, 70 ±10 lbs.
- New power steering belt–125 ±25 lbs.
- Used power steering belt–125 ±25 lbs.

11. Corolla drive belt tension:

- New alternator belt–175 ±5 lbs.
- Used alternator belt–115 ±20 lbs.
- New A/C belt–160 ±25 lbs.
- Used A/C belt–100 ±20 lbs.
- New power steering belt–125 ±25 lbs.
- Used power steering belt–80 ±20 lbs.

1.6L (4A-FE), 1.8L (7A-FE) and 2.2L (5S-FE) Engines

Alternator Belt

1. Disconnect the negative battery cable from the battery. On vehicles equipped with an air bag, wait at least 90 seconds before proceeding.

2. Loosen the pivot bolt and adjusting lock bolt.

3. Loosen the adjusting bolt and remove the drive belt.

4. Inspect the belt for cracks, glazing and other wear and/or damage. Replace the belt, as necessary.

To install:

5. Install the drive belt, making sure the belt is properly aligned on the pulleys.

6. If the old belt has been reinstalled, tighten the alternator adjusting bolt until the belt deflects 0.24–0.35 in. with approximately 22 lbs. pressure applied at a point midway between the pulleys, on the longest accessible belt span. If a new belt has been installed, tighten the alternator adjusting bolt until the belt deflects 0.20–0.31 in. with approximately 22 lbs. pressure applied at a point midway between the pulleys, on the longest accessible belt span.

7. Tighten the pivot bolt to 45 ft. lbs. (61 Nm) and the lock bolt to 14 ft. lbs. (19 Nm).

8. Connect the negative battery cable to the battery.

9. Run the engine and check operation.

A/C Belt

1. Disconnect the negative battery cable. On vehicles equipped with an air bag, wait at least 90 seconds before proceeding.

2. Remove the alternator belt.

3. Remove the windshield washer fluid reservoir tank, if necessary.

4. Loosen the idler pulley locknut.

5. Loosen the adjusting bolt for the idler pulley and remove the compressor drive belt.

6. Inspect the belt for cracks, glazing and other wear and/or damage. Replace the belt, as necessary.

To install:

7. Install the drive belt, making sure the belt is properly aligned on the pulleys.

8. If the old belt has been reinstalled, tighten the idler pulley adjusting bolt until the belt deflects 0.33–0.37 in. with approximately 22 lbs. pressure applied at a point midway between the pulleys, on the longest accessible belt span. If a new belt has been installed, tighten the idler pulley adjusting bolt until the belt de-

flects 0.24–0.28 in. with approximately 22 lbs. pressure applied at a point midway between the pulleys, on the longest accessible belt span.

9. Tighten the idler pulley locknut to 29 ft. lbs. (39 Nm).

10. Install the washer fluid reservoir tank, if removed.

11. Install the alternator drive belt.

12. Connect the negative battery cable to the battery.

13. Run the engine and check operation.

Power Steering Belt

1. Disconnect the negative battery cable from the battery. On vehicles equipped with an air bag, wait at least 90 seconds before proceeding.

2. Remove the alternator drive belt.

3. If equipped, remove the A/C compressor drive belt.

4. Loosen the two power steering pump mounting bolts and pivot the power steering pump.

5. Remove the belt from the pulleys.

6. Inspect the belt for cracks, glazing and other wear and/or damage. Replace the belt, as necessary.

To install:

7. Install the drive belt, making sure the belt is properly aligned on the pulleys.

8. Position a suitable tension gauge in the middle of the longest accessible belt span.

9. Adjust the belt tension by pulling back on the steering pump.

10. If the old belt is being reused, set the belt tension at 55–88 lbs. If a new belt is being installed, set the belt tension at 100–121 lbs.

11. When the belt tension is set, torque the power steering pump mounting bolts to 29 ft. lbs. (39 Nm).

12. If equipped, install the A/C compressor drive belt.

13. Install the alternator belt.

14. Connect the negative battery cable to the battery.

15. Run the engine and check operation.

2.0L (3S-GTE) Engine

Alternator and A/C Belt

1. Disconnect the negative battery cable. On vehicles equipped with an air bag, wait at least 90 seconds before proceeding.

2. Loosen the lock bolt and pivot bolt.

3. Loosen the belt by loosening the adjusting bolt at the alternator.

4. Remove the drive belt from its driven alternator and A/C units.

5. Visually check the belt for separation of the ribs, torn or worn ribs, or cracks in the inner ridges of the ribs. If necessary, replace the drive belt.

To install:

6. Position the drive belt over the component pulleys and drive pulley.

7. Tighten the adjusting bolt to tighten the belt.

8. Check the drive belt tension using a belt tension gauge.

9. On the Celica, tighten the belt to the following tension:
- With A/C–New belt to 175 ±5 lbs.
- Used belt to 115 ±20 lbs.
- Without A/C–New belt to 125 ±25 lbs.
- Used belt to 95 ±20 lbs.

10. On the MR2, tighten the belt to the following tension:
- New alternator belt–120 ±20 lbs.
- Used alternator belt–104 ±20 lbs.
- New A/C belt–160 ±25 lbs.
- Used A/C belt–100 ±20 lbs.

11. Tighten the pivot bolt to 40 ft. lbs. (54 Nm) and the lock bolt to 14 ft. lbs. (19 Nm).

NOTE: After installing the drive belt(s), check that the belt does not touch the bottom of the pulley groove (conventional type belts) or that it fits properly in the ribbed grooves (V-ribbed type belts).

12. Connect the negative battery cable.

13. If a new belt was installed, run the engine for approximately 5 minutes and then recheck the tension.

Power Steering

1. Disconnect the negative battery cable from the battery. On vehicles equipped with an air bag, wait at least 90 seconds before proceeding.

2. Remove the alternator–A/C belt from the engine.

3. Loosen the pivot bolt and adjusting bracket bolt for the power steering unit.

4. Remove the drive belt from the engine by pivoting the power steering unit.

To install:

5. Install the belt to the engine and tighten the belt by pulling back on the power steering unit.

6. Tighten the drive belt to the following specifications:
- New belt to 99–121 ft. lbs.
- Old belt to 44–77 ft. lbs.

7. Tighten the pivot and adjusting bracket bolt to 32 ft. lbs. (43 Nm).

8. Connect the negative battery cable to the battery.

2.2L (5S-FE), 3.0L (1MZ-FE and 3VZ-FE) Engines

Alternator and A/C

1. Disconnect the negative battery cable. On vehicles equipped with an air bag, wait at least 90 seconds before proceeding.

2. Loosen the lock bolt and pivot bolt.

3. Loosen the belt by loosening the adjusting bolt at the alternator.

4. Remove the drive belt from its driven alternator and A/C units.

5. Visually check the belt for separation of the ribs, torn or worn ribs, or cracks in the inner ridges of the ribs. If necessary, replace the drive belt.

To install:

6. Route the belt on the pulleys.

7. Tighten the adjusting bolt until the belt meets one of the following tensions:
- New alternator belt with A/C (5S-FE), 175 ±5 lbs.
- Used alternator belt with A/C (5S-FE), 130 ±10 lbs.
- New alternator belt without A/C (5S-FE), 125 ±25 lbs.
- Used alternator belt without A/C (5S-FE), 95 ±20 lbs.
- New alternator belt (3VZ-FE and 1MZ-FE), 175 ±5 lbs.
- Used alternator belt (3VZ-FE and 1MZ-FE), 115 ±20 lbs.

8. Tighten the alternator pivot bolt to 41 ft. lbs. (56 Nm) and the lock bolt to 13 ft. lbs. (18 Nm).

NOTE: After installing the drive belt(s), check that the belt does not touch the bottom of the pulley groove (conventional type belts) or that it fits properly in the ribbed grooves (V-ribbed type belts).

9. Connect the negative battery cable to the battery.

Power Steering

1. Disconnect the negative battery cable from the battery. On vehicles equipped with an air bag, wait at least 90 seconds before proceeding.

2. Remove the alternator–A/C belt from the engine.

3. If necessary, remove the right front wheel.

4. If necessary, remove the fender apron seal.

5. Loosen the pivot bolt and adjusting bracket bolt for the power steering unit.

6. Remove the belt from the engine by pivoting the power steering unit.

To install:

7. Route the belt on the pulleys.

8. Adjust the belt tension to the following tensions:
- New power steering belt (5S-FE), 125 ±25 lbs.
- Used power steering belt (5S-FE), 80 ±20 lbs.
- New power steering belt (3VZ-FE and 1MZ-FE), 150–185 lbs.
- Used power steering belt (3VZ-FE and 1MZ-FE), 115 ±20 lbs.

9. Tighten the power steering mounting bolts to 32 ft. lbs. (43 Nm).
10. Install the fender apron seal.
11. Install the wheel.
12. Lower the vehicle.
13. Install the alternator belt.

3.0L (2JZ-GE and 2JZ-GTE) Engines

1. Disconnect the negative battery cable from the battery. On vehicles equipped with an air bag, wait at least 90 seconds before proceeding.
2. Make a note of the belt routing.
3. If equipped with a manual transmission, remove the drive belt tensioner damper by removing the two nuts.
4. Turn the drive belt tensioner clockwise to release the tension.
5. Remove the drive belt.

To install:
6. Loosen the drive belt tensioner and position the belt over the pulleys.

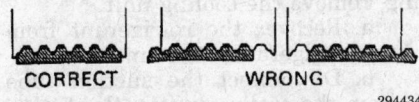

CORRECT   WRONG

294434

Properly align the belt on the pulley grooves

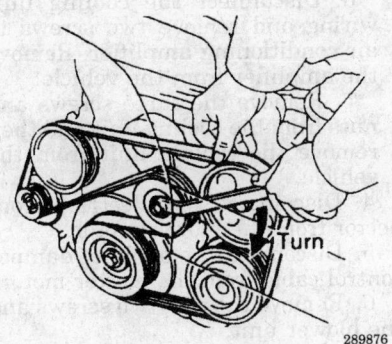

289876

Turn the drive belt tensioner clockwise to release the tension — 2JZ-GTE engine

7. If equipped with a manual transmission, install the drive belt tensioner damper and install the two nuts. Tighten the nut to 14 ft. lbs. (19 Nm).
8. If installing a used belt, ensure the arrow mark on the tensioner falls within the area **A** of the scale.
9. If installing a new belt, ensure the arrow mark on the tensioner falls within the area **B** of the scale.
10. If not as specified, replace the belt.

Starter

REMOVAL AND INSTALLATION

1.5L (3E-E and 5E-FE) Engines

Paseo and Tercel

1. Disconnect the negative battery cable. On vehicles equipped with an air bag, wait at least 90 seconds before proceeding.
2. If required, remove the transaxle cable and bracket from the transaxle.
3. For vehicles with distributorless ignition, disconnect the wire harness from the intake manifold support stay.
4. Remove the intake manifold support bracket.
5. Disconnect the connector wiring from the starter.
6. Remove the nut and disconnect the starter wire.
7. Unbolt the two starter mounting bolts and remove the starter.

To install:
8. Install the starter to the engine and torque the two bolts to 29 ft. lbs. (39 Nm).
9. Install the starter wiring to the starter and torque the nut to 78 inch lbs. (9 Nm).
10. Connect the starter connector to the starter.
11. Install the intake support stay to the engine and torque the bolt(s) and nut(s) to 13 ft. lbs. (18 Nm).
12. If removed, attach the transaxle cable and bracket to the transaxle.
13. If vehicle is equipped with distributorless ignition, connect the engine wire clamp to the intake manifold stay.
14. Connect the negative battery cable to the battery.

1.6L (4A-FE) and 1.8L (7A-FE) Engines

Celica and Corolla

1. Disconnect the negative battery cable. On vehicles equipped with an

air bag, wait at least 90 seconds before proceeding.
2. Disconnect the Intake Air Temperature (IAT) sensor from the air cleaner.
3. Loosen the air cleaner hose clamp bolt.
4. Disconnect the 4 air cleaner housing cover clips.
5. Disconnect the air cleaner hose from the throttle body and remove the air cleaner housing cover together with the air cleaner hose.
6. Disconnect the wire clamp for the starter motor.
7. Disconnect the starter motor electrical connector.
8. Remove the nut and the positive battery cable from the starter motor terminal.
9. Remove the 2 bolts and the starter motor from the vehicle.

To install:
10. Install the starter motor and tighten the 2 retaining bolts to 31 ft. lbs. (43 Nm).
11. Install the nut and the positive battery cable to the starter motor terminal.
12. Connect the starter motor electrical connector.
13. Connect the wire clamp for the starter motor.
14. Connect the air cleaner hose to the throttle body. Install the air cleaner housing cover and secure with the 4 clips.
15. Connect the IAT sensor to the air cleaner.
16. Tighten the air cleaner hose clamp bolt.
17. Connect the negative battery cable to the battery.
18. Check starter motor operation.

2.0L (3S-GTE) and 2.2L (5S-FE) Engines

Celica and MR2

1. Disconnect the battery, negative terminal first. Wait at least 90 seconds before performing any other work. Disconnect the positive cable.
2. On the MR2, remove left hand engine hood side panel.
3. Remove the air cleaner assembly.
4. On the Celica 3S-GTE engine, remove the two nuts holding the relay box and remove the relay box from the battery.
   a. Remove the lower cover from the relay box.
   b. Disconnect the fusible link cassette (tray) and the two engine harness connectors from the relay box.
   c. Remove the battery.

5. Disconnect and remove the connectors and wiring from the starter.

6. Remove the two starter mounting bolts and remove the starter.

To install:

7. Reinstall the starter and tighten the retaining bolts to 29 ft. lbs. (39 Nm).

8. Connect the wiring to the starter terminals.

9. If removed, install the battery.

10. If removed, install the engine relay box. Remember to connect the engine harnesses and secure the cassette within the box. Install the lower cover and mount the box with the two nuts.

11. Install the air cleaner base and the filter element.

12. Connect the air cleaner hose to the turbocharger, the PCV hose to the valve cover and the air hose to the air pipe.

13. Install the air cleaner top and air flow meter; connect the wiring for the air flow meter.

14. Install left hand engine hood side panel.

15. Connect the battery cables to the battery, negative cable last.

2.2L (5S-FE) and 3.0L (1MZ-FE and 3VZ-FE) Engines

Avalon and Camry

1. Disconnect the negative and positive battery cables. On vehicles equipped with an air bag, wait at least 90 seconds before proceeding.

2. Remove the battery from the vehicle by removing the battery clamp.

3. On models with cruise control, remove the actuator cover and disconnect the electrical connector.

4. Remove the three bolts and then lift out the cruise control actuator.

5. Disconnect the starter connector from the starter.

6. Remove the nut and disconnect the starter wire.

7. Support the starter by hand and remove the two mounting bolts.

8. Remove the starter from the transaxle.

To install:

9. Place the starter motor in the transaxle and support it by hand.

10. Install the two mounting bolts and torque them to 30 ft. lbs. (41 Nm).

11. Place the starter wire into position and tighten the nut.

12. Connect the electrical connector to the starter.

13. If removed, install the cruise control actuator and the three mounting bolts. Connect the electri-

cal connector to the actuator and install the actuator cover.

14. Install the battery and tighten down the battery clamp.

15. Connect the negative and positive battery cables.

16. Verify proper starter operation.

3.0L (2JZ-GE) and 3.0L (2JZ-GTE) Engines

Supra

1. Disconnect the negative battery cable. Wait at least 90 seconds before performing any other work.

2. Remove the rubber cap and nut at the starter; remove the starter wire.

3. Disconnect the electrical connector from the starter.

4. Remove the two starter bolts and engine wire clamp.

5. Remove the starter from the engine.

To install:

6. Place the engine wire clamp in place and install the starter. Tighten the mounting bolts to 29 ft. lbs. (39 Nm).

7. Connect the electrical connector to the starter.

8. Install the starter wire, nut, and the rubber cap.

9. Connect the negative battery cable to the battery.

10. Check the starter for proper operation.

CHASSIS ELECTRICAL

Blower Motor

REMOVAL AND INSTALLATION

Tercel

The blower motor is located under the dashboard on the far right side of the car. It is accessible from under the dashboard without removing the dash assembly.

1. Disconnect the negative battery cable. On vehicles equipped with an air bag, wait at least 90 seconds before proceeding.

2. Remove the three screws attaching the retainer.

3. Remove the glove box assembly.

4. Remove the air duct between the heater case and blower assembly.

5. Disconnect the blower motor wire connector at the motor case.

6. Disconnect the air source selector control cable at the blower assembly.

7. Loosen the two nuts and the bolt attaching the blower assembly, then remove the blower.

8. With the blower removed, check the case for any debris or signs of fan contact. Inspect the fan for wear spots, cracked blades or hub, loose retaining nut or poor alignment.

To install:

9. Place the blower in position, making sure it is properly aligned within the case. Install the two bolts and the nut and tighten them.

10. Connect the selector control cable at the blower assembly.

11. Connect the wire harness to the motor and install the ductwork between the heater case and the blower assembly.

12. Install the glove box assembly and install the retainer with its three screws. Reconnect battery and check operation of blower motor for all speeds and heater A/C system for proper operation.

Paseo

1. Disconnect the negative battery cable from the battery. On vehicles equipped with an air bag, wait at least 90 seconds before proceeding.

2. Disconnect the glove compartment and lower finish panel to gain access to the blower motor.

3. If equipped with air conditioning, remove the cooling unit.

   a. Recover the refrigerant from the refrigeration system.

   b. Disconnect the suction tube from the cooling unit outlet fitting and cap the opening immediately to prevent contamination.

   c. Disconnect the liquid tube from the cooling unit inlet fitting and cap the opening immediately to prevent contamination.

   d. Disconnect the cooling unit wiring and remove two screws to air conditioning amplifier. Remove the amplifier from the vehicle.

   e. Remove the three screws and nuts from the cooling unit and then remove the cooling unit from the vehicle.

4. Disconnect the electrical connector from the blower motor.

5. Disconnect the air inlet damper control cable from the blower motor.

6. Remove the nut, two screws and the blower unit.

To install:

7. Install the blower unit with the nut and two screws.

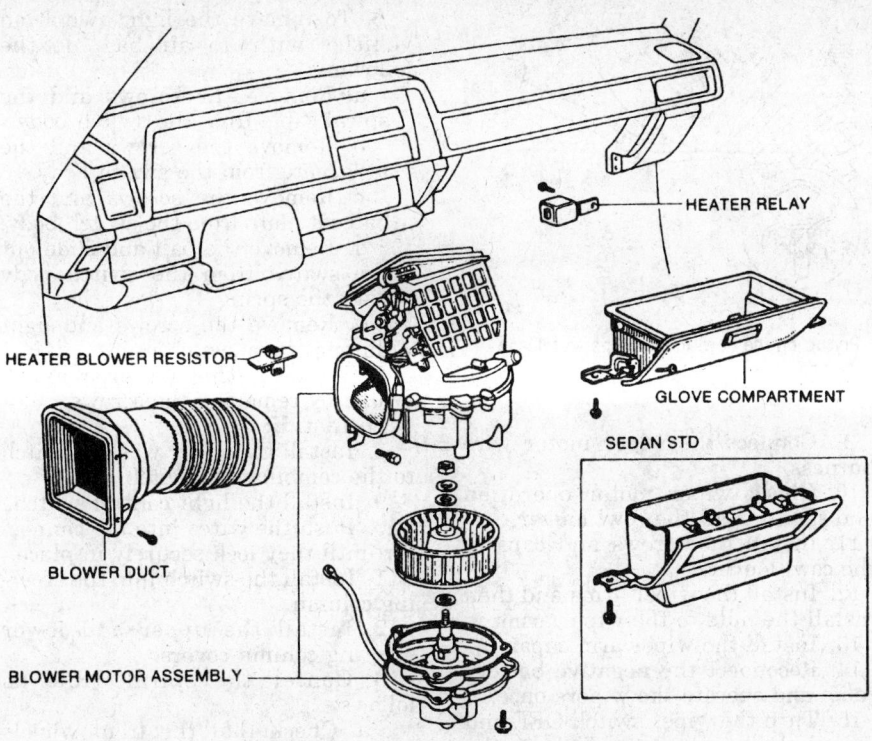

HEATER RELAY

HEATER BLOWER RESISTOR

GLOVE COMPARTMENT

SEDAN STD

BLOWER DUCT

BLOWER MOTOR ASSEMBLY

Typical early model heater blower assembly

122329

Exploded view of the blower motor components — Tercel shown

8. Connect the air inlet damper control cable to the blower unit.

9. Connect the electrical connector to the blower motor.

10. Install the air conditioning unit as follows:

a. Install the cooling unit with the three nuts and screws.

b. Install the A/C amplifier to the cooling unit with the two screws.

c. Connect the connector to the A/C amplifier.

d. Install the ground wire for the A/C system,

e. Connect the liquid tube to the cooling unit inlet fitting. Tighten the fitting to 10 ft. lbs. (14 Nm).

f. Connect the suction tube to the cooling unit outlet fitting. Tighten the fitting to 24 ft. lbs. (32 Nm).

g. If the evaporator was replaced, add compression oil to the compressor.

h. Evacuate the air from the refrigeration system.

11. Install the glove compartment and finish panel.

12. Connect the negative battery cable to the battery.

13. Charge the air conditioning system and check for leaks.

Celica, Corolla and MR2

1. Disconnect the negative battery cable from the battery. On vehicles equipped with an air bag, wait at least 90 seconds before proceeding.

——— **CAUTION** ———
The Air Bag System must be disarmed before removing the glove compartment. Failure to do so may cause accidental deployment of the air bag, resulting in unnecessary system repairs and/or personal injury.

2. If equipped, disable the air bag system.

3. Remove the glove compartment door finish plate inside the instrument panel box.

4. If equipped, pull up and disconnect the air bag connector.

5. Remove the glove compartment box by removing the screws.

6. On Celica, remove the left and right lower pad inserts.

7. Disconnect the connector from the blower motor.

8. Disconnect the air source selector control cable at the blower motor assembly.

9. Remove the three screws and the blower motor.

To install:

10. Install the blower motor with the three screws.

11. Connect the electrical connector to the blower motor.

12. Connect the air source selector control cable at the blower motor assembly.

13. Install the left and right lower pad inserts for the glove compartment.

14. Install the glove compartment box by installing the screws.

15. Connect the air bag connector.

16. Install the glove compartment door finish plate inside the instrument panel box.

17. Enable the air bag system.

18. Connect the negative battery cable to the battery.

Avalon and Camry

1. Disconnect the negative battery cable from the battery. On vehicles equipped with an air bag, wait at least 90 seconds before proceeding.

2. Remove the lower finish panel and the No. 2 undercover from the dashboard.

3. On Camry, remove the two screws to the connector bracket and remove the connector bracket.

4. Disconnect the electrical connector from the blower motor.

5. Remove the three screws from the blower motor and remove the blower motor.

To install:

6. Install the blower motor and install the three screws.

7. Connect the electrical connector to the blower motor.

8. Install the connector bracket and install the two screws.

9. Install the lower finish panel and the No. 2 undercover to the dashboard.

10. Connect the negative battery cable to the battery and check the operation of the blower motor.

Supra

1. Disconnect the negative battery cable from the battery. On vehicles equipped with an air bag, wait at least 90 seconds before proceeding.

2. Remove the glove box from the passenger compartment.

3. Remove the five screws and the brace around the glove box compartment area.

4. Remove the side air duct.

5. Remove the scuff plate from the passenger door.

6. Pull back the floor carpet on the passenger side.

7. Remove the ECM protector by removing the two nuts.

8. Disconnect the electrical connector to the blower motor.

9. Remove the blower motor by removing the three screws.

To install:

10. Install the blower motor to the vehicle and install the three screws.

11. Connect the electrical connector to the blower motor.

12. Install the ECM protector with the two nuts.

13. Replace the carpet to its original position.

14. Install the scuff plate.

15. Install the side air duct.

16. Install the glove compartment brace with the five screws.

17. Install the glove box to the vehicle.

18. Connect the negative battery cable to the battery.

19. Check the blower motor operation.

Windshield Wiper Motor

REMOVAL AND INSTALLATION

1. For easy removal of the wiper arms, operate the wipers and turn the ignition switch **OFF** when the wiper arms are about half way on the downward sweep.

2. Disconnect the negative battery cable from the battery. On vehicles equipped with an air bag, wait at least 90 seconds before proceeding.

3. Remove the nuts and the wiper arms.

4. If necessary, remove the cowl louver.

5. Disconnect the electrical connector to the wiper motor.

6. Remove the bolts to the wiper motor.

7. Disconnect the wiper link.

To install:

8. Install the wiper motor to the vehicle. Connect the linkage and secure the motor with the bolts.

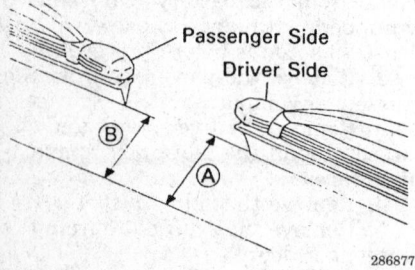

Passenger Side
Driver Side
Ⓑ
Ⓐ
286877

When in the PARK position, the wipers should rest at the same level when seated

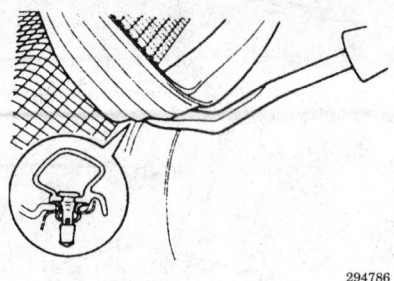

294786

Pry up on the cowl louver clips to release the unit

9. Connect the wiper motor wire harness.

10. Check wiper motor operation and then install the cowl louver.

11. Install the screws and caps to the cowl louver.

12. Install the wiper arms and then install the nuts to the wiper arms.

13. Install the wiper arm caps.

14. Reconnect the negative battery cable and operate the wipers once.

15. Turn the wiper switch OFF and install the wiper arms. Tighten the wiper arm nuts to 15–18 ft. lbs. (20–25 Nm).

16. Inspect the installed height of the wiper arms. The arms should be level when in the park position.

Combination Switch

REMOVAL AND INSTALLATION

Ball Type

The combination switch is composed of the turn signal, the headlight control, the dimmer switch, the wiper, and the washer switch.

──── **CAUTION** ────

The Air Bag System must be disarmed before performing this procedure. Failure to do so may cause accidental deployment of the air bag, resulting in unnecessary system repairs and/or personal injury.

1. Disconnect the negative battery cable. Remove the steering wheel.

2. Remove the upper and lower steering column shroud screws and the shrouds.

3. Remove the combination switch screws and the switch from the column.

4. Disconnect the electrical connectors from the combination switch.

5. Remove the light control switch by removing the screws.

6. To remove the light switch on vehicles with an air bag, do the following:

   a. Remove the screws and the spiral cable from the switch body.

   b. Remove the screws and the lock plate from the switch body.

   c. Remove the screws and the ball set plate from the switch body.

   d. Remove the ball and slide out the switch from the switch body with the spring.

   e. Remove the screws and light dimmer turn switch unit.

7. Remove the wiper washer switch by removing the screws.

To install:

8. Install the wiper washer switch to the combination switch.

9. Install the light control switch.

10. Push the wires into the connector until they lock securely in place.

11. Install the switch into the steering column.

12. Install the upper and lower steering column covers.

13. Center the spiral cable as follows:

   a. Check that the front wheels are facing straight ahead.

   b. Turn the cable counterclockwise by hand until it becomes harder to turn the cable.

   c. Rotate the cable clockwise about 2.5 turns to align the green mark.

14. Install the steering wheel and torque the nut to 25 ft. lbs. (34 Nm).

15. Connect the negative battery cable.

Screw-On Type

The combination switch incorporates the headlight switch, turn signal switch and the windshield wiper switch.

──── **CAUTION** ────

The air bag system must be disarmed before removing the combination switch. Failure to do so may cause accidental deployment, property damage, or personal injury.

1. Disconnect the negative battery cable. On vehicles equipped with an air bag, wait at least 90 seconds before proceeding.

2. Disable the air bag system.

3. Position the front wheels in a straight–ahead position.

4. Remove the instrument panel No. 1 undercover sub-assembly.

5. Remove the instrument panel lower finish panel and cluster finish panel.

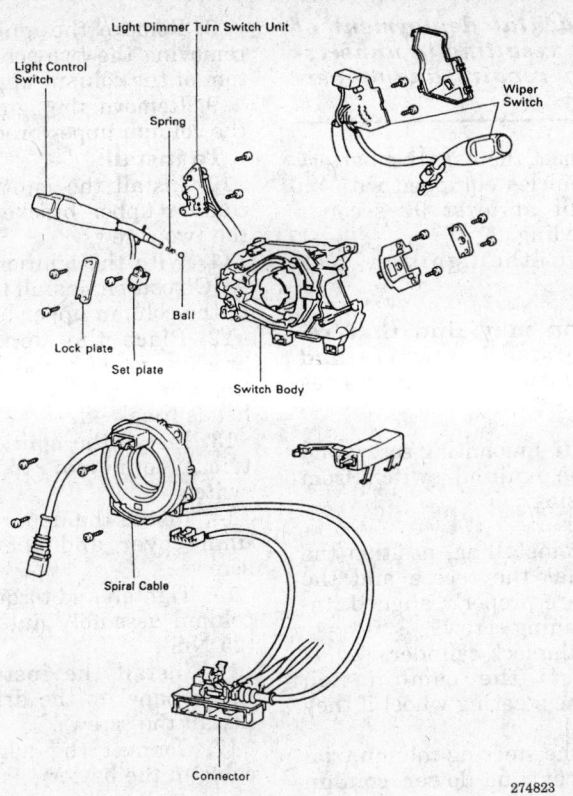

Exploded view of the Combination switch with air bag — Paseo

274823

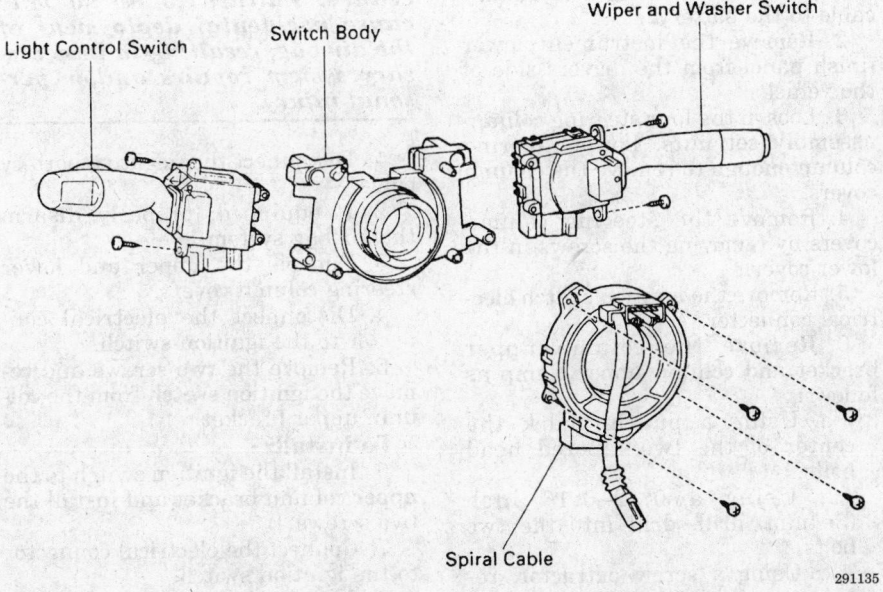

Exploded view of the combination switch — 1995–97 Tercel

291135

6. Remove the steering wheel center pad.

NOTE: When removing the wheel pad, take care not to pull the air bag harness connector. When storing the wheel pad, keep the upper surface of the pad facing upward. Since the air bag connector has a 2-stage lock, remove the 1st stage lock and disconnect the connector.

7. Remove the steering wheel assembly.

8. Remove the steering column covers. Remove the combination switch retaining screws. Disconnect the connectors and remove the combination switch assembly from the steering column.

9. Loosen the mounting screws and remove the turn signal switch from the switch body.

10. Separate the bracket from the wiper switch body. Remove the wiper switch from the switch body.

To install:

11. Install the wiper switch to the switch body and connect the mounting bracket.

12. Install the turn signal switch to the switch body and tighten the mounting screws.

13. Install the combination switch assembly to the steering column and tighten the mounting screws.

14. Connect the electrical connectors. Push in the terminals until they are securely locked in the connector lug.

NOTE: The spiral cable matchmarks must be aligned for correct installation with the front wheels in the straight head position.

15. Install both steering column covers.

16. Center the spiral cable as follows:

   a. Check that the front wheels are facing straight ahead.

   b. Turn the cable counterclockwise by hand until it becomes harder to turn the cable.

   c. Rotate the cable clockwise about three turns to align the red mark.

17. Install the steering wheel, using the proper tool.

18. Connect the air bag connector and replace the 1st stage lock.

19. Install the cluster finish panel. Install the steering wheel center pad and connect the wire connector.

20. Install the lower instrument trim panel and cover assembly.

21. Enable the air bag system.

22. Connect the negative battery cable.

Ignition Lock Cylinder

REMOVAL AND INSTALLATION

——————— CAUTION ———————

The Air Bag System must be disarmed before performing this procedure. Failure to do so may cause accidental deployment of the air bag, resulting in unnecessary system repairs and/or personal injury.

———————————————————

1. Disconnect the negative battery cable. On vehicles equipped with an air bag, wait at least 90 seconds before proceeding.
2. Unscrew the retaining screws and remove the upper and lower steering column covers.
3. Remove the retaining screws and remove the steering column trim.
4. Turn the ignition key to the **ACC** position.
5. Push the lock cylinder stop in with a small, round object (cotter pin, punch, etc.) and pull out the ignition key and the lock cylinder.

NOTE: You may find that removing the steering wheel and the combination switch makes the job easier.

To install:

6. Make sure that both the lock cylinder and the column lock are in the **ACC** position. Slide the cylinder into the lock housing until the stop tab engages the hole in the lock.
7. Make certain the stop tab is firmly seated in the slot. Turn the key to each switch position, checking for smoothness of motions and a positive feel. Remove and insert the key a few times, each time turning the key to each switch position.
8. If removed, install the combination switch and the steering wheel.
9. Install the steering column trim and the upper and lower column covers.
10. Connect the negative battery cable.

Ignition Switch

REMOVAL AND INSTALLATION

1993–94 Models

——————— CAUTION ———————

The Air Bag System must be disarmed before performing this procedure. Failure to do so may

cause accidental deployment of the air bag, resulting in unnecessary system repairs and/or personal injury.

———————————————————

1. Disconnect the negative battery cable. On vehicles equipped with an air bag, wait at least 90 seconds before proceeding.
2. Remove the ignition lock cylinder.

NOTE: You may find that removing the steering wheel and the combination switch makes the job easier.

3. Loosen the mounting screw and withdraw the ignition switch from the lock housing.

To install:

4. When reinstalling, position the switch so that the recess and the bracket tab are properly aligned. Install the retaining screw.
5. Install the lock cylinder.
6. Reinstall the combination switch and the steering wheel if they were removed.
7. Install the steering column trim and the upper and lower column covers.
8. Connect the negative battery cable.

1995–97 Except Corolla

1. Disconnect the negative battery cable to the battery.
2. Remove the instrument lower finish panel from the drivers side of the vehicle.
3. Loosen the four steering column assembly set nuts. Lower steering column enough to remove the column covers.
4. Remove the steering column covers by removing the screws in the lower cover.
5. Remove the ignition switch electrical connector.
6. Remove the column upper bracket and column upper clamp as follows:
   a. Using a punch, mark the center of the two tapered head bolts.
   b. Using a 0.12–0.16 inch (3–4mm) drill, drill into the two bolts.
   c. Using a screw extractor, remove the two bolts.
7. If replacing the key cylinder, follow the following steps:
   a. Place the ignition key at the **ACC** position.
   b. Push down the stop pin with a thin rod and pull out the key cylinder.

8. Remove the ignition switch by removing the two screws at the bottom of the column upper bracket.
9. Remove the ignition key from the column upper bracket.

To install:

10. Install the ignition key to the column upper bracket by installing the two screws.
11. With the ignition switch is the **ACC** position, install the ignition key to the column upper bracket.
12. Place the upper and lower bracket in place and install the two tapered head bolts until the bolt heads break off.
13. Install the ignition switch electrical connector to the ignition switch.
14. Install the upper and lower column cover and then install the screws.
15. Tighten and torque the steering column assembly nuts to 19 ft. lbs. (25 Nm).
16. Install the instrument lower finish panel to the drivers side and install the screws.
17. Connect the negative battery cable to the battery

1995–97 Corolla

——————— CAUTION ———————

The Air Bag System must be disarmed before performing this procedure. Failure to do so may cause accidental deployment of the air bag, resulting in unnecessary system repairs and/or personal injury.

———————————————————

1. Disconnect the negative battery cable.
2. If equipped, properly disarm the air bag system.
3. Remove the upper and lower steering column covers.
4. Disconnect the electrical connector to the ignition switch.
5. Remove the two screws and remove the ignition switch from the column upper bracket.

To install:

6. Install the ignition switch to the upper column bracket and install the two screws.
7. Connect the electrical connector to the ignition switch.
8. Install the upper and lower steering column covers.
9. If equipped, properly enable the air bag system.
10. Connect the negative battery cable.
11. Check ignition switch operation.

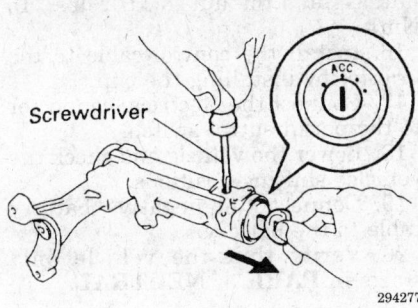

Push down the stop pin with a thin rod and pull out the key cylinder

Park/Neutral Safety Switch

REMOVAL AND INSTALLATION

Paseo and Tercel

1. Disconnect the negative battery cable to the battery. On vehicles equipped with an air bag, wait at least 90 seconds before proceeding.
2. Raise and safely support the vehicle.
3. Apply the parking brake fully and block the rear wheels.
4. Remove the clip holding the control cable to the bracket.

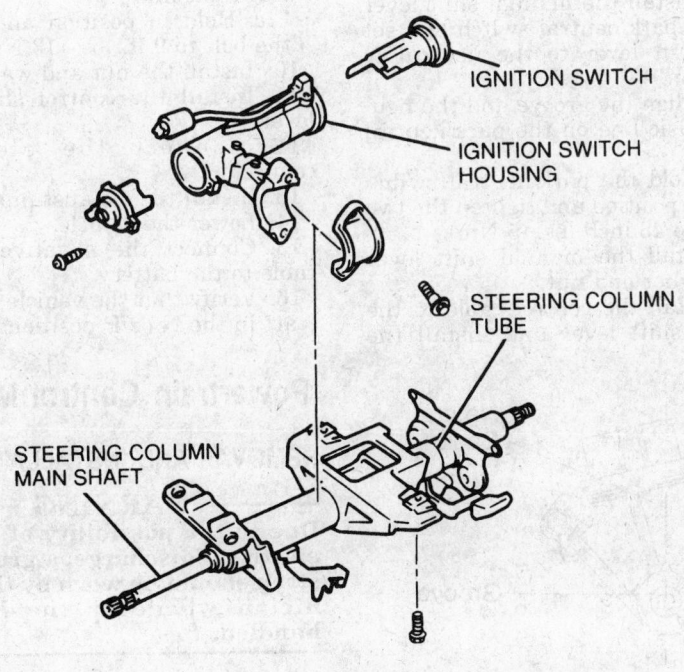

Exploded view of the ignition switch components — Corolla shown

IGNITION SWITCH

IGNITION SWITCH HOUSING

STEERING COLUMN TUBE

STEERING COLUMN MAIN SHAFT

5. Remove the locknut and disconnect the control cable from the manual shift lever.
6. Remove the nut, washer and the manual shift lever from the manual valve shaft.
7. Disconnect the electrical connector from the park/neutral position switch.
8. Using a screwdriver, bend the lock washer back away from the nut.
9. Remove the nut, lock washer and packing from the park/neutral switch.
10. Remove the two (2) bolts and the park/neutral position switch.

To install:
11. Install the park/neutral position switch to the manual valve shaft.
12. Install the packing, lock washer and nut. Tighten the nut to 61 inch lbs. (7 Nm).
13. Install the manual shift lever but do not install the nut at this time.
14. Turn the manual shift lever clockwise until it stops, then turn it counterclockwise three (3) notches.
15. Remove the manual shift lever and align the groove and neutral base line on the park/neutral switch.
16. After the neutral base line is aligned, install the two (2) park/neutral switch bolts and torque the bolts to 48 inch lbs. (5 Nm).

17. Using a screwdriver, bend the lock washer over the nut to secure the nut holding the park/neutral switch.
18. Install the manual shift lever with washer and nut.
19. Install the control cable to the manual shift lever and install the nut.
20. Install the control cable to the bracket by installing the clip.
21. Connect the electrical connector to the park/neutral switch.
22. Lower the vehicle and check the vehicles shifting positions.
23. Connect the negative battery cable to the battery.

Corolla and Celica

A245 and A246E Automatic Transaxles

1. Disconnect the negative battery cable. On vehicles equipped with an air bag, wait at least 90 seconds before proceeding.
2. Raise and safely support the vehicle.
3. Remove the engine undercover and disconnect the shift cable from the manual shift lever.
4. Remove the locknut, pry out the lock washer, and remove the manual shift lever.
5. Disconnect the park/neutral switch electrical connector.
6. Remove the two bolts and pull out the park/neutral switch.

To install:
7. Install the park/neutral switch to the manual valve shaft.
8. Install a new lock plate and tighten the nut. Tighten the nut to 108 inch lbs. (12 Nm).
9. Stake the nut with the lock plate.
10. Temporarily install the manual shift lever to the manual valve shaft.
11. Turn the lever counter clockwise until it stops, then turn it clockwise two notches. Remove the manual shift lever.
12. Adjust the park/neutral switch by aligning the groove and the neutral basic line. Install the switch bolts and torque to 48 inch lbs. (5 Nm).
13. Install the switch electrical connector.
14. Install the manual shift lever, tighten the nut, and install the shift control cable to the lever.
15. Install the engine undercover, safely lower the vehicle and install the negative battery cable.
16. Verify that the switch is operating correctly. The engine should only start when in the **P** or **N** position.

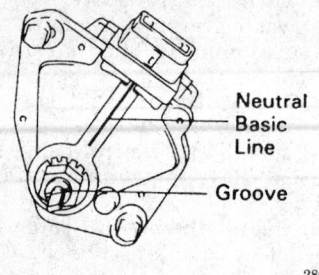

286762

The groove must be aligned with the neutral basic line (A245E) when installed — Corolla and Celica

A131L and A140E Automatic Transaxles

1. Disconnect the negative battery cable. On vehicles equipped with an air bag, wait at least 90 seconds before proceeding.

2. Raise and safely support the vehicle.

3. Remove the engine undercover and disconnect the control cable from the transaxle control shaft lever.

4. Disconnect the park/neutral switch electrical connector

5. Remove the two switch mounting bolts and remove the switch.

To install:

6. Install the park/neutral switch and loosely secure it with the two mounting bolts.

7. Set the shift selector to the **N** position. Align the groove and the neutral basic line. Hold the switch in position and torque the bolts to 48 inch lbs. (5 Nm).

8. Connect the switch electrical connector.

9. Install the shift control cable to the transaxle control shaft lever.

10. Install the engine undercover, safely lower the vehicle, and connect the negative battery cable.

11. Verify that the switch is operating correctly. The engine should only start when in the **P** or **N** position.

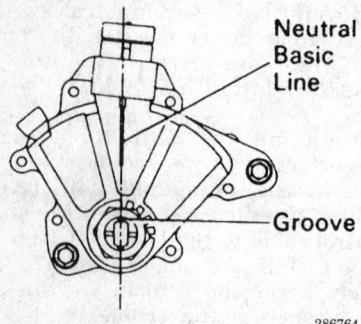

286764

Aligning the neutral base line and groove (A131L) — Corolla

Avalon, Camry and MR2

1. Disconnect the negative battery cable to the battery. On vehicles equipped with an air bag, wait at least 90 seconds before proceeding.

2. Apply the parking brake fully and block the rear wheels.

3. Disconnect the electrical connector from the park/neutral position switch.

4. Remove the locknut and disconnect the control cable from the manual shift lever.

5. Remove the clip from the holding the control cable to the bracket.

6. Remove the nut, washer and the manual shift lever from the manual valve shaft.

7. Using a flat bladed tool, bend the lock washer back away from the nut.

8. Remove the nut and lock washer from the park/neutral switch.

9. Remove the two bolts and the park/neutral position switch.

To install:

10. Install the park/neutral position switch to the manual valve shaft. Leave the two bolts loose at this time.

11. Install the lock washer and nut. Tighten the nut to 61 inch lbs. (7 Nm).

12. Using a flat bladed tool, bend the lock washer over the nut to secure the nut in place.

13. Adjust the park/neutral as follows:

 a. Install the manual shift lever to the park/neutral switch and set the shift lever to the neutral **N** position.

 b. Align the groove and the neutral basic line on the park/neutral switch.

 c. Hold the park/neutral switch in this position and tighten the two bolts to 48 inch lbs. (5 Nm).

14. Install the manual shift lever with washer and nut.

15. Install the control cable to the manual shift lever and install the

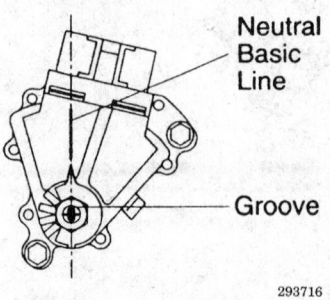

293716

Aligning the neutral base line and groove — Avalon and Camry

nut. Tighten the nut to 11 ft. lbs. (15 Nm).

16. Install the control cable to the bracket by installing the clip.

17. Connect the electrical connector to the park/neutral switch.

18. Lower the vehicle and check the vehicles shifting positions.

19. Connect the negative battery cable to the battery.

20. Verify that the vehicle only starts in **PARK** or **NEUTRAL**.

Supra

1. Disconnect the negative battery cable from the battery. On vehicles equipped with an air bag, wait at least 90 seconds before proceeding.

2. Raise and safely support the vehicle.

3. Remove the front exhaust pipe.

4. Disconnect the neutral start switch connector.

5. Remove the control shaft shift lever.

6. Pry off the washer and remove the nut.

7. Remove the bolt and pull out the neutral start switch.

To install:

8. Install the park/neutral switch and loosely install the bolt.

9. Adjust the switch as follows:

 a. Loosen the neutral start switch bolt and place the shift lever in the **N** position.

 b. Align the groove and the neutral basic line.

 c. Hold in position and tighten the bolt to 9 ft. lbs. (13 Nm).

10. Install the nut and washer.

11. Install the control shaft shift lever.

12. Connect the electrical connector.

13. Install the exhaust pipe.

14. Lower the vehicle.

15. Connect the negative battery cable to the battery.

16. Verify that the vehicle will only start in the **N** or **P** positions.

Powertrain Control Module

REMOVAL AND INSTALLATION

——— **WARNING** ———
Due to the possibility of a static electrical discharge, a grounding strap should be worn by the technician whenever the ECM is handled.

Paseo and Tercel

The engine control module is located behind the glove box on the Tercel

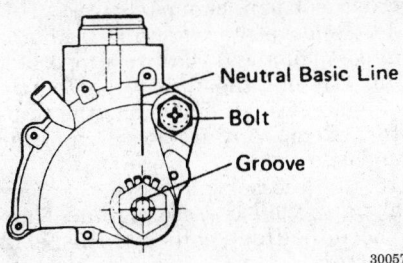

Neutral Basic Line
Bolt
Groove

300571

Park/Neutral switch groove and basic line adjustment — Supra

and 1996–97 Paseo. On the 1993–95 Paseo it is located behind the ash-tray/radio console.

1. Disconnect the negative battery cable. On vehicles equipped with an air bag, wait at least 90 seconds before proceeding.

2. Remove the screws and bolts to the glove compartment and then remove the glove compartment. The ECM is located behind the glove compartment.

3. On the 1993–95 Paseo, the ECM is located behind the center console. Remove the console.

4. Disconnect the ECM wire harness.

5. Remove the ECM securing bolt and nut.

6. Remove the ECM.

To install:

7. Position the ECM in place and secure it with the bolt and nut.

8. Connect the ECM wiring harness.

9. Install the glove compartment or center console assembly.

10. Connect the negative battery cable to the battery.

Corolla

The engine control module is located behind the passengers side carpet, on the console side.

1. Disconnect the negative battery cable. On vehicles equipped with an air bag, wait at least 90 seconds before proceeding.

2. Remove the floor mat bracket.

3. Disconnect the three connectors from the ECM.

4. Remove the mounting nuts or bolts and remove the ECM.

To install:

5. Install the ECM with the attaching nuts or bolts.

6. Reconnect the three electrical connectors.

7. Install the floor mat bracket and reconnect the negative battery cable.

Celica

The engine control module is located behind the drivers side carpet, on the console side.

1. Disconnect the negative battery cable. On vehicles equipped with an air bag, wait at least 90 seconds before proceeding.

2. Roll back the floor mat to gain access to the ECM. Remove the ECM cover.

3. Disconnect the 3 connectors from the ECM.

4. Remove the connector for the circuit opening relay

5. Remove the mounting nuts or bolts and remove the ECM.

6. Remove the brackets and the circuit opening relay from the ECM.

To install:

7. Secure the circuit opening relay and brackets to the ECM.

8. Install the ECM with the attaching nuts or bolts.

9. Connect the electrical connectors.

10. Install the ECM cover and replace the floor mats.

11. Connect the negative battery cable.

Avalon and Camry

The engine control module is located behind the glove box.

1. Disconnect the negative battery cable from the battery. On vehicles equipped with an air bag, wait at least 90 seconds before proceeding.

2. On the Avalon, remove the right front door scuff plate, right cowl side trim and the lower finish panel by removing the screws.

3. Remove the instrument lower panel.

4. Remove the instrument panel undercover.

5. Remove the glove compartment door by removing the three nuts at the bottom of the door.

6. Remove the glove compartment by removing the five screws.

7. Disconnect the ECM from its bracket by removing the nut and two screws.

8. Disconnect the electrical connectors from the ECM.

9. Remove the ECM from the vehicle.

To install:

10. Install the ECM to the vehicle.

11. Connect the electrical connector to the ECM.

12. Connect the ECM to its bracket and install the two screws and the nut.

13. Install the glove compartment and install the five screws.

14. Install the glove compartment door and install the three nuts.

15. Install the instrument panel undercover.

16. Install the instrument lower panel and install the screws.

17. On the Avalon, install the lower finish panel, cowl side trim and the front door scuff plate.

18. Connect the negative battery cable to the battery.

MR2

The engine control module is located in the luggage compartment.

1. Disconnect the negative battery cable from the battery. On vehicles equipped with an air bag, wait at least 90 seconds before proceeding.

2. Disconnect the front side of the luggage compartment floor mat from the partition panel to remove the ECM.

3. Disconnect the ECM wire harness.

4. Remove the ECM securing bolt or nut.

5. Remove the ECM.

To install:

NOTE: It is recommended that a grounding strap be worn whenever handling an ECM. The grounding strap will prevent a shock resulting from static electricity. A static electrical shock can severely damage the ECM.

6. Position the ECM in place and secure with nuts or bolts.

7. Connect the ECM wiring harness.

8. Install ECM cover and reconnect the luggage compartment floor mat to the partition panel.

9. Connect the negative battery cable.

Supra

The engine control module is mounted to the passengers side floorpan, beneath the carpet.

1. Disconnect the negative battery cable from the battery. On vehicles equipped with an air bag, wait at least 90 seconds before proceeding.

2. Remove the scuff plate from the right door.

3. Pull back the passenger side rug.

4. Remove the ECM protector by removing the two nuts.

5. Remove the nut and disconnect the ECM from the floor panel.

6. Fully loosen the bolt and disconnect the two ECM connectors.

7. Remove the ECM from the vehicle.

To install:

8. Connect the electrical connectors to the ECM and install the nut.

9. Connect the ECM to the vehicle floor and install the nut.

10. Install the ECM protector by installing the two bolts.

11. Replace the rug to its original position.

12. Install the scuff plate to the vehicle.

13. Connect the negative battery cable to the battery.

ENGINE COOLING

Radiator

REMOVAL AND INSTALLATION

Paseo and Tercel

1. Disconnect the cable at the negative battery terminal. On vehicles equipped with an air bag, wait at least 90 seconds before proceeding.

2. Remove the undercovers from the vehicle.

3. Drain the engine coolant.

4. Remove the air intake duct.

5. Remove the reservoir tank assembly.

6. Disconnect the No. 1 cooling fan connector.

7. If equipped with A/C, disconnect the No. 2 cooling fan connector.

8. Disconnect the upper and lower radiator hoses. Disconnect the lower radiator hose from the engine water inlet and the upper hose from the radiator.

9. If equipped with an automatic transmission, disconnect the two A/T oil cooler hoses from the radiator. Make sure to plug the transmission hoses to prevent oil from escaping.

10. Remove the two bolts and the two upper radiator supports.

11. Remove the radiator with the fan(s) from the engine compartment.

12. Remove the two lower radiator supports and the lower radiator hose.

13. Remove the No. 2 cooling fan from the radiator by removing the three bolts.

14. Remove the No. 1 cooling fan from the radiator by removing the two bolts.

To install:

15. Install the No. 1 cooling fan to the radiator by installing the two bolts.

16. Install the No. 2 cooling fan to the radiator by installing the three bolts.

17. Install the two lower radiator supports and the lower radiator hose. Make sure clamp for the radiator hose is around the radiator.

18. Install the radiator and fans in the engine compartment.

19. Install the two upper radiator supports and two bolts to the radiator and radiator support. Tighten the bolts to 9 ft. lbs. (12 Nm).

20. Clamp the two A/T oil cooler hoses to the radiator, if equipped.

21. Connect the upper and lower radiator hoses.

22. Install the reservoir tank assembly.

23. Connect the No. 1 cooling fan connector.

24. If equipped with A/C, connect the No. 2 cooling fan connector.

25. Install the engine undercovers and the intake air duct.

26. Fill the radiator with engine coolant.

27. Connect the battery negative cable and start the engine. Top off engine coolant when engine is hot.

28. Verify fan operation and check for leaks.

Avalon, Camry, Celica, Corolla and MR2

1. Disconnect the negative battery cable. On vehicles equipped with an air bag, wait at least 90 seconds before proceeding.

2. If necessary, remove the engine undercover.

3. Drain the engine coolant into a suitable container.

4. On the MR2, remove the upper radiator support seal.

5. On Canadian models, remove the two bolts and the Daytime Running Lamp (DRL) relay box.

6. On the Avalon, remove the battery hold down clamp and remove the battery from the vehicle. Remove the battery tray.

7. Disconnect the electric cooling fan connector.

8. Disconnect the upper radiator hose from the radiator.

9. Disconnect the lower hose from the water inlet pipe.

10. Disconnect the coolant reservoir hose.

11. If the vehicle is equipped with automatic transmission, disconnect the oil cooler hoses.

12. Disconnect the No. 1 cooling fan connector and wire clamp. If equipped with A/C, disconnect the No. 2 cooling fan connector.

13. On Celica, remove the two oxygen sensor wire clamps.

14. Remove the two upper radiator support bolts and the two supports.

15. Lift out the radiator assembly being careful not to damage the radiator. Remove the lower radiator supports.

16. If necessary, remove the four bolts and the No. 1 cooling fan. Then remove the three bolts and the No. 2 cooling fan.

To install:

17. Install the No. 2 cooling fan to the radiator with the three bolts and torque them to 43 inch lbs. (5 Nm).

18. Install the No. 1 cooling fan to the radiator with the four bolts and torque them to 43 inch lbs. (5 Nm).

19. Install the radiator lower supports and carefully lower the radiator assembly into position.

20. Install the two upper radiator supports with the support bolts. Tighten the support bolts to 9 ft. lbs. (12 Nm). Make sure that the rubber cushions are not distorted.

21. If detached, connect the No. 1 cooling fan connector and the wire clamp.

22. If equipped with A/C, connect the No. 2 cooling fan connector.

23. On the Celica, connect the two oxygen sensor wire clamps.

24. If the vehicle is equipped with automatic transmission, connect the oil cooler lines.

25. Connect the coolant reservoir hose.

26. Connect the lower radiator hose to the water inlet pipe.

27. Connect the upper radiator hose to the radiator.

28. Connect the electric cooling fan electrical connector.

29. On the MR2, install the radiator upper support seal.

30. Install the DRL relay box and secure it with the two mounting bolts.

31. If removed, install the battery tray, battery and clamp to the engine compartment.

32. Connect the negative battery cable.

33. Refill the cooling system with coolant and bleed the system. Check the cooling system for leaks and proper operation.

34. If the system is OK, install the engine undercover if removed.

Supra

1. Disconnect the negative battery cable. On vehicles equipped with an air bag, wait at least 90 seconds before proceeding.

2. Remove the battery and battery tray.

3. Raise and safely support the vehicle.

4. Remove the engine undercover.

5. Drain the engine coolant from the radiator.

6. Remove the No. 2 air tube.

7. Remove the No. 2 fan shroud.

8. Remove the air cleaner duct.

9. Remove the No. 5 air hose.

10. Remove the left hand headlight beam angle gauge by removing the screw.

11. Disconnect the reservoir inlet hose from the radiator.

12. Disconnect the upper and lower radiator hoses from the radiator.

13. If equipped with an automatic transmission, remove the two oil cooler hoses from the radiator and plug the ends to prevent fluid loss.

14. Disconnect the engine coolant temperature switch (for the electric cooling fan connector) and the wire harness.

15. Disconnect the electric cooling fan connector and wire harness.

16. Remove the upper radiator supports by removing the bolts.

17. Lift out the radiator assembly.

18. Remove the two lower radiator supports.

19. Remove the drain hose from the radiator.

20. Remove the engine coolant temperature switch from the radiator.

21. Remove the electric cooling fan from the radiator by removing the three bolts.

22. Remove the No. 1 fan shroud from the radiator by removing the four bolts.

To install:

23. Install the No. 1 fan shroud to the radiator and install the four bolts.

24. Connect the electric cooling fan to the radiator and install the three bolts.

25. Using a new O-ring, install the ECT switch to the radiator. Tighten the switch to 65 inch lbs. (7.4 Nm).

26. Install the drain hose to the radiator.

27. Install the lower radiator supports to the vehicle.

28. Install the radiator assembly to the vehicle.

29. Install the two upper radiator supports and install the bolts. Tighten the bolts to 11 ft. lbs. (15 Nm).

30. Connect the electric cooling fan connector and the wire harness.

31. Connect the ECT switch (for the cooling fan connector) and wire harness.

32. If equipped with an automatic transmission, connect the oil cooler hoses to the radiator.

33. Connect the lower and upper radiator hoses to the radiator.

34. Connect the reservoir inlet hose to the radiator.

35. Install the left hand headlight beam angle gauge and install the screw.

36. Install the No. 5 air hose.

37. Install the air cleaner duct.

38. Install the No. 2 fan shroud.

39. Install the No. 2 air tube.

40. Fill the engine coolant.

41. Install the battery and battery tray.

42. Install the engine undercover.

43. Lower the vehicle and start the engine. Check for coolant leaks, bleed the cooling system, and top off the radiator.

Water Pump

REMOVAL AND INSTALLATION

1.5L (3E-E) Engine

Tercel

1. Disconnect the negative battery cable. On vehicles equipped with an air bag, wait at least 90 seconds before proceeding.

2. Drain the engine coolant.

3. Remove the right side engine undercover.

4. Remove the HAC valve from the bracket, if equipped.

5. Remove the oil dipstick.

6. Loosen the water pump pulley mounting bolts.

7. Remove the alternator belt.

8. Remove the intake manifold bracket located behind the alternator.

9. Remove the No. 1 water by-pass hose, if equipped.

10. Remove the oil dipstick guide.

11. Remove the alternator and adjusting bar.

12. Disconnect the intake manifold water hose from the intake manifold, if equipped.

13. Remove the water pump pulley.

14. Remove the water inlet pipe mounting bolt from the cylinder block.

15. Remove the water pump attaching bolts and nuts. Remove the water pump assembly.

To install:

16. Scrape any remaining gasket material off the pump mating surface. Apply a 2–3mm (0.08–0.12 in.) bead of sealant to the groove in the pump.

17. Replace the O-ring on the water inlet pipe and lubricate the O-ring with a little soap and water. Install the pump assembly. Tighten the bolts to 13 ft. lbs. (17 Nm).

18. Connect the water inlet pipe to the cylinder block with a bolt.

19. Temporarily install the water pump pulley.

20. Connect the intake manifold water hose to the intake manifold, if equipped.

21. Install alternator and adjusting bar.

22. Replace the O-ring on the oil dipstick guide and install the assembly. Temporarily install the alternator adjusting bar and dipstick guide clamp bolt.

23. Connect the No. 1 water by-pass hose, if equipped.

24. Install the intake manifold bracket.

25. Install the alternator belt and adjust.

26. Install the oil dipstick.

27. Install the HAC valve to the bracket.

28. Install the right side engine undercover.

29. Refill the engine with coolant.

30. Connect the negative battery cable and start the engine.

31. Check for coolant leaks.

1.5L (5E-FE) Engine

Paseo and Tercel

1. Disconnect the negative battery cable. On vehicles equipped with an air bag, wait at least 90 seconds before proceeding.

2. Drain the engine coolant.

3. Remove the alternator.

4. For engines with distributorless ignition, remove the intake manifold stay bracket by disconnecting the wire clamps and removing the two nuts.

5. For engine with distributor ignition, remove the intake manifold stay bracket by removing the two nuts and two bolts.

6. Remove the water inlet pipe as follows:
    a. Disconnect the water inlet hose.
    b. Disconnect the heater hose.
    c. Disconnect the bypass hose.
    d. Remove the bolt, water inlet pipe, and O-ring.

7. Remove the oil dipstick guide.

8. Remove the alternator adjusting bar.

9. Remove the water pump attaching bolt and nuts. Remove the water pump assembly.

To install:

10. Scrape any remaining gasket material off the pump mating surface. Apply a 2–3mm (0.08–0.12 in.) bead of sealant to the groove in the pump.

11. Replace the O-ring on the water inlet pipe and lubricate the O-ring with a little soap and water. Install the pump assembly. Tighten the bolts to 13 ft. lbs. (17 Nm).

12. Replace the O-ring on the oil dipstick guide and install the assembly. Install the alternator adjusting bar and dipstick guide clamp bolt.

13. Connect the water inlet pipe to the cylinder block with a bolt. Tighten the bolt to 65 inch lbs. (7.5 Nm).

14. Connect the water by-pass, heater inlet and water inlet hoses.

15. For distributorless ignition engines, install the intake manifold bracket by installing the two bolts and the wire clamp. Tighten the bolts to 15 ft. lbs. (20 Nm).

16. For distributor ignition, install the intake manifold bracket by installing the two bolts. Tighten the bolts to 15 ft. lbs. (20 Nm).

17. Install the alternator and belt.

18. Refill the engine with coolant.

19. Connect the negative battery cable and start the engine.

20. Check for coolant leaks.

1.6L (4A-FE) and 1.8L (7A-FE) Engines

Celica and Corolla

1. Disconnect the negative battery cable. On vehicles equipped with an air bag, wait at least 90 seconds before proceeding.

2. Drain the engine coolant into a suitable container.

3. Remove the RH engine mounting insulator.

4. Remove No. 2 and No. 3 timing belt covers.

5. If equipped with power steering, safely raise and support the engine. Remove the hole cover and remove the two mounting bolts from the front engine mount insulator. Remove the nut and the through–bolt and remove the insulator.

6. If equipped with power steering, remove the electric cooling fan.

7. Remove the bolt and two nuts and remove the engine wire.

8. On the 7A-FE engine, disconnect crankshaft position sensor connector from the dipstick guide.

9. Remove the mounting bolt and pull out the dipstick guide and the dipstick.

10. Disconnect the water temperature sender gauge connector.

11. Remove the two nuts and the No. 2 water inlet from the water inlet hose.

12. Remove the three water pump bolts, the water pump, and the O-ring from the block.

To install:

13. Install a new O-ring on the block and install the water pump with the three bolts. Tighten the bolts to 10 ft. lbs. (14 Nm).

14. Connect the inlet hose to the water pump and install the water inlet No. 2 to the cylinder head with the two nuts. Tighten the nuts to 11 ft. lbs. (15 Nm).

15. Connect the water temperature sender gauge connector.

16. After applying a small amount of oil to the O-ring, install a new O-ring on the oil dipstick guide. Install the guide mounting bolt and tighten it to 82 inch lbs. (9 Nm).

17. On the 7A-FE engine, connect the crankshaft position sensor connector.

18. Connect the engine wire with the two nuts and the bolt.

19. If equipped with power steering, install the electric cooling fan.

20. If equipped with power steering, install the front mounting insulator through–bolt and nut. Tighten the nut to 64 ft. lbs. (87 Nm).

21. Install the engine insulator mounting bolts and torque the two bolts to 47 ft. lbs. (64 Nm). Install the hole cover and safely lower the engine.

22. Install the No. 2 and No. 3 timing belt covers.

23. Install the RH engine mounting insulator.

24. Refill the cooling system with coolant and connect the negative battery cable. Start the engine and bleed the cooling system. Check for cooling system leaks and proper system operation.

2.0L (3S-GTE) and 2.2L (5S-FE) Engines

Camry, Celica and MR2

1. Disconnect the negative battery cable. On vehicles equipped with an air bag, wait at least 90 seconds before proceeding.

2. Raise and safely support the vehicle.

3. Remove the right engine undercover.

4. Drain the engine coolant into a suitable container. Disconnect the lower radiator hose from the water outlet.

5. Remove the timing belt, timing belt tension spring, and the No. 2 idler pulley.

6. Remove the alternator, drive belt and the adjusting bar if necessary.

7. Remove the two nuts holding the water pump to the water bypass pipe and remove the three bolts in sequence.

8. Disconnect the water pump cover from the water bypass pipe and remove the water pump cover assembly.

9. Remove the gasket and two O-rings from the water pump and the bypass pipe.

10. Remove the water pump from the water pump cover by removing the three bolts in sequence.

To install:

11. Cleaned the gasket mating surfaces.

12. Install a new gasket and assemble the water pump to the water pump cover. Tighten the bolts to 78 inch lbs. (9 Nm) in proper sequence.

13. Install a new O-ring and gasket to the water pump cover and install a new O-ring on the water bypass pipe. Connect the water pump cover to the water bypass pipe, but do not install the nuts yet.

14. Install the water pump and tighten the three bolts in sequence. Tighten the bolts to 78 inch lbs. (9 Nm). Install the two nuts holding the water pump cover to the water bypass pipe and torque them to 82 inch lbs. (9 Nm).

15. Install the alternator drive belt adjusting bar with the bolt and torque the bolt to 13 ft. lbs. (18 Nm).

16. Install the No. 2 idler pulley and the timing belt tension spring.

17. Connect the lower radiator hose.

18. Install the timing belt.

19. Install the right undercover and safely lower the vehicle.

20. Fill the cooling system with coolant and connect the negative battery cable. Start the engine and bleed the cooling system. Check the cooling system for leaks and proper operation.

3.0L (3VZ-FE) Engine

Camry

1. Disconnect the negative battery cable from the battery. On vehicles equipped with an air bag, wait at least 90 seconds before proceeding.

2. Drain the engine coolant.

3. Disconnect the lower radiator hose at the water inlet.

4. Disconnect the timing belt from the water pump pulley.

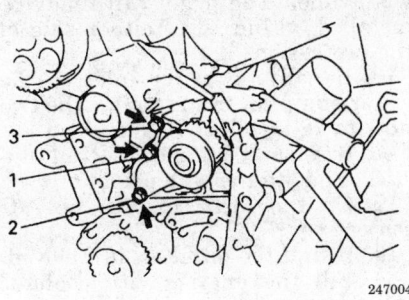

Install the three water pump bolts in this sequence — 5S-FE and 3S-GTE engines

247004

5. Remove the bolt holding the inlet pipe to the alternator belt adjusting bar and then remove the inlet pipe and O-ring.

6. Remove the water inlet and thermostat.

7. Remove the seven bolts and then pry off the water pump.

To install:

8. Scrape any remaining gasket material off the pump mating surface. Apply a 2–3mm (0.08–0.12 in.) bead of sealant to the groove in the pump and then install the pump. Tighten the bolts to 14 ft. lbs. (20 Nm).

9. Install the water inlet and thermostat.

10. Install a new O-ring to the water inlet pipe, coat it with soapy water and then connect the pipe to the inlet. Install the bolt holding the pipe to the adjusting bar and tighten it to 14 ft. lbs. (20 Nm).

11. Install the timing belt.

12. Connect the radiator hose and fill the engine with coolant.

3.0L (1MZ-FE) Engine

Avalon and Camry

1. Disconnect the negative battery cable from the battery. On vehicles equipped with an air bag, wait at least 90 seconds before proceeding.

2. Drain the engine coolant.

3. Remove the timing belt.

4. Mark the left and right camshaft pulleys with a touch of paint. Using SST tools 09249–63010 and 09960–10000 or equivalents, remove the bolts to the right and left camshaft pulleys. Remove the pulleys from the engine. Be sure not to mix up the pulleys.

5. Remove the No. 2 idler pulley by removing the bolt.

6. Disconnect the three clamps and engine wire from the rear timing belt cover.

7. Remove the six bolts holding the rear timing belt cover to the engine block.

8. Remove the four bolts and two nuts to the water pump.

9. Remove the water pump and the gasket from the engine.

To install:

10. Check that the water pump turns smoothly. Also check the air hole for coolant leakage.

11. Using a new gasket, apply liquid sealer to the gasket, water pump and engine block.

12. Install the gasket and pump to the engine and install the four bolts and two nuts. Tighten the nuts and bolts to 53 inch lbs. (6 Nm).

13. Install the rear timing belt cover and torque the six bolts to 74 inch lbs. (9 Nm).

14. Connect the engine wire with the three clamps to the rear timing belt cover.

15. Install the No. 2 idler pulley with the bolt. Tighten the bolt to 32 ft. lbs. (43 Nm). After tightening the bolt, make sure the idler pulley moves smoothly.

16. With the flange side **outward**, install the right hand camshaft pulley to the engine. Make sure to align the knock pin hole on the camshaft pulley with the knock pin on the camshaft. Using the same tools as removal, torque the camshaft bolt to 65 ft. lbs. (88 Nm).

17. With the flange side **inward**, install the left hand camshaft pulley to the engine. Make sure to align the knock-pin hole on the camshaft pulley with the knock pin on the camshaft. Using the same tools as removal, torque the camshaft bolt to 94 ft. lbs. (125 Nm).

18. Install the timing belt to the engine.

19. Fill the engine coolant.

20. Connect the negative battery cable to the battery and start the engine.

21. Top off the engine coolant and check for leaks.

3.0L (2JZ-GE and 2JZ-GTE) Engines

Supra

1. Disconnect the negative battery cable from the battery. On vehicles equipped with an air bag, wait at least 90 seconds before proceeding.

2. Remove the air cleaner and MAF meter assembly.

3. Remove the radiator assembly from the vehicle.

4. If equipped with manual transmission, remove the drive belt tensioner damper by removing the two nuts.

5. Loosen the four nuts holding the fan clutch to the water pump.

6. Loosen the drive belt tension by turning the drive belt tensioner clockwise. Remove the drive belt from the engine.

7. Remove the four nuts, the fan, fan clutch, and the water pump pulley.

8. Remove the water inlet, lower radiator hose assembly, and the thermostat.

9. Remove the timing belt.

10. Remove the alternator from the engine.

11. On turbo models disconnect the turbo water hoses from the water outlet.

12. Except for the California vehicles, remove the exhaust manifold heat insulator.

13. Remove the water outlet and No. 1 water bypass pipe.

14. Disconnect the No. 2 water bypass from the water pump by disconnecting the two nuts.

15. Disconnect the No. 3 turbo water hose from the water pump.

16. Remove the six bolts securing the water pump and remove the water pump from the engine. Make sure to replace the each bolt to its original position.

17. Clean the surface of the engine and remove the O-ring from the cylinder block.

To install:

18. Install the O-ring to the cylinder block.

19. Apply a thin layer of liquid sealant to the engine and water pump. Install a new gasket to the water pump.

20. Connect the water pump to the water bypass pipe. Do not install the nut at this time.

21. Install the water pump with the six bolts. Make sure to replace the bolts to their original positions. Tighten the bolts to 15 ft. lbs. (21 Nm).

22. Install the two nuts holding the No. 2 water bypass pipe to the water pump. Tighten the nuts to 15 ft. lbs. (21 Nm).

23. Connect the No. 3 turbo water hose to the water pump.

24. Install the water bypass outlet and No. 1 water bypass pipe.

25. Connect the turbo water hoses to the water outlet.

26. Install the alternator to the engine.

27. Except for California vehicle, install the exhaust manifold heat insulator.

28. Install the engine wire bracket with the bolt.

29. Install the timing belt.

30. Install the thermostat, water inlet, and the lower radiator hose assembly.

31. Install the water pump pulley, fan, fluid clutch assembly, and the drive belt. Tighten the fan nuts to 12 ft. lbs. (16 Nm).

32. If equipped with manual transmission, install the drive belt tensioner damper.

33. Install the radiator assembly to the vehicle.

34. Install the air cleaner and MAF meter assembly.

35. Install the No. 1 air hose.

36. Connect the negative battery cable to the battery.

37. Fill and bleed the cooling system.

38. Start the engine and check for leaks.

Thermostat

REMOVAL AND INSTALLATION

1.5L (3E-E and 5E-FE), 1.6L (4A-FE) and 1.8 (7A-FE) Engines

Celica, Corolla, Tercel and Paseo

1. Disconnect the negative battery cable from the battery. On vehicles equipped with an air bag, wait at least 90 seconds before proceeding.

2. Drain the cooling system.

3. On the 5E-FE engine, remove the air intake duct.

4. If necessary, remove the upper radiator or lower hose.

5. If equipped with a thermoswitch in the thermostat housing, unplug the lead wire.

6. Remove the thermostat housing by removing the two nuts.

7. Remove the gasket and thermostat from the engine.

To install:

8. Clean the gasket mating surfaces.

9. Be sure to use a new thermostat gasket. Install the thermostat with the spring inside the engine block. Align the juggle valve on the thermostat with the protrusion on the thermostat housing.

10. Install the thermostat housing and retaining bolts. On the 3E-E and 5E-FE engines, torque the nuts to 43 inch lbs. (5 Nm). On the 4A-FE and 7A-FE engines, torque the nuts to 82 inch lbs. (9 Nm).

11. Install the upper radiator or lower hose.

12. Connect the air intake duct.

13. Connect any wire connectors previously removed.

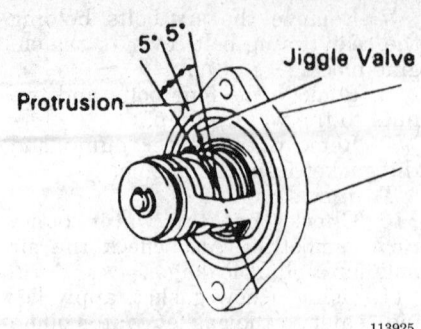

Align the juggle valve on the thermostat with the protrusion on the thermostat housing

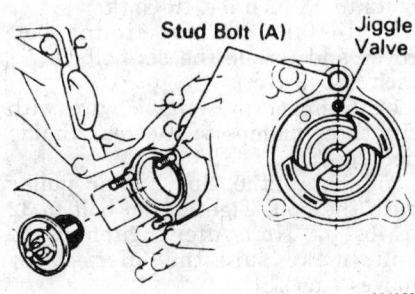

The jiggle valve must be aligned with the upper-most stud on the housing assembly

14. Connect the negative battery cable.

15. Fill and bleed the cooling system properly.

16. Start the engine and check for leaks.

2.0L (3S-GTE) and 2.2L (5S-FE) Engines

Camry, Celica and MR2

1. Disconnect the negative battery cable. On vehicles equipped with an air bag, wait at least 90 seconds before proceeding.

2. Drain the engine coolant into a suitable container.

3. On the MR2, disconnect the A/C compressor from the engine.

4. On the 1995–97 Celica, remove the oil filter.

5. On the 3S-GTE engine, disconnect the oil dipstick guide from the water inlet.

6. Remove the two water inlet nuts and remove the water inlet from the water pump cover.

7. Remove the thermostat and gasket.

To install:

8. Install a new gasket to the thermostat.

9. Align the jiggle valve of the thermostat with the protrusion of the water inlet. The jiggle valve may be installed within 5° of either side of the protrusion.

10. Install the water inlet to the water pump cover and torque the two nuts to 78 inch lbs. (9 Nm).

11. If removed, connect the oil dipstick guide to the water inlet.

12. On the MR2, install the A/C compressor.

13. Install the oil filter, if removed.

14. Fill the engine with coolant, check the oil level and add oil if necessary. Connect the negative battery cable and start the engine. Bleed the cooling system, check for leaks, and verify proper operation of the thermostat.

3.0L (1MZ-FE) Engine

Avalon and Camry

1. Disconnect the negative battery cable from the battery. On vehicles equipped with an air bag, wait at least 90 seconds before proceeding.

2. Drain the cooling system.

3. Remove the air cleaner cap assembly as follows:

  a. Disconnect the MAF sensor connector and wire clamp.

  b. Disconnect the accelerator cable clamp.

  c. Disconnect the PCV hose.

  d. Loosen the air cleaner hose clamp bolt.

  e. Disconnect the four air cleaner cap clips.

  f. Remove the air cleaner cap and MAF sensor together with the air cleaner hose.

4. Disconnect the heater hose.

5. Disconnect the engine wire from the thermostat housing and cylinder head by removing the two nuts.

6. Disconnect the two ECT switch connectors from the thermostat housing.

7. Remove the bolt holding the water inlet pipe to the cylinder head.

8. Disconnect the water inlet pipe from the thermostat housing and remove the O-ring.

9. Remove the three nuts and thermostat housing from the engine.

10. Remove the thermostat and gasket.

To install:

11. Clean the gasket mating surfaces.

12. Be sure to use a new thermostat and gasket. Install the thermostat with the spring inside the engine block. Align the jiggle valve on the thermostat with the stud bolt on the thermostat housing.

13. Install the thermostat housing and the three retaining nuts. Tighten the nuts to 69 inch lbs. (8.0 Nm).

14. Using a new O-ring, connect the water inlet pipe to the thermostat housing.

15. Install the bolt holding the water inlet pipe to the cylinder head. Tighten the bolt to 14 ft. lbs. (19.5 Nm).

16. Connect the ECT switch connectors.

17. Connect the engine wire to the thermostat housing and cylinder head with the two nuts.

18. Connect the heater hose.

19. Install the air cleaner cap assembly.

20. Connect all electrical connectors and all hoses.

21. Fill the engine coolant.

22. Connect the negative battery cable to the battery, start the engine, and bleed the cooling system.

23. Top off the engine coolant and check for leaks. Make sure the thermostat works properly.

3.0L (2JZ-GE and 2JZ-GTE) Engines

Supra

1. Disconnect the negative battery cable from the battery. On vehicles equipped with an air bag, wait at least 90 seconds before proceeding.

2. If necessary, remove the engine undercover.

3. With the engine cool, position a drain pan under the radiator drain cock and drain the cooling system.

4. Loosen the hose clamp and disconnect the lower radiator hose from radiator.

5. Remove the two nuts from the water inlet housing and remove the housing and from the water pump studs.

6. Remove the thermostat and then remove the gasket from the thermostat.

To install:

7. Make sure all the gasket surfaces are clean. Clean the inside of the inlet housing and the radiator hose connection with a rag.

8. Install a new gasket onto the thermostat.

9. Insert the thermostat into the housing.

10. Position the water inlet housing with the jiggle valve of the thermostat aligned with the protrusion of the water inlet; install the two nuts. Tighten the two nuts to 15 ft. lbs. (21 Nm).

11. Connect the lower radiator hose to the radiator and tighten the hose clamp.

12. Fill the cooling system with coolant.

13. Connect the negative battery cable to the battery.

14. Install the engine undercover.

15. Start the engine and inspect for leaks. Bleed the cooling system.

Electric Cooling Fan

REMOVAL AND INSTALLATION

Tercel and Paseo

Engine Cooling Fan

1. Disconnect the cable at the negative battery terminal. On vehicles equipped with an air bag, wait at least 90 seconds before proceeding.

2. Remove the engine undercovers, as required.

3. Drain the coolant.

4. Remove the air intake duct.

5. On the 3E-E engine, remove the exhaust manifold heat shroud.

6. On the 5E-FE engine, remove the reservoir tank assembly.

7. Disconnect the upper radiator hose from the radiator.

8. Tag and disconnect the cooling fan switch and oxygen sensor connectors.

9. Remove the mounting bolts and lift out the fan shroud/motor/fan assembly.

10. Unbolt the fan from the shaft, remove the motor mounting bolts and remove the motor from the shroud bracket.

To install:

11. Position the motor in the bracket and tighten the mounting bolts. Press the fan onto the shaft and tighten the bolt.

12. Install the shroud/fan assembly to the radiator.

13. Install the reservoir tank assembly on the 5E-FE engine.

14. Connect the electrical leads, upper radiator hose and refill radiator with coolant.

15. Install the engine undercovers, exhaust heat shroud (if removed), and the intake air duct.

16. Connect the battery negative cable and verify fan operation.

Condenser Cooling Fan

1. Disconnect the negative battery cable from the battery. On vehicles equipped with an air bag, wait at least 90 seconds before proceeding.

2. Remove the bolt and remove the air cleaner inlet.

3. Disconnect the electrical connector to the condenser fan.

4. Remove the three bolts and condenser fan.

To install:

5. Install the condenser fan and install the three bolts. Tighten the bolts to 61 inch lbs. (7 Nm).

6. Connect the condenser fan electrical connector.

7. Install the air cleaner inlet with the bolt.

8. Connect the negative battery cable to the battery.

Corolla

1. Disconnect the cable at the negative battery terminal. On vehicles equipped with an air bag, wait at least 90 seconds before proceeding.

2. Drain the coolant into a suitable container.

3. Remove the coolant reservoir tank.

4. Remove the upper radiator hose from the radiator.

5. Remove the cooling fan electrical connector from the fan shroud, and disconnect the electrical connector.

6. Remove the four mounting bolts and the cooling fan.

To install:

7. Reinstall the cooling fan and attach with the four bolts. Tighten the bolts to 43 inch lbs. (5 Nm).

8. Reconnect the cooling fan electrical connector and secure to the fan shroud.

9. Reinstall the upper radiator hose to the radiator.

10. Install the coolant reservoir tank.

11. Reconnect the cable at the negative battery terminal.

12. Refill and bleed the cooling system.

13. Check the cooling fan for proper operation.

Celica

Engine Cooling Fan

1. Disconnect the cable at the negative battery terminal. On vehicles equipped with an air bag, wait at least 90 seconds before proceeding.

2. Remove the engine undercovers, as required.

3. Drain the coolant into a suitable container.

4. Remove the battery if necessary.

5. For Canadian models, disconnect the relay box for the daytime running light system from the radiator.

6. Remove the upper radiator hose.

7. Remove the cooling fan electrical connector clamp and disconnect the electrical connector.

8. Remove the 4 mounting bolts and the cooling fan.

To install:

9. Reinstall the cooling fan and attach with the 4 bolts. Tighten the bolts to 43 inch lbs. (5 Nm).

10. Reconnect the cooling fan electrical connector and secure with the clamp.

11. Reinstall the upper radiator hose.

12. If removed, install the relay box for the daytime running light system.

13. Reconnect the cable at the negative battery terminal.

14. Refill the cooling system and bleed.

15. Check the cooling fan for proper operation.

Condenser Cooling Fan

1. Disconnect the cable at the negative battery terminal. On vehicles equipped with an air bag, wait at least 90 seconds before proceeding.

2. Drain the coolant into a suitable container.

3. Remove the cooling fan electrical connector.

4. Remove the oxygen sensor wire clamp.

5. Remove the three mounting bolts and the cooling fan.

To install:

6. Reinstall the cooling fan and attach with the three bolts. Tighten the bolts to 43 inch lbs. (5.0 Nm).

7. Reconnect the cooling fan electrical connector.

8. Reconnect the cable at the negative battery terminal.

9. Refill the cooling system and bleed.

10. Check the cooling fan for proper operation.

Avalon and Camry

No. 1 Cooling Fan

1. Disconnect the negative and positive battery cables from the battery. On vehicles equipped with an air bag, wait at least 90 seconds before proceeding.

2. For Canadian vehicles, disconnect the relay block from the battery hold down clamp.

3. On the Avalon, remove the battery clamp, battery, and tray from the engine compartment.

4. Disconnect the cruise control actuator from the body.

5. Disconnect the cruise control actuator wire from the clamp on the No. 1 fan shroud.

6. Disconnect the cooling fan electrical connector from the No. 1 fan.

7. Remove the four bolts and then remove the cooling fan from the radiator.

To install:

8. Install the No. 1 cooling fan to the radiator and install the four bolts. Tighten the bolts to 44 inch lbs. (5 Nm).

9. Connect the No. 1 cooling fan electrical connector.

10. Connect the cruise control actuator wire to the clamp on the No. 1 cooling fan shroud.

11. Connect the cruise control actuator to the body.

12. If removed, install the battery tray, battery, and clamp to the engine compartment.

13. For Canadian vehicles, connect the relay block to the battery hold down clamp.

14. Connect the negative and positive battery cables to the battery.

15. Check the operation of the cooling fan.

No. 2 Cooling Fan

1. Disconnect the negative battery cable from the battery. On vehicles equipped with an air bag, wait at least 90 seconds before proceeding.

2. Drain the engine coolant from the radiator.

3. Disconnect the upper radiator hose from the radiator.

4. Disconnect the No. 2 cooling fan electrical connector.

5. Remove the three (3) bolts and remove the cooling fan.

6. Remove the fan from the motor by removing the clip.

7. Disconnect the wire and connector from the fan shroud.

8. Remove the motor by removing the three screws.

To install:

9. Install the motor to the fan shroud and install the three (3) screws.

10. Connect the motor wire to the fan shroud.

11. Connect the fan to the motor and connect the clip. Install the clip from the side opposite the protrusion on the fan.

12. Install the No. 2 cooling fan to the radiator and install the three (3) bolts. Tighten the bolts to 44 inch lbs. (5 Nm).

13. Connect the No. 2 cooling fan electrical connector.

14. Connect the upper radiator hose to the radiator.

15. Fill the engine coolant.

16. Connect the negative battery cable to the battery and start the engine.

17. Top off the engine coolant and check for leaks.

18. Check the fan for proper operation.

MR2

1. Disconnect the cable at the negative battery terminal. On vehicles equipped with an air bag, wait at least 90 seconds before proceeding.

2. Remove the front luggage undercovers and upper radiator support, as required.

3. Tag and disconnect the cooling fan switch connectors.

4. Remove the mounting bolts and lift out the fan shroud/motor/fan assembly.

5. Unbolt the fan from the shaft, remove the motor mounting bolts and remove the motor from the shroud bracket.

To install:

6. Position the motor in the bracket and tighten the mounting bolts. Press the fan onto the shaft and tighten the bolt.

7. Install the shroud/fan assembly to the radiator.

8. Connect the electrical leads.

9. Install the radiator support and luggage undercovers.

10. Connect the battery negative cable.

11. Check the fan for proper operation.

Supra

1994 Models

1. Disconnect the negative battery cable. On vehicles equipped with an air bag, wait at least 90 seconds before proceeding.

2. Drain the coolant into a suitable container.

3. Remove the air cleaner case, as required.

4. Disconnect the upper radiator hose, as required.

5. Remove the 4 fan shroud bolts.

6. Remove the 4 bolts from the fluid coupling (fan clutch) flange.

7. Pull out the fan clutch along with the fan shroud.

8. Remove the fan from the fan clutch assembly.

To install:

9. Install the fan to the fan clutch assembly.

10. Install the fan clutch along with the fan shroud.

11. Install the 4 bolts to the fluid coupling (fan clutch) flange.

12. Install the 4 fan shroud bolts.

13. Install the upper radiator hose.

14. Install the air cleaner case.

15. Close the engine and radiator drain cock. Refill the coolant system.

16. Start the engine and check fan operation.

1995–97 Models

1. Disconnect the negative battery cable from the battery. On vehicles equipped with an air bag, wait at least 90 seconds before proceeding.
2. Raise and safely support the vehicle.
3. Remove the undercovers from the vehicle.
4. Disconnect the cooling fan connector.
5. Remove the three bolts to the cooling fan bracket and remove the assembly.
6. Remove the three screws and the fan blade from the cooling fan motor.
7. Disconnect the cooling fan connector from the fan bracket.
8. Remove the three screws and the fan motor from the fan bracket.
To install:
9. Install the fan to the fan bracket and install the three screws.
10. Connect the electrical connector to the fan bracket.
11. Connect the fan blade to the fan motor and install the three screws.
12. Install the fan assembly to the engine compartment and install the three bolts.
13. Connect the cooling fan connector.
14. Install the engine undercovers to the vehicle.
15. Connect the negative battery cable to the battery.
16. Lower the vehicle and check fan operation.

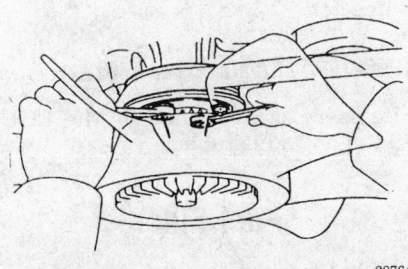

287646

Loosen the bolts holding the fan-clutch to the water pump — 1995–97 2JZ-GE and 2JZ-GTE engines

1995–97 Models With Belt Driven Fan

1. Disconnect the negative battery cable from the battery. On vehicles equipped with an air bag, wait at least 90 seconds before proceeding.
2. Loosen the four nuts holding the fan clutch to the water pump.
3. Remove the drive belt from the engine by turning the tensioner clockwise.
4. Remove the four nuts, the fan, and the fan clutch.
To install:
5. Install the fan clutch and fan to the engine and loosely install the four nuts.
6. Install the drive belt to the engine by turning the tensioner clockwise.
7. Tighten the four nuts for the fan clutch and torque the nuts to 12 ft. lbs. (16 Nm).
8. Connect the negative battery cable to the battery.

Cooling System Bleeding

After working on the cooling system, the system must be properly bled. Air trapped in the system will prevent proper filling and coolant circulation.

To bleed the system, start with the system cool, the radiator cap off and the radiator filled to about an inch below the filler neck.

1. Start the engine and run it at slightly above normal idle speed. This will insure adequate circulation. If air bubbles appear and the coolant level drops, fill the system with an antifreeze/water mixture to bring the level back to the proper level.
2. Run the engine this way until the thermostat opens. When this happens, coolant will move abruptly across the top of the radiator and the temperature of the radiator will suddenly rise.
3. At this point, air is often expelled and the level may drop quite a bit. Keep refilling the system until the level is near the top of the radiator and remains constant.
4. If the vehicle has an overflow tank, fill the radiator right up to the filler neck. Replace the radiator filler cap.

Fuel System Service Precautions

Safety is the most important factor when performing not only fuel system maintenance but any type of maintenance. Failure to conduct maintenance and repairs in a safe manner may result in serious personal injury or death. Maintenance and testing of the vehicle's fuel system components can be accomplished safely and effectively by adhering to the following rules and guidelines.

• To avoid the possibility of fire and personal injury, always disconnect the negative battery cable unless the repair or test procedure requires that battery voltage be applied.

• Always relieve the fuel system pressure prior to disconnecting any fuel system component (injector, fuel rail, pressure regulator, etc.), fitting or fuel line connection. Exercise extreme caution whenever relieving fuel system pressure to avoid exposing skin, face and eyes to fuel spray. Please be advised that fuel under pressure may penetrate the skin or any part of the body that it contacts.

• Always place a shop towel or cloth around the fitting or connection prior to loosening to absorb any excess fuel due to spillage. Ensure that all fuel spillage (should it occur) is quickly removed from engine surfaces. Ensure that all fuel soaked cloths or towels are deposited into a suitable waste container.

• Always keep a dry chemical (Class B) fire extinguisher near the work area.

• Do not allow fuel spray or fuel vapors to come into contact with a spark or open flame.

• Always use a backup wrench when loosening and tightening fuel line connection fittings. This will prevent unnecessary stress and torsion to fuel line piping. Always follow the proper torque specifications.

• Always replace worn fuel fitting O-rings with new. Do not substitute fuel hose or equivalent, where fuel pipe is installed.

Fuel System Pressure

RELIEVING

——————— CAUTION ———————

Failure to relieve fuel pressure before repairs or disassembly can cause serious personal injury and/or property damage. Fuel pressure is maintained within the fuel lines, even if the engine is OFF or has not been run in a period of time. This pressure must be safely relieved before any fuel-bearing line or component is loosened or removed. On vehicles equipped with inflatable restraints or air bag systems, wait at least 90 seconds after disconnecting the battery cable before performing any other work. The backup power will keep the restraint system energized for a period of time after the battery is disconnected.

1. Remove the fuse for the fuel pump.
2. Start the engine until the engine stalls.
3. Disconnect the negative battery terminal.
4. Place a catch-pan under the joint to be disconnected. A large quantity of fuel may be released when the joint is opened.
5. Wear eye or full face protection.
6. Place a shop towel over the area and slowly release the joint using a wrench of the correct size.
7. Allow the any fuel left in the line to bleed off slowly before fully disconnecting the joint.
8. Plug the opened lines immediately to prevent fuel spillage or the entry of dirt.
9. Dispose of the released fuel properly.
10. After connecting fuel lines, install the fuse for the fuel pump and start the engine.
11. Check for leaks and repair as needed.

Idle Speed

PRECAUTIONS

Not all tachometers are compatible with the Toyota electrical system. You are cautioned to check with the dealer and/or the tachometer manufacturer about suitability for your car. After a proper unit is selected,

you may have to make or buy an adapter to connect the tachometer test lead to the Toyota terminal.

——————— WARNING ———————

The electrical system and computer(s) are easily damaged through careless work habits. Since electricity travels at close to the speed of light, damage is instantaneous and expensive. Always observe the following rules when connecting the tachometer.

• Do not leave the ignition switch ON for more than 5 minutes when the engine is not running.
• The tachometer lead must only be connected to the correct terminal; accidental contact with another terminal can cause great damage.
• Once connected to the terminal, the tachometer lead and connectors must be protected against grounding to any metal surface of the car. It is highly recommended that any adapter have fully insulated connectors or be wrapped in dry cloth or tape for protection; otherwise, the igniter and/or ignition coil can be damaged.
• Never disconnect the battery cables while the engine is running.
• Make certain the engine and ignition wiring is connected properly before connecting the tachometer. It is particularly important that ground circuits be clean and tight.

ADJUSTMENT

1993–94 Models

1. Warm up the engine until it reaches normal operating temperature.
2. The air cleaner should be in place and all wires and vacuum hoses connected. All accessories should be **OFF**, the transmission in neutral and the drive wheels blocked.
3. Connect a tachometer to the engine.
4. Use a jumper wire to connect terminal TE_1 or T and terminal E_1 of the check connector.
5. Run the engine at 2,500 rpm for 2 minutes. This insures that the oxygen sensor and other sensors are fully up to temperature and stabilized.
6. Let the engine return to idle.
7. Set the idle speed by turning the idle adjusting screw to obtain the proper idle speed.
8. Disconnect the jumper wires and recheck idle.

9. If the idle is not at specification, reconnect jumper wires and adjust the idle speed adjusting screw on the throttle body.
10. Remove the jumper wire and check the idle speed again; it should now be the value in the Specification Chart.
11. If the idle is not at the specified value, start the engine and idle it for 30 seconds. Turn it off, then repeat the entire pattern (start — idle — off) several times. This should store the correct idle value within the Idle Speed Control unit (ISC) and the idle rpm will be at the correct setting.
12. Disconnect the tachometer and replace any covers or plugs which were removed from the service connector.

1995 MR2

1. Warm up the engine until it reaches normal operating temperature.
2. The air cleaner should be in place and all wires and vacuum hoses connected. All accessories should be OFF, the transmission in neutral, and the drive wheels blocked.
3. Connect a tachometer to the engine.
   a. Connect the tachometer test probe to the IG terminal of the DCL1.
4. Run the engine at 2,500 rpm for 2 minutes. This insures that the oxygen sensor and other sensors are fully up to temperature and stabilized.
5. Let the engine return to idle.
6. Set the idle speed by turning the idle adjusting screw to obtain the proper idle speed.
7. Disconnect the tachometer and replace any covers or plugs which were removed from the service connector.

1995–1997 Models Except MR2

Idle speed is controlled by the ECU and is not adjustable.

Fuel Filter

REMOVAL AND INSTALLATION

1. Disconnect the negative battery cable. On vehicles equipped with an air bag, wait at least 90 seconds before proceeding.
2. Unbolt the retaining screws and remove the protective shield for the fuel filter, if equipped.

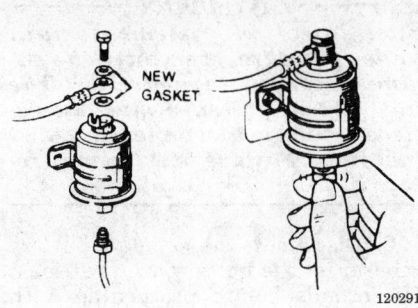

Always use new gaskets when replacing a fuel filter — Tercel shown

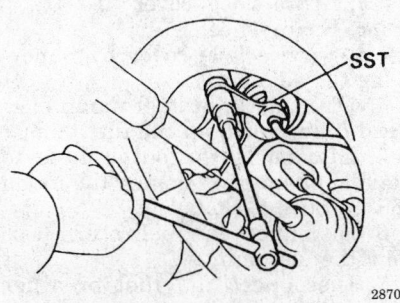

A line wrench with an extension may be needed to loosen the inlet line at the filter — 1995–97 Corolla

— CAUTION —

Fuel injection systems remain under pressure after the engine has been turned OFF. Properly relieve fuel pressure before disconnecting any fuel lines. Failure to do so may result in fire or personal injury.

3. Place a drain pan or plastic container under the fuel filter.
4. If necessary, remove the air cleaner hose and cap.
5. If necessary, remove the charcoal canister.
6. Slowly loosen the lower flare nut fitting until all the pressure is relieved and all the fuel is collected.
7. Loosen the union bolt on the upper portion of the filter and remove the banjo fitting and two metal gaskets. Discard the gaskets.
8. Loosen the fuel filter bracket bolt, remove the fuel line with the flared nut from the filter, and pull the filter from the mounting bracket.
To install:
9. Install a new fuel filter to the vehicle and tighten the bracket bolt.
10. Install the banjo fitting with a new metal gasket on each side and install the union bolt. Tighten the union bolt to 22 ft. lbs. (30 Nm).

11. Connect the flare nut to the lower connection. Tighten the flare nut to 22 ft. lbs. (30 Nm).
12. If removed, install the charcoal canister to the vehicle.
13. Install the air cleaner hose and cap.
14. If removed, install the protective shield.
15. Remove the drain pan and/or rags and connect the negative battery cable.
16. Start the engine and visually inspect the upper and lower connections for leaks.

Fuel Pump

REMOVAL AND INSTALLATION

— CAUTION —

The fuel system is under pressure. Release pressure slowly and contain spillage. Observe no smoking/no open flame precautions. Have a Class B-C (dry powder) fire extinguisher within arm's reach at all times.

Celica, Corolla, Paseo and Tercel

1. Relieve the fuel system pressure.
2. Disconnect the negative battery cable. On vehicles equipped with an air bag, wait at least 90 seconds before proceeding.
3. On the 3S-GTE engine, remove the fuel tank from the vehicle. Remove the access plate-to-fuel tank bolts, then, pull out the plate/fuel pump assembly.
4. On all other engines, remove the rear seat cushion and remove the 4 screws and the floor service hole cover.
   a. Disconnect the fuel pump sender and fuel pump connector.
   b. Disconnect the outlet pipe from the fuel pump bracket and

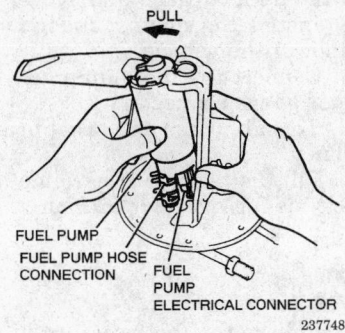

Pull the pump off the sender unit, the filter is still attached to the pump

disconnect the return hose from the pump bracket.
   c. Remove the 8 bolts and pull the fuel pump bracket assembly from the fuel tank.
5. Separate the fuel pump from the fuel pump bracket as follows:
   a. Pull off the lower side of the fuel pump from the pump bracket.
   b. Disconnect the fuel pump connector.
   c. Disconnect the fuel hose from the fuel pump and remove the rubber cushion from the pump.
   d. Remove the fuel filter from the pump by removing the small clip.
To install:
6. Install a new cushion to the fuel pump. Install a new fuel filter and a new clip to the fuel pump. Reconnect the fuel hose to the fuel pump, reconnect the fuel pump connector and reinstall the fuel pump to the bracket.
7. Install the fuel pump bracket assembly to the fuel tank using a new gasket and the 8 bolts. Tighten the bolts to 30 inch lbs. (3 Nm).
8. Connect the fuel return hose and the fuel outlet pipe to the fuel pump bracket.
9. Connect the fuel pump and fuel pump sender connector.
10. On the 3S-GTE engines, install the fuel tank.
11. Connect the negative battery cable, start the engine, and check for leaks and proper operation.
12. Install the floor service hole cover with the 4 screws and reinstall the rear seat cushion.

Avalon and Camry

1. Relieve the fuel system pressure.
2. Disconnect the negative battery cable from the battery. On vehicles equipped with an air bag, wait at least 90 seconds before proceeding.
3. On the 1993–94 5S-FE engines, position a suitable waste container under the fuel tank and drain the fuel from the tank. Next, remove the fuel tank.
4. On some models it will be necessary to remove the rear seat cushion.
5. Remove the floor service hole cover by removing the retaining screws.
6. Disconnect the electrical connector at the fuel pump assembly.
7. Disconnect the fuel outlet pipe from the fuel pump bracket.
8. Disconnect the return hose from the fuel pump bracket.
9. Disconnect the fuel pump bracket assembly from the fuel tank by removing the bolts.

10. Pull out the pump bracket assembly.

11. Remove the fuel pump from the fuel bracket.

To install:

12. Connect the fuel pump to the fuel bracket.

13. Install the pump bracket assembly.

14. Connect the fuel pump to the fuel tank by installing the bolts. Tighten the bolts to 35 inch lbs. (4 Nm).

15. Connect the return hose to the fuel pump bracket.

16. Connect the outlet pipe to the fuel pump bracket. Tighten to 21 ft. lbs. (28 Nm).

17. Install the service hole cover to the fuel tank and install the screws.

18. Connect the fuel pump connector.

19. Install the rear seat cushion.

20. Install the fuel tank.

21. Fill the fuel tank and check for leaks.

22. Connect the negative battery cable to the battery.

23. Check the operation of the fuel pump and check for leaks.

MR2

1. Disconnect the negative battery cable. On vehicles equipped with an air bag, wait at least 90 seconds before proceeding.

--- CAUTION ---

Fuel injection systems remain under pressure after the engine has been turned OFF. Properly relieve fuel pressure before disconnecting any fuel lines. Failure to do so may result in fire or personal injury.

2. Relieve the fuel system pressure.

3. Remove the ash receptacle retainer in the center console.

4. Disconnect the fuel pump and the fuel sender gauge connectors.

5. Raise and safely support the vehicle.

6. Remove the undercovers from the vehicle.

7. Remove the front luggage undercover.

8. Remove the fuel tank protectors.

9. Drain the fuel.

--- CAUTION ---

Do not allow fuel spray or fuel vapors to come in contact with a spark or open flame. Keep a dry chemical fire extinguisher nearby. Never store fuel in an open container due to the risk of fire or explosion.

10. If the vehicle has the 3S-GTE engine, remove the parking brake intermediate lever and the No. 1 center floor crossmember.

11. Remove the A/C pipes from the body.

12. Disconnect the radiator pipes from the body.

13. Disconnect the fuel hoses.
   a. Fuel inlet hose
   b. Fuel breather hose
   c. Fuel pump hose
   d. Fuel return hose
   e. Two evaporative vent hoses

14. Properly support the fuel tank before removing the tank band.

15. Remove the bolt, pin and the tank band.

16. Remove the bolts holding the No. 2 center floor crossmember.

17. Remove the fuel tank.

18. Remove the fuel pump and the gasket.

To install:

19. Install the fuel pump with a new gasket. Tighten the bolts to 3 ft. lbs. (3 Nm).

20. Install the fuel tank and tighten the tank band bolt to 10 ft. lbs. (13 Nm) for the 3S-GTE engine and 11 ft. lbs. (14 Nm) for the 5E-FE engine.

21. Install the No. 2 center floor crossmember and torque the bolts to 11 ft. lbs. (14 Nm) for the 3S-GTE engine and 10 ft. lbs. (13 Nm) for the 5E-FE engine.

22. Connect the fuel hoses.

23. Install the fuel tank heat insulators.

24. Connect the radiator pipes to the body.

25. Connect the A/C pipes to the body.

26. If removed, install the parking brake intermediate lever and the No. 1 center floor crossmember.

27. Install the fuel tank protectors and the front luggage undercover.

28. Install the engine undercovers and lower the vehicle.

29. Connect the fuel pump and the sender gauge connectors.

30. Install the ash receptacle retainer.

31. Fill the tank with gasoline and check the fuel pump operation.

Supra

1. Relieve the fuel pressure in the fuel line before disconnecting any fuel lines.

--- CAUTION ---

Fuel injection systems remain under pressure, even after the engine has been turned OFF. The fuel system pressure must be relieved before disconnecting any fuel lines. Failure to do so may result in fire and/or explosion.

2. Disconnect the negative battery cable from the battery. Wait at least 90 seconds before proceeding with any other work.

3. Open the hatchback and remove the following components:
   a. Floor carpet
   b. Spare wheel cover
   c. Spare wheel
   d. Service hole cover by removing the six nuts

4. Disconnect the fuel pump electrical connector from the fuel pump.

5. Disconnect the outlet hose to the fuel pump by removing the union bolt and two gaskets.

6. Disconnect the fuel return hose from the fuel pump.

7. Disconnect the fuel breather clamp.

8. Using SST 09808-14010 or equivalent, loosen the retainer to the fuel pump.

9. Disconnect the fuel return hose from the return port of the fuel pump bracket.

10. Remove the retainer, fuel pump, sender gauge assembly, and the gasket as a unit.

To install:

11. Install a new gasket to the fuel pump.

12. Insert the fuel pump and sender gauge assembly into the fuel tank.

13. Align the arrow marks of the fuel pump bracket and the fuel tank.

14. Using the same tool as removal, install and tighten the fuel pump retainer until the arrow mark on the retainer is within the lines on the fuel tank.

15. Check that the arrow marks of the fuel pump bracket and fuel tank are aligned.

16. Install the retainer clamp to the fuel pump.

17. Connect the fuel pump electrical connector to the fuel pump.

18. Connect the outlet hose with two new gaskets and the union bolt. Tighten the union bolt to 22 ft. lbs. (29 Nm).

19. Connect the fuel return hose to the fuel pump.

20. Connect the fuel breather hose to the fuel pump.

21. Connect the negative battery cable to the vehicle. Start the engine and check for fuel leaks.

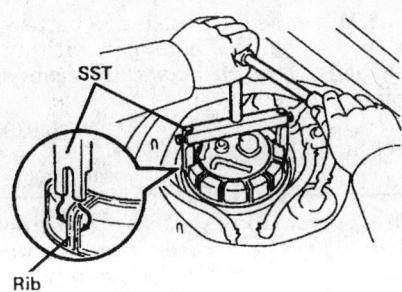

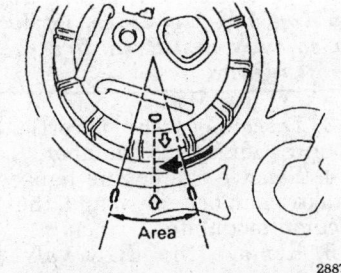

288790

Tighten the fuel pump retainer until the arrow mark on the retainer is within the lines on the fuel tank — Supra

22. Install the service hole cover with the six nuts.

23. Install the following components:
   a. Spare wheel
   b. Spare wheel cover
   c. Floor carpet

Fuel Injector

REMOVAL AND INSTALLATION

1.5L (3E-E) Engine

Tercel

> **CAUTION**
> *The fuel system is under pressure. Release pressure slowly and contain spillage. Observe no smoking/no open flame precautions. Have a Class B-C (dry powder) fire extinguisher within arm's reach at all times.*

1. Disconnect the negative battery cable. On vehicles equipped with an air bag, wait at least 90 seconds before proceeding.

2. Remove the gas cap and relieve fuel system pressure.

3. Disconnect the PCV hose from the valve cover.

4. Remove the air intake connector. Disconnect the accelerator cable and throttle cable.

5. Remove the vacuum sensing hose from the pressure regulator.

6. Remove the dash pot and link bracket.

7. Disconnect the fuel inlet and return hoses.

8. Place a towel or container under the cold start injector pipe. Loosen the two union bolts at the fuel line and remove the pipe with its gaskets.

9. Disconnect the injector electrical connections.

10. At the fuel delivery pipe (rail), remove the 2 bolts. Lift the delivery pipe and the injectors free of the engine.

11. Remove the four insulators and 2 collars from the cylinder head.

12. Pull the injectors free of the delivery pipe.

To install:

13. Before installing the injectors back into the fuel rail, install a NEW O-ring on each injector.

14. Coat each O-ring with a light coat of gasoline and alternately install the injectors into the delivery pipe, dark blue injector then a brown injector, dark blue injector, brown injector.

15. Make certain each injector can be smoothly rotated. If they do not rotate smoothly, the O-ring is not in its correct position.

16. Install the insulators into each injector hole. Place the 2 spacers on the delivery pipe mounting holes in the cylinder head.

17. Place the delivery pipe and injectors on the cylinder head and again check that the injectors rotate smoothly. Install the bolts and tighten them to 14 ft. lbs. (19 Nm).

18. Connect the electrical connectors to each injector.

19. Install two new gaskets and attach the inlet pipe and fuel union bolt. Tighten the bolt to 22 ft. lbs. (30 Nm). Install the mounting bolt.

20. Install new gaskets and connect the cold start injector pipe to the delivery pipe and cold start injector. Install the fuel line union bolts and tighten them to 13 ft. lbs. (17 Nm).

21. Connect the fuel return hose and the vacuum sensing hose to the pressure regulator. Attach the PCV hose to the valve cover.

22. Reconnect the accelerator cable, throttle cable and install the air intake connector. Install the dash pot and link bracket.

23. Connect the negative battery terminal. Start the engine and check for leaks.

> **CAUTION**
> *If there is a leak at any fitting, the line will be under pressure and the fuel may spray in a fine mist. This mist is extremely explosive. Shut the engine off immediately if any leakage is detected. Use rags to wrap the leaking fitting until the pressure diminishes and wipe up any fuel from the engine area.*

1.5L (5E-FE) Engine

Paseo and Tercel — Distributor Ignition

1. Relieve the fuel pressure.

2. Disconnect the negative battery cable from the battery. On vehicles equipped with an air bag, wait at least 90 seconds before proceeding.

3. Remove the PCV hose.

4. Disconnect the vacuum hose from the fuel pressure regulator.

5. Disconnect the fuel return hose and the delivery hose. Catch any fuel with a shop towel.

6. Remove the accelerator cable bracket and disconnect the four (4) injector connectors.

7. Remove the intake air chamber support bracket.

8. Remove the two (2) delivery pipe bolts and remove the delivery pipe with the injectors, being careful not to drop the injectors. Note the positions of the spacers for installation.

To install:

9. Install the injectors with new O-rings with the delivery pipe. Coat the new O-rings with gasoline. Make sure the injectors can be rotated by hand once they are installed.

10. Install the two (2) pipe bolts and torque them to 14 ft. lbs. (19 Nm).

11. Install the air intake chamber bracket and torque to 13 ft. lbs. (17 Nm).

12. Install the injector connectors in their original positions, then install the accelerator bracket.

13. Connect the inlet hose to the delivery pipe, using new gaskets, and torque the union bolt to 22 ft. lbs. (29 Nm).

14. Connect the return hose to the return pipe.

15. Connect the vacuum hose to the regulator.

16. Install the PCV hose and connect the negative battery cable.

17. Start the engine and check for leaks.

Paseo and Tercel — Distributorless Ignition

1. Relieve the fuel pressure.

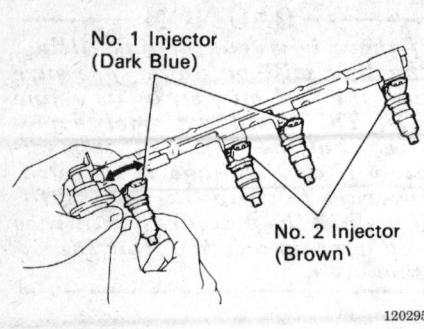

Make certain each injector can be smoothly rotated once installed

2. Disconnect the negative battery terminal. On vehicles equipped with an air bag, wait at least 90 seconds before proceeding.

3. Disconnect the vacuum hose from the fuel pressure regulator.

4. Disconnect the engine wire clamp from the No. 2 engine hanger.

5. Disconnect the fuel return hose and the delivery hose and catch the spilled fuel with a shop towel.

6. Disconnect the engine wire from the surge tank stay.

7. Disconnect the 4 injector connectors.

8. Remove the 2 delivery pipe bolts and remove the delivery pipe with the injectors, being careful not to drop the injectors. Note the positions of the spacers for installation.

To install:

9. Install the injectors with new O-rings with the delivery pipe. Coat the new O-rings with gasoline. Make sure the injectors can be rotated by hand once they are installed.

10. Install the two (2) pipe bolts and torque them to 14 ft. lbs. (19 Nm).

11. Install the air intake chamber bracket and torque to 13 ft. lbs. (17 Nm).

12. Install the injector connectors in their original positions, then install the accelerator bracket.

13. Connect the inlet hose to the delivery pipe, using new gaskets, and torque the union bolt to 22 ft. lbs. (29 Nm).

14. Connect the return hose to the return pipe.

15. Connect the vacuum hose to the regulator.

16. Connect the engine wire clamp to the No. 2 engine hanger.

17. Connect the negative battery cable.

18. Start the engine and check for leaks.

1.6L (4A-FE) and 1.8L (7A-FE) Engines

1993–95 Corolla and Celica

1. Disconnect the negative battery cable. On vehicles equipped with an air bag, wait at least 90 seconds before proceeding.

2. Relieve the fuel system pressure.

----- **CAUTION** -----
Fuel injection systems remain under pressure after the engine has been turned OFF. Properly relieve fuel pressure before disconnecting any fuel lines. Failure to do so may result in fire or personal injury.

3. Disconnect the throttle body from the air intake chamber.

4. Remove the engine hanger, air intake chamber stay, and the EGR vacuum modulator.

5. Remove the EGR valve and pipe.

6. Disconnect the following hoses:
• Vacuum sensing hose from the fuel pressure regulator
• PCV hose from the PCV valve
• PCV hose from the air intake chamber cover

7. Using a 6mm hexagon wrench, remove the 3 bolts, 2 nuts, and the air intake chamber cover and gasket.

8. Disconnect the injector connectors.

9. Remove the union bolt and two (2) gaskets and disconnect the inlet hose from the delivery pipe. Place a shop towel under the connection and slowly loosen the union bolt.

10. Disconnect the fuel return hose from the fuel pressure regulator.

11. Remove the two (2) bolts and the delivery pipe together with the four (4) injectors.

12. Remove the four (4) insulators and two (2) spacers from the intake manifold.

13. Pull out the four (4) injectors from the delivery pipe and remove the O-ring and the grommet from each injector.

To install:

14. Apply a light coat of gasoline to a new O-ring and install the O-rings to each of the four (4) injectors.

15. While turning the injector left and right, install them to the fuel delivery pipe. Make sure to position the connector upward.

16. Place four (4) new insulators and the two (2) spacers in position on the intake manifold. Place the four (4) injectors and the delivery pipe in position on the intake manifold.

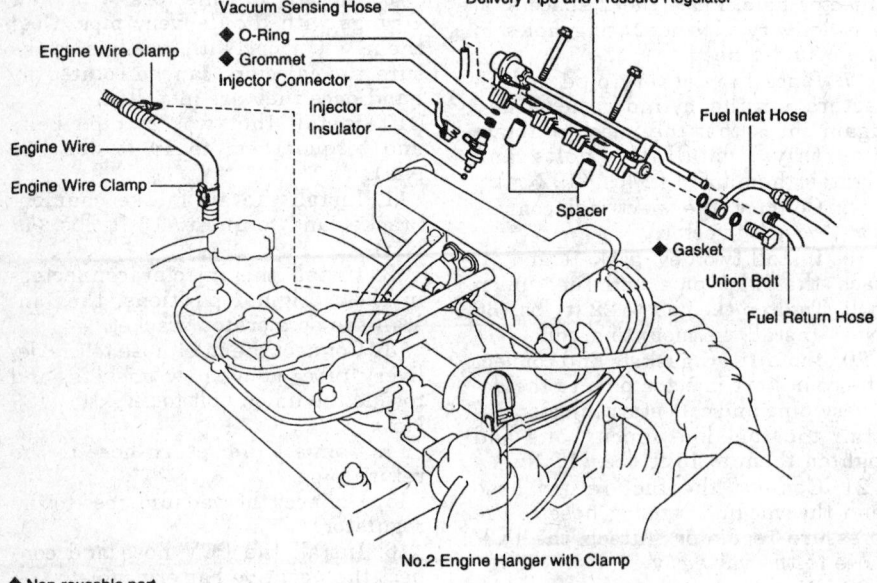

◆ Non-reusable part

Exploded view of the fuel injector and rail assembly — Paseo and Tercel 5E-FE engine

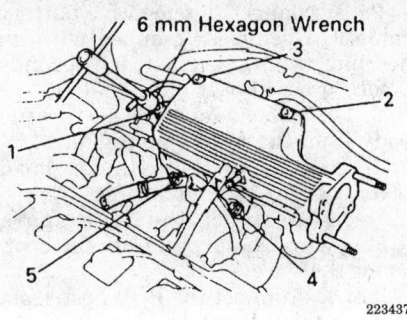

Removal and installation bolt sequence for the air intake chamber cover — 1993–95 7A-FE engine

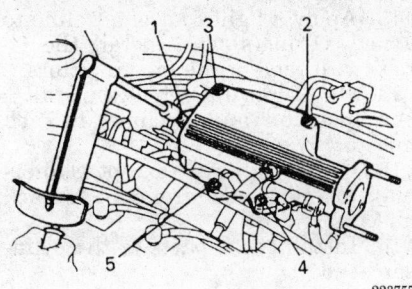

Tightening sequence for the air intake chamber — 1993–95 4A-FE engine

17. Temporarily install the two (2) bolts holding the delivery pipe to the intake manifold.

NOTE: Make sure that the injector rotates smoothly. If they do not the O-ring may be installed incorrectly.

18. Tighten the two (2) bolts and torque them to 11 ft. lbs. (15 Nm).
19. Connect the fuel return hose to the fuel pressure regulator.
20. Connect the fuel inlet hose to the fuel delivery pipe with two (2) new gaskets and the union bolt. Tighten the union bolt to 22 ft. lbs. (29 Nm).
21. Connect all of the injector connectors.
22. Using a 6mm hexagon wrench, install a new gasket and the air intake chamber cover with the 3 bolts and 2 nuts in sequence. Tighten the bolts and nuts to 14 ft. lbs. (19 Nm).
23. Connect the following hoses:
• Vacuum sensing hose from the fuel pressure regulator
• PCV hose from the PCV valve
• PCV hose from the air intake chamber cover
24. Install the EGR valve and pipe.
25. Install the engine hanger and air intake chamber stay.
26. Install the throttle body.
27. Connect the negative battery cable, start the engine, and check for leaks. Road test the vehicle for proper operation.

1996–97 Corolla and Celica

1. Remove the air cleaner hose and cap.
2. Relieve the fuel system pressure.
3. Disconnect the negative battery cable. On vehicles equipped with an air bag, wait at least 90 seconds before proceeding.

CAUTION

Fuel injection systems remain under pressure after the engine has been turned OFF. Properly relieve fuel pressure before disconnecting any fuel lines. Failure to do so may result in fire or personal injury.

---

4. Disconnect the following:
• The vacuum sensing hose from the fuel pressure regulator
• The PCV hose from the PCV valve
• The PCV hose from the air intake chamber cover
• The EGR hose from the EGR valve.
• The vacuum hose from the EGR valve.
• The EGR temperature sensor and EGR VSV electrical connectors.
5. Using a 6mm hexagon wrench, remove the three bolts, two nuts, and the air intake chamber cover and gasket.
6. Disconnect the injector connectors.
7. Remove the union bolt and two gaskets and disconnect the inlet hose from the delivery pipe. Place a shop towel under the connection to absorb the fuel and slowly loosen the union bolt.
8. Disconnect the fuel return hose from the fuel pressure regulator.
9. Remove the two bolts and the delivery pipe together with the four injectors.
10. Remove the four insulators and two spacers from the intake manifold.
11. Pull out the four injectors from the delivery pipe and remove the O-ring and the grommet from each injector.
To install:
12. Apply a light coat of gasoline to a new O-ring and install the O-rings to each of the four injectors.
13. While turning the injector left and right, install them to the fuel de-

livery pipe. Make sure to position the connector upward.
14. Place four new insulators and the two spacers in position on the intake manifold. Place the four injectors and the delivery pipe in position on the intake manifold.
15. Temporarily install the two bolts holding the delivery pipe to the intake manifold. Make sure that the injector rotates smoothly. If they do not, the O-ring may be installed incorrectly. Tighten the two bolts and torque them to 11 ft. lbs. (15 Nm).
16. Connect the fuel return hose to the fuel pressure regulator.
17. Connect the fuel inlet hose to the fuel delivery pipe with two new gaskets and the union bolt. Tighten the union bolt to 22 ft. lbs. (29 Nm).
18. Connect all of the injector connectors.
19. Using a 6mm hexagon wrench, install a new gasket and the air intake chamber cover. Tighten the three bolts and two nuts in sequence to 14 ft. lbs. (19 Nm).
20. Install the remaining components in the reverse order they were removed.
21. Install the air cleaner hose and cap.
22. Connect the negative battery cable and start the engine. Check for leaks and road test the vehicle for proper operation.

2.0L (3S-GTE) Engine

1. Disconnect the negative battery cable. On vehicles equipped with an air bag, wait at least 90 seconds before proceeding.
2. Remove the throttle body.
3. Remove the air cleaner, charcoal canister, EGR VSV and vacuum modulator.
4. Remove the EGR valve and pipe, cold start injector pipe and injector, water bypass hoses and air hose.
5. Disconnect the electrical connections from fuel injectors.
6. Disconnect the fuel inlet hose from the fuel filter.
7. Disconnect the fuel return hose from the fuel pressure regulator.
8. Remove the delivery pipe and fuel injector and regulator assembly (insulators, spacers, O-ring and grommet). Remove injectors from the delivery pipe.
To install:
9. Install the fuel inlet hose to the delivery pipe, tightening the bolt to 69 inch lbs. (7.8 Nm) and the union bolt to 22 ft. lbs. (29 Nm).
10. Install the injectors, together with the delivery pipe and fuel pres-

sure regulator. Tighten the 3 mounting bolts to 14 ft. lbs. (19 Nm).

11. Reconnect the fuel inlet hose and tighten to 22 ft. lbs. (29 Nm). Reconnect the brown No. 1 and 3 injectors and the gray No. 2 and 4 injector connectors.

12. Install the EGR valve and pipe, cold start injector pipe and injector, water bypass hoses and air hose.

13. Install the air cleaner, charcoal canister, EGR VSV and vacuum modulator.

14. Install the throttle body.

15. Connect the battery, start the engine and check for any fuel leaks.

2.2L (5S-FE) Engine

1993–95 Camry

1. Disconnect the negative battery cable. On vehicles equipped with an air bag, wait at least 90 seconds before proceeding.

—————— CAUTION ——————

Fuel injection systems remain under pressure after the engine has been turned OFF. Properly relieve fuel pressure before disconnecting any fuel lines. Failure to do so may result in fire or personal injury.

2. If equipped with A/T, disconnect the throttle cable from the throttle body.

3. Disconnect the accelerator cable from the throttle body.

4. Disconnect the throttle body from the intake manifold.

5. Remove the EGR valve.

6. Remove the bolts, wire bracket, No. 1 air intake chamber and manifold stays. Remove the intake manifold.

7. Disconnect the four injector connectors.

8. Relieve the fuel system pressure, loosen the pulsation damper, and disconnect the fuel inlet hose from the delivery pipe.

9. Remove the two bolts holding the delivery pipe to the cylinder head. Disconnect the delivery pipe from the four injectors and remove the delivery pipe. Remove the four injectors from the intake manifold.

10. Remove the four (except California) insulators and two spacers from the cylinder head.

11. On California vehicles, remove the two O-rings, insulator, and grommet from each injector.

12. On all vehicles (except California), remove the O-ring and grommet from each injector.

To install:

13. Apply a light coat of gasoline to a new O-ring(s) and install the O-ring(s) to each of the four injectors.

14. On California vehicles install a new insulator and grommet to each injector.

15. On all vehicles (except California), install a new grommet to each injector.

16. Install these parts to the cylinder head:

   a. Two spacers.

   b. Except California–four new insulators.

17. Place the delivery pipe between the cylinder head and the intake manifold. While turning the injector left and right install the four injectors to the delivery pipe. The injector connector should be facing upwards.

18. Temporarily install the two bolts holding the delivery pipe to the intake manifold.

NOTE: Make sure that the injector rotates smoothly. If they do not the O-ring may be installed incorrectly.

19. Tighten the two bolts holding the delivery pipe to 9 ft. lbs. (13 Nm).

20. Connect the fuel inlet hose to the delivery pipe with two new gaskets and the union bolt. Tighten the union bolt to 25 ft. lbs. (34 Nm). Be careful of the fuel inlet hose installation direction; the hole in the bottom of the delivery line is slightly smaller than the top. Tighten the fuel pulsation damper and connect the fuel return hose to the return pipe.

21. Connect the four injector connectors. The No. 1 & 3 injector connectors are brown and the No. 2 & 4 connectors are gray.

22. Connect the wire clamps to the wire brackets on the intake manifold.

23. Install a new gasket and the intake manifold. Uniformly tighten the bolts and nuts in several passes. Tighten to 14 ft. lbs. (19 Nm).

24. Install the No. 1 air intake chamber and the manifold stays.

25. Install the EGR valve.

26. If equipped, install the throttle cable to the throttle body.

27. Install the accelerator cable bracket with the two bolts and install the accelerator cable to the clamp on the intake manifold.

28. Install the throttle body.

29. Connect the negative battery cable, start the engine, and check for leaks. Road test the vehicle for proper operation.

Celica and 1996–97 Camry

1. Release the fuel system pressure.

2. Disconnect the negative battery cable. On vehicles equipped with an air bag, wait at least 90 seconds before proceeding.

3. On Celica, remove the throttle body from the intake manifold.

4. To remove the cylinder head cover, perform the following:

   a. Disconnect the four spark plug wires. note: pull from the boot, not the wire.

   b. Disconnect the PCV hose from the intake manifold.

   c. Disconnect the engine wire harness protector from the two mounting bolts of the No. 2 timing belt cover.

   d. Remove the four nuts & grommets, cylinder head cover, and the gasket.

NOTE: Cover the cylinder head with a shop cloth to prevent dirt from getting into the cylinder head. Arrange the grommets in order so they can be installed in their original position. This minimizes oil leakage due to grommet distortion.

5. Disconnect the vacuum sensing hose from the fuel pressure regulator.

6. On California vehicles, disconnect the air hose from the air assist system from the intake manifold port, and remove the air hose.

7. Disconnect the four injector connectors.

—————— CAUTION ——————

Fuel injection systems remain under pressure after the engine has been turned OFF. Properly relieve fuel pressure before disconnecting any fuel lines. Failure to do so may result in fire or personal injury.

8. Remove the union bolt and two gaskets, and disconnect the fuel inlet hose from the delivery pipe. Disconnect the fuel return hose from the return pipe.

9. Remove the two bolts holding the delivery pipe to the cylinder head. Disconnect the delivery pipe from the four injectors and remove the delivery pipe. Remove the four injectors from the intake manifold.

10. Remove the four insulators (except California) and two spacers from the cylinder head.

11. On California vehicles, remove the two O-rings, insulator, and grommet from each injector.

12. On all vehicles (except California), remove the O-ring and grommet from each injector.

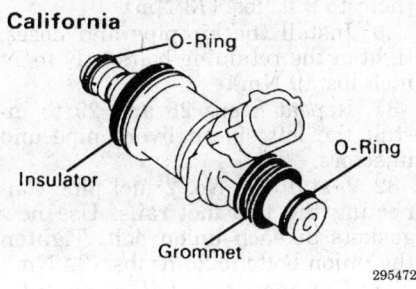

California

O-Ring

O-Ring

Insulator

Grommet

295472

California style injector, O-rings, grommet and insulator — 5S-FE engine

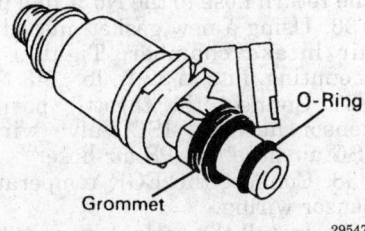

Except California

O-Ring

Grommet

295473

Fuel injector, O-ring and grommet (except California) — 5S-FE engine

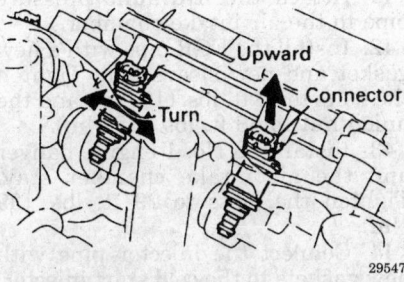

Upward

Turn

Connector

295475

While turning the injector left and right install the four injectors to the delivery pipe. The injector connector should be facing upwards — 5S-FE engine

To install:
13. Apply a light coat of gasoline to a new O-ring(s) and install the O-ring(s) to each of the four injectors.
14. On California vehicles install a new insulator to each injector.
15. On all vehicles, install a new grommet to each injector.
16. Install these parts to the cylinder head:
    a. Two spacers.
    b. Except California–four new insulators.
17. Place the delivery pipe between the cylinder head and the intake

manifold. While turning the injector left and right install the four injectors to the delivery pipe. The injector connector should be facing upwards.
18. Temporarily install the two bolts holding the delivery pipe to the intake manifold.

NOTE: Make sure that the injector rotates smoothly. If they do not the O-ring may be installed incorrectly.

19. Tighten the two bolts holding the delivery pipe to 9 ft. lbs. (13 Nm).
20. Connect the fuel inlet hose to the delivery pipe with two new gaskets and the union bolt. Tighten the union bolt to 25 ft. lbs. (34 Nm). Be careful of the fuel inlet hose installation direction; the hole in the bottom of the delivery line is slightly smaller than the top. Connect the fuel return hose to the return pipe.
21. Install the engine wire harness protector to the two brackets on the front side of the intake manifold.
22. Connect the four injector connectors. The No. 1 & 3 injector connectors are brown and the No. 2 & 4 connectors are gray.
23. Connect the vacuum sensing hose to the fuel pressure regulator. On California models, install the air hose for the air assist system to the intake manifold port.
24. Clean the cylinder head cover mating surface and install new sealant. Use Part No. 08826–00080 or equivalent. Install the gasket to the cylinder head cover and install with the four grommets and nuts. Tighten the nuts to 17 ft. lbs. (23 Nm).
25. Install the engine wire harness protector to the two mounting bolts of the No. 2 timing belt cover in reverse order of removal.
26. On the Celica, install the throttle body.
27. Connect the PCV hose to the intake manifold.
28. Connect the negative battery cable, start the engine, and check for

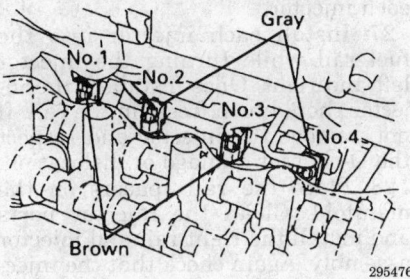

Gray

No.1

No.2

No.3

No.4

Brown

295476

Fuel injector connectors — Camry 5S-FE engine

leaks. Road test the vehicle for proper operation.

MR2
1. Relieve the fuel system pressure.
2. Disconnect the negative battery cable. On vehicles equipped with an air bag, wait at least 90 seconds before proceeding.
3. Remove the right engine hood side panel.
4. Remove the air cleaner hose.
5. Disconnect the accelerator cable from the throttle body.
6. If equipped with A/T, disconnect the throttle cable from the throttle body.
7. If equipped with cruise control, remove the cruise control actuator.
8. Disconnect the throttle body from the intake manifold.
9. Disconnect the engine wire protector.
10. Remove the air intake chamber stay.
11. Disconnect the vacuum sensing hose from the fuel pressure regulator and the brake booster vacuum hose from the intake manifold.

----CAUTION----
Fuel injection systems remain under pressure after the engine has been turned OFF. Properly relieve fuel pressure before disconnecting any fuel lines. Failure to do so may result in fire or personal injury.

12. Remove the bolt holding the fuel inlet pipe to the cylinder head and remove the inlet pipe and the return hose.
13. Remove the pulsation damper and gaskets.
14. Remove the delivery pipe and the injectors.
    a. Disconnect the four injector connectors.
    b. Remove the delivery pipe and the four injectors assembly from the right side of the engine.
    c. Remove the four insulators and the two spacers from the cylinder head.
    d. Pull out the injectors from the delivery pipe.
    e. Remove the O-ring and the grommet from each injector.
To install:
15. Install a new grommet to the injector and while turning the injector left and right, install it to the delivery pipes.
16. Place new insulators and spacers in position on the cylinder head.

17. Place the delivery pipe with the injectors in position on the cylinder head.

18. Temporarily install the bolts holding the delivery pipe and check that the injectors rotate smoothly.

19. Position the injector connector upward and tighten the fuel rail bolts to 9 ft. lbs. (13 Nm).

20. Install the fuel pulsation damper and gaskets.

21. Connect the fuel inlet pipe and the return hose. Tighten the union bolt to 25 ft. lbs. (34 Nm).

22. Connect the engine wire protector to the brackets on the front side of the intake manifold.

23. Install the clamps of the engine wire protector to each bolt.

24. Install the air intake chamber stay and torque the bolts to 31 ft. lbs. (42 Nm).

25. Connect the vacuum hose to the fuel pressure regulator and the brake booster vacuum hose to the intake manifold.

26. Install the throttle body.

27. Connect the accelerator cable to the throttle body.

28. If removed, connect the throttle cable to the throttle body.

29. If removed, connect the cruise control actuator.

30. Install the air cleaner hose.

31. Connect the negative battery cable.

32. Start the engine and check for fuel leaks.

33. Install the engine hood side panels.

3.0L (3VZ-FE) Engine

1. Disconnect the negative battery cable. On vehicles equipped with an air bag, wait at least 90 seconds before proceeding.

2. Drain the engine coolant.

3. Disconnect the accelerator cable from the throttle linkage.

4. If equipped with automatic transmission, disconnect the throttle cable from the throttle linkage.

5. Remove the air cleaner cap, air flow meter and the air cleaner hose as a unit.

6. Remove the two 5mm bolts holding the V-cover; remove the cover.

7. Disconnect the EGR temperature connector clamp from the set of emission control valves.

8. Label and remove the hoses from the fuel pressure control VSV. Disconnect the hoses from the IACV, disconnect the VSV wiring connectors and remove the emission control valve set.

9. Label and disconnect the brake booster vacuum hose, PS air hose, PCV hose and IACV vacuum hose.

10. Disconnect the two ground straps.

11. Remove the wiring connector from the cold start injector. Disconnect the fuel line from the cold start injector.

12. Remove the No. 1 engine hanger and the air intake chamber support.

13. Remove the EGR pipe.

14. Remove the bolt and disconnect the hydraulic pressure pipe from the air intake chamber.

15. Disconnect the 3 hoses at the air intake plenum, disconnect the two coolant by-pass hoses and disconnect the EGR temperature sensor connector if so equipped.

16. Disconnect the throttle position sensor connector. Detach the connector for the ISC valve and remove the air hoses from the ISC valve. Remove the PS air hose.

17. Remove the bolts and nuts holding the air plenum; remove the air plenum and gasket.

18. Disconnect the fuel return hoses from the No. 1 fuel pipe; then disconnect the fuel inlet hose from the filter.

19. Disconnect the wiring connectors from each injector.

20. Remove the No. 2 fuel pipe.

21. Remove the left delivery pipe or fuel rail; be careful not to drop the injectors during removal.

22. Remove the 3 injectors from the delivery pipe. Remove the rail spacers from the intake manifold.

23. Disconnect the two air hoses; remove the air pipe with the hoses attached.

24. Remove the right fuel rail and injectors. Take care not to drop an injector. Remove the injectors from the rail.

To install:

25. Install new grommets on each injector.

26. Apply a light coat of clean gasoline to new O-rings and install 2 on each injector.

27. Install each injector into the fuel rail while turning the injector left and right. Once installed, the injector should turn freely in the rail. If not, remove the injector and inspect the O-ring for damage or dislocation.

28. Place the rail spacers on the manifold. Clean the injector ports and install the right rail and injector assembly. Again check that the injectors turn freely in place.

29. Position the injector wiring connector upward. Install the bolts hold-ing the delivery pipe and tighten them to 9 ft. lbs. (13 Nm).

30. Install the air pipe and hoses; tighten the retaining bolts only to 74 inch lbs. (9 Nm).

31. Repeat Steps 28 and 29 to install the left side delivery pipe and injectors.

32. Install the No. 2 fuel pipe connecting the two fuel rails. Use new gaskets at each union bolt. Tighten the union bolts to 25 ft. lbs. (34 Nm).

33. Connect the IACV vacuum hose.

34. Attach the wiring connectors to their proper injectors.

35. Install the inlet hose to the fuel filter using new gaskets; tighten the bolt to 22 ft. lbs. (29 Nm). Connect the return hose to the No. 1 fuel pipe.

36. Using a new gasket, install the air intake chamber. Tighten the mounting nuts to 32 ft. lbs. (43 Nm).

37. Connect the throttle position sensor harness, ISC valve wiring, ISC air hose and PS air hose.

38. Connect the EGR temperature sensor wiring.

39. Install the coolant by-pass hose to the throttle body. Install the coolant by-pass hose to the EGR cooler.

40. Connect the vacuum hoses to the BVSV.

41. Attach the hydraulic pressure pipe to the air intake chamber.

42. Install the EGR pipe with a new gasket and new sleeve ball. Tighten the bolts to 13 ft. lbs. (18 Nm) and the union nut to 58 ft. lbs. (78 Nm).

43. Install the No. 1 engine hanger and the air intake chamber stay. Tighten the bolts to 29 ft. lbs. (39 Nm).

44. Connect the injector pipe with new gaskets to the cold start injector. Tighten the bolts to 11 ft. lbs. (15 Nm). Attach the cold start injector wiring connector.

45. Connect the 2 ground straps.

46. Connect the brake booster vacuum hose, PS air hose, PCV hose and the IACV vacuum hose.

47. Install the emission valve set and tighten the bolts. Connect the VSV connectors and attach the 2 vacuum hoses to the IACV VSV. Install the two hoses to the fuel pressure VSV. Connect the EGR temperature sensor connector clamp to the valve set.

48. Install the V-bank cover on the engine.

49. Install the air cleaner cover, air flow meter and air hose as a unit. Make certain the clips are correctly engaged.

50. Connect and adjust the throttle control cable if it was removed.

51. Connect the accelerator cable and adjust it as needed.

52. Refill the engine coolant.

53. Connect the negative battery cable.

54. Start the engine and check for leaks.

3.0L (1MZ-FE) Engine

1. Disconnect the negative battery cable from the battery. On vehicles equipped with an air bag, wait at least 90 seconds before proceeding.

2. Relieve the fuel pressure from the fuel lines.

───── **CAUTION** ─────

Fuel injection systems remain under pressure after the engine has been turned OFF. Properly relieve fuel pressure before disconnecting any fuel lines. Failure to do so may result in fire or personal injury.

3. Remove the air cleaner hose from the engine compartment.

4. Remove the V–bank cover from the engine.

5. Remove the air cleaner chamber assembly as follows:

a. Drain the engine coolant from the vehicle.

b. Disconnect the following connectors and cables:

- Accelerator cable
- A/T throttle cable
- TPS connector
- IAC valve connector
- EGR gas temperature sensor connector
- A/C idle up valve connector
- VSV connector for the ACIS
- VSV connector for the fuel pressure control
- For California vehicles, disconnect the VSV for the EVAP
- VSV connector for the EGR
- DLC1 from the bracket on the intake air control valve

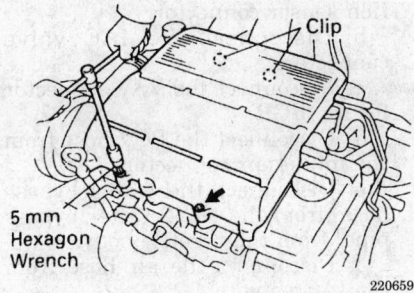

Using a 5mm wrench, remove the V-bank cover — 1MZ-FE engine

c. Remove the power steering pressure tube from the No. 1 engine hanger by removing the two bolts.

d. Disconnect the following hoses, clamps and cables:

- Brake booster vacuum hose from the intake air control valve for the ACIS
- PCV hose from the PCV valve on the right hand cylinder head
- Ground strap and cable from the air intake air control valve from the ACIS
- Ground cable from the air intake chamber
- Vacuum hose clamp from fuel pipe
- Two bypass hoses from the throttle body
- Air assist hose from the throttle body
- Two power steering air hoses to the air intake chamber.
- For California vehicles, remove the EVAP hose from the pipe on emission control valve set
- Two vacuum hoses from the pipes on the cylinder head rear plate
- Vacuum sensing hose from the fuel pressure regulator
- Engine wire clamp from emission control valve set
- Except California, remove the two EVAP hoses (from TVV) from the pipe on the throttle body
- Except California, remove the EVAP hose from hose clamp
- Except California, remove the EVAP hose (from charcoal canister) from the pipe on the throttle body

e. Remove the two bolts and the No. 1 engine hanger.

f. Remove the two bolts and the air intake chamber stay.

g. Remove the No. 2 EGR pipe and two gaskets by removing the four nuts.

h. Disconnect the hose from the VSV fro the EVAP.

i. Remove the two bolts, the two nuts, and the air intake chamber assembly, and the gasket.

6. Disconnect the fuel injector connectors.

7. Remove the air assist hoses and pipe.

8. Disconnect the fuel return hose from the No. 1 fuel pipe.

9. Disconnect the fuel inlet hose from the fuel filter. Catch any fuel leaking from the filter in a shop rag. Dispose of the rag properly.

10. Remove the delivery pipes and injectors from the engine as follows:

a. Loosen the two union bolts holding the No. 2 fuel pipe to the delivery pipes.

b. Disconnect the fuel return hose from the fuel pressure regulator.

c. Remove the union bolt and two gaskets for the right hand delivery pipe.

d. Remove the two bolts to the left hand delivery pipe and then remove the left hand delivery pipe, three injectors and the No. 2 fuel pipe as an assembly.

e. Remove the union bolt and two gaskets from the left hand delivery pipe. Disconnect the No. 2 fuel pipe from the left hand delivery pipe.

f. Remove the right hand delivery pipe by removing the three bolts. Remove the delivery pipe, injectors, and the fuel inlet hose as an assembly.

g. Remove the four spacers from the intake manifold.

h. Pull out the six injectors from the delivery pipes.

i. Remove the two O-rings and two grommets from each injector.

To install:

11. Install the injectors as follows:

a. Install two new grommets to each injector.

b. Apply a light coat of spindle oil or gasoline to the two O-rings and install them to each injector.

c. Install the injector into the delivery pipe by turning the injector back and forth. Install all the injectors into the delivery pipes. Make sure to position the injector electrical connector outward.

d. Place the four spacers in position on the intake manifold.

e. Place the right hand delivery pipe and the No. 1 fuel pipe together with the three injectors in position on the intake manifold.

f. Temporarily install the two bolts holding the right hand delivery pipe to the intake manifold.

g. Temporarily install the bolt holding the No. 1 fuel pipe to the intake manifold.

h. Place the left hand delivery pipe and the No. 2 fuel pipe together with the three injectors in position.

i. Connect the fuel return hose to the fuel pressure regulator.

j. Temporarily install the two bolts holding the left hand delivery pipe to the intake manifold.

k. Temporarily install the No. 2 fuel pipe to the left hand delivery pipe with the union bolts and two new gaskets.

l. Check that the injectors rotate smoothly. If the injectors do not rotate smoothly, the probable cause is incorrect installation of the O-rings. Replace the O-rings.

m. Tighten the four bolts holding the delivery pipes to the intake manifold. Tighten the bolts to 7 ft. lbs. (10 Nm).

n. Tighten the bolt holding the No. 1 fuel pipe to the intake manifold. Tighten the bolt to 14 ft. lbs. (20 Nm).

o. Tighten the two union bolts holding the No. 2 fuel pipe to the delivery pipes. Tighten the bolts to 24 ft. lbs. (33 Nm).

12. Connect the fuel inlet hose to the fuel filter by installing the union bolt. Use two new gaskets when installing the union bolt.

13. Connect the fuel return hose to the No. 1 fuel pipe. When routing the fuel return hose, pass the hose under the heater hoses.

14. Connect the air assist hoses to the intake manifold and then install the air assist pipe to the bracket on the No. 1 fuel pipe.

15. Connect the injector connectors.

16. Install the air intake chamber assembly.

17. Install the power steering pressure tube with the two nuts.

18. Connect the remaining components in the reverse order they were removed.

19. Fill the engine coolant.

20. Connect the negative battery cable to the battery and start the engine.

21. Check the engine coolant level and check for leaks.

2.0L (3S-GTE) Engine

1. Relieve the fuel system pressure.

2. Disconnect the negative battery cable. On vehicles equipped with an air bag, wait at least 90 seconds before proceeding.

3. Remove the throttle body.

4. Remove the cold start injector.

5. Remove the vacuum switching valve (VSV) and the EGR vacuum modulator.

6. Remove the vacuum hoses from the EGR vacuum modulator.

7. Remove the EGR valve.

8. Remove the water bypass hose and the air hose.

9. Disconnect the vacuum sensing hose from the vacuum pipe on the injector cover.

10. Disconnect the engine wire protector between the No. 3 timing belt cover and the cylinder head cover.

11. Disconnect the fuel injector connectors.

───── CAUTION ─────

Fuel injection systems remain under pressure after the engine has been turned OFF. Properly relieve fuel pressure before disconnecting any fuel lines. Failure to do so may result in fire or personal injury.

───────────────────

12. Disconnect the fuel return hose from the fuel pressure regulator.

13. Loosen the union bolt holding the fuel inlet hose to the delivery pipe.

14. Loosen the bolts holding the delivery pipe to the cylinder head and remove it.

15. Remove the insulators and spacers from the cylinder head.

16. Remove the injectors from the delivery pipe by using special tool 09268-74010 or equivalent.

To install:

17. Install a light coat of gasoline to a new O-ring and install it on the injector.

18. Install a new grommet to the injector and while turning the injector left and right, install it to the delivery pipes.

19. Place new insulators and spacers in position on the cylinder head.

20. Place the delivery pipe with the injectors in position on the cylinder head.

21. Temporarily install the bolts holding the delivery pipe and check that the injectors rotate smoothly.

22. Position the injector connector upward and tighten the fuel rail bolts to 9 ft. lbs. (13 Nm).

23. Connect the fuel inlet pipe and the return hose. Tighten the union bolt to 22 ft. lbs. (29 Nm).

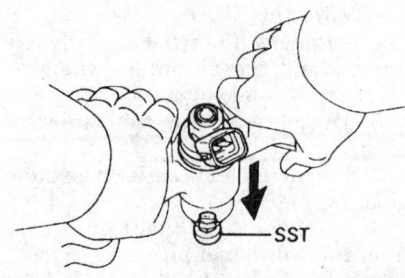

247561

Injector removal — 1995 MR2 3S-GTE engine

24. Connect the four injector connectors.

NOTE: The No. 1 and No. 3 injector connectors are brown, and the No. 2 and No. 4 connectors are gray.

25. Connect the remaining components in the reverse order they were removed.

26. Connect the negative battery cable.

27. Start the engine and check for fuel leaks.

3.0L (2JZ-GE) Engine

1. Relieve the fuel pressure from the fuel lines.

───── CAUTION ─────

Fuel injection systems remain under pressure even after the engine has been turned OFF. The fuel system pressure must be relieved before disconnecting any fuel lines. Failure to do so may result in fire and/or personal injury.

───────────────────

2. Disconnect the negative battery cable. Wait at least 90 seconds before performing any other work.

3. Drain the engine coolant.

4. Remove the throttle body along with the intake air connector.

5. Remove the VSV for the fuel pressure control.

6. Disconnect the EGR pipe by removing the two bolts and loosening the union nut.

7. Disconnect the EGR gas temperature sensor wiring connector.

8. Disconnect the connector at the No. 2 vacuum pipe and intake manifold.

9. Disconnect the throttle body bracket from the throttle body and cylinder head.

10. Remove the throttle body and intake air connector assembly as follows:

a. Disconnect the throttle position sensor connector.

b. Disconnect the IAC valve connector.

c. Disconnect the VSV connector for the EGR.

d. Disconnect the PCV hose from the intake air connector.

e. Disconnect the water bypass hose (from the No. 2 water bypass pipe) from the throttle body.

f. Disconnect the air hose from the IAC valve.

g. Remove the nut holding the VSV for the ACIS to the intake chamber.

h. Remove the four bolts and two nuts holding the intake air connector to the air intake chamber.

i. Disconnect the three vacuum hoses (from the No. 2 vacuum pipe) from the No. 1 vacuum pipe.

j. Disconnect the water bypass hose (from the water outlet) from the throttle body.

k. Disconnect the vacuum hose (from the actuator for the ACIS) from the No. 1 vacuum pipe.

l. Remove the throttle body, intake air connector assembly and gasket from the engine.

11. Remove the air intake chamber stays.

12. Disconnect the fuel inlet pipe from the delivery pipe.

13. Disconnect the fuel return pipe from the fuel pressure regulator.

14. Disconnect the fuel injector connectors.

15. Disconnect the actuator (for the ACIS) from the air intake chamber.

16. Remove the delivery pipe together with the six injectors by removing the three bolts.

17. Remove the three spacers from the intake manifold.

18. Pull out the six injectors from the delivery pipe.

19. Remove the two O-rings, insulator and grommet from each injector.

20. Remove the O-rings and grommet from each injector.

To install:

21. Install a new insulator and grommet to each injector.

22. Apply a light coat of gasoline to the two O-rings and install them to each injector.

23. While turning the injector back and forth, push the injector onto the delivery pipe. Install all six injectors.

24. Position the injector electrical connector outward.

25. Install the three spacers and except for California, install the six insulators to the intake manifold.

26. Install the three bolts to the delivery pipe.

27. Attach the six injectors and delivery pipe to the intake manifold.

28. Temporarily install the three bolts holding the delivery pipe to the intake manifold. Do not torque the bolts at this time.

29. Check that the injectors rotate smoothly and then position each injector connector upward.

30. Tighten the three injector bolts to 15 ft. lbs. (21 Nm).

31. Install the actuator with the two bolts. Tighten the bolt to 61 inch lbs. (6.8 Nm).

32. Connect the injector electrical connectors.

33. Connect the fuel inlet pipe to the delivery pipe and torque the union bolt to 30 ft. lbs. (41 Nm).

34. Connect the fuel return pipe to the fuel pressure regulator. Tighten the union bolt to 20 ft. lbs. (27 Nm).

35. Connect the vacuum sensing hose to the fuel pressure regulator.

36. Install the air intake chamber stays and torque the bolts and nuts to 13 ft. lbs. (18 Nm).

37. Install the throttle body and intake air connector assembly in the reverse order they were removed.

38. Install the throttle body bracket and torque to 15 ft. lbs. (21 Nm).

39. Install the No. 2 vacuum pipe and torque to 20 ft. lbs. (27 Nm).

40. Connect the EGR gas temperature sensor connector.

41. Install the EGR pipe and torque the two bolts to 20 ft. lbs. (27 Nm). Tighten the union nut for the EGR pipe to 47 ft. lbs. (64 Nm).

42. Install the VSV for the fuel pressure control.

43. Install the intake air connector pipe.

44. Connect the control cables to the throttle body.

45. Fill the engine coolant.

46. Connect the negative battery cable to the battery.

47. Start the engine and check for fuel leaks.

3.0L (2JZ-GTE) Engine

1. Remove the engine undercover.
2. Relieve the fuel pressure in the fuel line before disconnecting any fuel lines.

─────── **CAUTION** ───────
Fuel injection systems remain under pressure, even after the engine has been turned OFF. The fuel system pressure must be relieved before disconnecting any fuel lines. Failure to do so may result in fire and/or explosion.

3. Disconnect the negative battery cable to the battery. On vehicles equipped with an air bag, wait at least 90 seconds before proceeding.

4. Remove the throttle body as follows:

a. Drain the engine coolant from the radiator and engine.

b. Disconnect the following hose, cable and connectors from the throttle body:
   • Air hose
   • Accelerator cable
   • Cruise control actuator cable
   • Throttle position sensor connector

• Sub throttle position sensor connector
• Sub throttle actuator connector

c. Disconnect the throttle body from the air intake chamber by removing the two bolts and two nuts.

d. Remove the gasket from the throttle body.

e. Disconnect the following hoses from the throttle body and remove the throttle body from the engine:
   • EVAP hose
   • Water bypass hose (from the No. 4 water bypass pipe)
   • Power steering air hose

5. For vehicles with automatic transmissions, remove the transmission dipstick and guide.

6. Remove the oil dipstick and guide for the engine.

7. Remove the air intake chamber stay by removing the nut and bolts.

8. Disconnect the control cable bracket from the air intake chamber by removing the two bolts.

9. Disconnect the following connectors and hose:
   • IAC valve connector
   • Turbo pressure sensor connector
   • VSV connector for the fuel pressure control
   • VSV connector for the EGR valve

10. Remove the bolt and disconnect the engine wire protector from the vehicle body.

11. Disconnect the following hoses:
   • Disconnect the IAC valve pipe from the clamp on the cylinder head cover and disconnect the air hose form the IAC valve
   • Disconnect the air hose (from the air intake chamber) from the vacuum pipe on the IAC valve pipe
   • Air hose for the EGR from the valve pipe
   • PCV hose from the PCV valve
   • Vacuum sensing hose from the fuel pressure regulator
   • Water bypass hose (from the IAC valve) from the No. 4 water bypass pipe
   • EVAP hose (from the air intake chamber) from the vacuum pipe on the manifold stay
   • EVAP hose (from the vacuum pipe on the No. 4 water bypass pipe) from the No. 2 vacuum pipe
   • EVAP hose (from the charcoal canister) from the No. 2 vacuum pipe
   • Power steering air hose from the air intake chamber
   • Brake booster vacuum hose from the union on the air intake chamber

12. Disconnect the EGR gas temperature sensor connector from the No. 2 vacuum pipe and wiring connector.
13. Remove the EGR pipe.
14. Remove the No. 4 water bypass pipe by removing the two bolts.
15. Remove the intake manifold stay by removing the two bolts.
16. Remove the air intake chamber assembly as follows:
    a. Disconnect the ground cable from the intake manifold by removing the bolt.
    b. Remove the two bolts holding the engine wire protector to the intake manifold.
    c. Disconnect the two clamps for the engine wire protector from the brackets.
    d. Remove the five bolts, two nuts, and the engine wire bracket.
    e. Disconnect the air intake air chamber assembly from the intake manifold.
    f. Disconnect the water bypass hose from the IAC valve.
    g. Remove the gasket.
17. Disconnect the six injector connectors.
18. Disconnect the two camshaft position sensor connectors.
19. Disconnect the three engine wire clamps from the injector holders.
20. Disconnect the fuel inlet pipe from the delivery pipe.
21. Disconnect the fuel return pipe from the fuel pressure regulator.
22. Remove the delivery pipe and injectors as follows:
    a. Remove the six bolts and three injector holders.
    b. Remove the six insulators from the injectors.
    c. Remove the two bolts, delivery pipe, and the six injectors assembly.
    d. Remove the two spacers from the intake manifold.
    e. Remove the six insulators from the delivery pipe.

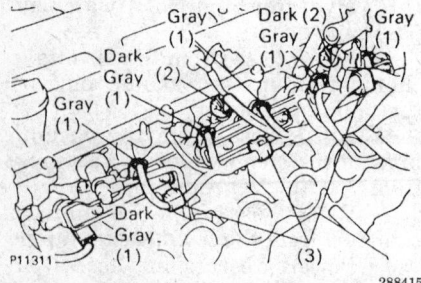

Fuel injector connectors — Supra 2JZ-GTE engine

To install:
23. Install the delivery pipe and injectors as follows:
    a. Install the six insulators to the delivery pipe.
    b. Install the two spacers to the intake manifold.
    c. Install the six injectors, delivery pipe, and the two bolts.
    d. Install the six insulators to the injectors.
    e. Install the three injector holders and six bolts.
24. Connect the fuel return pipe to the fuel pressure regulator and torque the union bolts to 20 ft. lbs. (27 Nm).
25. Connect the fuel inlet pipe to the delivery pipe and torque the union bolt to 30 ft. lbs. (41 Nm).
26. Connect the three engine wire clamps to the injector holders.
27. Install the two camshaft position sensor connectors.
28. Install the six injector connectors. The No. 1, 3, and 5 injector connectors are dark gray, and the No. 2, 4, and 6 injector connectors are gray.
29. Install the remaining components in the reverse order they were removed.
30. Fill engine coolant.
31. Install the engine undercover.
32. Connect the negative battery cable to the battery.
33. Start the engine and check for leaks. Bleed the cooling system.

ENGINE MECHANICAL

Engine Assembly

REMOVAL AND INSTALLATION

1.5L (3E-E) Engine

Tercel

NOTE: Each fuel and vacuum line must be tagged or labeled individually during disassembly for correct installation.

1. Disconnect the negative battery cable. On vehicles equipped with an air bag, wait at least 90 seconds before proceeding.
2. Drain the cooling system and save the coolant for reuse.
3. Drain the engine oil and the transmission oil.
4. With the help of an assistant, remove the hood (mark hood hinges for correct installation) from the car. Be careful not to damage the paint finish.
5. Remove the air cleaner assembly from the carburetor.
6. Disconnect the upper radiator hose from the engine and remove the overflow hose.
7. Remove the coolant hose at the cylinder head rear coolant pipe and remove the coolant hose at the thermostat housing.

NOTE: On the 3E-E engine, the fuel system is under pressure. Release pressure slowly and contain spillage. Observe no smoking/no open flame precautions. Have a Class B-C (dry powder) fire extinguisher within arm's reach at all times.

8. Remove the fuel hoses from the fuel pump.
9. Loosen the adjuster(s) and remove the alternator belt, the power steering and/or air conditioning drive belts depending on equipment.
10. Label and remove all wiring running to the motor. Be careful when unhooking wiring connectors; many have locking devices which must be released.
11. Label and disconnect vacuum hoses. Make sure your labels contain accurate information for reconnecting both ends of the hose. Make sure the labels will stay put on the hoses.
12. Disconnect the wiring at the transaxle.
13. Disconnect the speedometer cable at the transaxle.
14. Safely elevate and support the vehicle on jackstands.
15. Disconnect the exhaust pipe from the manifold. Be ready to deal with rusty hardware.
16. Disconnect the air hose at the converter pipe, if so equipped.
17. Loosen and remove the transaxle cooler lines at the radiator.
18. Remove the left and right undercovers (splash shields) under the car.
19. If so equipped, remove the power steering pump from its mounts and lay it aside. Leave the hoses attached. The pump may be hung on a piece of stiff wire to be kept out of the way.
20. If so equipped, remove the air conditioning compressor from its mounts and position out of the way. DO NOT loosen any hoses or fittings simply move the compressor out of the way. It may be hung from a piece of stiff wire to be kept out of the way.

21. Disconnect the cable and bracket from the transaxle.

22. Disconnect the steering knuckles at the lower control arms.

23. Have an assistant step on the brake pedal while you loosen the nuts and bolts holding the driveshafts to the transaxle. Let the disconnected shafts hang down clear of the transaxle. Remove the driveshaft assembly to the rear differential on 4-Wheel Drive vehicles.

24. Remove the flywheel cover. If equipped with an automatic transmission, remove the flexplate-to-torque converter bolts.

25. Disconnect the front and rear mounts at the crossmember by first removing the two bolt covers and removing the two bolts at each mount. Remove the center crossmember under the engine.

26. Remove the clip and washer holding the shift cable; disconnect the cable at the outer shift lever or outer selector lever.

27. Lower the vehicle to the ground. Remove the radiator and fan assembly.

28. Install the engine hoist to the lifting bracket on the engine. Keep the wiring harness in front of the chain. Draw tension on the hoist enough to support the engine but no more. Double check the hoist attachments before proceeding.

29. Remove the through-bolt to the right side engine mount.

30. Remove the left side transaxle mount bolt and remove the mount.

31. Lift the engine and transaxle assembly out of the engine compartment, proceeding slowly and watching for any interference. Pay particular attention to not damaging the right side engine mount, the power steering housing and the neutral safety switch. Make sure wiring, hoses and cables are clear of the engine.

32. Support the engine assembly on a suitable stand; do not allow it to remain on the hoist for any length of time.

33. To install lower the engine and transaxle into the car, paying attention to clearance and proper position. Raise and lower the vehicle as necessary to perform each service operation.

34. Install the left side transaxle mount and its bolt.

35. Install and tighten the right side through-bolt for the motor mount.

36. When the engine is securely mounted within the car, the lifting devices may be removed. Replace the radiator and fan assembly.

37. Safely elevate and support the vehicle on jackstands. Install the center crossmember and connect the shift cable.

38. Connect the front and rear mountings for the crossmember and reinstall the bolt covers.

39. Install the flywheel-to-torque converter bolts. Tighten them to the final specification in steps.

40. Replace the flywheel cover.

41. Connect the driveshafts to the transaxle. Install the driveshaft assembly to the rear differential on 4-Wheel Drive vehicles.

42. Reconnect the steering knuckles to the lower control arms.

43. Reattach the cable and bracket to the transaxle.

44. Depending on equipment, reinstall the power steering pump and/or the air conditioning compressor. Tighten the mounting bolts enough to hold the unit in place but no more; the belts will be installed later.

45. Install the left and right splash shields (undercovers).

46. Reconnect the transmission cooler lines to the radiator and connect the air hose at the converter pipe.

47. Connect the exhaust pipe to the manifold.

48. Lower the vehicle to the ground; reconnect the speedometer cable at the transaxle.

49. Paying close attention to proper routing and labeling, connect the wiring and vacuum hoses to the engine.

50. Install the drive belts (alternator, power steering and air conditioning) and make sure the belts are properly seated on the pulleys. Adjust the belts to the correct tension and tighten the bolts.

51. Connect the fuel hoses at the fuel pump. Use new clamps if necessary.

52. Attach the coolant hoses: at the thermostat housing, at the cylinder head rear pipe, at the overflow, and at the outlet for the upper hose. Insure that the hoses are firmly over the ports; use new clamps wherever needed.

53. Install the air cleaner assembly onto the carburetor.

54. Have an assistant help reinstall the hood. Make sure its is properly adjusted and secure.

55. Refill the transmission with the proper fluid.

56. Refill the engine oil to the proper level.

57. Refill the engine coolant with the proper amount of fluid.

58. Double check all installation items, paying particular attention to loose hoses or hanging wires, untightened nuts, poor routing of hoses and wires (too tight or rubbing) and tools left in the engine area.

59. Check all fluid levels. Bleed systems as necessary. Make all necessary adjustments. Start the engine. Check for any fluid leaks, road test the vehicle for proper operation.

1.5L (5E-FE) Engine

Tercel and Paseo

1. Disconnect the negative battery cable from the battery. On vehicles equipped with an air bag, wait at least 90 seconds before proceeding.

2. Remove the hood from the vehicle.

3. Remove the undercovers from the vehicle.

4. Drain the cooling system and remove the radiator.

5. If equipped with automatic transaxle, disconnect and plug the transaxle fluid lines and disconnect the accelerator cable.

6. Disconnect the throttle cable.

7. Drain the transaxle fluid and the engine oil.

8. Remove the air cleaner assembly and bracket.

9. Remove the charcoal canister.

CAUTION

To avoid personal injury, properly release the fuel pressure on any fuel injected model before disconnecting any fuel lines.

10. Disconnect the fuel return and inlet hoses.

11. Disconnect the speedometer cable.

12. Disconnect the idle up air hoses from the power steering air control valve.

13. Disconnect and label the oxygen sensor wire, the oil pressure switch wire, the coolant fan switch wire, the water temperature gauge wire, the back-up light switch and neutral safety switch wires.

14. Disconnect and tag any remaining wiring harnesses connected to the engine. Remove the wire harness from the engine.

15. Disconnect and tag the PCV hoses and any other vacuum hoses that prevent the removal of the engine.

16. Disconnect and tag the starter wires.

17. If equipped with cruise control, remove the actuator assembly.

18. Disconnect and tag the heater hoses.

19. If equipped with manual transaxle, remove the clutch release cylinder.

20. Remove the clips and washers, then disconnect the transaxle control cables.

21. If equipped with power steering, remove the power steering pump and position it aside. Do not disconnect the power steering hoses.

22. If equipped with air conditioning, remove the air conditioning compressor and position it aside. Leave the refrigerant lines connected.

23. Disconnect the front exhaust pipe.

24. Remove the halfshafts.

25. Support the engine/transaxle assembly properly.

26. Connect a suitable lifting device to the engine lifting hooks.

27. If equipped with manual transaxle, remove the rear mounting through-bolt and the rear mounting assembly. If equipped with automatic transaxle, remove the front mounting through-bolt and front mounting assembly.

28. Remove the right and left side mounting bolts and brackets.

29. Carefully lift the engine/transaxle assembly out of the vehicle.

30. Disconnect the starter from the engine/transaxle by removing the starter stay bracket and then the two bolts to the starter.

31. For automatic transaxles, disconnect the torque converter clutch mounting bolts.

32. Separate the engine from the transaxle by removing the bolts.

To install:

33. Attach the transaxle to the engine and install the bolts. On automatic transaxles, install the torque converter clutch and mounting bolts. Install the gray bolt first, then install the other five bolts. Torque the bolts to 20 ft. lbs. (27 Nm).

34. Install the starter with the two bolts and torque the bolts to 29 ft. lbs. (39 Nm).

35. Attach an engine sling to the engine hangers and lower the engine and transaxle assembly in the vehicle.

36. Tilt the transaxle downward, lower the engine and clear the left side mounting.

37. Keep the engine level and align the right side and left side mountings with the body bracket.

38. Attach the right side mounting insulator to the mounting bracket and the body and temporarily install the through-bolt, two bolts and nut.

39. Install the left side mounting bracket to the transaxle and mounting insulator with the 5 bolts and torque the bolts. Torque the bracket to transaxle bolts with the head marked (NT) to 47 ft. lbs. (64 Nm), and the bracket to insulator bolts with the bolt head marked (7T) to 35 ft. lbs. (48 Nm).

40. Connect the ground strap. Torque the ground strap to 35 ft. lbs. (49 Nm).

41. Install the rear mounting bracket to the transaxle with the 3 bolts (M/T) or 2 bolts (A/T) and torque to 35 ft. lbs. (48 Nm).

42. Install and torque the rear insulator through-bolt to 47 ft. lbs. (64 Nm).

43. Install the halfshafts.

44. Remove the engine sling from the engine.

45. Install the two bolts, nut and through-bolt of the right side mounting insulator. Torque the 2 bolts and nut to 47 ft. lbs. (64 Nm) and the through-bolt to 54 ft. lbs. (73 Nm).

46. Connect the front exhaust pipe.

47. If equipped with air conditioning, install the air conditioning compressor.

48. If equipped with power steering, install the power steering pump to the engine.

49. Install the transaxle control cables with the washers and clips.

50. If equipped with manual transaxle, install the clutch release cylinder. Torque the bolts to 9 ft. lbs. (12 Nm).

51. Connect the idle up air hoses to the power steering air control valve.

52. Connect the speedometer cable to the transaxle.

53. Connect the heater hoses.

54. If equipped with cruise control, connect the actuator assembly.

55. Connect the starter wires and nut.

56. Connect vacuum hoses to the engine.

57. Connect the oxygen sensor wire, oil pressure switch, coolant fan switch wire, water temp. gauge wire, back-up light switch, and the neutral safety switch wires.

58. Connect any other wiring to the engine.

59. Connect the fuel line hoses.

60. Connect the charcoal canister.

61. Install the air cleaner assembly and bracket.

62. Install the radiator. Connect the coolant hoses and on vehicles equipped with A/T, connect the cooling lines.

63. Refill all fluids to specifications.

64. Install the engine undercovers.

65. Install the hood.

66. Connect the negative battery cable to the battery.

67. Start the engine and check for leaks.

1.6L (4A-FE) and 1.8L (7A-FE) Engines
Corolla

1. Relieve the fuel system pressure.

2. Disconnect the negative battery cable. On vehicles equipped with an air bag, wait at least 90 seconds before proceeding.

3. Remove the battery.

4. Remove the hood.

5. Remove the engine undercover and then drain the engine coolant and oil.

6. Drain the transaxle assembly.

7. Disconnect the accelerator cable from the accelerator bracket.

8. If equipped with A/T, disconnect the throttle cable from the accelerator cable.

9. Remove the radiator assembly with the cooling fan.

10. Remove the air cleaner assembly.

11. Remove the two bolts and the coolant reservoir tank stay.

12. Disconnect the electrical connector, the hose, the mounting bolt, and remove the washer tank.

13. On models with cruise control, remove the actuator cover, unplug the connector, remove the three bolts, and then disconnect the actuator with the bracket.

14. Disconnect or remove the following components:

   a. The MAP sensor vacuum hose from the gas filter on the intake manifold

   b. The brake booster vacuum hose from the intake manifold

   c. With A/C: the A/C vacuum hose from the actuator

   d. With P/S: the air hose from the air pipe

   e. With A/C: the A/C actuator connector

15. Disconnect the following wires and connectors from the RH fender apron:

   a. The ground strap connector

   b. The MAP sensor connector

   c. With A/C: the A/C pressure switch

   d. The engine wire harness from the fender apron

16. Disconnect the DLC 1 connector and ground strap from the LH fender apron.

17. Remove the two bolts and remove the engine relay box.

18. Disconnect the four connectors from the engine relay box.

19. Disconnect the hose from the charcoal canister and remove the canister from the bracket.

20. Disconnect the two heater hoses from water inlet housing.

———— CAUTION ————

Fuel injection systems remain under pressure after the engine has been turned OFF. Properly relieve fuel pressure before disconnecting any fuel lines. Failure to do so may result in fire or personal injury.

21. Disconnect the fuel inlet and the fuel return hoses.

22. On M/T model, remove the three bolts and remove the clutch release cylinder without disconnecting the pipe.

23. Disconnect the transaxle control cable(s) from the transaxle.

24. Disconnect and remove the following components to disconnect the engine wire harness:

    a. The LH and RH front door scuff plate

    b. The lower finish panel

    c. The lower panel with the glove compartment

    d. The radio and center cluster finish panel

    e. The rear console box

    f. M/T: the shift lever knob an on A/T: the shifting hole bezel

    g. The lower center finish panel

    h. The floor carpet bracket

    i. The three ECM connectors and cowl wire connector

25. Pull out the wire harness from the cowl.

26. If equipped w/ A/C, disconnect the A/C compressor connector. Remove the drive belt and remove the four mounting bolts and the A/C compressor. Hint: securely hang the compressor out of the way.

27. Disconnect the front exhaust pipe from the exhaust manifold.

28. Remove the halfshafts from the transaxle.

29. If equipped with P/S, remove the drive belt and remove the two mounting bolts and the P/S pump. Hint: securely hang the pump out of the way.

30. Remove the engine mounting center member.

31. Remove the through–bolt and nut holding the mounting insulator to the mounting bracket.

32. Remove the engine and transaxle assembly from the vehicle.

33. Lift the engine and transaxle out of the vehicle slowly and carefully.

34. Remove the two bolts and the front engine mounting bracket.

35. Remove the three bolts (except the A131L A/T) or the two bolts on the A131L A/T; remove the mounting bracket.

36. Remove the starter and separate the transaxle from the engine.

To install:

37. Assemble the engine to the transaxle.

38. Install the starter.

39. Install the rear engine mounting bracket with two bolts on the A131L A/T or the three bolts (except the A131L A/T). The bolts are torqued to 57 ft. lbs. (77 Nm).

40. Install the front engine mounting bracket to the transaxle and torque the two bolts to 57 ft. lbs. (77 Nm).

41. Install the engine and transaxle assembly into the vehicle.

42. Install the engine mounting center member.

43. Tighten the through–bolt and nut holding the front engine mounting insulator to the mounting bracket. The torque is 64 ft. lbs. (87 Nm).

44. Install the halfshafts.

45. Install the front exhaust pipe.

46. If equipped with P/S, install the pump with the two bolts and torque to 29 ft. lbs. (39 Nm). Install the drive belt.

47. If equipped with A/C, install the compressor with the four mounting bolts and torque the bolts to 18 ft. lbs. (25 Nm). Install the drive belt and reconnect the connector.

48. To install and connect the engine wire harness, perform the following:

    a. Push the wire through the cowl

    b. Connect the three ECM connectors

    c. Attach the cowl wire connector

    d. Install the floor carpet bracket

    e. Install the center lower finish panel

    f. With A/T, install the shifting hole bezel, with M/T, install the shift lever knob

    g. Install the rear console box

    h. Install the center cluster finish panel and the radio

    i. Install the lower panel with the glove compartment door

    j. Install the RH and LH door scuff plate

    k. Install the lower finish panel

49. If equipped with M/T, install the clutch release cylinder with the tube and the three bolts.

50. Connect the transaxle control cable(s) to the transaxle.

51. Connect the fuel return hose and the fuel inlet hose. Torque the bolt to 22 ft. lbs. (29 Nm).

52. Connect the two heater hoses to the water inlet housing.

53. Connect the hose to the charcoal canister and install the canister to the bracket.

54. Connect the following wires and connectors on the LH fender apron:

    a. The four connectors to the engine relay box

    b. Install the engine relay box with the two bolts

    c. The DLC1 connector

    d. The connector from the fender apron

    e. The ground strap from the fender apron

55. Connect the following wires and connectors on the RH fender apron:

    a. The ground strap connector

    b. The MAP sensor connector

    c. With A/C, the A/C pressure switch

    d. The engine wire from the fender apron

56. Connect the following hoses and connectors:

    a. With A/C, the A/C actuator connector

    b. With P/S, the air hoses to the air pipe

    c. The vacuum hose from the MAP sensor to the gas filter to the intake chamber

    d. The brake booster vacuum hose to the air intake chamber

    e. With A/C, the A/C vacuum hose from the actuator

57. If equipped with cruise control, install the actuator and bracket with the three bolts. Connect the connector, connect the actuator cable to the actuator, and install the cover.

58. Install the connector and the vinyl hose. Install the washer tank with the bolt.

59. Install the coolant reservoir tank stay with the bolts.

60. Install the air cleaner assembly.

61. Install the radiator with the cooling fan.

62. If equipped with A/T, connect the throttle cable and adjust.

63. Install the accelerator cable and adjust it.

64. Fill the radiator with engine coolant.

65. Fill the engine with oil.

66. Fill the transaxle assembly with oil.

67. Connect the negative battery cable, start the engine, and check for leaks.

68. Perform engine adjustments, install the engine undercovers, and install the hood.

69. Road test the vehicle for proper operation.

70. Recheck all fluid levels.

Celica

1. Release the fuel system pressure.

2. Disconnect the negative battery cable. On vehicles equipped with an air bag, wait at least 90 seconds before proceeding.

3. Remove the battery.

4. Remove the hood.

5. Remove the engine undercover and then drain the engine coolant and oil.

6. Drain the oil from the transaxle.

7. Disconnect the accelerator cable from the throttle body, cable bracket, and the clamps.

8. Remove the air cleaner assembly.

9. Disconnect the cruise control actuator cable from the clamps.

10. Remove the radiator assembly.

---- CAUTION ----

Fuel injection systems remain under pressure after the engine has been turned OFF. Properly relieve fuel pressure before disconnecting any fuel lines. Failure to do so may result in fire or personal injury.

11. Disconnect or remove the following components:
- The MAP sensor vacuum hose from the gas filter on the intake manifold
- The P/S air hose from the intake manifold
- The P/S hose from the air the air pipe
- The brake booster vacuum hose from the intake manifold
- The A/C idle–up valve
- The A/C idle–up valve hose from the intake manifold
- The A/C idle–up valve hose from the air pipe
- The DLC1 from the bracket
- The engine wire harness from the bracket
- The ground cable from the body and the ground strap from the body
- The two heater hoses from the water outlet
- The heater hose from the water bypass pipe
- The fuel inlet hose from the fuel filter and the fuel return hose from the return pipe
- The EVAP hose from the charcoal canister

12. Disconnect the two connectors, remove the two relay box covers, and disconnect the engine wire harness from the engine compartment relay box.

13. To remove the engine wire harness from the vehicle cabin, remove or disconnect the following:
- The scuff plate
- The cowl side trim
- The finish panel from the lower instrument panel
- Remove the front side of the floor carpet
- The wire harness from the clamp of the ECM bracket
- The three ECM connectors
- The circuit opening relay connector
- The three connectors from the connectors on the bracket
- The A/C amplifier connector
- The MAP sensor connector
- The MAP sensor wire from the clamp on the bracket
- The wire clamp from the bracket
- The two nuts holding the engine wire harness to the cowl

14. Remove the front exhaust pipe.

15. Remove the halfshafts.

16. Remove the alternator drive belt.

17. Remove the A/C drive belt, disconnect the A/C compressor connector, remove the four bolts, and the A/C compressor. Do not disconnect the A/C lines and position the compressor safely out of the way.

18. Remove the drive belt and remove the four bolts that secure the P/S pump. Without disconnecting the lines, securely hang the pump out of the way.

19. Remove the two bolts and disconnect the A/C relay box from the body.

20. On M/T equipped vehicles, remove the bolt and disconnect the bracket from the transaxle. Remove the two mounting bolts and the clutch release cylinder from the transaxle.

21. Disconnect the transaxle control cable(s) from the transaxle.

22. On A/T equipped vehicles, remove the two bolts and disconnect the transaxle control cable from the engine mounting center member.

23. Remove the exhaust pipe support bracket.

24. To remove the engine mounting center member, remove the following components:
- The two dust covers from the rear side of the member
- The A/C pipe from the bracket
- The bolt and nut holding the front engine mounting bracket to the mounting insulator
- The bolt holding the rear engine mounting bracket to the insulator
- The bolt and two nuts holding the rear engine mounting insulator to the front suspension member
- The two bolts and the rear engine mounting bracket, and the center member with the rear mounting insulator.

25. Attach an engine chain hoist or suitable equivalent to the engine hangers. Remove the two bolts and nut and disconnect the LH engine mounting bracket from the mounting insulator.

26. Remove the through-bolt and the LH mounting insulator.

27. Disconnect the ground strap connector.

28. Remove the bolt, two nuts, and disconnect the RH engine mounting bracket from the mounting insulator.

29. Lift the engine and transaxle assembly from the vehicle slowly and carefully making sure that it is clear of all the wiring, cables, and hoses.

30. Separate the transaxle from the engine assembly.

To install:

31. Install the transaxle to the engine assembly.

32. Attach an engine chain hoist to the engine hangers and slowly lower the engine into the engine compartment. Tilt the transaxle downward, lower the engine and clear the LH body mounting.

33. Keeping the engine level, align the RH and the LH mounts with the body mounts and attach the RH engine mounting bracket to the mounting insulator. Temporarily install the three nuts.

34. Temporarily install the LH engine mounting insulator to the body with the through-bolt. Attach the LH engine mounting bracket to the mounting insulator and install the two bolts and nut. Torque the bolts and nut to 47 ft. lbs. (64 Nm).

35. Tighten the LH engine mounting through-bolt to the body. Torque the bolt to 54 ft. lbs. (73 Nm). Tighten the three nuts holding the RH mounting bracket to the insulator. Torque the 12mm nut to 21 ft. lbs. (28 Nm) and the 14mm nut to 38 ft. lbs. (52 Nm).

36. Connect the engine ground strap connector and remove the engine hoist.

37. To install the engine mounting center member, perform the following:
- Attach the center member together with the rear engine mounting insulator to the front suspension member

- Temporarily install the two bolts and nut holding the center member to the body
- Install the rear engine mounting bracket with the two bolts. Torque the bolts to 58 ft. lbs. (78 Nm
- Temporarily install the bolt and two nuts holding the rear engine mounting insulator to the front suspension member
- Temporarily install the bolt holding the rear engine mounting bracket to the insulator
- Temporarily install the bolt and nut holding the front engine mounting bracket to the insulator
- Torque the two bolts holding the center member to the body to 26 ft. lbs. (35 Nm)
- Torque the bolt and two nuts holding the rear mounting insulator to the front suspension member to 59 ft. lbs. (80 Nm)
- Torque the bolt holding the rear engine mounting bracket to the insulator to 65 ft. lbs. (88 Nm)
- Torque the bolt and nut holding the front engine mounting bracket to the insulator to 65 ft. lbs. (88 Nm
- Install the A/C pipe to the bracket and install the two dust covers to the center member

38. Install the exhaust pipe support bracket with the two bolts and the nut. Torque the bolts to 14 ft. lbs. (19 Nm).

39. Connect the transaxle control cable(s) to the transaxle.

40. On A/T vehicles, install the transaxle control cable to the engine mounting center member with the two clamps and two bolts.

41. On M/T vehicles, install the clutch release cylinder. Tighten the bolts to 9 ft. lbs. (12 Nm), then attach the bracket with the bolt.

42. Connect the A/C relay box to the body.

43. Install the P/S pump with the four bolts. Torque the 12mm bolts to 14 ft. lbs. (19 Nm) and torque the 14mm bolts to 29 ft. lbs. (39 Nm). Install the drive belt and torque the adjusting bolt to 29 ft. lbs. (39 Nm).

44. Install the A/C compressor with the four bolts and torque them to 18 ft. lbs. (25 Nm). Install the A/C drive belt with the adjusting bolt and tighten the idler pulley locknut to 29 ft. lbs. (39 Nm). Connect the connector.

45. Install the alternator drive belt.
46. Install the halfshafts.
47. Install the front exhaust pipe.

48. To install the engine wire harness to the vehicle cabin, perform the following:
- Push the harness through the cowl panel, install the retainer to the cowl with the two nuts and install the wire clamp to the bracket
- Connect the harness to the clamp on the ECM
- Connect the three ECM connectors and the circuit opening relay connector
- Connect the three connectors to the connectors on the bracket
- Connect the A/C amplifier connector
- Install the floor carpet, the lower instrument panel finish panel, the cowl side trim panel, and the scuff plate

49. Connect the engine wire harness with the two connectors to the engine compartment relay box and install the relay box covers.

50. Install and/or connect the following:
- The MAP sensor connector
- The MAP sensor wire to the clamp on the bracket
- The MAP sensor vacuum hose to the gas filter on the intake manifold
- The brake booster vacuum hose to the intake manifold
- The A/C idle–up valve connector
- The A/C idle–up valve hose to the intake manifold
- The A/C idle–up valve hose to the air pipe
- The DLC1 to the bracket
- The engine harness protector to the bracket
- The ground cable and the ground strap to the body
- The heater hose to the water outlet and the heater hose to the water bypass pipe
- Connect the fuel inlet hose to the fuel filter
- Connect the fuel inlet hose with two new gaskets and the union bolt–torque the union bolt to 22 ft. lbs. (30 Nm).
- Connect the fuel return hose to the return pipe and connect the EVAP hose to the charcoal canister
- Connect the P/S air hoses to the intake manifold and the air pipe.

51. Install the radiator assembly.
52. On the model equipped with cruise control, install the actuator cable to the clamps.
53. Install the accelerator cable to the throttle body, cable bracket, and the clamps.
54. Install the air cleaner assembly.
55. Install the battery tray and battery.
56. Install the hood.

57. Refill the transaxle assembly.
58. Fill the engine with oil and coolant. Connect the negative battery cable, start the engine, bleed the cooling system, and check for any leaks.
59. Install the engine undercover.
60. Road test the vehicle for any abnormal noise and verify proper operation.

2.0L (3S-GTE) Engine

MR2

1. Disconnect the negative battery cable. On vehicles equipped with an air bag, wait at least 90 seconds before proceeding.
2. Release the fuel pressure.

CAUTION

Fuel injection systems remain under pressure after the engine has been turned OFF. Properly relieve fuel pressure before disconnecting any fuel lines. Failure to do so may result in fire or personal injury.

---

3. Remove the engine hood.
4. Remove the engine hood side panels.
5. Remove the undercovers from the vehicle.
6. Drain the engine coolant.
7. Drain the engine oil.
8. Drain the transaxle oil.
9. Remove the upper suspension brace.
10. Remove the air cleaner.
11. Remove the No. 1 and No. 2 intake air connectors.
12. If equipped, remove the cruise control actuator and accelerator linkage.
13. If not equipped with cruise control, disconnect the accelerator cable from the throttle body.
14. Disconnect the ground strap connector.
15. Disconnect the brake booster vacuum hose.
16. Disconnect the data link connector 1 (DLC1) and map sensor.
17. Disconnect the A/C idle up valve and remove the solenoid resistor assembly.
18. Remove the water filler. Disconnect the water filler hose and the coolant reservoir hose.
19. Remove the charcoal canister.
20. Remove the engine relay box and disconnect the engine wire.
21. Disconnect the noise filter connectors, the igniter connector, and the high tension cord from the ignition coil.
22. Disconnect the two ECM connectors, the starter relay, the engine compartment electric cooling fan con-

nector, and pull the engine compartment wire from the luggage compartment.

23. Disconnect the starter cable from the starter terminal 30.

24. Disconnect the following hoses:

   a. Radiator hose from the water inlet

   b. Radiator hose from the water outlet

   c. Two heater water hoses

   d. Fuel inlet hose from the fuel filter

   e. Fuel return hose

25. Disconnect the transaxle control cables from transaxle.

26. Remove the tailpipe and the front exhaust pipe.

27. Remove the engine compartment electric cooling fan.

28. Disconnect the A/C compressor from engine.

29. Remove the Charge Air Cooler (CAC).

30. Remove the rear engine mounting insulator.

31. Remove the driveshafts.

32. Remove the rear suspension crossmember.

33. Remove the front engine mounting insulator and bracket.

34. Remove the right-hand engine mounting stay.

35. Remove the left-hand engine mounting stay.

36. Disconnect the lateral control rod from left-hand engine mounting insulator.

37. Disconnect the ground strap connector.

38. Remove the lateral control rod.

39. Attach the chain hoist to the engine hangers.

40. Remove the engine and transaxle assembly from the vehicle.

41. Remove the rear engine mounting bracket.

42. Remove the left-hand engine mounting bracket.

43. Remove the driveshaft bearing bracket.

To install:

44. Install the driveshaft bearing bracket. Torque the bolts to 47 ft. lbs. (64 Nm). Torque the bearing bracket stay bolts to 56 ft. lbs. (75 Nm).

45. Install the left-hand engine mounting bracket. Torque the bolts to 38 ft. lbs. (52 Nm).

46. Install the rear engine mounting bracket. Torque the bolts for the right-hand side to 29 ft. lbs. (39 Nm) and the bolts for the right rear to 57 ft. lbs. (77 Nm).

47. Install the engine and transaxle assembly in vehicle. Torque left-hand mounting insulator bolts to 47 ft. lbs. (64 Nm), the left-hand through-bolt

to 58 ft. lbs. (78 Nm). Torque the right-hand mounting insulator bolts to 38 ft. lbs. (52 Nm), right-hand through-bolt to 58 ft. lbs. (78 Nm).

48. Install the right-hand engine mounting stay. Torque the bolts to 54 ft. lbs. (73 Nm).

49. Install the lateral control rod to body. Torque the bolts to 26 ft. lbs. (35 Nm).

50. Install the lateral control rod and left-hand engine mounting stay. Torque the transaxle side to 18 ft. lbs. (25 Nm). Torque insulator side to 54 ft. lbs. (73 Nm). Tighten the A bolt to 27 ft. lbs. (37 Nm).

51. Connect the ground strap connector.

52. Install the front engine mounting bracket. Torque bolts holding mounting bracket to transaxle to 57 ft. lbs. (77 Nm).

53. Install the front engine mounting insulator. Torque the bolts to 47 ft. lbs. (64 Nm).

54. Install the rear suspension lower crossmember. Torque the bolts to 83 ft. lbs. (113 Nm).

55. Install the driveshafts.

56. Install the rear engine mounting insulator. Torque the three bolts holding the mounting bracket to 47 ft. lbs. (64 Nm). Torque the through-bolt to 64 ft. lbs. (87 Nm).

57. Tighten the front engine mounting through-bolt. Torque bolt to 71 ft. lbs. (96 Nm).

58. Install the CAC.

59. Install the A/C compressor.

60. Install the engine compartment electric cooling fan.

61. Install the front exhaust pipe.

62. Install the tailpipe.

63. Connect the transaxle control cables to transaxle.

64. Connect the following hoses:

   a. Radiator hose to the water inlet

   b. Radiator hose to the water outlet

   c. Two heater water hoses

   d. Fuel inlet hose to the fuel filter

   e. Fuel return hose

65. Connect the starter cable.

66. Connect the two ECM connectors, the starter relay connector, the engine compartment electric cooling fan connector, and the engine compartment wire connector. Push the engine wire into the luggage compartment.

67. Connect the noise filter connector, the igniter connector, and high tension cord to the ignition coil.

68. Connect the engine wire and install the engine relay box.

69. Install the charcoal canister.

70. Install the water filler. Connect the water filler hose and the coolant reservoir hose.

71. Install the solenoid resistor assembly and A/C idle up valve. Attach the solenoid resistor, fuel pump relay and fuel pump resistor connectors.

72. Connect the data link connector (DLC1) and map sensor.

73. Connect the brake booster vacuum hose.

74. Connect the ground strap connector.

75. If removed, install the cruise control actuator and accelerator linkage.

76. If equipped without cruise control, connect the accelerator cable to throttle body.

77. Install the No. 1 and No. 2 intake air connectors.

78. Install the air cleaner.

79. Install the upper suspension brace. Torque the bolt to 54 ft. lbs. (73 Nm) and the nut to 47 ft. lbs. (64 Nm).

80. Fill the engine coolant and oil.

81. Fill the transaxle with the appropriate fluid.

82. Connect the negative battery cable. Start the engine and check for leaks.

83. Perform all the necessary engine adjustments.

84. Install the engine undercovers.

85. Install the engine hood and hood side panels.

86. Road test the vehicle.

87. Recheck the coolant, oil, and the transaxle fluid levels.

2.2L (5S-FE) Engine

Camry

1. Disconnect the negative battery cable. On vehicles equipped with an air bag, wait at least 90 seconds before proceeding.

2. Remove the battery and the battery tray.

3. Remove the hood.

4. Remove the engine undercover and then drain the engine coolant and oil.

5. Disconnect the accelerator cable from the throttle body. On models with A/T, disconnect the throttle cable.

6. Remove the air cleaner assembly, resonator, and the air intake hose.

7. On models with cruise control, remove the actuator cover, unplug the connector, remove the three bolts, and then disconnect the actuator with the bracket.

8. Disconnect the ground strap at the battery carrier.

9. Remove the radiator and then disconnect the coolant reservoir hose.

10. Remove the washer tank and disconnect the electrical lead and hose.

11. Tag and disconnect the following:

a. The five connectors to the engine relay box

b. The igniter connector

c. The noise filter connector

d. The connector at the LH fender apron

e. The two ground straps from the LH and RH fender aprons

f. The data link connector (DLC1)

g. Disconnect the MAP sensor connector

12. Inside the vehicle, remove the dash panel undercover, the glove compartment door, the glove compartment, disconnect the cowl harness connectors and the two ECM connectors.

CAUTION

Fuel injection systems remain under pressure after the engine has been turned OFF. Properly relieve fuel pressure before disconnecting any fuel lines. Failure to do so may result in fire or personal injury.

13. Disconnect the heater hoses, the fuel return hose, and the fuel inlet hose.

14. On models with M/T, remove the starter and the clutch release cylinder. Don't disconnect the hydraulic line, simply hang the cylinder out of the way.

15. Disconnect the transaxle control cables at the transaxle.

16. Tag and disconnect all remaining vacuum hoses and connectors.

17. Remove the two nuts and pull out the engine wire from the cowl panel.

18. Without disconnecting the refrigerant lines, remove the A/C compressor and carefully position it out of the way.

19. Loosen the two bolts and disconnect the front exhaust pipe bracket. Use a deep 14mm socket and remove the three nuts attaching the front pipe to the manifold. Disconnect the front pipe from the exhaust manifold.

20. Remove the halfshafts.

21. Without disconnecting the hydraulic lines, remove the power steering pump and carefully position it aside.

22. Remove the three bolts (M/T) or four bolts (A/T) and then disconnect the left engine mounting insulator. Remove the access plugs, remove the three nuts and then remove the right rear engine mounting insulator. Remove the three bolts and disconnect the front right engine mounting insulator.

23. Attach an engine lifting device to the lift hooks. Remove the three bolts and disconnect the control rod. Slowly and carefully, lift the engine/transaxle assembly out of the engine compartment.

24. If equipped with A/T, remove the starter. Separate the engine assembly from the transaxle.

To install:

25. Connect the engine assembly to the transaxle. On vehicles equipped with A/T, install the starter.

26. Carefully lower the engine and transaxle assembly into the engine compartment. With the engine level and all the mounts aligned with their brackets, install the engine control rod. Tighten the three bolts, in the sequence to 47 ft. lbs. (64 Nm).

27. Connect the front right engine mount and tighten the bolts to 59 ft. lbs. (80 Nm). Connect the rear mount and tighten the nuts to 48 ft. lbs. (66 Nm). Don't forget the access plugs.

28. Connect the left mount and tighten the bolts (3 or 4) to 47 ft. lbs. (64 Nm).

29. Install the power steering pump and tighten the bolts to 31 ft. lbs. (43 Nm). Install the drive belt and connect the two air hoses to the air pipe.

30. Install the halfshafts.

31. Connect the front pipe to the manifold and tighten the new nuts to 46 ft. lbs. (62 Nm).

32. Install the A/C compressor and tighten the bolts to 20 ft. lbs. (27 Nm).

33. Feed the engine harness through the cowl and reattach the clamp to the cowl. Make the following connections:

a. The two ECM connectors

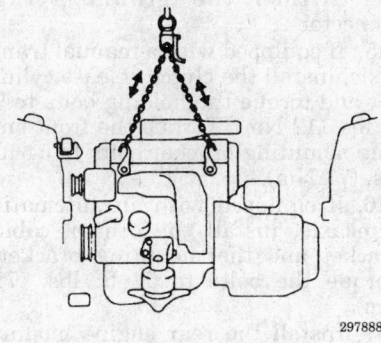

297888

Use a hoist to remove the engine assembly — Camry 5S-FE engine

b. The two cowl wire connectors

c. Install the glove compartment and door

d. Install the lower instrument panel and the undercover

34. Connect the vacuum hoses and the transaxle control cables.

35. On M/T vehicles, install the release cylinder and the starter.

36. Connect the fuel inlet hose and tighten it to 22 ft. lbs. (29 Nm). Connect the return hose and the two heater hoses.

37. Connect the following:

a. Attach the five connectors to the relay box

b. The connectors from the LH fender apron

c. Install the engine relay box

d. The igniter connector

e. On California models, the ignition coil connector

f. The noise filter connector

g. The two ground straps from the LH and RH fender apron

h. The data link connector (DLC1)

i. The MAP sensor connector.

38. Install the washer tank and connect the electrical lead and hose.

39. Install the coolant reservoir hose and the radiator.

40. If equipped with cruise control, install the actuator and bracket with the three bolts. Connect the actuator connector and install the cover.

41. Connect the ground strap to the battery carrier.

42. Install the air cleaner assembly.

43. On California models, connect the air hose to the air cleaner assembly and connect the air intake temperature sensor connector.

44. On vehicles equipped with A/T, connect and adjust the throttle cable.

45. Connect and adjust the accelerator cable.

46. Install the battery tray and battery.

47. Install the hood.

48. Fill the engine with oil and coolant. Connect the negative battery cable, start the engine, bleed the cooling system, and check for any leaks.

49. Install the engine undercover.

50. Road test the vehicle for any abnormal noise and verify proper operation.

MR2

1. Disconnect the negative battery cable. On vehicles equipped with an air bag, wait at least 90 seconds before proceeding.

2. Relieve the fuel system pressure.

----- CAUTION -----

Fuel injection systems remain under pressure after the engine has been turned OFF. Properly relieve fuel pressure before disconnecting any fuel lines. Failure to do so may result in fire or personal injury.

3. Remove the engine hood and hood side panels.

4. Remove the undercovers from the vehicle.

5. Drain the engine coolant.

6. Drain the engine oil.

7. Drain the transaxle oil.

8. Remove the air cleaner.

9. Disconnect the accelerator cable from the throttle body.

10. With cruise control, remove the cruise control actuator and the accelerator linkage.

11. Disconnect the ground strap connector.

12. Disconnect the brake booster vacuum hose.

13. Disconnect the Data Link Connector 1 (DLC1) and the MAP sensor.

14. Disconnect the A/C idle up valve.

15. Remove the water filler and disconnect the water filler and the coolant reservoir hoses.

16. Disconnect the three hoses and remove the charcoal canister.

17. Remove engine relay box and disconnect the engine wire.

18. Disconnect the noise filter and the igniter connectors.

19. Disconnect the two ECM connectors, starter relay connector, and the engine compartment wire connector from the luggage compartment.

20. Disconnect the starter cable from starter terminal 30.

21. Disconnect the following hoses:
   a. Radiator hose from the water inlet
   b. Radiator hose from the water outlet
   c. Two heater water hoses
   d. Fuel inlet hose from the fuel filter
   e. Fuel return hose

22. Disconnect transaxle control cables from transaxle.

23. If equipped with an automatic transaxle, disconnect the transaxle oil cooler hoses.

24. Remove the tailpipe.

25. Remove the front exhaust pipe.

26. Disconnect the A/C compressor from engine.

27. Remove the halfshafts.

28. Remove the front and rear engine mounting insulator with brackets.

29. Remove the right-hand engine mounting stay. On manual transaxles, remove the left-hand engine mounting stay.

30. Disconnect the lateral control rod from left-hand engine mounting insulator.

31. Disconnect the ground strap connector.

32. Remove the lateral control rod.

33. Attach the engine chain hoist to the engine hangers.

34. Remove the engine and transaxle assembly from the vehicle.

35. Remove the left-hand engine mounting bracket.

To install:

36. Install the left-hand engine mounting bracket to transaxle. Torque bolts to 38 ft. lbs. (52 Nm).

37. Install the engine and transaxle assembly in vehicle. Torque the left-hand mounting insulator bolt for the manual transaxle to:
   Bolt A — 47 ft. lbs. (64 Nm)
   Bolt B — 54 ft. lbs. (73 Nm)

38. Torque the left-hand mounting insulator bolt for the automatic transaxle to:
   Bolt A — 38 ft. lbs. (52 Nm)
   Bolt B — 47 ft. lbs. (63 Nm)
   Bolt C — 54 ft. lbs. (73 Nm)

39. Torque the through-bolts holding the left and right mounting insulators to the body to 58 ft. lbs. (78 Nm).

40. Install the right-hand engine mounting stay to 54 ft. lbs. (73 Nm).

41. Install the lateral control rod to body. Torque the bolts to 26 ft. lbs. (35 Nm).

42. If equipped with a manual transaxle, install the lateral control rod and left-hand engine mounting stay. Torque the transaxle side to 18 ft. lbs. (25 Nm). Torque the insulator side to 54 ft. lbs. (73 Nm). Tighten bolt A to 27 ft. lbs. (37 Nm).

43. If equipped with an automatic transaxle, torque the bolts for the lateral control rod to 27 ft. lbs. (37 Nm).

44. Attach the ground strap connector.

45. If equipped with a manual transaxle, install the clutch release cylinder and torque the holding bolts to 9 ft. lbs. (12 Nm). Torque the front engine mounting bracket bolts to 54 ft. lbs. (77 Nm).

46. If equipped with an automatic transaxle, install the control cable bracket and the mounting bracket. Torque the bolts to 54 ft. lbs. (77 Nm).

47. Install the rear engine mounting bracket. Torque bolts to 57 ft. lbs. (77 Nm).

48. Install the front engine mounting insulator. Torque bolts to 54 ft. lbs. (73 Nm).

49. Install the rear engine mounting insulator. Torque bolts to 47 ft. lbs. (64 Nm).

50. Install the through-bolt and torque to 64 ft. lbs. (87 Nm).

51. Tighten the front engine mounting through-bolt. Torque bolt to 70 ft. lbs. (96 Nm).

52. Install the halfshafts.

53. Install the A/C compressor.

54. Install the front exhaust pipe.

55. If equipped with an automatic transaxle, connect the oil cooler hoses.

56. Connect the transaxle control cables to transaxle.

57. Connect the following hoses:
   a. Radiator hose from the water inlet
   b. Radiator hose from the water outlet
   c. Two heater water hoses
   d. Fuel inlet hose from the fuel filter
   e. Fuel return hose

58. Connect the starter cable.

59. Connect the engine wire to luggage compartment and attach the following connectors:
   a. The two ECM connectors.
   b. The starter relay connector.
   c. The engine compartment wire connector.

60. Secure the connectors to the noise filter and igniter.

61. Connect the engine wire and install engine relay box.

62. Install the charcoal canister.

63. Install the water filler and connect the water filler and the coolant reservoir hoses.

64. Install the A/C idle up valve.

65. Connect the Data Link Connector 1 (DLC1) and MAP sensor.

66. Connect the brake booster vacuum hose.

67. Secure the ground strap connector.

68. If removed, install the cruise control actuator and the accelerator linkage.

69. Connect the accelerator cable to throttle body.

70. Install the air cleaner.

71. Fill the engine coolant and engine oil.

72. Fill the transaxle with oil.

73. Connect the negative battery cable.

74. Start the engine and check for leaks.

75. Perform all the necessary engine adjustments.

76. Install the engine undercovers.

77. Install the engine hood and hood side panels.
78. Perform a road test.
79. Recheck the engine coolant and oil levels.

Celica

1. Release the fuel system pressure.
2. Disconnect the negative battery cable. On vehicles equipped with an air bag, wait at least 90 seconds before proceeding.
3. Remove the battery.
4. Remove the hood.
5. Remove the engine undercover and then drain the engine coolant and oil.
6. Disconnect the accelerator cable from the throttle body, the cable bracket and clamps.
7. Remove the air cleaner assembly.
8. On models with cruise control, unplug the connector, remove the three bolts and then disconnect the actuator from the bracket.
9. Remove the radiator assembly.
10. Disconnect the following connectors, wire harness, cables, and hoses:
   a. The MAP sensor
   b. The MAP sensor wire from the clamp
   c. The MAP sensor vacuum hose from the gas filter on the intake manifold
   d. The brake booster vacuum hose from the intake manifold
   e. The DLC1 from the bracket
   f. The igniter connector
   g. On California models, the ignition coil connector
   h. On California models, the ignition coil high tension wire, the noise filter, the wire clamp from bracket, the ignition coil and igniter assembly, and disconnect the wire from the bracket
   i. The ground cable from the body
   j. The ground strap from the body
   k. The heater hose from the water outlet and bypass pipe

---- **CAUTION** ----
Fuel injection systems remain under pressure after the engine has been turned OFF. Properly relieve fuel pressure before disconnecting any fuel lines. Failure to do so may result in fire or personal injury.

l. The fuel inlet hose from the fuel filter and the fuel return hose from the return pipe

   m. The EVAP hose from the charcoal canister
11. Disconnect the two connectors, remove the two relay box covers, and disconnect the engine wire harness from the engine compartment relay box.
12. To remove the engine wire harness from the vehicle cabin, remove or disconnect the following:
   a. The scuff plate
   b. The cowl side trim
   c. The finish panel from the lower instrument panel
   d. Remove the front side of the floor carpet
   e. The wire harness from the clamp of the ECM bracket
   f. The three ECM connectors
   g. The circuit opening relay connector
   h. The three connectors from the connectors on the bracket
   i. The A/C amplifier connector
   j. The wire clamp from the bracket
   k. The two nuts holding the engine wire harness to the cowl
13. Remove the front exhaust pipe.
14. Remove the halfshafts.
15. Remove the alternator drive belt.
16. Disconnect the A/C compressor connector, remove the three bolts, remove the A/C compressor and suspend it securely out of the way.
17. Disconnect the two air hoses from the air tube, remove the drive belt, and remove the four bolts and the P/S pump. Securely hang the pump out of the way.
18. On M/T model, remove the starter.
19. On M/T, remove the clutch release cylinder and associated components.
20. Disconnect the transaxle control cable(s) from the transaxle.
21. On A/T model, remove the two bolts and disconnect the transaxle control cable from the engine mounting center member.
22. Remove the exhaust pipe support bracket.
23. To remove the engine mounting center member, remove the following components:
   a. The two dust covers from the rear side of the member
   b. The A/C pipe from the bracket
   c. The bolt and nut holding the front engine mounting bracket to the mounting insulator
   d. The bolt holding the rear engine mounting bracket to the insulator

   e. The bolt and two nuts holding the rear engine mounting insulator to the front suspension member
   f. The two bolts (A/T) or three bolts (M/T) and rear engine mounting bracket. Remove the center member with the rear mounting insulator.
24. Attach an engine chain hoist or suitable equivalent to the engine hangers. Remove the two nuts and bolt (M/T) or the two bolts (A/T), and disconnect the LH engine mount bracket from the mounting insulator.
25. Remove the through-bolt and the LH mounting insulator.
26. Disconnect the ground strap connector.
27. Remove the bolt, two nuts, and disconnect the RH engine mount bracket from the mounting insulator.
28. Remove the halfshaft bearing bracket.
29. Lift the engine and transaxle assembly from the vehicle slowly and carefully making sure that it is clear of all the wiring, cables and hoses. Separate the transaxle from the engine assembly.

To install:

30. Install the transaxle to the engine assembly and reattach the halfshaft bearing bracket with the three bolts. Torque the bolts to 47 ft. lbs. (64 Nm).
31. Attach an engine chain hoist to the engine hangers, and slowly lower the engine into the engine compartment. Tilt the transaxle downward, lower the engine and clear the LH body mounting.
32. Keeping the engine level, align the RH and the LH mounts with the body mounts, and attach the RH engine mount bracket to the mounting insulator, and temporarily install the bolt and two nuts.
33. Temporarily install the LH engine mounting insulator to the body with the through-bolt. Attach the LH engine mount bracket to the mounting insulator and install the two nuts and bolt (M/T) or the two bolts (A/T). Torque the bolts and nuts to 47 ft. lbs. (64 Nm).
34. Tighten the LH engine mounting through-bolt to the body. The torque is 54 ft. lbs. (73 Nm). Tighten the bolt and two nuts holding the RH mounting bracket to the insulator. Torque the bolt to 27 ft. lbs. (37 Nm) and the nut to 38 ft. lbs. (52 Nm).
35. Connect the engine ground strap connector and remove the engine hoist.

36. To install the engine mounting center member, perform the following:

a. Attach the center member together with the rear engine mounting insulator to the front suspension member

b. Temporarily install the two bolts holding the center member to the body

c. Install the rear engine mounting bracket with the two bolts (A/T) or the three bolts (M/T). The bolt torque is 58 ft. lbs. (79 Nm).

d. Temporarily install the bolt and two nuts holding the rear engine mounting insulator to the front suspension member

e. Temporarily install the bolt holding the rear engine mounting bracket to the insulator

f. Temporarily install the bolt and nut holding the front engine mounting bracket to the insulator

g. Torque the two bolts holding the center member to the body to 26 ft. lbs. (35 Nm).

h. Torque the bolt and two nuts holding the rear mounting insulator to the front suspension member to 59 ft. lbs. (80 Nm).

i. Torque the bolt holding the rear engine mounting bracket to the insulator to 65 ft. lbs. (88 Nm).

j. Torque the bolt and nut holding the front engine mounting bracket to the insulator to 65 ft. lbs. (88 Nm).

k. Install the A/C pipe to the bracket and install the two dust covers to the center member

37. Install the exhaust pipe support bracket with the two bolts and the nut. Torque the bolts to 14 ft. lbs. (19 Nm).

38. Connect the transaxle control cable(s) to the transaxle.

39. On the A/T vehicles, install the transaxle control cable to the engine mounting center member with the two clamps and two bolts.

40. On M/T model only, install and/or connect the following:

a. The clutch release cylinder–torque the two bolts to 9 ft. lbs. (12 Nm).

b. The bracket with the bolt and the tube to the bracket with the clamp

c. The tube with the clamp and bolt

d. The back-up switch connector

41. On M/T model only, install the starter, connect the starter wire and cable with the nut, and connect the starter connector.

42. Temporarily install the P/S pump with the four bolts. Tighten the three bolts (except the pivot bolt). The torque is 32 ft. lbs. (44 Nm). Install the drive belt and torque the adjusting bolt to 29 ft. lbs. (39 Nm).and the pivot bolt to 32 ft. lbs. (44 Nm). Connect the two hoses to the air tube.

43. Install the A/C compressor with the three bolts and torque them to 18 ft. lbs. (24 Nm) and connect the connector.

44. Install the alternator drive belt.

45. Install the halfshafts.

46. Install the front exhaust pipe.

47. To install the engine wire harness to the vehicle cabin, perform the following:

a. Push the harness through the cowl panel, install the retainer to the cowl with the two nuts and install the wire clamp to the bracket

b. Connect the harness to the clamp on the ECM

c. Connect the three ECM connectors and the circuit opening relay connector

d. Connect the three connectors to the connectors on the bracket

e. Connect the A/C amplifier connector

f. Install the floor carpet, the lower instrument panel finish panel, the cowl side trim panel, and the scuff plate

48. Connect the engine wire harness with the two connectors to the engine compartment relay box and install the relay box covers.

49. Install and/or connect the following:

a. The MAP sensor connector

b. The MAP sensor wire to the clamp on the bracket

c. The MAP sensor vacuum hose to the gas filter on the intake manifold

d. The brake booster vacuum hose to the intake manifold

e. The DLC1 to the bracket

f. The engine harness protector to the bracket

g. On California models, the engine harness clamp to the bracket and the ignition coil and igniter assembly with the three bolts, and install the harness to the bracket

h. The igniter connector

i. On California model, the ignition coil connector, high tension wire to the coil, and the noise filter

j. The ground cable and the ground strap to the body

k. The heater hose to the water outlet and the heater hose to the water bypass pipe

l. Connect the fuel inlet hose to the fuel filter

m. Connect the fuel inlet hose with two new gaskets and the union bolt–torque the union bolt to 22 ft. lbs. (30 Nm).

n. Connect the fuel return hose to the return pipe and connect the EVAP hose to the charcoal canister

50. Install the radiator assembly.

51. On models equipped with cruise control, install the actuator with the three bolts and connect the connector.

52. Install the accelerator cable to the throttle body, the cable bracket, and the clamps.

53. Install the air cleaner assembly.

54. Install the battery tray and battery.

55. Install the hood.

56. Fill the engine with oil and coolant. Connect the negative battery cable, start the engine, bleed the cooling system, and check for any leaks.

57. Install the engine undercover.

58. Road test the vehicle for any abnormal noise and verify proper operation.

3.0L (3VZ-FE) Engine

Camry

1. Disconnect the negative battery cable. On vehicles equipped with an air bag, wait at least 90 seconds before proceeding.

2. Remove the battery and its tray.

3. Remove the hood.

4. Remove the engine undercover and then drain the engine coolant and oil.

5. Disconnect the accelerator cable from the throttle body. On models with AT, its the throttle cable, not the accelerator.

6. Remove the air cleaner assembly, resonator and the air intake hose.

7. On models with cruise control, remove the actuator cover, unplug the connector, remove the three bolts and then disconnect the actuator with the bracket.

8. Disconnect the ground strap at the battery carrier.

9. Remove the radiator and then disconnect the coolant reservoir hose.

10. Remove the three washer tank mounting bolts, disconnect the connector and hose and then lift out the tank.

11. Tag and disconnect the:

a. Three connectors to the engine relay box

b. Two connectors from the left side fender apron

c. Igniter connector

d. Noise filter connector

e. Connector at the fender apron

f. Check connector

g. Ground strap at the right fender apron

h. Back-up light switch and speed sensor (models with MT)

12. Disconnect the heater hoses, fuel return hose and the fuel inlet hose. All of these may leak coolant or fuel, so have a container handy!

13. On models with MT, remove the starter and then remove the clutch release cylinder. Don't disconnect the hydraulic line, simply hang the cylinder out of the way.

14. Disconnect the transaxle control cables at the transaxle.

15. Tag and disconnect all remaining vacuum hoses.

16. Remove the undercover beneath the glove box. Remove the lower instrument panel, the glove box door and the box itself. Tag and disconnect the three ECU connectors, the five cowl wire connectors and the cooling fan ECU connector. Remove the two nuts and then pull the engine harness into the engine compartment.

17. Without disconnecting the refrigerant lines, remove the A/C compressor and hang it carefully out of the way.

18. Loosen the two bolts and disconnect the front exhaust pipe bracket. Use a deep 14mm socket and remove the three nuts attaching the front pipe to the manifold. Disconnect the pipe.

19. Remove the halfshafts.

20. Without disconnecting the hydraulic lines, remove the power steering pump and hang it aside; carefully! Disconnect the hydraulic cooling fan pressure hose.

21. Remove the three bolts (MT) or four bolts (AT) and then disconnect the left engine mounting insulator. Pop out the plugs, remove the four nuts and then remove the rear engine mounting insulator. Remove the four bolts and remove the mount absorber. Remove the three bolts and disconnect the front engine mounting insulator.

22. Attach an engine lifting device to the lift hooks. Remove the three bolts and disconnect the control rod. Slowly and carefully, lift the engine/transaxle assembly out of the engine compartment.

To install:

23. Carefully lower the engine into the engine compartment. With the engine level and all the mounts aligned with their brackets, install the engine control rod. Tighten the three bolts, in the sequence shown, to 47 ft. lbs. (64 Nm). Install the right side mounting stays and tighten the bolt

to 23 ft. lbs. (31 Nm) and the bolts to 46 ft. lbs. (62 Nm).

24. Connect the front engine mount and tighten the bolts to 59 ft. lbs. (80 Nm). Connect the engine mount absorber and tighten the bolts to 35 ft. lbs. (48 Nm). Connect the rear mount and tighten the nuts to 48 ft. lbs. (66 Nm). Don't forget the plugs.

25. Connect the left mount and tighten the bolts (3 or 4) to 47 ft. lbs. (64 Nm).

26. Install the power steering pump and tighten the bolts to 31 ft. lbs. (43 Nm).

27. Install the halfshafts. Connect the cooling fan pressure hose.

28. Connect the front pipe to the manifold and tighten the new nuts to 46 ft. lbs. (62 Nm), tighten the converter nuts to 32 ft. lbs. (43 Nm). Don't forget to install the bracket.

29. Install the A/C compressor and tighten the cylinder block bolts to 20 ft. lbs. (27 Nm); tighten the bracket bolts to 14 ft. lbs. (20 Nm).

30. Feed the engine harness through the cowl and reconnect it. Install the glovebox.

31. Connect the vacuum hoses and the transaxle control cables.

32. Install the release cylinder and the starter.

33. Connect the fuel inlet hose and tighten it to 22 ft. lbs. (29 Nm). Connect the return hose and the two heater hoses.

34. Reconnect all wires disconnected previously.

35. Install the washer tank and connect the electrical lead and hose.

36. Install the coolant reservoir hose and the radiator.

37. Connect the ground strap to the battery carrier and then install the cruise control actuator. Install the air cleaner assembly.

38. Connect the throttle/accelerator cable and adjust it.

39. Fill the engine with oil and coolant. Connect the battery cable, start the engine and check for any leaks.

3.0L (1MZ-FE) Engine

Camry and Avalon

1. Release the fuel system pressure.

2. Turn the ignition switch **OFF**. Disconnect the battery cables, negative cable first. On vehicles equipped with an air bag, wait at least 90 seconds before proceeding.

3. Matchmark the hood hinges and remove the hood. Remove the battery and battery tray.

4. Drain the engine oil and cooling system.

5. Disconnect the accelerator and throttle cables. Remove the cruise control actuator, if equipped.

6. Remove the air cleaner assembly, mass air flow meter and air cleaner hose.

7. Remove the radiator.

8. Remove the two bolts and disconnect the engine relay box. Disconnect the five connectors to the engine relay box.

9. Disconnect the following connectors:

a. Two igniter connectors

b. Noise filter connector

c. Connector from the left-hand fender apron

d. Disconnect the two ground straps and any other electrical connections keeping them from being removed.

10. Disconnect all vacuum hoses from the engine.

CAUTION
Fuel injection systems remain under pressure after the engine has been turned OFF. Properly relieve fuel pressure before disconnecting any fuel lines. Failure to do so may result in fire or personal injury.

11. Disconnect the fuel inlet and return hoses.

12. Disconnect the heater hoses.

13. Disconnect the transaxle control cable from the transaxle.

14. Remove the instrument panel undercover, the lower instrument panel and glove box assembly.

15. Disconnect the three ECM connectors, the five cowl wire connectors, and the cooling fan ECM connector. Push the engine wire through the cowl panel.

16. Remove the front exhaust pipe.

17. Remove the halfshafts from the vehicle.

18. Disconnect the power steering pressure tube.

19. Remove the power steering pump.

20. Remove the air conditioning compressor without disconnecting the hoses.

21. Remove the left-hand engine mounting insulator by removing the four bolts.

22. Remove the right-hand engine mounting insulator by removing the two hole plugs and then removing the four nuts.

23. Remove the four bolts to the engine mounting shock absorber and then remove the absorber.

24. Remove the front right engine mounting insulator by removing the three bolts.

25. Attach a hoist chain to the engine hangers.

26. Disconnect the coolant reservoir hose and remove the reservoir tank.

27. Remove right-side engine mounting stay bracket. Remove the engine control rod and bracket assembly.

NOTE: Make certain all wires, connectors and hoses are cleared from the engine.

28. Using an engine hoist, carefully lift the engine/transaxle assembly from the vehicle.

To install:

29. Carefully lower the engine position. Keep the engine level while aligning the engine mounts.

30. Install the engine control rod and bracket. Torque to 47 ft. lbs. (64 Nm).

31. Install the right engine mount stay bracket. Torque to 23 ft. lbs. (31 Nm).

32. Connect the engine ground straps. Install the coolant reservoir tank.

33. Install the front engine insulator. Torque to 48 ft. lbs. (66 Nm).

34. Install the engine mounting shock absorber. Torque to 35 ft. lbs. (48 Nm).

35. Install the left and right engine mounts. Torque to 48 ft. lbs. (66 Nm).

36. Install the power steering pump and air conditioning compressor.

37. Connect the power steering pressure tube.

38. Install the halfshafts and front exhaust pipe.

39. Push the engine wires through the cowl panel and connect all wires and connectors.

40. Connect the transaxle control cable to the transaxle.

41. Connect the fuel hoses and heater hoses.

42. Connect all vacuum hoses, wiring and connectors.

43. Install the radiator.

44. Install the cruise control actuator, if equipped. Connect the throttle cable and accelerator cable.

45. Install the Mass Air Flow meter, the air cleaner assembly, and air cleaner hose.

46. Fill the cooling system with the proper coolant/water mixture. Fill the engine with engine oil.

47. Install the battery tray and battery. Connect the battery cables; negative cable last.

48. Align the marks and install the hood.

49. Start the engine and check for leaks.

50. Perform a road test.

51. Recheck all fluid levels.

Celica

1. Turn the ignition switch **OFF**. Disconnect the negative battery cable and remove the battery. On vehicles equipped with an air bag, wait at least 90 seconds before proceeding.

2. Drain all coolant from the engine. Drain the turbocharger intercooler, if required.

3. Scribe matchmarks around the hinges and remove the hood.

4. Disconnect the accelerator cable at the throttle body. If equipped with automatic transaxle, disconnect the transaxle cooling lines. Remove the radiator assembly.

5. Disconnect the heater and intercooler hoses.

6. Relieve the fuel system pressure. Disconnect the fuel inlet line at the fuel filter and the return line at the return pipe.

7. Remove the cruise control actuator and bracket. Remove the air cleaner assembly.

8. Remove the clutch release cylinder and bracket without disconnecting the hydraulic line. Wire it out of the way.

9. Disconnect the speedometer and transaxle control cables. Remove the alternator.

10. Remove the air conditioning compressor without disconnecting the refrigerant lines and position it out of the way.

11. Tag and disconnect any wires, connectors and vacuum lines which might interfere with engine removal.

12. Raise the vehicle and support it safely. Drain the engine oil and remove the undercovers.

13. Remove the lower suspension crossmember. Remove the front halfshafts and the (mark before removing) driveshaft.

14. Remove the power steering pump and bracket without disconnecting the hydraulic lines and position it out of the way.

15. Disconnect the front exhaust pipe at the manifold and tailpipe and remove it.

16. Remove the engine mounting center member and lower the vehicle.

17. Unplug the 3 TCCS ECU connectors, remove the 2 screws and pull the connectors out through the firewall. Remove the power steering pump reservoir tank.

18. Attach an engine hoist chain to the lifting brackets on the engine. Remove the 2 bolts holding the right engine mount insulator to the mounting bracket.

19. Remove the 4 bolts holding the left engine mount insulator to the mounting bracket.

20. Carefully, remove the engine and transaxle assembly from the vehicle.

To install:

21. Install the removed engine/transaxle assembly in the vehicle. Install all engine mounts and brackets (mount isolator to mount bracket) in the correct locations. Tighten the right and left engine mount bracket bolts to 38 ft. lbs. (52 Nm).

22. When installing the engine mounting center member, tighten the outer bolts to 29 ft. lbs. (39 Nm), tighten the inner bolts to 38 ft. lbs. (52 Nm).

23. Raise and safely support the vehicle as necessary.

24. Install the power steering pump reservoir tank. Plug the 3 TCCS ECU connectors, install (pull the connectors in through the firewall) the retaining screws.

25. Install the complete exhaust system. Use new exhaust manifold gaskets if necessary.

26. Install the power steering pump and bracket with hydraulic lines attached.

27. Install the front halfshafts and the driveshaft. Install the lower suspension crossmember. When installing the lower suspension crossmember, tighten the outer bolts to 154 ft. lbs. (208 Nm), tighten the inner bolts to 29 ft. lbs. (39 Nm).

28. Install the engine undercovers as necessary. Connect any wires, connectors or vacuum lines that were disconnected for engine removal procedure.

29. Install the air conditioning compressor to the engine. Use caution when working near or around the air conditioning system.

30. Reconnect the speedometer cable and transaxle control cables. Install the alternator assembly.

31. Install the clutch release cylinder and bracket in the proper position.

32. Install the cruise control bracket, control actuator and air cleaner assembly.

33. Reconnect the fuel lines and accelerator at the throttle body.

34. Install the radiator assembly reconnect transaxle cooling lines if so equipped and all water hoses.

35. Install the battery and connect the battery cables. Install the engine hood in the correct location.

36. Refill all fluid levels with the correct fluid to the proper level. Bleed systems as necessary. Make all necessary adjustments. Start the engine.

Check for any fluid leaks, road test the vehicle for proper operation.

3.0L (2JZ-GE) and (2JZ-GTE) Engines

Supra

— CAUTION —

The Air Bag System must be disarmed before performing this procedure. Failure to do so may cause accidental deployment of the air bag, resulting in unnecessary system repairs and/or personal injury.

1. Turn the ignition switch **OFF**. Disconnect the negative battery cable from the battery. Do not start any work for at least 90 seconds to prevent accidental deployment of the air bag.
2. Remove the hood.
3. Remove the radiator assembly from the vehicle.
4. Relieve the fuel pressure from the fuel lines.

— CAUTION —

Fuel injection systems remain under pressure even after the engine has been turned OFF. The fuel system pressure must be relieved before disconnecting any fuel lines. Failure to do so may result in fire and/or personal injury.

5. Drain the engine oil.
6. Disconnect the accelerator cable and cruise control actuator.
7. Remove the air cleaner assembly, volume air flow meter, and the air intake hose.
8. Remove the drive belt by turning the tensioner clockwise.
9. Remove the fan, fan clutch, and the water pump pulley by removing the four nuts.
10. Remove the charcoal canister.
11. Disconnect the heater water hoses.
12. Disconnect the brake booster vacuum hose.
13. Disconnect the EVAP hose.
14. Disconnect the following the connectors and wires:
    • Noise filter connector
    • Ignition coil connector
    • Engine wire from the wire clamp
    • Rubber cap, nut and wire from the alternator
    • Engine room main wire
    • Igniter connector
    • Theft deterrent horn connector
    • Engine wire from the two wire clamps
    • Wire clamp and power steering solenoid valve connector
    • Ground strap from the cylinder block by removing the bolt
    • Rubber cap, nut, and the wire from the starter
15. Disconnect the fuel inlet hose from the engine by removing the union bolt and two gaskets. Suspend the hose union upward.
16. Disconnect the fuel return hose from the fuel return guide.
17. Disconnect the fuel return hose from the fuel return hose. Plug the hose end.
18. Remove the bolt and bracket and disconnect the engine wire from the intake manifold stay.
19. Disconnect the power steering pump.
20. Disconnect the power steering pressure tube from the engine by removing the two bolts.
21. Disconnect the A/C compressor without disconnecting the hoses.
22. Disconnect the engine wire from the cowl panel.
23. Disconnect the engine wire from the cabin as follows:
    a. Remove the scuff plate from the right door.
    b. Take out the front side of the floor carpet.
    c. Remove the two nuts and ECM protector.
    d. Remove the nut and disconnect the ECM from the floor panel.
    e. Disconnect the two connectors from the ECM.
    f. Disconnect the connector from the instrument panel wire.
    g. Disconnect the connector from the connector cassette.
    h. Pull out the engine wire from the cabin.
24. For vehicles equipped with an manual transmission, remove the upper console panel, shift lever boots, and holding bolts.
25. If equipped with manual transmission, disconnect the clutch release cylinder and the ground strap from the transmission.
26. Remove the No. 2 front exhaust pipe.
27. Remove the exhaust pipe heat insulator.
28. Remove the driveshaft.
29. For vehicles with automatic transmissions, disconnect the control rod from the shift lever by removing the nut.
30. Support the transmission with a jack.
31. Remove the rear support member by removing the eight bolts.
32. Attach the engine hoist chain to the engine and raise the engine slightly.

33. Remove the two nuts holding the engine front mounting insulators to the front suspension crossmember.
34. Lift the engine out of the vehicle. While lifting the engine, make sure the engine is clear of all wiring, hoses and cables.
35. Remove the oil dipstick guide from the transmission.
36. Disconnect the engine wire from the transmission.
37. Disconnect the starter connector, two bolts, engine wire bracket, and the starter.
38. For vehicles with automatic transmissions, remove the oil cooler tubes from the transmission.
39. For vehicle with automatic transmissions, remove the torque converter clutch mounting bolts.
40. Remove the six bolts from the transmission and remove the transmission from the engine.

To install:

41. Assemble the engine and transmission by installing the six bolts. Tighten the bolts as follows:
    • 14mm–29 ft. lbs. (39 Nm)
    • 17mm–43 ft. lbs. (72 Nm)
42. If equipped with an automatic transmission, install the torque converter clutch mounting bolts by first installing the gray bolt; then install the other five bolts. Tighten the bolts to 25 ft. lbs. (33 Nm).
43. If equipped with automatic transmission, install the oil cooler tubes to the transmission. Tighten the union nuts to 25 ft. lbs. (33 Nm).
44. Install the starter.
45. Connect the engine wire to the transmission.
46. If equipped with automatic transmission, install the oil dipstick guide and dipstick form the transmission.
47. Install the engine and transmission as an assembly to the vehicle. Keep slight tension on the engine until the mounting bolts and nuts are installed.
48. Install the two nuts holding the engine front mounting insulators to the front suspension crossmember. Tighten the nuts to 43 ft. lbs. (59 Nm).
49. Install the four bolts holding the support member to the body. Tighten the bolts to 19 ft. lbs. (25 Nm).
50. Install the four nuts holding the support member to the engine rear mounting insulator. Tighten the nuts to 10 ft. lbs. (13 Nm).
51. Remove the engine hoist from the engine.
52. Install the driveshaft.

53. For vehicles equipped with automatic transmissions, connect the transmission control rod as follows:

   a. Shift the shift lever to the **N** position.

   b. Fully turn the control shaft lever back and return two notches. The control shaft is now in the neutral position.

   c. Connect the control rod to the shift lever with the nut. Tighten the nut to 9 ft. lbs. (13 Nm).

54. Install the exhaust pipe heat insulator.

55. Install the No. 2 front exhaust pipe.

56. For vehicle with manual transmission, install the clutch release cylinder and ground strap. Tighten the clutch release cylinder bolts to 9 ft. lbs. (13 Nm) and the ground strap bolt to 27 ft. lbs. (37 Nm).

57. For vehicles with manual transmission, install the upper console panel, shift lever boots, and holding bolts.

58. Connect the engine wire to the cabin as follows:

   a. Push in the engine wire through the cowl panel.

   b. Connect the connector to the connector cassette.

   c. Connect the connector to the instrument panel wire connector.

   d. Connect the two connectors to the ECM.

   e. Insert the ECM bracket into the stay on the floor panel.

   f. Install the ECM with the nut.

   g. Install the ECM protector with the two nuts.

   h. Install the floor carpet.

   i. Install the scuff plate.

59. Connect the engine wire to the cowl panel.

60. Install the A/C compressor to the engine.

61. Install the power steering tube with the two clamp bolts.

62. Install the power steering pump as follows:

   a. Install the pump bracket with the two bolts. Tighten the lower bolt to 43 ft. lbs. (58 Nm) and the upper bolt to 29 ft. lbs. (39 Nm).

   b. Install the pump rear stay with the two bolts. Tighten the bolts to 29 ft. lbs. (39 Nm).

   c. Install the pump housing to the pump bracket.

   d. Connect the air hose to the No. 4 timing belt cover.

   e. Connect the air hose to the air intake chamber and secure the hose to the No. 4 timing belt cover.

   f. Install the front pump bracket with the two bolts. Tighten the bolts to 43 ft. lbs. (58 Nm).

   g. Install the plate washer and bolt to the power steering fluid pump.

63. Install the engine wire bracket.

64. Connect the fuel return hose to the fuel return pipe.

65. Install the fuel return hose to the clamp of the oil dipstick guide.

66. Install the fuel inlet hose with the two new gaskets and the union bolt. Tighten the bolt to 22 ft. lbs. (29 Nm).

67. Connect the wires and connectors.

68. Connect the EVAP hose.

69. Connect the brake booster vacuum hose.

70. Connect the heater hoses.

71. Install the charcoal canister.

72. Install the water pump pulley, fan, and the fan clutch.

73. Install the drive belt to the engine.

74. Install the air cleaner, VAF meter, and the intake air connector pipe assembly.

75. Connect the control cables to the throttle body.

76. Fill the engine with oil.

77. Install the radiator assembly.

78. Connect the negative battery cable to the battery.

79. Start the engine, bleed the cooling system, and check for leaks.

80. Install the hood.

81. Road test the vehicle and check all fluids.

Engine Mounts

REMOVAL AND INSTALLATION

1.5L (3E-E) Engine

Tercel

1. Disconnect the negative battery cable. On vehicles equipped with an air bag, wait at least 90 seconds before proceeding.

2. Raise and safely support the vehicle.

— CAUTION —
Make certain vehicle is properly supported when raising the engine.

3. Raise the engine slightly to take the tension off the motor engine mounts.

NOTE: When raising the engine to replace the engine mounts, DO NOT place the jack directly under the oil pan or the crankshaft torsional damper. Use a block of wood between the engine and the jack to prevent damage at the jacking point.

4. Loosen and remove the insulator mounting through-bolt and nut. Raise the engine sufficiently to remove the engine mount. Raising the engine assembly higher than necessary may cause damage.

5. Remove the engine mount mounting bolts and remove the mount.

To install:

6. Install the engine mount with the mounting bolts.

7. Slowly and safely lower the engine to install the engine mount through-bolt. Install the bolt and nut.

8. The engine mounts are to be tightened to the following torque specifications:

   a. LH mounting bracket to transaxle—36 ft. lbs. (49 Nm)

   b. LH insulator to bracket—35 ft. lbs. (47 Nm)

   c. RH mounting bracket to engine mounting bracket—47 ft. lbs. (64 Nm)

   d. RH mounting bracket to insulator nut—47 ft. lbs. (64 Nm)

   e. RH mounting insulator through-bolt—54 ft. lbs. (73 Nm)

   f. Rear mounting insulator to body—58 ft. lbs. (79 Nm)

   g. Rear mounting insulator through-bolt—47 ft. lbs. (64 Nm)

9. Safely lower and remove the engine jack.

10. Lower the vehicle and reconnect the negative battery cable.

1.5L (5E-FE) Engine

Tercel and Paseo — Right Mount

1. Disconnect the negative battery cable. On vehicles equipped with an air bag, wait at least 90 seconds before proceeding.

2. Raise and safely support the vehicle.

— CAUTION —
Make certain vehicle is properly supported when raising the engine.

3. Raise the engine slightly to take the tension off the motor engine mounts.

NOTE: When raising the engine to replace the engine mounts, DO NOT place the jack directly under the oil pan or the crankshaft torsional damper. Use a block of wood between the engine and the jack to prevent damage at the jacking point.

4. Remove the bolt and disconnect the ground strap from the engine mount.

5. Remove the engine mount through-bolt, nut and two bolts.

6. Remove the engine mount from the engine.

To install:

7. Install the engine mount to the bracket by installing the through-bolt. Do not torque the bolt at this time.

8. Install the engine mount to the engine and install the four nuts and two bolts.

9. Tighten the bolts as follows:
- Bolts and nut to 47 ft. lbs. (64 Nm)
- Through-bolt: 54 ft. lbs. (73 Nm)

10. Lower the engine and connect the negative battery cable.

Tercel and Paseo — Left Mount

1. Disconnect the negative battery cable.

2. Raise and safely support the vehicle.

─────── **CAUTION** ───────
Make certain vehicle is properly supported when raising the engine.

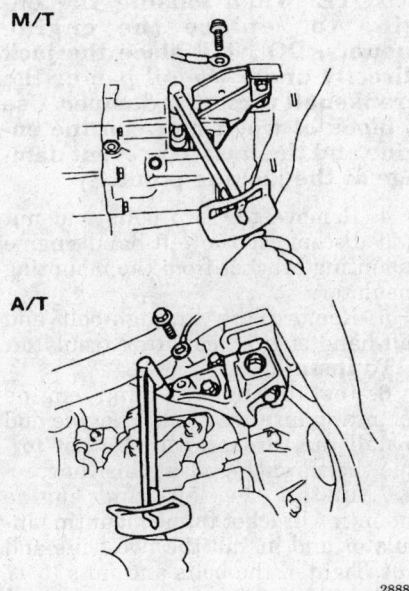

M/T

A/T

288811
LH engine mount — Paseo with 5E-FE engine

3. Raise the engine slightly to take the tension off the motor engine mounts.

NOTE: When raising the engine to replace the engine mounts, DO NOT place the jack directly under the oil pan or the crankshaft torsional damper. Use a block of wood between the engine and the jack to prevent damage at the jacking point.

4. Remove the bolt holding the engine ground to the left-hand bracket.

5. Remove the five bolts and disconnect the left-hand engine mounting bracket from the engine and transaxle.

To install:

6. Install the left-hand engine mounting bracket to the engine and transaxle and install the five bolts. Tighten the bolts to 47 ft. lbs. (64 Nm).

7. Connect the ground strap to the bracket and install the bolt. Tighten the bolt to 35 ft. lbs. (49 Nm).

8. Lower the engine and connect the negative battery cable.

Tercel and Paseo — Rear Mount

1. Disconnect the negative battery cable.

2. Raise and safely support the vehicle.

─────── **CAUTION** ───────
Make certain vehicle is properly supported when raising the engine.

3. Raise the engine slightly to take the tension off the motor engine mounts.

NOTE: When raising the engine to replace the engine mounts, DO NOT place the jack directly under the oil pan or the crankshaft torsional damper. Use a block of wood between the engine and the jack to prevent damage at the jacking point.

4. Remove the through-bolt nuts and the through-bolts.

5. Remove the engine mount bolts.

6. Raise the engine enough to remove the engine mounts.

NOTE: Only raise the engine enough to remove the mounts. Raising the engine too far could result in damage to some of the engine components.

To install:

7. Install the engine mounts and the engine bolts. Tighten the bolts to

going through the support brackets to 58 ft. lbs. (78 Nm) and the other bracket bolts to 67 ft. lbs. (90 Nm).

8. Install the through-bolt and torque the bolt to 54 ft. lbs. (73 Nm).

9. Lower the vehicle and connect the negative battery cable.

1.6L (4A-FE) and 1.8L (7A-FE) Engines

Corolla — Right Mount

1. Disconnect the negative battery cable. On vehicles equipped with an air bag, wait at least 90 seconds before proceeding.

2. Raise the engine slightly to take the tension off the motor engine mounts.

NOTE: When raising the engine to replace the engine mounts, DO NOT place the jack directly under the oil pan or the crankshaft torsional damper. Use a block of wood between the engine and the jack to prevent damage at the jacking point.

3. If equipped with cruise control, remove the following components:
   a. Remove the cruise control actuator cover.
   b. Disconnect the cable from the actuator.
   c. Disconnect the actuator connector.
   d. Remove the three bolts and disconnect the actuator with the bracket.
   e. Remove the three bolts and cruise control actuator bracket from the vehicle.

4. If equipped with A/C, remove the one bolt, one screw, and the pipe bracket from the engine mount.

5. If equipped with A/C, remove the bolt and damper.

6. Carefully place the A/C aside.

7. Remove the four mounting bolts, two nuts, and the rubber insulator from the engine mount.

To install:

8. Install the engine mount in the vehicle while carefully positioning the A/C line out of the way. Install the four mounting bolts and two nuts. Tighten as follows:
- Tighten the A bolts to 47 ft. lbs. (64 Nm)
- Tighten the B bolts to 38 ft. lbs. (52 Nm).
- Tighten the C bolt to 19 ft. lbs. (25 Nm).

9. If equipped with A/C, install the damper with the bolt.

10. If equipped with A/C, install the bracket with the bolt and screw.

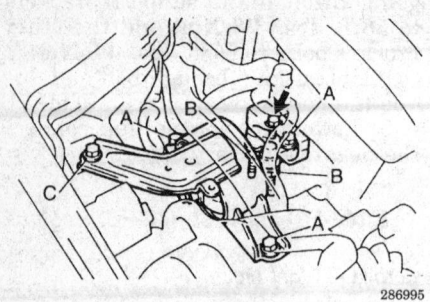

Tightening the RH mounting insulator — Corolla with 4A-FE and 7A-FE engines

286995

11. If equipped with cruise control, install the following:

    a. Install the actuator bracket and ground wire with the three bolts.

    b. Install the actuator with the bracket by installing the three bolts.

    c. Connect the actuator connector.

    d. Connect the cable to the actuator.

    e. Install the actuator cover.

12. Lower the engine and connect the negative battery cable to the battery.

Corolla — Left Mount

1. Disconnect the negative battery cable. On vehicles equipped with an air bag, wait at least 90 seconds before proceeding.

2. Raise the engine slightly to take the tension off the motor engine mounts.

NOTE: When raising the engine to replace the engine mounts, DO NOT place the jack directly under the oil pan or the crankshaft torsional damper. Use a block of wood between the engine and the jack to prevent damage at the jacking point.

3. Remove the two bolts to the mounting stay and remove the stay from the vehicle.

4. Remove the three bolts, through-bolt, and the left-hand insulator.

5. Remove the three bolts and the left-hand mounting bracket.

To install:

6. Install the left-hand mounting insulator to the body and temporarily install the through-bolt.

7. Connect the left-hand mounting insulator to the mounting bracket and install the three bolts. Tighten the bolts to 41 ft. lbs. (56 Nm).

8. Tighten the through-bolt to 64 ft. lbs. (87 Nm).

9. Install the mounting stay with the two bolts and torque the bolts to 15 ft. lbs. (21 Nm).

10. Lower the engine and connect the negative battery cable.

Corolla — Rear Mount

1. Disconnect the negative battery cable.

2. Raise the engine slightly to take the tension off the motor engine mounts.

NOTE: When raising the engine to replace the engine mounts, DO NOT place the jack directly under the oil pan or the crankshaft torsional damper. Use a block of wood between the engine and the jack to prevent damage at the jacking point.

3. Remove the through-bolt for the engine mount.

4. Remove the two bolts holding the mount insulator to the engine mounting center member.

5. Remove the rear engine mount insulator from the vehicle.

To install:

6. Install the rear engine mount insulator to the vehicle and install the two bolts. Tighten the bolts to 47 ft. lbs. (87 Nm).

7. Install the through-bolt for the mounting insulator and torque the bolt to 64 ft. lbs. (87 Nm).

8. Lower the engine, lower the vehicle, and connect the negative battery cable to the battery.

Celica — Right Mount

1. Disconnect the negative battery cable. On vehicles equipped with an air bag, wait at least 90 seconds before proceeding.

2. Raise and safely support the vehicle.

CAUTION

Make certain vehicle is properly supported when raising the engine.

3. Raise the engine slightly to take the tension off the engine mounts.

NOTE: When raising the engine to replace the engine mounts, DO NOT place the jack directly under the oil pan or the crankshaft torsional damper. Use a block of wood between the engine and the jack to prevent damage at the jacking point.

4. Remove the engine mount two bolts and four nuts.

5. Remove the engine mount from the engine.

6. Remove the engine mount from the bracket by removing the through-bolt.

To install:

7. Install the engine mount to the bracket by installing the through-bolt. Do not torque the bolt at this time.

8. Install the engine mount to the engine and install the four nuts and two bolts.

9. Tighten the bolts as follows:

- Bracket-to-insulator — 47 ft. lbs. (64 Nm)
- Through bolt — 54 ft. lbs. (73 Nm)
- Nut A — 38 ft. lbs. (52 Nm)
- Nut B — 21 ft. lbs. (28 Nm)

10. Lower the engine and connect the negative battery cable.

Celica — Left Mount

1. Disconnect the negative battery cable. On vehicles equipped with an air bag, wait at least 90 seconds before proceeding.

2. Raise and safely support the vehicle.

CAUTION

Make certain vehicle is properly supported when raising the engine.

3. Raise the engine slightly to take the tension off the motor engine mounts.

NOTE: When raising the engine to replace the engine mounts, DO NOT place the jack directly under the oil pan or the crankshaft torsional damper. Use a block of wood between the engine and the jack to prevent damage at the jacking point.

4. Remove the two bolts and nut and disconnect the left-hand engine mounting bracket from the mounting insulator.

5. Remove the through-bolt and left-hand engine mounting insulator.

To install:

6. Install the left-hand engine mounting insulator to the engine and install the through-bolt. Do not torque the through-bolt at this time.

7. Install the left-hand engine mounting bracket to the mounting insulator and install the two bolts and nut. Tighten the bolts and nuts to 47 ft. lbs. (64 Nm).

8. Torque the through-bolt for the left-hand mount to 54 ft. lbs. (52 Nm).

9. Lower the engine and connect the negative battery cable.

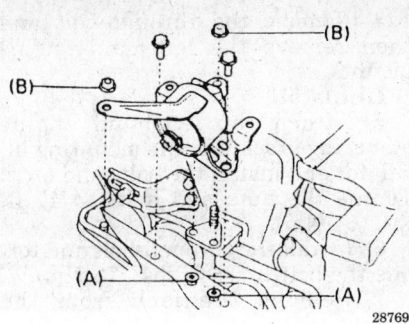

Right-hand engine mount — Celica with 7A-FE engine

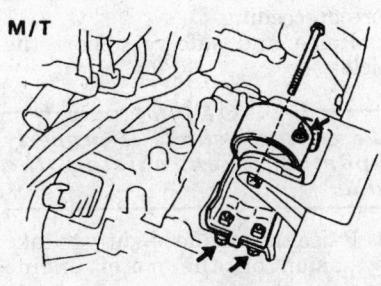

M/T

A/T

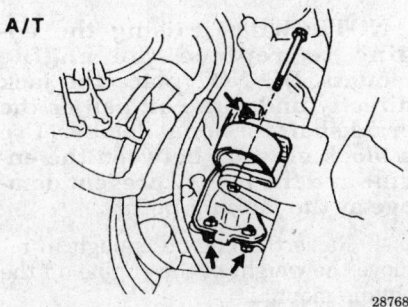

LH engine mount — Celica with 7A-FE engine

Celica — Front and Rear Mount

NOTE: The front engine mount is part of the center member. The engine mount insulator and center member are replaced as a unit.

1. Disconnect the negative battery cable. On vehicles equipped with an air bag, wait at least 90 seconds before proceeding.
2. Raise and safely support the vehicle.

— **CAUTION** —
Make certain vehicle is properly supported when raising the engine.

3. Raise the engine slightly to take the tension off the motor engine mounts.

NOTE: When raising the engine to replace the engine mounts, DO NOT place the jack directly under the oil pan or the crankshaft torsional damper. Use a block of wood between the engine and the jack to prevent damage at the jacking point.

4. Remove the two dust covers from the rear side of the center member.
5. Disconnect the A/C pipe from the bracket.
6. Remove the through-bolt to the front engine mount insulator.
7. Remove the through-bolt to the rear engine mount insulator.
8. Remove the two bolts and the rear engine mounting bracket.
9. Remove the two bolts holding the center member to the body and remove the center member and rear engine mounting insulator (front member is attached to center member).
   To install:
10. Connect the center member together with the rear engine mounting insulator to the front suspension member.
11. Install the two bolts holding center member to the body. Do not torque the bolts at this time.
12. Install the rear engine mounting bracket with the two bolts. Torque the bolts to 58 ft. lbs. (78 Nm).
13. Install the bolt and two nuts holding the rear engine mounting insulator to the front suspension member. Do not torque at this time.
14. Install the through-bolt to the rear engine mounting insulator. Do not torque at this time.
15. Install the through-bolt to the front engine mounting insulator. Do not torque at this time.
16. Torque the two bolts holding the center member to the body to 26 ft. lbs. (35 Nm).
17. Torque the bolt and two nuts holding the rear engine mounting insulator to the front suspension member to 59 ft. lbs. (80 Nm).
18. Torque the through-bolts for the rear and front insulators to 65 ft. lbs. (88 Nm).
19. Install the A/C pipe to the bracket.
20. Install the two dust covers.

2.0L (3S-GTE) Engine

Celica and MR2

1. Disconnect the negative battery cable. On vehicles equipped with an air bag, wait at least 90 seconds before proceeding.
2. Raise and safely support the vehicle.

— **CAUTION** —
Make certain vehicle is properly supported when raising the engine.

3. Raise the engine slightly to take the tension off the motor engine mounts.

NOTE: When raising the engine to replace the engine mounts, DO NOT place the jack directly under the oil pan or the crankshaft torsional damper. Use a block of wood between the engine and the jack to prevent damage at the jacking point.

4. Remove the through-bolt nuts and through-bolts.
5. Remove the engine mount bolts.
6. Raise the engine enough to remove the engine mounts.

NOTE: Only raise the engine enough to remove the mounts. Raising the engine too far could result in damage to some of the engine components.

To install:
7. Attach the RH mounting insulator to the mounting bracket and body. Temporarily install the through-bolt and two nuts.
8. Install the LH mounting bracket to the transaxle case with the three bolts, tighten the bolts to 38 ft. lbs. (52 Nm).
9. Attach the LH mounting insulator to the mounting bracket and body with that through-bolt and four bolts. Tighten the bolt to 47 ft. lbs. (63 Nm) and the through-bolt to 64 ft. lbs. (87 Nm).
10. Tighten the through-bolt and two nuts on the RH mounting insulator. Tighten the nut to 38 ft. lbs. (52 Nm) and the through-bolt to 64 ft. lbs. (87 Nm).
11. Remove the support from the engine.
12. Lower the vehicle and connect the negative battery cable.

2.2L (5S-FE) Engine

Celica — Right Mount

1. Disconnect the negative battery cable. On vehicles equipped with an

air bag, wait at least 90 seconds before proceeding.

2. Place a jack under the engine and relieve tension off the engine mount. Place a piece of wood between the engine and jack to prevent damaging the oil pan.

3. Disconnect and move the ground strap connector out of the way.

4. Remove the bolt and two nuts holding the mounting insulator to the mounting bracket.

5. Disconnect the power steering reservoir from the reservoir bracket.

6. Remove the bolt, nut and power steering bracket.

7. Remove the two bolts and the mounting insulator.

To install:

8. Install the engine mount to the vehicle and install the two bolts. Torque the bolts to 47 ft. lbs. (64 Nm).

9. Install the power steering bracket with the nut and bolt. Torque the nut and bolt to 21 ft. lbs. (28 Nm).

10. Install the bolt and two nuts to hold the mounting insulator to the mounting bracket. Torque the bolt to 27 ft. lbs. (37 Nm) and the nut to 38 ft. lbs. (52 Nm).

11. Connect the ground strap connector.

12. Connect the negative battery cable to the battery.

Celica — Front and Rear Mounts

1. Disconnect the negative battery cable. On vehicles equipped with an air bag, wait at least 90 seconds before proceeding.

2. Raise and safely support the vehicle with floor jacks.

3. Place a jack under the engine and relieve tension off the engine mount. Place a piece of wood between the engine and jack to prevent damaging the oil pan.

4. Remove the center centermember as follows:

   a. Disconnect the A/C pipe from the bracket on the centermember.

   b. Remove the two grommets from the sub-frame.

   c. Remove the set bolt and two nuts holding the engine rear mounting insulator to the centermember and sub-frame.

   d. Remove grommet from the centermember.

   e. Remove the two set bolts holding the front mounting insulator to the centermember.

   f. Remove the two bolts holding the centermember to the radiator support.

   g. Remove the centermember from the vehicle.

5. Remove the rear insulator and bracket as follows:

   a. Remove the through-bolt from the rear engine bracket.

   b. Remove the three (two for M/T) bolts and remove the bracket from the engine.

6. Remove the front insulator and bracket as follows:

   a. Remove the through-bolt from the front engine bracket.

   b. Remove the two bolts and remove the bracket from the engine.

To install:

7. Install the front bracket and insulator as follows:

   a. Install the bracket to the engine and install the two bolts. Torque the bolts to 57 ft. lbs. (77 Nm).

   b. Install the front insulator to the bracket and install the through-bolt. Torque the bolt to 64 ft. lbs. (87 Nm).

8. Install the rear bracket and insulator as follows:

   a. Install the bracket with the three (two for M/T) bolts. Torque the bolts to 57 ft. lbs. (64 Nm).

   b. Install the insulator to the bracket and torque the through-bolt to 64 ft. lbs. (87 Nm).

9. Install the center centermember as follows:

   a. Install the centermember and install the two bolts to hold the centermember to the radiator support. Torque the two bolts to 26 ft. lbs. (35 Nm).

   b. Install the two set bolts for the engine front mount and torque the bolts to 59 ft. lbs. (80 Nm).

   c. Install the grommet to the centermember.

   d. Install the set bolt and two nuts for the engine rear mount and torque to 59 ft. lbs. (80 Nm).

   e. Install the grommets to the sub-frame.

   f. Connect the A/C line to the bracket on the centermember.

10. Lower the vehicle and install the negative battery cable to the battery.

Celica — Left Mount

1. Disconnect the negative battery cable. On vehicles equipped with an air bag, wait at least 90 seconds before proceeding.

2. Place a jack under the engine and relieve tension off the engine mount. Place a piece of wood between the engine and jack to prevent damaging the oil pan.

3. Remove the two nuts and bolt (two bolts for A/T) and disconnect the left-hand engine mounting bracket from the left mounting insulator.

4. Remove the through-bolt and then remove the left-hand engine mount.

To install:

5. Attach the left-hand engine mounting bracket to the mounting insulator and install the bolts and nuts. Torque the nuts and bolts to 47 ft. lbs. (64 Nm).

6. Install the through-bolt and torque the bolt to 54 ft. lbs. (73 Nm).

7. Remove the jack from the engine.

8. Connect the negative battery cable to the battery.

Camry

1. Disconnect the negative battery cable. On vehicles equipped with an air bag, wait at least 90 seconds before proceeding.

2. Raise and safely support the vehicle.

—————— **CAUTION** ——————

Make certain vehicle is properly supported when raising the engine.

3. Raise the engine slightly to take the tension off the motor engine mounts.

NOTE: When raising the engine to replace the engine mounts, DO NOT place the jack directly under the oil pan or the crankshaft torsional damper. Use a block of wood between the engine and the jack to prevent damage at the jacking point.

4. Raise the engine enough to remove the weight of the engine off the engine mounts.

NOTE: Only raise the engine enough to remove the mounts. Raising the engine too far could result in damage to some of the engine components.

5. Disconnect the LH engine mounting insulator as follows:

   a. On M/T: remove the three bolts attaching the insulator to the transaxle

   b. On A/T: remove the four bolts attaching the insulator to the transaxle

6. Remove the two nuts holding the LH engine mounting insulator to the engine mounting member and remove the mount.

7. To remove the RR engine mounting insulator, remove the hole plugs and the three nuts and disconnect the insulator from the engine mounting member. Remove the four mounting bolts and the mounting insulator.

8. Remove the three bolts, and disconnect the RF engine mounting insulator from the sub frame. Remove the four bolts and nut attaching the RF mounting insulator bracket from the engine assembly and the manifold stay.

9. Remove the three bolts and remove the engine moving control rod.

To install:

10. Install the engine moving control rod with the three bolts and torque the bolts in sequence to 47 ft. lbs. (64 Nm).

11. Install the RF engine mounting insulator bracket with the four bolts and torque the bolts to 47 ft. lbs. (64 Nm). Install the manifold stay with the bolt and nut and torque to 31 ft. lbs. (42 Nm). Connect the mounting insulator to the sub frame with the three bolts and torque the bolts to:

    a. TMC made: 59 ft. lbs. (80 Nm)

    b. TMM made: 48 ft. lbs. (66 Nm)

12. Install the RR engine mounting insulator with the four bolts and torque the bolts to 47 ft. lbs. (64 Nm). Connect the insulator to the engine mounting member with the three nuts and torque the nuts to 48 ft. lbs. (66 Nm). Reinstall the hole plugs.

13. Install the LH engine mounting insulator as follows:

    a. On M/T: connect the insulator with the three bolts and torque to 47 ft. lbs. (64 Nm).

    b. On A/T: connect the insulator with the four bolts and torque to 47 ft. lbs. (64 Nm)

14. Install the two nuts attaching the LH engine mounting insulator to the engine mounting member.

15. Lower the engine.

16. Lower the vehicle and connect the negative battery cable.

MR2

1. Disconnect the negative battery cable. On vehicles equipped with an air bag, wait at least 90 seconds before proceeding.

2. Raise and safely support the vehicle.

———— **CAUTION** ————

Make certain vehicle is properly supported when raising the engine.

3. Raise the engine slightly to take the tension off the motor engine mounts.

NOTE: When raising the engine to replace the engine mounts, DO NOT place the jack directly under the oil pan or the

crankshaft torsional damper. Use a block of wood between the engine and the jack to prevent damage at the jacking point.

4. Remove the through-bolt nuts and the through-bolts.

5. Remove the engine mount bolts.

6. Raise the engine enough to remove the engine mounts.

NOTE: Only raise the engine enough to remove the mounts. Raising the engine too far could result in damage to some of the engine components.

To install:

7. Install the engine mounts and the engine bolts. Torque the mounting insulator through-bolts to:

    Right and left mounts — 58 ft. lbs. (78 Nm)

    Front mount — 70 ft. lbs. (96 Nm)

    Rear mount — 64 ft. lbs. (87 Nm)

8. Lower the vehicle and connect the negative battery cable.

3.0L (1MZ-FE and 3VZ-FE) Engines

Avalon and Camry — Front Mount

1. Disconnect the negative battery cable. On vehicles equipped with an air bag, wait at least 90 seconds before proceeding.

2. Raise and safely support the vehicle.

3. Raise the engine slightly with a suitable jacking device or engine support fixture.

4. Remove the four bolts attaching the mount to the engine.

5. Remove the bolts attaching the mount to the crossmember.

6. Remove the engine mount.

To install:

7. Position the mount next to the engine and install the mounting bolts. Tighten the bolts to 47 ft. lbs. (64 Nm).

8. Install the bolts that attach the mount to the crossmember. Torque the bolts to 59 ft. lbs. (80 Nm).

9. Remove the jack or engine support fixture and lower the vehicle to the floor.

Avalon and Camry — Rear Mount

1. Disconnect the negative battery cable. On vehicles equipped with an air bag, wait at least 90 seconds before proceeding.

2. Raise and safely support the vehicle.

3. Raise the engine slightly with a suitable jacking device or engine support fixture.

4. Remove the four nuts attaching the mount to the frame.

5. Remove the four bolts attaching the mount to the engine.

6. Remove the engine mount.

To install:

7. Position the mount next to the engine and install the four mounting bolts. Torque the bolts to 47 ft. lbs. (64 Nm).

8. Install the four nuts attaching the mount to the frame. Torque the nuts to 48 ft. lbs. (65 Nm).

9. Remove the jack or engine support fixture and lower the vehicle to the floor.

Left Mount

1. Disconnect the negative battery cable. On vehicles equipped with an air bag, wait at least 90 seconds before proceeding.

2. Raise and safely support the vehicle.

3. Raise the engine slightly with a suitable jacking device or engine support fixture.

4. Remove the two bolts that attach the mount to the frame.

5. Remove the four bolts that attach the mount to the transaxle.

6. Remove the engine mount.

To install:

7. Position the mount on the frame and install the four bolts that attach the mount to the transaxle. Torque the bolts to 47 ft. lbs. (64 Nm).

8. Lower the engine and install the two bolts that attach the mount to the frame. Torque the bolts to 47 ft. lbs. (64 Nm).

9. Remove the jack or engine support fixture and lower the vehicle to the floor.

Avalon and Camry — Engine Mount Shock Absorber

1. Disconnect the negative battery cable. On vehicles equipped with an air bag, wait at least 90 seconds before proceeding.

2. Raise and safely support the vehicle.

3. Raise the engine slightly with a suitable jacking device or engine support fixture.

4. Remove the four mounting bolts and the mount from the engine and frame.

To install:

5. Position the mount on the engine and frame, install the mounting bolts. Torque the bolts to 35 ft. lbs. (47 Nm).

6. Remove the jack or engine support fixture and lower the vehicle to the floor.

3.0L (2JZ-GE and 2JZ-GTE) Engines

Supra

1. Disconnect the negative battery cable. On vehicles equipped with an air bag, wait at least 90 seconds before proceeding.
2. Raise and safely support the vehicle.

——————— CAUTION ———————

Make certain vehicle is properly supported when raising the engine.

3. Raise the engine slightly to take the tension off the motor engine mounts.

NOTE: When raising the engine to replace the engine mounts, DO NOT place the jack directly under the oil pan or the crankshaft torsional damper. Use a block of wood between the engine and the jack to prevent damage at the jacking point.

4. Remove the through-bolt nuts and the through-bolts.
5. Remove the engine mount bolts.
6. Raise the engine enough to remove the engine mounts.

NOTE: Only raise the engine enough to remove the mounts. Raising the engine too far could result in damage to some of the engine components.

To install:

7. Install the engine mounts and the engine bolts. Torque the bolts for all the engine mounts to 43 ft. lbs. (59 Nm).
8. Lower the vehicle and connect the negative battery cable.

Cylinder Head

REMOVAL AND INSTALLATION

1.5L (3E-E) Engine

Tercel

1. Disconnect the negative battery cable. On vehicles equipped with an air bag, wait at least 90 seconds before proceeding.
2. Remove the right side under engine splash shield. Relieve the fuel pressure on fuel injected engines.
3. Drain the engine coolant from the radiator.
4. Remove the power steering pump and bracket (if equipped).
5. If equipped with A/C, but not power steering, remove the idler pulley/bracket.

6. Disconnect the radiator hoses. Disconnect the accelerator and throttle valve cable linkage.
7. Remove the timing belt and camshaft timing pulley.
8. Disconnect the heater inlet hose. Disconnect and plug the fuel lines. On the fuel injected engine remove the pulsation damper and disconnect the fuel inlet and return hoses.
9. Disconnect the power brake booster vacuum line from the intake manifold.
10. Disconnect the water inlet hose. Disconnect the intake manifold water hose from the intake manifold.
11. Tag (for identification) all vacuum lines, hoses and wires or harnesses to the intake manifold and cylinder head and disconnect them.
12. Remove the evaporative valve and cold enrichment valve if so equipped.
13. Disconnect the water bypass hoses from the carburetor if so equipped.
14. Disconnect the exhaust pipe from the exhaust manifold.
15. Remove the intake manifold stay bracket and ground strap. Remove the engine wire harness bracket clamp from the intake manifold.
16. Remove the cylinder head cover.
17. Loosen and remove the head mounting bolts gradually in three passes working from the ends of the cylinder head inward.
18. Lift the head straight up from the engine block.
19. Clean all gasket surfaces.
20. Service as necessary. Install in the reverse order of removal. Install a new cylinder head gasket. Torque the cylinder head mounting bolts in three progressive steps. Torque to 22 ft. lbs. (30 Nm) in the first pass. 36 ft. lbs. (49 Nm) in the second pass, and finally for the third pass, tighten the head bolts an additional 90° from the second pass. Tighten the bolts in sequence from the center of the head outwards (refer to the torque sequence illustration).
21. Adjust the valves by turning the crankshaft pulley until the groove in the pulley is aligned with the 0 mark on the timing belt cover. Check that No. 1 cylinder rocker arms are loose and No. 4 rocker arms are tight. If not, turn the engine one complete revolution and align the marks again. Adjust the valves to correct specifications. Turn one complete revolution and adjust the remaining valves. Complete installation, start the engine and check ignition timing

and carburetor adjustments. Road test for proper operation.

1.5L (5E-FE) Engine

Tercel and Paseo

1. Disconnect the negative battery cable. On vehicles equipped with an air bag, wait at least 90 seconds before proceeding.
2. Relieve the fuel pressure.
3. Remove the right engine undercover.
4. Drain the cooling system.
5. Disconnect the front exhaust pipe from the exhaust manifold by removing the two bolts and two compression springs.
6. Disconnect the accelerator and on vehicles equipped with automatic transmissions, disconnect the throttle cable.
7. Remove the air cleaner and air intake collector assembly.
8. Disconnect the fuel inlet and return hoses from the delivery pipe. Plug the hoses to prevent fuel leakage.
9. If equipped with power steering, remove the power steering pump and bracket. Do not disconnect the power steering hoses. Set and safely support the power steering pump aside.
10. Remove the ignition coils, spark plug wires, and spark plugs. Make sure to mark the position of the spark plug wires.
11. Tag and disconnect all electrical wire and vacuum hoses that interfere with removal of the cylinder head.
12. Remove the EGR pipe, EGR valve, and vacuum modulator.
13. Disconnect the radiator, water inlet and heater hoses.
14. Remove the water inlet and outlet housing.
15. Remove the throttle body assembly.
16. Remove the exhaust manifold.
17. Remove the fuel rail and injector assembly.
18. Remove the intake manifold.
19. Remove the valve cover.
20. Remove the No. 2 timing belt cover by removing the 4 bolts.
21. Remove the alternator belt, then the No. 3 timing belt cover from the No. 1 timing belt cover.
22. Turn the crankshaft pulley and align its groove with the timing mark 0 on the No. 1 timing belt cover.
23. Check that the hole of the camshaft timing pulley on the side with the 5E-FE mark is aligned with the timing mark on the No. 1 bearing cap. If the marks do not line up, turn the crankshaft pulley one (1) revolution and check marks.

24. Place matchmarks on the timing belt, loosen the No. 1 idler pulley and carefully remove the belt.
25. Remove the No. 2 idler pulley.

NOTE: Keep tension on the belt so it does not shift position and do not allow the crankshaft to rotate. This includes allowing the vehicle to roll while in gear. Do not allow anything to drop inside the belt cover, including dirt. The belt can be damaged. Do not let the belt come into contact with water, oil or grease.

26. Remove the camshaft timing pulley.
27. Remove the camshafts following the proper sequences and procedures.
28. Loosen the cylinder head bolts in several passes and in the reverse order of the installation sequence.
29. Remove the cylinder head from the engine. There are two different bolt lengths. Make note of there positions and replace the cylinder bolts in their original position.

—— **WARNING** ——
Failure to loosen the bolts as described can result in cylinder head warpage or cracking.

To install:
30. Clean the gasket mating surfaces using care not to damage the aluminum components, replace the gasket, then lower the cylinder head onto the engine. Make sure the dowel pins are aligned and no hoses or wires are between the head and cylinder block.

NOTE: The head bolts stretch and must be replaced once removed.

31. Lightly oil and place the 2 different size head bolts in their correct positions and tighten them in several passes in the proper sequence, evenly until arriving at a torque of 33 ft. lbs. (44 Nm).
32. Mark each bolt with a reference mark and tighten each bolt in sequence an additional 90 degrees.
33. Install the camshafts following the proper sequences and procedures.
34. Install the camshaft timing pulley in its original position and torque the bolt to 37 ft. lbs. (50 Nm).
35. Install the No. 2 idler pulley, tightening the bolt to 20 ft. lbs. (27 Nm).
36. Install the timing belt on the matchmarks and tension it properly.
37. Install the No. 3 timing belt cover, the alternator belt and the No. 2 timing belt cover with its gasket and 4 bolts.

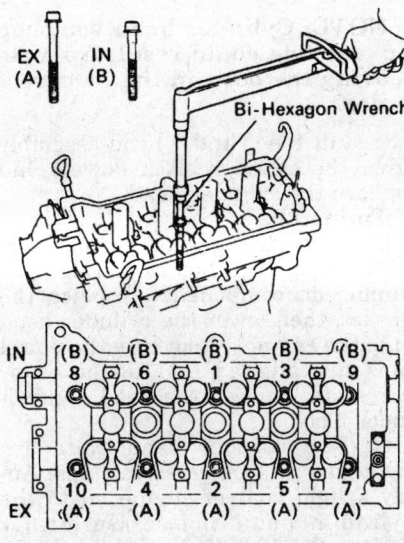

Cylinder head bolt tightening sequence — Tercel and Paseo with 5E-FE engine

Tightening head bolts — Tercel and Paseo with 5E-FE engine

38. Install the valve cover with the proper gasket and sealer and torque the nuts to 61 inch lbs. (7 Nm).
39. Install the intake manifold and torque the nuts and bolts evenly to 14 ft. lbs. (19 Nm).
40. Install the intake manifold stay with the bolt and nut. Torque the nut and bolt to 15 ft. lbs. (20 Nm).
41. Install the MAP sensor vacuum line and the brake booster.
42. Install the fuel injector rail assembly, using new grommets and O-rings. Lightly lubricate the O-rings with gasoline and check that the injectors can be rotated smoothly once pressed in.

43. Using a new gasket, install the exhaust manifold to the engine. Evenly tighten the nuts in several passes and then torque each nut to 35 ft. lbs. (48 Nm).
44. Place the exhaust manifold stay on the engine and exhaust manifold. Tighten the bolt and then the two (2) nuts to 29 ft. lbs. (40 Nm).
45. Install the exhaust manifold heat insulator with the three bolts and torque to 69 inch lbs. (8 Nm).
46. Install the water inlet and outlet housings.
47. Connect the radiator, heating and water inlet hoses.
48. Install the EGR valve, and vacuum modulator. Torque the EGR bolts to 13 ft. lbs. (18 Nm).
49. Install the EGR pipe and torque the union nut to 29 ft. lbs. (40 Nm) and the two nuts to 22 ft. lbs. (30 Nm).
50. Install the throttle body.
51. Reconnect all remaining electrical and vacuum hose fittings.
52. Install the ignition coils and spark plugs.
53. Install the power steering pump bracket. Torque the three bolts to 32 ft. lbs. (43 Nm). Install the power steering belt and adjust the belt.
54. Connect the fuel return hose and the union bolt for the fuel inlet hose. Torque the union bolt to 22 ft. lbs. (29 Nm).
55. Install the air cleaner assembly with the air intake connector.
56. Install and adjust the accelerator cable.
57. If equipped with A/T, install and adjust the throttle cable.
58. Using a new gasket, install the exhaust pipe using two springs and two bolts. Torque the bolts to 46 ft. lbs. (62 Nm).
59. Install the right-hand engine undercover.
60. Connect the battery cable, refill all fluids, and start the engine. Check the ignition timing and check for engine leaks.

1.6L (4A-FE) and 1.8L (7A-FE) Engines
Corolla
1. Release the fuel system pressure.
2. Disconnect the negative battery cable. On vehicles equipped with an air bag, wait at least 90 seconds before proceeding.
3. Drain the engine coolant into a suitable container.
4. Remove the air cleaner and cap assembly.
5. Disconnect the accelerator cable bracket from the throttle body.
6. Remove the alternator.

7. Disconnect the spark plug wires and remove the distributor assembly.

8. Disconnect the oxygen sensor connector and remove the front exhaust pipe.

9. Remove the exhaust manifold.

10. Disconnect the radiator inlet hose and remove the water outlet.

11. Disconnect the electrical connectors and hoses from the water inlet and the water inlet housing and remove the housing.

12. Disconnect the ground strap connector.

13. Disconnect all of the hoses from the air chamber.

14. If equipped with and EGR, remove the EGR VSV and remove the intake manifold stay.

15. Remove the air pipe.

16. Remove the throttle body assembly and remove the air intake chamber.

CAUTION

Fuel injection systems remain under pressure after the engine has been turned OFF. Properly relieve fuel pressure before disconnecting any fuel lines. Failure to do so may result in fire or personal injury.

17. Remove the fuel delivery pipe and the fuel injectors.

18. Disconnect the engine wire harness, if equipped with A/C, disconnect the A/C compressor connector. Disconnect the crankshaft position sensor connector.

19. Remove the intake manifold.

20. Safely support the engine assembly and raise the engine to remove the RH engine mounting insulator.

21. Remove the cylinder head cover and the spark plugs.

22. Remove the No. 3 and the No. 2 timing belt covers.

23. Set No. 1 cylinder to TDC/compression.

24. Place matchmarks on the camshaft timing pulley and timing belt. Remove the plug from the No. 1 timing belt cover, loosen the idler pulley and push the pulley as far left as possible. Remove the timing belt from the camshaft timing pulley.

25. Remove the camshaft timing pulley.

26. Remove the alternator bracket.

27. Remove the oil dipstick guide and the dipstick.

28. Remove the No. 2 water inlet.

29. Remove the intake and exhaust camshafts following the proper sequences and procedures.

30. Using an SST 09205–16010 tool or its equivalent, uniformly loosen the bolts in the reverse order of the installation sequence. Remove the 10 cylinder head bolts in several passes and in sequence.

NOTE: Cylinder head warpage or cracking could result from removing the bolts in the incorrect order.

31. Lift the cylinder head assembly from the cylinder block dowels and remove the cylinder head.

To install:

32. Clean the gasket mating surfaces using care not to damage the aluminum components, replace the gasket, then lower the cylinder head onto the engine. Make sure the dowel pins are aligned and no hoses or wires are between the head and cylinder block.

33. The cylinder head bolts are tightened in 3 progressive steps. Apply a light coat of engine oil to the cylinder head bolts. Uniformly tighten the 10 cylinder head bolts in several passes and in sequence.

NOTE: The cylinder head bolts are in lengths of 3.54 in. (90mm) and 4.25 in. (108mm). The 3.54 in. (90mm) bolts (A) are to be installed in the intake side of the cylinder head. The 4.25 in. (108mm) bolts (B) are to be installed in the exhaust manifold side of the cylinder head.

34. Mark the front of the cylinder head bolt with paint. tighten the cylinder head bolts by 90° in sequence. Tighten an additional 90° and make sure that the paint mark is now positioned toward the rear.

35. Install the intake and exhaust camshafts following the proper sequences and procedures.

36. Check and adjust valve clearance.

37. Install the No. 2 water inlet and connect the water inlet hose.

38. Install the oil dipstick guide, and the dipstick.

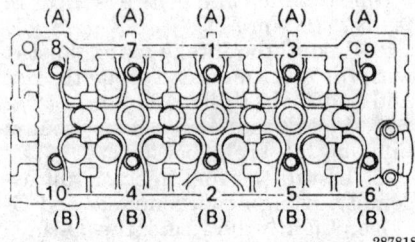

Cylinder head bolt positioning and tightening sequence — 4A-FE and 7A-FE engines

39. Install the alternator bracket.

40. Install the camshaft timing pulley.

41. Install the timing belt to the camshaft timing pulley aligning the matchmarks and properly tensioning the belt with the tensioning pulley. Loosen the pulley bolt ½ of a turn and turn the crankshaft clockwise 2 turns. Make sure that each pulley align with the timing marks.

42. Install the No. 2 and the No. 3 timing belt covers.

43. Install the spark plugs.

44. Install the semi-circular plug with sealant to the cylinder head and install the cylinder head cover with a new gasket to the cylinder head.

45. Install the RH engine mounting insulator and safely lower the engine.

46. Install the intake manifold and ground strap.

47. Connect the engine wire harness to the cylinder head cover and, if equipped with A/C, fasten the compressor connector. Also fasten the oil switch connector and the crankshaft position sensor connector.

48. Install the injectors and the delivery pipe.

49. Install the air intake chamber.

50. Install the throttle body assembly.

51. Install the air pipe.

52. Install the intake manifold stay and if equipped with an EGR, install the EGR VSV.

53. Connect all of the hoses to the air intake chamber.

54. Connect the ground strap connector.

55. Install the water inlet and the water inlet housing and connect all hoses and connectors.

56. Install the water outlet.

57. Install the exhaust manifold.

58. Install the front exhaust pipe and connect the oxygen sensor connectors.

59. Install the distributor and connect the spark plug wires.

60. Install the alternator.

61. Connect the accelerator cable bracket to the throttle body.

62. Install the air cleaner and cap assembly.

63. Connect the negative battery cable, fill the engine with coolant, start the engine, warm up, and check for leaks.

64. Install the RH engine undercover, check ignition timing, and road test for proper operation.

Celica

1. Release the fuel system pressure.

2. Disconnect the negative battery cable. On vehicles equipped with an

air bag, wait at least 90 seconds before proceeding.

3. Drain the engine coolant into a suitable container.

4. Remove the air cleaner and cap assembly.

5. Disconnect the accelerator cable bracket from the throttle body. Disconnect the throttle cable if equipped with A/T.

6. Remove the alternator.

7. If equipped with A/C, remove the A/C drive belt and idler pulley.

8. Remove the P/S pump adjusting bracket and drive belt.

9. Disconnect the spark plug wires and remove the distributor assembly.

10. Remove the front TWC.

11. Remove the exhaust manifold.

12. Disconnect the radiator inlet hose and remove the water outlet.

13. Disconnect the electrical connectors and hoses from the water inlet and the water inlet housing and remove the housing.

14. Disconnect all of the hoses from the air chamber.

15. If equipped with an EGR, remove the EGR VSV and remove the intake manifold stay.

16. Remove the air pipe.

17. Remove the throttle body assembly and remove the air intake chamber.

——— CAUTION ———
Fuel injection systems remain under pressure after the engine has been turned OFF. Properly relieve fuel pressure before disconnecting any fuel lines. Failure to do so may result in fire or personal injury.

18. Remove the fuel delivery pipe and the fuel injectors.

19. Disconnect the engine wire harness, if equipped with A/C, disconnect the A/C compressor connector. Disconnect the crankshaft position sensor connector.

20. Remove the intake manifold.

21. Remove the mounting bolt attaching the oil dipstick guide and the engine wire harness bracket.

22. Remove the No. 2 water inlet.

23. Remove the cylinder head cover and the spark plugs.

24. Remove the No. 3 and the No. 2 timing belt covers.

25. Set No. 1 cylinder to TDC/compression.

26. Place matchmarks on the camshaft timing pulley and timing belt. Remove the plug from the No. 1 timing belt cover, loosen the idler pulley and push the pulley as far left as possible. Remove the timing belt from the camshaft timing pulley.

27. Remove the camshaft timing pulley.

28. Remove the alternator bracket.

29. Remove the camshafts following the proper sequences and procedures.

30. Using an SST 09205–16010 tool or its equivalent, uniformly loosen the bolts in the reverse order of the installation sequence. Remove the 10 cylinder head bolts in several passes and in sequence.

NOTE: Cylinder head warpage or cracking could result from removing the bolts in the incorrect order.

31. Lift the cylinder head assembly from the cylinder block dowels and remove the cylinder head.

To install:

32. Clean the gasket mating surfaces using care not to damage the aluminum components, replace the gasket, then lower the cylinder head onto the engine. Make sure the dowel pins are aligned and no hoses or wires are between the head and cylinder block.

33. The cylinder head bolts are tightened in 3 progressive steps. Apply a light coat of engine oil to the cylinder head bolts. Uniformly tighten the 10 cylinder head bolts in several passes and in sequence. The cylinder head bolt torque is 22 ft. lbs. (29 Nm).

NOTE: The cylinder head bolts are in lengths of 3.54 in. (90mm) and 4.25 in. (108mm). The 3.54 in. (90mm) bolts (A) are to be installed in the intake side of the cylinder head. The 4.25 in. (108mm) bolts (B) are to be installed in the exhaust manifold side of the cylinder head.

34. Mark the front of the cylinder head bolt with paint. Tighten the cylinder head bolts by 90° in sequence. Tighten an additional 90° and make sure that the paint mark is now positioned toward the rear.

35. Install the camshafts following the proper sequences and procedures.

36. Check and adjust valve clearance.

37. Install the No. 2 water inlet and connect the water inlet hose.

38. Install the oil dipstick guide, and the dipstick.

39. Install the alternator bracket.

40. Install the camshaft timing pulley.

41. Install the timing belt to the camshaft timing pulley aligning the matchmarks and properly tensioning the belt with the tensioning pulley. Loosen the pulley bolt ½ of a turn and turn the crankshaft clockwise 2

turns. Make sure that each pulley align with the timing marks.

42. Install the No. 2 and the No. 3 timing belt covers.

43. Install the spark plugs.

44. Install the semi-circular plug with sealant to the cylinder head and install the cylinder head cover with a new gasket to the cylinder head.

45. Install the intake manifold and the ground strap. Torque the 7 bolts and 2 nuts to 14 ft. lbs. (19 Nm).

46. Install the intake manifold stay and connect the engine wire harness.

47. Install the injectors and the delivery pipe.

48. Install the air intake chamber.

49. Install the throttle body assembly.

50. Install the air pipe.

51. Install the intake manifold stay and if equipped with an EGR, install the EGR VSV.

52. Connect all of the hoses to the air intake chamber.

53. Install the water inlet and the water inlet housing and reconnect all hoses and connectors.

54. Install the water outlet.

55. Install the exhaust manifold.

56. Install the front TWC.

57. Install the front exhaust pipe.

58. Install the distributor and reconnect the spark plug wires.

59. Temporarily install the water pump pulley.

60. Install the P/S pump adjusting bracket and drive belt.

61. If equipped with A/C, install The A/C idler pulley and drive belt.

62. Install the alternator and drive belt.

63. Tighten the water pump pulley bolts.

64. Reconnect the accelerator cable bracket to the throttle body. Reconnect the throttle cable if equipped with A/T. If equipped with cruise control, install the cruise control actuator cable.

65. Install the air cleaner and cap assembly.

66. Reconnect the negative battery cable, fill the engine with coolant, start the engine, warm up, and check for leaks.

67. Install the RH engine undercover, check ignition timing, and road test for proper operation.

2.0L (3S-GTE) Engine

Celica

1. Disconnect the cable at the negative terminal of the battery. On vehicles equipped with an air bag, wait at least 90 seconds before proceeding.

2. Drain the engine coolant.

3. Tag and disconnect the ignition coil connector and the spark plug wire at the ignition coil. Remove the upper suspension brace. Remove the intercooler, if necessary for clearance.

4. On models with AT, disconnect the throttle cable with its bracket from the throttle body.

5. Disconnect the accelerator cable from the throttle body.

6. Remove the radiator overflow tank.

7. Remove the cruise control actuator and its bracket.

8. Disconnect the air flow meter connector. Remove the air cleaner cap clips. Loosen the hose clamp and remove the air cleaner hose and the air flow meter along with the air cleaner top. Lift out the filter element and then remove the air cleaner case.

9. Tag and disconnect the oxygen sensor lead. Remove the exhaust manifold heat insulator and disconnect or remove the front exhaust pipe.

10. Remove the alternator and its main bracket.

11. Raise the front of the vehicle and support it with safety stands. Remove the right front wheel.

12. Remove the right side engine undercover and remove the lower suspension crossmember.

13. Remove the turbocharger assembly and catalytic converter attaching exhaust clamps.

14. Remove the exhaust manifold stay and the EGR pipe. Unbolt the manifold and remove it along with the lower heat insulator.

15. Remove the distributor.

16. Tag and disconnect the oil pressure switch connector.

17. Tag and disconnect all electrical leads and vacuum hoses at the water outlet. Remove the upper radiator hoses, the heater outlet hose and the water bypass hose. Remove the water outlet.

18. Disconnect the heater inlet hose and the water bypass hose and then remove the water bypass pipe.

19. Disconnect the throttle position sensor lead, the ventilation hose, the air valve hose and any emission control vacuum hoses at the throttle body. Remove the four bolts and lift out the throttle body.

20. Remove the forward engine hanger and the No. 2 intake manifold stay.

21. Remove the EGR vacuum modulator.

22. Tag and disconnect any remaining vacuum hoses which may interfere with cylinder head removal.

23. Tag and disconnect the fuel injector electrical leads at the injector.

24. Disconnect the fuel inlet hose at the fuel filter. Disconnect the fuel return hose at the return pipe.

25. Remove the No. 1 and No. 3 intake manifold stays.

26. Tag and disconnect the two VSV connectors. Disconnect the two power steering vacuum hoses.

27. Remove the intake manifold and the air control valve.

28. Remove the fuel delivery pipe with the injectors attached. Pull the four injector insulators out of the injector holes in the cylinder head.

29. Remove the cylinder head cover. Remove the spark plugs.

30. Remove the No. 1 engine hanger.

31. Remove the power steering reservoir and position it out of the way with the hydraulic lines still attached.

32. Remove the camshaft timing pulleys. Remove the No. 1 idler pulley and tension spring.

33. Remove the bolt holding the No. 2 and No. 3 timing covers. Remove the four mounting bolts and remove the No. 3 timing cover.

34. Remove the camshafts following the proper sequences and procedures.

NOTE: When removing the camshaft bearing caps, keep them in the proper order.

35. Loosen and remove the cylinder head bolts in several stages with the proper socket tool, in the reverse order of the installation sequence. Remove the cylinder head.

To install:

36. Position the cylinder head onto the cylinder block with a new gasket. Lightly coat the cylinder head bolts with engine oil. Install them into the head and tighten them in several passes, in the sequence shown and to the following torque: 36 ft. lbs. (49 Nm), plus an additional 90 degrees.

NOTE: Place paint marks on the bolts right before tightening the additional 90 degrees to keep track of the angle tightened.

37. Install the camshafts following the proper sequences and procedures.

38. Coat the inside of a new oil seal with grease and carefully tap it onto the camshaft with a drift (SST No. 09223–50010) or equivalent.

39. Install the No. 3 timing belt cover.

40. Connect the idler pulley tension spring to the pulley and the pin on the cylinder head. Install the idler pulley onto the pivot pin, force it to the left as far as it will go and tighten

it. Make sure that the tension spring is not out of groove in the pin.

41. Install the camshaft timing pulleys and the timing belt, using the proper procedures.

42. Installation of the remaining components is in the reverse order of removal. Tighten the lower suspension crossmember end bolts to 154 ft. lbs. (208 Nm) and the center bolt to 29 ft. lbs. (39 Nm). Tighten the upper suspension brace bolts to 15 ft. lbs. (20 Nm) and the nuts to 47 ft. lbs. (64 Nm). Refill the engine with coolant an check the idle speed and ignition timing. Road test the vehicle for proper operation.

MR2

1. Release the fuel system pressure.

2. Disconnect the negative battery cable. On vehicles equipped with an air bag, wait at least 90 seconds before proceeding.

3. Drain the engine coolant.

4. Remove the undercovers from the vehicle.

5. Remove the engine hood side panels.

6. Remove the upper suspension brace.

7. Remove the No. 1 and No. 2 intake air connectors.

8. If equipped with cruise control, remove the cruise control actuator and the accelerator linkage.

9. If not equipped with cruise control, disconnect the accelerator cable from the throttle body.

10. Remove the air cleaner.

11. Disconnect the ground strap connector.

12. Disconnect the brake booster vacuum hose.

13. Disconnect the A/C idle–up air hoses.

   a. Air hose from the No. 2 air tube

   b. Air hose from the intake manifold

14. Remove the right front engine hanger.

15. Remove the engine compartment electric cooling fan.

16. Disconnect the A/C compressor from the engine.

17. Remove the Charge Air Cooler (CAC).

18. Remove the tailpipe.

19. Remove the front exhaust pipe.

20. Remove the front engine mounting insulator.

21. Remove the front engine mounting bracket and the clutch release cylinder.

22. Remove the catalytic converter assembly.

23. Remove the turbocharger.

24. Remove the throttle body.

━━━ **CAUTION** ━━━
Fuel injection systems remain under pressure after the engine has been turned OFF. Properly relieve fuel pressure before disconnecting any fuel lines. Failure to do so may result in fire or personal injury.

25. Remove the cold start injector pipe.

NOTE: Slowly loosen the union bolts. Put a suitable container under the injector pipe.

26. Disconnect the cold start injector connector.
27. Remove the distributor.
28. Remove the exhaust manifold.
29. Remove the No. 2 air tube.
   a. Air hose from the No. 1 air tube
   b. Air hose from the intake manifold.
30. Remove the left engine hanger.
31. Remove the EGR vacuum modulator and the VSV.
32. Disconnect the EGR and the vacuum hose from the EGR valve.
33. Remove the EGR valve and pipe.
34. Disconnect the ECT sensor, ECT gauge sender, and the cold start injector time switch connector. Remove the two bolts and the water outlet.
35. Disconnect the following hoses:
   a. Water filler hose
   b. Radiator hose
   c. Water bypass pipe hose from the IAC valve
   d. Heater water hose
   e. Water bypass hose from the water bypass pipe
   f. TVV vacuum hoses for the EVAP
36. Remove the oil pressure switch.
37. Remove the oil cooler.
38. Disconnect the following hoses:
   a. Water bypass hose from the cylinder block
   b. Water bypass hoses from the No. 1 air tube
   c. Air hose from the VSV for turbocharging pressure
   d. Heater water hose.
39. Remove the three bolts, two nuts, the water bypass pipe, gasket, and the O-ring.
40. Remove the intake manifold stays.
41. Remove the No. 1 air tube.
42. Remove the T-VIS vacuum tank, VSV for T-VIS, and the VSV for turbocharging pressure and bracket.

43. Disconnect the engine wire from the intake manifold and disconnect the electrical connectors.
44. Remove the intake manifold and T-VIS valve.
45. Remove the fuel delivery pipe assembly.
46. Remove the cylinder head cover.
47. Disconnect the timing belt from the camshaft timing pulleys.
48. Remove the camshaft timing pulleys.
49. Remove the No. 1 idler pulley.
50. Remove the No. 3 timing belt cover.
51. Support the timing belt, so that the meshing of the crankshaft timing pulley and the timing belt does not shift.
52. Remove the camshafts following the proper sequences and procedures.
53. Loosen and remove the cylinder head bolts in several stages with the proper socket tool, in the reverse order of the installation sequence. Remove the cylinder head.
To install:
54. Install the cylinder head with a new gasket.

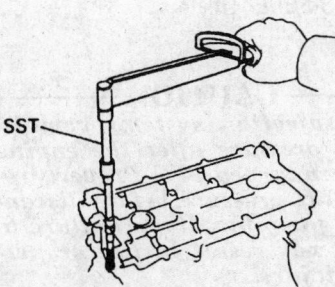

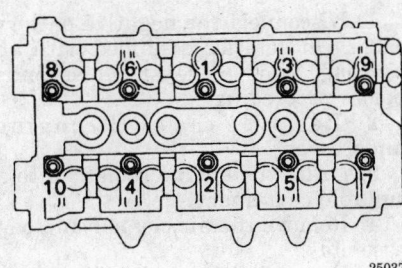

250378

Cylinder head bolts tightening sequence — 3S-GTE engine

55. Tighten the bolts in sequence in two progressive steps with a final torque of 36 ft. lbs. (49 Nm).

NOTE: Apply a light coat of engine oil to the threads and under the heads of the cylinder head bolts.

56. Install the camshafts following the proper sequences and procedures.
57. Adjust the valve clearance.
58. Install the No. 3 timing belt cover.
59. Install the No. 1 idler pulley and torque the bolt to 38 ft. lbs. (52 Nm). Make sure that the pulley moves smoothly.
60. Install the camshaft timing pulley.
   a. Using a wrench, turn and align the groove of the camshaft with the dot mark of the No. 1 camshaft bearing cap.
   b. Slide the timing pulley onto the camshaft facing the mark **S** upward.
   c. Align the pin holes of the camshaft and the timing pulley and insert the knock pin.
   d. Hold the hexagon wrench portion of the camshaft with a wrench and tighten the bolts. Torque to 43 ft. lbs. (59 Nm).
61. Connect the timing belt to the camshaft timing pulleys.
62. Install the cylinder head cover. Torque the bolts to 4 ft. lbs. (6 Nm).
63. Install the fuel delivery pipe assembly.
64. Install the T-VIS valve and the intake manifold. Torque the bolts to 14 ft. lbs. (19 Nm).
65. Install the engine wire to the intake manifold and connect the electrical connectors.
66. Install the T-VIS vacuum tank, VSV for T-VIS, and the VSV for the turbocharging pressure assembly.
67. Install the No. 1 air tube.
68. Install the intake manifold stays. Torque the bolts to 19 ft. lbs. (25 Nm).
69. Install the water bypass pipe with a new gasket and O-ring. Torque the nuts to 6 ft. lbs. (8 Nm).
70. Connect the following hoses:
   a. Water bypass hose to the cylinder block
   b. Water bypass hoses to the No. 1 air tube
   c. Air hose to the VSV for turbocharging pressure
   d. Heater water hose
71. Install the oil cooler.
72. Install the oil pressure switch.
73. Install the water outlet and housing. Torque the bolts to 29 ft. lbs. (39 Nm). Secure the fuel inlet and re-

turn line clamps with the water outlet bolts

74. Connect the ECT sensor, ECT gauge sender, and the cold start injector time switch connectors.

75. Connect the following hoses:
   a. Water filler hose
   b. Radiator hose
   c. Water bypass pipe hose from the IAC valve
   d. Heater water hose
   e. Water bypass hose to the water bypass pipe
   f. Two TVV vacuum hoses for EVAP

76. Install the EGR valve and pipe. Torque the bolts to the intake manifold to 14 ft. lbs. (19 Nm) and the bolts to the cylinder head to 19 ft. lbs. (25 Nm). Connect the vacuum hose to the EGR valve.

77. Install the EGR vacuum modulator and the VSV.

78. Install the left engine hanger. Torque the 12mm bolts to 9 ft. lbs. (13 Nm) and the 14mm bolts to 14 ft. lbs. (19 Nm).

79. Install the No. 2 air tube.

80. Install the distributor.

81. Install the exhaust manifold with a new gasket and torque the bolts to 38 ft. lbs. (52 Nm).

82. Install the cold start injector pipe and torque the bolts to 9 ft. lbs. (12 Nm). Connect the cold start injector connector.

83. Install the throttle body with a new gasket and torque the bolts to 14 ft. lbs. (19 Nm).

84. Install the turbocharger and stays.

85. Install the TWC with a new gasket. Torque the bolts to 22 ft. lbs. (29 Nm).

86. Install the clutch release cylinder and the front engine mounting bracket. Torque the clutch release cylinder bolts to 9 ft. lbs. (12 Nm) and the mounting bracket bolts to 57 ft. lbs. (77 Nm).

87. Install the front engine mounting insulator to the body and torque the bolts to 47 ft. lbs. (64 Nm).

88. Install the mounting insulator to the mounting bracket with the through-bolt and torque it to 71 ft. lbs. (96 Nm).

89. Install the front exhaust pipe. Torque the bolts to 46 ft. lbs. (62 Nm).

90. Install the tailpipe and tighten the front pipe to rear to 32 ft. lbs. (43 Nm). Tighten the front pipe to bracket to 14 ft. lbs. (19 Nm).

91. Install the Charge Air Cooler (CAC).

92. Install the A/C compressor and torque the mounting bolts to 18 ft. lbs. (25 Nm).

93. Install the electric cooling fan.

94. Install the right front engine hanger. Torque the bolts to the cylinder head to 29 ft. lbs. (39 Nm) and the bolts to the mounting bracket to 45 ft. lbs. (61 Nm).

95. Connect the A/C idle–up air hoses.

96. Connect the brake booster vacuum hose and the ground strap connector.

97. Install the air cleaner.

98. If removed, install the cruise control actuator and the accelerator linkage.

99. Connect the accelerator cable to the throttle body.

100. Install the No. 1 and the No. 2 intake air connectors.

101. Install the upper suspension brace and torque the bolt to 54 ft. lbs. (73 Nm) and the nut to 47 ft. lbs. (64 Nm).

102. Fill the engine with coolant.

103. Check the engine oil level and add oil as necessary.

104. Connect the negative battery cable.

105. Start the engine and check for leaks.

106. Adjust the ignition timing.

107. Install the engine hood side panels.

108. Install the engine undercovers.

109. Road test the vehicle.

2.2L (5S-FE) Engine

Celica

CAUTION

Fuel injection systems remain under pressure after the engine has been turned OFF. Properly relieve fuel pressure before disconnecting any fuel lines. Failure to do so may result in fire or personal injury.

1. Disconnect the negative battery cable. On vehicles equipped with an air bag, wait at least 90 seconds before proceeding.

2. Remove the RH engine undercover.

3. Drain the engine coolant into a suitable container.

4. Remove the air cleaner and cap assembly.

5. Remove the spark plug wires and the distributor assembly.

6. Remove the alternator.

7. Disconnect the sub oxygen and oxygen sensor connectors and remove the front exhaust pipe, the front TWC, and the exhaust manifold.

8. Disconnect the oil pressure switch connector.

9. Disconnect the sensor connectors and hoses from the water outlet. Remove the two nuts and gasket; remove the water outlet.

10. Disconnect the hoses; remove the heat protector and the water bypass pipe.

11. Remove the throttle body assembly.

12. Disconnect the vacuum hoses from the intake manifold and remove the A/C idle–up valve.

13. Remove the EGR valve, vacuum modulator, vacuum hoses and gasket.

14. Remove the intake manifold stay and disconnect the A/T throttle control cable.

15. Remove the air hoses from the air tube and remove the air tube assembly.

16. Except California vehicles, disconnect the sensing hoses and remove the vacuum pipe.

17. Disconnect the knock sensor connector, remove the bolt and the ground cable from the intake manifold.

18. On California vehicles, remove the VSV for fuel pressure control and EGR. On all vehicles (except California), remove the VSV for EGR.

19. Disconnect the PCV hose from the intake manifold and disconnect the A/T throttle control cable and bracket from the intake manifold.

20. Disconnect the engine wire harness from the starter bracket. Disconnect the VSS sensor connector and the wire harness protector from the LH side of the manifold.

21. Disconnect the fuel inlet hose from the delivery pipe and disconnect the fuel return hose from the return pipe.

22. Remove the six bolts, two nuts, and the intake manifold and gasket.

23. On California vehicles, remove the air hose for the air assist system.

24. Remove the fuel delivery pipe and the injectors.

25. Remove the timing belt from the camshaft timing pulley and remove the camshaft timing pulley.

26. Remove No. 1 idler pulley and tension spring.

27. Remove four bolts and the No. 3 timing belt cover.

NOTE: Support the timing belt, so that the meshing of the crankshaft timing pulley and the timing belt does not shift. Be careful not to drop anything inside the timing belt cover.

28. Remove the engine hangers and the alternator bracket.

29. Remove the oil pressure switch.

30. Remove the cylinder head cover. Remove the spark plug wire clamp, PCV valve and hoses from the cylinder head cover.

31. Remove the camshafts following the proper sequences and procedures.

32. Uniformly loosen and remove the cylinder head bolts in several passes and in the reverse order of the installation sequence. Lift the cylinder head from the cylinder block disengaging the cylinder head from the block dowel pins.

To install:

33. Clean the gasket mating surfaces using care not to damage the aluminum components, replace the gasket, then lower the cylinder head onto the engine. Make sure the dowel pins are aligned and no hoses or wires are between the head and cylinder block.

34. The cylinder head bolts are tightened in two progressive steps. Apply a light coat of engine oil to the cylinder head bolts. Uniformly tighten the 10 cylinder head bolts in several passes and in sequence. The torque for the head bolts is 36 ft. lbs. Mark the front of the cylinder head bolt with paint. Tighten the cylinder head bolts by 90° in sequence. Tighten an additional 90° and make sure that the paint mark is now positioned toward the rear.

35. Install the camshafts following the proper sequences and procedures.

36. Check and adjust valve clearance.

37. Install sealant to the two new semi-circular seals and install the seals to the cylinder head.

38. Install the PCV valve and hoses and the spark plug wire clamp to the cylinder head cover.

39. Install the cylinder head cover with a new gasket, the four grommets and nuts and uniformly tighten the nuts in several passes. The torque for the nuts is 17 ft. lbs. (23 Nm).

40. Install the oil pressure switch.

41. Install the alternator bracket and the engine hangers.

42. Install the No. 3 timing belt cover and temporarily install the No. 1 idler pulley and tension spring.

43. Install the camshaft timing pulley and install the timing belt to the pulley. Correctly tension the timing belt and make sure that the belt timing is correct.

44. Install the fuel injectors and the delivery pipe.

45. On California cars, install the air hose for the air assist system.

46. Install the intake manifold and insert the engine wire harness between the head and the intake mani-fold, install a new gaskets the six bolts and two nuts, and torque the intake manifold to 25 ft. lbs. (34 Nm)

47. Connect the fuel inlet and return hoses to the delivery and return pipes.

48. Install the engine wire protector and harness to the intake manifold and connect the four fuel injector connectors. Connect the VSS connector.

49. Install the cable bracket to the intake manifold and install the accelerator, A/T control cables and bracket.

50. Connect the PCV hose to the intake manifold.

51. On California vehicles, install the VSV assembly for fuel pressure control and EGR.

52. Except for California vehicles, install the VSV for EGR.

53. Connect the knock sensor connector and install the ground cable with the bolt.

54. Except for California vehicles, install the sensing hoses to the vacuum pipe and mount the pipe with the bolt.

55. Install the air tube and the bracket for the EGR. Connect the hoses to the air pipe.

56. Install the A/T throttle control cable to the clamp on the rear side of the intake manifold. Install the manifold stay and torque the bolt to 15 ft. lbs. (21 Nm) and torque the nut to 32 ft. lbs. (44 Nm).

57. Install the EGR valve and the vacuum modulator using a new gasket, connect the two vacuum hoses to the VSV for the EGR, and connect the EGR gas temperature sensor connector. Connect the EVAP hose to the charcoal canister.

58. Install the vacuum sensor hose to the gas filter and the brake booster vacuum hose to the intake manifold.

59. Install the A/C idle up valve.

60. Install the throttle body.

61. Connect the water bypass pipe to the water pump cover; install the water bypass pipe and connect the hoses to the water bypass pipe.

62. Install the water outlet with a new gasket and the two nuts and install the water hoses, vacuum hoses, and connect the sensor connectors.

63. Connect the oil pressure switch connector.

64. Assemble the TWC to the exhaust manifold and install the main oxygen sensor to the exhaust manifold. Install the sub oxygen sensor to the TWC.

65. Install the exhaust manifold to the cylinder head and connect the oxygen sensors connectors.

66. Install the front exhaust pipe and install the alternator with the drive belt.

67. Install the distributor assembly and connect the spark plug wires.

68. Install the air cleaner and cap assembly.

69. Connect the negative battery cable, fill the engine with coolant, start the engine, warm up, and check for leaks. Bleed the cooling system and top off coolant as necessary.

70. Install the RH engine undercover, check ignition timing, and road test the vehicle for proper operation.

Camry

1. Release the fuel system pressure.

2. Disconnect the negative battery cable. On vehicles equipped with an air bag, wait at least 90 seconds before proceeding.

3. Drain the engine coolant into a suitable container.

4. Disconnect the A/T throttle control cable and accelerator cable.

5. Remove the air cleaner and cap assembly.

6. Remove the alternator.

7. Remove the spark plug wires and the distributor assembly.

8. Disconnect the sub oxygen and oxygen sensor connectors and remove the front exhaust pipe, the front TWC, and the exhaust manifold.

9. Disconnect the oil pressure switch connector.

10. Disconnect the sensor connectors and hoses from the water outlet.

11. Remove the two nuts, water outlet and gasket.

12. Disconnect the hoses and remove the water bypass pipe.

13. Remove the throttle body assembly.

14. Remove the EGR valve, vacuum modulator, vacuum hoses assembly, and the gasket.

15. Disconnect the PCV hose from the intake manifold.

16. Remove the air hoses from the air tube and remove the air tube assembly.

17. Remove the air tube as follows:

   a. Remove the two power steering hoses from the air tube.

   b. Remove the power steering air hose from the intake manifold.

   c. Remove the air hose from the fuel pressure regulator.

   d. Remove the three bolts and air tube. Disconnect the ground cable and clamp bracket for the engine wire.

18. Remove the bolt and disconnect the ground cable from the intake manifold.

19. Disconnect the knock sensor connector.

20. Remove the bolt and disconnect the engine wire protector from the left-hand side of the intake manifold.

21. Disconnect the four injector connectors.

22. Disconnect the engine wire protector from the two brackets on the front side of the intake manifold.

23. Disconnect the engine wire protector from the two mounting bolts of the No. 2 timing belt cover.

24. Remove the four bolts, wire bracket, No. 1 air intake chamber and intake manifold stays.

25. Remove the six bolts and two nuts and remove the intake manifold and gasket.

--- **CAUTION** ---

Fuel injection systems remain under pressure after the engine has been turned OFF. Properly relieve fuel pressure before disconnecting any fuel lines. Failure to do so may result in fire or personal injury.

26. Remove the fuel delivery pipe and the injectors.

27. Remove the timing belt from the camshaft timing pulley. Remove the camshaft timing pulley.

NOTE: Support the timing belt, so that the meshing of the crankshaft timing pulley and the timing belt does not shift. Be careful not to drop anything inside the timing belt cover.

28. Remove No. 1 idler pulley and tension spring.

29. Remove four bolts and the No. 3 timing belt cover.

30. Remove the engine hangers and the alternator bracket.

31. Remove the oil pressure switch.

32. Remove the cylinder head cover.

33. Remove the camshafts following the proper sequences and procedures.

34. Uniformly loosen and remove the cylinder head bolts in several passes and in the reverse order of the installation sequence. Lift the cylinder head from the cylinder block disengaging the cylinder head from the block dowels.

To install:

35. Clean the gasket mating surfaces, but be careful not to damage the aluminum components, replace the gasket, then lower the cylinder head onto the engine. Make sure the dowel pins are aligned and no hoses or wires are between the head and cylinder block.

36. The cylinder head bolts are tightened in two progressive steps. Apply a light coat of engine oil to the cylinder head bolts. Uniformly tighten the 10 cylinder head bolts in several passes and in sequence. The torque for the head bolts is 36 ft. lbs. (49 Nm). Mark the front of the cylinder head bolt with paint. Tighten the cylinder head bolts an additional 90° in the proper sequence. The paint mark should now be 90° from the front.

37. Install the camshafts following the proper sequences and procedures.

38. Check and adjust valve clearance.

39. Install sealant to the two new semi-circular seal and install the seals to the cylinder head.

40. Install the cylinder head cover with a new gasket, the four grommets and nuts and uniformly tighten the nuts in several passes. The torque for the nuts is 17 ft. lbs. (23 Nm).

41. Install the oil pressure switch.

42. Install the alternator bracket and the engine hangers.

43. Install the No. 3 timing belt cover and temporarily install the No. 1 idler pulley and tension spring.

44. Install the camshaft timing pulley and install the timing belt to the pulley. Correctly tension the timing

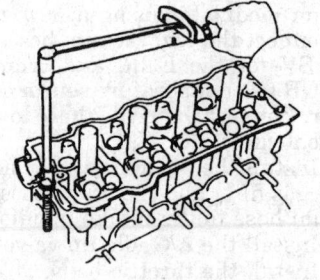

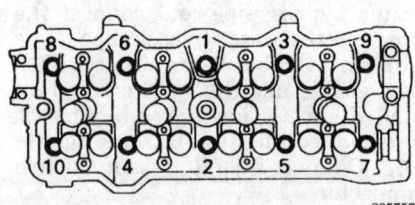

Cylinder head bolt tightening sequence — Camry, Celica and MR2 with 5S-FE engine

belt and make sure that the belt timing is correct.

45. Install the fuel injectors and the delivery pipe.

46. Install the intake manifold.

47. Install the engine wire protector and harness to the intake manifold and connect the four fuel injector connectors.

48. Install the air tube.

49. Connect the PCV hose to the intake manifold.

50. Install the EGR valve and the vacuum modulator using a new gasket, connect the two vacuum hoses to the VSV for the EGR, and connect the EGR gas temperature sensor connector.

51. Install the throttle body.

52. Connect the water bypass pipe to the water pump cover, install the water bypass pipe and reconnect the hoses to the water bypass pipe.

53. Install the water outlet with a new gasket and the two nuts. Torque the nuts to 11 ft. lbs. (15 Nm), Install the water hoses, vacuum hoses, and reconnect the sensor connectors.

54. Connect the oil pressure switch connector.

55. Install the exhaust manifold to the cylinder head. Uniformly tighten the nuts in several passes to 36 ft. lbs. (49 Nm).

56. Install the manifold stay with the bolt and nut. Torque the bolt and nut to 31 ft. lbs. (42 Nm).

57. Install the No. 1 manifold stay with the bolt and nut. Torque the bolt and nut to 31 ft. lbs. (42 Nm).

58. Install the manifold upper heat insulator with the four bolts.

59. Install the main oxygen sensor to the exhaust manifold. Install the sub oxygen sensor to the TWC.

60. Install the front exhaust pipe to the catalytic converter and torque the nuts to 46 ft. lbs. (62 Nm).

61. Install the alternator and drive belt.

62. Install the distributor assembly and reconnect the spark plug wires.

63. Install the air cleaner and cap assembly.

64. For vehicles with A/T, connect and adjust the throttle cable and accelerator cable.

65. Fill the engine with coolant.

66. Check all fluids.

67. Connect the negative battery cable to the battery.

68. Start the engine and check for leaks.

MR2

1. Release the fuel system pressure.

CAUTION

Fuel injection systems remain under pressure after the engine has been turned OFF. Properly relieve fuel pressure before disconnecting any fuel lines. Failure to do so may result in fire or personal injury.

2. Disconnect the negative battery cable. On vehicles equipped with an air bag, wait at least 90 seconds before proceeding.

3. Drain the engine coolant into a suitable container.

4. Remove the undercovers from the vehicle.

5. Remove the engine hood side panels.

6. Disconnect the accelerator cable from the throttle body.

7. If equipped with A/T, disconnect the throttle cable from the throttle body.

8. If equipped with cruise control, remove the cruise control actuator and the accelerator linkage.

9. Remove the air cleaner.

10. Disconnect the vacuum hoses from the MAP sensor, the brake booster, and the A/C idle up valve.

11. Remove the spark plug wires and the distributor assembly.

12. Remove the front exhaust pipe, the front TWC, and the exhaust manifold.

13. Remove the A/C compressor.

14. Remove the main oxygen sensor connector and on California models remove the sub-oxygen sensor connector.

15. Disconnect the oil pressure switch connector.

16. Disconnect the fuel inlet and return tubes from the water outlet housing.

17. Disconnect the ECT sensor and sender gauge connectors.

18. Disconnect the following hoses:
  a. Water filler hose.
  b. Radiator hose and water bypass hose.
  c. Heater water hose and the vacuum hoses from the upper and lower ports of the TVV.
  d. IAC valve water bypass pipe.

19. Remove the water bypass pipe as follows:
  a. Remove the bolt, two nuts, and the water bypass heat protector.
  b. Disconnect the IAC valve water bypass pipe, heater water hose, and the two oil cooler water bypass hoses.

20. Remove the throttle body assembly.

21. Remove the EGR valve, vacuum modulator, vacuum hoses, and the gasket.

22. Disconnect the vacuum hoses and remove the fuel pressure VSV.

23. Disconnect the knock sensor connector and remove the EGR VSV.

24. Remove the intake manifold.

25. Remove the delivery pipe and the four fuel injectors.

26. Remove the No. 2 timing belt cover, the timing belt, and the camshaft timing pulley.

27. Remove the No.1 idler pulley and the tension spring.

28. Remove four bolts and the No. 3 timing belt cover.

NOTE: Support the timing belt, so that the meshing of the crankshaft timing pulley and the timing belt does not shift. Be careful not to drop anything inside the timing belt cover.

29. Remove the engine hangers.

30. Remove the oil pressure switch.

31. Remove the cylinder head cover. Remove the spark plug wire clamp, PCV valve and hoses from the cylinder head cover.

32. Remove the camshafts following the proper sequences and procedures.

33. Uniformly loosen and remove the cylinder head bolts in several passes, in the reverse order of the installation sequence. Lift the cylinder head from the cylinder block, disengaging the cylinder head from the block dowels.

To install:

34. Clean the gasket mating surfaces, but be careful not to damage the aluminum components. Replace the gasket, then lower the cylinder head onto the engine. Make sure the dowel pins are aligned and no hoses or wires are between the head and cylinder block.

35. The cylinder head bolts are tightened in two progressive steps. Apply a light coat of engine oil to the cylinder head bolts. Uniformly tighten the 10 cylinder head bolts in several passes and in sequence. The torque for the head bolts is 36 ft. lbs. (49 Nm).

36. Mark the front of the cylinder head bolt with paint. Tighten the cylinder head bolts an additional 90° in the proper sequence. The paint mark should now be 90° from the front.

37. If removed, install the spark plug tubes.

38. Install the camshafts following the proper sequences and procedures.

39. Check and adjust valve clearance.

40. Install sealant to the two new semi-circular seal and install the seals to the cylinder head.

41. Install the PCV valve and hoses and the spark plug wire clamp to the cylinder head cover.

42. Install the cylinder head cover with a new gasket, the four grommets and nuts and uniformly tighten the nuts in several passes. The torque for the nuts is 17 ft. lbs. (23 Nm).

43. Install the engine hangers.

44. Install the oil pressure switch.

45. Install the No. 3 timing belt cover and temporarily install the No. 1 idler pulley and tension spring.

46. Install the camshaft timing pulley and install the timing belt to the pulley. Correctly tension the timing belt and make sure that the belt timing is correct.

47. Install the No. 2 timing belt cover.

48. Install the fuel injectors and the delivery pipe.

49. Connect the fuel inlet and return lines.

50. On California vehicles, install the air hose for the air assist system.

51. Install the intake manifold. Install a new gasket and uniformly tighten the bolts and nuts in several passes to 14 ft. lbs. (19 Nm).

52. Connect the PCV hose to the intake manifold.

53. Install the cable bracket to the intake manifold and install the accelerator, A/T control cables and bracket.

54. On California vehicles, install the VSV assembly for fuel pressure control and EGR.

55. On vehicles (except California), install the VSV for EGR.

56. Connect the knock sensor connector and install the ground cable with the bolt.

57. On vehicles (except California), install the sensing hoses to the vacuum pipe and mount the pipe with the bolt.

58. Install the air tube and the bracket for the EGR. Connect the hoses to the air pipe.

59. Install the A/T throttle control cable to the clamp on the rear side of the intake manifold. Install the manifold stay and torque the bolt to 15 ft. lbs. (21 Nm) and torque the nut to 32 ft. lbs. (44 Nm).

60. Install the EGR valve and the vacuum modulator using a new gasket, connect the two vacuum hoses to the VSV for the EGR, and connect the EGR gas temperature sensor connector. Connect the EVAP hose to the charcoal canister.

61. Install the throttle body.

62. Connect the water bypass pipe to the water pump cover, install the water bypass pipe and reconnect the hoses to the water bypass pipe.

63. Install the water outlet with a new gasket and the two nuts and install the water hoses, vacuum hoses, and reconnect the sensor connectors.

64. Connect the oil pressure switch connector.

65. Assemble the TWC to the exhaust manifold and torque to 31 ft. lbs. (42 Nm). Install the main oxygen sensor to the exhaust manifold and the sub oxygen sensor to the TWC.

66. Install the exhaust manifold to the cylinder head and uniformly tighten the nuts in several passes to 36 ft. lbs. (49 Nm).

67. Reconnect the oxygen sensors connectors.

68. Install the A/C compressor. Install the idler pulley bracket with the three bolts. Torque to:
Bolt A — 18 ft. lbs. (25 Nm)
Bolt B — 20 ft. lbs. (27 Nm)
Bolt C — 27 ft. lbs. (37 Nm)

69. Tighten the A/C compressor bolts to 18 ft. lbs. (25 Nm).

70. Install the front exhaust pipe and torque the new nuts to 46 ft. lbs. (62 Nm). Torque the bolts holding the front exhaust to the tailpipe to 32 ft. lbs. (43 Nm).

71. Install the distributor assembly and reconnect the spark plug wires.

72. Connect the ground strap.

73. Connect the following hoses:
a. MAP sensor hose to the intake manifold
b. A/C idle–up air hose (from port F) to the intake manifold
c. Brake booster vacuum hose to the intake manifold

74. Install the air cleaner assembly.

75. If removed, install the cruise control actuator and the accelerator linkage.

76. Connect the accelerator cable to the throttle body.

77. Connect the negative battery cable.

78. Fill the engine with coolant.

79. Start the engine, let it warm up, and check for leaks. Bleed the cooling system.

80. Check the ignition timing, and road test the vehicle for proper operation. Recheck the fluid levels.

81. Install the engine undercovers and hood side panels.

3.0L (3VZ-FE) Engine

Camry

1. Disconnect the negative battery cable. On vehicles equipped with an air bag, wait at least 90 seconds before proceeding.

2. Drain the cooling system.

3. On vehicles equipped with automatic transmission, disconnect the throttle cable and bracket from the throttle body.

4. Disconnect the accelerator cable and bracket from the throttle body and intake chamber.

5. On vehicles equipped with cruise control, remove the actuator, vacuum pump and bracket.

6. Remove the air cleaner hose. air flow meter and air cleaner cover.

7. Remove the alternator.

8. Remove the oil pressure gauge, engine hangers and alternator upper bracket.

9. Loosen the lug nuts on the right wheel and raise and support the vehicle safely.

10. Remove the right tire and wheel assembly.

11. Remove the right undercover.

12. Remove the front exhaust pipe.

13. Remove the distributor and the V-bank cover.

14. Disconnect the water temperature sender gauge connector, water temperature sensor connector, cold start injector time switch connector, upper radiator hose, water hoses, and the emission control vacuum hoses. Unbolt and remove the water outlet and gaskets.

15. Remove the water bypass pipe with O-rings and gasket.

16. Remove the EGR valve and vacuum modulator. Remove the exhaust crossover pipe.

17. Remove the throttle body.

18. Remove the cold start injector pipe.

19. Disconnect the air chamber hose, throttle body air hose and power steering hoses (if equipped). Remove the air tube.

20. Remove the intake manifold stay and disconnect the vacuum sensing hose. Remove the intake manifold and gasket. Purchase a new gasket.

21. Remove the fuel delivery pipe and the injectors.

22. Remove the rear cylinder head plate.

23. Remove the emission control valve set and the left side engine harness.

24. Remove the exhaust manifolds and the spark plugs.

25. Remove the oil dipstick.

26. Remove the timing belt, all camshaft timing pulleys and the No. 2 idler pulley.

27. Remove the No. 3 timing belt cover. Support the belt carefully so that the belt and pulley mesh does not shift.

28. Remove the cylinder head covers.

29. Remove the intake and exhaust camshafts from each head, following the proper sequences and procedures.

30. Remove the power steering pump bracket and the left side engine hanger.

31. Remove the two (one on each head) 8mm hex bolts. Loosen and remove the eight head bolts evenly, in three passes, in the reverse order of the installation sequence. Carefully lift the head from the engine and place it on wood blocks in a clean work area.

WARNING
If the cylinder head bolts are loosened out of sequence, warpage or cracking could result.

32. Remove the cylinder head gasket and purchase a new one (cylinder head gaskets must never be re-used). With a gasket scraper, carefully remove all the old gasket material from the cylinder head and engine block surfaces. Do not damage the block or cylinder head mounting surface.

To install:

33. Place the new cylinder head gasket onto the cylinder block. Place the cylinder head onto the gasket.

34. Coat the threads of the eight cylinder head bolts (12-sided) with clean engine oil and install the bolts into the cylinder head. Uniformly torque the bolts in three passes to an ultimate torque of 25 ft. lbs. (34 Nm), using the proper sequence. If any of the bolts do not meet the torque, replace it.

35. Mark the forward edge of each bolt with paint and then tighten each bolt an additional 90°, in the proper order. Now repeat the process once more, for an additional 90°. Check that each painted mark is now at a 180° angle to the front, facing the rear.

36. Coat the threads of the two remaining 8mm bolts with engine oil and install them. Tighten to 13 ft. lbs. (18 Nm).

37. Install the left engine hanger and tighten it to 27 ft. lbs. (37 Nm).

38. Install the power steering pump bracket.

39. Install the camshafts following the proper sequences and procedures.

40. Install the cylinder head covers and tighten the bolts to 52 inch lbs. (6 Nm).

41. Install the No. 3 timing belt cover and tighten the six bolts to 65 inch lbs. (7 Nm). Install the No. 2

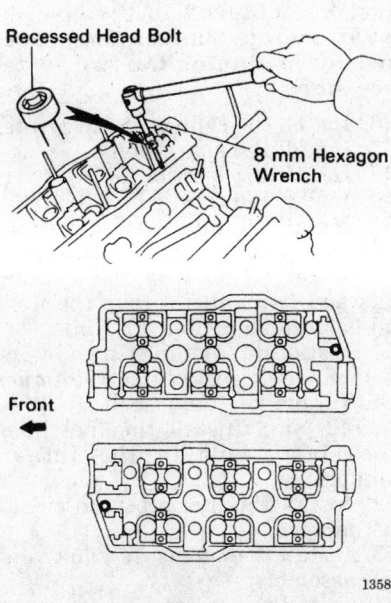

Recessed head bolts (2) — 3VZ-FE engine

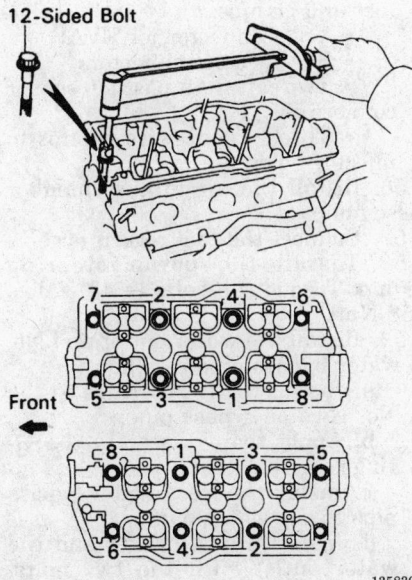

Cylinder head bolt tightening sequence — 3VZ-FE engine

idler pulley, the camshaft timing pulleys and the timing belt.

42. Install the spark plugs.

43. Install the right and left side exhaust manifolds and tighten them to 29 ft. lbs. (39 Nm).

44. Install the intake manifold and the No. 2 idler pulley bracket. Tighten all bolts to 13 ft. lbs. (18 Nm).

45. Install the cylinder head rear plate and the oil dipstick tube.

46. Install the water bypass outlet and tighten the bolts to 74 inch lbs. (8 Nm).

47. Install the injectors and delivery pipe. Tighten the bolts to 9 ft. lbs. (13 Nm).

48. Install the air pipe, the engine harness and the No. 1 EGR cooler. Tighten the pipe to 73 inch lbs. (8 Nm) and the cooler to 13 ft. lbs. (18 Nm).

49. Install the air intake chamber. Tighten the mounting bolts to 29 ft. lbs. (39 Nm).

50. Install the cold start injector. Install the distributor and the EGR assembly. Tighten the EGR bolts to 13 ft. lbs. (18 Nm).

51. Install the emission control valve set and tighten it to 73 inch lbs. (8 Nm).

52. Install the EGR pipe and tighten the bolt to 13 ft. lbs. (18 Nm) and the union nut to 58 ft. lbs. (78 Nm).

53. Install the throttle body and the ISC valve. Tighten both sets of bolts to 9 ft. lbs. (13 Nm).

54. Install the V-bank cover.

55. Install the front exhaust pipe and tighten the manifold nuts to 46 ft. lbs. (62 Nm), tighten the converter nuts to 32 ft. lbs. (43 Nm).

56. Install the alternator and adjust the drive belt tension.

57. Install the air cleaner hose.

58. If equipped, install the cruise control actuator and bracket.

59. Install and adjust the accelerator cable.

60. If equipped with automatic transaxle, connect and adjust the throttle cable.

61. Fill the cooling system to the proper level with a good brand of ethylene glycol coolant.

62. Connect the negative battery cable. Start the engine and check for leaks.

63. Adjust the valves and the ignition timing. Check the toe-in.

64. Road test the vehicle and check for unusual noise, shock, slippage, correct shift points and smooth operation.

65. Recheck the coolant and engine oil levels.

3.0L (2JZ-GTE) Engine

Supra

─────── **CAUTION** ───────
Fuel injection systems remain under pressure after the engine has been turned OFF. Properly relieve fuel pressure before disconnecting any fuel lines. Failure to do so may result in fire or personal injury.
───────────────────

1. Relieve the fuel pressure in the fuel line before disconnecting any fuel lines.

2. Disconnect the negative battery cable from the battery. On vehicles equipped with an air bag, wait at least 90 seconds before proceeding.

3. Drain the engine coolant from the engine and radiator.

4. Remove the turbocharger from the engine.

5. Remove the exhaust manifold by removing the 12 nuts and two gaskets.

6. If equipped with manual transmission, remove the drive belt tensioner damper by removing the two nuts.

7. Remove the drive belt by turning the tensioner clockwise.

8. Remove the water outlet and the No. 1 water bypass pipe.

9. Disconnect the power steering pump without disconnecting the hoses as follows:

   a. Disconnect the air hose from the throttle body.

   b. Disconnect the air hose from the air intake chamber.

   c. Remove the power steering pump housing from the pump bracket by removing the two bolts.

   d. Suspend the pump housing without disconnecting the hoses.

10. Disconnect the fuel return hose from the fuel return pipe. Plug the hose end.

11. Remove the air intake chamber assembly.

12. Disconnect the six injector connectors.

13. Disconnect the two camshaft position sensor connectors.

14. Disconnect the three engine wire clamps from the injector holders.

15. Disconnect the VSV connector for the EVAP.

16. Remove the two ground straps from the intake manifold by removing the two bolts.

17. Disconnect the engine wire protector from the intake manifold by removing the nut.

18. Remove the starter.

19. Remove the pressure tank and VSV assembly.

— CAUTION —

Fuel injection systems remain under pressure, even after the engine has been turned OFF. The fuel system pressure must be relieved before disconnecting any fuel lines. Failure to do so may result in fire and/or explosion.

20. Remove the fuel pressure pulsation damper.

21. Remove the fuel inlet pipe by disconnecting the union bolt and clamp bolt.

22. Remove the intake manifold and delivery pipe assembly by removing the four bolts and two nuts.

23. Remove the upper two timing belt covers (Nos. 2 and 3).

24. Remove the drive belt tensioner.

25. Set No. 1 cylinder to TDC/compression. Turn the crankshaft pulley clockwise to align the groove with the **0** mark on the lower (No. 1) timing belt cover. Check that the timing marks on the camshaft pulleys are aligned with the marks on the rear belt cover. If the camshaft marks do not align, turn the crankshaft another 360 degrees.

26. Alternately loosen the two bolts holding the timing belt tensioner. Remove the bolts and remove the tensioner.

27. Remove the timing belt from the camshaft pulleys. If the belt is to be reused, place matchmarks on the belt and gears before removing the belt. Mark the belt with an arrow to show direction of rotation.

28. Remove the ignition coils.

29. Remove the spark plugs.

30. Remove the cylinder head covers.

31. While holding each camshaft with a wrench, remove the camshaft bolts and remove the gears.

32. Remove the four bolts and lift out the No. 4 (inner) timing belt cover.

33. Remove the camshafts following the proper sequences and procedures.

34. Remove the cylinder head from the engine block as follows:

a. Using a 10mm bi-hexagon wrench, uniformly loosen and remove the 14 cylinder head bolts. Loosen the bolts in several passes and in the reverse order of the installation sequence.

b. Remove the 14 plate washers.

c. Lift the cylinder head fro the dowels on the cylinder block.

d. Place the head on wooden blocks on a bench.

35. Remove the engine hangers and the ground strap.

36. Remove the camshaft position sensors.

37. Remove the EGR cooler.

38. Remove the valve lifters and shims. Make sure to make a note to the positions of the lifters and shims. When installing the lifters and shims, install them in the same position as removal.

39. Using SST 09202–70010 or equivalent, compress the valve spring and remove the two keepers.

40. Remove the spring retainer, valve spring, valve and spring seat.

41. Using needle nose pliers, remove the oil seal.

To install:

42. Install new valve oil seals and assemble the cylinder head.

43. Install the engine hangers and the ground strap. Torque the mounting bolts to 29 ft. lbs. (39 Nm).

44. Install the camshaft position sensors and the EGR cooler. Torque the cooler and sensor mounting bolts to 78 inch lbs. (9 Nm).

45. Install the cylinder head as follows:

a. Clean the cylinder head and block.

b. Place a new cylinder head gasket in position on the cylinder block.

c. Place the cylinder head in position on the cylinder head gasket.

d. Lightly coat the head bolt threads and plate washers with engine oil. Install plate washers and bolts to the head.

e. Uniformly tighten the head bolts in several passes in the correct order. Torque the bolts to 25 ft. lbs. (34 Nm).

f. Mark the front (towards the front of the engine) of each bolt with a dot of paint. Following the correct order, tighten each bolt an additional 90 degrees. When complete, all the paint marks should face to the side of the engine.

g. Again following the correct order, tighten the bolts another 90 degrees of rotation. When complete, the paint marks should face the rear of the engine, exactly 180 degrees away from the original starting point. Correct bolt torque

is expressed as 25 ft. lbs. (34 Nm) + 90° + 90°.

NOTE: Correct bolt torque must be achieved in 3 steps; do not attempt to shorten the procedure by combining the two 90 degree steps.

46. Install the cylinder head covers.

47. Install the spark plugs.

48. Install the ignition coils.

49. Install the timing belt.

50. Install the intake manifold and delivery pipe assembly by installing a new gasket, engine wire and the four bolts and two nuts. Torque the nuts and bolts to 20 ft. lbs. (27 Nm).

51. Install the fuel inlet pipe by installing a new gasket and the union bolt. Torque the union bolt to 30 ft. lbs. (42 Nm). Install the fuel inlet pipe clamp bolt to the intake manifold.

52. Install the fuel pressure pulsation damper.

53. Install the pressure tank and VSV assembly.

54. Install the starter.

55. Connect the engine wire as follows:

a. Install the engine wire protector to the intake manifold with the nut.

b. Install the two ground straps to the intake manifold with the bolts.

c. Connect the following connectors and clamps:
- VSV connector for EVAP
- Six injector connectors
- Two camshaft position sensor connectors
- Three engine wire clamps to injector holders

56. Install the air intake chamber assembly.

57. Connect the fuel return hose

58. Install the power steering pump. Torque the bolts to 43 ft. lbs. (58 Nm).

59. Install the water outlet and No. 1 water bypass pipe as follows:

a. Install the two O-rings to the No. 1 water bypass pipe.

b. Apply soapy water to the O-rings.

c. Install the No. 1 water bypass pipe to the water pump.

d. Install a new gasket and the water outlet with the two bolts. Torque the bolts to 15 ft. lbs. (21 Nm).

e. Connect the ECT sensor and sender gauge connectors.

f. Connect the upper radiator hose to the water outlet.

60. Install the drive belt.

61. If equipped with manual transmission, install the drive belt ten-

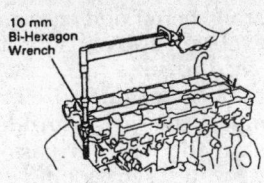

289708

Cylinder head bolt tightening sequence — 2JZ-GE and 2JZ-GTE engines

sioner damper. Torque the two nuts to 14 ft. lbs. (20 Nm).

62. Place two new gaskets to the cylinder head.

63. Install the exhaust manifold with 12 new nuts and tighten the nuts in several passes. Torque the bolts to 29 ft. lbs. (39 Nm).

64. Install the turbocharger.

65. Fill the engine with coolant.

66. Connect the negative battery cable to the battery.

67. Start the engine and check for leaks.

3.0L (2JZ-GE) Engine

Supra

1. Relieve the fuel pressure in the fuel lines.

— **CAUTION** —

Fuel injection systems remain under pressure after the engine has been turned OFF. Properly relieve fuel pressure before disconnecting any fuel lines. Failure to do so may result in fire or personal injury.

2. Disconnect the negative battery cable. Wait at least 90 seconds before performing any other work.

3. Drain the engine coolant.

4. Remove the undercovers from the vehicle.

5. Disconnect the accelerator, throttle control (A/T only) and cruise control cables from the throttle body.

6. Remove the air cleaner duct.

7. Remove the air cleaner, airflow meter, and the intake air pipe.

8. Remove the drive belt, the fan and fluid coupling, and the water pump pulley.

9. Remove the front exhaust pipe.

10. Except for California vehicles, remove the four nuts and the manifold heat insulator.

11. Disconnect the oxygen sensor connector(s).

12. Remove the exhaust manifolds and gaskets by removing the eight bolts.

13. Disconnect the power steering pump without disconnecting the hoses as follows:

a. Disconnect the air hose from the No. 4 timing belt cover.

b. Disconnect the air hose from the intake chamber.

c. Disconnect the power steering pump housing from the pump bracket by removing the two bolts.

d. Remove the two bolts and the pump rear stay.

14. Disconnect the brake booster vacuum hose.

15. Disconnect the EVAP hose.

16. Remove the throttle body and intake air connector assembly.

17. Remove the air intake chamber stays by removing the nut and bolt to each stay.

18. Remove the No. 2 vacuum pipe and VSV assembly.

19. Remove the No. 3 timing belt cover by removing the oil filler cap and six bolts.

20. Remove the cylinder head rear cover by removing the four bolts.

21. Disconnect the spark plug wires from the cylinder head covers.

22. Remove the distributor and wires from the engine.

23. Remove the spark plugs.

24. Remove the timing belt.

25. Remove the water bypass outlet and No. 1 bypass pipe.

26. Disconnect the fuel return hose from the fuel return pipe.

27. Disconnect the fuel return hose from the oil dipstick guide.

28. Remove the engine wire bracket from the intake manifold by removing the bolt.

29. Remove the oil dipstick and guide for the engine.

30. Remove the starter.

31. Remove the air intake chamber as follows:

a. Except for California vehicles, disconnect the vacuum sensing hose from the fuel pressure regulator.

b. Remove the bolt holding the engine wire protector to the air intake chamber.

c. Remove the five bolts, nut, air intake chamber, and the gasket.

32. Remove the vacuum control valve set by disconnecting the VSV connector and then removing the two nuts.

33. Disconnect the engine wire from the intake manifold as follows:

a. Remove the bolt and disconnect the engine wire bracket from the water pump.

b. Remove the two bolts and disconnect the two ground straps from the intake manifold.

c. Remove the two bolts and disconnect the two wire clamps from the intake manifold.

d. Disconnect the following connectors:

- Six injector connectors
- ECT sensor connector
- ECT sender gauge connector

e. Disconnect the engine wire protector from the intake manifold by removing the three nuts.

34. Remove the water outlet and No. 1 bypass hose assembly by removing the bolt and two nuts.

35. Remove the intake manifold stay by removing the two bolts.

36. Remove the fuel pressure pulsation damper.

37. Remove the fuel inlet pipe.

38. Remove the two nuts and six bolts to the intake manifold.

39. Remove the intake manifold and gaskets.

40. Remove the cylinder head covers (valve covers).

41. Remove the camshaft timing pulleys. Hold the hexagon portion of the camshaft with a wrench and remove the pulley mounting bolt and camshaft pulley.

42. Remove the No. 4 timing belt cover.

43. Remove the camshafts following the proper sequences and procedures.

44. Uniformly loosen the cylinder head bolts in several passes and in the reverse order of the installation sequence.

— **WARNING** —

The cylinder head may crack or warp if the correct order is not followed.

45. Remove the head from the engine. If prying is necessary, use a protected blade; take great care not to damage the mating surfaces of the head and block.

46. Clean the head and block of all gasket material. Take great care not to gouge or scratch the mating surfaces.

To install:

47. Place a new gasket on the cylinder block. Make sure it is positioned correctly. Install the cylinder head in position.

48. Lightly coat the head bolt threads and plate washers with engine oil. Install plate washers and bolts to the head.

49. Uniformly tighten the head bolts in several passes in the correct order. Torque the bolts to 25 ft. lbs. (34 Nm).

50. Mark the front (towards the front of the engine) of each bolt with a dot of paint. Following the correct order and tighten each bolt an additional 90 degrees. When complete, all the paint marks should face to the side of the engine.

51. Again following the correct order, tighten the bolts another 90 degrees of rotation. When complete, the paint marks should face the rear of the engine, exactly 180 degrees away from the original starting point. Correct bolt torque is expressed as 25 ft. lbs. (34 Nm) + 90° + 90°.

NOTE: Correct bolt torque must be achieved in 3 steps; do not attempt to shorten the procedure by combining the two 90 degree steps.

52. Coat the camshaft with engine oil and then position them in the cylinder head with the cam lobes and the knock pins in the correct position.

53. Position the No. 3 and No. 7 bearing caps in place, coat the bolt threads with oil and then uniformly and alternately tighten them temporarily.

54. Coat the new oil seals with multi-purpose grease and then slide them over the camshafts.

55. Clean the surfaces of the No. 1 bearing cap and cylinder head with cleaner. Apply seal packing to the No. 1 bearing cap.

56. Install the camshafts following the proper sequences and procedures.

57. Using SST tool 09316–60010 or equivalent, press the two oil seals in as far as it will go.

58. Rotate each camshaft until the forward straight (knock) pin is straight up. Loosen exhaust No. 1, 2, and 6 bearing cap bolts until they can be turned by hand; tighten, in several passes, to 14 ft. lbs. (20 Nm). Loosen intake No. 1, 2, and 5 and tighten, in several passes, to 14 ft. lbs. (20 Nm).

59. Turn each camshaft 1/3 of a revolution (120 degrees). Loosen exhaust Nos. 4 and 7 bearing cap bolts; tighten, in several passes, to 14 ft. lbs. (20 Nm). Loosen intake No. 4 and 6 bearing cap bolts; tighten, in several passes, to 14 ft. lbs. (20 Nm).

60. Turn each camshaft an additional 1/3 of a revolution, loosen exhaust bearing cap bolts Nos. 3 and 5 and then tighten them, in several passes, to 14 ft. lbs. (20 Nm). Loosen intake bearing cap bolts No. 3 and 7 and then tighten them, in several passes to 14 ft. lbs. (20 Nm).

61. Check and adjust the valve clearance.

62. Install the No. 4 timing belt cover. Torque the bolts to 78 inch lbs. (9 Nm).

63. Install the camshaft timing pulleys. Align the shaft pin with the pulley groove and slide the pulley on. Install the bolt temporarily. Hold the hex portion of the camshaft with a wrench and tighten the pulley bolt to 59 ft. lbs. (79 Nm).

64. Install the cylinder head covers.

65. Install the intake manifold with a new gasket. Tighten the bolts to 20 ft. lbs. (27 Nm).

66. Install the injectors and the delivery pipe. Tighten the bolts holding the pipe to the manifold to 20 ft. lbs. (27 Nm).

67. Install the fuel inlet pipe with two new gaskets and tighten the union bolt. Torque the union bolt to 30 ft. lbs. (42 Nm). Install the clamp bolt to the intake manifold.

68. Install the fuel pressure pulsation damper.

69. Install the intake manifold stay and torque the bolts to 29 ft. lbs. (39 Nm).

70. Install the water outlet and No. 1 bypass hose assembly.

71. Connect the engine wire as follows:

   a. Install the engine wire protector to the intake manifold with the three nuts.

   b. Connect the six injector connectors.

   c. connect the ECT sensor connector.

   d. Connect the ECT sender gauge connector.

   e. Install the two wire clamps to the intake manifold with the bolts.

   f. Install the two ground straps to the intake manifold with the bolts.

   g. Install the engine wire bracket to the water pump with the bolt.

72. Install the vacuum control valve set. Torque the bolts to 15 ft. lbs. (21 Nm). Connect the VSV connector.

73. Install the air intake chamber as follows:

   a. Install a new gasket and the intake chamber with the nut and five bolts. Torque the bolts and nut to 20 ft. lbs. (27 Nm).

   b. Install the bolt holding the engine wire protector to the air intake chamber.

   c. Except for California vehicles, connect the vacuum sensing hose to the fuel pressure regulator.

74. Install the starter.

75. Install the oil and transmission dipstick tubes. Always use a new O-ring on each tube.

76. Install the engine wire bracket.

77. Connect the fuel return hose.

78. Install the water bypass outlet and No. 1 water bypass pipe.

79. Install the timing belt as follows:

   a. Turn the crankshaft pulley and align its groove with the timing mark, **0** on the No. 1 timing belt cover.

   b. Align the timing marks on the camshaft timing gears and the No. 4 timing belt cover.

   c. Install the timing belt.

   d. Double check that all the timing marks for the crankshaft pulley and the camshaft gears are aligned as they were during disassembly.

   e. Set the timing belt tensioner:

   • Use a press to slowly push in the pushrod on the tensioner. This will require between 220–2200 pounds of pressure.

   • Align the holes of the pushrod and housing. Place a 1.5mm hex wrench through the holes to keep the pushrod retracted.

   • Release the press and install the dust boot onto the tensioner.

   f. Install the tensioner; alternately tighten the bolts to 20 ft. lbs. (26 Nm).

   g. Remove the hex wrench from the tensioner with a pair of pliers.

   h. Turn the crankshaft pulley two full turns clockwise. Check that each pulley's timing marks align correctly after the two turns. If any mark does not align, remove the timing belt and reinstall it.

80. Install the drive belt tensioner by installing the three bolts. Torque the bolts to 15 ft. lbs. (21 Nm).

81. Install the No. 2 timing belt cover.

82. Install the spark plugs.

83. Install the distributor and spark plug wires to the engine.

84. Connect the spark plug wires to the cylinder head covers.

85. Install the No. 3 timing belt cover.

86. Install the cylinder head rear cover.

87. Install the No. 2 vacuum pipe and VSV assembly.

88. Install the air intake chamber stays with the nut and bolt for each stay. Torque the bolt and nut to 13 ft. lbs. (18 Nm).

89. Install the throttle body and intake air connector assembly.

90. Connect the EVAP hose.

91. Connect the brake booster vacuum hose.

92. Install the power steering pump.

93. Install the exhaust manifolds.

94. Install the No. 2 front exhaust pipe.

95. Install the water pump pulley, fan, fluid coupling assembly, and the drive belt. Torque the four bolts for the pulley to 12 ft. lbs. (16 Nm).

96. Install the air cleaner, VAF meter, and the intake air connector pipe assembly.

97. Install the air cleaner duct.

98. Connect the cruise control cable, throttle control, and accelerator cables to the throttle body.

99. Fill the engine with coolant.

100. Connect the negative battery cable to the battery.

101. Start the engine and check for leaks.

102. Check the ignition timing.

103. Install the engine undercover.

104. Perform a road test.

3.0L (1MZ-FE) Engine

Camry and Avalon

— CAUTION —
Fuel injection systems remain under pressure even after the engine has been turned OFF. The fuel system pressure must be relieved before disconnecting any fuel lines. Failure to do so may result in fire and/or personal injury.

1. Relieve the fuel pressure.

2. Disconnect the negative battery cable. On vehicles equipped with an air bag, wait at least 90 seconds before proceeding.

3. Drain the cooling system.

4. Disconnect the accelerator cable and the throttle cable on vehicles equipped with an automatic transaxle.

5. Remove the air cleaner cover, air flow meter, and the air duct.

6. Remove the cruise control actuator and bracket, if equipped.

7. Disconnect the two engine ground straps.

8. Remove the right engine mounting support.

9. Disconnect the radiator hoses.

10. Disconnect the two heater hoses.

11. Disconnect and plug the fuel feed and return lines from the fuel rail assembly.

12. Disconnect and plug the pressure hose from the hydraulic motor.

13. Remove the V-bank cover.

14. Disconnect the following vacuum hoses:

   a. Fuel pressure control VSV

   b. Fuel pressure regulator

   c. Cylinder head rear plate

   d. Intake air control valve VSV

   e. EGR vacuum modulator

   f. EGR valve

15. Disconnect the following connectors:

   a. Intake air control valve

   b. Fuel pressure regulator

   c. EGR VSV

16. Remove the two nuts and the emission control valve set.

17. Disconnect the following hoses;

   a. Brake booster vacuum hose

   b. PCV hose

   c. Intake air control valve vacuum hose

18. Remove the data link connector from the mounting bracket.

19. Remove the two ground straps from the intake chamber.

20. Remove the hydraulic motor pressure hose from the intake chamber.

21. Remove the right oxygen sensor connector from the P/S pressure tube.

22. Remove the two nuts and the P/S pressure tube from the intake chamber.

23. Disconnect the two P/S air hoses.

24. Remove the engine hanger and the intake chamber support.

25. Remove the EGR pipe and gaskets.

26. Disconnect the following connectors;

   a. Throttle position sensor connector

   b. IAC valve connector

   c. EGR gas temperature connector

   d. A/C idle up connector

27. Disconnect the following vacuum hoses:

   a. Two vacuum hoses from the TVV

   b. Vacuum hose from the cylinder head rear plate

   c. Vacuum hose from the charcoal canister

28. Disconnect the air assist hose and the two water bypass hoses.

29. Remove the air intake chamber.

30. Disconnect the left engine wire harness and position it out of the way.

31. Remove the wire harness from the rear of the engine.

32. Disconnect the right engine wire harness and position it out of the way.

33. Remove the ignition coils and the spark plugs.

34. Remove the timing belt.

35. Remove the camshaft pulleys and the timing belt rear cover.

36. Remove the cylinder head rear plate.

37. Remove the water inlet pipe.

38. Remove the air assist hose and vacuum hose.

39. Remove the intake manifold and fuel rail assembly.

40. Remove the water outlet.

41. Remove the EGR pipe from the right exhaust manifold.

42. Remove the exhaust manifolds.

43. Remove the dipstick assembly and the P/S pump bracket.

44. Remove the valve covers and the camshaft position sensor.

45. Remove the camshafts following the proper sequences and procedures.

46. Make sure the engine is at or near ambient temperature and remove the two (one on each head) 8mm recessed hex bolts. Loosen and remove the 8 head bolts evenly, in 3 passes, in the reverse order of the installation sequence. Carefully lift the head from the engine; if it is necessary to pry the head loose, take great care not to damage the mating surfaces. Place the head on wood blocks in a clean work area.

NOTE: If the cylinder head bolts are loosened out of sequence, warpage or cracking could result.

47. Remove the cylinder head gasket. With a gasket scraper, carefully remove all the old gasket material from the cylinder head and engine block surfaces.

To install:

48. Place the new cylinder head gasket onto the cylinder block. Place the cylinder head onto the gasket.

49. Coat the threads of the 8 cylinder head bolts (12-sided) with clean engine oil and install the bolts into the cylinder head. Uniformly torque the bolts in sequence in three steps to an ultimate torque of 40 ft. lbs. (54 Nm), using the proper sequence. If any bolt does not meet the torque, replace it.

50. Mark the forward edge of each bolt with paint and then tighten each bolt, in proper sequence, an additional 90 degrees. Check that each painted mark is now at a 90 degrees angle to the front. The paint mark should have been applied to the bolt in the 9 o'clock position and should now be in the 12 o'clock position.

51. Coat the threads of the two remaining 8mm bolts with engine oil and install them. Tighten to 13 ft. lbs. (18 Nm).

52. Install the camshafts following the proper sequences and procedures.

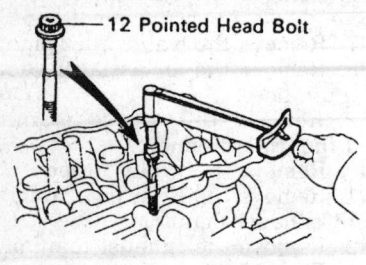

12 Pointed Head Bolt

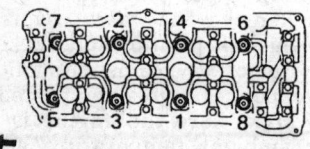

Front ◄

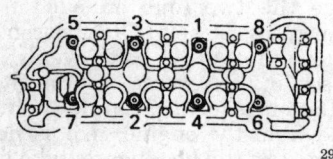

294418

Cylinder head tightening sequence — 1MZ-FE engine

53. Check and adjust the valves.
54. Apply sealant to the cylinder heads where the camshaft supports meet the cylinder heads.
55. Use new gaskets and install the cylinder head covers.
56. Install the dipstick and power steering pump bracket.
57. Install the exhaust manifolds. Torque the nuts to 36 ft. lbs. (49 Nm).
58. Install the EGR pipe to the right exhaust manifold.
59. Install the water outlet.
60. Install the intake manifold and the fuel rail assembly. Torque the intake manifold nuts and bolts to 11 ft. lbs. (15 Nm).
61. Install the air assist hose and the two water bypass hoses.
62. Install the water inlet pipe and the cylinder head rear plate.
63. Install the timing belt rear cover and the camshaft pulleys.
64. Install the timing belt.
65. Install the spark plugs and the ignition coils.
66. Install the right engine wire harness.
67. Install the wire harness to the rear of the engine.
68. Install the left engine wire harness.
69. Install the air intake chamber.
70. Use new gaskets and install the EGR pipe.

71. Connect the following vacuum hoses:
   a. The two TVV vacuum hoses.
   b. The vacuum hose to the rear cylinder head plate.
   c. Charcoal canister vacuum hose.
72. Connect the following electrical connectors:
   a. Throttle position sensor connector.
   b. IAC valve connector.
   c. EGR gas temperature connector.
   d. A/C idle up connector.
73. Install the engine hanger and the intake chamber support.
74. Connect the two P/S air hoses.
75. Install the P/S pressure tube to the intake chamber.
76. Install the oxygen sensor connector to the pressure tube.
77. Install the two ground straps to the intake chamber.
78. Install the data link connector to the bracket.
79. Connect the following hoses:
   a. Power brake booster vacuum hose.
   b. PCV hose.
   c. IAC valve vacuum hose.
80. Install the emission control valve set and related vacuum hoses and connectors.
81. Install the V-bank cover.
82. Connect the pressure hose to the hydraulic motor.
83. Connect the fuel lines to the fuel rail assembly.
84. Connect the heater and radiator hoses.
85. Install the right engine mounting support.
86. Connect the two engine ground straps.
87. Install the cruise control actuator and bracket.
88. Install the air cleaner, air flow meter, and air duct assembly.
89. Connect the accelerator cable and the throttle cable on vehicles equipped with an automatic transaxle.
90. Fill the cooling system to the proper level with coolant.
91. Connect the negative battery cable.
92. Start the engine and check for leaks. Bleed the air from the cooling system.
93. Adjust the ignition timing.
94. Road test the vehicle and check for unusual noise, shock, slippage, correct shift points and smooth operation.
95. Recheck the coolant and engine oil levels.

Valve Lash

ADJUSTMENT

1.5L (5E-FE) Engine

Tercel and Paseo

NOTE: Adjust the valve clearance when the engine is cold.

1. Disconnect the negative battery cable. On vehicles equipped with an air bag, wait at least 90 seconds before proceeding.
2. Remove the cylinder head covers.
3. Turn the crankshaft pulley and align its groove with the timing mark **0** of the No. 1 timing cover.
4. Check that the timing marks on the camshaft sprockets are in alignment with the marks on the No. 4 timing cover. If not, turn the crankshaft 1 complete revolution (360° degrees).
5. Measure the clearance between the valve lifter and the camshaft. Record the measurements on the intake valves No. 1 and 2. Measure the exhaust valves at No. 1 and 3.
   a. The intake valve clearance cold is 0.006–0.010 in. (0.15–0.25mm).
   b. The exhaust valve clearance cold is 0.012–0.016 in. (0.31–0.41mm).
6. Turn the crankshaft pulley one revolution (360°) and align the groove with the timing mark **0** of the No.1 timing belt cover.
7. Measure the clearance between the valve lifter and the camshaft. Record the measurements on the intake valves No. 3 and 4. Measure the exhaust valves at No. 2 and 4.
   a. The intake valve clearance cold is 0.006–0.010 in. (0.15–0.25mm).
   b. The exhaust valve clearance cold is 0.012–0.016 in. (0.31–0.41mm).
8. To adjust the valve clearance:
   a. Remove the adjusting shim and turn the crankshaft to position the cam lobe of the camshaft on the valve to be adjusted upward.
   b. Turn the valve lifter so that the notch is perpendicular to the camshaft and facing the spark plug side.
   c. Using SST 09248–55040 (valve lifter press) or equivalent, hold the camshaft in place.
   d. Using SST 09248–55040 (valve lifter press) or equivalent, press down the valve lifter and place SST 09248–05420 (valve lifter stopper) or equivalent be-

Adjusting Shim Selection Chart

Shim No.	New shim thickness mm (in.) Thickness	Shim No.	Thickness
02	2.500 (0.0984)	20	2.950 (0.1161)
04	2.550 (0.1004)	22	3.000 (0.1181)
06	2.600 (0.1024)	24	3.050 (0.1201)
08	2.650 (0.1043)	26	3.100 (0.1220)
10	2.700 (0.1063)	28	3.150 (0.1240)
12	2.750 (0.1083)	30	3.200 (0.1260)
14	2.800 (0.1102)	32	3.250 (0.1280)
16	2.850 (0.1122)	34	3.300 (0.1299)
18	2.900 (0.1142)		

7920eg01

Adjusting shim chart (intake and exhaust) — all 5E-FE and 3VZ-FE engines

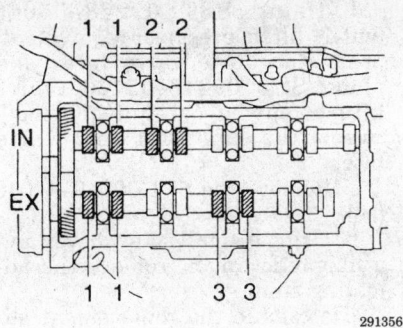

291356

Intake valves (1 and 2) and exhaust valves (1 and 3) — 5E-FE engine

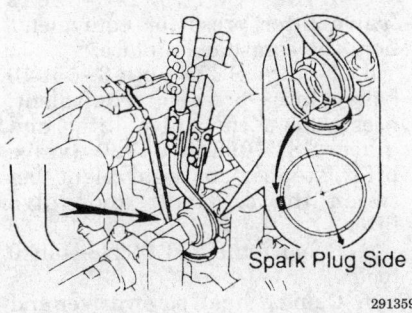

Spark Plug Side

291359

Common method of removing valve shims

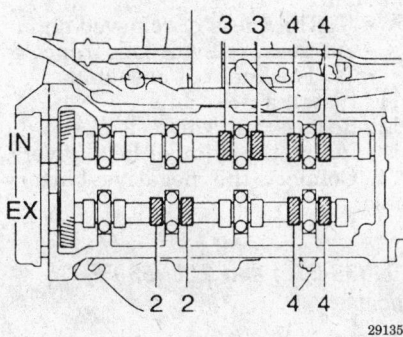

291357

Intake valves (3 and 4) and exhaust valves (2 and 4) — 5E-FE engine

Adjusting Shim Selection Chart

Shim No.	New shim thickness mm (in.) Thickness	Shim No.	Thickness
1	2.500 (0.0984)	10	2.950 (0.1161)
2	2.550 (0.1004)	11	3.000 (0.1181)
3	2.600 (0.1024)	12	3.050 (0.1201)
4	2.650 (0.1043)	13	3.100 (0.1220)
5	2.700 (0.1063)	14	3.150 (0.1240)
6	2.750 (0.1083)	15	3.200 (0.1260)
7	2.800 (0.1102)	16	3.250 (0.1280)
8	2.850 (0.1122)	17	3.300 (0.1299)
9	2.900 (0.1142)		

HINT: New shims have the thickness in millimeters imprinted on the face.

7920eg02

Adjusting shim chart (intake and exhaust) — all 4A-FE, 7A-FE, 5S-FE, 1MZ-FE, 2JZ-GE and 2JZ-GTE engines

tween the camshaft and valve lifter.

e. Remove the SST 09248–44040 tool.

f. Using a small screwdriver and a magnetic finger, remove the adjusting shim.

9. Determine the replacement adjusting shim size by either using the chart or the following formula:

- Intake–$N=T+A-0.008$ in. (0.20mm)
- Exhaust–$N=T+A-0.014$ in. (0.36mm)
- T=Thickness of removed shim
- A=measured valve clearance
- N=Thickness of new shim

10. Install a new shim.
11. Recheck the valve clearance.
12. Install the cylinder head covers.
13. Connect the negative battery cable.

1.6L (4A-FE) and 1.8L (7A-FE) Engine

Corolla and Celica

NOTE: Adjust the valve clearance when the engine is cold.

1. Disconnect the negative battery cable.
2. Remove the cylinder head covers.
3. Turn the crankshaft pulley and align its groove with the timing mark **0** of the No. 1 timing cover.
4. Check that the timing marks on the camshaft sprockets are in alignment with the marks on the No. 4 timing cover. If not, turn the crankshaft 1 complete revolution (360° degrees).
5. Measure the clearance between the valve lifter and the camshaft. Record the measurements on the intake valves No. 1 and 2. Measure the exhaust valves at No. 1 and 3.

a. The intake valve clearance cold is 0.006–0.010 in. (0.15–0.25mm).

b. The exhaust valve clearance cold is 0.010–0.014 in. (0.25–0.35mm).

6. Turn the crankshaft pulley one revolution (360°) and align the groove with the timing mark **0** of the No.1 timing belt cover.
7. Measure the clearance between the valve lifter and the camshaft. Record the measurements on the intake valves No. 3 and 4. Measure the exhaust valves at No. 2 and 4.

a. The intake valve clearance cold is 0.006–0.010 in. (0.15–0.25mm).

b. The exhaust valve clearance cold is 0.010–0.014 in. (0.25–0.35mm).

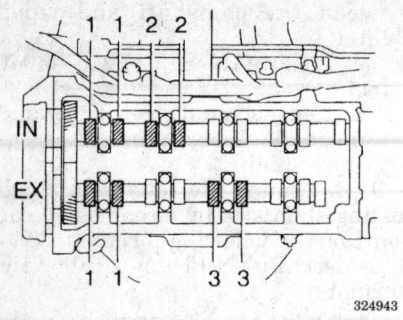

Intake valves (1 and 2) and exhaust valves (1 and 3) — Corolla and Celica 4A-FE and 7A-FE engines

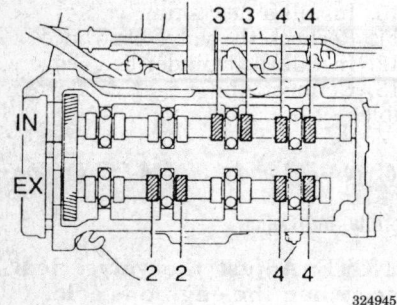

Intake valves (3 and 4) and exhaust valves (2 and 4) — Corolla and Celica 4A-FE and 7A-FE engines

8. To adjust the intake valve clearance:

a. Remove the intake camshaft.

b. Using a small screwdriver and a magnetic finger, remove the adjusting shim.

c. Determine the replacement adjusting shim size by either using the chart or the following formula:

- Intake: N=T+A-0.008 in. (0.20mm)
- T=Thickness of removed shim
- A=measured valve clearance
- N=Thickness of new shim

d. Install a new shim.

e. Install intake camshaft.

f. Recheck the valve clearance.

9. To adjust the exhaust valve clearance:

a. Turn the crankshaft to position the cam lobe of the camshaft on the valve to be adjusted, upward.

b. Turn the valve lifter so that the notch is perpendicular to the camshaft and facing the spark plug side.

c. Using SST 09248–55040 (valve lifter press) or equivalent, hold the camshaft in place.

d. Using SST 09248–55040 (valve lifter press) or equivalent, press down the valve lifter and place SST 09248–05420 (valve lifter stopper) or equivalent between the camshaft and valve lifter.

e. Remove the SST 09248–44040 tool.

f. Using a small screwdriver and a magnetic finger, remove the adjusting shim.

10. Determine the replacement adjusting shim size by either using the chart or the following formula:

- Exhaust–N=T+A-0.014 in. (0.36mm)
- T=Thickness of removed shim
- A=measured valve clearance
- N=Thickness of new shim

11. Install a new shim.

12. Recheck the valve clearance.

13. Install the cylinder head covers.

14. Connect the negative battery cable.

2.0L (3S-GTE) and 2.2L (5S-FE) Engines

Celica, Corolla and MR2

NOTE: Adjust the valve clearance when the engine is cold.

1. Disconnect the negative battery cable. On vehicles equipped with an air bag, wait at least 90 seconds before proceeding.

2. Remove the cylinder head covers.

3. Turn the crankshaft pulley and align its groove with the timing mark **0** of the No. 1 timing cover.

4. Check that the timing marks on the camshaft sprockets are in alignment with the marks on the No. 4 timing cover. If not, turn the crankshaft 1 complete revolution (360° degrees).

5. Measure the clearance between the valve lifter and the camshaft. Record the measurements on the intake valves No. 1 and 2. Measure the exhaust valves at No. 1 and 3.

a. The intake valve clearance cold is 0.007–0.011 in. (0.19–0.29mm).

b. The exhaust valve clearance cold is 0.011–0.015 in. (0.28–0.38mm).

6. Turn the crankshaft pulley one revolution (360°) and align the groove with the timing mark **0** of the No.1 timing belt cover.

7. Measure the clearance between the valve lifter and the camshaft. Record the measurements on the intake valves No. 3 and 4. Measure the exhaust valves at 2 and 4.

a. The intake valve clearance cold is 0.007–0.011 in. (0.19–0.29mm).

b. The exhaust valve clearance cold is 0.011–0.015 in. (0.29–0.38mm).

8. To adjust the valve clearance:

a. Turn the crankshaft to position the cam lobe of the camshaft on the valve to be adjusted, upward.

b. Turn the valve lifter so that the notch is perpendicular to the camshaft and facing the spark plug side.

c. Using SST 09248–55040 (valve lifter press) or equivalent, hold the camshaft in place.

d. Using SST 09248–55040 (valve lifter press) or equivalent, press down the valve lifter and place SST 09248–05420 (valve lifter stopper) or equivalent between the camshaft and valve lifter.

e. Remove the SST 09248–44040 tool.

f. Using a small screwdriver and a magnetic finger, remove the adjusting shim.

9. Determine the replacement adjusting shim size by either using the chart or the following formula:

- Intake–N=T+A-0.009 in. (24mm)
- Exhaust–N=T+A-0.013 in. (0.33mm)
- T=Thickness of removed shim
- A=measured valve clearance
- N=Thickness of new shim

10. Install a new shim.

11. Recheck the valve clearance.

12. Install the cylinder head covers.

13. Connect the negative battery cable.

3.0L (3VZ-FE) Engine

Camry

NOTE: Adjust and inspect the valve clearance when the engine is cold.

1. Disconnect the negative battery cable. On vehicles equipped with an air bag, wait at least 90 seconds before proceeding.

2. Drain the engine coolant.

3. Disconnect and remove the air intake tube.

4. Remove the cylinder head covers by removing the 6 nuts, seal washers and gaskets.

5. Use a wrench and turn the crankshaft until the notch in the pulley aligns with the timing mark **0** of

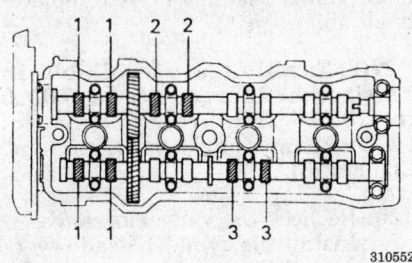

Intake valves (1 and 2) and exhaust valves (1 and 3) — Celica, Corolla and MR2 2.2L (5S-FE) engine

310552

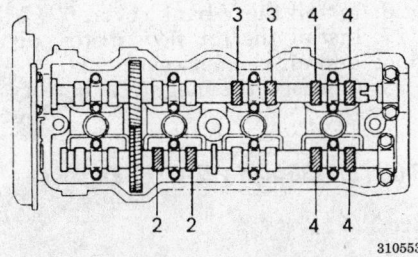

Intake valves (3 and 4) and exhaust valves (2 and 4) — Celica, Corolla and MR2 with 2.2L (5S-FE) engine

310553

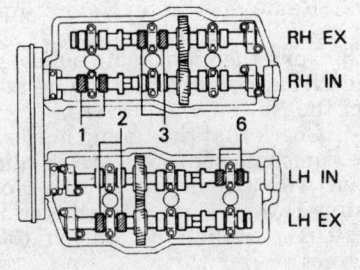

First pass valve adjustment — 3VZ-FE engine

136951

the No. 1 timing belt cover. This will insure that the engine is at TDC.

NOTE: Check that the valve lifters on the No. 1 (IN) cylinder are loose and that those on the No. 1 (EX) cylinder are tight. If not, turn the crankshaft one complete revolution (360 degrees) and then realign the mark.

6. Using a feeler gauge measure the clearance between the camshaft lobe and the valve lifter in the correct sequences (FIRST). This measure-

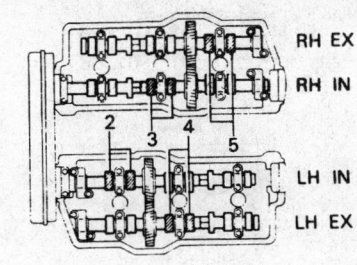

Second pass valve adjustment — 3VZ-FE engine

136952

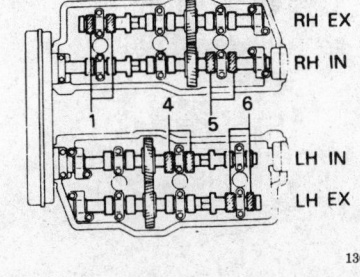

Third pass valve adjustment — 3VZ-FE engine

136953

ment should correspond to specification.

NOTE: If the measurement is within specifications, go on to the next step. If not, record the measurement taken for each individual valve. These measurements will be used later to determine the proper replacement adjusting shim.

7. Turn the crankshaft ⅔ of a revolution (240 degrees).
8. Now measure the clearance of the next set of valves (SECOND).

NOTE: If the measurement for this set of valves (and also the previous one) is within specifications, the procedure is finished. If not, record the measurements and then proceed to Step 9.

9. Turn the crankshaft a further ⅔ of a revolution (240 degrees).
10. Now measure the clearance of the next valves (THIRD).

NOTE: If the measurement for this set of valves (and also the previous 2) is within specifications, the procedure is finished. If not, record the measurements and then proceed to Step 11.

11. Turn the crankshaft to position the camshaft lobe of the valve to be adjusted faces upward.
12. Using a suitable tool, turn the valve lifter so the notch is easily accessible (toward the spark plug).
13. Install a valve lifter depressing tool between the camshaft lobe and the valve and then turn the handle so the tool presses down the valve lifter evenly.
14. Using a suitable tool and a magnet, remove the valve adjusting shim.
15. Measure the thickness of the old shim with a micrometer. Using this measurement and the clearance made earlier, determine what size replacement shim will be required in order to bring the valve clearance into specification.

NOTE: Replacement shims are available in 17 sizes, in increments of 0.0020 in. (0.05mm). Shim sizes are 0.0984–0.1299 in. (2.50–3.30mm).

16. The new replacement shim thickness may also be determined by a formula.
- T — Thickness of used shim
- A — Measured valve clearance
- N — Thickness of new shim
- Intake:

$N = T + (A - 0.18mm)$
- Exhaust:

$N = T + (A - 0.32mm)$

17. Select a new shim with a thickness as close as possible to the value arrived at in the formula. To convert millimeters to inches, multiply by 0.039. To convert inches to millimeters, multiply by 25.4.
18. Place the new shim on the valve lifter. Remove the special tool and then recheck the valve clearance.
19. Install the gasket onto the cylinder head cover. Install the head cover with the seal washers and nuts. Torque the nuts to 52 inch lbs. (6 Nm).
20. Install the intake tube.
21. Refill the engine with coolant and connect the battery cable.

3.0L (1MZ-FE) engine

Camry and Avalon

NOTE: Adjust the valve clearance when the engine is cold.

1. With the ignition switch in the **LOCK** position, disconnect the negative battery terminal. If equipped with an air bag system, wait at least 90 seconds or longer before performing any other work.
2. Disconnect the accelerator/throttle cable from the throttle linkage.

3. Remove the air cleaner cover, air flow meter, and air duct assembly.

4. Remove the V-bank cover.

5. Remove the emission control valve set.

6. Remove the air intake chamber.

7. Disconnect the engine harness from the injectors and the ignition coils.

8. Remove the ignition coils and keep them in order for reassembly.

9. Remove the spark plugs.

10. Remove the cylinder head covers.

11. Turn the crankshaft pulley and align its groove with the timing mark **0** of the No. 1 timing cover.

12. Check that the valve lifters on the No. 1 intake are loose and the No. 1 exhaust are tight. If not, turn the crankshaft one complete revolution (360 degrees).

NOTE: All measurements should be written down. These recorded measurements will need to be used in conjunction with a mathematical formula to determine the thickness of the replacement shims.

13. Measure the clearance between the valve lifters and the camshaft. Record the measurements on valves No. 1 and 6 intake; No. 2 and 3 exhaust.

   a. The intake valve clearance cold is 0.006–0.010 in. (0.15–0.25mm).

   b. The exhaust valve clearance cold is 0.010–0.014 in. (0.25–0.35mm).

14. Turn the crankshaft ⅔ of a revolution (240 degrees). Record the measurements on valves No. 2 and 3 intake; No. 4 and 5 exhaust.

15. Turn the crankshaft another ⅔ of a revolution. Record the measure-

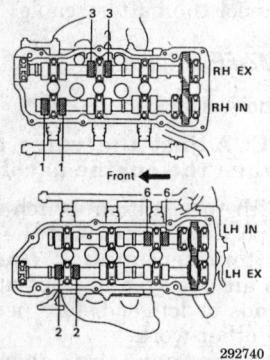

Adjust these valves–1st step — 1MZ-FE engine

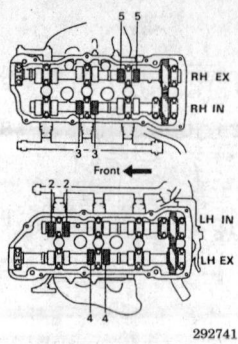

292741

Adjust these valves–2nd step — 1MZ-FE engine

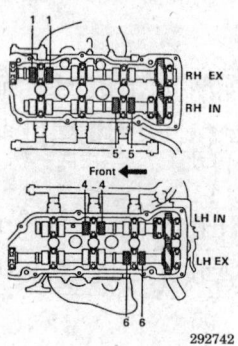

292742

Adjust these valves–3rd step — 1MZ-FE engine

ments on valves No. 4 and 5 intake; No. 1 and 6 exhaust.

16. Remove the adjusting shim by turning the crankshaft to position the cam lobe of the camshaft in the up position on the valve to be adjusted. Using a small thin flat bladed tool, turn the valve lifter so that the notches are perpendicular to the camshaft. Press down the valve lifter with SST 09248–55010 part A or equivalent. Place SST 09248–55010 part B between the camshaft and the valve lifter; remove part A.

17. Remove the adjusting shim with a magnet and a small screwdriver.

18. Determine the replacement adjusting shim size by either using the charts or the following formulas:

• Intake–N=T+(A-0.008 in. (0.020mm))
• Exhaust–N=T+(A-0.012 in. (0.30mm))
• T=Thickness of removed shim
• A=meassured valve clearance
• N=Thickness of new shim

19. Select a new shim with a thickness as close as possible to the calcu-

lated value. Install the new replacement shim.

NOTE: Shims are available in 17 sizes in increments of 0.0020 in. (0.050mm), from 0.0984 in. (2.500mm) to 0.1299 in. (3.300mm).

20. Recheck the valve clearance.

21. Install the cylinder head covers.

22. Install the spark plugs and the ignition coils.

23. Connect the engine wire harness to the injectors and the coils.

24. Install the intake chamber.

25. Install the emission control valve set.

26. Install the V-bank cover.

27. Install the air flow meter, air duct, and air cleaner cover.

28. Connect the negative battery cable.

3.0L (2JZ-GE and 2JZ-GTE) Engines

Supra

NOTE: Adjust the valve clearance when the engine is cold.

1. Disconnect the negative battery cable. On vehicles equipped with an air bag, wait at least 90 seconds before proceeding.

2. Remove the cylinder head covers.

3. Turn the crankshaft pulley and align its groove with the timing mark **0** of the No. 1 timing cover.

4. Check that the timing marks on the camshaft sprockets are in alignment with the marks on the No. 4 timing cover. If not, turn the crankshaft 1 complete revolution (360° degrees).

5. Measure the clearance between the valve lifter and the camshaft. Record the measurements on the intake valves No. 1, 2 and 4. Measure the exhaust valves at No. 1, 3 and 5.

   a. The intake valve clearance cold is 0.006–0.010 in. (0.15–0.25mm).

   b. The exhaust valve clearance cold is 0.010–0.014 in. (0.25–0.35mm).

6. Turn the crankshaft pulley one revolution (360°) and align the groove with the timing mark **0** of the No.1 timing belt cover.

7. Measure the clearance between the valve lifter and the camshaft. Record the measurements on the intake valves No. 3, 5 and 6. Measure the exhaust valves at No. 2, 4, and 6.

   a. The intake valve clearance cold is 0.006–0.010 in. (0.15–0.25mm).

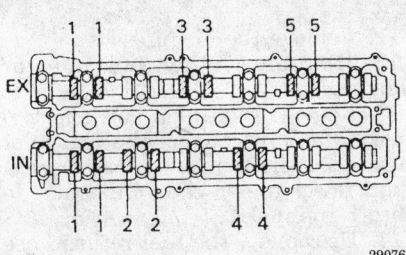

Valve clearance inspection (before turning crankshaft 360 degrees) — 2JZ-GE and 2JZ-GTE engines

290763

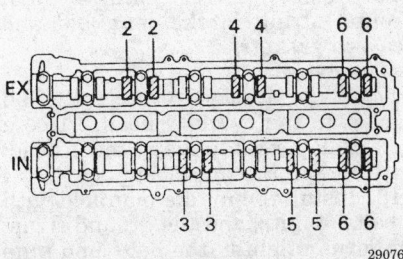

Valve clearance inspection (after turning crankshaft 360 degrees) — 2JZ-GE and 2JZ-GTE engines

290766

b. The exhaust valve clearance cold is 0.010–0.014 in. (0.25–0.35mm).

8. To adjust the valve clearance:

a. Remove the adjusting shim and turn the crankshaft to position the cam lobe of the camshaft on the adjusting valve upward.

b. Turn the valve lifter so that the notches are perpendicular to the camshaft.

c. Using SST 09248–55040 (valve lifter press) or equivalent, hold the camshaft in place.

d. Using SST 09248–55040 (valve lifter press) or equivalent, press down the valve lifter and place SST 09248–05420 (valve lifter stopper) or equivalent between the camshaft and valve lifter.

e. Remove the SST 09248–44040 tool.

f. Using a small screwdriver and a magnetic finger, remove the adjusting shim.

9. Determine the replacement adjusting shim size by either using the chart or the following formula:

- Intake–$N=T+(A-0.008$ in. (0.20mm))
- Exhaust–$N=T+(A-0.012$ in. (0.30mm))
- T=Thickness of removed shim
- A=meassured valve clearance
- N=Thickness of new shim

10. Recheck the valve clearance.

11. Install the cylinder head covers.

12. Connect the negative battery cable.

Rocker Arms

REMOVAL AND INSTALLATION

1.5L (3E-E) Engine

Tercel

1. Disconnect the negative battery cable. On vehicles equipped with an air bag, wait at least 90 seconds before proceeding.

2. Loosen the rocker arm adjusting screw locknuts and fully loosen the adjusting screw.

3. Remove the camshaft.

a. Uniformly loosen and remove bearing cap bolts in several posses, in the reverse order of the installation sequence.

b. Remove the camshaft, oil seal and camshaft bearing. Do not remove the distributor bearing cap.

c. Arrange the camshaft bearing caps in correct order.

4. Pull up on the top of the rocker arm spring while prying the spring with a suitable tool.

5. Remove the rocker arms and arrange them in order. Check the contact surface for any signs of pitting or wear. Replace if necessary.

To install:

6. Check that the adjusting screw is as shown and install a new spring to the rocker arm.

7. Press the bottom lip of the spring until it fits into the groove on the rocker arm pivot.

NOTE: Put the valve adjusting screw in the rocker arm pivot.

8. Pry the rocker spring clip onto the pivot. Pull the rocker arm up and down to check that there is spring tension and that the rocker does not rattle. Adjust the valves as necessary.

Intake Manifold

REMOVAL AND INSTALLATION

1.5L (3E-E) Engine

Tercel

1. Disconnect the negative battery cable. On vehicles equipped with an air bag, wait at least 90 seconds before proceeding.

2. If equipped with fuel injected engine, relieve the fuel pressure by removing fuel filler cap.

3. Drain the cooling system. Remove the air cleaner assembly.

4. Tag and disconnect all wires, hoses or cables that interfere with intake manifold removal.

5. Remove the necessary components in order to gain access to the intake manifold retaining bolts.

6. Remove the throttle body.

7. Disconnect the intake manifold water hoses.

8. Remove the intake manifold retaining bolts. Remove the intake manifold from the vehicle.

To install:

9. Using new gaskets, install the intake manifold and tighten the retaining bolts to 14 ft. lbs. (19 Nm) working from the center to the ends in steps.

10. Connect all necessary lines and hoses.

11. Connect all wire harnesses previously disconnected.

12. Install the throttle body.

13. Install the air cleaner assembly.

14. Fill the cooling system.

15. Connect the negative battery cable.

16. Start the engine, check for leaks and road test for proper operation.

1.5L (5E-FE) Engine

Tercel and Paseo

1. Disconnect the negative battery cable. On vehicles equipped with an air bag, wait at least 90 seconds before proceeding.

2. Drain the cooling system.

3. Disconnect the accelerator cable and if equipped with an automatic transmission, disconnect the throttle cable.

4. Remove the PCV hose.

5. Remove the air cleaner to throttle body hose.

6. Tag and disconnect all electrical wires and vacuum hoses that interfere with removal of the intake manifold.

7. Remove the EGR pipe from the intake manifold and EGR valve.

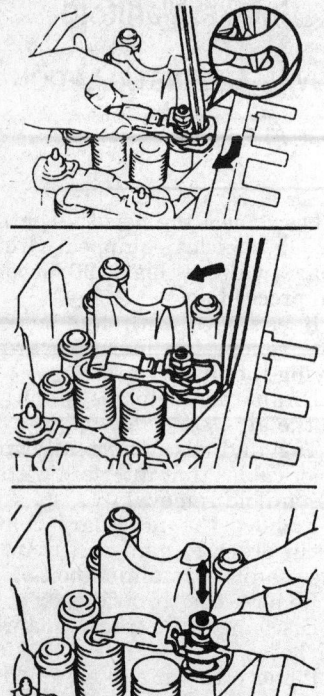

Rocker arm spring clip removal and installation — 3E-E engine

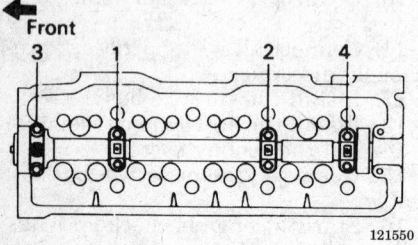

Bearing cap tightening sequence — 3E-E engine

8. Remove the throttle body assembly.

9. Disconnect the air intake chamber stay by removing the bolt and nut.

10. For vehicles with distributor ignition, disconnect the air pipe by removing the two bolts.

11. Disconnect the engine wire clamps from the intake manifold stay.

12. Remove the bolt(s) and nut(s) to the intake manifold stay. Remove the intake manifold stay from the engine.

13. Remove the intake manifold by removing two bolts and three nuts. Remove the intake manifold gasket.

To install:

14. Make sure the intake manifold surface is clean.

15. Using a new gasket, install the intake manifold and torque the nuts and bolts evenly to 14 ft. lbs. (19 Nm).

16. Install the intake manifold stay and torque the bolt(s) and nut(s) to 15 ft. lbs. (20 Nm).

17. For vehicles with distributor ignition, install the vacuum hoses and then install the air pipe. Torque the bolts to 48 inch lbs. (6 Nm).

18. Install the air intake chamber stay by installing the bolt and nut.

19. Install the throttle body, using a new gasket and tightening the nuts and bolts evenly to 9 ft. lbs. (13 Nm). Install all components to the throttle body.

20. Connect the EGR pipe to the intake manifold and EGR valve.

21. Connect all electrical wires and vacuum hoses removed from the intake manifold.

22. Install the air cleaner hose to the throttle body.

23. Refill all fluids.

24. Connect the negative battery cable, start the engine and check the ignition timing. Check for engine coolant leaks.

1.6L (4A-FE) and 1.8L (7A-FE) Engines

Celica and Corolla

1. Disconnect the negative battery cable. On vehicles equipped with an air bag, wait at least 90 seconds before proceeding.

— CAUTION —

Fuel injection systems remain under pressure after the engine has been turned OFF. Properly relieve fuel pressure before disconnecting any fuel lines. Failure to do so may result in fire or personal injury.

2. Drain the engine coolant into a suitable container.

3. Disconnect the throttle body from the air intake chamber.

4. Disconnect the ground strap connector.

5. Tag and disconnect the hoses from the intake chamber.

6. On vehicles with A/C remove the hose from the idle-up valve.

7. On vehicles with P/S remove the air hose from the air pipe.

8. Using a 6mm hexagon wrench, remove the 3 bolts, 2 nuts, and the air intake chamber cover and gasket.

9. If equipped with EGR, remove the EGR VSV.

10. Remove the intake manifold stay.

11. Remove the air pipe.

12. Disconnect the fuel injector connectors.

13. Remove the union bolt and two gaskets and disconnect the fuel inlet hose from the delivery pipe. Place a shop towel under the connection to absorb the fuel.

14. Disconnect the fuel return hose from the fuel pressure regulator and the air hose from the IAC valve to the air pipe.

15. Remove the two bolts and the delivery pipe together with the four injectors.

16. Remove the 7 bolts, 4 nuts, ground strap, intake manifold and the two gaskets.

To install:

17. Place a new intake manifold gasket to the cylinder head. Place a new EGR gasket to the cylinder head, with the protrusion facing down.

18. Install the intake manifold with 7 bolts, 4 nuts, and the ground strap. Uniformly tighten the bolts and nuts in several passes. Torque the **A** nuts to 9 ft. lbs. (13 Nm). Tighten all other nuts and bolts to 14 ft. lbs. (19 Nm).

19. Install the injectors and the delivery pipe.

20. If equipped with EGR, install the EGR VSV.

21. Install the air intake chamber cover with a new gasket.

22. Install the throttle body.

23. Install the air pipe and fuel inlet hose with the two bolts and nut. Install new gaskets on the fuel inlet hose and secure the clamp of the inlet hose to the intake manifold.

24. Connect the fuel return hose to the pressure regulator and the air hose from the IAC valve to the air pipe.

25. Install the intake manifold stay. Torque the 12mm head bolt to 14 ft. lbs. (19 Nm) and the 14mm head bolt to 29 ft. lbs. (39 Nm).

26. Connect the hoses to the intake chamber.

27. If equipped, install the air hoses the idle-up valve and the air pipe.

28. Connect the ground strap connector and connect the fuel injector connectors.

29. Refill the cooling system with coolant and connect the negative battery cable. Start the engine, check for leaks, and road test the vehicle for proper operation.

2.0L (3S-GTE) Engine

Celica and MR2

1. Relieve the fuel system pressure.

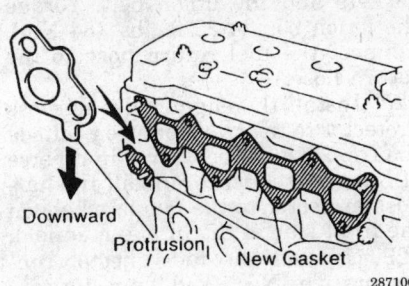

Installing the EGR and intake manifold gaskets — Celica and Corolla with 4A-FE and 7A-FE engines

---- **CAUTION** ----

Fuel injection systems remain under pressure after the engine has been turned OFF. Properly relieve fuel pressure before disconnecting any fuel lines. Failure to do so may result in fire or personal injury.

2. Disconnect the negative battery cable. On vehicles equipped with an air bag, wait at least 90 seconds before proceeding.

3. Drain the coolant. and remove the engine undercovers.

4. Remove the engine hood side panels the upper suspension brace.

5. Remove the No. 1 and No. 2 intake air connectors.

6. If equipped with cruise control, remove the cruise control actuator and the accelerator linkage.

7. Without cruise control, disconnect the accelerator cable from the throttle body.

8. Remove the air cleaner assembly and disconnect the ground strap connector.

9. Disconnect the brake booster vacuum hose the A/C idle up hoses.

10. Remove the turbocharger the throttle body.

11. Remove the cold start injector pipe and the injector connector.

12. Remove the distributor.

13. Remove the air hoses from the No. 1 air tube and the intake manifold. Remove the No. 2 air tube.

14. Remove the EGR vacuum modulator and the vacuum switching valve (VSV).

15. Remove the vacuum pipe.

16. Remove the hoses from the EGR valve. Remove the EGR valve and pipe assembly.

17. Disconnect the following:
   a. ECT sender gauge and the sensor connector.
   b. Cold start injector time switch connector.

c. Remove the bolts securing the fuel inlet and return hoses. Remove the hoses from the fuel rail.

18. Disconnect the following hoses:
   a. Water filler hose and radiator hose
   b. Water bypass pipe hose from the IAC valve
   c. Heater water hose and the water bypass hose from the bypass pipe
   d. Two TVV vacuum hoses for the EVAP

19. Remove the water outlet and housing and the No. 1 air tube and the water bypass pipe.

20. Remove the intake manifold stays.

21. Remove the T-VIS (Toyota variable induction system) vacuum tank, the VSV for the T-VIS, the VSV for the turbocharging pressure and the bracket.

22. Disconnect the two engine wire clamps from the mounting bolts on the No. 2 timing belt cover.

23. Disconnect the four injector connectors.

24. Remove the bolt and disconnect the ground strap.

25. Disconnect the knock sensor connector and the alternator wire and the connector.

26. Disconnect the vacuum sensing hose from the intake manifold.

27. Remove the two bolts and disconnect the engine wire from the intake manifold.

28. Remove the intake manifold and the T-VIS valve.

To install:

29. Install the T-VIS valve and the intake manifold and torque the bolts and nuts to 14 ft. lbs. (19 Nm).

30. Install the engine wire clamp to the intake manifold.

31. Connect the vacuum sensing hose to the intake manifold.

32. Connect the alternator wire, connector and the knock sensor connector.

33. Connect the ground strap with the bolt and the four injector connectors.

NOTE: The No. 1 and No. 3 injector connectors are brown, and the No. 2 and No. 4 injector connectors are gray.

34. Connect the engine wire clamps to the mounting bolts on the No. 2 timing belt cover.

35. Install the T-VIS vacuum tank, VSV for the T-VIS and the VSV for the turbocharging pressure assembly.

36. Install the intake manifold stays, torque the bolts to 19 ft. lbs. (25 Nm).

37. Install the No. 1 air tube and the water bypass pipe.

38. Install the water outlet and housing.

39. Connect these hoses:
   a. Water filler hose and the radiator hose
   b. Water bypass pipe hose from the IAC valve
   c. Heater water hose and the water bypass hose from the bypass pipe
   d. Two TVV vacuum hoses for the EVAP

40. Connect the following:
   a. ECT sender gauge and the sensor connector.
   b. Cold start injector time switch connector.
   c. Install the bolts securing the fuel inlet and return hoses. Install the fuel inlet and return hoses to the fuel rail.

41. Install the EGR valve and pipe. Torque the bolts to the manifold to 14 ft. lbs. (19 Nm) and the bolts to the cylinder head to 19 ft. lbs. (25 Nm).

42. Install the vacuum pipe.

43. Install the EGR vacuum modulator and the VSV.

44. Connect the EGR hoses to the EGR valve.

45. Install the No. 2 air tube. Connect the air hose to the No. 1 air tube and the intake manifold.

46. Install the distributor.

47. Connect the cold start injector connector and install the cold start injector pipe. Torque the bolts to 9 ft. lbs. (12 Nm).

48. Install the throttle body. Torque the bolts to 14 ft. lbs. (19 Nm).

49. Install the turbocharger. Torque the bolts:
   a. Turbocharger to the exhaust manifold: 47 ft. lbs. (64 Nm).
   b. Turbocharger to the oil pipe: 13 ft. lbs. (17 Nm).
   c. Oil pipe to the cylinder block: 38 ft. lbs. (51 Nm).
   d. Bracket of the oil pipe to the cylinder block: 32 ft. lbs. (43 Nm).

50. Connect the A/C idle up hoses and the brake booster vacuum hose.

51. Connect the ground strap connector and install the air cleaner.

52. For vehicles with cruise, install the cruise control actuator and the accelerator linkage.

53. For vehicles without cruise, connect the accelerator cable to the throttle body.

54. Install the No. 1 and No. 2 intake air connectors.

55. Install the upper suspension brace. Torque the bolts to 54 ft. lbs. (73 Nm) and the nuts to 47 ft. lbs. (64 Nm).

56. Fill the engine with coolant and install the engine hood side panels.

57. Install the engine undercovers and connect the negative battery cable.

58. Start the engine and check for leaks.

2.2L (5S-FE) Engine

Celica

1. Relieve the fuel system pressure.

2. Disconnect the negative battery cable. On vehicles equipped with an air bag, wait at least 90 seconds before proceeding.

3. Drain the engine coolant into a suitable container.

4. Disconnect and/or remove the following components:

a. The IAT sensor connector.

b. The high tension spark plug wire from air cleaner hose.

c. The accelerator cable from the clamp and the cruise control actuator cable from the clamps.

d. For California vehicles, disconnect the air hose for the idle up from the air cleaner hose.

e. The four clamps and the and the air cleaner cap from the air cleaner case.

f. The air cleaner hose from the throttle body.

g. The air cleaner cap and hose assembly.

5. Remove the throttle body.

6. Disconnect the vacuum sensor hose from the gas filter on the intake manifold.

7. Remove the brake booster vacuum line from the intake manifold.

8. If equipped with A/C, remove the A/C idle-up valve.

9. Disconnect the EGR temperature sensor connector from the bracket on the intake manifold, and disconnect the sensor connector from the wiring connector. Disconnect the EVAP hose from the charcoal canister and disconnect the hose clamp from the bracket on the air tube.

10. Disconnect the two vacuum hoses from the VSV for the EGR. Disconnect the vacuum modulator from the clamp on the intake manifold.

11. Loosen the union nut of the EGR pipe, and remove the two nuts, the EGR valve, vacuum modulator, vacuum hoses assembly and gasket.

12. Remove the bolt, nut, and the intake manifold stay.

13. Disconnect the A/T throttle control cable from the clamp on the rear side of the intake manifold.

14. Remove the following hoses:

• Two P/S air hose(s) from the air tube and the intake manifold

• Air hose from the fuel pressure regulator

15. Remove the two bolts, hose bracket (for EGR), and the air tube.

16. Remove the bolt and the vacuum pipe.

17. Disconnect the knock sensor connector. Remove the bolt and disconnect the ground cables from the intake manifold.

18. Remove the VSV assembly for fuel pressure control and EGR.

19. Disconnect the PCV hose, accelerator, A/T throttle control cables and bracket from the intake manifold.

20. Disconnect the engine wire harness protector from the bracket on the starter, disconnect the Vehicle Speed Sensor (VSS) connector, remove the bolt and disconnect the engine wire harness protector from the LH side of the intake manifold.

21. Disconnect the four fuel injector connectors and disconnect the engine wire harness protector from the two brackets on the front side of the intake manifold.

22. Disconnect the engine wire harness protector from the two mounting bolts of the No. 2 timing belt cover in sequence.

CAUTION
Fuel injection systems remain under pressure after the engine has been turned OFF. Properly relieve fuel pressure before disconnecting any fuel lines. Failure to do so may result in fire or personal injury.

23. Remove the union bolt and two gaskets and disconnect the fuel inlet hose from the delivery pipe. Disconnect the fuel return hose from the return pipe.

24. Remove the six bolts and two nuts and disconnect the engine wire harness between the intake manifold and the cylinder head. Remove the intake manifold and gasket.

To install:

25. Insert the intake manifold between the cylinder head and the firewall. Insert the engine wire harness between the intake manifold and the cylinder head.

26. Install a new gasket and the intake manifold with the six bolts and two nuts. Uniformly tighten the bolts and nuts in several passes. Torque them to 14 ft. lbs. (19 Nm).

27. Connect the fuel inlet pipe to the fuel delivery hose using two new gaskets and the union bolt. Torque the union bolt to 25 ft. lbs. (34 Nm). Connect the fuel return hose to the return hose.

28. Install the engine wire harness protector to the two mounting bolts of the No. 2 timing belt cover in reverse of removal sequence. Install the harness protector to the two brackets on the front side of the intake manifold.

29. Connect the fuel injector connectors. The No. 1 and 3 injector connectors are brown; the No. 2 and 4 connectors are gray.

30. Install the engine wire harness protector to the LH side of the intake manifold with the bolt and connect the VSS sensor connector.

31. Reinstall the engine wire harness protector to the starter bracket.

32. Install the cable bracket to the intake manifold with the two bolts, and install the accelerator and A/T throttle control cables to the four clamps.

33. Connect the PCV hose to the intake manifold.

34. Install the VSV for fuel pressure control and EGR.

35. Connect the knock sensor connector and install the ground cable to the intake manifold with the bolt.

36. Install the air tube and hose bracket (for EGR) with the two bolts and connect these hoses:

• The P/S air hose(s) to the air tube and the intake manifold

• The vacuum sensing hose to the fuel pressure regulator

37. Install the A/T throttle control cable to the clamp on the rear side of the intake manifold.

38. Install the intake manifold stay bolt with the bolt and nut. Torque the bolt to 15 ft. lbs. (21 Nm) and the nut to 32 ft. lbs. (44 Nm).

39. Install a new gasket, the EGR valve with the union bolt and two nuts. Torque the union bolt to 43 ft. lbs. (59 Nm) and the nuts to 9 ft. lbs. (13 Nm). Install the vacuum modulator to the clamp on the intake manifold.

40. Connect the vacuum hoses to the VSV for the EGR and install the hose clamp to the bracket on the air tube.

41. If equipped with A/C, install the A/C idle-up valve.

42. Connect the EGR gas temperature sensor connector and install the sensor connector to the bracket on the intake manifold. Connect the EVAP hose to the charcoal canister.

43. Install the vacuum sensor hose to the gas filter on the intake manifold and install the brake booster vacuum hose.

44. Install the throttle body assembly.

45. Install the air filter, the air cleaner cap and hose assembly, the air cleaner hose to the throttle body, and secure the air cleaner cap with the four clamps.

46. Install the accelerator and cruise control cables to the clamps. Connect the high tension spark plug wire to the air cleaner hose.

47. Connect the IAT sensor connector.

48. Refill the cooling system and connect the negative battery cable. Start the engine, check for leaks, and road test the vehicle for proper operation.

Camry

1. Disconnect the negative battery cable. On vehicles equipped with an air bag, wait at least 90 seconds before proceeding.

─────── **CAUTION** ───────

Fuel injection systems remain under pressure after the engine has been turned OFF. Properly relieve fuel pressure before disconnecting any fuel lines. Failure to do so may result in fire or personal injury.

2. Drain the engine coolant into a suitable container.

3. Disconnect the accelerator cable from the throttle body. If equipped with A/T, disconnect the throttle cable from the throttle body.

4. Disconnect the intake air temperature sensor connector

5. On California models, disconnect the air cleaner hose.

6. Loosen the air cleaner hose clamp bolt, disconnect the four air cleaner cap clips, disconnect the air hose from the throttle body, and remove the air cleaner cap together with the resonator and the air cleaner hose.

7. Tag and remove the electrical connections and hoses from the throttle body.

8. Remove the throttle body. Type A throttle bodies are secured with four bolts and Type B throttle bodies are secured with two bolts and two nuts.

9. Remove the vacuum hose bracket and the engine wire harness from the intake manifold.

10. Remove the EGR valve.

11. Remove the four bolts, the wire bracket, the No. 1 air intake chamber, and the manifold stays. Remove the six bolts, two nuts, the intake manifold, and the gasket.

To install:

12. Install the intake manifold to the cylinder head with a new gasket. Tighten the six bolts and the two nuts in several passes to 14 ft. lbs. (19 Nm). Install the two wire clamps to the wire brackets on the intake manifold.

13. Install the vacuum hose bracket and the engine wire harness.

14. Install the No. 1 air intake chamber and manifold stays with the four bolts. Torque the 14mm bolts to 31 ft. lbs. (42 Nm) and torque the 12mm bolts to 16 ft. lbs. (22 Nm).

15. Install the EGR valve.

16. Install the throttle body with a new gasket on the intake chamber. Connect the hoses and electrical connections to the throttle body.

NOTE: The protrusion on the gasket should be facing down and the water hose connections on the throttle body should also face down.

17. On type A throttle body, torque the four bolts to 14 ft. lbs. (19 Nm). Bolt A is 45mm in length and bolt B is 55mm

18. On type B throttle body, torque the two bolts and the two nuts to 14 ft. lbs. (19 Nm).

19. Make the following connections to the throttle body:
   • The PCV hose
   • The two vacuum hoses from the EGR modulator
   • The vacuum hose from the TVV (for EVAP)
   • The IAC valve connector
   • The throttle position sensor connector

20. Connect the air cleaner hose to the throttle body and install the air cleaner cap together with the resonator and the air cleaner hose. Connect the intake air temperature sensor connector.

21. On California models, connect the air hose to the air cleaner hose.

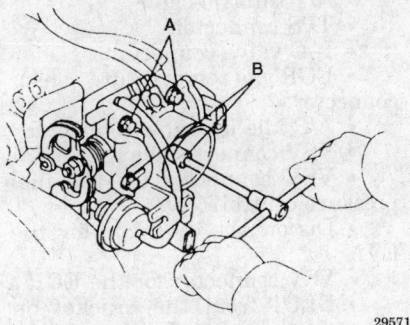

295716

Mounting bolt lengths for the throttle body — Camry with 5S-FE engine

22. If equipped with A/T, connect and adjust the throttle cable. Connect the accelerator cable.

23. Fill the cooling system with coolant and connect the negative battery cable. Start the engine, check for leaks, and road test the vehicle for proper operation.

MR2

1. Relieve the fuel system pressure.

2. Disconnect the negative battery cable. On vehicles equipped with an air bag, wait at least 90 seconds before proceeding.

3. Drain the coolant and remove the engine undercovers.

4. Remove the engine hood side panels.

5. Disconnect the accelerator cable from the throttle body.

6. If equipped with cruise control, remove the cruise control actuator and the accelerator linkage.

7. With A/T, disconnect the throttle cable from the throttle body.

8. Remove the air cleaner assembly.

9. Disconnect the ground strap connector.

10. Remove the distributor.

11. Tag and disconnect the MAP sensor, A/C idle-up and brake booster hoses from the intake manifold.

12. Remove the water outlet and the housing as follows:
   a. Disconnect the ECT sender gauge and the sensor connector
   b. Remove the bolt securing the fuel inlet and the fuel return line to the water outlet.
   c. Disconnect the water filler and radiator hoses.
   d. Disconnect the heater water and IAC water bypass hoses.
   e. Disconnect the vacuum hoses from the TVV.
   f. Remove the two bolts, the housing, and the gasket.

13. Remove the water bypass pipe.

─────── **CAUTION** ───────

Fuel injection systems remain under pressure after the engine has been turned OFF. Properly relieve fuel pressure before disconnecting any fuel lines. Failure to do so may result in fire or personal injury.

14. Remove the throttle body.

15. Remove the EGR valve and the vacuum modulator.

16. Remove the VSV for fuel pressure control.

17. Disconnect the knock sensor and the EGR VSV connectors.

18. Remove the EGR VSV.

19. Disconnect the alternator wire and the connector.

20. Remove the two bolts and the air intake chamber stay.

21. Disconnect the engine wire protector from the RH side of the intake manifold.

22. Disconnect the engine wire protector between the No. 3 timing belt cover and the cylinder head cover.

23. Remove the accelerator bracket and disconnect the PCV hose from the intake manifold.

24. Disconnect the engine wire protector from the brackets on the front side of the intake manifold.

25. Remove the two bolts and the intake manifold stay.

26. Remove the intake manifold.

To install:

27. Install the intake manifold and uniformly torque the bolts and nuts to 14 ft. lbs. (19 Nm).

28. Install the intake manifold stays, torque the 12mm bolts to 16 ft. lbs. (22 Nm) and the 14mm bolts to 31 ft. lbs. (42 Nm).

29. Install the engine wire clamp and the PCV hose to the intake manifold.

30. Install the accelerator bracket.

31. Install the two clamps and the engine wire protector.

32. Install the air intake chamber stay and torque the bolts to 31 ft. lbs. (42 Nm).

33. Connect the alternator wire and the connector.

34. Install the EGR VSV.

35. Connect the knock sensor and the EGR VSV connectors.

36. Install the VSV for the fuel pressure control.

37. Install the EGR valve and the vacuum modulator, torque the nut to 9 ft. lbs. (13 Nm) and the union nut to 43 ft. lbs. (59 Nm).

38. Install the throttle body, torque the bolts to 14 ft. lbs. (19 Nm).

39. Install the water bypass pipe. Tighten the bolts and nuts to 78 inch lbs. (9 Nm).

40. Install the water outlet and housing and torque the bolts to 11 ft. lbs. (15 Nm).

41. Perform the following steps:

a. Install the bolt that secures the fuel inlet and return lines.

b. Connect the ECT sensor and gauge sender connectors.

c. Connect the heater water and the IAC water bypass hoses.

d. Connect the radiator and water filler hoses.

e. Connect the vacuum hoses to the TVV.

42. Install the distributor.

43. Connect the MAP sensor, A/C idle-up and brake booster hoses to the intake manifold.

44. Connect the ground strap connector.

45. Install the air cleaner assembly.

46. If removed, install the cruise control actuator and the accelerator linkage.

47. If removed, connect the throttle and accelerator cables to the throttle body.

48. Fill the engine with coolant and install the engine hood side panels.

49. Install the engine undercovers and connect the negative battery cable.

50. Start the engine and check for leaks.

3.0L (1MZ-FE) Engine

Camry and Avalon

── **CAUTION** ──
Fuel injection systems remain under pressure even after the engine has been turned OFF. The fuel system pressure must be relieved before disconnecting any fuel lines. Failure to do so may result in fire and/or personal injury.

1. Relieve the fuel pressure from the fuel lines.

2. With the ignition switch in the **LOCK** position, disconnect the negative battery terminal. If equipped with an air bag system, wait at least 90 seconds or longer before performing any other work.

3. Remove the air cleaner hose from the engine compartment.

4. Remove the V–bank cover from the engine.

5. Remove the air cleaner chamber assembly as follows:

a. Drain the engine coolant from the vehicle.

b. Disconnect the following connectors and cables:
  • Accelerator cable
  • A/T throttle cable
  • TPS connector
  • IAC valve connector
  • EGR gas temperature sensor connector
  • A/C idle up valve connector
  • VSV connector for the ACIS
  • VSV connector for the fuel pressure control
  • Disconnect the VSV for the EVAP
  • VSV connector for the EGR
  • DLC1 from the bracket on the intake air control valve

c. Remove the power steering pressure tube from the No. 1 engine hanger by removing the two bolts.

d. Disconnect the following hoses, clamps, and cables:
  • Brake booster vacuum hose from the intake air control valve for the ACIS
  • PCV hose from the PCV valve on the right-hand cylinder head
  • Ground strap and cable from the air intake air control valve from the ACIS
  • Ground cable from the air intake chamber
  • Vacuum hose clamp from fuel pipe
  • Two bypass hoses from the throttle body
  • Two power steering air hoses to the air intake chamber.
  • Air assist hose from the throttle body
  • Remove the EVAP hose from the pipe on emission control valve set
  • Two vacuum hoses from the pipes on the cylinder head rear plate
  • Vacuum sensing hose from the fuel pressure regulator
  • Engine wire clamp from emission control valve set

e. Remove the two bolts and the No. 1 engine hanger.

f. Remove the two bolts and the air intake chamber stay.

g. Remove the No. 2 EGR pipe and two gaskets by removing the four nuts.

h. Disconnect the hose from the VSV from the EVAP.

i. Using an 8mm hexagon wrench, remove the two bolts. Remove the two nuts, the air intake chamber assembly and gasket.

6. Disconnect the fuel injector connectors.

7. Remove the air assist hoses and pipe.

8. Disconnect the fuel return hose from the No. 1 fuel pipe.

9. Disconnect the fuel inlet hose from the fuel filter. Catch any fuel leaking from the filter in a shop rag. Dispose of the rag properly.

10. Remove the delivery pipes and injectors from the engine as follows:

a. Loosen the two union bolts holding the No. 2 fuel pipe to the delivery pipes.

b. Disconnect the fuel return hose from the fuel pressure regulator.

c. Remove the union bolt and two gaskets for the right-hand delivery pipe.

d. Remove the two bolts to the left-hand delivery pipe and then remove the left-hand delivery pipe, three injectors, and the No. 2 fuel pipe as an assembly.

e. Remove the union bolt and two gaskets from the left-hand delivery pipe. Disconnect the No. 2 fuel pipe from the left-hand delivery pipe.

f. Remove the right-hand delivery pipe by removing the three bolts. Remove the delivery pipe, injectors, and the fuel inlet hose as an assembly.

g. Remove the four spacers from the intake manifold.

h. Pull out the six injectors from the delivery pipes.

i. Remove the two O-rings and two grommets from each injector.

11. Remove the heater hoses from the intake manifold.

12. Remove the nine bolts, two nuts, and two plate washers. Remove the intake manifold.

To install:

13. Thoroughly clean the intake manifold and cylinder head surfaces.

14. Using new gaskets, install the intake manifold with the two plate washers, nine bolts, and two nuts. Tighten the nuts and bolts to 11 ft. lbs. (15 Nm).

15. Install the heater hoses to the intake manifold.

16. Install the injectors as follows:

a. Install two new grommets to each injector.

b. Apply a light coat of spindle oil or gasoline to the two O-rings and install them to each injector.

c. Install the injector into the delivery pipe by turning the injector back and forth. Install all the injectors into the delivery pipes. Make sure to position the injector electrical connector outward.

d. Place the four spacers in position on the intake manifold.

e. Place the right-hand delivery pipe and the No. 1 fuel pipe together with the three injectors in position on the intake manifold.

f. Temporarily install the two bolts holding the right-hand delivery pipe to the intake manifold.

g. Temporarily install the bolt holding the No. 1 fuel pipe to the intake manifold.

h. Place the left-hand delivery pipe and the No. 2 fuel pipe together with the three injectors in position.

i. Connect the fuel return hose to the fuel pressure regulator.

j. Temporarily install the two bolts holding the left-hand delivery pipe to the intake manifold.

k. Temporarily install the No. 2 fuel pipe to the left and right-hand delivery pipes with the union bolts and two new gaskets.

l. Check that the injectors rotate smoothly. If the injectors do not rotate smoothly, the probable cause is incorrect installation of the O-rings. Replace the O-rings.

m. Tighten the four bolts holding the delivery pipes to the intake manifold. Torque the bolts to 7 ft. lbs. (10 Nm).

n. Tighten the bolt holding the No. 1 fuel pipe to the intake manifold. Torque the bolts to 14 ft. lbs. (20 Nm).

o. Tighten the two union bolts holding the No. 2 fuel pipe to the delivery pipes. Torque the bolts to 24 ft. lbs. (33 Nm).

17. Connect the fuel inlet hose to the fuel filter by installing the union bolt. Use two new gaskets when installing the union bolt.

18. Connect the fuel return hose to the No. 1 fuel pipe. When routing the fuel return hose, pass the hose under the heater hoses.

19. Connect the air assist hoses to the intake manifold and then install the air assist pipe to the bracket on the No. 1 fuel pipe.

20. Connect the injector connectors.

21. Install the air intake chamber assembly as follows:

a. Using an 8mm hexagon wrench, install the air intake chamber with a new gasket. Install the two bolts and two nuts. Uniformly tighten the bolts and nuts in several passes and then torque the bolts and nuts to 32 ft. lbs. (43 Nm).

b. Connect the hose to the VSV for the EVAP system.

c. Install the two new gaskets and No. 2 EGR pipe with the four nuts. Torque the nuts to 9 ft. lbs. (12 Nm).

d. Install the No. 1 engine hanger with the two bolts. Torque the bolts to 19 ft. lbs. (39 Nm).

e. Install the air intake chamber stay with the two bolts. Torque the bolts to 14 ft. lbs. (20 Nm).

f. Connect the hoses, clamp, and cables as follows:

• Brake booster vacuum hose to the intake air control valve for the ACIS.

• PCV hose to the PCV valve on the right-hand cylinder head.

• Ground strap and cable to the intake air control valve for the ACIS.

• Connect the ground cable and strap with the nut. Torque the nut to 10 ft. lbs. (15 Nm)

• Ground cable to air intake chamber

• Vacuum hose clamp to fuel pipe.

• Two water bypass hoses to the throttle body.

• Air assist hose to the throttle body.

• Two power steering air hoses to the air intake chamber.

• Connect the EVAP hose to the pipe on the emission control valve set

• Two vacuum hoses to the pipes on the cylinder head rear plate.

• Vacuum sensing hose to the fuel pressure regulator.

• Engine wire clamp to the emission control valve set.

22. Install the power steering pressure tube with the two nuts.

23. Connect the following connectors and cables:

• TPS sensor connector
• IAC valve connector
• EGR gas temperature sensor connector
• A/C idle up valve connector
• VSV connector for the ACIS
• VSV connector for the fuel pressure control
• For California vehicles, install the VSV connector for the EVAP
• VSV connector for the EGR
• DLC1 to the bracket on the intake air control valve
• Accelerator cable
• A/T throttle cable

24. Install the V-bank cover.

25. Install the air cleaner hose to the engine.

26. Fill the engine coolant.

27. Connect the negative battery cable to the battery and start the engine.

28. Check the engine coolant level and check for leaks.

3.0L (2JZ-GE) Engine

Supra

1. Disconnect the negative battery cable. Wait at least 90 seconds before proceeding with any other work.

2. Drain the engine coolant from the engine.

3. Remove the VSV connector.

4. Remove the vacuum sensing hose from the fuel pressure control and the air intake chamber.

5. For vehicles with automatic transmissions, remove the transmission dipstick and guide.

6. Remove the EGR pipe.

7. Disconnect the EGR gas temperature sensor connector from the No. 2 vacuum pipe and wiring connector.

8. Disconnect the No. 2 vacuum pipe from the air intake chamber and from the intake manifold by removing the two nuts.

9. Remove the air intake chamber as follows:

a. Remove the two nuts holding the throttle body bracket to the cylinder head.

b. Remove the intake air connector from the air intake chamber by removing the four bolts and two nuts.

c. Remove the bolt and disconnect the engine wire protector from the air intake chamber.

d. Disconnect the:
- Three vacuum hoses from the No. 1 vacuum pipe.
- Vacuum hose from the air intake chamber
- Power steering air hose from the air intake chamber
- Brake booster vacuum hose from the air intake chamber by removing the union bolt and two gaskets
- Vacuum hose (from the actuator for ACIS) from the No. 1 vacuum pipe
- Except for California vehicles, disconnect the vacuum sensing hose (from the fuel pressure regulator) from the air intake chamber

e. Loosen the two nuts holding the air intake chamber stays to the cylinder head

f. Remove the two bolts and disconnect the two air intake chamber stays from the air intake chamber.

g. Remove the air intake chamber by removing the nut and five bolts. Remove the gaskets.

10. Relieve the fuel pressure in the fuel line before disconnecting any fuel lines.

CAUTION

Fuel injection systems remain under pressure, even after the engine has been turned OFF. The fuel system pressure must be relieved before disconnecting any fuel lines. Failure to do so may result in fire and/or explosion.

11. Disconnect the fuel return hose.

12. Remove the oil dipstick and guide for the engine.

13. Disconnect the engine wire from the intake manifold as follows:

a. Remove the bolt and disconnect the engine wire bracket from the water pump.

b. Remove the two bolts and disconnect the two ground straps from the intake manifold.

c. Remove the two bolts and disconnect the two wire clamps from the intake manifold.

d. Disconnect the following connectors:
- Six injector connectors
- ECT sensor and the ECT sender gauge connectors

e. Disconnect the engine wire protector from the intake manifold by removing the three nuts.

14. Remove the intake manifold stay by removing the two bolts.

15. Remove the fuel inlet pipe as follows:

a. Remove the clamp bolt from the intake manifold.

b. Remove the union bolt and disconnect the fuel inlet pipe.

16. Remove the two nuts and six bolts to the intake manifold.

17. Remove the intake manifold and gaskets.

To install:

18. Inspect the contact surfaces of the head and manifold. Remove any traces of gasket material. Install a new gasket and then install the intake manifold. Tighten the bolts and nuts to 20 ft. lbs. (27 Nm).

19. Install the fuel inlet pipe as follows:

a. Install the inlet pipe with new gaskets. Torque the union bolt to 30 ft. lbs. (42 Nm).

b. Install the clamp bolt to the intake manifold.

20. Install the intake manifold stay and torque the bolts to 29 ft. lbs. (39 Nm).

21. Connect the engine wire as follows:

a. Install the engine wire protector to the intake manifold with the three nuts.

b. Connect the following connectors:
- Six injector connectors
- ECT sender gauge and the ECT sender gauge connectors

c. Install the two wire clamps to the intake manifold with the bolts.

d. Install the two ground straps to the intake manifold with the bolts.

e. Install the engine wire bracket to the water pump with the bolt.

22. Install the intake air chamber as follows:

a. Install a new gasket to the air intake chamber facing the protrusion on the gasket rearward.

b. Install the intake air chamber and install the nut and five bolts.

c. Install the two air intake chamber stays to the air intake chamber and install the two bolts. Torque the bolts to 13 ft. lbs. (18 Nm).

d. Tighten the two nuts holding the air intake chamber to the cylinder head. Torque the nuts to 13 ft. lbs. (18 Nm).

e. Connect the following hoses:
- Except California vehicles, connect the vacuum sensing hose (from the fuel pressure regulator) to the air intake chamber.
- Vacuum hose (from the actuator for ACIS) to the No. 1 vacuum pipe.
- Brake booster vacuum hose to the air intake chamber. Torque the union bolt to 22 ft. lbs. (29 Nm).
- Power steering air hose to the air intake chamber
- Vacuum hose (from the No. 2 vacuum pipe) to the air intake chamber
- Three vacuum hoses (from the No. 2 vacuum pipe) to the No. 1 vacuum pipe

f. Install the engine wire protector to the air intake chamber and install the bolt.

g. Install the four bolts and two nuts holding the intake air connector to the air intake chamber.

h. Install the throttle body bracket to the cylinder head and install the two nuts. Torque the nuts to 15 ft. lbs. (21 Nm).

23. Connect the No. 2 vacuum pipe to the air intake chamber and intake manifold. Torque the two nuts to 20 ft. lbs. (27 Nm).

24. Connect the EGR gas temperature sensor connector.

25. Install the EGR pipe. Torque the union nut to 47 ft. lbs. (64 Nm) and the two bolts to 20 ft. lbs. (27 Nm).

26. Install the oil dipstick guide and dipstick.

27. Connect the fuel return hose.

28. If equipped with automatic transmission, install the transmission oil dipstick guide and dipstick.

29. Install the VSV for the fuel pressure control.

30. Connect the negative battery cable to the battery.

31. Fill the cooling system, start the engine, and check for leaks. Bleed the cooling system.

32. Road test the vehicle for proper operation.

3.0L (2JZ-GTE) Engine

Supra

— **CAUTION** —
Fuel injection systems remain under pressure after the engine has been turned OFF. Properly relieve fuel pressure before disconnecting any fuel lines. Failure to do so may result in fire or personal injury.

1. Relieve fuel system pressure.
2. Disconnect the negative battery cable to the battery. On vehicles equipped with an air bag, wait at least 90 seconds before proceeding.
3. Remove the engine undercover.
4. Remove the throttle body as follows.
   a. Drain the engine coolant from the radiator and engine.
   b. Disconnect the following hose, cable and connectors from the throttle body:
   • Air hose
   • Accelerator and cruise control actuator cables
   • Throttle position sensor connector
   • Sub throttle position sensor connector
   • Sub throttle actuator connector
   c. Disconnect the throttle body from the air intake chamber by removing the two bolts and two nuts.
   d. Remove the gasket from the throttle body.
   e. Disconnect the following hoses from the throttle body and remove the throttle body from the engine:
   • EVAP hose
   • Water bypass hose (from the No. 4 water bypass pipe)
   • Power steering air hose
5. For vehicles with automatic transmissions, remove the transmission dipstick and guide.
6. Remove the oil dipstick and guide for the engine.
7. Remove the air intake chamber stay by removing the nut and bolts.
8. Disconnect the control cable bracket from the air intake chamber by removing the two bolts.
9. Disconnect the following connectors and hose:
   • IAC valve connector
   • Turbo pressure sensor connector
   • VSV connector for the fuel pressure control
   • VSV connector for the EGR valve

10. Remove the bolt and disconnect the engine wire protector from the vehicle body.
11. Disconnect the following hoses:
   • Disconnect the IAC valve pipe from the clamp on the cylinder head cover and disconnect the air hose form the IAC valve
   • Disconnect the air hose (from the air intake chamber) from the vacuum pipe on the IAC valve pipe
   • Air hose for the EGR from the valve pipe
   • PCV hose from the PCV valve
   • Vacuum sensing hose from the fuel pressure regulator
   • Water bypass hose (from the IAC valve) from the No. 4 water bypass pipe
   • EVAP hose (from the air intake chamber) from the vacuum pipe on the manifold stay
   • EVAP hose (from the vacuum pipe on the No. 4 water bypass pipe) from the No. 2 vacuum pipe
   • EVAP hose (from the charcoal canister) from the No. 2 vacuum pipe
   • Power steering air hose from the air intake chamber
   • Brake booster vacuum hose from the union on the air intake chamber
12. Disconnect the EGR gas temperature sensor connector from the No. 2 vacuum pipe and wiring connector.
13. Remove the EGR pipe.
14. Remove the No. 4 water bypass pipe by removing the two bolts.
15. Remove the intake manifold stay by removing the two bolts.
16. Remove the air intake chamber assembly as follows:
   a. Disconnect the ground cable from the intake manifold by removing the bolt.
   b. Remove the two bolts holding the engine wire protector to the intake manifold.
   c. Disconnect the two clamps for the engine wire protector from the brackets.
   d. Remove the five bolts, two nuts, and the engine wire bracket.
   e. Disconnect the air intake air chamber assembly from the intake manifold.
   f. Disconnect the water bypass hose from the IAC valve.
   g. Remove the gasket.
17. Disconnect the six injector connectors.
18. Disconnect the two camshaft position sensor connectors.
19. Disconnect the three engine wire clamps from the injector holders.

20. Disconnect the fuel inlet pipe from the delivery pipe.
21. Disconnect the fuel return pipe from the fuel pressure regulator.
22. Remove the intake manifold and delivery pipe assembly by removing the four bolts and two nuts.
To install:
23. Install a new gasket, the intake manifold and delivery pipe assembly with the four bolts and two nuts. Torque the six bolts and two nuts to 20 ft. lbs. (27 Nm).
24. Connect the fuel return pipe to the fuel pressure regulator and torque the union bolts to 20 ft. lbs. (27 Nm).
25. Connect the fuel inlet pipe to the delivery pipe and torque the union bolt to 30 ft. lbs. (41 Nm).
26. Connect the three engine wire clamps to the injector holders.
27. Install the two camshaft position sensor connectors.
28. Install the six injector connectors. The No. 1, 3, and 5 injector connectors are dark gray; the No. 2, 4, and 6 injector connectors are gray.
29. Install the air intake chamber assembly as follows:
   a. Install the gasket.
   b. Connect the water bypass hose to the IAC valve.
   c. Connect the air intake chamber assembly to the intake manifold and install the five bolts and two nuts. Tighten the nuts and bolts in several passes to 20 ft. lbs. (27 Nm).
   d. Connect the two clamps of the engine wire protector to the brackets.
   e. Install the two bolts to hold the engine wire protector to the intake manifold.
   f. Connect the ground cable to the intake manifold by installing the bolt.
30. Install the intake manifold stay and install the two bolts. Torque the bolts to 29 ft. lbs. (39 Nm).
31. Install the No. 4 water bypass pipe with the two bolts.
32. Install the EGR pipe as follows:
   a. Install the EGR pipe and gasket with the two bolts. Torque the bolts to 20 ft. lbs. (27 Nm).
   b. Install the union bolt holding the EGR pipe to the EGR valve and torque the union bolt to 47 ft. lbs. (64 Nm).
33. Connect the EGR gas temperature sensor connector.
34. Install the following hoses:
   • Brake booster vacuum hose to the union on the air intake chamber
   • Power steering air hose to the air intake chamber

- EVAP hose (from the charcoal canister) to the No. 2 vacuum pipe
- EVAP hose (from the vacuum pipe on the No. 4 water bypass pipe) to the No. 2 vacuum pipe
- EVAP hose (from the air intake chamber) to the vacuum pipe on the manifold stay
- Water bypass hose (from the IAC valve) to the No. 4 water bypass pipe
- Vacuum sensing hose to the fuel pressure regulator
- PCV hose to the PCV valve
- Air hose for the EGR to the valve pipe
- Air hose (from the air intake chamber) to the vacuum pipe on the IAC valve pipe
- Connect the IAC valve pipe to the clamp on the cylinder headcover
- Connect the air hose to the IAC valve

35. Connect the engine wire protector to the vehicle body with the bolt.
36. Connect the following connectors:
- VSV connector for the EGR valve and the fuel pressure control
- Turbo pressure sensor and the IAC valve connectors

37. Connect the control cable bracket to the intake chamber by installing the two bolts. Torque the bolts to 14 ft. lbs. (19 Nm).
38. Install the intake chamber stay by installing the bolt and nut. Torque the nut and bolt to 14 ft. lbs. (19 Nm).
39. Install the oil dipstick guide and dipstick to the engine. Use a new O-ring for the dipstick guide.
40. Install the transmission oil dipstick guide and dipstick to the transmission. Use a new O-ring for the dipstick guide.
41. Install the throttle body and install the following hoses:
- Power steering air hose
- Water bypass hose (from the cylinder head)
- Water bypass hose (from the No. 4 water bypass pipe)
- EVAP hose

42. Install the gasket and throttle body to the air intake chamber and install the two bolts and two nuts. Torque the nuts and bolts to 15 ft. lbs. (21 Nm).

43. Install the following hose, cables, and connectors to the throttle body:
- Sub throttle actuator connector
- Sub throttle position sensor connector
- Throttle position sensor connector
- Cruise control actuator and the accelerator cables
- Air hose

44. Fill engine coolant and install the engine undercover.
45. Connect the negative battery cable to the battery.

3.0L (3VZ-FE) Engine

Camry

1. Disconnect the negative battery cable. On vehicles equipped with an air bag, wait at least 90 seconds before proceeding.
2. Drain the engine coolant.
3. Disconnect the throttle/accelerator cable from the throttle body.
4. Disconnect the air cleaner hose at the air intake chamber and remove it.
5. Remove the V-bank cover.
6. Tag and disconnect all lines and hoses and then remove both the ISC valve and the throttle body.
7. Remove the EGR valve and vacuum modulator. Remove the distributor.
8. Remove the emission control valve set and then disconnect the left side engine harness.
9. Remove the cylinder head rear plate.
10. Remove the intake chamber stays, any wires and then remove the air intake chamber.
11. Remove the fuel injection delivery pipe and the injectors.
12. Remove the water outlet and the bypass outlet.
13. Remove the two bolts and the No. 2 idler pulley bracket stay. Remove the eight bolts and four nuts and then lift out the intake manifold.

To install:
14. Thoroughly clean the intake manifold and cylinder head surfaces. Using a machinist's straightedge and a feeler gauge, check the surface of the intake manifold for warpage. If the warpage is greater than 0.0039 in. (0.10mm), replace the intake manifold.
15. Place new gaskets onto the intake manifold and position the intake manifold between the cylinder heads. Tighten the nuts and bolts to 13 ft. lbs. (18 Nm). Tighten the No. 2 pulley bracket bolts to 13 ft. lbs. (18 Nm).

16. Install the water bypass outlet and tighten the bolts to 74 inch lbs. (8 Nm). Tighten the water outlet to 74 inch lbs. (8 Nm).
17. Install the injectors and delivery pipe.
18. Install the air intake chamber and tighten the two bolts and two nuts to 32 ft. lbs. (43 Nm); use an 8mm hex wrench. Install the chamber stays and tighten the mounting bolts to 29 ft. lbs. (39 Nm).
19. Install the distributor. Install the emission control valve set and tighten the two bolts to 73 inch lbs. (8 Nm).
20. Install the EGR valve and modulator (with new gaskets). Tighten the bolts and nuts to 13 ft. lbs. (18 Nm).
21. Place a new gasket onto the throttle body and attach the throttle body to the intake manifold with the four bolts. Tighten the bolts to 14 ft. lbs. (19 Nm).
22. Install the ISC valve and tighten it to 9 ft. lbs. (13 Nm).
23. Unplug and connect all hoses.
24. Connect the throttle position sensor and ISC valve connectors.
25. Connect the air cleaner hose and tighten the hose clamp.
26. Connect the throttle cable with bracket onto the throttle body. Install the return spring.
27. On vehicles equipped with automatic transaxle, connect the accelerator cable and adjust it.
28. Fill the cooling system to the proper level and connect the negative battery cable.
29. Start the engine and inspect for leaks.

Exhaust Manifold

REMOVAL AND INSTALLATION

1.5L (3E-E) Engine

Tercel

1. Disconnect the negative battery cable. On vehicles equipped with an air bag, wait at least 90 seconds before proceeding.
2. Remove the exhaust manifold heat insulator shield assembly.
3. Remove the necessary components in order to gain access to the exhaust manifold retaining nuts.
4. Disconnect the exhaust manifold bolts at the exhaust pipe. Disconnect the oxygen sensor electrical wire. It may be necessary to raise and support the vehicle safely before removing these bolts.

5. Remove the 6 exhaust manifold retaining nuts. Remove the exhaust manifold from the vehicle.

To install:

6. Place the **E** mark on the gasket outward.

7. Install the manifold and tighten the 6 retaining nuts to 38 ft. lbs. (51 Nm).

8. Connect the exhaust manifold to the exhaust pipe. Connect the oxygen sensor electrical wire.

9. Install the manifold heat shroud.

10. Connect the negative battery cable.

1.5L (5E-FE) Engine

Paseo

1. Disconnect the negative battery cable from the battery. On vehicles equipped with an air bag, wait at least 90 seconds before proceeding.

2. Tag and disconnect all electrical wires and vacuum hoses that interfere with removal of the exhaust manifold.

3. Remove the three bolts to the exhaust heat insulator and then remove the insulator from the engine.

4. Remove the exhaust pipe stay by removing the bolt and two nuts.

5. Disconnect the exhaust pipe from the manifold by removing the two bolts and two compression springs.

6. Remove the six nuts to the exhaust manifold and then remove the exhaust manifold.

To install:

7. Clean the gasket surfaces and install the exhaust manifold, tightening the six bolts to 35 ft. lbs. (47 Nm).

8. Install the exhaust stay to the engine and exhaust manifold. Torque the bolt and two nuts to 29 ft. lbs. (40 Nm).

9. Reconnect the exhaust manifold heat insulator with the three bolts. Torque the bolts to 69 inch lbs. (8 Nm).

10. Connect the exhaust pipe to the exhaust manifold with the two compression springs and two bolts. Torque the bolts to 46 ft. lbs. (62 Nm).

11. Connect all electrical wires and vacuum hoses that were disconnected for removal of the exhaust manifold.

12. Connect the battery cable, start the engine and check for exhaust leaks.

1.6L (4A-FE) and 1.8L (7A-FE) Engine

Corolla and Celica

1. Disconnect the negative battery cable. On vehicles equipped with an air bag, wait at least 90 seconds before proceeding.

2. Raise and safely support the vehicle.

3. Working from under the vehicle, remove the bolts holding the front exhaust pipe to the mounting bracket.

4. Using a 14mm deep socket wrench, remove the nuts and the gasket and disconnect the front exhaust pipe from the manifold.

5. Remove the main oxygen sensor connector.

6. Remove the bolts and the upper heat insulator.

7. Remove the nuts, the exhaust manifold, and the gasket.

8. Remove the bolts and lower heat insulator from the exhaust manifold.

To install:

9. Install the lower heat insulator to the exhaust manifold with the bolts.

10. Install a new gasket and the exhaust manifold with the nuts. Uniformly tighten the nuts in several passes. Torque the nuts to 25 ft. lbs. (34 Nm).

11. Install the upper heat insulator with the bolts.

12. Install the front exhaust pipe with a new gasket to the exhaust manifold. Install the nuts using a 14mm deep socket wrench. Torque the nuts to 46 ft. lbs. (62 Nm).

13. Secure the front exhaust pipe to the exhaust pipe bracket with the bolts.

14. Connect the main oxygen sensor connector.

15. Lower the vehicle safely and reconnect the negative battery cable.

16. Start the engine and make sure that there are no exhaust leaks.

2.0L (3S-GTE) Engine

Celica

1. Disconnect the negative battery cable. On vehicles equipped with an air bag, wait at least 90 seconds before proceeding.

2. Raise and support the vehicle safely as necessary. Remove the right-hand gravel shield from underneath the car.

3. Remove the exhaust pipe support stay. Unbolt (soak bolts with penetrating oil or equivalent) the exhaust pipe from the exhaust manifold flange.

4. Disconnect the oxygen sensor connector. Remove the upper heat insulator, if equipped.

5. Remove the manifold retaining bolts. Remove the exhaust manifold from the vehicle. If the manifold is being replaced, remove the catalytic converter.

6. Installation is the reverse of the removal procedure. Use a new gasket on exhaust manifold and exhaust pipe as required. Torque the retaining nuts evenly working from the center to the ends in steps to 38 ft. lbs. (52 Nm). Reconnect the battery cable. Start the engine and check for exhaust leaks.

MR2

1. Disconnect the negative battery cable. On vehicles equipped with an air bag, wait at least 90 seconds before proceeding.

2. Raise and safely support the vehicle.

3. Working from under the vehicle, remove the two bolts holding the front exhaust pipe to the mounting bracket.

4. Using a 14mm deep socket wrench, remove the three nuts and the gasket and disconnect the front exhaust pipe from the manifold.

5. Remove the front engine mounting insulator.

6. Remove the front engine mounting bracket and clutch release cylinder from the transaxle.

7. Remove the front TWC.

8. Remove the turbocharger.

— CAUTION —

Fuel injection systems remain under pressure after the engine has been turned OFF. Properly relieve fuel pressure before disconnecting any fuel lines. Failure to do so may result in fire or personal injury.

9. Remove the throttle body.

10. Remove the cold start injector pipe.

NOTE: Put a suitable container or shop towel under the injector pipe. Slowly loosen the union bolts.

11. Disconnect the cold start injector connector.

12. Remove the nuts and the exhaust manifold.

To install:

13. Install a new gasket and the exhaust manifold. Torque the nuts to 38 ft. lbs. (52 Nm).

14. Connect the cold start injector connector.

15. Install the cold start injector pipe. Torque the bolts to 9 ft. lbs. (12 Nm).

16. Install the throttle body.

17. Install the turbocharger.

18. Install the front TWC. Torque the bolts to 22 ft. lbs. (29 Nm).

19. Torque the right and left TWC stays to 43 ft. lbs. (59 Nm).

20. Install the clutch release cylinder and the front engine mounting bracket. Torque the clutch release bolts to 9 ft. lbs. (12 Nm).

21. Torque the holding mounting bracket bolts to the transaxle to 57 ft. lbs. (77 Nm).

22. Install the front engine mounting insulator. Torque the through-bolt to 71 ft. lbs. (96 Nm).

23. Install a new gasket with the front exhaust pipe and torque the nuts to 46 ft. lbs. (62 Nm).

24. Lower the vehicle safely and connect the negative battery cable.

25. Start the engine and make sure that there are no exhaust leaks.

2.2L (5S-FE) Engine

MR2

1. Disconnect the negative battery cable. On vehicles equipped with an air bag, wait at least 90 seconds before proceeding.

2. Raise and safely support the vehicle.

3. Working from under the vehicle, remove the two bolts holding the front exhaust pipe to the mounting bracket.

4. Remove the front exhaust pipe as follows:

   a. On California vehicles, remove the two bolts and the damper.

   b. Remove the two bolts holding the front exhaust pipe to the tailpipe bracket.

   c. Remove the two bolts holding the front exhaust pipe to the tailpipe.

   d. Using a 14mm deep socket wrench, remove the three nuts.

   e. Remove the front exhaust pipe and the two gaskets.

5. Disconnect the A/C compressor from the engine.

6. Remove the two bolts and lower the suspension brace.

7. Remove the clamp nut, and disconnect the two A/C pipes.

8. Loosen the idler pulley nut and the adjusting nut, and remove the drive belt.

9. Disconnect the A/C compressor connector.

10. Remove the idler pulley bracket and the A/C compressor.

11. Remove the upper manifold heat insulator.

12. Remove the exhaust manifold and the TWC assembly.

13. Disconnect the main oxygen sensor connector. On California vehicles, disconnect the sub-oxygen sensor connector.

To install:

14. Install the exhaust manifold and the front TWC assembly.

15. Install a new gasket, and torque the nuts to 36 ft. lbs. (49 Nm).

16. Install the TWC stay bolts and torque them to 31 ft. lbs. (42 Nm).

17. Install the upper manifold heat insulator.

18. Connect the main and the sub oxygen sensors connectors.

19. Install the A/C compressor.

20. Install the idler pulley bracket and torque the bolts to:

   Bolt A: 18 ft. lbs. (25 Nm)
   Bolt B: 20 ft. lbs. (27 Nm)
   Bolt C: 27 ft. lbs. (37 Nm)

21. Install the drive belt and tighten the idler pulley locknut.

22. Connect the A/C compressor connector.

23. Tighten the two bolts of the lower side of the A/C compressor to 18 ft. lbs. (25 Nm).

24. Install the A/C pipes.

25. Install the lower suspension brace and torque the bolts to 54 ft. lbs. (73 Nm).

26. Install the front exhaust pipe with a new gasket to the TWC. Install the three nuts using a 14mm deep socket wrench. Torque the nuts to 46 ft. lbs. (62 Nm).

27. Secure the front exhaust pipe to the exhaust pipe bracket with the two bolts. Torque the bolts to 14 ft. lbs. (19 Nm).

28. Install the two bolts securing the front pipe to the tailpipe. Torque the bolts to 32 ft. lbs. (43 Nm).

29. On California vehicles, install the damper and its two bolts. Torque the bolts to 15 ft. lbs. (21 Nm).

30. Connect the front exhaust pipe to the mounting bracket.

31. Lower the vehicle safely and connect the negative battery cable.

32. Start the engine and make sure that there are no exhaust leaks.

Celica and Camry

1. Disconnect the negative battery cable. On vehicles equipped with an air bag, wait at least 90 seconds before proceeding.

2. Raise and safely support the vehicle.

3. Working from under the vehicle, remove the bolts holding the front exhaust pipe to the mounting bracket.

4. Using a 14mm deep socket wrench, remove the nuts and the gasket and disconnect the front exhaust pipe from the manifold.

5. Disconnect the main oxygen sensor connector and the sub oxygen sensor connector.

6. Remove the bolt and nut and remove the LH exhaust manifold stay.

7. Remove the bolts and the upper manifold heat insulator.

8. Remove the bolts, nuts, and the RH exhaust manifold stay.

9. Remove the nuts, the exhaust manifold, and the three-way catalytic converter assembly.

10. Remove the nuts, oxygen sensor and gasket from the exhaust manifold.

11. Remove the sub oxygen sensor from the three-way catalytic converter.

12. Remove the bolts and the lower heat insulator from the exhaust manifold.

13. Remove the bolts and the three-way catalytic converter heat insulators.

14. Remove the bolts, nuts, TWC, gasket, retainer and cushion from the exhaust manifold.

To install:

15. Place the cushion, retainer, and a new gasket on the TWC and reinstall it to the exhaust manifold with the bolts and the nuts. Torque the bolts and nuts to 22 ft. lbs. (30 Nm).

16. Install the lower manifold heat insulator with the bolts and install the TWC heat insulators with the bolts.

17. Install the main oxygen sensor to the exhaust manifold with a new gasket and new nuts. Torque the nuts to 14 ft. lbs. (19 Nm).

18. Install the sub oxygen sensor to the front TWC and torque to 33 ft. lbs. (45 Nm).

19. Install a new gasket, the exhaust manifold and front TWC assembly to the engine with the nuts. Uniformly tighten the nuts in several passes. Torque the nuts to 36 ft. lbs. (49 Nm).

20. Install the RH exhaust manifold stay with the bolts and new nuts. Torque the bolts and nuts to 31 ft. lbs. (42 Nm).

21. Install the upper heat insulator with the bolts.

22. Install the LH exhaust manifold stay with the bolt and nut. Torque the bolt to 29 ft. lbs. (39 Nm) and torque the nut to 31 ft. lbs. (42 Nm).

23. Connect the main and the sub oxygen sensors connectors.

24. Install the front exhaust pipe with a new gasket to the TWC. Install the nuts using a 14mm deep socket wrench. Torque the nuts to 46 ft. lbs. (62 Nm).

25. Secure the front exhaust pipe to the exhaust pipe bracket with the bolts. Torque the bolts to 14 ft. lbs. (19 Nm).

26. Lower the vehicle safely and reconnect the negative battery cable.

27. Start the engine and make sure that there are no exhaust leaks.

3.0L (3VZ-FE) Engine

Camry

1. Disconnect the negative battery cable. On vehicles equipped with an air bag, wait at least 90 seconds before proceeding.

2. Raise the car, support it on safety stands and then remove the engine undercovers.

3. Remove the two front exhaust pipe stay bolts. Disconnect the front pipe from the center pipe and remove the gasket. Loosen the three nuts and then remove the front pipe.

4. Disconnect the O₂ sensor at the right side manifold. Remove the three mounting nuts and lift off the outside heat insulator.

5. Remove the six nuts and lift off the right side manifold and gasket.

6. Loosen the two nuts and bolt and lift off the left side heat insulator. Remove the six nuts and lift off the left side manifold and gaskets.

To install:

7. Scrape the mating surfaces of all old gasket material.

8. Install the right manifold with a new gasket. Tighten the nuts to 29 ft. lbs. (39 Nm). Install the outer insulator.

9. Use a new gasket and install the left manifold. Tighten the nuts to 29 ft. lbs. (39 Nm). Install the outer insulator.

10. Install the front exhaust pipe and tighten the manifold-to-pipe nuts to 46 ft. lbs. (62 Nm). Tighten the pipe-to-converter nuts to 32 ft. lbs. (43 Nm).

11. Connect the O₂ sensor, install the undercovers and then lower the car. Connect the battery cable.

3.0L (1MZ-FE) Engine

Camry and Avalon

1. Disconnect the negative battery cable from the battery.

2. Raise and safely support the vehicle.

3. Remove the engine undercovers from the vehicle.

4. From below the engine, disconnect the front exhaust pipe from the exhaust manifold by removing the two nuts.

5. If necessary, lower the vehicle to access the top of the engine.

6. Remove the EGR pipe from the exhaust manifold by removing the four nuts.

7. Disconnect the heated oxygen sensor connector to the right exhaust manifold.

8. Remove the exhaust manifold stay by removing the bolt and nut.

9. Remove the six nuts to the exhaust manifold and remove the exhaust manifold from the engine.

To install:

10. Using a new gasket, install the exhaust manifold to the engine and install the six nuts. Uniformly tighten and then torque the bolts to 36 ft. lbs. (49 Nm).

11. Install the exhaust manifold stay and install the bolt and nut. Torque the bolt and nut to 15 ft. lbs. (20 Nm).

12. Connect the heated oxygen sensor connector to the right exhaust manifold.

13. Using new gaskets, install the EGR pipe to the exhaust manifold and the engine. Torque the four nuts to 9 ft. lbs. (12 Nm).

14. Raise the vehicle and safely support.

15. Connect the front exhaust pipe to the exhaust manifold. Use a new gasket and torque the two nuts to 46 ft. lbs. (62 Nm).

16. Install the engine undercovers to the vehicle.

17. Lower the vehicle.

18. Connect the negative battery cable to the battery.

3.0L (2JZ-GTE) Engine

Supra

1. Disconnect the negative battery cable from the battery. On vehicles equipped with an air bag, wait at least 90 seconds before proceeding.

2. Drain the engine coolant from the engine.

3. Disconnect the cruise control actuator cable from the throttle body.

4. Remove the No. 1 air hose.

5. Remove the air cleaner and MAF meter assembly as follows:

  a. Remove the three bolts to the air assembly.

  b. loosen the hose clamp and disconnect the air hose from the intake air connector.

  c. Disconnect the MAF meter wire from the clamp on the air cleaner case.

  d. Disconnect the MAF meter connector and remove the air cleaner and MAF meter assembly.

6. Disconnect the theft deterrent horn from the body.

7. Raise and safely support the vehicle.

8. Remove the engine undercover.

9. Remove the front lower arm bracket stay by removing the two bolts, nut, and the plate washer.

10. Remove the front upper cross-member extension by removing the two bolts and two nuts.

11. Remove the No. 2 front exhaust pipe.

12. Remove the heat insulator for the No. 2 front exhaust pipe by removing the two bolts and two nuts.

13. If equipped with automatic transmission, disconnect the A/T oil cooler tubes from the engine.

14. Disconnect the engine wire protector from the body by removing the two bolts.

15. Disconnect the heater hose from the No. 3 water bypass pipe.

16. Disconnect the EVAP hose from the No. 1 vacuum pipe.

17. Disconnect the IAC valve pipe from the No. 2 air tube as follows:

  a. Disconnect the engine wire from the clamp.

  b. Disconnect the air hose (from the No. 1 vacuum pipe) from the IAC valve pipe.

  c. Disconnect the air hose from the No. 2 air tube.

  d. Disconnect the IAC valve pipe from the clamp.

18. Disconnect the No. 1 vacuum pipe from the air tubes as follows:

  a. Disconnect the VSV connector for the intake air control valve.

  b. Disconnect the VSV connector for the exhaust bypass valve.

  c. Disconnect the engine wire from the three clamps.

  d. Disconnect the following hoses:

  • Air hose from the No. 4 air tube
  • Air hose from the No. 1 air tube
  • Air hose (from the VSV for the waste gate valve) from the vacuum pipe
  • Air hose (from the VSV for the exhaust gas control valve) from the vacuum pipe
  • Vacuum hose (from the air bypass valve) from the No. 1 air tube
  • Two air hoses (from the VSV for exhaust bypass valve) from the vacuum pipe
  • Air hose (from the No. 2 air tube) from the vacuum pipe
  • Two air hoses (from the pressure tank) from the vacuum pipe

  e. Remove the three bolts and disconnect the vacuum pipe from the air tubes.

19. Remove the VSV assembly as follows:

a. Disconnect the air hose from the actuator for the waste gate valve.

b. Disconnect the air hose from the actuator for exhaust gas control valve.

c. Disconnect the air hose from the hose clamp.

d. Disconnect the engine wire from the wire clamp.

e. Remove the two bolts and disconnect the two VSV connectors.

f. Remove the VSV assembly.

20. Remove the air tubes and intake air connector as follows:

a. Disconnect the crankshaft position sensor connector from the clamp.

b. Disconnect the water bypass hose (from the water pump) from the No.1 turbo water pipe.

c. Disconnect the water bypass hose (from the water outlet) from the No. 1 turbo water pipe.

d. Disconnect the water bypass hose (from the water outlet) from the No. 2 turbo water pipe.

e. Remove the bolt and disconnect the No. 2 turbo water pipe from the No. 4 air tube.

f. Disconnect the No. 1 air tube from the No. 1 turbocharger by removing the two bolts.

g. Remove the two bolts holding the No. 4 air tube to the No. 1 turbocharger.

h. Disconnect the air hose from the No. 4 air tube.

i. Disconnect the air hose from the intake air connector.

j. Remove the No. 4 air tube and air bypass valve assembly.

k. Remove the intake air control valve and gasket by removing the two nuts.

l. Disconnect the air hose from the No. 2 air tube.

m. Disconnect the PCV hose from the No. 2 cylinder head cover.

n. Remove the intake air connector and No. 1 air tube assembly.

21. Remove the air inlet duct and cable bracket by removing the bolt and two nuts.

22. Remove the heat insulator for the turbocharger by removing the four bolts.

23. Remove the exhaust bypass pipe and gasket by removing the four nuts.

24. Remove the exhaust gas control valve stay by removing the nut and bolt.

25. Remove the main heated oxygen sensor by disconnecting the electrical connector and two nuts.

26. Remove the exhaust gas control valve by removing the three nuts.

27. Remove the No. 1 turbocharger stay by removing the nut and bolt.

28. Remove the No. 2 turbocharger stay by removing the nut and bolt.

29. Remove the No. 1 turbo oil pipe as follows:

a. Remove the union bolt holding the turbo oil pipe to the cylinder block. Remove the two gaskets.

b. Remove the two nuts and disconnect the turbo oil pipe from the turbocharger. Remove the gaskets.

c. Disconnect the turbo oil hose from the turbo oil outlet on the No. 1 oil pan. Remove the turbo oil pipe.

30. Remove the No. 2 turbo oil pipe as follows:

a. Remove the union bolt holding the turbo oil pipe to the cylinder block. Remove the two gaskets.

b. Disconnect the turbo oil pipe from the turbocharger by removing the two nuts. Remove the two gaskets.

c. Disconnect the turbo oil hose from the turbo oil outlet on the No. 1 oil pan and remove the turbo oil pipe.

31. Remove the turbochargers and turbine outlet elbow assembly as follows:

a. Disconnect the heater hose (from the No. 3 water bypass pipe) from the No. 2 water bypass pipe.

b. Disconnect the water bypass hose (from the No. 2 turbo water pipe) from the No. 2 water bypass pipe.

c. Remove the eight nuts holding the turbochargers to the exhaust manifold.

d. Remove the two turbochargers and turbine outlet elbow assembly.

e. Remove the two gaskets.

32. Remove the exhaust manifold by removing the 12 nuts and two gaskets.

To install:

33. Install the exhaust manifold with two new gaskets and install the nuts. Tighten the nuts in several passes and in sequence to 29 ft. lbs. (39 Nm).

34. Install the turbochargers and turbine outlet elbow assembly as follows:

a. Install two new gaskets.

b. Install the turbochargers and turbine outlet elbow assembly to the exhaust manifold.

c. Install eight new nuts and uniformly tighten the nuts in several passes. Torque the nuts to 40 ft. lbs. (54 Nm).

d. Connect the water bypass hose (from the No. 2 turbo water pipe) to the No. 2 water bypass pipe.

e. Connect the heater hose (from the No. 3 water bypass pipe) to the No. 2 water bypass pipe.

35. Install the No. 2 turbo oil pipe as follows:

a. Install the turbo oil pipe and connect the turbo oil hose to the turbo oil outlet on the No. 1 oil pan.

b. Using a new gasket, install the turbo oil pipe to the turbocharger by installing the two nuts. Torque the nuts to 15 ft. lbs. (21 Nm). Make sure to align the oil holes of the gasket and turbocharger housing.

c. Using two new gaskets, connect the union bolt to hold the turbo oil pipe to the cylinder block. Torque the union bolt to 29 ft. lbs. (39 Nm).

36. Install the No. 1 turbo oil pipe as follows:

a. Install the turbo oil pipe and connect the turbo oil hose to the turbo oil outlet on the No. 1 oil pan.

b. Using a new gasket, install the turbo oil pipe to the turbocharger by installing the two nuts. Torque the nuts to 15 ft. lbs. (21 Nm). Make sure to align the oil holes of the gasket and turbocharger housing.

c. Using two new gaskets, connect the union bolt to hold the turbo oil pipe to the cylinder block. Torque the bolts to 29 ft. lbs. (39 Nm).

37. Install the No. 2 turbocharger stay by installing the nut and bolt. Torque the nut and bolt to 32 ft. lbs. (43 Nm).

38. Install the No. 1 turbocharger stay by installing the nut and bolt. Torque the nut and bolt to 32 ft. lbs. (43 Nm).

39. Install the exhaust gas control valve by installing two new gaskets and the three nuts. Torque the three nuts to 51 ft. lbs. (69 Nm).

40. Install the main heated oxygen sensor by connecting the electrical connector and installing the two nuts. Torque the nuts to 14 ft. lbs. (20 Nm).

41. Install the exhaust gas control valve stay by installing the bolt and nut. Torque the bolt and nut to 32 ft. lbs. (43 Nm).

42. Install the exhaust bypass pipe by installing two new gaskets and four new nuts. Torque the nuts to 18 ft. lbs. (25 Nm).

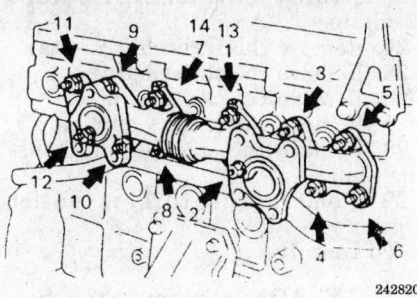

Exhaust manifold installation sequence — Supra with 2JZ-GTE engine

43. Install the heat insulator for the turbocharger by installing the four bolts.

44. Install the air inlet duct by installing the cable bracket, bolt, and two nuts.

45. Connect the air tubes and intake air connector as follows:

 a. Install the intake air connector and No. 1 air tube assembly.

 b. Connect the PCV hose to the No. 2 cylinder head cover.

 c. Connect the air hose to the No. 2 air tube.

 d. Install the intake air control valve and gasket by installing the two nuts. Torque the nut to 15 ft. lbs. (21 Nm).

 e. Install the No. 4 air tube and air bypass valve assembly.

 f. Connect the air hose to the intake air connector.

 g. Connect the air hose to the No. 4 air tube.

 h. Install the two bolts holding the No. 4 air tube to the No. 1 turbocharger. Torque the bolt to 15 ft. lbs. (21 Nm).

 i. Connect the No. 1 air tube to the No. 1 turbocharger and install the two bolts. Torque the bolts to 15 ft. lbs. (21 Nm).

 j. Install the No. 2 turbo water pipe to the No. 4 air tube by installing the bolt.

 k. Connect the water bypass hose (from the water outlet) to the No. 2 turbo water pipe.

 l. Connect the water hose (from the water outlet) to the No. 1 turbo water pipe.

 m. Connect the water bypass hose (from the water pump) to the No. 1 turbo water pipe.

 n. Connect the crankshaft position sensor connector to the clamp.

46. Install the VSV assembly as follows:

 a. Connect the VSV assembly and connect the two VSV connectors.

 b. Install the two bolts.

 c. Connect the engine wire to the wire clamp.

 d. Connect the air hose to the hose clamp.

 e. Connect the air hose to the actuator for the exhaust gas control valve.

 f. Connect the air hose to the actuator for the waste gate valve.

47. Connect the No. 1 vacuum pipe to the air tubes as follows:

 a. Install the vacuum pipe to the air tubes and install the three bolts.

 b. Connect the two air hoses (from the vacuum pressure tank) to the vacuum pipe.

 c. Connect the air hose to the VSV for the intake air control valve.

 d. Connect the air hose (from the No. 2 air tube) from the vacuum pipe.

 e. Connect the two air hoses (from the VSV for exhaust bypass valve) to the vacuum pipe.

 f. Connect the vacuum hose (from the air bypass valve) to the No. 1 air tube.

 g. Connect the air hose (from the VSV for exhaust gas control valve) to the vacuum pipe.

 h. Connect the air hose (from the VSV for waste gate valve) to the vacuum pipe.

 i. Connect the air hose to the No. 1 air tube.

 j. Connect the air hose to the No. 4 air tube.

 k. Connect the engine wire to the three clamps.

 l. Connect the VSV connector for the exhaust bypass valve.

 m. Connect the VSV connector for the intake air control valve.

48. Connect the IAC valve pipe to the No. 2 air tube as follows:

 a. Connect the IAC valve pipe to the clamp.

 b. Connect the air hose to the No. 2 air tube.

 c. Connect the air hose (from the No. 1 vacuum pipe) to the IAC valve pipe.

 d. Connect the engine wire to the clamp.

49. Connect the EVAP hose to the No. 1 vacuum pipe.

50. Connect the heater hose to the No. 3 water bypass pipe.

51. Connect the engine wire protector to the body by installing the two bolts.

52. Connect the A/T oil cooler tubes to the engine.

53. Install the heat insulator for the No. 2 front exhaust pipe by installing the two bolts and two nuts.

54. Using a new gasket, install the No. 2 front exhaust pipe and install the three nuts. Torque the nuts to 46 ft. lbs. (62 Nm).

55. Connect the front exhaust pipe to the No. 2 exhaust pipe and install the pipe support bracket and install the two bolts. Torque the bolts to 32 ft. lbs. (43 Nm).

56. Install the two bolts and nuts to hold the front exhaust pipe to the No. 2 front exhaust pipe. Torque the bolts and nuts to 43 ft. lbs. (58 Nm).

57. Install the upper front cross-member extension by installing the two bolts and two nuts. Torque the bolts to 22 ft. lbs. (29 Nm) and the nuts to 25 ft. lbs. (33 Nm).

58. Install the front lower arm bracket stay by installing the plate washer, nut, and two bolts. Torque the bolts to 33 ft. lbs. (44 Nm) and the nut to 43 ft. lbs. (59 Nm).

59. Safely lower the vehicle.

60. Connect the theft deterrent horn to the body.

61. Connect the air cleaner and MAF meter assembly as follows:

 a. Install the MAF meter assembly and air cleaner. Connect the MAF meter connector.

 b. Connect the MAF meter wire to the clamp on the air cleaner case.

 c. Connect the air hose to the intake air connector and tighten the clamp.

 d. Install the three bolts.

62. Install the air cleaner duct.

63. Install the No. 1 air hose.

64. Install the cruise control actuator cable to the throttle body.

65. Install the engine undercover.

66. Fill the engine coolant.

67. Check all fluids.

68. Connect the negative battery cable to the battery.

69. Start the engine, bleed the cooling system, and check for exhaust leaks.

3.0L (2JZ-GE) Engine

Supra

1. Disconnect the negative battery cable. Wait at least 90 seconds before performing any other work.

2. Raise the vehicle and support it safely.

3. Disconnect the two O_2 sensors at the exhaust manifolds.

4. If equipped, remove the four mounting nuts and lift off the outside heat insulator.

5. Remove the four nuts to disconnect the manifold from the front exhaust pipe. Loosen the mounting nuts to the exhaust manifolds and re-

move the two manifolds and the gasket.

To install:

6. Scrape the mating surfaces of all old gasket material.

7. Install the manifolds with a new gasket. Tighten the nuts to 29 ft. lbs. (39 Nm).

8. Use a new gasket and install the front exhaust pipe to the exhaust manifolds. Tighten the nuts to 46 ft. lbs. (62 Nm).

9. If equipped, install the outer insulator and tighten the nuts to 13 ft. lbs. (18 Nm).

10. Connect the O_2 sensors.

11. Install the undercovers.

12. Lower the vehicle.

13. Connect the battery cable.

14. Start the engine and check for exhaust leaks.

Turbocharger

REMOVAL AND INSTALLATION

2.0L (3S-GTE) Engine

Celica

1. Disconnect the negative battery cable. On vehicles equipped with an air bag, wait at least 90 seconds before proceeding.

2. Raise and support the vehicle safely as necessary to complete this service operation.

3. Drain the engine coolant.

4. Remove the air cleaner cap and engine undercovers.

5. Remove the suspension lower crossmember. Remove the front exhaust pipe.

6. Remove the engine mounting center member. Remove the front mounting insulator and bracket.

7. Remove the clutch release cylinder without disconnecting the tube. Remove the alternator.

8. Remove the idler pulley bracket and air conditioning compressor without disconnecting the hoses. Remove the catalytic converter.

9. Remove the intercooler cool air inlet and protector, then remove the intercooler.

10. Remove the oxygen sensor and remove the turbocharger heat insulators.

11. Disconnect all hoses from the turbocharger. Remove the turbocharger retaining stay. Disconnect all oil pipe lines and remove the turbocharger unit from the vehicle.

To install:

NOTE: After replacing the turbocharger assembly, pour or squirt new oil into the oil (about 20cc new oil) inlet and turn the impeller wheel by hand to splash (circulate) oil on the bearing.

12. Installation is the reverse of removal. Use new gaskets. Torque the turbocharger-to-exhaust manifold nuts to 47 ft. lbs. (64 Nm). Torque the oil pipe-to-turbocharger nuts to 13 ft. lbs. (17 Nm). Torque the oil pipe-to-block bolt to 38 ft. lbs. (51 Nm). Torque the oxygen sensor nuts to 33 ft. lbs. (44 Nm).

13. Refill the engine with the specified coolant to the correct level. Reconnect the battery cable and check oil level. Start engine check for any fluid leaks. Road test the vehicle for proper operation.

MR2

1. Disconnect the negative battery cable. On vehicles equipped with an air bag, wait at least 90 seconds before proceeding.

2. Drain the engine coolant.

3. Remove the undercovers from the vehicle.

4. Remove the left-hand engine hood side panel.

5. Remove the suspension upper brace.

6. Remove the air cleaner.

7. Remove the No. 1 and No. 2 air intake connectors.

8. If equipped with cruise control, disconnect the cruise control accelerator linkage from the body.

9. Remove the engine compartment electric cooling fan.

10. Disconnect the A/C compressor from the engine.

11. Remove the tail pipe and the front exhaust pipe.

12. Remove the air cleaner hose.

13. Remove the front engine mounting bracket.

14. Remove the front mounting insulator.

15. Remove the clutch release cylinder.

16. Remove the front TWC.

17. Disconnect the air bypass valve and the VTV from the clamp.

18. Remove the No. 4 air tube.

19. Remove the right front engine hanger.

20. Remove the turbocharger heat insulator.

21. Remove the heated oxygen sensor.

22. Remove the turbine outlet elbow heat insulators.

23. Disconnect the hoses.

a. Water hose from the water inlet housing

b. Water hose from the water bypass pipe

c. Air hose from the actuator

d. Turbo oil hose from the turbo oil pipe

24. Remove the turbocharger stay.

25. Remove the turbocharger.

26. Remove the turbo oil pipe.

27. Remove the turbo water pipe.

28. Remove the side bearing housing plate.

29. Remove the turbine outlet elbow.

To install:

NOTE: After replacing the turbocharger assembly pour 0.12 cu in. of new engine oil into the oil inlet and turn the impeller wheel by hand to splash oil on the bearing.

30. Install the turbine outlet elbow with a new gasket. Torque the nuts to 47 ft. lbs. (64 Nm).

31. Install the side bearing housing plate with a new gasket. Torque the nuts to 9 ft. lbs. (11 Nm).

32. Install the turbo water pipe with a new gasket. Torque the nuts to 9 ft. lbs. (11 Nm).

33. Install the turbo oil pipe.

34. Install the turbocharger with a new gasket.

a. Tighten the nuts holding the turbocharger to the exhaust manifold to 47 ft. lbs. (64 Nm).

b. Tighten the nuts holding the oil pipe to the turbocharger to 13 ft. lbs. (17 Nm).

c. Tighten the union bolt holding the oil pipe to the cylinder block to 38 ft. lbs. (51 Nm).

d. Tighten the bolt holding the bracket of the oil pipe to the cylinder block to 32 ft. lbs. (43 Nm).

35. Install the turbocharger stay attaching bolts. Torque bolt to the turbocharger to 51 ft. lbs. (69 Nm). Torque bolt to the cylinder head to 43 ft. lbs. (59 Nm).

36. Connect the hoses:

a. Water hose to the water inlet housing

b. Water hose to the water bypass pipe

c. Air hose to the actuator

d. Turbo oil hose to the turbo oil pipe

37. Install the heat insulators of the turbine outlet elbow.

38. Install the oxygen sensor and torque it to 33 ft. lbs. (44 Nm).

39. Install the turbocharger heat insulator.

40. Install the right front engine hanger. Torque the bolts to the cylinder head to 29 ft. lbs. (39 Nm) and the bolts to the mounting bracket to 45 ft. lbs. (61 Nm).

41. Install the No. 4 air tube. Torque the bolts to 14 ft. lbs. (19 Nm).

42. Install the air bypass valve. Torque the bolt to 14 ft. lbs. (19 Nm).
43. Install the air bypass hose and VTV clamps.
44. Connect the air bypass hoses.
45. Install the TWC and torque the bolts to 22 ft. lbs. (29 Nm).
   a. Install the right and left TWC stays and torque the bolts to 43 ft. lbs. (59 Nm)
46. Install the front mounting bracket and clutch release cylinder.
47. Install the front engine mounting insulator.
48. Install the air cleaner hose.
49. Install the tail pipe and the front exhaust pipe.
50. Install the A/C compressor and idler pulley bracket.
51. Install the engine compartment cooling fan.
52. If removed, install the cruise control accelerator linkage.
53. Install the No. 1 and No. 2 air intake connectors.
54. Install the air cleaner.
55. Install the suspension upper brace. Torque the bolt to 54 ft. lbs. (73 Nm) and the nut to 47 ft. lbs. (64 Nm).
56. Install the right engine hood side panel.
57. Fill the engine with coolant.
58. Connect the negative battery cable.
59. Start the engine and check for leaks.
60. Check the engine oil level.
61. Install the engine undercovers.

3.0L (2JZ-GTE) Engine

1. Disconnect the negative battery cable from the battery.
2. Drain the engine coolant from the engine.
3. Disconnect the cruise control actuator cable from the throttle body.
4. Remove the No. 1 air hose.
5. Remove the air cleaner and MAF meter assembly as follows:
   a. Remove the three bolts to the air assembly.
   b. loosen the hose clamp and disconnect the air hose from the intake air connector.
   c. Disconnect the MAF meter wire from the clamp on the air cleaner case.
   d. Disconnect the MAF meter connector and remove the air cleaner and MAF meter assembly.
6. Disconnect the theft deterrent horn from the body.
7. Raise and safely support the vehicle.
8. Remove the front lower arm bracket stay by removing the two bolts, nut, and plate washer.

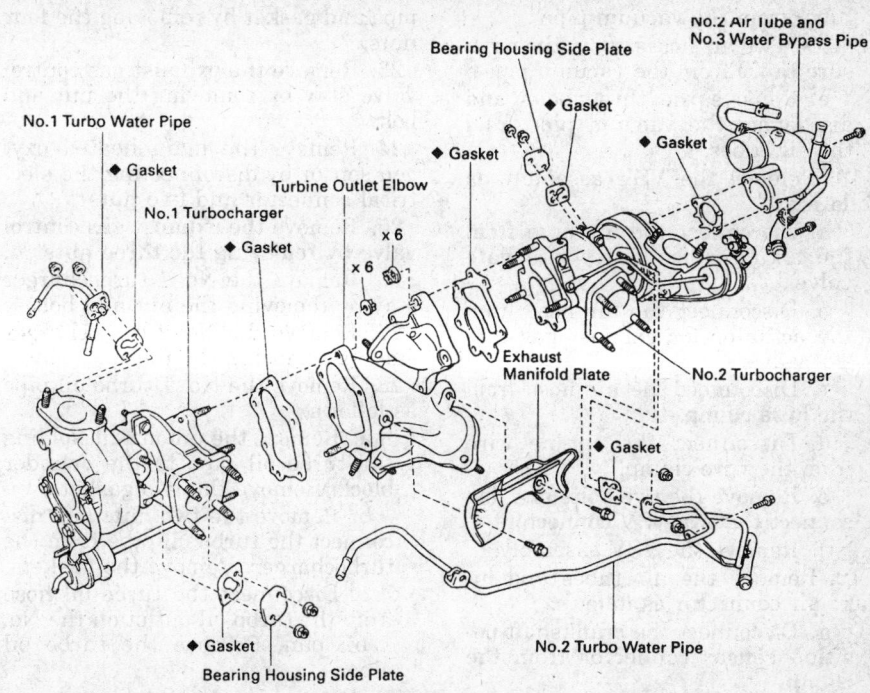

Turbocharger component assembly continued — Supra with 2JZ-GTE engine

◆ Non-reusable part

9. Remove the front upper cross-member extension by removing the two bolts and two nuts.
10. Remove the No. 2 front exhaust pipe as follows:
   a. Remove the front exhaust pipe to the No. 2 front exhaust pipe by removing the two bolts and nuts.
   b. Remove the pipe support bracket by removing the two bolts.
   c. Disconnect the front exhaust pipe from the No. 2 exhaust pipe and remove the gasket.
   d. Remove the three nuts and then remove the No. 2 front exhaust pipe and gasket.
11. Remove the heat insulator for the No. 2 front exhaust pipe by removing the two bolts and two nuts.
12. If equipped with automatic transmission, disconnect the A/T oil cooler tubes from the engine.
13. Disconnect the engine wire protector from the body by removing the two bolts.
14. Disconnect the heater hose from the No. 3 water bypass pipe.
15. Disconnect the EVAP hose from the No. 1 vacuum pipe.
16. Disconnect the IAC valve pipe from the No. 2 air tube as follows:
   a. Disconnect the engine wire from the clamp.
   b. Disconnect the air hose (from the No. 1 vacuum pipe) from the IAC valve pipe.
   c. Disconnect the air hose from the No. 2 air tube.
   d. Disconnect the IAC valve pipe from the clamp.
17. Disconnect the No. 1 vacuum pipe from the air tubes as follows:
   a. Disconnect the VSV connector for the intake air control valve.
   b. Disconnect the VSV connector fro the exhaust bypass valve.
   c. Disconnect the engine wire from the three clamps.
   d. Disconnect the following hoses:
   • Air hose from the No. 4 air tube
   • Air hose from the No. 1 air tube
   • Air hose (from the VSV for the waste gate valve) from the vacuum pipe
   • Air hose (from the VSV for the exhaust has control valve) from the vacuum pipe
   • Vacuum hose (from the air bypass valve) from the No. 1 air tube
   • Two air hoses (from the VSV for exhaust bypass valve) from the vacuum pipe

• Air hose (from the No. 2 air tube) from the vacuum pipe

• Two air hoses (from the pressure tank) from the vacuum pipe

e. Remove the three bolts and disconnect the vacuum pipe from the air tubes.

18. Remove the VSV assembly as follows:

a. Disconnect the air hose from the actuator for the waste gate valve.

b. Disconnect the air hose from the actuator for exhaust gas control valve.

c. Disconnect the air hose from the hose clamp.

d. Disconnect the engine wire from the wire clamp.

e. Remove the two bolts and disconnect the two VSV connectors.

f. Remove the VSV assembly.

19. Remove the air tubes and intake air connector as follows:

a. Disconnect the crankshaft position sensor connector from the clamp.

b. Disconnect the water bypass hose (from the water pump) from the No.1 turbo water pipe.

c. Disconnect the water bypass hose (from the water outlet) from the No. 1 turbo water pipe.

d. Disconnect the water bypass hose (from the water outlet) from the No. 2 turbo water pipe.

e. Remove the bolt and disconnect the No. 2 turbo water pipe from the No. 4 air tube.

f. Disconnect the No. 1 air tube from the No. 1 turbocharger by removing the two bolts.

g. Remove the two bolts holding the No. 4 air tube to the No. 1 turbocharger.

h. Disconnect the air hose from the No. 4 air tube.

i. Disconnect the air hose from the intake air connector.

j. Remove the No. 4 air tube and air bypass valve assembly.

k. Remove the intake air control valve and gasket by removing the two nuts.

l. Disconnect the air hose from the No. 2 air tube.

m. Disconnect the PCV hose from the No. 2 cylinder head cover.

n. Remove the intake air connector and No. 1 air tube assembly.

20. Remove the air inlet duct and cable bracket by removing the bolt and two nuts.

21. Remove the heat insulator for the turbocharger by removing the four bolts.

22. Remove the exhaust bypass pipe and gasket by removing the four nuts.

23. Remove the exhaust gas control valve stay by removing the nut and bolt.

24. Remove the main heated oxygen sensor by disconnecting the electrical connector and two nuts.

25. Remove the exhaust gas control valve by removing the three nuts.

26. Remove the No. 1 turbocharger stay by removing the nut and bolt.

27. Remove the No. 2 turbocharger stay by removing the nut and bolt.

28. Remove the No. 1 turbo oil pipe as follows:

a. Remove the union bolt holding the turbo oil pipe to the cylinder block. Remove the two gaskets.

b. Remove the two nuts and disconnect the turbo oil pipe from the turbocharger. Remove the gaskets.

c. Disconnect the turbo oil hose from the turbo oil outlet on the No. 1 oil pan. Remove the turbo oil pipe.

29. Remove the No. 2 turbo oil pipe as follows:

a. Remove the union bolt holding the turbo oil pipe to the cylinder block. Remove the two gaskets.

b. Disconnect the turbo oil pipe from the turbocharger by removing the two nuts. Remove the two gaskets.

c. Disconnect the turbo oil hose from the turbo oil outlet on the No. 1 oil pan and remove the turbo oil pipe.

30. Remove the turbochargers and turbine outlet elbow assembly as follows:

a. Disconnect the heater hose (from the No. 3 water bypass pipe) from the No. 2 water bypass pipe.

b. Disconnect the water bypass hose (from the No. 2 turbo water pipe) from the No. 2 water bypass pipe.

c. Remove the eight nuts holding the turbochargers to the exhaust manifold.

d. Remove the two turbochargers and turbine outlet elbow assembly.

e. Remove the two gaskets.

31. Remove the No. 1 vacuum pipe from the No. 2 turbocharger by disconnecting the two air hoses from the actuator.

32. Remove the No. 2 air tube and No. 3 water bypass pipe assembly from the No. 2 turbocharger by removing the two bolts and air tube.

33. Remove the exhaust manifold plate from the turbine outlet elbow.

34. Remove the No. 2 turbo water pipe from the No. 2 turbocharger by removing the two nuts.

35. Remove the bearing housing side plate from the No. 1 turbocharger by removing the two nuts.

36. Remove the No. 1 turbo water pipe from the No. 1 turbocharger by removing the two nuts.

37. Remove the bearing housing side plate from the No. 2 turbocharger by removing the two nuts.

38. Remove the No. 1 turbocharger from the turbine outlet elbow by removing the six nuts.

39. Remove the No. 2 turbocharger from the turbine outlet elbow by removing the six nuts.

To install:

40. Install the No. 2 turbocharger to the turbine outlet elbow by installing a new gaskets and six new nuts. Torque the nuts to 18 ft. lbs. (25 Nm).

41. Install the No. 1 turbocharger to the turbine outlet elbow by installing a new gasket and six new nuts. Torque the nuts to 18 ft. lbs. (25 Nm).

42. Install the bearing housing side plate to the No. 2 turbocharger by installing a new gasket and two nuts. Torque the nuts to 78 inch lbs. (9 Nm).

43. Install the No. 1 turbo water pipe to the No. 1 turbocharger by installing a new gasket and two nuts. Torque the nuts to 78 inch lbs. (9 Nm).

44. Install the bearing housing side plate to the No. 1 turbocharger by installing a new gasket and two nuts. Torque the nut to 78 inch lbs. (9 Nm).

45. Install the No. 2 turbo water pipe to the No. 2 turbocharger by installing a new gasket and two nuts. Torque the nuts to 78 inch lbs. (9 Nm).

46. Install the exhaust manifold plate to the turbine outlet elbow and install the two bolts.

47. Install the No. 2 air tube and No. 3 water bypass pipe assembly to the No. 2 turbocharger and install the gasket and two bolts. Torque the bolts to 15 ft. lbs. (21 Nm).

48. Install the No. 1 vacuum pipe to the No. 2 turbocharger by installing the two hoses.

49. Install the turbochargers and turbine outlet elbow assembly as follows:

a. Install two new gaskets.

b. Install the turbochargers and turbine outlet elbow assembly to the exhaust manifold.

c. Install eight new nuts and uniformly tighten the nuts in several passes. Torque the nuts to 40 ft. lbs. (54 Nm).

d. Connect the water bypass hose (from the No. 2 turbo water pipe) to the No. 2 water bypass pipe.

e. Connect the heater hose (from the No. 3 water bypass pipe) to the No. 2 water bypass pipe.

50. Install the No. 2 turbo oil pipe as follows:

a. Install the turbo oil pipe and connect the turbo oil hose to the turbo oil outlet on the No. 1 oil pan.

b. Using a new gasket, install the turbo oil pipe to the turbocharger by installing the two nuts. Torque the nuts to 15 ft. lbs. (21 Nm). Make sure to align the oil holes of the gasket and turbocharger housing.

c. Using two new gaskets, connect the union bolt to hold the turbo oil pipe to the cylinder block. Torque the union bolt to 29 ft. lbs. (39 Nm).

51. Install the No. 1 turbo oil pipe as follows:

a. Install the turbo oil pipe and connect the turbo oil hose to the turbo oil outlet on the No. 1 oil pan.

b. Using a new gasket, install the turbo oil pipe to the turbocharger by installing the two nuts. Torque the nuts to 15 ft. lbs. (21 Nm). Make sure to align the oil holes of the gasket and turbocharger housing.

c. Using two new gaskets, connect the union bolt to hold the turbo oil pipe to the cylinder block. Torque the bolts to 29 ft. lbs. (39 Nm).

52. Install the No. 2 turbocharger stay by installing the nut and bolt. Torque the nut and bolt to 32 ft. lbs. (43 Nm).

53. Install the No. 1 turbocharger stay by installing the nut and bolt. Torque the nut and bolt to 32 ft. lbs. (43 Nm).

54. Install the exhaust gas control valve by installing two new gaskets and the three nuts. Torque the three nuts to 51 ft. lbs. (69 Nm).

55. Install the main heated oxygen sensor by connecting the electrical connector and installing the two nuts. Torque the nuts to 14 ft. lbs. (20 Nm).

56. Install the exhaust gas control valve stay by installing the bolt and nut. Torque the bolt and nut to 32 ft. lbs. (43 Nm).

57. Install the exhaust bypass pipe by installing two new gaskets and four new nuts. Torque the nuts to 18 ft. lbs. (25 Nm).

58. Install the heat insulator for the turbocharger by installing the four bolts.

59. Install the air inlet duct by installing the cable bracket, bolt, and two nuts.

60. Connect the air tubes and intake air connector as follows:

a. Install the intake air connector and No. 1 air tube assembly.

b. Connect the PCV hose to the No. 2 cylinder head cover.

c. Connect the air hose to the No. 2 air tube.

d. Install the intake air control valve and gasket by installing the two nuts. Torque the nut to 15 ft. lbs. (21 Nm).

e. Install the No. 4 air tube and air bypass valve assembly.

f. Connect the air hose to the intake air connector.

g. Connect the air hose to the No. 4 air tube.

h. Install the two bolts holding the No. 4 air tube to the No. 1 turbocharger. Torque the bolt to 15 ft. lbs. (21 Nm).

i. Connect the No. 1 air tube to the No. 1 turbocharger and install the two bolts. Torque the bolts to 15 ft. lbs. (21 Nm).

j. Install the No. 2 turbo water pipe to the No. 4 air tube by installing the bolt.

k. Connect the water bypass hose (from the water outlet) to the No. 2 turbo water pipe.

l. Connect the water hose (from the water outlet) to the No. 1 turbo water pipe.

m. Connect the water bypass hose (from the water pump) to the No. 1 turbo water pipe.

n. Connect the crankshaft position sensor connector to the clamp.

61. Install the VSV assembly as follows:

a. Connect the VSV assembly and connect the two VSV connectors.

b. Install the two bolts.

c. Connect the engine wire to the wire clamp.

d. Connect the air hose to the hose clamp.

e. Connect the air hose to the actuator for the exhaust gas control valve.

f. Connect the air hose to the actuator for the waste gate valve.

62. Connect the No. 1 vacuum pipe to the air tubes as follows:

a. Install the vacuum pipe to the air tubes and install the three bolts.

b. Connect the two air hoses (from the vacuum pressure tank) to the vacuum pipe.

c. Connect the air hose to the VSV for the intake air control valve.

d. Connect the air hose (from the No. 2 air tube) from the vacuum pipe.

e. Connect the two air hoses (from the VSV for exhaust bypass valve) to the vacuum pipe.

f. Connect the vacuum hose (from the air bypass valve) to the No. 1 air tube.

g. Connect the air hose (from the VSV for exhaust gas control valve) to the vacuum pipe.

h. Connect the air hose (from the VSV for waste gate valve) to the vacuum pipe.

i. Connect the air hose to the No. 1 air tube.

j. Connect the air hose to the No. 4 air tube.

k. Connect the engine wire to the three clamps.

l. Connect the VSV connector for the exhaust bypass valve.

m. Connect the VSV connector for the intake air control valve.

63. Connect the IAC valve pipe to the No. 2 air tube as follows:

a. Connect the IAC valve pipe to the clamp.

b. Connect the air hose to the No. 2 air tube.

c. Connect the air hose (from the No. 1 vacuum pipe) to the IAC valve pipe.

d. Connect the engine wire to the clamp.

64. Connect the EVAP hose to the No. 1 vacuum pipe.

65. Connect the heater hose to the No. 3 water bypass pipe.

66. Connect the engine wire protector to the body by installing the two bolts.

67. Connect the A/T oil cooler tubes to the engine.

68. Install the heat insulator for the No. 2 front exhaust pipe by installing the two bolts and two nuts.

69. Install the No. 2 front exhaust pipe as follows:

a. Using a new gasket, install the No. 2 front exhaust pipe and install the three nuts. Torque the nuts to 46 ft. lbs. (62 Nm).

b. Connect the front exhaust pipe to the No. 2 exhaust pipe.

c. Install the pipe support bracket and install the two bolts. Torque the bolts to 32 ft. lbs. (43 Nm).

d. Install the two bolts and nuts to hold the front exhaust pipe to the No. 2 front exhaust pipe. Torque the bolts and nuts to 43 ft. lbs. (58 Nm).

70. Install the upper front cross-member extension by installing the two bolts and two nuts. Torque the bolts to 22 ft. lbs. (29 Nm) and the nuts to 25 ft. lbs. (33 Nm).

71. Install the front lower arm bracket stay by installing the plate washer, nut, and two bolts. Torque the bolts to 33 ft. lbs. (44 Nm) and the nut to 43 ft. lbs. (59 Nm).

72. Connect the theft deterrent horn to the body.

73. Connect the air cleaner and MAF meter assembly as follows:

a. Install the MAF meter assembly and air cleaner. Connect the MAF meter connector.

b. Connect the MAF meter wire to the clamp on the air cleaner case.

c. Connect the air hose to the intake air connector and tighten the clamp.

d. Install the three bolts.

74. Install the air cleaner duct.

75. Install the No. 1 air hose.

76. Install the cruise control actuator cable to the throttle body.

77. Install the engine undercover.

78. Fill the engine coolant.

79. Check all fluids.

80. Connect the negative battery cable to the battery.

Front Cover Seal

REMOVAL AND INSTALLATION

1.5L (3E-E) Engine

Tercel

1. Disconnect the negative battery cable. On vehicles equipped with an air bag, wait at least 90 seconds before proceeding.

2. Remove the timing belt.

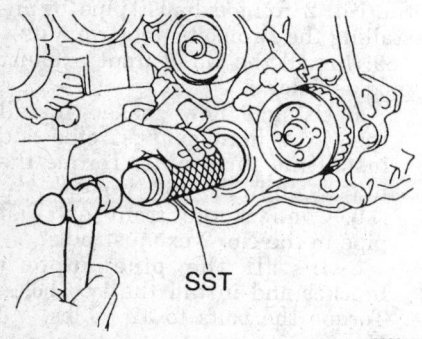

SST

288840

Typical method of installing the front cover seal

3. Remove the oil pan.

4. Remove the oil pump front cover seal using a suitable tool.

To install:

5. Install the seal into the front cover.

6. Install the timing belt and crankshaft pulleys.

7. Install the oil pan assembly.

8. Install the exhaust pipe and any front end components previously removed.

9. Install the engine undercover.

10. Lower the vehicle. Refill with engine oil. Start the engine and check for leaks.

Front Crankshaft Seal

REMOVAL AND INSTALLATION

1.5L (5E-FE) Engine

Tercel

NOTE: The front oil seal can be removed from the engine without removing the oil pump.

1. Disconnect the negative battery cable from the battery. On vehicles equipped with an air bag, wait at least 90 seconds before proceeding.

2. Remove the front covers and the timing belt.

3. Remove the crankshaft timing belt sprocket.

—————— **WARNING** ——————
When removing the front seal, be extremely careful not to damage the crankshaft.

4. Using a knife, cut off the oil seal lip.

5. Tape the end of a flat bladed tool to avoid damaging crankshaft. Pry out the oil seal using the taped end of the tool.

6. Inspect the oil seal riding surface on the crankshaft for signs of wear or damage.

To install:

7. Wipe the seal bore with a clean rag.

8. Apply multipurpose grease the lip of a new oil seal.

9. Drive the oil seal into place using tool SST 09309–37010 or an equivalent seal installer tool. Make sure the seal surface is flush with the oil pump case edge. Work from the front of the cover. Be extremely careful not to damage the seal.

10. Install the sprocket without disturbing the Woodruff key.

11. Install the timing belt and front covers.

12. Connect the negative battery cable to the battery.

13. Start the engine and check for leaks.

Corolla and Celica

1.6L (4A-FE) and 1.8L (7A-FE) Engines

1. Disconnect the negative battery cable. On vehicles equipped with an air bag, wait at least 90 seconds before proceeding.

2. Remove the timing belt and crankshaft sprocket.

3. Using a cutting tool, cut off the oil seal lip.

—————— **WARNING** ——————
When removing the oil seal, be careful not to damage the crankshaft or the oil pump housing.

4. Carefully pry out the oil seal from the oil pump housing.

To install:

5. Apply clean engine oil to the new oil seal.

6. Coat the inside of the new oil seal with multi-purpose grease.

7. Install the new oil seal to the oil pump using a suitable seal driver.

8. Install the timing belt and crankshaft sprocket.

9. Connect the negative battery cable. Check the engine oil level.

10. Run the engine and check for leaks.

2.0L (3S-GTE) and 2.2L (5S-FE) Engines

Camry, MR2 and Celica

1. Disconnect the negative battery cable. On vehicles equipped with an air bag, wait at least 90 seconds before proceeding.

NOTE: The front oil seal can be removed from the engine without removing the oil pump.

2. Disconnect the negative battery cable from the battery. On vehicles equipped with an air bag, wait at least 90 seconds before proceeding.

3. Remove the timing belt covers and the timing belt from the engine.

4. Using SST tool 09950–50010 or equivalent (crankshaft gear puller), remove the front crankshaft gear from the crankshaft. Make sure not to damage any part of the crankshaft.

5. Using a knife, cut off the oil seal lip.

6. Using a suitable tool, pry out the oil seal. Wrap the edge of the tool with a rag or tape to prevent damaging the crankshaft. Be careful not to damage the crankshaft.

To install:

7. Using a new seal, apply a thin layer of liquid sealer to the outside of the seal.

8. Apply multi purpose grease to the new oil seal lip.

9. Using SST tool 09223–00010 or equivalent (oil seal installer) and a hammer, tap in the oil seal until its surface is flush with the oil pump body edge.

10. Install the timing belt and the timing belt covers.

11. Install all other components and then connect the negative battery cable to the battery.

12. Start the engine and check for leaks.

3.0L (3VZ-FE) Engine

Camry

1. Disconnect the negative battery cable. On vehicles equipped with an air bag, wait at least 90 seconds before proceeding.

2. Drain the coolant and the engine oil.

3. Remove the timing belt and the crankshaft timing sprocket.

4. Using and sharp knife, cut off the oil seal lip.

5. Taking care not to damage the crankshaft, pry out the oil seal.

To install:

6. Coat the lip of the seal with multi-purpose grease.

7. Using the proper seal driver, tap in the new seal until it is flush with the oil pump case edge.

8. Install the crankshaft timing sprocket.

9. Install the timing belt.

10. Refill the engine with coolant and oil.

11. Connect the negative battery cable.

12. Start the engine and check for leaks.

3.0L (1MZ-FE) Engine

Camry

1. With the ignition switch in the **LOCK** position, disconnect the negative battery terminal. If equipped with an air bag system, wait at least 30 seconds or longer before performing any other work.

2. Remove the timing belt.

3. Remove the crankshaft timing gear.

4. Cut out the lip portion of the oil seal.

5. Tape the end of a suitable prybar to protect the crankshaft and carefully remove the oil seal.

― **WARNING** ―

Be careful not to damage the crankshaft sealing surface.

To install:

6. Apply multi-purpose grease to the lip of a new oil seal. Also apply a light coating of liquid sealant to the outside of the oil seal.

7. Lightly tap the oil seal with a hammer and installer 09309–37010 or equivalent, until its surface is flush with the oil pump case edge.

8. Install the crankshaft timing gear.

9. Install the timing belt.

10. Reconnect the negative battery cable.

3.0L (2JZ-GE and 2JZ-GTE) Engines

Supra

1. Remove the timing belt.

2. Remove the crankshaft timing pulley.

3. Cut the oil seal lip.

4. Tape the end of a small prying tool and remove the oil seal.

― **WARNING** ―

Be careful not to damage the crankshaft.

To install:

5. Apply multi-purpose grease to a new oil seal lip.

6. Lightly tap the oil seal with a hammer and installer 09316–60010, or equivalent, until its surface is flush with the oil pump case edge.

7. Install the crankshaft timing pulley.

8. Install the timing belt.

9. Start the engine and check for leaks.

Timing Belt, Sprockets, Tensioner and Front Cover

REMOVAL AND INSTALLATION

1.5L (3E-E) Engine

Tercel

1. Disconnect the negative battery cable. On vehicles equipped with an air bag, wait at least 90 seconds before proceeding.

2. Remove the right side engine undercover. Disconnect the accelerator and throttle cables, as required.

3. Remove the drive belts, alternator and alternator bracket. Remove the air cleaner and air intake collector assemblies and spark plugs on the fuel injected engine.

4. Raise the engine and remove the right side engine mounting insulator assembly.

5. Remove the cylinder head cover. Set the engine to TDC on the compression stroke. Remove the crankshaft pulley using the proper removal tool.

6. Remove both timing belt covers. Remove the timing belt guide. Remove the timing belt and the No. 1 idler pulley. If using the old belt matchmark it in the direction of engine rotation. Matchmark the pulleys.

7. Remove the tension spring. Remove the No. 2 idler pulley. Remove the crankshaft pulley, camshaft pulley and oil pump pulley using the proper tools.

To install:

8. Inspect the belt for defects. Replace as required. Inspect the idler pulleys and springs. Replace defective components as required.

9. Align and install the oil pump pulley. Torque the retaining bolt to 20 ft. lbs. (27 Nm).

10. To install the camshaft timing pulley, align the camshaft knock pin with the No. 1 bearing cap mark. Align the knock pin hole on the 3E-E mark side with the camshaft knock pin hole. Torque the retaining bolt to 37 ft. lbs. (51 Nm).

11. Install the crankshaft timing pulley and align the TDC marks on the oil pump body and the crankshaft timing pulley. Install the No. 1 idler pulley. Pry the idler pulley toward the left as far as it will go and temporarily tighten the retaining bolt.

12. Install the No. 2 idler pulley and torque the retaining bolt to 20 ft. lbs. (27 Nm). Install the timing belt. If reusing the old belt align it with the marks made during the removal procedure.

13. Inspect the valve timing and the belt tension by loosening the No. 1 idler pulley set bolt. Temporarily install the crankshaft pulley bolt and turn the crankshaft 2 complete revolutions in the clockwise direction.

14. Check that each pulley aligns with the proper markings. Torque the No. 1 idler pulley bolt to 13 ft. lbs. (17 Nm). Check for proper belt tension. The belt should deflect ¼ in. (6mm) at 4½ lbs. pressure. Install the belt guide.

15. Install the timing belt covers. Align and install the crankshaft pulley. Torque the retaining bolt to specifications.

16. Installation of the remaining components is the reverse of the removal procedure. Service components

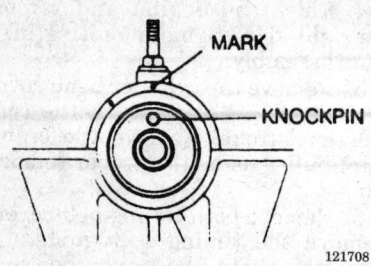

Align the camshaft knockpin with the mark in the bearing cap prior to installing the timing belt pulley — 3E-E engine

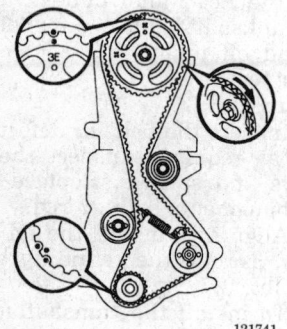

Timing belt and pulley alignment marks — 3E-E engine

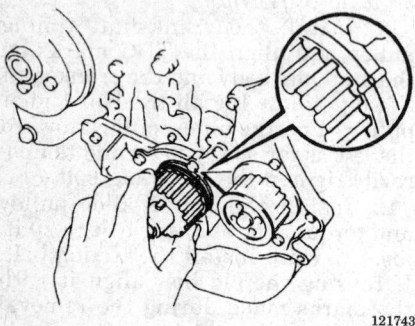

Aligning crankshaft pulley marks — 3E-E engine

as required. Make all necessary engine adjustments and road test for proper operation.

1.5L (5E-FE) Engine

Tercel and Paseo

1. Disconnect the negative battery cable. On vehicles equipped with an air bag, wait at least 90 seconds before proceeding.
2. Remove the right side engine undercover.
3. Remove the power steering and alternator drive belts.

4. Remove cruise control actuator, if equipped.
5. For vehicle with distributor ignition, mark the vacuum lines and remove VSV assembly.
6. For vehicles with distributorless ignition, remove the ground strap.
7. Raise the engine enough to remove weight from the right side engine mount and remove the mount.

NOTE: Be careful not to damage vacuum or electrical connections when raising the engine.

8. Remove the spark plugs.
9. Remove the No. 2 timing belt cover.
10. Rotate the engine clockwise until the crankshaft pulley is aligned with the **0** mark on the No. 1 timing belt cover. Verify the hole (5E-FE) in the camshaft timing pulley is aligned with the timing mark on the No. 1 bearing cap. If not as specified, rotate the crankshaft an additional 360°.
11. For vehicles with A/C and/or power steering, remove the four bolts to the No. 2 crankshaft pulley. Remove the No. 2 crankshaft pulley.
12. Using tools SST 09213–14010 and SST 09330–00021 or equivalent, remove the No. 1 crankshaft pulley bolt.
13. Using tools SST 09950–50010 or equivalent (crankshaft puller), remove the No. 1 crankshaft pulley from the crankshaft.
14. Remove the No. 3 timing belt cover (plug).
15. Remove the No. 1 timing belt cover and timing belt guide.
16. Place matchmarks on the timing belt on both sides of the cam and crankshaft gear timing marks. Also place an arrow on the top surface of the belt to indicate the direction of travel.
17. Using pliers, remove the tension spring.
18. Loosen the No. 1 idler pulley bolt and push the pulley to the left as

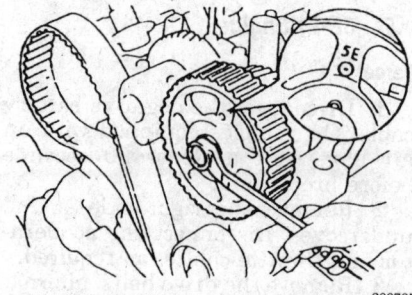

Ensure the match hole of the camshaft housing is aligned with that of the bearing cap — 5E-FE engine

far as it will go, then temporarily tighten the bolt.
19. Remove the belt.
20. Using tool SST 09960–10010 or equivalent, remove the bolt and camshaft timing pulley.
21. Remove the crankshaft timing sprocket. If the timing pulley cannot be removed by hand, use a plastic faced hammer. Do not pry on the gear.
22. Remove the No. 1 and No. 2 idler pulleys.

To install:

23. Install the No. 1 idler pulley and spring to the engine. Do not tighten the bolts at this time.
24. Torque the No. 2 idler pulley bolts to 20 ft. lbs. (28 Nm).
25. Install the camshaft timing sprocket and torque the bolt to 37 ft. lbs. (50 Nm). Make sure to align the camshaft knock pin with the knock pin groove on the pulley side with the 5E-FE marking.
26. Install the crankshaft sprocket, aligning the slot with the Woodruff key.
27. For vehicle with a distributor (distributor ignition), use the crankshaft bolt to turn the crankshaft until the timing marks on the sprocket and oil pump body align. This is setting the piston at TDC before the marks on the belt cover can be seen.
28. For vehicles with a crankshaft position sensor (distributorless ignition), use the crankshaft bolt to turn the crankshaft until the rotor side of the crankshaft position sensor faces inward.
29. Turn the camshaft and align the hole of the camshaft timing pulley with the timing mark of the bearing cap. The matchmarks, if using the old belt should line up. Place the belt over the crankshaft pulley and the idler pulleys.
30. Install the belt on the crankshaft gear (using matchmarks if reinstalling old belt) and install the timing belt guide with flange out.
31. Loosen the No. 1 idler pulley bolt until the pulley is moved slightly by the spring tension.
32. Turn the crankshaft pulley two revolutions from TDC to TDC.

NOTE: Always rotate the crankshaft clockwise.

33. Check that the pulleys align with the reference marks. If not, reinstall the belt.
34. When the timing is verified, torque the adjuster pulley (No. 1 idler pulley) to 13 ft. lbs. (18 Nm).
35. Install the No. 1 and No. 3 lower timing belt cover.

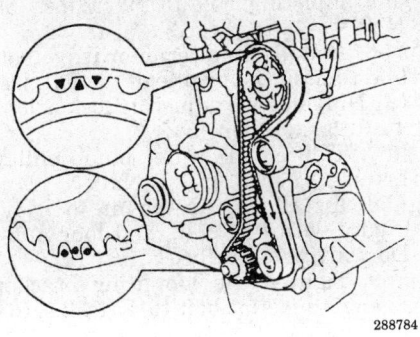

Mark the belt prior to removing — 5E-FE engine

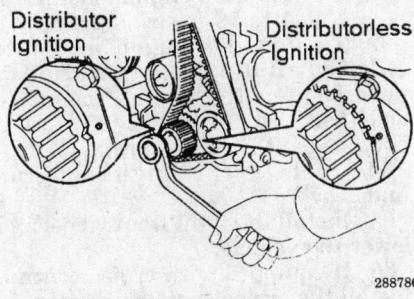

Distributor Ignition   Distributorless Ignition

Crankshaft gear timing marks (distributor and distributorless ignitions) — 5E-FE engine

36. Install the No. 1 crankshaft pulley and torque the pulley bolt to 112 ft. lbs. (152 Nm).

37. Torque the four No. 2 crankshaft pulley bolts to 14 ft. lbs. (19 Nm).

38. Install the No. 2 timing belt cover with the four bolts.

39. Install the right-hand engine mounting insulator. Torque bracket bolt to 47 ft. lbs. (64 Nm) and the through-bolt to 54 ft. lbs. (73 Nm).

40. For vehicles with distributorless ignition, connect the ground strap to the engine.

41. For vehicles with distributors, install the VSV assembly.

42. If equipped with cruise control, install the cruise control system.

43. Install the spark plugs to the engine.

44. Install the alternator and power steering drive belts.

45. Install the right-hand engine undercover.

46. Connect the negative battery cable to the battery.

1.6L (4A-FE) and 1.8L (7A-FE) Engines

Celica

1. Disconnect the negative battery cable. On vehicles equipped with an air bag, wait at least 90 seconds before proceeding.

2. If equipped with cruise control, remove the following.

 a. Remove the cruise control actuator cover and the cruise control electrical connector.

 b. Remove the cruise control cable and the three bolts and cruise control actuator from the vehicle.

3. Raise and support the vehicle safely.

4. Remove the right front wheel.

5. Remove the right front wheel housing cover.

6. Slightly jack up the engine.

7. Remove the right-hand engine mounting insulator by removing the two bolts and four nuts.

8. Loosen the water pump pulley bolts.

9. If equipped with A/C compressor, remove the A/C compressor and bracket.

10. Loosen the power steering pump pivot and lock bolts.

11. Push the power steering pump toward the engine and remove the drive belt.

12. Remove the water pump pulley by removing the four bolts.

13. Disconnect the wire and clamp from the alternator.

14. Remove the wire harness protector from the cylinder head by removing the two bolts.

15. Remove the spark plug wires.

16. Remove the spark plugs from the cylinder head.

17. Remove the two hose clamps and the PCV hoses from the cylinder head cover.

18. Remove the four cap nuts, the seal washers, the cylinder head cover, and the gasket from the cylinder head.

19. Turn the crankshaft to align the timing mark on crankshaft pulley at **0**, setting the piston in No. 1 cylinder at top dead center (TDC) on the compression stroke. Check that the valve lifters on the No. 1 cylinder are loose. If not, turn crankshaft pulley 1 complete revolution (360 degrees).

20. Remove the four bolts and coolant pump pulley from coolant pump.

21. Using SST 09213–54015, or equivalent, remove the crankshaft pulley bolt.

22. Remove the crankshaft pulley using a puller.

23. Remove the nine bolts and timing belt covers from the engine.

24. Slide the timing belt guide from crankshaft.

25. Set the camshaft and crankshaft timing sprockets to align the marks.

NOTE: Do not turn crankshaft or camshaft independently after removal of timing belt. Binding or damage to engine components could result. If timing belt is to be reused, mark timing belt with arrow showing direction of engine revolution. Put matchmarks where timing belt meets with crankshaft timing sprocket and camshaft timing sprocket to ensure installation in the same position.

26. Remove the timing belt tensioner bolt, tensioner, and the tension spring.

27. Remove the timing belt from camshaft and crankshaft timing sprockets.

— WARNING —

Do not bend, twist or turn the timing belt inside out. Do not allow the belt to come in contact with oil, coolant or steam.

28. Remove the camshaft timing sprocket bolt while holding the camshaft by the hexagonal wrench head section.

— WARNING —

Be careful not to damage the cylinder head when holding camshaft in place.

29. Remove crankshaft timing sprocket using 2 flat bladed prybars to pry off. Tape the end of the prybars to prevent damaging the crankshaft.

To install:

30. Check the camshaft and crankshaft timing sprockets to align the marks. Do not turn crankshaft or camshaft independently before installation of the timing belt or engine damage may occur.

31. Install the camshaft timing sprocket. Align the camshaft key with the groove on the sprocket and slide the sprocket on the camshaft; secure with the camshaft timing sprocket bolt. Hold the camshaft at hexagonal wrench head portion to prevent rotation and tighten the camshaft timing sprocket bolt to 43 ft. lbs. (59 Nm).

NOTE: Inspect the camshaft timing sprocket to ensure mark is still aligned as indicated.

32. Install the crankshaft timing sprocket. Turn the crankshaft and

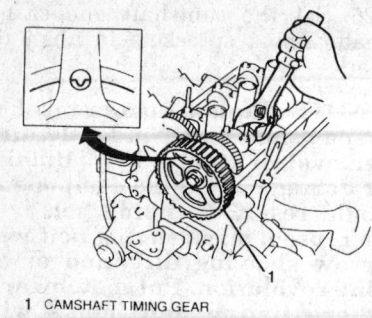

1 CAMSHAFT TIMING GEAR

287766

Aligning the camshaft timing sprocket timing marks — 7A-FE engine

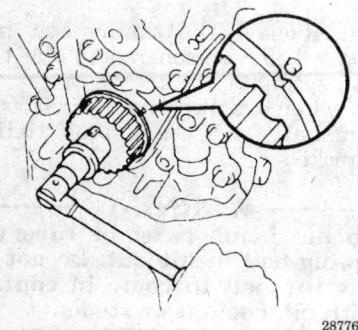

287768

Crankshaft gear timing position — 4A-FE and 7A-FE engines

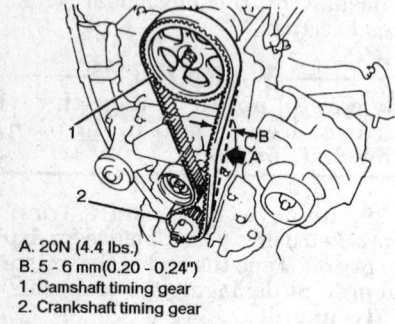

A. 20N (4.4 lbs.)
B. 5 - 6 mm(0.20 - 0.24")
1. Camshaft timing gear
2. Crankshaft timing gear

287765

Measuring timing belt deflection — 7A-FE engine

position the key groove of the crankshaft timing gear upward.

NOTE: Inspect the crankshaft timing sprocket to ensure mark is still aligned as indicated.

33. Install the timing belt tensioner and the tension spring. Pry the tensioner to the left as far as it will go and temporarily tighten the retaining bolt.

34. Install the timing belt. If reinstalling the old belt, observe the matchmarks made during removal. Make sure the belt is fully and squarely seated on the sprockets.

35. Loosen the retaining bolt for the timing belt tensioner and allow it to tension the belt.

36. Temporarily install the crankshaft pulley bolt and turn the crank clockwise 2 full revolutions from TDC to TDC. Insure that each timing mark realigns exactly.

37. Tighten the timing belt tensioner bolt to 27 ft. lbs. (37 Nm).

38. Measure the timing belt deflection at the **SIDE** point, looking for 5–6mm of deflection at 4.4 pounds of pressure. If the deflection is not correct, adjust with the timing belt tensioner.

39. Install the timing belt guide, with the cup side facing outward, onto the crankshaft and install the timing belt covers from the lowest to the highest. Tighten the nine cover bolts to 62 inch lbs. (7 Nm).

40. Install the crankshaft pulley to the crankshaft after aligning the pulley key with the slot on the pulley; secure with the bolt while holding the crankshaft pulley and tighten the pulley bolt to 87 ft. lbs. (118 Nm).

41. Install the new cylinder head cover gasket to the cylinder head.

42. Install new seal washers to the cylinder head cover.

43. Install the cylinder head cover to the cylinder head; secure with four cap nuts.

44. Tighten the cylinder head cap nuts to 53 inch lbs. (6 Nm).

45. Install two PCV hoses to the cylinder head cover; secure with hose clamps.

46. Install the spark plugs and the spark plug wires.

47. Install the engine wiring harness to the cylinder head cover.

48. Install the engine wiring harness cover; secure with two bolts.

49. Tighten the wiring harness cover bolts to 53 inch lbs. (6 Nm).

50. Connect the wire and clamp to the alternator.

51. Install the water pump pulley. Install the bolts but do not torque the bolts at this time.

52. Install the power steering pump pulley and pull the power steering pump back to tighten the drive belt. Torque the pivot and adjusting bolt to 29 ft. lbs. (39 Nm).

53. If equipped, install the A/C pipe mounting bracket and torque the bolts to 35 ft. lbs. (47 Nm).

54. Install the A/C compressor with the four bolts. Torque the bolts to 18 ft. lbs. (25 Nm) and connect the A/C compressor connector.

55. Install the drive belt with the adjusting bolt and install the idler pulley locknut and tighten the nut

and adjusting bolt to 29 ft. lbs. (39 Nm).

56. Install the alternator drive belt and tighten the pivot bolt to 45 ft. lbs. (61 Nm) and the adjusting lock bolt to 14 ft. lbs. (19 Nm).

57. Tighten the water pump pulley bolts to 82 inch. lbs. (9 Nm).

58. Install engine mount to body; secure with two bolts and four nuts. Do not tighten fully.

59. Tighten the mounting bracket to engine mount bolt to 47 ft. lbs. (64 Nm).

60. Tighten the mounting bracket to engine mount nuts to 38 ft. lbs. (52 Nm).

61. Tighten the engine mount to body bolt **A** to 19 ft. lbs. (25 Nm).

62. Tighten the engine mount to body bolts **B** to 19 ft. lbs. (25 Nm).

63. Tighten the engine mount to body bolt **C** to 19 ft. lbs. (25 Nm), if equipped with cruise control.

64. Install the right engine undercover.

65. Install the right front wheel and lower the vehicle.

66. If equipped with cruise control, install the cruise control actuator.

67. Install the negative battery cable.

68. Start the engine and check the ignition timing.

Corolla

1. Disconnect the negative battery cable.

2. Remove bolt, hose connector, electrical connector, and the windshield washer reservoir from the engine compartment.

3. If equipped with cruise control, remove the following.

   a. Remove the cruise control actuator cover.

   b. Remove the cruise control electrical connector.

   c. Remove the cruise control cable.

   d. Remove the three bolts and cruise control actuator from the vehicle.

4. Raise and support the vehicle safely.

5. Remove the right front wheel and tire assembly.

6. Remove the 10 bolts, two plastic clips, and the right front wheel housing.

7. Loosen the coolant pump pulley adjusting bolt, pivot bolt, and the lockbolt and remove the alternator/coolant pump drive belt.

8. Lower the vehicle.

9. If equipped with A/C, remove the A/C compressor without disconnecting the hoses. Set the A/C compressor aside.

10. Remove the four bolts and A/C compressor mounting bracket.

11. If equipped with power steering, loosen the power steering pump pivot and lock bolts.

12. Relocate the power steering pump toward the engine and remove the drive belt.

13. Remove the following electrical connectors: alternator electrical connector, alternator lead wire, and the oil pressure switch electrical connector.

14. Remove the two bolts and the engine wiring harness cover.

15. Remove the wiring harness from the cylinder head cover.

16. Remove the spark plug wires.

17. Remove the spark plugs from the cylinder head.

18. Remove the two hose clamps and the PCV hoses from the cylinder head cover.

19. Remove the four cap nuts, the seal washers, the cylinder head cover, and the gasket from the cylinder head.

20. Turn the crankshaft to align the timing mark on crankshaft pulley at **0**, setting the piston in No. 1 cylinder at Top Dead Center (TDC) on the compression stroke. Check that the valve lifters on the No. 1 cylinder are loose. If not, turn crankshaft pulley 1 complete revolution (360 degrees).

21. Remove the engine ground wire from the right fender apron.

22. Install a support under the engine.

23. Remove the engine mount.

24. Remove the four bolts and coolant pump pulley from coolant pump.

25. Remove the crankshaft pulley using a puller.

26. Remove the nine bolts and timing belt covers from the engine.

27. Slide the timing belt guide from crankshaft.

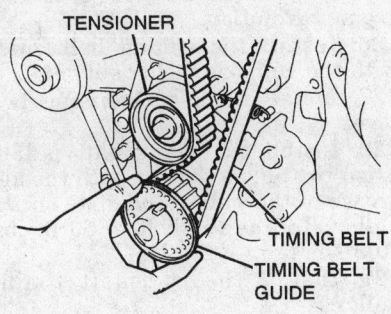

Removing the timing belt guide — 4A-FE engine

28. Set the camshaft and crankshaft timing sprockets to align the marks.

NOTE: Do not turn crankshaft or camshaft independently after removal of timing belt. Binding or damage to engine components could result. If the timing belt is to be reused, mark timing belt with arrow showing direction of engine revolution. Put matchmarks where timing belt meets with crankshaft timing sprocket and camshaft timing sprocket to ensure installation in the same position.

29. Remove the timing belt tensioner bolt, tensioner, and the tension spring.

30. Remove the timing belt from camshaft and crankshaft timing sprockets.

─────── **WARNING** ───────

Do not bend, twist or turn the timing belt inside out. Do not allow the belt to come in contact with oil, coolant or steam.

31. Remove the camshaft timing sprocket bolt while holding the camshaft by the hexagonal wrench head section.

─────── **WARNING** ───────

Be careful not to damage the cylinder head when holding camshaft in place.

32. Remove crankshaft timing sprocket using 2 flat bladed prybars to pry off. Tape the end of the prybars to prevent damaging the crankshaft.

To install:
33. Check the camshaft and crankshaft timing sprockets to align the marks. Do not turn crankshaft or camshaft independently before installation of the timing belt.

34. On 1993–95 models, install the crankshaft timing sprocket. For the 1.6L (4A-FE) engine, align the crankshaft gear timing mark with the timing mark on the oil pump. For the 1.8L (7A-FE) engine, turn the crankshaft and position the key groove of the crankshaft timing gear upward.

35. On 1996–97 models, install the camshaft timing sprocket as follows:

   a. Align the camshaft knock pin with the knock pin groove on the pulley side with the **K** mark and slide on the timing pulley.

   b. Hold the camshaft at hexagonal wrench head portion to prevent rotation and tighten the camshaft

timing sprocket bolt to 43 ft. lbs. (59 Nm).

NOTE: Inspect the camshaft timing sprocket to ensure mark is still aligned as indicated.

36. Install the crankshaft timing sprocket and align the crankshaft gear timing mark with the timing mark on the oil pump.

37. Reinstall the timing belt tensioner and the tension spring. Pry the tensioner to the left as far as it will go and temporarily tighten the retaining bolt.

38. Install the timing belt. If reinstalling the old belt, observe the matchmarks made during removal. Make sure the belt is fully and squarely seated on the sprockets.

39. Loosen the retaining bolt for the timing belt tensioner and allow it to tension the belt.

40. Temporarily install the crankshaft pulley bolt and turn the crank clockwise 2 full revolutions from TDC to TDC. Insure that each timing mark realigns exactly.

41. Tighten the timing belt tensioner bolt to 27 ft. lbs. (37 Nm).

42. Measure the timing belt deflection at the **SIDE** point, looking for 5–6mm of deflection at 4.4 pounds of pressure. If the deflection is not correct, adjust with the timing belt tensioner.

43. Install the timing belt guide, with the cup side facing outward, onto the crankshaft and install the timing belt covers from the lowest to the highest. Tighten the nine cover bolts to 62 inch lbs. (7 Nm).

44. Install the crankshaft pulley to the crankshaft after aligning the pulley key with the slot on the pulley; secure with the bolt while holding the crankshaft pulley and tighten the pulley bolt to 87 ft. lbs. (118 Nm).

45. Install engine mount to body; secure with four bolts. Do not tighten fully.

46. Raise and safely support the vehicle.

47. Install the mounting bracket(s) to the engine mount; secure with the bolt (2 if equipped with power steering) and two nuts.

48. Tighten the mounting bracket to engine mount bolt to 47 ft. lbs. (64 Nm).

49. Tighten the mounting bracket to engine mount nuts to 38 ft. lbs. (52 Nm).

50. Tighten the engine mount to body bolt **A** to 19 ft. lbs. (25 Nm).

51. Tighten the engine mount to body bolts **B** to 19 ft. lbs. (25 Nm).

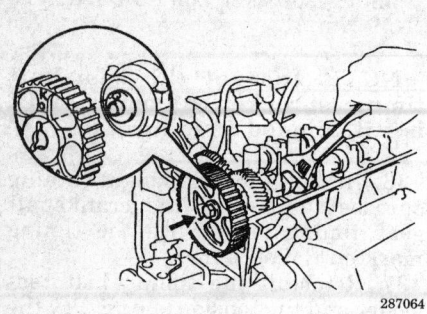

Camshaft pulley K mark — 4A-FE engine

287064

52. Tighten the engine mount to body bolt **C** to 19 ft. lbs. (25 Nm), if equipped with cruise control.

53. Install right-side wheel housing; secure with 10 bolts and two plastic clips.

54. Tighten the wheel housing bolts to 11 ft. lbs. (15 Nm).

55. Install the tire and wheel assembly.

56. Lower the vehicle.

57. Remove the support.

58. Install the A/C pipe bracket (2 pieces, if equipped, and secure with the bolt and screw.

59. Tighten the A/C pipe bracket bolt to 89 inch lbs. (10 Nm).

60. Install the engine ground wire to the right fender apron.

61. Install the new cylinder head cover gasket to the cylinder head.

62. Install new seal washers to the cylinder head cover.

63. Install the cylinder head cover to the cylinder head; secure with four cap nuts.

64. Tighten the cylinder head cap nuts to 53 inch lbs. (6 Nm).

65. Install two PCV hoses to the cylinder head cover; secure with hose clamps.

66. Install the secondary (spark plug) wires and the spark plugs.

67. Install the engine wiring harness to the cylinder head cover.

68. Install the engine wiring harness cover; secure with two bolts.

69. Tighten the wiring harness cover bolts to 53 inch lbs. (6 Nm).

70. Install the following electrical connectors: alternator electrical connector, alternator lead wire, and the oil pressure switch electrical connector.

71. If removed, install the power steering pump.

72. If removed, install the A/C compressor.

73. Adjust the power steering pump drive belt.

74. Adjust the A/C compressor drive belt.

75. If removed, install and adjust the cruise control actuator.

76. Install the windshield washer reservoir to fender; secure it with 1 bolt.

77. Tighten the windshield washer reservoir bolt to 53 inch lbs. (6 Nm).

78. Install the hose connector and electrical connector to the windshield washer reservoir.

79. Install the negative battery cable.

2.2L (5S-FE) Engine

Celica

1. Disconnect the negative battery cable. On vehicles equipped with an air bag, wait at least 90 seconds before proceeding.

2. Remove the alternator and alternator bracket.

3. Raise and support the vehicle safely. Remove the right tire and wheel assembly.

4. Remove the right side engine undercover.

5. Remove the power steering belt.

6. Disconnect the ground strap connector.

7. Raise the engine enough to move the right side engine mounting assembly.

8. Remove the spark plugs.

9. Remove the No. 2 timing cover.

10. Position the number one cylinder to TDC on the compression stroke by turning the crankshaft pulley and aligning its groove with the timing mark **0** of the No. 1 timing belt cover. Check that the hole of the camshaft timing pulley is aligned with the alignment mark of the bearing cap. If not, turn the crankshaft one revolution 360°.

11. Remove the timing belt from the camshaft timing pulley.

 a. If reusing the belt, place matchmarks on the timing belt and the camshaft pulley. Loosen the mount bolt of the No. 1 idler pulley and position the pulley toward the left as far as it will go. Tighten the bolt. Remove the belt from the camshaft pulley.

12. Remove the camshaft timing pulley.

 a. Using the SST Nos. 09249–63010 and 09960–10010, or equivalent, remove the bolt and the camshaft pulley.

13. Remove the crankshaft pulley.

 a. Using the SST Nos. 09213–54015 and 09330–00021, or equivalent, to hold the crankshaft pulley. Remove the pulley set bolt and remove the pulley using a puller.

14. Remove the No. 1 timing belt cover.

15. Remove the timing belt and the belt guide. If reusing the belt mark the belt and the crankshaft pulley in the direction of engine rotation and matchmark for correct installation.

16. Remove the No. 1 idler pulley and the tension spring.

17. Remove the No. 2 idler pulley.

18. Remove the crankshaft timing pulley. If the pulley cannot be removed by hand, use two prying tools with shop rags behind them to prevent damage to the engine.

19. Remove the oil pump pulley.

20. Inspect the belt for defects and replace as required. Inspect the idler pulleys and springs; replace the defective components as required.

 To install:

21. Align the cutouts of the oil pump pulley and shaft. Install the oil pump pulley and torque the retaining nut to 18 ft. lbs. (24 Nm).

22. Install the crankshaft timing pulley.

 a. Align the timing pulley set key with the key groove of the pulley.

 b. Slide on the timing pulley with the flange side facing inward.

23. Install the No. 2 idler pulley and torque the bolt to 31 ft. lbs. (42 Nm). Be sure that the pulley moves freely.

24. Temporarily install the No. 1 idler pulley and tension spring. Pry the pulley toward the left as far as it will go. Tighten the bolt. Make sure that the pulley rotates freely.

25. Temporarily install the timing belt.

 a. Using the crankshaft pulley bolt, turn the crankshaft and align the timing marks of the crankshaft timing pulley and the oil pump body.

 b. If reusing the old belt, align the marks made during removal, and install the belt with the arrow pointing in the direction of the engine revolution.

26. Install the timing belt guide with the cup side facing outward.

27. Install the No. 1 timing belt cover.

28. Install the crankshaft pulley. Align the pulley set key with the key groove of the pulley and slide on the pulley. Torque the bolt to 80 ft. lbs. (108 Nm).

29. Install the camshaft timing pulley.

 a. Align the camshaft knock pin with the knock pin groove of the pulley and slide on the timing pul-

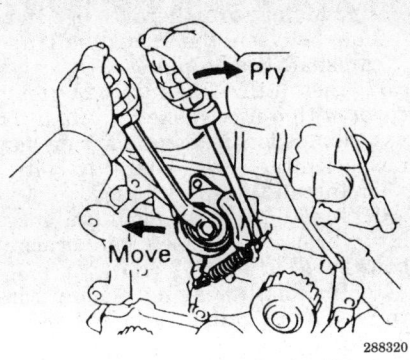

No. 1 idler pulley — 5S-FE engine

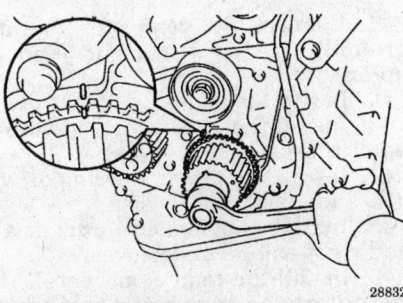

Crankshaft gear timing mark alignment (with oil pump body) — 5S-FE engine

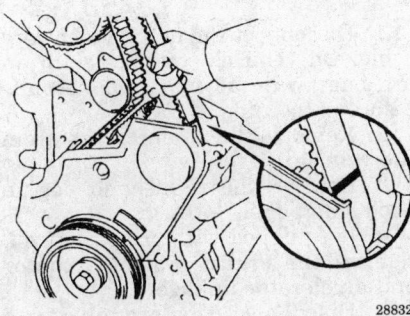

Aligning the timing belt mark with the No. 1 cover — 5S-FE engine

ley. Torque the bolt to 40 ft. lbs. (54 Nm).

30. With the No. 1 cylinder set at TDC on the compression stroke install the timing belt (all timing marks aligned). If reusing the belt, align with the marks made during the removal procedure.

   a. Turn the crankshaft pulley, and align its groove with the timing mark **0** of the No. 1 timing belt cover. Make sure the camshaft sprocket hole is aligned with the mark on the bearing cap.

31. Connect the timing belt to the camshaft timing pulley.

32. Check that the matchmark on the timing belt matches the end of the No. 1 timing belt cover.

33. Once the belt is installed be sure that there is tension between the crankshaft timing pulley and the camshaft pulley.

34. Check the valve timing.

   a. Loosen the No. 1 idler pulley mount bolt ½ turn. Turn the crankshaft pulley two revolutions from TDC in the clockwise direction. Always turn the crankshaft pulley clockwise.

   b. Make sure that the all the timing marks are aligned.

   c. Slowly turn the crankshaft pulley 1⅞ revolutions. Align its groove with the mark at 45° BTDC on the No. 1 timing belt cover for the No. 1 cylinder.

   d. Torque the No. 1 idler pulley mount bolt to 31 ft. lbs. (42 Nm).

35. Install the No. 2 timing belt cover.

   a. Install the upper gasket to the No. 1 timing belt cover.

   b. Disconnect the engine wire protector between the cylinder head cover and the No. 3 timing belt cover.

   c. Install the gasket to the timing belt cover.

   d. Install the belt cover.

   e. Install the engine wire protector to the two mounting bolts of the No. 2 timing belt cover. Install the right side first.

   f. Install the engine wire protector to the alternator bracket and adjusting bar.

36. Install the spark plugs.

37. Tighten the right engine mount bracket bolts to 38 ft. lbs. (52 Nm).

38. Lower the engine after installing the bracket.

39. Tighten the engine mount insulator bolt to 47 ft. lbs. (64 Nm). Tighten the through-bolt to 54 ft. lbs. (78 Nm).

40. Install the power steering reservoir bracket, torque the bolt to 21 ft. lbs. (28 Nm).

41. Install the power steering reservoir to the bracket. Torque the bolt to 27 ft. lbs. (37 Nm) and the nut to 38 ft. lbs. (52 Nm).

42. Connect the ground strap connector.

43. Install the power steering pump drive belt.

44. Install the front wheel.

45. Install the alternator.

46. Install the engine undercover.

47. Lower the vehicle and connect the negative battery cable.

48. Check all fluid levels and adjust all drive belts. Perform all necessary engine adjustments. Road test the vehicle for proper operation.

Camry

1. Disconnect the negative battery cable. On vehicles equipped with an air bag, wait at least 90 seconds before proceeding.

2. Remove the engine coolant reservoir tank.

3. Remove the alternator.

4. Remove the right front wheel and the fender apron seal.

5. Remove the power steering belt.

6. Slightly jack up the engine using a block of wood under the oil pan to prevent damage.

7. Remove the through-bolt, two nuts, and right-hand mounting bracket.

8. Remove the engine moving control rod.

9. Disconnect the connector from the ground wire on the right fender apron.

10. Remove the retaining bolts and remove the No. 2 engine mounting bracket.

11. Remove the spark plugs.

12. Remove the No. 2 timing belt cover.

13. Set the No. 1 cylinder to TDC of the compression stroke.

14. Turn the crankshaft pulley and align its groove with the timing mark **0** of the No. 1 timing belt cover. Check that the hole of the camshaft timing pulley is aligned with the timing mark of the bearing cap.

15. Remove the timing belt from the camshaft timing pulley.

NOTE: If the timing belt is to be reused, matchmark the timing belt to the timing pulleys and timing belt covers so the belt can be reinstalled in its original position. Also, be sure to mark an arrow on the belt to indicate which direction it was turning.

16. Hold the camshaft sprocket with a spanner wrench and remove the mounting bolt. Remove the camshaft pulley.

17. Remove the crankshaft pulley bolt and remove it by pulling it straight off the crankshaft.

18. Remove the No. 1 timing belt cover.

19. Remove the timing belt guide and the timing belt.

20. Remove the No. 1 idler pulley and tension spring.

21. Remove the No. 2 idler pulley.

22. Remove the crankshaft timing pulley.

23. Support the oil pump sprocket with a spanner wrench and remove

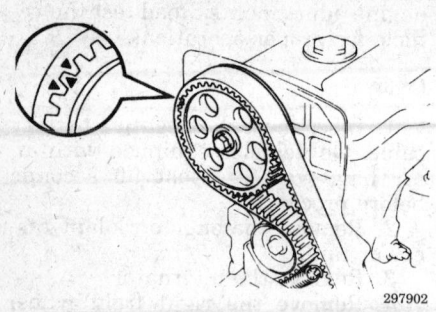

Matchmarking the timing belt to the camshaft sprocket — 5S-FE engine

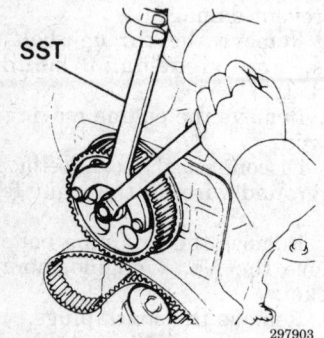

Removing the camshaft sprocket — 5S-FE engine

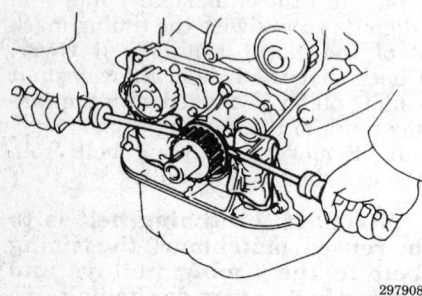

Removing the crankshaft sprocket — 5S-FE engine

the mounting bolt and remove the sprocket.

To install:

24. Install the oil pump pulley. Torque the nut to 21 ft. lbs. (28 Nm).

25. Install the crankshaft timing pulley. Align the pulley set key with the key groove of the pulley. Slide on the pulley facing the flange side inward.

26. Install the No. 2 idler pulley and tighten the mounting bolt to 31 ft. lbs. (42 Nm). Make sure that the pulley moves smoothly.

27. Install the No. 1 idler pulley with the bolt and the tension spring.

Pry the pulley toward the left as far as it will go and tighten the bolt. Make sure that the pulley moves smoothly.

28. Temporarily install the timing belt. Using the crankshaft pulley bolt, turn the crankshaft and position the key groove of the crankshaft timing pulley upward. The dot on the crankshaft pulley should align between the two dots on the timing belt.

29. Install the timing belt on the crankshaft timing pulley, oil pump pulley, No. 1 idler pulley, water pump pulley and the No. 2 idler pulley.

30. Install the timing belt guide.

NOTE: If the old timing belt is being reinstalled, make sure the directional arrow is facing in the original direction and that the belt and sprocket/cover matchmarks are properly aligned.

31. Install the lower (No. 1) timing belt cover and new gasket with the four bolts.

32. Align the crankshaft pulley set key with the pulley key groove. Install the pulley. Tighten the pulley bolt to 80 ft. lbs. (108 Nm).

33. Align the camshaft knock pin with the groove of the pulley, and slide the timing pulley onto the camshaft with the plate washer and set bolt.

34. Tighten the pulley set bolt to 40 ft. lbs. (54 Nm).

35. Turn the crankshaft pulley and align the **0** mark on the lower (No. 1) timing belt cover.

36. Finish installing the timing belt and check the valve timing as follows:

  a. If reusing the old timing belt, align the matchmarks that you made previously, and install the timing belt onto the camshaft pulley.

  b. Align the marks on the timing belt with the marks on the camshaft pulley.

  c. Loosen the No. 1 idler pulley set bolt ½ turn.

  d. Turn the crankshaft pulley two complete revolutions TDC to TDC. ALWAYS turn the crankshaft CLOCKWISE. Check that the pulleys are still in alignment with the timing marks.

  e. If the No. 1 idler pulley uses a green tension spring, slowly turn the crankshaft pulley 1⅞ revolutions, and align its groove with the mark at 45° BTDC (for the No. 1 cylinder) of the No. 1 timing belt cover.

  f. Tighten the No. 1 idler pulley set bolt to 31 ft. lbs. (42 Nm).

  g. Make sure there is belt tension between the crankshaft and camshaft timing pulleys.

37. Install the upper (No. 2) timing cover with a new gasket(s). Align the two clamps for the engine wiring harness with the cover mounting bolts.

38. Install the spark plugs.

39. Install the right mounting insulator bracket. Tighten the bracket bolts to 38 ft. lbs. (52 Nm).

40. Install the engine moving control rod and tighten the bolts to 47 ft. lbs. (64 Nm).

NOTE: Tighten the bracket bolts before tightening the control rod through-bolt.

41. Connect the connector to the ground wire on the right fender apron.

42. Lower the engine.

43. Install the alternator and alternator bracket.

44. Install the power steering drive belt and adjust the tension.

45. Install the fender apron seal and right engine undercover.

46. Install the right front wheel.

47. Start the engine and check the timing.

48. Connect the negative battery cable.

MR2

1. Disconnect the negative battery cable. On vehicles equipped with an air bag, wait at least 90 seconds before proceeding.

2. Remove the undercovers from the vehicle.

3. Remove the right-hand engine hood side panel.

4. If equipped with cruise control, remove the cruise control actuator and accelerator linkage.

5. Disconnect the ground strap connector.

6. Disconnect the brake booster vacuum hose.

7. Remove the A/C drive belt.

8. Remove the alternator drive belt.

9. Raise and safely support the rear of the vehicle.

10. Remove the right rear wheel.

11. Slightly jack up the engine. Use a block of wood between the jack and the oil pan to prevent damage.

12. Remove the right-hand engine mounting stay.

13. Remove the right-hand engine mounting insulator.

14. Remove the right-hand engine mounting bracket.

  a. Raise the engine as far as it will go, and remove the mounting bracket.

15. Remove the spark plugs.

16. Remove the No. 2 timing belt cover.

17. Set the No.1 cylinder to TDC compression stroke.

    a. Turn the crankshaft pulley and align its groove with the timing mark **0** of the No. 1 timing belt cover.

    b. Loosen the mounting bolt of the No. 1 idler pulley and shift the pulley toward the left as far as it will go, temporarily tighten it.

18. If re-using the old timing belt, matchmark the belt to the camshaft pulley. Place a matchmark on the timing belt to match the end of the No. 1 timing belt cover.

19. Remove the timing belt from the camshaft timing pulley.

    a. Loosen the mounting bolt of the No. 1 idler pulley and shift the pulley toward the left as far as it will go, and temporarily tighten it.

    b. Remove the timing belt from the camshaft timing pulley.

20. Using SST 09249–63010 and 09960–10010, or equivalent, remove the camshaft timing pulley.

21. Remove the crankshaft pulley.

22. Remove the No. 1 timing belt cover.

23. Remove the timing belt guide.

24. Remove the timing belt. If re-using the timing belt, draw a directional arrow on the timing belt in the direction of the engine revolution.

25. Remove the No.1 idler pulley and tension spring.

26. Remove the No. 2 idler pulley.

27. Remove the crankshaft timing pulley.

28. Using SST 09960–10010, or equivalent, remove the oil pump pulley.

To install:

29. Install the oil pump pulley and torque the nut to 18 ft. lbs. (24 Nm).

30. Install the crankshaft timing pulley as follows:

    a. Align the timing pulley set key with the key groove of the pulley.

    b. Slide on the timing pulley with the flange side facing inward.

31. Install the No. 2 idler pulley and torque the bolt to 31 ft. lbs. (42 Nm). Make sure that the pulley moves smoothly.

32. Temporarily install the No. 1 idler pulley and tension spring as follows:

    a. Align the bracket pin hole with the pivot pin. Install the tension spring.

    b. Pry the pulley toward the left as far as it will go and tighten the bolt.

    c. Verify that the pulley moves smoothly.

33. Temporarily install the timing belt as follows:

    a. Using the crankshaft pulley bolt, turn the crankshaft and align the timing marks of the crankshaft timing pulley and the oil pump body.

    b. Install the timing belt on the crankshaft timing pulley, oil pump pulley, No. 1 idler pulley, water pump pulley, and the No. 2 idler pulley.

34. If re-using the timing belt, align matchmarks.

35. Install the timing belt guide with the cup side facing outward.

36. Install the No.1 timing belt cover.

37. Install the crankshaft pulley as follows:

    a. Align the pulley set key with the key groove of the pulley, and slide on the pulley. Torque the bolt to 80 ft. lbs. (108 Nm).

38. Install camshaft timing pulley as follows:

    a. Align the camshaft knock pin with the knock pin groove of the pulley, and slide on the timing pulley.

    b. Install the pulley bolt and torque to 40 ft. lbs. (54 Nm).

39. Set No. 1 cylinder to TDC, compression stroke.

    a. Turn the crankshaft pulley and align its groove with the timing mark **0** of the No. 1 timing belt cover.

    b. Using SST 09960–10010, or equivalent, turn the camshaft and align the hole of the camshaft timing pulley with the timing mark of the bearing cap.

40. Finish installing the timing belt.

41. If re-using the timing belt, check that the matchmark on the timing belt matches the end of the No. 1 timing belt cover.

42. Check the valve timing.

    a. Loosen the No. 1 idler pulley bolt ½ turn.

    b. Slowly turn the crankshaft pulley clockwise two revolutions TDC to TDC.

    c. Check that each pulley aligns with the timing marks. If the marks do not align, remove the timing belt and reinstall it.

    d. Slowly turn the crankshaft pulley 1⅞ revolutions, and align its groove with the mark at 45° BTDC on the No. 1 timing belt cover. Always turn the crankshaft clockwise.

    e. Torque the mounting bolt of the No. 1 idler pulley to 31 ft. lbs. (42 Nm).

43. Install the No. 2 timing belt cover.

    a. Install the gasket to the No. 1 timing belt cover.

    b. Disconnect the engine wire protector between the No. 3 timing belt cover and the cylinder head cover.

    c. Install the belt cover.

    d. Install the two clamps of the engine wire protector to each bolt.

44. Install the spark plugs.

45. Install the right-hand engine mounting bracket. Torque the bolts to 45 ft. lbs. (61 Nm).

46. Install the right-hand engine mounting insulator. Torque nut to 38 ft. lbs. (52 Nm) and through-bolt to 58 ft. lbs. (78 Nm).

47. Install right-hand engine mounting stay. Torque bolt to 54 ft. lbs. (73 Nm).

48. Install the right rear tire.

49. Safely lower the vehicle.

50. Install the alternator drive belt.

51. Install the A/C drive belt.

52. Connect the ground strap connector.

53. Connect the brake booster hose.

54. If removed, install the cruise control actuator and accelerator linkage.

55. Install the right-hand engine hood side panel.

56. Install the engine undercovers.

57. Connect the negative battery cable.

58. Start the engine and check the ignition timing.

3.0L (3VZ-FE) Engine

Camry

1. Disconnect the cable from the negative battery terminal. On vehicles equipped with an air bag, wait at least 90 seconds before proceeding.

2. Remove the power steering pump reservoir and position it out of the way. Remove the right fender apron seal and then remove the alternator and power steering belts.

3. Remove the coolant reservoir hose, the washer tank and then the coolant overflow tank.

4. Remove the right side engine mount stays.

5. Position a piece of wood on a floor jack and then slide the jack under the oil pan. Raise the jack slightly until the pressure is off the engine mounts.

6. Remove the engine control rod.

7. Remove the spark plugs.

8. Remove the right side engine mounting bracket.

9. Remove the eight bolts and lift off the upper (No. 2) cover.

10. If reusing the belt paint matchmarks on the timing belt at all points where it meshes with the pulleys and the lower timing cover.

11. Set the No. 1 cylinder to TDC of the compression stroke and check that the timing marks on the camshaft timing pulleys are aligned with those on the No. 3 timing cover. If not, turn the engine one complete revolution (360°) and check again.

12. Remove the timing belt tensioner and the dust boot.

13. Turn the right camshaft pulley clockwise slightly to release tension and then remove the timing belt from the pulleys.

14. Use a spanner wrench to hold the pulley, loosen the set bolt and then remove the camshaft timing pulleys along with the knock pin. Be sure to keep track of which is which.

15. Remove the No. 2 idler pulley.

16. Remove the crankshaft pulley and then pull off the lower (No. 1) timing belt cover.

17. Remove the belt guide and remove the timing belt.

To install:

18. Install the belt on the sprocket and replace the belt guide.

19. Install the lower (No. 1) timing cover and tighten the bolts.

20. Align the crankshaft pulley set key with the key groove on the pulley and slide the pulley on. Tighten the bolt to 181 ft. lbs. (245 Nm).

21. Install the No. 2 idler pulley and tighten the bolt to 29 ft. lbs. (39 Nm). Check that the pulley moves smoothly.

22. Install the left camshaft pulley with the flange side outward. Align the knock pin hole in the camshaft with the knock pin groove on the pulley and then install the pin. Tighten the bolt to 80 ft. lbs. (108 Nm).

23. Set the No. 1 cylinder to TDC again. Turn the right camshaft until the knock pin hole is aligned with the timing mark on the No. 3 belt cover. Turn the left pulley until the marks on the pulley are aligned with the mark on the No. 3 timing cover.

24. Check that the mark on the belt matches with the edge of the lower cover. If not, shift it on the crank pulley until it does. Turn the left pulley clockwise a bit and align the mark on the timing belt with the timing mark on the pulley. Slide the belt over the left pulley. Now move the pulley until the marks on it align with the one on the No. 3 cover. There should be tension on the belt between the crank-

shaft pulley and the left camshaft pulley.

25. Align the installation mark on the timing belt with the mark on the right side camshaft pulley. Hang the belt over the pulley with the flange facing inward. Align the timing marks on the right pulley with the one on the No. 3 cover and slide the pulley onto the end of the camshaft. Move the pulley until the camshaft knock pin hole is aligned with the groove in the pulley and then install the knock pin. Tighten the bolt to 55 ft. lbs. (75 Nm).

26. Position a plate washer between the timing belt tensioner and the a block and then press in the pushrod until the holes are aligned between it and the housing. Slide a 1⁄2 in. Allen wrench through the hole to keep the pushrod set. Install the dust boot and then install the tensioner. Tighten the bolts to 20 ft. lbs. (26 Nm). Don't forget to pull out the Allen wrench!

27. Turn the crankshaft **clockwise** two complete revolutions and check that all marks are still in alignment. If they aren't, remove the timing belt and start over again.

28. Install the right engine mount bracket and tighten it to 30 ft. lbs. (39 Nm).

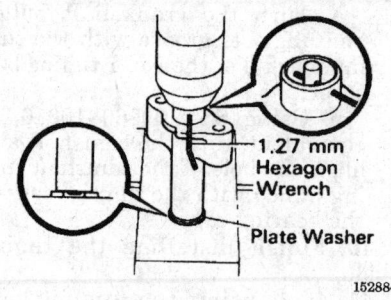

1.27 mm
Hexagon
Wrench

Plate Washer

152884

Setting the belt tensioner — 3VZ-FE engine

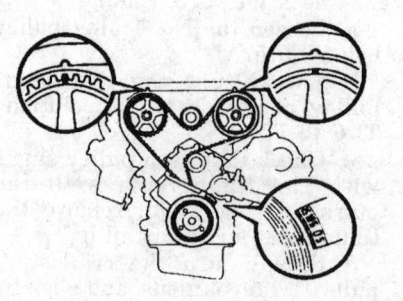

152885

Timing marks aligned with the belt installed — 3VZ-FE engine

29. Position a new gasket and then install the upper (No. 2) timing cover. refer to the illustration for bolt positioning.

30. Install the spark plugs.

31. Install the control rod and tighten the bolts to 47 ft. lbs. (64 Nm). Install the right stay and tighten it to 23 ft. lbs. (31 Nm).

32. Install and adjust the drive belts.

33. Install the fender apron seal and the wheel.

34. Install the No. 2 stay and tighten the bolt to 55 ft. lbs. (75 Nm), the nut to 46 ft. lbs. (62 Nm). Install the No. 3 stay and tighten it to 54 ft. lbs. (73 Nm).

35. Install the coolant overflow tank and the washer tank.

36. Install the power steering reservoir tank and the cruise control actuator.

37. Connect the battery cable, start the car and check for any leaks.

2.0L (3S-GTE) Engine

Celica

1. Disconnect the negative battery cable at the battery. On vehicles equipped with an air bag, wait at least 90 seconds before proceeding.

2. Raise the front of the vehicle and support it with safety stands. Remove the right front tire.

3. Remove the right side fender liner.

4. Remove the windshield washer and radiator reservoir tanks.

5. Remove the cruise control actuator, if so equipped. Remove the intercooler, throttle body, EGR vacuum modulator and vacuum solenoid valve, if necessary for clearance, using proper procedures.

6. Remove the power steering drive belt and then remove the pump itself. Position the pump out of the way with the hydraulic lines still attached.

7. Remove the alternator and its support bracket.

8. Jack up the engine slightly to remove the weight from the right engine mounting. Remove the right engine mounting from the vehicle.

9. Remove the No. 2 timing belt cover.

10. Set the No. 1 cylinder to TDC of the compression stroke by aligning the groove on the crank pulley with the "0" mark on the No. 1 timing belt cover. Check that the matchmarks on the two camshaft timing pulleys and the rear timing belt cover are aligned; if not, turn the crankshaft one complete revolution clockwise.

11. If the timing belt is to be re-used, draw a directional arrow on it and matchmark the belt to the two camshaft pulleys. Loosen the No. 1 idler pulley bolt and shift the pulley as far left as possible; tighten the set bolt. Remove the timing belt from the two camshaft pulleys.

12. Remove the crankshaft pulley and the lower timing belt cover.

13. Remove the timing belt guide and then remove the timing belt from the remaining pulleys. Be sure to have matchmarked the belt to the pulleys if it is to be reused.

14. Remove the No. 1 idler pulley and the tension spring, the No. 2 idler pulley and the oil pump pulley, if they are being replaced.

To install:

15. Install the oil pump pulley and tighten it to 26 ft. lbs. (35 Nm). Install the No. 2 idler pulley and tighten it to 32 ft. lbs. (43 Nm).

16. Install the No. 1 idler pulley, applying a thread lock compound on the threads of the pivot bolt. Install the plate washer and pulley with the pivot bolt, tightening it to 38 ft. lbs. (52 Nm). Check that the bracket moves freely.

17. Install the timing belt on all pulleys except the two camshaft pulleys. Make sure the matchmarks made earlier are in alignment, if re-using the old belt.

18. Install the timing belt guide with the cup side out.

19. Install the lower timing belt cover and then install the crankshaft pulley. Tighten it to 80 ft. lbs. (108 Nm).

20. Check that the No. 1 cylinder is at TDC of the compression stroke for the crankshaft. The crankshaft pulley groove should be aligned with the **0** mark on the lower timing belt cover.

21. Check that the camshaft pulley marks align with the marks on the valve cover. Install the timing belt on the two camshaft timing pulleys.

22. Set the timing belt tensioner by aligning the holes in the tensioner pushrod with the holes in the housing.

23. Using a press, slowly press the pushrod into the housing until the holes are exactly aligned.

24. Insert a 1.27mm hex wrench through the holes to retain the position of the pushrod.

25. Release the press.

26. Using a torque wrench, turn the No. 1 idler pulley bolt counterclockwise to 13 ft. lbs. (18 Nm), then temporarily install the timing belt tensioner with the 2 mounting bolts.

27. Slowly turn the crankshaft pulley 5/6 of a revolution and align the groove in the pulley with the 60° ATDC mark on the timing belt cover. Always turn the crankshaft in the clockwise direction.

28. Insert a 0.75 in. (1.90mm) feeler gauge between the tensioner body and the No. 1 idler pulley stopper.

29. Using a suitable torque wrench, turn the No. 1 idler pulley bolt counterclockwise to 13 ft. lbs. (18 Nm). Push on the tensioner and torque the tensioner retaining bolts to 15 ft. lbs. (21 Nm).

30. Remove the 1.27mm hex wrench from the tensioner.

31. Slowly turn the crankshaft pulley in the clockwise direction 1 revolution and align the groove in the pulley with the 60° ATDC mark on the timing belt cover. Always turn the crankshaft in the clockwise direction.

32. Turn the No. 1 idler pulley bolt counterclockwise to 13 ft. lbs. (18 Nm), as done previously.

33. Using a feeler gauge, check the clearance between the tensioner body and the No. 1 idler pulley stopper. The clearance should be 0.071–0.087 in. (1.8–2.2mm). If the clearance is not as specified, remove the tensioner and reinstall it.

34. Check for proper valve timing by rotating the crankshaft clockwise 2 revolutions from TDC to TDC. Check that each timing mark aligns.

35. Installation of the remaining components is in the reverse order of removal. Tighten the engine mount bracket bolts to 38 ft. lbs. (52 Nm). Tighten the engine mount bolt to 58 ft. lbs. (78 Nm) and the nuts to 38 ft. lbs. (52 Nm).

36. Check all fluid levels. Adjust all drive belts. Perform all necessary engine adjustments. Road test the vehicle for proper operation.

MR2

1. Disconnect the negative battery cable. On vehicles equipped with an air bag, wait at least 90 seconds before proceeding.

2. Remove the right-hand engine hood side panel.

3. Remove the upper suspension brace.

4. Remove the No. 1 and No. 2 intake air connectors.

5. If the vehicle has cruise control, remove the cruise control actuator and the accelerator linkage.

6. If the vehicle is equipped without cruise control, disconnect the accelerator cable from the throttle body.

7. Remove the engine compartment electric cooling fan.

8. Disconnect the A/C compressor from the engine.

9. Remove the Charge Air Cooler (CAC).

10. Remove the right rear wheel.

11. Remove the alternator drive belt.

12. Jack up the engine by using a block of wood under the oil pan to prevent damage.

13. Disconnect the ground strap connector.

14. Disconnect the brake booster vacuum hose.

15. Remove the right engine mounting stay.

16. Remove the right engine mounting insulator.

17. Remove the right engine mounting bracket.

18. Remove the throttle body.

19. Remove the cylinder head cover.

20. Remove the spark plugs.

21. Remove the No. 2 timing belt cover.

22. Set the No. 1 cylinder to TDC of the compression stroke by aligning the groove on the crank pulley with the **0** mark on the No. 1 timing belt cover. Check that the matchmarks on the two camshaft timing pulleys and the rear timing belt cover are aligned; if not, turn the crankshaft one complete revolution clockwise.

23. If the timing belt is to be re-used, draw a directional arrow on it and matchmark the belt to the two camshaft pulleys. Loosen the No. 1 idler pulley bolt and shift the pulley as far left as possible; tighten the set bolt. Remove the timing belt from the two camshaft pulleys.

24. Remove the timing belt from the camshaft timing pulleys.

25. Remove the camshaft timing pulleys.

26. Remove the crankshaft pulley.

27. Remove the No. 1 timing belt cover.

28. Remove the timing belt guide.

29. Remove the timing belt. When re-using the timing belt, place matchmarks on the timing belt and the crankshaft timing pulley.

30. Remove the No. 1 idler pulley.

31. Remove the No. 2 idler pulley.

32. Remove the crankshaft timing pulley.

33. Using SST 09960–10010 or equivalent, remove the oil pump pulley.

To install:

34. Install the oil pump pulley and tighten it to 26 ft. lbs. (35 Nm).

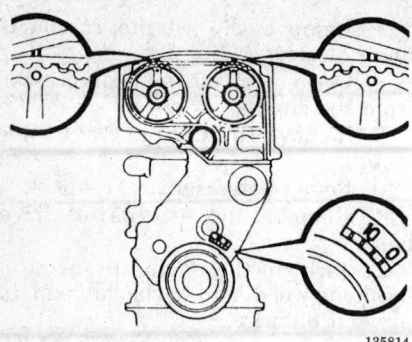

Making sure the crankshaft and camshaft matchmarks align — 3S-GTE engine

135814

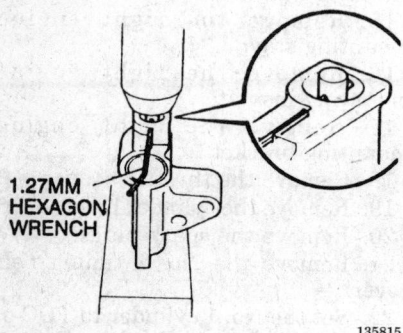

1.27MM HEXAGON WRENCH

Setting the timing belt tensioner with a press — 3S-GTE engine

135815

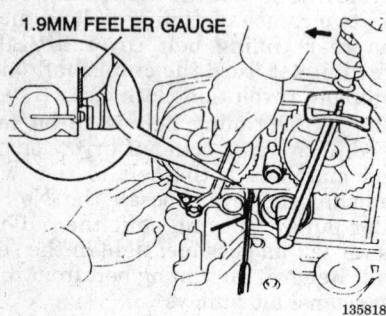

1.9MM FEELER GAUGE

Inserting a 0.075 in. (1.90mm) feeler gauge between the tensioner body and No. 1 idler pulley stopper — 3S-GTE engine

135818

35. Install the crankshaft timing pulley.

a. Align the timing pulley set key with the key groove of the pulley.

b. Slide on the timing with the flange side facing inward.

36. Install the No. 2 idler pulley and torque the bolt to 32 ft. lbs. (43 Nm). Make sure that the pulley moves smoothly.

37. Install the No. 1 idler pulley, applying a thread lock compound on the threads of the pivot bolt. Install the plate washer and pulley with the pivot bolt, tightening it to 38 ft. lbs.

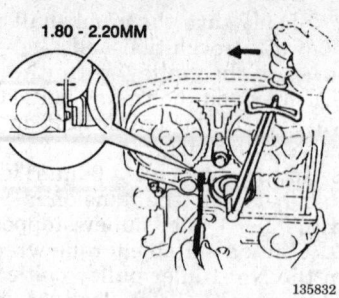

1.80 - 2.20MM

135832

Checking the clearance between the tensioner body and No. 1 idler pulley stopper with a feeler gauge after tightening the tensioner bolts — 3S-GTE engine

(52 Nm). Check that the pulley moves freely.

38. Temporarily install the timing belt. If reusing the old belt, make sure the matchmarks made earlier are in alignment.

39. Install the timing belt guide with the cup side facing out.

40. Install the lower timing belt cover.

41. Install the crankshaft pulley. Tighten it to 80 ft. lbs. (108 Nm).

42. Install the camshaft timing pulleys.

a. Align the groove of the camshaft with the dot mark of the No. 1 camshaft bearing cap.

b. Make sure the **S** mark on the camshaft pulleys is facing upward.

c. Align the pin holes of the camshaft and the timing pulley, and insert the knock pin.

d. Torque the bolts of the camshaft pulleys to 43 ft. lbs. (59 Nm).

43. Check that the No. 1 cylinder is at TDC of the compression stroke for the crankshaft. The crankshaft pulley groove should be aligned with the **0** mark on the lower timing belt cover.

44. Install the timing belt on the two camshaft timing pulleys. If re-using the timing belt align the matchmarks or the timing belt and the camshaft timing pulleys.

45. Set the timing belt tensioner by aligning the holes in the tensioner pushrod with the holes in the housing.

a. Using a press, slowly press the pushrod into the housing until the holes are exactly aligned.

b. Insert a 1.27mm hex wrench through the holes to retain the position of the pushrod.

c. Release the press.

46. Install the timing belt tensioner.

a. Using a torque wrench, turn the No. 1 idler pulley bolt counter-

clockwise to 13 ft. lbs. (18 Nm), then temporarily install the timing belt tensioner with the two mounting bolts.

b. Slowly turn the crankshaft pulley 5/6 of a revolution and align the groove in the pulley with the 60° ATDC mark on the timing belt cover. Always turn the crankshaft in the clockwise direction.

c. Insert a 0.075 in. (1.90mm) feeler gauge between the tensioner body and the No. 1 idler pulley stopper.

d. Using a suitable torque wrench, turn the No. 1 idler pulley bolt counterclockwise to 13 ft. lbs. (18 Nm). Push on the tensioner and torque the tensioner retaining bolts to 15 ft. lbs. (21 Nm).

e. Remove the 1.27mm hex wrench from the tensioner.

f. Remove the 1.90mm feeler gauge.

g. Slowly turn the crankshaft pulley in the clockwise direction one revolution and align the groove in the pulley with the 60° ATDC mark on the timing belt cover. Always turn the crankshaft in the clockwise direction.

h. Using a feeler gauge, check the specified clearance between the tensioner body and the No. 1 idler pulley stopper. Clearance should be 0.071–0.087 in. (1.80–2.20mm). If the clearance is not as specified, remove the tensioner and reinstall it.

47. Check for proper valve timing by rotating the crankshaft clockwise two revolutions from TDC to TDC. Check that each timing mark aligns.

48. Install the No. 2 timing belt cover.

49. Install the spark plugs.

50. Install the cylinder head cover, torque the screws to 4 ft. lbs. (6 Nm).

51. Install the throttle body.

52. Install the right engine mounting bracket. Torque the bolts to 45 ft. lbs. (61 Nm).

53. Install the right engine mounting insulator. Torque the nut to 38 ft. lbs. (52 Nm) and the through-bolt to 58 ft. lbs. (78 Nm).

54. Install the right engine mounting stay and torque the bolts to 54 ft. lbs. (73 Nm).

55. Connect the brake booster vacuum hose.

56. Connect the ground strap connector.

57. Install the alternator belt.

58. Install the right rear wheel.

59. Install the CAC.

60. Install the A/C compressor.

61. Install the engine compartment electric cooling fan.

62. Install the cruise control actuator and the accelerator linkage for cruise.

63. On vehicles without cruise control, connect the accelerator cable to the throttle body.

64. Install the No. 1 and No. 2 intake air connectors.

65. Install the upper suspension brace and torque the bolt to 54 ft. lbs. (73 Nm) and the nut to 47 ft. lbs. (64 Nm).

66. Install the right engine hood side panel.

67. Install the engine undercovers.

68. Check all fluid levels. Adjust all drive belts. Perform all necessary engine adjustments. Road test the vehicle for proper operation.

69. Check the ignition timing.

3.0L (2JZ-GTE and 2JZ-GE) Engines

Supra

1. Disconnect the negative battery cable. Wait at least 90 seconds before performing any other work.

2. Remove the battery and battery tray.

3. Remove the undercovers from the vehicle.

4. Drain the engine coolant and remove the radiator.

5. For vehicles with manual transmissions, remove the drive belt tensioner damper by removing the two nuts.

6. Remove the drive belt.

7. Remove the fan, the fluid coupling, and the water pump pulley.

8. Remove the upper two timing belt covers (Nos. 2 and 3).

9. Remove the drive belt tensioner.

10. Set No. 1 cylinder to TDC/compression. Turn the crankshaft pulley clockwise to align the groove with the **0** mark on the lower (No. 1) timing belt cover. Check that the timing marks on the camshaft pulleys are aligned with the marks on the rear belt cover. If the camshaft marks do not align, turn the crankshaft another 360 degrees.

11. Alternately loosen the two bolts holding the timing belt tensioner. Remove the bolts and remove the tensioner.

12. Remove the timing belt from the camshaft pulleys. If the belt is to be reused, place matchmarks on the belt and gears before removing the belt. Mark the belt with an arrow to show direction of rotation.

13. Using SST 09960-10010 or equivalent, remove the bolts for the camshaft timing gears.

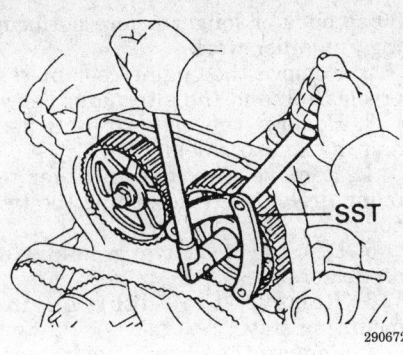

Removing the camshaft gears — 2JZ-GE and 2JZ-GTE engines

14. Remove the camshaft gears from the engine.

15. If necessary, disconnect the oil cooler tubes from the front of the engine by removing the two bolts and hose clamps.

16. Remove the crankshaft pulley by using tool 09330-0021, or equivalent, to hold the pulley and using tool 09213-70010, or equivalent, to remove the pulley bolt.

17. Remove the lower (No. 1) timing belt cover and the timing belt guide.

18. Remove the timing belt. If the belt is to be reused, protect it from contact with oil, grease, or fluids.

19. Remove the idler pulley by removing the pivot bolt and plate washer.

20. Remove the crankshaft timing gear as follows:

  a. Remove the bolt and timing belt plate.

  b. Remove the crankshaft timing pulley. It may be necessary to use a puller to remove the timing gear.

To install:

21. Install the crankshaft timing gear with the flange side inward. Install the timing belt plate with the bolt. Torque the bolt to 69 inch lbs. (8 Nm).

22. Install the idler pulley if it was removed. Place 2–3 drops of thread sealant (Loctite® 242, or equivalent)

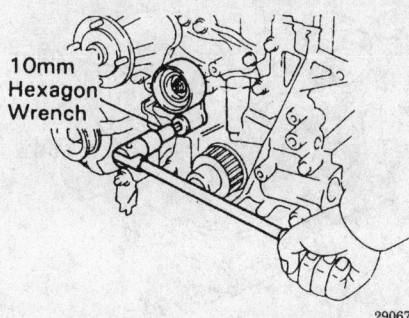

Timing belt idler pulley — 2JZ-GE and 2JZ-GTE engines

on the pivot bolt and tighten the bolt to 25 ft. lbs. (34 Nm).

23. Use the crankshaft pulley bolt to turn the crankshaft (clockwise) until the mark on the gear aligns with the oil pump body. Check all the pulleys and gears for cleanliness; remove any grease, oil or coolant. Install the timing belt onto the crankshaft gear and idler pulleys.

24. Install the timing belt guide with the cupped side facing outward.

25. Install the No. 1 timing belt cover.

26. Install the crankshaft pulley. Align the set key with the groove. Hold the pulley with the proper tool and tighten the pulley bolt to 239 ft. lbs. (324 Nm).

27. If equipped with automatic transmission, connect the oil cooler tubes with the clamps and two bolts.

28. Install the camshaft gears as follows:

  a. Align the camshaft knock pin with the groove on the gear and slide on the timing gear.

  b. Temporarily install the timing gear bolt.

  c. Using the same tools as removal, torque the camshaft gear bolts to 59 ft. lbs. (79 Nm).

  d. Turn the crankshaft pulley and align its groove with the timing mark, **0** on the No. 1 timing belt cover.

  e. Align the timing marks on the camshaft timing gears and the No. 4 timing belt cover.

29. Finish installing the timing belt.

30. Double check that all the timing marks for the crankshaft pulley and the camshaft gears are aligned as they were during disassembly.

31. Set the timing belt tensioner:

  a. Use a press to slowly push in the pushrod on the tensioner. This will require between 220 and 2200 pounds of pressure.

  b. Align the holes of the pushrod and housing. Place a 1.5mm hex wrench through the holes to keep the pushrod retracted.

  c. Release the press and install the dust boot onto the tensioner.

32. Install the tensioner; alternately tighten the bolts to 20 ft. lbs. (26 Nm).

33. Remove the hex wrench from the tensioner with a pair of pliers.

34. Turn the crankshaft pulley two full turns clockwise. Check that each pulley's timing marks align correctly after the two turns. If any mark does not align, remove the timing belt and reinstall it.

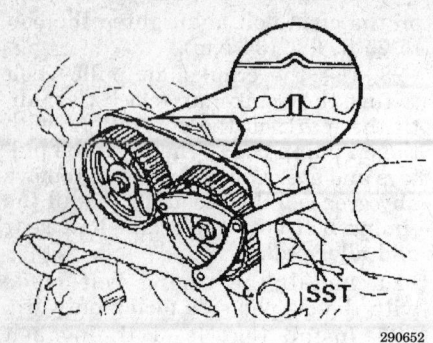

Aligning the camshaft pulleys — 2JZ-GE and 2JZ-GTE engines

290652

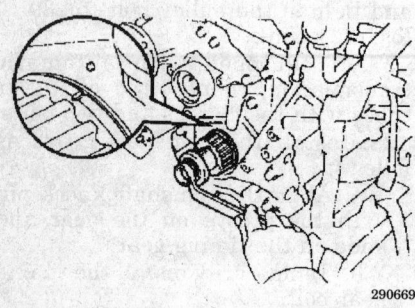

Aligning the crankshaft pulley timing mark — 2JZ-GE and 2JZ-GTE engines

290669

35. Install the drive belt tensioner and tighten the bolts to 15 ft. lbs. (21 Nm).

36. Install the Nos. 2 and 3 timing belt covers.

37. Install the water pump pulley, the fan with fluid coupling, and the drive belt.

38. If equipped with manual transmission, install the drive belt tensioner damper and torque the nuts to 14 ft. lbs. (20 Nm).

39. Install the radiator and connect the hoses and lines.

40. Install the battery tray and battery. Connect the battery cables.

41. Fill the coolant. Start the engine and check for leaks.

42. Bleed the cooling system and check the ignition timing.

43. Check the fluid levels, including the automatic transmission if so equipped.

44. Install the engine undercover. Road test the vehicle and recheck the coolant level.

3.0L (1MZ-FE) Engine

Camry and Avalon

1. With the ignition switch in the **LOCK** position, disconnect the negative battery terminal. If equipped with an air bag system, wait at least 30 seconds or longer before performing any other work.

2. Remove the engine coolant reservoir tank and the alternator belt.

3. Remove the right front wheel and the splash shield.

4. Remove the power steering pump drive belt by loosening the two bolts.

5. Disconnect the two ground wire connectors.

6. Remove the right engine mounting stay.

7. Remove the engine moving control rod and the No. 2 right engine mounting bracket.

8. Remove the No. 2 alternator bracket.

9. Using SST tool 09213–54015 and 09330–00021 or equivalent, remove the crankshaft pulley bolt.

10. Using SST tool 09950–50010 or equivalent (crankshaft pulley remover), remove the crankshaft pulley.

11. Remove the lower timing belt cover by removing four bolts.

12. Remove the No. 2 timing belt cover as follows:

   a. Remove the bolt and disconnect the engine wire protector from the No. 3 (rear) timing belt cover.

   b. Disconnect the engine wire protector clamp from the No. 3 timing belt cover.

   c. Remove the five bolts from the No. 2 timing belt cover.

   d. Remove the No. 2 cover from the engine.

13. Remove the right engine mounting bracket by removing the nut and two bolts.

14. Remove the crankshaft timing belt guide.

15. Temporarily install the crankshaft pulley bolt.

16. Turn the crankshaft and align the crankshaft timing pulley groove with the oil pump alignment mark. Always turn the engine clockwise.

17. Ensure the timing mark of the camshaft timing pulleys and rear

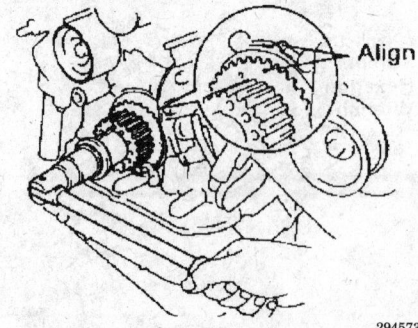

Aligning the crankshaft pulley mark to the oil pump mark — 1MZ-FE engine

294573

timing belt cover are aligned. If not, turn the engine over an additional 360 degrees.

18. Remove the crankshaft pulley bolt.

NOTE: If the belt is to be reused, align the installation marks on the belt to the marks on the pulleys. If the marks have worn off, make new ones.

19. Alternately loosen the two timing belt tensioner bolts. Remove the tensioner and dust boot.

20. Remove the timing belt.

To install:

21. Remove any oil or water from the pulleys.

22. Align the front mark of the timing belt with the dot mark of the crankshaft timing pulley.

23. Align the installation marks on the timing belt with the timing marks of the camshaft timing pulleys.

24. Install the timing belt in the following order:

   a. Crankshaft pulley
   b. Water pump pulley
   c. Left camshaft pulley
   d. No. 2 idler pulley
   e. Right camshaft pulley
   f. No. 1 idler pulley

25. Using a press, slowly press the timing belt tensioner until the holes of the pushrod and housing align. Insert a 1.27mm hexagon Allen wrench through the holes to preserve the setting position.

26. Install the dust boot to the tensioner.

27. Install the tensioner with the two bolts. Alternately tighten and then torque the bolts to 20 ft. lbs. (27 Nm). Remove the Allen wrench.

28. Turn the crankshaft clockwise and align the crankshaft timing pulley groove with the oil pump alignment mark.

29. Ensure the camshaft timing marks align with the timing marks on the rear timing belt cover.

30. Install the timing belt guide.

31. Install the right engine mounting bracket and torque the bolts to 21 ft. lbs. (28 Nm).

32. Install the upper timing belt cover with the five bolts. Torque the bolts to 74 inch lbs. (8 Nm).

33. Install the engine wire protector clamp to the No. 3 timing belt cover.

34. Install the engine wire protector to the No. 3 timing belt cover with the bolt.

35. Install the lower timing belt cover by installing the four bolts. Torque the bolts to 74 inch lbs. (8 Nm).

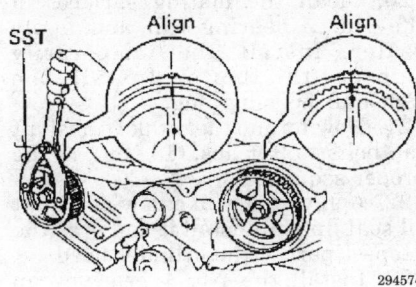

Aligning the camshaft marks — 1MZ-FE engine

294574

36. Install the crankshaft pulley. Torque the bolt to 159 ft. lbs. (215 Nm).

37. Install the No. 2 alternator bracket. Torque the nut to 21 ft. lbs. (28 Nm). Do not tighten the pivot bolt at this time.

38. Install the No. 2 right engine mounting bracket and the moving control rod.

39. Install the right engine mounting stay.

40. Connect the two ground wire connectors.

41. Install and adjust the drive belts.

42. Install the coolant reservoir.

43. Install the right front splash shield and wheel assembly.

44. Connect the negative battery cable.

45. Start the vehicle and check for any leaks.

46. Recheck the ignition timing.

Camshaft

REMOVAL AND INSTALLATION

1.5L (3E-E) Engine

Tercel

1. Disconnect the negative battery cable. On vehicles equipped with an air bag, wait at least 90 seconds before proceeding.

2. Remove the valve cover.

3. Drain the cooling system.

4. Loosen the water pump pulley bolts.

5. Remove the alternator belt.

6. Raise the vehicle and safely support it on jackstands.

7. Remove the power steering pivot bolt, if so equipped.

8. Remove the bolt which goes through both the upper and lower timing belt covers.

9. Lower the vehicle to the ground. Remove the power steering pump belt if so equipped.

10. Remove the water pump pulley.

11. Disconnect the upper radiator hose at the engine water outlet.

12. Label and disconnect vacuum hoses.

13. Remove the upper timing belt cover and its gasket.

14. At the distributor, label and disconnect the spark plug wires, the vacuum hoses and the electrical connections. Remove the distributor.

15. Disconnect the hoses at the fuel pump and remove the fuel pump.

16. Remove the distributor gear bolt.

17. Remove the rocker arm assembly.

18. Rotate the crankshaft clockwise and set the engine to TDC/compression on No. 1 cylinder. Make sure the rockers for cylinder No. 1 are loose. If not, rotate the crankshaft one full turn.

19. Matchmark and remove the timing belt from the camshaft pulley. Support the belt so that it doesn't change position on the crankshaft pulley.

20. Loosen the camshaft bearing cap bolts a little at a time and in the reverse order of the installation sequence. After removal, label each cap.

21. Remove the camshaft oil seal (at the pulley end) and then remove the camshaft by lifting it straight out of its bearings.

To install:

22. Coat all the bearing journals with engine oil and place the camshaft in the cylinder head.

23. Place bearing caps Nos. 2, 3 and 4 on each journal with the arrows pointing towards the front of the engine.

24. Apply multi-purpose grease to the inside of the oil seal. and apply liquid sealer on the outer circumference of the oil seal.

25. Install the seal in position, being very careful to get it straight. Do not allow the seal to cock or move out of place during installation.

26. Install the bearing cap for bearing No.1 and apply sealant to the lower ends of the of the seal.

27. Tighten each bearing cap little at a time and in the correct sequence. Tighten the bolts to 9 ft. lbs. (12 Nm).

28. Hold the camshaft with an adjustable wrench and tighten the drive gear to 22 ft. lbs. (30 Nm).

29. Install the timing belt.

30. Install the rocker arm assembly.

31. Install the fuel pump with a new gasket and connect the hoses.

32. Replace the distributor and connect its vacuum hoses, electrical connections and the spark plug wires.

33. Install the upper timing belt cover but don't install the bolt which holds both the upper and the lower covers yet.

34. Connect the vacuum hoses.

35. Connect the upper radiator hose to the water outlet.

36. Install the water pump pulley and tighten the bolts finger tight.

37. Install the power steering drive belt if so equipped.

38. Elevate the vehicle and safely support it on jackstands.

39. Install the bolt which connects the upper and lower timing belt covers.

40. Install the power steering pump pivot bolt and the belt; adjust the belt tension.

41. Lower the vehicle. Install the alternator belt and adjust it to the correct tension.

42. Tighten the water pump pulley bolts.

43. Refill the cooling system.

44. Install the valve cover with a new gasket.

1.5L (5E-FE) Engine

Tercel and Paseo

1. Disconnect the negative battery cable from the battery. On vehicles equipped with an air bag, wait at least 90 seconds before proceeding.

2. Remove the valve cover.

3. Remove the timing belt assembly.

4. Remove the camshaft timing sprocket.

NOTE: Due to the relatively small amount of camshaft thrust clearance, the camshaft must be kept level during removal. If the camshaft is not level on removal, the portion of the head receiving the thrust may crack or be damaged.

5. Set the intake camshaft so the service bolt holes of the intake camshaft gears are directly up.

6. Remove each front bearing cap of the intake and exhaust camshafts.

7. Secure the intake camshaft subgear to the main gear with a 6mm diameter bolt, 16–20mm long and with a pitch of 1.0mm Make sure that the torsional spring force of the subgear has been eliminated by the above operation.

8. In correct sequence, loosen and remove the 8 bolts of the 4 bearing caps of the exhaust camshaft and re-

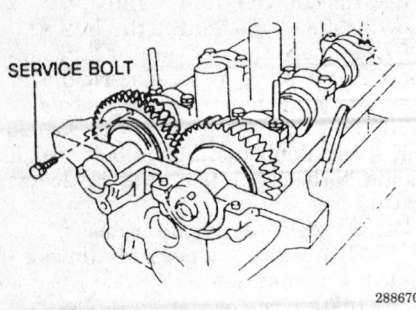

Securing intake camshaft sub-gear to main gear — 5E-FE engine

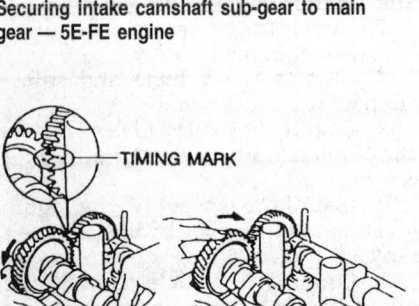

Exhaust and intake camshaft gear timing marks — 5E-FE engine

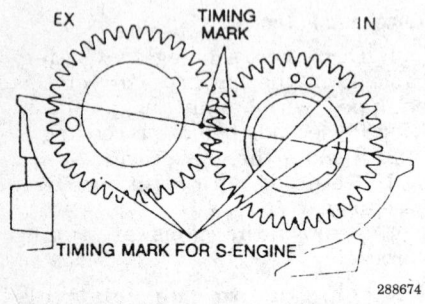

Correct exhaust and intake camshaft gear timing marks — 5E-FE engine

move the exhaust camshaft. If the camshaft is not being lifted out straight, reinstall the middle bearing cap and loosen it evenly to keep the camshaft straight.

9. In the reverse order of the installation sequence, loosen and remove the 8 bolts of the 4 bearing caps of the intake camshaft and remove the intake camshaft. If the camshaft is not being lifted out straight, reinstall the middle bearing cap and loosen it evenly to keep the camshaft straight.

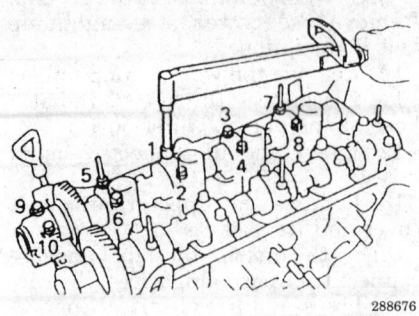

Intake bearing cap torque sequence — 5E-FE engine

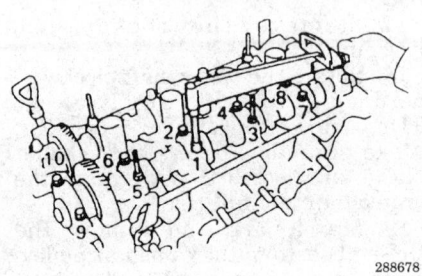

Exhaust bearing cap torque sequence — 5E-FE engine

10. Using a small screwdriver and a magnetic finger, remove the adjusting shim.

To install:

11. Install the valve shim to the engine.
12. Apply engine oil to the surface of the intake camshaft.
13. Place the intake camshaft on the cylinder head so the service bolt points directly up.
14. Install the 4 rearward bearing caps in their original order and temporarily tighten them evenly. Do not torque the bolts at this time.
15. Apply engine oil to the portion of the exhaust camshaft.
16. Engage the exhaust camshaft gear to the intake camshaft gear by matching the proper timing marks on each gear.
17. Roll down the exhaust camshaft onto the bearing journals while engaging the gears with each other.

NOTE: There are other marks present for the "S" engine. Do not use these marks.

18. Install the 4 rearward intake bearing caps in their original order and tighten evenly. Do not torque the bolts at this time.

19. Remove the service bolt.
20. Clean the mating surfaces of the No. 2 bearing cap and apply sealer. Install and temporarily tighten the bolts. Install the camshaft housing plug.
21. Now torque the intake camshaft cap bolts to 9 ft. lbs. (13 Nm), in the proper sequence.
22. Apply grease to a new camshaft oil seal lip and install it as far as the deepest part of the cylinder head.
23. Install the No. 1 bearing cap and temporarily tighten the bolts.
24. Now torque the exhaust camshaft bearing cap bolts to 9 ft. lbs. (13 Nm) evenly and in the proper sequence.
25. Turn the camshaft 1 revolution and check that the timing marks of the camshaft gears are aligned.
26. Check and adjust the valve clearance. Install the valve cover.
27. Install the camshaft timing sprocket and torque the bolt to 37 ft. lbs. (50 Nm).
28. Install the timing belt. start the engine and check for leaks.
29. Connect the negative battery cable to the battery.

1.6L (4A-FE) and 1.8L (7A-FE) Engines

Corolla and Celica

1. Disconnect the negative battery cable. On vehicles equipped with an air bag, wait at least 90 seconds before proceeding.
2. Disconnect the spark plug wires from the spark plugs. Be sure to make note of the proper firing order for easier installation.
3. Remove the valve cover.
4. Remove the timing belt covers.
5. Remove the timing belt and idler pulley.
6. Set the exhaust camshaft so that the knock pin is slightly above the cylinder head. This angle allows the No. 1 and No. 3 cylinder cam lobes of the intake camshaft to push their valve lifters evenly.
7. Remove the two bolts and the front bearing cap of the intake camshaft.
8. Secure the intake camshaft end gear to the sub-gear with a service bolt. The service bolt should match the following specifications:
   - Thread diameter: 6.0mm
   - Thread pitch: 1.0mm
   - Bolt length: 16mm
9. Uniformly loosen each intake camshaft bearing cap bolt in several passes in the reverse order of the installation sequence.

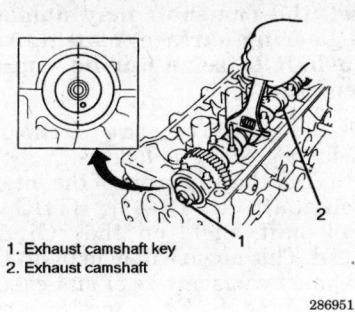

1. Exhaust camshaft key
2. Exhaust camshaft

286951

Positioning the exhaust camshaft for removal — Corolla and Celica with 4A-FE and 7A-FE engines

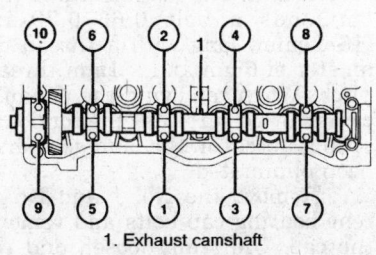

1. Exhaust camshaft

286948

Exhaust camshaft bearing cap tightening sequence — Corolla and Celica with 4A-FE and 7A-FE engines

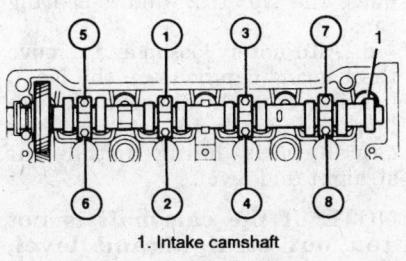

1. Intake camshaft

286946

Intake camshaft bearing cap tightening sequence — Corolla and Celica with 4A-FE and 7A-FE engines

———————— **WARNING** ————————

The camshaft must be held level while it is being removed. If the camshaft is not kept level, the portion of the cylinder head receiving the thrust may crack or become damaged. In turn, this could cause the camshaft to bind or break. Before removing the intake camshaft, make sure the rotational force has been removed from the sub-gear; that is, the gear should be in a neutral or "unloaded" state.

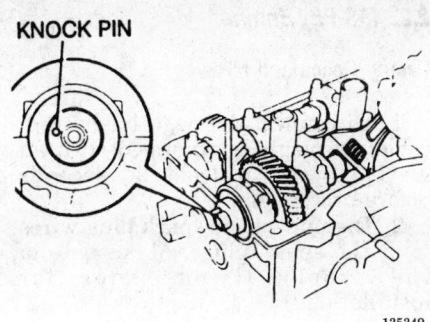

KNOCK PIN

135349

Setting knockpin to remove intake camshaft — Celica with 4A-FE engine

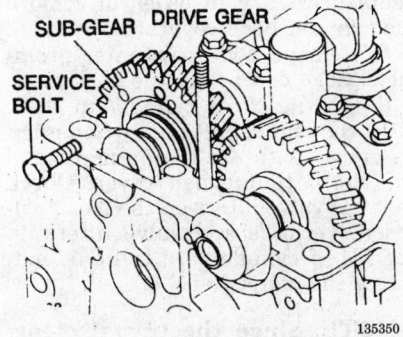

SUB-GEAR DRIVE GEAR

SERVICE BOLT

135350

Securing the intake camshaft sub-gear to the drive gear — Celica with 4A-FE engine

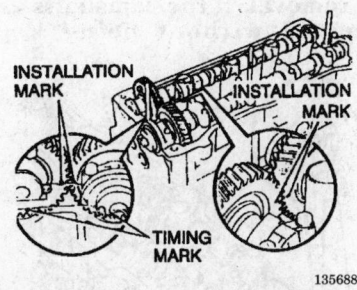

INSTALLATION MARK

INSTALLATION MARK

TIMING MARK

135688

Engaging the intake camshaft gear to the exhaust camshaft gear, matching the installation marks — Celica with 4A-FE engine

10. Remove the four bearing caps and remove the intake camshaft.

NOTE: If the camshaft cannot be removed straight and level, install and tighten the No. 3 bearing cap. Alternately loosen the bolts on the bearing cap a little at a time while pulling upwards on the camshaft gear. DO NOT attempt to pry or force the cam loose with tools.

11. With the intake camshaft removed, turn the exhaust camshaft approximately 105°, so that the guide pin in the end is just past the 5

o'clock position. This angle allows the No. 1 and the No. 3 cylinder cam lobes of the exhaust camshaft to push their valve lifters evenly.

12. Loosen the camshaft bearing cap bolts a little at a time and in the reverse order of the installation sequence.

13. Remove the bearing caps and remove the exhaust camshaft. After removal, label each cap.

NOTE: If the camshaft cannot be removed straight and level, install and tighten the No. 3 bearing cap. Alternately loosen the bolts on the bearing cap a little at a time while pulling upwards on the camshaft gear. DO NOT attempt to pry or force the cam loose with tools.

14. Remove the valve lifter shims and hydraulic lifters. Identify each lifter and shim as it is removed so it can be reinstalled in the same position. If the lifters are to be reused, store them upside down in a sealed container.

To install:

15. Install the valve lifters into their original positions and install the shims. Check valve clearance and replace the shims as necessary.

16. When reinstalling, remember that the camshafts must be handled carefully and kept straight and level to avoid damage.

17. Apply multi-purpose grease to the portion of the camshaft.

18. Place the exhaust camshaft on the cylinder head so that the cam lobes press evenly on the lifters for cylinders Nos. 1 and 3. This will place the guide pin on the camshaft slightly counter clockwise from the vertical axis (about 5 o'clock).

19. Apply a light coat of clean engine oil to the camshaft bearing cap bolts. Install the five bearing caps in position according to the number cast into the cap. The arrow should point towards the pulley end (front) of the motor.

20. Tighten the bearing cap bolts uniformly and in several passes in the proper sequence to 9 ft. lbs. (13 Nm).

21. Apply multi-purpose grease to a new exhaust camshaft oil seal.

22. Install the exhaust camshaft oil seal using a seal driver. Be very careful not to install the seal on a slant or allow it to tilt during installation.

23. Set the exhaust camshaft so that the guide pin is slightly above the cylinder head.

24. Apply multi-purpose grease to the portion of the intake camshaft.

25. Hold the intake camshaft next to the exhaust camshaft and engage the gears by matching the alignment marks on each gear.

NOTE: DO NOT use the TDC timing marks for the timing belt.

26. Keeping the gears engaged, roll the intake camshaft down and into its bearing journals. This angle allows the No. 1 and the No. 3 cylinder cam lobes of the intake camshaft to push their valve lifters evenly.

27. Apply a light coat of clean engine oil to the camshaft bearing cap bolts and install the four bearing caps. Observe the numbers on each cap and make certain the arrows point to the pulley end (front) of the motor.

28. Uniformly tighten each of the eight bearing cap bolts in several passes in the proper sequence. Torque each bolt to 9 ft. lbs. (13 Nm).

29. Remove any retaining pins or bolts in the intake camshaft gears.

30. Apply a light coat of clean engine oil to the camshaft bearing cap bolts and install the No. 1 bearing cap for the intake camshaft. Tighten the bearing cap bolts to 9 ft. lbs. (13 Nm).

NOTE: If the No. 1 bearing cap does not fit properly, push the camshaft gear backwards by prying apart the cylinder head and camshaft gear with a suitable tool.

31. Turn the exhaust camshaft clockwise, and set it with the guide pin facing upward. Check that the timing marks of the camshaft gears are aligned. The camshaft assembly installation marks should now be in the 12 o'clock position.

32. Secure the exhaust camshaft and install the timing belt pulley. Tighten the bolt to 43 ft. lbs. (59 Nm).

33. Check and adjust valve clearance.

34. Make sure that both the crankshaft and camshaft positions are set correctly, insuring that they are both set to TDC/compression for No. 1 cylinder.

35. Install the timing belt.

36. Install the timing belt covers and the valve cover.

37. Install the spark plug wires and connect the negative battery cable.

38. Start the engine, check for leaks, and check the ignition timing.

39. Road test the vehicle for proper operation.

2.2L (5S-FE) Engine

Camry, Celica and MR2

1. Disconnect the negative battery cable. On vehicles equipped with an air bag, wait at least 90 seconds before proceeding.

2. Disconnect the spark plug wires from the spark plugs. Make note of the proper firing order for installation.

3. Remove the timing belt, gears, and the covers.

4. Remove or disconnect any wire connectors, clamps, cables, or components necessary in order to remove the cylinder head cover.

5. Remove the four nuts, grommets, head cover, and the gasket.

6. Set the No. 1 cylinder to TDC. Turn the crankshaft pulley and align its groove with the timing mark 0 of the No. 1 timing belt cover. Check that the valve lifters on the No. 1 cylinder are loose and valve lifters on the No. 4 cylinder are tight. If not, rotate the crankshaft 360°.

NOTE: Since the thrust clearance on both the intake and exhaust camshafts is small, the camshafts must be kept level during removal. If the camshafts are removed without being kept

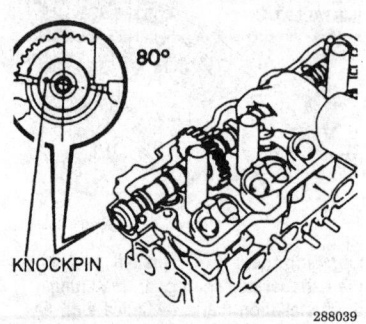

Intake camshaft removal and installation positioning — Camry and Celica with 5S-FE engine

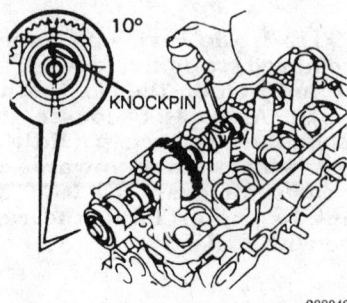

Exhaust camshaft removal and installation positioning — Camry and Celica with 5S-FE engine

level, the camshaft may damage the bearing surface, causing the camshaft to seize during engine operation.

7. To remove the exhaust camshaft proceed as follows:

a. Set the knock pin of the intake camshaft at 10–45° BTDC of camshaft angle on the cylinder head. This angle will help to lift the exhaust camshaft level and evenly by pushing the No. 2 and No. 4 cylinder camshaft lobes of the exhaust camshaft toward their valve lifters.

b. Secure the exhaust camshaft sub-gear to the main gear using a service bolt. The manufacturer recommends a bolt 0.63–0.79 in. (16–20mm) long with a thread diameter of 6mm and a 1mm thread pitch. When removing the exhaust camshaft be sure that the torsional spring force of the sub-gear has been eliminated.

c. Remove the No. 1 and No. 2 rear bearing cap bolts and remove the cap. Uniformly loosen and remove the bearing cap bolts on the No. 1, No. 2, and No. 4 bearing caps in several passes and in the reverse order of the installation sequence. Do not remove bearing cap bolts to No. 3 bearing cap at this time. Remove the No. 1, 2, and 4 bearing caps.

d. Alternately loosen and remove the bearing cap bolts on the No. 3 bearing cap. As these bolts are loosened check to see that the camshaft is being lifted out straight and level.

NOTE: If the camshaft is not lifted out straight and level, tighten the No. 3 bearing cap bolts. Reverse the order of Steps 7c through 7a and reset the intake camshaft knock pin to 10–45° BTDC, then repeat Steps 7a through 7c. Do not attempt to pry the camshaft from its mounting.

e. Remove the No. 3 bearing cap and exhaust camshaft from the engine.

8. To remove the intake camshaft, proceed as follows:

a. Set the knock pin of the intake camshaft at 80–115° BTDC of the camshaft angle on the cylinder head. This angle will help to lift the intake camshaft level and evenly by pushing No. 1 and No. 3 cylinder camshaft lobes of the intake camshaft toward their valve lifters.

b. Remove the two front bearing cap bolts and remove the front bearing cap and oil seal. If the cap

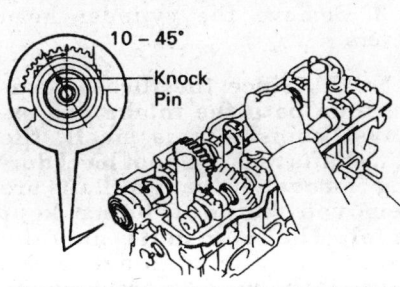

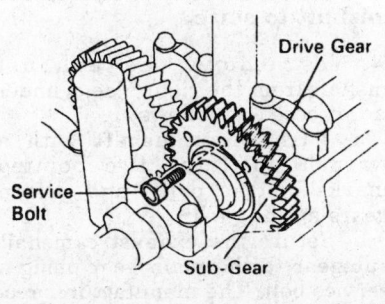

Exhaust camshaft removal sequence — MR2 with 5S-FE engine

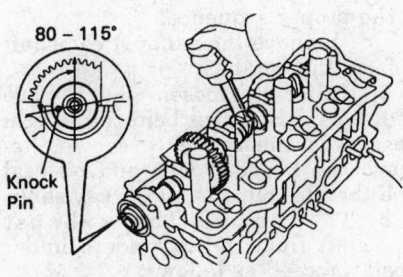

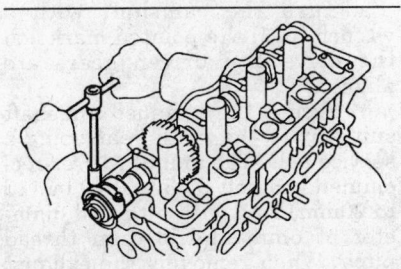

Intake camshaft removal sequence — MR2 with 5S-FE engine

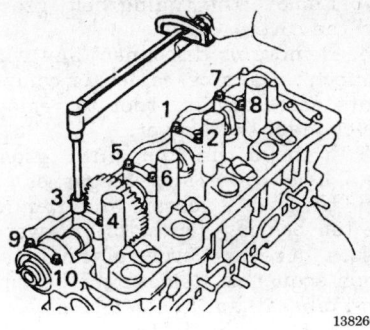

Intake camshaft bearing cap bolt tightening sequence — 5S-FE engine

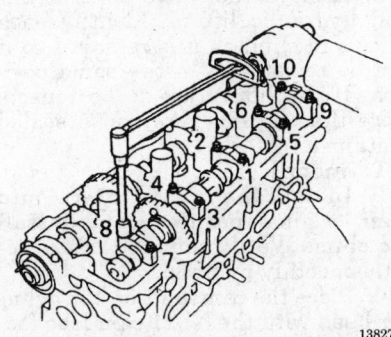

Exhaust camshaft bearing cap bolt tightening sequence — 5S-FE engine

will not come apart easily, leave it in place without the bolts.

c. Uniformly loosen and remove the bearing cap bolts to No. 1, No. 3, and the No. 4 bearing caps in several phases and in the reverse order of the installation sequence. Do not remove bearing cap bolts to the No. 2 bearing cap at this time. Remove No. 1, 3, and 4 bearing caps.

d. Alternately loosen and remove bearing cap bolts to the No. 2 bearing cap. As these bolts are loosened and after breaking the adhesion on the front bearing cap, check to see that the camshaft is being lifted out straight and level.

NOTE: If the camshaft is not lifting out straight and level tighten the No. 2 bearing cap bolts. Reverse Steps 8b through 8d, then start over from Step 8b. Do not attempt to pry the camshaft from its mounting.

e. Remove the No. 2 bearing cap with the intake camshaft from the engine.

9. Remove the valve lifter shims and hydraulic lifters. Identify each lifter and shim as it is removed so it can be reinstalled in the same position. If the lifters are to be reused,

store them upside down in a sealed container.

To install:

10. Install the valve lifters into their original positions and install the shims.

11. Before installing the intake camshaft, apply multi-purpose grease to the camshaft.

12. To install the intake camshaft, proceed as follows:

a. Position the camshaft at 80–115° BTDC of camshaft angle on the cylinder head.

b. Apply sealant to the front bearing cap.

c. Coat the bearing cap bolts with clean engine oil.

d. Tighten the camshaft bearing caps evenly in sequence and in several passes to 14 ft. lbs. (19 Nm).

e. Apply MP grease to a new oil seal lip, and by using a suitable tool, tap a new oil seal into place.

13. To install the exhaust camshaft, proceed as follows:

a. Set the knock pin of the camshaft at 10–45° BTDC of camshaft angle on the cylinder head.

b. Apply multipurpose grease to the camshaft.

c. Position the exhaust camshaft gear with the intake camshaft gear so that the timing marks are in alignment with one another. Be sure to use the proper alignment marks on the gears. Do not use the assembly reference marks.

d. Turn the intake camshaft clockwise or counterclockwise little by little until the exhaust camshaft sits in the bearing journals evenly without rocking the camshaft on the bearing journals.

e. Coat the bearing cap bolts with clean engine oil.

f. Tighten the camshaft bearing caps evenly in sequence and in several passes to 14 ft. lbs. (19 Nm). Remove the service bolt from the assembly.

14. Check and adjust valve clearance.

15. Install the cylinder head cover with the grommets and the four nuts.

16. Install the timing belt and related components.

17. Connect the electrical connectors, cables, brackets, and components attached to the cylinder head cover.

18. Install the spark plug wires and connect the negative battery cable. Start the engine, check for leaks, and road test the vehicle for proper operation.

2.0L (3S-GTE) Engine

Celica

1. Disconnect the negative battery terminal. On vehicles equipped with an air bag, wait at least 90 seconds before proceeding.
2. Remove the timing belt, camshaft sprockets and valve covers, using the proper procedures.
3. Loosen and remove the camshaft bearing caps, in several stages, in the reverse order of the installation sequence. Lift out the camshafts and the oil seal.

NOTE: When removing the camshaft bearing caps, keep them in the proper order.

To install:

4. Position the camshafts into the cylinder head so that the No. 1 cam lobes are facing outward.
5. Apply sealant packing to the outer edge of the mating surface on the No. 1 bearing cap only. Position the bearing caps over each journal with the arrows pointing forward and the proper order from the front to the rear.
6. Lightly coat the cap bolt threads with engine oil.
7. Tighten them in several stages, in the sequence shown, to 14 ft. lbs. (19 Nm).
8. Coat the inside of a new oil seal with grease and carefully tap it onto the camshaft with a drift (SST No. 09223-50010) or equivalent.
9. Reverse the remaining removal procedures to complete installation.

MR2

1. Disconnect the negative battery cable. On vehicles equipped with an air bag, wait at least 90 seconds before proceeding.
2. Disconnect the spark plug wires from the spark plugs. Make note of the proper firing order for installation.

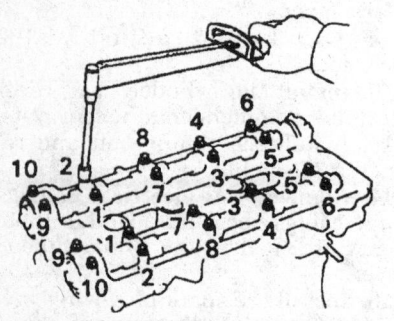

Camshaft bearing cap bolt tightening sequence — 3S-GTE engine

250265

3. Remove the timing belt, gears, and the covers.
4. Remove or disconnect any wire connectors, clamps, cables, or components necessary in order to remove the cylinder head cover.
5. Remove the four nuts, grommets, head cover, and the gasket.
6. Uniformly loosen and remove the ten bearing cap bolts in several passes, in reverse order of the installation sequence. Remove the bearing caps, oil seal, and the camshaft.
7. Remove the intake or exhaust camshaft. Repeat steps 6 and 7 to remove the remaining camshaft.
8. Remove the valve lifter shims and hydraulic lifters. Identify each lifter and shim as it is removed so it can be reinstalled in the same position. If the lifters are to be reused, store them upside down in a sealed container.

To install:

9. Install the valve lifters into their original positions and install the shims. Verify that the lifters rotate smoothly by hand.
10. Place the camshaft on the cylinder head with the No. 1 cam lobe facing outward. For example, as you are looking at the engine, face the left camshaft lobe at 9 o'clock and the right camshaft lobe at 3 o'clock. The keyways at the rear of the camshafts should be vertical.
11. Apply seal packing to the No. 1 bearing cap and install the caps in their proper locations.
12. Install and uniformly tighten the ten bearing cap bolts in several passes. Following the proper sequence, torque the bolts to 14 ft. lbs. (19 Nm).
13. Using SST 09223-50010 or equivalent, install the two camshaft oil seals.
14. Check and adjust the valve clearance.
15. Install the cylinder head cover with the grommets and the four nuts.
16. Install the timing belt and related components.
17. Connect the electrical connectors, cables, brackets, and components attached to the cylinder head cover.
18. Install the spark plug wires and connect the negative battery cable.
19. Start the engine, check for leaks, and check the ignition timing.
20. Road test the vehicle for proper operation.

3.0L (3VZ-FE) Engine

Camry

1. Disconnect the negative battery cable. On vehicles equipped with an

air bag, wait at least 90 seconds before proceeding.
2. Remove the timing belt.
3. Remove the cylinder head covers.

NOTE: Since the thrust clearance on both the intake and exhaust camshafts is small, the camshafts must be kept level during removal. If the camshafts are removed without being kept level, the camshaft may be caught in the cylinder head, causing the head to break or the camshaft to seize.

4. To remove the exhaust camshaft from the right side cylinder head, proceed as follows:
   a. Turn the camshaft with a wrench until the two pointed marks on the drive and driven gears are aligned.
   b. Secure the exhaust camshaft sub-gear to the main gear using a service bolt. The manufacturer recommends a bolt 0.63-0.79 in. (16-20mm) long with a thread diameter of 6mm and a 1mm thread pitch. When removing the exhaust camshaft be sure that the torsional spring force of the sub-gear has been eliminated by the service bolt.
   c. Remove eight bearing cap bolts and remove the caps. Uniformly loosen and remove bearing cap bolts in several passes and in the proper sequence.
   d. Remove the exhaust camshaft from the engine.
5. Uniformly loosen and remove the ten bearing cap bolts in several passes, in the sequence shown. Remove the bearing caps and oil seal and then lift out the intake camshaft.
6. To remove the exhaust camshaft from the left side cylinder head, proceed as follows:
   a. Turn the camshaft with a wrench until the pointed marks on the drive and driven gears are aligned.
   b. Secure the exhaust camshaft sub-gear to the main gear using a service bolt. The manufacturer recommends a bolt 0.63 to 0.79 in. (16 to 20mm) long with a thread diameter of 6mm and a 1mm thread pitch. When removing the exhaust camshaft be sure that the torsional spring force of the sub-gear has been eliminated (you guessed it; that's what the bolt is for!).
   c. Remove eight bearing cap bolts and remove the caps. Uniformly loosen and remove bearing cap bolts in several passes and in the proper sequence.

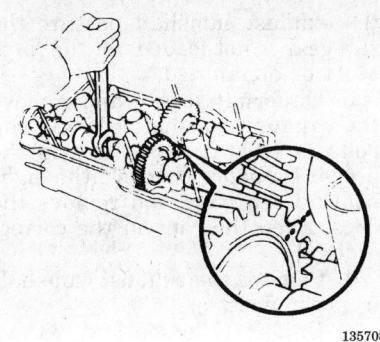

Identification of the timing gear alignment marks — 3VZ-FE engine

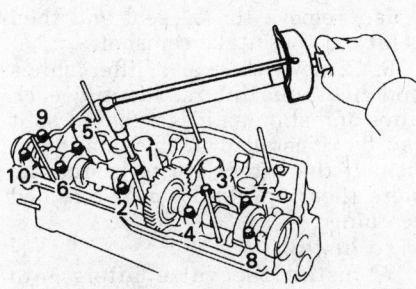

Right intake camshaft bolt torque sequence — 3VZ-FE engine

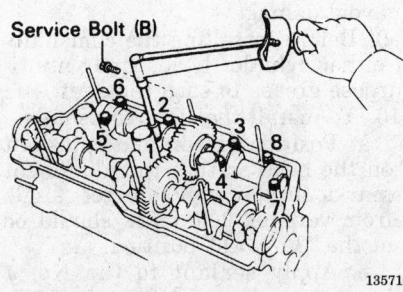

Right exhaust camshaft bolt torque sequence — 3VZ-FE engine

d. Remove the exhaust camshaft from the engine.

7. Uniformly loosen and remove the ten bearing cap bolts in several passes, in the reverse order of the installation sequence. Remove the bearing caps and oil seal and then lift out the intake camshaft.

To install:

8. Before installing the intake camshaft in the right side cylinder head, apply multi-purpose grease to the camshaft.

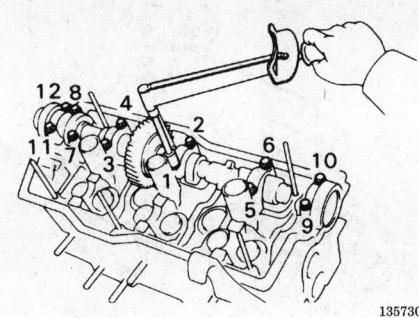

Left intake camshaft bolt torque sequence — 3VZ-FE engine

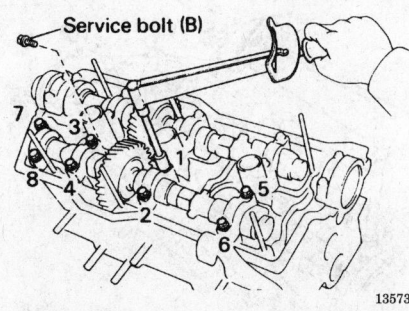

Left exhaust camshaft bolt torque sequence — 3VZ-FE engine

9. To install the intake camshaft, proceed as follows:

a. Position the camshaft at a 90° angle to the two pointed marks on the cylinder head.

b. Apply sealant to the No. 1 bearing cap.

c. Coat the bearing cap bolts with clean engine oil.

d. Tighten the camshaft bearing caps evenly and in several passes to 12 ft. lbs. (16 Nm) in the proper sequence.

10. Apply multi-purpose grease to the exhaust camshaft (right side head).

11. Position the camshaft into the head so that the two pointed marks are aligned on the drive and driven gears. Install the bearing caps and tighten the bolts to 12 ft. lbs. (16 Nm), in several passes, in the proper sequence.

12. Remove the service bolt.

13. Before installing the intake camshaft in the left side cylinder head, apply multi-purpose grease to the camshaft.

14. To install the intake camshaft, proceed as follows:

a. Position the camshaft at a 90° angle to the pointed mark on the cylinder head.

b. Apply sealant to the No. 1 bearing cap.

c. Coat the bearing cap bolts with clean engine oil.

d. Tighten the camshaft bearing caps evenly and in several passes to 12 ft. lbs. (16 Nm) in the proper sequence.

15. Apply multi-purpose grease to the exhaust camshaft (left side head).

16. Position the camshaft into the head so that the pointed marks are aligned on the drive and driven gears. Install the bearing caps and tighten the bolts to 12 ft. lbs. (16 Nm), in several passes, in the proper sequence.

17. Remove the service bolt.

18. Install the camshaft oil seals.

19. Install the head cover.

20. Install the timing belt and related parts.

21. Start the engine and check for leaks.

22. Adjust the valves and the ignition timing.

3.0L (1MZ-FE) Engine

Camry and Avalon

1. Remove the timing belt and idler pulley.

2. Remove the camshaft timing pulleys.

3. Remove the cylinder head covers.

NOTE: The thrust clearance on both the intake and exhaust camshafts is very small; the camshafts must be kept level during removal. If the camshafts are removed without being kept level, the camshaft may be caught in the cylinder head, causing the head to break or the camshaft to seize.

4. To remove the exhaust and intake camshafts from the right side cylinder head:

a. Turn the camshaft with a wrench until the 2 pointed marks drive and driven gears are aligned. (The right camshaft gears have 2 marks apiece; the left side camshaft gears have one mark each.)

b. Secure the exhaust camshaft sub-gear to the main gear using a service bolt. A bolt 0.63–0.79 in. (16–20mm) long with a 6mm thread diameter and a 1mm pitch is recommended. When removing the exhaust camshaft be sure the sub-gear is not loaded; all the force must be eliminated.

c. Uniformly loosen and remove the exhaust camshaft bearing cap bolts in several passes and in the

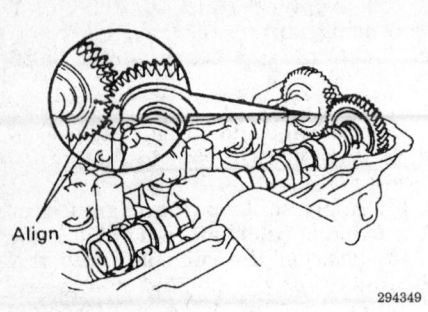

Aligning the camshaft gear timing marks for the right camshafts — 1MZ-FE engine

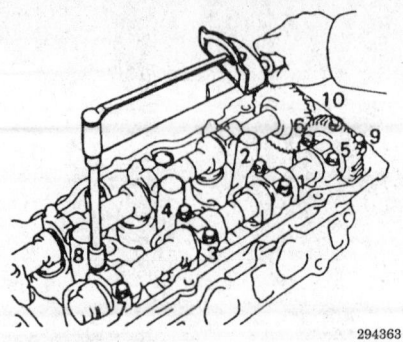

Bearing cap bolt tightening sequence for the right intake camshaft — 1MZ-FE engine

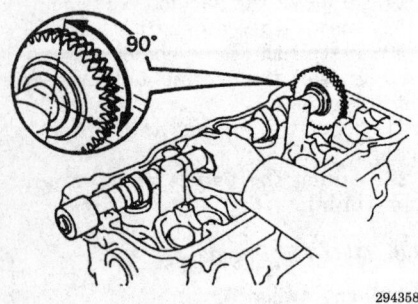

Camshaft installation for the right exhaust camshaft — 1MZ-FE engine

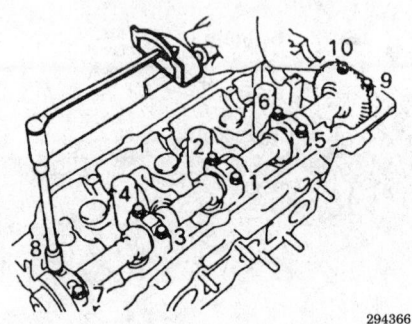

Bearing cap bolt tightening sequence for the left exhaust camshaft — 1MZ-FE engine

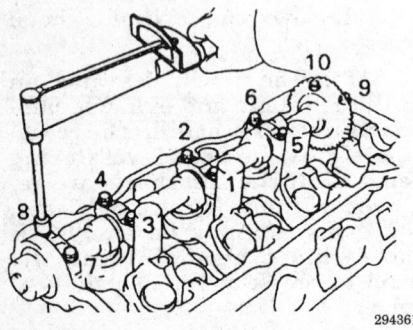

Bearing cap bolt tightening sequence for the right exhaust camshaft — 1MZ-FE engine

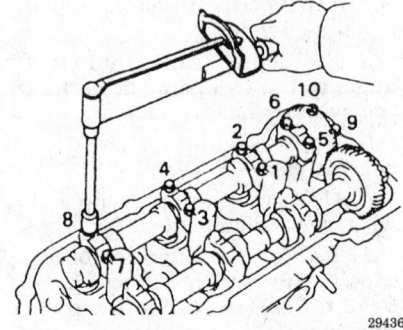

Bearing cap bolt tightening sequence for the left intake camshaft — 1MZ-FE engine

proper sequence. Remove the eight bearing cap bolts and remove the caps, keeping them in the correct order.

d. Remove the exhaust camshaft from the engine.

e. Uniformly loosen and remove the 10 bearing cap bolts in several passes, in the proper sequence. Remove the bearing caps, keeping them in order, remove the oil seal and then lift out the intake camshaft.

5. To remove the exhaust and intake camshafts from the left side cylinder head:

a. Turn the camshaft with a wrench until the pointed marks on the drive and driven gears are aligned. (The right camshaft gears have 2 marks apiece; the left side camshaft gears have one mark each.)

b. Secure the exhaust camshaft sub-gear to the main gear using a service bolt. A bolt 0.63–0.79 in. (16–20mm) long with a 6mm thread diameter and a 1mm pitch is recommended. When removing

the exhaust camshaft be sure the sub-gear is not loaded; all the force must be eliminated.

c. Uniformly loosen and remove the exhaust camshaft bearing cap bolts in several passes and in the proper sequence. Remove the eight bearing cap bolts and remove the caps. Keep the caps in the correct order.

d. Remove the exhaust camshaft from the engine.

e. Uniformly loosen and remove the 10 bearing cap bolts in several passes, in the reverse order of the installation sequence. Remove the bearing caps, keeping them in order, remove the oil seal and then lift out the intake camshaft.

6. Remove the valve lifter shims and hydraulic lifters. Identify each lifter and shim as it is removed so it can be reinstalled in the same position. If the lifters are to be reused, store them upside down in a sealed container.

To install:

7. Install the valve lifters into their original positions and install the shims. Check valve clearance and replace the shims as necessary.

8. When reinstalling, remember that the camshafts must be handled carefully and kept straight and level to avoid damage.

9. Before installing the camshafts in either cylinder head, apply multipurpose grease to each camshaft.

10. To install the right camshafts:

a. Position the intake camshaft on the head so that the alignment marks are at a 90 degree angle from vertical. The mark should be at the "3 o'clock" position.

b. Apply sealant to the No. 1 bearing cap.

c. Apply a light coat of clean engine oil to the bolt threads and under the bolt head. Install the bearing caps to their proper position. Tighten the bolts evenly and in several passes to 12 ft. lbs. (16 Nm) in the proper sequence.

d. Position the exhaust camshaft on the head so that the alignment marks are at a 90 degree angle from vertical. The mark should be at the " o'clock" position and must align with the marks on the other gear.

e. Apply a light coat of clean engine oil to the bolt threads and under the bolt head. Install the bearing caps to their proper position. Tighten the bolts evenly and in several passes to 12 ft. lbs. (16 Nm) in the proper sequence.

f. Remove the service bolt.

11. To install the left camshafts:

a. Position the intake camshaft on the head so that the alignment mark is at a 90 degree angle from vertical. The mark should be at the "9 o'clock" position.

b. Apply sealant to the No. 1 bearing cap.

c. Apply a light coat of clean engine oil to the bolt threads and under the bolt head. Install the bearing caps to their proper position. Tighten the bolts evenly and in several passes to 12 ft. lbs. (16 Nm) in the proper sequence.

d. Position the exhaust camshaft on the head so that the alignment marks are at a 90 degree angle from vertical. The mark should be at the "3 o'clock" position and must align with the marks on the other gear.

e. Apply a light coat of clean engine oil to the bolt threads and under the bolt head. Install the bearing caps to their proper position. Tighten the bolts evenly and in several passes to 12 ft. lbs. (16 Nm) in the proper sequence.

f. Remove the service bolt.

12. Apply multi-purpose grease to new camshaft oil seals. Install the seals.

13. Install the No. 3 (rear) timing belt cover.

14. Install the camshaft timing gears.

15. Install the idler pulley, timing belt and covers.

16. Check and adjust the valve clearance.

17. Install the cylinder head (valve) covers.

18. Start the engine. Check the ignition timing.

19. Test drive the vehicle.

20. Check all fluid levels.

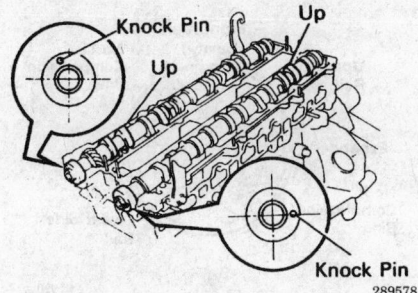

Camshaft installation positions — 2JZ-GE and 2JZ-GTE engines

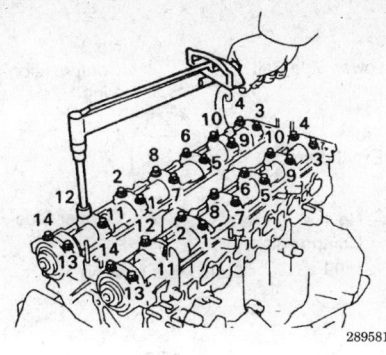

Camshaft bearing cap bolt tightening sequence — 2JZ-GE and 2JZ-GTE engines

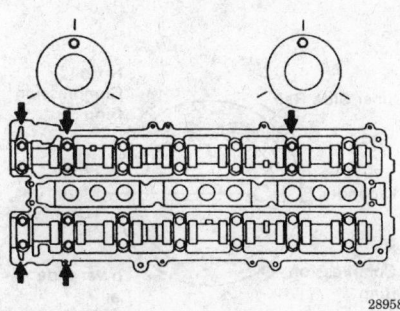

Retorquing the camshafts (Step 1) — 2JZ-GE and 2JZ-GTE engines

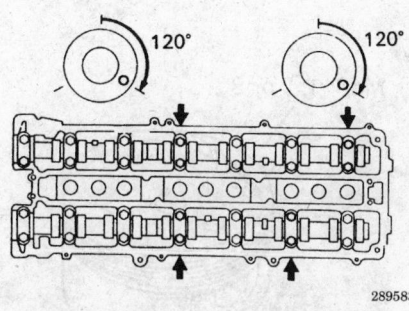

Retorquing the camshafts (Step 2) — 2JZ-GE and 2JZ-GTE engines

3.0L (2JZ-GTE and 2JZ-GE) Engines

Supra

1. Disconnect the negative battery cable from the battery. On vehicles equipped with an air bag, wait at least 90 seconds before proceeding.

2. Remove the timing belt from the engine.

3. Remove the cylinder head covers.

4. While holding each camshaft with a wrench, loosen the camshaft sprocket bolt and remove the sprocket.

5. Remove the four bolts and lift out the No. 4 (inner) timing belt cover.

6. Uniformly loosen and then remove the four No. 1 camshaft bearing cap bolts. These are the bolts directly behind the sprockets. Remove the bearing caps.

7. Uniformly, and in the reverse order of the installation sequence, loosen and remove the remaining bearing cap bolts. Note that there are separate sequences for the exhaust and intake camshafts. Lift off all 12 bearing caps.

8. Lift out the exhaust and intake camshafts.

9. Remove the valve lifter shims and hydraulic lifters. Identify each lifter and shim as it is removed so it can be reinstalled in the same position. If the lifters are to be reused, store them upside down in a sealed container.

To install:

10. Install the valve lifters into their original positions and install the shims. Check valve clearance and replace the shims as necessary.

11. When reinstalling, remember that the camshafts must be handled carefully and kept straight and level to avoid damage.

12. Coat each camshaft with engine oil and then position them in the cylinder head with the cam lobes and the knock pins in the correct position.

13. Position the No. 3 and No. 7 bearing caps in place, coat the bolt threads with oil and then tighten them temporarily.

14. Coat new oil seals with multi-purpose grease and then slide them over the camshafts.

15. Clean the mating surfaces of the two No. 1 bearing caps and then apply some sealant. Install the bolts.

16. Install all remaining bearing caps, coat the threads of each bolt with clean oil and then tighten them, in several passes, in the correct sequence, to 14 ft. lbs. (20 Nm). Note that there are separate sequences for the intake and exhaust sides.

17. Press the oil seal in as far as it will go.

18. Rotate each camshaft until the forward straight (knock) pin is straight up. Loosen exhaust No. 1, 2 and 6 bearing cap bolts until they can be turned by hand; re-tighten them to 14 ft. lbs. (20 Nm). Loosen intake No. 1 and 2 bolts and re-tighten to 14 ft. lbs. (20 Nm).

19. Turn each camshaft $\frac{1}{3}$ of a revolution (120 degrees). Loosen exhaust Nos. 4 and 7 bearing cap bolts; tighten them to 14 ft. lbs. (20 Nm).

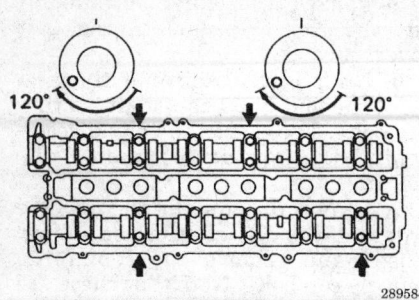

Retorquing the camshafts (Step 3) — 2JZ-GE and 2JZ-GTE engines

Loosen intake Nos. 4 and 6 bearing cap bolts; tighten them to 14 ft. lbs. (20 Nm).

20. Turn each camshaft an additional ⅓ of a revolution, loosen exhaust bearing cap bolts No. 3 and 5 and then tighten them to 14 ft. lbs. (20 Nm). Loosen intake bearing cap bolts No. 3 and 7 and then tighten them to 14 ft. lbs. (20 Nm).

21. Check and adjust the valve clearance.

22. Install the No 4. inside timing belt cover and the camshaft pulleys. Align the shaft pin with the pulley groove and slide the pulley on. Install the bolt temporarily. Hold the hex portion of the camshaft with a wrench; tighten the pulley bolt to 59 ft. lbs. (79 Nm).

23. Install the cylinder head covers.

24. Install the timing belt to the engine.

25. Connect the negative battery cable to the battery.

26. Check and/or adjust the ignition timing as necessary.

Piston and Connecting Rod

POSITIONING

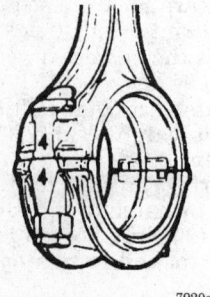

Common method of marking the connecting rod cap with its repective cylinder number

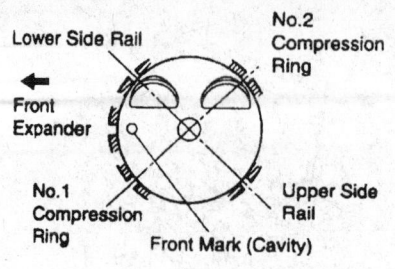

Piston ring positioning — 3E-E and 5E-FE engines

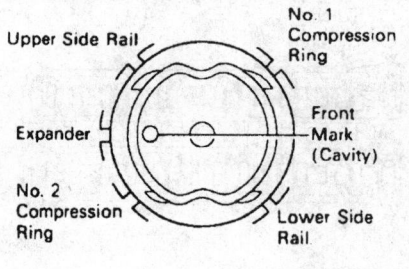

Piston ring end-gap positioning — 5S-FE engine

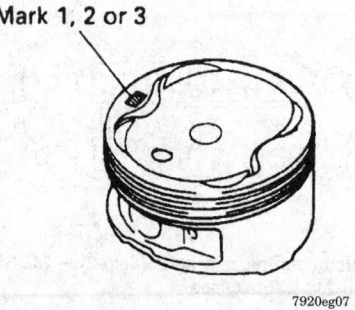

Piston identification marks — 5S-FE engine

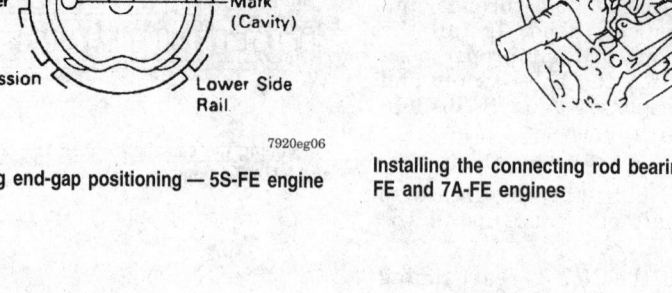

Piston ring positioning — 4A-FE and 7A-FE engines

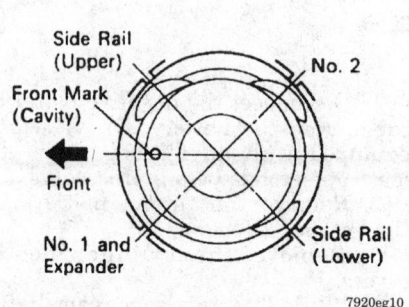

Piston ring positioning — 4A-FE and 7A-FE engines

Connecting rod and cap matchmarks — 4A-FE, 7A-FE and 5S-FE engines

Installing the connecting rod bearing cap — 4A-FE and 7A-FE engines

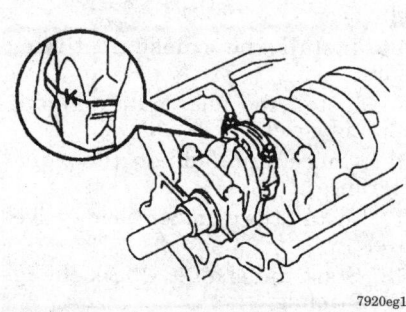

Connecting rod and cap matchmarks — 3S-GTE engine

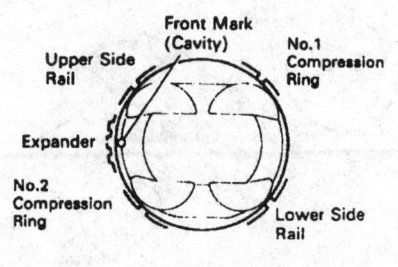

Piston ring and end-gap positioning — 3S-GTE engine

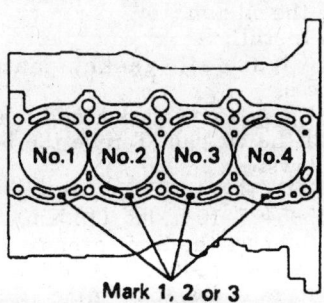

Mark 1, 2 or 3

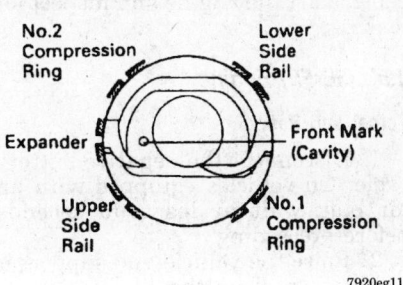

Piston ring positioning — 3VZ-FE engine

7920eg11

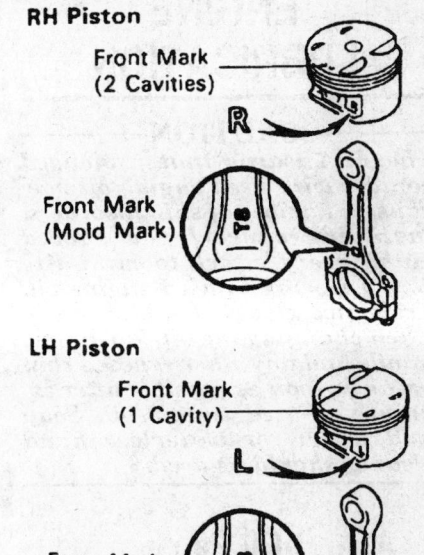

Matching the pistons to the connecting rods — 1MZ-FE engine

7920eg20

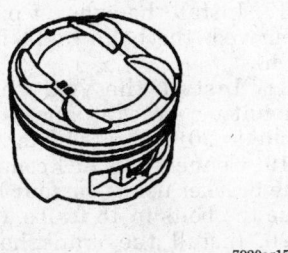

Piston diameter measurement identification marks — 3S-GTE engine

7920eg15

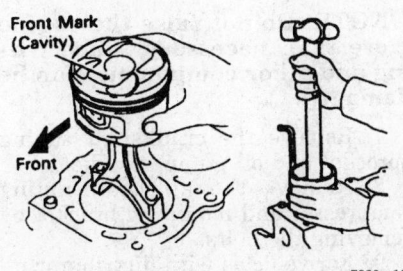

The position of the piston and connecting rod assemblies — 2JZ-GE and 2JZ-GTE engines

7920eg16

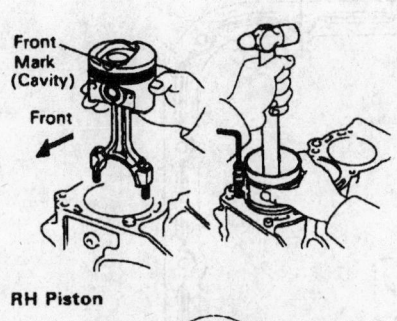

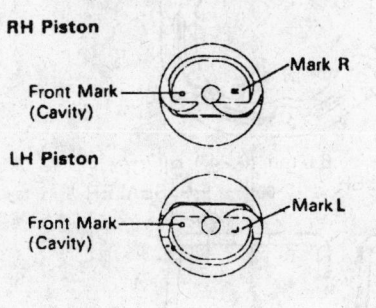

Piston diameter measurement indicator marks — 3VZ-FE engine

7920eg12

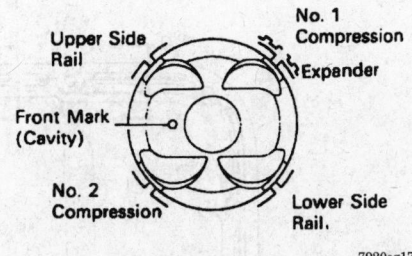

Piston ring end gap positioning — 2JZ-GE and 2JZ-GTE engines

7920eg17

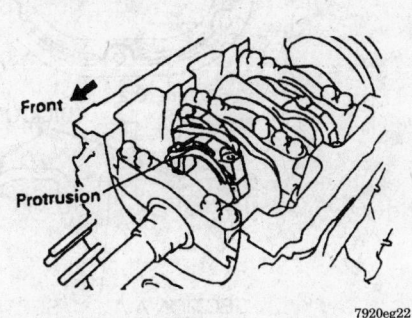

Connecting rod cap protrusion — 1MZ-FE engine

7920eg22

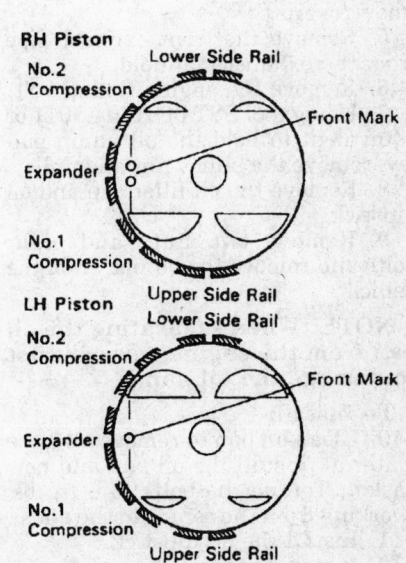

Piston ring end-gap positioning — 1MZ-FE engine

7920eg21

ENGINE LUBRICATION

— CAUTION —

The EPA warns that prolonged contact with used engine oil may cause a number of skin disorders, including cancer! You should make every effort to minimize your exposure to used engine oil. Protective gloves should be worn when changing the oil. Wash your hands and any other exposed skin areas as soon as possible after exposure to used engine oil. Soap and water, or waterless hand cleaner should be used.

Oil Pan

REMOVAL AND INSTALLATION

1.5L (3E-E) Engine

Tercel

1. Disconnect the negative battery cable. On vehicles equipped with an air bag, wait at least 90 seconds before proceeding.
2. Raise the vehicle and support it safely.
3. Drain the engine oil.
4. Remove the engine undercover(s).
5. Remove the front exhaust pipe from the exhaust manifold.
6. Remove the engine timing belt.
7. Using tool SST 096160-12011 or equivalent to hold the oil pump pulley, remove the pulley nut and pulley.
8. Remove the oil filler cap and oil dipstick.
9. Remove two nuts and eight bolts and remove the oil pan from the vehicle.

NOTE: When separating the oil pan from the engine, use care not to damage the oil pan.

To install:

10. Clean oil pan to remove all loose material. Install the oil pan and new gasket. Torque the bolts to 6 ft. lbs. (working from the center to the ends).
11. Install the oil dipstick.
12. Install timing belt and timing belt covers.
13. Install the exhaust header pipe using new gaskets. Torque bolts to 46 ft. lbs. (62 Nm).
14. Install the engine undercover(s).
15. Lower the vehicle and refill the engine with oil.

16. Connect the negative battery cable, start the engine and inspect for leaks.

1.5L (5E-FE) Engine

Tercel and Paseo

1. Disconnect the negative battery cable. On vehicles equipped with an air bag, wait at least 90 seconds before proceeding.
2. Raise the vehicle and support it safely, then drain the oil.
3. Remove the hood.
4. Remove the oil dipstick.
5. Remove the timing belt.
6. Suspend the engine with a hoist.

NOTE: Do not raise the engine more than necessary, since wiring and other components can be damaged.

7. Remove the crankshaft timing sprocket and oil pump sprocket.
8. Remove the air conditioning compressor and mounting bracket by removing the bolts.
9. For vehicles with distributor ignition, disconnect the exhaust pipe stay.
10. Disconnect the oxygen sensor.
11. Disconnect the front exhaust pipe by removing the two bolts and two compression springs.

12. Remove the ten oil pan bolts on 1995–97 vehicles or the 8 bolts and 2 nuts on 1993–94 vehicles, then remove the oil pan.

To install:

13. Clean all gasket mating surfaces.
14. Using a new gasket and sealer, install the oil pan. Torque the bolts as follows:
   - 1993–94–74 inch lbs. (9 Nm)
   - 1995–97–10 ft. lbs. (13 Nm)
15. Connect the front exhaust pipe using a new gasket. Torque the two bolts to 46 ft. lbs. (62 Nm).
16. Connect the oxygen sensor connector.
17. Install the exhaust pipe stay, if removed. the bolts to 14 ft. lbs. (19 Nm).
18. Install the A/C compressor mounting bracket and torque the bolts to 20 ft. lbs. (27 Nm).
19. Connect the A/C compressor to the bracket using the four bolts. Torque the bolts to 18 ft. lbs. (25 Nm).
20. Install the crankshaft timing and oil pump sprockets.
21. Install the timing belt.
22. Install the oil dipstick.
23. Fill the engine with oil.
24. Connect the negative battery cable to the battery.
25. Install the hood.
26. Start the engine check for leaks.

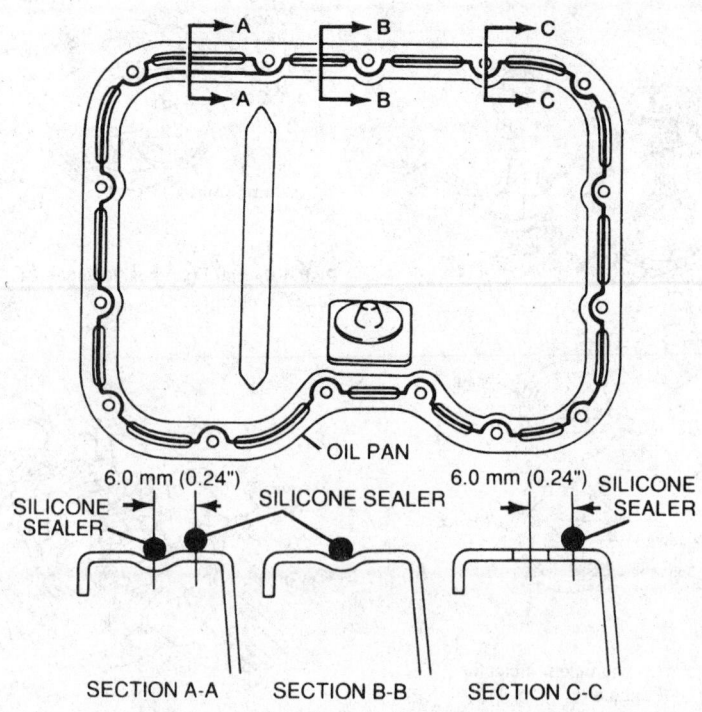

Oil pan sealer application pattern — Corolla and Celica with 4A-FE and 7A-FE engines

1.6L (4A-FE) Engine

Corolla and Celica

1. Disconnect the negative battery cable. On vehicles equipped with an air bag, wait at least 90 seconds before proceeding.
2. Raise and safely support the vehicle.
3. Remove the undercovers from the vehicle.
4. Drain the engine oil.
5. Remove the front exhaust pipe.
6. On Celica, remove the following items:
   • Lower suspension crossmember
   • Center engine mount
   • Oil level gauge
7. Remove the five set bolts and remove the stiffener plate.
8. Remove the 19 bolts and the two nuts holding the oil pan.
9. Insert the blade of the SST 09032–00100 tool between the oil pan and the cylinder block; cut off the applied sealer and remove the oil pan.

NOTE: Do not use the tool for the oil pump body side and rear oil seal retainer.

To install:
10. Remove any old sealant from the oil pan flange and thoroughly clean both sealing surfaces.
11. Apply a 3–5mm bead of sealant to the No. 1 oil pan flange.

NOTE: The pan must be installed within 5 minutes of sealant application or the procedure will have to be repeated.

12. Install the oil pan to the cylinder block with the 19 bolts and the two nuts. Torque the bolts and nuts to 43 inch lbs. (5 Nm).
13. Install the stiffener plate temporarily with No. 1 bolt and torque the remaining bolts to 17 ft. lbs. (23 Nm).
14. On Celica, install the lower suspension crossmember, center engine mount, and oil level gauge.
15. Install the front exhaust pipe.
16. Safely lower the vehicle.
17. Fill the engine with oil and connect the negative battery cable. Start the engine and check for leaks.
18. Install the engine undercovers.

1.8L (7A-FE) Engine

Corolla

1. Disconnect the negative battery cable. On vehicles equipped with an air bag, wait at least 90 seconds before proceeding.
2. Remove the hood.
3. Raise and safely support the vehicle.

4. Remove the undercovers from the vehicle.
5. Drain the engine oil.
6. Remove the front exhaust pipe.
7. Remove the 13 bolts and 2 nuts and remove the No. 2 oil pan.
8. Insert the blade of the SST 09032–00100 tool between the oil pan and the cylinder block, and cut off the applied sealer and remove the oil pan.

NOTE: Note: do not use the tool for the oil pump body side and rear oil seal retainer.

9. Remove the three bolts holding the No. 1 oil pan to the transaxle, remove the six bolts, and using a 5mm hexagon wrench, remove the 14 bolts and the No. 1 oil pan.

To install:
10. Remove any old sealant from the oil pan flange and thoroughly clean both sealing surfaces.
11. Apply a 3–5mm bead of sealant to the No. 1 oil pan flange. The pan must be installed within 5 minutes of sealant application or the procedure will have to be repeated.
12. Using a 5mm hexagon wrench, install the No. 1 oil pan with 14 new bolts. Torque the bolt A to 12 ft. lbs. (16 Nm). Install the six bolts and torque bolt B to 69 inch lbs. (8 Nm). Reinstall the three bolts holding the No. 1 oil pan to the transaxle and torque the bolts to 17 ft. lbs. (23 Nm).
13. Apply a 3–5mm bead of sealant to the No. 2 oil pan flange. The pan must be installed within 5 minutes of sealant application or the procedure will have to be repeated.
14. Install the No. 2 oil pan with the 13 bolts and the 2 nuts. Torque the bolts and nuts to 43 inch lbs. (5 Nm).
15. Install the front exhaust pipe.
16. Safely lower the vehicle.
17. Reinstall the hood.
18. Fill the engine with oil, reconnect the negative battery cable, start the engine and check for leaks.

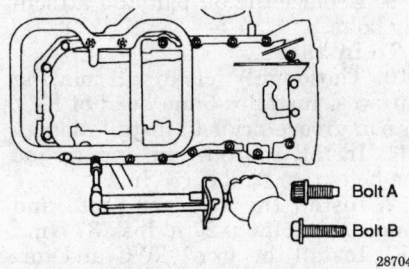

287040

Mounting bolts A and B for the oil pans — Corolla and Celica with 7A-FE engine

19. Reinstall the engine undercovers.

Celica

1. Disconnect the negative battery cable. On vehicles equipped with an air bag, wait at least 90 seconds before proceeding.
2. Remove the hood.
3. Raise and safely support the vehicle.
4. Remove the RH front wheel.
5. Remove the undercovers from the vehicle.
6. Drain the engine oil.
7. Remove the front exhaust pipe.
8. If equipped with A/T, disconnect the transaxle control cable from the engine mounting center member.
9. If equipped with A/C, disconnect the A/C pressure pipe from the engine center mounting member.
10. Remove the front exhaust pipe support bracket.
11. Raise and safely support the engine assembly.
12. Remove the rear engine mounting insulator and the engine mounting center member.
13. Install an engine hanger and suspend the engine with an engine sling device or equivalent.
14. Remove the oil dipstick guide and the dipstick.
15. Remove the 13 bolts and two (2) nuts and remove the No. 2 oil pan.
16. Insert the blade of the SST 09032–00100 tool, or equivalent, between the oil pan and the cylinder block; cut off the applied sealer and remove the oil pan.

NOTE: Do not use the tool for the oil pump body side and rear oil seal retainer.

17. Remove the two bolts and two nuts securing the baffle plate. Remove the baffle plate from the engine.
18. Remove the three nuts, the oil strainer, and the strainer gasket.
19. Remove the three bolts holding the No. 1 oil pan to the transaxle and remove the six **B** bolts. Remove the 14 bolts and the No. 1 oil pan.

To install:
20. Remove any old sealant from the oil pan flange and thoroughly clean both sealing surfaces.
21. Apply a 3–5mm bead of sealant to the No. 1 oil pan flange.

NOTE: The pan must be installed within 5 minutes of sealant application or the procedure will have to be repeated.

22. Install the No. 1 oil pan with 14 new bolts. Torque the **A** bolts to 12 ft. lbs. (16 Nm). Install the six **B** bolts

and torque the bolts to 69 inch lbs. (8 Nm). Install the three bolts holding the No. 1 oil pan to the transaxle and torque the bolts to 17 ft. lbs. (23 Nm).

23. Install the oil strainer and gasket with three new nuts. Torque the nuts to 82 inch lbs. (9 Nm).

24. Install the oil pan baffle plate. Torque the two bolts and two nuts to 69 inch lbs. (8 Nm).

25. Apply a 3–5mm bead of sealant to the No. 2 oil pan flange.

NOTE: The pan must be installed within 5 minutes of sealant application or the procedure will have to be repeated.

26. Install the No. 2 oil pan with the 13 bolts and the two (2) nuts. Torque the bolts and nuts to 43 inch lbs. (5 Nm).

27. Remove the engine sling device, safely lower the engine, remove the hanger, and install the rear engine mounting insulator and engine mounting center member.

28. Install the remaining components in the reverse order they were removed.

29. Fill the engine with oil and connect the negative battery cable. Start the engine and check for leaks.

30. Install the engine undercovers.

2.2L (5S-FE) Engine

Celica and Camry

1. Disconnect the negative battery cable. On vehicles equipped with an air bag, wait at least 90 seconds before proceeding.

2. Raise and safely support the vehicle.

3. Drain the engine oil and remove the engine undercovers.

4. Remove the front exhaust pipe.

5. Safely support the engine assembly and remove the engine mounting center member.

6. On Celica, remove the TWC as follows:
    a. Disconnect the sub oxygen sensor connector.
    b. Remove the RH exhaust manifold stay.
    c. Remove the three bolts, two nuts, the TWC with gasket, retainer, and the cushion.

7. Remove the rear end stiffener plate.

8. Remove the oil dipstick and remove the 17 bolts and two nuts attaching the oil pan to the engine.

9. Insert the blade of the SST 09032–00100 tool, or equivalent, between the oil pan and the cylinder

block; cut off the applied sealer and remove the oil pan.

NOTE: Do not use the tool for the oil pump body side and rear oil seal retainer.

To install:

10. Remove any old sealant from the oil pan flange and thoroughly clean both sealing surfaces.

11. Apply a 3–5mm bead of sealant to the oil pan flange.

NOTE: The pan must be installed within 5 minutes of sealant application or the procedure will have to be repeated.

12. Install the oil pan with the 17 bolts and two nuts. Uniformly tighten the bolts and nuts in several passes. Torque the bolts and nuts to 48 inch lbs. (5.4 Nm) and install the oil dipstick.

13. Install the rear end stiffener plate.

14. On Celica, install the TWC.

15. Install the engine mounting center member and safely lower the engine.

16. Install the front exhaust pipe.

17. Fill the engine with oil and connect the negative battery cable. Start the engine and check for leaks.

18. Recheck the engine oil level and reinstall the engine undercovers.

MR2

1. Disconnect the negative battery cable. On vehicles equipped with an air bag, wait at least 90 seconds before proceeding.

2. Drain the engine oil.

3. Remove the under engine covers.

4. Remove the right hand engine side panel.

5. Remove the front exhaust pipe.

6. Remove the stiffener plate.

7. Remove main TWC and TWC stay.

8. On the California vehicles, disconnect the sub-oxygen sensor connector.

9. Remove the oil pan and attaching bolts.

To install:

10. Thoroughly clean all mating surfaces. Install a 5mm bead of RTV in pan groove prior to installation.

11. Install the oil pan and torque the bolts to 4 ft. lbs. (5 Nm).

12. Install the stiffener plate and torque the bolts to 27 ft. lbs. (37 Nm).

13. Install the front TWC and torque the bolts to 21 ft. lbs. (29 Nm).

14. Install the TWC stay and torque the bolts to 31 ft. lbs. (42 Nm).

15. Install the front exhaust pipe.

16. Install the right hand engine side panel.

17. Connect the negative battery cable.

18. Fill with engine oil.

19. Start the engine and check for oil leaks.

20. Install the engine undercovers.

2.0L (3S-GTE) Engine

Celica

1. Disconnect the negative battery cable. On vehicles equipped with an air bag, wait at least 90 seconds before proceeding.

2. Raise the front of the vehicle and support it with jackstands or equivalent.

3. Drain the engine oil.

4. Remove the undercovers from the vehicle.

5. Disconnect the exhaust pipe from the exhaust manifold and related exhaust hangers as necessary.

6. Remove the lower suspension crossmember. Remove the center engine mount.

7. Remove the engine stiffener plate and the oil level gauge. Remove the oil pan retaining nuts and bolts, then remove the oil pan.

To install:

8. Apply a 5mm bead of RTV gasket material to the groove around the pan flange. Apply the oil within 3 minutes of application and tighten the mounting bolts and nuts to 4 ft. lbs. (5 Nm), working from the center to the ends in progressive steps.

9. Install the remaining components in the reverse order they were removed.

10. Fill with engine oil.

11. Start the engine and check for oil leaks.

MR2

1. Disconnect the negative battery cable. On vehicles equipped with an air bag, wait at least 90 seconds before proceeding.

2. Raise and safely support the vehicle.

3. Drain the engine oil.

4. Remove the under engine covers.

5. Remove the right hand engine side panel.

6. Remove the upper suspension brace.

7. Remove the air cleaner hose.

8. Remove the No. 1 and No. 2 intake air connectors.

9. If equipped, remove the cruise control actuator and accelerator linkage.

10. Remove the engine compartment electric cooling fan.

11. Disconnect the A/C compressor from the engine.
12. Remove the tail pipe.
13. Remove the front exhaust pipe.
14. Support the engine and remove the front engine mounting insulator.
15. Remove the front engine mounting bracket. If equipped, remove the clutch release cylinder.
16. Remove the stiffener plate.
17. Remove the main Three-Way Catalytic (TWC) converter stay.
18. Remove the oil pan and attaching bolts.

To install:

19. Thoroughly clean all the mating surfaces. Install a 5mm bead of RTV in pan groove prior to installation.
20. Install the oil pan and attaching bolts. Torque the bolts to 4 ft. lbs. (5.4 Nm).
21. Install the front TWC.
   a. Torque the three bolts to: 22 ft. lbs. (29 Nm).
   b. Torque the four bolts to the RH and LH stays to: 43 ft. lbs. (59 Nm)
22. Install the stiffener plate and torque the bolts to 27 ft. lbs. (37 Nm).
23. If removed, install the clutch release cylinder.
24. Install the front engine mounting bracket.

25. Install the front engine mounting insulator. Torque the nuts to 71 ft. lbs. (96 Nm).
26. Install the front exhaust pipe.
27. Install the tail pipe.
28. Install the A/C compressor. Torque the bolts to the following specifications:
   a. Bolt A: 18 ft. lbs. (25 Nm)
   b. Bolt B: 20 ft. lbs. (27 Nm)
   c. Bolt C: 27 ft. lbs. (37 Nm)
29. Install the engine compartment electric cooling fan.
30. If removed, install the cruise control actuator and the accelerator linkage.
31. Install the No. 1 and No. 2 intake air connectors.
32. Install the air cleaner hose.
33. Install the upper suspension brace. Torque the bolts to 54 ft. lbs. (73 Nm) and the nut to 47 ft. lbs. (64 Nm).
34. Install the right hand engine side panel.
35. Connect the negative battery cable.
36. Fill the engine with oil.
37. Start the engine and check for oil leaks.
38. Install the engine undercovers.
39. Safely lower the vehicle.

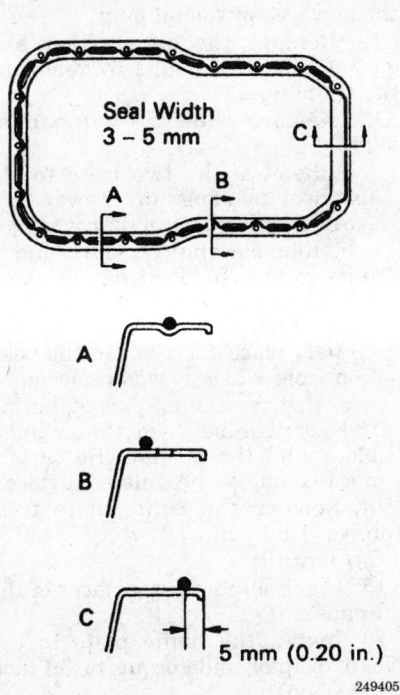

Oil pan sealing diagram — MR2 with 5S-FE and 3S-GTE engines

HINT: The stabilizer bar and drive shaft will be in the way while installing the exhaust pipe. Follow directions according to illustration.

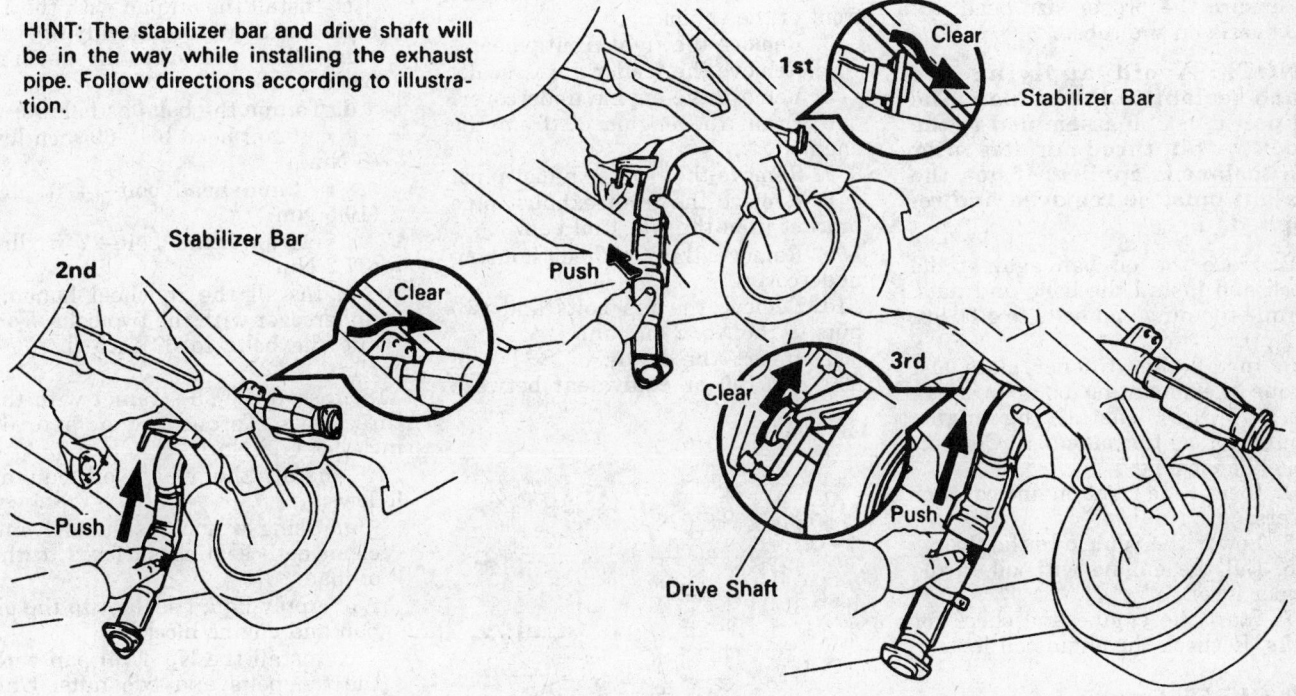

When removing or installing the front exhaust pipe position the pipe as shown — MR2 3S-GTE engine

3.0L (3VZ-FE) Engine

Camry

1. Disconnect the negative battery cable. On vehicles equipped with an air bag, wait at least 90 seconds before proceeding.
2. Raise and safely support the vehicle.
3. Drain the oil and remove the dipstick.
4. Remove the right engine undercover or covers.
5. Disconnect the exhaust pipe.
6. Remove the lower suspension crossmember, engine mounting center member and the stiffener plate.
7. Remove the fifteen bolts and four nuts from the oil pan flange.
8. Using SST No. 09032-00100 or equivalent (prybar), separate the oil pan from the cylinder block.
9. Lower the oil pan to the ground being careful not to damage the oil pan flange.

To install:

10. Using a gasket scraper and a wire brush, remove all the old gasket material from the oil pan and cylinder block gasket surfaces. Wipe the oil pan interior with a rag. Clean the contact surfaces with a non-residue type solvent.
11. Apply a thin bead of No. 102 sealant or equivalent to the oil pan. To ensure the proper size bead, cut the nozzle on the tube.

NOTE: Avoid applying too much sealant to the oil pan. The oil pan must be assembled to the block within three minutes after the sealant is applied. If not, the sealant must be removed and reapplied.

12. Place the oil pan against the block and install the bolts and nuts. Torque the nuts and bolts to 4 ft. lbs. (5 Nm).
13. Install the stiffener plate and torque the mounting bolts to 27 ft. lbs. (37 Nm). Install the engine mounting center member. Connect the exhaust pipe.
14. Install the right engine cover or covers.
15. Lower the vehicle to the floor.
16. Fill the engine with oil to the proper level.
17. Start the engine and check for leaks. Recheck the engine oil level.

3.0L (1MZ-FE) Engine

Camry and Avalon

1. Disconnect the negative battery cable from the battery. On vehicles

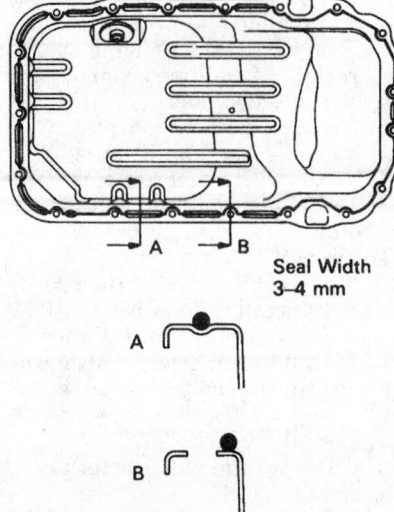

Seal Width 3–4 mm

135376

Apply sealant to the oil pan flange — 3VZ-FE engine

equipped with an air bag, wait at least 90 seconds before proceeding.
2. Raise and safely support the front of the vehicle.
3. Remove the right front wheel.
4. Remove the fender apron seal.
5. Remove the engine undercover.
6. Drain the engine oil from the engine.
7. Remove the front exhaust pipe.
8. Remove the front exhaust pipe bracket from the No. 1 oil pan.
9. Remove the flywheel housing undercover.
10. Remove the ten bolts and two nuts to the No. 2 oil pan.
11. Insert the blade of SST tool 09032–00100 or equivalent between

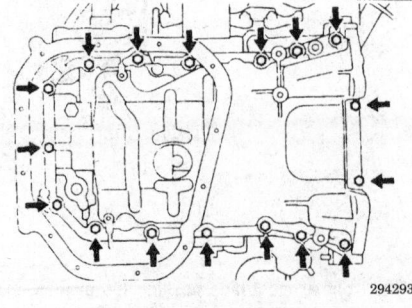

294293

No. 1 oil pan bolts — 1MZ-FE engine

the No. 1 and No. 2 oil pans. Clean the surfaces of the oil pans.
12. Remove the oil strainer and gasket from the engine by removing the three nuts.
13. Remove the No.1 oil pan as follows:

   a. Remove the two bolts to the flywheel housing undercover. Remove the flywheel undercover.

   b. Remove the 17 bolts and 2 nuts to the No. 1 oil pan. Make a note of the position of the each bolt. When replacing the bolts into the oil pan, place each bolt in the position from which it was removed.

   c. Remove the oil pan by prying the portions between the cylinder block and the oil pan. Be careful not to damage the contact surfaces.
14. Remove the baffle plate from the No. 1 oil pan.

To install:

15. Clean all mating surfaces of the oil pans.
16. Install the baffle plate to the No. 1 oil pan and torque to 69 inch lbs. (8 Nm).
17. Install the No. 1 oil pan as follows:

   a. Using a non residue solvent, clean both sealing surfaces to the oil pan.

   b. Apply liquid sealant to the oil pan and engine block.

   c. Install the oil pan with the 17 bolts and 2 nuts. Uniformly tighten the bolts and nuts in several passes.

   d. Torque the bolts as follows:
   • 10mm head bolt–69 inch lbs. (8 Nm)
   • 12mm head bolt–14 ft. lbs. (19.5 Nm)
   • 14mm head bolt–27 ft. lbs. (37.2 Nm)

   e. Install the flywheel housing undercover with the two bolts. Torque the bolts to 69 inch lbs. (7.8 Nm).
18. Install the oil strainer with the three nuts. Torque the nuts to 69 inch lbs. (7.8 Nm).
19. Install the No. 2 oil pan as follows:

   a. Using a non residue solvent, clean both sealing surfaces to the oil pan.

   b. Apply liquid sealant to the oil pan and engine block.

   c. Install the No. 2 oil pan with the ten bolts and two nuts. Uniformly tighten the bolts and nuts in several passes. Torque the bolts to 69 inch lbs. (7.8 Nm).
20. Install the flywheel housing undercover.

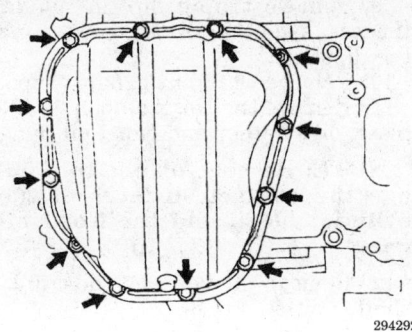

No. 2 oil pan bolts — 1MZ-FE engine

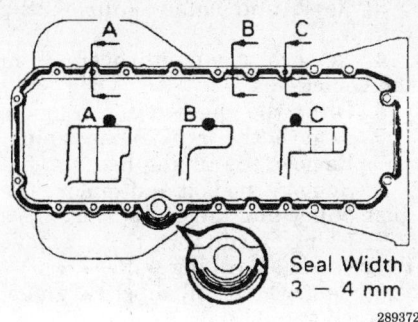

Upper oil pan sealant application — 2JZ-GE and 2JZ-GTE engines

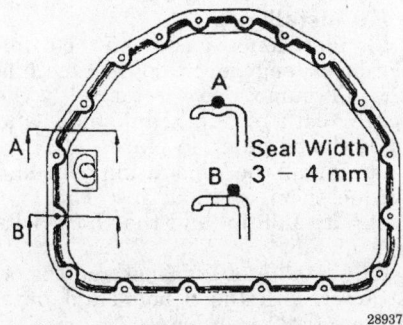

Lower oil pan sealant application — 2JZ-GE and 2JZ-GTE engines

21. Install the front exhaust pipe bracket to the No. 1 oil pan. Torque the bolts to 15 ft. lbs. (21 Nm).

22. Install the front exhaust pipe as follows:

a. Temporarily install the three new gaskets and the front exhaust pipe with the two bolts and six nuts.

b. Tighten the four nuts holding the exhaust manifolds to the front exhaust pipe. Torque the four nuts to 46 ft. lbs. (62 Nm).

c. Tighten the two bolts and two nuts holding the front exhaust pipe to the center exhaust pipe. Torque the bolts and nuts to 41 ft. lbs. (56 Nm).

d. Install the bracket with the two bolts and torque to 14 ft. lbs. (19 Nm).

e. Install the support stay with the two bolts and torque to 22 ft. lbs. (29 Nm).

23. Install the engine undercover.

24. Install the right fender apron seal.

25. Install the right front wheel and lower the vehicle.

26. Fill the engine with oil.

27. Start the engine and check for leaks.

3.0L (2JZ-GE and 2JZ-GTE) Engines

Supra

NOTE: The No. 1 oil pan can not be removed with the engine in the vehicle. The engine/transmission assembly must be removed. If only the No. 2 oil pan is being serviced, the engine/transmission assembly can remain in the vehicle.

1. Remove the engine/transmission assembly and then separate the transmission from the engine.

2. With the engine on a stand, remove the timing belt, the idler pulley and the crankshaft timing pulley.

3. Remove the oil dipstick and guide.

4. Disconnect the oil sensor lead, remove the four attaching bolts and lift off the oil level sensor. Be careful not to drop this sensor.

5. Remove the 14 bolts and two nuts and pry off the lower (No. 2) oil pan. Be careful not to damage the No. 1 pan while performing this procedure.

6. Remove the bolt and two nuts and drop down the oil strainer and gasket.

7. Remove the five bolts and two nuts and drop down the baffle plate.

8. On turbocharged engines, disconnect the turbo oil outlet pipe as follows:

a. Disconnect the two turbo oil outlet hoses.

b. Remove the two nuts, oil outlet pipe and gasket.

9. Remove the twenty two bolts and the carefully pry off the upper (No. 1) oil pan. Remove the O-ring from the cylinder block.

To install:

10. Position a new O-ring in the block and scrape off any old sealant. Apply sealant to the pan mating surface with a 1/8 inch (3–4mm) bead. Install the upper pan and tighten the

12mm bolts to 15 ft. lbs. (21 Nm) and the 14mm bolts to 29 ft. lbs. (39 Nm).

11. On turbocharged engines, install the turbo oil outlet pipe as follows:

a. Install the oil outlet pipe and gasket with the two nuts. Torque the nuts to 20 ft. lbs. (27 Nm).

b. Install the two turbo oil outlet hoses.

12. Install the baffle plate and oil strainer. Tighten them both to 78 inch lbs. (9 Nm).

13. Install the lower pan in the same manner as the upper pan and tighten the bolts to 78 inch lbs. (9 Nm).

14. Using a new gasket, install the oil level sensor and tighten it to 48 inch lbs. (5.4 Nm).

15. Install the oil dipstick and guide, the timing pulleys and belt, and reconnect the transmission to the engine.

16. Install the engine and transmission.

17. Refill all fluids.

18. Start the engine and check for leaks.

19. Road test the vehicle.

Oil Pump

REMOVAL AND INSTALLATION

1.5L (3E-E) Engine

1. Disconnect the negative battery cable. On vehicles equipped with an air bag, wait at least 90 seconds before proceeding.

2. Remove the right hand engine splash shield. Disconnect the exhaust pipe from the manifold. Remove the timing belt.

3. Drain the engine oil. Remove the oil pan, the oil strainer and the dipstick.

4. Remove the oil pump mounting bolts and the tensioner spring bracket. Remove the oil pump.

To install:

5. Install the pump with sealer and a new O-ring.

6. Install the timing belt and crankshaft pulleys.

7. Install the oil pan and dipstick assembly.

8. Install the exhaust pipe and any front end components previously removed.

9. Install the engine undercover.

10. Lower the vehicle. Refill with engine oil. Start the engine and check for leaks.

1.5L (5E-FE) Engine

1. Disconnect the negative battery terminal. On vehicles equipped with an air bag, wait at least 90 seconds before proceeding.
2. Remove the hood.
3. Raise the vehicle and support it safely, then drain the oil.
4. Remove the oil dipstick.
5. Remove the timing belt.
6. Suspend the engine with a hoist.

───── **WARNING** ─────
Do not raise the engine more than necessary, since wiring and other components can be damaged.

7. Remove the necessary items for access, then remove the oil pan.
8. Remove the three bolts and oil strainer with the O-ring.
9. Remove the pressure regulator valve.
10. Remove the oil pump by removing the nine bolts and tension spring bracket. Use a soft faced hammer to remove the oil pump.
11. Using a vise, remove the nut and oil pump pulley.
   To install:
12. Clean all gasket mating surfaces.
13. Install the oil pump pulley by first placing the driven rotors into the pump body with the marks facing the front.
14. Align the pulley and oil pump driveshaft.
15. Using a vise, install the oil pump pulley and nut. Torque the nut to 27 ft. lbs. (37 Nm).
16. Install the pressure regulator valve and torque to 22 ft. lbs. (30 Nm).
17. Using a new O-ring, install the oil strainer with the three bolts. Torque the bolts to 7 ft. lbs. (10 Nm).
18. Using a new gasket and sealer, install the oil pan.
19. Install all remaining components in the reverse order they were removed.
20. Fill the engine with oil.
21. Connect the negative battery cable to the battery.
22. Install the hood.
23. Start the engine check for leaks.

1.6L (4A-FE) Engine

Corolla and Celica

1. Disconnect the negative battery cable. On vehicles equipped with an air bag, wait at least 90 seconds before proceeding.
2. Remove the hood.

3. Raise and safely support the vehicle.
4. Remove the undercovers from the vehicle.
5. Drain the engine oil.
6. Remove the front exhaust pipe.
7. Remove the timing belt.
8. Remove the bolt and remove the idler pulley and tension spring.
9. Using a suitable tool, remove the crankshaft timing pulley.
10. Remove the oil dipstick guide, dipstick and oil pan.
11. Remove the 2 bolts and the 2 nuts, the oil strainer and gasket.
12. Remove the 7 mounting bolts, the oil pump, and the gasket.
   To install:
13. Place a new gasket on the cylinder block, engage the splined teeth of the oil pump drive rotor with the large teeth of the crankshaft, and slide the oil pump in place.
14. Install the 7 mounting bolts and torque them to 16 ft. lbs. The long bolts are 1.38 in. and the short bolts are 0.98 in.
15. Install a new gasket and the oil strainer with the 2 bolts and nuts. Torque to 82 inch lbs.
16. Install the oil pan.
17. Install the idler pulley and tension spring.
18. Install the timing belt.
19. Install the front exhaust pipe.
20. Safely lower the vehicle.
21. Install the oil dipstick guide and the dipstick.
22. Install the crankshaft timing pulley. Align the timing pulley set key with the groove of the pulley. Make sure the flange side of the pulley faces inward.
23. Install the hood.
24. Fill the engine with oil, reconnect the negative battery cable, start the engine and check for leaks.
25. Install the engine undercovers.

1.8L (7A-FE) Engine

Corolla

1. Disconnect the negative battery cable. On vehicles equipped with an air bag, wait at least 90 seconds before proceeding.
2. Remove the hood.
3. Raise and safely support the vehicle.
4. Remove the undercovers from the vehicle.
5. Drain the engine oil.
6. Remove the timing belt.
7. Remove the bolt and remove the idler pulley and tension spring.
8. Using a suitable tool, remove the crankshaft timing pulley.

9. Remove the oil dipstick guide, the crankshaft position sensor, and the dipstick.
10. Remove the front exhaust pipe.
11. Remove the No. 2 oil pan, baffle plate, oil strainer and No. 1 oil pan.

NOTE: Be careful not to damage the contact surfaces of the cylinder block and the No. 1 oil pan.

12. Remove the bolt and the crankshaft position sensor.
13. Remove the seven bolts and the oil pump.
   To install:
14. To install the oil pump, place a new gasket to the cylinder block. Engage the splined teeth of the oil pump drive rotor with the large teeth of the crankshaft, and slide the oil pump on.
15. Install the seven oil pump mounting bolts and torque them to 16 ft. lbs. The long bolts are 1.38 in. and the short bolts are 0.98 in.
16. Install the crankshaft position sensor.
17. Install the No. 1 oil pan, strainer, baffle plate and No. 2 oil pan.
18. Install a new O-ring to the dipstick guide, push the dipstick guide and the dipstick in together, and attach the crankshaft position sensor to the dipstick guide mounting bolt. Torque the bolt to 82 inch lbs. (9 Nm).
19. Install the idler pulley and the tensioner spring with the mounting bolt, do not tighten the bolt, but push the pulley as far left as possible.
20. Align the timing pulley set key with the groove of the pulley and slide the pulley on with the flange facing in.
21. Install the idler pulley and the tensioner spring with the mounting bolt, do not tighten the bolt, but push the pulley as far left as possible.
22. Install the timing belt.
23. Install the front exhaust pipe.
24. Safely lower the vehicle.
25. Reinstall the hood.
26. Fill the engine with oil, reconnect the negative battery cable, start the engine and check for leaks.
27. Reinstall the engine undercovers.

Celica

1. Disconnect the negative battery cable. On vehicles equipped with an air bag, wait at least 90 seconds before proceeding.
2. Remove the hood.
3. Raise and safely support the vehicle.
4. Remove the RH front wheel.

5. Remove the undercovers from the vehicle.

6. Drain the engine oil.

7. Remove the front exhaust pipe.

8. If equipped with A/T, disconnect the transaxle control cable from the engine mounting center member.

9. If equipped w/ A/C, disconnect the A/C pressure pipe from the engine mounting center member.

10. Remove the front exhaust pipe support bracket.

11. Raise and safely support the engine assembly.

12. Remove the rear engine mounting insulator and the engine mounting center member.

13. Install an engine hanger and suspend the engine with an engine sling device or equivalent.

14. Remove the timing belt.

15. Remove the bolt and remove the idler pulley and tension spring.

16. Using a suitable tool, remove the crankshaft timing pulley.

17. Remove the oil dipstick guide and the dipstick.

18. Remove the No. 2 oil pan, baffle plate, oil strainer, and the No. 1 oil pan.

19. Remove the oil pump 7 attaching bolts and by carefully tapping the oil pump body with a plastic tipped hammer, remove the oil pump and gasket.

To install:

20. Place a new gasket on the cylinder block, engage the spline teeth of the oil pump drive rotor with the large teeth of the crankshaft and slide the oil pump on.

21. Install the 7 oil pump mounting bolts and torque them to 16 ft. lbs. (21 Nm). The long bolts are 1.38 in. long and the short bolts are 0.98 in. long.

22. Install the No. 1 oil pan, oil strainer, baffle plate and the No. 2 oil pan.

23. Install the timing belt.

24. Remove the engine sling device, safely lower the engine, remove the hanger, and install the rear engine mounting insulator and engine mounting center member.

25. Install the front exhaust pipe support bracket.

26. If equipped with A/C, connect the A/C pressure pipe to engine mounting center member clamp.

27. If equipped with A/T, connect the transaxle control cable to the engine mounting center member.

28. Install the front exhaust pipe.

29. Install the RH front wheel and safely lower the vehicle.

30. Install the hood.

31. Fill the engine with oil, connect the negative battery cable, start the engine, and check for leaks.

32. Recheck the engine oil level and install the engine undercovers.

2.0L (3S-GTE) Engine

Celica

1. Disconnect the negative battery cable. On vehicles equipped with an air bag, wait at least 90 seconds before proceeding.

2. Raise and support the vehicle.

3. Drain the engine oil and remove the engine oil pan as outlined.

4. Remove oil strainer. Using a suitable engine hoist, support the engine (remove engine hood if necessary).

5. Remove the timing belt, idler pulley and crankshaft timing pulley dipstick guide tube, using the proper procedures.

6. Remove the oil pump retaining bolts (note location of retaining bolts as the bolts are different sizes). Remove the oil pump assembly.

To install:

7. Place a new gasket in position on the cylinder block. Engage the spline teeth of the oil pump drive rotor with the large teeth of the crankshaft and slide the oil pump on. Place the bolts in the proper positions according to length and torque the oil pump retaining bolts evenly.

8. Install the remaining components in the reverse order that they were removed.

9. Fill the engine with oil, connect the negative battery cable, start the engine, and check for leaks.

10. Recheck the engine oil level and install the engine undercovers.

2.2L (5S-FE) Engine

Celica

1. Disconnect the negative battery cable. On vehicles equipped with an air bag, wait at least 90 seconds before proceeding.

2. Raise and safely support the vehicle.

3. Drain the engine oil and remove the engine undercovers.

4. Remove the front exhaust pipe.

5. Safely support the engine assembly and remove the engine mounting center member.

6. Disconnect the sub oxygen sensor wiring. Remove the RH side exhaust manifold stay, TWC with gasket, retainer, and cushion.

7. Remove the rear end stiffener plate.

8. Remove the oil dipstick and oil pan.

9. Insert the blade of the SST 09032-00100 tool between the oil pan and the cylinder block; cut off the applied sealer and remove the oil pan.

NOTE: Do not use the tool for the oil pump body side and rear oil seal retainer.

10. Remove the oil strainer, baffle plate, and the gasket.

11. Safely support the engine with an engine sling or equivalent.

12. Remove the timing belt.

13. Remove the No. 2 idler pulley and the crankshaft timing pulley.

14. Using a suitable tool, remove the oil pump pulley.

15. Remove the 12 bolts and remove the oil pump and gasket.

To install:

16. Install the oil pump with a new gasket and the 12 mounting bolts. Uniformly tighten the oil pump bolts in several passes. Torque the bolts to 78 inch lbs. (9 Nm). Bolt A is 0.98 in. long and bolt B is 1.38 in. long.

17. Align the cut outs of the pulley and the shaft and slide on the oil pump pulley. Torque the pulley nut to 18 ft. lbs. (24 Nm).

18. Install the crankshaft timing pulley and the No. 2 idler pulley.

19. Install the timing belt.

20. Remove the engine sling or equivalent and safely lower the engine.

21. Install a new gasket, the oil strainer and baffle plate with the two bolts and two nuts. Torque the bolts and nuts to 48 inch lbs. (5 Nm).

22. Remove any old sealant from the oil pan flange and thoroughly clean both sealing surfaces. Apply a new bead (3–5mm) of sealant to the oil pan flange.

NOTE: The pan must be installed within 5 minutes of sealant application or the procedure will have to be repeated.

23. Install the oil pan and dipstick.

24. Install the rear end stiffener plate.

25. Install the cushion, retainer, and a new gasket to the front TWC. Install the TWC with the three bolts and two nuts. Torque the bolts and nuts to 21 ft. lbs. (29 Nm). Install the RH side exhaust manifold stay with the two bolts and two new nuts. Torque the bolts and nuts to 31 ft. lbs. (42 Nm) and connect the sub oxygen sensor connector.

26. Safely support the engine assembly and install the engine mounting center member.

27. Install the front exhaust pipe.

28. Lower the vehicle and fill the engine with oil. Connect the negative

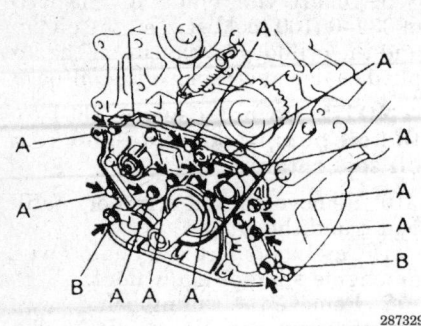

The oil pump mounting bolts — 5S-FE engine

287329

battery cable, start the engine, and check for leaks.

29. Recheck the engine oil level and install the engine undercovers.

Camry

1. Disconnect the negative battery cable. On vehicles equipped with an air bag, wait at least 90 seconds before proceeding.
2. Remove the hood.
3. Raise and safely support the vehicle.
4. Drain the engine oil.
5. Remove the front exhaust pipe.
6. Remove the rear end stiffener plate.
7. Remove the oil dipstick and oil pan.
8. Remove the oil strainer and gasket.
9. Carefully suspend the engine with a sling device or equivalent and remove the timing belt and pulleys.
10. If equipped, remove the crankshaft position sensor.
11. Remove the 12 mounting bolts, the oil pump, and the gasket.

To install:

12. Install a new gasket and the oil pump with the 12 bolts. Torque the bolts to 82 inch lbs. (9 Nm).

NOTE: The long bolts are 1.38 in. and all the others are 0.98 in.

13. Install the crankshaft position sensor, if removed.
14. Install the timing belt and pulleys, remove the engine sling device and safely lower the engine.
15. Install the oil strainer with a new gasket, then tighten to 48 inch lbs. (5 Nm).
16. Install the oil pan and dipstick.

NOTE: The pan must be installed within five minutes of sealant application or the procedure will have to be repeated.

17. Install the rear end stiffener plate and torque the three bolts to 27 ft. lbs. (37 Nm).

18. Install the front exhaust pipe.
19. Lower the vehicle and fill the engine with oil.

——— WARNING ———

Be sure to prime the oil pump prior to initial engine start-up or engine damage may occur.

20. Connect the negative battery cable, start the engine, and check for leaks.
21. Recheck the engine oil level and install the hood.

MR2

1. Disconnect the negative battery cable. On vehicles equipped with an air bag, wait at least 90 seconds before proceeding.
2. Drain the engine oil.
3. Remove the under engine covers and right hand engine side panel.
4. Remove the front exhaust pipe, TWC and the TWC stay.
5. Remove the oil pan and oil strainer.
6. Suspend the engine with sling devise.
7. Remove the timing belt, No. 2 idler pulley and crankshaft timing pulley.
8. Remove the oil pump pulley.
9. Remove the oil pump and attaching bolts.

To install:

NOTE: Lubricate the oil pump gears with clean engine oil prior to installing the oil pump.

10. Install the oil pump with new gasket and torque the bolts to 7 ft. lbs. (9 Nm).
11. Install the oil pump pulley.
12. Install the crankshaft timing pulley.
13. Install the No. 2 idler pulley.
14. Install the timing belt.
15. Remove the engine sling.
16. Install the oil strainer and oil pan.
17. Install the stiffener plate and torque the bolts to 27 ft. lbs. (37 Nm).
18. Install the Front TWC and torque the bolts to 21 ft. lbs. (29 Nm).
19. Torque the bolts for the TWC stay to 31 ft. lbs. (42 Nm).
20. Install the front exhaust pipe.
21. Install the right hand engine side panel.
22. Connect the negative battery cable.
23. Fill with engine oil.

NOTE: It is recommended to prime the oil pump prior to initial engine start-up. This can be done by disabling the ignition system and cranking the engine.

Once oil pressure is indicated, enable the ignition system and proceed with the installation procedure.

24. Start the engine and check for oil leaks.
25. Install the engine undercovers.
26. Verify that the oil pressure is within the proper specifications.

3.0L (3VZ-FE) Engine

Camry

1. Disconnect the negative battery cable. On vehicles equipped with an air bag, wait at least 90 seconds before proceeding.
2. Remove the hood
3. Raise and safely support the vehicle.
4. Drain the coolant and the engine oil.
5. Remove the front exhaust pipe.
6. Remove the stiffener plate.
7. Remove the oil pan, oil strainer, O-ring and oil pan baffle plate.
8. Lower the vehicle and remove the washer tank.
9. Remove the alternator and the A/C compressor with the hoses still connected. Also, remove the compressor bracket and the power steering pump adjusting bar.
10. Disconnect the lower radiator hose and water inlet pipe.
11. Suspend the engine with an engine lifting device.
12. Remove the timing belt.
13. Remove the No. 1 idler and the crankshaft timing pulleys.
14. Remove the nine oil pump retaining bolts.
15. With a soft-faced hammer or rubber mallet, tap the oil pump loose from the block.
16. Remove the oil pump gasket and replace a new one.

To install:

17. Clean the cylinder block and oil pump gasket contact surfaces.
18. Apply a 2-3mm bead of sealant to the oil pump. Install the pump within 5 minutes or the sealant must be removed and reapplied.
19. Position a new O-ring on the engine. Engage the spline teeth of the oil pump drive gear with the large teeth of the crankshaft and install the oil pump onto the block using the nine bolts.
20. Tighten the 12mm bolts (C and D) to 14 ft. lbs. (20 Nm) and the 14mm bolts (A and B) to 30 ft. lbs. (41 Nm).
21. Install the crankshaft timing pulley and the No. 1 idler pulley.
22. Install the timing belt and remove the engine lifting device.

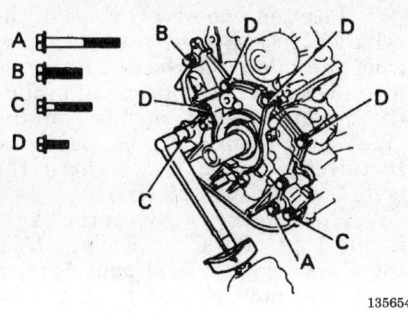

Oil pump bolts — 3VZ-FE engine

23. Use and new O-ring and install the water inlet pipe and connect the lower radiator hose.

24. Install the power steering belt adjusting bar, the A/C compressor bracket and the alternator.

25. Install the washer tank.

26. Raise the vehicle and install the baffle plate.

27. Place a new O-ring on the strainer pipe outlet and install the strainer.

28. Install the oil pan.

29. Install the stiffener plate and the front exhaust pipe.

30. Lower the vehicle to the floor.

31. Refill the engine coolant and the engine oil.

32. Connect the negative battery cable.

33. Start the engine and inspect for leaks.

34. Install the hood on the vehicle.

35. Recheck the engine oil level and coolant level.

3.0L (1MZ-FE) Engine

Camry and Avalon

1. Disconnect the negative battery cable from the battery. On vehicles equipped with an air bag, wait at least 90 seconds before proceeding.

2. Raise and safely support the front of the vehicle.

3. Remove the right front wheel.

4. Remove the fender apron seal.

5. Remove the engine undercover.

6. Drain the engine oil from the engine.

7. Remove the front exhaust pipe.

8. Remove the front exhaust pipe bracket from the No. 1 oil pan.

9. Remove the alternator drive belt from the engine.

10. Disconnect the A/C compressor from the engine.

11. Remove the power steering pump drive belt and adjusting strut.

12. Remove the timing belt and belt pulleys from the engine.

13. Remove the rear timing belt cover from the engine by removing the wire clamps and six bolts.

14. Remove the A/C compressor housing bracket by removing the three bolts.

15. Remove the No. 2 oil pan, oil strainer, No.1 oil pan and baffle plate.

16. Remove the crankshaft position sensor by removing the connector and bolt.

17. Remove the oil pump as follows:

   a. Remove the nine bolts. Make a note of the position of the each bolt. When replacing the bolts into the oil pump body, place each bolt in the position from which it was removed.

   b. Remove the oil pump body by prying between the oil pump and main bearing cap.

   c. Remove the O-ring from the cylinder block.

   d. Remove the plug, gasket, spring, and relief valve from the oil pump body.

   e. Remove the nine screws, pump body cover, drive, and driven rotors.

To install:

18. To install the oil pump:

   a. Install the driven rotors, drive, pump body cover, and then install the nine screws.

   b. Install the oil pump relief valve, spring, gasket, and the plug to the oil pump body.

   c. Place a new O-ring on the cylinder block.

   d. Using a non residue solvent, clean both sealing surfaces to the oil pump.

   e. Apply liquid sealant to the oil pump and engine block.

   f. Install the oil pump to the engine block. Make sure to engage the spline teeth of the oil pump drive gear with the large teeth of the crankshaft.

   g. Install the nine bolts to the oil pump and uniformly tighten the bolts in several passes. Torque the bolts as follows:

   • 10mm head—69 inch lbs. (8 Nm)

   • 12mm head—14 ft. lbs. (20 Nm)

19. Install the crankshaft position sensor and install the bolt. Torque the bolt to 69 inch lbs. (8 Nm).

20. Install the baffle plate to the No. oil pan and torque to 69 inch lbs. (8 Nm).

21. Install the No. 1 oil pan, oil strainer and No. 2 oil pan.

22. Install the A/C compressor housing bracket and three bolts. Tor-

que the three bolts to 18 ft. lbs. (25 Nm).

23. Install the rear timing belt cover with the six bolts and three wire clamps. Torque the bolts to 74 inch lbs. (9 Nm).

24. Install the timing belt pulleys.

25. Install the timing belt.

26. Install the adjusting strut and power steering drive belt as follows:

   a. Temporarily install the adjusting strut with the bolt and nut.

   b. Install the drive belt and then install the pivot and adjusting bolts. Torque the bolt and nut to 32 ft. lbs. (43 Nm).

27. Install the A/C compressor.

28. Install the alternator drive belt.

29. Install the front exhaust pipe bracket to the No. 1 oil pan. Torque the bolts to 15 ft. lbs. (21 Nm).

30. Install the front exhaust pipe.

31. Install the engine undercover.

32. Install the right fender apron seal.

33. Install the right front wheel and lower the vehicle.

34. Fill the engine with oil.

35. Connect the negative battery cable to the battery.

36. Start the engine and check for leaks.

37. Recheck the engine oil.

2.0L (3S-GTE) Engine

MR2

1. Disconnect the negative battery cable. On vehicles equipped with an air bag, wait at least 90 seconds before proceeding.

2. Drain the engine oil.

3. Remove the under engine covers.

4. Remove the right hand engine side panel.

5. Remove the upper suspension brace.

6. Remove the air cleaner hose.

7. Remove the No. 1 and No. 2 intake air connectors.

8. If equipped, remove the cruise control actuator and accelerator linkage.

9. Remove the engine compartment electric cooling fan.

10. Disconnect the A/C compressor from the engine.

11. Remove the tail pipe and front exhaust pipe.

12. Support the engine with a jack, but leave enough clearance that the oil pan can still be removed.

13. Remove the front engine mounting insulator.

14. Remove the front engine mounting bracket and clutch release cylinder.

15. Remove the stiffener plate.

16. Remove the main TWC stay.

17. Remove the oil pan and oil strainer.

18. Suspend the engine with a sling devise.

19. Remove the timing belt and No. 2 idler pulley.

20. Remove the crankshaft timing pulley.

21. Remove the oil pump pulley.

22. Remove the oil pump attaching bolts.

23. Using a plastic faced hammer, remove the oil pump by carefully tapping the oil pump body.

To install:

NOTE: Lubricate the oil pump gears with clean engine oil prior to installing the oil pump.

24. Install the oil pump with new gasket. Torque the bolts to 5 ft. lbs. (8 Nm).

25. Install the oil pump pulley.

26. Install the crankshaft timing pulley.

27. Install the No. 2 idler pulley.

28. Install the timing belt.

29. Remove the engine sling.

30. Install the oil strainer and oil pan.

31. Install the front TWC.

32. Install the stiffener plate and torque the bolts to 27 ft. lbs. (37 Nm).

33. Install the clutch release cylinder and front engine mounting bracket.

34. Install the front engine mounting insulator. Torque nuts to 71 ft. lbs. (96 Nm).

35. Install the front exhaust pipe and the tail pipe.

36. Install the A/C compressor.

37. Install the engine compartment electric cooling fan.

38. Install the cruise control actuator and accelerator linkage.

39. Install the No. 1 and No. 2 intake air connectors.

40. Install the air cleaner hose.

41. Install the upper suspension brace. Torque the bolts to 54 ft. lbs. (73 Nm) and the nut to 47 ft. lbs. (64 Nm).

42. Install the right hand engine side panel.

43. Connect the negative battery cable.

44. Fill with engine oil.

NOTE: It is recommended to prime the oil pump prior to initial engine start-up. This can be done by disabling the ignition system and cranking the engine. Once oil pressure is indicated, enable the ignition system and proceed with the installation procedure.

45. Start the engine and check for oil leaks.

46. Install the engine undercovers.

47. Verify that the oil pressure is within the proper specifications.

3.0L (2JZ-GE and 2JZ-GTE) Engines

1. Disconnect the negative battery cable. On vehicles equipped with an air bag, wait at least 90 seconds before proceeding.

2. Remove the engine and transmission.

3. Separate the transmission from the engine and mount the engine on a service stand.

4. Remove the timing belt.

5. Remove the idler pulley.

6. Remove the crankshaft timing pulley.

7. Remove the oil dipstick and tube, oil level sensor, No. 2 (lower) oil pan, oil strainer and oil baffle plate.

8. If the engine is turbocharged, remove the turbo oil outlet pipe by disconnecting the two turbo oil hoses and two nuts.

9. Remove the No.1 (upper) oil pan.

10. Remove the nine mounting bolts to the oil pump body. Carefully drive the pump off the cylinder block using a brass drift. Remove the two O-rings.

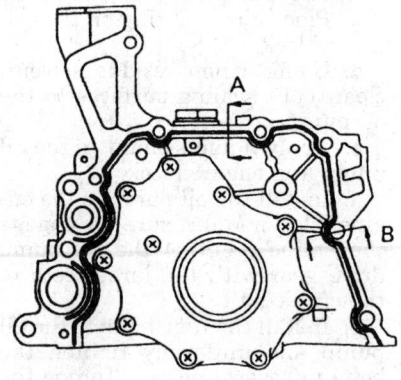

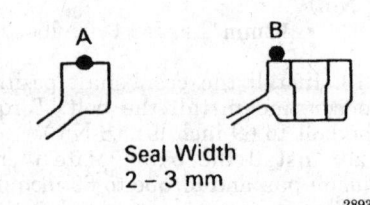

Seal Width
2 – 3 mm

289391

Oil pump sealant application — 2JZ-GE and 2JZ-GTE engines

To install:

11. Position two new O-rings in the cylinder block. Scrape any old sealant from the mating surfaces. Draw a 1/8 inch (3–4mm) bead of sealant around the pump mating surface, taking great care around the oil passages. Install the pump and tighten the bolts to 15 ft. lbs. (21 Nm).

12. Place a new O-ring on the block. Remove all of the old sealant from the block and No. 1 oil pan. Apply a bead of sealant around the No. 1 oil pan. Avoid excessive application. Install the No.1 oil pan and tighten the bolts with 12mm heads to 15 ft. lbs. (21 Nm). Tighten the bolts with 14mm heads to 29 ft. lbs. (39 Nm).

13. If removed, install the turbo oil outlet pipe by installing the two bolts and two hoses. Torque the bolts to 20 ft. lbs. (27 Nm).

14. Install the oil baffle plate, oil strainer, No. 2 oil pan, oil lever sensor and oil dipstick with a new O-ring.

15. Install the crankshaft and idler pulley.

16. Install the timing belt.

17. Connect the engine and transmission

18. Install the engine and transmission.

19. Fill all fluids.

20. Connect the negative battery cable.

21. Start the engine and check for leaks.

TRANSMISSION AND TRANSAXLE

Manual Transmission Assembly

REMOVAL AND INSTALLATION

Supra

1. Disconnect the negative battery cable. On vehicles equipped with an air bag, wait at least 90 seconds before proceeding.

2. Remove the fan shroud set bolts.

3. Remove the shift lever knob. Using a flat bladed tool, pry out the upper console panel. Remove the four mounting bolts. Remove the shift and select lever boot No. 1 and No. 2. Remove the four mounting bolts.

4. Raise and support the vehicle.

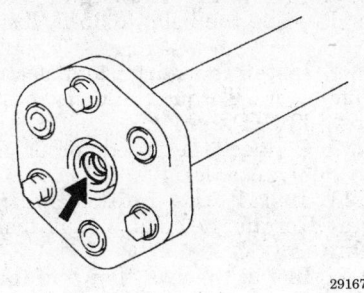

Apply grease to the flexible coupling center bushings — Supra

291677

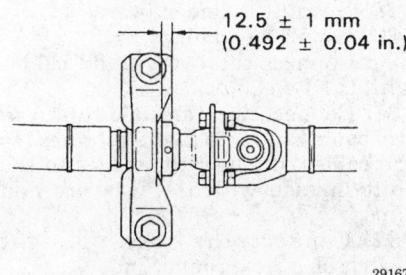

12.5 ± 1 mm
(0.492 ± 0.04 in.)

291678

Propeller shaft joint angle — Supra

5. Drain the transmission fluid.

6. Remove the oxygen sensor, exhaust front pipe and pipe support bracket.

7. Remove the exhaust center pipe. Remove the heat insulator.

8. Remove the center floor crossmember brace.

9. Matchmark and remove the driveshafts. Cap the end of the transmission to prevent leakage.

10. Remove the transmission lever bolt and nut, remove the shift lever.

11. Remove the two bolts and the clutch release cylinder. Remove the bolt, ground cable and flexible hose bracket.

12. Disconnect the starter connector.

13. Disconnect the back-up light switch and vehicle speed sensor connectors.

14. If equipped with a V160 transmission, remove the clutch cover set bolts and the service hole cover. Place matchmarks on the flywheel and clutch cover. Remove the six bolts.

15. Jack up the transmission slightly.

16. Remove the rear engine mounting member.

17. Lower the engine rear side and remove the two starter bolts and remove the starter.

18. Remove the remaining transmission mounting bolts. Lower the engine rear side and remove the transmission from the engine.

19. If equipped with a V160 transmission, remove the four bolts and the shift lever retainer from the transmission.

20. Remove the rear engine mounting from the transmission.

To install:

21. Install the rear engine mount to the transmission, torque the bolts to 18 ft. lbs. (25 Nm).

22. If equipped with a V160 transmission, install the shift lever retainer. Torque the bolts to 14 ft. lbs. (19 Nm). Torque the through-bolt and nut to 18 ft. lbs. (25 Nm).

23. Raise the engine front side, align the input spline with the clutch disc and install the transmission to the engine. Torque the mounting bolts to 53 ft. lbs. (72 Nm).

24. Install the starter, torque the bolts to 29 ft. lbs. (39 Nm).

25. Install the rear engine mounting member, torque the nuts to 10 ft. lbs. (13 Nm) and the bolts to 19 ft. lbs. (25 Nm).

26. If equipped with a V160 transmission, align matchmarks and then install the clutch cover set bolts, torque to 14 ft. lbs. (19 Nm). Install the service hole cover and torque the bolts to 9 ft. lbs. (12 Nm).

27. Connect the vehicle speed sensor and the back-up light switch connectors.

28. Connect the starter wire connector.

29. Install the clutch release cylinder, torque the bolts to 9 ft. lbs. (12 Nm). Install and torque the bolt with the clamp and ground cable to 53 ft. lbs. (72 Nm).

30. Install the transmission shift lever, torque the bolts to 14 ft. lbs. (19 Nm).

31. Apply grease to the flexible coupling centering bushings. Install the driveshafts.

32. Inspect the propeller shaft joint angle.

33. Install the center floor crossmember brace, torque the bolts to 9 ft. lbs. (13 Nm).

34. Install the heat insulator, torque the bolts to 48 inch lbs. (5 Nm).

35. Using new gaskets, install the center pipe. Torque the bolts to 14 ft. lbs. (19 Nm).

36. Using new gaskets, install the front pipe. Torque the bolts to 43 ft. lbs. (58 Nm).

37. Install the pipe support bracket and torque the bolts to 27 ft. lbs. (37 Nm).

38. Install the front pipe set bolts and nuts, torque to 32 ft. lbs. (43 Nm).

39. Install the oxygen sensor and cover using a new gasket. Torque to 13 ft. lbs. (18 Nm).

40. Fill the transmission with proper gear oil. Lower the vehicle.

41. Install the shift lever mount bolts, torque to 69 inch lbs. (8 Nm). Install the shift and select lever boot No. 1 and No. 2. Install the mounting bolts.

42. Fit the upper console panel to the console box with the four clips.

43. Install the shift lever knob.

44. Install the fan shroud set bolts.

45. Install the negative battery cable.

Transaxle Assembly

REMOVAL AND INSTALLATION

Tercel

1. Disconnect the negative battery cable, then the positive. On vehicles equipped with an air bag, wait at least 90 seconds before proceeding.

2. Remove the battery.

3. Remove the air cleaner case assembly with the air hose.

4. Remove the clutch release cylinder and tube clamp.

5. Remove the two bolts and ground cable from the transaxle side and engine left mounting bracket side.

6. Disconnect the back-up light switch electrical connector.

7. Remove the clips and washers that connect the shifter control cables to the transaxle.

8. Remove the retainers that attach the shifter control cables to the transaxle and position the cables out of the way.

9. Remove the upper transaxle attaching bolts.

10. Safely raise and support the vehicle.

11. Remove the front wheels.

12. Remove the engine undercovers, if equipped.

13. Drain the oil from the transaxle.

14. Remove the left and right halfshafts.

15. Remove the front exhaust pipe, if necessary.

16. Disconnect the speedometer cable from the transaxle.

17. Remove the starter electrical connectors and mounting bolts. Remove the starter from the engine.

18. Remove the rear engine mounting insulator and bracket.

19. Using a block of wood and a floor jack, place it under the engine's oil pan and support it.

20. Using a floor jack, position it under the transaxle and support its weight.

21. Remove the left engine mount.

22. Remove the transaxle attachment bolts from the engine.

23. Lower the left side of the engine and remove the transaxle from the vehicle.

To install:

24. Align the input shaft spline with the clutch disc and install the transaxle to the engine. Install the transaxle attaching bolts and torque the bolts to specification.

25. Install the left engine mounting bracket with the two bolts. Torque the bolts to 36 ft. lbs. (48 Nm).

26. Install the rear mounting insulator with the six bolts. Torque the bolts as follows:

- A — Body side bolt through brackets to 58 ft. lbs. (78 Nm)
- B — Through bolt to 47 ft. lbs. (64 Nm)
- C — Body side bolt though insulator to 67 ft. lbs. (90 Nm)

27. Remove the jacks from under the transaxle and engine.

28. Install the starter and electrical connectors. Torque the starter bolts to 29 ft. lbs. (39 Nm).

29. Connect the speedometer to the transaxle.

30. Install the front exhaust pipe, if removed.

31. Install the right and left halfshafts.

32. Fill the transaxle with 75W–90 or 80W–90 gear oil.

33. Install the undercovers, if removed.

34. Install the wheels.

35. Lower the vehicle.

36. Position the shift control cables and install the retainers, the washers and the clips.

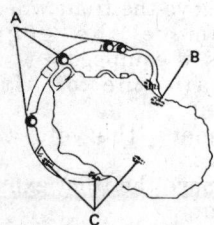

Torque A: 64 Nm (650 kgf.cm, 47 ft.lbf)
Torque B: 46 Nm (470 kgf.cm, 34 ft.lbf)
Torque C: 7 Nm (75 kgf.cm, 65 ft.lbf)

291015

Torque the transaxle mounting bolts to specification — Tercel

37. Connect the back-up light switch electrical connector.

38. Install the clutch release cylinder and tube clamp.

39. Install the air cleaner assembly with the air hose.

40. Connect the two ground cable to the transaxle.

41. Install the battery.

42. Road test the vehicle.

Paseo

1. Disconnect the negative battery cable. On vehicles equipped with an air bag, wait at least 90 seconds before proceeding.

2. Remove the air cleaner case assembly with the air hose.

3. Remove the clutch release cylinder and tube clamp.

4. Disconnect the backup light switch electrical connector.

5. Remove the transaxle control cable end clips and washers, then remove the retainer clips from the cables. Remove the transaxle shift control cables from the transaxle.

6. Remove the clutch release cylinder bracket and the ground cable.

7. Remove the bolts from the upper transaxle mount.

8. Raise the vehicle and support it safely.

9. Remove the engine undercovers and drain the transaxle fluid.

10. Disconnect the speedometer cable from the transaxle.

11. Disconnect both halfshafts from the transaxle.

12. Remove the engine rear mounting bracket.

13. Remove the starter assembly.

14. Support the engine and transaxle assembly using the proper equipment.

15. Disconnect the left engine mount.

16. Remove the bolts mounting the transaxle to the engine.

17. Carefully remove the transaxle assembly from the vehicle. Lower the engine left side to aid in the transaxle removal.

To install:

18. Align the input shaft spline with the clutch disc, and install the transaxle to the engine. Torque the bolts to the following specifications:

- Bolt A: 47 ft. lbs. (64 Nm)
- Bolt B: 34 ft. lbs. (46 Nm)
- Bolt C: 65 inch lbs. (7 Nm)
- Bolt D: 17 ft. lbs. (24 Nm)

19. Install the rear engine mounting brackets. Torque the bolts as follows:

- Bolt A: 67 ft. lbs. (90 Nm)
- Bolt B: 58 ft. lbs. (78 Nm)
- Bolt C: 47 ft. lbs. (64 Nm)

20. Install the left engine mount and torque the bolts to 35 ft. lbs. (48 Nm).

21. Install the starter and electrical connectors. Torque the bolts to 29 ft. lbs. (39 Nm).

22. Connect the speedometer cable to the transaxle.

23. Install the intake manifold stay. Torque the bolts to 14 ft. lbs. (20 Nm).

24. Install the sway bar and torque the bracket bolts to 14 ft. lbs. (20 Nm).

25. Install the front exhaust pipe.

26. Connect the halfshafts to the transaxle.

27. Fill the transaxle with gear oil.

28. Install the undercovers.

29. Lower the vehicle.

30. Connect the two ground cables with the two bolts.

31. Connect the transaxle shift control cables and install the retainers to the cables. Connect the cables to the shift linkage with the washers and clips.

32. Connect the back-up light switch electrical connector.

33. Install the clutch release cylinder and tube clamp.

34. Install the air cleaner assembly with the air hose.

35. Connect the negative battery cable.

36. Adjust the clutch and perform a road test of the vehicle.

Corolla

1. Disconnect the negative battery cable. On vehicles equipped with an air bag, wait at least 90 seconds before proceeding.

2. Remove the air cleaner case assembly with hose.

3. On 1993–95 models, remove the coolant reservoir tank.

4. Remove release cylinder tube bracket by removing the bolt.

5. Remove the clutch release cylinder by removing the two bolts.

6. Disconnect the back-up light switch connector.

7. Remove the ground cable.

8. Disconnect shift cables from the transaxle.

9. Disconnect the vehicle speed sensor connector or the speedometer cable.

10. Disconnect the engine wire clamps.

11. Remove the starter set bolt from the transaxle upper side.

12. Remove the two transaxle upper mounting bolts.

13. Remove the engine left mounting stay by removing the two bolts.

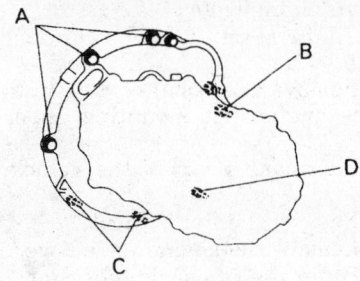

Torque A: 64 N·m (650 kgf·cm, 47 ft·lbf)
Torque B: 46 N·m (470 kgf·cm, 34 ft·lbf)
Torque C: 7 N·m (75 kgf·cm, 65 in.·lbf)
Torque D: 24 N·m (240 kgf·cm, 17 ft·lbf)

291073

Torque specifications for mounting transaxle to the engine — Paseo

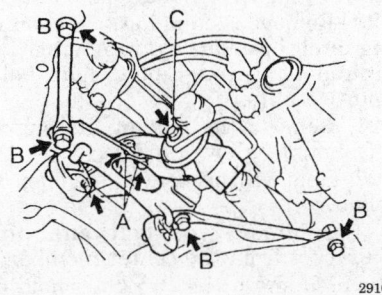

291075

Rear engine mounting bolt identification —
Paseo

14. Remove the engine left mounting set bolt from the rear side.
15. Install engine support fixture. Raise and safely support the vehicle.
16. Remove the front wheels. Remove the undercovers from the vehicle.
17. Drain the transaxle oil.
18. Disconnect the lower ball joint from the lower arm by removing the bolt and two nuts.
19. Remove the halfshafts.
20. Remove the front exhaust pipe.
21. Remove the hole cover.
22. Remove the engine front mounting set bolts.
23. Disconnect engine rear mounting by removing the three set nuts.
24. Place a jack under the engine center support member.
25. Remove the engine center support member by removing the eight bolts.
26. Remove the starter by disconnecting the electrical leads and removing the lower bolt.
27. For the 1.6L engine, remove stiffener plate.
28. Raise the transaxle and engine slightly with a jack.
29. For the 1.8L engine, remove the transaxle mounting bolts from the engine rear end plate side.

30. Remove the engine left mounting set bolts from the front side.
31. Remove the transaxle mounting bolts from the engine front side. Remove the transaxle mounting bolts from the engine rear side. Lower the engine left side and remove the transaxle from the engine.

To install:
32. Align the input shaft with the clutch disc and install the transaxle to the engine and torque the engine to transaxle bolts to 47 ft. lbs. (64 Nm) for 12mm bolts and 34 ft. lbs. (46 Nm) for 10mm bolts.
33. Raise the transaxle and engine slightly and install the left engine mounting. Install the left engine to transaxle mount bolts torque to 41 ft. lbs. (56 Nm).
34. For 1.8L engine, install the transaxle mounting bolts to the engine rear end plate side. Torque the bolts to 17 ft. lbs. (23 Nm).
35. For the 1.6L engine, install stiffener plate and torque the bolts to 17 ft. lbs. (23 Nm).
36. Install starter, lower bolt and connect the electrical connector to the starter. Torque the bolt to 29 ft. lbs. (39 Nm).
37. Install engine center support member and torque the support member to radiator support bolts to 45 ft. lbs. (61 Nm). Torque the support member to frame bolts to 152 ft. lbs. (206 Nm).
38. Connect the engine rear mounting torque bolts to 35 ft. lbs. (48 Nm).
39. Connect the engine front mounting and torque bolts to 47 ft. lbs. (64 Nm). Install hole covers.
40. Install front exhaust pipe.
41. Install the halfshafts.
42. Connect the lower ball joint to lower arm. Torque the bolt and nuts to 105 ft. lbs. (142 Nm).
43. Fill the transaxle with the correct gear oil.
44. Install undercovers.
45. Remove the engine support fixture.

46. Install front wheels and lower the vehicle.
47. Install the engine left mounting set bolt to the rear side. Torque the bolt to 41 ft. lbs. (56 Nm).
48. Install engine left mounting stay. Torque the bolt to 15 ft. lbs. (21 Nm).
49. Install the two transaxle upper side mounting bolts and torque to 29 ft. lbs. (39 Nm).
50. Install the starter set bolt to the transaxle upper side. Torque the bolt to 29 ft. lbs. (39 Nm).
51. Connect the engine wire clamps.
52. Connect the vehicle speed sensor connector or the speedometer cable.
53. Connect the transaxle shift cables and install ground cable.
54. Connect the back-up light switch connector.
55. Install release cylinder and release cylinder tube bracket. Torque the bolts to 9 ft. lbs. (12 Nm).
56. On 1993–95 models, install coolant reservoir tank.
57. Install air cleaner case assembly.
58. Connect the negative battery cable and check the front wheel alignment.
59. Road test the vehicle and check for abnormal noise and smooth shifting.

Celica

With C52 Transaxle

1. Disconnect the negative battery cable from the battery. On vehicles equipped with an air bag, wait at least 90 seconds before proceeding.
2. Remove the air cleaner case assembly with hose.
3. Remove release cylinder tube bracket by removing the bolt.
4. Remove the clutch release cylinder by removing the two bolts.
5. Disconnect the back-up light switch connector.
6. Remove the ground cable on the transaxle by removing the bolt.
7. Disconnect shift cables from the transaxle.
8. Disconnect the vehicle speed sensor connector or the speedometer cable.
9. Disconnect the engine wire clamps.
10. Remove the starter set bolt from the transaxle upper side.
11. Install a engine support fixture. Raise and safely support the vehicle.
12. Remove the front wheels.
13. Remove the undercovers from the vehicle.
14. Drain the transaxle oil.

15. Remove the halfshafts.
16. Remove the front exhaust pipe and support bracket.
17. Remove the starter.
18. Place a jack under the engine center support member.
19. Remove the engine center support member.
20. Disconnect engine rear mounting by removing the three set nuts.
21. Remove the engine front mounting bracket and insulator by removing the through-bolt and two set bolts.
22. Raise the transaxle and engine slightly with a jack.
23. Remove the engine left mounting bracket by removing the three set bolts.
24. Remove the transaxle mounting bolts from the engine rear end plate side.
25. Remove the transaxle case protector by removing the two bolts.
26. Lower the engine left side and remove the three upper transaxle bolts.
27. Remove the transaxle from the engine.

To install:
28. Connect the transaxle to the engine and raise the engine right side. Align the input shaft with the clutch disc and install the transaxle to the engine. Torque the three upper transaxle bolts to 47 ft. lbs. (64 Nm).
29. Install the transaxle case protector and torque the two bolts to 9 ft. lbs. (13 Nm).
30. Install and torque the four transaxle lower bolts as follows:
• Bolt A: 17 ft. lbs. (23 Nm)
• Bolt B: 34 ft. lbs. (46 Nm).
31. Raise the transaxle and engine slightly. Install the left engine mounting bracket to the engine left mounting insulator. Install the three set bolts and torque to 47 ft. lbs. (64 Nm).
32. Install the engine front mounting bracket and insulator and torque the two bracket bolts to 57 ft. lbs. (77

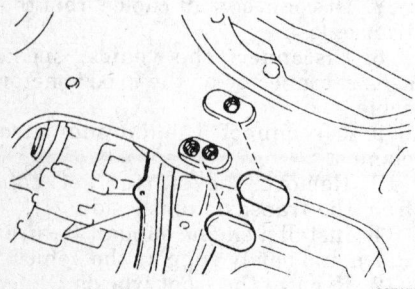

Rear mounting insulator set bolt — Celica C52

Nm). Torque the through-bolt to 64 ft. lbs. (87 Nm).
33. Install the engine rear mounting bracket and insulator and torque the bracket bolts to 57 ft. lbs. (77 Nm). Torque the through-bolt to 64 ft. lbs. (87 Nm).
34. Install engine center support member.
35. Install the starter.
36. Install front exhaust pipe and support bracket.
37. Install the halfshafts.
38. Fill the transaxle with the correct gear oil.
39. Install undercovers.
40. Remove the engine support fixture.
41. Install front wheels and lower the vehicle.
42. Install the starter set bolt to the transaxle upper side. Torque the bolt to 29 ft. lbs. (39 Nm).
43. Connect the engine wire clamps.
44. Connect the vehicle speed sensor connector or the speedometer cable.
45. Connect the transaxle shift cables and install ground cable.
46. Connect the back-up light switch connector.
47. Install the release cylinder.
48. Install air cleaner case assembly.
49. Connect the negative battery cable and check the front wheel alignment.
50. Road test the vehicle and check for abnormal noise and smooth shifting.

With S54 Transaxle

1. Disconnect the negative battery cable. On vehicles equipped with an air bag, wait at least 90 seconds before proceeding.
2. Remove the air cleaner assembly and battery.
3. Remove the cruise control actuator by disconnecting the connector and removing the three bolts. The cruise control actuator and bracket should be removed as an assembly.
4. Disconnect the wiring and remove the starter assembly by removing the two bolts.
5. Remove the clutch release cylinder and tube.
6. Disconnect the speedometer cable.
7. Disconnect the engine ground strap.
8. Disconnect the back-up light switch connector.
9. Disconnect the transaxle control cables by removing the four clips and washers.

10. Remove the three upper transaxle retaining bolts.
11. Install an engine support fixture.
12. Remove the bolt and two nuts for the engine left mounting upper side.
13. Raise and support the vehicle safely.
14. Remove the front wheel.
15. Remove the engine undercover.
16. Drain the transaxle lubricant.
17. Disconnect both halfshafts.
18. Remove the front exhaust pipe.
19. Remove the front engine mount and bracket by removing the grommet, through-bolt, and four bolts.
20. Remove the engine rear mounting and bracket by removing the grommets, through-bolt, four bolts and two nuts.
21. Remove the bolt of the engine left mounting lower side.
22. Remove the center member as follows:
a. Remove the bolt and pipe bracket from the center member.
b. Remove the two bolts holding the center member to the radiator support.
c. Remove the center member from the vehicle.
23. Remove the stiffener plate.
24. Using a transaxle jack, support the transaxle.
25. Remove the transaxle bolts, lower the left side of the engine and carefully ease the transaxle out of the engine compartment.

To install:
26. Align the input shaft spline with the clutch disc and carefully mate the transaxle to the engine. Torque the bolts as follows:
• Bolt A to 47 ft. lbs. (64 Nm)
• Bolt B to 34 ft. lbs. (46 Nm)
27. Install the stiffener plate. Torque the bolts as follows:
• Engine side bolts to 47 ft. lbs. (64 Nm)
• Transaxle side bolts to 15 ft. lbs. (21 Nm).
28. Install the center member as follows:
a. Install the center member to the vehicle.
b. Install the two front bolts to hold the center member to the radiator support. Torque the bolts to 38 ft. lbs. (52 Nm).
c. Install the pipe bracket with the bolt.
29. Install the engine left mounting bolt and torque to 47 ft. lbs. (64 Nm).
30. Install the engine rear mount and bracket by installing the four

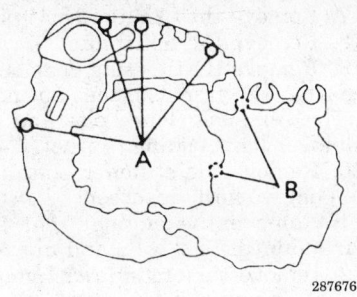

287676

Installing the upper transaxle bolts — Celica S54

bolts, two nuts and the through-bolt. Torque the bolts as follows:

• Transaxle side bolts to 57 ft. lbs. (77 Nm).

• Center member side bolts to 59 ft. lbs. (80 Nm)

• Through bolt to 64 ft. lbs. (87 Nm)

31. Install the engine front mount and bracket by installing the four bolts and through-bolt. Torque the bolts as follows:

• Transaxle side bolts to 57 ft. lbs. (77 Nm).

• Center member side bolts to 59 ft. lbs. (80 Nm)

• Through bolt to 64 ft. lbs. (87 Nm)

32. Install the grommets to the center member.

33. Install the front exhaust pipe.

34. Connect both halfshafts to the transaxle.

35. Fill the transaxle with gear oil.

36. Install the undercovers.

37. Install the front wheel and lower the vehicle.

38. Install the engine left mounting bolts of the upper side and torque to 47 ft. lbs. (63 Nm)

39. Connect the speedometer cable and the transaxle control cables.

40. Reconnect the engine ground strap.

41. Connect the back-up light switch connector.

42. Install the clutch release cylinder. Torque the bolts to 9 ft. lbs. (12 Nm).

43. Install the starter and reconnect the wiring. Torque the starter bolts to 29 ft. lbs. (39 Nm).

44. Install the cruise control actuator.

45. Install the air cleaner and the battery.

46. Connect the battery cables; positive first, negative last.

47. Bleed the clutch system.

48. Test drive the vehicle for shifting smoothness.

Camry

With S51 Transaxle

1. Disconnect the negative battery cable. On vehicles equipped with an air bag, wait at least 90 seconds before proceeding.

2. Remove the air cleaner assembly.

3. If equipped with cruise control, remove the cruise control actuator.

4. Remove the clutch release cylinder and tube clamp.

5. Remove the starter.

6. Disconnect the back-up light switch connector and ground strap.

7. Disconnect the wires clamp.

8. Remove the clips and washers that attach the transaxle control cables to the control levers. Remove the retaining clips and disconnect the transaxle control cables.

9. Disconnect the speed sensor connector.

10. Install a engine support fixture.

11. Tie the steering gear housing to the engine support fixture with a cord.

12. Raise the vehicle and support it safely.

13. Remove the engine undercovers and the front wheels.

14. Drain the fluid from the transaxle.

15. Remove the left and right halfshafts.

16. Disconnect the steering gear housing from the front suspension member as follows:

a. Remove the four bolts.

b. Remove the stabilizer bar bushing bracket.

c. Remove the two set bolts and nuts.

d. Disconnect the steering gear box from the suspension member and suspend it securely.

17. Remove the exhaust pipe.

18. Remove the stiffener plate.

19. Disconnect the engine front mounting from the suspension member by removing the two bolts.

20. Disconnect the engine rear mounting from the front suspension member by removing the two grommets and the three nuts.

21. With a transaxle jack and block of wood, raise the transaxle and engine slightly and disconnect the left engine mounting.

22. Disconnect the steering cooler pipe from the suspension member.

23. Remove the two fender liner set screws.

24. Disconnect the front suspension member as follows:

a. Remove the two bolts and four nuts located on the outside of the brackets.

b. Remove the four larger bolts holding the suspension member to the vehicle body.

c. Remove the two front lower braces, rear braces, and the front suspension member.

25. Remove the six transaxle mounting bolts from the engine.

26. Lower the left side of the engine and remove the transaxle.

27. Clean the mating surfaces of grease and dirt in preparation for reinstallation.

To install:

28. Move the transaxle into position so that the input shaft spline is aligned with the clutch disc.

29. Install the transaxle into the engine and secure with the lower mounting bolts. Torque the 10mm mounting bolts to 47 ft. lbs. and 12mm bolts to 34 ft. lbs.

30. Install the front suspension member to the vehicle and install the two front lower braces and rear lower braces. Install the four large bolts that hold the suspension member to the vehicle. Torque the bolts to 134 ft. lbs. (181 Nm).

31. Install the two outside bolts and four outside nuts. Torque the bolts to 24 ft. lbs. (32 Nm).

32. Install the two fender liner set screws.

33. Connect the steering cooler pipe to the suspension member.

34. Raise the transaxle and engine slightly with a jack and wooden block.

35. Install the engine left mounting as follows:

a. Install the engine left mounting and then install the three bolts. Torque the bolts to 38 ft. lbs. (52 Nm).

b. Install the two nuts and two grommets. Torque the nuts to 59 ft. lbs. (80 Nm).

36. Install the engine rear mounting to the front suspension member by installing the three nuts and two grommets. Torque the nuts to 59 ft. lbs. (80 Nm).

37. Install the engine front mounting to the suspension member and install the nut and two bolts. Torque the bolt to 59 ft. lbs. (80 Nm).

38. Install the stiffener plate and tighten the bolts to 27 ft. lbs. (37 Nm).

39. Install the exhaust pipe.

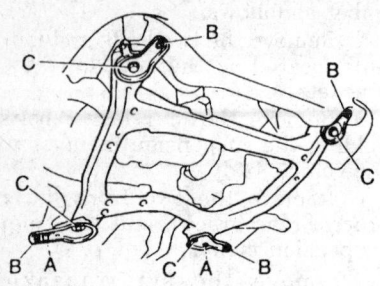

Bolt A: 32 N·m (330 kgf·cm, 24 ft·lbf)
Nut B: 36 N·m (370 kgf·cm, 27 ft·lbf)
Bolt C: 181 N·m (1,850 kgf·cm, 134 ft·lbf)

295232

Front suspension member — Camry S51

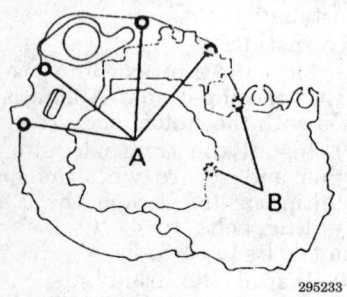

295233

Transaxle to engine bolts — Camry S51

40. Install the steering gear housing to the front suspension member as follows:

    a. Lower the steering gear housing onto the suspension member.

    b. Install the two set bolts and nuts. Torque the bolts and nuts to 134 ft. lbs. (181 Nm).

    c. Install the stabilizer bar bushing bracket.

    d. Install the four bolts and torque to (14 ft. lbs. (19 Nm).

41. Install the right and left halfshafts.

42. Install the engine undercovers.

43. Install the wheels and lower the vehicle.

44. Fill the transaxle with transaxle fluid.

45. Untie the steering gear housing from the engine support fixture.

46. Remove the engine support fixture from the vehicle.

47. Connect the vehicle speed sensor.

48. Connect the control cables by installing the washers and clips.

49. Connect the clamp that retains the wires to the transaxle.

50. Connect the backup light switch connector and ground cables.

51. Install the starter to the vehicle and install the two bolts. Torque the bolts to 29 ft. lbs. (39 Nm).

52. Install the pipe clamp and clutch release cylinder to the transaxle. Torque the bolts to 9 ft. lbs. (13 Nm).

53. Install the cruise control actuator.

54. Install the air cleaner case assembly.

55. Connect the negative battery cable to the battery.

With E53 Transaxle

1. Disconnect the negative battery cable. On vehicles equipped with an air bag, wait at least 90 seconds before proceeding.

2. Remove the air cleaner assembly with the air hose.

3. Remove the cruise control actuator as follows:

    a. Remove the cruise control actuator cover.

    b. Disconnect the connector and then remove the three bolts and the cruise control actuator. Remove the actuator and bracket as an assembly.

4. Remove the clutch release cylinder.

5. Remove the starter assembly by disconnecting the starter wiring and removing the two bolts.

6. Disconnect the back-up light switch electrical connector.

7. Remove the clutch release cylinder bracket.

8. Remove the ground straps.

9. Disconnect the control cables by removing clips and washers.

10. Remove the upper three (3) transaxle mounting bolts.

11. Disconnect the vehicle speed sensor connector.

12. Raise and support the vehicle safely. Use an engine support fixture to secure engine.

13. Remove the front wheels.

14. Tie the steering gear housing to the engine support fixture by cord or equivalent.

15. Remove the undercovers from the vehicle.

16. Drain the transaxle fluid.

17. Remove the exhaust front pipe.

18. Remove the halfshafts.

19. Remove the steering gear housing from the front suspension member by removing four bolts and the stabilizer bar bushing bracket.

20. Remove the stiffener plate and the engine shock absorber.

21. Remove the engine front and rear mounting set bolts and nut.

22. remove the engine rear mounting hole plugs and four nuts.

23. Remove the engine left mounting as follows:

    a. Raise the transaxle and engine slightly with a jack and wooden block.

    b. Remove the two hole plugs and nuts.

    c. Remove the three bolts and engine left mount.

24. Remove the steering cooler set bolts.

25. Remove the front suspension member as follows:

    a. Remove the left and right fender liner set screws.

    b. Remove the two bolts and four nuts from the outer suspension braces.

    c. Remove the inner four bolts to the suspension member and then remove the suspension braces and member.

26. Properly support the transaxle assembly. Remove the engine-to-transaxle bolts, lower the left side of the engine and carefully ease the transaxle out of the engine compartment.

To install:

27. Align the input shaft spline with the clutch disc and install the transaxle to the engine. Torque the two bolts to the left and right of the starter to 47 ft. lbs. (64 Nm). Torque the other two bolts to 34 ft. lbs. (46 Nm).

28. Install the front suspension member, rear lower brace, front lower brace and inner four bolts. Torque the bolts to 134 ft. lbs. (181 Nm).

29. Install the two bolts and four nuts to the outer braces and torque the bolts to 24 ft. lbs. (32 Nm). Torque the nuts to 27 ft. lbs. (36 Nm).

30. Install the left and right fender liner set screws.

31. Install the steering cooler pipe set bolts.

32. Install the engine left mounting and torque the three bolts to 47 ft. lbs. (64 Nm). Install and torque the two nuts to 59 ft. lbs. (80 Nm). Install the two hole plugs.

33. Install the torque the four nuts from the rear mounting set nuts. Tor-

que the nuts to 59 ft. lbs. (80 Nm). Install the two hole plugs.

34. Install the engine front mounting set bolts and nut. Torque the two bolts and nut to 59 ft. lbs. (80 Nm).

35. Install the engine shock absorber and torque the four bolts to 35 ft. lbs. (48 Nm).

36. Install the stiffener plate and torque the outside bolts to 13 ft. lbs. (18 Nm). Torque the inside bolts to 27 ft. lbs. (37 Nm).

37. Connect the steering gear housing to the front suspension member. Torque the two set bolts and nuts to 134 ft. lbs. (181 Nm).

38. Install the stabilizer bar bushing bracket and torque the four bolts to 14 ft. lbs. (19 Nm).

39. Install the halfshafts.

40. Install the exhaust front pipe.

41. Fill the transaxle with gear oil. Capacity is 4.2 liters (4.4 quarts.).

42. Install the engine undercovers and side covers.

43. Install the front wheel and lower the vehicle. Torque the wheel lug nuts to 76 ft. lbs. (103 Nm).

44. Connect the vehicle speed sensor connector.

45. Untie the steering gear housing from the engine support fixture. Remove the engine support fixture.

46. Connect the control cables with the washers and clips.

47. Install the ground cables and wire clamps.

48. Connect the backup light switch connector.

49. Install the clutch release cylinder bracket.

50. Install the clutch accumulator tube clamp and torque the two bolts to 14 ft. lbs. (20 Nm).

51. Install the starter.

52. Place the release cylinder.

53. Install the cruise control actuator with bracket. Install the actuator cover.

54. Install the air cleaner case assembly with the air hose.

55. Install the negative battery cable to the battery.

56. Inspect the front wheel alignment.

57. Perform a road test and check for abnormal noise and smooth shifting.

MR2

With S54 Transaxle

1. Disconnect the negative battery cable. On vehicles equipped with an air bag, wait at least 90 seconds before proceeding.

2. Remove the air cleaner housing assembly with the air hose.

3. Remove the left engine mounting stay by removing the two bolts.

4. Disconnect the electrical connectors to the starter and remove the two bolts. Remove the starter from the engine.

5. Disconnect the back-up light switch connector and ground cable.

6. Disconnect the vehicle speed sensor connector.

7. Remove the three transaxle upper side mounting bolts.

8. Raise the vehicle and support safely. Drain the transaxle fluid.

9. Remove the rear wheels and undercovers.

10. Disconnect the halfshafts from the vehicle.

11. Disconnect the shift cables from the transaxle by removing the two clips, two washers and two retainers.

12. Remove the front exhaust pipe.

13. Install an engine support fixture to the engine.

14. Remove the engine mounting front bracket and insulator by removing the through-bolt and nut, four bolts for the body side bracket, and two bolts for the transaxle side bracket.

15. Remove the clutch release cylinder from the transaxle by removing the two bolts.

16. Remove the engine mounting rear bracket and insulator by removing the through-bolt and nut, three bolts holding the bracket to the suspension member, and the three bolts holding the bracket to the transaxle.

17. Place a floor jack under the rear suspension crossmember.

18. Remove the rear suspension crossmember as follows:

   a. Disconnect the lower control arm from the suspension crossmember by removing the bolt.

   b. Remove the suspension arm from the rear axle hub by removing the bolt and nut.

   c. Disconnect the ABS speed sensor bracket from the suspension crossmember by removing the bolt.

   d. Remove the four outer bolts to the suspension crossmember and remove the suspension crossmember from the vehicle.

19. Raise the transaxle and engine slightly with a jack and wooden block.

20. Remove the three engine left-hand mounting set bolts.

21. Remove the engine stiffener plate by removing the three bolts.

22. Remove the rear end plate by removing the four bolts.

23. Place a suitable transaxle jack under the assembly, remove the bell

housing bolts and remove the transaxle from the vehicle.

To install:

24. Place a transaxle jack under the transaxle assembly.

25. Raise the vehicle and connect the transaxle to the engine. Install and torque the transaxle bolts as follows:
   • Bolt A to 47 ft. lbs. (64 Nm)
   • Bolt B to 34 ft. lbs. (46 Nm)

26. Install the rear end plate and install the four bolts. Torque the bolts to 82 inch lbs. (9 Nm).

27. Install the stiffener plate and install the three bolts.

28. Install the engine left hand mounting set bolts and torque the bolts to 47 ft. lbs. (64 Nm).

29. Install the rear suspension crossmember as follows:

   a. Raise the suspension member into place and install the outer bolts. Torque the bolts to 83 ft. lbs. (113 Nm).

   b. Connect the ABS speed sensor bracket to the suspension crossmember and install the bolt.

   c. Install the suspension arm to the rear axle hub and install the bolt and nut. Torque to 76 ft. lbs. (103 Nm).

   d. Install the lower control arm to the suspension crossmember and install the bolt. Torque the bolt to 98 ft. lbs. (132 Nm).

30. Install the engine mounting rear bracket and insulator. Torque the bolts as follows:
   • Transaxle side bracket bolts to 57 ft. lbs. (77 Nm)
   • Suspension crossmember bracket bolts to 47 ft. lbs. (64 Nm)
   • Through bolt to 64 ft. lbs. (87 Nm)

31. Install the clutch release cylinder to the transaxle and torque the two bolts to 9 ft. lbs. (12 Nm).

32. Install the engine mounting front bracket and insulator. Torque the bolts as follows:
   • Transaxle side bracket bolts to 57 ft. lbs. (77 Nm)
   • Body bracket bolts to 54 ft. lbs. (73 Nm)
   • Through bolt to 71 ft. lbs. (96 Nm)

33. Remove the engine support fixture from the engine.

34. Install the front exhaust pipe.

35. Connect the shift cables to the transaxle and install the retainers, washers, and two clips.

36. Connect the halfshafts to the vehicle.

37. Install the engine undercovers.

38. Install the rear wheels and lower the vehicle to the ground.

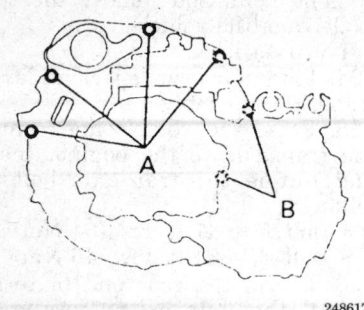

Installation of transaxle bolts — 1995 MR2 S54

39. Connect the vehicle speed sensor connector.

40. Connect the back-up light switch connector and ground cable.

41. Install the starter with the two bolts. Torque the two bolts to 29 ft. lbs. (39 Nm) and connect the electrical connectors to the starter.

42. Install the engine mounting stay by installing the two bolts. Torque the bolt to the body side to 54 ft. lbs. (73 Nm) and the transaxle side to 18 ft. lbs. (25 Nm).

43. Install the air cleaner housing assembly with the air hose.

44. Connect the negative battery cable, refill the transaxle and check for leaks.

With E153 Transaxle

1. Disconnect the negative battery cable. On vehicles equipped with an air bag, wait at least 90 seconds before proceeding.

2. Remove the strut bar by removing the nut and two bolts.

3. Remove the air cleaner housing assembly with the air hose.

4. Remove the left engine mounting stay by removing the two bolts.

5. Disconnect the electrical connectors to the starter and remove the two bolts. Remove the starter from the engine.

6. Disconnect the back-up light switch connector and ground cable.

7. Disconnect the vehicle speed sensor connector.

8. Remove the three transaxle upper side mounting bolts.

9. Raise the vehicle and support safely. Drain the transaxle fluid.

10. Remove the rear wheels and undercovers.

11. Disconnect the halfshafts from the vehicle.

12. Disconnect the shift cables from the transaxle by removing the two clips, two washers and two retainers.

13. Remove the front exhaust pipe.

14. Install an engine support fixture to the engine.

15. Remove the engine mounting front bracket and insulator by removing the through-bolt and nut, four bolts for the body side bracket, and two bolts for the transaxle side bracket.

16. Remove the clutch release cylinder from the transaxle by removing the two bolts.

17. Remove the engine mounting rear bracket and insulator by removing the through-bolt and nut, three bolts holding the bracket to the suspension member, and the three bolts holding the bracket to the transaxle.

18. Place a floor jack under the rear suspension crossmember.

19. Remove the rear suspension crossmember as follows:

   a. Disconnect the lower control arm from the suspension crossmember by removing the bolt.

   b. Remove the suspension arm from the rear axle hub by removing the bolt and nut.

   c. Disconnect the ABS speed sensor bracket from the suspension crossmember by removing the bolt.

   d. Remove the four outer bolts to the suspension crossmember and remove the suspension crossmember from the vehicle.

20. Raise the transaxle and engine slightly with a jack and wooden block.

21. Remove the three engine left hand mounting set bolts.

22. Remove the engine stiffener plate.

23. Remove the rear end plate by removing the four bolts.

24. Place a suitable transaxle jack under the assembly, remove the bell housing bolts and remove the transaxle from the vehicle.

To install:

25. Place a suitable transaxle jack under the transaxle assembly.

26. Raise the vehicle and connect the transaxle to the engine. Install

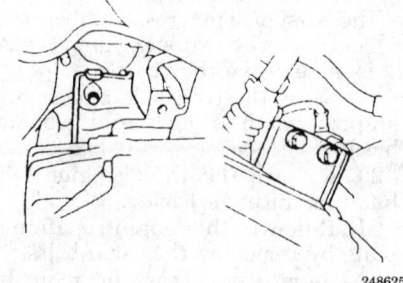

Engine left-hand mounting set bolts — 1995 MR2 E153 and S54

and torque the transaxle bolts as follows:

- Bolt A to 47 ft. lbs. (64 Nm)
- Bolt B to 34 ft. lbs. (46 Nm)

27. Install the rear end plate and install the four bolts. Torque the bolts to 82 inch lbs. (9 Nm).

28. Install the stiffener plate and install the three bolts.

29. Install the engine left hand mounting set bolts and torque the bolts to 47 ft. lbs. (64 Nm).

30. Install the rear suspension crossmember as follows:

   a. Raise the suspension member into place and install the outer bolts. Torque the bolts to 83 ft. lbs. (113 Nm).

   b. Connect the ABS speed sensor bracket to the suspension crossmember and install the bolt.

   c. Install the suspension arm to the rear axle hub and install the bolt and nut. Torque to 76 ft. lbs. (103 Nm).

   d. Install the lower control arm to the suspension crossmember and install the bolt. Torque the bolt to 98 ft. lbs. (132 Nm).

31. Install the engine mounting rear bracket and insulator. Torque the bolts as follows:

- Transaxle side bracket bolts to 57 ft. lbs. (77 Nm)
- Suspension crossmember bracket bolts to 47 ft. lbs. (64 Nm)
- Through bolt to 64 ft. lbs. (87 Nm).

32. Install the clutch release cylinder to the transaxle and torque the two bolts to 9 ft. lbs. (12 Nm).

33. Install the engine mounting front bracket and insulator. Torque the bolts as follows:

- Transaxle side bracket bolts to 57 ft. lbs. (77 Nm)
- Body bracket bolts to 54 ft. lbs. (73 Nm)
- Through bolt to 71 ft. lbs. (96 Nm).

34. Remove the engine support fixture from the engine.

35. Install the front exhaust pipe.

36. Connect the shift cables to the transaxle and install the retainers, washers, and two clips.

37. Connect the halfshafts to the vehicle.

38. Install the engine undercovers.

39. Install the rear wheels and lower the vehicle to the ground.

40. Connect the vehicle speed sensor connector.

41. Connect the back-up light switch connector and ground cable.

42. Install the starter with the two bolts. Torque the two bolts to 29 ft.

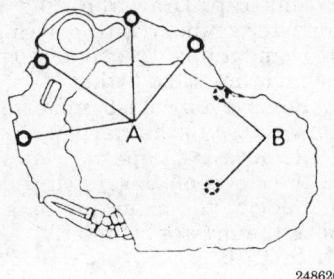

Installation of transaxle bolts — 1995
MR2 E153

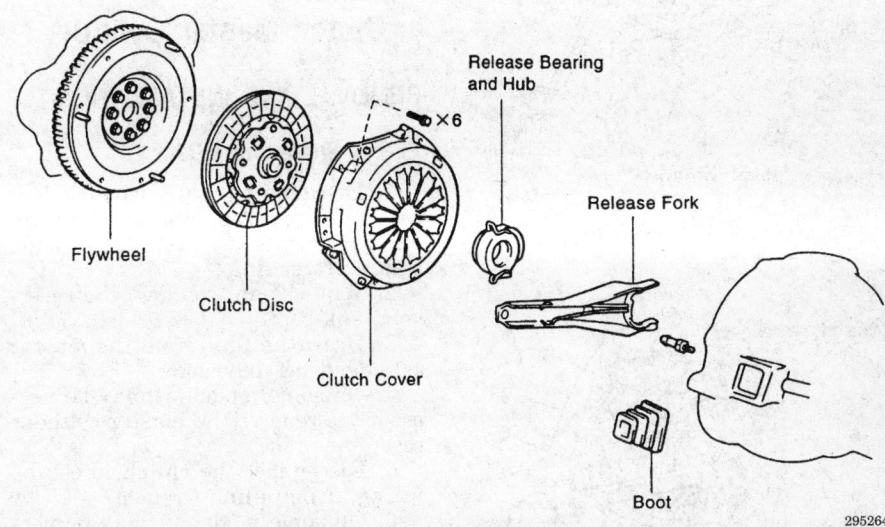

Clutch component assembly — Camry shown, others similar

lbs. (39 Nm) and connect the electrical connectors to the starter.

43. Install the engine mounting stay by installing the two bolts. Torque the bolt to the body side to 54 ft. lbs. (73 Nm) and the transaxle side to 18 ft. lbs. (25 Nm).

44. Install the air cleaner housing assembly with the air hose.

45. Install the strut bar by installing the nut and two bolts. Torque the bolt to 54 ft. lbs. (73 Nm) and the nut to 59 ft. lbs. (80 Nm).

46. Connect the negative battery cable, refill the transaxle and check for leaks.

Clutch Assembly

REMOVAL AND INSTALLATION

Paseo, Tercel, Supra and Camry

1. Disconnect the negative battery cable from the battery. On vehicles equipped with an air bag, wait at least 90 seconds before proceeding.

2. Raise and safely support the vehicle.

3. Remove the transaxle assembly from the vehicle.

4. Matchmark the clutch cover to the flywheel.

5. Loosen each set bolt one turn at a time until spring tension is released.

6. Once the tension on the springs are released, remove the clutch pressure plate retaining bolts.

7. Remove the clutch cover.

8. Remove the clutch disc.

9. Remove the retaining clip and withdraw the release bearing from the transaxle.

10. Remove the release fork and boot assembly.

To install:

11. Using a suitable clutch disc alignment tool, install the clutch disc onto the flywheel.

12. Position the clutch cover onto the flywheel and align the matchmarks.

13. Install the clutch cover retaining bolts. Torque the bolts in a criss-cross pattern to 14 ft. lbs. (19 Nm).

14. Lubricate the release fork pivot contact points and the release bearing, bearing hub and input shaft spline surfaces with a suitable molybdenum disulfide lithium based or multi-purpose grease.

15. Install the boot, release fork, hub and bearing assemblies.

16. Install the transaxle to the vehicle.

17. Lower the vehicle and check the clutch is working properly.

18. Connect the negative battery cable to the battery.

Corolla, Celica and MR2

NOTE: Do not allow grease or oil to get on any part of the disc, pressure plate, or flywheel surfaces.

1. Disconnect the negative battery cable. On vehicles equipped with an air bag, wait at least 90 seconds before proceeding.

2. Raise and safely support the vehicle.

3. Remove the transaxle.

4. Make matchmarks on the clutch cover (pressure plate) and flywheel so that the pressure plate can be returned to its original position during installation.

5. Unfasten the release fork bearing clips. Withdraw the release bearing hub, complete with the release bearing.

6. Remove the release fork and support.

7. Slowly unfasten the bolts which attach the pressure plate. Loosen each bolt one turn at a time until the spring tension is released.

— **CAUTION** —
If the bolts are released improperly the clutch assembly could fly apart, causing possible injury.

8. Separate the pressure plate from the clutch cover/spring assembly.

9. Inspect the disc, pressure plate and flywheel for damage and wear using a caliper to measure depth and width and a dial indicator to measure runout.

   a. The minimum clutch disc rivet head depth is 0.012 in. (0.3mm).

   b. The maximum clutch disc runout is 0.031 in. (0.8mm).

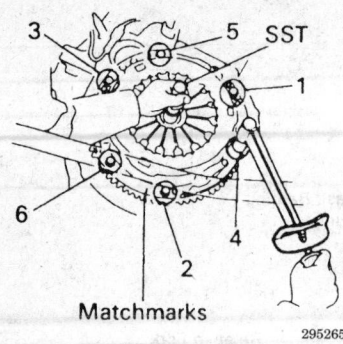

Torque sequence for the clutch cover — Camry

295265

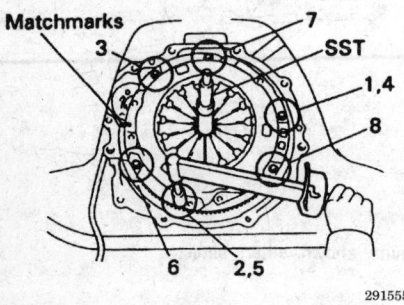

Pressure plate bolt tightening sequence — Supra

291555

c. The maximum pressure plate spring depth is 0.024 in. (0.6mm).

d. The maximum pressure plate spring width is 0.197 in. (5.0mm).

e. The maximum flywheel runout is 0.004 in. (0.1mm).

10. Replace or machine parts as necessary.

To install:

11. When reassembling, apply a thin coating of multipurpose grease to the release bearing hub and release fork contact points. Also, pack the groove inside the clutch hub with multipurpose grease and lubricate the pivot points of the release fork.

12. Align the matchmarks on the clutch cover and flywheel which were made during disassembly. Install the clutch disc and pressure plate assembly and tighten the retaining bolts just finger tight.

13. Center the clutch disc by using a clutch pilot tool or an old input shaft. Insert the pilot into the end of the input shaft front bearing, wiggle it gently to align the clutch disc and pressure plate and tighten the retaining bolts. The bolts should be tightened in 2 or 3 steps, gradually and evenly. Final bolt torque is 14 ft. lbs. (19 Nm).

14. Install the release bearing, fork, and the boot.

15. Install the transaxle and connect the negative battery cable.

16. Road test and check for proper clutch operation.

Clutch Master Cylinder

REMOVAL AND INSTALLATION

Except Supra and MR2

1. Disconnect the negative battery cable. On vehicles equipped with an air bag, wait at least 90 seconds before proceeding.

2. Wipe off and remove the reservoir tank cap.

3. Draw the fluid from the master cylinder with a syringe.

4. On some models it may be necessary to remove the master cylinder reservoir tank.

5. Disconnect the clutch line tube fitting, using a line wrench.

6. On some models it may be necessary to remove the lower instrument panel finish panel and disconnect the air duct from the panel, as required.

7. Remove the pedal return spring on the Celica and Corolla models.

8. Remove the clip and clevis pin with the spring washer, if equipped.

9. Remove the mounting nuts and remove the master cylinder.

To install:

10. Install the master cylinder and tighten the mounting nuts to specifications.

11. Connect the clutch line tube and tighten the fitting to specifications.

12. Connect the clevis and secure the clevis pin with the clip.

13. Install the pedal return spring.

14. Install the reservoir if removed and fill the system with brake fluid.

15. Bleed the clutch system.

16. Check for leaks.

17. Adjust the clutch pedal.

18. Connect the air duct to the lower instrument panel and install the panel, if previously removed.

19. Connect the negative battery cable.

MR2

1. Disconnect the negative battery terminal. On vehicles equipped with an air bag, wait at least 90 seconds before proceeding.

2. On the 1994 Supra, remove the ABS control relay and front suspension brace, if equipped and necessary for clearance.

3. On the MR2, remove the retractor control relay.

4. Wipe off and remove the reservoir tank cap. Draw the fluid from the master cylinder with a syringe.

5. Using a line wrench, disconnect the clutch line tube fitting.

6. Remove any dash undercovers to gain access to the clevis pin.

7. Remove the clip and clevis pin from the clevis on the clutch pedal.

8. Remove the mounting nuts and remove the master cylinder.

To install:

9. Install the master cylinder and tighten the mounting nuts to 9 ft. lbs. (13 Nm).

10. Connect the clevis and install the pin and clip.

11. Install any dash undercovers removed from the vehicle.

12. Connect the clutch line tube and tighten the fitting to 11 ft. lbs. (15 Nm).

13. Install the retractor control relay.

14. Fill the clutch master cylinder with brake fluid and bleed the system.

15. On the 1994 Supra, install the ABS control relay and front suspension brace, if removed.

16. Connect the negative battery cable to the battery.

Clutch Slave Cylinder

REMOVAL AND INSTALLATION

Except Supra and MR2

1. Disconnect the negative battery cable. On vehicles equipped with an air bag, wait at least 90 seconds before proceeding.

2. Raise and support the vehicle safely.

3. Remove the gravel shield, if equipped.

4. Disconnect the fluid line with a line wrench. Plug the line to prevent oil from dripping. Dispose of fluid properly.

5. Remove the slave cylinder retaining bolts.

6. Remove the clutch slave cylinder from the vehicle.

7. Reverse the removal procedure to install the slave cylinder.

8. Bleed the clutch hydraulic system and verify proper clutch operation.

MR2

1. Disconnect the negative battery cable. On vehicles equipped with an air bag, wait at least 90 seconds before proceeding.

2. Raise and support the vehicle safely.

3. Remove the engine undercover.

4. Disconnect the control cables from the transaxle.

5. Using SST 09751–36011 or equivalent (line wrench), disconnect the clutch line. Use a container to catch the brake fluid. Remove the retainer and disconnect the line from the transaxle.

6. Support the engine and transaxle.

7. From the engine side, remove the two engine front mounting bracket bolts.

8. Unbolt and remove the clutch slave cylinder.

To install:

9. Reverse the removal procedure to install the slave cylinder. Tighten the slave cylinder bolts to 9 ft. lbs. (12 Nm) and the engine mounting bracket bolts to 57 ft. lbs. (77 Nm).

10. Connect the hydraulic line and torque the line to 14 ft. lbs. (20 Nm).

11. Add clean fresh fluid. Bleed the clutch hydraulic system and verify proper clutch operation.

Supra

1. Disconnect the negative battery cable. On vehicles equipped with an air bag, wait at least 90 seconds before proceeding.

2. Raise and support the vehicle safely.

3. Disconnect the fluid line from the assembly. Use a container to catch any fluid. Dispose of the fluid properly.

4. Remove the three slave cylinder retaining bolts.

5. On some models, it may be necessary to remove the two bolts from the LH clutch housing cover.

6. Remove the clutch slave cylinder from the vehicle.

To install:

7. Install the slave cylinder and tighten the bolts to the following specifications:
- A — 9 ft. lbs. (12 Nm)
- B — 43 inch lbs. (5 Nm)

8. Connect the hydraulic line and tighten the union bolt to 11 ft. lbs. (15 Nm).

9. Fill the master cylinder with fluid.

10. Bleed the system and connect the negative battery cable.

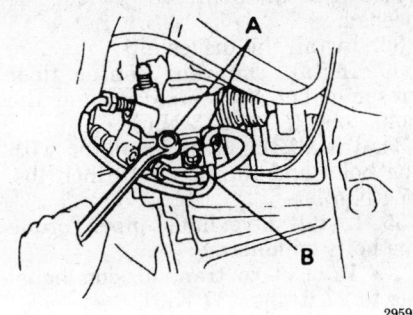

295937

Torquing the mounting bolts — Supra

Hydraulic Clutch System Bleeding

NOTE: If any maintenance on the clutch system was performed or the system is suspected of containing air, bleed the system. Use care; brake fluid will remove the paint from any surface. If the brake fluid spills onto any painted surface, wash it off immediately with soap and water.

1. Fill the clutch reservoir with brake fluid. Check the reservoir level frequently and add fluid as needed.

2. Connect one end of a vinyl tube to the bleeder plug on the slave cylinder and submerge the other end into a clear container half-filled with brake fluid.

3. Slowly pump the clutch pedal several times.

4. Have an assistant hold the clutch pedal down and loosen the bleeder plug until fluid and/or air starts to run out of the bleeder plug. Close the bleeder plug while the pedal is held to the floor.

NOTE: Do not allow the pedal to rise back-up while the bleeder is still open. If this happens, it will allow air to re-enter the slave cylinder and cause the clutch system not to work properly.

5. Repeat Steps 2 and 3 until all the air bubbles are removed from the system.

6. Tighten the bleeder plug when all the air is gone.

7. Refill the master cylinder to the proper level as required.

8. Check the system for leaks.

Automatic Transmission Assembly

REMOVAL AND INSTALLATION

Supra

1. Disconnect the negative battery cable from the battery. On vehicles equipped with an air bag, wait at least 90 seconds before proceeding.

2. Remove the transmission oil level gauge and filler pipe.

3. Raise and safely support the vehicle.

4. Remove the engine undercover.

5. Disconnect the oxygen sensor from the exhaust by removing the two nuts.

6. Remove the exhaust pipe.

7. Remove the heat insulator by removing the four bolts (normal roof) or six bolts (sport roof).

8. Remove the rear center floor crossmember brace.

9. Remove the driveshaft.

10. Disconnect the control rod from the shift lever by removing the nut.

11. Remove the shift control rod from the park/neutral position switch by removing the nut.

12. Disconnect the following:
- No. 1 vehicle speed sensor
- No. 2 vehicle speed sensor
- Solenoid wire
- Park/neutral position switch
- A/T fluid temperature sensor
- If equipped, O/D direct clutch speed sensor

13. Remove the starter electrical connector to the starter and then remove the nut and cable from the starter.

14. Remove the transmission oil cooler pipes as follows:
   a. Loosen the two oil cooler union nuts.
   b. Remove the center and rear oil cooler pipe brackets in front and in back of the lower control arm.
   c. Remove the front oil cooler pipe bracket next to the crankshaft pulley.
   d. Disconnect the two oil cooler pipes.

15. On turbocharged models, remove the intercooler pipe by loosening the two clamps and removing the two bolts.

16. Remove the torque converter clutch mounting bolts.

17. Support the transmission with a jack.

18. Remove the rear mounting by removing the four outer bolts.

19. Lower the transmission and remove the four wire harness clamps from the retainer.

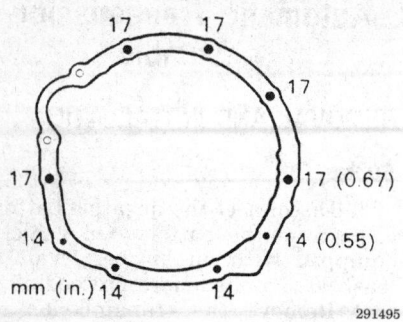

Transmission bolt identification — 1995–97 Supra

20. Remove the starter and transmission set bolts.

21. Remove the transmission from the engine and vehicle.

To install:

22. Install the transmission and starter and install the nine bolts. Torque the bolts as follows:
- 14mm head bolts to 27 ft. lbs. (37 Nm.)
- 17mm head bolts to 53 ft. lbs. (72 Nm)

23. Install the starter cable and nut and then install the electrical connector to the starter.

24. Install the four wire harness clamps to the retainer.

25. Install the rear mounting bolts and torque the bolts to 19 ft. lbs. (25 Nm).

26. Install the torque converter clutch mounting bolts.

27. On turbocharged models, install the intercooler pipe by installing the two bolts and tightening the two clamps.

28. Connect the transmission oil cooler pipes as follows.

   a. Install and connect the two oil cooler pipes. Connect the two oil cooler union nuts and torque to 25 ft. lbs. (34 Nm).

   b. Install the oil cooler pipe bracket next to the crankshaft pulley. Torque the bolt to 7 ft. lbs. (10 Nm).

   c. Connect the center and rear oil cooler pipe brackets and torque the bolts to 7 ft. lbs. (34 Nm).

29. Connect the following:
- A/T fluid temperature sensor
- Park/neutral position switch
- Solenoid wire
- No. 2 vehicle speed sensor
- No. 1 vehicle speed sensor
- O/D direct clutch speed sensor

30. Connect the shift control rod to the park/neutral position switch and torque the nut to 12 ft. lbs. (16 Nm).

31. Connect the shift control rod to the shift lever. Torque the nut to 9 ft. lbs. (13 Nm). Inspect and adjust the park/neutral position switch as needed.

32. Install the driveshaft.

33. Install the rear center floor crossmember brace and torque the bolts to 9 ft. lbs. 913 Nm).

34. Install the heat insulator with the bolts and torque to 48 inch lbs. (5.4 Nm).

35. Install the exhaust pipe. Torque the bolts as follows:
- Bracket to transmission housing to 27 ft. lbs. (37 Nm)
- No. 2 exhaust pipe to center exhaust pipe to 43 ft. lbs. (58 Nm)

36. Connect the oxygen sensor with the two nuts.

37. Install the engine undercover.

38. Install the transmission filler pipe and level gauge.

39. Connect the negative battery cable to the battery.

40. Check all fluids.

Automatic Transaxle Assembly

REMOVAL AND INSTALLATION

Tercel

1993–94 Models

1. Disconnect the negative battery cable. On vehicles equipped with an air bag, wait at least 90 seconds before proceeding.

2. Remove positive battery cable, battery hold down, and the battery.

3. Remove the air duct.

4. Disconnect the speedometer cable.

5. Remove the throttle cable from the throttle link.

6. Remove the starter,

7. Safely support and raise the vehicle and remove the engine undercovers.

8. Drain the fluid from the transaxle assembly and the differential.

9. Disconnect the oil cooler hose from the pipe and remove the engine undercover.

10. Remove the clips from the shift control cable and disconnect the cable.

11. Disconnect the right and left halfshafts. Suspend the halfshafts with a cord.

12. Remove the 3 bolts holding the front exhaust pipe to the engine, disconnect the oxygen sensor, and remove the 2 bolts from the rear exhaust pipe.

13. Disconnect the park/neutral switch connector.

14. Hold and turn the crankshaft pulley nut to gain access to and remove the 6 torque convertor mounting bolts.

15. Disconnect the bond cable from the LH mounting bracket. Remove the LH mounting bolts and the mount.

16. Support the engine and transaxle with 2 jacks or a chain block and jack. Remove the 3 bolts and remove the rear engine mounting bracket from the body.

17. Remove the transaxle mounting bolts and remove the transaxle from the engine.

To install:

18. Install the torque convertor to the transmission. If the torque convertor has been drained and flushed, Refill with ATF.

NOTE: Fluid type: ATF Dexron®II

19. Using calipers and a straight edge, measure from the installed surface of the torque convertor to the front surface of the transmission housing. The distance should be more than 13.4mm

20. Using a suitable jack, install the transaxle assembly to the engine aligning the 2 engine block knock pins with the convertor housing. Install 1 bolt.

21. Install the remaining transaxle mounting bolts and torque all the bolts to 47 ft. lbs.

22. Install the LH engine mounting bracket and torque the 2 bolts to 32 ft. lbs. Reconnect the bond cable with the bolt.

23. Install the rear engine mounting bracket and torque the 3 bolts to 54 ft. lbs.

24. Install the torque convertor mounting bolts. Install the gray bolt first, then the remaining 5 bolts. Torque the bolts to 13 ft. lbs.

25. Install the remaining components in the reverse order of removal.

26. Install and adjust the throttle cable and reconnect the speedometer cable.

27. Attach the air duct.

28. Install the battery and battery hold-down. Connect the battery cables.

29. Refill the transmission and differential.

NOTE: Use ATF Dexron®II

30. Road test the vehicle for proper operation and check for leaks.

1995–97 Models

1. Disconnect the negative battery cable. On vehicles equipped with an air bag, wait at least 90 seconds before proceeding.

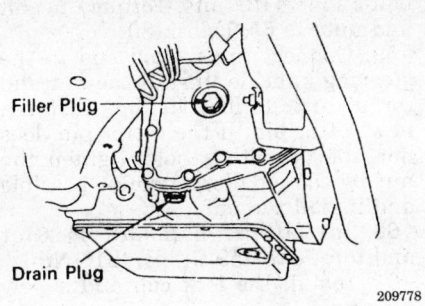

Differential drain and fill plugs — 1993–94 Tercel

209778

2. Remove positive battery cable, battery hold down, and the battery.

3. Remove the air cleaner assembly.

4. Remove the throttle cable from the throttle link.

5. Remove the ground cable and bracket from the transaxle.

6. Remove the upper side mounting bolts from the transaxle.

7. Disconnect the starter wiring.

8. Install an engine support device.

9. Safely raise and support the vehicle.

10. Remove the undercovers from the vehicle.

11. Drain the fluid from the transaxle assembly and the differential.

12. Disconnect the right and left halfshafts from the transaxle.

13. Disconnect the speedometer cable from the transaxle.

14. Remove the clip holding the shift control cable to the body.

15. Disconnect the shift control cable from the control lever by removing the nut.

16. Disconnect the park/neutral position switch electrical connector.

17. Disconnect the O/D solenoid connector, if equipped.

18. Disconnect the oil cooler hose from the transaxle.

19. Remove the starter assembly.

20. Remove the front exhaust pipe.

21. Remove the transaxle converter cover and turn the crankshaft to gain access to each converter bolt.

22. Hold the crankshaft pulley nut with a socket and breaker bar and remove the six converter bolts.

23. Support the engine and transaxle with two jacks or a chain block and jack. Remove the two bolts holding the left hand engine mounting bracket to the body.

24. Remove the rear mounting insulator through-bolt.

25. Remove the five bolts and the rear mounting insulator.

26. Remove the transaxle lower mounting bolts and remove the transaxle from the engine. The transaxle is removed from the bottom of the vehicle.

27. Remove the torque convertor from the transaxle.

To install:

28. Install the torque convertor to the transaxle. If the torque convertor has been drained and flushed, refill with ATF.

NOTE: Fluid type: ATF Dexron®II

29. Using a suitable jack, install the transaxle assembly to the engine aligning the two engine block knock pins with the convertor housing. Install two bolts, but do not torque at this time.

30. Install the remaining transaxle mounting bolts and torque the bolts to specifications.

31. Install the LH engine mounting bracket and torque the two bolts to 35 ft. lbs. (48 Nm).

32. Install the rear engine mounting bracket and torque the five bolts as follows:
• Outside bolts through brackets to 58 ft. lbs. (78 Nm)
• Inside bolts though rear mounting to 69 ft. lbs. (92 Nm)

33. Install the rear mounting insulator through-bolt and torque the bolt to 47 ft. lbs. (64 Nm).

34. Install the torque six convertor mounting bolts. Torque the bolts to 20 ft. lbs. (27 Nm).

35. Remove the jacks supporting the transaxle and install the torque converter cover.

36. Install the front exhaust pipe.

37. Install the starter assembly.

38. Connect the oil cooler hose to the transaxle.

39. If equipped, connect the O/D solenoid connector.

40. Connect the park/neutral switch electrical connector.

41. Install the shift control cable to the lever and secure with the nut.

42. Connect the shift control cable to the body by installing the clip.

43. Connect the speedometer cable to the transaxle.

44. Install the left and right halfshafts.

45. Install the engine undercovers.

46. Safely lower the vehicle.

47. Remove the engine support.

48. Install the ground cable and bracket to the transaxle.

49. Install the starter upper bolt. Torque the bolt to 29 ft. lbs. (39 Nm).

50. Install the transaxle upper side mounting bolts and torque the bolts to 47 ft. lbs. (64 Nm).

51. Install and adjust the throttle cable.

52. Install the air cleaner assembly.

53. Install the battery with holddown. Connect the battery cables.

54. Refill the transmission and differential.

NOTE: Use ATF Dexron®II

55. Road test the vehicle for proper operation and check for leaks.

Paseo

1. Disconnect the negative battery cable from the battery. On vehicles equipped with an air bag, wait at least 90 seconds before proceeding.

2. Remove the transaxle oil level gauge electrical connector.

3. Remove the air duct (leading from the front of the engine compartment to the air cleaner) by removing the two (2) bolts, nuts, and screw.

4. Remove the air cleaner upper cover, air cleaner, and lower case assembly.

5. Remove the throttle cable from the engine.

6. Remove all electrical connections on top of the transaxle and remove the wire harness.

290902

Lower transaxle mounting bolts (torque specifications) — 1995–97 Tercel

REMOVE TRANSAXLE MOUNTING BOLT
Torque:

Bolt A: 7.4 N·m (75 kgf·cm, 65 in.·lbf)
Bolt B: 64 N·m (650 kgf·cm, 47 ft·lbf)
Bolt C: 46 N·m (470 kgf·cm, 34 ft·lbf)

7. Remove the upper side bolt from the starter and the two (2) upper side mounting bolts.

8. Attach an engine chain hoist to the engine hangers.

9. Raise and safely support the vehicle.

10. Remove the engine undercovers from the vehicle.

11. Drain the fluid from the transaxle and differential by removing the drain plugs.

12. Remove the halfshaft.

13. Disconnect the tie rod end from the steering knuckle.

14. Disconnect the lower ball joint from the lower control arm.

15. Using a rubber mallet, disconnect the halfshaft from the steering knuckle.

16. Using a brass bar and hammer, drive out the halfshaft from the transaxle.

17. Remove the snaprings from the inboard joints of the halfshafts.

18. Pull out the clip from the shift cable.

19. Disconnect the solenoid connector from the transaxle.

20. Remove the locknut and disconnect the shift cable from the control lever.

21. Disconnect the connector from the park/neutral position switch.

22. Using a line wrench, loosen the nuts connecting the oil cooler tubes to the transaxle. Remove the oil cooler lines from the transaxle.

23. Remove the two (2) bolts and (2) clamps to the oil cooler tube and then remove the oil cooler.

24. If equipped, remove the ground wire from the transaxle.

25. Remove the two (2) bolts from the left engine mounting bracket.

26. Remove the starter by removing the two (2) nuts, electrical lines and the lower starter bolt.

27. Remove the speedometer cable from the transaxle.

28. Disconnect the No. 2 vehicle speed sensor.

29. Raise the transaxle and hold the transaxle with the a transaxle jack.

30. Remove the through-bolt from the rear mounting bracket.

31. Remove the hole plug and the three (3) cover plate bolts.

32. Turn the crankshaft to gain access to each torque converter bolt.

33. Hold the crankshaft pulley nut with a wrench and remove the six (6) converter bolts.

34. Remove the transaxle housing mounting bolts.

35. Remove the transaxle from the engine. Lower the rear of the engine slightly to help with the removal of the transaxle.

36. Remove the torque converter clutch from the transaxle.

To install:

37. Install the torque converter clutch in the transaxle.

NOTE: If the torque converter clutch has been drained and washed, refill with new ATF fluid.

38. Using a transaxle jack, install the transaxle to the engine. Make sure to align the two (2) knock pins of the engine block with the transaxle housing holes. Leave the jack supporting the transaxle until mounts are installed.

39. On 1993–95 models, install the transaxle housing mounting bolts and torque the bolts to 47 ft. lbs. (64 Nm).

40. On 1996–97 models, install the transaxle housing mounting bolts and torque the bolts as shown:
- Bolt A to 34 ft. lbs. (46 Nm).
- Bolt B to 66 inch lbs. (8.0 Nm).
- Bolt C to 29 ft. lbs. (39 Nm).
- Bolt D 18 ft. lbs. (25 Nm).
- Bolt E to 47 ft. lbs. (64 Nm).

41. Install all the torque converter clutch bolts to the transaxle. Torque the bolts to 13 ft. lbs. (18 Nm).

42. Install the hole plug and three (3) cover plate bolts.

43. Install the through-bolt to the rear mount bracket.

44. Release the transaxle jack.

45. Connect the vehicle speed sensor.

46. Install the speedometer cable to the transaxle.

47. Install the starter, then tighten the bolt to 29 ft. lbs. (39 Nm).

48. Install the two (2) bolts to the left engine mount side bracket. Torque the two (2) bolts to 32 ft. lbs. (43 Nm).

49. If removed, install the transaxle ground wire with the bolt. Torque the bolt to 32 ft. lbs. (43 Nm).

50. Install the oil cooler tubes to the transaxle.

51. Connect the electrical connector to the park/neutral position switch.

52. Install the shift cable to the control lever and install the locknut.

53. Install the clip keeping the shift cable to the vehicle body.

54. Connect the solenoid connector.

55. Install new snaprings to the inboard joints of the halfshafts.

56. Install the halfshafts to the transaxle.

57. Connect the halfshaft to the axle hub.

58. Connect the lower ball joint to the lower control arm by install the bolt and two (2) nuts. Torque the bolt and nuts to 59 ft. lbs. (80 Nm).

59. Connect the tie rod end to the steering knuckle. Install the nut and torque to 36 ft. lbs. (49 Nm). Install a new cotter pin. If the cotter pin does not line up with a hole, tighten the nut by the smallest amount possible and install the pin.

60. Install the halfshaft locknut and torque to 159 ft. lbs. (216 Nm).

61. Install the lock cap and a new cotter pin.

62. Install the engine undercovers.

63. Install the front wheels and lug nuts. Lower the vehicle.

64. Detach the engine chain hoist from the engine hangers.

65. Install the upper bolt to the starter and the two (2) upper side mounting bolts. Torque the starter bolt to 29 ft. lbs. (39 Nm) and the upper side mounting bolts to 47 ft. lbs. (64 Nm).

66. Install the wire harness.

67. Connect all electrical connectors on the top of the transaxle.

68. Install and adjust the throttle cable.

69. Install the air cleaner lower case, air cleaner element, and upper cover.

70. Install the air duct by installing the two (2) bolts, nuts, and screw.

71. Install the oil level gauge electrical connector.

72. Install the differential drain plug and the transaxle drain plug.

73. Fill the transaxle.

74. Connect the negative battery cable and start the engine.

75. Check for leaks and then check fluid level.

Corolla

1. Disconnect the negative battery cable. On vehicles equipped with an air bag, wait at least 90 seconds before proceeding.

2. Disconnect the negative battery cable from the transaxle.

3. Remove the transaxle level gauge.

4. If equipped with A245E transaxle, remove the reservoir tank and air cleaner assembly.

5. Remove the throttle cable from the bracket.

6. Remove the engine left mounting upper side bolts.

7. Remove the engine left mounting stay.

8. Remove the ground cable from the transaxle.

9. Disconnect wire harness clamp and throttle cable clamp.

10. If equipped with A245E transaxle, remove the starter upper side

RH Side View

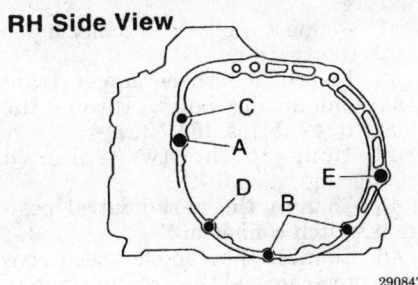

Transaxle bolts — Paseo

mounting bolt and the two transaxle mounting bolts from transaxle side.

11. Raise and safely support the vehicle.

12. Remove the undercovers from the vehicle.

13. Remove the left and right halfshafts.

14. Support the transaxle with a jack.

15. Remove the front exhaust pipe.

16. Install an engine support fixture.

17. Remove the suspension member by removing the grommet, 14 bolts, and 3 nuts.

18. Remove the starter.

19. Disconnect the vehicle speed sensor connector.

20. Disconnect the solenoid connector and park/neutral position switch connector. Remove the wire harness clamps.

21. Remove the nut from the manual shift lever and then disconnect the control cable from the bracket by removing the clip.

22. Loosen the two clips and disconnect the two oil cooler hoses.

23. Remove the transaxle filler tube.

24. If equipped with A131L transaxle, remove the stiffener plate.

25. If equipped with A245E transaxle, remove the converter cover.

26. Turn the crankshaft to gain access to the torque converter bolts.

27. Remove the six torque converter bolts.

28. Remove the five (A245E) or four (A131L) transaxle mounting bolts.

29. Disconnect the transaxle from the engine and lower the transaxle to the ground.

To install:

30. Raise the transaxle into position.

31. On vehicles equipped with the A245E transaxle, install the five

lower transaxle bolts. Torque the bolts as follows:

- Bolt A: 17 ft. lbs. (23 Nm)
- Bolt B: 18 ft. lbs. (25 Nm)
- Bolt C: 34 ft. lbs. (46 Nm)
- Bolt D: 47 ft. lbs. (64 Nm)

32. On vehicles equipped with the A131L transaxle, install the four lower transaxle bolts. Torque the bolts to 47 ft. lbs. (64 Nm).

33. Install the torque converter bolts to the transaxle and install the six bolts. Torque the bolt to 18 ft. lbs. (25 Nm).

34. On vehicles equipped with the A131L transaxle, install the stiffener plate and tighten the mounting bolts to 13 ft. lbs. (18 Nm).

35. On vehicles equipped with the A245E transaxle, install the torque converter cover.

36. Install the transaxle filler pipe.

37. Connect the two oil cooler hoses and replace the clips to their original positions.

38. Connect the control cable for the transaxle to the bracket and install the clip. Connect the control cable to the manual shaft lever by installing the nut.

39. Connect the solenoid connector and park/neutral position switch connector. Connect the wiring to the clamps.

40. Connect the vehicle speed sensor wiring.

41. Install the starter by installing the lower bolt and the electrical connector. Torque the lower bolt to 29 ft. lbs. (39 Nm).

42. Install the suspension member with the 14 bolts and 3 nuts. Torque the bolts and nuts as follows:

- Bolt A: 45 ft. lbs. (61 Nm)
- Bolt B: 47 ft. lbs. (64 Nm)
- Bolt C: 152 ft. lbs. (206 Nm)
- Bolt D: 152 ft. lbs. (206 Nm)
- Nuts: 42 ft. lbs. (57 Nm)

43. Once the suspension member is in place, remove the engine support fixture.

44. Install the front exhaust pipe.

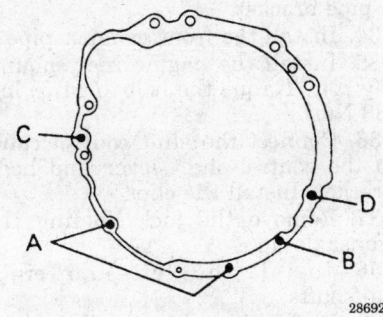

Transaxle bolts torque specifications — Corolla A245E transaxle

45. Remove the jack from the transaxle.

46. Install the left and right halfshafts.

47. Install the engine undercovers.

48. Lower the vehicle.

49. Install the two transaxle mounting bolts to the transaxle side.

50. Install the starter upper side mounting bolt. Torque the bolt to 29 ft. lbs. (39 Nm).

51. Connect the wire harness clamp and throttle cable clamp.

52. Install the ground cable and torque the bolt to 13 ft. lbs. (18 Nm).

53. Install the engine left mounting stay and torque the bolts to 15 ft. lbs. (21 Nm).

54. Install the engine left mounting upper side bolts and torque the bolts to 38 ft. lbs. (52 Nm).

55. Install the throttle cable.

56. Install the air cleaner and the reservoir tank, if removed.

57. Install the transaxle level gauge.

58. Fill the transaxle with fluid and adjust the throttle cable, shift cable and park/neutral position switch.

59. Connect the negative battery cable to the battery.

Celica

With A140E Transaxle

1. Disconnect the negative and positive battery cables from the battery. On vehicles equipped with an air bag, wait at least 90 seconds before proceeding.

2. Disconnect the throttle cable from the engine.

3. Remove the cruise control actuator by disconnecting the connector and removing the three bolts. The cruise control actuator and bracket should be removed as an assembly.

4. Remove the air cleaner assembly and battery.

5. Disconnect the vehicle speed sensor and the transaxle ground strap.

6. Remove the engine left mounting upper side bolt.

7. Disconnect the wiring and remove the starter assembly by removing the two bolts.

8. Disconnect the park/neutral position switch connector.

9. Disconnect the two solenoid connectors.

10. Remove the three upper transaxle retaining bolts.

11. Disconnect the transaxle oil cooler hoses by loosening the two clips.

12. Install a engine support fixture.

13. Raise and support the vehicle safely.

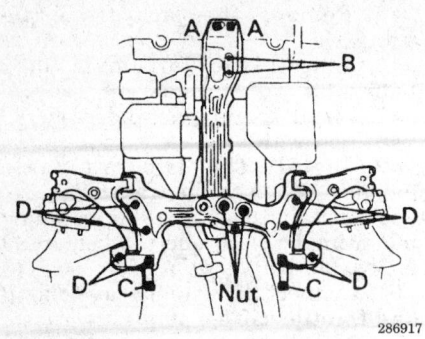

286917

Suspension member torque specifications — Corolla

14. Remove the front wheel.
15. Remove the engine undercover.
16. Drain the transaxle fluid.
17. Disconnect both halfshafts.
18. Support the transaxle with a jack.
19. Disconnect the shift control cable from the control shaft lever and body bracket.
20. Remove the engine rear mounting through-bolt.
21. Remove the front exhaust pipe.
22. Remove the suspension and center members as follows:
   a. Hold up the suspension and center members with a jack.
   b. Disconnect the air conditioner pipe bracket by removing the bolt.
   c. Remove the two shift cable mounting bolts and disconnect the cable from the suspension member.
   d. Remove the two power steering gear assembly set bolts and nuts.
   e. Remove the three grommets from the center crossmember.
   f. Remove the 13 bolts and 2 nuts holding the suspension and center crossmembers.
   g. Lower the jack and remove the crossmembers from the vehicle.
23. Remove the No. 1 manifold stay by removing the nut and bolt.
24. Remove the stiffener plate by removing the nut and six bolts.
25. Turn the crankshaft to gain access to the torque converter bolts. Remove all six bolts from the torque converter.
26. Remove the three bolts from the transaxle and lower the transaxle from the engine.
   To install:
27. Raise and connect the transaxle to the engine.
28. Install the three transaxle bolts to the engine and torque as follows:
   • 10mm bolt to 34 ft. lbs. (46 Nm)
   • 12mm bolt to 47 ft. lbs. (64 Nm)
29. Apply silicone to the transaxle torque converter bolts.

30. Turn the transaxle to install the six torque converter bolts. Torque the bolts to 18 ft. lbs. (25 Nm).
31. Install the stiffener plate and install the six bolts. Alternately tighten the six bolts and torque the bolts as follows:
   • 12mm bolts to 15 ft. lbs. (21 Nm)
   • 14mm bolts to 32 ft. lbs. (43 Nm)
32. Install the No. 1 manifold stay with the nut and bolt. Torque the bolt to 15 ft. lbs. (21 Nm) and the nut to 32 ft. lbs. (43 Nm).
33. Install the suspension member and center member as follows:
   a. Raise the suspension member into position and install the two bolts to hold the suspension to the body. Torque the bolts to 94 ft. lbs. (127 Nm).
   b. Install the three bolts to hold the rear of the lower control arms to the sub-frame and body. Torque the bolt that goes through the lower control arm to 123 ft. lbs. (167 Nm) and the other two bolts to 130 ft. lbs. (175 Nm). Install the bolts to both sides.
   c. Install the center member.
   d. Install the engine rear mount and bracket by installing the bolt and two nuts. Torque the as follows:
   • Two nuts to 59 ft. lbs. (80 Nm)
   • Bolt to 65 ft. lbs. (88 Nm)
   e. Install the engine front mount by installing the two bolts. Torque the bolts to 59 ft. lbs. (80 Nm).
   f. Install the two front bolts to connect the center mount to the radiator support. Torque the bolts to 26 ft. lbs. (35 Nm).
   g. Install the grommets to the center member.
   h. Install the two power steering gear assembly set bolts and nuts. Torque the nuts and bolts to 94 ft. lbs. (127 Nm).
   i. Install the two shift cable mounting bolts.
   j. Connect the air conditioner pipe bracket.
34. Install the front exhaust pipe.
35. Install the engine rear mounting bolt. Torque the bolt to 64 ft. lbs. (88 Nm).
36. Connect the shift control cable to the control shaft lever and body bracket. Install the clips.
37. Remove the jack holding the transaxle up.
38. Install the left and right halfshafts.
39. Install the engine undercovers.
40. Install the front wheel and lower the vehicle.

41. Remove the engine support fixture.
42. Connect the two oil cooler hoses with the two clips.
43. Install the three upper transaxle mounting bolts. Torque the bolts to 47 ft. lbs. (64 Nm).
44. Connect the two solenoid connectors.
45. Connect the park/neutral position switch connector.
46. Connect the vehicle speed sensor connector and the ground strap to the transaxle.
47. Install the starter with the two bolts. Torque the bolts to 29 ft. lbs. (39 Nm). Connect the starter electrical connectors.
48. Install the left mounting upper side bolt and torque the bolt to 47 ft. lbs. (64 Nm).
49. Install the air cleaner assembly.
50. Install the cruise control actuator.
51. Install the battery.
52. Install the adjust the throttle cable.
53. Adjust the shift control cable.
54. Adjust the park/neutral position switch.
55. Fill the transaxle with fluid.
56. Check all fluid levels.
57. Check front wheel alignment and ABS speed sensor signal.
58. Connect the positive and negative battery cable.

With A246E Transaxle
1. Disconnect the negative and positive battery cables from the battery. On vehicles equipped with an air bag, wait at least 90 seconds before proceeding.
2. Remove the transaxle oil level dipstick.
3. Disconnect the throttle cable from the engine.
4. Remove the air cleaner assembly and battery.
5. Remove the engine left mounting upper side bolt.
6. Remove the throttle cable mounting bolt from the transaxle.
7. Remove the ground cable bolt from the transaxle.
8. Remove the wire harness clamp mounting bolt.
9. Remove the starter upper bolt and the two transaxle mounting bolts.
10. Remove the transaxle filler pipe.
11. Disconnect the power steering oil cooler hoses by loosening the two clips.
12. Install a engine support fixture.
13. Support the steering gear housing securely from the engine support fixture.

14. Raise and support the vehicle safely.

15. Remove the engine undercover.

16. Drain the transaxle fluid.

17. Disconnect both halfshafts.

18. Support the transaxle with a jack.

19. Remove the engine rear mounting through-bolt.

20. Remove the front exhaust pipe.

21. Remove the suspension and center members as follows:

a. Hold up the suspension and center members with a jack.

b. Disconnect the air conditioner pipe bracket by removing the bolt.

c. Remove the two shift cable mounting bolts and disconnect the cable from the suspension member.

d. Remove the two power steering gear assembly set bolts and nuts.

e. Remove the three grommets from the center crossmember.

f. Remove the 13 bolts and 2 nuts holding the suspension and center crossmembers.

g. Lower the jack and remove the crossmembers from the vehicle.

22. Remove the nut and disconnect the connector and terminal from the starter.

23. Remove the bolt and starter from the engine.

24. Disconnect the vehicle speed sensor connector.

25. Disconnect the shift control cable from the control shaft lever and body bracket.

26. Disconnect the park/neutral position switch connector.

27. Disconnect the solenoid connectors.

28. Remove the converter clutch cover.

29. Turn the crankshaft to gain access to the torque converter bolts. Remove all six bolts from the torque converter.

30. Remove the two engine left mounting lower side bolts.

31. Remove the five lower bolts and lower the transaxle from the vehicle.

To install:

32. Raise and connect the transaxle to the engine.

33. Install the five transaxle bolts to the engine and torque as follows:

• Bolt A to 17 ft. lbs. (23 Nm)

• Bolt B to 34 ft. lbs. (46 Nm)

34. Apply silicone to the transaxle torque converter bolts.

35. Install the two engine left mounting lower side bolts. Torque the bolts to 47 ft. lbs. (64 Nm).

36. Turn the transaxle to install the six torque converter bolts. Torque the bolts to 18 ft. lbs. (25 Nm).

Transaxle lower bolts — Celica A246E

37. Install the converter clutch cover.

38. Connect the solenoid connectors.

39. Connect the park/neutral position switch connector.

40. Connect the shift control cable to the control shaft lever and body bracket. Install the clips.

41. Connect the vehicle speed sensor connector.

42. Install the starter to the engine and install the lower bolt. Torque the bolt to 29 ft. lbs. (39 Nm).

43. Install the connector and terminal to the starter and install the nut.

44. Install the suspension member and center member as follows:

a. Raise the suspension member into position and install the two bolts to hold the suspension to the body. Torque the bolts to 94 ft. lbs. (127 Nm).

b. Install the three bolts to hold the rear of the lower control arms to the sub-frame and body. Torque the bolt that goes through the lower control arm to 123 ft. lbs. (167 Nm) and the other two bolts to 130 ft. lbs. (175 Nm). Install the bolts to both sides.

c. Install the center member.

d. Install the engine rear mount and bracket by installing the bolt and two nuts. Torque the bolt and nuts as follows:

• Two nuts to 59 ft. lbs. (80 Nm)

• Bolt to 65 ft. lbs. (88 Nm)

e. Install the engine front mount by installing the two bolts. Torque the bolts to 59 ft. lbs. (80 Nm).

f. Install the two front bolts to connect the center mount to the radiator support. Torque the bolts to 26 ft. lbs. (35 Nm).

g. Install the grommets to the center member.

h. Install the two power steering gear assembly set bolts and nuts. Torque the nuts and bolts to 94 ft. lbs. (127 Nm).

i. Install the two shift cable mounting bolts.

j. Connect the air conditioner pipe bracket.

45. Install the front exhaust pipe.

46. Install the engine rear mounting through-bolt. Torque the bolt to 64 ft. lbs. (88 Nm).

47. Remove the jack holding the transaxle up.

48. Install the left and right halfshafts.

49. Install the engine undercovers.

50. Lower the vehicle.

51. Remove the engine support fixture and untie the steering gear assembly.

52. Connect the two oil cooler hoses with the two clips.

53. Install the transaxle filler pipe.

54. Install the upper transaxle mounting bolts. Torque the bolts to 47 ft. lbs. (64 Nm).

55. Install the starter upper bolt. Torque the bolt to 29 ft. lbs. (39 Nm).

56. Connect the ground cable to the transaxle and install the bolt.

57. Install the throttle cable and mounting bolt to the transaxle.

58. Install the left mounting upper side bolt and torque the bolt to 47 ft. lbs. (64 Nm).

59. Install the air cleaner assembly.

60. Install the battery.

61. Install and adjust the throttle cable.

62. Adjust the shift control cable.

63. Adjust the park/neutral position switch.

64. Fill the transaxle with fluid.

65. Check all fluid levels.

66. Check front wheel alignment and ABS speed sensor signal.

67. Connect the positive and negative battery cable.

Camry and Avalon

With A140E Transaxle

1. Disconnect the negative battery cable from the battery. On vehicles equipped with an air bag, wait at least 90 seconds before proceeding.

2. Remove the air cleaner assembly.

3. Disconnect the throttle cable from the throttle linkage.

4. Remove the cruise control actuator cover and disconnect the connector.

5. Disconnect the transaxle ground strap.

6. Disconnect the electrical connector from the vehicle speed sensor.

7. Remove the starter assembly.

8. Disconnect the park/neutral position switch connector.

9. Disconnect the solenoid connector.

10. Disconnect the oil cooler hoses from the transaxle. Plug the lines to prevent oil from spilling.

11. Disconnect the shift control cable by removing the clip and nut.

12. Remove the two front side engine mounting bolts.

13. Remove the two bolts and clamp from the front frame assembly.

14. Remove the three upper transaxle to engine bolt bolts.

15. Install an engine support fixture to the top of the engine.

16. Tie the steering gear housing to the engine support fixture by a cord or equivalent.

17. Raise and safely support the vehicle.

18. Remove the front wheels.

19. Remove the front exhaust pipe.

20. Drain the transaxle fluid from the transaxle and differential.

21. Remove the undercovers from the vehicle.

22. Remove the halfshafts from the transaxle.

23. Remove the front side engine mounting bolt.

24. Remove the three rear side engine mounting nuts.

25. Remove the four engine/transaxle mounting bolts.

26. Disconnect the steering gear housing from the frame as follows:

    a. Remove the four bolts and disconnect the stabilizer bar bushing bracket from the front frame assembly.

    b. Remove the two bolts and nuts from the steering gear housing.

27. Remove the front frame assembly as follows:

    a. Hold the front frame assembly with a jack.

    b. Remove the two set screws from the right and left fender liners.

    c. Remove the six bolts and four nuts.

    d. Remove the front frame assembly from the vehicle.

28. Support the transaxle with a jack and remove the three bolts to the stiffener plate. Remove the stiffener plate from the transaxle.

29. Remove the torque converter plate by removing the four bolts.

30. Remove the torque converter clutch mounting bolts. There are six bolts holding the torque converter clutch to the flywheel.

31. Remove the three transaxle to engine bolts. Make a note to the position of each bolt for installation.

32. Slowly lower the jack holding up the transaxle and remove the transaxle assembly from the engine.

To install:

33. Jack up the transaxle and connect the transaxle to the engine.

34. Install the three transaxle to engine bolts and torque as follows:

• 12mm head bolt: 47 ft. lbs. (64 Nm)

• 10mm head bolt: 34 ft. lbs. (46 Nm)

35. Using a sealer, install the six bolts to hold the torque converter to the flywheel.

36. Install the torque converter plate by installing the four bolts.

37. Install the stiffener plate and the three bolts. Torque the bolts to 27 ft. lbs. (37 Nm).

38. Install the front frame assembly as follows:

    a. Raise the front frame assembly with a jack.

    b. Install the four nuts and six bolts and torque as follows:

    • 19mm head bolt–134 ft. lbs. (181 Nm)

    • 12mm head bolt–24 ft. lbs. (32 Nm)

    • Nuts–27 ft. lbs. (36 Nm)

    c. Install the two set screws to the right and left fender liners.

39. Connect the steering gear housing to the frame by installing the two bolts and two nuts. Torque the bolts and nuts to 134 ft. lbs. (181 Nm).

40. Install the stabilizer bar bushing to the frame by installing the four bolts. Torque the bolts to 14 ft. lbs. (19 Nm).

41. Install the four engine transaxle mounting bolts and torque the bolts to 38 ft. lbs. (52 Nm).

42. Install the three rear side engine mounting nuts and torque the nuts to 48 ft. lbs. (66 Nm).

43. Install the front side engine mounting bolt and torque it to 59 ft. lbs. (80 Nm).

44. Install the halfshafts.

45. Install the engine undercovers.

46. Fill the transaxle with ATF DEXTRON®II.

47. Install the front exhaust pipe.

48. Install the front wheel and lower the vehicle.

49. Untie the steering gear housing from the engine support fixture.

50. Remove the engine support fixture.

51. Install the upper three transaxle to engine bolts and torque the bolts to 47 ft. lbs. (64 Nm).

52. Install the clamp and two bolts to the front frame assembly.

53. Install the two front side engine mounting bolts and torque the bolts to 59 ft. lbs. (80 Nm).

54. Install the shift control cable by installing the nut and clip. Torque the nut to 11 ft. lbs. (15 Nm) and adjust the shift cable as necessary.

55. Connect the transaxle oil cooler lines.

56. Connect the solenoid connector.

57. Install the park/neutral position switch connector.

58. Install the starter assembly.

59. Connect vehicle speed sensor.

60. Connect the ground cable.

61. Connect the connector for the cruise control actuator.

62. Install the cruise control actuator cover.

63. Install the throttle cable to the engine and torque to 11 ft. lbs. (15 Nm).

64. Install the air cleaner assembly.

65. Connect the negative battery cable to the battery.

66. Start the engine and check the transaxles fluid level.

67. Road test the vehicle.

With A541E Transaxle

1. Turn the ignition switch to the LOCK position and disconnect the negative battery cable. Wait at least 90 seconds or longer before doing any work on the vehicle.

2. Remove the battery.

3. Remove the air cleaner assembly.

4. Disconnect the throttle cable from the throttle body.

5. Remove the cruise control actuator cover and disconnect the connector, if equipped.

6. Remove the ground wire.

7. Remove the starter.

8. Disconnect speed sensor connectors, direct clutch speed sensor, and the park/neutral position switch connector on the transaxle.

9. Disconnect the solenoid connector on the transaxle.

10. Disconnect shift control cable.

11. Disconnect oil cooler hoses.

12. Remove the two front side transaxle mounting bolts.

13. Remove the two front engine mounting bolts.

14. Remove the oil cooler line mounting bolts from the front frame.

15. Remove the three upper transaxle to engine mounting bolts.

16. Install a engine support fixture. Tie steering gear housing to engine support fixture.

17. Raise and safely support the vehicle.

18. Drain the transaxle/differential fluid.

19. Remove the front wheels.

20. Remove the front exhaust pipe.

21. Remove the engine side covers and undercovers.

22. Disconnect both halfshafts.

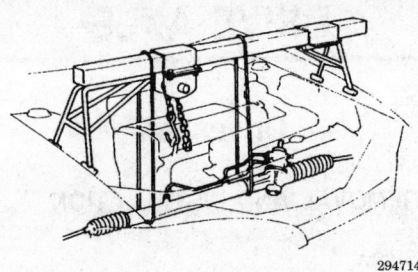

294714

Tie the steering rack to the engine support fixture components — Avalon and Camry with A541E transaxle

23. Remove the front side engine mounting nut.

24. Remove the rear side engine mounting bolts (remove hole plugs).

25. Remove the four left side transaxle mounting bolts.

26. Remove the steering gear housing.

27. Remove the front frame assembly.

28. Properly support the transaxle assembly.

29. Remove the rear end plate mounting bolts.

30. Remove the torque converter cover.

31. Remove the torque converter retaining bolts.

32. Remove the remaining transaxle mounting bolts.

33. Carefully remove the transaxle assembly from the vehicle.

To install:

34. Install the transaxle aligning the two dowel pins on the block with the converter housing. Torque the bolts as follows:
- 10mm bolts — 34 ft. lbs. (46 Nm)
- 12mm bolts — 47 ft. lbs. (64 Nm)

35. Coat the threads of the torque converter bolts with sealer. Install the bolts starting with the green bolt followed by the rest and torque the bolts evenly to 20 ft. lbs. (27 Nm).

36. Install the rear end plate and torque the bolts to 27 ft. lbs. (37 Nm).

37. Install the front frame assembly and torque the fasteners as follows:
- 12mm bolts — 24 ft. lbs. (32 Nm)
- 19mm bolts — 134 ft. lbs. (181 Nm)
- Nut — 27 ft. lbs. (36 Nm)

38. Install the two fender liner set screws.

39. Connect the steering gear to the frame and torque the bolts and nuts to 134 ft. lbs. (181 Nm).

40. Connect the sway bar brackets and toque the bolts to 14 ft. lbs. (19 Nm).

41. Install the left transaxle mounting bolts and torque them to 38 ft. lbs. (52 Nm).

42. Install the rear side mounting bolts and nuts and torque them to 48 ft. lbs. (66 Nm). Install the plugs.

43. Install the front engine mounting nut and torque it to 59 ft. lbs. (80 Nm).

44. Install the halfshafts.

45. Install the right and left engine side covers.

46. Install the lower engine cover.

47. Fill the transaxle/differential to the proper level with Dexron II® or equivalent.

48. Install the exhaust pipe and torque the nuts to 46 ft. lbs. (62 Nm). Connect the exhaust pipe to the converter and torque the nuts and bolts to 32 ft. lbs. (43 Nm). Always use new gaskets.

49. Install the wheel.

50. Lower the vehicle.

51. Remove the engine support.

52. Install the four upper transaxle mounting bolts and torque them to 47 ft. lbs. (64 Nm).

53. Install the oil cooler clamping bolts to the front frame.

54. Install the two front side engine mounting bolts and torque them to 59 ft. lbs. (80 Nm).

55. Install the two front side transaxle mounting bolts and torque them to 59 ft. lbs. (80 Nm).

56. Connect the oil cooler hoses.

57. Connect and adjust the shift control cable.

58. Connect the solenoid electrical connector.

59. Connect the park/neutral switch electrical connector.

60. Connect the speed sensor and the direct clutch speed sensor connectors.

61. Install the starter.

62. Connect the ground strap.

63. If equipped, install the cruise control actuator and cover.

64. Connect the throttle cable to the engine and torque the nuts to 11 ft. lbs. (15 Nm).

65. Install the air cleaner.

66. Install the battery and connect the battery cables.

67. Check the transaxle/differential fluid level.

68. Check the front wheel alignment.

MR2

1. Disconnect the negative battery cable. On vehicles equipped with an air bag, wait at least 90 seconds before proceeding.

2. Remove the air flow meter and the air cleaner hose.

3. Remove the level gauge and disconnect the breather hose.

4. Disconnect the throttle cable from the throttle linkage and the bracket.

5. Remove the starter assembly by disconnecting the electrical connectors and removing the two bolts.

6. Disconnect the No. 1 and No. 2 vehicle speed sensor connectors.

7. Remove the wire harness clamp bolt.

8. Remove the engine lateral control assembly.

9. Remove the three upper transaxle mounting bolts.

10. Raise and safely support the vehicle.

11. Remove the rear wheels.

12. Remove the engine undercover by removing the five bolts.

13. Drain the transaxle oil.

14. Remove the left and right driveshafts.

15. Disconnect the shift control cable by removing the bolt and clip.

16. Disconnect the oil cooler lines at the transaxle.

17. Disconnect the park/neutral position switch connector.

18. Remove the front exhaust pipe.

19. Install an engine support fixture to the engine.

20. Remove the engine mounting front bracket and insulator by removing the through-bolt and nut, four bolts for the body side bracket, and two bolts for the transaxle side bracket.

21. Remove the engine mounting rear bracket and insulator by removing the through-bolt and nut, three bolts holding the bracket to the suspension member, and the three bolts holding the bracket to the transaxle.

22. Place a floor jack under the rear suspension crossmember.

23. Remove the rear suspension crossmember as follows:

  a. Disconnect the lower control arm from the suspension crossmember by removing the bolt.

  b. Remove the suspension arm from the rear axle hub by removing the bolt and nut.

  c. Disconnect the ABS speed sensor bracket from the suspension crossmember by removing the bolt.

  d. Remove the four outer bolts to the suspension crossmember and remove the suspension crossmember from the vehicle.

24. Raise the transaxle and engine slightly with a jack and wooden block.

25. Remove the three engine left-hand mounting set bolts.

26. Remove the engine stiffener plate by removing the three bolts.

27. Remove the rear end-plate by removing the four bolts.

28. Turn the crankshaft to gain access to the torque converter clutch bolts. Remove the torque converter clutch mounting bolts.

29. Remove the transaxle mounting bolts from the engine.

30. Lower the engine left side and remove the transaxle from the engine.

To install:

31. Raise the transaxle into place and align the two knock pins on the engine with the converter housing.

32. Temporarily install all the bolts for the transaxle. After installing the bolts, torque the bolts as follows:
- Bolt A to 47 ft. lbs. (64 Nm)
- Bolt B to 34 ft. lbs. (46 Nm)
- Bolt C to 18 ft. lbs. (25 Nm)

33. Install the torque converter bolts. Install the gray bolt first and then the five black bolts. Torque the converter bolt to 20 ft. lbs. (27 Nm).

34. Install the rear end plate by installing the four bolts. Torque the bolts as follows:
- 10mm head bolts to 82 inch lbs. (9 Nm)
- 12mm head bolts to 14 ft. lbs. (19 Nm).

35. Install the stiffener plate with the three bolts. Torque the bolts to 27 ft. lbs. (37 Nm).

36. Install the engine left hand mounting set bolts and torque the bolts to 47 ft. lbs. (64 Nm).

37. Install the rear suspension crossmember as follows:

a. Raise the suspension member into place and install the outer bolts. Torque the bolts to 83 ft. lbs. (113 Nm).

b. Connect the ABS speed sensor bracket to the suspension crossmember and install the bolt.

c. Install the suspension arm to the rear axle hub and install the

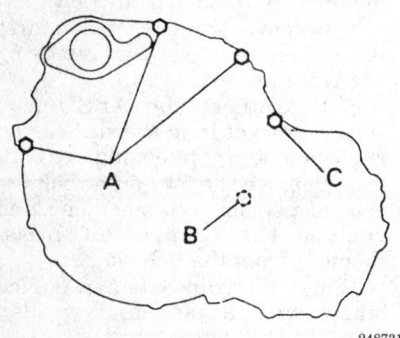

248731

Installation of transaxle bolts — MR2

bolt and nut. Torque to 76 ft. lbs. (103 Nm).

d. Install the lower control arm to the suspension crossmember and install the bolt. Torque the bolt to 98 ft. lbs. (132 Nm).

38. Install the engine mounting rear bracket and insulator. Torque the bolts as follows:
- Transaxle side-bracket bolts to 57 ft. lbs. (77 Nm)
- Suspension crossmember bracket bolts to 47 ft. lbs. (64 Nm)
- Through bolt to 64 ft. lbs. (87 Nm).

39. Install the engine mounting front bracket and insulator. Torque the bolts as follows:
- Transaxle side-bracket bolts to 57 ft. lbs. (77 Nm)
- Body bracket bolts to 54 ft. lbs. (73 Nm)
- Through bolt to 71 ft. lbs. (96 Nm).

40. Remove the engine support fixture from the engine.

41. Install the front exhaust pipe.

42. Connect the park/neutral position switch connector.

43. Connect the two oil cooler hoses.

44. Connect the shift cables to the transaxle and install the retainers, washers, and two clips.

45. Connect the halfshafts to the vehicle.

46. Install the engine undercovers.

47. Install the rear wheels and lower the vehicle to the ground.

48. Install the engine lateral control assembly as follows:

a. Install the lateral control assembly and install the six bolts to the lower bracket. Torque the bolts as follows:
- Bolt A to 54 ft. lbs. (35 Nm)
- Bolt B to 27 ft. lbs. (37 Nm)

b. Install the engine left hand mounting and install the four bolts. Torque the bolts as follows:
- Bolt A to 38 ft. lbs. (52 Nm)
- Bolt B to 47 ft. lbs. (63 Nm)
- Bolt C to 54 ft. lbs. (73 Nm)

49. Connect the No. 1 and No. 2 vehicle speed sensor connectors.

50. Install the wire harness clamp bolt.

51. Install the starter.

52. Install the throttle cable and adjust the cable if necessary.

53. Install the level gauge and connect the breather hose.

54. Install the air cleaner housing assembly with the air hose.

55. Fill the transaxle with fluid and check for leaks.

56. Connect the negative battery cable to the battery.

DRIVE AXLE

Driveshaft

REMOVAL AND INSTALLATION

Supra

1. Disconnect the negative battery cable. On vehicles equipped with an air bag, wait at least 90 seconds before proceeding.

2. Raise and safely support the vehicle.

3. Remove the exhaust pipe.

4. Remove the heat insulator by removing the four nuts.

5. Remove the center floor crossmember brace by removing the four (sport roof has six bolts) bolts.

6. On turbocharged models, loosen the driveshaft adjusting nut, using the proper tools (two of tool SST 09922–10010). Leave the nut hand tight.

7. Place matchmarks on the differential flange and the flexible coupling.

8. Remove the three bolts inserted from the differential side. Separate the flexible couplings from the transmission and the differential.

9. Remove the center support bearing set bolts and the adjusting washers. Some vehicles are not equipped with the adjusting washers.

NOTE: When removing the set bolts, support the center support bearing so the transmission and intermediate shaft, the driveshaft and differential remain in straight line.

10. On non turbocharged models, remove the driveshaft by pulling out toward the rear of the vehicle. Install SST 09325–20010 for M/T (09325–40010 for A/T) in the transmission to prevent oil leakage.

11. On turbocharged models, remove the driveshaft as follows:

a. Push the rear driveshaft forward to compress the driveshaft and pull out the driveshaft from the centering pin of the differential.

b. Remove the driveshaft by pulling out toward the rear of the vehicle.

To install:

12. Apply a suitable grease to the flexible coupling centering bushings.

13. Insert the driveshaft from the rear of the vehicle and connect the

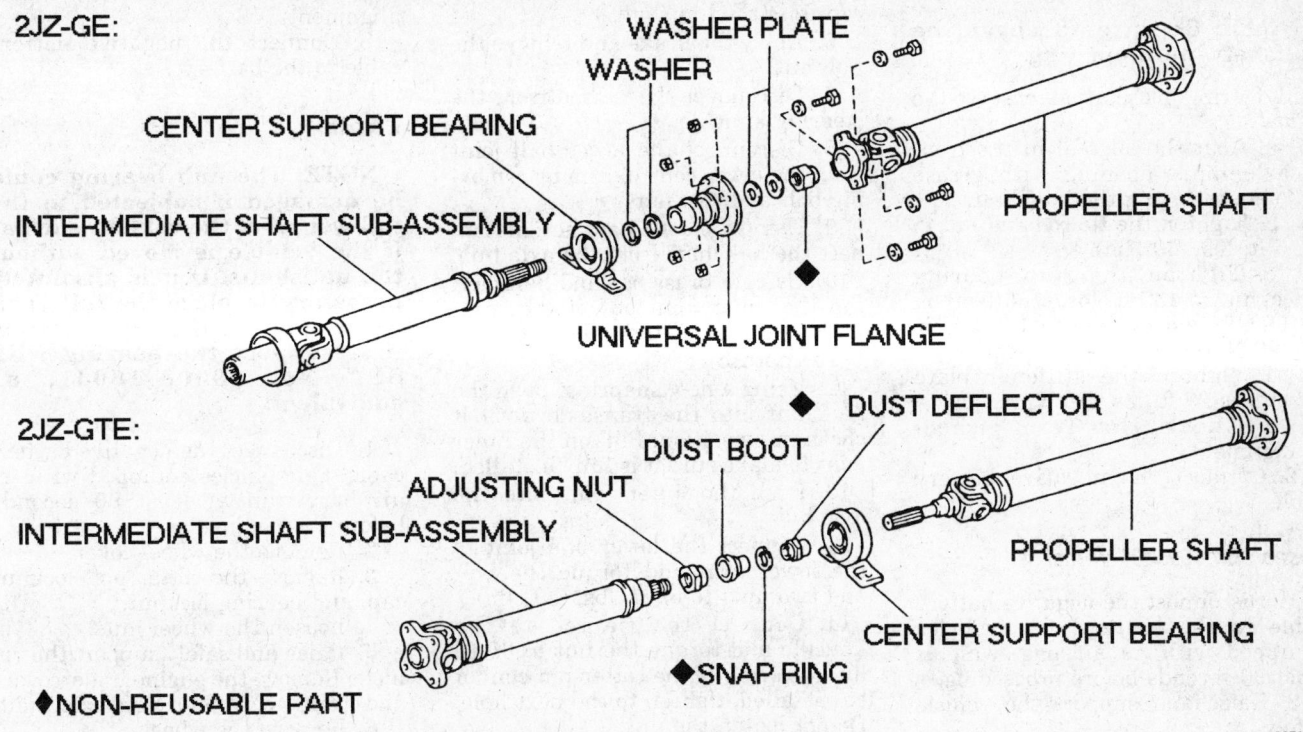

2JZ-GE:

WASHER PLATE
WASHER
CENTER SUPPORT BEARING
PROPELLER SHAFT
INTERMEDIATE SHAFT SUB-ASSEMBLY
UNIVERSAL JOINT FLANGE

2JZ-GTE:

DUST DEFLECTOR
DUST BOOT
ADJUSTING NUT
INTERMEDIATE SHAFT SUB-ASSEMBLY
PROPELLER SHAFT
CENTER SUPPORT BEARING
◆SNAP RING

◆NON-REUSABLE PART

287190

Driveshaft component identification — Supra

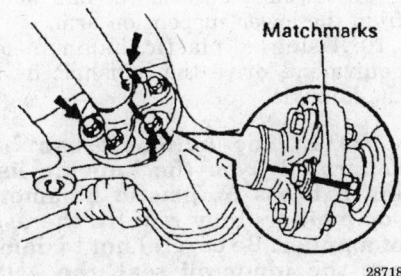

Matchmarks

287182

Matchmarking the driveshaft to the differential — Supra

driveshaft to the transmission and differential.

14. Temporarily, install the center support bearing set bolts with the adjusting washers. Use the adjusting washers that were removed.

15. On non turbocharged models, align the matchmarks and install the driveshaft to the differential. Insert the bolts from the differential side and tighten to 58 ft. lbs. (79 Nm).

16. On turbocharged models, install the driveshaft as follows:

   a. Align the matchmarks and connect the driveshaft to the transmission. Install the nuts and tighten to 41 ft. lbs. (56 Nm).

   b. Align the matchmarks and install the driveshaft to the differential. Insert the bolts from the differential side and tighten to 58 ft. lbs. (79 Nm).

17. Tighten the center bearing support bolts to 36 ft. lbs. (49 Nm).

18. Install the crossmember brace and tighten the bolts to 8 ft. lbs. (13 Nm).

19. Install the heat insulator and torque the bolts to 48 inch lbs. (5.4 Nm).

20. Install the exhaust pipe.

21. Connect the two oxygen sensors to the front exhaust pipe. Torque the sensor bolts to 34 ft. lbs. (44 Nm).

22. Lower the vehicle and connect the negative battery cable.

U-Joints

REMOVAL AND INSTALLATION

Supra

The driveshaft does not use a conventional style U-joint. Instead, flexible rubber couplings are used. These couplings are incorporated into the driveshaft and are non-serviceable.

Halfshaft

REMOVAL AND INSTALLATION

1993–94 Tercel

1. Disconnect the negative battery cable. On vehicles equipped with an air bag, wait at least 90 seconds before proceeding.

2. Raise the vehicle and support it safely.

3. Remove the cotter pin and locknut cap.

4. Loosen the bearing locknut.

5. Remove the brake caliper and position it aside. Remove the brake disc.

6. Remove the cotter pin and nut from the tie rod end. Using a suitable puller, disconnect the tie rod end from the steering knuckle.

7. Matchmark the lower strut mounting bracket where it attaches to the steering knuckle, remove the mounting bolts and disconnect the steering knuckle from the strut bracket.

8. Using a suitable puller, pull the axle hub off the outer halfshaft end.

9. Remove the stiffener plate from the left side of the transaxle assembly.

10. Using the proper tool, tap the halfshaft out of the transaxle casing.

To install:

NOTE: Be sure to cover the halfshaft input opening.

11. During installation, observe the following:

    a. Coat the oil seal in the transaxle input opening with grease before inserting the halfshaft.

    b. Tighten the tie rod end nut to 36 ft. lbs. (49 Nm).

    c. Tighten the hub bearing locknut to 137 ft. lbs. (186 Nm) on 1990–91 models and to 166 ft. lbs. (226 Nm) on 1992–94 models.

    d. Tighten the stiffener plate bolts to 29 ft. lbs. (39 Nm).

    e. Check the front wheel alignment.

12. Connect the negative battery cable.

Paseo and 1995–97 Tercel

1. Disconnect the negative battery cable to the battery. On vehicles equipped with an air bag, wait at least 90 seconds before proceeding.

2. Raise and support the vehicle safely.

3. Remove the left engine undercover and drain the transaxle.

4. If equipped with ABS brakes, disconnect the ABS speed sensor by removing bolt.

5. Remove the cotter pin and lock cap from the halfshaft.

6. Apply the brake and remove the hub nut.

7. Disconnect the tie rod from the steering knuckle.

8. Disconnect the lower ball joint from the lower control arm by removing bolt and two nuts.

9. Using a plastic hammer, disconnect the halfshaft from the axle hub.

10. Using a brass bar and hammer, tap the inner joint out of the transaxle and remove the halfshaft.

To install:

11. Using a new snapring, push the halfshaft into the transaxle until it clicks into position. Pull on the inner joint to make sure it is fully installed.

12. Push the outer joint into the axle hub.

13. Connect the lower ball joint to the lower arm and torque the bolt and two nuts to 59 ft. lbs. (80 Nm).

14. Connect the tie rod to the knuckle and torque the nut to 36 ft. lbs. (49 Nm). If the cotter pin cannot be installed, tighten to the next hole. Do not loosen the nut.

15. Install the hub nut and torque to 152 ft. lbs. (206 Nm). Install a new cotter pin.

16. Refill the transaxle. Install the undercover and the wheel.

17. Check the front wheel alignment.

18. Connect the negative battery cable to the battery.

Corolla

NOTE: The hub bearing could be damaged if subjected to the full weight of the vehicle, such as if the vehicle is moved without the halfshafts. If it is absolutely necessary to place the full vehicle weight on the hub bearing, first support the bearing with SST No. 09608–16041, or equivalent.

1. Disconnect the negative battery cable. On vehicles equipped with an air bag, wait at least 90 seconds before proceeding.

2. Remove the wheel cover.

3. Remove the cotter pin, locknut cap, and bearing locknut.

4. Loosen the wheel nuts.

5. Raise and safely support the vehicle. Remove the engine undercovers and drain gear oil or transaxle fluid.

6. Remove the wheel.

7. If equipped with ABS, remove the bolt and the speed sensor.

8. Separate the tie rod ball joint from the steering knuckle.

9. Disconnect the lower ball joint from the lower suspension arm.

10. Using a plastic hammer or equivalent, drive the halfshaft from the knuckle.

NOTE: The halfshaft can be separated from the knuckle using a brass or plastic hammer; some others may require the use of a puller. Be careful not to damage the inner oil seal, the ABS sensor rotor, or the halfshaft.

11. Using a suitable prying tool, pry the halfshaft from the transaxle. Remove the halfshaft from the vehicle.

To install:

12. Coat gear oil to the inboard joint tulip and position the snapring opening side facing downward.

13. Install the halfshaft into the transaxle. After installing the halfshaft to the transaxle, check that there is 0.08–0.12 in. (2–3mm) of axial play. Check that the halfshaft is making contact with the pinion shaft and that the halfshaft cannot be pulled out.

14. Install the halfshaft into the knuckle.

15. Install the lower suspension arm to the steering knuckle. Tighten the nuts and bolts to 105 ft. lbs. (142 Nm).

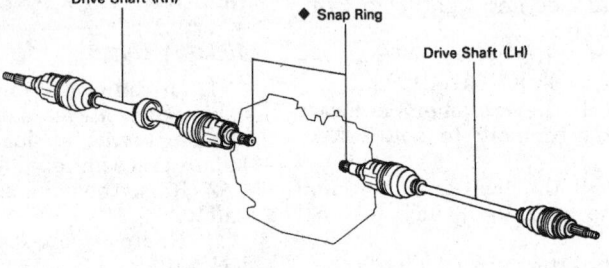

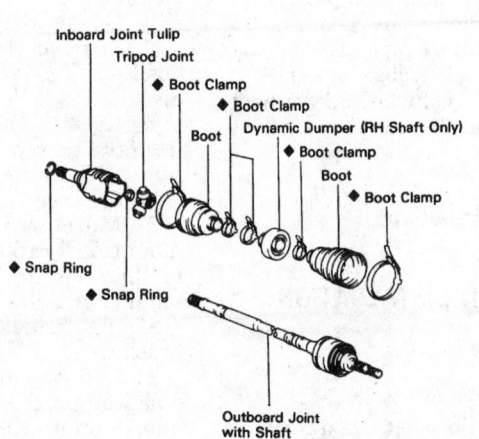

111051

Exploded view of the halfshaft assembly — 1993–94 Tercel

16. Install the tie rod end to the steering knuckle and tighten the nut to 36 ft. lbs. (49 Nm).

17. Install the ABS speed sensor with the attaching bolt, if removed.

18. Install the wheel.

19. Install the hub locknut and washer.

20. Lower the vehicle to the ground.

21. Connect the negative battery cable.

22. Tighten the wheel lug nuts to 76 ft. lbs. (103 Nm). Tighten the hub locknut to 159 ft. lbs. (216 Nm).

23. Install the locknut cap and a NEW cotter pin. Fill transaxle with gear oil or transaxle fluid if necessary.

24. Install engine cover. Check front wheel alignment and check ABS speed sensor signal.

Celica

NOTE: The hub bearing could be damaged if subjected to the full weight of the vehicle, such as if the vehicle is moved without the halfshafts. If it is absolutely necessary to place the full vehicle weight on the hub bearing, first support the bearing with SST No. 09608–16041, or equivalent.

1. Disconnect the negative battery cable. On vehicles equipped with an air bag, wait at least 90 seconds before proceeding.

2. Remove the wheel cover.

3. Remove the cotter pin, locknut cap, and the bearing locknut.

4. Loosen the wheel lug nuts.

5. Raise and safely support the vehicle. Remove the engine undercovers and drain the gear oil or transaxle fluid.

6. Remove the wheel.

7. Separate the tie rod ball joint from the steering knuckle.

8. Disconnect the stabilizer bar link from the lower suspension arm.

9. Disconnect the lower ball joint from the lower suspension arm.

10. Using a plastic hammer or equivalent, tap the halfshaft from the knuckle.

NOTE: Be careful not to damage the inner oil seal or the ABS sensor rotor on the halfshaft.

11. To remove the left side halfshaft, use a suitable prying tool and separate the halfshaft from the transaxle. Remove the halfshaft from the vehicle.

12. To remove the right side halfshaft perform the following steps:

   a. Remove the two (2) bolts of the center bearing bracket

   b. Pull the halfshaft out together with the center bearing case and the center halfshaft.

   c. Remove the center shaft with the RH halfshaft from the transaxle through the bearing bracket.

NOTE: Do not damage the oil seal lip.

To install:

13. Coat gear oil to the inboard joint tulip and position the snapring opening side facing downward.

14. To install the left side halfshaft, simply insert the halfshaft into the transaxle.

15. To install the right side halfshaft, insert the halfshaft with the bearing case and center shaft into the transaxle. Attach the center bearing case to and torque the two (2) bolts to 47 ft. lbs. (64 Nm).

16. After installing either halfshaft, check that there is 0.08–0.12 in. (2–3mm) of axial play. Check that the halfshaft is making contact with the pinion shaft and that the halfshaft cannot be pulled out.

17. Install the halfshaft into the knuckle.

18. Connect the lower suspension arm to the lower ball joint. Torque the ball joint bolt and nuts to 94 ft. lbs. (127 Nm).

19. Connect the tie rod end to the steering knuckle and tighten the nut to 36 ft. lbs. (49 Nm).

20. Install the stabilizer bar link to the lower suspension arm. Tighten the nuts to 33 ft. lbs. (44 Nm).

21. Install the wheel and tighten the wheel lugs to 76 ft. lbs.

22. Install the locknut and washer. Tighten the locknut to 159 ft. lbs. (216 Nm).

23. Lower the vehicle to the ground.

24. Connect the negative battery cable.

25. Install the locknut cap and a new cotter pin. Fill transaxle with gear oil or transaxle fluid if necessary.

26. Install engine cover, check front wheel alignment, and check ABS speed sensor signal.

Camry and Avalon

1. Disconnect the negative battery cable to the battery. On vehicles equipped with an air bag, wait at least 90 seconds before proceeding.

2. Raise and support the vehicle safely.

3. Remove the front wheel(s).

4. Remove the front fender apron seal.

5. Drain the transaxle.

6. Disconnect the tie rod end from the steering knuckle by removing the cotter pin and nut. Using tool SST 09628–62011 or equivalent, separate the tie rod from the steering knuckle.

7. Disconnect the stabilizer bar link from the lower control arm. Make note of the washers and cushions positions.

8. Disconnect the lower ball joint from the steering knuckle by removing the bolt and two nuts. Push down on the lower control arm and separate the steering knuckle from the ball joint.

9. Remove the cotter pin, lock cap and locknut holding the halfshaft to the steering knuckle.

10. Using a plastic hammer, disconnect the halfshaft from the steering knuckle.

11. Remove the left halfshaft from the transaxle as follows:

   a. Use a brass bar and hammer to tap the inner joint out of the transaxle.

   b. Remove the halfshaft.

   c. Once the halfshaft is removed from the vehicle, remove the snapring from the halfshaft.

12. Remove the right halfshaft from the transaxle as follows:

   a. Remove the bearing lockbolt. The lockbolt is located in the center of the halfshaft, near the dampener.

   b. Using snapring pliers, remove the snapring and pull the halfshaft from the transaxle.

To install:

13. To install the right halfshaft to the transaxle:

   a. Coat the side gear shaft and differential case sliding surface with gear oil.

   b. Using snapring pliers, install the snapring to the halfshaft.

   c. Install the halfshaft and the bearing lockbolt. Torque the lockbolt to 24 ft. lbs. (32 Nm).

14. To install the left halfshaft to the transaxle:

   a. Install a new snapring to the inner spline of the halfshaft.

   b. Coat the side gear shaft and differential case sliding surface with gear oil.

   c. Install the halfshaft to the transaxle with the snapring opening facing down. The halfshaft should click into place when installing.

   d. After installation of the halfshaft, check that the halfshaft cannot be removed by hand.

15. Connect the halfshaft to the steering knuckle, then install the

locknut. Torque the locknut to 217 ft. lbs. (294 Nm).

16. Install the lock cap and a new cotter pin to the halfshaft.

17. Connect the steering knuckle to the lower ball joint. Install the two nuts and bolt. Torque the nuts and bolt to 94 ft. lbs. (127 Nm).

18. Connect the stabilizer bar link to the lower control arm. Torque the nut to 29 ft. lbs. (39 Nm).

19. Connect the tie rod to the steering knuckle and torque the nut to 36 ft. lbs. (49 Nm). Install a new cotter pin to the tie rod end.

20. Install the front fender apron seal.

21. Install the wheel(s) and lower the vehicle. Torque the lug nuts to 76 ft. lbs. (103 Nm).

22. Refill the transaxle and check for leaks.

23. Connect the negative battery cable to the battery.

MR2

NOTE: After disconnecting halfshaft, work carefully so as not damage the sensor rotor serrations on the halfshaft.

1. Disconnect the negative battery cable. On vehicles equipped with an air bag, wait at least 90 seconds before proceeding.

2. Raise and support the vehicle safely.

3. Remove the wheels.

4. Remove the cotter pin, cap, and the locknut holding the halfshaft to the axle carrier.

5. Remove the transaxle gravel shield.

6. Drain the transaxle.

7. If removing the right side halfshaft, on vehicles equipped with 3S-GTE engines, perform the following steps:

   a. Place matchmarks on the halfshaft and side gear shaft.

   b. Loosen the six bolts attaching the inner end of the halfshaft to the center halfshaft.

8. Remove the brake caliper support by removing the two bolts.

9. Remove the brake rotor.

10. Disconnect the sway bar from the strut by removing the nut.

11. If equipped with ABS brakes, remove the ABS speed sensor from the axle carrier.

12. Disconnect the lower control arm from the axle carrier by removing the two bolts holding the ball joint to the axle carrier.

13. Disconnect the suspension arm from the axle carrier by removing the nut and bolt.

14. Using a rubber hammer, disconnect the halfshaft from the axle carrier.

15. Using a hammer and brass bar, drive out the left hand halfshaft from the transaxle.

16. On vehicles equipped with 5S-FE engines, remove the right halfshaft using a hammer and brass bar to drive out the halfshaft.

17. On vehicles equipped with 3S-GTE engines, remove the right halfshaft as follows:

   a. Using a hammer and prybar, remove the snapring from the bearing bracket.

   b. Remove the bolt from the bearing bracket.

   c. Remove the right hand halfshaft and center halfshaft as an assembly.

   d. Remove the halfshaft bearing bracket and bearing bracket stay by removing the nut and four bolts.

18. Using SST 09308–00010, or equivalent (oil seal puller), pull out the oil seals from the transaxle.

To install:

19. Using SST 09223–15020 and SST 09350–32014, or equivalent (seal installer), and a hammer, install a new left hand oil seal to the transaxle.

20. Coat the oil seal lips with multi-purpose grease.

21. Using SST 09316–60010 (seal installer), or equivalent, and a hammer, install a new right hand oil seal to the transaxle.

NOTE: Before installing the halfshaft to the transaxle, set the snapring opening side facing downward

22. On vehicles equipped with 3S-GTE engines, install the right hand halfshaft as follows:

   a. Install the bearing bracket to the engine and install the two bolts. Torque the bolts to 56 ft. lbs. (75 Nm).

   b. Install the bearing bracket stays to the bearing bracket and engine. Install the nut and two bolts and torque to 47 ft. lbs. (64 Nm).

   c. Install the right hand halfshaft (with the center shaft) to the transaxle.

   d. Install the bolt to the bearing bracket and torque to 24 ft. lbs. (32 Nm).

   e. Install the snapring to the bearing bracket.

23. To install right halfshafts in vehicles equipped with 5S-FE engines and all left side halfshafts, simply insert the joint end into the transaxle. After installation, check that the halfshafts cannot be removed by hand.

24. After installation, check that the halfshafts cannot be removed by hand.

25. Connect the halfshaft to the axle carrier.

26. Connect the suspension arm to the axle carrier and install the bolt and nut. Torque the bolt to 76 ft. lbs. (103 Nm).

27. Connect the lower control arm to the axle carrier and install the two bolts. Torque the bolts to 83 ft. lbs. (113 Nm).

28. If equipped with ABS brakes, install the ABS speed sensor to the axle carrier. Torque the bolt to 74 inch lbs. (9 Nm).

29. Connect the sway bar to the strut and install the bolt. Torque the bolt to 36 ft. lbs. (49 Nm).

30. Install the rotor.

31. Install the brake caliper support and install the two bolts. Torque the bolts to 43 ft. lbs. (56 Nm).

32. On vehicles equipped with 3S-GTE engines, torque the bolts to hold the right halfshaft to the center shaft to 48 ft. lbs. (65 Nm).

33. Install the locknut to the halfshaft and torque the nut to 217 ft. lbs. (294 Nm).

34. Install the lock cap and a new cotter pin to the halfshaft.

35. Fill the transaxle with the proper fluid.

36. Install the transaxle gravel shield.

37. Install the rear wheels and lower the vehicle.

38. Connect the negative battery cable.

Supra

1. Disconnect the negative battery cable. On vehicles equipped with an air bag, wait at least 90 seconds before proceeding.

2. Raise and safely support the vehicle.

3. Remove the rear tire and wheel assembly.

4. Remove the rear exhaust assembly.

5. Remove the cotter pin, locknut cap, and the locknut holding the halfshaft to the rear axle carrier.

6. Remove the lower suspension arm brace by removing the four bolts.

7. Place matchmarks on the halfshaft and the differential side gear shaft. Remove the hexagon bolts and washers with the proper tool.

8. Hold the inboard joint side of the halfshaft so the outboard joint side does not bend too much. Tap the end of the halfshaft with a rubber

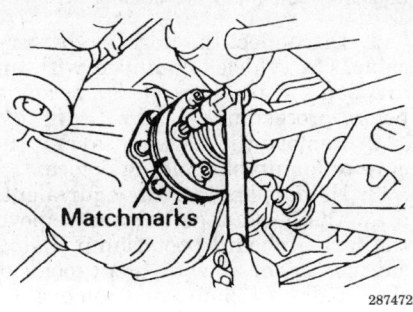

Matchmarks on the halfshaft and differential side gear — Supra

287472

mallet and disengage the halfshaft from the axle carrier.

9. Remove the halfshaft.

To install:

10. Insert the outboard joint side of the halfshaft and align the matchmarks on the side gear shaft and the halfshaft.

11. Coat the threads with clean oil and install the hexagon bolts. Torque the bolts to 61 ft. lbs. (83 Nm).

12. Install the lower suspension arm brace and torque the four bolts to 13 ft. lbs. (18 Nm).

13. Install the bearing locknut and torque the locknut to 213 ft. lbs. (289 Nm).

14. Install the locknut cap and install a new cotter pin.

15. Install the rear exhaust assembly.

16. Replace the rear tire and wheel assembly.

17. Lower the vehicle.

18. Connect the negative battery cable.

CV-Joint Boot

REPLACEMENT

Paseo and Tercel

1. Disconnect the negative battery cable from the battery. On vehicles equipped with an air bag, wait at least 90 seconds before proceeding.

2. Raise and support the vehicle safely.

3. Remove the halfshaft.

4. Remove the snapring from the end of the halfshaft.

--- **WARNING** ---

If equipped with ABS brakes, work carefully making sure not to damage the sensor rotor serrations on the halfshaft.

5. Mount the halfshaft in a suitable holding fixture. Measure the

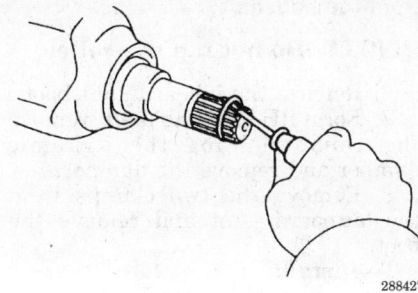

Removing the snapring from the end of the halfshaft — Paseo and Tercel

288427

length of the assembly and mark the positions of the boots.

6. Remove the inboard joint boot clamps and pull back the boot.

7. Place matchmarks on the inboard joint tulip and tripod.

8. Remove the inboard joint tulip from the halfshaft.

9. Remove the tripod joint snapring.

10. Place matchmarks on the halfshaft and tripod.

11. Using a brass punch or equivalent, remove the tripod joint from the halfshaft.

12. Remove inboard joint boot.

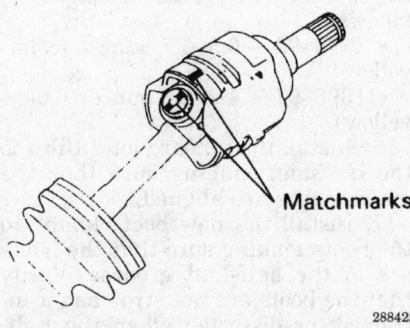

Placing matchmarks on inboard boot tulip and tripod — Paseo and Tercel

288425

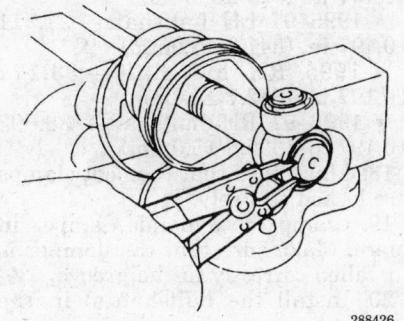

Removing the snapring to remove tripod — Paseo and Tercel

288426

13. For the right hand halfshaft only, remove the dynamic damper as follows:

   a. Remove the dynamic damper clamp.

   b. Slide out the dynamic damper. It may be necessary to press the dynamic damper off the halfshaft.

14. Remove the two boot clamps of the outboard joint boot.

15. Remove the outboard joint boot clamps and boot.

16. Using tool SST 09950–00020 and a press, press out the dust cover from the inboard joint tulip.

17. If equipped without ABS brakes, mount the outboard joint shaft in a soft jaw vise. Using a flat bladed tool and hammer, remove the dust deflector.

NOTE: Do not disassemble the outboard joint. The outboard joint and shaft are replaced as an assembly.

To install:

NOTE: Do not forget to install the inside boot clamps when installing the boots to the halfshaft.

18. If the vehicle does not have ABS brakes, mount the outboard joint shaft in a soft jaw vise. Using SST tool 09309–36010 or equivalent and a press, press a new dust deflector onto the halfshaft.

19. Using a press, press a new dust cover onto the halfshaft.

20. Temporarily install a new outboard joint boot to the halfshaft. Install the clamps with the boot, but do not tighten the clamps at this time.

21. Install a new damper to the right hand halfshaft.

22. Temporarily install a new inboard joint boot to the halfshaft. Install the clamps with the boot, but do not tighten the clamps at this time.

23. Install the tripod as follows:

   a. Place the beveled side of the tripod axial spline toward the outboard joint.

   b. Align the matchmarks placed on the tripod and halfshaft. Using a brass bar and hammer, tap the tripod onto the halfshaft.

   c. Using a snapring expander, install a new snapring to the end of the halfshaft.

24. Pack the outboard joint and boot with suitable grease. Use the grease supplied in a new boot kit. Do not tighten the clamps to the boot at this time.

25. Install the inboard joint to the front halfshaft as follows:

a. Pack the inboard tulip and boot with grease supplied with the boot kit.

b. Align the matchmarks placed on the tulip and tripod.

c. Install the inboard joint tulip to the halfshaft.

d. Temporarily install the boot to the inboard join tulip. Do not tighten the clamps at this time.

26. Assemble the boot clamps and dynamic damper clamp as follows:

a. Be sure the boots and damper clamp are on the shaft groove.

b. Set the driveshaft to standard length:

- Left hand side length–21.815 ±0.197 inch (554.1 ±5.0mm)
- Right hand side length–30.882 ±0.197 inch (784.4 ±5.0mm)

c. Secure the boot and damper clamps at this time.

NOTE: Use new boot retaining clamps and snaprings.

27. Make sure the length of the assembly equals the original length.

28. Install the snapring to the end of the halfshaft.

29. Install the halfshaft.

30. Fill the transaxle fluid and then install the engine undercover.

31. Install the wheel(s) and check the front wheel alignment.

32. Connect the negative battery cable to the battery.

Corolla

Toyota Type

The outboard joint is an integral part of the halfshaft and can't be removed. The halfshaft and outboard joint is replaced as a unit.

1. Disconnect the negative battery cable. On vehicles equipped with an air bag, wait at least 90 seconds before proceeding.

2. Remove the halfshaft(s) and secure to a suitable holding device.

3. Using a suitable tool, remove the two inboard joint boot clamps and slide the boot towards the outboard joint.

4. Place matchmarks on the tripod and inboard joint tulip. Do not punch the marks. Remove the inboard joint tulip from the halfshaft.

5. Using a snaping tool or equivalent, remove the snapring. Place matchmarks on the halfshaft and the tripod. Again, do not punch the marks. Using a brass drift and a

hammer, remove the tripod joint from the halfshaft.

NOTE: Do not tap the roller.

6. Remove the inboard joint boot.

7. For a RH side halfshaft, remove the clamp holding the dynamic damper and remove the damper.

8. Remove the two clamps from the outboard joint and remove the boot.

To install:

9. Temporarily install a new outboard boot and clamps to the halfshaft.

10. On the RH side drive axle, temporarily install the dynamic damper.

11. Temporarily install a new inboard boot to the halfshaft.

12. Place the beveled side of the tripod joint axial spline toward the outboard joint. Make sure to align the matchmarks made earlier. Using a hammer and brass drift, tap the tripod joint to the halfshaft.

NOTE: Do not tap the roller.

13. Install a new snapring.

14. Before installing the outboard boot, fill the joint with the grease supplied with the boot kit. Grease capacity: 4.2–4.6 ounces (color–black). Install the outboard boot.

15. Fill the inboard joint boot and joint with the grease supplied with the boot kit. The grease capacity is as follows:

- 1993–95 — 6.3–6.7 ounces (color-yellow)
- 1996–97 — 4.2–4.6 ounces (color-yellow)

16. Install the inboard joint tulip to the halfshaft making sure that the matchmarks are aligned.

17. Install the new boot clamps to the boots making sure that the boots are in the halfshaft grooves. Verify that the boots are not stretched, contracted, or distorted when the halfshaft is at standard length. The halfshaft standard length is:

- 1995 LH halfshaft — 20.799 ±0.197 in. (529 ±5.0mm)
- 1996–97 LH halfshaft — 21.311 ±0.197 in. (541.3 ±5.0mm)
- 1995 RH halfshaft — 33.177 ±0.197 in. (842.7 ±5.0mm)
- 1996–97 RH halfshaft — 33.693 ±0.197 in. (855.8 ±5.0mm)

18. Make sure that the boot clamps are locked securely.

19. Clamp the dynamic damper in place. Make sure that the damper is installed correctly in the groove.

20. Install the halfshaft(s) in the vehicle.

21. Connect the negative battery cable.

Saginaw Type — Inboard Joint

1. Disconnect the negative battery cable. On vehicles equipped with an air bag, wait at least 90 seconds before proceeding.

2. Remove the halfshaft(s) and secure to a suitable holding device.

3. Using pliers or an equivalent, draw the hooks together and remove the large inboard boot clamp. Using side cutters or an equivalent tool, cut the small boot clamp and remove it.

4. Place matchmarks on the inboard joint tulip, tripod and halfshaft. Do not punch the marks. Remove the inboard tulip from the halfshaft.

5. Using snapring pliers or an equivalent tool, remove the snapring.

6. Place matchmarks on the halfshaft and the tripod and pull out the tripod.

7. Remove the inboard boot.

To install:

8. Temporarily install the new boot and boot clamp for the inboard joint to the shaft.

9. Install a new clamp to the inboard joint boot small end and install it to the halfshaft.

10. Align the matchmarks and push the tripod joint onto the halfshaft by hand. Install a new snapring.

11. Pack the inboard tulip and boot with grease. Use the grease supplied with the boot kit. The grease capacity is: 8.1–8.8 ounces the color for (1995 black) and (1996–97 green).

12. Align the matchmarks and install the inboard tulip to the halfshaft. Install the boot to the inboard tulip.

13. Check that the boot is on the halfshaft groove. Make sure that the boot is not stretched, contracted, or distorted when the shaft is at standard length. The halfshaft standard length is:

- LH halfshaft — 21.268 in. (540.2mm)
- RH halfshaft — 33.756 in. (857.4mm)

14. Holding the boot large clamp near the closing hooks, use pliers to position the holes in the clamp's free end over the closing hook. Secure the clamp by drawing the closing hook together.

15. Secure the small clamp onto the boot by placing SST 09521–24010 or equivalent, on the clamp and pinching the clamp. Do not over tighten.

16. Use SST 09240–00020 or equivalent, to adjust the clamp clearance. The clearance should be 0.059 in. (1.5mm) or less.

17. Install the halfshaft(s) into the vehicle.

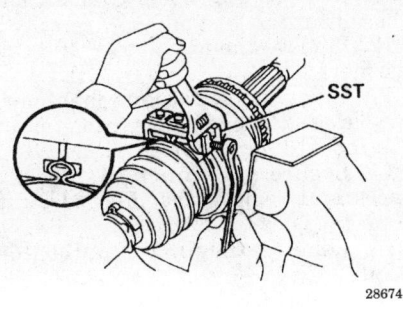

**Using the SST tool to secure the boot clamps
(Saginaw type) — Corolla**

286741

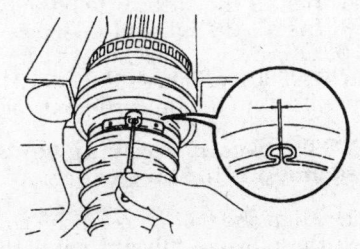

**Using the SST tool to measure the clamp
clearance (Saginaw type) — Corolla**

286742

Saginaw Type — Outboard Joint

1. Remove the halfshaft(s) and secure to a suitable holding device.
2. Using side cutters or an equivalent tool, cut the boot clamps and remove them.
3. Temporarily move the boot toward the inboard joint. Place matchmarks on the outboard joint's inner race and halfshaft.
4. Using snapring pliers or an equivalent tool, expand the snapring and push off the outboard joint. Remove the snapring and slide the outboard boot off.

To install:
5. Temporarily install the new outboard boot and clamps onto the halfshaft.
6. Install a new snapring to the outboard joint and align the matchmarks made during disassembly. Fill the outboard joint with grease and install the outboard joint to the halfshaft.
7. Before installing the outboard boot, fill the boot with grease. Use the grease supplied with the boot kit. The grease capacity is: 4.6–5.3 ounces the color for (1995 black) and (1996–97 green). Install the boot to the joint.

8. Check that the boot is on the halfshaft groove. Make sure that the boot is not stretched, contracted, or distorted when the shaft is at standard length. The halfshaft standard length is:
   a. LH halfshaft — 21.268 in. (540.2mm).
   b. RH halfshaft — 33.756 in. (857.4mm).
9. Secure the clamps onto the boot by placing SST 09521–24010 or equivalent, on the clamps and pinching the clamp. Do not over tighten.
10. Using SST 09240–00020 or equivalent, adjust the clearance of the clamps. The clearance should be as follows:
   a. Outboard large clamp — 0.047–0.102 in. (1.20mm–2.60mm).
   b. Outboard small clamp — 0.055–0.085 in. (1.40mm–2.15mm).
11. Install the halfshaft(s) into the vehicle.
12. Connect the negative battery cable.

Celica, Camry and Avalon

With 1.8L (7A-FE) Engine
The outboard joint is an integral part of the halfshaft and can't be removed. The halfshaft and outboard joint is replaced as a unit.
1. Disconnect the negative battery cable. On vehicles equipped with an air bag, wait at least 90 seconds before proceeding.
2. Remove the halfshaft(s) and secure it in a suitable holding device.
3. Using a suitable tool, remove the two inboard joint boot clamps and slide the boot towards the outboard joint.
4. Place matchmarks on the tripod and inboard joint tulip and center halfshaft. Do not punch the marks. Remove the inboard joint tulip from the halfshaft.
5. Using a snapring tool or equivalent, remove the snapring. Place matchmarks on the halfshaft and the tripod. Again, do not punch the marks. Using a brass drift and a hammer, remove the tripod joint from the halfshaft.

NOTE: Do not tap on the roller.

6. Remove the inboard joint boot. On the RH side halfshaft, remove the clamp holding the dynamic damper and remove the damper.
7. Remove the two clamps from the outboard joint and remove the boot.

To install:
8. Temporarily install a new outboard boot and boot clamps to the halfshaft.

9. On the RH side drive axle, temporarily install the dynamic damper.
10. Temporarily install a new inboard boot to the halfshaft.
11. Place the beveled side of the tripod joint axial spline toward the outboard joint. Make sure to align the matchmarks made earlier. Using a hammer and brass drift, tap the tripod joint to the halfshaft. Install a new snapring.

NOTE: Do not tap on the roller.

12. Before installing the outboard boot, fill the boot and the joint with the grease supplied with the boot kit. Grease capacity is: 4.2–4.6 ounces (color–black). Install the outboard boot.
13. Fill the inboard joint boot and joint with the grease supplied with the boot kit. The grease capacity is: 6.3–6.7 oz. (color–yellow).
14. Install the inboard joint tulip to the halfshaft making sure that the matchmarks are aligned. Install the inboard boot to the inboard joint
15. Install the new boot clamps to the boots making sure that the boots are in the halfshaft grooves, and that the boots are not stretched, contracted, or distorted when the halfshaft is at standard length. Make sure that the boot clamps are locked securely.
16. The halfshaft correct length is:
• LH shaft — 22.016 ±0.197 in.
• RH shaft–34.535 ±0.197 in.
17. Clamp the dynamic damper in place. Make sure that the damper is in the groove.
18. The correct distance measurement for the damper is:
• 15.453 in. ±0.197 in.
19. Install the halfshafts in the vehicle.
20. Connect the negative battery cable.

With 2.2L (5S-FE) Engine
1. Disconnect the negative battery cable from the battery. On vehicles equipped with an air bag, wait at least 90 seconds before proceeding.
2. Raise and support the vehicle safely.
3. Remove the wheel(s) from the vehicle.
4. Remove the halfshafts from the vehicle.
5. Mount the halfshaft in a suitable holding fixture. Measure the length of the assembly and mark the positions of the boots.
6. Remove the inboard and outboard joint boot clamps and pull back the boots.
7. Place matchmarks on the inboard joint tulip and tripod.

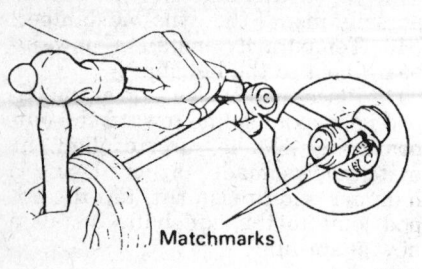

**Placing matchmarks on the tripod joint —
Celica**

287139

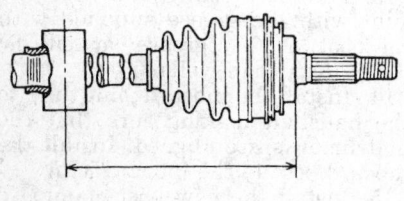

287140

**Measurement for RH halfshaft dynamic damper
installation position — Celica**

8. Remove the inboard joint tulip
from the halfshaft.

9. Remove the tripod joint outside
snapring.

10. Using a snapring expander,
temporarily slide the inside snapring
toward the outboard joint side.

11. Place matchmarks on the halfshaft and tripod.

12. Using a brass punch or
equivalent, remove the tripod joint
from the halfshaft.

13. Using a snapring expander, remove the inside snapring from the
halfshaft.

14. Slide the two boots off the
halfshaft.

15. For the left hand driveshaft, use
SST 09950–00020 and a press. Press
out the dust cover from the inboard
joint tulip.

To install:

16. For the left hand driveshaft, use
a press to install a new dust cover.

17. Install the outboard and inboard joint boots and new boot
clamps. Do not clamp the boot clamps
at this time.

18. Using a snapring expander, install the inside snapring to the
halfshaft.

19. Place the beveled side of the tripod joint axial spline toward the out-

board joint. Align the tripod
matchmarks and install the tripod to
the halfshaft with a brass bar and
hammer.

**NOTE: Do not tap the rollers of
the tripod.**

20. Using a snapring expander, install the outside snapring to the
halfshaft.

21. Pack the outboard joint and
boot with grease. Use the grease supplied in the boot kit.

22. Pack the inboard joint tulip boot
with grease.

23. Align the matchmarks placed
before removal and install the inboard joint tulip to the halfshaft.

24. Install the boot to the inboard
joint tulip.

25. Check that the boot is in the
shaft groove.

26. Ensure that the boot is not
stretched or contracted when the
driveshaft is at standard length.
Standard length is as follows:
 • Left hand side — 23.941 ±0.197
(608.1 ±5.0mm).
 • Right hand side — 34.102 ±0.197
866.2 ±5.0mm).

27. Bend the boot clamps and lock
the boots in place.

28. Install the halfshaft to the
vehicle.

29. Install the wheels to the
vehicle.

30. Lower the vehicle and install
the negative battery cable to the
battery.

With 3.0L (3VZ-FE and 1MZ-FE) Engines

1. Disconnect the negative battery
cable. On vehicles equipped with an
air bag, wait at least 90 seconds
before proceeding.

2. Raise and safely support the
vehicle.

3. Remove the front tire and wheel
assembly.

4. Remove the front halfshaft.

5. Remove the six bolts and three
washers and disconnect the center
axle shaft from the right axle shaft.
Remove the joint end cover gasket
from the axle shaft.

**NOTE: Do not compress the in-
board boot.**

6. Install nuts, bolts and washers
finger-tight to keep the inboard joint
together.

7. Remove the inboard and outboard CV-boot clamps.

8. Matchmark the inboard joint
and axle shaft.

9. Remove the snapring.

10. Remove the inboard joint from
the axle shaft, using the proper tool
to press out the joint.

11. Remove the inboard joint from
the inboard axle shaft.

12. Remove the nuts, bolts, and the
washers.

13. Using a prybar and hammer,
pry around the whole perimeter of
the inboard joint cover.

14. Secure the inner and outer
races and remove the inboard joint
from the cover.

15. Remove the inner and outer
boots.

To install:

16. Wrap vinyl tape around the
spline of the shaft to prevent damage
to the boot.

17. Pack the outer CV-boot with the
grease provided in the kit.

18. Install the outboard CV-boot.

19. Install the inboard CV-boot.

20. Apply seal packing
08826–00801 or Three Bond 1121 or
equivalent, to the inboard joint cover.

**NOTE: Avoid applying an ex-
cess amount to the surface.**

21. Align the bolt holes of the cover
with those of the inboard joint, then
insert the hexagon bolts.

22. Using a plastic faced hammer,
tap the rim of the inboard joint cover
into place. Repeat several times until
seated.

23. Install the bolts, nuts, and the
washers finger-tight to keep the joint
assembled.

24. Align the matchmarks placed
before removal.

25. Using a brass bar and hammer,
tap the inboard joint onto the axle
shaft.

**—————— WARNING ——————
Ensure the brass bar contacts the
inner race, not the cage.**

26. Remove the nuts, bolts, and the
washers.

27. Install a new snapring.

28. Pack grease into the inboard
boot and joint.

29. Install new clamps onto the CV-
boots. Ensure the boots are on the
shaft groove and are not stretched or
contracted when the driveshaft is at
the standard length of:
 • Right side — 17.949 inches
(455.0mm)
 • Left side — 17.953 inches
(455.9mm)

30. Pack grease into the center axle
shaft or side gear shaft.

31. Ensure there is not play in the
inboard and outboard joint.

32. Ensure the inboard joint slides
smoothly in the thrust direction.

33. Install the axle shaft.

34. Install the wheel.

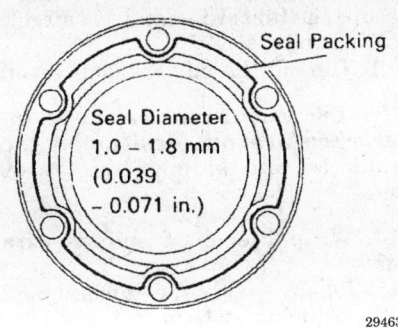

294631

Applying sealant to the inboard joint cover — Avalon and Camry with V6 engine

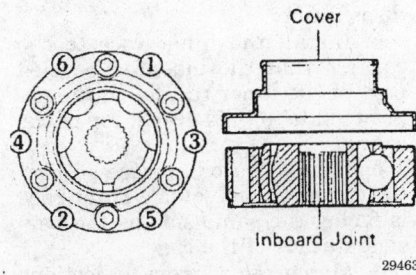

294633

Inboard joint cover installation sequence — Avalon and Camry with V6 engine

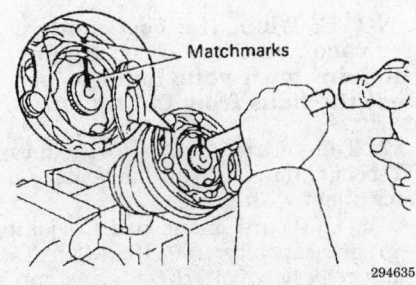

294635

Align matchmarks and install the inboard joint — Avalon and Camry with V6 engine

35. Lower the vehicle and connect the negative battery cable.

MR2

With 2.2L (5S-FE) Engine

The outboard joint is an integral part of the halfshaft and can't be removed. The halfshaft and outboard joint is replaced as a unit.

1. Disconnect the negative battery cable. On vehicles equipped with an air bag, wait at least 90 seconds before proceeding.

2. Remove the halfshaft(s) and secure it in a suitable holding device.

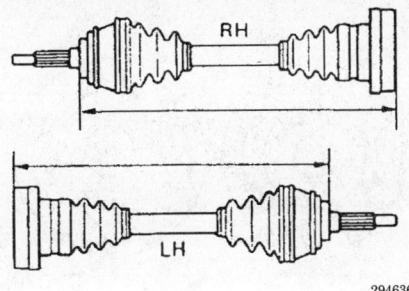

294636

Axle shaft measuring points — Avalon and Camry with V6 engines

3. Using a suitable tool, remove the two inboard joint boot clamps and slide the boot towards the outboard joint.

4. Place matchmarks on the tripod, inboard joint tulip, and the halfshaft. Do not punch the marks. Remove the inboard joint tulip from the halfshaft.

5. Using a snapring tool or equivalent, remove the outer snapring. Temporarily slide the inner snapring toward the outboard joint side. Place matchmarks on the halfshaft and the tripod. Again, do not punch the marks. Using a brass drift and a hammer, remove the tripod joint from the halfshaft.

NOTE: Do not tap on the roller.

6. Remove the inner snapring and remove the inboard joint boot. On the RH side halfshaft, remove the clamp holding the dynamic damper and remove the damper.

7. Remove the two clamps from the outboard joint and remove the boot.

To install:

8. Temporarily install a new outboard boot and boot clamps to the halfshaft.

9. On the RH side drive axle, temporarily install the dynamic damper and the clamp.

10. Temporarily install the inboard joint boot and clamps to the halfshaft.

11. Install a new inner snapring and place the beveled side of the tripod joint axial spline toward the outboard joint. Make sure to align the matchmarks made earlier. Using a hammer and brass drift, tap the tripod joint to the halfshaft. Install a new outer snapring. Make sure the inner and outer snaprings are seated correctly.

NOTE: Do not tap on the roller.

12. Before installing the outboard boot, fill the boot and the joint with

the grease supplied with the boot kit. Grease capacity is: 4.2–4.6 ounces (color–black). Install the outboard boot.

13. Fill the inboard joint boot and joint with the grease supplied with the boot kit. Grease capacity: 8.2–8.5 ounces (color–yellow).

14. Install the inboard joint tulip to the halfshaft making sure that the matchmarks are aligned. Install the inboard boot to the inboard joint.

15. Install the new boot clamps to the boots making sure that the boots are in the halfshaft grooves, and that the boots are not stretched, contracted, or distorted when the halfshaft is at standard length. Make sure that the boot clamps are locked securely.

16. The correct halfshaft length is:
- LH shaft — 21.272 ±0.197 in. (540.3 ±5.0mm)
- RH shaft — 32.732 ±0.197 in. (831.4 ±5.0mm)

17. Tighten the clamp that holds the dynamic damper in place. Make sure that the damper is in the groove.

NOTE: The correct distance measurement for the damper is 18.385 in. ±0.197 in.

18. Install the halfshafts in the vehicle and connect the negative battery cable.

With 2.0L (3S-GTE) Engine

The outboard joint is an integral part of the halfshaft and can't be removed. The halfshaft and outboard joint is replaced as a unit.

1. Disconnect the negative battery cable. On vehicles equipped with an air bag, wait at least 90 seconds before proceeding.

2. Remove the halfshafts from the vehicle.

3. Disconnect the side gear shaft or center halfshaft.

a. Remove the six bolts and washers.

b. After the center halfshaft is removed, use the bolts and washers to keep the inner joint together.

4. Remove the inboard joint and boot clamps.

a. Mount the inboard joint subassembly in a vise.

b. Remove the two inboard joint boot clamps.

c. Remove the inboard joint boot from the inboard joint cover.

5. Disassemble the inboard joint.

a. Place matchmarks on the inboard joint and the driveshaft. Do not punch marks.

b. Using a snapring expander, remove the snapring.

c. Using SST 09726–10010, or equivalent, a socket wrench and a press, remove the inboard joint from the driveshaft.

d. Remove the bolts, nuts and washers from the inboard joint. Pry out the inboard joint from the inboard joint cover.

NOTE: When lifting the inboard joint, hold onto the inner race and the outer race.

6. Remove the inboard and outboard joint boots.

7. Using a flat prying device and a hammer, tap out the dust cover.

To install:

8. Install the new dust cover using a press.

9. Align the matchmarks placed before disassembly. If the joint has become disassembled, assemble the inboard joint as follows:

a. Insert the socket wrench into the inner race.

b. Lift the outer race and cage, and insert the six balls.

c. Jiggle the outer race and cage to place the balls in their respective grooves.

d. Lower the outer race and cage so that they fit tightly with the inner race.

10. Install the new boots to the driveshaft. Wrap vinyl tape around the spline of the shaft to prevent damaging the boot.

11. Install the new inboard joint cover.

a. Apply sealant three bond # 1121, or equivalent, to the inboard joint cover.

b. Align the bolt holes of the cover with those of the inboard joint, then insert the hexagon bolts. Using a brass bar and hammer, tap the rim of the inboard joint cover into place.

c. Be sure to install and tighten the bolts in order and repeat the process several times. Use bolts, nuts and washers to keep the inboard joint together. Tighten the bolts by hand to avoid scratching the flange surface.

12. Install the inboard joint.

a. Align the matchmarks made prior to disassembly. Using a brass bar and hammer, tap the inboard joint onto the driveshaft.

b. Using a snapring expander, install a new snapring.

13. Pack the boots with grease and assemble the boots to the outboard and inboard joints.

14. Assemble boot clamps to both boots.

a. Make sure that the boot is on the shaft groove.

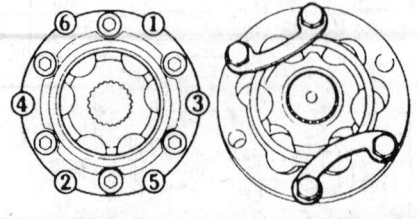

266237

Inboard bolt tightening pattern — 1995 MR2

b. Ensure that the boot is not stretched or contracted when the driveshaft is at standard length. Drive shaft length is 15.343 ±0.197 in. (388.37 ±5.0mm).

c. Bend the boot clamp and lock it.

15. Remove the hexagon bolts from the inboard joint cover and connect the side gear shaft or center halfshaft.

16. Install the halfshafts in the vehicle.

17. Connect the negative battery cable.

18. Test drive the vehicle and check the halfshaft for proper operation.

Supra

If the Outboard joint CV-boot is being replaced, the Inboard CV-joint and boot must be removed first.

1. Disconnect the negative battery cable. On vehicles equipped with an air bag, wait at least 90 seconds before proceeding.

2. Raise and safely support the vehicle.

3. Remove the rear tire and wheel assembly.

4. Remove the halfshaft.

5. Secure the halfshaft in a suitable holding fixture.

6. Wrap a shop towel around the speed sensor rotor to prevent damage.

7. Pry out the end cover.

8. Install nuts, bolts and washers finger-tight to keep the inboard joint together.

9. Remove the boot clamps from the inboard and outboard joint boots.

10. Matchmark the inboard joint and the halfshaft.

NOTE: Do not use punchmarks as the matchmarks.

11. Remove the snapring, using the proper pliers.

12. Press out the inboard joint from the halfshaft with the proper tools.

Secure the inboard joint in a suitable holding fixture.

13. Tap out the inboard joint cover.

NOTE: Ensure the cage and inner race are not positioned too much to one side of the outer race.

14. Wrap the shaft splines with tape.

15. Remove both the inboard and outboard joint boots.

To install:

16. If the joint has been disassembled:

a. Align the inboard joint matchmarks made prior to disassembly.

b. Install the inner race to the cage so that the indented beveled part of the inner race is on the opposite side to the beveled top of the cage.

c. Install the outer race so that the indented side of the outer race is facing the same side as the beveled surface of the cage.

d. Match the narrow projections of the inner race with the wide projections of the outer race.

e. Tilt the cage and the inner race to the side and insert the balls one by one.

NOTE: When the cage and inner race are tilted over, support the joint with your hand to prevent the balls from falling out.

17. Temporarily install the four CV-boot clamps. Install the boots and pack them with grease.

a. Outboard & inboard CV-joint grease capacity: 100–105g (3.5–3.7 ounces). Use all the grease supplied in the boot kit and thoroughly pack the ball contact surface inside the joint.

18. Apply seal packing 08826–00801, Three Bond 1121, or equivalent to the inboard joint cover as shown in the illustration.

NOTE: Avoid applying an excessive amount to the surface.

19. Remove the grease from the surface of the inboard joint facing the cover.

20. Align the bolt holes of the cover with those of the inboard joint, then insert the hexagon bolts.

21. Using a plastic tipped hammer, tap the inboard joint cover onto the joint. Strike the cover in the order shown and repeat several times.

22. Align the matchmarks for the joint and the halfshaft.

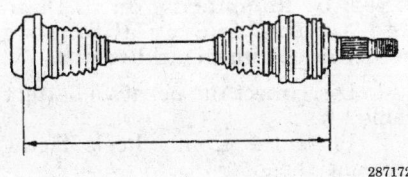

Halfshaft measuring points — Supra

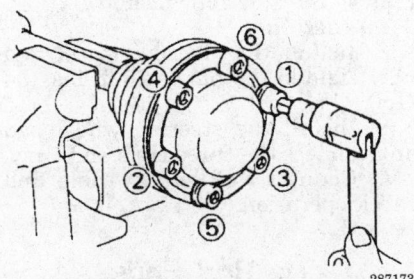

Cover tightening sequence — Supra

23. Using a brass bar and a hammer, tap the inboard joint onto the halfshaft.

NOTE: Ensure the brass bar is contacting the inner race and not the cage.

24. Install a new snapring.
25. Remove the hexagon bolts, nuts, and washers.
26. Position the boots and tighten the clamps.
27. Ensure the boots are not stretched with the halfshaft at normal length of:
  • Left halfshaft — 21.791 inches (553.5mm)
  • Left halfshaft (1995–97) M/T & 2JZ-GTE — 21.555 inches (547.5mm).
  • Right halfshaft — 23.602 inches (598.5mm)
28. Pack grease into the end cover. The end cover grease capacity is: 50–55g (1.8–1.9 ounces).
29. Wipe off any excess grease and install a new gasket on the end cover.
30. Align the bolt holes of the end cover with those of the inboard joint.
31. Install the 6 hexagon bolts and washers from the end cover side and 6 nuts to the boot side. Tighten the bolts in the following sequence several times.

32. Ensure the claw of the end cover touches the inboard joint.
33. Ensure the operation of the joint is smooth within the sliding region in the axial direction.

NOTE: If a large angle is used for the cross-groove type joint, the joint will feel like it is catching, but this does not indicate an abnormality.

34. Install the halfshaft.
35. Install the wheel.
36. Lower the vehicle and connect the negative battery cable.

STEERING

Air Bag

———— **CAUTION** ————
Some vehicles are equipped with an air bag system, also known as the Supplemental Inflatable Restraint (SIR) or Supplemental Restraint System (SRS). The system must be disabled before performing service on or around system components, steering column, instrument panel components, wiring and sensors. Failure to follow safety and disabling procedures could result in accidental air bag deployment, possible personal injury and unnecessary system repairs.

PRECAUTIONS

Several precautions must be observed when handling the inflator module to avoid accidental deployment and possible personal injury.
  • Never carry the inflator module by the wires or connector on the underside of the module.
  • When carrying a live inflator module, hold securely with both hands, and ensure that the bag and trim cover are pointed away.
  • Place the inflator module on a bench or other surface with the bag and trim cover facing up.
  • With the inflator module on the bench, never place anything on or close to the module which may be thrown in the event of an accidental deployment.

DISARMING

To avoid personal injury when working on vehicles equipped with an air bag, the negative battery cable must be disconnected and at least 90 seconds must elapse before working on the system. Failure to do so may result in deployment of the air bag.

Steering Wheel

REMOVAL AND INSTALLATION

With Air Bag

NOTE: Do not attempt to remove or install the steering wheel by hammering on it. Damage to the energy-absorbing steering column could result.

———— **CAUTION** ————
To avoid personal injury when working on air bag equipped vehicles, the system must be disarmed. When removing the air bag, take care not to pull the air bag wire harness. When carrying the wheel pad, carry it with the upper surface facing away. When storing it, keep the upper surface of the pad facing upward.

1. Disconnect the negative battery cable. Wait at least 90 seconds from the time the negative battery cable is disconnected to start any work.
2. Position the front wheels in a straight ahead position.
3. Remove the small covers behind the wheel. Loosen the Torx® screws until the groove along the screw circumference catches on the screw case.
4. Pull the wheel pad away from the steering wheel and disconnect the air bag connector.

NOTE: When removing the wheel pad, take care not to pull the air bag harness connector. When storing the wheel pad, keep the upper surface of the pad facing upward.

5. Disconnect the wire and remove the steering wheel retaining nut.
6. Place matchmarks on the steering wheel and mainshaft.
7. Remove the steering wheel with a suitable puller.
To install:
8. Check that the front wheels are facing straight ahead. Center the spiral cable. When centering the spiral cable be sure to use the following procedure:
  a. Check that the front wheels are pointing straight ahead.
  b. Turn the spiral cable counterclockwise by hand until it becomes harder to turn the cable. The spiral

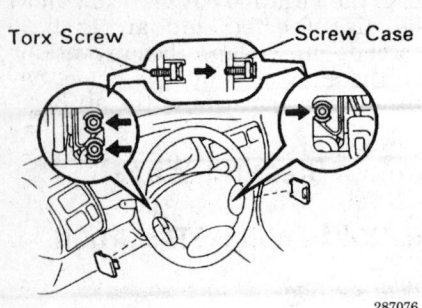

Air bag retaining screws

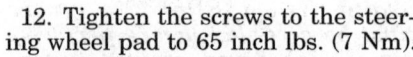

Airbag Connector

Correct / Wrong

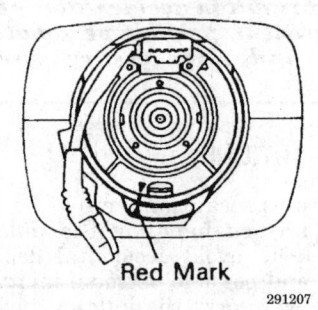

Storing the air bag

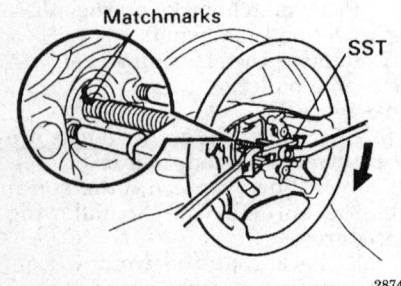

Matchmarks / SST

Common method of removing the steering wheel

cable will rotate approximately three turns to either the left or right of the center.

c. Then rotate the spiral cable clockwise approximately three turns to align the red mark.

9. Align the matchmarks and install the steering wheel. Tighten the nut to 26 ft. lbs. (35 Nm).

10. Engage the connector.

11. Install the steering wheel pad after confirming that the circumference groove of the screws is caught on the screw case.

NOTE: Make sure the wheel pad is installed to the specified torque. If the wheel pad has been dropped, or there are cracks, dents or other defects in the case or connector, replace the wheel pad with a new one. When installing the wheel pad, be sure the wires and connectors do not interfere with other parts and are not pinched between other parts.

12. Tighten the screws to the steering wheel pad to 65 inch lbs. (7 Nm).

13. Install the screw covers.

14. Connect the negative battery cable to the battery.

Red Mark

Aligning the spiral cable red mark

Green Mark

Alignment of spiral cable (green mark) — Tercel

Without Air Bag

NOTE: Do not attempt to remove or install the steering wheel by hammering on it. Damage to the energy-absorbing steering column could result.

1. Disconnect the negative battery cable.

2. Place the front wheels facing straight ahead.

3. Remove the screw located at the bottom of the steering pad.

4. Remove the three clips.

5. Remove the steering pad from the steering wheel.

6. Remove the steering wheel nut.

7. Using a steering wheel puller, remove the steering wheel.

To install:

8. Install the steering wheel and nut. Tighten the nut to 25 ft. lbs. (34 Nm).

9. Install the steering wheel pad and connect the three clips and crew.

10. Connect the battery cable and check operation.

Tie Rod Ends

REMOVAL AND INSTALLATION

1. Raise the front of the vehicle and support it safely. Remove the wheel.

2. Remove the cotter pin and nut holding the tie rod to the steering knuckle.

3. Using a tie rod separator, press the tie rod out of the knuckle.

NOTE: Use only the correct tool to separate the tie rod joint. Replace the boot if the rubber is cracked or ripped.

4. Matchmark the inner end of the tie rod to the end of the steering rack.

5. Loosen the locknut and remove the tie rod (count the number turns out to remove the tie rod) from the steering rack.

To install:

6. Install the tie rod end (count the same amount of turns in for correct installation) onto the rack ends and align the matchmarks made earlier.

7. Tighten the locknut to 41 ft. lbs. (56 Nm), except on 1994–97 Celica, Camry and Avalon. On these models, tighten the locknut to 55 ft. lbs. (74 Nm).

8. Connect the tie rod to the knuckle. Tighten the tie rod-to-steering knuckle nut to 36 ft. lbs. (49 Nm) and install a new cotter pin. Wrap the prongs of the cotter pin firmly around the flats of the nut.

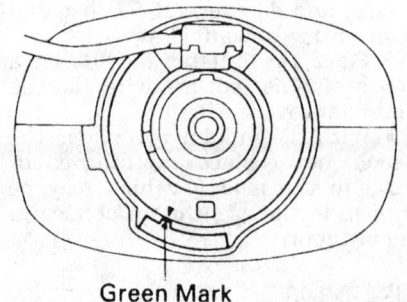

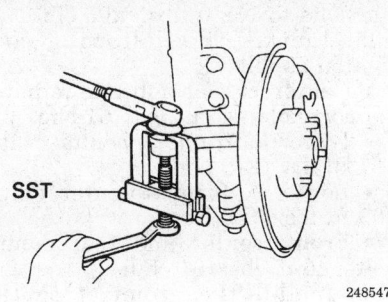

248547

Removing tie rod ends with a puller

9. If the vehicle is equipped with lubrication fittings, lubricate the suspension and chassis components.

10. Install the wheel and lower the vehicle to the ground. Check the front end alignment. The toe adjustment may have to be reset.

Manual Rack and Pinion

REMOVAL AND INSTALLATION

1. Position the front wheels straight ahead.

2. Disconnect the negative battery cable. On vehicles equipped with an air bag, wait at least 90 seconds before proceeding.

3. If equipped with an air bag, disable the system and secure the steering wheel.

4. Remove sliding yoke on rack and pinion assembly.

5. Raise and safely support the vehicle.

6. Remove the front wheels.

7. Disconnect the tie rod ends.

8. On MR2, remove front luggage undercover.

9. On Corolla, remove the engine rear mount insulator by removing the bolt and three nuts. Remove the engine rear mount bracket by removing the three bolts.

10. Remove rack and pinion assembly bracket bolts and rack and pinion assembly from the vehicle.

To install:

11. Install the rack and pinion assembly into the vehicle with the attaching bolts. Tighten the bolts to 43 ft. lbs. (58 Nm), except on MR2. On MR2, tighten the bolts to 32 ft. lbs. (42 Nm).

12. On MR2, install front luggage undercover and attaching bolts.

13. On Corolla, install the engine rear mount bracket and install the three bolts. Tighten the bolts to 57 ft. lbs. (77 Nm). Install the engine rear

mount insulator by installing the bolt and three nuts. Tighten the bolt to 64 ft. lbs. (87 Nm) and the nuts to 35 ft. lbs. (48 Nm).

14. Connect the tie rod ends to the knuckle arms.

15. Connect sliding yoke onto rack and pinion assembly. On Tercel, tighten the bolts to 19 ft. lbs. (26 Nm). On Corolla, tighten the lower bolt to 26 ft. lbs. (35 Nm) and the upper bolt to 20 ft. lbs. (27 Nm). On MR2, tighten the bolts to 26 ft. lbs. (35 Nm).

16. Install the front wheels.

17. Safely lower the vehicle.

18. Connect the negative battery cable.

19. Check the front end alignment. The toe adjustment may have to be reset.

Power Rack and Pinion

REMOVAL AND INSTALLATION

Tercel and Paseo

1. Position the front wheels straight ahead.

2. Disconnect the negative battery cable. On vehicles equipped with an air bag, wait at least 90 seconds before proceeding.

3. If equipped with an air bag, disable the system and secure the steering wheel.

4. Disconnect the tie rod ends from the knuckle arm using a tie rod separator or equivalent.

5. Remove the column hole cover. Matchmark the sliding yoke and control valve shaft for installation.

6. Loosen the upper bolt and disconnect the lower bolt to the control valve shaft. Slide the shaft upward and disconnect the control valve shaft from the steering rack.

7. Disconnect the oxygen sensor and the exhaust pipe.

8. If necessary for additional access, remove the stabilizer bar.

9. Remove the engine rear mount insulator bolts.

10. Disconnect the two vacuum hoses.

11. If equipped with a manual transaxle, disconnect the transmission control cables.

12. Disconnect the power steering hoses and drain the fluid into a container.

13. Disconnect the tube clamp.

14. Remove the two brackets and grommets.

15. Remove the housing-to-frame retaining bolts and remove the as-

sembly. Slide the housing out the left hand side of the vehicle.

To install:

16. Line up the steering splines, then install the unit.

17. Install the two grommets and brackets. Tighten the two bolts and nuts to 43 ft. lbs. (58 Nm).

18. Connect the pressure feed and return tubes and tighten to 26 ft. lbs. (36 Nm).

19. Connect the tube clamp and tighten the bolt to 9 ft. lbs. (13 Nm).

20. Connect the two vacuum hoses.

21. Install the rear brackets and tighten the bolts to 35 ft. lbs. (48 Nm).

22. Install the engine rear mount insulator. Tighten the through-bolt to 47 ft. lbs. (64 Nm) the support bracket bolts to 58 ft. lbs. (78 Nm).

23. If applicable, install the stabilizer bar to the vehicle.

24. Connect the sliding yoke to the control valve shaft. Tighten the bolts to 21 ft. lbs. (28 Nm).

25. Install the column hole cover with the three nuts and torque the nuts to 43 inch lbs. (5 Nm).

26. Connect the right and left tie rod ends.

27. Install the wheels to the vehicle and lower the vehicle.

28. Connect the negative battery cable to the battery.

29. Fill the power steering reservoir to specification and bleed the system.

30. Check the front end alignment. The toe adjustment may have to be reset.

Corolla

1. Position the front wheels straight ahead.

2. Disconnect the negative battery cable.

3. If equipped with an air bag, disable the system and secure the steering wheel.

4. Place a drain pan under the steering rack.

5. Remove the steering column hole cover by removing the five bolts.

6. Loosen the upper pinch bolt to the sliding yoke. Remove the lower pinch bolt at the pinion shaft.

7. Loosen the wheel lug nuts.

8. Raise and safely support the vehicle.

9. Remove both front wheels.

10. Remove the left and right engine undercovers.

11. Disconnect the left and right tie rod ends.

12. Install an engine support and tension it to support the engine without raising it.

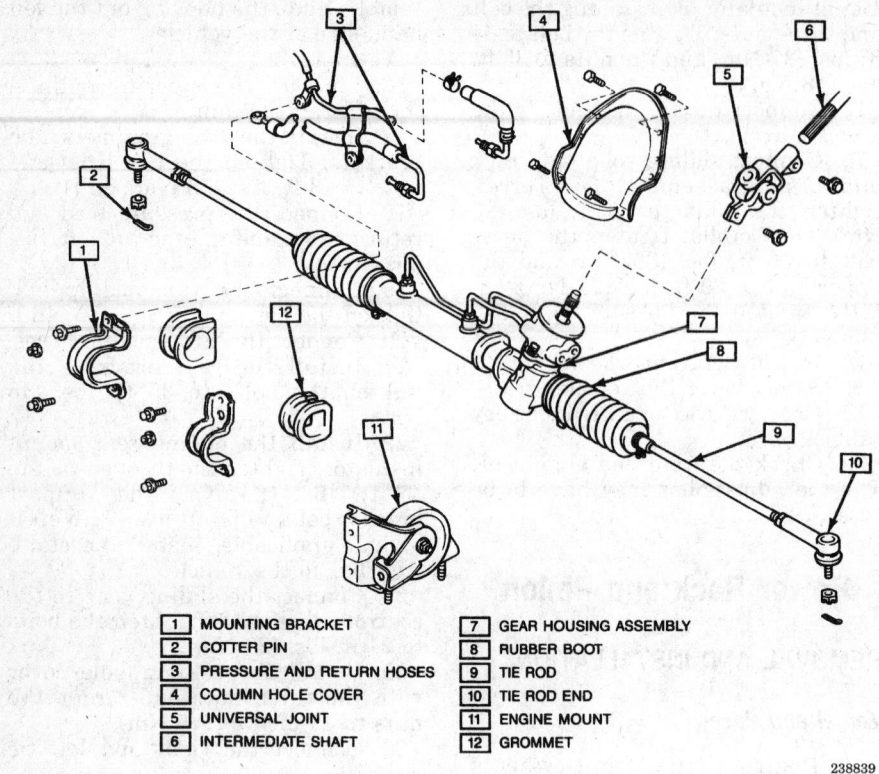

1	MOUNTING BRACKET	7	GEAR HOUSING ASSEMBLY
2	COTTER PIN	8	RUBBER BOOT
3	PRESSURE AND RETURN HOSES	9	TIE ROD
4	COLUMN HOLE COVER	10	TIE ROD END
5	UNIVERSAL JOINT	11	ENGINE MOUNT
6	INTERMEDIATE SHAFT	12	GROMMET

238839

Common power rack and pinion unit

—— CAUTION ——

The engine hoist is now in place and under tension. Use care when repositioning the vehicle and make necessary adjustments to the engine support.

13. Disconnect the lower control arms from the ball joints.

14. If equipped with a stabilizer bar, disconnect the stabilizer bar links from both lower control arms.

15. Remove the nut and three bolts and remove the right rear control arm bushing retaining bracket. Do this for both lower control arms.

16. Remove the stabilizer bar from the vehicle.

17. Remove the grommet in the crossmember.

18. Remove the bolt and four nuts holding in the middle of the crossmember.

19. Support the suspension crossmember with a jack.

20. Remove the six bolts from the outer side of the suspension crossmember.

21. Remove the suspension crossmember with the lower suspension arms.

22. Remove the exhaust front pipe support by removing the two bolts.

23. Remove the engine rear mount insulator by removing the bolt.

24. Remove the engine rear mount bracket by removing the three bolts.

25. Disconnect the pressure feed and return tubes.

26. Remove the two brackets and grommets to the power steering rack by removing the two bolts and nuts.

27. Slide the power steering gear assembly to the right side of the vehicle.

To install:

28. Install the power steering assembly.

29. Install the two grommets and brackets by installing the two bolts and two nuts. Tighten the nuts and bolts to 43 ft. lbs. (59 Nm).

30. Connect the pressure feed and return tubes and tighten the union nuts to 26 ft. lbs. (36 Nm).

31. Install the engine rear mount bracket by installing the three bolts. Tighten the bolts to 57 ft. lbs. (77 Nm).

32. Install the engine rear mount insulator by installing the bolt. Tighten the bolt to 64 ft. lbs. (87 Nm).

33. Install the exhaust front pipe support by installing the two bolts. Tighten the bolts to 14 ft. lbs. (19 Nm).

34. Raise the suspension crossmember with the lower control arms. Install the outer six bolts to hold the crossmember to the vehicle. Tighten the bolts to 152 ft. lbs. (206 Nm).

35. Install the following and tighten as follows:

• Center crossmember-to-radiator support bolts: 45 ft. lbs. (61 Nm).

• Lower A frame-to-center bolts: 161 ft. lbs. (218 Nm).

• Lower A frame-to-outer bolts: 109 ft. lbs. (147 Nm).

• Front, center and rear mount bolts: 45 ft. lbs. (61 Nm).

36. Install the grommet to the crossmember.

37. Install the stabilizer bar to the vehicle.

38. Install the lower control arm bushing retaining bracket and install the nut and three bolts. Do not tighten the bolts or nut at this time.

39. Connect the lower control arm to the lower ball joint by installing the bolt and two nuts. Tighten the bolt and nuts to 105 ft. lbs. (142 Nm). Connect both lower control arms to the ball joints.

40. Connect the stabilizer bar links to the lower control arms and tighten the nuts to 33 ft. lbs. (44 Nm).

41. Connect the sliding yoke to the pinion shaft. Install the lower bolt and tighten the bolt to 26 ft. lbs. (35 Nm). Tighten the upper bolt to the sliding yoke to 20 ft. lbs. (27 Nm).

42. Install the steering column hole cover by installing the five bolts. Tighten the bolts to 43 inch lbs. (5 Nm).

43. Install the left and right hand tie rod ends and tighten the nuts to 36 ft. lbs. (49 Nm).

44. Install the wheels and lower the vehicle to the ground.

45. Remove the engine support.

46. Lower the vehicle and stabilize the suspension by pushing up and down on the vehicle.

47. Tighten the rear lower control arm bracket bolts and nuts as follows:

• Control arm bracket bolts: 108 ft. lbs. (147 Nm)

• Stabilizer bar bracket bolt: 37 ft. lbs. (50 Nm)

• Bracket nut: 14 ft. lbs. (19 Nm)

48. Connect the negative battery cable to the battery.

49. Fill the power steering reservoir to specification and bleed the system.

50. Check the front end alignment. The toe adjustment may have to be reset.

Celica

1. Disconnect the negative battery cable. On vehicles equipped with an air bag, wait at least 90 seconds before proceeding.

2. Raise and safely support the front of the vehicle.

3. Remove the right and left hand engine undercovers.

4. Disconnect the left and right hand tie rod ends by removing the cotter pins and nuts. Separate the tie rod using a puller.

5. Remove the oxygen sensor.

6. Remove the front exhaust pipe and brackets.

7. Place matchmarks on the steering column intermediate shaft and the steering gear control valve shaft.

8. Loosen the upper bolt and remove the lower bolt to the intermediate shaft.

9. Disconnect the intermediate shaft from the control valve shaft.

10. Using a line wrench, disconnect the pressure feed and return tubes from the steering rack. Make sure to place a pan under the tubes to catch any fluid.

11. Disconnect the tube clamp bracket by removing the two bolts.

12. Support the engine and transaxle with a support fixture.

13. Disconnect the lower control arms from the lower ball joints by removing the bolt and two nuts on each side.

14. Remove the through-bolts to the front and rear mounting insulators.

15. Remove the three bolts holding the rear of the lower control arm to the sub-frame and body. Remove the bolts on both sides of the sub-frame.

16. Support the front sub-frame with a jack.

17. Remove the two bolts holding the sub-frame to the body. Lower the front sub-frame with the lower suspension arms and steering gear.

18. Remove the power steering gear assembly by removing the set bolts and nuts from the sub-frame.

To install:

19. Install the power steering gear assembly to the sub-frame by installing the two bolts and two nuts. Tighten to 94 ft. lbs. (127 Nm).

20. Raise the sub-frame into position and install the two bolts to hold the sub-frame to the body. Tighten the bolts to 94 ft. lbs. (127 Nm).

21. Install the three bolts to hold the rear of the lower control arms to the sub-frame and body. Tighten the bolt that goes through the lower control arm to 123 ft. lbs. (167 Nm) and the other two bolts to 130 ft. lbs. (175 Nm). Install the bolts to both sides.

22. Connect the two through-bolts to the front and rear mounting insulators and tighten to 64 ft. lbs. (88 Nm).

23. Connect the lower control arms to the lower ball joints by installing the bolt and two nuts on each side. Tighten the nuts and bolts to 94 ft. lbs. (127 Nm).

24. Remove the support from under the engine and transaxle.

25. Connect the pressure feed and return tubes to the steering rack. Tighten the tubes to 26 ft. lbs. (36 Nm).

26. Connect the tube clamp bracket and tighten the two bolts to 9 ft. lbs. (13 Nm).

27. For the intermediate shaft on the steering column, align the matchmarks on the intermediate shaft and control valve shaft.

28. Install the lower bolt and tighten the upper and lower bolts to 26 ft. lbs. (35 Nm).

29. Install the front exhaust pipe.

30. Install the oxygen sensor.

31. Connect the tie rod ends to the steering knuckles and install the nuts.

32. Install the right and left hand engine undercovers by installing the bolts.

33. Lower the vehicle.

34. Connect the negative battery cable to the battery.

35. Fill the power steering reservoir to specification and bleed the system.

36. Check the front end alignment. The toe adjustment may have to be reset.

Camry and Avalon

1. Position the front wheels straight ahead.

2. Disconnect the negative battery cable. On vehicles equipped with an air bag, wait at least 90 seconds before proceeding.

3. If equipped with an air bag, disable the system and secure the steering wheel.

4. Raise and support the vehicle safely. Remove the front wheels.

5. Remove the left and right front fender apron seals by removing the two bolts.

6. Remove the cotter pin and nut holding the steering knuckle to the tie rod end. Using a tie rod puller, disconnect the tie rod end from the steering knuckle.

7. Place matchmarks on the intermediate shaft and the control valve shaft.

8. Loosen the upper bolt and remove the lower bolt holding the control valve shaft to the intermediate shaft. Disconnect the intermediate shaft from steering rack housing.

9. Remove the nut to the tube clamp. Remove the clamp from the vehicle.

10. Disconnect the return line and the pressure line from the control valve housing. Use a small plastic container to catch the fluid.

11. Remove the four stabilizer bar bolts and two nuts. Position the stabilizer bar out of the way. Do not remove the bar from the vehicle.

12. If necessary, remove the rear engine mounting and bracket for additional clearance.

13. On the V6 engine, remove the oxygen sensor.

14. Remove the two steering gear mounting bolts and nuts. Remove the steering gear.

To install:

15. Position the steering gear on the vehicle and install the two mounting bolts and nuts. Tighten the nuts and bolts to 134 ft. lbs. (181 Nm).

16. On the V6, install the oxygen sensor.

17. If applicable, install the rear engine mounting bracket and tighten the retaining bolts to 38 ft. lbs. (52 Nm).

18. Install the stabilizer bar bolts and nuts.

19. Connect the tube clamp and tighten the nut to 7 ft. lbs. (10 Nm).

20. Install the intermediate shaft to the steering rack and tighten the retaining bolts to 26 ft. lbs. (35 Nm).

21. Connect the tie rods to the steering knuckles with the castellated nuts.

22. Install the front fender apron seals by installing the two bolts.

23. Install the front wheels and lower the vehicle.

24. Connect the negative battery cable to the battery.

25. Fill the power steering reservoir to specification and bleed the system.

26. Check the front end alignment. The toe adjustment may have to be reset.

MR2

1. Position the front wheels straight ahead.

2. Disconnect the negative battery cable. On vehicles equipped with an air bag, wait at least 90 seconds before proceeding.

3. If equipped with an air bag, disable the system and secure the steering wheel.

4. Raise and support the vehicle safely. Remove the front wheels.

5. Remove front luggage undercover.

6. Disconnect the right and left tie rod ends.

7. Remove the intermediate shaft dust seal cover by disconnecting the two clips.

8. Place matchmarks on the sliding yoke and the control valve shaft for the steering rack.

9. Loosen the upper bolt and remove the lower bolt for the sliding yoke.

10. Disconnect the sliding yoke from the steering rack.

11. Disconnect the hose from the return tube on the power steering gear assembly.

12. Remove the pressure feed tube on the rack and pinion assembly by removing the union bolt.

13. Remove rack and pinion assembly bracket bolts from the rack and pinion assembly.

14. Remove the rack and pinion unit from the vehicle.

To install:

15. Install the rack and pinion assembly into the vehicle and install the brackets and four bolts. Tighten bolts to 32 ft. lbs. (42 Nm).

16. Using a new gasket, connect the pressure tube to the rack and pinion assembly. Tighten the union bolt to 51 ft. lbs. (69 Nm).

NOTE: Make sure the stopper on the tube is touching the power steering gear body, then tighten the union bolt.

17. Connect the return hose to the power steering gear.

18. Connect the sliding yoke to the control valve shaft. Make sure to align the matchmarks on the sliding yoke and control valve shaft. Tighten the two bolts to 26 ft. lbs. (35 Nm).

19. Install the innershaft dust seal cover with the two clips.

20. Connect the tie rod ends to the steering knuckles and install the nuts.

21. Install front luggage undercover and attaching bolts.

22. Install the front wheels and lower the vehicle.

23. Connect the negative battery cable to the battery.

24. Fill the power steering reservoir to specification and bleed the system.

25. Check the front end alignment. The toe adjustment may have to be reset.

Supra

1. Position the front wheels straight ahead.

2. Disconnect the negative battery cable. On vehicles equipped with an air bag, wait at least 90 seconds before proceeding.

3. If equipped with an air bag, disable the system and secure the steering wheel.

4. Raise and safely support the vehicle. Remove the front wheels.

5. Remove the engine undercovers and front suspension member protection.

6. Place matchmarks on the universal joint and the control valve shaft.

7. Loosen the bolt on the upper side of the intermediate shaft.

8. Remove the bolt on the lower side of the intermediate shaft and disconnect the universal joint from the steering rack.

9. Disconnect and plug the hydraulic lines to the rack assembly by removing the union bolts and gaskets.

10. Disconnect the tie rod ends from the steering knuckles.

11. Disconnect the solenoid wiring from the rack and pinion unit.

12. Remove the bolts and brackets holding the steering rack to the frame.

13. Remove the rack assembly.

To install:

14. Install the rack. Tighten mounting bracket bolts to 55 ft. lbs. (75 Nm).

15. Connect the solenoid wiring.

16. Connect the tie rod ends to the steering knuckles.

17. Connect the fluid lines to the rack and pinion with new washers. Tighten the union bolts to 36 ft. lbs. (49 Nm).

18. Align the matchmarks on the universal joint and the control valve shaft. Tighten the upper and lower bolts to 26 ft. lbs. (35 Nm).

19. Install the front suspension member protection and the engine undercovers.

20. Install the front wheels to the vehicle and lower the vehicle to the ground.

21. Connect the negative battery cable to the battery.

22. Fill the power steering reservoir to specification and bleed the system.

23. Check the front end alignment. The toe adjustment may have to be reset.

Power Steering Pump

BLEEDING

Any time the power steering system has been opened or disassembled, the system must be bled to remove any air which may be trapped in the lines. Air will prevent the system from providing the correct pressures to the rack. The correct fluid level reading will not be obtained if the system is not bled.

1. With the engine running, turn the wheel all the way to the left and shut off the engine.

2. Add power steering fluid to the **COLD** mark on the indicator.

3. Start the engine and run at fast idle for about 15 seconds. Stop the engine and recheck the fluid level. Add to the **COLD** mark as needed.

4. Start the engine and bleed the system by turning the wheels from left to right 3 or 4 times.

5. Stop the engine and check the fluid level and condition. Fluid with air in it is a light tan color. This air must be eliminated from the system before normal operation can be obtained. Repeat Steps 3 and 4 until the correct fluid color and fluid level is obtained.

REMOVAL AND INSTALLATION

Tercel and Paseo

1. Disconnect the negative battery cable from the battery. On vehicles equipped with an air bag, wait at least 90 seconds before proceeding.

2. Loosen the adjusting bolt and the pivot bolt and remove the belt.

3. Pinch the clamp and remove the return hose.

4. Remove the bolt for the pressure line and discard the gasket.

NOTE: Tie the hose ends up high so the fluid cannot flow out of them. Drain or plug the pump to prevent fluid leakage.

5. Remove the adjusting and pivot bolts from the mounting braces.

6. Remove the pump from the engine.

To install:

7. Install the pump and related components.

8. Adjust the pump drive belt tension and tighten the mounting bolts to 32 ft. lbs. (43 Nm).

9. Install the pressure line bolt with a new gasket and tighten to 40 ft. lbs. (54 Nm).

10. Install the return hose with a new clamp.

11. Fill the reservoir with fluid. Bleed the air from the system.

12. Connect the negative battery cable to the battery.

13. Check for fluid leaks.

Corolla

1. Disconnect the negative battery cable from the battery. On vehicles equipped with an air bag, wait at least 90 seconds before proceeding.
2. Loosen the power steering pivot and adjusting bolts.
3. Swing the power steering pump toward the engine and remove the drive belt.
4. Loosen the power steering hose retaining clip and remove the return hose from the power steering pump.
5. Remove the pressure hose from the pump.
6. Remove the vacuum hoses from the air control valve of the power steering vane pump.
7. Remove the two mounting bolts from the steering pump.
8. Remove the two adjusting bracket set bolts and bracket.
9. Remove the power steering pump and bracket from the vehicle.

To install:
10. Install the power steering pump and bracket to the engine.
11. Install the adjusting bracket with the two bolts. Tighten the bolts to 29 ft. lbs. (39 Nm).
12. Install the two power steering pump set bolts. Do not tighten the bolts at this time.
13. Install and tighten the drive belt.
14. Tighten the two power steering bolts to 29 ft. lbs. (39 Nm).
15. Connect the two vacuum hoses.
16. Connect the pressure feed tube by connecting the union bolt over a new gasket. Tighten the union bolt to 40 ft. lbs. (54 Nm).
17. Refill and bleed the power steering system.

Celica

1.6L (4A-FE) and 1.8L (7A-FE) Engines

1. Disconnect the negative battery cable. On vehicles equipped with an air bag, wait at least 90 seconds before proceeding.
2. Raise and safely support the vehicle.
3. Loosen the pivot bolt and adjusting bolt to the power steering pump and remove the drive belt.
4. Place a drain pan below the power steering pump.
5. Remove the clamp from the fluid return hose and disconnect the hose from the pump.
6. Remove the union bolt and gasket for the pressure tube and disconnect the tube from the pump.
7. Remove the adjusting bolt from the power steering pump.

8. Remove the four bolts from the steering pump bracket and remove the assembly from the engine.
9. Remove the pump bracket from the steering pump by removing the pivot bolt.

To install:
10. Install the pump bracket to the steering pump by installing the pivot bolt. Do not tighten the bolt at this time.
11. Place the pump assembly in position and temporarily install the mounting bolts. Tighten the bolts as follows:
• 12mm head bolt to 14 ft. lbs. (19 Nm)
• 14mm head bolt to 29 ft. lbs. (39 Nm)
12. Install the adjusting bolt. Do not tighten the bolt at this time.
13. Connect the pressure feed tube with a new gasket and the union bolt. Tighten the bolt to 38 ft. lbs. (52 Nm).
14. Connect the return hose.
15. Install the drive belt. Adjust the belt to the proper tension. Tighten the adjusting bolt and pivot bolt to 29 ft. lbs. (39 Nm).
16. Install the right side engine undercover.
17. Lower the vehicle and connect the negative battery cable.
18. Fill the reservoir to the proper level with power steering fluid and bleed the system.

2.2L (5S-FE) Engine

1. Disconnect the negative battery cable. On vehicles equipped with an air bag, wait at least 90 seconds before proceeding.
2. Raise and safely support the vehicle.
3. Place a drain pan below the power steering pump.
4. If equipped with ABS brakes, remove the ABS speed sensor wiring harness clamp.
5. Remove the right hand halfshaft from the vehicle.
6. Disconnect the two vacuum hoses from the power steering pump.
7. Disconnect the return hose from the power steering hose.
8. Loosen the pivot bolt and adjusting bolt to the power steering pump and remove the drive belt.
9. Disconnect the power steering pressure feed tube from the rack and pinion as follows:
   a. Remove the two tube clamp bracket set bolts.
   b. Remove the three tube clamp brackets and grommets from the tube.
   c. Using a line wrench, disconnect the tube from the rack and pinion.

10. Remove the power steering pump assembly with the pressure feed tube and two brackets by removing the five bolts.
11. Remove the union bolt and gasket holding the pressure tube to the power steering pump. Disconnect the tube from the pump.

To install:
12. Install the pressure feed tube to the steering pump with the union bolt and a new gasket. Tighten the bolt to 38 ft. lbs. (52 Nm).
13. Install the power steering pump assembly with the pressure feed tube and two brackets. Install the five bolts but only tighten the three bolts for the power steering pump brackets. Do not tighten the pivot or adjusting bolt. Tighten the bolts to 32 ft. lbs. (43 Nm).
14. Connect the pressure feed tube to the steering rack as follows:
   a. Install the three tube clamp brackets and grommets.
   b. Tighten the two tube clamp bracket set bolts.
   c. Using a line wrench, connect the pressure feed tube to the steering rack. Tighten the tube to 26 ft. lbs. (36 Nm).
15. Install the adjust the power steering belt. Tighten the adjusting and pivot bolts to 32 ft. lbs. (43 Nm).
16. Connect the return hose.
17. Connect the two vacuum hoses.
18. Install the right halfshaft.
19. If equipped with ABS brakes, install the ABS speed sensor wiring harness clamp.
20. Install the engine undercover and connect the negative battery cable.
21. Bleed the power steering system.

Camry and Avalon

1. Disconnect the negative battery cable from the battery. On vehicles equipped with an air bag, wait at least 90 seconds before proceeding.
2. Remove the alternator drive belt.
3. Raise and safely support the vehicle.
4. Remove the right front fender apron seal by removing the two bolts.
5. Siphon the power steering fluid from the reservoir.
6. If applicable, disconnect the two vacuum hoses from the power steering pump.
7. Disconnect the return hose from the power steering vane pump.
8. Remove the pressure feed tube by removing the union bolt and gasket.

9. Loosen the pivot bolt and remove the adjusting bolt. The pivot bolt can not be removed.

10. Remove the power steering pump from the engine.

To install:

11. Install the power steering pump to the engine.

12. Install the adjusting bolt. Tighten the adjusting bolt and pivot bolt temporarily.

13. Adjust the power steering drive belt and then tighten the pivot and adjusting bolts to 32 ft. lbs. (43 Nm).

14. Connect the pressure feed tube with the union bolt and a new gasket. Tighten the union bolt to 38 ft. lbs. (51 Nm)

NOTE: Make sure the stopper of the tube is touching the front bracket when tightening the union bolt.

15. Connect the return hose to the power steering pump.

16. If applicable, connect the vacuum hoses to the pump.

17. Install the front fender apron seal and tighten the two bolts.

18. Lower the vehicle and refill the power steering reservoir.

19. Connect the negative battery cable to the battery.

20. Start the engine, check for leaks, and bleed the steering system.

MR2

1. Disconnect the negative battery cable. On vehicles equipped with an air bag, wait at least 90 seconds before proceeding.

2. Remove front luggage undercover.

3. Disconnect the return hose from the power steering vane pump.

4. Using a line wrench, disconnect the pressure tube from the power steering pump.

5. Remove rear pump stay by removing the two bolts. Remove the stay from the two connectors.

6. Unplug the two connectors from the power steering vane pump.

7. Remove the power steering unit from the engine by removing the two bolts, spacers, and bushings.

8. Remove the power steering pump shield.

To install:

9. Install the power steering pump shield.

10. Install the power steering pump with motor and attaching bolts/spacers. Tighten the bolts to 19 ft. lbs. (25 Nm).

11. Engage the two connectors to the power steering pump.

12. Connect return and pressure hoses to the pump. Tighten the pressure feed pump to 26 ft. lbs. (36 Nm).

13. Install rear pump stay to the two connectors. Install and tighten the two bolts for the pump stay to 13 ft. lbs. (18 Nm).

14. Install front luggage cover.

15. Connect the negative battery cable.

16. Fill the power steering fluid reservoir and bleed the system.

Supra

1. Remove the battery cover and the battery. On vehicles equipped with an air bag, wait at least 90 seconds before proceeding.

2. Raise and support the vehicle safely.

3. Remove the lower engine cover.

4. Remove the drive belt.

5. Remove the return hose from the power steering pump.

6. Disconnect the pressure feed tube from the pump assembly.

7. Remove the two power steering pump mounting bolts.

8. Remove the power steering pump assembly.

To install:

9. Install the pump and loosely install the bolts.

10. Install and tension the drive belt. Tighten the steering pump mounting bolts to 43 ft. lbs. (58 Nm).

11. Connect the return hose to the power steering pump.

12. Connect the hydraulic pressure tube to the pump with new sealing washers. Tighten the union bolt to 36 ft. lbs. (49 Nm).

13. Install the lower engine cover.

14. Lower the vehicle.

15. Install the battery and cover.

16. Refill and bleed the power steering system.

BRAKES

Anti-Lock Brake System Service

PRECAUTIONS

• Certain components within the Anti-Lock Brake System (ABS) are not intended to be serviced or repaired individually. Only those components with removal and installation procedures should be serviced.

• Do not use rubber hoses or other parts not specifically specified for and ABS system. When using repair kits, replace all parts included in the kit. Partial or incorrect repair may lead to functional problems and require the replacement of components.

• Lubricate rubber parts with clean, fresh brake fluid to ease assembly. Do not use lubricated shop air to clean parts; damage to rubber components may result.

• Use only specified brake fluid from an unopened container.

• If any hydraulic component or line is removed or replaced, it may be necessary to bleed the entire system.

• A clean repair area is essential. Always clean the reservoir and cap thoroughly before removing the cap. The slightest amount of dirt in the fluid may plug an orifice and impair the system function. Perform repairs after components have been thoroughly cleaned; use only denatured alcohol to clean components. Do not allow ABS components to come into contact with any substance containing mineral oil; this includes used shop rags.

• The Anti-Lock control unit is a microprocessor similar to other computer units in the vehicle. Ensure that the ignition switch is **OFF** before removing or installing controller harnesses. Avoid static electricity discharge at or near the controller.

• If any arc welding is to be done on the vehicle, the control unit should be unplugged before welding operations begin.

DEPRESSURIZING

Toyota vehicles use low pressure ABS systems. No special depressurizing procedures are required.

Master Cylinder

REMOVAL AND INSTALLATION

1. Disconnect the negative battery cable from the battery. On vehicles equipped with an air bag, wait at least 90 seconds before proceeding.

2. If necessary, remove the two bolts and four nuts attaching the suspension upper brace and remove the upper brace to gain access to the master cylinder. Temporarily fasten the shock absorber with the four nuts.

3. If necessary, remove the air cleaner assembly from the engine compartment.

4. If equipped with a Traction Control System (TRAC), perform the following procedure:

a. Connect a vinyl tube from a container to the bleeder plug of the TRAC actuator, then loosen the bleeder plug with the ignition in the **OFF** position.

b. Tighten the plug when the fluid stops flowing out.

— **WARNING** —

The fluid is under high pressure and could spray out with great force. Use caution when opening the bleeder plug.

5. Disconnect the electrical connector to the master cylinder.

6. Remove the fluid in the master cylinder with a suitable syringe.

NOTE: If brake fluid is spilled, wipe it up immediately and wash the area with water.

7. If equipped with TRAC, remove the hoses from the master cylinder.

8. Disconnect the hydraulic lines from the master cylinder. Plug the ends of the lines to prevent loss of fluid.

9. Unfasten the nuts and remove the master cylinder from the power brake unit.

To install:

NOTE: Never reuse brake fluid or leave the fluid uncovered for a period of time. Brake fluid will absorb moisture from the air and contaminated fluid can cause premature brake system failure.

10. Bench bleed the master cylinder.

11. Install the master cylinder. Tighten the hydraulic lines to 11 ft. lbs. (15 Nm) and the master cylinder mounting nuts to 9 ft. lbs. (13 Nm).

12. If equipped with TRAC, connect the hoses to the master cylinder.

13. Engage the electrical connector to the master cylinder.

14. Install the air cleaner and the suspension upper brace.

15. After installation is completed, fill and bleed the brake system.

16. If equipped, bleed the TRAC system.

17. Check for leaks in the system before starting the vehicle.

18. Check and adjust the brake pedal.

19. Connect the negative battery cable to the battery.

Brake Caliper

REMOVAL AND INSTALLATION

— **CAUTION** —

Brake pads may contain asbestos, which has been determined to be a cancer causing agent. Never clean the brake surfaces with compressed air! Avoid inhaling any dust from any brake surfaces. When cleaning brake surfaces, use a commercially available brake cleaning fluid.

1. Raise and support the vehicle safely.

2. Remove the wheels.

3. Disconnect the brake hose from the caliper. Plug the end of the hose to prevent loss of fluid.

4. Remove the bolts that attach the caliper to the torque plate. If applicable, hold the flats of the sliding pin with a wrench while loosening the caliper attaching bolts.

5. Lift up and remove the caliper assembly.

To install:

6. Grease the caliper slides and bolts with lithium grease or equivalent. Install the caliper and loosely install the bolts.

7. Hold the flats of the sliding pin with a wrench, then tighten the bolts. On front calipers, tighten single piston types to 18 ft. lbs. (25 Nm) and dual piston types to 25 ft. lbs. (34 Nm). Tighten the bolts on rear calipers to 14 ft. lbs. (20 Nm).

8. Connect the brake hose to the caliper, using two new washers.

NOTE: If applicable, insert the brake line securely into the lock hole in the caliper.

9. Fill the brake system to the proper level and bleed the brake system.

10. Install the tire and wheel assembly.

11. Add brake fluid to the reservoir to fill to the correct level. Check for leaks and proper brake operation.

12. Lower the vehicle to the ground.

— **WARNING** —

Before moving the vehicle, make sure to pump the brake pedal with short strokes to seat the pads against the rotors, otherwise the vehicle may have no braking.

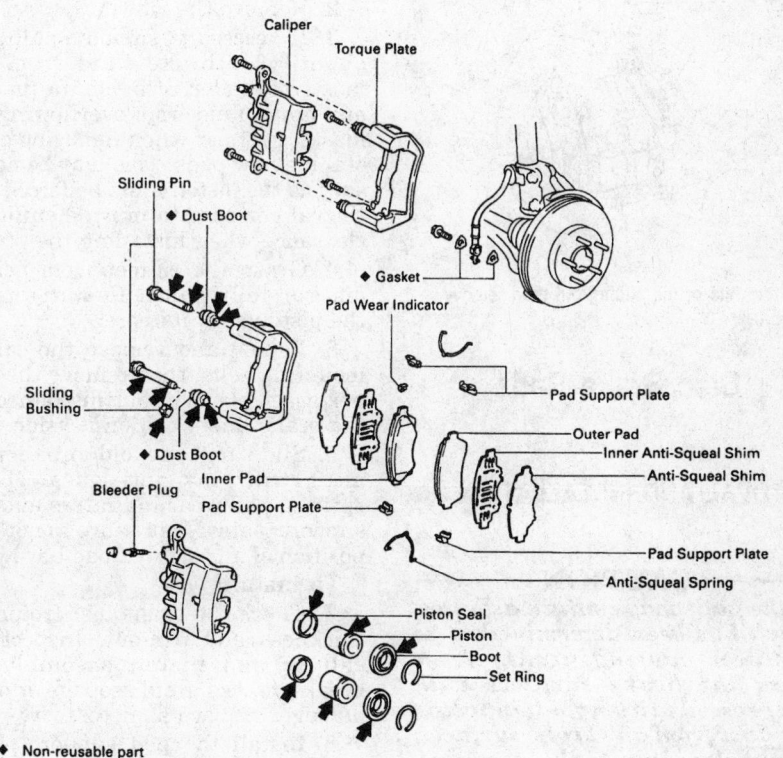

293013

Front brake assembly (two piston type) — Camry shown, others similar

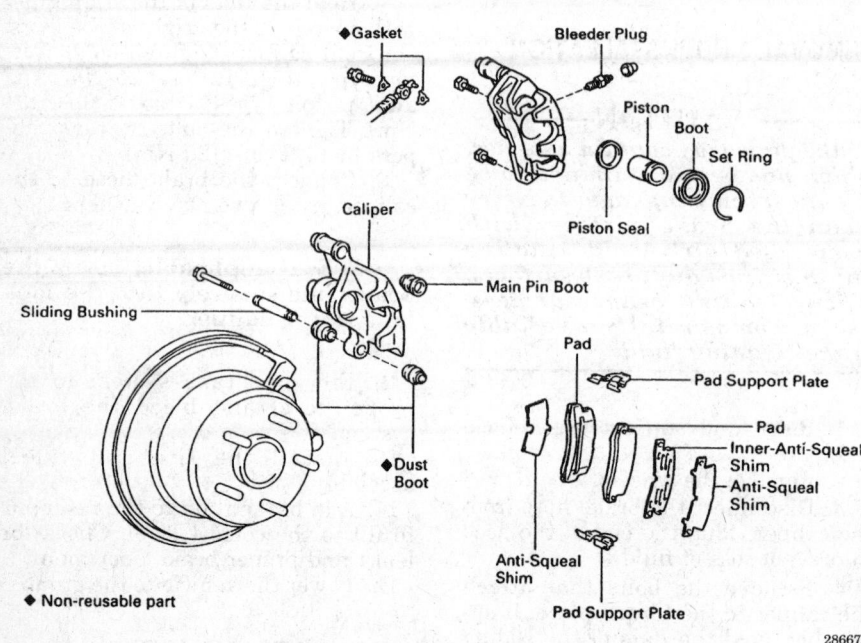

Rear disc brake components — Celica and Corolla shown, others similar

◆ Non-reusable part

the indicator plate is pointing in the direction of rotation.

10. Install the anti-squeal shims on the outside of each pad and then install the pad assemblies into the torque plate.

11. Position the caliper back down over the pads. If it won't fit, use a C-clamp or hammer handle and carefully force the piston into its bore, except for rear discs on Corolla. On Corolla, use tool SST 09719-14020 to rotate the piston clockwise while pressing it into the bore until it locks.

NOTE: On dual piston calipers, be sure to press both pistons in at the same time.

12. Install and tighten the caliper mounting bolts.

13. Install the wheels and lower the vehicle. Check the brake fluid level.

— **WARNING** —
Before moving the vehicle, make sure to pump the brake pedal with short strokes to seat the pads against the rotors, otherwise the vehicle may have no braking.

Brake Rotor

REMOVAL AND INSTALLATION

1. Loosen the wheel lugs slightly, then raise and safely support the car.
2. Remove the wheel(s) and temporarily attach the rotor disc with two of the wheel nuts.
3. Remove the bolts holding the caliper to the torque plate.
4. Remove the caliper from the torque plate and hang the caliper from a piece of wire.

— **WARNING** —
Do not allow the caliper to hang freely from the vehicle. Always support the caliper with a wire from the vehicle.

5. Unbolt and remove the torque plate.
6. Remove the two wheel nuts and pull the disc from the axle hub.
To install:
7. Position the new rotor disc onto the axle hub and reinstall the two wheel nuts temporarily.
8. Install the torque plate. Make sure the brake pads are seated correctly within the torque plate.
9. Install the caliper to the torque plate.
10. Remove the wheel nuts and install the wheels. Secure the wheel lugs.

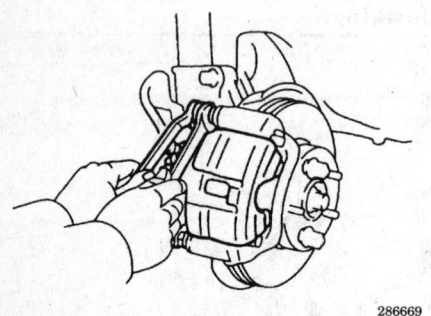

Hold the flats of the sliding pin then remove the bolts

Disc Brake Pads

REMOVAL AND INSTALLATION

— **CAUTION** —
Brake pads may contain asbestos, which has been determined to be a cancer causing agent. Never clean the brake surfaces with compressed air! Avoid inhaling any dust from any brake surfaces. When cleaning brake surfaces, use a commercially available brake cleaning fluid.

1. Raise and support the vehicle safely.
2. Remove the wheels.
3. If necessary, siphon a sufficient quantity of brake fluid from the master cylinder reservoir to prevent any brake fluid from overflowing the master cylinder when removing or installing new pads. This may be necessary as the piston must be forced into the caliper bore to provide sufficient clearance when installing the pads.
4. Grasp the caliper from behind and carefully pull it to start to seat the piston(s) in its bore.
5. Loosen and remove the caliper mounting bolts, then remove the caliper assembly, without disconnecting the brake line. Position it aside.
6. Slide out the old brake pads along with any anti-squeal shims, springs, pad wear indicators and pad support plates. Make sure to note the position of all assorted pad hardware.
To install:
7. Check the brake disc (rotor) for thickness and run-out. Inspect the caliper and piston assembly for breaks, cracks, fluid seepage or other damage. Replace as necessary.
8. Install the pad support plates into the torque plate.
9. Install the pad wear indicators onto the pads. Be sure the arrow on

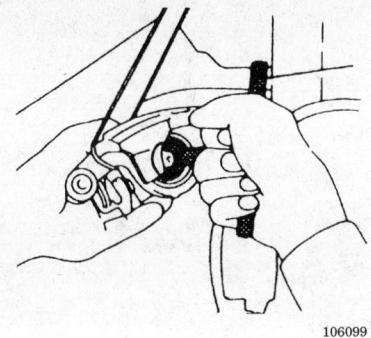

106099

Using the special tool to seat the rear brake caliper piston — Corolla

---
WARNING

Before moving the vehicle, make sure to pump the brake pedal with short strokes to seat the pads against the rotors, otherwise the vehicle may have no braking.

Brake Drums

REMOVAL AND INSTALLATION

Except Paseo and 1995–97 Tercel

1. Loosen the rear wheel lug nuts slightly. Release the parking brake.
2. Block the front wheels, raise the rear of the vehicle, and safely support it with jackstands.
3. Remove the wheel lug nuts and the wheel.
4. Remove the brake drum from the axle hub. If there is difficulty in removing the drum, insert a suitable tool through the hole in the rear of the backing plate, and hold the automatic adjusting lever away from the adjuster. Using another suitable tool at the same time, reduce the brake shoe adjuster by turning the adjusting wheel.
To install:
5. Install the brake drum and pull the parking brake lever all the way up until a clicking sound can no longer be heard.
6. Verify that the rear wheels will not turn. If the rear wheels turn, adjust the parking brake cable as necessary.
7. Release the parking brake and remove the brake drum. Measure the brake drum inside diameter and diameter of the brake shoes. Check that the difference between the diameters is the correct shoe clearance. Clearance is 0.024 in. (0.60mm).
8. If the brake shoe clearance is not correct, adjust the brake shoes

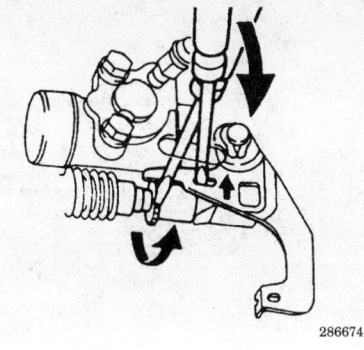

286674

Reducing the adjuster to remove the brake drum

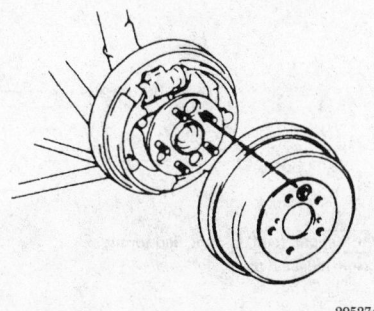

295274

Align the hole in the drum with the hub

until the clearance is correct. Install the drum.
9. Install the rear wheels, tighten the wheel lug nuts and lower the vehicle.
10. Retighten the wheel lug nuts and pump the brake pedal a few times before moving the vehicle.

Paseo and 1995–97 Tercel

1. Loosen the rear wheel lug nuts slightly. Release the parking brake.
2. Block the front wheels, raise the rear of the vehicle, and safely support it with jackstands.
3. Remove the wheel lug nuts and the wheel.
4. Remove the grease cap.
5. Remove the cotter pin and lock cap.
6. Remove the nut.
7. Remove the brake drum assembly with bearings, being careful not to drop the bearings. If there is difficulty in removing the drum, insert a suitable tool through the hole in the rear of the backing plate, and hold the automatic adjusting lever away from the adjuster. Using another suitable tool at the same time, reduce the brake shoe adjuster by turning the adjusting wheel.

To install:
8. Install the drum with inner bearing onto the spindle.
9. Install the outer bearing and washer.
10. Install the nut and tighten to 22 ft. lbs. (29 Nm) while turning the drum assembly several turns.
11. Loosen the nut until it can be turned freely by hand.
12. Measure the rotation frictional force of the oil seal.
13. Tighten the nut until preload is 0–2.6 lbs. (0–11.8 N) plus frictional force.
14. Install the lock cap, cotter pin, and the grease cap.
15. Pull the parking brake lever all the way up until a clicking sound can no longer be heard.
16. Verify that the rear wheels will not turn. If the rear wheels turn, adjust the parking brake cable as necessary.
17. Release the parking brake and remove the brake drum. Measure the brake drum inside diameter and diameter of the brake shoes. Check that the difference between the diameters is the correct shoe clearance. Clearance is 0.024 in. (0.60mm).
18. If the brake shoe clearance is not correct, adjust the brake shoes until the clearance is correct. Reinstall the drum.
19. Install the rear wheels, tighten the wheel lug nuts and lower the vehicle.
20. Retighten the wheel lug nuts and pump the brake pedal a few times before moving the vehicle.

Brake Shoes

REMOVAL AND INSTALLATION

Except Paseo and 1995–97 Tercel

1. Loosen the rear wheel lug nuts slightly. Release the parking brake.
2. Block the front wheels, raise the rear of the vehicle, and safely support it with jackstands.
3. Remove the wheel lug nuts and the wheel.
4. Remove the brake drum.

NOTE: Do not depress the brake pedal once the brake drum has been removed.

5. Carefully unhook the return spring from the leading (front) brake shoe. Grasp the hold-down spring pin with pliers and turn it until its in line with the slot in the hold-down spring. Remove the hold-down spring and the pin. Pull out the brake shoe and

unhook the anchor spring from the lower edge.

6. Remove the hold-down spring from the trailing (rear) shoe. Pull the shoe out with the adjuster strut, automatic adjuster assembly and springs attached and disconnect the parking brake cable. Unhook the return spring and then remove the adjusting strut. Remove the anchor spring.

7. Remove the adjusting strut. Unhook the adjusting lever spring from the rear shoe and then remove the automatic adjuster assembly by popping out the C-clip.

To install:

8. Inspect the shoes for signs of unusual wear or scoring.

9. Check the wheel cylinder for any sign of fluid seepage or frozen pistons.

10. Clean and inspect the brake backing plate and all other components. Lubricate the backing plate bosses and the anchor plate.

11. Mount the automatic adjuster assembly onto a new rear brake shoe. Make sure the C-clip fits properly. Connect the adjusting strut/return spring and then install the adjusting spring.

12. Connect the parking brake cable to the rear shoe and then position the shoe so the lower end rides in the anchor plate and the upper end is against the boot in the wheel cylinder. Install the pin and the hold-down spring. Rotate the pin so the crimped edge is held by the retainer.

13. Install the anchor spring between the front and rear shoes and then stretch the spring enough so the front shoe will fit as the rear did. Install the hold-down spring and pin. Connect the return spring/adjusting strut between the 2 shoes and connect it so it rides freely.

14. Install the drum.

15. Install the wheel and lower the vehicle. Check the level of brake fluid in the master cylinder, then test drive.

Paseo and 1995–97 Tercel

1. Loosen the rear wheel lug nuts slightly.

2. If the parking brake is applied, release the parking brake.

3. Block the front wheels, raise the rear of the vehicle, and safely support it with jackstands.

4. Remove the wheel lug nuts and the wheel.

5. Remove the drum.

NOTE: Do not depress the brake pedal once the brake drum has been removed.

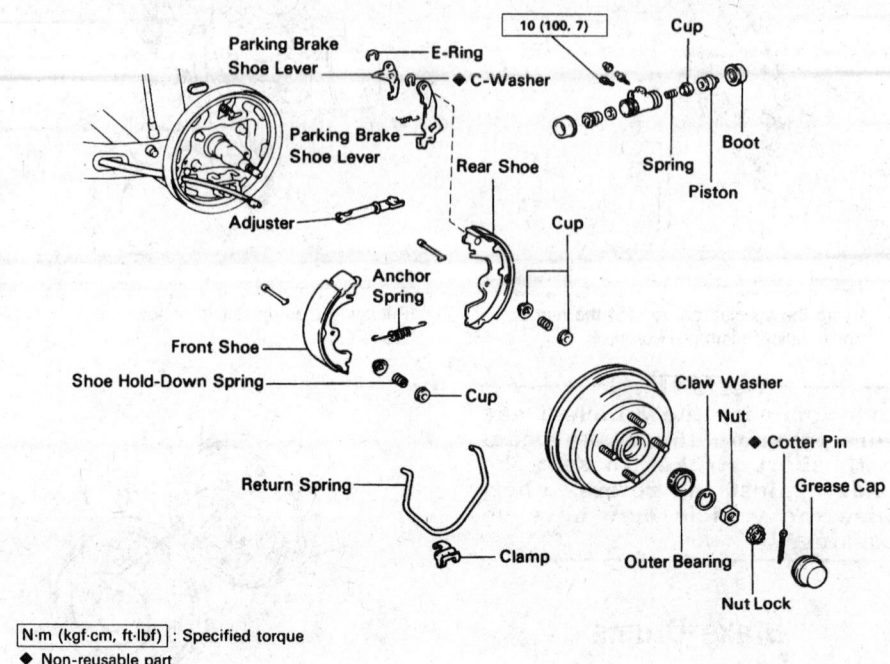

Brake drum and shoe assembly — Paseo and 1995–97 Tercel

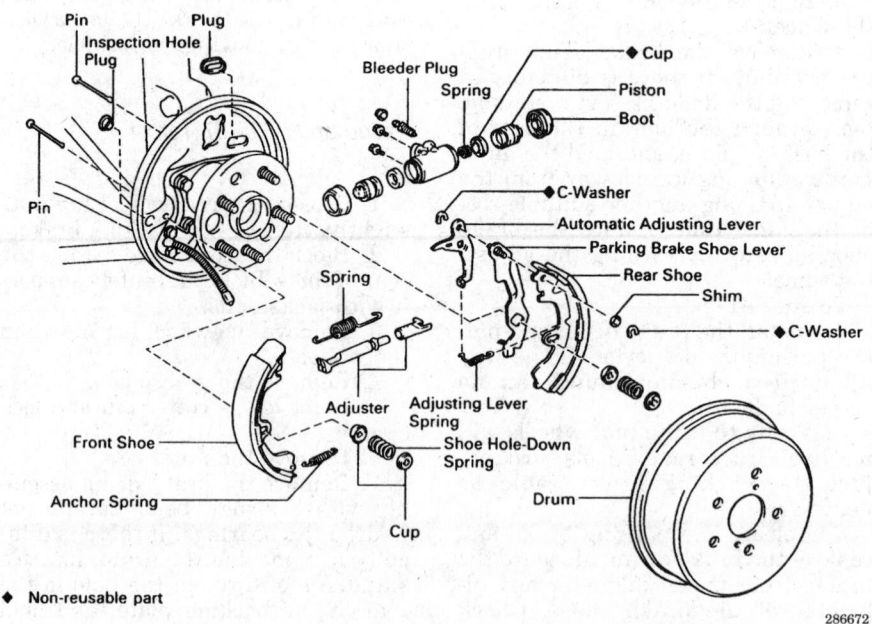

Rear drum brake components — except Paseo and 1995–97 Tercel

6. Carefully unhook the tension spring from the leading (front) brake shoe and remove the clamp.

7. Press the hold-down spring retainer in and turn the pin.

8. Remove the hold-down spring, retainers and the pin. Pull out the brake shoe and unhook the anchor spring from the lower edge.

9. Remove the hold-down spring from the trailing (rear) shoe. Pull the shoe out with the adjuster, automatic adjuster assembly and springs attached and disconnect the parking brake cable. Remove the tension/return and anchor springs from the rear shoe.

10. Unhook the adjusting lever spring from the rear shoe and then remove the automatic adjuster assembly by popping out the C-clip.

To install:

11. Inspect the shoes for signs of unusual wear or scoring.

12. Check the wheel cylinder for any sign of fluid seepage or frozen pistons.

13. Clean and inspect the brake backing plate and all other components. Lubricate the backing plate bosses and the anchor plate.

14. Mount the automatic adjuster assembly onto a new rear brake shoe. Make sure the C-clip fits properly. Connect the adjuster and install the spring.

15. Connect the parking brake cable to the rear shoe and then position the shoe so the lower end rides in the anchor plate and the upper end is against the boot in the wheel cylinder. Install the pin and the hold-down spring. Press the retainer down over the pin and rotate the pin so the crimped edge is held by the retainer. Install the anchor spring between the front and rear shoes and then stretch the spring enough so the front shoe will fit as the rear did. Install the hold-down spring, pin and retainer. Stretch the tension/return spring between the two shoes and connect it so it rides freely. Do not forget the return spring clamp.

16. Install the brake drum.

17. Install the rear wheels, tighten the wheel lug nuts, and lower the vehicle.

18. Check the level of brake fluid in the master cylinder.

19. Pump the brake pedal a few times before moving the vehicle. Adjust the rear brakes again if necessary.

Wheel Cylinder

REMOVAL AND INSTALLATION

1. Plug the master cylinder inlet to prevent hydraulic fluid from leaking.

2. Remove the brake drum.

3. Remove the brake shoe tension spring.

4. Working from behind the backing plate, disconnect the hydraulic line from the wheel cylinder and plug the opening.

5. Unfasten the screws or the clip retaining the wheel cylinder and withdraw the cylinder.

To install:

6. Insert the wheel cylinder into position and secure with clips or retaining bolts.

7. Connect the brake line to the wheel cylinder.

8. Tighten the retaining bolts to 7 ft. lbs. (10 Nm) and the line to 11 ft. lbs. (15 Nm).

9. Bleed the brake system after completing wheel cylinder, brake shoe and drum installation. Adjust the brakes.

Parking Brake Cable

ADJUSTMENT

Celica, Corolla, Paseo and Tercel

NOTE: The rear brake system, including brake linings, should be in good condition and properly adjusted prior to performing this adjustment.

1. Disconnect the negative battery cable from the battery.

2. Slowly pull the parking brake lever upward (without depressing the button on the end of it) while counting the number of notches required until the parking brake is applied.

3. The optimal setting is 5 to 8 clicks.

4. If the brake system requires adjustment, remove the console box and loosen the cable adjusting nut cap. The adjusting cap is located at the rear of the parking brake lever.

5. Take up the slack in the parking brake cable by rotating the adjusting nut with another open end wrench.

a. If the number of notches is less than specified, turn the nut counterclockwise.

b. If the number of notches is more than specified, turn the nut clockwise.

6. Tighten the adjusting cap, using care not to disturb the setting of the adjusting nut.

7. Check the rotation of the rear wheels to be sure that the brakes are not dragging with the parking brake released.

8. Install the console box.

9. Connect the negative battery cable to the battery.

1995–97 Avalon, Camry and Supra

1. Raise and support the vehicle safely.

2. Remove the rear wheels.

3. Temporarily install the hub nuts to hold the rotor on the vehicle.

4. Remove the brake rotor adjustment hole plug.

5. Turn the adjuster to expand the shoes until the rotor disc locks.

6. Return the adjuster 8 notches.

7. Install the plug.

8. Remove the lug nuts and then install the wheels.

9. Lower the vehicle.

MR2

Pull the parking brake lever all the way up and count the number of clicks. The parking brake lever should travel approximately 5–8 clicks.

1. Pull the parking brake lever all the way up and down two or three times. Then return the parking brake lever.

2. Depress the brake pedal several times.

3. Remove the fuel tank protector.

4. Loosen the adjusting nut and brake cable, and check that the parking brake crank touches the stopper pin.

5. Stretch the brake cable by turning the adjusting nut before the parking brake crank begins moving.

6. Tighten the adjusting nuts so that the equalizer is horizontal to the ground. Tighten the nuts to 12 ft. lbs. (16 Nm).

7. Install the fuel tank protector.

8. After adjusting the parking brake, confirm that the rear brakes are not dragging.

REMOVAL AND INSTALLATION

Front Cable

Pedal Operated

1. Remove the left inner kick panel to access the parking brake pedal assembly.

2. Remove the retainer clip from the pin on the pedal assembly and

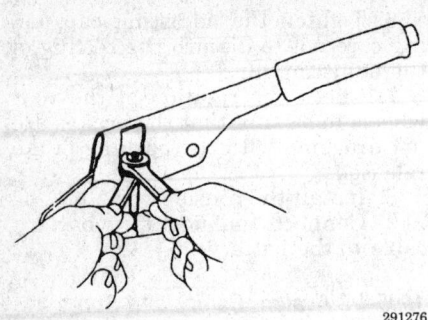

291276

Adjusting the parking brake cable — Celica, Corolla, Paseo and Tercel

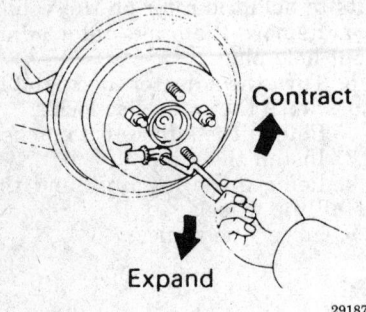

291873

Adjusting the parking brake shoes — Avalon, Camry and Supra

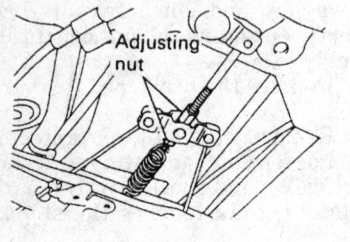

264874

Parking brake adjusting nut — MR2

withdraw the pin from the bracket and the eye of the cable.

3. Remove the spring clip holding the cable to the brake pedal assembly and remove the cable.

4. Disconnect the rear portion of the front parking brake wire and remove the cable from the vehicle.

To install:

5. Reconnect the rear portion of the brake cable.

6. Install the front of the brake cable to the brake pedal assembly and install the spring clip.

7. Install the pin through the pedal bracket and the eye of the

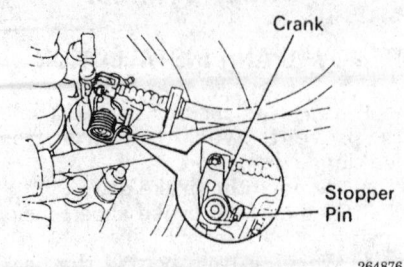

264876

The parking brake crank and the stopper pin — MR2

brake cable and secure with the retainer clip.

8. Reinstall the left inner kick panel.

9. Reconnect the negative battery cable.

Lever Operated

1. Raise and safely support the vehicle.

2. Remove the parking brake lever console box.

3. Remove the retainer nut located on the top of the lever shaft.

4. From under the vehicle disconnect the cable from the equalizer and extract the cable from the lever assembly. Remove the cable from the vehicle.

To install:

5. Install the cable to the equalizer.

6. Install the cable to the lever assembly and secure with the retaining nut. Tighten the retaining nut to 9 ft. lbs. (12 Nm).

7. Reinstall the brake lever console box.

8. Safely lower the vehicle and check for proper operation.

Rear Cable

1. Raise and safely support the vehicle.

2. Remove the tire and wheel assembly from the vehicle.

3. Working from underneath of the car, loosen the locknut on the parking brake cable equalizer.

4. Remove the brake drum or rotor from the vehicle.

5. If equipped, remove the rear brake shoe and disconnect the parking brake cable from the shoe.

6. Working on the shoe side of the backing plate, evenly compress the brake cable locking tines where the cable comes through the backing plate. At the same time, pull the cable free from behind the backing plate.

7. Remove the rear cable from the equalizer assembly and remove the cable from the vehicle.

To install:

8. Install the rear parking brake cable from behind the backing plate. Make sure all the locking tines are on the front side of the brake backing plate.

9. Install the rear brake cable to the equalizer.

10. Install the rear brake shoe lever to the rear brake cable and reinstall the brake shoe.

11. Install the brake drum or rotor, adjust the rear shoes, and adjust the brake cable. Tighten the equalizer locknut.

12. Install the wheel and tire assembly.

13. Safely lower the vehicle and road test for proper operation.

Brake System Bleeding

Start the bleeding procedure at the caliper or wheel cylinder the furthest from the master cylinder.

1. Connect a vinyl tube to the bleeder screw on the brake cylinder and submerge the other end of the tube in a transparent container half filled with clean brake fluid.

2. Pump the brake pedal several times and loosen the bleeder screw with the pedal held down.

3. When brake fluid stops coming out of the tube with the brake pedal held to the floor, tighten the bleeder screw and release the brake pedal.

4. Repeat Steps 2 and 3 until no air bubbles can be seen in the container.

5. Repeat the procedure for each wheel.

6. Check the level in the master cylinder. Add fluid as necessary.

Speed Sensor — ABS

REMOVAL AND INSTALLATION

Front Sensor

1. Disconnect the negative battery cable.

2. Raise and safely support the vehicle.

3. Remove the fender liner from the front fender.

4. Unplug the speed sensor connectors.

5. Remove the clamp bolts and clip holding the sensor harness.

6. Remove the bolt securing the speed sensor. Remove the speed sensor from the vehicle.

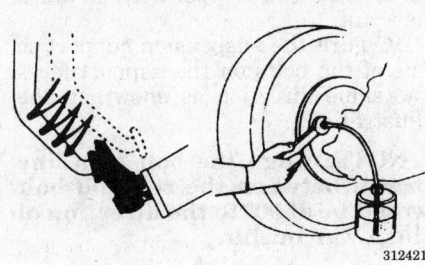

312421

Bleeding air from the wheel cylinder

To install:

7. Install the speed sensor. Tighten the bolt to 71 inch lbs. (8 Nm).

8. Tighten the clamp bolts holding the sensor harness to 44 inch lbs. (5 Nm).

9. Install the clip to hold the wire to the frame.

10. Engage the speed sensor connectors.

11. Install the fender liner and install the wheels.

12. Lower the vehicle.

13. Connect the negative battery cable.

Rear Sensor

1. Disconnect the negative battery cable.

2. Remove the rear seat cushion.

3. Unplug the sensor connector and pull out the sensor wire harness with the grommet.

4. Raise and safely support the vehicle.

5. Remove the clamp bolts securing the sensor wire harness to the control arm.

6. Remove the lockbolt from the axle beam and remove the speed sensor.

To install:

7. Install the speed sensor and tighten the lockbolt to 71 inch lbs. (8 Nm).

8. Install the clamp bolts to hold the sensor wire harness to the control arm. Tighten the bolts to 44 inch lbs. (5 Nm).

9. Lower the vehicle.

10. Engage the speed sensor connector and install the grommet.

11. Install the seat cushion.

12. Connect the negative battery cable.

FRONT SUSPENSION

Strut

REMOVAL AND INSTALLATION

Except Supra

1. Disconnect the negative battery cable. On vehicles equipped with an air bag, wait at least 90 seconds before proceeding.

2. Loosen the lug nuts.

3. Raise and support the vehicle safely.

————— **WARNING** —————
Do not support the weight of the vehicle on the suspension arm; the arm will deform under its weight.

4. Unfasten the lug nuts and remove the wheel.

5. Remove the bolt and disconnect the brake hose from the strut.

6. If equipped with ABS brakes, disconnect the wire harness from the strut.

7. Remove the bolts and disconnect the strut from the steering knuckle.

8. Remove the strut from the body.

9. Install a bolt and two nuts to the bracket at the lower portion of the strut shell and secure it in a vise.

10. Using a spring compressor tool SST 09727–30020 or equivalent, compress the coil spring.

————— **CAUTION** —————
This procedure requires the use of a spring compressor; it cannot be performed without one. If you do not have access to this special tool, do not attempt to disassemble the strut. The coil spring is retained under considerable pressure. It can exert enough force to cause serious injury! Exercise extreme caution when disassembling the strut.

11. Remove the dust cover and hold the spring seat so that it will not turn. Remove the nut on the top of the strut.

12. Remove the suspension support, bearing, dust seal, spring seat, spring, insulators and bumper.

To install:

13. To assemble the strut:

a. Install the spring bumper to piston.

b. Using a spring compressor, compress the spring.

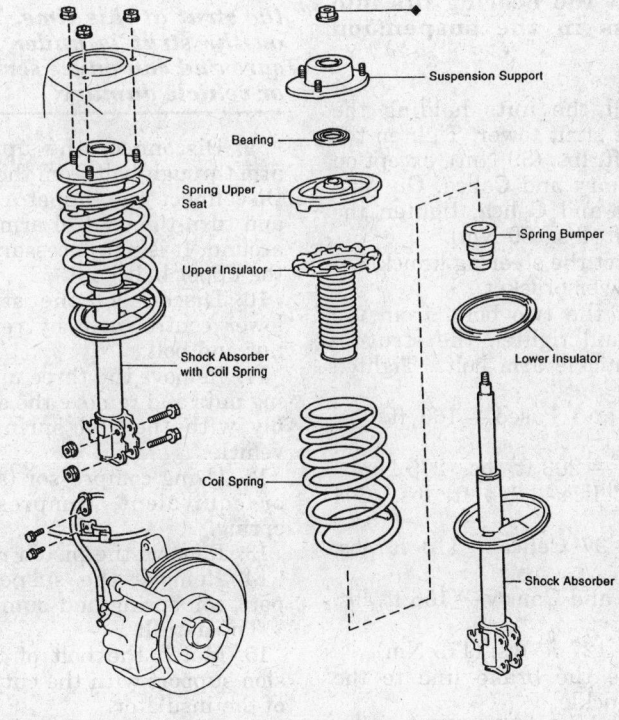

Suspension Support

Bearing

Spring Upper Seat

Upper Insulator

Shock Absorber with Coil Spring

Coil Spring

Spring Bumper

Lower Insulator

Shock Absorber

◆ Non-reusable part

292568

Common coil spring and strut component assembly — except Supra

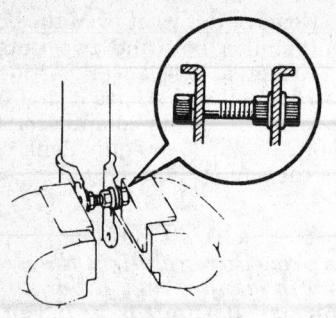

289026

Proper method of supporting the strut in a vise

c. Install the coil spring to the strut. Fit the lower end of the coil spring into the gap of the lower seat.

d. Install the spring seat with the insulator.

e. Install the dust seal on the spring seat.

f. Install the suspension support and tighten the new suspension nut to 35 ft. lbs. (47 Nm). After the nut has been tighten, release the compressor tool tension.

g. Pack multipurpose grease into the suspension support. Install the dustcover.

NOTE: Do not use an impact wrench to tighten the nut. Also, check that the bearing fits into the recess in the suspension support.

14. Install the nuts holding the strut to the strut tower. Tighten the nuts to 29 ft. lbs. (39 Nm), except on Avalon, Camry and Celica. On Avalon, Camry and Celica, tighten the nuts to 59 ft. lbs. (80 Nm).

15. Connect the steering knuckle to the strut lower bracket.

16. Insert the two bolts from the rear side and tighten the strut-to-steering knuckle arm bolts. Tighten as follows:

• Tercel and Paseo — 166 ft. lbs. (226 Nm)

• Corolla — 203 ft. lbs. (275 Nm)

• 1993 Celica — 224 ft. lbs. (304 Nm)

• 1994 — 97 Celica — 113 ft. lbs. (153 Nm)

• Avalon and Camry — 156 ft. lbs. (211 Nm)

• MR2 — 127 ft. lbs. (173 Nm)

17. Secure the brake line to the steering knuckle.

18. If equipped with ABS, secure the wire harness.

19. Install the wheel and lower the vehicle.

20. Connect the negative battery cable.

21. Have the front wheel alignment checked.

Supra

1. Disconnect the negative battery cable. On vehicles equipped with an air bag, wait at least 90 seconds before proceeding.

2. Raise and safely support the vehicle.

3. Remove the tire and wheel assembly.

4. Remove the brake caliper support bracket by removing the two bolts. Suspend it with a piece of wire.

5. Remove the fender apron, engine undercover, and the front fender wheel opening molding.

6. If removing the left side strut, disconnect the windshield washer tank.

7. Remove the bolt and disconnect the ABS speed sensor at the steering knuckle. Remove the three bolts and then disconnect the wire harness clamp in order to prevent the harness from being damaged when removing the through-bolt.

8. Remove the plug from the upper strut mount. Do not remove the center bolt.

——————— CAUTION ———————
Do not remove the center bolt to the strut at this time. The spring on the strut is under high pressure and can cause serious injury or vehicle damage.

9. Disconnect the upper control arm through-bolt from the sub-frame. Disconnect the upper control arm and turn the control arm completely around. It is not necessary to remove the upper ball joint.

10. Disconnect the strut at the lower control arm by removing the nut and bolt.

11. Remove the three upper mounting nuts and remove the strut assembly with the coil spring from the vehicle.

12. Using compressor 09727–30020 or equivalent, compress the coil spring.

13. Remove the piston rod locknut.

14. Remove the suspension support, coil spring and bumper.

To install:

15. Match the bolt of the suspension support with the cut out portion of the insulator.

16. Install the spring bumper.

17. Install the compressed coil spring. Match the end of the coil into the recess of the strut spring seat.

18. Install the suspension support to the rod and temporarily install a new nut.

19. Turn the suspension support so one of the bolts on the support faces the same direction as shown in the illustration.

NOTE: Align the bolt so a line drawn between the rod and bolt would be at 90° to the direction of the lower bushing.

20. Remove the spring compressor.

21. Install the strut and tighten the three upper strut mount nuts to 26 ft. lbs. (35 Nm). Tighten the middle nut to 22 ft. lbs. (29 Nm) and install the plug.

22. Connect the lower end of the strut to the lower control arm. Do not tighten the bolt at this time.

23. Install the upper control arm and install the through-bolt and nut. Do not tighten the bolt at this time.

24. Connect the speed sensor, wire harness, and the washer tank.

25. Install the fender apron and the engine undercover.

26. Install the caliper support bracket and tighten the bolts to 87 ft. lbs. (118 Nm).

27. Install the tire and wheel assembly.

28. Lower the vehicle.

29. Bounce the vehicle a few times to stabilize the suspension, then tighten the strut-to-lower arm bolt to 106 ft. lbs. (143 Nm). Tighten the upper arm to 121 ft. lbs. (164 Nm).

30. Connect the negative battery cable.

31. Have the front end alignment checked.

Upper Ball Joint

REMOVAL AND INSTALLATION

The upper ball joint (used only on the Supra) is an integral part of the upper arm and is not replaced separately. The upper ball joint replacement is accomplished by replacing the upper arm.

Lower Ball Joints

REMOVAL AND INSTALLATION

Except MR2

NOTE: On the Supra, the lower ball joint is not replaceable. If the lower ball joint is defective, replace the lower arm and ball joint as an assembly.

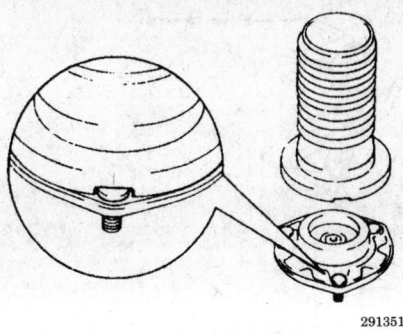

291351

Aligning the insulator to the support — Supra

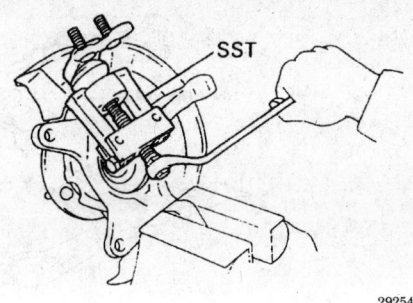

292540

Removing the ball joint from the knuckle

1. Disconnect the negative battery cable. On vehicles equipped with an air bag, wait at least 90 seconds before proceeding.
2. Raise and support the vehicle safely. Remove the front wheels.
3. Remove the cotter pin from the bearing locknut cap and then remove the cap.
4. Depress the brake pedal and loosen the axle nut.
5. Remove the brake caliper attaching hardware, position the caliper aside with the hydraulic line still attached and suspend it with a wire.
6. Remove the ABS speed sensor, if equipped.
7. Remove the brake rotor.
8. Loosen the two nuts holding the strut to the steering knuckle assembly. Do not remove at this time.
9. Remove the cotter pin and nut from the tie rod end. Using a tie rod end removal tool, separate the tie rod end from the steering knuckle.
10. Remove the ball joint bolt and two nuts to disconnect the steering knuckle from the lower arm.
11. Remove the two nuts and bolts holding the strut to the steering knuckle and separate the knuckle from the strut assembly.
12. Remove the axle nut and grasp the hub and knuckle assembly. With a plastic hammer tap the axle shaft to remove knuckle and hub.

NOTE: Cover the halfshaft boot with a shop rag to protect it from any damage.

13. Clamp the steering knuckle in a vise and remove the dust deflector. Remove the nut holding the steering knuckle to the ball joint. Press the ball joint out of the steering knuckle.
To install:
14. Reattach the ball joint to the steering knuckle. Tighten the ball joint-to-steering knuckle nut to 72 ft. lbs. (97 Nm) on Tercel and Paseo, 87 ft. lbs. (118 Nm) on Celica and Corolla or 90 ft. lbs. (123 Nm) on Avalon

and Camry. Install a new cotter pin. Drive the deflector shield onto the knuckle.
15. Install the knuckle and hub assembly to the axle and temporarily tighten the axle nut.
16. Connect the knuckle assembly to the lower strut bracket. Temporarily insert the mounting bolts from the rear and install the nuts.
17. Connect the lower ball joint to the lower arm and tighten the fasteners to 59 ft. lbs. (79 Nm) on Tercel and Paseo or 94 ft. lbs. (127 Nm) on Celica, Corolla, Avalon and Camry.
18. Connect the tie rod end to the knuckle.
19. Tighten the bolts on the lower side of the strut assembly.
20. If equipped, install the ABS speed sensor.
21. Install the brake disc and the caliper.
22. Tighten the axle nut.
23. Connect the negative battery cable.
24. Have the front end alignment checked.

MR2

1. Disconnect the negative battery cable. On vehicles equipped with an air bag, wait at least 90 seconds before proceeding.
2. Raise the vehicle and support it safely.
3. Remove the wheel.
4. Remove the cotter pin and castle nut from the lower ball joint.
5. Press the ball joint from the lower control arm.
6. Remove the two bolts and disconnect the ball joint from the steering knuckle.
To install:
7. Install the ball joint to the steering knuckle and install the two bolts. Tighten the bolts to 83 ft. lbs. (113 Nm).
8. Connect the lower control arm to the ball joint and tighten the castle

nut to 58 ft. lbs. (78 Nm). Install a new cotter pin.
9. Install the wheel and lower the vehicle.
10. Connect the negative battery cable.
11. Have the front end alignment checked.

Upper Control Arms

REMOVAL AND INSTALLATION

This procedure is applicable only for the Supra.
1. Disconnect the negative battery cable. On vehicles equipped with an air bag, wait at least 90 seconds before proceeding.
2. Raise the front of the vehicle and support it on safety stands.
3. Remove the wheel.
4. Remove the caliper support bracket by removing the two bolts. Leave the brake line connected and suspend it aside.
5. Remove the rotor.
6. Remove the front fender splash shield, fender liner, and wheel opening molding.
7. If removing the left side arm, remove the washer tank.
8. Remove the bolt and disconnect the ABS speed sensor from the steering knuckle. Remove the three bolts and disconnect the wire harness clamp.
9. Remove the cotter pin and the nut from the upper ball joint; press the upper ball joint from the knuckle.
10. Remove the through-bolt, nut, and the upper control arm.
To install:
11. Install the upper control arm. Connect the upper control arm to the sub-frame and install the through-bolt. Do not tighten the bolt at this time.

NOTE: The upper control arm mounting bolts are not tightened until the suspension has been assembled and vehicle is on the ground.

12. Install the ball joint to the knuckle and tighten the nut to 76 ft. lbs. (103 Nm). Install a new cotter pin.
13. Connect the wire harness and ABS speed sensor.
14. Install the washer tank, the fender liner, splash shield, and molding.
15. Install the rotor.
16. Install the caliper support bracket and tighten the bolts to 87 ft. lbs. (118 Nm).

17. Install the wheel.

18. Lower the vehicle.

19. Bounce the suspension several times to set the suspension.

20. Support the lower arm and tighten the upper control arm through-bolt and nut to 121 ft. lbs. (164 Nm).

21. Connect the negative battery cable.

22. Have the front wheel alignment checked.

Lower Control Arms

REMOVAL AND INSTALLATION

Tercel and Paseo

1. Disconnect the negative battery cable. On vehicles equipped with an air bag, wait at least 90 seconds before proceeding.

2. Raise and support the vehicle safely. Remove the wheels.

3. If equipped with a sway bar, disconnect the sway bar from the lower control arm.

4. Disconnect the lower arm from the lower ball joint by removing the bolt and 2 nuts.

5. Remove the lower arm by removing the 2 rear bracket bolts and the front through-bolt.

To install:

6. Install lower arm and temporarily install the 2 rear bracket bolts and one through-bolt.

7. Connect the lower arm to the lower ball joint and torque the bolt and nuts to 59 ft. lbs. (80 Nm).

8. If equipped with a sway bar, connect the sway bar to the lower control arm. Torque the nut to 13 ft. lbs. (18 Nm).

9. Install the wheels and lower the vehicle. Stabilize the suspension by bouncing the vehicle several times.

10. Raise the vehicle and support the body with stands.

11. Support the lower arm with a floor jack.

12. Torque the front side bolt to 105 ft. lbs. (142 Nm) and the rear bolt to 94 ft. lbs. (127 Nm).

13. Connect the negative battery cable.

14. Lower the vehicle and check the front end alignment.

Corolla

Except Left Side Control Arms with A/T

1. Disconnect the negative battery cable. On vehicles equipped with an air bag, wait at least 90 seconds before proceeding.

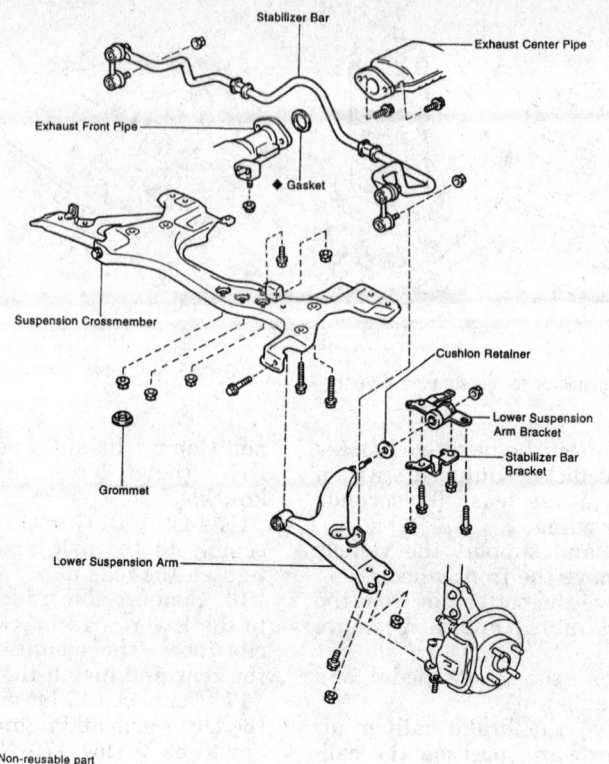

Lower control arm component assembly — Corolla

2. Raise and safely support the vehicle.

3. Remove the front wheel assembly.

4. If equipped with a stabilizer bar, disconnect the stabilizer bar link from the lower control arm by removing the nut.

5. Remove the bolt and nuts and separate the ball joints from the lower control arm.

6. Remove the nut and three bolts and remove the lower control arm bushing retaining bracket.

7. Remove the control arm-to-crossmember mounting bolt and remove the control arm from the crossmember.

To install:

8. Install the control arm to the suspension crossmember and install the front bolt. Do not tighten the bolt at this time.

9. Install the lower control arm bushing retaining bracket and install the nut and three bolts. Do not tighten the bolts or nut at this time.

10. Connect the lower control arm to the lower ball joint by installing the bolt and two nuts. Tighten the bolt and nuts to 105 ft. lbs. (142 Nm).

11. Connect the stabilizer bar link to the lower control arm and tighten the nut to 33 ft. lbs. (44 Nm).

12. Install the wheels and lower the vehicle to the ground.

13. Stabilize the suspension by pushing up and down the vehicle.

14. Tighten the front lower control arm bolt to 161 ft. lbs. (218 Nm).

15. Tighten the rear lower control arm bracket bolts and nuts as follows:

• Control arm bracket bolts — 108 ft. lbs. (147 Nm)

• Stabilizer bar bracket bolt — 37 ft. lbs. (50 Nm)

• Bracket nut — 14 ft. lbs. (19 Nm)

16. Connect the negative battery cable.

17. Have the wheel alignment checked.

Left Side Control Arms with A/T

NOTE: Both lower control arms and the suspension crossmember must be removed as unit when replacing the left hand control arm on a vehicle equipped with an automatic transaxle.

1. Disconnect the negative battery cable. On vehicles equipped with an air bag, wait at least 90 seconds before proceeding.

2. Raise and safely support the vehicle.

3. Remove the front wheels.

4. If equipped with a stabilizer bar, disconnect the stabilizer bar link from both lower control arms by removing the nut.

5. Remove the bolt and nuts and separate the ball joints from the lower control arms.

6. Remove the nut and three bolts and remove the right rear control arm bushing retaining bracket. Do this for both lower control arms.

7. Remove the stabilizer bar from the vehicle.

8. Remove the grommet in the crossmember.

9. Remove the bolt and four nuts holding in the middle of the crossmember.

10. Support the suspension crossmember with a jack.

11. Remove the six bolts from the outer side of the suspension crossmember.

12. Remove the suspension crossmember with the lower suspension arms.

13. Remove the bolt holding the front of the lower control arm to the suspension crossmember.

14. Disconnect the lower control arm from the suspension crossmember.

To install:

15. Install the lower control arm to the suspension crossmember.

16. Install the bolt to hold the lower control arm to the crossmember. Do not tighten the bolt at this time.

17. Raise the suspension crossmember with the lower control arms. Install the outer six bolts to hold the crossmember to the vehicle. Tighten the bolts to 152 ft. lbs. (206 Nm).

18. Install the center bolt and four nuts to the crossmember and tighten as follows:
- Nut A to 35 ft. lbs. (48 Nm).
- Bolt B and nut C to 45 ft. lbs. (61 Nm)

19. Install the grommet to the crossmember.

20. Install the stabilizer bar to the vehicle.

21. Install the lower control arm bushing retaining bracket and install the nut and three bolts. Do not tighten the bolts or nut at this time.

22. Connect the lower control arm to the lower ball joint by installing the bolt and two nuts. Tighten the bolt and nuts to 105 ft. lbs. (142 Nm). Connect both lower control arms to the ball joints.

23. Connect the stabilizer bar links to the lower control arms and tighten the nuts to 33 ft. lbs. (44 Nm).

24. Install the wheels and lower the vehicle to the ground.

25. Stabilize the suspension by pushing up and down the vehicle.

26. Tighten the front lower control arm bolt to 161 ft. lbs. (218 Nm).

27. Tighten the rear lower control arm bracket bolts and nuts as follows:
- Control arm bracket bolts — 108 ft. lbs. (147 Nm)
- Bolt B: Stabilizer bar bracket bolt — 37 ft. lbs. (50 Nm)
- Nut C : Bracket nut — 14 ft. lbs. (19 Nm)

28. Connect the negative battery cable.

29. Have the wheel alignment checked.

Celica

1. Disconnect the negative battery cable. On vehicles equipped with an air bag, wait at least 90 seconds before proceeding.

2. Raise the vehicle and support it safely. Remove the front wheels.

3. Remove the undercovers from the vehicle.

4. Remove halfshaft locknut. This is done by removing the cotter pin, lock cap and while applying the brakes.

5. Remove the cotter pin and nut from the tie rod end. Separate the tie rod from the steering knuckle using SST 09610–20012 or equivalent.

6. Disconnect the stabilizer bar link from lower control arm.

7. Disconnect lower control arm from lower ball joint by removing the bolt and two nuts.

8. Remove three bolts from the rear portion of the lower control arm.

9. Remove front sub-frame as follows:
    a. Support the transaxle with a jack.
    b. Remove the two exhaust pipe support bracket set bolts.
    c. Remove the two grommets and the center member set bolt.
    d. Remove the two center member set nuts.
    e. Remove the two power steering gear assembly set bolts and nut.
    f. Support the front sub-frame with a jack.
    g. Remove the two bolts attaching sub-frame to the body. One bolt is located above the left control arm and the other above the right control arm. Lower the front sub-frame with lower control arms.
    h. Remove both the stabilizer bar bushing retainer set bolts.

10. Remove the collar and bolt from the lower control arm. Remove lower control arm from the vehicle.

To install:

11. Put the lower control arm in place and install the bolt and collar.

12. Install the sub-frame as follows:
    a. Jack up the front sub-frame with the lower control arms.
    b. Install both stabilizer bar bushing retainers set bolts. Tighten to 14 ft. lbs. (19 Nm).
    c. Position the front sub-frame and tighten the two bolts that attach the sub-frame to the body. Tighten these two bolts to 94 ft. lbs. (127 Nm).
    d. Install the power steering gear assembly.
    e. Install the center member with the bolt and two nuts. Tighten to 59 ft. lbs. (80 Nm). Install the two grommets.
    f. Install the exhaust pipe support bracket with the two bolts. Tighten to 14 ft. lbs. (19 Nm).

13. Temporarily install the three bolts to hold the sub-frame to the rear of the lower control arm. Do not tighten completely at this time.

14. Connect the lower control arm to the lower ball joint. Tighten the bolt and two nuts to 94 ft. lbs. (127 Nm).

15. Connect stabilizer bar link to lower control arm. Torque to 33 ft. lbs. (44 Nm).

16. Connect the tie rod end to the steering knuckle. Tighten the nut 36 ft. lbs. (49 Nm). Install a new cotter pin.

17. Install the halfshaft locknut and tighten to 159 ft. lbs. (216 Nm). Install the lock cap and secure it with a new cotter pin.

18. Install the engine undercovers.

19. Install the front wheel and tighten to 76 ft. lbs. (103 Nm). Lower the vehicle.

20. Bounce the vehicle up and down several times to stabilize the suspension.

21. Tighten the bolts and nuts to lower control arm as follows:
- Bolt A — 116 ft. lbs. (157 Nm)
- Bolt B — 123 ft. lbs. (167 Nm)
- Bolt C — 130 ft. lbs. (175 Nm)

22. Have the front wheel alignment checked.

23. Connect the negative battery cable.

Camry and Avalon

1. Disconnect the negative battery cable. On vehicles equipped with an air bag, wait at least 90 seconds before proceeding.

2. Raise the vehicle and support it safely.

3. Remove the front wheel(s).

4. Remove the fender apron seal.

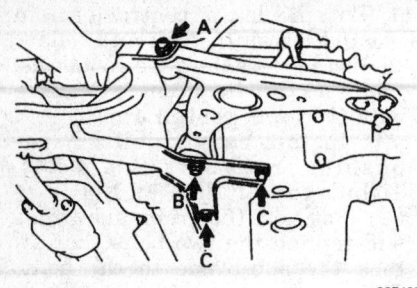

287485

A, B, and C bolt locations — 1994–97 Celica

5. Remove the cotter pin and lock cap from the halfshaft.

6. Have an assistant apply the brakes firmly; with the front brakes locked, remove the halfshaft locknut.

7. Disconnect and separate the tie rod end from the steering knuckle.

8. Remove the left and right stabilizer bar end brackets from the lower control arms.

9. Remove the bolt and two nuts and disconnect the lower arm from the ball joint.

10. Remove the halfshaft from the axle hub. Secure the shaft out of the way using wire.

─── WARNING ───
Be careful not to damage the shaft joint boot or ABS sensor rotor.

11. Remove the bolts from the front side of the lower control arm.

12. Remove the bolts and nuts from the rear side of the lower control arm and remove control arm from the vehicle.

To install:

13. Place the lower control arm onto the vehicle and temporarily install the rear mounting nut and bolt.

14. Install the lower control arm bushing stopper to the lower control arm shaft. Install the bolts on the front side of the control arm and tighten to 152 ft. lbs. (206 Nm), then tighten the bolts on the rear side to 152 ft. lbs. (206 Nm).

15. Install the halfshaft to the axle hub.

16. Connect the lower control arm to the lower ball joint and tighten the fasteners to 94 ft. lbs. (127 Nm).

17. Install both side stabilizer end brackets to the lower control arm and tighten to 41 ft. lbs. (56 Nm).

18. Connect the tie rod end to the steering knuckle.

19. Have a helper apply the brakes and install the halfshaft locknut. Tighten it to 217 ft. lbs. (294 Nm).

Install the lock cap and a new cotter pin.

20. Install front fender apron seal.

21. Install the front wheel(s).

22. Lower the vehicle to the floor and connect the negative battery cable.

MR2

1. Disconnect the negative battery cable. On vehicles equipped with an air bag, wait at least 90 seconds before proceeding.

2. Raise the vehicle and support it safely.

3. Remove the wheel.

4. Remove the cotter pin and castle nut from the ball joint.

5. Using SST 09610–20012 or equivalent (ball joint separator), press the lower control arm from the ball joint.

6. Remove the two nuts and disconnect the strut bar from the control arm.

7. Remove the lower control arm to body bolt and remove the control arm from the vehicle.

To install:

8. When installing the lower arm, position it in to strut bar and tighten the nuts finger-tight. Do the same thing with the control arm to body bolt.

9. Connect the control arm to the ball joint and tighten the castle nut to 58 ft. lbs. (78 Nm). Install a new cotter pin.

10. Tighten the strut bar to control arm bolts to 94 ft. lbs. (127 Nm).

11. Install the tires, lower the vehicle and bounce it several times to set the suspension.

12. Tighten the control arm-to-body bolt to 87 ft. lbs. (118 Nm). Connect the battery cable and check the wheel alignment.

Supra

1. Disconnect the negative battery cable. On vehicles equipped with an air bag, wait at least 90 seconds before proceeding.

2. Raise the front of the vehicle and support it on safety stands.

3. Remove the wheel and the engine undercover.

4. Remove the caliper support bracket from the vehicle by removing the two bolts. Support the caliper and bracket with a wire. Do not let the assembly hang from the brake line.

5. Remove the nut and disconnect the stabilizer bar from the lower control arm.

6. Remove the cotter pin and nut from the lower ball joint. Press the

lower ball joint out of the steering knuckle.

7. Disconnect the lower end of the strut by removing the nut and bolt.

8. Remove the nut, two bolts, and the front lower arm bracket stay.

9. Matchmark the front and rear adjustment cams to the body and then remove the nuts and adjusting cams.

10. Lift out the lower control arm.

11. Remove the bracket from the control arm by removing the two bolts.

To install:

12. Install the bracket to the lower control arm by installing the two bolts. Tighten the bolts to 38 ft. lbs. (52 Nm).

13. Install the lower control arm to the body and temporarily install the adjusting cams and nuts. Do not tighten the nuts at this time.

14. Connect the lower control arm to the knuckle and tighten the ball joint nut to 92 ft. lbs. (125 Nm). Install a new cotter pin.

15. Connect the strut to the arm and tighten the bolt and nut to 106 ft. lbs. (143 Nm).

16. Connect the stabilizer bar link and tighten the nut to 54 ft. lbs. (74 Nm).

17. Install the brake caliper support bracket to the vehicle and tighten the bolts to 87 ft. lbs. (118 Nm).

18. Install the wheel and lower the vehicle.

19. Bounce it several times to set the suspension.

20. Support the lower arm. Align the matchmarks on the adjusting cams and tighten the nuts to 166 ft. lbs. (226 Nm).

21. Connect the negative battery cable.

22. Have the wheel alignment checked.

Stabilizer Bar

REMOVAL AND INSTALLATION

1. Disconnect the negative battery cable. On vehicles equipped with an air bag, wait at least 90 seconds before proceeding.

2. Raise and safely support the vehicle.

3. Remove the front wheels.

4. Remove the stabilizer bar links.

5. Remove the stabilizer bar brackets.

6. If necessary, disconnect and remove the exhaust pipe.

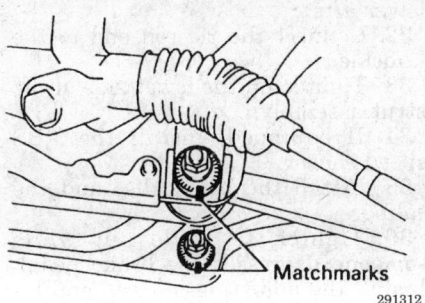

Matchmarks
291312

Matchmarking the adjustment cams — Supra

7. On Camry and Avalon, remove the rack and pinion mounting bolts. Secure it out of the way.

8. Remove the stabilizer bar from the vehicle.

To install:

9. Install the stabilizer bar bushings touching the line painted on the stabilizer bar.

10. Install the stabilizer bar to the vehicle. Do not tighten the bolts at this time.

11. On Camry and Avalon, install the rack and pinion unit.

12. Using a new gasket, install the exhaust pipe.

13. Lower the vehicle to the ground and connect the negative battery cable.

14. Tighten the bar brackets first, then the links as follows. On Paseo, tighten the brackets and links to 14 ft. lbs. (19 Nm). On Corolla, tighten the bolt next to the nut to 108 ft. lbs. (147 Nm), the bolts on the opposite side of the nut to 37 ft. lbs. (19 Nm) and the nut to 14 ft. lbs. (19 Nm). On 1993 Celica, tighten the brackets to 13 ft. lbs. (18 Nm) and the links to 26 ft. lbs. (35 Nm). On 1994–97 Celica, tighten the brackets and links to 33 ft. lbs. (44 Nm). On Camry and Avalon, tighten the brackets to 14 ft. lbs. (19 Nm) and the links to 47 ft. lbs. (64 Nm). On MR2, tighten the brackets and links to 47 ft. lbs. (64 Nm). On Supra, tighten the brackets to 13 ft. lbs. (18 Nm) and the links to 54 ft. lbs. (103 Nm).

Front Wheel Bearings

ADJUSTMENT

All models use a non-adjustable wheel bearing. To determine the condition of the wheel bearing, check the backlash in bearing shaft direction and the axle hub deviation. Maxi-

mum for backlash should be as follows:

- Corolla, Supra, Avalon and Camry — 0.0020 in. (0.05mm)
- 1994–97 Celica — 0.0031 in. (0.08mm)

Maximum axle hub deviation is:

- Corolla — 0.0028 inch (0.07mm)
- 1994–97 Celica — 0.0028 in. (0.07mm)
- Supra, Avalon and Camry — 0.0020 in. (0.05mm)

If the wheel bearing is out of specifications, replace the wheel bearing.

REMOVAL AND INSTALLATION

Except MR2 and Supra

1. Disconnect the negative battery cable from the battery. On vehicles equipped with an air bag, wait at least 90 seconds before proceeding.

2. Raise and support the vehicle safely. Remove the front wheels.

3. Remove the cotter pin from the axle nut cap and then remove the cap.

4. Depress the brake pedal and loosen the axle nut.

5. Remove the brake caliper attaching hardware, position the caliper aside with the hydraulic line still attached and suspend it with a wire.

6. Remove the ABS speed sensor, if equipped.

7. Remove the brake rotor.

8. Loosen the nuts on the lower side of the strut assembly. Do not remove at this time.

9. Remove the cotter pin and nut from the tie rod end. Using a tie rod end removal tool, separate the tie rod end from the steering knuckle.

10. Place matchmarks on the strut assembly lower mounting bracket and the camber adjustment cam.

11. Remove the ball joint bolt and two nuts and disconnect the steering knuckle from the lower control arm.

12. Remove the two nuts and bolts on the lower side of the strut mount

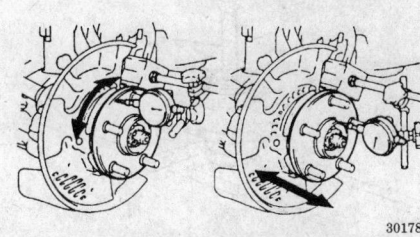

301789

Checking the wheel bearings

and separate the knuckle from the strut assembly.

13. Remove the axle nut and grasp the hub and knuckle assembly. With a plastic hammer tap the axle shaft to remove knuckle and hub.

NOTE: Cover the halfshaft boot with a shop rag to protect it from any damage.

14. Clamp the steering knuckle in a vise and remove the dust deflector. Remove the nut holding the steering knuckle to the ball joint. Press the ball joint out of the steering knuckle.

15. Remove the inner axle seal.

16. Using a Torx® wrench, remove the bolts securing the dust cover.

17. Using hub puller, remove the hub and backing plate from the steering knuckle.

18. Using a proper sized driver and a press, remove the inner hub race from the axle hub.

19. Using seal removal tool, remove the outer axle seal.

20. Using snapring pliers, remove the snapring from the inner side of the steering knuckle.

21. Using a proper sized driver and a press, remove the bearing from the steering knuckle. The bearing is pressed from the front of the steering knuckle and is removed through the back of the steering knuckle.

To install:

22. Using a proper sized driver and a press, install a new bearing to the steering knuckle.

23. Install the snapring to the steering knuckle using snapring pliers.

24. Using a seal driver and a hammer, install a new outer oil seal. Apply multipurpose grease to the oil seal lip.

25. Place the dust cover on the steering knuckle and tighten the bolts to 78 inch lbs. (9 Nm).

26. Using a press and a proper sized driver, install the axle hub to the steering knuckle.

27. Attach the ball joint to the steering knuckle. Install a new cotter pin.

28. Using a seal driver and a hammer, install a new inner oil seal. Apply multipurpose grease to the oil seal lip.

29. Install the knuckle and hub assembly to the axle and temporarily tighten the axle nut.

30. Connect the knuckle assembly to the lower strut bracket. Temporarily insert the mounting bolts from the rear and install the nuts making sure the matchmarks made earlier are in alignment.

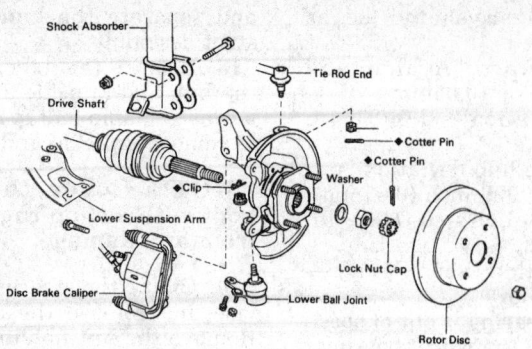

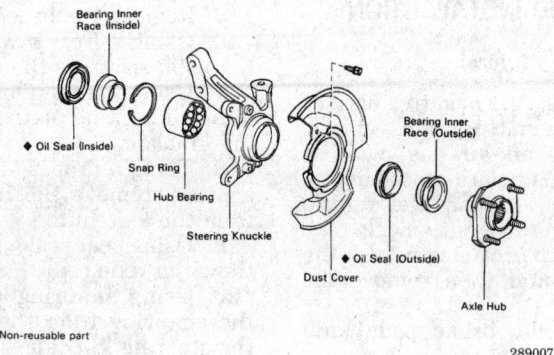

Steering knuckle and hub assembly — Tercel shown, all similar except MR2 and Supra

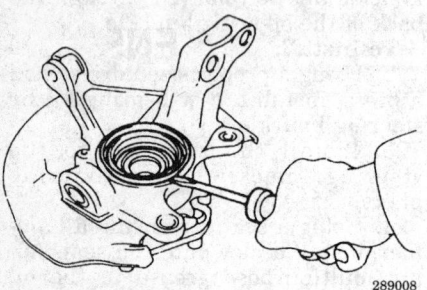

Removing the inner axle seal

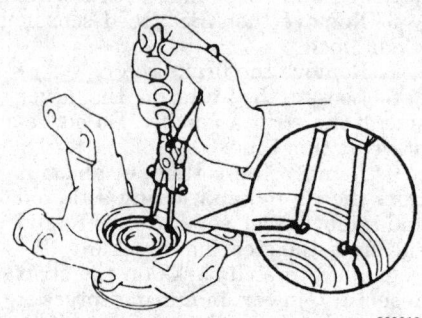

Removing the snapring

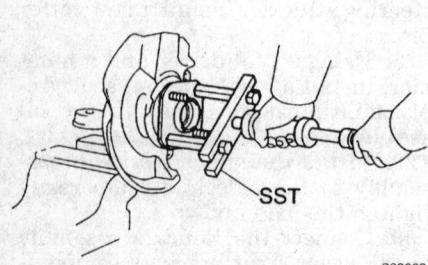

Removing the axle hub

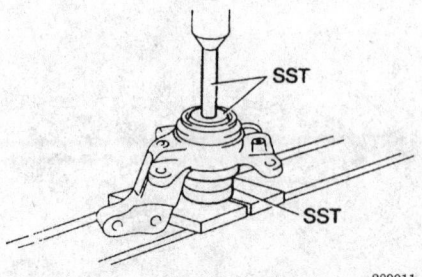

Removing the bearing from the steering knuckle

31. Connect the lower ball joint to lower arm.

32. Connect the tie rod end to the knuckle.

33. Tighten on the lower side of the strut assembly.

34. If equipped, install the ABS speed sensor.

35. Install the brake disc and the caliper.

36. Tighten the axle nut while someone depresses the brake pedal. Install the adjusting nut cap and insert a new cotter pin.

37. Install the wheels to the vehicle. Verify that the wheel turns freely.

38. Lower the vehicle. Connect the negative battery cable to the battery.

39. Have the wheel alignment checked.

MR2

1. Disconnect the negative battery cable. On vehicles equipped with an air bag, wait at least 90 seconds before proceeding.

2. Raise and support the vehicle safely. Remove the front wheels.

3. Remove the brake caliper attaching hardware, position the caliper aside with the hydraulic line still attached and suspend it with a wire.

4. Place matchmarks on the axle and hub and remove the brake rotor.

5. Remove the cotter pin and nut from the tie rod end. Using a tie rod end removal tool, separate the tie rod end from the steering knuckle.

6. Remove the bolt and pull out the ABS speed sensor.

7. Loosen the nuts and bolts holding the lower strut bracket to steering knuckle. Remove the nuts and washers but leave the bolts installed.

8. Remove the two bolts holding the ball joint to the steering knuckle.

9. Remove the two bolts from the lower strut bracket and remove the steering knuckle and hub assembly.

10. Clamp the steering knuckle in a soft jaw vise.

11. Using a prytool and hammer, remove the hub grease cap. Remove the O-ring.

12. Clamp the axle hub in a soft jaw vise.

13. Using a chisel and hammer, loosen the staked part of the locknut and remove the locknut from the hub.

14. If equipped with ABS brakes, remove the ABS speed sensor rotor. If not, remove the bearing inner spacer.

15. Using an axle hub puller, remove the axle hub from the steering knuckle.

16. Remove the brake dust cover by removing the four bolts.

17. Using a seal remover, remove the oil seal from the knuckle assembly.

18. Using a bearing puller, remove the hub bearing inner race from the axle hub.

19. Using snapring pliers, remove the snapring.

20. Place the removed inner race (outside) in the bearing.

21. Using a hammer and bearing driver, drive out the hub bearing from the steering knuckle.

To install:

22. Using a proper sized driver and a press, install a new axle hub bearing.

23. Install the snapring.

24. Rotate and insert the side lip of a new oil seal into a seal installer. Drive the oil seal into the steering knuckle. Apply multipurpose grease to the oil seal lip.

25. Install the brake dust cover and install the four bolts. Tighten the bolts to 74 inch lbs. (9 Nm).

26. Install the bearing outer and inner races into the axle bearing.

27. Using a hub installer, install the axle hub to the steering knuckle.

28. If equipped with ABS brakes, install the ABS speed sensor rotor. If not, install the bearing inner spacer.

29. Install a new locknut to hold the axle hub to the steering knuckle. Tighten the nut to 90 ft. lbs. (123 Nm).

30. Using a punch and hammer, stake the nut.

31. Install a new O-ring.

32. Install the hub grease cap.

33. Install the steering knuckle and hub to the lower strut mount. Insert the two bolts to hold the knuckle to the strut from the rear and temporarily tighten.

34. Raise the lower control arm and install the two bolts that attach the ball joint to the steering knuckle.

35. Connect the tie rod end to the knuckle.

36. If equipped, install the ABS speed sensor.

37. Tighten the steering knuckle-to-strut assembly bolts.

38. Install the brake disc, and bolt the caliper to the steering knuckle.

39. Install the wheels and lower the vehicle.

40. Connect the negative battery cable.

41. Have the front end alignment checked.

Supra

1. Disconnect the negative battery cable. On vehicles equipped with an

air bag, wait at least 90 seconds before proceeding.

2. Raise and safely support the vehicle.

3. Remove the front tire and wheel assembly.

4. Remove the brake caliper support bracket, leaving the brake line connected and support it using a piece of wire.

5. Remove the rotor by removing the two screws.

6. Disconnect the ABS speed sensor.

7. Remove the cotter pin and nut and disconnect the tie rod from the steering knuckle.

8. Remove the cotter pin and nut and disconnect the steering knuckle from the upper control arm.

9. Remove the clip and nut and press the knuckle off the lower control arm.

10. Remove the steering knuckle from the vehicle.

11. Pry the hub bearing cap from the steering knuckle. Using a hammer and chisel, loosen the staked part of the hub nut and remove it.

12. Remove the ABS sensor rotor.

13. Remove the four bolts and shift the brake dust shield toward the hub (outside).

14. Using a two armed puller, remove the axle hub from the knuckle.

15. With a puller, remove the inner bearing race from the axle hub. Pry out the oil seal.

16. Remove the bearing snapring, then position the inner race above the bearing on the inner side. Press the bearing out.

To install:

17. Press the bearing into the knuckle. If the inner race and balls come loose from the outer race, be sure to install them on the same side as before.

18. Install the snapring and inner race, then tap in a new oil seal until it is flush with the end surface of the knuckle.

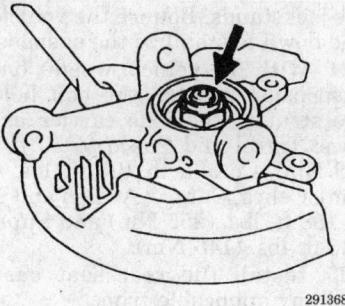

291368

Axle hub locknut — Supra

19. Install the brake dust cover and tighten the bolts to 74 inch lbs. (9 Nm).

20. Press the hub into the knuckle and install the speed sensor.

21. Install a new locknut and tighten it to 147 ft. lbs. (199 Nm). Stake the nut with a chisel. Tap the bearing cap into place.

22. Connect the knuckle to the upper control arm and tighten the nut to 76 ft. lbs. (103 Nm). Install a new cotter pin.

23. Connect the knuckle to the lower control arm and tighten the nut to 92 ft. lbs. (125 Nm). Install a new clip.

24. Connect the tie rod end to the steering knuckle with the nut.

25. Install the rotor by installing the two screws.

26. Install the caliper support bracket.

27. Connect the speed sensor to the knuckle.

28. Install the front wheel and lower the vehicle.

29. Connect the negative battery cable.

30. Have the front end alignment checked.

REAR SUSPENSION

Strut

REMOVAL AND INSTALLATION

1. Disconnect the negative battery cable from the battery. On vehicles equipped with an air bag, wait at least 90 seconds before proceeding.

2. Remove the rear seat cushion and any trim necessary to access the strut towers.

3. Loosen the lug nuts to the wheel on the side being serviced.

4. Raise and safely support the vehicle. Place jackstands under the vehicle frame.

5. Support the axle beam with a jack.

6. Remove the wheel on the side being serviced.

7. On Supra, remove the brake caliper support bracket by removing the two bolts. Leave the brake line connected and position it aside.

8. If equipped with ABS, disconnect the sensor wire from the strut.

9. If equipped, disconnect the stabilizer bar from the strut.

10. On Camry, disconnect the Load Sensing Proportioning Valve (LSPV) from the lower control arm by removing the bolt.

11. Loosen the fasteners securing the strut to the axle carrier. Do not remove the bolts at this time.

12. Support the axle carrier with a jack.

13. Disconnect the strut from the strut tower by the nuts.

--- CAUTION ---

Do not loosen the center nut on the top of the strut piston.

14. Remove the strut from the vehicle.

15. To disassemble, proceed as follows:

--- CAUTION ---

This procedure requires the use of a spring compressor; it cannot be performed without one! If you do not have access to this special tool, do not attempt to disassemble the strut. The coil springs are retained under considerable pressure. They can exert enough force to cause serious injury! Exercise extreme caution when disassembling the strut.

a. Place the strut assembly in a pipe vise or strut vise.

NOTE: Do not attempt to clamp the strut assembly in a flat jaw vise as this will result in damage to the strut tube.

b. Attach a spring compressor and compress the spring until the upper suspension support is free of any spring tension. Do not overcompress the spring.

c. Hold the upper support, then remove the nut on the end of the shock piston rod.

d. Remove the support, coil spring, insulator, and bumper.

16. Inspect the strut as follows:

a. Check the shock absorber by moving the piston shaft through its full range of travel. It should move smoothly and evenly throughout its entire travel without any trace of binding or notching.

b. Use a small straightedge to check the piston shaft for any bending or deformation.

c. Inspect the spring for any sign of deterioration or cracking. The waterproof coating on the coils should be intact to prevent rusting.

To install:

NOTE: Never reuse a self-locking nut. Always replace self-locking nuts and cotter pins as applicable.

17. Assemble the strut as follows:

a. Loosely assemble all components onto the strut assembly. Make sure the spring end aligns with the hollow in the lower seat.

b. Align the upper suspension support with the piston rod and install the support.

c. Align the suspension support with the strut lower bracket. This assures the spring will be properly seated top and bottom.

d. Compress the spring to expose the strut piston rod threads.

e. Install a new strut piston nut and tighten to the following: 1993–94 Tercel and Paseo to 40 ft. lbs. (54 Nm), 1995–96 Tercel and Paseo to 25 ft. lbs. (34 Nm), Corolla, Celica, Camry and Avalon to 36 ft. lbs. (49 Nm), MR2 to 54 ft. lbs. (73 Nm) and Supra to 20 ft. lbs. (27 Nm).

f. Remove the spring compressor. Make sure the paint mark on the upper support faces the outside of the strut.

18. Place the strut on the vehicle and install the nuts to hold the strut to the strut tower. Tighten the nuts to 29 ft. lbs. (39 Nm), except on MR2 and Supra. On MR2, tighten the nuts to 59 ft. lbs. (80 Nm). On Supra, tighten to 19 ft. lbs. (26 Nm).

19. Connect the strut to the axle carrier and install the bolt and nut. Do not tighten at this time.

20. Connect the stabilizer link to the strut.

21. On Camry, connect the load sensing proportioning valve to the control arm. Tighten to 94 ft. lbs. (130 Nm).

22. On Supra, install the brake caliper.

23. Install the wheel and remove the jackstands. Bounce the vehicle up and down to stabilize the suspension.

24. With the vehicle weight on the suspension, tighten the bolt holding the strut to the axle carrier as follows: Tercel and Paseo to 50 ft. lbs. (68 Nm), Corolla to 105 ft. lbs. (142 Nm), Celica, Camry, Avalon and MR2 to 188 ft. lbs. (255 Nm) and Supra to 106 ft. lbs. (143 Nm).

25. Install the rear seat cushion and any applicable trim.

26. Connect the negative battery cable to the battery.

Upper Control Arms

REMOVAL AND INSTALLATION

This procedure is applicable only to the Supra.

1. Disconnect the negative battery cable. On vehicles equipped with an air bag, wait at least 90 seconds before proceeding.

2. Raise and safely support the vehicle. Remove the tire and wheel assembly.

3. Remove the rear halfshaft.

4. Remove the brake caliper support bracket from the vehicle by removing the two bolts.

5. Remove the two bolts and the ABS wire harness clamp.

6. Remove the nut from the upper ball joint; press the upper ball joint out of the axle carrier.

7. Remove the upper mounting bolts and nuts. Remove the control arm from the vehicle.

To install:

8. Position the arm and tighten the upper bolts and nuts to 121 ft. lbs. (164 Nm).

NOTE: Ensure the tip of the nut lock is facing down.

9. Use a new nut and tighten the ball joint to 80 ft. lbs. (109 Nm).

10. Install the halfshaft.

11. Install the ABS clamp by installing the two bolts.

12. Install the brake caliper support bracket to the vehicle by installing the two bolts.

13. Install the wheel and lower the vehicle.

14. Connect the negative battery cable.

15. Have the vehicle alignment checked.

Lower Control Arms

REMOVAL AND INSTALLATION

Corolla

NOTE: There are two lower control arm bars on each side of the vehicle. The bar to the rear of the vehicle is referred to as No. 2 and the one towards the front of the vehicle is referred to as No. 1.

1. Disconnect the negative battery cable. On vehicles equipped with an air bag, wait at least 90 seconds before proceeding.

2. Loosen the lug nuts to the rear wheel.

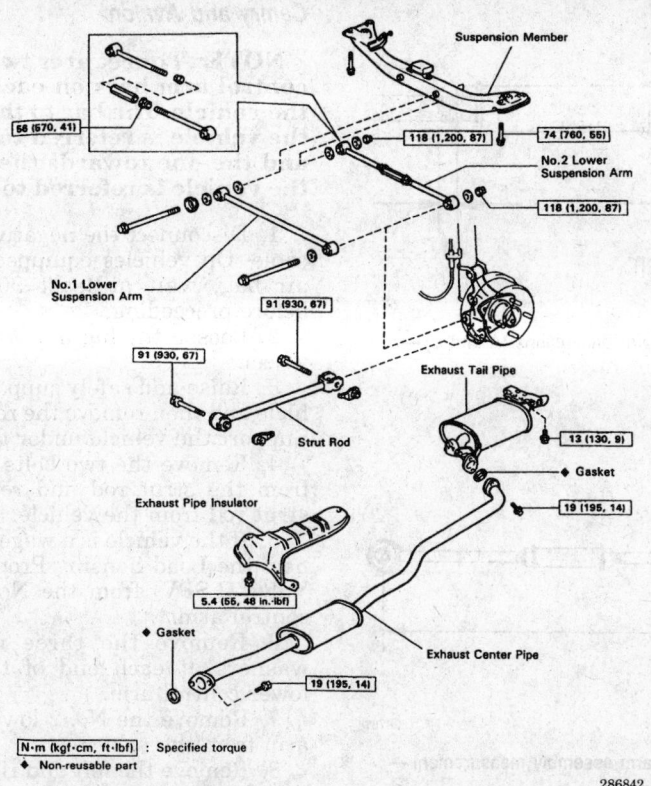

N·m (kgf·cm, ft·lbf) : Specified torque
◆ Non-reusable part

286842

Rear suspension assembly overview — Corolla shown, Celica similar

3. Raise and safely support the vehicle and then remove the rear wheel. Support the vehicle under the frame.

4. Remove the wheel on the side being serviced.

5. Remove the strut rod by removing the two bolts and two nuts.

6. Remove the two nuts and washers at each end of the No. 2 lower control arm.

7. Remove the No. 2 lower control arm from the vehicle.

8. Remove the exhaust center pipe and tail pipe from the vehicle.

9. Remove the four bolts to the exhaust tail pipe and then remove the exhaust pipe.

10. Disconnect the exhaust tail pipe from the bracket.

11. Remove the three bolts to the exhaust pipe insulator and then remove the exhaust pipe insulator.

12. Support the rear suspension crossmember with a jack.

13. Remove the six nuts to the suspension crossmember.

14. Lower the suspension crossmember enough to gain access to the No. 1 lower control arm bolts.

15. Remove the two bolts and the washers from the No. 1 lower control arm and then remove the control arm from the vehicle.

16. To disassemble the No. 2 lower control arm, loosen the two locknuts

and turn the adjusting tube. Remove the locknuts from the No. 2 control arm.

To install:

17. To assemble the No. 2 lower control arm, install the locknuts to the control arm. Turn the adjusting tube and assemble the No. 2 lower control arm.

18. Adjust the No. 2 lower control arm length by turning the adjusting tube. The arm length should be 19.69 ±0.059 inches (500.0 ±1.5mm) on 1993–95 models or 19.76 ±0.059 inches (502 ±1.5mm) on 1996–97 models.

NOTE: When adjusting the control arm length, try to adjust the arm so that the splines on each side of the locknuts are the same length. There should be a maximum difference (between A and B) of 0.12 in. (3mm).

19. Temporarily tighten the two locknuts. Make sure to tighten the nuts after the rear wheels are aligned.

20. Install the No. 1 lower control arm with the paint mark toward the inside of the vehicle.

21. Install the washers and through-bolts to the lower control arm.

22. Once the control arm is in place, lift the suspension crossmember and install the suspension crossmember support brackets and six nuts. Tighten the six suspension crossmember-to-body nuts to 55 ft. lbs. (74 Nm).

23. Install the exhaust pipe insulator with the three bolts. Tighten the bolts to 48 inch lbs. (5 Nm).

24. Install the exhaust tail pipe to the bracket. Install the four bolts and tighten to 9 ft. lbs. (13 Nm).

25. Install the exhaust center pipe and tail pipe.

26. Install the No. 2 lower suspension arm with the paint mark toward the inside of the vehicle. Install the washers and nuts but do not tighten completely at this point.

27. Install the strut rod and temporarily install the two bolts and nuts.

28. Install the rear wheel and lower the vehicle.

29. Bounce the vehicle up and down several times to stabilize the suspension.

30. Jack up the vehicle and support the body with stands.

31. Support the rear axle carrier with a jack.

32. Tighten the nuts of the lower control arms to 87 ft. lbs. (118 Nm).

33. Tighten the strut rod set bolts to 67 ft. lbs. (91 Nm).

34. Lower the vehicle.

35. Tighten the No. 2 lower suspension arm locknuts to 41 ft. lbs. (56 Nm).

36. Connect the negative battery cable.

37. Have the rear alignment checked.

Celica

NOTE: There are two lower control arm bars on each side of the vehicle. The bar to the rear of the vehicle is referred to as No. 2 and the one towards the front of the vehicle is referred to as No. 1.

1. Disconnect the negative battery cable. On vehicles equipped with an air bag, wait at least 90 seconds before proceeding.

2. Loosen the lug nuts to the rear wheel.

3. Raise and safely support the vehicle and then remove the rear wheel. Support the vehicle under the frame.

4. Remove the two nuts and washers at each end of the No. 2 lower control arm.

5. Remove the No. 2 lower control arm from the vehicle.

6. Support the rear suspension crossmember with a jack.

Tube Type

Bolt Type

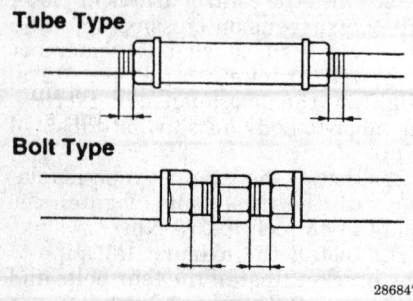

286847

Adjusting rod measurement (bolt and tube type) — Corolla

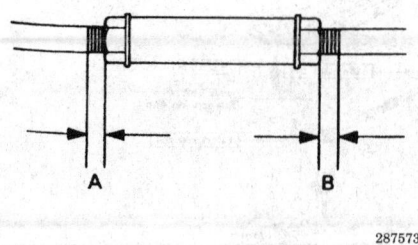

287573

No. 2 control arm dimensions (A and B) — Celica

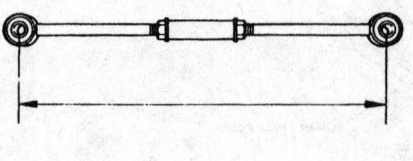

287574

No. 2 control arm assembly measurement — Celica

7. Remove the four outer nuts to the suspension crossmember.

8. Remove the two inner bolts to the suspension crossmember.

9. Lower the suspension crossmember enough to gain access to the No. 1 lower control arm bolts.

10. Remove the two bolts and the washers from the No. 1 lower control arm and then remove the control arm from the vehicle.

11. To disassemble the No. 2 lower control arm, loosen the two locknuts and turn the adjusting tube. Remove the locknuts from the No. 2 control arm.

To install:

12. Install the locknuts to the No. 2 control arm.

13. Turn the adjusting tube and assemble the No. 2 lower control arm.

14. Adjust the No. 2 lower control arm length by turning the adjusting tube. The arm length should be 19.46 ±0.059 inches (494.5 ±1.5mm).

NOTE: When adjusting the control arm length, try to adjust the arm so that the splines on each side of the locknuts are the same length. There should be a maximum difference (between A and B) of 0.12 in. (3mm).

15. Temporarily tighten the two locknuts. Make sure to tighten the nuts after the rear wheels are aligned.

16. Install the No. 1 lower control arm with the paint mark toward the outer rear of the vehicle.

17. Install the washers and through-bolts to the lower control arm.

18. Once the control arm is in place, lift the suspension crossmember and install the crossmember outer nuts. Tighten the four suspension crossmember-to-body nuts to 95 to 93 ft. lbs. (125 Nm) on 1994–95 models and

101 ft. lbs. (137 Nm) on 1996–97 models.

19. Tighten the inner two bolts for the crossmember to 14 ft. lbs. (20 Nm).

20. Install the No. 2 lower suspension arm with the paint mark toward the outer rear of the vehicle. Install the washers and nuts but do not tighten completely at this point.

21. Install the rear wheel and lower the vehicle.

22. Bounce the vehicle up and down several times to stabilize the suspension.

23. Jack up the vehicle and support the body with stands.

24. Support the rear axle carrier with a jack and remove the rear wheel assembly.

25. With the weight of the body on the axle, tighten the nuts of the lower control arms to 145 ft. lbs. (197 Nm).

26. Install the rear wheel and lower the vehicle.

27. Tighten the No. 2 lower suspension arm locknuts to 55 ft. lbs. (75 Nm).

28. Connect the negative battery cable.

29. Have the wheel alignment checked.

Camry and Avalon

NOTE: There are two lower control arm bars on each side of the vehicle. The bar to the rear of the vehicle is referred to as No. 2 and the one towards the front of the vehicle is referred to as No. 1.

1. Disconnect the negative battery cable. On vehicles equipped with an air bag, wait at least 90 seconds before proceeding.

2. Loosen the lug nuts to the rear wheel.

3. Raise and safely support the vehicle and then remove the rear wheel. Support the vehicle under the frame.

4. Remove the two bolts and nuts from the strut rod and remove the strut rod from the vehicle.

5. If the vehicle is a wagon, disconnect the Load Sensing Proportioning Valve (LSPV) from the No. 2 lower control arm.

6. Remove the three nuts and washers at each end of the No. 2 lower control arm.

7. Remove the No. 2 lower control arm from the vehicle.

8. Remove the left and right stabilizer bushing retainers.

9. Remove the exhaust center pipe and tailpipe from the vehicle.

10. Support the rear suspension crossmember with a jack.

11. Remove the six nuts to the suspension crossmember. Remove the left and right suspension crossmember lower supports.

12. Lower the suspension crossmember enough to gain access to the No. 1 lower control arm bolts.

13. Remove the two bolts and the washers from the No. 1 lower control arm and then remove the control arm from the vehicle.

14. To disassemble the No. 2 lower control arm, loosen the two locknuts and turn the adjusting tube. Remove the locknuts from the No. 2 control arm.

To install:

15. To assemble the No. 2 lower control arm, install the locknuts to the control arm. Turn the adjusting tube and assemble the No. 2 lower control arm. Adjust the No. 2 lower control arm length by turning the adjusting tube. The arm length should be 23.0 in. (584.2mm).

NOTE: When adjusting the control arm length, try to adjust the arm so that the splines on each side of the locknuts are the same length. There should be a maximum difference (between A and B) of 0.12 in. (3mm).

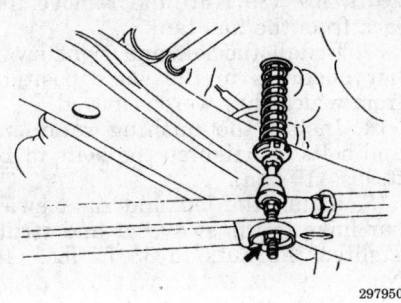

Removing the LSPV from the lower control arm — Camry and Avalon

16. Temporarily tighten the two locknuts. Make sure to tighten the nuts after the rear wheels are aligned.

17. Install the No. 1 lower control arm with the paint mark facing the rear of the vehicle.

18. Install the washers and through-bolts to the lower control arm.

19. Once the control arm is in place, lift the suspension crossmember and install the suspension crossmember support brackets and six nuts. Tighten the four suspension cross-member-to-body nuts to 83 ft. lbs.

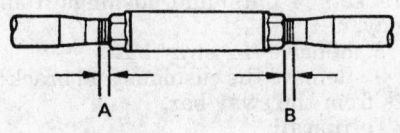

No. 2 control arm spline difference — Camry and Avalon

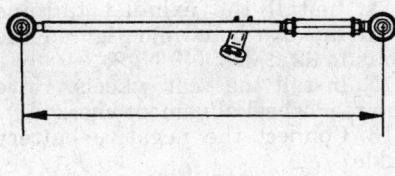

Measuring the No. 2 control arm length — Camry and Avalon

(113 Nm). Tighten the two front bracket bolts to 28 ft. lbs. (38 Nm).

20. Install the left and right stabilizer bushing retainers and tighten the bolts to 14 ft. lbs. (19 Nm).

21. Install the exhaust center pipe and tail pipe.

22. Install the No. 2 lower suspension arm with the paint mark facing rearward. Install the washers and nuts but do not tighten completely at this point.

23. If the vehicle is a wagon, install the load sensing proportioning valve to the No. 2 lower control arm. Tighten the nut to 9 ft. lbs. (13 Nm).

24. Install the strut rod and temporarily install the two bolts and nuts.

25. Tighten the nut on the outside of the lower control arms to 134 ft. lbs. (181 Nm).

26. Install the rear wheel and lower the vehicle.

27. Bounce the vehicle up and down several times to stabilize the suspension.

28. Jack up the vehicle and support the body with stands.

29. Support the rear crossmember with a jack.

30. Tighten the nuts on the inside of the lower control arm to 134 ft. lbs. (181 Nm).

31. Tighten the strut rod set bolts to 83 ft. lbs. (113 Nm).

32. Lower the vehicle and connect the negative battery cable

33. Tighten the No. 2 lower suspension arm locknuts to 41 ft. lbs. (56 Nm).

34. Have the wheel alignment checked.

MR2

1. Disconnect the negative battery cable. On vehicles equipped with an air bag, wait at least 90 seconds before proceeding.

2. Loosen the lug nuts to the rear wheel on the side being serviced.

3. Raise and safely support the vehicle. Place the jackstands under the frame of the body.

4. Remove the wheel from the side being serviced.

5. Remove the cotter pin and nut connecting the lower control arm to the ball joint.

6. Using a ball joint separator, disconnect the lower control arm from the ball joint.

7. Remove the strut rod nut and retainer from the lower control arm.

8. Remove the bolt holding the lower control arm to the body and remove the lower control arm from the vehicle.

9. Remove the strut rod cushion, collar, and the retainer from the lower control arm.

To install:

10. Install the strut rod retainer, cushion, and the collar to the strut rod.

11. Connect the lower control arm to the strut rod and install the strut rod nut. Do not tighten the nut at this time.

12. Install the lower control arm to the body and install the bolt and nut. Do not tighten at this time.

13. Connect the lower control arm to the ball joint and tighten the nut to 67 ft. lbs. (91 Nm).

14. Install a new cotter pin to the ball joint.

15. Install the wheel and lower the vehicle to the ground.

16. Bounce the vehicle up and down to stabilize the suspension.

17. With the weight of the vehicle on the suspension, tighten the strut rod nut to 87 ft. lbs. (118 Nm).

18. With the weight of the vehicle on the suspension, tighten the lower control arm holding nut to 98 ft. lbs. (132 Nm).

19. Connect the negative battery cable.

20. Have the wheel alignment checked.

Supra

NOTE: There are two lower control arm bars on each side of the vehicle. The bar to the rear of the vehicle is referred to as No. 2 and the one towards the front of the vehicle is referred to as No. 1.

1. Disconnect the negative battery cable. On vehicles equipped with an air bag, wait at least 90 seconds before proceeding.

2. Raise the rear of the vehicle and support it on safety stands.

3. Remove the wheel.

4. Remove the caliper support bracket by removing the two bolts. Leave the brake line connected and suspend it aside.

5. Remove the brake rotor.

6. Remove the bolt and nut and disconnect the strut rod from the axle carrier.

7. Remove the bolt and nut holding the strut rod to the body. Remove the strut rod from the vehicle.

8. Remove the nut and press the No. 1 control arm out of the axle carrier.

9. Place matchmarks on the adjusting cam and frame. Remove the cam and nut and lift out the No. 1 control arm.

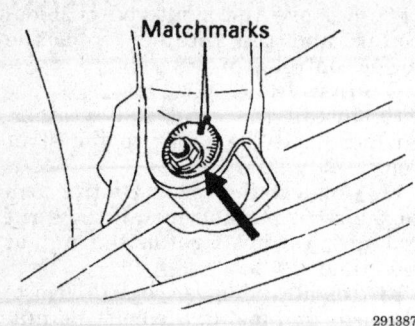

Matchmarking the No. 1 suspension arm adjusting cam to the body — Supra

291387

10. Disconnect the strut from the No. 2 lower control arm by removing the bolt and nut.

11. Disconnect the stabilizer bar link from the No. 2 lower control arm by removing the nut.

12. Remove the nut and press the No. 2 lower control arm out of the axle carrier.

13. Place matchmarks on the adjusting cam and frame for the No. 2 lower control arm. Remove the cam and nut and lift out the No. 2 arm.

To install:

14. Install the No. 2 arm to the axle carrier with a new nut and tighten it to 110 ft. lbs. (150 Nm).

15. Connect the No. 2 arm to the body and install the adjusting cam and nut, align the matchmarks. Do not tighten the bolts and nuts at this time.

16. Connect the strut to the lower arm by installing the bolt and nut. Do not tighten the bolt at this time.

17. Connect the stabilizer link.

18. Connect the No. 1 lower control arm to the axle carrier and tighten the new nut to 43 ft. lbs. (59 Nm).

19. Connect the No. 1 lower control arm to the body and install the adjusting cam and nut. Align the matchmarks. Do not tighten the bolts and nuts at this time.

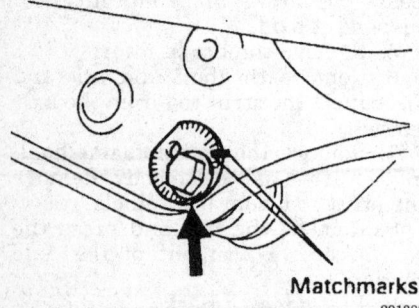

Matchmarks

Matchmarking the No. 2 suspension arm adjusting cam to the body — Supra

291388

20. Install the strut rod to the body and axle carrier. Do not tighten the bolts at this time.

21. Install the brake rotor.

22. Install the caliper support bracket.

23. Install the wheel and lower the vehicle.

24. Bounce the vehicle several times to set the suspension.

25. Support the axle carrier and tighten the strut rod bolts to 136 ft. lbs. (184 Nm).

26. Tighten the strut bolt to 106 ft. lbs. (143 Nm).

27. Align the adjusting cam matchmarks and tighten the nuts to 136 ft. lbs. (184 Nm).

28. Connect the negative battery cable.

29. Have the wheel alignment checked.

Sway Bar

REMOVAL AND INSTALLATION

Corolla

1. Disconnect the negative battery cable. On vehicles equipped with an air bag, wait at least 90 seconds before proceeding.

2. Loosen the lug nuts to the rear wheels.

3. Raise and safely support the vehicle and then remove the rear wheels.

4. Remove the brake hose from the strut using a suitable container to catch the fluid.

5. Disconnect the strut from the steering knuckle by removing the two nuts and bolts.

6. Disconnect the stabilizer bar links from both the strut and the sway bar by removing the nuts.

7. Remove the left and right bushing retainers holding the sway bar to the body. There are two bushing retainers and each has two bolts.

8. Remove the sway bar bushings.

9. Remove the fuel tank bands from vehicle as follows:

   a. Support the fuel tank with a jack and lower the fuel tank.

   b. Remove the two bolts and disconnect the left and right hand fuel tank bands.

   c. If necessary, lower the fuel tank slightly.

10. Remove the sway bar from the left hand side of the vehicle.

To install:

11. Install the sway bar to its normal position on the vehicle.

12. Install the fuel tank bands and install the bolts. Tighten the bolts to

29 ft. lbs. (39 Nm) and remove the jack from the fuel tank.

13. Install the left and right sway bar bushings in the same position from which they were removed.

14. Install the bushing retainers and bolts and tighten the bolts to 14 ft. lbs. (19 Nm).

15. Install the left and right sway bar links to the sway bar and strut. Tighten the nuts to 33 ft. lbs. (44 Nm).

16. Connect the strut to the steering knuckle and install the two bolts and nuts. Tighten to 105 ft. lbs. (142 Nm).

17. Connect the brake line to the strut and install the clip. Tighten the brake line to the brake hose to 11 ft. lbs. (15 Nm).

18. Install the rear wheels and lower the vehicle.

19. Connect the negative battery cable.

Celica

1. Disconnect the negative battery cable. On vehicles equipped with an air bag, wait at least 90 seconds before proceeding.

2. Raise and safely support the vehicle.

3. Remove the rear wheels.

4. Remove the exhaust tailpipe by removing the two bolts.

5. Remove the right hand fuel tank band by removing the bolt.

6. Remove sway bar links from the struts.

7. Remove the two sway bar brackets (4 nuts) and bushings from body.

8. Remove the sway bar.

9. Remove the cushions and brackets from the sway bar.

To install:

10. Install the sway bar in place.

11. Install the sway bushing and bracket. Align the marks and tighten the nuts to 14 ft. lbs. (19 Nm).

12. Install the sway link, tightening the nuts to 33 ft. lbs. (44 Nm).

13. Install the right hand fuel tank band with the bolt and tighten the bolt to 30 ft. lbs. (40 Nm).

14. Install the exhaust tailpipe with the two bolts and tighten the bolts to 32 ft. lbs. (40 Nm).

15. Install the rear wheels. Have the rear wheel alignment checked.

16. Connect the negative battery cable.

Avalon and Camry

1. Disconnect the negative battery cable. On vehicles equipped with an air bag, wait at least 90 seconds before proceeding.

2. Raise and support the vehicle safely. Remove the rear wheels.

3. Disconnect the stabilizer bar links from the struts. If the ball joint stud turns together with the nut, use a hexagon wrench to hold the stud.

4. Remove the four bolts, stabilizer brackets, and cushions.

5. Remove the stabilizer bar from the vehicle.

6. Remove the stabilizer bar links from the stabilizer bar.

7. Rotate the ball joint stud in all directions. If the movement is not smooth and free, replace the stabilizer link.

To install:

8. Install the stabilizer bar links to the stabilizer bar. Tighten the nuts to 47 ft. lbs. (64 Nm).

9. Install the stabilizer bar assembly. Tighten the stabilizer bar bracket retaining bolts to 14 ft. lbs. (19 Nm).

10. Connect the stabilizer bar links to the struts and tighten the stabilizer bar link nuts to 47 ft. lbs. (64 Nm).

11. Install the rear wheels.

12. Lower the vehicle and connect the negative battery cable.

MR2

1. Disconnect the negative battery cable. On vehicles equipped with an air bag, wait at least 90 seconds before proceeding.

2. Raise and support the rear axle housing and frame safely. Be sure to leave the jack under the rear axle.

3. Remove the nuts holding the sway bar links to the struts.

4. Remove the nuts holding the sway bar to the sway bar links.

5. If equipped with ABS brakes, remove the bolt and disconnect the ABS speed sensor bracket from the rear suspension crossmember.

6. Disconnect the rear axle carrier from the lower control arm by removing the two bolts.

7. Disconnect the suspension arm from the axle carrier by removing the nut and bolt.

8. Disconnect the exhaust pipe mounting from the rear suspension crossmember by removing the two bolts,

9. Disconnect the rear engine mount from the rear suspension member by removing the three bolts.

10. Support the rear suspension crossmember with a floor jack.

11. Remove the four suspension crossmember mounting bolts and lower the crossmember to the floor.

12. Remove the sway bar from the suspension crossmember by removing the two nuts and two bolts.

13. Remove the brackets and cushions from the sway bar.

To install:

14. Install the cushions and brackets to the sway bar.

15. Install the sway bar to the suspension crossmember and install the two nuts and two bolts. Tighten the nuts and bolts to 14 ft. lbs. (19 Nm).

16. Raise the suspension into place and install the four bolts for the crossmember. Tighten the bolts to 83 ft. lbs. (113 Nm).

17. Remove the jack from the crossmember.

18. Connect the rear engine mount to the rear suspension crossmember and install the three bolts. Tighten the bolts to 57 ft. lbs. (77 Nm).

19. Connect the exhaust pipe mounting to the rear suspension crossmember and install the two mounting bolts. Tighten the bolts to 15 ft. lbs. (21 Nm).

20. Connect the rear axle carrier to the suspension arm and install the nut and bolt. Do not tighten the nut and bolt at this time.

21. Connect the rear axle carrier to the lower control arm by installing the two bolts.

22. If equipped with ABS brakes, install the ABS speed sensor bracket to the rear suspension crossmember and install the bolt. Tighten the bolt to 48 inch lbs. (5 Nm).

23. Install the sway bar links and install the four nuts. Tighten the nuts to 36 ft. lbs. (49 Nm).

24. Install the wheels and lower the vehicle.

25. Bounce the vehicle up and down to stabilize the suspension.

26. Tighten the nut and bolt holding the rear axle carrier to the suspension arm to 76 ft. lbs. (103 Nm).

27. Connect the negative battery cable.

Supra

1. Disconnect the negative battery cable. On vehicles equipped with an air bag, wait at least 90 seconds before proceeding.

2. Raise and support the vehicle safely.

3. Disconnect the rear exhaust from the brackets.

4. Remove the stabilizer links. If the ball joint stud turns together with the nut, use a hexagon wrench to hold the stud.

5. Remove stabilizer bar brackets and cushions by removing the four bolts.

6. Remove the stabilizer bar from the vehicle.

To install:

7. Install the sway bar to the vehicle.

8. Install the stabilizer bar brackets and cushions. Install the four bolts and tighten the bolts to 23 ft. lbs. (31 Nm).

9. Connect the stabilizer bar links and tighten the nuts to 54 ft. lbs. (74 Nm).

10. Connect the exhaust to the brackets.

11. Lower the vehicle and connect the negative battery cable.

Wheel Bearings

ADJUSTMENT

Paseo and Tercel

1. Disconnect the negative battery cable. On vehicles equipped with an air bag, wait at least 90 seconds before proceeding.

2. Raise and support the vehicle safely.

3. Remove the rear wheels.

4. Remove the locknut cap and cotter pin. Remove the locknut.

5. Install the bearing locknut and tighten it to 22 ft. lbs. (29 Nm) while spinning the drum.

6. Spin the brake drum several times to snug down the bearing and then loosen the bearing locknut until it can be turned by hand.

NOTE: There must be absolutely no brake drag at this time.

7. Retighten the bearing locknut until there is a bearing preload of 0.9–2.2 lbs. (3.2–9.8 N) while turning the wheel. Measure with a spring scale hooked to one of the studs.

8. Install the locknut lock, a new cotter pin, and the cap. If the cotter pin hole does not align properly, align the holes by tightening the nut to the next hole. Do not loosen the nut.

9. Install the rear wheel and lower the vehicle.

10. Connect the negative battery cable.

Except Tercel and Paseo

Check the backlash in bearing shaft direction and the axle hub deviation. Maximum for backlash should be 0.0020 in. (0.05mm). Maximum axle hub deviation is 0.0028 in. (0.07mm), except on Supra, which is 0.020 in. (0.05mm).

NOTE: The wheel bearing is non-adjustable. If the wheel bearing is out of specifications, replace the wheel bearing.

REMOVAL AND INSTALLATION

Paseo and Tercel

1. Disconnect the negative battery cable. On vehicles equipped with an air bag, wait at least 90 seconds before proceeding.
2. Raise and support the vehicle safely.
3. Remove the rear wheels.
4. Remove the locknut cap and cotter pin. Remove the locknut.
5. Carefully pull off the brake drum along with the outer wheel bearing and thrust washer. Do not drop the bearing.
6. Pry the inner bearing oil seal out of the brake drum assembly and then remove the inner bearing.
7. Drive out the bearing races, as required.
To install:
8. Press the new outer bearing races into the brake drum and add a

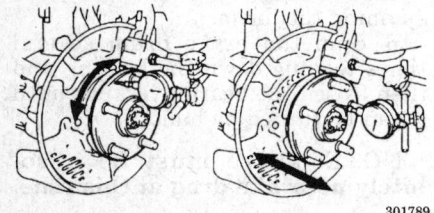

Checking wheel bearings, except Tercel and Paseo

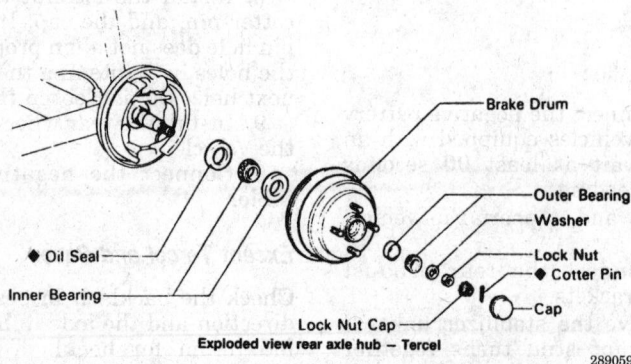

Rear wheel bearing assembly — Tercel shown, Paseo similar

liberal amount of bearing grease to the inside of the hub and the bearing cap.
9. Clean and pack the bearings with grease.
10. Position the inner bearing into the brake drum and then drive in a new oil seal to the original position. Lightly coat the seal with grease.
11. Position the brake drum onto the axle shaft. Install the outer bearing and position the thrust washer. Install the bearing locknut and tighten it to 22 ft. lbs. (29 Nm) while spinning the drum.
12. Spin the brake drum several times to snug down the bearing and then loosen the bearing locknut until it can be turned by hand.

NOTE: There must be absolutely no brake drag at this time.

13. Retighten the bearing locknut until there is a bearing preload of 0.9–2.2 lbs. (3.2–9.8 N) while turning the wheel. Measure with a spring scale hooked to one of the studs.
14. Install the locknut lock, a new cotter pin, and the cap. If the cotter pin hole does not align properly, align the holes by tightening the nut to the next hole. Do not loosen the nut.
15. Install the rear wheel and lower the vehicle.
16. Connect the negative battery cable.

Celica, Corolla, Camry and Avalon

1. Disconnect the negative battery cable. On vehicles equipped with an air bag, wait at least 90 seconds before proceeding.
2. Raise and safely support the vehicle.
3. Remove the wheel.
4. Remove the brake drum or rotor.
5. If equipped with ABS brakes, disconnect and remove the ABS wheel speed sensor.

6. Remove the four bolts securing the hub to the knuckle and remove the hub.
7. Remove the O-ring from the backing plate.
To install:
8. Install a new O-ring onto the backing plate. Coat the O-ring with multipurpose grease.
9. Install the hub to the knuckle with the mounting bolts and tighten to 59 ft. lbs. (80 Nm).
10. If equipped with ABS brakes, install the ABS wheel speed sensor.
11. Install the brake drum or rotor.
12. Install the wheel and lower the vehicle.
13. Connect the negative battery cable.
14. Have the wheel alignment checked.

MR2

1. Disconnect the negative battery cable. On vehicles equipped with an air bag, wait at least 90 seconds before proceeding.
2. Raise and safely support the vehicle.
3. Remove the rear tire and wheel assembly.
4. Remove the cotter pin and bearing locknut cap to the halfshaft.
5. With the parking brake pulled, remove the bearing locknut to the halfshaft.
6. Disconnect the brake caliper support bracket from the rear axle carrier and support it with a piece of wire.
7. Place matchmarks on the disc brake rotor and the axle hub. Remove the brake rotor.
8. Remove the nut and disconnect the sway bar link from the strut.
9. If equipped with ABS brakes, remove the ABS speed sensor from the axle carrier.
10. Disconnect the ball joint from the axle carrier by removing the two bolts.
11. Remove the suspension arm from the axle carrier by removing the bolt and nut.
12. Disconnect the axle carrier from the strut by removing the two nuts and two bolts.
13. Remove the axle carrier from the vehicle. Use a soft rubber hammer to remove the axle carrier from the halfshaft.
14. Clamp the rear axle carrier in a soft jaw vise.
15. Remove the dust deflector in the back of the axle carrier.
16. Using snapring pliers, remove the snapring.

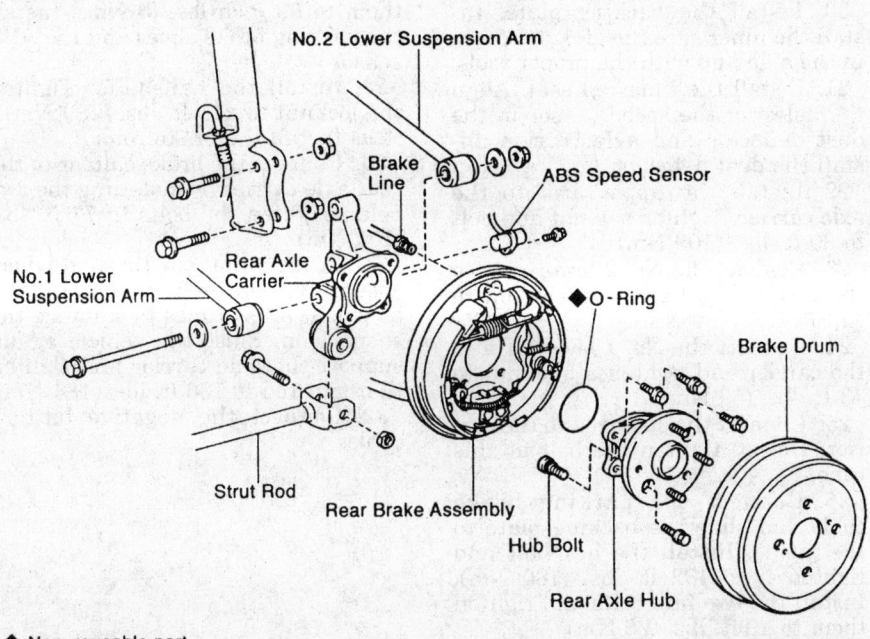

No.2 Lower Suspension Arm

Brake Line

ABS Speed Sensor

No.1 Lower Suspension Arm

Rear Axle Carrier

◆O - Ring

Brake Drum

Strut Rod

Rear Brake Assembly

Hub Bolt

Rear Axle Hub

◆ Non-reusable part

286839

Hub and wheel bearing component assembly — Corolla shown, Celica similar

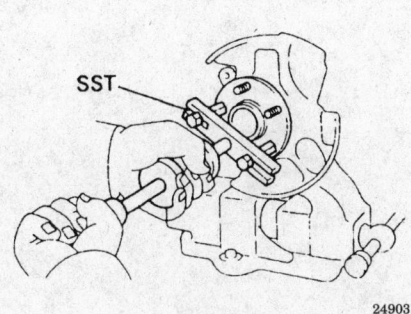

SST

249031

Removing the hub — MR2

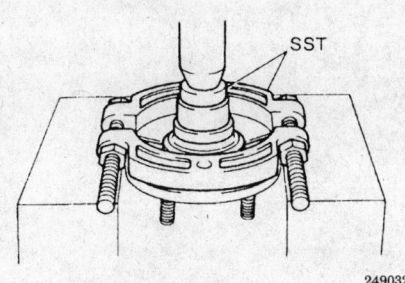

SST

249032

Removing the bearing inner race — MR2

17. Using a two arm puller, remove the axle hub from the axle carrier.

18. Remove the backing plate by removing the three bolts.

19. Using a bearing race puller, remove the inner race (outside) from the hub.

20. Place the removed inner race (outside) in the bearing and press out the bearing.

To install:

21. Using a proper sized bearing driver and a press, press a new bearing onto the axle hub.

NOTE: If the inner races come loose from the bearing outer race, be sure to install them on the same side as before.

22. Install the backing plate with the three bolts. Tighten the bolts to 73 inch lbs. (8 Nm).

23. Using a proper sized driver and a press, press the hub into the rear axle carrier.

24. Using snapring pliers, install a new snapring into the axle carrier.

25. Using a seal driver and a hammer, drive in a new dust deflector to the rear axle carrier.

NOTE: When installing the dust deflector, align the holes for the ABS speed sensor in the dust deflector and steering knuckle.

26. Install the axle carrier to the vehicle and install the two bolts and two nuts to hold the axle carrier to the strut. Hand-tighten the bolts.

27. Connect the suspension arm to the axle carrier and install the bolt and nut. Do not tighten the bolt and nut at this time.

28. Connect the ball joint to the axle carrier and install the two bolts. Tighten the bolts to 83 ft. lbs. (113 Nm).

29. If equipped with ABS brakes, install the ABS speed sensor to the axle carrier.

30. Connect the sway bar link to the strut and install the nut. Tighten the nut to 36 ft. lbs. (49 Nm).

31. Install the rotor to the axle carrier.

32. Install the brake caliper to the axle carrier and install the two bolts. Tighten the bolts to 43 ft. lbs. (59 Nm).

33. Install the halfshaft locknut to hold the axle carrier to the halfshaft. If equipped with a 5S-FE engine, tighten the nut to 159 ft. lbs. (216 Nm). If equipped with a 3S-GTE engine, tighten the nut to 217 ft. lbs. (294 Nm).

34. Install the halfshaft locknut cap and cotter pin.

35. Install the rear wheel and lower the vehicle.

36. Stabilize the vehicle and connect the negative battery cable.

37. With weight on the suspension, tighten the strut lower bolts to 127 ft. lbs. (173 Nm). Tighten the suspension arm bolt and nut to 76 ft. lbs. (103 Nm).

Supra

1. Disconnect the negative battery cable. On vehicles equipped with an air bag, wait at least 90 seconds before proceeding.

2. Raise and safely support the vehicle. Remove the rear tire and wheel assembly.

3. Disconnect the brake caliper support bracket from the rear axle carrier and support it with a piece of wire.

4. Place matchmarks on the disc brake rotor and the axle hub. Remove the brake rotor.

5. Remove the speed sensor.

6. Remove the rear halfshaft.

7. Remove the parking brake shoes.

8. Remove the two bolts at the parking brake cable. Remove the two hub bolts and the hex bolt. Slide the backing plate to the outside and disconnect the parking brake cable.

9. Disconnect the strut rod at the axle carrier.

10. Remove the nut and then press out the No. 1 lower suspension arm.

11. Remove the nut and then press out the No. 2 lower suspension arm.

12. Remove the nut and then press out the upper suspension arm. Remove the axle carrier.

13. Remove the dust deflector and pull out the oil seal.

14. Using a two arm puller, remove the axle hub from the carrier.

15. Remove the backing plate.

16. Press the inner race (outside) from the hub. Then, remove the oil seal and the snapring.

17. Place the inner race (outside) over the bearing and tap out the bearing and inner race (inside).

To install:

18. Install the bearing to the axle carrier.

NOTE: If the inner races come loose from the bearing outer race, be sure to install them on the same side as before.

19. Install the snapring, the inner race (outside), and a new oil seal.

20. Install the backing plate. Install the inner race (inside) and press in the axle hub with the proper tools.

21. Install the inner oil seal. Align the holes for the speed sensor in the dust deflector and axle carrier. Install the dust deflector.

22. Install the upper arm to the axle carrier. Tighten the nut and bolt to 80 ft. lbs. (109 Nm).

23. Connect the No. 2 lower arm to the carrier and tighten a new nut to 110 ft. lbs. (150 Nm).

24. Connect the No. 1 lower arm to the carrier and tighten a new nut to 43 ft. lbs. (59 Nm).

25. Connect the strut rod to the carrier. Do not tighten the bolt at this time.

26. Connect the parking brake cable and slide the backing plate to the inside. Install the hex bolt and tighten it to 132 ft. lbs. (180 Nm). Install the two hub bolts and tighten them to 19 ft. lbs. (26 Nm).

27. Install the two bolts at the parking brake cable and tighten them to 69 inch lbs. (8 Nm). Install the parking brake shoes and the ABS sensor.

28. Install the halfshafts. Tighten the locknut to 213 ft. lbs. (289 Nm).

29. Install the brake rotor.

30. Connect the brake caliper to the rear axle carrier by installing the two bolts. Tighten the bolts to 77 ft. lbs. (104 Nm).

31. Replace the rear tire and wheel assembly. Lower the vehicle and bounce it a few times to stabilize the suspension. Raise the vehicle again, support the axle carrier and tighten the strut rod to 136 ft. lbs. (184 Nm).

32. Connect the negative battery cable.

FIRING ORDERS

NOTE: To avoid confusion, always replace spark plug wires one at a time.

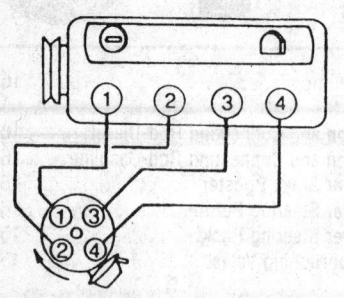

7913r001

4–Cylinder Engines
Engine Firing Order: 1–3–4–2
Distributor Rotation:
Counterclockwise

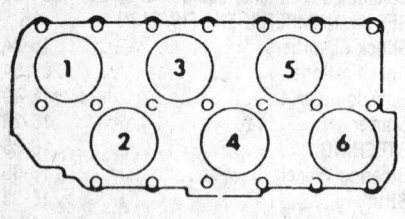

7913r002

6–Cylinder Engine
Engine Firing Order: 1–5–3–6–2–4
Distributor Rotation: Counterclockwise

SERIAL NUMBER IDENTIFICATION

Vehicle Identification Plate

All vehicles have an identification plate bearing the chassis number on the top of the dash board at the driver's side, visible through the windshield. The VIN indicates such information as model year, type, date of manufacture, etc. This information is more easily read from the Vehicle Identification Label in the luggage compartment. That label also provides engine and transaxle code numbers, paint, interior and option code numbers. Since the manufacturer sometimes makes production changes in mid-model year, this in-

formation is sometimes required when locating parts.

Engine Number

The diesel engine number with its 2 letter code is stamped on the block between the injection pump and the vacuum pump. On 4–cylinder gasoline engines, the engine number is stamped on the block just below the cylinder head, near the No. 2 or No. 3 spark plug. On 6–cylinder engines, the engine number is stamped into the front of the block just below the cylinder head gasket.

GASOLINE ENGINE MECHANICAL

NOTE: Disconnecting the negative battery cable on some vehicles may interfere with the functions of the on-board computer or security systems and may require reprogramming when the battery cable is reconnected.

Engine Assembly

REMOVAL AND INSTALLATION

Except Fox and Corrado SLC

NOTE: The engine and transaxle are lifted as an assembly from the vehicle.

——— CAUTION ———
Use care when disconnecting the fuel lines. Fuel under pressure may still be in the lines and, if sprayed, may cause fire or personal injury.

1. Disconnect the battery cables and remove the battery.
2. Open the fuel filler cap to relieve tank pressure and then relieve the fuel system pressure.
3. Remove the air intake duct between the fuel distributor and the throttle body. On Corrado, remove the air tubing from the G-charger and the intercooler. At the throttle body, pull back the accelerator cable clip and disconnect the cable from the ball. Loosen the accelerator cable locknut and remove the cable from the cylinder head cover.
4. Remove the radiator cap and set the heater temperature control to

full HOT. Place a pan under the thermostat housing and remove the thermostat flange to drain the coolant.
5. Remove the upper radiator hose and disconnect the wiring from the radiator fan motor and switches. Remove the mounting nuts or bolts and lift out the radiator and fan shroud as an assembly.
6. Except Corrado, at the front of the vehicle remove the apron, the trim and the grille. Disconnect the headlight electrical connectors and the hood release cable from the hood latch assembly.
7. Begin disconnecting electrical connections and vacuum lines, carefully labeling each one. Don't forget ground connections that are screwed to the body.

NOTE: If equipped with power steering, remove pump and reservoir and set them aside; do not disconnect the fluid lines. If equipped with air conditioning, remove the compressor and set it aside without disconnecting the lines.

8. On CIS fuel systems, much of the system can be removed as a unit without disconnecting fuel lines. Remove the injectors from their holes and protect them with caps. Remove the cold start injector and warm-up regulator, if equipped, and disconnect the fuel supply and return lines from the fuel distributor.
9. On vehicles with CIS fuel injection, unsnap the clips holding the air cleaner housing together and lift the fuel distributor/air sensor assembly from the vehicle with all the other fuel lines attached.

NOTE: If equipped with an automatic transaxle, place the selector lever in the P position.

10. On vehicles with cable shift linkage, disconnect the shift linkage cables and remove the clutch slave cylinder from the transaxle without disconnecting the line and set it aside.
11. On vehicles with rod shift linkage, remove the 2 rods with the plastic socket ends and unbolt the remaining linkage from the transaxle case as required. Disconnect the clutch cable, lift it from the case and set it aside.
12. Disconnect the electrical connectors from the starter, the backup light switch and the ground cable from the transaxle. Remove the speedometer cable from the transaxle and plug the hole in the case.
13. On vehicles with automatic transaxle, disconnect the cable from

the actuating lever and remove it from the bracket.

14. Attach an engine sling tool VW-2024A or equivalent, to the engine and attach the sling to a suitable lifting device. On the 16V engine, remove the idle stabilizer valve and the upper intake manifold to attach the sling.

15. Unbolt the exhaust pipe from the manifold or remove the spring clamps holding the exhaust pipe to the manifold and lower the pipe.

---- **CAUTION** ----

On some models, special tools are required for removing and installing the exhaust pipe-to-manifold spring clamps; VW3140/1 and /2 or equivalent. This is a set of different sized wedges for spreading the spring clamps in steps. The installed spring clamp has considerable tension and could cause damage or injury if not properly removed. Clamps with wedges installed are also under high tension and should be handled carefully.

16. Unbolt the halfshafts from the flanges and hang them from the body with wire.

17. Make sure everything is disconnected and unbolt the mounts. Remove the starter first and the front mount with it.

18. With all mounts unbolted, slightly lower the engine/transaxle assembly and tilt it towards the transaxle side. Then carefully lift the assembly from the vehicle.

To install:

19. Carefully install the engine/transaxle assembly and make sure all mounts are securely bolted to the engine/transaxle. Start all nuts and bolts that secure the mounts to the body but don't tighten them yet.

20. With all mounts installed and the engine safely in the vehicle, allow some slack in the lifting equipment. With the vehicle safely supported, shake the engine/transaxle as a unit to settle it in the mounts. Torque all mounting bolts, starting at the rear and working forward. Torque to 33 ft. lbs. (41 Nm) for 10mm bolts or 54 ft. lbs. (73 Nm) for 12mm bolts.

21. Install the starter and torque the bolts to 33 ft. lbs. (45 Nm).

22. Connect the halfshafts to the flanges and torque the bolts to 33 ft. lbs. (45 Nm).

23. Install the exhaust pipe and use new self-locking nuts to secure the flange. Torque the nuts to 30 ft. lbs. (40 Nm). If equipped with spring

clamps, the clamps can be used again.

24. Connect the shift linkage and install the clutch cable or slave cylinder. Adjust the clutch and shift linkage as required.

25. Install the fuel system components.

26. Install the air conditioning compressor and/or power steering pump, if equipped. Install and adjust the drive belts.

27. Connect the wiring and vacuum hoses.

28. Install the radiator, fan and heater hoses. Use a new O-ring on the thermostat and torque the thermostat housing bolts to 7 ft. lbs. (10 Nm).

29. Fill and bleed the cooling system. Check the adjustment of the accelerator cable.

30. Install any body parts that were removed.

Fox

The engine is lifted from the vehicle without the transaxle.

1. Disconnect the battery ground cable and remove the battery.

2. Open the heater valve and the cap on the coolant expansion tank. Drain the coolant by removing the bottom hose. Disconnect the electrical connector from the radiator cooling fan.

3. Remove the radiator and fan as an assembly.

4. If equipped with air conditioning, remove the compressor and condenser and place them aside without disconnecting any refrigerant lines.

5. Detach and label all the electrical wires and vacuum lines connecting the engine to the body.

6. Much of the fuel system can be removed as a unit without disconnecting fuel lines. Remove the injectors from their holes and protect them with caps. Remove the cold start injector and warm-up regulator, if equipped. Disconnect the throttle cable and remove the air duct. Place these aside without disconnecting the fuel lines.

7. Disconnect the speedometer cable from the transaxle and plug the hole. Detach the clutch cable.

8. Loosen the charcoal filter clamp and move the filter to the rear of the engine compartment.

9. Remove the upper engine-to-transaxle bolts.

10. Remove the left and right engine mounting nuts.

11. Remove the front engine stop and the starter.

12. Remove the clutch cover and the 2 lower engine-to-transaxle bolts.

13. Disconnect the exhaust pipe from the manifold at the flange. Then remove the bolt from the exhaust pipe support and remove the exhaust pipe from the manifold.

14. Install transaxle support bar VW-758/1 or equivalent, with slight preload. This is to hold the transaxle in place while the engine is out.

15. Install sling US-1105, or equivalent, on the engine lifting eyes located on the left side of the cylinder head.

16. Lift the engine until its weight is taken off the engine mounts.

17. Adjust the support bar to contact the transaxle.

18. Separate the engine and transaxle.

19. Carefully lift the engine out of the engine compartment so as not to damage the transaxle main shaft, clutch and body.

To install:

20. Lubricate the clutch release bearing and transaxle main shaft splines with MOS_2 grease or equivalent; do not lubricate the guide sleeve or the clutch release bearing.

21. Carefully guide the engine into the vehicle and attach to the transaxle while keeping weight off the motor mounts.

22. Remove the transaxle support bar and lower the engine onto the engine mounts.

23. The remainder of the installation is the reverse of the removal procedure. Torque the engine mounts and subframe bolts with the engine running at idle speed. This will minimize vibration.

24. Torque the following:
* Cold start valve, the radiator mount bolts and the engine-to-transaxle cover plate bolts — 7 ft. lbs. (10 Nm).
* Engine-to-transaxle bolts — 42 ft. lbs. (55 Nm).
* Engine mount bolts — 30 ft. lbs. (40 Nm).
* Engine stop-to-body block and exhaust pipe support bolts — 18 ft. lbs. (25 Nm).
* Exhaust pipe-to-manifold bolts — 22 ft. lbs. (30 Nm).
* Starter bolts — 18 ft. lbs. (25 Nm).

Corrado SLC

The engine and transaxle are removed as a unit and the job requires 2 people.

1. Remove the battery; disconnect the wiring and vacuum lines as re-

quired to remove the air cleaner housing air duct.

2. Remove the front grille, headlights and hood lock support.

3. Remove the front bumper:

a. Remove the front spoiler.

b. Remove the clips between the bumper skin and the engine mount.

c. Below the bracket on each side, remove the 2 engine mount bolts that hold the bracket.

d. Pull the bumper cover forward evenly from both sides. Two people are required.

4. Drain the coolant and remove the radiator.

5. On manual transaxle, remove the clutch slave cylinder and disconnect the shift cables and support bracket.

6. On automatic transaxle, remove the clip to disconnect the shift lever cable.

7. Thread a long 8 x 10mm bolt into the accessory drive belt tensioner to release the tension. Move the tensioner only as required to remove the belt from the steering pump and air conditioner compressor.

8. Without disconnecting any refrigerant lines, remove the air conditioner compressor from the engine and secure it to the body. Be careful not to kink the hoses.

9. Without disconnecting any hydraulic lines, remove the power steering pump and secure it to the body. Be careful not to kink the hoses.

10. Remove the various covers from the top of the engine.

11. Remove the distributor cap, ignition wires and guides as an assembly.

12. Disconnect the accelerator cable.

13. Disconnect the 42-pin main engine wiring connector.

14. Label and disconnect the alternator, starter and transaxle wiring.

15. Disconnect the heater hoses.

16. Disconnect the main vacuum lines from the throttle body and intake manifold.

17. Disconnect the fuel supply and return lines.

18. Disconnect the oxygen sensor wire. Disconnect the exhaust pipe from the manifold.

19. Remove the bolts to disconnect the axle shafts from the drive flanges. Support the axle shafts with wire, do not let them hang by the outer CV-joint.

20. Attach a chain hoist to the lifting points on the engine.

21. Disconnect the engine and transaxle mounts and carefully lift the engine and transaxle out of the vehicle.

To install:

22. Carefully fit the engine/transaxle assembly into place.

a. Be sure the tabs on the rubber motor mounts fit properly into the mount brackets on the engine.

b. Start all the mounting bolts but do not tighten them yet.

c. With the vehicle resting on its wheels, shake the engine/transaxle assembly to settle it in the mounts.

d. Torque the bolts that fit into the upper center hole on each mount to 44 ft. lbs. (60 Nm). These procedures are important to minimize vibration.

23. Connect the axle shafts and torque the bolts to 33 ft. lbs. (45 Nm).

24. Use new gaskets and self-locking nuts to connect the exhaust pipe. Torque to 30 ft. lbs. (40 Nm) and connect the oxygen sensor wires.

25. Install the air conditioner compressor and power steering pump; install the drive belt.

26. Connect all wiring, hoses and control cables.

27. Install the clutch slave cylinder, if equipped, and adjust the shift linkage as required.

28. Install the radiator, connect the hoses and refill the cooling system.

29. Install the bumper and torque the bracket bolts to 63 ft. lbs. (85 Nm). Install the hood lock support, headlights and grille.

30. Install the battery and air cleaner assembly.

31. After testing the engine and transmission, check the headlight adjustment.

Engine Mounts

REMOVAL AND INSTALLATION

Earlier vehicles have all rubber, not hydraulic mounts. These mounts are replaceable with the same type but they must be pressed in and out.

1. With the engine properly supported from above, remove the mount carrier.

2. On non-hydraulic mounts, note the position of the mount in the carrier before pressing the old mount out. The large air gap is always at the top.

3. Reinstall the carrier and mount and center the mount in the bracket on the frame while tightening the bolts.

ENGINE ALIGNMENT

If there is excessive engine vibration, before removing mounts, an engine alignment procedure may cure the problem. Loosen all the bolts that go into the rubber mounts themselves. With the vehicle safely supported, shake the engine/transaxle as a unit to settle it in the mounts. Retorque all mounting bolts, starting at the rear and working forward. If engine vibration is not reduced, check for torn rubber mounts. The mount at the timing belt end usually fails first.

Cylinder Head

REMOVAL AND INSTALLATION

Except VR6 Engine

1. Disconnect the negative battery cable.

NOTE: On some of the 16V models, removing the battery may make the job easier.

2. Open the radiator cap and remove the thermostat housing to drain the cooling system.

3. Disconnect the throttle cable. Label and disconnect all wiring and vacuum lines from the intake manifold. On the 16V engine, remove the upper half of the intake manifold.

4. On vehicles with CIS-E fuel injection, remove the injectors and the cold start valve without disconnecting the fuel lines and cap them. Secure all the lines aside.

5. On vehicles with Digifant fuel injection, the injectors and fuel rail assembly may be left on the head. Disconnect the fuel supply and return lines and the wiring connector for the injectors.

6. Disconnect the radiator and heater hoses.

7. Disconnect and label wiring for oil pressure and temperature sensors.

8. On vehicles with CIS-E fuel injection, remove the auxiliary air regulator from the intake manifold, if equipped.

9. Remove the distributor cap and wires. On 16V engines, remove the distributor with the cap and wires as an assembly.

10. Disconnect the exhaust pipe from the exhaust manifold. If the pipe is secured to the manifold with spring clamps, insert the wedge tools to remove the spring clamps and separate the pipe from the manifold.

—————— **CAUTION** ——————

Special tools are required for removing and installing the clamps; VW3140/1 and /2 or equivalent. This is a set of different sized wedges for spreading the spring clamps in steps. The installed spring clamp has considerable tension and could cause damage or injury if not properly removed. Clamps with wedges installed are also under high tension and should be handled carefully.

11. Remove the EGR pipe from the exhaust manifold, if equipped.

12. Remove the accessory drive belts and any accessory that is bolted to the head. On Corrado, a special clamping tool 3191, is required to remove the spring-loaded belt tensioner.

13. Turn the engine to TDC of No. 1 cylinder, if possible, and remove the cylinder head cover, timing belt cover and belt.

14. Loosen the cylinder head bolts in the reverse of the tightening sequence.

15. Remove the bolts and lift the head straight off.

To install:

16. Before reinstalling the head, check the flatness of the head and block in both width and length, then diagonally from each corner.

17. Install the new cylinder head gasket with the word TOP or OBEN facing upward; do not use any sealing compound.

18. Carefully fit the head in place and install the bolts in positions 8 and 10 in the torque sequence. These holes are smaller and will properly locate the gasket and cylinder head.

19. Install the remaining bolts. Torque the bolts in sequence in 3 steps: 29 ft. lbs. (39 Nm), 44 ft. lbs. (60 Nm) and an additional ½ turn. Two ¼ turns are allowed.

Cylinder head bolt torque sequence — 4–cylinder engine

7913r044

20. Install the camshaft drive belt and adjust the tension.

21. Connect the exhaust pipe to the manifold. Use new gaskets and self-locking nuts and torque to 18 ft. lbs. (25 Nm). On vehicles that use spring clamps, install the clamps and carefully remove the wedge tools.

22. Connect the EGR pipe, if equipped.

23. On 16V engines, install the distributor. Install the distributor cap and wires.

24. On vehicles with CIS-E fuel injection, install the auxiliary air regulator to the intake manifold, if equipped.

25. Connect wiring to the oil pressure and temperature sensors.

26. Install the ignition system components.

27. Connect the radiator and heater hoses.

28. Connect the throttle cable and all wiring and vacuum lines.

29. On vehicles with Digifant fuel injection, connect the fuel supply and return lines. Connect the wiring connector for the injectors.

30. Install the thermostat with a new O-ring. Torque the housing bolts to 7 ft. lbs. (10 Nm). Refill the cooling system.

31. On vehicles with CIS-E fuel injection, install the injectors and the cold start valve.

32. Install the accessory drive belts and adjust the tension.

33. On the 16V engine, install the upper half of the intake manifold. Torque the manifold retaining bolts to 18 ft. lbs. (25 Nm).

34. Connect all wiring and vacuum lines disconnected from the intake manifold. Connect the throttle cable.

35. Install the battery, if removed. Connect the negative and positive battery cables.

36. Refill and bleed the cooling system.

37. When everything has been properly installed and connected, be sure to change the oil and filter before starting the engine.

VR6 Engine

This procedure requires special tool 3268 or equivalent. This is a setting tool that holds the camshafts in the correct position for installing the timing chains. Before removing the cylinder head, make sure new bolts are available. The cylinder head bolts are made to stretch and cannot be used again.

1. Disconnect the battery cables and remove the battery.

2. Disconnect the wiring and vacuum lines as required to remove the air cleaner, air mass sensor and duct.

3. Open the radiator cap and remove the drain plug from the coolant pipe below the intake manifold to drain the cooling system.

4. Remove the engine trim cover. Remove the distributor cap, ignition wires and wire guide as an assembly.

5. Disconnect the throttle cable. Label and disconnect the wiring and vacuum lines from the intake manifold and remove the upper manifold.

6. The injectors and fuel rail assembly may be left on the manifold. Disconnect the fuel supply and return lines and the wiring connector for the injectors.

7. Disconnect the radiator and heater hoses.

8. Thread a long 8 x 10mm bolt into the accessory drive belt tensioner to release the tension. Move the tensioner only as required to remove the belt.

9. Remove the alternator and the belt tensioner.

10. Remove the heatshield and the bolts to disconnect the 2 piece exhaust manifold from the engine. Note the position of the gaskets.

11. Remove the distributor and the timing chain tensioner bolt from the timing chain cover.

12. Remove the cylinder head cover, upper timing chain cover and the retaining plate.

13. If possible, rotate the crankshaft to TDC of No. 1 piston. Clean the oil off the chain and sprockets and mark the direction of rotation for assembly.

14. Hold the camshafts at the flats with a 24mm wrench and remove the bolts to remove the sprockets and chain. Note the position of the distributor drive on the short camshaft.

NOTE: Do not use the setting tool to hold the camshafts when removing or installing the sprocket bolts. The camshafts and the tool will be damaged.

15. Carefully check to make sure all necessary wires, hoses and brackets and components have been removed.

16. Loosen the cylinder head bolts in the reverse of the torque sequence. Remove and discard the bolts.

17. Remove the cylinder head.

To install:

18. Carefully clean the old gasket material from the head and the block. Before reinstalling the head, check the flatness of the head and block in both width and length, then diagonally from each corner. Maximum al-

lowable distortion is 0.004 in. (0.1mm).

19. If the new head gasket already has sealant in the small holes at the timing chain end, remove the sealant. Apply a silicone sealer to the timing chain end and install the gasket onto the block with the word TOP or OBEN facing up.

20. Fit the cylinder head over the locating dowels and set the head onto the engine. Install new bolts and hand tighten them. Do not attempt to re-use the old bolts.

21. Torque the bolts in 3 steps:
- Step 1 — 29 ft. lbs. (40 Nm)
- Step 2 — 43 ft. lbs. (60 Nm)
- Step 3 — an additional ½ turn. Two ¼ turns are allowed

22. Make sure the crankshaft is at TDC on No. 1 piston. Install the setting tool to lock the camshafts in place, then install the timing chain and sprockets. Make sure they are positioned to rotate in the original direction.

23. Hold the camshaft with a 24mm wrench and install the sprocket bolt. Make sure the distributor drive is correctly positioned and torque the bolts to 74 ft. lbs. (100 Nm).

24. Install the tensioner shoe and temporarily install the upper timing chain cover. Install the tensioner bolt and remove the setting tool. Rotate the crankshaft 4 full turns and stop at TDC of No. 1 piston. The setting tool should fit into the camshafts.

25. Remove the tensioner bolt and upper timing chain cover again. Apply new sealant as required, install the cover and torque the bolts to 82 inch lbs. (10 Nm). Install the tensioner bolt and torque to 15 ft. lbs. (20 Nm).

26. Install the cylinder head cover.

27. Use new gaskets and install the intake and exhaust manifolds. Torque the nuts and bolts to 18 ft. lbs. (25 Nm).

28. Install the alternator belt and adjust tension.

29. Install the accessory drive belt and adjust tension.

30. Connect the radiator and heater hoses.

31. Install the injectors, fuel rail assembly and manifold. Connect the fuel supply and return lines. Connect the wiring connector for the injectors.

32. Install the upper manifold. Torque the manifold bolts to 18 ft. lbs. (25 Nm).

33. Connect the wiring and vacuum lines disconnected from the intake manifold. Connect the throttle cable.

34. Install the distributor cap, ignition wires and wire guide as an assembly. Install the engine trim cover.

35. Disconnect the battery cables and remove the battery.

36. Refill and bleed the cooling system.

37. When everything has been properly installed and connected, be sure to change the oil and filter before starting the engine.

Valve Lifters

REMOVAL AND INSTALLATION

1. Remove the camshaft(s).
2. The valve lifters can be easily lifted out of the head by hand. Place hydraulic lifters camshaft side down on a clean surface. Keep all lifters in order so they can be installed in the same position.
To install:
3. Make sure the engine is not at TDC of any cylinder.
4. Set the lifters in place and install the camshaft. Before running the engine, allow the lifters to bleed down for 30 minutes or the valves may hit the pistons.

Intake Manifold

REMOVAL AND INSTALLATION

1. Disconnect the negative battery cable. Remove the air duct from the throttle valve body and disconnect the accelerator cable.
2. On Corrado SLC, remove the engine trim cover. Remove the distributor cap, wires and wire guide as an assembly.
3. On Digifant fuel injection systems, remove the idle stabilizer valve, fuel pump pressure switch and the fuel injector wiring harness. Disconnect the fuel supply and return lines.
4. On CIS-E fuel injection systems, remove the auxiliary air regulator. Remove the fuel injectors and the cold start valve from the cylinder head without disconnecting the fuel lines.
5. Label and disconnect the vacuum hoses as required.
6. Label and disconnect any remaining wiring as required.
7. If equipped, disconnect the EGR pipe.
8. On 16V and VR6 engines, remove the bolts to remove the upper intake manifold.

9. Remove the bolts and remove the manifold from the cylinder head.
To install:
10. On 16V and VR6 engines, install the lower intake manifold to the cylinder head with a new gasket. Torque the bolts to 18 ft. lbs. (25 Nm).
11. On the VR6 engine, if the fuel injectors were removed, examine the injector O-rings and replace as required. Install the injectors and rail.
12. Use new gaskets and fit the manifold or the upper manifold in place. Torque the bolts to 18 ft. lbs. (25 Nm).
13. Connect fuel system hoses or install the injectors now to protect the system.
14. Connect all vacuum hoses and wiring.
15. If equipped, connect the EGR pipe.
16. Connect and adjust the throttle cable as required.
17. Install the remaining components and run the engine to check idle speed and ignition timing.

Exhaust Manifold

REMOVAL AND INSTALLATION

————— **CAUTION** —————
On some models, special tools are required for removing and installing the exhaust pipe-to-manifold spring clamps. Special tools VW3140/1 and /2, or equivalent, are a set of different sized wedges for spreading the spring clamps in steps. The installed spring clamp has considerable tension and could cause damage or injury if not properly removed. Clamps with wedges installed are under high spring pressure and should be handled carefully.

1. Disconnect the oxygen sensor wiring and remove any heatshields that may be in the way.
2. Remove the emissions sample tap and, if equipped, disconnect the EGR pipe from the exhaust manifold.
3. On models with manifold studs, remove the self-locking nuts and lower the exhaust pipe.
4. Expand the spring clamp by pushing the exhaust pipe to one side and insert the starter wedge into the clamp all the way up to the shoulder.
5. Push the pipe to the other side and install another wedge in the opposite clamp. Continue to work the pipe side to side while pushing the wedges into the clamps until the clamps are spread far enough to lift off easily.

---

6. Remove the self-locking nuts and remove the manifold. On the VR6 engine, the exhaust manifold is 2 sections. Note the position of the gaskets.

To install:

7. Installation is the reverse of removal. Use new gaskets and self-locking nuts and torque to 18 ft. lbs. (25 Nm).

8. If the exhaust pipe is bolted to the manifold, install a new gasket and use new self-locking nuts. Torque the nuts to 30 ft. lbs. (40 Nm).

9. If equipped with spring clamps, hold the pipe in position with a new gasket and install the clamps. Carefully remove the wedge tools.

Supercharger

REMOVAL AND INSTALLATION

Corrado

The early Corrado has a belt driven supercharger with an intercooler for supplying up to 11.6 psi (0.8 BAR) of boost. The belt is the same serpentine ribbed belt used to drive the other engine accessories. Belt tension is maintained with a spring-loaded automatic belt tensioner. Releasing the tension to remove the belt requires a special clamping tool, VW3191 or equivalent. The supercharger cannot be repaired; leaking or otherwise faulty units must be replaced.

1. Install clamping tool and compress the belt tensioner. Remove the belt from the supercharger pulley.

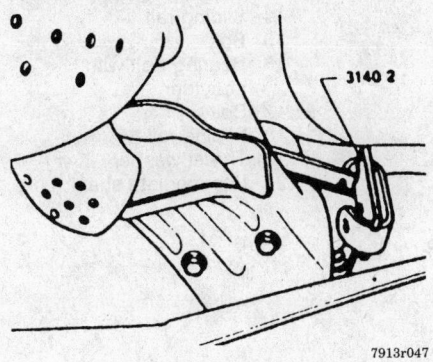

Exhaust pipe clamp removal tools

2. Remove the connector hose and silencer from the outlet side of the supercharger and remove the 2 upper inlet hoses.

3. Remove the front and rear mounting bolts and carefully lift the supercharger onto the top of the engine.

4. Allow the oil to drain back into the engine for a few minutes, then remove the oil lines and take the supercharger out of the vehicle.

5. Installation is the reverse of removal. Start the fittings for the oil lines but don't tighten them until the unit is bolted in place. Be sure to use new sealing rings. Torque the following:
- Supercharger mounting bolts — 18 ft. lbs. (25 Nm).
- Mounting bracket-to-engine bolt — 33 ft. lbs. (45 Nm)
- Oil line fittings — 11 ft. lbs. (15 Nm)

Timing Chain Cover

REMOVAL AND INSTALLATION

Corrado SLC

Only the upper timing chain cover can be removed with the engine in the vehicle. The flywheel must be removed to remove the lower cover. This cover also holds the rear main oil seal.

1. Remove the engine trim cover. Remove the distributor cap and wires with the wire guide as an assembly.

2. Remove the upper intake manifold and the cylinder head cover.

3. Remove the distributor and the chain tensioner bolt.

4. Remove the bolts to remove the upper timing chain cover.

5. Installation is the reverse of removal. Use new gaskets and torque the bolts to 82 inch lbs. (10 Nm).

Timing Chain

REMOVAL AND INSTALLATION

VR6 Engine

1. Remove the distributor cap, wires and wire guide as an assembly.

2. Remove the upper intake manifold.

3. Remove the cylinder head cover.

4. Remove the timing chain tensioner bolt and the upper timing chain cover.

5. Remove the transaxle and flywheel.

6. Rotate the crankshaft to TDC of No. 1 piston.

7. Mark the direction of travel on the upper camshaft drive chain before removing. Remove the tensioner shoe and the double row chain.

8. Remove the the lower chain tensioner and remove the single row chain.

To install:

9. Check the position of the crankshaft in reference to the intermediate shaft. The ground tooth of drive gear **A** must align with the bearing split; reposition it if necessary.

10. Install the single row chain in position marked during disassembly. Install the chain tensioner and retaining bolts. Torque the bolts to 7 ft. lbs. (10 Nm).

11. Install the double row chain in position marked during disassembly. Install the chain tensioner. Torque the tensior to 15 ft. lbs. (20 Nm).

NOTE: The marking on the intermediate shaft must align with notch B or C on the thrust washer.

12. Install the upper timing chain cover. Rotate the crankshaft 4 full turns and stop at TDC of No. 1 piston and check mark again.

13. Install the flywheel and transaxle.

14. If timing marks alignment checks good, install the cylinder head cover, upper intake manifold and ignition system components.

15. Connect the negative battery cable. Start the engine a check ignition timing.

Timing Belt Front Cover

REMOVAL AND INSTALLATION

Except Corrado SLC

1. Disconnect the negative battery cable.

2. Remove the accessory drive belts. On Corrado with a supercharger, install tool 3191 or equivalent, to compress the spring-loaded belt tensioner.

3. To remove the crankshaft accessory drive pulley, hold the center crankshaft sprocket bolt with a socket and loosen the pulley bolts.

4. The cover is now accessible. It comes off in 2 pieces; remove the upper ½ first. Take note of any special spacers or other hardware.

5. Installation is the reverse of removal.

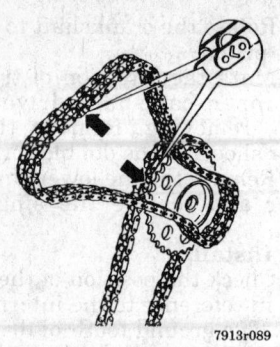

Mark the direction of the drive chains before removing — VR6 engine

7913r089

FRONT OIL SEAL REPLACEMENT

1. Remove the timing belt cover and the timing belt.
2. Remove the crankshaft sprocket.
3. Using a small prybar, pry the seal from the carrier or use the seal extractor tool VW–10–219 or equivalent, to pull out the seal.

NOTE: When removing the seal, be careful not to damage the carrier.

To install:
4. Lubricate the new seal lips and use the seal installation tool VW–10–203 or equivalent, to press the new seal into the carrier.
5. Install the crankshaft sprocket and torque the bolt to 133 ft. lbs. (180 Nm) for 12mm hex head bolts or 66 ft. lbs. (90 Nm) plus ½ turn for all 12 sided bolts.
6. Install the timing belt and check the ignition timing.

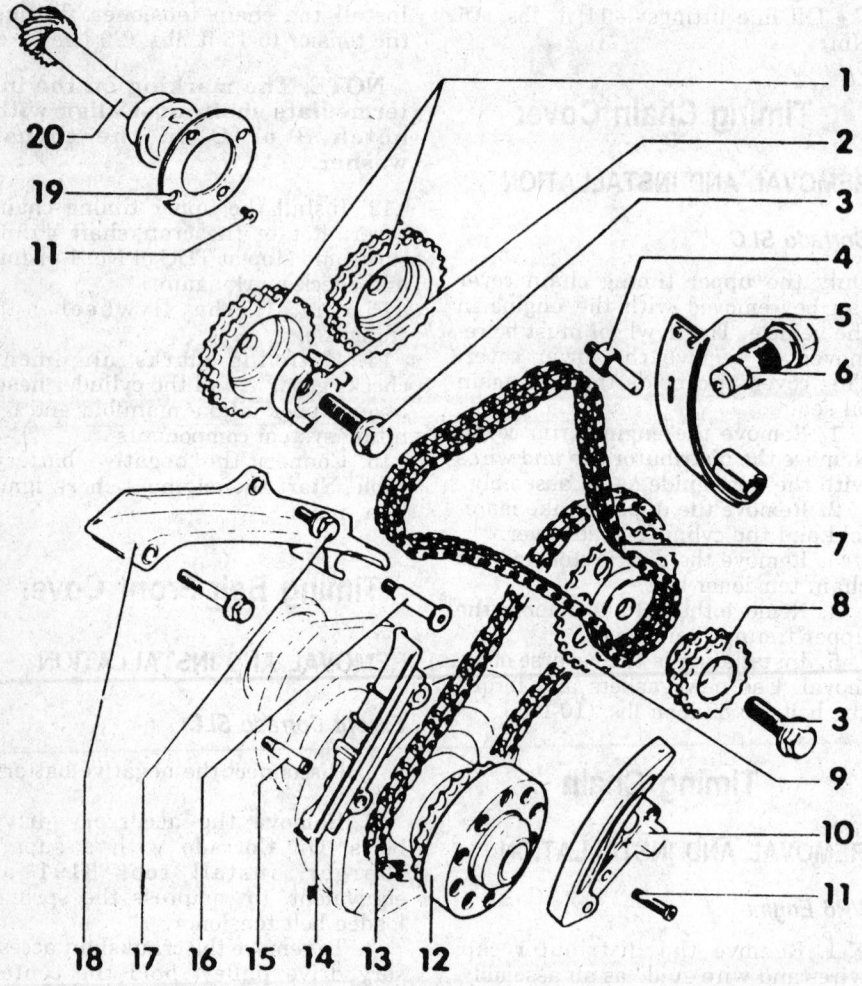

Removal and installation of the timing chains — VR6 engine

1 Camshaft chain sprocket
2 Distributor clutch
3 Bolt
4 Bearing bolt
5 Tension bar
6 Chain tensioner
7 Double row chain
8 Double row chain gear
9 Single row chain gear
10 Chain tensioner with bar
11 Bolt
12 Drive gear
13 Single row chain
14 Sliding rail
15 Bolt
16 Bearing bolt with shoulder
17 Bolt
18 Sliding rail
19 Thrust washer
20 Intermediate shaft

7913r090

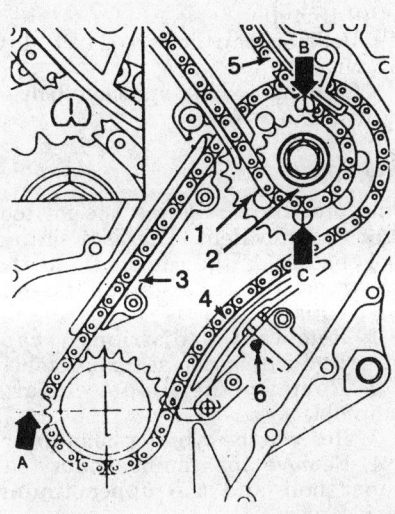

1 Single row chain gear
2 Double row chain gear
3 Sliding rail
4 Single row chain
5 Double row chain
6 Chain tensioner release screw

7913r091

Timing chain chain marks — VR6 engine

Timing Belt and Tensioner

REMOVAL AND INSTALLATION

NOTE: Do not turn the engine or camshaft with the camshaft drive belt removed. The pistons will contact the valves and cause internal engine damage.

1. Disconnect the negative battery cable and remove the accessory drive belts, crankshaft pulley and the timing belt cover(s).

2. Temporarily reinstall the crankshaft pulley bolt, if removed and turn the crankshaft to TDC of No. 1 piston. The mark on the camshaft sprocket should be aligned with the mark on the inner drive belt cover, if equipped, or the edge of the cylinder head.

3. On 8-valve engines, the notch on the crankshaft pulley should align with the dot on the intermediate shaft sprocket. With the distributor cap removed, the rotor should be pointing toward the No. 1 mark on the rim of the distributor housing.

4. Loosen the locknut on the tensioner pulley and turn the tensioner counterclockwise to relieve the tension on the timing belt.

5. Slide the timing belt from the sprockets.

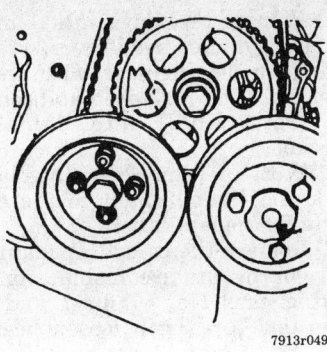

7913r049

Timing marks on the crankshaft pulley and the intermediate shaft sprocket

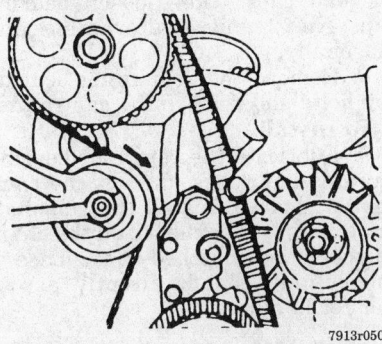

7913r050

Timing marks on camshaft sprockets: A — 16-valve engine, B — 8-valve engine

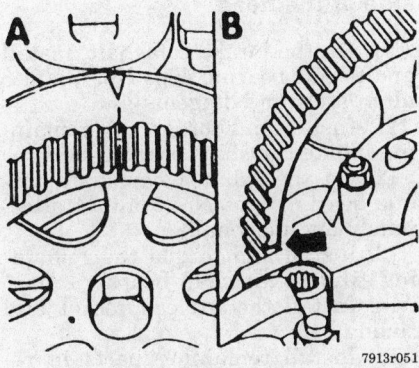

7913r051

Adjusting timing belt tension

To install:

6. Install the new timing belt and tension the belt so it can be twisted 90 degrees at the middle of its longest section, between the camshaft and intermediate sprockets.

7. Recheck the alignment of the timing marks and, if correct, turn the engine 2 full revolutions to return to TDC of No. 1 piston. Recheck belt tension and timing marks. Readjust as required. Torque the tensioner nut to 33 ft. lbs. (45 Nm).

8. Reinstall the belt cover and accessory drive belts.

9. When running the engine, there will be a growling noise that rises and falls with engine speed if the belt is too tight.

Timing Sprockets

REMOVAL AND INSTALLATION

Except VR6 Engine

1. Remove the timing belt covers and the timing belt. The crankshaft sprocket should slide off easily when the center bolt is removed. Don't lose the Woodruff key.

2. Remove the cylinder head cover.

3. Use a wrench to hold the camshaft on the flat section and remove the sprocket retaining bolt.

4. Gently pry or tap the sprocket off the shaft with a soft mallet. If the sprocket will not easily slide off the shaft, use a gear puller. Do not hammer on the sprocket or damage to the sprocket or bearings could occur.

5. When reinstalling the sprockets, torque the camshaft sprocket bolts to 58 ft. lbs. (80 Nm) and the crankshaft sprocket bolt to 66 ft. lbs. (90 Nm) plus ½ turn.

VR6 Engine

1. Disconnect the negative battery cable. Remove the distributor cap, wires and wire guide as an assembly.

2. Remove the upper intake manifold.

3. Remove the cylinder head cover.

4. Remove the timing chain tensioner bolt and the upper timing chain cover.

5. Rotate the crankshaft to TDC of No. 1 piston.

6. Mark the direction of travel on the upper camshaft drive chain. Remove the tensioner shoe and the chain.

7. Hold the camshafts at the flats with a 24mm wrench and remove the bolts to remove the sprockets. Note the position of the distributor drive on the short camshaft.

To install:

8. Hold the camshaft with a 24mm wrench and install the sprockets. Make sure the distributor drive is correctly positioned. Install the retaining bolts. Torque the bolts to 74 ft. lbs. (100 Nm).

9. Install the camshaft drive chain so the marks on the chain sprockets are matched at the base of the cylinder head, directly across from each other.

10. Install the timing chain tensioner. Install the tensioner bolt and torque to 15 ft. lbs. (20 Nm).

11. Install the upper timing chain cover and torque the bolts to 82 inch lbs. (10 Nm).

12. Install the cylinder head cover, upper intake manifold and ignition system components.

Camshaft

REMOVAL AND INSTALLATION

8-Valve Engine

1. Disconnect the negative battery cable. Remove the timing belt cover(s), the timing belt, camshaft sprocket and cylinder head cover.

2. Number the bearing caps from front to back. If the cap does not already have one, scribe an arrow pointing towards the front of the engine. The caps are offset and must be installed correctly. Factory numbers on the caps are not always on the same side.

3. Remove the front and rear bearing caps. Loosen the remaining bearing cap nuts diagonally, in several steps, starting from the outside caps near the ends of the head and working toward the center.

4. Remove the bearing caps and the camshaft.

To install:

5. Install a new oil seal and end plug in the cylinder head. Lubricate the camshaft bearing journals and lobes and set the camshaft in place.

6. Install the bearing caps in the correct position with the arrow pointing towards the front of the engine. Tighten the cap nuts diagonally and in several steps until they are torqued to 15 ft. lbs. (20 Nm). Do not over-torque.

7. Install the drive sprocket and torque the bolt to 58 ft. lbs. (80 Nm).

8. Align the timing marks, install the timing belt and adjust the tension.

9. On engines with hydraulic lifters, wait at least ½ hour after installing the camshaft before starting the engine to allow the lifters to leak down. Observe the following torques:
- Camshaft shaft end-play — 0.006 in. (0.15mm)
- Bearing cap bolts — 15 ft. lbs. (20 Nm)
- Camshaft sprocket bolt — 58 ft. lbs. (80 Nm)

16-Valve Engine

1. Remove the timing belt cover.

2. Remove the upper intake manifold and cylinder head cover.

3. Turn the engine to TDC on cylinder No. 1, then slacken and remove the timing belt and camshaft sprocket.

4. With a felt marker only, match-mark the timing chain to the camshafts for reinstallation.

5. Remove the camshaft chain.

6. On the intake camshaft, remove bearing caps No. 5 and 7 and the chain end cap. Then loosen bearing caps No. 6 and 8 alternately and diagonally.

7. On the exhaust camshaft, remove bearing caps No. 1 and 3 and the end caps. Then loosen bearing caps No. 2 and 4 alternately and diagonally.

8. Remove the remaining bearing cap bolts and remove the camshafts.

To install:

9. Lubricate the camshaft bearing journals and lobes and set the camshafts in place. Install the camshaft drive chain so the marks on the chain sprockets are matched at the base of the cylinder head, directly across from each other.

NOTE: When installing the bearing caps, make sure the notch points towards the intake side of the head.

10. On the intake camshaft, install and torque bearing caps No. 6 and 8 alternately and diagonally.

11. Install and torque the remaining intake camshaft bearing caps.

12. On the exhaust camshaft, torque bearing caps No. 2 and 4 alternately and diagonally.

13. Install and torque the remaining exhaust camshaft bearing caps.

14. Install the drive sprocket and timing belt.

15. Install remaining parts in reverse order of removal. Wait at least ½ hour after installing camshaft shafts before starting the engine to

7913r054

16-valve engine camshaft shaft alignment

allow the hydraulic lifters to leak down.
- Camshaft shaft end-play — 0.006 in. (0.15mm)
- Bearing cap bolts — 11 ft. lbs. (15 Nm)
- Camshaft shaft sprocket bolt — 48 ft. lbs. (65 Nm)

VR6 Engine

This procedure requires special tool 3268 or equivalent. This is a setting tool that holds the camshafts in the correct position for installing the timing chains.

1. Remove the distributor cap, wires and wire guide as an assembly.

2. Remove the upper intake manifold.

3. Remove the cylinder head cover.

4. Remove the timing chain tensioner bolt and the upper timing chain cover.

5. Rotate the crankshaft to TDC of No. 1 piston.

6. Mark the direction of travel on the upper camshaft drive chain. Remove the tensioner shoe and the chain.

7. Hold the camshafts at the flats with a 24mm wrench and remove the bolts to remove the sprockets. Note the position of the distributor drive on the short camshaft.

NOTE: Do not use the setting tool to hold the camshafts when removing or installing the sprocket bolts. The camshafts or the tool will be damaged.

8. On the long camshaft, remove the end bearing caps. Loosen the center cap nuts in a diagonal pattern 2 turns at a time until the valve springs are relieved. Remove the camshaft.

9. On the short camshaft, remove the center bearing cap and loosen the nuts on the end caps in a diagonal pattern 2 turns at a time. When the valve springs are relieved, remove the camshaft.

To install:

10. Lubricate the long camshaft and the cylinder head bearing surfaces and set the camshaft in place. Install bearing caps 3 and 5 and tighten the bolts 2 turns at a time in a diagonal pattern to draw the camshaft down against the valve springs.

11. Install the other bearing caps and torque all the nuts to 15 ft. lbs. (20 Nm).

12. Repeat the process with the short camshaft, using bearing caps 2 and 6 to draw the camshaft down against the springs.

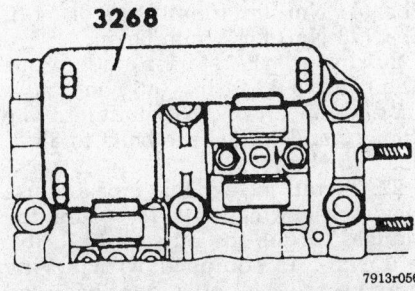

Hold the camshafts on the wrench flats to remove and install the sprocket bolts

3268

Camshaft setting tool holds the camshafts in place when installing the chain: do not use this tool to loosen or tighten the sprocket bolts

13. Hold the camshaft with a 24mm wrench and install the sprockets. Make sure the distributor drive is correctly positioned and torque the bolts to 74 ft. lbs. (100 Nm).

14. Make sure the crankshaft is at TDC on No. 1 piston. Install the setting tool and install the timing chain.

15. Install the tensioner shoe and temporarily install the upper timing chain cover. Install the tensioner bolt and remove the setting tool. Rotate the crankshaft 4 full turns and stop at TDC of No. 1 piston. The setting tool should fit into the camshafts.

16. Remove the tensioner bolt and upper timing chain cover again. Clean the old sealant off the cylinder head gasket and apply new sealant.

17. Install the upper timing chain cover and torque the bolts to 82 inch lbs. (10 Nm). Install the tensioner bolt and torque to 15 ft. lbs. (20 Nm).

18. Install the cylinder head cover, upper intake manifold and ignition system components.

Intermediate Shaft

REMOVAL AND INSTALLATION

Except VR6 Engine

1. Disconnect the negative battery cable. Remove the front cover upper and lower halves.

2. Remove the intermediate shaft sprocket.

3. Loosen and remove the 2 retaining screws and then remove the intermediate shaft mounting flange.

4. Carefully, slide the intermediate shaft out of the block. Turn the shaft, if necessary, to remove it. Inspect the gear on the intermediate shaft and replace it, if necessary.

To install:

5. Install the intermediate shaft to the block. Install new oil seal and O-ring in the mounting flange.

6. Install the mounting flange and retaining bolts. When installing the mounting flange be sure the oil return hole is at the bottom. Torque the retaining bolts to 18 ft. lbs. (25 Nm).

7. Install the front cover. Connect the negative battery cable.

VR6 Engine

1. Disconnect the negative battery cable. Remove the distributor cap, wires and wire guide as an assembly.

2. Remove the upper intake manifold.

3. Remove the cylinder head cover.

4. Remove the timing chain tensioner bolt and the upper timing chain cover.

5. Rotate the crankshaft to TDC of No. 1 piston.

6. Mark the direction of travel on the upper camshaft drive chain. Remove the tensioner shoe and the chain.

7. Hold the camshafts at the flats with a 24mm wrench and remove the bolts to remove the sprockets. Note the position of the distributor drive on the short camshaft.

8. With the intermediate shaft gear removed, remove the thrust washer retaining bolts. Pull the shaft out from the engine.

To install:

9. Install the intermediate shaft in the engine until the oil pump gear is completely engaged with the intermediate shaft gear.

10. Install the thrust washer and the retaining bolts. Torque the bolts to 7 ft. lbs. (10 Nm).

11. Hold the camshaft with a 24mm wrench and install the sprockets. Make sure the distributor drive is correctly positioned. Install the retaining bolts. Torque the bolts to 74 ft. lbs. (100 Nm).

12. Install the camshaft drive chain so the marks on the chain sprockets are matched at the base of the cylinder head, directly across from each other.

13. Install the timing chain tensioner. Install the tensioner bolt and torque to 15 ft. lbs. (20 Nm).

14. Install the upper timing chain cover and torque the bolts to 82 inch lbs. (10 Nm).

15. Install the cylinder head cover, upper intake manifold and ignition system components. Connect the negative battery cable.

Piston and Connecting Rod

POSITIONING

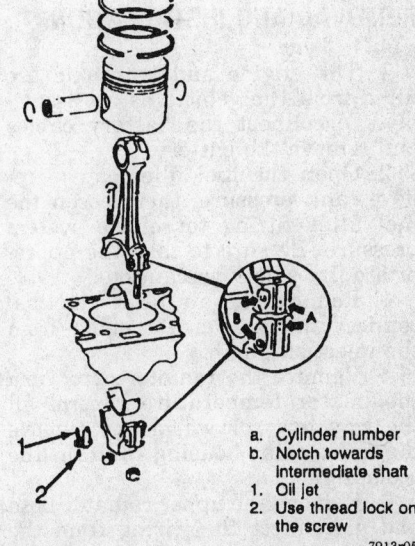

a. Cylinder number
b. Notch towards intermediate shaft
1. Oil jet
2. Use thread lock on the screw

7913r057

On all 4-cylinder engines, arrow on piston points towards the camshaft drive belt

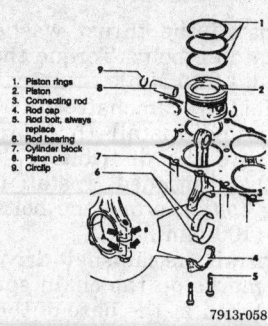

1. Piston rings
2. Piston
3. Connecting rod
4. Rod cap
5. Rod bolt, always replace
6. Rod bearing
7. Cylinder block
8. Piston pin
9. Circlip

7913r058

On VR6 engine, highest side of piston crown is toward center of engine block, connecting rod number is opposite the crown

DIESEL ENGINE MECHANICAL

NOTE: Disconnecting the negative battery cable on some vehicles may interfere with the functions of the on-board computer or security systems and may require reprogramming when the battery cable is reconnected.

Engine Assembly

REMOVAL AND INSTALLATION

1. The engine and transaxle are lifted from the vehicle as an assembly. Disconnect the battery cables and remove the battery.
2. Open the fuel filler cap to relieve tank pressure, then loosen the fuel filter fitting to relieve system pressure. Be sure to take the appropriate fire safety precautions.
3. Remove the air filter and disconnect the accelerator cable from the injection pump.
4. Remove the radiator cap. Turn the heater temperature control all the way towards warm and remove the thermostat housing to drain the coolant.
5. Remove the upper radiator hose and disconnect the wiring from the radiator fan motor and switches. Remove the mounting nuts or bolts and lift out the radiator and fan shroud as an assembly.
6. Begin disconnecting electrical connections and vacuum lines, carefully labeling each one. Don't forget ground connections that are screwed to the body.

7. If equipped with power steering, remove pump and secure it to the body. Do not disconnect the hydraulic lines. If equipped with air conditioning, remove the compressor and secure it aside without disconnecting the lines.
8. Disconnect the fuel inlet and outlet lines from the injection pump and plug the holes to keep the pump clean. Note the outlet fitting has a special orifice.
9. On turbocharged engines, disconnect the exhaust pipe and the oil lines from the turbocharger and cap the oil line fittings on the turbocharger. Unbolt the turbocharger and lift it out of the engine.
10. If equipped with an automatic transaxle, place the selector lever in **P** and disconnect the selector cable at the transaxle.
11. On manual transaxle shift linkage, remove the 2 rods with the plastic socket ends and unbolt the remaining linkage from the case as required. Disconnect the clutch cable, lift it out of the case and set it aside.
12. Disconnect the wiring from the starter, the backup light switch and the ground cable from the transaxle. Remove the speedometer cable from the transaxle and plug the hole in the case.
13. Attach an engine sling tool VW-2024A or equivalent, to the engine and attach the sling to a suitable lifting device.
14. Remove the nuts or spring clamps holding the exhaust pipe to the manifold or turbocharger.

CAUTION

On some models, special tools are required for removing and installing the exhaust pipe-to-manifold spring clamps; VW3140/1 and /2 or equivalent. This is a set of different sized wedges for spreading the spring clamps in steps. The installed spring clamp has considerable tension and could cause damage or injury if not properly removed. Clamps with wedges installed are also under high tension and should be handled carefully.

---

15. Unbolt the halfshafts from the flanges and hang them from the body with wire.
16. Make sure everything is disconnected and unbolt the mounts. Remove the starter first and the front mount with it.
17. With all mounts unbolted, slightly lower the engine/transaxle assembly and tilt it towards the tran-

saxle side. Then carefully lift the assembly out of the vehicle.

To install:

18. Carefully install the engine/transaxle assembly and make sure all mounts are securely bolted to the engine/transaxle. Start all nuts and bolts that secure the mounts to the body but don't tighten them yet.
19. With all mounts installed and the engine safely in the vehicle, allow some slack in the lifting equipment. With the vehicle safely supported, shake the engine/transaxle as a unit to settle it in the mounts. Torque all mounting bolts, starting at the rear and working forward. Torque to 33 ft. lbs. (41 Nm) for 10mm bolts or 54 ft. lbs. (73 Nm) for 12mm bolts.
20. Install the starter and torque the bolts to 33 ft. lbs. (45 Nm).
21. Connect the halfshafts to the flanges and torque the bolts to 33 ft. lbs. (45 Nm).
22. Install the exhaust pipe and use new self-locking nuts to secure the flange. Torque the nuts to 30 ft. lbs. (40 Nm). If equipped with spring clamps, the clamps can be used again.
23. Connect the shift linkage and the clutch cable, if equipped. Make any necessary adjustments.
24. Install the fuel injector lines and torque to 18 ft. lbs. (25 Nm). Be careful not to over torque the line nuts. If a line is damaged or clogged, replace all lines as a set.
25. Connect the inlet and outlet lines to the injector pump. Note the special outlet fitting has the word "OUT" printed on the top. Use new gaskets.
26. Install the air conditioning compressor and/or power steering pump, if equipped. Install and adjust the drive belts.
27. Connect the wiring and vacuum hoses.
28. Install the radiator, fan and heater hoses. Use a new O-ring on the thermostat and torque the thermostat housing bolts to 7 ft. lbs. (10 Nm).
29. Fill and bleed the cooling system. Check the adjustment of the accelerator cable.

Engine Mounts

REMOVAL AND INSTALLATION

1. Disconnect the negative battery cable.
2. Using an engine support fixture tool, center it on the cowl and attach it to the engine. Raise the engine

slightly to take the weight off of the engine mounts.

3. From the front of the engine, remove the engine mount bolts and the mount.

4. Inspect the engine mount for deterioration and replace it, if necessary.

To install:

5. To install, support the engine using a engine support fixture tool.

6. Install the engine mounts and the retaining bolts to the engine.

7. Torque all mounting bolts, starting at the rear and working forward. Torque to 33 ft. lbs. (41 Nm) for 10mm bolts or 54 ft. lbs. (73 Nm) for 12mm bolts.

ENGINE ALIGNMENT

After reinstalling all the mounts and mounting bolts, loosen all the bolts that go into the rubber mounts themselves. With the vehicle safely supported, shake the engine/transaxle as a unit to settle it in the mounts. Retorque all mounting bolts, starting at the rear and working forward. This procedure helps to minimize vibration.

Cylinder Head

REMOVAL AND INSTALLATION

NOTE: The cylinder head bolts on all diesel vehicles are stretch bolts and must be replaced when removed.

1. Disconnect the battery ground cable.

2. Remove the thermostat and drain the cooling system.

3. Remove the fuel lines from the injectors and the pump as an assembly. Put the lines where they will stay clean; protect the injector and pump fittings with caps.

4. Disconnect the radiator and heater hoses.

5. Disconnect all vacuum and electrical connections and carefully label for installation.

6. On turbocharged vehicles, unbolt the exhaust pipe and oil lines from the turbocharger and remove the turbocharger.

7. On non-turbocharged vehicles, remove the air cleaner and disconnect the exhaust pipe from the manifold.

--- **CAUTION** ---

On some models, special tools are required for removing and installing the exhaust pipe-to-mani-

fold spring clamps; VW3140/1 and /2 or equivalent. This is a set of different sized wedges for spreading the spring clamps in steps. The installed spring clamp has considerable tension and could cause damage or injury if not properly removed. Clamps with wedges installed are also under high tension and should be handled carefully.

---

8. Remove the cylinder head cover and camshaft drive belt cover.

9. Turn the engine to TDC of No. 1 cylinder, if possible, and remove the camshaft drive belt.

10. Remove the head bolts in the reverse order of installation sequence and lift the head out of the vehicle. The torque sequence is the same as for gasoline engines.

To install:

11. On these engines, the pistons actually project above the deck of the block. If the crankshaft and pistons are not to be removed, examine the old head gasket to see how many notches are on the edge near the oil return hole, between No. 2 and 3 cylinders. Replace the gasket with the same thickness.

12. If the pistons were removed or if the old gasket in not available, the piston height (pop up) must be measured to select the proper head gasket. Use a dial indicator or caliper to obtain the measurement.

- Pop-up on engines with solid lifters:
  - 0.026–0.031 in. (0.67–0.80mm)—1 notch
  - 0.032–0.035 in. (0.81–0.90mm)—2 notches
  - 0.036–0.040 in. (0.91–1.02mm)—3 notches
- Pop-up on engines with hydraulic lifters:
  - 0.026–0.034 in. (0.66–0.86mm)—1 notch
  - 0.034–0.035 in. (0.87–0.90mm)—2 notches
  - 0.036–0.040 in. (0.91–1.02mm)—3 notches

13. Install the new cylinder head gasket with the word TOP or OBEN facing upward. Do not use any sealing compound.

14. Turn the crankshaft to TDC of No. 1 cylinder, then back about 1/4 turn to bring all pistons about even.

15. Carefully lower the head on and install new head bolts into No. 8 and 10 first. These holes are smaller and will properly locate the gasket and cylinder head.

16. Install the remaining bolts and torque in the proper sequence in 3

steps: 29 ft. lbs. (40 Nm), 44 ft. lbs. (60 Nm), then a full 1/2 turn more. Two quarter turns are allowed.

17. Installation of the remaining parts is the reverse of removal, be sure to change the oil and filter. Install the camshaft drive belt and set injection pump timing.

18. Install the fuel injector lines and torque to 18 ft. lbs. (25 Nm). Be careful not to over torque the line nuts. If a line is damaged or clogged, replace all lines as a set.

19. After the engine has be run about 1000 miles, the cylinder head bolts must be re-torqued. Remove the cylinder head cover and turn each head bolt, in sequence, an additional 1/4 turn in 1 movement. This can be done on a cold or warm engine.

Valve Lifters

REMOVAL AND INSTALLATION

1. Remove the camshaft.

2. The valve lifters can be easily lifted out of the head by hand. Place hydraulic lifters camshaft side down on a clean surface. Keep all lifters in order so they can be installed in the same position.

To install:

3. Make sure the engine is not at TDC of any cylinder.

4. Set the lifters in place and carefully install the camshaft. Allow the lifters to bleed down for 30 minutes before turning the engine or the valves may hit the pistons.

5. Install the camshaft drive belt and adjust the belt tension and valve lash.

Valve Lash

ADJUSTMENT

All vehicles have hydraulic valve lifters and require no adjustment. On these vehicles there will be a sticker under the hood indicating hydraulic lifters.

Intake Manifold

REMOVAL AND INSTALLATION

1. Disconnect the hose and wiring from the blow-off valve.

2. Disconnect the air inlet hose.

3. Remove the bolts to remove the intake manifold.

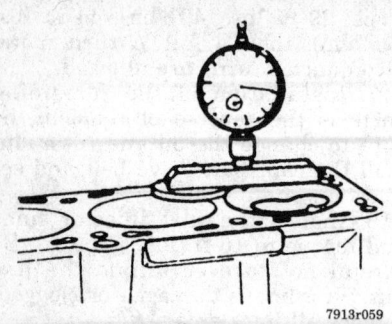

7913r059

Measure piston pop-up to determine required head gasket thickness — diesel engine

4. Installation is the reverse of removal. Use a new gasket and torque the bolts to 18 ft. lbs. (25 Nm).

Exhaust Manifold

REMOVAL AND INSTALLATION

――――― **CAUTION** ―――――

On some models, special tools are required for removing and installing the exhaust pipe-to-manifold spring clamps; VW3140/1 and /2 or equivalent. This is a set of different sized wedges for spreading the spring clamps in steps. The installed spring clamp has considerable tension and could cause damage or injury if not properly removed. Clamps with wedges installed are also under high tension and should be handled carefully.

1. Disconnect the negative battery cable and remove any heatshields that may be in the way.
2. On turbocharged engines, unbolt the exhaust pipe from the turbocharger outlet.
3. On non-turbocharged engines, expand the spring clamp by pushing the exhaust pipe to one side and insert the starter wedge into the clamp all the way up to the shoulder.
4. Push the pipe to the other side and install another wedge in the opposite clamp. Continue to work the pipe side to side while pushing the wedges into the clamps until the clamps are spread far enough to lift off easily.

――――― **CAUTION** ―――――

The removed spring clamps with wedges in them are under spring pressure and, if miss handled, could fly apart with enough force to cause serious injury. Store the removed clamps in a safe area where they won't be disturbed.

5. On turbocharged engines, remove the turbocharger oil lines and the turbocharger.
6. Remove the manifold locking nuts and lift the manifold off the head.
7. Installation is the reverse of removal. Use new gaskets and locking nuts and torque to 18 ft. lbs. (25 Nm).

Turbocharger

REMOVAL AND INSTALLATION

1. Disconnect the negative battery cable.
2. Remove the exhaust pipe from the turbocharger outlet.
3. Clean the oil supply fitting on the top of the turbocharger and remove the supply line and bracket.
4. Remove the inlet air hose.
5. Under the vehicle, remove the oil return line and the turbocharger mounting bracket.
6. Still underneath, remove the turbo-to-manifold bolts. Lift the turbocharger out from the top.
To install:
7. Installation is the reverse of removal. Before installing the oil supply line, fill the connection on the turbocharger with engine oil. Torque the following:
 • Turbocharger-to-exhaust manifold — 33 ft. lbs. (45 Nm)
 • Mounting bracket nuts — 18 ft. lbs. (25 Nm)
 • Turbocharger outlet nuts — 18 ft. lbs. (25 Nm)
 • Oil return line — 22 ft. lbs. (30 Nm)

Timing Belt Cover

REMOVAL AND INSTALLATION

1. Remove the accessory drive belts.
2. To remove the crankshaft accessory drive pulley, hold the center crankshaft sprocket bolt with a socket and loosen the pulley bolts.
3. The cover is now accessible. It comes off in 2 pieces, remove the upper half first. Take note of any special spacers or other hardware.
4. Installation is the reverse of removal. Torque the cover bolts to 87 inch lbs. (10 Nm).

Timing Belt and Tensioner

Timing belts are designed to last 60,000–75,000 miles. If the vehicle has been stored for long periods (2 years or more), the belt should be changed before returning the vehicle to service.

ADJUSTMENT

1. Disconnect the negative battery cable.
2. Remove the upper drive belt cover.
3. Strike the drive belt 1 time with a rubber hammer between the camshaft gear and injection pump gear.
4. Install and suitable belt tension gauge. Measure the belt tension between the camshaft and injection pump gear. Record the reading.
5. Turn the crankshaft 1 complete turn and measure the tension again. Compare the average of the 2 readings with specifications.
6. The belt tension specified value is 12–13.
7. If the belt tension is below specifications, turn the tensioner to the right. If belt tension is above specifications, turn the tensioner to the left.
8. Install the upper drive belt cover. Connect the negative battery cable.

REMOVAL AND INSTALLATION

Some special tools are required. A flat bar, VW tool 2065A or equivalent, is used to secure the camshaft in position. A pin, VW tool 2064 or equivalent, is used to fix the pump position while the timing belt is removed. The camshaft and pump work against spring pressure and will move out of position when the timing belt is removed. It is not difficult to find substitutes but do not remove the timing belt without these tools.

NOTE: Do not turn the engine or camshaft with the timing belt removed. The pistons will contact the valves and cause internal engine damage.

1. Disconnect the negative battery cable and remove the accessory drive belts, crankshaft pulley and the timing belt cover(s). Remove the camshaft cover and rubber plug at the back end of the camshaft.
2. Temporarily reinstall the crankshaft pulley bolt and turn the crankshaft to TDC of No. 1 piston. The mark on the camshaft sprocket should be aligned with the mark on the inner timing belt cover or the edge of the cylinder head.

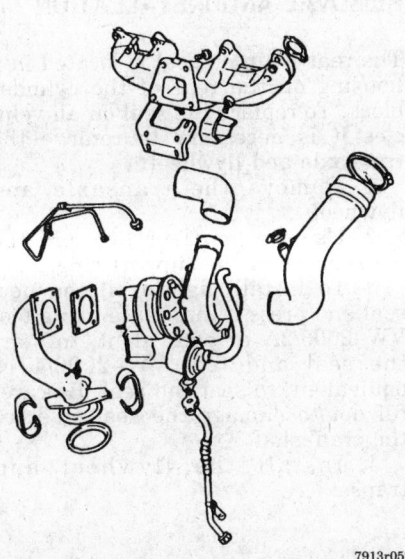

Diesel engine turbocharger and exhaust manifold

7913r052

3. With the engine at TDC, insert the bar into the slot at the back of the camshaft. The bar rests on the cylinder head to will hold the camshaft in position.

4. Insert the pin into the injection pump drive sprocket to hold the pump in position.

5. Loosen the locknut on the tensioner pulley and turn the tensioner counterclockwise to relieve the tension on the timing belt. Slide the timing belt from the sprockets.

To install:

6. Install the new timing belt and adjust the tension so the belt can be twisted 45 degrees at a point between the camshaft and pump sprockets.

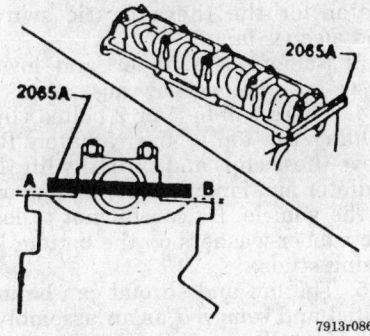

Locking the camshaft in TDC position using a special tool — diesel engine

7913r086

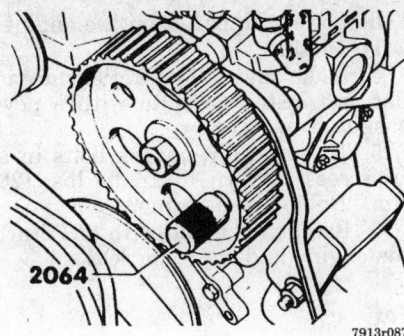

2064

7913r087

Locking the camshaft sprocket in TDC position using a special tool — diesel engine

Torque the tensioner nut to 33 ft. (45 Nm).

7. Remove the holding tools.

8. Turn the engine 2 full revolutions to return to TDC of No. 1 piston. Recheck belt tension and timing mark alignment, readjust as required.

9. Install the belt cover and accessory drive belts.

10. If the belt is too tight, there will be a growling noise that rises and falls with engine speed.

Timing Sprockets

REMOVAL AND INSTALLATION

NOTE: The 12-point crankshaft sprocket bolt is meant to be used 1 time only and must be replaced when removed.

1. Remove the timing belt covers and the timing belt. The crankshaft sprocket should slide off easily when the center bolt is removed. Don't lose the Woodruff key.

2. Remove the cylinder head cover.

3. Use a wrench to hold the camshaft on the flat section and remove the sprocket retaining bolt.

4. Gently pry or tap the sprocket off the shaft with a soft mallet. If the sprocket will not easily slide off the shaft, use a gear puller. Do not hammer on the sprocket or damage to the sprocket or bearings could occur.

5. Installation is the reverse of removal. On crankshaft sprocket bolts, oil the threads before installing the bolt. Torque the bolts as follows:

    a. Camshaft sprocket — 33 ft. lbs. (45 Nm).

    b. Crankshaft sprocket 12-point bolt — 66 ft. lbs. (90 Nm) plus ½ turn.

6. Install the timing belt, check valve timing, adjust the belt tension and install the covers.

Camshaft

REMOVAL AND INSTALLATION

1. Disconnect the negative battery cable. Remove the timing belt cover(s), the timing belt, cylinder head cover and the camshaft sprocket.

2. Number the bearing caps from front to back. If the cap does not already have one, scribe an arrow pointing towards the front of the engine. The caps are offset and must be installed correctly. Factory numbers on the caps are not always on the same side.

3. Remove the front and rear bearing caps. Loosen the remaining bearing cap nuts a little at a time to avoid bending the camshaft. Start from the outside caps near the ends of the head and work toward the center.

4. Remove the bearing caps and the camshaft.

To install:

5. Install a new oil seal and end plug in the cylinder head. Lubricate the camshaft bearing journals and lobes and set the camshaft in place.

6. Install the bearing caps in the correct position with the arrow pointing towards the front of the engine. Tighten the cap nuts diagonally and in several steps until they are torqued to 15 ft. lbs. (20 Nm). Do not over torque. Camshaft shaft end-play should be about 0.006 in. (0.15mm).

7. Install the drive sprocket and timing belt. Wait at least ½ hour after installing the camshaft before starting the engine to allow the lifters to leak down.

Intermediate Shaft

REMOVAL AND INSTALLATION

1. Disconnect the negative battery cable. Remove the front cover upper and lower halves.

2. Remove the intermediate shaft sprocket.

3. Loosen and remove the 2 retaining screws and then remove the intermediate shaft mounting flange.

4. Carefully, slide the intermediate shaft out of the block. Turn the shaft, if necessary, to remove it. Inspect the gear on the intermediate shaft and replace it, if necessary.

To install:

5. Install the intermediate shaft to the block. Install new oil seal and O-ring in the mounting flange.

6. Install the mounting flange and retaining bolts. When installing the

mounting flange be sure the oil return hole is at the bottom. Torque the retaining bolts to 18 ft. lbs. (25 Nm).

7. Install the front cover. Connect the negative battery cable.

Pistons and Connecting Rods

POSITIONING

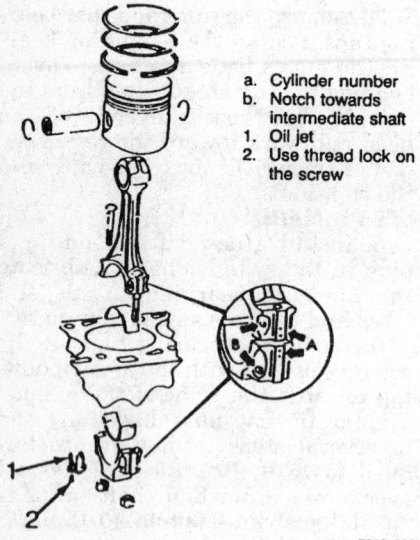

a. Cylinder number
b. Notch towards intermediate shaft
1. Oil jet
2. Use thread lock on the screw

7913r088

Piston and connecting rod positioning — diesel engine

ENGINE LUBRICATION

Oil Pan

REMOVAL AND INSTALLATION

Except Fox

The oil pan can be removed with the engine in the vehicle.

1. Raise and safely support the vehicle and drain the oil.
2. Loosen and remove the bolts retaining the oil pan.

3. Lower the pan from the engine.
To install:
4. Make sure the gasket surface is flat and install the pan with a new gasket.
5. Torque the retaining bolts in a crisscross pattern to 14 ft. lbs. (20 Nm). Do not over-torque.
6. Refill the engine with oil. Start the engine and check for leaks.

Fox

1. Raise and safely support the vehicle and drain the oil.
2. Support and slightly raise the engine from overhead with a suitable lifting device.
3. Gradually loosen the engine crossmember mounting bolts. Remove the left and right side engine mounts.
4. Carefully lower the crossmember from the vehicle.
5. Remove the oil pan retaining bolts and lower the pan from the vehicle.
To install:
6. Make sure the gasket surface is flat and install the pan and new gasket.
7. Torque the retaining bolts in a crisscross pattern to 14 ft. lbs. (20 Nm).
8. Install the crossmember and torque the crossmember-to-frame bolts to 42 ft. lbs. (57 Nm) and the engine mount bolts to 32 ft. lbs. (43 Nm).
9. Refill the engine with oil. Start the engine and check for leaks.

Oil Pump

REMOVAL AND INSTALLATION

1. Raise and safely support the vehicle and remove the oil pan.
2. Remove the mounting bolts and lower the pump from the engine.
3. Remove the bottom cover and disassemble the pump. The pressure relief valve is in the bottom cover.
4. Clean and inspect all parts for wear and replace as needed.
5. After reassembling the pump, prime it with oil and install in the reverse order of removal.
6. Observe the following torques:
• Oil pump bottom cover bolts — 7 ft. lbs. (10 Nm)
• Oil pump suction foot bolts — 7 ft. lbs. (10 Nm)
• Oil pump retaining bolts — 18 ft. lbs. (25 Nm)

Rear Main Bearing Oil Seal

REMOVAL AND INSTALLATION

The rear main oil seal is located in a housing on the rear of the cylinder block. To replace the seal on all vehicles it is necessary to remove the transaxle and flywheel.

1. Remove the transaxle and flywheel.
2. Using a small prybar, pry the old seal out of the support ring.
3. To install, lightly oil the new seal and press it into place using tool VW–2003/2A or equivalent, to start the seal and tool VW–2003/1 or equivalent, to seat the seal. Be careful not to damage the seal or score the crankshaft.
4. Install the flywheel and transaxle.

ENGINE COOLING

Radiator

REMOVAL AND INSTALLATION

Except Corrado SLC

NOTE: When replacing coolant/antifreeze, only a phosphate-free product must be used to help prevent damage to the water jacket sealing surfaces of the cylinder head. Other types of coolant may cause corrosion of the cooling system, thus leading to engine overheating and damage.

1. To drain the cooling system, remove the thermostat housing from under the water pump housing.
2. Disconnect the wiring on the radiator for the thermostatic switch and electric fan(s).
3. Remove the upper and lower hoses and the overflow hose.
4. There will be 1 or 2 bolted clips holding the top of the radiator. Remove these clips and carefully lift the radiator and fan assembly up and out of the vehicle. Be careful not to lose the rubber washers on the bottom locating studs.
5. The fan and shroud can be unbolted and removed as an assembly.
6. Installation is the reverse of removal. Torque the clip bolts to 7 ft. lbs. (10 Nm).

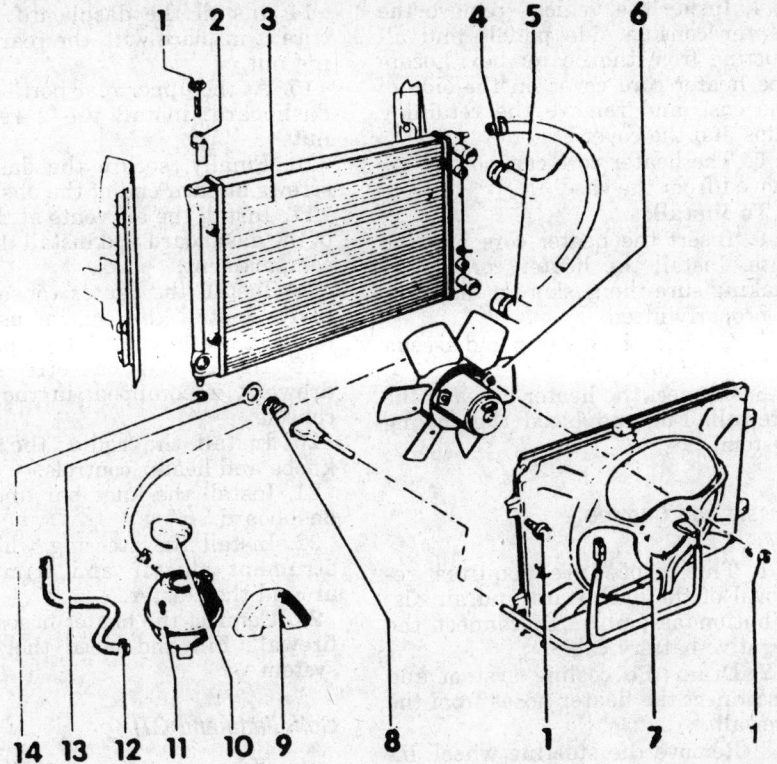

1. Bolt
2. Upper radiator mount
3. Radiator
4. Upper hose
5. Lower hose
6. Electric cooling fan
7. Fan shroud
8. Radiator fan thermo-switch
9. Cover
10. Pressure cap
11. Coolant expansion tank
12. To coolant hose
13. Sealing washer
14. Sealing washer

7913r023

Radiator and fan assembly — Jetta, GTI and Cabriolet

Corrado SLC

1. Read this entire procedure before starting work. Remove the battery.

2. Remove the front grille, headlights and hood lock support.

3. Remove the front bumper:

a. Remove the front spoiler.

b. Remove the clips between the bumper skin and the engine mount.

c. Below the bracket on each side, remove the 2 engine mount bolts that hold the bracket.

d. Pull the bumper cover forward evenly from both sides. Two people are required.

4. To drain the coolant, remove the drain plug from the coolant pipe below the intake manifold.

5. If equipped with air conditioning, remove the refrigerant hose clamps and pull the condenser forward as far as possible.

6. Remove the upper mount brackets and remove the radiator from the top.

To install:

7. Fit the radiator and fan assembly into place and install the brackets.

8. Secure the condenser in place and connect the fan wiring.

9. Connect the hoses and install the drain plug.

10. Before installing the body parts, fill the cooling system and make sure it does not leak.

11. Install the bumper and torque the bracket bolts to 63 ft. lbs. (85 Nm).

12. Install the remaining body parts and adjust the headlight aim as required.

Auxiliary Coolant Pump Switch

TESTING

Corrado SLC

At the thermostat housing, 2 switches and a plug, or on air conditioned vehicles, 3 switches are mounted in a row. The center yellow switch operates the cooling fans and the auxiliary electric coolant pump. When the ignition switch is **OFF** and the coolant temperature in the radiator is below about 145°F (63°C), but the coolant in the engine is above 215°F (101°C), this switch will close and the fan and pump will run.

TESTING

1. Turn the ignition switch **ON**, then **OFF** again.

2. At the thermostat housing, disconnect the wiring from the center yellow temperature switch.

3. Use a jumper wire to connect terminals **B** and **D**. The fan and the electric water pump should both run.

Heater Core

REMOVAL AND INSTALLATION

The heater core is contained in the fresh air/heater box located in the center of the dashboard. On air conditioned vehicles, the evaporator is also located in the heater box.

Fox

1. The entire dashboard and heater assembly must be removed. Disconnect the negative battery cable.

2. Drain the engine coolant or clamp the heater hoses.

3. Disconnect the heater hoses at the firewall and plug the core fittings.

4. Inside the vehicle, remove the knee bar, if equipped, the shifter han-

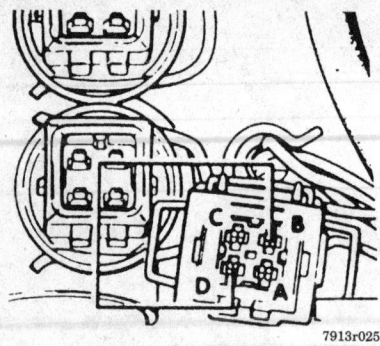

Jumper terminals B and D to run fans and auxiliary coolant pump — Corrado SLC

dle and boot. Remove the center console and the temperature controls from the dash.

5. Remove the left and right air distribution ducts.

6. In the engine compartment, remove the cowl cover and remove the air distribution housing cover.

7. Inside the vehicle, remove the lower housing retaining clips and remove the housing.

NOTE: If equipped with air conditioning, the heater box also contains the air conditioning system evaporator mounted in the lower housing cover. When removing the lower cover, place the cover and evaporator aside without removing the refrigerant lines.

8. Remove the bolts retaining the heater case and remove the case.

9. Remove the clips holding the case together and split the case. The heater core can now be removed.

To install:

10. Place the heater core into the case and reassemble the case.

11. Install the case into the vehicle. Attach the lower heater case cover to the heater case. Install the air distribution ducts and the control cables.

12. Install the center console. Install the shifter boot, handle and knee bar.

13. Reconnect the heater inlet hoses.

14. Install the air distribution housing cover and cowl and refill the cooling system.

Cabriolet

1. Disconnect the negative battery cable.

2. Drain the cooling system or clamp the heater hoses.

3. Disconnect the heater hoses at the firewall and plug the core fittings to prevent coolant leakage inside the vehicle.

4. Inside the vehicle, remove the center console side panels and all ducting from the heater box. Locate the heater core cover on the side of the case and remove the retaining clips and the cover.

5. The heater core can now be removed from the case.

To install:

6. Insert the heater core into the case. Install the heater core cover, making sure the gasket on the cover is properly fitted.

7. Connect the ducting and assemble the console.

8. Connect the heater hoses at the firewall. Fill and bleed the cooling system.

Passat and Corrado

1. This procedure requires removal of the dashboard and air distribution assembly. Disconnect the negative battery cable.

2. Drain the cooling system and disconnect the heater hoses from the firewall.

3. Remove the steering wheel, instrument cluster and trim panel around the cluster.

4. Remove the knee bar from below the dashboard.

5. Remove the cassette storage drawers, if equipped, from the center console. Pull the knobs off the heater controls and remove the radio.

6. Carefully pry out the heater control trim plate and remove the control assembly. Remove the center console trim plate.

7. Carefully pry out the air vents at the ends of the dashboard and remove the glove compartment.

8. Pry off the caps from the screws at each end of the dashboard and remove the bolts.

9. At the upper rear portion of the dashboard, remove the 2 nuts. These can only be seen with the aid of a mirror.

10. Remove the nut at the back of the center console and pull the dashboard slightly out to disconnect the remaining wires. Remove the dashboard.

11. Remove the air distribution assembly and remove the heater core.

To install:

12. If the retaining lugs are broken off, the heater core can be secured in place with screws. Install the heater core and the air distribution assembly. Make sure the seal is in good condition, replace, if necessary.

13. Prior to installing the dashboard, connect all wiring behind it.

14. Install the dashboard and secure it in place with the rear retaining nut.

15. At the upper rear portion of the dashboard, install the 2 retaining nuts.

16. Finally, secure the dash with screws at each end of the dashboard.

17. Install the air vents at the ends of the dashboard and install the glove compartment.

18. Install the center console trim plate. Install the control assembly and the heater control trim plate.

19. Install the cassette storage drawers, if equipped, in the center console.

20. Install the radio, the heater knobs and heater controls.

21. Install the knee bar under the dashboard.

22. Install the steering wheel, instrument cluster and trim panel around the cluster.

23. Connect the heater hoses to the firewall. Fill and bleed the cooling system.

Golf, Jetta and GTI

1. This procedure requires removal of the dashboard and air distribution assembly. Disconnect the negative battery cable.

2. Drain the cooling system and disconnect the heater hoses from the firewall.

3. Properly discharge the air conditioning system using freon recovery equipment.

4. Remove the gear shift knob and boot and remove the center console.

5. Remove the steering wheel.

6. Remove the knee bar from below the dashboard.

7. Remove the steering column support bracket and lower the column.

8. Pull the knobs off the heater controls and remove the control assembly and the radio.

9. Remove the headlight switch and switch blanks to gain access to the screws. Remove the instrument cluster and trim panel around the cluster.

10. Remove the glove compartment.

11. At the firewall, remove the plastic tray and remove the 2 nuts holding the top of the dashboard.

12. Remove the main fuse panel and disconnect the plugs at the back. Disconnect the ground wires.

13. Disconnect any remaining wiring from the dashboard and remove the 4 last screws. Remove the dashboard.

14. Disconnect the ducts and remove the heater housing. Remove the

screws and slide the heater core out of the housing.

To install:

15. Install the heater core and make sure the housing seals and gaskets are in good condition. Replace as necessary.

16. Connect the ducts to the heater housing.

17. Connect the wiring to the dashboard. Secure the dashboard in place with the 4 retaining screws; 1 at each end and 1 at each end of the instrument cluster area.

18. Connect the plugs at rear of the main fuse panel and secure it in place. Connect the ground wires.

19. At the firewall, install the plastic tray and install the 2 nuts securing the dashboard at the top.

20. Install the glove compartment.

21. Install the instrument cluster and trim panel around the cluster. Install the headlight switch and switch blanks.

22. Install the radio and control assembly. Install the knobs on the heater controls.

23. Install the steering column support bracket and lower the column.

24. Install the knee bar under the dashboard.

25. Install the steering wheel.

26. Install the center console. Install the gear shift boot and knob.

27. Evacuate and recharge the air conditioner. Fill and bleed the cooling system.

Water Pump

REMOVAL AND INSTALLATION

Except Corrado and Jetta Diesel

1. To drain the cooling system, remove the thermostat housing from under the water pump housing.

2. Raise and safely support the vehicle. Loosen but don't remove the bolts holding the pulley to the water pump.

3. Remove the timing belt cover.

4. Loosen the alternator and/or steering pump as required to remove the water pump drive belt.

5. Remove the water pump pulley. On some vehicles, the crankshaft pulley must also be removed by removing the bolts holding the pulley to the timing belt sprocket.

6. All the bolts are now accessible and the water pump can be removed from its housing.

To install:

7. Be sure to clean the housing before installing the new gasket. Install the pump into the housing and torque the pump-to-housing bolts to 7 ft. lbs. (10 Nm).

8. Install the water pump drive pulley and torque the bolts to 15 ft. lbs. (20 Nm). If the crankshaft drive pulley was removed, install it and torque the bolts to 15 ft. lbs. (20 Nm).

9. Adjust drive belt tension and install the thermostat and housing. Torque the bolts to 7 ft. lbs. (10 Nm).

Corrado

WITHOUT AIR CONDITIONING

The water pump is driven by the same belt that drives the alternator and supercharger. On vehicles with air conditioning, the water pump and steering pump use the same V-belt.

1. To drain the cooling system, remove the thermostat housing from under the water pump housing.

2. Loosen the water pump pulley bolts but don't remove the pulley yet.

3. Install the belt tensioner holding tool VW3191 or equivalent, to compress the tensioner and loosen the belt.

4. Loosen the power steering pump and remove its drive belt.

5. With the water pump pulley removed, the pump bolts are now accessible. Remove the pump.

To install:

6. Clean the housing and use a new gasket when installing the pump. Torque the bolts to 7 ft. lbs. (10 Nm).

7. The water pump and crankshaft pulleys must be aligned. Loosen the outer section bolts and turn the outer section relative to the inner section until the 2 pulleys are aligned; water pump pulley moves in and out. Torque the bolts to 18 ft. lbs. (25 Nm).

8. Remove the belt tensioner tool and complete the reassembly. Torque the thermostat housing bolts to 7 ft. lbs. (10 Nm) and fill the cooling system.

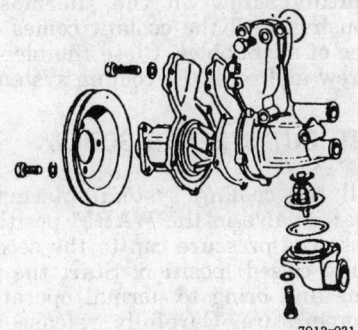

7913r031

Water pump and thermostat housing — all except Corrado and Jetta Diesel

WITH AIR CONDITIONING

1. To drain the cooling system, remove the thermostat housing from under the water pump housing.

2. Raise and safely support the vehicle.

3. Working under the vehicle, loosen but don't remove the bolts holding the pulley to the water pump.

4. Loosen the power steering pump and remove the drive belt.

5. Remove the water pump pulley and remove the pump.

To install:

6. Installation is the reverse of removal. Be sure to clean the pump housing before installing the pump with a new gasket. Torque the following:

 • Water pump-to-housing — 7 ft. lbs. (10 Nm)

 • Water pump drive pulley — 15 ft. lbs. (20 Nm)

 • Thermostat housing — 7 ft. lbs. (10 Nm)

 • Steering pump bolts — 18 ft. lbs. (25 Nm)

Corrado SLC

To remove the main coolant pump, the engine mounts must be unbolted and the engine must be lifted slightly. Do not jack the engine from below, use a hoist and lift from above. Also, a special wrench is required to remove the pump, tool VAG 1590.

1. Disconnect the negative battery cable and remove the plug in the coolant pipe below the intake manifold to drain the cooling system.

2. Disconnect the front exhaust pipe from the catalytic converter.

3. Thread a long 8 x 10mm bolt into the top of the belt tensioner. Tighten the bolt just enough to loosen and remove the accessory drive belt.

4. Remove the distributor cap and wires and the wire guide as an assembly. Attach a lifting yoke to the lifting eyes on the engine.

5. Remove the center bolt from each of the 3 engine mounts and lift the engine as required to gain access to the pump.

6. Use the special tool VAG 1590 to remove the coolant pump pulley bolts and remove the pulley.

7. Remove the bolts and push the engine slightly to the left to remove the pump and O-ring.

To install:

8. Install the pump with a new O-ring and torque the bolts to 15 ft. lbs. (20 Nm). Install the pulley.

9. Set the engine down on the mounts. Make sure the tabs on the front and rear engine mounts fit into the slot in the brackets on the engine.

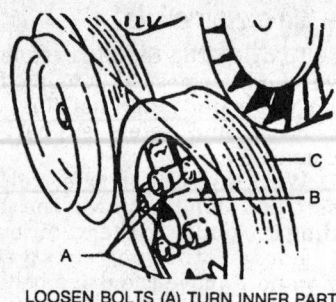

LOOSEN BOLTS (A) TURN INNER PART (B) TO ALIGN PULLEYS (C)

7913r032

Water pump pulley alignment — Corrado without air conditioning

Start all the mount bolts but do not tighten them yet.

10. Shake the engine to settle it in the mounts, then torque the bolts to 44 ft. lbs. (60 Nm).

11. Install the accessory drive belt and remove the bolt from the tensioner.

12. Use new gaskets and nuts and connect the exhaust pipe to the catalytic converter. Torque the nuts to 18 ft. lbs. (25 Nm).

13. Install the drain plug and refill the cooling system. The system holds 10.6 qts. (10L).

14. Connect the battery and run the engine to check for leaks.

Jetta Diesel

On some diesel engines, the belt tension is adjusted with shims between the outer and inner halves of the pulley. On others, the alternator swivels to adjust belt tension.

1. To drain the cooling system, remove the thermostat housing from under the water pump housing.

2. Raise and safely support the vehicle.

3. Working under the vehicle, loosen but don't remove the bolts holding the pulley to the water pump.

4. On vehicles with a movable alternator, loosen the alternator and remove the drive belt.

5. Remove the water pump pulley and remove the pump.

To install:

6. Installation is the reverse of removal. Be sure to clean the pump housing before installing the new gasket. Torque the following:

• Water pump-to-housing — 7 ft. lbs. (10 Nm)

• Water pump drive pulley — 15 ft. lbs. (20 Nm)

• Thermostat housing — 7 ft. lbs. (10 Nm)

• Alternator mounting bolts — 18 ft. lbs. (25 Nm)

Thermostat

REMOVAL AND INSTALLATION

Except Corrado SLC

The thermostat is the lowest point in the cooling system and is on the bottom of the water pump housing. Removing the thermostat is the only way to completely drain the coolant.

1. With a catch pan under the vehicle, loosen the bolts on the thermostat housing.

2. Remove the cap from the overflow bottle and allow the coolant to drain completely.

3. When the coolant is drained, remove the thermostat housing and clean both mating surfaces.

4. When installing, the thermostat spring goes up into the water pump housing and the new O-ring goes onto the thermostat. Torque the bolts to 7 ft. lbs. (10 Nm).

5. Refill the cooling system and check for leaks.

Corrado SLC

The thermostat housing is bolted to the flywheel end of the engine and includes the temperature sensor and 1 or 2 thermo-switches. The switches can be removed by removing the mounting clips. The cooling system drain plug is on the pipe leading from the housing to the other end of the engine. When removing any part from the thermostat housing, always replace the O-ring.

Cooling System Bleeding

WITH BLEEDER SCREW

Set the heat valve in the **WARM** position, start the engine and bring it to normal operating temperature. Run the engine at fast idle and open the venting screw on the thermostat housing until the coolant comes out free of air bubbles. Close the bleeder screw and refill the cooling system.

WITHOUT BLEEDER SCREW

Fill the cooling system, place the heater valve in the **WARM** position, close the pressure cap to the second (fully closed) position. Start the engine and bring to normal operating temperature. Carefully release the pressure cap to the first position and squeeze the upper and lower radiator hoses in a pumping action to allow

trapped air to escape through the radiator. Recheck the coolant level and close the pressure cap to its second position.

ENGINE ELECTRICAL

Distributor

REMOVAL

1. Disconnect the coil high tension wire and the connector plug at the distributor. Disconnect vacuum lines, if equipped.

2. Unsnap the cap retainer clips, and remove the cap and static shield as a unit.

3. At the front crankshaft pulley bolt, turn the engine to Top Dead Center (TDC) on No. 1 piston. Make a chalk or paint mark where the rotor points to the rim of the distributor; some vehicles already have a notch there. Also matchmark the distributor to the engine block or head.

4. Remove the bolt and distributor clamp and lift the distributor straight out.

INSTALLATION

Timing Not Disturbed

1. On some vehicles, the distributor engages its drive with an offset slot and is easy to reinstall in the reverse order of removal, even if the crankshaft or camshaft has been turned. Gently rotate the rotor while pushing the distributor into place. Install the hold-down bolt and adjust the ignition timing.

2. On engines with the drive gear on the distributor, make sure the engine is still at TDC and insert the distributor with the matchmarks aligned.

3. Install the hold-down clamp and bolt, connector plug, cap and static shield, and high tension wires.

4. Check and adjust the ignition timing.

Timing Disturbed

1. Rotate the crankshaft to TDC of No. 1 piston.

2. With a suitable tool, turn the oil pump drive so it is parallel with the crankshaft.

3. Install the rotor onto the distributor and align it with the No. 1 mark on the rim of the body.

4. Install the distributor, making sure the rotor still aligns with the mark when the distributor is all the way in.

5. With the distributor installed, install the hold-down clamp and bolt, connector plug, cap and static shield, and high tension wires.

6. Check and adjust the ignition timing.

Ignition Timing

ADJUSTMENT

NOTE: The manufacturer specifies timing, idle speed and CO value all be adjusted together. On some vehicles with Digifant engine management systems, these items are not adjustable. See the underhood sticker for details. There are 2 methods of checking ignition timing. One method is the use of a timing light, the other is the use of diagnostic tool VAG 1367 or equivalent.

1. Run the engine to normal operating temperature, stop engine and connect a tachometer. Using either a timing light or diagnostic tool VAG 1367 or equivalent, connect according to manufacturer's instructions.

2. If equipped, disconnect both plugs from the idle stabilizer and plug them together. On engines with Digifant engine management systems, with the ignition **ON** but the engine not running, verify that the idle stabilizer valve hums or buzzes. Do not disconnect any vacuum lines from the distributor.

3. Start the engine. On vehicles with Digifant engine management systems, disconnect the blue coolant temperature sensor plug.

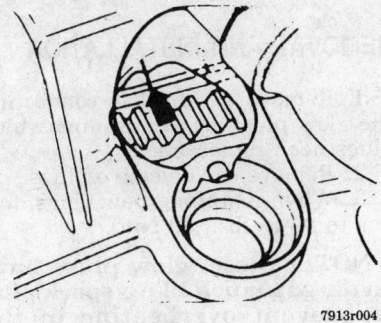

Ignition timing marks located on the flywheel

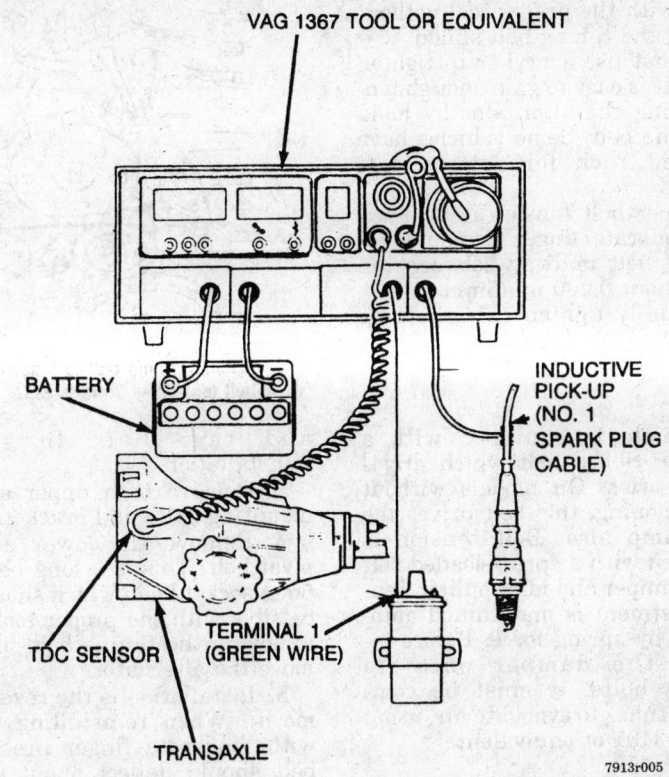

VAG 1367 TOOL OR EQUIVALENT

BATTERY

INDUCTIVE PICK-UP (NO. 1 SPARK PLUG CABLE)

TDC SENSOR

TERMINAL 1 (GREEN WIRE)

TRANSAXLE

7913r005

Using diagnostic tool VAG 1367 or equivalent to check ignition timing

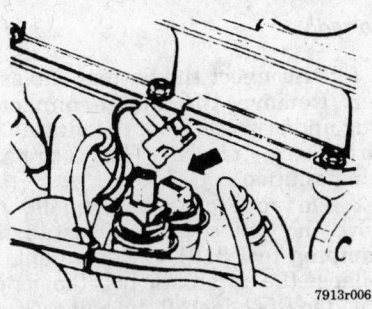

7913r006

For vehicles equipped with Digifant engine management systems, disconnect the blue coolant temperature sensor plug prior to checking ignition timing

4. Turn OFF all electrical equipment and set the idle speed.

5. If using a timing light, remove the timing mark cover from the top of the bell housing and, with the engine running, shine the timing light at the marks on the flywheel.

6. If using the diagnostic tool VAG 1367 or equivalent, observe the analog reading.

7. If adjustment is required, loosen the distributor clamp bolt and rotate the distributor as needed to set the correct timing degree. Stop the engine and reconnect plugs.

Alternator

PRECAUTIONS

• Before doing any work on any electrical system, always stop the engine and disconnect the battery cables.

• Disconnect the battery, engine control unit and ABS control unit, if equipped, before using electric welding equipment on the vehicle.

• Electronic parts and systems can be easily and permanently damaged through careless use of electric welding, charging, soldering or test equipment. Carefully follow manufacturer's instructions when using such equipment.

• If equipped with electronically theft-protected radios, obtain the security code before disconnecting the battery.

BELT TENSION ADJUSTMENT

Except Corrado

1. Loosen both upper alternator bracket bolts.

2. Loosen the lower alternator pivot bolt. This is a long bolt with a 6mm socket head which should be ac-

cessible with the proper tool without removing the timing belt shield.

3. Do not use a prybar to tighten the belt. It is easy to gain enough tension pulling the alternator by hand against the belt. Some vehicles have a toothed rack for setting belt tension.

4. Proper belt tension is attained when moderate finger pressure deflects the belt midway between the pulleys about 0.200 in. (5mm).

5. Securely tighten the mounting bolts.

Corrado

This vehicle is equipped with a serpentine ribbed belt which drives all accessories. On models without air conditioning, this belt drives the water pump also. Belt tension is maintained with a spring-loaded belt tension damper and idler pulley. Tension adjustment is maintained automatically by spring force. Before installing the damper onto its mounting bolts, it must be compressed 5 times to evacuate air, using VAG tool 3191 or equivalent.

Corrado SLC

This vehicle is equipped with a serpentine ribbed belt which drives all accessories. Belt tension is maintained with a spring-loaded idler pulley. Tension adjustment is maintained automatically by spring force. To loosen the tension, install a long 8 **x** 10mm bolt into the hole in the top of the tensioner. Tighten the bolt only as required to remove the belt from the alternator.

REMOVAL AND INSTALLATION

Except Corrado

1. Disconnect the battery cables.
2. Remove the multi-connector plug and/or wires from the alternator

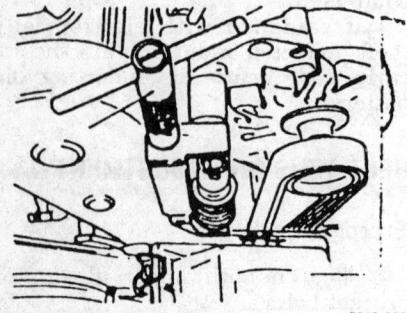

Belt tensioner compressing tool used — Corrado with supercharger

7913r008

Install 8x10mm bolt to loosen the belt tension — Corrado SLC

7913r009

and tag them for correct reinstallation.

3. Remove both upper alternator mounting bolts and bracket.

4. Remove the lower alternator pivot bolt. This is a long bolt with a 6mm socket head which should be accessible with the proper tool without removing the timing belt shield. Remove the alternator.

5. Installation is the reverse of removal. When reinstalling the belt, with moderate finger pressure the belt should deflect about 0.200 in. (5mm).

Corrado

1. Disconnect the battery cables.
2. Remove the multi-connector plug and/or wires from the alternator and tag them for correct reinstallation.
3. On supercharged models, remove the belt cover and install the clamping tool 3191 or equivalent, to collapse the automatic belt tensioner.
4. On SLC, install the bolt to move the spring tensioner and relieve the belt tension.
5. Remove the alternator bolts to lift the alternator from the vehicle.
6. Installation is the reverse of removal.

Voltage Regulator

REMOVAL AND INSTALLATION

The voltage regulator on all vehicles is mounted externally on the rear of the alternator. It can be removed without removal or disassembly of the alternator. The alternator field brushes are a part of the regulator and should project no less than 0.200 in. (5mm) from the regulator. If the brushes are not within specification, replace the regulator.

Starter

REMOVAL AND INSTALLATION

Bosch Starters

NOTE: On some vehicles, the same bolts hold an engine mount and the starter. Additionally some starter bolts hold the engine and transaxle together. Always support the weight of the engine when removing the starter.

1. Disconnect the battery ground cable.
2. Raise and safely support the vehicle.
3. Support the weight of the engine with tool 10–222 or equivalent. Do not jack up the oil pan.
4. Tag and disconnect the wires from the starter.
5. Remove starter mounting bolts and remove the starter.
6. Installation is the reverse of removal. Torque the mounting bolts to 33 ft. lbs. (45 Nm).

Mitsubishi Starters

1. Disconnect the battery ground cable.
2. Raise and safely support the vehicle.
3. Support the weight of the engine with tool 10–222 or equivalent. Do not jack up the oil pan.
4. Remove the engine/transaxle cover plate.
5. Unbolt and remove the starter side motor mount and carrier.
6. Disconnect and mark the starter wiring.
7. Remove starter mounting bolts and the starter.
8. Installation is the reverse of removal. Torque the mounting bolts to 33 ft. lbs. (45 Nm).

Diesel Glow Plugs

REMOVAL AND INSTALLATION

1. Remove the busbar connecting the glow plugs and determine which plugs need replacement.
2. Remove the defective plugs.
3. When installing new plugs, torque to 22 ft. lbs. (30 Nm).

NOTE: Diesel glow plugs have an air gap much like a spark plug to prevent overheating of the plug. Over-torquing the glow plug will close the gap and cause the plug to burn out.

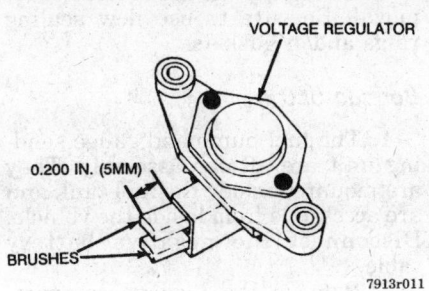

VOLTAGE REGULATOR

0.200 IN. (5MM)

BRUSHES

7913r011

Checking brush protrusion

TESTING

1. Disconnect the engine temperature sensor.

2. Connect a test light between No. 4 cylinder glow plug and ground. The glow plugs are connected by a flat, coated busbar, located near the bottom of the cylinder head.

3. Turn the ignition key **ON**; the test light should light, then go out after 10–30 seconds.

4. If there is no voltage, possible problems include a blown fuse, lack of power to or from the glow plug relay (check wiring) or the relay itself.

5. To test each plug individually, disconnect the wire and remove the busbar from the glow plugs.

6. Connect an ohmmeter to each glow plug connection or use a test light. Each plug must have continuity to ground. The engine will probably start with one defective glow plug, but it will produce excessive smoke.

EMISSION CONTROLS

Emission Warning Lamps

RESETTING

On models so equipped, the OXS warning light on the dash will turn on when it is time to replace the oxygen sensor. This is usually at 30,000 mile intervals. Under the hood near the wiper motor, is a black box with the speedometer cable connected to it. The mileage counter turns on the warning light. To reset the counter, find the white button on the box and push it in with a pen, listening (feeling) for the click.

GASOLINE FUEL SYSTEM

Fuel System Service Precautions

• Do not allow fuel spray or fuel vapors to come into contact with a heating element or open flame. Do not smoke while working on the fuel system.

• Always disconnect the negative battery cable unless the repair or test procedure requires that battery voltage be applied.

• Always relieve the fuel system pressure prior to disconnecting any fitting or fuel line connection.

• To control fuel spray when relieving system pressure, place a shop towel around the fitting prior to loosening to catch the spray. Ensure that all fuel spillage is quickly wiped up and that all fuel soaked rags are deposited into a proper fire safety container.

• Always keep a dry chemical (Class B) fire extinguisher near the work area.

• Always use a backup wrench when loosening and tightening fuel line fittings. Always follow the proper torque specifications.

• Do not re-use fuel system gaskets and O-rings, replace with new ones. Do not substitute fuel hose where fuel pipe is installed.

RELIEVING FUEL SYSTEM PRESSURE

On CIS systems, fuel pressure can be vented at the cold start injector line, either at the fuel distributor end or the injector end. Lay a rag over the fitting and use a socket or line wrench to crack the fitting.

On Digifant systems, pressure can be vented at the fuel pump switch in front of the throttle body. Lay a rag over the switch and loosen the clamp.

Fuel Tank

REMOVAL AND INSTALLATION

Except Fox

1. Disconnect the negative battery cable. Remove the access panel under the rear seat or in the luggage compartment and disconnect the gauge sending unit wiring and hoses.

2. Raise and safely support the vehicle and drain the fuel tank.

3. On Cabriolet, remove the right rear inner fender and disconnect the breather hose from the filler. Remove but do not disconnect the gravity valve.

4. Detach the fuel pump bracket from the body and lower the pump enough to disconnect the fuel hoses from the tank.

5. On Cabriolet, the rear axle must be dropped out of the way. Disconnect the brake hydraulic hoses at both sides of the rear axle.

6. Detach the rear axle from the body on both sides and let it hang on the parking brake cable guides.

7. Unhook the muffler supports and pull the large hose from the filler neck.

8. Support the tank, loosen the straps and carefully lower the tank out of the vehicle.

To install:

9. If a new tank is being installed, glue new foam strips to the tank in the same location as the old ones. Position the tank and connect the wiring and hoses to the sending unit.

10. Secure the tank in place with the straps and connect all hoses. Position the clamps so they do not contact the body.

11. Coat the tank with a rust protector or undercoating.

12. Install the rear axle, connect the hydraulic lines and bleed the brakes. Tighten the tank strap bolts to 17 ft. lbs. (23 Nm).

Fox

1. Disconnect the negative battery cable. Remove the access panel under the rear seat or in the luggage compartment and disconnect the gauge sending unit wiring and hoses.

2. Raise and safely support the vehicle; drain the fuel tank.

3. Disconnect the fuel filler hose.

4. Support the tank, loosen the straps and carefully lower the tank out of the vehicle.

5. Installation is the reverse of removal. Tighten the tank strap bolts to 17 ft. lbs. (23 Nm).

Fuel Filter

REMOVAL AND INSTALLATION

On the Digifant system, the fuel filter is a lifetime unit and only needs to be changed in the event of contamination. It is mounted under the vehicle, near the rear axle. The pump, accumulator, filter and reservoir are all part of a single assembly, but the filter can be removed separately.

On the CIS system, the fuel filter is mounted under the hood, sometimes on the fuel distributor. To make the job easier, open the clips holding the air filter housing and lift the whole assembly.

1. Disconnect the negative battery cable.
2. On vehicles with the Digifant system, raise and safely support the vehicle.
3. Relieve the fuel system pressure.
4. Remove the fuel lines, the mounting bracket nut and the filter.
5. Installation is the reverse of removal. Be sure to use the new sealing rings and torque the fuel lines to the filter to 14 ft. lbs. (20 Nm).

Electric Fuel Pump

TESTING

Fuel Pump Delivery

NOTE: Before testing the fuel pump, the battery must be fully charged and the fuel tank at least ¼ full.

1. Check the condition of the fuel filter.
2. Disconnect the high tension terminal from the ignition coil at the distributor and securely ground it.
3. Disconnect the fuel return fuel line and hold it in a measuring

container with a capacity of 1 qt. (1000cc).
4. Have an assistant run the starter for 30 seconds while watching the quantity of fuel delivered. The minimum allowable flow is 9/16 qt. (760cc) in 30 seconds.
5. If the flow is below specification, check the delivery of the fuel transfer pump, which is mounted with the gauge sending unit in the tank.
6. Under the rear seat or under the luggage compartment, remove the cover to expose the hoses and wires to the pump/sending unit.
7. Disconnect the output hose from the tank unit and plug it. Install a temporary fuel line and put the other end into the measuring container.
8. Have an assistant operate the starter for 10 seconds and measure the fuel delivered. The specification is about 10 oz. (300cc).
9. If the transfer pump is good, check for a dirty fuel filter, blocked lines or blocked fuel tank strainer, if equipped. If all of these are in good condition but the quantity measured is Step 4 is low, replace the main pump.

REMOVAL AND INSTALLATION

Except Corrado SLC

1. The main fuel pump is located under the vehicle in front of the rear axle or in front of the tank on the right side. Disconnect the negative battery cable.
2. Raise and safely support the vehicle.
3. Disconnect the electrical connector.
4. Relieve the fuel system pressure.
5. Remove the mounting bolts and the fuel pump.

6. Installation is the reverse of removal. Be sure to use new sealing rings and/or gaskets.

Corrado SLC

1. The fuel pump and gauge sending unit are all one assembly. They are mounted inside the fuel tank and are accessible from inside the vehicle. Disconnect the negative battery cable.
2. Remove the luggage compartment carpet and remove the access plate in the floor.
3. Label and disconnect the wiring and hoses.
4. Loosen and remove the flange nut and the O-ring.
5. Turn the pump counterclockwise to remove the pump and gauge sending unit.
6. Installation is the reverse of removal. Make sure the marks on the flange and fuel tank are aligned.

Transfer Pump

REMOVAL AND INSTALLATION

Except Corrado SLC

1. Disconnect the negative battery cable.
2. Under the rear seat or in the rear of the vehicle, pull back the carpet and remove the access plate from the floor (3 screws).
3. Disconnect the electrical connector and remove the fuel hoses from the sending unit.
4. Unscrew the cap and carefully lift the sending unit from the fuel tank. Note the orientation of the float in the tank.
5. Remove the transfer pump from the sending unit.
6. Installation is the reverse of removal. Be sure the float points the same way and use a new O-ring at the sending unit cap.

Fuel Injector

REMOVAL AND INSTALLATION

CIS-E

1. Relieve the pressure from the system.
2. Using a fuel injector removal tool, pry the injectors up out of the head. A spray lubricant can help release stuck injectors.
3. Hold the fuel line fitting with a line wrench and unscrew the injector.

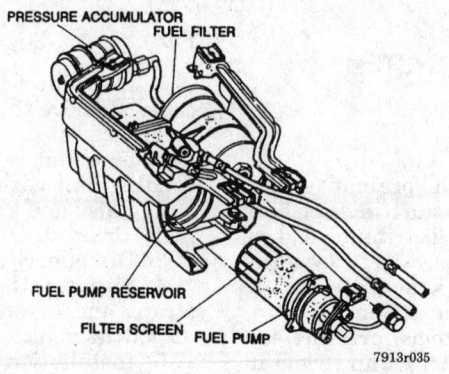

PRESSURE ACCUMULATOR
FUEL FILTER

FUEL PUMP RESERVOIR
FILTER SCREEN FUEL PUMP

7913r035

Fuel pump and reservoir assembly — Digifant system

4. Installation is the reverse of removal. Lightly lubricate the rubber rings.

Digifant

The electric injectors are held in place by the rail and cannot be removed separately.

Except Corrado SLC

1. Disconnect the negative battery cable.
2. Relieve the pressure from the fuel system.
3. Dismount the idle stabilizer valve and lay it aside.
4. Remove the intake manifold supports and cylinder head cover.
5. Unplug the wiring harness end connector and pry wiring guide away from the fuel distributor retainers.
6. Remove the fuel distributor retaining bolts and remove the rail, wiring guide and injectors as an assembly.
7. Installation is the reverse of removal.

Corrado SLC

1. Remove the ignition cap, wires and wire guide as an assembly.
2. Label and disconnect the wiring and hoses and remove the air intake duct with the mass air sensor.
3. Label and disconnect the wiring and hoses as required to remove the upper intake manifold.
4. Disconnect the wiring and the fuel supply and return hoses from the fuel rail. Disconnect the vacuum line from the pressure regulator.
5. Remove the bolts and remove the fuel rail and injectors as an assembly. It may be necessary to gently pry the injectors out of the ports.
To install:
6. Lightly lubricate the O-rings and fit the injectors into the ports. Secure the rail with the bolts.
7. Connect the wiring, fuel hoses and vacuum line.
8. Install the intake manifold with a new gasket and torque the bolts to 18 ft. lbs. (25 Nm).
9. Connect the wiring, hoses and control cables and install the air duct with the air mass sensor.
10. Install the ignition components and run the engine to check for leaks.

DRIVE AXLE

Halfshaft

REMOVAL AND INSTALLATION

NOTE: When loosening or tightening axle nuts, make sure the vehicle is on the ground. Axle nut torque is high enough that attempting to loosen it may cause the vehicle to fall off the jackstands.

1. With the vehicle on the ground, remove the front axle nut.
2. Raise and safely support vehicle and remove the front wheels.
3. Remove the socket head bolts retaining the halfshaft to the transaxle flange.
4. Separate the strut from the control arm:
   a. On Fox, matchmark the ball joint to the control arm and remove the nuts to disconnect the ball joint from the control arm.
   b. On Passat and Corrado, remove the bolts securing the ball joint to the control arm.
   c. On all other vehicles, remove the ball joint clamping bolt and push the control arm down, away from the ball joint.
5. Remove the transaxle side of the halfshaft from the drive flange and secure it out of the way. Do not let it hang unsupported.
6. Push the halfshaft out of the hub. A wheel puller may be required.
To install:
7. Fit the halfshaft to the drive flange and install the bolts. It is not necessary to torque them yet.
8. Apply a thread locking compound to the outer 1/4 in. of the spline. Slip the spline through the hub and loosely install a new axle nut.
9. Assemble the front suspension, being careful to align the matchmarks.
   a. On Fox, torque the ball joint-to-control arm nuts to 47 ft. lbs. (65 Nm).
   b. On Passat and Corrado, torque the ball joint bolts to 26 ft. lbs. (35 Nm).
   c. On all other models, torque the ball joint clamping bolt to 37 ft. lbs. (50 Nm).
10. Install the wheel and hold it to keep the axle from turning. Torque the inner axle bolts to 33 ft. lbs. (45 Nm).

11. With the vehicle on the ground, torque the axle nut:
   • Fox and Cabriolet — 175 ft. lbs. (240 Nm)
   • Golf, Jetta and Passat — 195 ft. lbs. (265 Nm)
   • Corrado with supercharger — 195 ft. lbs. (265 Nm)
   • Corrado SLC — 66 ft. lbs. (90 Nm) plus 45 degrees
12. Check and adjust the front wheel alignment.

CV-Joint/Boot

REMOVAL AND INSTALLATION

1. Raise and safely support the vehicle and remove the halfshaft.
2. Pry open and remove the boot clamps with a pair of wire cutters.
3. With the halfshaft securely clamped in a vise, the outer CV-joint and boot can be removed by sharply rapping out on the joint with a plastic hammer. The joint will snap off the circlip and slide off the axle.
4. To remove the inner joint, remove the circlip from the center and slide the joint and boot off the axle.
To install:
5. Always replace both circlips and make sure the CV-joint is clean before installation. Wrap a piece of black electrical tape around the shaft splines and slip the inner clamp and the boot onto the shaft.
6. Remove the tape and install the dished washer with the concave side out. On the outer joint, install the thrust washer and a new circlip.
7. To install the outer joint, place it onto the spline and carefully tap straight in on the end with a plastic hammer. The joint will click into place over the circlip.
8. To install the inner joint, slide it onto the spline and push in enough to allow the circlip to fit into the groove in the axle shaft.
9. Install the inner clamp on the boot and fill the boot with special CV-joint grease. Do not use any other type of grease.
10. On the inner joint, stick the gasket to the joint before installing the halfshaft into the vehicle. Install the outer boot clamp and install the halfshaft.

Front Steering Knuckle

REMOVAL AND INSTALLATION

On Fox, the strut must be removed but a spring compressor is not needed. On all models, the hub and bearing are pressed into the knuckle and the bearing cannot be reused once the hub has been removed.

NOTE: When loosening or tightening axle nuts, make sure the vehicle is on the ground. Axle nut torque is high enough that attempting to loosen it may cause the vehicle to fall off the jack stands.

1. With the vehicle on the ground, remove the front axle nut.

2. Raise and safely support the vehicle and remove the front wheels. On Fox, remove the strut.

3. Detach the brake line from the strut and remove the caliper. Hang it from the body with wire.

4. Remove the caliper carrier and brake rotor.

5. Remove the cotter pin and nut and press out the tie rod end. A small puller is required.

6. Remove the ball joint clamp bolt and push the control arm down to disengage the ball joint.

7. Front wheel camber is set with eccentric washers on the bolts holding the bearing housing to the strut. Clean and mark the position of these washers so they can be reinstalled in the same position.

8. Remove the bolts and take the knuckle and bearing housing off the strut.

To install:

9. Fit the knuckle to the strut and install the bolts. Align the marks and torque the nuts to 70 ft. lbs. (95 Nm).

10. Make sure the axle splines are clean and apply a bead of thread locking compound to the outer portion. Slide the axle into the hub and install a new axle nut. Do not torque it yet.

Camber adjusting eccentric washers

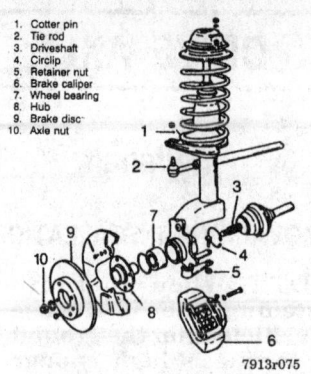

1. Cotter pin
2. Tie rod
3. Driveshaft
4. Circlip
5. Retainer nut
6. Brake caliper
7. Wheel bearing
8. Hub
9. Brake disc
10. Axle nut

7913r075

Front suspension components — Fox

11. Fit the lower ball joint in place and install the clamp bolt. Torque it to 37 ft. lbs. (50 Nm).

12. Connect the tie rod and torque the nut to 26 ft. lbs. (35 Nm), then tighten as required to install a new cotter pin.

13. Install the brake disc and caliper. Torque the carrier bolts to 92 ft. lbs. (125 Nm) and the caliper guide bolts to 26 ft. lbs. (35 Nm). Secure the brake line in place.

14. With the wheel installed and the vehicle on the ground, torque the axle nut:

- Cabriolet and Fox — 175 ft. lbs. (237 Nm)
- Passat, Corrado, Golf and Jetta — 195 ft. lbs. (265 Nm)

Front Wheel Bearing

REMOVAL AND INSTALLATION

1. Raise and safely support the vehicle and remove the strut or steering knuckle.

2. To remove the hub, support the strut or knuckle assembly in an arbor press with the hub facing down.

3. Use a proper size arbor that will fit through the bearing and press the hub out.

4. If the inner bearing race stayed on the hub, clamp the hub in a vise and use a bearing puller to remove it.

5. On the knuckle, remove the splash shield and internal snaprings from the bearing housing.

6. With the knuckle in the same pressing position, press the bearing out.

7. Clean the bearing housing and hub with a wire brush and inspect all

parts. Replace parts that have been distorted or discolored from heat. If the hub is not absolutely perfect where it contacts the inner bearing race, the new bearing will fail quickly.

To install:

8. The new bearing is pressed in from the hub side. Install the snapring and support the bearing housing on the press.

9. Using the old bearing as a press tool, press the new bearing into the housing up against the snapring. Make sure the press tool contacts only the outer race of the bearing.

10. Install the outer snapring and splash shield.

11. Support the inner race on the press and press the hub into the bearing. Make sure the inner race is supported or the bearing fail quickly.

12. Install the strut or knuckle and be sure to torque the axle nut before allowing the vehicle to roll.

Rear Axle Shafts/Stub Axles

REMOVAL AND INSTALLATION

1. Raise and safely support the vehicle and remove the rear wheels.

2. On drum brakes, insert a small pry tool through one of the wheel bolt holes and push the adjusting wedge up. On disc brakes, remove the caliper and carrier. Hang the caliper from the spring with wire.

3. Remove the grease cap, cotter pin, locknut, adjusting nut, thrust washer, wheel bearing and brake drum or disc.

4. On drum brakes, disconnect and plug the brake line.

5. Remove the brake backing plate, with the brakes attached and the stub axle.

To install:

6. Install the back plate and stub axle and torque the bolts:

- Stub axle/back plate on Golf and Jetta — 52 ft. lbs. (70 Nm)
- Stub axle/back plate on all others — 44 ft. lbs. (60 Nm)

7. When reinstalling the wheel bearing nut, the thrust washer must still move with a small pry tool. Don't forget to bleed the drum brakes.

8. On vehicles with rear disc brakes, torque the caliper bolts to 48 ft. lbs. (65 Nm).

MANUAL TRANSAXLE

Transaxle Assembly

REMOVAL AND INSTALLATION

Passat and Corrado

NOTE: If equipped with electronically theft-protected radio, obtain the security code before disconnecting the battery.

1. Disconnect the negative battery cable.
2. On Corrado with supercharger, remove the intercooler tubing.
3. Disconnect the backup light switch connector and the speedometer cable from the transaxle, plug the speedometer cable hole.
4. Remove the clutch slave cylinder without disconnecting the hydraulic line. Hang the cylinder from the body with wire.
5. On the cable shift linkage, remove the backup light switch bracket. Disconnect the cable from the relay lever but remove the gearshift lever with the cable still attached. Remove the cable support and set the cables aside.
6. If necessary, remove the intake hose from the air flow sensor.
7. Remove the upper transaxle-to-engine bolts.
8. Raise and safely support the vehicle and remove the front wheels. Connect the engine sling tool VW–10–222A or equivalent, to the loop in the cylinder head and just take the weight of the engine off the mounts. On 16V engine, the idle stabilizer valve must be removed to attach the tool. Do not try to support the engine from below.

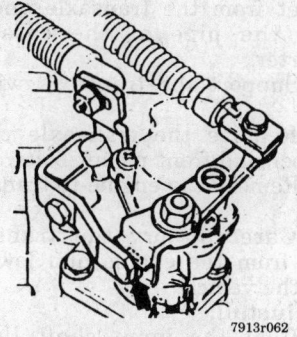

Cable shift linkage — Passat and Corrado transaxle; the relay lever is on the left

9. Remove the drain plug and drain the oil from the transaxle. Dispose of the oil properly.
10. Remove the starter and front mount.
11. Remove the 3 bolts from the right side mount, between engine and firewall.
12. Remove the large center bolt from the left side transaxle mount. On vehicles with ABS, this bolt can be reached by removing the cooling system overflow bottle.
13. Remove the radiator fan shroud and fan as an assembly.
14. Remove the long transaxle support bracket which connects the front and rear mounts on the left side.
15. Remove the heatshield for the right side inner CV-joint.
16. Disconnect the halfshafts from the output flanges and hang them from the body.
17. Remove the left rear transaxle mount. It may be necessary to push the engine/transaxle rearward to get the lower bolt out.
18. Lower the transaxle slightly.
19. Remove the bell housing cover and position a jack under the transaxle.
20. Remove the last transaxle-to-engine bolts and gently pry the transaxle away from the engine. Lower it carefully from the vehicle.

To install:

21. Press the clutch release lever towards the transaxle housing and secure it with a pin or 8mm bolt.
22. Coat the input shaft lightly with molybdenum grease and carefully fit the transaxle in place. If necessary, put the transaxle in any gear and turn an output flange to align the input shaft spline with the clutch spline.
23. Install the engine-to-transaxle bolts and torque to 59 ft. lbs. (80 Nm).
24. When installing the mounts to the transaxle, torque the left and rear bracket-to-transaxle bolts to 18 ft. lbs. (25 Nm). Torque the remaining mount-to-transaxle bolts to 44 ft. lbs. (60 Nm). Don't forget the balance weight. Install but do not torque the bolts that go into the rubber mounts.
25. Install the starter and front mount.
26. With all mounts installed and the transaxle safely in the vehicle, allow some slack in the lifting equipment. With the vehicle safely supported, shake the engine/transaxle as a unit to settle it in the mounts. Torque all mounting bolts, starting at the rear and working forward. Torque the bolts that go into the rubber

transaxle mounts to 44 ft. lbs. (60 Nm).
27. Install the halfshafts and torque the bolts to 33 ft. lbs. (45 Nm). Install the heatshield.
28. Remove the pin or bolt from the release lever and install the clutch slave cylinder. Torque the bolts to 18 ft. lbs. (25 Nm).
29. Lubricate the shift linkage lightly with molybdenum grease and install it. Torque the bolts to 18 ft. lbs. (25 Nm). Adjust the linkage as required.
30. Install the radiator fan assembly and connect the wiring.
31. Complete the installation and refill the transaxle with oil.

Jetta and Golf

NOTE: If equipped with electronically theft-protected radio, obtain the security code before disconnecting the battery.

1. Disconnect the negative battery cable.
2. Disconnect the backup light switch connector and the speedometer cable from the transaxle; plug the speedometer cable hole.
3. Remove the upper engine-to-transaxle bolts.
4. Remove the 3 right side engine mount bolts, between engine and firewall.
5. To disconnect the shift linkage, pry open the ball joint ends and remove the shift and relay shaft rods.
6. Remove the center bolt from the left transaxle mount.
7. Raise and safely support the vehicle and remove the front wheels. Connect the engine sling tool VW–10–222A or equivalent, to the loop in the cylinder head and just take the weight of the engine off the mounts. On 16V engine, the idle stabilizer valve must be removed to attach the tool. Do not try to support the engine from below.
8. Remove the drain plug and drain the oil from the transaxle. Dispose of the oil properly.
9. Remove the left inner fender liner.
10. Disconnect the halfshafts from the inner drive flanges and hang them from the body.
11. Remove the clutch cover plate and the small plate behind the right halfshaft flange.
12. Remove the starter and front engine mount.
13. Disconnect the clutch cable and remove it from the transaxle housing.
14. Remove the remaining transaxle mount bolts and mounts.

15. Place a jack under the transaxle and remove the last bolts holding it to the engine. Carefully pry the transaxle away from the engine and lower it from the vehicle.

To install:

16. Coat the input shaft lightly with molybdenum grease and carefully fit the transaxle in place. If necessary, put the transaxle in any gear and turn an output flange to align the input shaft spline with the clutch spline.

17. Install the engine-to-transaxle bolts and torque to 55 ft. lbs. (75 Nm).

18. When installing the mounts to the transaxle, torque the rear bracket-to-engine bolts and the transaxle support bolts to 18 ft. lbs. (25 Nm). Torque the left bracket-to-transaxle bolts to 25 ft. lbs. (35 Nm) and the remaining mounting bolts to 44 ft. lbs. (60 Nm). Install but do not torque the bolts that go into the rubber mounts.

19. Install the starter and front mount.

20. With all mounts installed and the transaxle safely in the vehicle, allow some slack in the lifting equipment. With the vehicle safely supported, shake the engine/transaxle as a unit to settle it in the mounts. Torque all mounting bolts, starting at the rear and working forward. Torque the bolts that go into the rubber mounts to 44 ft. lbs. (60 Nm).

21. Install the halfshafts and torque the bolts to 33 ft. lbs. (45 Nm). Install the clutch cover plates.

22. Connect the shift linkage and clutch cable and adjust as required.

23. Install the inner fender and complete the remaining installation. Refill the transaxle with oil.

Cabriolet

NOTE: If equipped with electronically theft-protected radio, obtain the security code before disconnecting the battery.

1. Disconnect the negative battery cable.

2. Disconnect the backup light switch connector and the speedometer cable from the transaxle, plug the speedometer cable hole.

3. Turn the engine to align the timing marks to TDC.

4. To disconnect the shift linkage, pry open the ball joint ends and remove both selector rods. Remove the pin, disconnect the relay rod and put the pin back in the hole on the rod for safe keeping.

5. Raise and safely support the vehicle and remove the front wheels.

Connect the engine sling tool VW-10-222A or equivalent, to the loop in the cylinder head and just take the weight of the engine off the mounts. On 16V engine, the idle stabilizer valve must be removed to attach the tool. Do not try to support the engine from below.

6. Remove the drain plug and drain the oil from the transaxle. Dispose of the oil properly.

7. Detach the clutch cable from the linkage and remove it from the transaxle case.

8. Remove the starter and front engine mount.

9. Remove the small cover behind the right halfshaft flange and remove the clutch cover plate.

10. Disconnect the halfshafts from the drive flanges and hang them up with wire.

11. Remove the long center bolt from the left side transaxle mount.

12. Remove the entire rear mount assembly from the body and differential housing.

13. Lower the engine hoist enough to let the left mount free of the body and remove the mount from the transaxle.

14. Place a transaxle support jack under the transaxle, remove all the transaxle-to-engine bolts and carefully pry the transaxle away from the engine. Lower the transaxle from under the vehicle.

To install:

15. Coat the input shaft lightly with molybdenum grease and carefully fit the transaxle in place. If necessary, put the transaxle in any gear and turn an output flange to align the input shaft spline with the clutch spline.

16. Install the engine-to-transaxle bolts and torque to 55 ft. lbs. (75 Nm).

17. When installing the mounts to the transaxle, torque the bolts to 33 ft. lbs. (45 Nm). Install but do not

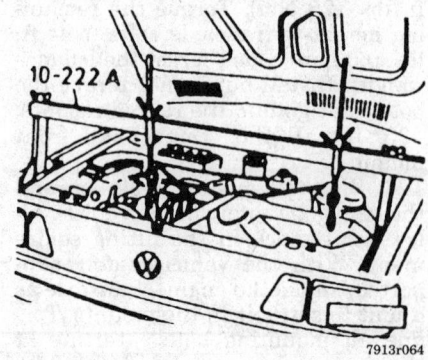

10-222 A

7913r064

Supporting the engine to remove the transaxle

torque the bolts that go into the rubber mounts.

18. Install the starter and front mount.

19. With all mounts installed and the transaxle safely in the vehicle, allow some slack in the lifting equipment. With the vehicle safely supported, shake the engine/transaxle as a unit to settle it in the mounts. Torque all mounting bolts, starting at the rear and working forward. Torque the bolts that go into the rubber mounts to 25 ft. lbs. (35 Nm). Torque the front mount bolts to 38 ft. lbs. (52 Nm).

20. Install the halfshafts and torque the bolts to 33 ft. lbs. (45 Nm). Install the clutch cover plates.

21. Connect the shift linkage and clutch cable and adjust as required.

22. Complete the remaining installation and refill the transaxle with oil.

Fox

NOTE: If equipped with electronically theft-protected radio, obtain the security code before disconnecting the battery.

1. Raise and safely support the vehicle and remove the front wheels.

2. Remove the drain plug and drain the oil from the transaxle. Dispose of the oil properly.

3. Disconnect the battery ground cable.

4. Disconnect the clutch cable.

5. Disconnect the exhaust pipe from the manifold.

6. Disconnect the speedometer cable and backup light switch.

7. Remove the bolt on the shift linkage, pry the control rod joint off and push the shift linkage coupling off the transaxle.

8. Detach the halfshafts from transaxle.

9. Remove the starter and clutch cover plate.

10. Remove the exhaust pipe bracket from the transaxle and remove the pipe at the catalytic converter.

11. Support the transaxle with a jack.

12. Remove the transaxle crossmember and front mount bolts.

13. Remove the engine-to-transaxle bolts.

14. Carefully pry the transaxle away from the engine and lower it from the vehicle.

To install:

15. Coat the input shaft lightly with molybdenum grease and carefully fit the transaxle in place. If necessary, put the transaxle in any gear

and turn an output flange to align the input shaft spline with the clutch spline.

16. Install the engine-to-transaxle bolts and torque to 40 ft. lbs. (55 Nm).

17. Torque the crossmember-to-body bolts to 47 ft. lbs. (65 Nm).

18. Install the rubber mount and torque the bracket bolts to 18 ft. lbs. (25 Nm) and the mount-to-body bolts to 80 ft. lbs. (110 Nm).

19. Connect the halfshafts to the drive flanges and torque the bolts to 33 ft. lbs. (45 Nm).

20. Install the remaining parts and adjust the clutch and shift linkage as required. Refill the transaxle with oil.

SHIFT LINKAGE ADJUSTMENT

Passat and Corrado

This procedure requires special tools VW 3193 and VW3192/1 or equivalent.

1. Put the transaxle in neutral, remove the shift knob and boot.

2. Loosen the nut and bolt connecting the cables to the shift levers so the cables move freely.

3. Loosen bolt **C** and install the adjusting tool.

4. Pivot the locating pin for the tool under the bearing plate and tighten nut **D**.

5. Push the shifter into the detent and all the way to the left and tighten the slide with bolt **E**.

6. Push the shifter all the way to the right, into the detent, and tighten bolt **C**.

7. At the other end of the cables, install the special wedge and pin so there is no play in the lever but the lever is not raised.

8. The linkage is now set in place. Tighten the cables to the levers and remove the tools to check shifter operation.

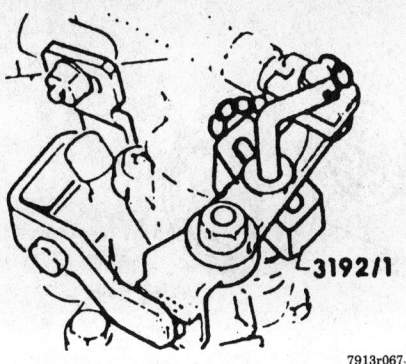

7913r067

Passat and Corrado adjusting wedge in place

Golf and Jetta

This procedure requires special tool VW 3104 or equivalent.

1. Put the transaxle in neutral.

2. Under the vehicle, loosen the clamp on the shifter rod so the shifter moves freely on the rod.

3. Remove the shifter knob and the boot.

4. Position the gauge alignment tool VW–3104 or equivalent, on the shifting mechanism and lock it in place.

5. Align the shift rod with the selector lever and torque the clamp to 19 ft. lbs. (26 Nm). The shifter linkage must not be under load during the adjustment.

6. Check shifter operation.

Cabriolet

1. Remove the shifter knob and boot.

2. Align the holes of the lever housing plate with the holes of the lever bearing plate. Check shifter operation.

3. If further adjustment is required, working under the vehicle remove the boot and loosen the shift rod clamp so the shifter moves easily on the rod.

4. Center the shift finger (fore and aft) in the lockout plate and move the

shifter so the finger is disengaged from the lock out by {169}/16 in. (15mm).

5. Tighten the rod clamp to 14 ft. lbs. (20 Nm) and check shifter operation. If operation is spongy or binding, readjust the lock out finger to ½ in. (13mm).

Fox

1. Shift into neutral.

2. Remove the gear shift lever knob and shift boot.

3. Loosen the clamp nuts and check that shift finger slides freely on the shift rod.

4. Move the gear shift lever to the right side, between 3rd and 4th gear position. The gear shift lever should remain perpendicular to the ball housing.

5. With the inner shift lever in neutral and the gear shift lever between 3rd and 4th gear, tighten the clamp nut.

6. Check the engagement of all gears, including reverse and make sure the gear shift lever moves freely.

CLUTCH

Clutch Assembly

REMOVAL AND INSTALLATION

Jetta, Golf and Cabriolet

1. Raise and safely support the vehicle and remove the transaxle.

2. Attach a toothed flywheel holder tool VW–558 or equivalent, to the flywheel and gradually loosen the flywheel-to-pressure plate bolts a few turns at a time. Use a crisscross pattern to prevent distortion.

3. Remove the flywheel and the clutch disc.

4. Use a small prybar to remove the release plate retaining ring. Remove the release plate.

To install:

5. Use new bolts to attach the pressure plate to the crankshaft. Use a thread locking compound and torque the bolts in a diagonal pattern to 72 ft. lbs. (100 Nm).

6. Lightly lubricate the clutch disc splines, release plate contact surface and pushrod socket with multi-purpose grease. Install the release plate, retaining ring and clutch disc.

7. Install a centering tool VW–547 or equivalent, to align the clutch disc.

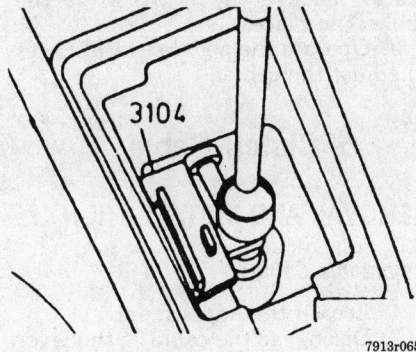

7913r066

Passat and Corrado shifter adjustment

7913r065

Golf and Jetta shifter adjusting tool

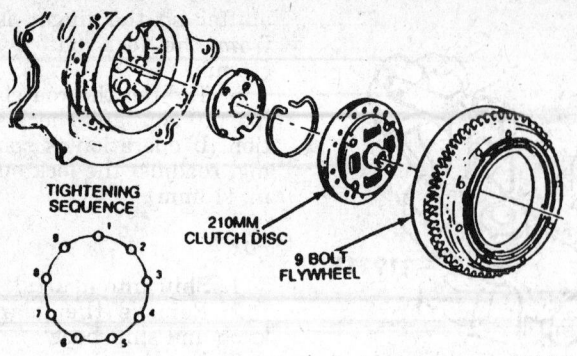

TIGHTENING SEQUENCE

210MM CLUTCH DISC

9 BOLT FLYWHEEL

7913r069

Clutch assembly — Golf, Jetta and Cabriolet

8. Install the flywheel, tightening the bolts 1–2 turns at a time in a crisscross pattern to prevent distortion. Torque the bolts to 14 ft. lbs. (20 Nm).

9. Remove the alignment tool, reinstall the transaxle and adjust the clutch cable.

Fox, Passat and Corrado

1. Raise and safely support the vehicle and remove the transaxle.

2. Matchmark the flywheel and pressure plate if the pressure plate is going to be reused.

3. Gradually loosen the pressure plate bolts 1–2 turns at a time in a crisscross pattern to prevent distortion.

4. Remove the pressure plate and disc.

5. Check the clutch disc for uneven or excessive lining wear. Examine the pressure plate for cracking, scorching or scoring. Replace any questionable components.

To install:

6. Install the clutch disc and pressure plate with the springs on the disc towards the plate. Use an alignment tool to keep the clutch disc centered.

7. Gradually tighten the pressure plate-to-flywheel bolts in a crisscross pattern. Tighten the bolts to 18 ft. lbs. (24 Nm).

8. Install the clutch release bearing.

9. Install the transaxle.

PEDAL HEIGHT/FREE-PLAY ADJUSTMENT

Hydraulic Clutch

If equipped with hydraulic clutch linkage, the slave cylinder has a bleeder screw to purge air from the system. The clutch pedal linkage rod is adjustable to maintain proper pedal height of 3/8 in. (10mm) above brake pedal.

Adjustable Clutch

On cable operated clutches with an adjustable cable, special tool US5043 or equivalent, is available to make it easier to determine proper adjustment. The tool is a simple go or no-go gauge, but proper adjustment can be accomplished without it.

1. Depress the clutch pedal several times.

2. Pull the cable adjusting sleeve up at the transaxle until resistance is felt and insert the gauge or measure the clearance.

3. Loosen the locknut and turn the adjusting sleeve until there is no free play at the gauge. Without the gauge, this distance should be 0.472 in. (12mm).

4. Tighten the locknut and operate the pedal several times. Recheck the adjustment.

Self-Adjusting Cables

1. If the original cable is being reinstalled, compress the spring and hold the cable in place on the transaxle. Another person is required to attach the cable to the clutch lever.

2. If a new cable is being installed, there is a strap holding the spring in place. Remove the strap after the cable is in place.

3. Operate the pedal several times to adjust the cable.

Clutch Cable

REMOVAL AND INSTALLATION

Adjustable Cable

1. Loosen the adjustment.

2. Disengage the cable at the lever arm, noting the placement of the parts.

3. Unhook the cable from the pedal and pull the cable from the firewall.

To install:

4. Grease the pedal end and install and connect the new cable. Adjust the pedal free-play.

Self-Adjusting Cable

1. Depress the pedal several times.

2. Compress the spring located under the boot at the top of the adjuster mechanism and remove the cable at the release lever, noting the placement of the parts.

3. Unhook the cable from the pedal and pull the cable from the firewall.

To install:

4. Grease the pedal end and install the new cable onto the pedal. Compress the spring and have a helper pull the cable down and install to the release lever.

5. If the adjuster spring is retained by a strap, remove the strap after cable installation.

6. Depress the clutch pedal several times to adjust the cable.

Clutch Master Cylinder

REMOVAL AND INSTALLATION

The clutch master cylinder is located on the firewall below the brake master cylinder. The clutch slave cylinder is located on top of the transaxle. The clutch master cylinder is supplied fluid from the brake fluid reservoir. Whenever any part of the system is removed or replaced the system must be bled to remove any air that may be in the lines.

1. Remove the windshield washer bottle.

2. Remove the pressure line from the rear of the clutch master cylinder and plug the fitting.

3. Disconnect the fluid supply hose from the brake fluid reservoir.

4. Inside the vehicle, disconnect the pushrod from the clutch pedal by removing the clip on the retaining pin.

5. Remove the 2 mounting nuts and remove the clutch master cylinder from the vehicle.

To install:

6. Insert the pushrod through the firewall, install new nuts and torque to 5 ft. lbs. (7 Nm). Pin the rod to the pedal and install the clip.

7. Connect the supply line to the brake master cylinder and install the

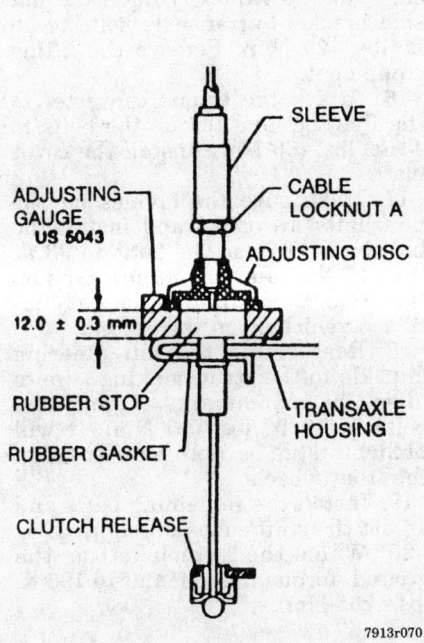

Clutch cable adjustment using gauge

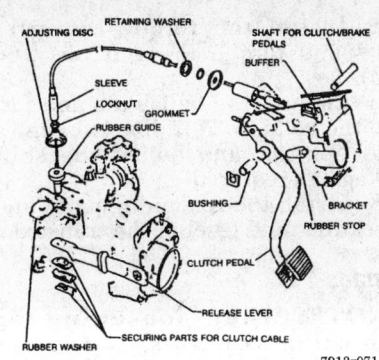

Clutch pedal and cable assembly

pressure line to the rear of the clutch master cylinder.

8. Fill the brake reservoir and bleed the clutch system.

Clutch Slave Cylinder

REMOVAL AND INSTALLATION

1. Raise and safely support the vehicle.
2. Disconnect and plug the pressure line to the slave cylinder.
3. Remove the slave cylinder by removing the spring pin and clip from the transaxle.

To install:

4. Align the slave cylinder on the transaxle housing and insert the spring pin and clip. Bolt the cylinder in place.
5. Connect the pressure line and lower the vehicle.
6. Fill the brake reservoir and bleed the system.

Hydraulic Clutch System Bleeding

1. The clutch and brakes share the same reservoir. Clean all dirt and grease from the cap to make sure no foreign substances enter the system.
2. Remove the cap and diaphragm and fill the reservoir to the top with the approved DOT 3 or 4 brake fluid. Fully loosen the bleed screw which is in the slave cylinder body next to the inlet connection.
3. At this point bubbles of air will appear at the bleed screw outlet. When the slave cylinder is full and a steady stream of fluid comes out of the slave cylinder bleeder, tighten the bleed screw.
4. Refill the reservoir and cap it. Exert a light load of about 20 lbs. to the slave cylinder piston by pushing the release lever towards the cylinder and loosen the bleed screw. Maintain a constant light load; fluid and any air that is left will be expelled through the bleed port. Tighten the bleed screw when a steady flow of fluid with no air is being expelled.
5. Fill the reservoir fluid level back to normal capacity and, if necessary repeat Step 4.
6. Exert a light load to the release lever but do not open the bleeder screw as the piston in the slave cylinder will move slowly down the bore. Repeat this operation 2–3 times; the fluid movement will force any air left in the system into the reservoir. The hydraulic system should now be fully bled.
7. Check the operation of the clutch hydraulic system and repeat this procedure, if necessary. Check the pushrod travel at the slave cylinder to insure the minimum travel is 0.57 in. (15mm).

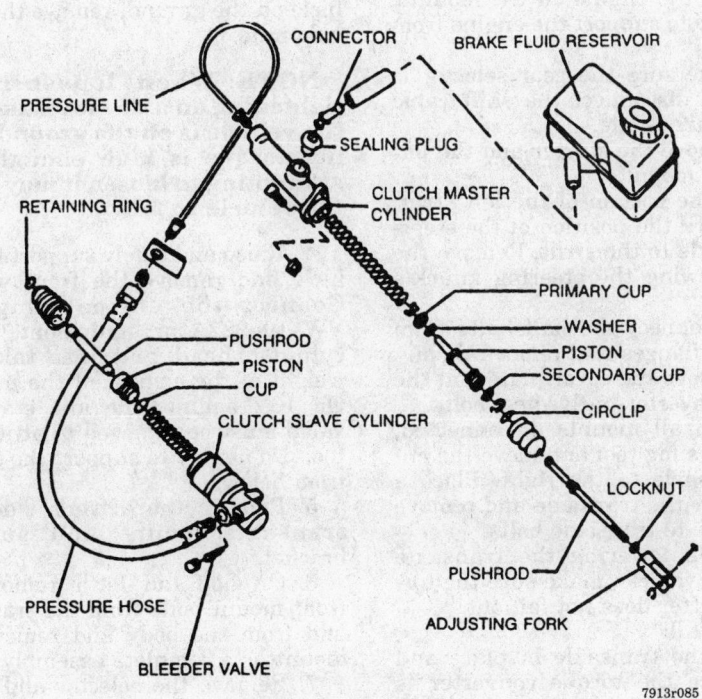

Hydraulic clutch components

AUTOMATIC TRANSAXLE

Transaxle Assembly

REMOVAL AND INSTALLATION

Passat

1. If equipped with electronically theft-protected radio, obtain the security code before disconnecting the battery.
2. Remove the battery and disconnect the wiring from the transaxle.
3. Remove the upper engine-to-transaxle bolts.
4. Raise and safely support the vehicle and remove the front wheels. Connect the engine sling tool VW-10-222A or equivalent, to the cylinder head and just take the weight of the engine off the mounts. On Passat, the idle stabilizer valve must be removed to attach the tool. Do not try to support the engine from below.
5. Put the shifter in **P** and disconnect the shift cable.
6. Clamp and remove the hoses at the transaxle cooler.
7. Remove the starter and the engine's left and right mounts.
8. Remove the skid plate and disconnect the halfshafts from the drive flanges. Hang them from the body with wire.
9. Remove the torque converter plate and turn the engine as needed to remove the torque converter-to-flywheel bolts.
10. Remove the remaining transaxle mounts and lower the hoist slightly.
11. Support the transaxle with a jack and remove the remaining engine-to-transaxle bolts. Be careful to secure the torque converter so it does not fall out of the transaxle.
12. Carefully lower the transaxle out of the vehicle.
To install:
13. Fit the transaxle into the vehicle and make sure the guide pins fit properly between the engine and transaxle. Install the bolts and torque the 12mm bolts to 59 ft. lbs. (80 Nm), the 10mm bolts to 44 ft. lbs. (60 Nm).
14. Install the transaxle mounts and torque the bolts to 44 ft. lbs. (60 Nm). Torque the left side bracket-to-transaxle bolts to 18 ft. lbs. (25 Nm).

15. Install the torque converter bolts and torque to 44 ft. lbs. (60 Nm).
16. Connect the halfshafts and torque the bolts to 33 ft. lbs. (45 Nm).
17. Connect and adjust the shift linkage as required.
18. Install the remaining parts and check the fluid level in the transaxle.

Corrado

NOTE: When loosening or tightening axle nuts, make sure the vehicle is on the ground. Axle nut torque is high enough that attempting to loosen it may cause the vehicle to fall off the support.

1. This vehicle is equipped with a theft protected radio. Obtain the security code and remove the battery.
2. With the vehicle on the ground, loosen the front axle nuts.
3. On supercharged models, remove the intercooler ducting. It may be necessary to remove the ducting for the brakes.
4. Disconnect the wiring from the transaxle.
5. Clamp the coolant hoses and disconnect them from the transaxle fluid intercooler.
6. Raise and safely support the vehicle and remove the front wheels. Connect the engine sling tool VW-10-222A or equivalent, to the cylinder head and just take the weight of the engine off the mounts. Do not try to support the engine from below.
7. Make sure the gear selector is in **P** and disconnect the shift cable from the transaxle.
8. Remove the starter and the left transaxle mount.
9. At the bottom of the left strut, matchmark the position of the steering knuckle to the strut. Remove the bolts to swing the steering knuckle down.
10. Disconnect the halfshafts from the drive flanges and remove them.
11. Remove the cover plate and the torque converter-to-flywheel bolts.
12. With all mounts disconnected, lower the sling tool and move the engine/transaxle to the right. Place a jack under the transaxle and remove the engine-to-transaxle bolts.
13. When lowering the transaxle out of the vehicle, make sure the torque converter does not fall out.
To install:
14. Fit the transaxle in place and make sure the torque converter is properly positioned when installing the bolts. Torque the 10mm bolts to 44 ft. lbs. (60 Nm) and the 12mm bolts to 59 ft. lbs. (80 Nm).

15. Install the mounts using the same torque values. Torque the left side bracket-to-transaxle bolts to 18 ft. lbs. (25 Nm). Remove the lifting equipment.
16. Attach the torque converter to the flywheel and torque the bolts to 44 ft. lbs. (60 Nm). Install the cover plate.
17. Make sure the splines on the halfshafts are clean and install the halfshafts. Torque the bolts to 33 ft. lbs. (45 Nm) and install new nuts on the axles. Do not torque the nuts until the vehicle is on the ground.
18. Reassemble the left steering knuckle to the strut, making sure to align the matchmarks. Torque the bolts to 70 ft. lbs. (95 Nm). It will probably still be necessary to align the front wheels.
19. Install the remaining parts and adjust the shift cable as required.
20. When the vehicle is on the ground, torque the axle nut to 195 ft. lbs. (265 Nm).

Golf, Jetta and Cabriolet

1. If equipped with electronically theft-protected radio, obtain the security code before disconnecting the battery.
2. Disconnect the battery and the speedometer drive and plug the hole in the transaxle.
3. On Golf and Jetta, with the vehicle on the ground, remove the front axle nuts.

NOTE: When loosening or tightening an axle nut, make sure the vehicle is on the ground. Axle nut torque is high enough that attempting to loosen it may cause the vehicle to fall.

4. Raise and safely support the vehicle and remove the front wheels. Connect the engine sling tool VW-10-222A or equivalent, to the cylinder head and just take the weight of the engine off the mounts. On 16V engine, the idle stabilizer valve must be removed to attach the tool. Do not try to support the engine from below.
5. Remove the driver's side rear transaxle mount and support bracket.
6. On Golf and Jetta, remove the front mount bolts from the transaxle and from the body and remove the mount as a complete assembly.
7. Remove the selector and accelerator cables from the transaxle lever but leave them attached to the bracket. Remove the bracket assembly to save the adjustment.

8. Unbolt the halfshafts from the drive flanges. On Golf and Jetta, the shafts must be removed, which may require separating the ball joints from the wheel bearing housing to gain the necessary clearance. Remove the ball joint clamping bolt.

9. Remove the heatshield and brackets and remove the starter. On Cabriolet, the front mount comes off with the starter.

10. Turn the engine as needed to remove the torque converter-to-flywheel bolts.

11. Remove the remaining transaxle mounts and, on Golf and Jetta, the subframe bolts and allow the subframe to hang free.

12. Support the transaxle with a jack and remove the remaining engine-to-transaxle bolts. Be careful to secure the torque converter so it does not fall out of the transaxle.

13. Carefully lower the transaxle from the vehicle.

To install:

14. When reinstalling, make sure the torque converter is fully seated on the pump shaft splines. The converter should be recessed into the bell housing and turn by hand. Keep checking that it still turns while drawing the engine and transaxle together with the bolts.

15. Install the engine-to-transaxle bolts and torque to 55 ft. lbs. (75 Nm).

16. Install all mount and subframe bolts before tightening any on them. Tighten the bolts starting at the rear and work forward. Torque the smaller bolts to 25 ft. lbs. (34 Nm) and the larger bolts to 58 ft. lbs. (80 Nm). Remove the lifting equipment when all mounts are installed.

17. Install the torque converter-to-flywheel bolts and torque them to 26 ft. lbs. (35 Nm).

18. Install the starter and torque the bolts to 14 ft. lbs. (20 Nm). Install the heatshields.

19. If the halfshafts were removed, make sure the splines are clean and apply a thread locking compound to the splines before sliding it into the hub. Connect the halfshafts to the drive flanges and torque the bolts to 37 ft. lbs. (50 Nm). Install new axle nuts but do not fully torque them until the vehicle is on the ground.

20. If removed, fit the ball joints to the control arm and torque the clamping bolt to 37 ft. lbs. (50 Nm).

21. Connect and adjust the shift linkage as required.

22. When assembly is complete and the vehicle is on its wheels, torque the axle nuts to 195 ft. lbs. (265 Nm).

SHIFT/THROTTLE LINKAGE ADJUSTMENT

1. With the engine warm and the gear selector in **P**, loosen the adjusting nut and disconnect the accelerator pedal cable from the transaxle.

2. On the intake plenum, loosen the nuts on the cable bracket and move the sleeve away from the throttle to take up any play. The throttle must remain closed.

3. Turn the nut on the throttle side of the bracket up to the bracket and tighten the other nut against the bracket. Be sure the throttle is still against its stop.

4. Reconnect the cable to the transaxle and have an assistant push the gas pedal to the floor.

5. Push the transaxle lever against the stop and turn the adjusting nut to remove all slack from the cable. Tighten the locknut, release the pedal and push it again to check adjustment.

FRONT SUSPENSION

MacPherson Strut

REMOVAL AND INSTALLATION

Except Fox

NOTE: When loosening or tightening axle nuts, make sure the vehicle is on the ground. Axle nut torque is high enough that attempting to loosen it may cause the vehicle to fall off the support.

1. With the vehicle on the ground, remove the front axle nut.

2. Raise and safely support the vehicle and remove the front wheels.

3. Detach the brake line from the strut and remove the caliper. Hang it from the body with wire.

4. Clean and matchmark the position of the strut to the wheel bearing housing for reassembly.

5. Remove the bolts and push the steering knuckle down away from the strut.

─── **CAUTION** ───
On Cabriolet, do not remove the large nut in the center of the top bearing. The spring will be released while still compressed.

6. On Cabriolet, remove the nuts holding the rubber strut bearing and lower the strut from the vehicle. On other vehicles, remove the large center nut to lower the strut from the vehicle.

To install:

7. Place the strut into the fender and install the nuts. On Cabriolet, torque the 3 nuts to 14 ft. lbs. (20 Nm). On all other models, torque the center nut to 44 ft. lbs. (60 Nm).

8. Fit the wheel bearing housing into the strut and torque the bolts to 70 ft. lbs. (95 Nm).

9. Install the brake caliper and torque the bolts to 44 ft. lbs. (60 Nm).

10. When assembly is complete and the vehicle is on the ground, torque the axle nut to 145 ft. lbs. (196 Nm) for M18 nut or 175 ft. lbs. (237 Nm) for M20 nut.

Fox

NOTE: When loosening or tightening axle nuts, make sure the vehicle is on the ground. Axle nut torque is high enough that attempting to loosen it may cause the vehicle to fall off the support.

1. With the vehicle on the ground, remove the front axle nut.

2. Raise and safely support the vehicle and remove the wheels.

3. Remove the brake caliper from the strut and hang from the body it with wire. Detach the brake line from the strut and remove the rotor.

4. At the tie rod end, remove the cotter pin and the castellated nut and remove the end from the strut with a puller.

5. Loosen the stabilizer bar bushings and detach the end from the strut being removed.

6. Remove the ball joint clamp bolt and push the control arm down to disengage the ball joint from the strut.

7. On some vehicles, the halfshaft spline is secured in the hub with thread sealer. The best way to remove it is to push it out with a wheel puller. Do not use heat; this will ruin the bearing. Pull the strut away from the halfshaft.

8. Remove the upper strut-to-fender retaining nut and lower the strut assembly down and out of the vehicle.

To install:

9. Install the upper end of the strut and torque the nut to 44 ft. lbs. (60 Nm).

10. Make sure the axle splines are clean and apply fresh thread sealer to the outer end. Insert the axle through the hub and install a new axle nut. Do not torque the nut until the vehicle is on the ground.

11. Fit the ball joint into the strut and torque the clamping bolt to 44 ft. lbs. (60 Nm).

12. Lightly lubricate the stabilizer arm bushings with silicone and install them. Torque the bolts to 15 ft. lbs. (20 Nm).

13. Install the brake calipers and torque the bolts to 44 ft. lbs. (60 Nm).

14. When the assembly is complete and the vehicle is on the ground, torque the axle nut to 170 ft. lbs. (230 Nm).

Lower Ball Joints

INSPECTION

1. To check the ball joint, raise and safely support the vehicle. Let the front wheels hang free.

2. Insert a prybar between the control arm and the ball joint clamping bolt. Be careful to not damage the ball joint boot.

3. Measure the play between the bottom of the ball joint and the clamping bolt with a caliper. Total must not exceed 0.100 in. (2.5mm).

REMOVAL AND INSTALLATION

1. Raise and safely support the vehicle, allowing the front wheels to hang. Remove the front wheels.

2. Matchmark the ball joint-to-control arm position.

3. Remove the ball joint clamping bolt.

4. Pry the lower control arm down to remove the ball joint from the strut.

5. Remove the ball joint-to-lower control arm retaining nuts and bolts or drill out the rivets with a ¼ in. (6mm) drill.

6. Remove the ball joint assembly.
To install:
7. Install the ball joint in the reverse order of removal. If no parts were installed other than the ball joint, align the matchmarks. No camber adjustment is necessary if this is done. Pull the ball joint into alignment with pliers. Tighten the 2 control arm-to-ball joint bolts to 47 ft. lbs. (64 Nm) and the ball joint clamping bolt to 44 ft. lbs. (60 Nm).

8. On all other vehicles, bolt the new ball joint in place. Torque the bolts to 18 ft. lbs. (25 Nm) and the ball joint clamping bolt to 37 ft. lbs. (50 Nm).

Lower Control Arm

REMOVAL AND INSTALLATION

NOTE: When removing the driver's side control arm on Cabriolet equipped with an automatic transaxle, it may be necessary to lift the engine/transaxle. First support the engine from above or below. Remove the front left engine mounting nut and bolt, remove the rear mount and raise the engine to expose the front control arm bolt.

1. Raise and safely support the vehicle and remove the wheels.

2. Remove the ball joint clamping bolt and pry the control arm down.

3. Remove the rubber bushings to unfasten the stabilizer bar.

4. Remove the control arm mounting bolts and remove the control arm.
To install:
5. Installation is the reverse of removal. Torque the following components:
• Fox control arm bushing bolts — 40 ft. lbs. (55 Nm)
• Cabriolet control arm bushing bolts — 50 ft. lbs. (68 Nm)
• All others: front bushing bolts — 96 ft. lbs. (130 Nm), rear bolts: 59 ft. lbs. (80 Nm)
• Stabilizer bar link rods — 18 ft. lbs. (25 Nm)
• Stabilizer bar bushing clamp bolts — 32 ft. lbs. (43 Nm)
• Ball joint clamping bolt — 37 ft. lbs. (50 Nm) for 8mm bolt or 44 ft. lbs. (60 Nm) for 10mm bolt

Sway Bar

REMOVAL AND INSTALLATION

1. Raise and safely support the vehicle.

2. Remove the front wheel and tire assemblies.

3. Disconnect the sway bar ends links from both lower control arms.

4. Remove the bolts retaining the sway bar mounting bushing brackets.

5. Remove the sway bar.

6. Installation is the reverse of the removal procedure. Torque the mounting bracket bolts and the end link nuts to 18 ft. lbs. (25 Nm).

REAR SUSPENSION

Shock Absorbers

REMOVAL AND INSTALLATION

NOTE: Do not remove both suspension struts at the same time or the axle beam will be hanging on the brake lines.

1. Working inside the vehicle, remove the cap from the top shock mount and note the way the washers and bushings installed.

2. Remove the upper strut-to-body bolts.

3. Slowly lift the vehicle until the wheels are slightly off the ground.

4. Unbolt the strut from the axle and carefully remove the strut from the vehicle. It may be necessary to press the axle down slightly when removing the strut.

5. Installation is the reverse of removal. Torque the strut-to-body bolts to 26 ft. lbs. (35 Nm) and the strut-to-axle bolts to 77 ft. lbs. (105 Nm).

Rear Wheel Bearings

REMOVAL AND INSTALLATION

Drum

1. Raise and safely support the vehicle and remove the rear wheels.

2. On drum brakes, insert a small pry tool through a wheel bolt hole and push up on the adjusting wedge to slacken the rear brake adjustment.

3. Remove the grease cap, cotter pin, locking ring, axle nut and thrust washer. Carefully remove the bearing and put all these parts where they will stay clean.

4. Before installing, pack the bearing. If any brake dust has fallen onto the axle, wipe off all the axle grease and put on new high temperature bearing grease.

5. Installation is reverse of removal.

Rotor

1. Raise and safely support the vehicle and remove the rear wheels.

2. Remove the brake caliper without disconnecting the hydraulic hose. Support the caliper so it does not hang by the hose.

3. Remove the brake disc.

4. Remove the hub cap and the nut and washer. The torque on the nut is very high, make sure the vehi-

cle is firmly supported and will not fall.

5. Remove the hub/bearing unit from the spindle. The bearing is a sealed unit pressed into the hub.

6. Installation is the reverse of the removal procedure. Torque the hub nut to 170 ft. lbs. (230 Nm).

ADJUSTMENT

1. When adjusting the bearing nut, the thrust washer must still be movable with light effort with a small pry tool.

2. When installing the locking ring, keep trying different positions of the ring on the nut until the cotter pin goes into the hole. Don't turn the nut to align the locking ring with the hole in the axle. Use a new cotter pin. Install the grease cap with a rubber hammer.

Rear Axle Assembly

REMOVAL AND INSTALLATION

1. Raise and safely support the vehicle and remove the rear wheels.

2. Remove the rear brake caliper or drum.

3. Disconnect the brake line and remove the caliper or back plate (with brakes attached) from the vehicle.

4. Disconnect the other end of the brake line; unclip the brake line and parking brake cable from the axle. Unhook the brake pressure regulator spring from the bracket.

5. Support one side of the axle beam so it does not fall and remove the lower shock mount bolts from both sides.

6. Unless it is the part being repaired, avoid removing the axle bushing brackets. Removing these will mean aligning the rear bushings upon reassembly.

7. Remove the bolt from the center of each bushing and lower the axle from the vehicle.

To install:

8. Install the axle but do not torque the bushing bolts yet. They should be torqued with the vehicle on the ground to properly align the bushings.

9. Install the brakes, connect the hydraulic line and bleed the brakes.

10. With the vehicle on the ground, torque the right side axle bushing bolt first, then pry the left side bushing slightly towards the center of the vehicle and torque the left side.

11. Torque the following:
- Passat axle bushing bolts — 52 ft. lbs. (70 Nm)
- All other axle bushing bolts — 44 ft. lbs. (60 Nm)
- Passat lower shock bolt — 77 ft. lbs. (105 Nm)
- Golf/Jetta and Corrado shock mount — 52 ft. lbs. (70 Nm)
- Cabriolet shock mount — 32 ft. lbs. (45 Nm)
- Fox shock mount — 52 ft. lbs. (70 Nm)

12. Connect the brake line at the axle side. Secure the brake line and parking brake cable at the axle.

13. Connect the brake pressure regulator spring to the bracket.

14. If equipped with disc brakes, install the caliper or back plate, with brakes attached. Connect the brake line to the caliper.

15. If equipped with drum brakes, install the brake drum.

16. Install the wheel assemblies.

STEERING

Steering Wheel

The air bag system is equipped with a backup power supply. The battery must be disconnected for more than 20 minutes before the power supply is fully discharged and the system is considered disarmed. A memory saver device will keep the power supply charged.

— CAUTION —
Failure to properly disarm the air bag system may result in accidental deployment of the air bag and possible personal injury.

REMOVAL AND INSTALLATION

With Air Bag

1. Disconnect battery, wait more than 20 minutes.

2. Remove the Torx head screws at the back of the steering wheel.

3. Carefully detach the air bag unit from the wheel and disconnect the wire at the center.

4. Place unit in a safe place, horn pad up.

5. Point the front wheels straight-ahead, remove the ignition key to lock the steering column and remove the nut and spring washer.

6. Mark the position of the wheel to the spline and pull the wheel straight off.

7. Installation is the reverse of removal. Torque the steering wheel nut to 30 ft. lbs. (40 Nm). When installing the air bag, use new Torx screws and tighten to 7.5 ft. lbs. (10 Nm). Do not over torque or the air bag may not function properly.

Without Air Bag

1. Remove the horn pad.

2. Remove the ignition key and turn the wheel until it locks.

3. Remove the center nut and matchmark the wheel to the splines.

4. Pull the wheel straight off.

5. Installation is the reverse of removal. Torque the center nut to 30 ft. lbs. (40 Nm).

Manual Steering Rack

ADJUSTMENT

On some vehicles, there is a rack and pinion free-play adjustment screw and locknut; however, this is not always accessible with the rack installed in the vehicle. Loosen the locknut and adjust the screw to allow smooth, non-binding movement of the rack.

REMOVAL AND INSTALLATION

1. Raise and safely support the vehicle.

2. Remove both front wheels and disengage both tie rod ends.

3. At the steering column, remove the boot clamp, push the boot towards the body and remove the clamp bolt from the universal joint.

4. Remove the rack mounting nuts and remove the rack from its mounts.

5. At this point on some vehicles, the rack cannot be removed from the body. Support the engine/transaxle and remove the subframe bolts or the rear transaxle mount and bracket to allow the rack to move towards the rear.

6. Installation is the reverse of removal. Torque the subframe bolts to 96 ft. lbs. (130 Nm).

Power Steering Rack

REMOVAL AND INSTALLATION

1. Raise and safely support the vehicle.

2. Remove both front wheels and disengage both tie rod ends.

3. Remove the low pressure (suction) hose from the pump and drain the system into a catch pan. Properly discard the fluid.

4. At the steering column, remove the boot clamp, push the boot towards the body and remove the clamp bolt from the universal joint.

5. On Cabriolet, remove the exhaust manifold and shift linkage bracket.

6. Remove the rack mounting clamp nuts and remove the clamps.

7. At this point on some vehicles, the rack cannot be removed from the body. Support the engine/transaxle and remove the subframe bolts to allow the rack to move towards the rear. On Cabriolet, remove the transaxle mount and bracket.

8. Disconnect the power steering hydraulic lines and remove the rack.
   To install:
9. Make sure the mounting bushings are in good condition. Fit the rack assembly into place and torque the clamp nuts to 22 ft. lbs. (32 Nm).

10. Install any subframe bolts that were removed.

11. Connect the hydraulic lines and install the steering column universal joint bolt.

12. Fill the system with new fluid and run the engine to check for leaks and bleed the system.

Power Steering Pump

REMOVAL AND INSTALLATION

1. Remove the suction hose and the pressure line from the pump, drain the fluid into a catch pan. Properly discard the fluid.

2. Loosen the tensioning bolt at the front of the tensioning bracket and remove the drive belt from the pump drive pulley.

3. Remove the pump mounting bolts and lift the pump from the vehicle.

4. To install, reverse the removal procedures. Torque the mounting bolts to 15 ft. lbs. (20 Nm). Tension the drive belt. Fill the reservoir with approved power steering fluid and bleed the system.

BELT ADJUSTMENT

To tension the drive belt, adjust the tensioner bolt so the belt will flex ½ in. (13.0mm) under light thumb pressure.

SYSTEM BLEEDING

1. With the wheels turned all the way to the left, add power steering fluid to the **COLD** mark on the fluid level indicator.

2. Start the engine and run at fast idle momentarily. Shut the engine **OFF** and recheck fluid level. If necessary; add fluid to bring level to the **COLD** mark.

3. Start the engine and bleed the system by turning the wheels from side to side without hitting the stops.

NOTE: Fluid with air in it has a light tan or red appearance.

4. Return the wheels to the center position and keep the engine running for 2–3 minutes.

5. Road test the vehicle and recheck the fluid level making sure it is at the **HOT** mark.

Tie Rod Ends

REMOVAL AND INSTALLATION

1. Raise and safely support the vehicle and remove the front wheels.

2. Remove the cotter pin and nut and press out the tie rod end. A small puller is required.

3. Hold the tie rod with a small pipe wrench or locking pliers and loosen the locking nut.

4. Back the nut away from the rod end far enough to mark the threads at the rod end with a crayon or chalk, then unscrew the end from the tie rod.
   To install:
5. When installing the new tie rod end, screw it onto the rod up to the mark on the threads. When tightening the locknut, hold the tie rod end and tighten the nut securely against it.

6. Reinstall the tie rod end into the steering knuckle and install the nut. Torque the nut to 22 ft. lbs. (30 Nm), then tighten as required to install a new cotter pin.

7. With the wheels on, lower the vehicle and roll it back and forth to settle the suspension. Check toe adjustment.

NOTE: On some vehicles, only the right tie rod is adjustable.

BRAKES

CAUTION

Vehicles equipped with anti-lock brakes have extremely high fluid pressure in the system at all times. Do not disconnect any hydraulic fittings or open a bleeder without properly relieving the system pressure. Failure to properly discharge the system pressure may result in severe personal injury.

Master Cylinder

REMOVAL AND INSTALLATION

Without ABS

1. Disconnect and plug the brake lines.

2. Disconnect the electrical plug from the sending unit for the low fluid switch.

3. Remove the 2 master cylinder mounting nuts and remove the master cylinder and reservoir.

4. The reservoir is held into the master cylinder by a press fit into rubber sealing plugs and can easily be pulled off. To reinstall, moisten the plugs with brake fluid and press it on.
   To install:
5. Position the master cylinder and reservoir assembly onto the mounting studs on the booster and install the washers and nuts. Tighten the nuts to 15 ft. lbs. (20 Nm).

6. Remove the plugs and connect the brake lines.

7. Fill the reservoir and bleed the entire brake system.

Proportioning Valve

REMOVAL AND INSTALLATION

1. Raise and safely support the vehicle.

2. Disconnect the spring and relieve the pressure by pushing the lever towards the axle.

3. Using a line wrench, loosen the lines to the proportioning valve.

4. Remove the retaining nuts and remove the proportioning valve from the frame.

5. Installation is the reverse of removal. Bleed the brake system.

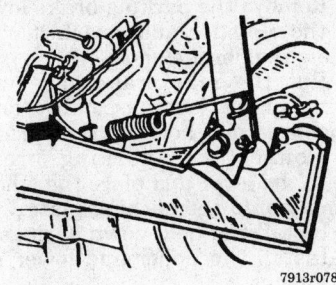

Relieving pressure at the proportioning valve — push the lever toward the axle

Power Brake Booster

REMOVAL AND INSTALLATION

Without ABS

1. Remove the master cylinder from the booster.
2. At the pedals, remove the clevis pin on the end of the booster pushrod by unclipping it and pulling it from the clevis.
3. Disconnect the vacuum hose from the booster.
4. Unbolt the booster; remove the 2 nuts under the dashboard or the 4 nuts holding the booster to its bracket. Remove the booster.
5. Installation is the reverse of removal. Install the master cylinder and bleed the system.

Brake Caliper

REMOVAL AND INSTALLATION

Front

1. Disconnect the negative battery cable. Draw off brake fluid with a suitable syringe.
2. Disconnect the hydraulic brake lines.
3. Raise and safely support the vehicle. Remove the front tire and wheel assembly.
4. Disconnect the hose union at the caliper. Plug the hose to prevent dirt from entering.
5. Remove the caliper mounting bolts and remove the caliper.
6. The installation is the reverse of the removal procedure. Bleed the brake system.
7. Torque the caliper bolts to 26 ft. lbs. (35 Nm).

Rear

1. Disconnect the negative battery cable. Draw off brake fluid with a suitable syringe.
2. Disconnect the hydraulic brake lines.
3. Raise and safely support the vehicle. Remove the rear tire and wheel assembly.
4. Remove the caliper mounting bolts and disconnect the brake pad wear indicator plug.
5. Remove the caliper assembly by pulling to the rear.
6. The installation is the reverse of the removal procedure. Bleed the brake system.
7. Torque the caliper bolts to 26 ft. lbs. (35 Nm).

Disc Brake Pads

REMOVAL AND INSTALLATION

Front

1. Raise and safely support the vehicle.
2. Remove the tire and wheel assembly.
3. Remove the caliper guide bolts and the spring clamp.
4. Turn up the caliper and remove the brake pads. The inner pad is located with a spring in the piston.

NOTE: The brake pads on both calipers on 1 axle should be replaced at the same time.

5. Lubricate the mounting pads with a suitable grease.
6. The installation is the reverse of the removal procedure. Bleed the brake system.

Rear

1. Raise and safely support the vehicle.
2. Remove the tire and wheel assembly.
3. Remove the caliper guide bolts and the spring clamp.
4. Turn up the caliper and remove the brake pads. The inner pad is located with a spring in the piston.

NOTE: The brake pads on both calipers on 1 axle should be replaced at the same time.

5. Lubricate the mounting pads with a suitable grease.
6. The installation is the reverse of the removal procedure. Bleed the brake system.

Brake Rotor

REMOVAL AND INSTALLATION

Front

1. Raise and safely support the vehicle. Remove the front tire and wheel assembly.
2. Disconnect the rubber grommet from the bracket, if equipped.
3. Disconnect and support the caliper, using a piece of wire.
4. Remove the mounting bolts and remove the brake rotor with the proper tool.
5. The installation is the reverse of the removal procedure.

Rear

1. Raise and safely support the vehicle. Remove the tire and wheel assembly.
2. Disconnect and support the caliper, using a piece of wire.
3. Remove the mounting bolts from the caliper support and remove the brake rotor.
4. Always replace both discs of the same axle.
5. The installation is the reverse of the removal procedure. Adjust the parking brake.
6. Torque the caliper support mounting bolts to 53 ft. lbs. (70 Nm).

Brake Drums

REMOVAL AND INSTALLATION

1. Raise and safely support vehicle and remove the rear wheels. Make certain the parking brake is released.
2. Insert a small pry tool through a wheel bolt hole and push up on the adjusting wedge to slacken the rear brake adjustment.
3. Remove the grease cap, cotter pin, locking ring, axle nut and thrust washer. Carefully remove the bearing and put all these parts where they will stay clean.
4. Carefully remove the drum.
5. Before installing, if any brake dust has fallen onto the axle, wipe off all the axle grease and apply a coat of new high temperature bearing grease. Install the parts in the reverse order of removal.

NOTE: When tightening the axle nut, the thrust washer must still be movable with a small pry tool. Spin the drum and check that the thrust washer can still be moved.

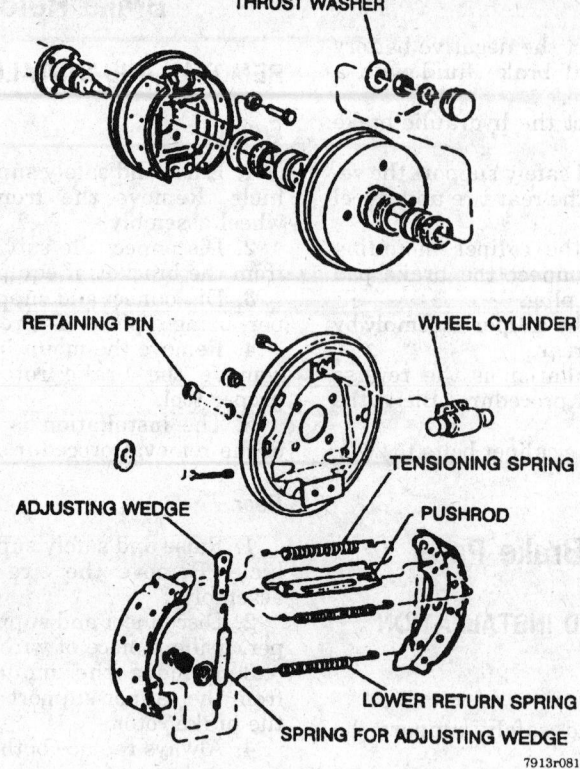

Rear brake assembly on drum brakes

6. When installing the locking ring, keep trying different positions of the ring on the nut until the cotter pin goes into the hole. Don't turn the nut to align the locking ring with the hole in the axle. Use a new cotter pin. Install the grease cap with a rubber hammer.

Brake Shoes

REMOVAL AND INSTALLATION

1. Raise and safely support the vehicle and remove the rear wheels. Make certain the parking brake is released.
2. Remove the rear brake drum.
3. Remove the spring retainers by holding the pin behind the back plate, push in on the retainer and turn it ¼ turn.
4. Remove the shoes from the back plate by pulling first 1 shoe, then the other against the upper spring and from its wheel cylinder slot. Detach the parking brake cable from the brake lever. The entire shoe assembly should now be free of the vehicle.
5. Carefully note the position of each spring, as spring shapes and positions have varied from vehicle to vehicle and year to year.

6. Clamp the pushrod in a vise and begin removing the springs, starting with the lower return spring, adjusting wedge spring, upper return spring and then the tensioning spring and adjusting wedge.
7. On most vehicles, the parking brake lever must be removed from the old shoes and reused. When new parts are purchased, don't forget the clip that holds the parking brake lever pin in place.

To install:
8. Check the wheel cylinder for frozen pistons or leaks. If any defects are found, replace the wheel cylinder.
9. Inspect the springs. If the springs are damaged or show signs of overheating they should be replaced. Indications of overheated springs are discoloration and distortion.
10. Inspect the brake drum and recondition or replace as necessary.
11. Clean the back plate and lubricate the shoe contact points with brake lubricant.
12. With the push rod clamped in a vise, attach the front brake shoe and tensioning spring.
13. Insert the adjusting wedge between the front shoe and pushrod so its lug is pointing toward the backing plate.

14. Remove the parking brake lever from the old shoe and attach it onto the new rear brake shoe.
15. Put the rear brake shoe and parking brake lever assembly onto the pushrod and hook up the spring.
16. Connect the parking brake cable to the lever and place the whole assembly onto the backing plate.
17. Install the hold-down springs.
18. Install the upper and lower return springs.
19. Install the adjusting wedge spring.
20. Center the brake shoes on the backing plate making sure the adjusting wedge is fully released (all the way up) before installing the drum.
21. Install the drum and wheel assembly.
22. Apply the brake pedal a few times to bring the brake shoe into adjustment.
23. If the wheel cylinder was replaced, bleed the system.
24. Road test the vehicle.

Wheel Cylinder

REMOVAL AND INSTALLATION

1. Raise and safely support the vehicle and remove the wheel, drum and brake shoes.
2. Loosen the brake line on the rear of the cylinder but do not pull the line away from the cylinder or it may bend.
3. Remove the bolts and lockwashers that attach the wheel cylinder to the backing plate and remove the cylinder.
4. Position the new wheel cylinder on the backing plate and install the cylinder attaching bolts and lockwashers. Torque to 6 ft. lbs. (8 Nm).
5. Attach the brake line.
6. Install the brakes and bleed the system.
7. Road test the vehicle.

Brake System Bleeding

The same procedure for bleeding the brake system may be used for vehicles with or without ABS brakes.

NOTE: Use only new DOT 4 brake fluid in all Volkswagen vehicles. Do not use silicone (DOT 5) fluid. Even the smallest traces can cause severe corrosion to the hydraulic system. All brake fluids are corrosive to paint.

1. On vehicles with power brakes, bleed brakes with the engine OFF and booster vacuum discharged; pump the pedal with the bleeders closed about 20 times until the pedal effort gets stiff.

2. Fill the fluid reservoir.

3. On vehicles with a rear brake pressure regulator at the rear axle, press the lever towards the rear axle when bleeding the brakes.

4. Connect a clear plastic tube to the bleeder valve at the right rear wheel, with the other end in a clean container.

5. Using either a power bleeder or an assistant pumping the pedal, open the bleeder valve until there are no air bubbles in the fluid stream. Be careful not to let the reservoir run out.

6. Repeat the procedure in sequence at the left rear, right front, left front: working farthest from the master cylinder to the nearest.

Anti-Lock Brake System Service

Vehicles with Anti-lock Brake Systems (ABS) have an electronic fault memory and an indicator light on the instrument panel. When the engine is first started, the light will go on to indicate the system is pressurizing and performing a self diagnostic check. After the system is at full pressure, the light will go out. If it remains lit, there is a fault in the system. The fault memory can only be accessed with the VW testers VAG 1551 or VAG 1598, or equivalent. Be sure to unplug the ABS control unit connector and ground before doing any electric welding on the vehicle.

------- CAUTION -------
The ABS modulator assembly is capable of self-pressurizing to more than 3000 psi. Serious injury may result if the brake service is attempted without disabling and depressurizing the system.

RELIEVING ANTI-LOCK BRAKE SYSTEM PRESSURE

With the ignition OFF, pump the brake pedal 25–35 times to depressurize the system. The system will recharge itself via the electric pump as soon as the ignition is turned ON. Disconnect the pump or the battery

to prevent unintended pressurization. The system can then be bled normally.

Modulator Assembly

REMOVAL AND INSTALLATION

1. Turn the ignition OFF and depress the brake pedal 25–35 times to depressurize the modulator assembly. Disconnect the pump or battery to prevent unintended pressurization.

2. Inside the vehicle, under the left rear seat for Passat, or behind the left kick panel for Corrado or near the right tail light for Golf and Jetta, locate and disconnect the ABS control unit and the ground connection.

3. Remove the brake fluid from the reservoir with a suction pump.

4. Disconnect the brake lines from the modulator assembly and protect its connections from contamination with suitable plugs.

5. Working inside the vehicle, remove the left shelf under the dash to gain access to the brake pedal linkage. Remove the clevis bolt and disconnect the pedal.

6. Remove the locknuts and remove the pressure modulator.

7. Installation is the reverse of removal. Use new locknuts and torque to 18 ft. lbs. (25 Nm). Refill the reservoir with new brake fluid and bleed the system.

Wheel Sensor

In addition to the pressure modulator and electronic control unit, the ABS system includes a wheel speed sensor. These sensors feed a speed signal to the control unit, which compares all the speed signals. Brake fluid pressure is modified as needed to prevent wheel lockup.

REMOVAL AND INSTALLATION

1. Raise and safely support the vehicle.

2. Remove the wheel and unbolt and remove the sensor from the wheel bearing housing.

3. The rotor portion of the sensor assembly is secured to the inside of the wheel hub. To remove the front rotor, the hub must be pressed out of the front wheel bearing.

4. On the rear wheels, the sensor is bolted to the stub axle just above

the axle beam mounting pad. The sensor rotor is pressed into the brake disc. To remove it:

    a. Remove the wheel bearing and the brake rotor.

    b. Insert a drift pin through the wheel bolt holes and gently tap the speed sensor rotor out a little bit at each hole, much like removing an inner wheel bearing race.

To install:

5. When reinstalling, use a suitable sleeve to drive the speed sensor rotor into the disc evenly. When the cover ring is installed, the distance from the ring to the splash shield should be 0.375 in. (9.5mm).

6. When reinstalling the sensor, use a dry lubricant on the sides of the sensor and torque the bolt to 7 ft. lbs. (10 Nm).

CONTROL UNIT

The ABS electronic control unit is located under the rear seat on the left side.

1. Disconnect the negative battery cable.

2. Remove the lower seat cushion.

3. Remove the control unit retaining screws. Disconnect the harness connector and remove the control unit.

4. Installation is the reverse order of the removal procedure.

5. Connect the negative battery cable.

CHASSIS ELECTRICAL

Air Bag

DISARMING

To disarm the air bag system, disconnect the negative battery cable for more than 20 minutes. This will allow the backup power supply capacitor to discharge. Do not use a memory-saver device or the power supply will remain charged. The air bag can then be removed and the battery connected for other electrical test work. Before installing the air bag, disconnect the battery and allow the power supply to discharge for more than 20 minutes.

PRECAUTIONS

• An air bag is an explosive device. Handle with extreme caution.

• Always disconnect the battery before beginning work on the air bag system and wait 20 minutes for the backup power supply to discharge.

• Do not use a computer memory-saver device. It will keep the backup power supply charged.

• Air bag components must not be repaired or opened. Always use new parts.

• Always place a removed air bag unit with the horn pad facing up. Put it in a safe place where it will not be disturbed.

• The air bag unit must not be exposed to grease, fluids, or cleaning agents.

• The air bag unit must not be exposed to temperatures above 194°F (90°C) at any time. Even the heat of a soldering iron can damage or ignite the charge.

• Any testing on the air bag system must be done with the air bag installed in the vehicle. Use only Volkswagen approved test equipment and procedures specified with that equipment's instruction manual.

• Storage and transport of air bags is subject to rules governing explosive devices and should be done only in the original package.

• Failure to follow proper safety precautions may result in personal injury through accidental firing of the air bag, or through failure of the air bag in an accident.

Heater Blower Motor

REMOVAL AND INSTALLATION

Except Cabriolet and Fox

The blower motor is located behind the glove box and it may be easier to remove the glove box to gain access to the motor. The series resistor is mounted on the motor.

1. Disconnect the wires at the blower motor.

2. At the blower motor flange near the cowl, disengage the retaining lug; pull down on the lug.

3. Turn the motor assembly clockwise to release it from its mount, then lower it from the plenum.

4. Installation is the reverse of removal.

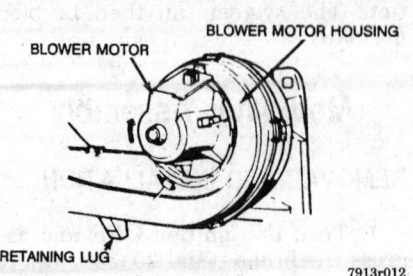

Removal and installation of the blower motor — except Cabriolet and Fox

Cabriolet and Fox

The blower motor and series resistor are reached from under the hood, just in front of the windshield.

1. Disconnect the negative battery cable.

2. Remove the clips and gasket holding the water deflector in place and remove the deflector.

3. To remove the plastic cover that is now visible. Some vehicles have fasteners which are accessed from both under the hood and under the dash. If after removing all screws, bolts or clips visible from above, the cover still won't lift off, check under the dash for more screws.

4. On vehicles with air conditioning, disconnect the linkage for the air distribution flaps. Remove the remaining plastic cover.

5. The blower and series resistor are now accessible. Remove the screws and the motor.

6. Installation is the reverse of removal. Be sure the seal around the motor is properly reinstalled.

Windshield Wiper Motor

REMOVAL AND INSTALLATION

When removing the wiper motor, leave the mounting frame in place. If possible, do not remove the wiper drive crank from the motor shaft.

1. Disconnect the negative battery cable and unplug the multi-connector from the wiper motor.

2. Disconnect the crank arm from the wiper arm assembly.

3. Remove the retaining nut and the crank arm from the wiper motor shaft.

4. Remove the motor mounting bolts and the motor from the vehicle.

To install:

5. Temporarily connect the multi-connector and run the motor. Turn the switch OFF and disconnect the power after the motor stops; it will stop in the park position.

6. Connect the crank arm to the wiper assembly and reverse the removal procedures.

Instrument Cluster

REMOVAL AND INSTALLATION

Except Corrado and Passat

1. Disconnect the negative battery cable and pull off the temperature control knobs and levers.

2. Unclip the heater control trim plate, separate the electrical connectors and remove the plate.

3. Remove the retaining screws and the instrument panel trim plate.

4. Remove the retaining screws and pull out the instrument panel.

5. Squeeze the clips on the speedometer cable head and remove the cable from the instrument cluster.

6. Disconnect all of the vacuum hose and the electrical connections.

7. Installation is the reverse of removal.

Corrado and Passat

1. Disconnect negative battery cable.

2. Peel off the horn button cover, starting at the bottom and remove the steering wheel.

3. Remove the screw trim caps, trim screws and cluster trim.

4. Unscrew the trip odometer reset button.

5. Remove the cover screws (1) and cover, then remove the cluster screws (2) and cluster.

6. Carefully disconnect the vacuum line and multi-point connector.

7. Installation is the reverse of removal.

Speedometer

REMOVAL AND INSTALLATION

1. Disconnect the negative battery cable.

2. Remove the instrument cluster from the instrument panel.

3. Remove the cluster lens and retainer from the cluster.

4. Remove the speedometer retaining screws from the rear of the cluster.

5. Remove the speedometer/odometer from the cluster.

6. Installation is the reverse order of the removal procedure. Connect battery negative cable.

FIRING ORDERS

NOTE: To avoid confusion, always replace spark plug wires one at a time.

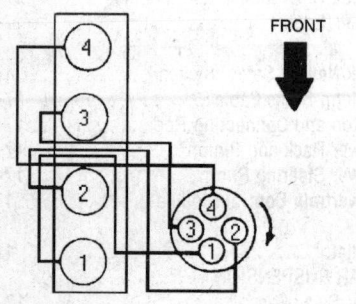

FRONT

185199

2.3L 4-Cyl. Engine
Firing Order: 1–3–4–2
Distributor Rotation: Clockwise

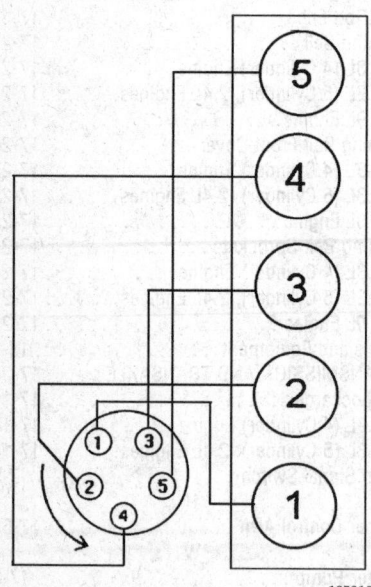

185316

2.3L and 2.4L 5-Cyl. Engines
Firing Order: 1–2–4–5–3
Distributor Rotation: Counterclockwise

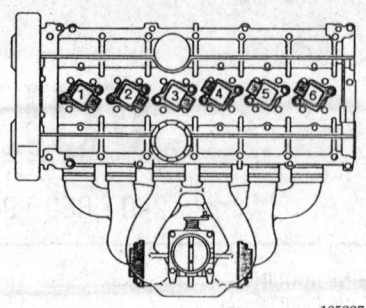

185227

2.9L Engine
Firing Order: 1–5–3–6–2–4
Direct Ignition: uses 1 coil at each cylinder

ENGINE ELECTRICAL

NOTE: Disconnecting the negative battery cable on some vehicles may interfere with the functions of the on-board computer systems and may require the computer to undergo a relearning process, once the negative battery cable is reconnected.

Distributor

REMOVAL AND INSTALLATION

2.3L (B230) 4-Cylinder Engine

1. Disconnect the negative battery cable.
2. Remove the protective cover and distributor cap.
3. Remove the rotor, dust cover and O-ring.
4. Remove the camshaft center bolt and rotor holder. Pull the distributor housing forward until the base rests against the rotor holder. Tap the distributor housing lightly to release the rotor holder from the shaft.
To install:
5. Install the distributor and tighten the rotor holder bolt to 52–66 ft. lbs. (70–90 Nm).
6. Install the distributor cap and protective cover. Connect the negative battery cable.

2.3L (B5234) and 2.4L (B5254) 5-Cylinder Engines

1. Disconnect the negative battery cable.
2. Disconnect the electrical leads to the distributor.

3. Remove the distributor cap. Mark the position of the rotor in relation to the cylinder head.
4. Remove the distributor retaining bolts and pull the distributor out.
To install:
5. Clean the cylinder head where the distributor installs.
6. Install the distributor with the scribe marks aligned and secure in place with the retaining bolts.
7. Install the distributor cap and connect the electrical leads.
8. Connect the negative battery cable.

Ignition Timing

ADJUSTMENT

The B230 engine is equipped with a Bosch LH 2.4 control system or Bosch LH Lambda control system.
The B6304 engine is equipped with a Motronic 4.4 control system. And the B5234 and B5254 engines are equipped with a Motronic 4.3 control system. Each control system adjusts the ignition timing continuously. Manual adjustment of the ignition timing is not possible.

Alternator

PRECAUTIONS

Several precautions must be observed with alternators to avoid damage to the unit.
• If the battery is removed for any reason, make sure it is reconnected with the correct polarity. Reversing the battery connections may result in damage to the 1-way rectifiers.
• When utilizing a booster battery as a starting aid, always connect the positive to positive terminals and the negative terminal from the booster battery to a good engine ground on the vehicle being started.
• Never use a fast charger as a booster to start vehicles.
• Disconnect the battery cables when charging the battery with a fast charger.
• Never attempt to polarize the alternator.
• Do not use test lights of more than 12 volts when checking diode continuity.
• Do not short across, or ground any of the alternator terminals.
• The polarity of the battery, alternator and regulator must be matched and considered before making any

electrical connections within the system.

• Never separate the alternator on an open circuit. Make sure all connections within the circuit are clean and tight.

• Disconnect the battery ground terminal when performing any service on electrical components.

• Disconnect the battery if arc welding is to be done on the vehicle.

REMOVAL AND INSTALLATION

1. Disconnect the negative battery cable.

2. Tag and disconnect the electrical leads on the alternator.

3. On 2.3L 4-cylinder and 2.9L 6-cylinder engines, remove the alternator drive belt.

4. On 2.3L and 2.4L 5-cylinder engines, remove the serpentine belt.

5. On 2.3L and 2.9L engines, remove the adjusting arm-to-alternator and the adjusting arm-to-engine bolts.

6. Remove the alternator mounting bolt(s) and slide the alternator out.

To install:

7. Position the alternator in place and secure with the mounting bolt.

8. On 2.3L 4-cylinder and 2.9L 6-cylinder engines, install the adjusting arm-to-engine and adjusting arm to alternator bolts.

9. Install the serpentine or alternator drive belt.

10. Connect the electrical lead to the alternator.

11. Connect the negative battery cable. Start the car and check the charging system.

Drive Belt

REMOVAL AND INSTALLATION

2.3L (B230) 4-Cylinder and 2.9L (B6304) 6-Cylinder Engines

1. Loosen the mount and adjusting bolts on the accessory, and move to the extreme loose position, generally by moving it toward the center of the engine.

2. Some belts run around a third pulley, which acts as an additional pivot in the belt's path. It may be possible to loosen the idler pulley as well as the main component.

3. Remove the belt.

4. Check the pulleys for dirt or built-up material which could affect belt contact.

NOTE: If a drive belt fails frequently, check for broken or bent brackets and replace the pulley as needed.

To install:

5. Carefully install the new belt. Adjust the belt tension. To check drive belt tension, push lightly on the belt midway between the pulleys. It should be possible to depress the drive belts $3/16$–$3/8$ in. (5–10mm) halfway between the pulleys.

6. Tighten the mounting bolts and recheck the tension.

2.3L (B5234) and 2.4L (B5254) 5-Cylinder Engines

1. Place a breaker bar or ratchet into the square cut hole in the belt tensioner bracket.

2. Pull against the tensioner spring to release tension on the belt.

3. Remove the belt from the alternator pulley.

To install:

4. Route the new belt around the crankshaft, A/C compressor, idler pulley, alternator, and air pump, in that order.

5. Move the tensioner and position the belt.

Starter

REMOVAL AND INSTALLATION

2.3L (B230) 4-Cylinder and 2.9L (B6304) 6-Cylinder Engines

1. Disconnect the negative battery cable at the battery.

2. Raise and support the vehicle safely.

3. Disconnect the leads from the starter motor.

4. Remove the bolts retaining the starter motor brace to the cylinder block, if required. Remove the bolts retaining the starter motor to the flywheel housing and lift it off.

To install:

5. Apply locking compound to the starter mounting bolt threads. Position the starter motor to the flywheel housing and install the retaining bolts finger-tight. Torque the bolts to approximately 25 ft. lbs. (34 Nm).

6. Connect the starter motor leads and the negative battery cable.

2.3L (B5234) and 2.4L (B5254) 5-Cylinder Engines

1. Disconnect the negative battery cable.

2. Raise and safely support the vehicle.

3. Disconnect all the electrical connections from the back of the starter. Remove the bracket and cover plate from the back and side of the starter. Remove the mounting bolts from the transmission, then remove the starter.

To install:

4. Clean the electrical connectors and transmission face where the starter is mounted. Install the starter and transmission attaching bolts loosely. Then install the rear bracket and side cover, and tighten the mounting bolts.

5. Connect all the electrical connections to the starter.

6. Lower the vehicle.

7. Connect the negative battery cable. Test the starter function.

CHASSIS ELECTRICAL

Blower Motor

REMOVAL AND INSTALLATION

240 Models

1. Disconnect the negative battery cable.

2. Remove the sound insulation and side panels on both sides of the radio, if equipped.

3. Remove the control panel and center console.

4. Remove the center air vents, cable and electrical connectors from the clock, glove compartment and air ducts for the center vents.

5. From the left and right sides, remove the air ducts and disconnect the vacuum hoses from the shutter actuators.

6. Fold back the floor mat, remove the rear floor duct screw and move the duct aside.

7. Remove the outer blower motor casing and blower motor wheel. It may be necessary to remove the support from under the glove compartment in order to remove the motor casing.

8. Disconnect the switch electrical leads and remove the switch.

9. Remove the inner blower motor casing, the vacuum hose from the rear floor shutter actuator, the electrical connector and the blower motor.

NOTE: Should the blower motor need to be replaced, a modified replacement unit is available. Certain modifications must be done and instructions are included with the new assembly.

To install:

10. Clean the heater housing of all dirt, leaves etc. before installation. Install the blower motor and connect the wiring and the vacuum hose to the rear floor shutter actuator. Install the inner blower motor casing.

11. Install the blower motor wheel and the outer blower motor casing.

12. Install the rear floor air duct and the floor mat.

13. At the left and right sides, connect the air ducts and the vacuum hoses to the shutter actuators.

14. Connect the electrical leads to the blower motor switch and fasten the switch to the center console.

15. Install the center air vents, cable and electrical connectors to the clock, glove compartment and air ducts for the center air vents.

16. Install the control panel and center console.

17. Install the sound insulation and side panels to both sides of the radio.

18. Reconnect the negative battery cable. Check the system for proper operation.

940 and 960 Models

NOTE: On some vehicles, a drum type fan is used which can be balanced by fitting steel clips to the outer edge. On most vehicles a hose is connected to the fan housing to supply cooling air to avoid damage to the fan motor assembly.

1. Disconnect the negative battery cable.

2. Remove the lower glove box panel.

3. If equipped with Automatic Climate Control (ACC), remove or disconnect the following;
    a. From the right side, remove the instep molding.
    b. Remove the panel from above the control unit, being careful not to damage it upon removal.
    c. Disconnect the electrical connector from the control unit.
    d. Remove the control unit bolts and the control unit. Remove the bracket bolts and bracket.

4. Disconnect the electrical connector from the blower motor.

5. Remove the ventilation pipe, the blower motor-to-housing screws and the blower motor.

To install:

6. Clean the heater housing of all dirt, leaves etc. before installation. Install the blower motor and the ventilation pipe.

7. Connect the harness to the blower motor.

8. For ACC systems;
    a. Install the control unit bracket and the control unit.
    b. Connect the electrical connector to the control unit.
    c. Install the panel above the control unit, the instep panel and the lower glove box panel.

9. Connect the negative battery cable and check the blower motor operation.

850 Models

1. Disconnect the negative battery cable.

2. Remove the following:
    • The three passenger's side soundproofing panel screws and panel
    • The four glove compartment screws and glove compartment
    • The four glove compartment door screws and door
    • Fan motor connector
    • Cable duct from the fan motor
    • Two connectors from the brackets
    • The four fan motor screws and fan motor

To install:

3. Install or connect the following in order:
    • New fan motor and attaching screws
    • Two connectors in the brackets
    • Cable duct to the fan motor

4. Install the glove compartment and door using the screws. Refit the soundproofing panel and screws.

5. Connect the negative battery cable and test the fan motor function.

Windshield Wiper Motor

REMOVAL AND INSTALLATION

Front

240 Models

1. Disconnect the negative battery cable.

2. Remove the right side kick panel and panel under the dashboard beneath the glove box.

3. Remove the defroster hoses and remove the glove box.

4. Remove the wiper arms.

5. Disconnect the wiper assembly and lift it out through the glove box opening.

To install:

6. Position the wiper motor in place.

7. Install the wiper arms, defroster hoses and glove box.

8. Install the right kick panel.

9. Connect the negative battery cable and test the wipers to make sure they function correctly.

940 and 960 Models

1. Disconnect the negative battery cable.

2. Remove the wiper arms and remove the rubber boot at the base of the arm.

3. Lift the hood to its uppermost position by pushing the catch on the hood hinges.

4. Release the washer hoses from the clips along the edge of the cowl. Unscrew the cowl retaining bolts. Older vehicles may use plastic clips instead of bolts to hold the cowl in place. Lower the hood to its normal position.

5. Remove the cowl by pulling it forward and then rotating the front edge upward. Close the hood.

6. Unscrew the bolts which hold the linkage assembly. One of the bolts is hidden beneath a rubber cap.

7. Remove the cover from the wiper motor, cut the cable tie and disconnect the plug to the motor.

8. Unscrew the spindle nut on the motor and the 3 bolts holding the motor to the linkage. Lift the motor out.

To install:

9. Install the motor and secure the wiring.

10. Connect the wiper linkage and check for any possible interference between the moving parts and the wiring. Always reinstall the wire tie on the motor harness.

11. Make sure the cowl is properly seated and install the bolts. Install the washer hoses in their clips, replace the rubber boots and install the wiper arms.

12. Connect the negative battery cable. Test the wiper system and make sure the arms return to the park position.

850 Models

1. Disconnect the negative battery cable.

2. Loosen the wiper arm nut. Remove the wiper arm by holding the nut with locking pliers and pulling the arm off. Remove the nut.

3. Remove wiper motor cover bolts and drain hoses. Lift the cover off.

4. Remove the wiper motor bolts and pull it free of the mounts. Disconnect the electrical connectors and remove the wiper assembly.

5. Mark the crank arm position and remove the nut from the wiper motor shaft. Remove the crank arm.

6. Remove the wiper motor bolts and pull the motor from the assembly.

To install:

7. Install the wiper motor in the assembly and fasten it with the bolts. Line up the crank marks and connect it to the motor. Install the nut with locking compound and torque the nut to 15 ft. lbs. (20 Nm).

8. Install the wiper assembly into the mounts and fasten with attaching bolts. Connect the electrical connectors.

9. Connect the negative battery cable. Test the wiper unit function.

10. Install the cover panel and drain hoses, fasten it with the attaching bolts.

11. Turn the ignition switch **ON** and run the wiper motor. Turn the wiper switch **OFF** and let the motor stop in the park position.

12. Install the wiper arms and nut hand-tight. Adjust the wiper arm position and torque the nut to 12 ft. lbs. (16 Nm).

Rear

NOTE: Refit the radio interference suppression unit if one is present.

1. Disconnect the negative battery cable.

2. Loosen the wiper arm nut a few turns. Remove the wiper arm by holding the nut with locking pliers and pulling the arm off. Remove the nut.

3. Remove the trunk panel. Remove the wiper motor assembly nuts and disconnect the electrical connectors. Lift the assembly out.

4. Remove the nut from the wiper motor shaft.

5. Remove the wiper motor.

To install:

6. Install the new wiper motor and fasten it with the mounting bolts. Install the shaft nut.

7. Install the wiper assembly and connect the electrical connectors. Fasten it with the attaching nuts and torque them to 7.5 ft. lbs. (10 Nm).

8. Install the trunk panel.

9. Connect the negative battery cable.

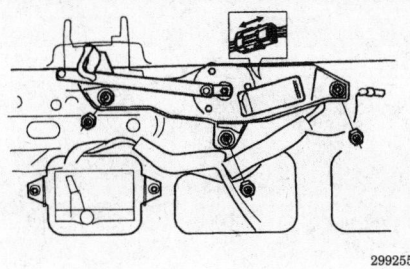

Removing/installing the rear wiper assembly

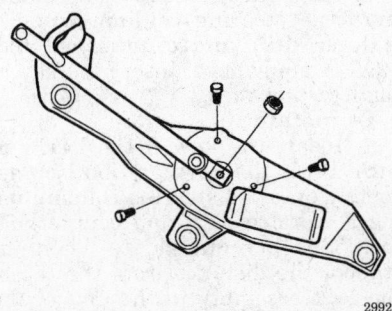

Disconnecting the rear wiper motor from the assembly

10. Run the motor and allow it to stop by itself when the switch is turned **OFF**.

11. Install the wiper arm and nut hand-tight.

12. Adjust the position of the arm so that it is parallel to the bottom of the rear window.

13. Torque the nut to 12 ft. lbs. (16 Nm).

Headlight Switch

REMOVAL AND INSTALLATION

1. Disconnect the negative battery cable.

2. On 240 models, do the following;

a. Disconnect the hose from the left side dash vent.

b. Carefully remove the trim below the vent, by removing the screws holding the vent in place. Remove the vent.

c. Using a very small screwdriver, loosen the setscrew holding the selector knob onto the shaft. If no setscrew is evident, simply pull on the knob until it comes free of the shaft.

3. On 940, 960 and 850 models, do the following;

a. Remove the selector knob by pulling it free of the shaft.

b. Remove the under dash panel.

4. Remove the shaft nut from the switch and remove the switch through the back of the dash. Remove the plug from the back of the switch assembly.

To install:

5. Connect the new switch to the wiring harness. Position the switch and secure with the shaft nut. Install the knob and to tighten the small setscrew, if equipped.

6. Install the remaining components.

7. Reconnect the battery and check the switch function.

Turn Signal Switch

REMOVAL AND INSTALLATION

— **CAUTION** —

On vehicles equipped with an Air Bag, the system must be properly disarmed and the Air Bag must be removed and properly stored. Unintended Air Bag deployment can cause serious or fatal injury.

1. Disconnect the negative battery cable.

2. Disarm the Air Bag, and remove the module if equipped.

3. If there is no Air Bag, remove the center pad from the steering wheel.

4. Remove the steering wheel, upper and lower steering column casings.

5. Remove the turn signal switch held to the column by two or three screws.

To install:

6. Fit the switch into place and install the screws. Check the position of all the wires so that nothing is pinched in the casings.

7. On 850 models, do the following;

a. Install and connect the wiper and/or turn signal switches and fasten them with the screws. Make sure the ground wire is under the mounting screws.

b. Install the stalks and fasten with the attaching screws.

8. Reinstall the column casings and then position the steering wheel with the matchmarks aligned.

9. If there is no Air Bag, install the steering wheel and center pad.

10. If equipped with an Air Bag carefully align the locking screw as

outlined in the Steering Wheel section. Install the steering wheel.

11. If equipped with an Air Bag, connect the module and install the Air Bag.

12. Connect the negative battery cable. Arm the Air Bag system.

Ignition Lock Cylinder

REMOVAL AND INSTALLATION

240 Models

— CAUTION —

On vehicles equipped with an Air Bag, the system must be properly disarmed and the Air Bag must be removed and properly stored. Unintended Air Bag deployment can cause serious or fatal injury.

1. Disconnect the negative battery cable.
2. Disarm the Air Bag, if equipped.
3. Remove the steering wheel.
4. Remove the steering column and mount it in a vice.
5. Break off the washers from the rear edge of the shearing bolts, then, using locking pliers, remove the shearing bolts.

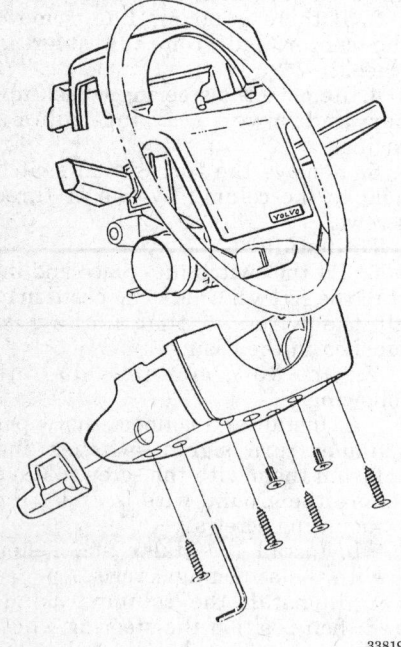

Steering column cover removal — 940 and 960 models

338197

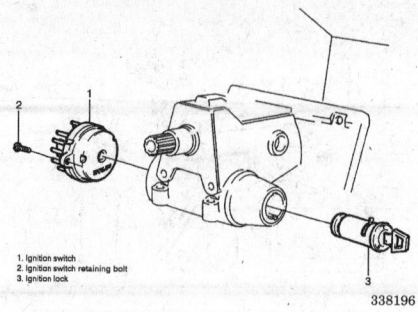

1. Ignition switch
2. Ignition switch retaining bolt
3. Ignition lock

338196

Ignition lock cylinder and switch removal — 940 and 960 models

6. Press the ignition lock assembly from the steering column, using a suitable drift and counterhold tool 5295 or equivalent. Insert the key in the lock and turn.

To install:

7. Insert the key in the lock and turn. Press the new ignition lock assembly onto the steering column, using a suitable drift and counterhold tool 5295 or equivalent. When installed, the distance from the top of the lock assembly to the end of the steering column above splined area should be 6 in. (152mm).

8. Remove the key from the lock. Turn the steering shaft and check that the lock barrel works.

9. Install the steering column.

10. Install the steering wheel.

11. Connect the negative battery cable. Arm the Air Bag, if equipped.

940 and 960 Models

— CAUTION —

On vehicles equipped with an Air Bag, the system must be properly disarmed and the Air Bag must be removed and properly stored. Unintended Air Bag deployment can cause serious or fatal injury.

1. Disconnect the negative battery cable. Disarm the Air Bag system.

2. Remove the steering wheel. Remove the combination switch.

3. Remove the steering column rake adjustment lever using a 0.12 in. (3mm) hex wrench.

4. Remove the parking plate around the steering tube (4 screws). Disconnect the ignition lock connector.

5. Turn the ignition switch to the **I** position. Use the shank of a 0.079 inch (2mm) drill bit to press down the tumblers in the cylinder. Remove the lock assembly.

To install:

6. Install the steering lock assembly. Turn the ignition switch to position **I** then press down the tumblers to install the lock assembly.

7. Install the combination switch assembly. Install the steering wheel.

8. Connect the negative battery cable, then the Air Bag system.

850 Models

— CAUTION —

On vehicles equipped with an Air Bag, the system must be properly disarmed and the Air Bag must be removed and properly stored. Unintended Air Bag deployment can cause serious or fatal injury.

1. Disconnect the negative battery cable. Disarm the Air Bag system.

2. Remove the steering wheel. Remove the upper and lower steering column covers. Remove the windshield wiper control unit.

3. Insert the ignition key into the lock cylinder and turn it to unlock the steering. Press the lock cylinder lug down using a 0.079 in. (2mm) pin or drill bit and remove the cylinder.

To install:

4. Insert the key into the new lock cylinder and turn it to the first position.

5. Press the lock lug down using the drill bit or pin and install the lock cylinder in the column.

6. Install the wiper control unit. Install the upper and lower steering column covers, making sure that no wires are pinched.

7. Install the steering wheel.

8. Connect the negative battery cable. Arm the Air Bag system.

Ignition Switch

REMOVAL AND INSTALLATION

— CAUTION —

When connecting the battery, make sure no one is in the vehicle in case of an SRS malfunction causing accidental Air Bag deployment.

1. Disconnect the negative battery cable.

2. Disarm the Air Bag system, if equipped.

3. Remove the steering column adjuster.

4. Remove upper and lower steering column covers.

5. Remove the turn signal stalk.

6. Disconnect the electrical connections to the ignition switch.

7. Remove the ignition switch retaining screws on the left hand side

of the column. Pull the switch from the column.

To install:

8. Slide the ignition switch into the steering column until it engages the lock cylinder.

9. Install the switch retaining screws and connect the electrical connections.

10. Install the turn signal stalk and steering column covers.

11. Install the steering wheel adjuster and set the position.

12. Connect the negative battery cable. Test switch function.

Park/Neutral Safety Switch

REMOVAL AND INSTALLATION

240, 940 and 960 Models

The start inhibitor (neutral safety switch) also serves to operate the reverse lights. The switch is found on the left side of the gear shift selector.

1. Disconnect the negative battery cable.

2. Remove the ashtray and panel in the center console.

3. Remove the faceplate with the gear position symbols.

4. Remove the start inhibitor switch. Open the connector and lift off the switch.

To install:

5. Install the switch and connect the wiring. Make sure that the tab on the selector lever enters the slot on the switch. Don't forget the prism which fits onto the top of the new switch.

6. Reinstall the holder and the shifter faceplate.

7. Install the panel and the ashtray in the center console.

8. Connect the negative battery cable.

850 Models

CAUTION

When connecting the battery, make sure that no one is in the vehicle in case of an SRS malfunction causing accidental Air Bag deployment.

1. Disconnect and remove the battery. Remove the air cleaner and ducting. Remove the battery tray and air cleaner bracket.

2. Disconnect the transmission cable from the rod arm and remove the arm. Remove the nut, washer and rubber seal from the position sensor.

3. Loosen the dipstick tube bracket and remove the two gear po-

sition sensor bolts. Carefully remove the sensor from the control shaft.

4. Using an angled pick remove the old shaft seal. Be careful not to damage the shaft.

To install:

5. Apply some automatic transmission fluid to the new shaft seal and install it using tool 999-5476 or equivalent.

6. Install the gear position sensor attaching bolts loosely. Install the rubber washer, lockwasher and nut on the end of the control shaft and tighten the nut. Adjust the position of the sensor.

7. Connect the dipstick tube bracket, making sure that the tube O-ring is in position. Torque the bolt to 17 ft. lbs. (24 Nm).

8. Connect the rod arm to the control shaft and torque the nut to 12 ft. lbs. (16 Nm). Connect the transmission cable to the rod arm and fasten it with the washer and lock clips. Apply grease sparingly to the rod arm pin.

9. Install the air filter bracket, battery tray, air cleaner assembly.

10. Connect the negative battery cable.

11. Clear any trouble codes if necessary. Check shifter function before driving vehicle.

Powertrain Control Module

REMOVAL AND INSTALLATION

2.3L (B230) 4-Cylinder Engine

The powertrain control module is located on the right side kick panel. In order to gain access to the powertrain control module, the passenger kick panel must be removed. Release the catch and unplug the connector from the control unit.

NOTE: Always turn the ignition switch OFF before disconnecting the powertrain control module wiring connectors.

2.9L (B6304) 6-Cylinder Engine

The powertrain control module is located behind the left side instrument panel. In order to gain access to the powertrain control module, the driver's side panel must be lowered. Release the catch and unplug the connector from the control unit.

NOTE: Always turn the ignition switch OFF before disconnecting the powertrain control module wiring connectors.

2.3L (B5234) and 2.4L (B5254) 5-Cylinder Engines

1. Disconnect the negative battery cable. Check that the power supply from the main relay has been interrupted by touching the Air Intake Charge (AIC) valve with fingertips. The valve should not vibrate.

2. Open the Electronic Control Module (ECM) box cover. Flip the keeper up and remove the powertrain control module.

To install:

3. Inspect the pins on the powertrain control module and the sockets in the base for damage. Carefully install the powertrain control module and push the keeper down.

4. Connect the negative battery cable. Turn the ignition switch **ON**. Connect the appropriate test equipment and check for fault codes.

ENGINE COOLING

Radiator

REMOVAL AND INSTALLATION

2.3L (B230) 4-Cylinder and 2.9L (B6304) 6-Cylinder Engine

NOTE: Perform this work only on a cold engine.

1. Disconnect the negative battery cable.

2. Set the heater control to MAX heat.

3. Remove the expansion tank cap. Open the draincock on the right-hand side of the engine block and the draincock on the radiator, and allow the coolant to empty into suitable containers.

4. Close the draincocks when the coolant is completely drained.

5. Remove the cooling fan shroud.

6. Disconnect the upper and lower radiator hoses.

7. On vehicles equipped with automatic transmissions, disconnect the transmission oil cooler lines at the radiator. Plug the lines immediately.

8. Remove the upper and lower radiator retainer bolts and lift the assembly from the vehicle.

To install:

9. Place the radiator in position and install the retaining bolts. Attach the fan shroud to the radiator.

10. On automatic transmission vehicles, connect the oil cooler lines.

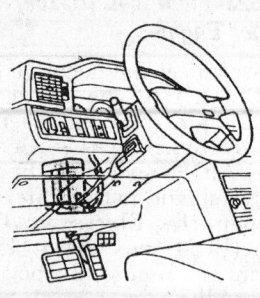

Location of powertrain control module — 2.9L engine

297461

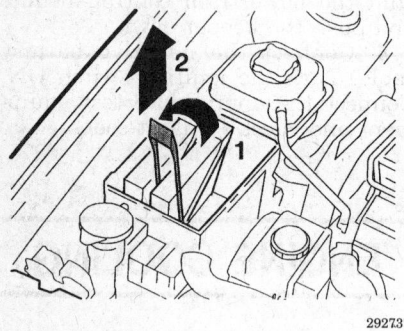

292737

Removing the control module — 2.3L and 2.4L 5-cylinder engines

11. Install the lower and upper radiator hoses.

12. Connect the expansion tank hose. Make sure that the overflow hose is clear of the fan and any sharp bends.

13. Fill the cooling system through the expansion tank, with a 50/50 solution of antifreeze and water.

14. Connect the negative battery cable. Run the engine until normal operating temperature is reached. Check for leaks. Top up the cooling system as required.

15. Check and top up the automatic transmission fluid level.

2.3L (B5234) and 2.4L (B5254) 5-Cylinder Engines

1. Remove the radiator expansion tank cap. Raise and safely support vehicle.

2. Remove the guard from below the radiator. Open the radiator draincock and drain the coolant into a suitable container. Close the draincock when empty.

3. Remove the electric cooling fan. Disconnect the upper and lower radiator hoses.

4. Remove the two upper bolts holding the A/C condenser to the ra-

diator and secure to keep the condenser from moving. Remove the lower condenser bolts to the radiator.

5. Remove the two radiator retaining bolts and pull the radiator out.

— WARNING —
Keep a firm hold on the radiator to avoid damage.

To install:

6. Install the radiator and attaching bolts. Torque to 17 ft. lbs. (30 Nm). Install the two lower A/C condenser bolts, then the two upper bolts and tighten.

7. Install the upper and lower radiator hoses and cooling fan. Connect the fan harness and install the splash guard.

8. Fill the cooling system through the expansion tank. The total capacity is about 7.6 qts. (7.2 liters). Run the engine until it is at operating temperature and top off if necessary. Check for leaks.

Water Pump

REMOVAL AND INSTALLATION

2.3 (B230) 4-Cylinder Engine

1. Disconnect the negative battery cable.

2. Set the heater control to MAX heat.

3. Remove the expansion tank cap. Open the draincock on the right-hand side of the engine block and on the radiator and drain the coolant into a suitable container.

4. Close the draincocks when the coolant is completely drained.

5. Remove the radiator shroud and fan

6. Remove the lower radiator hose at the water pump. If required, remove the retaining bolt for the coolant pipe beneath the exhaust manifold and pull the pipe rearward.

7. Remove the drive belts and water pump pulleys.

8. Remove the water pump bolts, washers and nuts. Remove the water pump assembly.

To install:

9. Clean the gasket contact surfaces thoroughly and use a new gasket and O-rings.

10. Install the water pump and tighten the bolts to 11-15 ft. lbs. (14-19 Nm). Install the coolant pipe and lower radiator hose. Install the accessory drive belts and water pump pulley.

11. Install the fan and shroud. Connect the negative battery cable.

12. Fill the coolant system with coolant. Start the engine and allow it to reach normal operating temperature. Check for leaks. Add coolant as necessary.

2.9L (B6304) 6-Cylinder Engine

1. Disconnect the negative battery cable.

2. Drain the cooling system by opening the draincock on the right side of the cylinder block.

3. Remove the timing belt.

4. Remove the water pump retaining bolts (7) and remove the water pump.

To install:

5. Before installing the water pump, clean the mating surfaces.

6. Install the water pump, using a new gasket. Tighten and torque the mounting bolts 15 ft. lbs. (20 Nm).

7. Install the timing belt.

8. Fill the cooling system. Connect the negative battery cable.

9. Start the engine and check for leaks.

2.3L (B5234) and 2.4L (B5254) 5-Cylinder Engines

1. Disconnect the negative battery cable.

2. Raise and safely support vehicle. Remove the splashguard from below the engine. Drain the cooling system.

3. Remove the following:
• Spark plug cover
• Fuel line clips
• Expansion tank
• Front timing cover
• Accessory belts

4. Remove the timing belt.

5. Remove the water pump and clean the cylinder block where the two mate.

To install:

6. Install the new water pump and gasket, and torque the bolts to 15 ft. lbs. (20 Nm).

7. Install the timing belt.

8. Install the following:
• The two fuel line clips
• Front timing cover
• Accessory belts
• Spark plug cover
• Vibration damper guard
• Wheel well panel
• Wheel

9. Connect the negative battery cable.

10. Fill the cooling system. Run the engine to normal operating temperature. Top off as necessary and check for leaks.

Thermostat

REMOVAL AND INSTALLATION

1. Disconnect the negative battery cable.
2. Drain the cooling system into a suitable container.
3. Disconnect the coolant hose attached to the thermostat housing.
4. Remove the thermostat housing retaining regular or Torx® bolts. Remove the thermostat housing, thermostat and gasket.

To install:

5. Before installing the thermostat, thoroughly clean the mating surfaces.
6. Fit a new gasket and place the thermostat into position.
7. Install the thermostat housing.
8. Fill the cooling system through the expansion tank.
9. Connect the negative battery cable. Start the engine and allow to reach normal operating temperature. Top up with coolant and check for leaks.

Electric Cooling Fan

REMOVAL AND INSTALLATION

2.3L (B230) 4-Cylinder and 2.9L (B6304) 6-Cylinder Engines

1. Disconnect the negative and positive battery cables. Remove the battery tray, as required.
2. Disconnect the harness on the crossmember. Undo the relay and disconnect the ground lead from the terminal on the right-hand wheel housing in the engine compartment.
3. Remove the fan shroud and cooling fan mounting bolts. Remove the fan assembly from the vehicle.

To install:

4. Position the fan in place and secure using bolts. Install the assembly in the engine compartment, and secure in place using the mounting bolts.
5. Connect the wire harness and relay.
6. Install the remaining components. Connect the battery cables.
7. Start the engine and check fan operation.

2.3L (B5234) and 2.4L (B5254) 5-Cylinder Engines

1. Disconnect the negative battery cable.

2. Remove the bolts holding the fan and shroud assembly. Lift the fan and shroud assembly up slightly and disconnect the fan connectors.
3. Remove the control module and air cleaner ducts. Remove the fan and shroud assembly from the engine compartment.
4. Remove the four bolts holding the cooling fan to the shroud. Remove the fan.

To install:

5. Install the fan in the shroud and attach with the bolts.
6. Position the fan and shroud in the engine compartment.
7. Install the relay holder, relays, and connectors.
8. Install or connect the air intake.
9. Install the fan shroud bolts and tighten.
10. Install the control module and air cleaner ducts.
11. Connect the negative battery cable.
12. Run engine and check fan function.

Cooling System Bleeding

—— **WARNING** ——
Do not loosen the expansion tank cap if the coolant is hot. Steam and boiling coolant can cause severe personal injury.

1. Set the heater control to maximum heat level.
2. Remove the expansion tank cap and fill to the proper level with coolant.
3. Run the engine to normal operating temperature.
4. Fill the expansion tank to the correct level and replace the cap.
5. Inspect the system for leaks.

FUEL SYSTEM

Fuel System Service Precautions

Safety is the most important factor when performing not only fuel system maintenance but any type of maintenance. Failure to conduct maintenance and repairs in a safe manner may result in serious personal injury or death. Maintenance and testing of the vehicle's fuel system components can be accomplished

safely and effectively by adhering to the following rules and guidelines.

• To avoid the possibility of fire and personal injury, always disconnect the negative battery cable unless the repair or test procedure requires that battery voltage be applied.

• Always relieve the fuel system pressure prior to disconnecting any fuel system component (injector, fuel rail, pressure regulator, etc.), fitting or fuel line connection. Exercise extreme caution whenever relieving fuel system pressure to avoid exposing skin, face and eyes to fuel spray. Please be advised that fuel under pressure may penetrate the skin or any part of the body that it contacts.

• Always place a shop towel or cloth around the fitting or connection prior to loosening to absorb any excess fuel due to spillage. Ensure that all fuel spillage (should it occur) is quickly removed from engine surfaces. Ensure that all fuel soaked cloths or towels are deposited into a suitable waste container.

• Always keep a dry chemical (Class B) fire extinguisher near the work area.

• Do not allow fuel spray or fuel vapors to come into contact with a spark or open flame.

• Always use a backup wrench when loosening and tightening fuel line connection fittings. This will prevent unnecessary stress and torsion to fuel line piping. Always follow the proper torque specifications.

• Always replace worn fuel fitting O-rings with new. Do not substitute fuel hose or equivalent, where fuel pipe is installed.

Fuel System Pressure

RELIEVING

1. Connect adapter 999-5484 or equivalent to fuel drainage unit 981-2270, 2273 and 2282 or suitable equivalent.
2. Remove protective cap from the valve on the fuel rail.
3. Connect the adapter in the locked or closed position to the valve on the fuel rail.
4. Start the fuel drainage unit.
5. Unlock or open the adapter valve.
6. Raise and safely support the vehicle.
7. Remove the fuel filter valve cap.
8. Connect vent hose 999-5480 to the upstream valve of the fuel filter.

9. Drain the system for approximately 2 minutes.

10. When the system is drained, disconnect vent hose and install the valve cap.

11. Lower the vehicle and disconnect the adapter from the fuel rail. Install the valve cap.

12. Install the protective cap for the fuel rail and throttle pulley cover.

Idle Speed

ADJUSTMENT

The engine control system maintains correct the idle speed under all speed/load conditions. Manual adjustment of the idle speed is not possible.

Idle Mixture

ADJUSTMENT

The engine control system maintains the correct idle mixture under all speed/load conditions. Manual adjustment of the idle mixture is not possible.

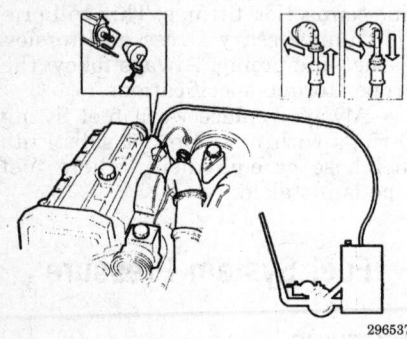

Connecting the adapter and drainage unit to the fuel rail — 2.9L engine

296537

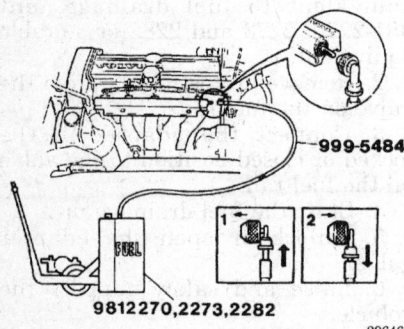

999-5484

9812270, 2273, 2282

296469

Connecting adapter and fuel drainage unit — 2.3L and 2.4L 5-cylinder engines

Fuel Filter

REMOVAL AND INSTALLATION

2.3L (B230) 4-Cylinder and 2.9L (B6304) 6-Cylinder Engines

— CAUTION —
The following procedure will produce fuel vapors and slight fuel spillage. Be sure that there is proper ventilation and take appropriate fire safety precautions.

The fuel filter is either on the left side of the firewall or next to the fuel pump near the left side of the fuel tank.

1. Disconnect the negative battery cable.

2. Relieve the fuel system pressure.

3. Remove the fuel filler cap.

4. Place a suitable container in position.

5. Hold a rag around the connections and loosen the fuel filter connections with flare wrenches.

6. Remove the clamp retaining the fuel filter to the bracket.

To install:

7. Transfer the bracket to the new filter.

8. Note the direction on the fuel filter and install the filter to the bracket.

9. Connect the fuel lines to the fuel filter. Check to ensure the copper seals are correctly installed.

10. Install the fuel filler cap.

11. Reconnect the negative battery cable.

2.3L (B5234) and 2.4L (B5254) 5-Cylinder Engines

1. Disconnect the negative battery cable.

2. Relieve the fuel system pressure.

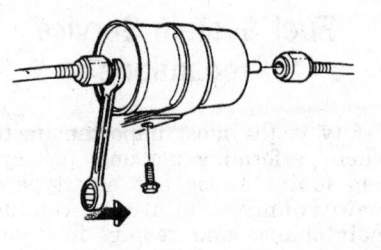

Fuel filter removal — 5-cylinder engines

293959

3. Raise and safely support vehicle.

4. Remove the quick disconnect couplers from the fuel filter. Use a 17mm open end wrench to push the sleeves back. Remove the bolt for the filter bracket and remove the filter.

To install:

5. Install the new filter in bracket and attach with bolt.

6. Push the quick-connectors onto both ends of the fuel filter.

7. Lower the vehicle. Connect the negative battery cable. Turn the ignition switch **ON** to pressurize the fuel system. Check for leaks.

Fuel Pump

REMOVAL AND INSTALLATION

2.3 (B230) 4-Cylinder and 2.9L (B6304) 6-Cylinder Engines

1. Disconnect the negative battery cable.

2. Raise and support the vehicle safely.

3. Relieve the fuel system pressure.

4. Remove the fuel tank.

5. Loosen the lock ring at the top of the fuel tank and remove the sending unit with the transfer pump attached. Note the direction of the float in the tank.

6. Remove the transfer pump from the sending unit.

To install:

7. Install the transfer pump on the sending unit. Install the sending unit in the fuel tank and tighten the lock-ring. Do not overtighten the lock ring, as the plastic threads on some fuel tanks are easily stripped.

8. Install the fuel tank in the vehicle.

9. Lower the vehicle.

10. Connect the negative battery cable, start the engine and check for leaks.

2.3L (B5234) and 2.4L (B5254) 5-Cylinder Engines

1. Disconnect the negative battery cable.

2. Relieve the fuel system pressure.

3. Tilt the rear seat forward and remove or fold back the trunk compartment carpet over the right-hand wheel well panel.

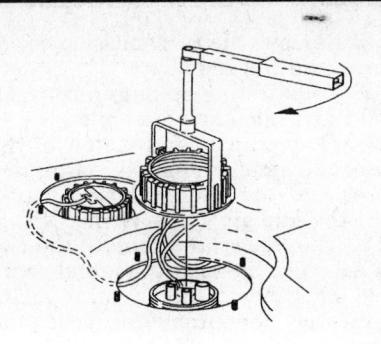

293969

Install the fuel pump retaining nut — 2.3L and 2.4L 5-cylinder engines

4. Disconnect the fuel pump electrical connections.

NOTE: Take note of the color markings on the hoses; colored tape should identify hose locations on the pump.

5. Disconnect the quick-connect couplers for the fuel delivery and return hoses.

6. Remove the pump unit by unscrewing the retaining nut using tool 999-5485 or equivalent. Lift the pump out carefully and remove the rubber seal. When lifting the pump out, do not grab the connections with pliers of any other sharp tools that might cause damage and result in fuel leakage.

To install:

—————— WARNING ——————
Install the retaining nut while the pump is removed, otherwise the tank connection may swell and the nut will be difficult to install

7. Install a new dry seal, making sure that it is seated properly. Lubricate the top and outer side of the seal with petroleum jelly.

8. Install the pump with the heater connection facing towards the right side of the vehicle.

9. Install the fuel pump retaining nut and torque it to 30 ft. lbs. (40 Nm) using tool 999-5484 or equivalent.

10. Apply a small amount of petroleum jelly to the delivery and return hose ends and install them on the pump. The delivery line is marked with yellow tape, which should be matched to the yellow marked pump outlet. Make sure that the quick-connectors are properly seated on the pump.

11. Connect the electrical connections, making sure that they are in the correct position. Install the panels and carpets.

12. Connect the negative battery cable.

13. Run the engine and check its function.

Fuel Injector

REMOVAL AND INSTALLATION

2.3L (B230) 4-Cylinder and 2.9L (B6304) 6-Cylinder Engines

1. Disconnect the negative battery cable.

2. Relieve the fuel system pressure.

3. Disconnect the fuel lines, electrical connectors and vacuum hose from the injection manifold and pressure regulator.

4. Unbolt the pressure regulator from the fuel rail bracket.

5. Remove the injector cover plate, if equipped.

6. Remove the fuel injection manifold retaining bolts. Remove the fuel injection manifold and injectors as one unit.

7. Secure the injection manifold in a suitable holding fixture and remove the fuel injectors.

To install:

8. Check the fuel injector O-rings; replace if necessary.

9. Coat the O-rings with petroleum jelly and install the fuel injectors to the injection manifold.

10. Install the fuel injection manifold and fuel injectors as one unit. Install and tighten the retaining bolts.

11. Install the injector cover plate, if equipped.

12. Connect the pressure regulator to the fuel injection manifold and then attach to bracket.

13. Connect the vacuum hose, fuel lines and electrical connectors.

14. Connect the negative battery cable.

2.3L (B5234) and 2.4L (B5254) 5-Cylinder Engines

1. Disconnect the negative battery cable.

2. Relieve the fuel system pressure.

3. Remove or disconnect the following components:
• upper air charge pipe
• fuel rail cover
• injector connectors
• fuel line clips

4. Install clamp tools 999-5533 or equivalents on injectors, press them on to make sure they are seated. Turn the adjusting screw until it

makes contact with the fuel rail and then another half turn.

5. Disconnect the vacuum hose from the pressure regulator. Remove the fuel rail retaining bolts and lift it off with the injectors.

NOTE: Handle the fuel injectors with care to avoid damaging the nozzles or needles. Make sure that the rubber dampers from the intake manifold are retained.

6. Remove the clamp from the injector to be replaced. Make sure that a spacer is in the fuel rail O-ring seat. Remove the injector.

To install:

7. Install the injector, lubricate the O-ring with petroleum jelly.

8. Install or connect the following:
• rubber damper on intake manifold
• fuel rail on the manifold
• new bolts in the fuel rail and torque to 7.5 ft. lbs. (10 Nm)
• fuel line clips
• injector connector with rubber seal
• fuel rail and throttle pulley covers
• upper air charge pipe

9. Connect the negative battery cable.

10. Run the vehicle and check for leaks.

ENGINE MECHANICAL

Engine Assembly

REMOVAL AND INSTALLATION

2.3L (B230) Engine

1. Disconnect the battery cables, negative lead first. Remove the battery.

2. If equipped with a manual transmission, remove the 4 retaining clips and lift up the shifter boot. Remove the snapring from the shifter.

3. Disconnect the windshield washer hose and engine compartment light wire. Scribe marks around the hood mount brackets on the underside of the hood for later alignment. Remove the hood.

4. Remove the overflow tank cap. Drain the cooling system.

5. Remove the upper and lower radiator hoses. Disconnect the overflow

hoses at the radiator. Disconnect the PCV hose at the cylinder head.

6. If equipped with an automatic transmission, disconnect the oil cooler lines at the radiator.

7. Remove the radiator and fan shroud.

8. Remove the air cleaner.

9. If equipped, disconnect the hoses at the air pump. Remove the air pump and drive belt.

10. Remove the vacuum pump and hoses. Disconnect the power brake booster vacuum hose.

11. Remove the power steering pump, drive belt and bracket. Position aside without disconnecting the hydraulic lines.

12. If equipped with A/C, remove the crankshaft pulley and compressor drive belt. Then install the pulley again for reference. Disconnect the air conditioning wiring and remove the compressor from the bracket. Position the compressor aside without disconnecting the hoses. Remove the bracket.

13. Disconnect the vacuum hoses from the engine. Disconnect the carbon canister hoses.

14. Disconnect the distributor wire connector, high tension lead, starter cables and the clutch cable clamp.

15. Disconnect the wiring harness at the voltage regulator. Disconnect the throttle cable at the pulley and the wire for the A/C at the manifold solenoid.

16. Remove the gas cap. Disconnect the fuel lines at the filter and return pipe.

17. At the firewall, disconnect the electrical connectors for the ballast resistor and relays. Disconnect the heater hoses.

18. Disconnect the micro-switch connectors at the intake manifold and all remaining harness connectors to the engine.

19. Drain the crankcase.

20. Remove the exhaust manifold flange retaining nuts. Loosen the exhaust pipe clamp bolts and remove the bracket for the front exhaust pipe mount.

21. From underneath, remove the front motor mount bolts.

22. If equipped with an automatic transmission, place the gear selector lever in **PARK** and disconnect the gear shift control rod from the transmission.

23. On manual transmission vehicles, disconnect the clutch controls, then drive out the pivot pin and remove the shifter from the control rod.

24. Disconnect the speedometer cable and the driveshaft from the transmission.

25. On overdrive equipped vehicles, disconnect the control wire from the shifter.

26. Raise and support the vehicle safely. Use a floor jack and a wooden block and support the weight of the engine beneath the transmission.

27. Remove the bolts for the rear transmission mount. Remove the transmission support crossmember.

28. Lift out the engine using the proper lifting equipment.

To install:

29. Install the engine assembly in the vehicle and tighten all engine mounting bolts. Install the transmission crossmember and remove the floor jack.

30. Install the driveshaft, speedometer cable, clutch controls (manual transmission) and gear selector mechanism.

31. Install the exhaust system.

32. Install the A/C compressor and related accessory drive units. Install all accessory drive belts. Install the vacuum pump.

33. Install the radiator and shroud. Install all vacuum, coolant and fuel lines and hoses. Connect all electrical connectors previously disconnected.

34. Install the hood, windshield wipers and battery with cables.

35. After installing the remaining components, fill the engine with oil, the radiator with coolant and the transmission with fluid.

36. Adjust the reverse lock clamp and the gear selector. Adjust the throttle valve/pulley, automatic transmission kick-down cable and link rod.

37. Start the engine and allow it to reach normal operating temperature. Check the ignition timing. Check for leaks.

2.9L (B6304) 6-Cylinder Engine

1. Disconnect the negative battery cables, negative first, then remove the battery.

2. Disconnect the ground lead connection to the body at the top of side member.

3. Remove the drive belt.

4. Remove the cooling fan.

5. Release the upper bolts and unfasten the connector at the relay in front of the battery. Disconnect the ground lead at the right-hand ground terminal.

6. Drain the cooling system.

7. Remove the upper and lower radiator hoses from the engine. Remove the radiator overflow hose.

8. Remove the transmission cooler lines from the radiator.

9. Remove the top nut on both left and right side engine mounts.

10. Disconnect and remove the large and small crankcase ventilation hoses and the idle air hose. Disconnect the idle air valve wiring.

11. Disconnect and remove the two EVAP valve hoses at the intake manifold. Disconnect the air mass meter connector, air preheater hose and throttle pulley cover.

12. Remove the servo pump mounting bolts.

13. Disconnect and remove the fuel return line at the regulator and fuel line at the bulkhead. Remove the throttle cable, cruise control vacuum hose and fuel line snap catches.

14. Remove the engine wiring harness cover and disconnect the harness. Disconnect the relay connector. Remove the harness duct retaining nuts.

15. Disconnect the heater hoses at the bulkhead, ECC hoses at the intake manifold and brake servo vacuum hose. Disconnect the timing pick up and camshaft sensor connectors.

16. Support the engine at the rear using engine removal tool assembly 5033, 5006, 5115, 5428 and 5429, or equivalent that will support the engine from above.

17. Remove the splashguard and air baffle under the engine.

18. Remove the radiator.

19. Drain the engine oil.

20. Disconnect the hose at the oil thermostat in the cylinder block.

21. Disconnect the A/C compressor wiring. Remove the compressor from the mount and set it aside without disconnecting the hoses.

22. Remove the exhaust pipe flanges at the manifold. Remove the lower section of the air preheater pipe and remove the exhaust pipe shield.

23. Remove the oil pipe connections at the gearbox. Plug the openings.

24. Remove the clips between the gear selector lever and control rod/reaction arm. Withdraw the rods from their mounting.

25. Disconnect and remove the oxygen sensor wiring.

NOTE: Before separating the driveshaft, mark the coupling halves for reassembly.

26. Disconnect the driveshaft and transmission support member.

27. Install engine lifting tool (2810 or equivalent) and adjust the lifting yoke to ensure the engine is balanced. Position the wiring harnesses so as to avoid damage when lifting.

28. Remove the engine and transmission assembly from the vehicle.

To install:

29. Install the engine into the vehicle, guiding the engine mounting into position.

30. Install the mounting nuts. Torque to 37 ft. lbs. (50 Nm).

31. Position a jack to support the transmission and disconnect the engine from the hoist.

32. Support the rear of the engine using support rails and lifting beam assembly. Remove the jack from beneath the transmission.

33. Using the transmission lifting tool (5972 or equivalent), raise the transmission. Tighten the bolted joints between the support member and side members. Tighten the transmission bump stop nut to 37 ft. lbs. (50 Nm).

34. Attach the control rod and reaction arm to the gear selector lever mounting. Install the locking clip.

35. Connect the oxygen sensor lead.

36. Install the driveshaft. Tighten the front and rear couplings, noting the marks made during removal.

37. Connect the preheater pipe to the exhaust pipe. Tighten the sump bolts.

38. Install the A/C compressor.

39. Reconnect the hoses to the oil cooler. Torque to 22 ft. lbs. (30 Nm).

40. Remove the lifting tools from rear of engine.

41. Install the heater hoses.

42. Install the timing pick up and camshaft position sensor connectors.

43. Connect the engine connector to the wiring harness at the left side wheel housing. Connect the relay and install the wiring duct retaining nuts.

44. Connect the fuel hoses, vacuum hoses and electrical connectors.

45. Install the throttle cable and throttle pulley cover.

46. Install the servo pump. Install the accessory drive belts.

47. Install the radiator and fan with the coolant hoses and transmission oil pipes.

48. Install the battery. Connect the lead at the right side wheel housing and battery positive leads.

49. Install the remaining components.

50. Fill the cooling system.

51. Connect the negative battery cable. Start the engine and check for leaks.

2.3L (B5234) and 2.4L (B5254) 5-Cylinder Engines

1. Disconnect the negative battery cable.

2. Remove the battery and tray.

3. Raise and safely support vehicle.

4. On vehicles with automatic transmissions remove the air baffle from below the engine.

5. Remove the radiator expansion cap. Drain the coolant into a suitable container.

6. Remove the front wheels and disconnect both track rods from the axle. Remove both ball joints from the control arm.

7. Remove the ABS/brake hose bracket bolt. Remove both halfshafts.

8. Remove the right side engine mount retainer bolts.

9. Remove the torque rod bolt in the gearbox

NOTE: Install transmission plugs in axle shaft holes to prevent fluid leakage.

10. Remove the front exhaust pipe lower nut and bolt from the bracket. Remove the two carriage bolts and skid plate. Disconnect the speedometer connection and remove the front and rear lower engine mount bolts. Lower the vehicle.

11. Remove the fresh air intake to the air cleaner, coil wires, throttle pulley cover and throttle cable from the pulley.

12. Tag and disconnect the throttle body inlet hose, idle air control valve, crankcase ventilation, preheat hoses and mass air flow sensor connector.

13. Disconnect the torque rod from the bracket and firewall.

14. Disconnect the ground strap from the firewall.

15. Unfasten the heated oxygen sensors and clips.

16. Remove the brake booster hose from the engine.

17. Remove the radiator and fan.

18. Remove the upper air charge pipe and fresh air intake from the radiator then disconnect the vacuum hoses to the turbocharger and EGR regulator. Protect the radiator with a piece plywood.

19. Remove the radiator, expansion tank and coolant hoses.

20. Remove the clutch slave cylinder retaining ring, if equipped. Make sure that the piston does not slip out. Remove the gear cable selector, after marking the position.

21. On automatic transmissions, mark the position and then remove the gear selector cable.

22. Remove the accessory drive belt

23. Remove the A/C compressor without disconnecting the lines and set it aside

24. Remove the starter.

25. Remove the fuel distribution manifold cover, injector covers, upper and lower fuel line clips and engine ground strap.

26. Install holders 999-5533 or equivalent on the injectors. Disconnect the fuel pressure regulator vacuum hose. Lift the fuel distribution manifold off and lay it aside.

NOTE: Make sure that the injectors and needles are not damaged.

27. Disconnect and remove the wiring harness from the engine.

28. Lift up the air pump and lay it to one side. Install engine lifting yoke 999-2810 and arm 999-5428, or equivalents, and connect to hoist. Remove the front engine mount when the engine/transmission is secured.

29. Lift the engine out of the vehicle. On vehicles with automatic transmissions remove the turbo oil cooler lines and valve from the right side of the oil sump.

To install:

30. Install the engine using the lifting yoke and arm. Adjust the yoke to properly balance the engine as it goes in.

31. Tighten the front engine mount nut.

32. Install the A/C compressor, air pump and shield.

33. Install the wiring harness and bracket.

34. Lubricate the O-rings on the fuel injectors and install. Tighten the fuel distribution manifold.

35. Install the fuel distribution manifold cover and upper and lower clips.

36. Install the accessory drive belt.

37. Install the radiator, expansion tank, cooling fan and hoses. If equipped, connect the upper and lower oil cooler hoses.

38. Install the starter motor.

39. If equipped with a manual transmission, install the clutch slave cylinder and clips.

40. On vehicles with manual transmissions, install the reverse light switch wiring, rubberized gear selector cable mount, and the gear selector cables to its mount on the transmission. They are color coded for easy installation.

41. On vehicles with automatic transmissions, connect the electrical connectors and install the gear selector cables.

NOTE: If the vehicle is equipped with cruise control, the vacuum hoses and motor wiring must be connected before the battery tray is installed.

42. Install the speedometer connector.

43. Install the front and rear engine mount bolts and torque to 37 ft. lbs. (50 Nm).

44. Instal a new torque rod bolt and nut in the transmission and torque to 26 ft. lbs. (35 Nm) plus 90 degrees

45. Install the halfshafts, followed by the ball joints.

46. Install the remaining engine components making sure all wire connections are made, and all vacuum lines are attached.

47. Lower the vehicle and tighten the three exhaust pipe nuts. Torque the bolts to 18 ft. lbs. (25 Nm)

48. Fill the radiator and crankcase with new anti-freeze and engine oil.

49. Start the engine and run it until the thermostat opens.

50. Check the engine for leaks and add fluids as necessary.

Engine Mounts

REMOVAL AND INSTALLATION

2.3L (B230) 4-Cylinder Engine

1. Disconnect the negative battery lead.

2. Using and engine lift, raise the engine slightly.

3. Remove the engine mount retaining nuts.

4. When replacing the left-side engine mount, cut the strap for the power steering hose.

To install:

5. Place the engine mount into position. On the left-side engine mount, attach the bracket between the intake manifold and engine mount. Install the strap for the power steering.

6. Install the lower engine mount on the front axle. Lower the engine and remove the lifting tools.

7. Connect the negative battery lead.

2.9L (B6304) 6-Cylinder Engine

1. Disconnect the negative battery cable.

2. Remove the engine mount top nut.

3. Use support bars, lifting yoke and lifting hook 5033, 5006, 5115 and 5186 or equivalent lifting equipment and raise the engine slightly.

4. Raise the vehicle and support it safely. Remove the splashguard under the engine.

5. Remove the mounting nuts and bolts.

6. On left-side engine mount, remove the mount and bracket towards the front.

7. On right-side engine mount, remove the mount and bracket towards the rear.

To install:

8. Assemble the mount on the bracket. Install the mount and bracket. Tighten the top mounting nut to 37 ft. lbs. (50 Nm).

9. Install the splashguard. Lower the vehicle and remove the lifting tools.

10. Connect the negative battery lead.

2.3L (B5234) and 2.4L (B5254) 5-Cylinder Engines

Transmission Mount

1. Remove the torque rod bolt from the engine. Loosen the bolt from the body mount and push it to one side.

2. Remove the nut from the engine mount.

3. Remove the splash guard below the engine.

4. Remove the torque rod bolt.

5. On automatic transaxle equipped models. remove the cooling fan.

6. Use a jack to lift the engine, and remove the engine mount bolt.

7. Remove the engine mount.

To install:

8. Raise the engine slightly to allow more space for installation, and install the engine mount.

NOTE: Make sure that the engine mount guide pin is correctly located when installing.

9. Install the engine mount bolt.

10. Install new lower torque rod bolts.

11. Install the engine mount nut.

12. Install the splash guard below the engine.

13. Install new upper torque rod bolts.

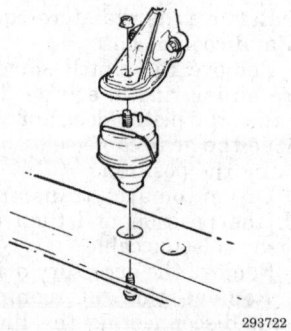

293722

**Engine mount and pin position —
2.3L and 2.4L 5-cylinder engines**

Rear Mount

1. Raise and safely support vehicle.

2. Remove the engine mount nut from top side of engine.

3. Remove the torque rod bolt from top side.

4. Remove the splash guard from below engine.

5. Remove the torque rod bolt from under engine.

6. Raise the engine a maximum of 1.18 in. (30mm). Remove the two engine mount bolts and loosen the third.

——— WARNING ———
Do not raise the engine more than 1.18 in. (30mm), as axle damage will result.

7. Remove the third engine mount bolt. Remove the engine mount by lifting on the collision shield and mount.

To install:

8. Install the engine mount and three mounting bolts to the transmission.

9. Install the engine mount bolt.

10. Install the new lower torque rod bolt. Torque to 37 ft. lbs. (50 Nm.) plus 90°.

11. Install the engine splash guard.

12. Install the new upper torque rod bolt. Torque to 37 ft. lbs. (50 Nm.) plus 90°.

13. Install the engine mount nut.

Cylinder Head

REMOVAL AND INSTALLATION

2.3L (B230) 4-Cylinder Engine

1. Disconnect the battery.

2. Remove the overflow tank cap and drain the coolant. Disconnect the upper radiator hose.

3. Remove the distributor cap and wires.

4. Remove the PCV hoses.

5. Remove the EGR valve and vacuum pump.

6. Remove the air pump, if equipped, and air injection manifold. Disconnect and remove all hoses to the turbocharger, if equipped. Plug all open hoses and holes immediately.

7. Remove the exhaust manifold and header pipe bracket.

8. Remove the intake manifold.

9. Remove the fuel injectors.

10. Remove the valve cover.

11. Remove the fan and shroud.

12. Remove the timing belt

Installing cylinder head with camshaft set at TDC: Make sure water pump O-ring sits correctly in groove — 2.3L 4-cylinder engine

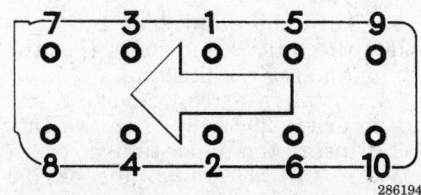

Cylinder head bolt torque sequence — 2.3L 4-cylinder engine

13. Loosen the cylinder head bolts by reversing the torque sequence. Remove the cylinder head.

To install:

14. Check the position of the crankshaft. No. 1 piston should be at TDC. Check the position of the camshaft for cylinder No. 1. Both lobes should be in such a position that if the head were installed, the valves would be closed.

15. Install the cylinder head gasket and the cylinder head. Ensure that the O-ring for the water pump is in place. Apply a light coat of oil to the head bolts and install.

16. Tighten the head bolts in three steps using the proper sequence.

   a. Step 1 — Tighten all bolts to 14 ft. lbs. (20 Nm)

   b. Step 2 — Tighten all bolts to 43 ft. lbs. (60 Nm)

   c. Step 3 — Angle tighten all bolts an additional 90 degrees

17. Install the timing belt.

18. Install the shroud and fan. Install the drive belts and pulleys.

19. Install the intake manifold, fuel injection system, throttle cable and valve covers.

20. Install the exhaust manifold and header pipe. Install the air pump assembly. If equipped with a turbo-

charger, install the turbocharger and related parts.

21. Install the EGR valve, vacuum pump, PCV hoses, distributor cap and wires, overflow tank and battery cables.

22. Fill the radiator with coolant, check the engine oil and transmission fluid. Start the engine and allow it to reach operating temperature. Check the timing.

2.9L (B6304) 6-Cylinder Engine

1. Disconnect the negative battery cable.

2. Drain the cooling system.

3. Remove the front exhaust pipe, heat shield and exhaust manifold(s).

4. Remove the coolant pipe bolts.

5. Remove the timing belt.

6. Remove the transmission mounting plate bolt.

7. Remove the air mass meter and intake hose.

8. Remove the throttle pulley cover, throttle cable and cable bracket.

9. Disconnect the throttle switch lead and vacuum hoses at throttle housing and cruise control servo.

10. Remove the intake manifold outer section.

11. Mark the positions and remove the ignition coils.

12. Mark the camshaft pulleys (intake and exhaust sides) and remove the pulleys, using holding tool 5199 or equivalent.

13. Remove the camshaft sensor, ground terminals and temperature sensor connector. Remove the coolant hose at rear.

14. Carefully tap the top half of the cylinder head upwards, using a soft mallet.

15. Tap the joint lugs and camshaft front ends. Remove the camshafts.

16. Remove the cylinder head bolts, starting at the outside and working inwards. Lift the cylinder head from the engine. Remove the gasket.

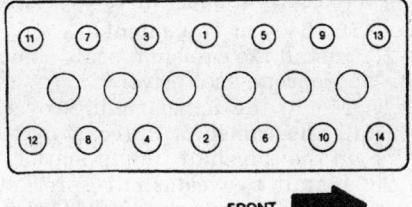

Tightening sequence for cylinder head bolts — 2.9L 6-cylinder engine

17. Clean and inspect the cylinder head and block mating surface.

To install:

18. Align the crankshaft timing mark by removing the starter motor and installing the crankshaft locking tool 5451 or equivalent. Turn the crankshaft until it is stopped by the tool.

19. Fit a new cylinder head gasket and install the bottom half of the cylinder head. Oil the cylinder head bolts; install and torque in sequence as follows:

   a. Stage 1 — 15 ft. lbs. (20 Nm)

   b. Stage 2 — 44 ft. lbs. (60 Nm)

   c. Stage 3 — angle tighten 130 degrees

20. Install new O-rings in the spark plug wells and oil the camshaft bearing seats.

21. Apply sealing compound (Part No. 1161059-9 or equivalent) to the upper section of the cylinder head.

NOTE: Do not allow any compound to penetrate the coolant or oil passages.

22. Install the camshaft.

23. Place the upper section of the cylinder head into position. Install the press tools (5454 or equivalent) and tighten against the lower section. Install the bolts, working from the inside outwards. Tighten to 13 ft. lbs. (17 Nm). Remove the tools.

24. Grease the camshaft front seal and tap the seal into place.

25. Place the upper timing cover into position. Install the camshaft pulleys while aligning the timing marks.

26. Temporarily install and tighten the pulley mounting bolts.

27. Remove the timing cover and install the mounting plate bolt.

28. Install the timing belt.

29. Loosen the camshaft pulley bolts and withdraw the tensioner locking pin. Insert the remaining camshaft pulley bolt. Hold the pulley using the counterhold tool 5199 or equivalent and tighten all bolts alternately to 15 ft. lbs. (20 Nm).

30. Remove the crankshaft locking tool. Install the protective plug and install the starter motor.

31. Install the upper timing cover.

32. Check that the timing marks on the crankshaft and camshaft pulleys are correctly aligned.

33. Install the camshaft sensor, ground terminals and temperature sensor connector. Install the coolant hose at rear.

34. Install the remaining components.

35. Change the engine oil. Fill the cooling system.

36. Connect the negative battery lead. Start the engine and check for leaks.

37. Recheck the cooling system level.

2.3L (B5234) and 2.4 (B5254) 5-Cylinder Engines

1. Disconnect the negative battery cable.

2. Raise and safely support vehicle. Remove the splash guard below the engine. Drain the coolant into a suitable container. Disconnect the exhaust pipe from the manifold.

3. Remove the exhaust manifold. Remove the timing belt. Disconnect the fuel distribution manifold and lift it and the injectors off to one side. Use 999-5533 holders or equivalent to separate them. Disconnect the two ground straps from the engine.

NOTE: Make sure that the injectors and needles are not damaged.

4. Remove the engine cooling fan.

5. Remove the intake manifold.

6. Remove the upper radiator hose from thermostat housing.

7. Remove the camshaft sprockets. Mark them intake or exhaust.

8. Remove the inner timing cover bolt.

9. Remove the air cleaner and hoses.

10. Remove the camshaft position sensor and damper.

11. Remove the distributor cap, wiring and rotor.

12. Remove the extension arm and brackets.

13. Working inwards from each end, loosen the bolts on the upper half of the cylinder head.

14. Gently tap the upper half with a soft mallet on the edges and front of the camshafts. Remove the bolts and upper half of the cylinder head.

15. Mark the camshafts and remove.

16. Remove the coolant pipe bolts. Remove the cylinder head bolts working inwards.

17. Remove the lower portion of the cylinder head and head gasket.

18. Clean all mating surfaces thoroughly.

——— WARNING ———
Do not use a metal scraper. Use a soft putty knife and gasket solvent cleaner with an exhaust fan. The surfaces must be totally clean to assure a tight seal.

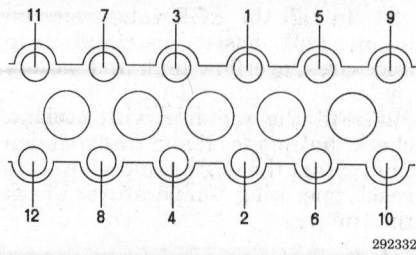

Lower half cylinder head torque order — 2.3L and 2.4L 5-cylinder engines

To install:

19. Align the crankshaft timing marks. Install crankshaft locking tool 999-5451 or equivalent and turn the crankshaft counterclockwise until it stops.

20. Install a new cylinder head gasket and the lower cylinder head. Apply a small amount of oil to the bolts.

21. Torque the lower cylinder head in three stages, starting on the inside and working outward as follows:
   a. 15 ft. lbs. (20 Nm)
   b. 44 ft. lbs. (60 Nm)
   c. angle tighten an additional 130° using an angle gauge

22. Install the coolant pipe using a new gasket. Replace the O-rings in the spark plug wells. Remove No. 1 and No. 5 spark plugs.

23. Using a roller, apply liquid gasket 161-059-9 or equivalent to the upper cylinder head.

NOTE: Make sure that no liquid gasket gets into the oil passages. Only a thin coating is required.

24. Install the camshafts and lock them in place using tools 999-5453 (front) and 999-5452 (rear) or equivalents.

25. Install the upper cylinder head. Pull the head down using press tools 999-5453 or 5454 (2) or equivalents. Torque the upper half working from the inside outward. Tighten to 13 ft. lbs. (17 Nm). Remove tools 999-5453 and 999-5454 or equivalents.

26. Install the camshaft seals using an appropriate seal driver.

27. Mount the upper timing cover. Install the camshaft sprockets and line up the camshaft timing marks.

28. Install two camshaft sprocket bolts furthest from the timing mark and tighten until they are just touching the sprocket. Remove the upper timing cover. Make sure that the remaining camshaft sprocket bolt hole is centered.

29. Install the tensioner pulley lever and torque to 18 ft. lbs. (25 Nm). Install the idler pulley and torque to 18 ft. lbs. (25 Nm).

30. Compress the tensioner fully with tool 999-5456 or equivalent and install a lock pin 2mm in diameter in the piston. If the tensioner leaks, has no resistance or will not compress, replace it.

31. Install the timing belt. Install the third camshaft sprocket bolt and torque the bolts to 15 ft. lbs. (20 Nm). Remove the tensioner lock pin.

32. Remove the crankshaft locking tool from the flywheel end of the block and install the plug in the hole. Install the starter motor. Remove the camshaft locking tool 999-5452 or its equivalent.

33. Turn the crankshaft two complete revolutions and check that the timing marks are lined up.

34. Install the rear camshaft seal using driver 999-5450 or equivalent.

35. Install the upper timing cover.

36. Install the remaining engine components

37. Connect the negative battery cable. Change the engine oil. Fill the cooling system.

38. Start the engine and run it until the thermostat opens, top it off as necessary.

39. Check the engine for leaks.

Lash Adjusters

BLEEDING

All Volvo engines are equipped with hydraulic valve lash adjusters and do not require bleeding.

Intake Manifold

REMOVAL AND INSTALLATION

2.3L (B230) 4-Cylinder Engine

1. Disconnect the negative battery cable. Remove the air cleaner and all necessary hoses.

2. Remove the PCV valve.

3. Disconnect the wiring and the fuel hose from the cold start injector. If necessary, remove the cold start injector.

4. Disconnect the wiring and the hoses at the auxiliary valve. If necessary, remove the auxiliary valve.

5. Remove the intake manifold brace.

6. Label and disconnect the vacuum hoses at the intake manifold.

7. Loosen the clamp for the rubber connecting pipe on the air-fuel control unit.

8. Remove the manifold bolts and manifold.

To install:

9. Clean the gasket mating surfaces thoroughly. Install the intake manifold, using new gaskets, and tighten the bolts to 15 ft. lbs. (20 Nm).

10. Install the intake manifold brace and the air-fuel control unit connecting pipe.

11. Install and connect the auxiliary valve, cold start injector and PCV valve.

12. Connect all vacuum hoses and electrical connectors.

13. Connect the negative battery cable, start the engine and bring it to normal operating temperature.

2.9L (B6304) 6-Cylinder Engine

1. Disconnect the negative battery lead.

2. Disconnect the harness at the air mass meter.

3. Disconnect the idle air valve wiring and air hose. Remove the flame trap holder and remove the intake hose.

4. Remove the throttle pulley cover.

5. Disconnect and remove the throttle switch wiring, throttle cable and bracket, cruise control vacuum servo and vacuum hoses at throttle the housing.

6. Remove the injector cover plate and distribution manifold retaining bolts (3).

7. Disconnect the pressure regulator vacuum hose and fuel line bracket.

8. Carefully lift out the injector and distribution manifold assembly.

9. Remove the air preheater hose. Remove the left and right side power stage connectors on the bottom of the manifold. Remove the manifold bottom mounting.

10. Disconnect the brake servo hose and vacuum hoses under the manifold.

11. Cut away the clamps securing the rubber sleeves between the manifold sections, and lift out the outer manifold section.

12. Remove the upper bolts and loosen the lower bolts. Remove the inner section of the manifold.

To install:

13. Install the inner section of the manifold, using a new gasket. Install the rubber sleeves on the inner section and lubricate the free ends with

petroleum jelly. Install the mounting bolts and torque to 15 ft. lbs. (20 Nm).

14. Route the wiring between the second and third branches of the outer manifold section. Place the manifold against the lower section and connect the crankcase ventilation hoses.

15. Insert the manifold branches in the rubber sleeves. Secure with new Oetiker clamps.

16. Tighten the manifold lower mounting. Reconnect the vacuum hoses, brake servo hose, power stage connectors and air preheater hose.

17. Inspect the injector O-rings. Lubricate with petroleum jelly.

18. Reconnect the fuel pressure regulator vacuum hose.

19. Press the fuel distribution manifold into position. Tighten the manifold.

20. Reconnect the injector harnesses and EGR vacuum hoses. Install the injector cover.

21. Install the throttle cable, throttle pulley cover and vacuum hoses (cruise control and throttle housing).

22. Install the cable bracket at the throttle pulley. Reconnect the PCV, idling valve wiring, air hose, air mass meter and throttle housing connector.

23. Connect the negative battery lead. Start the engine and check operation.

2.3L (B5234) and 2.4L (B5254) 5-Cylinder Engines

1. Disconnect the negative battery cable.

2. Remove the injector cover and throttle cable. Unfasten the connectors and clips from the injectors. Remove the two clips holding the fuel line. Remove the distribution manifold mounting bolts.

3. Install the five 999 5533 or equivalent injector holders and carefully pull the injectors off with the distribution manifold.

4. Disconnect the hose to the purge valve. Carefully lay the distribution manifold and injectors on the engine.

───── **WARNING** ─────
Make sure that the injectors and needles are not damaged.

5. Disconnect the throttle linkage from the pulley.

6. Disconnect the intake air hose to the throttle body.

7. Remove the multi-nipple.

8. Remove the throttle pulley with bracket and idle speed control valve

9. Remove the EGR hose clamp on turbo models.

10. Remove the pressure line to turbo instrumentation/EGR valve control.

11. Disconnect the vacuum hose.

12. Disconnect the brake booster hose.

13. Loosen the dipstick bracket and intake manifold lower bracket bolt. Loosen the lower intake manifold bolts several turns.

NOTE: The lower intake manifold bolts are not through-bolts.

14. Remove the upper intake manifold bolts. Remove the intake manifold by lifting it up 0.75 in. (20mm).

15. Make sure the mating surfaces of the cylinder head and intake manifold is clean.

To install:

16. Install a new intake gasket.

17. Install the intake manifold and upper bolts. Torque all bolts from inside to outside to 15 ft. lbs. (20 Nm).

18. Install the EGR valve (if equipped) with a new gasket.

19. Install the throttle body with a new gasket.

20. Install the throttle pulley and bracket.

21. Install the multi-nipple and connect the hoses.

22. Install the fuel distribution manifold.

23. Install the wiring and injector cover.

24. Install the remaining components.

25. Connect the negative battery cable.

26. Test run engine and check for leaks.

Exhaust Manifold

REMOVAL AND INSTALLATION

2.3L (B230) 4-Cylinder Engine

1. Disconnect the negative battery cable. Remove the air cleaner and all necessary hoses.

2. Remove the EGR valve pipe from the manifold.

3. Remove the exhaust pipe from the exhaust manifold.

4. Remove the manifold bolts and manifold.

To install:

5. Position and install the manifold using a new gasket. Torque the manifold bolts to 10–20 ft. lbs. (14–27 Nm).

6. Install the remaining components.

7. Connect the negative battery cable.

2.9L (B6304) 6-Cylinder Engine

1. Disconnect the negative battery lead.
2. Remove the exhaust pipe mounting nuts at the manifold joints.
3. Remove the heat shield retaining bolts and heat shield.
4. Remove the exhaust manifold mounting nuts. Remove the exhaust manifold and gasket.

To install:

5. Before installation, clean the manifold and cylinder head mating surfaces.
6. Fit a new gasket and place the exhaust manifold into position. Install the mount lifting lug on the studs between the 3rd and 4th exhaust branches. Torque the studs to 15 ft. lbs. (20 Nm) and mounting nuts to 18 ft. lbs. (25 Nm).
7. Install the heat shield to the rear manifold. Torque to 11 ft. lbs. (15 Nm).
8. Install the front exhaust pipe to manifold. Using thread locking compound, torque to 44 ft. lbs. (60 Nm).

NOTE: Loosen the joint at the catalytic converter and re-tighten to 18 ft. lbs. (25 Nm). This is necessary to prevent stresses in the system.

9. Connect the negative battery lead. Start the engine and check for leaks.

2.3L (B5234) and 2.4L (B5254) 5-Cylinder Engines

1. Disconnect the negative battery cable.
2. Raise and safely support the vehicle. Disconnect the exhaust pipe from the manifold by removing the nuts on the flanged joint. Remove the carriage bolts from the manifold.
3. Remove the two heat shields from the exhaust manifold. Remove the exhaust manifold bolts. Turn the manifold 90° to the right and lift it out from the top.

——————— **WARNING** ———————
When removing or installing the exhaust manifold, be careful not to damage the air conditioning pressure switch, if so equipped.

To install:

4. Check the gasket surface of the cylinder head, clean if necessary. Install the exhaust manifold using new gaskets. Line up the exhaust manifold with the pipe using the carriage bolts.

5. Install the exhaust manifold bolts using a locking compound on the threads. Torque the bolts to 18.5 ft. lbs. (25 Nm).
6. Install the heat shields. Tighten the carriage bolts using thread sealing compound. Torque the nuts to no more than 86 inch lbs. (10 Nm). Remember to install the springs and washers with the nuts.
7. Connect the negative battery cable.
8. Run the engine and check for leaks.

Turbocharger

REMOVAL AND INSTALLATION

2.3L (B230) 4-Cylinder Engine

1. Disconnect the negative battery cable.
2. Disconnect the expansion tank from the retainer. Remove the expansion tank retainer.
3. Remove preheater hose to the air cleaner. Remove the pipe and rubber bellows between the air/fuel control unit and the turbocharger unit. Pull out the crankcase ventilation hose from the pipe.
4. Remove the pipe and pipe connector between the turbocharger unit and the intake manifold.

NOTE: Cover the turbocharger intake and outlet ports to keep dirt out of the system.

5. Disconnect the exhaust pipe and place aside.
6. Disconnect the spark plug wires at the plugs.
7. Remove the upper heat shield. Remove the brace between the turbocharger unit and the manifold.
8. Remove the lower heat shield by removing the retaining screw under the manifold.
9. Remove the oil pipe clamp, retaining screws on the turbo unit and

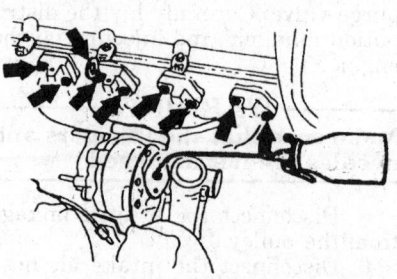

181701

Filling turbocharger inlet with oil, and tighten the bolts shown — 2.3L 4-cylinder engine

the pipe connection screw in the cylinder block under the manifold. Do not allow any dirt to enter the oil passages.
10. Remove the manifold retaining nuts and washers. Leave one nut in place to keep the manifold in position.
11. Remove the oil delivery pipe. Cover the opening on the turbo unit.
12. Disconnect the air/fuel control unit by loosening the clamps. Move the unit with the lower section of the air cleaner up to the right side wheel housing. Place a cover over the wheel housing as protection.
13. Remove the remaining nut and washer on the manifold. Lift the assembly forward and up. Remove the manifold gaskets. Disconnect the return oil pipe O-ring from the cylinder block.
14. Disconnect the turbocharger unit from the manifold.

To install:

15. Be sure to use a new gasket for the exhaust manifold and a new O-ring to the return oil pipe. Keep everything clean during assembly and use extreme care to keep dirt out of the various turbo inlet and outlet pipes and hoses.
16. Install the turbocharger on the exhaust manifold and tighten the bolts as follows:
   a. Step 1 — 0.7 ft. lbs. (0.9 Nm)
   b. Step 2 — 30 ft. lbs. (40 Nm)
   c. Step 3 — Tighten all bolts an additional 120 degrees (1/3 turn).
17. Install the exhaust manifold and turbocharger assembly on the engine. Connect all oil pipes from and to the turbocharger using new O-rings.
18. Install the air/fuel control unit and air cleaner. Install the heat shields, spark plug wires, exhaust pipes, preheater assembly and expansion tank. Connect the negative battery cable.
19. Disconnect the wire at terminal 15 (brown) of the ignition coil. Use the ignition key to turn the engine over for about 30 seconds. This circulates oil to the turbocharger, providing proper start-up lubrication.
20. Turn the ignition **OFF**, reconnect the coil wire, start the engine and allow it to idle for a few minutes prior to test driving.

2.3L (B5234) and 2.4L (B5254) 5-Cylinder Engines

1. Disconnect the negative battery cable.
2. Remove the heat shield from over the exhaust manifold.

3. Remove the upper air charge pipe and rubber hose from the turbo and move it to one side. Remove the fresh air intake hose and inner heat shield. Disconnect the upper turbo coolant return pipe and clamp off the hose, move it to the side. Disconnect the oil inlet pipe nipple.

4. Raise and safely support the vehicle. Remove or disconnect the following from under side:
  • clamp between the pipes
  • oil return pipe
  • exhaust pipe bracket and bolt
  • exhaust pipe to turbo nut
  • exhaust manifold to turbo nuts

5. From the top side remove the exhaust pipe to turbo nuts. Disconnect the coolant inlet pipe to the turbo. Remove the turbo/exhaust manifold nuts.

6. Disconnect the following hoses from the turbo:
  • red boost pressure
  • white bypass valve
  • yellow pressure regulator

7. Remove the turbo and the old pin bolts from the exhaust manifold.

To install:

8. Install new pin bolts with thread locking compound and torque to 15 ft. lbs. (20 Nm). Install the

turbo and connect the red, white, and yellow hoses to it.

9. Install the upper exhaust manifold nuts and tighten them lightly.

10. Working from under the vehicle, install the lower exhaust manifold nuts and torque them to 18 ft. (25 Nm). On the top side, torque the upper exhaust manifold nuts to 18 ft. lbs. (25 Nm). Torque the exhaust manifold/turbo nuts to 22 ft. lbs. (30 Nm) and check that they are mated properly.

11. Under the vehicle install the oil pipe, grease the O-ring. Install the exhaust pipe bracket bolt.

12. Lower the vehicle and install or connect the following:
  • oil inlet pipe
  • inlet and outlet coolant pipes (make sure the clamps are removed)
  • fresh air intake hose
  • inner heat shield
  • upper air charge pipe
  • outer heat shield

NOTE: Replace the copper coolant pipe and upper oil pipe washers.

13. Raise the vehicle and remove the clamp from coolant return hose.

14. Connect the negative battery cable.

15. Run the engine to check the boost pressure. Check oil and coolant levels.

NOTE: It may be necessary to reset a fault code after replacing the turbocharger.

Front Crankshaft Seal

REMOVAL AND INSTALLATION

2.3L (B230) 4-Cylinder Engine

1. Disconnect the negative battery cable.
2. Remove the cooling fan and shroud.
3. Remove the drive belts and water pump pulley.
4. Remove upper timing belt cover.
5. Set the crankshaft to TDC and remove the timing belt.

NOTE: Do not turn the crankshaft or camshaft. Pistons may strike valves.

6. Carefully pry loose the seal to be replaced. Do not damage the contact face.

To install:

7. Clean the contact faces. Lubricate the seal and seat, then press the seal into position.
8. Install the timing belt. Make sure the timing belt is correctly positioned.
9. Install the timing belt cover and vibration damper. Tighten the crankshaft center bolt to 45 ft. lbs. (60 Nm) plus an additional 60 degrees.
10. Install the drive belts and water pump pulley.
11. Install the cooling fan and shroud.
12. Connect the negative battery cable. Start the engine. Check for leaks and proper operation.

2.9L (B6304) 6-Cylinder Engine

1. Disconnect the negative battery cable.
2. Remove the timing belt.
3. Remove the crankshaft pulley, using a suitable puller.
4. Carefully pry out the old seal.

To install:

5. Before installing the new seal, thoroughly clean the crankshaft face.
6. Lubricate the new seal and tap the seal into place, using tool 5455 or equivalent.
7. Install the timing belt.
8. Connect the negative battery cable.

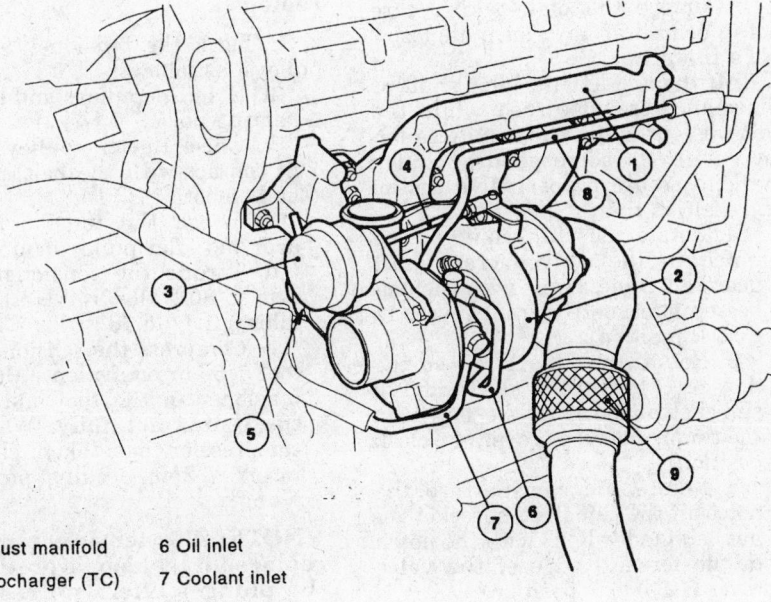

1 Exhaust manifold
2 Turbocharger (TC)
3 Pressure regulator
4 Link
5 Bypass valve
6 Oil inlet
7 Coolant inlet
8 Coolant return
9 Flexible joint (bellows type)

325043

Turbocharger and exhaust manifold — 2.3L and 2.4L 5-cylinder engines

2.3L (B5234) and 2.4L (B5254) 5-Cylinder Engines

1. Disconnect the negative battery cable.
2. Remove the fuel line clips.
3. Lift the coolant expansion tank and place it on top of the engine.
4. Remove the drive belts.
5. Remove the front timing cover.
6. Raise and safely support the vehicle.
7. Remove the right front wheel and loosen the inner fender liner. Remove the vibration damper guard and turn crankshaft pulley until all timing marks align.
8. Remove the timing belt.

WARNING

Do not turn the crankshaft or camshafts once the timing belt has been removed.

9. Install a universal puller so the claws pull against the bolts and not the sprocket. Pull the sprocket off.

WARNING

Make sure that the puller does not damage the sprocket teeth.

10. Remove the front seal using a groove cut chisel. Clean the mating surface.
To install:
11. Install the crankshaft timing belt sprocket using the nut and a spacer.
12. Install the timing belt.
13. Turn the crankshaft two complete revolutions and make sure the timing marks on the crankshaft and camshaft pulleys align properly.
14. Install the two fuel line clips.
15. Install the remaining components.
16. Install the wheel.
17. Connect the negative battery cable.
18. Test run the engine.

Timing Belt

REMOVAL AND INSTALLATION

2.3L (B230) 4-Cylinder Engine

1. Disconnect the negative battery cable.
2. Remove the drive belts.
3. Remove the cooling fan and shroud.
4. Remove the timing belt cover.
5. Remove the 6 retaining bolts and the crankshaft pulley.
6. To remove the tension from the belt, loosen the nut for the tensioner and press the idler roller back. The

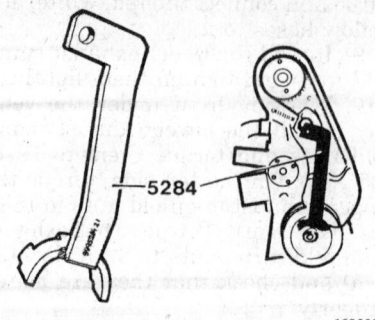

Removing vibration damper with special service tool 5284 — 2.3L 4-cylinder engine

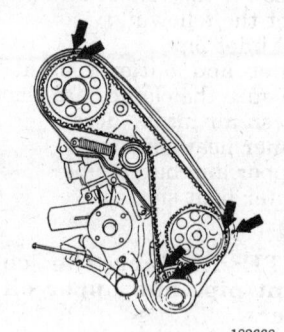

Proper timing marks alignment — 2.3L 4-cylinder engine

tension spring can be locked in this position by inserting the shank end of a 3mm drill through the pusher rod.
7. Remove the belt, taking care not to bend it at any sharp angles.
To install:
8. If the crankshaft, idler shaft or camshaft were disturbed while the belt was out, align each shaft as follows with its corresponding index mark to assure proper valve timing and ignition timing:
   a. Rotate the crankshaft so the notch in the convex crankshaft gear belt guide aligns with the embossed mark on the front cover (12 o'clock position).
   b. Rotate the idler shaft so the dot on the idler shaft drive sprocket aligns with the notch on the timing belt rear cover (4 o'clock position).
   c. Rotate the camshaft so the notch in the camshaft sprocket inner belt guide aligns with the notch in the forward edge of the valve cover (12 o'clock position).
9. Install the timing belt over the sprockets and then over the tensioner roller. Do not use any sharp tools. New belts have yellow marks. The 2 lines on the drive belt should fit to-

ward the crankshaft marks. The next mark should fit toward the intermediate shaft marks, etc.
10. Loosen the tensioner nut and let the spring tension automatically take up the slack. Tighten the tensioner nut to 37 ft. lbs. (51 Nm).
11. Rotate the crankshaft two full revolutions clockwise and make sure the timing marks align.
12. Install the drive belts, radiator fan and shroud. Connect the negative battery cable.

2.9L (B6304) 6-Cylinder Engine

1. Disconnect the negative battery cable.
2. Remove the drive belts.
3. Remove the timing belt cover.
4. Remove the splash guard, vibration damper guard and ignition coil cover.
5. Rotate the crankshaft clockwise, until the timing marks on the camshaft pulleys and timing belt mounting plate and crankshaft pulley/oil pump housing are aligned.
6. Remove the tensioner upper mounting bolts. Loosen the tensioner lower mounting bolt and twist the tensioner to free the plunger. Remove the lower mounting bolt and remove the tensioner.
7. Remove the timing belt.

NOTE: Do not rotate the crankshaft while the timing belt is removed.

8. Check the tensioner and idler pulleys, as follows:
   a. Spin the pulleys and listen for bearing noise.
   b. Check that the pulley surfaces in contact with the belt are clean and smooth.
   c. Check the tensioner pulley arm and idler pulley mountings.
   d. Torque the tensioner pulley arm to 30 ft. lbs. (40 Nm) and the idler pulley to 18 ft. lbs. (25 Nm).
   e. Compress the tensioner using tool 5456 or equivalent. Mount the tensioner in the tool and tighten the center nut fully. Wait until compression has taken place and insert a 2mm locking pin in the plunger.

NOTE: The tensioner must be replaced if leakage is observed or the plunger offers no resistance when depressed, or cannot be depressed.

To install:
9. Place the belt around the crankshaft pulley and right-side idler. Place the belt over the camshaft pulleys. Position the belt around the

water pump and press over the tensioner pulley.

10. Insert the tensioner mounting bolts. Torque to 18 ft. lbs. (25 Nm).

11. Remove the locking pin from the tensioner. Install the front timing belt cover.

NOTE: The lever bushing must be greased every time the belt is replace or the pulley is removed. Service the bushing, using the following procedure:

a. Remove the lever mounting bolt, tensioner pulley and sleeve.

b. Grease the surfaces of the bushing, bolt and sleeve, using Part No. 1161246-2 or equivalent.

c. Install the sleeve, tensioner pulley and lever mounting bolt.

d. Tighten the bolt to 30 ft. lbs. (40 Nm).

12. Turn the crankshaft two revolutions and check that the timing marks on the crankshaft and camshaft pulleys are correctly aligned.

13. Install the remaining components.

14. Connect the negative battery lead, start and check the engine operation.

2.3L (B5234) and 2.4L (B5254) 5-Cylinder Engines

1. Disconnect the negative battery cable.

2. Remove the coolant expansion tank and place it on top of the engine.

3. Remove the spark plug cover and drive belts.

4. Remove the timing belt cover.

5. Wait five minutes after lining up marks, then install gauge 998 8500 or equivalent between the exhaust camshaft and water pump. Read the gauge using a mirror, while

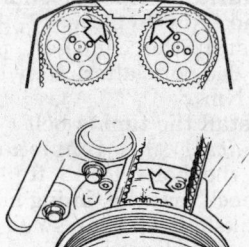

294501

Timing belt sprocket timing marks at TDC — 2.3L and 2.4L 5-cylinder and 2.9L 6-cylinder engines

still installed. For 23mm belts, the tension should be 2.7–4.0 units.

NOTE: If the belt tension is incorrect, the tensioner must be replaced.

6. Remove the upper tensioner bolt and loosen the lower bolt, turning the tensioner to free up the pulley.

7. Remove the lower bolt and the tensioner. Remove the timing belt.

To install:

8. Turn all the pulleys listening for bearing noise. Check to see that the contact surfaces are clean and smooth. Remove the tensioner pulley lever and idler pulley, lubricate the contact surfaces and bearing with grease. If the tensioner pulley lever or idler is seized replace it.

9. Install the tensioner pulley lever and idler pulley and torque to 18 ft. lbs. (25 Nm).

10. Compress the tensioner with tool 999 5456 or equivalent and insert a 0.079 in. (2.0mm) lock pin in the piston. If the tensioner leaks, has no resistance, or will not compress, replace it. Install the tensioner and torque to 18 ft. lbs. (25 Nm).

11. Install the timing belt in order:

a. Around the crankshaft sprocket.

b. Around the right idler pulley

c. Around the camshaft sprockets

d. Around the water pump

e. Onto the tensioner pulley

12. Pull the lock pin out from the tensioner and install the upper timing cover. Turn the crankshaft two complete revolutions and check to see that the timing marks on the crankshaft and camshaft pulleys are lined up.

13. Install the timing belt covers and the fuel line clips.

14. Install the accessory belts.

15. Install the vibration damper guard and the inner fender well.

16. Install the spark plug cover and any remaining components.

Timing Belt Sprockets

REMOVAL AND INSTALLATION

2.3L (B230) 4-Cylinder Engine

1. Disconnect the negative battery cable.

2. Remove the timing belt cover, followed by the timing belt and tensioner.

3. Hold the camshaft sprocket with tool 999-5199 or equivalent and remove the retainer bolt.

4. Remove the sprocket and guide plates from the camshaft.

— **WARNING** —
Do not rotate the crankshaft while the camshaft is removed from the cylinder head.

To install:

5. Install the camshaft guide plates and sprocket. Install the two bolts through the sprocket to lock it in place. Tighten the sprocket retainer bolt to 37 ft. lbs. (50 Nm).

6. Install the tensioner pulley and timing belt.

7. Install the remaining components.

8. Connect the negative battery cable.

2.9L (B6304) 6-Cylinder Engine

1. Disconnect the negative battery cable.

2. Remove the drive belts.

3. Remove the front timing belt cover followed by the timing belt.

NOTE: Do not rotate the crankshaft while the timing belt is removed.

4. Loosen and remove the camshaft retainer bolts.

5. Remove the camshaft pulleys, using holding tool 5199 or equivalent. Mark the pulleys intake and exhaust so they can be returned to their original side.

6. Remove the pulleys from the engine compartment.

To install:

7. Install the camshaft pulleys. Tighten the bolts alternately to 15 ft. lbs. (20 Nm).

8. Install the timing belt and cover.

9. Install the drive belts and the negative battery cable.

2.3L (B5234) and 2.4L (B5254) 5-Cylinder Engines

1. Disconnect the negative battery cable.

2. Remove the coolant expansion tank and place it on top of the engine.

3. Remove the timing belt cover and timing belt.

4. Hold the camshaft sprocket with tool 999-5199 or equivalent and remove the bolts. If both sprockets are being replaced, mark them for their respective cams.

5. Remove the sprockets from the camshaft(s).

To install:

6. Mount the upper timing cover. Install the camshaft sprocket aligning the marks. Install the two bolts

furthest from the mark. Tighten until they are just touching the sprocket. Remove the upper timing cover. Make sure that the remaining sprocket bolt hole is centered.

7. Turn all the pulleys listening for bearing noise. Check to see that the contact surfaces are clean and smooth. Remove the tensioner pulley lever and idler pulley, lubricate the contact surfaces and bearing with grease. If the tensioner pulley lever or idler is seized replace it.

8. Install the tensioner pulley and timing belt.

9. Install the third bolt on sprocket and torque the bolts to 15 ft. lbs. (20 Nm). Pull the lock pin out from the tensioner and install the upper timing cover. Turn the crankshaft two complete revolutions and check to see that the timing marks on the crankshaft and camshaft pulleys are lined up.

10. Install the remaining components.

11. Install the spark plug cover and any remaining components.

12. Connect the negative battery cable.

Timing Belt Front Cover

REMOVAL AND INSTALLATION

2.3L (B230) 4-Cylinder Engine

1. Disconnect the negative battery cable. Remove the cooling fan and shroud.

2. Remove the drive belts.

3. Remove the water pump pulley.

4. Remove the 4 retaining bolts and lift off the timing belt cover.

To install:

5. Clean all gasket mating surfaces thoroughly. Install the timing belt cover using a new gasket.

6. Install the water pump pulley, and all drive belts.

7. Install the fan and shroud. Connect the negative battery cable. Start the engine and check for leaks.

2.9L (B6304) 6-Cylinder Engine

1. Disconnect the negative battery lead.

2. Remove the drive belt.

3. Remove the lower timing belt cover, splash guard and vibration damper guard. Remove the ignition coil cover.

4. Remove the upper timing cover.

To install:

5. Install the upper timing belt cover. Install the ignition coil cover.

6. Install the lower timing belt cover, splash guard and vibration damper guard.

7. Install the drive belt.

8. Connect the negative battery lead.

2.3L (B5234) and 2.4L (B5254) 5-Cylinder Engines

1. Disconnect the negative battery cable. Remove the coolant expansion tank and place it on top of the engine.

2. Remove the spark plug cover.

3. Remove the drive belts.

4. Remove the fuel line clips.

5. Remove the right front wheel and loosen the inner fender well.

6. Remove the vibration damper guard and turn crankshaft pulley until the marks are lined up.

7. Remove the water pump pulley.

8. Remove the retaining bolts and lift off the timing belt cover.

To install:

9. Position the timing belt cover in place and secure with the retainer bolts.

10. Install the water pump pulley, followed by the drive belts.

11. Install the remaining components.

12. Connect the negative battery cable.

Camshaft

REMOVAL AND INSTALLATION

2.3L (B230) 4-Cylinder Engine

1. Disconnect the negative battery cable.

2. Remove the drive belts.

3. Remove the timing belt.

4. Remove the valve cover.

5. Remove the camshaft center bearing cap. Install camshaft press tool 5021 or equivalent over the center bearing journal to hold the camshaft in place while removing the other bearing caps.

6. Remove the 4 remaining bearing caps.

7. Remove the seal from the forward edge of the camshaft.

8. Release camshaft press tool and lift out the camshaft.

─────── **WARNING** ───────
Do not rotate the crankshaft while the camshaft is removed from the cylinder head.
───────────────────────

To install:

9. Apply sealant to the outer sealing surfaces of the front and rear caps. Lubricate and install the camshaft into position. The guide pin for the timing gear should face up.

10. Install the rear bearing cap. Slide the camshaft back and forth to check the camshaft end-play. End-play should be 0.004–0.016 in. (0.1–0.4mm).

11. Install the camshaft press tool. Install the camshaft seal. Lubricate and install the remaining caps. Tighten bolts to 14 ft. lbs. (20 Nm).

12. Lubricate the front seal and install, using tool 5025 or equivalent.

13. Install the camshaft gear and spacer washer. Install the timing belt.

14. Install the remaining components.

15. Connect the negative battery cable.

2.9L (B6304) 6-Cylinder Engine

1. Disconnect the negative battery cable.

2. Remove the drive belts.

3. Remove the timing belt.

NOTE: Do not turn the crankshaft while the belt is removed.

4. Remove the camshaft pulleys, using the holding tool 5199 or equivalent.

5. Remove the top half of the cylinder head. Tap the joint lugs and camshaft front ends lightly.

6. Remove the camshafts.

To install:

7. Lubricate the camshafts and bearing seats. Place the camshafts into position. Install the holding tool 5453 or equivalent to the front end and the locking tool 5452 or equivalent to the rear end of the cylinder head upper section.

8. Install the upper cylinder head section and tighten against the lower section, using the press tools 5454 or equivalent.

9. Install and tighten the retaining bolts to 13 ft. lbs. (17 Nm), starting from the inside and working outwards. Remove the tools.

10. Lubricate the camshaft front seals and tap into place.

11. Install the camshaft pulleys. Tighten the bolts alternately to 15 ft. lbs. (20 Nm).

12. Install the timing belt.

13. Install the tensioner and tighten the bolts to 18 ft. lbs. (25 Nm). Check that the timing marks on the crankshaft and camshaft pulleys are correctly aligned.

14. Install the remaining components.

15. Install the drive belts.

16. Connect the negative battery cable.

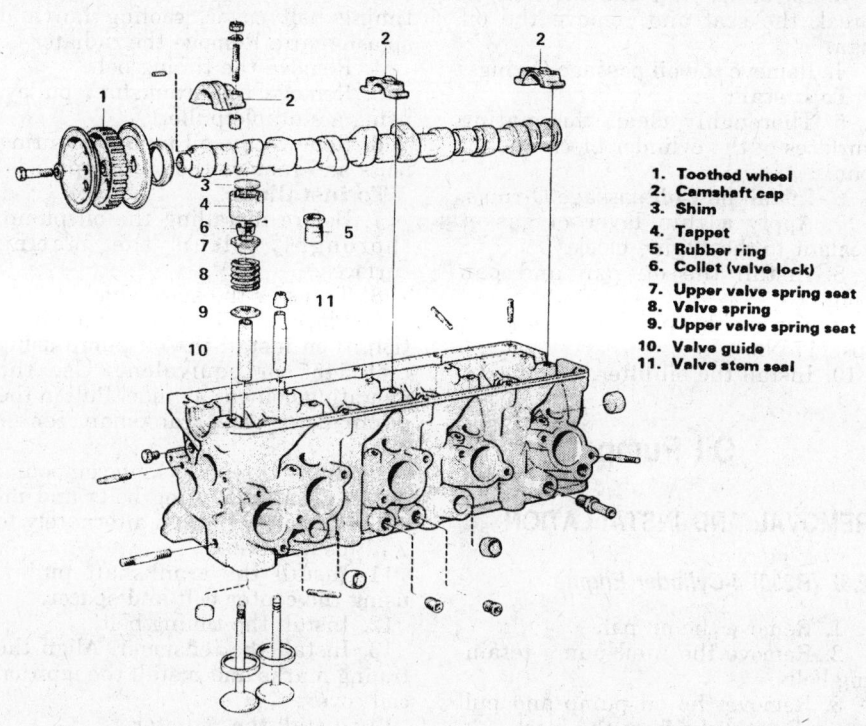

1. Toothed wheel
2. Camshaft cap
3. Shim
4. Tappet
5. Rubber ring
6. Collet (valve lock)
7. Upper valve spring seat
8. Valve spring
9. Upper valve spring seat
10. Valve guide
11. Valve stem seal

286192

Exploded view of cylinder head — 2.3L 4-cylinder engine

2.3L (B5234) and 2.4L (B5254) 5-Cylinder Engines

1. Disconnect the negative battery cable.
2. Remove the drive belt.
3. Remove the timing belt.
4. Remove the ignition coils cover.

NOTE: Do not turn the crankshaft while the belt is removed.

5. Remove the camshaft position sensor and shutter at the right rear of camshaft assembly. Remove the switch holder and shield at the left rear of assembly.
6. Remove the ignition coils. Mark their locations.
7. Remove the camshaft pulleys, using the holding tool 5199 or equivalent. Mark the pulleys intake and exhaust so they can be returned to their original side.
8. Remove the top half of the cylinder head. Tap the joint lugs and camshaft front ends lightly.
9. Remove the camshafts.
10. Thoroughly clean the mating surfaces between the upper and lower halves of the cylinder head.

— WARNING —
Do not use a metal scraper. Use a soft putty knife and gasket solvent cleaner with an exhaust fan. The surfaces must be totally clean to assure a tight seal.

To install:
11. Lubricate the camshafts and bearing seats. Place the camshafts into position.
12. Install the holding tool 5453 or equivalent to the front end and the locking tool 5452 or equivalent to the rear end of the cylinder head upper section.
13. Remove No. 1 and No. 5 spark plugs
14. Using a roller, apply liquid gasket 161 059-9 or equivalent to the upper half of the cylinder head.

NOTE: Make sure that no liquid gasket gets into the oil passages. Only a thin coating is required.

15. Install the upper cylinder head section and tighten against the lower section, using the press tools 5454 or equivalent.

16. Install and tighten the retaining bolts to 13 ft. lbs. (17 Nm), starting from the inside and working outwards. Remove the tools.
17. Lubricate the camshaft front seals and tap into place.
18. Mount the upper timing cover. Install the camshaft sprockets and line up the camshaft timing marks.
19. Install two camshaft sprocket bolts furthest from the timing mark and tighten until they are just touching the sprocket. Remove the upper timing cover. Make sure that the remaining camshaft sprocket bolt hole is centered.
20. Turn all the idler pulleys listening for bearing noise. Check to see that the contact surfaces are clean and smooth. Remove the tensioner pulley lever and idler pulley, lubricate the contact surfaces and bearing with grease. If the tensioner pulley lever or idler is seized, replace it.
21. Install the tensioner pulley lever and torque to 18 ft. lbs. (25 Nm). Install the idler pulley and torque to 18 ft. lbs. (25 Nm).
22. Compress the tensioner fully with tool 999 5456 or equivalent.
23. Install the timing belt.
24. Install the rear camshaft seal using drift 999 5450 or equivalent and press it carefully into position flush with the inner chamfer edge.
25. Install the remaining components
26. Connect the negative battery cable.
27. Start the engine and run it until the thermostat opens.
28. Check the engine for leaks.

Piston and Connecting Rod

POSITIONING

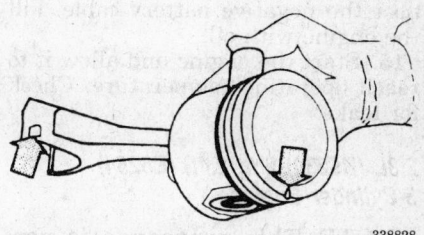

338828

Wrist pin installation and aligning the connecting rod and piston. Arrow indicates the front of the engine

ENGINE LUBRICATION

Oil Pan

REMOVAL AND INSTALLATION

2.3L (B230) 4-Cylinder and 2.9L (B6304) 6-Cylinder Engines

1. Disconnect the negative battery cable. Raise and support the vehicle safely.
2. Drain the engine oil.
3. Remove the splash guard, if equipped.
4. On 2.3L engines, perform the following steps;
    a. Remove the engine mount retaining nuts.
    b. Remove the lower bolt and loosen the top bolt on the steering column yoke.
    c. Slide the yoke assembly up on the steering shaft.
5. Raise and safely support the front of the engine.
6. Remove the retaining bolts for the front axle crossmember.
7. Remove the crossmember.
8. Remove the left engine mount.
9. Remove the pan support bracket.
10. Remove the pan bolts and remove the pan.
    To install:
11. Clean the gasket mating surfaces thoroughly. Install the oil pan and using new gaskets, tighten the bolts to 8 ft. lbs. (11 Nm).
12. Lower the engine and install all engine mounts. Install the front crossmember and install the bolts.
13. On 2.3L engines, install the yoke assembly on the steering shaft and tighten the bolts to 18 ft. lbs. (24 Nm).
14. Install the splash guard, if equipped. Lower the vehicle and connect the negative battery cable. Fill the engine with oil.
15. Start the engine and allow it to reach operating temperature. Check for leaks.

2.3L (B5234) and 2.4L (B5254) 5-Cylinder Engines

NOTE: This procedure is performed with engine removed from the car.

1. Remove the oil filter.
2. Remove the oil pan bolts.

3. Carefully tap the oil pan to break the seal and remove the oil pan.
4. Remove the oil passage O-rings.
To install:
5. Thoroughly clean the mating surfaces of the cylinder block and oil pan.
6. Install new oil passage O-rings.
7. Apply a thin layer of gasket sealant to the engine block.
8. Install the oil pan and pan bolts.
9. Torque the pan bolts to 12 ft. lbs. (17 Nm).
10. Install the oil filter.

Oil Pump

REMOVAL AND INSTALLATION

2.3L (B230) 4-Cylinder Engine

1. Remove the oil pan.
2. Remove the 2 oil pump retaining bolts.
3. Remove the oil pump and pull the delivery tube from the block.
To install:
4. Fit new sealing rings at either end of the delivery tube.
5. Install the pump with the delivery tube attached. Align the pipe to the block so that the seal does not become damaged. Tighten the two oil pump retaining bolts.
6. Attach the clamp for the oil trap drain hose to the oil pump bolts. Make sure the hose is securely clamped behind the oil pump shoulder. Do not shorten the hose.
7. Install the oil pan.
8. Fill the engine with oil. Start the vehicle and check the oil level.

2.9L (B6304) 6-Cylinder Engine

1. Disconnect the negative battery cable.
2. Drain the cooling system.

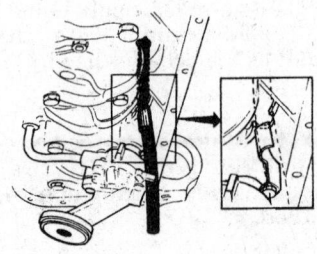

183542

Make sure hose from oil trap is securely clamped behind oil pump shoulder — 2.3L 4-cylinder engine

3. Remove the drive belts, front timing belt cover, cooling fan and splashguard. Remove the radiator.
4. Remove the timing belt.
5. Remove the crankshaft pulley, using a suitable puller.
6. Remove the oil pump mounting bolts and remove the oil pump.
To install:
7. Before installing the oil pump, thoroughly clean the mating surfaces.
8. Transfer the snow shield.
9. Place a new gasket into position, then install the oil pump using tool 5455 or equivalent. Use the mounting bolts as a guide. Pull in the pump using the crankshaft center nut.
10. Apply thread locking compound to the pump mounting bolts and install the bolts. Tighten alternately to 7 ft. lbs. (10 Nm).
11. Install the crankshaft pulley, using the center bolt and spacer.
12. Install the timing belt.
13. Install the tensioner. Align the timing marks and install the ignition coil cover.
14. Install the radiator.
15. Install the remaining components.
16. Connect the negative battery cable.
17. Fill the engine with oil.

2.3L (B5234) and 2.4L (B5254) 5-Cylinder Engines

The oil pump is on the front of the crankshaft.
1. Remove spark plug cover.
2. Remove the drive belts.
3. Remove the front timing cover and timing belt.
4. Raise and safely support the vehicle.

— WARNING —
Do not turn the crankshaft or camshafts once the timing belt has been removed.

5. Remove the crankshaft damper, using tool 999 5433 or equivalent to counterhold it from moving.
6. Remove the crankshaft sprocket.

— WARNING —
Make sure the puller does not damage the sprocket teeth.

7. Remove the old front seal using a groove cut chisel. Clean the mating surface where the seal lies.
8. Remove the four bolts retaining the oil pump. There are tabs on the oil pump housing located at the 6 o'clock and 11 o'clock positions.

9. Carefully pry out the oil pump using a groove cut chisel. Clean the surfaces where the pump mates to the engine.

To install:

10. Install the new oil pump using tool 999-5455 or equivalent using the bolts to guide it in. Use the crankshaft nut to press it in. Torque the bolts alternately to 7 ft. lbs. (10 Nm). Install the crankshaft timing belt sprocket using the nut and a spacer.

11. Install the timing belt and cover.

12. Install the drive belts.

13. Fill the engine with clean engine oil.

14. Start the engine and check for leaks.

TRANSMISSION AND TRANSAXLE

Manual Transmission Assembly

REMOVAL AND INSTALLATION

240 and 940 Models

1. Disconnect the battery. At the firewall, disconnect the back-up light connector.

2. Raise the front of the vehicle and install jackstands. Under the vehicle, loosen the setscrew and drive out the pin for the shifter rod. Disconnect the shift lever from the rod.

3. Drain the gear oil from the transmission's drain plug.

4. Inside the vehicle, pull up the shift boot. Remove the fork for the reverse gear detent. Remove the snapring and lift up the shifter. If overdrive-equipped, disconnect the engaging switch wire.

5. Disconnect the clutch controls and return spring at the fork.

6. Disconnect the exhaust pipe bracket from the flywheel cover. Remove the oil pan splash guard.

7. Using a floor jack and a block of wood, support the engine beneath the oil pan. Remove the transmission support crossmember.

8. Disconnect the driveshaft. Disconnect the speedometer cable.

9. Remove the starter.

—— CAUTION ——
Support the transmission weight with a second jack or hoist before removing. Do not allow the transmission to hang partially removed on the input shaft.

---

10. Support the transmission using another floor jack. Remove the flywheel housing-to-engine bolts and remove the transmission by pulling it straight back.

To install:

11. Prior to installation, inspect the condition of the clutch and throwout bearing. Replace the bearing if it is scored or has been noisy in operation.

12. Carefully fit the transmission into place and start all the mounting bolts. After all bolts are started, tighten to 30 ft. lbs. (39 Nm).

13. Reinstall the transmission crossmember. When secure, remove the jack from beneath the engine. Replace the splash guard and attach the exhaust bracket to the flywheel housing.

14. Install the starter.

15. Connect the driveshaft, the speedometer cable and if necessary, the overdrive wiring.

16. Fill the transmission with the correct type and amount of gear oil.

17. Reconnect the clutch controls and return spring to the fork.

18. Inside the vehicle, connect the shifter and the reverse gear detent fork. Connect the wiring for the overdrive switch and install the shift boot and cover.

19. Under the vehicle, connect the shifter rod to the shift lever. Don't forget to tighten the setscrew.

20. Connect the reverse light wiring and attach the negative battery cable.

Manual Transaxle Assembly

REMOVAL AND INSTALLATION

850 Models

1. With all four wheels on the ground, loosen the front axle shaft locknuts.

2. Place the transaxle in **NEUTRAL** and set the parking brake.

3. Disconnect and remove the battery, air cleaner air intake ducts. Remove the battery tray. On turbocharged models, disconnect the timing valve from air cleaner and the turbocharger air duct clamp and hose.

4. Disconnect the gear selector cables from the brackets and lever. Remove the selector link plate by tapping out the lock pin. Disconnect the reverse light switch connector.

5. On turbocharged models remove the control pulley cover. Disconnect the turbocharger inlet pipe and tie it back out of the way. Disconnect the upper coolant hose to the engine oil cooler.

6. Disconnect the clutch slave cylinder and remove the clip. Remove the ground strap from the transaxle.

7. Loosen the rear engine mount and splash guard nut. Remove the bolts connecting the engine, transaxle and starter. Disconnect the transaxle ground strap.

8. Disconnect the ground strap from the firewall. Remove the torque arm bolt.

9. Secure the engine from above with an engine support that rests on the inner edges of the engine compartment.

10. Lift the engine up slightly to take weight off the engine mounts.

11. Raise and safely support the vehicle and remove the wheels. Disconnect the ABS sensor from the left side axle shaft, but do not unfasten the connector.

12. Drain the gear oil from the transaxle.

13. Disconnect all brackets for the front brake lines and ABS wiring for both sides of the vehicle.

14. Remove the plastic inner fender liners on both sides. Remove and discard the axle shaft locknuts.

15. Separate the ball joint from control arm, being careful not to damage the boots. Disconnect the sway bar links on both sides.

16. Remove the mounting screws holding the cable to the front of the subframe and disconnect the cable from the subframe. Disconnect the carbon filter hoses.

17. Disconnect the exhaust pipe clamp behind the catalytic converter.

18. Remove the left and right halfshafts.

NOTE: Be careful not to damage the transaxle seal.

19. Install seal plugs in the transaxle.

20. Loosen the two right side subframe-to-body bolts approximately ½ in. (15mm). Remove the subframe-to-body bolts on the left side.

NOTE: Make sure the steering gear bolts come out of the subframe and the control arm is free of the axle shaft boot on the right side.

21. Remove the jack and let the frame hang down from the right side bolts.

22. Tie the left side of the steering gear to the left side frame rail for support. Remove the steering gear engine mount bolt and nut at the top of the mount and remove.

NOTE: Make sure the steering gear is properly secured so the lower steering shaft does not slide out of the steering column.

23. Disconnect the oxygen sensor wiring clamps from the cover, as well as the connector and wiring to the vehicle speed sensor. Remove the cover at the back of the engine and the mount from the transaxle.

24. Lower the engine and transaxle with the lifting hook.

─────── WARNING ───────
If the engine is lowered too far, the exhaust pipe will be crushed against the steering rack. Be careful not to pinch any wiring or hoses and be sure that the engine dipstick tube is free of the fan.

25. Remove the seven remaining transaxle-to-engine bolts. Pull the gearbox away from the engine. Lower the jack and move the transaxle away.

To install:

26. Secure the throwout bearing fork to the transaxle. Make sure the mating surfaces on the transaxle and engine are clean and that the dowel pins are in place on the engine.

NOTE: Do not grease the primary shaft or throwout bearing sleeve. Make sure there are no breaks in the clutch plate.

27. Lift the transaxle into place and mate to the engine. Install the seven bolts securing the engine and transaxle and tighten them a little at a time to draw the transaxle into place. Torque the bolts to 37 ft. lbs. (50 Nm) and remove the transaxle jack.

28. Lift the engine and transaxle up until the distance between the engine support beam and spark plug cover is 0.20 in. (5mm).

29. Install the rear transaxle mount and bolts. Torque the rear two bolts to 37 ft. lbs. (50 Nm), then remove the front bolt. Install the cover.

30. Install the engine mount by fitting its guide pin into the cover. Install a new nut and hand-tighten. Install the steering rack engine mount bolt but do not tighten. Remove the support for the steering gear.

31. Reconnect the oxygen sensor wiring and clamps on the cover. Install the vehicle speed sensor connector and wiring and connect the transaxle ground strap.

32. Install the subframe using new 4 x M14 bolts and apply grease to the threads. Starting on the left side, lift the frame with a jack. Mount the support brackets on both sides. Torque the frame bolts to 78 ft. lbs. (105 Nm), then tighten an additional 120 degrees. Torque the bracket bolts to 37 ft. lbs. (50 Nm). Remove the jack and repeat the procedure for the right side.

33. Remove the engine support tool and lifting eyelet from engine. Torque the engine mount nut to 37 ft. lbs. (50 Nm).

34. Install five new nuts on the steering rack and torque them to 37 ft. lbs. (50 Nm). Install the front engine mount nut, then torque the front and rear bolts to 37 ft. lbs. (50 Nm).

35. Install the torque rod mount on the transaxle using new bolts. On earlier vehicles equipped with M18 bolts, torque the bolts to 13 ft. lbs. (18 Nm) and then an additional 90 degrees. On later models with M10 bolts, torque to 26 ft. lbs. (35 Nm) and then an additional 40 degrees.

36. Install the oil line bracket bolts and torque to 19 ft. lbs. (25 Nm). Tighten the exhaust pipe clamp while rocking the pipe back and forth to seat it properly.

37. Install the right and left halfshafts.

NOTE: Make sure the transaxle axle seal and axle boot are not damaged.

38. Connect the control arms to the ball joints using new nuts.

39. Connect the brake line and ABS cable bracket on both sides. Install the ABS sensor on the axle shaft and clean if needed. Torque the sensor to 7.4 ft. lbs. (10 Nm).

40. Attach the cable pipe and carbon filter container to the subframe.

41. Install the front splash guard. Install the wheels.

42. Install the starter and torque the bolts to 30 ft. lbs. (40 Nm). Connect the cable conduit and oxygen sensor connectors. Install the dipstick tube with a new O-ring and torque the bolt to 19 ft. lbs. (25 Nm).

43. Connect the slave cylinder and clips.

44. Connect the reverse light connector, shift lever plate and pin. Install the cables and lubricate the levers, cables, washers and clips with grease.

45. Connect ground strip to the firewall. Install a new bolt and nut for the extension arm and torque rod.

46. Connect the oil cooler hose to the cooler, if equipped. Install the throttle body and cover over control pulley. Connect the intake manifold to the turbocharger.

47. Install the coolant expansion tank, battery tray, air cleaner and connectors. Connect the control valve to air cleaner on turbocharged models. Install the battery and attach leads.

48. Torque the axle shaft nut to 89 ft. lbs. (120 Nm), then tighten an additional 60 degrees. Lock the axle shaft nut by notching its flange into the axle shaft groove.

49. Fill the transaxle with the specified amount of oil. Reinstall the plug. Check the function of the clutch before driving.

Clutch Assembly

REMOVAL AND INSTALLATION

240 and 940 Models

1. Remove the transmission.
2. Scribe alignment marks on the clutch and flywheel. Slowly loosen the bolts holding the clutch to the flywheel in a diagonal pattern. Remove the bolts and lift off the clutch and pressure plate.
3. Inspect the pressure plate for heat damage, cracks, scoring or any other damage.
4. Place a ruler diagonally over the pressure plate friction surface and measure the distance between the straightedge of the ruler and the inner diameter of the pressure plate. This measurement must not be greater than 0.008 inch (0.2mm). In addition, there must be no clearance between the straightedge and the outer diameter of the pressure plate. This check should be made at several points.

To install:

5. Clean the pressure plate and flywheel with solvent to remove any traces of oil and wipe them clean with a cloth.

6. Position the clutch assembly with the longest side of the hub facing away from the engine. Fit it to the flywheel and align the bolt holes. Insert centering tool 5111 or equivalent or an input shaft from an old transmission of the same type, through the clutch assembly and flywheel. This centers the assembly and pilot bearing.

7. Install the clutch retaining bolts and tighten them in a diagonal pattern, a few turns at a time. After all the bolts are tightened, remove the centering tool.

8. Install the transmission.

850 Models

1. Remove the transaxle.

2. Lock the flywheel in position. Remove the six bolts retaining the pressure plate and disc, loosen them in rotation. Remove the pressure plate and disc.

3. Remove the throwout bearing from the sleeve and fork. Remove the fork and dust cap from the transaxle.

4. Clean and check the throwout bearing, It should rotate freely and quietly. Clean and check the fork for cracks and wear and that the dust cap is intact.

5. Check the pressure plate carefully for signs of overheating, cracks, scoring or other damage to the friction surface. Make sure that the diaphragm spring is not split or damaged. If any part of the pressure plate is damaged, it must be replaced.

6. Check the pressure plate for warpage by laying a straightedge across the contact surface and checking the distance with a feeler gauge. The maximum width is 0.008 in. (0.2mm).

NOTE: Warpage is permitted in one direction only.

7. Check the flywheel for cracks and heavy scoring. If it is damaged it must be replaced. Check the clutch disc for oil or dirt and clean if necessary.

To install:
8. Install the throwout bearing fork, and lubricate the ball joint with grease. Install the throwout bearing and dust cap. Secure the fork to the transaxle, so that it cannot be moved during installation.

9. Install the clutch disc and pressure plate using centering drift 999 5487 or equivalent clutch alignment tool. Install the clutch bolts, tightening them in rotation so that the clutch slides over the locating pins and lies evenly against the flywheel. Then torque the bolts to 18 ft. lbs. (25 Nm). Remove the centering drift.

NOTE: Do not apply any grease to the splines on the transaxle input shaft.

10. Install the transaxle and check the clutch operation.

Clutch Master Cylinder

REMOVAL AND INSTALLATION

240 and 940 Models

1. Disconnect the negative battery cable.

2. Disarm the Air Bag system, if equipped.

3. Drain the clutch reservoir with a bulb syringe. Be careful not to drip brake fluid on any painted surfaces.

4. Remove the underdash panel and remove the lockring and pin from the clutch pedal.

5. Remove the hose from the master cylinder. Use a clean jar to collect spillage.

6. Remove the retaining bolts and remove the master cylinder.

To install:
7. Position the master cylinder in place and secure with the retainer bolts. Make sure that the clearance (free-play) between the pushrod and piston is 0.040 in. (1mm).

8. Connect the hose to the master cylinder. Make sure the hose is correctly threaded and tight.

9. Connect the master cylinder to the clutch pedal, and install the underdash panel.

10. Top up the fluid and bleed the system.

11. Connect the negative battery cable. Arm the Air Bag.

850 Models

1. Disconnect the negative battery cable.

2. Disarm the arm bag system.

3. Remove the air cleaner, retaining bracket bolts, and move the cable tie to the drain hose to one side. Disconnect the hose from brake fluid reservoir. Place a shop towel under the clutch master cylinder to protect the paint. Disconnect the solid line from the cylinder and collect the brake fluid in a suitable container.

4. Remove the lower dash panel (on SRS models remove the kneeguard also), and fold back the carpeting. Disconnect the clutch master cylinder pushrod from the clutch pedal. Remove the master cylinder mounting nuts and bolts and remove the clutch master cylinder.

To install:
5. Attach the brake fluid supply hose to the new clutch master cylinder. Apply a small amount of silicone grease to the end of the pushrod.

6. Install the master cylinder, and fasten with nuts and bolts. Torque to 18 ft. lbs. (25 Nm). Connect the master cylinder pushrod and install

the clip holding the pushrod to the clutch pedal.

7. On cars with SRS install the kneeguard and torque the bolts to 15 ft. lbs. (20 Nm). Install the lower dash panel and return the carpet to its original position.

8. Connect the solid line, to the clutch master cylinder. Connect the hose to reservoir and drain hose with cable tie.

9. Fill the reservoir and bleed the system.

10. Install the bracket retaining bolts to the air cleaner, and the air cleaner.

11. Check the clutch function before driving the car.

12. Connect the negative battery cable.

13. Arm the arm bag system.

Clutch Slave Cylinder

REMOVAL AND INSTALLATION

240, 940 and 850 Models

1. Raise and safely support the vehicle.

2. Loosen the slave cylinder hose a half turn. Position a suitable container under the slave cylinder to collect any fluid spillage. Remove the slave cylinder retaining bolts and unscrew it from the hose.

To install:
3. Screw the new slave cylinder onto the hose, hand-tight. Mount the slave cylinder and fasten it with the attaching bolts. Make sure that it is properly seated. Tighten the hose, making sure that it is not twisted.

4. Fill the clutch master cylinder and bleed the system. Top off the clutch master cylinder as necessary.

5. Test the clutch function before driving the car.

Hydraulic Clutch System Bleeding

——— CAUTION ———
Use only DOT 4 brake fluid. Never reuse brake fluid. Always keep brake fluid well sealed in its original container.

1. Turn the ignition switch **OFF** and if equipped with ABS, remove the key to prevent accidental pump activation.

2. Raise and safely support the vehicle.

3. Clean the fluid reservoir filler cap. Remove the cap and fill the res-

ervoir completely. Replace the reservoir cap.

4. Depress the clutch pedal a few times to purge the air bubbles in the master cylinder.

NOTE: Repeat this step if multiple bleeding steps are needed. Check the reservoir fluid level during bleeding, add as necessary.

5. Connect a hose from a drain bottle to the nipple on the slave cylinder.

6. While the clutch pedal is depressed to the floor, open the bleed nipple.

7. Hold the pedal to the floor to allow brake fluid and air bubble to exist through the hose. Close the nipple.

8. Repeat this procedure until no air bubbles are visible in the escaping fluid.

9. Make sure the bleed nipple is tight.

10. Pump the clutch pedal a few times to build pressure in the system.

11. Check the fluid reservoir. The fluid level should not be above the MAX level.

12. Road test the vehicle.

Automatic Transmission Assembly

REMOVAL AND INSTALLATION

240, 940 and 960 Models

――――― CAUTION ―――――
If the vehicle has been driven within the last 3-5 hours, the transmission oil may still be hot. Use extreme care when draining the oil or handling components.

1. Disconnect the battery ground cable.

2. Place the gear selector in the **P** position. Disconnect the kickdown cable at the throttle pulley on the engine.

3. Disconnect the oil filler tube at the oil pan and drain the transmission oil.

4. Disconnect the control rod at the transmission lever and disconnect the reaction rod at the transmission housing.

5. On AW71 transmissions, disconnect the wire at the solenoid slightly to the rear of the transmission output flange.

6. Matchmark the transmission-to-driveshaft flange and unbolt the driveshaft.

7. Place a jack or transmission dolly under the transmission and support the unit. Remove the transmission crossmember assembly.

8. Disconnect the exhaust pipe at the joint and remove the exhaust pipe bracket from the exhaust pipe. Remove the rear engine mount with the exhaust pipe bracket.

9. Remove the starter motor.

10. Remove the cover plate at the torque converter housing.

11. Disconnect the oil cooler lines at the transmission.

12. Remove the upper bolts at the torque converter cover. Remove the oil filler tube.

NOTE: It is helpful to have another person steadying and guiding the transmission during the removal process.

13. Remove the lower bellhousing bolts.

14. Remove the bolts retaining the torque converter to the drive plate. Pry the torque converter back from the drive plate with a small prybar.

15. Slowly lower the transmission while pulling it back to clear the input shaft.

――――― WARNING ―――――
Do not tilt the transmission forward or the torque converter may fall out.

To install:

16. Install the two lower bolts in the casing as soon as the transmission is in place. On 2.3L engines, adjust the panel between the starter motor and torque converter casing and install the bolts for the starter.

17. Mount the oil filler tube at the oil pan but do not tighten the nut.

18. Install the tube bracket and the two upper bolts in the converter casing. Tighten the nut for the oil tube to 65 ft. lbs. (85 Nm).

19. Install the bolts for the torque converter. Hand-tighten the bolts first, then tighten in a crisscross pattern to 32 ft. lbs. (42 Nm).

20. Install the rear engine mount with the exhaust pipe bracket and reconnect the exhaust system.

21. Install the transmission crossmember. When it is securely bolted in place, the supporting jack may be removed.

22. Reinstall the driveshaft.

23. Making sure that both the transmission linkage and the shift selector in the vehicle are in the

PARK position. Attach the actuator rod and the reaction rod. Adjust the shift linkage as necessary.

24. On AW71 models, install and connect the wiring to the solenoid valve.

25. Connect the kickdown cable at the throttle pulley. Adjust the cable if necessary.

26. Fill the transmission with oil. Connect the negative battery cable.

27. Apply the parking brake. Start the engine and allow to idle. Move the selector lever through all gear positions.

28. Place the selector lever in **P**. Wait 2 minutes and check the fluid level. Fill, as required.

Automatic Transaxle Assembly

REMOVAL AND INSTALLATION

850 Models

1. Pull the steering wheel adjustment lever out and adjust the wheel in and up as far as it will go. Then lock it in position. Put the transaxle gear selector in **N** and set the parking brake.

2. Disconnect and remove the battery and air cleaner assembly. Remove the battery tray. On turbo models, disconnect the control valve from air cleaner and the air charge manifold clamp and hose. Also remove the turbocharger air cleaner intake.

3. Disconnect the transaxle cable and connector from the transaxle. Be careful not to damage the rubber seal. Remove the wiring harness and ground from the control system cover. On early models, disconnect the transaxle vent hose. On later models, disconnect the wiring and oxygen sensor from the transaxle brackets.

4. Disconnect the transaxle cooling lines from the quick disconnects and drain the transmission fluid.

5. Remove the dipstick and tube. On vehicles with an EGR valve, disconnect the hoses to the valve. On turbo models, remove the cover over the control pulley and disconnect the intake to the throttle body and pull it to one side so the throttle body is free. Seal all oil connections to prevent dirt from entering.

6. Remove the bolts connecting the engine and transaxle and starter. Disconnect the transaxle ground strap. Lift off the radiator overflow tank and let it hang.

7. Remove the torque rod extension arm bolt and swing it out of the way.

NOTE: It will be necessary to support the engine from above and still be able raise and lower the car.

8. Install lifting yoke 999-5534 or equivalent in place of the torque rod extension arm to lift from. Install support tool 999-5033 on the inside fender rail, lifting beam support 999-5006 or equivalent placing the beam directly over the eye let for lifting yoke. Install the lifting hook 999-5460 or equivalent and pull it up slightly until the load is taken off of the engine mounts. Measure the distance between the beam and spark plug cover and make note of it.

9. Raise and safely support the vehicle and remove the wheels. Disconnect the ABS sensor from the left side axle shaft but do not disconnect.

10. Disconnect all brackets for the brake lines and ABS wiring on both sides.

11. Remove the plastic inner fender wells on both sides. Remove the left and right side halfshafts.

12. Install a seal plug in the transaxle.

13. Remove all the splash guards. Separate the ball joints from the control arms, being careful not to damage the boots. Disconnect the sway bar links on both sides.

14. Remove the subframe cable mounting screws and disconnect them from the subframe. Disconnect the carbon filter and hoses from the subframe. Cut the wire tie and hang the holder on the body. Disconnect the exhaust pipe clamp behind the catalytic converter.

15. Remove the oil line bracket screws and torque rod holder mounting screws.

16. Back the engine mounting/steering gear bolt off one turn. Remove the five steering gear mounting nuts in the subframe.

17. Position a jack under the left-hand side of the subframe so that it is barely touching. Remove the subframe bracket bolts on the body. Back the 15mm bolts between the frame and body on the right-hand side several turns. Remove the bolts from the left.

NOTE: Make sure the steering gear bolts come out of the subframe.

18. Remove the jack and let the frame hang down from the right side bolts. Secure the end of the right side driveshaft on the oil lines.

19. Remove the steering gear engine mount bolt and nut at the top of the engine mount and remove the mount.

NOTE: Make sure the steering gear is properly secured so the lower steering shaft does not slide out of the steering column.

20. Disconnect the oxygen sensor wiring clamps from the cover and the connector and wiring to the vehicle speed sensor. Remove the cover at the back of the engine and the mount from the transaxle.

21. Lower the engine and transaxle with the lifting hook until the distance between the beam and spark plug cover is 12.6 in. (320mm).

--- **WARNING** ---
If the engine is lowered too far, the exhaust pipe will be crushed against the steering rack. Be careful not to pinch any wiring or hoses and be sure that the engine dipstick tube is free of the fan.

22. Install transaxle fixture 5463 on the transaxle jack or equivalent, using the torque rod mounting bolts to hold it in place. At the same time, fit tool 5463-1 support plate or equivalent and raise the jack so that it is making light contact.

23. Remove the six torque converter bolts using a TX50 Torx® socket. Remove the lower plastic nut and fold out the inner fender well on the right-hand side. Then turn the crankshaft with a socket and ratchet. Remove the seven bolts between the engine and transaxle.

24. Remove the torque converter bolts from the flywheel. Remove the transaxle making sure the torque converter comes out with it and does not slip off the shaft. Use the hole in the torque converter cover to press the torque converter in to keep it from sliding off.

--- **WARNING** ---
Do not pry against carrier plate rim, as damage may result.

To install:
25. Flush the oil lines with clean transmission fluid. Install the line and hose on the transaxle using new O-rings on the quick-connectors.

26. Install the hose on the upper transaxle cooler line catch pan under the return line.

27. Inspect all components before installing.

28. Apply a small amount of grease to the torque converter guide pin and install, making sure the converter is all the way into the transaxle. The distance between the cover and converter bolt flange should be 0.55 in. (14mm).

29. Install the transaxle securing in place with the seven bolts between the engine and transaxle. Tighten them to 37 ft. lbs. (50 Nm).

30. Install new torque converter bolts and torque to 22 ft. lbs. (30 Nm) using a TX50 Torx® socket.

NOTE: Remove the socket from the crankshaft.

31. Install the rear transaxle mount and three bolts. Torque the rear two bolts to 37 ft. lbs. (50 Nm), then remove the front bolt. Install the cover against the mount and torque the bolt to 37 ft. lbs. (50 Nm).

32. Install the engine mount guide pin into the cover. Install a new nut and hand-tighten. Install the steering rack engine mount bolt, but do not tighten.

33. Reconnect the oxygen sensor. Install the vehicle speed sensor connector and connect the transaxle ground strap.

34. Install the subframe using new bolts. Apply grease to the threads. Starting on the left, lift the frame with a transaxle jack. Mount the support brackets on both sides. Torque the frame bolts to 78 ft. lbs. (105 Nm) plus an additional 120 degrees. Torque the bracket bolts to 37 ft. lbs. (50 Nm). Remove the jack and repeat the procedure for the right side.

35. Install five new nuts on the steering rack and torque them to 37 ft. lbs. (50 Nm). Install the front engine mount nut then torque the front and rear bolts to 37 ft. lbs. (50 Nm).

36. Install the torque rod mount on the transaxle using new bolts, torque M18 bolts (early models) to 13 ft. lbs. (18 Nm) plus an additional 90 degrees or M10 bolts (later models) to 26 ft. lbs. (35 Nm) plus an additional 40 degrees.

37. Install the right and left halfshafts.

NOTE: Make sure the transaxle axle seal and axle boot are not damaged.

38. Connect the control arms to the ball joints using new nuts. Install the sway bar link using new nuts and torque to 37 ft. lbs. (50 Nm).

39. Attach the cable pipe and carbon filter container to the subframe. Tie the hoses with a strip clamp in subframe. Install engine splash guard on early models.

40. Install the front splash guard by pressing in the guides and installing the screws.

41. Install the five transaxle bolts on top side and torque to 37 ft. lbs. (50 Nm). Install the starter. Connect the cable conduit and oxygen sensor connectors. Install the dipstick tube with a new O-ring and torque to 19 ft. lbs. (25 Nm).

42. Connect the wiring harness, ground lead, transaxle connectors. Connect the transaxle vent and EGR hoses. Attach the transaxle cable and adjust.

43. Connect the ground strip to the firewall. Install a new bolt and nut for the extension arm and torque rod. Torque early model M8 bolts to 13 ft. lbs. (18 Nm) plus an additional 120 degrees. Torque later model M10 bolts to 26 ft. lbs. (35 Nm) plus an additional 90 degrees.

44. Install the remaining components.

45. Fill the transaxle with fluid. Check the fluid level after the engine has reached normal operating temperature to assure that it is correct.

DRIVELINE

Driveshaft

REMOVAL AND INSTALLATION

240, 940 and 960 Models

Without CV-Joints

1. Disconnect the negative battery cable.

2. Raise and safely support the vehicle.

3. Place the transmission in **NEUTRAL** and release the parking brake.

4. Mark the transmission and driveshaft to align when reassembling.

5. On models with plastic fuel tanks, loosen the fuel tank by removing one bolt and pushing the support aside.

6. Disconnect the driveshaft from the front and rear drive flange. Use tool 5244/5381, or the equivalent, for the U-joint flange nuts and bolts. Tool 5426, or equivalent, should be used to hold the rear CV-joint.

7. Remove the center support bearing retaining bolts.

8. While supporting the driveshaft, remove the bearing bracket mounting bolts. If washer or shims are present, the number should be noted.

NOTE: To preserve the balance of the driveshaft, exact replacement of any and all shims and washers should be done.

9. Remove the center support bearing bracket and remove the driveshaft, pulling toward the front of the vehicle.

To install:

10. A coating of grease should be applied to the splined section.

11. Check the splines for damage and check that the boot is in position.

12. Install the front of the driveshaft first, then the rear.

13. Install the bearing center support bracket and mounting bolts, but do not tighten the bolts.

14. Align the front drive flange markings and attach the driveshaft to the transmission drive flange. Install new nuts and bolts.

15. Tighten the flex joint bolts to 59 ft. lbs. (80 Nm) and the U-joint drive flange 10mm bolt to 37 ft. lbs. (50 Nm)

16. Models with 8mm U-joint drive flange bolts:
   a. Coat the drive flange with friction compound.
   b. Connect the driveshaft and lubricate the new bolts. Torque the bolts crosswise to 22 ft. lbs. (30 Nm), then an additional 60 degrees.

17. With the transmission in **PARK**, install the gear selector retaining pin and circlip.

18. Attach the rear drive flange, aligning the markings. Install new bolts and tighten to 37 ft. lbs. (50 Nm).

19. Install the center support bearing crossmember and tighten to 18 ft. lbs. (25 Nm).

20. With the center support bearing positioned correctly, tighten the four bearing bracket mounting bolts to 18 ft. lbs. (25 Nm).

21. Align the fuel tank support and install the mounting bolts to 15 ft. lbs. (20 Nm).

22. Check for proper clearance between the fuel tank, fuel lines and the driveshaft.

23. Lower the vehicle and connect the negative battery cable.

With CV-Joints

1. Disconnect the negative battery cable.

2. Raise and safely support the vehicle.

3. Place the transmission in **NEUTRAL** and release the parking brake.

4. Mark the transmission and driveshaft to align when reassembling. Mark the shaft tube in front of the rear joint boot.

5. On models with plastic fuel tanks, loosen the fuel tank by removing one bolt and pushing the support aside.

6. Disconnect the driveshaft from the front and rear drive flange. Tool 5426, or equivalent, should be used to hold the rear CV-joint.

7. Remove the center support bearing retaining bolts.

8. While supporting the driveshaft, remove the bearing bracket mounting bolts. If washer or shims are present, the number should be noted. To preserve the balance of the driveshaft, exact replacement of any and all shims and washers should be done.

9. Remove the center support bearing bracket and remove the driveshaft, pulling toward the front of the vehicle.

10. Mark the front and rear sections of the shaft in relation to each other. Record the distance between the sections. The proper distance is necessary for correct reinstallation.

11. Remove the setscrew from the shaft tube.

12. Remove the boot from the center bearing race, and separate the shafts.

To install:

13. A thin layer of penetrating oil should be applied to the splined section.

14. Check the splines for damage and check that the boot is in position.

15. Reassemble the shaft sections, lining up the markings and in accordance with the measured distance.

16. Tighten the shaft tube setscrew.

17. With the shaft compressed toward the forward position, the driveshaft should measure 61.12–61.14 in. (155–156cm). With the shaft extended toward the rear position, the driveshaft should measure 61.98–61.03 in. (157–158cm).

18. Apply thread sealant to the threads of the setscrew and tighten to 13 ft. lbs. (18 Nm).

19. Install the spline boot to the front section of the driveshaft.

20. Install the front of the driveshaft first, then the rear.

21. Install the bearing center support bracket and mounting bolts, but do not tighten the bolts.

22. Align the front drive flange markings and attach the driveshaft

to the transmission drive flange. Install new nuts and bolts.

23. Tighten the bolts to 37 ft. lbs. (50 Nm).

24. Attach the rear drive flange, aligning the markings. Install new bolts.

25. Tighten the bolts to 6 ft. lbs. (8 Nm), then tighten them to 22 ft. lbs. (30 Nm).

26. Install the center support bearing crossmember and tighten to 18 ft. lbs. (25 Nm).

27. With the center support bearing positioned correctly, tighten the four bearing bracket mounting bolts to 18 ft. lbs. (25 Nm).

28. Align the fuel tank support and install the mounting bolts to 15 ft. lbs. (20 Nm).

29. Check for proper clearance between the fuel tank, fuel lines and the driveshaft.

30. Lower the vehicle and connect the negative battery cable.

Halfshaft

REMOVAL AND INSTALLATION

940 and 960 Models

1. With the vehicle on all four wheels, loosen the large axle nut in the center of the wheel hub.

2. Raise and safely support the rear of the vehicle.

3. Remove the wheel and remove the axle nut.

4. Remove the eight bolts holding the upper and lower sections of the final drive subframe section.

5. Remove the bolts holding the halfshaft to the final drive unit (differential) and remove the shaft from the wheel bearing housing.

6. When the shaft is removed, inspect the rubber boots for any sign of splitting or cracking. A light coat of

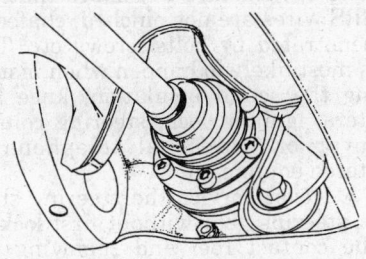

297650

Remove the halfshaft bolts — 940 and 960 models

silicone or vinyl protectant applied to a CV-boot will extend its life.

To install:

7. When reinstalling, fit the axle into the hub first, then position and secure the inboard end. Always use new, lightly oiled bolts and tighten them to 70 ft. lbs. (91 Nm). Do not tighten the axle nut yet.

8. Reinstall the lower section of the subframe section. Before tightening the eight mounting bolts, Install two long 12mm bolts or drift pins in the centering holes and align the panel. This is essential to insure correct wheel alignment when finished.

9. Tighten the eight mounting bolts to 52 ft. lbs. (68 Nm). plus an additional 30 degrees of rotation.

10. Use a new, lightly oiled axle nut and install it on the threaded end of the shaft. Tighten it until it is snug but do not attempt to apply final tightening.

11. Install the wheels. Lower the vehicle to the ground.

12. Apply the hand brake and tighten the axle nut to 103 ft. lbs. (134 Nm) plus an additional 60 degrees of rotation.

850 Models

1. With the vehicle sitting on all four wheels, loosen the axle shaft nut.

2. Raise and safely support the vehicle. Remove the wheels. Disconnect the ABS sensor from the halfshaft, but do not disconnect the harness.

3. Disconnect all brackets for brake lines and ABS wiring on both sides and let them hang.

4. Remove the axle nut. Push the end of the halfshaft from hub using a soft drift and a mallet.

5. Disconnect the sway bar from the link. Remove all splash guards.

6. Separate the ball joint from control arm, being careful not to damage the boots.

7. For the right side halfshaft, remove the bearing cap and pull the shaft out of the transmission while holding the strut out of the way. Install a seal plug in the transmission.

NOTE: Be careful not to damage the transmission seal.

8. For left side, remove the halfshaft by carefully prying between the transmission and the halfshaft. Hold the strut assembly out of the way. Install a seal plug in the transmission.

To install:

9. Install the right halfshaft and torque the bearing cap to 19 ft. lbs. (25 Nm). Install the splashguard.

NOTE: Make sure that the transmission axle seal and axle boot are not damaged.

10. Clean the ABS wheel if necessary. Apply metal adhesive to the halfshaft splines. Carefully press shaft in so that the lock ring engages with the differential gear. Check it by carefully pulling on the shaft joint housing. Install the axle nut and hand-tighten.

11. Connect the ball joints using new nuts. Install the sway bar link using new nuts.

12. Connect the brake line and ABS cable bracket on both sides. Install the ABS sensor on the halfshaft and clean it with a soft brush.

13. Install the wheels.

14. With all four wheels on the ground, torque the axle nut to 89 ft. lbs. (120 Nm) plus an additional 60 degrees. Lock the nut by staking its flange into the driveshaft groove.

CV-Joint Boot

REPLACEMENT

1. Raise and support the vehicle safely.

2. Remove the halfshaft from the vehicle. Support the assembly in a vise with soft jaws.

3. Remove the large boot clamp from the CV-joint and roll the boot back over the shaft.

4. Clean any excess grease. Spread the inner clip and separate it from its groove. Matchmark the axle shaft and tripod bearing for reassembly. Remove the wire ring bearing retainer from inside the outer race housing and remove the outer race.

5. Matchmark the tripod bearing and shaft. Remove the tripod bearing snapring and remove the tripod bearing from the shaft. It may be necessary to drive the tripod off the shaft with a brass punch.

6. Remove the small clamp and CV boot from the halfshaft.

To install:

7. The inner and outer CV-joint boots are different. To make sure the boot is being installed in the correct location, measure the outer diameter of the large end of the boot.

8. Wrap smooth electrical tape around the halfshaft spline to ease installation of the boot. Slide the

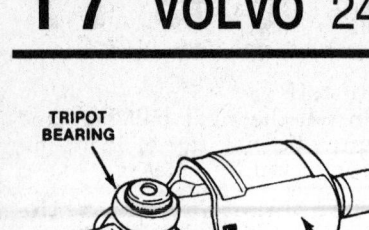

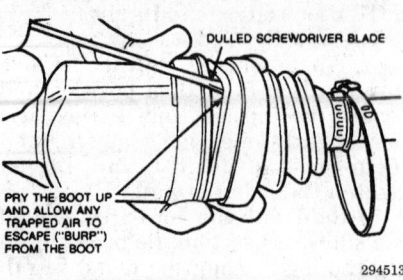

Release the air before installing the large clamp — 850 models

Alignment marks — 850 models

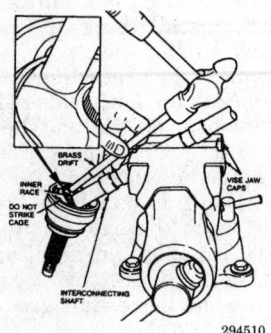

One sharp tap will disengage the CV-joint from the circlip — 850 models

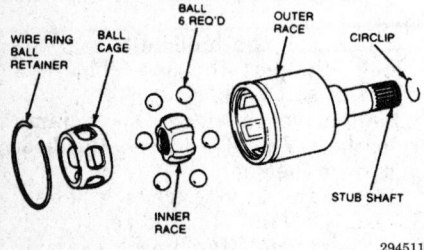

Outer CV-joint with retainer ring — 850 models

clamps and the outboard boot onto the shaft.

9. Before positioning the boot over the CV-joint, pack the CV-joint and the boot with grease. Be sure to use all of the grease in the pouch supplied with the boot kit.

10. Fit the boot into place on the CV-joint, making sure it is fully seated in the grooves in the shaft and outer race. Insert a suitable tool between the boot and the outer bearing race to allow trapped air to escape from the boot.

11. Install the boot clamps, wrapping them around the boots in the opposite direction of halfshaft rotation. Pull the clamps tight with a suitable tool and bend the locking tabs to secure in position. Make sure the band is tightened properly and the locking clip is secured.

12. Fit the inboard CV-joint boot and clamps onto the halfshaft and remove the electrical tape.

13. Install the tripod assembly on the halfshaft with the matchmarks aligned. Install the tripod retaining ring.

14. Fill the CV-joint outer race with CV-joint grease. Make sure the boot is also greased to prevent wear from internal contact of the ribs.

15. Install the outer race over the tripod joint with the matchmarks aligned and install the wire ring bearing retainer.

16. Fit the boot into place on the CV-joint, making sure it is fully seated in the grooves in the shaft and outer race. Insert a suitable tool between the boot and the outer bearing race to allow trapped air to escape from the boot.

17. Install the boot clamps, wrapping them around the boots in the opposite direction of halfshaft rotation. Pull the clamps tight with a suitable tool and bend the locking tabs to secure in position.

18. Work the CV-joint through its full range of travel at various angles. The joint should flex, extend, and compress smoothly. Wipe away any excess grease.

19. Install the halfshaft in the vehicle.

STEERING

Air Bag
— CAUTION —
Some vehicles are equipped with an Air Bag system, also known as the Supplemental Inflatable Restraint (SIR) or Supplemental Restraint System (SRS). The system must be disabled before performing service on or around system components, steering column, instrument panel components, wiring and sensors. Failure to follow safety and disabling procedures could result in accidental Air Bag deployment, possible personal injury and unnecessary system repairs.

PRECAUTIONS

Several precautions must be observed when handling the inflator module to avoid accidental deployment and possible personal injury.

• Never carry the inflator module by the wires or connector on the underside of the module.

• When carrying a live inflator module, hold securely with both hands, and ensure that the bag and trim cover are pointed away.

• Place the inflator module on a bench or other surface with the bag and trim cover facing up.

• With the inflator module on the bench, never place anything on or close to the module which may be thrown in the event of an accidental deployment.

• Before beginning work which could affect the SRS system, turn the ignition switch **OFF**, disconnect the negative battery cable and tape the cable end to avoid accidentally recharging the Air Bag power supply.

• When working around the instrument panel or steering column, take special care to ensure that the SRS wires are not pinched, chafed or penetrated by bolts/screws etc. This is most likely to happen when installing the sound insulation, knee bolsters, ignition lock, steering column cover or additional telephone or stereo equipment.

• Never service the steering shaft or steering gear without first locking the contact reel and removing the steering wheel.

• If it is necessary to connect the battery with the Air Bag removed, install the special tool 998 8695 or an equivalent resistor. This tool has the same resistance as the Air Bag assembly and will prevent setting a fault code in the SRS control unit.

• Do not connect an ohmmeter across the Air Bag connector terminals. The current in the meter is enough to fire the Air Bag explosive charge.

DISARMING

— WARNING —
Before beginning work on vehicles equipped with a Supplemental Restraint System (SRS), the system must be properly disarmed and the Air Bag must be removed and properly stored. Unintended Air Bag deployment can cause serious or fatal injury.

1. Disconnect the negative battery cable and tape the cable end. Allow sixty (60) seconds for the reserve power to dissipate.
2. Turn the ignition key to unlock the steering.
3. Turn the steering wheel slightly in order to reach the two Torx® bolts in back of the steering wheel. Unplug the connector and remove the Air Bag.

NOTE: Do not turn the ignition switch ON while the Air Bag assembly is removed. This will set a fault code in the SRS control unit.

4. Store the Air Bag module where it will not be disturbed. Set it face up and do not place anything on top that may become a projectile in case of accidental deployment.

— CAUTION —
When connecting the battery, make sure that no one is in the vehicle in case of an SRS malfunction causing accidental Air Bag deployment.

Steering Wheel

REMOVAL AND INSTALLATION
— CAUTION —
On vehicles equipped with an Air Bag, the system must be properly disarmed and the Air Bag must be removed and properly stored. Unintended Air Bag deployment can cause serious or fatal injury.

1. Disconnect the negative battery cable and tape the end to prevent ac-

cidental recharging of the Air Bag power supply.
2. Turn the steering wheel to the straight ahead position.
3. If equipped with an Air Bag, remove the TX30 Torx® screws from the rear of the steering wheel. Unplug the connector from the Air Bag module and lift it off. Place the module where it will not be disturbed. Lay it face up and do not place anything on top of the module.

NOTE: Do not turn ignition switch ON while Air Bag module is disconnected. This will register a fault code which will then need to be canceled.

4. Remove the bolt from the steering wheel center. Loosen the locking screw in the end of the plastic ribbon from parking hole in steering wheel. The screw should always remain in the plastic ribbon.
5. Set the lockscrew in the pin in the contact reel. This screw locks the contact reel in the zero position.

— WARNING —
Do not turn the wheel in this position. The pin will shear off and the whole contact reel will have to replaced.

6. Remove the steering wheel and pull the lead and plastic strip with the screw through the hole in the middle.
To install:
7. Fit the steering wheel onto the shaft.
8. If equipped with an Air Bag:
  a. Make the bolt only finger-tight at this time. Do not turn the wheel or contact pin will shear.
  b. Remove the locking-screw from the contact reel and install it in the parking hole.
  c. Torque the steering wheel bolt to 24 ft. lbs. (33 Nm).
9. Connect the Air Bag module wiring and fit the module into place.

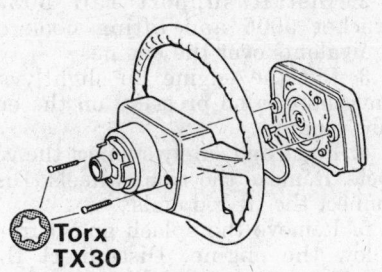

Torx TX30

299427

Removing the Air Bag assembly — 240, 940 and 960 models

Torque TX30 Torx® bolts to 72 inch lbs. (8 Nm). Make sure the cable is not pinched and the module is correctly mounted.

— CAUTION —
When connecting the battery, make sure that no one is in the vehicle in case of an Air Bag malfunction causing accidental Air Bag deployment.

10. Turn the ignition switch **ON** and connect the negative battery cable. If the SRS light is lit, there is a fault code present and the system will not function.

Tie Rod Ends

REMOVAL AND INSTALLATION

1. Raise and safely support the vehicle.
2. Remove the wheel.
3. Measure the length between the end of the tie rod and the steering rack housing and mark the position on the rack.
4. Disconnect the tie rod end from the steering arm using an appropriate puller.
5. Remove the tie rod end from the steering rack.
To install:
6. Install the tie rod end on the steering rack as per previous marking.
7. Torque the tie rod locknut to 52 ft. lbs. (70 Nm).
8. Connect the tie rod end to the steering arm.
9. Torque the ball stud nut.
  a. 850 models: torque to 52 ft. lbs. (70 Nm).
  b. 240, 940 and 960 models: torque to 44 ft. lbs. (60 Nm).
10. Install the wheels.
11. Lower the vehicle and check the front wheel alignment.

Power Rack and Pinion

REMOVAL AND INSTALLATION

240 Models

— CAUTION —
On vehicles with Air Bags, the front wheels must be pointing straight ahead with the steering wheel locked. If this is not done, the contact reel of the Air Bag system will reach its end position and deploy the Air Bag.

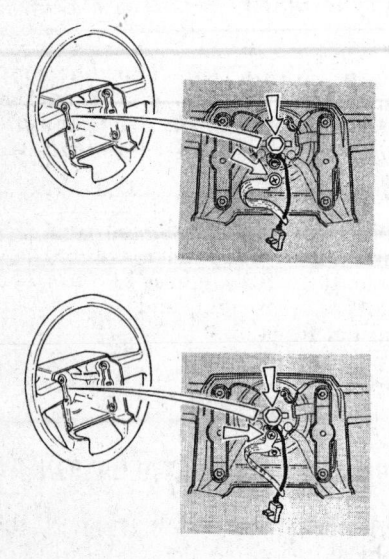

Locking contact reel in zero position — 240, 940 and 960 models

297735

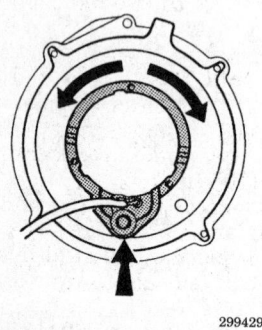

299429

Setting contact reel to zero position — 240, 940 and 960 models

1. Disconnect the negative battery cable. Disarm the Air Bag system, if equipped.

2. Raise and support the vehicle safely on jackstands. Remove the front wheels

3. From under the vehicle, remove the splashguard.

4. With the wheels pointed straight ahead, measure the length from one tie rod end to the steering rack housing.

5. Remove the steering column U-joint by scribing an alignment mark on the shaft. Remove the cotter pin

and loosen the nut and bolt securing the joint to the column shaft. Separate the joint and steering shaft.

6. Remove the tie rod ends.

7. With a drip pan placed below the hoses to the steering rack, tag then remove the hoses. Discard the copper sealing washers.

8. Remove the sway bar if equipped. Remove the two steering rack fixing bolts and nuts.

9. Lower the steering rack down from the vehicle frame and out.

To install:

10. Position the steering rack in the vehicle Secure the rack with the fixing nuts and bolts. Tighten the nuts and bolts to 32 ft. lbs. (44 Nm).

11. Install the sway bar if removed.

12. Install the hoses to the rack using new sealing washers. Tighten the hose bolts to 30 ft. lbs. (42 Nm).

13. Install the tie rod ends.

14. Connect the steering shaft and U-joint by aligning the scribe mark on the steering shaft with the mark on the U-joint. Tighten the retainer nut and bolt to 16 ft. lbs. (21 Nm). Install a new cotter pin.

15. Install the splashguard. Lower the vehicle, fill the steering reservoir and bleed the system

850 Models

--- **CAUTION** ---

On vehicles with Air Bags, the front wheels must be pointing straight ahead with the steering wheel locked. If this is not done, the contact reel of the Air Bag system will reach its end position and deploy the Air Bag.

NOTE: The front subframe must be lowered. The bolts cannot be used again once loosened: new subframe bolts are required.

1. Disconnect the negative battery cable. Disarm the Air Bag system, if equipped.

2. Install support rail 5033, bracket 5006 and lifting hook or equivalents over the engine.

3. Lift the engine up slightly so that there is no pressure on the engine mounts.

4. Raise and safely support the vehicle. Remove the front wheels. Disconnect the tie rod ends.

5. Remove the splash guard from below the engine. Disconnect the power steering fluid lines brackets and clamps from the front and rear of the subframe. Remove the five nuts holding the steering rack to the subframe.

6. Position a jack below the rear part of the subframe and remove the following:

• the four bolts holding the subframe to the body on both sides

• the two bolts and washers holding the bracket to the subframe

• loosen the front subframe bolts so the frame lowers 0.59–0.79 in. (15–20mm)

7. Lower the subframe using the jack, and place a spacer between the frame and the body at the rear edge so the frame will not pop up.

8. Position a catch pan under the steering rack and disconnect the power steering lines from the rack.

9. Remove the steering column joint bolt and press it up from the steering rack. Remove the bolt holding the rack to the engine mount.

10. Remove rack from the right side.

To install:

11. Transfer the thermal protection plate and center attachment mount, but do not tighten the mounting bolts. Install the protective plugs in the line connection holes.

12. Install the tie rod ends.

13. Install the steering rack from the right side and let it rest on the rear engine mount.

14. Raise the rack up on the right side so that it is straight in relation to the frame and torque the engine mount bolt to 37 ft. lbs. (50 Nm).

15. Connect the fluid lines and brackets loosely using new O-rings on the lines.

16. Align the fluid lines in the bracket and tighten them in the steering rack.

17. Fit the steering rack onto the steering shaft joint and torque the bolt to 15 ft. lbs. (20 Nm). Install the bolt lock clip.

18. Lift the rear of the subframe up using a jack and line up the steering rack mount bolts in the frame.

19. Install new subframe bolts loosely. Move the jack to the front and replace the bolts with new ones, but do not tighten. Torque the bolts on the left side of the subframe to 77 ft. lbs. (105 Nm) plus an additional 120 degrees. Torque the right side bolts the same way. Finally, torque the bracket bolts on both sides to 37 ft. lbs. (50 Nm).

20. Install new nuts on the steering rack and torque them to 37 ft. lbs. (50 Nm). Torque the steering rack center bolt to 59 ft. lbs. (80 Nm). Install the front and rear steering fluid line brackets and tighten them.

21. Install the engine splashguard below the engine.

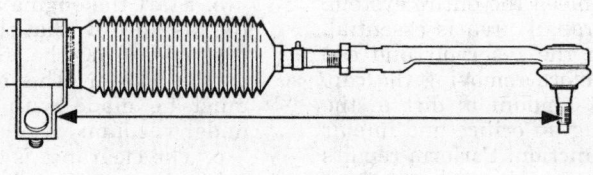

295596

Measuring tie rod distance from the rack housing

22. Install the wheels.
23. Fill the power steering fluid reservoir with fluid.
24. Check the fluid level once again. Lower the vehicle and check the toe-in.

Power Steering Pump

BLEEDING

1. Fill the reservoir with the proper type of fluid. Raise and support the vehicle safely. Place the transmission in **N** and apply the parking brake.
2. Start the engine and fill the reservoir as the level drops.
3. When the reservoir level has stopped dropping, slowly turn the steering wheel from lock to lock several times. Fill the reservoir if necessary.
4. Continue to turn the steering wheel slowly until the fluid in the reservoir is free of air bubbles.
5. Stop the engine and observe the oil level in the reservoir. If the oil level rises more than ¼ in. (6mm) past the level mark, air still remains in the system. Continue bleeding until the level rise is correct.
6. Lower the vehicle.

REMOVAL AND INSTALLATION

240 Models

1. Disconnect the negative battery cable.
2. Remove all dirt and grease from around the line connections at the pump.
3. Using a container to catch any fluid that might run out, disconnect the power steering lines and plug them to prevent dirt from entering the system.

4. Remove the tensioner locking screws on both sides of the pump and remove the drive belt.
5. Turn the pump up and remove the three bolts holding the bracket to the engine block. Remove the pump and bracket.
6. If the pump is being replaced with a new one, remove the nut and pulley from the old pump and transfer it to the new one. Separate the bracket and tensioner from the pump and install them loosely on the new pump.
To install:
7. Place the pump in position on the engine and install the retaining bolts and spacer.
8. Install the drive belt.
9. Adjust the belt tension and then tighten the nuts of the long bolts.
10. Use new copper washers, and reconnect the fluid lines to the pump.
11. Fill the reservoir with Type A or Dexron®III automatic transmission fluid and bleed the system.
12. Connect the negative battery cable.

940 and 960 Models

1. Disconnect the negative battery cable.
2. Remove the splash guard from under the engine.
3. Loosen the belt tensioner. Remove the mounting bracket and bolt.
4. Disconnect the lines at or near the pump. Depending on the type of pump, it may be necessary to disconnect the rubber hoses from the metal pipes instead of removing the lines at the pump body. Use a catch pan under the vehicle for spillage. Plug the lines and fittings immediately to avoid contamination.
5. Remove the large retaining bolt and remove the drive belt from the pump.
6. Lower the pump slightly and disconnect the filler hose from the

pump. Remove the pump from the vehicle.
7. If the pump is to be replaced with a new one, transfer the pulley, the mounting bracket and the washers to the new pump. Install the mounting bracket on the new pump; make sure the thick washer is between the bracket and the pump body. Install the pulley with the conical face of the washer must be to the outside.
To install:
8. Connect the filler hose to the pump.
9. Position the pump and install the retaining bolts loosely.
10. Install the mounting bracket and belt. Adjust the belt tension.
11. Tighten the lower retaining bolts.
12. Connect the fluid hoses to the pump. Use new copper washers and/or hose clamps. Tighten the banjo fittings to 31 ft. lbs. (40 Nm).
13. Connect the negative battery cable.
14. Fill the fluid reservoir and start the engine. Bleed the steering system.
15. Reinstall the splash guard.

850 Models

1. Disconnect the negative battery cable.
2. Remove the radiator reservoir cap. Open the radiator draincock on the left side under the radiator and drain out about 3.2 qts. (3 liters) of coolant. Disconnect the radiator hose from the thermostat housing.
3. Remove the oil hose holder from the dipstick tube and the air cooler hose from the control module box.
4. Using the proper sized ratchet, release the tension from the drive belt tensioner and remove the belt from the power steering pump.
5. Remove the long bolt and spacer from the plate. Loosen the pressure side hose a quarter turn and the lower plate mount nut a few turns.
6. Remove all of the pump mounting bolts. Lift the pump straight up and disconnect the pressure hose and old O-ring. Collect any fluid that spills. Carefully make a small cut in the end of the return line no longer than the mark on the hose itself. Remove the pump.
7. Raise and safely support the vehicle. Turn the steering wheel from lock to lock and collect the fluid from the lines. Make sure no oil gets into the alternator.

8. Place the old pump in a vice and remove the pulley using an appropriate puller.

To install:

9. Install the pulley on the new pump, using an appropriate pressing tool. Apply a small amount of oil to the shaft to ease the installation.

10. Install the pump and five mounting bolts, torque them to 18 ft. lbs. (25 Nm). Install the long bolt and cover plate spacer, torque it to 18 ft. lbs. (25 Nm). Tighten the lower attachment to 18 ft. lbs. (25 Nm).

11. Install the following:
• pressure hose with a new O-ring
• return hose
• hose bracket for power steering hoses to dipstick tube
• radiator hose
• pump drive belt
• cooling hose to the control module box

12. Fill the cooling system with coolant.

13. Fill the power steering pump reservoir with new fluid. Bleed the steering system.

14. Start the engine. Turn the wheel from lock to lock. Add more fluid as necessary.

15. Connect the negative battery cable.

BRAKES

Anti-Lock Brake System Service

PRECAUTIONS

• Certain components within the Anti-Lock Brake System (ABS) are not intended to be serviced or repaired individually. Only those components with removal and installation procedures should be serviced.

• Do not use rubber hoses or other parts not specifically specified for ABS system. When using repair kits, replace all parts included in the kit. Partial or incorrect repair may lead to functional problems and require the replacement of components.

• Lubricate rubber parts with clean, fresh brake fluid to ease assembly. Do not use lubricated shop air to clean parts; damage to rubber components may result.

• Use only specified brake fluid from an unopened container.

• If any hydraulic component or line is removed or replaced, it may be necessary to bleed the entire system.

• A clean repair area is essential. Always clean the reservoir and cap thoroughly before removing the cap. The slightest amount of dirt in the fluid may plug an orifice and impair the system function. Perform repairs after components have been thoroughly cleaned; use only denatured alcohol to clean components. Do not allow ABS components to come into contact with any substance containing mineral oil; this includes used shop rags.

• The Anti-Lock control unit is a microprocessor similar to other computer units in the vehicle. Ensure that the ignition switch is **OFF** before removing or installing controller harnesses. Avoid static electricity discharge at or near the controller.

• If any arc welding is to be done on the vehicle, the control unit should be unplugged before welding operations begin.

Master Cylinder

REMOVAL AND INSTALLATION

240 Models

1. Disconnect the negative battery cable and the low fluid indicator from the filler cap.

2. Disconnect the brake lines from the master cylinder and plug them immediately.

NOTE: If the vehicle has a hydraulic clutch, disconnect the line from the fluid reservoir. Plug and place it out of the way.

3. Remove the two nuts and lockwashers that attach the master cylinder to the brake booster and lift the assembly forward. Be careful not to spill any fluid on the fender.

— **WARNING** —
Do not depress the brake pedal while the master cylinder is removed.

To install:

4. A clearance of 0.004-0.040 in. (0.10–1.01mm) is required between the thrust rod and the primary piston with the master cylinder installed. To calculate and adjust the clearance:

a. Measure distance **A** between the face of the master cylinder

mounting flange and the center of the primary piston in the master cylinder. Use a depth caliper or depth micrometer.

b. Start the engine and measure the distance **B** that the thrust rod protrudes from the mounting face on the booster. This measurement must be made with the booster under vacuum.

c. The clearance is calculated by subtracting **B** from **A**.

d. Adjust the clearance by turning the adjusting nut on the thrust rod as required. Apply a few drops of thread locking compound to the adjusting nut.

5. Position the master cylinder and reservoir assembly onto the studs, install the washers and nuts and tighten to 17 ft. lbs. (22 Nm).

6. Remove the plugs and connect the brake lines.

7. Bleed the brake system.

8. Connect the negative battery cable and low fluid indicator.

940, 960 and 850 Models

1. Disconnect the negative battery cable and the low fluid indicator from the filler cap.

2. Place absorbent towels below the master cylinder to prevent paint damage from spillage.

3. Disconnect the brake lines from the master cylinder. Disconnect the clutch hose from the reservoir if equipped.

4. Remove the master cylinder mounting nuts.

5. Carefully lift the master cylinder off. Be careful not to spill any fluid.

To install:

6. Check and adjust the pushrod dimension according to the chart.

7. Install the new master cylinder and fasten it with the mounting nuts. Torque the nuts to 17 ft. lbs. (23 Nm).

8. Connect the brake lines and clutch hose if equipped.

9. Fill the master cylinder with approved brake fluid.

10. Bleed the brake system.

11. Depress the brake pedal to simulate hard braking for 30 seconds.

12. Check the master cylinder for signs of leakage.

NOTE: Depress the brake pedal 3 or 4 times, with the engine OFF. The brake pedal travel must not exceed 2.2 in. (55mm).

13. Connect the negative battery cable and the low fluid indicator.

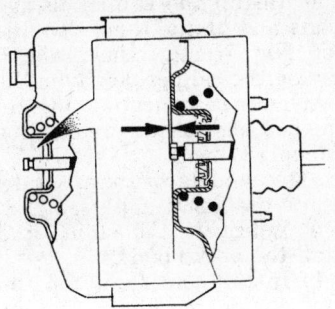

Dimension should be as follows in mm (in):

	Models to 1991 incl.		Models from 1992 on	
	10"	2×8"	10"	2×8"
Without ABS	2.5 (0.1)	2.5 (0.1)	2.5 (0.1)	2.5 (0.1)
With ABS	0.5 (0.02)	33.4* (1.31)	2.5 (0.1)	2.5 (0.1)

* This variant is provided with a spacer between the servo and master cylinder.

294497

Measuring the distance between the booster and pushrod — 940, 960 and 850 models

Brake Caliper

REMOVAL AND INSTALLATION

240 Models

Front

1. Raise and safely support the vehicle. Remove the front wheel.
2. Disconnect any ABS wires, if equipped.
3. Clean the caliper to ensure that no dirt gets into the brake line. Disconnect the brake lines, and remove the two caliper mounting bolts.

NOTE: If the brake line is seized to the caliper it should be replaced.

To install:

4. Install the caliper using new mounting bolts. Check the location of the caliper in relation to the disc.
   a. Use feeler gauges to check the distance between the disc and the caliper support stubs on both sides.
   b. The difference between the two measurements must not exceed 0.001 in. (0.25mm).
   c. Repeat the measurements using the upper and lower support stubs to check if the caliper is mounted parallel to the disc.
   d. If the caliper is not correctly aligned, shims can be used to adjust its position.
5. Install the brake pads and make sure that the disc can rotate freely between the pads.
6. Connect any ABS wires unfasten during removal.
7. Bleed the brake system.
8. Install the wheels.
9. Check the brake pedal function before driving vehicle.

Rear

1. Raise and safely support the vehicle. Remove the wheels.
2. Remove the guard plate, if equipped.

3. Disconnect any ABS wires, if equipped.
4. Disconnect the brake line from the rear axle and remove the mounting bolts for the caliper.
5. Disconnect the brake line from the caliper and lift the caliper off. Remove the brake pads.

To install:

6. Attach the brake line loosely to the caliper.
7. Install the caliper using new mounting bolts. Check the position of the caliper in relation to the disc.
   a. Use feeler gauges to check the distance between the disc and the caliper support stubs on both sides.
   b. The difference between the two measurements must not exceed 0.001 in. (0.25mm).
   c. Repeat the measurements using the upper and lower support stubs to check if the caliper is mounted parallel to the disc.
   d. If the caliper is not correctly aligned, shims can be used to adjust its position.
8. Tighten the bleed screw and line. Attach the brake line to the rear axle.
9. Install the brake pads.
10. Connect any ABS wires unfastened earlier.
11. Bleed the brake system.
12. Install the wheels.
13. Check brake function before driving vehicle.

940 and 960 Models

Front

1. Raise and safely support the vehicle. Remove the wheels.
2. Disconnect the ABS lead and brake hose from their clips. Clean the brake hose and line connection. Disconnect the hose from the line. Disconnect the hose from the caliper.
3. Remove the lower caliper guide pin bolt, swing the caliper up and remove the brake pads. Remove the cal-

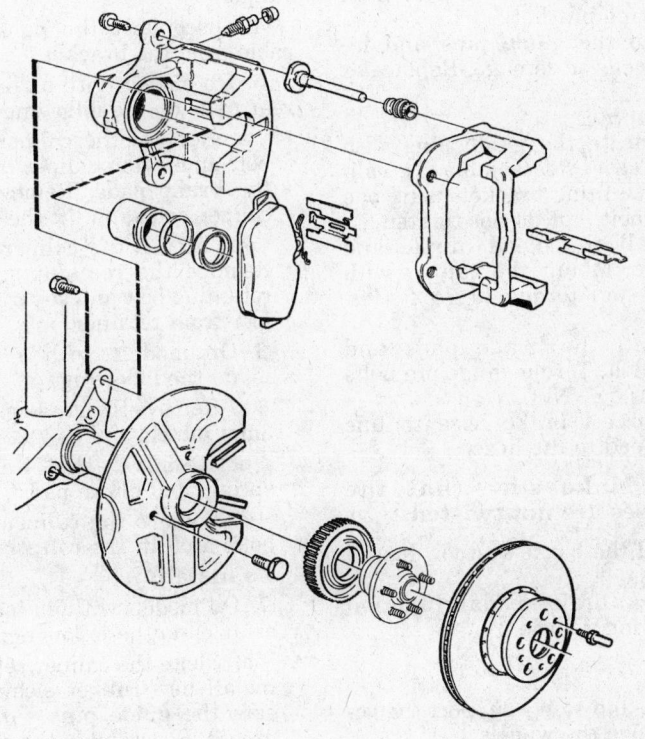

Exploded view of front brake components — 940 and 960 models

297579

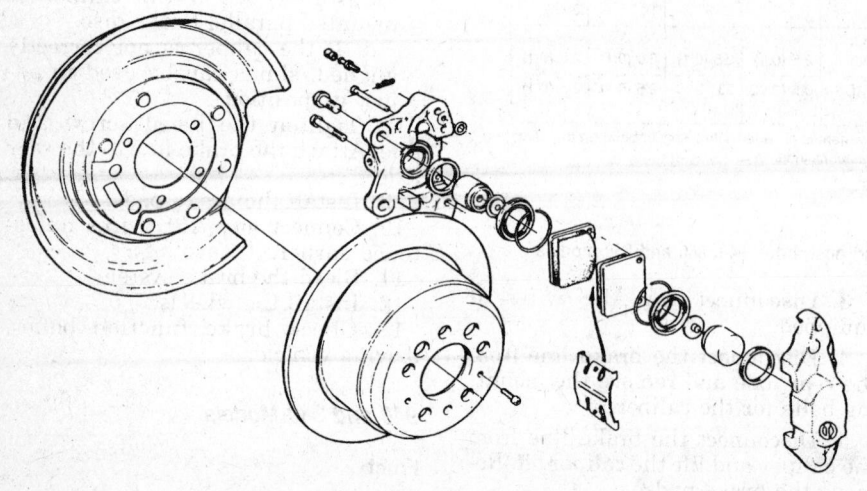

181960

Exploded view of Girling type rear brake components — 240 models

iper mounting bolts and lift the caliper off.

4. Remove the upper caliper guide pin bolt to separate the caliper from the mounting bracket.

5. Clean the guide pins and inspect for wear or damage. Replace as necessary.

To install:

6. Lubricate the guide pins with silicone grease. Reassemble the caliper and mounting bracket using one guide pin bolt, but do not tighten.

7. Install the bleed nipple and brake hose. Mount the caliper with new bolts and torque to 74 ft. lbs. (100 Nm).

8. Install the brake pads and guide pin bolt. Torque guide pin bolts to 20 ft. lbs. (27 Nm).

9. Reconnect brake hose to line and ABS lead to the hose.

NOTE: Make sure that the brake hoses are not twisted.

10. Bleed the brake system. Install the wheels.

11. Check brake pedal function before driving vehicle.

Rear

1. Raise and safely support the vehicle. Remove the wheels.

2. Disconnect any ABS wires, if equipped.

3. Clean the brake hose and line connection. Disconnect the hose from the line. Disconnect the hose from the caliper.

4. Disconnect the parking brake cable from the bracket.

5. On models with an independent rear axle, do the following;

a. Remove the caliper retaining bolt and swing caliper up. Remove the brake pads. Remove the two caliper bolts and lift the caliper off.

b. Mount the caliper in a vice. Remove the remaining guide pin retaining bolt and separate the caliper from retainer.

6. On models with a solid rear axle, do the following;

a. Remove the guide pins using a 3mm punch.

b. Remove the dampening spring, and brake pads.

c. Remove the caliper mounting bolts and lift the caliper off.

To install:

7. On models with an independent rear axle, do the following;

a. Clean the caliper retainer and install new rubber sleeves. Lubricate the guide pins with silicone grease. Reassemble the caliper and retainer using one retainer bolt, but do not tighten.

b. Install the caliper using new bolts and torque them to 43 ft. lbs. (58 Nm). Install the brake pads. Swing the caliper down and install the lower caliper guide pin retaining bolt and torque them both to 25 ft. lbs. (34 Nm).

8. On models with a solid rear axle, do the following;

a. Lubricate the shims and install the brake pads.

b. Install one fluid pin, and a new dampening plate. Then install the other guide pin.

9. Connect the brake hose, caliper and brake line.

10. Connect the parking brake cable to the bracket.

11. Connect any ABS wires, unfastened earlier.

NOTE: Make sure the hose is not twisted between the caliper and line.

12. Bleed the brake system using upper nipple.

13. Install the wheels.

850 Models

Front

1. Turn the ignition switch **OFF** and, if equipped with ABS, remove the key to prevent accidental pump activation.

2. Raise and safely support the vehicle and remove the wheel.

3. Disconnect any ABS wires, if equipped.

4. Loosen the brake hose a half turn. Remove the caliper bolts, lift the caliper off and unscrew the caliper from the hose. Drain the remaining brake fluid from the caliper. Remove the brake pads.

To install:

5. Grease the caliper bolts with lithium grease and insert them into the sleeves. Screw the caliper onto the brake hose. Install the brake pads and mount the caliper. Torque the caliper bolts to 22 ft. lbs. (30 Nm) and install the dust caps. Install the retaining clip.

6. Torque the brake hose to 12.6 ft. lbs. (18 Nm).

NOTE: Make sure that the brakes hose is not twisted.

7. Fill the master cylinder and bleed the brake system. Check the system for leaks and proper function.

8. Connect any ABS wires, removed earlier.

9. Install the wheels.

10. Check the brake function before driving the car.

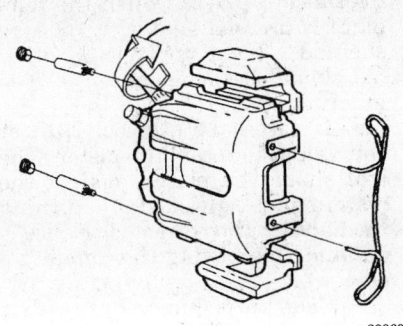

Brake caliper removal — 850 models

Rear

------ WARNING ------
Use only DOT 4 brake fluid. Never reuse brake fluid. Always keep brake fluid well sealed in its original container.

1. Turn the ignition switch **OFF** and, if equipped with ABS, remove the key to prevent accidental pump activation.
2. Raise and safely support the vehicle. Remove the wheel.
3. Disconnect any ABS wires, if equipped.
4. Clean the brake caliper thoroughly and remove the dust caps from the bleeder nipple.
5. Open the bleed nipple and lock the brake pedal the depressed position. Collect the brake fluid spillage in a suitable container.
6. Remove the brake pads and disconnect the brake pipes from the caliper.
7. Remove caliper mounting bolts and lift the caliper off. Drain the remaining brake fluid in the caliper into suitable container.
To install:
8. Clean the caliper mount where the caliper lies.
9. Install the caliper and new mounting bolts. Torque the bolts to 44 ft. lbs. (60 Nm).
10. Connect the brake line to the caliper and torque to 10 ft. lbs. (14 Nm).
11. Grease the brake pad shims with a thin layer of silicon grease. Put the shims on the back of the pads and install them in the caliper. Install one retaining pin, spring then the other retaining pin.
12. Fill the brake master cylinder and bleed the brake system.
13. Connect any ABS wires, unfastened earlier.
14. Install the wheels.

15. Lower the vehicle and check the brake function before driving.

Disc Brake Pads

REMOVAL AND INSTALLATION

240 Models
Front

1. Raise and safely support the vehicle.
2. Remove the front wheels.
3. Remove the spring clips and retaining pins. Remove the retaining springs and brake pads.

NOTE: If the brake pads are difficult to remove, tool 2917 or equivalent can be used to collapse the caliper pistons to ease removal.

4. Clean the caliper where the brake pads sit and inspect the dust caps for damage, and replace if necessary. Check the brake rotor surface for signs of wear, warping or variations in thickness.
5. Compress the caliper pistons using a large pair of pliers or a C - clamp.

NOTE: It may be necessary to remove some brake fluid from the reservoir when depressing the piston.

To install:
6. Install the brake pads, retaining springs, retaining pins, and spring clips.
7. Check the brake fluid level and pump the brake pedal several times. It may be necessary to bleed the brake system.
8. Install the wheels.
9. Check the brake pedal operation before driving the vehicle.

Rear Pads With ATE Caliper

1. Raise and safely support vehicle. Remove the rear wheels.
2. Drive the retaining pins out using a 3mm punch. Remove the retaining spring and brake pads.

NOTE: If the brake pads are difficult to remove tool 2917 or equivalent can be used to collapse the caliper pistons to ease removal.

3. Clean the brake caliper and inspect the dust caps for damage, replace if necessary. Check the rotor for signs of wear, warping or variations in thickness.

4. Compress the piston into the caliper using a large pair of pliers or tool 2809 or equivalent.

NOTE: It may be necessary to remove some fluid from the master cylinder, to prevent spilling when compressing the piston.

To install:
5. Caliper piston position must be checked before installation:
a. To check the position of the piston and prevent brake squeal, rotate the piston 20° in relation to the lower surface of the brake caliper.
b. Use tool 2919 or equivalent. The allowable tolerance is 18–22° when the template is pressed against one of the shoulders, the distance to the other (A) should be a maximum of 0.04 in. (1mm).
c. If necessary use tool 2918 or equivalent to rotate the piston. The tool should be placed against the piston, and tightened by turning the handle. The correct clearance is obtained by moving the handle up or down.
6. Install the brake pads and one retaining pin. Install a new retaining spring and the other retaining pin. Tap the retaining pins into place.
7. Check the brake fluid level and pump brake pedal repeatedly. It may be necessary to bleed the brake system.
8. Install the rear wheels. Check the brake pedal operation before driving the vehicle.

Rear Pads With Girling Calipers

1. Raise and safely support vehicle. Remove the rear wheels.
2. Remove the spring clips and retaining pins. Remove the retaining springs and brake pads.

NOTE: If the brake pads are difficult to remove tool 2917 or equivalent can be used to collapse the caliper pistons to ease removal.

3. Clean the brake caliper and inspect the dust caps for damage, replace if necessary. Check the rotor for signs of wear, warping or variations in thickness.
4. Compress the piston into the caliper using a large pair of pliers or tool 2809 or equivalent.

NOTE: It may be necessary to remove some fluid from the master cylinder, to prevent spilling when compressing the piston.

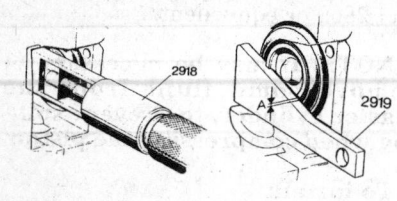

181374

Locating the ATE rear caliper piston using special tools — 240 models

To install:

5. Install the brake pads, retaining springs, retaining pins, and spring clips.

6. Check the brake fluid level and pump brake pedal repeatedly. It may be necessary to bleed the brake system.

7. Install the rear wheels. Check the brake pedal operation before driving the vehicle.

940 and 960 Models

Front

1. Raise and safely support the vehicle. Remove the wheels.

2. Remove the lower caliper guide pin bolt and swing the caliper upwards.

3. Remove the brake pads.

NOTE: Do not depress the brake pedal while pads are removed.

To install:

NOTE: The fluid level can rise in the reservoir when the piston is compressed.

4. Remove some brake fluid to prevent spillage. Air may be trapped in the dust seal of the piston. To avoid damage to the boot, it may be necessary to release the trapped air.

5. Press the piston back into the caliper.

6. Inspect the piston dust cap, if it is damaged, the caliper must be overhauled or replaced.

7. Check the disc brake surface for distortion or variation in thickness. Replace if not within specification.

8. Check to see that the metal guide plates are in position and install the pads. Check the guide pin boots for damage and replace them if necessary.

9. Swing the caliper down into position, being careful not to damage the guide pin boots. Torque the guide pin bolt to 20 ft. lbs. (27 Nm).

10. Check the reservoir fluid level and add as necessary. Operate the brake pedal repeatedly.

11. Install the wheels.

12. Test drive the vehicle.

Rear

1. Raise and safely support the vehicle. Remove the wheels.

2. On models with an independent rear axle, do the following;

    a. Remove the lower caliper guide bolt.

    b. Swing the caliper up and tie it back with a piece of wire. Remove the brake pads.

3. On models with solid rear axles, remove the brake pads in this sequence:

    a. Using a 3mm punch, drive out the guide pins.

    b. Remove spring plate.

    c. Remove the brake pads and shims.

NOTE: If the brake pads are difficult to remove, extractor 2917 or equivalent can be used to remove them.

NOTE: Do not operate the brake pedal while the pads are removed. Clean the caliper where the pads rest.

4. Inspect the piston dust boot for damage of dirt. If dirt has entered the caliper, it must be reconditioned or replaced.

5. Press the piston back into the caliper using large adjustable pliers, taking care not to damage the dust boot. If the piston is difficult to compress, this could indicate that oxidation has occurred. The caliper must be reconditioned or replaced in this case.

6. Inspect the rubber guide pin boots and replace if necessary. Check the disc brake surface for distortion or variation in thickness. Replace if needed

To install:

7. On models with an independent rear axle, install the new pads and lower the piston housing and torque the lower guide pin bolt to 25 ft. lbs. (34 Nm).

8. On models with a solid rear axle;

    a. Check the piston position with tool 2918 or equivalent to minimize brake noise.

    b. To check the position of the piston and prevent brake squeal, rotate the piston 20° in relation to the lower surface of the brake caliper.

    c. Use tool 2919 or equivalent to position the piston. The allowable

tolerance is 18–22° when the template is pressed against one of the shoulders, the distance to the other (A) should be a maximum of 0.04 in. (1mm).

    d. If necessary use tool 2918 or equivalent to rotate the piston. The tool should be placed against the piston and tightened by turning the handle. The correct clearance is obtained by moving the handle up or down.

    e. Coat anti-squeal shim with a thin coating of silicone grease on both sides. Then install the shims and brake pads in the order they where removed.

    f. Install one guide pin, a new spring plate, then the other guide pin.

9. Check the brake fluid reservoir level. Operate the brake pedal repeatedly.

10. Install the wheels.

850 Models

Front

1. Raise and safely support the vehicle. Remove the front wheels.

2. Carefully remove retaining spring, without bending. Remove the protective caps from the guide pin bolts. Using a 7mm Allen® key, remove the guide pin bolts.

3. Remove caliper from the carrier, then remove the brake pads. Hang the caliper from the spring so that the hose is not damaged.

——— WARNING ———
Do not depress the brake pedal while the pads are removed.

---

4. Clean the caliper carrier where the brake pads sit. Check the piston dust boot for damage or dirt. If the boot is damaged, the caliper should be overhauled or replaced. Check the brake rotor for signs of wear or damage. Check the guide pin bolt rubber sleeves for damage and replace if necessary.

To install:

5. Press the piston back into the caliper cylinder. Lubricate the caliper guide pin bolts with silicone grease. Insert the brake pads and slide the caliper on over them. Torque the guide pin bolts to 22 ft. lbs. (30 Nm) and replace the dust caps. Install the retaining spring.

6. Depress the brake pedal several times and check the brake fluid reservoir level.

7. Install the wheels.

8. Check brake pedal function before driving vehicle.

Rear

1. Raise and safely support the vehicle. Remove the wheels.

2. Drive out the retaining pins with a 3mm drift and remove the retaining spring.

3. Remove the brake pads and shims. If the pads are difficult to remove, use puller 999 2917 or equivalent to pull them out.

4. Clean the brake caliper surfaces where the pads lie. Check and clean the piston dust boot for dirt or damage and replace if necessary. Check the brake disc surface for damage, if it is warped or distorted, replace it or check the run-out.

5. Press the pistons back into their housing using adjustable pliers. Make sure that they are seated properly.

6. Grease the pad shims on both sides with a thin layer of silicone grease. Install the shims on the pads, Install the pads in the caliper.

7. Install one retaining pin, the retaining spring and then the other pin.

8. Depress the brake pedal several times and check the brake fluid reservoir level.

9. Install the wheels.

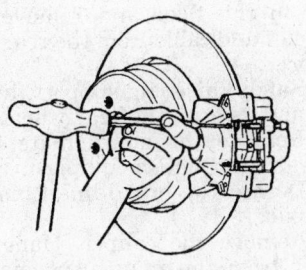

Removing the retaining pins — 850 models

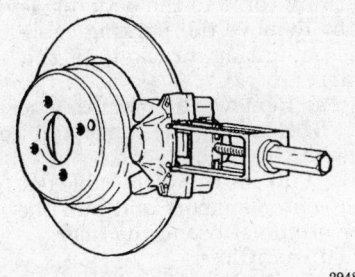

Remove the brake pads using special tool 999-2917 — 850 models

10. Check the brake function before driving the car.

Brake Rotor

REMOVAL AND INSTALLATION

240 Models

1. Raise and safely support vehicle. Mark the position of the wheel on the hub and remove the front wheel.

2. Remove the brake caliper and hang it from the spring with a piece of wire.

3. Remove the two screws securing the disc. Lift the disc off, it may be necessary to tap on the disc with a soft headed hammer.

To install:

4. Install the new disc, make sure the mating surfaces of the disc and hub are clean and dry.

5. Install the caliper using new mounting bolts. Check the location of the caliper in relation to the disc. Use feeler gauges to check the distance between the disc and the caliper support stubs on both sides. The difference between the two measurements must not exceed 0.001 in. (0.0013mm). Repeat the measurements using the upper and lower support stubs to check if the caliper is mounted parallel to the disc. If the caliper is not correctly aligned, shims can be used to adjust the caliper's position.

6. Install the brake pads and make sure that the disc can rotate freely between the pads. Install the disc securing screws.

7. Install the wheels.

8. Check brake pedal function before driving vehicle.

940 and 960 Models

Front

1. Raise and safely support the vehicle. Remove the wheels.

2. Remove the caliper and brake pads.

3. Hang the caliper from the spring, to avoid damaging the brake hose.

4. Remove the wheel pin guide and brake disc.

5. Clean the hub flange, remove the corrosion with a scraper and or wire brush.

6. Clean the ABS pick-up and toothed wheel using a soft brush.

To install:

7. Ensure that the mating surfaces on the hub and disc are clean. Install the disc with mark as close as possible to the hub mark.

8. Insert guide pin and tighten to 6 ft. lbs. (8 Nm). Reinstall the ring gauge and cross torque the lug nuts to 63 ft. lbs. (85 Nm).

9. Measure disc run-out. (Measure on the disc surface 0.60 in.(15mm) in from the edge.) Maximum run-out is 0.0023 in. (0.060mm). Remove the measuring equipment.

10. Install the caliper carrier using new mounting bolts, then install the caliper and brake pads.

11. Operate the brake pedal several times and check the fluid level.

12. Install the wheels.

13. Check brake function before driving vehicle. Limit hard braking whenever possible during the 500 miles (800 km) after pad replacement.

Rear

1. Raise and safely support vehicle. Mark the position of wheels on the hub and remove the wheels.

2. Remove the caliper and brake pads.

3. Hang the caliper on the spring to prevent damaging the brake hose.

4. Remove the locating pin and tap the disc off with a plastic mallet if stuck. Clean the axle shaft flange, remove all corrosion and dirt.

To install:

5. Install the new disc making sure that the friction surfaces on both sides are clean. Install the locating stud and torque to 6 ft. lbs. (8 Nm).

6. Install the caliper using new bolts, then install the brake pads.

7. Install the wheels.

850 Models

Front

1. Raise and safely support the vehicle. Remove the wheels.

2. Remove the brake caliper and brake pads.

3. Remove the carrier. Remove the guide pin bolt.

4. Remove the brake rotor and clean the hug flange of all corrosion and dirt.

5. Check the hub run-out by mounting gauge ring (Volvo tool 5419 from tool kit 5418) or equivalent.

6. Install the dial indicator on the spindle (using the caliper bracket bolt holes). Place the probe end to the gauge ring.

7. Turn hub slowly and identify the highest point. If the run-out exceeds 0.0007 in. (0.020mm), the hub must be replaced.

8. Remove the measuring equipment.

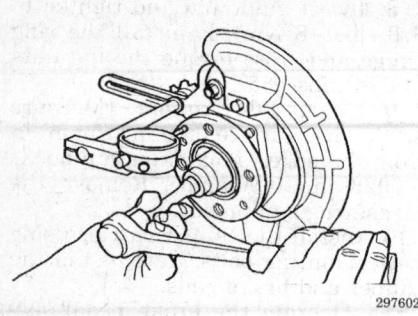

Checking hub run-out — 850, 940 and 960 models

To install:

9. Install the brake rotor. Tighten the guide pin bolt to 6 ft. lbs.

10. Install the caliper carrier using new bolts. Install the brake caliper and brake pads.

11. Depress the brake pedal several times and check the brake fluid reservoir.

12. Install the wheels.

13. Check the brake function before driving vehicle.

Rear

1. Raise and safely support the vehicle. Remove the wheels.

2. Remove the brake pads and shims.

3. Remove the two caliper and disconnect the brake line from its clip. Remove the brake caliper and hang it from the spring to prevent damaging the line.

4. Loosen the parking brake adjuster slightly so the disc a can be removed. Remove the guide pin bolt and disc.

To install:

5. Clean the face of the hub and check it for signs of damage. If the hub shows signs of damage, check the run-out by:

   a. Mounting a dial indicator and rotating the hub with sensor tip on the face of the hub.

   b. If the highest spot on the hub is higher than 0.007 in. (0.20mm), the hub must be replaced.

6. Carefully brush the ABS pulse wheel off with a soft brush.

7. Install the brake disc and guide pin bolt. Torque the bolt to 6 ft. lbs. (8 Nm).

8. Install the caliper and pads, using new mounting bolts. Install the brake line and mounting clip.

9. Depress the brake pedal several times and check the brake fluid reservoir. Adjust the parking brake.

10. Install the wheels.

11. Check the brake function before driving the vehicle.

Parking Brake Cable

ADJUSTMENT

1. The parking brake should be fully applied when the handle is moved no more than ten notches. If adjustment is required, adjust the brake shoes first.

2. For 850 models, full braking should be reached when the handle is raised about five notches. If cable adjustment is required:

   a. Remove the square cover plate under the arm rest in the middle console.

   b. Turn the adjustment screw so that full braking is obtained between the second and eighth notch.

   c. Replace the square cover plate.

3. Raise and safely support the vehicle and remove the rear wheels.

4. With the brake handle released, install a brake adjusting tool through the hole in the rotor between the studs.

5. Turn the adjuster wheel until the rotor will not turn.

6. Loosen the adjuster wheel 4–5 notches.

7. Make sure the rotor turns freely. If the shoes are binding inside the rotor even after loosening the adjuster more, remove the rotor to repair the problem.

8. Move the handle again: adequate braking power should be obtained at 3–7 notches with a normal pull force of approximately 65 lbs. (84 Nm). If cable adjustment is required, remove the access panel behind the brake handle and adjust through the opening in the rear console.

9. The yoke on top of the brake handle should be at right angles to the parking brake lever. If the yoke is out of alignment, lower the handle and turn the nuts at the cable ends to adjust. There should always be at

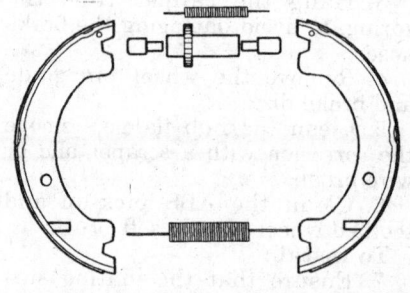

Parking brake shoes — 850 models

least 0.1 inch (2 mm) thread protruding.

10. Make sure the indicator light on the instrument panel illuminates when the brake is applied.

REMOVAL AND INSTALLATION

240 Models

1. Remove the two screws and the parking brake lever trim cover.

2. Disconnect the ashtray lamp connector.

3. Loosen the parking cable adjusting screw to release the cable tension.

NOTE: The parking brake cables crisscross under the vehicle. The cable attached to the driver side of the parking brake lever operates the passenger side rear parking brake and vise versa.

4. Loosen the locknut for the cable being removed.

5. Insert a small screw driver into the end of the cable to keep it from turning and remove the locknut.

6. Lift up the rear seat cushion and move the carpeting to one side.

7. Carefully remove the cable from under the carpet along the driveshaft tunnel.

8. Remove the clamps holding the cable to the floor and remove the grommet and cable from the rear seat support.

9. Safely raise and support the vehicle and remove the rear wheels.

10. Remove the guard plate from the rear caliper if so equipped.

11. Detach the brake line from the rear axle.

12. Remove the caliper. Hang the caliper from the rear suspension to avoid bending the brake line.

13. Remove the brake rotor, parking brake shoe retaining springs and parking brake shoes.

14. Press out the pin holding the parking cable to the actuator lever.

15. Remove the parking cable from the mounting bracket at the rear axle:

   a. Remove the mounting screw.

   b. Remove the cable and collar with the rubber seal.

16. Pull the complete cable out from the center support and pull the rubber grommet out of the floor.

To install:

17. Install the plastic collar through the mounting bracket and seat the rubber seal. Install the mounting screw.

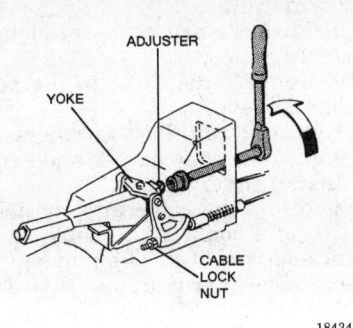

Adjust the parking brake as shown; 240, 940 and 960 models

18. Route the cable through the center support and floor panel.

NOTE: When the cables are installed, make certain they are crisscrossed with the cable from the driver side wheel under the cable from the passenger side wheel and they are inserted through the floor panel on the opposite side.

19. Lubricate the actuator lever and sliding surfaces of the backing plate for the parking brake shoes with a thin layer of heat resistant grease.
20. Position the end of the cable in the actuator lever and press in the retaining pin.
21. Push in the cable and position the actuator lever behind the rear axle flange.
22. Apply a thin layer of graphite grease to the sliding surfaces of the parking brake shoes. Install the shoes and retaining springs.
23. Install the upper return spring and adjuster.
24. Install the brake rotor and caliper using new mounting bolts.

NOTE: Make sure the caliper is properly positioned. The rotor should turn without contacting the brake pads.

25. Attach the brake line to the rear axle and install the guard plate onto the caliper if so equipped.
26. Adjust the parking brake shoes.
27. Reposition the carpeting and install the rear seat cushion.
28. Install the locknut until the threaded end of the cable goes through the nut at least 0.08 in. (2 mm). If necessary, insert a small screw driver into the end of the cable to keep it from turning.
29. Adjust the cable length with the locknuts until the parking brake lever yoke is square to the lever handle when the parking brake is applied.
30. Adjust the parking brake tension.
31. Reconnect the ashtray lamp and install the parking brake lever trim cover.
32. Install the rear wheels and lower the vehicle.

─────── **CAUTION** ───────
Make sure both foot brakes and parking brake are operating properly before test driving vehicle.
─────────────────────────

940 and 960 Models

1. If removing the left side cable:
    a. Remove the center console.
    b. Remove the rear seat cushion and fold the carpeting forward to access the cable where is enters the passenger compartment.
2. Remove the ashtray and holder from the rear of the center console.
3. Tap the spring sleeve free of the adjusting screw if necessary.
4. Back off the adjuster on the parking brake lever to release the tension on the parking brake cable.
5. On left side cables, bend aside the locking tab to release the cable from the parking brake lever. Remove the cable from the center console bracket. On 1995–97 models, loosen the fuel tank bolts and lower the fuel tank approximately ½ inch (10–15 mm). Remove the cable clamp in the floor of the vehicle.
6. Raise and safely support the vehicle. Remove the wheels
7. Remove the wheel, caliper and rotor. Hang the caliper from the coil spring with a wire. Unhook the rear return spring and remove the brake shoes.
8. Push out the pin holding the cable to the parking brake actuator lever. Remove the rubber boot from the backing plate and from the cable.
9. On left side cables,
    a. Pull out the cable from the backing plate and the equalizer on top of the rear axle.
    b. Remove the cable clamp on the torque rod bracket, above the driveshaft, and the cable. Carefully remove the cable from the grommet in the floor panel.
10. Remove the spring clip from the clevis pin. Remove the clevis pin and cable from the equalizer at the back of the differential housing.
11. On some vehicles, it may be necessary to remove 2 spring clips holding the cable to guides on the rear axle.
12. Remove the cable.
To install:
13. On left side cables;
    a. On 1995–97 models, carefully install the new cable over the fuel tank. Avoid damaging the plastic cable sleeve.
    b. Install the new cable through the grommet in the floor panel. Check that the grommet positioned correctly. Clamp the cable to the torque rod bracket. The cable should move easily in the clamp.
14. Install the cable through guide on the equalizer and through the backing plate. Make sure the boot is positioned correctly on the backing plate.
15. Install the cable and spring clips to hold the cable in the guides on the rear axle.
16. Tighten the fuel tank mounting bolts, if loosened earlier.
17. Inspect the rubber boot for wear or damage and replace if necessary. Install the boot onto the cable and position it through the hole in the backing plate.
18. Smear the contact surfaces of the brake levers with a thin layer of heat resistant graphite grease. Connect the cable to the lever and install the pin.

NOTE: The arrow stamped on the lever should be facing outwards and point upward.

19. Push the cable through and place the lever in position behind the rear axle flange.
20. Position the cable in the equalizer and install the clevis pin and spring clip.
21. Install the brake rotor, shoes, return spring, and caliper. Make sure the rotor turns freely. Install the wheel.
22. On left side cables, install the cable into the center console bracket. Attach the cable to the parking brake lever bend the locking tab into position to secure the cable.

NOTE: The vehicle must be on the ground before the parking brake can be adjusted.

23. Adjust the cable tension with the adjusting screw on the back of the lever.

─────── **CAUTION** ───────
Make sure both foot brakes and parking brake are operating properly before test driving vehicle.
─────────────────────────

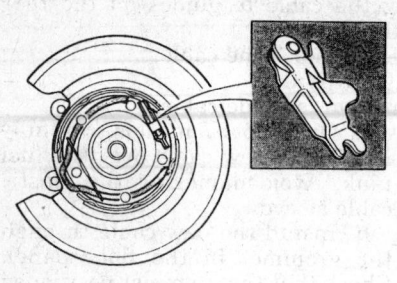

Cable expander position — 1993–94 940 and 960 models

850 Models

WARNING

On vehicles equipped with a Supplemental Restraint System (SRS), the system must be properly disarmed and the Air Bag must be removed and properly stored. Unintended Air Bag deployment can cause serious or fatal injury.

1. Remove the two cigarette lighter panel bolts. Pull the parking brake lever up and set shift lever in **NEUTRAL**.

2. With ignition in the **OFF** position, disconnect the negative battery cable.

3. To remove the console:

a. Remove the cover plates under the armrest and parking brake lever and remove the screws holding the console.

b. Unplug the electrical connectors in the front of the console.

c. Lift the back of the console up and push it forward so it clears the parking brake lever. Lift the console off.

4. Loosen the parking brake adjuster and disconnect the end of the cable from it.

5. Remove the rear seat and carpet keepers. Lift the carpet up and remove the cable retainers and grommet.

6. Raise and safely support vehicle and remove the rear wheels.

7. Remove the cable retainer bolt located in front of the rear wheel well. Disconnect the cable sleeve mount bolt and bracket to the rear wheel and pull the metal sleeve forward, if so equipped.

8. Disconnect the rear of the cable from the lever arm grip and metal sleeve from the cable. Remove the cable from below.

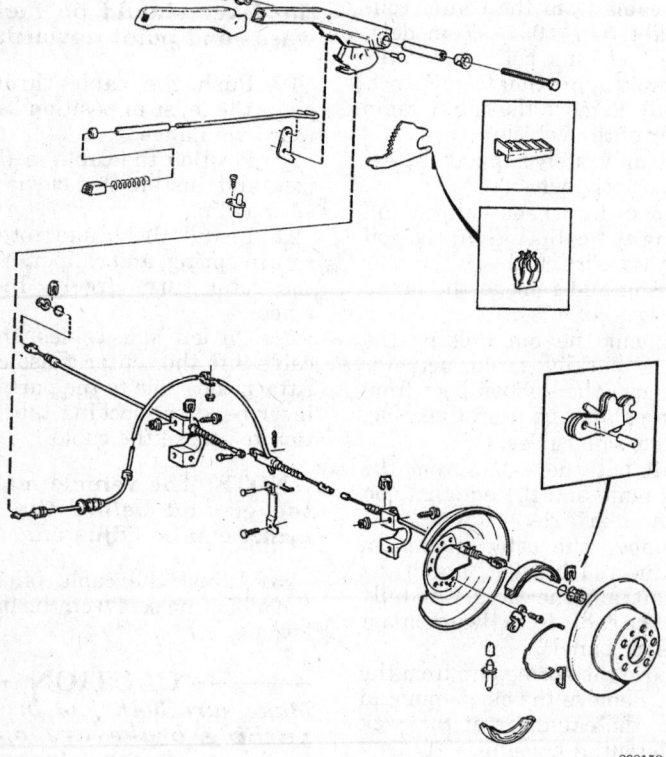

Parking brake assembly — 940 and 960 models

To install:

9. Push the new cable through the floor from below.

10. Connect the cable to the parking brake lever.

11. Install the floor retainers and grommet. Lay the carpet and secure it. Install the rear seat.

12. Insert the cable into the metal sleeve and connect the cable to the lever arm grip. Use pliers to squeeze the lever arm grip around the cable.

13. Mount the metal sleeve with bracket and bolt. Install the mounting bolt in front of the rear wheel well.

14. Install the rear wheels and tighten the lugs to 81 ft. lbs. (110 Nm).

15. Install the console.

CAUTION

When connecting the battery on vehicles equipped with an Air Bag, make sure no one is in the vehicle in case of an SRS malfunction causing accidental Air Bag deployment.

16. Adjust the brake cable tension.
17. Test both foot brake and parking brake operation before driving the vehicle.

Brake System Bleeding

PROCEDURE

CAUTION

Use only DOT 4 brake fluid. Never reuse brake fluid. Always keep brake fluid well sealed in its original container.

1. Turn the ignition switch **OFF** and if equipped with ABS, remove the key to prevent accidental pump activation.

2. Raise and safely support the vehicle.

3. Clean the brake fluid reservoir filler cap. Remove the cap and fill the brake fluid reservoir completely. Replace the reservoir cap.

4. Depress the brake pedal a few times to purge the air bubbles in the master cylinder. Repeat this step between bleeding each wheel. Check the reservoir fluid level during bleeding, add as necessary.

NOTE: On vehicles with two piston calipers or multi-link rear suspension, connect the bleeder hose to upper nipple when bleeding.

5. On all models except the 850, starting with the right rear wheel, re-

move the cap from the bleed nipple and connect the hose from the drain bottle.

6. On 850 models, starting with the left side front wheel, remove the cap from the bleed nipple and connect the hose from the drain bottle.

7. Open the bleed nipple.

8. Press the brake pedal to the floor and hold it there for 2 seconds before releasing. Repeat this 20–30 times or until no air bubbles are visible in the escaping fluid.

9. Tighten the bleed nipple.

10. Pump the brake pedal a few times to build pressure in the system.

11. Open the bleed nipple and close it while brake pedal is to the floor. Repeat this 3–5 times.

12. Tighten the front bleed nipples to 7.5 ft. lbs. (10 Nm) and rear to 3.3 ft. lbs. (4.5 Nm).

13. Remove the hose from the bleed nipple and put the cap back on.

14. On all models except the 850, repeat the procedure for the rest of the wheels in the following order:
- left rear wheel
- right front wheel
- left front wheel

15. On 850 models, repeat the procedure for the rest of the wheels in the following order:
- right front wheel
- left rear wheel
- right rear wheel

16. Check to make sure that there is no air in the brake system by depressing the brake pedal sharply 3–4 times with the engine off. The pedal travel should not exceed 2.2 inches (55 mm).

17. If the pedal travel is greater than 2.2 inches (55 mm), bleed and check the pedal pressure again.

18. Check the fluid reservoir. The fluid level should not be above the MAX level.

19. Test brake function.

Wheel Speed Sensor

REMOVAL AND INSTALLATION

1. Disconnect the negative battery cable.

2. Raise and support the vehicle on jackstands.

3. If removing a front speed sensor, remove the wheel from the side where the sensor is to be removed. If removing the rear speed sensor, the wheel does not have to be removed.

4. Disconnect the sensor harness running to the individual harness.

NOTE: The rear wheels share a common speed sensor.

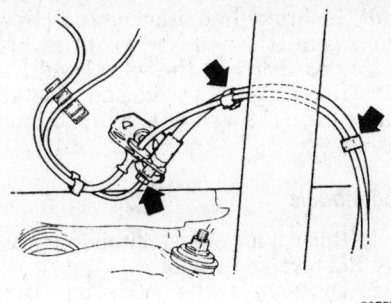

297578

ABS lead next to brake hose assembly — 940 and 960 models

5. Remove the retainer bolt securing the sensor to the front spindle or rear pulse wheel assembly.

6. Slide the sensor out. If reinstalling the original sensor, clean the pick-up end of the sensor.

To install:

7. Slide the sensor into position and secure in place with the retainer bolt.

8. Connect the sensor wire harness.

9. Install the wheel, if removed, and lower the car.

10. Connect the negative battery cable.

FRONT SUSPENSION

Strut

REMOVAL AND INSTALLATION

240, 940 and 960 Models

— CAUTION —
A coil spring compressor is required to remove the spring. Improper removal procedures may cause serious injury.

1. Raise and safely support vehicle. Remove the wheel.

2. Disconnect the tie rod end.

3. Place a floor jack under the control arm. Disconnect the sway bar from the link. Unbolt the brake lines from the bracket and detach them from the clips.

4. Remove the cover over the strut nut. Disconnect the coil wire and place it out of the way.

5. Hold the strut shaft with tool 5037 or equivalent and loosen the nut a few turns with tool 5036 or equivalent.

6. Mark the position of the upper mount in the housing, then remove the nuts and washers. Carefully lower the jack and pull the strut and spring out of the housing.

— WARNING —
Be careful not to damage the fender when removing the strut assembly. Use retaining hook 5045 or equivalent attached to the anti-sway bar to prevent it from falling.

7. To remove the spring:
  a. Attach spring compressor tool 5040 or equivalent to the spring. The two parts of the tool should be opposite each other and have three coils between the claws.
  b. Compress each side alternately until the strut is loose inside the spring.
  c. Hold the strut shaft with tool 5037 or equivalent and remove the nut with tool 5036 or equivalent and lift off the upper mount, spring retainer, spring and rubber bumper.

NOTE: On vehicles equipped with gas pressure struts, the bumper has been replaced by a rubber bellow and disc.

8. To remove the strut, unscrew the retaining nut and pull the shock insert out of the casing, using tool 5039 or equivalent for standard struts, or tool 5173 or equivalent for gas struts.

To install:

9. Insert the strut insert into the housing and torque the retaining nut to 119 ft. lbs. (160 Nm). Install the bumper or bellows and disc on the strut, making sure that the top of the bumper is lower than the top of the strut shaft.

10. Install the spring so the compressor tool bolt holes face upwards. Install the upper mount, washer and nut but do not tighten fully.

11. Remove the compressor loosening the bolts alternately and make sure that the ends of the spring fit correctly into the upper and lower plates.

12. Guide the strut assembly into the body.

13. Install the upper mount according to the earlier marking and torque to 30 ft. lbs. (40 Nm).

14. Torque the strut nut to 111 ft. lbs. (150 Nm) using socket 5036 and holder 5037 or equivalents. Press nut cover back on and connect the coil wire.

15. Install the sway bar link and tighten it until the distance between

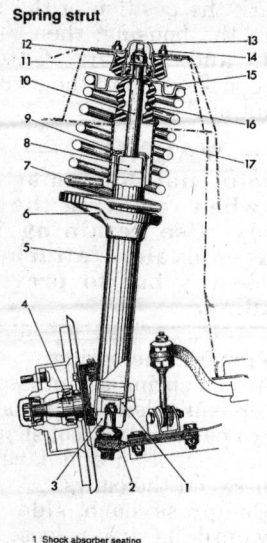

Spring strut

1. Shock absorber seating
2. Ball joint
3. Ball joint attachment
4. Wheel axle
5. Tube
6. Lower spring seating
7. Spring
8. Shock absorber
9. Shock absorber nut
10. Plunger rod, shock absorber
11. Upper mount
12. Wheel housing
13. Protective cap
14. Attaching nut, upper mount
15. Upper plate, spring
16. Rubber bump stop
17. Protective sleeve, shock absorber

190888

Spring strut assembly — 240 models

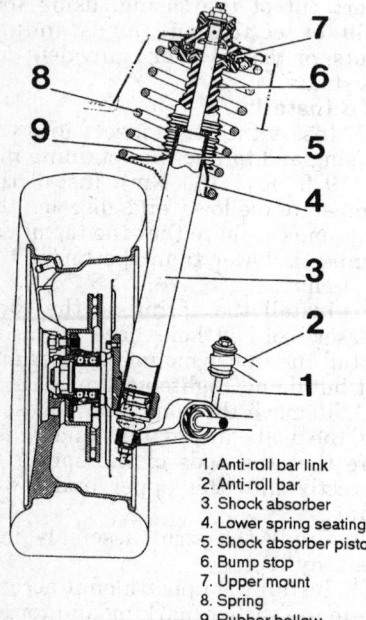

1. Anti-roll bar link
2. Anti-roll bar
3. Shock absorber
4. Lower spring seating
5. Shock absorber piston
6. Bump stop
7. Upper mount
8. Spring
9. Rubber bellow

186423

Spring strut assembly — 940 and 960 models

the washers is 1.65 in. (42mm). Install the brake line bracket and clips. Make sure that the brake lines are sitting correctly in the wheel well.

16. Install the tie rod end. Install the wheel. Lower the vehicle and test.

850 Models

1. Raise and safely support vehicle. Remove the wheel.

2. Disconnect the sway bar link from the strut. Remove the ABS sensor lead from the strut and brake bracket, but do not disconnect.

3. Install support tool 5466 or equivalent under the control arm.

—— **WARNING** ——
If this tool is not installed, the axle joint may be damaged from excessive downward pressure.

4. Remove the upper nuts attaching the strut attachment to the body.

5. Remove the two nuts and bolts holding the strut to the steering knuckle. Remove the spring and strut assembly.

6. Mount the spring and strut assembly in a vice and secure it. Install spring compressing tool 5407 or equivalent and alternately compress the spring.

7. Remove the bolt and washer from the strut attachment using socket 5467 and counterhold 5468 or equivalents.

8. Remove the strut nut using socket 5469 and counterhold 5468 or equivalents. Remove the spring seat, rubber stopper, boot, and check them for damage. Remove the compressed spring from the strut.

To install:

9. Compress the spring to about 12 in. (300mm) in length.

10. Install the rubber stopper.

11. Install the washer.

12. Install the spring (compressed).

13. Install the spring seat.

NOTE: Make sure that the spring ends are properly seated.

14. Install the strut nut and torque it to 52 ft. lbs. (70 Nm) using socket 5469 and counterhold 5468 or equivalents.

15. Install the spring attachment, washer, and nut and torque it to 52 ft. lbs. (70 Nm) using socket 5467 and counterhold 5468 or equivalents.

16. Slowly, alternately remove the spring compressor.

17. Install the spring and strut assembly in the spring housing and fasten it using new nuts and torque them to 18 ft. lbs. (25 Nm).

18. Connect the strut to the steering knuckle using new bolts and nuts.

19. Torque them to 48 ft. lbs. (65 Nm) and angle tighten 90°.

20. Connect the sway bar to the strut using new nuts.

21. Install the ABS sensor lead to the strut and brake pipe bracket. Remove the support tool.

22. Install the wheel.

Lower Ball Joints

REMOVAL AND INSTALLATION

240 Models

1. Raise the vehicle and support it safely.

2. Remove the wheel.

3. Remove the four bolts which retain the ball joint to the control arm and remove.

4. Remove the ball joint retaining nut and press the ball joint out of the strut.

To install:

NOTE: Ball joints are different for the right and left sides. It is therefore important that the correct ball joint is installed on each side.

5. Attach the ball joint and tighten to 44 ft. lbs. (60 Nm).

6. Install the ball joint to the strut using new locking bolts and washers. Tighten bolts to 17 ft. lbs. (23 Nm).

7. Install the ball joint to the control arm and torque to 85 ft. lbs. (115 Nm).

8. Install the wheel.

9. Lower the vehicle.

940 and 960 Models

1. Raise and safely support vehicle. Remove the wheel.

2. Disconnect the sway bar link to the control arm. Remove the cotter pin and nut from the ball joint stud.

3. Press the ball joint from the control arm using puller 5259 or equivalent. Make sure the puller is located directly in line with the stud and the rubber seal is not damaged.

4. Remove the bolts holding the ball joint to spring strut. Push the control arm down and remove the ball joint.

To install:

5. Install the new ball joint, using new bolts and apply Loctite® or equivalent to threads. Torque the bolts to 22 ft. lbs. (30 Nm). Check to see that the bolt heads are sitting flat

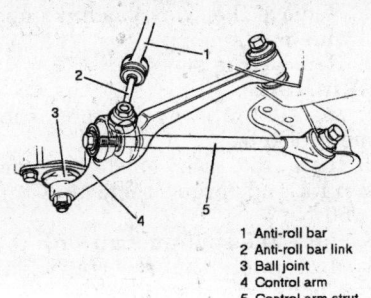

1 Anti-roll bar
2 Anti-roll bar link
3 Ball joint
4 Control arm
5 Control arm strut

297150

Front suspension assembly — 940 and 960 models

on the ball joint, then tighten them an additional 90°.

6. Slide the ball joint into the control arm. Install the nut and torque it to 44 ft. lbs. (60 Nm). Tighten and install a new cotter pin.

7. Attach the sway bar link to the control arm and torque to 63 ft. lbs. (85 Nm).

8. Install the wheel.

850 Models

1. Raise and safely support the vehicle.

2. Remove the wheel.

3. Remove the three nuts holding the ball joint to the lower control arm.

4. Remove the clamping bolt and nut from the steering knuckle where the ball joint is mounted.

5. Spread the ball joint apart and remove it from the hub housing.

To install:

6. Clean the control arm and steering knuckle where the new ball joint is fitted.

7. Install the new ball joint and clamping bolt and nut. Torque the bolt to 37 ft. lbs. (50 Nm).

8. Connect the ball joint to the lower control arm and fasten it with new nuts. Apply rustproofing compound to the nuts. Starting from inside, working outward, torque the nuts to 13 ft. lbs. (18 Nm) and then angle tighten 120°.

9. Lower the car.

Lower Control Arms

NOTE: On all models, always fully install the control arm and bounce the suspension several times before tightening the control arm mounting nuts or bolts.

REMOVAL AND INSTALLATION

240 Models

1. Raise the vehicle and support it safely. Remove the wheel.

2. Disconnect the sway bar link at the control arm.

3. Remove the control arm from the ball joint.

4. Remove the control arm rear attachment plate.

5. Remove the control arm front retaining bolt.

6. Remove the control arm.

To install:

7. If bushings are to be replaced, note that the right and left bushings are not interchangeable. The right side bushing should be turned so that the small slots point horizontally when installed.

8. Install the bracket onto the control arm. The nut should be tightened only enough to hold. The washer should be able to be turned after the nut is on.

9. Attach the control arm. Install the front retaining bolt and nut; tighten the nut only a few turns onto the bolt.

10. Guide the stabilizer link into position. Attach it loosely with its nut and bolt.

11. Install the ball joint and mount.

12. Install the rear bracket to the vehicle. Tighten the three bolts to 25–35 ft. lbs.

13. Tighten the stabilizer link.

14. Install the wheel.

15. Lower the vehicle. Bounce the front of the vehicle up and down. This "normalizes" the front suspension and allows the control arm to seek its final position.

16. Tighten the rear mount nut to 38–44 ft. lbs. (49–57 Nm). Tighten the front mount to 55 ft. lbs. (71 Nm).

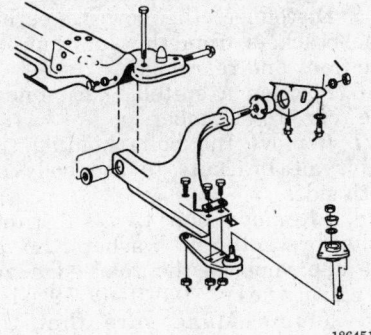

186451

Lower control arm details — 240 models

940 and 960 Models

1. Raise and safely support vehicle.

2. Remove the front wheel.

3. Remove the cotter pin and ball joint nut. Disconnect the sway bar link. Remove the control arm bolt and front bushing.

4. Remove the ball joint from the control arm using puller 5259 or equivalent. Remove the control arm from the cross member.

5. Press out the bushing with drift 2904 or equivalent.

To install:

6. Press in new bushing with drift 2904 or equivalent. Use disc 5240 or equivalent as a support. The disc recess should face upwards.

NOTE: Press the bushing in from the front side of the control arm.

7. Align control arm strut in control arm. Install control arm in crossmember but do not tighten fully.

8. Install the ball joint in the control arm, torque the nut to 44 ft. lbs. (60 Nm), and secure it with a new cotter pin.

9. Install the bushing, washer, and bolt for the control arm strut and torque it to 70 ft. lbs. (95 Nm). Attach the anti-sway link and torque it to 63 ft. lbs. (85 Nm).

10. Install the wheel.

11. Rock front end a few times. Torque control arm to 63 ft. lbs. (85 Nm).

850 Models

1. Raise and safely support vehicle.

2. Remove the three nuts holding the lower control arm to the ball joint.

3. Remove the bolts and nuts holding the lower control arm to the frame. Remove the lower control arm.

To install:

4. Clean the ball joint and frame where the lower control arm mates.

5. Install the lower control arm in the frame and attach with new bolts and nuts but do not tighten them yet.

6. Apply rustproofing compound to the lower control arm nuts. Connect the lower control arm to the ball joint using new nuts.

7. Torque the nuts to 13 ft. lbs. (18 Nm) and then angle tighten 120°.

8. Lower the car and bounce it up and down several times.

9. Torque the lower control arm subframe bolts to 48 ft. lbs. (65 Nm) and then angle tighten 120°.

Sway Bar

REMOVAL AND INSTALLATION

240 Models

1. Raise the vehicle and support it safely.
2. Remove the wheels.
3. Remove the underside splash guard panel, if equipped.
4. Remove the upper nut securing the anti-roll bar to the struts.
5. Remove the upper link nut on the opposite side.
6. Remove the bolts for the two retaining brackets and remove the bar.
7. If the link bushings are worn, remove the lower link bolts and remove the entire link. Inspect all the bushings for compression or elongation. Replace as required. The two U-shaped bushings from the front brackets are particularly prone to deforming.
   To install:
8. Reconnect the lower link to the arms, if removed.
9. Hold the bar in position and install the front brackets with bushings. Make sure the slot in the bushing faces forward.
10. Install the bar to the link on one side of the vehicle but do not tighten more than a few turns. Connect the bar to the link on the opposite side and install the bushings and nut.
11. Tighten each upper link nut until 1.65 inches (42mm) can be measured between the outer surfaces of the upper and lower washers.
12. Reinstall the underside splash panel, if required.
13. Install the wheel.
14. Lower the vehicle.

940 and 1993–94 960 Models

1. Raise and safely support vehicle.
2. Remove the splashguard from under the engine.
3. Disconnect the anti-sway bar from upper link mounts on both sides.
4. Remove the sway bar clamps on both sides and take the bar off.
   To install:
5. Install new rubber bushings on sway bar and attach to subframe with clamps.
6. Connect sway bar to upper link mounts and torque nut until the distance between bushing washers is 1.65 in. (42mm).
7. Install the splashguard under the engine.

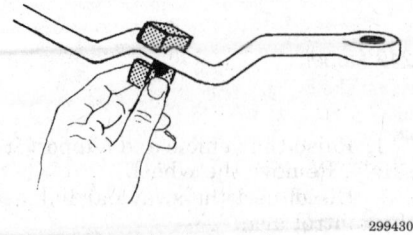

Sway bar bushing joint should face forward — 940 and 960 models

299430

8. Mount wheels as per previous marking and torque to 63 ft. lbs. (85 Nm).

1995–97 960 Models

1. Raise and safely support the vehicle.
2. Remove the wheels.
3. Remove the nuts connecting the sway bar to the axle beam.
4. Remove the bolts connecting the sway bar to the support arms.
5. Remove the sway bar.
   To install:
6. Attach the sway bar to the axle beam using new nuts. Torque the nuts to 15 ft. lbs. (20 Nm).
7. Connect the sway bar to the support arms using new bolts. Do not tighten the bolts.
8. Install the wheels. Lower the vehicle. Torque the support arm bolts to 63 ft. lbs. (80 Nm).

850 Models

1. Install support rails 5033, bracket 5006 an lifting hook 5115 or suitable equivalents. These make it possible to raise the engine in the vehicle.
2. Raise the engine slightly.
3. Raise and safely support the vehicle. Remove the under engine splashguard.
4. Remove the five nuts holding steering gear to the subframe.
5. Disconnect the power steering line brackets from the subframe at the front and rear edges.
6. Position a suitable jack under the rear crossmember.
7. Remove the bolts holding the subframe brackets to the body on both sides.
8. Remove the two subframe bolts, brackets and washers. Lower the subframe at the rear edge approximately 0.59–0.79 in. (15–20mm). Make sure that the steering gear bolts come away from the frame.

9. Remove the sway bar links and subframe brackets.
10. Remove the sway bar.
   To install:
11. Install the sway bar and subframe brackets.
12. Install the sway bar links using new nuts and torque them to 37 ft. lbs. (50 Nm).
13. Raise the subframe up with the jack and push the steering gear mount bolts into the frame.
14. Install the subframe brackets and new M14 bolts, but do not tighten fully.
15. Move the jack to the front edge of the frame and replace the bolts. Do not tighten fully.
16. First tighten the bolts on the left side on the frame to 77 ft. lbs. (105 Nm) and angle tighten 120°. Then do the same to the right side.
17. Tighten the bracket bolts to 37 ft. lbs. (50 Nm).
18. Install new attaching nuts to the steering gear and torque them to 37 ft. lbs. (50 Nm).
19. Tighten the power steering line brackets on the front and rear edges of the subframe.
20. Install the under engine splashguard.
21. Lower the vehicle and remove support rails, bracket and lifting hook.

Front Wheel Bearings

ADJUSTMENT

240, 940 and 960 Models

The front wheel bearings are not adjustable on the rear drive vehicles. If the lateral runout on the hub with the disc removed exceeds 0.0012 inch (0.030mm), the hub must be replaced.

850 Models

The front wheel bearings are not adjustable on the front drive vehicles. If the lateral runout on the hub with the disc removed exceeds 0.0007 inch (0.020mm), the hub must be replaced.

REMOVAL AND INSTALLATION

240, 940 and 960 Models

1. Raise and support the vehicle safely.
2. Remove the brake caliper. Hang the caliper out of the way with a piece of stiff wire. Do not let the caliper hang by the brake hose.

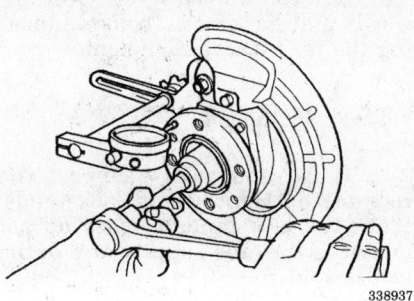

Checking hub runout — 240, 940 and 960 models

3. Pry off the grease cap. Remove the cotter pin and castle nut.

4. Remove the hub and brake disc assembly. Use a bearing puller 2722 or equivalent to remove the inner bearing if it stays on the spindle.

NOTE: If the vehicle is equipped with separate brake disc and hub, the guide pin and brake disc must be removed from the hub prior to bearing replacement.

5. Use a brass drift and carefully tap out the grease seal and inner bearing race.

6. Remove the outer bearing race, using a suitable handle and drift 2725 or equivalent.

To install:

7. Press in a new inner bearing race, using a suitable handle and drift 5005 or equivalent.

8. Press in a new outer bearing race, using a suitable handle and drift 2724 or equivalent.

9. Pack the wheel bearing between the cage and inner race with as much grease as possible. Also smear grease on the outer side of the bearing and bearing races inside the hub. Fill the space in the hub with grease to a diameter of the smallest ball races.

Removing the grease seal and inner bearing race from the brake disc with integrated hub

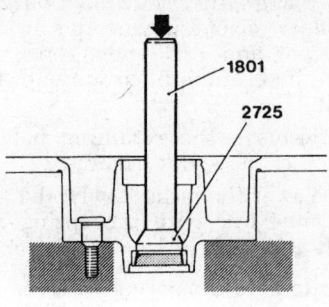

Removing the outer bearing race from the brake disc with integrated hub

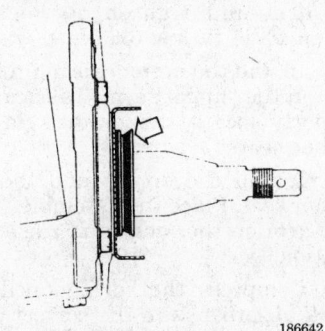

Pressing sealing ring onto the spindle from the vehicle with a separate hub and brake disc

10. On hubs with integrated brake disc, position the inner bearing seal in the hub and press the seal in so the edge lies in the same plane as the hub.

11. On hub with separate hub and brake disc:

   a. Press the sealing ring onto the spindle, making sure that the seal ring is square. The sealing ring lip should face outwards.

   b. Install the inner bearing in the hub. Press in the sealing washer.

12. Install the hub, outer race and castle nut.

NOTE: On vehicles with separate hub and brake disc, install the brake disc and guide pin.

13. To adjust the bearing pre-load:

   a. Spin the hub and simultaneously tighten the center nut to 42 ft. lbs. (57 Nm).

   b. Loosen the nut 1/2 turn, then tighten the nut by hand, approximately 1 ft. lb. (1.5 Nm).

   c. Install the cotter pin. If the pin hole in the spindle does not align with the pin hole in the nut, unscrew the nut slightly to the nearest pin hole.

   d. Install the protective cap.

14. Install the brake caliper.

15. Install the wheel. Lower the vehicle.

850 Models

1. Raise and safely support vehicle. Remove the wheel.

2. Disconnect the ABS sensor from the axle shaft but do not disconnect the sensor connector. Hang it out of the way.

3. Remove the caliper, carrier and rotor. Hang the caliper safely out of the way.

4. Remove the halfshaft.

5. Separate the ball joint from the control arm. Disconnect the sway bar link.

6. Remove the four Torx® E14 bolts retaining the hub. Remove the hub.

To install:

7. Clean the axle shaft and hub mating surfaces. Clean the ABS sensor with a soft brush.

8. Install the new hub and torque the bolts alternately to 33 ft. lbs. (45 Nm) plus an additional 60 degrees.

9. Insert the axle shaft into the hub and fit the splines. Tighten the new axle shaft nut by hand.

10. Connect the ball joint to the control arm using new nuts.

11. Connect the sway bar link.

12. Install the brake rotor, carrier and caliper.

13. Tighten the axle nut to 89 ft. lbs. (120 Nm) plus an additional 60 degrees, using tool 5461 or equivalent to counterhold. Lock the axle shaft nut using a chisel to tap the flange into the groove.

14. Clean the ABS sensor and its seat with a soft brush. Torque the sensor to 7 ft. lbs. (10 Nm).

15. Install the wheels. Lower the vehicle.

REAR SUSPENSION

Shock Absorber

REMOVAL AND INSTALLATION

All Models Except 850 and 960 Sedan

1. Raise and safely support the vehicle on jackstands.

2. Place a jack under the lower arm and lift the rear suspension to unload the shock.

3. Remove the lower nut and bolt securing the shock to the rear axle.

4. Lower the vehicle enough to remove the shock.

5. On 940 and 960 models, remove the rubber seal plug from the inner fender-well.

6. Remove the upper shock absorber through-bolt. Remove the old shock absorber.

To install:

7. Install the shock absorber in the upper mount and hand-tighten the through-bolt.

8. Raise the rear end and attach the shock using the lower mount bolt. Hand tighten the nut.

9. Torque the lower bolt to 63 ft. lbs. (85 Nm).

10. Remove the jack and lower the vehicle.

11. Torque the upper bolt to 63 ft. lbs. (85 Nm) and install the rubber plug, if removed.

1993–94 960 Sedan

NOTE: Multi-link suspensions require alignment any time the suspension components are disassembled. Installation tightening and torque procedures are critical to correct alignment.

1. Raise and support the vehicle safely on jackstands. Make sure the jackstands are placed as far forward as possible. Check that the rear jackstands will not interfere with the floor jack.

2. Remove the wheels. Remove the bolts holding the protective guard to the control arm, and remove the guard.

3. At the front of the arm, remove the two retaining bolts which hold the bracket (for the support arm) to the frame. Do not attempt to remove the through-bolt.

4. Remove the retaining bolt at the rear of the support arm.

5. Separate the rear end of the support arm from the wheel bearing housing.

6. Using either Volvo tool 5972 or two floor jacks, support the arm at the front and rear ends. Raise the jacks just enough to unload the shock.

7. Remove the retaining bolt at the top of the shock absorber.

8. Lower the jacks slowly; the arm will come free with the spring and shock attached.

9. Remove the spring with the upper and lower rubber seats. Unbolt the shock absorber from the arm.

To install:

10. Install the shock absorber on the arm and tighten the bottom mount to 41 ft. lbs. (56 Nm).

11. Install the bottom spring rubber seat on the support arm. Take care to properly locate the grooves in the rubber seat.

12. Install the spring and the top rubber seat. Place the assembled support arm on the jacks and raise into position.

13. Compress the spring until the shock absorber is in the correct position. The shock may be held in place temporarily with a drift in the hole. Insert the bolt, and tighten to 62 ft. lbs. (85 Nm).

14. Reinstall the mounting bolts at the front of the support arm bracket and tighten to 35 ft. lbs. (48 Nm). Tighten the large nut to 51 ft. lbs. (70 Nm) plus an additional 90 degrees of rotation.

15. At the rear of the support arm, tap the arm into place on the wheel bearing housing. Tighten the bolt to 44 ft. lbs. (60 Nm) plus an additional 90 degrees of rotation. Do not overtighten.

16. Reinstall the protective cover on the control arm. Install the wheel.

17. Lower the vehicle to the ground and final tighten the nuts to 62 ft. lbs. (85 Nm).

18. Roll the vehicle several feet forwards and backwards before adjusting the rear wheel alignment.

1995–97 960 Sedan

1. Raise and safely support the rear of the vehicle with jackstands. Place the jackstands so they do not interfere with the support arm of the vehicle.

2. Remove the wheel.

3. Use a floor jack and raise up the support arm.

4. Disconnect the shock absorber from the support arm.

5. Disconnect the shock absorber from the body. Remove shock absorber.

To install:

6. Install the shock absorber and connect it to the body using a new bolt. Torque the bolt to 59 ft. lbs. (80 Nm).

7. With the support arm raised, connect the shock absorber to the support arm and install a new bolt. Torque the bolt to 59 ft. lbs. (80 Nm).

8. Carefully lower the lift from the support arm.

9. Install the wheels.

10. Lower the vehicle.

850 Models

1. On early four door models, remove the plastic side panel, then fold the back seat forward and fold back the trunk carpet. Remove the support panel under the edge of the carpet and detach the side panels at the front and fold them over. Remove the back seat catch and panel mounting clip.

2. On later four door models, remove the support panel over the shock mount. Make small cuts in the panel if necessary to fold it up.

3. On five door models, remove the front floor panel bolts and pull the panel back to free it from the front mount. Remove the panel.

4. Raise and safely support the vehicle so the rear wheels are off the ground. Remove the two upper shock mount bolts.

5. Using a floor jack, press the trailing arm up to unload the shock absorber. Disconnect the shock absorber from the lower mount, then pull it off of the trailing arm.

6. Lower the vehicle and lift the shock assembly out from the top.

7. Check the shock upper mount bushing for damage and replace if necessary.

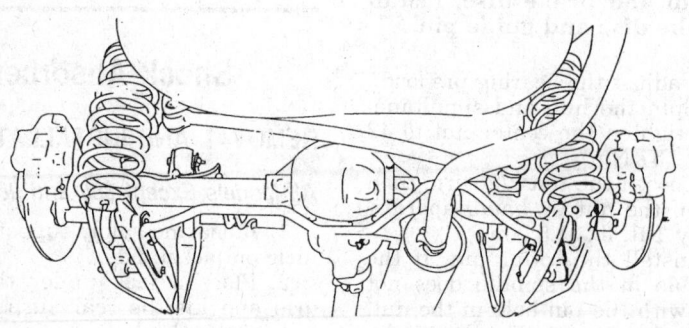

186689

Rear suspension components — 240 models

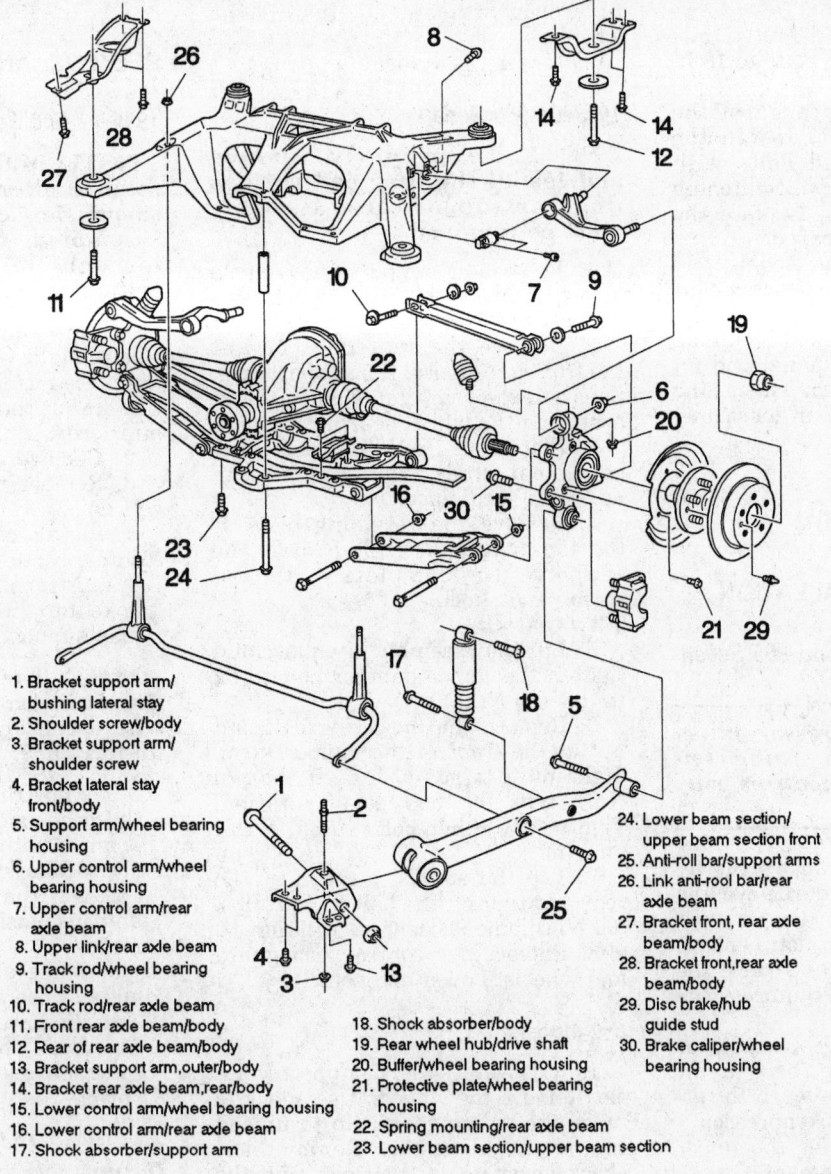

1. Bracket support arm/
   bushing lateral stay
2. Shoulder screw/body
3. Bracket support arm/
   shoulder screw
4. Bracket lateral stay
   front/body
5. Support arm/wheel bearing
   housing
6. Upper control arm/wheel
   bearing housing
7. Upper control arm/rear
   axle beam
8. Upper link/rear axle beam
9. Track rod/wheel bearing
   housing
10. Track rod/rear axle beam
11. Front rear axle beam/body
12. Rear of rear axle beam/body
13. Bracket support arm,outer/body
14. Bracket rear axle beam,rear/body
15. Lower control arm/wheel bearing housing
16. Lower control arm/rear axle beam
17. Shock absorber/support arm

18. Shock absorber/body
19. Rear wheel hub/drive shaft
20. Buffer/wheel bearing housing
21. Protective plate/wheel bearing
    housing
22. Spring mounting/rear axle beam
23. Lower beam section/upper beam section

24. Lower beam section/
    upper beam section front
25. Anti-roll bar/support arms
26. Link anti-rool bar/rear
    axle beam
27. Bracket front, rear axle
    beam/body
28. Bracket front,rear axle
    beam/body
29. Disc brake/hub
    guide stud
30. Brake caliper/wheel
    bearing housing

349467

Exploded view of the rear suspension assembly — 960 models with Mk II rear suspension

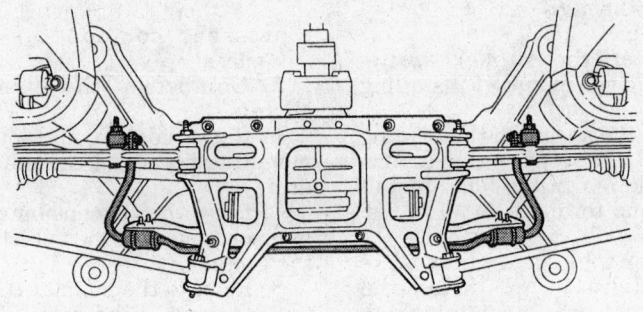

301369

Multi-link suspension rear anti-roll bar — 960 models

To install:

8. Install the upper mount on the new shock as follows:

• torque standard shock absorbers to 30 ft. lbs. (40 Nm)

• torque gas shock absorber M12 nuts to 30 ft. lbs. (40 Nm)

• torque gas shock absorber M10 nuts to 15 ft. lbs. (20 Nm) plus an additional 90 degrees

9. Install the shock absorber in the vehicle and turn the upper mount bolts a few turns.

10. Raise the vehicle and position the jack under the trailing arm and lift up. Connect the shock to the lower mount, making sure that the shock is seated correctly in the upper

mount. Torque the nut to 59 ft. lbs. (80 Nm).

11. Lower the vehicle and torque the upper shock mount bolts to 18 ft. lbs. (25 Nm).

12. Install the front edge of the panels using the clips. Install the back seat catches and bolts with thread locking compound, tightening to 15 ft. lbs. (20 Nm). Replace the cover plate and trunk carpet.

13. On later four door models, fold down the cover over the shock mount and install the carpeting.

14. On five door models, line up the front edge of the floor panel and install the bolt at the rear edge. Line up the panel with the rear floor panel and tighten the bolts.

Coil Spring

REMOVAL AND INSTALLATION

All Models Except 850 and 960 Sedan

——— CAUTION ———
A coil spring compressor is required to remove the spring. Improper removal procedures may cause serious injury.

1. Raise and safely support the vehicle on the frame and remove the rear wheels.

2. Place a hydraulic jack beneath the rear axle housing and raise the housing sufficiently to compress the spring.

3. Install the spring compressor and tighten. Make sure there are at least 3 coils of spring between the attachment points of the compressor.

4. Loosen the nuts for the upper and lower spring attachments.

5. Disconnect the shock absorber at the upper attachment. Lower the jack enough to remove the spring.

To install:

6. Make sure the coil spring is compressed.

7. Position the retaining bolt and inner washer for the upper attachment inside the spring. While holding the outer washer and rubber spacer to the upper body attachment, install the spring and inner washer to the upper attachment sandwiching the rubber spacer. Tighten the retaining bolt to 35 ft. lbs. (48 Nm).

8. Raise the jack and secure the bottom of the spring to the lower attachment with the washer and retaining bolt tightened to 63 ft. lbs. (85 Nm). Slowly remove the spring compressor.

9. Connect the shock absorber to its upper attachment. Install the wheels.

10. Lower the vehicle.

1993–94 960 Sedan

NOTE: To properly remove and install the rear coil springs the rear support arm assembly must be removed.

1. Raise and support the vehicle safely. Remove the rear wheels and support the arm guards.

2. Remove the retaining bolts at the front and rear of the support arm. Separate the rear end of the support arm from the wheel bearing housing.

3. Place a jack with fixture 5972 or equivalent, under the support arm and raise into place.

4. Remove the retaining bolts at the top of the shock and lower the support arm complete with the spring and shock.

To install:

5. Lift the assembly into place and tighten the upper damper bolt to 62 ft. lbs. (85 Nm).

6. Replace the mounting bolt and nut at the front of the support arm. Tighten the large nut to 51 ft. lbs. (70 Nm), plus 90 degrees of rotation. Tighten the other bolts to 35 ft. lbs. (48 Nm).

7. Tap the support arm in at the rear and tighten the bolt to 44 ft. lbs. (60 Nm), plus 90 degrees rotation.

8. Replace the control arm guard and wheels. Lower the vehicle.

850 Models

1. Raise and safely support vehicle. Remove the wheels. Use a jack to press the training arm up to unload the shock absorber. Remove the shock lower mount bolt and pull the shock off of its mount.

2. Lower the jack and remove the spring mounting nut. Remove the spring from the car.

To install:

3. Transfer the rubber spring spacer and lower mount if installing a new spring.

4. Install the spring in the trailing arm recess and center the mount washer guide pin in the hole. Install a new nut and torque it to 30 ft. lbs. (40 Nm).

5. Position the jack under the trailing and lift it up. Install the shock on the lower mount, making sure that the spring is correctly seated in the upper mount. Torque the shock nut to 59 ft. lbs. (80 Nm).

6. Install the wheels. Lower the car.

Leaf Spring

REMOVAL AND INSTALLATION

1995–97 960 Sedan

NOTE: Multi-Link suspensions require alignment any time the suspension components are disassembled. Installation tightening and torque procedures are critical to correct alignment.

1. Raise and support the vehicle safely. Make sure the front supports are placed as far forward as possible. Check that the rear supports will not interfere with the support arm.

2. Remove the rear wheels.

3. Remove the sway bar from the vehicle.

4. Install compression tool Kent-Moore® No. J-41470 (or equivalent):

   a. Attach one end of the tool yoke into the boss on one side of the axle support beam.

NOTE: There is a boss cast into the support on either side of where the spring is visible from underneath.

   b. Slide the tool into the boss on the other side of the axle support beam to fully engage the yoke.

   c. Make sure the tool is seated between the retaining tabs before any suspension components can be removed.

   d. Make sure the rollers on the ends of the tool are positioned inside the lower control arms.

   e. Turn the handle to raise the control arms and compress the leaf spring.

——— CAUTION ———
Do not tighten the tool after the pegs have contacted the axle support beam.

   f. Once the spring is compressed, components can be removed.

5. Compress the suspension slightly.

6. Removed the shock absorber mounting bolt from the rear support arm.

7. Press the suspension up to normal position, as if the vehicle were on the ground.

8. Remove the bolts at the front of the support arm bracket.

9. Remove the support arm-to-wheel bearing housing bolt.

10. Loosen the support arm bracket nut a few turns. Tap the support arm off the wheel bearing housing.

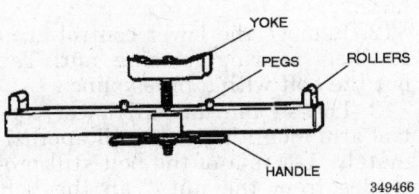

Leaf spring compression tool J-41470

11. Remove the support arm bracket nut and the support arm with the bracket attached. Do not remove the support arm-to-bracket through bolt.

12. Remove the track rod-to-wheel bearing housing bolt. Tap the track rod off the wheel bearing housing.

13. Remove the lower control arm-to-wheel bearing housing nut. Tap out the bolt with a brass punch.

14. Loosen the inboard lower control arm mounting nut until approximately. 1–2 mm of the bolt still protrudes from the nut. Tap the bolt with a brass punch until the nut contacts the lower control arm.

15. Repeat steps 6 through 14 to disassemble the opposite side of the suspension.

16. Lower the compression tool and control arms completely. Inspect the spring mountings in the lower control arms. Retain the mountings in the control arms for assembly.

17. Loosen all the lower rear axle support bolts 1 turn. Remove the 2 front bolts at the pinion flange. Remove the spacers.

18. Install rear axle retainer 5580 (or equivalent) to the front edge of the differential. Install rear axle retainer 5579 (or equivalent) to the rear edge of the differential.

19. Position a lift or jack against the lower axle support beam.

20. Remove the lower support beam mounting bolts and lower the support beam.

NOTE: Before removing the spring mounting plates, take note of the positioning. Installation must be in the same positions.

21. Remove the spring mounting plates and remove the spring.

22. Check all spring mountings for damage or wear.

To install:

23. Install the spring. Install the spring mounting plates using new bolts.

NOTE: The leaf spring is marked with a center line.

24. Make sure the spring is centered. Tighten the spring mounting plate bolts alternately to 37 ft. lbs. (50 Nm).

25. Position the lower axle support beam on a lift or jack.

26. Make sure the lower axle support beam contact surfaces are clean and that the lower differential bushing is in place.

27. Make sure the lower differential bushings are inside the guide flanges of the lower axle support beam.

28. Raise the lower axle support beam into position. Install new mounting bolts except at the pinion flange. Tighten the bolts until there is approximately 0.08–0.16 in. (2–4 mm) clearance between the support surfaces. Remove the lift or jack.

29. Remove the differential retainers 5579 and 5580 (or equivalents) from the differential.

30. Install the 2 front spacers between the upper and lower axle support beams. Install new bolts at the pinion flange.

31. Tighten all lower axle support bolts diagonally until snug. Torque the bolts diagonally to 59 ft. lbs. (80 Nm).

32. Install the compression tool. Install the spring into the mounting in the control arms.

33. Raise the control arms up to their normal position.

34. Install the lower control arm to the wheel bearing housing,

35. Install the track rod to the wheel bearing housing using a new nut and torque to 59 ft. lbs. (80 Nm).

36. Install the rear support arm using new bolts and nut. Do not tighten bolts and nut.

37. Torque the rear bolt and front nut to 59 ft. lbs. (80 Nm). Torque the front bolts to 37 ft. lbs. (50 Nm).

38. Install the shock absorbers.

39. Make sure the washers and bushings on the sway bar links are in place. Raise the sway bar into position and install.

40. Install the wheels.

41. Lower the vehicle.

42. Check wheel alignment.

Lower Control Arms

REMOVAL AND INSTALLATION

All Models Except 850 and 960 Sedan

1. Raise and safely support the vehicle on the frame and remove the rear wheels.

2. Place a hydraulic jack beneath the rear axle housing and raise the housing sufficiently to compress the spring and shock.

3. Install a compressor on the spring and tighten. Make sure there are at least 3 coils of spring between the attachment points of the compressor. Remove the lower shock attachment bolts and nut. Lower the rear axle and remove the spring.

4. Remove the sway bar from the trailing arms, if equipped.

5. Remove the axle-to frame through-bolt securing the trailing arm to the vehicle body.

6. Loosen and remove the through bolt securing the trailing arm to the rear axle. With the rear axle firmly supported, remove the trailing arm.

To install:

7. Install the control arm on the axle housing and tighten the bolt finger-tight.

8. Raise the assembly into position and install the axle-to-frame bolts finger-tight.

9. Install the shock and coil spring. Tighten all control arm bolts to 85 ft. lbs. (115 Nm).

10. Install the wheels. Lower the vehicle.

1993–94 960 Sedan

NOTE: Multi-Link suspensions require alignment any time the suspension components are disassembled. Installation tightening and torque procedures are critical to correct alignment.

1. Raise and support the vehicle safely. Make sure the front supports are placed as far forward as possible. Check that the rear supports will not interfere with the support arm.

2. Remove the wheels. Loosen and remove the bolts holding the protective guard to the arm and remove the guard.

3. At the front of the arm, remove the two retaining bolts which hold the bracket (for the control arm) to the frame. Do not attempt to remove the through-bolt.

4. Remove the retaining bolt at the rear of the control arm.

5. Separate the rear end of the control arm from the wheel bearing housing.

6. Using either Volvo tool 5972 or two floor jacks, support the arm at the front and rear ends. Raise the jacks just enough to relieve the tension on the shock absorber.

7. Remove the retaining bolt at the top of the shock absorber.

8. Lower the jacks slowly; the arm will come free with the spring and shock attached.

9. Unbolt the shock absorber from the arm, then unbolt and remove the bracket at the front of the arm. Take note of the relationship between the bracket and the arm: the bracket correctly mounts one way only.

To install:

10. Install the control arm bracket in the correct position and tighten the nut to 91 ft. lbs. (125 Nm) plus and additional 120 degrees of rotation.

11. Install the shock absorber on the arm and tighten the bottom mount to 41 ft. lbs. (56 Nm).

12. Install the bottom spring seat on the control arm. Take care to properly locate the grooves in the seat.

13. Install the spring and the top rubber seat. Place the assembled support arm on the jacks and raise into position.

14. Gently raise the jacks and compress the spring until the shock absorber is in the correct position. Install the shock.

15. Reinstall the mounting bolts at the front of the control arm bracket. Tighten the bolts to 35 ft. lbs. (48 Nm) and the large nut to 51 ft. lbs. (70 Nm) plus an additional 90 degrees of rotation.

16. At the rear of the control arm, tap the arm into place on the wheel bearing housing. Tighten the bolt to 44 ft. lbs. (60 Nm) plus an additional 90 degrees of rotation. Do not overtighten this fitting.

17. Reinstall the protective cover on the control arm. Install the wheel.

18. Lower the vehicle to the ground and final tighten the lugs to 62 ft. lbs. (85 Nm).

19. Roll the vehicle several feet forwards and backwards before adjusting the rear wheel alignment.

1995–97 960 Sedan

NOTE: Multi-Link suspensions require alignment any time the suspension components are dis- assembled. **Installation tightening and torque procedures are critical to correct alignment.**

1. Raise and support the vehicle safely. Make sure the front supports are placed as far forward as possible. Check that the rear supports will not interfere with the support arm.

2. Install compression tool Kent-Moore® No. J-41470 (or equivalent):

 a. Attach one end of the tool yoke into the boss on one side of the axle support beam.

NOTE: There is a boss cast into the support on either side of where the spring is visible from underneath.

 b. Slide the tool into the boss on the other side of the axle support beam to fully engage the yoke.

 c. Make sure the tool is seated between the retaining tabs before any suspension components can be removed.

 d. Make sure the rollers on the ends of the tool are positioned inside the lower control arms.

 e. Turn the handle to raise the control arms and compress the leaf spring.

CAUTION
Do not tighten the tool after the pegs have contacted the axle support beam.

 f. Once the spring is compressed, components can be removed.

3. Compress the suspension slightly.

4. Removed the shock absorber mounting bolt from the rear support arm.

5. Remove the sway bar-to-support arm bolts on both sides. Press the sway bar up away from the rear support arms.

6. Press the suspension up to normal position, as if the vehicle were on the ground.

7. Remove the bolts at the front of the support arm bracket.

8. Remove the support arm-to-wheel bearing housing bolt.

9. Loosen the support arm bracket nut a few turns. Tap the support arm off the wheel bearing housing.

10. Remove the support arm bracket nut and the support arm with the bracket attached. Do not remove the support arm-to-bracket through bolt.

11. Remove the brake caliper and hang it up with steel wire. Do not allow the caliper to hang by the brake hose.

12. Remove the lower control arm-to-wheel bearing housing nut. Tap out the bolt with a brass punch.

13. Loosen the inboard lower control arm mounting nut until approximately. 1–2 mm of the bolt still protrudes from the nut. Tap the bolt with a brass punch until the nut contacts the lower control arm.

14. Lower the compression tool and control arm completely. Inspect the spring mounting in the lower control arm. Retain the mountings in the control arm for assembly.

15. Remove the inboard lower control arm mounting nut. Tap out the bolt with a brass punch. Tap the lower control arm off the lower rear axle support beam.

16. Inspect the lower control arm bushings for damage or wear.

To install:

17. Install the lower control arm onto the lower rear axle support beam with a new nut and bolt. Do not tighten. Install the spring into the mounting in the control arm.

18. Use the compression tool to raise the control arms up to their normal position.

19. Install the lower control arm to the wheel bearing housing, using a new nut. Torque both ends of the control arm to 59 ft. lbs. (80 Nm).

20. Install the brake caliper with new bolts and torque to 44 ft. lbs. (60 Nm).

21. Install the rear support arm using new bolts and nut. Do not tighten bolts and nut.

22. Torque the rear bolt and front nut to 59 ft. lbs. (80 Nm). Torque the front bolts to 37 ft. lbs. (50 Nm).

23. Install the shock absorber:

 a. If equipped with standard shock absorbers, compress the shock by hand. Install a new bolt and torque to 59 ft. lbs. (80 Nm).

 b. If equipped with self-leveling shock absorbers, lower the compression tool. Install a new bolt and torque to 59 ft. lbs. (80 Nm).

24. Lower the sway bar into position and install new sway bar-to-support arm bolts, but do not tighten.

25. Install the wheels.

26. Lower the vehicle.

27. Torque sway bar bolts to 59 ft. lbs. (80 Nm).

28. Check wheel alignment and test drive vehicle.

Upper Control Arm

REMOVAL AND INSTALLATION

1993–94 960 Sedan

1. Raise and support the vehicle safely. Make sure the front supports are placed as far forward as possible.
2. Remove the wheels. Remove the brake caliper without disconnecting the brake hose. Tie it with wire out of the way. Do not allow it to hang by the hose.
3. Remove the bolt holding the lower support arm to the wheel bearing housing and tap the support arm loose.
4. Remove the nut and bolt holding the lower control arm (intermediate arm) to the wheel bearing housing.
5. Remove the bolt attaching the track rod to the wheel bearing housing. Use a small bearing puller to disconnect the track rod.
6. Remove the nut which holds the upper control arm to the wheel bearing housing. Collect and note the location of the spacers between the upper control arm and the bearing housing. They are alignment shims and must be reinstalled properly.
7. Remove the nut holding the rear upper control arm to the rear axle member (support).
8. At the front of the upper control arm, remove the nut and bolt which holds it to the rear axle member.
9. Use a pair of adjustable pliers to remove the control arm from the vehicle.
   To install:
10. Install the arm to the rear axle member and secure with nut and bolt. Install both the front and rear mounts.
11. Install the spacers at the wheel bearing housing, and install the nut holding the arm to the housing.
12. Inboard at the rear axle support, tighten the rear-most nut to 62 ft. lbs. (85 Nm). Tighten the front nut and bolt to 51 ft. lbs. (70 Nm) plus an additional 60 degrees of rotation.
13. Pull the top of the wheel bearing housing outwards away from the center of the vehicle. This is essential for correct wheel alignment. Tighten the upper control arm nut at the bearing housing to 84 ft. lbs. (115 Nm).
14. Pull the wheel bearing housing outward and install the lower control arm with the bolt and nut, but do not tighten it.
15. Pull the wheel bearing housing inwards towards the center of the ve-

hicle. Tighten the control arm nut to 37 ft. lbs. (50 Nm) plus and additional 90 degrees of rotation.
16. Reinstall the support arm; tighten the mount to 44 ft. lbs. (60 Nm) plus an additional 90 degrees of rotation.
17. Install the track rod and tighten to 62 ft. lbs. (85 Nm).
18. Install the brake caliper.
19. Install the wheels. Lower the vehicle to the ground.
20. Check and adjust the rear alignment if necessary.

1995–97 960 Sedan

1. Set the rear lift arms so that they do not interfere with support arm of the vehicle.
2. Raise and safely support the vehicle.
3. Remove the wheel.
4. Install compression tool J-41470 or a suitable equivalent. Apply light pressure to the link arm.
5. Disconnect the shock absorber from the support arm.
6. Remove the sway bar bolts from support arm on both sides.
7. Turn the compression tool up to its fully compressed position.
8. Remove the brake caliper and hang it safely out of the way so that the brake hose is not damaged.
9. Remove the nut connecting the upper control arm to the wheel bearing housing. Using a brass drift, tap the bolt out to release the upper control arm bushing from the wheel bearing housing.
10. Remove the bolt connecting the track rod to the wheel bearing housing. Separate the track rod from the wheel bearing housing.
11. Remove the nut connecting the lower control arm to the wheel bearing housing. Use a brass drift to tap out the bolt.
12. Separate the wheel bearing housing from the upper control arm. Allow the wheel bearing housing to rest on the lower control arm.
13. Remove the front and rear bolts from the upper control arm. Pull the upper control from the rear axle beam and remove it.
   To install:
14. Install the upper control arm in the rear axle beam and attach it with new front and rear bolts. Torque the bolts to 18 ft. lbs. (25 Nm).
15. Push the wheel bearing housing up and connect it to the upper control arm. Install a new attaching nut, but do not tighten.
16. Connect the support arm to the wheel bearing housing and install a

new attaching bolt, but do not tighten.
17. Connect the lower control arm to the wheel bearing housing. Install the through bolt and a new nut, but do not tighten.
18. Install the track rod and attaching bolt, but do not tighten.
19. Torque the attaching bolts and nuts in the following sequence:
   • nut-to-upper control arm, torque to 92 ft. lbs. (120 Nm)
   • nut-to-lower control arm, torque to 63 ft. lbs. (80 Nm)
   • bolts-to-support arm, torque to 63 ft. lbs. (80 Nm)
   • bolt-to-track rod, torque to 63 ft. lbs. (80 Nm)
20. Install the brake caliper and new attaching bolts. Torque the bolts to 47 ft. lbs. (60 Nm).
21. Install the shock absorber and a new attaching bolt. Torque the bolt to 63 ft. lbs. (80 Nm).
22. Release the pressure on the compression tool.
23. Connect the sway bar to the support arm using new bolts, but do not tighten.
24. Install the wheels and diagonally torque the lugs to 63 ft. lbs. (85 Nm).

Sway Bar

REMOVAL AND INSTALLATION

All Models Except 850 and 960 Sedan

1. Raise the vehicle and support it safely. Place the stands at the rear jacking points. If required, remove the wheels.
2. Use a floor jack to raise the rear axle just enough to unload the shock absorbers. Remove the lower shock retaining bolts.
3. Remove the nuts holding the sway bar to the brackets.
4. Remove the sway bar.
   To install:
5. When installing the sway bar, attach both the bracket nut and the lower shock retaining bolt hand-tight.
6. Once all four mounting points are snug, tighten the bracket nuts to 35 ft. lbs. (48 Nm) and the shock absorber bolts to 63 ft. lbs. (86 Nm).
7. Remove the jack from the axle. Install the wheels, if removed. Lower the vehicle to the ground.

1995–97 960 Sedan

1. Raise and safely support the vehicle. Remove the wheels.

2. Remove the nut and through-bolt securing the sway bar to the support arm.

3. Remove the nuts securing the bracket retainer to the axle beam.

4. Lower the sway bar.

To install:

5. Install the sway bar to the axle beam and secure in place using the nuts. Tighten the nuts to 15 ft. lbs.

6. Connect the sway bar to the support arm and attach the through-bolt and nut. Do not tighten at this time.

7. Install the wheels. Lower the vehicle.

8. With all four wheels on the ground, Tighten the support arm bolts to 63 ft. lbs. (80 Nm).

850 Models

1. Raise and safely support the vehicle. Remove the left-hand rubber silencer mount and hang it up as high as possible with a tie wrap.

2. Remove the outer transverse arm mount nut and bolt. Mark the position of the right side transverse arm mount in relation to the left-hand trailing arm hole. Punch a mark on the edge of the hole and remove the other mount bolt.

NOTE: It is important that this mark is made properly otherwise the toe in will be incorrect.

3. Remove the anti-sway bar mount bolts, then the bar.

To install:

4. Install the anti-sway bar using new nuts and bolts, but do not tighten completely.

5. Connect the transverse arm mount to the trailing arm with new bolts and nuts. Install the inner bolt first and line up the mark with the outer hole, tighten it enough to keep it in position. Then install the outer bolt and nut and torque to 37 ft. lbs. (50 Nm) and angle tighten 120°.

6. Torque the anti-sway bar bolts as follows:

• right side bolts to 37 ft. lbs. (50 Nm)

• left side forward bolt to 37 ft. lbs. (50 Nm) and angle tighten 90°

• left side rear bolt to 48 ft. lbs. (65 Nm) and angle tighten 90°

7. Cut the tie wrap holding the silencer and install the rubber mount. Lower the vehicle.

Wheel Bearings

ADJUSTMENT

The rear wheel bearings are sealed, pressed-in units, and no adjustment is possible.

REMOVAL AND INSTALLATION

All Models Except 850 and 960 Sedan

1. With the vehicle sitting on all four wheels, loosen the rear axle nut.

2. Raise and support the vehicle safely. Do not allow the lifting arms to interfere with the support arms.

3. Remove the wheels. Remove the brake caliper and use a piece of wire to hang the caliper out of the way.

4. Remove the brake disc and parking brake shoes.

5. Disconnect and remove the parking brake cable from the wheel bearing housing.

6. Remove the retaining bolt for the support arm at the housing. Tap the support arm loose.

7. Remove the nut and bolt holding the lower link arm to the housing.

8. Remove the retaining bolt for the track rod at the bearing housing and use a small claw-type puller to remove the track rod.

9. Remove the axle nut.

10. Remove the retaining nut for the upper link at the bearing housing. The wheel bearing housing can now be removed as a unit.

NOTE: There are shims between the bearing housing and the upper link arm. Collect them when the housing is removed.

11. Mount the housing assembly in a vise. Place counterhold tool 5340 or equivalent between the hub and bearing housing. Press out the hub with a proper sized drift.

12. Remove the circlip retaining the bearing in the wheel bearing housing and press the bearing out. Press against the inner race.

13. Use bearing puller 2722 or equivalent to pull the inner race off the hub.

To install:

14. Press the new bearing into the housing. Make sure the press tool contacts only the outer bearing race or the bearing will be damaged. Install the circlip.

15. Support the inner race and press the hub into the bearing. If the inner race is not properly supported, the bearing will be damaged.

16. Install the wheel bearing housing onto the housing and install the axle nut hand-tight.

17. Install the shims between the upper link and the wheel bearing housing and then install the retaining nut at the upper link.

18. Pull the wheel bearing housing outwards at the top and tighten the upper link arm nut to 85 ft. lbs. (110 Nm). This is essential to insure correct wheel alignment when completed.

19. Tilt the bearing housing outwards at the bottom as necessary to refit the lower link arm and its retaining bolt. When in place, pull the bottom of the bearing housing in towards the center of the vehicle and tighten the link arm to 36 ft. lbs. (47 Nm) plus an additional 90 degrees of rotation.

20. Install the support arm and its bolt.

21. Install the track rod and tighten to 63 ft. lbs. (82 Nm).

22. Reinstall the parking brake cable at the bearing housing.

23. Reinstall the parking brake shoes, the brake disc as marked and the brake caliper.

24. Install the wheels. Lower the vehicle.

25. With all four wheels on the ground, torque the axle nut to 103 ft. lbs. (134 Nm) plus an additional 60 degrees of rotation.

1993–96 940 and 960

1. With the vehicle sitting on all four wheels, loosen the rear axle nut.

2. Raise and support the vehicle safely. Do not allow the rear lifting arms to interfere with the support arms.

3. Remove the wheels. Remove the brake caliper and use a piece of wire to hang the caliper out of the way.

4. Mark the position of the brake disc relative to its small locating pin, then remove the disc. Remove the brake shoes.

5. Disconnect and remove the parking brake cable from the wheel bearing housing.

6. Remove the retaining bolt for the support arm at the housing. Tap the support arm loose.

7. Remove the nut and bolt holding the lower link arm to the housing.

8. Remove the retaining bolt for the track rod at the bearing housing and use a small claw-type puller to remove the track rod.

9. Remove the axle nut.

10. Remove the retaining nut for the upper link at the bearing hous-

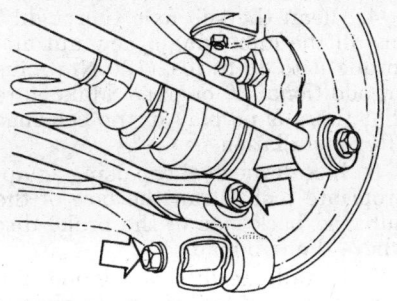

324608

Support arm bolt and lower link bolt

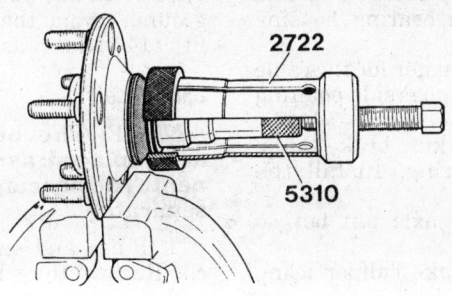

324610

Removing inner ring from hub

ing. The wheel bearing housing can now be removed unit.

NOTE: There are shims between the bearing housing and the upper link arm. Collect them when the housing is removed.

11. Mount the housing assembly in a vise. Place a counterhold tool 5340 or equivalent between the hub and bearing housing. Press out the hub with a proper sized drift.

12. Remove the circlip retaining the bearing in the wheel bearing housing and press the bearing out. Press against the inner race.

13. Use a bearing puller 2722 or equivalent to pull the inner race off the hub.

To install:

14. Press the new bearing into the bearing housing. Make sure the press tool contacts only the outer bearing race or the bearing will be damaged. Install the circlip.

15. Support the inner race and press the hub into the bearing. If the inner race is not properly supported, the bearing will be damaged.

16. Install the wheel bearing housing onto the halfshaft and install the axle nut hand-tight.

17. Install the shims between the upper link and the wheel bearing

housing and then install the retaining nut at the upper link.

18. Pull the wheel bearing housing outwards at the top and tighten the upper link arm nut to 85 ft. lbs. This is essential to insure correct wheel alignment when completed.

19. Tilt the bearing housing outwards at the bottom as necessary to refit the lower link arm and retaining bolt. When in place, pull the bottom of the bearing housing in towards the center of the vehicle and tighten the link arm to 36 ft. lbs. plus an additional 90 degrees of rotation.

20. Install the support arm and bolt. Tighten the nut to 44 ft. lbs. plus an additional 90 degrees of rotation.

21. Install the track rod and tighten to 63 ft. lbs.

22. Reinstall the parking brake cable at the bearing housing.

23. Reinstall the brake shoes, the brake disc as marked and the brake caliper. lbs.

24. Install the wheels. Lower the vehicle.

25. With all four wheels on the ground, torque the axle nut to 103 ft. lbs. plus an additional 60 degrees of rotation.

1995–96 960 Sedan

1. With the vehicle sitting on all four wheels, loosen the rear axle nut.

2. Raise and safely support vehicle. Position a lift or jackstands so they do not interfere with the suspension arms. Remove the wheel.

3. Use tool 999 5577 or equivalent to compress the suspension slightly against the spring.

4. Remove the damper bolt and pull the damper from the support arm. Remove the anti-sway bolts in both of the support arms.

5. Raise the suspension up into normal position. Remove the axle shaft nut. Remove the brake caliper and support it safely out of the way.

Mark the position of the brake disc and remove the disc.

6. Remove the parking brake shoes and disconnect the adjuster from the parking brake cable. Disconnect the parking brake cable from the wheel bearing housing and remove the nut for the upper link bushing. Tap the bushing bolt free of the wheel bearing housing.

7. Remove the bolt to the track rod and separate the rod from the wheel bearing housing. Remove the support arm bolt and tap if free of the bushing. Remove the link nut and tap the bolt out using a brass drift. Remove the wheel bearing housing.

NOTE: The bearing must be replaced any time the hub is pressed out.

8. Position the wheel bearing housing in a press so the hub can be pressed out. Using an appropriate sized drift, press the hub off.

9. Remove the circlip holding the bearing in the housing and press the bearing out.

10. Position the drift in the inner bearing race. Using the puller 999 2722 and counter hold 999 5310, pull the inner race out of the hub.

To install:

11. Properly support the bearing housing and press the new bearing in. Make sure the press tool contacts only the outer bearing race or the bearing will be damaged. Install the circlip.

12. Support the inner bearing race on the press table and press the hub into the bearing. Make sure the inner bearing race is supported or the bearing will be damaged.

13. Fit the wheel bearing housing to the upper link and driveshaft. Then install all the nuts and bolts before tightening any of them. Torque the nut for the upper link to 85 ft. lbs. (115 Nm) and all others to 63 ft. lbs. (80 Nm).

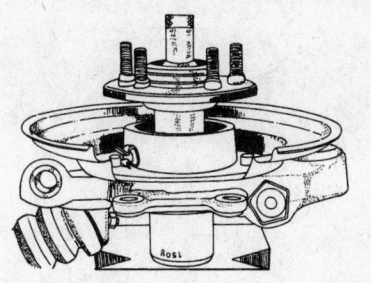

324609

Support the inner race when pressing hub into wheel bearing

14. Connect the parking brake cable to the wheel bearing housing and fasten with clip.

15. Install the adjuster for the cable with arrow on the upper side pointing up.

16. Install the parking brake shoes, retainers, and spring. Install the brake disc.

17. Install a new axle nut but do not tighten it yet.

18. Install the brake caliper using new bolts.

19. Install new anti-sway bolts on both sides, but do not tighten them fully.

20. Clean the face of the brake disc and the back side of the wheel where the two mate. Lubricate the guide pin with rust proofing compound.

21. Install the wheels.

22. Lower the car. Torque the anti-sway bar bolts on both sides to 63 ft. lbs. (80 Nm).

23. With all four wheels on the ground, torque the axle nut to 130 ft. lbs. (140 Nm) plus an additional 60°.

850 Models

NOTE: The bearing and hub are replaced as a single component. The bearing is not available separately.

1. Raise and safely support vehicle. Remove the wheels.

2. Remove the caliper and the brake line from the mounting clip. When the left brake caliper is removed, the three-way brake line connector mounting bolt must also be removed. Remove the caliper and hang it from the spring to prevent brake hose damage.

3. Back the parking brake adjustment off so the disc can be removed. Remove the disc. Remove the cap, hub nut and hub.

To install:

4. Clean the sub axle thoroughly. Install the hub using a new nut and torque it to 89 ft. lbs. (120 Nm) plus an additional 30 degrees. Make sure that there is no play in the bearings after installation.

5. Install the dust cap using an appropriate tool. Clean the face of the hub and back side of the brake disc where the two mate.

6. Install the disc and guide pin. Torque the pin to 7 ft. lbs. (9 Nm).

7. Adjust the parking brake shoes until the disc cannot be turned, then back it off four to six notches.

8. Install the brake caliper.

9. Install the brake line and mounting clips and the three-way connector on left-hand side, if applicable.

10. Install the wheels. Lower the vehicle.

WHAT TOOLS ARE NEEDED

Analyze the Needs

Nearly everybody needs some tools, whether they are fixing a kitchen sink, or overhauling the engine in the family car. As far as car repairs go, pliers and a can of oil won't get one very far down the path of do-it-yourself service. But, one doesn't have to equip the garage like the local service station either. Somewhere between these two extremes is a level that suits the average do-it-yourselfer. Just where that point is depends on one's ability and interest. The strategy is to match tools and equipment to the tasks one would like to tackle.

First, sort things out in a orderly manner. Think about the repair work on three levels: basic, average and advanced. Before purchasing any tools, sit down and determine the level of expertise, job needs to accomplish, and cost. Knowing what repairs one can or want to do is the most important step. Obviously, if all one intends to do is change the oil and spark plugs, one doesn't need many tools. If one plans some fairly extensive repair work, then the need is for a pretty complete collection of tools. Many expensive tools can be rented from automotive parts jobbers or tool rental centers. This allows many of us to do special repairs on an occasional basis.

BASIC AUTOMOTIVE TOOLS

Naturally, without the proper tools it is impossible to properly service a vehicle. It would be impossible to catalog each tool that is needed to perform every operation in this book. It would also be unwise for the amateur to rush out and buy an expensive set of tools on the theory that one or more may be needed at sometime.

The best approach is to proceed slowly, gathering together a good quality set of those tools that are used most frequently. Don't be misled by the low cost of bargain tools. It is far better to spend a little more for better quality. Forged wrenches, 6-point sockets and fine tooth ratchets are by far preferable to their less expensive counterparts. As any good mechanic can say, there are few worse experiences than trying to work on a truck with bad tools. Ones monetary savings will be far outweighed by frustration and mangled knuckles.

Certain tools, plus a basic ability to handle them, are required to get started. A basic tool set and a torque wrench, are good for a start. Begin accumulating those tools that are used most frequently (tools associated with routine maintenance/tune-up and engine repair). In addition to the normal assortment of screwdrivers and pliers, one should have the following tools for routine maintenance:

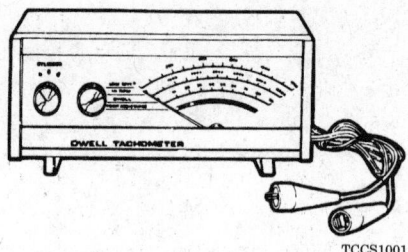

Dwell/tachometer unit (typical)

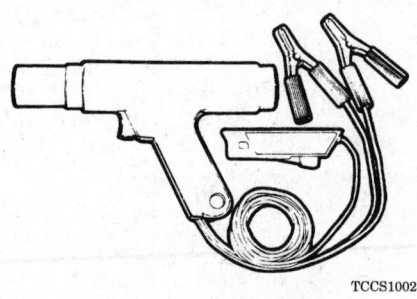

Inductive type timing light

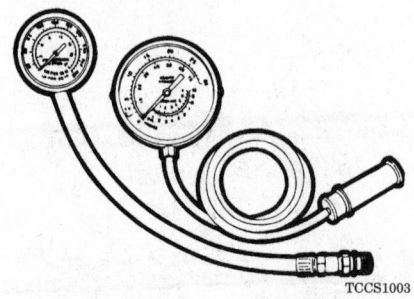

Compression gauge and a combination vacuum/fuel pressure test gauge

• Metric wrenches, sockets and combination open end/box end wrenches in sizes from 3–19mm, and a spark plug socket (5/8 inch or 16mm). If possible, buy various length socket drive extensions. One break in this department is that the metric sockets available in the U.S. will fit SAE ratchet handles and extensions one may already have (1/4 in., 3/8 in., and 1/2 in. drive).
• Jackstands for support.
• Oil filter wrench.
• Oil filler spout or funnel.
• Grease gun for chassis lubrication.
• Hydrometer or battery tester for checking the battery.
• A low flat pan for draining oil.
• Lots of rags for wiping up the inevitable mess.

In addition to the above items, there are several others that are not absolutely necessary, but handy to have around. These include oil-dry, a transmission fluid funnel and the usual supply of lubricants and fluids, although these can be purchased as needed. This is a basic list for routine maintenance, but only ones personal needs and desires can accurately determine a list of tools.

The second list of tools is for tune-ups. While these tools are slightly more sophisticated, they need not be outrageously expensive. There are several inexpensive tach/dwell meters on the market that are every bit as good for the average mechanic as a costly professional model. Just be sure that it goes to at least 1200–1500 rpm on the tach scale and that it works on 4 and 6-cylinder engines. A basic list of tune-up equipment could include:
• Tach/dwell meter.
• Spark plug wrench.
• Timing light (a DC light that works from the truck's battery is best).
• Wire spark plug gauge/adjusting tools.

Here again, be guided by ones own needs. In addition to these basic tools, there are several other tools and gauges which are useful. These include:
• A compression gauge. The screw-in type is slower to use, but eliminates the possibility of a faulty reading due to escaping pressure.
• A manifold vacuum gauge.

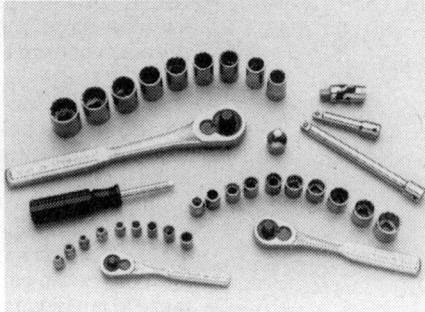

TCCS1200

All but the most basic procedure will require an assortment of ratchets and sockets

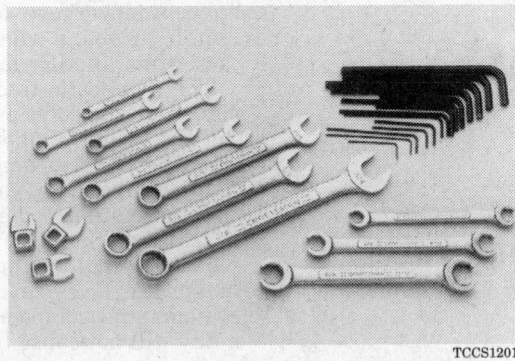

TCCS1201

In addition to ratchets, a good set of wrenches and hex keys will be necessary

• A test light.

• A DVOM digital volt-ohmmeter. This meter allows direct testing of electrical components and grounds.

As a final note, one may find a torque wrench necessary for most work. The beam type models are perfectly adequate, although the newer click (breakaway) type are more precise, and doesn't have to crane the neck to see a torque reading in awkward situations. The breakaway torque wrenches are more expensive and should be recalibrated periodically.

Correct tightening of bolts is an extremely important item on today's automobiles. The torque specification for each fastener will be given in the procedure whenever a specific torque value is required. An example of torque specifications are given, these values are only a guide, based upon fastener size:

Bolts marked 6T

6mm bolt/nut: 5–7 ft. lbs. (7–10 Nm)

8mm bolt/nut: 12–17 ft. lbs. (16–23 Nm)

10mm bolt/nut: 23–34 ft. lbs. (31–46 Nm)

12mm bolt/nut: 41–59 ft. lbs. (56–80 Nm)

14mm bolt/nut: 56–76 ft. lbs. (76–103 Nm)

Bolts marked 8T

6mm bolt/nut: 6–9 ft. lbs. (8–12 Nm)

8mm bolt/nut: 13–20 ft. lbs. (18–27 Nm)

10mm bolt/nut: 27–40 ft. lbs. (37–54 Nm)

12mm bolt/nut: 46–69 ft. lbs. (62–94 Nm)

14mm bolt/nut: 75–101 ft. lbs. (102–137 Nm)

Special Tools

Normally, special factory tools are avoided for repair procedures, since these many not be readily available for the do-it-yourself mechanic. When it is possible to perform the job with more commonly available tools, it will be pointed out, but occasionally, a special tool was designed to perform a specific function and should be used. Before substituting another tool, one should be convinced that neither safety nor the performance of the vehicle will be compromised.

Some special tools are available commercially from major tool manufacturers. Others can be purchased from dealers or from Kent-Moore Corporation, 29784 Little Mack, Roseville, Michigan 48066–2298. In Canada, contact Kent-Moore of Canada, Ltd., 2395 Cawthra Mississauga, Ontario, Canada L5A 3P2.

SPECIAL DIAGNOSTIC TOOLS

Frequent references to specific test equipment will be found in the text and in the diagnostic charts. This usually refers to scan tools used to communicate with electronic control units or special electronic testers. Among other features, scan tools combine many standard testers into a single device for quick and accurate circuit diagnosis. For many tests, a multimeter, test light, or other general test equipment can be substituted but the technician must be aware of the risk involved. The general test equipment may not be capable of safely testing the system or may generate incomplete or inaccurate test results. Some tests require activating system components and often this can only be done with scan tools or other special equipment.

Most test equipment is available through aftermarket tool manufacturers, but some can only be obtained through the vehicle manufacturer. Care should be taken that all test equipment being used is designed to diagnose that particular system accurately without damaging control modules or other components.

NOTE: When using special test equipment, the manufacturer's instructions provided with the tester should be read and clearly understood before attempting any test procedures.

Electrical Test Tools

ORGANIZED TROUBLESHOOTING

When diagnosing a specific problem, there are certain troubleshooting techniques that are standard:

1. Establish when the problem occurs. Does the problem appear only under certain conditions? Were there any noises, odors, or other unusual

A hydraulic floor jack and a set of jackstands are essential for lifting and supporting the vehicle

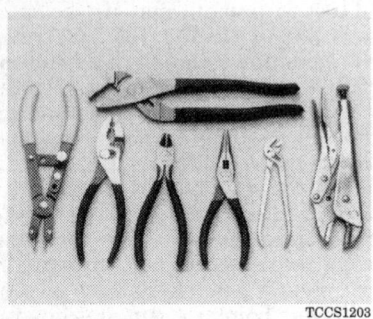

An assortment of pliers will be handy, especially for old rusted parts and stripped bolt heads

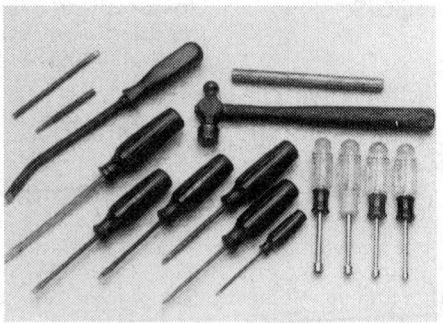

Various screwdrivers, a hammer, chisels and prybars are necessary to have in a toolbox

symptoms? Make notes on any symptoms found, including warning lights and trouble codes, if applicable.

2. Isolate the problem area. To do this, make some simple tests and observations; then eliminate the systems that are working properly. Check for obvious problems such as broken wires, split or disconnected vacuum hoses. Always check the obvious before assuming something complicated is the cause. Be suspicious of fuses, switches and connectors; wiring itself rarely fails.

3. Test for problems systematically to determine the cause once the problem area is isolated. Are all the components functioning properly? Is

there power going to electrical switches and motors? Is there vacuum at vacuum switches and/or actuators? Doing careful, systematic checks will often turn up most causes on the first inspection without wasting time checking components that have little or no relationship to the problem.

4. Test all repairs after the work is done to make sure that the problem is fixed. Some causes can be traced to more than 1 component, so a careful verification of repair work is important to pick up additional malfunctions that may cause a problem to reappear or a different problem to arise. A blown fuse, for example, is a simple problem that may require more than another fuse to repair.

The diagnostic tree charts are designed to help solve problems by leading the user through closely defined conditions and tests. Only the most likely components, vacuum and electrical circuits are checked for proper operation when troubleshooting a particular malfunction. By using the diagnostic trees to eliminate those systems and components which normally will not cause the condition described, a problem can be isolated within 1 or more systems or circuits without wasting time on unnecessary testing.

Experience has shown that most problems tend to be the result of a fairly simple and obvious cause, such as loose or corroded connectors; making careful inspection of components during testing is essential to quick and accurate troubleshooting. Frequent references to special test equipment will be found in the text and in the diagnosis charts. These devices or a compatible equivalent are necessary to perform some of the more complicated test procedures listed. Testers are available from a variety of aftermarket sources as well as from the vehicle manufacturer. Care should be taken that any test equipment being used is designed to diagnose that particular system accurately without damaging the computer control modules or components being tested.

NOTE: Pinpointing the exact cause of trouble in an electrical system can sometimes be accomplished only by the use of special test equipment. In addition to the information covered in this section, the manufacturer's instructions booklet provided with the tester should be read and clearly understood before attempting any test procedures.

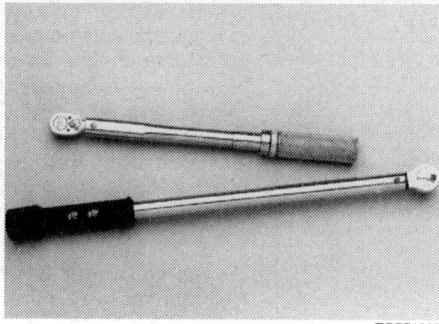

Many repairs will require the use of a torque wrench to assure the components are properly fastened

TCCS1205

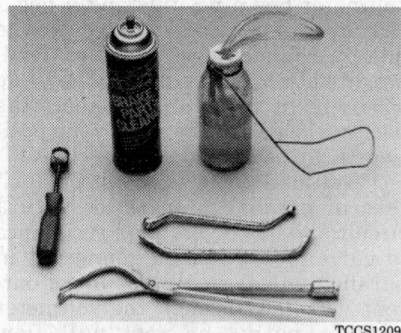

Although not always necessary, using specialized brake tools will save time

TCCS1209

Testers and Equipment

JUMPER WIRES

Jumper wires are simple, yet extremely valuable, pieces of test equipment. Jumper wires are merely wires that are used to bypass sections of a circuit. The simplest type of jumper wire is a length of multi-strand wire with an alligator clip at each end. Jumper wires are usually fabricated from lengths of standard automotive wire and whatever type of connector (alligator clip, spade connector or pin connector) is required for the vehicle being tested. Some jumper wires are made with 3 or more terminals coming from a common splice for special-purpose testing. In cramped, hard-to-reach areas it is advisable to have insulated boots over the jumper wire terminals in order to prevent accidental grounding and possible system damage.

Jumper wires are used primarily to locate open electrical circuits, on either the ground (-) side of the circuit or on the hot (+) side. If an electrical component fails to operate, connect the jumper wire between the component and a good ground. If the component operates only with the jumper installed, the ground circuit is open. If the ground circuit is good, but the component does not operate, the circuit between the power feed and component is open. Sometimes a fused jumper wire is connected directly from the battery to the hot terminal of the component, but first make sure the component uses 12 volts in operation.

By inserting an in-line fuse between a set of test leads, a fused jumper wire is created. A fused jumper wire can be used for bypassing open circuits. Use a 5 amp fuse to provide circuit protection.

NOTE: Never use jumpers made from wire that is of lighter gauge (smaller diameter) than used in the circuit under test. If the jumper wire is too small, it may overheat and possibly melt. Never use jumpers to bypass high-resistance loads (such as motors) in a circuit. Bypassing resistances, in effect, creates a short circuit which may cause damage and fire. Never use a jumper for anything other than temporary bypassing of components in a circuit, damage or fire could result.

TEST LIGHTS

12 Volt Test Light

The 12 volt test light is used to check circuits and components while electrical current is flowing through them. It is used for voltage and ground tests. Twelve volt test lights come in different styles, but all have 3 main parts; a ground clip, a probe and a light.

NOTE: Avoid piercing the insulation of any wire. While most probes are designed to pierce insulation, this can lead to corrosion or broken conductors within the wire. Trace the wire to a terminal that can be probed before piercing the insulation.

The most commonly used 12 volt test lights have pick-type probes. To use a 12 volt test light, connect the ground clip to a good ground and probe wherever necessary with the pick.

The wrap-around light is handy in hard to reach areas or where it is difficult to support a wire to push a probe pick into it. To use the wrap around light, hook the wire to be probed with the hook and pull the trigger. A small pick will be forced through the wire insulation into the wire core. Only use this type of test light as a last resort and do not use it on SRS or computer data lines.

NOTE: Never use a pick-type test light to probe wiring on computer controlled systems unless specifically instructed to do so. Any wire insulation that is pierced by the test light probe should be taped and sealed with silicone after testing.

The test light does not detect specific voltage amounts; it only detects

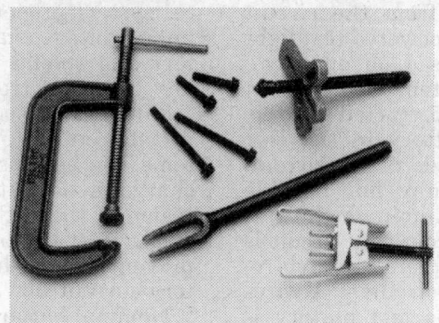

Various pullers, clamps and separator tools are needed for the repair of many components

TCCS1211

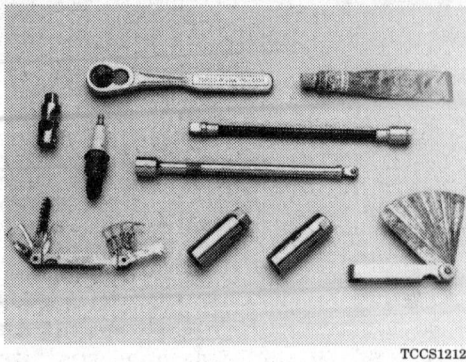

A variety of tools and gauges are needed for spark plug service

TCCS1212

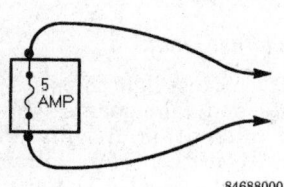

84688000

Schematic of a fused jumper wire

88513001

Jumper wires come in different gauges

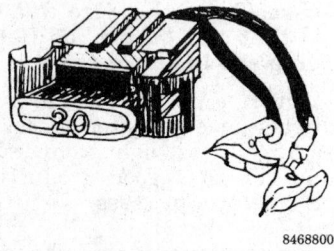

84688002

Fused jumper wire

that voltage is present. It is advisable before using the test light to touch its terminals across the battery posts to make sure the light is operating properly. Do not attempt to determine voltage by how brightly the tester glows; use a voltmeter if an exact reading is needed.

Use of a LED type test light is recommended for computer controlled circuits. A standard incandescent bulb test light can load the circuit causing a high current to flow and damage the components. An LED type test light will not load the circuit

and is safer to use in a computer controlled circuit.

Self-Powered Test Light

The self-powered test light usually contains a 1.5 volt penlight battery. One type is similar in design to the 12 volt test light. This type has both the battery and the light in the handle and pick-type probe tip. The second type has the light toward the open tip, so that the light illuminates the contact point. The self-powered test light is a dual-purpose piece of equipment. It can be used to test for either open or short circuits when power is isolated from the circuit (continuity test). A powered test light should never be used on any computer controlled system or component unless specifically instructed to do so.

The 1.5 volt battery in the test light does not provide much current. A weak battery may not provide enough power to illuminate the test light even when a complete circuit is made (especially if there are high resistances in the circuit). Always make sure that the test battery is strong. To check the battery, briefly touch the ground clip to the probe; if the light glows brightly, the battery

is strong enough for testing. Never use a self-powered test light to perform checks for opens or shorts when power is applied to the electrical system under test. The 12 volt vehicle power will quickly burn out the 1.5 volt light bulb in the test light.

VOLTMETER

A voltmeter is used to measure voltage at any point in a circuit, or to measure the voltage drop across any part of a circuit. Voltmeters usually have various scales on the meter dial and a selector switch to allow the selection of different test ranges. The voltmeter has a positive and a negative lead. To avoid damage to the meter, connect the negative lead, usually black, to the negative (-) side of circuit or to ground. Connect the positive lead, usually red, to the positive (+) or power side of the circuit.

A voltmeter can be connected either in parallel or in series with a circuit and has a very high resistance to current flow. When connected in parallel, only a small amount of current will flow through the voltmeter current path; the rest will flow through the normal current path and the circuit will work normally. When the voltmeter is connected in series with a circuit, only a small amount of current can flow through the circuit. The circuit will not work properly, but the voltmeter reading will show if the circuit is complete or not.

Available Voltage Measurement

Set the voltmeter selector switch to the 20V position and connect the meter negative lead to the negative post of the battery and connect the positive meter lead to the positive post of the battery. Read the voltage on the meter or digital display. A well-charged battery should register over 12 volts. If the meter reads below 11.5 volts, the battery power may be insufficient to operate the electrical system properly. This test determines voltage available from the battery and should be the first step in any electrical trouble diagnosis procedure. Many electrical problems, especially on computer controlled systems, can be caused by a low state of charge in the battery. Excessive corrosion at the battery cable terminals can cause a poor contact that will prevent proper charging and full battery current flow.

Nominal battery voltage is 12 volts but, when fully charged, should be about 13.2 volts. When the battery is supplying current to 1 or more cir-

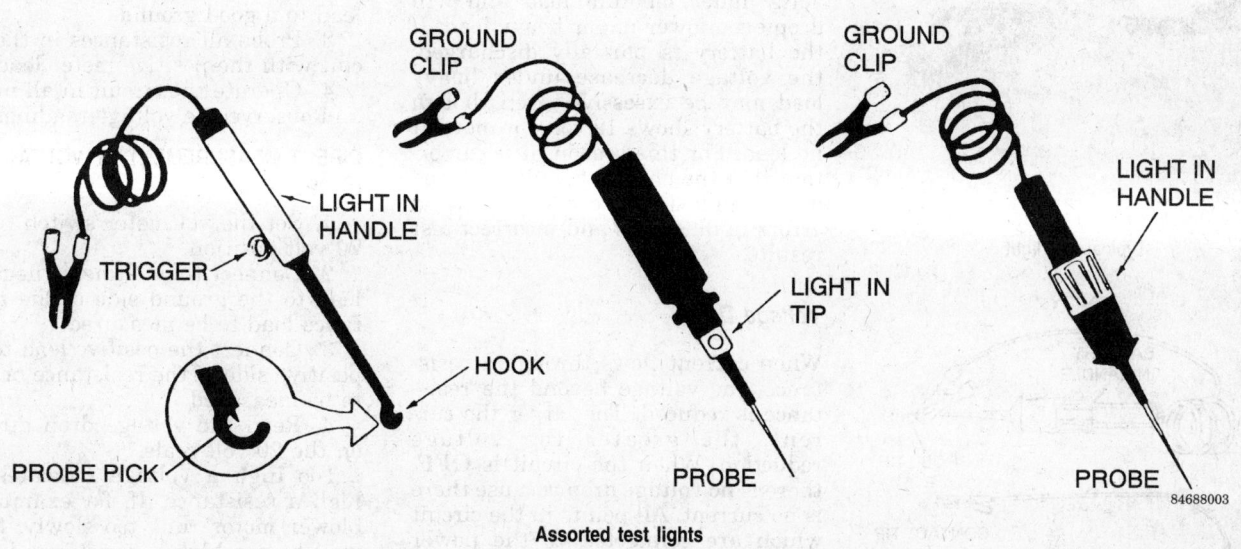

Assorted test lights

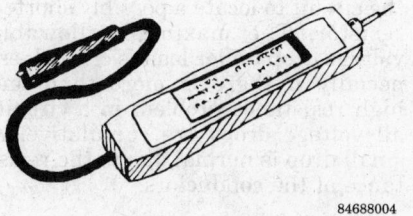

Logic probe type tester

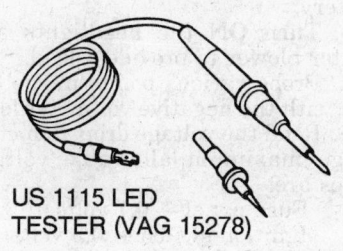

US 1115 LED TESTER (VAG 15278)

LED type test light for use on computer circuits

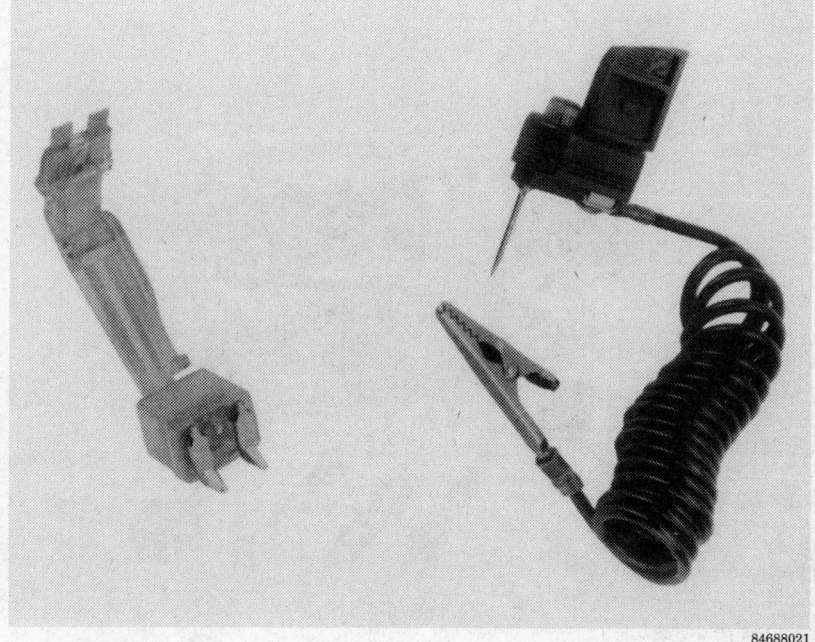

The device on the left is a fuse checker and the test light on the right is an LED type for use on computer circuits

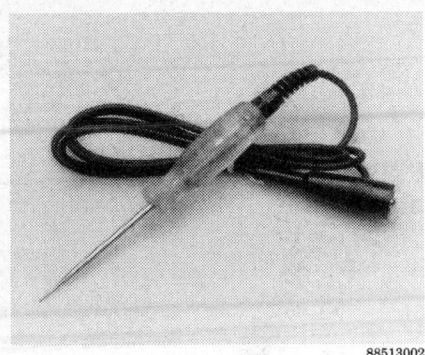

Typical test light

88513002

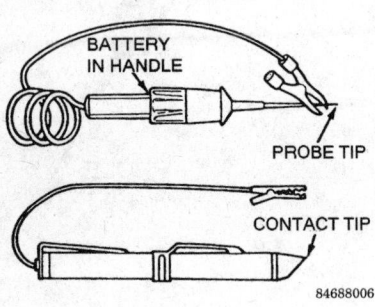

BATTERY IN HANDLE

PROBE TIP

CONTACT TIP

84688006

Types of self-powered test lights

cuits it is said to be under load. When everything is **OFF** the electrical system is under a no-load condition. A fully charged battery showing about 12.5 volts at no load may drop to 12 volts under medium load and will drop even lower under heavy load. If the battery is partially discharged, the voltage decrease under heavy load may be excessive, even though the battery shows 12 volts or more at no load. For this reason, it is important that the battery be fully charged during all testing procedures to avoid errors in diagnosis and incorrect test results.

Voltage Drop

When current flows through a resistance, the voltage beyond the resistance is reduced. The larger the current, the greater the voltage reduction. When the circuit is **OFF**, there is no voltage drop because there is no current. All points in the circuit which are connected to the power source are at the same voltage as the power source. In a long circuit with many connectors, a series of small, unwanted voltage drops due to corrosion at the connectors can add up to a total loss of voltage which impairs the operation of the loads in the circuit.

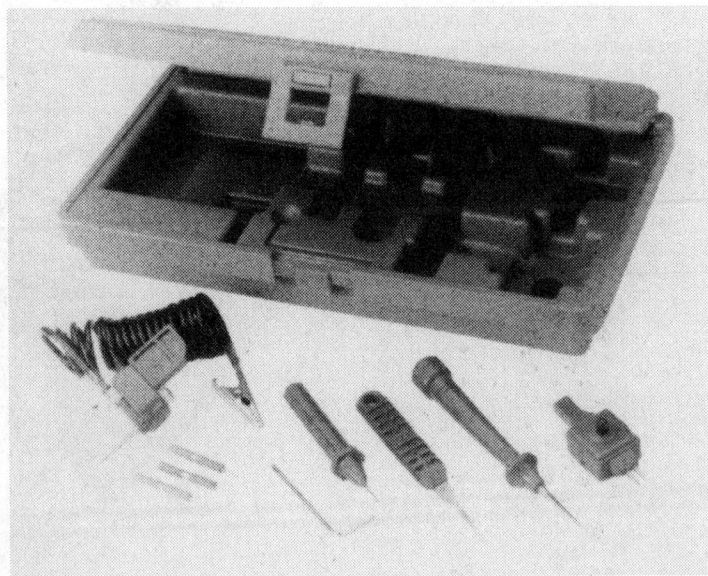

85438021

This computer circuit testing kit includes LED test lights that are safe for use on electronic circuits

INDIRECT COMPUTATION OF VOLTAGE DROPS

1. Set the voltmeter selector switch to the 20 volt position.
2. Connect the meter negative lead to a good ground.
3. Probe all resistances in the circuit with the positive meter lead.
4. Operate the circuit in all modes and observe the voltage readings.

DIRECT MEASUREMENT OF VOLTAGE DROPS

1. Set the voltmeter switch to the 20 volt position.
2. Connect the voltmeter negative lead to the ground side of the resistance load to be measured.
3. Connect the positive lead to the positive side of the resistance or load to be measured.
4. Read the voltage drop directly on the 20 volt scale.

Too high a voltage indicates too high a resistance. If, for example, a blower motor runs too slowly, there may be too high a resistance in the resistor pack. By taking voltage drop readings in all parts of the circuit, the problem can be isolated. Too low a voltage drop indicates too low a resistance. If, for example, a blower motor runs too fast in the **MED** and/or **LOW** position, the problem can be isolated to the resistor pack by taking voltage drop readings in all parts of the circuit to locate a possibly shorted resistor. The maximum allowable voltage drop under load is critical, especially if there is more than one high resistance problem in a circuit; all voltage drops are cumulative. A small drop is normal due to the resistance of the conductors.

High Resistance Testing

1. Set the voltmeter selector switch to the 2 volt position.
2. Connect the voltmeter positive lead to the positive post of the battery.
3. Turn **ON** the headlights and heater blower to provide a load.
4. Probe various points in the circuit with the negative voltmeter lead.
5. Read the voltage drop. Some average maximum allowable voltage drops are:

> Fuse panel — 0.7 volts
> Ignition switch — 0.5 volts
> Headlight switch — 0.7 volts
> Ignition coil (+) — 0.5 volts
> Any other load — 0.5–1.3 volts

NOTE: Voltage drops are all measured while a load is operating; without current flow, there will be no voltage drop.

OHMMETER

The ohmmeter is designed to read resistance (ohms) in a circuit or component. Although there are several different styles of ohmmeters, all will usually have a selector switch which permits the measurement of different ranges of resistance. Usually the selector switch allows the multiplication of the meter reading by 10, 100, 1000, 10,000, etc. A calibration knob allows the meter to be set at zero for accurate measurement. Since all ohmmeters are powered by an internal battery (usually 9 volts), the ohmmeter can be used as a self-powered test light. When the ohmmeter is connected, current from the ohmmeter flows through the circuit or component being tested. Since the ohmmeter's internal resistance and voltage are known values, the amount of current flow through the meter depends on the resistance of the circuit or component being tested.

The ohmmeter can be used to perform continuity tests for opens or shorts and to read actual resistance in a circuit. It should be noted that the ohmmeter is used to check the resistance of a component or wire while there is no voltage applied to the circuit. Current flow from an outside voltage source (such as the vehicle battery) can damage the ohmmeter, so the circuit or component should be isolated from the vehicle electrical system before any testing is done. Since the ohmmeter uses its own voltage source, either lead can be connected to any test point.

NOTE: When checking diodes or other solid state components, the ohmmeter leads can only be connected one way in order to measure current flow in a single direction. Make sure the positive (+) and negative (-) terminal connections are as described in the test procedures to verify the one-way diode operation.

When using the meter for continuity checks, do not be concerned with the actual resistance readings. Zero resistance, or any resistance reading, indicates continuity in the circuit. Infinite resistance indicates an open in the circuit. A high resistance reading where there should be none indicates a problem in the circuit. Checks for short circuits are made in the same manner as checks for open circuits except that the circuit must be isolated from both power and normal ground. Infinite resistance indicates no continuity to

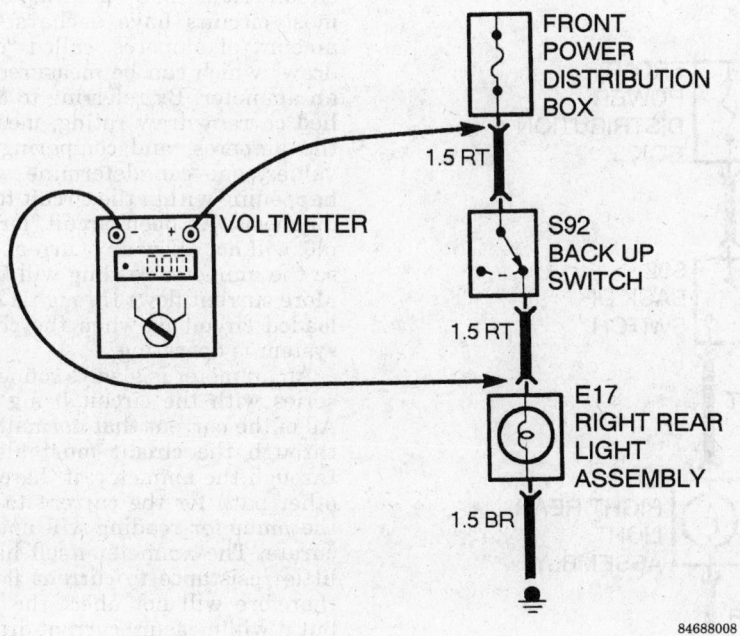

Measuring voltage at different points in the circuit

84688007

Checking for the voltage drop across a component in the circuit

84688008

ground, while zero resistance indicates a dead short to ground.

Resistance Measurement

The batteries in an ohmmeter may be affected by temperature and will weaken with age. The ohmmeter must be calibrated or "zeroed" before taking measurements. To zero the meter, place the selector switch in its lowest range and touch the 2 leads together. Turn the calibration knob

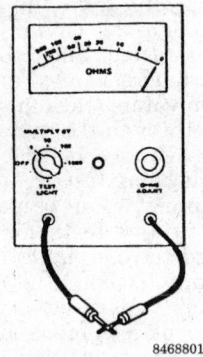

84688010

Zeroing the ohmmeter before using it

until the meter needle is exactly on zero.

NOTE: All analog (needle) type ohmmeters must be zeroed before use, but some digital ohmmeter models are automatically calibrated when the switch is turned ON. Self-calibrating digital ohmmeters do not have an adjusting knob, but it's a good idea to check for a zero readout before use by touching the leads together. All computer controlled systems require the use of a digital ohmmeter with at least 10 megohms impedance for testing. Before any test procedures are attempted, make sure the ohmmeter used is compatible with the electrical system, or damage to the on-board computer could result.

To measure resistance, first isolate the circuit from the vehicle power source by disconnecting the battery cables or the harness connector. Make sure the key is **OFF** when disconnecting any components or the battery. Where necessary, also isolate at least one side of the circuit to be checked to avoid reading parallel resistances. Parallel circuit resistances will always give a lower reading than the actual resistance of either of the branches. When measuring the resistance of parallel circuits, the total resistance will always be lower than the smallest resistance in the circuit. Connect the meter leads to both sides of the circuit (wire or component) and read the actual measured ohms on the meter scale. Make sure the selector switch is set to the proper ohm scale for the circuit being tested to avoid misreading the ohmmeter test value.

NOTE: Never use an ohmmeter with power applied to the circuit. Like the self-powered test light, the ohmmeter is designed to operate on its own power supply. The normal 12 volt automotive system could damage the meter.

AMMETERS

An ammeter measures the amount of current flowing through a circuit in units called amperes or amps. Amperes are units of electron flow which indicate how fast the electrons are flowing through the circuit. Since Ohm's Law dictates that current flow in a circuit is equal to the circuit voltage divided by the total circuit resistance, increasing voltage also increases the current level (amps). Likewise, any decrease in resistance will increase the amount of amps in a circuit. At normal operating voltage, most circuits have a characteristic amount of amperes, called "current draw" which can be measured using an ammeter. By referring to a specified current draw rating, measuring the amperes, and comparing the 2 values, one can determine what is happening within the circuit to aid in diagnosis. An open circuit, for example, will not allow any current to flow so the ammeter reading will be zero. More current flows through a heavily loaded circuit or when the charging system is operating.

An ammeter is always connected in series with the circuit being tested. All of the current that normally flows through the circuit must also flow through the ammeter; if there is any other path for the current to follow, the ammeter reading will not be accurate. The ammeter itself has very little resistance to current flow and therefore will not affect the circuit, but it will measure current draw only when the circuit is closed and electricity is flowing. Excessive current draw can blow fuses and/or drain the battery; a reduced current draw can cause motors to run slowly, lights to dim and other components to operate improperly.

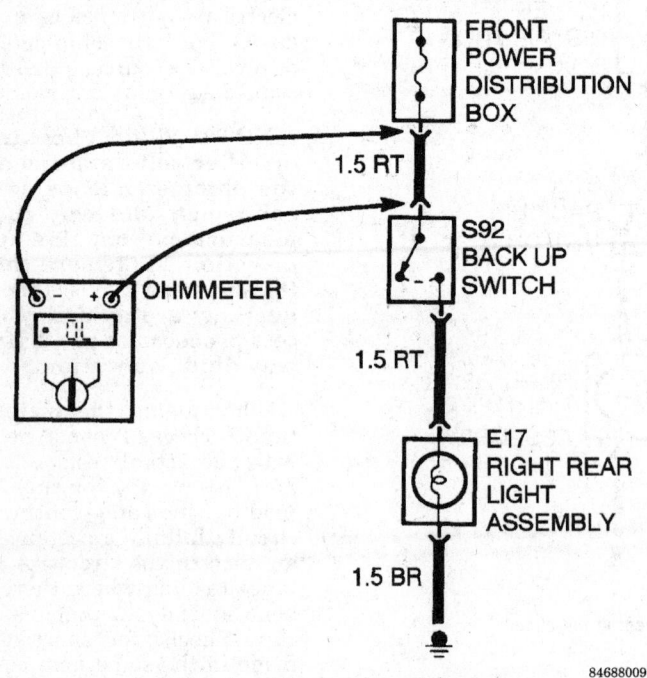

84688009

Using an ohmmeter to do a continuity test

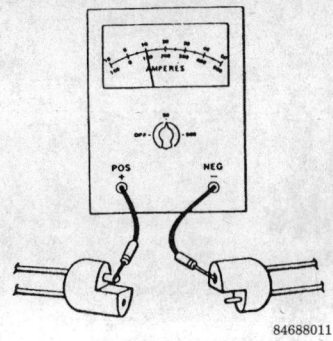

The ammeter is placed in line with
the circuit to be tested

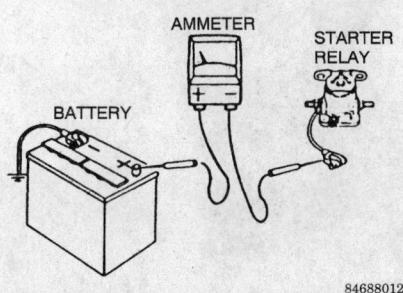

Checking the draw of the starter relay with an
ammeter

Different styles of multimeters allow a
choice of meter functions

DIGITAL VOLT-OHM METER (DVOM)

As its name implies, this tool combines a voltmeter and an ohmmeter into a single unit that has a digital display instead of a scale and pointer. The major advantage of a fully electronic meter is that there are no moving parts that require power to operate. Analog meters with an ultra light weight needle still require some power to move the needle. This limits the range and the features that can be built into the meter. Even the most basic DVOM can read a much greater range of voltage and resistance without imposing any load on the circuit being tested. It is usually the only equipment suitable for testing computer controlled circuits and is often the only test equipment needed.

Several additional features can be built into the same unit, such as circuitry for testing diodes and measuring AC voltage, AC and DC current, temperature, duty cycle, frequency, pulse width, dwell, and rpm. Some of the more sophisticated units also have storage capability, bar graph display, automatic shut-off, and can display the difference between two

readings. A top-of-the-line DVOM designed for automotive testing is probably the most useful and cost effective diagnostic tool available. Be sure to buy a unit with a high impedance, usually 10 megohms or higher.

Specialty Testers

FREQUENCY PROCESSOR

Some older DVOMs are not equipped to read frequency. There is at least one unit on the market that converts frequency signals to a millivolt signal that any DVOM can read. It is a simple box with input and output jacks and a "wake-up" circuit that automatically turns the unit on when needed. Its range of 10–5000 Hz makes it useful for checking rpm sensors, mass air flow sensors, Hall effect sensors and more. Instructions provided with the processor show how to interpret the readings.

BREAK-OUT BOX

The electronic Break-Out Box (BOB) is used to tap into the wiring of a control unit. The main connector to the electronic control unit is connected to the break-out box and an-

other wire harness is connected from the box to the control unit. The break-out box then allows the technician to access each circuit while it is operating without piercing the wire or causing damage to the connectors. All testing with the DVOM can be done safely at these terminals, eliminating the risk of damage due to backprobing at the control unit. Many times a break-out box is the only way to test a control unit function.

An Intelligent Break-Out Box (IBOB) connects to the vehicle diagnostic connector and has connector ports for a scan tool and/or a computer. On earlier electronic control units that do not generate a data stream, an IBOB will collect input/output data while the engine is running and present it to a scan tool or PC. Additionally, some manufacturers provide plastic overlays for the break-out box. This allows the box to be used on a variety of models; different overlays identify the changes in wire use or labeling. With the proper cable adaptors, an IBOB can be used with any engine, body or ABS control unit on any vehicle.

OSCILLOSCOPE

An oscilloscope is a voltmeter that presents a graphic picture of the voltage reading over time. Unlike a DVOM, it can show a voltage that exists for only a fraction of a second or occurs only at a specific time. Ignition oscilloscopes have been around for many years, but the latest generation of service bay oscilloscopes are more like those found in electronics labs. They can read voltages as small as one millivolt and can show a spike that occurs for as little as 10 nanoseconds (1 ns = of a second). Both the voltage and time scales are adjustable, so the same tool can be used to measure the fast, high voltage signal of the secondary ignition system and slow stable signals such as a temperature sensor. Another major feature of all oscilloscopes is an extremely high input impedance, meaning the oscilloscope imposes negligible current draw on the circuit being measured that might influence that measurement. Many times an oscilloscope is the only tool that can be used to measure low voltage, frequency, or duty cycle signals.

Like a timing light, an oscilloscope must be triggered. The trigger can be internal (automatic) or can come from an external source. On a multichannel oscilloscope, displaying the

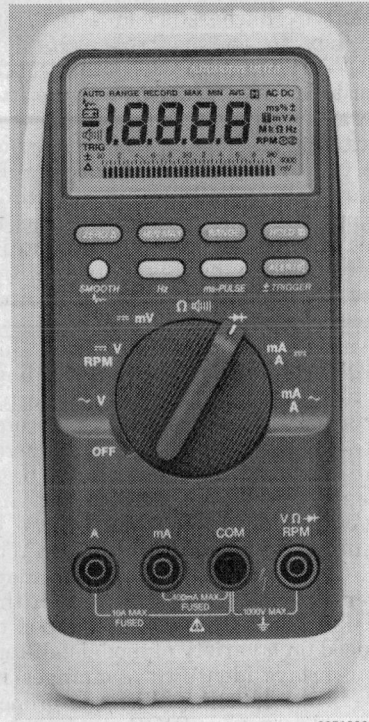

88513006

A good quality DVOM designed for automotive testing is the most useful diagnostic tool available

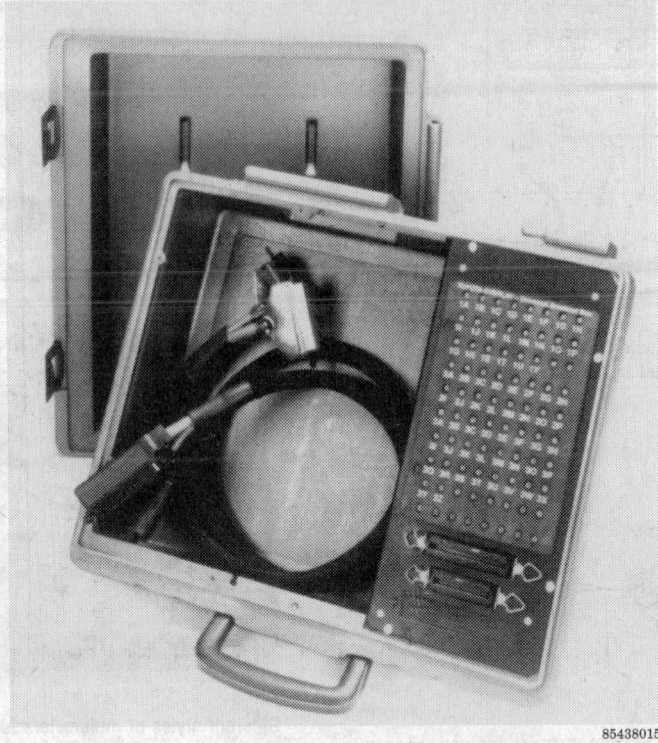

85438015

A break-out box makes it possible to tee into control unit circuits

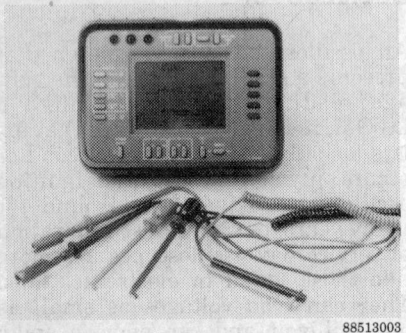

88513003

An oscilloscope shown with related testing probes

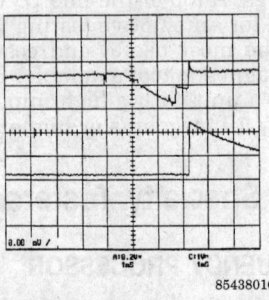

85438016

The oxygen sensor trace (top) shows a delayed cross-over coinciding with an injector pulse (bottom trace)

external trigger signal can show the timing of two events. For example, by taking the trigger from a suspected faulty fuel injector, it is possible to see the oxygen sensor signal only at the time of that injection event. The voltage level required to trigger the oscilloscope can also be adjusted, providing a simple method to look for low level or intermittent faults that may not set a code.

A digital oscilloscope converts the analog input signal to a digital form. A digital signal can be stored and played back by itself or along with another trace. Some units can also display the signal as numbers,

min/max values, change value and average value. If the oscilloscope is equipped with a computer port, the digitized traces and other data can also be down-loaded to save and/or print out. There is computer software available to aid organization and analysis of wave forms.

With its extremely fast sampling rate and graphic display, an oscilloscope can easily show a malfunction that occurs too fast for a voltmeter to show. For example, by adjusting the time sweep of the oscilloscope to show the full up-and-down stroke of a throttle position sensor, an intermit-

tent fault in the signal can be clearly shown. A voltmeter may also detect the fault but cannot change the display fast enough to show the resistance spike. A storage oscilloscope with min/max value capability can locate intermittent faults that even other oscilloscopes cannot.

Any oscilloscope used for automotive testing must be designed for use with automobiles. Standard lab oscilloscopes are usually not able to cope with the relatively harsh automotive electronic environment. Automotive oscilloscopes are available in a variety of types with a variety of features. Some are portable hand held models that operate on batteries or vehicle power. Even though they are small, the newest portable units include multi-trace and storage capabilities and are rugged enough to be used under the hood or on road tests. Larger more powerful models mounted in a console are often part of a top-of-the-line engine analyzer package. Most of the major diagnostic tool manufacturers produce at least one oscilloscope model.

As vehicles become more sophisticated and electronic controls become more powerful, an oscilloscope is fast becoming a necessary diagnostic tool. When the technician becomes proficient with an oscilloscope, many

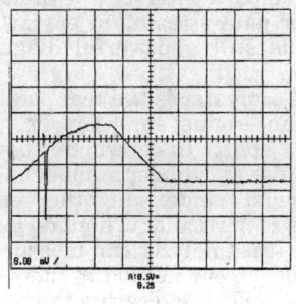

85438017

An intermittent fault in the throttle position sensor shows clearly on the scope trace

other diagnostic tools become unnecessary.

SCAN TOOL

This is the generic name for portable diagnostic equipment that communicates directly with an electronic control unit. The major vehicle manufacturers each have their own scan tool that is used by dealership technicians, such as Nissan's Consult and Honda's PGM Tester. Some of these are available through the dealer parts network or are sold outside the network under another name. Others such as Volkswagen's VAG 1551 are available only to authorized dealerships.

Scan tools are used to read and erase trouble codes stored in the control unit memory and to provide a direct data transfer link with the control unit's On Board Diagnostic (OBD) system. Reading the control unit memory through the scan tool is more complete than reading codes with the flashing light on the instrument panel. Some information is only available through the scan tool, such as the number of engine starts since the fault first appeared. Data transfer provides a real time display of control unit input/output signals.

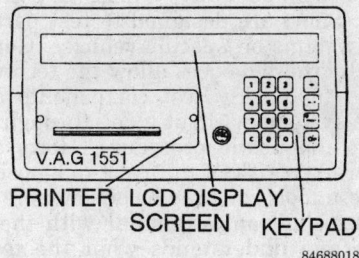

V.A.G 1551

PRINTER LCD DISPLAY
SCREEN   KEYPAD

84688018

Scan tools allow the testing of electronic control units

88513007

Aftermarket scan tool with a program module installed

Data such as the oxygen sensor reading or idle control motor duty cycle can be displayed while the engine is running. The scan tool can also be used as a volt/ohmmeter to check selected circuits without disconnecting them.

Some scan tools are designed to simulate sensor inputs to test the sensor circuit, the control unit and the output device. On many vehicles, the scan tool can communicate with every control unit on the vehicle through a single diagnostic connector. Some of the more advanced scan tools are equipped with a data memory to store test data during a road test. The test data is down loaded into a PC through an RS232 computer port, greatly increasing the processing power and expanding the amount of memory and information available.

Aftermarket scan tools require adaptors to match the different vehicle diagnostic connectors. A cartridge that plugs into the scan tool contains software needed for communicating with the different control units. The software is the tool's real power and is continuously evolving to enhance its capabilities. As the tool manufacturer's data base has grown, software now includes VIN-specific information that addresses some of the most

common trouble codes and driveability problems that don't always generate codes. Tests are menu-driven and many of the specifications are included right there in the program. Depending on the vehicle and the amount of computer control used, systems which may be viewed or investigated with a scan tool include:
1. Engine controllers/ECUs
2. Fuel/ignition systems
3. Electronic transmission control
4. Charging system
5. Suspension control functions
6. Anti-Lock brake system
7. Passive restraint system
8. Anti-theft system
9. Climate Control Systems

Body electrical systems including power locks, entertainment systems, sunroofs, and defoggers.

Aftermarket scan tools work well with most vehicles, but no scan tool can be used on all models and all are limited in their ability to communicate with European models. The Federal On-Board Diagnostic (OBD) specification requires all vehicles to have the same diagnostic connector and diagnostic trouble codes and to use the same data transfer language. This makes it possible to use a single scan tool to communicate with all engine control units from all manufacturers. Some manufacturers began production of OBD vehicles for the 1994 model year and all new vehicles must comply by the 1998 model year. As vehicle control units and scan tools become more powerful, data acquisition and test capabilities will also improve dramatically with each new generation of control unit and scan tool software.

EXHAUST GAS ANALYZERS

Exhaust gas analysis has long been an extremely valuable and versatile diagnostic tool. It can be used to troubleshoot fuel and ignition systems, locate vacuum leaks or EGR malfunctions, even diagnose mechanical problems such as worn valve guides. On most vehicles, it is the only way to accurately check air/fuel mixture.

The federal government regulates three exhaust gas components: hydrocarbons (HC), carbon monoxide (CO), and oxides of nitrogen (NOx). HC and CO are relatively easy to measure and have long been tested in states that require emissions inspections. Measuring carbon dioxide (CO_2) and oxygen (O_2) are also valuable diagnostic aids but NOx cannot be accurately measured at idle or no-

85438019

This type scan tool, used by dealer technicians, is available outside the dealer network and can scan all OBD II-spec vehicles. This one includes a built-in oscilloscope

load conditions. However as states enact tighter inspection and maintenance programs, service bay NOx analyzers and test procedures are being developed. For complete diagnostic and certification testing, a five-gas analyzer is required.

Most gas analyzers include a single tailpipe sample probe, sample pump and filtration system and a detector cell for each of the gasses being measured. Most stand-alone four-gas analyzers used for emissions inspections are equipped with a small microprocessor that includes testing and calibration programs, a self diagnostic program and a built-in printer. Other units are designed as part of a complete diagnostic station and are connected to a PC based computer, printer and monitor screen. There is at least one portable four-gas analyzer that is used along with the same manufacturer's scan tool. The scan tool's software guides the user through test procedures based on the gas and sensor readings.

There is also a series of small, hand-held oxygen and CO monitors available that do not require a sample pump and filter system. The measuring cell is built into the tailpipe probe and the monitor is battery operated. Since there is only a wire between the probe and the monitor,

these units can easily be used on a road test. They are also equipped with a memory that can record three minutes of test data. The CO monitor can be particularly useful for routine air/fuel adjustments. Each unit is available with its own display, with voltage outputs to use a DVOM display, or with computer ports and PC based software that includes test procedures.

All gas analyzers must be calibrated at least once per day. Industry standard calibration gasses are usually available through parts stores or tool outlets.

ENGINE ANALYZERS

A large, fully-equipped engine analyzer usually includes an oscilloscope, exhaust gas analyzer, vacuum and pressure sensors, a timing light and probe, electrical measuring equipment, and a computer. With a variety of electrical connections and a tail pipe probe, this analyzer can check the primary and secondary ignition systems, fuel injection controls and injectors, EGR systems, engine vacuum and compression, and the starting and charging systems. The computer can be used to read the engine control unit's data stream through the vehicle diagnostic connector.

Even on earlier vehicles with no sensors or data stream, an engine analyzer is still a powerful diagnostic tool.

The computer is the real power behind an engine analyzer. The computer's ability to determine the ignition pulse at cylinder number one can be used to index all other engine events to particular cylinders. For example, the analyzer can measure the starter current needed to move each piston to TDC, indicating the relative compression of each cylinder. The analyzer can also display spark plug firing voltage and duration on the oscilloscope. A spark plug that requires more voltage with less duration could indicate a faulty injector. The analyzer can detect and clearly show all the differences between cylinders. The computer can help the technician diagnose the data, determine the necessary repairs and even provide a print-out to present a clear explanation to the vehicle owner.

These analyzers are a major investment and are well supported by the manufacturer. They are frequently updated with a computer floppy disc that includes new vehicle information and test procedures. Some machines include CD-ROM equipment to read service manuals that are available on disc. They may also include a modem to communicate with the manufacturer or other computers via telephone. As vehicles and other shop equipment become more sophisticated, it should be possible to keep a computer based engine analyzer up to date and useful almost indefinitely.

Specific Test Equipment

There are many special diagnostic tools for testing individual components or systems, such as a Hall effect sensor, idle air control motor, fuel injectors, secondary ignition systems, and others. Most are designed for use on as many vehicles as possible. Some are designed to test parts or systems on specific vehicles. Generally these devices allow the technician to quickly test components or sub-systems without going through a long diagnostic procedure. However there is a risk of incorrect diagnosis. These tools can only be dependable if the technician is familiar with their use and understands what the test results really mean. A simple vacuum leak or loose connection may produce the same test result as a faulty component.

LEAK DETECTORS

A battery powered, hand-held vacuum leak detector uses a microphone and amplifier that detects noise in the ultrasonic range. Air moving through a vacuum leak will generate sound waves in the 40 kHz range, well above the range of human hearing. The detector will sound a beeper when a leak is found. Because of the high frequency sensed by the detector, it is not generally affected by normal engine or shop noises.

Leak detectors for air conditioning systems have a vacuum pump and probe to draw an air sample into the detecting cell. The cell detects halogen gas that is common to all air conditioning refrigerants. Most are capable of indicating the type of refrigerant in the system, as well as the rate of leakage. There are battery powered hand-held models and larger AC powered units suitable for mounting on an air conditioning service cart. The newest models are capable of detecting R134a and the sensitivity can be adjusted for possible background interference.

A combustible gas leak detector reacts to hydrocarbons present in fuels, exhaust gases, coolants, and lubricants. Models with adjustable sensitivity are typically used to look for fuel vapor leaks, head gasket leaks, and to measure the amount of exhaust leaking into the interior of a vehicle. With some imagination, this can be an extremely useful tool.

PYROMETER

A pyrometer measures a wide range of temperatures with a probe that only needs to touch the item being measured. As a general diagnostic tool, a hand held pyrometer can quickly locate hot or cold spots in a cooling system, a seized brake caliper, a dry bearing, test heater and A/C performance or even find a weak cylinder by measuring exhaust manifold runner temperatures. Most pyrometers are available with special probes for penetration and for measuring tire temperatures. There are even optical infrared non-contact pyrometers that measure temperature by the heat emission of a surface. This is useful as the surface to be measured does not have to actually touched with a probe. They can be calibrated quickly, have a very wide temperature range and can usually be switched to display Fahrenheit or Celsius degrees. With a little imagination, this can be an extremely useful tool.

IDLE AIR CONTROL TESTER

This is a kit used to isolate and test idle air control solenoids, motors, and signals. Some are made for use with a specific system, others include adaptors for use with many different vehicles. The device can activate solenoid valves and control motors to test the full range of motion with the engine not running. It can also be used to control idle speed for timing adjustment or other engine tests. Some can also check the control unit output signal to the idle air control motor. These functions can also be accomplished with scan tools but this tester can be faster and easier to use for some tests.

FUEL INJECTOR TESTER

This device can quickly check the coil resistance and current draw of an electric fuel injector while it is under load. Each injector is tested individually and the results are reported on a DVOM or oscilloscope. This information makes it possible to electronically check injector balance and detect intermittent faults. When used with equipment that measures fuel pressure and injection quantity, every function of the fuel system can be tested.

OXYGEN SENSOR TESTER

This kit usually includes a propane enrichment control valve, special connectors and test instructions, and the hose and fittings needed for connecting the valve to an intake manifold. The kit allows the technician to control air/fuel mixture and check the oxygen sensor response time. When the oxygen sensor is disconnected, forcing the control unit into open loop, sensor output voltage or resistance can be read with a DVOM. The instructions also include procedures for testing the control unit's response to the oxygen sensor signal.

SENSOR SIMULATOR

This device is used to take a sensor "out of the loop" and simulate its input signals to the control unit. It can simulate every type of voltage, resistance, and frequency signal one at a time to test the control unit's response to the input. The simulator can also measure any sensor output signal by back-probing the sensor connector. In addition to displaying the reading directly, some units can also output the reading to an oscilloscope, scan tool, or other diagnostic equipment.

POWER STEERING TESTER

A power steering tester can quickly confirm that the hydraulic system is functioning properly and indicate excess loads on the system due to mechanical malfunction in the suspension or steering linkage. The mechanical tester consists of a gauge, a heavy duty hose, and adapters for various models. Newer testers use an electronic pressure transducer that converts the pressure to an electrical signal. This allows the technician to road test variable effort power steering systems with a DVOM.

ELECTRONIC SIGHT GLASS

This device, used for troubleshooting air conditioning systems, includes two transducers and a battery operated meter. The transducers attach to the outside of the metal air conditioning lines without disconnecting the lines. While the A/C system is operating, the device ultrasonically detects bubbles in the refrigerant and uses an LED display to simulate a

88513008

An ultrasonic vacuum leak detector is unaffected by engine or shop noise

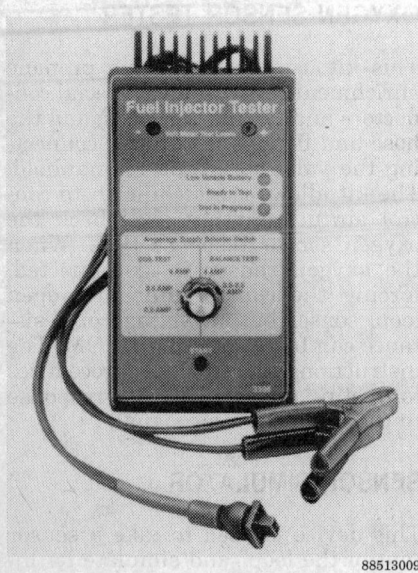

88513009

A fuel injector tester measures voltage drop while the injector is being activated

85438023

A power steering system testing gauge can quickly isolate hydraulic or linkage problems

sight glass, allowing the technician to actually "see" the bubbles in the system. The transducers can be fitted to any metal line at almost any point in the system, making it possible to test expansion valves and capillary tubes, or find other undesired restrictions.

ANTI-LOCK BRAKE TESTER

Anti-lock brake system control units are equipped with a self-diagnostic program that checks the system at engine start-up and de-activates the system if a fault is detected. Most control units also store diagnostic

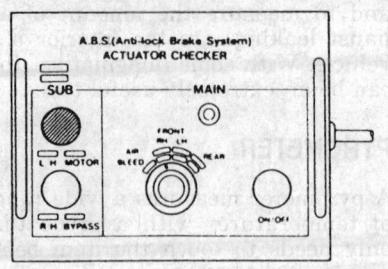

84688019

An ABS checker allows the testing of the anti-lock system

trouble codes. An ABS tester is basically a scan tool used to read and erase trouble codes and to provide a data link with the ABS control unit. The data link allows the tester to check sensor inputs and activate each solenoid and control valve to test the output system.

The testers for some of the early anti-lock braking systems are usually available only to dealers and function only with that manufacturer's vehicles. The aftermarket scan tool makers have developed the necessary adapters and software cartridges so that most scan tools used to communicate with engine control units can also be used on ABS control units. Unfortunately there are major differences in ABS designs and no scan tool is able to communicate with all ABS control units.

The most complete ABS test equipment is a software package, connector and interface pod that establishes a data link between the control unit and a PC-based computer in an engine analyzer or alignment station. This software uses the power of the computer to completely check the control unit and "bench test" every component of the system in about one minute. It is suitable for use with every anti-lock braking system on the market and the software can be updated as required to keep the system current.

Even without a scan tool, all of the system's sensors, solenoids, and actuators can still be individually checked with a DVOM and an oscilloscope. The trouble codes and other control unit functions can only be accessed with the correct scan tool or tester.

PASSIVE RESTRAINT TESTER

While there are some differences, the air bag system in most vehicles operates basically the same way. This has made it possible to develop a unit

that will test the circuits while they are fully connected and operating. It uses LEDs to read out circuit continuity, power supply, switch state and output device state. On some scan tools, such as the Nissan Consult, a software cartridge gives the tool the ability to test the passive restraint system.

—— CAUTION ——

If not familiar with disarming procedures and air bag system operation, do not attempt air bag system service. The air bag system must be disabled for some vehicle tests and before removing the steering wheel or dashboard air bag module. Failure to follow air bag safety procedures could result in accidental deployment and serious or fatal injury.

The air bag must function in an extremely short period of time after the crash sensor switches have closed. Most air bag systems include their own power supply. The squib that fires the gas generator charge must operate at low power levels to assure it will still fire if vehicle power is lost in the accident. This makes an air bag module quite sensitive to even small currents like static electricity. Even with such a "hair trigger", it is not difficult to test the system or handle an air bag module safely.

Air Bag Service Precautions

—— CAUTION ——

An air bag is an explosive device. Before beginning any air bag system service, disconnect and isolate the negative battery cable and backup power supply. Follow the procedures for disarming the air bag system exactly. Do not use any type of computer memory saver device. Do not use any self powered test equipment or test lights until the system is properly disabled. Failure to follow these precautions could result in accidental deployment of the air bag and possible serious or fatal injury.

- Disconnect and isolate the battery cable and any backup power supply when servicing the air bag system.
- Do not use a memory saver.
- When re-activating an air bag system, connect the power last and make sure no one is in the vehicle.
- Do not attempt to measure the resistance across the air bag module connectors with any type of ohmme-

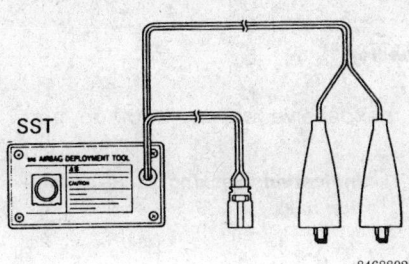

SST

84688020

An air bag deployment tool used when disposing of an air bag

ter. The ohmmeter battery can fire the air bag module.

• When working on air bag components in the passenger compartment, try to work away from where an air bag would deploy. Accidental deployment of the air bag against a body that is not in the proper position could result in severe or fatal injury.

• A removed air bag module must be placed away from sources of heat, sparks, or electricity, including static electricity.

• A removed air bag module must be placed with the cover pad facing up, so that accidental deployment will not launch the module into the air. Also be sure to carry the module with the cover pad facing away from the body.

• When the air bag module is removed, place it away from loose objects that would be thrown in the event of accidental deployment.

• When removing a steering column with the steering wheel attached, pay attention to where the air bag is aimed. Never stand the column on the face of the steering wheel. Lock the column to avoid damage to the clockspring.

• When handling an air bag which has been deployed, there may be a powdery residue which, though mostly talc, may irritate skin, eyes, and breathing passages. Wear gloves, glasses, and a dust mask while wrap-

ping the deployed air bag in a plastic bag for disposal.

• Sensor positioning is critical for proper system operation. If the sensor is in an area of vehicle damage, replace the sensor whether or not the air bag deployed. The proper torquing of the sensor retainers is critical.

• Any part of the air bag system found to be faulty must be replaced. No part can be repaired, not even the wiring.

• All diagnostic work is to be done with the air bag module(s) removed. Air bag simulators are available and can be installed for a full functional test of the control unit.

Mechanical Test Equipment

VACUUM GAUGE

Intake manifold vacuum is used to operate various systems and devices on all cars. To correctly diagnose and solve problems in vacuum control systems, a vacuum source is necessary for testing. In some cases, vacuum can be taken from the intake manifold when the engine is running, but vacuum is normally provided by a hand vacuum pump.

Most gauges are graduated in inches of Mercury (in. Hg), although a device called a manometer reads vacuum in inches of water (in. H_2O). The vacuum reading usually varies between 18 and 22 in. Hg at sea level. To test engine vacuum, the vacuum gauge must be connected to a source of manifold vacuum. Many engines have a plug in the intake manifold which can be removed and replaced with an adapter fitting. Connect the vacuum gauge to the fitting with a suitable rubber hose or, if no manifold plug is available, connect the vacuum gauge to any device using manifold vacuum, such as EGR valves, etc. The vacuum gauge can be used to determine the amount of vacuum reaching a component.

HAND VACUUM PUMP

Small, hand-held vacuum pumps come in a variety of designs and provide a source of vacuum for testing components without the engine operating. Most have a built-in vacuum gauge and allow a component to be tested without removing it from the vehicle. Operate the pump lever or plunger, applying the correct amount of vacuum required for the test. The level of vacuum in inches of Mercury (in. Hg) is indicated on the pump gauge. For some testing, an additional vacuum gauge may be necessary.

COMPRESSION GAUGE

A compression gauge measures the amount of pressure in pounds per square inch (psi) that a cylinder is producing. Some gauges have a hose that screws into the spark plug hole while others have a tapered rubber tip which is held by hand in the spark plug hole. Engine compression depends on the sealing ability of the rings, valves, head gasket and spark plug gaskets. If any of these parts are not sealing properly, compression will be lost and the power output of the engine will be reduced. The compression in each cylinder should be measured and the variation between cylinders should be noted. The engine should be cranked through 5 or 6 compression strokes while warm, with all plugs removed, ignition disabled and throttle valves wide open.

FUEL PRESSURE GAUGE

A fuel pressure gauge is required to test the operation of the fuel delivery and injection systems. Some systems also need a 3-way valve to check the fuel pressure in various modes of operation. Gauges may require special adapters for making fuel connections. Always observe the cautions outlined in the fuel system service section of the repair manual when working around any pressurized fuel system.

USING A VACUUM GAUGE

White needle = steady needle   Dark needle = drifting needle

The vacuum gauge is one of the most useful and easy-to-use diagnostic tools. It is inexpensive, easy to hook up, and provides valuable information about the condition of your engine.

Indication: *Normal engine in good condition*

Gauge reading: *Steady, from 17–22 in./Hg.*

Indication: *Sticking valve or ignition miss*

Gauge reading: *Needle fluctuates from 15–20 in./Hg. at idle*

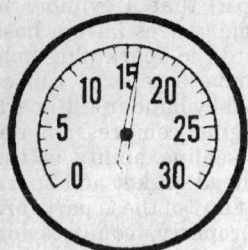

Indication: *Late ignition or valve timing, low compression, stuck throttle valve, leaking carburetor or manifold gasket.*

Gauge reading: *Low (15–20 in./Hg.) but steady*

Indication: *Improper carburetor adjustment, or minor intake leak at carburetor or manifold*

NOTE: *Bad fuel injector O-rings may also cause this reading.*

Gauge reading: *Drifting needle*

Indication: *Weak valve springs, worn valve stem guides, or leaky cylinder head gasket (vibrating excessively at all speeds).*

NOTE: *A plugged catalytic converter may also cause this reading.*

Gauge reading: *Needle fluctuates as engine speed increases*

Indication: *Burnt valve or improper valve clearance. The needle will drop when the defective valve operates.*

Gauge reading: *Steady needle, but drops regularly*

Indication: *Choked muffler or obstruction in system. Speed up the engine. Choked muffler will exhibit a slow drop of vacuum to zero.*

Gauge reading: *Gradual drop in reading at idle*

Indication: *Worn valve guides*

Gauge reading: *Needle vibrates excessively at idle, but steadies as engine speed increases*

88513010

A vacuum gauge is a good tool for diagnosing the general condition of an engine

BASIC MAINTENANCE AND TROUBLESHOOTING

19

WHERE TO START

Logical Diagnostic Procedures

Diagnosis of a driveability problem requires attention to detail and following the diagnostic procedures in the correct order. Resist the temptation to begin extensive testing before completing the preliminary diagnostic steps. The preliminary or visual inspection must be completed in detail before diagnosis begins. In many cases this will shorten diagnostic time and often cure the problem without the need for involved electronic testing.

There are two basic ways to check the vehicle engine for electronic problems. These are by symptom diagnosis and by the on-board computer self-diagnostic system. The first place to start is always the preliminary inspection. Intermittent problems are the most difficult to locate. If the problem is not present at the time of testing one may not be able to locate the fault.

PRELIMINARY INSPECTION

The visual inspection of all components is possibly the most critical step of diagnosis. A detailed examination of connectors, wiring and vacuum hoses can often lead to a repair without further diagnosis. Also, take into consideration if the vehicle has been serviced recently. Sometimes things get reconnected in the wrong place, or not at all. A careful inspector will check the undersides of hoses as well as the integrity of hard-to-reach hoses blocked by the air cleaner or other components. Correct routing for vacuum hoses can be ob-

Perform underhood inspection of all wiring and hoses

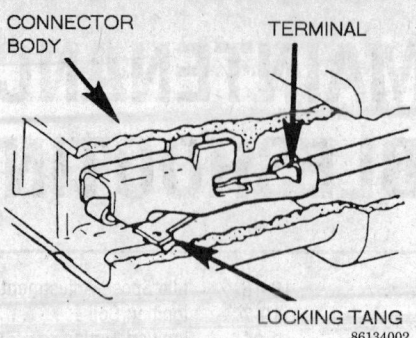

Check the individual terminals and wiring connectors for damage

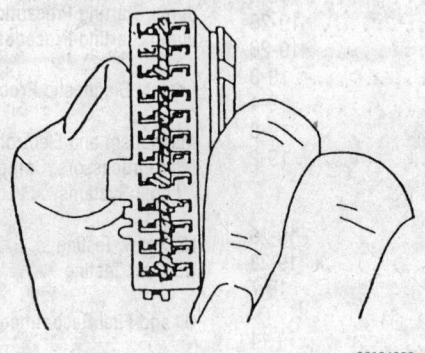

Inspect the connector terminals for damage

Check for damaged or broken wires

tained for from the vehicle service manual, or Vehicle Emission Control Information (VECI) label in the engine compartment of the vehicle. Wiring should be checked carefully for any sign of strain, burning, crimping or terminals pulled out from a connector.

Checking connectors at components or in harnesses is required; usually, pushing them together will reveal a loose fit. Also, check electrical connectors for corroded, bent, damaged, improperly seated pins, and bad wire crimps to terminals. Pay particular attention to ground

circuits, making sure they are not loose or corroded. Remember to inspect connectors and hose fittings at components not mounted on the engine, such as the evaporative canister or relays mounted on the fender aprons. Any component or wiring in the vicinity of a fluid leak or spillage should be given extra attention during inspection.

Additionally, inspect maintenance items such as belt condition and tension, battery charge and condition and the radiator cap carefully. Any of these very simple items may affect the system enough to set a fault code.

DIAGNOSIS BY SYMPTOM

Before the advent of the self-diagnostic system, diagnosis by symptom was the only method for investigation of an automotive problem. An attempt was made to solve problems by reviewing the symptoms and performing tests on suspected components until a defective component was located. The problem was then corrected and the vehicle checked for any other problems. This method is still used frequently today when a driveability complaint is made, but no code is set in the electronic control unit's memory.

When diagnosing by symptom the first step is to find out if the problem really exists. This may sound like a waste of time but one must be able to recreate the problem before beginning any testing. This is called an "operational check". Each operational check will give either a positive or negative answer (symptom). A positive answer is found when the check gives a positive result (the horn blows when the horn button is pressed). A negative answer is found when the check gives a negative result (the radio does not play when the knob is turned on). After performing several operational checks, a pattern may develop. This pattern is used in the next step of diagnosis to determine related symptoms.

In order to determine related symptoms, perform operational checks on circuits related to the problem circuit (the radio does not work and the dash lights do not go on). These checks can be made without the use of any test equipment. Simply follow the wires in the wiring harness or, if available, obtain a copy of the vehicle's specific wiring diagram. If one notices that the radio and the dash lights are on the same circuit, first check the radio to see if it works. Then check the dash lights. If neither the radio or dash lights work, this tells one that there is a problem in that circuit. Perform additional operational checks on that circuit and compile a list of symptoms.

When analyzing answers, a defect will always lie between a check which gave a positive answer and one which gives a negative answer. Look at the list of symptoms and try to determine probable areas to test. If one gets negative answers on related circuits, then maybe the problem is at the common junction. After one has determined what the symptoms are and where to go to look for defects, develop a plan for isolating the trouble. Ask a knowledgable automotive person which components frequently fail on the vehicle. Also notice which parts or components are easiest to reach and how can one accomplish the most by doing the least amount of checks.

A common way of diagnosis is to use the split-in-half technique. Each test that is made essentially splits the trouble area in half. By performing this technique several times the area where a problem is located becomes smaller and smaller until the problem can be isolated in a single wire or component. This area is most commonly between the two closest check points that produced a negative answer and a positive answer.

After the problem is located, perform the repair procedure. This may involve replacing a component, repairing a component or damaged wire, or making an adjustment.

NOTE: Never assume a component is defective until thoroughly tested it.

The final step is to make sure the complaint is corrected. Remember, the symptoms that one uncovers may lead to several problems that require separate repairs. Repeat the diagnosis and test procedures again and again until all negative symptoms are corrected.

DIAGNOSIS BY THE STORED TROUBLE CODE

When a fault code is detected, it appears as a flash of the CHECK ENGINE light on the instrument panel. This indicates that an abnormal signal in the system has been recognized by the ECM.

When diagnosing by code, the first step is to read any fault codes from the ECM. These codes will identify the area to perform more in-depth testing. After the fault codes have been read, proceed to test each of the components and component circuits indicated. Continue performing individual component tests until the failed component is located. Remember, fault codes do indicate the presence of a failure, but they do not identify the failed component directly.

SAFE VEHICLE SERVICING TIPS

It is virtually impossible to anticipate all of the hazards involved with automotive service, but care and common sense should prevent most accidents.

The rules of safety for mechanics range from "don't smoke around gasoline" to "use the proper tool for the job." The trick to avoiding injuries is to develop safe work habits and take every possible precaution.

Do's

• Do keep a fire extinguisher and first aid kit handy.

• Do wear safety glasses or safety goggles when cutting, drilling, grinding or prying. If one wears glasses for the sake of vision, wear safety goggles over regular glasses.

• Do shield ones eyes whenever working around the battery. Batteries contain sulfuric acid. In case of contact with the eyes or skin, flush the area with water or a mixture of water and baking soda, then get medical attention immediately.

• Do use safety stands for any under-vehicle service. Jacks are for raising vehicles; jackstands are for making sure the vehicle stays raised until one wants the vehicle to come down. Whenever the vehicle is raised, block the wheels remaining on the ground and set the parking brake.

• Do use adequate ventilation when working with chemicals. Asbestos dust from some worn brake linings can cause cancer.

• Do disconnect the negative battery cable when working on the vehicle.

• Do follow manufacturer's directions whenever working with potentially hazardous materials. Both

	Throttle Position Sensor	Coolant Temperature Sensor	MAP or MAF sensor	Air Temperature Sensor	Ignition Coil	Distributor	Spark Plug Wires	Fuel Filter	Air Filter	Vacuum Leak	Engine Mechanical	Knock Sensor / Spark Control	EGR System	Idle Control System	Camshaft sensor/Dist. pick-up	Oxygen Sensor	Ignition Module / Engine Computer	Torque Converter Clutch	PCV System
No Start	u	u	u		•	•		•			u	•			u	•	•		
Hard Start	•	•	•		•	•	•												u
Hesitation	•		•			•	•	•	•	•		•							
Stalling	•				•	•	•			•				•				•	u
Poor Idle	•				•	•	•			•			•	•					•
Dieseling							•					•	•						
Engine Lamp ON	•	•	•	•								•	•	•	•	•			
Knocks or Pings						•					•	•						•	
High Hydrocarbons	u	u	u		•	•	•	u	•	u	•	u	•			•			
Black Smoke	u	•	•	•									•						
Blue Smoke											•								
Poor Fuel Mileage	•	•	•	•	•	•	•	•					•			•		•	u
Lack of Power	•				•	•	•	•	•				•					•	u
Back fires		•			•	•	•									•		•	u
Runs Poor Cold	•				•	•	•			•			u	•				u	u
Runs Poor Hot	•	•	•		•	•	•			•			•	u	•			u	u
High speed surging		•	•					•										•	u

u: Although possible it is unlikely this component is at fault. A totally open or shorted circuit, or severe component fault may cause this condition.

8736MW01

DIAGNOSIS BY SYMPTOM — QUICK REFERENCE CHART

brake fluid and most types of anti-freeze are poisonous if taken internally.

• Do properly maintain ones tools. Loose hammerheads, mushroomed punches and chisels, frayed or poorly grounded electrical cords, excessively worn screwdrivers, spread wrenches (open end), cracked sockets, slipping ratchets, or faulty droplight sockets can cause accidents.

• Do use the proper size and type of tool for the job.

• Do, when possible, pull on a wrench handle rather than push on it, and adjust ones stance to prevent a fall.

• Do be sure that adjustable wrenches are tight on the nut or bolt and pulled so the force is on the fixed jaw.

• Do select a wrench or socket that fits the nut or bolt. The wrench or socket should sit straight, not cocked.

• Do strike squarely with a hammer. Avoid glancing blows.

• Do set the parking brake and block the wheels if work requires that the engine be running.

Don'ts

• Don't run an engine in a garage or anywhere else without proper ventilation — EVER! Carbon monoxide is poisonous and it is absorbed by the body faster than oxygen. It takes a long time to leave the human body, and one can build a deadly supply of it in the system by simply breathing in a little every day. One may not realize they are slowly being poisoned. Always use power vents, windows, fans or open the garage.

• Don't work around moving parts while wearing loose clothing. Short sleeves are much safer than long, loose sleeves. Hard-toed shoes with neoprene soles protect ones toes and give a better grip on slippery surfaces. Jewelry such as watches, fancy belt buckles, beads, or body adornment of any kind is not safe while working around a vehicle. Long hair should be hidden under a hat or cap.

• Don't use pockets for toolboxes. A fall or bump can drive a screwdriver deep into ones body. Even a wiping cloth hanging from the back pocket can wrap around a spinning shaft or fan.

• Don't smoke around gasoline, solvent or any flammable material.

• Don't smoke around the battery. When the battery is being charged, it gives off explosive hydrogen gas.

• Don't use gasoline to wash hands. There are excellent soaps available. Gasoline contains compounds which are hazardous to ones health and it removes natural oils from the skin so that bone dry hands will suck up oil and grease.

• Don't service the air conditioning system unless one is equipped with the necessary tools and training. The refrigerant is extremely cold, and when exposed to the air, will instantly freeze any surface it comes in contact with, including eyes. Keep refrigerant away from open flames; poisonous gas will be produced if refrigerant burns.

SERIAL NUMBER IDENTIFICATION

Vehicle Identification Number

The Vehicle Identification Number (VIN) is the number that will perhaps tell one every thing about the vehicle. The VIN number is somewhat like a Social Security Number in an automotive sense. The VIN is a standardized 17-digit number. Each digit of this number has a specific meaning or designation. Example: the 8th digit designates the engine code, the 10th digit designates the model year of the vehicle etc. The vehicle serial number is stamped on a plate fastened to the driver's side door pillar.

This number is usually located on the one of the fender aprons in the engine compartment (behind the wheel arch).

All models have the vehicle identification number stamped on a plate attached to the left side of the instrument panel. The plate is visible through the windshield.

The vehicle identification (model variation codes) may be interpreted as follows:

Using a Nissan/Infinity vehicle as an illustration, look at the VIN number (JN6H D2 1S*MW 000001), all models use a four letter prefix followed by the model designation (D21), then a four letter suffix (five on 1991 and later models), as shown in the illustration.

The serial number on all models is the new 17-digit format. The first three digits are the World Manufacturer Identification number. The next five digits are the Vehicle Description Section (same as the series identification number). The remaining nine digits are the production numbers.

NOTE: For specific identification of the vehicle see the Vehicle Identification Label on the vehicle. If the vehicle has be altered in some way or the engine or transmission has been changed the VIN may not coincide with the change that has been made. It a case like this, look for the serial number on the component to be sure the correct ordering of parts are made.

Engine Serial Number

The engine serial number consists of an engine series identification number followed by a six-digit production number. The number may be found in various places, depending upon the particular engine. Observe the following examples:

• Z24i Engine — the serial number is stamped on the left side of the cylinder block, below the No. 3 and No. 4 spark plugs.

• KA24E Engine — the serial number is stamped on the left side of the cylinder block, below the No. 2 and No. 3 spark plugs.

• VG30i and VG30E Engines — the serial number is stamped on the cylinder block, below the rear of the right side cylinder head.

NOTE: The illustrations given here are for a Nissan vehicle.

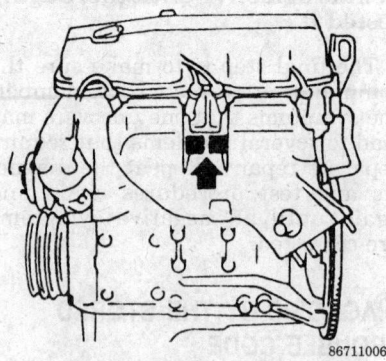

Typical engine serial number location

Transmission Serial Number

The transmission serial number is stamped on the front upper face of the transmission case on manual transmissions, or on the right side of

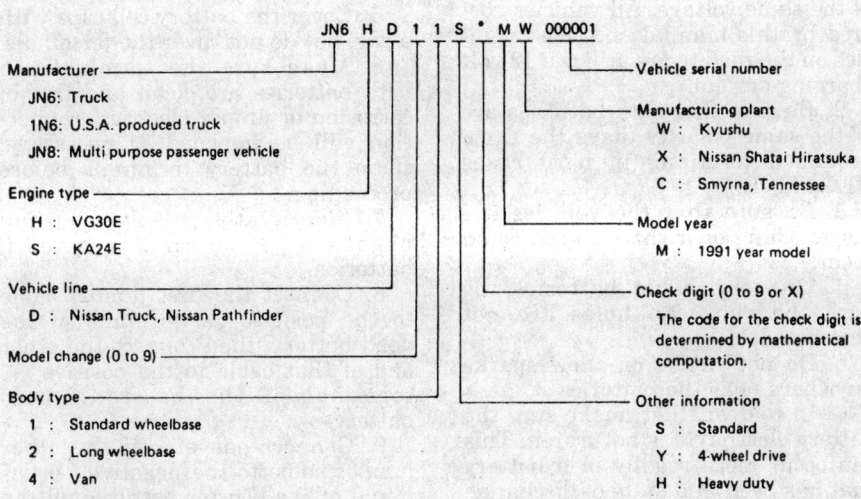

Manufacturer
  JN6 : Truck
  1N6 : U.S.A. produced truck
  JN8 : Multi purpose passenger vehicle

Engine type
  H : VG30E
  S : KA24E

Vehicle line
  D : Nissan Truck, Nissan Pathfinder

Model change (0 to 9)

Body type
  1 : Standard wheelbase
  2 : Long wheelbase
  4 : Van
  6 : King Cab, Wagon
  7 : 4-door wagon
  8 : 4-door van

JN6 H D 1 1 S ∗ M W 000001

Vehicle serial number

Manufacturing plant
  W : Kyushu
  X : Nissan Shatai Hiratsuka
  C : Smyrna, Tennessee

Model year
  M : 1991 year model

Check digit (0 to 9 or X)
  The code for the check digit is
  determined by mathematical
  computation.

Other information
  S : Standard
  Y : 4-wheel drive
  H : Heavy duty

86711004

Vehicle Identification Number (VIN) translation

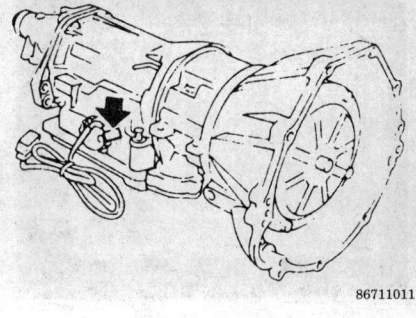

86711011

Typical automatic transmission serial number location

86801500

Vehicle Identification Number — visible through the windshield

86711131

Vehicle identification number is on the firewall-mounted label in the engine compartment

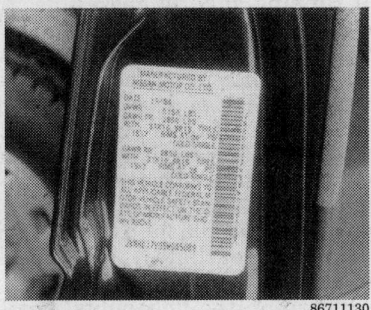

86711130

Manufacturer's label located in the door pillar area — note that the build date of vehicle is at the top

the transmission case on automatic transmissions.

NOTE: The illustrations given here are for a Nissan vehicle.

Vehicle Emissions Control Information (VECI) label

The Vehicle Emissions Control Information (VECI) label provides a wealth of information pertaining to the engine's Emission Control System. First, it identifies the engine's Cubic Inch Displacement (CID) and size in liter(s). It provides information for tune-up, such as spark plug gap, Ignition timing, idle/mixture and valve lash specifications. In some cases, it will provide a specific adjustment procedure as required. Some labels will also incorporate the vacuum routing of the engine's emission control system. Although the VECI label is very helpful to the person working on the vehicle, it should not be used as the main source for repair information. However, if there have not been any alterations to the engine and the Manual and sticker do not agree, use the Vehicle Emissions Control Information (VECI) sticker information, as it often reflects changes made during the production run.

Always keep in mind that the vehicle may have had an engine change at one time. If this is the case, one will have to identify the year and engine code in the vehicle. If the vehicle is missing the Emission Control Information (ECI) label, a new one can be ordered from a local dealer.

The Vehicle Emissions Control Information (VECI) label is usually located in the engine compartment. On some vehicles it will be found directly under the hood. Others may have it on the strut tower or radiator support.

86712037

Vehicle Emission Control Information (VECI) label located in the engine compartment

JUMP STARTING

Jump Starting a Dead Battery

Whenever a vehicle must be jump started, precautions must be followed in order to prevent the possibility of personal injury. Remember that batteries contain a small amount of explosive hydrogen gas which is a by-product of battery charging. Sparks should always be avoided when working around batteries, especially when attaching jumper cables. To minimize the possibility of accidental sparks, follow the procedure carefully.

WARNING

NEVER hook the batteries up in a series circuit or the entire electrical system will go up in smoke, especially the starter!

Vehicles equipped with a diesel engine utilize two 12 volt batteries, one on either side of the engine compartment. The batteries are connected in a parallel circuit (positive terminal to positive terminal, negative terminal to negative terminal). Hooking the batteries up in parallel circuit increases battery cranking power without increasing total battery voltage output. Output remains at 12 volts. On the other hand, hooking two 12 volt batteries up in a series circuit (positive terminal to negative terminal, positive terminal to negative terminal) increases total battery output to 24 volts (12 volts plus 12 volts).

Jump Starting Precautions

1. Be sure that both batteries are of the same voltage. All vehicles covered by this manual and most vehicles on the road today utilize a 12 volt charging system.
2. Be sure that both batteries are of the same polarity (have the same grounded terminal; in most cases NEGATIVE).
3. Be sure that the vehicles are not touching or a short circuit could occur.
4. On serviceable batteries, be sure the vent cap holes are not obstructed.
5. Do not smoke or allow sparks anywhere near the batteries.
6. In cold weather, make sure the battery electrolyte is not frozen. This can occur more readily in a battery that has been in a state of discharge.
7. Do not allow electrolyte to contact ones skin or clothing.

Jump Starting Procedure

1. Make sure that the voltages of the 2 batteries are the same. Most batteries and charging systems are of the 12 volt variety.
2. Pull the jumping vehicle (with the good battery) into a position so the jumper cables can reach the dead battery and that vehicle's engine. Make sure that the vehicles do NOT touch.
3. Place the transmissions of both vehicles in **NEUTRAL** or **PARK**, as applicable, then firmly set their parking brakes.

NOTE: If necessary for safety reasons, both vehicle's hazard lights may be operated throughout the entire procedure without significantly increasing the difficulty of jump starting the dead battery.

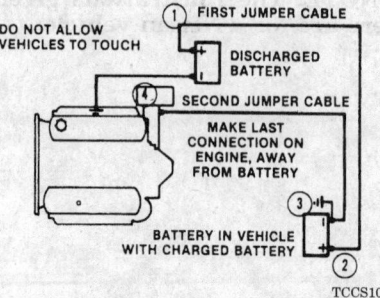

MAKE CONNECTIONS IN NUMERICAL ORDER

DO NOT ALLOW VEHICLES TO TOUCH

① FIRST JUMPER CABLE

DISCHARGED BATTERY

SECOND JUMPER CABLE

MAKE LAST CONNECTION ON ENGINE, AWAY FROM BATTERY

BATTERY IN VEHICLE WITH CHARGED BATTERY

TCCS1080

Connect the jumper cables to the batteries and engine in the order shown

4. Turn all lights and accessories **OFF** on both vehicles. Make sure the ignition switches on both vehicles are turned to the **OFF** position.
5. Cover the battery cell caps with a rag, but do not cover the terminals.
6. Make sure the terminals on both batteries are clean and free of corrosion or proper electrical connection will be impeded. If necessary, clean the battery terminals before proceeding.
7. Identify the positive (+) and negative (-) terminals on both batteries.
8. Connect the first jumper cable to the positive (+) terminal of the dead battery, then connect the other end of that cable to the positive (+) terminal of the booster (good) battery.
9. Connect one end of the other jumper cable to the negative (–) terminal of the booster battery and the other cable clamp to an engine bolt head, alternator bracket or other solid, metallic point on the dead battery's engine. Try to pick a ground on the engine that is positioned away from the battery, in order to minimize the possibility of the 2 clamps touching should one loosen during the procedure. DO NOT connect this clamp to the negative (–) terminal of the bad battery.

CAUTION
Be very careful to keep the jumper cables away from moving parts (cooling fan, belts, etc.) on both engines.

10. Check to make sure that the cables are routed away from any moving parts, then start the donor vehicle's engine. Run the engine at moderate speed for several minutes to allow the dead battery a chance to receive some initial charge.
11. With the donor vehicle's engine still running slightly above idle, try to start the vehicle with the dead battery. Crank the engine for no more than 10 seconds at a time and let the starter cool for at least 20 seconds between tries. If the vehicle does not start within 3 tries, it is likely that something else is also wrong.
12. Once the vehicle is started, allow it to run at idle for a few seconds to make sure that it is properly operating.
13. Turn on the headlights, heater blower and, if equipped, the rear defroster of both vehicles in order to reduce the severity of voltage spikes and subsequent risk of damage to the vehicles' electrical systems when the cables are disconnected.

14. Carefully disconnect the cables in the reverse order of connection. Start with the negative cable that is attached to the engine ground, then the negative cable on the donor battery. Disconnect the positive cable from the donor battery, then disconnect the positive cable from the formerly dead battery. Be careful when disconnecting the cables from the positive terminals not to allow the alligator clips to touch any metal on either vehicle or a short circuit and sparks will occur.

CHARGING SYSTEM

Alternator

The alternator converts the mechanical energy supplied by the drive belt into electrical energy by a process of electromagnetic induction. When the ignition switch is turned **ON**, current flows from the battery through the charging system light (or ammeter) to the voltage regulator, and finally to the alternator. When the engine is started, the drive belt turns the rotating field (rotor) in the stationary windings (stator), inducing alternating current. This alternating current is converted into usable direct current by the diode rectifier. Most of this current is used to charge the battery and to supply power for the vehicle's electrical accessories. A small part of this current is returned to the field windings of the alternator, enabling it to increase its power output. When the current in the field windings reaches a predetermined level, the voltage regulator grounds the circuit preventing any further increase. The cycle is continued so that the voltage supply remains constant.

All models use a 12-volt alternator. Amperage ratings vary according to the year and model. All models have an electronic, nonadjustable regulator, integral with the alternator.

ALTERNATOR PRECAUTIONS

To prevent damage to the alternator and regulator, the following precautionary measures must be taken when working with the electrical system:

• Never reverse the battery connections. Always visually check the battery polarity before any connections are made, to ensure that all connections correspond to the vehicle's battery ground polarity.

• Booster batteries must be connected properly. Make sure the positive cable of the booster battery is connected to the positive terminal of the battery which is getting the boost.

• Disconnect the battery cables before using a fast charger; the charger has a tendency to force current through the diodes in the opposite direction for which they were designed.

• Never use a fast charger as a booster for starting the vehicle.

• Never disconnect the voltage regulator while the engine is running, unless as noted for testing purposes.

• Do not ground the alternator output terminal.

• Do not operate the alternator on an open circuit with the field energized.

• Do not attempt to polarize the alternator.

• Disconnect the battery cables and remove the alternator before using an electric arc welder on the vehicle.

• Protect the alternator from excessive moisture. If the engine is to be steam cleaned, cover or remove the alternator.

REMOVAL & INSTALLATION

NOTE: On some models, the alternator is mounted very low on the engine. On these models, it may be necessary to remove the gravel shield and work from beneath the vehicle in order to gain access to the alternator.

1. Disconnect the negative battery cable.

2. On vehicles where the alternator can only be accessed from underneath the vehicle, raise the vehicle and support it safely with jackstands. Make sure the jackstands are at proper locations.

3. Remove the alternator pivot bolt. Push the alternator inward and remove the drive belt.

4. Pull back the rubber boots and disconnect the wiring from the back of the alternator.

5. Remove the alternator mounting bolt, then withdraw the alternator from its bracket.

To install:

6. Position the alternator in its mounting bracket, then lightly tighten the mounting and adjusting bolts.

7. Connect the electrical leads at the rear of the alternator, and return rubber boots to the proper position.

8. Adjust the belt tension.

9. Connect the negative battery cable.

10. Start the engine and perform a charging system voltage test to insure the charging system is putting out adequately.

One can make a basic check to see if the charging system is charging by using a voltmeter. Connect the voltmeter to the battery, battery voltage is approximately 12.6 volts. Start the engine and observe the voltmeter reading, it should read between 13.2–14.4 volts. If it remains at 12.6 volts the system is not charging.

Regulator

Regulators may be located internally to the the alternator or externally mounted on the firewall or inner fender panel depending on the vehicle. If faulty, it must be replaced; there are no adjustments which can be made.

REMOVAL & INSTALLATION

Internal Regulator

The electronic regulator is located inside the alternator. On some alternators the regulator is a simple bolt-on procedure, others may be soldered to the brush assembly. With a little knowledge of soldering one should be able to accomplish the job. The following procedure is for a regulator that is soldered to the brush assembly. The regulator is non-adjustable, and must be replaced together with the brush assembly, if faulty.

1. Remove the alternator.

2. Remove the through-bolts and separate the front cover from the stator housing.

3. Unsolder the wire connecting the diode plate to the brush at the brush terminal.

4. Remove the bolt retaining the diode plate to the rear cover.

5. Remove the nut securing the battery terminal bolt.

6. Lift the stator slightly, together with the diode plate, to gain access to the diode plate screw. Remove the screw.

7. Separate the stator and diode, then remove the brush and regulator assembly.

8. On assembly, apply soldering heat sparingly, carrying out the operation as quickly as possible, to avoid damage to the transistors and diodes.

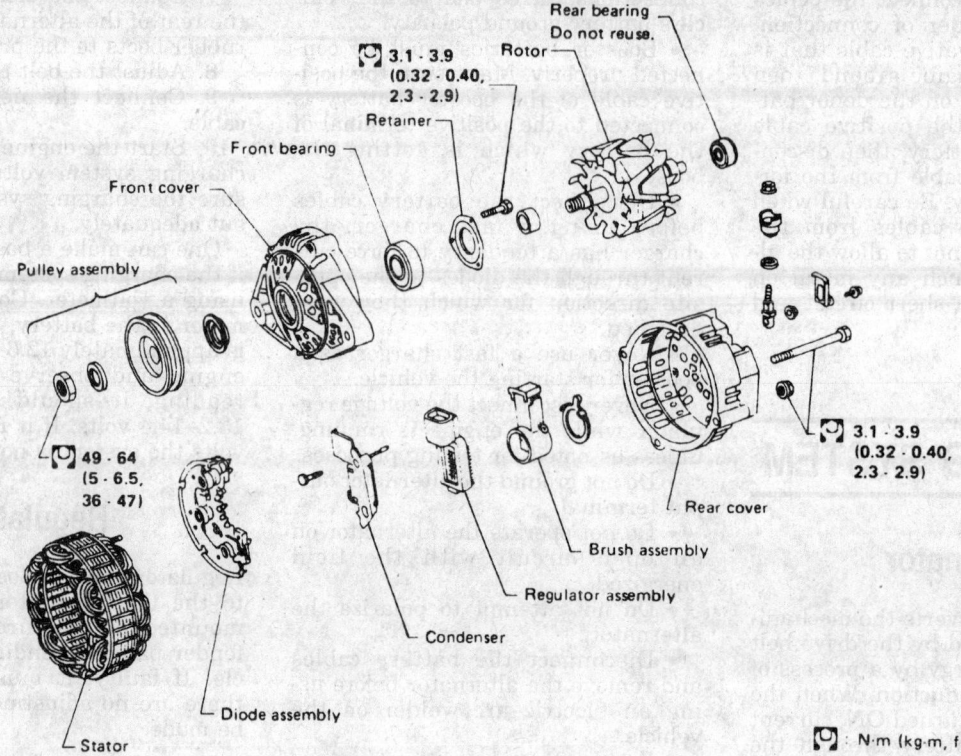

Exploded view of a typical alternator assembly

Before assembling the alternator halves, bend a piece of wire into an L-shape and slip it through the rear cover, next to the brushes. Use the wire to hold the brushes in a retracted position until the case halves are assembled. Remove the wire carefully, to prevent damage to the slip rings.

9. Install the alternator.

10. Start the engine and perform a charging system voltage test to insure charging system is putting out adequately.

One can make a basic check to see if the charging system is charging by using a voltmeter. Connect the voltmeter to the battery, battery voltage is approximately 12.6 volts. Start the engine and observe the voltmeter reading, it should read between 13.2–14.4 volts. If it remains at 12.6 volts the system is not charging.

External Regulator

Depending on the vehicle, the external regulator may be mounted on the firewall or inner fender panel. These regulators are much simpler to replace.

1. Disconnect the negative battery cable.

2. Locate the regulator.

3. Disconnect the harness connector from the regulator.

4. Remove the bolts holding the regulator to its mounting base and remove the regulator.

To install:

5. Install the regulator to the mounting base and secure it in place with the mounting screws.

6. Apply a coating of dielectric grease on the harness electrical connectors and connect.

7. Connect the negative battery cable.

8. Start the engine a perform a charging system voltage test to insure charging system is putting out adequately.

One can make a basic check to see if the charging system is charging by using a voltmeter. Connect the voltmeter to the battery, battery voltage is approximately 12.6 volts. Start the engine and observe the voltmeter reading, it should read between 13.2–14.4 volts. If it remains at 12.6 volts the system is not charging.

ADJUSTMENT

Voltage regulators on modern vehicles are not adjustable, most are electronic or even computer controlled.

STARTING SYSTEM

Starter

REMOVAL & INSTALLATION

1. Disconnect the negative battery cable at the battery, then disconnect the positive battery cable at the starter.

2. On vehicles where the starter can only be accessed from underneath the vehicle, observe the following caution:

——— CAUTION ———
Raise the vehicle and support safely with jackstands. Be sure to position the jackstands at proper frame locations.

3. On some vehicles it may be necessary to perform the following procedure:

 a. Remove the front right wheel.

 b. Remove the front gravel shield.

 c. Remove the exhaust manifold heat insulator.

 d. Remove the exhaust manifold.

 e. Detach the oil pressure switch connector.

Disconnect the cable attached the starter

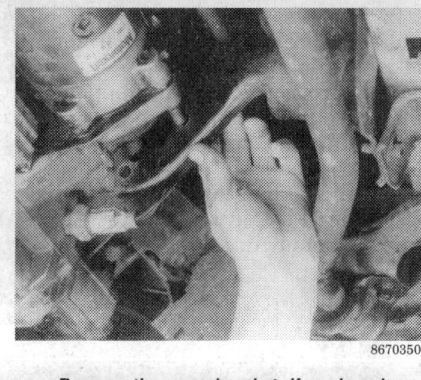

Remove the nose bracket, if equipped

Remove the starter bracket, if equipped

Remove the bolts holding the starter in place

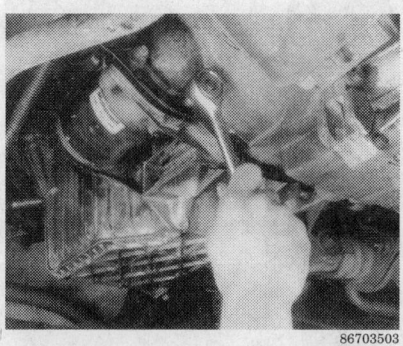

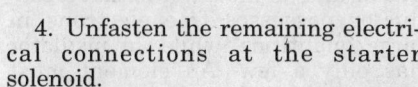

Loosen the holding the nose bracket, if equipped

Remove the the starter from the engine

4. Unfasten the remaining electrical connections at the starter solenoid.

5. Remove the two nuts holding the starter to the bell housing, then pull the starter toward the front of the vehicle and out.

To install:

6. Insert the starter into the bell housing, being sure that the starter drive is not jammed against the flywheel.

7. Tighten the attaching nuts and secure all electrical connections to the starter assembly.

8. Install all remaining components in reverse order of removal.

9. Reconnect the battery cables. If applicable, refill and check the oil level. Check the starter assembly for proper operation.

OVERHAUL

Solenoid Replacement

1. Remove the starter.
2. Unscrew the two solenoid switch (magnetic switch) retaining screws.

3. Remove the solenoid. In order to unhook the solenoid from the starter drive lever, lift it up at the same time that one is pulling it out of the starter housing.

4. Installation is in the reverse order of removal. Make sure that the solenoid switch is properly engaged with the drive lever before tightening the mounting screws.

Brush Replacement

NON-REDUCTION GEAR TYPE

1. Remove the starter.
2. Remove the solenoid (magnetic switch).
3. Unfasten the two end frame cap mounting bolts and remove the end frame cap.
4. Remove the O-ring and lock plate from the armature shaft groove, then slide the shims off the shaft.
5. Unfasten the two long housing screws (at the front of the starter) and carefully pull off the end plate.
6. Using a screwdriver, separate the brushes from the brush holder.
7. Slide the brush holder off of the armature shaft.
8. Crush the old brushes off of the copper braid and file away any remaining solder.
9. Fit the new brushes to the braid and spread the braid slightly.

NOTE: Use a soldering iron of at least 250 watts.

10. Using a light-grade solder, solder the brush to the braid. Grip the copper braid with flat pliers to prevent the solder from flowing down its length.
11. File off any extra solder and then repeat the procedure for the remaining three brushes.
12. Installation is in the reverse order of removal.

NOTE: When installing the brush holder, make sure that the brushes line up properly.

REDUCTION GEAR TYPE

1. Remove the starter and the solenoid.
2. Remove the through-bolts and the rear cover. The rear cover can be pried off with a small pry tool, but be careful not to damage the O-ring.
3. Separate the starter housing, armature, and brush holder from the center housing. They can be removed as an assembly.
4. Remove the positive side brush from its holder. The positive brush is insulated from the brush holder, and its lead wire is connected to the field coil.

5. Carefully lift the negative brush from the commutator and remove it from the holder.

6. Installation is in the reverse order of removal.

Starter Drive Replacement

NON-REDUCTION GEAR TYPE

1. With the starter motor removed from the vehicle, separate the solenoid from the starter.

2. Remove the two through-bolts and separate the gear case from the yoke housing.

3. Remove the pinion stopper clip and the pinion stopper.

4. Slide the starter drive off the armature shaft.

5. Install the starter drive and reassemble the starter in the reverse order of removal.

REDUCTION GEAR TYPE

1. Remove the starter.

2. Unfasten the solenoid and the shift lever.

3. Remove the bolts securing the center housing to the front cover and separate the parts.

4. Remove the gears and starter drive.

5. Installation is in the reverse order of removal.

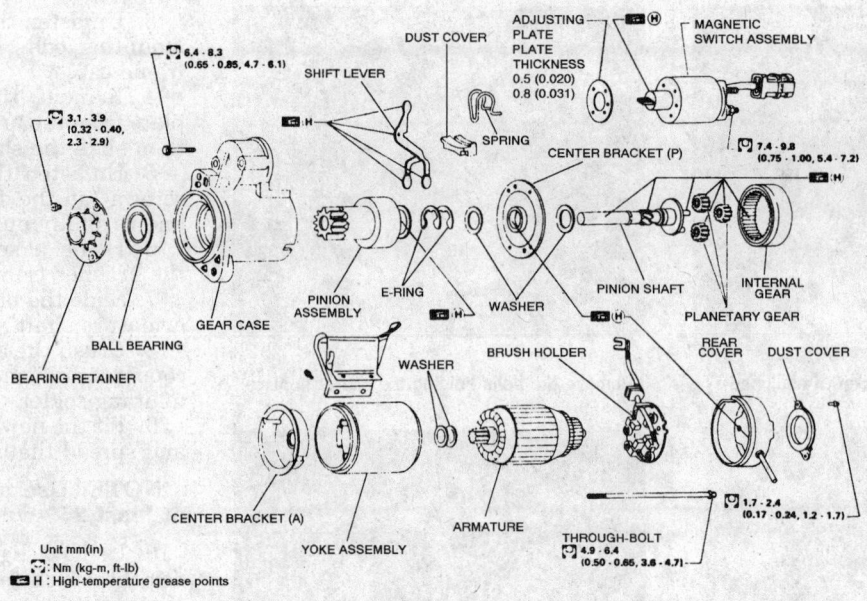

Exploded view of a reduction gear type starter — typical

FUNDAMENTALS OF ELECTRICITY

A good understanding of basic electrical theory and how circuits work is necessary to successfully perform the service and testing outlined in this manual. Therefore, this section should be read before attempting any diagnosis and repair.

All matter is made up of tiny particles called molecules. Each molecule is made up of two or more atoms. Atoms may be divided into even smaller particles called protons, neutrons and electrons. These particles are the same in all matter and differences in materials (hard or soft, conductive or non-conductive) occur only because of the number and arrangement of these particles. In other words, the protons, neutrons and electrons in a drop of water are the same as those in an ounce of lead, there are just more of them (arranged differently) in a lead molecule than in a water molecule. Protons and neutrons packed together form the nucleus of the atom, while electrons orbit around the nucleus much the same way as the planets of the solar system orbit around the sun.

The proton is a small positive natural charge of electricity, while the neutron has no electrical charge. The electron carries a negative charge equal to the positive charge of the proton. Every electrically neutral atom contains the same number of protons and electrons, the exact number of which determines the element. The only difference between a conductor and an insulator is that a conductor possesses free electrons in large quantities, while an insulator has only a few. An element must have very few free electrons to be a good insulator and vice-versa. When we speak of electricity, we're talking about these free electrons.

In a conductor, the movement of the free electrons is hindered by collisions with the adjoining atoms of the element (matter). This hindrance to movement is called **RESISTANCE** and it varies with different materials and temperatures. As temperature increases, the movement of the free electrons increases, causing more frequent collisions and therefore increasing resistance to the movement of the electrons. The number of colli-

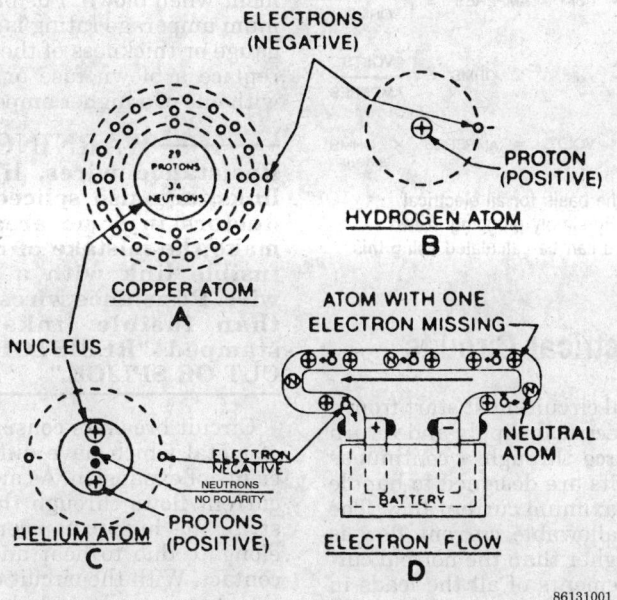

ATOMS AND ELECTRONS

Typical atoms of Copper (A), Hydrogen (B) and Helium (C). Electron flow in a battery circuit

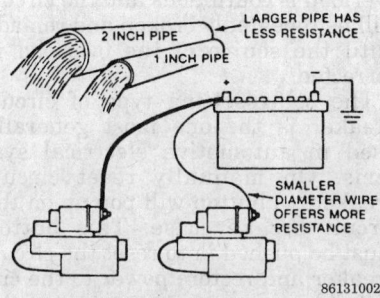

Electrical resistance can be compared to water flow through a pipe. The smaller the wire (pipe), the more resistance to the flow of electrons (water)

sions (resistance) also increases with the number of electrons flowing (current). Current is defined as the movement of electrons through a conductor such as a wire. In a conductor (such as copper) electrons can be caused to leave their atoms and move to other atoms. This flow is continuous in that every time an atom gives up an electron, it collects another one to take its place. This movement of electrons is called electric current and is measured in amperes. When 6.28 billion, billion electrons pass a certain point in the circuit in one second, the amount of current flow is called 1 ampere.

The force or pressure which causes electrons to flow in any conductor (such as a wire) is called **VOLTAGE**. It is measured in volts and is similar to the pressure that causes water to flow in a pipe. Voltage is the difference in electrical pressure measured between 2 different points in a circuit. In a 12 volt system, for example, the force measured between the two battery posts is 12 volts. Two important concepts are voltage potential and polarity. Voltage potential is the amount of voltage or electrical pressure at a certain point in the circuit with respect to another point. For example, if the voltage potential at one post of the 12 volt battery is 0, the voltage potential at the other post is 12 volts with respect to the first post. One post of the battery is said to be positive (+); the other post is negative (-) and the conventional direction of current flow is from positive to negative in an electrical circuit. It should be noted that the electron flow in the wire is opposite the current flow. In other words, when the circuit is energized, the current flows from positive to negative, but the electrons actually move from negative to positive. The voltage or pressure needed to produce a current flow in a circuit must be greater than the resistance present in the circuit. In other words, if the

voltage drop across the resistance is greater than or equal to the voltage input, the voltage potential will be zero — no voltage will flow through the circuit. Resistance to the flow of electrons is measured in ohms. One volt will cause 1 ampere to flow through a resistance of 1 ohm.

Units Of Electrical Measurement

There are 3 fundamental characteristics of a direct-current electrical circuit: volts, amperes and ohms.

VOLTAGE in a circuit controls the intensity with which the loads in the circuit operate. The brightness of a lamp, the heat of an electrical defroster, the speed of a motor are all directly proportional to the voltage, if the resistance in the circuit and/or mechanical load on electric motors remains constant. Voltage available from the battery is constant (normally 12 volts), but as it operates the various loads in the circuit, voltage decreases (drops).

AMPERE is the unit of measurement of current in an electrical circuit. One ampere is the quantity of current that will flow through a resistance of 1 ohm at a pressure of 1 volt. The amount of current that flows in a circuit is controlled by the voltage and the resistance in the circuit. Current flow is directly proportional to resistance. Thus, as voltage is increased or decreased, current is increased or decreased accordingly. Current is decreased as resistance is increased. However, current is also increased as resistance is decreased. With little or no resistance in a circuit, current is high.

OHM is the unit of measurement of resistance, represented by the Greek letter Omega (ω). One ohm is the resistance of a conductor through which a current of one ampere will flow at a pressure of one volt. Electrical resistance can be measured on an instrument called an ohmmeter. The loads (electrical devices) are the primary resistances in a circuit. Loads such as lamps, solenoids and electric heaters have a resistance that is essentially fixed; at a normal fixed voltage, they will draw a fixed current. Motors, on the other hand, do not have a fixed resistance. Increasing the mechanical load on a motor (such as might be caused by a misadjusted track in a power window system) will decrease the motor speed. The drop in motor rpm has the effect of reducing the internal resistance of the motor because the current draw of the motor varies directly with the mechanical load on the motor, although its actual resistance is unchanged. Thus, as the motor load increases, the current draw of the motor increases, and may increase up to the point where the motor stalls (cannot move the mechanical load).

Circuits are designed with the total resistance of the circuit taken into account. Troubles can arise when unwanted resistances enter into a circuit. If corrosion, dirt, grease, or any other contaminant occurs in places like switches, connectors and grounds, or if loose connections occur, resistances will develop in these areas. These resistances act like additional loads in the circuit and cause problems.

OHM'S LAW

Ohm's law is a statement of the relationship between the 3 fundamental characteristics of an electrical circuit. These rules apply to direct current (DC) only.

Ohm's law provides a means to make an accurate circuit analysis without actually seeing the circuit. If, for example, one wanted to check the condition of the rotor winding in a alternator whose specifications indicate that the field (rotor) current draw is normally 2.5 amperes at 12 volts, simply connect the rotor to a 12 volt battery and measure the current with an ammeter. If it measures about 2.5 amperes, the rotor winding can be assumed good.

An ohmmeter can be used to test components that have been removed from the vehicle in much the same manner as an ammeter. Since the voltage and the current of the rotor windings used as an earlier example are known, the resistance can be calculated using Ohms law. The formula would be ohms equals volts divided by amperes.

If the rotor resistance measures about 4.8 ohms when checked with an ohmmeter, the winding can be assumed good. By plugging in different specifications, additional circuit information can be determined such as current draw, etc.

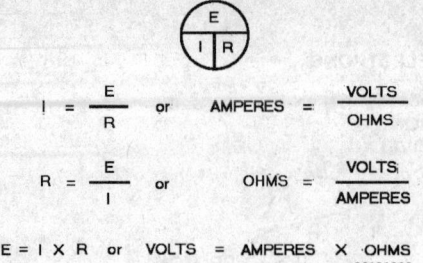

$$I = \frac{E}{R} \quad \text{or} \quad \text{AMPERES} = \frac{\text{VOLTS}}{\text{OHMS}}$$

$$R = \frac{E}{I} \quad \text{or} \quad \text{OHMS} = \frac{\text{VOLTS}}{\text{AMPERES}}$$

$$E = I \times R \quad \text{or} \quad \text{VOLTS} = \text{AMPERES} \times \text{OHMS}$$

86131003

Ohms Law is the basis for all electrical measurement. By simply plugging in two values, the third can be calculated using this formula

Electrical Circuits

An electrical circuit must start from a source of electrical supply and return to that source through a continuous path. Circuits are designed to handle a certain maximum current flow. The maximum allowable current flow is designed higher than the normal current requirements of all the loads in the circuit. Wire size, connections, insulation, etc., are designed to prevent undesirable voltage drop, overheating of conductors, arcing of contacts and other adverse effects. If the safe maximum current flow level is exceeded, damage to the circuit components will result; it is this condition that circuit protection devices are designed to prevent.

Protection devices are fuses, fusible links or circuit breakers designed to open or break the circuit quickly whenever an overload, such as a short circuit, occurs. By opening the circuit quickly, the circuit protection device prevents damage to the wiring, battery and other circuit components. Fuses and fusible links are designed to carry a preset maximum amount of current and to melt when that maximum is exceeded, while circuit breakers merely break the connection and may be manually reset.

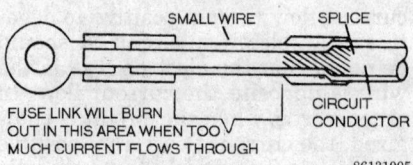

FUSE LINK WILL BURN OUT IN THIS AREA WHEN TOO MUCH CURRENT FLOWS THROUGH

SMALL WIRE   SPLICE

CIRCUIT CONDUCTOR

86131005

Typical fusible link wire

The maximum amperage rating of each fuse is marked on the fuse body and all contain a see-through portion that shows the break in the fuse element when blown. Fusible link maximum amperage rating is indicated by gauge or thickness of the wire. Never replace a blown fuse or fusible link with one of a higher amperage rating.

WARNING

Resistance wires, like fusible links, are also spliced into conductors in some areas. Do not make the mistake of replacing a fusible link with a resistance wire. Resistance wires are longer than fusible links and are stamped "RESISTOR-DO NOT CUT OR SPLICE."

Circuit breakers consist of 2 strips of metal which have different coefficients of expansion. As an overload or current flows through the bimetallic strip, the high-expansion metal will elongate due to heat and break the contact. With the circuit open, the bi-metal strip cools and shrinks, drawing the strip down until contact is re-established and current flows once again. In actual operation, the contact is broken very quickly if the overload is continuous and the circuit will be repeatedly broken and remade until the source of the overload is corrected.

The self-resetting type of circuit breaker is the one most generally used in automotive electrical systems. On manually reset circuit breakers, a button will pop up on the circuit breaker case. This button must be pushed in to reset the circuit breaker and restore power to the circuit. Always repair the source of the overload before resetting a circuit breaker or replacing a fuse or fusible link. When searching for overloads, keep in mind that the circuit protection devices protect only against overloads between the protection device and ground.

There are 2 basic types of circuit; series and parallel. In a series circuit, all of the elements are connected in chain fashion with the same amount of current passing through each element or load. No matter where an ammeter is connected in a series circuit, it will always read the same. The most important fact to remember about a series circuit is that the sum of the voltages across each element equals the source voltage. The total resistance of a series circuit is equal to the sum of the individual resistances within each element of the circuit. Using ohms law, one can deter-

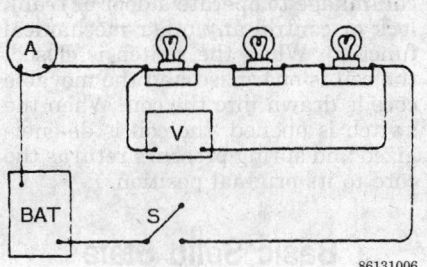

Example of a series circuit

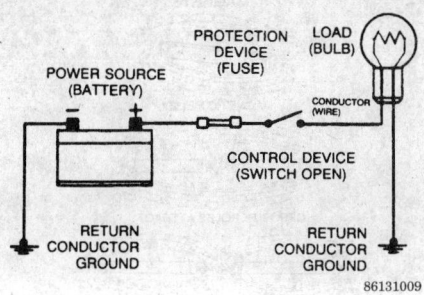

Typical circuit with all essential components

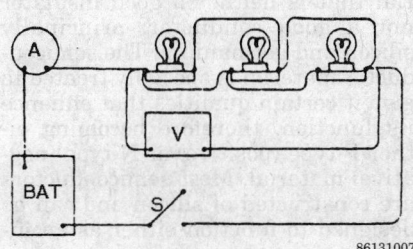

Example of a parallel circuit

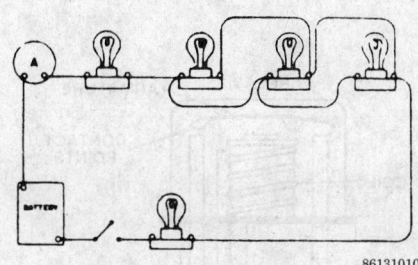

Example of a series-parallel circuit

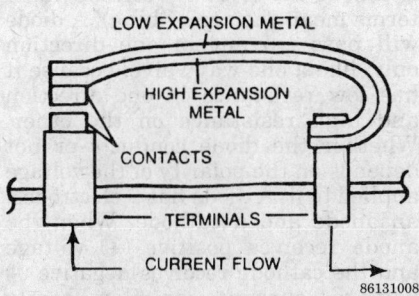

Typical circuit breaker construction

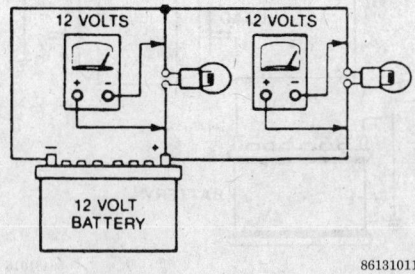

Voltage drop in a parallel circuit. Voltage drop across each lamp is 12 volts

A practical method of determining the resistance of a parallel circuit is to divide the product of the 2 resistances by the sum of 2 resistances at a time. Amperes through a parallel circuit is the sum of the amperes through the separate branches. Voltage across a parallel circuit is the same as the voltage across each branch.

By measuring the voltage drops the resistance of each element within the circuit is being measured. The greater the voltage drop, the greater the resistance. Voltage drop measurements are a common way of checking circuit resistances in automotive electrical systems. When part of a circuit develops excessive resistance (due to a bad connection) the element will show a higher than normal voltage drop. Normally, automotive wiring is selected to limit voltage drops to a few tenths of a volt. In parallel circuits, the total resistance is less than the sum of the individual resistances; because the current has 2 paths to take, the total resistance is lower.

Magnetism and Electromagnets

Electricity and magnetism are very closely associated because when electric current passes through a wire, a magnetic field is created around the wire. When a wire carrying electric current is wound into a coil, a magnetic field with North and South poles is created just like in a bar magnet. If an iron core is placed within the coil, the magnetic field becomes stronger because iron conducts magnetic lines much easier than air. This arrangement is called an electromagnet and is the basic principle behind the operation of such components as relays, buzzers and solenoids.

A relay is basically just a remote-controlled switch that uses a small amount of current to control the flow of a large amount of current. The simplest relay contains an electromagnetic coil in series with a voltage source (battery) and a switch. A movable armature made of some magnetic material pivots at one end and is held a small distance away from the electromagnet by a spring or the spring steel of the armature itself. A contact point, made of a good conductor, is attached to the free end of the armature with another contact point a small distance away. When the relay is switched on (energized), the

mine the voltage drop across each element in the circuit. If the total resistance and source voltage is known, the amount of current can be calculated. Once the amount of current (amperes) is known, values can be substituted in the Ohms law formula to calculate the voltage drop across each individual element in the series circuit. The individual voltage drops must add up to the same value as the source voltage.

A parallel circuit, unlike a series circuit, contains 2 or more branches, each branch a separate path independent of the others. The total current

draw from the voltage source is the sum of all the currents drawn by each branch. Each branch of a parallel circuit can be analyzed separately. The individual branches can be either simple circuits, series circuits or combinations of series-parallel circuits. Ohms law applies to parallel circuits just as it applies to series circuits, by considering each branch independently of the others. The most important thing to remember is that the voltage across each branch is the same as the source voltage. The current in any branch is that voltage divided by the resistance of the branch.

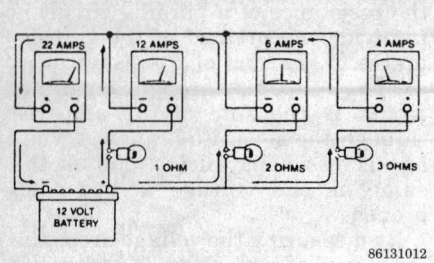

86131012

Total current in a parallel circuit: 4 + 6 +12 = 22 amps

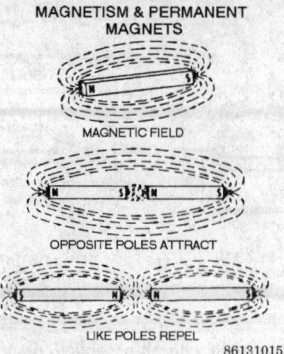

MAGNETISM & PERMANENT MAGNETS

MAGNETIC FIELD

OPPOSITE POLES ATTRACT

LIKE POLES REPEL

86131015

Magnetic field surrounding a bar magnet

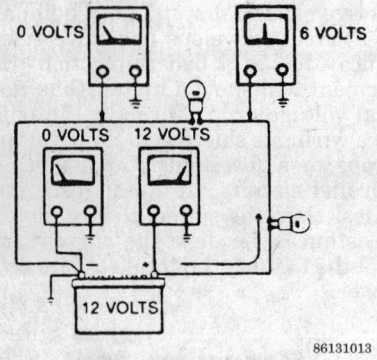

86131013

Voltage drop in a series circuit

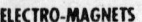

ELECTRO-MAGNETS

FORCE FIELD SURROUNDING A CURRENT CARRYING COIL
(WITHOUT IRON CORE)
ALL FORCE LINES ARE COMPLETE LOOPS

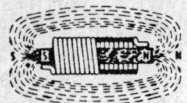

FORCE FIELD WITH SOFT IRON CORE
NOTE CONCENTRATION OF LINES IN IRON CORE

86131014

Magnetic field surrounding an electromagnet

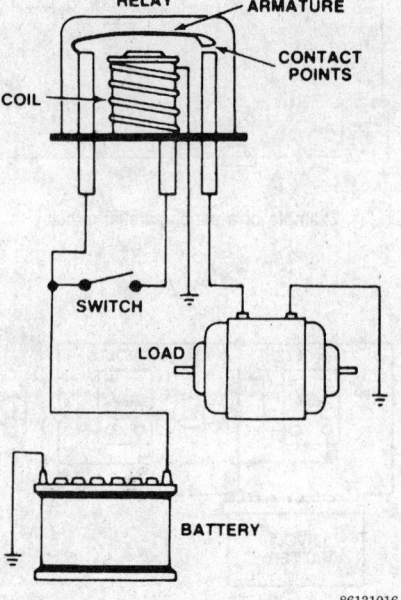

86131016

Typical relay circuit with basic components

magnetic field created by the current flow attracts the armature, bending it until the contact points meet, closing a circuit and allowing current to flow in the second circuit through the relay to the load the circuit operates. When the relay is switched off (de-energized), the armature springs back and opens the contact points, cutting off the current flow in the secondary, or controlled, circuit. Relays can be designed to be either open or closed when energized, depending on the type of circuit control a manufacturer requires.

A buzzer is similar to a relay, but its internal connections are different. When the switch is closed, the current flows through the normally closed contacts and energizes the coil. When the coil core becomes magnetized, it bends the armature down and breaks the circuit. As soon as the circuit is broken, the spring-loaded armature remakes the circuit and again energizes the coil. This cycle repeats rapidly to cause the buzzing sound.

A solenoid is constructed like a relay, except that its core is allowed to move, providing mechanical motion that can be used to actuate mechanical linkage to operate a door or trunk lock or control any other mechanical function. When the switch is closed, the coil is energized and the movable core is drawn into the coil. When the switch is opened, the coil is de-energized and spring pressure returns the core to its original position.

Basic Solid State

The term "solid state" refers to devices utilizing transistors, diodes and other components which are made from materials known as semiconductors. A semiconductor is a material that is neither a good insulator nor a good conductor; principally silicon and germanium. The semiconductor material is specially treated to give it certain qualities that enhance its function, therefore becoming either P-type (positive) or N-type (negative) material. Most semiconductors are constructed of silicon and can be designed to function either as an insulator or conductor.

DIODES

The simplest semiconductor function is that of the diode or rectifier (the 2 terms mean the same thing). A diode will pass current in one direction only, like a one-way valve, because it has low resistance in one direction and high resistance on the other. Whether the diode conducts or not depends on the polarity of the voltage applied to it. A diode has 2 electrodes, an anode and a cathode. When the anode receives positive (+) voltage and the cathode receives negative (-) voltage, current can flow easily through the diode. When the voltage is reversed, the diode becomes non-conducting and only allows a very slight amount of current to flow in the circuit. Because the semiconductor is not a perfect insulator, a small amount of reverse current leakage will occur, but the amount is usually too small to consider. The application of voltage to maintain the current flow described is called "forward bias."

A light-emitting diode (LED) is made of a particular type of crystal that glows when current is passed through it. LED's are used in display faces of many digital or electronic instrument clusters. LED's are usually arranged to display numbers (digital readout), but can be used to illuminate a variety of electronic graphic displays.

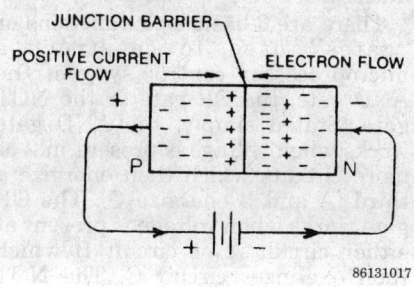

Diode with forward bias

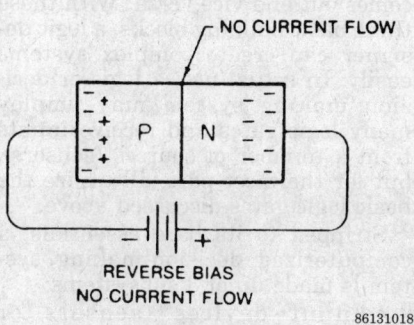

REVERSE BIAS
NO CURRENT FLOW

Diode with reverse bias

Like any other electrical device, diodes have certain ratings that must be observed and should not be exceeded. The forward current rating (or bias) indicates how much current can safely pass through the diode without causing damage or destroying it. Forward current rating is usually given in either amperes or milliamperes. The voltage drop across a diode remains constant regardless of the current flowing through it. Small diodes designed to carry low amounts of current need no special provision for dissipating the heat generated in any electrical device, but large current carrying diodes are usually mounted on heat sinks to keep the internal temperature from rising to the point where the silicon will melt and destroy the diode. When diodes are operated in a high ambient temperature environment, they must be de-rated to prevent failure.

Another diode specification is its peak inverse voltage rating. This value is the maximum amount of voltage the diode can safely handle when operating in the blocking mode. This value can be anywhere from 50–1000 volts, depending on the diode. If voltage amount is exceeded, it will damage the diode just as too

much forward current will. Most semiconductor failures are caused by excessive voltage or internal heat.

One can test a diode with a small battery and a lamp with the same voltage rating. With this arrangement one can find a bad diode and determine the polarity of a good one. A diode can fail and cause either a short or open circuit, but in either case it fails to function as a diode. Testing is simply a matter of connecting the test bulb first in one direction and then the other and making sure that current flows in one direction only. If the diode is shorted, the test bulb will remain on no matter how the light is connected.

TRANSISTORS

The transistor is an electrical device used to control voltage within a circuit. A transistor can be considered a "controllable diode" in that, in addition to passing or blocking current, the transistor can control the amount of current passing through it. Simple transistors are composed of 3 pieces of semiconductor material, P and N type, joined together and enclosed in a container. If 2 sections of P material and 1 section of N material are used, it is known as a PNP transistor; if the reverse is true, then it is known as an NPN transistor. The 2 types cannot be interchanged.

Most modern transistors are made from silicon (earlier transistors were made from germanium) and contain 3 elements; the emitter, the collector and the base. In addition to passing or blocking current, the transistor can control the amount of current passing through it and because of this can function as an amplifier or a switch. The collector and emitter form the main current-carrying circuit of the transistor. The amount of current that flows through the collec-

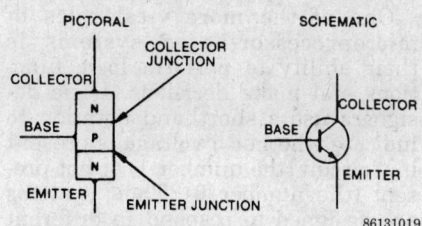

NPN transistor illustration (pictorial and schematic)

tor-emitter junction is controlled by the amount of current in the base circuit. Only a small amount of base-emitter current is necessary to control a large amount of collector-emitter current (the amplifier effect). In automotive applications, however, the transistor is used primarily as a switch.

When no current flows in the base-emitter junction, the collector-emitter circuit has a high resistance, like to open contacts of a relay. Almost no current flows through the circuit and transistor is considered OFF. By bypassing a small amount of current into the base circuit, the resistance is low, allowing current to flow through the circuit and turning the transistor ON. This condition is known as "saturation" and is reached when the base current reaches the maximum value designed into the transistor that allows current to flow. Depending on various factors, the transistor can turn ON and OFF (go from cutoff to saturation) in less than one millionth of a second.

Much of what was said about ratings for diodes applies to transistors, since they are constructed of the same materials. When transistors are required to handle relatively high currents, such as in voltage regulators or ignition systems, they are generally mounted on heat sinks in the same manner as diodes. They can be damaged or destroyed in the same manner if their voltage ratings are exceeded. A transistor can be checked for proper operation by measuring the resistance with an ohmmeter between the base-emitter terminals and then between the base-collector terminals. The forward resistance should be small, while the reverse resistance should be large. Compare the readings with those from a known good transistor. As a final check, measure the forward and reverse resistance between the collector and emitter terminals.

Integrated Circuits

The Integrated Circuit (IC) is an extremely sophisticated solid state device that consists of a silicone wafer (or chip) which has been doped, insulated and etched many times so that it contains an entire electrical circuit with transistors, diodes, conductors and capacitors miniaturized within each tiny chip. Integrated circuits are often referred to as "computers on a chip" and are largely responsible for the current boom in electronic control technology.

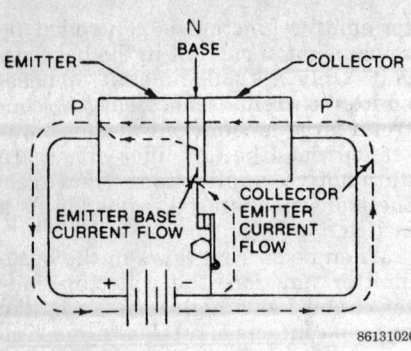

PNP transistor with base switch closed (base emitter and collector emitter current (flow)

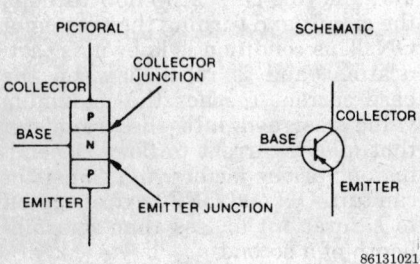

PNP transistor illustrations (pictorial and schematic)

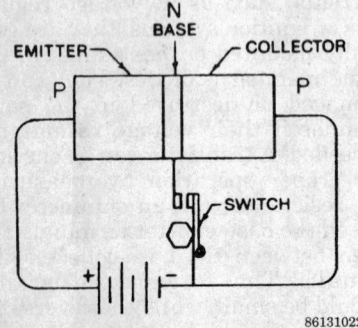

PNP transistor with base switch open (no current)

Microprocessors, Computers and Logic Systems

Mechanical or electromechanical control devices lack the precision necessary to meet the requirements of modern control standards. They do not have the ability to respond to a variety of input conditions common to anti-lock brakes, climate control and electronic suspension operation. To meet these requirements, manufacturers have gone to solid state logic

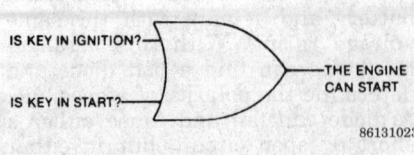

Typical two-input "OR" circuit operation

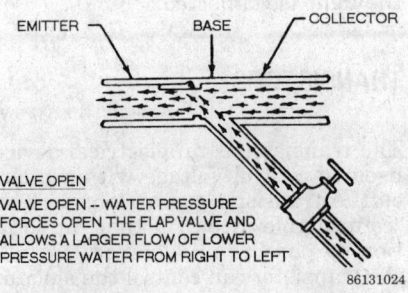

VALVE OPEN
VALVE OPEN -- WATER PRESSURE FORCES OPEN THE FLAP VALVE AND ALLOWS A LARGER FLOW OF LOWER PRESSURE WATER FROM RIGHT TO LEFT

Hydraulic analogy to transistor function is shown with the base circuit energized

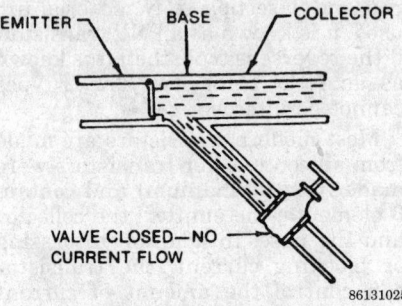

VALVE CLOSED—NO CURRENT FLOW

Hydraulic analogy to transistor function is shown with the base circuit off

systems and microprocessors to control the basic functions of suspension, brake and temperature control, as well as other systems and accessories.

One of the more vital roles of microprocessor-based systems is their ability to perform logic functions and make decisions. Logic designers use a shorthand notation to indicate whether a voltage is present in a circuit (the number 1) or not present (the number 0). Their systems are designed to respond in different ways depending on the output signal

(or the lack of it) from various control devices.

There are 3 basic logic functions or "gates" used to construct a microprocessor control system: the AND gate, the OR gate or the NOT gate. Stated simply, the AND gate works when voltage is present in 2 or more circuits which then energize a third (A and B energize C). The OR gate works when voltage is present at either circuit A or circuit B which then energizes circuit C. The NOT function is performed by a solid state device called an "inverter" which reverses the input from a circuit so that, if voltage is going in, no voltage comes out and vice versa. With these three basic building blocks, a logic designer can create complex systems easily. In actual use, a logic or decision making system may employ many logic gates and receive inputs from a number of sources (sensors), but for the most part, all utilize the basic logic gates discussed above.

Stripped to its bare essentials, a computerized decision-making system is made up of 3 subsystems:
• Input devices (sensors or switches)
• Logic circuits (computer control unit)
• Output devices (actuators or controls)

The input devices are usually nothing more than switches or sensors that provide a voltage signal to the control unit logic circuits that is read as a 1 or 0 (on or off) by the logic circuits. The output devices are anything from a warning light to solenoid-operated valves, motors, linkage, etc. In most cases, the logic circuits themselves lack sufficient output power to operate these devices directly. Instead, they operate some intermediate device such as a relay or power transistor which in turn operates the appropriate device or control. Many problems diagnosed as computer failures are really the result of a malfunctioning intermediate device like a relay. This must be kept in mind whenever troubleshooting any microprocessor-based control system.

As computer capacity is improved by the manufacturers, so does sensor technology. A few years ago, the onboard computer would receive a message from an engine sensor in a "go or no-go" form; for example the coolant temperature would either be above or below 150°F. Today's systems allow the same sensor to pass progressively more voltage as the engine warms up. The engine computer now knows

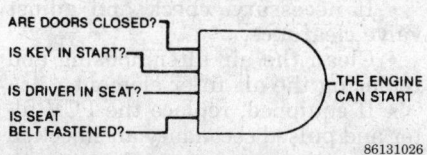

86131026

Multiple inputs "AND" operation in a typical automotive starting circuit

exactly what temperature the coolant is at all times. With this information the the computer can react to the changing voltage signal from the sensor instantly and control other engine functions based on engine warm-up or over heating conditions.

The logic systems discussed above are called "hardware" systems, because they consist only of the physical electronic components (gates, resistors, transistors, etc.). Hardware systems do not contain a program and are designed to perform specific or "dedicated" functions which cannot readily be changed. For many simple automotive control requirements,

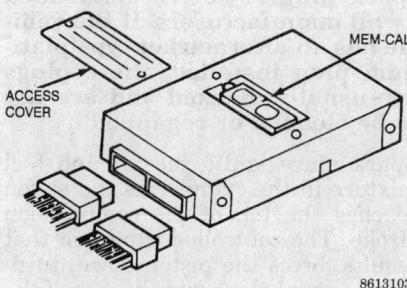

86131031

Typical General Motors engine control computer

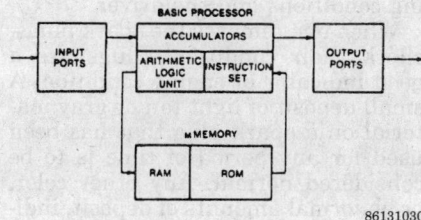

86131030

Schematic of typical microprocessor based on-board computer, showing essential common

such dedicated logic systems are perfectly adequate. When more complex logic functions are required, or where it may be desirable to alter these functions (such as from one model vehicle to another) a true computer system is used. A computer can be programmed through its software to perform many different functions and, if that program is stored on a separate integrated circuit chip called a ROM (Read Only Memory), it can be easily changed simply by plugging in a different ROM with the desired program. Most on-board automotive computers are designed with this capability. The on-board computer method of engine control offers the manufacturer a flexible method of responding to data from a variety of input devices and of controlling an equally large variety of output controls. The computer response can be changed quickly and easily by simply modifying its software program.

The microprocessor is the heart of the microcomputer. It is the thinking part of the computer system through which all the data from the various sensors passes. Within the microprocessor, data is acted upon, compared, manipulated or stored for future use. A microprocessor is not necessarily a microcomputer, but the differences between the 2 are becoming very minor. Originally, a microprocessor was a major part of a microcomputer, but nowadays microprocessors are being called "single-chip microcomputers". They contain all the essential elements to make them behave as a computer, including the most important ingredient–the program.

All computers require a program. In a general purpose computer, the program can be easily changed to allow different tasks to be performed. In a "dedicated" computer, such as most on-board automotive computers, the program isn't quite so easily altered. These automotive computers are designed to perform one or several specific tasks, such as maintaining the passenger compartment temperature at a specific, predetermined level. A program is what makes a computer smart; without a program a computer can do absolutely nothing. The term "software" refers to the computer's program that makes the hardware preform the function needed.

The software program is simply a listing in sequential order of the steps or commands necessary to make a computer perform the desired task. Before the computer can do an-

ything at all, the program must be fed into it by one of several possible methods. A computer can never be "smarter" than the person programming it, but it is a lot faster. Although it cannot perform any calculation or operation that the programmer himself cannot perform, its processing time is measured in millionths of a second.

Because a computer is limited to performing only those operations (instructions) programmed into its memory, the program must be broken down into a large number of very simple steps. Two different programmers can come up with 2 different programs, since there is usually more than one way to perform any task or solve a problem. In any computer, however, there is only so much memory space available, so an overly long or inefficient program may not fit into the memory. In addition to performing arithmetic functions (such as with a trip computer), a computer can also store data, look up data in a table and perform the logic functions previously discussed. A Random Access Memory (RAM) allows the computer to store bits of data temporarily while waiting to be acted upon by the program. It may also be used to store output data that is to be sent to an output device. Whatever data is stored in a RAM is lost when power is removed from the system by turning **OFF** the ignition key, for example.

Computers have another type of memory called a Read Only Memory (ROM) which is permanent. This memory is not lost when the power is removed from the system. Most programs for automotive computers are stored on a ROM memory chip. Data is usually in the form of a look-up table that saves computing time and program steps. For example, a computer designed to control the amount of distributor advance can have this information stored in a table. The information that determines distributor advance (engine rpm, manifold vacuum and temperature) is coded to produce the correct amount of distributor advance over a wide range of engine operating conditions. Instead of the computer computing the required advance, it simply looks it up in a pre-programmed table. However, not all electronic control functions can be handled in this manner; some must be computed. On an anti-lock brake system, for example, the computer must measure the rotation of each separate wheel and then calculate how much brake pressure to apply in order to prevent one wheel

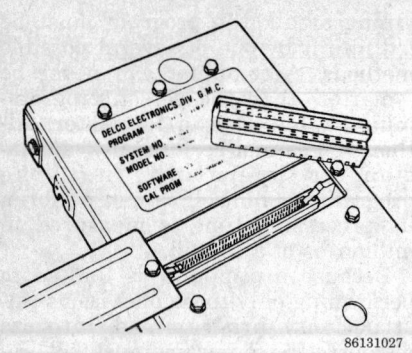

Electronic control module with Mem-Cal, used in General Motors vehicles

from locking up and causing a loss of control.

There are several ways of programming a ROM, but once programmed the ROM cannot be changed. If the ROM is made on the same chip that contains the microprocessor, the whole computer must be altered if a program change is needed. For this reason, a ROM is usually placed on a separate chip. Another type of memory is the Programmable Read Only Memory (PROM) that has the program "burned in" with the appropriate programming machine. Like the ROM, once a PROM has been programmed, it cannot be changed. The advantage of the PROM is that it can be produced in small quantities economically, since it is manufactured with a blank memory. Program changes for various vehicles can be made readily. There is still another type of memory called an EPROM (Erasable PROM) which can be erased and programmed many times. EPROM's are used only in research and development work, not on production vehicles.

Some automotive manufacturers refer to the engine controlling computer as an Electronic Control Module (ECM). The ECM contains the PROM necessary for all engine functions, it also contains a device that allows the fuel delivery function should other parts of the ECM become damaged. It has an access door in the ECM, like the PROM has.

Engines coupled to electronically controlled transmissions employ a Powertrain Control Module (PCM) to oversee both engine and transmission operation. This unit may be referred to as the PCM, the ECM/PCM or the PCM/TCM (Transmission Control Module). The integrated functions of engine and transmission control allow accurate gear selection and improved fuel economy.

For engine diagnostics, the PCM may be considered identical to an ECM system, although the combined unit will display additional codes relating to transmission function and components.

NOTE: When the term Powertrain Control Module (PCM) is used in this manual it will refer to the engine control computer regardless that it may be a Powertrain Control Module (PCM) or Electronic Control Module (ECM).

BASIC TUNE-UP PROCEDURES

In order to extract the best performance and economy from the engine, it is essential that it be properly tuned at regular intervals. A regular tune-up/inspection will keep the engine running smoothly and will prevent the annoying minor breakdowns and poor performance associated with an untuned engine.

A complete tune-up/inspection should generally be performed every 30,000 miles (48,000 km) or 24 months, whichever comes first. This interval should be halved (as a general rule of thumb) if the vehicle is operated under severe conditions, such as trailer towing, prolonged idling, continual stop and start driving, or if starting or running problems are noticed. It is assumed that the routine maintenance has been kept up, as this will have a decided effect on the results of a tune-up.

Some 1994 and newer vehicles are specified to go up to 100,000 miles between engine tune-ups, one can always refer the owners guide. A tune-

Because of the tangle of underhood wiring, ALWAYS tag/note wire locations before removal

up/inspection for all models should consists of the following items
• Inspect the drive belts.
• If necessary, check and adjust valve clearance.
• Clean the air filter housing and replacing the air filter element.
• If equipped, replace the PCV filter and pulsed secondary air injection filter.
• Inspect or replace the fuel filter assembly.
• Inspect all fuel and vapor lines.
• Check or replace the distributor cap, rotor and ignition wires.
• Replace the spark plugs and make all necessary engine adjustments.
• Always refer to the Maintenance Interval Chart for additional service information.

NOTE: If the tune-up specifications on the Vehicle Emission Control Information sticker in the engine compartment of the vehicle disagree with the Tune-Up Specifications, the figures on the sticker must be used. The sticker often reflects changes made during the production run.

Spark Plugs

NOTE: The platinum type spark plug is not recommended by all manufacturers. If the vehicle has an aftermarket type platinum plug installed, these plugs are usually marked and are not to be cleaned or regapped.

Spark plugs ignite the air and fuel mixture in the cylinder as the piston reaches the top of the compression stroke. The controlled explosion that results forces the piston down, turning the crankshaft and the rest of the drive train.

The average life of a spark plug is about 30,000 miles (48,000 km). This is, however, dependent on a number of factors: the mechanical condition of the engine, the type of fuel, the driving conditions and the driver.

When one removes the spark plugs, check their condition. Plugs are a good indicator of engine condition. A small deposit of light tan or gray material on a spark plug that has been used for any period of time is to be considered normal. Any other color, or abnormal amounts of deposit, indicates that there may be something wrong in the engine.

When a spark plug is functioning normally or, more accurately, when the plug is installed in an engine that is functioning properly, the plugs can be taken out, cleaned, regapped, and reinstalled in the engine without causing the engine any harm.

When, and if, a plug fouls and begins to misfire, one will have to investigate, correct the cause of the fouling, and either clean or replace the plug. There are several reasons why a spark plug will foul and one can learn which is at fault by just looking at the plug.

There are many spark plugs suitable for use in the engine and are offered in a number of different heat ranges. The amount of heat which the plug absorbs is determined by the length of the lower insulator. The longer the insulator the hotter the plug will operate; the shorter the insulator, the cooler it will operate. A spark plug that absorbs (or retains) little heat and remains too cool will accumulate deposits of lead, oil, and carbon, because it is not hot enough to burn them off. This leads to fouling and consequent misfiring. A spark plug that absorbs too much heat will have no deposits, but the electrodes will burn away quickly and, in some cases, pre-ignition may result. Pre-ignition occurs when the spark plug tips get so hot that they ignite the air/fuel mixture before the actual spark fires. This premature ignition will usually cause a pinging sound under conditions of low speed and heavy load. In severe cases, the heat may become high enough to start the air/fuel mixture burning throughout the combustion chamber rather than just to the front of the plug. In this case, the resultant explosion could be strong enough to damage pistons, rings, and valves.

In most cases the factory recommended heat range is correct; it is chosen to perform well under a wide range of operating conditions. However, if most of ones driving is long distance, high speed travel, one may want to install a spark plug one step colder than standard. If most of ones driving is of the short trip variety, when the engine may not always reach operating temperature, a hotter plug may help burn off the deposits normally accumulated under those conditions.

REMOVAL

NOTE: Some engines use two spark plugs per cylinder, be sure to replace them all.

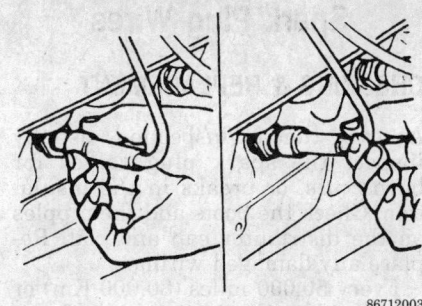

Twist and pull on the rubber boot to remove the spark plug wires; NEVER pull on the wire itself

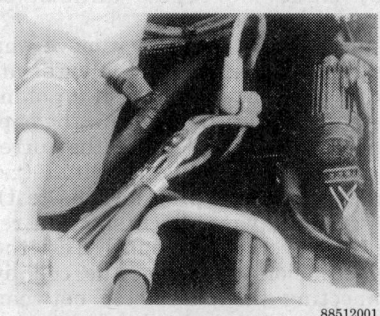

Using this special tool to remove the spark plug wire makes the job easier; NEVER pull on the wire itself

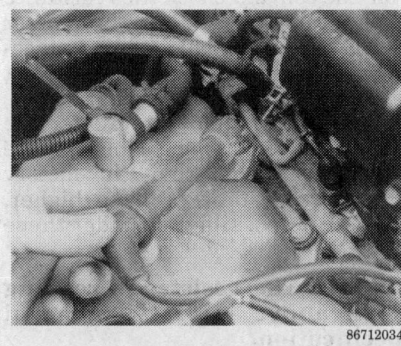

Mark and remove the spark plug wires one at a time to avoid a mix-up

1. Disconnect the negative battery cable.

NOTE: Always keep track of the spark plug cable routing and plug wire bracket locations.

2. Number the spark plug wires so that one won't cross them when they are reconnected.

3. Remove the wire from the end of the spark plug by grasping the rubber boot. If the boot sticks to the plug, remove it by twisting and pulling at the same time. DO NOT pull wire itself or it will damage the core.

4. Use a spark plug socket to loosen all of the plugs about two turns.

NOTE: Remove the spark plugs when the engine is cold, if possible, to prevent damage to the threads. If removal of the plugs is difficult, apply a few drops of penetrating oil or silicone spray to the area around the base of the plug, and allow it a few minutes to work.

5. If compressed air is available, apply it to the area around the spark plug holes. Otherwise, use a rag or a brush to clean the area. Be careful not to allow any foreign material to drop into the spark plug holes.

6. Remove the plugs by unscrewing them the rest of the way from the engine.

INSPECTION

Check the plugs for deposits and wear. If they are not going to be replaced, clean the plugs thoroughly. Remember that any kind of deposit will decrease the efficiency of the plug. Plugs can be cleaned on a spark plug cleaning machine, which can sometimes be found in service stations, or one can do an acceptable job of cleaning with a stiff brush. If the plugs are cleaned, the electrodes must be filed flat. Use an ignition points file, not an emery board or the like, which will leave deposits. The electrodes must be filed perfectly flat with sharp edges; rounded edges reduce the spark plug voltage by as much as 50%.

Check spark plug gap before installation. The ground electrode (the L-shaped one connected to the body of the plug) must be parallel to the center electrode and the specified size wire gauge (please refer to the Tune-Up Specifications chart for details) must pass between the electrodes with a slight drag.

NOTE: NEVER adjust the gap on a used platinum type spark plug.

Always check the gap on new plugs as they are not always set correctly at the factory. Do not use a flat feeler gauge when measuring the gap on a used plug, because the reading may be inaccurate. A wire type gapping tool is the best way to check the gap. Wire gapping tools usually have a bending tool attached. Use that to adjust the side electrode until the proper distance is obtained. Absolutely never attempt to bend the

86712035

Carefully unthread the spark plug from the cylinder head using the proper tools

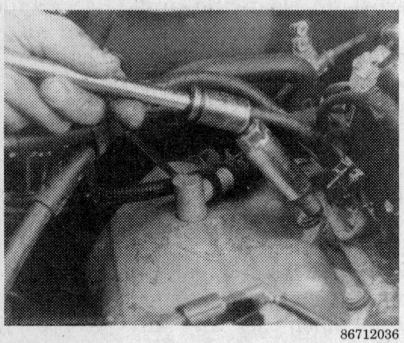

86712036

In this case, a universal joint made plug removal easier

center electrode. Also, be careful not to bend the side electrode too far or too often as it may weaken and break off within the engine, requiring removal of the cylinder head to retrieve it.

INSTALLATION

1. Lubricate the threads of the spark plugs with a drop of oil. Install the plugs and tighten them by hand first. Take care not to cross-thread them.
2. Tighten the spark plugs with a plug socket. Do not apply the same amount of force used for a bolt; just snug them in. If a torque wrench is available, tighten to specific specifications for the vehicle being worked on.
3. Install the wires on their respective plugs. Make sure the wires are firmly connected. One will be able to feel them click into place. Check the spark plug cable routing and always make sure the plug wires are in the correct plug wire bracket.
4. Connect the negative battery cable.

Spark Plug Wires

CHECKING & REPLACEMENT

At every tune-up/inspection, visually inspect the spark plug cables for burns cuts, or breaks in the insulation. Check the boots and the nipples on the distributor cap and coil. Replace any damaged wiring.

Every 50,000 miles (80,000 Km) or 60 months, the resistance of the wires should be checked with an ohmmeter. Wires with excessive resistance will cause misfiring, and may make the engine difficult to start in damp weather.

To check resistance, remove the distributor cap, leaving the wires attached. Connect one lead of an ohmmeter to an electrode within the cap; connect the other lead to the corresponding spark plug terminal (remove it from the plug for this test). Replace any wire which shows a resistance over 30,000 ohms. Test the high tension lead from the coil by connecting the ohmmeter between the center contact in the distributor cap and either of the primary terminals of the coil. If resistance is more than 25,000 ohms, remove the cable from the coil and check the resistance of the cable alone. Anything over 15,000 ohms is cause for replacement. It should be remembered that resistance is also a function of length; the longer the cable, the greater the resistance. Thus, if the cables on the vehicle are longer than the factory originals, resistance will be higher, and quite possibly outside these limits.

NOTE: The resistance reading given above is just a general specification.

When installing new cables, replace them one at a time to avoid mix-ups. Start by replacing the longest one first. Install the boot firmly over the spark plug. Route the wire over the same path as the original. Insert the nipple firmly into the tower on the cap or the coil. Check the spark plug cable routing and always make sure the plug wires are in the correct plug wire bracket.

Ignition Timing

Ignition timing is the measurement in degrees of crankshaft rotation of the instant the spark plug fires, in relation to the location of the piston (while the piston is on its compression stroke).

Although no periodic service is necessary, ignition timing can be adjusted by loosening the distributor locking device and turning the distributor in the engine.

Ideally, the air/fuel mixture in the cylinder will be ignited (by the spark plug) and just begin its rapid expansion as the piston passes top dead center (TDC) of the compression stroke. If this happens, the piston will be beginning the power stroke just as the compressed (by the movement of the piston) air/fuel mixture starts to expand. The expansion of the air/fuel mixture will then force the piston down on the power stroke and turn the crankshaft.

It takes a fraction of a second for the spark from the plug to completely ignite the mixture in the cylinder. Because of this, the spark plug must fire before the piston reaches TDC, if the mixture is to be completely ignited as the piston passes TDC. This measurement is given in degrees (of crankshaft rotation) Before the piston reaches Top Dead Center (BTDC). For example: if the ignition timing setting for an engine is seven (7°) BTDC, this means that the spark plug must fire at a time when the piston for that cylinder is 7° before top dead center of its compression stroke. However, this only holds true while ones engine is at idle speed.

As one accelerates from idle, the speed of the engine, in revolutions per minute (rpm), increases. The increase in rpm means that the pistons are now traveling up and down much faster. Because of this, the spark plugs will have to fire even sooner if the mixture is to be completely ignited as the piston passes TDC. To accomplish this, the ECU unit incorporates means to advance the timing of the spark as engine speed increases.

If ignition timing is set too far advanced (too far BTDC), the ignition and expansion of the air/fuel mixture in the cylinder will try to force the piston down the cylinder while it is still traveling upward. This causes engine "ping", a sound which resembles marbles being dropped into an empty tin can. If the ignition timing is too far retarded (after, or ATDC), the piston will have already started down on the power stroke when the air/fuel mixture ignites and expands. This will cause the piston to be forced down only a portion of its travel, resulting in poor engine performance and lack of power.

Ignition timing adjustment is checked with a timing light. This in-

GAP BRIDGED

IDENTIFIED BY DEPOSIT BUILD-UP CLOSING GAP BETWEEN ELECTRODES.

CAUSED BY OIL OR CARBON FOULING. REPLACE PLUG, OR, IF DEPOSITS ARE NOT EXCESSIVE THE PLUG CAN BE CLEANED.

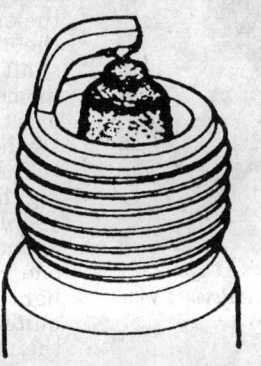

OIL FOULED

IDENTIFIED BY WET BLACK DEPOSITS ON THE INSULATOR SHELL BORE ELECTRODES.

CAUSED BY EXCESSIVE OIL ENTERING COMBUSTION CHAMBER THROUGH WORN RINGS AND PISTONS, EXCESSIVE CLEARANCE BETWEEN VALVE GUIDES AND STEMS, OR WORN OR LOOSE BEARINGS. CORRECT OIL PROBLEM. REPLACE THE PLUG.

CARBON FOULED

IDENTIFIED BY BLACK, DRY FLUFFY CARBON DEPOSITS ON INSULATOR TIPS, EXPOSED SHELL SURFACES AND ELECTRODES.

CAUSED BY TOO COLD A PLUG, WEAK IGNITION, DIRTY AIR CLEANER, DEFECTIVE FUEL PUMP, TOO RICH A FUEL MIXTURE, IMPROPERLY OPERATING HEAT RISER OR EXCESSIVE IDLING. CAN BE CLEANED.

NORMAL

IDENTIFIED BY LIGHT TAN OR GRAY DEPOSITS ON THE FIRING TIP.

PRE-IGNITION

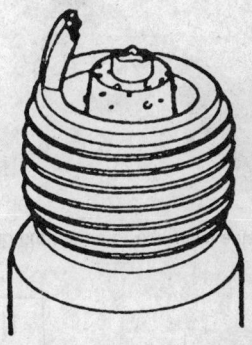

IDENTIFIED BY MELTED ELECTRODES AND POSSIBLY BLISTERED INSULATOR. METALIC DEPOSITS ON INSULATOR INDICATE ENGINE DAMAGE.

CAUSED BY WRONG TYPE OF FUEL, INCORRECT IGNITION TIMING OR ADVANCE, TOO HOT A PLUG, BURNT VALVES OR ENGINE OVERHEATING. REPLACE THE PLUG.

OVERHEATING

IDENTIFIED BY A WHITE OR LIGHT GRAY INSULATOR WITH SMALL BLACK OR GRAY BROWN SPOTS AND WITH BLUISH-BURNT APPEARANCE OF ELECTRODES.

CAUSED BY ENGINE OVER-HEATING, WRONG TYPE OF FUEL, LOOSE SPARK PLUGS, TOO HOT A PLUG, LOW FUEL PUMP PRESSURE OR INCORRECT IGNITION TIMING. REPLACE THE PLUG.

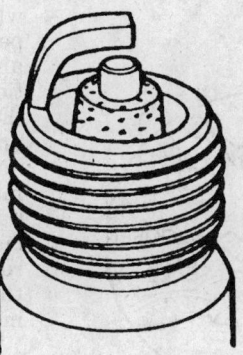

FUSED SPOT DEPOSIT

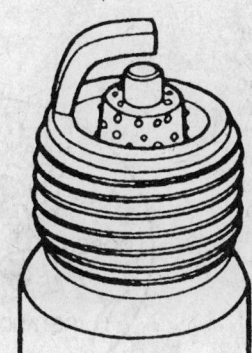

IDENTIFIED BY MELTED OR SPOTTY DEPOSITS RESEMBLING BUBBLES OR BLISTERS.

CAUSED BY SUDDEN ACCELERATION. CAN BE CLEANED IF NOT EXCESSIVE, OTHERWISE REPLACE PLUG.

TCCS2002

Inspect the spark plug to determine engine running conditions

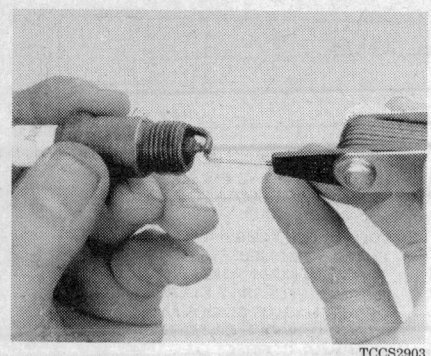

Check the spark plugs with a wire feeler gauge

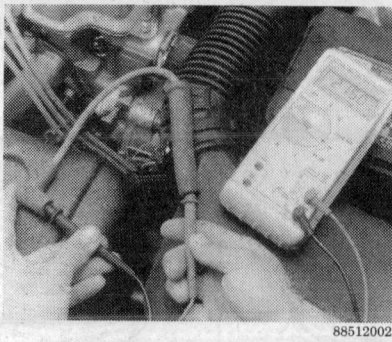

Checking individual plug wire resistance with a digital ohmmeter

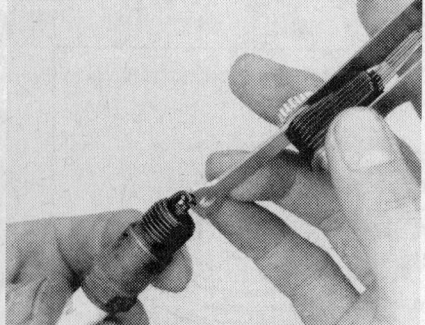

Bend the side electrode to adjust the gap

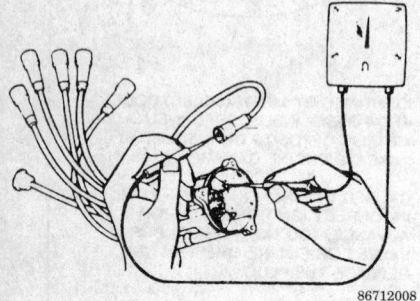

Checking plug wire resistance through the distributor cap with an ohmmeter

Checking and adjusting the ignition timing to specifications

low one to check the ignition timing. The timing light flashes every time the spark plug in the No. 1 cylinder of the engine fires. Since the flash from the timing light makes the crankshaft pulley seem stationary for a moment, one will be able to read the exact position of the piston in the No. 1 cylinder on the timing scale.

There are three basic types of timing lights available. The first is a simple neon bulb with two wire connections (one for the spark plug and one for the plug wire, connecting the light in series). This type of light is quite dim, and must be held closely to the marks to be seen, but it is inexpensive. The second type of light operates from the battery. Two alligator clips connect to the battery terminals, while a third wire connects to the spark plug with an adapter. This type of light is more expensive, but the xenon bulb provides a nice bright flash which can even be seen in sunlight. The third type replaces the battery source with 110 volt house current. Some timing lights have other functions built into them, such as dwell meters, tachometers, or remote starting switches. These are convenient, in that they reduce the tangle of wires under the hood, but may duplicate the functions of tools one already has.

For most vehicles, it is best to use a timing light with an inductive pickup. This pickup simply clamps onto the No. 1 plug wire, eliminating the adapter. It is not susceptible to crossfiring or false triggering, which may occur with a conventional light, due to the greater voltages produced by electronic ignition.

Idle Mixture Adjustment

Most vehicles today use a rather complex electronic fuel injection system which is regulated by a series of temperature, altitude (for California) and air flow sensors which feed information into an Electronic Control Unit (ECU). The control unit then relays an electronic signal to the injector nozzle(s), which allow(s) a predetermined amount of fuel into the combustion chamber. In this way all mixture control adjustments are regulated by the ECU, therefore on these vehicles no manual adjustments are necessary or possible.

Idle Speed Adjustment

Because of ECU control used on many vehicles today, no periodic ser-

strument is connected to the Number One (No. 1) spark plug of the engine. The timing light flashes every time an electrical current is sent from the distributor, through the No. 1 spark plug wire, to the spark plug. The crankshaft pulley and the front cover of the engine are marked with a timing pointer and a timing scale. When the timing pointer is aligned with the **0** mark on the timing scale, the piston for the No. 1 cylinder is at TDC of its compression stroke. With the engine running, and the timing light aimed at the timing pointer/scale, the flashes from the timing light will al-

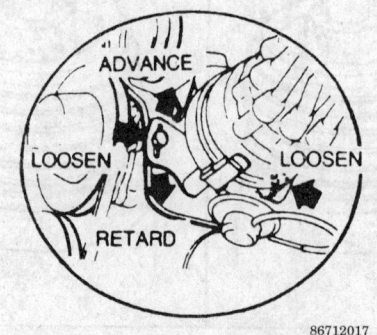

Adjust the ignition timing by rotating the distributor

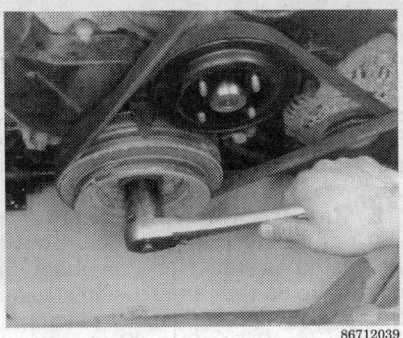

Timing marks on a 6-cylinder engine — note the fan was removed for a better view

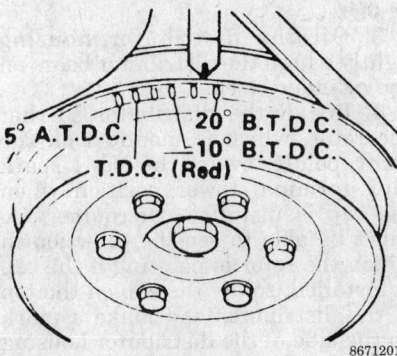

Typical timing marks

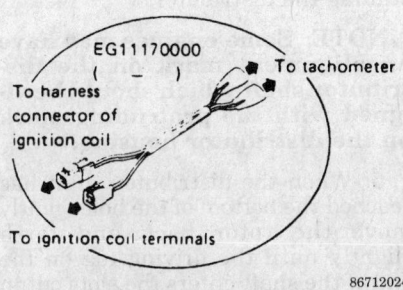

Some vehicles require a special harness for the tachometer connection to check idle speed

vice adjustments are necessary. If however the vehicle one is working with requires adjustment or an idle check, a general procedure is shown below. ALSO, always refer to the instructions or specifications found on the Vehicle Emission Control Information (VECI) label found underhood for additional or updated information which is applicable to a particular vehicle.

— CAUTION —
For manual transmission models, set parking brake and check idle speed in N position. For auto-

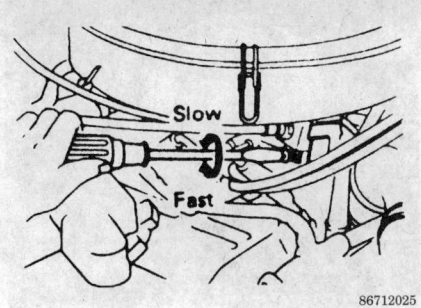

Idle speed adjustment — Carbureted engine

matic transmission equipped models, shifted into D for idle speed checks. When in DRIVE, the parking brake must be fully applied with both front and rear wheels chocked.

1. Turn **OFF** the: headlights, heater blower, air conditioning, and rear window defogger. If the vehicle has power steering, make sure the wheels are in the straight ahead position. The ignition timing must be correct to get an effective idle speed adjustment. Connect a tachometer (a special adapter harness may be needed) according to the instrument manufacturer's directions.

2. Start the engine and warm the engine so it reaches normal operating temperature. The water temperature indicator should be in the middle of the gauge.

— CAUTION —
NEVER run the engine in a closed garage. Always make sure there is proper ventilation to prevent carbon monoxide poisoning.

3. Run engine at 2000 rpm for about 2 minutes under no load.

4. Race the engine to 2000–3000 rpm a few times under no load and then allow it to return to idle speed.

5. Apply the parking brake securely. If equipped with an automatic, put the transmission into **DRIVE**.

6. Adjust the idle speed by turning the idle speed adjusting screw.

7. Turn the engine **OFF** and remove the tachometer. Road test for proper operation.

Distributor Cap

CHECKING & REPLACEMENT

Disconnect the negative battery cable. Individually disconnect each

ignition wire (one at a time) from the distributor cap and inspect the cap towers for corrosion build-up. Note, do not remove all of the spark plug wires from the cap, do this removing one wire, inspect the tower and wire contact then plug it back in. This will avoid getting the wires mixed up and out of the correct firing order. If all towers and spark plug wire contacts look good, make sure each wire is securely plug in its correct tower. Unfasten the retaining clips or unscrew the caps retaining screws and lift the cap off with the wires still attached. Inspect the underside of the distributor cap for cracks or carbon streaking between the contacts. Inspect the contacts for corrosion or wear. Replace the cap if any of the signs exists.

Sometimes water or condensation under the distributor cap will cause electrical current to short out between the contacts of the distributor cap or even wet ignition wires. If one has ever gone through a deep puddle of water and the engine stalls, chances are the ignition wires have become wet or the under the distributor cap. In this case it is possible to get the engine started by using a dry cloth and thoroughly drying the under side of the cap and the wires as best as possible.

Distributor

REMOVAL

1. Disconnect the negative battery cable.

2. Unfasten the retaining clips (only remove the coil wire if necessary) and lift the distributor cap straight off. It will be easier to install the distributor if the spark plug wiring is left connected to the cap. If the plug wires must be removed from the cap, mark their positions to aid in installation.

3. Remove the dust cover and mark the position of the rotor relative to the distributor body; then mark the position of the distributor body relative to the engine block.

4. Detach the harness assembly connector.

5. Remove the pinch-bolt and lift the distributor straight out, away from the engine. The rotor and body are marked so that they can be returned to the position from which they were removed. Do not turn or disturb the engine (unless absolutely necessary) after the distributor assembly has been removed.

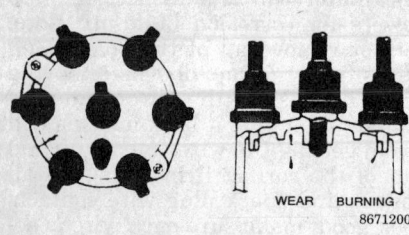

Check under the distributor cap for cracks; check the cable ends for wear

INSTALLATION

Timing Not Disturbed

1. Insert the distributor in the block and align all matchmarks made during removal.
2. Engage the distributor driven gear with the distributor drive.
3. Install the distributor clamp and secure it with the pinch-bolt.
4. Install the distributor cap and fasten the harness electrical connector.
5. If necessary, install the spark plug wires and coil wire.

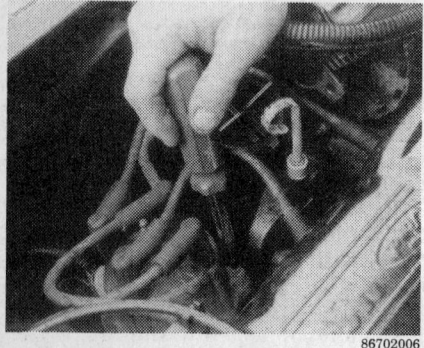

Unscrew the distributor cap retaining screws

Place the distributor cap and wires aside

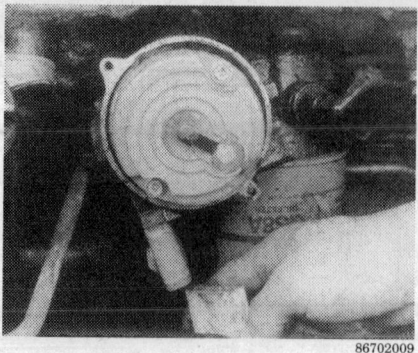

Unplug the distributor harness connector

Paint alignment marks on both the rotor cap and engine block

Remove the distributor retaining bolt

6. Start the engine. Check the timing and adjust it if necessary.

Timing Disturbed

This procedure gives a basic and simple way to install the distributor correctly if the engine was cranked while the distributor was out of the engine. Another reason this procedure may be helpful is if when the distributor was removed from the engine it was not marked for installa-

tion position as mentioned in the above procedure.

1. It is necessary to place the No. 1 cylinder in the firing position to correctly install the distributor. To locate this position, the ignition timing marks on the crankshaft front pulley can be used.
2. Remove the No. 1 cylinder spark plug. Turn the crankshaft until the piston in the No. 1 cylinder is moving up on the compression stroke. This can be determined by placing a thumb over the spark plug hole and feeling the air being forced out of the cylinder. Stop turning the crankshaft when the timing marks indicate **TDC** or **0**.
3. Oil the distributor housing lightly where the distributor bears on the cylinder block.
4. Install the distributor so that the rotor, which is mounted on the shaft, points toward the No. 1 spark plug terminal tower position when the cap is installed. Of course, one won't be able to see the direction in which the rotor is pointing if the cap is installed, so lay the cap on the top of the distributor and make a mark on the side of the distributor housing just below the No. 1 spark plug terminal. Make sure that the rotor points toward that mark when installing the distributor.

NOTE: Some engines may have an alignment mark on the distributor shaft which should be aligned with the protruding mark on the distributor housing.

5. When the distributor shaft has reached the bottom of the hole, gently move the rotor back and forth slightly until the driving lug on the end of the shaft enters the slots cut in the end of the oil pump shaft and the distributor assembly slides down into place.
6. Fasten the distributor hold-down bolt.
7. Install the spark plug into the No. 1 spark plug hole.
8. Install the distributor cap and engage the harness electrical connector.
9. If necessary, attach the spark plug wires and coil wire.
10. Start the engine. Check the timing and adjust it if necessary.

Ignition Coil

REMOVAL & INSTALLATION

1. Disconnect the negative battery cable.

Carefully remove the distributor

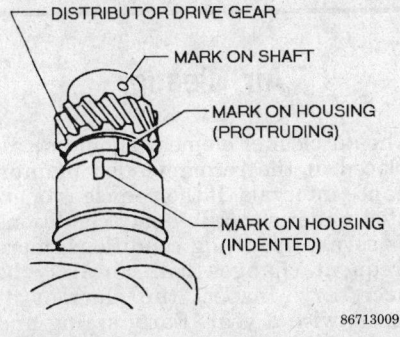

DISTRIBUTOR DRIVE GEAR
MARK ON SHAFT
MARK ON HOUSING (PROTRUDING)
MARK ON HOUSING (INDENTED)

Align the mark on the housing with the mark on the shaft

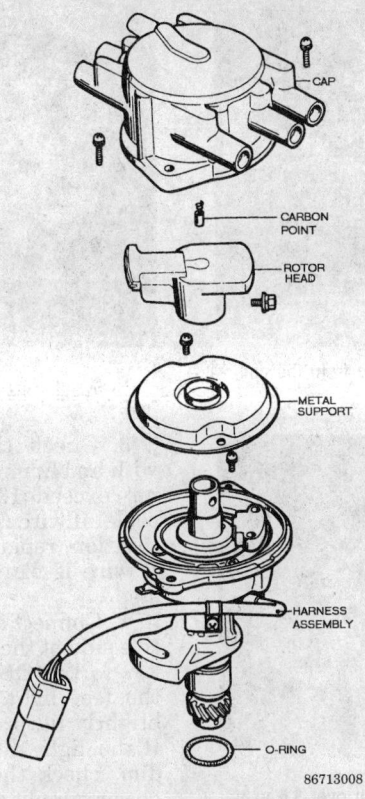

CAP
CARBON POINT
ROTOR HEAD
METAL SUPPORT
HARNESS ASSEMBLY
O-RING

Exploded view of the distributor assembly — typical 6-cylinder engine

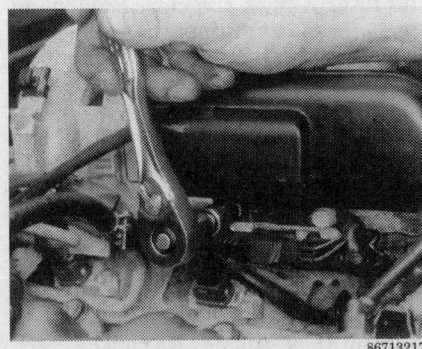

Remove the coil assembly mounting bolts

2. Remove the two mounting bolts and lift off the ignition coil.

3. Tag and disconnect all electrical leads at the coil.

4. Disconnect the coil high tension lead and remove the coil from the engine.

To install:

5. Connect all electrical leads to the coil as tagged.

6. Plug the high tension coil wire into the coil tower.

7. Install the coil in position and tighten the mounting bolts.

8. Connect the negative battery cable.

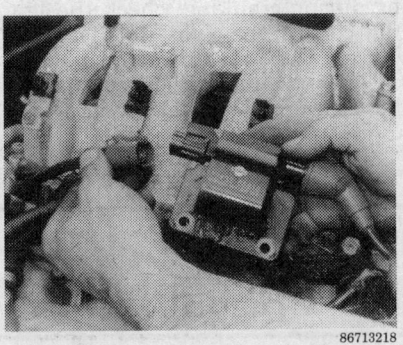

Disconnect the electrical connection from the coil assembly

ENGINE WILL NOT START

No Start Testing

1. Connect a voltmeter across the battery terminals. If battery voltage is not at least 12 volts, charge and test the battery before proceeding.

2. Turn the key to the **START** position and observe the voltmeter. If the engine turned over and battery voltage remained above 9.6 volts, go to next step. If the engine failed to crank and/or voltage was below 9.6 volts, proceed as follows:

　a. If the instrument panel lights dim, load test the battery, check the battery terminals and cables, test the starter motor and verify the engine turns.

　b. If the instrument panel lights do not dim, check the battery terminal connections, the ignition switch/wiring and the starter.

3. Using a spark tester, check for spark at two or more spark plugs. If okay go to next step, if not okay, perform No Spark Testing.

4. Cycle the ignition switch **ON** and **OFF**, several times, while listening for fuel pump operation. If fuel pump operates, proceed to next step. If fuel pump does not operate begin testing of the fuel pump circuit.

5. Verify adequate fuel in the tank, then connect a fuel pressure gauge and check fuel pressure. If fuel pressure is within specifications, proceed to next step, if not okay continue on checking the fuel pump and supply system.

6. Disconnect the fuel injector connector and connect a noid light to the wiring harness. Crank the engine,

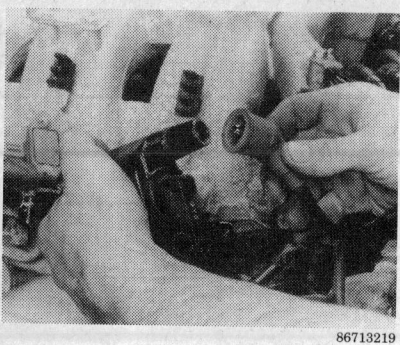

Remove the coil ignition wire from the coil assembly

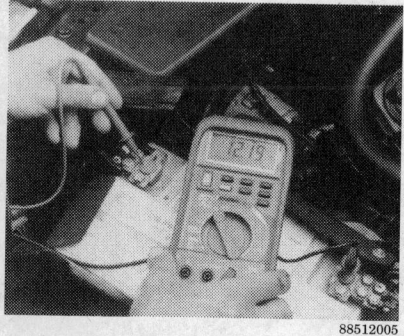

Battery voltage should remain over 9.6 volts, while the engine is cranking

while watching the light. Perform this test on at least two injectors before proceeding. If the light does not flash, go to the next step. If the light flashes, check the engine valve timing and overall mechanical condition of the engine. If okay, items such as; poor fuel quality, faulty injectors and computer controlled devices should be checked. Although these items are less likely, a shorted TPS or faulty coolant temperature sensor, are possibilities.

7. Check and verify the Malfunction Indicator Lamp (MIL) is operating properly. If the light does not operate, check the ECM and related wiring. If the MIL lamp is operational, check the injector wiring and circuitry.

No Spark Testing

1. Check for spark at two or more spark plugs. If spark does not exist go to next step, if spark is okay, check spark plugs, fuel system and engine mechanical condition.

2. Check for spark from the ignition coil wire. If spark does not exist go to next step, if spark is okay, check distributor cap, rotor and ignition wires.

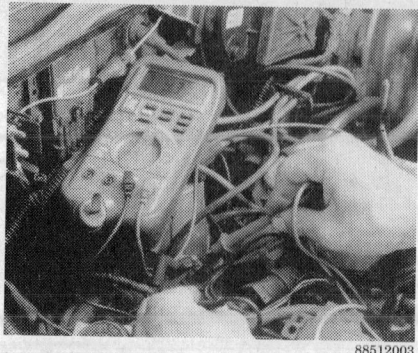

Testing the ignition coil resistance

3. Check the ignition coil wire with an ohmmeter. Resistance should not exceed 1000 ohms per inch of cable. If wire resistance exceeds specification, replace the wire and retest. If wire is okay, proceed to the next step.

4. Connect a test light to the negative side of the ignition coil. Turn the key to the **ON** position and observe the test light. If the light remains brightly lit, proceed to the next step. If the light did not light, or glowed dim, check the ignition switch and power supply circuit.

5. Observe the light while cranking the engine. If the light was flashing during cranking, check the ignition coil. If the light did not flash, verify the distributor rotates smoothly, then test the ignition module, pick-up coil or hall effect switch.

RELIEVING FUEL PRESSURE

The procedure is a generic procedure for most fuel injected vehicles. Carbureted vehicles may not use an electric fuel pump. Carbureted systems are lower pressure.

1. Disable the fuel pump by one of the following methods:
• Remove the fuel pump fuse.
• Remove the fuel pump relay.
• Locate and disconnect the fuel pump wiring.

NOTE: When removing the fuel pump fuse or relay to disable the fuel pump, it is important to make certain that the fuel injectors are not part of this circuit. If the injectors do not operate the residual fuel system pressure will not be relieved.

2. Start the engine and operate it until it stalls. Once the engine has stalled, crank the starter for an additional 10 seconds.

3. Place a rag over the connection in which one intends to disconnect and carefully separate the connections. Use the rag to absorb any remaining fuel.

PREVENTIVE MAINTENANCE

Air Cleaner

The air cleaner element should be replaced at the recommended maintenance intervals. If the vehicle is operated under severely dusty conditions or severe operating conditions, more frequent changes will certainly be necessary. Inspect the element at least twice a year. Early spring and early fall are good times for an inspection. Remove the element and check for any perforations or tears in the filter. Check the cleaner housing for signs of dirt or dust that may have leaked through the filter element or in through the snorkel tube. Position a droplight on one side of the element and look through the filter at the light. If no glow of light can be seen through the element material, replace the filter. If holes in the filter element are apparent, or signs of dirt seepage through the filter are evident, replace the filter.

REMOVAL & INSTALLATION

Air cleaners come a wide selection of shapes and sizes. Most common are either round or rectangular. In any event, air filter element replacement is usually pretty simple.

If the vehicle is equipped with a round type air cleaner, it probably has one or two wing nuts and/or some clips holding the air cleaner lid in place. Just remove the wing nuts and unclip the retaining clips. Lift the lid off and remove the element.

If the vehicle is equipped with a rectangular type air cleaner, unclip the retaining clips and lift off the air cleaner lid. remove the element from the assembly. Some these units may not be as easily accessible and may require removing a hose or two, but all in all it's pretty simple to do.

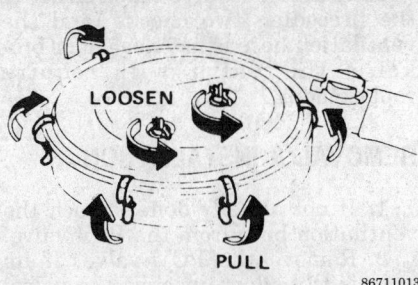

Removing the air cleaner filter element — round type

Removing the air cleaner filter element — rectangular

Air Cleaner Assembly (Housing)

1. Disconnect all hoses, ducts and vacuum tubes from the air cleaner assembly, after tagging them for easy identification.

2. Remove the top cover wing nuts and grommet (if so equipped). Most models also utilize four to five side clips to further secure the top of the assembly. Simply pull the wire tab and release the clip. On many vehicles, air cleaners are secured solely by means of clips (air box-to-cleaner housing). Remove the cover and lift out the filter element.

3. Remove any side mount brackets and/or retaining bolts, then lift off the air cleaner assembly.

4. Clean or replace the filter element as detailed previously. Wipe clean all surfaces of the air cleaner housing and cover. Check the condition of the mounting gasket and replace if it appears worn or broken.

5. Reposition the air cleaner assembly, then install the mounting bracket and/or bolts.

6. Reposition the filter element in the case and install the cover being careful not to overtighten the wingnut(s). On round-style cleaners, be certain that the arrows on the cover

lid and the snorkel match up properly.

NOTE: Filter elements on many engines have a TOP and BOTTOM side; be sure they are inserted correctly.

7. Reconnect all hoses, duct work and vacuum lines.

NOTE: Never operate the engine without the air filter element in place.

Air Cleaner Element

The air cleaner element can be replaced by removing the wingnut(s) and/or side clips, then removing the top cover as previously detailed.

Crankcase Ventilation Filter

Certain models may also utilize an air cleaner-mounted crankcase ventilation filter. If so, it should also be cleaned or replaced at the same time as the regular air filter element. To replace the filter, remove the air cleaner top cover and pull the filter from its housing on the side of the air cleaner assembly. Push a new filter into the housing and reinstall the cover. If the filter and plastic holder

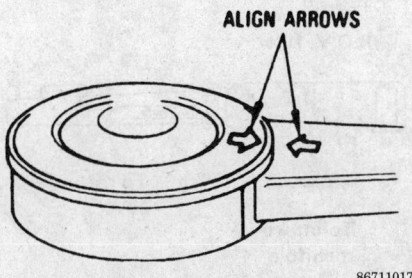

Many air cleaner assemblies have arrows on the housing and lid — always make sure they align

The air filter element may be cleaned with low pressure compressed air

need replacement, remove the clip mounting the feeder tube to the air cleaner housing, then remove the assembly from the air cleaner.

Fuel Filter

REMOVAL & INSTALLATION

———— CAUTION ————
NEVER SMOKE WHEN WORKING AROUND OR NEAR GASOLINE! MAKE SURE THAT THERE IS NO ACTIVE IGNITION SOURCE NEAR YOUR WORK AREA!

———— WARNING ————
Never attempt to remove the fuel filter without first relieving the fuel system pressure!

1. Release the fuel pressure from the fuel line as follows:
   a. Remove the fuel pump fuse at the fuse box.
   b. Start the engine.
   c. After the engine stalls, crank the engine two or three times to make sure that the fuel pressure is released.
   d. Turn the ignition switch **OFF** and reinstall the fuel pump fuse.
2. Loosen the hose clamps at the fuel inlet and outlet lines. Wrap a shop towel or absorbent rag around the filter, then slide each line off the filter nipples.
3. Remove the fuel filter and old hose clamps.
   To install:
4. Place new hose clamps on the fuel inlet and outlet lines.
5. Connect the fuel filter, being careful to observe the correct direction of flow, then tighten the hose clamps.
6. Start the engine and check for fuel leaks.

NOTE: Always use a high pressure-type fuel filter assembly. Do not use a synthetic resinous fuel filter.

PCV Valve

The PCV valve regulates crankcase ventilation during various engine operating conditions. At high vacuum (idle speed and partial load range) it will open slightly, and at low vacuum (full throttle) it will open fully. This causes vapor to be removed from the crankcase by the engine vacuum and

Loosen the air intake hose clamp before removing the filter element

Disconnect the air intake hose, being careful not to lose the retaining clamp

Unfasten the side retaining clamps so that the air filter housing can be opened

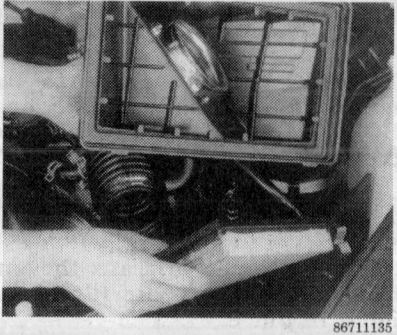

Remove the air filter element from the air filter housing

View of the air filter element. Make sure that the element is installed in the housing properly before fastening the side clamps

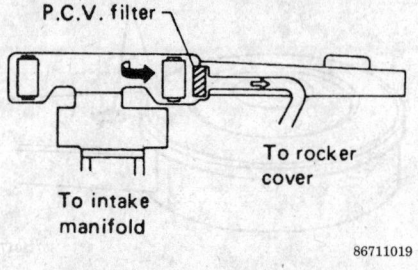

Crankcase ventilation filter replacement

then be sucked into the combustion chamber where it is burned.

NOTE: The PCV system will not function properly unless the oil filler cap is tightly sealed. Check the gasket on the cap and be certain it is not leaking. Replace the cap and/or gasket, if necessary, to ensure proper sealing.

TESTING

1. Check the ventilation hoses and lines for leaks or clogging. Clean or replace as necessary.
2. With the engine running at idle, locate the PCV valve in the cylinder head cover or intake manifold and remove the ventilation hose from the valve; a strong hissing sound should be heard as air passes through the valve.
3. With the engine still idling, place a finger over the valve; a strong vacuum should be felt.

4. If the PCV valve failed either of the preceding two checks (and the ventilation hose is not clogged or broken), the valve will require replacement.

REMOVAL & INSTALLATION

1. If not already done, detach the ventilation hose from the PCV valve.
2. Remove the PCV valve. If its base is threaded, unscrew the valve; otherwise, simply pull the valve from its retaining grommet.
   To install:
3. Depending on the type of valve, either screw in the replacement PCV valve or push it into its retaining grommet.
4. Slide the ventilation hose onto the end of the PCV valve.

Air Induction Valve Filter

REMOVAL & INSTALLATION

Regular maintenance for this component includes a check of the drive belt tension and replacement of the air pump air filter at the specified interval. The air filter case is located in the left front of the engine compartment on most models. To replace the air filter, simply unscrew the wing nut(s) securing the cover to the case, withdraw the old filter, install the new one, and reinstall the case.

Battery

NOTE: On a maintenance-free sealed battery, a built-in hydrometer or "eye" is used for checking the fluid level and specific gravity readings. If the battery is equipped with an eye, use it for checking the condition of the battery by observing the color of the eye. A green colored eye indicates good condition and a dark colored eye indicates the need for service. Replacement batteries could be either the sealed (maintenance-free) or non-sealed type.

FLUID LEVEL (EXCEPT MAINTENANCE-FREE SEALED BATTERIES)

Check the battery electrolyte level at least once a month, or more often in hot weather or during periods of extended operation. The level can be checked through the case on translu-

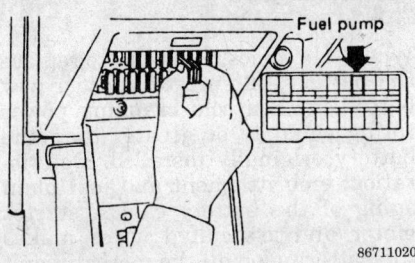

86711020

Remove the fuel pump fuse when releasing the fuel pressure — the fuse's location may vary in the box

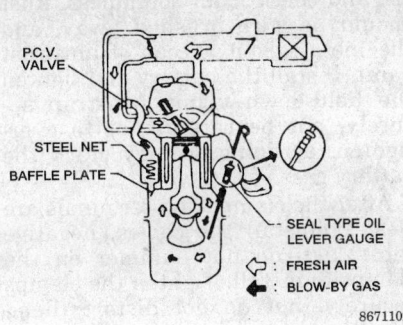

86711027

PCV valve location — typical 4-cylinder engine

86711137

Remove the fuel line hose clamp after releasing the fuel pressure

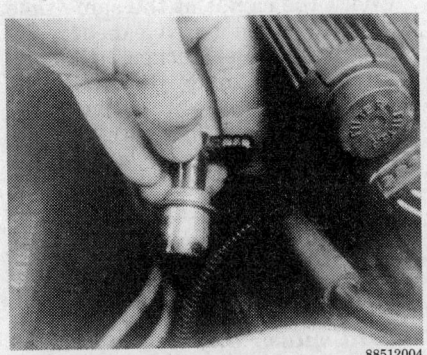

88512004

Removing the PCV valve

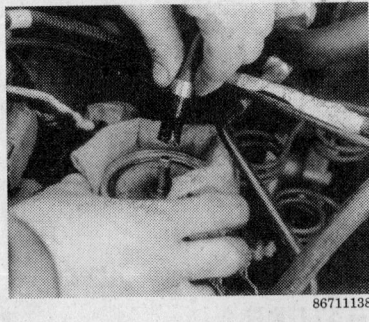

86711138

When removing the fuel line from the fuel filter, have a shop towel in position to catch any fuel that may spill from the filter

cent polypropylene batteries; the cell caps must be removed on other models. The electrolyte level in each cell should be kept filled to the bottom of the split ring inside, or to the line marked on the outside of the case.

If the level is low, add only distilled water, or colorless, odorless drinking water, through the opening until the level is correct. Each cell is completely separate from the others, so each must be checked and filled individually.

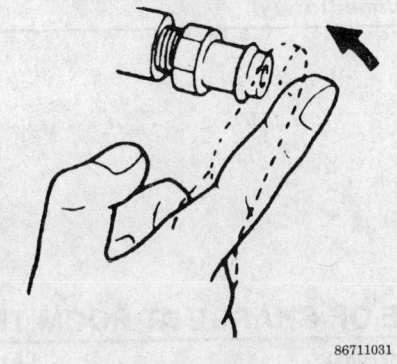

86711031

Checking the PCV valve

If water is added in freezing weather, the vehicle should be driven several miles to allow the water to mix with the electrolyte. Otherwise, the battery could freeze.

SPECIFIC GRAVITY (EXCEPT MAINTENANCE-FREE BATTERIES)

NOTE: On a maintenance-free sealed battery, a built-in eye is used for checking the specific gravity readings. Refer to the battery case for further instructions.

At least once a year, check the specific gravity of the battery. It should be 1.26–1.28 at room temperature.

The specific gravity can be checked with the use of a hydrometer, an inexpensive instrument available from many sources, including auto parts stores. The hydrometer has a squeeze bulb at one end and a nozzle at the other. Battery electrolyte is sucked into the hydrometer until the float is lifted from its seat. The specific gravity is then read by noting the position of the float. Generally, if after charging, the specific gravity of any two cells varies more than 50 points (0.50), the battery is bad and should be replaced.

It is not possible to check the specific gravity in this manner on sealed (maintenance-free) batteries. Instead, the indicator built into the top of the case must be relied on to display any signs of battery deterioration. On most batteries if the indicator is a light color, the battery can be assumed to be OK. If the indicator is a dark color, the specific gravity is low, and the battery should be charged or replaced. There should be specific notations on the battery to what color the indicator will be depending on the batteries state of charge.

CABLES AND CLAMPS

Once a year, the battery terminals and the cable clamps should be checked and cleaned, if necessary. Make sure that the ignition switch is turned to the OFF position. Loosen the clamps and remove the cables, negative cable first. On batteries with posts on top, the use of a puller specially made for this purpose is recommended. These are inexpensive, and available in most auto parts stores. Side terminal battery cables are secured with a bolt.

Clean the cable clamps and the battery terminal with a wire brush, until all corrosion, grease, etc., is removed and the metal is shiny. It is especially important to clean the inside of the clamp thoroughly, since a small deposit of foreign material or oxidation there will prevent a sound electrical connection and inhibit either starting or charging. Special tools are available for cleaning these parts, one type for top post batteries and another type for side terminal batteries.

Before installing the cables, loosen the battery hold-down clamp or strap, remove the battery and check the battery tray. Clear it of any deb-

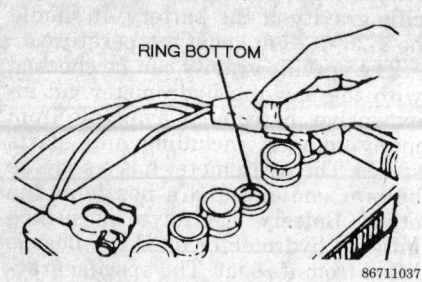

RING BOTTOM

86711037

Fill each battery cell to the bottom of the split ring with distilled water

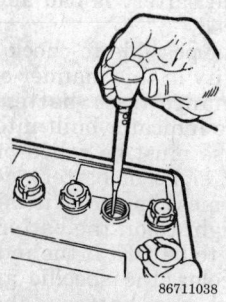

86711038

The specific gravity of the battery can be checked with a simple float-type hydrometer

ris, and check it for soundness. Rust should be wire brushed away, and the metal given a coat of anti-rust paint. Install the battery and tighten the hold-down clamp or strap securely, but be careful not to overtighten, as doing so may crack the battery case.

After the clamps and terminals are clean, reinstall the cables, negative cable last; do not hammer on the clamps to install. Tighten the clamps securely, but do not distort them. Give the clamps and terminals a thin external coat of grease after installation, to retard corrosion.

Check the cables at the same time that the terminals are cleaned. If the cable insulation is cracked or broken, or if the ends are frayed, the cable should be replaced with a new cable of the same length and gauge.

---- **CAUTION** ----

Keep flame or sparks away from the battery; it gives off explosive hydrogen gas! Battery electrolyte contains sulfuric acid! If one should splash any on the skin or in the eyes, flush the affected area with plenty of clear water. If it lands in the eyes, get medical help immediately!

REPLACEMENT

When it becomes necessary to replace the battery, be sure to select a new battery with a cold cranking power rating equal to or greater than the battery originally installed. Deterioration, embrittlement and just plain aging of the battery cables, starter motor and associated wires makes the battery's job all the more difficult in successive years. The slow increase in electrical resistance over time makes it prudent to install a new battery with a greater capacity than the old.

REMOVAL & INSTALLATION

1. Make sure the ignition switch is turned **OFF**.
2. Disconnect the negative battery cable from the terminal, then disconnect the positive cable. Special pullers are available to remove the clamps.

NOTE: To avoid sparks, always disconnect the negative cable first and reconnect it last.

3. Unscrew and remove the battery hold-down clamp.

BATTERY STATE OF CHARGE AT ROOM TEMPERATURE

Specific Gravity Reading	Charged Condition
1.260–1.280	Fully Charged
1.230–1.250	3/4 Charged
1.200–1.220	1/2 Charged
1.170–1.190	1/4 Charged
1.140–1.160	Almost no Charge
1.110–1.130	No Charge

86711039

Battery state of charge at room temperature — Generalized Specifications

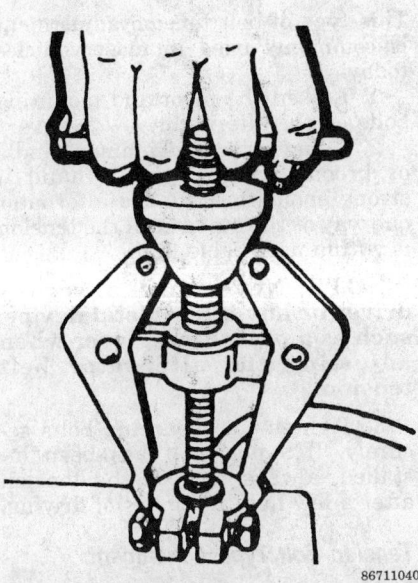

Special pullers are available to remove cable clamps

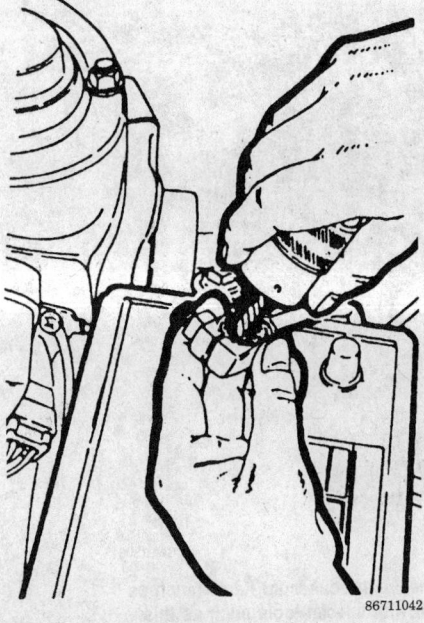

Clean the inside of the clamps with a wire brush or the special tool

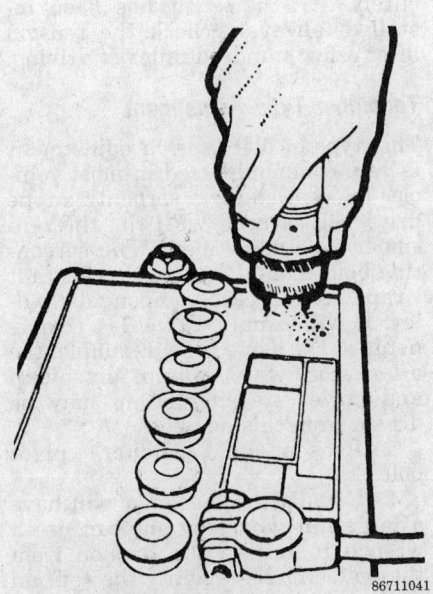

Clean the battery posts with a wire brush or the special tool shown

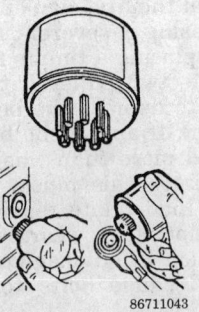

Special tools are also available for cleaning the posts and clamps of side terminal batteries

4. Remove the battery, being careful not to spill any of the acid.

NOTE: Spilled acid can be neutralized with a baking soda and water solution. If somehow one gets acid into the eyes, flush it out with lots of clean water and get to a doctor as quickly as possible.

To install:
5. Clean the battery posts thoroughly.

6. Clean the cable clamps using the special tools or a wire brush, both inside and out.
7. Install the battery, then fasten the hold-down clamp.
8. Connect the positive and then the negative cable. Do not hammer them into place. Coat the terminals with grease to prevent corrosion.

----- CAUTION -----
Make absolutely sure that the battery is connected properly before turning on the ignition switch. Reversed polarity can burn out the alternator and regulator in a matter of seconds.

Drive Belts

INSPECTION

Check the condition of the drive belts, and check the belt tension at least every 30,000 miles (48,000 km) or every 24 months.

Periodic inspection of the drive belts is important because of the following reasons; first of all, the drive belts drive various components such as the engine water pump, alternator, power steering pump and emission pump, etc.

Two of the components mention above play a vital part in keeping the engine running. They are the alternator and water pump. To give one a little example of how important drive belt inspection is picture this; suppose the alternator belt were to break due to wear or cracking, the alternator would be completely disabled and the battery would eventually go dead.

In case a like this one may find themselves sitting on the side of the road seeking someone to give a jump to get started. Not to mention, a possible tow job, battery charge and replacement of that drive belt that could have been detected during the inspection.

The water pump drive belt could even cause more severe complications, how about an excessively overheated engine. This could result in a very expensive engine repair, like a head gasket replacement, etc. So be sure to keep a good maintenance check on the drive belts.

1. Inspect the belts for signs of glazing or cracking. A glazed belt will be perfectly smooth from slippage, while a good belt will have a slight texture of fabric visible. Cracks will generally start at the inner edge of the belt and run outward. Replace

TCCS1206

Battery maintenance may be accomplished with household items (such as baking soda to neutralize spilled acid) or with special tools such as this post and terminal cleaner

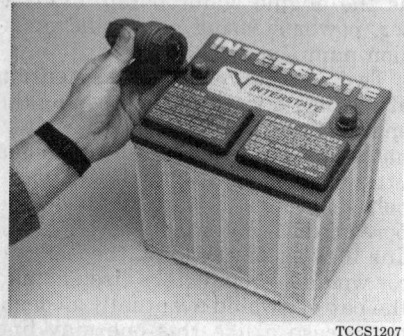

TCCS1207

The underside of this special battery tool has a wire brush to clean post terminals

TCCS1208

Place the tool over the terminals and twist to clean the post

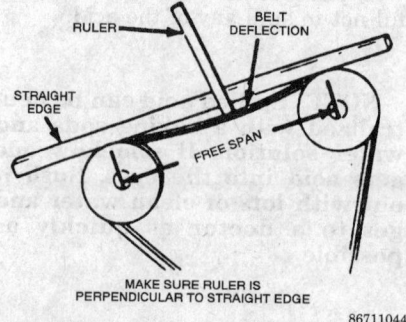

RULER
BELT DEFLECTION
STRAIGHT EDGE
FREE SPAN
MAKE SURE RULER IS PERPENDICULAR TO STRAIGHT EDGE

86711044

Measuring belt deflection with a straightedge and ruler

the belt at the first sign of cracking or if the glazing is severe.

2. By placing a thumb midway between the two pulleys, it should be possible to depress the belt ¼–½ in. (6–13mm). If any of the belts can be depressed more than this, or cannot be depressed this much, adjust the tension. Inadequate tension will result in slippage or wear, while excessive tension will damage pulley bearings and cause belts to fray and crack.

3. It's not a bad idea to replace all drive belts at 60,000 miles (96,000 km) or 48 months, regardless of their condition.

ADJUSTMENT

Pivot Type Adjustment

This type of belt tension adjustment is commonly used in most vehicles today.

1. Loosen the pivot and mounting bolts on the alternator.

2. Using a wooden hammer handle or broomstick (or even a hand if strong enough), move the alternator one way or the other until the tension is within acceptable limits.

NOTE: Never use a screwdriver or any other metal device, such as a prybar, as a lever when adjusting the alternator belt tension!

3. Tighten the mounting bolts securely. If a new belt has been installed, always recheck the tension after a few hundred miles of driving.

Tension Bolt Type Adjustment

Some belt tensions are adjusted by means of a tension adjusting bolt. This method of adjustment may use an idler pulley or the component being moved may slide on a bracket to increase or decrease belt tension.

1. Loosen the pivot bolt, then turn the adjusting bolt until proper tension is achieved.

2. Tighten the mounting bolts securely. If a new belt has been installed, always recheck the tension after a few hundred miles of driving.

Tensioner Type Adjustment

This type of belt tension adjustment is very commonly used in most vehicles today. Usually a serpentine type drive belt is used with the the tensioner type adjustment. The serpentine belt is one large single belt that wraps around each component's pulley. It will usually drive 3–4 components at the same time. Example: the alternator, water pump, air pump and power steering pump may be driven from this one belt.

1. Loosen the tensioner's pivot bolt.

2. Usually the tensioner will have a large nut where by one can use a wrench to relieve the tension from the belt. Moving against the tension of the tensioner will relieve the tension on the the belt. Allowing the tension of the tensioner to release will increase the tension on the belt within acceptable limits.

3. Tighten the pivot bolt securely. If a new belt has been installed, always recheck the tension after a few hundred miles of driving.

Timing Belt

INSPECTION

The timing belt is a bit more involved and a more critical service procedure. Although one can service the timing belt, remember that the correct procedures must be followed exactly. Be sure to have the correct repair manual for the vehicle. If possible, enlist the aid of someone who is experienced with timing belt replacement, it would be helpful.

NOTE: Do not bend or twist the timing belt. If the timing belt breaks while driving, or the crankshaft and/or camshaft are turned separately after the timing belt is removed, valves may strike the piston heads, causing engine damage. Make sure the timing belt and tensioner are clean and free from oil and water.

As a average rule, replace the timing belt at 60,000 miles (96,000 km).

Evaporative Canister

SERVICING

Check the evaporation control system, if so equipped, every 15,000 miles (24,000 km) or every 12 months. Check the fuel and vapor lines/hoses for proper connections, correct routing, and condition. Replace damaged or deteriorated parts as necessary.

To check the operation of the carbon canister purge control valve, disconnect the rubber hose between the canister control valve and the T-fitting at the T-fitting. Apply vacuum to the hose leading to the control valve. The vacuum condition should be maintained indefinitely. If the control valve leaks, remove the top cover of the valve and check for a dislocated or cracked diaphragm. If the diaphragm is damaged, a repair kit containing a new diaphragm, retainer, and spring is available and should be installed.

The carbon canister has a replaceable air filter in the bottom of the canister. The filter element should be checked once a year or every 15,000 miles (24,000 km); more frequently if the vehicle is operated in dusty areas. Replace the filter by pulling it out of the bottom of the canister and installing a new one.

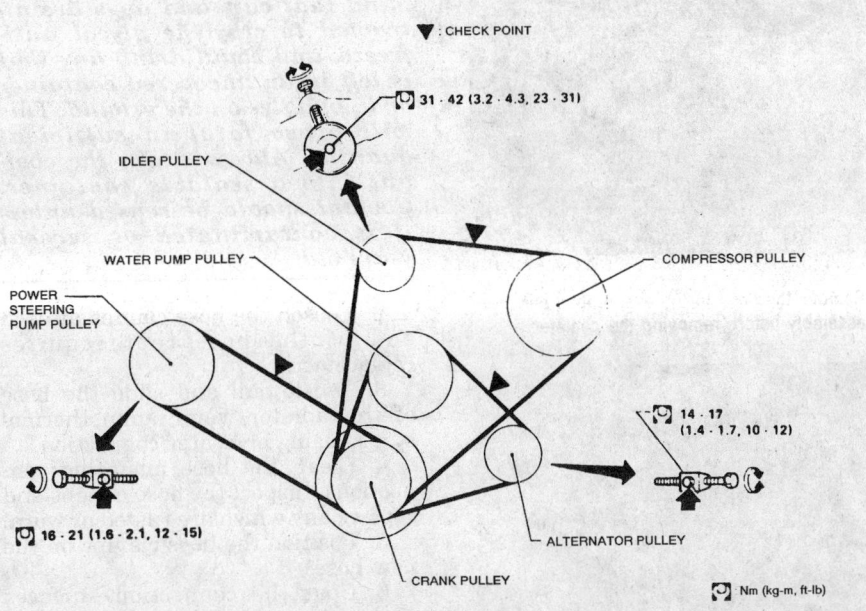

Drive belt tension inspection and adjustment points — typical

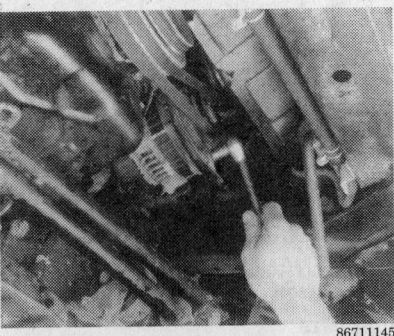

On some vehicles it is easier to access a component from underneath the vehicle

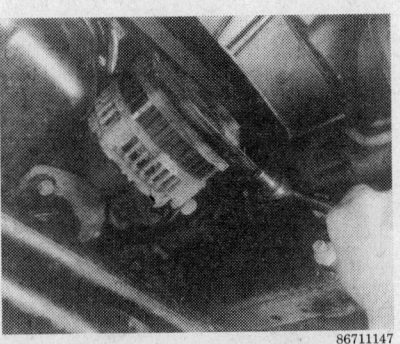

Use the adjusting bolt to vary tension on the belt

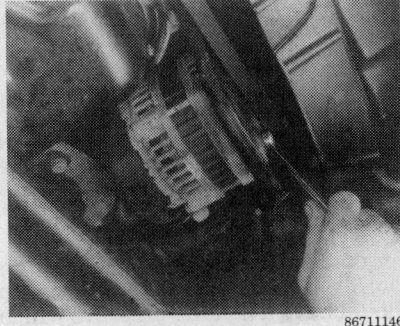

Loosen the alternator pivot bolt with a box wrench or a ratchet and socket

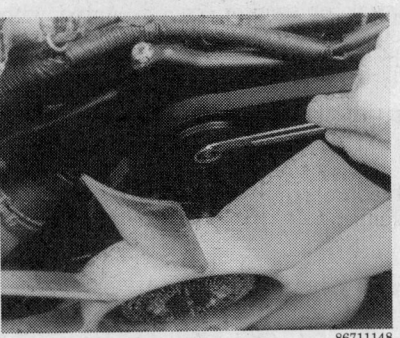

Loosen the locknut on the idler pulley before adjusting the belt

Turn the adjusting bolt until the correct belt tension is achieved

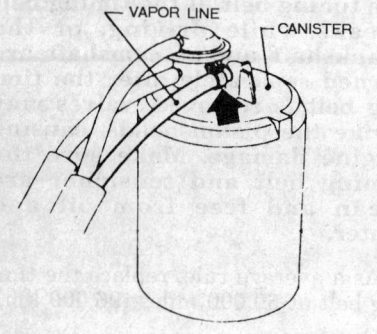

VAPOR LINE — CANISTER

Checking the evaporative canister

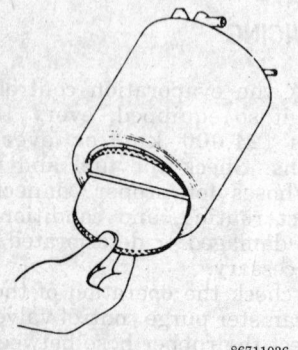

Replacing the evaporative canister filter

Hoses

INSPECTION

Inspect the condition of the radiator hoses, heater hoses and clamps periodically. Early spring and late fall are often good times to perform this, as well as other routine maintenance. Make sure the engine and cooling system are cold. Visually inspect for cracked, rotted or collapsed hoses,

Remove the lines to the evaporative canister assembly before removing the canister

Unfasten the evaporative canister assembly retaining clamp

Remove the evaporative canister assembly from the vehicle

and replace as necessary. Run a hand along the length of the hose. If a weak or swollen spot is noted when squeezing the hose wall, replace the hose.

REPLACEMENT

1. Drain the coolant into a suitable container (if the coolant is to be reused).

2. Loosen the hose clamps at each end of the hose that requires replacement.

3. Twist, pull and slide the hose off the radiator, water pump, thermostat housing or heater connection.

4. Clean the hose mounting connections. Inspect the hose clamps and replace any which are rusted or worn.

5. Position the hose clamps on the new hose.

6. Coat the connection surfaces with a water resistant sealer or equivalent and slide the hose into position. Make sure the hose clamps are located beyond the raised bead of the connector (if equipped) and centered in the clamping area of the connection.

7. Tighten the clamps evenly. Do not overtighten.

8. Refill the cooling system.

9. Start the engine and allow it to reach normal operating temperature. Check for coolant leaks, then top off the coolant level as necessary.

ADDITIONAL PREVENTIVE MAINTENANCE CHECKS

Antifreeze

In order to prevent heater core freeze-up during A/C operation, it is necessary to maintain permanent-type antifreeze protection of +15°F (-9°C) or lower. A reading of -15°F (-26°C) is ideal since this protection also supplies sufficient corrosion inhibitors for the protection of the engine cooling system.

NOTE: The same antifreeze should not be used longer than the manufacturer specifies.

Radiator Cap

For efficient operation of the vehicle's cooling system, the radiator cap should have a holding pressure which meets manufacturer's specifications. A cap which fails to hold the specified pressure should be replaced.

FLUIDS AND LUBRICANTS

Fluid Disposal

Used fluids such as engine oil, transmission fluid, antifreeze and brake fluid are hazardous wastes and must be disposed of properly. Before draining any fluids, consult with the local authorities; in many areas, waste oil, antifreeze, etc. are being accepted as a part of recycling programs. A number of service stations and auto parts stores are also accepting waste fluids for recycling.

Be sure of the recycling center's policies before draining any fluids, as many will not accept different fluids that have been mixed together, such as oil and antifreeze.

Oil and Fuel Recommendations

ENGINE OIL

The SAE (Society of Automotive Engineers) grade number indicates the viscosity of the engine oil (its resistance to flow at a given temperature). The lower the SAE grade number, the lighter the oil. For example, the mono-grade oils begin with SAE 5 weight, which is a thin, light oil, and continue in viscosity up to SAE 80 or 90 weight, which are heavy gear lubricants. These oils are also known as "straight weight," meaning they are of a single viscosity, and do not vary with engine temperature.

Multi-viscosity oils offer the important advantage of being adaptable to

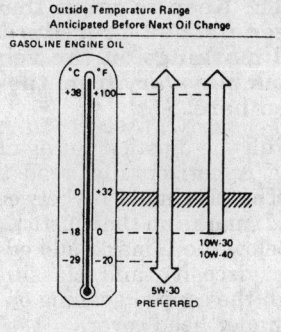

Engine oil viscosity chart — Typical

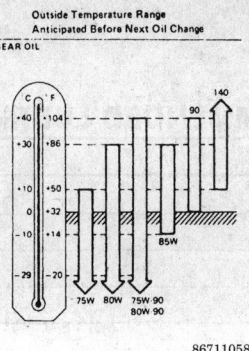

Gear oil viscosity chart — Typical

temperature extremes. These oils have designations such as 10W-40, 20W-50, etc. For example, 10W-40 means that in winter (the "W" in the designation) the oil acts like a thin 10 weight oil, allowing the engine to spin easily when cold and offering rapid lubrication. Once the engine has warmed up, however, the oil acts like a straight 40 weight, maintaining good lubrication and protection for the engine's internal components. A 20W-50 oil would therefore be slightly heavier than, and not as ideal, in cold weather as the 10W-40, but would offer better protection at higher rpm and temperatures because, when warm, it acts like a 50 weight oil. Whichever oil viscosity is chosen when changing the oil and filter, one is anticipating the temperatures the engine will be operating in until the oil is changed again. Refer to the oil viscosity chart that applies to the specific vehicle for oil recommendations according to temperature.

The API (American Petroleum Institute) designation indicates the classification of engine oil used under certain given operating conditions. Only oils designated for use "Service SG" should be used. Oils of the SG type perform a variety of functions inside the engine in addition to the basic function as a lubricant. Through a balanced system of metallic detergents and polymeric dispersants, the oil prevents the formation of high and low temperature deposits, and also keeps sludge and dirt particles in suspension. Acids, particularly sulfuric acid, as well as other by-products of combustion, are neutralized. Both the SAE grade number and the API designation can be found on the oil container.

Synthetic Oil

There are many excellent synthetic and fuel-efficient oils currently available that can provide better gas mileage, longer service life and, in some cases, better engine protection. These benefits do not come without a few hitches, however, the main one being the price of synthetic oils, which is three or four times the price per quart of conventional oil.

Synthetic oil is not for every vehicle and every type of driving, so one should consider ones engine's condition and the type of driving. Also, check the vehicle's warranty conditions regarding the use of synthetic oils.

Brand new engines and older, high mileage engines are not good candidates for synthetic oil. The synthetic oils are so slippery that they can prevent the proper break-in of new engines; most manufacturers recommend that one waits until the engine is properly broken in (3000 miles) before using synthetic oil. Older engines with wear have a different problem with synthetics: they "use" (consume during operation) more oil as they age. Slippery synthetic oils get past these worn parts easily. If the engine is using conventional oil, it will use synthetics much faster. Also, if the vehicle is leaking oil past old seals, there will be a much greater leak problem with synthetics.

Consider the type of driving. If most of the accumulated mileage is high speed, highway type driving, the more expensive synthetic oils may be a benefit. Extended highway driving gives the engine a chance to warm up, accumulating fewer acids in the oil, and putting less stress on the engine over the long run. Under these conditions, the oil change interval can be extended (as long as the oil filter can last the extended life of the oil) up to the advertised mileage claims of the synthetics. Vehicles with synthetic oils may show increased fuel economy in highway driving, due to less internal friction. However, many automotive experts agree that 50,000 miles (80,000 km) is too long to keep any oil in the engine.

Vehicles used under harsher circumstances, such as stop-and-go, city type driving, short trips, or extended idling, should be serviced more frequently. For the engines in these vehicles, the much greater cost of synthetic or fuel-efficient oils may not be worth the investment. Internal wear increases much quicker on these ve-

RECOMMENDED LUBRICANTS

Component	Lubricant
Engine oil	API SG
Coolant	Ethylene Glycol-based Antifreeze
Manual Transmission	API GL-4, SAE 75W-90
Automatic Transmission	ATF DEXRON®
Transfer Case	1989: API GL-4, SAE 75W-90
	1990-95: ATF DEXRON®
Differentials	API GL-5, SAE 80W-90
Limited Slip	Nissan-approved LSD
Master Cylinder	DOT 3, SAE J1703
Power Steering	ATF DEXRON®
Manual Steering	API GL-4, SAE 90W
Multi-Purpose Grease	NLGI #2
Free-Running Hub	Nissan-approved grease

86711059

Example of a Recommended Lubricants chart

hicles, causing greater oil consumption and leakage.

NOTE: The mixing of conventional and synthetic oils is possible but not recommended. Nondetergent or straight mineral oils must never be used in the engine.

FUEL

It is important to use fuel of the proper octane rating in ones vehicle. Octane rating is based on the quantity of anti-knock compounds added to the fuel, and also reflects the speed at which the gas will burn. The lower the octane rating, the faster it burns. The higher the octane, the slower the fuel will burn, and the greater the percentage of compounds in the fuel to prevent spark ping (knock), detonation and preignition (dieseling).

As the temperature of the engine increases, the air/fuel mixture exhibits a tendency to ignite before the spark plug is fired. If fuel of an octane rating too low for the engine is used, this will allow combustion to occur before the piston has completed its compression stroke, thereby creating a very high pressure very rapidly.

Fuel of the proper octane rating, for the compression ratio and ignition

timing of the vehicle, will slow the combustion process sufficiently to allow the spark plug enough time to ignite the mixture completely and smoothly. The use of super-premium fuel is no substitution for a properly tuned and maintained engine.

Light spark knock may be noticed when accelerating or driving up hills. The slight knocking may be considered normal (with 87 octane) because the maximum fuel economy is obtained under condition of occasional light spark knock. Gasoline with an octane rating higher than 87 may be used, but it is not necessary (in most cases) for proper operation.

NOTE: The engine's fuel requirement can change with time, mainly due to carbon buildup, which changes the compression ratio. If the engine pings, knocks or runs on, switch to a higher grade of fuel. Sometimes just changing brands may cure the problem.

OIL LEVEL CHECK

Every time one stops for fuel, check the engine oil as follows:

1. Park the vehicle on level ground.

NOTE: Although it is best for the engine to be at operating temperature, checking the oil immediately after stopping will lead to a false reading. Wait a few minutes after turning off the engine to allow the oil to drain back into the crankcase.

2. Open the hood and locate the dipstick. Pull the dipstick from its tube, wipe it clean and reinsert it.

NOTE: Keep in mind that this is a generalized procedure. The actual markings on the vehicle's dipstick may vary from those described here.

3. Pull the dipstick out again, and holding it horizontally, read the oil level. The oil should be between the **H** and **L** marks on the dipstick. If the oil is below the **L** mark, add oil of the proper viscosity and classification through the capped opening on top of the cylinder head cover.

4. Insert the dipstick and check the oil level again after adding any oil. Be careful not to overfill the

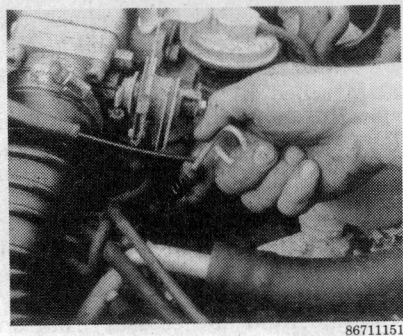

The oil dipstick in the engine compartment may be painted yellow on newer models

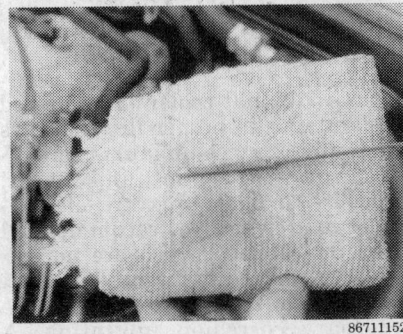

Check the oil dipstick for the correct level of engine oil — never overfill the engine oil

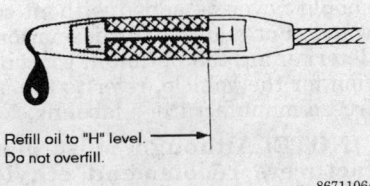

Refill oil to "H" level.
Do not overfill.

The engine oil level should be maintained between the L and H marks

crankcase. Approximately one quart of oil will raise the level from the **L** mark to the **H** mark. Excess oil will generally be consumed at an accelerated rate.

OIL AND FILTER CHANGE

NOTE: It may be a good idea to look under the vehicle, before starting any service procedure, to familiarize oneself with the necessary components and locations.

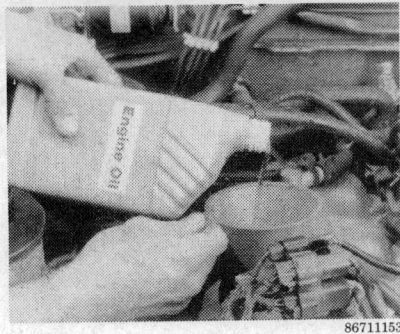

If the engine oil level is low, add engine oil, but do not overfill

The oil should be changed at least every 7500 miles (12,000 km) or every 6 months. Some manufacturers recommend changing the oil filter with every other oil change; we suggest that the filter be changed with **every** oil change. There is approximately 1 quart of dirty oil remaining in the old oil filter if it is not changed! A few dollars more every year seems a small price to pay for extended engine life — so change the filter every time one changes the oil!

CAUTION
Prolonged and repeated skin contact with used engine oil, with no effort to remove the oil, may be harmful. Always follow these simple precautions when handling used motor oil.

- Avoid prolonged skin contract with used motor oil
- Remove oil from skin by washing thoroughly with soap and water, or waterless hand cleaner. Do not use gasoline, thinners or other solvents
- Avoid prolonged skin contact with oil-soaked clothing

The mileage figures given are sample recommended intervals assuming normal driving and conditions. If the vehicle is being used under dusty, polluted or off-road conditions, change the oil and filter more frequently than specified. The same goes for vehicles driven in stop-and-go traffic or only for short distances. Always drain the oil after the engine has been running long enough to bring it to normal operating temperature. Hot oil will flow easier and more contaminants will be removed along with the oil than if it were drained cold. To change the oil and filter:

CAUTION
The EPA warns that prolonged contact with used engine oil may cause a number of skin disorders, including cancer! One should

make every effort to minimize the exposure to used engine oil. Protective gloves should be worn when changing the oil. Wash ones hands and any other exposed skin areas as soon as possible after exposure to used engine oil. Soap and water, or waterless hand cleaner should be used.

1. Run the engine until it reaches normal operating temperature.
2. Jack up the front of the vehicle and support it on safety stands.
3. Slide a drain pan of at least 6 quarts capacity under the oil pan.
4. Loosen the drain plug. Turn the plug out by hand. By keeping inward pressure on the plug as one unscrews it, oil won't escape past the threads, and it can be removed without being burned by hot oil.

CAUTION
The oil will be HOT! Be careful when removing the plug, so that it doesn't burn.

5. Allow the oil to drain completely. Clean and inspect the drain plug and oil pan sealing surface. If the plug is equipped with a removable gasket, also clean and inspect it.
6. Using a new plug gasket, if necessary, install the drain plug and tighten to correct specifications. Don't overtighten the plug; otherwise, a new pan or a replacement plug may have to be purchased.
7. Some engines will require the use of a oil filter strap wrench to remove the oil filter. Others may require a cap-type filter removal tool. Keep in mind that the filter is holding about one quart of dirty, hot oil.

NOTE: If the oil filter cannot be loosened by conventional methods, punch a hole through both sides near the mounting base of the filter and insert a punch, then turn to loosen the oil filter. After the oil filter is loosened, remove it from the engine with an oil filter wrench or equivalent.

8. Empty the old filter into the drain pan and properly dispose of the filter.
9. Using a clean rag, wipe off the filter adapter on the engine block. Be sure that the rag doesn't leave any lint which could clog an oil passage.
10. Coat the rubber gasket on the filter with fresh oil. Spin it onto the engine by hand; when the gasket touches the adapter surface, give it another 1/2–3/4 turn. Do not overtighten, or the gasket will leak.

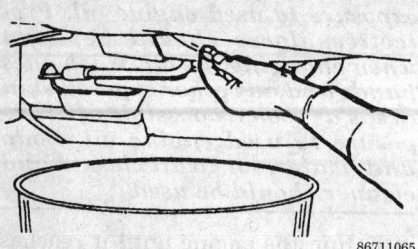

By keeping inward pressure on the drain plug as the plug is unscrewed, oil won't escape past the threads

86711065

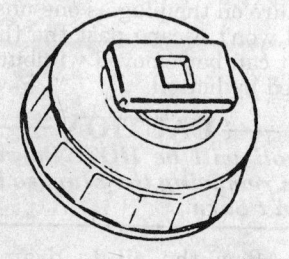

86711066

On some models, a cap-type oil filter removal tool works best

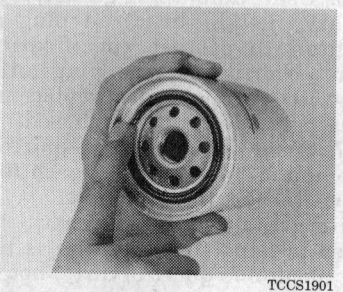

TCCS1901

Lubricate the gasket on the new filter with clean engine oil. A dry gasket may not make as good a seal, and could allow the filter to leak

11. Refill the engine with the correct amount of fresh oil.

12. Check the oil level on the dipstick. It is normal for the level to be a bit above the full mark. Start the engine and allow it to idle for a few minutes.

NOTE: Do not run the engine above idle speed until it has built up oil pressure, as indicated when the oil light goes out.

13. Shut off the engine and allow the oil to drain into the crankcase for a few minutes, then check the oil level. Check around the filter and

86711154

Removing the oil drain plug — do not over torque this drain plug upon installation

drain plug for any leaks and correct as necessary.

Power Steering Pump

Check the power steering fluid level every 6 months or 6000 miles (9600 km).

1. Park the vehicle on a level surface. Run the engine until normal operating temperature is reached.

2. Turn the steering all the way to the left and then all the way to the right several times. Center the steering wheel and shut off the engine.

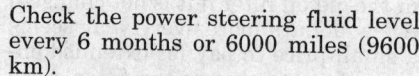

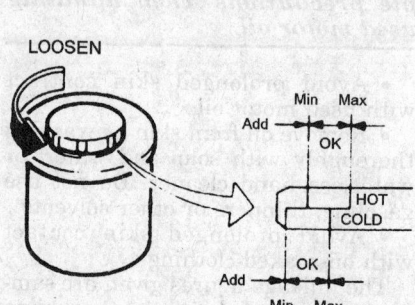

86711092

Checking the power steering fluid level

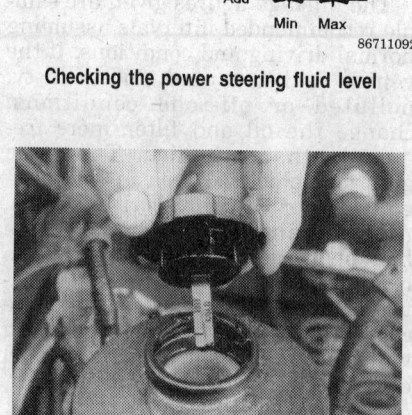

86711175

Remove the power steering cap to check the fluid level

3. Open the hood and check the power steering reservoir fluid level.

4. Remove the filler cap and wipe the attached dipstick clean.

5. Reinsert the dipstick and tighten the cap. Remove the dipstick and note the fluid level indicated on the dipstick.

6. The level should be at any point below the upper hash mark, but not below the lower hash mark (in the HOT or COLD ranges).

7. Add fluid as necessary, but do not overfill.

Cooling System

FLUID RECOMMENDATIONS

When additional coolant is required to maintain the proper level, always add a mixture of aluminum-compatible antifreeze/coolant and water. Typically, a 50/50 mixture of antifreeze and water is recommended (even for vehicles which are not exposed to cold winter temperatures), since this mixture also imparts the necessary corrosion inhibition. A greater concentration of antifreeze may be used, but the coolant mixture's level of protection actually lessens if too much antifreeze is used. Unless one is simply topping off the cooling system, straight antifreeze should never be added without some water. For additional information on determining the optimum concentration for the vehicle, refer to the antifreeze manufacturer's labeling.

NOTE: Although most manufacturers recommend ethylene glycol-based antifreeze (which has long been the prevalent type on the market), other types (such as propylene glycol) may also be suitable for use in the vehicle. Be sure to thoroughly read the alternative product's labeling to ensure compatibility before switching to a different formula. Check vehicle manufacturer's recommendations to be sure.

FLUID LEVEL CHECK

Dealing with the cooling system can be a tricky matter unless the proper precautions are observed. It is best to check the coolant level in the radiator when the engine is cold. This is done by checking the expansion tank. If coolant is visible above the **MIN** mark on the tank, the level is satisfactory. Always be certain that the filler caps on both the radiator and the reservoir are tightly closed.

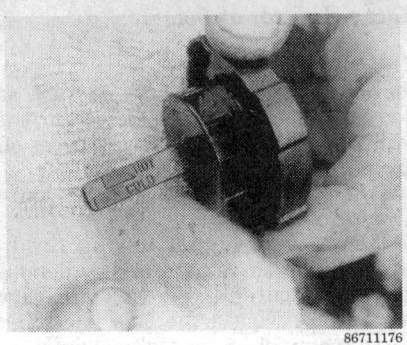

View of the power steering cap dipstick — note the hot and cold marks

Adding power steering fluid — use a funnel to avoid spills

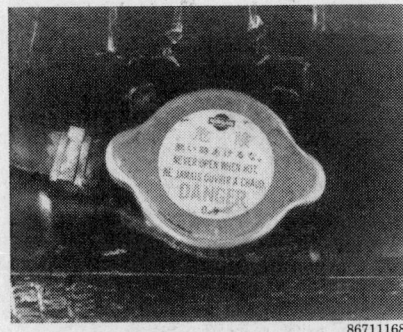

View of the radiator cap installed — never open when the engine is hot!

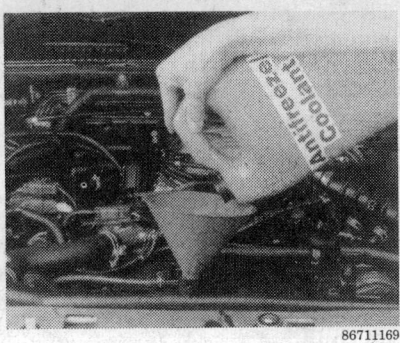

Add engine coolant to the radiator with a funnel to avoid spills

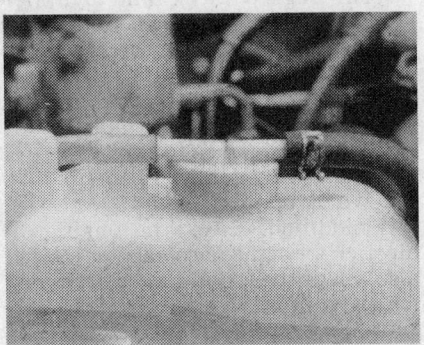

View of the coolant expansion tank

Remove the cap on the coolant expansion tank and add coolant to the proper level

In the event that the coolant level must be checked when the engine is warm or on engines without an expansion tank, place a thick rag over the radiator cap, then slowly turn the cap counterclockwise until it reaches the first detent. Allow all the hot steam to escape. This will allow the pressure in the system to drop gradually, preventing an explosion of hot coolant. When the hissing noise stops, remove the cap the rest of the way.

It's a good idea to check the coolant every time that one stops for fuel. If the coolant level is low, add equal amounts of suitable antifreeze and clean water. Fill the expansion tank to the **MAX** level. On models without an expansion tank, add coolant through the radiator filler neck.

NOTE: Never add cold coolant to a hot engine unless the engine is running, to avoid cracking the engine block.

Avoid using water that is known to have a high alkaline content or is very hard, except in emergency situations. Drain and flush the cooling system as soon as possible after using such water.

The radiator hoses and clamps and the radiator cap should be checked at the same time as the coolant level. Hoses which are brittle, cracked, or swollen should be replaced. Clamps should be checked for tightness (screwdriver-tight only)! Do not allow the clamp to cut into the hose or crush the fitting. The radiator cap gasket should be checked for any tears, cracks, swelling, or any signs of incorrect seating in the radiator neck.

DRAIN, REFILL AND FLUSH

———— **CAUTION** ————

When draining coolant, keep in mind that cats and dogs are attracted to ethylene glycol antifreeze, and could drink any that is left in an uncovered container or in puddles on the ground. This will prove fatal in sufficient quantity. Always drain the coolant into a sealable container. Coolant should be reused unless it is contaminated or two years old.

Complete draining and refilling of the cooling system at least once every two years will remove accumulated rust, scale and other deposits.

NOTE: Use a good quality antifreeze with water pump lubricants, rust inhibitors and other corrosion inhibitors along with acid neutralizers.

1. Drain the existing coolant as follows: Position suitable drain pans beneath the radiator and engine block. Open the radiator petcock and engine drain plug(s); there may be 1 or 2 drain plugs on the engine block depending on the type of engine. Another method of draining coolant is to disconnect the bottom radiator hose at the radiator outlet.

NOTE: If it is rusted or difficult to open, spray the radiator petcock with some penetrating lubricant.

2. Set the heater temperature controls to the full HOT position.
3. Close the petcock and tighten the drain plug(s) to correct specifications or reconnect the lower hose. Open the air relief plug, if so equipped, then fill the system with water.
4. Add a can of quality radiator flush. Be sure the flush is safe to use in engines having aluminum components.
5. Idle the engine until the upper radiator hose gets hot. Race it 2 or 3

Fluid level marks on the coolant recovery tank

Add engine coolant to the expansion tank with a funnel to avoid spills

times, then shut it **OFF**. Let the engine cool down.

6. Drain the system again.

7. Repeat this process until the drained water is clear and free of scale.

8. Close the petcock and drain plug(s) or, if applicable, connect the radiator hose.

9. If equipped with a coolant recovery system, flush the reservoir with water and leave empty.

NOTE: Always open the air relief plug before filling the cooling system, in order to bleed the trapped air. Only when the cooling system is bled properly can the correct amount of coolant be added to the system.

10. Determine the capacity of the cooling system. Add the appropriate ratio of quality aluminum-compatible antifreeze and water (normally a 50/50 mix) to provide the desired protection. With the air relief plug open, add the coolant mixture through the radiator filler neck until full, then close the bleeder plug and radiator cap.

11. Using the same concentration of clean antifreeze and water, fill the expansion tank to the **MAX** line, then cap the tank.

SYSTEM INSPECTION

Most permanent antifreeze/coolants have a colored dye added which makes the solution an excellent leak detector. When servicing the cooling system, check for leakage at:
- All hoses and hose connections
- Radiator seams, radiator core, and radiator draincock
- All engine block and cylinder head freeze (core) plugs, and drain plugs
- Edges of all cooling system gaskets (head gaskets, thermostat gasket)
- Transmission fluid cooler
- Heating system components
- Water pump

In addition, check the engine oil dipstick for signs of coolant in the oil; also, check the coolant in the radiator for signs of oil. Investigate and correct any indication of coolant leakage.

Check the Radiator Cap

While one is checking the coolant level, check the radiator cap for a worn or cracked gasket. If the cap doesn't seal properly, fluid will be lost and the engine will overheat. A worn cap should be replaced with a new one. The radiator cap must maintain pressure when the engine is running, or the cooling system will "boil over". The radiator cap also has a 2-way valve design to allow coolant to be drawn into the radiator from the coolant overflow tank. If this valve is not functioning properly, the vacuum cause in the system as the engine cools down can collapse and damage the hoses.

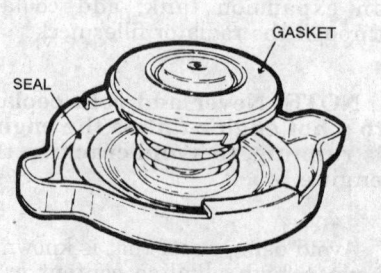

Check the radiator cap seal and gasket condition

Clean Radiator of Debris

Periodically clean any debris such as leaves, paper, insects, etc., from the radiator fins. Pick the large pieces off by hand. The smaller pieces can be washed away with water pressure from a hose.

Carefully straighten any bent radiator fins with a pair of needlenose pliers. Be careful, the fins are very soft. Don't wiggle the fins back and forth too much. Straighten them once and try not to move them again.

CHECKING SYSTEM PROTECTION

A 50/50 mix of antifreeze/coolant concentrate and water will usually provide the necessary protection. Freeze protection may be checked by using a cooling system hydrometer. Inexpensive hydrometers (floating ball types) may be obtained from a local department store (automotive section) or an auto supply store. Follow the directions packaged with the coolant hydrometer when checking protection.

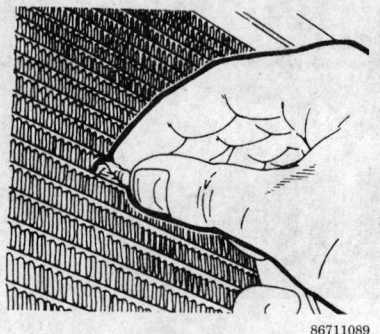

Clean the radiator fins of any debris which impedes air flow

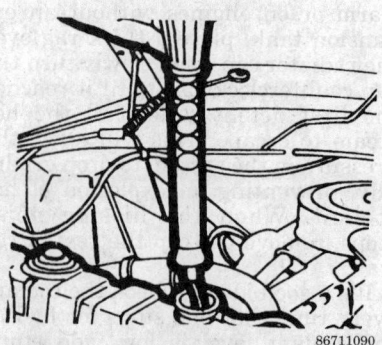

The freeze protection rating can be checked with an antifreeze tester

SPECIFICATIONS 20

ACURA

ENGINE IDENTIFICATION

Year	Model	Engine Displacement Liters (cc)	Engine Series (ID/VIN)	Fuel System	No. of Cylinders	Engine Type
1993	Integra	1.8 (1834)	B18A1	PGM-FI	4	DOHC
	Integra GSR	1.7 (1678)	B17A1	PGM-FI	4	DOHC
	Legend	3.2 (3206)	C32A1	PGM-FI	6	SOHC
	Vigor	2.5 (2451)	G25A1	PGM-FI	5	SOHC
	NSX	3.0 (2977)	C30A1	PGM-FI	6	DOHC
1994	Integra	1.8 (1834)	B18B1	PGM-FI	4	DOHC
	Integra GSR	1.8 (1797)	B18C1	PGM-FI	4	DOHC
	Legend	3.2 (3206)	C32A1	PGM-FI	6	SOHC
	Vigor	2.5 (2451)	G25A1	PGM-FI	5	SOHC
	NSX	3.0 (2977)	C30A1	PGM-FI	6	DOHC
1995	Integra	1.8 (1834)	B18B1	PGM-FI	4	DOHC
	Integra GSR	1.8 (1797)	B18C1	PGM-FI	4	DOHC
	Legend	3.2 (3206)	C32A1	PGM-FI	6	SOHC
	Vigor	2.5 (2451)	G25A1	PGM-FI	5	SOHC
	NSX	3.0 (2977)	C30A1	PGM-FI	6	DOHC
1996-97	Integra GSR	1.8 (1797)	B18C1	PGM-FI	4	DOHC
	Integra	1.8 (1834)	B18B1	PGM-FI	4	DOHC
	2.5TL	2.5 (2451)	G25A4	PGM-FI	5	SOHC
	NSX	3.0 (2977)	C30A1	PGM-FI	6	DOHC
	3.2TL	3.2 (3206)	C32A1	PGM-FI	6	SOHC

PGM-FI - Programmed fuel injection   SOHC - Single overhead camshaft   DOHC - Double overhead camshaft

GENERAL ENGINE SPECIFICATIONS

Year	Engine ID/VIN	Engine Displacement Liters (cc)	Fuel System Type	Net Horsepower @ rpm	Net Torque @ rpm (ft. lbs.)	Bore x Stroke (in.)	Compression Ratio	Oil Pressure @ rpm
1993	B18A1	1.8 (1834)	PGM-FI	140@6300	126@5000	3.19x3.50	9.2:1	50@3000
	B17A1	1.7 (1678)	PGM-FI	160@7600	117@7000	3.19x3.20	9.7:1	50@3000
	C32A1	3.2 (3206)	PGM-FI	200@5500	210@4500	3.54x3.31	9.6:1	50@3000
	C32A1	3.2 (3206)	PGM-FI	230@6200 ②	206@5000 ②	3.54x3.31	9.6:1	50@3000
	G25A1	2.5 (2451)	PGM-FI	176@6300	170@3900	3.35x3.40	9.0:1	50@3000
	C30A1	3.0 (2977)	PGM-FI	①	210@5300	3.54x3.07	10.2:1	50@3000
1994	B18B1	1.8 (1834)	PGM-FI	142@6300	127@5000	3.19x3.50	9.2:1	50@3000
	B18C1	1.8 (1797)	PGM-FI	170@7600	128@6200	3.19x3.43	10.0:1	50@3000
	C32A1	3.2 (3206)	PGM-FI	200@5500	210@4500	3.54x3.31	9.6:1	50@3000
	C32A1	3.2 (3206)	PGM-FI	230@6200 ②	206@5000 ②	3.54x3.31	9.6:1	50@3000
	G25A1	2.5 (2451)	PGM-FI	176@6300	170@3900	3.35x3.40	9.0:1	50@3000
	C30A1	3.0 (2977)	PGM-FI	①	210@5300	3.54x3.07	10.2:1	50@3000
1995	B18B1	1.8 (1834)	PGM-FI	142@6300	127@5000	3.19x3.50	9.2:1	50@3000
	B18C1	1.8 (1797)	PGM-FI	170@7600	128@6200	3.19x3.43	10.0:1	50@3000
	C32A1	3.2 (3206)	PGM-FI	200@5500	210@4500	3.54x3.31	9.6:1	50@3000
	C32A1	3.2 (3206)	PGM-FI	230@6200 ②	206@5000 ②	3.54x3.31	9.6:1	50@3000
	G25A1	2.5 (2451)	PGM-FI	176@6300	170@3900	3.35x3.40	9.0:1	50@3000
	C30A1	3.0 (2977)	PGM-FI	①	210@5300	3.54x3.07	10.2:1	50@3000

GENERAL ENGINE SPECIFICATIONS

Year	Engine ID/VIN	Engine Displacement Liters (cc)	Fuel System Type	Net Horsepower @ rpm	Net Torque @ rpm (ft. lbs.)	Bore x Stroke (in.)	Compression Ratio	Oil Pressure @ rpm
1996-97	B18C1	1.8 (1797)	PGM-FI	170@7600	128@6200	3.19x3.43	10.0:1	50@3000
	B18B1	1.8 (1834)	PGM-FI	142@6300	127@5200	3.19x3.50	9.2:1	50@3000
	G25A4	2.5 (2451)	PGM-FI	176@6300	170@3900	3.53x3.40	9.6:1	50@3000
	C30A1	3.0 (2977)	PGM-FI	①	210@5300	3.54x3.07	10.2:1	50@3000
	C32A1	3.2 (3206)	PGM-FI	200@5300	210@4500	3.54x3.31	9.6:1	50@3000

PGM-FI - Programmed fuel injection

① Manual transmission: 270@7100
Automatic transmission: 252@6600

② Legend Coupe and GS Sedan

GASOLINE ENGINE TUNE-UP SPECIFICATIONS

Year	Engine ID/VIN	Engine Displacement Liters (cc)	Spark Plugs Gap (in.)	Ignition Timing (deg.) MT	Ignition Timing (deg.) AT	Fuel Pump (psi)	Idle Speed (rpm) MT	Idle Speed (rpm) AT	Valve Clearance In.	Valve Clearance Ex.
1993	B18A1	1.8 (1834)	0.041	16B	16B	41-48 ①	700-800	700-800	0.003-0.005	0.006-0.008
	B17A1 ②	1.7 (1678)	0.041	16B	16B	48-56 ①	750-850	700-800	0.006-0.007	0.007-0.008
	C32A1	3.2 (3206)	0.041	15B	15B	44-51 ①	600-700	550-650	HYD	HYD
	G25A1	2.5 (2451)	0.043	15B	15B	43-50 ①	650-750	650-750	0.009-0.011	0.011-0.013
	C30A1	3.0 (2977)	0.041	15B	15B	47-53 ①	750-850	700-800 ④	0.006-0.007	0.007-0.008
1994	B18B1	1.8 (1834)	0.041	16B	16B	40-47 ①	700-800	700-800	0.003-0.005 ⑤	0.006-0.008 ⑤
	B18C1 ②	1.8 (1797)	0.049	16B	16B	48-55 ①	700-800	700-800	0.006-0.007 ⑤	0.007-0.008 ⑤
	C32A1	3.2 (3206)	0.043	15B	15B	44-51 ①	600-700	550-650	HYD	HYD
	C32A1 ③	3.2 (3206)	0.043	15B	15B	44-51 ①	630-730	580-680	HYD	HYD
	G25A1	2.5 (2451)	0.043	15B	15B	43-50 ①	650-750	650-750	0.009-0.011	0.011-0.013
	C30A1	3.0 (2977)	0.041	15B	15B	46-53 ①	750-850	700-800 ④	0.006-0.007	0.007-0.008
1995	B18B1	1.8 (1834)	0.041	16B	16B	40-47 ①	700-800	700-800	0.003-0.005 ⑤	0.006-0.008 ⑤
	B18C1 ②	1.8 (1797)	0.049	16B	16B	48-55 ①	700-800	700-800	0.006-0.007 ⑤	0.007-0.008 ⑤
	C32A1	3.2 (3206)	0.043	15B	15B	44-51 ①	600-700	550-650	HYD	HYD
	C32A1 ③	3.2 (3206)	0.043	15B	15B	44-51 ①	630-730	580-680	HYD	HYD
	G25A1	2.5 (2451)	0.043	15B	15B	43-50 ①	650-750	650-750	0.009-0.011	0.011-0.013
	C30A1	3.0 (2977)	0.041	15B	15B	46-53 ①	750-850	700-800 ④	0.006-0.007	0.007-0.008
1996-97	B18C1 ②	1.8 (1797)	0.051	16B	16B	48-55 ①	700-800	-	0.006-0.007	0.007-0.008
	B18B1	1.8 (1834)	0.039-0.043	16B	16B	40-47 ①	700-800	700-800 ④	0.003-0.005	0.006-0.008

GASOLINE ENGINE TUNE-UP SPECIFICATIONS

Year	Engine ID/VIN	Engine Displacement Liters (cc)	Spark Plugs Gap (in.)	Ignition Timing (deg.) MT	Ignition Timing (deg.) AT	Fuel Pump (psi)	Idle Speed (rpm) MT	Idle Speed (rpm) AT	Valve Clearance In.	Valve Clearance Ex.
1996-97	G25A4	2.5 (2451)	0.039-0.043	-	15B	43-50 ①	-	650-750	0.009-0.011	0.011-0.013
	C30A1	3.0 (2977)	0.043-0.047	15B	15B	47-53 ①	750-850	730 ④ 830	0.006-0.007	0.007-0.008
	C32A1	3.2 (3206)	0.039-0.043	-	15B	37-48 ①	-	600 700	HYD HYD	HYD HYD

NOTE: The Vehicle Emission Control Information label often reflects specification changes made during production. The label figures must be used if they differ from those in this chart.

B - Before top dead center

HYD - Hydraulic

① At idle, pressure regulator vacuum hose disconnected
② Integra GSR
③ Legend GS and Legend Coupe
④ In park or neutral
⑤ Specification is for a cold engine

CAPACITIES

Year	Model	Engine ID/VIN	Engine Displacement Liters (cc)	Engine Oil with Filter	Transmission (pts.) 5-Spd	Transmission (pts.) 6-Spd	Transmission (pts.) Auto.	Transfer Case (pts.)	Drive Axle Front (pts.)	Drive Axle Rear (pts.)	Fuel Tank (gal.)	Cooling System (qts.)
1993	Integra	B18A1	1.8 (1834)	4.0	4.6 ①	-	6.4 ①	-	-	-	13.2	②
	Integra	B17A1	1.7 (1678)	4.2	4.6 ①	-	6.4 ①	-	-	-	13.2	②
	Legend	C32A1	3.2 (3206)	5.0	-	4.8 ①	7.0 ①	-	2.2	-	18.0	7.9
	Vigor	G25A1	2.5 (2451)	4.5	3.8 ①	-	5.2 ①	-	1.9	-	17.2	6.3
	NSX	C30A1	3.0 (2977)	5.3	5.8 ①	-	6.2 ①	-	-	-	18.5	12.7
1994	Integra	B18B1	1.8 (1834)	4.0	4.6 ①	-	5.8 ①	-	-	-	13.2	③
	Integra	B18C1	1.8 (1797)	4.2	4.6 ①	-	5.8 ①	-	-	-	13.2	5.0
	Legend	C32A1	3.2 (3206)	4.8	4.8 ①	4.8 ①	7.0 ①	-	2.2	-	18.0	8.0
	Vigor	G25A1	2.5 (2451)	4.5	3.8 ①	-	5.2 ①	-	1.9	-	17.2	6.3
	NSX	C30A1	3.0 (2977)	5.3	5.8 ①	-	6.2 ①	-	-	-	18.5	12.7
1995	Integra	B18B1	1.8 (1834)	4.0	4.6 ①	-	5.8 ①	-	-	-	13.2	③
	Integra	B18C1	1.8 (1797)	4.2	4.6 ①	-	5.8 ①	-	-	-	13.2	5.0
	Legend	C32A1	3.2 (3206)	4.8	4.8 ①	4.8 ①	7.0 ①	-	2.2	-	18.0	8.0
	Vigor	G25A1	2.5 (2451)	4.5	3.8 ①	-	5.2 ①	-	1.9	-	17.2	6.3
	NSX	C30A1	3.0 (2977)	5.3	5.8 ①	-	6.2 ①	-	-	-	18.5	12.7
1996-97	Integra	B18C1	1.8 (1797)	4.2	4.6	-	-	-	-	-	13.2	5.0
	Integra	B18B1	1.8 (1834)	4.0	4.6	-	5.8	-	-	-	13.2	③
	2.5TL	G25A1	2.5 (2451)	4.5	-	-	5.8	-	-	-	17.2	5.5
	NSX	C30A1	3.0 (2977)	4.4	5.8	-	6.2	-	-	-	18.5	12.7
	3.2TL	C32A1	3.2 (3206)	5.0	-	-	7.0	-	-	-	17.2	3.4

NOTE: Capacities given are service, not overhaul capacities

① Transmission oil change capacity

② Automatic transmission: 5.2
Manual transmission: 5.3

③ Automatic transmission: 5.0
Manual transmission: 4.6

VALVE SPECIFICATIONS

Year	Engine ID/VIN	Engine Displacement Liters (cc)	Seat Angle (deg.)	Face Angle (deg.)	Spring Test Pressure (lbs. @ in.)	Spring Installed Height (in.)	Stem-to-Guide Clearance (in.)		Stem Diameter (in.)	
							Intake	Exhaust	Intake	Exhaust
1993	B18A1	1.8 (1834)	45	45	NA	NA	0.0010-0.0020	0.0020-0.0030	0.2591-0.2594	0.2579-0.2583
	B17A1	1.7 (1678)	45	45	NA	NA	0.0010-0.0022	0.0020-0.0031	0.2156-0.2159	0.2146-0.2150
	C32A1	3.2 (3206)	45	45	NA	NA	0.0027-0.0035	0.0039-0.0046	0.2157-0.2161	0.2146-0.2150
	G25A1	2.5 (2451)	45	45	NA	NA	0.0008-0.0018	0.0020-0.0030	0.2156-0.2159	0.2146-0.2150
	C30A1	3.0 (2977)	45	45	NA	NA	0.0010-0.0020	0.0020-0.0030	0.2156-0.2159	0.2146-0.2150
1994	B18B1	1.8 (1834)	45	45	NA	NA	0.0010-0.0020	0.0020-0.0030	0.2591-0.2594	0.2579-0.2583
	B18C1	1.8 (1797)	45	45	NA	NA	0.0010-0.0022	0.0020-0.0031	0.2156-0.2159	0.2146-0.2150
	C32A1	3.2 (3206)	45	45	NA	NA	0.0010-0.0020	0.0020-0.0030	0.2157-0.2161	0.2146-0.2150
	G25A1	2.5 (2451)	45	45	NA	NA	0.0008-0.0018	0.0020-0.0030	0.2156-0.2159	0.2146-0.2150
	C30A1	3.0 (2977)	45	45	NA	NA	0.0010-0.0020	0.0020-0.0030	0.2156-0.2159	0.2146-0.2150
1995	B18B1	1.8 (1834)	45	45	NA	NA	0.0010-0.0020	0.0020-0.0030	0.2591-0.2594	0.2579-0.2583
	B18C1	1.8 (1797)	45	45	NA	NA	0.0010-0.0022	0.0020-0.0031	0.2156-0.2159	0.2146-0.2150
	C32A1	3.2 (3206)	45	45	NA	NA	0.0010-0.0020	0.0020-0.0030	0.2157-0.2161	0.2146-0.2150
	G25A1	2.5 (2451)	45	45	NA	NA	0.0008-0.0018	0.0020-0.0030	0.2156-0.2159	0.2146-0.2150
	C30A1	3.0 (2977)	45	45	NA	NA	0.0010-0.0020	0.0020-0.0030	0.2156-0.2159	0.2146-0.2150
1996-97	B18C1	1.8 (1797)	45	45	NA	NA	0.0010-0.0022	0.0020-0.0031	0.2156-0.2159	0.2146-0.2150
	B18B1	1.8 (1834)	45	45	NA	NA	0.0010-0.0020	0.0020-0.0030	0.2591-0.2594	0.2579-0.2583
	G25A4	2.5 (2451)	45	45	NA	NA	0.0008-0.0018	0.0020-0.0030	0.2156-0.2159	0.2146-0.2150
	C30A1	3.0 (2977)	45	45	NA	NA	0.0010-0.0020	0.0020-0.0030	0.2156-0.2159	0.2146-0.2150
	C32A1	3.2 (3206)	45	45	NA	NA	0.0010-0.0020	0.0020-0.0030	0.2157-0.2181	0.2146-0.2150

NA - Not Available

TORQUE SPECIFICATIONS

All readings in ft. lbs.

Year	Engine ID/VIN	Engine Displacement Liters (cc)	Cylinder Head Bolts	Main Bearing Bolts	Rod Bearing Bolts	Crankshaft Damper Bolts	Flywheel Bolts	Manifold		Spark Plugs	Lug Nut
								Intake	Exhaust		
1993	B18A1	1.8 (1834)	③	⑧	⑫	130	②	17	23	13	80
	B17A1	1.7 (1678)	③	⑧	⑨	130	②	17	23	13	80
	G25A1	2.5 (2451)	72	⑫	24	181	②	16	23	13	80
	C32A1	3.2 (3206)	①	⑦	33	174	②	16	22	13	80

TORQUE SPECIFICATIONS
All readings in ft. lbs.

Year	Engine ID/VIN	Engine Displacement Liters (cc)	Cylinder Head Bolts	Main Bearing Bolts	Rod Bearing Bolts	Crankshaft Damper Bolts	Flywheel Bolts	Manifold Intake	Manifold Exhaust	Spark Plugs	Lug Nut
1993	C30A1	3.0 (2977)	56	④	⑤	⑥	②	16	25	13	80
1994	B18B1	1.8 (1834)	③	⑧	⑨	130	②	17	23	13	80
	B18C1	1.8 (1797)	③	⑩	⑪	130	②	17	23	13	80
	G25A1	2.5 (2451)	72	⑫	24	181	②	16	23	13	80
	C32A1	3.2 (3206)	①	⑦	33	174	②	16	22	13	80
	C30A1	3.0 (2977)	56	④	⑤	⑥	②	16	25	13	80
1995	B18B1	1.8 (1834)	③	⑧	⑨	130	②	17	23	13	80
	B18C1	1.8 (1797)	③	⑩	⑩	130	②	17	23	13	80
	G25A1	2.5 (2451)	72	⑫	24	181	②	16	23	13	80
	C32A1	3.2 (3206)	①	⑦	33	174	②	16	22	13	80
	C30A1	3.0 (2977)	56	④	⑤	⑥	②	16	25	13	80
1996-97	B18C1	1.8 (1797)	③	⑩	⑪	130	②	17	23	13	80
	B18B1	1.8 (1834)	③	56	⑨	130	②	17	23	13	80
	G25A4	2.5 (2451)	⑬	⑥	24	181	54	16	23	13	80
	C30A1	3.0 (2977)	56	⑭	⑤	181	②	16	25	13	80
	C32A1	3.2 (3206)	①	⑦	33	174	②	16	22	13	80

① Step 1: 29 ft. lbs.
Step 2: 56 ft. lbs.
② Manual transmission: 76 ft. lbs.
Automatic transmission: 54 ft. lbs.
③ Step 1: 22 ft. lbs.
Step 2: 61 ft. lbs.
④ Inner: 48 ft. lbs.
Outer: 29 ft. lbs.
⑤ Tighten to 14 ft. lbs. plus an additional 95 degrees
⑥ Step 1: Tighten to 203 ft. lbs.
Step 2: Loosen completely
Step 3: Tighten to 181 ft. lbs.

⑦ Inner: 57 ft. lbs.
Outer: 29 ft. lbs.
Side: 36 ft. lbs.
⑧ Step 1: 22 ft. lbs.
Step 2: 54 ft. lbs.
⑨ Step 1: 14 ft. lbs.
Step 2: 23 ft. lbs.
⑩ Step 1: 22 ft. lbs.
Step 2: Cap Nos. 1, 5: 56 ft. lbs.
Cap Nos. 2-4: 49 ft. lbs.
⑪ Step 1: 14 ft. lbs.
Step 2: 33 ft. lbs.

⑫ Step 1: 22 ft. lbs.
Step 2: 49 ft. lbs.
⑬ Step 1: 29 ft. lbs.
Step 2: 51 ft. lbs.
Step 3: 72 ft. lbs.
⑭ Cap bolts: 29 ft. lbs.
Cap bridge bolts: 48 ft. lbs.
Cap side bolts: 36 ft. lbs.

BRAKE SPECIFICATIONS
All measurements in inches unless noted

Year	Model		Master Cylinder Bore	Brake Disc Original Thickness	Brake Disc Minimum Thickness	Brake Disc Maximum Runout	Brake Drum Diameter Original Inside Diameter	Max. Wear Limit	Maximum Machine Diameter	Minimum Lining Thickness Front	Minimum Lining Thickness Rear
1993	Integra	F	NA	0.830	0.750	0.004	-	-	-	0.06	-
		R	NA	0.350	0.310	0.006	-	-	-	-	0.06
	Legend	F	NA	1.100	1.020	0.004	-	-	-	0.06	-
		R	NA	0.350	0.300	0.006	-	-	-	-	0.06
	Vigor	F	NA	0.910	0.830	0.004	-	-	-	0.06	-
		R	NA	0.390	0.310	0.004	-	-	-	-	0.06
	NSX	F	NA	1.100	1.020	0.004	-	-	-	0.06	-
		R	NA	0.830	0.750	0.004	-	-	-	-	0.06
1994	Integra	F	NA	0.830	0.750	0.004	-	-	-	0.06	-
		R	NA	0.350	0.310	0.006	-	-	-	-	0.06
	Legend	F	NA	1.100	1.020	0.004	-	-	-	0.06	-
		R	NA	0.350	0.300	0.006	-	-	-	-	0.06
	Vigor	F	NA	0.910	0.830	0.004	-	-	-	0.06	-
		R	NA	0.390	0.310	0.004	-	-	-	-	0.06
	NSX	F	NA	1.100	1.020	0.004	-	-	-	0.06	-
		R	NA	0.830	0.750	0.004	-	-	-	-	0.06
1995	Integra	F	NA	0.830	0.750	0.004				0.06	-
		R	NA	0.350	0.310	0.006				-	0.06

BRAKE SPECIFICATIONS
All measurements in inches unless noted

Year	Model		Master Cylinder Bore	Brake Disc Original Thickness	Brake Disc Minimum Thickness	Maximum Runout	Brake Drum Diameter Original Inside Diameter	Max. Wear Limit	Maximum Machine Diameter	Minimum Lining Thickness Front	Minimum Lining Thickness Rear
1995	Legend	F	NA	1.100	1.020	0.004	-	-	-	0.06	-
		R	NA	0.350	0.300	0.006	-	-	-	-	0.06
	Vigor	F	NA	0.910	0.830	0.004	-	-	-	0.06	-
		R	NA	0.390	0.310	0.004	-	-	-	-	0.06
	NSX	F	NA	1.100	1.020	0.004	-	-	-	0.06	-
		R	NA	0.830	0.750	0.004	-	-	-	-	0.06
1996-97	Integra	F	NA	0.830	0.750	0.004	-	-	-	0.06	-
		R	NA	0.350	0.310	0.004	-	-	-	-	0.06
	2.5TL	F	NA	0.910	0.830	0.004	-	-	-	0.06	-
		R	NA	0.350	0.300	0.004	-	-	-	-	0.06
	NSX	F	NA	1.100	1.020	0.004	-	-	-	0.06	-
		R	NA	0.830	0.750	0.004	-	-	-	-	0.06
	3.2TL	F	NA	0.910	0.830	0.004	-	-	-	0.06	-
		R	NA	0.350	0.300	0.004	-	-	-	-	0.06

NA - Not Available   F - Front   R - Rear

FREQUENT MAINTENANCE LABOR
ACURA

The following should be used as a guide when determining the amount of work required for a particular service if taken to a repair shop. In estimating how long a particular Frequent Maintenance Service item should take, please observe the following:
- **Chilton Time** is time that is based on field research and data supplied by the vehicle manufacturer.
- All labor time operations are given in hours and tenths of an hour.
- All labor operations, are to be used as a **guide**.

COOLING

Chilton Time

(G) Winterize Cooling System
Includes: Run engine to check for leaks, tighten all hose connections. Test radiator and pressure cap, drain radiator and engine block. Add antifreeze and refill system.
  1993-97 5

(G) Belt, Drive, Renew
Alternator
  1993-95 Integra
    wo/AC 6
    w/AC 7
  1993-95 Legend 4
  1993-94 Vigor 6
  1996 2.5TL 6
  1996 3.2TL 4
Power Steering
  1993 Integra 6
  1994-95 Integra 4
  1993-95 Legend 9
  1993-94 Vigor 6
  1996 2.5TL 6
  1996 3.2TL 6

Air Conditioning
  1993-95 Integra 4
  1993-95 Legend 6
  1993-94 Vigor 7
  1996 2.5TL 7
  1996 3.2TL 7

(G) Hoses, Radiator, Renew
Includes: Drain and refill cooling system as required.
  1993-97
    upper 7
    lower 7
    both 1.1

(G) Thermostat, Coolant, Renew
  1993-95 Integra 7
  1993-95 Legend 1.1
  1993-94 Vigor 4
  1996 2.5TL 4
  1996 3.2TL 1.1

FUEL

Chilton Time

(M) Air Cleaner, Service
  1993-97 2

(G) Filter, Fuel, Renew
  1993-97 5

BRAKES

(G) Bleed Brakes (Four Wheels)
Includes: Add fluid.
  1993-97 5

(M) Parking Brake, Adjust
  1993-97 4

LUBRICATION SERVICES

(M) Engine Oil & Filter, Renew
Includes: Inspect and correct all fluid levels.
  1993-97 3

FREQUENT MAINTENANCE LABOR (cont.)
ACURA

Chilton Time

(M) Lubricate Chassis, Change Oil & Filter
Includes: Inspect and correct all fluid levels.
1993-97 .6
Install grease fittings add 1

(M) Lubricate Chassis
Includes: Inspect and correct all fluid levels.
1993-97 .4
Install grease fittings add 1

WHEELS

(G) Wheel, Renew (One)
1993-97, one 5

Chilton Time

(G) Wheel, Balance
1993-97
  one 3
  each adtnl. 2
(G) Wheels, Rotate (All)
1993-97 5

ELECTRICAL

(G) Headlamps, Aim
1993-97 4
(G) Halogen Headlamp Bulb, Renew
1993-97, one 3
(G) High Mount Stop Lamp Bulb, Renew
1993-97 3
(G) License Lamp Bulb, Renew
1993-97, one or all 3

Chilton Time

(G) Park & Turn Signal Lamp Bulb or Lens, Renew
1993-97
  one 3
  each adtnl. 2
(G) Tail Lamp Lens or Bulb, Renew
1993-97
  one 3
  each adtnl. 1
(G) Horn, Renew
1993 Integra 7
1994-95 Integra 4
1993-95 Legend 3
1993-94 Vigor 3
1996 2.5TL, 3.2TL 3
(M) Terminals, Battery, Clean
1993-97 3

SCHEDULED MAINTENANCE INTERVALS
(ACURA 2.2CL, 2.5TL, 3.2TL, 3.5RL, INTEGRA, LEGEND, NSX & VIGOR)

TO BE SERVICED	TYPE OF SERVICE	VEHICLE MILEAGE INTERVAL (x1000)												
		7.5	15	22.5	30	37.5	45	52.5	60	67.5	75	82.5	90	97.5
Engine oil & filter	R	✓	✓	✓	✓	✓	✓	✓	✓	✓	✓	✓	✓	✓
Clutch release arm travel (1993 Integra)	S/I	✓	✓	✓	✓	✓	✓	✓	✓	✓	✓	✓	✓	✓
Front brake pads (Vigor & 1994 NSX)	S/I	✓	✓	✓	✓	✓	✓	✓	✓	✓	✓	✓	✓	✓
Rear brake discs, calipers & pads	S/I		✓		✓		✓		✓		✓		✓		
Rotate tires	S/I	✓	✓	✓	✓	✓	✓	✓	✓	✓	✓	✓	✓	✓
A/C filter (Legend)	R		✓		✓		✓		✓		✓		✓		
A/C filter (3.5RL)	R				✓				✓				✓		
Brake hoses & lines (including ABS)	S/I		✓		✓		✓		✓		✓		✓		

SCHEDULED MAINTENANCE INTERVALS
(ACURA 2.2CL, 2.5TL, 3.2TL, 3.5RL, INTEGRA, LEGEND, NSX & VIGOR) (Cont.)

TO BE SERVICED	TYPE OF SERVICE	VEHICLE MILEAGE INTERVAL (x1000)												
		7.5	15	22.5	30	37.5	45	52.5	60	67.5	75	82.5	90	97.5
Cooling system hoses & connections (2.2CL, 2.5TL, 3.2TL, 3.5RL, 1995 Integra & 1995-97 NSX)	S/I		✓		✓		✓		✓		✓		✓	
Cooling system hoses & connections (1993-94)	S/I				✓				✓				✓	
Driveshaft boots (2.2CL, 2.5TL, 3.2TL, 3.5RL & 1995 Integra)	S/I		✓		✓		✓		✓		✓		✓	
Exhaust pipe (after TWC converter) & muffler (1993-94)	S/I		✓		✓		✓		✓		✓		✓	
Exhaust system (2.2CL, 2.5TL, 3.2TL, 3.5RL, 1995 Integra & 1995-97 NSX)	S/I		✓		✓		✓		✓		✓		✓	
Front brake discs & calipers	S/I		✓		✓		✓		✓		✓		✓	
Front wheel alignment (1993-94)	S/I		✓		✓		✓		✓		✓		✓	
Fuel pipes, hoses & connections (2.2CL, 2.5TL, 3.2TL, 3.5RL, 1995 Integra & 1995-97 NSX)	S/I		✓		✓		✓		✓		✓		✓	
Fuel pipes, hoses & connections (1993-94)	S/I				✓				✓				✓	
Power steering system (1993-94)	S/I		✓		✓		✓		✓		✓		✓	
Suspension components (2.2CL, 2.5TL, 3.2TL, 3.5RL & 1995-97 NSX)	S/I		✓		✓		✓		✓		✓		✓	
Suspension mounting bolts	S/I		✓		✓		✓		✓		✓		✓	

SCHEDULED MAINTENANCE INTERVALS
(ACURA 2.2CL, 2.5TL, 3.2TL, 3.5RL, INTEGRA, LEGEND, NSX & VIGOR) (Cont.)

TO BE SERVICED	TYPE OF SERVICE	VEHICLE MILEAGE INTERVAL (x1000)												
		7.5	15	22.5	30	37.5	45	52.5	60	67.5	75	82.5	90	97.5
Tie rods, steering gear box & boots (2.2CL, 2.5TL, 3.2TL, 3.5RL, 1995 Integra & 1995-97 NSX)	S/I		✓		✓		✓		✓		✓		✓	
Steering operation, tie rod ends, steering gearbox & boots	S/I		✓		✓				✓				✓	
Valve clearance (2.5TL, 1993-94 Integra & NSX)	S/I		✓		✓		✓		✓		✓		✓	
Valve clearance (2.2CL, 1995 Integra & 1995-97 NSX)	S/I				✓				✓				✓	
Parking brake (2.2CL, 2.5TL, 3.2TL, 3.5RL, 1995 Integra & 1995-97 NSX)	S/I		✓		✓		✓		✓		✓		✓	
Parking brake (1993-94)	S/I		✓		✓				✓				✓	
Air cleaner element	R				✓				✓				✓	
Automatic transmission fluid	R				✓				✓				✓	
Brake fluid (including ABS) (Integra, Legend, Vigor, 2.5TL & 1993-95 NSX)	R				✓				✓				✓	
Brake fluid (including ABS) (3.5RL)	R								✓				✓	
Brake fluid (including ABS) (2.2CL, 3.2TL & 1996-97 NSX)	R						✓				✓			
Front differential oil (2.5TL, Legend & Vigor)	R				✓				✓				✓	

SCHEDULED MAINTENANCE INTERVALS
(ACURA 2.2CL, 2.5TL, 3.2TL, 3.5RL, INTEGRA, LEGEND, NSX & VIGOR) (Cont.)

TO BE SERVICED	TYPE OF SERVICE	VEHICLE MILEAGE INTERVAL (x1000)												
		7.5	15	22.5	30	37.5	45	52.5	60	67.5	75	82.5	90	97.5
Front differential fluid (3.2TL & 3.5RL)	R								✓				✓	
Manual transmission oil	R				✓				✓				✓	
ABS operation	S/I				✓				✓				✓	
Drive belt(s)	S/I				✓				✓				✓	
Exhaust pipe (before TWC converter) (1993-94)	S/I				✓				✓				✓	
Parking brake drums & linings (Legend)	S/I				✓				✓				✓	
Spark plugs (2.2CL, 2.5TL, Integra except GSR & Vigor)	R				✓				✓				✓	
Spark plugs (3.2TL, Integra GSR, Legend & NSX)	R								✓					
Spark plugs (3.5RL)①	R													
Engine coolant	R						✓				✓			
ABS high pressure hose (NSX)	R								✓					
Fuel filter	R								✓					
PCV valve	S/I								✓					
Timing belt (except as noted below)	R												✓	
Timing belt & timing balancer belt (2.2CL)	R												✓	
Timing belt & timing balancer belt (3.5RL)①	R													
Transmission fluid (2.2CL, 2.5TL, 3.2TL, 3.5RL & 1996-97 NSX)	R												✓	
Distributor, ignition cap & rotor (2.2CL, 2.5TL, Integra & Vigor)	S/I								✓					
Idle speed (2.2CL, 2.5TL, 3.2TL, Integra, Legend & NSX)	S/I								✓					

SCHEDULED MAINTENANCE INTERVALS
(ACURA 2.2CL, 2.5TL, 3.2TL, 3.5RL, INTEGRA, LEGEND, NSX & VIGOR) (Cont.)

TO BE SERVICED	TYPE OF SERVICE	VEHICLE MILEAGE INTERVAL (x1000)												
		7.5	15	22.5	30	37.5	45	52.5	60	67.5	75	82.5	90	97.5
Idle speed (3.5RL)②	S/I													
Ignition wires	S/I								✓					
TWC converter heat shield	S/I								✓					
Water pump	S/I												✓	
Water pump (3.5RL)②	S/I													

① Replace at 105,000 miles.
② Service or inspect at 105,000 miles.
R – Replace   S/I – Service or Inspect

FREQUENT OPERATION MAINTENANCE (SEVERE SERVICE)
If a vehicle is operated under any of the following conditions it is considered severe service:
- Extremely dusty areas.
- 50% or more of the vehicle operation is in 32°C (90°F) or higher temperatures, or constant operation in temperatures below 0°C (32°F).
- Prolonged idling (vehicle operation in stop and go traffic).
- Frequent short running periods (engine does not warm to normal operating temperatures).
- Police, taxi, delivery usage or trailer towing usage.

Clutch release arm travel (1993 Integra) - service or inspect every 3750 miles.
Oil & oil filter change – change every 3750 miles.
Brake hoses & lines (including ABS) (3.5RL) - service or inspect every 7500 miles.
Cooling system hoses & connections (3.5RL) - service or inspect every 7500 miles.
Driveshaft boots (2.2CL, 2.5TL, 3.2TL & 3.5RL) - check every 7500 miles.
Exhaust system (3.5RL) - check every 7500 miles.
Brake discs, calipers & pads - service or inspect every 7500 miles.
Fuel pipes, hoses & connections (3.5RL) - check every 7500 miles.
Power steering system - service or inspect every 7500 miles.
Suspension components - service or inspect every 7500 miles.
Tie rod ends, steering gear box & boots (2.2CL, 2.5TL, 3.2TL & 3.5RL) - service or inspect every 7500 miles.
Air cleaner element (1994-97 NSX) – service or inspect every 7500 miles.
Air cleaner element (except 1994-97 NSX) – service or inspect every 15,000 miles.
Front differential oil (2.5TL, 3.2TL, 3.5RL, Legend & Vigor) - replace every 15,000 miles.
Transmission oil (Legend, NSX & Vigor) - replace every 15,000 miles.
Transmission oil (2.2CL, 2.5TL, 3.2TL & 3.5RL) - replace every 30,000 miles.
Timing belt (2.5TL, 3.2TL & 1995 Integra) - replace every 60,000 miles.
Water pump (2.5TL & 3.2TL) - service or inspect every 60,000 miles.

AUDI

VEHICLE IDENTIFICATION CHART

Engine Code						Model Year	
Code	Liters	Cu. In. (cc)	Cyl.	Fuel Sys.	Eng. Mfg.	Code	Year
AAH	2.8	169 (2771)	6	MFI	Audi	P	1993
AAN	2.2	136 (2226)	5	MFI	Audi	R	1994
ABH	4.2	254 (4172)	8	MFI	Audi	S	1995
AFC	2.8	169 (2771)	6	MFI	Audi	T	1996
						V	1997

MFI - Multiport fuel injection

ENGINE IDENTIFICATION

Year	Model	Engine Displacement Liters (cc)	Engine Series (ID/VIN)	Fuel System	No. of Cylinders	Engine Type
1993	S4 Quattro	2.2 (2226)	AAN	MFI	5	DOHC-T
	90S	2.8 (2771)	AAH	MFI	6	SOHC
	90CS	2.8 (2771)	AAH	MFI	6	SOHC
	90CS Quattro	2.8 (2771)	AAH	MFI	6	SOHC
	100	2.8 (2771)	AAH	MFI	6	SOHC
	100S	2.8 (2771)	AAH	MFI	6	SOHC
	100CS	2.8 (2771)	AAH	MFI	6	SOHC
	100CS Quattro	2.8 (2771)	AAH	MFI	6	SOHC
	V8 Quattro	4.2 (4172)	ABH	MFI	8	DOHC
1994	S4 Quattro	2.2 (2226)	AAN	MFI	5	DOHC-T
	90S	2.8 (2771)	AAH	MFI	6	SOHC
	90CS	2.8 (2771)	AAH	MFI	6	SOHC
	90CS Quattro	2.8 (2771)	AAH	MFI	6	SOHC
	100	2.8 (2771)	AAH	MFI	6	SOHC
	100S	2.8 (2771)	AAH	MFI	6	SOHC
	100CS	2.8 (2771)	AAH	MFI	6	SOHC
	100CS Quattro	2.8 (2771)	AAH	MFI	6	SOHC
	Cabriolet	2.8 (2771)	AAH	MFI	6	SOHC
	V8 Quattro	4.2 (4172)	ABH	MFI	8	DOHC
1995	S6 Quattro	2.2 (2226)	AAN	MFI	5	DOHC-T
	90	2.8 (2771)	AAH	MFI	6	SOHC
	90	2.8 (2771)	AFC	MFI	6	SOHC
	90 Quattro	2.8 (2771)	AAH	MFI	6	SOHC
	Sport 90	2.8 (2771)	AAH	MFI	6	SOHC
	Sport 90	2.8 (2771)	AFC	MFI	6	SOHC
	Sport 90 Quattro	2.8 (2771)	AAH	MFI	6	SOHC
	A6	2.8 (2771)	AAH	MFI	6	SOHC
	A6	2.8 (2771)	AFC	MFI	6	SOHC
	A6 Quattro	2.8 (2771)	AAH	MFI	6	SOHC
	Cabriolet	2.8 (2771)	AAH	MFI	6	SOHC
	Cabriolet	2.8 (2771)	AFC	MFI	6	SOHC
1996-97	A4	2.8 (2771)	AFC	MFI	6	SOHC
	A6	2.8 (2771)	AFC	MFI	6	SOHC
	Cabriolet	2.8 (2771)	AFC	MFI	6	SOHC

MFI - Multiport fuel injection
SOHC - Single overhead camshaft

DOHC - Double overhead camshaft
DOHC-T - Double overhead camshaft, Turbocharged

GENERAL ENGINE SPECIFICATIONS

Year	Engine ID/VIN	Engine Displacement Liters (cc)	Fuel System Type	Net Horsepower @ rpm	Net Torque @ rpm (ft. lbs.)	Bore x Stroke (in.)	Compression Ratio	Oil Pressure @ rpm
1993	AAN	2.2 (2226)	MFI	227@5900	258@1950	3.19x3.40	9.3:1	29@2000
	AAH	2.8 (2771)	MFI	174@5500	184@3000	3.25x3.40	10.3:1	41@3000
	ABH	4.2 (4172)	MFI	280@5800	295@4000	3.29x3.62	10.6:1	29@2000
1994	AAN	2.2 (2226)	MFI	227@5900	258@1950	3.19x3.40	9.3:1	29@2000
	AAH	2.8 (2771)	MFI	174@5500	184@3000	3.25x3.40	10.3:1	41@3000
	ABH	4.2 (4172)	MFI	280@5800	295@4000	3.29x3.62	10.6:1	29@2000
1995	AAN	2.2 (2226)	MFI	227@5900	258@1950	3.19x3.40	9.3:1	29@2000
	AFC	2.8 (2771)	MFI	174@5500	184@3000	3.25x3.40	10.3:1	41@3000
	AAH	2.8 (2771)	MFI	174@5500	184@3000	3.25x3.40	10.0:1	41@3000
1996-97	AFC	2.8 (2771)	MFI	174@5500	180@3000	3.25x3.40	10.0:1	41@3000

MFI - Multiport fuel injection

GASOLINE ENGINE TUNE-UP SPECIFICATIONS

Year	Engine ID/VIN	Engine Displacement Liters (cc)	Spark Plugs Gap (in.)	Ignition Timing (deg.) MT	Ignition Timing (deg.) AT	Fuel Pump (psi)	Idle Speed (rpm) MT	Idle Speed (rpm) AT	Valve Clearance In.	Valve Clearance Ex.
1993	AAN	2.2 (2226)	0.024	①	①	58-61	770-830	770-830	HYD	HYD
	AAH	2.8 (2771)	0.039	①	①	55-61	700-800	700-800	HYD	HYD
	ABH	4.2 (4172)	0.032	①	①	58-62	690-750	710-770	HYD	HYD
1994	AAN	2.2 (2226)	0.024	①	①	58-61	770-830	770-830	HYD	HYD
	AAH	2.8 (2771)	0.039	①	①	55-61	700-800	700-800	HYD	HYD
	ABH	4.2 (4172)	0.032	①	①	58-62	690-750	710-770	HYD	HYD
1995	AAN	2.2 (2226)	0.024	①	①	58-61	770-830	770-830	HYD	HYD
	AFC	2.8 (2771)	0.039	①	①	52-61	650-750	650-750	HYD	HYD
	AAH	2.8 (2771)	0.039	①	①	55-61	700-800	700-800	HYD	HYD
1996-97	AFC	2.8 (2771)	0.039	①	①	46-55	650-750	650-750	HYD	HYD

NOTE: The Vehicle Emission Control Information label often reflects specification changes made during production. The label figures must be used if they differ from those in this chart.
HYD - Hydraulic
① The basic setting is controlled by the ECU and is not adjustable

CAPACITIES

Year	Model	Engine ID/VIN	Engine Displacement Liters (cc)	Engine Oil with Filter	Transmission (pts.) 4-Spd	Transmission (pts.) 5-Spd	Transmission (pts.) Auto.	Transfer Case (pts.)	Drive Axle Front (pts.)	Drive Axle Rear (pts.)	Fuel Tank (gal.)	Cooling System (qts.)
1993	S4 Quattro	AAN	2.2 (2226)	4.7	-	5.6	-	-	-	3.2	21.1	9.5
	90S	AAH	2.8 (2771)	5.3	-	5.0	①	-	-	-	17.4	11.6
	90CS	AAH	2.8 (2771)	5.3	-	5.0	①	-	-	-	17.4	11.6
	90CS Quattro	AAH	2.8 (2771)	5.3	-	5.0	-	-	-	2.0	21.1	11.6
	100	AAH	2.8 (2771)	5.3	-	5.0	③	-	④	2.0	21.1	8.4
	100S	AAH	2.8 (2771)	5.3	-	5.0	③	-	④	-	21.1	8.4
	100CS	AAH	2.8 (2771)	5.3	-	5.0	③	-	④	-	21.1	8.4
	100CS Quattro	AAH	2.8 (2771)	5.3	-	5.0	⑤	-	-	⑥	21.1	8.4
	V8 Quattro	ABH	4.2 (4172)	10.0	-	6.4	②	-	1.0	3.6	21.1	9.0
1994	S4 Quattro	AAN	2.2 (2226)	4.7	-	5.6	①	-	-	3.2	21.1	9.5
	90S	AAH	2.8 (2771)	5.3	-	5.0	①	-	-	-	17.4	11.6
	90CS	AAH	2.8 (2771)	5.3	-	5.0	①	-	-	-	17.4	11.6
	90CS Quattro	AAH	2.8 (2771)	5.3	-	5.0	-	-	-	2.0	16.9	11.6
	100	AAH	2.8 (2771)	5.3	-	5.0	③	-	④	-	21.1	8.4
	100S	AAH	2.8 (2771)	5.3	-	5.0	③	-	④	-	21.1	8.4
	100CS	AAH	2.8 (2771)	5.3	-	5.0	③	-	④	-	21.1	8.4
	100CS Quattro	AAH	2.8 (2771)	5.3	-	5.0	⑤	-	-	⑥	21.1	8.4
	Cabriolet	AAH	2.8 (2771)	5.3	-	-	①	-	-	-	17.4	11.6
	V8 Quattro	ABH	4.2 (4172)	10.0	-	6.4	②	-	1.0	3.6	21.1	9.0
1995	S6 Quattro	AAN	2.2 (2226)	4.7	-	5.6	-	-	-	3.2	21.1	9.5
	A6	AFC	2.8 (2771)	5.3	-	5.0	①	-	④	-	21.1	8.4
	A6	AAH	2.8 (2771)	5.3	-	5.0	①	-	④	-	21.1	8.4
	A6 Quattro	AFC	2.8 (2771)	5.3	-	5.0	⑤	-	-	⑥	21.1	8.4
	A6 Quattro	AAH	2.8 (2771)	5.3	-	5.0	⑤	-	-	⑥	21.1	8.4
	90/Sport 90	AFC	2.8 (2771)	5.3	-	5.0	①	-	-	-	17.4	11.6
	90/Sport 90	AAH	2.8 (2771)	5.3	-	5.0	①	-	-	-	17.4	11.6
	90/Sport 90 Quattro	AAH	2.8 (2771)	5.3	-	5.0	-	-	-	2.0	16.9	11.6
	Cabriolet	AFC	2.8 (2771)	5.3	-	-	⑩	-	-	-	17.4	11.6
	Cabriolet	AAH	2.8 (2771)	5.3	-	-	⑩	-	-	-	17.4	11.6
1996-97	A6	AFC	2.8 (2771)	5.3	-	5.0	③	-	④	-	21.1	8.4
	A6 Quattro	AFC	2.8 (2771)	5.3	-	5.0	⑤	-	1.5	⑥	21.1	8.4

CAPACITIES

Year	Model	Engine ID/VIN	Engine Displacement Liters (cc)	Engine Oil with Filter	Transmission (pts.) 4-Spd	5-Spd	Auto.	Transfer Case (pts.)	Drive Axle Front (pts.)	Rear (pts.)	Fuel Tank (gal.)	Cooling System (qts.)
1996-97	A4	AFC	2.8 (2771)	5.3	-	4.8	⑦,⑧	-	1.6	-	16.4	9.0
	A4 Quattro	AFC	2.8 (2771)	5.3	-	5.8	NA	-	NA	NA	15.6	9.0
	Cabriolet	AFC	2.8 (2771)	5.3	-	-	⑩	-	-	-	17.4	11.6

① Initial fill: 11.4 Change: 6.4
② Initial fill: 20.4 Change: 8.0
③ 097 transmission: Initial fill: 11.4 Change: 6.4 01K transmission: Initial fill: 14.8 Change: 5.8
④ 097 transmission: 2.0 01K transmission: 1.4
⑤ Initial fill: 14.8 Change: 5.8
⑥ Intermediate rear drive: 2.2 Final rear drive: 3.6
⑦ Initial fill: 19.0 Change: 5.4
⑧ Use Audi ATF only
⑨ Initial fill: 12.7 Change: 6.3
⑩ Initial Fill: 11.4 Change: NA

VALVE SPECIFICATIONS

Year	Engine ID/VIN	Engine Displacement Liters (cc)	Seat Angle (deg.)	Face Angle (deg.)	Spring Test Pressure (lbs. @ in.)	Spring Installed Height (in.)	Stem-to-Guide Clearance (in.) Intake		Stem Diameter (in.) Intake	Exhaust
1993	AAN	2.2 (2226)	45	45	NA	NA	0.039 ①	0.051 ①	0.2744	0.2732
	AAH	2.8 (2771)	NA	NA	NA	NA	0.039 ①	0.051 ①	NA	NA
	ABH	4.2 (4172)	45	45	NA	NA	0.039 ①	0.051 ①	NA	NA
1994	AAN	2.2 (2226)	45	45	NA	NA	0.039 ①	0.051 ①	0.2744	0.2732
	AAH	2.8 (2771)	45	45	NA	NA	0.039 ①	0.051 ①	NA	NA
	ABH	4.2 (4172)	45	45	NA	NA	0.039 ①	0.051 ①	NA	NA
1995	AAN	2.2 (2226)	45	45	NA	NA	0.039 ①	0.051 ①	0.2744	0.2732
	AAH	2.8 (2771)	45	45	NA	NA	0.039 ①	0.051 ①	NA	NA
	AFC	2.8 (2771)	45	45	NA	NA	0.039 ①	0.051 ①	NA	NA
1996-97	AFC	2.8 (2771)	45	45	NA	NA	0.039 ①	0.051 ①	NA	NA

NA - Not Available
① To measure Stem-to-Guide clearance, insert new valve into guide until end of valve is flush with end of guide. Use dial indicator to measure valve head movement. Specification given is for maximum wear.

TORQUE SPECIFICATIONS
All readings in ft. lbs.

Year	Engine ID/VIN	Engine Displacement Liters (cc)	Cylinder Head Bolts	Main Bearing Bolts	Rod Bearing Bolts	Crankshaft Damper Bolts	Flywheel Bolts	Manifold Intake	Exhaust	Spark Plugs	Lug Nut
1993	AAN	2.2 (2226)	①	48	②	③	②	15	18	28	89
	AAH	2.8 (2771)	⑥	⑪	②	⑫	⑨	15	18	22	89
	ABH	4.2 (4172)	①	NA	NA	③	62	④	18	22	81
1994	AAN	2.2 (2226)	①	48	②	③	②	15	18	28	89
	AAH	2.8 (2771)	⑥	⑪	②	⑫	⑨	15	18	22	89
	ABH	4.2 (4172)	①	NA	NA	③	62	④	18	22	81
1995	AAN	2.2 (2226)	①	48	②	③	②	15	18	28	89
	AAH	2.8 (2771)	⑥	⑪	②	⑫	⑨	15	18	22	89
	AFC	2.8 (2771)	⑥	⑪	②	⑫	⑨	15	18	22	89
1996-97	AFC	2.8 (2771)	⑥	⑪	②	⑫	⑨	15	18	22	89

NA - Not Available
① Step 1: 30 ft. lbs. Step 2: 44 ft. lbs. Step 3: +180 degrees
② Step 1: 22 ft. lbs. Step 2: +90 degrees
③ 258 ft. lbs. with special tool 332 ft. lbs. without special tool
④ Step 1: 11 ft. lbs. Step 2: 18 ft. lbs.
⑤ Step 1: 30 ft. lbs. Step 2: +180 degrees
⑥ Step 1: 44 ft. lbs. Step 2: +180 degrees

TORQUE SPECIFICATIONS
All readings in ft. lbs.

Year	Engine ID/VIN	Engine Displacement Liters (cc)	Cylinder Head Bolts	Main Bearing Bolts	Rod Bearing Bolts	Crankshaft Damper Bolts	Flywheel Bolts	Manifold Intake	Manifold Exhaust	Spark Plugs	Lug Nut

⑦ 44 ft. lbs. +90 degrees for automatic transmission
   30 ft. lbs. +180 degrees for manual transmission
⑧ Lower manifold to block: 7 ft. lbs.
   Lower manifold to upper manifold: 15 ft. lbs.
⑨ Step 1: 44 ft. lbs.
   Step 2: +90 degrees
⑩ Lower manifold to block: 7 ft. lbs.
   Lower manifold to upper manifold: 11 ft. lbs.

⑪ Main bearing bolts:
   Step 1: 44 ft. lbs.
   Step 2: Plus 180 degrees
   Cross bolts:
   Step 1: Hand tighten before installing main bearing ball
   Step 2: 18 ft. lbs.
⑫ Center bolt:
   Step 1: 148 ft. lbs.
   Step 2: Plus 180 degrees
   Outer bolts: 15 ft. lbs.

BRAKE SPECIFICATIONS
All measurements in inches unless noted

Year	Model		Master Cylinder Bore	Brake Disc Original Thickness	Brake Disc Minimum Thickness	Maximum Runout	Brake Drum Diameter Original Inside Diameter	Brake Drum Diameter Max. Wear Limit	Brake Drum Diameter Maximum Machine Diameter	Minimum Lining Thickness Front	Minimum Lining Thickness Rear
1993	S4		②	③	④	0.002	-	-	-	⑤	0.08
	90		0.937	⑥	⑦	0.002	-	-	-	⑧	⑧
	90	①	0.937	⑨	⑩	0.002	-	-	-	⑧	⑧
	100		0.937	⑥	⑪	0.002	-	-	-	⑤	⑧
	100	①	②	⑥	⑪	0.002	-	-	-	⑤	⑧
	V8		1.000	⑫	⑬	0.002	-	-	-	⑤	⑧
1994	S4		②	③	④	0.002	-	-	-	⑤	0.08
	90		0.937	⑥	⑦	0.002	-	-	-	⑧	⑧
	90	①	0.937	⑨	⑩	0.002	-	-	-	⑧	⑧
	100		0.937	⑥	⑪	0.002	-	-	-	⑤	⑧
	100	①	②	⑥	⑪	0.002	-	-	-	⑤	⑧
	V8		1.000	⑫	⑬	0.002	-	-	-	⑤	⑧
	Cabriolet		0.937	⑥	⑦	0.002	-	-	-	⑧	⑧
1995	S6		②	③	④	0.002	-	-	-	⑤	0.08
	90		0.937	⑥	⑦	0.002	-	-	-	⑧	⑧
	90	①	0.937	⑨	⑩	0.002	-	-	-	⑧	⑧
	A6		0.937	⑥	⑪	0.002	-	-	-	⑤	⑧
	A6	①	②	⑥	⑪	0.002	-	-	-	⑤	⑧
	Cabriolet		0.937	⑥	⑦	0.002	-	-	-	⑧	⑧
1996-97	A4		0.937	⑭	⑮	0.002	-	-	-	⑧	⑧
	A4	①	0.937	⑭	⑮	0.002	-	-	-	⑧	⑧
	A6		0.937	⑥	⑪	0.002	-	-	-	⑥	⑧
	A6	①	②	⑥	⑪	0.002	-	-	-	⑥	⑧
	Cabriolet		0.937	⑥	⑦	0.002	-	-	-	⑧	⑧

① Quattro
② Vacuum Brake Booster: 0.937
   Hydraulic Brake Booster: 1.000
③ Front: 1.18
   Rear: 0.79
④ Front: 1.10
   Rear: 0.71
⑤ When dashboard light illuminates
⑥ Front: 0.984
   Rear: 0.394

⑦ Front: 0.866
   Rear: 0.315
⑧ 0.28 including backing plate
⑨ Front: 0.984
   Rear: 0.984
⑩ Front: 0.905
   Rear: 0.905
⑪ Front: 0.905
   Rear: 0.315

⑫ Front: 0.984
   Rear: 0.787
⑬ Front: 0.905
   Rear: 0.709
⑭ Front:
   Teves/Ate Calipers:
   -Venetilated Disc: 0.984
   -Non-ventilated Disc: 0.590
   Lucas Calipers: 0.510
   Rear: 0.394

⑮ Front:
   Teves/Ate Calipers:
   -Venetilated Disc: 0.905
   -Non-ventilated Disc: 0.510
   Lucas Calipers: 0.430
   Rear: 0.315

FREQUENT MAINTENANCE LABOR
AUDI

The following should be used as a guide when determining the amount of work required for a particular service if taken to a repair shop. In estimating how long a particular Frequent Maintenance Service item should take, please observe the following:
- **Chilton Time** is time that is based on field research and data supplied by the vehicle manufacturer.
- All labor time operations are given in hours and tenths of an hour.
- All labor operations, are to be used as a **guide**.

COOLING

(G)Winter Cooling System
Includes: Run engine to check for leaks, tighten all hose connections. Test radiator and pressure cap, drain radiator and engine block. Add antifreeze and refill system.

	Chilton Time
1993-97	.5

(G) Belt, Serpentine Drive, Renew

1993-97	1.2

(G) Hoses, Radiator, Renew
Includes: Drain and refill cooling system as required.

1993-97 90/A4/Quattro	
upper	.6
lower	.9
1993-97 100/A6/Quattro	
upper	.6
lower	.9
1993-94 V8 Quattro	
one	.5
all	2.0
1993-97 S4/S6	
upper	.6
lower	.9
1995-97 Cabriolet	
upper	.6
lower	.9

(G) Thermostat, Coolant, Renew

	Chilton Time
1993-97 90/A4/Quattro	1.0
1993-97 100/A6/Quattro	1.0
1993-94 V8 Quattro	1.4
1993-97 S4/S6	1.0
1995-97 Cabriolet	1.0

BRAKES

(G) Bleed Brakes (Four Wheels)
Includes: Add fluid.

1993-97	.5

(G) Brakes, Adjust (Minor)
Includes: Adjust brakes, fill master cylinder.

1993-97	.7

(M) Parking Brake, Adjust

1993-97	.4

LUBRICATION SERVICES

(M) Engine Oil & Filter, Renew
Includes: Inspect and correct all fluid levels.

1993-97	.3

WHEELS

(G) Wheel, Renew (One)

	Chilton Time
1993-97	.5

(G) Wheel, Balance

1993-97	
one	.3
each adtnl.	.2

(G) Wheels, Rotate (All)

1993-97	.5

ELECTRICAL

(G) Headlamps, Aim

1993-97	.5

(G) Halogen Headlamp Bulb, Renew

1993-95 each	.3

(G) License Lamp Assy., Renew

1993-97	.2

(G) Parking Lamp Lens or Bulb, Renew

1993-97	.3

(G) Tail Lamp Lens or Bulb, Renew

1993-97	.3

(G) Horn, Renew

1993-97, one or both	.6

SCHEDULED MAINTENANCE INTERVALS
(AUDI 90, 100, S4, S6, A4, A6, A8, V8 QUATTRO & CABRIOLET)

TO BE SERVICED	TYPE OF SERVICE	VEHICLE MILEAGE INTERVAL (x1000)												
		7.5	15	22.5	30	37.5	45	52.5	60	67.5	75	82.5	90	97.5
Engine oil & filter②	R	✓	✓	✓	✓	✓	✓	✓	✓	✓	✓	✓	✓	✓
Automatic shiftlock operation	S/I	✓	✓	✓	✓	✓	✓	✓	✓	✓	✓	✓	✓	✓
Cooling system	S/I	✓	✓	✓	✓	✓	✓	✓	✓	✓	✓	✓	✓	✓
Passenger compartment air filter	R		✓		✓		✓		✓		✓		✓	
Automatic transmission fluid, filter & final drive	S/I		✓		✓		✓		✓		✓		✓	
Battery electrolyte level	S/I		✓		✓		✓		✓		✓		✓	
Brake system (brake pads & fluid level)	S/I		✓		✓		✓		✓		✓		✓	
Drive axle shaft boots	S/I		✓		✓		✓		✓		✓		✓	
Engine (check for leaks)	S/I		✓		✓		✓		✓		✓		✓	
Exhaust system	S/I		✓		✓		✓		✓		✓		✓	
Idle speed⑤	S/I		✓		✓		✓		✓		✓		✓	
Manual transmission oil	S/I		✓		✓		✓		✓		✓		✓	
ODB system check for codes③	S/I		✓		✓		✓		✓		✓		✓	
V-belts④	S/I			✓		✓				✓		✓		
Air cleaner element	R				✓				✓				✓	
Spark plugs	R				✓				✓				✓	
Power steering fluid level	S/I				✓				✓				✓	
Automatic transmission fluid⑦	R						✓						✓	
Timing belt (V8 Quattro)	S/I				✓				✓					
Timing belt	R												✓	
Brake fluid①	R													
Engine coolant (V8 Quattro)⑥	R								✓					
Front axle dust seals on ball joints & tie rod ends	S/I								✓					

SCHEDULED MAINTENANCE INTERVALS
(AUDI 90, 100, S4, S6, A4, A6, A8, V8 QUATTRO & CABRIOLET) (Cont.)

TO BE SERVICED	TYPE OF SERVICE	VEHICLE MILEAGE INTERVAL (x1000)												
		7.5	15	22.5	30	37.5	45	52.5	60	67.5	75	82.5	90	97.5
Poly-ribbed belt	R												✓	
Rotate tires	S/I	✓												

① Replace every 2 years regardless of mileage.
② Reset service interval display, if equipped.
③ If equipped.
④ Replace at 45,000 & 90,000 miles.
⑤ except California models.
⑥ Replace at mileage interval or every 2 years, whichever comes first.
⑦ A6, Cabriolet.
R – Replace   S/I – Service or Inspect

FREQUENT OPERATION MAINTENANCE (SEVERE SERVICE)
   If a vehicle is operated under any of the following conditions it is considered severe service:
- Extremely dusty areas.
- 50% or more of the vehicle operation is in 32°C (90°F) or higher temperatures, or constant operation in temperatures below 0°C (32°F).
- Prolonged idling (vehicle operation in stop and go traffic).
- Frequent short running periods (engine does not warm to normal operating temperatures).
- Police, taxi, delivery usage or trailer towing usage.
Oil & oil filter change – change every 3750 miles.
Air filter element – service or inspect every 15,000 miles.
Automatic transmission fluid - replace every 30,000 miles.

BMW

ENGINE IDENTIFICATION

Year	Model	Engine Displacement Liters (cc)	Engine Series (ID/VIN)	Fuel System	No. of Cylinders	Engine Type
1993	318i	1.8 (1796)	M42B18	M1.7	4	DOHC
	318iS	1.8 (1796)	M42B18	M1.7	4	DOHC
	325i	2.5 (2494)	M50B25	M3.1	6	DOHC
	325iC	2.5 (2494)	M20B25	M1.3	6	SOHC
	325iS	2.5 (2494)	M50B25	M3.1	6	DOHC
	525i	2.5 (2494)	M50B25	M3.1	6	DOHC
	525iT	2.5 (2494)	M50B25	M3.1	6	DOHC
	535i	3.4 (3430)	M30B35	M1.3	6	SOHC
	M5	3.6 (3535)	S38B36	M1.2	6	DOHC
	740i	4.0 (3982)	M60B40	M3.3	8	DOHC
	740iL	4.0 (3982)	M60B40	M3.3	8	DOHC
	750iL	5.0 (4988)	M70B50	M1.7	12	SOHC
	850Ci	5.0 (4988)	M70B50	M1.7	12	SOHC
1994	318i	1.8 (1796)	M42B18	M1.7	4	DOHC
	318iC	1.8 (1796)	M42B18	M1.7	4	DOHC
	318iS	1.8 (1796)	M42B18	M1.7	4	DOHC
	325i	2.5 (2494)	M50B25	M3.1	6	DOHC
	325iC	2.5 (2494)	M50B25	M3.1	6	DOHC
	325iS	2.5 (2494)	M50B25	M3.1	6	DOHC

ENGINE IDENTIFICATION

Year	Model	Engine Displacement Liters (cc)	Engine Series (ID/VIN)	Fuel System	No. of Cylinders	Engine Type
1994	525i	2.5 (2494)	M50B25	M3.1	6	DOHC
	525iT	2.5 (2494)	M50B25	M3.1	6	DOHC
	530i	3.0 (2997)	M60B30	M3.3	8	DOHC
	530iT	3.0 (2997)	M60B30	M3.3	8	DOHC
	540i	4.0 (3982)	M60B40	M3.3	8	DOHC
	730i	3.0 (2997)	M60B30	M3.3	8	DOHC
	740i	4.0 (3982)	M60B40	M3.3	8	DOHC
	740iL	4.0 (3982)	M60B40	M3.3	8	DOHC
	840Ci	4.0 (3982)	M60B40	M3.3	8	DOHC
	750iL	5.0 (4988)	M70B50	M1.7	12	SOHC
	850Ci	5.0 (4988)	M70B50	M1.7	12	SOHC
	850CSi	5.6 (5576)	S70B56	M1.7	12	SOHC
1995	318i	1.8 (1796)	M42B18	M1.7	4	DOHC
	318iS	1.8 (1796)	M42B18	M1.7	4	DOHC
	318iC	1.8 (1796)	M42B18	M1.7	4	DOHC
	325i	2.5 (2494)	M50B25	M3.1	6	DOHC
	325iS	2.5 (2494)	M50B25	M3.1	6	DOHC
	325iC	2.5 (2494)	M50B25	M3.1	6	DOHC
	525i	2.5 (2494)	M50B25	M3.1	6	DOHC
	525ti	2.5 (2494)	M50B25	M3.1	6	DOHC
	M3	3.0 (2990)	S50B30	M3.3	6	DOHC
	530i	3.0 (2997)	M60B30	M3.3	8	DOHC
	530ti	3.0 (2997)	M60B30	M3.3	8	DOHC
	540i	4.0 (3982)	M60B40	M3.3	8	DOHC
	740i	4.0 (3982)	M60B40	M3.3	8	DOHC
	740iL	4.0 (3982)	M60B40	M3.3	8	DOHC
	840Ci	4.0 (3982)	M60B40	M3.3	8	DOHC
	750iL	5.4 (5379)	M73B54	M5.2	12	SOHC
	850Ci	5.4 (5379)	M73B54	M5.2	12	SOHC
	850CSi	5.6 (5576)	S70B56	M3.3	12	SOHC
1996-97	318i	1.9 (1895)	M44B19	①	4	DOHC
	318iC	1.9 (1895)	M44B19	①	4	DOHC
	318is	1.9 (1895)	M44B19	①	4	DOHC
	318ti	1.9 (1895)	M44B19	①	4	DOHC
	Z3	1.9 (1895)	M44B19	M5.2	4	DOHC
	328i	2.8 (2793)	M52B28	②	6	DOHC
	328is	2.8 (2793)	M52B28	②	6	DOHC
	328iC	2.8 (2793)	M52B28	②	6	DOHC
	M3	3.2 (3152)	S52B32	M3.3	6	DOHC
	740iL	4.4 (4398)	M60B44	M5.2	8	DOHC
	750iL	5.4 (5379)	M73B54	M5.2	12	SOHC
	840Ci	4.4 (4398)	M60B44	M5.3	8	DOHC
	850Ci	5.4 (5379)	M73B54	M5.2	12	SOHC

DOHC - Double overhead camshaft
SOHC - Single overhead camshaft

① Bosch ML-Motronic w/knock control (2 sensors)
② Siemens MS 41.0 w/knock control (2 sensors)

GENERAL ENGINE SPECIFICATIONS

Year	Engine ID/VIN	Engine Displacement Liters (cc)	Fuel System Type	Net Horsepower @ rpm	Net Torque @ rpm (ft. lbs.)	Bore x Stroke (in.)	Compression Ratio	Oil Pressure @ rpm
1993	M42B18	1.8 (1796)	M1.7	134@6000	127@4600	3.31x3.19	10.0:1	18@idle
	M20B25	2.5 (2494)	M1.3	168@5800	164@4300	3.31x2.95	8.8:1	18@idle
	M50B25	2.5 (2494)	M3.1	189@5900	185@4200	3.31x2.95	10.5:1	28@idle
	M30B35	3.4 (3430)	M1.3	208@5700	225@4000	3.62x3.39	9.0:1	18@idle
	S38B36	3.6 (3535)	M1.2	311@6900	260@4750	3.68x3.39	10.0:1	18@idle
	M60B40	4.0 (3982)	M3.3	282@5800	295@4500	3.50x3.15	10.0:1	18@idle
	M70B50	5.0 (4988)	M1.7	296@5200	332@4100	3.31x2.95	8.8:1	18@idle
1994	M42B18	1.8 (1796)	M1.7	134@6000	127@4600	3.31x3.19	10.0:1	18@idle
	M50B25	2.5 (2494)	M3.1	189@5900	185@4200	3.31x2.95	10.5:1	28@idle
	M60B30	3.0 (2997)	M3.3	214@5800	214@4500	3.31x2.66	10.5:1	18@idle
	M60B40	4.0 (3982)	M3.3	282@5800	295@4500	3.50x3.15	10.0:1	18@idle
	M70B50	5.0 (4988)	M1.7	296@5200	332@4100	3.31x2.95	8.8:1	18@idle
	S70B56	5.6 (5576)	M1.7	375@5300	406@4000	3.39x3.15	9.8:1	18@idle
1995	M42B18	1.8 (1796)	M1.7	138@6000	129@4500	3.31x3.19	10.0:1	18@idle
	M50B25	2.5 (2494)	M3.1	189@5900	181@4200	3.31x2.95	10.5:1	28@idle
	S50B30	3.0 (2990)	M3.1	240@6000	225@4200	3.39x3.38	10.5:1	28@idle
	M60B30	3.0 (2997)	M3.3	215@5800	214@4500	2.94x2.66	10.5:1	18@idle
	M60B40	4.0 (3982)	M3.3	282@5800	295@4500	3.50x3.15	10.0:1	18@idle
	M73B54	5.4 (5379)	M5.2	322@5000	361@3900	3.35x3.11	10.0:1	18@idle
	S70B56	5.6 (5576)	M1.7	372@5300	402@4000	3.93x3.15	9.8:1	18@idle
1996-97	M44B19	1.9(1895)	N/A	N/A	N/A	N/A	N/A	N/A
	M52B28	2.8 (2793)	N/A	N/A	N/A	N/A	N/A	N/A
	S52B32	3.2 (3152)	N/A	240@6000	236@3800	3.40x3.53	N/A	N/A
	M60B44	4.4 (4398)	N/A	N/A	N/A	N/A	N/A	N/A
	M73B54	5.4 (5379)	M5.2	322@5000	361@3900	3.35x3.11	10.0:1	18@idle

GASOLINE ENGINE TUNE-UP SPECIFICATIONS

Year	Engine ID/VIN	Engine Displacement Liters (cc)	Spark Plugs Gap (in.)	Ignition Timing (deg.) MT	Ignition Timing (deg.) AT	Fuel Pump (psi)	Idle Speed (rpm) MT	Idle Speed (rpm) AT	Valve Clearance In.	Valve Clearance Ex.
1993	M42B18	1.8 (1796)	0.030	①	①	43	850	850	HYD	HYD
	M20B25	2.5 (2494)	0.030	①	①	43	760	760	0.010	0.010
	M50B25	2.5 (2494)	0.030	①	①	51	700	700	HYD	HYD
	M30B35	3.4 (3428)	0.030	①	①	43	800	800	0.012	0.012
	S38B36	3.6 (3535)	0.026	①	①	43	920	920	0.012	0.012
	M60B40	4.0 (3982)	0.030	①	①	51	600	600	HYD	HYD
	M70B50	5.0 (4988)	0.030	①	①	43	700	800	HYD	HYD
1994	M42B18	1.8 (1796)	0.030	①	①	43	850	850	HYD	HYD
	M50B25	2.5 (2494)	0.030	①	①	51	700	700	HYD	HYD
	M60B30	3.0 (2997)	0.030	①	①	51	600	600	HYD	HYD
	M60B40	4.0 (3982)	0.030	①	①	51	600	600	HYD	HYD
	M70B50	5.0 (4988)	0.030	①	①	43	700	800	HYD	HYD
	S70B56	5.6 (5576)	0.026	①	①	43	700	800	HYD	HYD
1995	M42B18	1.8 (1796)	0.030	①	①	43	850	850	HYD	HYD
	M50B25	2.5 (2494)	0.030	①	①	51	700	700	HYD	HYD
	S50B30	3.0 (2990)	0.030	①	①	51	700	700	HYD	HYD
	M60B30	3.0 (2997)	0.030	①	①	51	600	600	HYD	HYD
	M60B40	4.0 (3982)	0.030	①	①	51	600	600	HYD	HYD
	M73B54	5.4 (5379)	0.030	①	①	43	700	700	HYD	HYD
	S70B56	5.6 (5576)	0.030	①	①	43	700	800	HYD	HYD

GASOLINE ENGINE TUNE-UP SPECIFICATIONS

Year	Engine ID/VIN	Engine Displacement Liters (cc)	Spark Plugs Gap (in.)	Ignition Timing (deg.) MT	Ignition Timing (deg.) AT	Fuel Pump (psi)	Idle Speed (rpm) MT	Idle Speed (rpm) AT	Valve Clearance In.	Valve Clearance Ex.
1996-97	M44B19	1.9 (1895)	N/A	N/A	N/A	N/A	N/A	N/A	N/A	N/A
	M52B28	2.8 (2793)	N/A	N/A	N/A	N/A	N/A	N/A	N/A	N/A
	S52B32	3.2 (3152)	N/A	N/A	N/A	N/A	N/A	N/A	N/A	N/A
	M60B44	4.4 (4398)	N/A	N/A	N/A	N/A	N/A	N/A	N/A	N/A
	M73B54	5.4 (5379)	0.030	①	①	43	700	700	HYD	HYD

NOTE: The Vehicle Emission Control Information label often reflects specification changes made during production. The label figures must be used if they differ from those in this chart.

HYD - Hydraulic

① Computer-controlled; No adjustment or verification possible

CAPACITIES

Year	Model	Engine Displacement Liters (cc)	Engine ID/VIN	Engine Oil with Filter	Transmission (pts.) 4-Spd	Transmission (pts.) 5-Spd	Transmission (pts.) Auto.	Transfer Case (pts.)	Drive Axle Front (pts.)	Drive Axle Rear (pts.)	Fuel Tank (gal.)	Cooling System (qts.)
1993	318i	1.8 (1796)	M42B18	5.0	-	4.8	6.4	-	-	2.3	14.5	13.5
	318iS	1.8 (1796)	M42B18	5.0	-	2.2	6.4	-	-	2.3	14.5	13.5
	325i	2.5 (2494)	M50B25	6.9	-	2.6	6.4	-	-	3.6	16.6	11.0 ①
	325iC	2.5 (2494)	M20B25	6.9	-	2.6	6.4	-	-	3.6	16.6	11.0 ①
	325iS	2.5 (2494)	M50B25	6.9	-	2.6	6.4	-	-	3.6	16.6	11.0 ①
	525i	2.5 (2494)	M50B25	6.9	-	2.6	6.4	-	-	3.6	21.4	12.7 ①
	525iT	2.5 (2494)	M50B25	6.9	-	2.6	6.4	-	-	3.6	21.4	12.7 ①
	535i	3.4 (3430)	M30B35	6.1	-	2.6	6.4	-	-	4.0	21.4	12.7
	M5	3.6 (3353)	S38B36	6.1	-	2.6	-	-	-	4.0	21.4	14.0
	740i	4.0 (3982)	M60B40	7.9	-	-	11.6	-	-	3.7 ②	24.0	13.4
	740iL	4.0 (3982)	M60B40	7.9	-	-	11.6	-	-	3.7 ②	24.0	13.4
	750iL	5.0 (4988)	M70B50	7.9	-	-	6.4	-	-	3.7 ②	24.0	14.8
	850Ci	5.0 (4988)	M70B50	7.9	-	4.8	6.4	-	-	4.0	24.0	13.7
1994	318i	1.8 (1796)	M42B18	5.0	-	2.2	6.4	-	-	2.3	17.2	13.5
	318iC	1.8 (1796)	M42B18	5.0	-	2.2	6.4	-	-	2.3	16.4	13.5
	318iS	1.8 (1796)	M42B18	5.0	-	2.2	6.4	-	-	2.3	17.2	13.5
	325i	2.5 (2494)	M50B25	6.9	-	2.6	6.4	-	-	3.6	17.2	11.0 ①
	325iC	2.5 (2494)	M50B25	6.9	-	2.6	6.4	-	-	3.6	16.4	11.0 ①
	325iS	2.5 (2494)	M50B25	6.9	-	2.6	6.4	-	-	3.6	17.2	11.0 ①
	525i	2.5 (2494)	M50B25	6.9	-	2.6	6.4	-	-	3.6	21.1	12.7 ①
	525iT	2.5 (2494)	M50B25	6.9	-	2.6	6.4	-	-	3.6	21.1	12.7
	530i	3.0 (2997)	M60B30	7.9	-	2.6	6.4	-	-	4.0	21.1	12.7
	530iT	3.0 (2997)	M60B30	7.9	-	2.6	6.4	-	-	4.0	21.1	12.7
	540i	4.0 (3982)	M60B40	7.9	-	2.6	6.4	-	-	4.0	21.1	13.4
	730i	3.0 (2997)	M60B30	7.9	-	-	11.6	-	-	3.7 ②	21.5	13.4
	740i	4.0 (3982)	M60B40	7.9	-	-	11.6	-	-	3.7 ②	21.5	13.4
	740iL	4.0 (3982)	M60B40	7.9	-	-	11.6	-	-	3.7 ②	24.0	13.4
	840Ci	4.0 (3982)	M60B40	7.9	-	-	6.4	-	-	4.0	24.0	13.4
	750iL	5.0 (4988)	M70B50	7.9	-	-	6.4	-	-	3.7 ②	24.0	14.8
	850Ci	5.0 (4988)	M70B50	7.9	-	4.8	6.4	-	-	4.0	23.8	13.7
	850CSi	5.6 (5576)	S70B56	7.9	-	4.8	6.4	-	-	4.0	24.0	13.7
1995	318i	1.8 (1796)	M42B18	5.0	-	2.4	6.4	-	-	3.3	17.2	13.5
	318iS	1.8 (1796)	M42B18	5.0	-	2.4	6.4	-	-	3.3	17.2	13.5
	318iC	1.8 (1796)	M42B18	5.0	-	2.4	6.4	-	-	3.3	17.2	13.5
	325i	2.5 (2494)	M50B25	6.9	-	2.6	6.4	-	-	3.3	17.2	11.0
	325iS	2.5 (2494)	M50B25	6.9	-	2.6	6.4	-	-	3.3	17.2	11.0

CAPACITIES

Year	Model	Engine ID/VIN	Engine Displacement Liters (cc)	Engine Oil with Filter	Transmission (pts.)			Transfer Case (pts.)	Drive Axle		Fuel Tank (gal.)	Cooling System (qts.)
					4-Spd	5-Spd	Auto.		Front (pts.)	Rear (pts.)		
1995	325iC	2.5 (2494)	M50B25	6.9	-	2.6	6.4	-	-	3.3	17.2	11.0
	525i	2.5 (2494)	M50B25	6.9	-	2.6	6.4	-	-	3.3	21.1	12.7
	525ti	2.5 (2494)	M50B25	6.9	-	2.6	6.4	-	-	3.3	21.1	12.7
	M3	3.0 (2990)	S50B30	6.9	-	2.6	-	-	-	3.3	17.2	13.5
	530i	3.0 (2997)	M60B30	7.9	-	2.6	6.4	-	-	3.3	21.1	12.7
	530ti	3.0 (2997)	M60B30	7.9	-	2.6	6.4	-	-	3.3	21.1	12.7
	540i	4.0 (3982)	M60B40	7.9	-	2.6 ③	6.4	-	-	3.7	21.1	13.4
	740i	4.0 (3982)	M60B40	7.9	-	-	6.4	-	-	3.7	22.5	13.4
	740iL	4.0 (3982)	M60B40	7.9	-	-	6.4	-	-	3.7	25.1	13.4
	840Ci	4.0 (3982)	M60B40	7.9	-	-	6.4	-	-	3.7	23.8	13.4
	750iL	5.4 (5379)	M73B54	7.9	-	-	6.4	-	-	3.7	25.1	14.8
	850Ci	5.4 (5379)	M73B54	7.9	-	-	6.4	-	-	3.7	23.8	13.7
	850CSi	5.6 (5576)	S70B56	7.9	-	2.6	6.4	-	-	3.7	23.8	13.7
1996-97	318i	M44B19	1.9(1895)	N/A	N/A	N/A	N/A	N/A	N/A	N/A	N/A	N/A
	318iC	M44B19	1.9(1895)	N/A	N/A	N/A	N/A	N/A	N/A	N/A	N/A	N/A
	318is	M44B19	1.9(1895)	N/A	N/A	N/A	N/A	N/A	N/A	N/A	N/A	N/A
	318ti	M44B19	1.9(1895)	N/A	N/A	N/A	N/A	N/A	N/A	N/A	N/A	N/A
	Z3	M44B19	1.9(1895)	N/A	N/A	N/A	N/A	N/A	N/A	N/A	N/A	N/A
	328i	M52B28	2.8 (2793)	N/A	N/A	N/A	N/A	N/A	N/A	N/A	N/A	N/A
	328is	M52B28	2.8 (2793)	N/A	N/A	N/A	N/A	N/A	N/A	N/A	N/A	N/A
	328iC	M52B28	2.8 (2793)	N/A	N/A	N/A	N/A	N/A	N/A	N/A	N/A	N/A
	M3	S52B32	3.2 (3152)	N/A	N/A	N/A	N/A	N/A	N/A	N/A	N/A	N/A
	740iL	M60B44	4.4 (4398)	N/A	N/A	N/A	N/A	N/A	N/A	N/A	N/A	N/A
	750iL	M73B54	5.4 (5379)	7.9	-	-	6.4	-	-	3.7	25.1	14.8
	840Ci	M60B44	4.4 (4398)	N/A	N/A	N/A	N/A	N/A	N/A	N/A	N/A	N/A
	850Ci	M73B54	5.4 (5379)	7.9	-	-	6.4	-	-	3.7	23.8	13.7

① 11.0 without Air Conditioning
11.5 with Air Conditioning

② 3.6 with 7 bolt cover
4.0 wdith 8 bolt cover

③ 5 or 6 speed

VALVE SPECIFICATIONS

Year	Engine ID/VIN	Engine Displacement Liters (cc)	Seat Angle (deg.)	Face Angle (deg.)	Spring Test Pressure (lbs. @ in.)	Spring Installed Height (in.)	Stem-to-Guide Clearance (in.)		Stem Diameter (in.)	
							Intake	Exhaust	Intake	Exhaust
1993	M42B18	1.8 (1796)	45	NA	NA	NA	0.020 ①	0.020 ①	0.275	0.275
	M20B25	2.5 (2494)	45	NA	NA	NA	0.031 ①	0.031 ①	0.275	0.275
	M50B25	2.5 (2494)	45	NA	NA	NA	0.020 ①	0.020 ①	0.275	0.275
	M30B35	3.4 (3430)	45	NA	NA	NA	0.031 ①	0.031 ①	0.315	0.315
	S38B36	3.6 (3535)	45	NA	NA	NA	0.031 ①	0.031 ①	0.275	0.275
	M60B40	4.0 (3982)	45	NA	NA	NA	0.020 ①	0.020 ①	0.235	0.235
	M70B50	5.0 (4988)	45	NA	NA	NA	0.020 ①	0.020 ①	0.280	0.280
1994	M42B18	1.8 (1796)	45	NA	NA	NA	0.020 ①	0.020 ①	0.275	0.275
	M50B25	2.5 (2494)	45	NA	NA	NA	0.020 ①	0.020 ①	0.275	0.275
	M60B30	3.0 (2997)	45	NA	NA	NA	0.020 ①	0.020 ①	0.235	0.235
	M60B40	4.0 (3982)	45	NA	NA	NA	0.020 ①	0.020 ①	0.235	0.235
	M70B50	5.0 (4988)	45	NA	NA	NA	0.020 ①	0.020 ①	0.280	0.280
	S70B56	5.6 (5576)	45	NA	NA	NA	0.020 ①	0.020 ①	0.280	0.280
1995	M42B18	1.8 (1796)	45	NA	NA	NA	0.020 ①	0.020 ①	0.275	0.275
	M50B25	2.5 (2494)	45	NA	NA	NA	0.020 ①	0.020 ①	0.275	0.275
	S50B30	3.0 (2990)	45	NA	NA	NA	0.020 ①	0.020 ①	0.275	0.275

VALVE SPECIFICATIONS

Year	Engine ID/VIN	Engine Displacement Liters (cc)	Seat Angle (deg.)	Face Angle (deg.)	Spring Test Pressure (lbs. @ in.)	Spring Installed Height (in.)	Stem-to-Guide Clearance (in.)		Stem Diameter (in.)	
							Intake	Exhaust	Intake	Exhaust
1995	M60B30	3.0 (2997)	45	NA	NA	NA	0.020 ①	0.020 ①	0.235	0.235
	M60B40	4.0 (3982)	45	NA	NA	NA	0.020 ①	0.020 ①	0.235	0.235
	M73B54	5.4 (5379)	45	NA	NA	NA	0.020 ①	0.020 ①	0.280	0.280
	S70B56	5.6 (5576)	45	NA	NA	NA	0.021 ①	0.021 ①	0.280	0.280
1996-97	M44B19	1.9 (1895)	NA	NA	NA	NA	NA	NA	NA	NA
	M52B28	2.8 (2793)	NA	NA	NA	NA	NA	NA	NA	NA
	S52B32	3.2 (3152)	NA	NA	NA	NA	NA	NA	NA	NA
	M60B44	4.4 (4398)	NA	NA	NA	NA	NA	NA	NA	NA
	M73B54	5.4 (5379)	45	NA	NA	NA	0.020 ①	0.020 ①	0.280	0.280

NA - Not Available

① To measure Stem-to-Guide clearance, insert new valve into guide until end of valve is flush with end of guide. Use dial indicator to measure valve head movement. Specification given is for maximum wear.

TORQUE SPECIFICATIONS
All readings in ft. lbs.

Year	Engine ID/VIN	Engine Displacement Liters (cc)	Cylinder Head Bolts	Main Bearing Bolts	Rod Bearing Bolts	Crankshaft Damper Bolts	Flywheel Bolts	Manifold		Spark Plugs	Lug Nut
								Intake	Exhaust		
1993	M42B18	1.8 (1796)	①	②	③	217-231	90	⑬	⑬	14-22	65-79
	M20B25	2.5 (2494)	①	44	③	303	78	⑬	⑬	14-22	65-79
	M50B25	2.5 (2494)	①	②	③	303	78	⑬	⑬	14-22	65-79
	M30B35	3.4 (3430)	⑨	44	40	325	78	⑬	⑬	14-22	65-79
	S38B36	3.6 (3535)	⑧	②	⑩	⑫	78	⑬	⑬	14-22	65-79
	M60B40	4.0 (3982)	⑭	④	⑮	⑪	78	⑬	⑬	14-22	65-79
	M70B50	5.0 (4988)	⑦	④	③	⑪	78	⑬	⑬	14-22	65-79
1994	M42B18	1.8 (1796)	①	②	③	217-231	90	⑬	⑬	14-22	65-79
	M50B25	2.5 (2494)	①	②	③	303	78	⑬	⑬	14-22	65-79
	M60B30	3.0 (2997)	⑭	④	⑮	⑪	78	⑬	⑬	14-22	65-79
	M60B40	4.0 (3982)	⑭	④	⑮	⑪	78	⑬	⑬	14-22	65-79
	M70B50	5.0 (4988)	⑦	④	③	⑪	78	⑬	⑬	14-22	65-79
	S70B56	5.6 (4576)	⑦	④	③	⑪	78	⑬	⑬	14-22	65-79
1995	M42B18	1.8 (1796)	①	②	③	217-231	90	⑬	⑬	14-22	65-79
	M50B25	2.5 (2494)	①	②	③	303	78	⑬	⑬	14-22	65-79
	S50B30	3.0 (2990)	①	②	③	303	78	⑬	⑬	14-22	65-79
	M60B30	3.0 (2997)	⑤	⑥	⑮	⑪	78	⑬	⑬	14-22	65-79
	M60B40	4.0 (3982)	⑤	⑥	⑮	⑪	78	⑬	⑬	14-22	65-79
	M73B54	5.4 (5379)	⑦	⑥	③	⑪	78	⑬	⑬	14-22	65-79
	S70B56	5.6 (5576)	⑦	⑥	③	⑪	78	⑬	⑬	14-22	65-79
1996-97	M44B19	1.9 (1895)	N/A	N/A	N/A	N/A	N/A	N/A	N/A	N/A	N/A
	M52B28	2.8 (2793)	N/A	N/A	N/A	N/A	N/A	N/A	N/A	N/A	N/A
	S52B32	3.2 (3152)	N/A	N/A	N/A	N/A	N/A	N/A	N/A	N/A	N/A
	M60B44	4.4 (4398)	N/A	N/A	N/A	N/A	N/A	N/A	N/A	N/A	N/A
	M73B54	5.4 (5379)	⑦	⑥	③	⑪	78	⑬	⑬	14-22	65-79

① Step 1: 22 ft. lbs.
Step 2: Turn an additional 90 degrees
Step 3: Repeat Step 2
② Step 1: 14-18 ft. lbs.
Step 2: Turn an additional 50 degrees
③ Step 1: 17 ft. lbs.
Step 2: Turn an additional 70 degrees

④ Step 1: 15 ft. lbs.
Step 2: Turn an additional 70 degrees
⑤ Step 1: 22 ft. lbs.
Step 2: Turn an additional 80 degrees
Step 3: Repeat Step 2
⑥ Step 1: 14.5 ft. lbs.
Step 2: Turn an additional 47-53 degrees

⑦ Hex head bolts:
Step 1: 22 ft. lbs., No set time
Step 2: Turn an additional 120 degrees
Torx bolts:
Step 1: 22 ft. lbs., 15 minute settling time
Step 2: Turn an additional 120 degrees

TORQUE SPECIFICATIONS
All readings in ft. lbs.

Year	Engine ID/VIN	Engine Displacement Liters (cc)	Cylinder Head Bolts	Main Bearing Bolts	Rod Bearing Bolts	Crankshaft Damper Bolts	Flywheel Bolts	Manifold Intake	Manifold Exhaust	Spark Plugs	Lug Nut

⑧ Step 1: 15 ft. lbs.
  Step 2: Turn an additional 60 degrees
  Step 3: Turn an additional 70 degrees
⑨ Step 1: 44 ft. lbs., 15 minute settling time
  Step 2: 60 ft. lbs., 25 minute warm running time
  Step 3: Turn an additional 35 degrees
⑩ Step 1: 11 ft. lbs.
  Step 2: 22 ft. lbs.
  Step 3: Turn an additional 60 degrees

⑪ Step 1: 78 ft. lbs.
  Step 2: Turn an additional 60 degrees
  Step 3: Repeat Step 2
  Step 4: Turn an additional 30 degrees
⑫ Step 1: 110 ft. lbs.
  Step 2: Loosen bolts
  Step 3: 44 ft. lbs.
  Step 4: Turn an additional 60 degrees
  Step 5: Repeat Step 4
  Step 6: Turn an additional 30 degrees

⑬ M8 bolts: 16 ft. lbs.
  M7 bolts: 11 ft. lbs.
  M6 bolts: 7 ft. lbs.
⑭ Step 1: 22 ft. lbs.
  Step 2: Turn an additional 80 degrees
  Step 3: Repeat Step 2
⑮ Step 1: 3 ft. lbs.
  Step 2: 15 ft. lbs.
  Step 3: Turn an additional 80 degrees

BRAKE SPECIFICATIONS
All measurements in inches unless noted

Year	Model	Master Cylinder Bore	Brake Disc Original Thickness	Brake Disc Minimum Thickness	Brake Disc Maximum Runout	Brake Drum Diameter Original Inside Diameter	Brake Drum Diameter Max. Wear Limit	Brake Drum Diameter Maximum Machine Diameter	Minimum Lining Thickness Front	Minimum Lining Thickness Rear
1993	318i	NA	NA	⑦	0.008	-	-	-	0.079	0.079
	318iS	NA	NA	③	0.008	-	-	-	0.079	0.079
	325i	NA	NA	③	0.008	-	-	-	0.079	0.079
	325iC	NA	NA	②	0.008	-	-	-	0.079	0.079
	325iS	NA	NA	③	0.008	-	-	-	0.079	0.079
	525i	NA	NA	③	0.008	-	-	-	0.080	0.080
	525iT	NA	NA	③	0.008	-	-	-	0.080	0.080
	M5	NA	NA	④ ⑧	0.008	-	-	-	0.079	0.079
	540i	NA	NA	④	0.008	-	-	-	0.079	0.079
	740i	NA	NA	⑥	0.008	-	-	-	0.079	0.079
	750iL	NA	NA	⑥	0.008	-	-	-	0.079	0.079
	840Ci	NA	NA	⑥	0.008	-	-	-	0.079	0.079
	850Ci	NA	NA	⑥	0.008	-	-	-	0.079	0.079
1994	318i	NA	NA	⑦	0.008	-	-	-	0.079	0.079
	318iC	NA	NA	③	0.008	-	-	-	0.079	0.079
	318iS	NA	NA	③	0.008	-	-	-	0.079	0.079
	325i	NA	NA	③	0.008	-	-	-	0.079	0.079
	325iC	NA	NA	③	0.008	-	-	-	0.079	0.079
	325iS	NA	NA	③	0.008	-	-	-	0.079	0.079
	525i	NA	NA	③	0.008	-	-	-	0.080	0.080
	525iT	NA	NA	③	0.008	-	-	-	0.080	0.080
	530i	NA	NA	③	0.008	-	-	-	0.080	0.080
	530iT	NA	NA	③	0.008	-	-	-	0.080	0.080
	540i	NA	NA	④	0.008	-	-	-	0.079	0.079
	730i	NA	NA	⑥	0.008	-	-	-	0.079	0.079
	740i	NA	NA	⑥	0.008	-	-	-	0.079	0.079
	740iL	NA	NA	⑥	0.008	-	-	-	0.079	0.079
	750iL	NA	NA	⑥	0.008	-	-	-	0.079	0.079
	840Ci	NA	NA	⑥	0.008	-	-	-	0.079	0.079
	850Ci	NA	NA	⑥	0.008	-	-	-	0.079	0.079
	850CSi	NA	NA	⑥	0.008	-	-	-	0.079	0.079
1995	318i	NA	NA	①	0.008	-	-	-	0.079	0.079
	318iS	NA	NA	①	0.008	-	-	-	0.079	0.079
	325i	NA	NA	⑤	0.008	-	-	-	0.079	0.079

BRAKE SPECIFICATIONS
All measurements in inches unless noted

Year	Model	Master Cylinder Bore	Brake Disc Original Thickness	Brake Disc Minimum Thickness	Brake Disc Maximum Runout	Brake Drum Diameter Original Inside Diameter	Brake Drum Diameter Max. Wear Limit	Brake Drum Diameter Maximum Machine Diameter	Minimum Lining Thickness Front	Minimum Lining Thickness Rear
1995	325iS	NA	NA	⑤	0.008	-	-	-	0.079	0.079
	M3	NA	NA	⑤	0.008	-	-	-	0.079	0.079
	525i	NA	NA	⑤	0.008	-	-	-	0.079	0.079
	530i	NA	NA	⑤	0.008	-	-	-	0.079	0.079
	540i	NA	NA	⑨	0.008	-	-	-	0.079	0.079
	740i	NA	NA	⑩	0.008	-	-	-	0.079	0.079
	740iL	NA	NA	⑩	0.008	-	-	-	0.079	0.079
	750iL	NA	NA	⑩	0.008	-	-	-	0.079	0.079
	840Ci	NA	NA	⑩	0.008	-	-	-	0.079	0.079
	850Ci	NA	NA	⑩	0.008	-	-	-	0.079	0.079
	850CSi	NA	NA	⑩	0.008	-	-	-	0.079	0.079
1996-97	318i	NA	NA	NA	NA	NA	NA	NA	NA	NA
	318iC	NA	NA	NA	NA	NA	NA	NA	NA	NA
	318is	NA	NA	NA	NA	NA	NA	NA	NA	NA
	318ti	NA	NA	NA	NA	NA	NA	NA	NA	NA
	Z3	NA	NA	NA	NA	NA	NA	NA	NA	NA
	328i	NA	NA	NA	NA	NA	NA	NA	NA	NA
	328is	NA	NA	NA	NA	NA	NA	NA	NA	NA
	328iC	NA	NA	NA	NA	NA	NA	NA	NA	NA
	M3	NA	NA	NA	NA	NA	NA	NA	NA	NA
	740iL	NA	NA	NA	NA	NA	NA	NA	NA	NA
	750iL	NA	NA	⑩	0.008	-	-	-	0.079	0.079
	840Ci	NA	NA	NA	NA	NA	NA	NA	NA	NA
	850Ci	NA	NA	⑩	0.008	-	-	-	0.079	0.079

NA - Not Available
① Front: 0.409; Rear: 0.331
② Front: 0.906; Rear: 0.315
③ Front: 0.787; Rear: 0.315
④ Front: 1.039; Rear: 0.709
⑤ Front: 0.803; Rear: 0.331
⑥ Front: 1.102; Rear: 0.709
⑦ Front: 0.394; Rear: 0.315
⑧ Do not machine the front brake rotor
⑨ Front: 1.039; Rear: 0.724
⑩ Front: 1.118; Rear: 0.724

FREQUENT MAINTENANCE LABOR
BMW

The following should be used as a guide when determining the amount of work required for a particular service if taken to a repair shop. In estimating how long a particular Frequent Maintenance Service item should take, please observe the following:
- Chilton Time is time that is based on field research and data supplied by the vehicle manufacturer.
- All labor time operations are given in hours and tenths of an hour.
- All labor operations, are to be used as a guide.

COOLING

(G) Winterize Cooling System
Includes: Run engine to check for leaks, tighten all hose connections. Test radiator and pressure cap, drain radiator and engine block. Add antifreeze and refill system.
1993-97 5

(G) Belt, Serpentine Drive, Renew
1993-97 750iL 1.0

(G) Belt, Drive, Renew
1994-95 540i
one 1.0
each adtnl. 5
1994-95 740i, iL
one 1.0
each adtnl. 5
1993-94 750iL
one 1.0
each adtnl. 5

1994-97 840Ci
one 1.0
each adtnl. 5

(G) Hoses, Radiator, Renew
Includes: Drain and refill cooling system as required.
Upper
1993-97 5

FREQUENT MAINTENANCE LABOR (cont.)
BMW

Lower

	Chilton Time
1993-97 318i, is, iC	.7
1995-97 318ti	.7
1993-97 325i, 325is, iC	.7
1993-95 525i	.7
1993 535i, 535is	.7
1994-95 540i	1.2
1994-97 740i, iL	1.2
1993-97 750iL	1.0
1994-97 840Ci	1.2
1993-97 850i, Ci, CSi	1.0
1995-97 M3	.7

(G) Thermostat, Coolant, Renew

	Chilton Time
1993-97 318i, is, iC	1.3
1995-97 318ti	1.3
1993-97 325i, 325is	1.8
1993-97 325iC	1.9
1993-97 525i	1.6
1994-97 530i	1.3
1993 535i, 535is	1.5
1994-97 540i	1.3
1994 740i, iL	1.4
1995-97 740i, iL	1.9
1993-94 750iL	1.5
1995-97 750iL	2.9
1994-97 840Ci	1.3
1993-97 850i, Ci, CSi	
5.0L	1.1
5.4L	3.3
1995-97 M3	2.3

BRAKES

(G) Bleed Brakes (Four Wheels)
Includes: Add fluid.

	Chilton Time
1993-97 w/ABS	1.5

(M) Parking Brake, Adjust

	Chilton Time
1993-97 318i, is, iC	.6
1995-97 318ti	.6
1993-97 325i, 325is, iC	.6
1993-97 525i	1.0
1994-97 530i	1.0
1993 535i, 535is	.7
1994-97 540i	1.0
1994-97 740i, iL	.8
1993-97 750iL	.8
1994-97 840Ci	.8
1993-97 850i, Ci, CSi	.8
1995-97 M3	.6

LUBRICATION SERVICES

(M) Engine Oil & Filter, Renew
Includes: Inspect and correct all fluid levels.

	Chilton Time
1993-97	.6

ELECTRICAL

(G) Headlamps, Aim

	Chilton Time
1993-97	.6

(G) Halogen Headlamp Bulb, Renew

	Chilton Time
1993-97, each	.5

(G) High Mount Stop Lamp Bulb, Renew

	Chilton Time
1993-97	.2

(G) License Lamp Assy., Renew

	Chilton Time
1993-97, each	.5

(G) Parking Lamp Lens or Bulb, Renew

	Chilton Time
1993-97	.4

(G) Horn, Renew
Each

	Chilton Time
1993-97 318i, is, iC	.5
1995-97 318ti	.5
1993-97 325i, 325is, iC	.5
1993-97 525i	.5
1994-97 530i	.5
1993 535i, 535is	.5
1994-97 540i	.5
1994-95 740i, iL	.6
1993-94 750iL	.8
1995-97 750iL	.6
1994-97 840Ci	.5
1993-97 850i, Ci, CSi	.5
1995-97 M3	.5

(M) Terminals, Battery, Clean

	Chilton Time
1993-97	.3

20 SPECIFICATIONS

SCHEDULED MAINTENANCE INTERVALS
(1993 BMW 3 SERIES, 5 SERIES, 7 SERIES, 8 SERIES & M5)

Note: BMW does not rely solely on vehicle mileage to determine service intervals. An on-board diagnostic center, monitors engine operating conditions, along with mileage, to determine the most effective maintenance intervals. The information is then conveyed to the driver through the service indicator lights, located in the center of the instrument panel.

TO BE SERVICED	TYPE OF SERVICE	SERVICE INTERVALS			
		INITIAL 1200 MILES	OIL SERVICE	INSPECTION I	INSPECTION II
Oil level	S/I	✓			
Engine oil & filter	R①	✓②	✓	✓	✓
Engine air cleaner element	R③				✓
Spark plugs	R			⑤	✓
Fuel filter	R④				✓
Fuel, vapor lines & fuel cap	S/I	✓		✓	✓
Cooling system	S/I	✓		✓	✓
Exhaust pipe & muffler	S/I	✓		✓	✓
Catalytic converter & shielding	S/I	✓		✓	✓
Valve clearance	S/I	⑥		⑦	⑦
Throttle linkage	S/I			✓	✓
Engine (check for leakage)	S/I	✓			
Engine drive belts	S/I④				✓
Maintenance Indicators	RE		⑧	✓	✓
Engine coolant	R			⑨	⑨
Oxygen sensor	R⑩				
Timing belt	R⑪				✓
Intake air dust separators	S/I⑫				✓
Brake & clutch fluids ⑨	S/I			✓	✓
Brake pads & discs	S/I			✓	✓
Parking brake system	S/I			✓	✓
Power steering system	S/I			✓	✓
Rear axle oil	S/I			✓	✓
Steering play, suspension track rods, front axle joints, steering linkage & joint disc	S/I			✓	✓

20-28

SCHEDULED MAINTENANCE INTERVALS
(1993 BMW 3 SERIES, 5 SERIES, 7 SERIES, 8 SERIES & M5) (Cont.)

Note: BMW does not rely solely on vehicle mileage to determine service intervals. An on-board diagnostic center, monitors engine operating conditions, along with mileage, to determine the most effective maintenance intervals. The information is then conveyed to the driver through the service indicator lights, located in the center of the instrument panel.

TO BE SERVICED	TYPE OF SERVICE	SERVICE INTERVALS			
		INITIAL 1200 MILES	OIL SERVICE	INSPECTION I	INSPECTION II
Transmission fluid/oil	S/I			✓	✓⑭
Wheel centering hubs	S/I			✓	✓
Rear axle oil⑬	R		✓		✓
OBD system for codes	S/I	✓		✓	✓

R – Replace   S/I – Service or Inspect   RE – Reset

① On vehicle operated less than 6200 miles per year, more frequent service may be required.
② Service is not required for 318 & 325 models with E36 chassis.
③ Replace more frequently if vehicle is operated in dusty conditions.
④ Recommend service for California models, required for all other models.
⑤ Required service for M20 engines only.
⑥ Required service for M20 & M30 engines only.
⑦ Required service for M20, M30 & S38 engines only.
⑧ Reset the oil service indicator lights only.
⑨ Replace every 2 years with inspection service.
⑩ Replace at the following mileage intervals: 525I, 525iT, 325I & 325iS - replace every 100,000 miles. Other models: replace every 60,000 miles.
⑪ Replace every second inspection II or every 4 years, whichever occurs first.
⑫ Required service for 850 models only.
⑬ At first oil service, then at each inspection II.
⑭ Change fluid (A/T) or oil (M/T) at inspection II.

SCHEDULED MAINTENANCE INTERVALS
(1994-97 BMW 3 SERIES, 5 SERIES, 7 SERIES, 8 SERIES, M3 & Z3)

Note: BMW does not rely solely on vehicle mileage to determine service intervals. An on-board diagnostic center, monitors engine operating conditions, along with mileage, to determine the most effective maintenance intervals. The information is then conveyed to the driver through the service indicator lights, located in the center of the instrument panel.

TO BE SERVICED	TYPE OF SERVICE	SERVICE INTERVALS			
		INITIAL 1200 MILES	OIL SERVICE	INSPECTION I	INSPECTION II
Oil level	S/I	✓			
Engine oil	R	✓①			
Engine oil & filter	R②		✓	✓	✓
Engine air cleaner element	R③				✓
Spark plugs	R				✓
Fuel filter	R④				✓
Fuel, vapor lines & fuel cap	S/I	✓		✓	✓
Cooling system	S/I	✓		✓	✓
Exhaust pipe & muffler	S/I	✓		✓	✓

SCHEDULED MAINTENANCE INTERVALS
(1994-97 BMW 3 SERIES, 5 SERIES, 7 SERIES, 8 SERIES, M3 & Z3) (Cont.)

Note: BMW does not rely solely on vehicle mileage to determine service intervals. An on-board diagnostic center, monitors engine operating conditions, along with mileage, to determine the most effective maintenance intervals. The information is then conveyed to the driver through the service indicator lights, located in the center of the instrument panel.

TO BE SERVICED	TYPE OF SERVICE	SERVICE INTERVALS			
		INITIAL 1200 MILES	OIL SERVICE	INSPECTION I	INSPECTION II
Catalytic converter & shielding	S/I	✓		✓	✓
Throttle linkage	S/I			✓	✓
Engine (check for leakage)	S/I	✓			
Engine drive belts	S/I				✓
Maintenance Indicators	RE		⑤	✓	✓
Engine coolant	R			⑥	⑥
Oxygen sensor	R ⑦				
Intake air dust separators	S/I ⑧				✓
Brake & clutch fluids ⑥	S/I			✓	✓
Brake pads & discs	S/I			✓	✓
Parking brake system	S/I			✓	✓
Power steering system	S/I			✓	✓
Rear axle oil	S/I			✓	✓
Steering play, suspension track rods, front axle joints, steering linkage & joint disc	S/I			✓	✓
Transmission fluid/oil	S/I			✓	✓ ⑩
Wheel centering hubs	S/I			✓	✓
Rear axle oil ⑨	R		✓		✓
OBD system for codes	S/I	✓		✓	✓

R – Replace   S/I – Service or Inspect   RE – Reset

① Service is not required for 318 & 325 models.
② On vehicle operated less than 6200 miles per year, more frequent service may be required.
③ Replace more frequently if vehicle is operated in dusty conditions.
④ Recommend service for California models, required for all other models.
⑤ Reset the oil service indicator lights only.
⑥ Replace every 2 years with inspection service.
⑦ Replace every 100,000 miles on all models.
⑧ Required service for 850 models only.
⑨ At first oil service, then at each inspection II.
⑩ Change fluid (A/T) or oil (M/T) at inspection II.

CHRYSLER IMPORT

VEHICLE IDENTIFICATION CHART

Engine Code						Model Year	
Code	Liters	Cu. In. (cc)	Cyl.	Fuel Sys.	Eng. Mfg.	Code	Year
A	1.5	90 (1468)	4	MFI	Mitsubishi	P	1993
C	1.8	112 (1834)	4	MFI	Mitsubishi	R	1994
G	2.4	146 (2350)	4	MFI	Mitsubishi	S	1995
						T	1996
						V	1997

MFI - Multiport fuel injection

ENGINE IDENTIFICATION

Year	Model	Engine Displacement Liters (cc)	Engine Series (ID/VIN)	Fuel System	No. of Cylinders	Engine Type
1993	Colt	1.5 (1468)	A	MFI	4	SOHC
	Colt	1.8 (1834)	C	MFI	4	SOHC
	Colt Vista	1.8 (1834)	C	MFI	4	SOHC
	Colt Vista	2.4 (2351)	G	MFI	4	SOHC
	Summit	1.5 (1468)	A	MFI	4	SOHC
	Summit	1.8 (1834)	C	MFI	4	SOHC
	Summit Wagon	1.8 (1834)	C	MFI	4	SOHC
	Summit Wagon	2.4 (2350)	G	MFI	4	SOHC
1994	Colt	1.5 (1468)	A	MFI	4	SOHC
	Colt	1.8 (1834)	C	MFI	4	SOHC
	Colt Vista	1.8 (1834)	C	MFI	4	SOHC
	Colt Vista	2.4 (2351)	G	MFI	4	SOHC
	Summit	1.5 (1468)	A	MFI	4	SOHC
	Summit	1.8 (1834)	C	MFI	4	SOHC
	Summit Wagon	1.8 (1834)	C	MFI	4	SOHC
	Summit Wagon	2.4 (2350)	G	MFI	4	SOHC
1995	Summit	1.5 (1468)	A	MFI	4	SOHC
	Summit	1.8 (1834)	C	MFI	4	SOHC
	Summit Wagon	1.8 (1834)	C	MFI	4	SOHC
	Summit Wagon	2.3 (2351)	G	MFI	4	SOHC
1996-97	Summit	1.5 (1468)	A	MFI	4	SOHC
	Summit	1.8 (1834)	C	MFI	4	SOHC
	Summit Wagon	1.8 (1834)	C	MFI	4	SOHC
	Summit Wagon	2.4 (2351)	G	MFI	4	SOHC

MFI - Multipoint fuel injection
SOHC - Single overhead camshaft

GENERAL ENGINE SPECIFICATIONS

Year	Engine ID/VIN	Engine Displacement Liters (cc)	Fuel System Type	Net Horsepower @ rpm	Net Torque @ rpm (ft. lbs.)	Bore x Stroke (in.)	Compression Ratio	Oil Pressure @ rpm
1993	A	1.5 (1468)	MFI	92@6000	93@3000	2.97x3.23	9.2:1	50-64@700
	C	1.8 (1834)	MFI	113@6000	116@4500	3.19x3.50	9.5:1	56.9@2000
	G	2.4 (2351)	MFI	136@5500	145@4250	3.41x3.94	8.5:1	42.7@2000
1994	A	1.5 (1468)	MFI	92@6000	93@3000	2.97x3.23	9.2:1	54@2000
	C	1.8 (1834)	MFI	113@6000	116@4500	3.19x3.50	9.5:1	56.9@2000
	G	2.4 (2351)	MFI	136@5500	145@4250	3.41x3.94	9.5:1	42.7@2000
1995	A	1.5 (1468)	MFI	92@6000	93@3000	2.97x3.23	9.2:1	①
	C	1.8 (1834)	MFI	113@6000 ②	116@4500	3.19X3.50	9.5:1	①
	G	2.4 (2351)	MFI	136@5500	145@4250	3.41x3.94	9.5:1	①
1996-97	A	1.5 (1468)	MFI	92@6000	93@3000	2.97x3.23	9.2:1	①
	C	1.8 (1834)	MFI	113@6000 ②	116@4500	3.19x3.50	9.5:1	①
	G	2.4 (2351)	MFI	136@5500	145@4250	3.41x3.94	9.5:1	①

MFI - Multiport fuel injection   ① 11.4 psi or more at curb idle speed   ② Manual: 119@6000

GASOLINE ENGINE TUNE-UP SPECIFICATIONS

Year	Engine ID/VIN	Engine Displacement Liters (cc)	Spark Plugs Gap (in.)	Ignition Timing (deg.) MT	AT	Fuel Pump (psi)	Idle Speed (rpm) MT	AT	Valve Clearance In.	Ex.
1993	A	1.5 (1468)	0.039-0.043	5B	5B	47.6	700	750	0.006 ②	0.01
	C	1.8 (1834)	0.039-0.043	5B	5B	47.6	700	650	HYD	HYD
	G	2.4 (2351)	0.039-0.043	5B	5B	47.6	700	650	HYD	HYD
1994	A	1.5 (1468)	0.039-0.043	5B	5B	38 ①	750	750	0.008 ②	0.01 ②
	C	1.8 (1834)	0.039-0.043	5B	5B	38 ①	750	750	0.008 ②	0.012 ②
	G	2.4 (2351)	0.039-0.043	5B	5B	38	750	750	HYD	HYD
1995	A	1.5 (1468)	0.039-0.043	5B	5B	38 ①	750	750	0.008	0.010
	C	1.8 (1834)	0.039-0.043	5B	5B	38 ①	700	700	0.008	0.012
	G	2.4 (2351)	0.039-0.043	5B	5B	38 ①	750	750	HYD	HYD
1996-97	A	1.5 (1468)	0.039-0.043	5B	5B	38 ①	750	70	0.008	0.010
	C	1.8 (1934)	0.039-0.043	5B	5B	38 ①	700	700	0.008	0.012
	G	2.4 (2351)	0.039-0.043	5B	5B	38 ①	750	750	HYD	HYD

NOTE: The Vehicle Emission Control Information label often reflects specification changes made during production. The label figures must be used if they differ from those in this chart.
B - Before top dead center
HYD - Hydraulic
① Pressure at idle with vacuum applied to fuel pressure regulator
② Adjustment should be made with engine cold

CAPACITIES

Year	Model	Engine ID/VIN	Engine Displacement Liters (cc)	Engine Oil with Filter	Transmission (pts.) 4-Spd	5-Spd	Auto.	Transfer Case (pts.)	Drive Axle Front (pts.)	Rear (pts.)	Fuel Tank (gal.)	Cooling System (qts.)
1993	Colt	A	1.5 (1468)	3.5	-	3.8	12.6	-	-	-	13.2	5.3
	Colt	C	1.8 (1834)	4.0	-	3.8	12.6	-	-	-	13.2	6.3
	Colt Vista	C	1.8 (1834)	4.0	-	①	②	1.5	-	1.5	14.5	6.3
	Colt Vista	G	2.4 (2351)	4.5	-	4.8	②	1.2	-	1.5	14.5	6.8
	Summit	A	1.5 (1468)	3.0	-	3.8	13.0	-	-	-	13.2	5.3
	Summit	C	1.8 (1834)	4.0	-	3.8	13.0	-	-	-	13.2	6.3
	Summit Wagon	C	1.8 (1934)	4.0	-	3.8	13.0	1.25	-	1.5	14.5	6.3
	Summit Wagon	G	2.4 (2350)	4.1	-	4.8	13.0	1.25	-	1.5	16.0	6.8

CAPACITIES

Year	Model	Engine ID/VIN	Engine Displacement Liters (cc)	Engine Oil with Filter	Transmission (pts.) 4-Spd	5-Spd	Auto.	Transfer Case (pts.)	Drive Axle Front (pts.)	Rear (pts.)	Fuel Tank (gal.)	Cooling System (qts.)
1994	Colt	A	1.5 (1468)	3.5	-	3.8	12.6	-	-	-	13.2	5.3
	Colt	C	1.8 (1834)	4.0	-	3.8	12.6	-	-	-	13.2	6.3
	Colt Vista	C	1.8 (1834)	4.0	①	②		1.2	-	1.5	14.5	6.3
	Colt Vista	G	2.4 (2351)	4.5	-	4.8	②	1.2	-	1.5	14.5	6.8
	Summit	A	1.5 (1468)	3.0	-	3.8	13.0	-	-	-	13.2	5.3
	Summit	C	1.8 (1834)	4.0	-	3.8	13.0	-	-	-	13.2	6.3
	Summit Wagon	C	1.8 (1834)	4.0	-	3.8	13.0	1.25	-	1.5	14.5	6.3
	Summit Wagon	G	2.4 (2350)	4.1	-	4.8	13.0	1.25	-	1.5	14.5	6.8
1995	Summit	A	1.5 (1468)	3.7	-	3.8	12.6	-	-	-	13.2	5.3
	Summit	C	1.8 (1834)	4.2	-	3.8	12.6	-	-	-	13.2	6.3
	Summit Wagon	C	1.8 (1834)	4.0	-	①	③	1.20	-	1.5	14.5	6.3
	Summit Wagon	G	2.4 (2351)	4.5	-	4.8	③	1.20	-	1.5	14.5	6.8
1996-97	Summit	A	1.5 (1468)	3.7	-	3.8	12.6	-	-	-	13.2	5.3
	Summit	C	1.8 (1834)	4.2	-	3.8	12.6	-	-	-	13.2	6.3
	Summit Wagon	C	1.8 (1834)	4.0	-	①	③	1.20	-	1.5	14.5	6.3
	Summit Wagon	G	2.4 (2351)	4.5	-	3.8	③	1.20	-	1.5	14.5	6.8

① 2WD models: 3.8 pts.; 4WD models: 4.8 pts.
② 2WD models: 6.4 qts.; 4WD models: 6.9 qts.
③ 2WD models: 12.8 qts.; 4WD models: 13.8 qts.

VALVE SPECIFICATIONS

Year	Engine ID/VIN	Engine Displacement Liters (cc)	Seat Angle (deg.)	Face Angle (deg.)	Spring Test Pressure (lbs. @ in.)	Spring Installed Height (in.)	Stem-to-Guide Clearance (in.) Intake	Exhaust	Stem Diameter (in.) Intake	Exhaust
1993	A	1.5 (1468)	44-44.5	45-45.5	①	②	0.0008-0.0020	0.0020-0.0035	0.2585-0.2591	0.2571-0.2579
	C	1.8 (1834)	43.5-44	45-45.5	③	④	0.0008-0.0016	0.0012-0.0024	0.2350-0.2354	0.2350-0.2354
	G	2.4 (2351)	43.5-44	45-45.5	⑤	1.740	0.0008-0.0020	0.0012-0.0028	0.2350-0.2354	0.2343-0.2350
1994	A	1.5 (1468)	44-44.5	45-45.5	①	②	0.0008-0.0020	0.0020-0.0035	0.2585-0.2591	0.2571-0.2579
	C	1.8 (1834)	43.5-44	45-45.5	③	⑥	0.0008-0.0016	0.0012-0.0024	0.2350-0.2354	0.2343-0.2350
	G	2.4 (2351)	43.5-44	45-45.5	⑤	1.740	0.0008-0.0020	0.0012-0.0028	0.2350-0.2354	0.2343-0.2350
1995	A	1.5 (1468)	44-44.5	45-45.5	①	②	0.0008-0.0020	0.0020-0.0035	0.2585-0.2591	0.2571-0.2579
	C	1.8 (1834)	43.5-44	45-45.5	③	⑥	0.0008-0.0016	0.0012-0.0024	0.2350-0.2354	0.2343-0.2350
	G	2.4 (2351)	43.5-44	45-45.5	⑤	1.740	0.0008-0.0020	0.0012-0.0028	0.2350-0.2354	0.2343-0.2350
1996-97	A	1.5 (1468)	44	45-45.5	①		0.0008-0.0039	0.0020-0.0059	0.2587-0.2591	0.2571-0.2579
	C	1.8 (1834)	43.5-44	45-45.5	132	⑧	0.0008-0.0039	0.0012-0.0059	0.2350-0.2354	0.2343-0.2350
	G	2.4 (2351)	43-44	45-45.5	60	⑧ 1.740	0.0008-0.0039	0.0012-0.0059	0.2350-0.2354	0.2343-0.2350

① Intake: 51@ installed height; Exhaust: 64@ installed height
② Intake free length: 1.776-1.815; Exhaust free length: 1.803-1.843
③ 132@ installed height
④ Free length: 1.898-1.937
⑤ 60@installed height
⑥ Free length: 1.965-2.004
⑦ 73 @ installed height
⑧ At installed height

TORQUE SPECIFICATIONS
All readings in ft. lbs.

Year	Engine ID/VIN	Engine Displacement Liters (cc)	Cylinder Head Bolts	Main Bearing Bolts	Rod Bearing Bolts	Crankshaft Damper Bolts	Flywheel Bolts	Manifold Intake	Manifold Exhaust	Spark Plugs	Lug Nut
1993	A	1.5 (1468)	①	47-51	36-38	51-72	94-101	11-14	11-14	15-21	65-80
	C	1.8 (1834)	④	⑤	⑥	134	72	13	②	15-21	65-80
	G	2.4 (2351)	③	⑤	⑥	-	98	13	13	18	65-80
1994	A	1.5 (1468)	①	38	14.5	62	94-101	13	13	15-21	65-80
	C	1.8 (1834)	④	⑤	⑥	134	72	14	②	15-21	65-80
	G	2.4 (2351)	③	⑤	14.5	-	98	13	13	18	65-80
1995	A	1.5 (1468)	①	38	14.5	62	94-101	13	13	15-21	65-80
	C	1.8 (1834)	④	⑤	⑥	134	72	14	②	15-21	65-80
	G	2.4 (2351)	③	⑤	14.5	-	98	13	13	18	65-80
1996-97	A	1.5 (1468)	53	38	⑥	62	98	13	13	15-21	65-80
	C	1.8 (1834)	④	⑤	⑥	134	72	14	11-14 ⑦	18	65-80
	G	2.4 (2351)	③	⑤	⑥	-	98	13	18-22	18	65-80

① Cold: 51-54 ft. lbs.; Hot: 58-61 ft. lbs.
② Upper bolts: 11-14 ft. lbs. Lower bolts: 22 ft. lbs.
③ Step 1: 58 ft. lbs. Step 2: Fully loosen Step 3: 14 ft. lbs. Step 4: +90 degrees Step 5: +90 degrees
④ Step 1: 54 ft. lbs. Step 2: Fully loosen all bolts Step 3: 14.5 ft. lbs. Step 4: +90 degrees Step 5: +90 degrees
⑤ Step 1: 18 ft. lbs. Step 2: +90 degrees
⑥ Step 1: 14.5 ft. lbs. Step 2: +90 degrees
⑦ Lower exhaust manifold nuts: 22 ft. lbs.

BRAKE SPECIFICATIONS
All measurements in inches unless noted

Year	Model		Master Cylinder Bore	Brake Disc Original Thickness	Brake Disc Minimum Thickness	Maximum Runout	Brake Drum Diameter Original Inside Diameter	Max. Wear Limit	Maximum Machine Diameter	Minimum Lining Thickness Front	Minimum Lining Thickness Rear
1993	Colt	⑦	0.812	0.512	0.450	0.003	7.00	7.20	NA	0.080	0.040
	Colt	②	⑧	0.710	0.450	0.003	8.00	8.10	NA	0.080	0.040
	Colt	③	-	0.468	0.650	0.003	-	-	-	-	0.040
	Colt Vista	④	0.945	0.882	0.003	⑤	⑥	NA	0.080	0.040	
	Colt Vista	③	-	0.394	0.331	0.003	-	-	-	-	0.040
	Summit	①	0.813	0.510	0.449	0.006	7.10	7.20	NA	0.080	0.040
	Summit	②	0.875	0.710	0.646	0.006	7.10	7.20	NA	0.080	0.040
	Summit Wagon	⑨	⑩	⑪	0.003	⑤	⑥	NA	0.080	0.040	
1994	Colt	⑦	0.812	0.512	0.450	0.003	7.00	7.20	NA	0.080	0.040
	Colt	②	⑧	0.710	0.045	0.003	8.00	8.10	NA	0.080	0.040
	Colt	③	-	0.394	0.331	0.003	-	-	-	-	0.040
	Colt Vista	④	0.945	0.882	0.003	⑤	⑥	NA	0.080	0.040	
	Colt Vista	③	-	0.394	0.331	0.003	-	-	-	-	0.080
	Summit	①	0.813	0.510	0.449	0.003	7.10	7.20	NA	0.080	0.040
	Summit	②	⑬	0.710 ⑭	0.646 ⑮	0.003	8.00	8.10	NA	0.080	0.040 ⑫
	Summit Wagon	④	⑩	⑪	0.003	⑤	⑥	NA	0.080	0.040 ⑫	
1995	Summit	①	0.813	0.510	0.449	0.003	7.10	7.20	NA	0.080	0.040
	Summit	②	⑬	0.710 ⑭	0.646 ⑮	0.003	8.00	8.10	NA	0.080	0.040 ⑫
	Summit Wagon	④	⑩	⑪	0.003	⑤	⑥	NA	0.080	0.040 ⑫	
1996-97	Summit	⑦	0.813	0.710 ⑯	0.449	0.003	8.00	8.10	NA	0.080	0.040
	Summit	②	⑧	0.710 ⑭	0.646 ⑮	0.003	8.00	8.10	NA	0.080	0.040 ⑫
	Summit Wagon	④	⑩	⑪	0.003	⑤	⑥	NA	0.080	0.040 ⑫	

NA - Not Available
① Hatchback
② Sedan
③ Rear disc
④ With ABS: 1.000; Without ABS: 0.9375
⑤ 8" drum: 7.992; 9" drum: 9.000
⑥ 8" drum: 8.071; 9" drum: 9.079
⑦ Coupe
⑧ With ABS: 0.937; Without ABS: 0.812
⑨ With ABS: 15/16 Without ABS: 1
⑩ Front: 0.945 Rear: 0.394
⑪ Front: 0.882 Rear: 0.331
⑫ Rear disc brakes: 0.080
⑬ Master cylinder bore: 0.813 With ABS: 0.938
⑭ Rear: 0.390
⑮ Rear: 0.330
⑯ Solid front disc: 0.510

FREQUENT MAINTENANCE LABOR
CHRYSLER IMPORTS
SUMMIT

The following should be used as a guide when determining the amount of work required for a particular service if taken to a repair shop. In estimating how long a particular Frequent Maintenance Service item should take, please observe the following:
- **Chilton Time** is time that is based on field research and data supplied by the vehicle manufacturer.
- All labor time operations are given in hours and tenths of an hour.
- All labor operations, are to be used as a **guide**.

COOLING

(G) Winterize Cooling System
Includes: Run engine to check for leaks, tighten all hose connections. Test radiator and pressure cap, drain radiator and engine block. Add antifreeze and refill system.
1993-97 5

(G) Belt, Drive, Renew
1993-97
  serpentine
    4 cyl. (.3) 5
    V6 (.5) 8
  V-belt
    one (.3) 4
    each adtnl. 1
w/AC add 1

(G) Belt, Drive, Adjust
1993-97
  one 3
  each adtnl. 1

(G) Hoses, Radiator, Renew
Includes: Drain and refill cooling system as required.
1993-97 Summit
  upper (.3) 4
  lower (.4) 5

(G) Thermostat, Coolant, Renew
1993-97 Summit (.3) 4

FUEL

(M) Air Cleaner, Service
1993-97 3

(G) Filter, Fuel, Renew
1993-97 Summit
  in tank (1.2) 1.5
  in line (.5) 7

BRAKES

(G) Bleed Brakes (Four Wheels)
Includes: Add fluid.
1993-97 (.4) 5

(G) Brakes, Adjust (Minor)
Includes: Adjust brakes, fill master cylinder.
1993-97 Summit 4

(M) Parking Brake, Adjust
1993-97 Summit 4

LUBRICATION SERVICES

(M) Engine Oil & Filter, Renew
Includes: Inspect and correct all fluid levels.
1993-97 (.3) 4

(M) Lubricate Chassis
Includes: Inspect and correct all fluid levels.
1993-97 4
Install grease fittings, add 1

WHEELS

(G) Wheel, Renew (One)
1993-97 5

(G) Wheel, Balance
1993-97
  one 3
  each adtnl. 2

(G) Wheels, Rotate (All)
1993-97 5

ELECTRICAL

(G) Headlamps, Aim
1993-97
  two 4
  four 6

(G) High Mount Stop Lamp and/or Lens, Renew
1993-97 (.2) 4

(G) License Lamp Assy., Renew
1993-97 Summit Wagon (.4) 6
1993-97 Summit (.2) 4

(G) Tail Lamp Assy., Renew
1993-97 (.3) 4

(G) Turn Signal & Parking Lamp Assy., Renew
1993-97 Summit (.2) 3

(G) Horn, Renew
1993-97 Summit (.6) 8

(M) Terminals, Battery, Clean
1993-97 3

FREQUENT MAINTENANCE LABOR
CHRYSLER IMPORTS
COLT/VISTA

The following should be used as a guide when determining the amount of work required for a particular service if taken to a repair shop. In estimating how long a particular Frequent Maintenance Service item should take, please observe the following:
- **Chilton Time** is time that is based on field research and data supplied by the vehicle manufacturer.
- All labor time operations are given in hours and tenths of an hour.
- All labor operations, are to be used as a **guide**.

COOLING

(G) Winterize Cooling System
Includes: Run engine to check for leaks, tighten all hose connections. Test radiator and pressure cap, drain radiator and engine block. Add antifreeze and refill system.
1993-94 .5

(G) Belt, Drive, Renew
1993-94
Alternator or PS 4
AC
   SOHC 3

(G) Belt, Drive, Adjust
1993-94
one 3
each adtnl. 1

(G) Hoses, Radiator, Renew
Includes: Drain and refill cooling system as required.
1993-94
upper 4
lower .5

(G) Thermostat, Coolant, Renew
1993-94 .5

FUEL

(M) Air Cleaner, Service
1993-94 2

(G) Filter, Fuel, Renew
1993-94 Colt
  in line 7
  in tank 1.7
1993-94 Vista
  in line 4
  in tank 1.7

BRAKES

(G) Bleed Brakes (Four Wheels)
Includes: Add fluid.
1993-94 .5

(G) Brakes, Adjust (Minor)
Includes: Adjust brakes, fill master cylinder.
1993-94 .4

(M) Parking Brake, Adjust
1993-94 .4

LUBRICATION SERVICES

(M) Engine Oil & Filter, Renew
Includes: Inspect and correct all fluid levels.
1993-94 .3

(M) Lubricate Chassis, Change Oil & Filter
Includes: Inspect and correct all fluid levels.
1993-94 .6
Install grease fittings add 1

(M) Lubricate Chassis
Includes: Inspect and correct all fluid levels.
1993-94 .4
Install grease fittings add 1

WHEELS

(G) Wheel, Renew (One)
1993-94 .5

(G) Wheel, Balance
1993-94
one 3
each adtnl. 2

(G) Wheels, Rotate (All)
1993-94 .5

ELECTRICAL

(G) Headlamps, Aim
1993-94
two 4
four 6

(G) Parking Lamp Lens or Bulb, Renew
1993-94 .3

(G) Tail Lamp Lens or Bulb, Renew
1993-94 .3

(G) Horn, Renew
1993-94 Colt 8
1993-94 Vista 3

(M) Terminals, Battery, Clean
1993-94 .3

SCHEDULED MAINTENANCE INTERVALS
(CHRYSLER COLT, SUMMIT, SUMMIT WAGON & VISTA WAGON)

TO BE SERVICED	TYPE OF SERVICE	VEHICLE MILEAGE INTERVAL (x1000)												
		7.5	15	22.5	30	37.5	45	52.5	60	67.5	75	82.5	90	97.5
Engine oil & filter	R	✓	✓	✓	✓	✓	✓	✓	✓	✓	✓	✓	✓	✓
Coolant level, hoses & clamps	S/I	✓	✓	✓	✓	✓		✓	✓	✓	✓	✓	✓	✓
Rotate tires	S/I	✓	✓	✓	✓	✓	✓	✓	✓	✓	✓	✓	✓	✓
Brake hoses	S/I		✓		✓		✓		✓		✓		✓	
Disc brake pads	S/I		✓		✓		✓		✓		✓		✓	
Drive shaft boots	S/I		✓		✓		✓		✓		✓		✓	
Valve clearances (1.5L & 1.8L)	S/I		✓		✓		✓		✓		✓		✓	
Air cleaner element	R				✓				✓				✓	
Engine Coolant	R				✓				✓				✓	
Spark plugs	R				✓				✓				✓	
Automatic transaxle fluid & filter①	R				✓				✓				✓	
Ball joints & steering linkage seals	S/I				✓				✓				✓	
Drive belt(s)	S/I				✓				✓				✓	
Exhaust system	S/I				✓				✓				✓	
Fuel hoses	S/I				✓				✓				✓	
Manual transaxle oil	S/I				✓				✓				✓	
Rear axle oil (Colt Vista & Summit Wagon)	S/I				✓				✓				✓	
Rear drum brake linings & rear wheel cylinders	S/I				✓				✓				✓	
Ignition cables	R								✓					
Timing belt	R								✓					
Distributor cap & rotor	S/I								✓					
EVAP system (except canister)	S/I								✓					
Fuel system	S/I								✓					

① Automatic transaxle fluid & filter - service or inspect every 15,000 miles.
R – Replace   S/I – Service or Inspect

SCHEDULED MAINTENANCE INTERVALS
(CHRYSLER COLT, SUMMIT, SUMMIT WAGON & VISTA WAGON) (Cont.)

FREQUENT OPERATION MAINTENANCE (SEVERE SERVICE)

If a vehicle is operated under any of the following conditions it is considered severe service:
- Extremely dusty areas.
- 50% or more of the vehicle operation is in 32°C (90°F) or higher temperatures, or constant operation in temperatures below 0°C (32°F).
- Prolonged idling (vehicle operation in stop and go traffic).
- Frequent short running periods (engine does not warm to normal operating temperatures).
- Police, taxi, delivery usage or trailer towing usage.

Oil & oil filter change – change every 3000 miles.
Disc brake pads - service or inspect every 7500 miles (1993-95) or every 6000 miles (1996-97).
Air cleaner element – service or inspect every 15,000 miles.
Rear drum brake linings & rear wheel cylinders - service or inspect every 15,000 miles.
Spark plugs - replace every 15,000 miles.

HONDA

ENGINE IDENTIFICATION

Year	Model		Engine Displacement Liters (cc)	Engine Series (ID/VIN)		Fuel System	No. of Cylinders	Engine Type
1993	Civic		1.5 (1493)	D15B7		MFI	4	SOHC 16V
	Civic		1.5 (1493)	B15B8		MFI	4	SOHC 8V
	Civic		1.5 (1493)	D15Z1	①	MFI	4	SOHC 16V
	Civic		1.6 (1590)	D16Z6	⑤	MFI	4	SOHC 16V
	del Sol		1.5 (1493)	D15B7		MFI	4	SOHC 16V
	del Sol		1.6 (1590)	D16Z6	⑤	MFI	4	SOHC 16V
	Accord DX/LX	②	2.2 (2156)	F22A1		MFI	4	SOHC 16V
	Accord EX/SE	③	2.2 (2156)	F22A6		MFI	4	SOHC 16V
	Prelude S		2.2 (2156)	F22A1		MFI	4	SOHC 16V
	Prelude Si VTEC		2.2 (2157)	H22A1	⑤	MFI	4	DOHC 16V
	Prelude Si	⑥	2.3 (2259)	H23A1		MFI	4	DOHC 16V
1994	Civic		1.5 (1493)	D15B7		MFI	4	SOHC 16V
	Civic		1.5 (1493)	D15B8		MFI	4	SOHC 8V
	Civic		1.5 (1493)	D15Z1	①	MFI	4	SOHC 16V
	Civic		1.6 (1590)	D16Z6	⑤	MFI	4	SOHC 16V
	del Sol		1.5 (1493)	D15B7		MFI	4	SOHC 16V
	del Sol		1.6 (1590)	D16Z6	⑤	MFI	4	SOHC 16V
	del Sol		1.6 (1595)	B16A3	⑤	MFI	4	DOHC 16V
	Accord EX	④	2.2 (2156)	F22B1	⑤	MFI	4	SOHC 16V
	Accord DX/LX	②	2.2 (2156)	F22B2		MFI	4	SOHC 16V
	Prelude S		2.2 (2156)	F22A1		MFI	4	SOHC 16V
	Prelude Si VTEC		2.2 (2157)	H22A1	⑤	MFI	4	DOHC 16V
	Prelude Si	⑥	2.3 (2259)	H23A1		MFI	4	DOHC 16V
1995	Civic		1.5 (1493)	D15B7		MFI	4	SOHC 16V
	Civic		1.5 (1493)	D15B8		MFI	4	SOHC 8V
	Civic		1.5 (1493)	D15Z1	①	MFI	4	SOHC 16V
	Civic		1.6 (1590)	D16Z6	⑤	MFI	4	SOHC 16V
	del Sol		1.5 (1493)	D15B7		MFI	4	SOHC 16V

ENGINE IDENTIFICATION

Year	Model		Engine Displacement Liters (cc)	Engine Series (ID/VIN)	Fuel System	No. of Cylinders	Engine Type
1995	del Sol		1.6 (1590)	D16Z6 ⑤	MFI	4	SOHC 16V
	del Sol		1.6 (1595)	B16A3 ⑤	MFI	4	DOHC 16V
	Accord EX	④	2.2 (2156)	F22B1 ⑤	MFI	4	SOHC 16V
	Accord DX/LX	②	2.2 (2156)	F22B2	MFI	4	SOHC 16V
	Accord V-6		2.7 (2675)	C27A4	MFI	6	SOHC 24V
	Prelude S		2.2 (2156)	F22A1	MFI	4	SOHC 16V
	Prelude Si VTEC		2.2 (2157)	H22A1 ⑤	MFI	4	DOHC 16V
	Prelude Si	⑥	2.3 (2259)	H23A1	MFI	4	DOHC 16V
1996-97	Civic		1.6 (1590)	D16Y7	MFI	4	SOHC 16V
	Civic		1.6 (1590)	D16Y8 ⑤	MFI	4	SOHC 8V
	Civic		1.6 (1590)	D16Y5 ①	MFI	4	SOHC 16V
	del Sol		1.6 (1590)	D16Y7	MFI	4	SOHC 16V
	del Sol		1.6 (1590)	D16Y8 ⑤	MFI	4	SOHC 16V
	del Sol		1.6 (1595)	B16A2	MFI	4	DOHC 16V
	Accord EX		2.2 (2156)	F22B1	MFI	4	SOHC 16V
	Accord DX/LX		2.2 (2156)	F22B2	MFI	4	SOHC 16V
	Accord V-6		2.7 (2675)	C27A4	MFI	6	SOHC 24V
	Prelude S		2.2 (2156)	F22A1	MFI	4	SOHC 16V
	Prelude Si VTEC		2.2 (2157)	H22A1	MFI	4	DOHC 16V
	Prelude Si		2.3 (2259)	H23A1	MFI	4	DOHC 16V

MFI- Multiport Fuel Injection
DOHC- Dual Overhead Camshafts
SOHC- Single Overhead Camshaft

① VTEC-E
② Canadian models LX/EX
③ Canadian models EX/R/SE

④ Canadian models EX-R
⑤ VTEC
⑥ Including 4 wheel steering

GENERAL ENGINE SPECIFICATIONS

Year	Engine ID/VIN		Engine Displacement Liters (cc)	Fuel System Type	Net Horsepower @ rpm	Net Torque @ rpm (ft. lbs.)	Bore x Stroke (in.)	Compression Ratio	Oil Pressure @ rpm
1993	D15B7		1.5 (1493)	MFI	102@5900	98@5000	2.95x3.33	9.2:1	50@3000
	D15B8		1.5 (1493)	MFI	70@5000	91@2000	2.95x3.33	9.1:1	50@3000
	D15Z1		1.5 (1493)	MFI	92@5500	97@4500	2.95x3.33	9.3:1	50@3000
	D16Z6		1.6 (1590)	MFI	125@6600	106@5200	2.95x3.54	9.2:1	50@3000
	F22A1	①	2.2 (2156)	MFI	125@5200	137@4000	3.35x3.74	8.8:1	50@3000
	F22A1	②	2.2 (2156)	MFI	135@5200	142@4000	3.35x3.74	8.8:1	50@3000
	F22A6		2.2 (2156)	MFI	140@5600	142@4000	3.35x3.74	8.8:1	50@3000
	H22A1		2.2 (2157)	MFI	190@6800	158@5300	3.43x3.57	10.0:1	50@3000
	H23A1		2.3 (2259)	MFI	160@5800	156@4500	3.43x3.74	9.8:1	50@3000
1994	D15B7		1.5 (1493)	MFI	102@5900	98@5000	2.95x3.33	9.2:1	50@3000
	D15B8		1.5 (1493)	MFI	70@5000	91@2000	2.95x3.33	9.1:1	50@3000
	D15Z1		1.5 (1493)	MFI	92@5500	97@4500	2.95x3.33	9.3:1	50@3000
	D16Z6		1.6 (1590)	MFI	125@6600	106@5200	2.95x3.54	9.2:1	50@3000
	B16A3		1.6 (1595)	MFI	160@7600	111@7000	3.19x3.05	10.2:1	50@3000
	F22B1		2.2 (2156)	MFI	145@5500	147@4500	3.35x3.74	8.8:1	50@3000
	F22B2		2.2 (2156)	MFI	130@5300	139@4200	3.35x3.74	8.8:1	50@3000
	F22A1		2.2 (2156)	MFI	135@5200	142@4000	3.35x3.74	8.8:1	50@3000
	H22A1		2.2 (2157)	MFI	190@6800	158@5500	3.43x3.57	10.0:1	50@3000
	H23A1		2.3 (2259)	MFI	160@5800	156@4500	3.43x3.74	9.8:1	50@3000
1995	D15B7		1.5 (1493)	MFI	102@5900	98@5000	2.95x3.33	9.2:1	50@3000
	D15B8		1.5 (1493)	MFI	70@5000	91@2000	2.95x3.33	9.1:1	50@3000
	D15Z1		1.5 (1493)	MFI	92@5500	97@4500	2.95x3.33	9.3:1	50@3000

GENERAL ENGINE SPECIFICATIONS

Year	Engine ID/VIN	Engine Displacement Liters (cc)	Fuel System Type	Net Horsepower @ rpm	Net Torque @ rpm (ft. lbs.)	Bore x Stroke (in.)	Compression Ratio	Oil Pressure @ rpm
1995	D16Z6	1.6 (1590)	MFI	125@6600	106@5200	2.95x3.54	9.2:1	50@3000
	B16A3	1.6 (1595)	MFI	160@7600	111@7000	3.19x3.05	10.2:1	50@3000
	F22B1	2.2 (2156)	MFI	145@5500	147@4500	3.35x3.74	8.8:1	50@3000
	F22B2	2.2 (2156)	MFI	130@5300	139@4200	3.35x3.74	8.8:1	50@3000
	F22A1	2.2 (2156)	MFI	135@5200	142@4000	3.35x3.74	8.8:1	50@3000
	H22A1	2.2 (2157)	MFI	190@6800	158@5500	3.43x3.57	10.0:1	50@3000
	H23A1	2.3 (2259)	MFI	160@5800	156@4500	3.43x3.74	9.8:1	50@3000
	C27A4	2.7 (2675)	MFI	170@5600	165@4500	3.43x2.95	9.0:1	63@3000
1996-97	D16Y7	1.6 (1590)	MFI	106@6200	103@4600	2.95x3.54	9.4:1	50@3000
	D16Y8	1.6 (1590)	MFI	127@6600	107@5500	2.95x3.54	9.4:1	50@3000
	D16Y5	1.6 (1590)	MFI	115@6300	104@5400	2.95x3.54	9.6:1	50@3000
	B16A2	1.6 (1595)	MFI	160@7600	111@7000	3.19x3.05	10.2:1	50@3000
	F22B1	2.2 (2156)	MFI	145@5500	147@4500	3.35x3.74	8.8:1	50@3000
	F22B2	2.2 (2156)	MFI	130@5300	139@4200	3.35x3.74	8.8:1	50@3000
	F22A1	2.2 (2156)	MFI	135@5200	142@4000	3.35x3.74	8.8:1	50@3000
	H22A1	2.2 (2157)	MFI	190@6800	158@5500	3.43x3.57	10.0:1	50@3000
	H23A1	2.3 (2259)	MFI	160@5800	156@4500	3.43x3.74	9.8:1	50@3000
	C27A4	2.7 (2675)	MFI	170@5600	165@4500	3.43x2.95	9.0:1	63@3000

MFI - Multiport fuel injection   ① Accord   ② Prelude

GASOLINE ENGINE TUNE-UP SPECIFICATIONS

Year	Engine ID/VIN	Engine Displacement Liters (cc)	Spark Plugs Gap (in.)	Ignition Timing (deg.) MT	Ignition Timing (deg.) AT	Fuel Pump (psi)	Idle Speed (rpm) MT	Idle Speed (rpm) AT	Valve Clearance In.	Valve Clearance Ex.
1993	D15B7	1.5 (1493)	0.039-0.043	16B	16B	31-38	620-720	650-750	0.007-0.009	0.009-0.011
	D15B8	1.5 (1493)	0.039-0.043	12B	-	31-38	620-720	-	0.007-0.009	0.009-0.011
	D15Z1	1.5 (1493)	0.039-0.043	16B	-	31-38	550-650	-	0.007-0.009	0.009-0.011
	D16Z6	1.6 (1590)	0.039-0.043	16B	16B	31-38	620-720	650-750	0.007-0.009	0.009-0.011
	F22A1	2.2 (2156)	0.039-0.043	15B	15B	30-38 ①	650-750	650-750	0.009-0.011	0.011-0.013
	F22A6	2.2 (2156)	0.039-0.043	15B	15B	30-38	650-750	650-750	0.009-0.011	0.011-0.013
	H22A1	2.2 (2157)	0.039-0.043	15B	-	24-31	650-750	-	0.006-0.007	0.007-0.008
	H23A1	2.3 (2259)	0.039-0.043	15B	15B	28-35	650-750	650-750	0.003-0.004	0.006-0.007
1994	D15B7	1.5 (1493)	0.039-0.043	16B	16B	31-38	620-720	650-750	0.007-0.009	0.009-0.011
	D15B8	1.5 (1493)	0.039-0.043	12B	-	31-38	620-720	-	0.007-0.009	0.009-0.011
	D15Z1	1.5 (1493)	0.039-0.043	16B	-	31-38	550-650	-	0.007-0.009	0.009-0.011
	D16Z6	1.6 (1590)	0.039-0.043	16B	16B	31-38	620-720	650-750	0.007-0.009	0.009-0.011

GASOLINE ENGINE TUNE-UP SPECIFICATIONS

Year	Engine ID/VIN	Engine Displacement Liters (cc)	Spark Plugs Gap (in.)	Ignition Timing (deg.) MT	AT	Fuel Pump (psi)	Idle Speed (rpm) MT	AT	Valve Clearance In.	Ex.
1994	B16A3	1.6 (1595)	0.047-0.051	16B	-	31-38	650-750	-	0.006-0.007	0.007-0.008
	F22B1	2.2 (2156)	0.039-0.043	15B	15B	30-37	650-750	650-750	0.009-0.011	0.011-0.013
	F22B2	2.2 (2156)	0.039-0.043	15B	15B	30-37	650-750	650-750	0.009-0.011	0.011-0.013
	F22A1	2.2 (2156)	0.039-0.043	15B	15B	28-35	650-750	650-750	0.009-0.011	0.011-0.013
	H22A1	2.2 (2157)	0.039-0.043	15B	-	24-31	650-750	-	0.006-0.007	0.007-0.008
	H23A1	2.3 (2259)	0.039-0.043	15B	15B	28-35	650-750	650-750	0.003-0.004	0.006-0.007
1995	D15B7	1.5 (1493)	0.039-0.043	16B	16B	31-38	620-720 ②	650-750 ②	0.007-0.009	0.009-0.011
	D15B8	1.5 (1493)	0.039-0.043	12B	-	31-38	620-720	650-750	0.007-0.009	0.009-0.011
	D15Z1	1.5 (1493)	0.039-0.043	16B	-	31-38	550-650 ④	-	0.007-0.009	0.009-0.011
	D16Z6	1.6 (1590)	0.039-0.043	16B	16B	31-38	620-720 ③	650-750 ②	0.007-0.009	0.009-0.011
	B16A3	1.6 (1595)	0.047-0.051	16B	-	31-38	650-750 ②	-	0.006-0.007	0.007-0.008
	F22B1	2.2 (2156)	0.039-0.043	15B	15B	30-37	650-750	650-750	0.009-0.011	0.011-0.013
	F22B2	2.2 (2156)	0.039-0.043	15B	15B	30-37	650-750	650-750	0.009-0.011	0.011-0.013
	F22A1	2.2 (2156)	0.039-0.043	15B	15B	28-35	650-750	650-750	0.009-0.011	0.011-0.013
	H22A1	2.2 (2157)	0.039-0.043	15B	-	24-31	650-750	-	0.006-0.007	0.007-0.008
	H23A1	2.3 (2259)	0.039-0.043	15B	15B	28-35	650-750	650-750	0.003-0.004	0.006-0.007
	C27A4	2.7 (2675)	0.039-0.043	-	15B	30-37	-	650-750	0.009-0.011	0.011-0.013
1996-97	D16Y7	1.6 (1590)	0.039-0.043	12B	12B	28-36	620-720 ②	650-750 ②	0.007-0.009	0.009-0.011
	D16Y8	1.6 (1590)	0.039-0.043	12B	12B	28-36	620-720 ②	650-750 ②	0.007-0.009	0.009-0.011
	D16Y5	1.6 (1590)	0.039-0.043	12B	12B	28-36	620-720 ②	650-750 ②	0.007-0.009	0.009-0.011
	B16A2	1.6 (1595)	0.047-0.051	16B	-	31-38	650-750	-	0.006-0.007	0.007-0.008
	F22B1	2.2 (2156)	0.039-0.043	15B	15B	30-37	650-750	650-750	0.009-0.011	0.011-0.013
	F22B2	2.2 (2156)	0.039-0.043	15B	15B	30-37	650-750	650-750	0.009-0.011	0.011-0.013
	F22A1	2.2 (2156)	0.039-0.043	15B	15B	28-35	650-750	650-750	0.009-0.011	0.011-0.013
	H22A1	2.2 (2157)	0.039-0.043	15B	-	24-31	650-750	-	0.006-0.007	0.007-0.008

GASOLINE ENGINE TUNE-UP SPECIFICATIONS

Year	Engine ID/VIN	Engine Displacement Liters (cc)	Spark Plugs Gap (in.)	Ignition Timing (deg.) MT	Ignition Timing (deg.) AT	Fuel Pump (psi)	Idle Speed (rpm) MT	Idle Speed (rpm) AT	Valve Clearance In.	Valve Clearance Ex.
1996-97	H23A1	2.3 (2259)	0.039-0.043	15B	15B	28-35	650-750	650-750	0.003-0.004	0.006-0.007
	C27A4	2.7 (2675)	0.039-0.043	-	15B	30-37	-	650-750	0.009-0.011	0.011-0.013

NOTE: The Vehicle Emission Control Information label often reflects specification changes made during production. The label figures must be used if they differ from those in this chart.

B - Before top dead center

① Prelude: 28-35
② Canada: 700-800
③ Canadian Civic: 650-750
  Canadian del Sol: 700-800
④ Canada: 650-750

CAPACITIES

Year	Model	Engine ID/VIN	Engine Displacement Liters (cc)	Engine Oil with Filter	Transmission (pts.) 4-Spd	Transmission (pts.) 5-Spd	Transmission (pts.) Auto.	Transfer Case (pts.)	Drive Axle Front (pts.)	Drive Axle Rear (pts.)	Fuel Tank (gal.)	Cooling System (qts.)
1993	Civic	D15B7	1.5 (1493)	3.5	-	3.8	5.6	-	-	-	11.9	①
	Civic	D15B8	1.5 (1493)	3.5	-	3.8	-	-	-	-	11.9	3.8
	Civic	D15Z1	1.5 (1493)	3.5	-	3.8	-	-	-	-	11.9	3.7
	Civic	D16Z6	1.6 (1590)	3.5	-	3.8	5.6	-	-	-	11.9	①
	del Sol	D15B7	1.5 (1493)	3.5	-	3.8	5.6	-	-	-	11.9	①
	del Sol	D16Z6	1.6 (1590)	3.5	-	3.8	5.6	-	-	-	11.9	①
	Accord DX/LX ③	F22A1	2.2 (2156)	4.0	-	4.0	5.0	-	-	-	17.0	②
	Accord EX/SE ④	F22A6	2.2 (2156)	4.0	-	4.0	5.0	-	-	-	17.0	②
	Prelude S	F22A1	2.2 (2156)	4.0	-	4.0	5.0	-	-	-	15.9	①
	Prelude VTEC	H22A1	2.2 (2157)	5.1	-	4.0	-	-	-	-	15.9	4.4
	Prelude Si	H23A1	2.3 (2259)	4.5	-	4.0	5.0	-	-	-	15.9	①
1994	Civic	D15B7	1.5 (1493)	3.5	-	3.8	5.6	-	-	-	11.9	①
	Civic	D15B8	1.5 (1493)	3.5	-	3.8	-	-	-	-	10.0	①
	Civic	D15Z1	1.5 (1493)	3.5	-	3.8	-	-	-	-	10.0	3.8
	Civic	D16Z6	1.5 (1493)	3.5	-	3.8	5.6	-	-	-	11.9	3.7
	del Sol	D15B7	1.5 (1493)	3.5	-	3.8	5.6	-	-	-	11.9	①
	del Sol	D16Z6	1.6 (1590)	3.5	-	3.8	5.6	-	-	-	11.9	①
	del Sol	B16A3	1.6 (1595)	4.2	-	4.6	-	-	-	-	11.9	4.1
	Accord EX ⑤	F22B1	2.2 (2156)	4.0	-	4.0	5.0	-	-	-	17.0	②
	Accord DX/LX ③	F22B2	2.2 (2156)	4.0	-	4.0	5.0	-	-	-	17.0	②
	Prelude S	F22A1	2.2 (2156)	4.0	-	4.0	5.0	-	-	-	15.9	①
	Prelude VTEC	H22A1	2.2 (2157)	5.1	-	4.0	-	-	-	-	15.9	4.6
	Prelude Si	H23A1	2.3 (2259)	4.5	-	4.0	5.0	-	-	-	15.9	①
1995	Civic	D15B7	1.5 (1493)	3.5	-	3.8	5.8	-	-	-	11.9	①
	Civic	D15B8	1.5 (1493)	3.5	-	3.8	-	-	-	-	11.9	①
	Civic	D15Z1	1.5 (1493)	3.5	-	3.8	-	-	-	-	11.9	3.8
	Civic	D16Z6	1.6 (1590)	3.5	-	3.8	5.8	-	-	-	11.9	3.7
	del Sol	D15B7	1.5 (1493)	3.5	-	4.0	5.6	-	-	-	11.9	①
	del Sol	D16Z6	1.6 (1590)	3.5	-	4.0	5.6	-	-	-	11.9	①
	del Sol	B16A3	1.6 (1595)	4.2	-	4.8	-	-	-	-	11.9	4.1
	Accord EX ⑤	F22B1	2.2 (2156)	4.5	-	4.0	5.0	-	-	-	17.0	②
	Accord DX/LX ③	F22B2	2.2 (2156)	4.0	-	4.0	5.0	-	-	-	17.0	②
	Accord V-6	C27A4	2.7 (2675)	4.6	-	-	6.2	-	-	-	17.0	7.2
	Prelude S	F22A1	2.2 (2156)	4.0	-	4.0	5.0	-	-	-	15.9	①
	Prelude VTEC	H22A1	2.2 (2157)	5.1	-	4.0	-	-	-	-	15.9	4.6
	Prelude Si	H23A1	2.3 (2259)	4.5	-	4.0	5.0	-	-	-	15.9	①

CAPACITIES

Year	Model	Engine ID/VIN	Engine Displacement Liters (cc)	Engine Oil with Filter	Transmission (pts.) 4-Spd	5-Spd	Auto.	Transfer Case (pts.)	Drive Axle Front (pts.)	Rear (pts.)	Fuel Tank (gal.)	Cooling System (qts.)
1996-97	Civic	D16Y7	1.6 (1590)	3.5	-	3.8	5.8	-	-	-	11.9	4.4
	Civic	D16Y8	1.6 (1590)	3.5	-	3.8	5.8	-	-	-	11.9	4.3
	Civic	D16Y5	1.6 (1590)	3.5	-	3.8	5.8 ⑥	-	-	-	11.9	4.5
	del Sol	D16Y7	1.6 (1590)	3.5	-	4.0	5.6	-	-	-	11.9	②
	del Sol	D16Y8	1.6 (1590)	3.5	-	4.0	5.6	-	-	-	11.9	8.2
	del Sol	B16A2	1.6 (1595)	4.2	-	4.8	-	-	-	-	11.9	4.1
	Accord EX	F22B1	2.2 (2156)	4.5	-	4.0	5.0	-	-	-	17.0	②
	Accord DX/LX	F22B2	2.2 (2156)	4.0	-	4.0	5.0	-	-	-	17.0	②
	Accord V-6	C27A4	2.7 (2675)	4.6	-	-	6.2	-	-	-	17.0	7.2
	Prelude S	F22A1	2.2 (2156)	4.0	-	4.0	5.0	-	-	-	15.9	①
	Prelude VTEC	H22A1	2.2 (2157)	5.1	-	4.0	-	-	-	-	15.9	4.6
	Prelude Si	H23A1	2.3 (2259)	4.5	-	4.0	5.0	-	-	-	15.9	①

NOTE: Capacities given are service, not overhaul capacities

① Automatic transaxle: 4.0   ② Automatic transaxle: 5.6   ③ Canadian models: LX/EX   ⑤ Canadian models EX-R
   Manual transaxle: 3.8   Manual transaxle: 5.7   ④ Canadian models EX-R/SE   ⑥ Continuous variable transaxle: 8.2

VALVE SPECIFICATIONS

Year	Engine ID/VIN	Engine Displacement Liters (cc)	Seat Angle (deg.)	Face Angle (deg.)	Spring Test Pressure (lbs. @ in.)	Spring Installed Height (in.)	Stem-to-Guide Clearance (in.) Intake	Exhaust	Stem Diameter (in.) Intake	Exhaust
1993	D15B7	1.5 (1493)	45	45	NA	NA	0.0008-0.0020	0.0020-0.0030	0.2157-0.2161	0.2134-0.2150
	D15B8	1.5 (1493)	45	45	NA	NA	0.0010-0.0030	0.0020-0.0050	0.2146-0.2161	0.2134-0.2150
	D15Z1	1.5 (1493)	45	45	NA	NA	0.0010-0.0030	0.0020-0.0050	0.2146-0.2161	0.2134-0.2150
	D16Z6	1.6 (1590)	45	45	NA	NA	0.0008-0.0020	0.0020-0.0030	0.2157-0.2161	0.2134-0.2150
	F22A1	2.2 (2156)	45	45	NA	NA	0.0008-0.0030	0.0022-0.0050	0.2148-0.2163	0.2134-0.2150
	F22A6	2.2 (2156)	45	45	NA	NA	0.0008-0.0030	0.0022-0.0050	0.2148-0.2163	0.2134-0.2150
	H22A1	2.2 (2157)	45	45	NA	NA	0.0010-0.0030	0.0020-0.0040	0.2144-0.2159	0.2144-0.2159
	H23A1	2.3 92259)	45	45	NA	NA	0.0010-0.0030	0.0020-0.0040	0.2580-0.2594	0.2570-0.2583
1994	D15B7	1.5 (1493)	45	45	NA	NA	0.0010-0.0030	0.0020-0.0050	0.2146-0.2161	0.2134-0.2150
	D15B8	1.5 (1493)	45	45	NA	NA	0.0010-0.0030	0.0020-0.0050	0.2146-0.2161	0.2134-0.2150
	D15Z1	1.5 (1493)	45	45	NA	NA	0.0010-0.0030	0.0020-0.0050	0.2146-0.2161	0.2134-0.2150
	D16Z6	1.6 (1590)	45	45	NA	NA	0.0010-0.0030	0.0020-0.0050	0.2146-0.2161	0.2134-0.2150
	B16A3	1.6 (1595)	45	45	NA	NA	0.0010-0.0030	0.0020-0.0040	0.2144-0.2459	0.2134-0.2150
	F22B1	2.2 (2156)	45	45	NA	NA	0.0008-0.0030	0.0022-0.0050	0.2148-0.2163	0.2134-0.2150

VALVE SPECIFICATIONS

Year	Engine ID/VIN	Engine Displacement Liters (cc)	Seat Angle (deg.)	Face Angle (deg.)	Spring Test Pressure (lbs. @ in.)	Spring Installed Height (in.)	Stem-to-Guide Clearance (in.)		Stem Diameter (in.)	
							Intake	Exhaust	Intake	Exhaust
1994	F22B2	2.2 (2156)	45	45	NA	NA	0.0008-0.0018	0.0022-0.0031	0.2159-0.2163	0.2146-0.2150
	F22A1	2.2 (2156)	45	45	NA	NA	0.0008-0.0030	0.0022-0.0050	0.2148-0.2163	0.2134-0.2150
	H22A1	2.2 (2157)	45	45	NA	NA	0.0010-0.0030	0.0020-0.0040	0.2144-0.2159	0.2144-0.2159
	H23A1	2.3 (2259)	45	45	NA	NA	0.0010-0.0030	0.0020-0.0040	0.2580-0.2594	0.2570-0.2583
1995	D15B7	1.5 (1493)	45	45	NA	NA	0.0010-0.0030	0.0020-0.0040	0.2150-0.2160	0.2130-0.2150
	D15B8	1.5 (1493)	45	45	NA	NA	0.0010-0.0030	0.0020-0.0040	0.2150-0.2160	0.2130-0.2150
	D15Z1	1.5 (1493)	45	45	NA	NA	0.0010-0.0030	0.0020-0.0050	0.2150-0.2160	0.2130-0.2150
	D16Z6	1.6 (1590)	45	45	NA	NA	0.0010-0.0030	0.0020-0.0050	0.2150-0.2160	0.2130-0.2150
	B16A3	1.6 (1595)	45	45	NA	NA	0.0010-0.0030	0.0020-0.0040	0.2144-0.2459	0.2134-0.2150
	F22B1	2.2 (2156)	45	45	NA	NA	0.0008-0.0030	0.0022-0.0050	0.2148-0.2163	0.2134-0.2150
	F22B2	2.2 (2156)	45	45	NA	NA	0.0008-0.0030	0.0022-0.0050	0.2148-0.2163	0.2134-0.2150
	F22A1	2.2 (2156)	45	45	NA	NA	0.0008-0.0030	0.0022-0.0050	0.2148-0.2163	0.2134-0.2150
	H22A1	2.2 (2157)	45	45	NA	NA	0.0010-0.0030	0.0020-0.0040	0.2144-0.2459	0.2144-0.2459
	H23A1	2.3 (2259)	45	45	NA	NA	0.0010-0.0030	0.0020-0.0040	0.2580-0.2594	0.2570-0.2583
	C27A4	2.7 (2675)	45	45	NA	NA	0.0008-0.0030	0.0020-0.0040	0.2580-0.2594	0.2570-0.2583
1996-97	D16Y7	1.6 (1590)	45	45	NA	NA	0.0010-0.0020	0.0020-0.0030	0.2157-0.2161	0.2146-0.2150
	D16Y8	1.6 (1590)	45	45	NA	NA	0.0010-0.0020	0.0020-0.0030	0.2157-0.2161	0.2146-0.2150
	D16Y5	1.6 (1590)	45	45	NA	NA	0.0010-0.0020	0.0020-0.0030	0.2157-0.2161	0.2146-0.2150
	B16A2	1.6 (1595)	45	45	NA	NA	0.0010-0.0022	0.0020-0.0031	0.2156-0.2159	0.2146-0.2150
	F22B1	2.2 (2156)	45	45	NA	NA	0.0008-0.0030	0.0020-0.0050	0.2148-0.2163	0.2134-0.2150
	F22B2	2.2 (2156)	45	45	NA	NA	0.0008-0.0030	0.0020-0.0050	0.2148-0.2163	0.2134-0.2150
	F22A1	2.2 (2156)	45	45	NA	NA	0.0008-0.0018	0.0022-0.0050	0.2159-0.2163	0.2146-0.2150
	H22A1	2.2 (2156)	45	45	NA	NA	0.0010-0.0022	0.0020-0.0031	0.2156-0.2159	0.2156-0.2159
	H23A1	2.2 (2157)	45	45	NA	NA	0.0008-0.0018	0.0020-0.0030	0.2591-0.2594	0.2579-0.2583
	C27A4	2.7 (2675)	45	45	NA	NA	0.0008-0.0030	0.0020-0.0040	0.2580-0.2594	0.2570-0.2583

NA - Not Available

TORQUE SPECIFICATIONS
All readings in ft. lbs.

Year	Engine ID/VIN	Engine Displacement Liters (cc)	Cylinder Head Bolts	Main Bearing Bolts	Rod Bearing Bolts	Crankshaft Damper Bolts	Flywheel Bolts	Manifold Intake	Manifold Exhaust	Spark Plugs	Lug Nut
1993	D15B7	1.5 (1493)	⑧	⑪	23	134 ②	87 ①	17	23	13	80
	D15B8	1.5 (1493)	⑧	⑪	23	134 ②	87 ①	17	23	13	80
	D15Z1	1.5 (1493)	⑨	⑪	23	134 ②	87 ①	17	23	13	80
	D16Z6	1.6 (1590)	⑨	⑫	23	134 ②	87 ①	17	23	13	80
	F22A1	2.2 (2156)	⑦	⑬	34	159 ②	76 ①	16	23	13	80
	F22A6	2.2 (2156)	⑦	⑬	34	159 ②	76 ①	16	23	13	80
	H22A1	2.2 (2157)	⑦	⑬	34	159 ②	76 ①	16	23	13	80
	H23A1	2.3 (2259)	⑦	⑬	34	159 ②	76 ①	16	23	13	80
1994	D15B7	1.5 (1493)	⑧	⑪	23	134	87 ①	17	23	13	80
	D15B8	1.5 (1493)	⑧	⑪	23	134	87 ①	17	23	13	80
	D15Z1	1.5 (1493)	⑨	⑪	23	134	87 ①	17	23	13	80
	D16Z6	1.6 (1590)	⑨	⑫	23	134	87 ①	17	23	13	80
	B16A3	1.6 (1595)	⑩	⑬	30	130	76 ①	17	23	13	80
	F22B1	2.2 (2156)	③	⑨	34	181	76 ①	16	23	13	80
	F22B2	2.2 (2156)	③	⑨	34	181	76 ①	16	23	13	80
	F22A1	2.2 (2156)	③	⑨	34	181	76 ①	16	23	13	80
	H22A1	2.2 (2157)	③	⑨	34	181	76 ①	16	23	13	80
	H23A1	2.3 (2259)	③	⑨	34	181	76 ①	16	23	13	80
1995	D15B7	1.5 (1493)	⑧	⑪	23	134	87 ①	17	23	13	80
	D15B8	1.5 (1493)	⑧	⑪	23	134	87	17	23	13	80
	D15Z1	1.5 (1493)	⑨	⑪	23	134	87	17	23	13	80
	D16Z6	1.6 (1590)	⑨	⑫	23	134	87 ①	17	23	13	80
	B16A3	1.6 (1595)	⑩	⑬	30	130	87	17	23	13	80
	F22B1	2.2 (2156)	③	⑨	34	181	76 ①	16	23	13	80
	F22B2	2.2 (2156)	③	⑨	34	181	76 ①	16	23	13	80
	F22A1	2.2 (2156)	③	⑨	34	181	76 ①	16	23	13	80
	H22A1	2.2 (2157)	③	⑨	34	181	76	16	23	13	80
	H23A1	2.3 (2259)	③	⑨	34	181	76 ①	16	23	13	80
	C27A4	2.7 (2675)	⑭	⑥	33	181	54	16	22	13	80
1996-97	D16Y7	1.6 (1590)	⑮	⑫	23	134	87	17	23	13	80
	D16Y8	1.6 (1590)	⑮	⑫	23	134	87	17	23	13	80
	D16Y5	1.6 (1590)	⑮	⑫	23	134	87	17	23	13	80
	B16A2	1.6 (1595)	⑩	⑬	30	130	76	17	23	13	80
	F22B1	2.2 (2156)	③	⑨	34	181	⑤	16	23	13	80
	F22B2	2.2 (2156)	③	⑨	34	181	⑤	16	23	13	80
	F22A1	2.2 (2156)	③	⑨	34	181	76	16	23	13	80
	H22A1	2.2 (2157)	⑨	⑨	34	181	76	16	23	13	80
	H23A1	2.3 (2259)	⑨	⑨	34	181	76	16	23	13	80
	C27A4	2.7 (2675)	⑭	⑥	33	181	54	16	22	13	80

NOTE: Dip main bearing bolts and crankshaft damper bolt in clean engine oil

① Automatic transaxle: 54 ft. lbs.
② Dip bolts in clean engine oil
③ Step 1: 29 ft. lbs.
   Step 2: 51 ft. lbs.
   Step 3: 72 ft. lbs.
④ (Not Used)
⑤ (Not Used)

⑥ Inner: 48 ft. lbs.
   Outer: 29 ft. lbs.
   Side: 36 ft. lbs.
⑦ Step 1: 22 ft. lbs.
   Step 2:
   Prelude: 72 ft. lbs.
   Except Prelude: 78 ft. lbs.

⑧ Step 1: 22 ft. lbs.
   Step 2: 47 ft. lbs.
⑨ Step 1: 22 ft. lbs.
   Step 2: 53 ft. lbs.
⑩ Step 1: 22 ft. lbs.
   Step 2: 61 ft. lbs.

⑪ Step 1: 18 ft. lbs.
   Step 2: 33 ft. lbs.
⑫ Step 1: 18 ft. lbs.
   Step 2: 38 ft. lbs.
⑬ Step 1: 18 ft. lbs.
   Step 2: 54 ft. lbs.
⑭ Step 1: 29 ft. lbs.
   Step 2: 58 ft. lbs.

⑮ Cylinder head bolts to be tightened in four steps:
   Step 1: 14 ft. lbs.
   Step 2: 36 ft. lbs.
   Step 3: 49 ft. lbs.
   Step 4: Bolts 1-2, tighten an additional 49 ft. lb

BRAKE SPECIFICATIONS
All measurements in inches unless noted

Year	Model		Master Cylinder Bore	Brake Disc Original Thickness	Minimum Thickness	Maximum Runout	Brake Drum Diameter Original Inside Diameter	Max. Wear Limit	Maximum Machine Diameter	Minimum Lining Thickness Front	Rear
1993	Civic	F	NA	0.830	0.750	0.004	7.09 ①	7.13 ①	7.13 ①	0.060	0.080 ③
		R	NA	0.350	0.310	0.004	7.87 ②	7.91 ②	7.91 ②	-	-
	del Sol	F	NA	0.830	0.750	0.004	7.09	7.13	7.13	0.060	0.080 ③
		R	NA	0.350	0.310	0.004	-	-	-	-	-
	Accord	F	NA	0.910 ④	0.830 ⑤	0.004	8.66	8.70	8.70	0.060	0.080 ③
		R	NA	0.390	0.310	0.004	-	-	-	-	-
	Prelude	F	NA	0.910	0.830	0.004	-	-	-	0.060	0.060
		R	NA	0.390	0.320	0.004	-	-	-	-	-
1994	Civic	F	NA	0.830	0.750	0.004	7.09 ①	7.13 ①	7.13 ①	0.060	0.080 ③
		R	NA	0.350	0.310	0.004	7.87 ②	7.91 ②	7.91 ②	-	-
	del Sol	F	NA	0.830	0.750	0.004	7.09	7.13	7.13	0.060	0.080 ③
		R	NA	0.350	0.310	0.004	-	-	-	-	-
	Accord	F	NA	0.910 ④	0.830 ⑤	0.004	8.66	8.70	8.7	0.060	0.080 ③
		R	NA	0.400	0.310	0.004	-	-	-	-	-
	Prelude	F	NA	0.910	0.830	0.004	-	-	-	0.060	0.060
		R	NA	0.390	0.320	0.004	-	-	-	-	-
1995	Civic	F	NA	0.830	0.750	0.004	-	-	-	0.060	-
		R	-	0.350	0.310	0.004	⑥	⑦	⑦	-	0.080 ③
	del Sol	F	NA	0.830	0.750	0.004	-	-	-	0.060	-
		R	-	0.350	0.310	0.004	7.09	7.13	7.13	-	0.080 ③
	Accord	F	NA	0.910 ④	0.830 ⑤	0.004	-	-	-	0.060	-
		R	-	0.400	0.310	0.004	8.66	8.70	8.70	-	0.080 ③
	Prelude	F	NA	0.910	0.830	0.004	-	-	-	0.060	-
		R	-	0.390	0.320	0.004	-	-	-	-	0.060
1996-97	Civic	F	NA	0.840	0.750	0.004	-	-	-	0.060	-
		R	-	-	-	-	7.87	7.91	7.91	-	0.080
	del Sol	F	NA	0.830	0.750	0.004	-	-	-	0.060	-
		R	-	0.350	0.310	0.004	7.09	7.13	7.13	-	0.080
	Accord	F	NA	0.910 ④	0.830 ⑤	0.004	-	-	-	0.060	-
		R	-	0.400	0.310	0.004	8.66	8.70	8.70	-	0.080
	Prelude	F	NA	0.910	0.830	0.004	-	-	-	0.060	-
		R	-	0.390	0.320	0.004	-	-	-	-	0.060

NA - Not Available
F - Front
R - Rear
① Manual trans. except 1.6L Coupe
② Automatic trans. and 1.6L Coupe with manual trans.
③ Rear disc brakes: 0.060
④ Wagon and V6 models: 0.990
⑤ Wagon and V6 models: 0.910
⑥ 7.09: Cars with manual trans.; Except Coupe 1.6L manual trans.
   7.87: Cars with automatic trans., Coupe 1.6L manual trans.
⑦ 7.13: Cars with manual trans.; Except Coupe 1.6L manual trans.
   7.91: Cars with automatic trans.; Coupe 1.6L manual trans.

FREQUENT MAINTENANCE LABOR
HONDA

The following should be used as a guide when determining the amount of work required for a particular service if taken to a repair shop. In estimating how long a particular Frequent Maintenance Service item should take, please observe the following:
- **Chilton Time** is time that is based on field research and data supplied by the vehicle manufacturer.
- All labor time operations are given in hours and tenths of an hour.
- All labor operations, are to be used as a **guide**.

COOLING

(G) Winterize Cooling System
Includes: Run engine to check for leaks, tighten all hose connections. Test radiator and pressure cap, drain radiator and engine block. Add antifreeze and refill system.

	Chilton Time
1993-97	.5

(G) Hoses, Radiator, Renew
Includes: Drain and refill cooling system as required.

1993-97	
upper	.5
lower	.6
by-pass	.6

(G) Thermostat, Coolant, Renew

1993-97 Accord	
Four	.8
V6	1.1
1993-97 del Sol	.8
1993-97 Civic	.8
1995-97 Odyssey	.8
1993-97 Prelude	.8

FUEL

(G) Filter, Fuel, Renew

1993 Accord	.7
1994-97 Accord	1.0
1993-97 Civic	.5
1993-97 del Sol	.5
1995-97 Odyssey	.5
1993-97 Prelude	.5

BRAKES

(G) Bleed Brakes (Four Wheels)
Includes: Add fluid.

	Chilton Time
1993-97	.5

(G) Brakes, Adjust (Minor)
Includes: Adjust brakes, fill master cylinder.

1993-97	.4

(M) Parking Brake, Adjust

1993-97	.4

LUBRICATION SERVICES

(M) Engine Oil & Filter, Renew
Includes: Inspect and correct all fluid levels.

1993-97	.3

(M) Lubricate Chassis, Change Oil & Filter
Includes: Inspect and correct all fluid levels.

1993-97	.6
Install grease fittings add	.1

(M) Lubricate Chassis
Includes: Inspect and correct all fluid levels.

1993-97	.4
Install grease fittings add	.1

WHEELS

(G) Wheel, Renew (One)

1993-97	.5

(G) Wheel, Balance

	Chilton Time
1993-97	
one	.3
each adtnl.	.2

(G) Wheels, Rotate (All)

1993-97	.5

ELECTRICAL

(G) Headlamps, Aim

1993-97 Accord	.4
1993-97 del Sol	.6
1993-97 Civic	.6
1995-97 Odyssey	.4
1993-97 Prelude	.4

(G) Halogen Headlamp Bulb, Renew

1993-97, each	.4

(G) Parking Lamp Lens or Bulb, Renew

1993-97, one	.3

(G) Tail Lamp Lens or Bulb, Renew

1993-97 Accord	.5
1993-97 del Sol	.2
1993-97 Civic	.2
1995-97 Odyssey	.2
1993-97 Prelude	.5

(G) Horn, Renew

1993 Accord	1.3
1994-97 Accord	.5
1993-97 del Sol	1.0
1993-97 Civic	1.0
1995-97 Odyssey	.3
1993-97 Prelude	.7

(M) Terminals, Battery, Clean

1993-97	.3

SCHEDULED MAINTENANCE INTERVALS
(HONDA ACCORD, CIVIC, DEL SOL, & PRELUDE)

TO BE SERVICED	TYPE OF SERVICE	7.5	15	22.5	30	37.5	45	52.5	60	67.5	75	82.5	90	97.5
		VEHICLE MILEAGE INTERVAL (x1000)												
Engine oil & filter	R	✓	✓	✓	✓	✓	✓	✓	✓	✓	✓	✓	✓	✓
Front brake pads	S/I	✓	✓	✓	✓	✓	✓	✓	✓	✓	✓	✓	✓	✓
Rotate tires	S/I	✓	✓	✓	✓	✓	✓	✓	✓	✓	✓	✓	✓	✓
Brake hoses & lines (including ABS)	S/I		✓		✓		✓		✓		✓		✓	
Cooling system, hoses & connections (1995-97)	S/I		✓		✓		✓		✓		✓		✓	
Cooling system hoses & connections (1993-94)	S/I				✓				✓				✓	
Driveshaft boots	S/I		✓		✓		✓		✓		✓		✓	
Exhaust pipe (after TWC converter) & muffler (1993-94)	S/I		✓		✓		✓		✓		✓		✓	
Exhaust pipe (before TWC converter) (1993-94)	S/I				✓				✓				✓	
Exhaust system (1995-97)	S/I		✓		✓		✓		✓		✓		✓	
Front brake discs & calipers	S/I		✓		✓		✓		✓		✓		✓	
Front wheel alignment	S/I		✓		✓		✓		✓		✓		✓	
Front & rear wheel alignment (Prelude w/4WS)	S/I		✓		✓		✓		✓		✓		✓	
Fuel pipes, hoses & connections (1995-97)	S/I		✓		✓		✓		✓		✓		✓	
Fuel pipes, hoses & connections (1993-94)	S/I				✓				✓				✓	
Parking brake adjustment	S/I		✓		✓		✓		✓		✓		✓	
Power steering system	S/I		✓		✓		✓		✓		✓		✓	
Rear brake discs, calipers & pads	S/I		✓		✓		✓		✓		✓		✓	
Steering operation, tie rod ends, steering gearbox, boots & rear actuator (1993-94 Prelude w/4WS)	S/I		✓		✓				✓				✓	

SCHEDULED MAINTENANCE INTERVALS
(HONDA ACCORD, CIVIC, DEL SOL, & PRELUDE) (Cont.)

TO BE SERVICED	TYPE OF SERVICE	VEHICLE MILEAGE INTERVAL (x1000)												
		7.5	15	22.5	30	37.5	45	52.5	60	67.5	75	82.5	90	97.5
Steering operation, tie rod ends, steering gearbox & boots (1993-94)	S/I		✓		✓				✓				✓	
Suspension components	S/I		✓		✓		✓		✓		✓		✓	
Suspension mounting bolts	S/I		✓		✓		✓		✓		✓		✓	
Tie rods, steering gear box & boots (1995-97)	S/I		✓		✓		✓		✓		✓		✓	
Valve clearance (1993-94, 1995 Civic & 1995-97 Prelude VTEC)	S/I		✓		✓		✓		✓		✓		✓	
Valve clearance (1995-97 Accord L4, 1996-97 Civic, 1995 Del Sol non-VTEC & 1995-97 Prelude non-VTEC)	S/I				✓				✓				✓	
Valve clearance (1996-97 Del Sol VTEC)	S/I				✓				✓				✓	
Valve clearance (1996-97 Del Sol non-VTEC)	S/I				✓									
Parking brake	S/I		✓		✓				✓				✓	
Air cleaner element	R				✓				✓				✓	
Transmission fluid (1996-97 Civic CVT)	R				✓		✓		✓		✓		✓	
Transmission fluid (A/T or M/T) (except as noted below)	R				✓				✓				✓	
Transmission fluid (1996-97 Prelude L4 & Del Sol A/T or M/T)	R												✓	
Brake fluid (including ABS) (all 1993-95 & 1996-97 Accord V6)	R				✓				✓				✓	
Brake fluid (including ABS) (1996-97 Accord L4, Civic, Del Sol & Prelude)	R						✓						✓	

SCHEDULED MAINTENANCE INTERVALS
(HONDA ACCORD, CIVIC, DEL SOL, & PRELUDE) (Cont.)

TO BE SERVICED	TYPE OF SERVICE	VEHICLE MILEAGE INTERVAL (x1000)												
		7.5	15	22.5	30	37.5	45	52.5	60	67.5	75	82.5	90	97.5
Spark plugs (non-VTEC)	R				✓				✓				✓	
Spark plugs (VTEC)	R								✓					
ABS operation	S/I				✓				✓				✓	
Alternator drive belt	S/I				✓				✓				✓	
Power steering pump belt	S/I				✓				✓				✓	
Rear brake drums, wheel cylinders & linings (except Prelude)	S/I				✓				✓				✓	
Engine coolant	R						✓				✓			
ABS high pressure hose	R								✓					
Fuel filter	R								✓					
Timing belt	R												✓	
Timing balancer belt	R												✓	
Distributor, ignition cap & rotor	S/I								✓					
Idle speed	S/I								✓					
Ignition wires	S/I								✓					
PCV valve	S/I								✓					
TWC converter heat shield	S/I								✓					
Water pump	S/I												✓	

R – Replace   S/I – Service or Inspect

FREQUENT OPERATION MAINTENANCE (SEVERE SERVICE)

If a vehicle is operated under any of the following conditions it is considered severe service:
- Extremely dusty areas.
- 50% or more of the vehicle operation is in 32°C (90°F) or higher temperatures, or constant operation in temperatures below 0°C (32°F).
- Prolonged idling (vehicle operation in stop and go traffic).
- Frequent short running periods (engine does not warm to normal operating temperatures).
- Police, taxi, delivery usage or trailer towing usage.

Oil & oil filter change – change every 3750 miles.
Driveshaft boots - service or inspect every 7500 miles.
Front brake discs & calipers, & rear brake discs, calipers & pads - service or inspect every 7500 miles.
Power steering system - service or inspect every 7500 miles.
Suspension components - service or inspect every 7500 miles.
Tie rods, steering gear box & boots - service or inspect every 7500 miles.
Air cleaner element – service or inspect every 15,000 miles.
Transmission fluid (all 1993-95, 1996-97 Accord V6 & 1996-97 Civic CVT) - replace every 15,000 miles.
Transmission fluid (1996-97 Accord L4, Civic, Del Sol & Prelude) - replace every 30,000 miles.
Timing balancer belt - replace every 60,000 miles.
Timing belt - replace every 60,000 miles.
Water pump - service or inspect every 60,000 miles.

HYUNDAI

ENGINE IDENTIFICATION

Year	Model		Engine Displacement Liters (cc)	Engine Series (ID/VIN)	Fuel System	No. of Cylinders	Engine Type
1993	Excel	①	1.5 (1468)	M	2 bbl	4	SOHC
	Excel		1.5 (1468)	J	MFI	4	SOHC
	Scoupe		1.5 (1495)	N	MFI	4	SOHC
	Elantra		1.6 (1596)	R	MFI	4	DOHC
	Elantra		1.8 (1836)	M	MFI	4	DOHC
	Sonata		2.0 (1997)	F	MFI	4	DOHC
	Sonata		3.0 (2972)	T	MFI	6	SOHC
1994	Excel		1.5 (1468)	J	MFI	4	SOHC
	Scoupe		1.5 (1495)	N	MFI	4	SOHC
	Elantra		1.6 (1596)	R	MFI	4	DOHC
	Elantra		1.8 (1836)	M	MFI	4	DOHC
	Sonata		2.0 (1997)	F	MFI	4	DOHC
	Sonata		3.0 (2972)	T	MFI	6	SOHC
1995	Accent		1.5 (1495)	K	MFI	4	SOHC
	Scoupe		1.5 (1495)	N	MFI	4	SOHC
	Elantra		1.6 (1596)	R	MFI	4	DOHC
	Elantra		1.8 (1796)	M	MFI	4	DOHC
	Sonata		2.0 (1997)	P	MFI	4	DOHC
	Sonata		3.0 (2972)	T	MFI	6	SOHC
1996-97	Accent		1.5 (1495)	K	MFI	4	SOHC
	Accent		1.5 (1495)	K	MFI	4	DOHC
	Elantra		1.8 (1796)	M	MFI	4	DOHC
	Sonata		2.0 (1997)	F	MFI	4	DOHC
	Sonata		3.0 (2972)	K	MFI	6	SOHC

MFI - Multiport fuel injection
2 bbl - 2 barrel carburetor

SOHC - Single overhead camshaft
DOHC - Double overhead camshaft

① Canada only

GENERAL ENGINE SPECIFICATIONS

Year	Engine ID/VIN		Engine Displacement Liters (cc)	Fuel System Type	Net Horsepower @ rpm	Net Torque @ rpm (ft. lbs.)	Bore x Stroke (in.)	Compression Ratio	Oil Pressure @ rpm
1993	M		1.5 (1468)	2 bbl	81@5500	91@3000	2.97x3.23	9.4:1	12@750
	J		1.5 (1468)	MFI	81@5500	91@3000	2.97x3.23	9.4:1	12@750
	N		1.5 (1495)	MFI	92@5500	97@4500	2.97x3.29	10.0:1	21@800
	N	①	1.5 (1495)	MFI	115@5500	123@4500	2.97x3.29	7.5:1	21@800
	R		1.6 (1596)	MFI	113@6000	102@5000	3.24x2.95	9.2:1	12@750
	M	②	1.8 (1836)	MFI	124@6000	116@5000	3.17x3.46	9.2:1	12@750
	F		2.0 (1997)	MFI	128@6000	121@5000	3.35x3.46	9.0:1	12@750
	T		3.0 (2972)	MFI	142@5000	168@2500	3.59x2.99	8.9:1	30-80@2000
1994	J		1.5 (1468)	MFI	81@5500	91@3000	2.97x3.23	9.4:1	12@750
	N		1.5 (1495)	MFI	92@5500	97@4500	2.97x3.29	10.0:1	21@800
	N	①	1.5 (1495)	MFI	115@5500	123@4500	2.97x3.29	7.5:1	21@800
	R		1.6 (1596)	MFI	113@6000	102@5000	3.24x2.95	9.2:1	12@750
	M	②	1.8 (1836)	MFI	124@6000	116@5000	3.17x3.46	9.2:1	12@750
	F		2.0 (1997)	MFI	128@6000	121@5000	3.35x3.46	9.0:1	12@750
	T		3.0 (2972)	MFI	142@5000	168@2500	3.59x2.99	8.9:1	30-80@2000

GENERAL ENGINE SPECIFICATIONS

Year	Engine ID/VIN		Engine Displacement Liters (cc)	Fuel System Type	Net Horsepower @ rpm	Net Torque @ rpm (ft. lbs.)	Bore x Stroke (in.)	Com-pression Ratio	Oil Pressure @ rpm
1995	K		1.5 (1495)	MFI	92@5500	96@3000	2.97x3.29	10.0:1	21@idle
	N		1.5 (1495)	MFI	92@5500	97@4500	2.97x3.29	10.0:1	21@idle
	N	①	1.5 (1495)	MFI	115@5500	123@4000	2.97x3.29	7.5:1	21@idle
	R		1.6 (1596)	MFI	113@6000	102@5000	3.24X2.95	9.2:1	11@idle
	M		1.8 (1796)	MFI	124@6000	116@4500	3.17X3.46	9.2:1	11@idle
	P		2.0 (1997)	MFI	137@6000	129@4000	3.35X3.46	9.0:1	11@idle
	T		3.0 (2972)	MFI	142@5000	168@2500	3.59X2.99	8.9:1	11@idle
1996-97	K		1.5 (1495)	MFI	92@5500	97@4000	2.97x3.29	10.0:1	21@idle
	K	③	1.5 (1495)	MFI	105@6000	101@4500	2.97x3.29	9.5:1	21@idle
	M		1.8 (1796)	MFI	124@6000	116@4500	3.23x3.46	10.0:1	24@idle
	F		2.0 (1997)	MFI	137@6000	129@4000	3.35x3.46	9.0:1	11@idle
	K		3.0 (2972)	MFI	142@5000	168@2500	3.59x2.99	8.9:1	11@idle

2 bbl - 2 barrel carburetor

MFI - Multiport fuel injection

① Scoupe Turbocharged

② Elantra

③ Double overhead camshaft

GASOLINE ENGINE TUNE-UP SPECIFICATIONS

Year	Engine ID/VIN		Engine Displacement Liters (cc)	Spark Plugs Gap (in.)	Ignition Timing (deg.)		Fuel Pump (psi)	Idle Speed (rpm)		Valve Clearance	
					MT	AT		MT	AT	In.	Ex.
1993	M		1.5 (1468)	0.039-0.043	5B	5B	2.8-3.6	700	700	0.006	0.010
	J		1.5 (1468)	0.039-0.043	5B	5B	48	700	700	0.006	0.010
	N		1.5 (1495)	0.039-0.043	9B	9B	43	800	800	0.007	0.009
	N	①	1.5 (1495)	0.031-0.035	9B	9B	43	800	800	0.007	0.009
	R		1.6 (1596)	0.039-0.043	5B	5B	48	750	750	HYD	HYD
	M	②	1.8 (1836)	0.039-0.043	5B	5B	48	700	700	HYD	HYD
	F		2.0 (1997)	0.039-0.043	5B	5B	48	750	750	HYD	HYD
	T		3.0 (2972)	0.039-0.043	12B	12B	48	750	750	HYD	HYD
1994	J		1.5 (1468)	0.039-0.043	5B	5B	48	700	700	0.006	0.010
	N		1.5 (1495)	0.039-0.041	9B	9B	43	800	800	0.007	0.009
	N	①	1.5 (1495)	0.031-0.035	9B	9B	43	800	800	0.007	0.009
	R		1.6 (1596)	0.039-0.043	5B	5B	48	750	750	HYD	HYD
	M	②	1.8 (1836)	0.039-0.043	5B	5B	48	700	700	HYD	HYD
	F		2.0 (1997)	0.039-0.043	5B	5B	48	750	750	HYD	HYD
	T		3.0 (2972)	0.039-0.043	12B	12B	48	750	750	HYD	HYD
1995	K		1.5 (1495)	0.039-0.043	11B	11B	43	800	800	HYD	HYD
	N		1.5 (1495)	0.039-0.043	10B	10B	43	800	800	HYD	HYD
	N	①	1.5 (1495)	0.031-0.034	10B	10B	43	800	800	HYD	HYD
	R		1.6 (1596)	0.039-0.043	5B	5B	48	750	750	HYD	HYD
	M		1.8 (1796)	0.039-0.043	5B	5B	48	700	700	HYD	HYD
	P		2.0 (1997)	0.039-0.043	5B	5B	48	750	750	HYD	HYD
	T		3.0 (2972)	0.039-0.043	5B	5B	48	750	750	HYD	HYD
1996-97	K		1.5 (1495)	0.039-0.043	9B	9B	43	800	800	HYD	HYD
	K	③	1.5 (1495)	0.039-0.043	9B	9B	43	800	800	HYD	HYD
	M		1.8 (1796)	0.039-0.043	10B	10B	48	800	800	HYD	HYD
	F		2.0 (1997)	0.039-0.043	5B	5B	48	750	750	HYD	HYD
	K		3.0 (2972)	0.039-0.043	5B	5B	48	750	750	HYD	HYD

NOTE: The Vehicle Emission Control Information label often reflects specification changes made during production. The label figures must be used if they differ from those in this chart.

B - Before top dead center

HYD - Hydraulic

① Scoupe Turbocharged

② Elantra

③ Double overhead camshaft

CAPACITIES

Year	Model	Engine ID/VIN	Engine Displacement Liters (cc)	Engine Oil with Filter	Transmission (pts.) 4-Spd	Transmission (pts.) 5-Spd	Transmission (pts.) Auto.	Transfer Case (pts.)	Drive Axle Front (pts.)	Drive Axle Rear (pts.)	Fuel Tank (gal.)	Cooling System (qts.)
1993	Excel	M	1.5 (1468)	3.6	4.4	4.4	12.2	-	-	-	10.6 ①	5.6
	Excel	J	1.5 (1468)	3.6	4.4	4.4	12.2	-	-	-	10.6 ①	5.6
	Scoupe	N	1.5 (1495)	3.4	-	4.4	12.8	-	-	-	11.9	5.6
	Elantra	R	1.6 (1596)	4.6	-	3.8	12.8	-	-	-	13.7	5.4
	Elantra	M	1.8 (1836)	4.6	-	3.8	12.8	-	-	-	13.7	5.4
	Sonata	F	2.0 (1997)	4.0	-	4.0	12.8	-	-	-	17.2	7.7
	Sonata	T	3.0 (2972)	4.0	-	-	12.3	-	-	-	17.2	7.4
1994	Excel	J	1.5 (1468)	3.6	3.6	3.8	12.2	-	-	-	11.9	5.6
	Scoupe	N	1.5 (1495)	3.4	-	4.4	12.2	-	-	-	11.9	5.6
	Elantra	R	1.6 (1596)	4.6	-	3.8	12.8	-	-	-	13.7	5.4
	Elantra	M	1.8 (1836)	4.6	-	3.8	12.8	-	-	-	13.7	5.4
	Sonata	F	2.0 (1997)	4.0	-	4.0	12.8	-	-	-	17.2	7.7
	Sonata	T	3.0 (2972)	4.0	-	-	12.3	-	-	-	17.2	7.4
1995	Accent	K	1.5 (1495)	3.5	-	4.6	12.8	-	-	-	11.9	5.8
	Scoupe	N	1.5 (1495)	3.5	-	5.0	12.2	-	-	-	11.9	10.2
	Elantra	R	1.6 (1596)	4.7	-	3.8	12.8	-	-	-	13.8	5.4
	Elantra	M	1.8 (1796)	4.7	-	3.8	12.8	-	-	-	13.8	5.4
	Sonata	P	2.0 (1997)	4.0	-	3.8	12.8	-	-	-	17.2	7.7
	Sonata	T	3.0 (2972)	4.3	-	-	15.8	-	-	-	17.2	9.0
1996-97	Accent	K	1.5 (1495)	3.5	-	5.0	13.6	-	-	-	11.9	5.8
	Elantra	M	1.8 (1796)	4.2	-	3.8	12.8	-	-	-	14.5	6.3
	Sonata	F	2.0 (1997)	4.0	-	3.8	12.8	-	-	-	17.2	7.7
	Sonata	K	3.0 (2972)	4.3	-	-	15.8	-	-	-	17.2	9.0

① Optional 13.2 gal. tank

VALVE SPECIFICATIONS

Year	Engine ID/VIN	Engine Displacement Liters (cc)	Seat Angle (deg.)	Face Angle (deg.)	Spring Test Pressure (lbs. @ in.)	Spring Installed Height (in.)	Stem-to-Guide Clearance (in.) Intake	Stem-to-Guide Clearance (in.) Exhaust	Stem Diameter (in.) Intake	Stem Diameter (in.) Exhaust
1993	M	1.5 (1468)	45	45	53@1.07 ①	1.417 ③	0.0012-0.0024	0.0020-0.0035	0.2598	0.2598
	J	1.5 (1468)	45	45	53@1.07 ①	1.417 ③	0.0012-0.0024	0.0020-0.0035	0.2598	0.2598
	N	1.5 (1495)	45	45	53@1.259	1.259	0.0012-0.0024	0.0020-0.0031	0.2362	0.2362
	R	1.6 (1596)	44-44.5	45-45.5	66@1.575	①	0.0008-0.0019	0.0020-0.0033	0.2585-0.2591	0.2571-0.2579
	M	1.8 (1836)	44-44.5	45-45.5	66@1.575	①	0.0008-0.0019	0.0020-0.0033	0.2585-0.2591	0.2571-0.2579
	F	2.0 (1997)	44-44.5	45-45.5	66@1.575	①	0.0008-0.0019	0.0020-0.0033	0.2585	0.2591
	T	3.0 (2972)	44-44.5	45-45.5	74@1.591	1.590	0.0012-0.0024	0.0020-0.0033	0.3150	0.3134
1994	J	1.5 (1468)	45	45	53@1.07 ①	1.417 ②	0.0012-0.0024	0.0020-0.0035	0.2598	0.2598
	N	1.5 (1495)	45	45	53@1.259	1.259	0.0012-0.0024	0.0020-0.0031	0.2362	0.2362
	R	1.6 (1596)	44-44.5	45-45.5	66@1.575	③	0.0008-0.0019	0.0020-0.0033	0.2585-0.2591	0.2571-0.2579

VALVE SPECIFICATIONS

Year	Engine ID/VIN	Engine Displacement Liters (cc)	Seat Angle (deg.)	Face Angle (deg.)	Spring Test Pressure (lbs. @ in.)	Spring Installed Height (in.)	Stem-to-Guide Clearance (in.)		Stem Diameter (in.)	
							Intake	Exhaust	Intake	Exhaust
1994	M	1.8 (1836)	44-44.5	45-45.5	66@1.575	③	0.0008-0.0019	0.0020-0.0033	0.2585-0.2591	0.2571-0.2579
	F	2.0 (1997)	44-44.5	45-45.5	66@1.575	③	0.0008-0.0019	0.0020-0.0033	0.2585	0.2591
	T	3.0 (2972)	44-44.5	45-45.5	74@1.591	1.590	0.0012-0.0024	0.0020-0.0035	0.3150	0.3134
1995	K	1.5 (1495)	45	45-45.3	54.5@1.358	1.358	0.0012-0.0024	0.0014-0.0026	0.2340-0.2350	0.2340-0.2350
	N	1.5 (1495)	45	45	54.5@1.358	1.358	0.0012-0.0024	0.0014-0.0026	0.2360	0.2360
	R	1.6 (1596)	44-44.5	45-45.5	66@1.575	1.575	0.0008-0.0019	0.0020-0.0033	0.2585-0.2591	0.2571-0.2579
	M	1.8 (1796)	44-44.5	45-45.5	66@1.575	1.575	0.0008-0.0019	0.0020-0.0033	0.2585-0.2591	0.2571-0.2579
	P	2.0 (1997)	44-44.5	45-45.5	66@1.575	1.575	0.0008-0.0019	0.0020-0.0033	0.2585-0.2591	0.2571-0.2579
	T	3.0 (2972)	44-44.5	45-45.5	74@1.591	1.591	0.0012-0.0024	0.0020-0.0035	0.3143-0.3148	0.3126-0.3134
1996-97	K	1.5 (1495)	45	45-45.3	54.5@1.358	1.358	0.0012-0.0024	0.0014-0.0026	0.2340-0.2350	0.2340-0.2350
	K ④	1.5 (1495)	45	45	54.5@1.358	1.358	0.0012-0.0024	0.0020-0.0031	0.2344-0.2350	0.2337-0.2343
	M	1.8 (1796)	45	45-45.5	56@1.456	1.358	0.0008-0.0019	0.0020-0.0033	0.2348-0.2354	0.2334-0.2342
	F	2.0 (1997)	44-44.5	45-45.5	66@1.575	1.575	0.0008-0.0019	0.0020-0.0033	0.2585-0.2591	0.2571-0.2591
	K	3.0 (2972)	44-44.5	45-45.5	74@1.591	1.591	0.0012-0.0024	0.0020-0.0035	0.3143-0.3148	0.3126-0.3134

① Jet valve: 7.7@0.846   ② Jet valve: 0.846   ③ Free length: 1.902   ④ Double overhead camshaft

TORQUE SPECIFICATIONS
All readings in ft. lbs.

Year	Engine ID/VIN	Engine Displacement Liters (cc)	Cylinder Head Bolts	Main Bearing Bolts	Rod Bearing Bolts	Crankshaft Damper Bolts	Flywheel Bolts	Manifold		Spark Plugs	Lug Nut
								Intake	Exhaust		
1993	M	1.5 (1468)	①	36-39	23-25	51-72	94-101	11-14	12-14	18	80
	J	1.5 (1468)	①	36-39	23025	51-72	94-101	11-14	12-14	18	80
	N	1.5 (1495)	①	39-43	25-27	137-145	94-101	12-14	12-14	18	80
	R	1.6 (1596)	②	47-51	36-38	80-94	94-101	③	18-22	15-21	80
	M	1.8 (1836)	76-83	47-51	36-38	80-94	94-101	③	18-22	15-21	80
	F	2.0 (1997)	76-83	47-51	38	80-94	94-101	22	18-22	14-22	80
	T	3.0 (2972)	②	55-61	36-38	109-115	65-70	11-14	11-16	18	80
1994	J	1.5 (1468)	①	36-39	23-25	51-72	94-101	11-14	12-14	18	80
	N	1.5 (1495)	①	39-43	25-27	137-145	94-101	12-14	12-14	18	80
	R	1.6 (1596)	②	47-51	36-38	80-94	94-101	③	18-22	15-21	80
	M	1.8 (1836)	76-83	47-51	36-38	80-94	94-101	③	18-22	15-21	80
	F	2.0 (1997)	76-83	47-51	38	80-94	94-101	22	18-22	14-22	80
	T	3.0 (2972)	②	55-61	36-38	109-115	65-70	11-14	11-16	18	80

TORQUE SPECIFICATIONS
All readings in ft. lbs.

Year	Engine ID/VIN	Engine Displacement Liters (cc)	Cylinder Head Bolts	Main Bearing Bolts	Rod Bearing Bolts	Crankshaft Damper Bolts	Flywheel Bolts	Manifold Intake	Manifold Exhaust	Spark Plugs	Lug Nut
1995	K	1.5 (1495)	①	40-44	25-28	110-118	94-101	11-14	11-14	18	80
	N	1.5 (1495)	①	40-43	25-27	103-110	94-101	11-14	11-14	18	80
	N ④	1.5 (1495)	①	40-43	25-27	103-110	94-101	11-14	18-20	18	80
	R	1.6 (1596)	76-83 ⑤	47-51	36-38	80-94	94-101	③	18-22	18-21	80
	M	1.8 (1796)	76-83 ⑤	47-51	36-38	80-94	94-101	③	18-22	18-21	80
	F	2.0 (1997)	②	55-61	36-38	108-116	94-101	③	18-22	18-21	80
	T	3.0 (2972)	②	55-61	36-38	109-115	94-101	11-14	11-16	18-21	80
1996-97	K	1.5 (1495)	①	40-44	25-28	110-118	94-101	11-14	11-14	18	80
	K ⑦	1.5 (1495)	①	40-43	23-26	103-110	88-96	13-18	22-30	18	80
	M	1.8 (1796)	⑥	47-51	⑧	125-133	88-95	13-18	22-30	18-21	80
	F	2.0 (1997)	②	55-61	36-38	108-116	94-101	③	18-22	18-21	80
	K	3.0 (2972)	②	55-61	36-38	109-115	94-101	11-14	11-16	18-21	80

① Cold: 51-54 ft. lbs.
Warm: 58-61 ft. lbs.
② Cold: 65-72 ft. lbs.
Warm: 72-80 ft. lbs.
③ M8 bolt: 11-14 ft. lbs.
Except M8 bolt: 22-30 ft. lbs.
④ Turbocharged
⑤ Cold engine
⑥ Step 1:
M10 bolts: 22 ft. lbs.
M12 bolts: 26 ft. lbs.
Step 2: Plus 60-65 degrees
Step 3: Plus 60-65 degrees
⑦ Double overhead camshaft
⑧ Step 1: 20-24 ft. lbs.
Step 2: Plus 60-65 degrees

BRAKE SPECIFICATIONS
All measurements in inches unless noted

Year	Model	Master Cylinder Bore	Brake Disc Original Thickness	Brake Disc Minimum Thickness	Brake Disc Maximum Runout	Brake Drum Original Inside Diameter	Brake Drum Max. Wear Limit	Brake Drum Maximum Machine Diameter	Minimum Lining Thickness Front	Minimum Lining Thickness Rear
1993	Excel	0.875	0.750	0.669	0.006	7.100	7.200	7.165	0.039	0.039
	Sonata	1.024	0.866	0.787	0.004	8.858	8.958	8.936	0.079	0.031
	Sonata ①	-	0.472	0.413	0.005	-	-	-	-	0.031
	Scoupe	0.875	0.750	0.669	0.006	7.100	7.200	7.165	0.039	0.031
	Elantra	0.875	0.866	0.787	0.006	8.000	8.100	8.079	0.079	0.059
1994	Excel	0.875	0.750	0.669	0.006	7.100	7.200	7.165	0.039	0.039
	Sonata	1.024	0.866	0.787	0.004	8.858	8.958	8.936	0.079	0.031
	Sonata ①	-	0.472	0.413	0.005	-	-	-	-	0.031
	Scoupe	0.875	0.750	0.669	0.006	7.100	7.200	7.165	0.039	0.031
	Elantra	0.875	0.866	0.787	0.006	8.000	8.100	8.079	0.079	0.059
1995	Accent	0.813	0.750	0.669	0.002	7.09	7.20	7.17	0.039	0.039
	Elantra	0.875	0.866	0.787	0.002	8.00	8.11	8.08	0.079	0.059
	Scoupe	0.875	0.750	0.669	0.002	7.09	7.20	7.17	0.039	0.031
	Sonata	1.000	0.866	0.787	0.004	9.00	9.11	9.08	0.079	0.059
	Sonata ①	-	0.470	0.413	0.004	-	-	-	-	0.079
1996-97	Accent	0.813	0.750	0.669	0.002	7.09	7.20	7.17	0.039	0.039
	Elantra	0.875	0.866	0.787	0.002	8.00	8.11	9.08	0.079	0.059
	Sonata	1.000	0.866	0.787	0.004	9.00	9.11	9.08	0.079	0.059
	Sonata ①	-	0.470	0.413	0.004	-	-	-	-	0.079

① Specifications for models with rear disc brakes

FREQUENT MAINTENANCE LABOR
HYUNDAI

The following should be used as a guide when determining the amount of work required for a particular service if taken to a repair shop. In estimating how long a particular Frequent Maintenance Service item should take, please observe the following:
- **Chilton Time** is time that is based on field research and data supplied by the vehicle manufacturer.
- All labor time operations are given in hours and tenths of an hour.
- All labor operations, are to be used as a **guide**.

COOLING

(G) Winterize Cooling System
Includes: Run engine to check for leaks, tighten all hose connections. Test radiator and pressure cap, drain radiator and engine block. Add antifreeze and refill system.
All models 5

(G) Belt, Drive, Renew
All models 4
w/AC, add 1
w/PS, add 1
w/DOHC, add 2

(G) Belt, Drive, Adjust
All models
one 3
each adtnl. 1

(G) Hoses, Radiator, Renew
Includes: Drain and refill cooling system as required.
All models
upper or lower, each 5

(G) Thermostat, Coolant, Renew
1993-94 Excel 6
1995 Accent 6
1993-95 Elantra 6
1993-95 Sonata 6
1993-95 Scoupe 6

FUEL

(M) Air Cleaner, Service
All models 2

(G) Filter, Fuel, Renew
1993-94 Excel 9
1995 Accent 9
1993-95 Elantra 9
1993-95 Sonata
 4 cyl. 7
 V6 9
1993-95 Scoupe 9

BRAKES

(G) Bleed Brakes (Four Wheels)
Includes: Add fluid.
All models 6

(G) Brakes, Adjust (Minor)
Includes: Adjust brakes, fill master cylinder.
All models, two wheels 5

(M) Parking Brake, Adjust
All models 3

LUBRICATION SERVICES

(M) Engine Oil & Filter, Renew
Includes: Inspect and correct all fluid levels.
All models 6

WHEELS

(G) Wheel, Renew (One)
1993-94 Excel 8
1995 Accent 8
1993-95 Elantra 8
1993-95 Sonata 8
1993-95 Scoupe 8

(G) Wheel, Balance
All models
one 3
each adtnl. 2

(G) Wheels, Rotate (All)
All models 5

ELECTRICAL

(G) Headlamps, Aim
All models, two 5

(G) Parking Lamp Lens or Bulb, Renew
All models 3

(G) Horn, Renew
All models 3

(M) Terminals, Battery, Clean
All models 3

SCHEDULED MAINTENANCE INTERVALS
(HYUNDAI ACCENT, ELANTRA, EXCEL, SCOUPE & SONATA)

TO BE SERVICED	TYPE OF SERVICE	7.5	15	22.5	30	37.5	45	52.5	60	67.5	75	82.5	90	97.5
Engine oil & filter	R	✓	✓	✓	✓	✓	✓	✓	✓	✓	✓	✓	✓	✓
Automatic transaxle fluid	S/I		✓		✓		✓		✓		✓		✓	
Brake pads, calipers & rotors	S/I		✓		✓		✓		✓		✓		✓	
Brake hoses & lines	S/I		✓		✓		✓		✓		✓		✓	
Driveshafts & boots (except 1993 Sonata)	S/I		✓		✓		✓		✓		✓		✓	
Driveshafts & boots (1993 Sonata)	S/I				✓		✓		✓		✓		✓	
Valve clearance (Excel & Scoupe)	S/I		✓		✓		✓				✓		✓	
Wheel bearing grease (1993 Sonata)	S/I		✓		✓				✓				✓	
Wheel bearing grease (except 1993 Sonata)	S/I				✓				✓				✓	
Air cleaner filter	R				✓				✓				✓	
Automatic transaxle fluid & filter	R				✓				✓				✓	
Brake fluid	R				✓				✓				✓	
Engine Coolant	R				✓				✓				✓	
Fuel hose, vapor hose & fuel filler cap	S/I							✓						
Spark plugs	R				✓				✓				✓	
Spark plugs (Sonata 3.0L V6)	R								✓					
Bolts & nuts on chassis & body (Accent)	S/I				✓				✓				✓	
Drive belts	S/I				✓				✓				✓	
Exhaust pipe connections, muffler & suspension bolts	S/I				✓				✓				✓	
Manual transaxle oil	S/I				✓				✓				✓	
Rear brake drums, linings & parking brake	S/I				✓				✓				✓	
Steering gear rack, linkage & boots	S/I				✓				✓				✓	
Suspension ball joints & dust covers (Accent)	S/I				✓				✓				✓	

VEHICLE MILEAGE INTERVAL (x1000)

SCHEDULED MAINTENANCE INTERVALS
(HYUNDAI ACCENT, ELANTRA, EXCEL, SCOUPE & SONATA) (Cont.)

TO BE SERVICED	TYPE OF SERVICE	VEHICLE MILEAGE INTERVAL (x1000)												
		7.5	15	22.5	30	37.5	45	52.5	60	67.5	75	82.5	90	97.5
Timing belt (1996-97 Accent & Elantra)	S/I				✓				✓				✓	
Timing belt (except 1996-97 Accent & Elantra)	R								✓					
Fuel filter	R							✓						
Crankcase emission control system (carburetor)	S/I								✓					
Fuel lines & connections	S/I								✓					
Vacuum & crankcase ventilation hoses	S/I								✓					

R – Replace   S/I – Service or Inspect

FREQUENT OPERATION MAINTENANCE (SEVERE SERVICE)
  If a vehicle is operated under any of the following conditions it is considered severe service:
- Extremely dusty areas.
- 50% or more of the vehicle operation is in 32°C (90°F) or higher temperatures, or constant operation in temperatures below 0°C (32°F).
- Prolonged idling (vehicle operation in stop and go traffic).
- Frequent short running periods (engine does not warm to normal operating temperatures).
- Police, taxi, delivery usage or trailer towing usage.
Oil & oil filter change – change every 3000 miles.
Brake pads, calipers & rotors - service or inspect every 7500 miles.
Driveshaft boots - service or inspect every 7500 miles.
Steering gear rack, linkage & boots - service or inspect every 7500 miles.
Air cleaner filter – service or inspect every 15,000 miles.
Automatic transaxle fluid & filter - replace every 15,000 miles.
Rear brake drums & linings - service or inspect every 15,000 miles.
Spark plugs - replace every 24,000 miles.
Crankcase emission control system (carburetor) - service or inspect every 30,000 miles.

INFINITI

ENGINE IDENTIFICATION

Year	Model	Engine Displacement Liters (cc)	Engine Series (ID/VIN)	Fuel System	No. of Cylinders	Engine Type
1993	G20	2.0 (1998)	SR20DE (C)	MFI	4	DOHC
	J30	3.0 (2960)	VG30DE (A)	MFI	6	DOHC
	Q45	4.5 (4494)	VH45DE (N)	MFI	8	DOHC
1994	G20	2.0 (1998)	SR20DE (C)	MFI	4	DOHC
	J30	3.0 (2960)	VG30DE (A)	MFI	6	DOHC
	Q45	4.5 (4494)	VH45DE (N)	MFI	8	DOHC

ENGINE IDENTIFICATION

Year	Model	Engine Displacement Liters (cc)	Engine Series (ID/VIN)	Fuel System	No. of Cylinders	Engine Type
1995	G20	2.0 (1998)	SR20DE (C)	MFI	4	DOHC
	J30	3.0 (2960)	VG30DE (A)	MFI	6	DOHC
	Q45	4.5 (4494)	VH45DE (N)	MFI	8	DOHC
1996-97	G20	2.0 (1998)	SR20DE (C)	MFI	4	DOHC
	J30	3.0 (2960)	VG30DE (A)	MFI	6	DOHC
	I30	3.0 (2988)	VQ30DE (C)	MFI	6	DOHC
	Q45	4.5 (4494)	VH45DE (N)	MFI	8	DOHC

MFI - Multiport fuel injection   DOHC - Double overhead camshaft

GENERAL ENGINE SPECIFICATIONS

Year	Engine ID/VIN	Engine Displacement Liters (cc)	Fuel System Type	Net Horsepower @ rpm	Net Torque @ rpm (ft. lbs.)	Bore x Stroke (in.)	Compression Ratio	Oil Pressure @ rpm
1993	C	2.0 (1998)	MFI	140@6400	132@4800	3.39x3.39	9.5:1	46-57@3200
	A	3.0 (2960)	MFI	210@6400	193@4800	3.43x3.27	10.5:1	51-65@3000
	N	4.5 (4494)	MFI	278@6000	292@4000	3.66x3.26	10.2:1	67-81@3000
1994	C	2.0 (1998)	MFI	140@6400	132@4800	3.39x3.39	9.5:1	46-57@3200
	A	3.0 (2960)	MFI	210@64C0	193@4800	3.43x3.27	10.5:1	51-65@3000
	N	4.5 (4494)	MFI	278@6000	292@4000	3.66x3.26	10.2:1	67-81@3000
1995	C	2.0 (1998)	MFI	140@6400	132@4800	3.39x3.39	9.5:1	46-57@3200
	A	3.0 (2960)	MFI	210@6400	193@4800	3.43x3.27	10.5:1	51-65@3000
	N	4.5 (4494)	MFI	278@6000	292@4000	3.66x3.26	10.2:1	67-81@3000
1996-97	C	2.0 (1998)	MFI	140@6400	132@4800	3.39x3.39	9.5:1	46-57@3200
	A	3.0 (2960)	MFI	210@6400	193@4800	3.43x3.27	10.5:1	51-65@3000
	C	3.0 (2988)	MFI	190@5600	205@4000	3.66x2.89	10.1:1	63-80@3000
	N	4.5 (4494)	MFI	278@6000	292@4000	3.66x3.26	10.2:1	67-81@3000

MFI - Multiport fuel injection

GASOLINE ENGINE TUNE-UP SPECIFICATIONS

Year	Engine ID/VIN	Engine Displacement Liters (cc)	Spark Plugs Gap (in.)	Ignition Timing (deg.) MT	Ignition Timing (deg.) AT	Fuel Pump (psi)	Idle Speed (rpm) MT	Idle Speed (rpm) AT	Valve Clearance In.	Valve Clearance Ex.
1993	C	2.0 (1998)	0.039-0.041	15B	15B	34 ①	800	800	HYD	HYD
	A	3.0 (2960)	0.039-0.041	-	15B	34 ①	-	720	HYD	HYD
	N	4.5 (4494)	0.039-0.041	-	15B	34 ①	-	650	HYD	HYD
1994	C	2.0 (1998)	0.039-0.041	15B	15B	34 ①	800	800	HYD	HYD
	A	3.0 (2960)	0.039-0.041	-	15B	34 ①	-	720	HYD	HYD
	N	4.5 (4494)	0.039-0.041	-	15B	34 ①	-	650	HYD	HYD
1995	C	2.0 (1998)	0.039-0.041	15B	15B	34 ①	800	800	HYD	HYD
	A	3.0 (2960)	0.039-0.041	-	15B	34 ①	-	720	HYD	HYD
	N	4.5 (4494)	0.039-0.041	-	15B	34 ①	-	650	HYD	HYD
1996-97	C	2.0 (1998)	0.039-0.041	15B	15B	34 ①	800	800	HYD	HYD
	A	3.0 (2960)	0.039-0.041	-	15B	34 ①	-	720	HYD	HYD
	C	3.0 (2988)	0.039-0.043	15B	15B	34 ①	625	700	HYD	HYD
	N	4.5 (4494)	0.039-0.041	-	15B	34 ①	-	650	HYD	HYD

NOTE: The Vehicle Emission Control Information label often reflects specification changes made during production. The label figures must be used if they differ from those in this chart.

B - Before top dead center   HYD - Hydraulic   ① 43 psi with regulator vacuum hose disconnected

CAPACITIES

Year	Model	Engine ID/VIN	Engine Displacement Liters (cc)	Engine Oil with Filter	Transmission (pts.)			Transfer Case (pts.)	Drive Axle		Fuel Tank (gal.)	Cooling System (qts.)
					4-Spd	5-Spd	Auto.		Front (pts.)	Rear (pts.)		
1993	G20	C	2.0 (1998)	3.60	-	7.60	7.40	-	-	-	15.9	①
	J30	A	3.0 (2960)	4.50	-	-	8.75	-	-	3.20	19.0	9.75
	Q45	N	4.5 (4494)	6.35	-	-	10.8	-	-	3.20	22.5	10.9
1994	G20	C	2.0 (1998)	3.60	-	7.60	7.40	-	-	-	15.9	①
	J30	A	3.0 (2960)	4.50	-	-	8.75	-	-	3.20	19.0	9.75
	Q45	N	4.5 (4494)	6.35	-	-	10.8	-	-	3.20	22.5	10.9
1995	G20	C	2.0 (1998)	3.60	-	7.60	7.40	-	-	-	15.9	①
	J30	A	3.0 (2960)	4.50	-	-	8.75	-	-	3.20	19.0	9.75
	Q45	N	4.5 (4494)	6.35	-	-	10.8	-	-	3.20	22.5	10.9
1996-97	G20	C	2.0 (1998)	3.60	-	7.6	14.8	-	-	-	15.9	①
	J30	A	3.0 (2960)	4.50	-	-	17.5	-	-	3.20	19.0	9.8
	I30	C	3.0 (2988)	4.25	-	②	19.8	-	-	-	18.5	9.0
	Q45	N	4.5 (4494)	6.35	-	-	22.2	-	-	2.80	22.5	10.9

① With automatic transaxle: 6.9 qts.
   With manual transaxle: 6.5 qts.

② RSF50V: 9.13-9.50
   RSF50A: 9.50-10.13

VALVE SPECIFICATIONS

Year	Engine ID/VIN	Engine Displacement Liters (cc)	Seat Angle (deg.)	Face Angle (deg.)	Spring Test Pressure (lbs. @ in.)	Spring Installed Height (in.)	Stem-to-Guide Clearance (in.)		Stem Diameter (in.)	
							Intake	Exhaust	Intake	Exhaust
1993	C	2.0 (1998)	45.25-45.75	44.85-45.10	①	1.943	0.0008-0.0021	0.0016-0.0029	0.2348-0.2354	0.2341-0.2346
	A	3.0 (2960)	45.25-45.75	45	120.6@1.043	1.697	0.0008-0.0021	0.0016-0.0029	0.2348-0.2354	0.2341-0.2346
	N	4.5 (4494)	45.25-45.75	44.85-45.10	120.4@1.055	1.862	0.0011-0.0020	0.0014-0.0020	0.2743-0.2744	0.3134-0.3136
1994	C	2.0 (1998)	45.25-45.75	44.85-45.10	①	1.943	0.0008-0.0021	0.0016-0.0029	0.2348-0.2354	0.2341-0.2346
	A	3.0 (2960)	45.25-45.75	45	120.6@1.043	1.697	0.0008-0.0021	0.0016-0.0029	0.2348-0.2354	0.2341-0.2346
	N	4.5 (4494)	45.25-45.75	44.85-45.10	120.4@1.055	1.862	0.0011-0.0020	0.0014-0.0020	0.2743-0.2744	0.3134-0.3136
1995	C	2.0 (1998)	45.25-45.75	44.85-45.10	127.9-144.3@1.181	1.943 ②	0.0008-0.0021	0.0016-0.0029	0.2348-0.2354	0.2341-0.2346
	A	3.0 (2960)	45.25-45.75	45	120.6@1.043	1.697 ②	0.0008-0.0021	0.0016-0.0029	0.2348-0.2354	0.2341-0.2346
	N	4.5 (4494)	45.25-45.75	44.85-45.10	120.4@1.055	1.862 ②	0.0011-0.0020	0.0014-0.0020	0.2743-0.2744	0.3134-0.3136
1996-97	C	2.0 (1998)	45.25-45.75	44.85-45.10	127.9-144.3@1.181	1.943 ②	0.0008-0.0021	0.0016-0.0029	0.2348-0.2354	0.2341-0.2346
	A	3.0 (2960)	45.25-45.75	45	120.6@1.043	1.697 ②	0.0008-0.0021	0.0016-0.0029	0.2348-0.2354	0.2341-0.2346
	C	3.0 (2988)	45.25-45.75	NA	120.1@1.085	1.845 ②	0.0008-0.0021	0.0016-0.0029	0.2348-0.2354	0.2341-0.2346
	N	4.5 (4494)	45.25-45.75	44.85-45.10	120.4@1.055	1.946 ②	0.0011-0.0020	0.0014-0.0020	0.2743-0.2744	0.3134-0.3136

① Inner: 57.3@0.984
   Outer: 117.7@1.181

② Free height

TORQUE SPECIFICATIONS
All readings in ft. lbs.

Year	Engine ID/VIN	Engine Displacement Liters (cc)	Cylinder Head Bolts	Main Bearing Bolts	Rod Bearing Bolts	Crankshaft Damper Bolts	Flywheel Bolts	Manifold Intake	Manifold Exhaust	Spark Plugs	Lug Nut
1993	C	2.0 (1998)	④	⑤	③	105-112	61-69	13-15	27-35	14-22	72-87
	A	3.0 (2960)	⑥	67-74	⑦	159-174	61-69	12-15	17-20	14-22	72-87
	N	4.5 (4494)	①	②	③	260-275 ✓	61-69	12-15	20-23	14-22	72-87
1994	C	2.0 (1998)	④	⑤	③	105-112	61-69	13-15	27-35	14-22	72-87
	A	3.0 (2960)	⑥	67-74	⑦	159-174	61-69	12-15	17-20	14-22	72-87
	N	4.5 (4494)	①	②	③	260-275	61-69	12-15	20-23	14-22	72-87
1995	C	2.0 (1998)	④	⑤	③	105-112	61-69	13-15	27-35	14-22	72-87
	A	3.0 (2960)	⑥	67-74	⑦	159-174	61-69	12-15	17-20	14-22	72-87
	N	4.5 (4494)	①	②	③	260-275	61-69	12-15	20-23	14-22	72-87
1996-97	C	2.0 (1998)	④	⑤	③	105-112	61-69	13-15	27-35	14-22	72-87
	A	3.0 (2960)	⑥	67-74	⑦	159-174	61-69	12-15	17-20	14-22	72-87
	C	3.0 (2988)	⑧	⑨	⑩	⑪	61-69	20-23 ⑫	21-24	14-22	72-87
	N	4.5 (4494)	①	⑬	⑦	260-275	61-69	12-15	20-23	14-22	72-87

① Step 1: 22 ft. lbs.
  Step 2: 69 ft. lbs.
  Step 3: Loosen bolts completely
  Step 4: 18-25 ft. lbs.
  Step 5: Tighten an additional 90-95 degrees or 69-72 ft. lbs.
② See text
③ Step 1: 10-12 ft. lbs.
  Step 2: Tighten an additional 60-65 degrees or 28-33 ft. lbs.
④ Step 1: 29 ft. lbs.
  Step 2: 58 ft. lbs.
  Step 3: Loosen bolts completely
  Step 4: 25-33 ft. lbs.
  Step 5: Tighten an additional 90-100 degrees
  Step 6: Repeat Step 5.

⑤ Step 1: 24-28 ft. lbs.
  Step 2: Tighten an additional 45-50 degrees or 54-61 ft. lbs.
⑥ Step 1: 29 ft. lbs.
  Step 2: 90 ft. lbs.
  Step 3: Loosen bolts completely
  Step 4: 25-33 ft. lbs.
  Step 5: Tighten an additional 70-75 degrees or 90 ft. lbs.
⑦ Step 1: 10-12 ft. lbs.
  Step 2: Tighten an additional 60-65 degrees or 43-48 ft. lbs.
⑧ Step 1: 72 ft. lbs.
  Step 2: Loosen bolts completely
  Step 3: 25-33 ft. lbs.
  Step 4: Tighten an additional 90-95 degrees
  Step 5: Repeat Step 4

⑨ Step 1: Shift crankshaft to align the bearing beam
  Step 2: Tighten all bolts to 24-28 ft. lbs.
  Step 3: Tighten an additional 90-95 degrees
⑩ Step 1: Tighten all nuts to 15 ft. lbs.
  Step 2: Tighten an additional 90-95 degrees
⑪ Step 1: 29-36 ft. lbs.
  Step 2: Tighten an additional 60-66 degrees
⑫ Intake collector: 13-16 ft. lbs.
⑬ Step 1: Shift crankshaft back and forth to seat bearing caps
  Step 2: Tighten inner cap bolts to 27-31 ft. lbs.
  Step 3: Tighten outer cap bolts to 20-24 ft. lbs.
  Step 4: Tighten No. 1-3, 5 inner cap bolts an additional 60 degrees
  Step 5: Tighten No. 4 inner cap bolt and additional 35 degrees
  Step 6: Tighten outer cap bolts an additional 35 degrees
  Step 7: Tighten bearing cap side bolts to 34-38 ft. lbs.

BRAKE SPECIFICATIONS
All measurements in inches unless noted

Year	Model		Master Cylinder Bore	Brake Disc Original Thickness	Brake Disc Minimum Thickness	Brake Disc Maximum Runout	Brake Drum Diameter Original Inside Diameter	Brake Drum Diameter Max. Wear Limit	Brake Drum Diameter Maximum Machine Diameter	Minimum Lining Thickness Front	Minimum Lining Thickness Rear
1993	G20	F	0.937	0.870	0.787	0.0028	-	-	-	0.079	-
		R	-	0.350	0.310	0.0028	-	-	-	-	0.079
	J30	F	1.000	1.100	1.024	0.0028	-	-	-	0.079	-
		R	-	0.630	0.551	0.0028	-	-	-	-	0.079
	Q45	F	1.625	1.100	1.024	0.0028	-	-	-	0.079	-
		R	-	0.350	0.315	0.0028	-	-	-	-	0.079
1994	G20	F	0.937	0.870	0.787	0.0028	-	-	-	0.079	-
		R	-	0.350	0.310	0.0028	-	-	-	-	0.079
	J30	F	1.000	1.100	1.024	0.0028	-	-	-	0.079	-
		R	-	0.630	0.551	0.0028	-	-	-	-	0.079
	Q45	F	1.625	1.100	1.024	0.0028	-	-	-	0.079	-
		R	-	0.350	0.315	0.0028	-	-	-	-	0.079
1995	G20	F	0.937	0.870	0.787	0.003	-	-	-	0.079	-
		R	-	0.350	0.310	0.003	-	-	-	-	0.059
	J30	F	1.000	1.100	1.024	0.003	-	-	-	0.079	-
		R	-	0.630	0.551	0.006	-	-	-	-	0.079

BRAKE SPECIFICATIONS
All measurements in inches unless noted

Year	Model		Master Cylinder Bore	Brake Disc Original Thickness	Brake Disc Minimum Thickness	Brake Disc Maximum Runout	Brake Drum Diameter Original Inside Diameter	Brake Drum Diameter Max. Wear Limit	Brake Drum Diameter Maximum Machine Diameter	Minimum Lining Thickness Front	Minimum Lining Thickness Rear
1995	Q45	F	1.625	1.000	1.024	0.003	-	-	-	0.079	-
		R	-	0.350	0.315	0.003	-	-	-	-	0.079
1996-97	G20	F	0.937	0.870	0.787	0.003	-	-	-	0.079	-
		R	-	0.350	0.310	0.003	-	-	-	-	0.059
	I30	F	0.937	0.870	0.787	0.003	-	-	-	0.079	-
		R	-	0.350	0.310	0.003	-	-	-	-	0.059
	J30	F	1.000	1.100	1.024	0.003	-	-	-	0.079	-
		R	-	0.630	0.551	0.006	-	-	-	-	0.079
	Q45	F	1.000	1.100	1.024	0.003	-	-	-	0.079	-
		R	-	0.350	0.315	0.003	-	-	-	-	0.079

F - Front
R - Rear

FREQUENT MAINTENANCE LABOR
INFINITI

The following should be used as a guide when determining the amount of work required for a particular service if taken to a repair shop. In estimating how long a particular Frequent Maintenance Service item should take, please observe the following:
- **Chilton Time** is time that is based on field research and data supplied by the vehicle manufacturer.
- All labor time operations are given in hours and tenths of an hour.
- All labor operations, are to be used as a **guide**.

COOLING

Chilton Time

(G) Winterize Cooling System
Includes: Run engine to check for leaks, tighten all hose connections. Test radiator and pressure cap, drain radiator and engine block. Add antifreeze and refill system.
1993-97 .5

(G) Belt, Drive, Renew
1993-97 G20
   one 6
   two 1.0
1993-97 J30
   one 6
   two 7
   three 1.0
1993-97 Q45
   one 6
   two 7
   three 8
1996 I30
   one 7
   two 9

(G) Belt, Drive, Adjust
1993-97
   one 4
   each adtnl. 1

Chilton Time

(G) Hoses, Radiator, Renew
Includes: Drain and refill cooling system as required.
Upper
   1993-97 8
Lower
   1993-97 Q45 . . . . 1.4
   1993-97 G20 8
   1993-97 J30 7
   1996 I30 . . . . 1.1

(G) Thermostat, Coolant, Renew
   1993-97 Q45 . . . . 1.0
   1993-97 G20 . . . . 1.0
   1993-97 J30 . . . . 2.2
   1996 I30 . . . . 1.0

FUEL

(M) Air Cleaner, Service
   1993-97 2

(G) Filter, Fuel, Renew
In Tank
   1993-94 G20 . . . . 1.1
In Line
   1993-97 G20, Q45, J30 4
   1996 I30 9

Chilton Time

BRAKES

(G) Bleed Brakes (Four Wheels)
Includes: Add fluid.
   1993-97 8

(M) Parking Brake, Adjust
   1993-97 4

LUBRICATION SERVICES

(M) Engine Oil & Filter, Renew
Includes: Inspect and correct all fluid levels.
   1993-97 Q45 6
   1993-97 G20 4
   1993-97 J30 4
   1996 I30 4

WHEELS

(G) Wheel, Renew (One)
   1993-97 G20, I30 5

(G) Wheel, Balance
1993-97 G20, I30
   one 3
   each adtnl. 2

FREQUENT MAINTENANCE LABOR (cont.)
INFINITI

(G) Wheels, Rotate (All)
1993-97 G20, I30 . . . . .5 *(Chilton Time)*

ELECTRICAL

(G) Headlamps, Aim
1993-97
two .4
four .6

(G) Halogen Headlamp Bulb, Renew *(Chilton Time)*
One
1993-97 Q45 6
1993-97 G20 3
1993-97 J30 3
1996 I30 3

(G) Rear Combination Lamp Assy., Renew
One
1993-97 Q45 8
1993-97 G20 . . . . 1.4
1993-97 J30 4
1996 I30 . . . . 1.4

(G) Rear Combination Lamp Bulb, Renew *(Chilton Time)*
1993-97
one .3
each adtnl. 1

(G) Horn, Renew
1993-97 Q45 9
1993-97 G20 5
1993-97 J30 5
1996 I30 5

(M) Terminals, Battery, Clean
1993-97 3

SCHEDULED MAINTENANCE INTERVALS
(INFINITI G20, I30, J30 & Q45)

TO BE SERVICED	TYPE OF SERVICE	7.5	15	22.5	30	37.5	45	52.5	60	67.5	75	82.5	90	97.5
Engine oil & filter	R	✓	✓	✓	✓	✓	✓	✓	✓	✓	✓	✓	✓	✓
Automatic transaxle fluid	S/I		✓		✓		✓		✓		✓		✓	
Brake lines & cables	S/I		✓		✓		✓		✓		✓		✓	
Brake pads & discs	S/I		✓		✓		✓		✓		✓		✓	
Differential gear oil (J30 & Q45)	S/I		✓		✓		✓		✓		✓		✓	
Driveshaft boots (I30 & G20)	S/I		✓		✓		✓		✓		✓		✓	
Full-active suspension fluid (Q45)①	S/I		✓		✓		✓		✓		✓		✓	
Manual transaxle oil (G20 & I30)	S/I		✓		✓		✓		✓		✓		✓	
Air filter element	R				✓				✓				✓	
Exhaust system	S/I				✓				✓				✓	
Fuel lines	S/I				✓				✓				✓	
Steering gear linkage axle & suspension parts	S/I				✓				✓				✓	
SUPER HICAS linkage (J30 & Q45)	S/I			✓					✓				✓	

VEHICLE MILEAGE INTERVAL (x1000)

SCHEDULED MAINTENANCE INTERVALS
(INFINITI G20, I30, J30 & Q45) (Cont.)

TO BE SERVICED	TYPE OF SERVICE	VEHICLE MILEAGE INTERVAL (x1000)												
		7.5	15	22.5	30	37.5	45	52.5	60	67.5	75	82.5	90	97.5
Vapor lines	S/I				✓				✓				✓	
Engine Coolant	R								✓				✓	
Spark plugs	R								✓					
Timing belt	R								✓					
Drive belts	S/I								✓					

① Replace at 60,000 miles (if not previously replaced).
R – Replace   S/I – Service or Inspect

FREQUENT OPERATION MAINTENANCE (SEVERE SERVICE)

If a vehicle is operated under any of the following conditions it is considered severe service:
- Extremely dusty areas.
- 50% or more of the vehicle operation is in 32°C (90°F) or higher temperatures, or constant operation in temperatures below 0°C (32°F).
- Prolonged idling (vehicle operation in stop and go traffic).
- Frequent short running periods (engine does not warm to normal operating temperatures).
- Police, taxi, delivery usage or trailer towing usage.

Oil & oil filter change – change every 3750 miles.
Brake pads & discs - service or inspect every 7500 miles.
Driveshaft boots (G20 & I30) - service or inspect every 7500 miles.
Exhaust system - service or inspect every 7500 miles.
Steering gear, linkage, axle & suspension parts - service or inspect every 7500 miles.
Steering linkage ball joints & front suspension ball joints - service or inspect every 7500 miles.
SUPER HICAS linkage (J30 & Q45) - service or inspect every 7500 miles.

LEXUS

ENGINE IDENTIFICATION

Year	Model	Engine Displacement Liters (cc)	Engine Series (ID/VIN)	Fuel System	No. of Cylinders	Engine Type
1993	GS300	3.0 (2997)	2JZ-GE	SFI	6	DOHC
	ES300	3.0 (2959)	3VZ-FE	SFI	6	DOHC
	SC300	3.0 (2997)	2JZ-GE	SFI	6	DOHC
	LS400	4.0 (3969)	1UZ-FE	MFI	8	DOHC
	SC400	4.0 (3969)	1UZ-FE	MFI	8	DOHC
1994	GS300	3.0 (2997)	2JZ-GE	SFI	6	DOHC
	ES300	3.0 (2995)	1MZ-FE	SFI	6	DOHC
	SC300	3.0 (2997)	2JZ-GE	SFI	6	DOHC
	LS400	4.0 (3969)	1UZ-FE	MFI	8	DOHC
	SC400	4.0 (3969)	1UZ-FE	MFI	8	DOHC
1995	ES300	3.0 (2959)	1MZ-FE	MFI	6	DOHC
	GS300	3.0 (2997)	2JZ-GE	MFI	6	DOHC
	SC300	3.0 (2997)	2JZ-GE	MFI	6	DOHC
	LS400	4.0 (3969)	1UZ-FE	MFI	8	DOHC

ENGINE IDENTIFICATION

Year	Model	Engine Displacement Liters (cc)	Engine Series (ID/VIN)	Fuel System	No. of Cylinders	Engine Type
1995	SC400	4.0 (3969)	1UZ-FE	MFI	8	DOHC
1996-97	ES300	3.0 (2995)	1MZ-FE	MFI	6	DOHC
	GS300	3.0 (2997)	2JZ-GE	MFI	6	DOHC
	SC300	3.0 (2997)	2JZ-GE	MFI	6	DOHC
	LS400	4.0 (3969)	1UZ-FE	MFI	8	DOHC
	SC400	4.0 (3969)	1UZ-FE	MFI	8	DOHC

EFI - Electronic fuel injection
MFI - Multiport fuel injection

DOHC - Double overhead camshaft
SFI - Sequential Fuel Injection

GENERAL ENGINE SPECIFICATIONS

Year	Engine ID/VIN	Engine Displacement Liters (cc)	Fuel System Type	Net Horsepower @ rpm	Net Torque @ rpm (ft. lbs.)	Bore x Stroke (in.)	Compression Ratio	Oil Pressure @ rpm
1993	3VZ-FE	3.0 (2959)	SFI	185@5200	195@4400	3.44x3.23	9.6:1	43-78@3000
	2JZ-GE	3.0 (2997)	SFI	225@6000	210@4800	3.39x3.39	10.2:1	43-78@3000
	1UZ-FE	4.0 (3969)	MFI	250@5600	260@4400	3.44x3.25	10.0:1	36-71@3000
1994	1MZ-FE	3.0 (2995)	SFI	188@5200	203@4400	3.44x3.27	10.5:1	43-78@3000
	2JZ-GE	3.0 (2997)	SFI	225@6000	210@4800	3.39x3.39	10.2:1	43-78@3000
	1UZ-FE	4.0 (3969)	MFI	250@5600	260@4400	3.44x3.25	10.0:1	36-71@3000
1995	1MZ-FE	3.0 (2995)	MFI	188@5200	203@4400	3.44x3.27	10.5:1	43-78@3000
	2JZ-GE	3.0 (2997)	MFI	220@5800	210@4800	3.39x3.39	10.0:1	47-84@3000
	1UZ-FE ①	4.0 (3969)	MFI	250@5600	260@4400	3.44x3.25	10.0:1	43-85@3000
	1UZ-FE ②	4.0 (3969)	MFI	260@5300	270@4500	3.45x3.25	10.4:1	43-85@3000
1996-97	1MZ-FE	3.0 (2995)	MFI	188@5200	203@4400	3.44x3.27	10.5:1	43-78@3000
	2JZ-GE	3.0 (2997)	MFI	220@5800	210@4800	3.39x3.39	10.0:1	47-84@3000
	1UZ-FE	4.0 (3969)	MFI	260@5300	270@4500	3.45x3.25	10.4:1	43-85@3000

MFI - Multiport fuel injection
SFI - Sequential Fuel Injection

① SC400
② LS400

GASOLINE ENGINE TUNE-UP SPECIFICATIONS

Year	Engine ID/VIN	Engine Displacement Liters (cc)	Spark Plugs Gap (in.)	Ignition Timing (deg.) MT	Ignition Timing (deg.) AT	Fuel Pump (psi)	Idle Speed (rpm) MT	Idle Speed (rpm) AT	Valve Clearance In.	Valve Clearance Ex.
1993	3VZ-FE	3.0 (2959)	0.043	10B ①	10B ①	38-44	600-700	650-750	0.005-0.009	0.011-0.015
	2JZ-GE	3.0 (2997)	0.043	10B ①	10B ①	38-44	650-750	600-700	0.006-0.010	0.010-0.014
	1UZ-FE	4.0 (3969)	0.043	-	10B ①	38-44	-	600-700	0.006-0.010	0.010-0.014
1994	1MZ-FE	3.0 (2995)	0.043	-	10B ①	38-44	-	650-750	0.006-0.010	0.010-0.014
	2JZ-GE	3.0 (2997)	0.043	10B ①	10B ①	38-44	650-750	600-700	0.006-0.010	0.010-0.014
	1UZ-FE	4.0 (3969)	0.043	-	10B ①	38-44	-	600-700	0.006-0.010	0.010-0.014
1995	1MZ-FE	3.0 (2995)	0.043	-	10B ①	38-44	-	650-750	0.006-0.010	0.010-0.014

GASOLINE ENGINE TUNE-UP SPECIFICATIONS

Year	Engine ID/VIN	Engine Displacement Liters (cc)	Spark Plugs Gap (In.)	Ignition Timing (deg.) MT	Ignition Timing (deg.) AT	Fuel Pump (psi)	Idle Speed (rpm) MT	Idle Speed (rpm) AT	Valve Clearance In.	Valve Clearance Ex.
1995	2JZ-GE	3.0 (2997)	0.043	10B ①	10B ①	38-44	650-750	650-750	0.006-0.010	0.010-0.014
	1UZ-FE	4.0 (3969)	0.043	-	10B ①	38-44	-	600-700	0.006-0.010	0.010-0.014
1996-97	1MZ-FE	3.0 (2995)	0.043	-	10B ①	38-44	-	650-750	0.006-0.010	0.010-0.014
	2JZ-GE	3.0 (2997)	0.043	10B ①	10B ①	38-44	650-750	650-750	0.006-0.010	0.010-0.014
	1UZ-FE	4.0 (3969)	0.043	-	10B ①	38-44	-	600-700	0.006-0.010	0.010-0.014

NOTE: The Vehicle Emission Control Information label often reflects specification changes made during production. The label figures must be used if they differ from those in this chart.

B - Before top dead center

① Terminals TE1 and E1 of check connector must be connected

CAPACITIES

Year	Model	Engine ID/VIN	Engine Displacement Liters (cc)	Engine Oil with Filter	Transmission (pts.) 4-Spd	Transmission (pts.) 5-Spd	Transmission (pts.) Auto.	Drive Axle Front (pts.)	Drive Axle Rear (pts.)	Fuel Tank (gal.)	Cooling System (qts.)
1993	GS300	2JZ-GE	3.0 (2997)	5.7	-	-	4.0	-	2.8	21.1	7.9
	ES300	3VZ-FE	3.0 (2959)	4.5	-	8.8	6.6	1.6	-	18.5	9.0
	SC300	2JZ-GE	3.0 (2997)	5.1	-	5.4	3.4	-	2.8	20.6	8.9
	LS400	1UZ-FE	4.0 (3969)	5.6	-	-	4.0	-	2.8	22.5	11.2
	SC400	1UZ-FE	4.0 (3969)	5.1	-	-	4.0	-	2.8	20.6	11.4
1994	GS300	2JZ-GE	3.0 (2997)	5.7	-	-	4.0	-	2.8	21.1	7.9
	ES300	1MZ-FE	3.0 (2995)	5.1	-	-	7.1	1.6	-	18.5	9.2
	SC300	2JZ-GE	3.0 (2997)	5.1	-	5.4	3.4	-	2.8	20.6	8.9
	LS400	1UZ-FE	4.0 (3969)	5.6	-	-	4.0	-	2.8	22.5	11.4
	SC400	1UZ-FE	4.0 (3969)	5.1	-	-	4.0	-	2.8	20.6	12.0
1995	ES300	1MZ-FE	3.0 (2995)	5.1	-	-	7.4	1.8	-	18.5	9.2
	GS300	2JZ-GE	3.0 (2997)	5.7	-	-	4.0	-	2.8	21.1	7.9
	SC300	2JZ-GE	3.0 (2997)	5.1	-	5.4	3.4	-	2.8	20.6	8.9
	LS400	1UZ-FE	4.0 (3969)	5.6	-	-	4.0	-	2.8	22.5	11.4
	SC400	1UZ-FE	4.0 (3969)	5.6	-	-	4.0	-	2.8	20.6	12.0
1996-97	ES300	1MZ-FE	3.0 (2995)	5.1	-	-	7.4	1.8	-	18.5	9.2
	GS300	2JZ-GE	3.0 (2997)	5.7	-	-	4.0	-	2.8	21.1	7.9
	SC300	2JZ-GE	3.0 (2997)	5.1	-	5.4	3.4	-	2.8	20.6	8.9
	LS400	1UZ-FE	4.0 (3969)	5.6	-	-	4.0	-	2.8	22.5	11.4
	SC400	1UZ-FE	4.0 (3969)	5.6	-	-	4.0	-	2.8	20.6	12.0

VALVE SPECIFICATIONS

Year	Engine ID/VIN	Engine Displacement Liters (cc)	Seat Angle (deg.)	Face Angle (deg.)	Spring Test Pressure (lbs. @ in.)	Spring Installed Height (in.)	Stem-to-Guide Clearance (in.) Intake	Exhaust	Stem Diameter (in.) Intake	Exhaust
1993	3VZ-FE	3.0 (2959)	NA	44.5	38.4-42.4 @ 1.311	1.630 ②	0.0010-0.0024	0.0012-0.0026	0.2350-0.2356	0.2348-0.2354
	2JZ-GE	3.0 (2997)	NA	44.5	41.9-46.3 @ 1.358	① ② ③	0.0010-0.0024	0.0012-0.0026	0.2350-0.2356	0.2348-0.2354
	1UZ-FE	4.0 (3969)	NA	44.5	41.9-46.3 @ 1.295	1.717 ②	0.0010-0.0024	0.0012-0.0026	0.2350-0.2356	0.2348-0.2354
1994	1MZ-FE	3.0 (2995)	NA	44.5	41.9-46.3 @ 1.331	1.791 ②	0.0010-0.0024	0.0012-0.0026	0.2154-0.2159	0.2152-0.2157
	2JZ-GE	3.0 (2997)	NA	44.5	41.9-46.3 @ 1.358	② ③	0.0010-0.0024	0.0012-0.0026	0.2350-0.2356	0.2348-0.2354
	1UZ-FE	4.0 (3969)	NA	44.5	41.9-46.3 @ 1.295	1.717 ②	0.0010-0.0024	0.0012-0.0026	0.2350-0.2356	0.2348-0.2354
1995	1MZ-FE	3.0 (2995)	NA	44.5	41.9-46.3 @ 1.331	1.791 ②	0.0010-0.0024	0.0012-0.0026	0.2154-0.2159	0.2152-0.2157
	2JZ-GE	3.0 (2997)	NA	44.5	41.9-46.3 @ 1.358	② ③	0.0010-0.0024	0.0012-0.0026	0.2350-0.2356	0.2348-0.2354
	1UZ-FE ④	4.0 (3969)	NA	44.5	41.9-46.3 @ 1.295	1.717 ②	0.0010-0.0024	0.0012-0.0026	0.2350-0.2356	0.2348-0.2354
	1UZ-FE ⑤	4.0 (3969)	NA	44.5	41.9-46.3 @ 1.295	2.039 ②	0.0010-0.0024	0.0012-0.0026	0.2350-0.2356	0.2348-0.2354
1996-97	1MZ-FE	3.0 (2995)	NA	44.5	41.9-46.3 @ 1.331	1.791 ②	0.0010-0.0024	0.0012-0.0026	0.2154-0.2159	0.2152-0.2157
	2JZ-GE	3.0 (2997)	NA	44.5	41.9-46.3 @ 1.358	1.642 ②	0.0010-0.0024	0.0012-0.0026	0.2350-0.2356	0.2348-0.2354
	1UZ-FE	4.0 (3969)	NA	44.5	41.9-46.3 @ 1.295	2.039 ②	0.0010-0.0024	0.0012-0.0026	0.2350-0.2356	0.2348-0.2354

NA - Not Available
① Pink mark: 1.6433
  White mark: 1.6339
② Free length
③ Blue - 1.6433
  Yellow - 1.6417
④ SC400
⑤ LS400

TORQUE SPECIFICATIONS
All readings in ft. lbs.

Year	Engine ID/VIN	Engine Displacement Liters (cc)	Cylinder Head Bolts	Main Bearing Bolts	Rod Bearing Bolts	Crankshaft Damper Bolts	Flywheel Bolts	Manifold Intake	Exhaust	Spark Plugs	Lug Nut
1993	3VZ-FE	3.0 (2959)	①	45 ②	⑨	181	61	13	29	13	76
	2JZ-GE	3.0 (2997)	①	33 ②	22 ②	239	③	15	29	13	76
	1UZ-FE	4.0 (3969)	⑦	20 ②	⑨	181	72	13	29	13	76
1994	1MZ-FE	3.0 (2959)	⑤,⑥	④	⑨	159	61	11	29	13	76
	2JZ-GE	3.0 (2997)	①	33 ②	22 ②	239	③	15	29	13	76
	1UZ-FE	4.0 (3969)	⑦	20 ②	⑨	181	72	13	29	13	76
1995	1MZ-FE	3.0 (2995)	⑧	④	⑨	159	61	11	36	13	76
	2JZ-GE	3.0 (2997)	①	⑩	⑪	239	⑫	15	29	13	76
	1UZ-FE ⑬	4.0 (3969)	⑦	⑭	⑨	181	72	13	29	13	76
	1UZ-FE ⑮	4.0 (3969)	⑦	⑭	⑨	181	61	13	32	13	76
1996-97	1MZ-FE	3.0 (2995)	⑧	④	⑦	159	61	11	36	13	76

TORQUE SPECIFICATIONS
All readings in ft. lbs.

Year	Engine ID/VIN	Engine Displacement Liters (cc)	Cylinder Head Bolts	Main Bearing Bolts	Rod Bearing Bolts	Crankshaft Damper Bolts	Flywheel Bolts	Manifold Intake	Manifold Exhaust	Spark Plugs	Lug Nut
1996-97	2JZ-GE	3.0 (2997)	①	⑩	⑪	239	⑫	20	29	13	76
	1UZ-FE	4.0 (3969)	⑦	⑭	⑨	181	61	13	32	13	76

① Step 1: 25 ft. lbs.
  Step 2: Tighten an additional 90 degrees
  Step 3: Tighten an additional 90 degrees
② Plus an additional 90 degrees
③ Automatic Transmission - 72 ft. lbs.
  Manual Transmission
    Step 1: 36 ft. lbs.
    Step 2: Plus an additional 90 degrees
④ 6-point bolts - 20 ft. lbs.
  12-point bolts
    Step 1: 16 ft. lbs.
    Step 2: Plus an additional 90 degrees

⑤ Step 1: Tighten to 40 ft. lbs.
  Step 2: Plus an additional 90 degrees
⑥ Recessed head bolt tighten to 13 ft. lbs.
⑦ Step 1: Tighten to 29 ft. lbs.
  Step 2: Plus an additional 90 degrees
⑧ Head bolt:
  Step 1: 40 ft. lbs.
  Step 2: Plus 90 degrees
  Recessed head bolt: 13 ft. lbs.
⑨ Step 1: 18 ft. lbs.
  Step 2: Plus 90 degrees
⑩ Step 1: 33 ft. lbs.
  Step 2: Plus 90 degrees

⑪ Step 1: 22 ft. lbs.
  Step 2: Plus 90 degrees
⑫ Driveplate: 61 ft. lbs.
  Flywheel:
  Step 1: 36 ft. lbs.
  Step 2: Plus 90 degrees
⑬ SC400
⑭ Nuts:
  Step 1: 20 ft. lbs.
  Step 2: Plus 90 degrees
  Bolts: 36 ft. lbs.
⑮ LS400

BRAKE SPECIFICATIONS
All measurements in inches unless noted

Year	Model		Master Cylinder Bore	Brake Disc Original Thickness	Brake Disc Minimum Thickness	Brake Disc Maximum Runout	Brake Drum Diameter Original Inside Diameter	Brake Drum Diameter Max. Wear Limit	Brake Drum Diameter Maximum Machine Diameter	Minimum Lining Thickness Front	Minimum Lining Thickness Rear
1993	ES300	F	NA	1.102	1.063	0.0020	-	-	-	0.039	-
		R	NA	0.394	0.354	0.0059	-	-	-	-	0.039
	SC300	F	NA	1.102	1.024	0.0020	-	-	-	0.039	-
		R	NA	0.630	0.591	0.0020	-	-	-	-	0.039
	GS300	F	NA	1.260	1.181	0.0020	-	-	-	0.039	-
		R	NA	0.630	0.591	0.0020	-	-	-	-	0.039
	LS400	F	NA	1.102	1.024	0.0020	-	-	-	0.039	-
		R	NA	0.630	0.591	0.0020	-	-	-	-	0.039
	SC400	F	NA	1.260	1.181	0.0020	-	-	-	0.039	-
		R	NA	0.630	0.591	0.0020	-	-	-	-	0.039
1994	ES300	F	NA	1.102	1.063	0.0020	-	-	-	0.039	-
		R	NA	0.394	0.354	0.0059	-	-	-	-	0.039
	SC300	F	NA	1.102	1.024	0.0020	-	-	-	0.039	-
		R	NA	0.630	0.591	0.0020	-	-	-	-	0.039
	GS300	F	NA	1.260	1.181	0.0020	-	-	-	0.039	-
		R	NA	0.630	0.591	0.0020	-	-	-	-	0.039
	LS400	F	NA	1.102	1.024	0.0020	-	-	-	0.039	-
		R	NA	0.630	0.591	0.0020	-	-	-	-	0.039
	SC400	F	NA	1.260	1.181	0.0020	-	-	-	0.039	-
		R	NA	0.630	0.591	0.0020	-	-	-	-	0.039
1995	ES300	F	NA	1.102	1.024	0.0020	-	-	-	0.039	-
		R	NA	0.394	0.354	0.0059	-	-	-	-	0.039
	GS300	F	NA	1.260	1.181	0.0020	-	-	-	0.039	-
		R	NA	0.630	0.591	0.0020	-	-	-	-	0.039
	LS400	F	NA	1.102	1.024	0.0020	-	-	-	0.012	-
		R	NA	0.630	0.591	0.0020	-	-	-	-	0.098
	SC300	F	NA	1.102	1.024	0.0020	-	-	-	0.039	-
		R	NA	0.630	0.591	0.0020	-	-	-	-	0.039

BRAKE SPECIFICATIONS
All measurements in inches unless noted

Year	Model		Master Cylinder Bore	Brake Disc			Brake Drum Diameter			Minimum Lining Thickness	
				Original Thickness	Minimum Thickness	Maximum Runout	Original Inside Diameter	Max. Wear Limit	Maximum Machine Diameter	Front	Rear
1995	SC400	F	NA	1.260	1.181	0.0020	-	-	-	0.039	-
		R	NA	0.630	0.591	0.0020	-	-	-	-	0.039
1996-97	ES300	F	NA	1.102	1.024	0.0020	-	-	-	0.039	-
		R	NA	0.394	0.354	0.0059	-	-	-	-	0.039
	SC300	F	NA	1.102	1.024	0.0020	-	-	-	0.039	-
		R	NA	0.630	0.591	0.0020	-	-	-	-	0.039
	GS300	F	NA	1.260	1.181	0.0020	-	-	-	0.039	-
		R	NA	0.630	0.591	0.0020	-	-	-	-	0.039
	LS400	F	NA	1.102	1.024	0.0020	-	-	-	0.118	-
		R	NA	0.630	0.591	0.0020	-	-	-	-	0.098
	SC400	F	NA	1.260	1.181	0.0020	-	-	-	0.039	-
		R	NA	0.630	0.591	0.0020	-	-	-	-	0.039

NA - Not Available
F - Front
R - Rear

FREQUENT MAINTENANCE LABOR
LEXUS

The following should be used as a guide when determining the amount of work required for a particular service if taken to a repair shop. In estimating how long a particular Frequent Maintenance Service item should take, please observe the following:
- **Chilton Time** is time that is based on field research and data supplied by the vehicle manufacturer.
- All labor time operations are given in hours and tenths of an hour.
- All labor operations, are to be used as a **guide**.

COOLING

(G) Winterize Cooling System
Includes: Run engine to check for leaks, tighten all hose connections. Test radiator and pressure cap, drain radiator and engine block. Add antifreeze and refill system.
1993-97 5

(G) Belt, Fan Drive, Renew
1993-95 ES250, ES300 4

(G) Belt, Drive, Adjust
1993-95 ES250, ES300 3

(G) Hoses, Radiator, Renew
Includes: Drain and refill cooling system as required.
1993-97 SC300, GS300
    upper 8
    lower 1.2
1993-97 ES300
    upper 8
    lower 1.3

1993-94 LS400
    upper 1.0
    lower 9
1995-97 LS400
    each 1.2
1993-97 SC400
    upper 1.2
    lower 1.1

(G) Thermostat, Coolant, Renew
1993-97
    Six 8
    V6 .... 1.0
    V8 9

BRAKES

(G) Bleed Brakes (Four Wheels)
Includes: Add fluid.
1993-97
    one circuit 5
    both circuits 8

(M) Parking Brake, Adjust
1993-97 6

LUBRICATION SERVICES

(M) Engine Oil & Filter, Renew
Includes: Inspect and correct all fluid levels.
1993-97
    Six 3
    V6 3
    V8 5

WHEELS

(G) Wheel, Renew (One)
1993-97
    ES250, ES300 5

(G) Wheel, Balance
1993-97
    ES250, ES300
      one 3
      each adtnl. 2

FREQUENT MAINTENANCE LABOR (cont.)
LEXUS

	Chilton Time
(G) Wheels, Rotate (All)	
1993-97	
ES250, ES300	.5

ELECTRICAL

	Chilton Time
(G) Headlamps, Aim	
1993-97	
two	.4
four	.6

	Chilton Time
(G) Halogen Headlamp Bulb, Renew	
1993-97	.3
(G) Parking Lamp Lens or Bulb, Renew	
1993-97 LS400	.3
1993-97 ES300	.6
1993-97 SC300, SC400	1.2
1994-97 GS300	.3
(G) Rear Combination Lamp Bulb, Renew	
1993-97 LS400	.4
1993-97 ES250, ES300	.4

	Chilton Time
1993-97 SC300, SC400	.7
1994-97 GS300	.4
(G) Horn, Renew	
1993-97 LS400	.4
1993-97 ES300	.7
1993-97 SC300, SC400	.6
1994-97 GS300	.4
(M) Terminals, Battery, Clean	
1993-97	.3

SCHEDULED MAINTENANCE INTERVALS
(LEXUS ES300, SC300, LS400, SC400 & GS300)

TO BE SERVICED	TYPE OF SERVICE	7.5	15	22.5	30	37.5	45	52.5	60	67.5	75	82.5	90	97.5
Engine oil & filter	R	✓	✓	✓	✓	✓	✓	✓	✓	✓	✓	✓	✓	✓
Air conditioning filter (LS400)②	S/I	✓	✓	✓	✓	✓	✓	✓	✓	✓	✓	✓	✓	✓
Automatic transmission fluid & filter	S/I		✓		✓		✓		✓		✓		✓	
Ball joints & dust covers	S/I		✓		✓		✓		✓		✓		✓	
Bolts & nuts on chassis & body	S/I		✓		✓		✓		✓		✓		✓	
Brake fluid①	S/I		✓		✓		✓		✓		✓		✓	
Brake line pipes & hoses	S/I		✓		✓		✓		✓		✓		✓	
Brake linings & drums	S/I		✓		✓		✓		✓		✓		✓	
Brake pads & discs (front & rear)	S/I		✓		✓		✓		✓		✓		✓	
Differential oil	S/I		✓		✓		✓		✓		✓		✓	
Driveshaft boots (ES300)	S/I		✓		✓		✓		✓		✓		✓	
Manual transmission oil (SC300 & SC400)	S/I		✓		✓		✓		✓		✓		✓	
Steering gear housing oil	S/I		✓		✓		✓		✓		✓		✓	

SCHEDULED MAINTENANCE INTERVALS
(LEXUS ES300, SC300, LS400, SC400 & GS300) (Cont.)

TO BE SERVICED	TYPE OF SERVICE	VEHICLE MILEAGE INTERVAL (x1000)												
		7.5	15	22.5	30	37.5	45	52.5	60	67.5	75	82.5	90	97.5
Steering linkage	S/I		✓		✓		✓		✓		✓		✓	
Air filter	R				✓				✓				✓	
Exhaust pipes & mountings	S/I				✓				✓				✓	
Fuel lines & connections	S/I				✓				✓				✓	
Engine Coolant	R						✓					✓		
Fuel tank cap gasket	R								✓					
Spark plugs	R								✓					
Charcoal canister	S/I								✓					
Drive belts	S/I								✓					
Valve clearance (1995-97)	S/I								✓					

① Replace every 30,000 miles (unless previously replaced).
② Replace every 15,000 miles.
R – Replace   S/I – Service or Inspect

FREQUENT OPERATION MAINTENANCE (SEVERE SERVICE)
   If a vehicle is operated under any of the following conditions it is considered severe service:
- **Extremely dusty areas.**
- **50% or more of the vehicle operation is in 32°C (90°F) or higher temperatures, or constant operation in temperatures below 0°C (32°F).**
- **Prolonged idling (vehicle operation in stop and go traffic).**
- **Frequent short running periods (engine does not warm to normal operating temperatures).**
- **Police, taxi, delivery usage or trailer towing usage.**
Oil & oil filter change – change every 3750 miles.
Ball joints & dust covers - service or inspect every 7500 miles.
Bolts & nuts on chassis & body - service or inspect every 7500 miles.
Brake linings & drums - service or inspect every 7500 miles.
Brake pads & discs (front & rear) - service or inspect every 7500 miles.
Driveshaft boots (ES300) - service or inspect every 7500 miles.
Steering linkage - service or inspect every 7500 miles.
Air filter - service or inspect every 15,000 miles.
Automatic transmission fluid & filter - replace every 15,000 miles.
Differential oil - replace every 15,000 miles.
Exhaust pipes & mountings - service or inspect every 15,000 miles.
Manual transmission oil - replace every 15,000 miles.
Drive belts - service or inspect at 60,000 miles & every 7500 miles thereafter.
Timing belt - replace every 60,000 miles.

MAZDA

ENGINE IDENTIFICATION

Year	Model	Engine Displacement Liters (cc)	Engine Series (ID/VIN)	Fuel System	No. of Cylinders	Engine Type
1993	323	1.6 (1597)	B6E	EFI	4	SOHC
	626	2.0 (1991)	FS	EFI	4	DOHC
	626	2.5 (2496)	KL	EFI	6	DOHC
	929	3.0 (2954)	JE-ZE	EFI	6	DOHC
	Miata	1.6 (1597)	B6ZE	EFI	4	DOHC
	MX3	1.6 (1597)	B6E	EFI	4	SOHC
	MX3	1.8 (1844)	K8D	EFI	6	DOHC
	MX6	2.0 (1991)	FS	EFI	4	DOHC
	MX6	2.5 (2497)	KL	EFI	6	DOHC
	Protege	1.8 (1839)	BPE	EFI	4	SOHC
	Protege	1.8 (1839)	BPD	EFI	4	DOHC
	RX7	1.3 (1308)	13B	EFI	-	Rotary Turbo
1994	323	1.6 (1597)	B6E	EFI	4	SOHC
	626	2.0 (1991)	FS	EFI	4	DOHC
	626	2.5 (2497)	KL	EFI	6	DOHC
	929	3.0 (2954)	JE-ZE	EFI	6	DOHC
	Miata	1.8 (1839)	BPD	EFI	4	DOHC
	MX3	1.6 (1597)	B6ZE	EFI	4	DOHC
	MX3	1.8 (1844)	K8D	EFI	6	DOHC
	MX6	2.0 (1991)	FS	EFI	4	DOHC
	MX6	2.5 (2497)	KL	EFI	6	DOHC
	Protege	1.8 (1839)	BPE	EFI	4	SOHC
	Protege LX	1.8 (1839)	BPD	EFI	4	DOHC
	RX7	1.3 (1308)	13B	EFI	-	Rotary Turbo
1995	323	1.6 (1597)	B6E	EFI	4	SOHC
	626	2.0 (1991)	FS	EFI	4	DOHC
	626	2.5 (2497)	KL	EFI	6	DOHC
	929	3.0 (2954)	JE-ZE	EFI	6	DOHC
	Miata	1.8 (1839)	BPD	EFI	4	DOHC
	Millenia	2.5 (2497)	KLD	EFI	6	DOHC
	Millenia S	2.3 (2255)	KJS	EFI	6	DOHC
	MX3	1.6 (1597)	B6ZE	EFI	4	DOHC
	MX3	1.8 (1844)	K8D	EFI	6	DOHC
	MX6	2.0 (1991)	FS	EFI	4	DOHC
	MX6	2.5 (2497)	KL	EFI	6	DOHC
	Protege	1.8 (1839)	Z5D	EFI	4	SOHC
	Protege LX	1.8 (1839)	BPD	EFI	4	DOHC
	RX7	1.3 (1308)	13B	EFI	-	Rotary Turbo
1996-97	626 DX	2.0 (1991)	FS	EFI	4	DOHC
	626 ES	2.5 (2497)	KL	EFI	6	DOHC
	626 LX	2.0 (1991)	FS	EFI	4	DOHC
	626 LX-V6	2.5 (2497)	KL	EFI	6	DOHC
	Miata	1.8 (1839)	BPD	EFI	4	DOHC
	Millenia	2.5 (2497)	KLD	EFI	6	DOHC
	Millenia S	2.3 (2255)	KJS	EFI	6	DOHC
	MX6	2.0 (1991)	FS	EFI	4	DOHC
	MX6 LS	2.5 (2497)	KL	EFI	6	DOHC
	Protege DX	1.5 (1489)	Z5D	EFI	4	DOHC

ENGINE IDENTIFICATION

Year	Model	Engine Displacement Liters (cc)	Engine Series (ID/VIN)	Fuel System	No. of Cylinders	Engine Type
1996-97	Protege ES	1.8 (1839)	BPD	EFI	4	DOHC
	Protege LX	1.5 (1489)	Z5D	EFI	4	DOHC

EFI - Electronic fuel injection   SOHC - Single overhead camshaft   DOHC - Double overhead camshaft

GENERAL ENGINE SPECIFICATIONS

Year	Engine ID/VIN	Engine Displacement Liters (cc)	Fuel System Type	Net Horsepower @ rpm	Net Torque @ rpm (ft. lbs.)	Bore x Stroke (in.)	Compression Ratio	Oil Pressure @ rpm
1993	13B	1.3 (1308)	EFI Turbo	255 @ 6500	217 @ 5000	3.07x3.29	⑥	43-57@3000
	B6E	1.6 (1597)	EFI	④	⑤	3.07x3.29	⑥	43-57@3000
	B6ZE	1.6 (1597)	EFI	①	②	3.07x3.29	③	43-57@3000
	BPD	1.8 (1839)	EFI	125@6500	114@4500	3.27x3.35	9.0:1	43-57@3000
	BPE	1.8 (1839)	EFI	103@5500	111@4000	3.27x3.35	8.9:1	43-57@3000
	FS	2.0 (1991)	EFI	118@5500	127@4500	3.27x3.62	9.0:1	57-71@3000
	JE-ZE	3.0 (2954)	EFI	195@5750	200@3500	3.54x3.05	9.2:1	53-75@3000
	K8D	1.8 (1844)	EFI	130@6500	115@4500	2.95x2.74	9.2:1	48-71@3000
	KL	2.5 (2497)	EFI	164@5600	160@4800	3.33x2.92	9.2:1	49-71@3000
1994	13B	1.3 (1308)	EFI Turbo	255 @ 6500	217 @ 5000	3.07x3.29	⑥	43-57@3000
	B6E	1.6 (1597)	EFI	82@5000 ⑦	92@2500 ⑧	3.07x3.29	9.3:1	43-57@3000
	B6ZE	1.6 (1597)	EFI	105@6200	100@3600	3.07x3.29	9.0:1	43-57@3000
	BPD	1.8 (1839)	EFI	126@6500	114@4500	3.27x3.35	9.0:1	43-57@3000
	BPE	1.8 (1839)	EFI	103@5500	111@4000	3.27x3.35	8.9:1	43-57@3000
	FS	2.0 (1991)	EFI	118@5500 ⑩	127@4500 ⑫	3.27x3.62	9.0:1	57-71@3000
	JE-ZE	3.0 (2954)	EFI	195@5750	200@3500	3.54x3.05	9.2:1	53-75@3000
	K8D	1.8 (1844)	EFI	130@6500 ⑨	115@4500	2.95x2.74	9.2:1	48-71@3000
	KL	2.5 (2497)	EFI	164@5600 ⑪	160@4800 ⑬	3.33x2.92	9.2:1	49-71@3000
1995	13B	1.3 (1308)	EFI Turbo	255 @ 6500	217 @ 5000	3.07x3.29	⑥	43-57@3000
	B6E	1.6 (1597)	EFI	82@5000 ⑦	92@2500 ⑧	3.07x3.29	9.3:1	43-57@3000
	B6ZE	1.6 (1597)	EFI	105@6200	100@3600	3.07x3.29	9.0:1	43-57@3000
	BPD	1.8 (1839)	EFI	126@6500	114@4500	3.27x3.35	9.0:1	43-57@3000
	BPE	1.8 (1839)	EFI	103@5500	111@4000	3.27x3.35	8.9:1	43-57@3000
	FS	2.0 (1991)	EFI	118@5500 ⑩	127@4500 ⑫	3.27x3.62	9.0:1	57-71@3000
	JE-ZE	3.0 (2954)	EFI	195@5750	200@3500	3.54x3.05	9.2:1	53-75@3000
	K8D	1.8 (1844)	EFI	130@6500 ⑨	115@4500	2.95x2.74	9.2:1	48-71@3000
	KJS	2.3 (2255)	EFI	210@5300	210@3500	3.16x2.92	10.0:1	44-66@3000
	KL	2.5 (2497)	EFI	164@5600 ⑪	160@4800 ⑬	3.33x2.92	9.2:1	49-71@3000
	KLD	2.5 (2497)	EFI	170@5800	160@4800	3.33x2.92	9.2:1	49-71@3000
	Z5D	1.5 (1489)	EFI	⑭	96@4000	2.96x3.29	9.4:1	43-57@3000
1996-97	BPD	1.8 (1839)	EFI	⑮	⑯	3.27x3.35	9.0:1	43-57@3000
	FS	2.0 (1991)	EFI	118@5500	127@4500	3.27x3.62	9.0:1	57-71@3000
	KJS	2.3 (2255)	EFI	210@5300	210@3500	3.16x2.92	10.0:1	44-66@3000
	KL	2.5 (2497)	EFI	164@5500	⑬	3.33x2.92	9.2:1	49-71@3000
	KLD	2.5 (2497)	EFI	170@5800	160@4800	3.33x2.92	9.2:1	49-71@3000
	Z5D	1.5 (1489)	EFI	⑭	96@4000	2.96x3.29	9.4:1	43-57@3000

EFI - Electronic fuel injection
2 BBL - 2 barrel carburetor
① Manual transmission: 116@6500
   Automatic transmission: 105@6000
② Manual transmission: 100@5500
   Automatic transmission: 100@4000
③ Manual transmission: 9.4:1
   Automatic transmission: 9.0:1
④ 323: 82@5000
   MX3: 88@5000

⑤ 323: 92@2500
   MX3: 98@4000
⑥ 323: 9.3:1
   MX3: 9.0:1
⑦ California: 88@5000
⑧ California: 98@4000
⑨ California: 128@6000
⑩ California: 114@5500
⑪ California: 160@5500
⑫ California: 129@4500

⑬ California and New York: 156@5000
   Except California and New York: 160@4800
⑭ California and New York: 90@5500
   Except California and New York: 92@5500
⑮ Protege: 122@6000, Miata: 133@6500
⑯ Protege: 117@4000, Miata: 114@5000

GASOLINE ENGINE TUNE-UP SPECIFICATIONS

Year	Engine ID/VIN	Engine Displacement Liters (cc)	Spark Plugs Gap (in.)	Ignition Timing (deg.) MT	Ignition Timing (deg.) AT	Fuel Pump (psi)	Idle Speed (rpm) MT	Idle Speed (rpm) AT	Valve Clearance In.	Valve Clearance Ex.
1993	13B	1.3 (1308)	0.056	(10)	(10)	NA	750	750	NA	NA
	B6E	1.6 (1597)	0.041	6-8B (2),(6)	6-8B (2),(6)	30-38 (1)	750 (2),(4)	750 (2),(4)	HYD	HYD
	B6ZE	1.6 (1597)	0.041	9-11B (3)	7-9B (3)	30-38 (1)	750	750	HYD	HYD
	BPD	1.8 (1839)	0.041	9-11B (2)	9-11B (2)	30-38 (1)	750 (2),(4)	750 (2),(4)	HYD	HYD
	BPE	1.8 (1839)	0.041	4-6B (2)	4-6B (2)	30-38 (1)	750 (2),(5)	750 (2),(5)	HYD	HYD
	FS	2.0 (1991)	0.041	11-13B (2)	11-13B (2)	30-38 (1)	700 (2)	700 (2)	HYD	HYD
	JE-ZE	3.0 (2954)	0.041	11-13B (2)	11-13B (2)	30-38 (1)	700 (5)	700 (5)	HYD	HYD
	K8D	1.8 (1844)	0.041	9-11B (2)	9-11B (2)	30-38 (1)	670 (2)	670 (2)	HYD	HYD
	KL	2.5 (2497)	0.041	9-11B (2)	9-11B (2)	30-38 (1)	670 (2)	650 (2)	HYD	HYD
1994	13B	1.3 (1308)	0.056	(10)	(10)	NA	750	750	NA	NA
	B6E	1.6 (1597)	0.041	6-8B (2)	6-8B (2)	30-38 (1)	750 (2),(4)	750 (2),(4)	HYD	HYD
	B6ZE	1.6 (1597)	0.041	9-11B (2)	9-11B (2)	30-38 (1)	700 (3)	750 (3)	HYD	HYD
	BPD	1.8 (1839)	0.041	9-11B (2),(7)	9-11B (2),(7)	30-38 (1)	750 (2),(4)	750 (2),(4)	HYD	HYD
	BPE	1.8 (1839)	0.041	4-6B (2)	4-6B (2)	30-38 (1)	750 (2),(4)	750 (2),(4)	HYD	HYD
	FS	2.0 (1991)	0.041	11-13B (2)	11-13B (2)	30-38 (1)	700 (2)	700 (2)	HYD	HYD
	JE-ZE	3.0 (2954)	0.041	11-13B (2)	11-13B (2)	30-38 (1)	700 (2),(5)	700 (2),(5)	HYD	HYD
	K8D	1.8 (1844)	0.041	9-11B (2)	9-11B (2)	30-38 (1)	700 (2)	750 (2)	HYD	HYD
	KL	2.5 (2497)	0.041	9-11B (2)	9-11B (2)	30-38 (1)	670 (2)	650 (2)	HYD	HYD
1995	13B	1.3 (1308)	0.056	(10)	(10)	NA	750	750	NA	NA
	B6E	1.6 (1597)	0.041	6-8B (2)	6-8B (2)	30-38 (1)	750 (2),(4)	750 (2),(4)	HYD	HYD
	B6ZE	1.6 (1597)	0.041	9-11B (2)	9-11B (2)	30-38 (1)	700 (3)	750 (3)	HYD	HYD
	BPD	1.8 (1839)	0.041	9-11B (2),(7)	9-11B (2),(7)	30-38 (1)	750 (2),(4)	750 (2),(4)	HYD	HYD
	BPE	1.8 (1839)	0.041	4-6B (2)	4-6B (2)	30-38 (1)	750 (2),(4)	750 (2),(4)	HYD	HYD
	FS	2.0 (1991)	0.041	11-13B (2)	11-13B (2)	30-38 (1)	700	700	HYD	HYD
	JE-ZE	3.0 (2954)	0.041	11-13B (2)	11-13B (2)	30-38 (1)	700 (2),(5)	700 (2),(5)	HYD	HYD
	K8D	1.8 (1844)	0.041	9-11B (2)	9-11B (2)	30-38 (1)	700 (3)	750 (3)	HYD	HYD
	KJS	2.3 (2255)	0.030	6B (2)	6B (2)	39-48 (1)	670	650	0.011	0.011
	KL	2.5 (2497)	0.041	9-11B (2)	9-11B (2)	30-38 (1)	670 (2)	650 (2)	HYD	HYD
	KLD	2.5 (2497)	0.041	9-11B (2)	9-11B (2)	39-45 (1)	670	650	HYD	HYD
	Z5D	1.5 (1489)	0.041	6-18B	6-18B	29-34 (1)	650-750	700-800	0.011	0.011
1996-97	BPD	1.8 (1839)	0.041	(2), (9)	(2), (9)	39-45 (1)	(2), (8)	(2), (8)	HYD	HYD
	FS	2.0 (1991)	0.041	11-13B (2)	6-18B (2)	37-46 (1)	650-750 (2)	650-750 (2)	HYD	HYD
	KJS	2.3 (2255)	0.030	6B (2)	6B (2)	39-48 (1)	600-700	600-700	0.011	0.011
	KL	2.5 (2497)	0.041	9-11B (2)	9-11B (2)	39-45 (1)	600-700 (2)	600-700 (2)	HYD	HYD
	KLD	2.5 (2497)	0.041	9-11B (2)	9-11B (2)	39-45 (1)	600-700	600-700	HYD	HYD
	Z5D	1.5 (1489)	0.041	6-18B	6-18B	29-34 (1)	650-750	700-800	0.011	0.011

NOTE: The Vehicle Emission Control Information label often reflects specification changes made during production. The label figures must be used if they differ from those in this chart.

B - Before top dead center

HYD - Hydraulic

① Pressure indicated is with gauge in-line, regulator vacuum hose connected and engine idling
② Data link connector terminal 10 grounded
③ With system selector test switch on self test
④ Canadian vehicles: With parking brake applied

⑤ Vehicle in park
⑥ MX3: 9-11 before top dead center
⑦ Miata/MX5: 10 before top dead center
⑧ Protege: 700-800
Miata: 800-900

⑨ Protege: 0
Miata: 9-11B
⑩ Electronic Distributor: 5A Leading, 20A Trailing

CAPACITIES

Year	Model	Engine ID/VIN	Engine Displacement Liters (cc)	Engine Oil with Filter	Transmission (pts.) 4-Spd	5-Spd	Auto.	Transfer Case (pts.)	Drive Axle Front (pts.)	Rear (pts.)	Fuel Tank (gal.)	Cooling System (qts.)
1993	323	B6E	1.6 (1597)	3.6	-	5.6	13.4	-	②	-	13.2	5.3 ①
	626	FS	2.0 (1991)	3.9	-	5.8	18.6	-	②	-	15.5	7.4
	626	KL	2.5 (2496)	5.2	-	5.8	18.6	-	②	-	15.5	7.9
	929	JE-ZE	3.0 (2954)	5.7	-	-	18.2	-	-	2.8	18.5	③
	Miata	B6ZE	1.6 (1597)	3.8	-	4.2	14.2	-	-	1.4	11.9	6.3
	MX3	B6E	1.6 (1597)	3.6	-	5.6	13.4	-	②	-	13.2	6.3
	MX3	K8D	1.8 (1844)	5.2	-	5.7	12.2	-	②	-	13.2	7.9
	MX6	FS	2.0 (1991)	3.9	-	5.8	18.6	-	②	-	15.5	7.4
	MX6	KL	2.5 (2496)	5.2	-	5.8	18.6	-	②	-	15.5	7.9
	Protege	BPE	1.8 (1839)	4.2	-	5.6	13.4	-	②	-	14.5	5.3 ①
	Protege	BPD	1.8 (1839)	4.2	-	7.1	13.4	-	②	-	14.5	5.3 ①
	RX7	13B	1.3 (1308)	5.2 ④	-	5.2	18.2	-	-	2.8	20.1	9.3 ①
1994	323	B6E	1.6 (1597)	3.6	-	5.6	13.4	-	②	-	13.2	5.3 ①
	626	FS	2.0 (1991)	3.9	-	5.8	14.6	-	②	-	15.5	7.4
	626	KL	2.5 (2496)	5.2	-	5.8	18.6	-	②	-	15.5	7.9
	929	JE-ZE	3.0 (2954)	5.6	-	-	18.2	-	-	2.8	18.5	③
	Miata	BPD	1.8 (1839)	3.8	-	4.2	15.4	-	-	2.1	12.7	6.3
	MX3	B6ZE	1.6 (1597)	3.2	-	5.6	12.2	-	②	-	13.2	6.3
	MX3	K8D	1.8 (1844)	5.1	-	5.6	12.2	-	②	-	13.2	7.9
	MX6	FS	2.0 (1991)	3.9	-	5.8	14.6	-	②	-	15.5	7.4
	MX6	KL	2.5 (2496)	5.2	-	5.8	18.6	-	②	-	15.5	7.9
	Protege	BPE	1.8 (1839)	3.8	-	5.6	13.4	-	②	-	14.5	5.3 ①
	Protege LX	BPD	1.8 (1839)	3.8	-	6.6	13.4	-	②	-	14.5	5.3 ①
	RX7	13B	1.3 (1308)	5.6	-	5.2	18.2	-	-	2.6	⑤	9.3 ①
1995	323	B6E	1.6 (1597)	3.6	-	5.6	13.4	-	②	-	13.2	5.3 ①
	626	FS	2.0 (1991)	3.7	-	5.8	17.6	-	②		15.5	7.4
	626	KL	2.5 (2497)	4.2	-	5.8	18.6	-	②		15.5	7.9
	929	JE	3.0 (2954)	5.3	-	-	18.2	-	-	2.8	18.5	③
	Miata	BP	1.8 (1839)	4.0	-	4.2	15.4	-	-	2.1	12.7	6.3
	Millenia	KL	2.5 (2497)	4.2	-	-	18.6	-	②		18.0	7.4
	Millenia S	KJ	2.3 (2255)	4.3	-	-	18.6	-	②		18.0	7.4
	MX3	B6	1.6 (1598)	3.5	-	5.6	12.2	-	②	-	13.2	6.3
	MX3	K8	1.8 (1844)	5.2	-	5.6	12.2	-	②	-	13.2	7.9
	MX6	FS	2.0 (1991)	3.7	-	5.8	17.6	-	②		15.5	7.4
	MX6	KL	2.5 (2497)	4.2	-	5.8	18.6	-	②		15.5	7.9
	Protege	Z5	1.5 (1498)	3.7	-	5.6	10.4	-	②	-	14.5 ④	6.3 ①
	Protege LX	BP	1.8 (1839)	4.0	-	5.6	10.4	-	②	-	14.5 ④	6.3 ①
	RX7	13B	1.3 (1308)	4.0	-	5.2	18.2	-	-	2.7	⑤	9.2 ①
1996-97	626 DX	FS	2.0 (1991)	3.7	-	5.8	18.4	-	②	-	15.5	7.4
	626 ES	KL	2.5 (2497)	4.2	-	5.8	18.6	-	②	-	15.5	7.9
	626 LX	FS	2.0 (1991)	3.7	-	5.8	18.4	-	②	-	15.5	7.4
	626 LX-V6	KL	2.5 (2497)	4.2	-	5.8	18.6	-	②	-	15.5	7.9
	Miata	BP	1.8 (1839)	4.0	-	4.2	13.5	-	②	2.1	12.7	6.3
	Millenia	KL	2.5 (2497)	4.2	-	-	16.9	-	②	-	18.0	7.4
	Millenia S	KJ	2.3 (2255)	4.3	-	-	16.9	-	②	-	18.0	7.4
	MX6	FS	2.0 (1991)	3.7	-	5.8	18.4	-	②	-	15.5	7.4
	MX6	KL	2.5 (2497)	4.2	-	5.8	18.6	-	②	-	15.5	7.9
	Protege DX	Z5	1.5 (1498)	3.7	-	5.6	11.3	-	②	-	14.5 ⑥	6.3
	Protege ES	BP	1.8 (1839)	4.0	-	5.6	11.3	-	②	-	14.5 ⑥	6.3
	Protege LX	Z5	1.5 (1498)	3.7	-	5.6	11.3	-	②	-	14.5 ⑥	6.3

① Automatic transmission: 6.3 qts.
② Included in transaxle
③ With heater: 9.9 qts.
Without heater: 9.3 qts.
④ With RI package: 5.7 qts.
⑤ Manual transmission: 20.1 gals.
Automatic transmission: 18.5 gals.
⑥ California and New York: 13.2 gals.

VALVE SPECIFICATIONS

Year	Engine ID/VIN	Engine Displacement Liters (cc)	Seat Angle (deg.)	Face Angle (deg.)	Maximum out of Square (in.)	Spring Free Length (in.)	Stem-to-Guide Clearance (in.)		Stem Diameter (in.)	
							Intake	Exhaust	Intake	Exhaust
1993	B6E	1.6 (1597)	45	45	0.060	1.7188	0.0010-0.0024	0.0011-0.0026	0.2744-0.2750	0.2742-0.2748
	B6E ⑥	1.6 (1597)	45	45	③	⑦	0.0010-0.0024	0.0012-0.0026	0.2350-0.2356	0.2348-0.2354
	B6ZE	1.6 (1597)	45	45	①	②	0.0010-0.0024	0.0012-0.0026	0.2350-0.2356	0.2348-0.2354
	BPD	1.8 (1839)	45	45	0.064	1.821	0.0010-0.0024	0.0012-0.0026	0.2350-0.2356	0.2348-0.2354
	BPE	1.8 (1839)	45	45	③	④	0.0010-0.0024	0.0011-0.0026	0.2350-0.2356	0.2348-0.2354
	FS	2.2 (2184)	45	45	0.061	⑤	0.0010-0.0024	0.0012-0.0026	0.2351-0.2356	0.2349-0.2354
	JE-ZE	3.0 (2954)	45	45	0.060	1.720	0.0010-0.0024	0.0012-0.0026	0.2350-0.2356	0.2348-0.2354
	K8D	1.8 (1844)	45	45	0.064	1.847	0.0010-0.0024	0.0012-0.0026	0.2350-0.2356	0.2348-0.2354
	KL	2.5 (2496)	45	45	0.064	1.437	0.0010-0.0023	0.0012-0.0025	0.2351-0.2356	0.2351-0.2356
1994	B6E	1.6 (1597)	45	45	0.060	⑧	0.0010-0.0023	0.0012-0.0025	⑨	⑩
	B6ZE	1.6 (1597)	45	45	①	②	0.0010-0.0024	0.0012-0.0026	0.2350-0.2356	0.2348-0.2354
	BPD	1.8 (1839)	45	45	0.064	1.821	0.0008-0.0024	0.0012-0.0023	0.2350-0.2356	0.2348-0.2354
	BPE	1.8 (1839)	45	45	③	④	0.0010-0.0023	0.0012-0.0025	0.2350-0.2356	0.2348-0.2354
	FS	2.0 (1991)	45	45	0.061	1.437	0.0010-0.0024	0.0012-0.0026	0.2351-0.2356	0.2349-0.2354
	JE-ZE	3.0 (2954)	45	45	0.060	1.720	0.0010-0.0024	0.0012-0.0026	0.2350-0.2356	0.2348-0.2354
	K8D	1.8 (1844)	45	45	0.064	1.847	0.0010-0.0024	0.0012-0.0026	0.2350-0.2356	0.2348-0.2354
	KL	2.5 (2497)	45	45	0.064	1.437	0.0010-0.0023	0.0012-0.0025	0.2351-0.2356	0.2351-0.2356
1995	B6	1.6 (1597)	45	45	0.0670	1.570	0.0010-0.0023	0.0012-0.0025	0.2351-0.2356	0.2349-0.2354
	BPD	1.8 (1839)	45	45	0.0640	1.560	0.0010-0.0023	0.0012-0.0025	0.2351-0.2356	0.2349-0.2354
	FS	2.0 (1991)	45	45	0.0610	1.437	0.0010-0.0023	0.0012-0.0025	0.2351-0.2356	0.2349-0.2354
	JE-ZE	3.0 (2954)	45	45	0.0600	1.723	0.0010-0.0023	0.0012-0.0025	0.2351-0.2356	0.2349-0.2354
	K8D	1.8 (1844)	45	45	0.064	1.847	0.0010-0.0024	0.0012-0.0026	0.2350-0.2356	0.2348-0.2354
	KJS	2.3 (2255)	NA	NA	NA	NA	NA	NA	NA	NA
	KL	2.5 (2497)	45	45	0.0642	1.437	0.0010-0.0023	0.0012-0.0025	0.2351-0.2356	0.2349-0.2354
	Z5D	1.5 (1489)	45	45	0.0520	1.240	0.0010-0.0023	0.0012-0.0025	0.2154-0.2159	0.2152-0.2157

VALVE SPECIFICATIONS

Year	Engine ID/VIN	Engine Displacement Liters (cc)	Seat Angle (deg.)	Face Angle (deg.)	Maximum out of Square (in.)	Spring Free Length (in.)	Stem-to-Guide Clearance (in.)		Stem Diameter (in.)	
							Intake	Exhaust	Intake	Exhaust
1996-97	BPD	1.8 (1839)	45	45	50.2-2.57@ 1.56	1.560 ⑪	0.0010-0.0023	0.0012-0.0025	0.2351-0.2356	0.2349-0.2354
	FS	2.0 (1991)	45	45	39.7-44.9@ 1.44	1.732	0.0010-0.0023	0.0012-0.0025	0.2351-0.2356	0.2349-0.2354
	KJS	2.3 (2255)	NA	45	42-46@1.41	1.413	0.0010-0.0023	0.0012-0.0025	0.2351-0.2356	0.2349-0.2354
	KL ⑬	2.5 (2497)	45	45	52.3-59.1@ 1.52	1.847 ⑫	0.0010-0.0023	0.0012-0.0025	0.2351-0.2356	0.2349-0.2354
	KLD ⑭	2.5 (2497)	45	45	⑮	⑯	0.0010-0.0023	0.0012-0.0025	0.2351-0.2356	0.2349-0.2354
	Z5D	1.5 (1489)	45	45	29-32@1.24	1.240	0.0010-0.0023	0.0012-0.0025	0.2154-0.2159	0.2152-0.2157

NA - Not Available

① Intake: 0.0661 Exhaust: 0.0665
② Intake: 1.850-1.890 Exhaust: 1.862-1.902
③ Intake: 0.063 Exhaust: 0.060
④ Intake: 1.815 Exhaust: 1.717
⑤ Intake: 1.902-1.949 Exhaust: 1.937-1.984
⑥ MX3
⑦ Intake: 1.816 Exhaust: 1.687
⑧ Intake: California: 1.816 Except California: 1.719 Exhaust: California: 1.687 Except California: 1.719
⑨ California: 0.2351-0.2356 Except California: 0.2745-0.2749
⑩ California: 0.2349-0.2354 Except California: 0.2743-0.2748
⑪ Free length: 1.821
⑫ Free length only
⑬ 626 and MX6
⑭ Millenia
⑮ Intake: 53-60@1.46 Exhaust: 52-59@1.52
⑯ Intake: 1.457 Exhaust: 1.524

TORQUE SPECIFICATIONS
All readings in ft. lbs.

Year	Engine ID/VIN	Engine Displacement Liters (cc)	Cylinder Head Bolts	Main Bearing Bolts	Rod Bearing Bolts	Crankshaft Damper Bolts	Flywheel Bolts	Manifold Intake	Manifold Exhaust	Spark Plugs	Lug Nut
1993	13B	1.3 (1308)	NA	NA	NA	NA	NA	12-16	48-57	NA	NA
	B6E	1.6 (1597)	56-60	40-43	⑤	116-123	71-76	14-19	12-17	11-17	65-87
	B6ZE	1.6 (1597)	56-60	40-43	37-40	116-123	71-76	14-19	28-34	11-17	65-87
	BPD	1.8 (1839)	56-60	40-43	35-37	116-123	71-76	14-19	28-34	11-17	65-87
	BPE	1.8 (1839)	56-60	40-43	36-38	116-123	71-76	14-19	12-17	11-17	65-87
	FS	2.0 (1991)	⑨	⑩	⑪	116-123	71-76	14-18	④	11-16	65-87
	JE-ZE	3.0 (2954)	①	②	③	116-123	76-81	14-19	16-21	10-13	65-87
	K8D	1.8 (1844)	⑧	⑥	⑦	116-123	45-50	14-19	14-19	11-17	65-87
	KL	2.5 (2496)	⑧	⑥	⑪	116-123	45-49	14-18	14-18	11-16	65-87
1994	13B	1.3 (1308)	NA	NA	NA	NA	NA	12-16	48-57	NA	NA
	B6E	1.6 (1597)	56-60	40-43	⑤	116-123	71-76	14-19	12-17	11-17	65-87
	B6ZE	1.6 (1597)	56-60	40-43	37-40	116-123	71-76	14-19	28-34	11-17	65-87
	BPD	1.8 (1839)	56-60	40-43	35-37	116-123	71-76	14-19	28-34	11-17	65-87
	BPE	1.8 (1839)	56-60	40-43	36-38	116-123	71-76	14-19	12-17	11-17	65-87
	FS	2.0 (1991)	⑨	⑩	⑪	116-123	71-76	14-18	④	11-16	65-87
	JE-ZE	3.0 (2954)	①	②	③	116-123	76-81	14-19	16-21	10-13	65-87
	K8D	1.8 (1844)	⑧	⑥	⑦	116-123	45-50	14-19	14-19	11-17	65-87
	KL	2.5 (2496)	⑧	⑥	⑪	116-123	45-49	14-18	14-18	11-16	65-87
1995	13B	1.3 (1308)	NA	NA	NA	NA	NA	12-16	48-57	NA	NA
	B6E	1.6 (1597)	56-60	40-43	⑤	116-123	71-76	14-19	12-17	11-17	65-87
	B6ZE	1.6 (1597)	56-60	40-43	37-40	116-123	71-76	14-19	28-34	11-17	65-87
	BPD	1.8 (1839)	56-60	40-43	35-37	116-123	71-76	14-19	28-34	11-17	65-87
	BPE	1.8 (1839)	56-60	40-43	36-38	116-123	71-76	14-19	12-17	11-17	65-87
	FS	2.0 (1991)	⑨	⑩	⑪	116-123	71-76	14-18	④	11-16	65-87

TORQUE SPECIFICATIONS
All readings in ft. lbs.

Year	Engine ID/VIN	Engine Displacement Liters (cc)	Cylinder Head Bolts	Main Bearing Bolts	Rod Bearing Bolts	Crankshaft Damper Bolts	Flywheel Bolts	Manifold Intake	Manifold Exhaust	Spark Plugs	Lug Nut
1995	JE-ZE	3.0 (2954)	①	②	③	116-123	76-81	14-19	16-21	10-13	65-87
	K8D	1.8 (1844)	⑧	⑥	⑦	116-123	45-50	14-19	14-19	11-17	65-87
	KJS	2.3 (2255)	⑧	NA	NA	116-122	45-49	14-18	12-17	11-16	65-87
	KL	2.5 (2496)	⑧	⑥	⑪	116-123	45-49	14-18	14-18	11-16	65-87
	KLD	2.5 (2496)	⑧	⑥	⑪	116-122	45-49	14-18	14-18	11-16	65-87
	Z5D	1.5 (1489)	⑨	40-43	22-25	116-122	71-76	14-18	14-16	11-16	65-87
1996-97	BPD	1.8 (1839)	56-60	40-43	35-36	116-122	71-76	14-18	29-34	11-16	65-87
	FS	2.0 (1991)	⑨	⑩	⑪	116-122	71-76	14-18	15-20	11-16	65-87
	KJS	2.3 (2255)	⑧	⑫	⑪	116-122	45-49	14-18	14-18	11-16	65-87
	KL	2.5 (2496)	⑧	⑫	⑪	116-122	45-49	14-18	14-18	11-16	65-87
	KLD	2.5 (2496)	⑧	⑥	⑪	116-122	45-49	14-18	14-18	11-16	65-87
	Z5D	1.5 (1489)	⑨	40-43	22-25	116-122	71-76	14-18	14-16	11-16	65-87

① Step 1: 14 ft. lbs.
Step 2: Turn each bolt 90 degrees
Step 3: Repeat Step 2

② Step 1: 14 ft. lbs.
Step 2: Turn each bolt 90 degrees
Step 3: Turn each bolt 45 degrees

③ Step 1: 22 ft. lbs.
Step 2: Turn each nut 90 degrees

④ Nuts: 15-20 ft. lbs.
Bolts: 12-16 ft. lbs.

⑤ 323: 35-38 ft. lbs.
MX3: 35-37 ft. lbs.

⑥ Step 1: Inner bolts: 17-18 ft. lbs.; Outer bolts: 13-15 ft. lbs.
Step 2: Inner bolt Nos. 1-3: Turn each bolt 70 degrees
Step 3: Inner bolt No. 4: Turn each bolt 80 degrees
Step 4: Outer bolts: Turn each bolt 60 degrees
Step 5: Repeat Step 2

⑦ Step 1: 16-19 ft. lbs.
Step 2: Turn each bolt 90 degrees
Step 3: Repeat Step 2

⑧ Step 1: 17-19 ft. lbs.
Step 2: Turn each bolt 90 degrees
Step 3: Repeat Step 2

⑨ Step 1: 12.7-16.2 ft. lbs.
Step 2: Turn each bolt 90 degrees
Step 3: Repeat Step 2

⑩ Step 1: 12.7-16.2 ft. lbs.
Step 2: Turn each bolt 90 degrees

⑪ Step 1: 16.3-19.8 ft. lbs.
Step 2: Plus 90 degrees

⑫ Step 1: Inner bolts: 17-19 ft. lbs.
Step 2: Outer bolts: 13.5-15.5 ft. lbs.
Step 3: Inner bolt Nos. 1-3: Plus 70 degrees
Step 4: Inner bolt No. 4: Plus 80 degrees
Step 5: Outer bolts: Plus 60 degrees
Step 6: Repeat Steps 3-5

BRAKE SPECIFICATIONS
All measurements in inches unless noted

Year	Model		Master Cylinder Bore	Brake Disc Original Thickness	Brake Disc Minimum Thickness	Maximum Runout	Brake Drum Diameter Original Inside Diameter	Brake Drum Diameter Max. Wear Limit	Brake Drum Diameter Maximum Machine Diameter	Minimum Lining Thickness Front	Minimum Lining Thickness Rear
1993	323	F	0.875	0.870	0.790	0.004	-	-	-	0.080	-
		R	-	0.350	0.280	0.004	7.87	7.91	NA	-	0.040
	Protege	F	0.875	0.870	0.790	0.004	-	-	-	0.080	-
		R	-	0.350	0.280	0.004	7.87	7.91	NA	-	0.040
	Miata	F	0.875	0.710	0.630	0.004	-	-	-	0.040	-
		R	-	0.350	0.280	0.004	-	-	-	-	0.040
	MX3	F	①	0.870	0.790	0.004	-	-	-	0.080	-
		R	-	0.350	0.310	0.004	7.87	7.91	NA	-	0.040
	626	F	0.937	0.940	0.870	0.004	-	-	-	0.040	-
		R	-	0.390	0.310	0.004	9.00	9.06	NA	-	0.040
	MX6	F	0.937	0.940	0.870	0.004	-	-	-	0.040	-
		R	-	0.390	0.310	0.004	9.00	9.06	NA	-	0.040
	929	F	NA	0.940	0.870	0.004	-	-	-	0.040	-
		R	-	0.710	0.630	0.004	-	-	-	-	0.040
	RX7	F	NA	0.870	0.790	0.004	-	-	-	0.040	-
		R	-	②	③	0.004	-	-	-	-	0.040
1994	323	F	0.875	0.870	0.790	0.004	-	-	-	0.080	-
		R	-	0.350	0.280	0.004	7.90	7.91	NA	-	0.040

BRAKE SPECIFICATIONS
All measurements in inches unless noted

Year	Model		Master Cylinder Bore	Brake Disc Original Thickness	Brake Disc Minimum Thickness	Maximum Runout	Brake Drum Diameter Original Inside Diameter	Brake Drum Diameter Max. Wear Limit	Brake Drum Diameter Maximum Machine Diameter	Minimum Lining Thickness Front	Minimum Lining Thickness Rear
1994	Protege	F	0.875	0.870	0.790	0.004	-	-	-	0.080	-
		R	-	0.350	0.280	0.004	7.90	7.91	NA	-	0.040
	Miata	F	0.875	0.790	0.710	0.004	-	-	-	0.040	-
		R	-	0.350	0.310	0.004	-	-	-	-	0.040
	MX3	F	①	0.870	0.790	0.004	-	-	-	0.080	-
		R	-	0.350	0.310	0.004	7.87	7.93	NA	-	0.040
	626	F	0.937	0.940	0.870	0.004	-	-	-	0.040	-
		R	-	0.390	0.310	0.004	9.00	NA	NA	-	0.040
	MX6	F	0.937	0.940	0.870	0.004	-	-	-	0.040	-
		R	-	0.390	0.310	0.004	9.00	NA	NA	-	0.040
	929	F	NA	0.940	0.870	0.004	-	-	-	0.040	-
		R	-	0.710	0.630	0.004	-	-	-	-	0.040
	RX7	F	NA	0.870	0.790	0.004	-	-	-	0.040	-
		R	-	0.790	0.710	0.004	-	-	-	-	0.040
1995	RX7	F	NA	0.870	0.790	0.004	-	-	-	0.040	-
		R	-	0.790	0.710	0.004	-	-	-	-	0.040
	Protege	F	0.937	0.870	0.790	0.002	-	-	-	0.080	-
		R	-	0.354	0.276	0.002	7.87	7.91	NA	-	0.040
	MX3	F	0.875	0.870	0.790	0.004	-	-	-	0.080	-
		R	-	0.354	0.310	0.004	7.87	7.93	NA	-	0.040
	Miata	F	0.875	0.790	0.710	0.004	-	-	-	0.040	-
		R	-	0.350	0.310	0.004	-	-	-	-	0.040
	626	F	0.937	0.940	0.870	0.004	-	-	-	0.080	-
		R	-	0.390	0.310	0.004	9.00	NA	NA	-	0.040
	MX6	F	0.937	0.940	0.870	0.004	-	-	-	0.080	-
		R	-	0.390	0.310	0.004	9.00	NA	NA	-	0.040
	929	F	NA	0.950	0.870	0.004	-	-	-	0.040	-
		R	-	0.710	0.630	0.004	-	-	-	-	0.040
	Millenia	F	1.000	1.100	1.020	0.004	-	-	-	0.080	-
		R	-	0.370	0.290	0.004	-	-	-	-	0.080
1996	Protege	F	0.937	0.870	0.790	0.002	-	-	-	0.040	-
		R	-	0.354	0.276	0.002	7.87	7.91	NA	-	0.040
	Miata	F	0.875	0.790	0.710	0.004	-	-	-	0.040	-
		R	-	0.350	0.310	0.004	-	-	-	-	0.040
	626	F	0.937	0.940	0.870	0.002	-	-	-	0.080	-
		R	-	0.390	0.310	0.002	9.00	NA	9.06	-	0.040
	MX6	F	0.937	0.940	0.870	0.002	-	-	-	0.080	-
		R	-	0.390	0.310	0.002	9.00	NA	9.06	-	0.040
	Millenia	F	1.000	1.100	1.020	0.002	-	-	-	0.080	-
		R	-	0.370	0.290	0.002	-	-	-	-	0.080

NA - Not Available
F - Front
R - Rear

① 1.6L engine: 0.875
   1.8L engine: 0.937

② Ventilated disc: 0.790
   Solid disc: 0.390

③ Ventilated disc: 0.710
   Solid disc: 0.310

FREQUENT MAINTENANCE LABOR
MAZDA

The following should be used as a guide when determining the amount of work required for a particular service if taken to a repair shop. In estimating how long a particular Frequent Maintenance Service item should take, please observe the following:
- **Chilton Time** is time that is based on field research and data supplied by the vehicle manufacturer.
- All labor time operations are given in hours and tenths of an hour.
- All labor operations, are to be used as a **guide**.

COOLING

(G) Winterize Cooling System
Includes: Run engine to check for leaks, tighten all hose connections. Test radiator and pressure cap, drain radiator and engine block. Add antifreeze and refill system.
1993-97 5

(G) Belt, Fan Drive, Renew
1993-97 929 6
w/AC add 1

(G) Hoses, Radiator, Renew
Includes: Drain and refill cooling system as required.
1993-97 MX-3
  upper 6
  lower 9
1993 MX-5 Miata
  upper 3
  lower 5
1994-97 MX-5 Miata
  upper 9
  lower 1.0
1995-97 Millenia
  upper 9
  lower
    KJ 1.2
    KL 1.0
1993-94 323, Protege
  upper 3
  lower 5
1995-97 323, Protege
  upper 1.0
  lower 1.2
1993-97 626, MX-6
  upper 8
  lower 1.0
1993-97 929, each 1.0

1993-97 RX-7
  upper 1.1
  lower 1.0

(G) Thermostat, Coolant, Renew
1993-97 MX-3 . . . . 1.3
1993-97 MX-5 Miata 1.0
1995-97 Millenia
  KJ 1.2
  KL 1.1
1993-97 323, Protege 1.0
1993-97 626, MX-6
  4 cyl. 1.5
  w/AC add 3
  V6 . . . . 1.4
1993-97 929 . . . . 1.8
1993-97 RX-7 . . . . 2.5

FUEL

(M) Air Cleaner, Service
1993-97 2

(G) Filter, Fuel, Renew
Piston Engine
  1993-97 3
Rotary Engine
  1993-97 RX-7 6

BRAKES

(G) Bleed Brakes (Four Wheels)
Includes: Add fluid.
  1993-97 8
w/ABS, add 3

(G) Brakes, Adjust (Minor)
Includes: Adjust brakes, fill master cylinder.
  1993-97, two wheels 4

(M) Parking Brake, Adjust
1993-97 3

LUBRICATION SERVICES

(M) Engine Oil & Filter, Renew
Includes: Inspect and correct all fluid levels.
  1993-97 3

WHEELS

(G) Wheel, Renew (One)
1993-97, one 5

(G) Wheel, Balance
1993-97
  one .3
  each adtnl. 2

(G) Wheels, Rotate (All)
1993-97 5

ELECTRICAL

(G) Headlamps, Aim
1993-97
  one side 3
  both sides 5

(G) Halogen Headlamp Bulb, Renew
1993-97 each 2

(G) License Lamp Assy., Renew
1993-97 MX-5 Miata 4
1995-97 Millenia 9
1993-97 323, Protege 4
1993-97 626, MX-6 3
1993-97 929 8
1993-97 RX-7 3

(G) Rear Combination Lamp Bulb, Renew
1993-97, each 3

(G) Horn, Renew
1993-97
  one .4
  each adtnl. 1

SCHEDULED MAINTENANCE INTERVALS
(MAZDA 323, 626, 929, MX3, MX6, RX-7, MIATA, MILLENIA & PROTEGE)

TO BE SERVICED	TYPE OF SERVICE	VEHICLE MILEAGE INTERVAL (x1000)												
		7.5	15	22.5	30	37.5	45	52.5	60	67.5	75	82.5	90	97.5
Engine oil & filter④	R	✓	✓	✓	✓	✓	✓	✓	✓	✓	✓	✓	✓	✓
Air cleaner element	R				✓				✓				✓	
Engine coolant③	R				✓				✓				✓	
Spark plugs (except Millenia KJ engine)	R				✓				✓				✓	
Spark plugs (Millenia KJ engine)	R								✓					
Automatic transaxle fluid	S/I				✓				✓				✓	
Bolts & nuts on chassis & body	S/I				✓				✓				✓	
Brake lines, hoses & connections	S/I				✓				✓				✓	
Cooling system	S/I				✓				✓				✓	
Disc brakes	S/I				✓				✓				✓	
Drive belts (Millenia①)	S/I				✓				✓				✓	
Drive shaft dust boots	S/I				✓				✓				✓	
Exhaust system heat shield	S/I				✓				✓				✓	
Front & rear suspension ball joints	S/I				✓				✓				✓	
Fuel lines & hoses	S/I				✓				✓				✓	
Idle speed⑤	S/I				✓				✓				✓	
Steering operation & linkages	S/I				✓				✓				✓	
Engine timing belt②	R								✓					
Fuel filter	R								✓					
Manual transmission oil (Miata)	R								✓					
Hose & tube for emission	S/I								✓					

① (Millenia KJ engine) - replace every 105,000 miles.
② (Calif.) - inspect every 30,000 miles & replace at 105,000 miles (if not replaced previously).
③ (Millenia) - replace initially at 45,000 miles, & every 30,000 miles thereafter.
④ (RX-7) - change every 5000 miles.
⑤ (RX-7 & 929) - check every 15,000 miles.
R – Replace   S/I – Service or Inspect

SCHEDULED MAINTENANCE INTERVALS
(MAZDA 323, 626, 929, MX3, MX6, RX-7, MIATA, MILLENIA & PROTEGE) (Cont.)

FREQUENT OPERATION MAINTENANCE (SEVERE SERVICE)

If a vehicle is operated under any of the following conditions it is considered severe service:
- Extremely dusty areas.
- 50% or more of the vehicle operation is in 32°C (90°F) or higher temperatures, or constant operation in temperatures below 0°C (32°F).
- Prolonged idling (vehicle operation in stop and go traffic).
- Frequent short running periods (engine does not warm to normal operating temperatures).
- Police, taxi, delivery usage or trailer towing usage.

Oil & oil filter change – change every 5000 miles.
Oil & oil filter change (Puerto Rico) - change every 3000 miles.
Air cleaner element – service or inspect every 15,000 miles.
Automatic transaxle fluid - service or inspect every 15,000 miles.
Bolts & nuts on chassis & body - tighten every 15,000 miles.
Disc brakes - service or inspect every 15,000 miles.

MERCEDES-BENZ

VEHICLE IDENTIFICATION CHART

		Engine Code					Model Year	
Code	Liters	Cu. In. (cc)	Cyl.	Fuel Sys.	Eng. Mfg.		Code	Year
102.985	2.3	139 (2299)	4	CIS-E	MB		P	1993
103.942	2.6	158 (2599)	6	CIS-E	MB		R	1994
103.985	3.0	180 (2960)	6	CIS-E	MB		S	1995
104.941	2.8	171 (2799)	6	HFM	MB		T	1996
104.941 ①	3.6	220 (3606)	6	HFM	MB		V	1997
104.942	2.8	171 (2799)	6	HFM	MB			
104.981	3.0	180 (2962)	6	LH	MB			
104.990	3.2	195 (3199)	6	LH	MB			
104.991	3.2	195 (3199)	6	HFM	MB			
104.992	3.2	195 (3199)	6	HFM	MB			
104.994	3.2	195 (3199)	6	HFM	MB			
104.995	3.2	195 (3199)	6	HFM	MB			
111.961	2.2	134 (2199)	4	HFM	MB			
119.970	5.0	303 (4973)	8	LH	MB			
119.970	5.0	303 (4973)	8	LH	MB			
119.971	4.2	256 (4196)	8	LH	MB			
119.972	5.0	303 (4973)	8	LH	MB			
119.974	5.0	303 (4973)	8	LH	MB			
119.975	4.2	256 (4196)	8	LH	MB			
119.980	5.0	303 (4973)	8	ME	MB			
119.981	4.2	256 (5196)	8	ME	MB			
119.982	5.0	303 (4973)	8	ME	MB			
120.980	6.0	365 (5987)	12	LH	MB			
120.981	6.0	365 (5987)	12	LH	MB			
120.982	6.0	365 (5987)	12	ME	MB			
120.983	6.0	365 (5987)	12	ME	MB			
602.962	2.5	152 (2497)	5	EDS	MB			
603.971	3.5	210 (3449)	6	EDS	MB			

VEHICLE IDENTIFICATION CHART

		Engine Code					Model Year	
Code	Liters	Cu. In. (cc)	Cyl.	Fuel Sys.	Eng. Mfg.		Code	Year
606.910	3.0	183 (2996)	6	EDS	MB			
606.912	3.0	183 (2996)	6	EDS	MB			

CIS-E - Continuous Injection System with electronic controls
EDS - Electronic Diesel System
HFM - Multiport Fuel Injection and Ignition System
ME - Multiport Fuel Injection and Ignition System
① C36

ENGINE IDENTIFICATION

Year	Model	Engine Displacement Liters (cc)	Engine Series (ID/VIN)	Fuel System	No. of Cylinders	Engine Type
1993	190E 2.3	2.3 (2299)	102.985	CIS-E	4	SOHC
	300D 2.5	2.5 (2497)	602.962	EDS	5	SOHC
	190E 2.6	2.6 (2599)	103.942	CIS-E	6	SOHC
	300E 2.8	2.8 (2799)	104.942	HFM	6	DOHC
	300CE 4M	3.0 (2960)	103.985	CIS-E	6	SOHC
	300E 4M	3.0 (2960)	103.985	CIS-E	6	SOHC
	300SL	3.0 (2960)	104.981	LH	6	DOHC
	300CE	3.2 (3199)	104.992	HFM	6	DOHC
	300E	3.2 (3199)	104.992	HFM	6	DOHC
	300SE	3.2 (3199)	104.990	LH	6	DOHC
	300TE	3.2 (3199)	104.992	HFM	6	DOHC
	300SD	3.5 (3449)	603.971	EDS	6	SOHC
	400E	4.2 (4196)	119.975	LH	8	DOHC
	400SEL	4.2 (5196)	119.971	LH	8	DOHC
	500E	5.0 (4973)	119.974	LH	8	DOHC
	500SEC	5.0 (4973)	119.970	LH	8	DOHC
	500SEL	5.0 (4973)	119.970	LH	8	DOHC
	500SL	5.0 (4973)	119.972	LH	8	DOHC
	600SEC	6.0 (5987)	120.980	LH	12	DOHC
	600SEL	6.0 (5987)	120.980	LH	12	DOHC
	600SL	6.0 (5987)	120.981	LH	12	DOHC
1994	C220	2.2 (2199)	111.961	HFM	4	DOHC
	C280	2.8 (2799)	104.941	HFM	6	DOHC
	E320	3.2 (3199)	104.992	HFM	6	DOHC
	S320	3.2 (3199)	104.994	HFM	6	DOHC
	SL320	3.2 (3199)	104.991	HFM	6	DOHC
	S350	3.5 (3449)	603.971	EDS	6	SOHC
	E420	4.2 (4196)	119.975	LH	8	DOHC
	S420	4.2 (5196)	119.971	LH	8	DOHC
	E500	5.0 (4973)	119.974	LH	8	DOHC
	S500	5.0 (4973)	119.970	LH	8	DOHC
	SL500	5.0 (4973)	119.972	LH	8	DOHC
	S600	6.0 (5987)	120.980	LH	12	DOHC
	SL600	6.0 (5987)	120.981	LH	12	DOHC
1995	C220	2.2 (2199)	111.961	HFM	4	DOHC
	C280	2.8 (2799)	104.941	HFM	6	DOHC
	E320	3.2 (3199)	104.992	HFM	6	DOHC
	S320	3.2 (3199)	104.994	HFM	6	DOHC
	SL320	3.2 (3199)	104.991	HFM	6	DOHC

ENGINE IDENTIFICATION

Year	Model	Engine Displacement Liters (cc)	Engine Series (ID/VIN)	Fuel System	No. of Cylinders	Engine Type
1995	S350	3.5 (3449)	603.971	EDS	6	SOHC
	E420	4.2 (4196)	119.975	LH	8	DOHC
	S420	4.2 (5196)	119.971	LH	8	DOHC
	E500	5.0 (4973)	119.974	LH	8	DOHC
	S500	5.0 (4973)	119.970	LH	8	DOHC
	SL500	5.0 (4973)	119.972	LH	8	DOHC
	S600	6.0 (5987)	120.980	LH	12	DOHC
	SL600	6.0 (5987)	120.981	LH	12	DOHC
1996-97	C220	2.2 (2199)	111.961	HFM	4	DOHC
	C280	2.8 (2799)	104.941	HFM	6	DOHC
	C36	2.8 (2799)	104.941	HFM	6	DOHC
	E300	3.0 (2996)	606.912	EDS	6	DOHC
	E320	3.2 (3199)	104.995	HFM	6	DOHC
	S320	3.2 (3199)	104.994	HFM	6	DOHC
	SL320	3.2 (2199)	104.991	HFM	6	DOHC
	S420	4.2 (5196)	119.981	ME	8	DOHC
	S500	5.0 (4973)	119.980	ME	8	DOHC
	SL500	5.0 (4973)	119.982	ME	8	DOHC
	S600	6.0 (5987)	120.982	ME	12	DOHC
	SL600	6.0 (5987)	120.983	ME	12	DOHC

CIS-E - Continuous Injection System with electronic controls
EDS - Electronic Diesel System
HFM - Multiport Fuel Injection

SOHC - Single overhead camshaft
DOHC - Double overhead camshaft
ME - Multiport Fuel Injection

GENERAL ENGINE SPECIFICATIONS

Year	Engine ID/VIN	Engine Displacement Liters (cc)	Fuel System Type	Net Horsepower @ rpm	Net Torque @ rpm (ft. lbs.)	Bore x Stroke (in.)	Compression Ratio	Oil Pressure @ rpm
1993	102.985	2.3 (2299)	CIS-E	130@5100	146@3500	3.76x3.16	9.0:1	①
	602.962	2.5 (2497)	EDS	121@4600	165@2400	3.43x3.31	22.0:1	①
	103.942	2.6 (2599)	CIS-E	158@5800	162@4600	3.26x3.16	9.2:1	①
	104.942	2.8 (2799)	HFM	194@5500	199@3750	3.54x2.89	10.0:1	69.6@2000
	103.985	3.0 (2960)	CIS-E	177@5700	188@4400	3.48x3.16	9.2:1	①
	104.981	3.0 (2960)	LH	228@6300	201@4600	3.48x3.16	10.0:1	①
	104.990	3.2 (3199)	LH	228@5800	229@4100	3.54x3.30	10.0:1	①
	104.992	3.2 (3199)	HFM	217@5500	229@3750	3.54x3.30	10.0:1	69.6@2000
	603.971	3.5 (3449)	EDS	148@4000	232@2000	3.50x3.60	22.0:1	①
	119.975	4.2 (4196)	LH	275@5700	295@3900	3.62x3.11	11.0:1	23.2-72.5@2000
	119.971	4.2 (5196)	LH	275@5700	302@3900	3.62x3.11	11.0:1	23.2-72.5@2000
	119.970	5.0 (4973)	LH	315@5600	347@3900	3.80x3.35	10.0:1	23.2-72.5@2000
	119.972	5.0 (4973)	LH	315@5600	347@3900	3.80x3.35	10.0:1	23.2-72.5@2000
	119.974	5.0 (4973)	LH	315@5600	347@3900	3.80x3.35	10.0:1	23.2-72.5@2000
	120.980	6.0 (5987)	LH	389@5200	420@3800	3.50x3.16	10.0:1	①
	120.981	6.0 (5987)	LH	389@5200	420@3800	3.50x3.16	10.0:1	①
1994	111.961	2.2 (2199)	HFM	148@5500	155@4000	3.54x3.41	9.8:1	43.5-58@2000
	104.941	2.8 (2799)	HFM	194@5500	199@3750	3.54x2.89	10.0:1	69.6@2000
	104.991	3.2 (3199)	HFM	229@5600	232@3750	3.54x3.30	10.0:1	69.6@2000
	104.992	3.2 (3199)	HFM	217@5500	229@3750	3.54x3.30	10.0:1	69.6@2000
	104.994	3.2 (3199)	HFM	228@5600	229@3750	3.54x3.30	10.0:1	69.6@2000
	603.971	3.5 (3449)	EDS	148@4000	232@2000	3.50x3.60	22.0:1	①
	119.971	3.5 (3449)	EDS	275@5700	302@3900	3.62x3.11	11.0:1	23.2-72.5@2000
	119.975	4.2 (5196)	LH	275@5700	295@3900	3.62x3.11	11.0:1	23.2-72.5@2000

GENERAL ENGINE SPECIFICATIONS

Year	Engine ID/VIN	Engine Displacement Liters (cc)	Fuel System Type	Net Horsepower @ rpm	Net Torque @ rpm (ft. lbs.)	Bore x Stroke (in.)	Compression Ratio	Oil Pressure @ rpm
1994	119.970	5.0 (4973)	LH	315@5600	345@3900	3.80x3.35	10.0:1	23.2-72.5@2000
	119.972	5.0 (4973)	LH	315@5600	345@3900	3.80x3.35	10.0:1	23.2-72.5@2000
	119.974	5.0 (4973)	LH	315@5600	347@3900	3.80x3.35	10.0:1	23.2-72.5@2000
	120.980	6.0 (5987)	LH	389@5200	421@3800	3.50x3.16	10.0:1	①
	120.981	6.0 (5987)	LH	389@5200	420@3800	3.50x3.16	10.0:1	①
1995	111.961	2.2 (2199)	HFM	148@5500	155@4000	3.54x3.41	9.8:1	43.5-58@2000
	104.941	2.8 (2799)	HFM	194@5500	199@3750	3.54x2.89	10.0:1	69.6@2000
	104.991	3.2 (3199)	HFM	229@5600	232@3750	3.54x3.30	10.0:1	69.6@2000
	104.992	3.2 (3199)	HFM	217@5500	229@3750	3.54x3.30	10.0:1	69.6@2000
	104.994	3.2 (3199)	HFM	228@5600	229@3750	3.54x3.30	10.0:1	69.6@2000
	603.971	3.5 (3449)	EDS	148@4000	232@2000	3.50x3.60	22.0:1	①
	119.971	3.5 (3449)	EDS	275@5700	302@3900	3.62x3.11	11.0:1	23.2-72.5@2000
	119.975	4.2 (5196)	LH	275@5700	295@3900	3.62x3.11	11.0:1	23.2-72.5@2000
	119.970	5.0 (4973)	LH	315@5600	345@3900	3.80x3.35	10.0:1	23.2-72.5@2000
	119.972	5.0 (4973)	LH	315@5600	345@3900	3.80x3.35	10.0:1	23.2-72.5@2000
	119.974	5.0 (4973)	LH	315@5600	347@3900	3.80x3.35	10.0:1	23.2-72.5@2000
	120.980	6.0 (5987)	LH	389@5200	421@3800	3.50x3.16	10.0:1	①
	120.981	6.0 (5987)	LH	389@5200	420@3800	3.50x3.16	10.0:1	①
1996-97	111.961	2.2 (2199)	HFM	148@5500	155@4000	3.54x3.41	9.8:1	43.5-58@2000
	104.941	2.8 (2799)	HFM	194@5500	199@3750	3.54x2.89	10.0:1	69.6@2000
	606.912	3.0 (2966)	EDS	134@5000	155@2600	3.43x3.31	22.0:1	①
	104.991	3.2 (3199)	HFM	229@5600	232@3750	3.54x3.30	10.0:1	69.6@2000
	104.994	3.2 (3199)	HFM	228@5600	232@3750	3.54x3.30	10.0:1	69.6@2000
	104.995	3.2 (3199)	HFM	217@5500	229@5750	3.54x3.30	10.0:1	NA
	104.941 ②	3.6 (3606)	HFM	268@5750	280@4000	3.64x3.58	10.5:1	NA
	119.981	4.2 (5196)	ME	275@5700	295@3900	3.62x3.11	11.0:1	23.2-72.5@2000
	119.980	5.0 (4973)	ME	315@5600	347@3900	3.80x3.35	10.0:1	23.2-72.5@2000
	119.982	5.0 (4973)	ME	315@5600	345@3900	3.80x3.35	10.0:1	23.2-72.5@2000
	120.982	6.0 (5987)	ME	389@5200	420@3800	3.50x3.16	10.0:1	①
	120.983	6.0 (5987)	ME	389@5200	420@3800	3.50x3.16	10.0:1	①

CIS-E - Continuous Injection System with electronic controls
EDS - Electronic Diesel System
HFM - Multiport Fuel Injection and Ignition System
ME - Multiport Fuel Injection and Ignition System
NA - Not Available
① With engine at operating temperature, oil pressure should be a minimum of 43.5 psi @ idle. When engine is accelerated, oil pressure should increase immediately and reach a minimum pressure of 43.5 psi @ 3000 rpm.
② C36

GASOLINE ENGINE TUNE-UP SPECIFICATIONS

Year	Engine ID/VIN	Engine Displacement Liters (cc)	Spark Plugs Gap (in.)	Ignition Timing (deg.) MT	Ignition Timing (deg.) AT	Fuel Pump (psi)	Idle Speed (rpm) MT	Idle Speed (rpm) AT	Valve Clearance In.	Valve Clearance Ex.
1993	102.985	2.3 (2299)	0.031	8-12B	8-12B	77-80	700-800	700-800	HYD	HYD
	103.942	2.6 (2599)	0.031	7-11B	7-11B	77-80	650-750	650-750	HYD	HYD
	104.942	2.8 (2799)	0.031	-	①	③	-	650-750	HYD	HYD
	103.985	3.0 (2960)	0.031	-	6-11B	77-80	-	650-750	HYD	HYD
	104.981	3.0 (2960)	0.031	6-10B	6-10B	③	650-750	650-750	HYD	HYD
	104.990	3.2 (3199)	0.031	-	①	③	-	650-750	HYD	HYD
	104.992	3.2 (3199)	0.031	-	①	③	-	650-750	HYD	HYD
	119.975	4.2 (4196)	0.031	-	①	③	-	600-750	HYD	HYD
	119.971	4.2 (5196)	0.031	-	①	③	-	600-750	HYD	HYD

GASOLINE ENGINE TUNE-UP SPECIFICATIONS

Year	Engine ID/VIN	Engine Displacement Liters (cc)	Spark Plugs Gap (in.)	Ignition Timing (deg.) MT	Ignition Timing (deg.) AT	Fuel Pump (psi)	Idle Speed (rpm) MT	Idle Speed (rpm) AT	Valve Clearance In.	Valve Clearance Ex.
1993	119.970	5.0 (4973)	0.031	-	①	③	-	600-750	HYD	HYD
	119.972	5.0 (4973)	0.031	-	①	③	-	600-750	HYD	HYD
	119.974	5.0 (4973)	0.031	-	①	③	-	600-750	HYD	HYD
	120.980	6.0 (5987)	0.031	-	①	③	-	600-750	HYD	HYD
	120.981	6.0 (5987)	0.031	-	①	③	-	600-750	HYD	HYD
1994	111.961	2.2 (2199)	0.031	-	①	③	-	700-800	HYD	HYD
	104.941	2.8 (2799)	0.031	-	7-11B	③	-	650-750	HYD	HYD
	104.991	3.2 (3199)	0.031	-	①	③	-	650-750	HYD	HYD
	104.992	3.2 (3199)	0.031	-	①	③	-	650-750	HYD	HYD
	104.994	3.2 (3199)	0.031	-	①	③	-	650-750	HYD	HYD
	119.975	4.2 (4196)	0.031	-	①	③	-	600-750	HYD	HYD
	119.971	4.2 (5196)	0.031	-	①	③	-	600-750	HYD	HYD
	119.970	5.0 (4973)	0.031	-	①	③	-	600-750	HYD	HYD
	119.972	5.0 (4973)	0.031	-	①	③	-	600-750	HYD	HYD
	119.974	5.0 (4973)	0.031	-	①	③	-	600-750	HYD	HYD
	120.980	6.0 (5987)	0.031	-	①	③	-	600-750	HYD	HYD
	120.981	6.0 (5987)	0.031	-	①	③	-	600-750	HYD	HYD
1995	111.961	2.2 (2199)	0.031	-	①	③	-	700-800	HYD	HYD
	104.941	2.8 (2799)	0.031	-	7-11B	③	-	650-750	HYD	HYD
	104.991	3.2 (3199)	0.031	-	①	③	-	650-750	HYD	HYD
	104.992	3.2 (3199)	0.031	-	①	③	-	650-750	HYD	HYD
	104.994	3.2 (3199)	0.031	-	①	③	-	650-750	HYD	HYD
	119.975	4.2 (4196)	0.031	-	①	③	-	600-750	HYD	HYD
	119.971	4.2 (5196)	0.031	-	①	③	-	600-750	HYD	HYD
	119.970	5.0 (4973)	0.031	-	①	③	-	600-750	HYD	HYD
	119.972	5.0 (4973)	0.031	-	①	③	-	600-750	HYD	HYD
	119.974	5.0 (4973)	0.031	-	①	③	-	600-750	HYD	HYD
	120.980	6.0 (5987)	0.031	-	①	③	-	600-750	HYD	HYD
	120.981	6.0 (5987)	0.031	-	①	③	-	600-750	HYD	HYD
1996-97	111.961	2.2 (2199)	0.031	-	①	③	-	700-800	HYD	HYD
	104.941	2.8 (2799)	0.031	-	7-11B	③	-	650-750	HYD	HYD
	104.991	3.2 (3199)	0.031	-	①	③	-	650-750	HYD	HYD
	104.994	3.2 (3199)	0.031	-	①	③	-	650-750	HYD	HYD
	104.995	3.2 (3199)	0.031	-	①	③	-	650-750	HYD	HYD
	104.941 ②	3.6 (3606)	0.031	-	①	③	-	650-750	HYD	HYD
	119.981	4.2 (4196)	0.031	-	①	③	-	600-750	HYD	HYD
	119.980	5.0 (4973)	0.031	-	①	③	-	600-750	HYD	HYD
	119.982	5.0 (4973)	0.031	-	①	③	-	600-750	HYD	HYD
	120.982	6.0 (5987)	0.031	-	①	③	-	600-750	HYD	HYD
	120.983	6.0 (5987)	0.031	-	①	③	-	600-750	HYD	HYD

NOTE: The Vehicle Emission Control Information label often reflects specification changes made during production. The label figures must be used if they differ from those in this chart.

B - Before top dead center

HYD - Hydraulic

① Timing controlled by engine control module. Adjustment is not possible.

② C36

③ 46-52 psi without vacuum applied
53-61 psi with vacuum applied
36 psi retention after 30 minutes

DIESEL ENGINE TUNE-UP SPECIFICATIONS

Year	Engine ID/VIN	Engine Displacement cu. in. (cc)	Valve Clearance Intake (in.)	Valve Clearance Exhaust (in.)	Intake Valve Opens (deg.)	Injection Pump Setting (deg.)	Injection Nozzle Pressure (psi) New	Injection Nozzle Pressure (psi) Used	Idle Speed (rpm)	Cranking Compression Pressure (psi)
1993	602.962	2.5 (2497)	HYD	HYD	12A	14-15A	1957-2102	1740	660-760	377-464
	603.971	3.5 (3449)	HYD	HYD	12A	13.5-14.5A	1667-1812	1450	560-660	377-464

DIESEL ENGINE TUNE-UP SPECIFICATIONS

Year	Engine ID/VIN	Engine Displacement cu. in. (cc)	Valve Clearance Intake (in.)	Valve Clearance Exhaust (in.)	Intake Valve Opens (deg.)	Injection Pump Setting (deg.)	Injection Nozzle Pressure (psi) New	Injection Nozzle Pressure (psi) Used	Idle Speed (rpm)	Cranking Compression Pressure (psi)
1994	603.971	3.5 (3449)	HYD	HYD	12A	13.5-14.5A	1667-1812	1450	560-660	377-464
1995	603.971	3.5 (3449)	HYD	HYD	12A	13.5-14.5A	1667-1812	1450	560-660	377-464
1996-97	606.912	3.0 (2966)	HYD	HYD	NA	NA	NA	NA	NA	NA

NOTE: The Vehicle Emission Control Information label often reflects specification changes made during production. The label figures must be used if they differ from those in this chart.
HYD - Hydraulic
A - At top dead center

CAPACITIES

Year	Model	Engine ID/VIN	Engine Displacement Liters (cc)	Engine Oil with Filter	Trans 4-Spd	Trans 5-Spd	Trans Auto.	Transfer Case (pts.)	Drive Axle Front (pts.)	Drive Axle Rear (pts.)	Fuel Tank (gal.)	Cooling System (qts.)
1993	190E 2.3	102.985	2.3 (2299)	5.3	-	3.2	12.7	-	-	2.3	14.5	8.9
	300D 2.5	602.962	2.5 (2497)	7.0	-	-	12.7	-	-	2.3	18.5	9.5
	190E 2.6	103.942	2.6 (2599)	6.3	-	3.2	12.7	-	-	2.3	14.5	①
	300E 2.8	104.942	2.8 (2799)	7.9	-	-	12.7	-	-	2.7	18.5	9.5
	300CE 4M	103.985	3.0 (2960)	7.0	-	-	13.1	1.3	2.1	2.3	18.5	①
	300E 4M	103.985	3.0 (2960)	7.0	-	-	13.1	1.3	2.1	2.3	18.5	①
	300SL	104.981	3.0 (2960)	7.9	-	3.4	13.1	-	-	2.7	21.1	12.1
	300CE	104.992	3.2 (3199)	7.9	-	-	13.1	-	-	2.7	18.5	9.5
	300E	104.992	3.2 (3199)	7.9	-	-	13.1	-	-	2.7	18.5	9.5
	300SE	104.990	3.2 (3199)	7.9	-	-	13.1	-	-	2.7	26.4	15.3
	300TE	103.992	3.2 (3199)	7.9	-	-	13.1	-	-	2.7	19.0	9.5
	300SD	603.971	3.5 (3449)	8.5	-	-	13.1	-	-	2.7	26.4	10.6
	400E	119.975	4.2 (4196)	8.5	-	-	16.3	-	-	2.7	23.7	13.2
	400SEL	119.971	4.2 (5196)	8.5	-	-	16.3	-	-	2.7	26.4	17.4
	500E	119.974	5.0 (4973)	8.5	-	-	16.3	-	-	2.7	23.7	13.2
	500SEC	119.970	5.0 (4973)	8.5	-	-	16.3	-	-	2.9	26.4	17.4
	500SEL	119.970	5.0 (4973)	8.5	-	-	16.3	-	-	2.9	26.4	17.4
	500SL	119.972	5.0 (4973)	8.9	-	-	16.3	-	-	2.9	21.1	15.9
	600SEC	120.980	6.0 (5987)	10.0	-	-	16.3	-	-	2.9	26.4	19.6
	600SEL	120.980	6.0 (5987)	10.0	-	-	16.3	-	-	2.9	26.4	19.6
	600SL	120.981	6.0 (5987)	10.6	-	-	16.3	-	-	2.9	21.1	21.1
1994	C220	111.961	2.2 (2199)	6.1	-	-	11.7	-	-	2.3	16.4	8.8
	C280	104.941	2.8 (2799)	7.9	-	-	11.7	-	-	2.3	16.4	10.5
	E320	104.992	3.2 (3199)	7.9	-	-	13.1	-	-	2.7	18.5	9.5
	S320	104.994	3.2 (3199)	7.9	-	-	13.1	-	-	2.7	26.4	15.3
	SL320	104.991	3.2 (3199)	7.9	-	-	13.1	-	-	2.7	21.1	11.6
	S350	603.971	3.5 (3449)	8.5	-	-	13.1	-	-	2.7	26.4	10.6
	E420	119.975	4.2 (4196)	8.5	-	-	16.3	-	-	2.7	23.7	13.2
	S420	119.971	4.2 (5196)	8.5	-	-	16.3	-	-	2.7	26.4	17.4
	E500	119.974	5.0 (4973)	8.5	-	-	16.3	-	-	2.7	23.7	13.2
	S500	119.970	5.0 (4973)	8.5	-	-	16.3	-	-	2.9	26.4	17.4
	SL500	119.972	5.0 (4973)	8.9	-	-	16.3	-	-	2.9	21.1	15.9
	S600	120.980	6.0 (5987)	10.0	-	-	16.3	-	-	2.9	26.4	19.6
	SL600	120.981	6.0 (5987)	10.6	-	-	16.3	-	-	2.9	21.1	21.1
1995	C220	111.961	2.2 (2199)	6.1	-	-	11.7	-	-	2.3	16.4	8.8
	C280	104.941	2.8 (2799)	7.9	-	-	11.7	-	-	2.3	16.4	10.5

CAPACITIES

Year	Model	Engine ID/VIN	Engine Displacement Liters (cc)	Engine Oil with Filter	Transmission (pts.) 4-Spd	Transmission (pts.) 5-Spd	Transmission (pts.) Auto.	Transfer Case (pts.)	Drive Axle Front (pts.)	Drive Axle Rear (pts.)	Fuel Tank (gal.)	Cooling System (qts.)
1995	E320	104.992	3.2 (3199)	7.9	-	-	13.1	-	-	2.7	18.5	9.5
	S320	104.994	3.2 (3199)	7.9	-	-	13.1	-	-	2.7	26.4	15.3
	SL320	104.991	3.2 (3199)	7.9	-	-	13.1	-	-	2.7	21.1	11.6
	S350	603.971	3.5 (3449)	8.5	-	-	13.1	-	-	2.7	26.4	10.6
	E420	119.975	4.2 (4196)	8.5	-	-	16.3	-	-	2.7	23.7	13.2
	S420	119.971	4.2 (5196)	8.5	-	-	16.3	-	-	2.7	26.4	17.4
	E500	119.974	5.0 (4973)	8.5	-	-	16.3	-	-	2.7	23.7	13.2
	S500	119.970	5.0 (4973)	8.5	-	-	16.3	-	-	2.9	26.4	17.4
	SL500	119.972	5.0 (4973)	8.9	-	-	16.3	-	-	2.9	21.1	15.9
	S600	120.980	6.0 (5987)	10.0	-	-	16.3	-	-	2.9	26.4	19.6
	SL600	120.981	6.0 (5987)	10.6	-	-	16.3	-	-	2.9	21.1	21.1
1996-97	C220	111.961	2.2 (2199)	6.2	-	-	11.7	-	-	2.3	16.4	8.8
	C280	104.941	2.8 (2799)	7.9	-	-	11.7	-	-	2.3	16.4	10.5
	E300	606.912	3.0 (2996)	NA	-	-	NA	-	-	NA	17.2	NA
	E320	104.995	3.2 (3199)	NA	-	-	NA	-	-	NA	21.1	NA
	S320	104.994	3.2 (3199)	7.9	-	-	13.1	-	-	2.7	26.4	15.3
	SL320	104.991	3.2 (3199)	7.9	-	-	13.1	-	-	2.7	21.1	11.6
	C36	104.941	3.6 (3606)	7.9	-	-	NA	-	-	NA	16.4	7.9
	S420	119.981	4.2 (5196)	8.5	-	-	16.3	-	-	2.7	26.4	17.4
	S500	119.980	5.0 (4973)	8.5	-	-	16.3	-	-	2.9	26.4	17.4
	SL500	119.982	5.0 (4973)	8.9	-	-	16.3	-	-	2.9	21.1	15.9
	S600	120.982	6.0 (5987)	10.0	-	-	16.3	-	-	2.9	26.4	19.6
	SL600	120.983	6.0 (5987)	10.6	-	-	16.3	-	-	2.9	21.1	21.1

① With A/C or automatic climate control: 10.0
Without A/C or automatic climate control: 9.5

VALVE SPECIFICATIONS

Year	Engine ID/VIN	Engine Displacement Liters (cc)	Seat Angle (deg.)	Face Angle (deg.)	Spring Test Pressure (lbs. @ in.)	Spring Installed Height (in.)	Stem-to-Guide Clearance (in.) Intake	Stem-to-Guide Clearance (in.) Exhaust	Stem Diameter (in.) Intake	Stem Diameter (in.) Exhaust
1993	102.985	2.3 (2299)	45	45.25	232-250@ 1.196	NA	NA	NA	0.3129- 0.3137	0.3518- 0.3527
	602.962	2.5 (2497)	45	45.25	①	NA	NA	NA	0.3131- 0.3137	0.3521- 0.3527
	103.942	2.6 (2599)	45	45	NA	NA	NA	NA	0.3131- 0.3137	0.3521- 0.3527
	104.942	2.8 (2799)	45	45	NA	NA	NA	NA	0.2755	0.2755
	103.985	3.0 (2960)	45	45	NA	NA	NA	NA	0.3131- 0.3137	0.3521- 0.3527
	104.981	3.0 (2960)	45	45	NA	NA	NA	NA	②	②
	104.990	3.2 (3199)	45	45	NA	NA	NA	NA	③	③
	104.992	3.2 (3199)	45	45	NA	NA	NA	NA	0.2755	0.2755
	603.971	3.5 (3449)	45	45.25	①	NA	NA	NA	0.3131- 0.3137	0.3521- 0.3527
	119.975	4.2 (4196)	NA	NA	NA	NA	NA	NA	0.2755	0.2755
	119.971	4.2 (5196)	NA	NA	NA	NA	NA	NA	0.2755	0.2755
	119.970	5.0 (4973)	NA	NA	NA	NA	NA	NA	0.2755	0.2755
	119.972	5.0 (4973)	NA	NA	NA	NA	NA	NA	0.2755	0.2755

VALVE SPECIFICATIONS

Year	Engine ID/VIN	Engine Displacement Liters (cc)	Seat Angle (deg.)	Face Angle (deg.)	Spring Test Pressure (lbs. @ in.)	Spring Installed Height (in.)	Stem-to-Guide Clearance (in.)		Stem Diameter (in.)	
							Intake	Exhaust	Intake	Exhaust
1993	119.974	5.0 (4973)	NA	NA	NA	NA	NA	NA	0.2755	0.2755
	120.980	6.0 (5987)	NA	NA	NA	NA	NA	NA	NA	NA
	120.981	6.0 (5987)	NA	NA	NA	NA	NA	NA	NA	NA
1994	111.961	2.2 (2199)	NA	NA	NA	NA	NA	NA	NA	NA
	104.941	2.8 (2799)	45	45	NA	NA	NA	NA	0.2755	0.2755
	104.991	3.2 (3199)	45	45	NA	NA	NA	NA	0.2755	0.2755
	104.992	3.2 (3199)	45	45	NA	NA	NA	NA	0.2755	0.2755
	104.994	3.2 (3199)	45	45	NA	NA	NA	NA	0.2755	0.2755
	603.971	3.5 (3449	45	45.25	①	NA	NA	NA	0.3131-0.3137	0.3521-0.3527
	119.975	4.2 (4196)	NA	NA	NA	NA	NA	NA	0.2755	0.2755
	119.971	4.2 (5196)	NA	NA	NA	NA	NA	NA	0.2755	0.2755
	119.970	5.0 (4973)	NA	NA	NA	NA	NA	NA	0.2755	0.2755
	119.972	5.0 (4973)	NA	NA	NA	NA	NA	NA	0.2755	0.2755
	119.974	5.0 (4973)	NA	NA	NA	NA	NA	NA	0.2755	0.2755
	120.980	6.0 (5987)	NA	NA	NA	NA	NA	NA	NA	NA
	120.981	6.0 (5987)	NA	NA	NA	NA	NA	NA	NA	NA
1995	111.961	2.2 (2199)	NA	NA	NA	NA	NA	NA	NA	NA
	104.941	2.8 (2799)	45	45	NA	NA	NA	NA	0.2755	0.2755
	104.991	3.2 (3199)	45	45	NA	NA	NA	NA	0.2755	0.2755
	104.992	3.2 (3199)	45	45	NA	NA	NA	NA	0.2755	0.2755
	104.994	3.2 (3199)	45	45	NA	NA	NA	NA	0.2755	0.2755
	603.971	3.5 (3449	45	45.25	①	NA	NA	NA	0.3131-0.3137	0.3521-0.3527
	119.975	4.2 (4196)	NA	NA	NA	NA	NA	NA	0.2755	0.2755
	119.971	4.2 (5196)	NA	NA	NA	NA	NA	NA	0.2755	0.2755
	119.970	5.0 (4973)	NA	NA	NA	NA	NA	NA	0.2755	0.2755
	119.972	5.0 (4973)	NA	NA	NA	NA	NA	NA	0.2755	0.2755
	119.974	5.0 (4973)	NA	NA	NA	NA	NA	NA	0.2755	0.2755
	120.980	6.0 (5987)	NA	NA	NA	NA	NA	NA	NA	NA
	120.981	6.0 (5987)	NA	NA	NA	NA	NA	NA	NA	NA
1996-97	111.961	2.2 (2199)	NA	NA	NA	NA	NA	NA	NA	NA
	104.941	2.8 (2799)	45	45	NA	NA	NA	NA	0.2755	0.2755
	606.912	3.0 (2996)	NA	NA	NA	NA	NA	NA	NA	NA
	104.991	3.2 (3199)	45	45	NA	NA	NA	NA	0.2755	0.2755
	104.994	3.2 (3199)	45	45	NA	NA	NA	NA	0.2755	0.2755
	104.995	3.2 (3199)	45	45	NA	NA	NA	NA	0.2755	0.2755
	104.941 ④	3.6 (3606)	NA	NA	NA	NA	NA	NA	NA	NA
	119.981	4.2 (4196)	NA	NA	NA	NA	NA	NA	0.2755	0.2755
	119.980	5.0 (4973)	NA	NA	NA	NA	NA	NA	0.2755	0.2755
	119.982	5.0 (4973)	NA	NA	NA	NA	NA	NA	0.2755	0.2755
	120.982	6.0 (5987)	NA	NA	NA	NA	NA	NA	NA	NA
	120.983	6.0 (5987)	NA	NA	NA	NA	NA	NA	NA	NA

NA - Not Available
① 1st version: 159.6-177.5@ 1.062 in.
   2nd version: 152.8-166.3@ 1.062 in.

② 1st version: Intake - 0.3149, Exhaust - 0.3543
   2nd version: Intake - 0.2755, Exhaust - 0.3149
   3rd version: Intake - 0.2755, Exhaust - 0.2755

③ 1st version: Intake - 0.2755, Exhaust - 0.3149
   2nd version: Intake - 0.2755, Exhaust - 0.2755
④ C36

TORQUE SPECIFICATIONS
All readings in ft. lbs.

Year	Engine ID/VIN	Engine Displacement Liters (cc)	Cylinder Head Bolts	Main Bearing Bolts	Rod Bearing Bolts	Crankshaft Damper Bolts	Flywheel Bolts	Manifold Intake	Manifold Exhaust	Spark Plugs	Lug Nut
1993	102.985	2.3 (2299)	①	⑦	⑩	⑤	⑬	18.4	22.1	③	81
	602.962	2.5 (2497)	⑥	⑦	⑫	⑤	⑮	18.4	18.4	③	81
	103.942	2.6 (2599)	②	⑧	⑩	221	⑬	18.4	22.1	③	81
	104.942	2.8 (2799)	①	⑧	⑩	⑲	⑬	18.4	29.5	③	81
	103.985	3.0 (2960)	②	⑧	⑩	221	⑬	18.4	22.1	③	81
	104.981	3.0 (2962)	①	⑧	⑩	⑲	⑬	18.4	29.5	③	81
	104.990	3.2 (3199)	①	⑧	⑩	⑲	⑬	18.4	29.5	③	95
	104.992	3.2 (3199)	①	⑧	⑩	⑲	⑬	18.4	29.5	③	81
	603.971	3.5 (3449)	⑥	⑦	⑪	⑤	⑮	18.4	18.4	③	95
	119.975	4.2 (4196)	①	⑨	⑭	295	⑯	18.4	22.1	③	95
	119.971	4.2 (5196)	①	⑨	⑭	295	⑯	18.4	22.1	③	81
	119.970	5.0 (4973)	①	⑨	⑭	295	⑯	18.4	22.1	③	95
	119.972	5.0 (4973)	①	⑨	⑭	295	⑯	18.4	22.1	③	81
	119.974	5.0 (4973)	①	⑨	⑭	295	⑯	18.4	22.1	③	81
	120.980	6.0 (5987)	⑱	⑨	⑭	295	⑯	18.4	29.5	③	95
	120.981	6.0 (5987)	⑱	⑨	⑭	295	⑯	18.4	29.5	③	81
1994	111.961	2.2 (2199)	①	⑧	⑩	221	⑩	15	29.5	③	81
	104.941	2.8 (2799)	①	⑧	⑩	⑰	⑬	18.4	29.5	③	81
	104.991	3.2 (3199)	①	⑧	⑩	⑰	⑬	18.4	29.5	③	81
	104.992	3.2 (3199)	①	⑧	⑩	⑰	⑬	18.4	29.5	③	81
	104.994	3.2 (3199)	①	⑧	⑩	⑰	⑬	18.4	29.5	③	95
	603.971	3.5 (3449)	⑥	⑳	⑪	⑰	⑮	18.4	18.4	③	95
	119.975	4.2 (4196)	①	⑨	⑭	295	⑯	18.4	22.1	③	81
	119.971	4.2 (5196)	①	⑨	⑭	295	⑯	18.4	22.1	③	95
	119.970	5.0 (4973)	①	⑨	⑭	295	⑯	18.4	22.1	③	95
	119.972	5.0 (4973)	①	⑨	⑭	295	⑯	18.4	22.1	③	81
	119.974	5.0 (4973)	①	⑨	⑭	295	⑯	18.4	22.1	③	81
	120.980	6.0 (5987)	⑱	⑨	⑭	295	⑯	18.4	29.5	③	95
	120.981	6.0 (5987)	⑱	⑨	⑭	295	⑯	18.4	29.5	③	81
1995	111.961	2.2 (2199)	①	⑧	⑩	221	⑩	15	29.5	③	81
	104.941	2.8 (2799)	①	⑧	⑩	⑰	⑬	18.4	29.5	③	81
	104.991	3.2 (3199)	①	⑧	⑩	⑰	⑬	18.4	29.5	③	81
	104.992	3.2 (3199)	①	⑧	⑩	⑰	⑬	18.4	29.5	③	81
	104.994	3.2 (3199)	①	⑧	⑩	⑰	⑬	18.4	29.5	③	95
	603.971	3.5 (3449)	⑥	⑳	⑪	⑰	⑮	18.4	18.4	③	95
	119.975	4.2 (4196)	①	⑨	⑭	295	⑯	18.4	22.1	③	81
	119.971	4.2 (5196)	①	⑨	⑭	295	⑯	18.4	22.1	③	95
	119.970	5.0 (4973)	①	⑨	⑭	295	⑯	18.4	22.1	③	95
	119.972	5.0 (4973)	①	⑨	⑭	295	⑯	18.4	22.1	③	81
	119.974	5.0 (4973)	①	⑨	⑭	295	⑯	18.4	22.1	③	81
	120.980	6.0 (5987)	⑱	⑨	⑭	295	⑯	18.4	29.5	③	95
	120.981	6.0 (5987)	⑱	⑨	⑭	295	⑯	18.4	29.5	③	81
1996-97	111.961	2.2 (2199)	②	⑧	⑩	300	⑩	14.7	29.5	③	81
	104.941	2.8 (2799)	①	⑧	⑩	⑰	⑬	18.4	29.5	③	81
	606.912	3.0 (2996)	NA	NA	NA	NA	NA	NA	NA	NA	NA
	104.991	3.2 (3199)	①	⑧	⑩	⑰	⑬	18.4	29.5	③	81
	104.994	3.2 (3199)	①	⑧	⑩	⑰	⑬	18.4	29.5	③	110
	104.995	3.2 (3199)	NA	NA	NA	NA	NA	NA	NA	NA	NA
	104.941 ④	3.6 (3606)	NA	NA	NA	NA	NA	NA	NA	NA	NA

TORQUE SPECIFICATIONS
All readings in ft. lbs.

Year	Engine ID/VIN	Engine Displacement Liters (cc)	Cylinder Head Bolts	Main Bearing Bolts	Rod Bearing Bolts	Crankshaft Damper Bolts	Flywheel Bolts	Manifold Intake	Manifold Exhaust	Spark Plugs	Lug Nut
1996-97	119.981	4.2 (4196)	①	⑨	⑭	400	⑯	18.4	22.1	③	81
	119.980	5.0 (4973)	①	⑨	⑭	400	⑯	18.4	22.1	③	110
	119.982	5.0 (4973)	①	⑨	⑭	400	⑯	18.4	22.1	③	81
	120.982	6.0 (5987)	⑱	⑨	⑭	400	⑯	18.4	29.5	③	110
	120.983	6.0 (5987)	⑱	⑨	⑭	400	⑯	18.4	29.5	③	81

① Step 1: 40.5 ft. lbs.
   Step 2: + 90 degrees
   Step 3: + 90 degrees
   Step 4: M8 bolts: 18.4 ft. lbs.
② Step 1: 51.6 ft. lbs.
   Step 2: + 90 degrees
   Step 3: + 90 degrees
③ Spark plug with conical seat: 7.3-14.7 ft. lbs.
   Spark plug with flat seat: 14.7-22.1 ft. lbs.
④ C36
⑤ Without stretch bolt: 221 ft. lbs.
   With stretch bolt:
   Step 1: 147.5 ft. lbs.
   Step 2: + 90 degrees
   Step 3: + 90 degrees
⑥ Step 1: 11 ft. lbs.
   Step 2: 25.8 ft. lbs.
   Step 3: + 90 degrees
   Step 4: Wait ten minutes
   Step 5: + 90 degrees
   Step 6: M8 bolts: 18.4 ft. lbs.

⑦ M11 bolts:
   Step 1: 40.5 ft. lbs.
   Step 2: + 90-100 degrees
   M12 bolts: 66.3 ft. lbs.
⑧ Step 1: 40.5 ft. lbs.
   Step 2: + 90-100 degrees
⑨ M8 bolts: 22.1 ft. lbs.
   M10 bolts: 36.8 ft. lbs.
⑩ Step 1: 22.1 ft. lbs.
   Step 2: + 90-100 degrees
⑪ Step 1: 29.5 ft. lbs.
   Step 2: + 90-100 degrees
⑫ Step 1: 29.5 ft. lbs.
   Step 2: + 90-100 degrees
⑬ Without dual mass flywheel:
   Step 1: 22.1 ft. lbs.
   Step 2: + 90-100 degrees
   With dual mass flywheel:
   Step 1: 29.5 ft. lbs.
   Step 2: + 90-100 degrees

⑭ Step 1: 33.1 ft. lbs.
   Step 2: + 90-100 degrees
⑮ Without dual mass flywheel:
   Step 1: 25.8 ft. lbs.
   Step 2: + 90-100 degrees
   With dual mass flywheel:
   Step 1: 29.5 ft. lbs.
   Step 2: + 90-100 degrees
⑯ Step 1: 25.8 ft. lbs.
   Step 2: + 90-100 degrees
⑰ Step 1: 147.5 ft. lbs.
   Step 2: + 90 degrees
⑱ Step 1: 40.5 ft. lbs.
   Step 2: + 90 degrees
   Step 3: + 90 degrees
⑲ From 8/93 with 8.8 stretch bolt:
   Step 1: 147.5 ft. lbs.
   Step 2: + 90 degrees
   Without stretch bolt: 400 ft. lbs.

⑳ M11 bolts:
   Step 1: 40.5 ft. lbs.
   Step 2: + 90-100 degrees
⑱ Step 1: 40.5 ft. lbs.
   Step 2: + 90 degrees
   Step 3: + 90 degrees
⑲ From 8/93 with 8.8 stretch bolt:
   Step 1: 147.5 ft. lbs.
   Step 2: + 90 degrees
   Without stretch bolt: 400 ft. lbs.
⑳ M11 bolts:
   Step 1: 40.5 ft. lbs.
   Step 2: + 90-100 degrees

BRAKE SPECIFICATIONS
All measurements in inches unless noted

Year	Model		Master Cylinder Bore	Brake Disc Original Thickness	Brake Disc Minimum Thickness	Brake Disc Maximum Runout	Brake Drum Diameter Original Inside Diameter	Brake Drum Diameter Max. Wear Limit	Brake Drum Diameter Maximum Machine Diameter	Minimum Lining Thickness Front	Minimum Lining Thickness Rear
1993	190E 2.3	F	②	0.866	0.763	0.0047	-	-	-	0.078	-
		R		0.354	0.287	0.0059				-	0.078
	190E 2.6	F	①	0.866	0.763	0.0047	-	-	-	0.078	-
		R		0.354	0.287	0.0059				-	0.078
	300CE	F	①	0.984	0.881	0.0047	-	-	-	0.078	-
		R		0.354	0.287	0.0059				-	0.078
	300TE 4M	F	①	0.866	0.763	0.0047	-	-	-	0.078	-
		R		0.354	0.287	0.0059				-	0.078
	300D 2.5	F	①	0.866	0.763	0.0047	-	-	-	0.078	-
		R		0.354	0.287	0.0059				-	0.078
	300E	F	①	0.984	0.881	0.0047	-	-	-	0.078	-
		R		0.354	0.287	0.0059				-	0.078
	300E 2.8	F	①	0.866	0.763	0.0047	-	-	-	0.078	-
		R		0.354	0.287	0.0059				-	0.078
	300E 4M	F	①	0.866	0.763	0.0047	-	-	-	0.078	-
		R		0.354	0.287	0.0059				-	0.078
	300SD	F	③	④	⑤	0.0031	-	-	-	0.078	-
		R		0.472	0.385	0.0039				-	0.078
	300SE	F	③	④	⑤	0.0031	-	-	-	0.078	-
		R		0.472	0.385	0.0039				-	0.078

BRAKE SPECIFICATIONS
All measurements in inches unless noted

Year	Model		Master Cylinder Bore	Brake Disc Original Thickness	Brake Disc Minimum Thickness	Brake Disc Maximum Runout	Brake Drum Diameter Original Inside Diameter	Brake Drum Diameter Max. Wear Limit	Brake Drum Diameter Maximum Machine Diameter	Minimum Lining Thickness Front	Minimum Lining Thickness Rear
1993	300SL	F	③	1.102	0.999	0.0047	-	-	-	0.078	
		R		0.354	0.287	0.0059				-	0.078
	300TE	F	①	0.984	0.881	0.0047	-	-	-	0.078	-
		R		0.787	0.685	0.0059				-	0.078
	400E	F	③	0.984	0.881	0.0047				0.078	-
		R		0.944	0.842	0.0059				-	0.078
	400SEL	F	⑦	④	⑤	0.0031			-	0.078	-
		R		0.866	0.763	0.0039				-	0.078
	500E	F	③	1.181	1.102	0.0031	-	-	-	0.078	-
		R		0.944	0.842	0.0059				-	0.078
	500SEC	F	⑦	④	⑤	0.0031	-	-	-	0.078	-
		R		0.866	0.763	0.0039				-	0.078
	500SEL	F	⑦	④	⑤	0.0031	-	-	-	0.078	-
		R		0.866	0.763	0.0039				-	0.078
	500SL	F	③	1.102	0.999	0.0047	-	-	-	0.078	-
		R		0.354	0.287	0.0059				-	0.078
	600SEC	F	⑦	④	⑤	0.0031	-	-	-	0.078	-
		R		0.866	0.763	0.0039				-	0.078
	600SEL	F	⑦	④	⑤	0.0031	-	-	-	0.078	-
		R		0.866	0.763	0.0039				-	0.078
	600SL	F	⑦	1.181	1.102	0.0031	-	-	-	0.078	-
		R		0.866	0.763	0.0039				-	0.078
1994	C220	F	①	0.866	0.763	0.0047	-	-	-	0.078	-
		R		0.354	0.287	0.0059				-	0.078
	C280	F	①	0.866	0.763	0.0047	-	-	-	0.078	-
		R		0.354	0.287	0.0059				-	0.078
	E320	F	①	0.984 ⑥	0.881 ⑥	0.0047	-	-	-	0.078	-
		R		0.354	0.287	0.0059				-	0.078
	E420	F	③	0.984	0.881	0.0047	-	-	-	0.078	-
		R		0.944	0.842	0.0059				-	0.078
	E500	F	③	1.181	1.102	0.0031	-	-	-	0.078	-
		R		0.944	0.842	0.0059				-	0.078
	S320	F	③	1.102 ④	0.999 ⑤	0.0031	-	-	-	0.078	-
		R		0.472	0.385	0.0039				-	0.078
	S350	F	③	1.102 ④	0.999 ⑤	0.0031	-	-	-	0.078	-
		R		0.472	0.385	0.0039				-	0.078
	S420	F	⑦	1.102 ④	0.999 ⑤	0.0031	-	-	-	0.078	-
		R		0.866	0.763	0.0039				-	0.078
	S500	F	⑦	1.102 ④	0.999 ⑤	0.0031	-	-	-	0.078	-
		R		0.866	0.763	0.0039				-	0.078
	S600	F	⑦	1.102 ④	0.999 ⑤	0.0031	-	-	-	0.078	-
		R		0.866	0.763	0.0039				-	0.078
	SL320	F	③	1.102	0.999	0.0047	-	-	-	0.078	-
		R		0.354	0.287	0.0059				-	0.078
	SL500	F	③	1.102	0.999	0.0047	-	-	-	0.078	-
		R		0.354	0.287	0.0059				-	0.078
	SL600	F	⑦	1.181	1.102	0.0031	-	-	-	0.078	-
		R		0.866	0.763	0.0039				-	0.078
1995	C220	F	①	0.866	0.763	0.0047	-	-	-	0.078	-
		R		0.354	0.287	0.0059				-	0.078
	C280	F	①	0.866	0.763	0.0047	-	-	-	0.078	-
		R		0.354	0.287	0.0059				-	0.078

BRAKE SPECIFICATIONS
All measurements in inches unless noted

Year	Model		Master Cylinder Bore	Brake Disc Original Thickness	Brake Disc Minimum Thickness	Maximum Runout	Brake Drum Diameter Original Inside Diameter	Max. Wear Limit	Maximum Machine Diameter	Minimum Lining Thickness Front	Minimum Lining Thickness Rear
1995	E320	F	①	0.984 ⑥	0.881 ⑥	0.0047	-	-	-	0.078	-
		R		0.354	0.287	0.0059				-	0.078
	E420	F	③	0.984	0.881	0.0047	-	-	-	0.078	-
		R		0.944	0.842	0.0059				-	0.078
	E500	F	③	1.181	1.102	0.0031	-	-	-	0.078	-
		R		0.944	0.842	0.0059				-	0.078
	S320	F	③	1.102 ④	0.999 ⑤	0.0031	-	-	-	0.078	-
		R		0.472	0.385	0.0039				-	0.078
	S350	F	③	1.102 ④	0.999 ⑤	0.0031	-	-	-	0.078	-
		R		0.472	0.385	0.0039				-	0.078
	S420	F	⑦	1.102 ④	0.999 ⑤	0.0031	-	-	-	0.078	-
		R		0.866	0.763	0.0039				-	0.078
	S500	F	⑦	1.102 ④	0.999 ⑤	0.0031	-	-	-	0.078	-
		R		0.866	0.763	0.0039				-	0.078
	S600	F	⑦	1.102 ④	0.999 ⑤	0.0031	-	-	-	0.078	-
		R		0.866	0.763	0.0039				-	0.078
	SL320	F	③	1.102	0.999	0.0047	-	-	-	0.078	-
		R		0.354	0.287	0.0059				-	0.078
	SL500	F	③	1.102	0.999	0.0047	-	-	-	0.078	-
		R		0.354	0.287	0.0059				-	0.078
	SL600	F	⑦	1.181	1.102	0.0031	-	-	-	0.078	-
		R		0.866	0.763	0.0039				-	0.078
1996-97	C220	F	①	0.866	0.763	0.0047	-	-	-	0.078	-
		R	-	0.354	0.287	0.0059	-	-	-	-	0.078
	C280	F	①	0.866	0.763	0.0047	-	-	-	0.078	-
		R	-	0.354	0.287	0.0059	-	-	-	-	NA
	C36	F	①	1.181	1.102	0.0031	-	-	-	0.078	-
		R	-	0.945	0.842	0.0059	-	-	-	-	0.078
	E300		NA	NA	NA	NA	NA	NA	NA	NA	NA
	E320		NA	NA	NA	NA	NA	NA	NA	NA	NA
	S320	F	③	1.102 ④	0.999 ⑤	0.0031	-	-	-	0.078	-
		R	-	0.472	0.385	0.0039	-	-	-	-	0.078
	SL320	F	③	1.102	0.999	0.0047	-	-	-	0.078	-
		R	-	0.866	0.287	0.0059	-	-	-	-	0.078
	S420	F	⑦	1.102 ④	0.999 ⑤	0.0031	-	-	-	0.078	-
		R	-	0.866	0.763	0.0039	-	-	-	-	0.078
	S500	F	⑦	1.102 ④	0.999 ⑤	0.0031	-	-	-	0.078	-
		R	-	0.866	0.763	0.0039	-	-	-	-	0.078
	SL500	F	③	1.102	0.999	0.0047	-	-	-	0.078	-
		R	-	0.354	0.287	0.0059	-	-	-	-	0.078
	S600	F	⑦	1.102 ④	0.999 ⑤	0.0031	-	-	-	0.078	-
		R	-	0.866	0.763	0.0039	-	-	-	-	0.078
	SL600	F	⑦	1.181	1.102	0.0031	-	-	-	0.078	-
		R	-	0.866	0.763	0.0039	-	-	-	-	0.078

NA - Not Available
F - Front
R - Rear

① Stepped bore:
   Step 1: 0.938
   Step 2: 0.750

② Stepped bore:
   Step 1: 0.875
   Step 2: 0.688

③ Stepped bore:
   Step 1: 1.000
   Step 2: 0.750

④ With 2 piston fixed caliper: 1.102
   With 4 piston fixed caliper: 1.181

⑤ With 2 piston fixed caliper: .999
   With 4 piston fixed caliper: 1.078

⑥ Wagon:
   Original Thickness: 0.984 - 0.787
   Minimum Thickness 0.881 - 0.686

⑦ Stepped bore:
   Step 1: 1.063
   Step 2: 1.000

FREQUENT MAINTENANCE LABOR
MERCEDES-BENZ

The following should be used as a guide when determining the amount of work required for a particular service if taken to a repair shop. In estimating how long a particular Frequent Maintenance Service item should take, please observe the following:
- **Chilton Time** is time that is based on field research and data supplied by the vehicle manufacturer.
- All labor time operations are given in hours and tenths of an hour.
- All labor operations, are to be used as a **guide**.

COOLING

(G) Belt, Serpentine Drive, Renew

	Chilton Time
1993 190D, 190E	.6
1994-97 C220	.5
1994-97 C280	.9
1995-97 C36 AMG	.9
1993 300D	.6
1995-97 E300D	.9
1993 300SD	.6
1994-95 S350TD	.6
1993 300E	.9
1993 300CE, 300TE	.9
1993 300SE	.9
1993 300SL	.6
1994-95 E320	.9
1994-97 S320	1.2
1993 400E, 400SEL	.8
1994-97 E420	.8
1993-94 500E, E500	.6
1993 500SEC	.6
1993 500SEL	.6
1993 500SL	.8
1994-97 SL500	.6
1993 600SEC, 600SEL	1.5
1993 600SL	1.6
1994-97 S600	1.5
1994-97 SL600	1.6

(G) Belt, Drive, Adjust

	Chilton Time
1993 190D, 190E	
one	.2
each adtnl.	.1
1993 300D	
one	.2
each adtnl.	.1
1993 300E	
one	.2
each adtnl.	.1
1993 300CE, 300TE, 300SE	
one	.2
each adtnl.	.1
1993 300SL	
one	.2
each adtnl.	.1
one	.2
each adtnl.	.1

(G) Hoses, Radiator, Renew
Includes: Drain and refill cooling system as required.

	Chilton Time
Upper	
1993-97	.9
Lower	
1993 190D, 190E	1.1
1994-97 C220	.6
1994-97 C280	.6
1995-97 C36 AMG	1.1
1993 300D	1.1
1993 300SD	.9
1993 300E	1.1
1993 300CE, 300TE	1.1
1993 300SE	1.1
1993 300SL	1.1
1993 400E, 400SEL	.9
1994-97 E420	.9
1993-94 500E	.9
1993 500SEC	.9
1993 500SEL	.9
1993 500SL	1.1
1993 600SEL	.9
1993 600SEC, 600SL	.9

(G) Thermostat, Coolant, Renew

	Chilton Time
1993 190D, 190E	.8
1994-97 C220	.6
1994-97 C280	.6
1995-97 C36 AMG	.6
1993 300D	.8
1995-97 E300D	.8
1993 300E	.8
1993 300SD	.9
1993 300CE, 300TE	1.8
1993 300SE	.8
1993 300SL	.8
1994-95 E320	.8
1994-97 S320	.8
1994-95 S350TD	.9
1993 400E, 400SEL	2.0
1994-97 E420	2.0
1993-94 500E, E500	2.0
1993 500SL	1.2
1994-97 SL500	1.2
1993 500SEC	1.4
1993 500SEL	1.4
1993 600SEC, 600SEL	1.8
1993 600SL	1.8
1994-97 S600	1.8
1994-97 SL600	1.8

FUEL

(M) Air Cleaner, Service

	Chilton Time
1993-97, each	.3

(G) Filter, Fuel, Renew

	Chilton Time
1993 190E	.6
1994-97 C220	.6
1994-97 C280	.6
1995-97 C36 AMG	.6
1993 300CE, 300TE	.3
1993 300E	.6
1993 300SE	.9
1993 300SL	.6
1994-95 E320	.6
1994-97 S320	.9
1993 400E	.6
1993 400SEL	.8
1994-97 E420	.6
1993-94 500E, E500	.6
1993 500SL	.9
1994-97 SL500	.9
1993 500SEL	.9
1993 500SEC	.9
1993 600SEC, 600SEL	.9
1993 600SL	.9
1994-97 S600	.9
1994-97 SL600	.9

BRAKES

(G) Bleed Brakes (Four Wheels)
Includes: Add fluid.

	Chilton Time
1993-97	.6
w/ABS add	.5

(M) Parking Brake, Adjust

	Chilton Time
1993-97	.5

LUBRICATION SERVICES

(M) Engine Oil & Filter, Renew
Includes: Inspect and correct all fluid levels.

	Chilton Time
1993-97	.5

ELECTRICAL

(G) Horn, Renew

	Chilton Time
1993-97	.6

(M) Terminals, Battery, Clean

	Chilton Time
1993-97	.3

SCHEDULED MAINTENANCE INTERVALS
(MERCEDES-BENZ 190, 220, 280, 300, 320, 350, 400, 420, 500, 560, 600 & C36)

TO BE SERVICED	TYPE OF SERVICE	VEHICLE MILEAGE INTERVAL (x1000)												
		7.5	15	22.5	30	37.5	45	52.5	60	67.5	75	82.5	90	97.5
Engine oil & filter	R	✓	✓	✓	✓	✓	✓	✓	✓	✓	✓	✓	✓	✓
Automatic transmission fluid & filter	S/I	✓	✓	✓	✓	✓	✓	✓	✓	✓	✓	✓	✓	✓
Brake calipers	S/I	✓	✓	✓	✓	✓	✓	✓	✓	✓	✓	✓	✓	✓
Brake pads & brake discs (front & rear)	S/I	✓	✓	✓	✓	✓	✓	✓	✓	✓	✓	✓	✓	✓
Chassis lubrication	S/I	✓	✓	✓	✓	✓	✓	✓	✓	✓	✓	✓	✓	✓
Lubricate throttle linkage	S/I	✓	✓	✓	✓	✓	✓	✓	✓	✓	✓	✓	✓	✓
Dust caps	S/I		✓		✓		✓		✓		✓		✓	
Front axle ball joints	S/I		✓		✓		✓		✓		✓		✓	
Manual transmission oil	S/I		✓		✓		✓		✓		✓		✓	
Parking brake	S/I		✓		✓		✓		✓		✓		✓	
Poly-V-belt	S/I		✓		✓		✓		✓		✓		✓	
Power steering	S/I		✓		✓		✓		✓		✓		✓	
Rear axle oil	S/I		✓		✓		✓		✓		✓		✓	
Rotate tires	S/I		✓		✓		✓		✓		✓		✓	
Steering gear mounting bolts	S/I		✓		✓		✓		✓		✓		✓	
Steering mechanism (tie rods & idler arm)	S/I		✓		✓		✓		✓		✓		✓	
Air filter element	R				✓				✓				✓	
Brake fluid	R				✓				✓				✓	
Spark plugs	R				✓				✓				✓	
Injection timing	S/I				✓				✓				✓	
Soft top & roll bar hydraulic system	S/I				✓				✓				✓	
Engine Coolant	R						✓						✓	
Fuel filter	R								✓					
Fuel pre-filter	R								✓					
Drive shaft flexible couplings	S/I								✓					

R – Replace   S/I – Service or Inspect

SCHEDULED MAINTENANCE INTERVALS
(MERCEDES-BENZ 190, 220, 280, 300, 320, 350, 400, 420, 500, 560, 600 & C36) (Cont.)

FREQUENT OPERATION MAINTENANCE (SEVERE SERVICE)

If a vehicle is operated under any of the following conditions it is considered severe service:
- Extremely dusty areas.
- 50% or more of the vehicle operation is in 32°C (90°F) or higher temperatures, or constant operation in temperatures below 0°C (32°F).
- Prolonged idling (vehicle operation in stop and go traffic).
- Frequent short running periods (engine does not warm to normal operating temperatures).
- Police, taxi, delivery usage or trailer towing usage.

Oil & oil filter change – change every 3750 miles.
Air filter element – service or inspect every 15,000 miles.

MITSUBISHI

ENGINE IDENTIFICATION

Year	Model	Engine Displacement Liters (cc)	Engine Series (ID/VIN)	Fuel System	No. of Cylinders	Engine Type
1993	Precis	1.5 (1468)	G4DJ	MFI	4	SOHC
	Mirage	1.5 (1468)	4G15	MFI	4	SOHC
	Mirage	1.8 (1834)	4G93	MFI	4	SOHC
	Eclipse	1.8 (1755)	4G37	MFI	4	SOHC
	Eclipse Turbo	2.0 (1997)	4G63	MFI	4	DOHC
	Galant	2.0 (1997)	4G63	MFI	4	SOHC
	Galant	2.0 (1997)	4G63	MFI	4	DOHC
	3000GT	3.0 (2972)	6G72	MFI	6	DOHC
	3000GT Turbo	3.0 (2972)	6G72	MFI	6	DOHC
	Diamante	3.0 (2972)	6G72	MFI	6	SOHC
	Diamante	3.0 (2972)	6G72	MFI	6	DOHC
1994	Precis	1.5 (1468)	G4DJ	MFI	4	SOHC
	Mirage	1.5 (1468)	4G15	MFI	4	SOHC
	Mirage	1.8 (1834)	4G93	MFI	4	SOHC
	Eclipse	1.8 (1755)	4G37	MFI	4	SOHC
	Eclipse Turbo	2.0 (1997)	4G63	MFI	4	DOHC
	Galant	2.4 (2350)	4G64	MFI	4	SOHC
	Galant	2.4 (2350)	4G64	MFI	4	DOHC
	3000GT	3.0 (2972)	6G72	MFI	6	DOHC
	3000GT Turbo	3.0 (2972)	6G72	MFI	6	DOHC
	Diamante	3.0 (2972)	6G72	MFI	6	SOHC
	Diamante	3.0 (2972)	6G72	MFI	6	DOHC
1995	Mirage	1.5 (1468)	4G15	MFI	4	SOHC
	Mirage	1.8 (1834)	4G93	MFI	4	SOHC
	Eclipse	2.0 (1996)	420A	MFI	4	DOHC
	Eclipse	2.0 (1997)	4G63	MFI	4	DOHC
	Galant	2.4 (2350)	4G64	MFI	4	SOHC
	Diamante	3.0 (2972)	6G72	MFI	6	SOHC
	Diamante	3.0 (2972)	6G72	MFI	6	DOHC
	3000GT	3.0 (2972)	6G72	MFI	6	DOHC

ENGINE IDENTIFICATION

Year	Model	Engine Displacement Liters (cc)	Engine Series (ID/VIN)	Fuel System	No. of Cylinders	Engine Type
1996-97	Mirage	1.5 (1468)	4G15	MFI	4	SOHC
	Mirage	1.8 (1834)	4G93	MFI	4	SOHC
	Eclipse	2.0 (1996)	420A	MFI	4	DOHC
	Eclipse	2.0 (1997)	4G63	MFI	4	DOHC
	Eclipse	2.4 (2350)	4G64	MFI	4	SOHC
	Galant	2.4 (2350)	4G64	MFI	4	SOHC
	Diamante	3.0 (2972)	6G72	MFI	6	SOHC
	Diamante	3.0 (2972)	6G72	MFI	6	DOHC
	3000GT	3.0 (2972)	6G72	MFI	6	DOHC

MFI - Multiport fuel injection
SOHC - Single overhead camshaft
DOHC - Double overhead camshaft

GENERAL ENGINE SPECIFICATIONS

Year	Engine ID/VIN	Engine Displacement Liters (cc)	Fuel System Type	Net Horsepower @ rpm	Net Torque @ rpm (ft. lbs.)	Bore x Stroke (in.)	Compression Ratio	Oil Pressure @ rpm
1993	G4DJ	1.5 (1468)	MFI	81@5500	91@3000	2.97x3.23	9.4:1	54@2000
	4G15	1.5 (1468)	MFI	92@6000	93@3000	2.97x3.23	9.2:1	54@2000
	4G37	1.8 (1755)	MFI	92@5000	105@3500	3.17x3.39	9.0:1	41@2000
	4G93	1.8 (1834)	MFI	113@6000	116@4500	3.19x3.50	9.5:1	41@2000
	4G63 [1],[6]	2.0 (1997)	MFI	121@6000	120@4750	3.35x3.46	8.5:1	41@2000
	4G63 [1],[7]	2.0 (1997)	MFI	121@6000	120@4750	3.35x3.46	9.5:1	41@2000
	4G63 [2]	2.0 (1997)	MFI	135@6000	125@3000	3.35x3.46	9.0:1	41@2000
	4G63	2.0 (1997)	MFI	[4]	[5]	3.35x3.46	7.8:1	41@2000
	6G72 [1]	3.0 (2972)	MFI	175@5500	185@3000	3.59x2.99	10.0:1	30-80@2000
	6G72 [2]	3.0 (2972)	MFI	[8]	[9]	3.59x2.99	10.0:1	30-80@2000
	6G72 [3]	3.0 (2972)	MFI	300@6000	307@2500	3.59x2.99	8.0:1	30-80@2000
1994	G4DJ	1.5 (1468)	MFI	81@5500	91@3000	2.97x3.23	9.4:1	54@2000
	4G15	1.5 (1468)	MFI	92@6000	93@3000	2.97x3.23	9.2:1	54@2000
	4G37	1.8 (1755)	MFI	92@5000	105@3500	3.17x3.39	9.0:1	41@2000
	4G93	1.8 (1834)	MFI	113@6000	116@4500	3.19x3.50	9.5:1	41@2000
	4G63	2.0 (1997)	MFI	135@6000	125@3000	3.35x3.46	9.0:1	41@2000
	4G63 [3]	2.0 (1997)	MFI	[4]	[5]	3.35x3.46	7.8:1	41@2000
	4G64 [2]	2.4 (2350)	MFI	160@6000	160@4250	3.41x3.94	10.0:1	30-80@2000
	6G72 [1]	3.0 (2972)	MFI	175@5500	185@3000	3.59x2.99	10.0:1	30-80@2000
	6G72 [2]	3.0 (2972)	MFI	[8]	[10]	3.59x2.99	10.0:1	30-80@2000
	6G72 [3]	3.0 (2972)	MFI	320@6000	315@2500	3.59x2.99	8.0:1	30-80@2000
1995	4G15	1.5 (1468)	MFI	92@6000	93@3000	2.97x3.23	9.2:1	54@2000
	4G93	1.8 (1834)	MFI	[11]	116@4500	3.19x3.50	9.5:1	41@2000
	420A	2.0 (1996)	MFI	140@6000	130@4800	3.44x3.27	9.6:1	11@idle
	4G63	2.0 (1997)	MFI	[12]	[13]	3.35x3.46	8.5:1	11@idle
	4G64	2.4 (2350)	MFI	[14]	148@3000	3.41x3.94	9.5:1	41@2000
	6G72 [15]	3.0 (2972)	MFI	175@5500	185@3000	3.59x2.99	10.0:1	30-80@2000
	6G72 [16]	3.0 (2972)	MFI	202@6000	201@3500	3.59x2.99	10.0:1	30-80@2000
	6G72 [17]	3.0 (2972)	MFI	[18]	[19]	3.59x2.99	9.0:1	30-80@2000
	6G72 [20]	3.0 (2972)	MFI	[21]	205@4500	3.59x2.99	10.0:1	30-80@2000
	6G72 [22]	3.0 (2972)	MFI	320@6000	315@2500	3.59x2.99	8.0:1	30-80@2000
	6G74	3.5 (3497)	MFI	214@5000	228@3000	3.66x3.38	9.5:1	30-80@2000
1996-97	4G15	1.5 (1468)	MFI	92@6000	93@3000	2.97x3.23	9.2:1	54@2000
	4G93	1.8 (1834)	MFI	[11]	116@4500	3.19x3.50	9.5:1	41@2000
	420A	2.0 (1996)	MFI	140@6000	130@4800	3.44x3.27	9.6:1	11@idle
	4G63	2.0 (1997)	MFI	[12]	[13]	3.35x3.46	8.5:1	11@idle
	4G64	2.4 (2350)	MFI	[14]	148@3000	3.41x3.94	9.5:1	41@2000
	6G72 [15]	3.0 (2972)	MFI	175@5500	185@3000	3.59x2.99	10.0:1	30-80@2000
	6G72 [16]	3.0 (2972)	MFI	202@6000	201@3500	3.59x2.99	10.0:1	30-80@2000

GENERAL ENGINE SPECIFICATIONS

Year	Engine ID/VIN	Engine Displacement Liters (cc)	Fuel System Type	Net Horsepower @ rpm	Net Torque @ rpm (ft. lbs.)	Bore x Stroke (in.)	Compression Ratio	Oil Pressure @ rpm
1996-97	6G72 ⑰	3.0 (2972)	MFI	⑱	⑲	3.59x2.99	9.0:1	30-80@2000
	6G72 ⑳	3.0 (2972)	MFI	㉑	205@4500	3.59x2.99	10.0:1	30-80@2000
	6G72 ㉒	3.0 (2972)	MFI	320@6000	315@2500	3.59x2.99	8.0:1	30-80@2000
	6G74	3.5 (3497)	MFI	214@5000	228@3000	3.66x3.38	9.5:1	30-80@2000

MFI - Multiport fuel injection
BBL - Barrel carburetor
① Single overhead camshaft
② Double overhead camshaft
③ Turbocharged
④ Manual transmission: 195@6000
   Automatic transmission: 180@5500
⑤ Manual transmission: 203@3500
   Automatic transmission: 195@3000
⑥ 2 valves per cylinder

⑦ 4 valves per cylinder
⑧ 3000GT: 222@6000
   Diamante: 202@6000
⑨ 3000GT: 201@4500
   Diamante: 199@3000
⑩ 3000GT: 205@4500
   Diamante: 201@3500
⑪ California: 111@6000
   Except California: 113@6000

⑫ Manual transaxle: 210@6000
   Automatic transaxle: 205@6000
⑬ Manual transaxle: 214@3000
   Automatic transaxle: 220@3000
⑭ California: 138@5500
   Except California: 141@5500
⑮ Diamante SOHC, 2 valves per cylinder
⑯ Diamante DOHC, 4 valves per cylinder
⑰ Montero SOHC, 4 valves per cylinder

⑱ California: 168@5500
   Except California: 177@5500
⑲ California: 183@4500
   Except California: 188@4500
⑳ 3000GT DOHC, 4 valves per cylinder
㉑ California: 218@6000
   Except California: 222@6000
㉒ 3000GT Spyder DOHC, 4 valves
   per cylinder, Twin Turbochargers

GASOLINE ENGINE TUNE-UP SPECIFICATIONS

Year	Engine ID/VIN	Engine Displacement Liters (cc)	Spark Plugs Gap (in.)	Ignition Timing (deg.) MT	Ignition Timing (deg.) AT	Fuel Pump (psi)	Idle Speed (rpm) MT	Idle Speed (rpm) AT	Valve Clearance In.	Valve Clearance Ex.
1993	G4DJ	1.5 (1468)	0.039-0.043	5B	5B	39	700	700	0.006	0.010
	4G15	1.5 (1468)	0.039-0.043	5B	5B	38	750	750	0.008	0.010
	4G37	1.8 (1755)	0.039-0.043	5B	5B	38	700	700	HYD	HYD
	4G93	1.8 (1834)	0.039-0.043	5B	5B	38	750 ②	750 ②	0.008	0.012
	4G63	2.0 (1997)	0.039-0.043	5B	5B	38	750	750	HYD	HYD
	4G63	2.0 (1997)	0.028-0.031	5B	5B	①	750	750	HYD	HYD
	6G72	3.0 (2972)	0.039-0.043	5B	5B	38	700	700	HYD	HYD
	6G72	3.0 (2972)	0.039-0.043	5B	5B	38	700	700	HYD	HYD
	6G72	3.0 (2972)	0.039-0.043	5B	5B	34	700	700	HYD	HYD
1994	G4DJ	1.5 (1468)	0.039-0.043	5B	5B	39	700	700	0.006	0.010
	4G15	1.5 (1468)	0.039-0.043	5B	5B	38	750	750	0.008	0.010
	4G37	1.8 (1755)	0.039-0.043	5B	5B	38	700	700	HYD	HYD
	4G93	1.8 (1834)	0.039-0.043	5B	5B	38	750	750	0.008	0.012
	4G63	2.0 (1997)	0.039-0.043	5B	5B	38	750	750	HYD	HYD
	4G63 ③	2.0 (1997)	0.028-0.031	5B	5B	①	750	750	HYD	HYD
	4G64 ⑦	2.4 (2350)	0.039-0.043	5B	5B	38	750	750	HYD	HYD
	4G64 ⑧	2.4 (2350)	0.039-0.043	5B	5B	38	800	800	HYD	HYD
	6G72 ⑦	3.0 (2972)	0.039-0.043	5B	5B	38	700	700	HYD	HYD
	6G72 ⑧	3.0 (2972)	0.039-0.043	5B	5B	38	700	700	HYD	HYD
	6G72 ③	3.0 (2972)	0.039-0.043	5B	5B	34	700	700	HYD	HYD
1995	G4DJ	1.5 (1468)	0.039-0.043	5B	5B	39	700	700	0.006	0.010
	4G15	1.5 (1468)	0.039-0.043	5B	5B	38	750	750	0.008	0.010
	4G37	1.8 (1755)	0.039-0.043	5B	5B	38	700	700	HYD	HYD
	4G93	1.8 (1834)	0.039-0.043	5B	5B	38	750	750	0.008	0.012
	4G63	2.0 (1997)	0.039-0.043	5B	5B	38	750	750	HYD	HYD
	4G63 ③	2.0 (1997)	0.028-0.031	5B	5B	①	750	750	HYD	HYD
	4G64 ⑦	2.4 (2350)	0.039-0.043	5B	5B	38	750	750	HYD	HYD
	4G64 ⑧	2.4 (2350)	0.039-0.043	5B	5B	38	800	800	HYD	HYD
	6G72 ⑦	3.0 (2972)	0.039-0.043	5B	5B	38	700	700	HYD	HYD
	6G72 ⑧	3.0 (2972)	0.039-0.043	5B	5B	38	700	700	HYD	HYD
	6G72 ③	3.0 (2972)	0.039-0.043	5B	5B	34	700	700	HYD	HYD
1996-97	4G15	1.5 (1468)	0.039-0.043	5B	5B	38 ④	650-850	650-850	0.008 ⑤	0.010 ⑤
	4G93	1.8 (1834)	0.039-0.043	5B	5B	38 ④	600-800	650-850	0.008 ⑤	0.012 ⑤
	420A	2.0 (1996)	0.033-0.038	12B	12B	38 ④	700-900	700-900	HYD ⑤	HYD ⑤

GASOLINE ENGINE TUNE-UP SPECIFICATIONS

Year	Engine ID/VIN	Engine Displacement Liters (cc)	Spark Plugs Gap (in.)	Ignition Timing (deg.) MT	Ignition Timing (deg.) AT	Fuel Pump (psi)	Idle Speed (rpm) MT	Idle Speed (rpm) AT	Valve Clearance In.	Valve Clearance Ex.
1996-97	4G63 ③	2.0 (1997)	0.028-0.031	5B	5B	38 ④	650-850	650-850	HYD ⑤	HYD ⑤
	4G64	2.4 (2350)	0.039-0.043	5B	5B	38 ④	650-850	650-850	HYD ⑤	HYD ⑤
	6G72	3.0 (2972)	0.039-0.043	5B	5B	38 ④	600-800	600-800	HYD ⑤	HYD ⑤
	6G72	3.0 (2972)	0.039-0.043	5B	5B	38 ④	600-800	600-800	HYD ⑤	HYD ⑤
	6G72 ③	3.0 (2972)	0.039-0.043	5B	5B	34 ④	600-800	600-800	HYD ⑤	HYD ⑤
	6G74	3.5 (3497)	0.039-0.043	5B	5B	47-53 ⑥	600-800	600-800	HYD ⑤	HYD ⑤

NOTE: The Vehicle Emission Control Information label often reflects specification changes made during production. The label figures must be used if they differ from those in this chart.

B - Before top dead center
HYD - Hydraulic
① Manual transmission: 36
Automatic transmission: 43
② California: 700
③ Turbocharged
④ With vacuum hose connected
⑤ Hot engine
⑥ With vacuum hose disconnected
⑦ Single Overhead Camshaft
⑧ Double Overhead Camshaft

CAPACITIES

Year	Model	Engine ID/VIN	Engine Displacement Liters (cc)	Engine Oil with Filter	Transmission (pts.) 4-Spd	Transmission (pts.) 5-Spd	Transmission (pts.) Auto.	Transfer Case (pts.)	Drive Axle Front (pts.)	Drive Axle Rear (pts.)	Fuel Tank (gal.)	Cooling System (qts.)
1993	Precis	G4DJ	1.5 (1468)	3.6	3.6	3.8	12.8	NA	NA	NA	11.9	5.6
	Mirage	4G15	1.5 (1468)	3.5	-	3.8	12.6	NA	NA	NA	13.2	5.3
	Mirage	4G93	1.8 (1834)	4.0	-	3.8	12.6	NA	NA	NA	13.2	5.3
	Eclipse	4G37	1.8 (1755)	4.1	-	3.8	12.8	NA	NA	NA	15.9	6.6
	Eclipse	4G63	2.0 (1997)	②	-	⑧	③	1.2	NA	1.48	15.9	7.6
	Galant	4G63	2.0 (1997)	⑩	-	⑨	④	1.2	NA	1.48	⑤	7.6
	3000GT	6G72	3.0 (2972)	⑥	-	①	15.8	⑦	NA	2.3	19.8	8.5
	Diamante	6G72	3.0 (2972)	4.5	-	-	15.8	NA	NA	2.3	NA	8.5
1994	Precis	G4DJ	1.5 (1468)	3.6	3.6	3.8	12.8	NA	NA	NA	11.9	5.6
	Mirage	4G15	1.5 (1468)	3.5	-	3.8	12.6	NA	NA	NA	13.2	5.3
	Mirage	4G93	1.8 (1834)	4.0	-	3.8	12.6	NA	NA	NA	13.2	5.3
	Eclipse	4G37	1.8 (1755)	4.1	-	3.8	12.8	NA	NA	NA	15.9	6.6
	Eclipse	4G63	2.0 (1997)	②	-	⑧	③	1.2	NA	1.48	15.9	7.6
	Galant	4G64	2.4 (2350)	4.7	-	4.4	⑪	NA	NA	NA	16.9	7.4
	3000GT	6G72	3.0 (1972)	⑥	-	①	15.8	⑦	NA	2.3	19.8	8.5
	Diamante	6G72	3.0 (1972)	4.5	-	-	15.8	NA	NA	NA	19.0	8.5
1995	Precis	G4DJ	1.5 (1468)	3.6	3.6	3.8	12.8	NA	NA	NA	11.9	5.6
	Mirage	4G15	1.5 (1468)	3.5	-	3.8	12.6	NA	NA	NA	13.2	5.3
	Mirage	4G93	1.8 (1834)	4.0	-	3.8	12.6	NA	NA	NA	13.2	5.3
	Eclipse	4G37	1.8 (1755)	4.1	-	3.8	12.8	NA	NA	NA	15.9	6.6
	Eclipse	4G63	2.0 (1997)	②	-	⑧	③	1.2	NA	1.48	15.9	7.6
	Galant	4G64	2.4 (2350)	4.7	-	4.4	⑪	NA	NA	NA	16.9	7.4
	3000GT	6G72	3.0 (1972)	⑥	-	①	15.8	⑦	NA	2.3	19.8	8.5
	Diamante	6G72	3.0 (1972)	4.5	-	-	15.8	NA	NA	NA	19.0	8.5
1996-97	Mirage	4G15	1.5 (1468)	3.7	-	3.8	12.6	NA	NA	NA	13.2	5.3
	Mirage	4G93	1.8 (1834)	4.2	-	3.8	12.6	NA	NA	NA	13.2	6.3
	Eclipse	420A	2.0 (1996)	4.5	-	4.2	18.2	NA	NA	NA	17.0	7.4
	Eclipse	4G63	2.0 (1997)	4.5	-	⑬	14.2	1.2	NA	1.48	17.0	7.4
	Eclipse	4G64	2.4 (2350)	4.5	-	4.2	12.8	NA	NA	NA	17.0	14.8
	Galant	4G64	2.4 (2350)	4.5	-	4.6	12.6	NA	NA	NA	16.9	7.4

CAPACITIES

| Year | Model | Engine ID/VIN | Engine Displacement Liters (cc) | Engine Oil with Filter | Transmission (pts.) | | | Transfer Case (pts.) | Drive Axle | | Fuel Tank (gal.) | Cooling System (qts.) |
					4-Spd	5-Spd	Auto.		Front (pts.)	Rear (pts.)		
1996-97	Diamante	6G72	3.0 (2972)	4.5	-	-	15.8	NA	NA	NA	19.0	8.5
	3000GT	6G72	3.0 (2972)	4.5	-	①	15.8	⑭	NA	2.3	19.8	8.5

NA - Not Available

① FWD: 4.8 pts.
   AWD: 5.0 pts.
② Without Turbocharger: 4.6 qts.
   With Turbocharger: 5.1 qts.
③ Without Turbocharger: 12.8 pts.
   With Turbocharger: 14.8 pts.

④ FWD: 12.8 pts.
   AWD: 13.8 pts.
⑤ FWD: 15.9 gals.
   AWD: 16.4 gals.
⑥ Without Turbocharger: 4.5 qts
   With Turbocharger: 4.9 qts.
⑦ 5 speed: 0.58 pts.
   6 speed: 1.26 pts.

⑧ Without Turbocharger: 3.8 pts.
   With Turbocharger: 4.9 pts.
⑨ Single overhead camshaft: 3.8 pts.
   Double overhead camshaft: 4.8 pts.
⑩ FWD: 3.8 pts.
   AWD: 4.8 pts.
⑪ Single overhead camshaft: 12.6 pts.
   Double overhead camshaft: 15.8 pts.

⑫ Single overhead camshaft: 4.2 qts.
   Double overhead camshaft, without Turbocharger: 4.7 qts.
   Double overhead camshaft, with Turbocharger: 5.2 qts.
⑬ FWD: 4.2 qts.
   AWD: 4.6 qts.
⑭ M/T: 0.58 pts.
   A/T: 0.64 pts.

VALVE SPECIFICATIONS

| Year | Engine ID/VIN | Engine Displacement Liters (cc) | Seat Angle (deg.) | Face Angle (deg.) | Spring Test Pressure (lbs. @ in.) | Spring Installed Height (in.) | Stem-to-Guide Clearance (in.) | | Stem Diameter (in.) | |
							Intake	Exhaust	Intake	Exhaust
1993	G4DJ	1.5 (1468)	44-44.5	45-45.5	③	1.57	0.001-0.002	0.002-0.004	0.260	0.256
	4G15	1.5 (1468)	44-44.5	45-45.5	③	1.57	0.001-0.002	0.002-0.004	0.260	0.256
	4G37	1.8 (1755)	44-44.5	45-45.5	68@1.47	1.47	0.001-0.002	0.002-0.004	0.315	0.315
	4G93	1.8 (1834)	44-44.5	45-45.5	49@1.74	1.74	0.001-0.002	0.001-0.002	0.236	0.236
	4G63	2.0 (1997)	44-44.5	45-45.5	73@1.59	1.59	0.001-0.002	0.002-0.004	0.315	0.311
	4G63	2.0 (1997)	44-44.5	45-45.5	60@1.74	1.74	0.001-0.002	0.002-0.003	0.236	0.232
	4G63 ②	2.0 (1997)	44-44.5	45-45.5	66@1.57	1.57	0.001-0.002	0.002-0.004	0.260	0.256
	6G72	3.0 (2972)	44-44.5	45-45.5	72.5@1.59	1.59	0.001-0.002	0.002-0.004	0.315	0.311
	6G72 ②	3.0 (2972)	44-44.5	45-45.5	52.9@1.49	1.49	0.001-0.002	0.002-0.004	0.260	0.256
1994	G4DJ	1.5 (1468)	44-44.5	45-45.5	③	1.57	0.001-0.002	0.002-0.004	0.260	0.256
	4G15	1.5 (1468)	44-44.5	45-45.5	③	1.57	0.001-0.002	0.002-0.004	0.260	0.256
	4G37	1.8 (1755)	44-44.5	45-45.5	68@1.47	1.47	0.001-0.002	0.002-0.004	0.315	0.315
	4G93	1.8 (1834)	44-44.5	45-45.5	49@1.74	1.74	0.001-0.002	0.001-0.002	0.236	0.236
	4G63	2.0 (1997)	44-44.5	45-45.5	66@1.57	1.57	0.001-0.002	0.002-0.004	0.260	0.256
	4G64 ①	2.4 (2350)	44-44.5	45-45.5	60@1.74	1.74	0.001-0.002	0.002-0.003	0.236	0.232
	4G64 ②	2.4 (2350)	44-44.5	45-45.5	54@1.57	1.57	0.001-0.002	0.002-0.004	0.260	0.256
	6G72	3.0 (2972)	44-44.5	45-45.5	72.5@1.59	1.59	0.001-0.002	0.002-0.004	0.315	0.311

VALVE SPECIFICATIONS

Year	Engine ID/VIN	Engine Displacement Liters (cc)	Seat Angle (deg.)	Face Angle (deg.)	Spring Test Pressure (lbs. @ in.)	Spring Installed Height (in.)	Stem-to-Guide Clearance (in.) Intake	Stem-to-Guide Clearance (in.) Exhaust	Stem Diameter (in.) Intake	Stem Diameter (in.) Exhaust
1994	6G72 ②	3.0 (2972)	44-44.5	45-45.5	52.9@1.49	1.49	0.001-0.002	0.002-0.004	0.260	0.256
1995	G4DJ	1.5 (1468)	44-44.5	45-45.5	③	1.57	0.001-0.002	0.002-0.004	0.260	0.256
	4G15	1.5 (1468)	44-44.5	45-45.5	③	1.57	0.001-0.002	0.002-0.004	0.260	0.256
	4G37	1.8 (1755)	44-44.5	45-45.5	68@1.47	1.47	0.001-0.002	0.002-0.004	0.315	0.315
	4G93	1.8 (1834)	44-44.5	45-45.5	49@1.74	1.74	0.001-0.002	0.001-0.002	0.236	0.236
	4G63	2.0 (1997)	44-44.5	45-45.5	66@1.57	1.57	0.001-0.002	0.002-0.004	0.260	0.256
	4G64 ①	2.4 (2350)	44-44.5	45-45.5	60@1.74	1.74	0.001-0.002	0.002-0.003	0.236	0.232
	4G64 ②	2.4 (2350)	44-44.5	45-45.5	54@1.57	1.57	0.001-0.002	0.002-0.004	0.260	0.256
	6G72	3.0 (2972)	44-44.5	45-45.5	72.5@1.59	1.59	0.001-0.002	0.002-0.004	0.315	0.311
	6G72 ②	3.0 (2972)	44-44.5	45-45.5	52.9@1.49	1.49	0.001-0.002	0.002-0.004	0.260	0.256
1996-97	4G15	1.5 (1468)	44-44.5	45-45.5	③	1.57	0.0008-0.0020	0.0020-0.0035	0.260	0.256
	4G93	1.8 (1834)	44-44.5	45-45.5	49@1.74	1.74	0.0008-0.0016	0.0012-0.0024	0.236	0.236
	420A	2.0 (1996)	45	45-45.5	38@1.50	1.50	0.0019-0.0026	0.0029-0.0037	0.234	0.233
	4G63	2.0 (1997)	44-44.5	45-45.5	54@1.57	1.57	0.0008-0.0020	0.0020-0.0035	0.260	0.256
	4G64 ④	2.4 (2350)	44-44.5	45-45.5	73@1.59	1.59	0.0008-0.0020	0.0020-0.0035	0.315	0.311
	4G64 ⑤	2.4 (2350)	44-44.5	45-45.5	60@1.74	1.74	0.0008-0.0020	0.0008-0.0028	0.236	0.232
	6G72 ⑥	3.0 (2972)	44-44.5	45-45.5	72.5@1.59	1.59	0.0012-0.0024	0.0020-0.0035	0.315	0.311
	6G72 ⑦	3.0 (2972)	44-44.5	45-45.5	60@1.74	1.74	0.0008-0.0020	0.0016-0.0028	0.236	0.236
	6G72 ⑧	3.0 (2972)	44-44.5	45-45.5	52.9@1.49	1.49	0.0008-0.0020	0.0020-0.0035	0.260	0.256
	6G74	3.5 (3497)	44-44.5	45-45.5	52.9@1.49	1.49	0.0008-0.0020	0.0020-0.0035	0.260	0.256

① Single overhead camshaft
② Double overhead camshaft
③ Intake: 51 @ 1.57
　Exhaust: 64 @ 1.57
④ 8 valve SOHC
⑤ 16 valve SOHC
⑥ 12 valve SOHC
⑦ 24 valve SOHC
⑧ 24 valve DOHC

TORQUE SPECIFICATIONS
All readings in ft. lbs.

Year	Engine ID/VIN	Engine Displacement Liters (cc)	Cylinder Head Bolts	Main Bearing Bolts	Rod Bearing Bolts	Crankshaft Damper Bolts	Flywheel Bolts	Manifold Intake	Manifold Exhaust	Spark Plugs	Lug Nut
1993	G4DJ	1.5 (1468)	51-54	36-39	23-25	51-72	94-101	11-14	11-14	18	65-80
	4G15	1.5 (1468)	53	38	14.5 ①	61	98	13	13	18	65-80
	4G37	1.8 (1755)	53	38	25	87	98	13	13	18	87-101
	4G93	1.8 (1834)	⑦	18 ①	14.5 ①	134	72	14	③	18	65-80
	4G63	2.0 (1997)	⑦	18 ①	14.5 ①	87	98	26	20	18	⑨
	6G72 ④	3.0 (2972)	80	57	38	136	54	10	14	18	②
	6G72 ⑤	3.0 (2972)	80	67	38	136	54	10	14	18	②
	6G72 ⑥	3.0 (2972)	91 ⑧	54	38	136	54	10	14	18	②
1994	G4DJ	1.5 (1468)	51-54	36-39	23-25	51-72	94-101	11-14	11-14	18	65-80
	4G15	1.5 (1468)	53	38	14.5 ①	61	98	13	13	18	65-80
	4G37	1.8 (1755)	53	38	25	87	98	13	13	18	87-101
	4G93	1.8 (1834)	⑦	18 ①	14.5 ①	134	72	14	③	18	65-80
	4G63	2.0 (1997)	⑦	18 ①	14.5 ①	87	98	26	20	18	87-101
	4G93	1.8 (1834)	⑦	18 ①	14.5 ①	134	72	14	③	18	65-80
	4G93	1.8 (1834)	⑦	18 ①	14.5 ①	134	72	14	③	18	65-80
	4G93	1.8 (1834)	⑦	18 ①	14.5 ①	134	72	14	③	18	65-80
	4G93	1.8 (1834)	⑦	18 ①	14.5 ①	134	72	14	③	18	65-80
	4G93	1.8 (1834)	⑦	18 ①	14.5 ①	134	72	14	③	18	65-80
1995	G4DJ	1.5 (1468)	51-54	36-39	23-25	51-72	94-101	11-14	11-14	18	65-80
	4G15	1.5 (1468)	53	38	14.5 ①	61	98	13	13	18	65-80
	4G37	1.8 (1755)	53	38	25	87	98	13	13	18	87-101
	4G93	1.8 (1834)	⑦	18 ①	14.5 ①	134	72	14	③	18	65-80
	4G63	2.0 (1997)	⑦	18 ①	14.5 ①	87	98	26	20	18	87-101
	4G93	1.8 (1834)	⑦	18 ①	14.5 ①	134	72	14	③	18	65-80
	4G93	1.8 (1834)	⑦	18 ①	14.5 ①	134	72	14	③	18	65-80
	4G93	1.8 (1834)	⑦	18 ①	14.5 ①	134	72	14	③	18	65-80
	4G93	1.8 (1834)	⑦	18 ①	14.5 ①	134	72	14	③	18	65-80
	4G93	1.8 (1834)	⑦	18 ①	14.5 ①	134	72	14	③	18	65-80
1996-97	4G15	1.5 (1468)	53	38	14.5 ①	83	98	13	13	18	65-80
	4G93	1.8 (1834)	⑦	18 ①	14.5 ①	134	72	14	③	18	65-80
	420A	2.0 (1996)	⑩	55	20 ①	45	94-101	17	17	18	65-80
	4G63	2.0 (1997)	⑦	18 ①	14.5 ①	87	98	26	⑪	18	65-80
	4G64	2.4 (2350)	⑦	14.5 ①	14.5 ①	87	98	13	⑭	18	⑨
	6G72 ⑫	3.0 (2972)	80	57	38	136	54	10	14	18	⑨
	6G72 ⑬	3.0 (2972)	80	67	38	134	54	16	22	18	⑨
	6G72 ⑤	3.0 (2972)	80	67	38	134	54	10	33	18	⑨

① Torque to specification plus an additional 1/4 turn
② 3000GT: 87-101 ft. lbs.
   Diamante: 65-80 ft. lbs.
③ M10 bolts: 22-30 ft. lbs.
   M8 bolts: 11-14 ft. lbs.
④ Single overhead camshaft
⑤ Double overhead camshaft
⑥ Turbocharged

⑦ Step 1: 54 ft. lbs.
   Step 2: Loosen and retorque to 14.5 ft. lbs.
   Step 3: Torque an additional 1/4 turn
   Step 4: Torque an additional 1/4 turn
⑧ Torque to specification, loosen and retorque to value
⑨ Diamante and Galant: 65-80 ft. lbs.
   Eclipse and 3000GT: 85-100 ft. lbs.

⑩ Step 1:
   Bolts 1-6: 24 ft. lbs.
   Bolts 7-10: 20 ft. lbs.
   Step 2:
   Bolts 1-6: 48 ft. lbs.
   Bolts 7-10: 20 ft. lbs.
   Step 3: Repeat Step 2
   Step 4: Turn all bolts 1/4 turn

⑪ Bolts: 14 ft. lbs.
   Nuts: 26 ft. lbs.
⑫ 12 valve, SOHC
⑬ 24 valve, SOHC
⑭ M8 bolts: 20 ft. lbs.
   M10 bolts: 22 ft. lbs.

BRAKE SPECIFICATIONS
All measurements in inches unless noted

Year	Model		Master Cylinder Bore	Brake Disc			Brake Drum Diameter			Minimum Lining Thickness	
				Original Thickness	Minimum Thickness	Maximum Runout	Original Inside Diameter	Max. Wear Limit	Maximum Machine Diameter	Front	Rear
1993	Diamante	F	1.000	0.940	0.880	0.003	-	-	-	0.079	-
		R	1.000	0.710	0.650	0.003	-	-	-	-	0.079
	Eclipse ①	F	0.875 ②	0.940	0.880	0.003	-	-	-	0.079	-
		R	-	0.390	0.330	0.003	-	-	-	-	0.079
	Eclipse ③	F	1.000	0.940	0.880	0.003	-	-	-	0.079	-
		R	-	0.390	0.003	0.003	-	-	-	-	0.079
	Galant ④	F	0.875	0.940	0.880	0.003	-	-	-	0.079	-
		R	-	-	-	0.003	-	9.00	-	-	0.079
	Galant ⑤	F	0.938 ⑥	0.940	0.880	0.003	-	-	-	0.079	-
		R	-	0.390	0.330	0.003	-	9.00	-	-	0.079 ⑦
	Mirage	F	0.813 ②	⑧	⑨	0.003	-	⑪	-	0.079	-
		R	-	0.370	0.330	0.003	-	-	-	-	0.079 ⑦
	Precis		0.875	0.750	0.669	0.006	7.10	7.17	-	0.039	0.039
	3000GT ⑪	F	1.000 ⑫	0.940	0.880	0.003	-	-	-	0.079	-
		R	-	0.710	0.650	0.003	-	-	-	-	0.079
	3000GT ⑬	F	1.060	1.180	1.120	0.003	-	-	-	0.079	-
		R	-	0.790	0.720	0.003	-	-	-	-	0.079
1994	Diamante	F	1.000	0.940	0.880	0.003	-	-	-	0.079	-
		R	-	0.710	0.650	0.003	-	-	-	-	0.079
	Eclipse ①	F	0.875 ②	0.940	0.880	0.003	-	-	-	0.079	-
		R	-	0.390	0.330	0.003	-	-	-	-	0.079
	Eclipse ③	F	1.000	0.940	0.880	0.003	-	-	-	0.079	-
		R	-	0.390	0.330	0.003	-	-	-	-	0.079
	Galant ⑤	F	0.938 ⑥	0.940	0.880	0.003	-	-	-	0.079	-
		R	-	0.390	0.330	0.003	-	9.00	-	-	0.079 ⑦
	Mirage	F	0.813 ②	⑧	⑨	0.003	-	-	-	0.079	-
		R	-	0.370	0.330	0.003	-	⑥	-	-	0.079 ⑦
	Precis		0.875	0.750	0.669	0.006	7.10	-	7.17	0.039	0.039
	3000GT ⑪	F	1.000 ⑫	0.940	0.880	0.003	-	-	-	0.079	-
		R	-	0.710	0.650	0.003	-	-	-	-	0.079
	3000GT ⑩	F	1.060	1.180	1.120	0.003	-	-	-	0.079	-
		R	-	0.790	0.720	0.003	-	-	-	-	0.079
1995	Diamante	F	1.000	0.940	0.880	0.003	-	-	-	0.079	-
		R	-	0.710	0.650	0.003	-	-	-	-	0.079
	Eclipse	F	0.938 ⑭	0.940	0.880	0.003	-	-	-	0.079	⑦
		R	-	⑮	⑯	0.003	8.00	-	8.10	-	0.079
	Galant ⑤	F	0.938 ⑥	0.940	0.880	0.003	-	-	-	0.079	-
		R	-	0.390	0.330	0.003	-	9.00	-	-	0.079 ⑦
	Mirage	F	0.813 ⑰	⑧	⑨	0.003	-	-	-	0.079	-
		R	-	0.390	0.330	0.003	8.00	-	8.10	-	0.079 ⑦
	3000GT ⑪	F	1.000 ⑫	0.940	0.880	0.003	-	-	-	0.079	-
		R	-	0.710	0.650	0.003	-	-	-	-	0.079
	3000GT ⑩	F	1.060	1.180	1.120	0.003	-	-	-	0.079	-
		R	-	0.790	0.720	0.003	-	-	-	-	0.079
1996-97	Diamante	F	1.000	0.940	0.880	0.003	-	-	-	0.079	-
		R	-	0.710	0.650	0.003	-	-	-	-	0.079
	Eclipse	F	0.938 ⑭	0.940	0.880	0.003	-	-	-	0.079	-
		R	-	⑮	⑯	0.003	9.00	-	9.10	-	⑦

BRAKE SPECIFICATIONS
All measurements in inches unless noted

Year	Model		Master Cylinder Bore	Brake Disc			Brake Drum Diameter			Minimum Lining Thickness	
				Original Thickness	Minimum Thickness	Maximum Runout	Original Inside Diameter	Max. Wear Limit	Maximum Machine Diameter	Front	Rear
1996-97	Galant ⑤	F	0.938 ⑥	0.940	0.880	0.003	-	-	-	0.079	-
		R	-	0.390	0.330	0.003	-	9.00	-	-	0.079 ⑦
	Mirage	F	0.813 ⑰	⑧	⑨	0.003	-	-	-	0.079	-
		R	-	0.390	0.330	0.003	8.00	-	8.10	-	⑦
	3000GT ⑪	F	1.000 ⑫	0.940	0.880	0.003	-	-	-	0.079	-
		R	-	0.710	0.650	0.003	-	-	-	-	0.079
	3000GT ⑬	F	1.060	1.180	1.120	0.003	-	-	-	0.079	-
		R	-	0.790	0.720	0.003	-	-	-	-	0.079

① Non-Turbocharged
② With ABS: 0.938
③ Turbocharged
④ Single overhead camshaft
⑤ Double overhead camshaft

⑥ With ABS: 1.000
⑦ Drum shoe: 0.040
⑧ Front solid: 0.510
   Front ventilated: 0.710
⑨ Front solid: 0.450
   Front ventilated: 0.650

⑩ 7 inch drum: 7.20
   8 inch drum: 8.10
⑪ FWD
⑫ With ABS: 1.060
⑬ AWD
⑭ With ABS or AWD: 1.000

⑮ AWD: 0.790
   FWD: 0.390
⑯ AWD: 0.720
   FWD: 0.330
⑰ 4 door models with ABS: 0.938

FREQUENT MAINTENANCE LABOR
MITSUBISHI

The following should be used as a guide when determining the amount of work required for a particular service if taken to a repair shop.
In estimating how long a particular Frequent Maintenance Service item should take, please observe the following:
- **Chilton Time** is time that is based on field research and data supplied by the vehicle manufacturer.
- All labor time operations are given in hours and tenths of an hour.
- All labor operations, are to be used as a **guide**.

COOLING

(G) Winterize Cooling System
Includes: Run engine to check for leaks, tighten all hose connections. Test radiator and pressure cap, drain radiator and engine block. Add antifreeze and refill system.
1993-97 5

(G) Belt, Drive, Renew
1993-97
  V belt 5
  Serpentine belt 7

(G) Belt, Drive, Adjust
1993-97
  one 3
  each adtnl. 2

(G) Hoses, Radiator, Renew
Includes: Drain and refill cooling system as required.
Upper
  1993-94 Precis 4
  1993-97 Mirage 6
  1993 Galant 4
  1994-97 Galant 6

	Chilton Time
1993-94 Eclipse	.4
1995-97 Eclipse	.6
1993-97 Diamante	.6
1993-97 3000GT	.6
w/Turbo add	.3

Lower
  1993-94 Precis 5
  1993-97 Mirage 1.3
  1993 Galant 6
  1994-97 Galant 8
  1993-97 Eclipse 8
  1993-97 Diamante 6
  1993-97 3000GT 1.0
  w/Turbo add 3

(G) Thermostat, Coolant, Renew
1993-94 Precis 4
1993-97 Eclipse 6
1993-97 Mirage, Galant 6
1993-97 Diamante, 3000GT 6

FUEL

(M) Air Cleaner, Service
1993-97 5

(G) Filter, Fuel, Renew
1993-94 Precis 5
1993-97 Mirage
  in line 8
  in tank 2.0
1993 Galant
  in line 4
  in tank 7
1994-97 Galant
  in line 7
  in tank 1.2
1993-94 Eclipse
  in line 4
  in tank 8
1995-97 Eclipse
  in tank 1.2
1993-97 Diamante
  in line 5
  in tank
    Sedan 2.6
    w/4 WS add 2
    Wagon 1.1
1993-97 3000GT
  in line 9
  w/Turbo add 3
  in tank 8

FREQUENT MAINTENANCE LABOR (cont.)
MITSUBISHI

(G) Carburetor, Adjust (On Car)

1993-94 6

BRAKES

(G) Bleed Brakes (Four Wheels)
Includes: Add fluid.

1993-97 5
w/ABS add 4

(G) Brakes, Adjust (Minor)
Includes: Adjust brakes, fill master cylinder.

1993-97, two wheels 4

(M) Parking Brake, Adjust

1993-97 5

LUBRICATION SERVICES

(M) Engine Oil & Filter, Renew
Includes: Inspect and correct all fluid levels.

1993-97 3

(M) Lubricate Chassis, Change Oil & Filter
Includes: Inspect and correct all fluid levels.

1993-97 6
Install grease fittings add 1

(M) Lubricate Chassis, Change Oil & Filter
Includes: Inspect and correct all fluid levels.

Install grease fittings add 1

(M) Lubricate Chassis
Includes: Inspect and correct all fluid levels.

1993-97 4
Install grease fittings add 1

WHEELS

(G) Wheel, Renew (One)

1993-97 5

(G) Wheel, Balance

1993-97
one .3
each adtnl. 2

(G) Wheels, Rotate (All)

1993-97 5

ELECTRICAL

(G) Headlamps, Aim

1993-97
two .4
four .6

(G) Halogen Headlamp Bulb, Renew

1993-97 3000GT 6
1993-97 Eclipse 5
1993-97 Mirage, Galant 5
1993-97 Diamante 5

(G) License Lamp Assy., Renew

1993-94 Precis 4
1993-97 Mirage 3
1993-97 Galant 4
1993-97 Eclipse 3
1993-97 Diamante, 3000GT 3

(G) Park & Turn Signal Lamp Assy., Renew

1993-97 3

(G) Horn, Renew

1993-94 Precis 3
1993-97 Mirage 7
1993 Galant 4
1994-97 Galant 7
1993-94 Eclipse 4
1995-97 Eclipse 7
1993-97 Diamante 4
1993-97 3000GT 3

(M) Terminals, Battery, Clean

1993-97 3

SCHEDULED MAINTENANCE INTERVALS
(MITSUBISHI DIAMANTE, ECLIPSE, EXPO, GALANT, MIRAGE, PRECIS & 3000GT)

TO BE SERVICED	TYPE OF SERVICE	VEHICLE MILEAGE INTERVAL (x1000)												
		7.5	15	22.5	30	37.5	45	52.5	60	67.5	75	82.5	90	97.5
Engine oil & filter③	R	✓	✓	✓	✓	✓	✓	✓	✓	✓	✓	✓	✓	✓
Automatic transaxle fluid & filter	S/I		✓		✓		✓		✓		✓		✓	
Brake hoses	S/I		✓		✓		✓		✓		✓		✓	
Disc brake pads	S/I		✓		✓		✓		✓		✓		✓	
Driveshaft boots	S/I		✓		✓		✓		✓		✓		✓	
Valve clearance (Mirage)	S/I		✓		✓		✓		✓		✓		✓	
Valve clearance (Expo 1.8L)	S/I		✓		✓		✓		✓		✓		✓	
Air cleaner element	R				✓				✓				✓	
Engine coolant	R				✓				✓				✓	
Spark plugs (except Diamante & 3000GT w/platinum tip)	R				✓				✓				✓	
Spark plugs (Diamante & 3000GT w/platinum tip)	R								✓					
Ball joints & steering linkage seals	S/I				✓				✓				✓	
Drive belt(s)	S/I				✓				✓				✓	
Exhaust system	S/I				✓				✓				✓	
Fuel hoses	S/I				✓				✓				✓	
Manual transaxle oil (Mirage)	S/I				✓				✓				✓	
Manual transaxle oil (including transfer) (Eclipse, Expo & 3000GT)	S/I				✓				✓				✓	
Manual transaxle oil (Galant)	S/I				✓				✓				✓	
Rear axle oil (Eclipse, Expo AWD & 3000GT AWD)②	S/I				✓				✓				✓	
Rear drum brake linings & rear wheel cylinders (Eclipse, Expo, Galant & Mirage)	S/I				✓				✓				✓	
Ignition cables	R								✓					

SCHEDULED MAINTENANCE INTERVALS
(MITSUBISHI DIAMANTE, ECLIPSE, EXPO, GALANT, MIRAGE, PRECIS & 3000GT) (Cont.)

TO BE SERVICED	TYPE OF SERVICE	VEHICLE MILEAGE INTERVAL (x1000)												
		7.5	15	22.5	30	37.5	45	52.5	60	67.5	75	82.5	90	97.5
Timing belt(s)	R								✓					
Distributor cap & rotor (except 3000GT)	S/I								✓					
EVAP system (except canister)	S/I								✓					
Fuel system (tank, pipe line, connection & fuel tank filler tube cap)	S/I								✓					

① Replace every 30,000 miles.
② With LSD - replace every 30,000 miles.
③ 3000GT turbo - replace every 5000 miles.
R – Replace   S/I – Service or Inspect

FREQUENT OPERATION MAINTENANCE (SEVERE SERVICE)
   If a vehicle is operated under any of the following conditions it is considered severe service:
- Extremely dusty areas.
- 50% or more of the vehicle operation is in 32°C (90°F) or higher temperatures, or constant operation in temperatures below 0°C (32°F).
- Prolonged idling (vehicle operation in stop and go traffic).
- Frequent short running periods (engine does not warm to normal operating temperatures).
- Police, taxi, delivery usage or trailer towing usage.
Oil & oil filter change – change every 3000 miles.
Disc brake pads - service or inspect every 6000 miles.
Air filter element – service or inspect every 15,000 miles.
Automatic transaxle fluid & filter - replace every 15,000 miles.
Rear drum brake linings & rear wheel cylinders (Eclipse, Expo, Galant & Mirage)
Spark plugs (except Diamante & 3000GT w/platinum tip) - replace every 15,000 miles.
Manual transaxle oil (including transfer (Expo, Galant, Mirage & 3000GT) - replace every 30,000 miles.

NISSAN

ENGINE IDENTIFICATION

Year	Model	Engine Displacement Liters (cc)	Engine Series (ID/VIN)	Fuel System	No. of Cylinders	Engine Type
1993	Sentra	1.6 (1597)	GA16DE	MFI	4	DOHC
	Sentra	2.0 (1998)	SR20DE	MFI	4	DOHC
	Altima	2.4 (2389)	KA24DE	MFI	4	DOHC
	240SX	2.4 (2389)	KA24DE	MFI	4	DOHC
	300ZX	3.0 (2960)	VG30DE	MFI	6	DOHC
	300ZX	3.0 (2960)	VG30DETT	MFI	6	DOHC
	Maxima	3.0 (2960)	VG30E	MFI	6	SOHC
	Maxima	3.0 (2960)	VE30DE	MFI	6	DOHC
1994	Sentra	1.6 (1597)	GA16DE	MFI	4	DOHC
	Sentra	2.0 (1998)	SR20DE	MFI	4	DOHC
	Altima	2.4 (2389)	KA24DE	MFI	4	DOHC
	240SX	2.4 (2389)	KA24DE	MFI	4	DOHC

20 SPECIFICATIONS

ENGINE IDENTIFICATION

Year	Model	Engine Displacement Liters (cc)	Engine Series (ID/VIN)	Fuel System	No. of Cylinders	Engine Type
1994	300ZX	3.0 (2960)	VG30DE	MFI	6	DOHC
	300ZX	3.0 (2960)	VG30DETT	MFI	6	DOHC
	Maxima	3.0 (2960)	VG30E	MFI	6	SOHC
	Maxima	3.0 (2960)	VE30DE	MFI	6	DOHC
1995	Sentra	1.6 (1597)	GA16DE	MFI	4	DOHC
	Sentra	2.0 (1998)	SR20DE	MFI	4	DOHC
	Altima	2.4 (2389)	KA24DE	MFI	4	DOHC
	200SX	1.6 (1597)	GA16DE	MFI	4	DOHC
	200SX	2.0 (1998)	SR20DE	MFI	4	DOHC
	240SX	2.4 (2389)	KA24DE	MFI	4	DOHC
	300ZX	3.0 (2960)	VG30DE	MFI	6	DOHC
	300ZX	3.0 (2960)	VG30DETT	MFI	6	DOHC
	Maxima	3.0 (2960)	VG30E	MFI	6	SOHC
	Maxima	3.0 (2960)	VE30DE	MFI	6	DOHC
1996-97	Sentra	1.6 (1597)	GA16DE	MFI	4	DOHC
	200SX	1.6 (1597)	GA16DE	MFI	4	DOHC
	200SX	2.0 (1998)	SR20DE	MFI	4	DOHC
	Altima	2.4 (2389)	KA24DE	MFI	4	DOHC
	240SX	2.4 (2389)	KA24DE	MFI	4	DOHC
	300ZX	3.0 (2960)	VG30DE	MFI	6	DOHC
	300ZX	3.0 (2960)	VG30DETT	MFI	6	DOHC
	Maxima	3.0 (2988)	VQ30DE	MFI	6	DOHC

MFI - Multiport fuel injection
DOHC - Double overhead camshaft
SOHC - Single overhead camshaft

GENERAL ENGINE SPECIFICATIONS

Year	Engine ID/VIN	Engine Displacement Liters (cc)	Fuel System Type	Net Horsepower @ rpm	Net Torque @ rpm (ft. lbs.)	Bore x Stroke (in.)	Compression Ratio	Oil Pressure @ rpm
1993	GA16DE	1.6 (1597)	MFI	110@6000	108@4000	2.99x3.46	9.5:1	50@3000
	SR20DE	2.0 (1998)	MFI	140@6400	132@4800	3.39x3.39	9.5:1	46@3200
	KA24DE	2.4 (2389)	MFI	①	③	3.50x3.78	10.5:1	60@3000
	VG30E	3.0 (2960)	MFI	160@5200	182@2800	3.43x3.27	9.0:1	53@3200
	VE30DE	3.0 (2960)	MFI	190@5600	190@4000	3.43x3.27	10.0:1	51@3000
	VG30DE	3.0 (2960)	MFI	222@6400	198@4800	3.43x3.27	10.5:1	51@3000
	VG30DETT	3.0 (2960)	MFI	②	283@3600	3.43x3.27	8.5:1	51@3000
1994	GA16DE	1.6 (1597)	MFI	110@6000	108@4000	2.99x3.46	9.5:1	50@3000
	SR20DE	2.0 (1998)	MFI	140@6400	132@4800	3.39x3.39	9.5:1	46@3200
	KA24DE	2.4 (2389)	MFI	①	③	3.50x3.78	④	60@3000
	VG30E	3.0 (2960)	MFI	160@5200	182@2800	3.43x3.27	9.0:1	53@3200
	VE30DE	3.0 (2960)	MFI	190@5600	190@4000	3.43x3.27	10.0:1	51@3000
	VG30DE	3.0 (2960)	MFI	222@6400	198@4800	3.43x3.27	10.5:1	51@3000
	VG30DETT	3.0 (2960)	MFI	②	283@3600	3.43x3.27	8.5:1	51@3000
1995	GA16DE	1.6 (1597)	MFI	110@6000	108@4000	2.99x3.46	9.5:1	50@3000
	SR20DE	2.0 (1998)	MFI	140@6400	132@4800	3.39x3.39	9.5:1	46@3200
	KA24DE	2.4 (2389)	MFI	①	③	3.50x3.78	④	60@3000
	VG30E	3.0 (2960)	MFI	160@5200	182@2800	3.43x3.27	9.0:1	53@3200
	VE30DE	3.0 (2960)	MFI	190@5600	190@4000	3.43x3.27	10.0:1	51@3000
	VG30DE	3.0 (2960)	MFI	222@6400	198@4800	3.43x3.27	10.5:1	51@3000
	VG30DETT	3.0 (2960)	MFI	②	283@3600	3.43x3.27	8.5:1	51@3000

GENERAL ENGINE SPECIFICATIONS

Year	Engine ID/VIN	Engine Displacement Liters (cc)	Fuel System Type	Net Horsepower @ rpm	Net Torque @ rpm (ft. lbs.)	Bore x Stroke (in.)	Compression Ratio	Oil Pressure @ rpm
1996-97	GA16DE	1.6 (1597)	MFI	115@6000	108@4000	2.99x3.46	9.9:1	50@3000
	SR20DE	2.0 (1998)	MFI	140@6400	132@4800	3.39x3.39	9.5:1	46@3200
	KA24DE	2.4 (2389)	MFI	①	③	3.50x3.78	④	60@3000
	VG30DE	3.0 (2960)	MFI	222@6400	198@4800	3.43x3.27	10.5:1	60@3000
	VG30DETT	3.0 (2960)	MFI	300@6400	283@3600	3.43x3.27	8.5:1	51@3000
	VQ30DE	3.0 (2988)	MFI	190@5600	205@4000	3.66x2.89	10.0:1	63@3000

MFI - Mutliport fuel injection
① 240SX: 155@5600
   Altima: 150@5600
② Manual transmission: 300@6400
   Automatic transmission: 280@6400
③ 240SX: 160@4400
   Altima: 154@4400
④ 240SX: 9.5:1
   Altima: 9.2:1

GASOLINE ENGINE TUNE-UP SPECIFICATIONS

Year	Engine ID/VIN	Engine Displacement Liters (cc)	Spark Plugs Gap (in.)	Ignition Timing (deg.) MT	AT	Fuel Pump (psi)	Idle Speed (rpm) MT	AT	Valve Clearance In.	Ex.
1993	GA16DE	1.6 (1597)	0.041	10B	10B	36 ①	700	800 ②	0.015 ③	0.016 ③
	SR20DE	2.0 (1998)	0.033 ④	15B	15B	36 ①	800	800 ②	HYD	HYD
	KA24DE	2.4 (2389)	0.041	20B	20B	33 ①	700	700 ②	0.014 ③	0.015 ③
	VG30E	3.0 (2960)	0.041	15B	15B	33 ①	750	750 ②	HYD	HYD
	VE30DE	3.0 (2960)	0.041	15B	15B	36 ①	750	750 ②	HYD	HYD
	VG30DE	3.0 (2960)	0.041	15B	15B	36 ①	700	770 ②	HYD	HYD
	VG30DETT	3.0 (2960)	0.041	15B	15B	36 ①	700	750 ②	HYD	HYD
1994	GA16DE	1.6 (1597)	0.041	10B	10B	36 ①	700	800 ②	0.015 ③	0.016 ③
	SR20DE	2.0 (1998)	0.033 ④	15B	15B	36 ①	800	800 ②	HYD	HYD
	KA24DE	2.4 (2389)	0.041	20B	20B	33 ①	700	700 ②	0.014 ③	0.015 ③
	VG30E	3.0 (2960)	0.041	15B	15B	33 ①	750	750 ②	HYD	HYD
	VE30DE	3.0 (2960)	0.041	15B	15B	36 ①	750	750 ②	HYD	HYD
	VG30DE	3.0 (2960)	0.041	15B	15B	36 ①	700	770 ②	HYD	HYD
	VG30DETT	3.0 (2960)	0.041	15B	15B	36 ①	700	750 ②	HYD	HYD
1995	GA16DE	1.6 (1597)	0.041	10B	10B	36 ①	700	800 ②	0.015 ③	0.016 ③
	SR20DE	2.0 (1998)	0.033 ④	15B	15B	36 ①	800	800 ②	HYD	HYD
	KA24DE	2.4 (2389)	0.041	20B	20B	33 ①	700	700 ②	0.014 ③	0.015 ③
	VG30E	3.0 (2960)	0.041	15B	15B	33 ①	750	750 ②	HYD	HYD
	VE30DE	3.0 (2960)	0.041	15B	15B	36 ①	750	750 ②	HYD	HYD
	VG30DE	3.0 (2960)	0.041	15B	15B	36 ①	700	770 ②	HYD	HYD
	VG30DETT	3.0 (2960)	0.041	15B	15B	36 ①	700	750 ②	HYD	HYD
1996-97	GA16DE	1.6 (1597)	0.041	8B	8B	36 ①	625 ⑤	725 ②	0.015 ③	0.016 ③
	SR20DE	2.0 (1998)	0.033	15B	15B	36 ①	800	800 ②	HYD	HYD
	KA24DE	2.4 (2389)	0.041 ⑥	20B	20B	33 ①	650	650 ②	0.015 ③	0.015 ③
	VG30DE	3.0 (2960)	0.041 ⑥	10B	10B	36 ①	700	770 ②	HYD	HYD
	VG30DETT	3.0 (2960)	0.041 ⑥	15B	15B	36 ①	700 ⑤	750 ②	HYD	HYD
	VQ30DE	3.0 (2988)	0.041 ⑥	15B	15B	34 ①	650	700 ②	0.014 ③	0.015 ③

NOTE: The Vehicle Emission Control Information label often reflects specification changes made during production. The label figures must be used if they differ from those in this chart.
B - Before top dead center
HYD - Hydraulic
① 1 System pressure at idle with vacuum hose connected; should increase to 43 psi when disconnected
② 2 Automatic transmission in neutral
③ Engine warm
④ Conventional - .033
   Platinum - .041
⑤ Canada: 750
⑥ Do not check or adjust gap on platinum-tipped spark plugs

CAPACITIES

Year	Model	Engine ID/VIN	Engine Displacement Liters (cc)	Engine Oil with Filter (qts.)	Transmission (pts.) 4-Spd	5-Spd	Auto.	Drive Axle Front (pts.)	Rear (pts.)	Fuel Tank (gal.)	Cooling System (qts.)
1993	Sentra	GA16DE	1.6 (1597)	3.4	5.9	6.1	14.7	-	-	13.0	5.6
	Sentra	SR20DE	2.0 (1998)	3.4	-	7.5	14.7	-	-	13.0	6.0
	Altima	KA24DE	2.4 (2389)	4.1	-	10.0	20.0	-	-	16.0	8.2
	240SX	KA24DE	2.4 (2389)	3.8	-	5.1	17.5	-	①	16.0	7.1
	300ZX	VG30DE	3.0 (2960)	3.6	-	5.9	17.5	-	3.1	19.0	9.5
	300ZX	VG30DETT	3.0 (2960)	3.6	-	5.9	17.2	-	3.9	19.0	9.5
	Maxima	VG30E	3.0 (2960)	4.1	-	10.0	15.5	-	-	18.0	9.3
	Maxima	VE30DE	3.0 (2960)	4.0	-	10.0	18.0	-	-	18.0	11.2
1994	Sentra	GA16DE	1.6 (1597)	3.4	-	6.1	14.8	-	-	13.0	5.6
	Sentra	SR20DE	2.0 (1998)	3.4	-	7.5	14.8	-	-	13.0	6.0
	Altima	KA24DE	2.4 (2389)	4.1	-	10.0	20.0	-	-	16.0	8.2
	240SX	KA24DE	2.4 (2389)	3.8	-	5.1	17.5	-	①	16.0	7.1
	300ZX	VG30DE	3.0 (2960)	3.6	-	5.9	17.5	-	3.1	19.0	9.5
	300ZX	VG30DETT	3.0 (2960)	3.6	-	5.9	17.3	-	3.9	19.0	9.5
	Maxima	VG30E	3.0 (2960)	4.1	-	4.8	15.5	-	-	18.0	9.3
	Maxima	VE30DE	3.0 (2960)	4.0	-	4.8	20.3	-	-	18.0	11.2
1995	Sentra	GA16DE	1.6 (1597)	3.4	-	④	14.8	-	-	13.2	②
	Sentra	SR20DE	2.0 (1998)	3.4	-	7.5	14.8	-	-	13.0	6.0
	200SX	GA16DE	1.6 (1597)	3.4	-	8.2	14.8	-	-	13.2	②
	200SX	SR20DE	2.0 (1998)	3.6	-	8.2	14.8	-	-	13.2	6.5
	Altima	KA24DE	2.4 (2389)	4.1	-	10.0	20.0	-	-	15.9	8.3
	240SX	KA24DE	2.4 (2389)	4.0	-	5.1	17.5	-	①	17.2	7.3
	300ZX	VG30DE	3.0 (2960)	3.6	-	5.9	17.6	-	2.8	③	9.5
	300ZX	VG30DETT	3.0 (2960)	3.6	-	5.9	17.2	-	3.9	18.7	9.5
	Maxima	VG30E	3.0 (2960)	4.1	-	4.8	15.5	-	-	18.0	9.3
	Maxima	VE30DE	3.0 (2960)	4.6	-	9.5	20.0	-	-	18.5	9.4
1996-97	Sentra	GA16DE	1.6 (1597)	3.5	-	④	14.8	-	-	13.2	⑤
	Sentra	SR20DE	2.0 (1998)	3.4	-	7.5	14.8	-	-	13.0	6.0
	200SX	GA16DE	1.6 (1597)	3.5	-	④	14.8	-	-	13.2	⑤
	200SX	SR20DE	2.0 (1998)	3.6	-	④	14.8	-	-	13.2	6.5
	Altima	KA24DE	2.4 (2389)	4.1	-	10.0	20.0	-	-	15.9	8.3
	240SX	KA24DE	2.4 (2389)	4.0	-	5.3	17.5	-	①	17.2	7.3
	300ZX	VG30DE	3.0 (2960)	3.6	-	5.9	17.6	-	2.8	③	9.5
	300ZX	VG30DETT	3.0 (2960)	3.6	-	5.9	17.2	-	3.9	18.7	9.5
	Maxima	VQ30DE	3.0 (2988)	4.3	-	9.5	20.0	-	-	18.5	9.4

① With limited slip: 3.1 pts.
   Standard: 2.8 pts.
② GA16DE with MT: 5.4 qts.
   GA16DE with AT: 5.6 qts.
③ 2 seaters and 2+2: 18.7 gals.
   Convertible: 18.2 gals.
④ RS5F31A: 6.5 pts.
   RS5F32V: 8.0 pts.
⑤ GA16DE with MT: 5.5 qts.
   GA16DE with AT: 6.0 qts.

VALVE SPECIFICATIONS

Year	Engine ID/VIN	Engine Displacement Liters (cc)	Seat Angle (deg.)	Face Angle (deg.)	Spring Test Pressure (lbs. @ in.)	Spring Installed Height (in.)	Stem-to-Guide Clearance (in.) Intake	Exhaust	Stem Diameter (in.) Intake	Exhaust
1993	GA16DE	1.6 (1597)	45	45.25-45.75	77@0.9945	NA	0.0008-0.0020	0.0016-0.0028	0.2152-0.2157	0.2144-0.2150
	SR20DE	2.0 (1998)	45	45.25-45.75	144@1.181	NA	0.0008-0.0021	0.0016-0.0029	0.2348-0.2354	0.2341-0.2346
	KA24DE	2.4 (2389)	45	45.25-45.75	123@1.024	NA	0.0008-0.0021	0.0016-0.0029	0.2742-0.2748	0.2734-0.2740

VALVE SPECIFICATIONS

Year	Engine ID/VIN	Engine Displacement Liters (cc)	Seat Angle (deg.)	Face Angle (deg.)	Spring Test Pressure (lbs. @ in.)	Spring Installed Height (in.)	Stem-to-Guide Clearance (in.) Intake	Stem-to-Guide Clearance (in.) Exhaust	Stem Diameter (in.) Intake	Stem Diameter (in.) Exhaust
1993	VG30E	3.0 (2960)	45	45.25-45.75	①	NA	0.0008-0.0021	0.0016-0.0029	0.2742-0.2748	0.3136-0.3138
	VE30DE	3.0 (2960)	45	45.25-45.75	120@1.043	NA	0.0008-0.0021	0.0016-0.0029	0.2352-0.2354	0.2344-0.2346
	VG30DE	3.0 (2960)	45	45.25-45.75	120@1.043	NA	0.0008-0.0021	0.0016-0.0029	0.2348-0.2354	0.2341-0.2346
	VG30DETT	3.0 (2960)	45	45.25-45.75	120@1.043	NA	0.0008-0.0021	0.0016-0.0029	0.2348-0.2354	0.2341-0.2346
1994	GA16DE	1.6 (1597)	45	45.25-45.75	77@0.9945	NA	0.0008-0.0020	0.0016-0.0028	0.2152-0.2157	0.2144-0.2150
	SR20DE	2.0 (1998)	45	45.25-45.75	144@1.181	NA	0.0008-0.0021	0.0016-0.0029	0.2348-0.2354	0.2341-0.2346
	KA24DE	2.4 (2389)	45	45.25-45.75	106@1.026	NA	0.0008-0.0021	0.0016-0.0029	0.2742-0.2748	0.2734-0.2740
	VG30E	3.0 (2960)	45	45.25-45.75	①	NA	0.0008-0.0021	0.0016-0.0029	0.2742-0.2748	0.3136-0.3138
	VE30DE	3.0 (2960)	45	45.25-45.75	②	NA	0.0008-0.0021	0.0016-0.0029	0.2352-0.2354	0.2344-0.2346
	VG30DE	3.0 (2960)	45	45.25-45.75	②	NA	0.0008-0.0021	0.0016-0.0029	0.2348-0.2354	0.2341-0.2346
	VG30DETT	3.0 (2960)	45	45.25-45.75	②	NA	0.0008-0.0021	0.0016-0.0029	0.2348-0.2354	0.2341-0.2346
1995	GA16DE	1.6 (1597)	45	45.25-45.75	77@0.9945	NA	0.0008-0.0020	0.0016-0.0028	0.2152-0.2157	0.2144-0.2150
	SR20DE	2.0 (1998)	45	45.25-45.75	144@1.181	NA	0.0008-0.0021	0.0016-0.0029	0.2348-0.2354	0.2341-0.2346
	KA24DE	2.4 (2389)	45	45.25-45.75	106@1.026	NA	0.0008-0.0021	0.0016-0.0029	0.2742-0.2748	0.2734-0.2740
	VG30E	3.0 (2960)	45	45.25-45.75	①	NA	0.0008-0.0021	0.0016-0.0029	0.2742-0.2748	0.3136-0.3138
	VE30DE	3.0 (2960)	45	45.25-45.75	②	NA	0.0008-0.0021	0.0016-0.0029	0.2352-0.2354	0.2344-0.2346
	VG30DE	3.0 (2960)	45	45.25-45.75	②	NA	0.0008-0.0021	0.0016-0.0029	0.2348-0.2354	0.2341-0.2346
	VG30DETT	3.0 (2960)	45	45.25-45.75	②	NA	0.0008-0.0021	0.0016-0.0029	0.2348-0.2354	0.2341-0.2346
1996-97	GA16DE	1.6 (1597)	45	45.25-45.75	77@0.995	NA	0.0008-0.0020	0.0016-0.0028	0.2152-0.2157	0.2144-0.2150
	SR20DE	2.0 (1998)	45	45.25-45.75	137@1.181	NA	0.0008-0.0021	0.0016-0.0029	0.2348-0.2354	0.2341-0.2346
	KA24DE	2.4 (2389)	45	45.25-45.75	123@1.024	NA	0.0008-0.0021	0.0016-0.0029	0.2742-0.2748	0.2734-0.2740
	VG30DE	3.0 (2960)	45	45.25-45.75	120@1.043	NA	0.0008-0.0021	0.0016-0.0029	0.2348-0.2354	0.2341-0.2346
	VQ30DE	3.0 (2988)	45	45.25-45.75	102@1.085	NA	0.0008-0.0021	0.0016-0.0029	0.2348-0.2354	0.2341-0.2346
	VG30DETT	3.0 (2960)	45	45.25-45.75	120@1.043	NA	0.0008-0.0021	0.0016-0.0029	0.2348-0.2354	0.2341-0.2346

NA - Not Available

① Inner: 57.3 @ 0.984
Outer: 117.7 @ 1.181

② Maxima: 120 @ 1.059
300ZX: 120 @ 1.043

TORQUE SPECIFICATIONS
All readings in ft. lbs.

Year	Engine ID/VIN	Engine Displacement Liters (cc)	Cylinder Head Bolts	Main Bearing Bolts	Rod Bearing Bolts	Crankshaft Damper Bolts	Flywheel Bolts	Manifold Intake	Manifold Exhaust	Spark Plugs	Lug Nut
1993	GA16DE	1.6 (1597)	①	34-38	②	98-112	③	14	14	18	75
	SR20DE	2.0 (1998)	④	⑤	⑥	105-112	61-69	14	30	18	75
	KA24DE	2.4 (2389)	⑦	34-41	⑥	105-112	105-112	14	32	18	75
	VG30E	3.0 (2960)	⑪	67-74	⑥	90-98	72-80	⑫	15	18	75
	VE30DE	3.0 (2960)	③	67-74	④	123-130	61-69	⑫	㉓	18	75
	VG30DE	3.0 (2960)	⑬	67-74	⑥	123-130	61-69	⑮	18	18	75
	VG30DETT	3.0 (2960)	⑬	67-74	⑭	159-174	61-69	⑮	22	18	75
1994	GA16DE	1.6 (1597)	①	34-38	②	98-112	③	14	14	18	75
	SR20DE	2.0 (1998)	④	⑤	⑥	105-112	61-69	14	30	18	75
	KA24DE	2.4 (2389)	⑦	34-41	⑥	105-112	105-112	14	32	18	75
	VG30E	3.0 (2960)	⑪	67-74	⑥	90-98	72-80	⑫	15	18	75
	VE30DE	3.0 (2960)	③	67-74	④	123-130	61-69	⑫	㉓	18	75
	VG30DE	3.0 (2960)	⑬	67-74	⑥	123-130	61-69	⑮	18	18	75
	VG30DETT	3.0 (2960)	⑬	67-74	⑭	159-174	61-69	⑮	22	18	75
1995	GA16DE	1.6 (1597)	①	34-38	②	98-112	③	14	19	18	79
	SR20DE	2.0 (1998)	④	⑤	⑥	105-112	61-69	14	30	18	79
	KA24DE	2.4 (2389)	⑦	34-41	⑥	105-112	105-112	14	32	18	80
	VG30E	3.0 (2960)	⑪	67-74	⑥	90-98	72-80	⑫	15	18	75
	VE30DE	3.0 (2960)	③	67-74	④	123-130	61-69	⑫	㉓	18	75
	VG30DE	3.0 (2960)	⑲	67-74	⑯	159-174	61-69	⑳	22	18	80
	VG30DETT	3.0 (2960)	⑲	67-74	㉑	159-174	61-69	⑳	22	18	80
1996-97	GA16DE	1.6 (1597)	①	34-38	②	98-112	③	14	19	18	79
	SR20DE	2.0 (1998)	④	⑨	⑥	105-112	61-69	14	30	18	79
	KA24DE	2.4 (2389)	⑦	34-41	⑥	105-112	105-112	14	32	18	80
	VG30DE	3.0 (2960)	⑧	67-74	⑩	159-174	61-69	⑯	22	18	80
	VG30DETT	3.0 (2960)	⑧	67-74	⑩	159-174	61-69	⑯	22	18	80
	VQ30DE	3.0 (2988)	⑰	⑱	⑲	㉒	61-69	㉑	23	18	80

① Bolt Nos. 1-10:
 Step 1: 22 ft. lbs.
 Step 2: 43 ft. lbs.
 Step 3: Loosen completely then retorque to 22 ft. lbs.
 Step 4: 43 ft. lbs. or an additional 50-55 degrees
 Bolt Nos. 11-15: Torque last, to 72 inch lbs.
② Step 1: 12 ft. lbs.
 Step 2: 19 ft. lbs. or an additional 35-40 degrees
③ Manual transmission: 61-69 ft. lbs.
 Automatic transmission: 69-76 ft. lbs.
④ Step 1: 29 ft. lbs.
 Step 2: 58 ft. lbs.
 Step 3: Loosen completely then retorque to 30 ft. lbs.
 Step 4: Turn each bolt, in sequence, an additional 90-100 degrees
 Step 5: Repeat Step 4
⑤ Step 1: 28 ft. lbs.
 Step 2: 54-61 ft. lbs. or an additional 45-50 degrees
⑥ 12 ft. lbs. plus an additional 60-65 degrees
⑦ Step 1: 22 ft. lbs.
 Step 2: 58 ft. lbs.
 Step 3: Loosen completely then retorque to 22 ft. lbs.
 Step 4: 58 ft. lbs. or an additional 80-85 degrees

⑧ Step 1: 72 ft. lbs.
 Step 2: Loosen completely
 Step 3: 25-30 ft. lbs.
 Step 4: Turn all bolts 90-95 degrees
 Step 5: Repeat Step 4
⑨ Step 1: 20-24 ft. lbs.
 Step 2: 75-80 degrees
 Step 3: Loosen completely and retorque to 24-28 ft. lbs.
 Step 4: 45-50 degree turn
⑩ Step 1: 24-28 ft. lbs.
 Step 2: Turn all bolts 90-95 degrees
⑪ Step 1: 22 ft. lbs.
 Step 2: 43 ft. lbs.
 Step 3: Loosen completely then retorque to 22 ft. lbs.
 Step 4: 40-47 ft. lbs. or an additional 60-65 degrees
⑫ Step 1: Tighten nuts and bolts to 3 ft. lbs.
 Step 2: Tighten bolts to 12-14 ft. lbs; nuts to 17-20 ft. lbs.
 Step 3: Repeat Step 2
⑬ Step 1: 29 ft. lbs.
 Step 2: 90 ft. lbs.
 Step 3: Loosen completely then retorque to 29 ft. lbs.
 Step 4: 90 ft. lbs. or an additional 70 degrees

⑭ Step 1: 12 ft. lbs.
 Step 2: 43-48 ft. lbs. or an additional 60-65 degrees
⑮ 14 ft. lbs., in two steps
⑯ Step 1: 14-15 ft. lbs.
 Step 2: Turn nuts 90-95 degrees
⑰ Step 1: 29-36 ft. lbs.
 Step 2: Plus 60-65 degrees
⑱ Step 1: 3.6-7.2 ft. lbs.
 Step 2: 20-23 ft. lbs.
⑲ Step 1: 29 ft. lbs.
 Step 2: 90 ft. lbs.
 Step 3: Loosen completely and retorque to 25-33 ft. lbs.
 Step 4: Plus 90 ft. lbs. or 70 degrees
 Step 5: Tighten two bolts marked with an "X" to 7-9 ft. lbs.
⑳ Step 1: Nuts and bolts: 2.2-3.6 ft. lbs.
 Step 2: Nuts: 17-20 ft. lbs.; Bolts: 12-14 ft. lbs.
㉑ Step 1: 10-12 ft. lbs.
 Step 2: 43-48 ft. lbs. or an additional 60-65 degrees
㉒ Step 1: 2.2-3.6 ft. lbs.
 Step 2: 9-12 ft. lbs.
㉓ 18 ft. lbs., in two steps

BRAKE SPECIFICATIONS
All measurements in inches unless noted

Year	Model	Master Cylinder Bore	Brake Disc Original Thickness	Brake Disc Minimum Thickness	Maximum Runout	Brake Drum Diameter Original Inside Diameter	Brake Drum Diameter Max. Wear Limit	Brake Drum Diameter Maximum Machine Diameter	Minimum Lining Thickness Front	Minimum Lining Thickness Rear
1993	240SX	①	NA	②	0.003	-	-	-	0.079	0.059
	300ZX	③	NA	④	0.003	-	-	⑤	0.079	0.079
	Maxima	⑪	NA	⑫	0.003	9.000	-	9.060	0.079	⑧
	Sentra	⑨	NA	⑩	0.003	7.090	-	7.130	0.079	⑧
	Altima	⑪	NA	⑫	0.003	9.000	-	9.060	0.079	⑧
1994	240SX	①	NA	②	0.003	-	-	-	0.079	0.059
	300ZX	③	NA	④	0.003	-	-	⑤	0.079	0.079
	Maxima	⑪	NA	⑫	0.003	9.000	-	9.060	0.079	⑧
	Sentra	⑨	NA	⑩	0.003	7.090	-	7.130	0.079	⑧
	Altima	⑪	NA	⑫	0.003	9.000	-	9.060	0.079	⑧
1995	200SX	0.813	⑬	⑭	0.003	7.09	7.13	7.13	0.079	0.059
	240SX	①	NA	②	0.003	-	-	-	0.079	0.059
	300ZX	③	NA	④	0.003	-	-	⑤	0.079	0.079
	Maxima	⑪	NA	⑫	0.003	9.000	-	9.060	0.079	⑧
	Sentra	⑨	NA	⑩	0.003	7.090	-	7.130	0.079	⑧
	Altima	⑪	NA	⑫	0.003	9.000	-	9.060	0.079	⑧
1996-97	Sentra	0.813	⑬	⑭	0.003	7.09	7.13	7.13	0.079	0.059
	200SX	0.813	⑬	⑭	0.003	7.09	7.13	7.13	0.079	0.059
	240SX	⑮	⑯	⑰	0.003	-	-	-	0.079	0.079
	300ZX	1.000	⑱	⑲	0.003	-	-	6.81 ⑤	0.079	0.079
	Maxima	0.937	⑦	⑫	0.003	-	-	-	0.079	0.059
	Altima	⑪	⑥	⑫	0.003	9.00	NA	9.06	0.079	0.059

NA - Not Available

① With CL22VB front brake and M23 booster: 0.875
   With CL25VA front brake and M195T booster: 0.9375
② Front, without ABS: 0.709
   Front, with ABS: 0.787
   Rear: 0.315
③ Without ABS: 0.9375
   With ABS: 1.0625
④ Front: 1.102
   Rear: 0.630
⑤ Parking brake drum: 6.81
⑥ Front: 0.870
   Rear: 0.390

⑦ Front: 0.870
   Rear: 0.350
⑧ Disc brake: 0.079
   Drum brake: 0.059
⑨ Sentra XE and GXE models with 1.6L engine: 0.8125
   Sentra STD with 1.6L engine: 0.7500
   All other models: 0.8750
⑩ Front: AD22VF 0.945
   Front: AD18VE 0.630
   Rear: 0.236
⑪ With ABS: 1.000
   Without ABS: 0.9375
⑫ Front: 0.787
   Rear: 0.315

⑬ Front: 0.710
   Rear: 0.280
⑭ Front: 0.630
   Rear: 0.236
⑮ Manual transaxle without ABS: 0.875
   Automatic transaxle or all with ABS: 0.937
⑯ Front: 0.790
   Rear: 0.350
⑰ Front: 0.710
   Rear: 0.310
⑱ Front: 1.180
   Rear: 0.710
⑲ Front: 1.100
   Rear: 0.630

FREQUENT MAINTENANCE LABOR
NISSAN

The following should be used as a guide when determining the amount of work required for a particular service if taken to a repair shop. In estimating how long a particular Frequent Maintenance Service item should take, please observe the following:
- **Chilton Time** is time that is based on field research and data supplied by the vehicle manufacturer.
- All labor time operations are given in hours and tenths of an hour.
- All labor operations, are to be used as a **guide**.

COOLING

(G) Winterize Cooling System
Includes: Run engine to check for leaks, tighten all hose connections. Test radiator and pressure cap, drain radiator and engine block. Add antifreeze and refill system.
1993-97 5

(G) Belt, Drive, Renew
1993-97 7
  each adtnl. belt, add 4

(G) Belt, Drive, Adjust
1993-97
  one 4
  each adtnl. 1

(G) Hoses, Radiator, Renew
Includes: Drain and refill cooling system as required.
1993-97, each 6

(G) Thermostat, Coolant, Renew
1993-94 Maxima
  SOHC 1.0
  DOHC 1.8
1995-97 Maxima 1.0
1993-97 Sentra/NX 1.0
1993-97 Altima 8
1995-97 200SX 1.0

1993-97 240SX 1.2
1993-97 300ZX 2.2

FUEL

(M) Air Cleaner, Service
1993-97 fuel injected 4

(G) Filter, Fuel, Renew
1993-97 gas 5

BRAKES

(G) Bleed Brakes (Four Wheels)
Includes: Add fluid.
1993-97 6

(M) Parking Brake, Adjust
1993-97 4

LUBRICATION SERVICES

(M) Engine Oil & Filter, Renew
Includes: Inspect and correct all fluid levels.
1993-97 4

WHEELS

(G) Wheel, Renew (One)
1993-97 (FWD) 5

(G) Wheel, Balance
1993-97 (FWD)
  one 3
  each adtnl. 2

(G) Wheels, Rotate (All)
1993-97 (FWD) 5

ELECTRICAL

(G) Headlamps, Aim
1993-97
  two 4
  four 6

(G) Halogen Headlamp Bulb, Renew
1993-97, each 3

(G) Park & Turn Signal Lamp Bulb or Lens, Renew
1993-97, each 3

(G) Rear Combination Lamp Bulb, Renew
1993-97, each 3

(G) Horn, Renew
1993-97 Maxima 4
1993-97 Sentra/NX 4
1993-97 Altima 4
1995-97 200SX 4
1993-97 240SX 4
1993-97 300ZX 4

SCHEDULED MAINTENANCE INTERVALS
(NISSAN 200SX, 240SX, 300ZX, ALTIMA, MAXIMA & SENTRA/NX)

TO BE SERVICED	TYPE OF SERVICE	VEHICLE MILEAGE INTERVAL (x1000)												
		7.5	15	22.5	30	37.5	45	52.5	60	67.5	75	82.5	90	97.5
Engine oil & filter①	R	✓	✓	✓	✓	✓	✓	✓	✓	✓	✓	✓	✓	✓
Brake lines & cables	S/I		✓		✓		✓		✓		✓		✓	
Brake pads, discs, drums & linings	S/I		✓		✓		✓		✓		✓		✓	
Differential gear oil (240SX & 300ZX)	S/I		✓		✓		✓		✓		✓		✓	
Driveshaft boots (Altima, Maxima, Sentra/NX & 200SX)	S/I		✓		✓		✓		✓		✓		✓	
Exhaust system (300ZX)	S/I		✓		✓		✓		✓		✓		✓	

SCHEDULED MAINTENANCE INTERVALS
(NISSAN 200SX, 240SX, 300ZX, ALTIMA, MAXIMA & SENTRA/NX) (Cont.)

TO BE SERVICED	TYPE OF SERVICE	VEHICLE MILEAGE INTERVAL (x1000)												
		7.5	15	22.5	30	37.5	45	52.5	60	67.5	75	82.5	90	97.5
Exhaust system (except 300ZX)	S/I				✓				✓				✓	
Transmission or transaxle oil	S/I		✓		✓		✓		✓		✓		✓	
Air cleaner filter	R				✓				✓				✓	
Spark plugs (except below)	R				✓				✓				✓	
Spark plugs (platinum tip) (Sentra/NX, 200SX SR20DE, Maxima VE30DE, 300ZX & 1995-97 240SX)	R								✓					
Idle RPM (Sentra/NX & 200SX GA16DE)	S/I				✓				✓				✓	
Steering gear & linkage, axle & suspension parts	S/I				✓				✓				✓	
SUPER HICAS linkage (300ZX turbo & 1993-94 240SX)	S/I				✓				✓				✓	
Engine Coolant	R								✓					
Timing belt (1993-94 Maxima VE30DE & 1993-95 300ZX)	R								✓					
Drive belts	S/I								✓					
Fuel lines	S/I								✓					
Vapor lines	S/I								✓					

① 300ZX turbo - replace every 5000 miles.
R – Replace   S/I – Service or Inspect

FREQUENT OPERATION MAINTENANCE (SEVERE SERVICE)

If a vehicle is operated under any of the following conditions it is considered severe service:
- Extremely dusty areas.
- 50% or more of the vehicle operation is in 32°C (90°F) or higher temperatures, or constant operation in temperatures below 0°C (32°F).
- Prolonged idling (vehicle operation in stop and go traffic).
- Frequent short running periods (engine does not warm to normal operating temperatures).
- Police, taxi, delivery usage or trailer towing usage.

Oil & oil filter change (300ZX turbo) – change every 3000 miles.
Oil & oil filter change (except 300ZX turbo) – change every 3750 miles.
Brake pads & discs - service or inspect every 7500 miles.
Driveshaft boots (Altima, Maxima, Sentra/NX & 200SX) - service or inspect every 7500 miles.
Exhaust system - service or inspect every 7500 miles.
Steering gear & linkage, axle & suspension parts - service or inspect every 7500 miles.
Steering linkage ball joints & front suspension ball joints - service or inspect every 7500 miles.
SUPER HICAS linkage (300ZX turbo & 1993-94 240SX) - service or inspect every 7500 miles.
Air cleaner filter - service or inspect every 15,000 miles.

PORSCHE

VEHICLE IDENTIFICATION CHART

Code	Liters	Cu. In. (cc)	Cyl.	Fuel Sys.	Eng. Mfg.
M28/49, 50	5.4	329 (5397)	8	LH-Jetronic	Porsche
M44/43, 44	3.0	182 (2990)	4	DME	Porsche
M64/01, 02	3.6	220 (3602)	6	DME	Porsche
M64/07, 08	3.6	220 (3602)	6	DME	Porsche
M64/23, 24	3.6	220 (3600)	6	DME	Porsche
M64/50	3.6	220 (3602)	6	DME	Porsche
M64/60	3.6	220 (3600)	6	DME	Porsche

Model Year Code	Year
P	1993
R	1994
S	1995
T	1996
V	1997

DME - Digital motor electronic

ENGINE IDENTIFICATION

Year	Model	Engine Displacement Liters (cc)	Engine Series (ID/VIN)	Fuel System	No. of Cylinders	Engine Type
1993	968	3.0 (2990)	M44/43, 44	DME	4	DOHC
	911 Carrera 2/4	3.6 (3600)	M64/01, 02	DME	6	SOHC
	928 GTS	5.4 (5397)	M28/49, 50	LH	8	DOHC
1994	968	3.0 (2990)	M44/43, 44	DME	4	DOHC
	911 Carrera 2/4	3.6 (3600)	M64/01, 02	DME	6	SOHC
	911 Turbo	3.6 (3600)	M64/50	KE	6	SOHC
	928 GTS	5.4 (5397)	M28/49, 50	LH	8	DOHC
1995	968	3.0 (2990)	M44/43, 44	DME	4	DOHC
	911 Carrera	3.6 (3600)	M64/07, 08	DME	6	SOHC
	911 Carrera 4	3.6 (3600)	M64/07	DME	6	SOHC
	911 Turbo	3.6 (3600)	M64/50	KE	6	SOHC
	928 GTS	5.4 (5397)	M28/49, 50	LH-Jetronic	8	DOHC
1996-97	911 Carrera	3.6 (3600)	M64/23, 24	DME	6	SOHC
	911 Carrera 4	3.6 (3600)	M64/23	DME	6	SOHC
	911 Turbo	3.6 (3600)	M64/60	DME	6	SOHC

DME - Digital motor electronic
DOHC - Double overhead camshaft
KE - Bosch electronic CIS
LH - Bosch air flow-controlled
SOHC - Single overhead camshaft

GENERAL ENGINE SPECIFICATIONS

Year	Engine ID/VIN	Engine Displacement Liters (cc)	Fuel System Type	Net Horsepower @ rpm	Net Torque @ rpm (ft. lbs.)	Bore x Stroke (in.)	Compression Ratio	Oil Pressure @ rpm
1993	M44/43, 44	3.0 (2990)	DME	236@6200	225@4100	4.09x3.46	11.0:1	44@3000
	M64/01, 02	3.6 (3600)	DME	247@6100	228@4800	3.94x3.01	11.0:1	73@5000
	M28/49, 50	5.4 (5397)	LH	345@5700	369@4250	3.94x3.38	10.4:1	73@5000
1994	M44/43, 44	3.0 (2990)	DME	236@6200	225@4100	4.09x3.46	11.0:1	44@3000
	M64/01, 02	3.6 (3600)	DME	247@6100	228@4800	3.94x3.01	11.0:1	44@3000
	M64/50	3.6 (3600)	KE	355@5500	383@4500	3.94x3.01	7.5:1	73@5000
	M28/49, 50	5.4 (5397)	LH	345@5700	369@4250	3.94x3.38	10.4:1	73@5000

GENERAL ENGINE SPECIFICATIONS

Year	Engine ID/VIN	Engine Displacement Liters (cc)	Fuel System Type	Net Horsepower @ rpm	Net Torque @ rpm (ft. lbs.)	Bore x Stroke (in.)	Compression Ratio	Oil Pressure @ rpm
1995	M44/43, 44	3.0 (2990)	DME	236@6200	225@4100	4.09x3.46	11.0:1	44@3000
	M64/07, 08	3.6 (3600)	DME	272@6100	243@4800	3.94x3.01	11.3:1	73@5000
	M64/50	3.6 (3600)	KE	355@5500	383@4500	3.94x3.01	7.5:1	73@5000
	M28/49, 50	5.4 (5397)	LH-Jetronic	345@5700	369@4250	3.94x3.38	10.4:1	73@5000
1996-97	M64/23, 24	3.6 (3600)	DME	282@6300	250@5250	3.94x3.01	11.3:1	95@5000
	M64/60	3.6 (3600)	DME	400@5700	400@4500	3.94x3.01	8.0:1	95@5000

DME - Digital motor electronic   KE - Bosch electronic CIS   LH - Bosch air flow-controlled

GASOLINE ENGINE TUNE-UP SPECIFICATIONS

Year	Engine ID/VIN	Engine Displacement Liters (cc)	Spark Plugs Gap (in.)	Ignition Timing (deg.) MT	Ignition Timing (deg.) AT	Fuel Pump (psi)	Idle Speed (rpm) MT	Idle Speed (rpm) AT	Valve Clearance In.	Valve Clearance Ex.
1993	M44/43, 44	3.0 (2990)	0.028	10B	10B	53-59	840	880	HYD	HYD
	M64/01, 02	3.6 (3600)	0.032	-	-	45-52	880	880	0.004	0.004
	M28/49, 50	5.4 (5397)	0.028	10B	10B	75-80	675	700	HYD	HYD
1994	M44/43, 44	3.0 (2990)	0.028	10B	10B	53-59	840	880	HYD	HYD
	M64/50	3.6 (3600)	0.031	-	-	29-58	950	-	0.004	0.004
	M64/01, 02	3.6 (3600)	0.032	-	-	45-52	880	880	0.004	0.004
	M28/49, 50	5.4 (5397)	0.028	10B	10B	75-80	700	700	HYD	HYD
1995	M44/43, 44	3.0 (2990)	0.028	10B	10B	53-59	840	880	HYD	HYD
	M64/50	3.6 (3600)	0.031	-	-	29-58	950	-	0.004	0.004
	M64/07, 08	3.6 (3600)	0.032	-	-	45-52	880	880	0.004	0.004
	M28/49, 50	5.4 (5397)	0.028	10B	10B	75-80	700	700	HYD	HYD
1996-97	M64/23, 24	3.6 (3600)	0.026	①	①	53-59	800	750	HYD	HYD
	M64/60	3.6 (3600)	0.026	①	①	53-59	800	-	HYD	HYD

NOTE: The Vehicle Emission Control Information label often reflects specification changes made during production. The label figures must be used if they differ from those in this chart.
B - Before top dead center
HYD - Hydraulic
TDC - Top Dead Center
① Not adjustable

CAPACITIES

Year	Model	Engine ID/VIN	Engine Displacement Liters (cc)	Engine Oil with Filter	Transmission (pts.) 4-Spd	Transmission (pts.) 5-Spd	Transmission (pts.) Auto.	Transfer Case (pts.)	Drive Axle Front (pts.)	Drive Axle Rear (pts.)	Fuel Tank (gal.)	Cooling System (qts.)
1993	968	M44/43, 44	3.0 (2990)	7.4	-	5.8	14.8	-	-	1.4	19.6	8.5
	911 Carrera 2	M64/01, 02	3.6 (3600)	12.1	-	7.5	7.4	-	-	-	20.3	-
	911 Carrera 4	M64/01	3.6 (3600)	12.1	-	8.0	-	-	2.5	-	20.3	-
	928 GTS	M28/49, 50	5.4 (5397)	8.0	-	8.5	19.6	-	-	2.0	22.7	16.9
1994	968	M44/43, 44	3.0 (2990)	7.4	-	5.8	14.8	-	-	1.4	19.6	8.5
	911 Turbo	M64/50	3.6 (3600)	13.8	-	-	-	-	-	①	20.3	-
	911 Carrera 2	M64/01, 02	3.6 (3600)	12.1	-	7.5	7.4	-	-	-	20.3	-
	911 Carrera 4	M64/01, 02	3.6 (3600)	12.1	-	8.0	-	-	2.5	①	20.3	-
	928 GTS	M28/49, 50	5.4 (5397)	8.0	-	8.5	19.6	-	-	2.0	22.7	16.9
1995	968	M44/43, 44	3.0 (2990)	7.4	-	5.8	14.8	-	-	1.4	19.6	8.5
	911 Turbo	M64/50	3.6 (3600)	13.8	-	-	-	-	-	①	20.3	-
	911 Carrera 2	M64/07, 08	3.6 (3600)	12.1	-	7.5	7.4	-	-	-	20.3	-

CAPACITIES

Year	Model	Engine ID/VIN	Engine Displacement Liters (cc)	Engine Oil with Filter	Transmission (pts.) 4-Spd	5-Spd	Auto.	Transfer Case (pts.)	Drive Axle Front (pts.)	Rear (pts.)	Fuel Tank (gal.)	Cooling System (qts.)
1995	911 Carrera 4	M64/07, 08	3.6 (3600)	12.1	-	8.0	-	-	2.5	①	20.3	-
	928 GTS	M28/49, 50	5.4 (5397)	8.0	-	8.5	19.6	-	-	2.0	22.7	16.9
1996-97	911 Carrera	M64/23	3.6 (3600)	10.0	-	3.75 ②	-	-	-	-	16.0 ③	-
	911 Carrera	M64/24	3.6 (3600)	10.0	-	-	9.5	-	-	-	16.0 ③	-
	911 Carrera 4	M64/23	3.6 (3600)	10.0	-	3.75 ②	-	-	-	1.0	16.0 ③	-
	911 Turbo	M64/60	3.6 (3600)	10.0	-	3.75 ②	-	-	1.2	-	16.0 ③	-

NA - Not Available
① Manual transaxle is included in transmission
② 6 speed manual transaxle
③ Optional 20.2 gals.

VALVE SPECIFICATIONS

Year	Engine ID/VIN	Engine Displacement Liters (cc)	Seat Angle (deg.)	Face Angle (deg.)	Spring Test Pressure (lbs. @ in.)	Spring Installed Height (in.)	Stem-to-Guide Clearance (in.) Intake	Exhaust	Stem Diameter (in.) Intake	Exhaust
1993	M44/43, 44	3.0 (2990)	45	45	-	①	0.032	0.032	0.275	0.274
	M64/01, 02	3.6 (3600)	45	45	-	④	0.002	0.003	0.352	0.351
	M28/49, 50	5.4 (5397)	45	45	-	⑤	0.032	0.032	0.352	0.351
1994	M44/43, 44	3.0 (2990)	45	45	-	①	0.032	0.032	0.264	0.273
	M64/50	3.6 (3600)	45	45	②	③	0.030-0.060	0.050-0.080	0.275	0.274
	M64/01, 02	3.6 (3600)	45	45	-	④	0.002	0.003	0.353	0.352
	M28/49, 50	5.4 (5397)	45	45	-	⑤	0.032	0.032	0.352	0.351
1995	M44/43, 44	3.0 (2990)	45	45	-	①	0.032 ⑥	0.032 ⑥	0.274	0.273
	M64/07, 08	3.6 (3602)	45	45	-	④	0.032 ⑥	0.032 ⑥	0.275	0.274
	M64/50	3.6 (3600)	45	45	②	③	0.030-0.060	0.050-0.080	0.353	0.352
	M28/49, 50	5.4 (5397)	45	45	-	⑤	0.032 ⑥	0.032 ⑥	0.353	0.352
1996-97	M64/23, 24	3.6 (3600)	45	45	-	④	0.032 ⑥	0.032 ⑥	0.274	0.273
	M64/60	3.6 (3600)	45	45	-	④	0.032 ⑥	0.032 ⑥	0.353	0.352

① Intake: 1.496 Exhaust: 1.467
② Intake: 176@1.21 Exhaust: 165@1.25
③ Intake: 1.378 Exhaust: 1.398
④ Intake: 1.358 Exhaust: 1.319
⑤ Intake: 1.398 Exhaust: 1.358
⑥ With valve 0.39 in. off seat, rock back and forth with dial indicator set perpendicular to valve stem on valve head. Maximum dial indicator runout: 0.032 in.

TORQUE SPECIFICATIONS
All readings in ft. lbs.

Year	Engine ID/VIN	Engine Displacement Liters (cc)	Cylinder Head Bolts	Main Bearing Bolts	Rod Bearing Bolts	Crankshaft Damper Bolts	Flywheel Bolts	Manifold Intake	Exhaust	Spark Plugs	Lug Nut
1993	M44/43, 44	3.0 (2990)	①	②	③	210	④	15	15	18-22	96
	M64/01, 02	3.6 (3600)	⑦	⑨	⑤	173	④	15	15	18-22	96
	M28/49, 50	5.4 (5397)	⑤	⑥	③	218	④	15	15	18-22	96
1994	M44/43, 44	3.0 (2990)	①	②	③	215	④	15	15	18-22	96
	M64/50	3.6 (3600)	⑦	25	⑧	123	④	18	14-17	18-22	96
	M64/01, 02	3.6 (3600)	⑦	⑨	⑤	173	④	15	15	18-22	96
	M28/49, 50	5.4 (5397)	⑤	⑥	③	218	④	15	15	18-22	96

TORQUE SPECIFICATIONS
All readings in ft. lbs.

Year	Engine ID/VIN	Engine Displacement Liters (cc)	Cylinder Head Bolts	Main Bearing Bolts	Rod Bearing Bolts	Crankshaft Damper Bolts	Flywheel Bolts	Manifold Intake	Manifold Exhaust	Spark Plugs	Lug Nut
1995	M44/43, 44	3.0 (2990)	①	②	③	210	④	15	15	18-22	96
	M64/07, 08	3.6 (3602)	⑦	⑨	⑤	173	④	15	15	18-22	96
	M28/49, 50	5.4 (5397)	⑤	⑥	③	218	④	15	15	18-22	96
1996-97	M64/23, 24	3.6 (3600)	⑦	⑨	①	173	④	15	15	18-22	96
	M64/60	3.6 (3600)	⑦	⑨	①	173	④	15	15	18-22	96

① Step 1: 15 ft. lbs.
Step 2: Turn an additional 60 degrees
Step 3: Turn an additional 90 degrees

② M12 bolts:
Step 1: 22 ft. lbs.
Step 2: Turn an additional 60 degrees
M10 bolts:
Step 1: 15 ft. lbs.
Step 2: 37 ft. lbs.
M8 bolts: 15 ft. lbs.
M6 bolts: 7 ft. lbs.

③ Step 1: 18 ft. lbs.
Step 2: Turn an additional 90 degrees

④ Step 1: 29 ft. lbs.
Step 2: 66 ft. lbs.

⑤ Step 1: 15 ft. lbs.
Step 2: Turn an additional 90 degrees
Step 3: Turn an additional 90 degrees

⑥ M12 bolts:
Step 1: 22 ft. lbs.
Step 2: 41 ft. lbs.
Step 3: 59 ft. lbs.
M10 bolts:
Step 1: 15 ft. lbs.
Steo 2: 41 ft. lbs.
M8 bolts:
Step 1: 11 ft. lbs.
Step 2: 15 ft. lbs.

⑦ Step 1: 11 ft. lbs.
Step 2: Turn an additional 90 degrees

⑧ Step 1: 14 ft. lbs.
Step 2: Turn an additional 90 degrees

⑨ M10 bolts: 29 ft. lbs.
M8 bolts: 17 ft. lbs.

BRAKE SPECIFICATIONS
All measurements in inches unless noted

Year	Model		Master Cylinder Bore	Brake Disc Original Thickness	Brake Disc Minimum Thickness	Brake Disc Maximum Runout	Brake Drum Diameter Original Inside Diameter	Max. Wear Limit	Maximum Machine Diameter	Minimum Lining Thickness Front	Minimum Lining Thickness Rear
1993	911 Carrera 2/4	F	0.937 ⑤	1.102 ①	1.024 ③	0.004	-	-	-	0.079	0.079
		R	0.937 ⑤	0.945 ②	0.866 ④	0.004	-	-	-	0.079	0.079
	928 GTS	F	0.937	1.260	1.181	0.004	-	-	-	0.079	0.079
		R	0.813	0.945	0.866	0.004	-	-	-	0.079	0.079
	968	F	0.937	1.102 ⑥	1.024 ⑦	0.004	-	-	-	0.079	0.079
		R	0.813	0.945	0.866	0.004	-	-	-	0.079	0.079
1994	911 Carrera 2/4	F	0.937 ⑤	1.102 ①	1.024 ③	0.004	-	-	-	0.079	0.079
		R	0.937 ⑤	0.945 ②	0.866 ④	0.004	-	-	-	0.079	0.079
	911 Turbo	F	0.937	1.260	1.181	0.004	-	-	-	0.079	0.079
		R	0.937	1.102	1.024	0.004	-	-	-	0.079	0.079
	928 GTS	F	0.937	1.260	1.181	0.004	-	-	-	0.079	0.079
		R	0.813	0.945	0.866	0.004	-	-	-	0.079	0.079
	968	F	0.937	1.102 ⑥	1.024 ⑦	0.004	-	-	-	0.079	0.079
		R	0.813	0.945	0.866	0.004	-	-	-	0.079	0.079
1995	911 Carrera	F	0.940	1.102	1.024	0.004	-	-	-	0.079	-
		R	0.940	0.950	0.866	0.004	-	-	-	-	0.079
	911 Carrera 4	F	0.940	1.102	1.024	0.004	-	-	-	0.079	-
		R	0.940	0.950	0.866	0.004	-	-	-	-	0.079
	928 GTS	F	0.937	1.260	1.181	0.004	-	-	-	0.079	-
		R	0.937	0.945	0.866	0.004	-	-	-	-	0.079
	968	F	0.937	1.260	1.181	0.004	-	-	-	0.079	-
		R	0.937	0.945	0.866	0.004	-	-	-	-	0.079
1996-97	911 Carrera	F	0.940	1.102	1.024	0.004	-	-	-	0.079	-
		R	0.940	0.950	0.866	0.004	-	-	-	-	0.079
	911 Carrera 4	F	0.940	1.102	1.024	0.004	-	-	-	0.079	-
		R	0.940	0.950	0.866	0.004	-	-	-	-	0.079

BRAKE SPECIFICATIONS
All measurements in inches unless noted

Year	Model		Master Cylinder Bore	Brake Disc Original Thickness	Brake Disc Minimum Thickness	Brake Disc Maximum Runout	Brake Drum Diameter Original Inside Diameter	Brake Drum Diameter Max. Wear Limit	Brake Drum Diameter Maximum Machine Diameter	Minimum Lining Thickness Front	Minimum Lining Thickness Rear
1996-97	911 Turbo	F	1.000	1.102	1.024	0.004	-	-	-	0.079	-
		R	1.000	0.950	0.866	0.004	-	-	-	-	0.079

F - Front
R - Rear

① 911 Carrera 2 Turbo - Look option use 1.260
② 911 Carrera 2/4 Turbo - Look option use 1.102
③ 911 Carrera 2 Turbo - Look option use 1.181

④ 911 Carrera 2/4 Turbo - Look option use 1.024
⑤ 911 Carrera 2 without Turbo - Look option and 911 with RS America use 0.813
⑥ With sport running gear use 1.260
⑦ 7 With sport running gear use 1.181

FREQUENT MAINTENANCE LABOR
PORSCHE 911

The following should be used as a guide when determining the amount of work required for a particular service if taken to a repair shop. In estimating how long a particular Frequent Maintenance Service item should take, please observe the following:
- **Chilton Time** is time that is based on field research and data supplied by the vehicle manufacturer.
- All labor time operations are given in hours and tenths of an hour.
- All labor operations, are to be used as a **guide**.

COOLING

	Chilton Time
(G) Belt, Fan Drive, Renew	
1993-97 911	
wo/AC	.9
w/AC	1.2

FUEL

	Chilton Time
(G) Filter, Fuel, Renew	
1993-94 911 RS	.9
1993-94 911 Carrera 2/4	.9
1995-97 911 Carrera (993)	1.2

BRAKES

	Chilton Time
(G) Bleed Brakes (Four Wheels)	
Includes: Add fluid.	
1993-94 911 RS	2.0
1993-94 911 Carrera 2/4	2.0
1995-97 911 Carrera (993)	2.0

LUBRICATION SERVICES

	Chilton Time
(M) Engine Oil & Filter, Renew	
Includes: Inspect and correct all fluid levels.	
1993-97	.3

ELECTRICAL

	Chilton Time
(G) Headlamps, Aim	
1993-97	.3

	Chilton Time
(G) Halogen Headlamp Bulb, Renew	
1993-97	.6
(G) Parking Lamp Lens or Bulb, Renew	
1993-97	.6
(G) Tail Lamp Lens or Bulb, Renew	
1993-97	.3
(G) Horn, Renew	
1993-94 911 RS	
one	.8
both	.9
1993-94 911 Carrera 2/4	
one	.8
both	.9
1995-97 911 Carrera (993)	
one	.8
both	.9

FREQUENT MAINTENANCE LABOR
PORSCHE 928, 968

The following should be used as a guide when determining the amount of work required for a particular service if taken to a repair shop. In estimating how long a particular Frequent Maintenance Service item should take, please observe the following:

- **Chilton Time** is time that is based on field research and data supplied by the vehicle manufacturer.
- All labor time operations are given in hours and tenths of an hour.
- All labor operations, are to be used as a **guide**.

COOLING

(G) Winterize Cooling System
Includes: Run engine to check for leaks, tighten all hose connections. Test radiator and pressure cap, drain radiator and engine block. Add antifreeze and refill system.
1993-95 5

(G) Belt, Serpentine Drive, Renew
1993-95 968 . . . . 1.0

(G) Belt, Drive, Renew
1993-94 928S4/GT 1.0

(G) Hoses, Radiator, Renew
Includes: Drain and refill cooling system as required.
1993-95 928, 944, 968
one 1.5

(G) Thermostat, Coolant, Renew
1993-94 928 6
1993-95 968 . . . . 1.4

BRAKES

(G) Bleed Brakes (Four Wheels)
Includes: Add fluid.
1993-95 6

(M) Parking Brake, Adjust
1993-95 7

LUBRICATION SERVICES

(M) Engine Oil & Filter, Renew
Includes: Inspect and correct all fluid levels.
1993-95 3

ELECTRICAL

(G) Headlamps, Aim
1993-95
two .4
four .6

(G) License Lamp Assy., Renew
1993-95 968
one or all 3

(G) Parking Lamp Lens or Bulb, Renew
1993-95 3

(G) Tail Lamp Lens or Bulb, Renew
1993-95 3

(G) Horn, Renew
1993-94 928 5
1993-95 968
one 1.4
both 1.5

Chilton Time

SCHEDULED MAINTENANCE INTERVALS
(PORSCHE 911, 928 & 968)

TO BE SERVICED	TYPE OF SERVICE	VEHICLE MILEAGE INTERVAL (x1000)												
		7.5	15	22.5	30	37.5	45	52.5	60	67.5	75	82.5	90	97.5
Engine oil & filter	R	✓	✓	✓	✓	✓	✓	✓	✓	✓	✓	✓	✓	✓
Battery	S/I		✓		✓		✓				✓		✓	
Brake system	S/I		✓		✓		✓		✓		✓		✓	
Chassis lubrication	S/I		✓		✓		✓		✓		✓		✓	
Clutch adjustment	S/I		✓		✓		✓		✓		✓		✓	
Clutch hydraulic system	S/I		✓		✓		✓		✓		✓		✓	
Coolant level, hoses & connections	S/I		✓		✓		✓		✓		✓		✓	
Crankcase ventilation system	S/I		✓		✓		✓		✓		✓		✓	

SCHEDULED MAINTENANCE INTERVALS
(PORSCHE 911, 928 & 968) (Cont.)

TO BE SERVICED	TYPE OF SERVICE	VEHICLE MILEAGE INTERVAL (x1000)												
		7.5	15	22.5	30	37.5	45	52.5	60	67.5	75	82.5	90	97.5
Drive belts	S/I		✓		✓		✓		✓		✓		✓	
Fuel system	S/I		✓		✓		✓		✓		✓		✓	
Intake air system	S/I		✓		✓		✓		✓		✓		✓	
Parking brake system	S/I		✓		✓		✓		✓		✓		✓	
Steering system	S/I		✓		✓		✓		✓		✓		✓	
Suspension components & wheel bearings	S/I		✓		✓		✓		✓		✓		✓	
Throttle linkage	S/I		✓		✓		✓		✓		✓		✓	
Transmission fluid level	S/I		✓		✓		✓		✓		✓		✓	
Valve clearances	S/I		✓		✓		✓		✓		✓		✓	
Air filter element	R				✓				✓				✓	
Automatic transmission fluid & filter	R				✓				✓				✓	
Auxiliary air pump filter element	R				✓				✓				✓	
Brake fluid	R				✓				✓				✓	
Engine Coolant	R				✓				✓				✓	
Fuel filter	R				✓				✓				✓	
Limited slip differential fluid	R				✓				✓				✓	
Spark plugs	R				✓				✓				✓	
Automatic transmission differential oil	R								✓					
Axle drive fluid	R								✓					
Camshaft timing belt	R								✓					

R – Replace   S/I – Service or Inspect

FREQUENT OPERATION MAINTENANCE (SEVERE SERVICE)
If a vehicle is operated under any of the following conditions it is considered severe service:
- Extremely dusty areas.
- 50% or more of the vehicle operation is in 32°C (90°F) or higher temperatures, or constant operation in temperatures below 0°C (32°F).
- Prolonged idling (vehicle operation in stop and go traffic).
- Frequent short running periods (engine does not warm to normal operating temperatures).
- Police, taxi, delivery usage or trailer towing usage.
Oil & oil filter change – change every 3750 miles.
Air filter element – service or inspect every 15,000 miles.

SAAB

VEHICLE IDENTIFICATION CHART

		Engine Code					
Code	Liters	Cu. In. (cc)	Cyl.	Fuel Sys.	Eng. Mfg.		
B	2.3	140 (2290)	I4	MFI	Saab		
E	2.1	129 (2119)	I4	MFI	Saab		
L	2.0	121 (1985)	I4	MFI-Turbo	Saab		
M	2.3	140 (2290)	I4	MFI-Turbo	Saab		
N	2.0	121 (1985)	I4	MFI-Turbo	Saab		
R	2.3	140 (2290)	I4	MFI-Turbo	Saab		
U	2.3	140 (2290)	I4	MFI-Turbo	Saab		
V	2.5	152 (2498)	V6	MFI	GM		
W	3.0	180 (2961)	V6	MFI	GM		

Model Year	
Code	Year
P	1993
R	1994
S	1995
T	1996
V	1997

MFI - Multiport fuel injection

ENGINE IDENTIFICATION

Year	Model	Engine Displacement Liters (cc)	Engine Series (ID/VIN)	Fuel System	No. of Cylinders	Engine Type
1993	900	2.0 (1985)	B202L/L	MFI-Turbo	4	DOHC
	900	2.1 (2119)	B212I/E	MFI	4	DOHC
	9000	2.3 (2290)	B234I/B	MFI	4	DOHC
	9000	2.3 (2290)	B234L/M	MFI-Turbo	4	DOHC
1994	900	2.0 (1985)	B204L/N	MFI-Turbo	4	DOHC
	900	2.3 (2290)	B234I/B	MFI	4	DOHC
	900	2.5 (2498)	B258I/V	MFI	6	DOHC
	9000	2.3 (2290)	B234I/B	MFI	4	DOHC
	9000	2.3 (2290)	B234L/M	MFI-Turbo	4	DOHC
	9000	2.3 (2290)	B234R/R	MFI-Turbo	4	DOHC
	9000	2.3 (2290)	B234E/U	MFI-Turbo	4	DOHC
1995	900	2.0 (1985)	B204L/N	MFI-Turbo	4	DOHC
	900	2.3 (2290)	B234I/B	MFI	4	DOHC
	900	2.5 (2498)	B258I/V	MFI	6	DOHC
	9000	2.3 (2290)	B234I/B	MFI	4	DOHC
	9000	2.3 (2290)	B234L/M	MFI-Turbo	4	DOHC
	9000	2.3 (2290)	B234R/R	MFI-Turbo	4	DOHC
	9000	2.3 (2290)	B234E/U	MFI-Turbo	4	DOHC
1996-97	900	2.0 (1985)	B204L/N	MFI-Turbo	4	DOHC
	900	2.3 (2290)	B234I/B	MFI	4	DOHC
	900	2.5 (2498)	B258I/V	MFI	6	DOHC
	9000	2.3 (2290)	B234L/M	MFI-Turbo	4	DOHC
	9000	2.3 (2290)	B234R/R	MFI-Turbo	4	DOHC
	9000	2.3 (2290)	B234E/U	MFI-Turbo	4	DOHC
	9000	3.0 (2961)	B308I/W	MFI	6	DOHC

MFI - Multiport fuel injection
DOHC - Double overhead camshaft

GENERAL ENGINE SPECIFICATIONS

Year	Engine ID/VIN	Engine Displacement Liters (cc)	Fuel System Type	Net Horsepower @ rpm	Net Torque @ rpm (ft. lbs.)	Bore x Stroke (in.)	Compression Ratio	Oil Pressure @ rpm
1993	B202L/L	2.0 (1985)	MFI-Turbo	160@5500 ①	188@3000 ②	3.54x3.07	9.0:1	39@2000
	B212I/E	2.1 (2119)	MFI	140@6000	133@2900	3.66x3.07	10.1:1	39@2000
	B234I/B	2.3 (2290)	MFI	150@5500	157@3800	3.54x3.54	10.0:1	39@2000
	B234L/M	2.3 (2290)	MFI-Turbo	200@5000	244@2000 ③	3.54x3.54	8.5:1	39@2000
1994	B204L/N	2.0 (1985)	MFI-Turbo	185@5500	194@2100	3.54x3.07	9.2:1	39@2000
	B234I/B	2.3 (2290)	MFI	150@5700	155@4300	3.54x3.54	10.5:1	39@2000
	B234L/M	2.3 (2290)	MFI-Turbo	200@5500 ④	240@1800 ⑤	3.54x3.54	9.25:1	39@2000
	B234R/R	2.3 (2290)	MFI-Turbo	225@5500	253@1800	3.54x3.54	9.25:1	39@2000
	B234E/U	2.3 (2290)	MFI-Turbo	170@5700	192@3200	3.54x3.54	9.25:1	39@2000
	B258I/V	2.5 (2498)	MFI	170@5900	167@4200	3.21x3.13	10.8:1	NA
1995	B204L/N	2.0 (1985)	MFI-Turbo	185@5500	194@2100	3.54x3.07	9.2:1	39@2000
	B234I/B	2.3 (2290)	MFI	150@5700	155@4300	3.54x3.54	10.5:1	39@2000
	B234L/M	2.3 (2290)	MFI-Turbo	200@5500 ④	240@1800 ⑤	3.54x3.54	9.25:1	39@2000
	B234R/R	2.3 (2290)	MFI-Turbo	225@5500	253@1800	3.54x3.54	9.25:1	39@2000
	B234E/U	2.3 (2290)	MFI-Turbo	170@5700	192@3200	3.54x3.54	9.25:1	39@2000
	B258I/V	2.5 (2498)	MFI	170@5900	167@4200	3.21x3.13	10.8:1	NA
1996-97	B204L/N	2.0 (1985)	MFI-Turbo	185@5500	194@2100	3.54x3.07	9.2:1	39@2000
	B234I/B	2.3 (2290)	MFI	150@5700	155@4300	3.54x3.54	10.5:1	39@2000
	B234E/U	2.3 (2290)	MFI-Turbo	170@5700	192@3200	3.54x3.54	9.25:1	39@2000
	B234R/R	2.3 (2290)	MFI-Turbo	225@5500	153@1800	3.54x3.54	9.25:1	39@2000
	B234L/M	2.3 (2290)	MFI-Turbo	200@5500	240@1800	3.54x3.54	9.25:1	39@2000
	B258I/V	2.5 (2498)	MFI	170@5900	167@4200	3.21x3.13	10.8:1	NA
	B308I/W	3.0 (2961)	MFI	210@6200	200@3300	3.38x3.34	10.8:1	NA

MFI - Multiport fuel injection
NA - Not Available
① 900 SPG: 175@5500
② 900 SPG: 195@3000
③ 9000 with automatic transaxle: 222@2000
④ 9000 Aero, manual: 225@5500
⑤ 9000 Aero, manual: 258@1950

GASOLINE ENGINE TUNE-UP SPECIFICATIONS

Year	Engine ID/VIN	Engine Displacement Liters (cc)	Spark Plugs Gap (in.)	Ignition Timing (deg.) MT	Ignition Timing (deg.) AT	Fuel Pump (psi)	Idle Speed (rpm) MT	Idle Speed (rpm) AT	Valve Clearance In.	Valve Clearance Ex.
1993	B202L/L	2.0 (1985)	0.024-0.028	16B	16B	40 ①	850	850	HYD	HYD
	B212I/E	2.1 (2119)	0.024-0.028	14B	14B	41 ①	850	850	HYD	HYD
	B234I/B	2.3 (2290)	0.023-0.027	②	②	43 ①	850	850	HYD	HYD
	B234L/M	2.3 (2290)	0.023-0.027	②	②	40 ①	850	850	HYD	HYD
1994	B204L/N	2.0 (1985)	0.039	②	②	43 ①	850	850	HYD	HYD
	B234I/B	2.3 (2290)	0.024	②	②	43 ①	850	850	HYD	HYD
	B234L/M	2.3 (2290)	0.024	②	②	43 ①	850	850	HYD	HYD
	B234R/R	2.3 (2290)	0.024	②	②	43 ①	850	850	HYD	HYD
	B234E/U	2.3 (2290)	0.024	②	②	43 ①	850	850	HYD	HYD
	B258I/V	2.5 (2498)	0.031	②	②	43 ①	800	800	HYD	HYD
1995	B204L/N	2.0 (1985)	0.039	②	②	43 ①	850	850	HYD	HYD
	B234I/B	2.3 (2290)	0.024	②	②	43 ①	850	850	HYD	HYD
	B234L/M	2.3 (2290)	0.024	②	②	43 ①	850	850	HYD	HYD
	B234R/R	2.3 (2290)	0.024	②	②	43 ①	850	850	HYD	HYD
	B234E/U	2.3 (2290)	0.024	②	②	43 ①	850	850	HYD	HYD
	B258I/V	2.5 (2498)	0.031	②	②	43 ①	800	800	HYD	HYD
1996-97	B204L/N	2.0 (1985)	0.039	②	②	43 ①	850	850	HYD	HYD
	B234I/B	2.3 (2290)	0.024	②	②	43 ①	850	850	HYD	HYD

GASOLINE ENGINE TUNE-UP SPECIFICATIONS

Year	Engine ID/VIN	Engine Displacement Liters (cc)	Spark Plugs Gap (in.)	Ignition Timing (deg.) MT	Ignition Timing (deg.) AT	Fuel Pump (psi)	Idle Speed (rpm) MT	Idle Speed (rpm) AT	Valve Clearance In.	Valve Clearance Ex.
1996-97	B234L/M	2.3 (2290)	0.040	②	②	43 ①	850	850	HYD	HYD
	B234R/R	2.3 (2290)	0.040	②	②	43 ①	850	850	HYD	HYD
	B234E/U	2.3 (2290)	0.040	②	②	43 ①	850	850	HYD	HYD
	B258I/V	2.5 (2498)	0.031	②	②	43 ①	800	800	HYD	HYD
	B308I/W	3.0 (2961)	0.031	②	②	44 ①	800	800	HYD	HYD

NOTE: The Vehicle Emission Control Information label often reflects specification changes made during production. The label figures must be used if they differ from those in this chart.

B - Before top dead center
HYD - Hydraulic
① Fuel line pressure regulator before the control pressure regulator
② Pre-programmed in ECU and cannot be adjusted

CAPACITIES

Year	Model	Engine ID/VIN	Engine Displacement Liters (cc)	Engine Oil with Filter (qts.)	Transmission (pts.) 4-Spd	Transmission (pts.) 5-Spd	Transmission (pts.) Auto.	Transfer Case (pts.)	Drive Axle Front (pts.)	Drive Axle Rear (pts.)	Fuel Tank (gal.)	Cooling System (qts.)
1993	900	B202L/L	2.0 (1985)	4.5	-	6.0	17.0	-	2.6	-	18.0	10.5
	900	B212I/E	2.1 (2119)	4.2	-	6.0	17.0	-	2.6	-	18.0	10.5
	9000	B234I/B	2.3 (2290)	4.5	-	5.3	16.5	-	-	-	17.4	9.5
	9000	B234L/M	2.3 (2290)	4.5	-	5.3	16.5	-	-	-	17.4	9.5
1994	900	B204L/N	2.0 (1985)	5.5	-	3.6	14.8	-	-	-	18.0	8.7
	900	B234I/B	2.3 (2290)	5.1	-	3.6	14.8	-	-	-	18.0	8.7
	900	B258I/V	2.5 (2498)	4.6	-	3.6	14.8	-	-	-	18.0	8.2
	9000	B234I/B	2.3 (2290)	4.5	-	5.3	16.5	-	-	-	17.4	9.5
	9000	B234L/M	2.3 (2290)	4.5	-	5.3	16.5	-	-	-	17.4	9.5
	9000	B234R/R	2.3 (2290)	4.5	-	5.3	16.5	-	-	-	17.4	9.5
	9000	B234E/U	2.3 (2290)	5.1	-	7.2	14.0	-	-	-	17.4	9.5
1995	900	B204L/N	2.0 (1985)	5.5	-	3.6	14.8	-	-	-	18.0	8.7
	900	B234I/B	2.3 (2290)	5.1	-	3.6	14.8	-	-	-	18.0	8.7
	900	B258I/V	2.5 (2498)	4.6	-	3.6	14.8	-	-	-	18.0	8.2
	9000	B234I/B	2.3 (2290)	4.5	-	5.3	16.5	-	-	-	17.4	9.5
	9000	B234L/M	2.3 (2290)	4.5	-	5.3	16.5	-	-	-	17.4	9.5
	9000	B234R/R	2.3 (2290)	4.5	-	5.3	16.5	-	-	-	17.4	9.5
	9000	B234E/U	2.3 (2290)	5.1	-	7.2	14.0	-	-	-	17.4	9.5
1996-97	900	B204L/N	2.0 (1985)	4.1	-	7.2	13.6	-	-	-	18.0	8.7
	900	B234I/B	2.3 (2290)	4.1	-	7.2	13.6	-	-	-	18.0	8.7
	900	B258I/V	2.5 (2498)	4.6	-	-	13.6	-	-	-	18.0	8.2
	9000	B234L/M	2.3 (2290)	5.1	-	7.2	14.0	-	-	-	17.4	9.5
	9000	B234R/R	2.3 (2290)	5.1	-	7.2	-	-	-	-	17.4	9.5
	9000	B234E/U	2.3 (2290)	5.1	-	7.2	14.0	-	-	-	17.4	9.5
	9000	B308I/W	3.0 (2961)	4.9	-	-	14.0	-	-	-	17.4	9.0

VALVE SPECIFICATIONS

Year	Engine ID/VIN	Engine Displacement Liters (cc)	Seat Angle (deg.)	Face Angle (deg.)	Spring Test Pressure (lbs. @ in.)	Spring Installed Height (in.)	Stem-to-Guide Clearance (in.)		Stem Diameter (in.)	
							Intake	Exhaust	Intake	Exhaust
1993	B202L/L	2.0 (1985)	45	44.5	131-141@ 1.12	1.46	0.020	0.020	0.2740-0.2746	0.2738-0.2748
	B212I/E	2.1 (2119)	45	44.5	131-141@ 1.12	1.46	0.020	0.020	0.2740-0.2746	0.2738-0.2748
	B234I/B	2.3 (2290)	45	44.5	131-141@ 1.12	1.46	0.020	0.020	0.2740-0.2746	0.2738-0.2748
	B234L/M	2.3 (2290)	45	44.5	138-141@ 1.12	1.46	0.020	0.020	0.2740-0.2746	0.2738-0.2748
1994	B204L/N	2.0 (1985)	45	44.5	138-150@ 1.12	1.46	0.020	0.020	0.2740-0.2746	0.2738-0.2748
	B234I/B	2.3 (2290)	45	44.5	138-150@ 1.12	1.46	0.020	0.020	0.2740-0.2746	0.2738-0.2748
	B234L/M	2.3 (2290)	45	44.5	138-141@ 1.12	1.46	0.020	0.020	0.2740-0.2746	0.2738-0.2748
	B234R/R	2.3 (2290)	45	44.5	138-141@ 1.12	1.46	0.020	0.020	0.2740-0.2746	0.2738-0.2748
	B234E/U	2.3 (2290)	45	44.5	138-141@ 1.12	1.46	0.020	0.020	0.2740-0.2746	0.2738-0.2748
	B258I/V	2.5 (2498)	45	45.3	142-153@ 0.94	NA	NA	NA	0.2344-0.2350	0.2340-0.2346
1995	B204L/N	2.0 (1985)	45	44.5	138-150@ 1.12	1.46	0.020	0.020	0.2740-0.2746	0.2738-0.2748
	B234I/B	2.3 (2290)	45	44.5	138-150@ 1.12	1.46	0.020	0.020	0.2740-0.2746	0.2738-0.2748
	B234L/M	2.3 (2290)	45	44.5	138-141@ 1.12	1.46	0.020	0.020	0.2740-0.2746	0.2738-0.2748
	B234R/R	2.3 (2290)	45	44.5	138-141@ 1.12	1.46	0.020	0.020	0.2740-0.2746	0.2738-0.2748
	B234E/U	2.3 (2290)	45	44.5	138-141@ 1.12	1.46	0.020	0.020	0.2740-0.2746	0.2738-0.2748
	B258I/V	2.5 (2498)	45	45.3	142-153@ 0.94	NA	NA	NA	0.2344-0.2350	0.2340-0.2346
1996-97	B204L/N	2.0 (1985)	45	44.5	138-150@ 1.12	1.46	0.020	0.020	0.2740-0.2746	0.2738-0.2748
	B234I/B	2.3 (2290)	45	44.5	138-150@ 1.12	1.46	0.020	0.020	0.2740-0.2746	0.2738-0.2748
	B234L/M	2.3 (2290)	45	44.5	138-141@ 1.12	1.46	0.020	0.020	0.2740-0.2746	0.2738-0.2748
	B234R/R	2.3 (2290)	45	44.5	138-141@ 1.12	1.46	0.020	0.020	0.2740-0.2746	0.2738-0.2748
	B234E/U	2.3 (2290)	45	44.5	138-141@ 1.12	1.46	0.020	0.020	0.2740-0.2746	0.2738-0.2748
	B258I/V	2.5 (2498)	45	45.3	142-153@ 0.94	NA	NA	NA	0.2344-0.2350	0.2340-0.2346
	B308I/W	3.0 (2961)	45	45.3	142-153@ 0.94	NA	NA	NA	0.2344-0.2350	0.2340-0.2346

NA - Not Available

TORQUE SPECIFICATIONS
All readings in ft. lbs.

Year	Engine ID/VIN	Engine Displacement Liters (cc)	Cylinder Head Bolts	Main Bearing Bolts	Rod Bearing Bolts	Crankshaft Damper Bolts	Flywheel Bolts	Manifold Intake	Manifold Exhaust	Spark Plugs	Lug Nut
1993	B202L/L	2.0 (1985)	①	81	41	129	②	16	19	20	80-90
	B212I/E	2.1 (2119)	①	81	41	129	②	16	13	20	80-90
	B234I/B	2.3 (2290)	①	81	35	129	63	16	13	20	80-90
	B234L/M	2.3 (2290)	①	81	35	129	63	16	19	20	80-90
1994	B204L/N	2.0 (1985)	③	81	35	130	59	16	19	20	80-90
	B234I/B	2.3 (2290)	③	81	35	130	59	16	13	20	80-90
	B234L/M	2.3 (2290)	③	81	35	130	59	16	19	20	80-90
	B234R/R	2.3 (2290)	③	81	35	130	59	16	19	20	80-90
	B234E/U	2.3 (2290)	①	81	35	130	59	16	19	20	80-90
	B258I/V	2.5 (2498)	④	⑤	⑥	⑦	⑧	15	15	19	80-90
1995	B204L/N	2.0 (1985)	③	81	35	130	59	16	19	20	80-90
	B234I/B	2.3 (2290)	③	81	35	130	59	16	13	20	80-90
	B234L/M	2.3 (2290)	③	81	35	130	59	16	19	20	80-90
	B234R/R	2.3 (2290)	③	81	35	130	59	16	19	20	80-90
	B234E/U	2.3 (2290)	①	81	35	130	59	16	19	20	80-90
	B258I/V	2.5 (2498)	④	⑤	⑥	⑦	⑧	15	15	19	80-90
1996-97	B204L/N	2.0 (1985)	①	81	35	130	59	16	19	20	80-90
	B234I/B	2.3 (2290)	①	81	35	130	59	16	13	20	80-90
	B234L/M	2.3 (2290)	①	81	35	130	59	16	19	20	80-90
	B234R/R	2.3 (2290)	①	81	35	130	59	16	19	20	80-90
	B234E/U	2.3 (2290)	①	81	35	130	59	16	19	20	80-90
	B258I/V	2.5 (2498)	④	⑤	⑥	⑦	⑧	15	15	19	80-90
	B308I/W	3.0 (2961)	④	⑤	⑥	⑦	⑧	15	15	19	80-90

① Step 1: 44 ft. lbs.
Step 2: 59 ft. lbs. If 17mm head bolts, torque to 70 ft. lbs.
Step 3: Tighten each bolt an additional 90 degrees

② Step 1: 44 ft. lbs.
Step 2: Tighten 19mm head bolts to 63 ft. lbs.

③ Step 1: 44 ft. lbs.
Step 2: 59 ft. lbs.
Step 3: Tighten each bolt an additional 90 degrees

④ Step 1: 19 ft. lbs.
Step 2: Tighten each bolt an additional 90 degrees
Step 3: Repeat Step 2
Step 4: Repeat Step 2

⑤ Step 1: 37 ft. lbs.
Step 2: Tighten each bolt an additional 60 degrees
Step 3: Tighten each bolt an additional 15 degrees

⑥ Step 1: 26 ft. lbs.
Step 2: Tighten each bolt an additional 45 degrees
Step 3: Tighten each bolt an additional 15 degrees

⑦ Step 1: 185 ft. lbs.
Step 2: Turn bolt an additional 45 degrees

⑧ Step 1: 48 ft. lbs.
Step 2: Turn bolt an additional 30 degrees

BRAKE SPECIFICATIONS
All measurements in inches unless noted

Year	Model		Master Cylinder Bore	Brake Disc Original Thickness	Brake Disc Minimum Thickness	Brake Disc Maximum Runout	Brake Drum Diameter Original Inside Diameter	Brake Drum Diameter Max. Wear Limit	Brake Drum Diameter Maximum Machine Diameter	Minimum Lining Thickness Front	Minimum Lining Thickness Rear
1993	900	F	0.870	0.870	0.790	0.003	-	-	-	0.160	0.160
		R		0.350	0.290	0.003					
	9000	F	0.870	0.980	0.900	0.003	-	-	-	0.160	0.160
		R		0.350	0.290	0.003					
1994	900	F	0.940	0.940	0.870	0.002	-	-	-	0.200	0.200
		R		0.390	0.310	0.003					
	9000	F	0.870	0.980	0.900	0.003	-	-	-	0.160	0.160
		R		0.350	0.290	0.003					
1995	900	F	0.940	0.940	0.870	0.002	-	-	-	0.200	0.200
		R		0.390	0.310	0.003					
	9000	F	0.870	0.980	0.900	0.003	-	-	-	0.160	0.160
		R		0.350	0.290	0.003					

BRAKE SPECIFICATIONS
All measurements in inches unless noted

Year	Model		Master Cylinder Bore	Brake Disc			Brake Drum Diameter			Minimum Lining Thickness	
				Original Thickness	Minimum Thickness	Maximum Runout	Original Inside Diameter	Max. Wear Limit	Maximum Machine Diameter	Front	Rear
1996-97	900	F	0.940	0.940	0.870	0.002	-	-	-	0.200	-
		R	-	0.390	0.310	0.003	-	-	-	-	0.200
	9000	F	0.870	0.980	0.900	0.003	-	-	-	0.160	-
		R	-	0.350	0.290	0.003	-	-	-	-	0.160

FREQUENT MAINTENANCE LABOR
SAAB

The following should be used as a guide when determining the amount of work required for a particular service if taken to a repair shop. In estimating how long a particular Frequent Maintenance Service item should take, please observe the following:
- **Chilton Time** is time that is based on field research and data supplied by the vehicle manufacturer.
- All labor time operations are given in hours and tenths of an hour.
- All labor operations, are to be used as a **guide**.

COOLING

(G) Winterize Cooling System
Includes: Run engine to check for leaks, tighten all hose connections. Test radiator and pressure cap, drain radiator and engine block. Add antifreeze and refill system.
1993-97 .5

(G) Belt, Serpentine Drive, Renew
1993 900 4
1994-97 900
  4 cyl. 4
  V6 5
1993-97 9000
  2.3L .7

(G) Hoses, Radiator, Renew
Includes: Drain and refill cooling system as required.
1993 900 5
1994-97 900
  upper 3
  lower 5
1993-97 9000
  upper 5
  lower 4

(G) Thermostat, Coolant, Renew
1993 900 6
1994-97 900
  4 cyl. 9
  V6 . . . . 1.6
1993-97 9000
  2.3L 7

BRAKES

(G) Bleed Brakes (Four Wheels)
Includes: Add fluid.
1993-97 4
w/ABS add 3

(M) Parking Brake, Adjust
1993-97 900 3
1993-97 9000 3

LUBRICATION SERVICES

(M) Engine Oil & Filter, Renew
Includes: Inspect and correct all fluid levels.
1993-97 3

WHEELS

(G) Wheel, Renew (One)
1993-97 5

(G) Wheel, Balance
1993-97
  one 3
  each adtnl. 2

(G) Wheels, Rotate (All)
1993-97 5

ELECTRICAL

(G) Headlamps, Aim
1993-97 4

(G) License Lamp Bulb, Renew
1993-97, each 2

(G) Parking Lamp Lens or Bulb, Renew
1993-97 3

(G) Tail Lamp Lens or Bulb, Renew
1993 900
  3 door 3
  2 or 4 door 5
1994-97 900 3
1993-97 9000 3

(G) Horn, Renew
1993-97, each 4

SCHEDULED MAINTENANCE INTERVALS
(SAAB 900 & 9000)

TO BE SERVICED	TYPE OF SERVICE	VEHICLE MILEAGE INTERVAL (x1000)												
		5	10	15	20	25	30	35	40	45	50	55	60	65
Engine oil & filter	R	✓	✓	✓	✓	✓	✓	✓	✓	✓	✓	✓	✓	✓
Battery electrolyte level	S/I	✓		✓		✓		✓		✓		✓		✓
Brake fluid①	S/I	✓		✓		✓		✓		✓		✓		✓
Brake lines & hoses	S/I	✓		✓		✓		✓		✓		✓		✓
Brake pads & discs	S/I	✓		✓		✓		✓		✓		✓		✓
Drive belts	S/I	✓		✓		✓		✓		✓		✓		✓
Engine coolant strength	S/I	✓		✓		✓		✓		✓		✓		✓
Exhaust system	S/I	✓		✓		✓		✓		✓		✓		✓
Final drive oil level (900 A/T)	S/I	✓		✓		✓		✓		✓		✓		✓
Gearbox oil	S/I	✓		✓		✓		✓		✓		✓		✓
Outer & inner drive joint boots	S/I	✓		✓		✓		✓		✓		✓		✓
Rotate tires (front to rear)	S/I	✓		✓		✓		✓		✓		✓		✓
Automatic transmission fluid & filter	R	✓						✓						✓
Air cleaner element	R							✓						✓
Engine coolant	R							✓						✓
Spark plugs	R							✓						✓
Ventilation air filter	R							✓						✓
Ball joint clearance	S/I							✓						✓
Engine cooling system, hoses & cap	S/I	✓						✓						
Front wheel alignment	S/I							✓						✓
Fuel lines	S/I							✓						✓
Shock absorbers & bushings	S/I							✓						✓
Fuel filter	R													✓
Power steering fluid	R							✓						
Crankcase ventilation & vacuum lines	S/I													✓

SCHEDULED MAINTENANCE INTERVALS
(SAAB 900 & 9000) (Cont.)

TO BE SERVICED	TYPE OF SERVICE	VEHICLE MILEAGE INTERVAL (x1000)												
		5	10	15	20	25	30	35	40	45	50	55	60	65
Distributor cap & rotor	S/I													✓
EVAP system	S/I													✓
Front suspension, rear axle mountings	S/I	✓												
Parking brake adjustment	S/I	✓												
Spark plug wires	S/I													✓

① replace every 35,000 miles.

R – Replace   S/I – Service or Inspect

FREQUENT OPERATION MAINTENANCE (SEVERE SERVICE)

If a vehicle is operated under any of the following conditions it is considered severe service:
- Extremely dusty areas.
- 50% or more of the vehicle operation is in 32°C (90°F) or higher temperatures, or constant operation in temperatures below 0°C (32°F).
- Prolonged idling (vehicle operation in stop and go traffic).
- Frequent short running periods (engine does not warm to normal operating temperatures).
- Police, taxi, delivery usage or trailer towing usage.

Oil & oil filter change – change every 2500 miles.

Air filter element – service or inspect every 15,000 miles.

SUBARU

VEHICLE IDENTIFICATION CHART

		Engine Code						Model Year	
Code	Liters	Cu. In. (cc)	Cyl.	Fuel Sys.	Eng. Mfg.		Code	Year
2		1.8	111 (1829)	4	MFI	Subaru		P	1993
3		3.3	202 (3318)	6	MFI	Subaru		R	1994
4	①	1.8	108 (1781)	4	SFI	Subaru		S	1995
5	②	1.8	108 (1781)	4	MFI	Subaru		T	1996
6		2.2	135 (2212)	4	MFI	Subaru		V	1997
7		1.2	72 (1189)	3	MFI	Subaru			
8	②	1.2	72 (1189)	3	MFI	Subaru			

MFI - Multiport fuel injection   ① 2WD

SFI - Sequential fuel injection   ② 4WD

ENGINE IDENTIFICATION

Year	Model		Engine Displacement Liters (cc)	Engine Series (ID/VIN)	Fuel System	No. of Cylinders	Engine Type
1993	Justy	①	1.2 (1189)	7	MFI	3	SOHC
	Justy	②	1.2 (1189)	8	MFI	3	SOHC
	Loyale	①	1.8 (1781)	4	SFI	4	SOHC
	Loyale	②	1.8 (1781)	5	SFI	4	SOHC
	Impreza		1.8 (1829)	2	MFI	4	SOHC
	Legacy		2.2 (2212)	6	MFI	4	SOHC
	SVX		3.3 (3318)	3	MFI	6	DOHC
1994	Justy		1.2 (1189)	7	MFI	3	SOHC
	Loyale	①	1.8 (1781)	4	SFI	4	SOHC
	Loyale	②	1.8 (1781)	5	SFI	4	SOHC
	Impreza		1.8 (1829)	2	MFI	4	SOHC
	Legacy		2.2 (2212)	6	MFI	4	SOHC
	SVX		3.3 (3318)	3	MFI	6	DOHC
1995	Impreza		1.8 (1829)	2	MFI	4	SOHC
	Impreza		2.2 (2212)	6	MFI	4	SOHC
	Legacy		2.2 (2212)	6	MFI	4	SOHC
	SVX		3.3 (3318)	3	MFI	6	DOHC
1996-97	Impreza		1.8 (1829)	2	MFI	4	SOHC
	Impreza		2.2 (2212)	6	MFI	4	SOHC
	Legacy		2.2 (2212)	6	MFI	4	SOHC
	SVX		3.3 (3318)	3	MFI	6	DOHC

DOHC - Double overhead camshaft
MFI - Multiport fuel injection
SFI - Sequential fuel injection
SOHC - Single overhead camshaft
① 2WD
② 4WD

GENERAL ENGINE SPECIFICATIONS

Year	Engine ID/VIN		Engine Displacement Liters (cc)	Fuel System Type	Net Horsepower @ rpm	Net Torque @ rpm (ft. lbs.)	Bore x Stroke (in.)	Compression Ratio	Oil Pressure @ rpm	
1993	7		1.2 (1189)	MFI	73@5600	71@2800	3.07x3.27	9.1:1	30@1500	①
	8		1.2 (1189)	MFI	73@5600	71@2800	3.07x3.27	9.1:1	30@1500	①
	4		1.8 (1781)	SFI	90@5200	101@2800	3.62x2.64	9.5:1	14@550	①
	5		1.8 (1781)	SFI	90@5200	101@2800	3.62x2.64	9.5:1	14@550	①
	2		1.8 (1829)	MFI	110@5600	110@4400	3.46x2.95	9.5:1	14@600	①
	6	②	2.2 (2212)	MFI	130@4500	137@2400	3.82x2.95	9.5:1	14@600	①
	6	③	2.2 (2212)	MFI	160@5600	181@2400	3.82x2.95	8.0:1	14@600	①
	3		3.3 (3318)	MFI	230@5400	228@4400	3.82x2.95	10.1:1	14@600	①
1994	7		1.2 (1189)	MFI	73@5600	71@2800	3.07x3.27	9.1:1	30@1500	①
	2		1.8 (1829)	MFI	110@5600	110@4400	3.46x2.95	9.5:1	14@600	①
	4		1.8 (1781)	SFI	90@5200	101@2800	3.62X2.64	9.5:1	14@550	①
	5		1.8 (1781)	SFI	90@5200	101@2800	3.62X2.64	9.5:1	14@550	①
	6	②	2.2 (2212)	MFI	130@5600	137@2400	3.82x2.95	9.5:1	14@600	①
	6	③	2.2 (2212)	MFI	160@5600	181@2400	3.82x2.95	8.0:1	14@600	①
	3		3.3 (3318)	MFI	230@5400	228@4400	3.82x2.95	10.1:1	14@600	①
1995	2		1.8 (1829)	MFI	110@5600	110@4400	3.46x2.95	9.5:1	14@600	①
	6		2.2 (2212)	MFI	135@5400	140@4400	3.82x2.95	9.5:1	14@600	①
	3		3.3 (3318)	MFI	230@5400	228@4400	3..82x2.95	10.1:1	14@600	①
1996-97	2		1.8 (1829)	MFI	110@5600	110@4400	3.46x2.95	9.5:1	14@600	①
	6		2.2 (2212)	MFI	135@5400	140@4400	3.82x2.95	9.5:1	14@600	①
	3		3.3 (3318)	MFI	230@5400	228@4400	3..82x2.95	10.1:1	14@600	①

MFI - Multiport fuel injection
SFI - Sequential fuel injection
① Test with engine at normal operating temperature
② Non-Turbo
③ Turbo

GASOLINE ENGINE TUNE-UP SPECIFICATIONS

Year	Engine ID/VIN	Engine Displacement Liters (cc)	Spark Plugs Gap (in.)	Ignition Timing (deg.) MT	AT	Fuel Pump (psi)	Idle Speed (rpm) MT	AT	Valve Clearance In.	Ex.
1993	7	1.2 (1189)	0.039-0.043	5B	5B	43	650-750	650-750	④	⑤
	8	1.2 (1189)	0.039-0.043	5B	5B	43	650-750	650-750	④	⑤
	4	1.8 (1781)	0.039-0.043	20B	20B	38-50	700	700	HYD	HYD
	5	1.8 (1781)	0.039-0.043	20B	20B	38-50	700	700	HYD	HYD
	2	1.8 (1829)	0.039-0.043	20B	20B	36	700	700	HYD	HYD
	6 ①	2.2 (2212)	0.039-0.043	20B	20B	36	③	③	HYD	HYD
	6 ②	2.2 (2212)	0.039-0.043	15B	15B	43	③	③	HYD	HYD
	3	3.3 (3318)	0.039-0.043	-	20B	43	-	610	HYD	HYD
1994	7	1.2 (1189)	0.039-0.043	5B	5B	43	650-750	650-750	④	⑤
	4	1.8 (1781)	0.039-0.043	20B	20B	38-50	700	700	HYD	HYD
	5	1.8 (1781)	0.039-0.043	20B	20B	38-50	700	700	HYD	HYD
	2	1.8 (1829)	0.039-0.043	20B	20B	36	700	700	HYD	HYD
	6 ①	2.2 (2212)	0.039-0.043	20B	20B	36	③	③	HYD	HYD
	6 ②	2.2 (2212)	0.039-0.043	15B	15B	43	③	③	HYD	HYD
	3	3.3 (3318)	0.039-0.043	-	20B	43	-	610	HYD	HYD
1995	2	1.8 (1829)	0.039-0.043	20B	20B	36	700	700	HYD	HYD
	6	2.2 (2212)	0.039-0.043	20B	20B	36	③	③	HYD	HYD
	3	3.3 (3318)	0.039-0.043	-	20B	43	-	610	HYD	HYD
1996-97	2	1.8 (1829)	0.039-0.043	20B	20B	36	700	700	HYD	HYD
	6	2.2 (2212)	0.039-0.043	20B	20B	36	③	③	HYD	HYD
	3	3.3 (3318)	0.039-0.043	-	20B	43	-	610	HYD	HYD

NOTE: The Vehicle Emission Control Information lable often reflects specification changes made juring production. The lable fugures must be used if they differ from those in this chart.

B - Before top dead center

HYD - Hudraulic

① Non-Turbo
② Turbo

③ With A/C off: 600-800
   With A/C on: 800-900

④ 0.0051-0.0067
⑤ 0.0091-0.0106

CAPACITIES

Year	Model	Engine ID/VIN	Engine Displacement Liters (cc)	Engine Oil with Filter	Transmission (pts.) 4-Spd	5-Spd	Auto.	Transfer Case (pts.)	Drive Axle Front (pts.)	Rear (pts.)	Fuel Tank (gal.)	Cooling System (qts.)
1993	Justy	7	1.2 (1189)	3.0	-	5.0	7.0	-	-	1.6	13.2	6.2
	Justy	8	1.2 (1189)	3.0	-	7.2	8.8	-	-	-	9.2	②
	Loyale	4	1.8 (1781)	4.2	-	5.4	13.2	2.6	-	-	15.9	5.8
	Loyale	5	1.8 (1781)	4.2	-	5.4	13.2	2.6	-	1.6	15.9	5.8
	Impreza	2	1.8 (1829)	4.0	-	5.4	16.8	2.6	-	1.6	13.2	6.2
	Legacy	6	2.2 (2212)	4.8	-	①	17.6	3.0	-	1.6	15.9	③
	SVX	3	3.3 (3318)	6.3	-	-	20.8	2.8	-	1.8	18.5	7.4
1994	Justy	7	1.2 (1189)	3.0	-	5.0	-	-	-	-	9.2	5.2
	Loyale	4	1.8 (1781)	4.2	-	5.4	13.2	2.6	-	-	15.9	5.8
	Loyale	5	1.8 (1781)	4.2	-	5.4	13.2	2.6	-	-	15.9	5.8
	Impreza	2	1.8 (1829)	4.0	-	5.4	16.8	2.6	-	1.6	13.2	6.2
	Legacy	6	2.2 (2212)	4.8	-	①	17.6	3.0	-	1.6	15.9	②
	SVX	3	3.3 (3318)	6.3	-	-	20.8	2.8	-	1.8	18.5	7.4
1995	Impreza	2	1.8 (1829)	4.2	-	④	17.5	2.6	-	1.6	13.2	6.2
	Impreza	6	2.2 (2212)	4.4	-	④	17.5	-	-	NA	13.2	5.8
	Legacy	6	2.2 (2212)	4.4	-	⑤	15.0	3.0	-	1.6	15.9	6.3
	SVX	3	3.3 (3318)	6.3	-	-	20.0	2.8	-	1.8	18.5	7.4
1996-97	Impreza	2	1.8 (1829)	4.2	-	④	17.5	2.6	-	1.6	13.2	6.2
	Impreza	6	2.2 (2212)	4.4	-	④	17.5	-	-	NA	13.2	5.8

CAPACITIES

Year	Model	Engine ID/VIN	Engine Displacement Liters (cc)	Engine Oil with Filter	Transmission (pts.) 4-Spd	Transmission (pts.) 5-Spd	Transmission (pts.) Auto.	Transfer Case (pts.)	Drive Axle Front (pts.)	Drive Axle Rear (pts.)	Fuel Tank (gal.)	Cooling System (qts.)
1996-97	Legacy	6	2.2 (2212)	4.4	-	⑤	15.0	3.0	-	1.6	15.9	6.3
	SVX	3	3.3 (3318)	6.3	-	-	20.0	2.8	-	1.8	18.5	7.4

① 2WD: 7.0   4WD: 7.4
② 2 barrel carburetor: 4.9   Multiport fuel injection: 5.2
③ 2WD: 6.2   4WD: 7.4
④ FWD: 5.5   AWD: 7.4
⑤ FWD: 7.0   AWD: 7.6

VALVE SPECIFICATIONS

Year	Engine ID/VIN	Engine Displacement Liters (cc)	Seat Angle (deg.)	Face Angle (deg.)	Spring Test Pressure (lbs. @ in.)	Spring Installed Height (in.)	Stem-to-Guide Clearance (in.) Intake	Stem-to-Guide Clearance (in.) Exhaust	Stem Diameter (in.) Intake	Stem Diameter (in.) Exhaust
1993	7	1.2 (1189)	90	90	1.248@ 113-129	1.248	0.0008-0.0020	0.0016-0.0028	0.2742-0.2748	0.2734-0.2740
	8	1.2 (1189)	90	90	1.248@ 113-129	1.248	0.0008-0.0020	0.0016-0.0028	0.2742-0.2748	0.2734-0.2740
	4	1.8 (1781)	90	90	①	②	0.0014-0.0059	0.0016-0.0059	0.2736-0.2742	0.2734-0.2740
	5	1.8 (1781)	90	90	①	②	0.0014-0.0059	0.0016-0.0059	0.2736-0.2742	0.2734-0.2740
	2	1.8 (1829)	90	90	1.150@ 90-103	1.150	0.0014-0.0059	0.0016-0.0059	0.2343-0.2348	0.2341-0.2346
	6	2.2 (2212)	90	90	1.150@ 90-103	1.150	0.0014-0.0059	0.0016-0.0059	0.2343-0.2348	0.2341-0.2346
	3	3.3 (3318)	90	90	③	④	0.0012-0.0039	0.0016-0.0059	0.2344-0.2350	0.2341-0.2346
1994	7	1.2 (1189)	90	90	1.248@ 113-129	1.248	0.0008-0.0020	0.0016-0.0028	0.2742-0.2748	0.2734-0.2740
	4	1.8 (1781)	90	90	①	②	0.0014-0.0059	0.0016-0.0059	0.2736-0.2742	0.2734-0.2740
	5	1.8 (1781)	90	90	①	②	0.0014-0.0059	0.0016-0.0059	0.2736-0.2742	0.2734-0.2740
	2	1.8 (1829)	90	90	1.150@ 90-103	1.150	0.0014-0.0059	0.0016-0.0059	0.2343-0.2348	0.2341-0.2346
	6	2.2 (2212)	90	90	1.150@ 90-103	1.150	0.0014-0.0059	0.0016-0.0059	0.2343-0.2348	0.2341-0.2346
	3	3.3 (3318)	90	90	③	④	0.0012-0.0039	0.0016-0.0059	0.2344-0.2350	0.2341-0.2346
1995	2	1.8 (1829)	90	90	1.150@ 90-103	1.150	0.0014-0.0059	0.0016-0.0059	0.2343-0.2348	0.2341-0.2346
	6	2.2 (2212)	90	90	1.150@ 90-103	1.150	0.0014-0.0059	0.0016-0.0059	0.2343-0.2348	0.2341-0.2346
	3	3.3 (3318)	90	90	③	④	0.0012-0.0039	0.0016-0.0059	0.2344-0.2350	0.2341-0.2346
1996-97	2	1.8 (1829)	90	90	1.150@ 90-103	1.150	0.0014-0.0059	0.0016-0.0059	0.2343-0.2348	0.2341-0.2346
	6	2.2 (2212)	90	90	1.150@ 90-103	1.150	0.0014-0.0059	0.0016-0.0059	0.2343-0.2348	0.2341-0.2346
	3	3.3 (3318)	90	90	③	④	0.0012-0.0039	0.0016-0.0059	0.2344-0.2350	0.2341-0.2346

① Inner spring: 1.122@46-51   Outer spring: 1.240@101-105
② Inner spring: 1.122   Outer spring: 1.240
③ Inner spring: 0.772@33-37   Outer spring: 0.831@70-80
④ Inner spring: 0.772   Outer spring: 0.831

TORQUE SPECIFICATIONS
All readings in ft. lbs.

Year	Engine ID/VIN	Engine Displacement Liters (cc)	Cylinder Head Bolts	Main Bearing Bolts	Rod Bearing Bolts	Crankshaft Damper Bolts	Flywheel Bolts	Manifold Intake	Manifold Exhaust	Spark Plugs	Lug Nut
1993	7	1.2 (1189)	③	30-35	29-33	58-72	65-71	14-22	14-21	13-17	58-72
	8	1.2 (1189)	③	30-35	29-33	58-72	65-71	14-22	14-21	13-17	58-72
	4	1.8 (1781)	①	⑤	29-31	66-79	51-55	13-16	-	13-17	58-72
	5	1.8 (1781)	①	⑤	29-31	66-79	51-55	13-16	-	13-17	58-72
	2	1.8 (1829)	⑩	⑥	32-34	69-76	51-55	7-11	-	13-17	58-72
	6	2.2 (2212)	②	⑥	32-34	69-76	51-55	⑦	-	13-17	58-72
	3	3.3 (3318)	⑪	22 ⑨	32-34	108-123	51-55	17-20	-	13-17	58-72
1994	7	1.2 (1189)	③	30-35	29-33	58-72	65-71	14-22	14-21	13-17	58-72
	4	1.8 (1781)	①	⑤	29-31	66-79	51-55	13-16	-	13-17	58-72
	5	1.8 (1781)	①	⑤	29-31	66-79	51-55	13-16	-	13-17	58-72
	2	1.8 (1829)	⑩	⑥	32-34	69-76	51-55	7-11	-	13-17	58-72
	6	2.2 (2212)	②	⑥	32-34	69-76	51-55	⑦	-	13-17	58-72
	3	3.3 (33180)	⑧	22 ⑨	32-34	108-123	51-55	17-20	-	13-17	58-72
1995	2	1.8 (1829)	⑩	⑥	32-34	69-76	51-55	7-11	-	13-17	58-72
	6	2.2 (2212)	⑩	⑥	32-34	69-76	51-55	⑦	-	13-17	58-72
	3	3.3 (3318)	⑪	22 ⑨	32-34	108-123	51-55	17-20	-	13-17	58-72
1996-97	2	1.8 (1829)	⑩	⑥	32-34	69-76	51-55	7-11	-	13-17	58-72
	6	2.2 (2212)	⑩	⑥	32-34	69-76	51-55	⑦	-	13-17	58-72
	3	3.3 (3318)	⑪	22 ⑨	32-34	108-123	51-55	17-20	-	13-17	58-72

① Step 1: 22 ft. lbs.
Step 2: 43 ft. lbs.
Step 3: 47 ft. lbs.

② Step 1: 22 ft. lbs.
Step 2: 51 ft. lbs.
Step 3: Loosen all bolts 180 degrees
Step 4: Repeat Step 3
Step 5: Bolts 1-2:
Non-Turbo: 25 ft. lbs.
Turbo: 27 ft. lbs.
Step 6: Bolts 3-6:
Non-Turbo: 11 ft. lbs.
Turbo: 14 ft. lbs.
Step 7: Tighten all bolts 80-90 degrees

③ Step 1: 29 ft. lbs.
Step 2: 54 ft. lbs.
Step 3: Loosen all bolts 90 degrees
Step 4: 51-57 ft. lbs.

④ Step 1: 29 ft. lbs.
Step 2: 47 ft. lbs.
Step 3: Loosen all bolts 90 degrees
Step 4: 44-50 ft. lbs.

⑤ Engine block connecting bolts:
Short bolts: 17-20 ft. lbs.
Long bolts: 29-35 ft. lbs.

⑥ Engine block connecting bolts:
Short bolts: 17-20 ft. lbs.
Long bolts: 33-37 ft. lbs.

⑦ Long bolts: 4.3-5.1 ft. lbs.
Short bolts: 21-25 ft. lbs.

⑧ Step 1: 22 ft. lbs.
Step 2: 51 ft. lbs.
Step 3: Loosen all bolts 180 degrees
Step 4: Repeat Step 3
Step 5: 20 ft. lbs.
Step 6: Bolts 1-4: 80-90 degrees
Step 7: Bolts 5-8: 33 ft. lbs.
Step 8: Tighten all bolts 80-90 degrees

⑨ Engine block connecting bolts

⑩ Step 1: 22 ft. lbs.
Step 2: 51 ft. lbs.
Step 3: Loosen all bolts 180 degrees
Step 4: Repeat Step 3
Step 5: Bolts 1-2: 25 ft. lbs.
Step 6: Bolts 3-6: 11 ft. lbs.
Step 7: Tighten all bolts 80-90 degrees

⑪ Step 1: 22 ft. lbs.
Step 2: 51 ft. lbs.
Step 3: Loosen all bolts 180 degrees
Step 4: Repeat Step 3
Step 5: 20 ft. lbs.
Step 6: Bolts 1-4: 80-90 degrees
Step 7: Bolts 5-8: 33 ft. lbs.

BRAKE SPECIFICATIONS
All measurements in inches unless noted

Year	Model	Master Cylinder Bore	Brake Disc Original Thickness	Brake Disc Minimum Thickness	Maximum Runout	Brake Drum Diameter Original Inside Diameter	Brake Drum Diameter Max. Wear Limit	Maximum Machine Diameter	Minimum Lining Thickness Front	Minimum Lining Thickness Rear
1993	Impreza	⑯	⑱	⑲	⑳	㉑	㉒	-	0.295 ⑫	⑬
	Justy	①	0.709	0.610	0.0059	7.09	7.17	-	⑪	0.067
	Legacy	⑰	④	⑦	0.0039	3.36 ⑩	-	-	0.295 ⑫	⑬
	Loyale	③	0.710	0.630	0.0039	7.09	7.17	-	0.295 ⑫	0.059
	SVX	1.063	⑭	⑮	0.0039	7.48 ⑩	7.52 ⑩	-	0.295 ⑫	⑬
1994	Impreza	⑯	⑱	⑲	⑳	㉑	㉒	-	0.295	⑬
	Justy	①	0.709	0.610	0.0059	7.09	7.17	-	⑪	0.067
	Legacy	⑰	④	⑦	0.0039	3.36 ⑩	-	-	0.295 ⑫	⑬
	Loyale	③	0.710	0.630	0.0039	7.09	7.17	-	0.295 ⑫	0.059
	SVX	1.063	⑭	⑮	0.0039	7.48 ⑩	7.52 ⑩	-	0.295 ⑫	⑬

BRAKE SPECIFICATIONS
All measurements in inches unless noted

Year	Model	Master Cylinder Bore	Brake Disc Original Thickness	Brake Disc Minimum Thickness	Maximum Runout	Brake Drum Diameter Original Inside Diameter	Brake Drum Diameter Max. Wear Limit	Brake Drum Diameter Maximum Machine Diameter	Minimum Lining Thickness Front	Minimum Lining Thickness Rear
1995	Impreza	⑯	⑱	⑲	⑳	㉑	㉒	-	0.295 ⑫	⑬
	Legacy	②	④	⑦	0.0039	6.69 ⑩	-	-	0.295 ⑫	⑬
	SVX	1.063	⑭	⑮	0.0039	7.48 ⑩	7.52 ⑩	-	0.295 ⑫	⑬
1996-97	Impreza	⑯	⑱	⑲	⑳	㉑	㉒	-	0.295 ⑫	⑬
	Legacy	②	④	⑦	0.0039	6.69 ⑩	-	-	0.295 ⑫	⑬
	SVX	1.063	⑭	⑮	0.0039	7.48 ⑩	7.52 ⑩	-	0.295 ⑫	⑬

① Small diameter: 0.875
Large diameter: 1.000
② Without ABS: 1.000
With ABS: 1.059
③ Small diameter: 0.813
Large diameter: 1.000
④ Front: 0.940
Rear Non-Turbo: 0.390
Rear Turbo: 0.710
⑤ Front: 0.710
Rear: 0.390
⑥ Front: 0.870
Rear: 0.390

⑦ Front: 0.870
Rear Non-Turbo: 0.335
Rear Turbo: 0.650
⑧ Front: 0.630
Rear: 0.335
Front: 0.787
Rear: 0.335
⑨ Front: 0.787
Rear: 0.335
⑩ Parking brake drum
⑪ Includes metal backing:
DL: 0.295; GL: 0.315

⑫ Includes metal backing
⑬ Includes metal backing on pads:
Pads: 0.256; Shoes: 0.059
⑭ Front: 1.100
Rear: 0.390
⑮ Front: 1.020
Rear: 0.335
⑯ LS: 1.000
Except LS: 0.9374
⑰ Without ABS: 1.000
With ABS: 1.059

⑱ Front 2WD L and L+: 0.710
Front except 2WD L and L+: 0.940
Rear: 0.390
⑲ Front 2WD L and L+: 0.630
Front except 2WD L and L+: 0.870
Rear: 0.335
⑳ Front: 0.0030
Rear: 0.0039
㉑ Brake: 9.00
Parking brake: 6.69
㉒ Brake: 9.079
Parking brake: 6.73

FREQUENT MAINTENANCE LABOR
SUBARU

The following should be used as a guide when determining the amount of work required for a particular service if taken to a repair shop. In estimating how long a particular Frequent Maintenance Service item should take, please observe the following:
- **Chilton Time** is time that is based on field research and data supplied by the vehicle manufacturer.
- All labor time operations are given in hours and tenths of an hour.
- All labor operations, are to be used as a **guide**.

COOLING

(G) Winterize Cooling System

Includes: Run engine to check for leaks, tighten all hose connections. Test radiator and pressure cap, drain radiator and engine block. Add antifreeze and refill system.

1993-97 5

(G) Belt, Drive, Renew

1993-97 4
w/AC add 2
w/PS add 2

(G) Belt, Drive, Adjust

1993-97 3

(G) Hoses, Radiator, Renew

Includes: Drain and refill cooling system as required.

1993-97
upper 3
lower 4

(G) Thermostat, Coolant, Renew

1993-97 6
w/MFI add 4

FUEL

(M) Air Cleaner, Service

1993-97 2

(G) Filter, Fuel, Renew

1993-97 w/FI 4

BRAKES

(G) Bleed Brakes (Four Wheels)

Includes: Add fluid.

1993-97 5
w/ABS add 5

(G) Brakes, Adjust (Minor)

Includes: Adjust brakes, fill master cylinder.

1993-97, two wheels 4

(M) Parking Brake, Adjust

1993-97 4

LUBRICATION SERVICES

(M) Engine Oil & Filter, Renew

Includes: Inspect and correct all fluid levels.

1993-97 3

WHEELS

(G) Wheel, Renew (One)

1993-97 5

(G) Wheel, Balance

1993-97
one 3
each adtnl. 2

(G) Wheels, Rotate (All)

1993-97 5

FREQUENT MAINTENANCE LABOR (cont.)
SUBARU

	Chilton Time
ELECTRICAL	
(G) Headlamps, Aim	
1993-97	.4
(G) Halogen Headlamp Bulb, Renew	
1993-97	.3

	Chilton Time
(G) License Lamp Assy., Renew	
1993-97	.4
(G) Parking Lamp Lens or Bulb, Renew	
1993-97	.3
(G) Tail Lamp Lens or Bulb, Renew	
1993-97	.3

	Chilton Time
(G) Horn, Renew	
1993-97 Impreza	.4
1993-94 Justy	.5
1993-97 Legacy	.3
1993-94 Loyale	
one	.5
each adtnl.	.3
1993-97 SVX	
one or both	.5

SCHEDULED MAINTENANCE INTERVALS
(SUBARU JUSTY, LEGACY, LOYALE, SVX & IMPREZA)

TO BE SERVICED	TYPE OF SERVICE	VEHICLE MILEAGE INTERVAL (x1000)												
		7.5	15	22.5	30	37.5	45	52.5	60	67.5	75	82.5	90	97.5
Engine oil & filter	R	✓	✓	✓	✓	✓	✓	✓	✓	✓	✓	✓	✓	✓
Brake lines	S/I		✓		✓		✓		✓		✓		✓	
Clutch & hill holder system	S/I		✓		✓		✓		✓		✓		✓	
Disc brake pads & discs, front & rear axle boots & axle shaft joint portions	S/I		✓		✓		✓		✓		✓		✓	
Parking brake	S/I		✓		✓		✓		✓		✓		✓	
Steering & suspension	S/I		✓		✓		✓		✓		✓		✓	
Air filter element	R				✓				✓				✓	
Engine Coolant	R				✓				✓				✓	
Fuel filter	R				✓				✓				✓	
Spark plugs (except below)	R				✓				✓				✓	
Spark plugs (SVX & 1996-97 Legacy)	R								✓					
Automatic transmission fluid & filter	S/I				✓				✓				✓	
Brake fluid	S/I				✓				✓				✓	
Brake linings & drums	S/I				✓				✓				✓	
Camshaft drive belt①	S/I				✓				✓				✓	

SCHEDULED MAINTENANCE INTERVALS
(SUBARU JUSTY, LEGACY, LOYALE, SVX & IMPREZA) (Cont.)

TO BE SERVICED	TYPE OF SERVICE	7.5	15	22.5	30	37.5	45	52.5	60	67.5	75	82.5	90	97.5
		VEHICLE MILEAGE INTERVAL (x1000)												
Coolant level, hoses & clamps	S/I				✓				✓				✓	
Drive belts	S/I				✓				✓				✓	
Fuel system, hoses & connections	S/I				✓				✓				✓	
Differential gear oil (front & rear) (SVX)	S/I				✓								✓	
Transmission and/or differential gear oil (except SVX)	S/I				✓								✓	
Front & rear wheel bearing repack	S/I								✓					

① Non-California vehicles - replace every 60,000 miles.
R – Replace   S/I – Service or Inspect

FREQUENT OPERATION MAINTENANCE (SEVERE SERVICE)
If a vehicle is operated under any of the following conditions it is considered severe service:
- Extremely dusty areas.
- 50% or more of the vehicle operation is in 32°C (90°F) or higher temperatures, or constant operation in temperatures below 0°C (32°F).
- Prolonged idling (vehicle operation in stop and go traffic).
- Frequent short running periods (engine does not warm to normal operating temperatures).
- Police, taxi, delivery usage or trailer towing usage.
Oil & oil filter change – change every 3750 miles.
Clutch & hill holder system - service or inspect every 7500 miles.
Disc brake pads & discs, front & rear axle boots & axle shaft joint portions - service or inspect every 7500 miles.
Steering & suspension - service or inspect every 7500 miles.
Air filter element – service or inspect every 15,000 miles.
Automatic transmission fluid – service or inspect every 15,000 miles.
Brake linings & drums – service or inspect every 15,000 miles.
Coolant level, hoses & clamps - service or inspect every 15,000 miles.
Differential gear oil (front & rear) (SVX) - service or inspect every 15,000 miles.
Drive belts - service or inspect every 15,000 miles.
Transmission/differential gear oil (except SVX) - service or inspect every 15,000 miles.
Front & rear wheel bearing repack - service or inspect every 30,000 miles.

SUZUKI

ENGINE IDENTIFICATION

Year	Model	Engine Displacement Liters (cc)	Engine Series (ID/VIN)	Fuel System	No. of Cylinders	Engine Type
1993	Swift	1.3 (1298)	3	EFI	4	SOHC
	Swift GT	1.3 (1298)	3	MFI	4	DOHC
1994	Swift	1.3 (1298)	3	TFI	4	SOHC
	Swift GT	1.3 (1298)	3	MFI	4	DOHC

20 SPECIFICATIONS

ENGINE IDENTIFICATION

Year	Model	Engine Displacement Liters (cc)	Engine Series (ID/VIN)	Fuel System	No. of Cylinders	Engine Type
1995	Swift	1.3 (1298)	3	TFI	4	SOHC
1996-97	Swift	1.3 (1298)	2	TFI	4	SOHC
	Esteem	1.6 (1590)	3	MFI	4	SOHC

EFI - Electronic fuel injection
MFI - Multiport fuel injection
TFI - Throttle body fuel injection

SOHC - Single overhead camshaft
DOHC - Double overhead camshaft

GENERAL ENGINE SPECIFICATIONS

Year	Engine ID/VIN	Engine Displacement Liters (cc)	Fuel System Type	Net Horsepower @ rpm	Net Torque @ rpm (ft. lbs.)	Bore x Stroke (in.)	Compression Ratio	Oil Pressure @ rpm
1993	3 ①	1.3 (1298)	MFI	100@6500	83@5000	2.91x2.97	10.0:1	54-68@3000
	3	1.3 (1298)	EFI	70@6000	74@3500	2.91x2.97	9.5:1	47-61@3000
1994	3 ①	1.3 (1298)	MFI	100@6500	83@5000	2.91x2.97	10.0:1	54-68@3000
	3	1.3 (1298)	TFI	70@6000	74@3500	2.91x2.97	9.5:1	47-61@3000
1995	3 ②	1.3 (1298)	TFI	70@6000	74@3500	2.91x2.97	9.5:1	47-61@3000
1996-97	2	1.3 (1298)	TFI	70@5500	74@3000	2.91x2.97	9.5:1	47-61@3000
	3 ③	1.6 (1590)	MFI	98@6000	94@3200	2.95x3.54	9.5:1	47-61@4000

EFI - Electronic fuel injection
MFI - Multiport fuel injection
TFI - Throttle body fuel injection
① GT models
② Swift
③ Esteem

GASOLINE ENGINE TUNE-UP SPECIFICATIONS

Year	Engine ID/VIN	Engine Displacement Liters (cc)	Spark Plugs Gap (in.)	Ignition Timing (deg.) MT	AT	Fuel Pump (psi)	Idle Speed (rpm) MT	AT	Valve Clearance In.	Ex.
1993	3 ①	1.3 (1298)	0.029	6B	6B	26-30	850	850	HYD	HYD
	3	1.3 (1298)	0.029	6B	6B	13-20	850	850	②	②
1994	3 ①	1.3 (1298)	0.029	6B	6B	26-30	850	850	HYD	HYD
	3	1.3 (1298)	0.029	6B	6B	13-20	850	850	②	②
1995	3	1.3 (1298)	0.029	5B	5B	13-20 ③	750	850	HYD	HYD
1996-97	2	1.3 (1298)	0.029	5B	5B	13-20 ③	750	850	HYD	HYD
	3	1.6 (1590)	0.029	5B	5B	28-34 ③	750-800	750-800	④	④

NOTE: The Vehicle Emission Control Information label often reflects specification changes made during production. The label figures must be used if they differ from those in this chart.
HYD - Hydraulic
B - Before top dead center
① GT models
② Specifications for:
   Hot engine: Intake- 0.009 - 0.011; Exhaust 0.010 - 0.012
   Cold engine: Intake- 0.0051-0.0067; Exhaust 0.0063-0.0079
③ At idle
④ When cold: 0.005-0.007
   When hot: 0.007-0.008

CAPACITIES

Year	Model	Engine ID/VIN	Engine Displacement Liters (cc)	Engine Oil with Filter (qts)	Transmission (pts.) 4-Spd	5-Spd	Auto.	Transfer Case (pts.)	Drive Axle Front (pts.)	Rear (pts.)	Fuel Tank (gal.)	Cooling System (qts.)
1993	Swift	3	1.3 (1298)	3.7	-	4.2	10.4 ①	-	-	-	10.6	5.0
1994	Swift	3	1.3 (1298)	3.5	-	5.0	10.4 ①	-	-	-	10.6	4.9

20-138

CAPACITIES

Year	Model	Engine ID/VIN	Engine Displacement Liters (cc)	Engine Oil with Filter (qts)	Transmission (pts.) 4-Spd	5-Spd	Auto.	Transfer Case (pts.)	Drive Axle Front (pts.)	Rear (pts.)	Fuel Tank (gal.)	Cooling System (qts.)
1995	Swift	3	1.3 (1298)	3.5	-	5.0	10.4 ①	-	-	-	10.6	4.9
1996-97	Swift	2	1.3 (1298)	3.3	-	5.0	10.4 ①	-	-	-	10.6	②
	Esteem	3	1.6 (1590)	3.3	-	5.0	10.4 ①	-	-	-	13.5	③

① Specification for automatic transaxle is after complete overhaul. Drain and fill will be less
② Manual transmission: 4.8 qts.
  Automatic transmission: 4.9 qts.
③ Manual transmission: 4.8 qts.
  Automatic transmission: 4.7 qts.

VALVE SPECIFICATIONS

Year	Engine ID/VIN	Engine Displacement Liters (cc)	Seat Angle (deg.)	Face Angle (deg.)	Spring Test Pressure (lbs. @ in.)	Spring Installed Height (in.)	Stem-to-Guide Clearance (in.) Intake	Exhaust	Stem Diameter (in.) Intake	Exhaust
1993	3 ①	1.3 (1298)	45	45	55-64@1.63	1.9409	0.0008-0.0019	0.0014-0.0025	0.2742-0.2748	0.2737-0.2742
	3	1.3 (1298)	45	45	61-74@1.67	2.0079	0.0008-0.0018	0.0014-0.0024	0.2152-0.2157	0.2146-0.2151
1994	3 ①	1.3 (1298)	45	45	55-64@1.63	1.9409	0.0008-0.0019	0.0014-0.0025	0.2742-0.2748	0.2737-0.2742
	3	1.3 (1298)	45	45	61-74@1.67	2.0079	0.0008-0.0018	0.0014-0.0024	0.2152-0.2157	0.2146-0.2151
1995	3	1.3 (1298)	45	45	55-64@1.63	1.941	0.0008-0.0019	0.0014-0.0025	0.2742-0.2748	0.2737-0.2742
1996-97	2	1.3 (1298)	45	45	55-64@1.63	1.941	0.0008-0.0019	0.0014-0.0025	0.2742-0.2748	0.2737-0.2742
	3	1.6 (1590)	45	45	24-28@1.24	1.450	0.0008-0.0018	0.0018-0.0028	0.2152-0.2157	0.2142-0.2148

① Swift GT DOHC engine

TORQUE SPECIFICATIONS
All readings in ft. lbs.

Year	Engine ID/VIN	Engine Displacement Liters (cc)	Cylinder Head Bolts	Main Bearing Bolts	Rod Bearing Bolts	Crankshaft Damper Bolts	Flywheel Bolts	Manifold Intake	Exhaust	Spark Plugs	Lug Nut
1993	3 ①	1.3 (1298)	48-51	36-41	24-26	76-83 ②	50-52	13-20	13-20	15-22	36-50
	3	1.3 (1298)	51-54	36-41	24-26	76-83 ②	50-52	13-20	13-20	15-22	36-50
1994	3 ①	1.3 (1298)	48-51	36-41	24-26	76-83 ②	50-52	13-20	13-20	15-22	36-50
	3	1.3 (1298)	51-54	36-41	24-26	76-83 ②	50-52	13-20	13-20	15-22	36-50
1995	3	1.3 (1298)	51-54	36-41	24-26	76-83 ②	41-47	13-20	13-20	14-21	36-57
1996-97	2	1.3 (1298)	51-54	36-41	24-26	76-83 ②	41-47	13-20	13-20	14-21	36-57
	3	1.6 (1590)	48-51	36-41	24-26	76-83 ②	57	13-20	13-20	14-21	58-80

① GT models
② Specification shown is for Crankshaft timing sprocket nut

BRAKE SPECIFICATIONS
All measurements in inches unless noted

Year	Model		Master Cylinder Bore	Brake Disc			Brake Drum Diameter			Minimum Lining Thickness			
				Original Thickness	Minimum Thickness	Maximum Runout	Original Inside Diameter	Max. Wear Limit	Maximum Machine Diameter	Front		Rear	
1993	Swift	②	NA	0.670	0.590	0.004	7.09	7.16	7.16	0.315	④	0.110	④
	Swift	①	NA	⑤	⑥	0.004	-	-	-	0.315	④	0.236	④
	Swift	③	NA	0.670	0.590	0.004	7.87	7.95	7.95	0.315	④	0.110	④
1994	Swift	②	NA	0.670	0.590	0.004	7.09	7.16	7.16	0.315	④	0.110	④
	Swift	①	NA	⑤	⑥	0.004	-	-	-	0.315	④	0.236	④
	Swift	③	NA	0.670	0.590	0.004	7.87	7.95	7.95	0.315	④	0.110	④
1995	Swift	②	NA	0.670	0.670	0.004	7.09	7.87	7.87	0.236	④	0.110	④
	Swift	③	NA	0.670	0.670	0.004	7.16	7.95	7.95	0.236	④	0.110	④
1996-97	Swift	②	NA	0.670	0.670	0.004	7.09	7.87	7.87	0.236	④	0.110	④
	Swift	③	NA	0.670	0.670	0.004	7.16	7.95	7.95	0.236	④	0.110	④
	Esteem		NA	0.790	0.710	0.004	7.87	7.95	7.95	0.240	④	0.110	④

NA - Not Available
① GT models
② Hatchback except GT
③ Sedan models
④ Measurement is for lining and backing
⑤ Front - 0.730 Rear - 0.394
⑥ Front - 0.650 Rear - 0.315

FREQUENT MAINTENANCE LABOR
SUZUKI

The following should be used as a guide when determining the amount of work required for a particular service if taken to a repair shop. In estimating how long a particular Frequent Maintenance Service item should take, please observe the following:
- **Chilton Time** is time that is based on field research and data supplied by the vehicle manufacturer.
- All labor time operations are given in hours and tenths of an hour.
- All labor operations, are to be used as a **guide**.

COOLING
(G) Winterize Cooling System
Includes: Run engine to check for leaks, tighten all hose connections. Test radiator and pressure cap, drain radiator and engine block. Add antifreeze and refill system.
1993-97 5

(G) Hoses, Radiator, Renew
Includes: Drain and refill cooling system as required.
1993-97
  upper 5
  lower 7

(G) Thermostat, Coolant, Renew
1995-97 Esteem 6
1993-95 Samurai 4
1993-97 Swift 6
1993-97 Sidekick, X-90 5

FUEL
(G) Filter, Fuel, Renew
w/Fuel Injection
1995-97 Esteem 6

1993-97 Samurai, Sidekick,
  X-90 7
1993-97 Swift 8

BRAKES
(G) Bleed Brakes (Four Wheels)
Includes: Add fluid.
1993-97 8
w/ABS add 2

(G) Brakes, Adjust (Minor)
Includes: Adjust brakes, fill master cylinder.
1993-97 4

(M) Parking Brake, Adjust
1993-97 5

LUBRICATION SERVICES
(M) Engine Oil & Filter, Renew
Includes: Inspect and correct all fluid levels.
1993-97 3

WHEELS
(G) Wheel, Renew (One)
1993-97 Esteem, Swift 5

(G) Wheel, Balance
1993-97 Esteem, Swift
  one .3
  each adtnl. 2

(G) Wheels, Rotate (All)
1993-97 Esteem, Swift 5

ELECTRICAL
(G) Headlamps, Aim
1993-97 4

(G) Halogen Headlamp Bulb, Renew
1993-97 3

(G) License Lamp Assy., Renew
1995-97 Esteem 3
1993-95 Samurai 5
1993-97 Sidekick, X-90 4
1993-97 Swift 3

(G) Rear Combination Lamp Assy., Renew
1995-97 Esteem 3
1993-95 Samurai 4
1993-97 Sidekick, X-90 3
1993-94 Swift 6
1995-97 Swift 3

FREQUENT MAINTENANCE LABOR (cont.)
SUZUKI

	Chilton Time		Chilton Time		Chilton Time
*(G) Tail Lamp Lens or Bulb, Renew 1993-97	.3	(G) Horn, Renew 1993-97	.3	(M) Terminals, Battery, Clean 1993-97	.3

SCHEDULED MAINTENANCE INTERVALS
(SUZUKI SWIFT & ESTEEM)

TO BE SERVICED	TYPE OF SERVICE	VEHICLE MILEAGE INTERVAL (x1000)												
		7.5	15	22.5	30	37.5	45	52.5	60	67.5	75	82.5	90	97.5
Engine oil & filter	R	✓	✓	✓	✓	✓	✓	✓	✓	✓	✓	✓	✓	✓
Automatic transmission fluid & filter③	S/I	✓	✓	✓	✓	✓	✓	✓	✓	✓	✓	✓	✓	✓
Clutch pedal free travel	S/I	✓	✓	✓	✓	✓	✓	✓	✓	✓	✓	✓	✓	✓
Drive axle boots	S/I	✓	✓	✓	✓	✓	✓	✓	✓	✓	✓	✓	✓	✓
Gear shift control lever/shift operation	S/I	✓	✓	✓	✓	✓	✓	✓	✓	✓	✓	✓	✓	✓
Inspect & rotate tires	S/I	✓	✓	✓	✓	✓	✓	✓	✓	✓	✓	✓	✓	✓
Manual transmission oil②	S/I	✓	✓	✓	✓	✓	✓	✓	✓	✓	✓	✓	✓	✓
Power steering system	S/I	✓	✓	✓	✓	✓	✓	✓	✓	✓	✓	✓	✓	✓
Suspension system	S/I	✓	✓	✓	✓	✓	✓	✓	✓	✓	✓	✓	✓	✓
Brake discs, pads, drums & shoes	S/I	✓		✓		✓		✓		✓		✓		✓
Brake hoses, pipes, brake lever & cable	S/I	✓		✓		✓		✓		✓		✓		✓
Brake fluid①	S/I		✓		✓		✓		✓		✓		✓	
Brake pedal	S/I		✓		✓		✓		✓		✓		✓	
Cooling system, hoses & connections	S/I		✓		✓		✓		✓		✓		✓	
Fuel tank, cap & lines	S/I		✓		✓		✓		✓		✓		✓	
Valve lash (clearance)	S/I			✓			✓			✓			✓	
Air cleaner filter element	R				✓				✓				✓	
Engine Coolant	R				✓				✓				✓	

SCHEDULED MAINTENANCE INTERVALS
(SUZUKI SWIFT & ESTEEM) (Cont.)

TO BE SERVICED	TYPE OF SERVICE	VEHICLE MILEAGE INTERVAL (x1000)											
		7.5	15	22.5	30	37.5	45	52.5	60	67.5	75	82.5	90	97.5
Spark plugs	R				✓				✓				✓	
Drive belts	S/I				✓				✓				✓	
Exhaust system	S/I				✓				✓				✓	
Automatic transmission fluid hose	R						✓						✓	
Camshaft timing belt	R								✓					
Ignition wiring	S/I								✓					

① Replace every 60,000 miles.   ② Replace every 15,000 miles.   ③ Replace every 100,000 miles.
R – Replace   S/I – Service or Inspect

FREQUENT OPERATION MAINTENANCE (SEVERE SERVICE)
If a vehicle is operated under any of the following conditions it is considered severe service:
- Extremely dusty areas.
- 50% or more of the vehicle operation is in 32°C (90°F) or higher temperatures, or constant operation in temperatures below 0°C (32°F).
- Prolonged idling (vehicle operation in stop and go traffic).
- Frequent short running periods (engine does not warm to normal operating temperatures).
- Police, taxi, delivery usage or trailer towing usage.
Oil & oil filter change – change every 3000 miles.
Brake discs, pads, drums & shoes - service or inspect initially at 3000 miles, 6000 miles, & every 12,000 miles thereafter.
Brake hoses & pipes - service or inspect initially at 3000 miles, 6000 miles, & every 12,000 miles thereafter.
Air cleaner filter element - service or inspect every 3000 miles & replace every 30,000 miles (if not replaced previously).
Automatic transmission fluid & filter - service or inspect every 6000 miles & replace every 15,000 miles (if not replaced previously).
Clutch pedal free travel - service or inspect every 6000 miles.
Gear shift control lever/shift operation - service or inspect every 6000 miles.
Inspect & rotate tires - service or inspect every 6000 miles.
Manual transmission oil - service or inspect every 6000 miles & replace every 12,000 miles (if not replaced previously).
Power steering system - service or inspect every 6000 miles.
Steering system - service or inspect every 6000 miles.
Suspension system - service or inspect every 6000 miles.
Drive belts - service or inspect every 15,000 miles.
Exhaust system - service or inspect every 15,000 miles.

TOYOTA

ENGINE IDENTIFICATION

Year	Model	Engine Displacement Liters (cc)	Engine Series (ID/VIN)	Fuel System	No. of Cylinders	Engine Type
1993	Tercel	1.5 (1457)	3E-E	EFI	4	SOHC
	Paseo	1.5 (1495)	5E-FE	EFI	4	DOHC
	Corolla	1.6 (1587)	4A-FE	EFI	4	DOHC
	Corolla	1.8 (1762)	7A-FE	EFI	4	DOHC
	Celica	1.6 (1587)	4A-FE	EFI	4	DOHC
	Celica	2.0 (1998) ①	3S-GTE	EFI	4	DOHC

ENGINE IDENTIFICATION

Year	Model	Engine Displacement Liters (cc)	Engine Series (ID/VIN)	Fuel System	No. of Cylinders	Engine Type
1993	Celica	2.2 (2164)	5S-FE	EFI	4	DOHC
	MR2	2.0 (1998) ①	3S-GTE	EFI	4	DOHC
	MR2	2.2 (2164)	5S-FE	EFI	4	DOHC
	Camry	2.2 (2164)	5S-FE	EFI	4	DOHC
	Camry	3.0 (2952)	3VZ-FE	EFI	6	DOHC
1994	Tercel	1.5 (1457)	3E-E	EFI	4	SOHC
	Paseo	1.5 (1495)	5E-FE	EFI	4	DOHC
	Corolla	1.6 (1587)	4A-FE	EFI	4	DOHC
	Corolla	1.8 (1762)	7A-FE	EFI	4	DOHC
	Celica	1.8 (1762)	7A-FE	EFI	4	DOHC
	Celica	2.2 (2164)	5S-FE	EFI	4	DOHC
	MR2	2.0 (1998) ①	3S-GTE	EFI	4	DOHC
	MR2	2.2 (2164)	5S-FE	EFI	4	DOHC
	Camry	2.2 (2164)	5S-FE	EFI	4	DOHC
	Camry	3.0 (2952)	1MZ-FE	EFI	6	DOHC
	Avalon	3.0 (2952)	1MZ-FE	EFI	6	DOHC
	Supra	3.0 (2997)	2JZ-GE	EFI	6	DOHC
	Supra	3.0 (2997) ②	2JZ-GTE	EFI	6	DOHC
1995	Tercel	1.5 (1457)	3E-E	EFI	4	SOHC
	Paseo	1.5 (1495)	5E-FE	EFI	4	DOHC
	Corolla	1.6 (1587)	4A-FE	EFI	4	DOHC
	Corolla	1.8 (1762)	7A-FE	EFI	4	DOHC
	Celica	1.8 (1762)	7A-FE	EFI	4	DOHC
	Celica	2.2 (2164)	5S-FE	EFI	4	DOHC
	MR2	2.0 (1998) ①	3S-GTE	EFI	4	DOHC
	MR2	2.2 (2164)	5S-FE	EFI	4	DOHC
	Camry	2.2 (2164)	5S-FE	EFI	4	DOHC
	Camry	3.0 (2952)	1MZ-FE	EFI	6	DOHC
	Supra	3.0 (2997)	2JZ-GE	EFI	6	DOHC
	Supra	3.0 (2997) ②	2JZ-GTE	EFI	6	DOHC
1996-97	Paseo	1.5 (1497)	5E-FE	EFI	4	DOHC
	Tercel	1.5 (1497)	5E-FE	EFI	4	DOHC
	Corolla	1.6 (1587)	4A-FE	EFI	4	DOHC
	Corolla	1.8 (1762)	7A-FE	EFI	4	DOHC
	Celica	1.8 (1762)	7A-FE	EFI	4	DOHC
	Celica	2.2 (2164)	5S-FE	EFI	4	DOHC
	Camry	2.2 (2164)	5S-FE	EFI	4	DOHC
	Camry	3.0 (2995)	1MZ-FE	EFI	6	DOHC
	Avalon	3.0 (2995)	1MZ-FE	EFI	6	DOHC
	Supra	3.0 (2997)	2JZ-GE	EFI	6	DOHC
	Supra	3.0 (2997) ①	2JZ-GTE	EFI	6	DOHC

EFI - Electronic fuel injection
SOHC - Single overhead camshaft
DOHC - Double overhead camshaft

① Turbocharged
② Twin Turbocharged

20 SPECIFICATIONS

GENERAL ENGINE SPECIFICATIONS

Year	Engine ID/VIN	Engine Displacement Liters (cc)	Fuel System Type	Net Horsepower @ rpm	Net Torque @ rpm (ft. lbs.)	Bore x Stroke (in.)	Compression Ratio	Oil Pressure @ rpm	
1993	3E-E	1.5 (1456)	EFI	82@5200	89@4400	2.87x3.43	9.3:1	4.3	①
	5E-FE	1.5 (1495)	EFI	100@6400	91@3200	2.91x3.43	9.4:1	4.3	①
	4A-FE	1.6 (1587)	EFI	102@5800	101@4800	3.19x3.03	9.5:1	4.3	①
	7A-FE	1.8 (1762)	EFI	115@5600	115@2800	31.9x3.37	9.5:1	4.3	①
	3S-GTE	2.0 (1998)	EFI	200@6000	200@3200	3.39x3.39	8.8:1	4.3	①
	5S-FE	2.2 (2164)	EFI	135@3400	145@4400	3.43x3.58	9.5:1	4.3	①
	3VZ-FE	3.0 (2952)	EFI	185@5200	195@4400	3.44x3.23	9.6:1	4.3	①
1994	3E-E	1.5 (1456)	EFI	82@5200	89@4400	2.87x3.43	9.3:1	4.3	①
	5E-FE	1.5 (1495)	EFI	100@6400	91@3200	2.91x3.43	9.4:1	4.3	①
	4A-FE	1.6 (1587)	EFI	102@5800	101@4800	3.19x3.03	9.5:1	4.3	①
	7A-FE	1.8 (1762)	EFI	115@5600	115@2800	3.19x3.37	9.5:1	4.3	①
	3S-GTE	2.0 (1998)	EFI	200@6000	200@3200	3.39x3.39	8.8:1	4.3	①
	5S-FE	2.2 (2164)	EFI	135@5400	145@4400	3.43x3.58	9.5:1	4.3	①
	2JZ-GE	3.0 (2997)	EFI	220@5800	210@4800	3.39x3.39	10.0:1	36-71@3000	
	2JZ-GTE	3.0 (2997)	EFI	320@5600	315@4000	3.39x3.39	8.5:1	36-71@3000	
	1MZ-FE	3.0 (2952)	EFI	185@5200	195@4400	3.45x3.45	9.6:1	4.3	①
1995	3E-E	1.5 (1456)	EFI	82@5200	89@4400	2.87x3.43	9.3:1	4.3	①
	5E-FE	1.5 (1495)	EFI	100@6400	91@3200	2.91x3.43	9.4:1	4.3	①
	4A-FE	1.6 (1587)	EFI	102@5800	101@4800	3.19x3.03	9.5:1	4.3	①
	7A-FE	1.8 (1762)	EFI	115@5600	115@2800	3.19x3.37	9.5:1	4.3	①
	3S-GTE	2.0 (1998)	EFI	200@6000	200@3200	3.39x3.39	8.8:1	4.3	①
	5S-FE	2.2 (2164)	EFI	135@5400	145@4400	3.43x3.58	9.5:1	4.3	①
	2JZ-GE	3.0 (2997)	EFI	220@5800	210@4800	3.39x3.39	10.0:1	36-71@3000	
	2JZ-GTE	3.0 (2997)	EFI	320@5600	315@4000	3.39x3.39	8.5:1	36-71@3000	
	1MZ-FE	3.0 (2952)	EFI	185@5200	195@4400	3.45x3.45	9.6:1	4.3	①
1996-97	5E-FE	1.5 (1497)	EFI	93@5400	100@4400	2.91x3.43	9.4:1	43	①
	4A-FE	1.6 (1587)	EFI	100@5600	105@4400	3.19x3.03	9.5:1	43	①
	7A-FE	1.8 (1762)	EFI	105@5600	117@2800	3.19x3.37	9.5:1	43	①
	5S-FE	2.2 (2164)	EFI	130@5400	145@4400	3.43x3.58	9.5:1	43	①
	1MZ-FE	3.0 (2952)	EFI	192@5200	210@4400	3.44x3.27	10.5:1	43	①
	2JZ-GE	3.0 (2997)	EFI	220@5800	210@4800	3.39x3.39	10.0:1	36-71@3000	
	2JZ-GTE	3.0 (2997)	EFI	320@5600	315@4000	3.39x3.39	8.5:1	36-71@3000	

EFI - Electronic fuel injection   ① At idle

GASOLINE ENGINE TUNE-UP SPECIFICATIONS

Year	Engine ID/VIN	Engine Displacement Liters (cc)	Spark Plugs Gap (in.)	Ignition Timing (deg.) MT	Ignition Timing (deg.) AT	Fuel Pump (psi)	Idle Speed (rpm) MT	Idle Speed (rpm) AT	Valve Clearance In.	Valve Clearance Ex.
1993	3E-E	1.5 (1456)	0.043	10B	10B	41-45	750	800	0.008	0.008
	5E-FE	1.5 (1495)	0.043	10B	10B	41-42	750	750	0.006-0.010	0.012-0.016
	4A-FE	1.6 (1587)	0.031	10B	10B	38-44	①	①	0.006-0.010	0.008-0.012
	7A-FE	1.8 (1762)	0.031	10B	10B	38-44	800	800	0.006-0.010	0.008-0.012
	3S-GTE	2.0 (1998)	0.031	10B	10B	33-38	750-850	750-850	0.006-0.010	④
	5S-FE	2.2 (2164)	0.043	10B	10B	38-44	②	③	0.007-0.011	0.011-0.015

20-144

GASOLINE ENGINE TUNE-UP SPECIFICATIONS

Year	Engine ID/VIN	Engine Displacement Liters (cc)	Spark Plugs Gap (in.)	Ignition Timing (deg.) MT	Ignition Timing (deg.) AT	Fuel Pump (psi)	Idle Speed (rpm) MT	Idle Speed (rpm) AT	Valve Clearance In.	Valve Clearance Ex.
1993	3VZ-FE	3.0 (2952)	0.043	10B	10B	38-44	650-750	650-750	0.005-0.009	0.011-0.015
1994	3E-E	1.5 (1456)	0.043	10B	10B	41-45	750	800	0.008	0.008
	5E-FE	1.5 (1495)	0.043	10B	10B	41-42	750	750	0.012-0.010	0.012-0.016
	4A-FE	1.6 (1587)	0.031	10B	10B	38-44	①	①	0.006-0.010	0.008-0.012
	7A-FE	1.8 (1762)	0.031	10B	10B	38-44	800	800	0.006-0.010	0.008-0.012
	3S-GTE	2.0 (1998)	0.031	10B	10B	33-38	750-850	750-850	0.006-0.010	④
	5S-FE	2.2 (2164)	0.043	10B	10B	38-44	②	③	0.007-0.011	0.011-0.015
	2JZ-GE	3.0 (2997)	0.031	⑤	⑤	33-40	650-750	650-750	0.006-0.010	0.010-0.014
	2JZ-GTE	3.0 (2997)	0.031	⑤	⑤	33-40	600-700	600-700	0.006-0.010	0.010-0.014
	1MZ-FE	3.0 (2952)	0.043	⑤	⑤	38-44	650-750	650-750	0.006-0.010	0.010-0.014
1995	3E-E	1.5 (1456)	0.043	10B	10B	41-45	750	800	0.008	0.008
	5E-FE	1.5 (1495)	0.043	10B	10B	41-42	750	750	0.012-0.010	0.012-0.016
	4A-FE	1.6 (1587)	0.031	10B	10B	38-44	①	①	0.006-0.010	0.008-0.012
	7A-FE	1.8 (1762)	0.031	10B	10B	38-44	800	800	0.006-0.010	0.008-0.012
	3S-GTE	2.0 (1998)	0.031	10B	10B	33-38	750-850	750-850	0.006-0.010	④
	5S-FE	2.2 (2164)	0.043	10B	10B	38-44	②	③	0.007-0.011	0.011-0.015
	2JZ-GE	3.0 (2997)	0.031	⑤	⑤	33-40	650-750	650-750	0.006-0.010	0.010-0.014
	2JZ-GTE	3.0 (2997)	0.031	⑤	⑤	33-40	600-700	600-700	0.006-0.010	0.010-0.014
	1MZ-FE	3.0 (2952)	0.043	⑤	⑤	38-44	650-750	650-750	0.006-0.010	0.010-0.014
1996-97	5E-FE	1.5 (1497)	0.043	⑤	⑤	41-42	700-800	750	0.006-0.010	0.012-0.016
	4A-FE	1.6 (1587)	0.031	⑤	⑤	38-44	650-750	650-750	0.006-0.010	0.010-0.014
	7A-FE	1.8 (1762)	0.031	⑤	⑤	38-44	650-750	650-750	0.006-0.010	0.010-0.014
	5S-FE	2.2 (2164)	0.043	⑤	⑤	38-44	700-800	700-800	0.007-0.011	0.011-0.015
	1MZ-FE	3.0 (2952)	0.043	-	10B	38-44	650-750	650-750	0.006-0.010	0.010-0.014

GASOLINE ENGINE TUNE-UP SPECIFICATIONS

Year	Engine ID/VIN	Engine Displacement Liters (cc)	Spark Plugs Gap (in.)	Ignition Timing (deg.) MT	Ignition Timing (deg.) AT	Fuel Pump (psi)	Idle Speed (rpm) MT	Idle Speed (rpm) AT	Valve Clearance In.	Valve Clearance Ex.
1996-97	2JZ-GE	3.0 (2997)	0.031	⑤	⑤	33-40	650-750	650-750	0.006-0.010	0.010-0.014
	2JZ-GTE	3.0 (2997)	0.031	⑤	⑤	33-40	600-700	600-700	0.006-0.010	0.010-0.014

NOTE: The Vehicle Emission Control Information label often reflects specification changes made during production. The label figures must be used if they differ from those in this chart.

B - Before top dead center

① 2WD Federal and Canada: 700
2WD California and 4WD: 800

② USA: 750
Canada: 850

③ USA: 700
Canada: 750

④ MR2: 0.008-0.012
Celica: 0.011-0.015

⑤ 10B at idle, with terminal TE1 and E1 connected of DLC1

CAPACITIES

Year	Model	Engine ID/VIN	Engine Displacement Liters (cc)	Engine Oil with Filter	Transmission (pts.) 4-Spd	Transmission (pts.) 5-Spd	Transmission (pts.) Auto.	Transfer Case (pts.)	Drive Axle Front (pts.)	Drive Axle Rear (pts.)	Fuel Tank (gal.)	Cooling System (qts.)
1993	Tercel	3E-E	1.5 (1547)	3.4	5.0	5.0	5.2	-	3.0	-	11.9	②
	Paseo	5E-FE	1.5 (1495)	3.4	-	5.0	6.6	-	3.0	-	11.9	⑤
	Corolla	4A-FE	1.6 (1587)	3.4	-	5.4	5.2	-	-	-	13.2	6.5
	Corolla	7A-FE	1.8 (1762)	3.4	-	5.4	5.2	-	-	-	13.2	6.6
	Camry	5S-FE	2.2 (2164)	4.3	-	5.4	5.2	-	3.4	-	18.5	6.7
	Camry	3VZ-FE	3.0 (2952)	4.5	-	8.8	6.2	-	3.4	-	18.5	9.0
	Celica	4A-FE	1.6 (1587)	3.4	-	5.4	5.2	-	3.4	-	⑥	③
	Celica	3S-GTE	2.0 (1998)	4.1	-	5.4	5.2	④	3.4	2.4	15.9	③
	Celica	5S-FE	2.2 (2164)	①	-	11.0	-	-	3.4	-	15.9	6.9
	MR2	3S-GTE	2.0 (1998)	4.1	-	8.8	7.0	-	-	3.4	14.3	14.4
	MR2	5S-FE	2.2 (2164)	4.5	-	5.4	7.0	-	-	3.4	14.3	13.7
1994	Tercel	3E-E	1.5 (1457)	3.4	5.0	5.0	5.2	-	3.0	-	11.9	②
	Paseo	5E-FE	1.5 (1495)	3.4	-	5.0	6.6	-	3.0	-	11.9	⑤
	Corolla	4A-FE	1.6 (1587)	3.4	-	5.4	5.2	-	-	-	13.2	6.5
	Corolla	7A-FE	1.8 (1762)	3.4	-	5.4	5.2	-	-	-	13.2	6.6
	Camry	5S-FE	2.2 (2164)	4.3	-	5.4	5.2	-	3.4	-	18.5	6.7
	Camry	1MZ-FE	3.0 (2952)	4.5	-	8.8	6.2	-	3.4	-	18.5	9.0
	Celica	4AFE	1.6 (1587)	3.4	-	5.4	5.2	-	3.4	-	⑥	③
	Celica	5S-FE	2.2 (2164)	①	-	11.0	-	-	3.4	-	15.9	6.9
	MR2	3S-GTE	2.0 (1998)	4.1	-	8.8	7.0	-	-	3.4	14.3	14.4
	MR2	5S-FE	2.2 (2164)	4.5	-	5.4	7.0	-	-	3.4	14.3	13.7
	Supra	2JZ-GE	3.0 (2997)	5.5	-	5.4	15.2	-	-	2.9	18.5	⑦
	Supra	2JZ-GTE	3.0 (2997)	5.3	-	5.4	17.4	-	-	2.9	18.5	⑧
1995	Tercel	3E-E	1.5 (1457)	3.4	5.0	5.0	5.2	-	3.0	-	11.9	②
	Paseo	5E-FE	1.5 (1495)	3.4	-	5.0	6.6	-	3.0	-	11.9	⑤
	Corolla	4A-FE	1.6 (1587)	3.4	-	5.4	5.2	-	-	-	13.2	6.5
	Corolla	7A-FE	1.8 (1762)	3.4	-	5.4	5.2	-	-	-	13.2	6.6
	Camry	5S-FE	2.2 (2164)	4.3	-	5.4	5.2	-	3.4	-	18.5	6.7
	Camry	1MZ-FE	3.0 (2952)	4.5	-	8.8	6.2	-	3.4	-	18.5	9.0
	Avalon	1MZ-FE	3.0 (2952)	5.0	-	-	7.4	-	1.8	-	18.5	9.8
	Celica	4AFE	1.6 (1587)	3.4	-	5.4	5.2	-	3.4	-	⑥	③
	Celica	5S-FE	2.2 (2164)	①	-	11.0	-	-	3.4	-	15.9	6.9
	MR2	3S-GTE	2.0 (1998)	4.1	-	8.8	7.0	-	-	3.4	14.3	14.4
	MR2	5S-FE	2.2 (2164)	4.5	-	5.4	7.0	-	-	3.4	14.3	13.7
	Supra	2JZ-GE	3.0 (2997)	5.5	-	5.4	15.2	-	-	2.9	18.5	⑦
	Supra	2JZ-GTE	3.0 (2997)	5.3	-	5.4	17.4	-	-	2.9	18.5	⑧
1996-97	Tercel	5E-FE	1.5 (1497)	3.0	5.0	5.0	⑨	-	3.0 ⑩	-	11.9	②
	Paseo	5E-FE	1.5 (1497)	3.0	-	4.0	6.6	-	-	-	11.9	⑤

CAPACITIES

Year	Model	Engine ID/VIN	Engine Displacement Liters (cc)	Engine Oil with Filter	Transmission (pts.) 4-Spd	5-Spd	Auto.	Transfer Case (pts.)	Drive Axle Front (pts.)	Rear (pts.)	Fuel Tank (gal.)	Cooling System (qts.)
1996-97	Corolla	4A-FE	1.6 (1587)	3.2	-	4.0	5.2	-	3.0	-	13.2	⑪
	Corolla	7A-FE	1.8 (1762)	3.9	-	4.0	6.6	-	-	-	13.2	⑭
	Camry	5S-FE	2.2 (2164)	3.8	-	5.4	5.2	-	3.4	-	18.5	6.7
	Camry	1MZ-FE	3.0 (2995)	5.0	-	-	7.4	-	1.8	-	18.5	9.8
	Celica	7A-FE	1.8 (1762)	3.9	-	4.0	6.6	-	-	-	15.9	⑬
	Celica	5S-FE	2.2 (2164)	4.1	-	5.4	5.2	-	3.4	-	15.9	⑫
	Avalon	1MZ-FE	3.0 (2995)	5.0	-	-	7.4	-	1.8	-	18.5	9.8
	Supra	2JZ-GE	3.0 (2997)	5.5	-	5.4	3.4	-	-	2.9	18.5	⑦
	Supra	2JZ-GTE	3.0 (2997)	5.3	-	3.8	4.0	-	-	2.9	18.5	⑧

① With oil cooler: 4.4
   Without oil cooler: 4.3
② Manual transmission: 5.2
   Automatic transmission: 5.7
③ Manual transmission: 5.5
   Automatic transmission: 5.8
④ Included with transmission

⑤ Manual transmission: 5.3
   Automatic transmission: 5.7
⑥ 2WD: 15.9; 4WD: 18.0
⑦ Manual transmission: 7.7
   Automatic transmission: 8.8
⑧ Manual transmission: 10.0
   Automatic transmission: 9.9

⑨ A132L transmission: 5.2
   A242L transmission: 6.6
⑩ A132L transmission
⑪ M/T with Nippodenso radiator: 5.6
   A/T with Nippodenso radiator: 6.2
   M/T with Harrison radiator: 6.3
   A/T with Harrison radiator: 6.2

⑫ Manual trans.: 7.1
   Automatic trans.: 7.5
⑬ Manual trans.: 6.4
   Automatic trans.: 7.0
⑭ M/T with Nippodenso radiator: 5.8
   A/T with Nippodenso radiator: 6.2
   M/T with Harrison radiator: 6.6
   A/T with Harrison radiator: 6.4

VALVE SPECIFICATIONS

Year	Engine ID/VIN	Engine Displacement Liters (cc)	Seat Angle (deg.)	Face Angle (deg.)	Spring Test Pressure (lbs. @ in.)	Spring Installed Height (in.)	Stem-to-Guide Clearance (in.) Intake	Exhaust	Stem Diameter (in.) Intake	Exhaust
1993	3E-E	1.5 (1456)	45	44.5	35.1	1.384	0.0010-0.0024	0.0012-0.0026	0.2350-0.2356	0.2348-0.2354
	5E-FE	1.5 (1495)	45	44.5	37-37	1.252	0.0010-0.0024	0.0012-0.0026	0.2350-0.2356	0.2348-0.2354
	4A-FE	1.6 (1587)	45	44.5	37.3	1.248	0.0010-0.0024	0.0012-0.0026	0.2350-0.2356	0.2348-0.2354
	7A-FE	1.8 (1762)	45	44.5	37.3	1.248	0.0010-0.0023	0.0012-0.0025	0.2346-0.2352	0.2344-0.2350
	3S-GTE	2.0 (1998)	45	45.5	53.1	1.354	0.0010-0.0023	0.0012-0.0025	0.2346-0.2352	0.2344-0.2350
	5S-FE	2.2 (2164)	45	45.5	42.5	1.366	0.0010-0.0024	0.0012-0.0026	0.2350-0.2356	0.2348-0.2354
	3VZ-FE	3.0 (2952)	45	44.5	38-42	1.311	0.0010-0.0024	0.0012-0.0026	0.2350-0.2356	0.2348-0.2354
1994	3E-E	1.5 (1456)	45	44.5	35.1	1.384	0.0010-0.0024	0.0012-0.0026	0.2350-0.2356	0.2348-0.2354
	5E-FE	1.5 (1495)	45	44.5	37-37	1.252	0.0010-0.0024	0.0012-0.0026	0.2350-0.2356	0.2348-0.2354
	4A-FE	1.6 (1587)	45	44.5	37.3	1.248	0.0010-0.0024	0.0012-0.0026	0.2350-0.2356	0.2348-0.2354
	7A-FE	1.8 (1762)	45	44.5	37.3	1.248	0.0010-0.0023	0.0012-0.0025	0.2346-0.2352	0.2344-0.2350
	3S-GTE	2.0 (1998)	45	45.5	53.1	1.354	0.0010-0.0023	0.0012-0.0025	0.2346-0.2352	0.2344-0.2350
	5S-FE	2.2 (2164)	45	45.5	42.5	1.366	0.0010-0.0024	0.0012-0.0026	0.2350-0.2356	0.2348-0.2354
	1MZ-FE	3.0 (2952)	45	44.5	42-46	1.331	0.0010-0.0024	0.0012-0.0026	0.2154-0.2159	0.2152-0.2157

VALVE SPECIFICATIONS

Year	Engine ID/VIN	Engine Displacement Liters (cc)	Seat Angle (deg.)	Face Angle (deg.)	Spring Test Pressure (lbs. @ in.)	Spring Installed Height (in.)	Stem-to-Guide Clearance (in.)		Stem Diameter (in.)	
							Intake	Exhaust	Intake	Exhaust
1994	2JZ-GE	3.0 (2997)	45	44.5	42-46@ 1.358	1.358	0.0010-0.0024	0.0012-0.0026	0.2350-0.2356	0.2348-0.2358
	2JZ-GTE	3.0 (2997)	45	44.5	42-46@ 1.358	1.358	0.0010-0.0024	0.0012-0.0026	0.2350-0.2356	0.2348-0.2358
1995	3E-E	1.5 (1456)	45	44.5	35.1	1.384	0.0010-0.0024	0.0012-0.0026	0.2350-0.2356	0.2348-0.2354
	5E-FE	1.5 (1495)	45	44.5	37-37	1.252	0.0010-0.0024	0.0012-0.0026	0.2350-0.2356	0.2348-0.2354
	4A-FE	1.6 (1587)	45	44.5	37.3	1.248	0.0010-0.0024	0.0012-0.0026	0.2350-0.2356	0.2348-0.2354
	7A-FE	1.8 (1762)	45	44.5	37.3	1.248	0.0010-0.0023	0.0012-0.0025	0.2346-0.2352	0.2344-0.2350
	3S-GTE	2.0 (1998)	45	45.5	53.1	1.354	0.0010-0.0023	0.0012-0.0025	0.2346-0.2352	0.2344-0.2350
	5S-FE	2.2 (2164)	45	45.5	42.5	1.366	0.0010-0.0024	0.0012-0.0026	0.2350-0.2356	0.2348-0.2354
	1MZ-FE	3.0 (2952)	45	44.5	42-46	1.331	0.0010-0.0024	0.0012-0.0026	0.2154-0.2159	0.2152-0.2157
	2JZ-GE	3.0 (2997)	45	44.5	42-46@ 1.358	1.358	0.0010-0.0024	0.0012-0.0026	0.2350-0.2356	0.2348-0.2358
	2JZ-GTE	3.0 (2997)	45	44.5	42-46@ 1.358	1.358	0.0010-0.0024	0.0012-0.0026	0.2350-0.2356	0.2348-0.2358
1996-97	5E-FE	1.5 (1497)	45	44.5	33.3-36.8@ 1.252	1.252	0.0010-0.0024	0.0012-0.0026	0.2350-0.2356	0.2348-0.2354
	4A-FE	1.6 (1587)	45	44.5	35.5-39.0@ 1.248	1.248	0.0010-0.0024	0.0012-0.0026	0.2350-0.2356	0.2348-0.2354
	7A-FE	1.8 (1762)	45	44.5	35.5-39.0@ 1.248	1.248	0.0010-0.0024	0.0012-0.0025	0.2346-0.2352	0.2344-0.2350
	5S-FE	2.2 (2164)	45	44.5	36.8-42.5@ 1.366	1.366	0.0010-0.0024	0.0012-0.0026	0.2350-0.2356	0.2348-0.2354
	1MZ-FE	3.0 (2995)	45	44.5	41.9-46.3@ 1.331	1.331	0.0010-0.0024	0.0012-0.0026	0.2154-0.2159	0.2152-0.2157
	2JZ-GE	3.0 (2997)	45	44.5	42-46@ 1.358	1.358	0.0010-0.0024	0.0012-0.0026	0.2350-0.2356	0.2348-0.2358
	2JZ-GTE	3.0 (2997)	45	44.5	42-46@ 1.358	1.358	0.0010-0.0024	0.0012-0.0026	0.2350-0.2356	0.2348-0.2358

TORQUE SPECIFICATIONS

All readings in ft. lbs.

Year	Engine ID/VIN	Engine Displacement Liters (cc)	Cylinder Head Bolts	Main Bearing Bolts	Rod Bearing Bolts	Crankshaft Damper Bolts	Flywheel Bolts	Manifold Intake	Manifold Exhaust	Spark Plugs	Lug Nut
1993	3E-E	1.5 (1456)	①	42	29	112	88	14	38	13	76
	5E-FE	1.5 (1495)	③	42	29	112	65	14	35	13	76
	4A-FE	1.6 (1587)	44	44	36	87	58	14	18	13	76
	7A-FE	1.8 (1762)	②	44	⑦	87	⑨	14	29	13	76
	3S-GTE	2.0 (1998)	③	43	49	80	80	14	38	13	76
	5S-FE	2.2 (2164)	③	43	④	80	⑤	14	36	13	76
	3VZ-FE	3.0 (2952)	⑧	④	⑥	181	61	13	29	13	76

TORQUE SPECIFICATIONS
All readings in ft. lbs.

Year	Engine ID/VIN	Engine Displacement Liters (cc)	Cylinder Head Bolts	Main Bearing Bolts	Rod Bearing Bolts	Crankshaft Damper Bolts	Flywheel Bolts	Manifold Intake	Manifold Exhaust	Spark Plugs	Lug Nut
1994	3E-E	1.5 (1456)	①	42	29	112	88	14	38	13	76
	5E-FE	1.5 (1495)	③	42	29	112	65	14	35	13	76
	4A-FE	1.6 (1587)	44	44	36	87	58	14	18	13	76
	7A-FE	1.8 (1762)	②	44	⑦	87	⑨	14	29	13	76
	3S-GTE	2.0 (1998)	③	43	49	80	80	14	38	13	76
	2JZ-GE	3.0 (2997)	⑧	⑫	⑦	239	⑭	15	29	13	76
	2JZ-GTE	3.0 (2997)	⑧	⑫	⑦	239	⑭	20	29	13	76
	5S-FE	2.2 (2164)	③	43	④	80	⑤	14	36	13	76
	1MZ-FE	3.0 (2952)	⑩	⑪	④	159	30	14	36	13	76
1995	3E-E	1.5 (1456)	①	42	29	112	88	14	38	13	76
	5E-FE	1.5 (1495)	③	42	29	112	65	14	35	13	76
	4A-FE	1.6 (1587)	44	44	36	87	58	14	18	13	76
	7A-FE	1.8 (1762)	②	44	⑦	87	⑨	14	29	13	76
	3S-GTE	2.0 (1998)	③	43	49	80	80	14	38	13	76
	5S-FE	2.2 (2164)	③	43	④	80	⑤	14	36	13	76
	2JZ-GE	3.0 (2997)	⑧	⑫	⑦	239	⑭	20	29	13	76
	2JZ-GTE	3.0 (2997)	⑧	⑫	⑦	239	⑭	20	29	13	76
	1MZ-FE	3.0 (2952)	⑩	⑪	④	159	30	14	36	13	76
1996-97	5E-FE	1.5 (1497)	⑫	42	29	112	65	14	35	13	76
	4A-FE	1.6 (1587)	②	44	⑦	87	⑨	14	25	14	76
	7A-FE	1.8 (1762)	②	44	④	87	⑨	14	25	13	76
	5S-FE	2.2 (2164)	③	43	④	80	⑤	14	36	13	76
	1MZ-FE	3.0 (2995)	⑪	⑬	④	159	61	11	36	13	76
	2JZ-GE	3.0 (2997)	⑧	⑫	⑦	239	⑭	20	29	13	76
	2JZ-GTE	3.0 (2997)	⑧	⑫	⑦	239	⑭	20	29	13	76

① Step 1: 22 ft. lbs.
Step 2: 36 ft. lbs.
Step 3: 90 degree turn
② Step 1: 22 ft. lbs.
Step 2: 90 degree turn
Step 3: 90 degree turn
③ Step 1: 36 ft. lbs.
Step 2: 90 degree turn

④ Step 1: 18 ft. lbs.
Step 2: 90 degree turn
⑤ Manual transmission: 65 ft. lbs.
Automatic transmission: 61 ft. lbs.
⑥ Step 1: 45 ft. lbs.
Step 2: 90 degree turn
⑦ Step 1: 22 ft. lbs.
Step 2: 90 degree turn

⑧ Step 1: 25 ft. lbs.
Step 2: 90 degree turn
Step 3: 90 degree turn
Recessed head bolt: 13 ft. lbs.
⑨ Manual transmission: 58 ft. lbs.
Automatic transmission: 47 ft. lbs.
⑩ Step 1: 40 ft. lbs.
Step 2: 90 degree turn
Recessed head bolt: 13 ft. lbs.

⑪ Step 1: 16 ft. lbs.
Step 2: 90 degree turn
⑫ Step 1: 33 ft. lbs.
Step 2: 90 degree turn
⑬ 12 pointed head: 16 ft. lbs. + 90 degrees
6 pointed head: 20 ft. lbs.
⑭ Manual trans.: 36 ft. lbs. + 90 degrees
Automatic trans.: 61 ft. lbs.

BRAKE SPECIFICATIONS
All measurements in inches unless noted

Year	Model		Master Cylinder Bore	Brake Disc Original Thickness	Brake Disc Minimum Thickness	Brake Disc Maximum Runout	Brake Drum Diameter Original Inside Diameter	Brake Drum Diameter Max. Wear Limit	Brake Drum Diameter Maximum Machine Diameter	Minimum Lining Thickness Front	Minimum Lining Thickness Rear
1993	Camry	F	NA	1.102	1.024	0.0020	-	-	-	0.039	-
		R	NA	0.394	0.354	0.0059	9.00	-	9.08	-	0.039
	Celica	F	NA	0.984	0.906	0.0028	-	-	-	0.039	-
		R	NA	0.394	0.354	0.0059	7.87	-	7.91	-	0.039
	Corolla		NA	0.709	0.669	0.0035	7.87	-	7.91	0.039	0.039
	MR2	F	NA	①	②	0.0028	-	-	-	0.039	-
		R	NA	③	④	0.0039	-	-	-	-	0.039
	Paseo		NA	0.710	0.670	0.0035	7.09	-	7.13	0.039	0.039
	Tercel		NA	0.709	0.669	0.0035	7.09	-	7.13	0.039	0.039

BRAKE SPECIFICATIONS
All measurements in inches unless noted

Year	Model		Master Cylinder Bore	Brake Disc Original Thickness	Brake Disc Minimum Thickness	Brake Disc Maximum Runout	Brake Drum Diameter Original Inside Diameter	Brake Drum Diameter Max. Wear Limit	Brake Drum Diameter Maximum Machine Diameter	Minimum Lining Thickness Front	Minimum Lining Thickness Rear
1994	Camry	F	NA	1.102	1.024	0.0020	-	-	-	0.039	-
		R	NA	0.394	0.354	0.0059	9.00	-	9.08	-	0.039
	Celica	F	NA	0.984	0.906	0.0028	-	-	-	0.039	-
		R	NA	0.394	0.354	0.0059	7.87	-	7.91	-	0.039
	Corolla		NA	0.709	0.669	0.0035	7.87	-	7.91	0.039	0.039
	MR2	F	NA	①	②	0.0028	-	-	-	0.039	-
		R	NA	③	④	0.0039	-	-	-	-	0.039
	Paseo		NA	0.710	0.670	0.0035	7.09	-	7.13	0.039	0.039
	Supra	F	NA	⑤	⑥	0.0020	-	-	-	0.039	-
		R	NA	0.630	0.591	-	7.48	-	7.52	-	0.039
	Tercel		NA	0.709	0.669	0.0035	7.09	-	7.13	0.039	0.039
1995	Avalon	F	NA	1.102	1.024	0.0020	-	-	-	0.039	-
		R	NA	0.354	0.315	0.0059	-	-	6.73	-	0.039
	Camry	F	NA	1.102	1.024	0.0020	-	-	9.08	0.039	-
		R	NA	0.354	0.315	0.0059	9.00	-	6.73	-	0.039
	Celica	F	NA	⑦	⑧	0.0020	-	-	7.91	0.039	-
		R	NA	0.394	0.354	0.0059	7.87	-	6.73	-	0.039
	Corolla		NA	0.866	0.787	0.0020	7.87	-	7.91	0.039	0.039
	MR2	F	NA	⑨	②	0.0020	-	-	-	0.039	-
		R	NA	③	④	0.0039	-	-	-	-	0.039
	Supra	F	NA	⑤	⑥	0.0020	-	-	-	0.039	-
		R	NA	0.630	0.591	-	7.48	-	7.52	-	0.039
	Paseo		NA	0.709	0.669	0.0035	7.09	-	7.13	0.039	0.039
	Tercel		NA	0.709	0.669	0.0035	7.09	-	7.13	0.039	0.039
1996-97	Avalon	F	NA	1.102	1.024	0.0020	-	-	-	0.039	-
		R	NA	0.354	0.315	0.0059	-	-	-	-	0.039
	Camry	F	NA	1.102	1.024	0.0020	-	-	-	0.039	-
		R	NA	0.354	0.315	0.0059	9.00	-	9.08	-	0.039
	Celica	F	NA	⑦	⑧	0.0020	-	-	-	0.039	-
		R	NA	0.394	0.354	0.0059	7.87	-	7.91	-	0.039
	Corolla		NA	0.866	0.787	0.0020	7.87	-	7.91	0.039	0.039
	Paseo		NA	0.709	0.669	0.0035	7.09	-	7.13	0.039	0.039
	Supra	F	NA	⑤	⑥	0.0020	-	-	-	0.039	-
		R	NA	0.630	0.591	-	7.48	-	7.52	-	0.039
	Tercel		NA	0.709	0.669	0.0035	7.09	-	7.13	0.039	0.039

NA - Not Available
F - Front
R - Rear

① 3S-GTE engine: 1.181
5S-FE engine: 0.630

② 3S-GTE engine: 1.102
5S-FE engine: 0.945

③ 3S-GTE engine: 0.866
5S-FE engine: 0.630

④ 3S-GTE engine: 0.8287
5S-FE engine: 0.591

⑤ 2JZ-GTE engine: 1.181
2JZ-GE engine: 1.260

⑥ 2JZ-GTE engine: 1.102
2JZ-GE engine: 1.181

⑦ 7A-FE engine: 0.984
5S-FE engine: 1.102

⑧ 7A-FE engine: 0.906
5S-FE engine: 1.024

⑨ 3S-GTE engine: 1.181
5S-FE engine: 0.984

FREQUENT MAINTENANCE LABOR
TOYOTA

The following should be used as a guide when determining the amount of work required for a particular service if taken to a repair shop. In estimating how long a particular Frequent Maintenance Service item should take, please observe the following:
- **Chilton Time** is time that is based on field research and data supplied by the vehicle manufacturer.
- All labor time operations are given in hours and tenths of an hour.
- All labor operations, are to be used as a **guide**.

COOLING

(G) Winterize Cooling System

Includes: Run engine to check for leaks, tighten all hose connections. Test radiator and pressure cap, drain radiator and engine block. Add antifreeze and refill system.

	Chilton Time
1993-97	.5

(G) Belt, Drive, Renew

1995-97 Avalon	.4
1993-97 Camry	.4
1993 Celica	.4
1994-97 Celica	.5
1993-95 Corolla	.4
1993-95 MR2	.7
1993-95 Paseo	.7
1996 Paseo	.6
1994-97 Supra	
wo/Turbo	.3
w/Turbo	.6
1993-94 Tercel	.7
1995-97 Tercel	.6

(G) Belt, Drive, Adjust

1993-97	
one	.3
each adtnl.	.1

(G) Hoses, Radiator, Renew

Includes: Drain and refill cooling system as required.

Upper
1995-97 Avalon	.8
1993-97 Camry	.8
1993-97 Celica	.8
1993-97 Corolla FWD	.8
1993-97 Paseo	.9
1994-97 Supra	
wo/Turbo	.6
w/Turbo	.8
1993-97 Tercel	.8

Lower
1995-97 Avalon	1.3
1993-97 Camry	
Four	1.0
V6	1.3
1993 Celica	1.0
1994-97 Celica	.8
1993-95 Corolla FWD	.9
1993-97 Paseo	1.0
1994-97 Supra	
wo/Turbo	.8
w/Turbo	1.0
1993-94 Tercel	1.0
1995-97 Tercel	1.1

(G) Thermostat, Coolant, Renew

1995-97 Avalon	1.1
1993-97 Camry	
Four	1.1
V6	.9
1993 Celica	
wo/Turbo	1.1
w/Turbo	1.4
1994-97 Celica	.8
1993-97 Corolla FWD	1.0
1993-95 MR2	
wo/Turbo	1.1
w/Turbo	1.8
1993-97 Paseo	1.0
1994-97 Supra	
wo/Turbo	.9
w/Turbo	1.8
1993-97 Tercel	.9

FUEL

(M) Air Cleaner, Service

1993-97	.3

(G) Filter, Fuel, Renew

1995-97 Avalon	.7
1993-97 Camry	
Four	.6
V6	.7
1993 Celica	.6
1994-97 Celica	1.1
1993-94 Corolla FWD	.6
1995-97 Corolla FWD	1.0
1993-95 MR2	1.0
1993-95 Paseo	.7
1996 Paseo	1.0
1993-97 Supra	.8
1993-94 Tercel	.6
1995-97 Tercel	.9

BRAKES

(G) Bleed Brakes (Four Wheels)

Includes: Add fluid.

1993-97	
two wheels	.6
four wheels	1.0

(G) Brakes, Adjust (Minor)

Includes: Adjust brakes, fill master cylinder.

1993-97, two wheels	.4

(M) Parking Brake, Adjust

1995-97 Avalon	.6
1993-97 Camry	.7
1993 Celica	.5

1994-97 Celica	.6
1993-97 Corolla FWD	.8
1993-95 MR2	.3
1993-97 Paseo	.6
1993-97 Supra	.8
1993-97 Tercel	.6

LUBRICATION SERVICES

(M) Engine Oil & Filter, Renew

Includes: Inspect and correct all fluid levels.

1993-97	.3

WHEELS

(G) Wheel, Renew (One)

1993-97	.5

(G) Wheel, Balance

1993-97	
one	.3
each adtnl.	.2

(G) Wheels, Rotate (All)

1993-97	.5

ELECTRICAL

(G) Headlamps, Aim

1993-97	
two	.4
four	.6

(G) License Lamp Assy., Renew

1995-97 Avalon	.4
1993-97 Camry	.6
1993-97 Celica	.6
1993-97 Corolla FWD	.3
1993-95 MR2	.4
1993-97 Paseo	.3
1994-97 Supra	
wo/sport roof	1.9
w/sport roof	2.3
1993-94 Tercel	
Sedan	.3
1995-97 Tercel	.4

(G) Parking Lamp Lens or Bulb, Renew

1995-97 Avalon	.3
1993-97 Camry	.3
1993-97 Celica	.3
1993-97 Corolla FWD	.3
1993-95 MR2	.3
1994-97 Supra	.3
1993-97 Tercel	.3

FREQUENT MAINTENANCE LABOR (cont.)
TOYOTA

	Chilton Time
(G) Rear Combination Lamp Bulb, Renew	
1995-97 Avalon	.3
1993-97 Camry	.4
1993 Celica	.3
1994-97 Celica	.4
1993-97 Corolla FWD	.5
1993-95 MR2	.5
1993-95 Paseo	.4

	Chilton Time
1996 Paseo	.3
1993-97 Supra	.3
1993-97 Tercel	.4
(G) Horn, Renew	
1995-97 Avalon	.6
1993-97 Camry	.6
1993 Celica	.3
1994-97 Celica	.4

	Chilton Time
1993-97 Corolla FWD	.3
1993-95 MR2	2.2
1993-97 Paseo	.3
1994-97 Supra	.3
1993-97 Tercel Sedan	.3
(M) Terminals, Battery, Clean	
1993-97	.3

SCHEDULED MAINTENANCE INTERVALS
(TOYOTA AVALON, CAMRY, CELICA, COROLLA, MR2, PASEO, SUPRA & TERCEL)

TO BE SERVICED	TYPE OF SERVICE	VEHICLE MILEAGE INTERVAL (x1000)												
		7.5	15	22.5	30	37.5	45	52.5	60	67.5	75	82.5	90	97.5
Engine oil & filter②	R	✓	✓	✓	✓	✓	✓	✓	✓	✓	✓	✓	✓	✓
Idle speed (Paseo)	S/I	✓		✓		✓		✓		✓		✓		✓
Drive belts	S/I								✓	✓	✓	✓	✓	✓
Automatic transaxle fluid & filter	S/I		✓		✓		✓		✓		✓		✓	
Ball joints & dust covers	S/I		✓		✓		✓		✓		✓		✓	
Bolts & nuts on body & chassis	S/I			✓		✓		✓		✓		✓		
Brake line pipes & hoses	S/I		✓		✓		✓		✓		✓		✓	
Brake linings & drums (except MR2)	S/I		✓		✓		✓		✓		✓		✓	
Brake pads & discs (front & rear if equipped)	S/I		✓		✓		✓		✓		✓		✓	
Differential oil (Camry, Celica, Corolla & Supra①)	S/I		✓		✓		✓		✓		✓		✓	
Drive shaft boots (except Supra)	S/I		✓		✓		✓		✓		✓		✓	
Manual transaxle oil	S/I		✓		✓		✓		✓		✓		✓	
Steering gear housing oil	S/I		✓		✓		✓		✓		✓		✓	
Steering linkage	S/I		✓		✓		✓		✓		✓		✓	

SCHEDULED MAINTENANCE INTERVALS
(TOYOTA AVALON, CAMRY, CELICA, COROLLA, MR2, PASEO, SUPRA & TERCEL) (Cont.)

TO BE SERVICED	TYPE OF SERVICE	VEHICLE MILEAGE INTERVAL (x1000)												
		7.5	15	22.5	30	37.5	45	52.5	60	67.5	75	82.5	90	97.5
Air filter	R				✓				✓				✓	
Rear wheel bearings (Paseo & Tercel)	R				✓				✓				✓	
Spark plugs (Corolla, Paseo & Tercel)	R				✓				✓				✓	
Spark plugs (platinum tip) (Avalon, Camry, Celica, MR2 & Supra)	R								✓					
Exhaust system	S/I				✓				✓				✓	
Fuel lines & connections	S/I				✓				✓				✓	
Valve clearance	S/I				✓				✓				✓	
Engine Coolant	R						✓				✓			
Fuel tank cap gasket	R								✓					
Charcoal canister	S/I								✓					

① Supra w/LSD - replace every 30,000 miles.
② MR2 3S-GTE & Supra 3JZ-GTE - change every 5000 miles.
R – Replace   S/I – Service or Inspect

FREQUENT OPERATION MAINTENANCE (SEVERE SERVICE)

If a vehicle is operated under any of the following conditions it is considered severe service:
- **Extremely dusty areas.**
- **50% or more of the vehicle operation is in 32°C (90°F) or higher temperatures, or constant operation in temperatures below 0°C (32°F).**
- **Prolonged idling (vehicle operation in stop and go traffic).**
- **Frequent short running periods (engine does not warm to normal operating temperatures).**
- **Police, taxi, delivery usage or trailer towing usage.**

Oil & oil filter change – change every 6000 miles.
Oil & oil filter change (MR2 3S-GTE & Supra 3JZ-GTE) – change every 2500 miles.
Bolts & nuts on chassis & body - tighten every 7500 miles.
Ball joints & dust covers - service or inspect every 12,000 miles.
Brake linings & drums (except MR2) - service or inspect every 12,000 miles.
Brake pads & discs (front & rear if equipped) - service or inspect every 12,000 miles.
Drive shaft boots (except Supra) - service or inspect every 12,000 miles.
Steering linkage - service or inspect every 12,000 miles.
Air filter - service or inspect every 15,000 miles.
Exhaust system - service or inspect every 15,000 miles.
Timing belt - replace every 60,000 miles.

VOLKSWAGEN

ENGINE IDENTIFICATION

Year	Model	Engine Displacement Liters (cc)	Engine Series (ID/VIN)	Fuel System	No. of Cylinders	Engine Type
1993	Cabriolet	1.8 (1780)	2H	Digifant	4	SOHC
	Fox	1.8 (1780)	ABG	Digifant	4	SOHC
	Passat	2.0 (1984)	9A	CIS-E/Motronic	4	DOHC
	Passat GLX	2.8 (2792)	AAA	Motronic	6	DOHC
	Golf III	2.0 (1984)	ABA	Motronic	4	SOHC
	GTI	2.0 (1984)	ABA	Motronic	4	SOHC
	Jetta III	2.0 (1984)	ABA	Motronic	4	SOHC
	Jetta III GLX	2.8 (2792)	AAA	Motronic	6	DOHC
	Corrado SLC	2.8 (2792)	AAA	Motronic	6	DOHC
1994	Passat	2.0 (1984)	9A	CIS-E/Motronic	4	DOHC
	Passat GLX	2.8 (2792)	AAA	Motronic	6	DOHC
	Golf III	2.0 (1984)	ABA	Motronic	4	SOHC
	Jetta III	2.0 (1984)	ABA	Motronic	4	SOHC
	Jetta III GLX	2.8 (2792)	AAA	Motronic	6	DOHC
	Corrado SLC	2.8 (2792)	AAA	Motronic	6	DOHC
1995	Golf III	1.8 (1780)	ACC	Mono Motronic	4	SOHC
	Golf III	2.0 (1984)	ABA	Motronic	4	SOHC
	Golf III	2.8 (2782)	AAA	Motronic	6	DOHC
	Cabrio	2.0 (1984)	ABA	Motronic	4	SOHC
	Jetta III	1.8 (1780)	ACC	Mono Motronic	4	SOHC
	Jetta III	2.0 (1984)	ABA	Motronic	4	SOHC
	Jetta III	2.8 (2782)	AAA	Motronic	6	DOHC
	GTI	2.0 (1984)	ABA	Motronic	4	SOHC
	Passat	2.8 (2782)	AAA	Motronic	6	DOHC
1996-97	Golf	1.8 (1780)	ACC	Mono Motronic	4	SOHC
	Golf	2.0 (1984)	ABA	Motronic	4	SOHC
	Golf	2.8 (2782)	AAA	Motronic	6	DOHC
	Cabrio	2.0 (1984)	ABA	Motronic	4	SOHC
	GTI	2.0 (1984)	ABA	Motronic	4	SOHC
	GTI	2.8 (2782)	AAA	Motronic	6	DOHC
	Jetta	2.0 (1984)	ABA	Motronic	4	SOHC
	Jetta	2.8 (2782)	AAA	Motronic	6	DOHC
	Passat	2.8 (2782)	AAA	Motronic	6	DOHC

DSL - Diesel
CIS-E - Continuous injection system with electronic controls

SOHC - Single overhead camshaft
DOHC - Double overhead camshaft

GENERAL ENGINE SPECIFICATIONS

Year	Engine ID/VIN	Engine Displacement Liters (cc)	Fuel System Type	Net Horsepower @ rpm	Net Torque @ rpm (ft. lbs.)	Bore x Stroke (in.)	Compression Ratio	Oil Pressure @ rpm
1993	2H	1.8 (1780)	Digifant	98@5500	100@3000	3.19x3.40	8.5:1	28@2000
	ABG	1.8 (1780)	Digifant	81@5500	93@3250	3.19x3.40	9.0:1	28@2000
	ABA	2.0 (1984)	Motronic	115@5400	122@3200	3.25x3.65	10.0:1	28@2000
	9A	2.0 (1984)	CIS-E/Motronic	134@5800	133@4400	3.25x3.65	10.8:1	28@2000
	AAA	2.8 (2790)	Motronic	178@5800	177@4200	3.20x3.52	10.0:1	28@2000

GENERAL ENGINE SPECIFICATIONS

Year	Engine ID/VIN	Engine Displacement Liters (cc)	Fuel System Type	Net Horsepower @ rpm	Net Torque @ rpm (ft. lbs.)	Bore x Stroke (in.)	Compression Ratio	Oil Pressure @ rpm
1994	ABA	2.0 (1984)	Motronic	115@5400	122@3200	3.25x3.65	10.0:1	28@2000
	9A	2.0 (1984)	CIS-E	134@5800	133@4400	3.25x3.65	10.8:1	28@2000
	AAA	2.8 (2792)	Motronic	178@5800	177@4200	3.20x3.52	10.0:1	28@2000
1995	ACC	1.8 (1780)	Mono Motronic	90@5000	106@2500	3.19x3.40	9.0:1	29@2000
	ABA	2.0 (1984)	Motronic	115@5400	122@3200	3.25x3.65	10.4:1	29@2000
	AAA	2.8 (2782)	Motronic	178@5800	177@4200	3.19x3.56	10.0:1	29@2000
1996-97	ACC	1.8 (1780)	Mono Motronic	90@5000	106@2500	3.19x3.40	9.0:1	29@2000
	ABA	2.0 (1984)	Motronic	115@5400	122@3200	3.25x3.65	10.4:1	29@2000
	AAA	2.8 (2782)	Motronic	178@5800	177@4200	3.19x3.56	10.0:1	29@2000

DSL - Diesel
CIS-E - Continuous injection system with electronic controls

GASOLINE ENGINE TUNE-UP SPECIFICATIONS

Year	Engine ID/VIN	Engine Displacement Liters (cc)	Spark Plugs Gap (in.)	Ignition Timing (deg.) MT	Ignition Timing (deg.) AT	Fuel Pump (psi)	Idle Speed (rpm) MT	Idle Speed (rpm) AT	Valve Clearance In.	Valve Clearance Ex.
1993	2H	1.8 (1780)	0.028-0.036	6B	6B	36 ①	700-900	700-900	HYD	HYD
	ABG	1.8 (1780)	0.028-0.032	6B	6B	36 ①	800-1000	800-1000	HYD	HYD
	ABA	2.0 (1984)	0.024	NA	NA	36 ①	800-880	800-880	HYD	HYD
	9A	2.0 (1984)	0.028-0.032	6B	6B	②	800-1000	800-1000	HYD	HYD
	AAA	2.8 (2792)	0.027-0.032	6B	6B	52 ①	650-750	650-750	HYD	HYD
1994	ABA	2.0 (1984)	0.024	NA	NA	36 ①	800-880	800-880	HYD	HYD
	9A	2.0 (1984)	0.028-0.032	6B	6B	②	800-1000	800-1000	HYD	HYD
	AAA	2.8 (2792)	0.027-0.032	6B	6B	52 ①	650-750	650-750	HYD	HYD
1995	ACC	1.8 (1780)	0.030	5-7B	5-7B	17.4	800-1000	800-1000	HYD	HYD
	ABA	2.0 (1984)	0.024	5-7B	5-7B	43.5	800-880	800-880	HYD	HYD
	AAA	2.8 (2782)	0.028	5-7B	5-7B	58	650-750	650-750	HYD	HYD
1996-97	ACC	1.8 (1780)	0.030	5-7B	5-7B	17.4	800-1000	800-1000	HYD	HYD
	ABA	2.0 (1984)	0.024	5-7B	5-7B	43.5	800-880	800-880	HYD	HYD
	AAA	2.8 (2782)	0.028	5-7B	5-7B	58	650-750	650-750	HYD	HYD

NOTE: The Vehicle Emission Control Information label often reflects specification changes made during production. The label figures must be used if they differ from those in this chart.

B - Before top dead center
HYD - Hydraulic
① System pressure at idle
② System pressure: 88-96 psi
  Differential pressure, regulator wiring disconnected: 4-7 psi lower
  Differential pressure, regulator wiring connected and ignition switch on: 19-23 psi lower

CAPACITIES

Year	Model	Engine ID/VIN	Engine Displacement Liters (cc)	Engine Oil with Filter	Transmission (pts.) 4-Spd	Transmission (pts.) 5-Spd	Transmission (pts.) Auto.	Drive Axle Front (pts.)	Drive Axle Rear (pts.)	Fuel Tank (gal.)	Cooling System (qts.)
1993	Cabriolet	2H	1.8 (1780)	4.3	-	4.2	6.4	①	①	13.7	5.1
	Fox	ABG	1.8 (1780)	3.7	3.6	4.4	-	-	-	12.4	6.4 ②
	Golf III	ABA	2.0 (1984)	4.3	-	4.2	6.4	①	①	14.5	6.5
	GTI	ABA	2.0 (1984)	4.3	-	4.2	6.4	①	①	14.5	6.5
	Jetta III	ABA	2.0 (1984)	4.3	-	4.2	6.4	①	①	14.5	6.5

CAPACITIES

Year	Model	Engine ID/VIN	Engine Displacement Liters (cc)	Engine Oil with Filter	Transmission (pts.) 4-Spd	5-Spd	Auto.	Drive Axle Front (pts.)	Rear (pts.)	Fuel Tank (gal.)	Cooling System (qts.)
1993	Jetta III GLX	AAA	2.8 (2792)	7.3	-	4.2	6.4	①	①	14.5	9.6
	Passat	9A	2.0 (1984)	4.3	-	4.2	6.4	①	①	18.5	6.8
	Passat GLX	AAA	2.8 (2792)	7.4	-	4.2	6.4	①	①	18.5	10.5
	Corrado SLC	AAA	2.8 (2792)	6.3	-	4.2	6.0	①	①	18.5	10.6
1994	Passat	9A	2.0 (1984)	4.3	-	4.2	6.4	①	①	18.5	6.8
	Passat GLX	AAA	2.8 (2792)	7.4	-	4.2	6.4	①	①	18.5	10.5
	Golf III	ABA	2.0 (1984)	4.3	-	4.2	6.4	①	①	14.5	6.5
	Jetta III	ABA	2.0 (1984)	4.3	-	4.2	6.4	①	①	14.5	6.5
	Jetta III GLX	AAA	2.8 (2792)	7.3	-	4.2	6.4	①	①	14.5	9.6
	Corrado SLC	AAA	2.8 (2792)	6.3	-	4.2	6.0	①	①	18.5	10.6
1995	Cabrio	ABA	2.0 (1984)	4.3	-	4.2	11.8	-	-	14.5	6.5
	Golf III	ACC	1.8 (1780)	4.2	-	4.2	11.8	-	-	14.5	6.5
	Golf III	ABA	2.0 (1984)	4.3	-	4.2	11.8	-	-	14.5	6.5
	Golf III	AAA	2.8 (2782)	7.4	-	4.2	11.8	-	-	14.5	9.5
	GTI	AAA	2.8 (2782)	7.4	-	4.2	11.8	-	-	14.5	9.5
	Jetta III	ACC	1.8 (1780)	4.2	-	4.2	11.8	-	-	14.5	6.5
	Jetta III	ABA	2.0 (1984)	4.3	-	4.2	11.8	-	-	14.5	6.5
	Jetta III	AAA	2.8 (2782)	7.4	-	4.2	11.8	-	-	14.5	9.5
	Passat	AAA	2.8 (2782)	7.4	-	4.2	11.8	-	-	18.5	9.5
1996-97	Cabrio	ABA	2.0 (1984)	4.3	-	4.2	11.8	-	-	14.5	6.5
	Golf	ACC	1.8 (1780)	4.2	-	4.2	11.8	-	-	14.5	6.5
	Golf	ABA	2.0 (1984)	4.3	-	4.2	11.8	-	-	14.5	6.5
	Golf	AAA	2.8 (2782)	7.4	-	4.2	11.8	-	-	14.5	9.5
	GTI	ABA	2.0 (1984)	4.3	-	4.2	11.8	-	-	14.5	6.5
	GTI	AAA	2.8 (2782)	7.4	-	4.2	11.8	-	-	14.5	9.5
	Jetta	ABA	2.0 (1984)	4.3	-	4.2	11.8	-	-	14.5	6.5
	Jetta	AAA	2.8 (2782)	7.4	-	4.2	11.8	-	-	14.5	9.5
	Passat	AAA	2.8 (2782)	7.4	-	4.2	11.8	-	-	18.5	9.5

① Final drive unit of automatic transaxle: 0.8 qts. ② With air conditioning: 6.9 qts.

VALVE SPECIFICATIONS

Year	Engine ID/VIN	Engine Displacement Liters (cc)	Seat Angle (deg.)	Face Angle (deg.)	Spring Test Pressure (lbs. @ in.)	Spring Installed Height (in.)	Stem-to-Guide Clearance (in.) Intake	Exhaust	Stem Diameter (in.) Intake	Exhaust
1993	2H	1.8 (1780)	45	45	NA	NA	0.039	0.051	0.3138	0.3130
	ABG	1.8 (1780)	45	45	NA	NA	0.039	0.059	0.3138	0.3130
	9A	2.0 (1984)	45	45	NA	NA	0.039	0.051	0.2744	0.2732
	ABA	2.0 (1984)	45	45	NA	NA	0.039	0.051	0.2744	0.2736
	AAA	2.8 (2792)	45	45	NA	NA	0.039	0.051	0.2744	0.2736
1994	9A	2.0 (1984)	45	45	NA	NA	0.039	0.051	0.2744	0.2732
	ABA	2.0 (1984)	45	45	NA	NA	0.039	0.051	0.2744	0.2736
	AAA	2.8 (2792)	45	45	NA	NA	0.039	0.051	0.2744	0.2736
1995	ACC	1.8 (1780)	45	45	NA	NA	0.039	0.051	0.3130	0.3130
	ABA	2.0 (1984)	45	45	NA	NA	0.039	0.051	0.2744	0.2736
	AAA	2.8 (2782)	45	45	NA	NA	0.039	0.051	0.2744	0.2736
1996-97	ACC	1.8 (1780)	45	45	NA	NA	0.039	0.051	0.3138	0.3130
	ABA	2.0 (1984)	45	45	NA	NA	0.039	0.051	0.2744	0.2736
	AAA	2.8 (2782)	45	45	NA	NA	0.039	0.051	0.2744	0.2736

NA - Not Available

TORQUE SPECIFICATIONS
All readings in ft. lbs.

Year	Engine ID/VIN	Engine Displacement Liters (cc)	Cylinder Head Bolts	Main Bearing Bolts	Rod Bearing Bolts	Crankshaft Damper Bolts	Flywheel Bolts	Manifold Intake	Manifold Exhaust	Spark Plugs	Lug Nut
1993	2H	1.8 (1780)	①	48	②	③	④	18	18	15	81
	ABG	1.8 (1780)	①	48	②	③	⑦	18	18	15	81
	9A	2.0 (1984)	①	48	②	③	④	15	18	18	81
	ABA	2.0 (1984)	①	48	②	③	⑨	15	18	18	81
	AAA	2.8 (2792)	①	48	②	⑤	⑨	18	⑥	18	81
1994	9A	2.0 (1984)	①	48	②	③	④	15	18	18	81
	ABA	2.0 (1984)	①	48	②	③	⑨	15	18	18	81
	AAA	2.8 (2792)	①	48	②	⑤	⑨	18	⑥	18	81
1995	ACC	1.8 (1780)	①	48	②	⑩	⑨	22	22	18	81
	ABA	2.0 (1984)	①	48	⑤	⑩	⑨	15	15	22	81
	AAA	2.8 (2782)	①	48	②	⑤	⑨	18	18	22	81
1996-97	ACC	1.8 (1780)	①	48	②	⑩	⑨	22	22	18	81
	ABA	2.0 (1984)	①	48	⑤	⑩	⑨	15	15	22	81
	AAA	2.8 (2782)	①	48	②	⑤	⑨	18	18	22	81

① Torque in four steps: Use new bolts on AAA and ABG engines
    Step 1: 30 ft. lbs.
    Step 2: 44 ft. lbs.
    Step 3: + 90 degrees
    Step 4: + 90 degrees
② Torque to 22 ft. lbs. plus an additional 90 degrees
③ Damper to crankshaft sprocket - 15 ft. lbs.
    Crankshaft sprocket to crankshaft:
    Six sided bolt with washer: 137 ft. lbs.
    Allen bolt with washer: 111 ft. lbs.
    Twelve-sided bolt with collar: 66 ft. lbs. plus an additional 180 degrees. Use new bolt.
④ Flywheel to manual transaxle pressure plate: 15 ft. lbs.
    Pressure plate to crankshaft: 44 ft. lbs. plus an additional 90 degrees. Use new bolts.
    Automatic transaxle drive plate to flywheel: 22 ft. lbs. If spacing not as specified,
    correct it and tighten in three steps:
    Step 1: 22 ft. lbs.
    Step 2: 44 ft. lbs.
    Step 3: plus an additional 90 degrees
    On Passat GLX, perform steps 1 and 3 only. Use new bolts.

⑤ 74 ft. lbs. plus an additional 1/4 turn. Use new bolts.
⑥ 18 ft. lbs. Use new nuts.
⑦ 74 ft. lbs. Use new bolts
⑧ 22 ft. lbs. plus an additional 1/2 turn Use new bolts.
⑨ 44 ft. lbs. plus an additional 90 degrees;
    except Passat GLX: 52 ft. lbs. plus 90 degrees.
⑩ Step 1: 66 ft. lbs.
    Step 2: Plus 90 degrees

BRAKE SPECIFICATIONS
All measurements in inches unless noted

Year	Model			Master Cylinder Bore	Brake Disc Original Thickness	Brake Disc Minimum Thickness	Maximum Runout	Brake Drum Diameter Original Inside Diameter	Max. Wear Limit	Maximum Machine Diameter	Minimum Lining Thickness Front	Minimum Lining Thickness Rear
1993	Cabriolet	①	F	0.820	0.472	0.393	0.002	-	-	-	0.276	-
			R	-	0.394	0.319	0.002	7.09	7.13	7.11	-	③
	Cabriolet	②	F	0.820	0.787	0.708	0.002	-	-	-	0.276	-
			R	-	-	-	0.002	7.09	7.13	7.11	-	③
	Corrado		F	④	0.866	0.787	0.002	-	-	-	⑤	-
			R	-	0.394	0.315	0.002	-	-	-	-	⑥
	Fox	①	F	0.820	0.472	0.393	0.002	-	-	-	0.276	-
			R	-	0.394	0.319	0.002	7.09	7.13	7.11	-	③
	Fox	②	F	0.820	0.787	0.708	0.002	-	-	-	0.276	-
			R	-	-	-	0.002	7.09	7.13	7.11	-	③

BRAKE SPECIFICATIONS
All measurements in inches unless noted

Year	Model			Master Cylinder Bore	Brake Disc Original Thickness	Brake Disc Minimum Thickness	Maximum Runout	Brake Drum Diameter Original Inside Diameter	Max. Wear Limit	Maximum Machine Diameter	Minimum Lining Thickness Front	Minimum Lining Thickness Rear
1993	Golf III	①	F	NA	0.472	0.393	0.002	-	-	-	⑤	-
			R	-	0.394	0.319	0.002	7.87	7.91	NA	-	③
	Golf III	②	F	NA	0.787	0.708	0.002	-	-	-	⑤	-
			R	-	-	-	0.002	7.87	7.91	NA	-	③
	Jetta III	①	F	NA	0.472	0.393	0.002	-	-	-	⑤	-
			R	-	0.394	0.319	0.002	7.87	7.91	NA	-	③
	Jetta III	②	F	NA	0.787	0.708	0.002	-	-	-	⑤	-
			R	-	-	-	0.002	7.87	7.91	NA	-	③
	Passat		F	④	0.866	0.787	0.002	-	-	-	⑤	-
			R	-	0.394	0.315	0.002	-	-	-	-	⑥
1994	Corrado		F	④	0.866	0.787	0.002	-	-	-	⑤	-
			R	-	0.394	0.315	0.002	-	-	-	-	⑥
	Golf III	①	F	NA	0.472	0.393	0.002	-	-	-	⑤	-
			R	-	0.394	0.319	0.002	7.87	7.91	NA	-	③
	Golf III	②	F	NA	0.787	0.708	0.002	-	-	-	⑤	-
			R	-	-	-	0.002	7.87	7.91	NA	-	③
	Jetta III	①	F	NA	0.472	0.393	0.002	-	-	-	⑤	-
			R	-	0.394	0.319	0.002	7.87	7.91	NA	-	③
	Jetta III	②	F	NA	0.787	0.708	0.002	-	-	-	⑤	-
			R	-	-	-	0.002	7.87	7.91	NA	-	③
	Passat		F	④	0.866	0.787	0.002	-	-	-	⑤	-
			R	-	0.394	0.315	0.002	-	-	-	-	⑥
1995	Cabrio		F	0.874	0.790	0.709	0.002	-	-	-	⑥	-
			R	-	0.390	0.315	0.002	-	-	-	-	⑥
	Golf III	⑦	F	0.874	0.790	0.709	0.002	-	-	-	⑥	-
			R	-	0.390	0.315	0.002	-	-	-	-	⑥
	Golf III	⑧	F	0.874	0.870	0.787	0.002	-	-	-	⑥	-
			R	-	0.390	0.315	0.002	-	-	-	-	⑥
	GTI		F	0.874	0.870	0.787	0.002	-	-	-	⑥	-
			R	-	0.390	0.315	0.002	-	-	-	-	⑥
	Jetta III	⑦	F	0.874	0.790	0.709	0.002	-	-	-	⑥	-
			R	-	0.390	0.315	0.002	-	-	-	-	⑥
	Jetta III	⑨	F	0.874	0.870	0.787	0.002	-	-	-	⑥	-
			R	-	0.390	0.315	0.002	-	-	-	-	⑥
	Passat		F	0.874	0.870	0.787	0.002	-	-	-	⑥	-
			R	-	0.390	0.315	0.002	-	-	-	-	⑥
1996-97	Cabrio		F	0.874	0.790	0.709	0.002	-	-	-	⑥	-
			R	-	0.390	0.315	0.002	-	-	-	-	⑥
	Golf	⑩	F	0.874	0.790	0.709	0.002	-	-	-	⑥	-
			R	-	0.390	0.315	0.002	-	-	-	-	⑥
	Golf	⑧	F	0.874	0.870	0.787	0.002	-	-	-	⑥	-
			R	-	0.390	0.315	0.002	-	-	-	-	⑥
	GTI	⑪	F	0.874	0.790	0.709	0.002	-	-	-	⑥	-
			R	-	0.390	0.315	0.002	-	-	-	-	⑥
	GTI	⑫	F	0.874	0.870	0.787	0.002	-	-	-	⑥	-
			R	-	0.390	0.315	0.002	-	-	-	-	⑥
	Jetta	⑦	F	0.874	0.790	0.709	0.002	-	-	-	⑥	-
			R	-	0.390	0.315	0.002	-	-	-	-	⑥

BRAKE SPECIFICATIONS
All measurements in inches unless noted

Year	Model			Master Cylinder Bore	Brake Disc			Brake Drum Diameter			Minimum Lining Thickness	
					Original Thickness	Minimum Thickness	Maximum Runout	Original Inside Diameter	Max. Wear Limit	Maximum Machine Diameter	Front	Rear
1996-97	Jetta ⑨	F		0.874	0.870	0.787	0.002	-	-	-	⑥	-
		R		-	0.390	0.315	0.002	-	-	-	-	⑥
	Passat	F		0.874	0.870	0.787	0.002	-	-	-	⑥	-
		R		-	0.390	0.315	0.002	-	-	-	-	⑥

NA - Not Available
① With solid disc brakes
② With vented disc brakes
③ Disc brake pads: 0.276
   Drum brake shoes: 0.098

④ Without ABS: 0.874
   With ABS: 0.813
⑤ 0.433 including backing plate in.
⑥ 0.275 including backing plate in.
⑦ GL, GLS models

⑧ VR6 model
⑨ GLX model
⑩ GL, GLX models
⑪ 2.0L engine
⑫ 2.8L engine

FREQUENT MAINTENANCE LABOR
VOLKSWAGEN

The following should be used as a guide when determining the amount of work required for a particular service if taken to a repair shop. In estimating how long a particular Frequent Maintenance Service item should take, please observe the following:
- **Chilton Time** is time that is based on field research and data supplied by the vehicle manufacturer.
- All labor time operations are given in hours and tenths of an hour.
- All labor operations, are to be used as a **guide**.

COOLING

(G) Winterize Cooling System
Includes: Run engine to check for leaks, tighten all hose connections. Test radiator and pressure cap, drain radiator and engine block. Add antifreeze and refill system.
1993-97 5

(G) Belt, Serpentine Drive, Renew
1993-94 Corrado
  4 cyl. 1.1
  V6 . . . . 1.2
1995-97 Cabrio 8
1993-97 Jetta, Golf
  4 cyl. .8
  V6 . . . . 1.0
1993-97 Passat 8

(G) Hoses, Radiator, Renew
Includes: Drain and refill cooling system as required.
1993-97
  one 6
  both 8

(G) Thermostat, Coolant, Renew
1993 Cabriolet
  wo/AC 8
  w/AC 1.2
1995-97 Cabrio 8

1993-94 Corrado
  wo/AC 1.2
  w/AC 1.6
1993 Fox
  wo/AC 1.2
  w/AC 2.5
1993-97 Jetta, Golf
  wo/AC 1.2
  w/AC 1.6
1993-97 Passat
  wo/AC 1.2
  w/AC 1.6

BRAKES

(G) Bleed Brakes (Four Wheels)
Includes: Add fluid.
1993-97 5

(G) Brakes, Adjust (Minor)
Includes: Adjust brakes, fill master cylinder.
1993-97 7

LUBRICATION SERVICES

(M) Engine Oil & Filter, Renew
Includes: Inspect and correct all fluid levels.
1993-97
  gas 3
  diesel 8

WHEELS

(G) Wheel, Renew (One)
1993-97 5

(G) Wheel, Balance
1993-97
  one 3
  each adtnl. 2

(G) Wheels, Rotate (All)
1993-97 5

ELECTRICAL

(G) Headlamps, Aim
1993-97
  two 4
  four 6

(G) Halogen Headlamp Bulb, Renew
1993-97 4

(G) High Mount Stop Lamp Bulb, Renew
1993 Cabriolet 3
1995-97 Cabrio 3
1993-94 Corrado 3
1993-97 Jetta, Golf 3
1993-97 Passat 3

(G) Parking Lamp Lens or Bulb, Renew
1993-97 3

(G) Horn, Renew
1993-97, each 2

SCHEDULED MAINTENANCE INTERVALS
(VOLKSWAGEN CABRIO, CABRIOLET, CORRADO, FOX, GOLF, JETTA, PASSAT & GTI)

TO BE SERVICED	TYPE OF SERVICE	VEHICLE MILEAGE INTERVAL (x1000)												
		7.5	15	22.5	30	37.5	45	52.5	60	67.5	75	82.5	90	97.5
Engine oil & filter	R	✓	✓	✓	✓	✓	✓	✓	✓	✓	✓	✓	✓	✓
Brake pad thickness①	R	✓	✓	✓	✓	✓	✓	✓	✓	✓	✓	✓	✓	✓
A/T final drive fluid level	S/I		✓		✓		✓		✓		✓		✓	
Battery	S/I		✓		✓		✓		✓		✓		✓	
Brake system	S/I		✓		✓		✓		✓		✓		✓	
Clutch (Cabriolet)	S/I		✓		✓		✓		✓		✓		✓	
Cooling system	S/I		✓		✓		✓		✓		✓		✓	
Driveshaft boots	S/I		✓		✓		✓		✓		✓		✓	
Engine (check for leaks)	S/I		✓		✓		✓		✓		✓		✓	
Engine coolant level	S/I		✓		✓		✓		✓		✓		✓	
Exhaust system	S/I		✓		✓		✓		✓		✓		✓	
Fuel system	S/I		✓		✓		✓		✓		✓		✓	
Idle speed (gasoline)	S/I		✓		✓		✓		✓		✓		✓	
Idle speed (diesel)	S/I				✓				✓				✓	
Intake air system	S/I		✓		✓		✓		✓		✓		✓	
OBD system - check for codes	S/I		✓		✓		✓		✓		✓		✓	
Power steering fluid level	S/I		✓		✓		✓		✓		✓		✓	
Steering system	S/I		✓		✓		✓		✓		✓		✓	
Timing belt (1993-94 16V gasoline)②	S/I		✓		✓		✓		✓		✓		✓	
Timing belt (diesel)	S/I				✓				✓				✓	
Transaxle fluid level	S/I		✓		✓		✓		✓		✓		✓	
Water separator (diesel)	S/I		✓		✓		✓		✓		✓		✓	
Air filter element	R				✓				✓				✓	
Engine coolant	R				✓				✓				✓	
Fuel filter (diesel)	R				✓				✓				✓	
Spark plugs (w/o supercharger)	R				✓				✓				✓	
Spark plugs (w/supercharger)	R						✓							

SCHEDULED MAINTENANCE INTERVALS
(VOLKSWAGEN CABRIO, CABRIOLET, CORRADO, FOX, GOLF, JETTA, PASSAT & GTI)
(Cont.)

TO BE SERVICED	TYPE OF SERVICE	VEHICLE MILEAGE INTERVAL (x1000)												
		7.5	15	22.5	30	37.5	45	52.5	60	67.5	75	82.5	90	97.5
Passenger compartment air filter	R				✓				✓				✓	
Drive belts	S/I				✓				✓				✓	
Dust seals on ball joints, tie rod ends & tie rods	S/I				✓				✓				✓	
Brake fluid③	R													

① Diesel
② Replace every 60,000 miles.
③ Replace every two years regardless of mileage.
R – Replace   S/I – Service or Inspect

FREQUENT OPERATION MAINTENANCE (SEVERE SERVICE)
If a vehicle is operated under any of the following conditions it is considered severe service:
- Extremely dusty areas.
- 50% or more of the vehicle operation is in 32°C (90°F) or higher temperatures, or constant operation in temperatures below 0°C (32°F).
- Prolonged idling (vehicle operation in stop and go traffic).
- Frequent short running periods (engine does not warm to normal operating temperatures).
- Police, taxi, delivery usage or trailer towing usage.
Oil & oil filter change – change every 3750 miles.
Air filter element – service or inspect every 15,000 miles.
Automatic transaxle fluid & filter - replace every 30,000 miles.

VOLVO

ENGINE IDENTIFICATION

Year	Model		Engine Displacement Liters (cc)	Engine Series (ID/VIN)	Fuel System	No. of Cylinders	Engine Type	
1993	240		2.3 (2316)	B230F/88	EFI	4	SOHC	
	940		2.3 (2316)	B230F/88	EFI	4	SOHC	
	940	②	2.3 (2316)	B230FT/87	EFI	4	SOHC	
	850		2.3 (2435)	B5254FS/55	EFI	5	DOHC	③
	960		2.9 (2922)	B6304F/95	EFI	6	DOHC	①
1994	940		2.3 (2316)	B230F/88	EFI	4	SOHC	
	940	②	2.3 (2316)	B230FT/87	EFI	4	SOHC	
	850	②	2.3 (2319)	B5234T/57	EFI	5	DOHC	③
	850		2.4 (2435)	B5254S/55	EFI	5	DOHC	③
	960		2.9 (2922)	B6304F/95	EFI	6	DOHC	①

ENGINE IDENTIFICATION

Year	Model	Engine Displacement Liters (cc)	Engine Series (ID/VIN)	Fuel System	No. of Cylinders	Engine Type	
1995	940 Turbo	2.3 (2316)	B230FT/86 or 87	EFI	4	SOHC	
	940 Sedan	2.3 (2316)	B230F/88	EFI	4	SOHC	
	940 Wagon	2.3 (2316)	B20F/88	EFI	4	SOHC	
	T-5R	2.3 (2319)	B5234T5/58	EFI	5	DOHC	③
	850 Turbo	2.3 (2319)	B5234T/57	EFI	5	DOHC	③
	850	2.4 (2435)	B5254S/55	EFI	5	DOHC	③
	850 GLT	2.4 (2435)	B5254S/55	EFI	5	DOHC	③
	960 Sedan	2.9 (2922)	B6304F/95 ④	EFI	6	DOHC	①
	960 Wagon	2.9 (2922)	B6304F/95 ④	EFI	6	DOHC	①
1996-97	850 Turbo	2.3 (2319)	B5234T/57	EFI	5	DOHC	③
	850 R	2.3 (2319)	B5234FT/58	EFI	5	DOHC	③
	850	2.4 (2435)	B5254S/55	EFI	5	DOHC	③
	960	2.9 (2922)	B6304F/96	EFI	6	DOHC	①

NOTE: Since all 850 engines are now in principle F-engines, this symbol is no longer needed. Engine are now denoted with S or T.

B - Petrol (gasoline)
F - Fuel-injected with catalytic converter   EFI - Electronic fuel injection   ① 24 valve engine   ④ 95 with air pump
S - Normally-aspirated   SOHC - Single overhead camshaft   ② Turbocharged   96 without air pump
T - Turbocharged engine   DOHC - Double overhead camshaft   ③ 20 valve engine

GENERAL ENGINE SPECIFICATIONS

Year	Engine ID/VIN	Engine Displacement Liters (cc)	Fuel System Type	Net Horsepower @ rpm	Net Torque @ rpm (ft. lbs.)	Bore x Stroke (in.)	Compression Ratio	Oil Pressure @ rpm
1993	B230F/88	2.3 (2316)	EFI	114@5400	136@2150	3.78x3.15	9.8:1	35-85@2000
	B230FT/87	2.3 (2316)	EFI	162@4800	195@3450	3.78x3.15	8.7:1	35-85@2000
	B5254FS/55	2.4 (2435)	EFI	168@6200	162@3300	3.27x3.54	10.5:1	49.8@4000
	B6304F/95	2.9 (2922)	EFI	201@6000	197@4300	3.27x3.54	10.7:1	36@2000
1994	B230F/88	2.3 (2316)	EFI	114@5400	136@2150	3.78X3.15	9.8:1	35-85@2000
	B230FT/87	2.3 (2316)	EFI	162@4800	195@3450	3.78x3.15	8.7:1	35-85@2000
	B5234T/57	2.3 (2319)	EFI	222@5200	221@2100	3.19X3.54	8.5:1	49.8@4000
	B5254S/55	2.4 (2435)	EFI	168@6200	162@3300	3.27X3.54	10.5:1	49.8@4000
	B6304F/95	2.9 (2922)	EFI	201@6000	197@4300	3.27x3.54	10.7:1	36@2000
1995	B230F/88	2.3 (2316)	EFI	114@5400	136@2150	3.78x3.15	9.8:1	35-85@2000
	B230FT/86 or 87	2.3 (2316)	EFI	162@4800	195@3450	3.78x3.15	8.7:1	35-85@2000
	B5234T/57	2.3 (2319)	EFI	222@5200	221@2100	3.19x3.54	8.5:1	49.8@4000
	B5234T5/58	2.3 (2319)	EFI	240@5600	221@2100	3.19x3.54	8.5:1	49.8@4000
	B5254S/55	2.4 (2435)	EFI	168@6200	162@3300	3.27x3.54	10.5:1	49.8@4000
	B6304F/95	2.9 (2922)	EFI	181@5200	199@4100	3.27x3.54	10.7:1	36@2000
1996-97	B5234T/57	2.3 (2319)	EFI	222@5280	221@2000	3.19x3.54	8.5:1	49.8@4000
	B5234FT/58	2.3 (2319)	EFI	240@5600	221@2100	3.19x3.54	8.5:1	49.8@4000
	B5254S/55	2.4 (2435)	EFI	168@6200	162@3300	3.27x3.54	10.5:1	49.8@4000
	B6304F/96	2.9 (2922)	EFI	181@5200	199@4100	3.27x3.54	10.7:1	36@2000

EFI - Electronic fuel injection

GASOLINE ENGINE TUNE-UP SPECIFICATIONS

Year	Engine ID/VIN	Engine Displacement Liters (cc)	Spark Plugs Gap (in.)	Ignition Timing (deg.) MT	Ignition Timing (deg.) AT	Fuel Pump (psi)	Idle Speed (rpm) MT	Idle Speed (rpm) AT	Valve Clearance In.	Valve Clearance Ex.
1993	B230FT/87	2.3 (2316)	0.030	-	12B	43	-	775	0.014-0.018	0.014-0.018
	B230F/88	2.3 (2316)	0.028	-	12B	43	-	775	0.014-0.018	0.014-0.018
	B5234T/57	2.3 (2319)	0.028	-	6B	58	-	800-900	HYD	HYD
	B5254S/55	2.4 (2435)	0.028	-	10B	43	-	750-850	HYD	HYD
	B6304F/95	2.9 (2922)	0.026	-	16B	43	-	700-800	HYD	HYD
1994	B230FT/87	2.3 (2316)	0.030	-	12B	43	-	775	0.014-0.018	0.014-0.018
	B230F/88	2.3 (2316)	0.028	-	12B	43	-	775	0.014-0.018	0.014-0.018
	B5234T/57	2.3 (2319)	0.028	-	6B	58	-	800-900	HYD	HYD
	B5254S/55	2.4 (2435)	0.028	-	10B	43	-	750-850	HYD	HYD
	B6304F/95	2.9 (2922)	0.026	-	16B	43	-	700-800	HYD	HYD
1995	B230FT/86 or 87	2.3 (2316)	0.030	-	12B	43	-	775	0.014-0.018	0.014-0.018
	B230F/88	2.3 (2316)	0.028	-	12B	43	-	775	0.014-0.018	0.014-0.018
	B5234T/57	2.3 (2319)	0.028	-	6B	58	-	800-900	HYD	HYD
	B5234T5/58	2.3 (2319)	0.028	-	6B	58	-	800-900	HYD	HYD
	B5254S/55	2.4 (2435)	0.028	-	10B	43	-	750-850	HYD	HYD
	B6304F/95	2.9 (2922)	0.026	-	16B	43	-	700-800	HYD	HYD
1996-97	B5234T/57	2.3 (2319)	0.028-0.032	-	4-6B	58	-	850	HYD	HYD
	B5234FT/58	2.3 (2319)	0.028-0.032	-	4-6B	58	-	850	HYD	HYD
	B5254S/55	2.4 (2435)	0.028-0.032	3-7B	3-7B	43.5	-	850	HYD	HYD
	B6304F/96	2.9 (2922)	0.028-0.032	-	7-11B	43.5	-	700-800	HYD	HYD

NOTE: The Vehicle Emission Control Information label often reflects specification changes made during production. The label figures must be used if they differ from those in this chart.

HYD - Hydraulic

B - Before top dead center

CAPACITIES

Year	Model	Engine ID/VIN	Engine Displacement Liters (cc)	Engine Oil with Filter	Transmission (pts.) 4-Spd	Transmission (pts.) 5-Spd	Transmission (pts.) Auto.	Transfer Case (pts.)	Drive Axle Front (pts.)	Drive Axle Rear (pts.)	Fuel Tank (gal.)	Cooling System (qts.)
1993	240	B230F	2.3 (2316)	4.7	-	2.7	14.0	-	-	①	15.8	10.0
	940	B230F	2.3 (2316)	4.7	-	-	14.0	-	-	①	19.8	10.0
	940	B230FT	2.3 (2316)	4.7 ②	-	-	14.0	-	-	①	19.8	10.0
	850	B5254FS	2.4 (2435)	5.6	-	4.4	8.4	-	-	-	19.3	7.6
	960	B6304F	2.9 (2922)	6.0	-	-	③	-	-	①	⑤	11.3
1994	940	B230F	2.3 (2316)	4.7	-	-	14.0	-	-	①	19.8	10.0
	940	B230FT	2.3 (2316)	4.7 ②	-	-	14.0	-	-	①	19.8	10.0
	850	B5234T	2.3 (2319)	5.6 ④	-	4.4	8.4	-	-	-	19.3	7.4
	850	B5254S	2.4 (2435)	5.6	-	4.4	8.4	-	-	-	19.3	7.6
	960	B6304F	2.9 (2922)	6.0	-	-	③	-	-	-	⑤	11.3
1995	940 Turbo	B230FT	2.3 (2316)	4.7 ②	-	-	14.0	-	-	①	19.8	10.0
	940	B230F	2.3 (2316)	4.7	-	-	14.0	-	-	①	19.8	10.0
	960 Sedan	B6304F	2.9 (2922)	6.0	-	-	16.4	-	-	-	19.8	10.0
	960 Wagon	B6304F	2.9 (2922)	6.0	-	-	16.4	-	-	-	21.8	10.0

CAPACITIES

Year	Model	Engine ID/VIN	Engine Displacement Liters (cc)	Engine Oil with Filter	Transmission (pts.) 4-Spd	5-Spd	Auto.	Transfer Case (pts.)	Drive Axle Front (pts.)	Rear (pts.)	Fuel Tank (gal.)	Cooling System (qts.)
1995	850 Turbo	B5234T	2.3 (2319)	5.6 ④	-	-	16.0	-	-	-	19.3	7.6
	T5R	B5234T5	2.3 (2319)	5.6 ④	-	-	16.0	-	-	-	19.3	7.6
	850	B5254S	2.4 (2435)	5.6	-	4.4	16.0	-	-	-	19.3	7.8
	850 GLT	B5254S	2.4 (2435)	5.6	-	4.4	16.0	-	-	-	19.3	7.8
1996-97	850 Turbo	B5234T	2.3 (2319)	5.6 ④	-	4.4	8.4	-	-	-	19.3	7.4
	850R	B5234FT	2.3 (2319)	5.6 ④	-	4.4	8.4	-	-	-	19.3	7.4
	850	B5254S	2.4 (2435)	5.6	-	4.4	8.4	-	-	-	19.3	7.6
	960	B6304F	2.9 (2922)	6.0	-	-	③	-	-	-	⑤	11.3

① 1030 axle: 2.8 pts.
1031 axle: 3.4 pts.
② Add 0.6 qts. if oil cooler has been drained
③ Oil change (draining only): 6.3 pts.
④ If engine is completely drained, add 0.95 qts. for oil cooler
⑤ Sedan: 21.8
Wagon: 19.8

VALVE SPECIFICATIONS

Year	Engine ID/VIN	Engine Displacement Liters (cc)	Seat Angle (deg.)	Face Angle (deg.)	Spring Test Pressure (lbs. @ in.)	Spring Installed Height (in.)	Stem-to-Guide Clearance (in.) Intake	Exhaust	Stem Diameter (in.) Intake	Exhaust
1993	B230F	2.3 (2316)	45	44.5	158@1.08	1.79	0.0012-0.0024	0.0024-0.0036	0.3132-0.3138	0.3128-0.3134
	B230FT	2.3 (2316)	45	44.5	158@1.08	1.79	0.0012-0.0024	0.0024-0.0036	0.3132-0.3138	0.3128-0.3134
	B5254FS	2.4 (2435)	45	NA	①	NA	0.0012-0.0024	0.0012-0.0024	0.2738-0.2750	0.2734-0.2746
	B6304F	2.9 (2922)	45.25	45.5	①	NA	0.0012-0.0024	0.0012-0.0024	0.2738-0.2744	0.2738-0.2744
1994	B230F	2.3 (2316)	45	44.5	158@1.08	1.79	0.0012-0.0024	0.0024-0.0036	0.3132-0.3138	0.3128-0.3134
	B230FT	2.3 (2316)	45	44.5	158@1.08	1.79	0.0012-0.0024	0.0024-0.0036	0.3132-0.3138	0.3128-0.3134
	B5234T	2.3 (2319)	45	NA	①	NA	0.0012-0.0024	0.0012-0.0024	0.2738-0.2750	0.2734-0.2746
	B5254S	2.3 (2319)	45	NA	①	NA	0.0012-0.0024	0.0012-0.0024	0.2738-0.2750	0.2734-0.2746
	B6304F	2.9 (2922)	45.25	45.5	①	NA	0.0012-0.0024	0.0012-0.0024	0.2738-0.2744	0.2738-0.2744
1995	B230F	2.3 (2316)	45	44.5	158@1.08	1.79	0.0012-0.0024	0.0024-0.0036	0.3132-0.3138	0.3128-0.3134
	B230FT	2.3 (2316)	45	44.5	158@1.08	1.79	0.0012-0.0024	0.0024-0.0036	0.3132-0.3138	0.3128-0.3134
	B5234T	2.3 (2319)	45	NA	①	NA	0.0012-0.0024	0.0012-0.0024	0.2738-0.2750	0.2734-0.2746
	B5234T5	2.3 (2319)	45	NA	①	NA	0.0012-0.0024	0.0012-0.0024	0.2738-0.2750	0.2734-0.2746
	B5254S	2.3 (2319)	45	NA	①	NA	0.0012-0.0024	0.0012-0.0024	0.2738-0.2750	0.2734-0.2746
	B6304F	2.9 (2922)	45.25	45.5	①	NA	0.0012-0.0024	0.0012-0.0024	0.2738-0.2744	0.2738-0.2744

VALVE SPECIFICATIONS

Year	Engine ID/VIN	Engine Displacement Liters (cc)	Seat Angle (deg.)	Face Angle (deg.)	Spring Test Pressure (lbs. @ in.)	Spring Installed Height (in.)	Stem-to-Guide Clearance (in.)		Stem Diameter (in.)	
							Intake	Exhaust	Intake	Exhaust
1996-97	B5234T	2.3 (2319)	45	44.5	①	NA	0.0012-0.0024	0.0012-0.0024	0.2738-0.2750	0.2734-0.2746
	B5234FT	2.3 (2319)	45	44.5	①	NA	0.0012-0.0024	0.0012-0.0024	0.2738-0.2750	0.2734-0.2746
	B5254S	2.4 (2435)	45	44.5	①	NA	0.0012-0.0024	0.0012-0.0024	0.2738-0.2750	0.2734-0.2746
	B6304F	2.9 (2922)	45	44.5	①	NA	0.0012-0.0024	0.0012-0.0024	0.2738-0.2744	0.2738-0.2744

NOTE: Exhaust valves for turbocharged engines are stellite-coated and must not be machined. They may be ground against the valve seat.

NA - Not Available

① 150@1.00
61@1.34

TORQUE SPECIFICATIONS
All readings in ft. lbs.

Year	Engine ID/VIN	Engine Displacement Liters (cc)	Cylinder Head Bolts	Main Bearing Bolts	Rod Bearing Bolts	Crankshaft Damper Bolts	Flywheel Bolts	Manifold		Spark Plugs	Lug Nut
								Intake	Exhaust		
1993	B230F	2.3 (2316)	①	80	②	③	51	12	12	18	④
	B230FT	2.3 (2316)	①	80	②	③	51	12	12	18	④
	B5254FS	2.4 (2435)	⑨	⑧	⑥	133	⑦	15	18	18	④
	B6304F	2.9 (2922)	⑨	⑧	⑥	221	⑦	15	18	18	④
1994	B230F	2.3 (2316)	①	80	②	③	51	12	12	18	④
	B230FT	2.3 (2316)	①	80	②	③	51	12	12	18	④
	B5234T	2.3 (2319)	⑨	⑧	⑥	133	⑦	15	18	18	81
	B5254S	2.4 (2435)	⑨	⑧	⑥	133	⑦	15	18	18	81
	B6304F	2.9 (2922)	⑨	⑧	⑥	221	⑦	15	18	18	④
1995	B230FT	2.3 (2316)	①	80	②	③	51	12	12	18	④
	B230F	2.3 (2316)	①	80	②	③	51	12	12	18	④
	B5234T	2.3 (2319)	⑨	⑧	⑥	133	⑤	15	18	18	81
	B5234T5	2.3 (2319)	⑨	⑧	⑥	133	⑤	15	18	18	81
	B5254S	2.4 (2435)	⑨	⑧	⑥	133	⑤	15	18	18	81
	B6304F	2.9 (2922)	⑨	⑧	⑥	221	⑤	15	18	18	④
1996-97	B5234T	2.3 (2319)	①	⑧	⑥	133	⑦	15	18	18	81
	B5234FT	2.3 (2319)	①	⑧	⑥	133	⑦	15	18	18	81
	B5254S	2.4 (2435)	①	⑧	⑥	133	⑦	15	18	18	81
	B6304F	2.9 (2922)	①	⑧	⑥	221	⑦	15	18	18	④

① Step 1: 14 ft. lbs.
Step 2: 43 ft. lbs.
Step 3: + 90 degrees

② Step 1: 14 ft. lbs.
Step 2: + 90 degrees

③ Step 1: 43 ft. lbs.
Step 2: + 90 degrees

④ Torque lugs in a diagonal pattern
P20: 85 ft. lbs. (115nm)
P70/90: 63 ft. lbs. (85nm)

⑤ Step 1: 33 ft. lbs.
Step 2: + 65 degrees

⑥ Step 1: 15 ft. lbs.
Step 2: + 90 degrees

⑦ Step 1: 33 ft. lbs.
Step 2: + 50 degrees

⑧ Tighten cylinder block, intermediate section, in stages:
Step 1: M10 bolts: 15 ft. lbs. (20mm)
Step 2: M10 bolts: 33 ft. lbs. (45mm)
Step 3: M8 bolts: 18 ft. lbs. (25mm)
Step 4: M7 bolts: 13 ft. lbs. (17mm)
Step 5: M10 bolts: + 90 degrees

⑨ Step 1: 15 ft. lbs.
Step 2: 44 ft. lbs.
Step 3: + 130 degrees

Bolts should be tightened in sequence from center towards the ends

BRAKE SPECIFICATIONS
All measurements in inches unless noted

Year	Model		Master Cylinder Bore	Brake Disc Original Thickness	Brake Disc Minimum Thickness	Maximum Runout	Brake Drum Diameter Original Inside Diameter	Brake Drum Diameter Max. Wear Limit	Brake Drum Diameter Maximum Machine Diameter	Minimum Lining Thickness Front	Minimum Lining Thickness Rear
1993	240 Series	F	0.880	①	②	0.004	-	-	-	0.125	-
		R	-	0.380	0.330	0.004	-	-	-	-	0.075
	850 Series	F	0.937	1.024	0.906	0.002	-	-	-	0.125	-
		R	-	0.378	0.330	0.002	-	-	-	-	0.075
	940 Series	F	0.937	③	④	0.002	-	-	-	0.125	-
		R	-	⑤	⑥	0.002	-	-	-	-	0.075
	960 Series	F	0.937	③	④	0.002	-	-	-	0.125	-
		R	-	⑤	⑥	0.002	-	-	-	-	0.075
1994	850 Series	F	0.937	1.024	0.906	0.002	-	-	-	0.125	-
		R	-	0.378	0.330	0.002	-	-	-	-	0.075
	940 Series	F	0.937	③	④	0.002	-	-	-	0.125	-
		R	-	⑤	⑥	0.002	-	-	-	-	0.075
	960 Series	F	0.937	③	④	0.002	-	-	-	0.125	-
		R	-	⑤	⑥	0.002	-	-	-	-	0.075
1995	850 Series	F	0.937	1.024	0.906	0.002	-	-	-	0.125	0.075
		R	-	0.378	0.331	0.003	-	-	-	-	-
	T-5R	F	0.937	1.024	0.906	0.002	-	-	-	0.125	0.075
		R	-	0.378	0.331	0.003	-	-	-	-	-
	960 Series ⑦	F	0.937	0.866	0.787	0.002	-	-	-	0.125	0.075
		R	-	0.378	0.331	0.002	-	-	-	-	-
	960 Series ⑧	F	0.937	1.024	0.906	0.002	-	-	-	0.125	0.075
		R	-	0.394	0.315	0.002	-	-	-	-	-
	940 Series ⑦	F	0.937	0.866	0.787	0.002	-	-	-	0.125	0.075
		R	-	0.378	0.331	0.002	-	-	-	-	-
	940 Series ⑧	F	0.937	1.024	0.906	0.002	-	-	-	0.125	0.075
		R	-	0.394	0.315	0.002	-	-	-	-	-
1996-97	850 Series	F	0.937	1.024	0.906	0.002	-	-	-	0.125	-
		R	-	0.378	0.330	0.002	-	-	-	-	0.075
	960 Series	F	0.937	③	④	0.002	-	-	-	0.125	-
		R	-	⑤	⑥	0.002	-	-	-	-	0.075

① With solid front rotors: 0.563
With vented front rotors: 0.945

② With solid front rotors: 0.500
With vented front rotors: 0.803

③ With standard vented rotor: 0.866
With heavy duty vented rotor: 1.024

④ With standard vented rotor: 0.787
With heavy duty vented rotor: 0.905

⑤ With standard rear axle: 0.378
With Multi-link rear axle: 0.394

⑥ With standard rear axle: 0.330
With Multi-link rear axle: 0.315

⑦ With heavy duty jumbo rotors

⑧ Rear disc brakes with multi-link rear axle

FREQUENT MAINTENANCE LABOR
VOLVO

The following should be used as a guide when determining the amount of work required for a particular service if taken to a repair shop. In estimating how long a particular Frequent Maintenance Service item should take, please observe the following:
- **Chilton Time** is time that is based on field research and data supplied by the vehicle manufacturer.
- All labor time operations are given in hours and tenths of an hour.
- All labor operations, are to be used as a **guide**.

COOLING

(G) Winterize Cooling System
Includes: Run engine to check for leaks, tighten all hose connections. Test radiator and pressure cap, drain radiator and engine block. Add antifreeze and refill system.

	Chilton Time
1993-97	.5

(G) Belt, Serpentine Drive, Renew

	Chilton Time
1993-97 850	.3
1993-97 960	.3

(G) Belt, Drive, Renew

1993-97	
one	.7
each adtnl.	.1

(G) Belt, Drive, Adjust

1993-97	
one	.3
each adtnl.	.1

(G) Hoses, Radiator, Renew
Includes: Drain and refill cooling system as required.

1993-97	
upper	.4
lower	.5

(G) Thermostat, Coolant, Renew

1993-97	
4, 5 cyl.	.5
6 cyl.	.7

FUEL

(M) Air Cleaner, Service

	Chilton Time
1993-97	.2

(G) Filter, Fuel, Renew

1993-97	.6

BRAKES

(G) Bleed Brakes (Four Wheels)
Includes: Add fluid.

1993 240	.8
1993-97 850	.6
1993-97 940, 960	.6
w/ABS add	.2

(M) Parking Brake, Adjust

1993-97	.3

LUBRICATION SERVICES

(M) Engine Oil & Filter, Renew
Includes: Inspect and correct all fluid levels.

1993-97	.3

WHEELS

(G) Wheel, Renew (One)

	Chilton Time
1993-97 850	.5

(G) Wheel, Balance

1993-97 850	
one	.3
each adtnl.	.2

(G) Wheels, Rotate (All)

1993-97 850	.5

ELECTRICAL

(G) Headlamps, Aim

1993-97	
two	.4
four	.6

(G) Halogen Headlamp Bulb, Renew

1993 240	.4
1993-97 850	.4
1993-97 940, 960	.3

(G) Tail Lamp Lens or Bulb, Renew

1993-97, each	.2

(G) Horn, Renew

1993-97, each	.2

SCHEDULED MAINTENANCE INTERVALS
(VOLVO 240, 850, 940 & 960)

TO BE SERVICED	TYPE OF SERVICE	VEHICLE MILEAGE INTERVAL (x1000)												
		5	10	20	30	40	50	60	70	80	90	100	110	120
Engine oil & filter①	R	✓	✓	✓	✓	✓	✓	✓	✓	✓	✓	✓	✓	✓
Automatic transmission fluid⑥	S/I	✓	✓	✓	✓	✓	✓	✓	✓	✓	✓	✓	✓	✓
Fluid levels (all)	S/I	✓	✓	✓	✓	✓	✓	✓	✓	✓	✓	✓	✓	✓
Rotate tires	S/I		✓	✓	✓	✓	✓	✓	✓	✓	✓	✓	✓	✓
Automatic transmission shift control	S/I		✓	✓	✓	✓	✓	✓	✓	✓	✓	✓	✓	✓
Brake pads & parking brake	S/I		✓	✓	✓	✓	✓	✓	✓	✓	✓	✓	✓	✓
Driveshaft boots (850)	S/I		✓	✓	✓	✓	✓	✓	✓	✓	✓	✓	✓	✓
Engine & transmission (check for leaks)	S/I		✓	✓	✓	✓	✓	✓	✓	✓	✓	✓	✓	✓
Exhaust system	S/I		✓	✓	✓	✓	✓	✓	✓	✓	✓	✓	✓	✓
Grease link arm stops (1994-97 850)	S/I		✓	✓	✓	✓	✓	✓	✓	✓	✓	✓	✓	✓
Reset service reminder①	S/I		✓	✓	✓	✓	✓	✓	✓	✓	✓	✓	✓	✓
Driveshaft, U-joints	S/I					✓	✓	✓	✓	✓	✓	✓	✓	✓
Driveshaft joints (850)	S/I							✓	✓	✓	✓	✓	✓	✓
Clutch	S/I			✓		✓		✓		✓		✓		✓
Kickdown cable (240/940)	S/I			✓		✓		✓		✓		✓		✓
Brake & fuel lines & hoses	S/I					✓		✓		✓		✓		✓
Steering & suspension	S/I					✓		✓		✓		✓		✓
Air cleaner filter	R				✓			✓			✓			✓
Spark plugs	R				✓			✓			✓			✓
Timing belt (1993 960)	R				✓			✓			✓			✓
Timing belt (1994-95 960)	R						✓					✓		
Timing belt (850 & 1996-97 960)	R								✓					

SCHEDULED MAINTENANCE INTERVALS
(VOLVO 240, 850, 940 & 960) (Cont.)

TO BE SERVICED	TYPE OF SERVICE	VEHICLE MILEAGE INTERVAL (x1000)												
		5	10	20	30	40	50	60	70	80	90	100	110	120
Timing belt (B230F, FT)	R						✓					✓		
Timing belt (B230F, FT	S/I							✓					✓	
Timing belt (B230FD)	S/I		✓										✓	
Timing belt (B230FD)	R											✓		
Timing gear belt tensioner pivot bearing (1993 960)	S/I				✓			✓			✓			✓
Timing gear belt tensioner pivot bearing (1993 850)	R						✓					✓		
Valve clearance (240/740)	S/I				✓			✓			✓			✓
Drive belt tensioner (850)	S/I				✓			✓			✓			✓
Drive belts⑤	S/I				✓			✓			✓			✓
Fuel line filter④	R							✓						✓
EGR system	S/I							✓				✓		
PCV nipple (orifice)/hoses	S/I							✓				✓		
Check suspension torques③	S/I		✓											
Brake fluid②	R													

① Perform operation every 5000 miles on turbocharged models.
② Replace every 2 years or 30,000 miles, whichever comes first under normal conditions, more frequently in mountainous areas or moist climates.
③ Except 850 shown. 850 (1993) - perform at 1500 miles.
④ 1993-95 shown; replace every 100,000 miles (1996-97)
⑤ 850, 960: replace at 60,000 miles.
⑥ Replace as follows: 1993 240/940 - every 20,000 miles; 1994 940 - every 40,000 miles.
R – Replace   S/I – Service or Inspect

FREQUENT OPERATION MAINTENANCE (SEVERE SERVICE)
If a vehicle is operated under any of the following conditions it is considered severe service:
- Extremely dusty areas.
- 50% or more of the vehicle operation is in 32°C (90°F) or higher temperatures, or constant operation in temperatures below 0°C (32°F).
- Prolonged idling (vehicle operation in stop and go traffic).
- Frequent short running periods (engine does not warm to normal operating temperatures).
- Police, taxi, delivery usage or trailer towing usage.
Oil & oil filter change – (all models) change every 5000 miles.
Air filter element – service or inspect every 15,000 miles.

—NOTES—

TECHNICAL SERVICE BULLETINS 21

TECHNICAL SERVICE BULLETINS

What is a TSB?

All vehicle manufacturers experience occasional problems with one or more of their model lines, requiring that fixes be made after the vehicle is sold to the customer. Manufacturers therefore issue Technical Service Bulletins (TSBs) to inform and to suggest certain repairs or component replacements. These fixes may cover a variety of issues including: safety, general maintenance, part replacement, engine driveability improvements or general repairs. If the item at issue is a noted safety related problem, it is likely that the manufacturer will also issue SAFETY RECALL CAMPAIGN notices.

NOTE: The major difference between a TSB and a Recall Campaign is that the manufacturer wants the repairs performed to ALL vehicles affected by a recall, in order to prevent a problem (often safety related) from occurring. TSBs, on the other hand, are issued to help the dealership service facility cope with a problem (usually non-safety related) that may occur to SOME vehicles. The repair or change may not be necessary if the component or system in question never develops a problem.

The TSBs and Recall Campaigns notices are sent directly to the dealership repair facility. Safety Recall Campaign notices are also sent to the vehicle owners, but in case of a sale, transfer of title or other circumstances, the owner may not receive the actual notice.

All of these factory notifications provide a description of the problem, the vehicles which are affected by the problem, how the problem is fixed and whether or not the problem is warranty-related.

Federal law also requires that the general public have access to TSBs and Recalls. You can obtain copies of any authorized bulletins from various sources including the manufacturer service information groups or distributors, federal or state government agencies dealing in transportation or publication, electronic based professional information systems such as Chilton On Disc, or even a cooperative dealer service department.

We have provided the following examples of bulletins so that you can see the kind of information that a TSB might provide. You will also note that each bulletin is numbered, providing a valuable way to access this information. See the index for the list of bulletins applying to your vehicle which was current at time of publication.

Using the TSB Index

The TSB index in this manual is divided into groups of charts covering each manufacturer. Each group contains separate charts for the various vehicle sub-systems or service categories.

The charts we have provided contain 4 columns. The first column, **MODELS**, provides you with a coded listing of what actual vehicle nameplates from that manufacturer are affected by the bulletin. The model listing is a numerical code, explained in the footer (the bottom) of each chart.

The second column, **YEAR**, lists the model year(s) of those nameplates which are affected. For example, 92–94 would include all 1992, 1993 and 1994 models carrying the nameplate listed in the first column. Similarly, 94–94 would only affect vehicles built for the 1994 model year.

The third column, **TSB#**, was the last revised (if more than one was published) part number or code to retrieve that particular service bulletin.

Finally, the fourth column, **DESCRIPTION**, provides a brief idea of what the bulletin is about

(new components, procedures, specifications, or possibly just dealer network service policy revisions).

To determine if there are any bulletins for your vehicle within a certain category:

1. Locate the group containing charts for your vehicle's manufacturer.

2. Next, turn to the page(s) covering the specific system you are curious about (such as Heating, Air Conditioning, Ventilation, Defogger or Lighting, Horns, Turn Signals, Steering Column).

3. Once you have reached the chart for the category, the next step is to determine if any bulletins apply to the particular model on which you are working. Determine what number represents your model using the footer at the bottom of the chart. Scan the **MODELS** column for that number.

4. Whatever matches you find are service bulletins that *could* apply to your vehicle. If the **YEAR** and **DESCRIPTION** columns also match your model and problem, then you can use the **TSB#** to help attain the bulletin in question.

5. If you do not find a number match, check the chart footer again to see if any codes such as 1=All or 2=Most are used for that manufacturer. Sometimes components or procedures are shared by all or the majority of a manufacturer's model lines. In this case, a bulletin may apply to your model as well, but only if the **YEAR** and **DESCRIPTIONS** still match.

NOTE: Keep in mind that even if your model and year match a bulletin, it does not necessarily mean that the repair is required for your vehicle. A TSB (not a Safety Recall) repair should be performed ONLY if the problem (not just the symptom) exists on your vehicle. Remember that a TSB lists the probable cause of a symptom, but it is often not the only possible cause.

Technical Service Information

BODY
BO92-009
DECEMBER 4, 1992
ES 300

Title **ES 300 RATTLE NOISE FROM SUN VISOR**   Page 1 of 1

Some ES 300 vehicles may have a rattling noise caused by excessive clearance between the plastic sun visor shaft roller and the sun visor holder. To correct this condition in production, a rib has been added to the sun visor holder as follows:

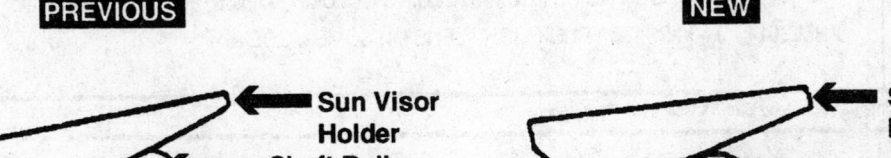

PRODUCTION EFFECTIVE:

	VIN	PRODUCTION DATE
FROM:	JT8VK13T-N0014939	09/91

PART NUMBER INFORMATION:

Previous	New	Description
74348-33010-02	NO CHANGE	Visor Holder, Taupe
74348-33010-04	NO CHANGE	Visor Holder, Blue
74348-33010-08	NO CHANGE	Visor Holder, Ivory
74348-33010-13	NO CHANGE	Visor Holder, Gray

FIELD-FIX PROCEDURE:

On vehicles with a rattling noise from the sun visor holder, a section of felt from the Squeak & Rattle Repair Kit **(Part number 08231-00801)** should be cut to the following dimensions. Remove release sheet from back side of felt and inserted into the sun visor holder as shown below:

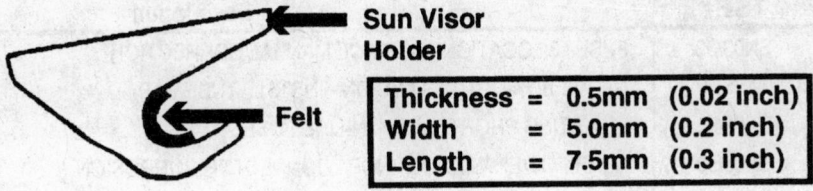

Thickness	=	0.5mm	(0.02 inch)
Width	=	5.0mm	(0.2 inch)
Length	=	7.5mm	(0.3 inch)

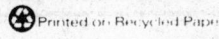 Printed on Recycled Paper

Lexus Supports ASE Certification

7920tsb2

A typical Lexus TSB for interior trim noises

21 TECHNICAL SERVICE BULLETINS

ACURA

Lighting, Horns, Turn Signals, Steering Column

Models	Year	TSB#	Description
4	94 - 94	95-001	GROAN NOISE FROM STEERING COLUMN - DPMS
3,7	94 - 94	AHM/B51	MISSING TRUNK LIGHT LENS ON SOME MODELS

Model Legend: 1= All 2= Most 3= Integra 4= Legend 5= NSX Coupe 6= TL 7= Vigor

Instruments, Dash Cluster, Warning Lights, Mirrors

Models	Year	TSB#	Description
4	94 - 94	95-003	CRUISE CONTROL WILL NOT ENGAGE
4	94 - 94	SN04/94	CRUISE CONTROL WILL NOT SET - TIPS
3	94 - 95	95-009	OUTSIDE MIRROR VIBRATES
4,5,7	91 - 93	92-039	REARVIEW MIRROR REINSTALLATION PROCEDURE
4	91 - 93	SN09/93	SEDAN DOOR MIRROR SEALS AVAILABLE SEPARATELY
1	Up - 93	SN03/93	WARN LITE/OIL - STAYS ON BELOW FREEZING - PRECAUTION

Model Legend: 1= All 2= Most 3= Integra 4= Legend 5= NSX Coupe 6= TL 7= Vigor

Chassis Electrical, Wiring Harness, Fuses-Circuit Breakers, Wipers, Window Motors

Models	Year	TSB#	Description
3	93 - 93	SN03/93	FUSE 13 LOCATION - SERVICE MANUAL CORRECTION
4,3	91 - 94	92-026 SEP '92	GLOVE BOX LIGHT STAYS ON - INSTALL TUBE CAP
1	Up - 96	SN05/95	ISOLATING SHORTS WITH IN-LINE FUSES - TIPS
4	91 - 93	SN09/93	REAR POWER WINDOW INOP - CONNECTOR CORROSION
4	91 - 93	93-021	REAR WINDOW INOP - CORRODED TERMINALS - PROCEDURE

Model Legend: 1= All 2= Most 3= Integra 4= Legend 5= NSX Coupe 6= TL 7= Vigor

Auxiliary Equipment, Jacks, Trailer Hitches, Towing

Models	Year	TSB#	Description
4	91 - 93	93-009	CELLULAR PHONE - HANDSET CORD DAMAGED - DIAG/REPAIR
2	Up - 94	SN02/94	IN-DASH CELLULAR PHONE - HANDS FREE OPS EXPLAINED
3	94 - 94	94-004	IN-DASH PHONE TROUBLESHOOTING - ETM SUPPLEMENT
4	91 - 93	93-023	INTERMITTENT POWER TO PHONE - STARTER INTERFERENCE
4,7,5	91 - 95	94-012	OUT OF WARRANTY TELEPHONE REPAIR
7	94 - 94	AHM/B06	WING SPOILER - CHECK VIN BEFORE INSTALL BRAKE SENSOR

Model Legend: 1= All 2= Most 3= Integra 4= Legend 5= NSX Coupe 6= TL 7= Vigor

Heating, Air Conditioning, Ventilation, Defogger

Models	Year	TSB#	Description
1	94 - 94	SN02/94	A/C - R-134A SYS WON'T CHARGE - CHECK SCHRADER VALVE
1	Up - 95	SN03/95	A/C - R134A O-RING LUBRICATION
4,5	93 - 93	92-027	A/C - R134A REFRIGERANT SYSTEM
1	92 - 93	SN10/92	A/C - STOPS COOLING INTERMITTENTLY - SERVICE TIP
3	94 - 94	SN09/94	A/C INOP - CORROSION IN C151
3	94 - 95	95-013	A/C INOP - ENG SPLASH SHIELD KNOCKS OFF A/C BELT
3	94 - 94	AHM/B49	A/C KITS - COMPRESSOR EXCHANGE PROGRAM
1	Up - 94	SN04/94	ND REPL'MENT A/C COMPRESSOR - XFER MANIFOLD - SPECS
4	92 - 94	SN08/94	ODOR - CLEAN EVAP - REPLACE FILTER
4,5	93 - 93	SN05/93	USE ONLY ND-8 REFRIGERANT OIL

Model Legend: 1= All 2= Most 3= Integra 4= Legend 5= NSX Coupe 6= TL 7= Vigor

Entertainment Devices, Stereo, Radio, Etc.

Models	Year	TSB#	Description
3	93 - 93	AHM/B08	ANTENNA (POWER) - NUT LOOSE - TORQUE PROCEDURE
7	Up - 94	SN02/94	ANTENNA OPERATES WITH RADIO OFF - CHECK BATTERY
1	Up - 93	SN10/92	AUDIO - SERVICE BULLETIN FILING SYSTEM
2	Up - 93	92-018	CD CHANGER - MAGAZINE WON'T EJECT - CORRECTION
3	94 - 94	SN04/94	CD INSTALLATION - CORRECT DIN CONNECTOR POSITION
1	Up - 94	SN09/93	INSTALLING CD CHANGER MAY REQUIRE CONTROL HEAD
4	91 - 95	94-015	RADIO - ACURA/BOSE AMPLIFIER REMANUFACTURING PROGRAM
4	91 - 95	91-045 SEP`92	RADIO - ERRATIC REMOTE AUDIO CONTROL OPERATION - REV
1	86 - 94	88-009 FEB`93	RADIO - OUT-OF-WARRANTY REPAIR
1	86 - 93	88-009 JUL`92	RADIO - OUT-OF-WARRANTY REPAIR
3,7	90 - 95	91-053 OCT`92	RADIO - POOR AM RECEPTION/NOISE - REVISED
4,3,5,7	86 - 93	91-053 JUN`92	RADIO - POOR AM RECEPTION/POPPING NOISE - REVISED
4,5	91 - 94	94-011	RADIO - POOR RECEPTION OR INTERFERENCE
4,7	Up - 93	SN03/93	RADIO RECEPTION PROBLEM - CHECK FOR GOLD EMBLEM KIT
1	Up - 93	87-015 MAY`92	RADIO/TAPE/CD PLAYER - EXCHANGE PROGRAM PROCEDURES
2	86 - 94	SN09/94	REPLACEMENT ANTENNA MOTORS WITHOUT ANTENNA TUBES
3	94 - 94	SN09/93	TRIPLE FUNCTION AUDIO (RADIO/CD/CASS) EXPLAINED

Model Legend: 1= All 2= Most 3= Integra 4= Legend 5= NSX Coupe 6= TL 7= Vigor

Seats, Belts, Interior Trim, Carpets, Air Bags

Models	Year	TSB#	Description
4,7	91 - 93	91-019 MAR`91	AIRBAG/PASSENGERS - ADJUSTMENT PROCEDURES - REVISED
4	92 - 93	SN05/93	BELT PRETENSIONERS TERMINALS WRONG - S/M CORRECTION
4	93 - 93	SN03/93	BELT/FRONT SHOULDER - ANCHOR ADJUSTER RATTLE - PROC
4	92 - 94	94-007	BEVERAGE HOLDER DOES NOT OPEN OR CLOSE PROPERLY
3	94 - 95	SN08/94	CUP HOLDER STICKS - SPILLED DRINKS - DISSASSEMBLY
3	90 - 93	92-033	CUPHOLDER IN OPTIONAL ARMREST - BROKEN - REPLACE
4	91 - 93	SN09/93	DRIVER SEAT ROCK - CHECK SCREWS BEFORE REPLACE BASE
4	93 - 95	94-020	NOISE/BUZZ FROM SEAT BELT SHOULDER ANCHOR
4	91 - 93	SN05/93	POWER SEAT CONTROL UNIT - HINTS ON REMOVING
4	91 - 93	92-017 JAN`93	SEAT (DRIVER) - ROCKS ON MOUNTS - REPAIR PROC - REV
1	Up - 94	91-050 JAN`92	SEAT BELT (3 POINT ACTIVE) - SLOW/PARTIAL RETRACT
1	Up - 95	AHM/B19	SEAT BELT BUCKLE CAMPAIGN CORRECTION - NO 3 POINT
1	Up - 95	AHM/B21	SEAT BELT BUCKLE CAMPAIGN UPDATE
2	92 - 94	93-015	SEAT BELT TONGUE STOPPER BUTTON BROKEN - REV PART
5,4	Up - 93	SN03/93	SEAT BELTS/FRONT - REPLACE AFTER AIRBAG DEPLOYMENT
4	91 - 93	SN03/93	SEAT/DRIVER - SLIDE MOTOR RUNS BUT SEAT DOESN'T MOVE
1	Up - 94	SN04/94	SRS - TORX BOLT REPLACEMENT CAUTION
6	95 - 96	95-012	TWEETER GRILLE FABRIC PULLS LOOSE

Model Legend: 1= All 2= Most 3= Integra 4= Legend 5= NSX Coupe 6= TL 7= Vigor

Glass, Doors, Hood, Decklid, Tailgate, Liftgate, Locks

Models	Year	TSB#	Description
4	93 - 93	AHM/B33	DOOR - BLACK STAINS ON LEFT FRONT DOOR OPENING - FIX
7	92 - 93	93-001	DOOR - UNLOCKING & OPENING SETS OFF SECURITY ALARM
3	90 - 93	93-011	DOOR GLASS CHATTER - IMPROPER REGULATOR POSITION
3	90 - 93	SN12/92	DOOR GLASS/FRONT - CHATTERS/GOING UP & DOWN - REPAIR
4	91 - 93	SN03/93	DOOR LOCK KNOB STICKS UP TOO HIGH/MAY BREAK - INFO
3	94 - 94	93-018	DOOR WILL NOT UNLOCK WITH KEY - INSTALL SPACER
4	91 - 94	91-012 FEB`92	FRONT DOOR GLASS - WIND NOISE - CAUSE/CORRECT -REV 2
3	94 - 94	93-014	HATCH - LATCH RATTLES ON ROUGH ROADS - DIAG/REPAIR
3	94 - 94	93-014 AUG`93	HATCH - LATCH RATTLES ON ROUGH ROADS - REVISED
3	94 - 94	94-009	HATCH HARD TO CLOSE AND LATCH
3	94 - 95	94-009 AUG`94	HATCH HARD TO CLOSE AND LATCH - REVISED
3	94 - 94	SN09/93	HATCH HITS GLASS INSTALLING SPOILER - PRECAUTIONS
4	91 - 95	SN03/95	MOONROOF ADJUSTMENT SHIMS
3	94 - 95	94-017	MOONROOF CHATTER WHILE OPENING OR CLOSING
3	94 - 95	94-016	NOISE/RATTLE FROM PARTIALLY OPEN WINDOW
3	94 - 94	94-005	POWER DOOR LOCKS CYCLE FROM LOCKED TO UNLOCKED
4	91 - 93	SN09/93	REAR POWER WINDOW INOP - CONNECTOR CORROSION
5	91 - 93	93-004	WINDOW REGULATOR POP NOISE ON RAISING/LOWERING GLASS
3	94 - 94	94-002	WINDOW SPEED IS TOO SLOW OR STOPS HALF WAY DOWN
5	91 - 93	93-013	WINDSHIELD - UPPER MOULDING SHRINKS/LEAVES GAP - FIX
5	91 - 93	93-013 AUG`93	WINDSHIELD - UPPER MOULDING SHRINKS/LEAVES GAP - REV

Model Legend: 1= All 2= Most 3= Integra 4= Legend 5= NSX Coupe 6= TL 7= Vigor

Finishes, Body Structure, Frame, Bumpers

Models	Year	TSB#	Description
4	91 - 94	94-006	FRONT BUMPER FACE PULLS LOOSE
4	95 - 95	94-021	GAP IN FORNT DOOR SILL MOLDING AT B-PILLAR
4	91 - 93	93-005	MOULDING/QUARTER WINDOW - LOOSE OR MISSING - REPAIR
1	95 - 95	94-014	PAINT CODES
1	93 - 93	92-038	PAINT CODES - 1993 MODELS
1	94 - 94	93-022	PAINT CODES - 1994 MODELS
4	91 - 93	SN05/93	ROOF RATTLE - BEND MOONROOF CABLE TUBES

Model Legend: 1= All 2= Most 3= Integra 4= Legend 5= NSX Coupe 6= TL 7= Vigor

AUDI

Lighting, Horns, Turn Signals, Steering Column

Models	Year	TSB#	Description
8	96 - 96	57/96-01	IGNITION LOCK CYLINDER REPLACE PROCEDURE
5	92 - 93	92/92-03	HEADLIGHTS - WEAK SPRAY FROM WASHER SYSTEM - SERVICE
5	92 - 93	94/91-02	TURN SIGNALS/FRONT - LOOSE ASSY/ADJUST DURING PDI
5	92 - 93	94/92-01A	HEADLIGHTS - UNEVEN W/HOOD - PDI CORRECTION PROC
2	Up - 93	94/93-01	IGNITION LOCK CYLINDER - REMOVE/INSTALL - S/M REV
9	95 - 95	94/95-01	FOG LIGHTS - MODIFICATION TO WORK ONLY W/ LOW BEAMS
4	93 - 93	96/92-05	BULBS (IP SWITCHES) - DETERMINING IF REPLACEABLE
5,12,11	92 - 93	96/93-01	BULBS (IP SWITCHES) - REPLACEMENT INFORMATION
8	96 - 96	96/95-09	REAR READING LAMP LOOSE

Model Legend: 1= All 2= Most 3= 80 4= 90 5= 100 6= 200 7= 90 Quattro 8= A4, A4 Quattro 9=A6, A6 Quattro 10= Cabriolet 11= 4000S, Quattro, V8 12= S4 13= S6

Instruments, Dash Cluster, Warning Lights, Mirrors

Models	Year	TSB#	Description
4,8,9,10	95 - 96	27/95-02	CRUISE CNTRL - NOT HOLD SET SPEED - INTERMITTENT
5,12	93 - 94	70/93-01	ADDITIONAL I/P ATTACHMENT - WITH PASSENGER AIRBAG
7	93 - 93	4/92-03	INSTRUMENTS IN CONSOLE - CHANGES (QUATTRO SPORT)
11	92 - 93	4/93-01	OUTSIDE AIR TEMP DISPLAY - TRANSFERRING TO NEW I/C
9	95 - 95	4/94-04	IC - CODING/TESTING - SRI INFO
1	Up - 96	4/95-03	SPEEDO REPLACEMENT - COMPLY WITH REGS
9,13	95 - 95	4/95-01	IC - CODING/TESTING - SRI INFO - REV
8	96 - 96	4/95-08	FUEL GAUGE DOES NOT REGISTER FULL
8	96 - 96	4/96-01	TACH STICKS - A/C READING GOES TO CELSIUS
4	93 - 93	96/92-03	MIRROR (HEATED OUTSIDE) - CHANGES
11	93 - 93	96/92-04	MIRROR (HEATED OUTSIDE) - CHANGES (V8)
5,9,13	95 - 95	TROUB/94-01D	CRUISE CONTROL SYSTEM TROUBLESHOOTING

Model Legend: 1= All 2= Most 3= 80 4= 90 5= 100 6= 200 7= 90 Quattro 8= A4, A4 Quattro 9=A6, A6 Quattro 10= Cabriolet 11= 4000S, Quattro, V8 12= S4 13= S6

Chassis Electrical, Wiring Harness, Fuses-Circuit Breakers, Wipers, Window Motors

Models	Year	TSB#	Description
4,10	93 - 96	37/96-01	CLICK NOISE - SHIFT LOCK SOLENOID
5Q	92 - 93	38/92-01	WIRING HARNESS (01F A/T) - REPLACEMENT PROCEDURES
9,13	95 - 95	4/95-02	CIGARETTE LIGHTER INOPERATIVE
5	92 - 94	91/94-04	FUSE S-14 BLOWS (TELEPHONE) - MOISTURE
1	Up - 93	92/92-01	WIPER BLADES SMEARING/CHATTERING - PDI REPAIR - REV
1	Up - 96	92/95-01	WIPER BLADES SKIP
1	Up - 95	97/94-01	ELECTRICAL CONTACT ENHANCER - STABILANT 22A
4,10	94 - 94	97/95-01	FUSE S12 BLOWN - NO START - A/T - NEW SWITCH
1	Up - 95	97/94-01	ELECTRICAL CONTACT ENHANCER - STABILANT 22A - REV
4,10	95 - 95	WD/95-01	WIRING DIAGRAM - A/T 01N
4	93 - 93	WD93-04	WIRING DIAGRAM - MAIN
4	93 - 93	WD93-05	WIRING DIAGRAM - AUTOMATIC TRANSMISSION
4	93 - 93	WD93-06	WIRING DIAGRAM - MINI CHECK SYSTEM
4	93 - 93	WD93-07	WIRING DIAGRAM - ABS QUATTRO
4	93 - 93	WD93-08	WIRING DIAGRAM - MANUAL AIR CONDITIONING
4	93 - 93	WD93-09	WIRING DIAGRAM - AUTOMATIC CLIMATE CONTROL
4	93 - 93	WD93-10	WIRING DIAGRAM - POWER SUNROOF
4	93 - 93	WD93-11	WIRING DIAGRAM - POWER DRIVER'S SEAT W/0 MEMORY
4	93 - 93	WD93-12	WIRING DIAGRAM - AM/FM STEREO RADIO W/6 SPEAKERS
4	93 - 93	WD93-13	WIRING DIAGRAM - HEATED FRONT SEATS
4	93 - 93	WD93-14	WIRING DIAGRAM - HEATED DOOR LOCKS
4	93 - 93	WD93-15	WIRING DIAGRAM - DAYTIME RUNNING LIGHTS
10	96 - 96	WD95-02	WIRING DIAGRAM - MAIN WIRING
10	96 - 96	WD95-03	WIRING DIAGRAM - ABS
10	96 - 96	WD95-04	WIRING DIAGRAM - MANUAL A/C WITH A/T
9	96 - 96	WD95-05	WIRING DIAGRAM - AUTOMATIC CLIMATE CONTROL
10	96 - 96	WD95-05	WIRING DIAGRAM - INTEGRAL SAFETY SEAT
10	96 - 96	WD95-06	WIRING DIAGRAM - AUTOMATIC TRANSMISSION
10	96 - 96	WD95-07	WIRING DIAGRAM - AM/FM STEREO RADIO
9,13	96 - 96	WD95-26	WIRING DIAGRAM - POWER SUNROOF

Model Legend: 1= All 2= Most 3= 80 4= 90 5= 100 6= 200 7= 90 Quattro 8= A4, A4 Quattro 9=A6, A6 Quattro 10= Cabriolet 11= 4000S, Quattro 12= S4 13= S6

Auxiliary Equipment, Jacks, Trailer Hitches, Towing

Models	Year	TSB#	Description
5,12	92 - 94	91/94-01	TELEPHONE - ASSEMBLY/REPAIRING/TROUBLESHOOTING
2	4 - 94	91/94-02	TELEPHONE - ASSEMBLY/REPAIRING/TROUBLESHOOTING - V8
5	93 - 94	91/94-05	TELEPHONE INSTALLATION WITH BOSE SOUND SYSTEM
1	Up - 95	91/94-06	HIGH PITCH WHISTLE OVER 55 MPH - REAR PHONE ANTENNA
9,13	95 - 95	91/95-01	CELLULAR PHONE INSTALLATION INSTRUCTIONS
9	96 - 96	91/95-07	PHONE INSTALLATION INSTRUCTIONS
9	95 - 96	91/96-02	CELLULAR TELEPHONE INSTALL INSTRUCTIONS

Model Legend: 1= All 2= Most 3= 80 4= 90 5= 100 6= 200 7= 90 Quattro 8= A4, A4 Quattro 9=A6, A6 Quattro 10= Cabriolet 11= 4000S, Quattro, V8 12= S4 13= S6

Heating, Air Conditioning, Ventilation, Defogger

Models	Year	TSB#	Description
5,12,11,4,7	92 - 93	01/92-01	A/C - OBD CONTROL HEAD DIAG INFO - S/M ADDITION
4,7	Up - 95	01/95-01	A/C VACUUM LEAK - HOSE ROUTING
5	92 - 93	87/92-03	CLIMATE CONTROL HEAD - ERRATIC ILLUMINATION/REPLACE
4	93 - 93	87/93-01	CONDENSATION IN VENTILATION SYSTEM & FOOTWELL
3,4,5,12,11	91 - 93	87/93-02	CONDENSATION IN VENTILATION SYSTEM & FOOTWELL
4	93 - 93	87/93-03	A/C - R-134a SYS SERVICING/REPAIR - S/M SUPPLEMENT
11	93 - 93	87/93-04	A/C - R-134a SYS SERVICING/REPAIR - S/M SUPPLEMENT
5,12	93 - 93	87/93-05	R-134A A/C SYSTEM SERVICE/REPAIR - S/M SUPPLEMENT
4	93 - 93	87/94-01	A/C COMPRESSOR INOP OR INTERMIT - POSS CODE 01270
9,13	95 - 95	87/95-01	CLIMATE CONTROL MALFUNCTIONS - PROCEDURE/TIPS
9,13	95 - 95	87/95-01	CLIMATE CONTROL MALFUNCTIONS - PROCEDURE/TIPS
1	86 - 93	87/95-03	A/C SYSTEM - RETROFITTING R12 TO R134A
5,9,12,13	93 - 96	87/95-04	A/C REFRIGERANT SYS (R134A) - CAPACITY - S/M REV
9	96 - 96	87/96-01	A/C COMPRESSORS - NIPPONDENSO - PROCEDURES - S/M REV
8	96 - 96	87/96-02	A/C COMPRESSORS - NIPPONDENSO - PROCEDURES - S/M REV
8,9	96 - 96	87/96-03	A/C COMP - NIPPONDENSO - IDENT & CODING
8	96 - 96	4/96-01	A/C READING GOES TO CELSIUS WHEN ENG IS SHUT OFF

Model Legend: 1= All 2= Most 3= 80 4= 90 5= 100 6= 200 7= 90 Quattro 8= A4, A4 Quattro 9=A6, A6 Quattro 10= Cabriolet 11= 4000S, Quattro, V8 12= S4 13= S6

Entertainment Devices, Stereo, Radio, Etc.

Models	Year	TSB#	Description
5,12,11	92 - 93	91/93-01	CD CHANGER - BLAUPUNKT - T/S DIFFERENT FROM ALPINE
10	96 - 96	91/95-02	DELTA RADIO - NEW - INSTALL/REPAIR - S/M REV
9	96 - 96	91/95-04	DELTA RADIO - NEW - INSTALL/REPAIR - S/M REV
8	96 - 96	91/95-05	CD PLAYER - INSTALL INSTRUCTIONS
8	96 - 96	91/95-06	CELLULAR TELEPHONE - INSTALL INSTRUCTIONS
8	96 - 96	91/95-08	RADIO - VOLUME GOES TO MAX WHEN ADJUSTED
8	96 - 96	91/95-09	CD CHANGER SKIPS
8	96 - 96	91/95-10	BUZZ SOUND FROM FRONT SPEAKERS
9	96 - 96	91/95-11	PHONE DIFFICULT TO RETRIEVE FROM CRADLE
8	96 - 96	91/95-06	CELLULAR TELEPHONE - INSTALL INSTRUCTIONS - REV

Model Legend: 1= All 2= Most 3= 80 4= 90 5= 100 6= 200 7= 90 Quattro 8= A4, A4 Quattro 9=A6, A6 Quattro 10= Cabriolet 11= 4000S, Quattro, V8 12= S4 13= S6

Seats, Belts, Interior Trim, Carpets, Air Bags

Models	Year	TSB#	Description
10	96 - 96	01/95-05	AIRBAG V REPAIR PROCEDURES - S/M UPDATE
9	96 - 96	01/95-06	AIRBAG V REPAIR PROCEDURES - S/M UPDATE
4	93 - 93	57/92-08	SEAT (REAR) - BACKREST LOCKING SYSTEM PROCEDURES
5	92 - 94	57/94-01	PROTECTIVE PADDING - DOORS - REMOVE/INSTALL PROCED
4,10	94 - 95	57/94-02	PROTECTIVE PADDING -SIDE/DOORS - REMOVE/INSTALL PROC
1	94 - 95	68/94-02	AIR BAG RETAINING CLIP DISCONTINUED
8	96 - 96	68/95-01	REAR ASHTRY LID BINDS
10	96 - 96	69/95-01	AIRBAG - SINGLE PIN CONNECTOR DELETED
9	96 - 96	69/95-02	AIRBAG - SINGLE PIN CONNECTOR DELETED
10	96 - 96	69/95-03	SPORT SEAT - BELT & PYROTECHNIC UNIT R & R - S/M REV
10	96 - 96	69/95-04	SPORT SEAT MICRO SWITCH - R & R PROCED - S/M REV
5,12	93 - 94	70/93-01	ADDITIONAL I/P ATTACHMENT - WITH PASSENGER AIRBAG
4	94 - 94	70/94-01	ARMREST REMOVE/INSTALL PROCEDURE
8	96 - 96	70/95-02	HATSHELF TRIM (BEADING) LOOSE
5	92 - 93	74/92-01	SEATS (FRONT) - SQUEAKING NOISE - PDI INSPECT/REPAIR
4	93 - 93	74/93-02	HEAD RESTRAINT UPHOLSTERY COVERS - R&I - S/M SUPPL
11	4 - 93	74/93-04	ARMREST W/PHONE WON'T HOLD OPEN - NEW RATCHET MECHAN
4	93 - 94	96/93-02	AIRBAG II SYSTEM - ASSEMBLY/DIAGNOSTICS - S/M UPDATE
4,10	93 - 95	96/95-01	AIRBAG - DTC 00588/01025 - PROCEDURE
9	95 - 95	96/95-02	AIRBAG III REPAIR PROCEDURES - S/M SUPP
4	95 - 95	96/95-03	AIRBAG III REPAIR PROCEDURES - S/M SUPP

Model Legend: 1= All 2= Most 3= 80 4= 90 5= 100 6= 200 7= 90 Quattro 8= A4, A4 Quattro 9=A6, A6 Quattro 10= Cabriolet 11= 4000S, Quattro, V8 12= S4 13= S6

Glass, Doors, Hood, Decklid, Tailgate, Liftgate, Locks

Models	Year	TSB#	Description
8,9	96 - 96	28/95-02	INTERIOR PROFILE KEYS AND LOCKS - REPLACEMENT
8,9	96 - 96	28/95-02	SIDEWINDER KEYS AND LOCKS - INFO
5	92 - 93	55/91-08	DECK LID (REAR) - DIFFICULT TO CLOSE - PDI REPAIR
5	92 - 93	57/92-02	KEY CODE LOCATION - SERVICE INFORMATION
5	92 - 93	57/92-03	LOCKING SYS/INFRARED - OPERATION/TROUBLESHOOTING REV
5	92 - 93	57/92-05	DOOR - DIFFICULT TO CLOSE - PDI CHECKING/ADJUSTING
4	93 - 93	57/92-06	DOOR - DIFFICULT TO CLOSE - PDI CHECKING/ADJUSTING
5	92 - 93	57/92-07	DOOR WINDOW - WIND NOISE - PDI CHECK/WINDOW SEALS
4	93 - 93	57/92-08	LOCKS - CENTRAL LOCKING SYSTEM/LOCATIONS & FUNCTIONS
5,12,11,4	92 - 93	57/92-09	LOCKS - CENTRAL LOCKING SYSTEM CHANGES
9,13	95 - 95	57/94-10	SAFETY FEATURE FOR UNLOCKING DOORS FROM INSIDE
8	96 - 96	57/96-01	TRUNK LOCK CYLINDER REPLACE PROCEDURE
8	96 - 96	57/96-02	REPLACEMENT KEYS
9	96 - 96	28/95-03	SIDEWINDER KEYS AND LOCKS - INFO - REV
5	92 - 93	60/93-02	PRE-SELECT SUNROOF DOES NOT STOP IN CORRECT POSITION
10	94 - 95	61/94-01	CONVERTIBLE TOP LATCH RELEASES INTERMITTENTLY
10	94 - 95	64/94-02	WINDOW RATTLE WITH WINDOWS UP & CONV TOP DOWN
4	93 - 93	70/92-04	WINDOW (REAR) - LOCK-OUT SWITCH R&R - PDI REPAIR
4	93 - 93	70/92-07	DOOR (FRONT) - REMOVING/INSTALLING TRIM PANEL
2	Up - 93	94/93-01	IGNITION LOCK CYLINDER - REMOVE/INSTALL - S/M REV
4,7	93 - 93	96/92-02	LOCKING (INFRARED) - COMPONENTS & TROUBLESHOOTING

Model Legend: 1= All 2= Most 3= 80 4= 90 5= 100 6= 200 7= 90 Quattro 8= A4, A4 Quattro 9=A6, A6 Quattro 10= Cabriolet 11= 4000S, Quattro, V8 12= S4 13= S6

Finishes, Body Structure, Frame, Bumpers

Models	Year	TSB#	Description
4,7	93 - 93	40/93-01	SUBFRAME MOUNTING BOLT/WASHER NEW - SPECS & TORQUE
1	Up - 94	50/93-02	INDUSTRIAL FALLOUT - NEUTRALIZATION & REMOVAL
1	Up - 94	50/93-02	INDUSTRIAL FALLOUT - NEUTRALIZATION & REMOVAL - REV
10	94 - 94	50/94-04	BODY DIMENSION CHARTS
1	95 - 95	50/94-07	PAINT - REDUCING SWIRL MARKS & FINE SCRATCHES
5	92 - 93	68/92-05	B-PILLAR TRIM - NOISY WHILE DRIVING - PDI REPAIR

Model Legend: 1= All 2= Most 3= 80 4= 90 5= 100 6= 200 7= 90 Quattro 8= A4, A4 Quattro 9=A6, A6 Quattro 10= Cabriolet 11= 4000S, Quattro, V8 12= S4 13= S6

BMW

Lighting, Horns, Turn Signals, Steering Column

Models	Year	TSB#	Description
4	93 - 95	63 01 94	10-WATT REAR LIGHT BULB FAILURE
1	Up - 95	61 03 94	BRAKE LIGHT SWITCH - NEW - APPLICATIONS
3,4,6,7	Up - 93	63 03 88	EMERGENCY FLASHERS - SELF-ACTIVATE - DIAG/CORRECTION
3	92 - 94	63 04 93	FOGLIGHT LENSES CRACKED - POSSIBLE THERMAL SHOCK
3	94 - 94	63 03 93	GLOVEBOX LIGHT STAYS ON - NEW PART - CONVERTIBLE
6	95 - 96	63 01 95	HEADLIGHT AIMING PROCEDURE
6	93 - 93	63 01 92	HEADLIGHTS (LOW BEAM) - NEW GAS DISCHARGE XENON SYS
7	Up - 93	63 04 92	HEADLIGHTS - AIMING PROBLEM - CORRECTION
3	92 - 93	04 35 92	HEADLIGHTS - SPECIAL ADAPTERS NEEDED FOR ALIGNMENT
3	Up - 93	63 01 93	LIGHT/ENG COMPART (UNDERHOOD) - INFORMATION
1	Up - 94	61 06 94	MALFUNCTION OF TURN SIGNAL & HAZARD FLASHER SYS
4	93 - 94	63 05 93	THIRD BRAKELIGHT BULB AVAILABLE - TOURING
6	93 - 93	63 02 93	XENON HEADLIGHTS - COLOR/FUSES & WARRANTY INFO

Model Legend: 1= All 2= Most 3= 3 Series 4= 5 Series 5= 6 Series 6= 7 Series 7= 8 Series

Instruments, Dash Cluster, Warning Lights, Mirrors

Models	Year	TSB#	Description
3	94 - 95	65 02 95	CRUISE CONTROL SET SPEED DROPS SLIGHTLY
3	Up - 93	62 12 91 DEC`92	FUEL GAUGE APPEARS TO BE STUCK/LEVEL SENDER STICKING
3	92 - 95	62 01 93 OCT`93	FUEL GAUGE DISPLAYS INACCURATE FUEL TANK LEVEL
3	92 - 94	62 01 93 JUN`93	GAUGE (FUEL) - INACCURATE - DIAGNOSIS & REPAIR
3	92 - 93	62 01 93	GAUGE (FUEL) - INACCURATE OR DROPS DURING CRANKING
3	92 - 93	62 01 93 FEB`93	GAUGE (FUEL) - INACCURATE OR DROPS DURING CRANKING
3	Up - 93	62 12 91 MAR`93	GAUGE/FUEL - APPEARS TO BE STUCK AT 1/2 - DIAG - REV
3	Up - 93	62 12 91 APR`93	GAUGE/FUEL - APPEARS TO BE STUCK AT 1/2 - DIAG - REV
3	92 - 93	62 01 93 APR`93	GAUGE/FUEL - INACCURATE/DROPS DURING CRANKING - REV
6,4	Up - 93	62 06 92	IC - ERRONEOUS CHECK CONTROL FAULT DISPLAYS - REPAIR
3	92 - 93	62 03 93	IC - SEAT BELT CHIME ACTIVATING IRREGULARITIES/INFO
3	93 - 93	62 07 92	IC - SOFTWEAR & ACOUSTIC TRANSMITTER IMPROVEMENTS
3	92 - 94	62 01 94	INSTRUMENT CLUSTER HANDLING INSTRUCTIONS
4,6	93 - 93	62 02 93 FEB`93	IP - "TRANS PROGRAM" MESSAGE - REV - SERVICE ACTION
4,6	93 - 93	62 02 93	IP DISPLAY - "TRANS PROGRAM" MESSAGE DEFINED/IGN OFF
3	92 - 93	63 02 92 FEB`92	ODOMETER DISPLAY/MID BACKLIGHT FLICKER - PROC - REV
3	Up - 95	62 02 95	REPLACEMENT IC - TRANSFER OF ODOMETER READING
4,6	93 - 93	62 03 94	TURN SIGNAL INDICATOR (ACOUSTIC RELAY) FAILURE

Model Legend: 1= All 2= Most 3= 3 Series 4= 5 Series 5= 6 Series 6= 7 Series 7= 8 Series

Chassis Electrical, Wiring Harness, Fuses-Circuit Breakers, Wipers, Window Motors

Models	Year	TSB#	Description
3	92 - 95	12 03 94 JUL`94	B+ LEAD IN ENG COMP - M50 W/ ASC +T - SERVICE ACTION
6	88 - 95	12 04 94 AUG`94	B+ LEAD RUBS FUEL RETURN LINE - M60 - SERVICE ACTION
3,4,6	88 - 94	61 04 93	CONSOLE WINDOW SWITCHES - LIQUID (COFFEE) CONTAMINAT
1	Up - 93	04 31 92	ELEC CONNECTOR KIT #4 - REPLACEMENT PARTS AVAILABLE
3	92 - 94	PARTS/II-60	MAIN WIRE HARNESS (E36) IDENT & REPLACEMENT GUIDE
6	93 - 93	61 07 93	POWER DISTRIBUTION BOX LABEL INCORRECT - 750iL
4	93 - 94	61 04 94	REPEAT FAILURE OF FUSE #20 - PARKED CAR VENTING
4	90 - 94	61 05 94	SQUEAK - WINDSHIELD WIPER CONSOLE
3	92 - 93	51 03 93	WINDOW REGULATOR & MOTOR - REPLACEMENT INFORMATION
3,4,6,7	Up - 93	61 08 89 JUN`92	WINDSHIELD WIPER CHATTER - SPECIAL TOOL MODIFIED
1	Up - 95	61 02 94	WIPERS NOISY OR DO NOT CLEAN PROPERLY - CLEANER

Model Legend: 1= All 2= Most 3= 3 Series 4= 5 Series 5= 6 Series 6= 7 Series 7= 8 Series

Auxiliary Equipment, Jacks, Trailer Hitches, Towing

Models	Year	TSB#	Description
1	Up - 94	65 07 93	AFTERMARKET PORTABLE PHONES - INTERFERENCE
7	94 - 94	PARTS/IV-83	CELLULAR PHONE & CD CHANGER INSTALLATION PROCEDURES
3	92 - 93	65 23 92	CELLULAR PHONE - FM RADIO INTERFERENCE - CORRECTION
1	Up - 93	65 01 93	CELLULAR PHONE - ANTENNA MAST WHISTLE - CORRECTION
3,4,6	92 - 94	02 05 93 MAR`93	CELLULAR PHONE/MOTOROLA - INSTALLATION KITS
2	92 - 94	02 05 93 JUN`93	CELLULAR PHONE/MOTOROLA - INSTALLATION KITS
2	92 - 95	02 05 93 NOV`93	CELLULAR PHONE/MOTOROLA - INSTALLATION KITS - UPDATE
1	94 - 95	65 04 94	CMT 2000VR PHONE NUMBER MEMORY #60 TO #99
6	95 - 95	65 05 94	CMT 3000VR CELLULAR TELEPHONE HANDSET DISPLAY
6	95 - 96	71 01 95	JACK REPLACEMENT - CAMPAIGN
3	92 - 94	51 04 94	M SPOILER W/INTEGRAL BRAKE LIGHT - INSTALL REV
6	95 - 95	66 02 95	MISSING UNIVERSAL GARAGE DOOR OPENER (UGDO)

Model Legend: 1= All 2= Most 3= 3 Series 4= 5 Series 5= 6 Series 6= 7 Series 7= 8 Series

Heating, Air Conditioning, Ventilation, Defogger

Models	Year	TSB#	Description
3	93 - 93	64 08 93	A/C (R134a) - WHISTLES/HOWLS WHEN COMP CYCLES - FIX
1	Up - 97	00 20 91 JUN`91	A/C - CHLOROFLUOROCARBONS (CFC) LEGISLATION
2	92 - 93	64 20 92	A/C - MICROFILTER INTRODUCTIONS - INFORMATION
3	Up - 93	64 15 92	A/C - NEW COMPRESSOR PART - REPLACEMENT INFORMATION
1	Up - 94	64 02 92 FEB`92	A/C - NON-APPROVED REFRIGERANTS
1	Up - 93	64 01 90 APR`90	A/C - REFRIGERANT FILL CAPACITY & INFO AVAILABLE
7,6,4	92 - 93	64 17 92	A/C - SMALL FLAKES FROM VENTS - COATED EVAPORATOR
7,6,4,3	92 - 93	64 17 92 MAR`93	A/C - SMALL FLAKES OUT VENTS/COATED EVAPORATOR - REV
3	Up - 93	64 06 93	A/C - SMALL FLAKES/MUSTY ODOR FROM VENTS - SERV PROC
1	Up - 93	64 13 92	A/C - SYS PRESSURES R-12/R-13a - SUPERCEDES ALL INFO
6	Up - 93	64 03 93	A/C - WHISTLE WHEN SUNROOF OR WINDOW SLIGHTLY OPEN
1	Up - 94	PARTS/II-59	A/C COMPRESSER - REDISIGNED BUSHING ASSEMBLY
3,4	Up - 93	64 01 93	A/C DRIVE BELT - CHECKING TENSION W/SPECIAL TOOL
1	Up - 95	64 11 93	A/C OIL CHARGE PROCEDURE - R-12 & R-134A
3	95 - 95	64 02 95	AIR DISTRIBUTION KNOB BINDS - SERVICE ACTION
1	Up - 94	64 02 94	AVAILABILITY OF R-12 A/C COMPRESSORS
4	92 - 93	64 12 93	DROP IN HEATER TEMP DURING EXTENDED DRIVE - NEW FAN
3,4,6,7	Up - 93	64 05 93	HEATER - DELAYED OUTPUT - CORRECTION & PARTS INFO
3	92 - 93	64 16 92	HEATER HOSE - WORN - RUBS INTAKE MANIFOLD - CAMPAIGN
3,4,6	93 - 94	64 01 94	HONK/HOOT - A/C EXPANSION VALVE
3	Up - 93	64 04 93	IHKR - RECIRC CONTROL STUCK ON OR OFF - CORRECTION
3	Up - 93	64 02 93	IHKR - SUDDEN LOSS OF HEAT OR AIRFLOW - CORRECTION
3	Up - 93	64 02 93 APR`93	IHKR - SUDDEN LOSS OF HEAT OR AIRFLOW - CORRECTION
3,4	94 - 95	64 03 94	IHKR FAULT CODES 34/61
3	94 - 94	64 05 94	RATCHET/CLICK NOISE FROM A/C AFTER TURN OFF - M42
3	92 - 93	64 10 93 NOV`93	REAR WINDOW DEFOGGER INOP/POOR RADIO - REVISED
3	92 - 93	64 10 93	REAR WINDOW DEFOGGER INOP/POOR RADIO - WIRES REVERSE
1	Up - 94	64 09 93	REPLACEMENT OF RECEIVER/DRYER - WHEN TO DO IT
2	Up - 94	64 01 95	RETROFITTING R12 A/C WITH R134A

Model Legend: 1= All 2= Most 3= 3 Series 4= 5 Series 5= 6 Series 6= 7 Series 7= 8 Series

21 TECHNICAL SERVICE BULLETINS

Electronic Devices, Computers, PROMS, Sensors

Models	Year	TSB#	Description
3,7	91 - 94	34 03 93	ABS SENSOR LEAD FAULTS - REPAIR PROCEDURE
3,7	91 - 94	34 03 93 SEP`93	ABS SENSOR LEAD FAULTS - REPAIR PROCEDURE
3	94 - 94	65 02 94	ALPINE ALARM/ZKE IV SYSTEM INCOMPATIBILITY - PROCED
1	93 - 94	65 01 94	ANTI-THEFT SYS (DWA & ALPINE) CHANGES
1	94 - 94	61 01 94 MAR`94	ANTI-THEFT SYS - CHANGES - REVISED
1	94 - 94	61 01 94	ANTI-THEFT SYS - IMMOBILIZING CIRCUIT ADDED
3,4,7	94 - 94	72 01 93	CENTRAL ACTIVATION MODULE EXPLAINED - ZAE-SRS
3,4,7	94 - 95	72 01 93 NOV`93	CENTRAL AIRBAG ELECT MODULE (SRS) - NEW - INFO
4	89 - 95	62 05 94	CHECK CONTROL MESSAGE "TRANS PROGRAM" - PROCEDURE
1	Up - 93	IDC015/92-01	CONTROL MODULES - UPDATED BMW SERV TESTER SOFTWARE
4	93 - 93	13 04 94	COOLANT TEMP SENSOR (NTC) REPLACE - SERVICE ACTION
3	92 - 95	12 05 94	DIAGN PLUG GROUND PIN 19 DAMAGE - SERVICE ACTION
4	95 - 95	12 06 95	DME EPROM UPDATE - M60 W/ M/T
6,7	Up - 95	12 04 95	DME M5.2 CONTROL MODULES - SERVICE ACTION
4,6,7	Up - 95	12 09 94R	DRIVEABILITY - EPROM UPDATE - M60 - SERVICE ACTION
4,7	94 - 95	65 03 95	DWA IV FAILS TO ARM WHEN DOUBLE-LOCKING
3,4	93 - 93	12 01 93	ECM - FAULT CODE 13 - SOFTWARE DISK X ERROR DEFINED
3	92 - 93	SUPP: 12 08 91	ECM - SERVICE ACTIONS 65/70/74 - PROCEDURES
4	93 - 94	12 02 93	ECM RELAY SELF-ACTIVATION - M60 W/ AGS - NEW ECM
1	Up - 93	13 02 90 MAR`91	ECM/ CROSS REF LIST OF ALL KNOWN COMPATIBLE US CODES
1	95 - 96	61 02 95	ELECTRONIC IMMOBILIZATION SYS (EWS II) PARTS INFO
1	95 - 95	61 01 95	ELECTRONIC IMMOBILIZATION SYSTEMS (EWS II) CAUTION
1	95 - 95	61 01 95	ELECTRONIC IMMOBILIZATION SYSTEMS (EWS II) INFO
6,7	Up - 93	12 04 92	EML - NEW PROC TO DETERMINE "TYPE" OF FAULT CODE 09
4,6	88 - 93	62 06 92 DEC`92	ERRONEOUS CHECK CONTROL FAULT DISPLAYS - REPLACE LKM
3	92 - 93	62 01 93 APR`93	FAULT CODE 10 - INFO - FUEL GAUGE DROP DURING CRANK
4,6	93 - 93	64 07 93	FAULT CODE 38/AUXILIARY WATER PUMP - INFORMATION
3	Up - 93	64 04 93	FAULT CODE 52/IHKR CONTROL MODULE - CORRECTION
3	95 - 95	62 04 94	FAULT CODES 5/13 IN BC V
3	92 - 95	34 01 94	IMPAIRED DIAGNOSTICS - IMPLAUSIBLE FAULT CODES
3	92 - 95	34 01 94 SEP`94	IMPAIRED DIAGNOSTICS - IMPLAUSIBLE FAULT CODES - REV
4	91 - 93	13 04 94 AUG`94	INCORRECT COOLANT TEMP SENSORS - M5 - SERVICE ACTION
3,4	89 - 94	13 03 93 JUL`93	MIL ON - ECM CODE 12/73/105
3,4	93 - 95	12 01 94 JAN`94	MIL ON - O2 SENSOR MALFUNCTION - EPROM - M50 - REV
3,4	93 - 95	12 07 94 JAN`95	MIL ON - O2 SENSOR MALFUNCTION - EPROM - M50 - REV2
3,4	93 - 93	12 01 94	MIL ON - O2 SENSOR MALFUNCTION - EPROM UPDATE - M50
3,4,6,7	Up - 93	63 03 88	MODULE/CRASH CONTROL - R&R - FLASHERS SELF-ACTIVATE
6,4	Up - 93	62 06 92	MODULE/LAMP CONTROL - REPLACE - FALSE IC DISPLAYS
3	Up - 93	64 02 93 APR`93	NEW CONTROL MODULE SOFTWARE (IHKR)- SUDDEN HEAT LOSS

Model Legend: 1= All 2= Most 3= 3 Series 4= 5 Series 5= 6 Series 6= 7 Series 7= 8 Series

Entertainment Devices, Stereo, Radio, Etc.

Models	Year	TSB#	Description
4	94 - 94	65 10 93	ACCESSORY CD CHANGER INOP - CHECK RADIO SOFTWARE
4	Up - 93	65 22 92	CD CHANGER - DIFFICULT TO INSTALL (TOURING VEHICLES)
7	94 - 94	65 14 93	CD CHANGER/PHONE INSTALLATION - 840CI
7	94 - 94	65 14 93 FEB`94	CD CHANGER/PHONE INSTALLATION - 840CI - REVISED
1	Up - 93	65 09 86 NOV`86	CD PLAYER - SHIPPING PROCEDURES - PRECAUTIONS
1	Up - 94	65 09 93	COMPACT DISCS - CARE AND CLEANING
3	92 - 94	65 08 93	CONVERTIBLE CD CHANGER INOP - REPAIR PROCEDURES
2	92 - 93	65 04 95	CRACKLE/STATIC IN DIVERSITY ANTENNA RADIO RECEPTION
1	Up - 95	65 03 94	HANDS FREE AUDIO - TELEPHONE DUAL COIL SPEAKER DIAGN
3	92 - 93	64 10 93 NOV`93	POOR RADIO RECEPTION/REAR DEFOG INOP - REVISED
3	92 - 93	64 10 93	POOR RADIO RECEPTION/REAR DEFOG INOP - WIRES REVERSE
3	93 - 93	65 02 93	RADIO - THEFT CODE NOT RETAINED IN COLD WEATHER
3	93 - 94	65 12 93	RADIO INCOMPATIBLE WITH CD PREWIRING - EARLY 318iA
1	Up - 93	65 05 93	RADIO/BAVARIA BUSINESS RDS - USA COMPATIBILITY INFO
3	92 - 93	65 23 92	RADIO/FM - INTERFERENCE FROM CELLULAR PHONE - REPAIR
1	93 - 93	65 21 92	RADIOS - NEW WITH SPECIAL FEATURES - IDENTIFICATION
1	93 - 93	65 21 92 DEC`92	RADIOS - NEW/SPECIAL FEATURES - IDENTIFICATION - REV
1	93 - 94	65 21 92 FEB`93	RADIOS - NEW/SPECIAL FEATURES - IDENTIFICATION - REV
1	93 - 94	65 21 92 MAY`93	RADIOS - NEW/SPECIAL FEATURES - IDENTIFICATION - REV
4	93 - 94	65 13 93	RATTLE/BUZZ NOISE - REAR HEADLINER BASS SPEAKERS

Model Legend: 1= All 2= Most 3= 3 Series 4= 5 Series 5= 6 Series 6= 7 Series 7= 8 Series

Seats, Belts, Interior Trim, Carpets, Air Bags

Models	Year	TSB#	Description
1	Up - 95	72 01 95	AIRBAG - SERVICING ZAE SYSTEMS - TIPS/CAUTIONS
3,4,7	94 - 95	72 01 93 NOV`93	CENTRAL AIRBAG ELECT MODULE (SRS) - NEW - INFO
3	94 - 95	51 08 94	DASH WARPING ABOVE CENTER VENT
3	94 - 95	51 08 94 MAR'94	DASH WARPING ABOVE CENTER VENT - REV
4	95 - 95	52 04 95	FRONT SEAT BACKREST PIVOT - SERVICE ACTION
3	92 - 96	54 04 95	HEADLINER PLASTIC PANELS BREAK - CONVERTIBLE
3	92 - 93	51 01 92	HEADLINER SAGS - HIGH SPEEDS W/SUNROOF OPEN - REPAIR
3	94 - 94	72 01 94	MODIFICATION TO FRONT RESTRAINT SYSTEM - 325i/A 4-DR
4,6	94 - 95	52 02 95	POWER SEAT & MIRROR MEMORY MALFUNCTIONS
3	92 - 93	72 04 91 NOV`91	PYROTECHNIC SAFETY BELT TENSIONER FOR CONVERTIBLE
3	Up - 95	52 03 95	SEAT BACK RELEASE DIFFICULT TO OPERATE
3	92 - 93	52 01 93	SEAT CUSHION - CREAK OR SQUEAK NOISES - CORRECTION
4	Up - 93	52 04 92	SEATS/ACCESSORY - DO NOT MOUNT IN CARGO AREA (525i)
1	Up - 95	72 01 88 MAY`88	SRS (AIR BAG) - COLLISION REPAIR GUIDELINES
1	Up - 93	32 01 91	SRS - CHECKING AIRBAG SYSTEMS - INFORMATION REVISED
67	93 - 93	72 01 92	SRS - NOW FOR FRONT PASSENGER'S & DRIVER'S SIDE
3,4,7	94 - 94	72 01 93	ZAE-SRS - CENTRAL ACTIVATION MODULE EXPLAINED

Model Legend: 1= All 2= Most 3= 3 Series 4= 5 Series 5= 6 Series 6= 7 Series 7= 8 Series

Glass, Doors, Hood, Decklid, Tailgate, Liftgate, Locks

Models	Year	TSB#	Description
3	91 - 93	51 09 93	CENTRAL LOCK SYS FAILS - INTERNAL FUSE - REPL MODULE
3	84 - 93	51 09 93 SEP`93	CENTRAL LOCKING SYSTEM FAILURE - REPAIR PROCEDURE
3,4	89 - 94	51 01 94	CENTRAL LOCKING SYSTEM T/S TIPS
3,4	92 - 93	51 11 93	CENTRAL LOCKING SYSTEM TROUBLESHOOTING
3	92 - 95	54 06 95	CONV STORAGE LID PROTECTIVE CAP - SERVICE ACTION
3	Up - 95	54 01 93 MAR'93	CONV TOP MAT'L PULLS AWAY FROM REAR WINDOW TRIM -REV
3	92 - 95	54 02 95	CONV TOP MAT'L STICKS OUT NEAR C-PILLAR
3	92 - 93	54 07 92	CONVERTIBLE - INSTALLATION OF AMCO BRAND SOFT TOPS
3	92 - 93	54 04 93	CONVERTIBLE E-M TOP - INITIALIZATION PROCEDURE
3	92 - 95	54 04 94	CONVERTIBLE HEADLINER SAGS
3	93 - 94	54 01 93A	CONVERTIBLE TOP (E36) - IMPROVEMENTS/OPERATION/INFO
3	Up - 93	54 06 92	CONVERTIBLE TOP - DAMAGE FROM WIRE TIE - CORRECTION
3	92 - 94	41 02 94	CONVERTIBLE TOP STORAGE COMP LID COVERING
3	92 - 94	41 02 94 AUG`94	CONVERTIBLE TOP STORAGE COMP LID COVERING - REVISED
3	Up - 94	54 02 93A	CONVERTIBLE/E36 - ROLLOVER PROTECTIONS SYS AVAILABLE
3	92 - 94	51 08 93	DAMAGED DOOR PANEL UPPER RETAINING HOLES - INST CLIP
3	92 - 94	51 14 93	DOOR WINDOWS DO NOT DROP WHEN OPENING DOOR - SWITCH
3	92 - 94	54 03 94	E-M TOP FAILS WHEN LOWERING - REV PARTS
3	92 - 96	54 05 95	HEADLINER GETS CAUGHT BY STORAGE LID COVER - CONVERT
3	92 - 93	51 15 93	INCREASED WINDOW THICKNESS AND GUIDES - PROCEDURE
3	92 - 94	51 10 93	INNER DOOR PANEL LOOSE/NOISE - NEW CLIPS & RING -4DR
3	92 - 93	54 01 94	RATTLE - SUNROOF WIND DEFLECTOR
3	Up - 93	54 07 93	REPLACE TOP ACTUATING ROD - CONVERTIBLE - CAMPAIGN
3	Up - 93	54 06 93	REPLACE TOP STORAGE LID CABLES - CONV - CAMPAIGN
3	Up - 93	54 01 93	SOFT TOP - MATERIAL PULLS AWAY FROM REAR WINDOW TRIM
3	91 - 94	54 05 93	SOFT TOP MATERIAL RUBS ON TENSION SPRING MOUNT -CONV
3	92 - 94	54 02 94	SQUEAKING FROM CONVERTIBLE TOP SEALS
4	Up - 93	54 05 92	SUNROOF (DUAL LID) - UPDATED REPAIR INSTRUCTIONS
3	92 - 96	54 03 95	SUNROOF CASSETTE INSTALL INSTRUCTIONS
3,6	96 - 96	54 08 95	SUNROOF LID MOUNTING REVISED
3,4	92 - 94	51 06 93	TRUNK LID ACTUATOR/LINKAGE ROD - PROPER INSTALL -REV
3,4	Up - 93	51 06 93	TRUNK LID ACTUATOR/LINKAGE ROD - PROPER INSTALLATION
3	92 - 94	51 02 94	WATER SPILLS INTO TRUNK - CONVERTIBLE
4	93 - 93	51 10 92	WINDOW (REAR HATCH) - ATTACHMENT METHOD MODIFIED
3	94 - 94	51 05 94	WINDOW GLASS DROPS SLIGHTLY AFTER CLOSING - 4-DR
3	92 - 93	51 03 92	WINDOW STOP ADDED - LIMITS UPWARD TRAVEL/CONVERTIBLE
3	Up - 93	51 14 91 JAN`93	WINDOW/ONE-TOUCH OPERATION - MALFUNCTIONS - 2ND REV
3	92 - 94	51 14 91 APR`93	WINDOW/ONE-TOUCH OPERATION - MALFUNCTIONS - 3RD REV
3	92 - 93	51 05 92	WINDOW/ONE-TOUCH OPERATION - MALFUNCTIONS - REVISED
3	Up - 93	54 02 93	WINDOW/REAR - MARKED BY LINKAGE COVER (EM TOP) - FIX
3	Up - 93	51 02 93	WINDOW/REAR QUARTER - SEALS TEAR AT THE BOTTOM - FIX

Model Legend: 1= All 2= Most 3= 3 Series 4= 5 Series 5= 6 Series 6= 7 Series 7= 8 Series

Finishes, Body Structure, Frame, Bumpers

Models	Year	TSB#	Description
1	Up - 93	99 02 92	PAINT - COLOR PROGRAM AND FORMULA OVERVIEW
1	Up - 93	99 01 88	PAINT DAMAGE - ACID RAIN/AIRBORNE CONTAMINANTS - FIX
1	94 - 94	99 02 94	PAINT FORMULAS - 1994 MODELS
1	93 - 94	99 03 94	PAINT SEALER
3,6	92 - 96	99 01 95	PAINTING PROCEDURE FOR ALUMINUM COMPONENTS
4,6	88 - 94	51 07 93	RAIN GUTTER MOLDING TRIM TAPE DEFORMED - REMOVE INST
3	92 - 93	51 05 93	REAR WING KIT - INSTALLATION INSTRUCTIONS AVAILABLE
1	Up - 95	41 01 95	STRUCTURAL ADHESIVES - RECOMMENDATIONS
4,6	94 - 94	99 04 94	TINTED PRIMER-FILLERS IN ENG COMPARTMENT & TRUNK
3	92 - 94	41 04 94	WATER LEAK INTO DRIVER'S SIDE FOOTWELL

Model Legend: 1= All 2= Most 3= 3 Series 4= 5 Series 5= 6 Series 6= 7 Series 7= 8 Series

HONDA

Lighting, Horns, Turn Signals, Steering Column

Models	Year	TSB#	Description
5	93 - 94	94-033	BURNED OUT AUXILIARY LIGHT BULBS
3	94 - 94	AHM/A80	GLOVE BOX LIGHT STAYS ON - ADJUST STRIKER
7	92 - 93	93-024	LIGHT/CEILING - INTERMITTENT OR NO WORK - SERV PROC
1	92 - 95	95-024	STEER COLUMN - SRS - CABLE REEL REMOVAL CAUTIONS
3	92 - 93	AHM/A14	STEER COLUMN SQUEAK WHEN TURNING - CAUSE/CORRECTION
3	92 - 93	AHM/A15	STEER COLUMN SQUEAK WHEN TURNING - CAUSE/CORRECTION
3	94 - 94	AHM/A71	TRUNK OPEN WARN LIGHT STAYS ON - ADJUST TRUNK LID
3	90 - 94	91-008 MAR`91	TURN SIGNAL - BUZZ NOISE DURING LEVER OPERATION -REV

Model Legend: 1= All 2= Most 3= Accord 4= Civic 5= del Sol 6= Odyssey 7= Prelude 8= Passport

Instruments, Dash Cluster, Warning Lights, Mirrors

Models	Year	TSB#	Description
6	95 - 95	95-010	BUZZ NOISE IN IP NEAR SPEEDOMETER
4	92 - 93	94-031	BUZZ/RATTLE/KNOCK NOISE IN DASH ABOVE 4000 RPM
4	92 - 94	SN08/94	CHIRP NOISE FROM SIDE(S) OF DASHBOARD - PROCEDURE
4	91 - 93	SN02/93	CRUISE CONTROL ACTUATOR ASSY TEST - S/M CORRECTION
4	92 - 95	92-050 DEC'92	DASH - WATER LEAK FROM UNDER CORNER - DIAG - REV
4	92 - 95	92-050R MAY'95	DASH - WATER LEAK FROM UNDER CORNER - DIAG - REV
4	92 - 93	92-050	DASH - WATER LEAK FROM UNDER CORNER - DIAG/REPAIR
6	95 - 95	95-027	DASHBOARD COMPARTMENT LATCH COVER COMES OFF
3	94 - 94	94-021	ENGINE TEMP GAUGE ALWAYS READS HOT - REVISED GAUGE
3	94 - 95	95-013	FUEL GAUGE DOES NOT READ FULL - SENDING UNIT
6	95 - 95	95-019	FUEL GAUGE READS LESS THAN FULL
8	94 - 94	SN06/94	ILLUMINATION CONTROLLER TEST - S/M CORRECTION
3,4,7	90 - 93	92-022 JAN`93	MIRROR (DOOR) - WIND NOISE CORRECTION - 2ND REVISION
3,4,7	90 - 93	92-022 JUL`92	MIRROR (DOOR) - WIND NOISE CORRECTION - REVISION
7,4	92 - 93	92-047	MIRROR/REARVIEW - REINSTALLATION PROCEDURE
7,4	92 - 95	92-047 NOV`92	MIRROR/REARVIEW - REINSTALLATION PROCEDURE - REV
3	93 - 93	93-017	MIRRORS (PWR DOOR) - CAUSE POPPING NOISE IN RADIO
3	94 - 94	93-049	NOISE/CREAK IN INSTRUMENT PANEL - PROCEDURES
3	94 - 95	93-049 NOV`93	NOISE/CREAK IN INSTRUMENT PANEL - PROCEDURES - REV
3,7	90 - 94	93-046	REARVIEW MIRROR TENSION ADJUSTMENT PROCEDURE
3	94 - 94	93-045 OCT`93	TRUNK OPEN INDICATOR LIGHT STAYS ON - PROC REVISED
3	94 - 94	93-045	TRUNK OPEN INDICATOR LIGHT STAYS ON - PROCEDURE

Model Legend: 1= All 2= Most 3= Accord 4= Civic 5= del Sol 6= Odyssey 7= Prelude 8= Passport

Chassis Electrical, Wiring Harness, Fuses-Circuit Breakers, Wipers, Window Motors

Models	Year	TSB#	Description
1	Up - 95	95-023	CONNECTOR TERMINAL REPLACEMENT - KIT/INSTRUCTIONS
4	94 - 94	SN08/94	CREAK NOISE FROM POWER WINDOW - REMOVE SPRING
3	94 - 94	AHM/A74	DEALER CHECK OF POWER WINDOW OPERATION & REGULATOR
4	92 - 93	SN02/94	DRIVER DOOR HARNESS CONNECTOR MISLABELED -CORRECTION
3	94 - 94	93-043	ELECT SEAT HEIGHT ADJ INOP - MOVE HARNESS
1	Up - 96	SN05/95	ISOLATING SHORTS WITH IN-LINE FUSES - TIPS
6	95 - 95	95-012	LOOSE CIGARETTE LIGHTER SOCKET - PDI
4	92 - 93	93-051	NOISE/INOP DOOR LOCKS/SPEAKER - HARNESS CONNECTOR
3	94 - 94	AHM/A94	POWER SEAT WIRING COUPLER DISCONNECT - REROUTE WIRE
4	95 - 95	SN12/94	POWER WINDOWS/MIRRORS INOP - CONNECTOR

Model Legend: 1= All 2= Most 3= Accord 4= Civic 5= del Sol 6= Odyssey 7= Prelude 8= Passport

Auxiliary Equipment, Jacks, Trailer Hitches, Towing

Models	Year	TSB#	Description
3,7,6	94 - 95	95-008	CELLULAR PHONE - OUT-OF-WARRANTY REPAIR
3,7	94 - 94	SN02/94	IN-DASH CELLULAR PHONE - HANDS FREE OPS EXPLAINED
3,7	Up - 95	94-014	IN-DASH CELLULAR PHONE T/S - ETM SUPPLEMENT
3,7	94 - 94	SN07/94	PHONE PROGRAMMING - BE QUICK - TIPS
1	Up - 95	SN05/95	REPLACING DAMAGED IN-DASH PHONE LENSES
6	95 - 95	AHM/A50	ROOF RACK BASKET & SKI ATTACHMENTS - REV KIT
4	94 - 95	AHM/A40	SPOILER INSTALL - WIRING CAUTION

Model Legend: 1= All 2= Most 3= Accord 4= Civic 5= del Sol 6= Odyssey 7= Prelude 8= Passport

Heating, Air Conditioning, Ventilation, Defogger

Models	Year	TSB#	Description
4	Up - 93	SN11/92	A/C - COMPRESSOR CLUTCH BELT NOISES - DIAG/REPAIR
1	Up - 94	SN02/94	A/C - R-134A SYS WON'T CHARGE - CHECK SCHRADER VALVE
1	Up - 95	SN03/95	A/C - R134A O-RING LUBRICATION
4	92 - 93	SN01/93	A/C - SWITCH BUTTON INSTALLATION - PRECAUTION
4	95 - 95	SN12/94	A/C COMPRESSOR FAN INOP - CONNECTOR
4	92 - 94	SN07/94	A/C INOP - CHECK HARNESS CONNECTORS
3	94 - 94	AHM/A69	A/C POOR PERFORMANCE - ADJUST CONTROL CABLE
3	94 - 94	93-037	A/C POOR PERFORMANCE - CONTROL CABLES
3	94 - 94	93-037 SEP`93	A/C POOR PERFORMANCE - CONTROL CABLES - REVISED
3	94 - 94	93-037 JAN`94	A/C POOR PERFORMANCE - CONTROL CABLES - REVISION 2
3	94 - 94	93-037 SEP`94	A/C POOR PERFORMANCE - CONTROL CABLES - REVISION 2
4	92 - 94	SN02/94	AIR MIX CABLE REPLACEMENT PROCEDURE
3,4,7,5	92 - 94	95-011	BLOWER MOTOR NOISE
8	94 - 94	94-002	CONDENSER FAN MOUNTING SCREWS BREAK - UPDATED SCREWS
1	Up - 93	95-020	CONVERTING R-12 A/C SYSTEMS TO R-134A
3	94 - 95	95-018	HEATER CONTROL PANEL INDICATORS DO NOT LIGHT
3	94 - 94	SN02/94	INSTALLING CONDENSER BRACKET MAY PINCH HOOD CABLE
5	93 - 93	94-008	MOISTURE BETW REAR WINDOW & SEAL/REAR DEFOG PROBLEMS
4	92 - 93	93-038	R134A KIT ('94 CIVIC) FITS '92-93 MODELS
4,5	92 - 94	94-016	TEMPERATURE CONTROL LEVER HARD TO MOVE
4,5	92 - 94	94-016 JUN'94	TEMPERATURE CONTROL LEVER HARD TO MOVE - REV
6	95 - 95	95-015	TEMPERATURE LEVER IS HARD TO MOVE

Model Legend: 1= All 2= Most 3= Accord 4= Civic 5= del Sol 6= Odyssey 7= Prelude 8= Passport

Entertainment Devices, Stereo, Radio, Etc.

Models	Year	TSB#	Description
1	Up - 93	87-014 JUL`92	AUDIO COMPONENTS - OUT-OF-WARRANTY REPAIR INFO
1	Up - 95	89-029 FEB'93	AUDIO UNIT - EXCHANGE PROGRAM PROC - REV
7	92 - 94	94-018	BUZZ IN DRIVER DOOR WHEN AUDIO TURNED UP
1	Up - 93	92-026	CD CHANGER MAGAZINE - WON'T EJECT - CORRECTION
1	Up - 93	92-026 OCT`92	CD CHANGER MAGAZINE WON'T EJECT-REPLACEMENT P/N REV
3	Up - 93	SN09/93	INSTALLING CD CHANGER WILL REQUIRE CONTROL HEAD
4	92 - 93	92-051 DEC'92	LOOSE FRONT SPEAKER COVER - REV
1	Up - 95	SN08/94	PHONE ESN CHANGES EXPLAINED
3	93 - 93	SN08/94	POP NOISE FROM SPEAKERS OR BLOWN FUSE 35
1	Up - 94	SN06/94	RADIO - ENTERING ANTI-THEFT CODES - TIPS
3	93 - 93	94-036	RADIO - HONDA/BOSE AMPLIFIER REMANUFACTURING PROGRAM
1	82 - 94	87-014 JAN`93	RADIO - OUT-OF-WARRANTY REPAIR INFORMATION - REV
3,7	Up - 93	SN03/93	RADIO - POOR RECEPTION - GOLD EMBLEM CHECK
7,3,5	88 - 93	92-036	RADIO - POOR RECEPTION ON AM BAND - DIAGNOSIS/REPAIR
3,7	90 - 93	94-011	RADIO - POOR RECEPTION OR INTERFERENCE
1	Up - 93	SN11/92	RADIO/1000 SERIES - POOR SOUND - CHECK FADER CONTROL
1	Up - 93	89-029 MAY`92	RADIO/TAPE/CD PLAYER - EXCHANGE PROGRAM PROC - REV

Model Legend: 1= All 2= Most 3= Accord 4= Civic 5= del Sol 6= Odyssey 7= Prelude 8= Passport

Seats, Belts, Interior Trim, Carpets, Air Bags

Models	Year	TSB#	Description
3	94 - 94	SN09/93	A/T GEAR INDICATOR PANEL TURNS WHITE - PROCEDURES
7	Up - 93	SN02/93	AIRBAG - ADJUSTMENT TO FIT FLUSH WITH DASHBOARD
3,4,7	92 - 93	93-033	BROKEN SEAT BELT STOPPER BUTTON - REVISED BUTTON
3	94 - 95	94-040	BUZZ NOISE FROM REAR SHELF
3	94 - 95	94-039	BUZZ NOISE FROM SEAT BELT ANCHOR ON B-PILLAR
3	94 - 95	94-039 NOV`94	BUZZ NOISE FROM SEAT BELT ANCHOR ON B-PILLAR - REV
7	88 - 93	93-012	CENTER CONSOLE - SQUEAK NOISE - DIAGNOSIS/REPAIR
3,7	92 - 94	94-017	CLICK/CREAK NOISE - DRIVER'S SEAT TRACK
7	94 - 94	94-015	CUP HOLDER DOES NOT STAY CLOSED - REVISED LATCH PIN
3	94 - 95	SN05/95	CUP HOLDER LID WON'T LATCH
4	92 - 95	92-048 APR'93	CUP HOLDER OPENS BY ITSELF OR HARD TO OPEN - REV
7	93 - 93	94-003	CUPHOLDER IN ACCESSORY ARMREST BROKEN
1	Up - 95	SN05/95	DEFOGGER GRILLE REPAIR KIT
3	94 - 94	93-043	ELECT SEAT HEIGHT ADJ INOP - MOVE HARNESS
6	95 - 95	95-016	FRONT DOOR PANELS CREAK
4	94 - 95	95-022	GAP IN HEADREST SUPPORT COVER
4	93 - 93	93-007	HEAD RESTRAINT - LOWER SEAM LOOSE - REPAIR (COUPE)
4	92 - 94	93-040	HOLES IN FLOOR CARPET - TEARS AT LEFT FOOT WELL AREA
3	94 - 94	SN07/94	MATTED ARMRESTS - USE STEAMER
7	94 - 95	95-006	NOISE/JINGLING - LEFT SIDE OF DASH - SPEAKER GRILLE
4	92 - 95	95-026	RATTLE/NOISE FROM REAR OF HEADLINER
3	90 - 93	93-020	REAR SEAT BELT RETRACTOR SENSOR - R&R/DEALER CMPGN
7	92 - 93	SN03/93	REAR SHELF AREA - CLICK NOISE - SERVICE INFORMATION
4	92 - 93	93-014	SEAT BACK/REAR - CAN'T BE UNLATCHED W/KEY OR LOOP
1	Up - 93	91-030 DEC`91	SEAT BELT - SLOW/INCOMPLETE RETRACT - REV PROCEDURE
3	94 - 94	SN09/93	SEAT(SHOULDER) BELT ANCHOR BUZZ - PROCEDURES
3	94 - 95	95-002	SQUEAK NOISE FROM INTERIOR DOOR PANEL
1	92 - 95	95-024	SRS - CABLE REEL REMOVAL - HOLDING FIXTURE/CAUTIONS
3,5	95 - 95	SN03/95	SRS - DTC 5-1 TROUBLESHOOTING CHART - S/M REVISION
3	95 - 95	SN12/94	SRS TROUBLESHOOTING - TIPS/PARTS
5	93 - 94	94-006	SUN VISOR DOES NOT STAY UP - INSTALL WASHER
4	93 - 95	95-021	SUNSHADE RATTLE
3	94 - 94	94-032	TONNEAU COVER CREAK NOISE - WAGON
6	95 - 95	SN05/95	UNDER SEAT DRAWER WON'T FIT BENCH SEAT
7	92 - 94	94-044	WARPED FRONT CONSOLE

Model Legend: 1= All 2= Most 3= Accord 4= Civic 5= del Sol 6= Odyssey 7= Prelude 8= Passport

Glass, Doors, Hood, Decklid, Tailgate, Liftgate, Locks

Models	Year	TSB#	Description
7	92 - 94	94-018	BUZZ IN DRIVER DOOR W/ AUDIO TURNED UP - LOCK CNTRL
3	94 - 95	95-001	BUZZ NOISE INSIDE FRONT DOOR HANDLE
3	94 - 94	94-024	CLUNK NOISE FROM WINDOW GLASS WHEN LOWERING
3	94 - 95	SN03/95	CRACKING NOISE FROM WINDSHIELD
3,4,7	90 - 93	92-022 JAN`93	DOOR (FRONT) - WIND NOISE CORRECTION - 2ND REVISION
4	92 - 95	92-021 JUN'92	DOOR (REAR) - WEATHERSTRIP NOISE/DETACHED - REV
4	Up - 93	SN11/92	DOOR CHECKER BRACKETS - NOW AVAILABLE W/INSTRUCTIONS
3	94 - 94	93-041 OCT`93	DOOR HANDLE SEAL DAMAGED - REVISED
3	94 - 95	93-041 MAR'94	DOOR HANDLE SEAL DAMAGED - REVISED
3	94 - 94	93-050	DOOR WEATHERSTRIP OUT OF POSITION
3,4,7	90 - 93	92-022 JUL`92	DOOR/FRONT - WIND NOISE CORRECTION AROUND MIRROR/REV
5	93 - 93	93-025	DOORS - RATTLE/SCRAPE NOISE ON BUMPS/ROUGH ROADS
3	90 - 93	93-005 FEB`93	DOORS HARD TO OPEN OR CLOSE
3	90 - 93	93-005	DOORS/WINDOWS - HARD TO OPEN FROM SUN/RAIN - REPAIR
5	93 - 94	94-009	INSIDE DOOR HANDLE STAYS OPEN
4	88 - 93	SN03/93	KEY BLANKS - LENGTH INFORMATION
5	93 - 93	94-008	MOISTURE BETW REAR WINDOW & SEAL OR GLASS BREAKS
7	92 - 94	92-014 & 92-028	NOISE FROM SUNROOF OR ROOF AREA
1	94 - 94	AHM/A77	PDI CHECK - MISSING TRUNK LIGHT LENSE
3	94 - 94	94-019	POWER DOOR LOCKS CYCLE FROM LOCKED TO UNLOCKED
3	94 - 94	AHM/A94	POWER SEAT WIRING COUPLER DISCONNECT - REROUTE WIRE
8	94 - 94	SN01/94	RATTLE - TAILGATE AND SPARE TIRE CARRIER
4	92 - 94	94-037	RATTLE FROM INTERIOR FRONT DOOR PANEL
3	94 - 94	AHM/A93	RATTLE FROM MOONROOF - CHECK HINGE RIVET TENSION
4	93 - 94	95-004	RATTLE NOISE IN REAR - TAILGATE - 3-DOOR
3	94 - 94	93-047	RATTLE/NOISE FROM MOONROOF WIND DEFLECTOR - ADJUST
3	94 - 94	93-041	REAR DOOR HANDLE SEAL DAMAGED
5	93 - 95	93-004 SEP`93	ROOF AREA WATER LEAKS - TESTING/REPAIR PROC - REV
3	90 - 93	91-017 MAY`91	SUNROOF - RATTLE WHEN PARTIALLY OPEN - REVISED
7	92 - 94	SN07/94	SUNROOF SQUEAKS REVISITED
8	94 - 95	SN12/94	TAILGATE RELEASE INOP - TIPS
8	94 - 94	SN01/94	TAILGATE RELEASE OPERATION EXPLAINED
4	92 - 95	93-002 JAN`93	TRUNK - HARD TO OPEN WITH KEY - CORRECTION - REVISED
4	92 - 93	93-002	TRUNK - HARD TO OPEN WITH KEY - CORRECTIVE ACTION
3	94 - 94	94-038	TRUNK DIFFICULT TO CLOSE
3	94 - 94	93-039	WATER LEAK AT TOP OF DOORS
5	93 - 93	AHM/A20	WATER LEAK CHECK INFORMATION
3	94 - 94	94-028	WIND NOISE AT FRONT EDGE OF DOOR
3	94 - 94	94-001	WIND NOISE AT FRONT EDGE OF WINDOW - OUTSIDE MIRROR
3	94 - 94	93-042	WIND NOISE AT THE DOOR MIRROR - APPLY EPT SEALER

Model Legend: 1= All 2= Most 3= Accord 4= Civic 5= del Sol 6= Odyssey 7= Prelude 8= Passport

Glass, Doors, Hood, Decklid, Tailgate, Liftgate, Locks

Models	Year	TSB#	Description
3	94 - 94	94-027	WIND NOISE/WHISTLE AT A-PILLAR
3	95 - 95	95-032	WINDNOISE AT TOP OF WINDSHIELD
7	92 - 93	SN03/93	WINDOW & GLASS BRACKETS (REAR) CONTACT/CLICK NOISE
3,7	90 - 93	91-010 NOV`92	WINDOW - RATTLES WHEN HALF OPEN/CORRECTION - 2ND REV
3,7	90 - 93	91-010 AUG`93	WINDOW - RATTLES WHEN HALF OPEN/CORRECTION - 3RD REV
5	93 - 93	93-004 FEB`93	WINDOW AREA WATER LEAKS - TESTING/REPAIR PROC - REV
5	93 - 93	93-004R JUN`93	WINDOW AREA WATER LEAKS - TESTING/REPAIR PROC - REV
5	93 - 93	93-004	WINDOW AREA WATER LEAKS - TESTING/REPAIR PROCEDURES
4	93 - 93	94-030	WINDOW BINDS - GLASS OUT OF CHANNEL - PROCEDURE
3	93 - 93	93-027	WINDOW RUN CHANNEL FLARED/WINDNOISE - SERVICE PROC
5	93 - 93	93-026	WINDOW/SIDE - CHATTERS WHILE GOING UP OR DOWN
7,3,4	88 - 93	91-001 NOV`91	WINDOWS CREAK - RAISING/LOWERING - PARTS INFO - REV

Model Legend: 1= All 2= Most 3= Accord 4= Civic 5= del Sol 6= Odyssey 7= Prelude 8= Passport

Finishes, Body Structure, Frame, Bumpers

Models	Year	TSB#	Description
1	93 - 93	87-006 APR`89	BODY - PDI SHIPPING WAX REMOVAL/SOLVENTS - REVISION
3	94 - 94	95-003	BUZZ NOISE IN B-PILLAR
7	92 - 94	92-014 & 92-028	NOISE FROM SUNROOF OR ROOF AREA
3	95 - 95	94-042	NOISE/RATTLE BETWEEN WINDSHIELD & MOONROOF
1	94 - 94	93-036	PAINT CODES
1	95 - 95	94-035 OCT'94	PAINT CODES - 1995 MODELS
8	95 - 95	95-009	PAINT CODES - 1995 PASSPORT
1	93 - 93	92-045	PAINT CODES AND DESCRIPTIONS
8	94 - 94	93-053	PASSPORT PAINT CODES
3	94 - 94	SN09/93	RAIL DUST REPAIR REMINDER SEE 88-035
5	93 - 93	93-023 JUN`93	RATTLE/CREAK/CHATTER NOISE FROM ROOF - PROCEDURES
5	93 - 94	93-023 JUN`94	RATTLE/CREAK/CHATTER NOISE FROM ROOF - PROCEDURES
5	93 - 94	93-023 SEP`93	RATTLE/CREAK/CHATTER NOISE FROM ROOF - REVISED
5	93 - 93	93-023	ROOF - RATTLE WHEN DRIVING ON ROUGH ROADS
5	93 - 93	93-023	ROOF - RATTLE/SQUEAK ON ROUGH ROADS - SERV INFO
5	93 - 93	93-004 FEB`93	ROOF AREA WATER LEAKS - TESTING/REPAIR PROC - REV
5	Up - 94	93-004R JUN`93	ROOF AREA WATER LEAKS - TESTING/REPAIR PROC - REV
5	93 - 95	93-004 SEP`93	ROOF AREA WATER LEAKS - TESTING/REPAIR PROC - REV
5	93 - 93	93-004	ROOF AREA WATER LEAKS - TESTING/REPAIR PROCEDURES
5	93 - 94	93-023	ROOF NOISE - ADD'L STEPS & REVISIONS
3	94 - 95	94-034	ROOF RACK END COVERS FALL OFF - WAGON
4	92 - 95	SN08/94	SPOILER GAPS TO BUMPER - PROCEDURE
8	94 - 94	93-052	TRANSIT COATING REMOVAL
3	93 - 93	AHM/A28	TRIM STUD/TRUNK SILL - COMES OFF - REPAIR PROCEDURE
5	93 - 94	94-026	WATER LEAK AT ANTENNA INTO TRUNK
5	93 - 93	AHM/A20	WATER LEAK CHECK INFORMATION
3	90 - 93	93-005	WEATHERSTRIP - R&R - REPAIR STUCK DOORS & WINDOWS
5	Up - 93	SN03/93	WEATHERSTRIP/ROOF SIDE - SEALING INFORMATION

Model Legend: 1= All 2= Most 3= Accord 4= Civic 5= del Sol 6= Odyssey 7= Prelude 8= Passport

HYUNDAI

Lighting, Horns, Turn Signals, Steering Column

Models	Year	TSB#	Description
1	Up - 95	95-90-009	BRAKE LIGHT SWITCH - MEASURE & ADJ PROCEDURE/SPECS
3	95 - 96	96-90-005	HEADLAMP FLASHER TURNS OFF SLOWLY IN COLD WEATHER
3	95 - 95	95-01-001	HEADLAMP SWITCH REPLACEMENT INST - CAMPAIGN
7	95 - 95	95-90-008	HIGH MOUNT STOP LAMP ASMBLY REV -BULB REPLACE PROCED
5	93 - 94	93-90-015	IGNITION KEY LOCK CONTROLS (NEW) - A/T VEHICLES
7	95 - 95	94-90-016	INOP INTERIOR/CLUSTER LAMPS - DOOR SWITCH GROUNDING
6	Up - 93	93-90-013	NEW FRONT TURN SIGNAL LAMP INSTALL PROCEDURE - REV
6	Up - 93	93-90-013	NEW FRONT TURN SIGNAL LAMP INSTALLATION PROCEDURE
7,4	95 - 95	94-90-005	TURN SIGNAL INDICATOR DOES NOT CANCEL

Model Legend: 1= All 2= Most 3= Accent 4= Elantra 5= Excel 6= Scoupe 7= Sonata

Instruments, Dash Cluster, Warning Lights, Mirrors

Models	Year	TSB#	Description
5	92 - 93	93-80-008	AIR VENT DUCT/LOWER CRASH PAD PANEL - MODIFIED
1	Up - 95	95-90-009	CRUISE CONTROL - BRAKE LIGHT SWITCH ADJUSTMENT SPECS
4	93 - 94	93-36-008	CRUISE CONTROL - OPERATION PROCEDURE
4	92 - 93	93-36-015	CRUISE CONTROL CABLE ADJUSTMENT PROCEDURE
4,7	Up - 95	94-90-008	CRUISE CONTROL DIAGNOSTIC CODES - RETRIEVAL/CHART
7	95 - 95	95-90-006	FUEL GAUGE ACCURACY FIELD FIX
6	93 - 93	92-30-005	GAUGE/FUEL PRESSURE - INSPECT CONNECTION VALVE/LEAKS
6	Up - 94	94-36-003	IP - TURBO LED INDICATORS - PRESSURE VALUES
6,7	Up - 95	95-90-002	IP SWITCH BULB REPLACEMENT PROCEDURE
7	95 - 95	94-80-005	MODIFICATION OF OUTSIDE MIRROR ASSEMBLY
1	Up - 95	94-80-008	OUTSIDE MIRROR BASE PLATE - REPLACEMENT PROCEDURE

Model Legend: 1= All 2= Most 3= Accent 4= Elantra 5= Excel 6= Scoupe 7= Sonata

Chassis Electrical, Wiring Harness, Fuses-Circuit Breakers, Wipers, Window Motors

Models	Year	TSB#	Description
6	93 - 93	93-90-001	CIRCUIT/FOG LAMP - ADDED TO TURBO - SCHEMATIC
6	93 - 93	93-90-001	CIRCUIT/FOG LAMP - SCHEMATIC (TURBO) - REVISED
5	94 - 94	94-90-006	COOLING SYSTEM CIRCUIT MODIFICATION - S/M REVISION
4	94 - 94	94-90-007	HARNESS CHANGE - CLOCK POWER SUPPLY - REV SCHEMATIC
4	94 - 94	94-90-002	HORN RELAY LOCATION CHANGED
5	93 - 94	93-90-015	IGNITION KEY LOCK CONTROLS (NEW) - A/T VEHICLES
7	92 - 93	93-90-016	INTERMITTENT WIPER SPEED VARIES WITH ENGINE RPM
4	92 - 93	94-90-017	PWR WINDOW NOISE - MOTOR BREATHER HOSE ADJUSTMENT
7	89 - 93	93-90-005	RELAY/POWER WINDOW - CHANGED TO INCORPORATE A DIODE
5	93 - 94	93-36-011	SPARK PLUG CABLE ROUTING CHANGE
4,7	94 - 94	94-90-012	SUNROOF MOTOR INSTALLATION - TIPS
6	93 - 93	92-00-003	TOWING (TRAILER) - NOT RECOMMENDED - O/M CORRECTION
4	96 - 96	96-90-003	UNDERHOOD FUSE/RELAY BOX POWER CONNECTOR - STORAGE
1	Up - 94	94-90-001	WINDSHIELD WASHER CONNECTOR - IMPROVED SHAPE
7	93 - 93	93-90-008	WINDSHIELD WASHER FILTER ADDITION
7	95 - 95	95-36-004	WIRING DIAGRAM - STARTER RELAY ADDED
3	95 - 95	95-01-003	WIRING HARNESS ROUTING - ECM - M/T ONLY - CAMPAIGN

Model Legend: 1= All 2= Most 3= Accent 4= Elantra 5= Excel 6= Scoupe 7= Sonata

Auxiliary Equipment, Jacks, Trailer Hitches, Towing

Models	Year	TSB#	Description
3	95 - 96	95-00-004	TOWING GUIDELINES - BEHIND RECREATIONAL VEHICLE

Model Legend: 1= All 2= Most 3= Accent 4= Elantra 5= Excel 6= Scoupe 7= Sonata

Heating, Air Conditioning, Ventilation, Defogger

Models	Year	TSB#	Description
3	95 - 95	95-97-004	A/C - COMPRESSOR DISENGAGEMENT DURING ACCEL
3	95 - 95	95-97-004	A/C - COMPRESSOR DISENGAGEMENT DURING ACCEL - REV
5,7,4	94 - 94	94-97-002	A/C - DEALER INSTALLED UNITS MUST BE R-134A
5,4,6	94 - 94	93-97-003	A/C - R-134A REFRIGERANT - PROCEDURES & TOOLS
4	92 - 93	93-20-007	A/C - REFRIG TEMP SENSOR SPEC CHANGES - S/M UPDATE
1	95 - 96	95-97-002	A/C COMPRESSOR OPERATION
6	94 - 94	94-01-010	A/C DISCHARGE HOSE - SERVICE CAMPAIGN
7	93 - 93	93-97-001	A/C LO PRESSURE SWITCH - SPEC CHANGE/IMPROVE COOLING
4	94 - 94	94-01-012	ADD REFRIGERANT OIL TO R-134A SYS - CAMPAIGN
4	94 - 94	94-01-012	ADD REFRIGERANT OIL TO R-134A SYS - CAMPAIGN - REV
5	94 - 94	94-90-006	COOLING SYSTEM CIRCUIT MODIFICATION - S/M REVISION
4,6	94 - 94	94-97-003	EVAPORATOR TEMPERATURE SWITCH CHANGED
1	Up - 93	92-20-009	HEATER - NO HEAT/FAULT CODE 21 - THERMOSTAT STICKING
1	Up - 94	93-97-004	MISTING FROM VENTS WITH A/C ON - EXPLAINED
7	Up - 93	93-60-003	ODOR/NOISE/FOG - WATER LEAK THROUGH PLENUM DUCT
6	Up - 94	95-90-001	REAR DEFROSTER - EXTENSION OF "ON" TIME
7	Up - 93	93-97-002	TEMPERATURE CONTROL KNOB REINFORCED
1	Up - 94	94-97-001	WINDSHIELD DEFROST/DEFOG SLOW

Model Legend: 1= All 2= Most 3= Accent 4= Elantra 5= Excel 6= Scoupe 7= Sonata

Entertainment Devices, Stereo, Radio, Etc.

Models	Year	TSB#	Description
3,4	96 - 96	96-90-004	AM/FM-CD INSTALLATION - KIT REQ'D
4,5	93 - 93	93-90-018	ANTENNA GROUND TERMINAL ADDED - IMPROVED RECEPTION
1	Up - 95	SUPP: 91-90-021	ANTI-THEFT RADIO - CANCELLATION OF LOCK-OUT CODES
3	95 - 95	94-90-015	AUDIO - FRONT/REAR SPEAKER REPLACEMENT PROCEDURE
1	Up - 93	92-00-002	AUDIO LOCKOUT CODES/KEY CODES - DIRECT ACCESS/DCS
7	95 - 95	95-90-004	AUDIO SYSTEM STEERING WHEEL REMOTE CONTROL
2	90 - 95	96-90-002	CD PLAYER ERROR CODES
5	94 - 94	94-90-010	RADIO - ADDING REAR SPEAKERS - PROCEDURE
5	Up - 95	94-90-014	RADIO APPLICATIONS & CONVERSIONS
1	95 - 95	95-90-003	RADIO CHASSIS NUMBER STICKER LOCATION
4,6	94 - 94	94-90-003	RADIO WHINE NOISE IN REAR SPEAKERS - 820
3	Up - 96	95-60-007	REAR SPEAKER INSTALLATION - 3-DR
7	Up - 93	93-90-007	SPEAKER BUZZING/FRONT - FIELD FIX (MED/HIGH RADIOS)

Model Legend: 1= All 2= Most 3= Accent 4= Elantra 5= Excel 6= Scoupe 7= Sonata

Seats, Belts, Interior Trim, Carpets, Air Bags

Models	Year	TSB#	Description
3,4,7	Up - 95	94-90-018	AIR BAG REPAIR KIT PART NUMBERS
1	96 - 96	96-90-001	AIRBAG (SRS) DIAGNOSTIC CODES (DTC)
3	95 - 95	95-90-010	AIRBAG - LOW PRESSURE SENSOR ELIMINATED
3	Up - 96	95-80-002	CONSOLE ARMREST - REV HARDWARE & INSTALL INSTRUCTION
5	94 - 94	94-80-006	COVER ELIMINATED - CHILD SEAT RESTRAINT HOOK - REV
5	93 - 93	94-80-006	COVER ELIMINATED - CHILD SEAT RESTRAINT HOOK HOLDER
6	Up - 93	93-80-014	DOOR SCUFF TRIM MOUNTING CHANGE - INSTALL PROCEDURES
7	93 - 93	93-80-016	GLOVE BOX RATTLE
7	91 - 93	94-80-002	MAP POCKET REPLACEMENT
7	89 - 94	94-01-002	PASSIVE SEAT BELT - LUBRICATION & DIAGN - CAMPAIGN
5	90 - 93	93-80-018	REAR SEAT BACK RATTLE - NEW STRIKER
3	95 - 96	95-80-001	REAR SEAT CUSHION MOUNTING CHANGE
7	93 - 93	93-80-003	SEAT BELT GUIDE REPLACEMENT
4	92 - 93	94-80-003	SQUEAK/RATTLE FROM REAR PACKAGE TRAY
3,4,7	95 - 96	96-90-006	SRS (AIR BAG) CONTROL MODULE CHANGE TO 30-PIN
4	94 - 95	95-01-004	SRS SYSTEM TERMINAL HOLDER - CAMPAIGN
4	Up - 96	95-60-006	SUNROOF SUNSHADE SLIDES OPEN WHILE DRIVING
6	Up - 93	93-80-013	SUNVISOR MATERIAL CHANGE - VINYL SQUEAKS
5,7	93 - 93	93-80-006	TRIM PANEL/REAR DOOR/DRIVER SIDE - ASH TRAY DELETED

Model Legend: 1= All 2= Most 3= Accent 4= Elantra 5= Excel 6= Scoupe 7= Sonata

Glass, Doors, Hood, Decklid, Tailgate, Liftgate, Locks

Models	Year	TSB#	Description
7	94 - 94	94-90-013	CHANGE OF MOUNTING METHOD FOR DOOR SWITCH
5	Up - 95	94-80-007	DOOR CHECKER NOISE - PROCEDURE
4	93 - 93	93-80-010	DOOR OPENING WEATHERSTRIP INSTALLATION
4	92 - 93	93-80-009	DOOR TRIM CHANGED AT UPPER SEAT BELT OPENING
1	94 - 94	94-80-001	FUEL FILLER DOOR EMERGENCY RELEASE HANDLE CHANGE
1	Up - 93	92-00-002	KEY CODES/AUDIO LOCKOUT CODES - DIRECT ACCESS/DCS
1	Up - 95	95-00-001	KEY CUTTING CODES
7	95 - 95	94-00-002	KEY CUTTING INFORMATION
7	89 - 93	94-80-004	RATTLE -SUNROOF DRAIN HOSE
7	95 - 95	94-00-001	THREE KEYS PROVIDED - ONE IS A VALET KEY
1	Up - 95	95-60-002	WATER LEAK DETECTION TIPS
4,5,6	Up - 94	93-80-015	WEATHERSTRIP - DOOR OPENING - INSTALLATION PROCEDURE
3	95 - 96	96-60-001	WINDOW GLASS GRIP REPAIR PROCEDURE

Model Legend: 1= All 2= Most 3= Accent 4= Elantra 5= Excel 6= Scoupe 7= Sonata

Finishes, Body Structure, Frame, Bumpers

Models	Year	TSB#	Description
6	93 - 93	93-80-017	BACK PANEL MOLDING CRACKING - FIELD FIX
5	93 - 93	92-60-002	BODY - DIMENSIONS - S/M CORRECTIONS
5	90 - 93	93-60-002	BODY DIMENSIONS - S/M CORRECTIONS
6	Up - 95	94-60-002	C-PILLAR COVER INSTALLATION - REV PART
7	93 - 93	93-60-001	CHASSIS - PLUG INSTALLATION INFORMATION (PDI)
4	92 - 93	95-60-001	COWL TOP COVER MOUNTING CHANGE - REVISED CLIPS
7	95 - 95	94-60-001	DRIP RAIL FRONT MOULDING MOUNTING CLIP ADDED
6	92 - 93	93-80-005	MOULDING (REAR WINDOW SIDE/UPPER) - SHAPE CHANGES
1	92 - 93	92-99-003	PAINT - APPLICATION OF ANTI-CORROSION TREATMENT
1	93 - 93	92-99-004	PAINT - EXTERIOR COLOR CODES AND APPLICATIONS
1	95 - 95	95-99-002	PAINT - EXTERIOR PLASTIC COMPONENT IDENT/PROCEDURES
1	Up - 93	93-99-002	PAINT - PEELING (CLEARCOAT DEGRADATION) - CORRECTION
1	96 - 96	95-99-003	PAINT CODES
1	95 - 95	94-99-004	PAINT CODES & APPLICATIONS - 1995 MODELS
1	94 - 94	93-99-003	PAINT CODES - 1994 MODELS
1	96 - 96	95-99-003	PAINT CODES - UPDATE
1	93 - 95	93-99-003	PAINT CODES REVISED - 1994 MODELS & 1995 SONATA
1	95 - 95	94-99-004	PAINT COLOR CODES AND APPLICATIONS - REVISED
1	93 - 95	93-99-001	PAINT/EXTERIOR - PROPER CARE - ENVIRO DAMAGE - REV
1	Up - 96	94-99-002	PAINT/EXTERIOR - PROPER CARE - ENVIRO DAMAGE - REV2
1	93 - 93	93-99-001	PAINT/EXTERIOR - PROPER CARE - ENVIRONMENTAL DAMAGE
1	94 - 95	94-99-003	PLASTIC COMPONENT IDENT FOR PROPER PAINTING PROCED
6	93 - 93	93-80-012	REAR BUMPER ADJUSTMENT PROCEDURE
4	95 - 95	95-60-005	REAR SPOILER ASSIST SPRING MOUNTING HOLE LOCATION
4	95 - 95	95-60-004	SPOILER INSTALLATION INSTRUCTIONS - SE MODEL
1	Up - 95	95-60-002	WATER LEAK DETECTION TIPS
6	Up - 95	95-60-003	WATER LEAKS INTO TRUNK NEAR WHEEL WELL
7,5	93 - 93	93-80-004	WEATHERSTRIP/DOOR OUTSIDE BELT - REPLACEMENT PROC

Model Legend: 1= All 2= Most 3= Accent 4= Elantra 5= Excel 6= Scoupe 7= Sonata

INFINITI

Lighting, Horns, Turn Signals, Steering Column

Models	Year	TSB#	Description
5	93 - 93	FC0008670	HEADLAMP - LOW BEAM INOPERABLE - WIRING REPAIR INFO
5	93 - 93	AM93-002	HEADLAMP AIMING SPECS - S/M REVISION
3	91 - 93	FC0008687	HEADLIGHTS - TURNING ON BLOWS FUSE FOR INT LIGHTS
5	93 - 93	FC0008578	HORN/THEFT WARNING - CAN'T FIND UNDER HOOD/INFO
7	90 - 93	FC0008592	LIGHT/DRIVER'S DOOR KEY - INOP/DOME LITE DOESN'T DIM

Model Legend: 1= All 2= Most 3= G20 4= I30 5= J30 6= M30 7= Q45

Instruments, Dash Cluster, Warning Lights, Mirrors

Models	Year	TSB#	Description
7	91 - 93	FC0008266	ASCD LIGHT IN DASH FLASHES/SYS INOP INTERMITTENTLY
5,7	93 - 94	EL93-012	CLOCK ILLUMINATION DIM AT NIGHT - PROCEDURES
5	93 - 93	FC0008553	DISPLAY/RADIO & A/C - GOES OFF INTERMITTENTLY - PROC
7	91 - 94	FC0008597	LIGHT/ACTIVE SUSP - ON/SELF DIAG CODE 41 - SERV PROC
5	93 - 93	FC0008263	MIRROR/LEFT - WIND NOISE AT 60-65 MPH - KIT INFO
5	93 - 93	FC0008661	SPEEDO (AT STEADY SPD OF 60 MPH) MOVES TO 75 & BACK
7	90 - 94	FC0008549	TCS LIGHT ACTIVATES/MOTOR NOISE/SELF-DIAG CODE 18
5	93 - 93	FC0008642	WARN LIGHT/ABS - CONTINUOUS ON - WIRING REPAIR
3	91 - 93	BR93-001	WARN LIGHT/ANTI-LOCK - STAYS ON/WIRE HARNESS DAMAGE
7	90 - 93	FC0008280	WARN LITE/BRAKE - ALWAYS ON - WHEEL SENSOR WIRE INFO
5	93 - 93	FC0008150	WARNING LAMP (SEAT BELT) - FLASHES - FIELD FIX

Model Legend: 1= All 2= Most 3= G20 4= I30 5= J30 6= M30 7= Q45

Chassis Electrical, Wiring Harness, Fuses-Circuit Breakers, Wipers, Window Motors

Models	Year	TSB#	Description
5	93 - 93	EL94-007	ACCESSORY POWER LOSS - IGNITION KEY STICKS
3	91 - 93	FC0008687	FUSE FOR INT LIGHTS BLOWS WHEN HEADLIGHTS TURNED ON
1	Up - 94	EF93-005	GREEN OR BLUE (TYPE 1M) RELAY CAUTION - DAMAGE ECM
5	93 - 93	FC0008510	NO START/CRANK - CLICK BEHIND DASH - BAD GROUND
7	94 - 94	FC0008825	NOISE/CLICK - REGULATOR - COUNTERMEASURE PART
1	Up - 96	EC95-005	RELAY CAUTION - FUEL PUMP & A/C
7	94 - 94	FC0008716	SEAT & WINDOWS 25 AMP FUSIBLE LINK BLOWS
5	93 - 93	FC0008749	STALL/NO RESTART - FUEL PUMP FUSE BLOWN - WIRE SHORT
5	93 - 94	EL93-016	STARTER HOLD RELAY CIRCUIT CHANGES EXPLAINED
7	94 - 94	EL94-008	STARTER RELAY REPLACEMENT - USE NEW HARNESS
5	93 - 93	FC0008390	WARN LIGHTS - BRAKE/STOP/TAIL/CHARGE STAY ON
5	93 - 93	FC0008305	WINDSHIELD WIPERS - ERRATIC OPERATION - REPAIR INFO
3	92 - 93	FC0008151	WIPERS - STAY ON WITH SWITCH OFF - FIELD FIX
3	91 - 93	FC0007844	WIPERS ERRATIC/JUMP ACROSS WINDSHIELD/WON'T PARK
5	93 - 93	FC0008662	WIPERS/RADIO/CLIMATE CONTROL INTERMIT OP - IGN PROC
7	94 - 94	AM94-001	WIRING DIAG - ANTI-DAZZLING MIRROR - S/M ADDITION

Model Legend: 1= All 2= Most 3= G20 4= I30 5= J30 6= M30 7= Q45

Auxiliary Equipment, Jacks, Trailer Hitches, Towing

Models	Year	TSB#	Description
5	93 - 93	FC0008210	CELLULAR PHONE - HANDS FREE MODE INOPERATIVE - INFO
5,7	94 - 94	BF93-025	CELLULAR PHONE - METALLIC WIDOW TINTS CAUSE PROBLEMS
5	93 - 93	EL92-009	CELLULAR PHONE - STATIC/POOR RECEPTION/DROPPED CALLS
7	94 - 94	EL93-002	CELLULAR PHONE ANTENNA - SERVICE INFORMATION
5	93 - 94	EL93-011	CELLULAR PHONE INTERFERENCE - PROCEDURES
7	94 - 94	EL93-006	CELLULAR PHONE MICROPHONE - ECHOING IN EARLY MODELS
7	94 - 94	FC0008536	CELLULAR PHONE WON'T COME ON AFTER INSTALL - REPAIR
5	93 - 93	EL92-007	CELLULAR TELEPHONE ANTENNA - INFORMATION
5	93 - 93	FC0007591	COMPACT COMMUNICATOR PHONE - PROGRAMMING INFORMATION
7	94 - 94	FC0008800	PHONE CONTROLS ON STEERING WHEEL INOP
7	94 - 94	FC0008443	PHONE ECHO - NEW MICROPHONE
5,7	91 - 94	FC0008840	POOR CELLULAR PHONE PERFORM - METALALIC WINDOW TINT

Model Legend: 1= All 2= Most 3= G20 4= I30 5= J30 6= M30 7= Q45

Heating, Air Conditioning, Ventilation, Defogger

Models	Year	TSB#	Description
5	93 - 93	FC0008553	A/C & RADIO DISPLAY - GOES OFF INTERMITTENTLY - PROC
5	93 - 93	FC0007673	A/C - BLOWER RUNS WITH A/C IN OFF POSITION - REPAIR
7	92 - 94	FC0008691	A/C - BLOWS WARM AIR - PERFORMANCE TEST INFO
5	93 - 93	FC0008301	A/C - COMPRESSOR INOP - SENSOR RESISTANCE INFO
5	93 - 93	FC0008282	A/C - COMPRESSOR INOP - WIRING INFORMATION
7	90 - 94	HA94-004	A/C - POOR/NO COOLING - DIAGNOSTIC TIPS
1	Up - 96	EL95-003	A/C - R134A CHARGE AMOUNTS/PAG LUBRICANT CHART
1	Up - 96	HA95-005	A/C - R134A CHARGE AMOUNTS/PAG LUBRICANT CHART
7	94 - 94	FC0008766	A/C AIR FLOW REDUCED - ADDED SRS AIR BAG - NORMAL
7	90 - 94	FC0008532	A/C BLOWS WARM AFTER A MINUTE - EXPANSION VALVE
3	93 - 93	HA93-001	A/C CHANGES FOR R134A REFRIGERANT - DETAILS
1	Up - 96	HA95-004	A/C COMPRESSOR LEAK/NOISE DIAGNOSIS
7	94 - 94	FC0008799	A/C DISPLAY IN CELSIUS ONLY - CANADIAN UNIT
5	93 - 93	FC0008300	A/C FAN STAYS ON HIGH SPEED IN AUTO MODE - SERV PROC
1	Up - 96	HA94-003	A/C LUBRICANT/OIL - SPECS
1	90 - 94	HA94-003	A/C LUBRICANT/OIL SPECIFICATIONS
5	93 - 93	FC0008384	A/C ON - FLUTTER NOISE UNDER DASH RIGHT SIDE
5	93 - 94	HA94-007	A/C ON AUTO - IDLE SPEED FLUCTUATION BELOW 56 DEG F
1	Up - 95	HA95-002	A/C PARTS - PROPER INSTALLATION PROCEDURE
3	91 - 93	FC0008706	A/C POOR COOL - WARN LIGHTS ON - BUZZ BEHIND DASH
1	Up - 95	HA95-001	A/C REFRIGERANT LEAKS - DETECTION PROCEDURES
5	93 - 93	FC0008381	A/C SHUTS OFF - ERRATIC/INTERMITENT DISPLAY
5	93 - 93	HA94-001	A/C TEMPERATURE FLUCTUATION
5	93 - 93	FC0008579	ATC - RECIRCULATION TIMER/PRODUCTION CHANGE INFO
5	93 - 94	HA94-002	ATC ERROR CODE 52
7	90 - 94	FC0008548	BLOWER MOTOR INOP/FUSES BLOW ON SPEEDS 3 & 4 - PROC
5	93 - 93	FC0008662	CLIMATE CONTROL/WIPERS/RADIO INTERMIT OP - IGN PROC
7	94 - 94	HA94-005	NOISE - A/C COMPRESSOR
5	93 - 93	HA93-005	ODOR - ATC RECIRCULATION TIMER LOGIC - RETROFIT PART
1	Up - 94	HA94-006	R134 RETROFIT PROCEDURES
1	Up - 94	HA93-003	R134A - KENT-MOORE ACR4 MACHINE USAGE INFORMATION

Model Legend: 1= All 2= Most 3= G20 4= I30 5= J30 6= M30 7= Q45

Electronic Devices, Computers, PROMS, Sensors

Models	Year	TSB#	Description
5	93 - 93	FC0007673	A/C CNTRL UNIT/REPLACE - BLOWER RUNS IN OFF POSITION
5	93 - 93	FC0008423	A/T LIGHT ON - NO CODES - ECU RUST - SUNROOF DRAIN
3	93 - 93	FC0008628	ABS CODES 41/61/63 STORED/ABS LIGHT ON - SERV PROC
7	90 - 94	FC0008745	ABS LIGHT ON - NO TCS LIGHT - SHORT IN HARNESS
7	94 - 95	EL94-004	ADP - AUTOMATIC DRIVE POSITIONER MEMORY SETTING
1	Up - 94	FC0008529	ALARM (FALSE) - AFTERMARKET PHONE
5	93 - 93	FC0008368	ALARM/DOOR LOCK - ERRATIC OPERATION/SELF GENER LOCK
7	94 - 94	EL93-014	AUTOMATIC DRIVE POSITIONER RESET PROCEDURE
7	90 - 94	FC0008552	AUTOMATIC DRIVE POSITIONER SYSTEM INOP - RESET PROC
7	90 - 94	FC0008624	CONSULT SHOWS INJECTOR LEAK/INJ CHECK OUT OK - SERV
1	90 - 95	FC0010108	DRIVEABILITY - ECM CONNECTOR NOT FULLY INSERTED
1	Up - 95	EC95-003	ECCS SYSTEM TESTING - PRECAUTIONS TO AVOID DAMAGE
1	91 - 93	EF93-001	ECM ID/VENDOR SERIAL # & CROSS-REFERENCES TO INF P/N
7	91 - 94	FC0008651	ECM/NEW-LACK OF PERFORM/CODE 34 KNOCK SENSOR CIRCUIT
5	93 - 93	FC0008652	ECU - #4 INJECTOR PIN #112 POOR CONTACT/ROUGH IDLE
2	90 - 94	EF94-008	EGINE CONTROL MODULE (ECM) IDENTIFICATION
7	94 - 94	AM93-005	EL (LAN) TROUBLE DIAGNOSES - S/M REVISION
7	91 - 93	FC0008396	FAS WARNING LIGHT ON - CODE 41
5	94 - 95	AM94-003	FUEL PRESSURE REGULATOR VACUUM RELIEF SYS CHANGES
1	Up - 94	EF93-005	GREEN OR BLUE (TYPE 1M) RELAY CAUTION - DAMAGE ECM
4	96 - 96	EL95-001	INTEGRATED HOMELINK TRANSMITTER SYSTEM
4	96 - 96	EL95-005	IVMS (LAN) TROUBLE DIAGNOSES SYSTEM - S/M REV
3,5,7	Up - 96	HA95-005	KEYLESS ENTRY - BATTERY REPLACE & RE-PROG PROCEDURE
5	93 - 93	FC0008370	KEYLESS ENTRY - CAN'T PROGRAM MORE THAN ONE REMOTE
5	93 - 93	FC0008371	KEYLESS ENTRY - LIMITED RANGE - ANTENNA CONNECTION
3,5,7	93 - 94	EL94-003	KEYLESS ENTRY REMOTE - BATTERY REPLACEMENT
5	93 - 93	FC0008367	KEYLESS ENTRY/ALARM INOP - IGN SWITCH ASSEMBLY
7	94 - 94	EL93-010	LAN SELF DIAGNOSIS RESULTS
7	94 - 95	EC95-004	LOW/ROUGH IDLE - ECM
3	94 - 95	EF95-003	MIL ON - DTC P1336
7	94 - 94	FC0008535	NEW SENSOR (STOP/TAIL LAMP)- DID INDICATES LAMP INOP
3	94 - 94	AT94-005	OBD-II - "A/T INITIAL START" TROUBLE CODE
3	94 - 94	EF94-006	OBD-II - ENG CONTROL SYS - SERVICING CAUTIONS
1	95 - 95	AT95-003	OBD-II ECM CONNECTOR BOLT - NEW - SPECS
3	94 - 94	EF94-003	POWERTRAIN ON BOARD DIAGNOSTICS (OBD II) INFO
3	94 - 94	EF94-007	REAR O2 SENSOR TORQUE PRECAUTION - SPECS
5	93 - 93	BF92-008	REMOTE CONTROLLER - TRANSMITTER PROGRAMMING
3	93 - 93	FC0008821	REMOTE LOCK INOP - REPLACE - MID-YEAR CHANGE
3,5,7	93 - 94	EL93-015	REMOTE LOCKS - TRANSMITTER PROGRAMMING PROCEDURE
5	93 - 93	FC0008448	SEAT BEL LIGHT ON - INTERMITTENT - NO CODES

Model Legend: 1= All 2= Most 3= G20 4= I30 5= J30 6= M30 7= Q45

Entertainment Devices, Stereo, Radio, Etc.

Models	Year	TSB#	Description
1	Up - 93	EL92-006	ANTENNA (POWER) - INOPERATIVE OR IMPROPER OPERATION
2	Up - 93	EL93-001	CASSETTE TAPE PLAYER - MAINTENANCE GUIDELINES
7	94 - 94	EL93-007	CD AUTO CHANGER - UNIT REPLACEMENT - ACCESSORY KITS
7	94 - 94	FC0008527	CD JAM - EXCHANGE PROGRAM
5	93 - 93	EL92-008	CD PLAYER - AFTERMARKET AUTOCHANGER INSTALL WARNING
1	Up - 94	EL93-005	CD PLAYER FAULTS - AMBIENT TEMP FACTOR
4	96 - 96	EL95-002	CELLULAR PHONE HANDS FREE OPERATION
5	93 - 93	EL93-008	CLARION CD/CASSETTE/RADIO - ERROR MESSAGES EXPLAINED
4	Up - 96	EL95-004	POOR RADIO RECEPTION - LOOSE ANTENNA GROUND
1	Up - 94	EL92-006	POWER ANTENNA ROD REPLACEMENT - REVISED
1	Up - 95	EL94-002	POWER ANTENNA ROD REPLACEMENT - REVISED
5	93 - 93	FC0008553	RADIO & A/C DISPLAY - GOES OFF INTERMITTENTLY - PROC
7	94 - 94	EL93-003	RADIO HEAD UNIT - POPPING NOISE - SERVICE PROCEDURE
7	94 - 94	FC0008403	RADIO SPEAKER POP NOISE - CAPACITORS BACKWARDS
5	93 - 93	FC0008389	RADIO TURNS ITSELF ON THEN OFF - ANTENNA UP AND DOWN
5	93 - 93	FC0008662	RADIO/WIPERS/CLIMATE CONTROL INTERMIT OP - IGN PROC
1	Up - 96	GI95-008	REMAN BOSE SPEAKERS FOR OUT-OR-WARRANTY REPAIRS
5,7	94 - 95	EL95-006	SONY CD AUTOCHANGER DIAGNOSTIC INFO

Model Legend: 1= All 2= Most 3= G20 4= I30 5= J30 6= M30 7= Q45

Seats, Belts, Interior Trim, Carpets, Air Bags

Models	Year	TSB#	Description
2	Up - 93	BF92-009	AIR BAG - SRS DIAGNOSIS AFTER A COLLISION
7,5	95 - 95	RS95-001	AIR BAG WARNING LIGHT ON - TIPS
5	93 - 93	BF93-011	AIR BAG/PASSENGER - P/Ns AND SERVICE PROCEDURES
5	93 - 95	BT95-002	B-PILLAR LOWER FINISHER LOOSE
5	93 - 95	BT95-002	B-PILLAR LOWER FINISHER LOOSE - REV
6	Up - 94	BF94-005	CENTER CONSOLE LID LATCH REPLACEMENT
4	96 - 96	BT95-010	CENTER CONSOLE LID LATCH REPLACEMENT
5	93 - 95	RS95-002	FRONT DOOR FINISHER DAMAGE FROM SEAT BELT
3,5,7	93 - 93	BF94-009	FRONT SEAT BELT EXTENDERS - CAUTION/APPLICATIONS
7	94 - 94	FC0008645	HEADREST - WON'T SET IN MEMORY IN LAN SYS - REPAIR
7	90 - 94	FC0008819	ODOR/PASS SIDE CARPET WET - COWL DRAIN
5	93 - 95	BT95-001	RATTLE NOISE - FRONT SEAT
5	93 - 95	BT95-001	RATTLE NOISE - FRONT SEAT
5	93 - 94	BF94-016	REAR PARCEL SHELF FINISHER SAGS -ALIGNMENT PROCEDURE
5	93 - 93	FC0008638	REAR SEAT AREA RATTLE - NO WINDOW LOCATING PIN FELT
5	93 - 93	BF93-002	SEAT (REAR) - SEAT BACK RATTLE - DIAGNOSIS/REPAIR
5	93 - 93	EL92-004	SEAT - RECLINER SWITCH BUTTON REVISED - CAMPAIGN
1	Up - 93	BF92-016	SEAT BELT EXTENDERS - P/N's AND APPLICATIONS
5	93 - 93	BF93-008	SEAT BELT PRE-TENSIONERS - ON-BOARD SELF-DIAGNOSIS
3	95 - 95	GI95-007	SEAT BELT/CHILD RESTRAINT - OWNERS MANUAL SUPP
5	93 - 93	BF93-007	SEAT SIDE FINISHER - MATERIAL DEFECT/REPLACE
3	91 - 93	BF92-010	SHOULDER BELT - IMPROVED OPERATION - REVISED
7	Up - 94	BF94-002	SUNVISOR VANITY MIRROR REPAIRS

Model Legend: 1= All 2= Most 3= G20 4= I30 5= J30 6= M30 7= Q45

Glass, Doors, Hood, Decklid, Tailgate, Liftgate, Locks

Models	Year	TSB#	Description
5	93 - 93	FC0008368	ALARM/DOOR LOCK - ERRATIC OPERATION/SELF GENER LOCK
1	Up - 94	BF93-014	DIAGNOSING WATER LEAKS
5	93 - 94	BF94-012	DOOR LOCK/KEY REMINDER - DIAGNOSIS & SERVICE
3	91 - 93	FC0007845	LT DOOR (OPENED W/KEY) WON'T OPEN RT DOOR - REPAIR
5,7	94 - 94	BF93-025	METALLIC WIDOW TINTS CAUSE PROBLEM/PHONE/DEFROST/ETC
5	93 - 93	BF94-001	NOISE/CREAK - SUNROOF - ADJUSTMENT
5	93 - 95	BT95-003	POWER DOOR LOCKS - INTERMITTENT/ERRATIC OPS
3,5,7	93 - 94	EL93-015	REMOTE LOCKS - TRANSMITTER PROGRAMMING PROCEDURE
3	Up - 93	BF93-006	SUNROOF - WIND NOISE/WHISTLE ABOVE 35 MPH - SERVICE
5	93 - 93	FC0008297	SUNROOF - WON'T CLOSE BUT MOTOR RUNS - SERVICE PROC
5	93 - 93	FC0008205	TRUNK LID - INOPERATIVE - FIELD FIX
5	93 - 93	FC0008249	TRUNK LID - WON'T STAY CLOSED - SERVICE PROCEDURE
3,5	92 - 94	EL94-009	TRUNK LID ACTUATOR REPLACEMENT
5	93 - 93	FC0008415	TRUNK LID RELEASE SWITCH INOP
3	Up - 93	BF93-017	WATER LEAK (BODY) DIAGNOSIS - REFERS TO BF93-014
5	Up - 93	BF93-018	WATER LEAK (BODY) DIAGNOSIS - REFERS TO BF93-014
7	Up - 94	BF93-015	WATER LEAK TROUBLE DIAGNOSIS - SEAL LOCATIONS
7	90 - 93	FC0008633	WATER LEAKS ONTO DRIVER'S SIDE CARPET - REPAIR PROC
7	94 - 94	BF94-007	WINDOW GLASS RETAINING BUTTON REPLACEMENT
5	93 - 93	FC0008584	WINDOWS - TINT COLOR/PRODUCTION CHANGE INFO

Model Legend: 1= All 2= Most 3= G20 4= I30 5= J30 6= M30 7= Q45

Finishes, Body Structure, Frame, Bumpers

Models	Year	TSB#	Description
1	Up - 95	BF94-015	3-COAT PEARLESCENT PAINT REFINISHING
7,6,3,5	90 - 93	BF92-005	BUMPER FACIA - IDENTIFICATION/REPAIR/REFINISHING
1	95 - 95	BT95-004	BUMPER FACIA REPAIR PROCEDURES
1	Up - 95	BF94-014	BUMPER REFINISHING TECHNIQUES
5,7	93 - 94	BF93-013C	CLEAR COAT (NCLC) PAINT FINISH - REFINISH PRODUCTS
5	93 - 94	BF93-023	CROSS-LINKING CLEAR COAT - SERVICE INFORMATION
1	Up - 94	BF93-014	DIAGNOSING WATER LEAKS
5	93 - 93	BF93-001	FENDER INNER PANEL - RATTLES IN UPPER DASH/DOOR AREA
4	96 - 96	BT95-004	FRONT LICENSE PLATE BRACKET INSTALLATION CAUTION
7,5	93 - 94	BF93-013B	NEW CLEAR COAT PAINT FINISH - MATERIALS UPDATE
1	94 - 95	BT95-006	NEW CROSS LINK CLEAR COAT - CODES & INFO
7,5	93 - 95	BF94-004	NEW CROSS-LINKING CLEAR COAT (NCLC) POLISH PROCEDURE
3	95 - 95	BF95-006	NEW PAINT GUARD FILM - RAPGARD F5A
1	95 - 95	BT95-008	NEW PAINT GUARD FILM INFO
1	94 - 94	BF93-024	PAINT CODES FOR 1994 MODELS
1	Up - 94	BF94-004	PAINT CONTAMINATION IDENTIFICATION AND REPAIR
7	93 - 94	BF93-013	PAINT FINISH - NEW CROSS LINK CLEAR COAT (NCLC)
7,5	93 - 94	BF93-013A	PAINT FINISH - NEW CROSS LINK CLEAR COAT - 2ND REV
7,5	93 - 94	BF93-013	PAINT FINISH - NEW CROSS LINK CLEAR COAT/NCLC - REV
1	95 - 95	BF94-013	PAINT FORMULA CODES FOR 1995 MODELS
7,5	93 - 94	BF93-022	PAINT GUARD FILM (PGF) WITH CROSS-LINKING CLEAR COAT
2	93 - 94	BF93-019	PAINT GUARD FILM - CLEAN UP PROCEDURES DURING PDI
1	93 - 93	BF93-009	PAINT GUARD FILM REMOVAL
1	93 - 93	BF92-011	PAINT- CODES/COLORS & REFINISH FORMULA CODE NUMBERS
1	96 - 96	BT95-007	REFINISH PAINT CODES
7	94 - 94	BF94-008	REMOVING WHITE STAINS IN CLEARCOAT
1	Up - 94	BF93-021	VEHICLE PAINT MAINTENANCE & STORAGE - FALLOUT
3	Up - 93	BF93-017	WATER LEAK (BODY) DIAGNOSIS - REFERS TO BF93-014
5	Up - 93	BF93-018	WATER LEAK (BODY) DIAGNOSIS - REFERS TO BF93-014
7	90 - 93	FC0008633	WATER LEAKS ONTO DRIVER'S SIDE CARPET - REPAIR PROC
1	Up - 95	BT95-005	WATERBORNE REFINISH PAINTS

Model Legend: 1= All 2= Most 3= G20 4= I30 5= J30 6= M30 7= Q45

ISUZU

Instruments, Dash Cluster, Warning Lights, Mirrors

Models	Year	TSB#	Description
2	88 - 93	93-04-004	WARNING UNIT/LOW FUEL - INSPECTION - S/M CORRECTION

Model Legend: 1= All 2= Most 3= Impulse 4= Stylus

Chassis Electrical, Wiring Harness, Fuses-Circuit Breakers, Wipers, Window Motors

Models	Year	TSB#	Description
1	Up - 96	95-14-004	ELECT TERMINAL KIT AVAILABLE

Model Legend: 1= All 2= Most 3= Impulse 4= Stylus

Heating, Air Conditioning, Ventilation, Defogger

Models	Year	TSB#	Description
1	93 - 93	93-12-002	R-12 A/C SYSTEM WARNING LABEL

Model Legend: 1= All 2= Most 3= Impulse 4= Stylus

Seats, Belts, Interior Trim, Carpets, Air Bags

Models	Year	TSB#	Description
1	Up - 95	95-11-003	SEAT BELT EXTENDER PROGRAM

Model Legend: 1= All 2= Most 3= Impulse 4= Stylus

Glass, Doors, Hood, Decklid, Tailgate, Liftgate, Locks

Models	Year	TSB#	Description
1	93 - 93	93-14-004	LABEL ADHESIVE REMOVAL/ DON'T SCRAPE GLASS - REMOVER

Model Legend: 1= All 2= Most 3= Impulse 4= Stylus

Finishes, Body Structure, Frame, Bumpers

Models	Year	TSB#	Description
1	93 - 93	92-11-010	PAINT COLORS AND CODES - 1993 MODELS

Model Legend: 1= All 2= Most 3= Impulse 4= Stylus

JAGUAR

Lighting, Horns, Turn Signals, Steering Column

Models	Year	TSB#	Description
3	95 - 95	57-19	STEER WHEEL - REPL LEATHER WITH WOOD -SERVICE ACTION
3	95 - 95	86-136	HEADLIGHTS - ADAPTER REQ'D FOR AIMING

Model Legend: 1= All 2= Most 3= Sedan Range 4= XJ12 5= XJ6 6= XJR-S 7=XJS

Instruments, Dash Cluster, Warning Lights, Mirrors

Models	Year	TSB#	Description
3,7	94 - 94	14-Apr	MIRRORS - MEMORY SEAT/AUTO DIP FUNCTIONS DELETED
3	93 - 93	76-88	DASH LINER - RETENTION - SPECIAL FASTENERS
3	94 - 94	86-131	INOP - DRIVER POWER SEAT OR MIRROR FUNCTIONS
3	94 - 94	86-131	INOP - POWER MIRROR OR DRIVER SEAT FUNCTIONS - REV
3	94 - 94	86-131R 10/94	INOP - POWER MIRROR OR DRIVER SEAT FUNCTIONS - REV
3	95 - 95	86-137	IP ILLUM EXPLAINED - MASTER SW & SPEED CONTROL
5	93 - 93	88-18	FUEL GUAGE NOT READING FULL - RECALIBRATE USING JDS

Model Legend: 1= All 2= Most 3= Sedan Range 4= XJ12 5= XJ6 6= XJR-S 7=XJS

21 TECHNICAL SERVICE BULLETINS

Chassis Electrical, Wiring Harness, Fuses-Circuit Breakers, Wipers, Window Motors

Models	Year	TSB#	Description
3	92 - 93	84-10	NOISE/JUDDER - WIPER BLADE - REVISED WIPER ARM
7	94 - 95	84-11	HEATED WINDSHIELD WASHER JETS - SERVICE ACTION S447
7	94 - 95	84-12	HEATED WINDSHIELD WASHER JETS - SERVICE ACTION S447
7	94 - 95	84-12	HEATED WINDSHIELD WASHER JETS - SRVC ACTION S447 REV
3	93 - 94	86-124	SEAT LUMBAR PUMP CIRCUIT DIAGRAMS - CORRECTION
7	Up - 93	86-122	WINDOW LIFT MOTOR - NEW PARTS & REPAIR PROCEDURE
3	95 - 95	86-138	FUSES BLOW - POWER SEAT - REVISED RATING

Model Legend: 1= All 2= Most 3= Sedan Range 4= XJ12 5= XJ6 6= XJR-S 7=XJS

Auxiliary Equipment, Jacks, Trailer Hitches, Towing

Models	Year	TSB#	Description
3	94 - 94	86-128	CELLULAR PHONE - CONNECTOR/WIRING INFORMATION
3	95 - 95	86-142	CELLULAR PHONE INSTALL INFO
3	94 - 95	86-144	CELLULAR PHONE TRANSCEIVER MULTIPLUG RELOCATED
3	95 - 95	86-144	CELLULAR PHONE TRANSCEIVER MULTIPLUG RELOCATED -REV

Model Legend: 1= All 2= Most 3= Sedan Range 4= XJ12 5= XJ6 6= XJR-S 7=XJS

Heating, Air Conditioning, Ventilation, Defogger

Models	Year	TSB#	Description
1	88 - 93	82-30	BLOWER MOTORS - NOISY ON ACCEL OR BRAKING - REV INFO
3	93 - 93	82-32	A/C - R134A REFRIGERANT INTRODUCTION
3	93 - 93	82-33	A/C - R134A REFRIG SYS - BUTYL O-RING SEALS USAGE
3	93 - 93	82-35	A/C R134A CONNECTIONS - TORQUE SPECS - S/M REVISION
3	93 - 93	82-36	A/C VACUUM RESERVOIR - NEW LOCATION
3	93 - 93	82-37	A/C DIAGNOSTICS FLOW CHART
3	91 - 94	82-38	A/C CONTROL PANEL VIBRATES OR INTERMITTENT OPERATION
1	Up - 94	82-39	R-134A RETROFIT KITS AVAILABLE
3	95 - 95	82-39R	A/C SYS - NEW FEATURES - SCHRADER VALVES & TORQUE
3	95 - 95	82-40	A/C SYS CHANGES - SCHRADER VALVES ADDED
4	95 - 95	82-41	A/C SUCTION HOSE UNION BOLT - SERVICE ACTION S621

Model Legend: 1= All 2= Most 3= Sedan Range 4= XJ12 5= XJ6 6= XJR-S 7=XJS

Entertainment Devices, Stereo, Radio, Etc.

Models	Year	TSB#	Description
3	93 - 93	86-125	RADIO - FREQUENT RE-CODING - SECURE ANTENNA CABLE
3	95 - 95	86-147	SOUND SYSTEM - TWEETER SPECIFICATION
3	95 - 95	86-149	RADIO SELECTABLE FEATURES EXPLAINED
1	Up - 95	86-150	CD AUTO CHANGER - CD BUTTON INOP

Model Legend: 1= All 2= Most 3= Sedan Range 4= XJ12 5= XJ6 6= XJR-S 7=XJS

Seats, Belts, Interior Trim, Carpets, Air Bags

Models	Year	TSB#	Description
3	95 - 95	76-106	DISCOLORED LEATHER TRIM - SERVICE ACTION S609
3	Up - 94	76-92	RATTLE - FRONT SEAT BELT HEIGHT ADJUSTER (ACTIVE)
3	93 - 94	86-124	SEAT LUMBAR PUMP CIRCUIT DIAGRAMS - CORRECTION
3	94 - 94	86-131	INOP - POWER MIRROR OR DRIVER SEAT FUNCTIONS
3	94 - 94	86-131	INOP - POWER MIRROR OR DRIVER SEAT FUNCTIONS - REV
3	94 - 94	86-131R 10/94	INOP - POWER MIRROR OR DRIVER SEAT FUNCTIONS - REV

Model Legend: 1= All 2= Most 3= Sedan Range 4= XJ12 5= XJ6 6= XJR-S 7=XJS

Glass, Doors, Hood, Decklid, Tailgate, Liftgate, Locks

Models	Year	TSB#	Description
1	Up - 93	Jun-32	KEYS - PROPER USAGE AND FUNCTION EXPLAINED
3	95 - 95	76-100	HOOD ORNAMENT - INSTALLING ON REPLACEMENT HOODS
3	95 - 95	76-101	TRUNK RELEASE BUTTON STICKS - SERVICE ACTION S614
7	95 - 95	76-103	WATER LEAK - CONVERTIBLE TOP - PROCEDURE
4	94 - 95	76-105	HOOD LINER MODIFICATION - SERVICE ACTION
7	Up - 94	76-93	CONVERTIBLE TOP REPLACEMENT INFO/PARTS
3	94 - 94	76-94	WIND NOISE AROUND WINDSHIELD - PROCEDURE
7	94 - 95	76-96	HOOD PANEL DIFFICULT TO LATCH - SERVICE ACTION S448
1	Up - 94	84-11	WINDSHIELD SMEAR FROM WIPER - CLEANING PASTE

Model Legend: 1= All 2= Most 3= Sedan Range 4= XJ12 5= XJ6 6= XJR-S 7=XJS

Finishes, Body Structure, Frame, Bumpers

Models	Year	TSB#	Description
3	95 - 95	18-44	PLENUM CHAMBER DRAIN VALVE DISTORTED - SRVC ACTION
6	95 - 95	76-102	NOISE DURING TIGHT TURNS - FRONT WHEEL ARCH LINER
1	93 - 93	76-85	PAINT - COLORS/CODES INFORMATION
3	95 - 95	76-97	ROCKER PANEL BLACKOUT AREA - SERVICE ACTION S616
3	95 - 95	76-97	ROCKER PANEL BLACKOUT AREA - SERVICE ACTION S616-REV
7,3	94 - 95	76-98	PAINT COLORS - PRIMER COLOR-KEYED TO FINISH COAT

Model Legend: 1= All 2= Most 3= Sedan Range 4= XJ12 5= XJ6 6= XJR-S 7=XJS

LEXUS

Lighting, Horns, Turn Signals, Steering Column

Models	Year	TSB#	Description
4	96 - 96	BO001-96	HEADLAMP ADJUSTMENT - RECTANGULAR ADJUSTER - S/M REV

Model Legend: 1= All 2= Most 3= ES250 4= ES300 5= GS300 6= LS400-S 7= SC300 8= SC400

Instruments, Dash Cluster, Warning Lights, Mirrors

Models	Year	TSB#	Description
6	93 - 93	EL92-003	IP - WARNING INDICATOR LENSES BUCKLED - CORRECTION
7,8	Up - 95	BO94-003	IP REGISTER ASSEMBLY REPLACEMENT PROCEDURE
7,8	Up - 93	BO92-008	MIRROR (INNER REAR VIEW) - SUPPLY PARTS CHANGED

Model Legend: 1= All 2= Most 3= ES250 4= ES300 5= GS300 6= LS400-S 7= SC300 8= SC400

Chassis Electrical, Wiring Harness, Fuses-Circuit Breakers, Wipers, Window Motors

Models	Year	TSB#	Description
7,8	Up - 93	EL93-003	COMBINATION SWITCH CONNECTOR - NEW DOUBLE LOCKING
6	93 - 93	EL92-004	CONNECTOR/LOW INSERTION FORCE (LIF) - SERVICE PROC
6	95 - 95	EL95-005	FUSE CHANGE - A/C CONTROL UNIT POWER SOURCE
5	93 - 93	EL93-001	FUSES (DOME AND ECU-B) - INSTALLATION AT PDS
1	Up - 96	BO95-004	WIPER INSERT REPLACEMENT METHOD
1	Up - 95	EL95-004	WIPER MOTOR CIRCUIT BREAKER INSPECTION PROCEDURE
7,8,5,4	Up - 95	BO94-004	WIRE TYPE WIDOW REGULATOR ASSEMBLY AVAILABLE
7,8,5,4	Up - 95	BO94-004	WIRE TYPE WIDOW REGULATOR ASSEMBLY AVAILABLE - REV
6	93 - 93	PG003-96	WIRING ROUTING - S/M REVISIONS AVAILABLE

Model Legend: 1= All 2= Most 3= ES250 4= ES300 5= GS300 6= LS400-S 7= SC300 8= SC400

Auxiliary Equipment, Jacks, Trailer Hitches, Towing

Models	Year	TSB#	Description
6	92 - 93	AX92-002	CELLULAR PHONE MODULE/NEW - HANDS-FREE MODE IMPROVED
6	95 - 95	EG95-001	HEAVY DUTY RADIATOR FOR TOWING AVAILABLE
1	Up - 94	AU93-002	MOBILE COMMUNICATION EQUIP - INSTALL PRECAUTIONS
6	95 - 95	AX95-001	MUDGUARD INSTALLATION PROCEDURES

Model Legend: 1= All 2= Most 3= ES250 4= ES300 5= GS300 6= LS400-S 7= SC300 8= SC400

Heating, Air Conditioning, Ventilation, Defogger

Models	Year	TSB#	Description
3	Up - 93	AC92-002	A/C - DRONE NOISE AT IDLE - DIAGNOSIS/REPAIR KITS
1	Up - 93	AC93-003	A/C - INSPECTION/DIAG/REPAIR OF LOW REFRIG LEVEL
1	Up - 93	AC92-003	A/C - O-RING COLOR CHANGES DUE TO NEW REFRIGERANTS
1	Up - 94	AC93-004	A/C COMPRESSOR OIL APPLICATIONS
5	Up - 96	AC002-96	A/C CONTROL ASSEMBLY SERVICE PARTS
6	95 - 95	EL95-005	A/C CONTROL UNIT POWER SOURCE CHANGED
1	Up - 95	AC95-001	A/C REFRIGERANT LEAKS - SERVICE TIPS
6	93 - 93	AC92-004	A/C REFRIGERANT PRESSURE CONVERSION - METRIC TO PSIG
5	93 - 93	PG003-96	A/C SYSTEM - S/M REVISIONS AVAILABLE
6,5	Up - 93	AC93-001	A/C W/HFC134a REFRIGERANT - NEW COMPONENTS & O-RINGS
1	Up - 93	AC94-001	ALTERNATIVE REFRIGERANTS FOR R12
1	Up - 96	AC001-96	AMBIENT TEMP DISPLAYS -22 DEG F - PROCEDURE
1	93 - 93	AC93-002	HFC134a REFRIGERANT - HANDLING PRECAUTIONS

Model Legend: 1= All 2= Most 3= ES250 4= ES300 5= GS300 6= LS400-S 7= SC300 8= SC400

Entertainment Devices, Stereo, Radio, Etc.

Models	Year	TSB#	Description
7,8	92 - 94	EL95-006	AM RADIO - STATIC NOISE ON WEAK STATIONS
6	93 - 93	AU93-001	CD CHANGER - "OPEN" BUTTON IMPROVEMENT
7,8	92 - 93	AU94-001	CD CHANGER INOP/SKIP/DISPLAYS ERROR - PROCEDURE
6	93 - 93	AU93-003	PIONEER SOUND SYS - IMPROVED BASS - WOOFER AMPLIFIER
1	Up - 94	AU94-002	RADIO MAIN AMPLIFIER INOP - REVISED AMP

Model Legend: 1= All 2= Most 3= ES250 4= ES300 5= GS300 6= LS400-S 7= SC300 8= SC400

Seats, Belts, Interior Trim, Carpets, Air Bags

Models	Year	TSB#	Description
7,8,6	93 - 93	EL93-004	AIRBAG/PASSENGER - NEW WIRE HARNESS CONNECTOR DESIGN
4	Up - 93	BO92-007	FLOOR CARPET/PASSENGER'S - PULLS DOWN - CORRECTION
4,5	Up - 95	EL95-001	FRONT PASS AIRBAG ASSEMBLY - REPL PROCEDURE REVISED
7,8	Up - 95	BO95-002	FRONT SEAT BACK PROTECTOR CHANGED
6	95 - 95	BO94-005	POP/SQUEAK NOISE - FRONT SEAT CUSHION
6	95 - 95	BO95-005	RATTLE/CREAK NOISE FROM SUNVISOR
1	Up - 93	BO93-005	SEAT BELT EXTENDER - AVAILABLE AT NO COST - PROC
1	Up - 96	BO95-006	SEAT BELT EXTENDERS
1	Up - 94	BO94-001	SEAT BELT EXTENDERS AVAILABLE - NO COST
7,8	93 - 93	BO93-003	SEAT/FRONT - CUSHION SHIELD WEAR IMPROVED
4	93 - 93	BO92-010	SEAT/FRONT - HEADREST SUPPORT RATTLES - MODIFICATION
1	Up - 93	EL93-002	SRS - BACKUP POWER SOURCE - SERVICE INFORMATION
7,8	94 - 94	PG95-001	SRS ASSEMBLY CONNECTOR - S/M REV
4	92 - 93	BO92-009	SUN VISOR - RATTLES - DIAGNOSIS AND REPAIR
5	94 - 94	BO94-002	SUNSHADE TRIM KNOB REDESIGNED

Model Legend: 1= All 2= Most 3= ES250 4= ES300 5= GS300 6= LS400-S 7= SC300 8= SC400

Glass, Doors, Hood, Decklid, Tailgate, Liftgate, Locks

Models	Year	TSB#	Description
4	93 - 93	BO93-002	DOOR ARMREST PLUG - MODIFICATION
6	92 - 93	BO93-008	MOONROOF - WATER LEAKAGE - REPAIR PROCEDURE
7,8	Up - 95	BO95-003	MOONROOF PANEL SERVICE TIPS
4	Up - 96	BO95-007	MOONROOF PANEL SERVICE TIPS
4	Up - 96	BO95-008	MOONROOF PANEL WIND NOISE
7,8	Up - 95	BO95-001	POP NOISE OR DOOR CONTACTS FENDER - MODIFIED HINGE
7,8	93 - 93	BO93-001	WINDOW GLASS - CREAKS/POPS RAISING/LOWERING - REPAIR

Model Legend: 1= All 2= Most 3= ES250 4= ES300 5= GS300 6= LS400-S 7= SC300 8= SC400

Finishes, Body Structure, Frame, Bumpers

Models	Year	TSB#	Description
5	93 - 93	BO93-006	BODY - HEADLAMP ASM INSTALLATION HOLES - DIMENSIONS
5	94 - 94	PG93-010	BUMPER - FRONT - S/M REVISION
7,8	94 - 94	PG95-001	FRONT BUMPER COMPONENTS - S/M REV
5	Up - 93	BO93-007	LICENSE PLATES/FRONT & REAR - INSTALLATION PROCEDURE
6	93 - 93	AX92-003	MUDGUARD KIT - INSTALLATION INSTRUCTIONS
1	95 - 95	PA94-002	PAINT & REFINISH CODES - 1995 MODELS
1	96 - 96	PA002-96	PAINT & REFINISH FORMULA CODES
7,8	93 - 93	PA93-001	PAINT - ANTI-CHIP PRIMER APPLIED TO FRONT OF HOOD
1	94 - 94	PA94-001	PAINT AND REFINISH CODES - 1994 MODELS
1	93 - 93	PA93-002	PAINT AND REFINISH CODES - UPDATED
1	93 - 93	PA92-004	PAINT AND REFINISH CODES FOR 1993 VEHICLES
6,5,4	95 - 95	PA95-001	PAINT COLORS ON CLADDING AND BUMPERS
1	Up - 93	PA92-003	PAINT GUIDE AND REPAIR PROCEDURES
1	95 - 96	PA95-002	PAINT REPAIRS ON POLYURETHANE BUMPERS
8	93 - 93	PG93-002	SPOILER/FRONT-TRANSIT PROTECTION SPACER-PDS REMOVAL
4	Up - 94	BO93-009	WATER LEAK - PASSENGER SIDE FLOOR

Model Legend: 1= All 2= Most 3= ES250 4= ES300 5= GS300 6= LS400-S 7= SC300 8= SC400

MAZDA

Lighting, Horns, Turn Signals, Steering Column

Models	Year	TSB#	Description
6	95 - 95	T-001/95	A/T POSITION INDICATOR LIGHT FAILURE
3	95 - 95	T-003/95	COURTESY LIGHT LENS REMOVAL
4,9	93 - 95	TIPS12/94	DOME LAMP INOP - INTERMITTENT - TIPS
4,9	93 - 94	T-008/94	DOOR KEY CYL NOT LIT AND/OR CPU FAILS - HARNESS
6	95 - 95	N-004/95	HARD EFFORT TO UNLOCK STEERING LOCK SYSTEM
3	95 - 95	S-012/95	INBOARD COMBINATION LIGHT REPLACE PROCEDURE
9	93 - 93	S-003/93	LAMP (HMSL) - RATTLE NOISE REPAIR (W/O OEM SPOILER)
9	93 - 95	T-002/95	LAMP - SPOILER HIGH MOUNT STOP - SERVICE PROCEDURE
5	93 - 93	Z-002/93	LIGHTING - ELECTRICAL TROUBLESHOOTING MAN - H/L REV
5	93 - 93	B-005/93	LIGHTS - "AUTO LIGHT OFF" SYS OPER CHANGES - REVISED
5	93 - 93	B-005/93	LIGHTS - "AUTO LIGHT OFF" SYSTEM OPERATION CHANGES
6	95 - 95	N-002/95	STEERING COLUMN - AUTO-TILT INOPERATIVE
6	95 - 95	TIPS08/94	STEERING WHEEL LOCKED - IGNITION KEY WILL NOT TURN
3	95 - 95	TIPS09/94	TAIL LIGHT LENS MOISTURE MAY BE NORMAL - TIP

Model Legend: 1= All 2= Most 3= 323 4= 626 5= 929 6= Millenia 7= MX-3, MX-3GS 8= MX5, Miata 9= MX-6

Instruments, Dash Cluster, Warning Lights, Mirrors

Models	Year	TSB#	Description
7	93 - 93	S-036/93	ASHTRAY - LID DOES NOT OPEN FULLY - REPAIR PROCEDURE
4,9	93 - 94	S-032/94	ASHTRAY WILL NOT OPEN OR OPENS UNINTENTIALLY
9,4	93 - 93	Z-016/93	CRUISE CONTROL - ACTUATOR R&I - BETM CORRECTION
6	95 - 95	T-006/94	CRUISE CONTROL SURGING - KL ENGINE - MOD PART
10	93 - 95	TIPS04/95	DOOR MIRROR VIBRATION
4,9	93 - 94	TIPS06/94	GLOVE BOX DOOR RATTLE
5	92 - 93	K-017/92	HOLD INDICATER LIGHT FLASHES - FALSE CODE 57 SHOWN
7	92 - 93	S-017/93	I/P - CRACKING NOISE WITH A/C ON AFTER SIT IN SUN
6	95 - 95	T-006/95	INFO DISPLAY HARD TO READ WITH HEADLIGHTS ON
4,9	93 - 94	S-033/94	REAR VIEW MIRROR FALLS OFF
4,9	93 - 94	S-033/94	REAR VIEW MIRROR FALLS OFF - REV
10	93 - 94	S-020/95	VIBRATION - DOOR MIRROR
4	93 - 94	S-012/94	WIND NOISE NEAR SIDE MIRRORS & A-PILLAR

Model Legend: 1= All 2= Most 3= 323 4= 626 5= 929 6= Millenia 7= MX-3, MX-3GS 8= MX5, Miata 9= MX-6 10= MX7

Chassis Electrical, Wiring Harness, Fuses-Circuit Breakers, Wipers, Window Motors

Models	Year	TSB#	Description
4	93 - 95	TIPS09/94	DEAD BATTERY/ALARM PROBLEMS - TRUNK LAMP SWITCH -TIP
4,9	93 - 94	T-008/94	DOOR KEY CYL NOT LIT AND/OR CPU FAILS - HARNESS
4,9	93 - 94	S-052/93	DOOR KEY ILLUMINATION INOP - BULB/SWITCH/HARNESS
8	90 - 94	T-006/93	INSTALLATION OF "ROOM" FUSE -TAPE DECK REJECTS TAPES
3	90 - 94	TIPS06/94	RATTLE OR WINDOW HARD TO OPEN - REGULATOR BOLTS
6	95 - 95	T-004/94	SPARE FUSE LOCATIONS
4	93 - 93	TIPS08/94	WINDOW REGULATOR - ADAPTING '94 STYLE TO '93 MODELS
8	90 - 93	S-038/93	WINDOW REGULATOR HARD OPERATION - SPACER
8	92 - 94	S-011/94	WINDOW WILL NOT OPEN FULLY - ROTATE MOTOR
4,9	93 - 93	Z-004/92	WIRING DIAGRAM (ABS) - CORRECTION

Model Legend: 1= All 2= Most 3= 323 4= 626 5= 929 6= Millenia 7= MX-3, MX-3GS 8= MX5, Miata 9= MX-6

Heating, Air Conditioning, Ventilation, Defogger

Models	Year	TSB#	Description
1	93 - 94	U-002/93	A/C (R-12 CFC SYS) - WARNING LABEL INFORMATION
4,9	93 - 95	U-007/95	A/C - INSUFFICIENT COOLING - RELAY FAILURE
2	88 - 93	U-016/92	A/C - O-RING INSTALLATION PROCEDURE
2	Up - 94	U-003/93	A/C - R134A SERVICE PRECAUTIONS & PARTS COMPARISON
2	Up - 94	U-003/93	A/C - R134A SERVICE PRECAUTIONS/PARTS REVISED
2	94 - 94	U-002/94	A/C 134A SIGHT GLASS ELIMINATED - INSPECT PROCEDURE
6	95 - 95	U-003/94	A/C EVAP FREEZING - RELOCATE TEMP SENSOR
6	95 - 95	U-003/94	A/C EVAP FREEZING - RELOCATE TEMP SENSOR - REVISED
4,9	93 - 95	TIPS10/94	A/C EVAP HOSE LEAK - TIPS
1	88 - 94	U-001/94	A/C O-RING REPLACEMENT - APPLICATIONS
1	88 - 94	U-001/94	A/C O-RING REPLACEMENT - APPLICATIONS - REVISED
3	95 - 95	TIPS09/94	A/C POOR PERF - COMPRESSOR BOLTS - TIP
6	95 - 95	TIPS08/94	A/C POOR PERF - WATER LEAK ONTO PASS SIDE FLOOR
7	92 - 93	U-017/92	A/C POOR PERFORMANCE - MODIFIED SUB-HARNESS
2	Up - 95	U-001/95	A/C RECEIVER DRIER REPLACEMENT CRITERIA
3	95 - 95	U-002/95	BLOWER MOTOR BRUSH NOISE
3	95 - 95	S-008/95	BLOWER MOTOR NOISE - DEAD LEAVES
9,4	93 - 93	T-020/92	DEFROSTER (REAR WINDOW) - OPERATION CLARIFIED
4,9	93 - 93	U-003/95	HEATER UNIT CASE SERVICE - PART MODIFICATION
4	93 - 95	TIPS04/95	HVAC - SWING LOUVER INOP - OPERATION EXPLAINED
3	95 - 95	T-010/95	REAR DEFROSTER & INDICATOR LIGHT INOP - SWITCH
5	91 - 94	T-001/94	UNINTENDED REAR DEFROSTER OPERATION
5	92 - 93	T-001/94	UNINTENDED REAR DEFROSTER OPERATION
4,9	93 - 95	U-004/95	WHISTLE NOISE WHEN A/C ENGAGES - EXPANSION VALVE

Model Legend: 1= All 2= Most 3= 323 4= 626 5= 929 6= Millenia 7= MX-3, MX-3GS 8= MX5, Miata 9= MX-6

Entertainment Devices, Stereo, Radio, Etc.

Models	Year	TSB#	Description
5,9,4,10	92 - 93	T-017/92	ANTENNA MAST - REPLACEMENT & SERVICE INFORMATION
6	95 - 95	T-003/94	AUDIO "BLANK/SKIP" FUNCTION EXPLAINED
7	95 - 95	T-007/95	AUDIO SPECIFICATIONS
10	94 - 94	TIPS10/94	AUDIO SYSTEM SPEAKER INTERMITTENT - GROUND
8	95 - 95	TIPS03/95	CD AUDIO UNIT CHANGES
6	95 - 95	TIPS06/94	CD PLAYER - RANDOM FEATURE EXPLAINED
8	Up - 94	TIPS10/94	CD PLAYER SKIPS - EXCHANGE UNIT
7	95 - 95	T-007/95	MISMATCHED AUDIO UNITS
2	Up - 95	T-005/95	POWER ANTENNA MAST REPLACEMENT
9	95 - 95	TIPS09/94	RADIO - NEW FMS - RESETTING AFTER POWER LOSS - TIP
10	93 - 93	T-018/92	RADIO W/ CD PLAYER - REMOVAL TIP & TOOLS
3	90 - 94	T-007/93	REAR SPEAKERS INOP/INTERMITTENT
8	90 - 94	T-006/93	TAPE DECK REJECTS TAPE AFTER INSTALLING "ROOM" FUSE

Model Legend: 1= All 2= Most 3= 323 4= 626 5= 929 6= Millenia 7= MX-3, MX-3GS 8= MX5, Miata 9= MX-6 10= MX7

Seats, Belts, Interior Trim, Carpets, Air Bags

Models	Year	TSB#	Description
9,4	93 - 93	T-002/93	AIR BAG - NEW SAFETY FEATURE - SERVICE INFORMATION
7	93 - 93	S-036/93	ASHTRAY - LID DOES NOT OPEN FULLY - REPAIR PROCEDURE
6	95 - 95	S-027/94	ASSIST HANDLE (CEILING MOUNT) CAP DETACHES
5	92 - 93	S-028/93	ASSIST HANDLE - CAP COMES OFF - REPAIR PROCEDURES
7	92 - 93	S-018/93	CARPET COMES OUT OF SCUFF PLATE/UNSIGHTLY APPEARANCE
4	93 - 94	S-030/94	CENTER CONSOLE CUP HOLDER - RETAINING CLIP POPS OUT
6	95 - 95	S-039/94	CUPHOLDER OPERATION - MODIFIED HOLDER
4,9	93 - 94	S-028/94	DOOR INTERIOR TRIM - REMOVE/INSTALL PROCEDURES
6	95 - 95	TIPS03/95	DOOR TRIM PANEL REMOVAL CAUTIONS
6	95 - 95	TIPS09/94	FRONT DOOR TRIM PANEL - REMOVE/INSTALL TIPS
5	92 - 94	S-006/94	FRONT DOOR TRIM SPLIT OR WARPED
4	93 - 93	S-035/93	HEADLINER - RATTLE FROM REAR - REPAIR PROCEDURE
4	93 - 94	S-029/94	HEADLINER RATTLE - SUNROOF AREA
10	93 - 93	S-056/92	INTERIOR TRIM - BLACK FINISH PEELING - REPAIR PROC
3,7	90 - 95	S-009/95	LUGGAGE COMP COVER HINGE HOLDER BREAKAGE
5	92 - 93	S-050/93	NOISE/RATTLE FROM CENTER CONSOLE - LOOSE CONSOLE LID
5	92 - 93	S-050/93	NOISE/RATTLE FROM CENTER CONSOLE - REVISED
5	92 - 95	TIPS12/94	RATTLE - REAR PACKAGE TRIM (PARCEL SHELF)
9	93 - 93	S-034/93	REAR PACKAGE TRAY RATTLE - REPAIR PROCEDURE
7	92 - 93	S-053/93	REAR SEAT CUSHION SEPARATES AT SEAMS - NEW MATERIAL
7	92 - 93	S-053/93	REAR SEAT CUSHION SEPARATES AT SEAMS - REVISED
6	95 - 95	S-023/94	ROOF INSULATOR PEELING OFF
6	95 - 95	S-023/94	ROOF INSULATOR PEELING OFF - REVISED
7	92 - 93	S-005/93	SEAT (FRONT) - TRIM MATERIAL SNAGGING/LOOKS WORN OUT
10,8	93 - 94	S-001/94	SEAT BELT LABEL COVER DROPS - INSPECT BELTS
6	95 - 95	S-002/95	SNAP NOISE - BASE OF A PILLAR IN DASH
4,9	93 - 94	S-035/94	SQUEAK IN MANUAL DRIVER'S SEAT
3	90 - 93	S-008/94	WATER LEAK THROUGH FIREWALL INTO PASS AREA- REVISED
3	90 - 93	S-054/93	WATER LEAK THROUGH FIREWALL INTO PASSENGER AREA- REV
3	90 - 93	S-054/93	WATER LEAK THROUGH FIREWALL NEAR HEATER PIPING -SEAL

Model Legend: 1= All 2= Most 3= 323 4= 626 5= 929 6= Millenia 7= MX-3, MX-3GS 8= MX5, Miata 9= MX-6 10= MX7

Glass, Doors, Hood, Decklid, Tailgate, Liftgate, Locks

Models	Year	TSB#	Description
4	93 - 94	S-018/94	ADJUSTMENT OF B-PILLAR DOOR SASH
4	93 - 94	S-031/94	B-PILLAR WEATHERSTRIP WEAR/DAMAGE
3	95 - 95	S-005/95	BUMP NOISE WHEN OPENING SUNROOF
3	95 - 95	S-015/95	CHATTER NOISE - FRONT POWER WINDOW
8	90 - 94	S-044/93	CONVERTIBLE TOP ZIPPER REPLACEMENT PROCEDURES
5	92 - 93	S-030/93	CRACKED DOOR GRIP - FRONT & REAR - REVISED
3	95 - 95	S-001/95	CRACKING NOISE AT B & C PILLAR
4,9	93 - 94	TIPS07/94	DOOR & TRUNK KEY CYLINDERS HARD TO MOVE
10	93 - 93	S-026/93	DOOR - OUTER HANDLE RATTLES - DIAGNOSIS/REPAIR
3	95 - 95	TIPS08/94	DOOR FRAME SEAMING WELT HANGS DOWN
9	93 - 93	S-040/92	DOOR GLASS - RATTLE - REPAIR PROCEDURE/WARRANTY INFO
4	93 - 94	S-013/94	DOOR GLASS COMES OFF TRACK
10	93 - 93	S-033/93	DOOR GRIP (PASSENGER SIDE) - MAY BREAK - REPAIR PROC
5	92 - 93	S-030/93	DOOR GRIP/SWITCH PANEL - MAY CRACK - DIAG & REPAIR
4,9	93 - 94	S-052/93	DOOR KEY ILLUMINATION INOP - BULB/SWITCH/HARNESS
4,9	93 - 93	T-003/93	DOOR LOCK CYL - WATER ENTRY - THEFT DETERRENT INOP
5	92 - 93	S-023/93	DOOR WINDOWS (FRONT/REAR) - THUMP WHEN OPENING - FIX
5	92 - 93	S-023/93	DOOR WINDOWS (FRONT/REAR) - THUMP WHEN OPENING - REV
10	93 - 93	S-002/93	DOOR WINDOWS - WIND NOISE - REPAIR INFORMATION
5	92 - 93	S-048/93	DRAGGING/STICKING SUNROOF - REVISED
6	95 - 95	S-018/95	FRONT DOOR WEATHERSTRIP OUT OF POSITION
5	92 - 94	S-051/93	FRONT DOOR WEATHERSTRIP TWISTED - NEW DESIGN STRIP
4,5,7,9	92 - 94	S-021/94	FROZEN LOCK CYLINDERS (DOOR OR HATCH) - PROCED - REV
4,5,7,9	92 - 94	S-021/94	FROZEN LOCK CYLINDERS (DOOR OR HATCH) - PROCEDURE
7	92 - 93	S-004/94	GLASS RUN CHANNEL WRINKLES AT CORNER
10	93 - 93	S-057/92	HOOD - SQUEAKING SOUND - REPAIR PROCEDURE
9	93 - 93	S-042/92	HOOD - VIBRATION - INSPECTION & REPAIR PROCEDURE
10	93 - 95	TIPS02/95	HOOD RELEASE INOP OR STRONG EFFORT
4	93 - 93	S-037/94	INSTALLING '94 WINDOW REGULATOR IN '93 MODEL
9,4	93 - 93	K-018/92	KEY - CAN'T BE REMOVED FROM IGN SWITCH - DIAG/REPAIR
1	Up - 95	AD-005/94	KEY REPLACEMENT INFO - CALIFORNIA ONLY
6	95 - 95	AD-001/94	KEY REPLACEMENT PROGRAM
6	95 - 95	AD-001/94	KEY REPLACEMENT PROGRAM - REVISED
5	92 - 93	S-048/93	NOISE/CHATTER - SUNROOF - NEW POSITION PLATE WASHER
7	93 - 93	S-047/93	NOISE/RATTLE - REAR HATCH - MODIFIED HINGES
3	95 - 95	TIPS12/94	OUTER DOOR HANDLE LOOSE - SELF-LOCKING NUTS
3	95 - 95	TIPS12/94	POWER DOOR LOCKS - OPERATION EXPLAINED
6	95 - 95	TIPS08/94	RATTLE - SUNROOF FRAME - PROCEDURE
3	95 - 95	S-021/95	RATTLE NOISE FROM HOOD
4	93 - 95	S-010/95	RATTLE NOISE IN SUNROOF

Model Legend: 1= All 2= Most 3= 323 4= 626 5= 929 6= Millenia 7= MX-3, MX-3GS 8= MX5, Miata 9= MX-6 10= MX7

Glass, Doors, Hood, Decklid, Tailgate, Liftgate, Locks

Models	Year	TSB#	Description
3	90 - 94	S-026/94	REAR DOOR WEATHERSTRIP BREAKAGE
10	93 - 93	S-010/93	REAR HATCH - HINGE NOISY OVER BUMPS - REPAIR PROC
8	Up - 93	S-024/93	SOFT TOP PLASTIC WINDOWS - PROTECTION PROCEDURES
8	90 - 93	S-007/94	SOFT TOP VELCRO TAPE PEELS OFF
5	92 - 94	S-006/94	SPLIT/WARPED FRONT DOOR TRIM - MODIFIED TRIM PANEL
9	93 - 94	TIPS06/94	SQUEAK - WINDOW TO WEATHERSTRIP - TIPS
5	92 - 93	S-009/94	SQUEAK NOISE - OPEN/CLOSE DOOR - NEW DOOR CHECKER
10	93 - 93	S-010/94	SQUEAK NOISE - OPEN/CLOSE DOOR - NEW DOOR CHECKER
4	93 - 94	S-046/93	SUNROOF BINDING - PROCEDURES
5	92 - 94	S-013/95	SUNROOF CHATTER NOISE WHEN USING TILT FUNCTION
9	93 - 93	S-045/92	TRUNK - WATER ENTRY - HINGE GASKET MISSING
5	92 - 94	TIPS10/94	WATER/WIND LEAK - WEATHERSTRIP & GLASS ADJUST
4	93 - 94	S-012/94	WIND NOISE NEAR A-PILLAR AND SIDE MIRRORS
3	90 - 94	S-022/93	WINDOW REGULATOR - DIFFICULT TO OPERATE - REVISED
3	90 - 94	S-036/94	WINDOW REGULATOR - DIFFICULT TO OPERATE - REVISED
8	92 - 94	S-011/94	WINDOW WILL NOT OPEN FULLY - ROTATE MOTOR
10	93 - 93	S-031/93	WINDOW/POWER - OPERATION NOISE - DIAG/CORRECTION
9	93 - 93	S-027/93	WINDOWS (DRIVER/PASSENGER) - RATTLE/SQUEAK - REPAIR
5	92 - 93	S-014/93	WINDOWS - SMEARED AFTER OPERATION - DIAGNOSIS/REPAIR
9	93 - 93	S-016/93	WINDOWS - SQUEAK WHEN CLOSING - DIAG & REPAIR PROC
9	93 - 93	S-016/93	WINDOWS SQUEAK WHEN CLOSING - DIAG/REPAIR PROC - REV
10	93 - 93	R-014/92	WINDSHIELD - WHISTLING NOISE FROM MOULDING - REPAIR
7	92 - 93	S-017/93	WINDSHIELD AREA CRACKING NOISE - REPAIR PROCEDURE

Model Legend: 1= All 2= Most 3= 323 4= 626 5= 929 6= Millenia 7= MX-3, MX-3GS 8= MX5, Miata 9= MX-6 10= MX7

Finishes, Body Structure, Frame, Bumpers

Models	Year	TSB#	Description
5	93 - 93	S-015/92	BODY - NEW PROTECTIVE VEHICLE FILM - HANDLING PROC
8	90 - 95	S-034/94	CLOGGED SIDE SILL DRAIN HOLES
3	95 - 95	S-004/95	COLOR COATING ON RADIATOR SHROUD ELIMINATED
4,9	93 - 93	S-041/92	LICENSE PLATE BRACKET - INSTALLATION PROCEDURE
2	Up - 95	TIPS10/94	LICENSE PLATE FRAME INSTALLATION - TIP
4,9	93 - 93	S-054/92	LICENSE PLATE/FRONT - BRACKET FALLS OUT OF POSITION
4,9	93 - 94	S-020/94	LOOSE FRONT AIR DAM SKIRT
4,9	93 - 94	S-020/94	LOOSE FRONT AIR DAM SKIRT - REVISED
4,9	93 - 94	S-020/94R MAY`94	LOOSE FRONT AIR DAM SKIRT - REVISED
5	92 - 93	S-014/93	MOULDINGS (BELTLINE) - REPLACE - SMEARING WINDOWS
3	95 - 95	TIPS07/94	PAINT CODES
4	95 - 95	TIPS09/94	ROOF MOULDING POPS UP NEAR WINDSHIELD/WINDOW - TIP
1	92 - 93	S-026/92	TRANSIT COATING REMOVAL - SPECIAL SOLUTION
1	95 - 95	TIPS02/95	TRANSIT COATING REMOVAL TIPS
3	90 - 93	S-008/94	WATER LEAK INTO PASS AREA NEAR HEATER PIPE - REVISED
3	90 - 93	S-054/93	WATER LEAK THROUGH FIREWALL INTO PASSENGER AREA
3	90 - 93	S-054/93	WATER LEAK THROUGH FIREWALL NEAR HEATER PIPING - REV

Model Legend: 1= All 2= Most 3= 323 4= 626 5= 929 6= Millenia 7= MX-3, MX-3GS 8= MX5, Miata 9= MX-6

MERCEDES-BENZ

Lighting, Horns, Turn Signals, Steering Column

Models	Year	TSB#	Description
479	92 - 94	82/81	BOND HEADLAMP LENS TO HOUSING - SERVICE ACTION
479	94 - 94	82/60	HEADLAMP/FOGLAMP ADJUSTMENT PROCEDURES
7	95 - 95	Jan-95	MOISTURE IN HORN - VENT SYSTEM CHANGED
7	95 - 95	Aug-95	TRUNK LAMP ACTUATING LEVER DELETED

Model Legend: 1= All 2= Most 3=SL to 1989 4= 1986 and up D, E, CE, TE 5= 1981-89 SD, SE 6= 1990 and up SL 7= 1992 and up SD,SE 8= 190 Series 9= 202 Series

21 TECHNICAL SERVICE BULLETINS

Instruments, Dash Cluster, Warning Lights, Mirrors

Models	Year	TSB#	Description
9	Up - 95	Jun-95	ADJUSTING ODOMETER ON REPLACEMENT IC
2	Up - 93	30/05	CRUISE CONTROL - IDENTIFYING PROPER DIAGNOSTIC PROC
9	Up - 94	RC#94-0421	CRUISE CONTROL LINKAGE - RECALL - REVISED KIT
4	95 - 95	21/95	CRUISE CONTROL SURGES - TIPS
679	Up - 94	AM#661	I/C REPLACEMENT - PARTS INFO - ODOMETER SETTINGS
6	Up - 94	14/95	IC - OUTSIDE TEMP GAUGE AVAILABLE - ANALOG ODOMETER
8	95 - 95	AM#513	IC CLOCK RESETS ITSELF - IMPROVED IC
7	Up - 94	Apr-95	LCD IC INSTALLED - ALTERNATOR LAMP STAYS LIT
9	94 - 94	RC#94-0421	MAINT FREE CRUISE CONTROL LINKAGE - RECALL
9	94 - 94	RC#94-0421	MAINT FREE CRUISE CONTROL LINKAGE - RECALL REVISED
7	92 - 93	REF: 00/51	MIRROR (OUTSIDE REARVIEW) - VALUE COMMITMENT PROGRAM
7	Up - 93	00/284A	MIRRORS (REAR VANITY/EXTERIOR) - WORK INSTRUCTIONS
3,457	Up - 93	54/39	SPEEDOMETER/ELECTRONIC - CHECKING PROC PRIOR TO R&R
4	Up - 93	AM#009	WARN LAMP (ABS) - ERRATIC OPER - FAULTY RELAY (HFM)

Model Legend: 1= All 2= Most 3= SL to 1989 4= 1986 and up D, E, CE, TE 5= 1981-89 SD, SE 6= 1990 and up SL 7= 1992 and up SD,SE 8= 190 Series 9= 202 Series

Chassis Electrical, Wiring Harness, Fuses-Circuit Breakers, Wipers, Window Motors

Models	Year	TSB#	Description
4	Up - 93	30/11	HARNESS (EA/CC/ISC ACTUATOR) - ROUTING CAUTION
48	Up - 93	42/293	HARNESS RE-ROUTING - ABS MIL ON AFTER ENG START
7	Up - 93	AM#546	REAR WINDOW LIFTER - IMPROVED PART PREVENTS BREAKAGE

Model Legend: 1= All 2= Most 3= SL to 1989 4= 1986 and up D, E, CE, TE 5= 1981-89 SD, SE 6= 1990 and up SL 7= 1992 and up SD,SE 8= 190 Series 9= 202 Series

Auxiliary Equipment, Jacks, Trailer Hitches, Towing

Models	Year	TSB#	Description
49	Up - 94	AM#147	CELLULAR PHONE - DROPPED CALLS
1	Up - 93	AM#120	CELLULAR PHONE - WARRANTY PARTS REPLACEMENT

Model Legend: 1= All 2= Most 3= SL to 1989 4= 1986 and up D, E, CE, TE 5= 1981-89 SD, SE 6= 1990 and up SL 7= 1992 and up SD,SE 8= 190 Series 9= 202 Series

Heating, Air Conditioning, Ventilation, Defogger

Models	Year	TSB#	Description
9	95 - 95	RC#95-0801	A/C PUSHBUTTON CONTROL MODULE SOFTWARE - RECALL
7	Up - 95	MEMO/GRP 83	A/C PUSHBUTTON MODULES - HIGH BLOWER SPEED - REV MOD
7	93 - 94	83/79	A/C SYSTEM MODIFICATIONS EXPLAINED
8	95 - 95	AM#058	ACC PUCHBUTTON MODULE CHANGES SETTINGS
9	95 - 95	17/95	ACC PUSHBUTTON MODULE REVISED
7	Up - 95	Nov-95	AUXILIARY FANS INOP - MODIFIED RELAY
1	Up - 94	83/81A	CONVERTING R12 TO R134A - INFO ON RETROFIT

Model Legend: 1= All 2= Most 3= SL to 1989 4= 1986 and up D, E, CE, TE 5= 1981-89 SD, SE 6= 1990 and up SL 7= 1992 and up SD,SE 8= 190 Series 9= 202 Series

Entertainment Devices, Stereo, Radio, Etc.

Models	Year	TSB#	Description
1	92 - 93	AM#580	RADIO - POOR AM RECEPTION/DOWNTOWN AREAS - REPAIR
47	94 - 94	AM#560	RADIO REPLACEMENT - ERRATIC OPERATION - REVISED
47	94 - 94	AM#560	RADIO REPLACEMENT - WON'T TURN ON OR ON BY ITSELF
7	Up - 93	82/66	RADIO/CD CHANGER - MAGAZINE WARPED - INFORMATION
1	Up - 93	82/68	RADIOS/BECKER - LIQUID INTRUSION/POST-WTY VEHICLES
7	93 - 93	AM#350	SPEAKERS (DOOR) - DAMAGED FROM DOOR PANEL R&I - INFO

Model Legend: 1= All 2= Most 3= SL to 1989 4= 1986 and up D, E, CE, TE 5= 1981-89 SD, SE 6= 1990 and up SL 7= 1992 and up SD,SE 8= 190 Series 9= 202 Series

Seats, Belts, Interior Trim, Carpets, Air Bags

Models	Year	TSB#	Description
1	Up - 93	91/48	AIRBAG - UNIT REPLACEMENT EXTENDED FROM 10 TO 15 YRS
7	94 - 94	91/52	DRIVER'S SEAT BACK "PLAY" - REVISED PRODUCTION/PART
9	95 - 95	16/95	FRONT SEAT HEAD RESTRAINT - OPERATION CHANGED
7	93 - 94	91/26	GOLD STAINS ON PARCHMENT LEATHER SEAT COVERS
7	93 - 94	91/26	GOLD STAINS ON PARCHMENT LEATHER SEAT COVERS - REV
6	94 - 94	AM#117	HEADLINER IMPROVEMENTS - COLOR CHANGES

Model Legend: 1= All 2= Most 3= SL to 1989 4= 1986 and up D, E, CE, TE 5= 1981-89 SD, SE 6= 1990 and up SL 7= 1992 and up SD,SE 8= 190 Series 9= 202 Series

Glass, Doors, Hood, Decklid, Tailgate, Liftgate, Locks

Models	Year	TSB#	Description
7	Up - 95	AM#381	BOSCH CENTRAL LOCKING (PSE) PUMP FAILS - PROCEDURE
6	95 - 95	AM#044	CONVERTIBLE TOP - NEW CONTROL MODULE
9	Up - 95	88/35	DEFLECTOR FOR ENG HOOD SEFETY CATCH - REPL PROCEDURE
7	93 - 93	AM#350	DOOR PANELS - R&I CAUTION - POSSIBLE SPEAKER DAMAGE
7	Up - 93	72/51	DOORS - WIND NOISE MODIFICATIONS & IMMEDIATE REPAIR
7	92 - 93	72/51	DOORS - WIND NOISE MODIFICATIONS/REPAIR - REVISED
7	Up - 93	72/51	DOORS - WIND NOISE REPAIR/TOOL P/N CORRECTION - REV
67	93 - 93	AM#340	LOCK/STEERING - MODIFIED TO IMPROVE THEFT PROTECTION
7	93 - 93	AM#339	LOCK/TRUNK - IMPROVED TONGUE/GRIP HANDLE ASSY - INFO
4	Up - 93	77/31	MODIFIED C-PILLAR MOUNTING FOR SOFT TOP CANVASS
6	95 - 95	77/35	POWER SOFT TOP - MODIFIED PARTS
7	Up - 93	AM#546	REAR WINDOW LIFTER - IMPROVED PART PREVENTS BREAKAGE
4	Up - 95	77/11	REPLACING SOFT TOP FABRIC BOW ACTUATING LEVERS
9	Up - 95	May-95	SLIDING ROOF REMOVAL - BUSHING BREAKS
9	94 - 94	RC#94-0551	SLIDING/POP UP ROOF BRACKETS - SERVICE CAMPAIGN
4	94 - 95	77/11	SOFT TOP ACTUATING LEVER REPLACEMENT
4	94 - 95	77/11	SOFT TOP ACTUATING LEVER REPLACEMENT - REV
6	Up - 95	77/24 & 77/26	SOFT TOP COVER WEAR - DIAGN & REPAIR
6	95 - 95	16/95	SOFT TOP MANUAL OPERATION - CHANGES
6	94 - 94	AM#117	SOFTTOP IMPROVEMENTS - COLOR CHANGES
7	93 - 93	AM#077	TRUNK LID HANDLE - WON'T EXTEND WHEN OPENING LID
4	Up - 93	AM#609	WINDOW/REAR HEATED - INOP - WIRE DAMAGE - REPAIR
7	93 - 93	AM#338	WINDOWS (FRONT/REAR) - LOWERING FEATURES CHANGED
7	Up - 93	.67/03	WINDSHIELD - CRACKS AT BOTTOM - DIAGNOSIS/CORRECTION
7	Up - 93	67/03	WINDSHIELD - CRACKS AT BOTTOM - PARTS/INFO REVISION

Model Legend: 1= All 2= Most 3= SL to 1989 4= 1986 and up D, E, CE, TE 5= 1981-89 SD, SE 6= 1990 and up SL 7= 1992 and up SD,SE 8= 190 Series 9= 202 Series

Finishes, Body Structure, Frame, Bumpers

Models	Year	TSB#	Description
1	95 - 95	00/40A	ANTI-CORROSION COATING FOR ENGINE COMPARTMENT
7	93 - 93	AM#077	LICENSE PLATE FRAME - OBSTRUCTING TRUNK LID HANDLE
9	94 - 94	97/03	WATER LEAK INTO TRUNK - SEAL SPOT WELD
6	Up - 93	77/07A	WIND SCREEN - RATTLING NOISE WHEN FOLDED - REVISION

Model Legend: 1= All 2= Most 3= SL to 1989 4= 1986 and up D, E, CE, TE 5= 1981-89 SD, SE 6= 1990 and up SL 7= 1992 and up SD,SE 8= 190 Series 9= 202 Series

MITSUBISHI

Lighting, Horns, Turn Signals, Steering Column

Models	Year	TSB#	Description
4	92 - 93	93-54-001	HEADLIGHT - AIMING PROCEDURE CHANGES - S/M SUPPL
6	94 - 95	94-54-003	HEADLIGHTS DIM ON DECEL - ECM MODULATOR KIT
2	Up - 93	93-54-003	HORN - UNNECESSARY R&R - ADJUSTMENT PROCEDURE
3,4	94 - 95	95-37A-001	HORN BUTTON REPLACEMENT - IMPROVED PARTS

Model Legend: 1= All 2= Most 3= 300GT 4= Diamante 5= Eclipse 6= Galant 7= Expo LRV 8= Mirage 9= Precis 10= Spyder

Instruments, Dash Cluster, Warning Lights, Mirrors

Models	Year	TSB#	Description
3	91 - 93	93-22-002	SPEEDOMETER - READS HIGHER THAN ACTUAL SPEED - FIX

Model Legend: 1= All 2= Most 3= 300GT 4= Diamante 5= Eclipse 6= Galant 7= Expo LRV 8= Mirage 9= Precis 10= Spyder

Chassis Electrical, Wiring Harness, Fuses-Circuit Breakers, Wipers, Window Motors

Models	Year	TSB#	Description
9	92 - 93	93-54-007	FUSE AND RELAY INFORMATION - S/M REVISION

Model Legend: 1= All 2= Most 3= 300GT 4= Diamante 5= Eclipse 6= Galant 7= Expo LRV 8= Mirage 9= Precis 10= Spyder

Heating, Air Conditioning, Ventilation, Defogger

Models	Year	TSB#	Description
1	Up - 93	93-55-001	A/C - MUSTY ODOR EMITTED - CAUSE/CORRECTION
8	93 - 93	92-55-005	A/C - SYSTEM CHECK TO ENSURE PROPER FUNCTIONING
1	94 - 94	93-55-001	A/C COMPRESSOR OIL REQUIREMENTS - R134A SYSTEMS
6	94 - 94	93-55-004	A/C INOP - BROKEN WIRE AT B-13 CONNECTOR - PROCEDURE
1	Up - 94	94-55-001	A/C LEAK CHECKING PROCEDURES
6	94 - 94	93-55-006	A/C REFRIGERANT TEMP SWITCH - S/M REVISION
6	94 - 95	95-55-002	A/C WHOOPING/RESONATING NOISE - REV EXP VALVE
8	95 - 95	95-55-001	TEMPERATURE CONTROL KNOB INOPERATIVE

Model Legend: 1= All 2= Most 3= 300GT 4= Diamante 5= Eclipse 6= Galant 7= Expo LRV 8= Mirage 9= Precis 10= Spyder

Entertainment Devices, Stereo, Radio, Etc.

Models	Year	TSB#	Description
4,3	93 - 93	92-54-006	AUDIO - SERVICE PROCEDURES FOR NEW ANTI-THEFT SYSTEM
4,3	93 - 93	92-54-010	AUDIO RX-367 W/REAR CD - HIGH PITCHED WHINE - REPAIR
8	93 - 93	93-00-001	AUDIO SYS (DEALER INSTALLED) - NECESSARY HARDWARE
4,3	93 - 93	93-54-006	CD CHANGER - CHARACTERISTICS AND TROUBLESHOOTING

Model Legend: 1= All 2= Most 3= 300GT 4= Diamante 5= Eclipse 6= Galant 7= Expo LRV 8= Mirage 9= Precis 10= Spyder

Seats, Belts, Interior Trim, Carpets, Air Bags

Models	Year	TSB#	Description
6	94 - 94	94-52A-002	FRONT SEAT TRACK ADJUSTMENT PROCEDURES
5	95 - 95	95-52A-001	HEADLINER SAG
4,3,7	93 - 93	93-52A-001	PROTECTIVE COVERINGS/INTERIOR - DESCRIPTION/REMOVAL
3	91 - 93	94-52A-001	RATTLE/LOOSE - REAR CARGO COVER - NEW RETAINING CLIP
4	93 - 93	SR-93-002	REAR SEAT BELT LOWER ANCHOR BOLT RE-TORQUE - RECALL
6,8,5,7	89 - 93	92-52-003	SHOULDER BELT - NORMAL MOVEMENT/DIAGNOSIS INFO
4,3	92 - 93	93-52B-001	SRS - CHANGES AFFECTING SERVICING & DIAG CODE UPDATE
2	90 - 94	94-52B-001	SRS - INSPECTION/REPLACEMENT PROCEDURE - S/M UPDATE
6	94 - 95	95-52B-001	SRS TROUBLESHOOTING - S/M REV
5	95 - 95	94-52A-003	TAILGATE INTERIOR TRIM - IMPROVED FIT

Model Legend: 1= All 2= Most 3= 300GT 4= Diamante 5= Eclipse 6= Galant 7= Expo LRV 8= Mirage 9= Precis 10= Spyder

Glass, Doors, Hood, Decklid, Tailgate, Liftgate, Locks

Models	Year	TSB#	Description
5	90 - 94	93-42B-003	DOOR AND WINDOW ADJUSTMENT PROCEDURES
3	91 - 93	93-42B-004	DOOR GLASS ADJUSTMENT PROCEDURE
5	95 - 95	95-42A-004	DOOR GLASS ADJUSTMENT PROCEDURE
6	94 - 94	94-42B-003	DOOR LOCKS/UNLOCKS BY SELF - REV ACTUATOR
6	94 - 94	93-42B-005	DOOR RATTLE REPAIR PROCEDURE
8	93 - 93	92-00-008	KEYS - NEW BLANKS AND CUTTING EQUIPMENT DEFINED
2	94 - 94	93-00-004	PWR DOOR LOCK/KEYLESS ENTRY "TWO TURN" SYS EXPLAINED
6	94 - 94	94-42B-001	REAR DOOR WEATHERSTRIP TEARING - PREVENTION
10	95 - 96	95-42A-001	RETRACTABLE HARDTOP REPAIR PROCEDURE
3	92 - 93	93-42B-002	SUNROOF - REPAIR PROCEDURES - S/M SUPPLEMENT
5	95 - 95	95-42A-003	SUNROOF/SUNSHADE INOP OR NOISY
3	92 - 93	93-42B-001	TILT-TOP SUNROOF - WATER LEAK & LOOSE COVER REPAIRS

Model Legend: 1= All 2= Most 3= 300GT 4= Diamante 5= Eclipse 6= Galant 7= Expo LRV 8= Mirage 9= Precis 10= Spyder

Finishes, Body Structure, Frame, Bumpers

Models	Year	TSB#	Description
8	93 - 93	92-00-002	BODY PROTECTIVE SHIPPING FILM - KANSAI RAPGARD F-3
7	92 - 93	93-51-002	BUMPERS - SLIDE/GUIDE BRACKET ELIMINATION - INFO
4,9	94 - 94	93-51-005	EXTERIOR COLOR CODES
2	93 - 93	92-51-004	PAINT - EXTERIOR COLOR CODES AND APPLICATIONS
2	94 - 94	93-51-004	PAINT - EXTERIOR COLOR CODES AND APPLICATIONS
4	93 - 93	92-51-005	PAINT - EXTERIOR COLOR CODES AND APPLICATIONS (WGN)
6	94 - 94	93-51-001	PAINT - EXTERIOR COLOR CODES/MANUFACTURERS' NUMBERS
1	95 - 95	94-51-001	PAINT CODES & APPLICATIONS - 1995 MODELS
6,5	95 - 95	95-51-001	PAINT SPOTTING (WHITENING)
4,3,7	93 - 93	93-52A-001	PROTECTIVE COVERINGS/EXTERIOR - DESCRIPTION/REMOVAL
7	92 - 93	93-51-003	ROOF RACK FADING/PAINT PEELING - CAUSE/CORRECTION
2	93 - 94	93-52A-001	VEHICLE PROTECTIVE COVERINGS - REVISED

Model Legend: 1= All 2= Most 3= 300GT 4= Diamante 5= Eclipse 6= Galant 7= Expo LRV 8= Mirage 9= Precis 10= Spyder

NISSAN

Lighting, Horns, Turn Signals, Steering Column

Models	Year	TSB#	Description
7	93 - 94	EL94-005	COMBINATION SWITCH REPLACEMENT CAUTION
8	95 - 95	EL94-016	FOG LIGHT INSTALLATION CAUTION
7	93 - 93	FC0008482	HEADLIGHT/RIGHT - BURNS OUT REPEATEDLY & FUSE BLOWS
7	91 - 93	FC0008437	HEADLIGHT/RIGHT SIDE - FUSE BLOWS W/HIGH BEAM ON
11	93 - 93	FC0008569	HEADLIGHTS - APPEAR DIM OR OUT OF ADJUSTMENT - INFO
7	93 - 93	FC0008424	HEADLIGHTS - INOPERATIVE - POOR GROUND CONNECTION
11,4	95 - 95	EL95-005	HORN ACTIVATES RANDOMLY
11	Up - 93	EL93-011	LAMP (CHMSL) - LENS REPLACEMENT INFORMATION
7	93 - 93	BF93-008	LAMP/HIGH MOUNT STOP - DIFFICULT TO REMOVE - PROC
3	89 - 93	FC0007849	LAMP/TRUNK - STAYS ON - DISCHARGES BAT IN 2/3 DAYS
4,3	92 - 93	FC0008212	LIGHT/INT DOME - STAYS ON - WIRING INFORMATION
7	93 - 93	FC0008480	TURN SIGNAL - INOPERATIVE - POOR TERMINAL CONNECTION

Model Legend: 1= All 2= Most 3= 240SX 4= 200SX 5= 280ZX 6= 300ZX 7= Altima 8= Maxima 9= NX1600, 2000 Coupe 10= Pulsar 11= Sentra Wagon/Coupe 12= Stanza

Instruments, Dash Cluster, Warning Lights, Mirrors

Models	Year	TSB#	Description
3	89 - 93	FC0007783	CEL - COMES ON WHILE DRIVING - EGR VALVE IS OK - FIX
11	91 - 93	FC0007809	CEL LIGHT ON - CODE 32/EGR SYS MALFUNCTION - REPAIR
11	90 - 93	FC0007810	CEL ON - CODE 45 (INJ LEAK) SET FALSELY - WIRE FIX
8	92 - 93	FC0008494	CRUISE CONTROL & INDICATOR LIGHT - INOP - FIELD FIX
7	93 - 93	FC0008456	DASH - BUZZING/SOUNDS LIKE ENG RELATED NOISE - FIX
7	93 - 93	FC0008454	DASH - RATTLE/BUZZ DURING ACCEL IN ALL GEARS - FIX
7	93 - 93	BF93-019	DASH PANEL/LOWER - REMOVAL PROCEDURE TO AVOID DAMAGE
1	Up - 94	FC0008618	DIGITAL DASH INOP ON REPLACEMENT UNIT - RESET ODOM
11,4	95 - 95	EL95-005	FUEL GAUGE NOT INDICATING FULL TANK
11,9	91 - 93	EL93-017	GAUGE (FUEL) - ERRATIC OPERATION - SERVICE PROCEDURE
7	93 - 93	FC0008352	GAUGE (FUEL) - FUEL LEVEL INCONSISTENT - EXPLANATION
11,9	91 - 93	EL93-017	GAUGE/FUEL - ERRATIC OPERATION - SERVICE PROC - REV
7	93 - 93	EF93-005	GAUGE/FUEL - INACCURATE/OUT OF CALIBR - REPAIR PROC
7	93 - 94	EF93-005	GAUGE/FUEL - INACCURATE/OUT OF CALIBR - REVISION
7	93 - 93	FC0008383	H.U.D. DISPLAY/SPEEDOMETER - FADE OUT INTERMITTENTLY
11	93 - 93	EL93-018	INDICATOR LIGHT/FOG LAMP - GLOWS WHEN OFF - WIRING
7	93 - 93	FC0008120	LOWER PANEL/STEER COLUMN - BREAKS - INSTALLING RADIO
7	93 - 93	FC0008637	SEAT BELT MOTOR INOP/AIR BAG LIGHT ON/10A FUSE BLOWS
7	93 - 93	FC0008484	SPEEDOMETER - INOPERATIVE/OTHER GAUGES OK - WIRING
7	93 - 93	FC0008176	SPEEDOMETER/HEAD-UP DISPLAY - INOP - CEL ON/CODE 14
7	93 - 93	FC0008382	SPEEDOMETER/ODOMETER - INACCURATE READINGS - REPAIR
8	89 - 93	BF93-010	SQUEAKS AND RATTLES - LOCATIONS AND REPAIRS
3	89 - 93	BF93-016	SQUEAKS AND RATTLES - LOCATIONS AND REPAIRS
6	90 - 93	BF93-017	SQUEAKS AND RATTLES - LOCATIONS AND REPAIRS
7	93 - 93	FC0008394	TACHOMETER - INTERMITTENTLY INOP - WIRE REPAIR
6	90 - 93	FC0008531	WARN LITE (ABS) - ON UNTIL CAR IS DRIVEN A FEW MILES
7	93 - 93	FC0008267	WARN LITE/OIL - ON/OIL PRESSURE SEND UNIT OK - FIX
3	89 - 93	FC0008530	WARNING LIGHT (ABS) - COMES ON AFTER 1/2 HR DRIVING
8	91 - 93	FC0008444	WARNING LIGHT (AIR BAG) - WIRING REPAIR
7	93 - 93	FC0008499	WARNING LIGHT (SEAT BELT) - CONSTANT ON - FIELD FIX
8	92 - 93	FC0007738	WARNING LIGHT (SRS) - BLINKING - DIAGNOSIS/REPAIR
11	91 - 93	EL93-019	WARNING LIGHT/ABS - STAYS ON AFTER ENG STARTS - DIAG

Model Legend: 1= All 2= Most 3= 240SX 4= 200SX 5= 280ZX 6= 300ZX 7= Altima 8= Maxima 9= NX1600, 2000 Coupe 10= Pulsar 11= Sentra Wagon/Coupe 12= Stanza

Chassis Electrical, Wiring Harness, Fuses-Circuit Breakers, Wipers, Window Motors

Models	Year	TSB#	Description
8	89 - 93	FC0008235	ECCS RELAY - REPLACE - ENG CRANKS BUT WON'T START
7	93 - 93	FC0008372	ELEC ACCES INOP - WIPERS/BLOWER/WARN LITES/WINDOWS
1	Up - 94	EL93-027	ELECTRICAL SYSTEM DIAGNOSTIC PROCEDURES
3	92 - 93	FC0007861	FUSE BLOWS - INTERIOR LT/CLOCK/RADIO/ANTENNA INOP
7	93 - 93	FC0008399	FUSE BLOWS-COMB METER/SHIFT LOCK/BACK-UP LITES INOP
1	Up - 95	EC95-014	GREEN OR BLUE (FUEL PUMP & A/C) RELAY CAUTION
7	95 - 95	EL95-014	MIL ON - BLOWN FUSE 25 OR 26 - POSS SHIFT CONCERNS
3	89 - 93	BF93-025	REAR WASHER NOZZLE GASKET LEAK - REPLACE (HATCHBACK)
1	Up - 94	EF93-016	RELAY TYPE 1M (GREEN OR BLUE) - SERVICE CAUTIONS
7	93 - 93	FC0008637	SEAT BELT MOTOR INOP/AIR BAG LIGHT ON/10A FUSE BLOWS
8	90 - 93	FC0008324	SWITCH/LEFT RR WINDOW - R&R TO CORRECT BATTERY DRAIN
1	Up - 93	EL92-011	WIPER BLADES - MAINTENANCE/SERVICE INFORMATION
7	93 - 93	FC0008321	WIRING-ABS LITE ON-WINDOWS/LOCKS/ANTENNA/CRUISE INOP

Model Legend: 1= All 2= Most 3= 240SX 4= 200SX 5= 280ZX 6= 300ZX 7= Altima 8= Maxima 9= NX1600, 2000 Coupe 10= Pulsar 11= Sentra Wagon/Coupe 12= Stanza

Auxiliary Equipment, Jacks, Trailer Hitches, Towing

Models	Year	TSB#	Description
11	93 - 93	BF93-037	ACCESSORY ARMREST NOT FOR CARS WITH AIR BAG

Model Legend: 1= All 2= Most 3= 240SX 4= 200SX 5= 280ZX 6= 300ZX 7= Altima 8= Maxima 9= NX1600, 2000 Coupe 10= Pulsar 11= Sentra Wagon/Coupe 12= Stanza

Heating, Air Conditioning, Ventilation, Defogger

Models	Year	TSB#	Description
1	93 - 93	HA93-010	A/C (R134a SYS) - SERVICE PARTS ORDERING INFORMATION
11	93 - 93	FC0008565	A/C - BLOWS WARM AIR - INSPECT/REPAIR INFORMATION
8	93 - 93	HA92-004	A/C - CHANGES/SPECIFIC DETAILS FOR R-13a REGRIGERANT
8	93 - 93	FC0008514	A/C - DASH VENT TEMP VARIES TOO MUCH - FIELD FIX
8	89 - 93	FC0008542	A/C - KNOCK NOISE W/FAN ON LOW SPEED - SERV INFO
2	Up - 94	HA94-001	A/C - LUBRICANTS & OILS
11	91 - 93	FC0008353	A/C - NO COOL - THERMO CNTRL/ECM/COOLANT TEMP SENSOR
11	91 - 93	FC0007815	A/C - NOT COLD ENOUGH/CONDENSER FANS INOP - WIRE FIX
7	93 - 93	FC0008481	A/C - NOT COOLING/NO POWER TO COMP - WIRING REPAIR
8	93 - 93	HA93-013	A/C - POOR PERF/VENT TEMP VARIES TOO MUCH - REPAIR
8	93 - 93	HA93-013	A/C - POOR PERF/VENT TEMP VARIES TOO MUCH - REVISED
7	93 - 93	HA94-004	A/C - POOR PERFORMANCE - EVAPORATOR BY-PASS
7	93 - 93	HA93-006	A/C - POOR PERFORMANCE/EVAPORATOR ICING UP - REPAIR
7	93 - 93	HA93-011	A/C - POOR PERFORMANCE/RELAYS MISSING - SERV CMPGN
11	91 - 93	FC0007836	A/C - RECIRCULATION LEVER BREAKS - DIAGNOSIS/REPAIR
7	93 - 93	FC0008458	A/C - WATER LEAKS ON FLOOR WHEN A/C IS ON - REPAIR
7,11	Up - 93	HA93-014	A/C - WHITE FLAKES FROM VENTS - NEW EVAPORATOR
7	93 - 93	HA93-003	A/C BUTTON - INSTALLATION CLARIFIED
11	94 - 94	REF: HA93-001	A/C CHANGES FOR R134A REFRIGERANT
6	94 - 94	HA93-023	A/C CHANGES FOR R134A REFRIGERANT
1	93 - 93	HA93-001	A/C CHANGES FOR R134A REFRIGERANT - DETAILS
7	93 - 93	FC0008250	A/C COMPRESSOR INOP/RADIATOR FAN ALWAYS ON - SERVICE
1	Up - 95	HA95-011	A/C COMPRESSOR LEAK/NOISE DIAGNOSIS
11	91 - 93	FC0008598	A/C COMPRESSOR NOT CYCLING ON - LOOSE WIRE
11	93 - 93	HA94-002	A/C INOP - COMPRESSOR CHECK VALVE
11	91 - 93	FC0008600	A/C INOP AFTER ENGINE WARMS - ECCS ECU
1	Up - 96	HA95-017	A/C LUBRICANT/OIL - SPECS
7	93 - 93	FC0008634	A/C NOT COLD - COMPRESSOR NOT ENGAGED - WIRE
7	93 - 93	FC0008580	A/C NOT COLD - PINHOLE IN TOP SEAM OF RECEIVER DRYER
1	Up - 95	HA95-004	A/C PARTS - PROPER INSTALLATION PROCEDURE
8	93 - 93	FC0008586	A/C POOR COOL - MOVE TIP OF THERMISTER
3	93 - 93	HA93-016	A/C R134A SYS CHANGES - DIAGNOSIS AND PROCEDURES
8	95 - 95	HA95-001	A/C REFRIG CAPACITY SPEC - S/M REV
7	Up - 95	FC0010087	A/C REFRIGERANT LEAK - TIPS
8	92 - 93	HA93-002	A/C SERVICE ON SRS VEHICLES - REMOVING COOLING UNIT
7	93 - 93	HA92-005	A/C SYSTEM (DEALER & PORT INSTALLED) - SERVICE INFO
11,9	93 - 93	HA92-013	A/C SYSTEM - COMPRESSOR CHANGE FOR 1993 MODEL YEAR
3	93 - 94	HA95-006	A/C SYSTEM WITHOUT THERMO CONTROL UNIT - INFO
11	91 - 93	FC0008078	A/C VENT - DRIPS CONDENSATION ON DRIVER'S FEET - FIX

Model Legend: 1= All 2= Most 3= 240SX 4= 200SX 5= 280ZX 6= 300ZX 7= Altima 8= Maxima 9= NX1600, 2000 Coupe 10= Pulsar 11= Sentra Wagon/Coupe 12= Stanza

Heating, Air Conditioning, Ventilation, Defogger

Models	Year	TSB#	Description
6	93 - 93	HA93-018	ATC RECIRCULATION TIMER LOGIC EXPLAINED
7	93 - 93	HA93-007	DEFROSTER - MODIFICATION TO IMPROVE OPERATION
7	93 - 95	HA95-002	FRESH AIR/HEATER PERFORMANCE - OPERATION EXPLAINED
11	93 - 93	FC0008697	HEAT TEMP CONTROL LEVER MOVES TO CENTER BY ITSELF
1	Up - 93	FC0008303	HEATER - PERFORMANCE INFORMATION
8	93 - 93	HA93-017	HEATER AIRFLOW OPERATION - OWNERS MANUAL INCORRECT
7	93 - 93	HA93-012	HEATER AND/OR A/C INOPERATIVE
7	93 - 93	HA93-012	HEATER AND/OR A/C INOPERATIVE - REV
7	93 - 93	HA93-012	HEATER AND/OR A/C INOPERATIVE - REVISED
7	93 - 93	FC0008366	HEATER INOP - NO OUTPUT - CLICK NOISE FROM DASH
11	91 - 93	FC0008405	HEATER-A/C - AIR INTAKE CONTROL LEVER BINDING/BROKEN
7	94 - 95	HA94-007	INSTALLING A/C - DRAIN HOSE LEAKAGE - ECM DAMAGE
1	Up - 96	HA95-013	R134A CHARGE AMOUNTS/PAG LUBRICATION CHART
1	Up - 95	HA95-003	REFRIGERANT LEAK DETECTION PROCEDURE
2	Up - 94	HA94-005	SRVC PROCED - R134A RETROFITTED A/C SYS (EXC QUEST)
1	Up - 95	HA94-003	THERMOSTATS AND HEATER PERFORMANCE

Model Legend: 1= All 2= Most 3= 240SX 4= 200SX 5= 280ZX 6= 300ZX 7= Altima 8= Maxima 9= NX1600, 2000 Coupe 10= Pulsar 11= Sentra Wagon/Coupe 12= Stanza

Electronic Devices, Computers, PROMS, Sensors

Models	Year	TSB#	Description
8	92 - 93	94V-007	AIR BAG - SENSOR REPLACEMENT - RECALL
5,6	93 - 93	FC0008270	ALARM SYS - SOUNDS INTERMITTENTLY FOR NO REASON- FIX
7	93 - 93	FC0008225	CODE 21/ROUGH IDLE/STALLS ON STOP - SERVICE INFO
11	93 - 93	FC0008233	CODE 35/EGR TEMP SENSOR - CEL ON - SERVICE PROCEDURE
8	89 - 94	FC0008812	CODES 11/21 - POOR DRIVEABILITY - CHECK CPS
2	88 - 93	EF93-012	CODES 32/45 DIAGNOSIS/REPAIR - CALIFORNIA EMISSIONS
8	92 - 94	EF95-002	CONSULT/CHECKER BOX CONNECTOR USAGE - VG30E
8	89 - 93	FC0008347	CONTROL UNIT (ECCS) - CODE 55 - LATE/HARD A/T SHIFTS
7	93 - 93	FC0008502	CONTROL UNIT/ECCS - R&R - CORRECT WARM ENG NO START
7	93 - 93	FC0008501	CONTROL UNITS (ECCS/FUEL PUMP) - R&R - NO START COND
3	93 - 93	EF93-015	DIAGNOSTIC TESTS - S/M SUPP FOR CONVERTIBLE
1	90 - 94	FC0008541	DIAGNOSTIC/SERVICE INFORMATION - IDLE FLUCTUATES
7	94 - 94	GI93-007	DIAGNOSTICS (VARIOUS SYSTEMS) - S/M IMPROVEMENTS
7	94 - 95	HA94-007	DRIVEABILITY - ECM DAMAGE FROM A/C DRAIN HOSE LEAK
2	Up - 95	EF95-005	DRIVEABILITY AND/OR DTC 45 (INJECTOR LEAK)
1	Up - 95	EC95-011	ECCS SYSTEM TESTING/DIAGNOSIS - PRECAUTIONS
8	92 - 93	EF93-006	ECM ASSEMBLY - R&R - LACK OF POWER AT WOT CORRECTION
1	77 - 93	EF93-013	ECM IDENTIFICATION & CROSS REFERENCE
1	75 - 95	EF94-015	ECM IDENTIFICATION & CROSS REFERENCES
1	91 - 95	EF94-011	ECM INSTALLATION - SEATING/SECURING THE CONNECTOR
3	91 - 93	FC0008326	ECU - R&R - DAMAGE FROM SHORTED WIRE - NO START COND
1	Up - 94	EL93-028	ELECTRONIC CONTROL UNIT (ECU) - OPERATION EXPLAINED
8	90 - 93	FC0008622	ENGINE CRANKS/NO START - COOLANT DAMAGED ECU
8	95 - 95	EC95-095	HARD/SLOW CRANKING - DIAGN - ECM
3	95 - 95	EF94-013	IDLE SPEED SPECIFICATIONS - S/M REVISION
8	96 - 96	EL95-015	IVMS (LAN) TROUBLE DIAGNOSES SYSTEM - S/M REV AVAIL
7	93 - 93	EL93-022	KEYLESS ENTRY - HARNESS & COMPONENT TESTING W/TESTER
8,6,3,11,4	95 - 95	EL95-011	KEYLESS ENTRY REMOTE - BATTERY REPLACE/RE-PROGRAM
7	95 - 95	EF95-004	MIL ON - CODE P1336 - PROCEDURE
8	95 - 95	EF95-008	MIL ON - DTC P0340/P1335/P1336/P0335
8	95 - 95	EC95-012	MIL ON - DTC P0443 - CAL ONLY - DIAGNOSTICS
3	95 - 95	EF95-010	MIL ON - DTC P1336
7	93 - 93	FC0008198	MODULE/AIR BAG - CONSULT UNABLE TO ACCESS DIAGNOSIS
1	95 - 95	EC95-009	OBD II ECM CONNECTOR & BOLT - NEW - PROCEDURE
8,3	95 - 95	AT94-005	OBD-II - "A/T INITIAL START" TROUBLE CODE
3,8	95 - 95	EF94-009	OBD-II - ENG CONTROL SYS SERVICE CAUTIONS
3,8	95 - 95	EF94-006	OBD-II - POWERTRAIN ON BOARD DIAGNOSTICS OVERVIEW
3,8	95 - 95	EF94-012	REAR O2 SENSOR TORQUE PRECAUTION/SPECS

Model Legend: 1= All 2= Most 3= 240SX 4= 200SX 5= 280ZX 6= 300ZX 7= Altima 8= Maxima 9= NX1600, 2000 Coupe 10= Pulsar 11= Sentra Wagon/Coupe 12= Stanza

Electronic Devices, Computers, PROMS, Sensors

Models	Year	TSB#	Description
8	95 - 95	EL94-015	REMOTE POWER LOCK/SECURITY SYS OPERATION EXPLAINED
7	93 - 94	PI95-003	SECURITY SYS (VSS) - INSTALLATION TIPS
7	95 - 95	EL95-001	SECURITY SYS (VSS) TROUBLE DIAGNOSIS GUIDE
7	95 - 95	EL95-001	SECURITY SYS (VSS) TROUBLE DIAGNOSIS GUIDE - REV
11	91 - 93	FC0007809	SENSOR (EXH GAS TEMP) - REPLACED TO ELIMINATE CEL ON
7	93 - 93	FC0008351	SENSOR (TPS) - R&R TO CORRECT HARSH/LATE A/T SHIFTS
7	93 - 93	FC0008250	SENSOR/COOLANT TEMP - COOLANT LEAK - SERVICE PROC
7	93 - 93	FC0008331	SENSOR/ENG COOLANT TEMP - R&R - HARD COLD STARTS FIX
3	89 - 93	FC0007783	SENSOR/EXH GAS TEMP - REPLACE - CEL LIGHT ON/DRIVING
8	92 - 93	FC0008494	SENSOR/SPEED - REPLACED TO CORRECT INOP CRUISE CNTRL
8	95 - 95	BT95-003	SRS FLASHING LIGHT - DIAGNOSTIC FAULT TREE
11	91 - 93	FC0007817	TIME CONTROL UNIT-REPLACE-KEY WARN BUZZER STAYS ON
6	90 - 93	EL93-029	TWS/THEFT WARN SYS - INOP/INTERMITTENT - PROCEDURES
5,6	93 - 93	FC0008208	VEHICLE ALARM SOUNDS BY ITSELF - WIRING REPAIR
7	93 - 93	FC0008126	VSS - INOPERATIVE - DIAGNOSIS AND REPAIR
11,4	95 - 95	EL95-009	VSS DASH LAMP FUNCTION EXPLAINED

Model Legend: 1= All 2= Most 3= 240SX 4= 200SX 5= 280ZX 6= 300ZX 7= Altima 8= Maxima 9= NX1600, 2000 Coupe 10= Pulsar 11= Sentra Wagon/Coupe 12= Stanza

Entertainment Devices, Stereo, Radio, Etc.

Models	Year	TSB#	Description
1	Up - 93	EL93-009	ANTENNA/POWER - ROD RELACEMENT PROCEDURE
1	Up - 96	EL94-004	ANTENNA/POWER - ROD RELACEMENT PROCEDURE - REV2
1	Up - 94	EL93-009	ANTENNA/POWER - ROD RELACEMENT PROCEDURE - REVISED
2	Up - 93	EL94-007	AUDIO HEADUNIT MOUNTING SCREWS - IMPROVED
3,8	95 - 95	PI94-001	AUDIO SYS DATA SHEETS - SPECIFICATIONS
1	Up - 94	EL94-009	BOSE SPEAKER EXCHANGE - REPLACE AS UNIT WITH AMP
8,11,10,9	89 - 93	EL93-013	CASSETTE - SIMULTANEOUS FAST FORWARD/PLAY - REPAIR
2	Up - 93	EL93-004	CASSETTE TAPE PLAYER - MAINTENANCE GUIDELINES
1	Up - 93	EL92-009	CD AUTO CHANGER/DIAGNOSING AFTERMARKET INSTALLATION
2	93 - 93	EL92-014	CD PLAYER - CHANGE IN AUDIO UNIT HARNESS CONNECTOR
6	93 - 93	FC0008436	CD PLAYER INOP AFTER NEW INSTALLATION - FIELD FIX
8	95 - 95	EL94-010	CD PLAYER INSTALL - REVISION TO FACTORY PRE-WIRING
8,3	95 - 95	EL94-014	CD PLAYER INSTALLATION - USE VEHICLE'S HARNESS
7,8	93 - 93	EL93-026	CLARION CD/CASSETTE/RADIO ERROR MESSAGES EXPLAINED
8	95 - 95	EL95-013	POOR RADIO RECEPTION - LOOSE ANTENNA GROUND
1	Up - 93	EL92-011	POWER ANTENNA - MAINTENANCE/SERVICE INFORMATION
1	93 - 93	EL93-007	RADIO - IDENTIFICATION INFORMATION
2	95 - 95	EL94-012	RADIO CONNECTOR MODIFIED - ADAPTER HARNESS
1	Up - 96	PI95-013	REMAN BOSE SPEAKERS AVAILABLE
7	93 - 93	FC0008637	SEAT BELT MOTOR INOP/SIR LITE ON/CLOCK-RADIO MEMORY

Model Legend: 1= All 2= Most 3= 240SX 4= 200SX 5= 280ZX 6= 300ZX 7= Altima 8= Maxima 9= NX1600, 2000 Coupe 10= Pulsar 11= Sentra Wagon/Coupe 12= Stanza

Seats, Belts, Interior Trim, Carpets, Air Bags

Models	Year	TSB#	Description
11	93 - 93	BF93-037	ACCESSORY ARMREST NOT FOR CARS WITH AIR BAG
8	92 - 93	94V-007	AIR BAG - SENSOR REPLACEMENT - RECALL
8	92 - 93	FC0007738	AIR BAG - WARN LIGHT BLINKS - OPEN CIRCUIT IN MODULE
7	93 - 93	FC0008198	AIR BAG LIGHT ON/UNABLE TO ACCESS MODULE DIAGNOSIS
7	93 - 93	BF93-006	ARM REST (ACCESSORY) - MOUNTING INSTRUCTIONS
4	95 - 95	BT95-004	C-PILLAR FINISHER LIFTING
8	95 - 96	BT95-007	CENTER CONSOLE LID LATCH REPLACEMENT
8	93 - 94	BF94-001	CHILD RESTRAINT INSTALLATION NOTICE
11,9	91 - 93	BF93-038	FRONT FLOORMAT RETAINING HOOK REPLACEMENT
1	94 - 94	BF94-009	FRONT SEAT BELT EXTENDERS - CAUTION/APPLICATIONS
8,11,3	95 - 95	BF95-001	FRONT SEAT BELT EXTENDERS AVAILABLE
3	89 - 93	BF93-025	HEADLINER WATER ENTRY VIA REAR WASHER NOZZLE (HATCH)
7	93 - 93	GI92-004	HEEL MAT/FOOT PAD - ADHESIVE RESIDUE - CLEANING PROC
11	91 - 93	RS95-004	IMPROVED MOTORIZED SEAT BELT RETRACTOR
7	93 - 93	RS95-004	IMPROVED MOTORIZED SEAT BELT RETRACTOR
6	93 - 93	BF92-035	INTERIOR FRONT GARNISH LOOSE - REPAIR (CONVERTIBLE)
9	Up - 94	BF94-002	PARCEL SHELF HINGE REPAIR - REVISED
9	Up - 94	BF94-002	PARCEL SHELF HINGE REPAIR PROCEDURE
9	Up - 94	BF94-002	PARCEL SHELF HINGE REPAIR PROCEDURE - REVISED
6	90 - 93	BF93-012	POCKET COVER LOOSE (DRIVER'S DOOR) - SERV PROCEDURE
7	93 - 93	BF94-027	SEAT BELT EXTENDERS AVAILABLE
7	93 - 93	FC0008637	SEAT BELT MOTOR INOP/AIR BAG LIGHT ON/10A FUSE BLOWS
2	Up - 93	BF93-034	SEAT BELTS - EXTENDERS AVAILABLE - P/N INFO
11,9	91 - 93	BF93-015	SQUEAKS & RATTLES REPAIR PROCEDURE
1	Up - 94	BF94-020	SRS - AIR BAG SYSTEM DIAGNOSTIC PROCEDURES
1	Up - 93	BF92-022	SRS - CLARIFICATION OF INSPECTION/REPAIR PROCEDURES
8	95 - 95	BT95-003	SRS FLASHING LIGHT - DIAGNOSTIC FAULT TREE
7,11	95 - 95	RS95-001	SRS SELF DIAGNOSIS LOGIC CHANGES
11,4	95 - 95	PI95-011	SUN VISOR CONTACTS INSIDE REAR VIEW MIRROR
1	Up - 94	GI93-009	TWO POINT MOTORIZED SHOULDER BELT - LOCKUP EXPLAINED
7	93 - 93	FC0008229	WATERLEAK TO FLOOR AREA - INSPECT & REPAIR PROCEDURE
7	94 - 94	GI93-007	WIRING DIAGRAMS - S/M IMPROVEMENTS

Model Legend: 1= All 2= Most 3= 240SX 4= 200SX 5= 280ZX 6= 300ZX 7= Altima 8= Maxima 9= NX1600, 2000 Coupe 10= Pulsar 11= Sentra Wagon/Coupe 12= Stanza

Glass, Doors, Hood, Decklid, Tailgate, Liftgate, Locks

Models	Year	TSB#	Description
7	93 - 93	BF93-043	BLACK SEALANT OR SCRATCHES ON SUNROOF
6	93 - 93	BF92-024	CONVERTIBLE - HEADER CORNER TRIM MODIFICATION
6,3	92 - 93	BF92-030	CONVERTIBLE CAR WASH-REAR WINDOW CLEANING/POLISHING
6	93 - 93	BF93-013	CONVERTIBLE CENTER PILLAR - REPLACEMENT INFORMATION
6	93 - 93	BF92-024	CONVERTIBLE HEADER CORNER TRIM - R&R PROC - CAMPAIGN
6,3	Up - 93	BF93-022	CONVERTIBLE TOP - FACILITIES FOR SPECIALTY REPAIRS
6,3	Up - 93	BF93-022	CONVERTIBLE TOP - FACILITIES/SPECIALTY REPAIRS - REV
6	Up - 94	BF94-016	CONVERTIBLE TOP STORAGE LID HANDLES CRACK
7	93 - 93	BF93-021	DOOR - TRIM PANEL MAP POCKETS PEELING - SERVICE INFO
7	93 - 93	BF93-009	DOOR - TRIM PANEL REMOVAL - SERVICE PROCEDURE
8	93 - 93	FC0008152	DOOR LOCKS/POWER - LOCK BY THEMSELVES - DIAG/REPAIR
1	92 - 93	EL92-010	DOOR LOCKS/POWER - LOGIC REFERENCE CHART
7	93 - 93	BF93-024	DOORS/WINDOWS - SQUEAK DIAGNOSIS & CORRECTION
11,4	95 - 95	BT95-006	FRONT WINDOW MISALIGNMENT
11	95 - 95	BT95-001	FRONT WINDOW WILL NOT GO COMPLETELY DOWN
11,9	92 - 93	HA92-011	HOOD - SEAL ELIMINATED IN VEHICLES WITH A/C
5,6	90 - 93	FC0008575	HOOD RELEASE INOP/CABLE BROKEN - SERVICE PROCEDURE
11,4	95 - 95	EL95-010	POWER DOOR LOCK SWITCH - INTERMITTENT OPS
11,4	95 - 95	EL95-006	POWER WINDOW INOP - REV MAIN SWITCH
7	93 - 93	FC0008696	RATTLE/CREAK NOISE RIGHT B-PILLAR - HARD ACCEL/DECEL
8	95 - 95	BF94-025	REAR WINDOW RUBBER MOULDING REPLACEMENT PROCEDURE
11	94 - 94	P3139	SEALING DOOR HINGES - SERVICE CAMPAIGN RECALL
11,9	91 - 93	BF93-015	SQUEAKS & RATTLES REPAIR PROCEDURE
8	89 - 93	BF93-010	SQUEAKS AND RATTLES - LOCATIONS AND REPAIRS
3	89 - 93	BF93-016	SQUEAKS AND RATTLES - LOCATIONS AND REPAIRS
6	90 - 93	BF93-017	SQUEAKS AND RATTLES - LOCATIONS AND REPAIRS
7	93 - 93	FC0008426	SUNROOF - INOPERATIVE - POOR GROUND CONNECTION
7	93 - 93	FC0008567	SUNROOF STICKS WHEN OPENING - INFORMATION/PROCEDURE
7	93 - 93	BF93-044	SUNROOF WATER LEAK - REPAIR PROCEDURE
7	93 - 93	FC0008551	TRUNK - KNOCK NOISE OVER ROUGH ROADS/CONNECTOR INFO
7	93 - 93	FC0008459	TRUNK AREA (RIGHT) - RATTLE OVER BUMPS - FIELD FIX
11	Up - 94	BF94-005	TRUNK TORSION BARS - STIFFER VERSION AVAILABLE
6	90 - 93	BF93-042	WATER LEAK - T-BAR ROOF - SEALANTS & PROCEDURES
7	93 - 93	REF: BF93-023	WATER LEAK TROUBLE DIAGNOSIS
8	89 - 93	REF: BF93-023	WATER LEAKS TROUBLE DIAGNOSIS
6	90 - 93	REF: BF93-023	WATER LEAKS TROUBLE DIAGNOSIS
11	91 - 93	REF: BF93-023	WATER LEAKS TROUBLE DIAGNOSIS
3	89 - 93	REF: BF93-023	WATER LEAKS TROUBLE DIAGNOSIS

Model Legend: 1= All 2= Most 3= 240SX 4= 200SX 5= 280ZX 6= 300ZX 7= Altima 8= Maxima 9= NX1600, 2000 Coupe 10= Pulsar 11= Sentra Wagon/Coupe 12= Stanza

Glass, Doors, Hood, Decklid, Tailgate, Liftgate, Locks

Models	Year	TSB#	Description
7	93 - 94	BF94-026	WIND NOISE AROUND FRONT DOOR
8	95 - 95	BF95-003	WIND NOISE AROUND FRONT DOOR
11,4	95 - 95	BT95-005	WIND NOISE FIELD FIX PROCEDURE
11,4	95 - 95	NTB95-052	WIND NOISE FIELD FIX PROCEDURE - REV
7	93 - 93	FC0008500	WINDOW (RIGHT POWER) - WON'T GO UP - FIELD FIX
8	89 - 93	FC0008184	WINDOW/POWER - INTERMITTENT SELF-OPERATING - REPAIR
7	93 - 93	EL93-010	WINDOW/POWER - PROTECTING SWITCHES DURING REMOVAL
7	93 - 93	FC0008230	WINDOWS/POWER - OPEN SELVES WHILE DRIVING - PROC

Model Legend: 1= All 2= Most 3= 240SX 4= 200SX 5= 280ZX 6= 300ZX 7= Altima 8= Maxima 9= NX1600, 2000 Coupe 10= Pulsar 11= Sentra Wagon/Coupe 12= Stanza

Finishes, Body Structure, Frame, Bumpers

Models	Year	TSB#	Description
8,3,6	Up - 95	PI94-007	3-COAT PEARLESCENT PAINT REFINISHING - MATCHING TIPS
1	Up - 93	BF93-023	BODY - DIAGNOSING WATER LEAKS W/VAC PRESSURE DEVICE
1	Up - 95	PI94-003	BUMPER REFINISHING TECHNIQUES
6	94 - 94	BF93-045	CLEAR COAT (NCLC) PAINT FINISH INFORMATION
11	91 - 93	FC0007864	COWL WATER LEAK - PASSENGER SIDE UNDER GLOVE BOX/FIX
7	93 - 93	BF93-001	LICENSE PLATE (FRONT) - PDI INSTALLATION PROCEDURES
6	94 - 94	BF93-045	NEW CLEAR COAT PAINT (NCLC) - APPROVED MANUFACTURERS
3,6,8	94 - 95	BF94-003	NEW CROSS-LINKING CLEAR COAT (NCLC) POLISH PROCEDURE
7,11	95 - 95	PI95-001	NEW PAINT GUARD FILM
8,3	95 - 95	PI95-009	NEW PAINT GUARD FILM
6	95 - 95	PI95-012	NEW PAINT GUARD FILM
11	91 - 93	BF94-007	NOISE/POPPING - AT B PILLAR/DOOR STRIKER - 2-DR
1	Up - 93	BF93-014	PAINT - CONTAMINATION IDENTIFICATION & REPAIR PROC
1	Up - 93	BF93-033	PAINT - MAINTENANCE/STORAGE RECOMMENDATIONS
6	93 - 93	FC0008410	PAINT - NEW CLEARCOAT - SUPPLIERS/REPAIR PROCEDURES
3,6	94 - 95	PI95-004	PAINT - NEW CROSS LINK CLEAR COAT (NCLC)
7	95 - 95	U4007	PAINT CODE LABEL - CAMPAIGN
1	94 - 94	BF93-039	PAINT CODE NUMBERS - 1994 MODELS
1	95 - 95	PI94-005	PAINT CODES - 1995 MODELS
3,8	95 - 95	BF94-014	PAINT CODES AND CLEAR COAT (NCLC) IDENTIFICATION
1	93 - 93	BF92-023	PAINT CODES/COLOR NAMES/REFINISH PAINT CODES INFO
8	95 - 95	PI94-004	PAINT COLOR CODE IDENTIFICATION
1	Up - 94	BF94-003	PAINT CONTAMINATION IDENTIFICATION AND REPAIR
2	Up - 93	BF92-019	PAINT GUARD COATING - REMOVER/SERVICE INFORMATION
1	93 - 93	BF93-011	PAINT GUARD COATING REMOVER - NEW FORMULA/NO AMMONIA
2	93 - 94	BF93-032	PAINT GUARD FILM - CLEAN UP PROCEDURES DURING PDI
1	94 - 95	BF94-010	PAINT GUARD FILM - TIMING OF REMOVAL

Model Legend: 1= All 2= Most 3= 240SX 4= 200SX 5= 280ZX 6= 300ZX 7= Altima 8= Maxima 9= NX1600, 2000 Coupe 10= Pulsar 11= Sentra Wagon/Coupe 12= Stanza

Finishes, Body Structure, Frame, Bumpers

Models	Year	TSB#	Description
1	94 - 95	BF94-010	PAINT GUARD FILM - TIMING OF REMOVAL - REVISED
2	93 - 93	BF93-007	PAINT GUARD FILM REMOVAL
7,11	95 - 95	PI95-010	PAINT GUARD FILM SWELLING - PROCEDURE
2	95 - 95	PI95-002	POLYURETHANE & POLYPROPYLENE BUMPER FACIA REPAIR
8	95 - 95	BF95-002	RATTLE/NOISE - SPARE TIRE COVER
1	96 - 96	PI95-015	REFINISH PAINT FORMULA CODES
9	Up - 93	BF92-034	SPLASH GUARD - AVAILABLE/SERVICE INFORMATION
7,11	94 - 94	BF93-047	UNDERBODY BLACK BITUMIN WAX DISCONTINUED
11	90 - 93	BF94-024	WATER LEAK INTO FRONT PASSENGER FLOOR AREA
7	93 - 93	REF: BF93-023	WATER LEAK TROUBLE DIAGNOSIS
8	89 - 93	REF: BF93-023	WATER LEAKS TROUBLE DIAGNOSIS
6	90 - 93	REF: BF93-023	WATER LEAKS TROUBLE DIAGNOSIS
11	91 - 93	REF: BF93-023	WATER LEAKS TROUBLE DIAGNOSIS
3	89 - 93	REF: BF93-023	WATER LEAKS TROUBLE DIAGNOSIS
1	Up - 95	PI95-007	WATERBORNE REFINISH PAINTS - APPROVED SUPPLIERS
11,4	95 - 95	BT95-005	WIND NOISE FIELD FIX PROCEDURE
11,4	95 - 95	NTB95-052	WIND NOISE FIELD FIX PROCEDURE - REV

Model Legend: 1= All 2= Most 3= 240SX 4= 200SX 5= 280ZX 6= 300ZX 7= Altima 8= Maxima 9= NX1600, 2000 Coupe 10= Pulsar 11= Sentra Wagon/Coupe 12= Stanza

PORSCHE

Lighting, Horns, Turn Signals, Steering Column

Models	Year	TSB#	Description
1	Up - 93	06/9205	STEERING COLUMN NOISE - REPAIR (4/6/8 CYL W/AIR BAG)

Model Legend: 1= All 2= Most 3= 911 4= 924 5= 928 6= 944 7= 968

Instruments, Dash Cluster, Warning Lights, Mirrors

Models	Year	TSB#	Description
3	94 - 94	09/9307	GAUGE/BOOST - ONLY READS 0.7 BAR AT MAX BOOST - INFO
3	94 - 94	09/9307 5/18/93	GAUGE/BOOST - ONLY READS 0.7 BAR AT MAX BOOST - REV
5	89 - 93	09/9302	IC (BACKLIGHTED) - READING ELAPSED MILEAGE - PROC
7	93 - 94	09/9402	OIL LEVEL WARNING LIGHT ON - PROCEDURE
3	95 - 96	09/9503	REPLACING LARGE INSTRUMENT FOR OIL PRESS/TEMP/WARN

Model Legend: 1= All 2= Most 3= 911 4= 924 5= 928 6= 944 7= 968

Chassis Electrical, Wiring Harness, Fuses-Circuit Breakers, Wipers, Window Motors

Models	Year	TSB#	Description
7	92 - 94	09/9309	ALARM/DEAD BATTERY - DOOR SWITCH/GLOVE BOX LIGHT
1	Up - 93	09/9203	ELEC CONDITIONS/COMPONENT INTERMITTENT OP - PROC
3	93 - 93	09/9306	FUSES - REMOVAL DURING TRANSPORTATION/STORAGE - INFO
7	94 - 94	09/9502	WIRING - ECU TERMINAL 15 - S/M REV

Model Legend: 1= All 2= Most 3= 911 4= 924 5= 928 6= 944 7= 968

Auxiliary Equipment, Jacks, Trailer Hitches, Towing

Models	Year	TSB#	Description
3	87 - 93	09/9301	CELLULAR PHONE (FUJITSU) - INSTALLATION PROCEDURES

Model Legend: 1= All 2= Most 3= 911 4= 924 5= 928 6= 944 7= 968

Heating, Air Conditioning, Ventilation, Defogger

Models	Year	TSB#	Description
3,4,6,7	Up - 94	01/9305	A/C COOLING FAN MOTOR - INSTALL WITH VENT HOLE DOWN
3	95 - 95	06/9503	HEATED REAR WINDOW - INFO UPDATE
1	Up - 93	08/9205 8/4/92	A/C SYSTEM - NEW VERSION O-RINGS MUST BE USED - REV
1	93 - 93	08/9206	A/C - USE OF R-134a CFC FREE REFRIGERANT - INFO
1	Up - 93	08/8604	A/C - OIL FILLING CAPACITIES - REVISED
7	92 - 93	08/9301	A/C - BELT SQUEAK AFTER COLD START - SERV PROC/TOOL
3	89 - 95	08/9401	A/C REFRIG LOSS - SEAL PIPE CONNECTIONS ON EXPAN VLV
3	89 - 94	08/9402	INOP REAR HEATER BLOWER MOTOR - RESISTOR

Model Legend: 1= All 2= Most 3= 911 4= 924 5= 928 6= 944 7= 968

Entertainment Devices, Stereo, Radio, Etc.

Models	Year	TSB#	Description
1	Up - 93	09/9204	ALPINE RADIOS - REPAIR/REPLACEMENT INFORMATION
1	Up - 93	09/9305	CD CHANGER - DISC JAMMED - DIAGNOSIS & CORRECTION
1	Up - 93	09/9305 03/23/93	CD CHANGER - DISC JAMMED - REVISED
3	95 - 95	06/9506	COMPACT DISC TRAYS - INSTALLATION
3	95 - 96	09/9504	DIGITAL SOUND PROCESSING SYSTEM - INSTALLATION
3	95 - 95	09/9401	HI-FI SOUND SYSTEM M490 - CARRERA
1	89 - 94	09/9308	RADIO INSTAL - DISCCHARGED BATTERY/ALARM MALFUNCTION

Model Legend: 1= All 2= Most 3= 911 4= 924 5= 928 6= 944 7= 968

Seats, Belts, Interior Trim, Carpets, Air Bags

Models	Year	TSB#	Description
3	Up - 93	06/9001	AIR BAG - NOTIFICATION OF DEPLOYMENT & SERVICE INFO
1	92 - 93	06/9305	AIR BAG SYSTEM - FAULT DIAGNOSIS - SPECIAL TOOL
3	95 - 95	06/9404	LOCKING AIRBAG CONTROL CIRCUIT
3	95 - 95	07/9501	NOISE - PASS AIRBAG COVER
3,5	Up - 94	07/9301	SEAT RAIL ADJUSTMENT - MORE LEG ROOM

Model Legend: 1= All 2= Most 3= 911 4= 924 5= 928 6= 944 7= 968

Glass, Doors, Hood, Decklid, Tailgate, Liftgate, Locks

Models	Year	TSB#	Description
3	95 - 95	06/9504	CABRIOLET TOP - INSUFFICIENT CLEARANCE
3	86 - 94	06/9502	CABRIOLET TOP INSTALL - TENSION CABLE - TIPS
3	95 - 96	05/9401	CARE OF INSIDE PAINTED SURFACES OF FRONT & REAR LID
3	95 - 95	05/9502	DOOR LOCK CYLINDER COVER
3	94 - 94	06/9401	GUIDE PINS FOR FOLDING TOP - SPEEDSTER
4,6,7	Up - 94	05/9302	HATCH - WATER ENTRY THRU REAR LID LATCH ASSY - PROC
3	95 - 95	06/9501	INSTALLING CABRIO REAR WINDOW - HOT GLUE TIMER
7	92 - 93	06/9301	LIFTING ROOF PANEL PROTRUDES - GASKET & CUT MODIFIED
7	92 - 93	06/9301 1/14/93	LIFTING ROOF PANEL PROTRUDES - GASKET & CUT MODIFIED
1	Up - 95	06/9403	PROPER MIXING OF WINDSHIELD CEMENT
3	95 - 95	06/9406	PUNCTURES/CHAFING - CABRIOLET TOP - PROCEDURE
6,7	Up - 93	06/9304	REAR LID SIDE PIECE BREAKS - MODIFIED PARTS
3	Up - 93	06/9303	VENT WINDOW FRAM SUPPORT BRAKES - MODIFIED PARTS
3	93 - 93	06/9206	WINDOW - HEIGHT ADJUSTMENT PROCEDURE CHANGED

Model Legend: 1= All 2= Most 3= 911 4= 924 5= 928 6= 944 7= 968

Finishes, Body Structure, Frame, Bumpers

Models	Year	TSB#	Description
3	94 - 94	00/9301	BODY - COVERS FOR REAR TRANSPORT TIE DOWNS
3	Up - 93	05/9301	GRILL/VENTILATION - MOUNTING/REPAIR INFORMATION
3	95 - 95	05/9501	PAINT DAMAGE AROUND WINDOWS - REPAIR PROCEDURE
3	95 - 96	06/9507	FITTING STAINLESS STEEL DOOR SILL COVERS
1	94 - 94	WAR/9316	PAINT CODES - 1994 MODELS
1	93 - 93	X/9302	PAINT/BLACK METALLIC - INCORRECT PAINT CODE STICKER

Model Legend: 1= All 2= Most 3= 911 4= 924 5= 928 6= 944 7= 968

SAAB

Lighting, Horns, Turn Signals, Steering Column

Models	Year	TSB#	Description
2	93 - 93	08/94-0487	BRAKE LIGHT SWITCH REPLACEMENT - RECALL 288
2	95 - 95	11/94-0520	DELETING DAYTIME RUNNING LIGHTS AT CUSTOMER REQUEST
3	94 - 95	11/94-0521	FITTING WIRING FOR DAYTIME RUNNING LIGHTS
3	94 - 94	MEMO10/18/94	IGNITION SWITCH ILLUMINATION DELETED
2	94 - 94	06/94-0473	IMPROPERLY WIRED BRAKE LIGHTS - RECALL 286
1	Up - 93	ST01/93	LIGHT (MAP) - BLOWS FUSE - CORRECTION
3	94 - 94	03/94-0453	STIFF/BINDING/NOISE IN COLUMN - EXCEPT CONVERTIBLE
3	94 - 94	ST04/94	TURN SIGNALS WON'T CANCEL - TIP

Model Legend: 1= All 2= Most 3= 900 4= 9000

Instruments, Dash Cluster, Warning Lights, Mirrors

Models	Year	TSB#	Description
3	93 - 93	ST01/93	CLOCK - INOPERATIVE OR ERRATIC - DIAGNOSTIC INFO
3	94 - 94	08/94-0489	CONTROLS - PEDALS - RAISING KIT
2	Up - 94	11/93-0411	CRUISE CONTROL BALL CHAIN - IMPROVED CLIP
2	95 - 95	01/95-0538	EDU ILLUMINATION COMPLAINTS
3	94 - 94	05/94-0468	GLOVEBOX LID REPLACEMENT - SERVICE DIRECTIVE
2	94 - 94	10/93-0399	HEADLIGHT INDICATOR LAMP - OWNER'S MANUAL REVISION
2	93 - 94	MEMO08/03/94	INACCURATE SCC SPEED WARNING - POOR EDU ILLUMINATION
2	93 - 93	ST01/93	PICTOGRAM ON W/BRAKE PEDAL DEPRESSED-S/M CORRECTION
2	95 - 95	08/95-0607	RATTLE/SQUEAK NOISE IN DASHBOARD
3	94 - 94	11/93-0406	SAAB INSTRUMENT DISPLAY (SID) EXPLAINED
2	94 - 94	03/94-0449	SURGE - INCORRECTLY WIRED CRUISE CONTROL

Model Legend: 1= All 2= Most 3= 900 4= 9000

Chassis Electrical, Wiring Harness, Fuses-Circuit Breakers, Wipers, Window Motors

Models	Year	TSB#	Description
3	94 - 94	01/95-0544	ADJUSTABLE WINDSHIELD WASHER NOZZLES - INSTRUCTIONS
3	94 - 95	05/95-0581	ADJUSTABLE WINDSHIELD WASHER NOZZLES - SERVICE PARTS
3	Up - 93	06/93-0355	BINDING IGNITION SWITCH CONTACTS
2	93 - 93	08/94-0487	BRAKE LIGHT SWITCH REPLACEMENT - RECALL 288
3	94 - 94	11/93-0414	CONNECTOR FOR MOBILE PHONE - WIRING DIAGRAM
2	93 - 93	05/94-0461	ENGINE WIRING HARNESS POSITION - SERVICE DIRECTIVE
2	89 - 94	12/93-0415	FAN RESISTOR (2 SPEED FAN) NOW REPLACEABLE
2	Up - 93	12/92-0278	FUSE BOX - SECURING DISTRIBUTION TERMINALS
2	92 - 93	11/92-0275	FUSES/WINDOW MOTOR - MAY BLOW - ADDITIONAL PDI STEP
3	93 - 93	10/93-0384	IGNITION RELAY (NEW) & WIRING DIAGRAM - S/M REVISION
3	93 - 93	10/93-0384	IGNITION RELAY (NEW)/WIRING DIAG - S/M REVISION -REV
3	94 - 94	10/93-0397	MAXI FUSES - LOCATIONS/SPECS - S/M REVISION
2	Up - 95	11/94-0522	POOR/IMPROPER POWER WINDOW OPS - TIPS
2	94 - 94	09/93-0378	REAR DOOR ELECT WINDOW SWITCHES - SERVICE DIRECTIVE
2	89 - 93	06/92-0218	RELAYS (RADIATOR FAN) - STANDARD VS TIME-DELAY
2	94 - 94	02/94-0437	RELAYS - REPLACE DEFECTIVE - SERVICE DIRECTIVE
2	93 - 93	03/93-0312	RELAYS/WINDSHIELD WIPER - ADDITIONAL PDI STEP
2	92 - 93	01/93-0299	WINDOW MOTOR FUSES - ADDITIONAL PDI STEP
2	93 - 93	ST01/93	WIPER (REAR) - TIGHTENING OF RETAINING NUT
2	Up - 93	05/93-0336	WIPERS - BLADE CARE
3	94 - 94	05/94-0468	WIRE HARNESS PINCHED - SERVICE DIRECTIVE - REVISED
3	94 - 94	05/94-0468	WIRE HARNESS PINCHED AT DOOR SILL -SERVICE DIRECTIVE
2	94 - 95	07/95-0593	WIRING HARNESS CHAFING - CAMPAIGN 418

Model Legend: 1= All 2= Most 3= 900 4= 9000

Auxiliary Equipment, Jacks, Trailer Hitches, Towing

Models	Year	TSB#	Description
1	Up - 93	REF: 12/91-0169	CELLULAR PHONE (NOKIA) - WARRANTY PROCEDURE INFO
3	94 - 94	11/93-0414	CONNECTOR FOR MOBILE PHONE - WIRING DIAGRAM
3	94 - 94	10/93-0390	LIFTING AND JACKING POINTS - 1994 HATCHBACK MODELS
3	94 - 95	10/94-0509	TRAILER HITCH KIT W/ WIRING HARNESS AVAILABLE

Model Legend: 1= All 2= Most 3= 900 4= 9000

Heating, Air Conditioning, Ventilation, Defogger

Models	Year	TSB#	Description
3	95 - 95	MEMO07/29/94	A/C INOP - CONVERTIBLE
3	94 - 95	08/94-0480	A/C POOR PERFORMANCE - DIAGNOSTICS
3	93 - 93	ST07/93	A/C REMAINS ON/KEY OFF - CHECK RELAY
3	94 - 94	ST04/94	CENTER VENT INOP - REMOVAL TIP
2	Up - 93	04/93-0328	DEFOGGER/REAR - CRACKED WIRES/RADIO INTERFERENCE/FIX
2	90 - 94	ST07/93	FAN SPEED - OVERHEATED FAN CONTROL UNIT
3	94 - 94	05/94-0468	HEAT & A/C CAPACITY - SERVICE DIRECTIVE - REVISION
2	92 - 94	10/94-0513	HIGH FREQ NOISE FROM CLIMATE CONTROL UNIT
3	94 - 94	05/94-0468	IMPROVING HEAT & A/C CAPACITY - SERVICE DIRECTIVE
2	92 - 94	03/95-0557	POOR PERFORMANCE - REAR DEFOGGER
3	94 - 95	05/95-0571	WINDSHIELD FOGGING - PROCEDURE

Model Legend: 1= All 2= Most 3= 900 4= 9000

Entertainment Devices, Stereo, Radio, Etc.

Models	Year	TSB#	Description
2	93 - 93	09/92-0254	ANTENNA - CHECKING ADJUSTMENT - ADDED PDI STEP
2	95 - 95	MEMO02/13/95	AUDIO SYS CONNECTOR HOUSINGS
3	94 - 94	10/93-0395	AUDIO SYSTEM - NEW ANTI-THEFT RADIO INFORMATION
2	Up - 93	12/92-0289	CD AUTOCHANGER - COVER AVAILABLE/PREVENT WATER ENTRY
3	94 - 94	03/94-0440	CD AUTOCHANGER - INSTALL INSTRUCTIONS SUPPLEMENT
2	Up - 94	ST11/93	CD AUTOCHANGER INSTALLATION - TECH TIP
2	Up - 95	MEMO08/24/94	CD AUTOCHANGER SKIPS - TIPS
2	95 - 95	MEMO02/02/95	INSTALLING 6-DISC CD AUTOCHANGER - ADAPTER
3	95 - 95	06/95-0589	NOISE FROM REAR SPEAKERS
2	95 - 95	MEMO10/04/94	PRESTIGE AUDIO SYSTEM INFO
2	Up - 93	04/93-0328	RADIO - POOR RECEPTION - REAR WINDOW HEATING WIRES
3	94 - 94	MEMO06/03/94	REMOVING REAR SPEAKER GRILLES - CAUTION
2	Up - 93	ST07/93	REPLACEMENT ANTENNA MAST

Model Legend: 1= All 2= Most 3= 900 4= 9000

Seats, Belts, Interior Trim, Carpets, Air Bags

Models	Year	TSB#	Description
2	89 - 94	08/94-0494	AIR BAG (SRS) - DTC'S - SEPARATE WIRING HARNESS
3	93 - 94	08/94-0493	AIR BAG (SRS) - SEPARATE WIRING HARNESS AVAILABLE
1	93 - 93	09/92-0235	FLOOR MATS (PLUSH) - STANDARD EQUIP - INSTALL PROC
3	94 - 95	05/95-0586	INTEG CHILD SAFETY SEAT BACKREST REPLACE PROCEDURE
3	94 - 94	10/94-0506	MANUAL FRONT SEAT RAIL SPRING RODS - RECALL 290
3	94 - 94	08/94-0488	MOUNTING WELDS - SEAT HEIGHT ADJUSTER - RECALL 287
3	94 - 94	05/94-0468	NOISE FROM A-PILLAR AREA (5-DOOR) -SERVICE DIRECTIVE
2	94 - 94	12/93-0420	NOISE/SQUEAK - PASSENGER AIR BAG - WIRING HARNESS
2	93 - 93	04/94-0452	REPLACEMENT AIRBAG - DIFFERENT APPEARANCE & P/N
2	88 - 93	12/92-0290	SEAT BELTS - EXTENDED LENGTH - INFORMATION
2	Up - 93	05/93-0342	SEAT CUSHION/REAR - DIFFICULT TO LATCH - SERV PROC
3	94 - 95	01/95-0536	SEAT RECLINER MECHANISM - RECALL 293

Model Legend: 1= All 2= Most 3= 900 4= 9000

Glass, Doors, Hood, Decklid, Tailgate, Liftgate, Locks

Models	Year	TSB#	Description
3	95 - 95	09/94-0499	ADJUSTMENT OF TONNEAU COVER LATCHES - S/M REV
2	Up - 94	06/93-0344	ATTACHING DOOR SEALS - REPLACE CLIPS WITH TAPE
3	95 - 95	MEMO08/18/94	AVOID MANUAL RAISING OF CONVERTIBLE TOP
3	94 - 94	ST11/93	CENTRAL LOCK FEATURE - TECH TIP
2	Up - 94	ST07/93	CLAX 200-S FOR CLEANING CONVERTIBLE TOPS
3	95 - 95	03/95-0554	CONV TOP STACK MECHANISM - IMPROVEMENTS
3	95 - 95	MEMO08/19/94	CONVERTIBLE - SUMMARY OF TECH ASSIST CALLS
3	86 - 93	10/91-0138	CONVERTIBLE TOP - MINIMIZING TOP FABRIC WEAR - CMPGN
3	Up - 93	07/89-1156	CONVERTIBLE TOP HYDRAULICS - CHECKING FLUID LEVEL
3	95 - 95	MEMO08/11/94	CONVERTIBLE TOP REPAIRS - TIPS & PARTS
3	94 - 94	08/94-0484	FRONT DOOR WINDOW SEAL LOOSE
2	Up - 93	03/93-0309	HATCH/REAR - GAS SPRING REPLACEMENT INFORMATION
3	95 - 95	05/95-0572	HEADLINER DRAGS/CATCHES - CONVERTIBLE
3	90 - 93	02/93-0304	HOOD/BULKHEAD AREA - SQUEAK/RUB NOISES/NEW HOOD SEAL
2	Up - 93	ST07/93	IMPROVED DOOR SWITCH - CORROSION PROTECTED
3	94 - 94	10/93-0398	LIGHTED KEY - SERVICING BULB/BATTERY/CONTACTS
3	94 - 94	MEMO07/08/94	LOOSE DOOR GLASS MOLDING
2	93 - 93	03/93-0315	LUGGAGE COMPARTMENT - WATER ENTRY - REPAIR CAMPAIGN
3	86 - 94	03/93-0316	MINIMIZE TOP FABRIC WEAR - CONVERTIBLE
2	Up - 95	11/94-0523	MODIFIED/RELOCATED HOOD STOPS
2	Up - 94	08/94-0485	NEW ADHESIVE FOR GLASS INSTALLATION
2	93 - 93	09/93-0373	NEW HOOD SEAL - FIX WINDNOISE/DAMAGED MOUNTING LUGS
2	Up - 94	ST07/93	NEW PLASTIC DOOR STOP SEAL - STOPS SQUEAKING - 4-DR
3	94 - 94	10/94-0507	PORES IN SHEET METAL SEAM - LH REAR DOOR
2	Up - 95	01/95-0543	RAIN/WASHER FLUID RUNS OVER REAR WINDOW - CD MODELS
3	94 - 94	12/94-0527	REAR SIDE DOOR BEAM SUPPORT BRACKET - RECALL 291
2	93 - 93	11/93-0409	TAILGATE INFRASONIC VIBRATION - RUBBER SUPPORTS
2	93 - 93	12/92-0279	TRUNK CARPET WET - LUGGAGE COMPARTMENT WATER LEAK
3	Up - 93	06/93-0350	WATER & WIND LEAKS - CONVERTIBLE
3	95 - 95	MEMO10/14/94	WATER LEAK AT WINDSHIELD FRAME - CONVERTIBLE
2	93 - 94	10/93-0383	WHISTLING NOISE FROM WINDSHIELD - APPLY SEALANT
3	94 - 94	08/94-0486	WIND NOISE - WINDSHIELD/A-PILLAR
3	93 - 93	07/93-0362	WINDOWS (REAR) - FRAMES RUSTING - DIAG/REPAIR
2	Up - 93	04/93-0321	WINDOWS - BINDING/SLOW OPERATION - CORRECTION PROC
2	Up - 93	ST07/93	WINDSHIELD WHISTLING NOISE AT 75 MPH+ - DIAGNOSTICS

Model Legend: 1= All 2= Most 3= 900 4= 9000

Finishes, Body Structure, Frame, Bumpers

Models	Year	TSB#	Description
2	86 - 93	04/93-0323	BODY & BODYWORK REPAIRS - S/M CORRECTION
2	93 - 93	10/92-0260	BODY (EXTERIOR) - GALVANIZED SHEET METAL USAGE
2	93 - 93	04/93-0325	BUMPER (REAR) - REMOVING/REFITTING (AERO MODEL)
2	93 - 93	03/93-0315	C-PILLAR VENT OUTLET - WATER ENTRY - REPAIR CAMPAIGN
3	95 - 95	06/95-0590	CREAK NOISE - RT HAND A-PILLAR - 5-DR
2	Up - 93	06/93-0351	FRONT SPOILER INSTALLATION
2	Up - 93	11/92-0267	GRILLE - WHISTLE NOISE/45-65 MPH - DIAG/REPAIR - REV
2	Up - 93	11/92-0267	GRILLE - WHISTLE NOISE/45-65 MPH - DIAGNOSIS/REPAIR
2	93 - 93	05/93-0338	HOOD PROTECTIVE STRIP AVAILABLE FOR DEALER PURCHASE
2	93 - 93	09/92-0255	LICENSE PLATE - ADJUSTING SCREWS - ADDED PDI STEP
3	94 - 94	10/93-0390	LIFTING AND JACKING POINTS - 1994 HATCHBACK MODELS
2	Up - 93	ST07/92	MOULDINGS/WHEEL ARCH - INSTALLATION DIFFICULT - TIPS
2	Up - 93	05/91-0107	PAINT (TOUCH-UP) - P/N's - DESCRIPTION - COLOR CODES
1	93 - 93	09/92-0251	PAINT - COLOR NAMES/SAAB CODES/SUPPLIERS
1	95 - 95	09/94-0500	PAINT CODES - 1995 MODELS
1	94 - 94	09/93-0377	PAINT COLORS - 1994 MODELS
1	94 - 94	09/93-0377	PAINT COLORS AND CODES
2	93 - 93	11/93-0404	PAINT FINISH - STONE GUARD DISCONTINUED MID-YEAR
2	Up - 93	06/93-0354	REPAIRING MINOR CRACKS IN REAR WHEEL HOUSINGS
2	90 - 93	06/92-0213	ROOF SEAL - NEW DESIGN TO REDUCE SQUEAKS - REVISION
2	93 - 93	04/93-0324	SILL PLATE (DOOR) - REMOVING/REFITTING (AERO MODEL)
2	93 - 93	10/92-0257	SPOILER (FRONT) - HARDWARE & INSTALLATION PROCEDURES
2	Up - 93	05/93-0339	SPOILER/BRIDGE TYPE - AVAILABLE FOR CS MODEL

Model Legend: 1= All 2= Most 3= 900 4= 9000

SUBARU

Lighting, Horns, Turn Signals, Steering Column

Models	Year	TSB#	Description
2	95 - 95	SH03/95	FOG LAMP KIT INSTALLATION TIP
5	Up - 94	SH04/94	HEADLIGHTS DIM AT IDLE - MAY BE NORMAL
7	Up - 94	01-136-93	STEER COLUMN INSPECTION AFTER COLLISION - PROCEDURE
9	92 - 93	08-032-94	TURN SIGNAL - PREMATURE CANCELLATION - PROCEDURE

Model Legend: 1= All 2= Most 3= DL 4= GL 5= Impreza 6= Justy 7= Legacy 8= Loyale 9= SVX

Instruments, Dash Cluster, Warning Lights, Mirrors

Models	Year	TSB#	Description
2	Up - 95	SH04/95	CRUISE CONTROL AND USING SELECT MONITOR
7	95 - 95	SH04/95	CRUISE CONTROL DIAGNOSTICS - S/M REV
7,5	Up - 94	16-060-94	SPEEDO CABLE ADAPTER REMOVE/INSTALL -4EAT - CAUTIONS
7	93 - 93	SH04/93	WARN LITE (SRS AIRBAG) - STAYS ON - CHECK CONNECTORS

Model Legend: 1= All 2= Most 3= DL 4= GL 5= Impreza 6= Justy 7= Legacy 8= Loyale 9= SVX

Chassis Electrical, Wiring Harness, Fuses-Circuit Breakers, Wipers, Window Motors

Models	Year	TSB#	Description
6	Up - 93	SH01/93	CONNECTORS/D-CHECK/READ MEMORY - ECU/ECVT CLARIFIED
7	95 - 95	SH07/94	ELECTRICAL COMPONENT LOCATION QUICK REFERENCE LIST
1	Up - 93	SH06/92	WIPER BLADE/WINDSHIELD - STREAKING/SMEARING - REPAIR
1	Up - 93	01-134-93	WIPER BLADES - SERV PROC FOR WINDSHIELD STREAKING
7	95 - 95	SH07/94	WIRING DIAGRAM - SECURITY SYSTEM CONTROL UNIT
5	93 - 93	18-014-93	WIRING DIAGRAMS - ABS & A/T CONTROL - S/M REVISION
5	93 - 93	18-016-93	WIRING DIAGRAMS/FUSES - S/M REVISION

Model Legend: 1= All 2= Most 3= DL 4= GL 5= Impreza 6= Justy 7= Legacy 8= Loyale 9= SVX

Heating, Air Conditioning, Ventilation, Defogger

Models	Year	TSB#	Description
7	Up - 93	SH04/93	A/C (ZEXEL) - CHIRP W/COMP ENGAGED OR WITH A/C CYCLE
1	93 - 93	SH03/93	A/C - A TSB REGARDING R134a SYSTEM WILL BE RELEASED
1	Up - 93	SH08/92	A/C - EVAPORATOR ODOR - DIAGNOSIS/REPAIR - CLEANER
1	Up - 94	10-065-94	A/C - PROPER REFRIGERANT - CHART
1	Up - 94	10-066-94	A/C - R-12 & R-134A HANDLING PROCEDURES
2	Up - 94	SH06-08/94	A/C - SPECIAL SUPPLEMENT - T/S TIPS
9	93 - 93	10-063-93	A/C - SYS CHANGES/SERV EQUIP/HANDLING R134a REFRIG
6	87 - 93	SH07/92	A/C KIT INSTALLATION - TECH INFO & WIRING DIAGRAMS
5	95 - 95	SH10/94	BUBBLES FROM DASH VENTS AFTER WASHING
7	92 - 94	10-067-94	CLICK NOISE FROM HEATER MODE DOOR ACTUATOR
5	93 - 93	10-064-93	HEATER VENT DOOR BINDS OR LEAKS AIR
1	Up - 95	07-050-93	REAR DEFOGGER GRID OR METAL TAB REPAIR
7	93 - 93	02-089-94	STUMBLE - INTERMITTENT WHEN A/C CYCLES

Model Legend: 1= All 2= Most 3= DL 4= GL 5= Impreza 6= Justy 7= Legacy 8= Loyale 9= SVX

Entertainment Devices, Stereo, Radio, Etc.

Models	Year	TSB#	Description
1	Up - 93	15-096-92	ANTENNA/POWER - MAST STUCK/BINDING - SERVICE PROC
9	Up - 93	SH06/93	AUDIO ASSEMBLY - REMOVAL INFORMATION

Model Legend: 1= All 2= Most 3= DL 4= GL 5= Impreza 6= Justy 7= Legacy 8= Loyale 9= SVX

Seats, Belts, Interior Trim, Carpets, Air Bags

Models	Year	TSB#	Description
6	Up - 93	12-058-93	FRONT SEAT TRACKS - ADJUSTMENT PROCEDURE
7	93 - 93	SH03/94	HEADLINER DROOPS
2	Up - 94	SH04/94	PASSIVE RESTRAINT IMPROPER OPERATION - TIPS
5	93 - 93	12-061-93	REAR LUGGAGE COVER WON'T STAY ENGAGED - WAGON
2	Up - 95	17-02-95	SRS - AIRBAG SERVICE PROCEDURES
1	Up - 95	17-001-92	SRS AIR BAG SYS - INSPECTION & SERVICING - REVISED

Model Legend: 1= All 2= Most 3= DL 4= GL 5= Impreza 6= Justy 7= Legacy 8= Loyale 9= SVX

Glass, Doors, Hood, Decklid, Tailgate, Liftgate, Locks

Models	Year	TSB#	Description
2	Up - 94	SH04/94	DOOR/IGNITION KEY CODE LOCATION
5	94 - 94	12-060-93	KEY CODE LOCATION (DOOR/IGN) - FOR DUPLICATING KEYS
9	Up - 93	12-059-93	MOULDING/WINDSHIELD - DISTORTED - REPAIR PROCEDURE
9	Up - 93	12-059-93	MOULDING/WINDSHIELD - FLAT RATE INFO - REVISED

Model Legend: 1= All 2= Most 3= DL 4= GL 5= Impreza 6= Justy 7= Legacy 8= Loyale 9= SVX

Finishes, Body Structure, Frame, Bumpers

Models	Year	TSB#	Description
5	93 - 93	SH03/93	MOULDINGS/SIDE GUARD - SLIGHT SEPARATION - INFO
1	93 - 93	13-062-92	PAINT CODES
1	94 - 94	13-064-94	PAINT CODES - 1994 MODELS
7,5	96 - 96	13-067-95	PAINT CODES - TWO-TONE
7,5	96 - 96	13-067-95	PAINT CODES - TWO-TONE
7,5	96 - 96	13-067-95	PAINT CODES - TWO-TONE - REVISED
7,5	Up - 95	13-066-94	PAINT CODES - TWO-TONE PAINT
1	93 - 93	13-063-93	PAINT CODES AND COLOR NAMES
1	95 - 95	13-065-94	PAINT FORMULA INFORMATION
1	96 - 96	13-068-95	PAINT FORMULA INFORMATION
1	Up - 94	01-135-93	REMOVING PROTECTIVE TRANSIT COATING

Model Legend: 1= All 2= Most 3= DL 4= GL 5= Impreza 6= Justy 7= Legacy 8= Loyale 9= SVX

SUZUKI

Lighting, Horns, Turn Signals, Steering Column

Models	Year	TSB#	Description
4	89 - 93	TS02-14	INT DOME LIGHT/WARNING BUZZER TEST PROC - S/M REV

Model Legend: 1= All 2= Most 3= Esteem 4= Swift 5= X-90

Instruments, Dash Cluster, Warning Lights, Mirrors

Models	Year	TSB#	Description
5	Up - 96	TS04-22	FUEL GAUGE ACCURACY - PARTS MODIFICATION
3	95 - 96	TS07-02	INACCURATE FUEL GAUGE - MODIFIED GAUGE/BREATHER PIPE

Model Legend: 1= All 2= Most 3= Esteem 4= Swift 5= X-90

Chassis Electrical, Wiring Harness, Fuses-Circuit Breakers, Wipers, Window Motors

Models	Year	TSB#	Description
1	Up - 93	SA01-01/93	ELECTRICAL TERMINAL KIT - NEW - MANDATORY PARTS
1	94 - 94	TS00-07	STOP LIGHT SWITCH MODIFIED - ADJUST PROCEDURES
4	93 - 93	TS02-15	WIRING DIAGRAM CORRECTIONS - W/D MANUAL MODIFICATION

Model Legend: 1= All 2= Most 3= Esteem 4= Swift 5= X-90

Heating, Air Conditioning, Ventilation, Defogger

Models	Year	TSB#	Description
4	95 - 95	TS07-09	A/C - COMPRESSOR BRACKET BOLT - A/C MANUAL REV
4	95 - 95	TS07-09	A/C - COMPRESSOR BRACKET BOLT - A/C MANUAL REV - REV
4	95 - 95	TS07-08	A/C - HOSE TO CONDENSER TORQUE SPEC - A/C MANUAL REV
4	94 - 94	TS07-06	A/C - MODIFICATION/CHANGE IN SYSTEM AND S/M
1	95 - 95	TS00-11	A/C - RECOMMENDED OIL FOR R134A SYSTEMS
4	91 - 93	TS07-04	A/C COMPRESSOR - MODIFICATIONS AND SPECIAL TOOLS
1	Up - 94	TS00-01	AUTHORIZED A/C REFRIGERANT - FREON R-12
1	Up - 96	TS00-01 06031	AUTHORIZED A/C REFRIGERANT - R-12 OR R134A - REV

Model Legend: 1= All 2= Most 3= Esteem 4= Swift 5= X-90

Seats, Belts, Interior Trim, Carpets, Air Bags

Models	Year	TSB#	Description
1	95 - 95	TS00-10R	AIR BAG SYSTEM ENABLE/DISABLE PROCEDURE
1	Up - 95	TS01-10	AIR BAG SYSTEM ENABLE/DISABLE PROCEDURE
1	95 - 96	TS00-10R	AIR BAG SYSTEM ENABLE/DISABLE PROCEDURE - REV
3	96 - 96	TS07-03	SDM & AIR BAG SYS DIAGNOSTICS - S/M REV

Model Legend: 1= All 2= Most 3= Esteem 4= Swift 5= X-90

Glass, Doors, Hood, Decklid, Tailgate, Liftgate, Locks

Models	Year	TSB#	Description
4	89 - 93	TS01-12	HOOD STRIKER - SERVICE CAMPAIGN RECALL
1	Up - 93	SA01-07/87	KEY CODES - AVAILABILITY & PROCEDURE TO BE FOLLOWED

Model Legend: 1= All 2= Most 3= Esteem 4= Swift 5= X-90 6= Sidekick

Finishes, Body Structure, Frame, Bumpers

Models	Year	TSB#	Description
4	92 - 93	TS01-10	BUMPER - BOLTS MODIFIED IN PRODUCTION/NO INTERCHANGE
1	Up - 94	TS00-06	BUMPER - PRECAUTIONS WHEN PAINTING REPLACEMENT PART
4	89 - 93	TS01-02	PAINT - CODE IDENTIFICATION PLATE LOCATION
4	89 - 95	TS01-02R	PAINT CODES
5	89 - 96	TS1-02	PAINT CODES
4	89 - 94	TS01-02R	PAINT CODES - IDENTIFICATION PLATE LOCATION
3	95 - 96	TS01-01	PAINT CODES - LOCATION OF PAINT CODE ID PLATE
5	96 - 96	TS01-01	PAINT CODES - LOCATION OF PAINT CODE ID PLATE
4	89 - 96	TS01-02	PAINT CODES - LOCATION OF PAINT CODE ID PLATE
1	Up - 94	TS00-06R	REPLACEMENT BUMPER PAINTING PRECAUTIONS
1	Up - 94	TS00-08	UNDERHOOD (PRIMER) PAINT COLOR CODES
1	Up - 96	TS0-08	UNDERHOOD (PRIMER) PAINT COLOR CODES - REV
1	Up - 95	TS00-08	UNDERHOOD (PRIMER) PAINT COLOR CODES - REV
1	Up - 96	TS00-08R	UNDERHOOD (PRIMER) PAINT COLOR CODES - REV

Model Legend: 1= All 2= Most 3= Esteem 4= Swift 5= X-90

TOYOTA

Lighting, Horns, Turn Signals, Steering Column

Models	Year	TSB#	Description
5	92 - 93	ST93-001	STEERING COLUMN & SHAFT - NEW PARTS/REPAIR PROCEDURE

Model Legend: 1= All 2= Most 3= Celica 4= Avalon 5= Camry 6= Corolla 7= MR2 8= Paseo 9= Supra 10= Tercel

Instruments, Dash Cluster, Warning Lights, Mirrors

Models	Year	TSB#	Description
10	95 - 95	AX94-005	ACCESSORY RIGHT HAND SIDE MIRROR INSTALLATION
6	93 - 95	BO95-004	INSTRUMENT PANEL REGISTER SUB-ASSEMBLY AVAILABLE
5	92 - 93	EL93-002	SPEEDOMETER MODIFIED - CLICK/TICK NOISE REPAIR (WGN)

Model Legend: 1= All 2= Most 3= Celica 4= Avalon 5= Camry 6= Corolla 7= MR2 8= Paseo 9= Supra 10= Tercel

Chassis Electrical, Wiring Harness, Fuses-Circuit Breakers, Wipers, Window Motors

Models	Year	TSB#	Description
5	Up - 95	BO95-009	FRONT DOOR WINDOW REGULATOR - IMPROVED PART
6	Up - 95	BO94-018	FRONT DOOR WINDOW REGULATOR - REV PARTS
9	93 - 93	EL93-003	FUSES-REMOVED FOR TRANSPORT/STORAGE-REINSTALL AT PDI
6,3,5,9	Up - 95	BO94-015	POWER WINDOW - NEW WIRE TYPE REGULATOR SUBASSEMBLY
6,3,5,9	Up - 96	BO94-015	POWER WINDOW - WIRE TYPE REGULATOR SUBASSEMBLY - REV
3	94 - 94	EL93-005	SHORT PIN IN RELAY BLOCK 2 - CORRECT POSITION
6	Up - 95	BO95-002	WINDSHIELD WASHER BOTTLE - IMPROVED STRENGTH
6	Up - 96	BO95-002	WINDSHIELD WASHER BOTTLE - IMPROVED STRENGTH - REV
1	Up - 96	EL95-004	WIPER MOTOR CKT BREAKER INSPECT PROCEDURE
6	93 - 94	SSC#R03	WIRE HARNESS CONNECTOR - SERVICE CAMPAIGN

Model Legend: 1= All 2= Most 3= Celica 4= Avalon 5= Camry 6= Corolla 7= MR2 8= Paseo 9= Supra 10= Tercel

Auxiliary Equipment, Jacks, Trailer Hitches, Towing

Models	Year	TSB#	Description
1	Up - 94	AU93-001	MOBILE COMMUNICATION EQUIP - INSTALL PRECAUTIONS
10	95 - 95	AX95-002	MUDGUARD INSTALLATION INSTRUCTIONS
4	95 - 95	AX95-003	MUDGUARD INSTALLATION INSTRUCTIONS
6	92 - 94	AX94-004	ROOF RACK WIND NOISE REPAIRS - WAGON
5	Up - 95	AX94-003	SPOILER - LAMP FAILURE SENSOR REPLACEMENT
5	Up - 95	AX94-003	SPOILER - LAMP FAILURE SENSOR REPLACEMENT - REVISED

Model Legend: 1= All 2= Most 3= Celica 4= Avalon 5= Camry 6= Corolla 7= MR2 8= Paseo 9= Supra 10= Tercel

Heating, Air Conditioning, Ventilation, Defogger

Models	Year	TSB#	Description
10,8	91 - 93	AC93-001	A/C (PIO/DIO) - KIT INSTALLATION PRECAUTIONS
2	89 - 93	SSC#P01	A/C - EXPANSION VALVE MALFUNCTIONS - SPECIAL CMPGN
1	Up - 94	SUPP: SSC#P01	A/C - EXPANSION VALVE REPLACEMENT/WARRANTY INFO
1	Up - 93	AC93-003	A/C - INSPECTION/DIAG/REPAIR OF LOW REFRIG LEVEL
1	92 - 93	AC92-004	A/C - O-RING CHANGED DUE TO R134a REFRIGERANT
1	Up - 96	AC002-96	A/C COMPRESSOR MAINTENANCE FOR STORED VEHICLES
1	Up - 94	AC93-005	A/C COMPRESSOR OIL APPLICATIONS
8,10	Up - 94	AC94-002	A/C POOR PERF - NEW HARNESS CONNECTOR & ROUTING
1	Up - 94	AC94-001	ALTERNATIVE REFRIGERANTS FOR R12
4	Up - 96	AC001-96	AMBIENT TEMP DISPLAYS -22 DEG F - PROCEDURE
1	93 - 94	PG93-003	CAUTION LABELS/EPA REQUIRED - A/C SYS W/REFRIG R-12
6	Up - 95	AC95-002	HEATER CONTROL KNOB RETENTION IMPROVEMENT
5	Up - 93	AC93-006	INOP A/C MAIN SWITCH - MODIFIED PART - DURABILITY
1	Up - 95	AC95-001	REFRIGERANT LEAK DETECTION - SERVICE HINTS

Model Legend: 1= All 2= Most 3= Celica 4= Avalon 5= Camry 6= Corolla 7= MR2 8= Paseo 9= Supra 10= Tercel

Entertainment Devices, Stereo, Radio, Etc.

Models	Year	TSB#	Description
1	92 - 93	AU92-002	CASSETTE PLAYER WON'T PLAY - DIAGNOSIS/REPAIR
2	92 - 93	AU94-001	CD CHANGER INOP/SKIP/DISPLAYS ERROR - PROCEDURE
4	95 - 95	AU001-96	DELCO RADIO FM HIGH PITCHED TONE
5,11	Up - 95	SSC#P03B	RADIO - SSC PO3 - CAMPAIGN UPDATE
5,11	Up - 95	SSC#P03	RADIO - SSC PO3 - CAMPAIGN UPDATE - EXCHANGE ONLY
3	94 - 94	AU94-002	RADIO ILLUMINATION MODIFIED FOR DIMMING

Model Legend: 1= All 2= Most 3= Celica 4= Avalon 5= Camry 6= Corolla 7= MR2 8= Paseo 9= Supra 10= Tercel

Seats, Belts, Interior Trim, Carpets, Air Bags

Models	Year	TSB#	Description
5	95 - 95	EL95-002	FRONT PASS AIRBAG - NEW REPLACEMENT PROCEDURE
6	Up - 94	BO94-005	REAR SEAT BOLSTER COVER LOOSE - CLIPS AVAILABLE
11	Up - 94	BO94-003	S SHAPE CLIPS FOR REINSTALLING SEAT COVERS AVAIL
2	93 - 93	SSC#N06	SEAT BELT - INCORRECT TONGUE PLATE SIZE - RECALL
6	94 - 94	SUPP: SSC#R02	SEAT BELT BUCKLE ASSEMBLY - CAMPAIGN SUPPLEMENT
6	94 - 94	SSC#R02	SEAT BELT BUCKLE ASSEMBLY - SPECIAL SERVICE CAMPAIGN
1	85 - 94	BO94-006	SEAT BELT EXTENDER - APPLICATIONS & PROCEDURES
1	83 - 93	BO93-006	SEAT BELTS - EXTENDERS AVAIL THRU SPECIAL PROGRAM
1	Up - 93	EL93-001	SRS - SERVICE INFORMATION PRECAUTION
4	95 - 96	BO005-96	SUNVISOR RATTLE NOISE
5	Up - 94	BO93-010	WATER LEAK - PASSENGER SIDE FLOOR
5	94 - 94	AX94-001	WOOD DASH/IP TRIM REPAIR INSTRUCTIONS

Model Legend: 1= All 2= Most 3= Celica 4= Avalon 5= Camry 6= Corolla 7= MR2 8= Paseo 9= Supra 10= Tercel

Glass, Doors, Hood, Decklid, Tailgate, Liftgate, Locks

Models	Year	TSB#	Description
3	90 - 93	BO93-001	CONVERTIBLE COMPONENTS - SERV PROC - S/M REVISIONS
6	93 - 93	BO92-011	DOOR ARM REST COVER RETENTION - IMPROVEMENT
11	92 - 93	BO92-007	DOORS-HINGE/HINGE PILLAR/CHECK ARM STRENGTH CHANGES
10	95 - 96	BO003-96	FRONT DOOR WINDOW REGULATOR OPERATION
6	93 - 93	PG92-006	KEY CODE & CUTTER - TOTALLY NEW FOR COROLLA WAGON
5	95 - 96	BO004-96	MOONROOF PANEL WIND NOISE
5	95 - 95	BO94-016	QUARTER WINDOW UPPER/REAR MOLDINGS LOOSE - WAGON
5	96 - 96	BO006-96	RATTLE NOISE FROM REAR WINDOW
6	94 - 95	BO94-014	REAR DOOR LOCK OPERATION IMPROVEMENT
9	Up - 96	BO95-010	REMOVABLE ROOF - LOCK STRIPS - REV PARTS
4	95 - 95	BO95-003	SLOW RETURN OF INSIDE DOOR HANDLES
5	92 - 96	AX95-007	TRUNK LID TORSION ROD ADJUSTMENT
5	92 - 95	BO94-013	WATER LEAK INTO TRUNK AFTER OPENING LID
6	Up - 94	BO94-008	WIND NOISE - A-PILLAR/WINDSHIELD MOLDING
5	Up - 94	BO94-009	WIND NOISE - REAR DOOR WINDOW
10	95 - 95	BO95-005	WIND NOISE - WINDSHIELD MOLDING
4	95 - 95	BO95-006	WIND NOISE - WINDSHIELD MOLDING
10	93 - 95	BO94-011	WIND NOISE - WINDSHIELD MOULDING
10,8	93 - 94	BO94-004	WINDSHIELD GLASS AND MOUNTING SPACERS - CHANGES

Model Legend: 1= All 2= Most 3= Celica 4= Avalon 5= Camry 6= Corolla 7= MR2 8= Paseo 9= Supra 10= Tercel

Finishes, Body Structure, Frame, Bumpers

Models	Year	TSB#	Description
6	93 - 93	BO93-002	FENDER/FRONT - REPLACEMENT SUPPLY PARTS INFO
9	93 - 93	BO93-004	LICENSE PLATE (FRONT/REAR) - INSTALLATION PROCEDURE
3	94 - 94	BO93-009	LICENSE PLATE INSTALLATION PROCEDURES
9	93 - 93	AX93-003	MUDGARD/FRONT - INSTALLATION INSTRUCTIONS
10	Up - 93	AC92-001	MUDGUARD - KIT AVAILABLE - INSTALL INSTRUCTIONS REV
10	93 - 93	AC92-001	MUDGUARD - KIT AVAILABLE - INSTALL INSTRUCTIONS REV
7	93 - 93	AX92-004	MUDGUARD - KIT AVAILABLE - INSTALLATION INSTRUCTIONS
6	93 - 93	AX92-005	MUDGUARD - KIT AVAILABLE - INSTALLATION INSTRUCTIONS
8	96 - 96	AX95-005	MUDGUARD INSTALLATION
1	93 - 93	PA92-001	PAINT & REFINISH FORMULA CODES
9	93 - 93	PA93-001	PAINT & REFINISH FORMULAS - CODES & COMPANIES LIST
7,6,11	93 - 93	BO93-003	PAINT - ANTI-CHIP PRIMER APPLIED TO SPECIAL AREAS
1	93 - 93	BO92-005	PAINT - NEW VEHICLE WASHING SCHEDULE FOR PROTECTION
1	95 - 95	PA94-001	PAINT AND REFINISH CODES - 1995 MODELS
1	94 - 94	PA93-002	PAINT AND REFINISH CODES - OEM AND REFURBISH
1	96 - 96	PA001-96	PAINT AND REFINISH FORMULA CODES
1	Up - 96	PA95-001	PAINT REPAIRS ON POLYURETHANE BUMPER
3	Up - 95	BO95-001	RATTLE NOISE LH QUARTER PANEL - FUEL LID CABLE
5	Up - 95	BO94-017	TRUNK FINISH PLATE LOOSE - REV CLIPS
5	Up - 94	BO93-010	WATER LEAK - PASSENGER SIDE FLOOR
4,5	95 - 96	PA95-002	WATER-BORNE PAINT COLOR INTRODUCTION

Model Legend: 1= All 2= Most 3= Celica 4= Avalon 5= Camry 6= Corolla 7= MR2 8= Paseo 9= Supra 10= Tercel

VOLKSWAGEN

Lighting, Horns, Turn Signals, Steering Column

Models	Year	TSB#	Description
6,8	95 - 95	97/95-04	DAYTIME RUNNING LIGHTS - RESISTOR WIRE INSTALLATION
7,8	93 - 93	94/93-01	HEADLIGHT ASSEMBLY - S/M ADDITION
6,8	96 - 96	94/95-01	HEADLIGHT BULB - LONG LIFE
6,8	Up - 94	48/93-05	HORN PAD ASSEMBLY REMOVAL PROCEDURES
6,8	94 - 94	90/94-01	HORN SYSTEM INFO FOR AIRBAG EQUIPPED VEHICLES
6,7,8,3	96 - 96	94/95-02	IGNITION/STARTER SWITCH - POWER SUPPLY TO RADIO
6,7,8	96 - 96	96/96-01	LIGHT SWITCH - NEW FUNCTION - PARK/HEADLIGHTS/FOG
6,8	93 - 94	96/93-01	REAR ASHTRAY LIGHT - NO WIRING HARNESS
4	93 - 94	94/93-03	STEERING COLUMN ASM/COMPONENTS - R&I (A/T) - S/M REV
9,4	90 - 93	94/90-02	TURN SIGNAL & HEADLIGHT DIMMER/FLASHER - INOP - REV

Model Legend: 1= All 2= Most 3= Cabriolet 4= Corrado 5= Fox 6= Golf 7= GTI 8= Jetta 9= Passat 10=Scirocco

Instruments, Dash Cluster, Warning Lights, Mirrors

Models	Year	TSB#	Description
9,4	90 - 93	90/93-04	CLOCK/DIGITAL (MFA DISPLAY) - LOSES TIME - SERVICE
5	93 - 93	80/92-01	CONTROL LEVER (TEMP) - HARD OP IN HEATER MODE - REV
5	93 - 93	80/92-01	CONTROL LEVER (TEMPERATURE) - HARD OP IN HEATER MODE
5	93 - 93	80/92-03	CONTROL LEVER/TEMP - HARD OP/HEATER MODE - 2ND REV
9	90 - 94	27/93-06	CRUISE CONTROL TROUBLESHOOTING USING VAG 1466 (A/T)
9	90 - 94	27/93-07	CRUISE CONTROL TROUBLESHOOTING USING VAG 1466 (M/T)
6,7,8,3	93 - 95	96/94-01	CRUISE CONTROL VACUUM VENT VALVES - S/M UPDATE
6,8	95 - 95	90/95-04	FUEL GAUGE INACCURATE
6,8	95 - 95	90/95-06	FUEL GAUGE INACCURATE - REV
9,4	91 - 93	90/93-05	GAUGE (FUEL/ENG COOLANT TEMP) - INOP OR INTERMITTENT
9,4	91 - 93	90/93-05	GAUGE (FUEL/ENG COOLANT TEMP) INOP/INTERMIT - REV
6,7,8	95 - 95	90/95-02	IC - BRAKE CONTROL LIGHT MALFUNCTION
9	95 - 95	90/95-01	IC - LOSS OF MEMORY - REVISED PC BOARD
9	95 - 95	90/95-01	IC - LOSS OF MEMORY - REVISED PC BOARD - REV
6,8,7,3	93 - 95	90/94-04	IC - REMOVE/INSTALL/REPAIR - S/M REVISION
6,7,8	93 - 95	90/95-07	IC ILLUMINATION INTERMITTENT - DRL VEHICLES
9	93 - 93	91/93-05	MFA - MULTI-FUNCTION INDICATOR TROUBLESHOOTING - REV
4	93 - 94	91/93-04	MULTI-FUNCTION INDICATOR - TROUBLESHOOTING - REVISED
4	93 - 94	91/93-04	MULTI-FUNCTION INDICATOR - TROUBLESHOOTING INFO
9	93 - 94	91/93-05	MULTI-FUNCTION INDICATOR - TROUBLESHOOTING INFO
6,7,8	93 - 93	90/93-02	SERVICE REMINDER INDICATOR (SRI) - S/M ADDITION
1	Up - 95	90/95-05	SPEEDO REPLACEMENT - COMPLY WITH REGULATIONS
5	Up - 94	66/93-04	WIND NOISE - EXTERIOR REAR VIEW MIRRORS
6,8	93 - 94	66/94-02	WINDNOISE - OUTSIDE REAR VIEW MIRROR
4	92 - 94	WD92-21	WIRING DIAGRAM - POWER MIRRORS

Model Legend: 1= All 2= Most 3= Cabriolet 4= Corrado 5= Fox 6= Golf 7= GTI 8= Jetta 9= Passat 10=Scirocco

Chassis Electrical, Wiring Harness, Fuses-Circuit Breakers, Wipers, Window Motors

Models	Year	TSB#	Description
1	Up - 93	91/93-02	ELECTRICAL AFTERMARKT EQUIPMENT - PRECAUTIONS
6,7,8,3	Up - 95	97/94-02	ELECTRICAL CONTACT ENHANCER - STABILANT 22A
9	95 - 95	97/94-03	ELECTRICAL CONTACT ENHANCER - STABILANT 22A
1	Up - 95	97/94-04	ELECTRICAL CONTACT ENHANCER - STABILANT 22A
6,7,8,3	Up - 95	97/94-02	ELECTRICAL CONTACT ENHANCER - STABILANT 22A - REV
9	95 - 95	97/94-03	ELECTRICAL CONTACT ENHANCER - STABILANT 22A - REV
1	Up - 95	97/94-04	ELECTRICAL CONTACT ENHANCER - STABILANT 22A - REV
6,8	93 - 95	57/94-05	FRONT ELECTRIC WINDOW REGULATOR SERVICE - S/M SUPP
1	90 - 94	97/93-02	HARNESS CONNECT TERMINALS MUST MATCH SENSOR (GOLD)
6,8	93 - 93	97/93-01	NOISE - DAYTIME RUNNING LIGHT RELAY
9	93 - 95	28/95-07	POWER ACCESSORY CIRCUIT - INTERMITTENT OPERATION
6,7,8,3	95 - 96	96/95-03	POWER WINDOWS CHILD SAFETY SYS - PROGRAMMING
6,8	96 - 96	92/95-04	PROGRAMMABLE WINDSHIELD WIPERS
6,7,8,3	95 - 96	96/95-01	PWR WINDOWS CHILD SAFETY SYS PROGRAMMING - S/M REV
9	92 - 93	CCC:SW	REPLACE POWER WINDOW REGULATORS - CAMPAIGN
4	Up - 93	97/92-02	TERMINALS (ECM) - ID AND LOCATIONS - S/M UPDATE
9	Up - 93	97/92-03	TERMINALS (ECM) - ID AND LOCATIONS - S/M UPDATE
1	Up - 96	69/95-03	USING STABILANT 22A ON ELECT CONN OF AIR BAG SYS
1	76 - 95	92/95-02	WIPER BLADES JERK OR SKIP
6,8,7	93 - 95	92/95-01	WIPER FUSE BLOWS - PROCEDURE
9,4,3	92 - 93	92/92-02	WIPER/WINDSHIELD - INFINITELY VARIABLE INTERMIT OPER
9	93 - 93	WD93-10	WIRING DIAGRAM - A/C (2.0L 16V WITH M/T)
4	92 - 93	WD92-04	WIRING DIAGRAM - AIR CONDITIONING
4	94 - 94	WD93-25	WIRING DIAGRAM - AIR CONDITIONING
4	95 - 95	WD95-01	WIRING DIAGRAM - AIRBAG
9	93 - 93	WD93-23	WIRING DIAGRAM - FOG LIGHTS
4	92 - 93	WD92-05	WIRING DIAGRAM - HEATED SEATS
4	94 - 94	WD93-21	WIRING DIAGRAM - MAIN
3	Up - 95	WD95-02	WIRING DIAGRAM - SHIFT LOCK
4	94 - 94	WD93-24	WIRING DIAGRAM - STEREO RADIO
6,7,8,3	93 - 96	WD95-09	WIRING DIAGRAMS - S/M REV
9	95 - 95	LIT/V94-37	WIRING DIAGRAMS - S/M REVISION
9	95 - 95	SI/02A	WIRING DIAGRAMS - S/M REVISION
6,8,7,3	93 - 95	SI/06	WIRING DIAGRAMS - S/M REVISION
9	95 - 95	SI/01	WIRING DIAGRAMS - T/S AND COMPONENT LOCATIONS
6,8	93 - 94	SI/04	WIRING DIAGRAMS - T/S AND COMPONENT LOCATIONS - REV
9	95 - 95	SI/04A	WIRING DIAGRAMS - UPDATE
9	95 - 95	SI/06	WIRING DIAGRAMS/TROUBLESHOOT/COMPONENT LOC

Model Legend: 1= All 2= Most 3= Cabriolet 4= Corrado 5= Fox 6= Golf 7= GTI 8= Jetta 9= Passat 10=Scirocco

Chassis Electrical, Wiring Harness, Fuses-Circuit Breakers, Wipers, Window Motors

Models	Year	TSB#	Description
9	95 - 96	SI/07	WIRING DIAGRAMS/TROUBLESHOOT/COMPONENT LOC
9	95 - 96	SI/08	WIRING DIAGRAMS/TROUBLESHOOT/COMPONENT LOC
6,7,8,3	93 - 96	SI/10	WIRING DIAGRAMS/TROUBLESHOOT/COMPONENT LOC
9	95 - 95	SI/02A	WIRING DIAGRAMS/TROUBLESHOOT/COMPONENT LOC - REVISED
9	95 - 95	SI/03B	WIRING DIAGRAMS/TROUBLESHOOT/COMPONENT LOC - REVISED
6,7,8,3	93 - 95	SI/05	WIRING DIAGRAMS/TROUBLESHOOTING/LOCATIONS

Model Legend: 1= All 2= Most 3= Cabriolet 4= Corrado 5= Fox 6= Golf 7= GTI 8= Jetta 9= Passat 10=Scirocco

Auxiliary Equipment, Jacks, Trailer Hitches, Towing

Models	Year	TSB#	Description
6,7,8,3	93 - 96	63/95-01	TOWING EYE (LENS) COVER REMOVAL - S/M REV

Model Legend: 1= All 2= Most 3= Cabriolet 4= Corrado 5= Fox 6= Golf 7= GTI 8= Jetta 9= Passat 10=Scirocco

Heating, Air Conditioning, Ventilation, Defogger

Models	Year	TSB#	Description
9	91 - 93	80/92-02	A/C - DUST AND POLLEN FILTER INSTALLATION - PARTS
9,4	93 - 94	87/93-05	A/C - R134a SYSTEM - HISSING SOUND - SERVICE
4	93 - 95	87/93-05	A/C - R134a SYSTEM - HISSING SOUND - SERVICE - REV
9	93 - 93	WD92-36	A/C - WIRING DIAGRAM (2.0L 16V WITH A/T)
9	93 - 93	WD92-34	A/C - WIRING DIAGRAM (2.8L VR6)
9	93 - 93	WD92-38	A/C - WIRING DIAGRAM (ECO DSL - 1.9L/CANADA ONLY)
9	93 - 93	WD92-37	A/C - WIRING DIAGRAM (SYNCRO - 1.8L G60/CANADA ONLY)
6,8	93 - 95	87/94-03	A/C COMP MOUNT BRACKET INST - 1.9L DIESEL - S/M REV
3	90 - 93	87/92-01	A/C COMPRESSOR BELT - ADJUSTING PROCEDURE - REVISED
3	90 - 93	87/92-01	A/C COMPRESSOR BELT - ADJUSTING PROCEDURE/TOOLS
9	95 - 95	87/94-04	A/C COMPRESSOR BRACKET - 1.9L CANADA ONLY - S/M REV
6,8,3	93 - 95	87/94-05	A/C CUT OFF THERMAL SWITCH - 2.0L - S/M UPDATE
6,8	Up - 93	87/93-03	A/C DISCHARGE LINE RUBS ON FAN OR FAN SHROUD
2	93 - 95	87/94-02	A/C POOR PERFORMANCE - VARIABLE DISPLACEMENT COMP
3	96 - 96	00/95-03	COOLANT SPEC - S/M REV
6,8	Up - 94	85/94-01	POOR AIR DIST TO FLOOR VENTS/KNOB BINDING
6,7,8	93 - 96	85/95-01	POOR AIR FLOW FROM DUCT - LEFT SIDE OF IP
9,4	93 - 93	87/93-01	R-134A COMPONENTS - REPLACING & CAPABILITIES
6,7,8	93 - 93	87/93-02	R-134A COMPONENTS - REPLACING & CAPABILITIES
6,7,8,4	93 - 94	87/93-01	R-134A COMPONENTS - REPLACING & CAPABILITIES - REV
9	93 - 95	87/93-05	R-134A EXPANSION VALVE NOISY
9,4	93 - 94	87/93-05	R134A EXPANSION VALVE NOISY - NEW VALVE (REVISED)
9	Up - 93	80/93-02	SIDE WINDOWS FOG UP DURING HEATER OPERATION

Model Legend: 1= All 2= Most 3= Cabriolet 4= Corrado 5= Fox 6= Golf 7= GTI 8= Jetta 9= Passat 10=Scirocco

21 TECHNICAL SERVICE BULLETINS

Entertainment Devices, Stereo, Radio, Etc.

Models	Year	TSB#	Description
6,8	93 - 93	91/93-06	ANTENNA LEAD - INSUFFICIENT SLACK TO REMOVE RADIO
9,4	92 - 93	91/93-01A	CASSETTE RADIO - DELUXE & PREMIUM - GEN INFO
1	Up - 95	91/95-03	CD CHANGER SKIPS
3	95 - 95	91/94-03	CD CHANGER SKIPS ON ROUGH ROADS
3	95 - 95	91/94-03	CD CHANGER SKIPS ON ROUGH ROADS - REVISED
1	Up - 93	WAR/V-93-13	HEIDELBERG LOCKED-UP RADIOS/OUT OF WARR UNLOCK INFO
6,8	96 - 96	91/96-02	NEW FIXED ANTENNA REMOVE/INSTALL INST - S/M REV
1	Up - 96	91/95-04	PANASONIC CD CHANGER NOISE - REMOVE ADD'L GROUND
9	95 - 95	91/94-04	POOR OR NO RADIO RECEPTION
9	95 - 95	91/95-02	POOR OR NO RADIO RECEPTION - REV
6,7,8,3	95 - 96	91/95-06	RADIO - PREMIUM STEREO - NEW FUNCTIONS
1	Up - 94	91/94-02	RADIO DIAGNOSIS INSTRUCTIONS
3	95 - 95	91/95-01	RADIO SPEAKERS - REAR - REMOVE/INSTALL - S/M UPDATE
6,8	93 - 93	90/93-08	RADIO WILL NOT RELEASE FROM IP
8	96 - 96	91/96-01	REAR SPEAKERS IN PARCEL SHELF - REMOVE/INST -S/M REV
6,7,8	93 - 93	91/93-03	STEREO SYSTEM - GEN INFO/ASSEM/REPAIR - S/M ADDITION
4	94 - 94	WD94-02	WIRING DIAGRAM - RADIO WITH CD CHANGER

Model Legend: 1= All 2= Most 3= Cabriolet 4= Corrado 5= Fox 6= Golf 7= GTI 8= Jetta 9= Passat 10=Scirocco

Seats, Belts, Interior Trim, Carpets, Air Bags

Models	Year	TSB#	Description
1	Up - 95	68/94-04	AIR BAG DEPLOYMENT - REPORT INCIDENT
1	Up - 96	69/95-03	AIR BAG SYS - USE OF STABILANT 22A ON ELECT CONNECT
6,7,8	95 - 96	68/95-06	AIRBAG CONTROL MODULE/WIRING HARNESS - NEW - S/M REV
6,7,8	94 - 95	01/95-03	AIRBAG DIAGNOSTICS - VAG 1551 - S/M REV
3	95 - 95	01/95-04	AIRBAG DIAGNOSTICS - VAG 1551 - S/M REV
6,8	94 - 95	68/94-06	AIRBAG ECM SAFETY LOCK CONNECTOR REPLACEMENT
3	95 - 95	68/94-07	AIRBAG ECM SAFETY LOCK CONNECTOR REPLACEMENT
9	95 - 95	68/94-08	AIRBAG ECM SAFETY LOCK CONNECTOR REPLACEMENT
6,8	94 - 94	01/94-01	AIRBAG OBD - ON BOARD DIAGNOSTIC/ASSEMBLY/COMPONENTS
9	95 - 96	01/95-09	AIRBAG SYS - DRIVER/PASS - ON BOARD DIAGN - S/M REV
3	95 - 96	01/95-10	AIRBAG SYS - DRIVER/PASS - ON BOARD DIAGN - S/M REV
6,7,8	95 - 96	01/96-01	AIRBAG SYS - DRIVER/PASS - ON BOARD DIAGN - S/M REV
6,7,8	94 - 95	69/95-01	AIRBAG UNITS - REPLACING AFTER ACCIDENT
3	95 - 95	69/95-02	AIRBAG UNITS - REPLACING AFTER ACCIDENT
2	85 - 94	69/95-04	CHILD SAFETY SEATS - ON FRONT PASS SEAT - INSTALL
6,8	93 - 94	68/94-02	CHILD SEAT - INSTALLING ON FRONT SEAT - INSTRUCTIONS
6,7,8	Up - 93	70/93-02	DOOR PANEL REMOVAL PROCEDURES
1	Up - 94	70/93-04	FABRIC/VINYL/PLASTIC CLEANING PROCEDURES
6,7,8	96 - 96	68/96-01	NOISE IN REAR CENTER CONSOLE
6,8	93 - 95	68/94-05	REAR CHILD RESTRAINT ANCHORAGE - INSTALL - S/M SUPP
2	80 - 96	69/95-05	REAR CHILD RESTRAINT ANCHORAGES
8	96 - 96	70/96-01	REAR PARCEL SHELF/REAR DOOR TRIM SVC - S/M REV
6,7,8	93 - 93	70/93-01	ROOF TRIM/HEADLINER REMOVAL - S/M ADDITION
6,7,8,3	93 - 96	68/95-05	SUN VISOR INBOARD CLIP CRACKS

Model Legend: 1= All 2= Most 3= Cabriolet 4= Corrado 5= Fox 6= Golf 7= GTI 8= Jetta 9= Passat 10=Scirocco

Glass, Doors, Hood, Decklid, Tailgate, Liftgate, Locks

Models	Year	TSB#	Description
9,4	85 - 95	64/95-01	ADHESIVE MAT'L FOR FLUSH BONDED WINDOWS - S/M REV
6,7,8	Up - 93	57/93-04	CENTRAL LOCK SYS - PNEUMATIC OPERATION TROUBLESHOOT
6,7,8	93 - 95	57/93-04	CENTRAL LOCK SYS - PNEUMATIC OPERATION TROUBLESHOOT
6,7,8	96 - 96	57/95-09	CENTRAL LOCKING - SWITCH IN IP - S/M REV
9	90 - 93	57/93-01	CENTRAL LOCKING SYS - TROUBLESHOOTING - S/M ADDITION
6,8	Up - 95	57/94-02	CENTRAL LOCKING SYSTEM TROUBLESHOOTING
6,8	94 - 94	57/93-06	CENTRAL LOCKING SYSTEM VACUUM LINE MODIFICATION
9	93 - 93	57/92-02	DOOR HANDLE (RIGHT SIDE) - MODIFICATIONS
9	93 - 93	57/92-03	DOOR LOCK - REMOVING/INSTALLING PROCEDURES
9	95 - 96	57/96-01	DOOR LOCK CYLINDER REPAIR - S/M REV
4	93 - 93	57/93-03	DOOR LOCK W/FREEPLAY ADJUSTMENT -REMOVING/INSTALLING
5	93 - 94	57/93-05	DOORS - WIND NOISE - SEAL ON TOP OF DOORS TOO SHORT
9	95 - 95	50/94-01A	ELIMINATING WINDNOISES
6,7,8	93 - 95	57/95-01	EXTERIOR DOOR LOCKS FREEZE
3	95 - 95	57/95-02	EXTERIOR DOOR LOCKS FREEZE
3	95 - 95	57/95-02	EXTERIOR DOOR LOCKS FREEZE
6,8,7	93 - 95	57/95-01	EXTERIOR DOOR LOCKS FREEZE - REVISED
6,7,8	93 - 95	57/95-06	FRICTION NOISE - FRONT/REAR DOOR SEALS
6,7,8	93 - 96	57/95-06	FRICTION NOISE - FRONT/REAR DOOR SEALS - REV
9	92 - 93	57/93-02	FRONT DOOR LOCK - UNABLE TO UNLOCK
6,8	93 - 93	55/93-03	HOOD - UNABLE TO OPEN WITH RELEASE HANDLE -MAKE TOOL
6,7,8	93 - 95	55/95-01	HOOD LOCK FASTENER - TORQUE REVISION
8	93 - 94	55/93-02	LOCK (TRUNK) - CORRECTION PROCEDURE FOR DEFICIENCIES
6,8,7	Up - 95	60/95-01	REMOVING/REPAIRING SUNROOF CABLE ASSEMBLIES
9	93 - 93	60/93-02	SUNROOF - DOES NOT CLOSE W/KEY AS STATED IN O/M
6,8,7	93 - 95	60/94-05	SUNROOF NOISE AND VIBRATION WHEN OPEN
6,8,7	93 - 95	60/94-03	SUNROOF POP NOISE/JAM WHILE OPENING
9	95 - 95	60/94-04	SUNROOF POP NOISE/JAM WHILE OPENING
3	95 - 95	61/94-01	TOP BOOT COVER LOCKING SNAPS BROKEN
8	96 - 96	55/96-01	TRUNK LID - ADJUSTABLE STOP - PROCEDURE
8	Up - 94	55/94-01	TRUNK LID CONTACTS TAIL LIGHT LENSES
8	93 - 94	55/93-02	TRUNK LOCK - CORRECTION PROCED FOR DEFICIENCIES -REV
6,7,8	93 - 95	57/95-07	WIND NOISE AT A-PILLAR - DOOR GLASS SEAL
3	95 - 95	57/95-03	WIND NOISE OR WATER LEAK - A PILLAR DOOR SEAL
6,7,8	Up - 93	60/93-01	WINDNOISE - SUNROOF SEAL ADJUSTMENT
3	95 - 95	57/94-01	WINDNOISE/WATER LEAK - DOOR WINDOW SEAL
6,7,8	93 - 93	64/93-01	WINDOW GLASS INSTALL INSTRUCTIONS - S/M ADDITION
9	Up - 94	64/92-01	WINDOWS/REAR POWER - ERRATIC OPER - REPAIR - REV

Model Legend: 1= All 2= Most 3= Cabriolet 4= Corrado 5= Fox 6= Golf 7= GTI 8= Jetta 9= Passat 10=Scirocco

Finishes, Body Structure, Frame, Bumpers

Models	Year	TSB#	Description
6,7,8	93 - 93	40/93-01	CRACKING NOISE WHEN BRAKES APPLIED - SUBFRAME
6,8	93 - 94	66/93-03	FRONT & REAR VW EMBLEMS PEEL OR DISCOLOR - REPLACE
9	95 - 95	66/94-03	FRONT BUMPER LICENSE PLATE BRACKET INSTALLATION
6,8	Up - 93	66/93-02	GRILLE (RADIATOR) - REMOVAL PROCEDURE PRECAUTIONS
6,7,8	95 - 95	50/95-02	INCORRECT PAINT LABELS - FLUSH RED
1	Up - 94	02/92-02	INDUSTRIAL FALLOUT REMOVAL
1	95 - 95	50/94-03	PAINT - REDUCING SWIRL MARKS & FINE SCRATCHES
6,7,8	95 - 95	50/95-01	PAINT SURFACES - DELETION OF CLEAR COAT
8	96 - 96	66/96-01	RADIATOR GRILLE - REMOVE/INSTALL - S/M REV
6,7,8	95 - 96	50/95-03	RAPGARD REMOVAL - PAINT SWELLING REPAIR
6,7	96 - 96	66/96-02	REAR LICENSE PLATE TRIM CRACKED
9	92 - 93	66/92-01	ROOF RAIL - REMOVE/INSTALL - S/M ADDITION
6,8	93 - 94	50/94-01	WATER BORNE/CLEAR COATED UNICOLOR PAINTS
6,8	94 - 94	51/93-02	WATER LEAKS - SEAL ENGINE BULKHEAD AT PDI

Model Legend: 1= All 2= Most 3= Cabriolet 4= Corrado 5= Fox 6= Golf 7= GTI 8= Jetta 9= Passat 10=Scirocco

VOLVO

Lighting, Horns, Turn Signals, Steering Column

Models	Year	TSB#	Description
6	95 - 95	35/907	BEAM ALIGNMENT - SPOILER MOUNTED FOGLIGHTS
5	95 - 95	DCS/83-05	COURTESY LIGHTS ON WITH DOORS CLOSED
5	94 - 94	DCS/36-01	DAYTIME RUNNING LIGHTS - INFO
5	94 - 94	35/804	HEADLAMP REPLACEMENT/ADJUST PROCEDURE - TWIN BULBS
5	Up - 93	39/801	HEADLAMP WASHER/WIPER MOTOR - RADIO INTERFERENCE/FIX
6	95 - 95	DCS/36-03	NO RETURN OF DIRECTION INDICATOR
5	94 - 94	DCS/64-03	NOISE/CLICK/RUB IN STEERING WHEEL - HORN CONTACTS
6	95 - 95	DCS/36-04	POOR SOUND FROM ALARM HORN - TAB
5	94 - 94	35/803	REAR LIGHT REPLACEMENT PROCEDURES - 850 WAGON
5	95 - 95	35/806	SPOILER MOUNTED FOGLIGHTS - BEAM ADJUSTMENT INSTR
5	93 - 94	64/802	SPRING SOUND FROM AIRBAG STEERING WHEEL
6	92 - 94	64/905	SPRING SOUND FROM AIRBAG STEERING WHEEL

Model Legend: 1= All 2= Most 3= 200 Series 4= 700 Series 5= 800 Series 6= 900 Series

Instruments, Dash Cluster, Warning Lights, Mirrors

Models	Year	TSB#	Description
5	93 - 93	38/806	BULB FAILURE WARNING SENSOR FOR TRAILER DISCONTINUED
3	86 - 94	38/203	COOLANT TEMPERATURE GAUGE - FALSE READING
3	90 - 93	27/202	CRUISE CONTROL - NOISE FROM VAC PUMP - REPAIR PROC
5	92 - 96	27/802	CRUISE CONTROL WILL NOT ENGAGE
5	Up - 93	DCS/88-01	DASH COVER - R&R FOR COSMETIC REASONS CLARIFIED
6	95 - 95	DCS/88-02	GLOVE BOX RATTLE
5	93 - 93	38/004	I/P PLASTIC LENS/COMBINED - FLOAT LINES EXPLAINED
5	93 - 93	38/802	IC (COMBINED) - COMPONENTS R&R - PARTS & PROCEDURES
5	Up - 93	DCS/38-02	INSTRUMENT (COMBINED) - DTC 2-3-1 - DIAG PROCEDURE
5	Up - 93	REF: 38/802	INSTRUMENT CIRCUIT BOARD R&R - LAMP INFORMATION
5	94 - 94	38/807	IP - NEW YAZAKI COMBINED INSTRUMENT - REMOVE/INSTALL
3,4,6	92 - 93	84/203-703	MIRROR (OUTSIDE) - GLASS R&R - NEW METHOD
3,4,5,6	Up - 93	38/114	ODOMETER - REPLACEMENT/STICKER INFORMATION
5	95 - 95	38/811	READING DTC'S - NEW VDO IC
5	95 - 95	38/111 APR'95	READING DTC'S - NEW VDO IC - REV
5	93 - 93	38/113	WARNING SYS (OIL LEVEL) - COMPONENTS & OPERATION
5	93 - 94	38/113 FEB`93	WARNING SYS (OIL LEVEL) - COMPONENTS & OPERATION REV
5	92 - 93	DCS 2/1/93	WARNING SYS/OIL LEVEL - DESCRIPTION/INFORMATION

Model Legend: 1= All 2= Most 3= 200 Series 4= 700 Series 5= 800 Series 6= 900 Series

Chassis Electrical, Wiring Harness, Fuses-Circuit Breakers, Wipers, Window Motors

Models	Year	TSB#	Description
5	94 - 94	37/806	ACCESSORY WIRING CONNECTIONS UNDER DASHBOARD
6	95 - 95	DCS/36-02	BOUNCING REAR WIPER BLADE - IMPROVED BLADES
6	95 - 95	37/912	CONNECTORS FOR ACCESSORIES BELOW DASHBOARD
5	93 - 93	37/809	DASH WIRE HARNESS REPL - ADD'L REQ'D STEPS
5	Up - 94	DCS/83-01	FRONT WINDOW REGULATOR - PDI TEST
5	93 - 94	37/804	FUSABLE LINK CABLE TO ELECT COOL FAN - REPL PROCEDUR
3	80 - 93	37/011 JUL`88	HARNESS (IGN CABLE)/NEW - USAGE/OLDER VEHICLES - REV
5	92 - 95	36/810	IMPROVED WASHER JET FOR WINDSHIELD
5	92 - 95	36/811	LEAK - REAR WASHER JET - PARTS/PROCEDURE
3	75 - 94	37/204	MEASURING VOLTAGE DROP ACROSS GRND CONNECT - PROCED
5	94 - 94	37/803	MEASURING VOLTAGE DROP ACROSS GRND CONNECT - PROCED
4,6	Up - 94	37/909	MEASURING VOLTAGE DROP ACROSS GRND CONNECT - PROCED
3	85 - 93	37/207	NEW CABLE TERMINAL AS A REPLACEMENT PART
6	95 - 95	37/911	PHONE ACCESSORY CONNECTOR UNDER DASHBOARD
6	95 - 96	37/911	PHONE ACCESSORY CONNECTOR UNDER DASHBOARD - REV
5	Up - 95	36/808	REAR WASHER JET ADJUSTMENT
3	93 - 93	DCS/28-07	SHIFT LOCK RELAY - IMPROVED PART
6	93 - 94	37/908	SRS WIRING & CONNECTOR - CARDBOARD GUIDE IN TUNNEL
5	94 - 96	35/807	SWITCH WITH REPLACEABLE ILLUMINATION BULB
6	95 - 96	35/908	SWITCH WITH REPLACEABLE ILLUMINATION BULB
6	93 - 93	DCS/37-02	TIGHTENING OF GROUND SCREW ON FRONT SIDE MEMBER
5	93 - 94	36/804	WIPER ASSEMBLY INSTALLATION PROCEDURES - S/M REVISED
5	93 - 94	36/804 OCT`93	WIPER ASSEMBLY INSTALLATION PROCEDURES - S/M REVISED
3	Up - 93	36/204	WIPER MOTOR (WINDSHIELD) - GROUNDING & REPLACEMENT
5	94 - 94	36/802	WIPER MOTOR - WGN TAILGATE - REPLACEMENT PROCEDURE
5	92 - 96	36/809	WIPER MOTOR LINK ARM ADJUSTMENT PROCED - REV 2
5	92 - 95	36/809	WIPER MOTOR LINK ARM ADJUSTMENT PROCED - REVISED
3,4,6	93 - 93	37/701	WIRING - SEALING SPLICES WITH NEW CRIMPING SLEEVE
5	93 - 93	37/801	WIRING - SEALING SPLICES WITH NEW CRIMPING SLEEVE
6	93 - 93	39/727	WIRING DIAGRAMS - 1993 MODIFIED ELECTRICAL SYSTEMS
6	93 - 94	88/920X	WIRING DIAGRAMS - S/M REVISION
5	Up - 96	37/808	WIRING TO FRONT WHEEL SENSOR - REPL PROCED/PARTS
4,6	83 - 94	85/913	WIRING/ROUTING - FRONT SEAT

Model Legend: 1= All 2= Most 3= 200 Series 4= 700 Series 5= 800 Series 6= 900 Series

Auxiliary Equipment, Jacks, Trailer Hitches, Towing

Models	Year	TSB#	Description
5	95 - 95	89/002	JACK REPLACEMENT - RECALL 72
6	95 - 95	37/911	PHONE ACCESSORY CONNECTOR UNDER DASHBOARD
6	95 - 96	37/911	PHONE ACCESSORY CONNECTOR UNDER DASHBOARD - REV

Model Legend: 1= All 2= Most 3= 200 Series 4= 700 Series 5= 800 Series 6= 900 Series

Heating, Air Conditioning, Ventilation, Defogger

Models	Year	TSB#	Description
6	93 - 96	87/933	A/C - COMP/CLUTCH/BELT PULLEY REPL PROCED - 940 -REV
6	93 - 96	87/949	A/C - COMP/CLUTCH/BELT PULLEY REPL PROCED - 960 -REV
6	93 - 96	87/933	A/C - COMP/CLUTCH/BELT PULLEY REPLACE PROCEDURE -940
6	93 - 96	87/949	A/C - COMP/CLUTCH/BELT PULLEY REPLACE PROCEDURE -960
3	75 - 94	87/217	A/C - CONVERT TO R134A - RETROFIT PROCEDURES - REV2
3	75 - 94	87/217 AUG`93	A/C - CONVERTING TO R134A - RETROFIT PROCEDURES -REV
5	92 - 94	87/814X	A/C - DUCT TEMP SENSORS - S/M REVISION
6	93 - 94	87/945	A/C - HIGH NOISE LEVEL ON COMPRESSOR - MODIFIED HOSE
6	92 - 95	87/954	A/C - IMPROVED PERFORMANCE DURING IDLING
6	92 - 95	87/954	A/C - IMPROVED PERFORMANCE DURING IDLING - REV
4,6	92 - 93	DCS/87-01	A/C - POOR PERF SITTING AT IDLE/HIGH AMB - REPAIR
4,6	92 - 94	DCS/87-01 JUL`93	A/C - POOR PERF SITTING AT IDLE/HIGH AMB - REVISED
4,6	92 - 94	DCS/87-07	A/C COMPRESSORS CYCLING PRESSURES & FAULT TRACING
4,5,6	Up - 94	87/722	A/C CONDENSATE LEAK - AIR DUCT OVER GLOVE COMPARTM'T
4,6	92 - 94	28/913	A/C CYCLES - MOMENTARY IDLE SPD DECREASE - MFI
4,5,6	Up - 95	87/948 JAN`95	A/C EVAPORATOR ODOR TREATMENT
6	93 - 96	87/953	A/C POOR PERF AT IDLE - PRESSURE SWITCH OPERATION
6	93 - 96	87/953	A/C POOR PERF AT IDLE - PRESSURE SWITCH OPS - REV
5	92 - 94	87/816X	A/C TROUBLESHOOTING - S/M REVISION
5	Up - 95	87/811	A/C- STEEL LOW PRESS SWITCH - NEW TORQUE - S/M REV
4,6	Up - 93	87/930 MAR`93	BLOWER MOTOR - REPLACEMENT - PARTS & PROC - 2ND REV
4,6	Up - 93	87/707	BLOWER MOTOR - REPLACEMENT - PARTS AND PROC - REV
4,6	Up - 94	87/940	CLIMATE CONTROL REC MOD - IMPROVE RECIRCULATION
5	94 - 94	DCS/87-05	CLIMATE CONTROL SYS PRODUCTION CHANGES
5	92 - 95	87/820	ECC CONTROL MODULE - REPLACING FRONT PANEL
5	92 - 95	87/817	ECC CONTROL MODULE INTERFERENCE
5	92 - 95	87/817	ECC CONTROL MODULE INTERFERENCE - SUPPLEMENT
5	92 - 95	87/819	EVAPORATOR ODOR TREATMENT
4,6	Up - 95	87/948	EVAPORATOR ODOR TREATMENT
4,6	Up - 95	87/948	EVAPORATOR ODOR TREATMENT - REV
5	92 - 95	87/819 NOV`94	EVAPORATOR ODOR TREATMENT - REV
5	92 - 95	87/819	EVAPORATOR ODOR TREATMENT - REV
6	Up - 95	87/948	EVAPORATOR ODOR TREATMENT - REV
5	92 - 95	87/819 MAY'95	EVAPORATOR ODOR TREATMENT - REV2
4,6	Up - 95	87/948 NOV`94	EVAPORATOR ODOR TREATMENT - REVISED
4,6	92 - 93	87/728	HEAT - IMPROVED FRONT FLOOR AIR DISTRIBUTION
4,6	Up - 93	87/012 MAR`88	HEAT - MODIFY TO IMPROVE FLOOR HEAT DISTRIBUTION
5	93 - 95	87/813	HEATER HOSES - NEW SEAL/QUICK COUPLING AT FIREWALL

Model Legend: 1= All 2= Most 3= 200 Series 4= 700 Series 5= 800 Series 6= 900 Series

Heating, Air Conditioning, Ventilation, Defogger

Models	Year	TSB#	Description
3	75 - 93	87/219	HEATER UNIT DRAINAGE HOSE CLEANING PROCEDURE
3	75 - 95	87/219 FEB`94	HEATER UNIT DRAINAGE HOSE CLEANING PROCEDURE - REV
6	93 - 94	87/944	LIQUID REFRIGERANT IN A/C COMPRESSOR
4,6	88 - 94	87/942	ODOR FROM EVAP - BLOWER FAN CONTROL MODULE
4,6	88 - 95	87/942	ODOR FROM EVAP - BLOWER FAN CONTROL MODULE - REV
6	88 - 95	87/955	ODOR FROM EVAP - BLOWER FAN CONTROL MODULE - REV2
6	95 - 95	87/950	PROPER POSITIONING OF A/C HOSE
4,6	Up - 93	87/941	REPLACEMENT OF CONDENSER/PRESSURE PIPE IN A/C SYSTEM

Model Legend: 1= All 2= Most 3= 200 Series 4= 700 Series 5= 800 Series 6= 900 Series

Entertainment Devices, Stereo, Radio, Etc.

Models	Year	TSB#	Description
3	92 - 93	39/209	LOUDSPEAKER GRILLE (REPLACEMENT) IN PARCEL SHELF
3	91 - 93	39/211	TAPE PLAYER INTERFERENCE - RE-ROUTE LEADS
3,4,6	Up - 93	39/113	PWR ANTENNA - DIAGN PROCEDURE BEFORE REPLACING - REV
5	Up - 93	39/801	RADIO - INTERFERENCE ON FM/AM BAND - DIAG/REPAIR
5	94 - 94	39/803	SIDE WINDOW ANTENNA & BOOSTER - DIAGN & REPAIR - WGN
5	94 - 95	39/803	SIDE WINDOW ANTENNA/BOOSTER - DIAGN/REPAIR -WGN- REV
5	93 - 94	39/805	POWER ANTENNA MALFUNCTION - CLEANING AND REPAIR
5	93 - 94	39/806	INSTALLATION OF EXTERNAL MAST ANTENNA - WAGON
5	Up - 96	39/808	RADIO SC-815 SRC SWITCH
6	93 - 93	39/929	RADIO/ANTI-THEFT - CODE-INHIBIT LITERATURE INFO REV
6	93 - 94	39/930	INTEGRAL WINDOW ANTENNA - DIAGNOSTIC PROCEDURE
6	91 - 94	39/934	RADIO INTERFERENCE WHEN PLAYING TAPE
6	Up - 96	39/939	RADIO SC-815 SRC SWITCH
4	Up - 93	80/906	ANTENNA COVER - REPLACEMENT PARTS MODIFICATION (780)
6	94 - 94	DCS/39-01	NOISE/CD CHANGER - INSTALL IN RT REAR WHEEL HOUSING
6	Up - 94	DCS/39-02	CD CHANGER DIAGNOSIS CODE E-01

Model Legend: 1= All 2= Most 3= 200 Series 4= 700 Series 5= 800 Series 6= 900 Series

Seats, Belts, Interior Trim, Carpets, Air Bags

Models	Year	TSB#	Description
6	93 - 94	37/908	SRS WIRING & CONNECTOR - CARDBOARD GUIDE IN TUNNEL
4	Up - 93	80/906	SEAT/PARCEL SHELF/TRIM - MODIFY REPLACE PARTS (780)
4,6	91 - 94	83/910	REPLACEMENT OF UPHOLSTERY FABRIC ON SUNROOF PANEL
6	91 - 93	85/001	ARMREST/CHILD BOOSTER SEAT TORQUE SPEC - S/M REV
5	95 - 95	85/010	FRONT SEAT BELT CATCH - 850 - RECALL 75
6	92 - 93	85/117	SEAT/POWER - UPGRADE TO OPERATE WITH IGNITION OFF
6	93 - 93	85/708	SEAT/REAR (NEW) - NEW COMPONENT R&R (949/960 WAGONS)
5	Up - 93	85/803	HEATED FRONT SEATS/THERMOSTAT REPLACEMENT - S/M SUPP
5	93 - 94	85/803 JAN`93	HEATED FRONT SEAT/THERMOSTAT REPLCM'T -S/M SUPP(REV)
5	93 - 93	85/805	POWER SEAT WITH MEMORY - FAULT TRACING/CALIBRATION
5	93 - 93	85/805 SEP`92	POWER SEAT/MEMORY - FAULT TRACING/CALIBRATION - REV
5	93 - 94	85/806	MODIFYING HEADLINER FOR ECC IN-CAR TEMP SENSOR
5	93 - 94	85/805	POWER SEAT WITH MEMORY - FAULT TRACING/CALIBRAT -REV
5	94 - 94	85/808	NEW MOUNTING CLIPS FOR BACKREST SUPPORT HOOKS - WGN
5	94 - 94	85/809	REPLACEMENT OF MOUNTING CLIPS ON SECURITY COVER
5	93 - 95	85/810	FLOOR CARPET REPLACEMENT PROCEDURE
5	93 - 95	85/811	INTEGRAL CHILD SEAT - REPLACING FOLDING CABLES
5	Up - 95	85/813	REINFORCEMENT FOR MOUNTING FOR CARPETED FLOOR MAT
5	93 - 95	85/816	SQUEAK FROM PARCEL SHELF - RUBBER COATED STUDS
5	93 - 95	85/817	SERVICE UPHOLSTERY FOR LEATHER SEAT BACK
5	93 - 95	85/818	INTEGRAL CHILD SEAT - LUBRICATING FOLDING MECHANISM
5	93 - 96	85/819	POWER SEAT DIAGNOSTICS
6	93 - 93	85/912	POWER SEAT/MEMORY-FAULT TRACING/CALIBRATION MODIFIED
6	92 - 94	85/912	POWER SEAT WITH MEMORY - FAULT TRACING/CALIBRAT -REV
4,6	82 - 95	85/917	NEW HEIGHT ADJUSTER FOR POWER SEATS - INSTALL INST
4,6	95 - 95	85/913	WIRING/ROUTING - FRONT SEAT - NEW HEATING PAD
6	Up - 95	85/920	INTEGRATED CHILD SEAT - LUBRICATION PROCEDURE
6	91 - 96	85/921	NOISE FROM MANUAL FRONT SEATS
6	96 - 96	85/922	POWER SEAT DIAGNOSTICS
6	92 - 96	85/922	POWER SEAT DIAGNOSTICS - REV
6	95 - 95	85/918	WIRING/ROUTING - FRONT SEAT - NEW HEATING PAD - REV
1	75 - 94	88/001 JUL`94	SEAT BELT EXTENDERS - INFORMATION - REVISION TWO
3,4,5,6	Up - 93	88/001 JUN`92	SEAT BELT EXTENDERS - AVAILABLE FOR FRONT SEATS ONLY
1	75 - 94	88/001 NOV`92	SEAT BELT EXTENDERS - INFORMATION - REVISED
6	95 - 95	88/006	AIR BAG MODULE - 960 - RECALL
6	94 - 94	88/008	INTERIOR TRIM - 960 - SERVICE ACTION
4,6	92 - 93	88/009	SEAT BELT WEBBING GUIDE D-RING - RECALL 74
5	93 - 93	88/101	AIR BAG/PASSENGER SIDE - REPLACEMENT - S/M REV

Model Legend: 1= All 2= Most 3= 200 Series 4= 700 Series 5= 800 Series 6= 900 Series

Seats, Belts, Interior Trim, Carpets, Air Bags

Models	Year	TSB#	Description
3	93 - 95	88/205	DRILLING OUT AIRBAG SCREW IN STEER WHEEL
5	93 - 94	88/804	NOISE FROM INTERIOR FITTINGS - DIAGNOSIS & REPPAIR
4,6	92 - 94	88/908	REPLACING HEAD RESTRAINT WITH DETACHABLE CUSHION
6	95 - 95	88/912	SRS (AIR BAG) CHANGES - CABLING/SENSOR/CODE 127
6	94 - 94	88/918	SRS AIR BAG - PASSENGER SIDE - REPLACEMENT INSTRUCT
4,6	91 - 94	83-910	REPLACEMENT OF UPHOLSTERY FABRIC ON SUNROOF PANEL
5	Up - 95	DCS/81-01	KNOCK/RATTLE UNDER FRONT SEAT
5	95 - 95	DCS/88-03	SEAT CAUTIONS - SIDE IMPACT AIR BAG

Model Legend: 1= All 2= Most 3= 200 Series 4= 700 Series 5= 800 Series 6= 900 Series

Glass, Doors, Hood, Decklid, Tailgate, Liftgate, Locks

Models	Year	TSB#	Description
4,6	Up - 93	36/908	DOOR SWITCH/NEW-TYPE - AVAILABLE AS REPLACEMENT PART
6	96 - 96	37/914	CENTRAL LOCKING - MODIFIED HARNESS - 965 ONLY
3	93 - 93	43/206	KEY LOCK CABLE REPLACEMENT - 240 WITH A/T
4	Up - 93	80/906	DOORS/SIDE & FUEL FILLER- MODIFY REPLACE PARTS (780)
5	92 - 96	81/807	DOOR STOP MOUNTING REPAIR PROCEDURE
5	94 - 94	82/802	HOOD RELEASE ADJUSTMENT PROCEDURES
5	93 - 96	82/803	ADJUSTING HOOD LATCH
6	93 - 93	83/103	GLASS SUNROOF - ADJUSTMENT PROC - WIND NOISE REPAIR
4,6	83 - 93	83/104	SUNROOF - DRAIN TUBE CLAMP ADJUSTMENT/REDUCE NOISE
3	Up - 93	83/202	SUNROOF - NEW WEATHERSTRIP AVAILABLE FOR WIND NOISE
3	Up - 93	83/203	SUNROOF - WATER LEAKAGE - REPAIR PROCEDURES
5	92 - 93	83/802	WINDOW WINDER UPPER STOP/ELEC - ADJUSTMENT FOR NOISE
5	92 - 93	83/803	TRUNK LID - ADJUSTMENT OF RUBBER STOPS
5	92 - 94	83/804	DOOR LOCK SWITCHES - NOW STOCKED AS R&R PARTS - PROC
5	93 - 93	83/805	TRUNK LOCK & LOCK CYL - LINKAGE ADJUSTMENT - S/M REV
5	93 - 94	83/808	GLASS SUNROOF WEATHERSTRIP INSTALL PROCEDURE
5	93 - 94	83/809	DOOR CLOSING FORCE - REDUCING WITH SPECIAL SHIM
5	93 - 95	DCS/83-02	RATTLE FROM TAILGATE LOCK
5	92 - 95	83/804	DOOR LOCK SWITCHES - NOW STOCKED AS R&R PARTS - REV
5	93 - 95	83/815	NOISE FROM DOOR STOPS - PROCEDURE
5	93 - 95	83/817	GREASING DOOR LOCKS AND STRIKER PLATES
5	Up - 95	83/818	PAINT SCRATCH - ADJ REAR DOOR CHILD-PROOF LOCKS

Model Legend: 1= All 2= Most 3= 200 Series 4= 700 Series 5= 800 Series 6= 900 Series

Glass, Doors, Hood, Decklid, Tailgate, Liftgate, Locks

Models	Year	TSB#	Description
5	93 - 95	83/819	RATTLE FROM SUNROOF - REPAIRS
6	Up - 93	83/705	TRUNK LOCK - REPLACEMENT/ADJUSTMENT PROCEDURES - REV
6	95 - 95	83/913	INNER DOOR LOCK/DOOR OPENER REPLACEMENT - S/M REV
6	95 - 95	83/914	OUTER DOOR OPENER LINK - NEW TYPE
6	95 - 95	83/916	INNER DOOR OPENER - SPRING REPLACEMENT - S/M REV
6	94 - 94	84/010	REAR WINDOW - 960 W/SANDSTONE PAINT - RECALL
1	93 - 96	84/011	WINDSHIELD REPLACEMENT - SRS - PASS AIRBAG VEHICLES
3	75 - 95	84/204	WINDSHIELD REPLACEMENT WITH PUR BONDED W/S
5	93 - 93	84/801	WINDSHIELD - REPLACEMENT PARTS AND PROCEDURES (850)
5	93 - 93	84/801 OCT'92	WINDSHIELD - REPLACEMENT PARTS & PROC (850) - REV
5	93 - 94	84/809	PUR BONDED WINDOWS - REPLACEMENT PROCEDURES
5	93 - 95	84/811	TAILGATE ADJUSTMENT & WEATHERSTRIP REPLACEMENT - WGN
6	93 - 94	84/904	WINDOWS/REAR SIDE - NOW INSTALLED WITH PUR ADHESIVE
6	94 - 94	84/907	PUR BONDED WINDSHIELD REPLACEMENT PROCEDURES - 940
4,6	Up - 94	84/908	PUR BONDED WINDOWS - REPLACEMENT PROCEDURES
5	Up - 95	88/809	REMOVING GLASS SPLINTERS FROM DEFROST/VENT DUCTS
6	Up - 95	88/921	REMOVING GLASS SPLINTERS FROM DEFROST/VENT DUCTS
5	Up - 95	89/804	TAILGATE SPOILER INSTALLATION INST - SUPPLEMENT
5	Up - 95	DCS/36-05	BUZZ/RATTLE LEFT A-PILLAR - TIPS
6	94 - 94	84-907	PUR BONDED WINDSHIELD REPLACEMENT PROCEDURES - REV
5	Up - 94	DCS/83-01	FRONT WINDOW REGULATOR - PDI TEST
5	Up - 95	DCS/83-02	RATTLE/SQUEAK FROM TAILGATE LOCK - 5-DOOR
1	Up - 95	DCS/83-03	AFTERMARKET ACCESSORY SUNROOF - PRECAUTIONS
6	95 - 95	DCS/83-06	PAINT CHIP ON LEFT REAR DOOR MOLDING
5	95 - 95	DCS/85-01	MANUAL PASS SEAT - LUMBAR/HEIGHT CONTROLS DELETED
5	Up - 94	DCS/89-02	THIRD SEAT & LOCKRING IN REAR SEAT BACKREST

Model Legend: 1= All 2= Most 3= 200 Series 4= 700 Series 5= 800 Series 6= 900 Series

Finishes, Body Structure, Frame, Bumpers

Models	Year	TSB#	Description
5	93 - 93	10/801	BODYWORK JACKING POINTS FOR LIFT PADS AND SHOP JACKS
5	94 - 95	17/808	RAPGARD-F PAINTWORK PROTECTION REMOVAL
5	95 - 95	17/808 JUL`94	RAPGARD-F PAINTWORK PROTECTION
6	94 - 95	17/910	RAPGARD-F PAINTWORK PROTECTION REMOVAL
6	95 - 95	17/910 JUL`94	RAPGARD-F PAINTWORK PROTECTION
5	93 - 93	21/808	FENDER/FRONT - LINER EXTENDED - DRIVE BELT NOISE FIX
5	93 - 94	21/813	NOISE (DRIVE BELT) - NEW RIGHT FRONT FENDER LINER
5	93 - 94	21/813 MAR`94	NOISE (DRIVE BELT) - NEW RT FRONT FENDER LINER - REV
5	92 - 96	21/825	REPLACEMENT OF SUB-FRAME/RIGHT ENGINE PAD
5	93 - 93	80/801	PAINT REPAIR WORK - PLASTIC EXTERIOR TRIM
5	93 - 93	80/803	BUMPER COVER - BASE-SOLID PAINT & CLEAR LACQUER USED
5	94 - 95	80/807	MODIFIED PAINT CODES - FIFTH DIGIT
5	Up - 96	80/810	REPAIR PAINTING WITH VOLVO 3-COAT PEARL
4,6	93 - 93	80/905 APR`91	PAINT/BASE-SOLID & CLEAR LACQUER - REPAIR INFO
4	Up - 93	80/906	FENDER LINER/FRONT - MODIFY REPLACEMENT PARTS (780)
6	94 - 94	80/909	3-COAT PEARL FINISH TOUCH-UP PROCEDURES
6	94 - 94	80/909 FEB`94	3 COAT PEARL FINISH TOUCH UP PROCEDURE
6	94 - 95	80/910	MODIFIED PAINT CODES - FIFTH DIGIT
1	Up - 95	81/001	PAINT - APPROVED PRODUCTS
3	75 - 93	81/207	REAR FLOOR PANEL/CROSSMEMBER/WELLS REPLACEMENT - WGN
3	86 - 94	81/208	FRONT PANEL MODIFIED - PREPARATION PROCEDURES
3	Up - 93	81/209	SURGE ARRESTORS W/ ARC WELDING NOT PERMITTED
5	93 - 93	81/802	PILLARS - SOUND INSULATION WHEN REPLACING BODY PARTS
5	92 - 95	81/805	WELDING CLIP MOUNTINGS FOR ROOF MOLDING
5	Up - 95	81/806	SURGE ARRESTORS W/ ARC WELDING NOT PERMITTED
6	95 - 95	81/912	THRESHOLD PLATE REPLACEMENT
4,6	Up - 95	81/913	SURGE ARRESTORS W/ ARC WELDING NOT PERMITTED
4,6	Up - 93	83/708	BULKHEAD JOINT/WIPER WELL COVER WHISTLE - SEAL PROC
5	93 - 96	83/821	DAMAGE TO PAINT BY TAILGATE HANDLE - 855
5	94 - 94	DCS/84-01	ATTACHING WAIST TRIM TO DOORS - PROCEDURES
5	93 - 95	84/812	NEW PLACEMENT FOR EMBLEM
5	94 - 94	86/803	NEW BUMPERS - REPLACEMENT PROCEDURE -
6	95 - 95	86/902	BUMPER WRAPAROUND TRIM MOLDING REPLACEMENT PROCEDURE
5	93 - 95	89/803	WHISTLE NOISE UNDER CAR - RUBBER PLUGS
5	94 - 94	DCS/84-01	BODY SIDE MOLDING MAY COME LOOSE - PROCEDURE

Model Legend: 1= All 2= Most 3= 200 Series 4= 700 Series 5= 800 Series 6= 900 Series

---NOTES---